# CHILTON®

## EUROPEAN
## SERVICE MANUAL
## 2008 EDITION
## Audi
## BMW
## Mercedes–Benz
## MINI
## Saab
## Volkswagen
## Volvo

CENGAGE
Learning™

Australia • Brazil • Japan • Korea • Mexico • Singapore • Spain • United Kingdom • United States

**CHILTON®**
**European Service Manual**
**2008 Edition**
**Audi, BMW, Mercedes-Benz, MINI,
Saab, Volkswagen, Volvo**

**Vice President,
Technology & Trades Professional
Business Unit:**
Gregory L. Clayton

**Publisher,
Technology & Trades Professional
Business Unit:**
David Koontz

**Director of Marketing:**
Beth A. Lutz

**Marketing Manager:**
Jennifer Stall

**Marketing Assistant:**
Rachael Conover

**Production Director:**
Carolyn Miller

**Editorial Assistant:**
Jason Yager

**Production Manager:**
Andrew Crouth

**Publishing Coordinator:**
Paula Baillie

**Sr. Content Project Manager:**
Elizabeth C. Hough

**Managing Editor:**
Terry L. Blomquist

**Editors:**
Dennis Bailey
Ken Burdette
Sherry Burdette
Tim Crain
Jacques Gordon
John Howard
Eugene F. Hannon, Jr.
Maureen Lazarz
Doug Lee
James R. Marotta
Ryan Price
Christine Sheeky

**Graphical Designer:**
Melinda Possinger

For more information contact:
Cengage Learning
Executive Woods
5 Maxwell Drive, PO Box 8007,
Clifton Park, NY 12065-8007
Visit us at **www.chilton.cengage.com**
Visit our corporate website at **www.cengage.com**
For permission to use material from
the text or product, contact us by
Tel. (800) 730-2214
Fax (800) 730-2215
**www.cengage.com/permissions**

Cengage Learning products are represented in Canada by Nelson Education, Ltd.

ISBN 10: 1-4283-2220-5
ISBN 13: 978-1-4283-2220-2
ISSN: 1939-621X

**NOTICE TO THE READER**

Publisher does not warrant or guarantee any of the products described herein or perform any independent analysis in connection with any of the product information contained herein. Publisher does not assume, and expressly disclaims, any obligation to obtain and include information other than that provided to it by the manufacturer.

The reader is expressly warned to consider and adopt all safety precautions that might be indicated by the activities herein and to avoid all potential hazards. By following the instructions contained herein, the reader willingly assumes all risks in connection with such instructions.

The publisher makes no representation or warranties of any kind, including but not limited to, the warranties of fitness for particular purpose or merchantability, nor are any such representations implied with respect to the material set forth herein, and the publisher takes no responsibility with respect to such material. The publisher shall not be liable for any special, consequential, or exemplary damages resulting, in whole or part, from the readers' use of, or reliance upon, this material.

Printed in the United States of America
1 2 3 4 5 xx 13 12 11 10 09 08

# Table of Contents

# Model Index

| Model | Section No. | Model | Section No. | Model | Section No. |
|---|---|---|---|---|---|
| 3 Series | 3-1 | C Class | 7-1 | **R** | |
| 5 Series | 3-1 | Cooper | 9-1 | Rabbit | 14-1 |
| 7 Series | 3-1 | Cooper S | 9-1 | **S** | |
| 9-2x | 11-1 | **E** | | S40 | 17-1 |
| 9-3 | 11-1 | E Class | 7-1 | S60 | 17-1 |
| 9-5 | 11-1 | **G** | | S80 | 17-1 |
| 9-7x | 12-1 | Golf | 14-1 | S Class | 7-1 |
| **A** | | GTI | 14-1 | **T** | |
| A3 | 1-1 | **J** | | Touareg | 15-1 |
| A4 | 1-1 | Jetta | 14-1 | TT | 1-1 |
| A4 Avant | 1-1 | **M** | | TT Quattro | 1-1 |
| A6 | 1-1 | ML350 | 8-1 | **V** | |
| A6 Avant | 1-1 | ML500 | 8-1 | V50 | 17-1 |
| A6 Quattro | 1-1 | ML550 | 8-1 | V70 | 17-1 |
| A8 | 1-1 | **N** | | **X** | |
| A8 Quattro | 1-1 | New Beetle | 14-1 | X3 | 4-1 |
| **C** | | **P** | | X5 | 5-1 |
| C30 | 17-1 | Passat | 14-1 | XC70 | 17-1 |
| C70 | 17-1 | | | XC90 | 18-1 |

# USING THIS INFORMATION

## Organization

To find where a particular model section or procedure is located, look in the Table of Contents. Main topics are listed with the page number on which they may be found. Following the main topics is an alphabetical listing of all of the procedures within the section and their page numbers.

## Manufacturer and Model Coverage

This product covers 2007–2008 European models that are produced in sufficient quantities to warrant coverage, and which have technical content available from the vehicle manufacturers before our publication date. Although this information is as complete as possible at the time of publication, some manufacturers may make changes which cannot be included here. While striving for total accuracy, the publisher cannot assume responsibility for any errors, changes, or omissions that may occur in the compilation of this data.

## Part Numbers & Special Tools

Part numbers and special tools are recommended by the publisher and vehicle manufacturer to perform specific jobs. Before substituting any part or tool for the one recommended, you must be completely satisfied that neither your personal safety, nor the performance of the vehicle will be endangered.

# ACKNOWLEDGEMENT

The publisher would like to express appreciation to the following vehicle manufacturers for their assistance in producing this manual. No further reproduction or distribution of the material in this manual is allowed without the expressed written permission of the vehicle manufacturers and the publisher. Audi of America, Inc., BMW of North America, LLC, Mercedes-Benz USA LLC, Mini USA, Volkswagen of America, Inc., Volvo Cars of North America, LLC.

# PRECAUTIONS

Before servicing any vehicle, please be sure to read all of the following precautions, which deal with personal safety, prevention of component damage, and important points to take into consideration when servicing a motor vehicle:

• Always wear safety glasses or goggles when drilling, cutting, grinding or prying.

• Steel-toed work shoes should be worn when working with heavy parts. Pockets should not be used for carrying tools. A slip or fall can drive a screwdriver into your body.

• Work surfaces, including tools and the floor should be kept clean of grease, oil or other slippery material.

• When working around moving parts, don't wear loose clothing. Long hair should be tied back under a hat or cap, or in a hair net.

• Always use tools only for the purpose for which they were designed. Never pry with a screwdriver.

• Keep a fire extinguisher and first aid kit handy.

• Always properly support the vehicle with approved stands or lift.

• Always have adequate ventilation when working with chemicals or hazardous material.

• Carbon monoxide is colorless, odorless and dangerous. If it is necessary to operate the engine with vehicle in a closed area such as a garage, always use an exhaust collector to vent the exhaust gases outside the closed area.

• When draining coolant, keep in mind that small children and some pets are attracted by ethylene glycol antifreeze, and are quite likely to drink any left in an open container, or in puddles on the ground. This will prove fatal in sufficient quantity. Always drain the coolant into a sealable container.

• To avoid personal injury, do not remove the coolant pressure relief cap while the engine is operating or hot. The cooling system is under pressure; steam and hot liquid can come out forcefully when the cap is loosened slightly. Failure to follow these instructions may result in personal injury. The coolant must be recovered in a suitable, clean container for reuse. If the coolant is contaminated it must be recycled or disposed of correctly.

• When carrying out maintenance on the starting system be aware that heavy gauge leads are connected directly to the battery. Make sure the protective caps are in place when maintenance is completed. Failure to follow these instructions may result in personal injury.

• Do not remove any part of the engine emission control system. Operating the engine without the engine emission control system will reduce fuel economy and engine ventilation. This will weaken engine performance and shorten engine life. It is also a violation of Federal law.

• Due to environmental concerns, when the air conditioning system is drained, the refrigerant must be collected using refrigerant recovery/recycling equipment. Federal law requires that refrigerant be recovered into appropriate recovery equipment and the process be conducted by qualified technicians who have been certified by an approved organization, such as MACS, ASI, etc. Use of a recovery machine dedicated to the appropriate refrigerant is necessary to reduce the possibility of oil and refrigerant incompatibility concerns. Refer to the instructions provided by the equipment manufacturer when removing refrigerant from or charging the air conditioning system.

• Always disconnect the battery ground when working on or around the electrical system.

• Batteries contain sulfuric acid. Avoid contact with skin, eyes, or clothing. Also, shield your eyes when working near batteries to protect against possible splashing of the acid solution. In case of acid contact with skin or eyes, flush immediately with water for a minimum of 15 minutes and get prompt medical attention. If acid is swallowed, call a physician immediately. Failure to follow these instructions may result in personal injury.

• Batteries normally produce explosive gases. Therefore, do not allow flames, sparks or lighted substances to come near the battery. When charging or working near a battery, always shield your face and protect your eyes. Always provide ventilation. Failure to follow these instructions may result in personal injury.

• When lifting a battery, excessive pressure on the end walls could cause acid to spew through the vent caps, resulting in personal injury, damage to the vehicle or battery. Lift with a battery carrier or with your hands on opposite corners. Failure to follow these instructions may result in personal injury.

• Observe all applicable safety precautions when working around fuel. Whenever servicing the fuel system, always work in a well-ventilated area. Do not allow fuel spray or vapors to come in contact with a spark, open flame, or excessive heat (a hot drop light, for example). Keep a dry chemical fire extinguisher near the work area. Always keep fuel in a container specifically designed for fuel storage; also, always properly seal fuel containers to avoid the possibility of fire or explosion. Do not smoke or carry lighted tobacco or open flame of any type when working on or near any fuel-related components.

• Fuel injection systems often remain pressurized, even after the engine has been turned OFF. The fuel system pressure must be relieved before disconnecting any fuel lines. Failure to do so may result in fire and/or personal injury.

• The evaporative emissions system contains fuel vapor and condensed fuel vapor. Although not present in large quantities, it still presents the danger of explosion or fire. Disconnect the battery ground cable from the battery to minimize the possibility of an electrical spark occurring, possibly causing a fire or explosion if fuel vapor or liquid fuel is present in the area. Failure to follow these instructions can result in personal injury.

• The EPA warns that prolonged contact with used engine oil may cause a number of skin disorders, including cancer! You should make every effort to minimize your exposure to used engine oil. Protective gloves should be worn when changing oil. Wash your hands and any other exposed skin areas as soon as possible after exposure to used engine oil. Soap and water, or waterless hand cleaner should be used.

• Some vehicles are equipped with an air bag system, often referred to as a Supple-mental Restraint System (SRS) or Supplemental Inflatable Restraint (SIR) system. The system must be disabled before performing service on or around system components, steering column, instrument panel components, wiring and sensors. Failure to follow safety and disabling procedures could result in accidental air bag deployment, possible personal injury and unnecessary system repairs.

• Always wear safety goggles when working with, or around, the air bag system. When carrying a non-deployed air bag, be sure the bag and trim cover are pointed away from your body. When placing a non-deployed air bag on a work surface, always face the bag and trim cover upward, away from the surface. This will reduce the motion of the module if it is accidentally deployed.

• Electronic modules are sensitive to electrical charges. The ABS module can be damaged if exposed to these charges.

• Brake pads and shoes may contain asbestos, which has been determined to be a cancer-causing agent. Never clean brake surfaces with compressed air. Avoid inhaling brake dust. Clean all brake surfaces with a commercially available brake cleaning fluid.

• When replacing brake pads, shoes, discs or drums, replace them as complete axle sets.

• When servicing drum brakes, disassemble and assemble one side at a time, leaving the remaining side intact for reference.

• Brake fluid often contains polyglycol ethers and polyglycols. Avoid contact with the eyes and wash your hands thoroughly after handling brake fluid. If you do get brake fluid in your eyes, flush your eyes with clean, running water for 15 minutes. If eye irritation persists, or if you have taken brake fluid internally, immediately seek medical assistance.

• Clean, high quality brake fluid from a sealed container is essential to the safe and proper operation of the brake system. You should always buy the correct type of brake fluid for your vehicle. If the brake fluid becomes contaminated, completely flush the system with new fluid. Never reuse any brake fluid. Any brake fluid that is removed from the system should be discarded. Also, do not allow any brake fluid to come in contact with a painted or plastic surface; it will damage the paint.

• Never operate the engine without the proper amount and type of engine oil; doing so will result in severe engine damage.

• Timing belt maintenance is extremely important! Many models utilize an interference-type, non-freewheeling engine. If the timing belt breaks, the valves in the cylinder head may strike the pistons, causing potentially serious (also time-consuming and expensive) engine damage.

• Disconnecting the negative battery cable on some vehicles may interfere with the functions of the on-board computer system (s) and may require the computer to undergo a relearning process once the negative battery cable is reconnected.

• Steering and suspension fasteners are critical parts because they affect performance of vital components and systems and their failure can result in major service expense. They must be replaced with the same grade or part number or an equivalent part if replacement is necessary. Do not use a replacement part of lesser quality or substitute design. Torque values must be used as specified during reassembly to ensure proper retention of these parts.

# AUDI

**1**

A3 • A4 • A4 Avant • A6 • A6 Avant • A6 Quattro • A8 • A8 Quattro • TT • TT Quattro

## SPECIFICATIONS AND MAINTENANCE CHARTS

### ENGINE AND VEHICLE IDENTIFICATION

| Engine | | | | | | | Model Year | |
|---|---|---|---|---|---|---|---|---|
| Code | Liters (cc) | Cu. In. | Cyl. | Fuel Sys. | Engine Type | Eng. Mfg. | Code ① | Year |
| AMB | 1.8 (1781) | 107 | 4 | MFI-Turbo | DOHC | Audi | 6 | 2006 |
| AWP | 1.8 (1781) | 107 | 4 | MFI-Turbo | DOHC | Audi | 7 | 2007 |
| BEA | 1.8 (1781) | 107 | 4 | MFI-Turbo | DOHC | Audi | 8 | 2008 |
| AXX | 2.0 (1998) | 121 | 4 | FSI-Turbo | DOHC | Audi | | |
| BPY | 2.0 (1998) | 121 | 4 | FSI-Turbo | DOHC | Audi | | |
| BPG | 2.0 (1998) | 121 | 4 | FSI-Turbo | DOHC | Audi | | |
| BGN | 3.0 (2976) | 183 | 6 | MFI | DOHC | Audi | | |
| BHE | 3.2 (3200) | 195 | 6 | MFI | DOHC | Audi | | |
| BUB | 3.2 (3200) | 195 | 6 | MFI | DOHC | Audi | | |
| BKH | 3.2 (3200) | 190 | 6 | FSI | DOHC | Audi | | |
| BNK | 4.2 (4172) | 255 | 8 | MFI | DOHC | Audi | | |

MFI: Multi-point Fuel Injection

FSI: Direct Fuel Injection

DOHC: Double Overhead Camshaft

① 10th digit of the Vehicle Identification Number (VIN)

22205_AUDI_C0001

## GENERAL ENGINE SPECIFICATIONS

| Year | Model | Engine Displacement Liters | Engine ID | Net Horsepower @ rpm | Net Torque@rpm (ft. lbs.) | Bore x Stroke (in.) | Com- pression Ratio | Oil Pressure @ rpm |
|---|---|---|---|---|---|---|---|---|
| 2006 | A3 | 2.0 | AXX | 200@5,100-600 | 207@1800-5000 | 3.25x3.65 | 10.3:1 | 39-65@2000 |
| | A3 | 2.0 | BPY | 200@5,100-600 | 207@1800-5000 | 3.25x3.65 | 10.3:1 | 39-65@2000 |
| | A3 | 3.2 | BUB | 250@6,300 | 236@2800-3200 | 3.31x3.78 | 11.3:1 | 39-65@2000 |
| | A4 | 2.0 | BPG | 200@5,100-600 | 207@1800-5000 | 3.25x3.65 | 10.5:1 | 43@2000 |
| | A4 | 3.2 | BKH | 255@6500 | 243@3250 | 3.33x3.65 | 12.5:1 | 43@2000 |
| | A6 | 3.2 | BKH | 255@6500 | 243@3250 | 3.33x3.65 | 12.5:1 | 43@2000 |
| | A6 | 4.2 | BNK | 335@6500 | 310@3500 | 3.33x3.65 | 11.0:1 | 29@2000 |
| | TT | 1.8 | AWP | 132@5500 | 235@1950 | 3.18x3.48 | 9.5:1 | 51-65@2000 |
| | TT | 1.8 | BEA | 225@5900 | 220@5900 | 3.18x3.40 | 9.0:1 | 51-65@3000 |
| | TT Quattro | 3.2 | BHE | 184@6300 | 320@2800 | 3.31x3.78 | 11.3:1 | 44@2000 |
| | A8 | 4.2 | BNK | 335@6500 | 310@3500 | 3.33x3.65 | 11.0:1 | 29@2000 |
| 2007 | A3 | 2.0 | BPY | 200@5,100-600 | 207@1800-5000 | 3.25x3.65 | 10.3:1 | 39-65@2000 |
| | A3 | 3.2 | BUB | 250@6,300 | 236@2800-3200 | 3.31x3.78 | 11.3:1 | 39-65@2000 |
| | A4 | 2.0 | BPG | 200@5,100-600 | 207@1800-5000 | 3.25x3.65 | 10.5:1 | 43@2000 |
| | A4 | 3.2 | BKH | 255@6500 | 243@3250 | 3.33x3.65 | 12.5:1 | 43@2000 |
| | A6 | 3.2 | BKH | 255@6500 | 243@3250 | 3.33x3.65 | 12.5:1 | 43@2000 |
| | A6 | 4.2 | BNK | 335@6500 | 310@3500 | 3.33x3.65 | 11.0:1 | 29@2000 |
| | A8 | 4.2 | BNK | 335@6500 | 310@3500 | 3.33x3.65 | 11.0:1 | 29@2000 |
| 2008 | A3 | 2.0 | BPY | 200@5,100-600 | 207@1800-5000 | 3.25x3.65 | 10.3:1 | 39-65@2000 |
| | A3 | 3.2 | BUB | 250@6,300 | 236@2800-3200 | 3.31x3.78 | 11.3:1 | 39-65@2000 |
| | A4 | 2.0 | BPG | 200@5,100-600 | 207@1800-5000 | 3.25x3.65 | 10.5:1 | 43@2000 |
| | A4 | 3.2 | BKH | 255@6500 | 243@3250 | 3.33x3.65 | 12.5:1 | 43@2000 |
| | A6 | 3.2 | BKH | 255@6500 | 243@3250 | 3.33x3.65 | 12.5:1 | 43@2000 |
| | A6 | 4.2 | BNK | 335@6500 | 310@3500 | 3.33x3.65 | 11.0:1 | 29@2000 |
| | TT | 1.8 | BEA | 225@5900 | 220@5900 | 3.18x3.40 | 9.0:1 | 51-65@3000 |
| | TT Quattro | 3.2 | BUB | 184@6300 | 320@2800 | 3.31x3.78 | 11.3:1 | 44@2000 |
| | A8 | 4.2 | BNK | 335@6500 | 310@3500 | 3.33x3.65 | 11.0:1 | 29@2000 |

MFI: Multi-point Fuel Injection

22205_AUDI_C0002

## ENGINE TUNE-UP SPECIFICATIONS

| Year | Engine Displacement Liters | Engine ID/VIN | Spark Plug Gap (in.) | Ignition Timing (deg.) | | Fuel Pump (psi) | Idle Speed (rpm) | | Valve Clearance | |
|------|------|------|------|------|------|------|------|------|------|------|
| | | | | MT | AT | | MT | AT | In. | Ex. |
| **2006** | 1.8 | BEA | 0.039 | ① | ① | 36 | 700-820 | 700-820 | HYD | HYD |
| | 1.8 | AWP | 0.039 | ① | ① | 36 | 700-820 | 700-820 | HYD | HYD |
| | 2.0 | AXX | 0.039 | ① | ① | 87 | 640-800 | 640-800 | HYD | HYD |
| | 2.0 | BPG | 0.039 | ① | ① | 87 | 640-800 | 640-800 | HYD | HYD |
| | 2.0 | BPY | 0.039 | ① | ① | 87 | 640-800 | 640-800 | HYD | HYD |
| | 3.2 | BUB | 0.039 | ① | ① | 40 | 600-700 | 600-700 | HYD | HYD |
| | 3.2 | BHE | 0.039 | ① | ① | 46-55 | 650-750 | 650-750 | HYD | HYD |
| | 3.2 | BKH | 0.039 | ① | ① | 87 | 650-750 | 650-750 | HYD | HYD |
| | 4.2 | BNK | 0.039 | ① | ① | 87 | 650-750 | 650-750 | HYD | HYD |
| **2007** | 2.0 | BPG | 0.039 | ① | ① | 87 | 640-800 | 640-800 | HYD | HYD |
| | 2.0 | BPY | 0.039 | ① | ① | 87 | 640-800 | 640-800 | HYD | HYD |
| | 3.2 | BUB | 0.039 | ① | ① | 40 | 600-700 | 600-700 | HYD | HYD |
| | 3.2 | BKH | 0.039 | ① | ① | 87 | 650-750 | 650-750 | HYD | HYD |
| | 4.2 | BNK | 0.039 | ① | ① | 87 | 650-750 | 650-750 | HYD | HYD |
| **2008** | 1.8 | BEA | 0.039 | ① | ① | 36 | 700-820 | 700-820 | HYD | HYD |
| | 2.0 | BPY | 0.039 | ① | ① | 87 | 640-800 | 640-800 | HYD | HYD |
| | 2.0 | BPG | 0.039 | ① | ① | 87 | 640-800 | 640-800 | HYD | HYD |
| | 3.2 | BUB | 0.039 | ① | ① | 40 | 600-700 | 600-700 | HYD | HYD |
| | 3.2 | BHE | 0.039 | ① | ① | 46-55 | 650-750 | 650-750 | HYD | HYD |
| | 3.2 | BKH | 0.039 | ① | ① | 87 | 650-750 | 650-750 | HYD | HYD |
| | 4.2 | BNK | 0.039 | ① | ① | 87 | 650-750 | 650-750 | HYD | HYD |

NOTE: The Vehicle Emission Control Information label reflects specification changes made during production and must be used if different from this chart.

NOTE: Fuel pump pressure specifications with the fuel pressure regulator vacuum hose attached.

HYD: Hydraulic

① The basic setting is controlled by the ECU and is not adjustable

22205_AUDI_C0003

## CAPACITIES

| Year | Model | Engine Displacement Liters | Engine ID/VIN | Engine Oil with Filter (qts) | Transmission (pts.) | | Drive Axle | | Fuel Tank (gal.) | Cooling System (qts.) |
|------|-------|------|------|------|------|------|------|------|------|------|
| | | | | | 5-Spd | Auto | Front (pts.) | Rear (pts.) | | |
| **2006** | A3 | 2.0 | AXX | 4.8 | 4.8 | 12.6 | NA | NA | 14.5 | 8.5 |
| | A3 | 2.0 | BPY | 4.8 | 4.8 | 12.6 | NA | NA | 14.5 | 8.5 |
| | A3 | 3.2 | BUB | NA | NA | NA | NA | NA | 14.5 | 7.4 |
| | A4 | 2.0 | BPG | 4.8 | 7.4 | 6.4 | 1.6 | 3.8 | 16.5 | 7.4 |
| | A4 | 3.2 | BKH | 8.6 | 7.4 | 6.4 | 1.6 | NA | 16.6 | — |
| | A6 | 3.2 | BKH | 8.6 | 7.4 | 4.6 | NA | — | 31.1 | 8.5 |
| | A6 | 4.2 | BNK | 8.0 | — | 8.0 | — | 3.2 | 21.1 | 9.5 |
| | TT | 1.8 | AWP | 4.8 | ① | 14.8 | — | — | 16.3 | 5.2 |
| | TT | 1.8 | BEA | 4.8 | ① | 14.8 | — | — | 14.5 | 5.2 |
| | TT Quattro | 3.2 | BHE | 5.8 | ① | ② | — | — | — | — |
| | A8 | 4.2 | BNK | 8.0 | — | 8.0 | — | 3.2 | 21.1 | 9.5 |
| **2007** | A3 | 2.0 | BPY | 4.8 | 4.8 | 12.6 | NA | NA | 14.5 | 8.5 |
| | A3 | 3.2 | BUB | NA | NA | NA | NA | NA | 14.5 | 7.4 |
| | A4 | 2.0 | BPG | 4.8 | 7.4 | 6.4 | 1.6 | 3.8 | 16.5 | 7.4 |
| | A4 | 3.2 | BKH | 8.6 | 7.4 | 6.4 | 1.6 | NA | 16.6 | — |
| | A6 | 3.2 | BKH | 8.6 | 7.4 | 4.6 | NA | — | 31.1 | 8.5 |
| | A6 | 4.2 | BNK | 8.0 | — | 8.0 | — | 3.2 | 21.1 | 9.5 |
| | A8 | 4.2 | BNK | 8.0 | — | 8.0 | — | 3.2 | 21.1 | 9.5 |
| **2008** | A3 | 2.0 | BPY | 4.8 | 4.8 | 12.6 | NA | NA | 14.5 | 8.5 |
| | A3 | 3.2 | BUB | NA | NA | NA | NA | NA | 14.5 | 7.4 |
| | A4 | 2.0 | BPG | 4.8 | 7.4 | 6.4 | 1.6 | 3.8 | 16.5 | 7.4 |
| | A4 | 3.2 | BKH | 8.6 | 7.4 | 6.4 | 1.6 | NA | 16.6 | — |
| | A6 | 3.2 | BKH | 8.6 | 7.4 | 4.6 | NA | — | 31.1 | 8.5 |
| | A6 | 4.2 | BNK | 8.0 | — | 8.0 | — | 3.2 | 21.1 | 9.5 |
| | TT | 1.8 | BEA | 4.8 | ① | 14.8 | — | — | 14.5 | 5.2 |
| | TT Quattro | 3.2 | BHE | 5.8 | ① | ② | — | — | — | — |
| | A8 | 4.2 | BNK | 8.0 | — | 8.0 | — | 3.2 | 21.1 | 9.5 |

NOTE: All capacities are approximate. Add fluid gradually and ensure a proper fluid level is obtained.

NA: Not Available

① Front Wheel Drive: 4.2 pts.
   All Wheel Drive: 5.4 pts.

② Initial Fill: 15,2 pts.
   Refill/Change: 12.7 pts.

## VALVE SPECIFICATIONS

| Year | Engine Displacement Liters | Engine ID/VIN | Seat Angle (deg.) | Face Angle (deg.) | Spring Test Pressure (lbs. @ in.) | Spring Installed Height (in.) | Stem-to-Guide Clearance (in.) | | Stem Diameter (in.) | |
|---|---|---|---|---|---|---|---|---|---|---|
| | | | | | | | Intake | Exhaust | Intake | Exhaust |
| **2006** | 1.8 | AWP | 45 | 45 | NA | NA | 0.031 ① | 0.031 ① | NA | NA |
| | 1.8 | BEA | 45 | 45 | NA | NA | 0.031 ① | 0.031 ① | NA | NA |
| | 2.0 | AXX | 45 | 45 | NA | NA | 0.031 ① | 0.031 ① | NA | NA |
| | 2.0 | BPY | 45 | 45 | NA | NA | 0.031 ① | 0.031 ① | NA | NA |
| | 2.0 | BPG | 45 | 45 | NA | NA | 0.031 ① | 0.031 ① | NA | NA |
| | 3.2 | BUB | 45 | 45 | NA | NA | 0.031 ① | 0.031 ① | 0.2346 | 0.2343 |
| | 3.2 | BHE | 45 | 45 | NA | NA | 0.031 ① | 0.031 ① | 0.2346 | 0.2343 |
| | 3.2 | BKH | 45 | 45 | NA | NA | 0.031 ① | 0.031 ① | 0.2346 | 0.2343 |
| | 4.2 | BNK | 45 | 45 | NA | NA | 0.031 ① | 0.031 ① | 0.2346-0.2350 | 0.2339-0.2343 |
| **2007** | 2.0 | BPY | 45 | 45 | NA | NA | 0.031 ① | 0.031 ① | NA | NA |
| | 2.0 | BPG | 45 | 45 | NA | NA | 0.031 ① | 0.031 ① | NA | NA |
| | 3.2 | BUB | 45 | 45 | NA | NA | 0.031 ① | 0.031 ① | 0.2346 | 0.2343 |
| | 3.2 | BHE | 45 | 45 | NA | NA | 0.031 ① | 0.031 ① | 0.2346 | 0.2343 |
| | 3.2 | BKH | 45 | 45 | NA | NA | 0.031 ① | 0.031 ① | 0.2346 | 0.2343 |
| | 4.2 | BNK | 45 | 45 | NA | NA | 0.031 ① | 0.031 ① | 0.2346-0.2350 | 0.2339-0.2343 |
| **2008** | 1.8 | BEA | 45 | 45 | NA | NA | 0.031 ① | 0.031 ① | NA | NA |
| | 2.0 | BPY | 45 | 45 | NA | NA | 0.031 ① | 0.031 ① | NA | NA |
| | 2.0 | BPG | 45 | 45 | NA | NA | 0.031 ① | 0.031 ① | NA | NA |
| | 3.2 | BUB | 45 | 45 | NA | NA | 0.031 ① | 0.031 ① | 0.2346 | 0.2343 |
| | 3.2 | BKH | 45 | 45 | NA | NA | 0.031 ① | 0.031 ① | 0.2346 | 0.2343 |
| | 4.2 | BNK | 45 | 45 | NA | NA | 0.031 ① | 0.031 ① | 0.2346-0.2350 | 0.2339-0.2343 |

NA: Not Available

① To measure: Insert a new valve into guide with end of valve flush with end of guide. Use a dial indicator to measure axial valve head movement.

22205_AUDI_C0006

## CRANKSHAFT AND CONNECTING ROD SPECIFICATIONS
All measurements are given in inches.

| Year | Engine Size Liters | Engine ID/VIN | Crankshaft | | | | Connecting Rod | | |
|------|------|------|------|------|------|------|------|------|------|
| | | | Main Brg. Journal Dia. | Main Brg. Oil Clearance | Shaft End-play | Thrust on No. | Journal Diameter | Oil Clearance | Side Clearance |
| 2006 | 1.8 | AWP | 2.1251-2.1259 | 0.0004-0.0015 | 0.0028-0.0083 | 3 | 1.8802-1.8819 | 0.0004-0.002 | 0.0040-0.014 |
| | 1.8 | BEA | 2.1251-2.1259 | 0.0004-0.0015 | 0.0028-0.0083 | 3 | 1.8802-1.8819 | 0.0004-0.002 | 0.0040-0.014 |
| | 2.0 | AXX | NA | NA | NA | NA | NA | NA | NA |
| | 2.0 | BPY | NA | NA | NA | NA | NA | NA | NA |
| | 2.0 | BPG | NA | NA | NA | NA | NA | NA | NA |
| | 3.2 | BUB | 2.3606-2.3613 | 0.0008-0.0024 | 0.0003-0.009 | 5 | 2.1243-2.1251 | 0.0008-0.0028 | NA |
| | 3.2 | BHE | 2.3606-2.3613 | 0.0008-0.0024 | 0.0003-0.009 | 5 | 2.1243-2.1251 | 0.0008-0.0028 | NA |
| | 3.2 | BKH | 2.3606-2.3613 | 0.0008-0.0024 | 0.0003-0.009 | 5 | 2.1243-2.1251 | 0.0008-0.0028 | NA |
| | 4.2 | BNK | NA | NA | NA | NA | NA | NA | NA |
| 2007 | 2.0 | BPY | NA | NA | NA | NA | NA | NA | NA |
| | 2.0 | BPG | NA | NA | NA | NA | NA | NA | NA |
| | 3.2 | BUB | 2.3606-2.3613 | 0.0008-0.0024 | 0.0003-0.009 | 5 | 2.1243-2.1251 | 0.0008-0.0028 | NA |
| | 3.2 | BHE | 2.3606-2.3613 | 0.0008-0.0024 | 0.0003-0.009 | 5 | 2.1243-2.1251 | 0.0008-0.0028 | NA |
| | 3.2 | BKH | 2.3606-2.3613 | 0.0008-0.0024 | 0.0003-0.009 | 5 | 2.1243-2.1251 | 0.0008-0.0028 | NA |
| | 4.2 | BNK | NA | NA | NA | NA | NA | NA | NA |
| 2008 | 1.8 | BEA | 2.1251-2.1259 | 0.0004-0.0015 | 0.0028-0.0083 | 3 | 1.8802-1.8819 | 0.0004-0.002 | 0.0040-0.014 |
| | 2.0 | BPY | NA | NA | NA | NA | NA | NA | NA |
| | 2.0 | BPG | NA | NA | NA | NA | NA | NA | NA |
| | 3.2 | BUB | 2.3606-2.3613 | 0.0008-0.0024 | 0.0003-0.009 | 5 | 2.1243-2.1251 | 0.0008-0.0028 | NA |
| | 3.2 | BHE | 2.3606-2.3613 | 0.0008-0.0024 | 0.0003-0.009 | 5 | 2.1243-2.1251 | 0.0008-0.0028 | NA |
| | 3.2 | BKH | 2.3606-2.3613 | 0.0008-0.0024 | 0.0003-0.009 | 5 | 2.1243-2.1251 | 0.0008-0.0028 | NA |
| | 4.2 | BNK | NA | NA | NA | NA | NA | NA | NA |

NA: Not Available

22205_AUDI_C0005

## PISTON AND RING SPECIFICATIONS

All measurements are given in inches

| Year | Engine Size Liters | Engine ID/VIN | Piston Clearance | Ring Gap | | | Ring Side Clearance | | |
| | | | | Top Compression | Bottom Compression | Oil Control | Top Compression | Bottom Compression | Oil Control |
|---|---|---|---|---|---|---|---|---|---|
| **2006** | 1.8 | AWP | 0.0005-0.0011 | 0.006-0.0157 | 0.006-0.0157 | 0.0098-0.0197 | 0.0028 | 0.0008-0.0028 | 0.0008-0.0023 |
| | 1.8 | BEA | 0.0005-0.0011 | 0.006-0.0157 | 0.006-0.0157 | 0.0098-0.0197 | 0.0028 | 0.0008-0.0028 | 0.0008-0.0023 |
| | 2.0 | AXX | NA | NA | NA | NA | NA | NA | NA |
| | 2.0 | BPY | NA | NA | NA | NA | NA | NA | NA |
| | 2.0 | BPG | NA | NA | NA | NA | NA | NA | NA |
| | 3.2 | BUB | NA | 0.0078-0.0157 | 0.0078-0.0157 | 0.0098-0.0197 | 0.0016-0.0035 | 0.0012-0.0023 | 0.0008-0.0023 |
| | 3.2 | BHE | NA | 0.0078-0.0157 | 0.0078-0.0157 | 0.0098-0.0197 | 0.0016-0.0035 | 0.0012-0.0023 | 0.0008-0.0023 |
| | 3.2 | BKH | NA | 0.0078-0.0157 | 0.0078-0.0157 | 0.0098-0.0197 | 0.0016-0.0035 | 0.0012-0.0023 | 0.0008-0.0023 |
| | 4.2 | BNK | NA | NA | NA | NA | NA | NA | NA |
| **2007** | 2.0 | BPY | NA | NA | NA | NA | NA | NA | NA |
| | 2.0 | BPG | NA | NA | NA | NA | NA | NA | NA |
| | 3.2 | BUB | NA | 0.0078-0.0157 | 0.0078-0.0157 | 0.0098-0.0197 | 0.0016-0.0035 | 0.0012-0.0023 | 0.0008-0.0023 |
| | 3.2 | BHE | NA | 0.0078-0.0157 | 0.0078-0.0157 | 0.0098-0.0197 | 0.0016-0.0035 | 0.0012-0.0023 | 0.0008-0.0023 |
| | 3.2 | BKH | NA | 0.0078-0.0157 | 0.0078-0.0157 | 0.0098-0.0197 | 0.0016-0.0035 | 0.0012-0.0023 | 0.0008-0.0023 |
| | 4.2 | BNK | NA | NA | NA | NA | NA | NA | NA |
| **2008** | 1.8 | BEA | 0.0005-0.0011 | 0.006-0.0157 | 0.006-0.0157 | 0.0098-0.0197 | 0.0028 | 0.0008-0.0028 | 0.0008-0.0023 |
| | 2.0 | BPY | NA | NA | NA | NA | NA | NA | NA |
| | 2.0 | BPG | NA | NA | NA | NA | NA | NA | NA |
| | 3.2 | BUB | NA | 0.0078-0.0157 | 0.0078-0.0157 | 0.0098-0.0197 | 0.0016-0.0035 | 0.0012-0.0023 | 0.0008-0.0023 |
| | 3.2 | BHE | NA | 0.0078-0.0157 | 0.0078-0.0157 | 0.0098-0.0197 | 0.0016-0.0035 | 0.0012-0.0023 | 0.0008-0.0023 |
| | 3.2 | BKH | NA | 0.0078-0.0157 | 0.0078-0.0157 | 0.0098-0.0197 | 0.0016-0.0035 | 0.0012-0.0023 | 0.0008-0.0023 |
| | 4.2 | BNK | NA | NA | NA | NA | NA | NA | NA |

NA: Not Available.

22205_AUDI_C0007

## TORQUE SPECIFICATIONS
All readings in ft. lbs.

| Year | Engine Size Liters | Engine ID/VIN | Cylinder Head Bolts | Main Bearing Bolts | Rod Bearing Bolts | Crankshaft Damper Bolts | Flywheel Bolts | Manifold Intake | Manifold Exhaust | Spark Plugs | Oil Pan Drain Plug |
|---|---|---|---|---|---|---|---|---|---|---|---|
| 2006 | 1.8 | AWP | ① | ② | ③ | ④ | ⑤ | 7 | 18 | 22 | 22 |
| | 1.8 | BEA | ⑥ | ② | ③ | ⑦ | ⑧ | 7 | 18 | 22 | 22 |
| | 2.0 | AXX | ⑥ | ② | ③ | NA | ⑧ | 9 | 18 | 22 | 22 |
| | 2.0 | BPY | ⑥ | ② | ③ | NA | ⑧ | 9 | 18 | 22 | 22 |
| | 2.0 | BPG | ⑥ | ② | ③ | NA | ⑧ | 9 | 18 | 22 | 22 |
| | 3.2 | BUB | ⑨ | ⑩ | ③ | NA | ⑧ | 9 | 18 | 22 | 22 |
| | 3.2 | BHE | ⑨ | ⑩ | ③ | ⑪ | ⑤ | 10 | 17 | 22 | 22 |
| | 3.2 | BKH | ⑥ | ⑩ | ③ | ⑪ | — | 10 | 17 | 22 | 22 |
| | 4.2 | BNK | ⑫ | NA | NA | 30 | ⑤ | 7 | 18 | 22 | 36 |
| 2007 | 2.0 | BPY | ⑥ | ② | ③ | NA | ⑧ | 9 | 18 | 22 | 22 |
| | 2.0 | BPG | ⑥ | ② | ③ | NA | ⑧ | 9 | 18 | 22 | 22 |
| | 3.2 | BUB | ⑨ | ⑩ | ③ | NA | ⑧ | 9 | 18 | 22 | 22 |
| | 3.2 | BHE | ⑨ | ⑩ | ③ | ⑪ | ⑤ | 10 | 17 | 22 | 22 |
| | 3.2 | BKH | ⑥ | ⑩ | ③ | ⑪ | — | 10 | 17 | 22 | 22 |
| | 4.2 | BNK | ⑫ | NA | NA | 30 | ⑤ | 7 | 18 | 22 | 36 |
| 2008 | 1.8 | BEA | ⑥ | ② | ③ | ⑦ | ⑧ | 7 | 18 | 22 | 22 |
| | 2.0 | BPY | ⑥ | ② | ③ | NA | ⑧ | 9 | 18 | 22 | 22 |
| | 2.0 | BPG | ⑥ | ② | ③ | NA | ⑧ | 9 | 18 | 22 | 22 |
| | 3.2 | BUB | ⑨ | ⑩ | ③ | NA | ⑧ | 9 | 18 | 22 | 22 |
| | 3.2 | BKH | ⑥ | ⑩ | ③ | ⑪ | — | 10 | 17 | 22 | 22 |
| | 4.2 | BNK | ⑫ | NA | NA | 30 | ⑤ | 7 | 18 | 22 | 36 |

NA: Not Available

① Step 1: 30 ft. lbs.
　Step 2: Plus 90 degrees

② Step 1: 48 ft. lbs.
　Step 2: Plus 90 degrees

③ Step 1: 22 ft. lbs.
　Step 2: 90 degrees

④ Center Bolt, installed with oil:
　Step 1: 148 ft. lbs.
　Step 2: 180 degrees
　Damper Bolts: 15 ft. lbs.

⑤ Step 1: 44 ft. lbs.
　Step 2: Plus 90 degrees

⑥ Step 1:29 ft. lbs.
　Step 2: Plus 90 degrees
　Step 3: Plus 90 degrees

⑦ Step 1:74 ft. lbs.
　Step 2: Plus 90 degrees

⑧ 22.5 mm Bolt: 44 ft. lbs. Plus 90 degrees
　35 & 43 mm Bolts: 44 ft.lbs. Plus 180 degrees

⑨ Step 1: 22 ft. lbs.
　Step 2: 37 ft. lbs.
　Step 3: Plus 90 degrees
　Step 4: Plus 90 degrees

⑩ Step 1: 22 ft. lbs.
　Step 2: Plus 90 degrees

⑪ Step 1: 74 ft. lbs.
　Step 2: Plus 90 degrees

⑫ Step 1: 22 ft. lbs.
　Step 2: 44 ft. lbs.
　Step 3: Plus 90 degrees
　Step 4: Plus 90 degrees

22205_AUDI_C0008

## WHEEL ALIGNMENT

| Year | Model | | Caster Range (+/-Deg.) | Caster Preferred Setting (Deg.) | Camber Range (+/-Deg.) | Camber Preferred Setting (Deg.) | Toe Setting (in.) |
|---|---|---|---|---|---|---|---|
| **2006** | A3 | F | — | — | 0.30 | 0.30 | 0.16 +/- 0.03 |
| | Standard Suspension | R | — | — | 0.20 | 0.20 | 0.50 +/- 0.16 |
| | A3 | F | — | — | 0.41 | -0.78 | 0.16 +/- 0.03 |
| | Sport Suspension | R | — | — | 0.20 | -1.17 | 0.50 +/- 0.16 |
| | A3 | F | — | — | 0.42 | -0.35 | 0.16 +/- 0.03 |
| | Heavy Duty Suspension | R | — | — | 0.42 | -0.35 | 0.50 +/- 0.16 |
| | A4 | F | — | — | 0.42 | -0.50 | 0.16 +/- 0.03 |
| | Standard Suspension | R | — | — | 0.50 | -1.17 | 0.50 +/- 0.16 |
| | A4 | F | — | — | 0.42 | -0.78 | 0.16 +/- 0.03 |
| | Sport Suspension | R | — | — | 0.50 | -1.17 | 0.50 +/- 0.16 |
| | A4 | F | — | — | 0.42 | -0.35 | 0.16 +/- 0.03 |
| | Heavy Duty Suspension | R | — | — | 0.42 | -0.35 | 0.50 +/- 0.16 |
| | A6 ① | F | — | — | 0.42 | -0.83 | 0.18 +/- 0.03 |
| | FWD | R | — | — | 0.33 | -1.50 | 0.13 +/- 0.08 |
| | A6 ② | F | — | — | 0.42 | -1.09 | 0.18 +/- 0.03 |
| | FWD | R | — | — | 0.33 | -1.50 | 0.13 +/- 0.08 |
| | A6 ③ | F | — | — | 0.42 | -0.58 | 0.18 +/- 0.03 |
| | FWD | R | — | — | 0.33 | -1.50 | 0.18 +/- 0.03 |
| | A6 ① | F | — | — | 0.42 | -0.83 | 0.16 +/- 0.03 |
| | AWD | R | — | — | 0.50 | -0.67 | 0.18 +/- 0.03 |
| | A6 ② | F | — | — | 0.42 | -0.83 | 0.16 +/- 0.03 |
| | AWD | R | — | — | 0.50 | -0.67 | 018 +/- 0.03 |
| | A6 ③ | F | — | — | 0.42 | -0.83 | 0.16 +/- 0.03 |
| | AWD | R | — | — | 0.50 | -0.67 | 0.18 +/- 0.03 |
| | A6 ① | F | — | — | 0.42 | -1.09 | 0.18 +/- 0.03 |
| | Quattro | R | — | — | 0.50 | -0.67 | 0.33 +/- 0.21 |
| | A6 ② | F | — | — | 0.42 | -1.08 | 0.18 +/- 0.03 |
| | Quattro | R | — | — | 0.50 | -1.00 | 0.33 +/- 0.21 |
| | A6 ③ | F | — | — | 0.42 | -0.83 | 0.18 +/- 0.03 |
| | Quattro | R | — | — | 0.50 | -0.67 | 0.23 +/- 0.21 |
| | TT | F | — | — | ④ | — | 0.08 +/- 0.08 |
| | FWD Std. Suspension | R | — | — | 0.33 | -0.20 | 0.20 +/- 0.01 |
| | TT | F | — | — | 0.50 | -0.75 | 0.08 +/- 0.08 |
| | FWD Sport Suspension | R | — | — | 0.20 | -2.00 | 0.20 +/- 0.05 |
| | TT | F | — | — | ⑤ | — | 0.08 +/- 0.08 |
| | AWD | R | — | — | ⑥⑦ | ⑥⑦ | ⑧ |
| **2007** | A3 | F | — | — | 0.30 | 0.30 | 0.16 +/- 0.03 |
| | Standard Suspension | R | — | — | 0.20 | 0.20 | 0.50 +/- 0.16 |
| | A3 | F | — | — | 0.41 | -0.78 | 0.16 +/- 0.03 |
| | Sport Suspension | R | — | — | 0.20 | -1.17 | 0.50 +/- 0.16 |
| | A3 | F | — | — | 0.42 | -0.35 | 0.16 +/- 0.03 |
| | Heavy Duty Suspension | R | — | — | 0.42 | -0.35 | 0.50 +/- 0.16 |
| | A4 | F | — | — | 0.42 | -0.50 | 0.16 +/- 0.03 |
| | Standard Suspension | R | — | — | 0.50 | -1.17 | 0.50 +/- 0.16 |
| | A4 | F | — | — | 0.42 | -0.78 | 0.16 +/- 0.03 |
| | Sport Suspension | R | — | — | 0.50 | -1.17 | 0.50 +/- 0.16 |
| | A4 | F | — | — | 0.42 | -0.35 | 0.16 +/- 0.03 |
| | Heavy Duty Suspension | R | — | — | 0.42 | -0.35 | 0.50 +/- 0.16 |
| | A6 ① | F | — | — | 0.42 | -0.83 | 0.18 +/- 0.03 |
| | FWD | R | — | — | 0.33 | -1.50 | 0.13 +/- 0.08 |
| | A6 ② | F | — | — | 0.42 | -1.09 | 0.18 +/- 0.03 |
| | FWD | R | — | — | 0.33 | -1.50 | 0.13 +/- 0.08 |
| | A6 ③ | F | — | — | 0.42 | -0.58 | 0.18 +/- 0.03 |
| | FWD | R | — | — | 0.33 | -1.50 | 0.18 +/- 0.03 |

## WHEEL ALIGNMENT

| Year | Model | | Caster Range (+/-Deg.) | Caster Preferred Setting (Deg.) | Camber Range (+/-Deg.) | Camber Preferred Setting (Deg.) | Toe Setting (in.) |
|------|-------|---|---|---|---|---|---|
| 2007 Cont. | A6 ① | F | — | — | 0.42 | -0.83 | 0.16 +/- 0.03 |
| | AWD | R | — | — | 0.50 | -0.67 | 0.18 +/- 0.03 |
| | A6 ② | F | — | — | 0.42 | -0.83 | 0.16 +/- 0.03 |
| | AWD | R | — | — | 0.50 | -0.67 | 018 +/- 0.03 |
| | A6 ③ | F | — | — | 0.42 | -0.83 | 0.16 +/- 0.03 |
| | AWD | R | — | — | 0.50 | -0.67 | 0.18 +/- 0.03 |
| | A6 ① | F | — | — | 0.42 | -1.09 | 0.18 +/- 0.03 |
| | Quattro | R | — | — | 0.50 | -0.67 | 0.33 +/- 0.21 |
| | A6 ② | F | — | — | 0.42 | -1.08 | 0.18 +/- 0.03 |
| | Quattro | R | — | — | 0.50 | -1.00 | 0.33 +/- 0.21 |
| | A6 ③ | F | — | — | 0.42 | -0.83 | 0.18 +/- 0.03 |
| | Quattro | R | — | — | 0.50 | -0.67 | 0.23 +/- 0.21 |
| | TT | F | — | — | ④ | — | 0.08 +/- 0.08 |
| | FWD Std. Suspension | R | — | — | 0.33 | -0.20 | 0.20 +/- 0.01 |
| | TT | F | — | — | 0.50 | -0.75 | 0.08 +/- 0.08 |
| | FWD Sport Suspension | R | — | — | 0.20 | -2.00 | 0.20 +/- 0.05 |
| | TT | F | — | — | ⑤ | — | 0.08 +/- 0.08 |
| | AWD | R | — | — | ⑥⑦ | ⑥⑦ | ⑧ |
| 2008 | A3 | F | — | — | 0.30 | 0.30 | 0.16 +/- 0.03 |
| | Standard Suspension | R | — | — | 0.20 | 0.20 | 0.50 +/- 0.16 |
| | A3 | F | — | — | 0.41 | -0.78 | 0.16 +/- 0.03 |
| | Sport Suspension | R | — | — | 0.20 | -1.17 | 0.50 +/- 0.16 |
| | A3 | F | — | — | 0.42 | -0.35 | 0.16 +/- 0.03 |
| | Heavy Duty Suspension | R | — | — | 0.42 | -0.35 | 0.50 +/- 0.16 |
| | A4 | F | — | — | 0.42 | -0.50 | 0.16 +/- 0.03 |
| | Standard Suspension | R | — | — | 0.50 | -1.17 | 0.50 +/- 0.16 |
| | A4 | F | — | — | 0.42 | -0.78 | 0.16 +/- 0.03 |
| | Sport Suspension | R | — | — | 0.50 | -1.17 | 0.50 +/- 0.16 |
| | A4 | F | — | — | 0.42 | -0.35 | 0.16 +/- 0.03 |
| | Heavy Duty Suspension | R | — | — | 0.42 | -0.35 | 0.50 +/- 0.16 |
| | A6 ① | F | — | — | 0.42 | -0.83 | 0.18 +/- 0.03 |
| | FWD | R | — | — | 0.33 | -1.50 | 0.13 +/- 0.08 |
| | A6 ② | F | — | — | 0.42 | -1.09 | 0.18 +/- 0.03 |
| | FWD | R | — | — | 0.33 | -1.50 | 0.13 +/- 0.08 |
| | A6 ③ | F | — | — | 0.42 | -0.58 | 0.18 +/- 0.03 |
| | FWD | R | — | — | 0.33 | -1.50 | 0.18 +/- 0.03 |
| | A6 ① | F | — | — | 0.42 | -0.83 | 0.16 +/- 0.03 |
| | AWD | R | — | — | 0.50 | -0.67 | 0.18 +/- 0.03 |
| | A6 ② | F | — | — | 0.42 | -0.83 | 0.16 +/- 0.03 |
| | AWD | R | — | — | 0.50 | -0.67 | 018 +/- 0.03 |
| | A6 ③ | F | — | — | 0.42 | -0.83 | 0.16 +/- 0.03 |
| | AWD | R | — | — | 0.50 | -0.67 | 0.18 +/- 0.03 |
| | A6 ① | F | — | — | 0.42 | -1.09 | 0.18 +/- 0.03 |
| | Quattro | R | — | — | 0.50 | -0.67 | 0.33 +/- 0.21 |
| | A6 ② | F | — | — | 0.42 | -1.08 | 0.18 +/- 0.03 |
| | Quattro | R | — | — | 0.50 | -1.00 | 0.33 +/- 0.21 |
| | A6 ③ | F | — | — | 0.42 | -0.83 | 0.18 +/- 0.03 |
| | Quattro | R | — | — | 0.50 | -0.67 | 0.23 +/- 0.21 |
| | TT | F | — | — | ④ | — | 0.08 +/- 0.08 |
| | FWD Std. Suspension | R | — | — | 0.33 | -0.20 | 0.20 +/- 0.01 |
| | TT | F | — | — | 0.50 | -0.75 | 0.08 +/- 0.08 |
| | FWD Sport Suspension | R | — | — | 0.20 | -2.00 | 0.20 +/- 0.05 |
| | TT | F | — | — | ⑤ | — | 0.08 +/- 0.08 |
| | AWD | R | — | — | ⑥⑦ | ⑥⑦ | ⑧ |

## WHEEL ALIGNMENT
### Footnotes

① With standard suspension

② With sport suspension

③ With heavy duty suspension

④ Rear camber at standing vehicle height of 13.8", measured from centerline of rear wheel to bottom of rear wheel opening.

⑤ Range is -0.75 to +0.50 degrees for FWD and AWD models.

⑥ With standard suspension, rear camber at standing vehicle height of 13.9", measured from centerline of rear wheel to bottom of rear wheel opening.

⑦ With sport suspension, rear camber at standing vehicle height of 13.1", measured from centerline of rear wheel to bottom of rear wheel opening.

⑧ Preferred setting is 0.075 in., with a range of +0.075 in. to -0.05 in.

22205_AUDI_C0011

## TIRE, WHEEL AND BALL JOINT SPECIFICATIONS

| Year | Model | OEM Tires | | Tire Pressures (psi) | | Wheel Size | Ball Joint Inspection | Lug Nut (ft. lbs.) |
| | | Standard | Optional | Front | Rear | | | |
| --- | --- | --- | --- | --- | --- | --- | --- | --- |
| 2006 | A8 | 225/60HR16 | 225/45R18 245/45R18 | 34 | 34 | Std: 7-J Opt: 8-J | NS | 120 |
| | A6 | 205/55R16 | 235/45R17 | 34 | 34 | 7-J | NS | 120 |
| | A4 | 205/65HR15 | 215/55HR16 | 34 | 34 | 7-J | NS | 120 |
| | A3 | 205/65HR15 | 215/55HR16 | 34 | 34 | 7-J | NS | 120 |
| | TT | 205/55WR16 | 225/45YR17 | 34 | 34 | 7-J | NS | 120 |
| 2007 | A8 | 225/60HR16 | 225/45R18 245/45R18 | 34 | 34 | Std: 7-J Opt: 8-J | NS | 120 |
| | A6 | 205/55R16 | 235/45R17 | 34 | 34 | 7-J | NS | 120 |
| | A4 | 205/65HR15 | 215/55HR16 | 34 | 34 | 7-J | NS | 120 |
| | A3 | 205/65HR15 | 215/55HR16 | 34 | 34 | 7-J | NS | 120 |
| | TT | 205/55WR16 | 225/45YR17 | 34 | 34 | 7-J | NS | 120 |
| 2008 | A8 | 225/60HR16 | 225/45R18 245/45R18 | 34 | 34 | Std: 7-J Opt: 8-J | NS | 120 |
| | A6 | 205/55R16 | 235/45R17 | 34 | 34 | 7-J | NS | 120 |
| | A4 | 205/65HR15 | 215/55HR16 | 34 | 34 | 7-J | NS | 120 |
| | A3 | 205/65HR15 | 215/55HR16 | 34 | 34 | 7-J | NS | 120 |
| | TT | 205/55WR16 | 225/45YR17 | 34 | 34 | 7-J | NS | 120 |

OEM: Original Equipment Manufacturer

PSI: Pounds Per Square Inch

STD: Standard

OPT: Optional

NS: Not Specified by manufacturer

22205_AUDI_C0012

## BRAKE SPECIFICATIONS
All measurements in inches unless noted

| Year | Model | | Brake Disc | | | Minimum Lining Thickness | | Brake Caliper | |
| | | | Original Thickness | Minimum Thickness | Maximum Runout | Front | Rear | Bracket Bolts (ft. lbs.) | Mounting Bolts (Nm.) |
|------|-------|---|----|----|----|----|----|----|----|
| 2006 | A4 Sedan | F | ① | ② | 0.002 | ③ | 0.275 | 144 | 18 |
| | and Avant | R | 0.394 | 0.315 | 0.002 | — | 0.275 | 103 | 25 |
| | A3 Sedan | F | ④ | ⑤ | 0.002 | ⑥ | 0.023 | 144 | 22 |
| | and Avant | R | ⑦ | ⑧ | 0.002 | — | 0.023 | 66 | 25 |
| | A6 Sedan | F | 0.984 | 0.906 | 0.002 | 0.078 | — | 144 | 18 |
| | and Avant | R | 0.394 | 0.315 | 0.002 | — | 0.079 | 103 | 25 |
| | A8 | F | 0.984 | 0.905 | 0.002 | 0.078 | — | — | 18 |
| | | R | 0.390 | 0.315 | 0.002 | — | 0.079 | ⑨ | 26 |
| | TT | F | ⑩ | ⑪ | 0.002 | 0.551 | — | 92 | 20 |
| | | R | ⑫ | ⑬ | 0.002 | — | 0.079 | 47 | 25 |
| 2007 | A4 Sedan | F | ① | ② | 0.002 | ③ | 0.275 | 144 | 18 |
| | and Avant | R | 0.394 | 0.315 | 0.002 | — | 0.275 | 103 | 25 |
| | A3 Sedan | F | ④ | ⑤ | 0.002 | ⑥ | 0.023 | 144 | 22 |
| | and Avant | R | ⑦ | ⑧ | 0.002 | — | 0.023 | 66 | 25 |
| | A6 Sedan | F | 0.984 | 0.906 | 0.002 | 0.078 | — | 144 | 18 |
| | and Avant | R | 0.394 | 0.315 | 0.002 | — | 0.079 | 103 | 25 |
| | A8 | F | 0.984 | 0.905 | 0.002 | 0.078 | — | — | 18 |
| | | R | 0.390 | 0.315 | 0.002 | — | 0.079 | ⑨ | 26 |
| | TT | F | ⑩ | ⑪ | 0.002 | 0.551 | — | 92 | 20 |
| | | R | ⑫ | ⑬ | 0.002 | — | 0.079 | 47 | 25 |
| 2008 | A4 Sedan | F | ① | ② | 0.002 | ③ | 0.275 | 144 | 18 |
| | and Avant | R | 0.394 | 0.315 | 0.002 | — | 0.275 | 103 | 25 |
| | A3 Sedan | F | ④ | ⑤ | 0.002 | ⑥ | 0.023 | 144 | 22 |
| | and Avant | R | ⑦ | ⑧ | 0.002 | — | 0.023 | 66 | 25 |
| | A6 Sedan | F | 0.984 | 0.906 | 0.002 | 0.078 | — | 144 | 18 |
| | and Avant | R | 0.394 | 0.315 | 0.002 | — | 0.079 | 103 | 25 |
| | A8 | F | 0.984 | 0.905 | 0.002 | 0.078 | — | — | 18 |
| | | R | 0.390 | 0.315 | 0.002 | — | 0.079 | ⑨ | 26 |
| | TT | F | ⑩ | ⑪ | 0.002 | 0.551 | — | 92 | 20 |
| | | R | ⑫ | ⑬ | 0.002 | — | 0.079 | 47 | 25 |

① Teves/Ate Calipers:
  Ventilated Disc: 0.984 inches
  Non-ventilated Disc: 0.590 inches
  Lucas Calipers:
  Ventilated Disc: 0.870 inches
  Non-ventilated Disc: 0.590 inches
  Double Piston Calipers:
  Ventilated Disc: 1.180 inches

② Teves/Ate Calipers:
  Venetilated Disc: 0.905 inches
  Non-ventilated Disc: 0.510
  Lucas Calipers:
  Venetilated Disc: 0.790 inches
  Non-ventilated Disc: 0.430 inches
  Double Piston Calipers:
  Venetilated Disc: 1.100 inches

③ Teves/Ate and Lucas Calipers: 0.080 inches
  Double Piston Calipers: 0.12 inches

④ C54 15 inch :
  FN 3 15 inch
  FN 3 16 inch
  FNR-G60 16 inch
  FNR-G60 17 inch
  Brembo 18 inch

⑤ C54: 0.790 inch
  FN 3 15in.: 0.980 inch
  FN 3 16 in: 0.980 inch
  FNR-G60 16 in: 1.180 inc
  FNR-G60 17 in: 1.180 inch
  Brembo 18 inch: 1.340 inch

⑥ C54: 0.063 inch
  FN 3 15in.: 0.066 inch
  FN 3 16 in: 0.066 inch
  FNR-G60 16 in: 0.066 inc
  FNR-G60 17 in: 0.066 inch
  Brembo 18 inch: not avaialable

⑦ C38 15 inch
  C38 16 inch
  C41 16 inch
  C43 16 inch
  C43 18 inch

⑧ C38 15 inch: 0.31 inch
  C38 16 inch: 0.39 inch
  C41 16 inch: 0.39 inch
  C43 16 inch: 0.79 inch
  C43 18 inch: 0.79 inch

⑨ Steel wheel carrier: 44 ft. lbs.
  Aluminum wheel carrier: 51 ft. lbs. Plus 90 deg.

⑩ Exc. P/N 1ZT: 0.984 inches
  P/N 1ZT: 1.26 inches

⑪ Exc. P/N 1ZT: 0.905 inches
  P/N 1ZT: 1.181 inches

⑫ Non-ventilated Disc: 0.354 inches
  Ventilated Disc: 0.866 inches

⑬ Non-ventilated Disc: 0.276 inches
  Ventilated Disc: 0.787 inches

## SCHEDULED MAINTENANCE INTERVALS
### Audi—TT, A3, A4, A6 and A8

| TO BE SERVICED | TYPE OF SERVICE | VEHICLE MILEAGE INTERVAL (x1000) | | | | | | | | | | | | |
|---|---|---|---|---|---|---|---|---|---|---|---|---|---|---|
| | | 5 | 10 | 20 | 30 | 40 | 50 | 60 | 70 | 80 | 90 | 100 | 105 | 120 |
| Engine oil & filter ① | R | ✓ | ✓ | ✓ | ✓ | ✓ | ✓ | ✓ | ✓ | ✓ | ✓ | ✓ | ✓ | ✓ |
| Service reminder reset | S/I | ✓ | ✓ | ✓ | ✓ | ✓ | ✓ | ✓ | ✓ | ✓ | ✓ | ✓ | ✓ | ✓ |
| Rotate wheels | S/I | ✓ | | | | | | | | | | | | |
| Fluid levels | S/I | ✓ | ✓ | ✓ | ✓ | ✓ | ✓ | ✓ | ✓ | ✓ | ✓ | ✓ | ✓ | ✓ |
| Auto shift lock | S/I | ✓ | ✓ | ✓ | ✓ | ✓ | ✓ | ✓ | ✓ | ✓ | ✓ | ✓ | ✓ | ✓ |
| Brake system | S/I | ✓ | ✓ | ✓ | ✓ | ✓ | ✓ | ✓ | ✓ | ✓ | ✓ | ✓ | ✓ | ✓ |
| Brake fluid ① | R | | | | | | | | | | | | | |
| M/T shift & clutch interlock | S/I | ✓ | ✓ | ✓ | ✓ | ✓ | ✓ | ✓ | ✓ | ✓ | ✓ | ✓ | ✓ | ✓ |
| Cooling system | S/I | | ✓ | ✓ | ✓ | ✓ | ✓ | ✓ | ✓ | ✓ | ✓ | ✓ | ✓ | ✓ |
| Exhaust system | S/I | | ✓ | ✓ | ✓ | ✓ | ✓ | ✓ | ✓ | ✓ | ✓ | ✓ | ✓ | ✓ |
| ODB check for codes | S/I | | ✓ | ✓ | ✓ | ✓ | ✓ | ✓ | ✓ | ✓ | ✓ | ✓ | ✓ | ✓ |
| Door hinges - lubricate | S/I | | ✓ | ✓ | ✓ | ✓ | ✓ | ✓ | ✓ | ✓ | ✓ | ✓ | ✓ | ✓ |
| Battery level | S/I | | ✓ | ✓ | ✓ | ✓ | ✓ | ✓ | ✓ | ✓ | ✓ | ✓ | ✓ | ✓ |
| Windshield washer fluid | S/I | | ✓ | ✓ | ✓ | ✓ | ✓ | ✓ | ✓ | ✓ | ✓ | ✓ | ✓ | ✓ |
| Tire condition & pressure | S/I | | ✓ | ✓ | ✓ | ✓ | ✓ | ✓ | ✓ | ✓ | ✓ | ✓ | ✓ | ✓ |
| Drive shaft boots | S/I | | ✓ | ✓ | ✓ | ✓ | ✓ | ✓ | ✓ | ✓ | ✓ | ✓ | ✓ | ✓ |
| Road test | S/I | | ✓ | ✓ | ✓ | ✓ | ✓ | ✓ | ✓ | ✓ | ✓ | ✓ | ✓ | ✓ |
| Lighting | S/I | | ✓ | ✓ | ✓ | ✓ | ✓ | ✓ | ✓ | ✓ | ✓ | ✓ | ✓ | ✓ |
| Engine for leaks | S/I | | | ✓ | | ✓ | | ✓ | | ✓ | | ✓ | | ✓ |
| Front axle dust seals on ball joints & tie rod ends | S/I | | | ✓ | | ✓ | | ✓ | | ✓ | | ✓ | | ✓ |
| Transmission for leaks | S/I | | | ✓ | | ✓ | | ✓ | | ✓ | | ✓ | | ✓ |
| Multronic trans fluid | R | | | | | ✓ | | | | ✓ | | | | ✓ |
| MT final drive fluid | S/I | | | ✓ | | ✓ | | ✓ | | ✓ | | ✓ | | ✓ |
| AT final drive fluid | S/I | | | | | ✓ | | | | ✓ | | | | ✓ |
| Dust/pollen filter | R | | | ✓ | | ✓ | | ✓ | | ✓ | | ✓ | | ✓ |
| Sliding roof rails | S/I | | | ✓ | | ✓ | | ✓ | | ✓ | | ✓ | | ✓ |
| PS fluid | S/I | | | | | ✓ | | | | ✓ | | | | ✓ |
| Air cleaner element | R | | | | | ✓ | | | | ✓ | | | | ✓ |
| Spark plugs | R | | | | | ✓ | | | | ✓ | | | | ✓ |
| Serpentine belt | R | | | | | | | | | ✓ | | | | |
| Timing belt 1.8L Turbo ② | R | | | | | | | | | | | | | ✓ |
| Timing belt 2.0L, 3.2L & 4.2L | R | | | | | | | | | | | | ✓ | |

R: Replace     S/I: Service or Inspect

① Replace fluid every two years, regardless of mileage.

② 1.8L Turbo, exc. TT models; on TT models with 1.8L Turbo & 3.2L V6, replace every 105K miles.

22205_AUDI_C0014

## PRECAUTIONS

Before servicing any vehicle, please be sure to read all of the following precautions, which deal with personal safety, prevention of component damage, and important points to take into consideration when servicing a motor vehicle:

• Never open, service or drain the radiator or cooling system when the engine is hot; serious burns can occur from the steam and hot coolant.

• Observe all applicable safety precautions when working around fuel. Whenever servicing the fuel system, always work in a well-ventilated area. Do not allow fuel spray or vapors to come in contact with a spark, open flame, or excessive heat (a hot drop light, for example). Keep a dry chemical fire extinguisher near the work area. Always keep fuel in a container specifically designed for fuel storage; also, always properly seal fuel containers to avoid the possibility of fire or explosion. Refer to the additional fuel system precautions later in this section.

• Fuel injection systems often remain pressurized, even after the engine has been turned **OFF** . The fuel system pressure must be relieved before disconnecting any fuel lines. Failure to do so may result in fire and/or personal injury.

• Brake fluid often contains polyglycol ethers and polyglycols. Avoid contact with the eyes and wash your hands thoroughly after handling brake fluid. If you do get brake fluid in your eyes, flush your eyes with clean, running water for 15 minutes. If eye irritation persists, or if you have taken

brake fluid internally, IMMEDIATELY seek medical assistance.

• The EPA warns that prolonged contact with used engine oil may cause a number of skin disorders, including cancer. You should make every effort to minimize your exposure to used engine oil. Protective gloves should be worn when changing oil. Wash your hands and any other exposed skin areas as soon as possible after exposure to used engine oil. Soap and water, or waterless hand cleaner should be used.

• All new vehicles are now equipped with an air bag system, often referred to as a Supplemental Restraint System (SRS) or Supplemental Inflatable Restraint (SIR) system. The system must be disabled before performing service on or around system components, steering column, instrument panel components, wiring and sensors. Failure to follow safety and disabling procedures could result in accidental air bag deployment, possible personal injury and unnecessary system repairs.

• Always wear safety goggles when working with, or around, the air bag system. When carrying a non-deployed air bag, be sure the bag and trim cover are pointed away from your body. When placing a non-deployed air bag on a work surface, always face the bag and trim cover upward, away from the surface. This will reduce the motion of the module if it is accidentally deployed. Refer to the additional air bag system precautions later in this section.

• Clean, high quality brake fluid from a sealed container is essential to the safe and

proper operation of the brake system. You should always buy the correct type of brake fluid for your vehicle. If the brake fluid becomes contaminated, completely flush the system with new fluid. Never reuse any brake fluid. Any brake fluid that is removed from the system should be discarded. Also, do not allow any brake fluid to come in contact with a painted surface; it will damage the paint.

• Never operate the engine without the proper amount and type of engine oil; doing so WILL result in severe engine damage.

• Timing belt maintenance is extremely important. Many models utilize an interference-type, non-freewheeling engine. If the timing belt breaks, the valves in the cylinder head may strike the pistons, causing potentially serious (also time-consuming and expensive) engine damage. Refer to the maintenance interval charts for the recommended replacement interval for the timing belt, and to the timing belt section for belt replacement and inspection.

• Disconnecting the negative battery cable on some vehicles may interfere with the functions of the on-board computer system(s) and may require the computer to undergo a relearning process once the negative battery cable is reconnected.

• When servicing drum brakes, only disassemble and assemble one side at a time, leaving the remaining side intact for reference.

• Only an MVAC-trained, EPA-certified automotive technician should service the air conditioning system or its components.

## BRAKES

### GENERAL INFORMATION

*PRECAUTIONS*

• Certain components within the ABS system are not intended to be serviced or repaired individually.

• Do not use rubber hoses or other parts not specifically specified for and ABS system. When using repair kits, replace all parts included in the kit. Partial or incorrect repair may lead to functional problems and require the replacement of components.

• Lubricate rubber parts with clean,

fresh brake fluid to ease assembly. Do not use shop air to clean parts; damage to rubber components may result.

• Use only DOT 3 brake fluid from an unopened container.

• If any hydraulic component or line is removed or replaced, it may be necessary to bleed the entire system.

• A clean repair area is essential. Always clean the reservoir and cap thoroughly before removing the cap. The slightest amount of dirt in the fluid may plug an orifice and impair the system function. Perform repairs after components have been thor-

## ANTI-LOCK BRAKE SYSTEM (ABS)

oughly cleaned; use only denatured alcohol to clean components. Do not allow ABS components to come into contact with any substance containing mineral oil; this includes used shop rags.

• The Anti-Lock control unit is a microprocessor similar to other computer units in the vehicle. Ensure that the ignition switch is **OFF** before removing or installing controller harnesses. Avoid static electricity discharge at or near the controller.

• If any arc welding is to be done on the vehicle, the control unit should be unplugged before welding operations begin.

## BRAKES · BLEEDING THE BRAKE SYSTEM

### BLEEDING PROCEDURE

*BLEEDING PROCEDURE*

#### ❋❋ CAUTION

Brake fluid will damage painted surfaces. Be careful not to spill any on painted surfaces. If it is spilled, wipe it off immediately.

➡ Keep the fluid level in the reserve tank at 3/4 full or more during the air bleeding.

➡ Begin air bleeding with the brake caliper that is furthest from the master cylinder.

1. Before servicing the vehicle, refer to the Precautions Section.
2. Remove the bleeder cap on the brake caliper, and attach a vinyl tube to the bleeder screw.
3. Place the other end of the vinyl tube in a clear container and fill the container with fluid during air bleeding.
4. Working with two people, one should pump the brake pedal several times and depress and hold the pedal down.
5. While the brake pedal is depressed, the other should loosen the bleeder screw using the SST, drain out any fluid containing air bubbles, and tighten the bleeder screw.
6. Repeat until no air bubbles are seen.
7. Perform air bleeding as described in the above procedures for all brake calipers.

## BRAKES · FRONT DISC BRAKES

#### ❋❋ CAUTION

Dust and dirt accumulating on brake parts during normal use may contain asbestos fibers from production or aftermarket brake linings. Breathing excessive concentrations of asbestos fibers can cause serious bodily harm. Exercise care when servicing brake parts. Do not sand or grind brake lining unless equipment used is designed to contain the dust residue.

Do not clean brake parts with compressed air or by dry brushing. Cleaning should be done by dampening the brake components with a fine mist of water, then wiping the brake components clean with a dampened cloth. Dispose of cloth and all residue containing asbestos fibers in an impermeable container with the appropriate label. Follow practices prescribed by the Occupational Safety and Health Administration (OSHA) and the Environmental Protection Agency (EPA) for the handling, processing, and disposing of dust or debris that may contain asbestos fibers.

### BRAKE CALIPER

*REMOVAL & INSTALLATION*

*See Figures 1 through 6.*

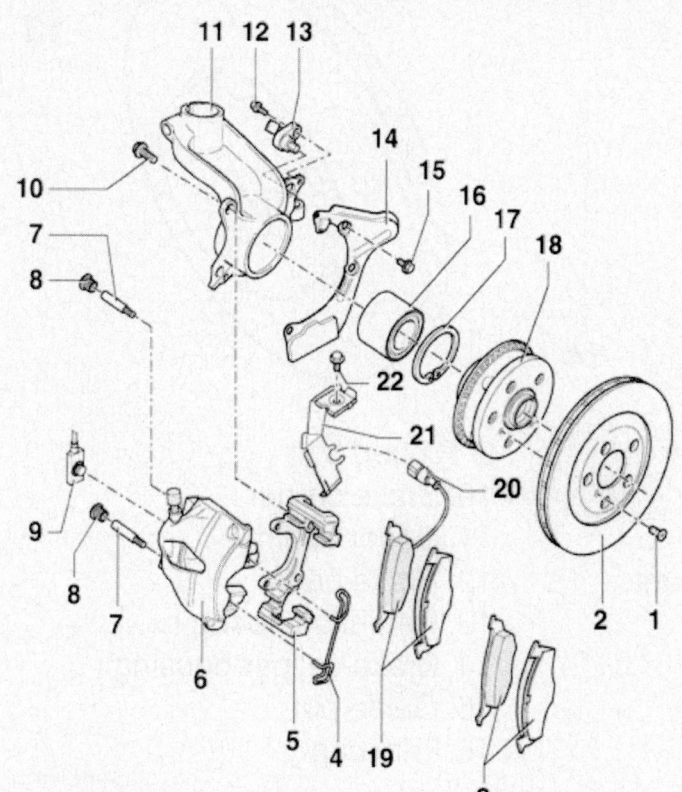

1. Philips screw
2. Brake disc
3. Brake pads
4. Retaining spring
5. Brake carrier
6. Brake caliper
7. Guide pin
8. Cap
9. Brake hose
10. Ribbed bolt
11. Wheel bearing housing
12. Hex head bolt
13. Speed sensor
14. Backing plate
15. Bolt
16. Wheel bearing
17. Circlip
18. Wheel hub and rotor
19. Brake pads
20. Wiring connector
21. Bracket
22. Self tapping screw

2348-TTTT-G25

Fig. 1 Exploded view of front brake components—TT models

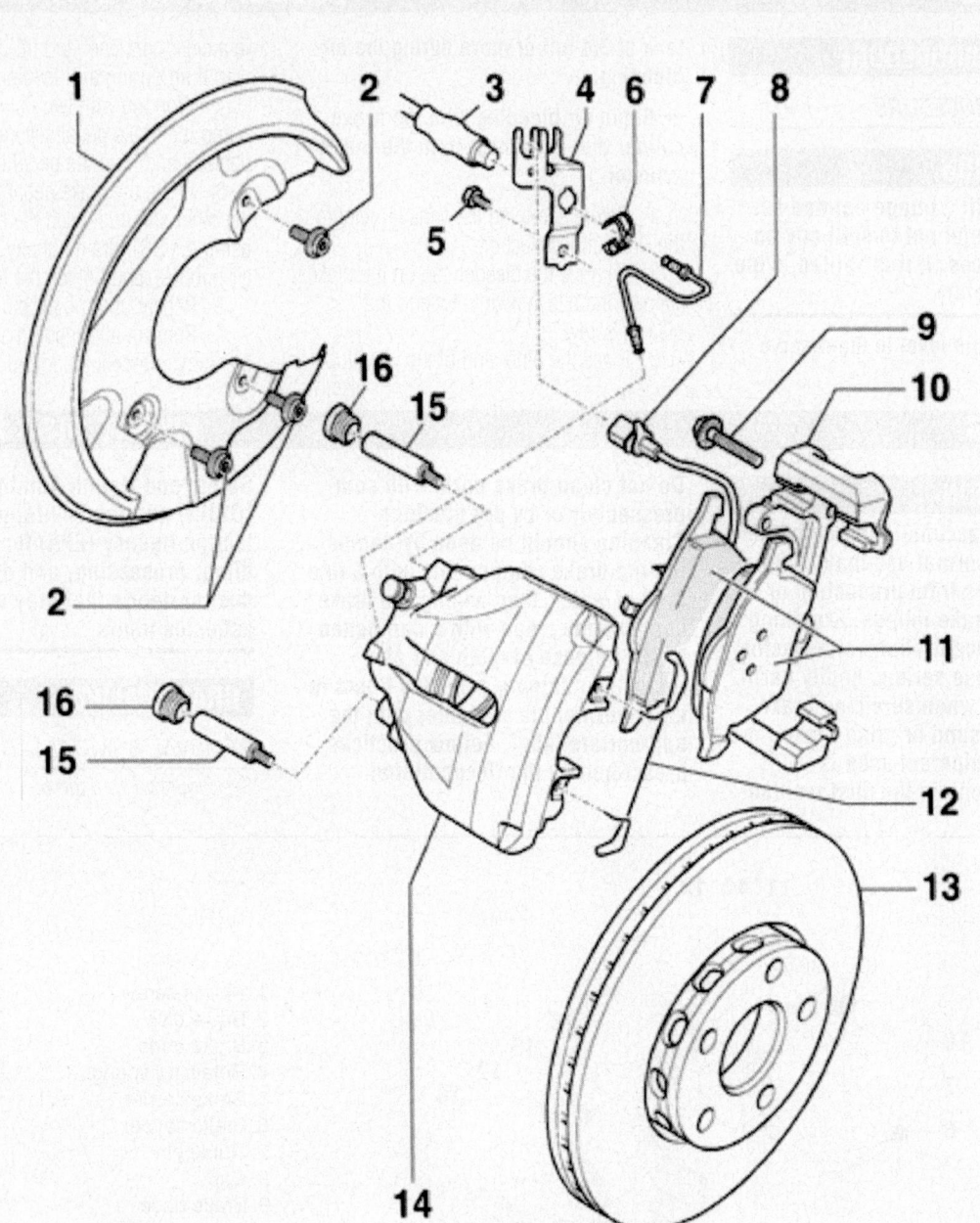

1. Brake disc cover
2. Flange bolt
3. Brake hose
4. Brake hose mount
5. Hex bolt
6. Spring clip
7. Brake line
8. Connector
9. Ribbed bolt
10. Brake carrier
11. Upper air line
12. Brake pads
13. Retention spring
14. Brake caliper housing
15. Guide pin
16. Trim cap

2348-A4A4-G23

**Fig. 2 Exploded view of FN3 type front brake components—A4 and A6 shown**

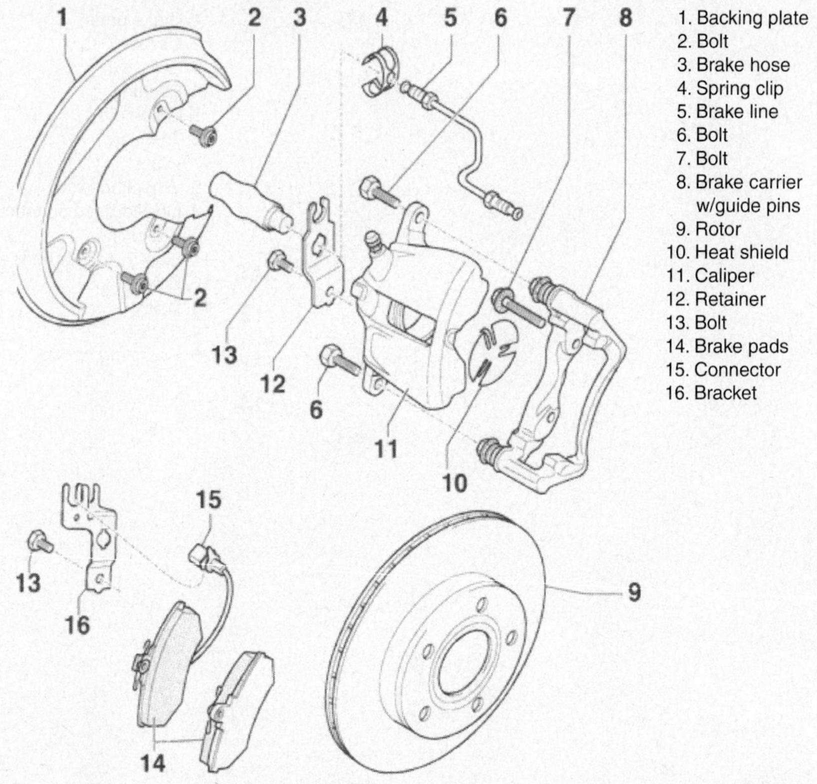

1. Backing plate
2. Bolt
3. Brake hose
4. Spring clip
5. Brake line
6. Bolt
7. Bolt
8. Brake carrier
   w/guide pins
9. Rotor
10. Heat shield
11. Caliper
12. Retainer
13. Bolt
14. Brake pads
15. Connector
16. Bracket

67200-A4A4-G04

**Fig. 3 Exploded view of C54 type front brake components—A4 models**

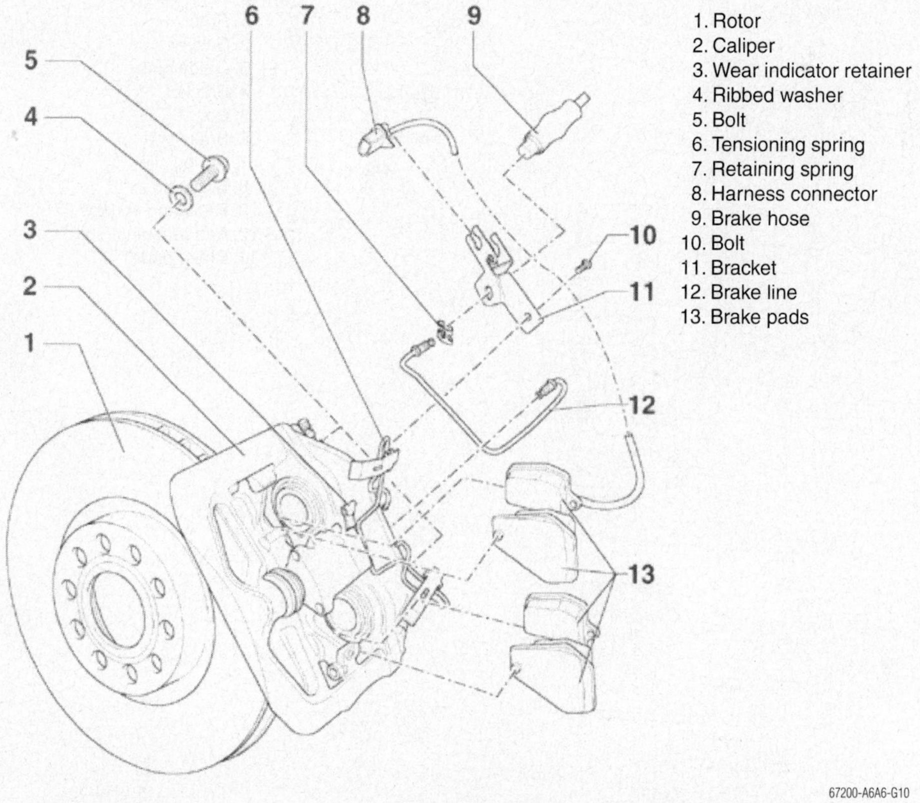

1. Rotor
2. Caliper
3. Wear indicator retainer
4. Ribbed washer
5. Bolt
6. Tensioning spring
7. Retaining spring
8. Harness connector
9. Brake hose
10. Bolt
11. Bracket
12. Brake line
13. Brake pads

67200-A6A6-G10

**Fig. 4 Exploded view of HP-2 type (aluminum caliper) front brake components—A6 shown; others similar**

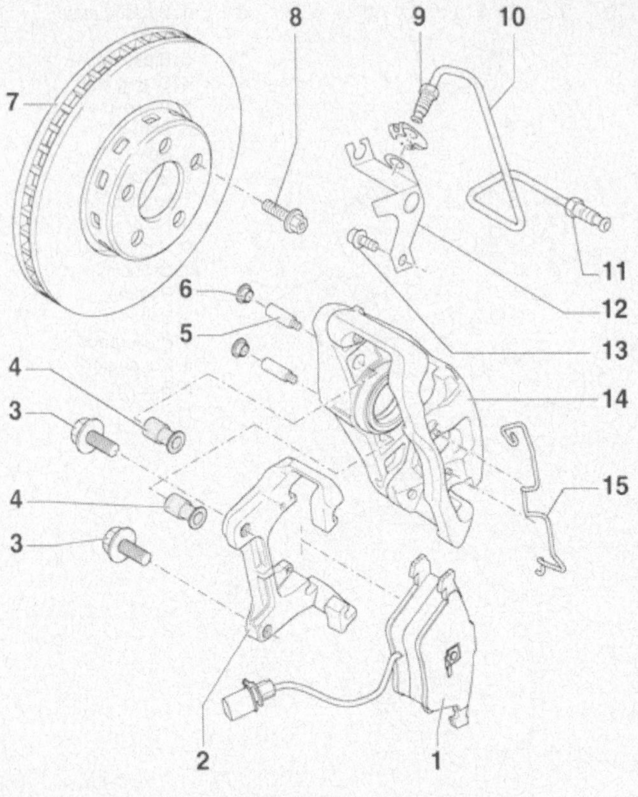

1. Brake pads
2. Carrier
3. Bolt
4. Bushing
5. Guide pin
6. Trim cap
7. Rotor
8. Wheel bolts
9. Line-to-hose connection
10. Brake line
11. Line-to-caliper connection
12. Bracket
13. Bolt
14. Caliper
15. Retaining spring

67200-A6A6-G11

**Fig. 5 Exploded view of FNR-60 type front brake components—A4 and A6 shown**

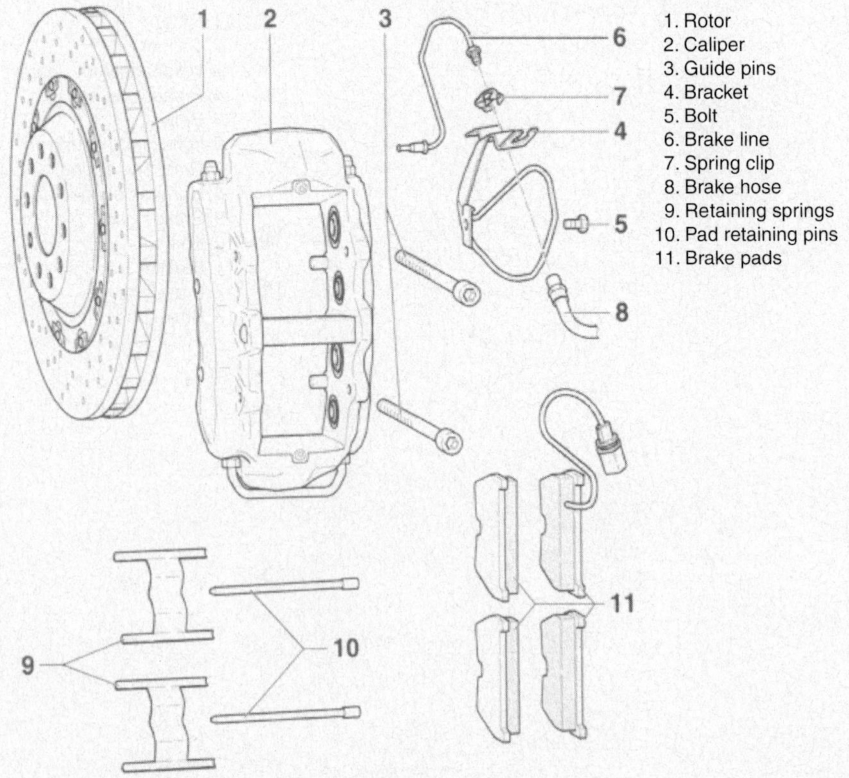

1. Rotor
2. Caliper
3. Guide pins
4. Bracket
5. Bolt
6. Brake line
7. Spring clip
8. Brake hose
9. Retaining springs
10. Pad retaining pins
11. Brake pads

67200-A6A6-G12

**Fig. 6 Exploded view of Brembo 18" type front brake components—A6 shown; others similar**

➡ **Procedures may vary slightly, depending on vehicle or brake type. Refer to appropriate illustration for additional assistance.**

1. Before servicing the vehicle, refer to the precautions section.

2. Remove the wheels.

3. Loosen the hydraulic line at the caliper, then remove the caliper from the carrier. With guide pin calipers, be sure to hold the pin with a back-up wrench when removing the caliper bolts.

4. Remove the caliper from the hydraulic line.

5. The carrier can be removed by removing the 2 bolts.

### To install:

6. If removed, install the carrier. Torque the carrier bolts to 92 ft. lbs. (125 Nm).

7. Thread the caliper onto the hydraulic line and hand-tighten it. Fit the caliper into place on the carrier.

8. Torque the caliper bolts (guide pins) as follows:

- On TT models: 21 ft. lbs. (28 Nm)
- On A4 and A6 models with FN3 type brakes: (25 Nm)
- On All A3 models, A4 and A6 models with FNR-G60 type brakes: (30 Nm)
- On A6 models with HP2 (aluminum caliper) type brakes: (190 Nm)
- On A6 models with 18" Brembo type brakes: (110 Nm)

9. Torque the brake line connection to (15 Nm).

10. Tighten the hydraulic line and bleed the brakes.

### DISC BRAKE PADS

#### REMOVAL & INSTALLATION

1. Before servicing the vehicle, refer to the precautions section.

2. Remove the front wheels.

3. Hold the lower guide pin with an open wrench and remove the bolt securing the caliper to the guide pin.

4. Pivot the caliper up on the upper guide pin and slide the pads straight out to remove them.

### To install:

5. Compress the caliper piston into the bore.

6. Fit the new pads into the carrier and pivot the caliper into place.

7. The original bolts are micro-encapsulated with a thread locking compound. Install a new bolt or clean the old bolt and apply a thread-locking compound.

8. When tightening the caliper bolts (guide pins) torque them as follows:

- On TT models: 21 ft. lbs. (28 Nm)
- On all A3 models, A4 and A6 models with FN3 type brakes: (25 Nm)
- On A4 and A6 models with FNR-G60 type brakes: (30 Nm)
- On A6 models with HP2 (aluminum caliper) type brakes: (190 Nm)
- On A6 models with 18" Brembo type brakes: (110 Nm)

9. Install the wheels.

## BRAKES

## REAR DISC BRAKES

### ✳ CAUTION

**Dust and dirt accumulating on brake parts during normal use may contain asbestos fibers from production or aftermarket brake linings. Breathing excessive concentrations of asbestos fibers can cause serious bodily harm. Exercise care when servicing brake parts. Do not sand or grind brake lining unless equipment used is designed to contain the dust residue. Do not clean brake parts with compressed air or by dry brushing. Cleaning should be done by dampening the brake components with a fine mist of water, then wiping the brake components clean with a dampened cloth. Dispose of cloth and all residue containing asbestos fibers in an impermeable container with the appropriate label. Follow practices prescribed by the Occupational Safety and Health Administration (OSHA) and the Environmental Protection Agency (EPA) for the handling, processing, and disposing of dust or debris that may contain asbestos fibers.**

### BRAKE CALIPER

#### REMOVAL & INSTALLATION

*See Figures 7 through 9.*

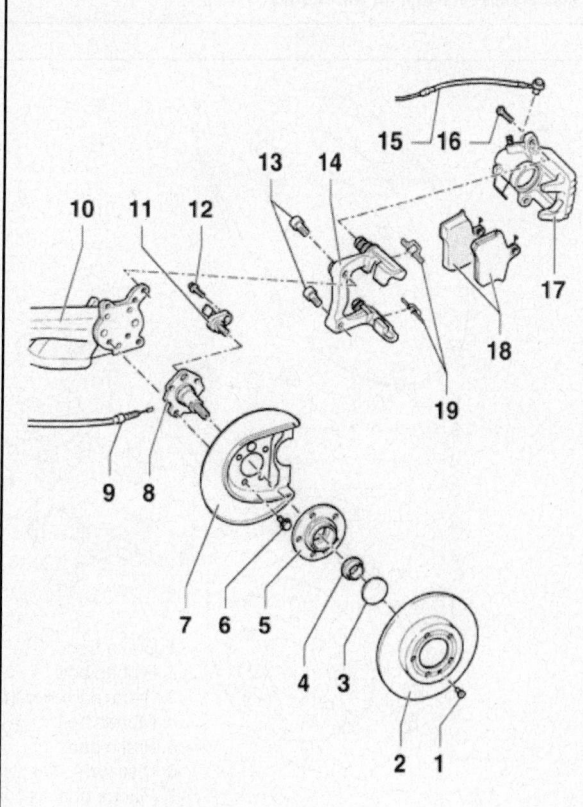

1. Philips screw
2. Brake disc
3. Grease cap
4. 12 point nut
5. Wheel hub
6. Bolt
7. Backing plate
8. Stub axle
9. Parking brake cable
10. Axle beam
11. Speed sensor
12. Hex head bolt
13. Hex head bolt
14. Brake carrier
15. Brake hose/line
16. Self-locking bolt
17. Caliper
18. Brake pads
19. Retaining springs

2348-TTTT-G26

**Fig. 7 Exploded view of rear disc brakes—TT with front wheel drive**

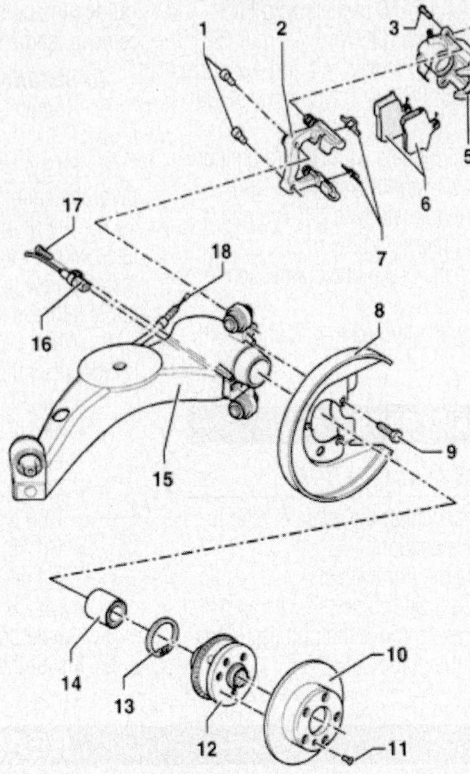

1. Hex head bolt
2. Brake carrier
3. Bolt
4. Brake hose/line
5. Caliper
6. Brake pads
7. Retaining springs
8. Backing plate
9. Bolt
10. Brake disc
11. Philips screw
12. Wheel hub with roto
13. Circlip
14. Wheel bearing
15. Trailing arm
16. Speed sensor
17. Hex head bolt
18. Parking brake cable

2348-TTTT-G27

**Fig. 8  Exploded view of rear disc brakes—TT with all wheel drive**

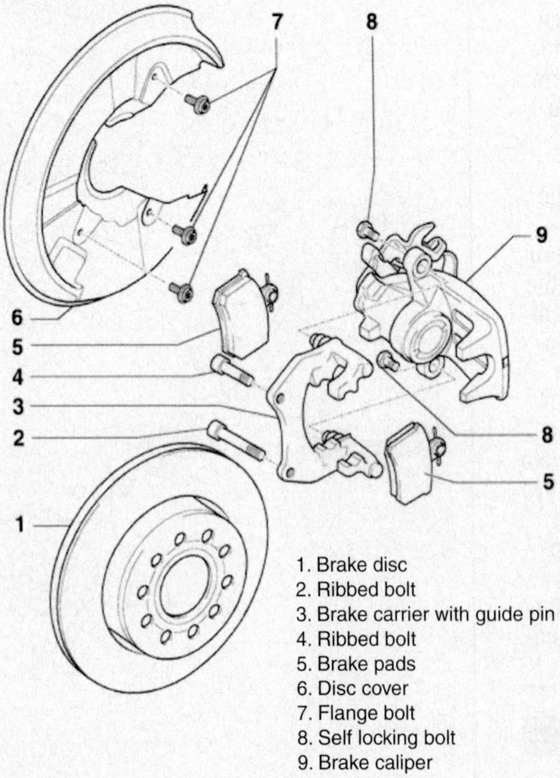

1. Brake disc
2. Ribbed bolt
3. Brake carrier with guide pin
4. Ribbed bolt
5. Brake pads
6. Disc cover
7. Flange bolt
8. Self locking bolt
9. Brake caliper

2348-A4A4-G24

**Fig. 9  Exploded view of rear disc brakes—A4 shown; A6 and A8 similar**

1. Before servicing the vehicle, refer to the precautions section.

2. Turn the ignition switch **OFF** and pump the brake pedal 25– 35 times to relieve the system pressure.

3. Remove the wheels.

4. Disconnect the parking brake cable.

5. Loosen the hydraulic line.

6. Use a back-up wrench to hold the guide pins and remove the caliper bolts.

7. Lift the caliper off the carrier and unscrew it from the hydraulic line.

***To install:***

8. Thread the caliper onto the hydraulic line and hand-tighten it. Fit the caliper into

place on the carrier. Torque the bolts to 26 ft. lbs. (35 Nm).

9. Bleed the brakes.

10. Install the wheels.

## DISC BRAKE PADS

### REMOVAL & INSTALLATION

1. Before servicing the vehicle, refer to the precautions section.

2. Remove the rear wheels.

3. Remove the parking brake cable clip from the caliper. Disconnect the parking brake cable.

4. Hold the guide pin with a back-up wrench and remove the upper mounting bolt from the brake caliper.

5. Swing the caliper downward and remove the brake pads.

***To install:***

6. Retract the piston into the housing by rotating the piston clockwise.

7. Install the new brake pads onto the pad carrier.

8. Install the caliper to the pad carrier using a new self locking bolt or a thread locking compound and torque to 26 ft. lbs. (35 Nm).

9. Attach the hand brake cable to the caliper.

10. Check the parking brake operation and adjust the cable if necessary.

11. Install the wheels.

# CHASSIS ELECTRICAL

## GENERAL INFORMATION

### ✳✳ CAUTION

**These vehicles are equipped with an air bag system. The system must be disarmed before performing service on, or around, system components, the steering column, instrument panel components, wiring and sensors. Failure to follow the safety precautions and the disarming procedure could result in accidental air bag deployment, possible injury and unnecessary system repairs.**

### SERVICE PRECAUTIONS

Disconnect and isolate the battery negative cable before beginning any airbag system component diagnosis, testing, removal, or installation procedures. Allow system capacitor to discharge for two minutes before beginning any component service. This will disable the airbag system. Failure to disable the airbag system may result in accidental airbag deployment, personal injury, or death.

Do not place an intact undeployed airbag face down on a solid surface. The airbag will propel into the air if accidentally deployed and may result in personal injury or death.

When carrying or handling an undeployed airbag, the trim side (face) of the airbag should be pointing towards the body to minimize possibility of injury if accidental deployment occurs. Failure to do this may result in personal injury or death.

Replace airbag system components with OEM replacement parts. Substitute parts

# AIR BAG (SUPPLEMENTAL RESTRAINT SYSTEM)

may appear interchangeable, but internal differences may result in inferior occupant protection. Failure to do so may result in occupant personal injury or death.

Wear safety glasses, rubber gloves, and long sleeved clothing when cleaning powder residue from vehicle after an airbag deployment. Powder residue emitted from a deployed airbag can cause skin irritation. Flush affected area with cool water if irritation is experienced. If nasal or throat irritation is experienced, exit the vehicle for fresh air until the irritation ceases. If irritation continues, see a physician.

Do not use a replacement airbag that is not in the original packaging. This may result in improper deployment, personal injury, or death.

The factory installed fasteners, screws and bolts used to fasten airbag components have a special coating and are specifically designed for the airbag system. Do not use substitute fasteners. Use only original equipment fasteners listed in the parts catalog when fastener replacement is required.

During, and following, any child restraint anchor service, due to impact event or vehicle repair, carefully inspect all mounting hardware, tether straps, and anchors for proper installation, operation, or damage. If a child restraint anchor is found damaged in any way, the anchor must be replaced. Failure to do this may result in personal injury or death.

Deployed and non-deployed airbags may or may not have live pyrotechnic material within the airbag inflator.

Do not dispose of driver/passenger/curtain airbags or seat belt tensioners unless you are sure of complete deployment. Refer

to the Hazardous Substance Control System for proper disposal.

Dispose of deployed airbags and tensioners consistent with state, provincial, local, and federal regulations.

After any airbag component testing or service, do not connect the battery negative cable. Personal injury or death may result if the system test is not performed first.

If the vehicle is equipped with the Occupant Classification System (OCS), do not connect the battery negative cable before performing the OCS Verification Test using the scan tool and the appropriate diagnostic information. Personal injury or death may result if the system test is not performed properly.

Never replace both the Occupant Restraint Controller (ORC) and the Occupant Classification Module (OCM) at the same time. If both require replacement, replace one, then perform the Airbag System test before replacing the other.

Both the ORC and the OCM store Occupant Classification System (OCS) calibration data, which they transfer to one another when one of them is replaced. If both are replaced at the same time, an irreversible fault will be set in both modules and the OCS may malfunction and cause personal injury or death.

If equipped with OCS, the Seat Weight Sensor is a sensitive, calibrated unit and must be handled carefully. Do not drop or handle roughly. If dropped or damaged, replace with another sensor. Failure to do so may result in occupant injury or death.

If equipped with OCS, the front passenger seat must be handled carefully as well. When removing the seat, be careful when setting on floor not to drop. If dropped, the sensor may be inoperative, could result in occupant injury, or possibly death.

If equipped with OCS, when the passenger front seat is on the floor, no one should sit in the front passenger seat. This uneven force may damage the sensing ability of the seat weight sensors. If sat on and damaged,

the sensor may be inoperative, could result in occupant injury, or possibly death.

### DISARMING THE SYSTEM

1. Before servicing the vehicle, refer to the precautions section.
2. Disconnect and shield the negative battery cable.

3. Wait at least 5 minutes before servicing the vehicle.

### ARMING THE SYSTEM

1. Before servicing the vehicle, refer to the precautions section.
2. Remove the shield.
3. Connect the negative battery cable.

## DRIVETRAIN

### AUTOMATIC TRANSAXLE ASSEMBLY

#### REMOVAL & INSTALLATION

**A3 Models—Front Wheel Drive**

*See Figures 10 through 12.*

1. Before servicing the vehicle, refer to the precautions section.
2. Place selector lever in position "P".
3. Disconnect the electrical harness connector at Mass Air Flow (MAF) sensor.
4. Unclip air intake nozzle.
5. Remove the engine cover.
6. Disconnect the battery cables, negative cable first.
7. Remove the battery.
8. Remove the battery carrier.
9. Clamp off coolant hoses of transmission fluid cooler using hose clamps and remove them from transmission fluid cooler. Seal transmission fluid cooler using plugs.
10. Remove the securing plate.
11. Press circlip upward and remove it. Press selector lever cable off from selector shaft lever and pull it out of mounting bracket.

➡ **Do not bend or kink the selector lever cable.**

12. Disconnect the electrical connector.
13. Remove the cable tie for protective sleeve.
14. Flip protective sleeve back and unscrew B+ cable from solenoid switch.

#### ✳✳ CAUTION
**Never touch the contacts in the transmission connector, under any circumstances, because static discharge could destroy the control module along with the mechatronics unit. Touch a vehicle ground with your hand to discharge static.**

15. Twist bayonet connection of 20-pin connector to release it from transmission, and disconnect.
16. Unbolt ground cable.

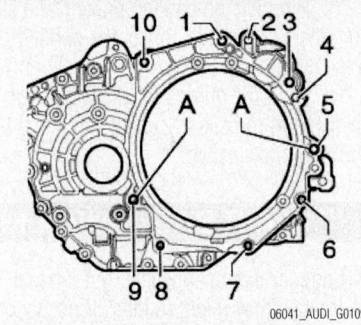

**Fig. 10 Transmission bolt locations— Front wheel drive A3 with automatic transmission**

17. Unhook connector and wiring harnesses from bracket on transmission.
18. Remove the bolts starter.
19. Remove the engine/transmission assembly bolts which are accessible from above.
20. Set engine support tool onto bolted flanges of fender.
21. Hook spindle of engine support tool into left engine lifting eye.
22. Pretension engine using spindle, but do not lift.
23. Remove the both front wheels.
24. Remove the center sound insulation fasteners.
25. Remove the left sound insulation fasteners.
26. Remove the air duct hose to charge

**Fig. 11 Attach an engine support tool— Front wheel drive A3 with automatic transmission**

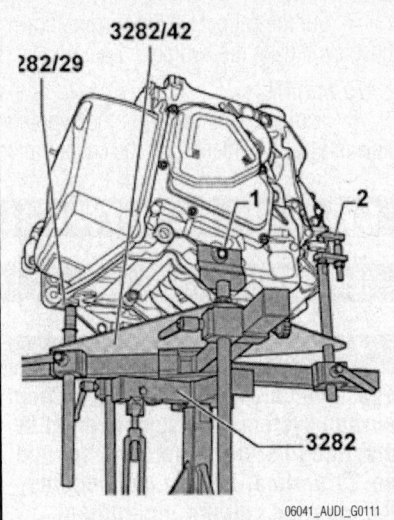

**Fig. 12 Remove the transmission with a suitable jack and attachments—Front wheel drive A3 with automatic transmission**

air cooler , if equipped, remove shield at bottom of transmission.

27. Remove the both nuts that secure bracket on front side of transmission oil pan.

➡ **The threaded studs are welded to the front of the transmission oil pan.**

28. Press off the bracket from threaded studs of transmission oil pan and lay it aside up top, so that it will not get caught up when lowering transmission.
29. Remove the heat shield for right drive axle.

➡ **To prevent damage, decoupling element of the front exhaust pipe must never be angled by more than 10 degrees.**

30. Loosen the nuts of clamping sleeve.
31. To disconnect exhaust system, slide the clamping sleeve towards the rear.
32. Remove the bracket for exhaust system.
33. Remove the bolt from exhaust system bracket.

34. Remove the bolts and remove pendulum support.

35. Disconnect the oil Level thermal sensor connector and unclip the bracket at the sub frame.

36. Unbolt the drive axles from the transmission flange and support them with wire.

37. Remove the bolt and pull right flange shaft off from transmission.

38. Disconnect the left front level control system sensor connector.

39. Remove the bolt and the sub frame.

➡ **When removing the control arm console, the locating fixture T10096 must be used, otherwise a vehicle alignment will need to be performed afterwards.**

40. Remove the left control arm console from body.

41. Remove the bolts and remove transmission support.

42. Lower the engine/transmission assembly, using spindle of engine support tool 100 to 110mm between transmission housing and transmission mounting.

43. To remove transmission, transmission support 3282 is equipped with adjustment plate 3282/42 and positioned on engine/transmission assembly jack.

44. Align arms of transmission support to correspond with holes in adjustment plate 3282/42.

45. Screw in support elements as shown on adjustment plate 3282/42.

46. Place engine/transmission assembly jack under vehicle, with the arrow on adjustment plate 3282/42 pointing to front of vehicle.

47. Align transmission support 3282 parallel to transmission.

48. Secure support element to transmission with bolt.

49. Screw stud 3282/29 into transmission.

50. Insert support element into transmission and secure by tightening nut.

51. Support transmission from underneath by raising engine/transmission assembly jack .

52. Remove the remaining engine/transmission assembly bolts.

53. Press transmission off from alignment sleeves.

54. Pull transmission off from engine.

55. Carefully lower transmission using engine/transmission assembly jack .

56. Use spindles of transmission support 3282 to adjust position of transmission while lowering.

### To install:

57. Replace self-locking nuts and bolts.

58. Replace bolts that are tightened to torque as well as O-rings and gaskets.

59. Clean the input shaft splines and the hub splines, remove corrosion and apply only a very thin coating of grease onto the splines. Excess grease must be removed.

60. Check if the alignment sleeves for centering engine/transmission assembly are in cylinder block; if necessary, install alignment sleeves.

61. Replace seal for right flange shaft

62. Make sure that the intermediate plate is hooked into sealing flange of engine and is pushed onto alignment sleeves.

63. Carefully raise transmission using engine/transmission assembly jack and move it into installation position using transmission support 3282.

64. Fasten transmission to engine. Tighten the M12 bolts to 80 Nm and the M10 bolts to 40 Nm.

65. Install the mountings as follows:

   a. Position transmission support between transmission and support arm of transmission mount.

   b. Fasten transmission support to transmission using new bolts and tighten to 40 Nm plus 90 degrees.

   c. Use spindle of engine support tool to lift transmission up to support arm of transmission mounting.

   d. Temporarily screw in bolts hand-tight.

   e. Remove the transmission support 3282 from transmission.

   f. Install the left control arm console.

   g. Install the sub frame and tighten the bolts to 110 Nm plus 90 degrees.

66. Install the left front level control system sensor, screw in bolt and connect electrical connector.

67. Install the flange.

68. Install the drive axles.

69. Fasten pendulum supports to transmission first and then to subframe bolts.

70. Install the exhaust system and align it free of tension.

71. Install the starter

72. Check adjustment of assembly mounting. Fasten bolts of assembly mounting for transmission to 60 Nm plus an additional 90 degrees.

73. Remove the engine support tool from engine.

➡ **Do not bend or kink the selector lever cable.**

74. Install the selector lever cable to mounting bracket and to selector shaft lever.

75. Install the remaining components in the reverse order of removal.

### A3 Models—All Wheel Drive

*See Figures 13 through 18.*

1. Before servicing the vehicle, refer to the precautions section.

2. Remove the molded tool insert under luggage compartment floor covering.

3. Remove the battery compartment cover.

4. Remove the molded insert over battery.

5. Disconnect the negative battery cable.

6. Remove the vacuum hose on air duct hose.

7. Disconnect the air guide hose at throttle valve control module.

8. Disconnect the electrical harness connector at Mass Air Flow (MAF) sensor.

9. Remove the upper part of air filter housing.

10. Remove the filter insert.

11. Remove the air duct cover.

12. Unclip air duct.

13. Unscrew bottom part of air filter housing.

14. Remove the air filter housing bolts.

15. Place selector lever in position "P".

16. Press lock washer upward and remove it. Press off selector lever cable from lever/selector shaft in direction of and set it on top. Do not loosen bolt.

17. Remove the both bolts and remove the selector cable. Do not kink selector lever cable.

18. Remove the starter.

19. Mark coolant hoses to transmission fluid cooler to prevent mix-ups when installing.

20. Disconnect the transmission fluid cooler hoses with hose clamps from transmission fluid cooler. Plug transmission fluid cooler.

21. Remove the all engine/transmission assembly bolts that can be reached from top.

22. Secure a shackle /12 at right rear engine lifting eye.

23. Position engine support tool with adapters on to bolted flanges of fender.

24. Disconnect the any hoses and cables in area of engine support tool lifting eyes on engine.

25. Engage spindles in lifting eyes or at shackles.

26. Slightly pre-tension engine/transmission assembly on spindle.

27. Remove the center noise insulation fasteners.

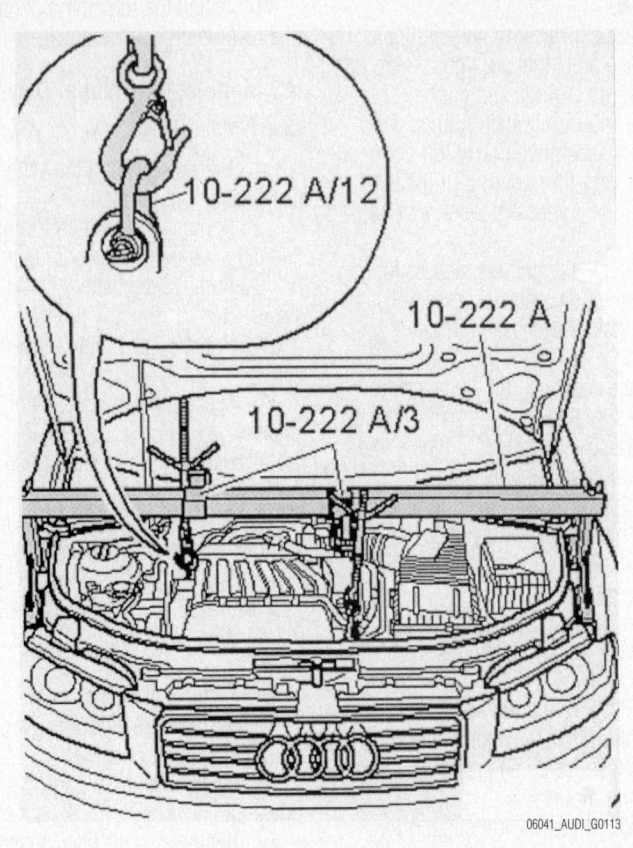

**Fig. 13 Attaching the engine support device to the engine—All wheel drive A3 with automatic transmission**

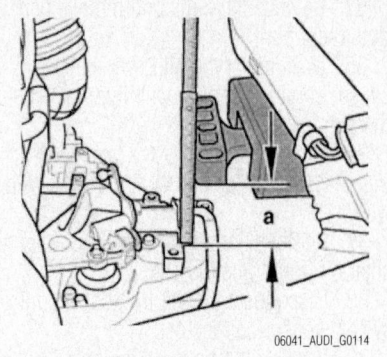

**Fig. 14 The distance between transmission housing and transmission bearing must be 100 to 110 mm—All wheel drive A3 with automatic transmission**

28. Remove the left and right noise insulation fasteners.

29. Remove the both front wheels.

30. If installed, remove guard plate at bottom of transmission.

31. Remove the bracket for exhaust system.

32. Unclip bracket for electrical wiring to oil level thermal sensor at subframe. Remove the bolts and bracket.

33. Remove the nut and press bracket upward so that it does not hang when transmission is lowered.

➡ **Bracket is secured at front of transmission fluid pan.**

34. Disconnect the exhaust system at double clamp. Push clamping sleeve toward rear.

35. Tie exhaust system in position.

➡ **Flex joint of exhaust system must not be bent more than 10 degrees, otherwise it may be damaged.**

36. Remove the right axle shaft heat shield from bevel box.

37. Remove the drive axles from transmission flange.

38. Secure the axles with wire.

39. Mark the position of flexible disc and bevel box flange in relation to each other.

40. Remove the drive shaft flexible disc at bevel box.

➡ **When loosening bolts, provide support at drive shaft flange/final drive with a suitable lever. Press drive axle horizontally toward rear as far as possible.**

### ⁜ CAUTION

**The sealing ring in drive shaft flange must not be damaged when removing transmission. If the seal is damaged, the drive axle must be replaced. Be aware that loosening pedal support/subframe bolts, the engine/transmission assembly moves forward slightly.**

41. Remove the screws and press assembly forward slightly and then detach drive shaft from bevel box.

42. Set down drive shaft above bevel box and secure it in position.

43. If equipped, remove the left front level control system sensor.

44. Remove the sub frame.

45. Remove the left control arm console from the body.

46. Remove the bolts and at bevel gear transfer case bracket.

47. Remove the bracket for bevel gear transfer case.

48. Remove the bolts and remove transmission console.

49. Lower engine/transmission assembly at spindle a short distance.

50. The distance between transmission housing and transmission bearing must be 100 to 110 mm.

51. To remove transmission, the transmission support 3282 is equipped with adjustment plate 3282/42 and positioned on engine/transmission assembly jack .

52. Align arms of transmission support to correspond with holes in adjustment plate 3282/42.

53. Screw in support elements as shown on adjustment plate 3282/42.

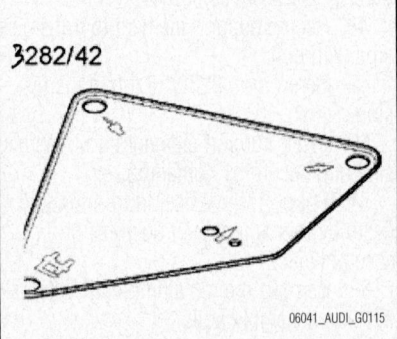

**Fig. 15 To remove transmission, the transmission support 3282 is equipped with adjustment plate 3282/42 and positioned on engine/transmission assembly jack—All wheel drive A3 with automatic transmission**

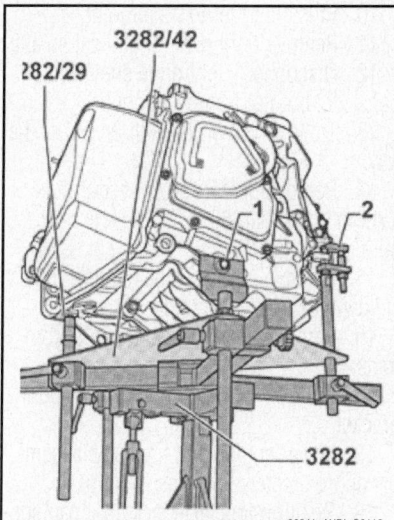

**Fig. 16 Screw in support elements as shown on adjustment plate 3282/42—All wheel drive A3 with automatic transmission**

54. Place engine/transmission assembly jack under vehicle, arrow symbol on adjustment plate 3282/42 points in direction of travel/vehicle.

55. Align transmission support 3282 parallel to transmission.

56. Secure support element to transmission with bolt.

57. Screw bolt 3282/29 into transmission.

58. Insert support element into transmission and secure by tightening nut.

59. Support transmission from below by lifting engine/transmission assembly jack .

60. Remove the remaining engine/transmission assembly connecting bolts.

61. Press off transmission from alignment sleeves.

62. Detach transmission from engine, starting on bevel box side and then at front.

63. Carefully lower transmission over engine/transmission assembly jack , making sure no wires are pinched.

64. Use spindles of transmission support 3282 to adjust position of transmission while lowering.

***To install:***

65. Replace self-locking nuts and bolts.

66. Replace bolts that are tightened to torque as well as O-rings and gaskets.

67. Clean the input shaft splines and on a used clutch plate, the hub splines, remove corrosion and apply only a very thin coating of grease onto the splines. Then move clutch plate back and forth on input shaft until the hub moves freely on the shaft. Excess grease must be removed.

68. If the transmission is being replaced, transfer over the transmission shift lever.

69. Check if the alignment sleeves for centering engine/transmission assembly are in the cylinder block, and install if necessary. If alignment sleeves are not installed, difficulty in shifting, clutch problems and possibly noises from the transmission (loose gear chatter) may develop.

70. Make sure that the intermediate plate is hooked into sealing flange and is pushed onto alignment sleeves.

71. Check release bearing for wear. If necessary, replace slave cylinder with release

72. Carefully raise transmission and move it into installation position.

73. Align transmission to engine and install. Tighten M12 bolts to 80 Nm and the M10 bolts to 40 Nm.

　a. Use spindle of engine support tool to lift transmission up to support arm of transmission mounting.

　b. Temporarily screw in bolts hand-tight.

　c. Remove the transmission support 3282 from transmission.

74. Tighten the mounting bracket to transmission to 20 Nm plus 90 degrees.

75. Tighten the transmission to console to 40 Nm plus 90 degrees.

**✳✳ CAUTION**

**Before installing the bolts, the transmission console and transmission mount support arm must be absolutely parallel, otherwise the threads will be damaged. If necessary, use a floor jack to raise the rear of the transmission. Remove the engine support tool only when all sub frame mounting bolts are tightened to the torque specifications and the sub frame is installed.**

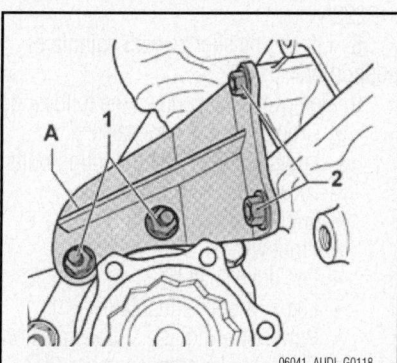

**Fig. 17 Bevel box bracket bolt locations—All wheel drive A3 with automatic transmission**

76. Install the bolts and hand-tight for securing bevel box bracket as follows:

　a. Step 1: Tighten nuts (1) first to 3 Nm.

　b. Step 1: Tighten nuts (2) to 35 Nm.

　c. Step 1: Tighten nuts to (1) 45 Nm.

77. Install the pendulum support to transmission bolts to 40 Nm plus 90 degrees.

78. Install the pendulum support to sub frame bolts to 110 Nm plus 90 degrees.

79. Install the transmission mount to transmission console bolts to 60 Nm plus 90 degrees.

80. Install the left front level control system sensor, screw in bolt and connect electrical connector.

81. Install the flange.

82. Install the drive axles.

83. Fasten pendulum supports to transmission first and then to sub frame bolts.

84. Install the exhaust system and align it free of tension.

85. Install the starter

86. Check adjustment of assembly mounting. Fasten bolts of assembly mounting for transmission to 60 Nm plus an additional 90 degrees.

87. Remove the engine support tool from engine.

➡ **Do not bend or kink the selector lever cable.**

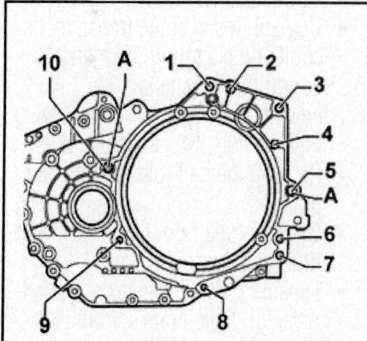

**Securing engine/transmission**

|  |  | Nm |
|---|---|---|
| 1, 3, 6 | M12x55 | 80 |
| 5 | M12x65 | 80 |
| 7, 8 | M10x50 | 40 |
| 9 | M10x45 | 40 |
| 10 | M12x70 | 80 |
| A | Alignment sleeves for centering | |

**Fig. 18 Transmission bolt locations and torque values—All wheel drive A3 with automatic transmission**

88. Install the selector lever cable to mounting bracket and to selector shaft lever.

89. Install the remaining components in the reverse order of removal.

## A4 and A6 Models

1. Before servicing the vehicle, refer to the precautions section.

2. Remove or disconnect the following:
   - Negative battery cable
   - Engine cover
   - Air cleaner duct
   - Air cleaner assembly
   - Upper engine-to-transaxle bolts

3. Using an engine support tool, secure it to the engine and the vehicle.

4. Remove or disconnect the following:
   - Front wheels
   - Transmission range switch, connector and bracket
   - Heat shields for drive shafts
   - Drive shaft from transmission flanges
   - Speed sensor connectors (mark for reinstallation)
   - Both oxygen sensors (mark left and right for reinstallation)
   - Both top bolts at the front of the engine
   - Starter
   - Torque converter-to-driveplate bolts through the starter opening
   - Torque converter cover plate
   - Coolant hoses at the transmission cooler by clamping them off

5. Support the halfshafts

6. Remove or disconnect the following:
   - Inner halfshaft-to-transaxle bolts
   - Remove the ball joint and the support
   - Oil filler tube from the oil pan and drain the fluid
   - Exhaust pipe-to-transaxle bracket
   - Selector cable bracket from the transaxle
   - Selector cable circlip and the cable at the transaxle shift lever
   - Accelerator cable bracket and the cable from the operating lever
   - Center bolt, from the transaxle mount.

7. Using the engine support tool, lift the engine slightly.

8. Remove the throttle cable bracket bolts and the bracket.

9. Support the transaxle and lift it slightly.

10. Remove or disconnect the following:
    - Lower transaxle-to-engine bolts
    - Transaxle from the engine

➡ **Be sure to secure the torque converter.**

### To install:

11. Installation is the reverse of the removal procedure, while using the following torque values:
    - Transaxle flange bolts: 41 ft. lbs. (56 Nm)
    - Subframe bolts: 52 ft. lbs. (71 Nm)
    - Transaxle mount center bolt: 30 ft. lbs. (40 Nm)
    - Torque converter bolts: 63 ft. lbs. (85 Nm)
    - Halfshaft bolts: 33 ft. lbs. (45 Nm)
    - Ball joint bolts: 48 ft. lbs. (65 Nm)
    - Starter motor bolts: 48 ft. lbs. (65 Nm)

## TT Models

1. Before servicing the vehicle, refer to the precautions section.

2. Place gearshift in neutral.

3. Before servicing the vehicle, refer to the precautions section.

4. Place transaxle in neutral.

5. Remove or disconnect the following:
   - Negative battery cable
   - Engine covers
   - Engine top cover over cylinder head
   - Hoses and connectors from air filter housing
   - Air filter housing
   - Battery and battery carrier
   - Selector lever off selector lever shaft
   - Selector lever cable
   - Multi-function switch connector
   - Cables and connectors from starter
   - Starter
   - ATF cooler hoses after clamping off

6. Install Engine Support bracket 10-222A and supports 10-222A1.

7. Install Retainer 3180 to right engine lifting eye and attach it to support bracket 10-222A.

8. Lift engine slightly with spindle of support bracket 1022A.

9. Remove or disconnect the following:
   - Ground cable at transaxle
   - Engine to transaxle mounting bolts accessible from above
   - Engine undercover
   - Front wheels
   - Left drive shaft
   - Left side sound insulator
   - Power steering line at transmission
   - Front tubular cross member (if equipped)
   - Cross brace at subframe
   - Sway brace

10. Press out the right ball joint.

11. Remove right drive shaft heat shield.

12. Disconnect right drive shaft from transaxle, but not from wheel hub

13. Tie the right drive shaft up out of the way.

14. Remove six torque converter nuts, turning crankshaft 60° each time to access nuts.

15. Remove exhaust pipe clamps and suspend exhaust pipe under vehicle.

16. Remove electrical connectors from transaxle.

17. Remove transaxle console support bracket.

18. Lower engine and transaxle assembly approximately 2 inches (50mm).

19. Position transaxle jack so it can support full weight of transaxle.

20. Remove the engine to transaxle mounting bolts.

21. With the aid of an assistant slide transaxle back slightly until torque converter can be secured in place, then lower transaxle and move away from vehicle.

### To install:

22. Installation is the reverse of the removal procedure, while using the following torque values:
    - Transaxle to engine M12 bolts: 59 ft. lbs. (80 Nm)
    - Transaxle to engine M10 bolts: 33 ft. lbs. (45 Nm)
    - Starter bolts: 59 ft. lbs. (80 Nm)
    - Subframe bolts: 74 ft. lbs. (100 Nm) plus 90°
    - Sway brace side bolts: 15 ft. lbs. (20 Nm) plus 90°
    - Sway brace bottom bolts: 30 ft. lbs. (40 Nm) plus 90°

## MANUAL TRANSAXLE ASSEMBLY

### REMOVAL & INSTALLATION

#### A3 Models

*See Figures 19 through 21.*

1. Before servicing the vehicle, refer to the precautions section.

2. Obtain the radio anti-theft code.

3. Disconnect the negative battery cable.

4. Remove the engine cover.

5. Remove the battery and battery carrier.

6. Remove the circlips and at both cables.

7. Pull cable retainers with cables off from transmission shift lever and relay lever.

8. Remove the cable mounting bracket

from transmission and tie it up off to side with cables.

9. Unfasten the ground cable from upper starter bolt.

10. Remove the connector and wiring from starter.

➡**During the following procedures, make sure that brake fluid does not leak onto the starter or the transmission. If this does occur, these areas must be cleaned thoroughly.**

11. Pull clamp for line/hose assembly out to the stop.

12. Pull pressure line off from bleeder/slave cylinder and seal it off.

### ❋❋ CAUTION

**Do not press clutch pedal after line/hose assembly has been removed.**

13. Remove the bolts at engine/transmission assembly flange.

14. Remove the upper bolt on starter.

15. Disconnect the any hoses and cables in the area of engine lifting eyes on engine.

16. Install engine support tool.

17. Pretension engine with the support tool, but do not lift.

18. With the vehicle weight still on the wheels, loosen left front collar (drive axle) bolt a maximum of 90 degrees, otherwise wheel bearing will be damaged.

19. Loosen the left front wheel bolts.

20. Remove the left front wheel.

21. Remove the sound insulation.

22. Remove the front part of left front wheel housing liner.

23. If equipped, remove left front level control system sensor.

24. Disconnect the electrical connector for back-up light switch.

25. Remove the nut and bracket with electrical lines from lower starter bolt.

26. Remove the starter motor.

27. Separate air duct hose/pressure pipe connection.

28. Remove the exhaust system bracket from subframe.

29. Separate exhaust system at clamping sleeve and tie up front exhaust pipe.

➡**To prevent damage, decoupling element of the front exhaust pipe must never be kinked by more than 10 degrees.**

30. Disconnect the oil level thermal sensor.

31. Remove the shield for right drive axle.

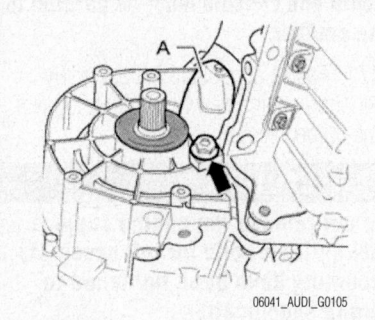

**Fig. 19 Remove the bolt and small cover plate—A3 2.0L engine with manual transmission**

32. Unbolt right drive axle from flange shaft on transmission.

33. Completely remove left drive axle

34. Remove the bolt and small cover plate.

35. Remove the bolts and remove pendulum support.

36. Before removing subframe, match mark its position.

37. Remove the subframe .

38. Remove the left console with the control arm.

39. Remove the bolts from assembly mounting on transmission.

40. Lower transmission the engine support tool by 60 mm.

41. Unfasten the console from transmission.

42. Pull off circlip and remove relay lever.

43. Remove the lower engine/transmission assembly bolts.

44. Loosen the engine transmission bolts and leave them screwed in hand-tight.

45. Attach transmission support 3282 for removal of the transmission using adjustment plate 3282/33.

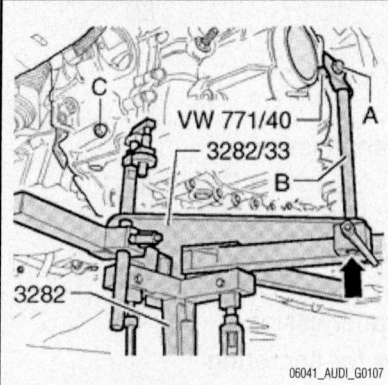

**Fig. 20 Secure the adapter VW 771/40 in the threaded bore of the transmission housing—A3 2.0L engine with manual transmission**

46. Place transmission support 3282 in the engine/transmission assembly jack V.A.G 1383A.

47. Align the arms of transmission support according to holes in adjustment plate.

48. Screw the support elements onto adjustment plate as shown.

49. Instead of the support element, screw in the stud 3282/29.

50. Place engine/transmission assembly jack under vehicle, with arrow symbol on adjustment plate pointing to front of vehicle.

51. Align adjustment plate and transmission parallel to one another.

52. Secure the adapter VW 771/40 in the threaded bore of the transmission housing, as shown in the accompanying illustration.
Screw stud 3282/29 into bore for securing bolt of pendulum support on transmission.

53. Secure transmission on transmission support 3282 using an M10X M20 bolt. The drift should thereby sit flush at bottom with guide of transmission support 3282.

54. Remove the last connecting bolt for engine/transmission assembly.

55. Press transmission off from engine locating sleeves.

56. Using spindles of the transmission support 3282, screw up the transmission in the area of the differential.

57. Tilt transmission toward left via the spindles of the transmission support 3282.

58. Carefully lower transmission.

***To install:***

59. Replace self-locking nuts and bolts.

60. Replace bolts that are tightened to torque as well as O-rings and gaskets.

61. Clean the input shaft splines and on a used clutch plate, the hub splines, remove corrosion and apply only a very thin coating of grease onto the splines. Then move clutch plate back and forth on input shaft until the hub moves freely on the shaft. Excess grease must be removed.

62. If the transmission is being replaced, transfer over the transmission shift lever.

63. Check if the alignment sleeves for centering engine/transmission assembly are in the cylinder block, and install if necessary. If alignment sleeves are not installed, difficulty in shifting, clutch problems and possibly noises from the transmission (loose gear chatter) may develop.

64. Make sure that the intermediate plate is hooked into sealing flange and is pushed onto alignment sleeves.

65. Check release bearing for wear. If necessary, replace slave cylinder with release

66. Carefully raise transmission and move it into installation position.

67. Align transmission to engine and install. Tighten M12 bolts to 80 Nm and the M10 bolts to 40 Nm.

68. Insert relay lever into selector mechanism and clip in circlip.

69. Install the console onto transmission with new hex bolts to 60 Nm plus 90 degrees

70. Align engine/transmission assembly in installed position. To do this, lift until console is in complete contact with left assembly mounting.

➡ To prevent the threads in the console from being damaged, the transmission mount and console must be parallel to one another.

71. Fasten console to left assembly mounting , using new hex bolts to 60 Nm plus 90 degrees.

### ❊ CAUTION

**Do not remove the engine support tool until all bolts for left assembly mounting have been tightened to torque specification.**

72. Install the small cover plate for flywheel.

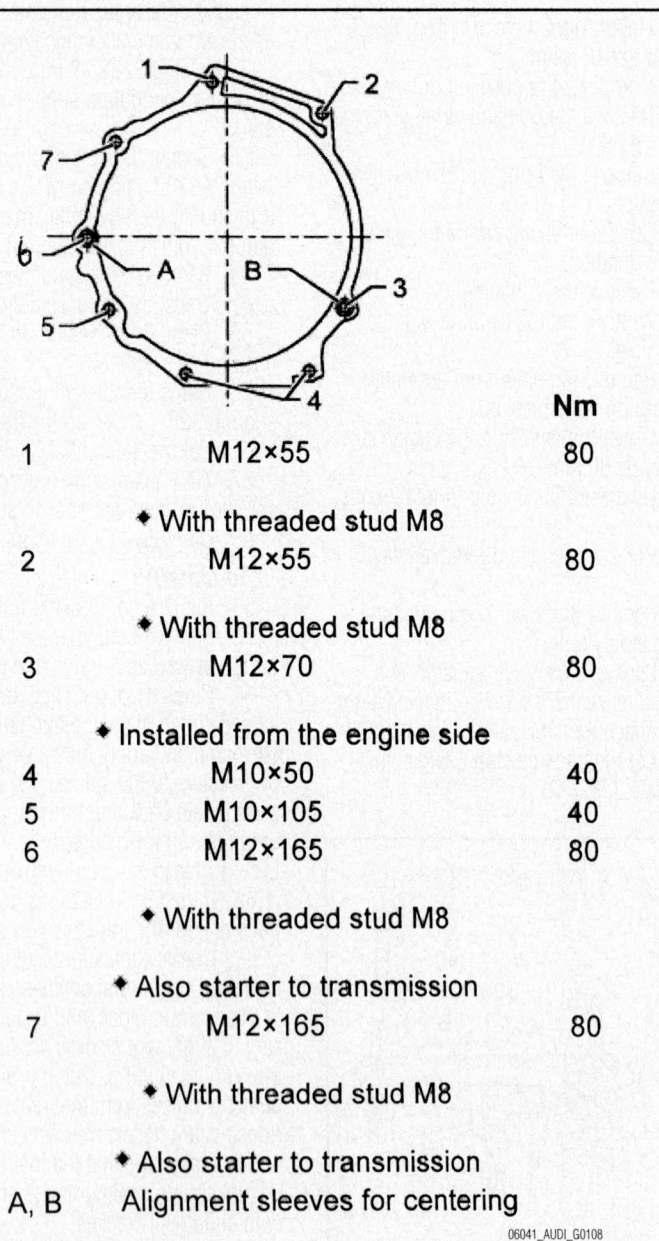

| | | Nm |
|---|---|---|
| 1 | M12×55 | 80 |
| | ✦ With threaded stud M8 | |
| 2 | M12×55 | 80 |
| | ✦ With threaded stud M8 | |
| 3 | M12×70 | 80 |
| | ✦ Installed from the engine side | |
| 4 | M10×50 | 40 |
| 5 | M10×105 | 40 |
| 6 | M12×165 | 80 |
| | ✦ With threaded stud M8 | |
| | ✦ Also starter to transmission | |
| 7 | M12×165 | 80 |
| | ✦ With threaded stud M8 | |
| | ✦ Also starter to transmission | |
| A, B | Alignment sleeves for centering | |

06041_AUDI_G0108

Fig. 21 Transmission bolt location and their torque values—A3 2.0L engine with manual transmission

73. Install left and right drive axles to transmission

74. Install the shield for right drive axle.

75. Install the subframe/control arm.

76. Attach steering gear to sub frame.

77. Fasten pendulum supports to transmission first and then to subframe bolts , and use new bolts to fasten. Replace the bolts, tighten the outer bolt to 100 Nm plus 90 degrees and the two inner bolts to 40 Nm plus 90 degrees.

78. Check adjustment of subframe mounting, adjust if necessary

79. Install the remaining components in the reverse order of removal.

80. Bleed the clutch system .

81. Check the gear oil level in manual transmission.

### A4 and A6 Models

1. Before servicing the vehicle, refer to the precautions section.

2. Remove or disconnect the following:
- Negative battery cable
- Intake hose
- Expansion tank (move aside)
- Oxygen sensors
- Upper engine-to-transaxle mounting bolts
- Engine undercover and bracket
- Front exhaust pipe with the catalytic converter
- Driveshaft for All Wheel Drive (AWD) models only
- Starter
- Shift rod and joint bolt at the transaxle and separate from the rear of the shift rod
- Shift rod
- Vehicle Speed Sensor (VSS) connector
- Backup light switch connector
- Driveshaft from transaxle and rest on heat shield

3. Support the transaxle, using a transmission/transaxle jack.

4. Remove or disconnect the following:
- Transaxle mount heat shield
- Right mount at the transaxle
- Left mount with the bushings
- Left and right halfshafts
- Axle heat shield
- Remaining engine-to-transaxle mounting bolts

5. Pry the transaxle off the dowel sleeves and carefully lower the transaxle until the slave cylinder is accessible, approx. 6 inches. (15cm).

6. Remove or disconnect the following:

- Clutch slave cylinder from the transaxle with the hydraulic line attached
- Transaxle

### To install:

7. Installation is the reverse of the removal procedure, while using the following torque values:

- Transaxle flange M12 bolts: 48 ft. lbs. (65 Nm)
- Transaxle flange M10 bolts: 33 ft. lbs. (45 Nm)
- Transaxle flange M8 bolts: 18 ft. lbs. (25 Nm)
- Slave cylinder bolts: 15 ft. lbs. (20 Nm)
- Transaxle mount bolts: 30 ft. lbs. (40 Nm)
- Shift rod bolts: 15 ft. lbs. (20 Nm)
- Pushrod to transaxle: 30 ft. lbs. (40 Nm)
- Driveshaft to transaxle: 41 ft. lbs. (55 Nm)
- Exhaust pipe clamp: 30 ft. lbs. (40 Nm)
- Axle shaft to drive flange: M8 bolts 30 ft. lbs. (40 Nm); M10 bolts 52 ft. lbs. (70 Nm)

### TT Models

*See Figure 22.*

1. Before servicing the vehicle, refer to the precautions section.
2. Place gearshift in neutral.
3. Remove or disconnect the following:
- Negative battery cable
- Engine covers
- Engine top cover over cylinder head
- Hoses and connectors from air filter housing
- Air filter housing
- Battery and battery carrier
- Cable mounting bracket at transaxle
- Shift cables at transaxle
- Speed sensor connector
- Ground cable
- Cables and connectors from starter
- Upper starter mounting bolts
- Mark rotation direction of accessory belt, then remove belt
- Air duct at lower right cross member
- Front Wheels
- Engine undercover
- Charge air cooler lines
- Power steering pump pulley
- Clamp off power steering hoses
- Pressure pipe at power steering pump
- Power steering pipe from transaxle

- Starter
- Reverse light connector
- Clutch cylinder hydraulic line
- Exhaust pipe clamps
- Loosen exhaust system and suspend under vehicle
- Mark position of driveshaft relative to flexible coupling
- Driveshaft from flexible coupling
- Press front driveshaft tube as far back as possible
- Sway brace support under vehicle
- Steering box bolts

4. Place a transaxle jack under transaxle.
5. Remove the subframe bolts.
6. Lower the subframe and leave it resting on the ball joints.
7. Suspend steering box up out the way.
8. Disconnect the drive axles and place on subframe.
9. Pull the subframe back and tie it under vehicle.
10. Remove right drive axle heat shield bolted to bevel box.
11. Remove the bevel box.
12. Remove the flywheel heat shield behind the bevel box.
13. Install Engine Support bracket 10-222A and supports 10-222A1.
14. Install Retainer 3180 to right engine lifting eye and attach it to support bracket 10-222A.
15. Lift engine slightly with spindle of support bracket 1022A.
16. Remove the left side transaxle support bolts.
17. Lower the engine slightly.
18. Position transaxle jack so it can support full weight of transaxle.
19. Remove the engine to transaxle mounting bolts.
20. Slide transaxle back and then lower away from vehicle.

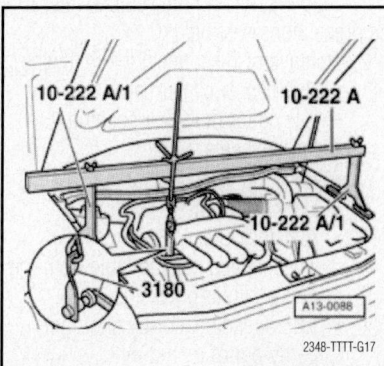

**Fig. 22 Attaching engine support and lifting tools—Audi TT 1.8L 4-Cyl engine**

### To install:

21. Installation is the reverse of the removal procedure, while using the following torque values:

- Cable mounting bracket: 17 ft. lbs. (23 Nm)
- Transaxle flange M12 bolts: 59 ft. lbs. (80 Nm)
- Transaxle flange M10 bolts: 30 ft. lbs. (40 Nm)
- Slave cylinder bolts: 18 ft. lbs. (25 Nm)
- Starter bolts: 48 ft. lbs. (65 Nm)
- Steering box bolts: 15 ft. lbs. (20 Nm) plus 90°
- Subframe bolts: 74 ft. lbs. (100 Nm) plus90°
- Sway brace side bolts: 15 ft. lbs. (20 Nm) plus 90°
- Sway brace bottom bolts: 30 ft. lbs. (40 Nm) plus 90°

## CLUTCH

### REMOVAL & INSTALLATION

1. Before servicing the vehicle, refer to the precautions section.
2. Remove or disconnect the following:
- Negative battery cable
- Transaxle

➡**If the pressure plate is to be reused, matchmark its relationship to the flywheel.**

3. Using a Flywheel Locking tool, lock the flywheel.
4. Remove or disconnect the following:
- Pressure plate from the flywheel by loosening the bolts alternately, a little at a time, to prevent warpage
- Clutch disc

### To install:

5. Install or connect the following:
- Clutch with the driven plate on the pressure plate so the spring cage is facing the pressure plate.
6. Hold the clutch assembly against the flywheel, aligning the matchmarks and the dowel pins on the flywheel with the pressure plate. Insert an alignment shaft tool through the pressure plate and the driven plate into the crankshaft pilot bearing.
7. Install the pressure plate to the flywheel. Torque the bolts evenly in a diagonal pattern, to avoid distortion. Tighten all models except A3 to 18 ft. lbs. (24 Nm). On A3 models, tighten the M6 bolts to 13 Nm or the M8 bolts to 20 Nm.
8. Remove the alignment shaft.

## ✴ WARNING

**The clutch release bearing in the front of the transaxle should be checked before reassembly. It is retained by 2 springs.**

9. Install or connect the following:
- Transaxle
- Negative battery cable

### *BLEEDING*

The clutch system can be bled using a pressure bleeder. Follow the instructions that come with the pressure bleeder for the proper pressure bleeding procedure. The maximum line pressure while pressure bleeding must not exceed 36 psi (248 kPa).

1. Before servicing the vehicle, refer to the precautions section.
2. To bleed the system perform the following:
   a. Top off the hydraulic fluid reservoir using a fluid that meets the standards of the vehicle's hydraulic system.
   b. Open the clutch slave cylinder bleed screw and press the clutch pedal to the floor and hold the pedal down.
   c. Close the clutch slave cylinder bleed screw.
   d. Release the clutch pedal.
   e. Check the hydraulic fluid level and top off as necessary.
3. Repeat the above steps until the discharged fluid is clean and no air bubbles appear during the bleeding process

## FRONT HALFSHAFT

### *REMOVAL & INSTALLATION*

#### A3 Models

1. Before servicing the vehicle, refer to the precautions section.
2. Remove the sound insulation.
3. Remove the drive axle bolts on wheel side.
4. Remove the wheel.
5. Remove the hex flange bolts.
6. Remove the hex nuts on ball joint.
7. Remove the hex nut from bracket for level control system sensor if equipped with automatic vertical headlight aim control on left side of vehicle.
8. Disengage control arm from ball joint.
9. Swing suspension strut outward, at the same time press drive axle out of wheel bearing unit using a brass drift if necessary
10. Support the axle with and pull drive axle out of splines and remove.

### *To install:*

11. Installation is the reverse of removal, please note the following:
   a. Replace gasket at transmission bolt.
   b. Install the drive axle on transmission side first. Tighten hex flange bolts to 20 Nm.

#### A6 Models

1. Before servicing the vehicle, refer to the precautions section.
2. Remove or disconnect the following:
- Halfshaft nut
- Front wheels
- Hex bolt for outer joint of drive axle
- Anti-lock Brake System (ABS) speed sensor by sliding it partly out of its mount
- Halfshaft from the transaxle flange
3. Press the halfshaft upward toward the front of the vehicle.
4. Turn the steering to full lock.
5. Remove the halfshaft.

### *To install:*

➡️**If equipped, replace the inner CV-joint gasket.**

- Halfshaft into the wheel hub
- Halfshaft-to-transaxle flange. Torque the bolts to 33 ft. lbs. (45 Nm) for the M8 bolts and to 59 ft. lbs. (80 Nm) for the M10 bolts
- ABS speed sensor
- Halfshaft nut. Torque it to 148 ft. lbs. (200 Nm) plus an additional¼ (90 degree) turn
- Hex bolt for outer joint of drive axle. Torque M14 bolts to 85 ft. lbs. (115 Nm) and M16 bolts to (190 Nm).
- Front wheels

#### A4 Models

*See Figure 23.*

1. Before servicing the vehicle, refer to the precautions section.
2. Remove or disconnect the following:
- Hub cap or center cap
- Hex collar bolt by loosening it
- Front wheels
- Halfshaft-to-transaxle flange bolts
- Hex collar bolt
- Anti-lock Brake System (ABS) wheel speed sensor cable from the brake caliper bracket
- ABS speed sensor by sliding it partly out of its mount
- Remove nut/bolt No. 1, as shown
3. Pull both arms up and out of the swing arm

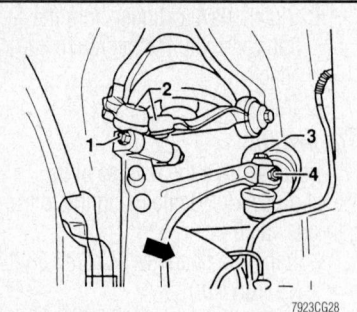

**Fig. 23 Loosen nut (1), remove the hex bolt and pull both arms (2) upward and out—A4 models**

## ✴ WARNING

**The slots in the swing arm must not be widened. Do not loosen the bolts No. 3 and 4, otherwise the axle geometry must be checked.**

4. Tilt the swing arm out and to the rear of the vehicle
5. Remove the halfshaft

### *To install:*

6. Install or connect the following:
- Halfshaft into the wheel hub
- Swing arm bolt. Torque it to 30 ft. lbs. (40 Nm)
- Halfshaft-to-transaxle flange. Torque the bolts to 30 ft. lbs. (40 Nm) for the M8 bolts and to 57 ft. lbs. (77 Nm) for the M10 bolts
- ABS wheel speed sensor
- Sensor cable into the caliper bracket
- Front wheels
7. Tighten the axle bolt as follows:
- M14 bolt: 85 ft. lbs. (115 Nm) plus an additional¼ (90 degree) turn
- M16 bolt: 140 ft. lbs. (190 Nm) plus an additional¼ (90degree) turn

## REAR HALFSHAFT

### *REMOVAL & INSTALLATION*

#### A6 Models

*See Figure 24.*

1. Before servicing the vehicle, refer to the precautions section.
2. Remove or disconnect the following:
- Halfshaft bolt
- Rear wheel
- Brake caliper without disconnecting the hydraulic line and support it on a wire
- Brake rotor
- Inner halfshaft flange bolts

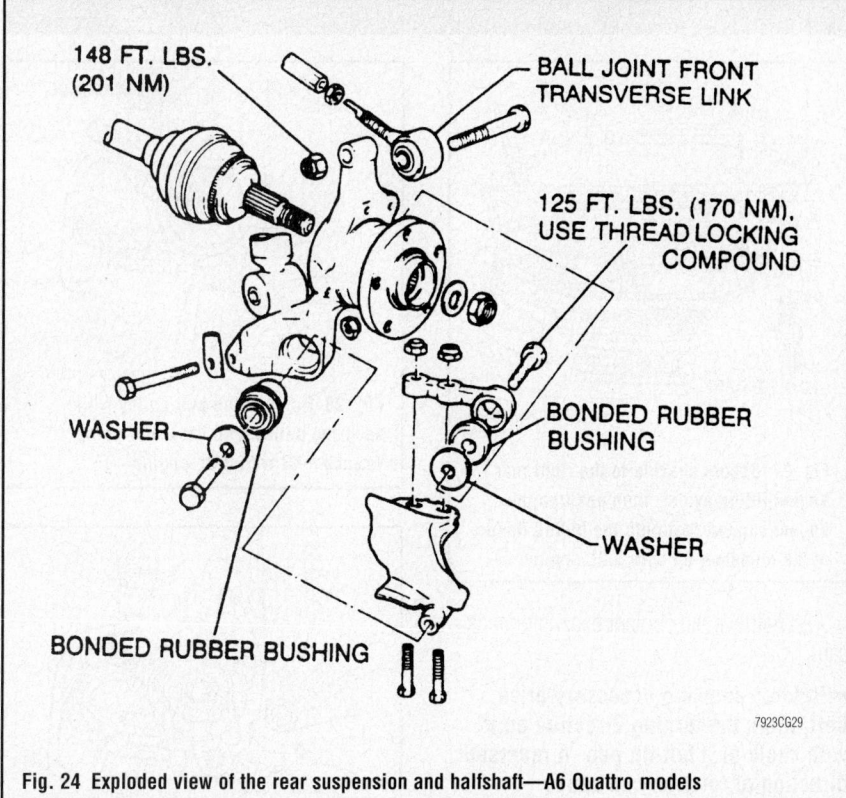

**Fig. 24 Exploded view of the rear suspension and halfshaft—A6 Quattro models**

148 FT. LBS. (201 NM)

BALL JOINT FRONT TRANSVERSE LINK

125 FT. LBS. (170 NM). USE THREAD LOCKING COMPOUND

BONDED RUBBER BUSHING

WASHER

WASHER

BONDED RUBBER BUSHING

7923CG29

3. Support the halfshaft.

4. Remove or disconnect the following:
- Fuel tank cover plate and/or inner CV-joint heat shield
- Anti-lock Brake System (ABS) speed sensor by sliding it partly out of its mount
- Lower strut bolt
- Transverse link from the wheel bearing housing
- Halfshaft by pressing down on wheel bearing housing

5. Clean the halfshaft splines of any grease, dirt or locking compound.

**To install:**

6. Use a new inner flange gasket and reverse the removal procedures.

7. Torque the halfshaft flange bolts as follows:
- M8 bolts: 33 ft. lbs. (45 Nm)
- M10 bolts: 59 ft. lbs. (80 Nm)

8. Install or connect the following:
- Brake caliper. Torque the bolts to 48 ft. lbs. (65 Nm)

➡**Adjustment of parking brake may be necessary.**

- Halfshaft bolt and washer assembly. Torque the bolt to 148 ft. lbs. (200 Nm) plus an additional ¼ (90 degree) turn
- ABS speed sensor
- Rear wheels

### A4 Models

*See Figure 25.*

1. Before servicing the vehicle, refer to the precautions section.

2. Remove or disconnect the following:
- Halfshaft bolt
- Rear wheel
- Anti-lock Brake System (ABS) speed sensor by sliding it partly out of its mount
- CV-joint from the final drive

3. Loosen the sway bar link mounting bolt at the wheel bearing housing.

4. Loosen the upper control arm mounting bolt at the wheel bearing housing.

5. Remove or disconnect the following:
- The center and rear mufflers, if servicing the left halfshaft
- Halfshaft from the final drive and the wheel bearing housing

**To install:**

6. Install or connect the following:
- Halfshaft into the wheel bearing and the final drive
- Upper control arm. Torque the bolt to 52 ft. lbs. (70 Nm) plus a ¼ (90 degree) turn
- Sway bar link. Torque the bolt to 37 ft. lbs. (50 Nm)
- Center and rear mufflers, if removed
- CV-joint to the final drive. Torque the bolts to 30 ft. lbs. (40 Nm)
- ABS wheel speed sensor
- Halfshaft bolt. Torque the bolt to 85 ft. lbs. (115 Nm) plus ¼ (90 degree) turn
- Rear wheel

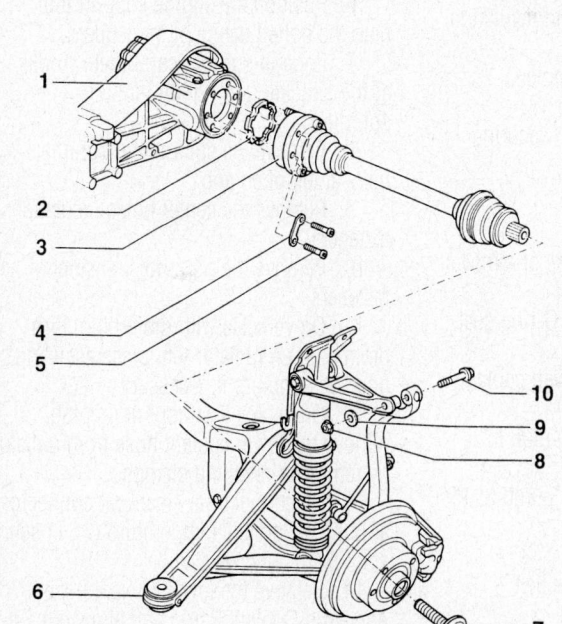

1. Rear final drive
2. Gasket
3. Halfshaft
4. Spacer plate
5. Halfshaft retaining bolts
6. Subframe
7. Collar bolt
8. Self-locking nut
9. Washer
10. Halfshaft retaining bolt

7923CG30

**Fig. 25 Exploded view of the rear halfshaft and related component mounting—A4 Quattro models**

## ENGINE COOLING

### WATER PUMP

*REMOVAL & INSTALLATION*

**A3 Models**

*2.0L Engine*

See Figure 26.

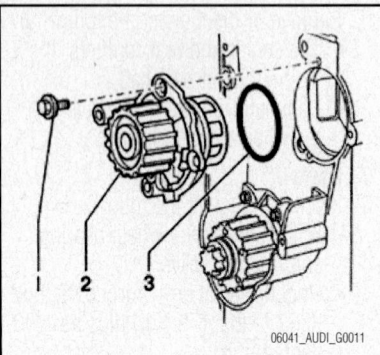

06041_AUDI_G0011

**Fig. 26 Exploded view of the water pump assembly—A3 with 2.0L engine**

1. Before servicing the vehicle, refer to the precautions section.
2. Remove the coolant expansion tank cap.
3. Disconnect the electrical connector.
4. Disconnect the intake connection using spring-type clip pliers.
5. Remove the engine cover.
6. Remove the center sound insulation fasteners.
7. Drain and recycle the coolant.
8. Remove the timing belt.
9. Remove the water pump securing bolts and remove water pump.
10. Remove the O-ring.

**To install:**

11. Installation is the reverse of removal, please note the following:
    a. Clean and/or smooth O-ring sealing surface.
    b. Moisten new O-ring with coolant.
    c. Insert water pump.
    d. Make sure the sealing plug in housing points downward.
    e. Tighten bolts of water pump to 15 Nm.
    f. Install the timing belt.
    g. Install accessory drive belt
12. Fill the cooling system.

*3.2L Engine*

See Figures 27 through 29.

1. Before servicing the vehicle, refer to the precautions section.

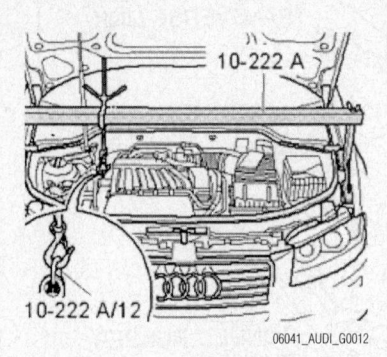

10-222 A

10-222 A/12

06041_AUDI_G0012

**Fig. 27 Secure shackle to the right rear engine lifting eyelet, then position the engine support tool onto the bolted flange of the fenders—A3 with 3.2L engine**

2. Remove the coolant expansion tank cap.

➡**Before removing accessory drive belt, mark the turning direction on it with chalk or a felt tip pen. A reversed direction of rotation can cause damage to the belt under operating conditions.**

3. Remove the accessory drive belt.
4. Disconnect the air guide hose at Throttle Valve Control Module.
5. Secure shackle to the right rear engine lifting eyelet.
6. Position the engine support tool onto the bolted flange of the fenders.
7. Engage spindle carabineer hooks at the shackle and lightly tension spindle.
8. Remove the subframe mounting bolts at left of engine.
9. Remove the center noise insulation fasteners.
10. Remove the right noise insulation fasteners.
11. On vehicles with drain plug, turn drain plug on radiator left, place assisting hose on supports if necessary.
12. On vehicles without drain plug, remove the lower coolant hose from radiator by removing retaining clamps.
13. Disconnect the electrical connector at Engine Coolant Temperature (ECT) sensor.
14. Remove the lower coolant hose to After-Run Coolant Pump and allow remaining coolant to drain.
15. Remove the pendulum support by removing bolts.
16. Unclip electrical wiring bracket to Oil Level Thermal sensor.

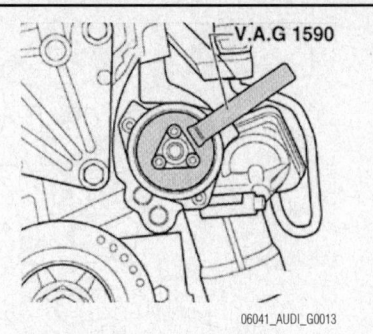

V.A.G 1590

06041_AUDI_G0013

**Fig. 28 Remove the belt pulley while securing pulley with the water pump wrench—A3 with 3.2L engine**

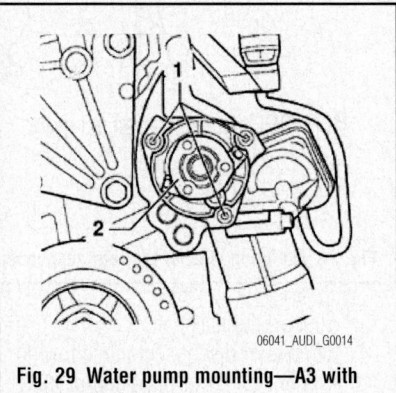

06041_AUDI_G0014

**Fig. 29 Water pump mounting—A3 with 3.2L engine**

17. Lower the engine at the spindle about 60 mm.
18. Remove the belt pulley while securing pulley with the water pump wrench.

➡**If engine is removed, the belt pulley must not be removed. Water pump bolts are then accessible through belt pulley holes.**

19. Remove the bolts on water pump and swing the water pump forward.

**To install:**

20. Installation is the reverse of removal, please note the following:

➡**Replace O-ring.**

    a. Clean or smooth O-ring sealing surface.
    b. Coat new O-ring with coolant additive.
    c. Install the water pump, tighten bolts to 8 Nm.
21. Install the belt pulley, tighten the bolts to 20 Nm.
22. Install the accessory drive belt .
23. Fill the cooling system.
    a. Adjust engine mounts.

**A4 Models**

*See Figures 30 through 32.*

➡ **The coolant pump is bolted to the brackets for the alternator, power steering pump and cooling fan. To gain access to the front of the engine, the front bumper must be removed and the hood lock carrier assembly moved forward to the service position.**

1. Before servicing the vehicle, refer to the precautions section.

2. Drain the engine coolant.

3. Turn the ignition switch to the **OFF** position.

4. Remove or disconnect the following:
   • Negative battery cable
   • Front bumper assembly and move the front inner fender lock carrier assembly forward to the lock carrier service position, on A4 models

5. With the front bumper removed, move the lock carrier assembly into the service position as follows:

   a. Release the 3 quick-release screws on the front of the lower engine splash shield.

   b. Unbolt the air guide between the lock carrier and the air filter assembly.

   c. Remove the 2 bolts that attach the lock carrier assembly to the side of the front fender and the 2 front bolts that mount the carrier to the top of the fender.

   d. Remove the right side top outer bumper energy absorbing strut mounting bolt and install support tool 3369 into the top outer threaded holes on both the left and right sides of the bumper energy absorbing strut mounting surface.

➡ **The 2 front bumper-to-bumper energy absorbing strut fasteners can be substituted for support tool 3369.**

   e. Remove the remaining bumper energy absorbing strut mounting bolts.

   f. Remove the remaining 2 bolts that attach the lock carrier assembly to the top of the front fenders. Pull the lock carrier assembly forward until the rearmost bolt holes of the carrier align with the front-most threaded mounting points on the top of the front fender. Install the 2 carrier mounting bolts through the carrier into threaded mounting points of front fender to secure the lock carrier assembly in this position.

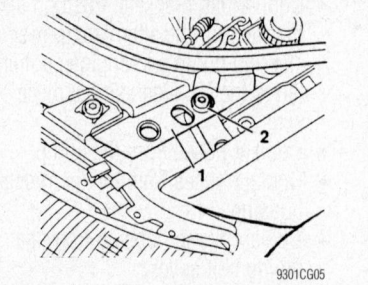

Fig. 31 The hood lock carrier assembly (1) secured in the service position with the mounting fastener (2) installed—A4 models

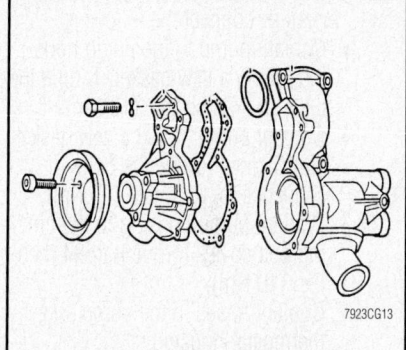

Fig. 32 Exploded view of the water pump, housing and related components

➡ **The 2 front bumper mounting fasteners can be substituted for Support tool 3369.**

   g. Remove the remaining bumper energy absorbing strut bolts and pull the lock carrier out to the stop.

   h. To secure the lock carrier, install the appropriate M6 bolts into the rear of the lock carrier and fender.

6. Remove or disconnect the following:
   • Accessory drive belt
   • Cooling fan
   • Lower engine slash shield

7. Loosen the clamps for the coolant hoses at the water pump.

8. Remove or disconnect the following:
   • Intake air duct between the intake manifold and the charge air cooler
   • Alternator mounting bolts and slide it forward
   • Alternator wiring

9. Unbolt the following supports and brackets for the alternator, power steering pump and engine cooling fan:
   • Intake manifold support
   • Support for the engine torque bracket
   • Brace to the cylinder block
   • Alternator brackets
   • Power steering pump brackets

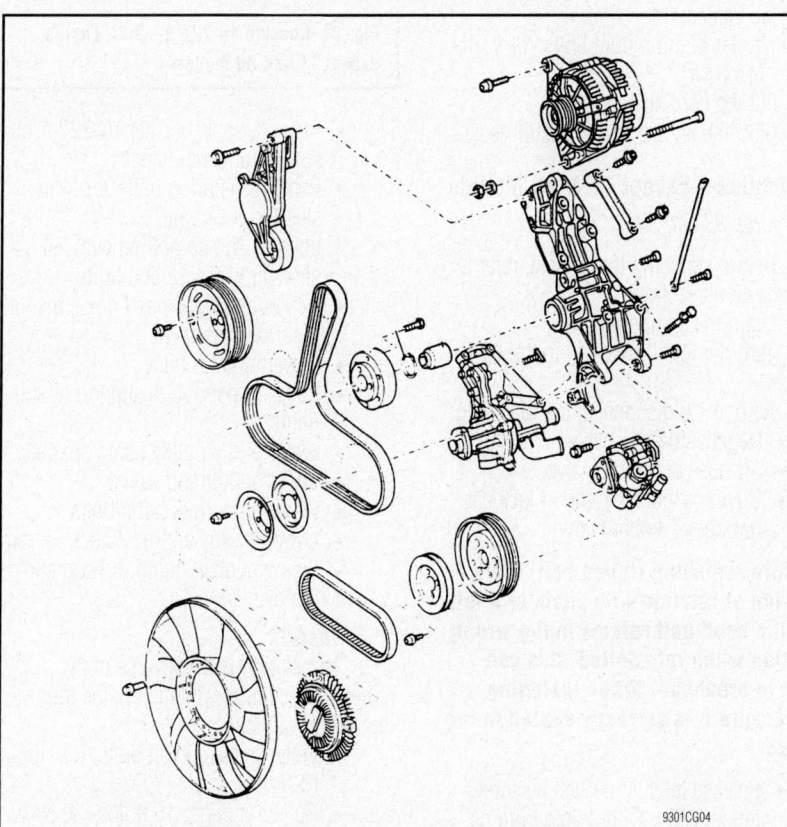

Fig. 30 The water pump, alternator and power steering pump all mount to the same engine bracket

- Cooling fan brackets. Position the brackets for the alternator, power steering pump and engine cooling fan to the left side using a piece of wire
- Coolant hoses from the pump
- Coolant hoses from the thermostat housing
- Coolant pump housing from the timing belt cover
- Coolant pump
- Impeller housing from the pump housing

10.  Clean all gasket and O-ring sealing surfaces.

### To install:

11.  Install or connect the following:
- Coolant pump to the pump housing, using a new gasket. Torque the bolts to 84 inch lbs. (10 Nm).
- Coolant pump, using a new gasket and O-ring. Torque the bolts in sequence to 18 ft. lbs. (25 Nm).
- Coolant pump housing to the timing belt cover. Torque it to 84 inch lbs. (10 Nm).
- Coolant hoses to the pump and thermostat housing
- Brackets. Torque the bolts to 18 ft. lbs. (25 Nm)
- Alternator and wire connectors
- Air intake duct between the intake manifold and the charge air cooler

12.  The remaining steps are the reverse of the removal procedure noting the following items:
- Fill the engine with coolant
- Verify that the key is in the **OFF** position before connecting the battery
- Fully close all power windows to stop, operate all window switches for at least 1 second in the close direction to activate the one touch opening/closing function
- After installing the lock carrier, check the wiring for proper routing near the cooling fan

### TT Models

1.  Before servicing the vehicle, refer to the precautions section.
2.  Drain the engine coolant.
3.  Turn the ignition switch to the **OFF** position.
4.  Remove or disconnect the following:
- Negative battery cable
- Engine cover
- Air duct at bottom of right-hand cross member
- Center and right-hand noise insulation under engine

- Accessory drive belt
- Vacuum hoses from carbon canister and throttle valve
- Coolant expansion tank and hoses
- Power steering reservoir, leaving hoses connected
- Upper and center timing belt covers

5.  Set crankshaft to TDC of No. 1 cylinder by turning central bolt on crankshaft sprocket in direction of rotation.
6.  Insert a M5 x 55 stud into the belt tensioner. Screw hex nut onto stud using large washer.
7.  Tension the piston of the tensioner only until it can be secured with a locking pin such as an Allen key or split pin.
8.  Remove the timing belt from camshaft sprocket.
9.  The vibration damper and bottom timing belt guard do not have to be removed.
10.  The timing belt should be left in position on the crankshaft sprocket.
11.  Cover timing belt with a cloth to protect it from coolant before removing coolant pump.
- Lower air duct
- Water pump

### To install:

12.  Installation is the reverse of the removal procedure, while using the following torque values:
- Water pump mount bolts: 11 ft. lbs. (15 Nm).
13.  Fill the cooling system.
14.  Start the engine and check for leaks.

### 3.2L Engine—Except TT & A6 Models

*See Figures 33 and 34.*

1.  Before servicing the vehicle, refer to the precautions section.
2.  Drain the engine coolant.
3.  Turn the ignition switch to the **OFF** position.
4.  Remove or disconnect the following:
- Negative battery cable
- Air hose at throttle valve
- Covers (right side and at lock carrier above coolant tank)

➡ **Before removing ribbed belt, mark direction of rotation with chalk or a felt pen. If a used belt rotates in the wrong direction when reinstalled, this can result in breakage. When installing belt, ensure it is correctly seated in the pulleys.**

- Release tension on belt by screwing an M8 x 40 bolt into bore on tensioning element and remove belt
- Attach shackle 10-222 A/12 or equivalent to right rear lifting eye

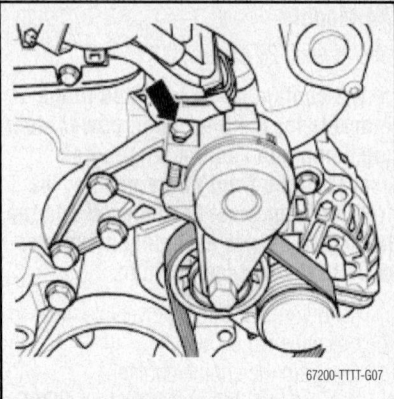

**Fig. 33 Releasing tension on drive belt— 3.2L Engine, except TT and A6 models**

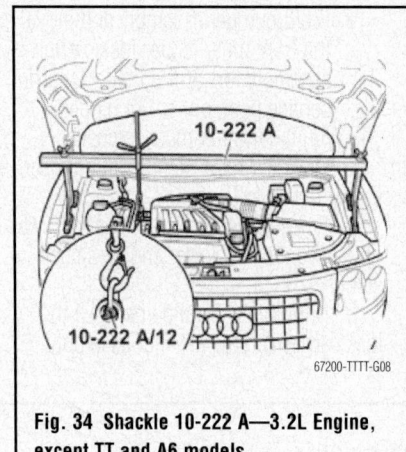

**Fig. 34 Shackle 10-222 A—3.2L Engine, except TT and A6 models**

- Place support bracket 10-222 A on surface of panel flange
- Remove securing bolt for power steering reservoir
- Loosen, but do not remove, left side engine mounting bolts
- Center and right-hand noise insulation under engine
- Lower radiator hose
- Bolts for pendulum support at subframe
- Water pump pulley using spanner VAG 1590 or equivalent
- Water pump mounting bolts
- Lower engine approximately 65 mm
- Swing coolant pump forward and remove

### To install:

5.  Installation is the reverse of the removal procedure, while using the following torque values:
- Water pump mount bolts: 6 ft. lbs. (8 Nm)
- Pulley to pump 15 ft. lbs (20 Nm)
6.  Fill the cooling system.
7.  Start the engine and check for leaks.

### A6 and A8 Models

#### 3.2L Engine

1. Before servicing the vehicle, refer to the precautions section.
2. Drain the cooling system.
3. Remove or disconnect the following:
   - Negative battery cable
   - Water pump

#### To install:

4. Install or connect the following:

---

- Water pump. Tighten the bolts to 89 inch lbs. (10 Nm).
- Negative battery cable
5. Fill the cooling system.
6. Start the engine and check for leaks.

#### 4.2L Engine

1. Before servicing the vehicle, refer to the precautions section.
2. Drain the cooling system.
3. Remove or disconnect the following:
   - Negative battery cable

---

- Timing belt
- Water pump

#### To install:

4. Install or connect the following:
   - Water pump. Tighten the bolts to 89 inch lbs. (10 Nm).
   - Timing belt
   - Negative battery cable
5. Fill the cooling system.
6. Start the engine and check for leaks.

---

## ENGINE ELECTRICAL

## CHARGING SYSTEM

### ALTERNATOR

#### REMOVAL & INSTALLATION

#### A3 Models

##### 2.0L Engine

1. Before servicing the vehicle, refer to the precautions section.
2. Disconnect the negative battery cable.
3. Disconnect the electrical connection at Mass Air Flow (MAF) sensor.
4. Remove the engine cover.
5. Lift the EVAP canister with the hoses still connected out of bracket and set aside.

➡**Before removing accessory drive belt, mark direction of rotation. Belt damage will result if not reinstalled in proper direction.**

6. Remove the accessory drive belt tension by turning tensioner in direction of.
   Lock tensioner in place using mandrel T10050 A as illustrated.
7. Remove the accessory drive belt.
8. Remove the bolts and remove tensioner.
9. Remove the bolts and remove alternator from bracket.
10. Swing alternator to right.
11. Disconnect the electrical connection.
12. Remove the B+ terminal nut and terminal eyelet.
13. Remove the cable retainer.
14. Remove the alternator.
15. Installation is the reverse order of removal.
16. Tighten the alternator to bracket bolts and belt tensioner to bracket bolts to 23 Nm.

##### 3.2L Engine

1. Before servicing the vehicle, refer to the precautions section.
2. Disconnect the negative battery cable.

---

3. Remove the right front wheel.
4. Remove the noise insulation.
5. Remove the cover for air duct and the air duct.
6. Pull off vacuum hose at air intake hose.
7. Remove the air hose from Throttle Valve Control Module.
8. Disconnect the electrical connection at Mass Air Flow (MAF) sensor.
9. Remove the top section of air cleaner housing.
10. Remove the filter element.
11. Remove the screws and remove bottom section of air cleaner housing.
12. Separate secondary air hose at point indicated.
13. Set aside air hose to secondary air pump.
14. Remove the top screws at radiator cowl.

➡**Before removing accessory drive belt, mark direction of rotation. Belt damage will result if not reinstalled in proper direction.**

15. Remove the accessory drive belt.
16. Move aside and retain lower coolant hose at radiator cowl.
17. Disconnect the electrical connection at Secondary Air Injection (AIR) Pump Motor set aside harness.
18. Remove the bolt and remove bracket for coolant pipe.
19. Remove the bolts.
20. Remove the bolt and remove secondary air pump with bracket.
21. Disconnect the electrical connections for coolant fan.
22. Remove the lower screws for radiator cowl.
23. Pull out radiator cowl with both fans from below.
24. Remove the electrical connection for a/c clutch and set aside wiring.
25. Remove the bolts for air conditioner

---

compressor and remove compressor with the hoses still attached from bracket.
26. Disconnect the electrical connection.
27. Remove the B+ terminal nut and terminal eyelet.
28. Remove the upper idler roller.
29. Loosen the bolts and.

➡**The alternator can only be removed with upper bolt still in place. If alternator is difficult to remove from mounting, reinstall the mounting bolts loosely and carefully strike bolt heads with a rubber mallet in order to loosen the mounting threaded bushings.**

30. Pull alternator away from bracket.
31. Remove the alternator from lower left engine compartment.
32. Installation is the reverse order of removal.
33. Tighten the alternator to bracket bolts to 23 Nm.
34. Tighten the tensioning roller bracket bolt to 50 Nm.
35. Tighten the A/C compressor bracket bolt to 25 Nm.

#### A6 and A8 Models

##### 3.2L Engine

1. Before servicing the vehicle, refer to the precautions section.
2. Remove or disconnect the following:
   - Air intake channel
   - Engine undercover
   - Accessory drive belt
   - Hose to charge air cooler
   - Alternator wiring harness
   - A/T and A/C lines
   - Alternator bolts
   - Alternator

##### To install:

3. Install or connect the following:
   - Alternator. Tighten the 8mm bolt to 16 ft. lbs. (22 Nm) and the 10mm bolts to 33 ft. lbs. (45 Nm).

- A/T and A/C lines
- Alternator wiring harness
- Hose to charge air cooler
- Accessory drive belt
- Engine undercover
- Air intake channel
- Negative battery cable

### 4.2L Engine

1. Before servicing the vehicle, refer to the precautions section.

2. Remove or disconnect the following:

- Negative battery cable
- Air cleaner assembly (A8)
- Engine undercover
- Alternator air duct hose
- Alternator wiring harness
- Accessory drive belt
- Alternator bolts
- Alternator

### To install:

3. Install or connect the following:

- Alternator. Tighten the alternator to engine bolts to 22 Nm, tighten the tensioner roller to bracket bolts to 22 Nm.
- Accessory drive belt
- Alternator wiring harness. Tighten the **B** terminal to 12 ft. lbs. (16 Nm) and the **D** to 9 Nm.
- Alternator air duct hose
- Engine undercover
- Air cleaner assembly
- Negative battery cable

### TT Models

### 1.8L Engine

1. Before servicing the vehicle, refer to the precautions section.

### ❊❊ CAUTION

**Before beginning repairs on the electrical system:**

- Obtain the anti-theft radio security code.
- Switch off all electrical components.
- Switch ignition off and remove the ignition key.
- Disconnect negative battery terminal.

2. Remove the engine cover and cover in front of the intake manifold.

3. Release the fasteners and remove the left lock carrier cover.

4. Release the fasteners and remove the right lock carrier cover and expansion tank cover.

5. Before removing drive belt, mark direction of rotation. Belt damage will result if not reinstalled in proper direction.

6. Remove the drive belt.

7. Remove the secondary air system pipe from brackets at engine.

8. Remove the bolts and pull off holder from guide tube for dipstick.

9. On the underside of the holder, disconnect all electrical connections.

10. Disconnect the vacuum hose to solenoid valves at the intake manifold.

11. Set aside the holder with hoses still connected.

12. Pull the air hoses off of the secondary air pump.

13. Disconnect electrical connector.

14. Remove the nut from alternator terminal B+ and remove the connector eyelet from terminal.

15. Remove the harness retainer at alternator.

16. Remove the bolts and the alternator.

### To install:

17. Installation is the reverse of removal, please note the following:

18. Press the sleeves into alternator mounting to aid in installation.

19. Before installing the drive belt, make sure that all subassemblies (alternator, a/c compressor etc.) are securely mounted and turn freely. When installing belt, ensure correct seating in the belt pulleys!

20. Note running direction of drive belt as previously marked and route as illustrated.

21. Tighten the alternator mounting bolts to 16 ft. lbs. (23 Nm).

22. Start engine and check for proper operation.

### 3.2L Engine

1. Before servicing the vehicle, refer to the precautions section.

### ❊❊ CAUTION

**Before beginning repairs on the electrical system:**

- Obtain the anti-theft radio security code.

- Switch off all electrical components.
- Switch ignition off and remove the ignition key.
- Disconnect negative battery terminal.

2. Disconnect the negative battery cable.

3. Release the fastener and remove the cover.

4. Before removing drive belt, mark direction of rotation. Belt damage will result if not reinstalled in proper direction.

5. Thread M8 x 40 bolt into the tensioner enough to relieve drive belt tension.

6. Remove the drive belt.

7. Remove the upper bolt.

8. Remove the screws and the center noise insulation panel.

9. Remove the screws and the right and left noise insulation.

10. Remove the bolts and the cross brace.

11. Remove the bolts from the power steering pump and position the pump aside with the lines attached.

12. Disconnect A/C compressor electrical connector, remove the bolts and position aside with the hoses attached.

13. Remove the alternator lower bolt.

14. Remove the alternator from bracket leaving the wiring connected.

15. Remove the harness retainer at alternator.

16. Remove the nut from alternator terminal B+ and the connector eyelet from terminal.

17. Disconnect the electrical connector and remove the alternator.

### To install:

18. Installation is the reverse of removal, please note the following:

19. Before installing the drive belt, make sure that all subassemblies (alternator, a/c compressor etc.) are securely mounted and turn freely. When installing belt, ensure correct seating in the belt pulleys!

20. Note running direction of drive belt as previously marked and route as illustrated.

21. Tighten the alternator mounting bolts to 16 ft. lbs. (23 Nm).

22. Start engine and check for proper operation.

## ENGINE ELECTRICAL                                                    IGNITION SYSTEM

### FIRING ORDER

*See Figures 35 through 37.*

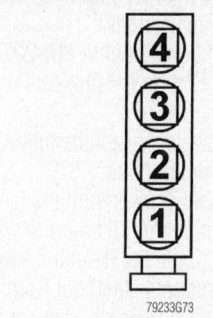

**Fig. 35 1.8L & 2.0L Engines–Firing Order 1-3-4-2–Distributorless ignition system (one coil on each cylinder)**

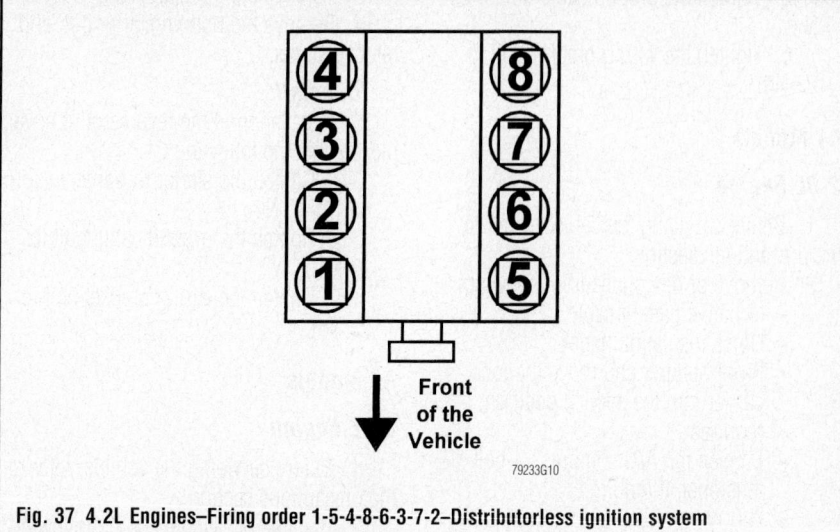

**Fig. 37 4.2L Engines–Firing order 1-5-4-8-6-3-7-2–Distributorless ignition system**

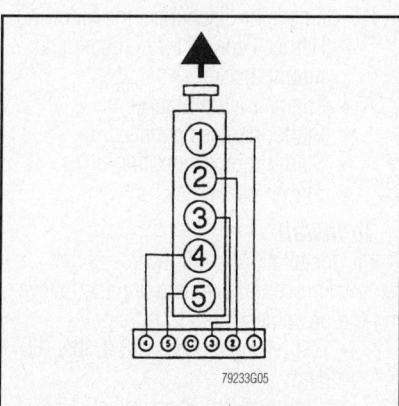

**Fig. 36 3.2L Engines–Firing order 1-4-3-6-2-5**

### IGNITION COIL

#### REMOVAL & INSTALLATION

See Spark Plugs.

### IGNITION TIMING

All vehicles in this section are equipped with distributorless ignition systems. No adjustments are possible.

### SPARK PLUGS

#### REMOVAL & INSTALLATION

1. Before servicing the vehicle, refer to the precautions in the beginning of this section.
2. Disconnect the coil harness connectors.
3. Remove the ignition coils.
4. Remove the spark plugs.

**To install:**

5. Install the spark plugs and tighten as follows:
   - 4 cylinder engines: 22 ft. lbs. (30 Nm)
   - 6 cylinder engines: 15 ft. lbs. (20 Nm)
6. Install the ignition coils.
7. Connect the coil wiring harness connectors.

## ENGINE ELECTRICAL                                                    STARTING SYSTEM

### STARTER

#### REMOVAL & INSTALLATION

#### A3 Models

##### 2.0L Engine

1. Before servicing the vehicle, refer to the precautions section.
2. Disconnect the negative battery cable.
3. Disconnect the electrical connection at Mass Air Flow (MAF) sensor.
4. Remove the engine cover.
5. Remove the cable tie for protective sleeve.
6. Disconnect the electrical connections.
7. Remove the top starter bolt.
8. Remove the bottom starter bolt and remove starter.

**To install:**

9. Installation is the reverse of removal, please note the following:
   a. Tighten the starter to transmission bolt to 80 Nm.
   b. Tighten the ground cable to starter bolt to 23 Nm.
   c. Tighten the wiring bracket bolt to 23 Nm.

##### 3.2L Engine

1. Before servicing the vehicle, refer to the precautions section.
2. Disconnect the negative battery cable.
3. Pull off vacuum hose at air intake hose.
4. Remove the air hose from Throttle Valve Control Module.
5. Disconnect electrical connection at Mass Air Flow (MAF) sensor.
6. Remove the top section of air cleaner housing.
7. Remove the filter element.
8. Pull off cover for air duct.
9. Unclip air duct.
10. Remove the screws and remove bottom section of air cleaner housing.
11. Remove the screws and remove bracket for air cleaner housing.
12. Disconnect the electrical connections.
13. Remove the top starter bolt.
14. Remove the bottom starter bolt and remove starter.

**To install:**

15. Installation is the reverse of removal, please note the following:

a. Tighten the starter to transmission bolt to 40 Nm.

b. Tighten the ground cable bolt to 23 Nm.

c. Tighten the wiring bracket bolt to 23 Nm.

## A4 Models

### 2.0L Engine

1. Before servicing the vehicle, refer to the precautions section.

2. Remove or disconnect the following:
- Negative battery cable
- Noise insulation panel
- Front bumper and move the lock carrier into the service position, if required
- Loosen the A/C compressor belt tensioner, if required
- A/C compressor and move it aside, if required
- Starter electrical connectors
- Starter

### To install:

3. Install or connect the following:
- Starter motor. Torque the mounting bolts to 48 ft. lbs. (65 Nm).
- Starter electrical connectors
- A/C compressor and torque the bolts to 18 ft. lbs. (25 Nm), if removed
- A/C belt and torque the tensioner bolt to 15 ft. lbs. (20 Nm), if removed
- Front bumper, if removed
- Noise insulation panel
- Negative battery cable

4. Enter the radio code and the preset frequencies.

### 3.2L Engine

1. Before servicing the vehicle, refer to the precautions section.

2. Disconnect the negative battery cable.

3. Pull off vacuum hose at air intake hose.

4. Remove the air hose from Throttle Valve Control Module.

5. Disconnect electrical connection at Mass Air Flow (MAF) sensor.

6. Remove the top section of air cleaner housing.

7. Remove the filter element.

8. Pull off cover for air duct.

9. Unclip air duct.

10. Remove the screws and remove bottom section of air cleaner housing.

11. Remove the screws and remove bracket for air cleaner housing.

12. Disconnect the electrical connections.

13. Remove the top starter bolt.

14. Remove the bottom starter bolt and remove starter.

### To install:

15. Installation is the reverse of removal, please note the following:

a. Tighten the starter to transmission bolt to 40 Nm.

b. Tighten the ground cable bolt to 23 Nm.

c. Tighten the wiring bracket bolt to 23 Nm.

## A6 Models

### 3.2L Engine

1. Before servicing the vehicle, refer to the precautions section.

2. Remove or disconnect the following:
- Negative battery cable
- Noise insulation panel
- Starter retainer at engine mount
- Electrical connection at engine mount
- Starter ground wire
- Starter wiring connectors
- Rotate starter downward while turning clockwise to remove

### To install:

3. Install or connect the following:
- Starter and torque the bolts to 48 ft. lbs. (65 Nm)
- Starter wiring connectors
- Ground wire
- Electrical connection
- Starter retainer
- Noise insulation panel
- Negative battery cable

### 4.2L Engine

1. Before servicing the vehicle, refer to the precautions section.

2. Relieve the fuel system pressure.

3. Remove or disconnect the following:
- Negative battery cable
- Engine cover
- Intake air duct between the air filter and the throttle valve part
- Air-induction ports between the air filter housing and the metering unit
- Fan shroud and the electric fan
- Engine undercover
- Alternator air duct hose
- Alternator wiring harness
- Accessory drive belt
- Alternator bolts
- Alternator
- The coverings over the strut tower and the air filter housing

4. Place the support device 10-222A with adapter for support device 10-222A/4 onto the screws of the strut tower.

5. Unscrew the fuel line from the pressure regulator.

6. Insert support device 3180 on the transmission side into the engine hoisting ring and screw it in securely.

7. Insert additional hook A (10-222A/2) into the hoisting ring on the pressure regulator.

8. Remove or disconnect the following:
- Engine mount bolts
- Attachment screws from the torque converter bearing on the right front

9. Use the spindles to hoist the engine until the engine mounts stand out from the subframe.

➡**Be careful not to let the throttle valve part damage the soundproofing material on the bulkhead.**

10. Remove or disconnect the following:
- Harness connector at the engine mount bracket
- Engine mount bracket
- Starter harness connectors
- Starter motor mounting bolts
- Starter motor

### To install:

11. Installation is the reverse of the removal procedure, while using the following torque values:
- Starter motor bolts: 48 ft. lbs. (65 Nm)
- Starter motor **B+** terminal: 12 ft. lbs. (16 Nm)

## A8

### 3.2L Engine

1. Before servicing the vehicle, refer to the precautions section.

2. Remove or disconnect the following:
- Negative battery cable
- Noise insulation panel
- Starter retainer at engine mount
- Electrical connection at engine mount
- Starter ground wire
- Starter wiring connectors
- Rotate starter downward while turning clockwise to remove

### To install:

3. Install or connect the following:
- Starter and torque the bolts to 48 ft. lbs. (65 Nm)
- Starter wiring connectors
- Ground wire
- Electrical connection
- Starter retainer

- Noise insulation panel
- Negative battery cable

### 4.2L Engine

1. Before servicing the vehicle, refer to the precautions section.
2. Relieve the fuel system pressure.
3. Remove or disconnect the following:
   - Negative battery cable
   - Engine cover
   - Intake air duct between the air filter and the throttle valve part
   - Air-induction ports between the air filter housing and the metering unit
   - Fan shroud and the electric fan
   - Engine undercover
   - Alternator air duct hose
   - Alternator wiring harness
   - Accessory drive belt
   - Alternator bolts
   - Alternator
   - The coverings over the strut tower and the air filter housing
4. Place the support device 10-222A with adapter for support device 10-222A/4 onto the screws of the strut tower.
5. Unscrew the fuel line from the pressure regulator.
6. Insert support device 3180 on the transmission side into the engine hoisting ring and screw it in securely.
7. Insert additional hook A (10-222A/2) into the hoisting ring on the pressure regulator.
8. Remove or disconnect the following:
   - Engine mount bolts
   - Attachment screws from the torque converter bearing on the right front

9. Use the spindles to hoist the engine until the engine mounts stand out from the subframe.

➡**Be careful not to let the throttle valve part damage the soundproofing material on the bulkhead.**

10. Remove or disconnect the following:
    - Harness connector at the engine mount bracket
    - Engine mount bracket
    - Starter harness connectors
    - Starter motor mounting bolts
    - Starter motor

#### To install:
11. Installation is the reverse of the removal procedure, while using the following torque values:
    - Starter motor bolts: 48 ft. lbs. (65 Nm)
    - Starter motor **B+** terminal: 12 ft. lbs. (16 Nm)

### TT Models

#### 1.8L Engine

1. Before servicing the vehicle, refer to the precautions section.
2. Remove or disconnect the following:
   - Battery
   - Battery platform
   - Air cleaner housing bolts on battery platform
   - On Roadsters, remove support at front cross member
   - Cable retainers
   - Starter electrical connectors
   - Starter

#### To install:
3. Install or connect the following:
   - Starter motor. Torque the mounting bolts to 48 ft. lbs. (65 Nm)
   - Starter electrical connectors
   - Front cross member support
   - Air cleaner housing
   - Battery platform
   - Battery cables
4. Enter the radio code and the preset frequencies.

#### 3.2L Engine

1. Before servicing the vehicle, refer to the precautions section.
2. Remove or disconnect the following:
   - Battery
   - Battery platform
   - Air cleaner housing bolts on battery platform
   - On Roadsters, remove support at front cross member
   - Cable retainers
   - Starter electrical connectors
   - Starter

#### To install:
3. Install or connect the following:
   - Starter motor. Torque the mounting bolts to 48 ft. lbs. (65 Nm)
   - Starter electrical connectors
   - Front cross member support
   - Air cleaner housing
   - Battery platform
   - Battery cables
4. Enter the radio code and the preset frequencies.

## ENGINE MECHANICAL

➡**Disconnecting the negative battery cable may interfere with the functions of the on board computer systems and may require the computer to undergo a relearning process, once the negative battery cable is reconnected.**

### ACCESSORY DRIVE BELTS

#### ACCESSORY BELT ROUTING

See Figures 38 through 41.

#### INSPECTION

Inspect the drive belt for signs of glazing or cracking. A glazed belt will be perfectly smooth from slippage, while a good belt will have a slight texture of fabric visible. Cracks will usually start at the inner edge of the belt and run outward. All worn or damaged drive belts should be replaced immediately.

2348-TTTT-G20

**Fig. 38 Accessory drive belt routing—1.8L Engine**

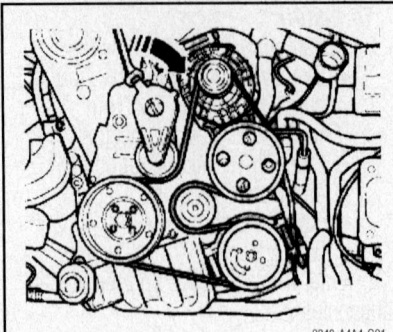

Fig. 39 Accessory drive belt routing—2.0L Engine

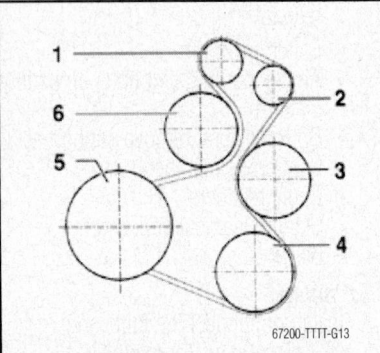

Fig. 40 Accessory drive belt routing—3.2L Engine

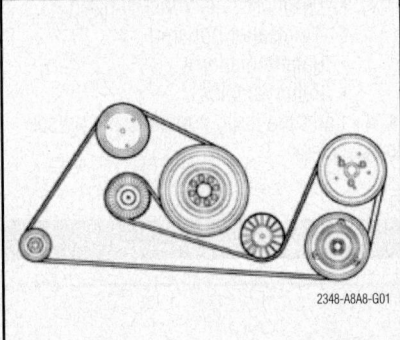

Fig. 41 Accessory drive belt routing—4.2L Engine

## REMOVAL & INSTALLATION

1. Before servicing the vehicle, refer to the precautions in the beginning of this section.

2. Pivot the belt tensioner pulley and remove the belt from the alternator pulley.

3. Remove the center and right engine undercovers.

4. Remove the belt from the remaining pulleys.

5. Installation is the reverse of the removal procedure.

## CAMSHAFT AND VALVE LIFTERS

### REMOVAL & INSTALLATION

#### 1.8L Engine

*See Figures 42 through 46.*

1. Before servicing the vehicle, refer to the precautions section.

2. Turn the ignition switch to the **OFF** position.

3. Disconnect the negative battery cable.

4. On A4 models, place the lock carrier into the service position.

5. Remove or disconnect the following:
   • Accessory drive belts
   • Engine covers
   • Timing belt upper cover

6. Turn the crankshaft, in the direction of rotation (clockwise), until the No. 1 cylinder is at Top Dead Center (TDC).

7. Remove or disconnect the following:
   • Timing belt tensioner by loosening it using Torx® wrench T45
   • Belt from the camshaft gear by pushing the tensioner downward
   • Torx® bolt and swing the tensioner assembly bracket forward
   • Timing belt
   • Valve cover
   • Cam gear retaining bolt by loosening it using retainer tool 3036

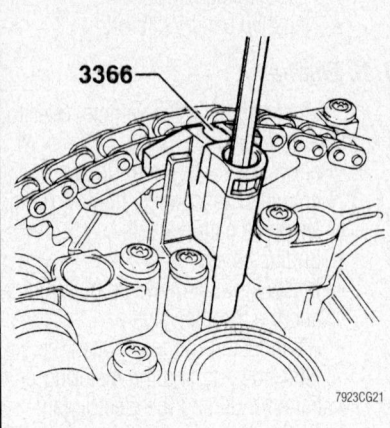

Fig. 43 Do not overtighten the chain tensioner tool 3366, it can be damaged

   • Camshaft gear
   • Camshaft Position (CMP) housing sensor and shutter wheel
   • Hydraulic chain tensioner by securing it with bracket tensioner tool 3366

8. Verify that the camshafts are at TDC for the No. 1 cylinder. Both camshaft markings must align with arrows on the bearing caps.

9. Clean the drive chain and the cam chain gears opposite both arrows on the bearing caps. Matchmark the installed position using paint.

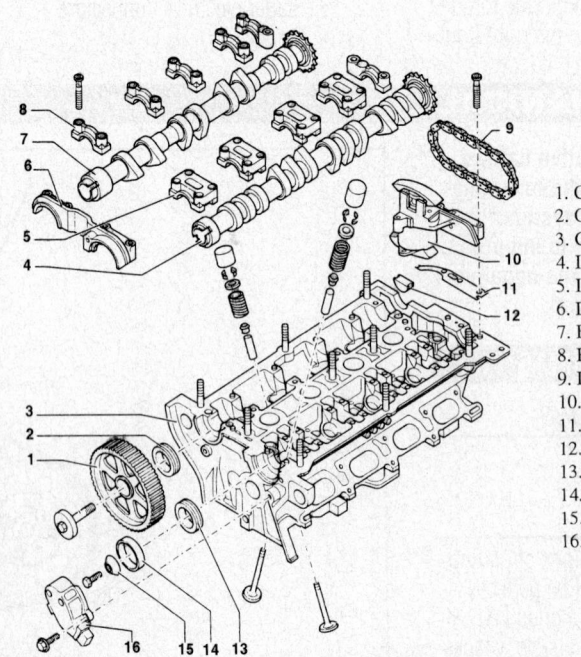

1. Camshaft gear
2. Oil seal
3. Cylinder head
4. Intake camshaft
5. Intake camshaft bearing cap
6. Double bearing cap
7. Exhaust camshaft
8. Exhaust camshaft bearing cap
9. Drive chain
10. Hydraulic chain tensioner
11. Rubber/metal seal
12. Gasket
13. Oil seal
14. Shutter wheel for the CMP
15. Washer
16. CMP sensor housing

Fig. 42 Exploded view of the camshaft mounting and related components—1.8L engine

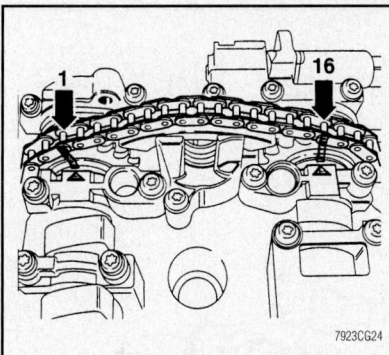

**Fig. 44 To ensure proper installation, matchmark the chain-to-camshaft position**

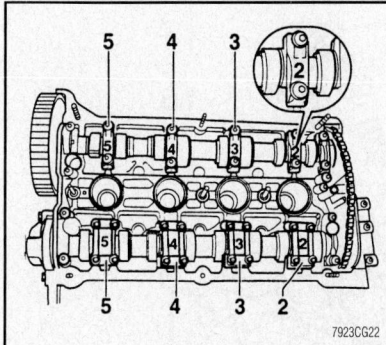

**Fig. 45 Camshaft bearing cap identification**

➡**The distance between the 2 arrows/paint marks is equivalent to 16 drive chain rollers and the notch on the exhaust camshaft is slightly offset inward toward the drive chain roller.**

10. Remove or disconnect the following:
- Bearing caps No. 3 and 5 from the intake and exhaust camshafts
- Double bearing cap
- Both bearing caps from the chain gears on the intake and exhaust camshafts
- Hydraulic chain tensioner retaining bolts

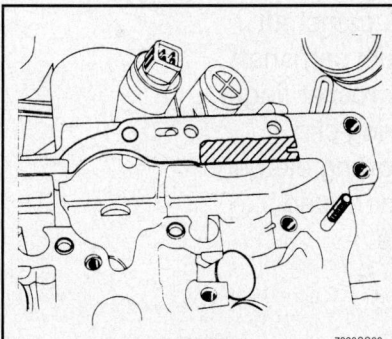

**Fig. 46 To ensure a proper seal, be sure to apply sealant to the hatched area**

- Intake and exhaust manifold bearing caps No. 2 and 4 by loosening them in an alternating and diagonal sequence
- Camshafts with the hydraulic chain tensioner

*To install:*

11. Replace the rubber/metal chain tensioner gasket and apply sealant to the hatched area, as shown.
12. Install or connect the following:
- Drive chain on the camshaft

➡**If installing the old chain, align the paint marks with the camshaft marks. If installing a new chain, the distance between the notches A and B on the camshafts must equal the distance between 16 drive chain rollers.**

- Hydraulic chain tensioner by sliding it between the drive chains
- Camshafts with the chain tensioner lubricated with engine oil into the cylinder head

➡**When installing the bearing caps, verify the markings on the caps are readable from the intake side of the cylinder head.**

- Intake and exhaust camshafts bearing caps No. 2 and 4. Torque them in an alternating diagonal sequence to 89 inch lbs. (10 Nm).
- Both the intake and exhaust camshafts bearing caps on the chain sprockets. Torque the bolts to 89 inch lbs. (10 Nm).

13. Verify the correct positions of the camshafts.
14. Remove the bracket tensioner.
15. Install or connect the following:
- Cylinder head-to-double bearing cap mating surface by lightly coating it with sealant. Torque the remaining bearing caps to 89 inch lbs. (10 Nm)
- Camshaft gear. Torque the bolt to 48 ft. lbs. (65 Nm)
- CMP shutter wheel and housing cover
- Valve cover

16. Align the camshaft gear and the vibration damper with the TDC markings.
17. Install or connect the following:
- Timing belt
- Accessory drive belts and the engine cover
- Lock carrier
- Negative battery cable

18. Fully close all power windows to stop, operate all window switches for at

least 1 second in the close direction to activate the one-touch opening/closing function

**✳✳ CAUTION**

**After installing the lifters or the camshaft(s), the engine must NOT be started for at least 30 minutes. Otherwise the valves could strike the pistons. Rotate the engine by hand, at least 2 revolutions, to ensure that the valves do not strike the pistons.**

19. Check the oil level before starting the engine.
20. Adjust the headlights.

### 3.2L Engine—Except A3 & A6

*See Figures 47 through 55.*

1. Before servicing the vehicle, refer to the precautions section.
2. Turn the ignition switch to the **OFF** position.
3. Disconnect the negative battery cable.
4. Drain coolant
5. Remove or disconnect the following:
- Intake manifold
- Air cleaner housing
- Fuse box
- Carrier
- Coolant hoses
- Bracket for secondary air injection hose

➡**When removing the thermostat housing, push the coolant pipe toward the coolant pump to prevent it from also being removed**

- Thermostat housing
- Throttle valve control module
- Camshaft position sensors
- Camshaft adjustment valves
- Fuel lines
- Cylinder head cover removing bolts in sequence
- Insert camshaft plate T10058 or equivalent into grooves on both camshafts
- Bolt for timing chain tensioner
- Upper timing chain cover.

➡**Be careful not to damage the head gasket when removing the upper timing chain cover**

- Guide rail
- Remove camshaft plate T10058

➡**The camshaft plate must not be in place when loosening or tightening the bolts for the camshaft adjuster**

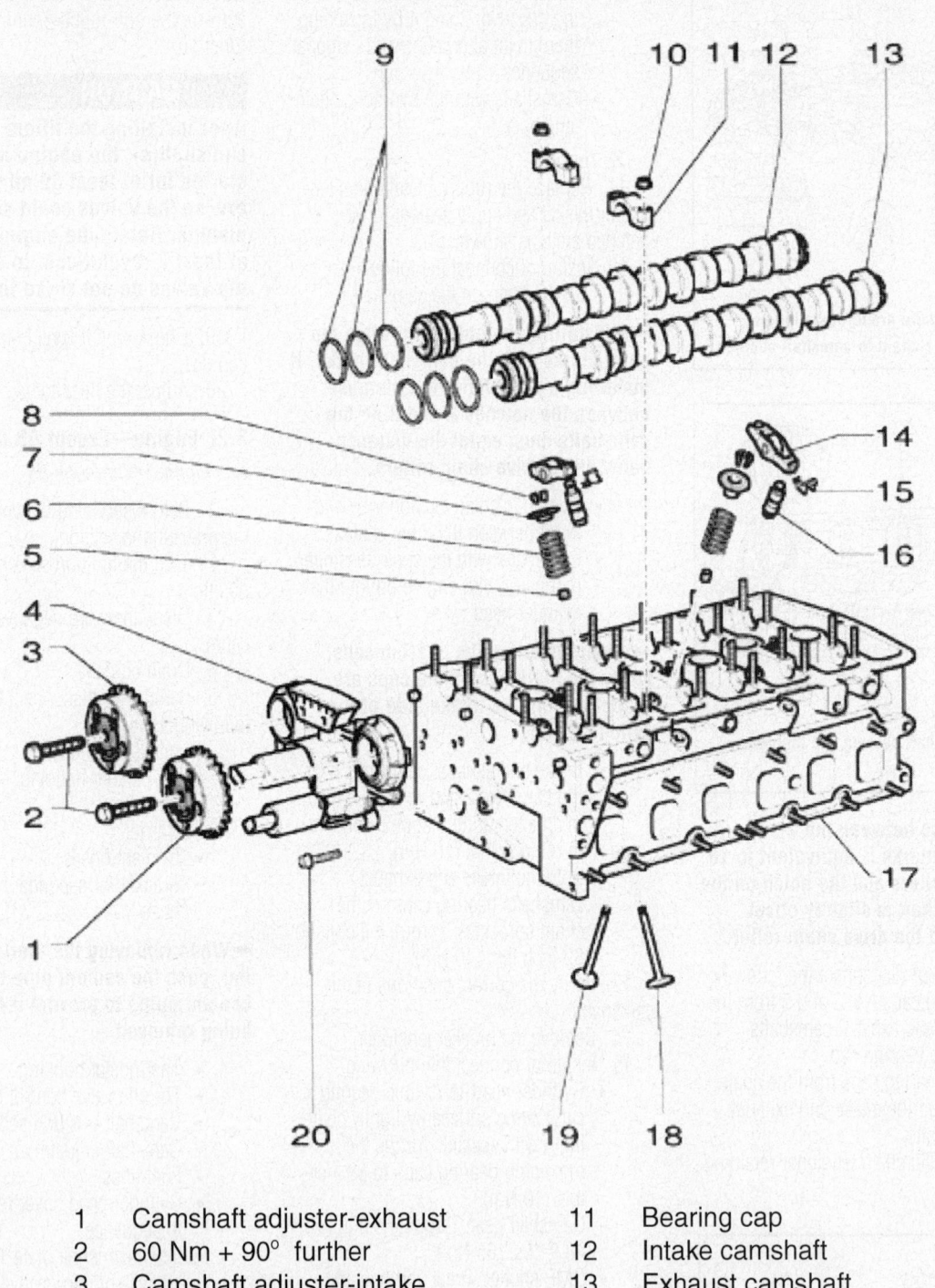

| | | | |
|---|---|---|---|
| 1 | Camshaft adjuster-exhaust | 11 | Bearing cap |
| 2 | 60 Nm + 90° further | 12 | Intake camshaft |
| 3 | Camshaft adjuster-intake | 13 | Exhaust camshaft |
| 4 | Valve timing housing | 14 | Roller rocker finger |
| 5 | Valve stem oil seal | 15 | Securing clip |
| 6 | Valve spring | 16 | Supporting element |
| 7 | Valve spring plate | 17 | Cylinder head |
| 8 | Valve keepers | 18 | Valves |
| 9 | Camshaft seals | 19 | 8 Nm |
| 10 | 5 Nm + 45° | | |

67200-TTTT-G01

**Fig. 47 Exploded view of the camshaft mounting and related components—3.2L engine, except A3 and A6**

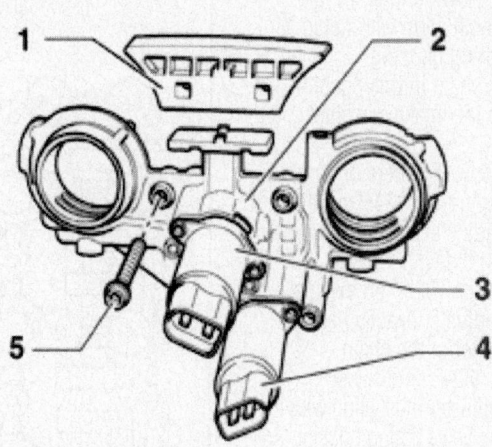

1 - Guide rail; clipped onto valve timing housing

2 - Valve timing housing

3 - Valve -1- for camshaft adjustment

4 - Camshaft adjustment valve 1 (exhaust)

5 - 8 Nm

67200-TTTT-G03

**Fig. 48 Valve timing housing—3.2L engine, except A3 and A6**

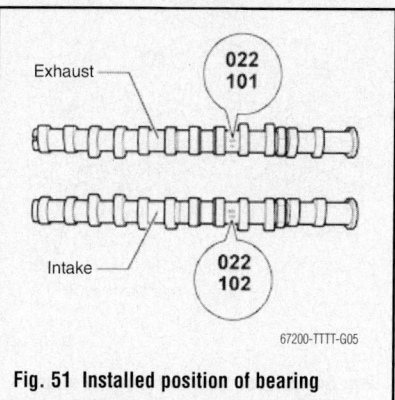

67200-TTTT-G05

**Fig. 51 Installed position of bearing caps—3.2L engine, except A3 and A6**

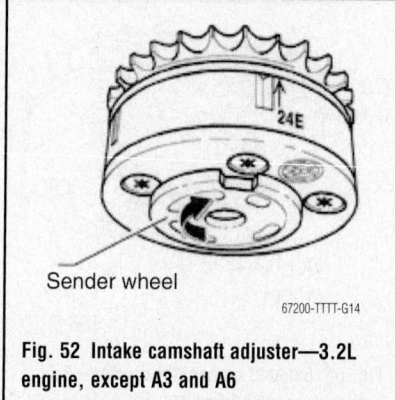

Sender wheel

67200-TTTT-G14

**Fig. 52 Intake camshaft adjuster—3.2L engine, except A3 and A6**

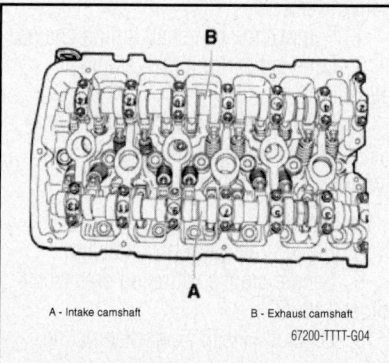

A - Intake camshaft    B - Exhaust camshaft

67200-TTTT-G04

**Fig. 49 Intake and exhaust camshafts—3.2L engine, except A3 and A6**

- Camshaft adjuster, intake side
- Camshaft adjuster, exhaust side
- Timing chain from camshafts and set aside
- Valve timing housing

6. Remove intake camshaft as follows

- Unbolt bearing caps 1 and 13
- Unbolt bearing caps 3 and 11
- Unbolt bearing cap 7
- Loosen bearing caps 5 and 9 alternately in a diagonal sequence and remove

7. Remove exhaust camshaft as follows

- Unbolt bearing caps 2 and 14
- Unbolt bearing caps 4 and 12
- Unbolt bearing cap 8
- Loosen bearing caps 6 and 10 alternately in a diagonal sequence and remove

8. Remove camshafts

**To install:**

➡Pay attention to the camshaft identification located between cam pairs for cylinder number 4 and 5.

- Exhaust camshaft identification 022–index 101
- Intake camshaft identification 022–index 102

9. Install camshafts so that they are located at TDC for cylinder number 1

- It must be possible to insert camshaft plate T10058 or equivalent in both camshaft grooves

10. Install bearing caps, pointed section "A" of bearing cap faces outward

11. Tighten intake camshaft as follows:

- Caps 5 and 9 alternately in a diagonal sequence and then tighten to final torque
- Caps 1 and 13
- Cap 7
- Caps 3 and 11

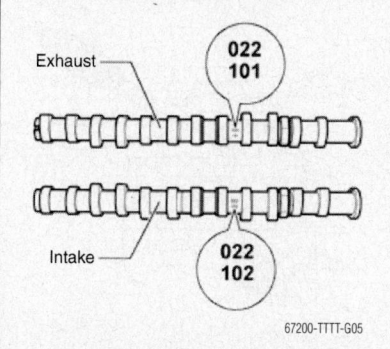

Exhaust    022 101
Intake    022 102

67200-TTTT-G05

**Fig. 50 Intake and exhaust camshaft identification—3.2L engine, except A3 and A6**

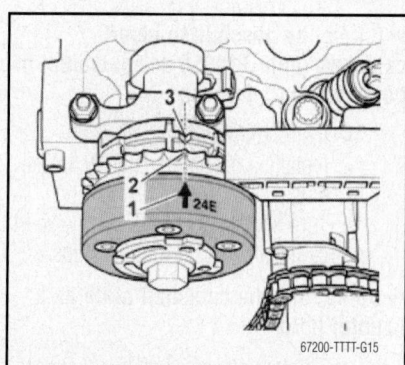

67200-TTTT-G15

**Fig. 53 Intake camshaft adjuster alignment—3.2L engine, except A3 and A6**

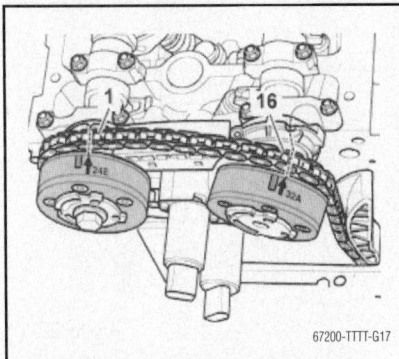

**Fig. 54 Exhaust camshaft adjuster alignment—3.2L engine, except A3 and A6**

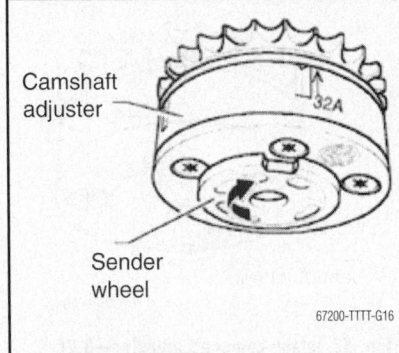

**Fig. 55 Exhaust camshaft adjuster—3.2L engine, except A3 and A6**

12. Tighten exhaust camshaft as follows:
- Caps 6 and 10 alternately in a diagonal sequence and then tighten to final torque
- Caps 2 and 14
- Cap 8
- Caps 4 and 12
- Bearing cap to cylinder head 44 in. lbs (5Nm) + 45° (45° = 1/8 turn)

13. Install valve timing housing
- Timing housing 71 in. lbs (8 Nm)

➡**Maximum Camshaft end play = .004 (.10 mm)**

➡**It must be possible to insert camshaft plate T10058 or equivalent in both camshaft grooves**

- Install guide rail
- Install camshaft adjuster for intake side
- Turn sender wheel "1" clockwise in camshaft adjuster until it stops

➡**Do not use the camshaft plate as a counter hold**

- Tighten intake camshaft adjuster to intake camshaft in this position finger tight

- Inscription "24E" and tooth just behind it must align with notch "3" in valve timing housing
- Lay timing chain in this position tautly onto the intake camshaft adjuster
- Count exactly 16 rollers of timing chain starting from "24E" on the camshaft adjuster moving towards the exhaust side
- Place tooth on "32A" on exhaust camshaft adjuster exactly behind the 16th roller of the chain
- Inscription "32A" on exhaust camshaft adjuster must align with the notch in the valve timing housing
- Turn sender wheel "2" clockwise in exhaust camshaft adjuster until it stops
- Tighten exhaust camshaft adjuster to exhaust camshaft in this position finger tight

➡**When turning the crankshaft it is necessary to press the guide rail firmly against the timing chain to prevent the timing chain from slipping**

- Turn the crankshaft 2 revolutions in normal direction and return to TDC
- Insert camshaft plate T10058 in both camshaft grooves. If camshaft plate cannot be inserted, repeat adjustment of valve timing

➡**If valve timing is correct, torque camshaft adjuster bolts to 44 ft. lbs (60 Nm) + 90 degrees (1/4 turn) install new bolts**

- Install upper timing chain cover
- Tighten bolts in following sequence
- Timing cover to engine block initially to 44 inch lbs (5 Nm)
- Bottom of timing cover at head gasket 17 ft. lbs (23 Nm)
- Timing cover to engine block to 88 inch lbs (10 Nm)

14. The remainder of the installation is the reverse of the removal procedure

➡**The engine must not be started for 30 minutes after installing the camshafts. The hydraulic valve compensation elements have to settle otherwise valves will strike pistons. After 30 minutes, turn the engine carefully at least 2 rotations to ensure that none of the valves make contact when the starter is operated**

### 3.2L Engine—A6 Models

*See Figures 56 through 62.*

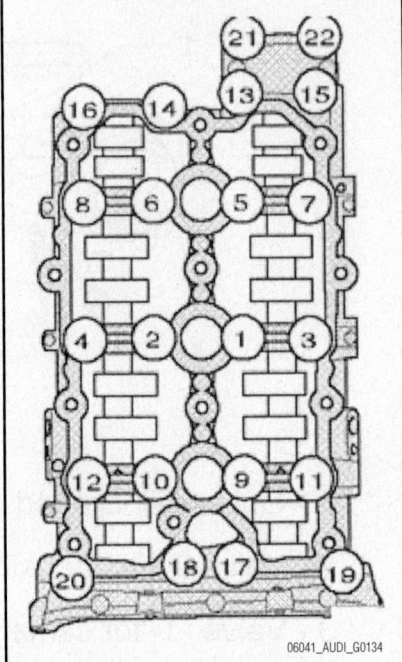

**Fig. 56 Bearing bracket loosening sequence—A6 with 3.2L engine**

1. Before servicing the vehicle, refer to the precautions section.
2. Remove the cylinder head covers.
3. Remove the left and right timing chain covers.
4. Remove the camshaft timing chains.
5. Loosen bearing bracket bolts in sequence.
6. Remove the bearing bracket.
7. Mark camshafts and their related components and remove.

### To install:

8. Replace all gaskets and seals
9. Secure crankshaft using crankshaft holder T40059 .
10. Carefully clean all gasket mating surfaces.
11. Oil the camshaft journal surfaces.

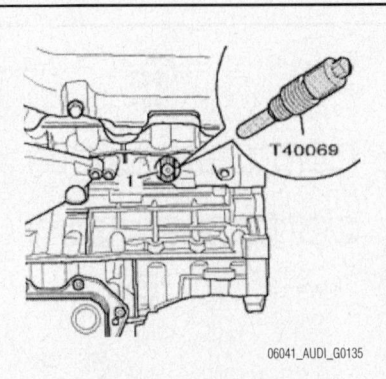

**Fig. 57 Secure crankshaft using crankshaft holder T40059—A6 with 3.2L engine**

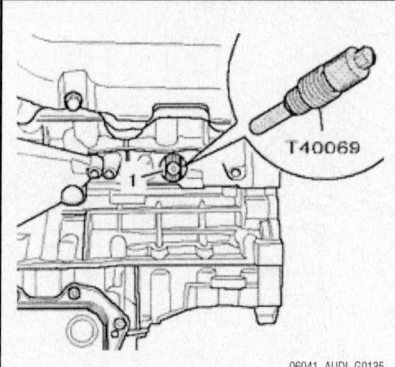

06041_AUDI_G0135

**Fig. 58 The placement of camshafts must be exactly within the axial bearings (arrows) of the bearing bracket. The ends of the piston rings (1 and 2) must face upward or downward, never sideways—A6 with 3.2L engine**

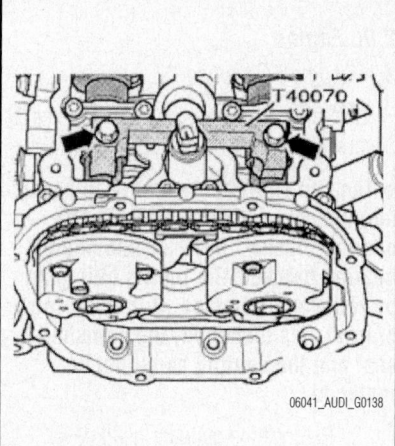

06041_AUDI_G0138

**Fig. 60 Mount camshaft locating tool T40070—A6 with 3.2L engine**

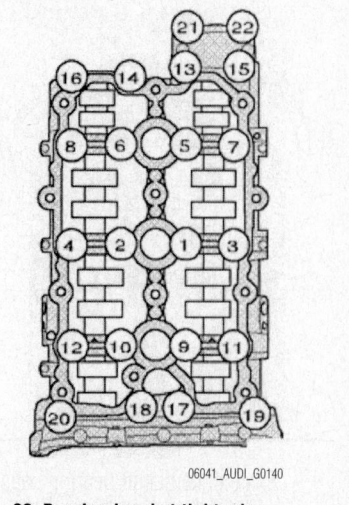

06041_AUDI_G0140

**Fig. 62 Bearing bracket tightening sequence—A6 with 3.2L engine**

12. Place bearing bracket onto a soft cloth on bench.

13. Set the camshafts into bearing bracket.

14. The placement of camshafts must be exactly within the axial bearings of the bearing bracket.

15. The ends of the piston rings must face upward or downward, and must never face sideways.

16. Turn over the bearing bracket with installed camshafts, holding camshafts tight within bearing bracket.

17. Turn camshafts until threaded holes face upward.

18. Make sure that the camshafts are still positioned exactly within the axial bearings of the bearing bracket.

19. Mount camshaft locating tool T40070 as shown and tighten bolts to 20 Nm.

20. Apply a 1mm bead of sealant of the bearing bracket as illustrated.

21. The grooves of sealing surface must be completely filled with sealant.

22. The sealant beads must be 1.5 to 2.0 mm above sealing surface.

➡**The sealant beads must be applied exactly as instructed, otherwise excess sealant could enter the camshaft bearings.**

23. Place bearing bracket immediately onto cylinder head.

➡**Placing the bearing bracket in place and tightening it should occur without interruption, since the sealant begins to harden immediately.**

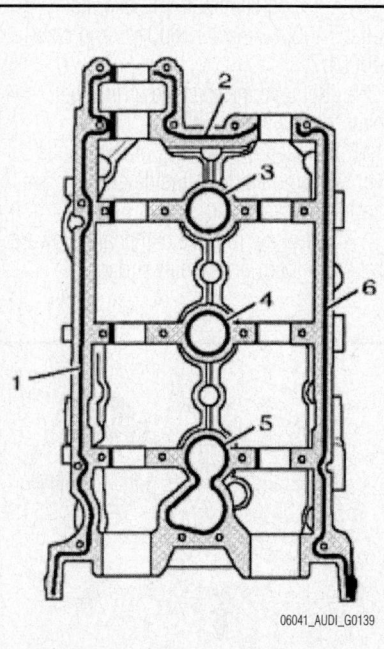

06041_AUDI_G0139

**Fig. 61 Apply a 1mm bead of sealant of the bearing bracket at the locations indicated—A6 with 3.2L engine**

➡**After installing bearing bracket, sealant must dry for approx. 30 minutes.**

24. Hand-tighten bearing bracket bolts equally, in sequence.

25. The bearing bracket must be in contact with the entire contact surface of the cylinder head.

26. Fasten bearing bracket bolts in sequence to 9 Nm until they stop.

27. Install the remaining components in the reverse order of removal.

➡**After installing the camshafts, the engine may not be started for approx. 30 minutes. The hydraulic equalization elements must seat themselves (otherwise the valves will hit the pistons).**

28. After working on the valve gear, carefully rotate engine by hand at least 2 full revolutions to ensure that valves do not strike the pistons when starting.

### 4.2L Engine

*See Figure 63.*

1. Before servicing the vehicle, refer to the precautions section.

2. Remove or disconnect the following:
- Negative battery cable
- Timing belt
- Cylinder head cover
- Camshaft sprockets
- Exhaust camshaft intermediate flange and rear-most bearing cap for camshaft position sensor housing
- Exhaust camshaft bearing cap in front of drive chain, and bearing caps 2 and 3

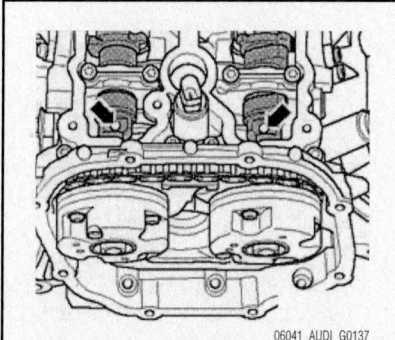

06041_AUDI_G0137

**Fig. 59 Turn camshafts until threaded holes face upward—A6 with 3.2L engine**

**Fig. 63 Camshaft alignment—A8**

- Exhaust camshaft bearing caps 1 and 4, alternating in diagonal sequence
- Intake camshaft bearing cap 6 and 7
- Intake camshaft bearing caps 5 and 8, alternating in diagonal sequence
- Intake and exhaust camshafts
- Hydraulic valve lifters

➥**Keep all valvetrain components in order for assembly.**

*To install:*

3. Install the hydraulic valve lifters in their original positions.

4. Install camshafts with drive chain so that markings on sprockets are aligned with each other (arrows).

➥**When installing the bearing caps, make sure that the stamped numbers appear upright when viewed from the intake side.**

➥**The contact surface of both outer bearing caps should be coated lightly with AMV 174 003 01.**

5. Install the intake camshaft bearing caps as follows:

a. Step 1: Tighten bearing caps 5 and 8 evenly, alternating in diagonal sequence to 11 ft. lbs. (15 Nm).

b. Step 2: Tighten the remaining bearing caps to 11 ft. lbs. (15 Nm).

6. Install the exhaust camshaft bearing caps as follows:

a. Step 1: Tighten bearing caps 1 and 4 evenly, alternating in diagonal sequence to 11 ft. lbs. (15 Nm).

b. Step 2: Tighten the remaining bearing caps to 11 ft. lbs. (15 Nm).

7. Install the camshaft sprockets and tighten the bolts to 41 ft. lbs. (55 Nm).

8. The remainder of the installation is the reverse of the removal procedure.

## A3 Models

### 2.0L Engine

*See Figures 64 through 79.*

1. Before servicing the vehicle, refer to the precautions section.

➥**The camshaft bearings are integrated into the cylinder head and into the bearing bracket. Before removing the bearing bracket, the timing belt must be relieved of tension. If the bearing bracket was loosened, the camshaft seal and the sealing cap must be replaced.**

2. Remove the cylinder head cover

3. Remove the camshaft adjuster as follows:

4. Remove the high-pressure pump

5. Remove the cylinder head cover

6. Remove the camshaft adjuster cover.

7. Move marking on camshaft sprocket so that it is across from the marking on the timing belt cover. The notches of camshafts now face toward each other vertically.

8. Install the camshaft locating tool T10252 and secure it.

9. Loosen the bolt from camshaft adjuster using socket T40080.

10. Compress chain tensioner and secure it with locking pin T10115.

11. Remove the bolt from camshaft adjuster and remove it together with chain.

12. Remove the timing belt.

13. Loosen the camshaft adjuster.

14. Pull off camshaft sprocket using puller T40001 , claw T40001/6 and claw T40001/7.

15. Remove the rear the timing belt cover from cylinder head.

16. Loosen the bearing bracket bolts evenly from outside to inside and remove bearing bracket.

17. Carefully remove camshafts upward and place them on a clean surface.

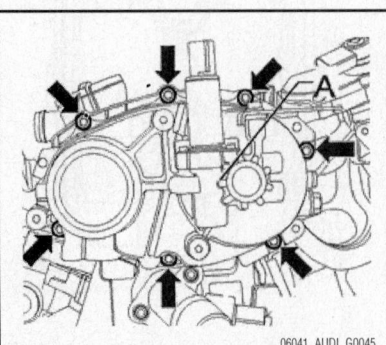

**Fig. 64 Remove the camshaft adjuster cover—A3 with 2.0L engine**

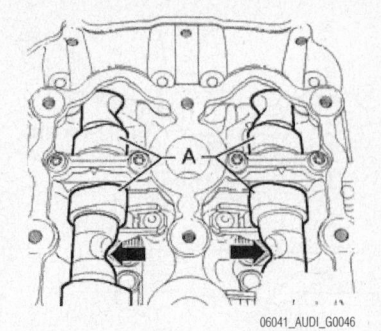

**Fig. 65 Move marking on camshaft sprocket so that it is across from the marking on the timing belt cover. The notches of camshafts now face toward each other vertically—A3 with 2.0L engine**

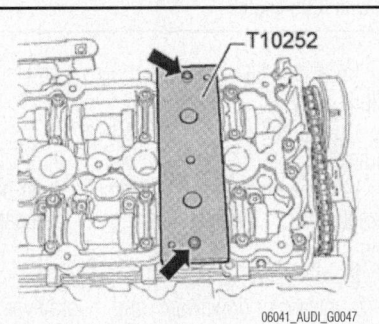

**Fig. 66 Install the camshaft locating tool T10252 and secure it—A3 with 2.0L engine**

18. Remove the old sealant from bearing bracket groove and from sealing surfaces.

19. Do not allow dirt or sealant residue to get into cylinder head.

*To install:*

➥**The sealing surfaces must be free of oil and grease.**

➥**The pistons must not be positioned at TDC.**

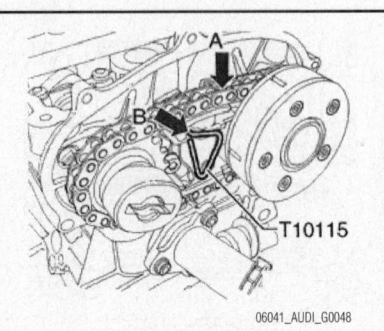

**Fig. 67 Compress chain tensioner and secure it with locking pin T10115—A3 with 2.0L engine**

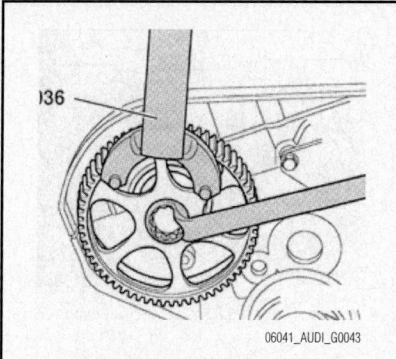

**Fig. 68 Loosen the camshaft adjuster—A3 with 2.0L engine**

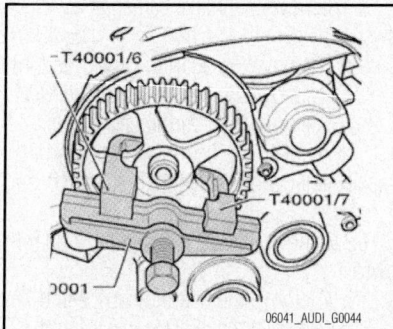

**Fig. 69 Pull off camshaft sprocket using puller T40001, claw T40001/6 and claw T40001/7—A3 with 2.0L engine**

20. Make sure that all roller rocker levers are positioned correctly on the valve stem ends.

21. Oil the journal surfaces of camshafts.

22. Carefully place camshafts into camshaft bearings of cylinder head. The cam lobes of cylinder 4 must face toward one another.

23. Apply an even, light sealant bead into clean bearing bracket groove. Spread sealant evenly on sealing surface.

➡**The sealant must not be applied too thickly. Wipe off excess sealant with a lint-free rag, if necessary.**

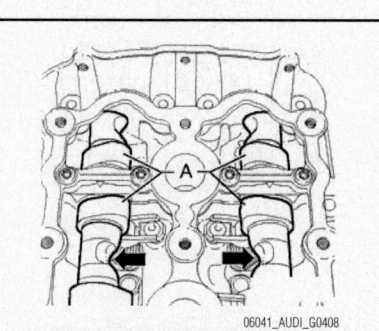

**Fig. 70 The cam lobes of cylinder 4 must face toward one another—A3 with 2.0L engine**

**Fig. 71 Set bearing bracket in place so that it can get by the EGR valve—A3 with 2.0L engine**

➡**Install the bearing bracket and tighten it right away, since the sealant begins to harden immediately when the sealing surfaces come into contact.**

24. Set bearing bracket in place so that it can get by the EGR valve.

25. Fasten bolts in using several passes using an inside to outside pattern.

26. Tighten bolts in sequence to 8 Nm.

27. Drive in sealing cap with the thrust piece 3202 about 1 to 2 mm deep.

28. Install the camshaft.

29. Replace the exhaust camshaft seal as follows:

 a. Thread special tool 2085/2 of the seal puller into camshaft by hand up to stop.

 b. Unscrew inner portion of seal puller two rotations or about 3 mm from outer portion and secure with knurled-head screw.

 c. Lubricate threaded head of seal puller, place against seal, and with strong force screw into seal as far as possible.

 d. Loosen the knurled bolt and turn inner part against camshaft until seal is removed.

 e. Secure seal remover in a vise at the flat spots.

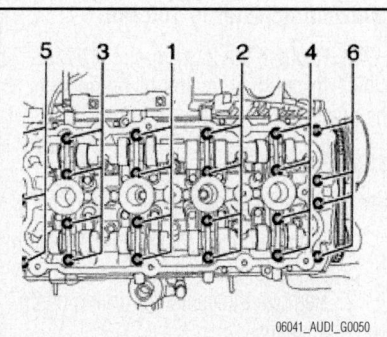

**Fig. 72 Camshaft bolt tightening sequence—A3 with 2.0L engine**

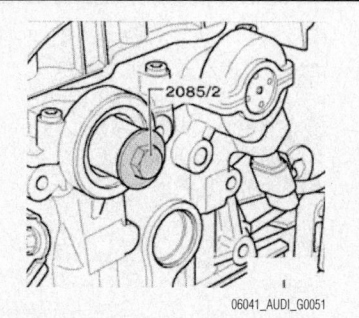

**Fig. 73 Thread special tool 2085/2 of the seal puller into camshaft by hand up to stop—A3 with 2.0L engine**

 f. Remove the seal with pliers

 g. Do not oil sealing lip of seal.

 h. Position guide sleeve T10070/1 of polydrive bit and drive socket T10070 on end of camshaft.

 i. Slide seal over guide sleeve onto camshaft end.

 j. Remove the guide sleeve.

 k. Press in seal with pressure sleeve T10071/3 and bolt T10071/4 from Poly-drive Bit And Drive Socket T10070 up to stop.

30. Install the camshaft sprocket.

➡**The thin rib of the camshaft sprocket points outward and the cyl. 1 TDC mark should be visible.**

31. Install the bolt for camshaft sprocket using retainer 3036 and tighten to 65 Nm.

➡**By turning the camshaft, the valves could contact the stationary pistons at TDC. Therefore, the pistons must not be at TDC. Valves/pistons may be damaged.**

32. Install the timing belt.

33. Camshaft sprocket to camshaft.

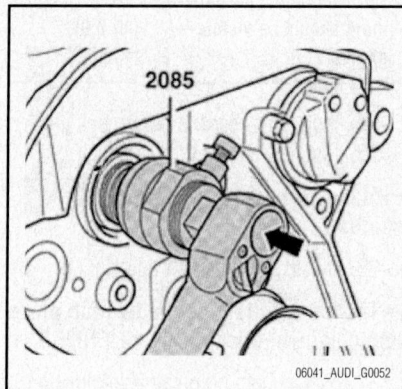

**Fig. 74 Lubricate threaded head of seal puller, place against seal, and with strong force screw into seal as far as possible—A3 with 2.0L engine**

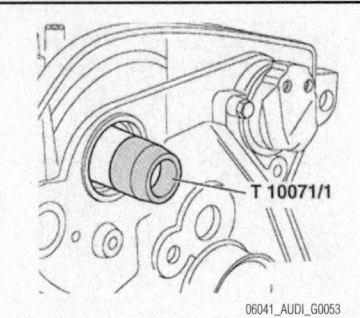

Fig. 75 Position guide sleeve T10070/1 of polydrive bit and drive socket T10070 on end of camshaft—A3 with 2.0L engine

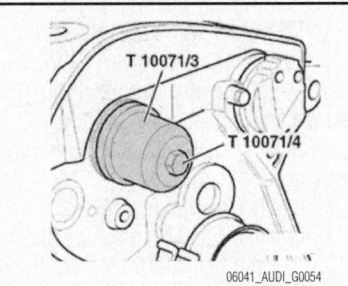

Fig. 76 Press in seal with pressure sleeve T10071/3 and bolt T10071/4 from Polydrive Bit And Drive Socket T10070 up to stop—A3 with 2.0L engine

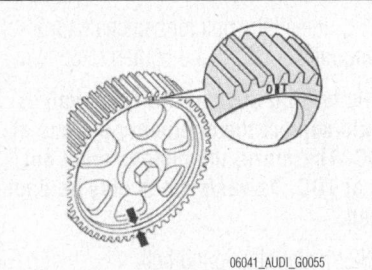

Fig. 77 The thin rib of the camshaft sprocket points outward and the cyl. 1 TDC mark should be visible—A3 with 2.0L engine

34. Install the rear the timing belt cover.
35. Insert fitted key into camshaft.

➡**Make sure fitted key is properly seated.**

36. Install the camshaft adjuster

➡**The camshafts must be fixed in place using camshaft locating tool T10252.**

37. Place chain onto camshaft adjuster.
38. Hold camshaft adjuster in front of exhaust camshaft so that the notch and pin are positioned across from one another. In this position, place chain at top onto chain sprocket of intake manifold.

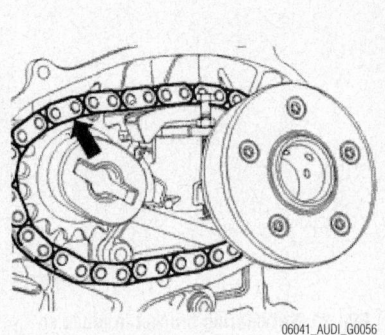

Fig. 78 Hold camshaft adjuster in front of exhaust camshaft so that the notch and pin are positioned across from one another. In this position, place chain at top onto chain sprocket of intake manifold—A3 with 2.0L engine

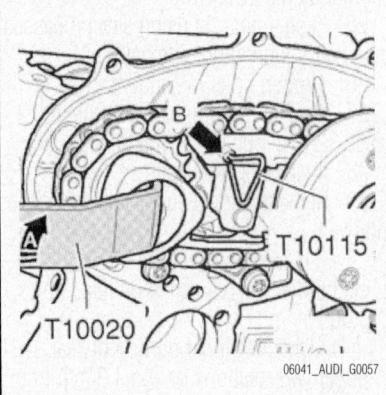

Fig. 79 Slowly turn intake camshaft using the timing belt tensioner tool T10020 until camshaft adjuster fits onto camshaft—A3 with 2.0L engine

39. Slowly turn intake camshaft using the timing belt tensioner tool T10020 until camshaft adjuster fits onto camshaft.

➡**If the pin does not fit into the notch, remove the chain and then re-install the chain again. Tighten bolt of camshaft adjuster to 100 Nm.**

40. Remove the locking pin T10115.
41. Install the remaining components in the reverse of removal.

### 3.2L Engine
*See Figures 80 through 85.*

1. Before servicing the vehicle, refer to the precautions section.
2. Remove the camshaft timing chain from camshafts.
3. Remove the control housing.
4. Carefully pull control housing off camshaft seals.

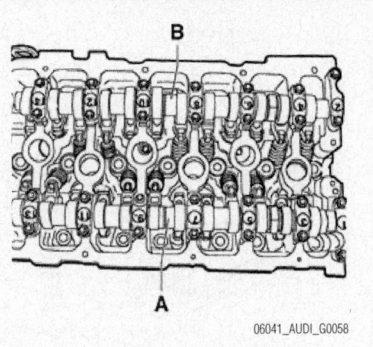

Fig. 80 Camshaft bearing cap location—A3 with 3.2L engine

5. Remove the intake camshaft as follows:
   a. Remove the bearing caps 1 and 13.
   b. Remove the bearing caps 3 and 11.
   c. Remove the bearing cap 7.
   d. Loosen the bearing caps 5 and 9 alternately and in diagonal sequence, and remove.
6. Remove the exhaust camshaft as follows:
   a. Remove the bearing caps 2 and 14.
   b. Remove the bearing caps 4 and 12.
   c. Remove the bearing cap 8.
   d. Loosen the bearing caps 6 and 10 alternately and in diagonal sequence, and remove.
7. Carefully remove camshafts and place on a clean surface.

### *To install:*

➡**When installing camshafts, crankshaft must not be at TDC with any piston. Valves and/or pistons may be damaged.**

➡**After installing the camshafts, the engine may not be started for about 30 minutes. The hydraulic equalization elements must seat themselves (otherwise the valves will crash into the pistons). After working on the valvetrain**

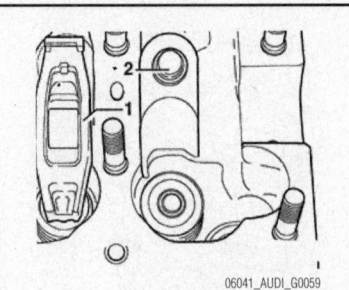

Fig. 81 Install the rocker lever so that it rests against valve stem ends and is clipped into support element—A3 with 3.2L engine

**and lifters, carefully rotate the crank-shaft by hand at least 2 full revolutions before starting to be sure that valves do not strike the pistons.**

8. Insert support element in cylinder head.

9. Install the rocker lever so that it rests against valve stem ends and is clipped into support element.

10. Oil the journal surfaces of camshafts.

11. Place respective camshaft into camshaft bearings of cylinder head.

12. Note the camshaft identification between cylinder 4 and cylinder 5 cam pair:

   a. A–Exhaust camshaft–Identification 022–Index 101.

   b. B–Intake camshaft–Identification 022–Index 102.

13. Insert camshafts into cylinder head in such a way that they are set at TDC.

14. It must be possible to insert camshaft bar T10058 into both shaft grooves.

15. Observe installed position of bearing caps as follows:

   a. Points of intake and exhaust camshaft bearing caps face outward.

   b. Identification on bearing caps is legible when read from intake side.

16. Lightly coat bearing cap 7 and 8 contact surface with adhesive lubricating paste.

17. Install the intake camshaft as follows:

   a. Fasten bearing caps 5 and 9 alternately and diagonally and then tighten to 5 Nm plus an additional 45 degree turn.

   b. Fasten bearing caps 1 and 13 to 5 Nm plus an additional 45 degree turn.

   c. Fasten bearing cap 7 to 5 Nm plus an additional 45 degree turn.

   d. Fasten bearing caps 3 and 11 to 5 Nm plus an additional 45 degree turn.

18. Install the exhaust camshaft as follows:

   a. Fasten bearing caps 6 and 10 alternately and diagonally and then tighten to 5 Nm plus an additional 45 degree turn.

   b. Fasten bearing caps 2 and 14 to 5 Nm plus an additional 45 degree turn.

   c. Fasten bearing cap 8 to 5 Nm plus an additional 45 degree turn.

   d. Fasten bearing caps 4 and 12 to 5 Nm plus an additional 45 degree turn.

19. Position camshafts in cylinder head to TDC. It must be possible to insert camshaft bar T10058 into both shaft grooves.

20. Unclip screen at backside of control housing and remove any contaminants.

21. Lightly oil camshaft seal contact surfaces in control housing.

22. Lightly oil sealing ring contact surfaces of camshafts and carefully slide control housing over camshaft sealing rings.

23. Tighten control housing.

24. Install the camshaft timing chain.

## CRANKSHAFT FRONT SEAL

### REMOVAL & INSTALLATION

**TT Models**

*1.8L Engine*

1. Before servicing the vehicle, refer to the precautions section.

2. Turn the ignition switch to the **OFF** position.

3. Remove or disconnect the following:
- Negative battery cable
- Accessory drive belt
- Timing belt
- Lower air duct

4. Hold the crankshaft timing belt gear using tool 3099 and remove crankshaft sprocket
- Crankshaft oil seal

***To install:***

5. Install or connect the following:
- New oil seal lubricated with engine oil using a Seal Driver until it is flush
- Crankshaft sprocket: Torque to 66 ft. lbs. (90 Nm) plus 90 degrees
- Lower air duct
- Timing belt
- Accessory drive belt
- Negative battery cable

### 3.2L Engine—Except TT and A6

*See Figure 86.*

1. Before servicing the vehicle, refer to the precautions section.

2. Turn the ignition switch to the **OFF** position.

3. Remove or disconnect the following:
- Negative battery cable
- Cover over coolant expansion tank

➡**Before removing ribbed belt, mark direction of rotation with chalk or a felt pen. If a used belt rotates in the wrong direction when reinstalled, this can result in breakage. When installing belt, ensure it is correctly seated in the pulleys.**

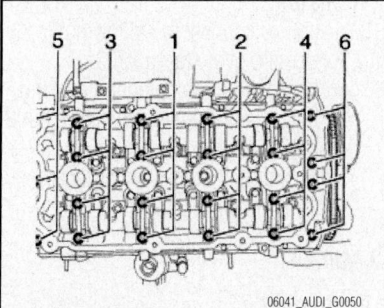

06041_AUDI_G0050

**Fig. 82 Note the camshaft identification between cylinder 4 and cylinder 5 cam pair—A3 with 3.2L engine**

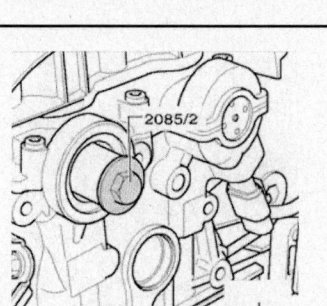

06041_AUDI_G0051

**Fig. 83 Points of intake and exhaust camshaft bearing caps face outward—A3 with 3.2L engine**

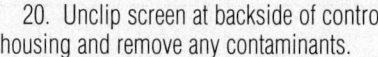

06041_AUDI_G0052

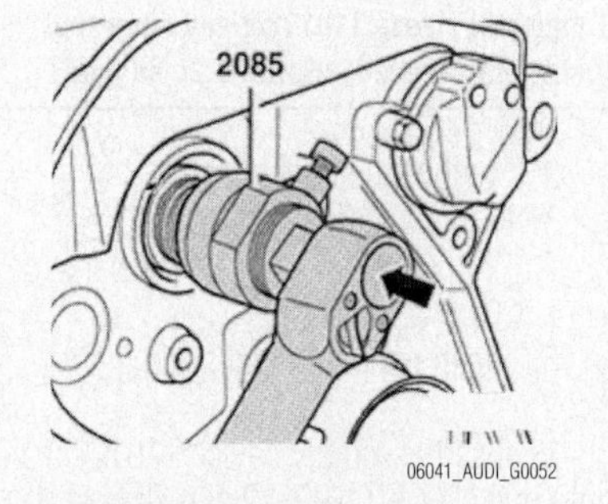

**Fig. 84 Unclip screen at backside of control housing and remove any contaminants—A3 with 3.2L engine**

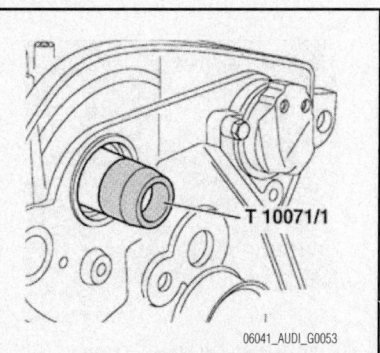

06041_AUDI_G0053

**Fig. 85 Tighten the control housing (indicated by arrows)—A3 with 3.2L engine**

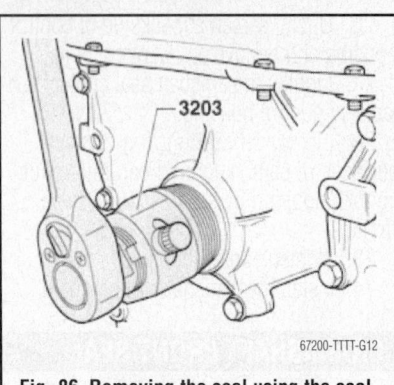

Fig. 86 Removing the seal using the seal remover—3.2L engine, except TT and A6

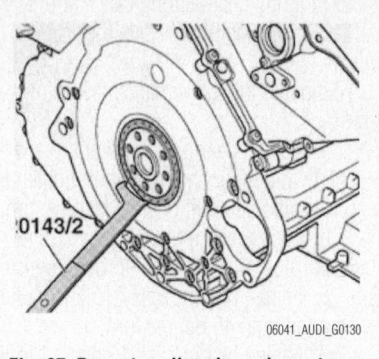

Fig. 87 Pry out sealing ring using extractor lever T20143/2—A6 with 3.2L engine

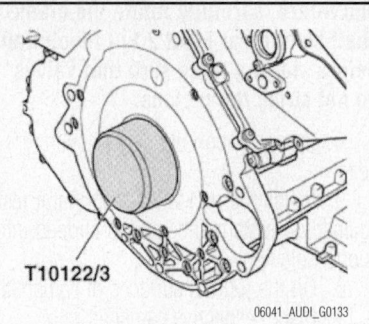

Fig. 90 Press sealing ring in evenly and flush, using pressure sleeve T10122/3—A6 with 3.2L engine

- Release tension on belt by screwing an M8 x 40 bolt into bore on tensioning element and remove belt
- Center and right sound insulation from under vehicle
- Remove bolt for vibration damper using counter hold tool T10059 or equivalent
- Unscrew inner section of oil seal extractor 3203 3 turns and lock with knurled screw
- Lubricate the threaded head of oil seal extractor, put into position, exert firm pressure and screw into oil seal as far as possible
- Loosen the knurled screw and turn inner part against the crankshaft until the oil seal is pulled out

### To install:

→Do not lubricate the oil seal before installing

4. Place guide sleeve T10053/1 or equivalent on crankshaft journal
5. Push oil seal carefully over the guide sleeve onto the crankshaft journal as far as possible and remove the guide sleeve
6. Press the oil seal in as far as possible using the old securing bolt for the vibration damper and installation sleeve from 3266 or equivalent
7. Install or connect the following:
- Vibration damper with new securing bolt and tighten to 74 ft. lbs (100 Nm) + 90 degrees (1/4 turn)
- Accessory drive belt
- Sound insulation
- Cover over coolant expansion tank
- Negative battery cable

### A6 and A8 Models

#### 3.2L Engine

See Figures 87 through 90.

1. Before servicing the vehicle, refer to the precautions section.

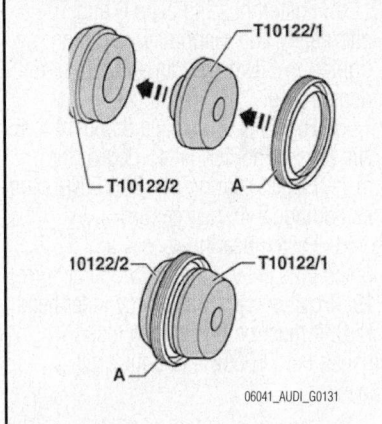

Fig. 88 Insert assembly device T10122/1 onto pull sleeve T10122/2 and slide seal onto pull sleeve—A6 with 3.2L engine

2. Remove the transmission
3. Remove the drive plate
4. Pry out sealing ring using extractor lever T20143/2.
5. Clean operating and sealing surfaces.
6. Insert assembly device T10122/1 onto pull sleeve T10122/2 and slide seal onto pull sleeve.
7. Remove the assembly device T10122/1.

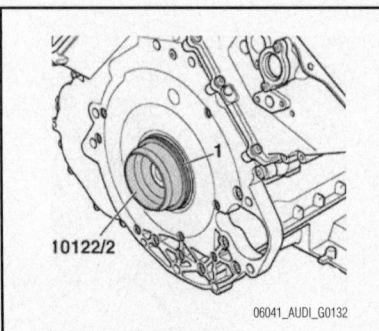

Fig. 89 Install pull sleeve T10122/2 with sealing ring onto crankshaft—A6 with 3.2L engine

### To install:

8. Install pull sleeve T10122/2 with sealing ring onto crankshaft.
9. Press sealing ring in evenly and flush, using pressure sleeve T10122/3.
10. Install drive plate.
11. Install transmission.

#### 4.2L Engine

1. Before servicing the vehicle, refer to the precautions section.
2. Remove or disconnect the following:
- Timing belt
- Timing belt crankshaft sprocket
- Oil seal using seal remover 3203

### To install:

3. Install or connect the following:
- Oil seal over sleeve 3202
- Using vibration damper mounting bolt, press in seal with sleeve 3202 until flush
- Timing belt crankshaft sprocket
- Timing belt

### A3 Models

#### 2.0L Engine

See Figures 91 through 95.

1. Before servicing the vehicle, refer to the precautions section.

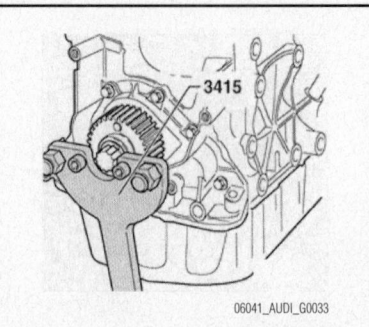

Fig. 91 Lock the timing belt sprocket with counter support 3415—A3 with 2.0L engine

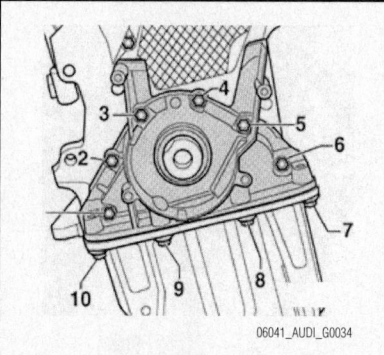

**Fig. 92 Location of the sealing flange bolts—A3 with 2.0L engine**

2. Remove the accessory drive belt.

3. Remove the timing belt.

4. Remove the timing belt crankshaft sprocket. To do this, lock the timing belt sprocket with counter support 3415.

5. Remove the bolts.

6. Pry off and remove front sealing flange.

7. Drive out seal from removed flange.

***To install:***

➡Lay a rag over the open portion of the oil pan.

**Carefully remove any residual sealant from cylinder block and oil pan.**

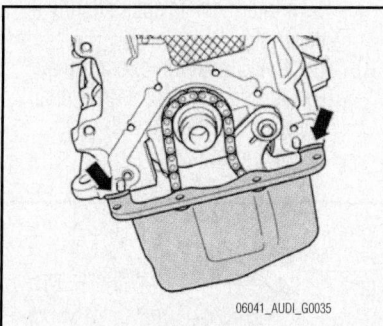

**Fig. 93 Apply a 2 to 3mm bead of sealant to edge between cylinder block and oil pan—A3 with 2.0L engine**

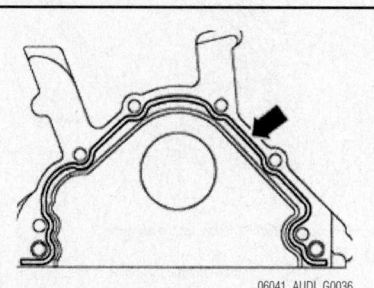

**Fig. 94 Apply a 2 to 3mm bead of sealant to sealing surface of sealing flange—A3 with 2.0L engine**

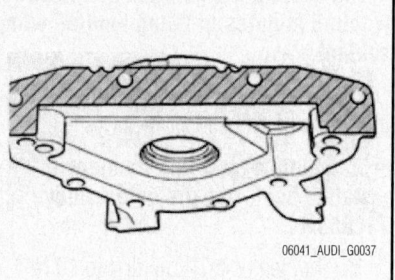

**Fig. 95 Lightly coat lower sealing surface of sealing flange with sealant—A3 with 2.0L engine**

8. Remove the residual sealant from sealing flange e.g. with a rotating plastic brush.

9. Clean sealing surfaces, they must be free of oil and grease.

10. Apply a 2 to 3mm bead of sealant to edge between cylinder block and oil pan.

11. Apply a 2 to 3mm bead of sealant to sealing surface of sealing flange.

➡**The sealant bead may not be thicker than 3 mm, otherwise excess sealant could enter the oil pan and clog the oil intake tube.**

12. Lightly coat lower sealing surface of sealing flange with sealant.

➡**The sealing flange must be installed within 5 minutes of being applied with sealant.**

13. Carefully push sealing flange onto locating pins on cylinder block.

➡**To install the sealing flange with installed seal, use the guide sleeve T10053/1.**

14. Tighten bolts 1 to 6 first to 15 Nm and then bolts 7 through 10 to 15 Nm.

15. Install the crankshaft seal on the belt pulley side.

16. Replace central bolt for the timing belt sprocket.

➡**There must be no oil on contact surface between the timing belt sprocket and crankshaft. Do not oil bolt for crankshaft the timing belt sprocket.**

17. Install the crankshaft the timing belt sprocket, counter-holding the timing belt sprocket with counter support 3415.

18. Install the remaining components in the reverse order of removal.

### 3.2L Engine

*See Figures 96 through 100.*

1. Before servicing the vehicle, refer to the precautions section.

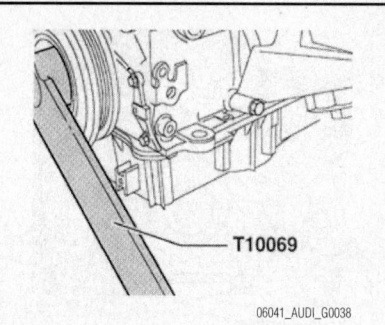

**Fig. 96 Remove the bolt from vibration damper, using counter-holder tool T10059 hold it—A3 with 3.2L engine**

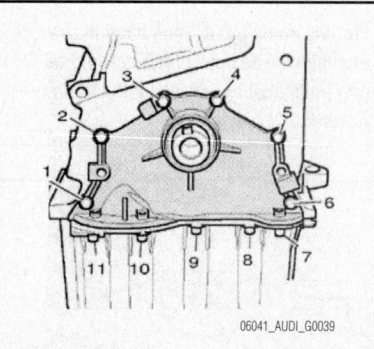

**Fig. 97 Front sealing flange bolt locations—A3 with 3.2L engine**

2. Before removing accessory drive belt, mark the turning direction on it with chalk or a felt tip pen.

3. Remove the center noise insulation fasteners.

4. Remove the right noise insulation fasteners.

5. Remove the accessory drive belt.

6. Remove the bolt from vibration damper, using counter-holder tool T10059 to hold it.

7. Remove the bolts, then pry off and remove front sealing flange.

8. Drive out the seal from the removed flange.

***To install:***

➡Lay a rag over the open portion of the oil pan. Carefully remove any residual sealant from cylinder block and oil pan.

9. Remove the residual sealant from sealing flange e.g. with a rotating plastic brush.

10. Clean sealing surfaces, they must be free of oil and grease.

11. Apply a 2 to 3mm bead of sealant to edge between cylinder block and oil pan.

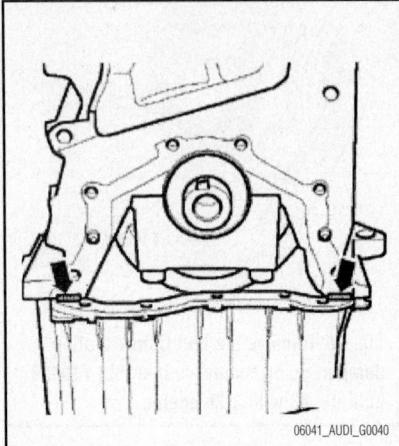

**Fig. 98 Apply a 2 to 3mm bead of sealant to edge between cylinder block and oil pan (indicated by arrows in illustration)—A3 with 3.2L engine**

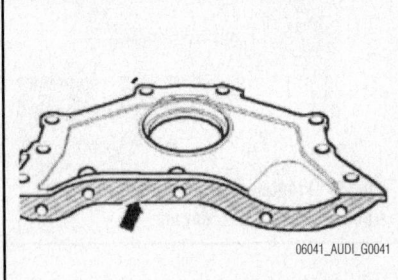

**Fig. 99 Apply a 2 to 3mm bead of sealant to surface of sealing flange—A3 with 3.2L engine**

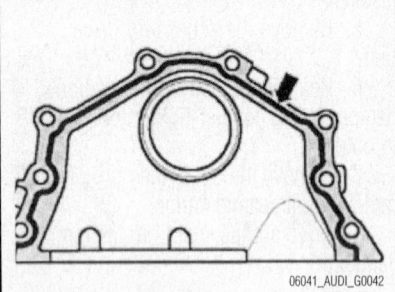

**Fig. 100 Lightly coat lower sealing surface of sealing flange with sealant—A3 with 3.2L engine**

12. Apply a 2 to 3mm bead of sealant to surface of sealing flange.

➡**The sealant bead may not be thicker than 3 mm, otherwise excess sealant could enter the oil pan and clog the oil intake tube.**

13. Lightly coat lower sealing surface of sealing flange with sealant.

➡**The sealing flange must be installed within 5 minutes of being applied with sealant.**

14. Carefully push sealing flange onto locating pins on cylinder block.

➡**To install the sealing flange with installed seal, use the guide sleeve T10053/1.**

15. Tighten bolts 1 to 6 first to 10 Nm and then bolts 7 through 10 to 10 Nm.

16. Install the crankshaft seal on the belt pulley side.

17. Replace central bolt for the timing belt sprocket.

➡**There must be no oil on contact surface between the timing belt sprocket and crankshaft. Do not oil bolt for crankshaft the timing belt sprocket.**

18. Install the crankshaft the timing belt sprocket, counter-holding the timing belt sprocket with counter support 3415.

19. Install the remaining components in the reverse order of removal.

### 3.2L Engine

#### TT

*See Figures 101 through 104.*

1. Disconnect the negative battery cable.
2. Remove engine
3. Remove flywheel
4. Remove intake manifold
5. Remove cylinder head cover
6. Remove thermostat housing
7. Remove oil pan
8. Remove or disconnect the following:
   - Bolt for timing chain tensioner
   - Upper timing chain cover. Be careful not to damage the head gasket when removing
   - Lower timing chain cover. Be careful not to damage the head gasket when removing
   - Crankshaft oil seal (rear main)

➡**Mark rotation direction of timing chains with a colored arrow before removing. Do not mark the chains using a center punch, notch, or anything similar**

   - Guide rail
   - Timing chain from sprocket for intermediate shaft and from behind projection of head gasket
   - Loosen bolt on intermediate shaft sprocket approximately 1 turn by counter holding the vibration damper with T10059 or equivalent

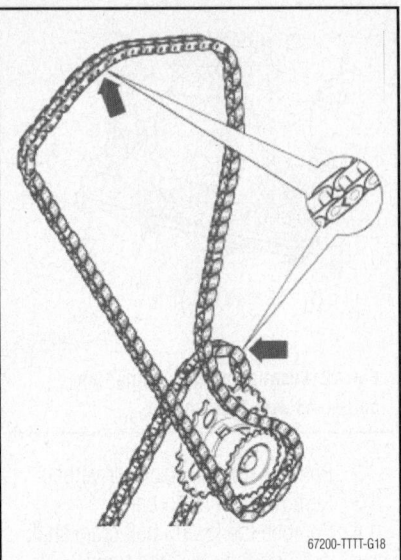

**Fig. 101 Mark timing chain rotation—3.2L engine**

   - Loosen valve gear chain tensioner
   - Remove bolt from intermediate shaft gear and removing chain together with gears

#### To install:

9. Install chain together with intermediate shaft sprocket
   - Sprocket can only be installed on intermediate shaft in one position. Turn intermediate shaft slightly if necessary
10. Check TDC position of crankshaft
   - Ground down tooth of drive chain sprocket must align with main bearing joint (arrow A)

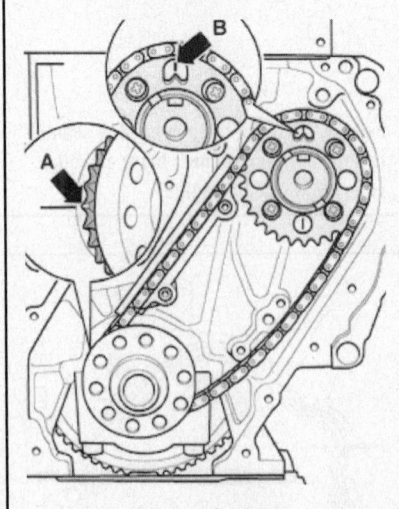

A. Main bearing joint
B. Thrust washer groove

**Fig. 102 Timing gear alignment—3.2L engine**

- Notch on intermediate shaft sprocket must align with groove on thrust washer for intermediate shaft (arrow B)

11. Install small sprocket on intermediate shaft and install bolt finger tight

12. Release the locking spline in timing chain tensioner with a small screwdriver and press the guide rail against the chain tensioner. Secure in this position

13. Tighten intermediate shaft securing bolt by counter holding at vibration damper with T10059

14. Install timing chain as follows:
- Yellow chain link "1" must align with marking "2" on the small chain sprocket for the intermediate shaft
- Copper chain link "4" must align with arrows "3" on camshaft adjusters and notches "5" on valve timing housing
- There must be 16 chain rollers between intake and exhaust camshaft adjusters

15. Install guide rail
16. Remove camshaft plate T10058

➡ **When turning the crankshaft you must press the guide rail firmly against the timing chain to prevent the chain from slipping**

17. Turn the crankshaft 2 revolutions in normal direction until timing mark is at TDC

➡ **The marked chain links will now be in an undefined position**

18. Insert camshaft plate T10058 in both camshaft grooves. If camshaft plate cannot be inserted, repeat adjustment of valve timing

19. Install lower timing cover. Tighten bolts in diagonal sequence and in stages

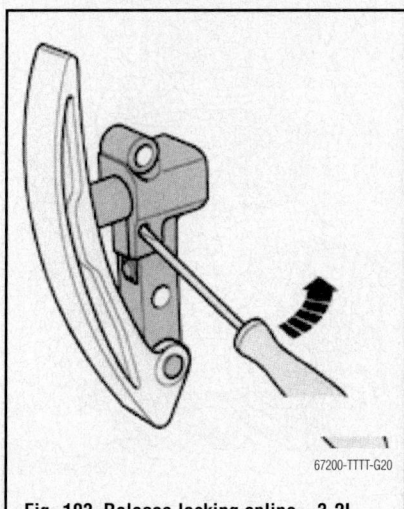

**Fig. 103 Release locking spline—3.2L engine**

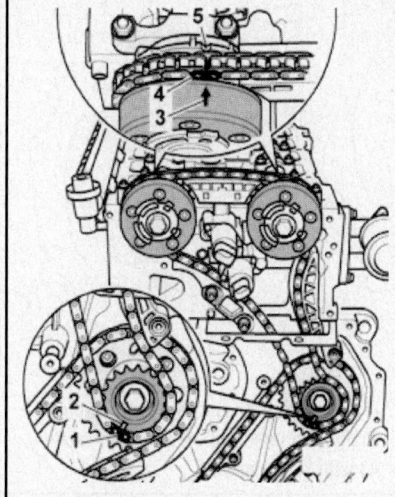

1. Yellow chain link
2. Marking
3. Camshaft adjusters
4. Copper colored chain link
5. Notches on valve timing housing

**Fig. 104 Camshaft timing alignment—3.2L engine**

20. Install upper timing cover and tighten bolts in following sequence:
- Timing cover to engine block initially to 44 inch lbs (5 Nm)
- Bottom of timing cover at head gasket 17 ft. lbs (23 Nm)
- Timing cover to engine block to 88 inch lbs (10 Nm)
21. Secure timing chain tensioner
22. The remainder of the installation is the reverse of the removal procedure

➡ **Torque Specs**

- Pins for guide rail to cylinder block 88 inch lbs (10 Nm)
- Chain tensioner to cylinder block 71 inch lbs (8 Nm)
- Chain sprockets to intermediate shaft 44 ft. lbs (60 Nm) + 90 degrees (1/4/ turn) Install new bolt
- Camshaft adjusters to camshaft 44 ft. lbs (60 Nm) + 90 degrees (1/4/ turn) Install new bolt
- Lower timing chain cover to cylinder block 88 inch lbs (10 Nm)

### A3

*See Figures 105 through 116.*

1. Before servicing the vehicle, refer to the precautions section.
2. Remove the engine.
3. Remove the transmission and flywheel.
4. Remove the intake manifold.
5. Remove the valve cover.

6. Remove the coolant thermostat housing.
7. Remove the oil pan.
8. Remove the transmission.
9. Set crankshaft to TDC mark by turning crankshaft on vibration damper bolt in a clockwise direction of engine rotation. At the same time, camshaft bar T10058 must engage in both shaft grooves. Turn crankshaft 1 additional turn if necessary.
10. Disconnect all necessary electrical connections.
11. Remove the camshaft timing chain tensioner.
12. Remove the bolts and upper timing chain cover horizontally from cylinder head. When doing so, ensure cylinder head bolts are not damaged.
13. Remove the bolts and remove lower timing chain cover horizontally from cylinder block. When doing so, ensure cylinder head bolt is not damaged.
14. Press crankshaft seal on the timing chain side out of the cover.
15. Identify timing chain direction of rotation with arrow mark before removing, using paint.

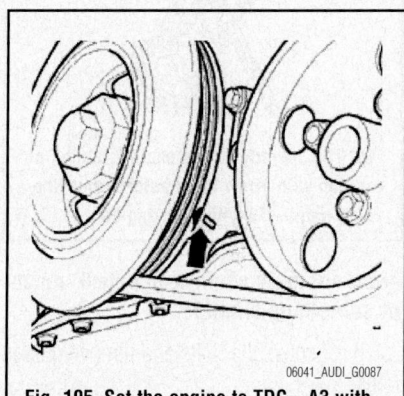

**Fig. 105 Set the engine to TDC—A3 with 3.2L engine**

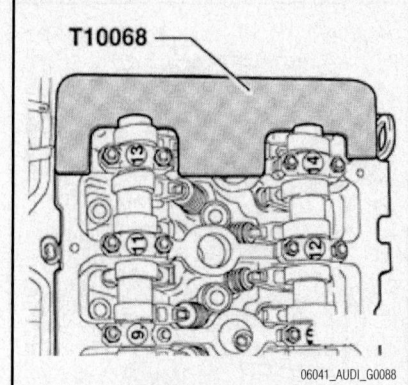

**Fig. 106 When at TDC the camshaft bar T10068 must engage in both shaft grooves—A3 with 3.2L engine**

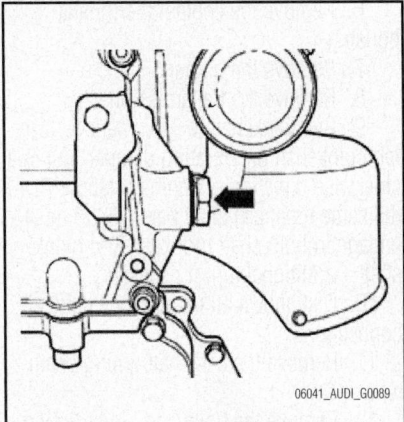

**Fig. 107 Remove the camshaft timing chain tensioner—A3 with 3.2L engine**

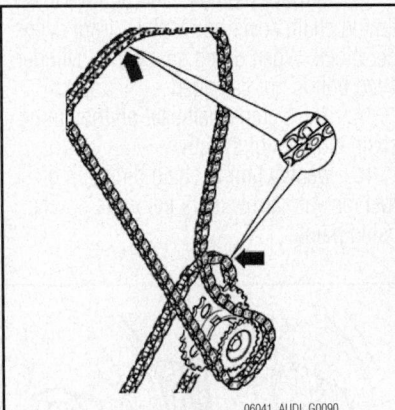

**Fig. 108 Identify timing chain direction of rotation with arrow mark before removing, using paint—A3 with 3.2L engine**

➡**Do not mark chain with punch, notch or something similar!**

16. Remove the bolts and remove guide track.

17. Remove the camshaft timing chain from intermediate shaft sprocket.

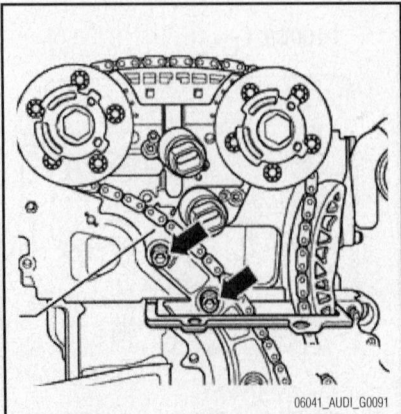

**Fig. 109 Remove the bolts and remove guide track—A3 with 3.2L engine**

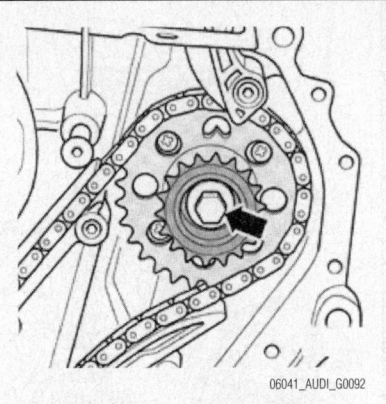

**Fig. 110 Remove the camshaft timing chain from intermediate shaft sprocket— A3 with 3.2L engine**

18. Carefully remove camshaft timing chain behind cylinder head seal rib.

19. Loosen the chain sprocket bolt on intermediate shaft about 1 turn while holding the vibration damper.

20. Remove the timing drive chain tensioner.

21. Remove the bolt and remove drive chain together with chain sprockets from intermediate shaft.

### To install:

➡**When reinstalling a used timing mechanism drive chain, note running direction identification mark.**

22. Install the drive chain together with large sprocket.

23. Note the location of sprocket, it can only be installed in one position in intermediate shaft. Turn intermediate shaft slightly if necessary.

24. Verify the crankshaft is at the TDC position as follows:

　　a. The milled drive chain sprocket tooth must align with bearing joint.

　　b. The tab on intermediate shaft sprocket must align with notch on intermediate shaft thrust washer.

25. Install the small sprocket on intermediate shaft as follows:

　　a. It can only be installed in one position on intermediate shaft.

　　b. Replace intermediate shaft sprocket securing bolt.

　　c. Hand tighten bolt.

　　d. Release locking splines in chain tensioner with a small screwdriver and press tensioning rail against chain tensioner.

　　e. Tighten chain tensioner in this position.

　　f. Tighten sprocket securing bolt while holding the vibration damper to 60 Nm plus an additional 60 degrees.

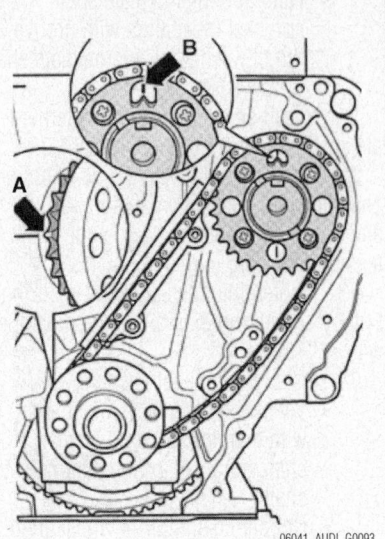

**Fig. 111 The milled drive chain sprocket tooth must align with bearing joint (A) and the tab on intermediate shaft sprocket (B) must align with notch on intermediate shaft thrust washer—A3 with 3.2L engine**

26. Route camshaft timing chain as follows:

　　a. Yellow link must align with marking on smaller intermediate chain sprocket.

　　b. Both copper-colored links must align with arrows at camshaft adjusters and notch on control housing. 16 chain rollers must now lie between arrows.

27. Insert guide track and tighten bolts.

28. Remove the camshaft bar T10058.

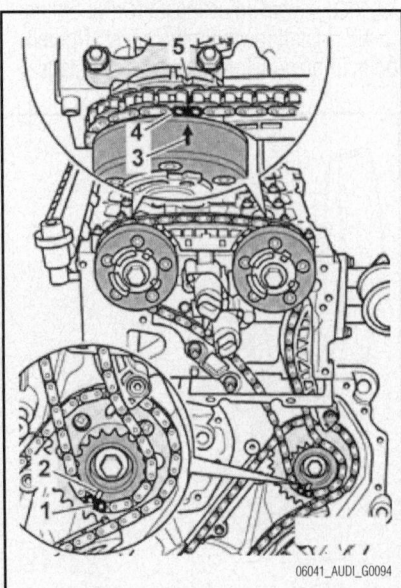

**Fig. 112 Timing chain routing—A3 with 3.2L engine**

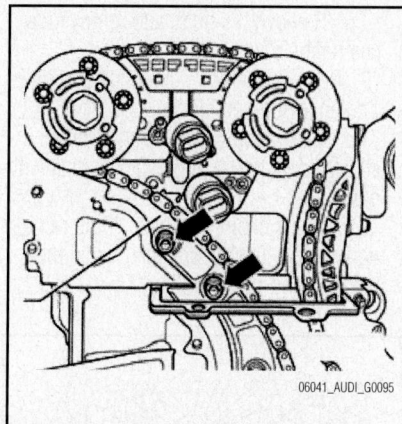

**Fig. 113 16 chain rollers must now lie between arrows—A3 with 3.2L engine**

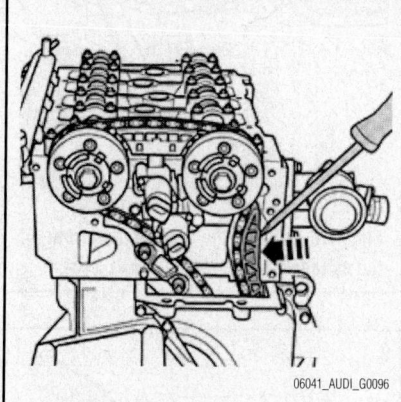

**Fig. 114 Insert guide track and tighten bolts—A3 with 3.2L engine**

➡**When turning crankshaft, tensioning rail must be pressed strongly with hand (instead of chain tensioner) against camshaft timing chain to prevent chain from jumping off.**

29. Set crankshaft to TDC and at the same time, the camshaft bar T10058 must engage in both shaft grooves.

30. Tighten the camshaft timing adjuster to camshafts to 60 Nm plus 90 degrees

31. Tighten the camshaft timing chain tensioner to cylinder head to 40 Nm

➡**Marked links, now take on an undefined position.**

32. If camshaft bar cannot be inserted, repeat valve timing adjustment.

33. Clean old sealant from the 3mm holes in cylinder head seal.

➡**With the cylinder head installed only half of the holes in the cylinder head gasket are visible.**

34. Clean sealing surfaces at engine and both covers; they must be free of oil and grease.

35. Check whether lower timing chain cover alignment pins are inserted in cylinder block.

36. Lightly coat the lower timing chain cover sealing surfaces with sealing paste AMV 188 001 02.

➡**The lower timing chain cover must be installed within 5 minutes of applying sealing paste.**

37. Tighten the lower timing chain cover bolts diagonally in stages.

38. Fill the 3mm holes in cylinder head seal using black sealant AMV 174 04 01.

39. Replace seal in upper timing chain cover

40. Check whether alignment sleeves and are inserted in upper timing chain cover.

41. Insert a new gasket and a new seal.

42. Lightly coat clean upper timing chain cover sealing surfaces with sealing paste AMV 188 001 02.

43. Set upper timing chain cover in place and tighten bolts as follows:
   a. Step 1: Tighten bolts 1 and 2 to 5 Nm.
   b. Step 2: Tighten bolts 3 to 23 Nm.

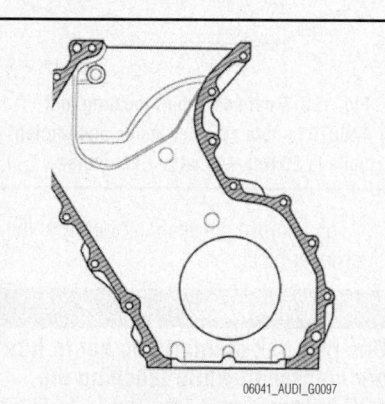

**Fig. 115 Lightly coat the lower timing chain cover sealing surfaces with sealing paste—A3 with 3.2L engine**

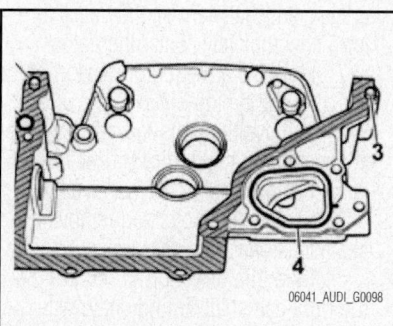

**Fig. 116 Lightly coat the upper timing chain cover sealing surfaces with sealing paste—A3 with 3.2L engine**

   c. Step 3: Tighten bolts 1 and 2 to 10 Nm.
   d. Step 1: Tighten camshaft timing chain tensioner.
   e. Step 1: Turn crankshaft 2 times in direction of engine rotation and check valve timing again.

44. Install the sealing ring for crankshaft on the timing belt side.

45. Install the cylinder head cover.

46. Install the coolant thermostat housing.

47. Install the oil pan.

48. Install the flywheel.

49. Install the transmission and engine.

### A6

*See Figures 117 through 138.*

1. Before servicing the vehicle, refer to the precautions section.

2. Drain engine oil.

3. Remove the transmission

4. Remove the drive plate.

5. Remove the left and right timing chain covers as follows:

6. Remove the rear engine covers as follows:
   a. Remove the coolant expansion tank.
   b. Disconnect electrical connection from Engine Coolant Level (ECL) switch at bottom of coolant reservoir and set aside coolant reservoir with coolant hoses still connected.
   c. Disconnect electrical harness connectors.
   d. Remove the nut.
   e. Remove the retainer for connection.
   f. Remove the double-bolt lying beneath.
   g. Remove the retainer for connections.
   h. Remove the bolts and separate electrical connections at ignition coils.
   i. Remove the bolts and remove the left timing chain cover.
   j. Disconnect check valve from connection at air duct hose.
   k. Remove the air duct hose.
   l. Disconnect electrical connector.
   m. Remove the bolt and remove retainer for connection.
   n. Remove the bolts and remove right timing chain cover.

7. Remove the lower timing chain cover as follows:
   a. Remove the cap for oil filter housing.
   b. Remove the oil filter element.
   c. Disconnect electrical harness connector from oil pressure switch.
   d. Remove the oil pressure switch.

e. Remove the oil filter housing.

f. Remove the bolts and remove lower timing chain cover.

8. Remove the camshaft timing chains as follows:

a. Remove the cylinder head cover.

b. Insert guide pin of adapter T40058 so that large diameter points to engine the small diameter should point to the adapter.

c. Using socket T40058, rotate crankshaft in a clockwise direction of engine rotation to TDC.

d. The threaded holes in camshafts must face upward.

e. Mount camshaft locating tool T40070 to both cylinder heads and tighten bolts to 20 Nm.

f. The camshaft locating tool T40070 is correctly positioned when the holes for the cylinder head bolts remain free.

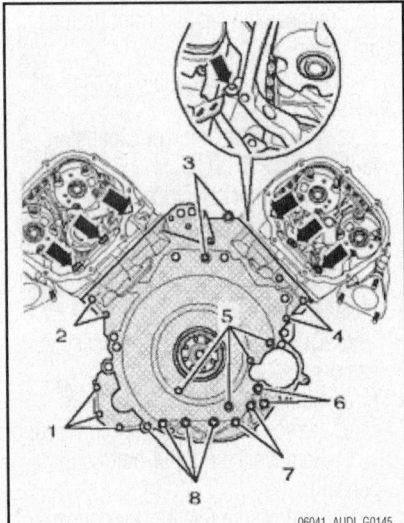

**Fig. 117 Remove the bolts and remove lower timing chain cover—A6 with 3.2L engine**

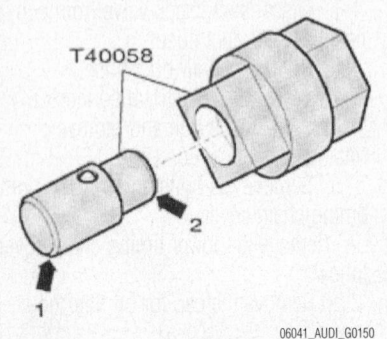

**Fig. 118 Insert guide pin of adapter T40058 so that large diameter points to engine the small diameter should point to the adapter—A6 with 3.2L engine**

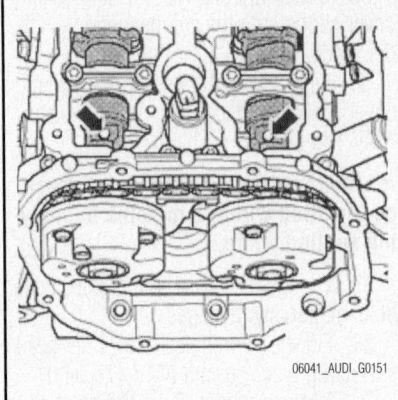

**Fig. 119 The threaded holes in camshafts must face upward—A6 with 3.2L engine**

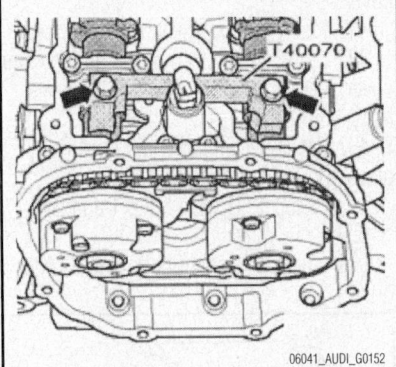

**Fig. 120 Mount camshaft locating tool T40070 to both cylinder heads and tighten bolts to 20 Nm—A6 with 3.2L engine**

g. Remove the sealing plug from the cylinder block.

### ✳✳ CAUTION

**Due to a risk of injury, do not to turn the crankshaft while touching the TDC hole.**

h. Screw the crankshaft holder T40059 into hole using a torque wrench to a torque value of 10 Nm, if necessary rotate crankshaft very slightly back and forth to completely center the holder.

i. Mark the direction of rotation of the left camshaft timing chain with paint.

j. Remove the bolts for the camshaft adjuster using multipoint socket T10035.

k. Remove the both camshaft adjusters.

l. Remove the bolts and remove the chain tensioner.

m. Mark the direction of rotation of the right camshaft timing chain with paint.

n. Remove the bolts for the camshaft adjuster using multipoint socket T10035.

o. Remove the both camshaft adjusters.

p. Remove the bolts and remove the chain tensioner.

9. Remove the drive chain for oil pump and balance shaft as follows:

a. Mark running direction of drive chain for oil pump and balance shaft with paint.

b. Press chain tensioner guide rail in downwards and secure chain tensioner with locking pin T40071.

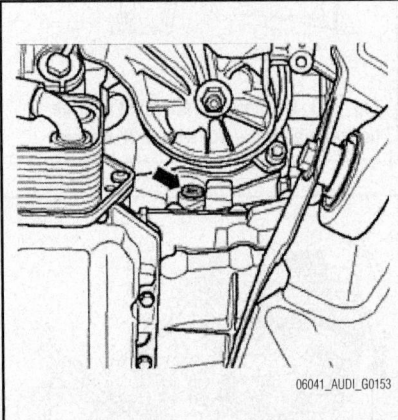

**Fig. 121 Remove the sealing plug from the cylinder block—A6 with 3.2L engine**

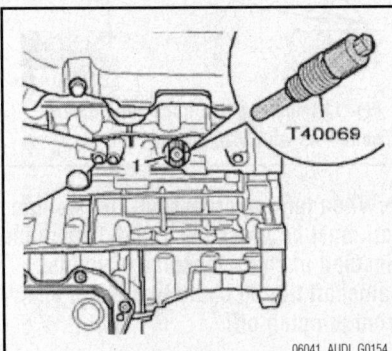

**Fig. 122 Screw the crankshaft holder T40059 into hole using a torque wrench to a torque value of 10 Nm—A6 with 3.2L engine**

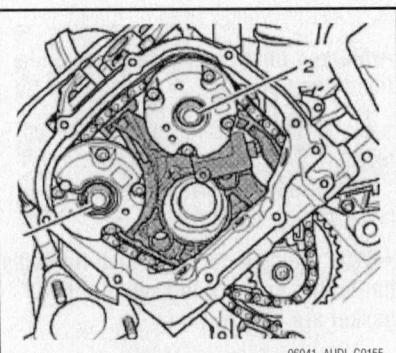

**Fig. 123 Remove the bolts for the left camshaft adjuster—A6 with 3.2L engine**

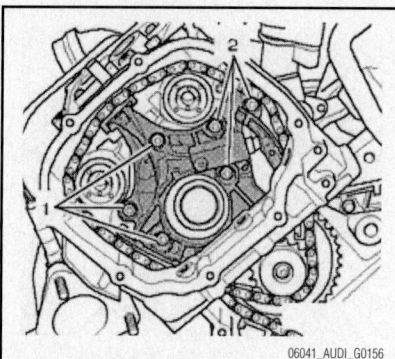

**Fig. 124 Remove the bolts for the left chain tensioner—A6 with 3.2L engine**

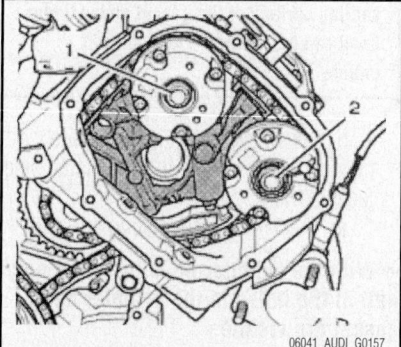

**Fig. 125 Remove the bolts for the right camshaft adjuster—A6 with 3.2L engine**

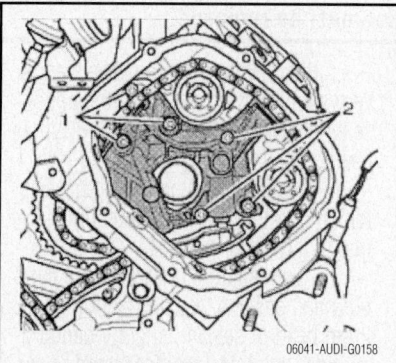

**Fig. 126 Remove the bolts for the right chain tensioner—A6 with 3.2L engine**

c. Remove the bolts and remove chain tensioner, balance shaft sprocket and chain.

### To install:

10. Install drive chain for oil pump and balance shaft as follows:

    a. Secure crankshaft in TDC position using crankshaft holder T40059.

    b. Mount chain tensioner with chain and balance shaft sprocket.

    c. Secure balance shaft with 8 mm diameter drill bit in TDC position. The slots in balance shaft sprocket must be at the middle position in relation to the

threaded holes of the balance shaft. If necessary, adjust chain by one tooth.

    d. Tighten chain tensioner bolts.

    e. Loosely screw in bolts for sprocket.

    f. The chain sprocket must still be able to be rotated on balance shaft and must not tip.

    g. Remove the locking pin T40071 to release the chain tensioner.

    h. Press against chain tensioner guide rail with a screwdriver and fasten the bolts for sprocket.

    i. Tighten the chain tensioner to cylinder block to 6 Nm plus an additional 45 degrees and the balance shaft sprocket to balance weight to 15 Nm plus an additional 90 degrees.

    j. Remove the drill from the balance shaft.

    k. Install a new sealing plug in the block and tighten to 14 Nm.

11. Install the camshaft timing chains as follows:

➡ **Always replace bolts that are tightened to torque as well as O-rings and gaskets.**

➡ **When turning camshaft, crankshaft must not be at TDC for any cylinder. Valves and/or pistons may be damaged.**

    a. The drive chain for the timing mechanism installed.

    b. Secure crankshaft in TDC position using crankshaft holder T40059 and the camshaft locating tool T40070 must be mounted on both cylinder heads and fastened to 20 Nm.

    c. Fully relieve tension of guide rail for left and right camshaft timing chain tensioner. The chain tensioner piston must be driven out completely, releasing the retainer.

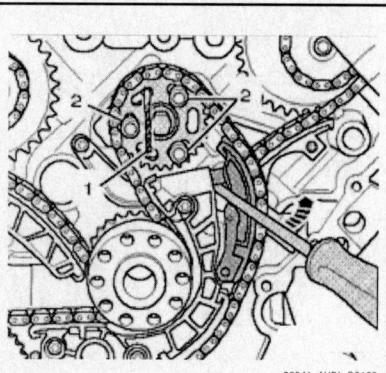

**Fig. 127 Secure balance shaft with 8 mm diameter drill bit in TDC position—A6 with 3.2L engine**

    d. Press guide rail of the left and right camshaft timing chain inward up to the stop and secure chain tensioner with the locking pin T40071.

    e. Place a new gasket onto the rear of the chain tensioner.

    f. Install the set chain tensioner in place on the left cylinder head and install camshaft timing chain, as illustrated.

    g. Install the bolts indicated by the numbers 1 and 2 on the illustration. Replace the camshaft bolts.

    h. Both camshaft adjusters must be able to rotate on camshaft and must not tip.

    i. Set chain tensioner in place on right cylinder head and install camshaft timing chain, as illustrated.

    j. Install the bolts indicated by the numbers 1 and 2 on the illustration. Replace the camshaft bolts.

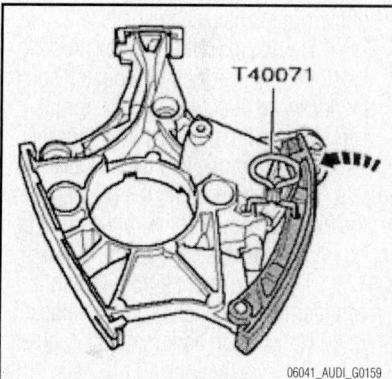

**Fig. 128 Press guide rail of the left and right camshaft timing chain inward up to the stop and secure chain tensioner with the locking pin T40071—A6 with 3.2L engine**

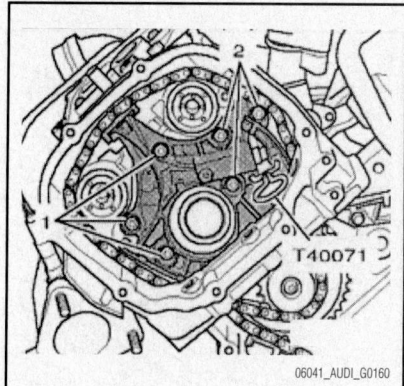

**Fig. 129 Install the set chain tensioner in place on the left cylinder head and install camshaft timing chain. Install the bolts indicated by the numbers 1 and 2—A6 with 3.2L engine**

k. Both camshaft adjusters must be able to rotate on camshaft and must not tip.

l. Set counter-holder T10172 with pin T10172/2 in place on camshaft adjuster of the left intake camshaft.

m. Hold camshaft timing chain pre-tensioned by pressing on counter-holder in direction of–arrow -.

n. Simultaneously, pre-torque camshaft bolt using multipoint socket T10035 and torque wrench. Tighten to 40 Nm.

o. Continue holding the pretension on intake camshaft and pre-torque bolt on the exhaust camshaft and tighten to 40 Nm.

p. Tighten the camshaft bolts on left cylinder head to 80 Nm plus an additional 90 degrees

q. Set counter-holder T10172 with pin T10172/2 in place on camshaft adjuster of the right exhaust camshaft.

r. Hold camshaft timing chain pre-tensioned by pressing on counter-holder in clockwise direction and pre-torque camshaft bolt using multipoint socket T10035 and torque wrench to 40 Nm.

s. Continue holding the pretension on the exhaust camshaft and pre-torque bolt on intake camshaft to 40 Nm.

t. Tighten camshaft bolts on right cylinder head to final torque specification to 80 Nm plus an additional 90 degrees.

u. Remove the locking pin T40071 to release chain tensioner.

v. Remove the camshaft locators T40070 on both cylinder heads.

w. Remove the crankshaft holder T40059.

x. Using key T40058 turn crankshaft two complete rotations in clockwise

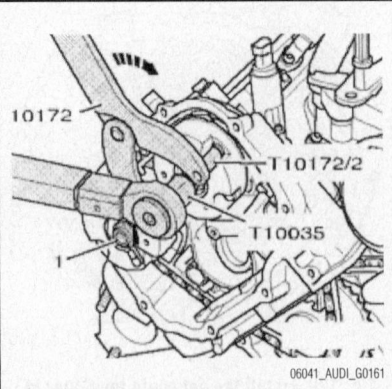

**Fig. 130 Set counter-holder T10172 with pin T10172/2 in place on camshaft adjuster of the left intake camshaft—A6 with 3.2L engine**

direction of engine rotation until crank-shaft stands at TDC again.

➡ **If rotated unintentionally beyond TDC, turn back crankshaft again approx. 30° and set to TDC again.**

y. The threaded holes in the camshafts must face upward.

z. Mount camshaft locating tools T40070 to both cylinder heads and tighten bolts to 20 Nm.

aa. The camshaft locating tool T40070 is correctly positioned when the holes for the cylinder head bolts remain free.

bb. Screw the crankshaft holder T40059 directly into the hole. The crank-shaft holder T40059 must engage in locating hole of crankshaft, otherwise repeat adjustment.

cc. Remove the camshaft locating tools on both cylinder heads.

dd. Remove the crankshaft holder.

ee. Screw sealing plug of TDC-marking with new seal into cylinder block.

12. Install the lower timing chain cover as follows:

a. Replace all gaskets, seals and O-rings.

b. Pull the alignment bushing out of top right of cylinder block.

c. Chamfer the alignment bushing with a file, as shown in illustration. Dimension **X** should be 6.5 mm and dimension **Y** should be 8 mm.

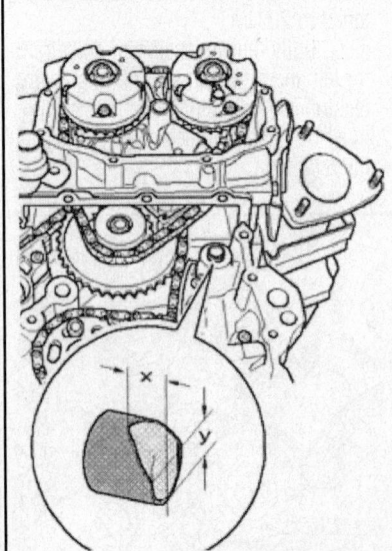

**Fig. 131 Chamfer the alignment bushing with a file, as shown. Dimension X should be 6.5 mm and dimension Y should be 8 mm—A6 with 3.2L engine**

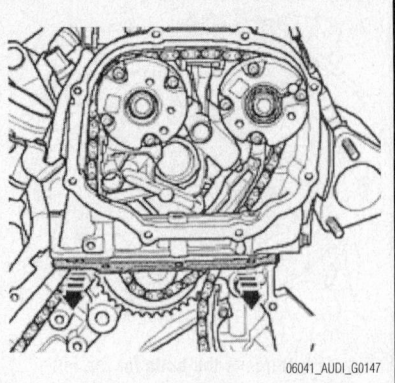

**Fig. 132 Bend cylinder head gaskets ends very slightly downward until the upper sealing surface of the gasket and cylinder head can be cleaned—A6 with 3.2L engine**

d. Install alignment bushing into cylinder block so that the chamfered side faces upward.

e. Clean all gasket mating surfaces.

➡ **With the cylinder head installed only half of the holes in the cylinder head gasket are visible.**

❄❄ **CAUTION**

**The cylinder head gasket must not be kinked. A kinked cylinder head gasket must be replaced.**

f. Bend cylinder head gaskets ends very slightly downward until the upper sealing surface of the gasket and cylinder head can be cleaned.

g. Clean both cylinder head gaskets, top and bottom, so they are completely free of any oil or grease.

h. Coat sealing surfaces of the cylinder head gaskets, top and bottom, with a 2mm bead of sealant, slightly bending cylinder head gaskets downward again. A feeler gauge can be used to coat the surface between the cylinder head and gasket.

➡ **The covers for timing chain must be installed within 5 minutes after applying sealant.**

i. Apply a 2mm bead of sealant as shown to the lower timing chain cover. The groove of sealing surface must be completely filled with sealant and the sealant bead must be 1.5 to 2.0 mm above sealing surface.

j. Install the lower timing chain cover in place, guiding cover at an angle from below onto sealing surface of the cylinder block and cylinder head.

k. Insert bolts indicated by arrows on the illustration with locking compound and tighten to 5 Nm.

l. Tighten bolts 1 through 8 using a diagonal sequence to 10 Nm.

m. Tighten the bolts indicated by arrows in the illustration to 10 Nm.

n. Tighten bolts 6, 7 and 8 to 22 Nm.

13. Install the rear engine covers as follows:

a. Clean all gasket mating surfaces.

➥**Covers for the timing chain must be installed within 5 minutes after applying sealant.**

b. Apply a 1mm bead of sealant as illustrated to the left timing chain cover. The groove of sealing surface must be completely filled with sealant and the sealant bead must be 1.5 to 2.0 mm above sealing surface.

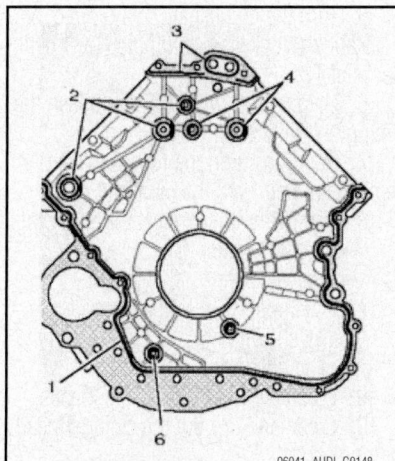

Fig. 133 Apply a 2mm bead of sealant to the locations illustrated on the lower timing chain cover—A6 with 3.2L engine

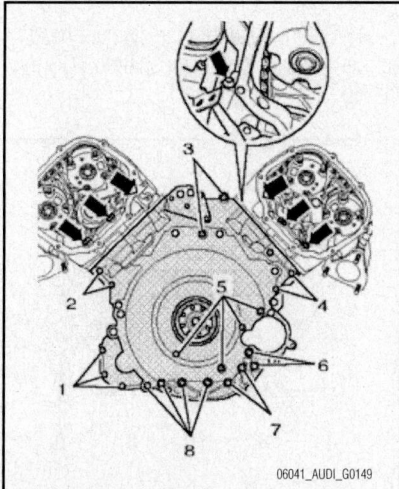

Fig. 134 Lower timing chain cover bolt locations—A6 with 3.2L engine

c. Install the left timing chain cover and tighten bolts in the sequence illustrated to 9 Nm.

d. Apply a 1mm bead of sealant as illustrated to the right timing chain cover. The groove of sealing surface must be completely filled with sealant and the sealant bead must be 1.5 to 2.0 mm above sealing surface.

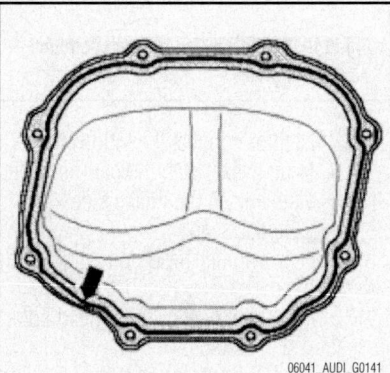

Fig. 135 Apply a 1mm bead of sealant as shown to the left timing chain cover—A6 with 3.2L engine

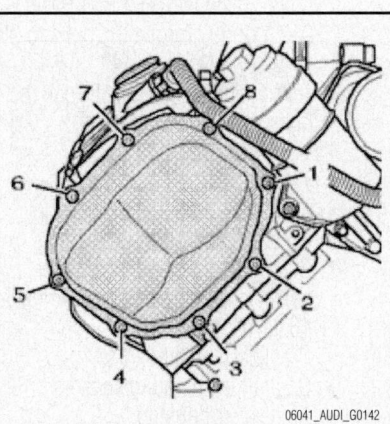

Fig. 136 Left timing chain cover bolt locations—A6 with 3.2L engine

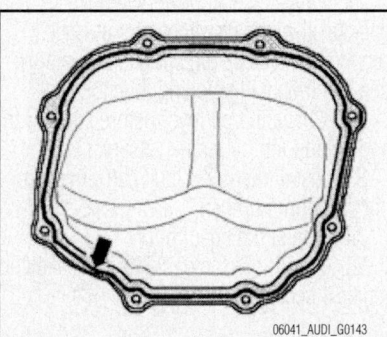

Fig. 137 Apply a 1mm bead of sealant as shown to the right timing chain cover—A6 with 3.2L engine

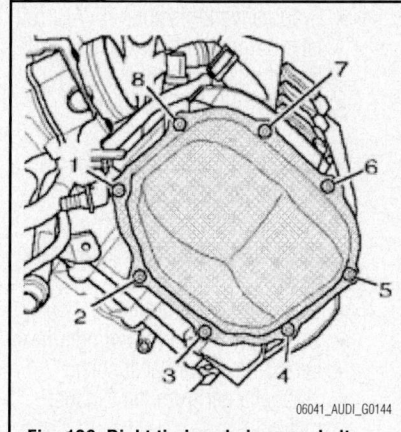

Fig. 138 Right timing chain cover bolt locations—A6 with 3.2L engine

e. Install the right timing chain cover and tighten bolts in the sequence illustrated to 9 Nm.

## CYLINDER HEAD

➥**Before removing or installing the cylinder head, align the engine timing marks at Top Dead Center (TDC). Rotate the crankshaft mark away about ¼ turn Before Top Dead Center (BTDC). This will prevent the valves from hitting the piston heads. Be sure to turn the crankshaft to the proper position after cylinder head installation.**

### ✵✵ CAUTION

**Cylinder head removal should not be attempted unless the engine is cold.**

*REMOVAL & INSTALLATION*

**A4 Models**

*2.0L Engine*

*See Figures 139 and 140.*

1. Before servicing the vehicle, refer to the precautions section.

2. Remove the front bumper if necessary for access.

3. Place the hood lock carrier into the service position.

4. Turn the ignition switch to the **OFF** position.

5. Remove or disconnect the following:
- Negative battery cable
- Accessory drive belt
- Cooling fan

6. Drain the engine coolant.

7. Remove or disconnect the following:
- Intake manifold
- Accessory drive belts
- Wastegate bypass regulator valve

- Evaporative Emissions (EVAP) canister purge regulator valve
- Power outage stage
- Mass Air Flow (MAF) sensor
- Air cleaner housing
- Engine Temperature Control (ETC) and the temperature II sensor harness connector
- All connections from the cylinder head
- Crankcase breather line
- Oil supply line at the cylinder head
- Exhaust manifold heat shield
- Turbocharger from the exhaust manifold
- Coolant hose to the heat exchanger at the rear of the cylinder head
- Upper timing belt cover

8. Turn the crankshaft, in the direction of rotation (clockwise), until the No. 1 cylinder is at TDC.

9. Using Torx® wrench T45, loosen the timing belt tensioner.

10. Push down on the tensioner and remove the belt from the camshaft gear.

11. Remove or disconnect the following:
- Torx® bolt and swing the tensioner assembly bracket forward
- Valve cover
- Cylinder head bolts, in sequence, as shown
- Cylinder head

12. Clean the gasket mating surfaces.

13. Clean and dry out the cylinder head bolt holes.

### To install:

➡**Always replace the cylinder head bolts, self-locking nuts, bolts, gaskets and O-rings.**

14. Before installing the cylinder head, set the crankshaft and camshaft to TDC for the No. 1 cylinder.

15. Loosen the turbocharger support bracket to reduce the likelihood of any tension while installing the cylinder head.

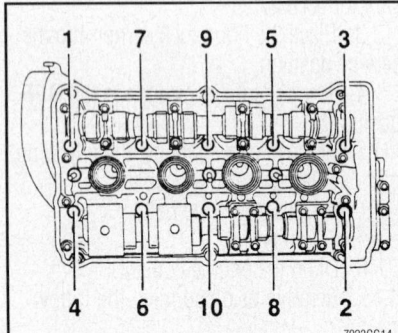

Fig. 139 Cylinder head bolt removal sequence—A4 2.0L engine

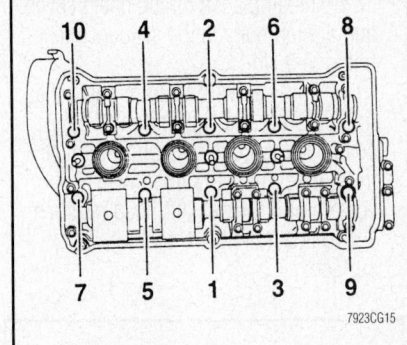

Fig. 140 Cylinder head torque sequence—A4 2.0L engine

16. Install or connect the following:
- Head gasket with the part number visible from the intake side
- Cylinder head
- New cylinder head bolts, tighten by hand

17. Tighten the new cylinder head bolts in sequence in 2 steps:
   a. Step 1: 44 ft. lbs. (60 Nm)
   b. Step 2: Plus 180 degrees

18. Install or connect the following:
- Turbocharger to the exhaust manifold using new gaskets and the bolts coated with Hot Bolt Paste G 052 112 A3. Torque the bolts to 26 ft. lbs. (35 Nm).
- Turbo support bracket. Torque the bolts to 33 ft. lbs. (40 Nm).
- Valve cover
- Timing belt
- Accessory drive belts
- Exhaust manifold heat shield
- Oil supply lines to the cylinder head. Torque the retaining straps to 15 ft. lbs. (20 Nm).
- Crankcase breather
- Coolant temperature sensors
- Air cleaner housing

19. Fill the engine with coolant and bleed, if necessary.

20. Connect the negative battery cable.

21. Fully close all power windows to stop, operate all window switches for at least 1 second in the close direction to activate the one touch opening/closing function.

22. Check the oil level before starting the engine and top off, as necessary.

23. Install the hood lock carrier assembly and front bumper.

24. Adjust the headlights.

25. Start the vehicle, check for leaks and repair if necessary.

## TT Models

### 1.8L Engine

*See Figures 141 and 142.*

1. Before servicing the vehicle, refer to the precautions section.

2. Turn the ignition switch to the **OFF** position.

3. Drain the engine coolant.

4. Remove or disconnect the following:
- Negative battery cable
- Intake manifold
- Coolant line at thermostat
- Connectors on ignition coils
- All connections from the cylinder head
- Oil return line retainer
- Upper air line to turbocharger
- Exhaust manifold heat shield
- Turbocharger from the exhaust manifold
- Accessory drive belt
- Coolant expansion tank with hoses
- Power steering reservoir, leaving hoses attached
- Upper timing belt cover

5. Turn the crankshaft, in the direction of rotation (clockwise), until the No. 1 cylinder is at TDC.

6. Using Torx® wrench T45, loosen the timing belt tensioner.

7. Push down on the tensioner and remove the belt from the camshaft gear.

8. Remove or disconnect the following:
- Ignition coils
- Valve cover
- Cylinder head bolts, in sequence, as shown
- Cylinder head

9. Clean the gasket mating surfaces.

10. Clean and dry out the cylinder head bolt holes.

### To install:

➡**Always replace the cylinder head bolts, self-locking nuts, bolts, gaskets and O-rings.**

11. Before installing the cylinder head, set the crankshaft and camshaft to TDC for the No. 1 cylinder.

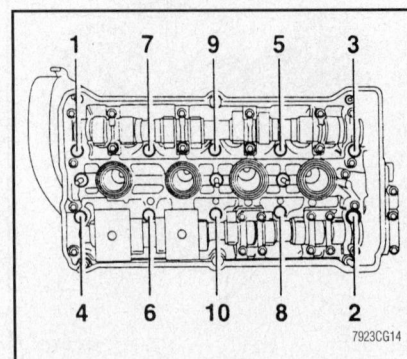

Fig. 141 Cylinder head bolt removal sequence—TT 1.8L engine

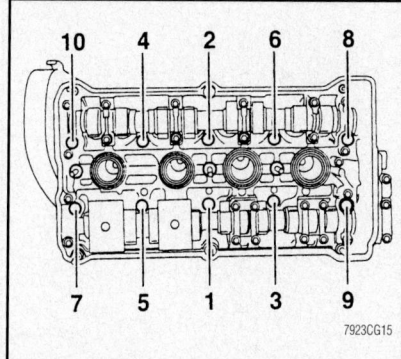

**Fig. 142 Cylinder head torque sequence— TT 1.8L engine**

12. Loosen the turbocharger support bracket to reduce the likelihood of any tension while installing the cylinder head.

13. Install or connect the following:
- Head gasket with the part number visible from the intake side
- Cylinder head
- New cylinder head bolts, tighten by hand

14. Tighten the new cylinder head bolts in sequence in 2 steps:
a. Step 1: 30 ft. lbs. (40 Nm)
b. Step 2: Plus 180 degrees

15. Install or connect the following:
- Turbocharger to the exhaust manifold using new gaskets and the bolts coated with Hot Bolt Paste G 052 112 A3. Torque the bolts to 26 ft. lbs. (35 Nm).
- Turbo support bracket. Torque the bolts to 33 ft. lbs. (40 Nm).
- Timing belt.
- Power steering reservoir
- Coolant expansion tank
- Accessory drive belt
- Exhaust manifold heat shield
- Upper air line to turbocharger
- Oil return line retainer
- All connections from the cylinder head
- Connectors on ignition coils
- Coolant line at thermostat
- Intake manifold

16. Fill the engine with coolant and bleed, if necessary.

17. Connect the negative battery cable.

18. Fully close all power windows to stop, operate all window switches for at least 1 second in the close direction to activate the one touch opening/closing function.

19. Check the oil level before starting the engine and top off, as necessary.

20. Start the vehicle, check for leaks and repair if necessary.

### 3.2L Engine

*See Figures 143 and 144.*

1. Before servicing the vehicle, refer to the precautions section.

2. Turn the ignition switch to the **OFF** position.

3. Drain the engine coolant.

4. Remove or disconnect the following:
- Negative battery cable

➡ **Before removing ribbed belt, mark direction of rotation with chalk or a felt pen. If a used belt rotates in the wrong direction when reinstalled, this can result in breakage. When installing belt, ensure it is correctly seated in the pulleys.**

- Release tension on belt by screwing an M8 x 40 bolt into bore on tensioning element and remove belt
- Upper mounting bolt for the alternator
- Upper mounting bolts for radiator cowl
- Secondary air hose
- Center and right-hand noise insulation under engine
- Cross piece
- Power steering pump from bracket, hang pump from lock carrier with hoses still attached
- Secondary air injection pump with bracket
- Lower radiator cowl bolts and remove the radiator cowl with fans

➡ **The air conditioning compressor can only be removed from the bracket with the upper securing bolt still inserted. To accomplish this, press the engine simultaneously to the left side. If necessary, loosen all bolts of the assembly mounting.**

- Mounting bolt for air conditioning compressor
- Secure the air conditioning compressor together with the hoses at front longitudinal member (hoses remain connected)
- Alternator
- Oil dipstick tube
- Vacuum reservoir
- Mounting bracket for alternator and air conditioning compressor
- Bolts at front of intake manifold
- Attach puller T10095A or equivalent to ignition coils and remove coils
- Body brace
- Air intake hose

- Crankcase breather hose
- Throttle body
- Hose bracket
- Rear intake mounting bolts, swing intake manifold forward
- Cylinder head cover, remove bolts in sequence
- Timing chain from camshafts
- Exhaust pipe from manifold
- Remove cylinder head bolts in sequence indicated
- Attach lifting shackle 3033 or equivalent and carefully lift off cylinder head

➡ **Replace cylinder head bolts. Replace self locking nuts and bolts that are tightened by turning to a specified angle**

5. Set camshafts to TDC for number 1 cylinder
- It must be possible to insert camshaft plate T10058 in both camshaft grooves

6. Install cylinder head gasket and cylinder head

➡ **The long cylinder head bolts are inserted in the middle bores of the cylinder head**

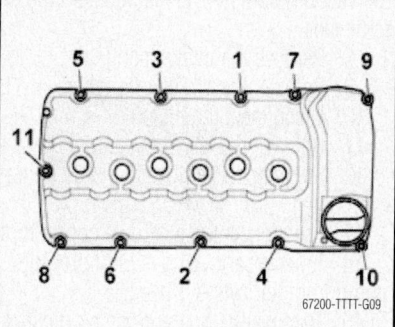

**Fig. 143 Cylinder head cover bolt removal sequence—TT 3.2L engine**

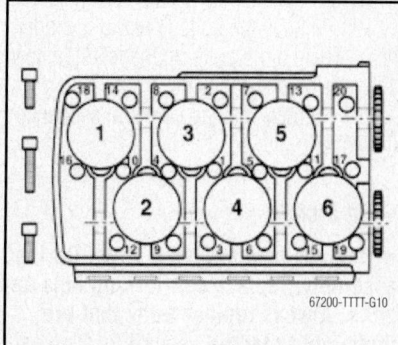

**Fig. 144 Cylinder head bolt removal and tightening sequence—TT 3.2L engine**

7. Tighten cylinder head bolts in 4 stages using the following sequence:
- 1$^{st}$ stage 22 ft. lbs (30 Nm)
- 2$^{nd}$ stage 37 ft. lbs (50 Nm)
- 3$^{rd}$ stage 90° (1/4 turn) further
- 4$^{th}$ stage 90° (1/4 turn) further

8. Install or connect the following
- Timing chain
- Cylinder head cover
- Intake manifold

9. Fill the engine with coolant and bleed, if necessary.

10. Connect the negative battery cable.

11. Check the oil level before starting the engine and top off, as necessary.

12. Start the vehicle, check for leaks and repair if necessary.

### A6 and A8 Models

#### 3.2L Engine

*See Figures 145 through 147.*

1. Before servicing the vehicle, refer to the precautions section.
2. Drain the cooling system.
3. Remove the front exhaust pipe.
4. Remove the drive belt.
5. Remove the front coolant pipe.
6. Remove the power-steering pump.
7. To remove the right cylinder head, the vacuum pump for brake booster must be removed.
8. Remove the intake manifold.
9. Remove the oil filter housing.
10. Disconnect the fuel line.
11. Remove the low-pressure line.
12. If equipped, remove the large lifting eye.
13. Disconnect the electrical connectors.
14. If necessary, remove high pressure pump from left cylinder head.
15. Remove the bolt and the oil dipstick tube.
16. Remove the cylinder head cover.
17. Remove the left and right timing chain covers.
18. Remove the camshaft timing chains.
19. Remove the bolts at rear of cylinder head. There are 3 bolts on the left side head and four on the right side head.
20. Remove the head bolts in sequence illustrated.
21. Carefully remove cylinder head.

#### To install:

➡Replace cylinder head bolts. During assembly, replace self-locking nuts and bolts. Always replace bolts that are tightened to torque as well as O-rings and gaskets.

22. Clean the gasket mating surfaces.

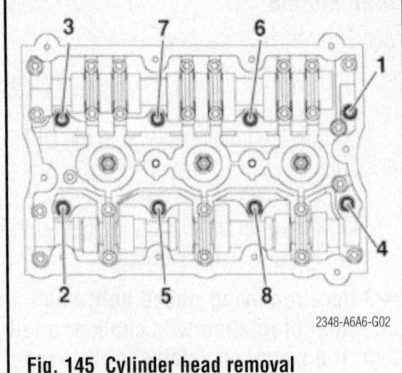

**Fig. 145 Cylinder head removal sequence—A6 with 3.2L engine**

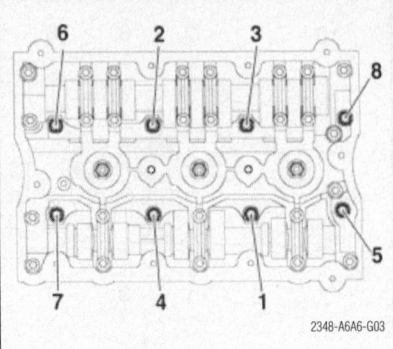

**Fig. 146 Cylinder head torque sequence—A6 with 3.2L engine**

➡Only unpack new cylinder head gasket immediately prior to installation. Handle gasket carefully. Damages to the silicone layer and in areas of recesses may result in leaks.

23. Install the cylinder head gasket onto guide sleeves. Marking "oben" (top) or part number must face toward cylinder head.

➡After installing a replacement cylinder head with camshafts installed, oil contact surfaces between roller cam followers and cam lubricating surfaces after installing cylinder head.

24. Do not remove plastic bases protecting exposed valves until immediately before installing cylinder head.

➡When replacing the cylinder head or cylinder head gasket, coolant must be completely replaced.

➡After working on the valvetrain, carefully rotate engine by hand at least 2 full revolutions to ensure that valves do not strike the pistons when starting. Change contaminated engine oil.

25. Before installing cylinder head, set crankshaft and camshafts to TDC setting, mounting camshaft locating tool T40070 on both cylinder heads and tightening to 20 Nm.

26. The camshaft locating tool T40070 is correctly positioned when the holes for cylinder head bolts remain free. Crankshaft holder T40059 must be screwed into crankshaft.

27. Position cylinder head gasket.

28. Pay close attention to centering pins in cylinder block.

29. Pay attention to installation position of cylinder head gasket, marking "oben" (top) or part number must face toward cylinder head.

30. Install the cylinder head.

31. Insert new cylinder head bolts and tighten by hand.

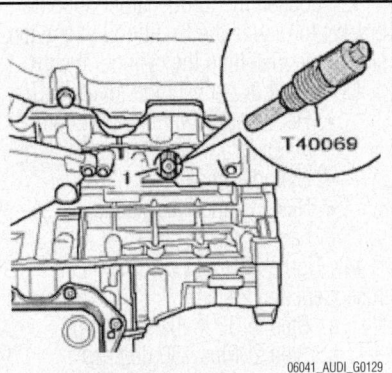

**Fig. 147 Sub frame bolt locations The camshaft locating tool T40070 is correctly positioned when the holes for cylinder head bolts remain free—A6 with the 3.2L engine**

32. Tighten cylinder head in sequence in 3 steps as follows:
33. Step1: tighten to 40 Nm.
   a. Step 2: Tighten an additional 90 degree.
   b. Step 3: Tighten an additional 90 degree.
34. Tighten the bolts at the rear of the heads to 9 Nm.
35. Installation of the remaining components is the reverse order of removal.
36. Check the oil level.
37. Replace the coolant.
38. Fill and bleed the power steering system.

#### 4.2L Engine

*See Figure 148.*

1. Before servicing the vehicle, refer to the precautions section.
2. Drain the cooling system.
3. Relieve the fuel system pressure.
4. Remove or disconnect the following:
- Negative battery cable
- Dipstick from LH cylinder head
- Intake duct for air cleaner

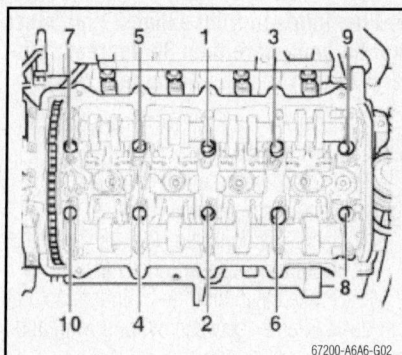

**Fig. 148 Cylinder head torque sequence— 4.2L Engines**

67200-A6A6-G02

- Fuel supply and return lines
- Upper bolt between exhaust manifold and exhaust pipe
- Oxygen sensor connectors
- Exhaust pipe from LH manifold (loosen double clamps rearward)
- Upper LH and RH timing belt guards
- Timing belt from camshafts (without moving position)
- Lower timing belt guard
- Connectors, pipes and coolant lines (mark for position reference)
- Cylinder head covers
- Left and right knock sensor connectors
- Fuel injector harness connectors
- Fuel rail, carefully remove along with fuel injectors, and place down to one side
- Coolant line at rear between cylinder heads
- Intake manifold
- Cylinder head bolts, in sequence, following reverse order of tightening sequence
- Cylinder head

***To install:***

5. Install the cylinder head with a new gasket and tighten the bolts in sequence as follows:

   a. Step 1: 26 ft. lbs. (35 Nm)
   b. Step 2: 44 ft. lbs. (60 Nm)
   c. Step 3: Plus 90 degrees
   d. Step 4: Plus 90 degrees

6. The remainder of the installation is the reverse of the removal procedure.

### A3 Models

#### 2.0L Engine

*See Figures 149 through 153.*

1. Before servicing the vehicle, refer to the precautions section.
2. Disconnect the electrical connector.
3. Remove the engine cover upward and out in direction of

4. Remove the coolant expansion tank cap.
5. Remove the center sound insulation fasteners.
6. Remove the right sound insulation fasteners.
7. Drain and recycle the coolant.
8. Remove the wiper arms.
9. Unclip spray nozzles and push spray nozzles, with lines still attached, back through opening and into plenum chamber.
10. Remove the rubber seal for plenum chamber cover.
11. Remove the plenum chamber cover.
12. Loosen up the rear engine wiring harness at partition for plenum chamber.
13. Remove the electrical connector for oxygen sensor before catalytic converter from bracket, disconnect it and lay it aside.
14. Remove the partition for plenum chamber.
15. Disconnect the coolant line to coolant expansion tank.
16. Disconnect the coolant hose.
17. Disconnect the coolant hose.
18. Unscrew the ground wire.
19. Disconnect the electrical connector.
20. Disconnect the coolant hoses.
21. Remove the intake manifold
22. Remove the catalytic converter with front exhaust pipe
23. Remove the right driveshaft
24. Remove the heat shield for right driveshaft.
25. Pull air duct off from charge air cooler.
26. Unfasten the air duct from turbocharger.
27. Disconnect the connector and loosen up cable.
28. Unscrew bolts and remove air duct.
29. Disconnect the oil supply line for turbocharger at cylinder block.
30. Remove the oil supply line for turbocharger.
31. Disconnect the coolant supply line for turbocharger at cylinder block.
32. Unscrew coolant supply line to turbocharger.
33. Remove the coolant supply line for turbocharger at cylinder block.
34. Remove the oil return line at turbocharger.
35. Remove the bolts and remove turbocharger support.
36. Remove the timing belt.
37. Loosen the bolt for camshaft sprocket using retainer 3036.
38. Pull off camshaft sprocket using two-arm puller T40001 and claws T40001/6 and T40001/7.

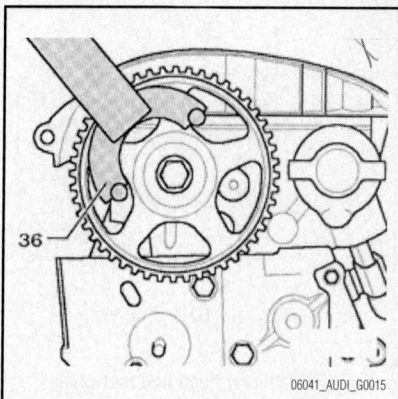

**Fig. 149 Loosen the bolt for camshaft sprocket using retainer 3036—A3 with 2.0L engine**

06041_AUDI_G0015

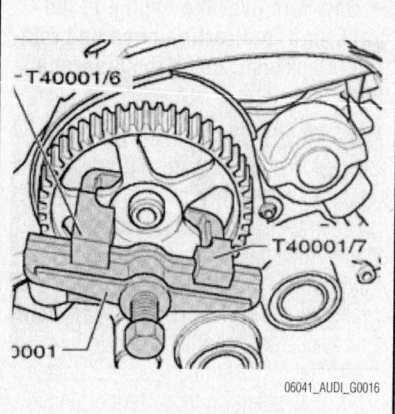

**Fig. 150 Pull off camshaft sprocket using two-arm puller T40001 and claws T40001/6 and T40001/7—A3 with 2.0L engine**

06041_AUDI_G0016

39. Remove the rear the timing belt housing from cylinder head.
40. Remove the cylinder head cover.
41. Loosen the cylinder head bolts in the sequence illustrated.

➡️**Verify that all hose and line connections between engine, transmission and body have been disconnected.**

42. Remove the cylinder head.

***To install:***

➡️**Use new cylinder head bolts.**

43. Always replace self-locking nuts, bolts that have been tightened to tightening torque as well as gaskets and O-rings.
44. Clean all gasket mating surfaces.
45. Only unpack new cylinder head gasket immediately prior to installation. Handle gasket carefully. Damages to the silicone layer and in areas of recesses may result in leaks.
46. There must be no oil or coolant in the blind holes for the cylinder head bolts in

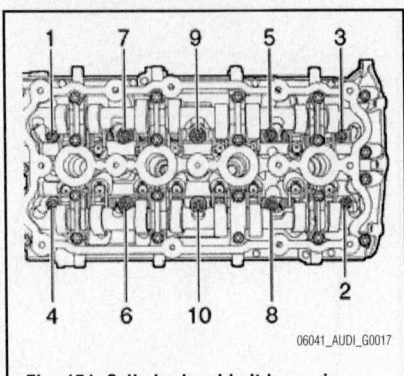

**Fig. 151 Cylinder head bolt loosening sequence—A3 with 2.0L engine**

the cylinder block. High-temperature lubricant Parts Catalog

➡ **Only turn over the engine at the crankshaft in direction of engine rotation (clockwise) using the crankshaft bolt.**

47. Move marking on the camshaft sprocket so that it is across from the marking on the timing belt cover. The notches of camshafts should now face toward each other vertically.

48. If the crankshaft was turned in the meantime: set cyl1 to TDC and turn crankshaft back again slightly.

49. Set cylinder head gasket in place.

50. Pay attention to centering pins in cylinder block.

51. Observe the installed position of the cylinder head gasket marking. The replacement part numbers must be legible from the intake side.

52. Install the cylinder head in place.

53. Insert cylinder head bolts and tighten by hand.

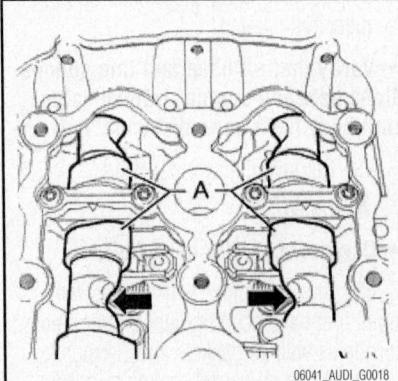

**Fig. 152 Move marking on the camshaft sprocket so that it is across from the marking on the timing belt cover. The notches of camshafts should now face toward each other vertically—A3 with 2.0L engine**

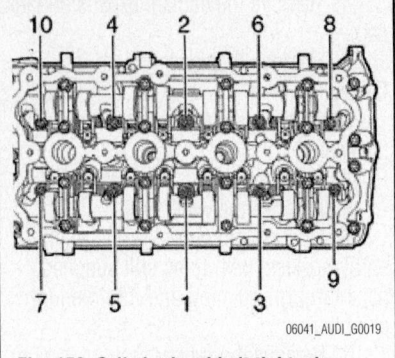

**Fig. 153 Cylinder head bolt tightening sequence—A3 with 2.0L engine**

54. Tighten the cylinder head in the sequence illustrated as follows:

   a. Step 1: tighten to 40 Nm.

   b. Step 2: tighten and additional 180 degrees.

55. Install the cylinder head cover and tighten to 10 Nm.

56. Install the wiper arms.

57. Install the timing belt.

58. Install the accessory drive belt.

59. Install the intake manifold.

60. Connect the negative battery cable.

61. Check the engine oil level and replenish as needed.

62. Fill the cooling system.

> ❋❋ **CAUTION**
>
> **Do not use a battery charger for starting assistance! There is the risk that the vehicle control modules could be damaged.**

### 3.2L Engine

*See Figures 154 through 158.*

1. Before servicing the vehicle, refer to the precautions section.

2. Drain the cooling system.

3. Remove the intake manifold.

4. Remove the coolant hoses to heater core at the bulkhead.

5. Disconnect the following hose connections:

   a. Coolant hose to reservoir.

   b. Vacuum hose to Evaporative Emission (EVAP) canister.

   c. Vacuum hose to leak detection pump

   d. Fuel supply line.

   e. Coolant hose to reservoir.

   f. Disconnect the coolant hoses.

6. Remove the wiring connections with coolant and vacuum hoses.

7. Remove the camshaft timing chain from camshafts.

➡ **Flex joints in front exhaust pipe must not be bent more than 30 degrees, otherwise they may be damaged.**

8. Remove the front exhaust pipe from exhaust manifolds.

9. Loosen and remove cylinder head bolts in sequence shown.

10. Install lifting tackle 3033 as illustrated.

11. Carefully raise cylinder head.

12. Place clean cloth in cylinders so that no dirt can enter between cylinder wall and pistons.

*To install:*

➡ **Use new cylinder head bolts.**

➡ **Always replace self-locking nuts, bolts that have been tightened to tightening torque as well as gaskets and O-rings.**

13. Clean the gasket mating surfaces.

14. Only unpack new cylinder head gasket immediately prior to installation. Handle gasket carefully. Damage in silicon layer and recessed area lead to leakage. The plastic protectors installed to protect the open valves must only be removed immediately before installing the cylinder head.

15. When installing a cylinder head other than the original cylinder head, the

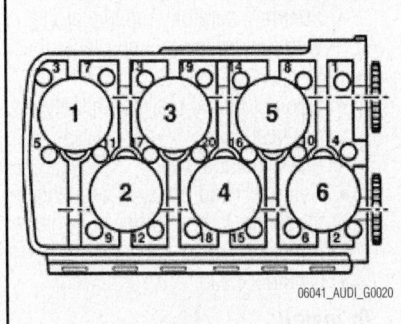

**Fig. 154 Cylinder head bolt loosening sequence—A3 with 3.2L engine**

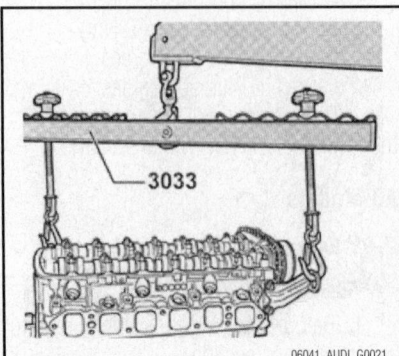

**Fig. 155 Install lifting tackle 3033 as shown—A3 with 3.2L engine**

contact surfaces between bearing elements, roller rocker levers and cam contact surfaces must be oiled before installation.

➡**There must be no oil or coolant in the blind holes for the cylinder head bolts in the cylinder block.**

16. Position camshafts at TDC marking before positioning cylinder head. It must be possible to insert camshaft bar T10058 into both shaft grooves.

17. Make sure the crankshaft is positioned at TDC.

18. Install cylinder head gasket. Check the centering pins in cylinder block. Check the cylinder head seal location marks, the part number must be visible from intake side.

19. Install the cylinder head and hand tighten the bolts.

➡**The longer cylinder head bolts must be inserted in the middle holes of cylinder head.**

20. Tighten the bolts in sequence as follows:

    a. Step 1: tighten to 30 Nm.
    b. Step 2: tighten to 50 Nm.

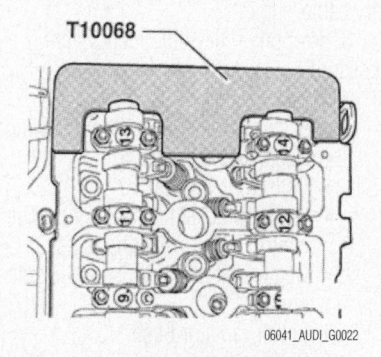

Fig. 156 Position camshafts at TDC marking before positioning cylinder head. It must be possible to insert camshaft bar T10068 into both shaft grooves—A3 with 3.2L engine

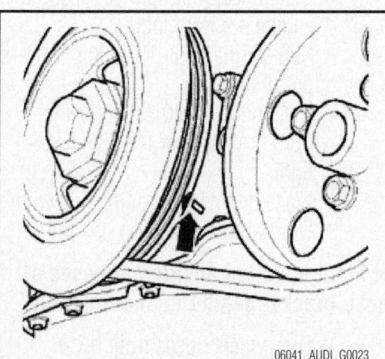

Fig. 157 Make sure the crankshaft is positioned at TDC—A3 with 3.2L engine

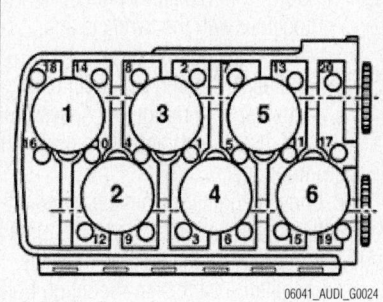

Fig. 158 Cylinder head torque sequence—A3 with 3.2L engine

    c. Step 3: Using a Torx key tighten an additional 90 degrees.
    d. Step 4: Using a Torx key tighten an additional 90 degrees.

21. Install the camshaft timing chain.
22. Install the cylinder head cover
23. Install the intake manifold
24. Fill the cooling system.
25. Connect the negative battery cable.

## ENGINE ASSEMBLY

### REMOVAL & INSTALLATION

#### TT Models

##### 1.8L Engine

*See Figure 159.*

1. Before servicing the vehicle, refer to the precautions section.

2. Turn the ignition switch to the **OFF** position, then disconnect the negative battery cable.

3. Properly relieve the fuel system pressure.

4. Drain the engine coolant.

5. Remove or disconnect the following:

- Negative battery cable
- Engine cover from cylinder head cover
- Cover in front of intake manifold
- Air duct at bottom of right-hand longitudinal member
- Noise insulation in center and on left and right sides
- Pendulum support
- Both drive axles from transmission flange
- Move left-hand drive axle toward rear of vehicle and secure in place by tying to anti-roll bar
- Front exhaust pipe

➡**Do not bend the flexible connection on front exhaust pipe more than 10 degrees or it may be damaged**

- Secure support rails to subframe with M8 x 25 bolt. Also secure engine support device to support rails
- Press engine forward
- Loosen right-hand 12-point nut for drive axle. When doing this, vehicle must be on its wheels
- Unbolt wheel
- Press out and remove drive axle
- Air hose and wiring harness from air mass meter
- Vacuum hose from air recirculation valve
- Hose from pressure control valve for crankcase breather
- Connector from charge pressure control valve
- Hose from charge pressure control valve
- Hose from solenoid valve to turbocharger
- Solenoid charge pressure control valve from air intake hose and place it to one side on engine
- Hose between activated charcoal filter and turbocharger
- Retainer off turbocharger connection and detach air intake hose
- Battery and platform
- Ground wire on engine/transmission flange
- Wires from starter and retainer on starter and move wires to one side
- Connector for electrical change-over valve on underside of bracket
- Bracket from intake manifold and disengage dipstick tube
- Alternator wiring harness
- A/C compressor wiring harness
- Coolant hoses
- Connectors from knock sensors, speed sensor and oil pressure switch
- Vacuum pipe from intake manifold.
- Connector for intake air temperature sensor and throttle valve control part (under throttle valve control part)
- Air hose from throttle valve control part
- Vacuum hose (for activated charcoal filter) from throttle valve control part
- Connector from Hall sensor
- Vacuum pipe from vacuum reservoir
- Vacuum reservoir and detach it from the bracket
- Connectors from injectors and unclip support bar from fuel rail

- Ground wire between ignition coils 1 and 2
- Connectors from ignition coils
- Connector from coolant temperature sensor and speedometer sensor
- Mark threaded sections of both selector cables with a permanent marker pen so that they can be installed in the same positions later
- Pull collar toward ball joint to compress selector cable spring, then turn collar clockwise to lock in position
- Disengage both threaded sections
- Unclip clutch slave cylinder hose from selector cable support bracket

➡ **Do not depress the clutch or open the pipe system.**

- Fuel supply line and fuel return line at connection point

➡ **Check the colors of connectors when connecting fuel supply and return lines**

➡ **Before removing ribbed belt, mark direction of rotation with chalk or a felt pen. If a used belt rotates in the wrong direction when reinstalled, this can result in breakage. When installing belt, ensure it is correctly seated in the pulleys.**

- Loosen ribbed belt by turning tensioning element clockwise and remove belt

➡ **Tensioning element can be locked in position with a suitable punch.**

- Connecting pipe between both left and right crossmembers
- Right cross member
- Connectors at both starter and transmission and starter positive cable
- Power steering fluid cooling line from transmission
- Open hose clamp at lower left of radiator and completely remove cooling line
- Pulley from power steering pump (counterhold with socket wrench)
- Place clean container under pump. Open spring clips from intake hose and remove hose.
- Pump bolts from bracket
- Connector and hoses at secondary air pump
- Secondary air pump
- Two radiator fan connectors on the radiator cowl (bottom left)

- Cowl from radiator (4 bolts) together with the two fans and remove by pulling down
- Unbolt A/C compressor, lift clear together with refrigerant hoses (do not disconnect) and tie in place on hood lock

6. Loosen engine/transmission mounting bolts (about two turns only) on engine side. Do not remove the bolts completely. Loosen engine/transmission mounting bolts (about two turns only) on transmission side. Do not remove the bolts completely.

7. Raise vehicle

8. Bolt engine bracket T10012, "or equivalent" to cylinder block with securing nut and bolt (M10 x 25/8.8). Tighten to about 15 ft. lb. (20 Nm).

9. Install engine/transmission jack to engine bracket and raise engine/transmission slightly

10. Unbolt engine/transmission mounting on the engine side

11. Unbolt engine/transmission mounting on the transmission side

12. Carefully lower the engine/transmission assembly

### To install:

13. Installation is the reverse of the removal procedure, while using the torque values below. Note the following engine mounting adjustment procedure.

14. Adjust the engine mounting so that distance "a" (on the right-hand mounting) is about 0.51" (13 mm). It should be possible to insert a 0.47" (12 mm) flat metal bar without any difficulty.

15. To move the engine within the engine console, first loosen bolts "b" on the left and right engine supports by about 2 turns each

16. If the gap is too narrow or too wide, proceed as follows: Remove air hose from air mass meter and disconnect electrical connectors.

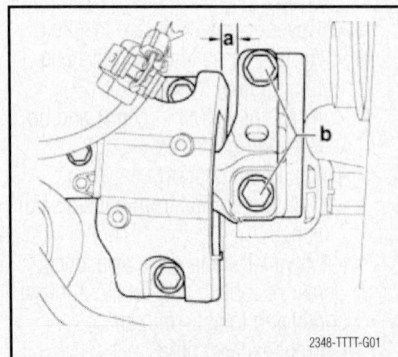

**Fig. 159 Adjusting engine mounting— 1.8L engine TT**

2348-TTTT-G01

17. Remove air cleaner housing

18. Raise engine slightly, taking up the weight evenly on both sides

19. Loosen bolts "b" on left and right-hand support arms by about two turns each

20. Insert a metal bar between engine console and support arm, then move engine until gap "a" measures 0.51" (13 mm)

21. Tighten (by hand) all four bolts and left and right-hand support arms, then final tighten all four bolts to 63 ft. lbs. (85 Nm.)

- M8 bolts: 15 ft. lbs. (205 Nm)
- M10 bolts: 33 ft. lbs. (45 Nm)
- M12 bolts: 48 ft. lbs. (65 Nm)
- Drive shaft-to-transmission bolts: 30 ft. lbs. (40 Nm)
- Engine support-to-engine console bolts: 63 ft. lbs. (85 Nm)
- Transmission support-to-transmission console bolts: 44 ft. lbs. (60 Nm), plus 90 degrees

### 3.2L Engine

*See Figure 160.*

1. Before servicing the vehicle, refer to the precautions section.

2. Turn the ignition switch to the **OFF** position, then disconnect the negative battery cable.

3. Properly relieve the fuel system pressure.

4. Drain the engine coolant.

5. Remove or disconnect the following:
- Negative battery cable
- Both front wheels
- Unbolt bracket for headlight range control in left wheel housing
- Center, left and right sound insulation
- Lower radiator hose
- Body brace
- Right side bracket for body brace
- Air cleaner housing
- Fuse box and carrier
- Wiper arms
- Plenum chamber cover
- Disconnect multi pin connector at engine control module
- Pull engine control module forward out of bracket
- Disconnect electrical connector from bracket near left side headlight
- Place engine control module with wiring harness on top of engine

➡ **Mark installed position of hoses at hose bracket behind intake manifold**

- Disconnect hoses from hose bracket and remove bracket
- Fuel lines

➡️**Before removing ribbed belt, mark direction of rotation with chalk or a felt pen. If a used belt rotates in the wrong direction when reinstalled, this can result in breakage. When installing belt, ensure it is correctly seated in the pulleys.**

- Release tension on belt by screwing a bolt into bore on tensioning element and remove belt
- Upper radiator hose
- Upper securing bolts for radiator cowl
- Bracket and selector cable from transmission
- Loosen, but do not remove bolts on mounting assembly on engine and transmission
- Cross piece
- Power steering pump
- Secondary air injection pump and bracket
- Lower securing bolts for radiator cowl
- Radiator cowl with both fans

➡️**The air conditioning compressor can only be removed from the bracket with the upper securing bolt still inserted. To accomplish this, press the engine simultaneously to the left side. If necessary, loosen all bolts of the assembly mounting.**

- Mounting bolt for air conditioning compressor
- Secure the air conditioning compressor together with the hoses at front longitudinal member (hoses remain connected)
- Alternator
- Catalytic converter
- Front exhaust pipe

➡️**When loosening and removing the axle shaft, the wheel bearing must NOT be under load (vehicle must not be standing on its wheels)**

- Right and left axle shafts
- Unbolt flexible coupling for driveshaft, push driveshaft as far back as possible

➡️**The oil seal in the flange of the driveshaft must NOT be damaged. The driveshaft must be replaced if the oil seal is damaged**

- Pendulum support
6. Attach engine bracket T10102 or equivalent to cylinder block
7. Install engine/transmission jack to engine bracket and raise assembly slightly

8. Remove engine/transmission mounting bolts
9. Pull engine/transmission assembly as far forward as possible and carefully lower

### To install:

10. Installation is the reverse of the removal procedure, while using the torque values below. Note the following engine mounting adjustment procedure.
11. Adjust the engine mounting so that distance "X" (on the right-hand mounting) is about 0.51" (13 mm). It should be possible to insert a 0.47" (12 mm) flat metal bar without any difficulty.
12. To move the engine within the engine console, first loosen bolts on the left and right engine supports by about 2 turns each
13. If the gap is too narrow or too wide, proceed as follows: Remove air hose from air mass meter and disconnect electrical connectors.
14. Remove air cleaner housing
15. Raise engine slightly, taking up the weight evenly on both sides
16. Loosen bolts on left and right-hand support arms by about two turns each
17. Insert a metal bar between engine console and support arm, then move engine until gap "a" measures 0.51" (13 mm)
18. Tighten (by hand) all four bolts and left and right-hand support arms, then final tighten all four bolts to 63 ft. lbs. (85 Nm)
19. Transmission to engine block
   - M12 bolts 59 ft. lbs (80 Nm)
   - M10 bolts 30 ft. lbs (40 Nm)
20. Other bolts/nuts
   - M6 7 ft. lbs (10 Nm)
   - M8 15 ft. lbs (20 Nm)

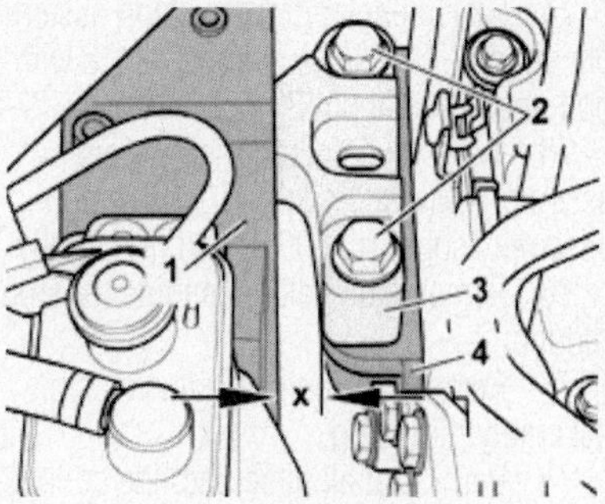

1. Engine mount
2. Bolts
3. Support arm
4. Engine support

67200-TTTT-G11

**Fig. 160 Adjusting engine mounting—3.2L engine TT**

- M10 33 ft. lbs (45 Nm)
- M12 48 ft. lbs (65 Nm)
- Flexible coupling to bevel box 44 ft. lbs (60 Nm)
- Exhaust pipe to manifold 30 ft. lbs (40 Nm)
- Exhaust pipe to catalytic converter 18 ft. lbs (25 Nm)
- Pendulum support to transmission 30 ft. lbs (40 Nm) + 90° (1/4 turn)
- Pendulum support to subframe 15 ft. lbs (20 Nm) + 90° (1/4 turn)

### A3 Models

#### 2.0L Engine

*See Figures 161 through 163.*

1. Before servicing the vehicle, refer to the precautions section.

➡️**The engine is removed downward together with the transmission.**

2. Disconnect the electrical connector and unclip intake connection.
3. Remove the engine cover upward and out in direction of.
4. Remove the electrical connector for oxygen sensor before catalytic converter from the bracket, disconnect it and lay it aside.
5. Remove the coolant expansion tank cap.
6. Remove the cover from above battery, by pressing release button.
7. Disconnect the negative battery cable.
8. Remove the battery.
9. Remove the battery carrier.
10. Pry off caps on wiper arms with a screwdriver and remove nuts.
11. Pull off and remove wiper arms from wiper axles.
12. Unclip spray nozzles.
13. Push spray nozzles, with the lines still attached, back through the opening and into the plenum chamber.
14. Remove the rubber seal for plenum chamber cover.
15. Remove the plenum chamber cover.
16. Loosen up the rear engine wiring harness at partition for plenum chamber.
17. Remove the partition for plenum chamber.
18. Remove the Engine Control Module (ECM).
19. Disconnect the electrical connector for the engine wiring harness.
20. Pull up guide for engine wiring harness.
21. Open the wiring guide bracket.
22. Remove the engine wiring harness to control module from the wiring guide.

23. Disconnect the electrical connector for back-up light switch.

24. Remove the ground wire.

25. Push the cover towards the rear.

26. Remove the wires at the starter.

27. Remove the cover from above the battery, by pressing the buttons.

28. Remove the electrical harness from alternator.

29. Remove the front wheels.

30. Remove the center sound insulation fasteners.

31. Remove the left and right sound insulation fasteners.

32. Drain the coolant.

33. Disconnect the upper coolant hose from the radiator.

34. Disconnect the coolant hoses to the heat exchanger.

35. Remove the nuts for front exhaust pipe/turbocharger that can be reached from above.

36. Disconnect the vacuum hose from brake booster.

37. Unclip lock washers and at both cables. Remove the cable retainers from transmission selector lever and relay lever.

38. Remove the cable mounting bracket from the transmission and lay aside.

### ✳✳ CAUTION

**Do not operate the clutch pedal anymore after the hose for the slave cylinder has been disconnected.**

39. Pull out the clip to stop and pull off the slave cylinder hose line.

40. Relieve the fuel system pressure.

41. Disconnect the fuel line and lay it aside.

42. Disconnect the EVAP line and lay it aside.

43. Disconnect the vacuum line to EVAP canister.

44. Remove the EVAP canister upwards.

45. Disconnect the coolant hoses.

46. Discharge air conditioning system

47. Remove the refrigerant line at compressor.

48. Unscrew bolts of engine mountings about 2 turns.

49. Unscrew bolts of transmission mountings about 2 turns.

50. Disconnect the electrical connector.

51. Open cable retainer. and loosen up any electrical wiring.

52. Remove the bolts from above.

53. Remove the air duct. for the charge air cooler.

54. Disconnect the electrical connector.

55. Remove the bolts and remove fan shroud.

56. Disconnect the connector at left longitudinal member and loosen up cable.

57. Pull air duct off from charge air cooler.

58. Disconnect the refrigerant line from condenser.

59. Disconnect the connector from Charge Air Pressure sensor

60. Disconnect the electrical connector at oil level sensor.

61. Disconnect the electrical connector at after-run water pump motor and loosen up wiring.

62. Remove the heat shield for right drive axle.

63. Remove the left and right drive axle.

64. Remove the right cover for vehicle floor.

65. Disconnect the connector for oxygen sensor after catalytic converter and loosen up cable.

66. Remove the remaining nuts for front exhaust pipe/turbocharger from below.

67. Remove the supports for front exhaust pipe with catalytic converter, by unscrewing bolts.

➡**Decoupling elements in front exhaust pipes must not be bent more than 10° as damage may occur.**

68. Disconnect the exhaust system at clamping sleeve.

69. Remove the cross member for vehicle floor.

70. Remove the bracket for exhaust system.

71. Remove the catalytic converter with front exhaust pipe.

72. Remove the bolts and remove pendulum support.

73. Remove the auxiliary water pump bolts.

74. Fasten engine/transmission assembly support T10012 to cylinder block with bolt and nut to about 20 Nm.

75. Insert an engine/transmission assembly jack into engine/transmission assembly support T10012 and lift slightly.

76. Remove the bolts at engine assembly mounting.

77. Remove the bolts at transmission assembly mounting.

78. Verify that all hose and line connections between engine, transmission and body have been disconnected.

79. While lowering, carefully guide the engine/transmission assembly in order to prevent damage.

80. Pull engine/transmission assembly as far forward as possible and slowly lower it.

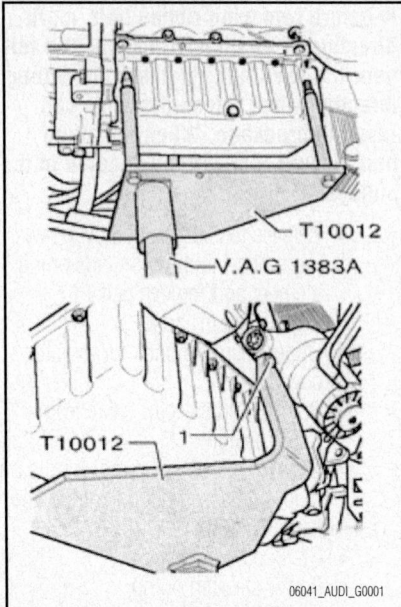

Fig. 161 Fasten engine/transmission assembly support T10012 to cylinder block—A3 with 2.0L engine

### To install:

81. Installation is in the reverse order of removal, please note the following:

a. Check whether dowel sleeves for centralizing engine/transmission assembly are in cylinder block, install if necessary.

b. Make sure that the intermediate plate is hooked in at the sealing flange and is pushed onto alignment sleeves.

c. Check clutch release bearing for wear and replace if necessary.

➡**Torque specifications only apply to lightly greased, oiled, phosphated or blackened nuts and bolts. additional lubricants, such as engine or transmission oil are permissible, although lubricants containing graphite are not.**

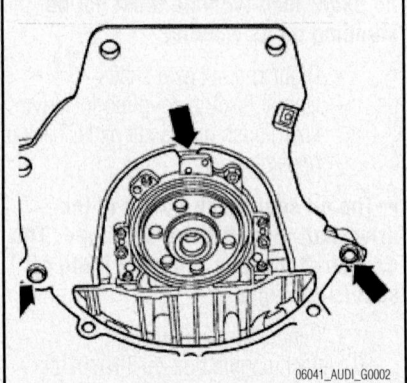

Fig. 162 Check whether dowel sleeves for centralizing engine/transmission assembly are in cylinder block—A3 with 2.0L engine

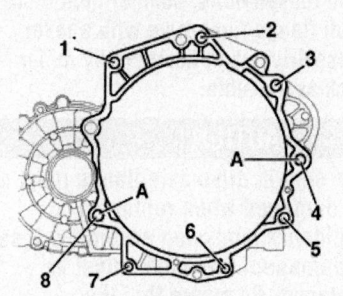

Engine/transmission, fastening

| Pos. | Bolt | Nm |
|------|------|-----|
| 1, 2 | M12×65 | 80 |
| 3 *, 4 * | M12×135 | 80 |
| 5 to 7 | M10×60 | 40 |
| 8 | M12×90 | 65 |

\* Bolt with threaded pin M8

06041_AUDI_G0003

**Fig. 163 Transmission-to-engine bolts torque values—A3 with 2.0L engine**

d. Do not use any degreased parts.

e. Tighten the M12 X 65 and M12 X 135 bolts to 80 Nm, the M12 X 90 bolts to 65 Nm and the M10 X 60 bolts 40 Nm.

➡**The bolts are first tightened to final torque after the engine mounts have been adjusted.**

f. Remove the engine/transmission assembly support T10012 from engine.

g. Fasten pendulum supports to transmission and to sub frame. Tighten pendulum to transmission bolts to 40 Nm plus an additional 90 degrees. Tighten the pendulum to sub frame bolts to 100 Nm plus 90 degrees.

**✴✴ CAUTION**

**Do not use a battery charger for starting assistance! There is the risk that the vehicle control modules could be damaged.**

h. Tighten all remaining M6 bolts to 10 Nm, M8 bolts to 20 Nm, M10 bolts to 45 Nm and M12 bolts to 65 Nm.

i. Tighten the A/C compressor to accessory assembly bracket to 25 Nm.

j. Tighten the ground strap to transmission to 23 Nm.

### 3.2L Engine

*See Figures 164 through 170.*

1. Before servicing the vehicle, refer to the precautions section.

➡**The engine is removed downward together with the transmission.**

2. Remove the tool formed insert under luggage compartment floor covering.

3. Remove the cover for battery compartment

4. Remove the formed insert from over battery.

5. Disconnect the negative battery cable.

6. Remove the coolant expansion tank cap.

7. Remove the both front wheels.

8. Remove the center noise insulation fasteners.

9. Remove the left and right noise insulation fasteners.

10. Disconnect the electrical connector at Engine Coolant Temperature (ECT) sensor.

11. On vehicles with drain plug, turn the drain plug left on radiator, place assisting hose on supports if necessary.

12. Remove the lower coolant hose from radiator by removing retaining clamps and drain the coolant.

13. If equipped, remove vacuum hose to air guide hose.

14. Disconnect the air guide hose at Throttle Valve Control Module.

15. Disconnect the Mass Air Flow (MAF) sensor connector.

16. Remove the upper part of air filter housing.

17. Remove the filter element.

18. Remove the air duct cover by releasing retaining clasps on sides.

19. Remove the lower part of air filter housing.

20. Remove the air filter housing bracket.

21. Disconnect the electrical connector for Heated Oxygen sensors (HO2S)

22. Move the electrical wiring to oxygen sensors aside.

23. Pry off the wiper arms caps using a screwdriver.

24. Loosen the wiper arm nuts using several turns.

25. Loosen the wiper arms by gently rocking wiper arm.

26. Remove the nuts and remove wiper arms.

27. Unclip spray nozzles. Push the spray nozzle with lines still attached back through opening and into plenum chamber.

28. Remove the rubber seal for plenum chamber cover.

29. Remove the plenum chamber cover.

30. Loosen up the rear engine wiring harness at the partition for plenum chamber.

31. Remove the partition for plenum chamber.

32. Remove the Engine Control Module (ECM)

33. Separate electrical connector for engine wiring harness.

34. Release pass-through for engine wiring harness and pull off upward.

35. On vehicles with the older with E-box version, press the retaining clamps in and remove the engine compartment E-box cover.

36. On vehicles with the newer with E-box version, press in both latches and remove engine compartment E-box cover.

37. Remove the terminal 30 wiring at engine compartment E-box. Open the wiring bracket.

38. Loosen up the electrical connector and disconnect, open underlying wiring guide bracket.

39. Remove the engine wiring harness to control module from wiring.

40. Disconnect the Secondary Air Injection hose at position indicated by.

41. Loosen up the air hose to the secondary air injection pump.

42. Disconnect the electrical harness connectors.

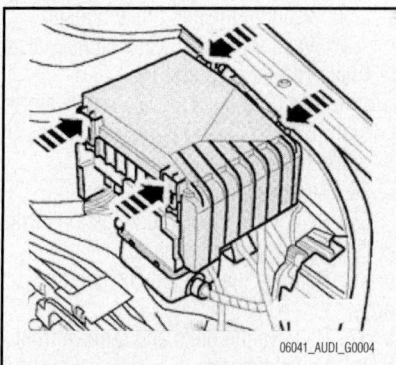

06041_AUDI_G0004

**Fig. 164 View of the older E-box version—A3 with 3.2L engine**

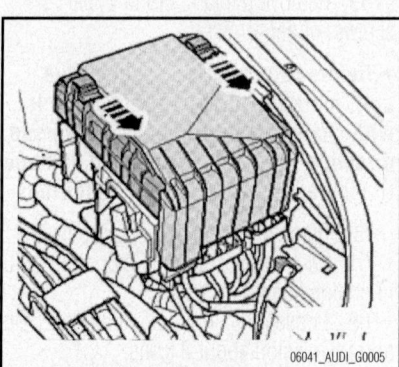

06041_AUDI_G0005

**Fig. 165 View of the newer E-box version—A3 with 3.2L engine**

43. Remove the protective boot and remove electrical wiring at starter solenoid switch.

44. Remove the ground wire.

45. Remove the bolts for bracket.

46. Remove the retaining clip upward and remove selector lever cable from transmission.

47. Remove the coolant hoses to heater core on bulkhead.

48. Disconnect the coolant hoses.

49. Disconnect the electrical connection at throttle valve control module.

50. Remove the Throttle Valve Control Module from the intake manifold.

51. Properly relieve the fuel system pressure.

52. Disconnect the following hose connections:

    a. Coolant hose to reservoir.

    b. Vacuum hose to Evaporative Emission (EVAP) canister .

    c. Vacuum hose to leak detection pump.

    d. Fuel supply line.

    e. Coolant hose to reservoir.

53. Remove the following hoses and electrical connections:

    a. Electrical connector to Evaporative Emission (EVAP) canister Purge Regulator Valve.

    b. Vacuum hose to brake booster.

    c. Vacuum hose to exhaust flap, also to leak detection pump.

    d. Vacuum hose to evaporative emission (EVAP) canister purge regulator valve.

54. Remove the wiring connection.

55. Remove the wiring connections with coolant and vacuum hoses.

56. Remove the Heated Oxygen sensors.

57. Remove the bolts and remove front exhaust pipe support.

58. Remove the front exhaust pipe nuts and exhaust manifolds.

59. Remove the fan shroud upper screws—upper arrows -.

➡**Before removing accessory drive belt, mark the turning direction on it with chalk or a felt tip pen. A reversed direction of rotation can cause damage to the belt under operating conditions.**

60. Remove the accessory drive belt.

61. Loosen the subframe mounting bolts to engine about 2 turns.

62. Loosen the subframe mounting bolts to transmission about 2 turns.

63. Remove the cover for vehicle floor on the right.

64. Disconnect the electrical connector to Oxygen sensors (O2S) behind the Three Way Catalytic Converter (TWC).

65. Remove the connector coupling from bracket and loosen up wiring to the oxygen sensors.

66. Remove the front vehicle floor cross member by removing the retainers.

➡**Flex joints in front exhaust pipe must not be bent more than 30 degrees, otherwise they may be damaged.**

67. Separate exhaust system at clamping sleeve.

68. Remove the exhaust system bracket and remove exhaust system.

69. Loosen up the lower coolant hose at fan shroud.

70. Disconnect the electrical connector at Secondary Air Injection (AIR) Pump Motor and loosen up electrical wiring.

71. Remove the coolant pipe bracket to transmission oil cooler.

72. Remove the bolts.

73. Loosen the bolt and remove Secondary Air Injection (AIR) pump with bracket.

74. Disconnect the lower coolant fan electrical connectors at fan shroud.

75. Remove the lower fan shroud screws.

76. Remove the fan shroud with both fans downward.

77. Disconnect the solenoid clutch electrical connector at A/C compressor and loosen up electrical wiring.

78. Remove the A/C compressor bolts.

79. Secure A/C compressor with coolant hoses attached at front of longmember.

80. Remove the electrical wire on alternator.

81. Disconnect the electrical connector.

82. Remove the upper idler roller.

83. Remove the alternator bolts and.

➡**Alternator can only be removed from bracket with upper securing bolt still in place.**

84. Remove the alternator from accessory assembly bracket.

85. Remove the alternator toward lower left.

86. Disconnect the electrical connector to Secondary Air Injection (AIR) Solenoid Valve.

87. Have a helper press foot brake and loosen the collar bolt at right drive axle.

88. Remove the right drive axle heat shield from the bevel box.

89. Remove the right drive axle.

90. Unbolt left drive axle from transmission flange shaft.

91. Remove the drive shaft flexible disc at bevel box.

➡**To loosen bolts, counter hold at drive shaft flange/final drive with a lever. Press drive shaft horizontally as far back as possible.**

**✖✖ CAUTION**

**The seal in drive axle flange must not be damaged when removing engine/transmission assembly. If seal is damaged, drive axle must be replaced. Be aware that the engine/transmission assembly swings forward slightly when pendulum support/subframe bolts are loosened.**

92. Remove the bolts and remove pendulum support.

93. Disconnect the electrical connector at Oil Level Thermal sensor.

94. Unclip bracket for electrical wire to Oil Level Thermal sensor at subframe.

95. Disconnect the electrical connector at After-Run Coolant Pump.

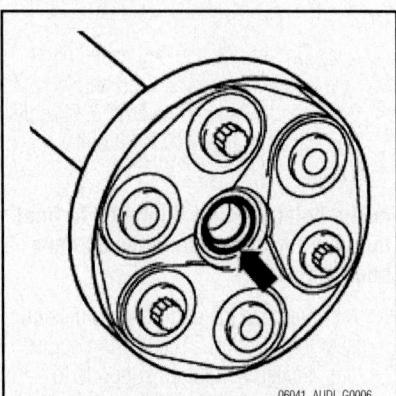

06041_AUDI_G0006

**Fig. 166 To loosen bolts, counter hold at drive shaft flange/final drive with a lever—A3 with 3.2L engine**

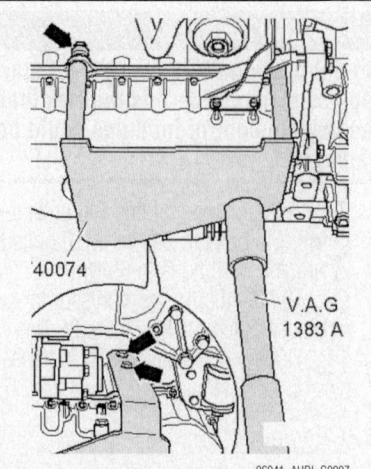

06041_AUDI_G0007

**Fig. 167 Bolt engine mount T40074 to cylinder block and tighten to 20 Nm—A3 with 3.2L engine**

96. Remove the After-Run Coolant Pump downward with coolant lines attached from rubber bracket loops. Spray rubber loops with silicone-free lubricant, if necessary.

➡**The water pump remains with coolant hoses connected at engine.**

97. Remove the bolts and accessory assembly bracket.

98. Bolt engine mount T40074 to cylinder block and tighten to 20 Nm.

99. Insert engine and transmission jack at engine mount T40074 and raise engine/transmission assembly slightly.

100. Remove bolts for the assembly mounting.

101. Remove the subframe mounting bolts at the engine.

102. Remove the subframe mounting bolts at the transmission.

➡**Verify that all hose and line connections between engine, transmission and body have been disconnected.**

103. While lowering, carefully guide the engine/transmission assembly in order to prevent damages.

104. Pull engine/transmission assembly as far forward as possible while slowly lowering it.

***To install:***

105. Installation is the reverse of removal, please note the following:

   a. Always replace self-locking nuts, bolts that have been tightened to tightening torque as well as gaskets and O-rings.

   b. Make sure alignment sleeves for engine to transmission are installed in cylinder block.

   c. A pilot needle bearing must be installed in the crankshaft in engines for vehicles with DSG transmission.

   d. Use new bolts when securing the transmission to engine.

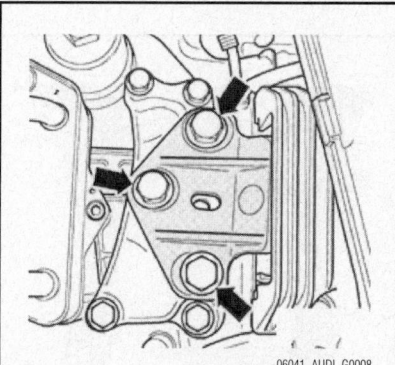

**Fig. 168 Remove the sub frame mounting bolts at the transmission—A3 with 3.2L engine**

➡**Torque specifications only apply to lightly greased, oiled, phosphated or blackened nuts and bolts. Additional lubricants, such as engine or transmission oil are permissible, although lubricants containing graphite are not.**

   e. Tighten bolts position 1 on the transmission to engine illustration to 80 Nm.

   f. Tighten bolts position 2 on the transmission to engine illustration to 40 Nm.

   g. Tighten bolts position 3 and 6 on the transmission to engine illustration to 80 Nm.

   h. Tighten bolts position 4 and 9 on the transmission to engine illustration to 40 Nm.

   i. Tighten bolts position 5 on the transmission to engine illustration to 80 Nm.

   j. Tighten bolts position 7 and 8 on the transmission to engine illustration to 40 Nm.

   k. Tighten bolts position 10 on the transmission to engine illustration to 80 Nm.

   l. Install the bevel box bracket, tighten the bolts (1) to 3 Nm, then bolts (2) to 35

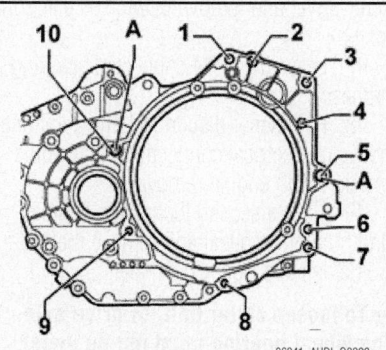

**Fig. 169 Transmission to engine bolt locations—A3 with 3.2L engine**

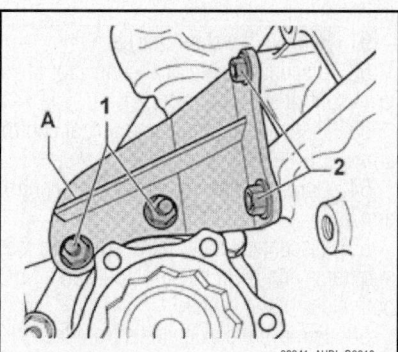

**Fig. 170 Bevel box bolt locations—A3 with 3.2L engine**

Nm and finally bolts (1) to 45 Nm. Refer to the illustration for bolt locations.

   m. Guide engine/transmission assembly into the body and hand tighten engine support bolts to engine console.

   n. Tighten transmission support bolts to transmission console.

   o. Remove the engine/transmission assembly support from engine.

   p. Press engine/transmission assembly toward bulkhead, when doing so bevel box pins must be carefully guided into drive shaft flange.

   q. Fasten pendulum supports to transmission and to sub frame. Tighten pendulum to transmission bolts to 40 Nm plus an additional 90 degrees. Tighten the pendulum to sub frame bolts to 100 Nm plus 90 degrees.

   r. Fasten drive shaft with flexible disc to bevel box flange to 60 Nm.

   s. To make it easier to install the alternator, drive the bushings for the retaining bolts back slightly.

**✴✴ CAUTION**

**Do not use a battery charger for starting assistance, there is the risk that the vehicle control modules could be damaged.**

   t. Tighten the front exhaust pipe to exhaust manifold to 40 Nm with new fasteners.

**✴✴ CAUTION**

**Do not use a battery charger for starting assistance! There is the risk that the vehicle control modules could be damaged.**

   u. Tighten all remaining M6 bolts to 10 Nm, M7 bolts to 15 Nm, M8 bolts to 20 Nm, M10 bolts to 45 Nm and M12 bolts to 65 Nm.

   v. Tighten the ground strap to transmission to 22 Nm.

### A6 Models

***3.2L Engine***

*See Figure 171.*

1. Before servicing the vehicle, refer to the precautions section.

**✴✴ CAUTION**

**Before removing the engine, secure the vehicle to prevent it from tipping over. The luggage compartment must be empty.**

2. Remove the luggage compartment floor trim.

➡ **So that the front wheels can still be turned with the battery disconnected, the battery must only be disconnected with the ignition key inserted.**

3. Disconnect the negative battery cable.
4. Remove the front engine cover.
5. Remove the rear engine cover.
6. Open cap of coolant expansion tank after the engine has fully cooled.
7. Remove the drive belt.
8. Remove the both front wheels.

➡ **Secure the brake rotors with wheel bolts.**

9. Loosen quick-release fasteners and remove sound insulation.
10. Place a drip tray under the engine.
11. Remove the drain plug on coolant thermostat housing and drain coolant from engine.
12. Disconnect lower coolant hose from radiator and drain radiator coolant.
13. Disconnect the A/C compressor clutch connector.
14. Unbolt the A/C compressor and move to one side with the lines attached.
15. Disconnect the power steering hoses, unbolt the pump and move aside.
16. If equipped with an auxiliary heater, remove the left front fender liner.
17. Remove the muffler and corrugated exhaust pipe.
18. Remove the bolts and the torque support.
19. Disconnect the transmission cooler lines.
20. Remove the ground strap from the longitudinal member.
21. Disconnect vacuum hose for brake booster at bulkhead.
22. Disconnect the hoses from the heater core.
23. Disconnect the hoses at the expansion tank and remove the tank.
24. Disconnect electrical connector to Engine Coolant Level (ECL) warning switch at bottom of coolant expansion tank.
25. Disconnect the hose at the coolant line.
26. Disconnect the hoses from the radiator.
27. Disconnect the vacuum hose from the leak detection pump.
28. Disconnect fuel line.
29. Remove the air duct and the upper air filter housing.
30. Remove the lower air filter housing.
31. Disconnect vacuum hose to EVAP canister.

32. Remove the nuts and the right bracket for harness connectors from bulkhead.
33. Remove the ground wire on left strut tower.
34. Remove the nuts and left bracket for harness connectors from bulkhead.
35. Remove the rubber seal for plenum chamber cover.
36. Remove the plenum chamber cover.
37. Remove the wiper arms and the cowl grill
38. Remove the strut tower brace if equipped.
39. Remove the E box cover and box electrical connections.
40. Remove the bolts and remove cover from electronics box at left in engine compartment.
41. Release retaining tabs and remove the fuse holder and the relay carrier.
42. Remove the bolt for electrical wiring connection
43. Disconnect all electrical connections at rear on connector strip.
44. Disengage engine wiring harness at electronics box and move the wiring aside.
45. Remove the bolts and remove cover from electronics box at right in engine compartment.
46. Carefully pry off retaining clip with a screwdriver and remove ECM from electronics box.
47. Leave the ECM connected at wiring harness.
48. Remove or disconnect any additional hoses or electrical connections that would interfere with engine removal.
49. Have a second technician depress and hold brake pedal and remove the collar bolt at left and right axle shafts.

➡ **To loosen collar bolt for drive axle, the wheel bearing must not be under load (vehicle must not be standing on its wheels).**

50. Remove the bolts for stabilizer bar uniformly on both sides.
51. Remove the stabilizer bar.
52. Disconnect electrical connector at level control system sensor.
53. Disconnect connecting link at control arm.
54. Unbolt suspension strut from control arm.
55. Disconnect the control arm from the sub frame and pivot the arms outwards on both sides of the vehicle.
56. Remove the axle shafts from the steering knuckle.
57. Unbolt the transverse beam from the sub frame.

58. Unhook shift cable.
59. Remove the transverse beam.
60. Remove the bolts at transmission/driveshaft flange.
61. Remove the left and right axle shafts.
62. Position a lift table such as VAS 6131 with support set VAS 6131/10 and VAS 6131/11.
63. Remove the bolts at tunnel cross member.
64. Remove the bolts at the sub frame, mark installation position of sub frame and of both engine bearing consoles to long members using a felt-tip marker.
65. Remove the bolts in a diagonal sequence using multiple passes.
66. Make sure that all hoses and lines between engine, transmission, sub frame and body have been disconnected.
67. While lowering, carefully guide engine/transmission subassembly with sub frame out of engine compartment in order to prevent damage.
68. Slowly lower engine/transmission subassembly.

**To install:**

69. Tighten the engine to transmission bolts as follows:
   a. M12 bolts to 65 Nm.
   b. M10 X 150 bolts to 65 Nm.
   c. M10 X 70 bolts to 45 Nm.
70. Tighten the sub frame bolts as follows, refer to the illustration for bolt locations:
71. Position 1 : 50 Nm.
72. Position 2 : 150 Nm.
73. Position 3 : 150 Nm.
74. Position 4 : 75 Nm.
75. Install the remaining components in the reverse order of removal, please note the following specifications:
   a. Tighten all M6 bolts to 9 Nm.
   b. Tighten all M8 bolts to 20 Nm.
   c. Tighten all M10 bolts to 40 Nm.
   d. Tighten all M12 bolts to 65 Nm.

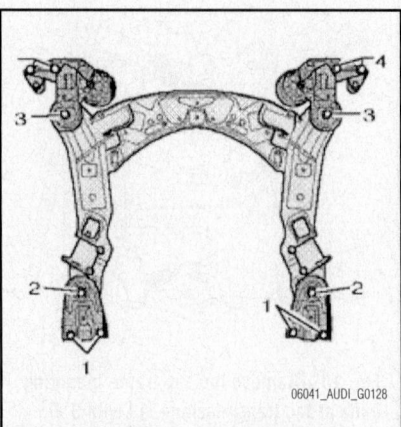

06041_AUDI_G0128

**Fig. 171 Sub frame bolt locations—A6**

e. Tighten the drive plate to torque to 85 Nm using new bolts.

f. Tighten the engine support to cylinder block to 40 Nm.

g. Tighten the engine mount console to longitudinal member to 75 Nm.

h. Tighten the tunnel cross member to body to 40 Nm.

i. Tighten the pressure line to the power steering pump to 47 Nm.

j. Tighten the fuel supply line to 22 Nm.

k. Tighten the torque bracket to engine to 40 Nm.

l. Tighten the torque support stop to lock carrier to 23 Nm.

### 4.2L Engine

1. Before servicing the vehicle, refer to the precautions section.

**�֍֍ CAUTION**

**Before removing the engine, secure the vehicle to prevent it from tipping over. The luggage compartment must be empty.**

2. Remove the luggage compartment floor trim.

➡**So that the front wheels can still be turned with the battery disconnected, the battery must only be disconnected with the ignition key inserted.**

3. Disconnect the negative battery cable.
4. Remove the front engine cover.
5. Remove the rear engine cover.
6. Open cap of coolant expansion tank after the engine has fully cooled.
7. Remove the both front wheels.

➡**Secure the brake rotors with wheel bolts.**

8. Loosen quick-release fasteners and remove sound insulation.
9. Place a drip tray under the engine.
10. Remove the drain plug on coolant thermostat housing and drain coolant from engine.
11. Disconnect lower coolant hose from radiator and drain radiator coolant.
12. Remove the bumper.
13. Remove the lock carrier and attachments.
14. Disconnect connector from the secondary air injection pump and move the wiring aside.
15. Disconnect the following hoses:
    a. Ventilation hose.
    b. Variable intake manifold hose.
    c. Secondary Air Injection (AIR) pump.

d. Disconnect coolant hose from the thermostat housing.
    e. Coolant expansion tank.
    f. Coolant line.
16. Remove the coolant expansion tank.
17. Disconnect electrical connector to Engine Coolant Level (ECL) warning switch at the bottom of coolant expansion tank.
18. Clamp off hose of power steering fluid reservoir using hose clamps up to 25mm diameter.
19. Disconnect hose for power steering .
20. Unclip fuel hose.
21. Disconnect electrical connector from mass air flow (MAF) sensor and separate the sensor from air intake hose.
22. Remove the upper part of air filter housing with mass air flow (MAF) sensor.
23. Separate hose for secondary air injection (AIR) system.
24. Disconnect electrical connector from the air flap positioner.
25. Remove the pin from spreader clip and remove lower part of air filter housing.
26. Pull resonator up and unclip it.
27. Properly relieve the fuel system pressure.
28. Disconnect fuel line.
29. Disconnect vacuum hose for brake booster at bulkhead.
30. Remove the intake hose.
31. Disconnect vacuum hose to EVAP canister.
32. Remove the ground wire on right strut tower.
33. Remove the nuts and the right bracket for harness connectors from bulkhead.
34. Remove the ground wire on left strut tower.
35. Remove the nuts and left bracket for harness connectors from bulkhead.
36. Remove the rubber seal for plenum chamber cover.
37. Remove the plenum chamber cover.
38. Remove the wiper arms and the cowl grill.
39. Remove the bolts and remove cover from electronics box at left in engine compartment.
40. Release retaining tabs and remove the fuse holder and the relay carrier.
41. Remove the bolt for electrical wiring connection.
42. Disconnect all electrical connections at rear on connector strip.
43. Disengage engine wiring harness at electronics box and move the wiring aside.
44. Remove the bolts and remove cover from electronics box at right in engine compartment.

45. Carefully pry off retaining clip with a screwdriver and remove ECM from electronics box.
46. Leave the ECM connected at wiring harness.
47. Remove or disconnect any additional hoses or electrical connections that would interfere with engine removal.
48. Remove the bolts and remove front transverse beam.
49. Remove the heat shield for the driveshaft.
50. Remove the bolts at transmission/driveshaft flange.
51. Push driveshaft toward rear final drive the Constant velocity (CV) joints can move axially.
52. Tie driveshaft up and to side against heat shield
53. Remove the heat shield for left driveshaft.
54. Remove the left and right axle shafts from transmission.
55. Have a second technician depress and hold brake pedal and remove the collar bolt at left and right axle shafts.

➡**To loosen collar bolt for drive axle, the wheel bearing must not be under load (vehicle must not be standing on its wheels).**

56. Remove the bolts for stabilizer bar uniformly on both sides.
57. Remove the stabilizer bar.
58. Disconnect electrical connector at the level control system sensor.
59. Disconnect the connecting link at the control arm.
60. Unbolt the suspension strut from the control arm.
61. Loosen the nut of the guide link ball joint until it is flush with end of thread.

➡**Ensure that the ball joint protective boot does not become damaged when pressing off. Make sure that both lever arms of ball joint puller are positioned parallel to one another while increasing force.**

62. Secure ball joint puller against falling.
63. Using ball joint puller T40043 press guide link ball joint off of tapered seat.
64. Remove the nut and remove ball joint from guide link.
65. Loosen nut of ball joint at control arm until it is flush with end of thread.

➡**Make sure that the ball joint protective boot does not become damaged when pressing off. Make sure that both lever arms of ball joint puller are positioned parallel to one another while increasing force.**

66. Using ball joint puller T40042 press control arm ball joint off from tapered seat.

67. Remove the nut and remove ball joint from control arm.

68. Unbolt guide link and control arm at sub frame and remove them.

69. Perform these removal procedures on both sides of the vehicle.

70. Remove the bolts for support bracket from transmission.

71. Unhook shift cable.

72. Position a lift table such as VAS 6131 with support set VAS 6131/10 and VAS 6131/11.

73. Remove the bolts at tunnel cross member.

74. Remove the bolts at the sub frame, mark installation position of sub frame and of both engine bearing consoles to long members using a felt-tip marker.

75. Remove the bolts in a diagonal sequence using multiple passes.

76. Make sure that all hoses and lines between engine, transmission, sub frame and body have been disconnected.

77. While lowering, carefully guide engine/transmission subassembly with sub frame out of engine compartment in order to prevent damage.

78. Slowly lower engine/transmission subassembly.

### To install:

79. Tighten the engine to transmission bolts as follows:
   a. M12 bolts to 65 Nm.
   b. M10 X 150 bolts to 65 Nm.
   c. M10 X 70 bolts to 45 Nm.

80. Tighten the sub frame bolts as follows, refer to the illustration for bolt locations:

81. Position 1: 50 Nm.

82. Position 2: 150 Nm.

83. Position 3: 150 Nm.

84. Position 4: 75 Nm.
   a. Tighten all M6 bolts to 9 Nm.
   b. Tighten all M8 bolts to 20 Nm.
   c. Tighten all M10 bolts to 40 Nm.
   d. Tighten all M12 bolts to 65 Nm.
   e. Tighten the right engine support to cylinder block to 42 Nm.
   f. Tighten the engine mount to the engine support and to the engine console to 23 Nm.
   g. Tighten the pressure line to the power steering pump to 47 Nm.
   h. Tighten the transmission mount to sub frame to 23 Nm.
   i. Tighten the selector lever heat shield to 22 Nm.
   j. Tighten the exhaust pipe straps to 25 Nm.
   k. Tighten the heat shield to the transmission to 25 Nm.
   l. Tighten the fuel supply line to 25 Nm.
   m. Tighten the chassis ground wire to 25 Nm.
   n. Tighten the torque support stop to lock carrier to 28 Nm.

### A8

➡The engine is removed from the front together with the transmission.

➡All tie wraps that are loosened or removed in order to remove the engine must be replaced or reinstalled to the same place when the engine is reinstalled.

➡The battery is located in the luggage compartment to the right and under the cover.

1. Before servicing the vehicle, refer to the precautions section.
2. Drain the cooling system.
3. Relieve the fuel system pressure.
4. Drain the transaxle fluid.
5. Remove or disconnect the following:
- Negative battery cable
- Intake air grille
- Noise insulation panel
- Front bumper
- Intake air duct between air cleaner and throttle body
- Engine cover
- Coolant hose clamps at right-side belt guard
- Electronics box cover by removing seven bolts
- All connectors from Engine Control Module (ECM) and Transmission Control Module (TCM) at bulkhead
- Cruise control module with relay and fuse bracket
- Sealing strip between engine compartment and plenum chamber
- Wiring harness from plenum panel, remove spacer sleeves, expose wiring and place down on engine
- Vacuum line to brake booster unit
- Vacuum line at cruise control module
- Accelerator pedal cable
- Coolant lines to heater core (supply and return) at vent valves
- Heated Oxygen (HO$_2$S) sensor connectors on left and right sides and push out of holder
- Ignition power output stage
- Top part of air cleaner.
- Harness connector at Mass Air Flow (MAF) sensor
- Coolant vent line to expansion tank at radiator
- Coolant supply line at expansion tank
- Hydraulic reservoir
- Vacuum line for intake manifold change-over valve, located at front by left headlight
- Left and right knock sensor connectors at fuel rail
- Harness connectors at fuel injectors
- Fuel rail with fuel injectors, and place down to side
- Exhaust manifold to front exhaust pipe on left and right sides
- Crossmember
- Three Way Catalytic Converter (TWC)
- Exhaust system at retaining loop and remove
- Exhaust pipes (left and right) from exhaust manifold
- TWC bracket along with spring
- TWC along with front pipe
- Heat shield above three way catalytic converter
- Heat shield for selector lever cable at transmission pan

6. For all-wheel drive vehicles, mark alignment of driveshaft to transmission output, disconnect the driveshaft, and install support for the driveshaft.

7. For all vehicles, remove or disconnect the following:
- Selector lever cable at bracket
- Left and right drive axles
- Left drive axle shield
- Starter cable connector at junction box on right side long member
- Junction box cover
- Electrical connectors
- Starter cable in junction box and at bracket
- Generator air guide
- Transmission oil cooler
- Lower radiator hose
- Engine Coolant Temperature (ECT) sensor above hose
- Torque support (note washers)
- Radiator air guide on left and right sides
- High pressure switch connector
- Left and right headlight connectors
- Left headlight
- Outside temperature sensor at bottom in front of radiator, cut open cable clip, and expose cable
- Cooling loop for power steering fluid and place to side
- A/C condenser and tie up to side (to relieve strain on lines and radiator)

### ✳✳ CAUTION

**Disconnect A/C refrigerant line brackets and support points only. DO NOT open the air conditioning refrigerant circuit. Avoid damage from bending. Refrigerant lines kink easily.**

- Coolant fan wiring
- Hood supports at both front fenders
- Lock carrier bolts located below hood support (note washers)
- Impact absorbers for bumper on left and right sides
- Lock carrier along with radiator
- Accessory drive belt
- Hydraulic pump belt pulley (note washers)
- Hydraulic pump from mounting bracket and place down at side member
- A/C compressor, lines remains connected
- Torque support with bonded rubber bushing at left-front of engine

8. Raise transmission using tool VAG 1383.
- Right and left transmission mounts
- Left transmission support

9. Lower and remove transmission hoist.
- Left and right engine mounts
- Both gas-filled struts at hood

10. Position hood in vertical position and support with auxiliary tool.

11. Install brackets VAG 3180 on both rear engine lifting points.

12. Attach lifting device 3033 to both brackets evenly.

13. Slide in assembly crane VAG 1202 A (500 kg), and attach lifting device 3033.

14. Raise engine carefully and pull out toward front.

#### To install:

15. Installation is the reverse of the removal procedure, while using the following torque values:
- A/C compressor bracket to engine: 18 ft. lbs. (25 Nm)
- A/C compressor to bracket: 18 ft. lbs. (25 Nm)
- Drive axles to flange shafts: 33. ft. lbs. (45 Nm)
- Engine support (right front) to engine: 33 ft. lbs. (45 Nm)
- Engine support to body: 37 ft. lbs. (50 Nm)
- Generator to engine 8mm bolt: 18 ft. lbs. (25 Nm)
- Generator to engine 10mm bolt: 30 ft. lbs. (40 Nm)
- Exhaust manifold to cylinder head: 18 ft. lbs. (25 Nm)

- Exhaust pipe to exhaust manifold: 30 ft. lbs. (40 Nm)
- Hydraulic pump bracket to engine: 18 ft. lbs. (25 Nm)

### EXHAUST MANIFOLD

#### REMOVAL & INSTALLATION

### A4 Models

#### 2.0L Engine

1. Before servicing the vehicle, refer to the precautions section.

2. Remove or disconnect the following:
- Cylinder head cover
- Intake manifold cover
- Turbocharger to charcoal filter hose
- Upper air pipe and heat shield
- Turbocharger
- Exhaust manifold nuts
- Exhaust manifold

3. Clean the gasket mounting surfaces.

#### To install:

4. Install or connect the following:
- Exhaust manifold using a new gasket. Torque the nuts to 18 ft. lbs. (25 Nm).
- Turbocharger. Torque the bolts to 25 ft. lbs. (35 Nm)
- Upper air pipe and heat shield. Torque the bolts to 89 inch lbs. (10 Nm).
- Turbocharger to charcoal filter hose
- Intake manifold cover
- Cylinder head cover

### TT Models

#### 1.8L Engine

1. Before servicing the vehicle, refer to the precautions section.

2. Remove or disconnect the following:
- Negative battery cable.
- Cylinder head and intake manifold covers
- Noise insulation under vehicle
- Right drive axle heat shield
- Air hoses
- Upper air line bracket
- Turbocharger from exhaust manifold
- Cylinder head heat shield
- Exhaust manifold

3. Clean all gasket mating surfaces.

#### To install:

4. Installation is the reverse of the removal procedure, using the following torque values:
- Manifold using a new gasket. Torque the nuts to 18 ft. lbs. (25 Nm)
- Manifold heat shield: 84 inch lbs. (10 Nm)

- Drive shaft heat shield: 26 ft. lbs. (35 Nm)

#### 3.2L Engine—Except TT and A6

1. Before servicing the vehicle, refer to the precautions section.

2. Remove or disconnect the following:
- Negative battery cable
- Intake manifold
- Heat shield
- Remove 7 nuts and washers on both exhaust manifolds
- Pull exhaust manifolds towards the rear and remove

#### To install:

3. Installation is the reverse of the removal procedure, using the following torque values:
- Exhaust manifold to cylinder head 17 ft. lbs (23 Nm)
- Exhaust pipe to manifold 29 ft. lbs (40 NM)

### A6 and A8 Models 4.2L Engine

#### Both Manifolds

➡**This procedure requires engine removal. See ENGINE ASSEMBLY section.**

1. Before servicing the vehicle, refer to the precautions section.

2. Remove or disconnect the following:
- Negative battery cable
- Dipstick tube (LH manifold only)
- Exhaust system from the manifold
- Heat shield
- Manifold nuts
- Exhaust manifold

3. Clean all gasket mating surfaces.

#### To install:

4. Install or connect the following:
- Exhaust manifold using a new gasket. Torque the nuts to 18 ft. lbs. (25 Nm)
- Heat shield
- Exhaust system to the manifold using a new gasket
- Dipstick tube
- Negative battery cable

5. Reinstall engine assembly.

6. Start the vehicle, check for leaks and repair if necessary.

### TURBOCHARGER

#### REMOVAL & INSTALLATION

### TT Models

#### 1.8L Engine

*See Figures 172 and 173.*

1. Before servicing the vehicle, refer to the precautions section.

2. Drain coolant

3. Drain engine oil

4. Remove or disconnect the follow-ing:

- Negative battery cable
- Engine cover
- Noise insulation under engine
- Heat shield for right drive axle
- Right drive axle
- Front exhaust pipe

➡**The exhaust flex pipe may be dam-aged if bent more than 10 degrees.**

- Air hose between upper air line and lower air line
- Bracket for upper air line
- Oil return line
- Turbocharger bracket from cylinder block
- Bracket for coolant return line
- Coolant supply line from cylinder block
- Hose from pressure unit for charge air pressure control
- Turbocharger support bracket
- Oil return line
- Air hoses
- Pressure control valve
- Connector from charge air pressure control valve
- Vacuum line from air recirculation valve
- Air intake hose from air cleaner
- Pull locking clip out of tur-bocharger connection and take off air intake hose
- Coolant return hose from the Y connection on the right next to the cylinder head
- Heat shield on back of cylinder head
- Turbocharger from exhaust mani-fold
- Exhaust manifold
- Retainer piece for oil supply line from turbocharger
- Oil supply line and coolant return line
- Turbocharger

**To install:**

5. Installation is the reverse of the removal procedure, while using the follow-ing notes and torque values:

6. Bolt turbocharger bracket onto tur-bocharger but do not tighten bolts

7. Position turbocharger against engine from below, then tighten bolts by hand to secure bracket to cylinder block

8. Tighten bolts for oil supply line

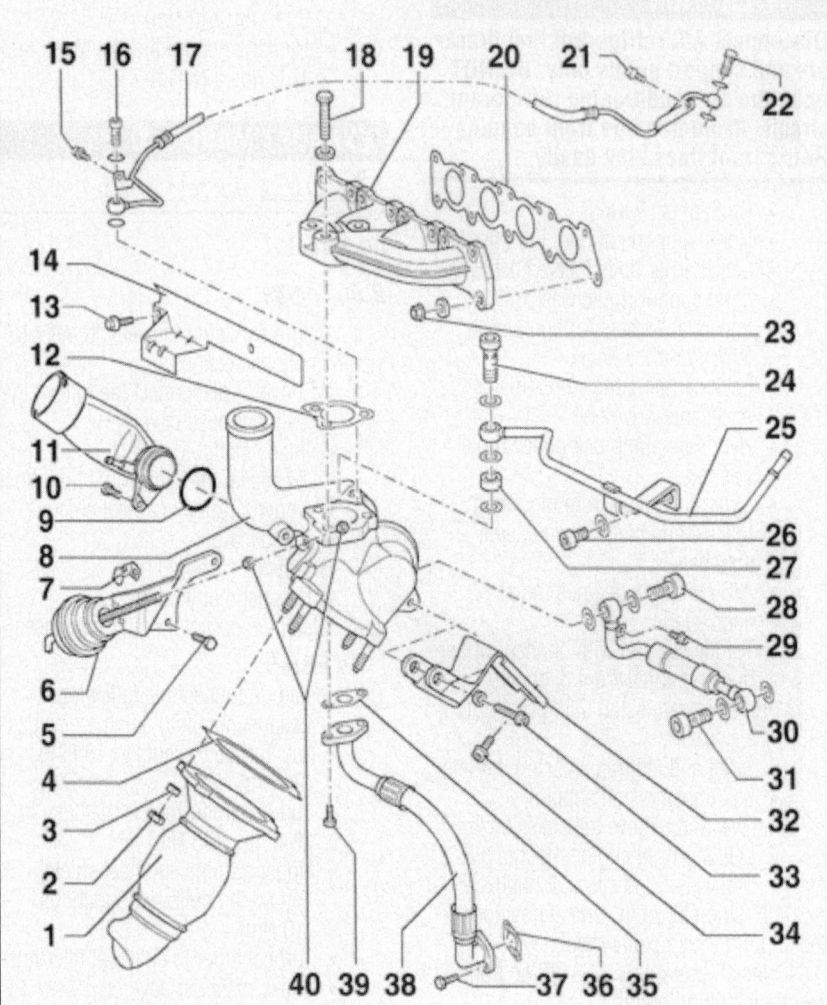

1. Front exhaust pipe
2. Nut
3. Spacer sleeve
4. Gasket
5. Bolt
6. Pressure unit
7. Securing clip
8. Turbocharger
9. O-ring
10. Bolt
11. Connection
12. Gasket
13. Bolt
14. Heat shield
15. Bolt
16. Banjo bolt
17. Oil supply line
18. Bolt
19. Exhaust manifold
20. Gasket

21. Bolt
22. Banjo bolt
23. Nut
24. Banjo bolt
25. Coolant return line
26. Bolt
27. Spacer sleeve
28. Banjo bolt
29. Bolt
30. Coolant supply line
31. Banjo bolt
32. Bracket
33. Bolt
34. Bolt
35. Gasket
36. Gasket
37. Bolt
38. Oil return line
39. Bolt
40. Nut

2348-TTTT-G04

**Fig. 172 Exploded view of the turbocharger and related components—1.8L engine TT**

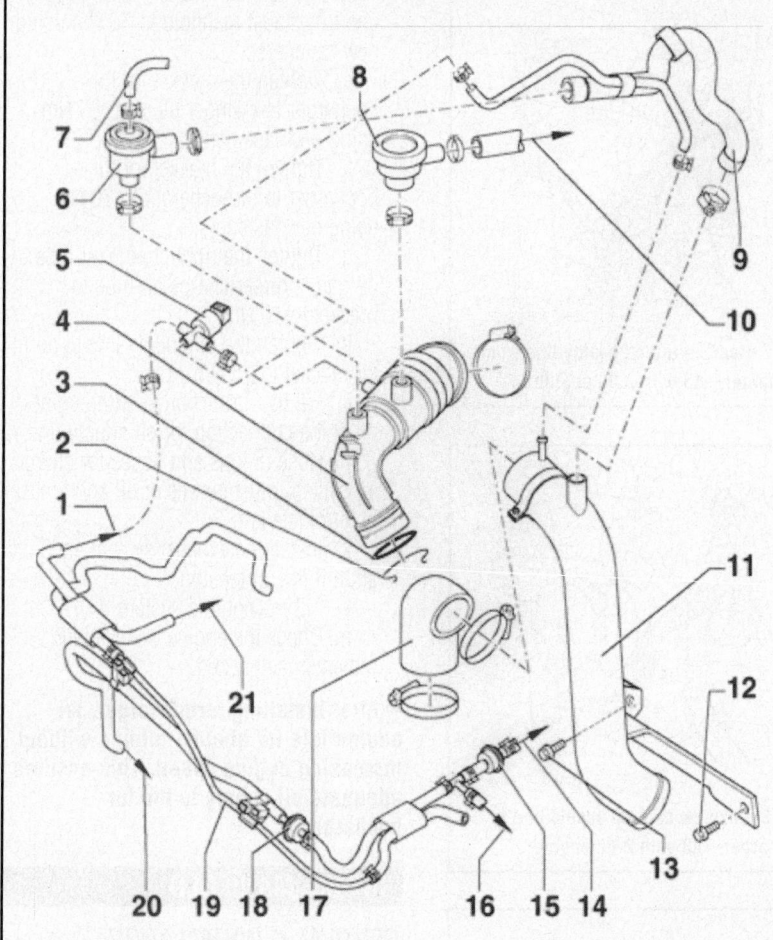

1. To charge air control valve
2. Locking clip
3. O-ring
4. Air intake hose
5. Solenoid valve
6. Air pressure bypass valve
7. Air recirculation valve hose
8. Pressure regulating valve
9. Hose
10. To crankcase breather
11. Upper air line
12. Bolt
13. Bolt
14. To throttle valve housing
15. Non return valve
16. To EVAP canister
17. Hose
18. Non return valve
19. Coolant line
20. To turbocharger
21. To pressure unit for air pressure valve

2348-TTTT-G05

**Fig. 173 Exploded view of the vacuum hoses and connections related to the turbocharger—1.8L engine TT**

9. Install coolant return line with spacer sleeve and bolt it to the turbocharger

10. Unbolt turbocharger bracket from cylinder block again

11. Install exhaust manifold

12. Bolt turbocharger to exhaust manifold

13. Install or connect the following:
- Coolant supply banjo fitting: 26 ft. lbs. (35 Nm)
- Turbocharger exhaust manifold bolts: 18 ft. lbs. (25 Nm)
- Turbo support bracket bolts: 22 ft. lbs. (30 Nm)
- Turbo return line: 26 ft. lbs. (35 Nm)

### A3 Models

#### 2.0L Engine

*See Figures 174 through 179.*

1. Before servicing the vehicle, refer to the precautions section.

2. Disconnect the intake connection using spring-type clip pliers.

3. Remove the engine cover upward and out in direction of.

4. Remove the coolant expansion tank cap.

5. Drain and recycle the coolant.

6. Remove the wiper arms.

7. Unclip spray nozzles and push spray nozzles, with lines still attached, back through opening and into plenum chamber.

8. Remove the rubber seal for plenum chamber cover.

9. Remove the plenum chamber cover.

10. Loosen up the rear engine wiring harness at partition for plenum chamber.

11. Remove the electrical connector for the Heated Oxygen sensor (HO2S) and Oxygen sensor (O2S) located before the catalytic converter from the bracket, disconnect it and lay it aside.

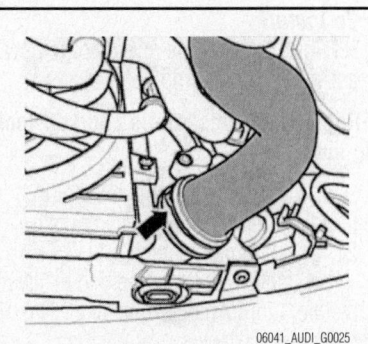

06041_AUDI_G0025

**Fig. 174 Pull air duct off from charge air cooler—A3 with 2.0L engine**

06041_AUDI_G0026

**Fig. 175 Unfasten the air duct from turbocharger—A3 with 2.0L engine**

12. Remove the partition for plenum chamber.

13. Disconnect the connectors from ignition coils.

14. Remove the heat shield.

15. Disconnect the coolant line to coolant expansion tank.

16. Disconnect the coolant hoses.

17. Unfasten the coolant line.

18. Unfasten the crankcase ventilation line with heat shield from turbocharger.

19. Pull off crankcase ventilation line from cylinder head cover and remove it.

20. Disconnect the EVAP line to turbocharger from cylinder head cover.

21. Unscrew oil supply line from turbocharger.

22. Remove the catalytic converter with front exhaust pipe

23. Remove the center sound insulation fasteners.

24. Remove the right sound insulation fasteners.

25. Remove the right drive axle.

26. Remove the heat shield for right drive axle.

27. Pull air duct off from charge air cooler.

28. Unfasten the air duct from turbocharger.

29. Unscrew bolts and remove air duct.

30. Remove the oil supply line from turbocharger.

31. Unscrew coolant supply line to turbocharger.

32. Remove the oil return line at turbocharger.

33. Remove the bolts and remove turbocharger support.

34. Disconnect the coolant pipe.

35. Remove the coolant pipe from cylinder head.

36. Remove the upper nuts.

37. Remove the turbocharger/exhaust manifold upward and out.

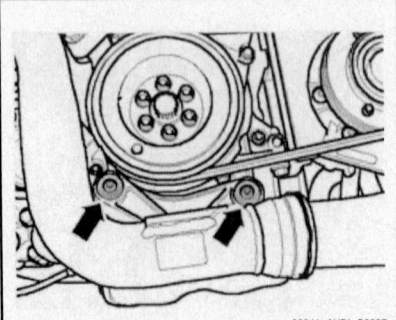

**Fig. 176 Unscrew bolts and remove air duct—A3 with 2.0L engine**

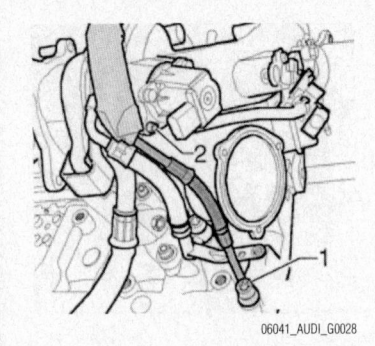

**Fig. 177 Remove the oil supply line from turbocharger—A3 with 2.0L engine**

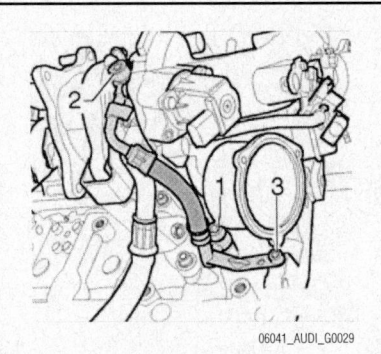

**Fig. 178 Unscrew coolant supply line to turbocharger—A3 with 2.0L engine**

**Fig. 179 Remove the turbocharger support—A3 with 2.0L engine**

### To install:

38. Installation is the reverse of removal, please note the following:

➡ **Replace gaskets, seals and self-locking nuts.**

a. Tighten the exhaust manifold/turbocharger to cylinder head nuts to 21 Nm using new fasteners.

b. Tighten the heat shield to exhaust manifold/turbocharger nuts to 20 Nm using new fasteners.

c. Tighten the oil supply line to turbocharger to 30 Nm.

d. Tighten the bracket for oil supply line to exhaust manifold to 20 Nm using new fasteners.

e. Tighten the bracket for turbocharger to cylinder block to 30 Nm using new fasteners.

f. Tighten the bracket for turbocharger to turbocharger to 30 Nm using new fasteners.

g. Tighten the right charge air tube to oil pan and rear charge air tube to bracket to 10 Nm.

h. Tighten the oil supply line to oil filter bracket to 25 Nm.

i. Fill the turbocharger with engine oil at the connection for oil supply line.

j. Hose unions and hoses for charge air system must be free of oil and grease before installing.

k. Install the exhaust system and align it free of tension

l. Fill the cooling system.

m. Check the engine oil level and replenish as needed.

➡ **After installing turbocharger, let engine idle for about 1 minute without increasing engine speed. This ensures adequate oil supply to the turbocharger.**

## INTAKE MANIFOLD

### REMOVAL & INSTALLATION

#### A4 Models

##### 2.0L Engine

1. Before servicing the vehicle, refer to the precautions section.

2. Drain the engine coolant.

3. Properly relieve the fuel system pressure.

4. Turn the ignition switch to the **OFF**.

5. Remove or disconnect the following:

- Negative battery cable
- Engine covers
- Hose for the Leak Detection Pump (LDP)
- Accelerator pedal cable from the throttle valve control module
- Air guide hose from the throttle valve control module
- Vacuum line from the Evaporative Emissions (EVAP) canister
- Brake booster vacuum hose
- Intake Air Temperature (IAT) sensor and the throttle valve control module
- Camshaft Position (CMP) sensor wiring harness connector
- Fuel rail with the injectors

- Coolant hoses from to the intake manifold
- Crankcase breather hose at the intake manifold
- Intake manifold brace and the oil dipstick
- Manifold at the mounting flange
6. Clean all gasket surfaces.

### To install:

7. Install or connect the following:
- Intake manifold using new gaskets. Torque the fasteners to 89 inch lbs. (10 Nm).
- Manifold brace. Torque the bolts to 15 ft. lbs. (20 Nm).
- Dipstick
- Crankcase breather and coolant hoses to the intake manifold
- Fuel injector sealing O-ring
- Fuel rail with the injectors. Torque the bolts to 89 inch lbs. (10 Nm).
- CMP and the IAT sensors to the throttle valve control module
- Brake booster and EVAP canister vacuum hoses
- Air guide hose and the accelerator pedal cable to the throttle valve control module
- Hose for the LDP

8. Top off the engine coolant and bleed, if necessary.
9. Connect the negative battery cable.
10. Fully close all power windows to stop, operate all window switches for at least 1 second in the close direction to activate the one-touch opening/closing function.
11. Check the oil level before starting the engine and top off, as necessary.

### TT Models

#### 1.8L Engine

1. Before servicing the vehicle, refer to the precautions section.
2. Drain the engine coolant.
3. Properly relieve the fuel system pressure.
4. Turn the ignition switch to the **OFF**.
5. Remove or disconnect the following:
- Negative battery cable
- Engine cover
- IAT and throttle control valve connectors
- Throttle control valve air hose
- Vacuum hoses
- Hall sensor connector
- Fuel injector connectors
- Support bar from fuel rail
- Electrical change over valve connector

- Dipstick tube bracket
- Release tabs from fuel supply and return lines
- PCV hose
- Intake manifold support
- Intake manifold

### To install:

6. Installation is the reverse of the removal procedure, using the following torque values:
- Fuel rail bolts: 84 inch lbs. (10 Nm)
- Intake manifold bolts: 84 inch lbs. (10 Nm)

#### 3.2L Engine

1. Before servicing the vehicle, refer to the precautions section.
2. Drain the engine coolant.
3. Properly relieve the fuel system pressure.
4. Turn the ignition switch to the **OFF**.
5. Remove or disconnect the following:
- Negative battery cable
- Right and left covers at lock carrier, above main fuse box and coolant expansion tank

➡**Before removing ribbed belt, mark direction of rotation with chalk or a felt pen. If a used belt rotates in the wrong direction when reinstalled, this can result in breakage. When installing belt, ensure it is correctly seated in the pulleys.**

- Release tension on belt by screwing an M8 x 40 bolt into bore on tensioning element and remove belt
- Upper mounting bolt for the alternator
- Upper mounting bolts for radiator cowl
- Secondary air hose
- Center and right-hand noise insulation under engine
- Cross piece
- Power steering pump from bracket, hang pump from lock carrier with hoses still attached
- Secondary air injection pump with bracket
- Lower radiator cowl bolts and remove the radiator cowl with fans

➡**The air conditioning compressor can only be removed from the bracket with the upper securing bolt still inserted. To accomplish this, press the engine simultaneously to the left side. If necessary, loosen all bolts of the assembly mounting.**

- Mounting bolt for air conditioning compressor
- Secure the air conditioning compressor together with the hoses at front longitudinal member (hoses remain connected)
- Alternator
- Oil dipstick tube
- Vacuum reservoir
- Mounting bracket for alternator and air conditioning compressor
- Bolts at front of intake manifold
- Attach puller T10095A or equivalent to ignition coils and remove coils
- Body brace
- Air intake hose
- Crankcase breather hose
- Throttle body
- Hose bracket
- Rear intake mounting bolts, swing intake manifold forward
- Fuel rail with fuel injectors
- Swing back to installed position and remove manifold

### To install:

6. Installation is the reverse of the removal procedure, using the following torque values:
- Intake manifold to cylinder head 115 inch lbs. (13 Nm)
- Bracket for water pump to cylinder block 15 ft. lbs. (20 Nm)
- Intake manifold support to intake manifold 15 ft. lbs. (20 Nm)
- Throttle body to intake manifold 88 inch lbs. (10 Nm)

### A6 and A8 Models

#### 3.2L Engine

1. Before servicing the vehicle, refer to the precautions section.
2. Remove the engine covers.
3. Disconnect the check valve from connection at air duct hose.
4. Remove the air duct hose.
5. Disconnect the electrical harness connectors from the change-over valve for intake manifold flap and the Intake Manifold Tuning (IMT) valve position sensor.
6. Remove and necessary vacuum lines and hoses.
7. Disconnect the following electrical connectors.
   a. Camshaft adjustment valve.
   b. Throttle valve control module.
   c. Evaporative emission (EVAP) canister purge regulator valve.
   d. Intake Air Temperature (IAT) sensor G42 / Manifold Absolute Pressure (MAP) sensor.

e. Intake Manifold Runner Control (IMRC) valve.

8. Remove the Evaporative Emission (EVAP) canister purge regulator valve from the throttle valve control module.

9. Disconnect the crankcase breather hose at pressure regulator valve.

10. Remove the bolts and the upper intake manifold.

➡**Plug the intake ports of the cylinder head with clean rags.**

11. Remove the high-pressure line, thereby removing bolts and union nuts—1 to 7–.

12. Disconnect the following electrical connectors:

   a. Intake Manifold Runner Position sensor.
   b. Fuel Pressure sensor.

13. Remove the nuts and bolts and remove lower intake manifold with fuel rail.

➡**Plug the intake ports of the cylinder head with clean rags.**

### To install:

14. Replace the lower manifold gaskets and O-rings.

15. Tighten the lower manifold bolts to 9 Nm using several passes in a diagonal sequence.

16. Replace the upper manifold gaskets and O-rings.

17. Tighten the upper manifold bolts to 6 Nm using several passes in a diagonal sequence.

18. Tighten the high pressure line to 25 Nm at the pump and fuel rail.

19. Install the remaining components in the reverse order of removal.

### 4.2L Engine

1. Before servicing the vehicle, refer to the precautions section.

2. Relieve the fuel system pressure.

3. Remove or disconnect the following:
- Negative battery cable
- Noise insulation panel
- Intake air grille in bumper
- Accessory drive belt
- Timing belt
- Vacuum line to brake booster unit
- Vacuum line at cruise control module
- Accelerator pedal cable
- Coolant lines to heater core (supply and return to vent valves)
- Left and right knock sensor connectors at fuel rail and remove cable ties (4x)
- Fuel injector harness connectors

- Fuel rail, carefully remove along with fuel injectors, and place down to one side
- Intake manifold

### To install:

4. Install or connect the following:
- Intake manifold. Tighten the fasteners to 15 ft. lbs. (20 Nm).
- Fuel rail. Tighten the bolts to 84 inch lbs. (10 Nm).
- Fuel injector harness connectors
- Left and right knock sensor connectors at fuel rail
- Coolant lines to heater core (supply and return to vent valves)
- Accelerator pedal cable
- Vacuum line at cruise control module
- Vacuum line to brake booster unit
- Timing belt
- Accessory drive belt
- Intake air grille in bumper
- Noise insulation panel
- Negative battery cable

### A3 Models

#### 2.0L Engine

1. Before servicing the vehicle, refer to the precautions section.

2. Relieve the fuel system pressure.

3. Remove the engine cover.

4. Disconnect the all necessary electrical harness connectors.

5. Disconnect the vacuum hose between intake manifold and vacuum pump from intake manifold.

6. Disconnect the hose connections from valve cover.

7. Loosen the fuel supply line and hose connection from EVAP canister and cap off open lines.

8. Open both fuel lines at high-pressure pump.

9. Disconnect the electrical connector from Fuel Pressure sensor.

10. Remove the oil dipstick and remove bolt for oil dipstick guide tube.

11. Remove the bolt for oil line from intake manifold.

12. Open hose clamps and pull intake hose off from Throttle Valve Control Module and intake manifold.

13. Remove the nut from intake manifold bracket, if necessary, remove Throttle Valve Control Module.

14. Separate oil dipstick guide tube at separation point and remove bolt from intake manifold bracket.

15. Remove the bolts from intake manifold.

16. Carefully pull intake manifold with fuel rail off from cylinder head.

➡**The fuel injectors could remain stuck in the fuel rail.**

### To install:

17. Installation is the reverse of removal, please note the following:

   a. Intake manifold bolts are tightened to 9 Nm
   b. Intake manifold support bolts are tightened to 23 Nm
   c. Intake manifold support nuts are tightened to 10 Nm
   d. Fuel supply line to pump are tightened to 27 Nm
   e. New fuel return line to pump are tightened to 17 Nm

### 3.2L Engine

*See Figures 180 and 181.*

1. Remove the molded insert for tools under luggage compartment floor cover.

2. Unscrew cover for battery compartment.

3. Remove the molded insert over battery.

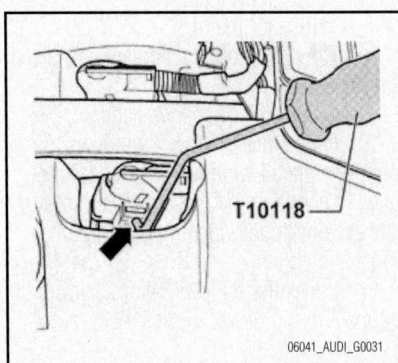

06041_AUDI_G0031

**Fig. 180 Disconnect the electrical connectors from the ignition coils—A3 with 3.2L engine**

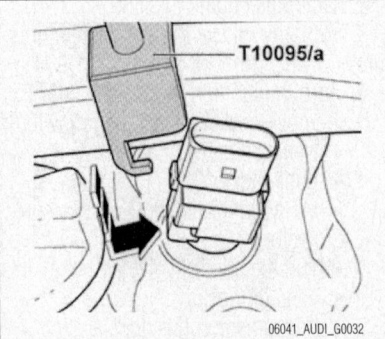

06041_AUDI_G0032

**Fig. 181 Attach Puller for ignition coil tool T10095A on ignition coils and pull out ignition coils in succession—A3 with 3.2L engine**

4. Disconnect the negative battery cable.

5. Remove the noise insulation.

6. Disconnect the vacuum hose from air duct hose.

7. Disconnect the air duct hose from Throttle Valve Control Module.

8. Disconnect the electrical harness connector at Mass Air Flow (MAF) sensor.

9. Unbolt upper section of air filter housing.

10. Remove the filter element.

11. Unbolt lower section of air duct.

12. Unbolt lower section of air filter housing.

13. Disconnect the secondary air hose at position indicated by.

14. Loosen up the air hose to Secondary Air Injection (AIR) pump.

15. Disconnect the electrical harness connectors and.

16. Disconnect the crankcase ventilation hose.

17. Unscrew Throttle Valve Control Module from the intake manifold and lay it aside with coolant hoses connected.

18. Disconnect the electrical connectors from the ignition coils.

19. Attach Puller for ignition coil tool T10095A on ignition coils as shown in the illustration and pull out ignition coils in succession.

20. Remove the ignition wiring harness strip and set it aside.

21. Disconnect the vacuum hoses at rear from intake manifold.

22. Remove the line union from intake manifold and place it in the rear with hoses connected.

23. Remove the bolts for rear intake manifold mount.

24. Unscrew top mounting bolts for fan shroud.

25. Remove the dipstick.

➡**Mark direction of rotation of accessory drive belt using chalk or felt-tip marker before removing. Reversing the direction of rotation of a run-in belt can destroy the belt**

26. Remove the accessory drive belt.

27. Loosen up the coolant hose at bottom on fan shroud.

28. Disconnect the electrical harness connector at Secondary Air Injection (AIR) pump motor.

29. Unscrew bracket for coolant pipe.

30. Remove the bolts.

31. Loosen the bolt and remove Secondary Air Injection (AIR) pump with bracket.

32. Disconnect the electrical harness connectors for coolant fans at bottom on fan shroud.

33. Unscrew mounting bolts for fan shroud at the bottom.

34. Pull out fan shroud downward with both fans.

35. Disconnect the electrical harness connector for A/C clutch on A/C compressor and loosen up electrical wire.

36. Remove the bolts for A/C compressor. Securely tie the A/C compressor with connected coolant hoses at front on the longitudinal member.

37. Unscrew electrical wire on alternator.

38. Disconnect the electrical harness connector.

39. Remove the upper idler roller.

40. Remove the mounting bolts and for alternator.

➡**The alternator can be removed from bracket only with the upper mounting bolt still installed.**

41. Remove the alternator with electrical wires connected from bracket for assemblies.

42. Remove the alternator downward and to the left.

43. Disconnect the electrical harness connector at the after-run coolant pump.

44. Remove the bolts and remove bracket for assemblies.

45. Pull the after-run coolant pump with coolant hoses connected, downward out of rubber loops of retainer. Spray rubber loops with silicon-free lubricant if necessary.

➡**The water pump remains on the engine with the coolant hoses connected.**

46. Disengage harness connector for Engine Speed (RPM) sensor from bracket on oil dipstick guide tube.

47. Unscrew bolt and pull out oil dipstick guide tube.

48. Unscrew bolt and pull off vacuum reservoir from intake manifold.

49. Set aside vacuum reservoir with lines still connected.

50. Remove the bolts at front on intake manifold.

51. Disconnect the vacuum line from actuator for intake manifold change-over.

➡**Protect intake manifold from damage with a clean cloth.**

52. Swivel intake manifold forward and then pull slightly toward left.

➡**Seal intake channels in cylinder head with clean rags or foam pieces so that small pieces cannot fall in.**

53. Remove the bracket for right fuel line at cylinder head cover.

54. Expose electrical wiring harness on cylinder head cover.

55. Remove the bolts.

56. Pull off fuel rail pipe with fuel injectors from cylinder head and set aside with fuel line connected.

57. Swivel intake manifold back into installation position and remove.

### To install:

58. Installation is the reverse of removal, please note the following:

➡**Replace all gaskets.**

a. Intake manifold to cylinder head retainers are tightened to 13 Nm.

b. Intake manifold support to intake manifold are tightened to 20 Nm.

c. Throttle Valve Control Module to intake manifold retainers are tightened to 10 Nm.

## OIL PAN

### REMOVAL & INSTALLATION

#### A4 Models

##### 2.0L Engine

See Figures 182 through 184.

1. Before servicing the vehicle, refer to the precautions section.

2. Drain the engine oil.

3. Remove or disconnect the following:

- Negative battery cable
- Engine undercover
- Accessory drive belts and the air conditioning belt tension pulley
- Torque support stop and side brace
- Starter wiring
- Hose from the turbocharger at the air guide tube in the lock carrier
- Bottom nuts from the lower engine mount
- Top engine cover

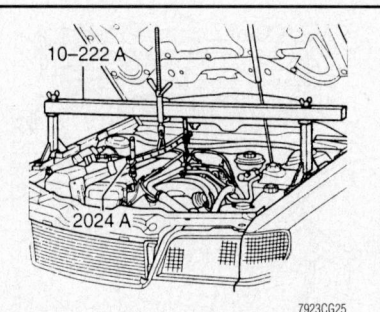

Fig. 182 The engine must be supported, because the subframe mounting bolts must be loosened—2.0L engine

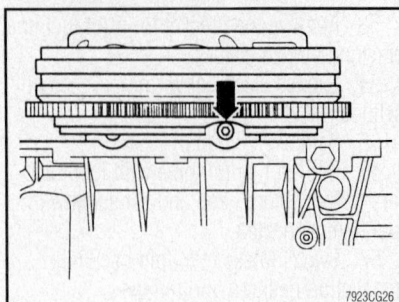

**Fig. 183 If equipped with a manual transaxle, align the flywheel as shown to remove the rear oil pan bolts—2.0L engine**

4. Install the Engine Support Bridge tool 10-222A and the Engine Sling tool 2024A; then, lift the engine as far as possible.

5. Support the subframe.

6. Remove or disconnect the following:
   - Front bolts No. 2 and 3
   - Bolt No. 1 from the subframe

7. Slowly lower the subframe.

8. If equipped with a manual transaxle, loosen the left transaxle mount nut until it is aligned with the lower edge of the bolt (approx. 4 turns).

9. If equipped with an automatic transaxle, loosen the rear bolt for the left transaxle mount several turns, then remove the front bolt for the transaxle mount.

10. At the right transaxle mount, loosen the rear bolt mount several turns and remove the front bolt.

➡ **If equipped with a manual transaxle, both of the rear bolts on the oil pan can be accessed through the opening on the flywheel. Turn the flywheel as needed.**

11. Remove the oil pan.

12. Clean all gasket mating surfaces.

### To install:

13. Apply sealant to the front and rear contact areas of the oil seal carriers.

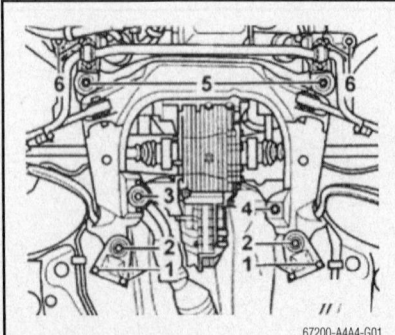

**Fig. 184 Subframe bolt tightening sequence—2.0L engine**

14. Install or connect the following:
   - New oil pan gasket
   - Oil pan

15. Tighten the oil pan bolts as follows:
   - Oil pan-to-engine bolts: 44 inch lbs. (5 Nm)
   - M10 bolts between the oil pan and engine: 33 ft. lbs. (45 Nm)
   - M6 bolts between the oil pan and engine: 84 inch lbs. (10 Nm)

16. Tighten the subframe bolts/nuts using the illustration as follows:
   - Bolts 2 and 5: 81 ft. lbs. (110 Nm) plus a ¼ (90 degree) turn
   - Bolt 6: 55 ft. lbs. (75 Nm)
   - Bolt 1: 17 ft. lbs. (23 Nm)
   - Nuts 3 and 4: 30 ft. lbs. (40 Nm)

17. Install or connect the following:
   - Subframe-to-transaxle. Torque the nuts to 17 ft. lbs. (23 Nm)
   - Engine mount-to-subframe. Torque the nuts to 18 ft. lbs. (25 Nm)
   - Turbocharger air hose
   - Starter wiring
   - Torque support stop and brace. Torque the fasteners to 18 ft. lbs. (25 Nm)
   - Air conditioning belt tensioner
   - Accessory drive belts
   - Negative battery cable

18. Fill the engine with oil and check the level.

19. Start the vehicle and check for leaks, then recheck the engine oil level.

20. Install or connect the following:
   - Engine cover
   - Undercover

## TT Models

### 1.8L Engine

1. Before servicing the vehicle, refer to the precautions section.

2. Disconnect the battery negative cable.

3. Drain engine oil

4. Remove or disconnect the following:
   - Engine undercover
   - Oil return line
   - Oil pan

### To install:

5. Apply silicone sealant to oil pan sealing surface

6. Install or connect the following:
   - Oil pan
   - Tighten the oil pan bolts to engine diagonally to 84 inch lbs., (10 Nm)
   - Tighten the oil pan bolts to transmission to 33 ft. lbs. (45 Nm)
   - Oil return line
   - Engine undercover
   - Negative battery cable

## 3.2L Engine—Except A3 and A6

1. Before servicing the vehicle, refer to the precautions section.

2. Disconnect the battery negative cable.

3. Drain engine oil

4. Remove or disconnect the following:
   - Center and right sound insulation
   - Oil thermal sensor
   - Secondary air injection pump with bracket
   - Oil pan

### To install:

5. Installation is the reverse of the removal procedure, while using the following torque values:
   - Tighten the oil pan bolts in diagonal sequence initially to 44 inch lbs. (5 Nm)
   - Tighten bolts for sump/transmission to 30 ft. lbs. (40 Nm) Use new bolts
   - Tighten the oil pan bolts in diagonal sequence to 105 inch lbs. (12 Nm)

## A6 and A8 Models

### 3.2L Engine

1. Before servicing the vehicle, refer to the precautions section.

2. Disconnect the battery negative cable.

3. Remove the engine undercover sound insulation.

4. Drain engine oil

5. Remove or disconnect the following:
   - Oil thermal sensor
   - Oil pan

### To install:

6. Installation is the reverse of the removal procedure, while using the following torque values:
   - Tighten the oil pan bolts in diagonal sequence initially to 9 Nm
   - Tighten the oil pan drain plug to 30 Nm.

### 4.2L Engine

1. Before servicing the vehicle, refer to the precautions section.

2. Remove the engine from the vehicle and mount it on a suitable workstand.

3. Remove or disconnect the following:
   - Lower oil pan
   - Upper oil pan
   - Honeycomb baffle

### To install:

4. Installation is the reverse of the removal procedure, while using the following torque values:

- Upper oil pan bolts: 89 inch lbs. (10 Nm)
- Lower oil pan bolts: 89 inch lbs. (10 Nm)

### A3 Models

#### 2.0L Engine

*See Figures 185 and 186.*

1. Before servicing the vehicle, refer to the precautions section.
2. Disconnect the negative battery cable.
3. Disconnect the electrical connector.
4. Disconnect the intake connection using spring-type clip pliers.
5. Remove the engine cover upward and out in direction of.
6. Remove the center sound insulation fasteners.
7. Remove the right sound insulation fasteners.
8. Remove the air duct from charge air cooler.
9. Unfasten the air duct from turbocharger.
10. Remove the bolts and remove air duct.
11. Remove the air duct for charge air cooler.

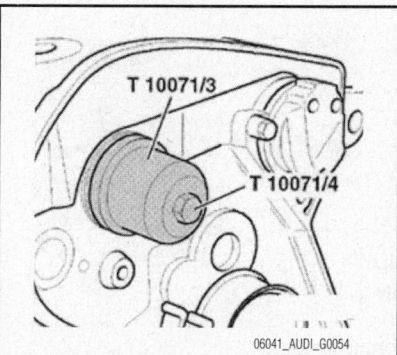

06041_AUDI_G0054

**Fig. 185 Oil pan bolt loosening/tightening sequence—A3 with 2.0L engine**

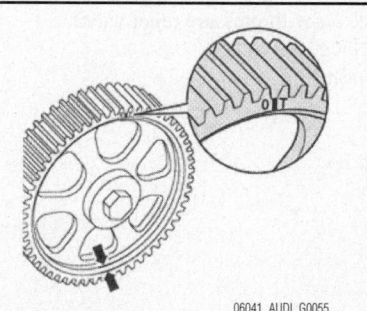

06041_AUDI_G0055

**Fig. 186 Apply 2 to 3mm bead of silicone sealant to the sealing surface of oil pan—A3 with 2.0L engine**

12. Disconnect the charge air pressure sensor connector.
13. Remove the air duct.
14. Disconnect the oil Level thermal sensor connector.
15. Remove the oil return line from turbocharger.
16. Drain the engine oil.
17. Remove the bolts for oil pan/transmission.
18. Loosen the bolts in a diagonal sequence.
19. Remove the oil pan, if necessary loosen by applying light strikes with a rubber mallet.

**To install:**

➡**The oil pan must be installed within 5 minutes after applying the silicone sealant.**

20. Clean the oil pan mating surfaces.
21. Apply 2 to 3mm bead of silicone sealant as illustrated to the sealing surface of oil pan.

➡**The sealant bead may not be thicker than specified, otherwise excess sealant could enter the oil pan and clog the oil intake tube.**

22. Apply a bead of sealant at the rear area of the sealing flange as shown in the illustration.
23. Install the oil pan immediately and tighten bolts in sequence as follows:
   a. Step 1: Bolts 1 through 20 to 5 Nm.
   b. Step 2: Oil pan to transmission bolts to 40 Nm.
   c. Step 3: Bolts 1 through 20 to 15 Nm.

➡**Make sure that the oil pan is positioned flush with the cylinder block on the flywheel side.**

➡**After installing oil pan, allow sealant to dry for about 30 minutes, then fill the crankcase with oil.**

24. Check and adjust the oil level.
25. Install the remaining components in the reverse order of removal.

#### 3.2L Engine

*See Figures 187 and 188.*

1. Before servicing the vehicle, refer to the precautions section.
2. Disconnect the negative battery cable.
3. Disconnect the secondary air injection hose.
4. Remove the center noise insulation fasteners.

5. Remove the right noise insulation fasteners.
6. Drain the engine oil.
7. Disconnect the lower coolant hose at fan shroud.
8. Disconnect the electrical connector at Secondary Air Injection (AIR) pump.
9. Remove the coolant pipe bracket to transmission oil cooler.
10. Remove the bolts.
11. Loosen the bolt and remove Secondary Air Injection (AIR) pump with bracket.
12. Disconnect the oil level thermal sensor connector.
13. Remove the bolts for oil pan/transmission.
14. Loosen the oil pan bolts diagonally and remove.
15. Remove the oil pan, and if necessary loosen by applying light strikes with a rubber mallet.

**To install:**

➡**The oil pan must be installed within 5 minutes after applying the silicone sealant.**

16. Clean the oil pan mating surfaces.
17. Apply 2 to 3mm bead of silicone sealant as illustrated to the sealing surface of oil pan.

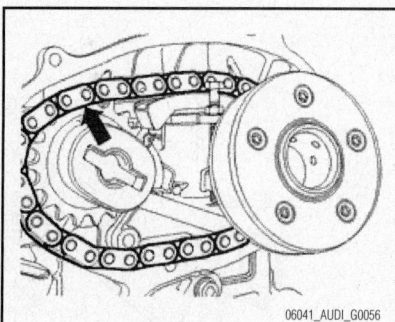

06041_AUDI_G0056

**Fig. 187 Apply 2 to 3mm bead of silicone sealant to the sealing surface of oil pan—A3 with 3.2L engine**

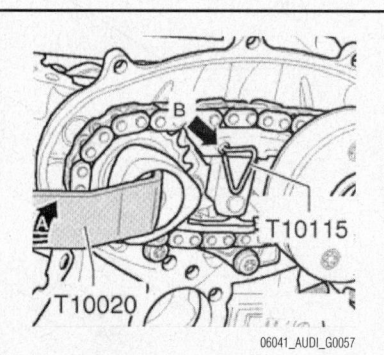

06041_AUDI_G0057

**Fig. 188 Oil pan bolt location—A3 with 3.2L engine**

➡The sealant bead may not be thicker than specified, otherwise excess sealant could enter the oil pan and clog the oil intake tube.

18. Install the oil pan immediately and tighten bolts in sequence as follows:
    a. Step 1: Oil pan bolts to 5 Nm.
    b. Step 2: New oil pan to transmission bolts to 40 Nm.
    c. Step 3: Oil pan bolts to 12 Nm.

➡Make sure that the oil pan is positioned flush with the cylinder block on the flywheel side.

➡After installing oil pan, allow sealant to dry for about 30 minutes, then fill the crankcase with oil.

19. Check and adjust the oil level.
20. Install the remaining components in the reverse order of removal.

## OIL PUMP

### REMOVAL & INSTALLATION

#### A4 Models

##### 2.0L Engine

*See Figure 189.*

1. Before servicing the vehicle, refer to the precautions section.
2. Drain the engine oil.
3. Remove or disconnect the following:
   - Oil pan
   - Baffle plate
   - Oil pump-to-engine bolts
   - Oil pump by pressing down on the subframe

#### To install:

4. Install or connect the following:
   - Oil pump by pressing down on the subframe. Torque the Oil pump-to-engine bolts to 18 ft. lbs. (25 Nm)
   - Baffle plate
   - Oil pan
5. Fill the engine with clean oil.
6. Start the vehicle, check for leaks and repair if necessary.

#### TT Models

##### 1.8L Engine

1. Before servicing the vehicle, refer to the precautions section.
2. Drain the engine oil.
3. Remove or disconnect the following:
   - Oil pan

   - Oil pump
   - Chain sprocket

#### To install:

4. Install or connect the following:
   - Dowel sleeves at top of oil pump
   - Chain sprocket. Torque the sprocket bolts to 11 ft. lbs. (15 Nm).
   - Oil pump. Torque the Oil pump-to-engine bolts to 18 ft. lbs. (25 Nm).
   - Oil pan
5. Fill the engine with clean oil.
6. Start the vehicle, check for leaks and repair if necessary.

#### 3.2L Engine

##### Except A3 and A6

1. Before servicing the vehicle, refer to the precautions section.
2. Disconnect the battery negative cable.
3. Drain engine oil
4. Remove or disconnect the following:
   - Center and right sound insulation
   - Oil thermal sensor
   - Secondary air injection pump with bracket
   - Oil pan
   - Oil pump

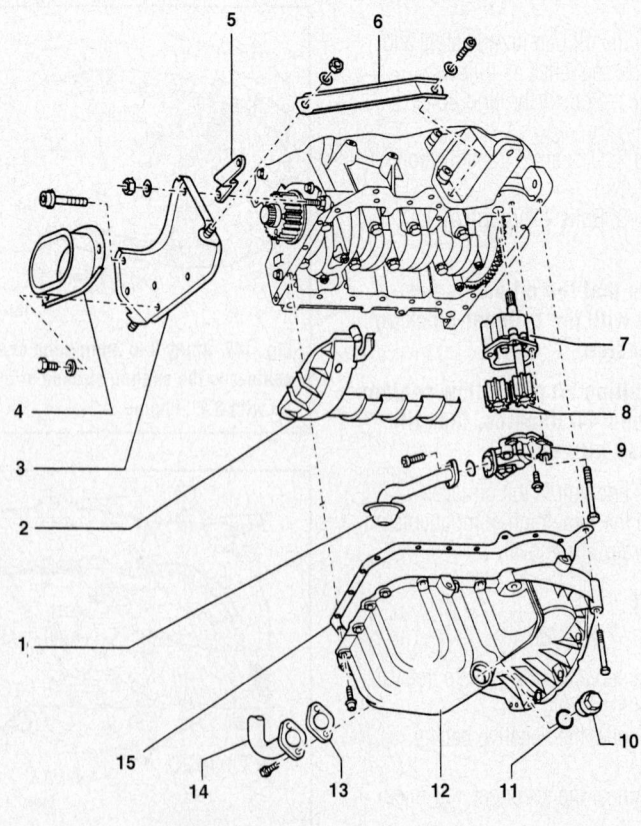

1. Suction pipe
2. Baffle plate
3. Bracket
4. Stop for torque support
5. Brace
6. Side brace
7. Oil pump housing
8. Gears
9. Oil pump cover with pressure relief valve
10. Oil drain plug
11. Sealing washer
12. Oil pan
13. Gasket
14. Oil return line
15. Gasket

**Fig. 189 Exploded view of the oil pan and pump—2.0Lengine A4**

7923CG27

**To install:**

5. Installation is the reverse of the removal procedure, while using the following torque values:

- Oil pump to engine block 17 ft. lbs (23 Nm)
- Suction tube 70 inch lbs (8 Nm)
- Tighten the oil pan bolts in diagonal sequence initially to 44 inch lbs. (5 Nm)
- Tighten bolts for sump/transmission to 30 ft. lbs. (40 Nm) Use new bolts
- Tighten the oil pan bolts in diagonal sequence to 105 inch lbs. (12 Nm)

### A6

1. Before servicing the vehicle, refer to the precautions section.
2. Drain the engine oil.
3. Remove the oil pan.
4. Remove the pump and pickup tube bolts and remove the pump assembly.

**To install:**

5. Installation is the reverse of removal, tighten the oil pump bolts to 20 Nm.

### 4.2L Engine

1. Before servicing the vehicle, refer to the precautions section.
2. Drain the engine oil.
3. Remove or disconnect the following:

- Engine oil dipstick
- Lower oil pan
- Oil pump

**To install:**

4. Installation is the reverse of the removal procedure, while using the following torque values:

- Oil pump bolts: 18 ft. lbs. (25 Nm)
- Lower oil pan bolts: 89 inch lbs. (10 Nm)

### A3 Models

#### 2.0L Engine

*See Figures 190 through 194.*

1. Before servicing the vehicle, refer to the precautions section.
2. Remove the oil pan

➡**Only rotate the engine at the crankshaft in a clockwise direction of rotation.**

3. Set camshaft gear to marking for TDC by turning crankshaft. The marking on the camshaft sprocket must align with marks on the timing belt cover.
4. Pull off the chain guard. If necessary, release retaining tabs through openings with a small screwdriver.
5. Loosen the bolt from oil pump chain

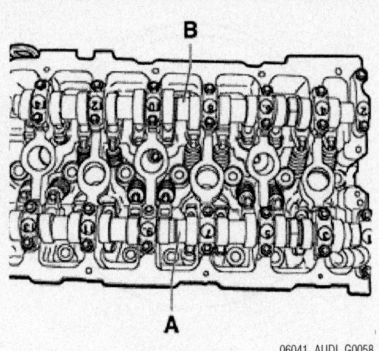

Fig. 190 Oil pan bolt location Set camshaft gear to marking for TDC by turning crankshaft. The marking on the camshaft sprocket must align with marks on the timing belt cover—A3 with 2.0L engine

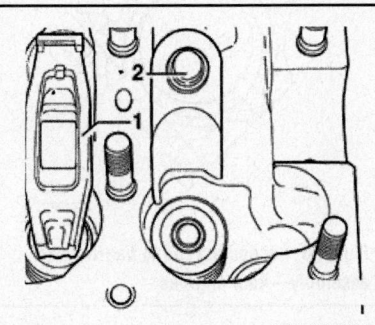

Fig. 191 Loosen the bolt from oil pump chain sprocket while holding the central bolt for the vibration damper—A3 with 2.0L engine

sprocket while holding the central bolt for the vibration damper.

6. Relieve tension on chain rail with a screwdriver and secure it with a 3mm Allen wrench.
7. Remove the oil pump chain sprocket and unhook chain at balance shaft drive.
8. Remove the oil baffle.

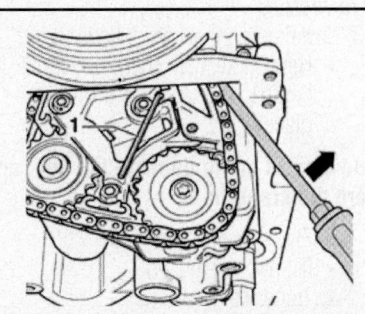

Fig. 192 Relieve tension on chain rail with a screwdriver and secure it with an 3mm Allen wrench—A3 with 2.0L engine

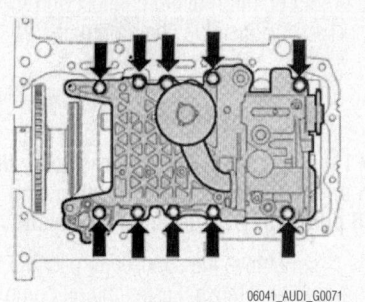

Fig. 193 Loosen the bolts from balance shaft housing from outside to inside and remove it—A3 with 2.0L engine

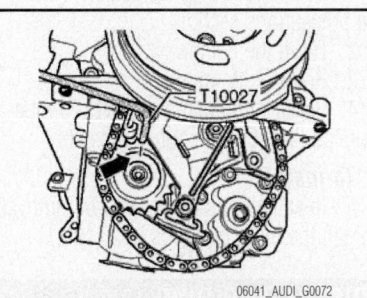

Fig. 194 Turn the crankshaft clockwise the marking on camshaft sprocket and marking on crankshaft must be positioned at TDC for cyl. 1—A3 with 2.0L engine

9. Loosen the bolts from balance shaft housing from outside to inside and remove it.

**To install:**

10. Installation is the reverse of removal, please note the following:

a. Replace all bolts of balance shaft housing

b. Install the balance shaft housing with oil pump and intermediate plate. Fasten bolts from inside to outside to 15 Nm plus an additional 90 degree turn.

➡**Observe the different lengths of bolts and pay attention to the centering sleeves.**

c. Install the oil baffle, by inserting tabs into balance shaft housing and tightening bolt to 21 Nm and bolts to 9 Nm.

d. Set engine to TDC at cylinder1, by turning crankshaft in engine clockwise direction at the central bolt of the crankshaft the timing belt sprocket. Marking on camshaft sprocket and marking on crankshaft must be positioned at TDC for cyl. 1.

e. Position the marking on the balance shaft chain sprocket across from locating hole. Secure the chain sprocket in this position using connecting pin T10027. Place the chain onto chain sprocket of balance shaft.

    f. Install the oil pump chain sprocket and tighten bolt by hand.

➡**The oil pump chain sprocket only fits in one position. When installing, the oil pump may be turned exclusively.**

    g. Remove the connecting pin T10027 and Allen wrench. Fasten the oil pump chain sprocket. To 20 Nm plus and additional 90 degrees.

### 3.2L Engine

1. Before servicing the vehicle, refer to the precautions section.
2. Drain the engine oil.
3. Remove the oil pan.
4. Remove the pump and pickup tube bolts and remove the pump assembly.

**To install:**

5. Installation is the reverse of removal, tighten the oil pump bolts to 23 Nm.

## PISTON AND RING

### POSITIONING

*See Figures 195 through 198.*

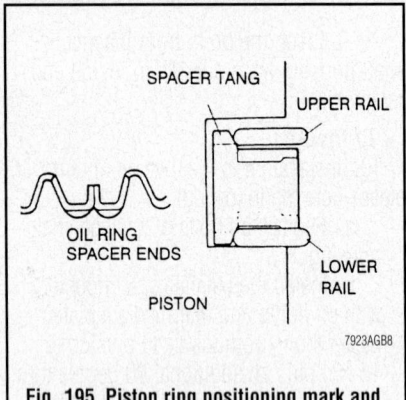

**Fig. 195 Piston ring positioning mark and location—Audi engines**

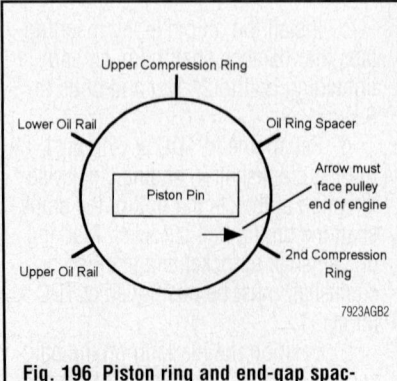

**Fig. 196 Piston ring and end-gap spacing—Audi engines**

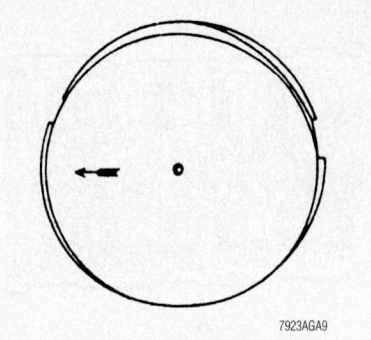

**Fig. 197 Arrow on the piston crown must face the front of the engine—Audi engines**

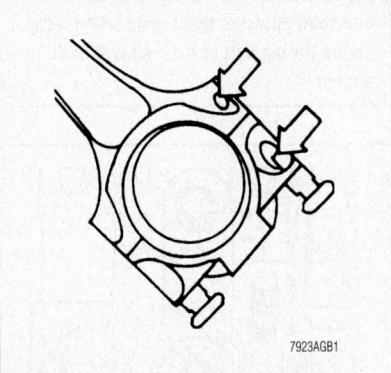

**Fig. 198 Connecting rod to bearing cap assembly—Audi engines**

## REAR MAIN SEAL

### REMOVAL & INSTALLATION

#### Except A3 and A8

1. Before servicing the vehicle, refer to the precautions section.
2. Remove or disconnect the following:
   - Negative battery cable
   - Transaxle
   - Oil pan if needed to access seal
   - Flywheel/flexplate assembly
   - Oil seal by prying it out of the housing.

**To install:**

3. Install or connect the following:
   - New oil seal by coating it with engine oil and press it into place

➡**Be careful not to damage the seal or score the crankshaft.**

   - Flywheel/flexplate
   - Oil pan if removed
   - Transaxle
   - Negative battery cable

#### A3 Models

*See Figures 199 through 203.*

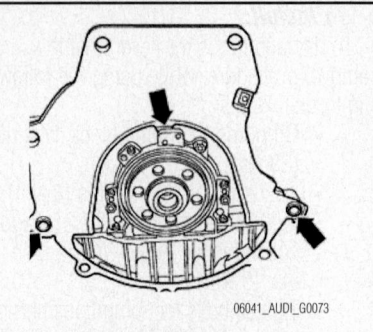

**Fig. 199 Disconnect the intermediate plate at sealing flange and at the alignment sleeves—A3 engine**

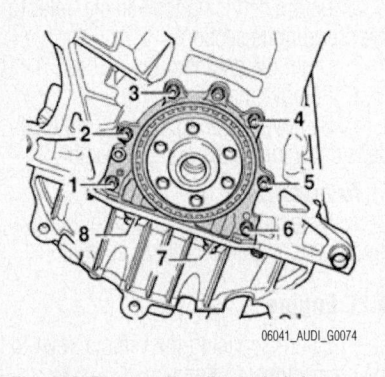

**Fig. 200 Rear sealing flange bolt location—A3**

1. Before servicing the vehicle, refer to the precautions section.
2. Remove the transmission.
3. Remove the flywheel.
4. Disconnect the intermediate plate at sealing flange and at the alignment sleeves.
5. Remove the bolts and the rear sealing flange.

**To install:**

➡**Lay a rag over the open portion of the oil pan.**

6. Clean the gasket mating surfaces.

**Fig. 201 Apply a thin bead of sealant to edge between cylinder block and oil pan—A3**

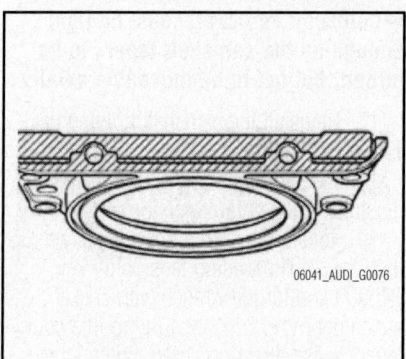

**Fig. 202 Lightly coat lower sealing surface of sealing flange with sealant—A3**

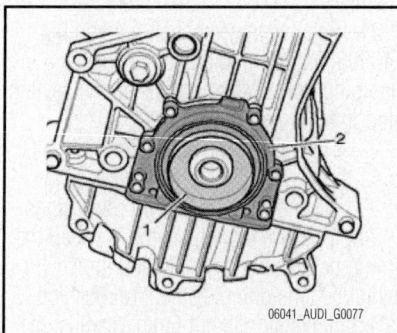

**Fig. 203 Push the sealing flange together with the guide sleeve, which is included with replacement part onto crankshaft—A3**

7. Apply a thin bead of sealant to edge between cylinder block and oil pan.

8. Lightly coat lower sealing surface of sealing flange with sealant.

➡**The sealing flange must be installed within 5 minutes of sealant being applied.**

9. Push the sealing flange together with the guide sleeve, which is included with replacement part onto crankshaft.

10. Carefully push the sealing flange onto the locating pins on cylinder block.

11. Fasten the bolts to 15 Nm.

12. Install the remaining components in the reverse order of removal.

**A8**

1. Before servicing the vehicle, refer to the precautions section.

2. Remove the transaxle from the vehicle.

3. Remove the flexplate from the crankshaft.

4. Use extraction hook 10-221 and spacer, approx. 6 mm (1/4 in.) thick, and pry out oil seal.

*To install:*

5. Using tool 2003 and drive plate mounting bolts, press in oil seal until fully seated.

6. Install the flexplate. Tighten the bolts to 44 ft. lbs. Plus 90 degrees.

7. Install the transaxle to the vehicle.

8. Start the engine and check for leaks.

## TIMING BELT, FRONT COVER AND SEAL

*REMOVAL & INSTALLATION*

**TT Models**

*See Figure 204.*

1. Disconnect the negative battery cable.

2. From under the vehicle, remove the splash shield and right side noise insulation panel.

3. Remove the intake air duct at the bottom of the right cross member.

4. Remove the accessory drive belt and belt tensioner.

5. Remove coolant expansion tank and power steering reservoir with hoses attached and place to one side.

6. Disconnect throttle valve and charcoal canister vacuum hoses.

7. Disconnect electrical connectors from coolant tank and charcoal canister.

8. Remove the upper timing belt cover.

9. Install Engine Support bracket 10-222A and supports 10-222A1.

10. Install Retainer 3180 to right engine lifting eye and attach it to support bracket 10-222A.

11. Lift engine slightly with spindle of support bracket 1022A.

➡**If reusing the timing belt, mark its rotational direction so it may be installed in its original position.**

12. Using the center bolt, rotate the crankshaft in the direction of engine rotation to position the No. 1 cylinder at Top Dead Center (TDC) of its compression stroke.

13. Remove the damper pulley-to-crankshaft bolts and the damper.

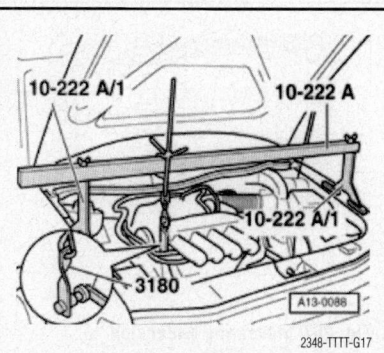

**Fig. 204 Attaching engine support and lifting tools—Audi TT 1.8L 4-Cyl engine**

14. Remove the center and lower timing belt covers.

15. Unbolt front engine support from engine console and engine console from body.

16. Disconnect connector between engine console and engine support.

17. Remove engine console.

18. Raise or lower the engine enough to remove engine support bolts. Do not remove support.

19. Using a Torx Wrench T45, loosen the timing belt tensioner, push the tensioner downward and remove timing belt.

*To install:*

20. Align the camshaft sprocket timing mark with the cylinder head cover mark.

21. Install the timing belt on the crankshaft sprocket with the arrow facing the rotational direction.

22. Install the lower timing belt cover. Torque the bolts to 84 inch lbs. (10 Nm).

23. Using a bolt, secure the damper/belt pulley on the crankshaft. Torque the bolt to 18 ft. lbs. (25 Nm).

24. Align the crankshaft damper/belt pulley with the housing timing mark so that the No. 1 cylinder is at TDC of its compression stroke.

25. Install the timing belt on the water pump, tensioning pulley and finally the camshaft sprocket.

26. Rotate the crankshaft 2 complete rotations in the running direction, and then set to TDC on no. 1 cylinder.

27. Check that timing belt is properly aligned.

28. Tighten engine support bolts to engine. Torque the bolts to 33 ft. lbs. (45 Nm).

29. Install engine console to body. Torque the bolts to 40 ft. lbs. (30 Nm) plus 90 degrees.

30. Install engine console to engine support. Torque the bolts to 44 ft. lbs. (60 Nm) plus 90 degrees.

31. Install the center and upper timing belt covers. Torque the bolts to 84 inch lbs. (10 Nm).

32. Replace the remaining components by reversing the removal procedures.

33. Install the negative battery cable last.

34. Test drive the vehicle.

**4.2L Engine**

*See Figures 205 through 207.*

1. Before servicing the vehicle, refer to the precautions section.

2. Drain the cooling system.

➡ **The battery is located in the luggage compartment, right side, under a cover.**

3. Remove or disconnect the following:
- Negative battery cable
- Engine cover
- Cylinder head covers
- Spring clips at timing belt cover
- Timing belt cover
- Sound insulation from rear of timing cover
- Accessory drive belt

4. Turn crankshaft to TDC by hand to align timing marks

5. Check position of camshafts; ensure larger holes in securing plates on sprockets align with each other and are facing inward (if not, turn crankshaft one revolution).

6. Remove seal plug from left side of cylinder block and check that the TDC drilling in the crankshaft is aligned in the hole. Then screw in a clamping bolt (3242) to hold this position.

7. Remove or disconnect the following:
- Serpentine drive belt tensioner
- Vibration damper
- Torque support

➡ **Mark the direction of rotation of the timing belt before removing.**

➡ **The timing belt tensioner is oil-filled. To compress the tensioner, apply a slow, constant pressure.**

8. With an 8mm Allen key, turn the timing belt tensioner in a counterclockwise direction until the tensioner is compressed far enough to insert a holding pin T40011 into the drilled hole and into the plunger.

9. Install a camshaft holding bar T40005 onto the securing plated of the two camshafts.

10. Loosen two camshaft bolts about five turns, and then remove the camshaft holding bar.

11. With a puller, remove the camshaft sprocket from the camshaft.

12. Loosen the timing belt tensioning idler and remove the timing belt with the camshaft sprocket for bank 1—4. Do not change position of belt on sprocket.

***To install:***

13. Ensure all pulleys, timing belt idlers and wheels are installed. Torque bolts as follows:
- Idler wheel: 33 ft. lbs. (45 Nm)
- Tensioning roller: 15 ft. lbs. (20 Nm)
- Pulley to crankshaft: 15 ft. lbs. (20 Nm)
- Timing belt tensioner: 7 ft. lbs. (10 Nm)

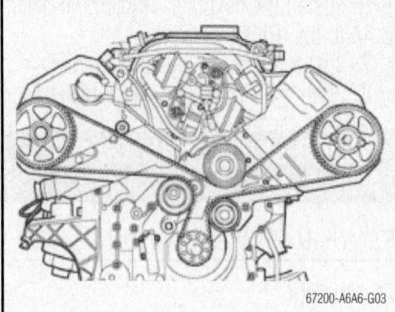

Fig. 205 Showing timing belt routing— 4.2L engines

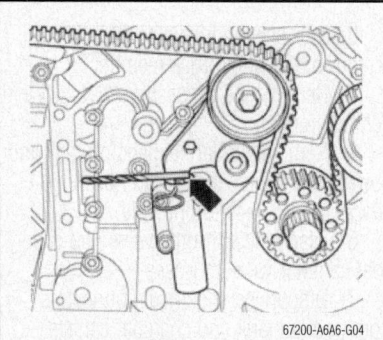

Fig. 206 Inserting drill bit between tensioner and hydraulic piston—4.2L engines

- Eccentric wheel: 16 ft. lbs. (22 Nm)
- Center bolt to crankshaft: 148 ft. lbs. (200 Nm), plus 180 degrees

14. First, install the timing belt onto the crankshaft sprocket, idler wheel for tensioner, tensioning roller, camshaft sprocket for bank 5—8, coolant pump, and damper wheel.

15. Then, take camshaft sprocket for cylinder bank 1—4 and install timing belt and bolt sprocket onto camshaft.

16. Lightly secure both camshaft sprockets with securing plates by tightening hand tight.

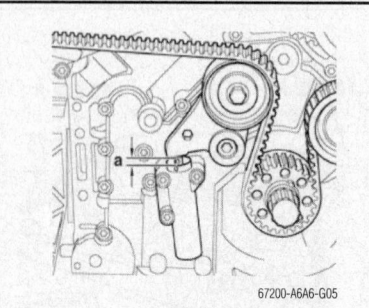

Fig. 207 Measuring dimension "a" between timing belt tensioning lever and hydraulic tensioning element—4.2L engines

➡ **Camshaft sprockets must be tight enough on the camshaft tapers to be turned, but not to be moveable axially.**

17. Reinstall the camshaft holding bar T40005.

18. Insert a 5mm drill bit between the tensioner and hydraulic piston.

19. Tension the timing belt initially to 35 inch lbs. (4 Nm), using tensioning key T40009 and torque wrench with socket attachment VAS 5122 by turning in a counterclockwise direction. Then, tighten the eccentric wheel to 16 ft. lbs. (22 Nm).

20. Using the 8mm Allen key, turn the timing belt tensioning lever to the left until the insert pin T40011 can be removed.

21. Remove the camshaft holding bar T40005, 5mm drill bit, and crankshaft clamping bolt 3242 and reinstall the sealing plug in the hole.

22. Tighten the camshaft sprocket bolts to 41 ft. lbs. (55 Nm).

23. Turn the crankshaft two revolutions.

24. Measure the dimension between the timing belt tensioning lever and the hydraulic tensioning element (dimension "a"). If dimension is not 5mm, readjust the tensioning roller as needed.

25. Measure the clearance between the timing belt guard and outer surface of timing belt (at both cam sprockets). The clearance should be 1mm. If not, adjust cover until clearance is set.

26. Install all remaining components in reverse of removal procedure.

### A8 Models

1. Before servicing the vehicle, refer to the precautions section.

2. Drain the cooling system.

➡ **The battery is located in the luggage compartment, right side, under a cover.**

3. Remove or disconnect the following:
- Negative battery cable
- Noise insulation panel
- Intake air grille from bumper
- Accessory drive belt
- Engine cover
- Coolant hose clamp on right belt guard
- Air intake duct between air cleaner and throttle body
- Air shroud for viscous fan and electric coolant fan on radiator (upper-left)
- Viscous fan. Counterhold with 3212 two-hole nut driver.

➡ **The viscous fan has a left-hand thread; loosen by turning clockwise.**

- Viscous fan with air shroud, from above
- Engine support mount (right front)
- Coolant hoses at engine
- Loosen coolant hose on upper-right of radiator, and turn hose to right
- Loosen center bolt for vibration damper approx. 1 turn. Counterhold using special tool 3197.

➡**Loosening and removing the vibration damper with the camshaft drive belt sprocket is only necessary if the belt is to be replaced, or if other engine disassembly requires it.**

➡**When the center bolt has been loosened, it must be replaced.**

4. Rotate crankshaft to align TDC marks on vibration damper.

5. Remove Camshaft Position (CMP) sensor housing at rear of left cylinder head.

6. If camshaft position sensor is not positioned behind sensor plate window, rotate crankshaft 360.

7. Remove camshaft flange at rear of right cylinder head.

8. Remove or disconnect the following:

- Harness connector from switch for intake manifold change-over valve - N156-
- Belt guard
- Cap at guide pulley for ribbed belt at right belt guard
- Top part of air cleaner housing
- Right belt guard mounting bolts (6X)
- Right belt guard
- Hold camshaft sprocket with holder 3036 and loosen camshaft sprockets bolts
- Camshaft Position (CMP) sensor plate and housing

9. Install camshaft locking device 3341 at rear of each cylinder head, and tighten. If necessary, rotate camshaft slightly to allow locking device to engage.

### ❉❉ CAUTION

**The camshaft locking devices 3341 are not to be used as counterholds when loosening or tightening the camshaft sprocket bolts.**

10. Using two-hole nut turner, e.g. Matra V/159, loosen eccentric tensioning roller and turn to lowest point.

11. Compress drive belt damper by hand and remove tensioning roller.

➡**Mark the running direction of the belt with a felt pen or equivalent. Reversed running direction can lead to damage.**

12. Remove belt from camshaft sprockets.

13. Remove 4 screws fastening vibration damper to drive belt sprocket, remove center bolt, and remove vibration damper.

14. Remove camshaft drive belt.

### *To install:*

15. Place belt over crankshaft sprocket and install vibration damper on crankshaft.

16. By hand, screw in 4 bolts that connect vibration damper to sprocket.

17. Lightly oil thread and bolt contact surfaces of center bolt.

18. Thread in new center bolt and tighten by hand. Counterhold with special tool 3197.

➡**Always replace center bolt.**

19. Place belt in position over all sprockets, guide pulleys and water pump pulley.

20. Push belt over tensioning roller with eccentric insert and tighten nut.

➡**Eccentric insert must still be able to turn.**

21. Tighten 4 mounting bolts connecting vibration damper and crankshaft sprocket to 18 ft. lbs. (25 Nm).

22. Turn eccentric insert of tensioning roller clockwise using commercial two-hole nut turner (e.g. Matra V/159).

23. Adjust position of tensioning roller until damper length is within specified range.

24. Install or connect the following:

- Engine cold: 5.35–5.47 inches (136-139 mm)
- Engine warm: 4.96-5.08 inches (126–129 mm)

25. Tighten tensioning roller bolt to 18 ft. lbs. (25 Nm).

26. Remove camshaft locking device from each camshaft (rear of cylinder head).

27. Install housing behind camshaft position sensor and tighten to 15 ft. lbs. (20 Nm).

28. Crank engine at least two revolutions by hand and check timing belt alignment.

29. Tighten crankshaft center bolt as follows:

- With special tool 2079: 258 ft. lbs. (350 Nm)
- Without special tool 2079: 332 ft. lbs. (450 Nm)

30. The remaining installation of the camshaft drive belt is the reverse of removal.

### A3 Models

#### *2.0L Engine*

*See Figures 208 through 216.*

1. Remove the engine cover.

2. Open the coolant expansion tank cap after the engine is cooled.

3. Remove the center sound insulation.

4. Remove the right sound insulation.

5. Drain the cooling system.

6. Relieve the fuel system pressure.

7. Disconnect the fuel line.

8. Disconnect the EVAP lines .

9. Remove the EVAP canister.

10. Loosen coolant hose from retainer.

11. Remove the bracket for EVAP the filter.

12. Unbolt filler tube for windshield washer fluid reservoir.

13. Remove the retaining clip, disconnect the electrical connector and remove the coolant expansion tank.

14. Disconnect the coolant pipe.

15. Remove the accessory drive belt and remove locking pin T10050A .

16. Install an engine support tool such as 10–222 A onto fender bolting edge. The spindles should be positioned forward. Hook the carabiner hooks of spindles with shackles 10–222 A /12 into lifting eyes.

17. Remove the bolts from assembly mount/engine support remove the assembly mount completely.

### ❉❉ CAUTION

**When lifting the engine using the engine support tool , make sure that no components/hoses are damaged, overstressed or torn off.**

18. Lift engine using engine support tool far enough so that both upper bolts of engine support can be loosened and removed.

Set camshaft gear to marking for TDC by turning crankshaft. The marking on the camshaft sprocket must align with the mark on timing belt cover.

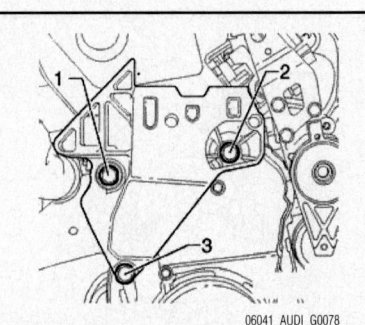

06041_AUDI_G0078

**Fig. 208 Lift the engine using the support tool and remove the upper engine support bolts—A3 with 2.0L engine**

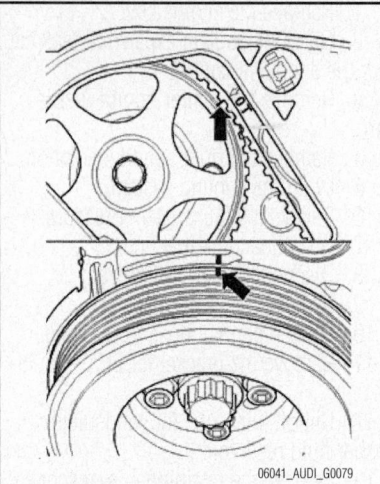

**Fig. 209 Align the marks as shown by setting the engine to TDC—A3 with 2.0L engine**

19. Remove the vibration damper/belt pulley.

20. Remove the bolts from lower timing belt cover.

21. Loosen the lower bolts and remove the engine support.

22. Remove the remaining bolts from the timing belt cover and remove timing belt cover from engine.

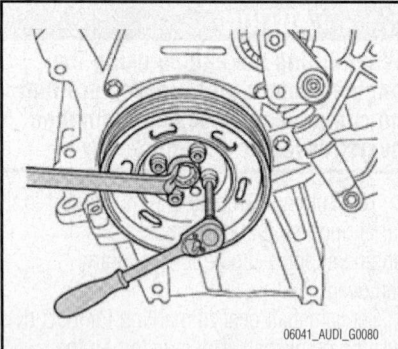

**Fig. 210 Remove the vibration damper pulley—A3 with 2.0L engine**

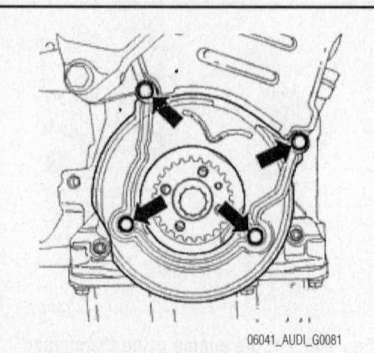

**Fig. 211 Remove the lower timing belt cover—A3 with 2.0L engine**

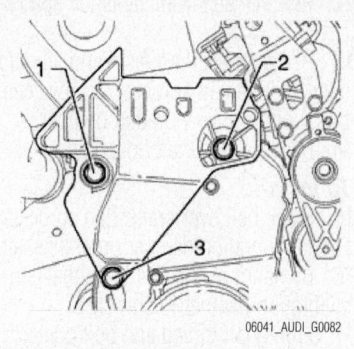

**Fig. 212 Remove the lower bolts and the engine support—A3 with 2.0L engine**

23. Mark the direction of rotation of timing belt.

24. Loosen tensioning roller and remove the timing belt.

25. Turn crankshaft back slightly.

### To install:

➡ **When turning the camshaft, the crankshaft must not be at TDC as the valves and/or pistons may be damaged.**

26. Place the timing belt onto the crankshaft sprocket making sure of the belt direction of rotation.

27. Secure lower timing belt cover with two lower bolts.

28. Install the vibration damper/belt pulley with new bolts. Tighten to 10 Nm plus an additional 90 degrees.

29. Set the crankshaft and camshaft to TDC at cylinder 1.

30. Route the timing belt as follows:
   a. Tension roller.
   b. Camshaft sprocket.
   c. Coolant pump.
   d. Idler pulley.

➡ **Make sure the tensioning roller is correctly seated in the cylinder head.**

31. Tension timing belt by turning hex key on the eccentric toward right in a clockwise direction until notch is positioned above tab.

32. Relieve tension on timing belt again.

33. Now tension timing belt until notch and tab are positioned across from each other and tighten the nut to 25 Nm.

34. Turn the crankshaft 2 revolutions further in direction of engine rotation until engine is positioned at TDC again. When doing so, it is important that the last 45 degrees is turned without interruption.

35. Check timing belt tension again. The tab and notch are positioned across from each other.

36. Check valve timing again.

**Fig. 213 Remove the timing belt cover—A3 with 2.0L engine**

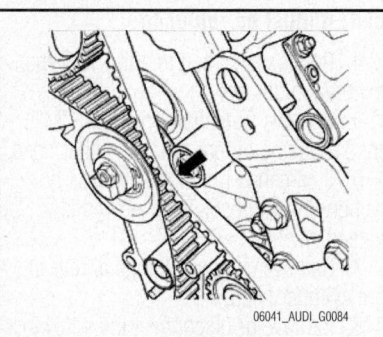

**Fig. 214 Install the belt over the tension roller, camshaft sprocket, water pump and idler puller—A3 with 2.0L engine**

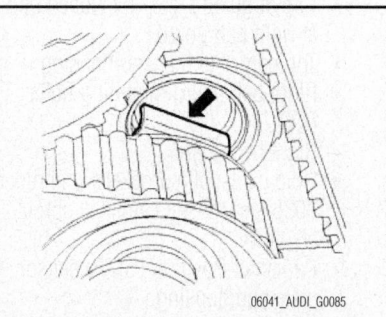

**Fig. 215 Make sure the tension roller is seated correctly in the cylinder head—A3 with 2.0L engine**

37. If the marks do not align, repeat the valve timing adjustment procedure.

38. If the marks align, install timing belt guard.

➡ **Observe the different lengths of bolts. The lower bolt is approx. 25 mm shorter than the upper bolts.**

39. Insert the lower bolt into the engine support.

40. Install engine support from below onto cylinder block and hand-tighten bolt.

41. Lift engine using engine support tool so that both upper bolts can be inserted. Tighten the bolts to 45 Nm.

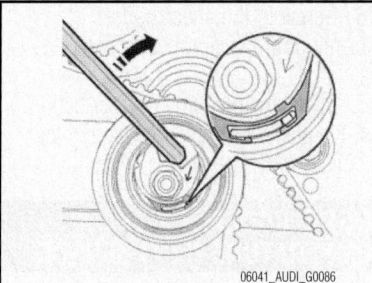

**Fig. 216 Tension the timing belt by turning the hex key clockwise until the notch is positioned above the tab—A3 with 2.0L engine**

42. Lower engine to installed position.
43. Install assembly mount for engine

completely. The clearance between engine support and right longitudinal member should be 16mm. The casting edge of engine support must be parallel to the support arm.

44. Install the engine assembly mount to engine support bringing contact surfaces into contact using engine support tool. Tighten to 60 Nm plus an additional 90 degrees.

45. Remove the engine support bridge 10-222A .

46. Install the remaining components in the reverse order of removal, please note the following:

a. Tighten the timing belt tensioning roller to cylinder head to 23 Nm.

b. Tighten the lower and middle timing belt covers to cylinder block, use a locking compound and tighten to 10 Nm.

c. Tighten the engine support to cylinder block to 45 Nm.

d. Tighten the engine console to body, using new bolts to 40 Nm plus an additional 90 degrees.

e. Tighten the bracket to body/engine console, using new bolts to 20 Nm plus an additional 90 degrees.

## VALVE LASH

### ADJUSTMENT

Audi engines are equipped with hydraulic lash adjusters. No adjustment is possible.

# ENGINE PERFORMANCE & EMISSION CONTROL

## CAMSHAFT POSITION (CMP) SENSOR

### LOCATION

#### 1.8L Engines

*See Figure 217.*

The Camshaft Position (CMP) Sensor is mounted on the cylinder head, front, beneath timing belt cover.

**Fig. 217 Camshaft Position sensor location for the 1.8L engine**

#### 3.2L Engines

*See Figures 218 and 219.*

The Camshaft Position (CMP) Sensor is mounted in the left side of the engine compartment to the left of the oil fill cap.

The Camshaft Position (CMP) Sensor 2 is mounted in the left side of the engine compartment with a black connector.

**Fig. 218 Camshaft Position sensor location for the 3.2L engine**

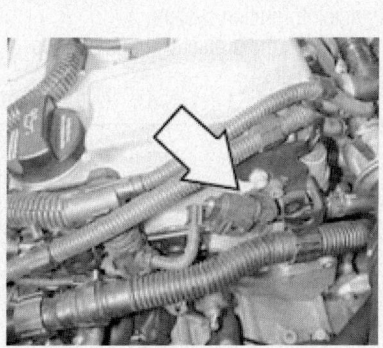

**Fig. 219 Camshaft Position sensor 2 location for the 3.2L engine**

#### 4.2L Engines

*See Figure 220.*

The Camshaft Position (CMP) Sensor is located on the rear of the left cylinder head.

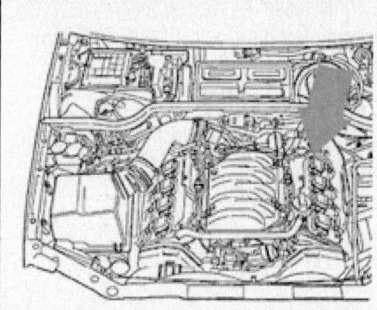

**Fig. 220 Camshaft Position sensor location for the 4.2L engine**

## ELECTRONIC CONTROL MODULE (ECM)

### LOCATION

*See Figure 221.*

The Motronic Engine Control Module is located in the center of the plenum.

**Fig. 221 Motronic Engine Control Module location for all engines**

*REMOVAL & INSTALLATION*

*See Figures 222 through 226.*

1. Remove the windshield wiper arms, the plenum chamber cover and the plenum chamber bulkhead.

2. Cut the heads of shear bolts so that two parallel surfaces remain (arrows).

3. Remove the bolts with locking pliers.

4. Insert a screwdriver between the protective housing and the retaining plate.

5. Use screwdriver to pry the protective housing upward, and then pull it off sideways from the retaining plate.

6. Release and pull off the front connector from the Motronic Engine Control Module (ECM).

7. Pry the retaining tab slightly open.

8. Then, push the Motronic Engine Control Module (ECM) out of retainer.

9. Release the rear connector from the Motronic Engine Control Module (ECM) and pull it off.

### To install:

10. Connect the rear connector to the Motronic Engine Control Module (ECM) and lock it into place.

11. Push the Motronic Engine Control Module (ECM) onto retaining plate.

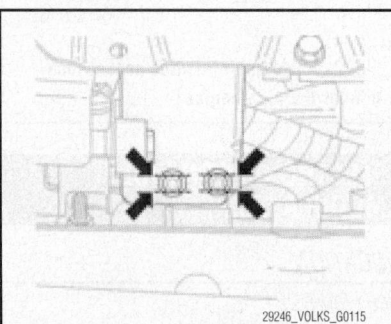

**Fig. 222 Cut the heads of shear bolts so that two parallel surfaces remain (arrows).**

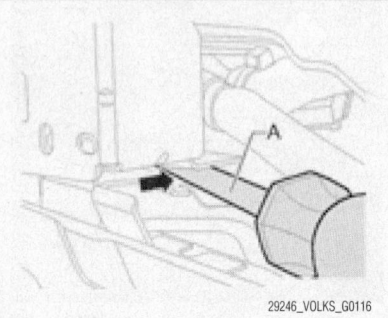

**Fig. 223 Insert a screwdriver (A) between protective housing and retaining plate (arrow).**

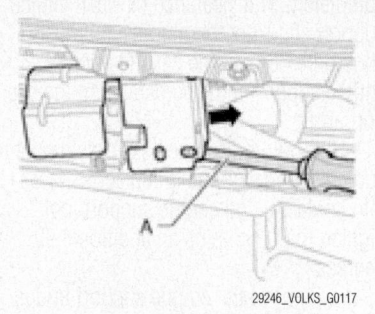

**Fig. 224 Use screwdriver (A) to pry protective housing upward, and then pull it off sideways from retaining plate (arrow)**

**Fig. 225 Release and pull off front connector (1) from ECM and pry retaining tab (2) slightly open.**

12. Connect the front connector to the Motronic Engine Control Module (ECM) and lock it into place.

13. Check whether the new control module identification matches the old control module identification.

14. Perform function "Replacing control module" using the vehicle diagnosis and service information system.

15. Push the protective housing onto retaining plate.

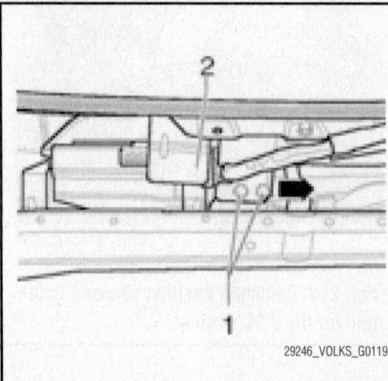

**Fig. 226 Push protective housing (2) onto retaining plate, and screw in shear bolts (1) uniformly until the bolt heads begin to shear.**

16. Screw in the shear bolts uniformly until the bolt heads begin to shear.

17. Install the plenum chamber bulkhead, the plenum chamber cover and the windshield wiper arms.

## ENGINE COOLANT TEMPERATURE (ECT) SENSOR

*LOCATION*

### 1.8L and 3.2L Engines

*See Figure 227.*

**Fig. 227 Engine Coolant Temperature sensor location for the 1.8L and 3.2L engines**

The Engine Coolant Temperature Sensor is located in the coolant flange to the left of the cylinder head and is integral with the ECT sensor.

### 4.2L Engines

*See Figure 228.*

The Engine Coolant Temperature Sensor is located on the back of the right cylinder head and is integral with the ECT sensor.

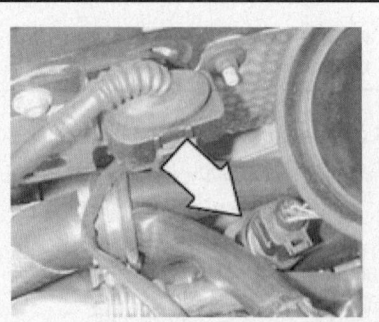

**Fig. 228 Engine Coolant Temperature sensor location for the 4.2L engines**

## HEATED OXYGEN (HO2S) SENSOR

### LOCATION

**1.8L Engines**

*See Figure 229.*

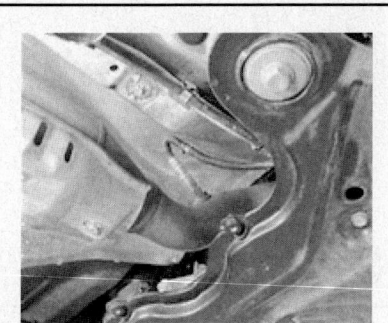

29246_AUDI_G0012

**Fig. 229 Heated Oxygen sensor location for the 1.8L engine**

The Heated Oxygen Sensor is located on the exhaust system, in front of the three-way catalytic converter. It is integral with the Oxygen Sensor Heater.

**3.2L Engines**

*See Figure 230.*

The Heated Oxygen Sensor is located behind the engine in the left and right front exhaust pipes. It is integral with the Oxygen Sensor Heater.

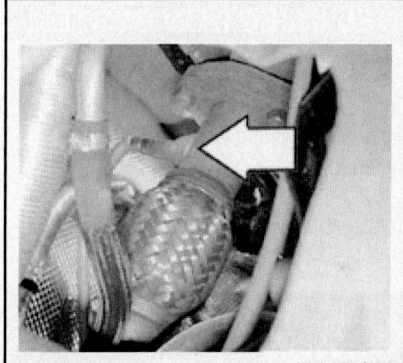

29246_AUDI_G0013

**Fig. 230 Heated Oxygen sensor location for the 3.2L engine**

**4.2L Engines**

The Heated Oxygen Sensor is located behind the exhaust manifold in the left and right exhaust pipes. It is integral with the Oxygen Sensor Heater. (no image available)

## INTAKE AIR TEMPERATURE (IAT) SENSOR

### LOCATION

The Intake Air Temperature (IAT) Sensor is located on the intake manifold, in front of the engine compartment.

## KNOCK SENSOR (KS)

### LOCATION

**1.8L and 2.0L Engines**

*See Figure 231.*

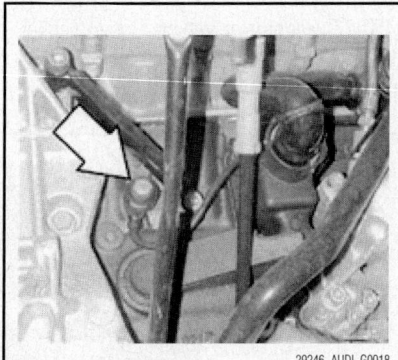

29246_AUDI_G0018

**Fig. 231 Knock Sensor location**

The Knock Sensor (KS) 1 is located in front of the engine block on the passenger's side.

The Knock Sensor (KS) 2 is located in front of the engine block on the driver's side behind the oil filter housing.

**3.2L and 4.2L Engines**

*See Figures 232 and 233.*

The Knock Sensor (KS) 1 is located below the intake manifold on the right rear cylinder head (shown with the intake manifold removed).

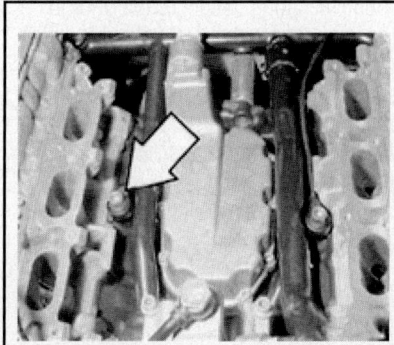

29246_AUDI_G0049

**Fig. 232 Knock Sensor 1 location**

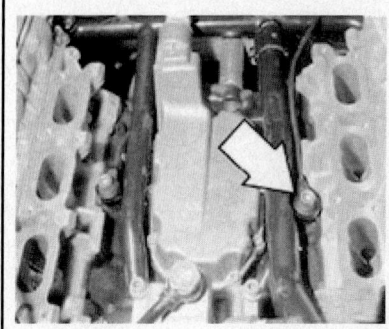

29246_AUDI_G0050

**Fig. 233 Knock Sensor 2 location**

The Knock Sensor (KS) 2 is located below the intake manifold on the left rear cylinder head (shown with the intake manifold removed).

## MANIFOLD ABSOLUTE PRESSURE (MAP) SENSOR

### LOCATION

**All Engines**

*See Figure 234.*

The Manifold Absolute Pressure Sensor is located below the passenger side of the headlight assembly in the outlet flange of the intercooler.

29246_AUDI_G0020

**Fig. 234 Manifold Absolute Pressure sensor location for all engines**

## MASS AIR FLOW (MAF) SENSOR

### LOCATION

**1.8L Engines**

*See Figure 235.*

The Mass Air Flow Sensor is located in the left rear of the engine compartment, in the intake air duct, to the right of the air cleaner housing.

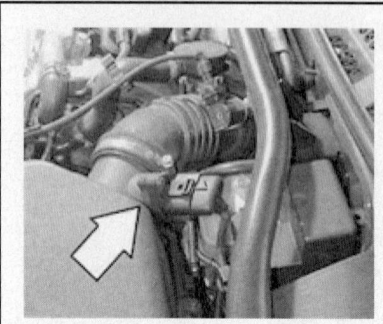

**Fig. 235 Mass Air Flow sensor location for the 1.8L engine**

### 3.2L Engines

*See Figure 236.*

The Mass Air Flow Sensor is located in the left rear of the engine compartment, near the intake air duct.

**Fig. 236 Mass Air Flow sensor location for the 3.2L engine**

## THROTTLE POSITION SENSOR (TPS)

### LOCATION

*See Figure 237.*

The Throttle Position (TP) Sensor is located on the bulkhead, above the accelerator pedal.

## VEHICLE SPEED SENSOR (VSS)

### LOCATION

#### All Engines

*See Figure 238.*

The Speedometer Vehicle Speed Sensor (VSS) is located in the transmission housing on the passenger side output flange.

**Fig. 238 Speedometer Vehicle Speed Sensor location for all engines**

**Fig. 237 Throttle Position Sensor location**

## FUEL

## GASOLINE FUEL INJECTION SYSTEM

### FUEL SYSTEM SERVICE PRECAUTIONS

Safety is the most important factor when performing not only fuel system maintenance but any type of maintenance. Failure to conduct maintenance and repairs in a safe manner may result in serious personal injury or death. Maintenance and testing of the vehicle's fuel system components can be accomplished safely and effectively by adhering to the following rules and guidelines.

• To avoid the possibility of fire and personal injury, always disconnect the negative battery cable unless the repair or test procedure requires that battery voltage be applied.

• Always relieve the fuel system pressure prior to disconnecting any fuel system component (injector, fuel rail, pressure regulator, etc.), fitting or fuel line connection.

Exercise extreme caution whenever relieving fuel system pressure to avoid exposing skin, face and eyes to fuel spray. Please be advised that fuel under pressure may penetrate the skin or any part of the body that it contacts.

• Always place a shop towel or cloth around the fitting or connection prior to loosening to absorb any excess fuel due to spillage. Ensure that all fuel spillage (should it occur) is quickly removed from engine surfaces. Ensure that all fuel soaked cloths or towels are deposited into a suitable waste container.

• Always keep a dry chemical (Class B) fire extinguisher near the work area.

• Do not allow fuel spray or fuel vapors to come into contact with a spark or open flame.

• Always use a back-up wrench when loosening and tightening fuel line connection fittings. This will prevent unnecessary stress and torsion to fuel line piping.

• Always replace worn fuel fitting O-rings with new Do not substitute fuel hose or equivalent where fuel pipe is installed.

Before servicing the vehicle, make sure to also refer to the precautions in the beginning of this section as well.

### RELIEVING FUEL SYSTEM PRESSURE

The fuel injection system operates under high pressure. This makes it necessary to first relieve the system of pressure before servicing. The pressurized fuel, when released, may ignite or cause personal injury.

**✴✴ CAUTION**

**The fuel injection system remains under pressure after the engine has been turned OFF. Properly relieve**

**fuel pressure before disconnecting any fuel lines. Failure to do so may result in fire or personal injury.**

1. Before servicing the vehicle, refer to the precautions section.
2. Remove or disconnect the following:
   - Power to the fuel pump by removing the relay or the fuel pump fuse. The fuse can be removed to stop the fuel pump from running. With the engine operating at idle, wait until the engine stalls from fuel starvation.
3. Switch the ignition **OFF** and remove the negative battery cable.
4. Slowly and carefully open the fuel tank filler cap for a brief moment and reinstall.
5. Carefully loosen the fuel line on the control pressure regulator or component to be serviced.
6. Wrap a clean rag around the connection, while loosening, to catch any fuel.
7. After service is complete, discard the fuel soaked rag in the proper manner and reconnect negative battery cable, relay or fuses.

## FUEL FILTER

*REMOVAL & INSTALLATION*

Most vehicles use a fuel filter mounted under the vehicle, below the fuel tank or in the engine compartment. An arrow on the filter indicates fuel flow direction. Use care not to mix up fuel supply or return lines. Fuel pressure applied to the return side of the system will cause damage.

1. Before servicing the vehicle, refer to the precautions section.
2. Properly relieve the residual fuel pressure.
3. Remove or disconnect the following:
   - Negative battery cable
   - Fuel lines leading into and out of the fuel filter.
   - Filter retaining bracket
   - Fuel filter

*To install:*
4. Install or connect the following:
   - New fuel filter in the bracket
   - Fuel filter bracket

➡**Be sure the arrows are pointing in the direction of the fuel flow**

   - Fuel lines
   - Negative battery cable
5. Start the engine, check for leaks and repair if necessary.

## FUEL INJECTORS

*REMOVAL & INSTALLATION*

### Except A3

1. Before servicing the vehicle, refer to the precautions section.
2. Relieve the fuel system pressure.
3. Remove or disconnect the following:
   - Negative battery cable
   - Engine under cover, if applicable
   - Fuel lines
   - Fuel injector connectors
   - Fuel pressure line
   - Fuel supply rail with injectors attached
   - Fuel injectors from the supply rail

*To install:*
4. Install or connect the following:
   - Fuel injectors to the fuel supply rail using new O-rings
   - Fuel supply rail with injectors attached. Tighten the bolts to 89 inch lbs. (10 Nm)
   - Fuel injector connectors
   - Fuel pressure line
   - Fuel lines
   - Engine under cover, if applicable
   - Negative battery cable
5. Start the engine, check for leaks and repair if necessary.

### A3 Models

#### 2.0L Engine

*See Figures 239 through 243.*

1. Before servicing the vehicle, refer to the precautions section.
2. Remove the intake manifold and fuel rail .
3. Carefully pull fuel injectors out of fuel rail.

➡**With fuel injector installed, the radial adjustment is clipped into the support ring. To remove the fuel injector, the support ring must be separated from the fuel injector in order to install the puller T10133/2 into the indentation on the fuel injector.**

4. Cover open intake channels with a clean rag.
5. Disconnect the electrical harness connector at fuel injector that is to be removed.
6. Using a screwdriver, bend the retaining tabs of the radial adjustment aside and pull support ring off from fuel injector.
7. Screw together slide hammer T10133/3 (1) with puller T10133/2 (2). Then, guide puller T10133/2 into the groove

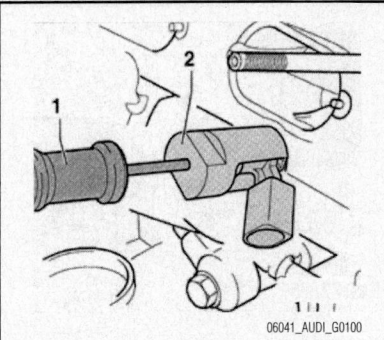

**Fig. 239 Screw together slide hammer T10133/3 (1) with puller T10133/2 (2). Then, guide puller T10133/2 into the groove on the fuel injector and carefully drive the fuel injector out—A3 with 2.0L engine**

on the fuel injector and carefully drive the fuel injector out. By doing this, it is possible that the radial adjustment will be damaged by breaking the retaining tabs. These must be replaced when re-installing the fuel injector.

➡**The combustion chamber seal must always be replaced before re-installing the high-pressure fuel injector.**

8. Clean off the old Teflon seal, make sure not to damage the groove and rib in groove base.

➡**If the groove is damaged, the fuel injector must be replaced.**

*To install:*
9. Before installing new Teflon seal, the seal groove and shaft of the fuel injector must be clean.
10. Place assembly cone T10133/5 with new Teflon seal onto fuel injector.
11. Using the assembly sleeve T10133/6, push Teflon seal further onto assembly cone T10133/5 until the Teflon seal is engaged in seal groove. Do not use any lubricants.
12. By pushing the Teflon seal onto the

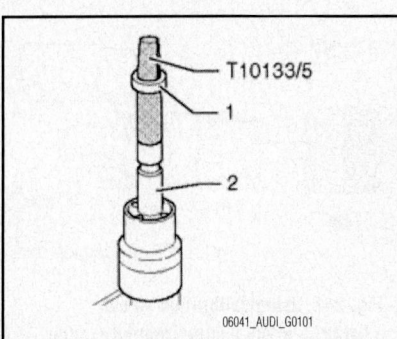

**Fig. 240 Place assembly cone T10133/5 with new Teflon seal onto fuel injector— A3 with 2.0L engine**

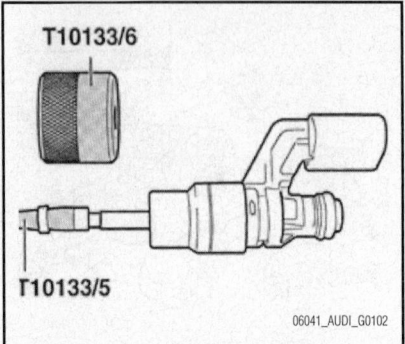

**Fig. 241 Using the assembly sleeve T10133/6, push Teflon seal further onto assembly cone T10133/5 until the Teflon seal is engaged in seal groove—A3 with 2.0L engine**

fuel injector, the Teflon seal is expanded. For this reason, the Teflon seal must be contracted again, once it has been pushed on as follows:

a. Using calibration sleeve T10133/7, a 180 degree rotating motion and light pressure, push calibration sleeve T10133/7 over the fuel injector up to stop. Pull calibration sleeve T10133/7 off again, rotating it in the opposite direction.

b. Using calibration sleeve T10133/8, 180 degree rotating motion and light pressure, push calibration sleeve T10133/8 over the fuel injector up to stop. Pull calibration sleeve T10133/8 off again, rotating it in the opposite direction.

13. Replace O-ring on the fuel injector and on spacer sleeve. Moisten O-ring with clean engine oil before installing.

➡ **The Teflon seal must not get oil on it.**

14. Clean the bores for the high-pressure fuel injectors in the cylinder head before installing the fuel injectors.

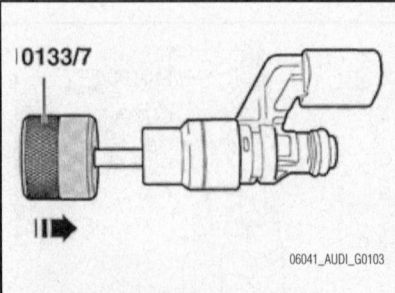

**Fig. 242 Using calibration sleeve T10133/7, a 180 degree rotating motion and light pressure, push calibration sleeve T10133/7 over the fuel injector up to stop—A3 with 2.0L engine**

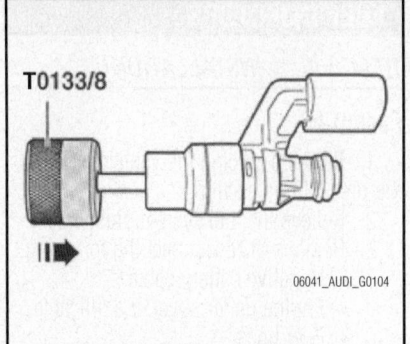

**Fig. 243 Using calibration sleeve T10133/8, 180 degree rotating motion and light pressure, push calibration sleeve T10133/8 over the fuel injector up to a stop—A3 with 2.0L engine**

➡ **An open intake valve could possibly hinder the cleaning process. In this case, turn the engine further by hand using a socket wrench at the crankshaft.**

15. Equip the fuel injector again with support ring and clip the radial adjustment in at the support ring.

16. Using the assembly drift T10133/9, press the fuel injector into its bore in the cylinder head until it stops. Make sure fuel injectors are correctly positioned in cylinder head.

17. Install the remaining components in the reverse order of removal.

### 3.2L Engine

1. Before servicing the vehicle, refer to the precautions section.
2. Relieve the fuel system pressure.
3. Disconnect the fuel line.
4. Remove the intake manifold and fuel rail.
5. Pull off retaining clip and remove the relevant fuel injector.

### To install:

6. Installation is the reverse of removal, please note the following:

a. Replace O-rings at all opened connections. Never pull off the plastic cap from valve head to replace the front O-ring always lift O-ring over the plastic cap.

b. Lightly lubricate O-rings with clean engine oil.

c. Make sure fuel injectors are in place and seated correctly

d. Check clip for unobstructed fitting.

e. Place fuel rail with secured fuel injectors onto intake manifold, and apply uniform pressure to press it in and tighten the rail bolts to 10 Nm.

## FUEL PUMP

### REMOVAL & INSTALLATION

#### Except A3 Models

*See Figures 244 and 245.*

The fuel pump is located in the fuel tank. It is recommended that the fuel tank not be filled more than ⅓ full. If necessary the fuel must be drained using an approved fuel cart.

1. Before servicing the vehicle, refer to the precautions section.
2. Properly relieve fuel pressure.
3. Remove or disconnect the following:
   - Negative battery cable
   - Inspection cover by lifting the cargo area trim
4. Mark the fuel pump/gauge sending unit assembly supply and return lines for reassembly.

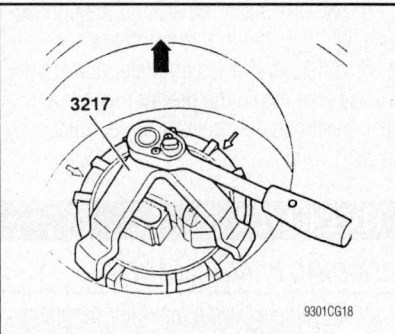

**Fig. 244 Tool 3217 is used to loosen and tighten the union nut on the fuel tank for access to the fuel pump and fuel level sending unit**

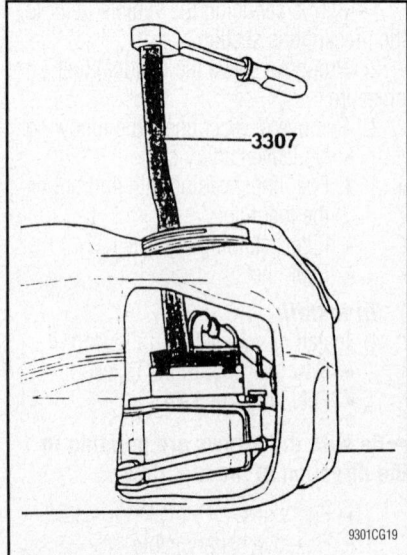

**Fig. 245 Tool 3307 is used to remove and install the fuel pump in the fuel tank**

5. Remove or disconnect the following:
- Electrical connector
- Union nut by loosening it using tool 3217
- Fuel pump/gauge sending unit assembly by matchmarking it
- Level connector and the fuel return line

6. Using tool 3307, turn the fuel pump module to the left (counterclockwise) about 15 degrees.

7. Remove or disconnect the following:
- Fuel pump supply hose
- Electrical connectors
- Fuel pump

***To install:***

8. The installation is reverse of the removal procedure.

9. Install or connect the following:
- Flange cover using a new O-ring lubricated the O-ring with fuel
- Union nut using tool 3217. Torque it to 59 ft. lbs. (80 Nm)
- Negative battery cable

10. Start the vehicle, check for leaks and repair if necessary.

### A3 Models

*See Figure 246.*

1. Before servicing the vehicle, refer to the precautions section.

➡**To remove fuel pump, the fuel tank may only be a maximum of one third filled, drain the tank as needed.**

2. Remove the spare wheel and formed insert for tool kit from under luggage compartment floor covering.

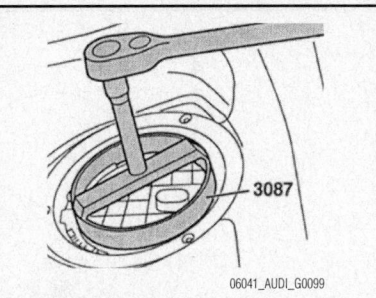

06041_AUDI_G0099

**Fig. 246 Remove/tighten the locking ring using wrench 3087—A3**

3. Disconnect the negative battery cable.

4. Remove the rear seat bench.

5. Unclip cover retainers for right sealing flange.

6. Disconnect the electrical harness connector at locking flange.

7. Mark fuel supply line and fuel return line.

8. Pull both lines from locking flange, by pressing release buttons.

9. Remove the locking ring using wrench 3087.

10. Pull the right sealing flange and fuel pump seal out from opening in fuel tank and lay aside with lines still attached.

11. Through the fuel tank opening, separate suction jet pipe from suction jet pump, by pressing release button.

12. Disconnect the fuel delivery line by pressing the release.

13. Remove the fuel pump.

➡**When removing fuel pump, be sure not to bend floater arm of fuel level sensor.**

***To install:***

14. Installation is the reverse of removal, please note the following:

a. When inserting fuel pump, be sure not to bend fuel level sensor.

b. Ensure fuel hoses are seated securely.

c. Insert fuel pump into fuel tank with locking flange set aside.

d. Connect the fuel lines until a click is heard.

e. Install the new locking flange seal dry.

f. Insert guide of locking flange into guide at fuel pump, observing spring while doing so.

g. Press locking flange down against spring pressure and bring it into installation position.

➡**The tab on locking flange faces toward the arrow marking and must lie between tabs and on fuel tank.**

h. Tighten locking ring to 145 Nm.

i. Install the rear seat bench.

### IDLE SPEED

*ADJUSTMENT*

Idle speed is maintained by the Powertrain Control Module (PCM). No adjustment is necessary or possible.

### THROTTLE BODY

*REMOVAL & INSTALLATION*

See Intake Manifold procedures.

## HEATING & AIR CONDITIONING SYSTEM

### HEATER CORE

*REMOVAL & INSTALLATION*

**A4 and TT Models**

*See Figures 247 through 252.*

1. Perform the Output Diagnostic Test Mode (DTM) using the VAG 1551 function 03 by performing the following procedure:
- Once the air flow flap closes, cancel the Output DTM by pressing the "C" button.

➡**If equipped with power seats, move the seats as far rearward as possible. Also, obtain the anti-theft radio coding and the preset radio stations from the owner.**

2. Discharge and recover the air conditioning system refrigerant.

3. Remove or disconnect the following:
- Air plenum cover, the water guide and the dust/pollen filter, located at the right side
- Negative battery cable
- Positive battery cable and remove the battery
- Coolant recovery tank cap to relieve the pressure from the system

4. Drain the cooling system into a clean container for reuse.

5. Label and disconnect or clamp off the heater hoses from (at) the heater core.

6. Place a container under the right heater core tube, induce pressurized air to the left tube and blow compressed into the tube to drain excess coolant.

7. Remove or disconnect the following:
- Air conditioning system vacuum supply hose and attach it at the heater core inlet/outlet
- Heater core-to-chassis boot
- Refrigerant lines-to-evaporator core clamp bolt, disconnect the refrigerant line clamp and discard the O-rings. Plug the openings to prevent contamination.
- Evaporator core-to-chassis boot
- Low pressure switch electrical connector and secure it to the evaporator fixture

8. Remove the glove box, the driver's side lower shelf and the instruments panel center section by removing or disconnecting the following:

- 5 glove box-to-instrument panel bolts and the glove box
- 3 clips, remove the 3 stowage compartment bolts and the stowage compartment at the driver's side
- Rear console
- 3 the knobs, remove the 4 center instrument panel trim bolts, the 2 screws and the trim
- Radio
- 4 front console-to-instrument panel bolts
- Trim cover and the console-to-chassis nut at the left side of the front console
- Pedal assembly at the driver's side

9. Place the front wheels in the straight-ahead position.

10. Remove the driver's side air bag module by removing or disconnecting the following:
- Air bag module-to-steering wheel screws using a T30 Torx bit; the screws are located on both sides of the steering wheel
- Air bag module and disconnect the electrical connector
- Place the air bag module in a safe place with the front facing upward.

11. Remove the steering wheel by removing or disconnecting the following:
- Steering wheel nut
- Horn and air bag electrical connectors
- 4 carrier unit-to-steering wheel bolts and the carrier unit
- Steering wheel

12. At the steering column-to-instrument panel connection, secure the steering column wire to keep the steering column from sliding apart.

13. Remove or disconnect the following:
- Steering column-to-steering gear bolt
- Electronic box electrical connectors located in the left air plenum chamber
- All instrument panel-to-chassis electrical connectors

14. Remove the instrument panel by removing or disconnecting the following:
- Instrument panel-to-chassis bolts
- Any necessary electrical connectors
- Instrument panel

15. Remove or disconnect the following:
- Ducts, heater/air conditioning housing assembly-to-chassis bolts and the assembly
- Heater core-to-heater/air conditioning housing screws
- Press the heater core-to-heater housing catches and remove the

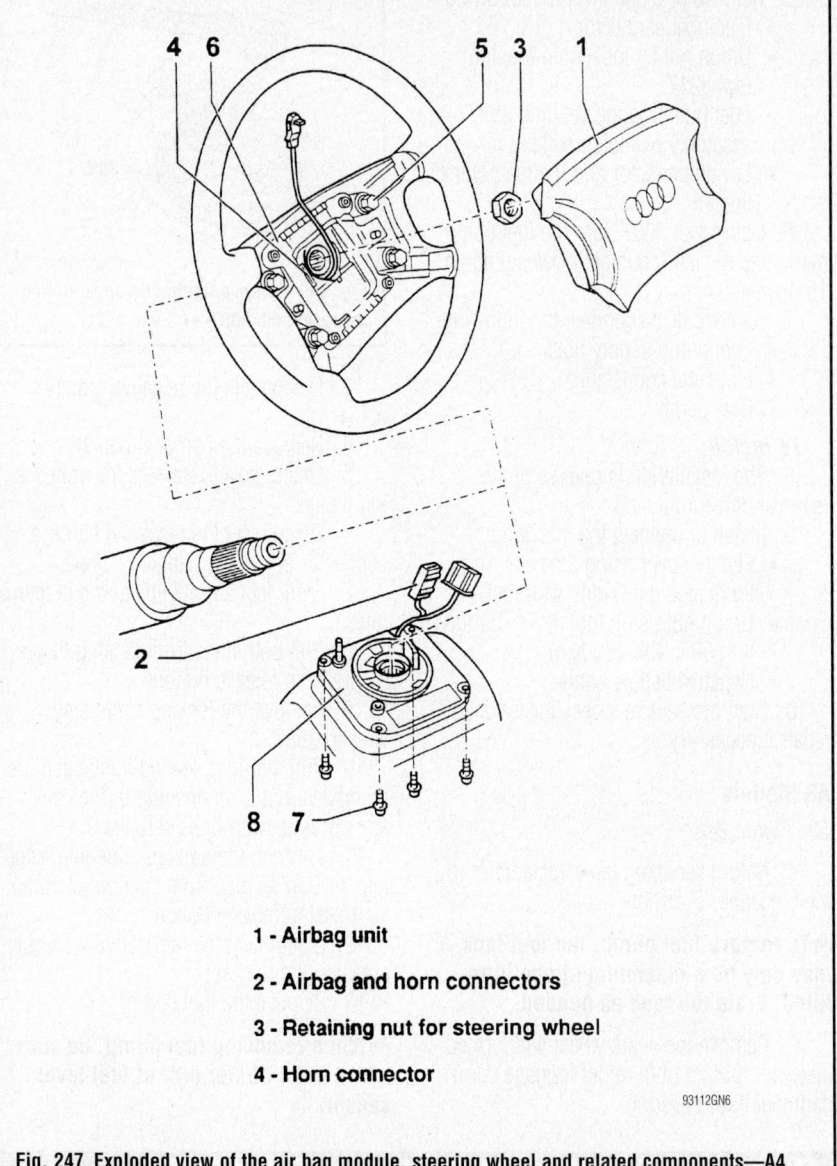

1 - Airbag unit

2 - Airbag and horn connectors

3 - Retaining nut for steering wheel

4 - Horn connector

93112GN6

**Fig. 247 Exploded view of the air bag module, steering wheel and related components—A4**

heater core from the heater/air conditioning housing

**To install:**

16. Install or connect the following:
- Heater core to the heater/air conditioning housing and press the heater core into heater housing until the latches catch
- Heater core-to-heater/air conditioning housing screws
- Heater/air conditioning housing assembly and connect the ducts

17. Install the instrument panel by installing or connecting the following:
- Instrument panel
- Any necessary electrical connectors
- Instrument panel-to-chassis bolts and torque the bolts to 44 inch lbs. (5 Nm)

- All instrument panel-to-chassis electrical connectors
- Electronic box electrical connectors located in the left air plenum chamber
- Steering column-to-steering gear bolt

18. Install the steering wheel by installing or connecting the following:
- Steering wheel
- Carrier unit and the 4 carrier unit-to-steering wheel bolts
- Horn and air bag electrical connectors
- Steering wheel nut

19. Install the driver's side air bag module by installing or connecting the following:
- Air bag module and connect the electrical connector

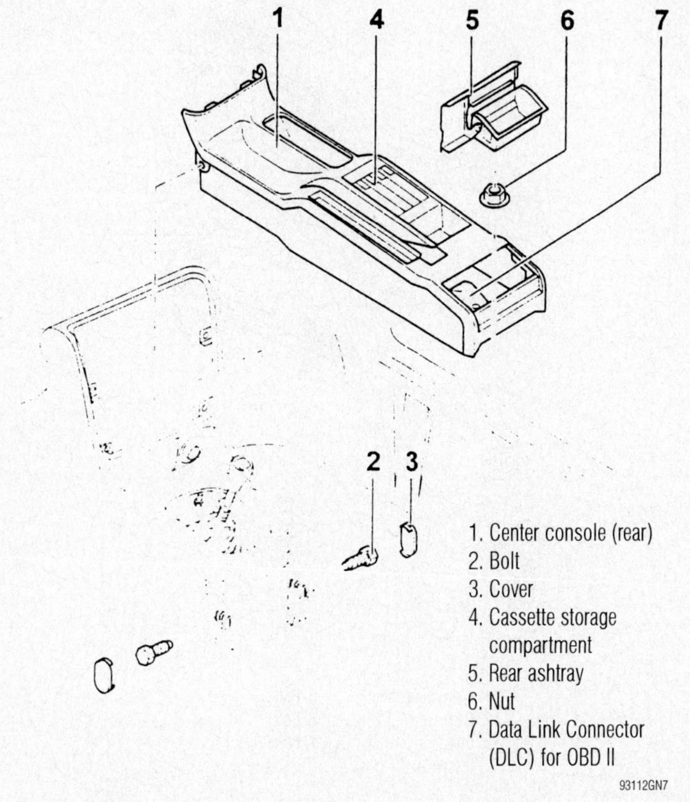
1
4
5
6
7
2
3
1. Center console (rear)
2. Bolt
3. Cover
4. Cassette storage compartment
5. Rear ashtray
6. Nut
7. Data Link Connector (DLC) for OBD II
93112GN7

Fig. 248 Exploded view of the rear console assembly—A4

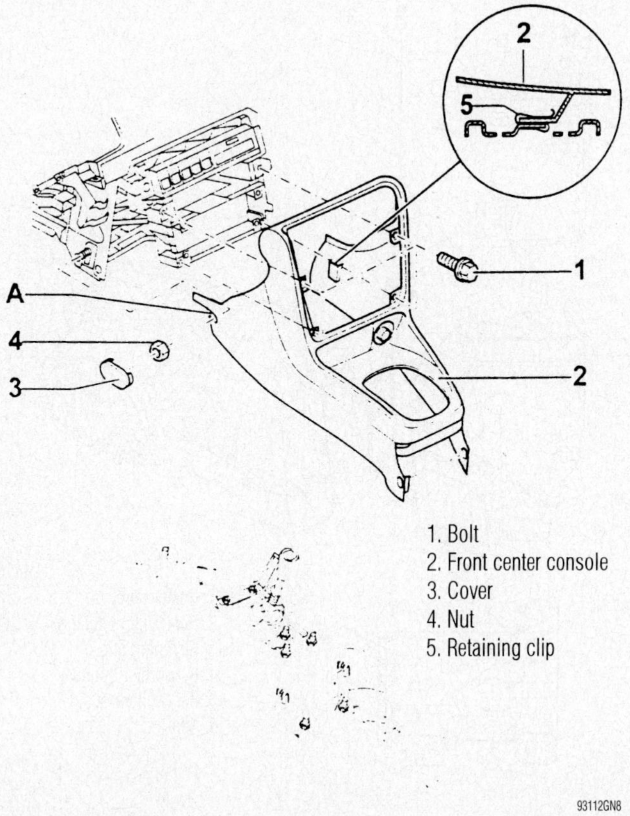
2
5
1
A
4
3
2
1. Bolt
2. Front center console
3. Cover
4. Nut
5. Retaining clip
93112GN8

Fig. 249 Exploded view of the front console assembly—A4

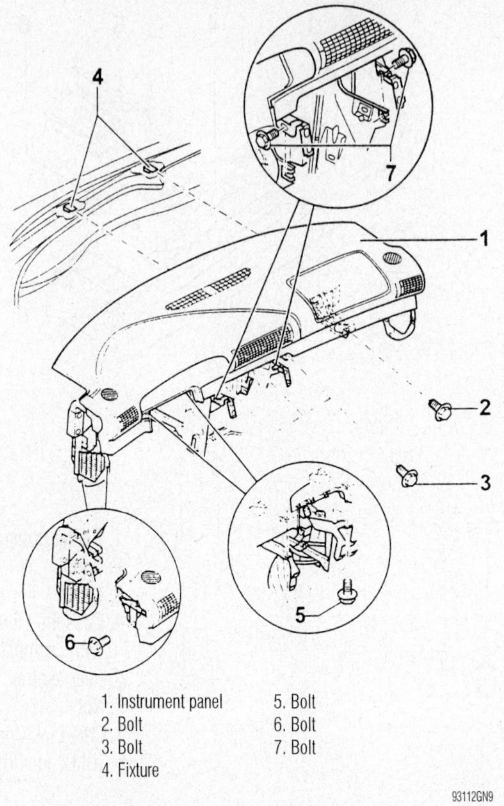
4
7
1
2
3
5
6
1. Instrument panel
2. Bolt
3. Bolt
4. Fixture
5. Bolt
6. Bolt
7. Bolt
93112GN9

**Fig. 250 Exploded view of the instrument panel assembly—A4**

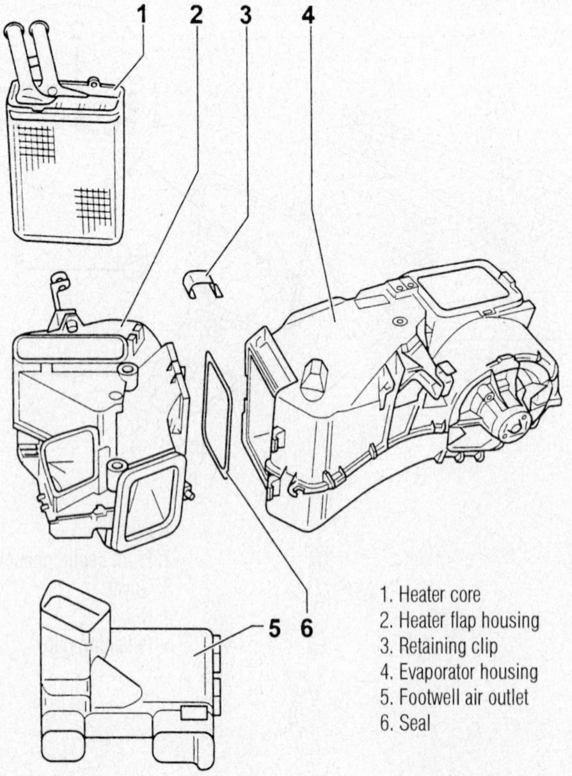
1
2
3
4
5
6
1. Heater core
2. Heater flap housing
3. Retaining clip
4. Evaporator housing
5. Footwell air outlet
6. Seal
93112GN1

**Fig. 251 Exploded view of the heater core, heater/air conditioning housing and related components—A4**

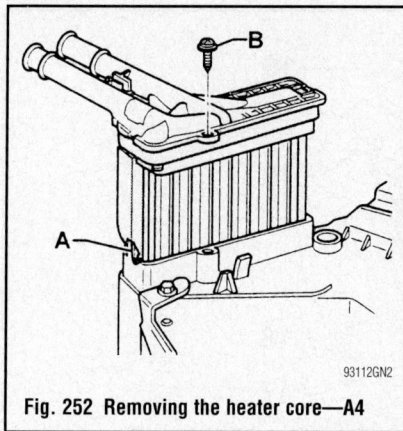

**Fig. 252 Removing the heater core—A4**

93112GN2

- Torque the air bag module-to-steering wheel screws to 53 inch lbs. (6 Nm) using a T30 Torx bit; the screws are located on both sides of the steering wheel
- Pedal assembly at the driver's side

20. Install the glove box, the driver's side lower shelf and the instruments panel center section by installing or connecting the following:
  - Console-to-chassis nut and the trim cover at the left side of the front console. Torque the nut to 31 inch lbs. (3.5 Nm)
  - 4 front console-to-instrument panel bolts and torque the bolts to 44 inch lbs. (5 Nm)
  - Radio
  - 4 center instrument panel trim, the 4 center instrument panel trim bolts, the 2 screws and the 3 the knobs. Torque the bolts to 44 inch lbs. (5 Nm) and the screws to 31 inch lbs. (3.5 Nm)
  - Rear console
  - Stowage compartment, the 3 stowage compartment bolts and engage the 3 clips. Torque the bolts to 44 inch lbs. (5 Nm)
  - Glove box and the 5 glove box-to-instrument panel bolts; then, torque the bolts to 44 inch lbs. (5 Nm)
21. Install or connect the following:
  - Low pressure switch electrical connector
  - Evaporator core-to-chassis boot.
  - Refrigerant lines-to-evaporator core clamp and connect the refrigerant line clamp bolt making sure to use new O-rings
  - Heater core-to-chassis boot
  - Air conditioning system vacuum supply hose
  - Heater hoses to the heater core
22. Refill the cooling system.
23. Install the coolant recovery tank cap.

24. Install the battery and connect the positive (+) battery cable.
25. Connect the negative (−) battery cable.
26. At the right side, install the dust/pollen filter, the water guide and the air plenum cover.
27. Evacuate, charge and leak test the air conditioning system.
28. Operate the engine to normal operating temperature; then, check the climate control operation and check for leaks.

### A6 Models

➡If equipped with power seats, move the seats as far rearward as possible. Also, acquire the anti-theft radio coding from the owner.

1. Remove or disconnect the following:
  - Negative battery cable
  - Positive battery cable and remove the battery
  - Coolant recovery tank cap to relieve the pressure from the system
  - Coolant
  - Heater hoses from (at) the heater core
  - Heater core tubes-to-chassis grommet

➡If desired, instead of draining cooling system, disconnect heater hoses from tubes in engine compartment and apply compressed air to upper tube and catch coolant from heater core coming out of bottom tube.

2. Remove the storage compartment on driver's side.
3. From inside passenger compartment, remove bolts and brackets from heater core tubes.
4. Slightly pull the heater core from the A/C-heater housing.
5. Pull two coolant lines out of the heater core.
6. Remove the heater core.
7. If the heater core cannot be pulled out of the A/C housing, loosen the instrument panel cross member by removing or disconnecting the following:
  - Bolt at base of windshield (above meter cluster)
  - Loosen RH panel cross member bolts
  - Two bolts from bottom of instrument panel brace
  - ECM cover
  - Pedal bracket from cross member
  - LH instrument panel cover (push in steering column as far as possible)
  - Loosen 2 LH instrument panel nuts

8. Pull back instrument panel and cross member enough to allow heater core to be removed.

### To install:

9. Installation is the reverse of the removal procedure, noting the following:
  - Install new clamps and O-rings at all connections
  - Before installing storage compartment on driver's side, check for cooling system leaks
  - Ensure WOT stop of accelerator pedal is incorrect position

### A8 Models

#### Passenger's Side

See Figures 253 and 254.

1. Disconnect the negative battery cable.
2. Turn the ignition switch OFF.
3. Remove the cowl panel.
4. Drain the cooling system into the clean container for reuse.
5. Remove or disconnect the following:
  - Engine-to-pump valve unit heater hoses. Drain and plug the openings
  - Electrical connectors from the pump valve unit
6. Remove the reinforcement plate (plenum) by removing or disconnecting the following:
  - Intake hose from the air filter
  - Sound proofing mat at the front wall of the plenum
  - Reinforcement plate (plenum)-to-chassis nuts/bolts and the plenum
  - In the plenum, loosen the heater hose holder screw about 2 turns
7. Remove the knee bar by removing or disconnecting the following:
  - Open the glove box
  - 2 knee bar trim screws
  - 4 knee bar screws
  - Pull out the knee bar and disengage the latch
  - Electrical connector and remove the knee bar
8. Remove or disconnect the following:
  - Left support bolts and the retainer bolts; then, remove the left support and the retainer
  - Center console side trim.
  - Footwell air outlet
9. Place an absorbent cover on the floor of the car, under the heater core to catch any spilt coolant.
  - Passenger's side heater hose clamps, move the heater hoses toward the plenum and discard the O-rings

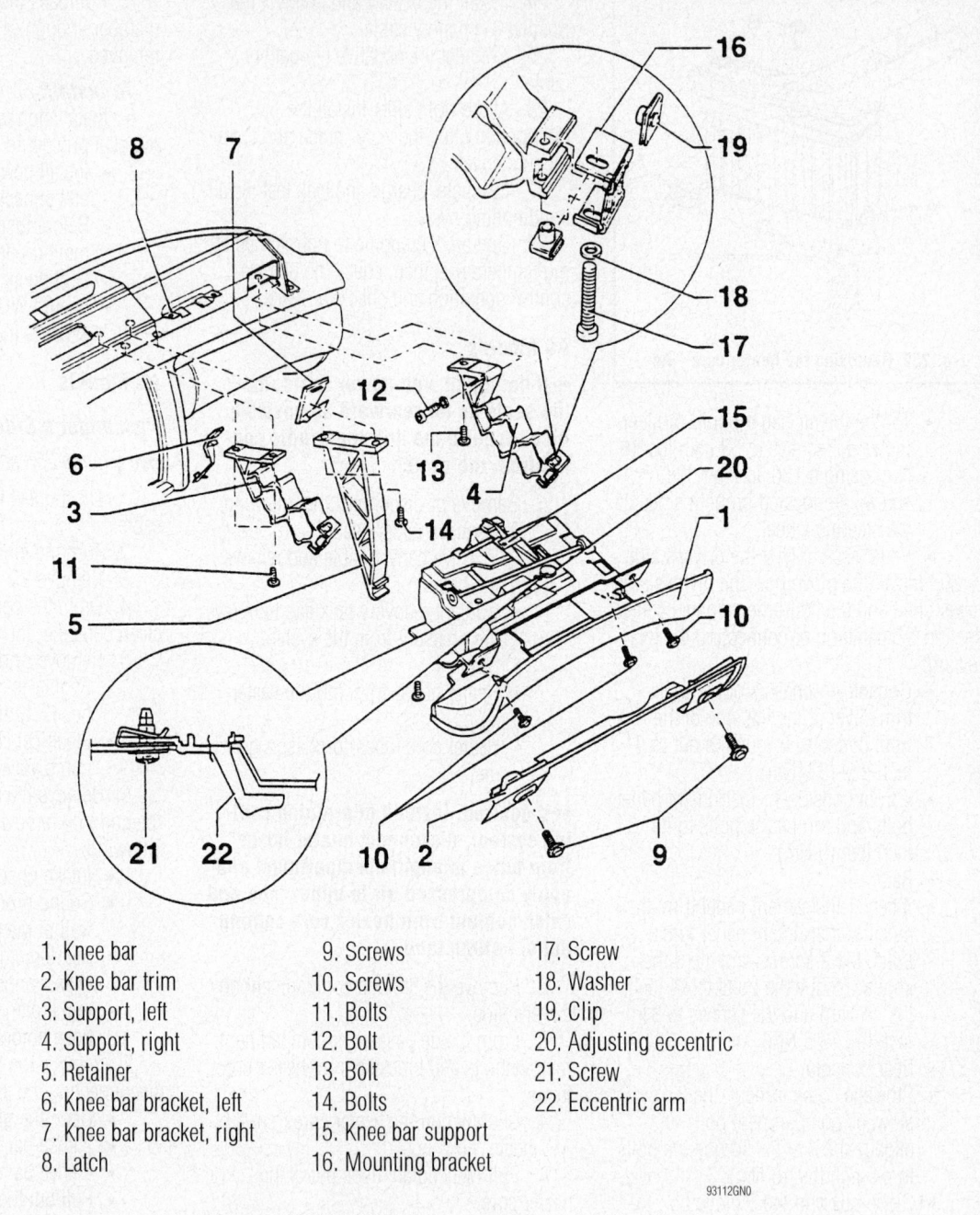

1. Knee bar
2. Knee bar trim
3. Support, left
4. Support, right
5. Retainer
6. Knee bar bracket, left
7. Knee bar bracket, right
8. Latch
9. Screws
10. Screws
11. Bolts
12. Bolt
13. Bolt
14. Bolts
15. Knee bar support
16. Mounting bracket
17. Screw
18. Washer
19. Clip
20. Adjusting eccentric
21. Screw
22. Eccentric arm

93112GN0

**Fig. 253 Exploded view of the glove box, knee bar assembly and related components—A8**

- Heater core-to-heater case clamp screws and the clamp
- Heater core

### To install:

10. Install or connect the following:
- Heater core
- Heater core-to-heater case clamp and the clamp screws
- Heater hoses to the plenum (using O-rings) and install the passenger's side heater hose clamps
- Footwell air outlet
- Center console side trim

11. Install the knee bar by installing or connecting the following:
- Left support and the retainer and the left support bolts and the retainer bolts; then, torque the bolts to 16 ft. lbs. (22 Nm)
- Electrical connector and install the knee bar
- Knee bar and engage the latch
- 4 knee bar screws and torque to 22 inch lbs. (2.5 Nm)
- 2 knee bar trim screws and torque to 22 inch lbs. (2.5 Nm)

- Tighten the heater hose holder screw in the plenum

12. Install the reinforcement plate (plenum) by installing or connecting the following:
- Reinforcement plate (plenum) and the plenum-to-chassis nuts/bolts
- Sound proofing mat at the front wall of the plenum
- Intake hose to the air filter
- Electrical connectors to the pump valve unit

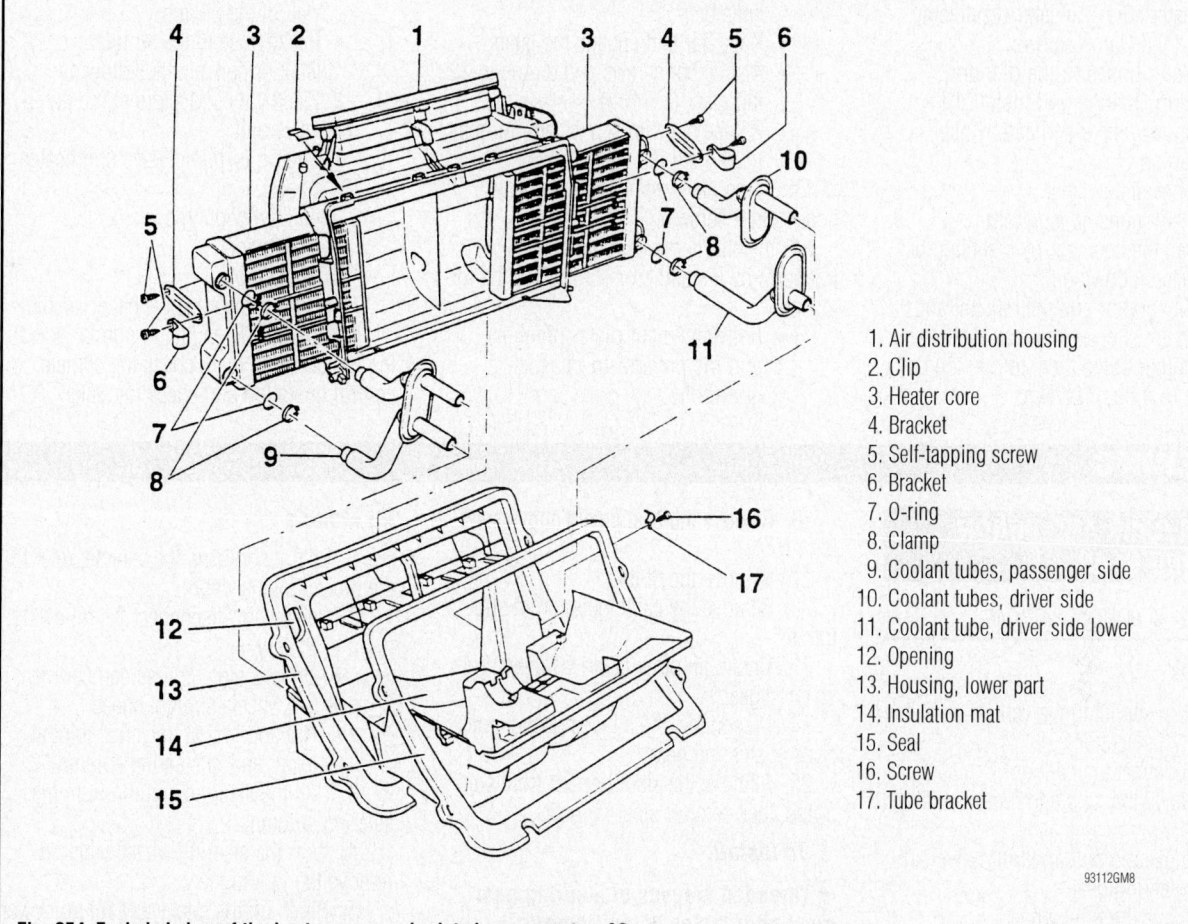

1. Air distribution housing
2. Clip
3. Heater core
4. Bracket
5. Self-tapping screw
6. Bracket
7. O-ring
8. Clamp
9. Coolant tubes, passenger side
10. Coolant tubes, driver side
11. Coolant tube, driver side lower
12. Opening
13. Housing, lower part
14. Insulation mat
15. Seal
16. Screw
17. Tube bracket

93112GM8

**Fig. 254 Exploded view of the heater cores and related components—A8**

- Engine-to-pump valve unit heater hoses
13. Refill the cooling system.
14. Install the cowl panel.
15. Connect the negative battery cable.

### Driver's Side

1. Disconnect the negative battery cable.
2. Turn the ignition switch OFF.
3. Remove the windshield.
4. Remove the cowl panel.
5. Drain the cooling system into the clean container for reuse.
6. Remove or disconnect the following:
   - Engine-to-pump valve unit heater hoses. Drain and plug the openings
   - Electrical connectors from the pump valve unit
7. Remove the reinforcement plate (plenum) by removing or disconnecting the following:
   - Remove the intake hose from the air filter.
   - At the front wall of the plenum, remove the sound proofing mat.
   - Remove the reinforcement plate

(plenum)-to-chassis nuts/bolts and the plenum
8. Remove or disconnect the following:
   - Loosen the heater hose holder screw in the plenum about 2 turns
9. Remove the knee bar by removing or disconnecting the following:
   - Open the glove box
   - 2 knee bar trim screws
   - 4 knee bar screws
   - Pull out the knee bar and disengage the latch
   - Electrical connector and remove the knee bar
10. Remove or disconnect the following:
    - Left support bolts and the retainer bolts; then, remove the left support and the retainer
    - Center console side trim
    - Passenger's footwell air outlet
    - Passenger's side heater hose clamps, move the heater hoses toward the plenum and discard the O-rings
    - Passenger's side heater core-to-heater case clamp screws and the clamp

- Passengers side heater core
- Driver's side shelf and center console's side trim
- Loosen the heater hose holder screws about 2 turns at the driver's side heater core
- Driver's side footwell air outlet
- Driver's side heater hose clamps, move the heater hoses toward the plenum and discard the O-rings
- Driver's side heater core through the passenger's side of the heater housing

### To install:

11. Install or connect the following:
    - Driver's side heater core through the passenger's side of the heater housing
    - Move the heater hoses away from the plenum and install the driver's side heater hose clamps using new O-rings
    - Driver's side footwell air outlet
    - Heater hose holder screws at the driver's side heater core
    - Driver's side shelf and center console's side trim

- Passenger's side heater core
- Heater core-to-heater case clamp and the clamp screws
- Heater hoses to the plenum using O-rings and install the passenger's side heater hose clamps
- Footwell air outlet
- Center console side trim

12. Install the knee bar by installing or connecting the following:

- Left support and the retainer and the left support bolts and the retainer bolts; then, torque the bolts to 16 ft. lbs. (22 Nm)

- Electrical connector and install the knee bar
- Knee bar and engage the latch
- 4 knee bar screws and torque to 22 inch lbs. (2.5 Nm)
- 2 knee bar trim screws and torque to 22 inch lbs. (2.5 Nm)

13. In the plenum, tighten the heater hose holder screw.

14. Install the reinforcement plate (plenum) by installing or connecting the following:

- Reinforcement plate (plenum) and the plenum-to-chassis nuts/bolts

- Sound proofing mat at the front wall of the plenum
- Intake hose to the air filter.

15. Install or connect the following:

- Electrical connectors to the pump valve unit
- Engine-to-pump valve unit heater hoses

16. Refill the cooling system.

- Cowl panel
- Windshield

17. Connect the negative battery cable.

18. Operate the engine to normal operating temperature; then, check the climate control operation and check for leaks.

# STEERING

## POWER RACK AND PINION STEERING GEAR

### REMOVAL & INSTALLATION

### A3 Models

1. Before servicing the vehicle, refer to the precautions section.

2. Disconnect the battery.

3. Remove the nuts and remove foot well trim.

4. Remove the bolt and pull universal joint from steering gear.

5. Remove the front wheels.

6. Loosen the nut from tie rod end but do not remove completely.

7. Press off tie rod end from wheel bearing housing using Ball Joint Puller 3287A.

8. Remove the sound insulation at bottom.

9. Remove the coupling rod from stabilizer bar.

10. Remove the bolts.

11. Remove the bolts and remove pendulum support from transmission.

12. Remove the bracket for exhaust system from subframe.

13. Remove the bolts for heat shield.

14. Remove the heat shield from subframe.

15. Loosen the bolts for steering gear and stabilizer bar.

16. Secure sub frame and consoles.

17. Position engine/transmission assembly jack under subframe.

18. Place wooden block or similar between engine/transmission assembly jack and subframe.

19. Loosen the bolts and lower sub frame and consoles slightly. Make sure all electrical connections are out of the way.

20. Remove the heat shield above steering gear.

21. Remove the bolts.

22. Remove the cable guide from subframe.

23. Unclip all other cable fasteners from steering gear.

24. Disconnect the all electrical connections at steering gear.

25. Remove the stabilizer bar from subframe, then remove steering gear.

### To install:

➡ **Threaded sleeves of steering gear must sit in holes of subframe.**

➡ **Coat seal on steering gear with lubricant, e.g. lubricating soap, before installing.**

26. After attaching steering gear to drive axle, make sure that seal on steering gear makes contact with mounting plate without kinking and seals opening to foot well properly. Otherwise, this may result in water leaks and/or noise.

27. Before positioning bolts for subframe, position steering gear on subframe and fasten bolts for steering gear and stabilizer bar.

28. Bolt on heat shields with steering gear.

29. Bolt on sub frame and consoles with body.

30. Bolt on steering gear with sub frame.

31. Bolt on ball joint with control arm. Always use new bolts.

➡ **Make sure that boot is not damaged or twisted.**

32. Bolt on universal joint with steering gear.

33. After installation, position of steering wheel must be checked with a road test.

### A6 Models

1. Before servicing the vehicle, refer to the precautions section.

2. Remove or disconnect the following:

- Battery
- Driver's side storage compartment
- Set wheels straight ahead
- Nut and bolt at universal joint at lower end of steering column

3. Pull the steering column from the steering column.

4. With the steering wheel centered, remove the ignition key.

5. Pinch off the intake and return power steering hoses.

6. Remove the front wheels.

7. Remove the tie rod end from the steering knuckle connection.

8. Remove the steering column opening cover from the dash panel.

9. If necessary, remove the left exhaust pipe.

10. Drain the power steering fluid by removing the banjo bolts from the steering gear.

11. Remove the steering gear mounting bolts from the cross member and remove the steering gear through the left wheel housing.

### To install:

12. Steering gear into the vehicle. Temporarily install mounting bolts from above, but do not tighten.

13. Install the single mounting bolt from below and tighten.

14. Install the return hose and tighten the banjo bolt to 37 ft. lbs. (50 Nm).

15. Screw of the pressure hose and tighten the banjo fitting to 30 ft. lbs. (40 Nm).

16. Install the ignition key and release the steering lock.

17. Center the steering wheel, then install the universal joint to the steering gear pinion.

18. Install the eccentric bolt and tighten in a counterclockwise direction.

19. Install and tighten a new self-locking nut on the eccentric bolt. Torque the nut to 22 ft. lbs. (30 Nm).

20. Install the driver's side storage compartment.

21. Install the dash panel cover.

22. Fully tighten steering gear mounting bolts in cross member.

23. Remove the power steering hose pinchers.

24. Install the front wheels and battery.

25. Refill the power steering fluid.

26. Start the engine, cycle the steering wheel from lock to lock. Turn the engine off and check and top up the power steering fluid.

## A4 Models

*See Figure 256.*

1. Before servicing the vehicle, refer to the precautions section.

2. Remove or disconnect the following:
- Battery
- Battery box
- Steering column U-joint bolt

3. Release the eccentric by turning the Torx® T50 bolt clockwise, then remove the bolt.

➡**Before removing the steering column from the steering gear, secure the steering column with safety wire.**

### ✳✳ WARNING

**Be sure to lock the steering wheel, otherwise the air bag unit coil spring may be damaged.**

4. Lock the steering wheel in the center position and do not move during the repairs.

➡**The splines between the top and bottom part of the steering column must not be separated.**

5. Move the U-joint down and out of the way.

6. Using hose clamps tool 3094, pinch off the suction and return lines to the steering gear.

7. Remove or disconnect the following:
- Front wheels
- Left and right tie rods
- Tie rod opening cover

➡**Place a drip tray under the vehicle to catch any residual power steering fluid.**

- Banjo bolts for the steering gear suction and return hydraulic hoses

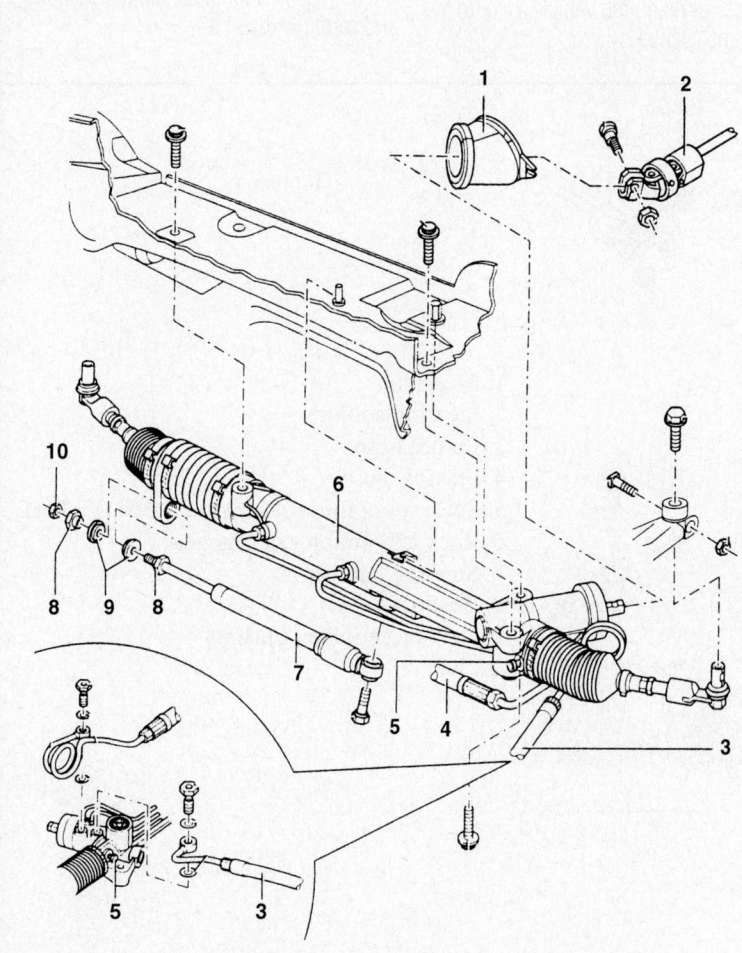

1. Boot seal
2. Steering column
3. Return hose
4. Flexible hose
5. Screw plug for centering the steering wheel
6. Rack and pinion steering gear
7. Steering damper
8. Bushing
9. Two-piece rubber bushing
10. Nut

7923CG33

**Fig. 256 Exploded view of the steering gear mounting—A4 models**

- Steering gear mounting bolts
- Steering gear through the left side wheel opening

### To install:

8. Remove the screw plug to lock the steering gear in the center position with locking tool VAG 1907 and torque to 13 ft. lbs. (18 Nm)

9. Install or connect the following:
- Steering gear through the left side wheel opening. Torque bolt No. 3 to 48 ft. lbs. (65 Nm) and bolts No. 1 and 2 to 48 ft. lbs. (65 Nm).
- Power steering gear hoses using new sealing gaskets. Torque the return hose banjo bolt to 37 ft. lbs. (50 Nm) and the suction hose banjo bolt to 30 ft. lbs. (40 Nm).
- Left and right tie rods. Torque the bolts to 33 ft. lbs. (45 Nm).
- Tie rod opening cover
- U-joint to the steering gear and the Torx® adjusting bolt by turning it clockwise

10. Remove the locking tool VAG 1907.

11. Install or connect the following:
- Screw plug. Torque it to 13 ft. lbs. (18 Nm)
- Adjusting bolt. Torque the nut to 30 ft. lbs. (40 Nm)

12. Remove the steering wheel lock

13. Remove the Hose Clamp tools 3094 and check the hydraulic fluid

14. Install or connect the following:
- Battery tray
- Battery

> ❋❋ **WARNING**
>
> **If the hydraulic fluid requires being topped of, use only an approved fluid, otherwise internal damage may occur.**

15. Start the vehicle and check for leaks.

16. Check and/or adjust the wheel alignment.

### TT Models

*See Figure 257.*

1. Before servicing the vehicle, refer to the precautions section.

2. Secure the front wheel in a straight ahead position.

3. Remove or disconnect the following:
- Cover in back of brake pedal
- Steering column U-joint bolt
- Engine undercover
- Hydraulic hose bracket on left sub-frame
- Servo reservoir suction hose
- Servo reservoir return line
- Front wheels
- Left and right tie rods

➡ **Place a drip tray under the vehicle to catch any residual power steering fluid.**

- Banjo bolts for the steering gear suction and return hydraulic hoses
- Sway brace support on transaxle side

4. Place a suitable jack under steering gear subframe.

5. Remove the bolts and screws from subframe.

6. Pry off subframe using a tire iron and carefully lower it.

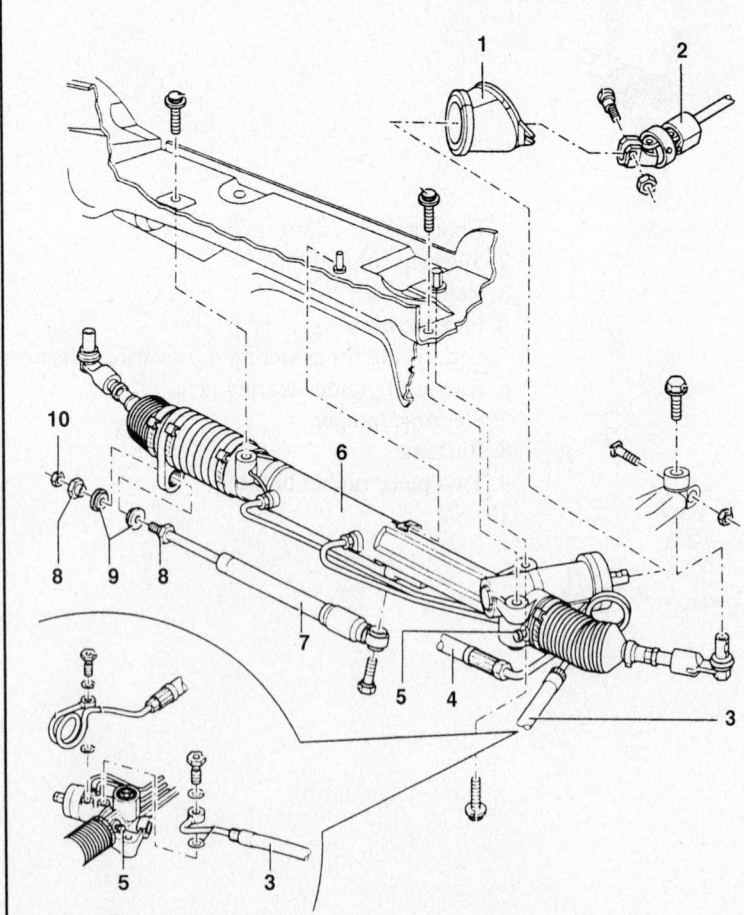

1. Boot seal
2. Steering column
3. Return hose
4. Flexible hose
5. Screw plug for centering the steering wheel
6. Rack and pinion steering gear
7. Steering damper
8. Bushing
9. Two-piece rubber bushing
10. Nut

7923CG33

**Fig. 257 Exploded view of the steering gear mounting—A4 models**

7. Unbolt steering gear return line from clamp and rotary slide valve housing.
  • Steering gear mounting bolts
  • Steering gear through the left side wheel opening

**To install:**

8. Remove the screw plug to lock the steering gear in the center position with locking tool VAG 1907 and torque to 13 ft. lbs. (18 Nm)

9. Install or connect the following:
  • Steering gear with guide sleeve into subframe and hand-tighten the bolts.
  • Return line to clamp on steering gear
  • Return line with new gaskets to rotary slide valve housing

10. Check sealing cuff at the steering pinion for proper seating.

11. Raise the subframe and insert steering pinion into hole on underside of vehicle. Bolt on subframe with old bolts. Torque subframe bolts to: 74 ft. lbs. (100 Nm) plus90°. Torque steering gear bolts to 15 ft. lbs. (20 Nm).

➡**Replace old bolts of subframe when aligning the vehicle.**

12. Install or connect the following:
  • Expansion hose with new gaskets to rotary slide valve housing
  • Sway brace support and exhaust system bracket. Torque sway brace side bolts: 15 ft. lbs. (20 Nm) plus 90°, sway brace bottom bolts: 30 ft. lbs. (40 Nm) plus 90°
  • Tie rods in steering knuckle and bolt on. If joint bolt also turns when tightening, counterhold with

internal Torx screw (T40). Torque the bolts to 33 ft. lbs. (45 Nm).
  • Return line to servo fluid reservoir
  • Engine undercover.

13. Slide universal joint onto steering pinion and secure with new bolts. Torque the bolts to: 22 ft. lbs. (30 Nm).

14. Install cover behind brake pedal.

**⁕⁂ WARNING**

**If the hydraulic fluid requires being topped of, use only an approved fluid, otherwise internal damage may occur.**

15. Start the vehicle and check for leaks.

16. Check and/or adjust the wheel alignment.

---

## SUSPENSION                                    FRONT SUSPENSION

### COIL SPRING

*REMOVAL & INSTALLATION*

*See Figures 258 and 259.*

1. Before servicing the vehicle, refer to the precautions section.

2. Remove the strut from the vehicle.

3. Clamp the spring compressor tool VAG 1752/2, in a vise.

4. Install the strut into the spring compressor.

5. Pry off the mounting bolt cap.

6. Compress the coil spring and remove the self-locking nut from the piston rod.

7. Matchmark the position of the spring retainer and spring mount.

8. Remove or disconnect the following:
  • Spring seat and related components
  • Strut from the spring compressor

9. Release the tension on the coil spring.

10. Remove the spring out of the compressor.

**To install:**

11. Install the spring into the compressor.

12. Compress the spring and insert the strut through the spring.

13. Install or connect the following:
  • Spring seat and related components in the reverse order as they were removed by aligning the matchmarks
  • New self-locking nut
  • Mounting bolt cap

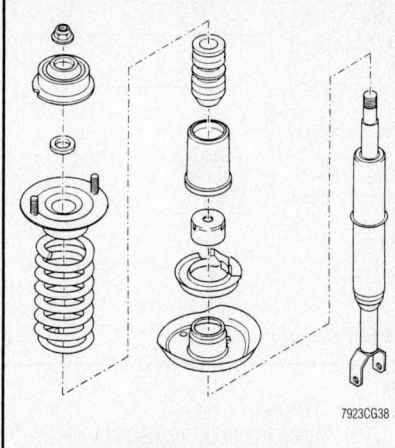

**Fig. 258 Exploded view of the front strut—A4 model**

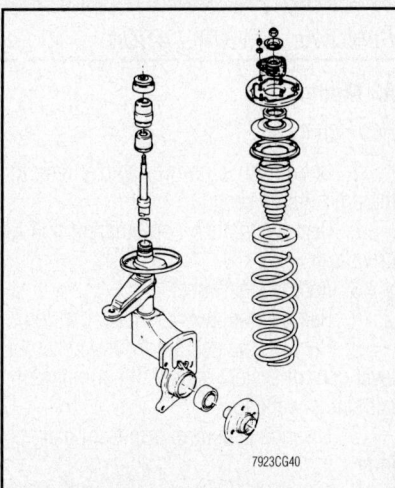

**Fig. 259 Exploded view of the front strut—except A4 model**

14. Release the spring compressor.

15. Install the strut into the vehicle.

**A4 Models**

*See Figure 260.*

The Audi A4 front suspension is equipped with 2 separate upper ball joints that are not replaceable, therefore the upper link (front or rear) must be replaced. To remove this link, perform the following procedures.

1. Before servicing the vehicle, refer to the precautions section.

2. Remove or disconnect the following:
  • Front wheels
  • Pinch bolt and pull both control arms upward and out

3. Cover the steering gear boot.

4. Remove or disconnect the following:
  • Guide link ball joint and press off the joint
  • Anti-lock Brake System (ABS) wheel speed sensor wire from the brake caliper bracket

5. Support the suspension from excessive rebound travel.

6. Remove or disconnect the following:
  • Lower strut bolt and swing the wheel bearing housing aside
  • Rubber grommets from the plenum chamber
  • Upper strut-to-body nuts
  • Strut together with the mounting bracket

7. Clamp the strut in a vise with the protective jaw covers.

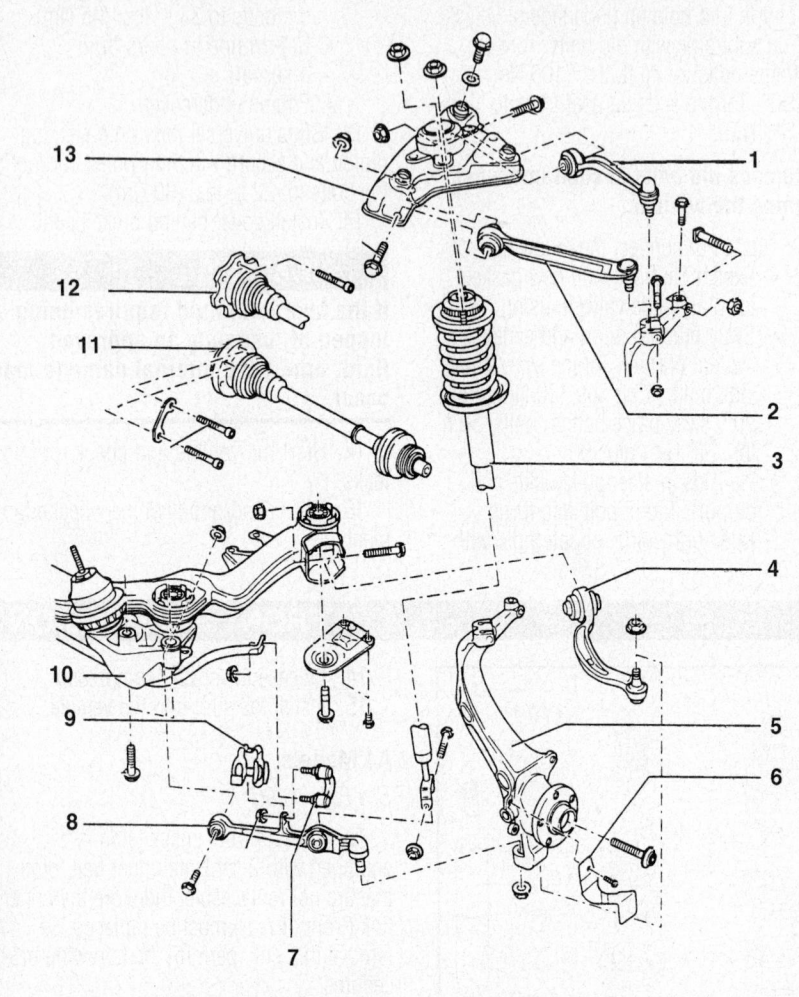

1. Upper link, rear
2. Upper link, front
3. Suspension strut
4. Guide link
5. Wheel bearing housing
6. Splash shield
7. Connecting link
8. Lower track control link
9. Clamp
10. Subframe
11. Halfshaft w/CV joint
12. Halfshaft w/triple-rotor joint
13. Mounting bracket

7923CG41

**Fig. 260 Exploded view of the front suspension—A4 and A6 models**

8. Remove or disconnect the following:
• Upper link bolts and detach both of the links
• Bracket-to-strut nuts, then separate

**To install:**
9. Install or connect the following:
• Brackets and links, as shown. Torque the bracket-to-strut mounting nuts to 15 ft. lbs. (20 Nm)
• Links by aligning them, as shown. Torque to 37 ft. lbs. (50 Nm) plus a ¼ (90 degree) turn
• Strut with mounting bracket. Torque the upper strut-to-body nuts to 48 ft. lbs. (75 Nm)
• Lower strut bolt. Torque it to 66 ft. lbs. (90 Nm)
• Nut on the ball joint. Torque to 74 ft. lbs. (100 Nm)
• Upper links to the wheel bearing housing. Torque the pinch bolt to 30 ft. lbs. (40 Nm)
• ABS wiring to the brake caliper bracket

• Wheels
10. Check the front suspension alignment.

### LOWER BALL JOINT

*REMOVAL & INSTALLATION*

#### A3 Models

*See Figure 261.*

1. Before servicing the vehicle, refer to the precautions section.
2. Remove the hex-combination bolt for drive axle
3. Remove the wheel.
4. Remove the hex nuts from ball joint.
5. Remove the hex nut from bracket for level control system sensor, if equipped on left side of vehicle.
6. Disengage control arm from ball joint.
7. Press constant velocity joint out of wheel bearing unit using a brass drift.

8. Swing suspension strut outward, at the same time guide drive axle out of wheel bearing.

➡**Drive axle must not hang down, otherwise inner joint will be damaged by over bending. Tie up drive axle at body using wire.**

9. Position ball joint puller 3287 A at ball joint as shown in the accompanying illustration.
10. Press ball joint out of wheel bearing housing.

**To install:**
11. Installation is the reverse order of removal, please note the following:
a. Install the new self-locking nut, while holding with internal Torx T40. Tighten the ball joint nuts to 75 Nm.

➡**Instead of insertion tool V.A.G 1332/10 other commercially available 18mm ring-insertion tools may be used.**

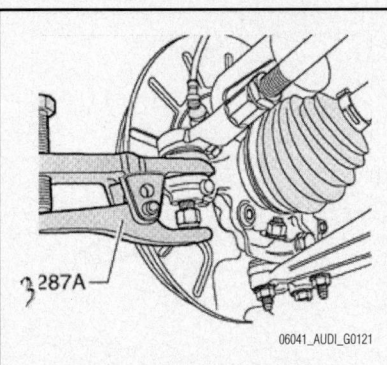

**Fig. 261 Press ball joint out of wheel bearing housing using puller 3287—A3**

b. Guide drive axle into wheel hub splines, tighten the bolt to 200 Nm plus 180 degrees.

### A4 and A6 Models

➡The A4 and A6 models are equipped with 2 lower ball joints that are not serviceable. The control arms must be replaced if a joint is worn. The lower track control link ball joint stud faces down, and the guide link ball joint stud faces up.

#### Lower Track Control Link

1. Before servicing the vehicle, refer to the precautions section.
2. Remove or disconnect the following:
   • Front wheels
   • Nut from the lower track control link
3. Press the ball joint out of the tapered seat.
4. Support the wheel bearing housing to prevent excessive rebound travel in the suspension.
5. Remove or disconnect the following:
   • Stabilizer link and lower strut mounting bolt
   • Lower track control link-to-subframe attaching bolt
   • Lower track control link

#### To install:

6. Install or connect the following:
   • Lower track control link
   • Subframe attaching bolt. Torque the bolt to 74 ft. lbs. (100 Nm)
7. Install or connect the following:
   • Stabilizer link. Torque the upper bolt to 30 ft. lbs. (40 Nm) plus ¼ (90 degree) turn and the lower bolt to 74 ft. lbs. (100 Nm)
8. Load the suspension and torque the subframe bolt to 59 ft. lbs. (80 Nm) plus ¼ (90 degree) turn
9. Front wheels

10. Check and/or adjust the front suspension alignment.

#### Lower Guide Link

*See Figure 262.*

1. Before servicing the vehicle, refer to the precautions section.
2. Remove or disconnect the following:
   • Front wheels
   • Nut from the lower guide link joint and press the joint from the wheel bearing housing
3. Loosen lower guide link-to-subframe attaching bolt

➡**The subframe must be lowered at the rear to remove the lower guide link-to-subframe attaching bolt.**

4. Loosen the rear subframe support plate bolts and subframe bolts.
5. Remove or disconnect the following:
   • Lower guide link-to-subframe bolt
   • Link from the vehicle

#### To install:

6. Install or connect the following:
   • Link into the vehicle
   • Guide link-to-subframe mounting bolt
7. Torque the support plate bolts as follows:
   • Bolt type **A**: 18 ft. lbs. (25 Nm)
   • Bolt type **B**: 55 ft. lbs. (75 Nm)
8. Install or connect the following:
   • New subframe bolts. Torque the bolts to 81 ft. lbs. (110 Nm) plus a ¼ (90 degree) turn
   • Joint end into the wheel bearing housing. Torque the nut to 74 ft. lbs. (100 Nm)
9. Load the suspension. Torque the lower guide link-to-subframe attaching bolt to 66 ft. lbs. (90 Nm) plus ¼ (90 degree) turn.

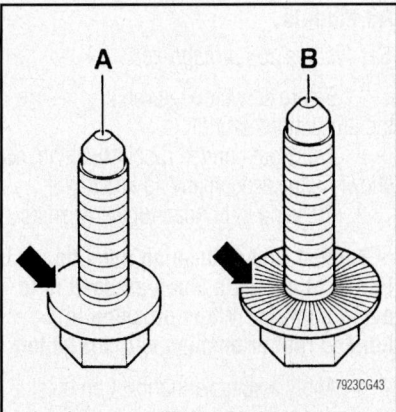

**Fig. 262 Subframe support bracket bolt identification—A4 model**

10. Install the front wheels.
11. Check and/or adjust the front suspension alignment.

#### TT Models

1. Before servicing the vehicle, refer to the precautions section.
2. Remove appropriate wheel.
3. Mark the installation positions of the ball joint nuts.
4. Remove upper and lower ball joint nuts.
5. Press ball joint from control arm.

#### To install:

6. Install ball joint into control arm and install retaining nuts to marked positions as removed. Torque the bolts to 55 ft. lbs. (75 Nm).
7. Install ball joint into wheel bearing housing using new self locking nut. Torque the bolts to 55 ft. lbs. (75 Nm).
8. Install wheel.

### LOWER CONTROL ARM

*REMOVAL & INSTALLATION*

#### A3 Models

*See Figure 263.*

1. Before servicing the vehicle, refer to the precautions section.
2. Remove the wheel.
3. Remove the lower sound insulation
4. Remove the coupling rod of left front level control system sensor from control arm, if installed.
5. Remove the bolts.
6. Pull out wheel bearing housing with lower ball joint out of control arm.
7. Set position for mounting bracket.
8. Replace bolt for left side and the right side by using locating pins T10096 and tighten locating pins to 20 Nm.

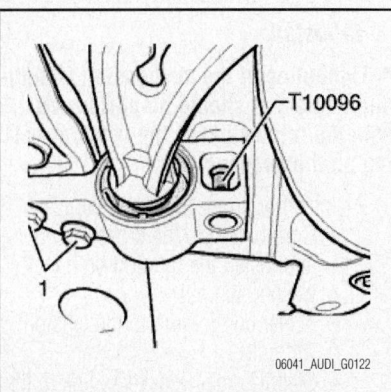

**Fig. 263 Replace bolt for left side and the right side by using locating pins T10096—A3**

**✱✱ CAUTION**

**Locating pins T10096 must not be tightened to more than 20 Nm, otherwise thread of locating pins can be damaged.**

9. Remove the item for left vehicle side and item for right vehicle side.
10. Remove the bolts.
11. Remove the control arms.

*To install:*

12. Installation is the reverse order of removal, please note the following:

   a. Insert control arm with mounting bracket into subframe.

   b. Position but do not tighten yet.

   c. Tighten the mounting bracket bolts to 50 Nm plus 90 degrees.

   d. Tighten the lower control arm to spindle bolts to 75 Nm

   e. Replace locating pins T10096 with new bolt and tighten to 70 Nm plus 90 degrees.

13. Bolt the control arm to the ball joint.
14. Tighten control arm to the console in the curb weight position.
15. Install the remaining components in the reverse order of removal.

### Except A3 Models

1. Remove or disconnect the following:
   - Front wheels
   - Cotter pin and castle nut from the tie rod end
   - Tie rod end from the steering knuckle
   - Cotter pin and castle nut from the ball joint stud
   - Knuckle from the ball joint
   - Stabilizer bar link from the lower arm
   - Nuts and bolts that connect the lower control arm
   - Lower control arm from the vehicle

*To install:*

➡ **Tightening of the suspension component fasteners should be performed with the full weight of the vehicle resting on the suspension.**

2. Install or connect the following:
   - Control arm to the cross member and install the through bolt, lockwasher and nut
   - Lower control arm to the tension rod nuts
   - Stabilizer bar transverse link to the lower arm and tighten the link
   - Knuckle to the lower ball joint.
   - Tie rod end to the knuckle. Torque

the castle nut to 22–36 ft. lbs. (29–49 Nm)
   - New cotter pin
   - Front wheels

3. Tighten all nuts and bolts to specification
   - Check and/or adjust the wheel alignment.

### CONTROL ARM BUSHING REPLACEMENT

### Except A3

#### Front Bushings

1. Before servicing the vehicle, refer to the precautions section.
2. Remove the lower control arm from the vehicle.
3. Press the front bushing out of the control arm.

*To install:*

4. Lubricate the front bushing with soap and press into the control arm.
5. Install the control arm to the vehicle.
6. Check and/or adjust the wheel alignment.

#### Rear Bushings

1. Before servicing the vehicle, refer to the precautions section.
2. Remove or disconnect the following:
   - Rear wheel
   - Rear control arm
   - Press the rear control arm bushing out

*To install:*

3. Lube the rear bushing with soap
4. Install or connect the following:
   - Rear bushing
   - Rear control arm
   - Front wheel

5. Check and/or adjust the wheel alignment.

### A3 Models

*See Figures 264 through 268.*

1. Before servicing the vehicle, refer to the precautions section.
2. Press out bonded rubber bushing, as shown in the accompanying illustration.
3. Pressing in bonded rubber bushing

➡ **Bonded rubber bushing must be positioned at an angle when pressed in to avoid damage. When pressing in, bonded rubber bushing will straighten.**

4. Apply a suitable lubricant on outside of bonded rubber bushing.
5. Position bonded rubber bushing at an angle in direction of control arm. The lip

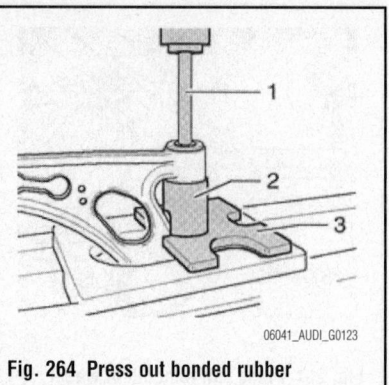

**Fig. 264 Press out bonded rubber bushing—A3**

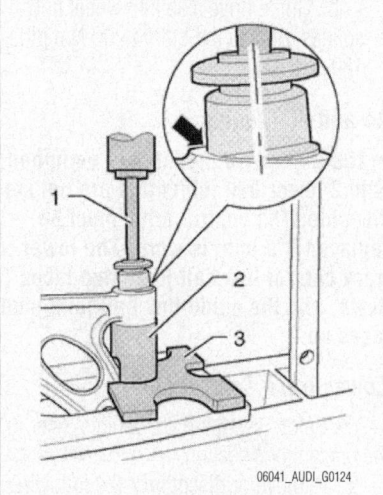

**Fig. 265 Position bonded rubber bushing at an angle in direction of control arm. The lip (arrow) must slide into hole—A3**

must slide into hole as shown in the accompanying illustration.

6. Press in bonded rubber bushing until core and hole of control arm are at same level.
7. Press back bushing slightly in control arm.
8. The dimensions and must be the same as shown.

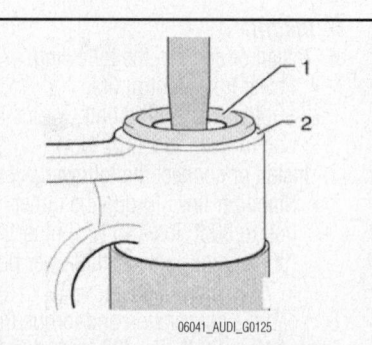

**Fig. 266 Press in bonded rubber bushing until core and hole of control arm are at same level—A3**

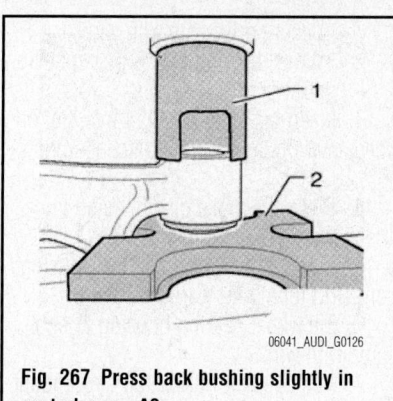

**Fig. 267 Press back bushing slightly in control arm—A3**

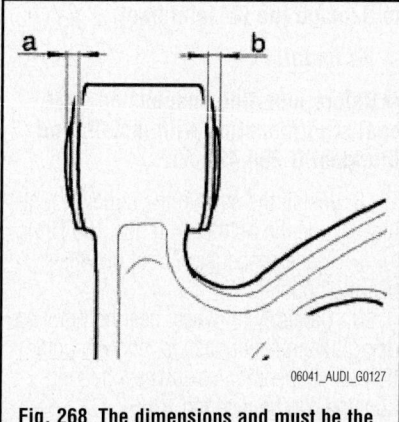

**Fig. 268 The dimensions and must be the same as shown—A3**

## MACPHERSON STRUT

### REMOVAL & INSTALLATION

### A3 Models

*See Figures 269 and 270.*

1. Before servicing the vehicle, refer to the precautions section.
2. Loosen the hex-combination bolt for drive axle.
3. Remove the wheel.
4. Remove the upper coupling rod hex nut from suspension strut.

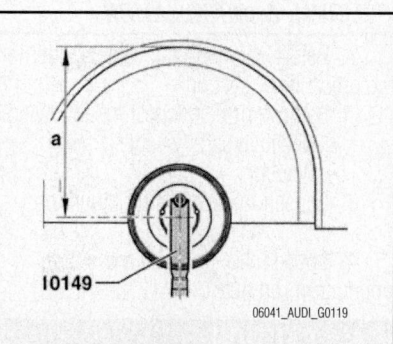

**Fig. 269 Secure Engine/transmission jack with wheel hub support T10149 to wheel hub using a wheel bolt—A3**

5. Disengage wheel speed sensor wire from suspension strut.
6. Remove the bolts.
7. Pull out wheel bearing housing with lower ball joint out of control arm.
8. Fasten drive axle to body with wire.

➡ **Drive axle must not hang down, otherwise inner joint will be damaged by over bending.**

9. Bolt ball joint to the control arm.
10. Secure Engine/transmission jack with wheel hub support T10149 to wheel hub using a wheel bolt.
11. Remove the wheel bearing housing/suspension strut bolt connection
12. Insert Spreader 3424 into slot of wheel bearing housing.
13. Turn ratchet 90 degrees and remove from spreader 3424.
14. Press brake disc by hand in direction of suspension strut. Otherwise strut tube may be canted in hole of wheel bearing housing.
15. Pull off wheel bearing housing downward from strut tube using engine/transmission jack and lower until strut tube hangs free.
16. Tie wheel bearing housing with wire at console/subframe.
17. Remove the Engine/transmission jack under wheel bearing housing.
18. Remove the wiper arms and plenum chamber cover
19. Remove the hex bolts for top strut mounting and remove strut.

### To install:

20. Insert strut, one of two markings must point in direction of travel.
21. Tighten hex bolts for top strut mounting to 15 Nm plus 90 degrees.
22. Secure Engine/transmission jack with Wheel hub support T10149 to wheel hub using wheel bolt.
23. Place suspension strut on wheel bearing housing.

**Fig. 270 Insert strut, one of two markings must point in direction of travel—A3**

24. Remove the wire on wheel bearing housing.
25. Install the remaining components in the reverse order of removal, please note the following:
    a. Tighten the strut to wheel bearing housing to bolt/nut to 70 Nm plus an additional 90 degrees.
    b. Tighten the drive axle bolt to 200 Nm plus an additional 180 degrees.

### A4 Models

1. Before servicing the vehicle, refer to the precautions section.
2. Remove or disconnect the following:
    - Front wheels
    - Rubber grommets from the plenum chamber
    - Upper strut-to-body mounting nuts
    - Anti-lock Brake System (ABS) wheel speed sensor wire from the brake caliper bracket
    - Upper control arm pinch bolt and both upper control links
    - Guide link ball joint, by swiveling the wheel bearing housing aside
    - Lower strut mounting bolt

➡ **When removing the strut, be sure not to damage the CV-joint boot.**

    - Strut

### To install:

➡ **The bonded rubber bushing can only turned to a limited extent. The bolted connections between the suspension strut and the lower track control links should therefore only be tightened when the vehicle is standing on the ground.**

3. Install or connect the following:
    - Strut by positioning it so that the spring hole plate faces the middle of the vehicle. Torque the bolt to 66 ft. lbs. (90 Nm)
    - Upper control links to the wheel bearing housing. Torque the pinch bolt to 30 ft. lbs. (40 Nm)

➡ **It may be necessary to hold the ball joint stud with a 4mm hex wrench.**

    - Ball joint. Torque the nut to 74 ft. lbs. (100 Nm)
    - ABS wheel speed sensor wire into the brake caliper holder
    - Upper strut-to-body nuts. Torque them to 15 ft. lbs. (20 Nm)
    - Rubber grommets into the plenum chamber
    - Front wheels
4. Test drive the vehicle.

5. Check and/or adjust the front alignment.

## A6 Models

*See Figure 271.*

1. Before servicing the vehicle, refer to the precautions section.

2. Remove or disconnect the following:
- Plenum chamber cover
- Battery
- Front wheels
- Rubber grommets in plenum
- Support lower control arm with a strong jack
- Top mounting nuts for strut through plenum openings
- ABS sensor cable from bracket on caliper
- Steering knuckle upper mounting nut (innermost nut only; do not loosen outer bolts)
- Guide link from wheel bearing housing

### ✳✳ WARNING

**Do not damage CV boot during this process.**

- Lower strut mounting bolt
- Strut assembly from vehicle.

### To install:

3. Installation is the reverse of the removal procedures.

4. Install or connect the following:
- Lower strut mounting nut to 66 ft. lbs. (90 Nm)
- Upper links in wheel bearing housing nuts to 30 ft. lbs. (40 Nm)
- Joint bolt nut to 74 ft. lbs. (100 Nm)
- Strut upper mounting bolts (new) to 15 ft. lbs. (20 Nm)

## TT Models

1. Before servicing the vehicle, refer to the precautions section.

2. Remove or disconnect the following:
- Front wheels
- Coupling rod from suspension strut on both sides
- Spring clip from bracket and detach brake hose
- Speed sensor wiring
- Noise insulation panel
- Drive axle from flange shaft/transmission

- Secure drive axle with wire
- Wheel bearing housing/suspension strut bolt connection

3. Insert special tool 3424 into slot on wheel bearing housing and ratchet around 90°.

4. Press the brake disc by hand in direction of suspension strut

5. Remove the wheel bearing housing from strut tube downward.

6. Tie wheel bearing housing to subframe with wire

7. Remove hex-nut for top strut mounting .

➡ **When removing the strut, be sure not to damage the CV-joint boot.**

### To install:

➡ **Before inserting suspension strut, coat strut mounting with installation lubricant G 294 421 A1.**

8. Install the strut to the upper mounting. Torque the bolt to 44 ft. lbs. (60 Nm).

9. Place suspension strut on wheel bearing housing.

10. Carefully lift wheel bearing housing using transmission jack far enough until bolt for suspension strut/wheel bearing housing can be inserted.

11. Press the brake disc by hand in direction of the suspension strut.

12. Remove spreader 3424 from wheel bearing housing.

13. Strut must be installed up to stop in wheel carrier. Torque the bolt to 44 ft. lbs. (60 Nm) plus 90°.

14. Install the ball joint. Torque the bolt to 55 ft. lb. (75) Nm).

15. Reverse the removal procedure for the remaining components.

16. Test drive the vehicle.

17. Check and/or adjust the front alignment.

### UPPER CONTROL ARM

#### REMOVAL & INSTALLATION

1. Before servicing the vehicle, refer to the precautions section.

2. Remove or disconnect the following:
- Negative battery cable
- Wheel

3. Loosen the upper strut mounting nuts.

4. Loosen, but do not remove, the upper strut rod nut.

### ✳✳ CAUTION

**DO NOT completely remove the upper strut nut at this time.**

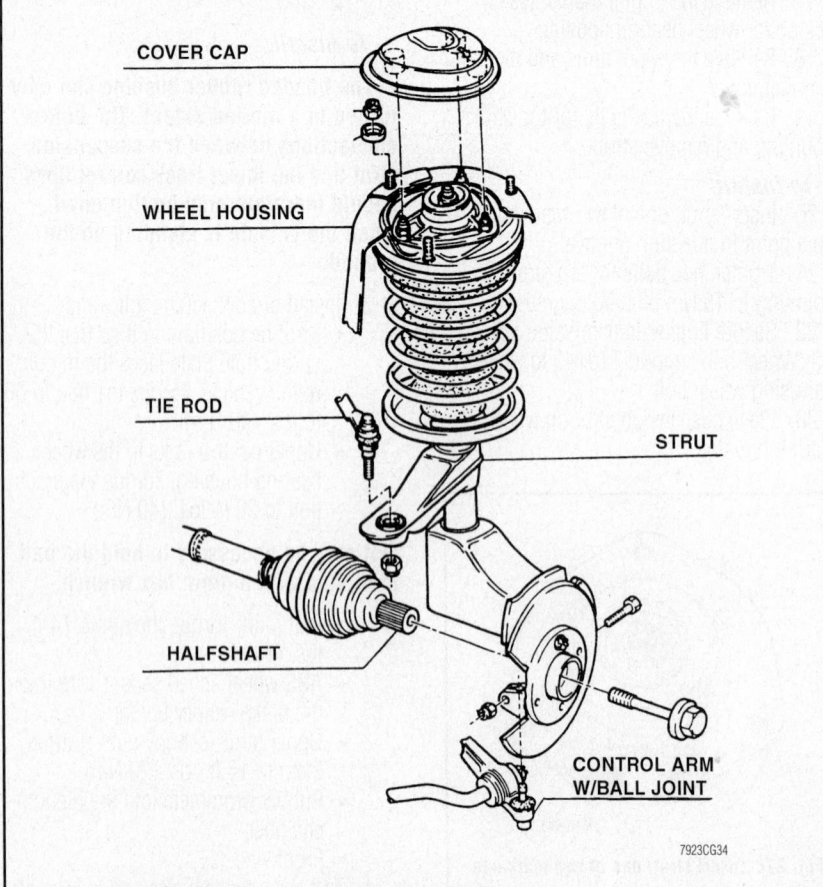

COVER CAP

WHEEL HOUSING

TIE ROD

STRUT

HALFSHAFT

CONTROL ARM
W/BALL JOINT

7923CG34

**Fig. 271 Exploded view of the front strut mounting—A6**

5. Remove or disconnect the following:
- Brake caliper, leaving the line attached and secure it out of the way
- Anti-lock Brake System (ABS) speed sensor and harness, if applicable
- Cotter pin and nut from the upper control arm
- Upper control arm from the steering knuckle
- Stabilizer bar from the link, if applicable
- Cotter pin and nut from the lower control arm
- Strut
- Upper strut mounting nuts
- Strut
- Upper control arm

***To install:***

6. Install or connect the following:
- Upper suspension arm
- Strut. Torque the upper nuts to 42 ft. lbs. (56 Nm)
- Strut to the lower arm
- Stabilizer bar bracket
- Stabilizer bar to the link
- Upper suspension arm to the steering knuckle. Torque the nut to 64 ft. lbs. (87 Nm)
- New cotter pin
- ABS speed sensor. Torque the bolt to 69 inch lbs. (8 Nm)
- Brake caliper
- Front wheel

7. Bounce the vehicle several times to stabilize the suspension.
8. Tighten the lower strut bolt
9. Check and/or adjust the front wheel alignment.

### CONTROL ARM BUSHING REPLACEMENT

The upper control arm bushings are serviced with the control arm as an assembly.

## WHEEL BEARINGS

### REMOVAL & INSTALLATION

### A3 Models

1. Before servicing the vehicle, refer to the precautions section.
2. Remove the hex-combination bolt for drive axle.
3. Remove the wheel.
4. Remove the brake caliper and hang on body using tie wire.
5. Remove the ABS wheel speed sensor
6. Remove the brake disc.
7. Loosen the nut from tie rod end but do not remove completely. Press off tie rod end from wheel bearing housing using Ball Joint Puller 3287A.
8. Remove the wheel bearing housing/suspension strut bolt connection
9. Insert Spreader 3424 into slot of wheel bearing housing.
10. Turn ratchet around 90 degrees and remove from the ratchet from spreader 3424.
11. Remove the lower control arm to spindle bolts.
12. Guide control arm out of wheel bearing housing with ball joint.
13. Remove the wheel bearing housing from strut tube and pull driveshaft out of wheel hub.

➡**Drive axle must not hang down, otherwise inner joint will be damaged by over bending.**

14. Fasten drive axle to body with wire.
15. Remove the wheel bearing housing with ball joint.

***To install:***

16. Installation is the reverse order of removal, please note the following:
   a. Tighten the NEW wheel bearing housing bolts to 70 Nm plus 90 degrees.
   b. Tighten the tie rod end nut to 20 Nm plus 90 degrees.
   c. Tighten the lower control arm to spindle bolts to 75 Nm
   d. Tighten the drive axle bolts to 200 Nm plus 180 degrees.

### A4 and A6 Models

*See Figures 272 and 273.*

1. Before servicing the vehicle, refer to the precautions section.
2. Loosen the halfshaft retaining bolt.
3. Remove or disconnect the following:
- Front wheel, then reinstall all 5 wheel bolts at this time
- Anti-lock Brake System (ABS) wheel speed sensor
- Caliper (suspend by wire to body)
- Rotor
- Backing plate bolts

4. Pull ABS wheel speed sensor from wheel bearing housing
5. Loosen the mounting nuts for the lower guide and track links.
6. Remove ABS cable grommet and clips and feed the cable through the opening.
7. Remove the tie rod end from the wheel bearing housing (use care to not damage the CV joint boot).
8. Remove the mounting nuts for the lower guide and track links and press out the joints

➡**If equipped with self-leveling control (for headlight aiming), disconnect or remove actuating arm.**

9. Remove the upper mounting nuts for the wheel bearing housing and lift out both links. Swing the wheel bearing housing toward the rear while pulling the drive axle end out of the wheel hub.
10. Unscrew the nut from the joint bolt of the track control link and remove the wheel bearing housing.
11. Place the wheel bearing housing on a press.

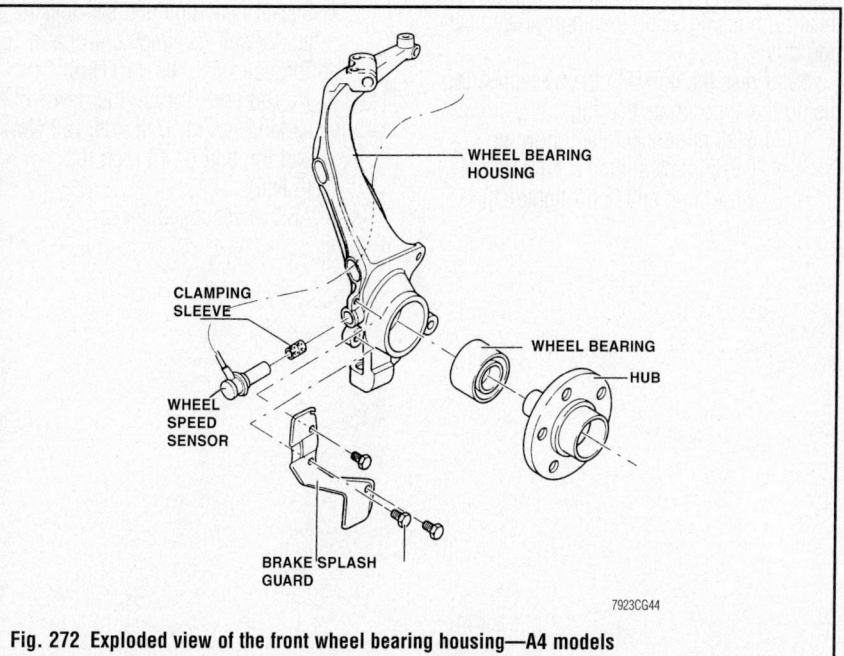

**Fig. 272 Exploded view of the front wheel bearing housing—A4 models**

WHEEL BEARING HOUSING

CLAMPING SLEEVE

WHEEL BEARING

HUB

WHEEL SPEED SENSOR

BRAKE SPLASH GUARD

7923CG44

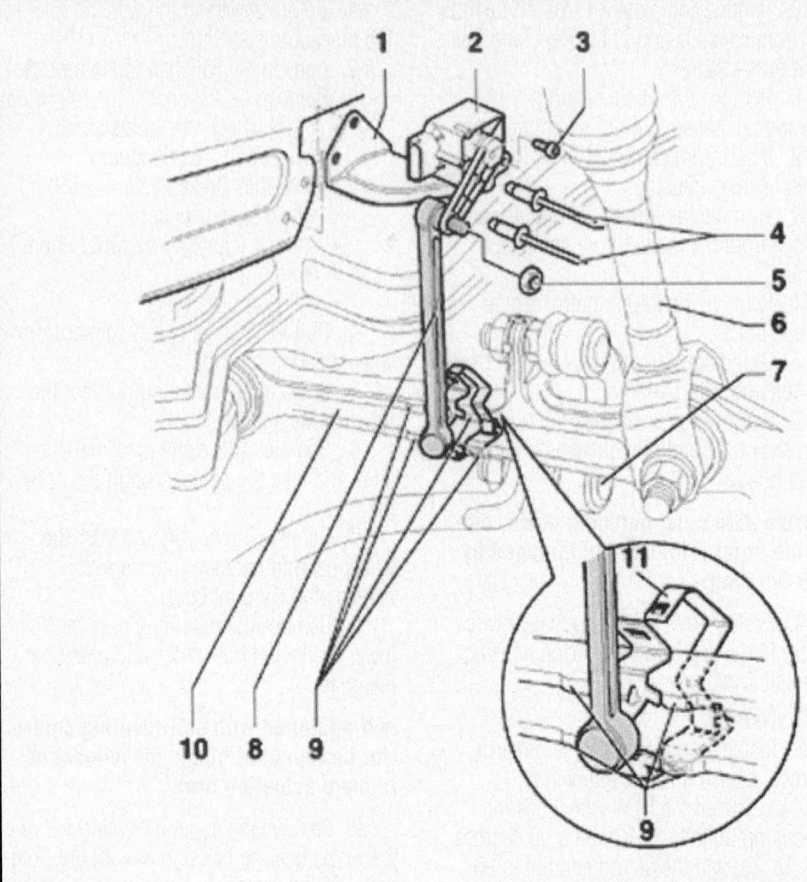

1. Bracket
2. LF level control system sensor
3. Bolt
4. Rivets
5. Nut
6. Front strut
7. Stabilizer bar link
8. Actuating arm
9. Alignment marks for mounting
10. Lower front link
11. Clip

67200-A6A6-G07

**Fig. 273 Showing the self-leveling control assembly on front suspension—A6 models (optional equipment)**

12. Drive out the hub with the wheel bearing.

13. Using a bearing separator and press, drive hub out of the bearing.

### To install:

14. Press the new wheel bearing into the bearing housing using the appropriate bearing driver.

15. Press the hub into the wheel bearing using the appropriate bearing driver.

16. Install or connect the following:
- CV-joint by sliding it through the wheel hub and hand-tighten the new nut

- Lower track control and guide link. Torque the new self-locking nut to 74 ft. lbs. (100 Nm)
- Both of the upper link ball joints into the wheel bearing. Torque the pinch bolt to 30 ft. lbs. (40 Nm)
- Socket-head bolt and self-locking nut for self-leveling control arm. Torque to 7 ft. lbs. (10 Nm)
- Tie rod end. Torque the new self-locking nut to 37 ft. lbs. (50 Nm) and the bolt to 44 inch lbs. (5 Nm)
- ABS wheel speed sensor

- Brake splash guard. Torque the bolts to 84 inch lbs. (10 Nm)
- Brake rotor
- Bake caliper. Torque the bolt to 89 ft. lbs. (120 Nm)
- Front wheel

17. Tighten the halfshaft retaining bolt as follows:
- M14 bolt: 85 ft. lbs. (115 Nm) plus ½ (180 degree) turn
- M16 bolt: 140 ft. lbs. (190 Nm) plus ½ (180 degree) turn

18. Check and/or adjust the front alignment, if necessary.

## SUSPENSION

### COIL SPRING

*REMOVAL & INSTALLATION*

*See Figures 274 and 275.*

1. Before servicing the vehicle, refer to the precautions section.
2. Remove the strut from the vehicle.
3. Clamp the spring compressor tool VAG 1752/2, in a vise.
4. Install the strut into the spring compressor.
5. Pry off the mounting bolt cap.
6. Compress the coil spring and remove the self-locking nut from the piston rod.
7. Matchmark the position of the spring retainer and spring mount.
8. Remove or disconnect the following:
   • Spring seat and related components
   • Strut from the spring compressor
9. Release the tension on the coil spring.
10. Remove the spring out of the compressor.

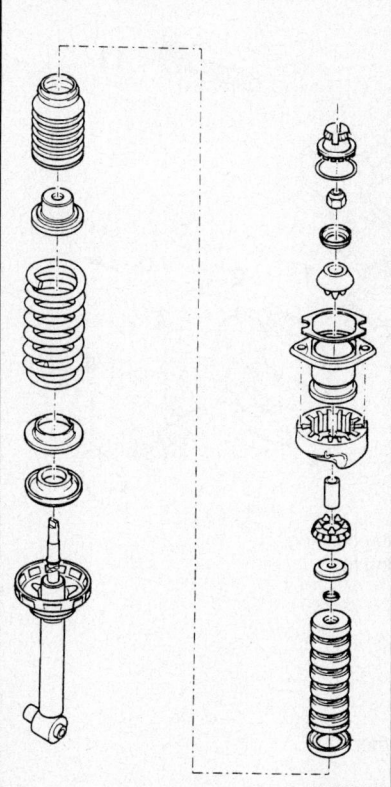

**Fig. 274 Exploded view of the rear strut— A4 model**

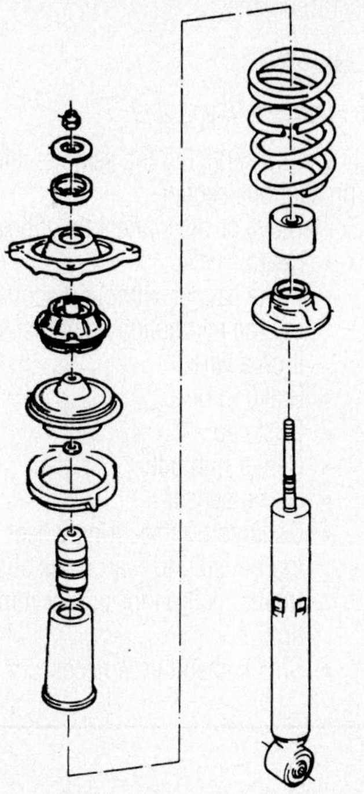

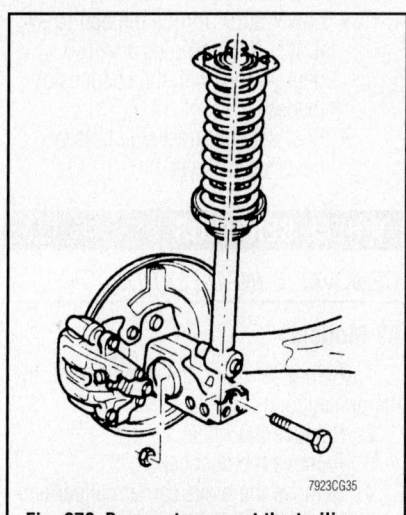

**Fig. 275 Exploded view of the rear strut— except A4 model**

*To install:*

11. Install the spring into the compressor.
12. Compress the spring and insert the strut through the spring.
13. Install or connect the following:
    • Spring seat and related components in the reverse order as they were removed by aligning the matchmarks
    • New self-locking nut
    • Mounting bolt cap
14. Release the spring compressor.
15. Install the strut into the vehicle.

### STRUT & SPRING ASSEMBLY

*REMOVAL & INSTALLATION*

**A3 Models**

1. Before servicing the vehicle, refer to the precautions section.
2. Remove the wheel.
3. Measure dimension from center of wheel to lower edge of wheel housing
4. Remove the wheel housing liner.
5. Remove the coil spring as follows:
   a. Compress the coil spring until it can be removed. Using spring compressor V.A.G 1752/1 and spring holders V.A.G 1752/3.
   b. Remove the spring.

6. Remove the upper strut bolts.
7. Remove the lower bolt.
8. Remove the shock absorber.

*To install:*

9. Installation is the reverse order of removal, please note the following:
   a. The end of the spring must rest against stop of lower spring support.
   b. Install the spring together with spring support.
   c. The spring support has two pins at bottom.
   d. Insert the pins into the bores of lower control arm.
   e. Insert the upper spring support into the upper end of spring.
   f. Release the spring. Position the upper spring support on lug of body while doing so.
   g. Tighten the upper bolts to 50 Nm plus an additional 45 degrees.
   h. Tighten the lower bolt to 180 Nm.

**A4 Models**

*See Figures 276 and 277.*

1. Before servicing the vehicle, refer to the precautions section.
2. Support the trailing arms.
3. Remove or disconnect the following:
   • Lower strut mounting bolt
   • Rear seat backrest side bolster cover or backrest to access the upper mounting
   • Upper strut-to-body mounting nuts

➡**In addition to the bolted connection, the strut is also attached to the body by 4 retaining lugs.**

**Fig. 276 Be sure to support the trailing arm before removing the lower strut mounting bolt—A4 models**

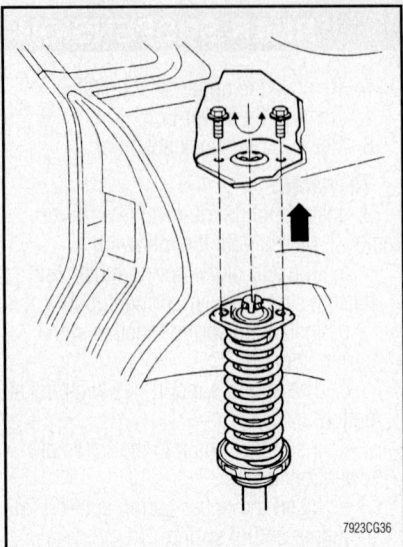

**Fig. 277 After loosening the 2 attaching bolts, rotate the upper strut mount to disengage the strut from the vehicle—A4 models**

4. Turn the strut until the retaining lugs are positioned above the recesses, then pull the strut downward out of its mount.

### To install:

➡The bonded rubber bushing can only turned to a limited extent. The bolted connections between the suspension strut and the rear axle should therefore only be tightened when the vehicle is standing on the ground.

5. Install or connect the following:
  • Strut by engaging the retaining lugs
  • Upper strut-to-body. Torque the nuts to 18 ft. lbs. (25 Nm)
  • Lower strut. Torque the bolt to 37 ft. lbs. (50 Nm) plus a ¼ (90 degree) turn with the suspension loaded
  • Rear seat backrest side bolster cover or backrest

### WHEEL BEARINGS

#### REMOVAL & INSTALLATION

#### A3 Models

1. Before servicing the vehicle, refer to the precautions section.
2. Remove the wheel.
3. Remove the dust cap.
4. Remove the brake carrier caliper and support to one side with a piece of wire.
5. Remove the brake rotor.
6. Remove the nut and wheel bearing.

### To install:

7. Installation is the reverse order of removal, tighten the bearing nut to 180 Nm plus 180 degrees:

#### A4 Models

*See Figures 278 and 279.*

1. Before servicing the vehicle, refer to the precautions section.
2. Remove or disconnect the following:
  • Rear wheel
  • Brake caliper, without disconnecting the hydraulic line and suspend it on a wire.
  • Brake rotor
  • Dust cap
  • Center hub bolt
  • Speed sensor
  • Collared bolt securing wheel bearing housing to rear shock absorber (mark position for proper reinstallation)
  • Shock absorber, if necessary

  • Wheel bearing housing
  • Wheel hub

➡The wheel hub assembly and wheel bearing are an integral unit. The wheel bearing cannot be replaced separately. If the wheel hub assembly can be pulled from the wheel bearing housing by hand, it is okay and can be reinstalled. If not okay, wheel hub assembly must be replaced.

### To install:

3. Clean and inspect mating surfaces for bearing races.
4. Install in reverse of the removal procedure, noting the following:

➡If shock absorber was removed, torque upper mounting bolts to 27 ft. lbs. (36 Nm).

  • Center hub bolt to 140 ft. lbs. (190 Nm), plus an additional 180° with wheel raised off ground; when

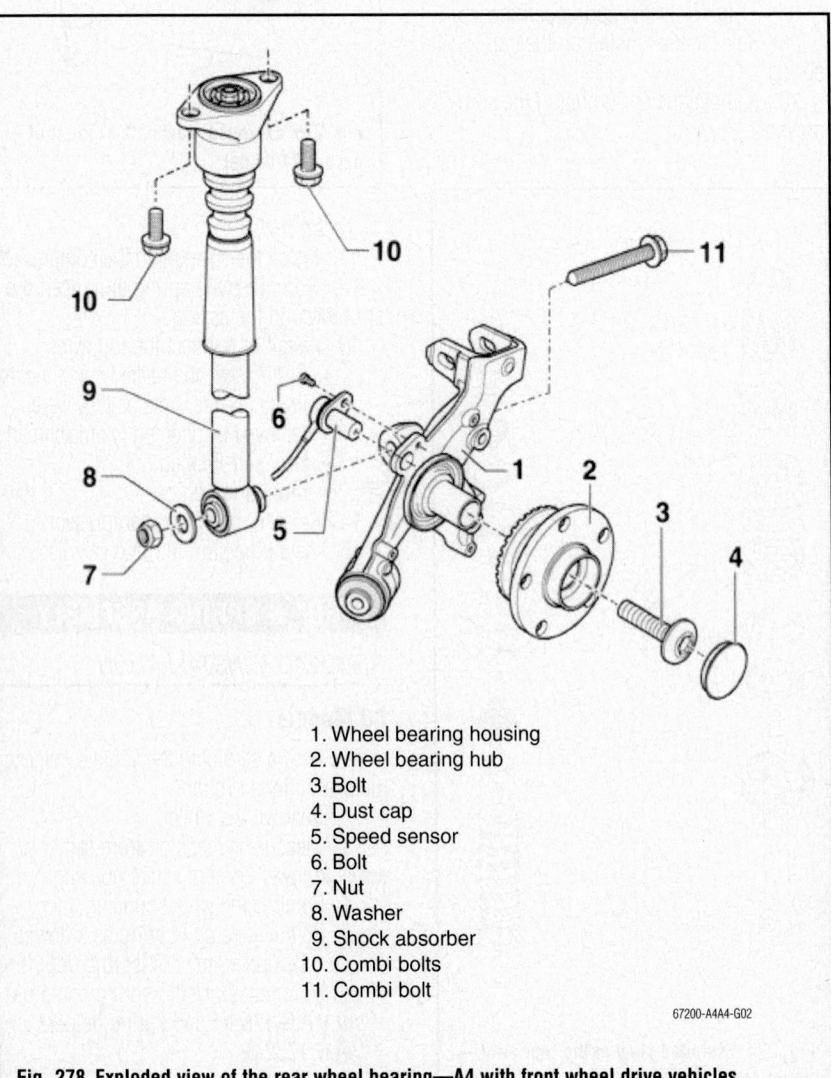

1. Wheel bearing housing
2. Wheel bearing hub
3. Bolt
4. Dust cap
5. Speed sensor
6. Bolt
7. Nut
8. Washer
9. Shock absorber
10. Combi bolts
11. Combi bolt

67200-A4A4-G02

**Fig. 278 Exploded view of the rear wheel bearing—A4 with front wheel drive vehicles**

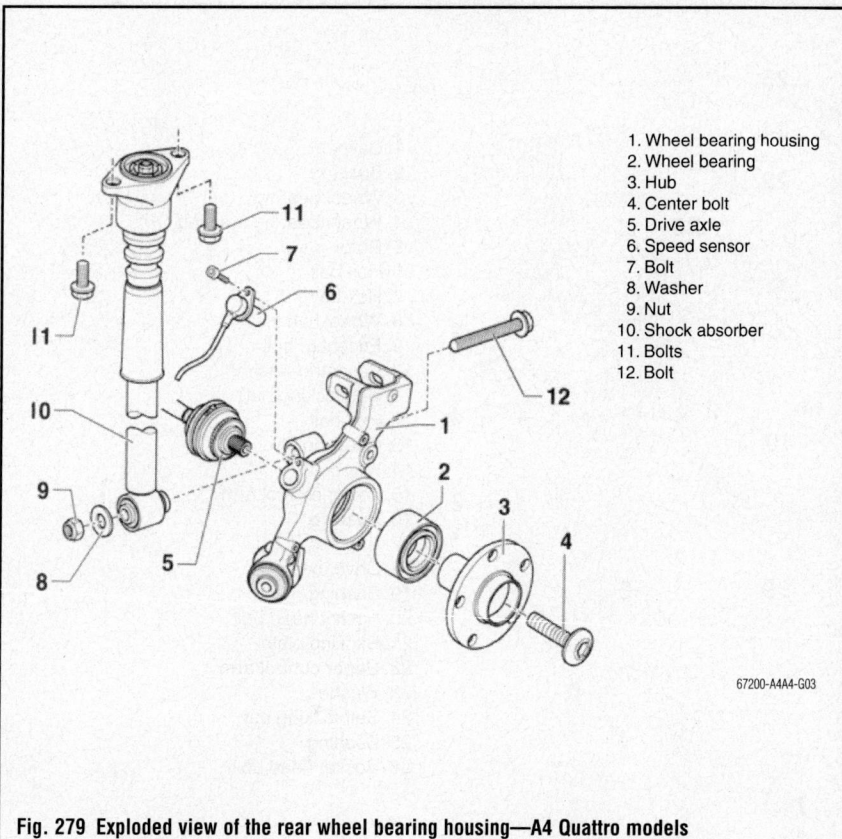

1. Wheel bearing housing
2. Wheel bearing
3. Hub
4. Center bolt
5. Drive axle
6. Speed sensor
7. Bolt
8. Washer
9. Nut
10. Shock absorber
11. Bolts
12. Bolt

67200-A4A4-G03

**Fig. 279 Exploded view of the rear wheel bearing housing—A4 Quattro models**

3. Installation is the reverse of the removal procedure, noting the following:
- Wheel hub/bearing bolts: torque to 44 ft. lbs. (60 Nm)
- Brake caliper bolts: torque to 26 ft. lbs. (35 Nm)
- Wheel bolts: torque to 89 ft. lbs. (120 Nm)

### A6 Quattro Models

*See Figure 281.*

1. Before servicing the vehicle, refer to the precautions section.
2. Remove or disconnect the following:
- Wheel assembly
- Caliper (hang by wire from body)
- Backing plate
- Brake rotor
- ABS wheel speed sensor from the wheel bearing housing
- Coupling rod bolt
- Tie rod bolt (mark position of eccentric bolt before removal)
- Control arm bolt
- Wheel bearing housing from drive axle

3. Remove the wheel bearing by pressing it from the hub.

vehicle is returned to ground, loosen bolt by 90°
- Wheel bearing housing to shock absorber new bolt. Torque self-locking nut to 118 ft. lbs. (160 Nm), plus an additional 90°, with vehicle standing on ground.

5. Check the brakes for proper operation.

### A6 FWD Models

*See Figure 280.*

1. Before servicing the vehicle, refer to the precautions section.
2. Remove or disconnect the following:
- Wheel(s)
- Brake caliper (suspend by wire from body)
- Rotor
- Wheel speed sensor bracket
- Speed sensor
- Wheel hub/bearing assembly bolts
- Wheel hub/bearing assembly with backing plate

### To install:

➡ **Wheel hub and bearing assembly is an integral unit. The wheel bearing cannot be replaced separately. If the wheel hub and bearing assembly is replaced, also replace the wheel speed sensor.**

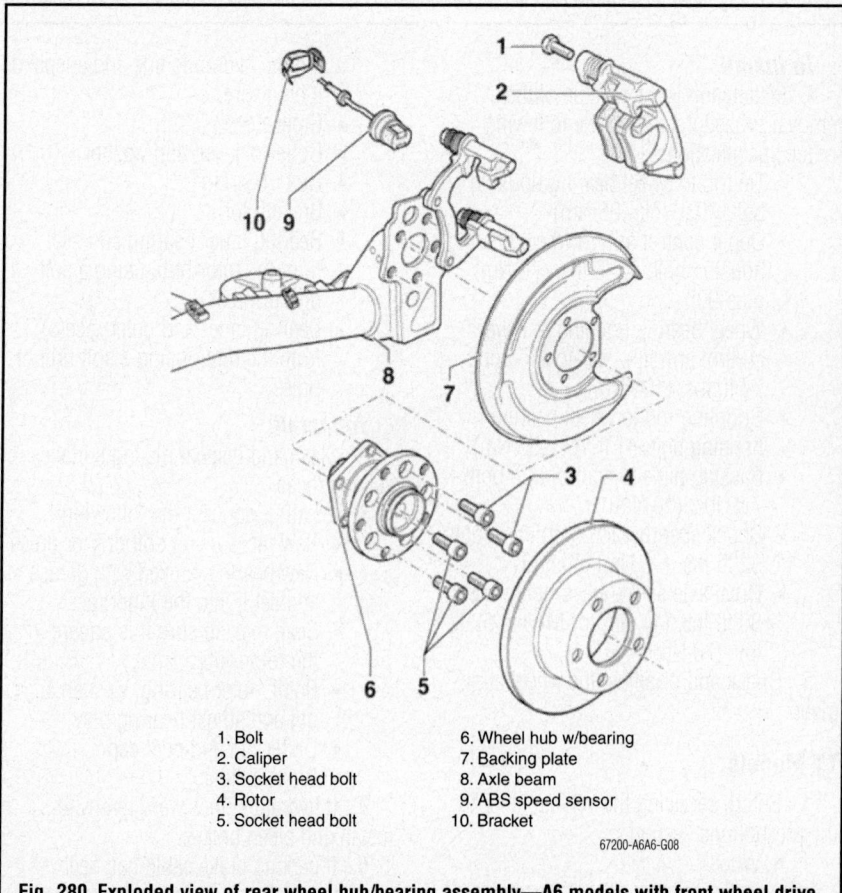

1. Bolt
2. Caliper
3. Socket head bolt
4. Rotor
5. Socket head bolt
6. Wheel hub w/bearing
7. Backing plate
8. Axle beam
9. ABS speed sensor
10. Bracket

67200-A6A6-G08

**Fig. 280 Exploded view of rear wheel hub/bearing assembly—A6 models with front wheel drive**

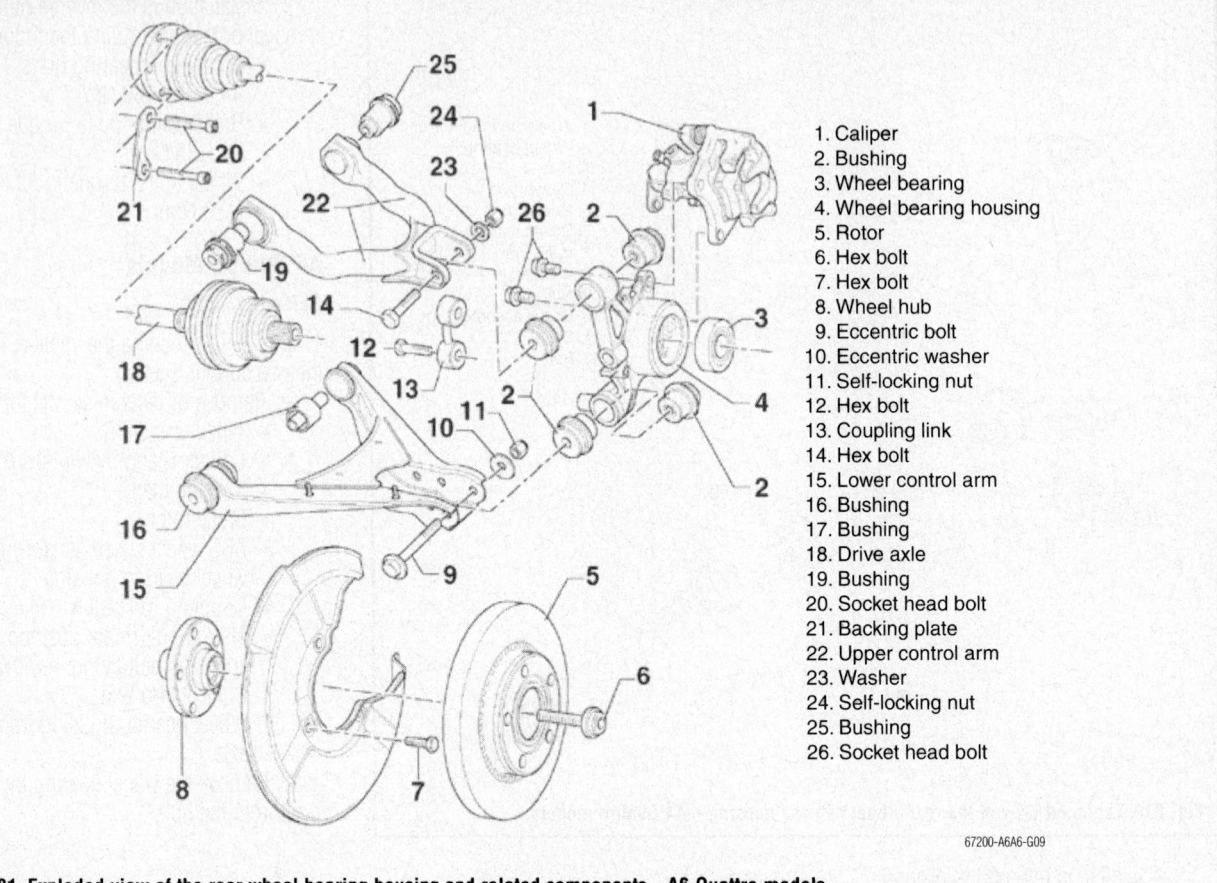

1. Caliper
2. Bushing
3. Wheel bearing
4. Wheel bearing housing
5. Rotor
6. Hex bolt
7. Hex bolt
8. Wheel hub
9. Eccentric bolt
10. Eccentric washer
11. Self-locking nut
12. Hex bolt
13. Coupling link
14. Hex bolt
15. Lower control arm
16. Bushing
17. Bushing
18. Drive axle
19. Bushing
20. Socket head bolt
21. Backing plate
22. Upper control arm
23. Washer
24. Self-locking nut
25. Bushing
26. Socket head bolt

67200-A6A6-G09

**Fig. 281 Exploded view of the rear wheel bearing housing and related components—A6 Quattro models**

## To install:

4. Installation is the reverse of the removal procedure, noting the following torque specifications:

- Tie rod to wheel bearing housing bolt: 70 ft. lbs. (95 Nm)
- Upper control arm to wheel bearing housing bolt: 52 ft. lbs. (70 Nm) plus 90°
- Wheel bearing housing to lower control arm new self-locking nut: 74 ft. lbs. (100 Nm)
- Coupling rod to wheel bearing housing bolt: 41 ft. lbs. (55 Nm)
- Backing plate to control arm bolt: 7 ft. lbs. (10 Nm)
- Shock absorber to control arm bolt: 52 ft. lbs. (70 Nm) plus 90°
- Drive axle socket head bolts: M8 to 30 ft. lbs. (40 Nm) or M10 to 52 ft. lbs. (70 Nm)

5. Check and/or adjust the wheel alignment.

### TT Models

1. Before servicing the vehicle, refer to the precautions section.
- Wheel
- Brake caliper, without disconnect-ing the hydraulic line and suspend it on a wire.
- Grease cap
- Cotter pin, nut and washer
- Outer bearing
- Brake rotor
- Bearing inner bearing and seal from the rotor hub, using a soft drift or press
- Bearing inner and outer race(s) from the rotor, using a soft drift or press

### To install:

2. Clean and inspect mating surfaces for bearing races.

3. Install or connect the following:
- New races using soft drift or press
- New bearing packed with grease and set it into the inner race
- Seal, making sure it is square in the rotor hub
- Rotor, outer bearing, washer, and nut and adjust bearing play
- Cotter pin and dust cap
- Brake caliper

4. If hydraulic lines were removed, install and bleed brakes.

5. If parking brake cable has been remove, install and adjust as necessary.

6. Install the wheel.

7. Check the brakes for proper operation.

## ADJUSTMENT

### Non-Quattro Models

1. Before servicing the vehicle, refer to the precautions section.

2. Remove or disconnect the following:
- Grease cap
- Cotter pin and the locking nut

3. While turning the wheel, so the wheel bearing does not jam, tighten the adjusting nut firmly.

4. Back the nut off slightly. The nut is properly adjusted when it is possible to pry the thrust washer side to side with some drag but using light pressure on the tool.

5. Install the locking nut and a new cotter pin.

6. When installing the cap, be sure it is securely in place.

### Quattro Models

The wheel bearings are sealed; no adjustment is necessary or possible. If the bearings are found to be loose or noisy, they must be replaced.

# AUDI

## Diagnostic Trouble Codes

## DIAGNOSTIC TROUBLE CODES

### OBD II VEHICLE APPLICATIONS

*AUDI*

**A4**

2007–2008
- 2.0L V6 MPI Engine Codes: BPG, BWT
- 3.2L V6 MPI . . . . . Engine Codes: BKH

**A6**

2007–2008
- 3.2L V6 MPI . . . . . Engine Codes: BKH
- 4.2L V8 MPI  Engine Codes: BNK, BVJ
- 5.2L V8 MPI . . . . . Engine Codes: BXA

*GAS ENGINE TROUBLE CODE LIST*

### Gas Engine OBD II Trouble Code List (P0xxx Codes)

| DTC | Trouble Code Title, Conditions & Possible Causes |
|---|---|
| **DTC: P0010**<br>**1T CCM, MIL: Yes**<br>**Years: 2007, 2008**<br>**Models:** A4, A6<br>**Engines:** All<br>**Transmissions:** All | **"A" Camshaft Position Actuator Circuit (Bank 1) Conditions:**<br>Key on or engine running; and the ECM detected an unexpected high voltage or low voltage condition on the camshaft position sensor. The relative position between the camshaft and crankshaft needs to be optimal so the engine has better torque, fuel economy and emissions.<br>**Note: The camshaft adjustment is load- and RPM dependant. The electrical camshaft adjustment valve 1 switches oil pressure onto camshaft adjuster (mechanical adjustment mechanism), which adjusts the camshaft.**<br>**Possible Causes:**<br>&bull; Fuel pump has failed<br>&bull; Actuator circuit is open<br>&bull; ECM has failed<br>&bull; Battery voltage below 11.5 volts<br>&bull; Position actuator circuit may short to B+ or Ground |
| **DTC: P0011**<br>**1T CCM, MIL: Yes**<br>**Years: 2007, 2008**<br>**Models:** A4, A6<br>**Engines:** All<br>**Transmissions:** All | **"A" Camshaft Position Timing Over-Advanced (Bank 1) Conditions:**<br>Engine started and driven at an engine speed of more than 400 RPM; and the ECM detected the camshaft timing exceeded the maximum calibrated advance value, or the camshaft remained in an advanced position during the CCM test. The valve timing did not change from the current valve timing or it remained fixed during the testing.<br>**Note: The camshaft adjustment is load- and RPM dependant. The electrical camshaft adjustment valve 1 switches oil pressure onto camshaft adjuster (mechanical adjustment mechanism), which adjusts the camshaft.**<br>**Possible Causes:**<br>&bull; Fuel pump has failed<br>&bull; CPS circuit is open, shorted to ground or shorted to power<br>&bull; ECM has failed<br>&bull; Battery voltage below 11.5 volts<br>&bull; Position actuator circuit may short to B+ or Ground<br>&bull; Camshaft timing improperly set, or continuous oil flow to the VCT piston chamber<br>&bull; Camshaft advance mechanism (the VCT unit) is sticking or binding mechanically<br>&bull; VCT solenoid valve is stuck in open position |
| **DTC: P0012**<br>**1T CCM, MIL: Yes**<br>**Years: 2007, 2008**<br>**Models:** A4, A6<br>**Engines:** All<br>**Transmissions:** All | **"A" Camshaft Position Over-Retarded (Bank 1) Conditions:**<br>Engine started and driven at an engine speed of more than 400 RPM; and the ECM detected the camshaft timing exceeded the minimu calibrated retarded value, or the camshaft remained in an retarded position during the CCM test. The valve timing did not change from the current valve timing or it remained fixed during the testing.<br>**Note: The camshaft adjustment is load- and RPM dependant. The electrical camshaft adjustment valve 1 switches oil pressure onto camshaft adjuster (mechanical adjustment mechanism), which adjusts the camshaft.**<br>**Possible Causes:**<br>&bull; Fuel pump has failed<br>&bull; CPS circuit is open, shorted to ground or shorted to power<br>&bull; ECM has failed<br>&bull; Battery voltage below 11.5 volts<br>&bull; Position actuator circuit may short to B+ or Ground<br>&bull; Camshaft timing improperly set, or continuous oil flow to the VCT piston chamber<br>&bull; Camshaft advance mechanism (the VCT unit) is sticking or binding mechanically<br>&bull; VCT solenoid valve is stuck in open position |
| **DTC: P0013**<br>**1T CCM, MIL: Yes**<br>**Years: 2007, 2008**<br>**Models:** A4, A6<br>**Engines:** All<br>**Transmissions:** All | **"B" Camshaft Position Actuator Circuit (Bank 1) Conditions:**<br>Key on or engine running; and the ECM detected an unexpected high voltage or low voltage condition on the camshaft position sensor. The relative position between the camshaft and crankshaft needs to be optimal so the engine has better torque, fuel economy and emissions.<br>**Note: The camshaft adjustment is load- and RPM dependant. The electrical camshaft adjustment valve 1 switches oil pressure onto camshaft adjuster (mechanical adjustment mechanism), which adjusts the camshaft.**<br>**Possible Causes:**<br>&bull; Fuel pump has failed<br>&bull; ECM has failed<br>&bull; Battery voltage below 11.5 volts<br>&bull; Position actuator circuit may short to B+ or Ground |

| DTC | Trouble Code Title, Conditions & Possible Causes |
|---|---|
| **DTC: P0014**<br>**1T CCM, MIL: Yes**<br>**Years: 2007, 2008**<br>**Models:** A4, A6<br>**Engines:** All<br>**Transmissions:** All | **"B" Camshaft Position Timing Over-Advanced (Bank 1) Conditions:**<br>Engine started and driven at an engine speed of more than 400 RPM; and the ECM detected the camshaft timing exceeded the maximum calibrated advance value, or the camshaft remained in an advanced position during the CCM test. The valve timing did not change from the current valve timing or it remained fixed during the testing.<br>**Note: The camshaft adjustment is load- and RPM dependant. The electrical camshaft adjustment valve 1 switches oil pressure onto camshaft adjuster (mechanical adjustment mechanism), which adjusts the camshaft.**<br>**Possible Causes:**<br>• Fuel pump has failed<br>• CPS circuit is open, shorted to ground or shorted to power<br>• ECM has failed<br>• Battery voltage below 11.5 volts<br>• Position actuator circuit may short to B+ or Ground<br>• Camshaft timing improperly set, or continuous oil flow to the VCT piston chamber<br>• Camshaft advance mechanism (the VCT unit) is sticking or binding mechanically<br>• VCT solenoid valve is stuck in open position |
| **DTC: P0015**<br>**1T CCM, MIL: Yes**<br>**Years: 2007, 2008**<br>**Models:** A4, A6<br>**Engines:** All<br>**Transmissions:** All | **"B" Camshaft Position Over-Retarded (Bank 1) Conditions:**<br>Engine started and driven at an engine speed of more than 400 RPM; and the ECM detected the camshaft timing exceeded the minimu calibrated retarded value, or the camshaft remained in an retarted position during the CCM test. The valve timing did not change from the current valve timing or it remained fixed during the testing.<br>**Note: The camshaft adjustment is load- and RPM dependant. The electrical camshaft adjustment valve 1 switches oil pressure onto camshaft adjuster (mechanical adjustment mechanism), which adjusts the camshaft.**<br>**Possible Causes:**<br>• Fuel pump has failed<br>• CPS circuit is open, shorted to ground or shorted to power<br>• ECM has failed<br>• Battery voltage below 11.5 volts<br>• Position actuator circuit may short to B+ or Ground<br>• Camshaft timing improperly set, or continuous oil flow to the VCT piston chamber<br>• Camshaft advance mechanism (the VCT unit) is sticking or binding mechanically<br>• VCT solenoid valve is stuck in open position |
| **DTC: P0020**<br>**2T CCM, MIL: Yes**<br>**Years: 2007, 2008**<br>**Models:** A4, A6<br>**Engines:** All<br>**Transmissions:** All | **"A" Camshaft Position Timing Over-Advanced (Bank 2) Conditions:**<br>Engine started and driven at an engine speed of more than 400 RPM; and the ECM detected the camshaft timing exceeded the maximum calibrated advance value, or the camshaft remained in an advanced position during the CCM test. The valve timing did not change from the current valve timing or it remained fixed during the testing.<br>**Possible Causes:**<br>• Fuel pump has failed<br>• CPS circuit is open, shorted to ground or shorted to power<br>• ECM has failed<br>• Battery voltage below 11.5 volts<br>• Position actuator circuit may short to B+ or Ground<br>• Camshaft timing improperly set, or continuous oil flow to the VCT piston chamber<br>• Camshaft advance mechanism (the VCT unit) is sticking or binding mechanically<br>• VCT solenoid valve is stuck in open position |
| **DTC: P0021**<br>**2T CCM, MIL: Yes**<br>**Years: 2007, 2008**<br>**Models:** A4, A6<br>**Engines:** All<br>**Transmissions:** All | **"A" Camshaft Position Actuator Circuit (Bank 2) Conditions:**<br>Key on or engine running; and the ECM detected an unexpected high voltage or low voltage condition on the camshaft position sensor. The relative position between the camshaft and crankshaft needs to be optimal so the engine has better torque, fuel economy and emissions.<br>**Possible Causes:**<br>• Fuel pump has failed<br>• Actuator circuit is open, shorted to ground or shorted to power<br>• ECM has failed<br>• Battery voltage below 11.5 volts<br>• Position actuator circuit may short to B+ or Ground |

| DTC | Trouble Code Title, Conditions & Possible Causes |
|---|---|
| **DTC: P0022**<br>**2T CCM, MIL: Yes**<br>**Years: 2007, 2008**<br>**Models:** A4, A6<br>**Engines:** All<br>**Transmissions:** All | **"A" Camshaft Position Over-Retarded (Bank 2) Conditions:**<br>Engine started and driven at an engine speed of more than 400 RPM; and the ECM detected the camshaft timing exceeded the minimu calibrated retarded value, or the camshaft remained in an retarted position during the CCM test. The valve timing did not change from the current valve timing or it remained fixed during the testing.<br>**Possible Causes:**<br>• Fuel pump has failed<br>• CPS circuit is open, shorted to ground or shorted to power<br>• ECM has failed<br>• Battery voltage below 11.5 volts<br>• Position actuator circuit may short to B+ or Ground<br>• Camshaft timing improperly set, or continuous oil flow to the VCT piston chamber<br>• Camshaft advance mechanism (the VCT unit) is sticking or binding mechanically<br>• VCT solenoid valve is stuck in open position |
| **DTC: P0023**<br>**2T CCM, MIL: Yes**<br>**Years: 2007, 2008**<br>**Models:** A4, A6<br>**Engines:** All<br>**Transmissions:** All | **"B" Camshaft Position Actuator Circuit (Bank 2) Conditions:**<br>Key on or engine running; and the ECM detected an unexpected high voltage or low voltage condition on the camshaft position sensor. The relative position between the camshaft and crankshaft needs to be optimal so the engine has better torque, fuel economy and emissions.<br>**Possible Causes:**<br>• Fuel pump has failed<br>• Actuator circuit is open, shorted to ground or shorted to power<br>• ECM has failed<br>• Battery voltage below 11.5 volts<br>• Position actuator circuit may short to B+ or Ground |
| **DTC: P0024**<br>**2T CCM, MIL: Yes**<br>**Years: 2007, 2008**<br>**Models:** A4, A6<br>**Engines:** All<br>**Transmissions:** All | **"B" Camshaft Position Timing Over-Advanced (Bank 2) Conditions:**<br>Engine started and driven at an engine speed of more than 400 RPM; and the ECM detected the camshaft timing exceeded the maximum calibrated advance value, or the camshaft remained in an advanced position during the CCM test. The valve timing did not change from the current valve timing or it remained fixed during the testing.<br>**Possible Causes:**<br>• Fuel pump has failed<br>• CPS circuit is open, shorted to ground or shorted to power<br>• ECM has failed<br>• Battery voltage below 11.5 volts<br>• Position actuator circuit may short to B+ or Ground<br>• Camshaft timing improperly set, or continuous oil flow to the VCT piston chamber<br>• Camshaft advance mechanism (the VCT unit) is sticking or binding mechanically<br>• VCT solenoid valve is stuck in open position |
| **DTC: P0025**<br>**2T CCM, MIL: Yes**<br>**Years: 2007, 2008**<br>**Models:** A4, A6<br>**Engines:** All<br>**Transmissions:** All | **"B" Camshaft Position Over-Retarded (Bank 2) Conditions:**<br>Engine started and driven at an engine speed of more than 400 RPM; and the ECM detected the camshaft timing exceeded the minimu calibrated retarded value, or the camshaft remained in an retarted position during the CCM test. The valve timing did not change from the current valve timing or it remained fixed during the testing.<br>**Possible Causes:**<br>• Fuel pump has failed<br>• CPS circuit is open, shorted to ground or shorted to power<br>• ECM has failed<br>• Battery voltage below 11.5 volts<br>• Position actuator circuit may short to B+ or Ground<br>• Camshaft timing improperly set, or continuous oil flow to the VCT piston chamber<br>• Camshaft advance mechanism (the VCT unit) is sticking or binding mechanically<br>• VCT solenoid valve is stuck in open position |

| DTC | Trouble Code Title, Conditions & Possible Causes |
|---|---|
| **DTC: P0030**<br>**2T CCM, MIL: Yes**<br>**Years:** 2007, 2008<br>**Models:** A4, A6<br>**Engines:** All<br>**Transmissions:** All | **HO2S Heater (Bank 1 Sensor 1) Control Circuit Malfunction Conditions:**<br>Engine started, battery voltage must be at least 11.5v, all electrical components must be off, the ground between the engine and the chassis must be well connected, the exhaust system must be properly sealed between the catalytic converter and the cylinder head, the coolant temperature must be 80 degrees Celsius, and the oxygen sensor heater for oxygen sensor before the catalytic converter must be properly functioning. The ECM detected the HO2S signal was in a negative voltage range referred to as "character shift downward". This code sets when the HO2S signal remains in a low state (usually less than 156 mv). In effect, it does not switch properly between 0.1v and 1.1v in closed loop operation.<br>**Possible Causes:**<br>• HO2S is contaminated (due to presence of silicone in fuel)<br>• HO2S signal and ground circuit wires crossed in wiring harness<br>• HO2S signal circuit is shorted to sensor or chassis ground<br>• HO2S element has failed (internal short condition)<br>• ECM has failed |
| **DTC: P0031**<br>**2T CCM, MIL: Yes**<br>**Years:** 2007, 2008<br>**Models:** A4, A6<br>**Engines:** All<br>**Transmissions:** All | **HO2S Heater (Bank 1 Sensor 1) Circuit Low Input Conditions:**<br>Engine started, battery voltage must be at least 11.5v, all electrical components must be off, the ground between the engine and the chassis must be well connected, the exhaust system must be properly sealed between the catalytic converter and the cylinder head, the coolant temperature must be 80 degrees Celsius, and the oxygen sensor heater for oxygen sensor before the catalytic converter must be properly functioning. The ECM detected the HO2S signal was in a negative voltage range referred to as "character shift downward". This code sets when the HO2S signal remains in a low state. In effect, it does not switch properly in the closed loop operation. The HO2S (before the three-way catalytic converter) has a short circuit to ground that has lasted longer than 200 seconds<br>**Possible Causes:**<br>• HO2S is contaminated (due to presence of silicone in fuel)<br>• HO2S signal and ground circuit wires crossed in wiring harness<br>• HO2S signal circuit is shorted to sensor or chassis ground<br>• HO2S element has failed (internal short condition)<br>• ECM has failed |
| **DTC: P0032**<br>**2T CCM, MIL: Yes**<br>**Years:** 2007, 2008<br>**Models:** A4, A6<br>**Engines:** All<br>**Transmissions:** All | **HO2S Heater (Bank 1 Sensor 1) Circuit High Input Conditions:**<br>Engine started, battery voltage must be at least 11.5v, all electrical components must be off, the ground between the engine and the chassis must be well connected, the exhaust system must be properly sealed between the catalytic converter and the cylinder head, the coolant temperature must be 80 degrees Celsius, and the oxygen sensor heater for oxygen sensor before the catalytic converter must be properly functioning. The ECM detected the HO2S signal remained in a high state.<br>**Note: The HO2S signal circuit may be shorted to the heater power circuit due to tracking inside of the HO2S connector. Remove the connector and visually inspect the connector for signs of oil or water.**<br>**Possible Causes:**<br>• HO2S signal shorted to heater power circuit inside connector<br>• HO2S signal circuit shorted to ground or to system voltage<br>• ECM has failed |
| **DTC: P0036**<br>**2T CCM, MIL: Yes**<br>**Years:** 2007, 2008<br>**Models:** A4, A6<br>**Engines:** All<br>**Transmissions:** All | **HO2S Heater (Bank 1 Sensor 2) Control Circuit Malfunction Conditions:**<br>Engine started, battery voltage must be at least 11.5v, all electrical components must be off, the ground between the engine and the chassis must be well connected, the exhaust system must be properly sealed between the catalytic converter and the cylinder head, the coolant temperature must be 80 degrees Celsius, and the oxygen sensor heater for oxygen sensor before the catalytic converter must be properly functioning. The ECM detected the HO2S signal was in a negative voltage range referred to as "character shift downward". This code sets when the HO2S signal remains in a low state.<br>**Possible Causes:**<br>• HO2S is contaminated (due to presence of silicone in fuel)<br>• HO2S signal and ground circuit wires crossed in wiring harness<br>• HO2S signal circuit is shorted to sensor or chassis ground<br>• HO2S element has failed (internal short condition)<br>• ECM has failed |

| DTC | Trouble Code Title, Conditions & Possible Causes |
|---|---|
| **DTC: P0037**<br>**2T CCM, MIL: Yes**<br>**Years: 2007, 2008**<br>**Models:** A4, A6<br>**Engines:** All<br>**Transmissions:** All | **HO2S Heater (Bank 1 Sensor 2) Circuit Low Input Conditions:**<br>Engine started, battery voltage must be at least 11.5v, all electrical components must be off, the ground between the engine and the chassis must be well connected, the exhaust system must be properly sealed between the catalytic converter and the cylinder head, the coolant temperature must be 80 degrees Celsius, and the oxygen sensor heater for oxygen sensor before the catalytic converter must be properly functioning. The ECM detected the HO2S signal was in a negative voltage range referred to as "character shift downward". This code sets when the HO2S signal remains in a low state. In effect, it does not switch properly in the closed loop operation. The HO2S (before the three-way catalytic converter) has a short circuit to ground that has lasted longer than 200 seconds<br>**Possible Causes:**<br>• HO2S is contaminated (due to presence of silicone in fuel)<br>• HO2S signal and ground circuit wires crossed in wiring harness<br>• HO2S signal circuit is shorted to sensor or chassis ground<br>• HO2S element has failed (internal short condition)<br>• ECM has failed |
| **DTC: P0038**<br>**2T CCM, MIL: Yes**<br>**Years: 2007, 2008**<br>**Models:** A4, A6<br>**Engines:** All<br>**Transmissions:** All | **HO2S Heater (Bank 1 Sensor 2) Circuit High Input Conditions:**<br>Engine started, battery voltage must be at least 11.5v, all electrical components must be off, the ground between the engine and the chassis must be well connected, the exhaust system must be properly sealed between the catalytic converter and the cylinder head, the coolant temperature must be 80 degrees Celsius, and the oxygen sensor heater for oxygen sensor before the catalytic converter must be properly functioning. The ECM detected the HO2S signal remained in a high state.<br>**Note: The HO2S signal circuit may be shorted to the heater power circuit due to tracking inside of the HO2S connector. Remove the connector and visually inspect the connector for signs of oil or water.**<br>**Possible Causes:**<br>• HO2S signal shorted to heater power circuit inside connector<br>• HO2S signal circuit shorted to ground or to system voltage<br>• ECM has failed |
| **DTC: P0040**<br>**2T CCM, MIL: Yes**<br>**Years: 2007, 2008**<br>**Models:** A4, A6<br>**Engines:** All<br>**Transmissions:** All | **O2 Sensor Signals Swapped (Bank 1 Sensor 1/Bank 2 Sensor 1) Conditions:**<br>Engine started, battery voltage must be at least 11.5v, all electrical components must be off, the ground between the engine and the chassis must be well connected, the exhaust system must be properly sealed between the catalytic converter and the cylinder head, and the coolant temperature must be 80 degrees Celsius. The ECM detected the O2 signals were mixed and reading implausible results from both.<br>**Possible Causes:**<br>• HO2S-11 and HO2S-21 harness connectors are swapped<br>• HO2S-11 and HO2S-21 wiring is crossed inside the harness<br>• HO2S-11 and HO2S-21 wires are crossed at 104-pin connector<br>• Connector coding and color mixed with correct catalytic converter |
| **DTC: P0041**<br>**2T CCM, MIL: Yes**<br>**Years: 2007, 2008**<br>**Models:** A4, A6<br>**Engines:** All<br>**Transmissions:** All | **O2 Sensor Signals Swapped (Bank 1 Sensor 2/Bank 2 Sensor 2) Conditions:**<br>Engine started, battery voltage must be at least 11.5v, all electrical components must be off, the ground between the engine and the chassis must be well connected, the exhaust system must be properly sealed between the catalytic converter and the cylinder head, and the coolant temperature must be 80 degrees Celsius. The ECM detected the O2 signals were mixed and reading implausible results from both.<br>**Possible Causes:**<br>• HO2S-12 and HO2S-22 harness connectors are swapped<br>• HO2S-12 and HO2S-22 wiring is crossed inside the harness<br>• HO2S-12 and HO2S-22 wires are crossed at 104-pin connector<br>• Connector coding and color mixed with correct catalytic converter |
| **DTC: P0050**<br>**2T CCM, MIL: Yes**<br>**Years: 2007, 2008**<br>**Models:** A4, A6<br>**Engines:** All<br>**Transmissions:** All | **HO2S Heater (Bank 2 Sensor 1) Control Circuit Malfunction Conditions:**<br>Engine started, battery voltage must be at least 11.5v, all electrical components must be off, the ground between the engine and the chassis must be well connected, the exhaust system must be properly sealed between the catalytic converter and the cylinder head, and the coolant temperature must be 80 degrees Celsius. The ECM detected the HO2S signal was in a negative voltage range referred to as "character shift downward".<br>**Possible Causes:**<br>• HO2S is contaminated (due to presence of silicone in fuel)<br>• HO2S signal and ground circuit wires crossed in wiring harness<br>• HO2S signal circuit is shorted to sensor or chassis ground<br>• HO2S element has failed (internal short condition)<br>• ECM has failed |

| DTC | Trouble Code Title, Conditions & Possible Causes |
|---|---|
| **DTC: P0051**<br>**2T CCM, MIL: Yes**<br>**Years: 2007, 2008**<br>**Models:** A4, A6<br>**Engines:** All<br>**Transmissions:** All | **HO2S Heater (Bank 2 Sensor 1) Circuit Low Input Conditions:**<br>Engine started, battery voltage must be at least 11.5v, all electrical components must be off, the ground between the engine and the chassis must be well connected, the exhaust system must be properly sealed between the catalytic converter and the cylinder head, and the coolant temperature must be 80 degrees Celsius. The ECM detected the HO2S signal was in a negative voltage range referred to as "character shift downward". This code sets when the HO2S signal remains in a low state. In effect, it does not switch properly in the closed loop operation. The HO2S (before the three-way catalytic converter) has a short circuit to ground that has lasted longer than a specified time.<br>**Possible Causes:**<br>• HO2S is contaminated (due to presence of silicone in fuel)<br>• HO2S signal and ground circuit wires crossed in wiring harness<br>• HO2S signal circuit is shorted to sensor or chassis ground<br>• HO2S element has failed (internal short condition)<br>• ECM has failed |
| **DTC: P0052**<br>**2T CCM, MIL: Yes**<br>**Years: 2007, 2008**<br>**Models:** A4, A6<br>**Engines:** All<br>**Transmissions:** All | **HO2S Heater (Bank 2 Sensor 1) Circuit High Input Conditions:**<br>Engine started, battery voltage must be at least 11.5v, all electrical components must be off, the ground between the engine and the chassis must be well connected, the exhaust system must be properly sealed between the catalytic converter and the cylinder head, and the coolant temperature must be 80 degrees Celsius. The ECM detected the HO2S signal was in a negative voltage range referred to as "character shift downward". This code sets when the HO2S signal remains in a low state. In effect, it does not switch properly in the closed loop operation. The HO2S (before the three-way catalytic converter) has a short circuit to ground that has lasted longer than a specified time.<br>**Possible Causes:**<br>• HO2S is contaminated (due to presence of silicone in fuel)<br>• HO2S signal and ground circuit wires crossed in wiring harness<br>• HO2S signal circuit is shorted to sensor or chassis ground<br>• HO2S element has failed (internal short condition)<br>• ECM has failed |
| **DTC: P0056**<br>**2T CCM, MIL: Yes**<br>**Years: 2007, 2008**<br>**Models:** A4, A6<br>**Engines:** All<br>**Transmissions:** All | **HO2S Heater (Bank 2 Sensor 2) Circuit High Input Conditions:**<br>Engine started, battery voltage must be at least 11.5v, all electrical components must be off, the ground between the engine and the chassis must be well connected, the exhaust system must be properly sealed between the catalytic converter and the cylinder head, and the coolant temperature must be 80 degrees Celsius. The ECM detected the HO2S signal remained in a high state.<br>**Note: The HO2S signal circuit may be shorted to the heater power circuit due to tracking inside of the HO2S connector. Remove the connector and visually inspect the connector for signs of oil or water.**<br>**Possible Causes:**<br>• HO2S signal shorted to heater power circuit inside connector<br>• HO2S signal circuit shorted to ground or to system voltage<br>• ECM has failed |
| **DTC: P0057**<br>**2T CCM, MIL: Yes**<br>**Years: 2007, 2008**<br>**Models:** A4, A6<br>**Engines:** All<br>**Transmissions:** All | **HO2S Heater (Bank 2 Sensor 2) Control Circuit Malfunction Conditions:**<br>Engine started, battery voltage must be at least 11.5v, all electrical components must be off, the ground between the engine and the chassis must be well connected, the exhaust system must be properly sealed between the catalytic converter and the cylinder head, and the coolant temperature must be 80 degrees Celsius. The ECM detected the HO2S signal was in a negative voltage range referred to as "character shift downward".<br>**Possible Causes:**<br>• HO2S is contaminated (due to presence of silicone in fuel)<br>• HO2S signal and ground circuit wires crossed in wiring harness<br>• HO2S signal circuit is shorted to sensor or chassis ground<br>• HO2S element has failed (internal short condition)<br>• ECM has failed |
| **DTC: P0058**<br>**2T CCM, MIL: Yes**<br>**Years: 2007, 2008**<br>**Models:** A4, A6<br>**Engines:** All<br>**Transmissions:** All | **HO2S Heater (Bank 2 Sensor 2) Circuit Low Input Conditions:**<br>Engine started, battery voltage must be at least 11.5v, all electrical components must be off, the ground between the engine and the chassis must be well connected, the exhaust system must be properly sealed between the catalytic converter and the cylinder head, and the coolant temperature must be 80 degrees Celsius. The ECM detected the HO2S signal was in a negative voltage range referred to as "character shift downward". This code sets when the HO2S signal remains in a low state. In effect, it does not switch properly in the closed loop operation. The HO2S (before the three-way catalytic converter) has a short circuit to ground that has lasted longer than a specified time.<br>**Possible Causes:**<br>• HO2S is contaminated (due to presence of silicone in fuel)<br>• HO2S signal and ground circuit wires crossed in wiring harness<br>• HO2S signal circuit is shorted to sensor or chassis ground<br>• HO2S element has failed (internal short condition)<br>• ECM has failed |

| DTC | Trouble Code Title, Conditions & Possible Causes |
|---|---|
| **DTC: P0087**<br>**2T CCM, MIL: Yes**<br>**Years: 2007, 2008**<br>**Models:** A6<br>**Engines:** 3.2L<br>**Transmissions:** All | **Fuel Rail/System Pressure Too Low Conditions**<br>Engine started, battery voltage must be at least 11.5v, all electrical components must be off, the ground between the engine and the chassis must be well connected, the exhaust system must be properly sealed between the catalytic converter and the cylinder head, and the coolant temperature must be 80 degrees Celsius. The ECM detected that the system's fuel pressure has fallen below the accepted normal calibrated value.<br>**Possible Causes:**<br>• Fuel Pressure Regulator Valve faulty<br>• Fuel Pressure Sensor faulty<br>• Fuel Pump (FP) Control Module faulty<br>• Fuel pump faulty<br>• Low fuel |
| **DTC: P0088**<br>**2T CCM, MIL: Yes**<br>**Years:**2007, 2008<br>**Models:** A6<br>**Engines:** 3.2L<br>**Transmissions:** All | **Fuel Rail/System Pressure Too High Conditions**<br>Engine started, battery voltage must be at least 11.5v, all electrical components must be off, the ground between the engine and the chassis must be well connected, the exhaust system must be properly sealed between the catalytic converter and the cylinder head, and the coolant temperature must be 80 degrees Celsius. The ECM detected that the system's fuel pressure has risen above the accepted normal calibrated value.<br>**Possible Causes:**<br>• Fuel Pressure Regulator Valve faulty<br>• Fuel Pressure Sensor faulty<br>• Fuel Pump (FP) Control Module faulty<br>• Fuel pump faulty |
| **DTC: P0089**<br>**2T CCM, MIL: Yes**<br>**Years: 2007, 2008**<br>**Models:** A6<br>**Engines:** 3.2L<br>**Transmissions:** All | **Fuel Pressure Regulator Range/Performance Conditions**<br>Engine started, battery voltage must be at least 11.5v, all electrical components must be off, the ground between the engine and the chassis must be well connected, the exhaust system must be properly sealed between the catalytic converter and the cylinder head, and the coolant temperature must be 80 degrees Celsius. The ECM detected that the system's fuel pressure sensor is providing a signal that is either outside the accepted normal values or is not receiving a signal at all.<br>**Possible Causes:**<br>• Fuel Pressure Regulator Valve faulty<br>• Fuel Pressure Sensor faulty<br>• Fuel Pump (FP) Control Module faulty<br>• Fuel pump faulty |
| **DTC: P0101**<br>**2T CCM, MIL: Yes**<br>**Years: 2007, 2008**<br>**Models:** A4, A6<br>**Engines:** All<br>**Transmissions:** All | **Mass or Volume Air Flow Circuit Range/Performance Conditions**<br>Engine running, with the system voltage more than 11.0v, and the temperature must be at least 185-degrees (F) and all electrical equipment (A/C, lights, etc) must be off. The ECM has detected that the MAF signal was out of a calculated range with the engine (or undetectable) for a certain period of time.<br>**Possible Causes:**<br>• Mass air flow (MAF) sensor has failed or is damaged<br>• ECM has failed<br>• Signal and ground wires of Mass Air Flow (MAF) sensor has short circuited |
| **DTC: P0102**<br>**1T CCM, MIL: Yes**<br>**Years: 2007, 2008**<br>**Models:** A4, A6<br>**Engines:** All<br>**Transmissions:** All | **MAF Sensor Circuit Low Input Conditions:**<br>Key on, engine started, and the ECM detected the MAF sensor signal was less than the minimum calibrated value. The engine temperature must beat least 185-degrees (F) and all electrical equipment (A/C, lights, etc) must be off. The ECM has detected that the MAF signal was less than the required minimum.<br>**Possible Causes:**<br>• Check for leaks between MAF sensor and throttle valve control module<br>• Voltage supply faulty.<br>• Sensor power circuit open from fuel pump relay to MAF sensor<br>• Sensor signal circuit open (may be disconnected) from ECM and MAF<br>• Faulty ground cable resistance between connector terminal 1 and Ground<br>• MAF Sensor malfunction |
| **DTC: P0103**<br>**1T CCM, MIL: Yes**<br>**Years: 2007, 2008**<br>**Models:** A4, A6<br>**Engines:** All<br>**Transmissions:** All | **MAF Sensor Circuit High Input Conditions:**<br>Key on, engine started, and the ECM detected the MAF sensor signal was more than the minimum calibrated value. The engine temperature must beat least 185-degrees (F) and all electrical equipment (A/C, lights, etc) must be off. The ECM has detected that the MAF signal was more than the required minimum.<br>**Possible Causes:**<br>• Check for leaks between MAF sensor and throttle valve control module<br>• Voltage supply faulty.<br>• Sensor power circuit open from fuel pump relay to MAF sensor<br>• Sensor signal circuit open (may be disconnected) from ECM and MAF<br>• Faulty ground cable resistance between connector terminal 1 and Ground<br>• MAF Sensor malfunction |

| DTC | Trouble Code Title, Conditions & Possible Causes |
|---|---|
| **DTC: P0106**<br>**2T CCM, MIL: Yes**<br>**Years:** 2007, 2008<br>**Models:** A4, A6<br>**Engines:** All<br>**Transmissions:** All | **Manifold Absolute Pressure/Barometric Pressure Sensor Circuit Performance Conditions:**<br>Engine started, the temperature must beat least 185-degrees (F) and all electrical equipment (A/C, lights, etc) must be off. The ECM detected the BARO sensor was out of range during the CCM test. The BARO sensor signal should be in 4.5v.<br>**Possible Causes:**<br>• Sensor has deteriorated (response time too slow) or has failed<br>• MAP sensor signal circuit is shorted to ground<br>• MAP sensor circuit (5v) is open<br>• MAP sensor is damaged or it has failed<br>• BARO sensor signal circuit is shorted to ground<br>• BARO sensor circuit (5v) is open<br>• BARO sensor is damaged or it has failed<br>• ECM is not connected properly<br>• ECM has failed |
| **DTC: P0107**<br>**1T CCM, MIL: Yes**<br>**Years:** 2007, 2008<br>**Models:** A4, A6<br>**Engines:** 3.2L<br>**Transmissions:** All | **Manifold Absolute Pressure/Barometric Pressure Sensor Circuit Low Input Conditions:**<br>Engine started, the temperature must beat least 185-degrees (F) and all electrical equipment (A/C, lights, etc) must be off. The ECM detected the BARO sensor was out of range during the CCM test. The BARO sensor signal should be in 4.5v. The BARO sensor is a variable capacitance unit used to detect altitude.<br>**Possible Causes:**<br>• Sensor has deteriorated (response time too slow) or has failed<br>• MAP sensor signal circuit is shorted to ground<br>• MAP sensor circuit (5v) is open<br>• MAP sensor is damaged or it has failed<br>• BARO sensor signal circuit is shorted to ground<br>• BARO sensor circuit (5v) is open<br>• BARO sensor is damaged or it has failed<br>• ECM is not connected properly<br>• ECM has failed |
| **DTC: P0108**<br>**1T CCM, MIL: Yes**<br>**Years:** 2007, 2008<br>**Models:** A4, A6<br>**Engines:** 3.2L<br>**Transmissions:** All | **Manifold Absolute Pressure/Barometric Sensor Circuit High Input Conditions:**<br>Engine started, the temperature must beat least 185-degrees (F) and all electrical equipment (A/C, lights, etc) must be off. The ECM detected the BARO sensor was out of range during the CCM test. The BARO sensor signal should be in 4.5v. The BARO sensor is a variable capacitance unit used to detect altitude.<br>**Possible Causes:**<br>• Sensor has deteriorated (response time too slow) or has failed<br>• MAP sensor signal circuit is shorted to ground<br>• MAP sensor circuit (5v) is open<br>• MAP sensor is damaged or it has failed<br>• BARO sensor signal circuit is shorted to ground<br>• BARO sensor circuit (5v) is open<br>• BARO sensor is damaged or it has failed<br>• ECM is not connected properly<br>• ECM has failed |
| **DTC: P0111**<br>**2T CCM, MIL: Yes**<br>**Years:** 2007, 2008<br>**Models:** A4, A6<br>**Engines:** 3.2L<br>**Transmissions:** All: All | **Intake Air Temperature Sensor Circuit Low Input Conditions:**<br>Key on or engine running, the temperature must beat least 185-degrees (F) and all electrical equipment (A/C, lights, etc) must be off; and the ECM detected the IAT sensor signal was less than the self-test minimum. This is a thermistor-type sensor with a variable resistance that changes when exposed to different temperatures. This means: the higher the temperature, the lower the resistance value.<br>**Possible Causes:**<br>• IAT sensor signal circuit is grounded (check wiring & connector)<br>• Resistance value between sockets 33 and 36 out of range<br>• IAT sensor has an open circuit<br>• IAT sensor is damaged or it has failed<br>• ECM has failed |

| DTC | Trouble Code Title, Conditions & Possible Causes |
|---|---|
| **DTC: P0112**<br>**1T CCM, MIL: Yes**<br>**Years:** 2007, 2008<br>**Models:** A4, A6<br>**Engines:** All<br>**Transmissions:** All | **Intake Air Temperature Sensor Circuit Low Input Conditions:**<br>Key on or Engine running, the temperature must beat least 185-degrees (F) and all electrical equipment (A/C, lights, etc) must be off; and the ECM detected the IAT sensor signal was less than the self-test minimum. This is a thermistor-type sensor with a variable resistance that changes when exposed to different temperatures. This means: the higher the temperature, the lower the resistance value.<br>**Possible Causes:**<br>• IAT sensor signal circuit is grounded (check wiring & connector)<br>• Resistance value between sockets 33 and 36 out of range<br>• IAT sensor has an open circuit<br>• IAT sensor is damaged or it has failed<br>• ECM has failed |
| **DTC: P0113**<br>**1T CCM, MIL: Yes**<br>**Years:** 2007, 2008<br>**Models:** A4, A6<br>**Engines:** All<br>**Transmissions:** All | **Intake Air Temperature Sensor Circuit High Input Conditions:**<br>Key on or engine running, the temperature must beat least 185-degrees (F) and all electrical equipment (A/C, lights, etc) must be off; and the ECM detected the IAT sensor signal was more than the self-test maximum. This is a thermistor-type sensor with a variable resistance that changes when exposed to different temperatures. This means: the higher the temperature, the lower the resistance value.<br>**Possible Causes:**<br>• IAT sensor signal circuit is open (inspect wiring & connector)<br>• IAT sensor signal circuit is shorted<br>• Resistance value between sockets 33 and 36 out of range<br>• IAT sensor is damaged or it has failed<br>• ECM has failed |
| **DTC: P0116**<br>**2T CCM, MIL: Yes**<br>**Years:** 2007, 2008<br>**Models:** A4, A6<br>**Engines:** All<br>**Transmissions:** All | **ECT Sensor / CHT Sensor Signal Range/Performance Conditions:**<br>Engine started (cold), battery voltage must be 11.5, and all equipment must be off. The ECM detected the ECT sensor exceeded the required calibrated value, or the engine is at idle and doesn't reach operating temperature quickly enough; the Catalyst, Fuel System, HO2S and Misfire Monitor did not complete, or the timer expired. Testing completion of procedure, the engine's temperature must rise uniformly during idle.<br>**Possible Causes:**<br>• Check for low coolant level or incorrect coolant mixture<br>• ECM detects a short circuit wiring in the ECT<br>• CHT sensor is out-of-calibration or it has failed<br>• ECT sensor is out-of-calibration or it has failed |
| **DTC: P0117**<br>**1T CCM, MIL: Yes**<br>**Years:** 2007, 2008<br>**Models:** A4, A6<br>**Engines:** All<br>**Transmissions:** All | **ECT Sensor Circuit Low Input Conditions:**<br>Engine started (cold), battery voltage must be 11.5, and all equipment must be off. The ECM detected the ECT sensor signal was less than the self-test minimum. This is a thermistor-type sensor with a variable resistance that changes when exposed to different temperatures<br>**Possible Causes:**<br>• ECT sensor signal circuit is grounded in the wiring harness<br>• ECT sensor doesn't react to changes in temperature<br>• ECT sensor is damaged or the ECM has failed |
| **DTC: P0118**<br>**1T CCM, MIL: Yes**<br>**Years:** 2007, 2008<br>**Models:** A4, A6<br>**Engines:** All<br>**Transmissions:** All | **ECT Sensor Circuit High Input Conditions:**<br>Engine started (cold), battery voltage must be 11.5, and all equipment must be off. The ECM detected the ECT sensor signal was more than the self-test maximum. This is a thermistor-type sensor with a variable resistance that changes when exposed to different temperatures<br>**Possible Causes:**<br>• ECT sensor signal circuit is open (inspect wiring & connector)<br>• ECT sensor signal circuit is shorted to ground<br>• ECT sensor is damaged or it has failed<br>• ECM has failed |
| **DTC: P0121**<br>**1T CCM, MIL: Yes**<br>**Years:** 2007, 2008<br>**Models:** A4, A6<br>**Engines:** All<br>**Transmissions:** All | **Throttle/Pedal Position Sensor Signal Range/Performance Conditions:**<br>Engine started; then immediately following a condition where the engine was running under at off-idle, the ECM detected the TP sensor signal indicated the throttle did not return to its previous closed position during the Rationality test.<br>**Possible Causes:**<br>• Throttle plate is binding, dirty or sticking<br>• Throttle valve is damaged or dirty<br>• Throttle valve control module is faulty<br>• TP sensor signal circuit open (inspect wiring & connector)<br>• TP sensor ground circuit open (inspect wiring & connector)<br>• TP sensor and/or control module is damaged or has failed<br>• MAF sensor signal is damaged, has failed or a short is present |

| DTC | Trouble Code Title, Conditions & Possible Causes |
|---|---|
| **DTC: P0122**<br>**1T CCM, MIL: Yes**<br>**Years: 2007, 2008**<br>**Models:** A4, A6<br>**Engines:** All<br>**Transmissions:** All | **Throttle/Pedal Position Sensor Circuit Low Input Conditions:**<br>Engine started, at idle, the temperature must be at least 80 degrees Celsius. The throttle position sensor supplies implausible signal to the ECM.<br>**Possible Causes:**<br>• TP sensor signal circuit open (inspect wiring & connector)<br>• TP sensor signal shorted to ground (inspect wiring & connector)<br>• TP sensor is damaged or has failed<br>• Throttle control module's voltage supply is shorted or open<br>• ECM has failed |
| **DTC: P0123**<br>**1T CCM, MIL: Yes**<br>**Years: 2007, 2008**<br>**Models:** A4, A6<br>**Engines:** All<br>**Transmissions:** All | **TP Sensor Circuit High Input Conditions:**<br>Engine started, at idle, the temperature must be at least 80 degrees Celsius. The ECM detected the TP sensor signal was more than the self-test maximum during testing.<br>**Possible Causes:**<br>• TP sensor not seated correctly in housing (may be damaged)<br>• TP sensor signal is circuit shorted to ground or system voltage<br>• TP sensor ground circuit is open (check the wiring harness)<br>• TP sensor and/or ECM has failed |
| **DTC: P0130**<br>**2T OBD/O2S1, MIL: Yes**<br>**Years: 2007, 2008**<br>**Models:** A4, A6<br>**Engines:** All<br>**Transmissions:** All | **O2 Sensor Circuit Bank 1 Sensor 1 Conditions:**<br>Engine running, battery voltage 11.5, all electrical components off, ground between engine and chassis well connected and the exhaust system must be properly sealed between catalytic converter and the cylinder head. The ECM detected the HO2S signal was implausible or not detected.<br>**Possible Causes:**<br>• Oxygen sensor heater for oxygen sensor (HO2S) before catalytic converter is faulty<br>• HO2S is contaminated (due to presence of silicone in fuel)<br>• HO2S signal and ground circuit wires crossed in wiring harness<br>• HO2S signal circuit is shorted to sensor or chassis ground<br>• HO2S element before the catalytic converter has failed (internal short condition)<br>• Leaks present in the exhaust manifold or exhaust pipes<br>• ECM has failed |
| **DTC: P0131**<br>**2T CCM, MIL: Yes**<br>**Years: 2007, 2008**<br>**Models:** A4, A6<br>**Engines:** All<br>**Transmissions:** All | **HO2S (Bank 1 Sensor 1) Circuit Low Input Conditions:**<br>Engine running, battery voltage 11.5, all electrical components off, ground between engine and chassis well connected and the exhaust system must be properly sealed between catalytic converter and the cylinder head. The ECM detected the HO2S signal was in a negative voltage range referred to as "character shift downward". This code sets when the HO2S signal remains in a low state for a measured period of time. In effect, it does not switch properly in the closed loop operation.<br>**Possible Causes:**<br>• HO2S is contaminated (due to presence of silicone in fuel)<br>• HO2S signal and ground circuit wires crossed in wiring harness<br>• HO2S signal circuit is shorted to sensor or chassis ground<br>• HO2S element has failed (internal short condition)<br>• Leaks present in the exhaust manifold or exhaust pipes<br>• ECM has failed |
| **DTC: P0132**<br>**2T CCM, MIL: Yes**<br>**Years: 2007, 2008**<br>**Models:** A4, A6<br>**Engines:** All<br>**Transmissions:** All | **HO2S (Bank 1 Sensor 1) Circuit High Input Conditions:**<br>Engine running, battery voltage 11.5, all electrical components off, ground between engine and chassis well connected and the exhaust system must be properly sealed between catalytic converter and the cylinder head. The ECM detected the HO2S signal was in a high state. This code sets when the HO2S signal remains in a high state for a measured period of time. In effect, it does not switch properly in the closed loop operation.<br>**Note: The HO2S signal circuit may be shorted to the heater power circuit due to tracking inside of the HO2S connector. Remove the connector and visually inspect the connector for signs of oil or water.**<br>**Possible Causes:**<br>• HO2S is contaminated (due to presence of silicone in fuel)<br>• HO2S signal and ground circuit wires crossed in wiring harness<br>• HO2S signal circuit is shorted to sensor or chassis ground<br>• HO2S element has failed (internal short condition)<br>• Leaks present in the exhaust manifold or exhaust pipes<br>• ECM has failed |

| DTC | Trouble Code Title, Conditions & Possible Causes |
|---|---|
| **DTC: P0133**<br>**2T OBD/O2S1, MIL: Yes**<br>**Years:** 2007, 2008<br>**Models:** A4, A6<br>**Engines:** All<br>**Transmissions:** All | **HO2S (Bank 1 Sensor 1) Circuit Slow Response Conditions:**<br>Engine running, battery voltage 11.5, all electrical components off, ground between engine and chassis well connected and the exhaust system must be properly sealed between catalytic converter and the cylinder head. The ECM detected the HO2S amplitude and frequency were out of the normal range (e.g., the HO2S rich to lean switch) during the HO2S Monitor test.<br>**Possible Causes:**<br>• HO2S before the three-way catalytic converter is contaminated (due to presence of silicone in fuel); Run the engine for three minutes at 3500 RPM as a self-cleaning effect<br>• HO2S signal circuit open<br>• Leaks present in the exhaust manifold or exhaust pipes<br>• HO2S is damaged or has failed<br>• ECM has failed |
| **DTC: P0134**<br>**2T OBD/O2S1, MIL: Yes**<br>**Years:** 2007, 2008<br>**Models:** A4, A6<br>**Engines:** All<br>**Transmissions:** All | **HO2S (Bank 1 Sensor 1) Circuit No Activity Conditions:**<br>Engine running, battery voltage 11.5, all electrical components off, ground between engine and chassis well connected and the exhaust system must be properly sealed between catalytic converter and the cylinder head. The ECM detected the HO2S signal failed to meet the maximum or minimum voltage levels (i.e., it failed the voltage range check).<br>**Possible Causes:**<br>• Leaks present in the exhaust manifold or exhaust pipes<br>• HO2S signal wire and ground wire crossed in connector (voltage jumps)<br>• HO2S element is fuel contaminated or has failed<br>• ECM has failed |
| **DTC: P0135**<br>**2T OBD/O2S1, MIL: Yes**<br>**Years:** 2007, 2008<br>**Models:** A4, A6<br>**Engines:** All<br>**Transmissions:** All | **HO2S (Bank 1 Sensor 1) Heater Circuit Malfunction Conditions:**<br>Engine running, battery voltage 11.5, all electrical components off, ground between engine and chassis well connected and the exhaust system must be properly sealed between catalytic converter and the cylinder head. The ECM detected an unexpected voltage condition, or it detected excessive current draw in the heater circuit during the CCM test.<br>**Possible Causes:**<br>• HO2S heater power circuit is open or heater ground circuit open<br>• HO2S signal tracking (due to oil or moisture in the connector)<br>• HO2S is damaged or has failed<br>• ECM has failed |
| **DTC: P0136**<br>**2T OBD/O2S1, MIL: Yes**<br>**Years:** 2007, 2008<br>**Models:** A4, A6<br>**Engines:** All<br>**Transmissions:** All | **HO2S (Bank 1 Sensor 2) Circuit Malfunction Conditions:**<br>Engine running, battery voltage 11.5, all electrical components off, ground between engine and chassis well connected and the exhaust system must be properly sealed between catalytic converter and the cylinder head. The ECM detected the HO2S signal failed to meet the maximum or minimum voltage levels (i.e., it failed the voltage range check).<br>**Possible Causes:**<br>• Leaks present in the exhaust manifold or exhaust pipes<br>• HO2S signal wire and ground wire crossed in connector<br>• HO2S element is fuel contaminated or has failed<br>• ECM has failed |
| **DTC: P0137**<br>**2T CCM, MIL: Yes**<br>**Years:** 2007, 2008<br>**Models:** A4, A6<br>**Engines:** All<br>**Transmissions:** All | **HO2S (Bank 1 Sensor 2) Circuit Low Input Conditions:**<br>Engine running, battery voltage 11.5, all electrical components off, ground between engine and chassis well connected and the exhaust system must be properly sealed between catalytic converter and the cylinder head. The ECM detected the HO2S signal remained in a high state.<br>**Note: The HO2S signal circuit may be shorted to the heater power circuit due to "tracking inside of the HO2S connector. Remove the connector and visually inspect the connector for signs of oil or water.**<br>**Possible Causes:**<br>• HO2S signal shorted to heater power circuit in the connector<br>• HO2S signal circuit shorted to ground (for more than 200 seconds) or to system voltage<br>• ECM has failed |
| **DTC: P0138**<br>**2T CCM, MIL: Yes**<br>**Years:** 2007, 2008<br>**Models:** A4, A6<br>**Engines:** All<br>**Transmissions:** All | **HO2S (Bank 1 Sensor 2) Circuit High Input Conditions:**<br>Engine running, battery voltage 11.5, all electrical components off, ground between engine and chassis well connected and the exhaust system must be properly sealed between catalytic converter and the cylinder head. The ECM detected the HO2S signal remained in a high state.<br>**Note: The HO2S signal circuit may be shorted to the heater power circuit due to "tracking inside of the HO2S connector. Remove the connector and visually inspect the connector for signs of oil or water.**<br>**Possible Causes:**<br>• HO2S signal shorted to heater power circuit in the positive connector<br>• HO2S signal circuit shorted to ground or to system voltage<br>• HO2S has failed<br>• ECM has failed |

| DTC | Trouble Code Title, Conditions & Possible Causes |
|---|---|
| **DTC: P0139**<br>**2T CCM, MIL: Yes**<br>**Years:** 2007, 2008<br>**Models:** A4, A6<br>**Engines:** All<br>**Transmissions:** All | **HO2S (Bank 1 Sensor 2) Slow Response Conditions:**<br>Engine running, battery voltage 11.5, all electrical components off, ground between engine and chassis well connected and the exhaust system must be properly sealed between catalytic converter and the cylinder head. The ECM detected the HO2S amplitude and frequency were out of the normal range during the HO2S Monitor test.<br>**Possible Causes:**<br>• HO2S signal shorted to heater power circuit in the connector<br>• HO2S signal circuit shorted to VREF or to system voltage<br>• ECM has failed |
| **DTC: P0140**<br>**2T OBD/O2S1, MIL: Yes**<br>**Years:** 2007, 2008<br>**Models:** A4, A6<br>**Engines:** All<br>**Transmissions:** All | **HO2S (Bank 1 Sensor 2) No Activity Conditions:**<br>Engine running, battery voltage 11.5, all electrical components off, ground between engine and chassis well connected and the exhaust system must be properly sealed between catalytic converter and the cylinder head. The ECM detected the HO2S signal failed to meet the maximum or minimum voltage levels (i.e., it failed the voltage range check).<br>**Possible Causes:**<br>• HO2S before the three-way catalytic converter is contaminated (due to presence of silicone in fuel); Run the engine for three minutes at 3500 RPM as a self-cleaning effect<br>• Leaks present in the exhaust manifold or exhaust pipes<br>• HO2S signal wire and ground wire crossed in connector (voltage jumps)<br>• HO2S element is contaminated or has failed<br>• ECM has failed |
| **DTC: P0141**<br>**2T OBD/O2S1, MIL: Yes**<br>**Years:** 2007, 2008<br>**Models:** A4, A6<br>**Engines:** All<br>**Transmissions:** All | **HO2S (Bank 1 Sensor 2) Malfunction Conditions:**<br>Engine running, battery voltage 11.5, all electrical components off, ground between engine and chassis well connected and the exhaust system must be properly sealed between catalytic converter and the cylinder head. The ECM detected the HO2S signal failed to meet the maximum or minimum voltage levels (i.e., it failed the voltage range check).<br>**Possible Causes:**<br>• Leaks present in the exhaust manifold or exhaust pipes<br>• HO2S signal wire and ground wire crossed in connector<br>• HO2S element is fuel contaminated or has failed<br>• ECM has failed |
| **DTC: P0150**<br>**2T CCM, MIL: Yes**<br>**Years:** 2007, 2008<br>**Models:** A4, A6<br>**Engines:** All<br>**Transmissions:** All | **HO2S (Bank 2 Sensor 1) Circuit Malfunction Conditions:**<br>Engine running, battery voltage 11.5, all electrical components off, ground between engine and chassis well connected and the exhaust system must be properly sealed between catalytic converter and the cylinder head. The ECM detected the HO2S signal failed to meet the maximum or minimum voltage levels (i.e., it failed the voltage range check).<br>**Possible Causes:**<br>• Leaks present in the exhaust manifold or exhaust pipes<br>• HO2S signal wire and ground wire crossed in connector<br>• HO2S element is fuel contaminated or has failed<br>• ECM has failed |
| **DTC: P0151**<br>**2T CCM, MIL: Yes**<br>**Years:** 2007, 2008<br>**Models:** A4, A6<br>**Engines:** All<br>**Transmissions:** All | **HO2S (Bank 2 Sensor 1) Low Input Conditions:**<br>Engine running, battery voltage 11.5, all electrical components off, ground between engine and chassis well connected and the exhaust system must be properly sealed between catalytic converter and the cylinder head. The ECM detected the HO2S signal remained in a high state.<br>**Note: The HO2S signal circuit may be shorted to the heater power circuit due to "tracking inside of the HO2S connector. Remove the connector and visually inspect the connector for signs of oil or water.**<br>**Possible Causes:**<br>• HO2S is contaminated (due to presence of silicone in fuel)<br>• HO2S signal tracking (due to oil or moisture in the connector)<br>• HO2S signal circuit is open or shorted to VREF<br>• ECM has failed |
| **DTC: P0152**<br>**2T CCM, MIL: Yes**<br>**Years:** 2007, 2008<br>**Models:** A4, A6<br>**Engines:** All<br>**Transmissions:** All | **HO2S (Bank 2 Sensor 1) Circuit High Input Conditions:**<br>Engine running, battery voltage 11.5, all electrical components off, ground between engine and chassis well connected and the exhaust system must be properly sealed between catalytic converter and the cylinder head. The ECM detected the HO2S signal remained in a high state (more than 1.5v).<br>**Note: The HO2S signal circuit may be shorted to the heater power circuit due to "tracking inside of the HO2S connector. Remove the connector and visually inspect the connector for signs of oil or water.**<br>**Possible Causes:**<br>• HO2S is contaminated (due to presence of silicone in fuel)<br>• HO2S signal tracking (due to oil or moisture in the connector)<br>• HO2S signal circuit is open or shorted to VREF<br>• ECM has failed |

| DTC | Trouble Code Title, Conditions & Possible Causes |
|---|---|
| **DTC: P0153**<br>**2T CCM, MIL: Yes**<br>**Years:** 2007, 2008<br>**Models:** A4, A6<br>**Engines:** All<br>**Transmissions:** All | **HO2S (Bank 2 Sensor 1) Circuit Slow Response Conditions:**<br>Engine running, battery voltage 11.5, all electrical components off, ground between engine and chassis well connected and the exhaust system must be properly sealed between catalytic converter and the cylinder head. The the ECM detected the HO2S amplitude and frequency were out of the normal range during the HO2S Monitor test.<br>**Possible Causes:**<br>• HO2S is contaminated (due to presence of silicone in fuel)<br>• Leaks present in the exhaust manifold or exhaust pipes<br>• HO2S is damaged or has failed<br>• ECM has failed |
| **DTC: P0154**<br>**2T OBD/O2S1, MIL: Yes**<br>**Years:** 2007, 2008<br>**Models:** A4, A6<br>**Engines:** All<br>**Transmissions:** All | **HO2S (Bank 2 Sensor 1) Circuit No Activity Conditions:**<br>Engine running, battery voltage 11.5, all electrical components off, ground between engine and chassis well connected and the exhaust system must be properly sealed between catalytic converter and the cylinder head. The ECM detected the HO2S signal failed to meet the maximum or minimum voltage (i.e., it failed the voltage check).<br>**Possible Causes:**<br>• Leaks present in the exhaust manifold or exhaust pipes<br>• HO2S signal wire and ground wire crossed in connector<br>• HO2S element is fuel contaminated or has failed<br>• ECM has failed |
| **DTC: P0155**<br>**2T OBD/O2S1, MIL: Yes**<br>**Years:** 2007, 2008<br>**Models:** A4, A6<br>**Engines:** All<br>**Transmissions:** All | **HO2S (Bank 2 Sensor 1) Heater Circuit Malfunction Conditions:**<br>Engine running, battery voltage 11.5, all electrical components off, ground between engine and chassis well connected and the exhaust system must be properly sealed between catalytic converter and the cylinder head. The ECM detected an open or shorted condition, or excessive current draw in the heater circuit.<br>**Possible Causes:**<br>• HO2S heater power circuit is open<br>• HO2S heater ground circuit is open<br>• HO2S signal tracking (due to oil or moisture in the connector)<br>• HO2S is damaged or has failed<br>• ECM has failed |
| **DTC: P0156**<br>**2T OBD/O2S1, MIL: Yes**<br>**Years:** 2007, 2008<br>**Models:** A4, A6<br>**Engines:** All<br>**Transmissions:** All | **HO2S (Bank 2 Sensor 2) Circuit No Activity Conditions:**<br>Engine running, battery voltage 11.5, all electrical components off, ground between engine and chassis well connected and the exhaust system must be properly sealed between catalytic converter and the cylinder head. The ECM detected the HO2S signal failed to meet the maximum or minimum voltage (i.e., it failed the voltage check).<br>**Possible Causes:**<br>• Leaks present in the exhaust manifold or exhaust pipes<br>• HO2S signal wire and ground wire crossed in connector<br>• HO2S element is fuel contaminated or has failed<br>• ECM has failed |
| **DTC: P0157**<br>**2T OBD/O2S1, MIL: Yes**<br>**Years:** 2007, 2008<br>**Models:** A4, A6<br>**Engines:** All<br>**Transmissions:** All | **HO2S (Bank 2 Sensor 2) Circuit Low Voltage Conditions:**<br>Engine running, battery voltage 11.5, all electrical components off, ground between engine and chassis well connected and the exhaust system must be properly sealed between catalytic converter and the cylinder head. The ECM detected the HO2S signal remained in a high state.<br>**Note: The HO2S signal circuit may be shorted to the heater power circuit due to "tracking inside of the HO2S connector. Remove the connector and visually inspect the connector for signs of oil or water**<br>**Possible Causes:**<br>• HO2S is contaminated (due to presence of silicone in fuel)<br>• HO2S signal tracking (due to oil or moisture in the connector)<br>• HO2S signal circuit is open or shorted to VREF<br>• ECM has failed |
| **DTC: P0158**<br>**2T OBD/O2S1, MIL: Yes**<br>**Years:** 2007, 2008<br>**Models:** A4, A6<br>**Engines:** All<br>**Transmissions:** All | **HO2S (Bank 2 Sensor 2) Circuit High Input Conditions:**<br>Engine running, battery voltage 11.5, all electrical components off, ground between engine and chassis well connected and the exhaust system must be properly sealed between catalytic converter and the cylinder head. The ECM detected the HO2S signal remained in a high state (i.e., more than 1.5v).<br>**Note: The HO2S signal circuit may be shorted to the heater power circuit due to "tracking inside of the HO2S connector. Remove the connector and visually inspect the connector for signs of oil or water.**<br>**Possible Causes:**<br>• HO2S signal shorted to the heater power circuit (due to oil or moisture in the connector)<br>• HO2S signal circuit shorted to VREF or to system voltage<br>• ECM has failed |

| DTC | Trouble Code Title, Conditions & Possible Causes |
|---|---|
| **DTC: P0159**<br>**2T OBD/O2S1, MIL: Yes**<br>**Years:** 2007, 2008<br>**Models:** A4, A6<br>**Engines:** All<br>**Transmissions:** All | **HO2S (Bank 2 Sensor 2) Circuit Slow Response Conditions:**<br>Engine running, battery voltage 11.5, all electrical components off, ground between engine and chassis well connected and the exhaust system must be properly sealed between catalytic converter and the cylinder head. The ECM detected the HO2S amplitude and frequency were out of the normal range during the HO2S Monitor test.<br>**Possible Causes:**<br>• HO2S is contaminated (due to presence of silicone in fuel)<br>• Leaks present in the exhaust manifold or exhaust pipes<br>• HO2S is damaged or has failed<br>• ECM has failed |
| **DTC: P0160**<br>**2T OBD/O2S1, MIL: Yes**<br>**Years:** 2007, 2008<br>**Models:** A4, A6<br>**Engines:** All<br>**Transmissions:** All | **HO2S (Bank 2 Sensor 2) Circuit No Activity Detected Conditions:**<br>Engine running, battery voltage 11.5, all electrical components off, ground between engine and chassis well connected and the exhaust system must be properly sealed between catalytic converter and the cylinder head. The ECM detected the HO2S signal failed to meet the maximum or minimum voltage (i.e., it failed the voltage check).<br>**Possible Causes:**<br>• Leaks present in the exhaust manifold or exhaust pipes<br>• HO2S signal wire and ground wire crossed in connector<br>• HO2S element is fuel contaminated or has failed<br>• ECM has failed |
| **DTC: P0161**<br>**2T OBD/O2S1, MIL: Yes**<br>**Years:** 2007, 2008<br>**Models:** A4, A6<br>**Engines:** All<br>**Transmissions:** All | **HO2S (Bank 2 Sensor 2) Heater Circuit Malfunction Conditions:**<br>Engine running, battery voltage 11.5, all electrical components off, ground between engine and chassis well connected and the exhaust system must be properly sealed between catalytic converter and the cylinder head. The the ECM detected an open or shorted condition, or excessive current draw in the heater circuit.<br>**Possible Causes:**<br>• HO2S heater power circuit or the heater ground circuit is open<br>• HO2S signal tracking (due to oil or moisture in the connector)<br>• HO2S has failed, or the ECM has failed |
| **DTC: P0170**<br>**2T CCM, MIL: Yes**<br>**Years:** 2007, 2008<br>**Models:** A4, A6<br>**Engines:** All<br>**Transmissions:** All | **Fuel System Malfunction (Cylinder Bank 1) Conditions:**<br>The engine is running in a closed loop at a stable engine speed, and the ECM detected the lean or rich fuel trim correction valve was more than or less than a calibrated limit.<br>**Possible Causes:**<br>• Air leaks after the MAF sensor, or leaks in the PCV system<br>• Exhaust leaks before or near where the HO2S is mounted<br>• Fuel injector(s) restricted or not supplying enough fuel<br>• Fuel system not supplying enough fuel during high fuel demand conditions (e.g., the fuel pump may not supply enough fuel)<br>• Leaking EGR gasket, or leaking EGR valve diaphragm<br>• MAF sensor dirty (causes ECM to underestimate airflow)<br>• Vehicle running out of fuel or engine oil dip stick not seated |
| **DTC: P0171**<br>**2T CCM, MIL: Yes**<br>**Years:** 2007, 2008<br>**Models:** A4, A6<br>**Engines:** All<br>**Transmissions:** All | **Fuel System Too Lean (Cylinder Bank 1) Conditions:**<br>Key on or engine running, all electrical components off and coolant temperature at least 80 degrees Celsius; and the ECM detected the Bank 1 Adaptive Fuel Control System reached its rich correction limit (a lean A/F condition).<br>**Possible Causes:**<br>• Air leaks after the MAF sensor, or leaks in the PCV system<br>• Exhaust leaks before or near where the HO2S is mounted<br>• Fuel injector(s) restricted or not supplying enough fuel<br>• Fuel pump not supplying enough fuel during high fuel demand conditions<br>• Leaking EGR gasket, or leaking EGR valve diaphragm<br>• MAF sensor dirty (causes ECM to underestimate airflow)<br>• Vehicle running out of fuel or engine oil dip stick not seated |
| **DTC: P0172**<br>**2T CCM, MIL: Yes**<br>**Years:** 2007, 2008<br>**Models:** A4, A6<br>**Engines:** All<br>**Transmissions:** All | **Fuel System Too Rich (Cylinder Bank 1) Conditions:**<br>Key on or engine running, all electrical components off and coolant temperature at least 80 degrees Celsius; and the ECM detected the Bank 1 Adaptive Fuel Control System reached its rich correction limit (a rich A/F condition).<br>**Possible Causes:**<br>• Camshaft timing is incorrect, or the engine has an oil overfill condition<br>• EVAP vapor recovery system failure (may be pulling vacuum)<br>• Fuel pressure regulator is damaged or leaking<br>• HO2S element is contaminated with alcohol or water<br>• MAF or MAP sensor values are incorrect or out-of-range<br>• One of more fuel injectors is leaking |

| DTC | Trouble Code Title, Conditions & Possible Causes |
|---|---|
| **DTC: P0173**<br>**2T CCM, MIL: Yes**<br>**Years:** 1995, 1996<br>**Models:** A4, A6<br>**Engines:** All<br>**Transmissions:** All | **Fuel System Malfunction (Cylinder Bank 1) Conditions:**<br>Key on or engine running, all electrical components off and coolant temperature at least 80 degrees Celsius; and the ECM detected the Bank 1 Fuel Control System experienced a implausible signal<br>**Possible Causes:**<br>• Air leaks after the MAF sensor, or leaks in the PCV system<br>• Exhaust leaks before or near where the HO2S is mounted<br>• Fuel injector(s) restricted or not supplying enough fuel<br>• Fuel system not supplying enough fuel during high fuel demand conditions (e.g., the fuel pump may not supply enough fuel)<br>• Leaking EGR gasket, or leaking EGR valve diaphragm<br>• MAF sensor dirty (causes ECM to underestimate airflow)<br>• Vehicle running out of fuel or engine oil dip stick not seated |
| **DTC: P0174**<br>**2T CCM, MIL: Yes**<br>**Years:** 2007, 2008<br>**Models:** A4, A6<br>**Engines:** All<br>**Transmissions:** All | **Fuel System Too Lean (Cylinder Bank 2) Conditions:**<br>Key on or engine running, all electrical components off and coolant temperature at least 80 degrees Celsius; and the ECM detected the Bank 2 Fuel Control System reached its lean correction limit<br>**Possible Causes:**<br>• Air leaks after the MAF sensor, or leaks in the PCV system<br>• Exhaust leaks before or near where the HO2S is mounted<br>• Fuel injector(s) restricted or not supplying enough fuel<br>• Fuel pump not supplying enough fuel during high fuel demand conditions<br>• Leaking EGR gasket, or leaking EGR valve diaphragm<br>• MAF sensor dirty (causes ECM to underestimate airflow)<br>• Vehicle running out of fuel or engine oil dip stick not seated |
| **DTC: P0175**<br>**2T CCM, MIL: Yes**<br>**Years:** 2007, 2008<br>**Models:** A4, A6<br>**Engines:** All<br>**Transmissions:** All | **Fuel System Too Rich (Cylinder Bank 2) Conditions:**<br>Key on or engine running, all electrical components off and coolant temperature at least 80 degrees Celsius; and the ECM detected the Bank 2 Adaptive Fuel Control System reached its rich correction limit (a rich A/F condition).<br>**Possible Causes:**<br>• Air leaks after the MAF sensor, or leaks in the PCV system<br>• Exhaust leaks before or near where the HO2S is mounted<br>• Fuel injector(s) restricted or not supplying enough fuel<br>• Fuel pump not supplying enough fuel during high fuel demand conditions<br>• Leaking EGR gasket, or leaking EGR valve diaphragm<br>• MAF sensor dirty (causes ECM to underestimate airflow)<br>• Vehicle running out of fuel or engine oil dip stick not seated |
| **DTC: P0190**<br>**1T CCM, MIL: Yes**<br>**Years:** 2007, 2008<br>**Models:** A6<br>**Engines:** 3.2L<br>**Transmissions:** All | **Fuel Rail Pressure Sensor Circuit Conditions**<br>Key on or engine running, all electrical components off and coolant temperature at least 80 degrees Celsius; and the ECM detected the fuel rail pressure sensor signal was outside the required voltage parameters in the self-test.<br>**Possible Causes:**<br>• Fuel Pressure Regulator Valve faulty<br>• Fuel Pressure Sensor faulty<br>• Fuel Pump (FP) Control Module faulty<br>• Fuel pump faulty |
| **DTC: P0192**<br>**1T CCM, MIL: Yes**<br>**Years:** 2007, 2008<br>**Models:** A6<br>**Engines:** 3.2L<br>**Transmissions:** All | **Fuel Rail Pressure Sensor Circuit Low Conditions**<br>Key on or engine running, all electrical components off and coolant temperature at least 80 degrees Celsius; and the ECM detected the fuel rail pressure sensor signal was below the required voltage in the self-test.<br>**Possible Causes:**<br>• Fuel Pressure Regulator Valve faulty<br>• Fuel Pressure Sensor faulty<br>• Fuel Pump (FP) Control Module faulty<br>• Fuel pump faulty |
| **DTC: P0201**<br>**1T CCM, MIL: Yes**<br>**Years:** 2007, 2008<br>**Models:** A4, A6<br>**Engines:** All<br>**Transmissions:** All | **Cylinder 1 Injector Circuit Malfunction Conditions:**<br>Engine started, and the ECM detected the fuel injector "1" control circuit was in a high state when it should have been low, or in a low state when it should have been high (wiring harness & injector okay).<br>**Possible Causes:**<br>• Injector 1 connector is damaged, open or shorted<br>• Injector 1 control circuit is open, shorted to ground or to power<br>• ECM has failed (the injector driver circuit may be damaged) |

| DTC | Trouble Code Title, Conditions & Possible Causes |
|---|---|
| **DTC: P0202**<br>**1T CCM, MIL: Yes**<br>**Years: 2007, 2008**<br>**Models:** A4, A6<br>**Engines:** All<br>**Transmissions:** All | **Cylinder 2 Injector Circuit Malfunction Conditions:**<br>Engine started, and the ECM detected the fuel injector "2" control circuit was in a high state when it should have been low, or in a low state when it should have been high (wiring harness & injector okay).<br>**Possible Causes:**<br> • Injector 2 connector is damaged, open or shorted<br> • Injector 2 control circuit is open, shorted to ground or to power<br> • ECM has failed (the injector driver circuit may be damaged) |
| **DTC: P0203**<br>**1T CCM, MIL: Yes**<br>**Years: 2007, 2008**<br>**Models:** A4, A6<br>**Engines:** All<br>**Transmissions:** All | **Cylinder 3 Injector Circuit Malfunction Conditions:**<br>Engine started, and the ECM detected the fuel injector "3" control circuit was in a high state when it should have been low, or in a low state when it should have been high (wiring harness & injector okay).<br>**Possible Causes:**<br> • Injector 3 connector is damaged, open or shorted<br> • Injector 3 control circuit is open, shorted to ground or to power<br> • ECM has failed (the injector driver circuit may be damaged) |
| **DTC: P0204**<br>**1T CCM, MIL: Yes**<br>**Years: 2007, 2008**<br>**Models:** A4, A6<br>**Engines:** All<br>**Transmissions:** All | **Cylinder 4 Injector Circuit Malfunction Conditions:**<br>Engine started, and the ECM detected the fuel injector "4" control circuit was in a high state when it should have been low, or in a low state when it should have been high (wiring harness & injector okay).<br>**Possible Causes:**<br> • Injector 4 connector is damaged, open or shorted<br> • Injector 4 control circuit is open, shorted to ground or to power<br> • ECM has failed (the injector driver circuit may be damaged) |
| **DTC: P0205**<br>**1T CCM, MIL: Yes**<br>**Years: 2007, 2008**<br>**Models:** A4, A6<br>**Engines:** All<br>**Transmissions:** All | **Cylinder 5 Injector Circuit Malfunction Conditions:**<br>Engine started, and the ECM detected the fuel injector "5" control circuit was in a high state when it should have been low, or in a low state when it should have been high (wiring harness & injector okay).<br>**Possible Causes:**<br> • Injector 5 connector is damaged, open or shorted<br> • Injector 5 control circuit is open, shorted to ground or to power<br> • ECM has failed (the injector driver circuit may be damaged) |
| **DTC: P0206**<br>**1T CCM, MIL: Yes**<br>**Years: 2007, 2008**<br>**Models:** A4, A6<br>**Engines:** All<br>**Transmissions:** All | **Cylinder 6 Injector Circuit Malfunction Conditions:**<br>Engine started, and the ECM detected the fuel injector control circuit was in a high state when it should have been low, or in a low state when it should have been high (wiring harness & injector okay).<br>**Possible Causes:**<br> • Injector 6 connector is damaged, open or shorted<br> • Injector 6 control circuit is open, shorted to ground or to power<br> • ECM has failed (the injector driver circuit may be damaged) |
| **DTC: P0207**<br>**1T CCM, MIL: Yes**<br>**Years: 2007, 2008**<br>**Models:** A6<br>**Engines:** All<br>**Transmissions:** All | **Cylinder 7 Injector Circuit Malfunction Conditions:**<br>Engine started, and the ECM detected the fuel injector "7" control circuit was in a high state when it should have been low, or in a low state when it should have been high (wiring harness & injector okay).<br>**Note: Monitor the INJIF PID Fault "flags" with the Scan Tool. The appropriate INJF PID "flag" will read Yes when this code is set.**<br>**Possible Causes:**<br> • Injector 7 connector is damaged, open or shorted<br> • Injector 7 control circuit is open, shorted to ground or to power<br> • ECM has failed (the injector driver circuit may be damaged) |
| **DTC: P0208**<br>**1T CCM, MIL: Yes**<br>**Years: 2007, 2008**<br>**Models:** A6<br>**Engines:** All<br>**Transmissions:** All | **Cylinder 8 Injector Circuit Malfunction Conditions:**<br>Engine started, and the ECM detected the fuel injector "8" control circuit was in a high state when it should have been low, or in a low state when it should have been high (wiring harness & injector okay).<br>**Note: Monitor the INJIF PID Fault "flags" with the Scan Tool. The appropriate INJF PID "flag" will read Yes when this code is set.**<br>**Possible Causes:**<br> • Injector 8 connector is damaged, open or shorted<br> • Injector 8 control circuit is open, shorted to ground or to power<br> • ECM has failed (the injector driver circuit may be damaged) |
| **DTC: P0219**<br>**1T CCM, MIL: Yes**<br>**Years:** 2007, 2008<br>**Models:** A6<br>**Engines:** 3.2L, 4.2L<br>**Transmissions:** All | **Engine Over-Speed Condition Conditions:**<br>Engine started, and the ECM determined the vehicle had been driven in a manner that caused the engine to over-speed, and to exceed the engine speed calibration limit stored in memory.<br>**Possible Causes:**<br> • Engine operated in the wrong transmission gear position<br> • Excessive engine speed with gear selector in Neutral position<br> • Wheel slippage due to wet, muddy or snowing conditions |

| DTC | Trouble Code Title, Conditions & Possible Causes |
|---|---|
| **DTC: P0221**<br>**1T CCM, MIL: Yes**<br>**Years: 2007, 2008**<br>**Models:** A4, A6<br>**Engines:** All<br>**Transmissions:** All | **Throttle Position Sensor 'B' Signal Performance Conditions:**<br>Engine started, battery voltage at least 11.5v, all electrical components off, ground connections between engine and chassis well connected, coolant temperature at least 80-degrees Celicius and the throttle valve must not be damaged or dirty; and the ECM detected the TP Sensor 'B' circuit was out of its normal operating range during a condition with the throttle wide open, or with it completely closed. The throttle valve activation occurs via an electric motor (throttle drive) in the throttle valve control module. It is activated by the ECM according to specifications of the two sensors, Throttle Position Sensor and Accelerator Pedal Position Sensor 2. Slowly depress accelerator pedal up to Wide Open Throttle (WOT) stop while observing the percentage display on the PID data function of the scan tool. The percentage display must increase uniformly.<br>**Possible Causes:**<br>• Throttle body is damaged<br>• Throttle linkage is binding or sticking<br>• ETC TP Sensor 'B' signal circuit to the ECM is open<br>• ETC TP Sensor 'B' ground circuit is open<br>• ETC TP Sensor 'B' is damaged or it has failed |
| **DTC: P0222**<br>**1T CCM, MIL: Yes**<br>**Years: 2007, 2008**<br>**Models:** A4, A6<br>**Engines:** All<br>**Transmissions:** All | **Throttle Position Sensor 'B' Circuit Low Input Conditions:**<br>Engine started, battery voltage at least 11.5v, all electrical components off, ground connections between engine and chassis well connected, coolant temperature at least 80-degrees Celicius and the throttle valve must not be damaged or dirty; and the ECM detected the TP Sensor 'B' circuit was out of its normal operating range during a condition with the throttle wide open, or with it completely closed. The throttle valve activation occurs via an electric motor (throttle drive) in the throttle valve control module. It is activated by the ECM according to specifications of the two sensors, Throttle Position Sensor and Accelerator Pedal Position Sensor 2. Slowly depress accelerator pedal up to Wide Open Throttle (WOT) stop while observing the percentage display on the PID data function of the scan tool. The percentage display must increase uniformly.<br>**Possible Causes:**<br>• ETC TP Sensor 'B' connector is damaged or shorted<br>• ETC TP Sensor 'B' signal circuit is shorted to ground<br>• ETC TP Sensor 'B' is damaged or it has failed<br>• ECM has failed |
| **DTC: P0223**<br>**1T CCM, MIL: Yes**<br>**Years: 2007, 2008**<br>**Models:** A4, A6<br>**Engines:** All<br>**Transmissions:** All | **Throttle Position Sensor 'B' Circuit High Input Conditions:**<br>Engine started, battery voltage at least 11.5v, all electrical components off, ground connections between engine and chassis well connected, coolant temperature at least 80-degrees Celicius and the throttle valve must not be damaged or dirty; and the ECM detected the TP Sensor 'B' circuit was out of its normal operating range during a condition with the throttle wide open, or with it completely closed. The throttle valve activation occurs via an electric motor (throttle drive) in the throttle valve control module. It is activated by the ECM according to specifications of the two sensors, Throttle Position Sensor and Accelerator Pedal Position Sensor 2. Slowly depress accelerator pedal up to Wide Open Throttle (WOT) stop while observing the percentage display on the PID data function of the scan tool. The percentage display must increase uniformly.<br>**Possible Causes:**<br>• ETC TP Sensor 'B' connector is damaged or open<br>• ETC TP Sensor 'B' signal circuit is open<br>• ETC TP Sensor 'B' signal circuit is shorted to VREF (5v)<br>• ETC TP Sensor 'B' is damaged or it has failed |
| **DTC: P0230**<br>**1T CCM, MIL: Yes**<br>**Years: 2007, 2008**<br>**Models:** A4, A6<br>**Engines:** All<br>**Transmissions:** All | **Fuel Pump Primary Circuit Malfunction Conditions:**<br>Engine started, battery voltage at least 11.5v, all electrical components off, ground connections between engine and chassis well connected, coolant temperature at least 80-degrees Celicius. The ECM detected high current in fuel pump or fuel shutoff valve (FSV) circuit, or it detected voltage with the valve off, or it did not detect voltage on the circuit. The circuit is used to energize the fuel pump relay at key on or while running. Fuel pressure value should be 3000 to 5000 kPa at idle.<br>**Possible Causes:**<br>• FP or FSV circuit is open or shorted<br>• Fuel pump relay VPWR circuit open<br>• Fuel pump relay is damaged or has failed<br>• Fuel pressure sensor has failed<br>• Fuel pump control module is faulty<br>• ECM has failed |

| DTC | Trouble Code Title, Conditions & Possible Causes |
|---|---|
| **DTC: P0234**<br>**1T CCM, MIL: Yes**<br>**Years:** 1997, 1998, 1999, 2000, 2001, 2002, 2003, 2004, 2005, 2006<br>**Models:** A4<br>**Engines:** All<br>**Transmissions:** All | **Turbo/Supercharger Overboost Condition Conditions:**<br>Engine started, battery voltage at least 11.5v, all electrical components off, ground connections between engine and chassis well connected, coolant temperature at least 80-degrees Celicius. The ECM detected an operating condition that could harm the engine or automatic transmission.<br>**Possible Causes:**<br>• Ignition misfire condition exceeds the calibrated threshold<br>• Knock sensor circuit has failed, or excessive knock detected<br>• Low speed fuel pump relay not switching properly<br>• Transmission oil temperature beyond the calibrated threshold<br>• Shaft bearing of charge pressure regulator valve in turbocharger is blocked |
| **DTC: P0261**<br>**2T CCM, MIL: Yes**<br>**Years:** 2007, 2008<br>**Models:** A4, A6<br>**Engines:** All<br>**Transmissions:** All | **Cylinder 1 Injector Circuit Low Input/Short to Ground Conditions:**<br>Key on or engine running, fuses in the instrument panel and the E-box in the engine compartment must be functioning, and the ground connections between the engine ad the chassis must be well connected; and the ECM detected an unexpected voltage condition on the injector circuit<br>**Possible Causes:**<br>• Injector 1 control circuit is open<br>• Injector 1 power circuit (B+) is open<br>• Injector 1 control circuit is shorted to chassis ground<br>• Injector 1 is damaged or has failed<br>• ECM is not connected or has failed |
| **DTC: P0262**<br>**2T CCM, MIL: Yes**<br>**Years:** 2007, 2008<br>**Models:** A4, A6<br>**Engines:** All<br>**Transmissions:** All | **Cylinder 1 Injector Circuit Low Input/Short to B+ Conditions:**<br>Key on or engine running, fuses in the instrument panel and the E-box in the engine compartment must be functioning, and the ground connections between the engine ad the chassis must be well connected; and the ECM detected an unexpected voltage condition on the injector circuit<br>**Possible Causes:**<br>• Injector control circuit is open<br>• Injector power circuit (B+) is open<br>• Injector control circuit is shorted to chassis ground<br>• Injector is damaged or has failed<br>• ECM is not connected or has failed<br>• Fuel pump relay has failed<br>• Fuel injectors may have malfunctioned<br>• Faulty engine speed sensor |
| **DTC: P0264**<br>**2T CCM, MIL: Yes**<br>**Years:** 2007, 2008<br>**Models:** A4, A6<br>**Engines:** All<br>**Transmissions:** All | **Cylinder 2 Injector Circuit Low Input/Short to Ground Conditions:**<br>Key on or engine running, fuses in the instrument panel and the E-box in the engine compartment must be functioning, and the ground connections between the engine ad the chassis must be well connected; and the ECM detected an unexpected voltage condition on the injector circuit<br>**Possible Causes:**<br>• Injector control circuit is open<br>• Injector power circuit (B+) is open<br>• Injector control circuit is shorted to chassis ground<br>• Injector is damaged or has failed<br>• ECM is not connected or has failed<br>• Fuel pump relay has failed<br>• Fuel injectors may have malfunctioned<br>• Faulty engine speed sensor |
| **DTC: P0265**<br>**2T CCM, MIL: Yes**<br>**Years:** 2007, 2008<br>**Models:** A4, A6<br>**Engines:** All<br>**Transmissions:** All | **Cylinder 2 Injector Circuit Low Input/Short to B+ Conditions:**<br>Key on or engine running, fuses in the instrument panel and the E-box in the engine compartment must be functioning, and the ground connections between the engine ad the chassis must be well connected; and the ECM detected an unexpected voltage condition on the injector circuit<br>**Possible Causes:**<br>• Injector control circuit is open<br>• Injector power circuit (B+) is open<br>• Injector control circuit is shorted to chassis ground<br>• Injector is damaged or has failed<br>• ECM is not connected or has failed<br>• Fuel pump relay has failed<br>• Fuel injectors may have malfunctioned<br>• Faulty engine speed sensor |

| DTC | Trouble Code Title, Conditions & Possible Causes |
|---|---|
| **DTC: P0267**<br>**2T CCM, MIL: Yes**<br>**Years: 2007, 2008**<br>**Models:** A4, A6<br>**Engines:** All<br>**Transmissions:** All | **Cylinder 3 Injector Circuit Low Input/Short to Ground Conditions:**<br>Key on or engine running, fuses in the instrument panel and the E-box in the engine compartment must be functioning, and the ground connections between the engine ad the chassis must be well connected; and the ECM detected an unexpected voltage condition on the injector circuit<br>**Possible Causes:**<br>• Injector control circuit is open<br>• Injector power circuit (B+) is open<br>• Injector control circuit is shorted to chassis ground<br>• Injector is damaged or has failed<br>• ECM is not connected or has failed<br>• Fuel pump relay has failed<br>• Fuel injectors may have malfunctioned<br>• Faulty engine speed sensor |
| **DTC: P0268**<br>**2T CCM, MIL: Yes**<br>**Years: 2007, 2008**<br>**Models:** A4, A6<br>**Engines:** All<br>**Transmissions:** All | **Cylinder 3 Injector Circuit Low Input/Short to B+ Conditions:**<br>Key on or engine running, fuses in the instrument panel and the E-box in the engine compartment must be functioning, and the ground connections between the engine ad the chassis must be well connected; and the ECM detected an unexpected voltage condition on the injector circuit<br>**Possible Causes:**<br>• Injector control circuit is open<br>• Injector power circuit (B+) is open<br>• Injector control circuit is shorted to chassis ground<br>• Injector is damaged or has failed<br>• ECM is not connected or has failed<br>• Fuel pump relay has failed<br>• Fuel injectors may have malfunctioned<br>• Faulty engine speed sensor |
| **DTC: P0270**<br>**2T CCM, MIL: Yes**<br>**Years: 2007, 2008**<br>**Models:** A4, A6<br>**Engines:** All<br>**Transmissions:** All | **Cylinder 4 Injector Circuit Low Input/Short to Ground Conditions:**<br>Key on or engine running, fuses in the instrument panel and the E-box in the engine compartment must be functioning, and the ground connections between the engine ad the chassis must be well connected; and the ECM detected an unexpected voltage condition on the injector circuit<br>**Possible Causes:**<br>• Injector control circuit is open<br>• Injector power circuit (B+) is open<br>• Injector control circuit is shorted to chassis ground<br>• Injector is damaged or has failed<br>• ECM is not connected or has failed<br>• Fuel pump relay has failed<br>• Fuel injectors may have malfunctioned<br>• Faulty engine speed sensor |
| **DTC: P0271**<br>**2T CCM, MIL: Yes**<br>**Years: 2007, 2008**<br>**Models:** A4, A6<br>**Engines:** All<br>**Transmissions:** All | **Cylinder 4 Injector Circuit Low Input/Short to B+ Conditions:**<br>Key on or engine running, fuses in the instrument panel and the E-box in the engine compartment must be functioning, and the ground connections between the engine ad the chassis must be well connected; and the ECM detected an unexpected voltage condition on the injector circuit<br>**Possible Causes:**<br>• Injector control circuit is open<br>• Injector power circuit (B+) is open<br>• Injector control circuit is shorted to chassis ground<br>• Injector is damaged or has failed<br>• ECM is not connected or has failed<br>• Fuel pump relay has failed<br>• Fuel injectors may have malfunctioned<br>• Faulty engine speed sensor |

| DTC | Trouble Code Title, Conditions & Possible Causes |
|---|---|
| **DTC: P0274**<br>**2T CCM, MIL: Yes**<br>**Years: 2007, 2008**<br>**Models:** A4, A6<br>**Engines:** All<br>**Transmissions:** All | **Cylinder 5 Injector Circuit Low Input/Short to Ground Conditions:**<br>Key on or engine running, fuses in the instrument panel and the E-box in the engine compartment must be functioning, and the ground connections between the engine ad the chassis must be well connected; and the ECM detected an unexpected voltage condition on the injector circuit<br>**Possible Causes:**<br>• Injector control circuit is open<br>• Injector power circuit (B+) is open<br>• Injector control circuit is shorted to chassis ground<br>• Injector is damaged or has failed<br>• ECM is not connected or has failed<br>• Fuel pump relay has failed<br>• Fuel injectors may have malfunctioned<br>• Faulty engine speed sensor |
| **DTC: P0274**<br>**2T CCM, MIL: Yes**<br>**Years: 2007, 2008**<br>**Models:** A4, A6<br>**Engines:** All<br>**Transmissions:** All | **Cylinder 5 Injector Circuit Low Input/Short to B+ Conditions:**<br>Key on or engine running, fuses in the instrument panel and the E-box in the engine compartment must be functioning, and the ground connections between the engine ad the chassis must be well connected; and the ECM detected an unexpected voltage condition on the injector circuit<br>**Possible Causes:**<br>• Injector control circuit is open<br>• Injector power circuit (B+) is open<br>• Injector control circuit is shorted to chassis ground<br>• Injector is damaged or has failed<br>• ECM is not connected or has failed<br>• Fuel pump relay has failed<br>• Fuel injectors may have malfunctioned<br>• Faulty engine speed sensor |
| **DTC: P0276**<br>**2T CCM, MIL: Yes**<br>**Years: 2007, 2008**<br>**Models:** A4, A6<br>**Engines:** All<br>**Transmissions:** All | **Cylinder 6 Injector Circuit Low Input/Short to Ground Conditions:**<br>Key on or engine running, fuses in the instrument panel and the E-box in the engine compartment must be functioning, and the ground connections between the engine ad the chassis must be well connected; and the ECM detected an unexpected voltage condition on the injector circuit<br>**Possible Causes:**<br>• Injector control circuit is open<br>• Injector power circuit (B+) is open<br>• Injector control circuit is shorted to chassis ground<br>• Injector is damaged or has failed<br>• ECM is not connected or has failed<br>• Fuel pump relay has failed<br>• Fuel injectors may have malfunctioned<br>• Faulty engine speed sensor |
| **DTC: P0277**<br>**2T CCM, MIL: Yes**<br>**Years: 2007, 2008**<br>**Models:** A4, A6<br>**Engines:** All<br>**Transmissions:** All | **Cylinder 6 Injector Circuit Low Input/Short to B+ Conditions:**<br>Key on or engine running, fuses in the instrument panel and the E-box in the engine compartment must be functioning, and the ground connections between the engine ad the chassis must be well connected; and the ECM detected an unexpected voltage condition on the injector circuit<br>**Possible Causes:**<br>• Injector control circuit is open<br>• Injector power circuit (B+) is open<br>• Injector control circuit is shorted to chassis ground<br>• Injector is damaged or has failed<br>• ECM is not connected or has failed<br>• Fuel pump relay has failed<br>• Fuel injectors may have malfunctioned<br>• Faulty engine speed sensor |

| DTC | Trouble Code Title, Conditions & Possible Causes |
|---|---|
| **DTC: P0279**<br>**2T CCM, MIL: Yes**<br>**Years: 2007, 2008**<br>**Models:** A6<br>**Engines:** All<br>**Transmissions:** All | **Cylinder 7 Injector Circuit Low Input/Short to Ground Conditions:**<br>Key on or engine running, fuses in the instrument panel and the E-box in the engine compartment must be functioning, and the ground connections between the engine ad the chassis must be well connected; and the ECM detected an unexpected voltage condition on the injector circuit<br>**Possible Causes:**<br>• Injector control circuit is open<br>• Injector power circuit (B+) is open<br>• Injector control circuit is shorted to chassis ground<br>• Injector is damaged or has failed<br>• ECM is not connected or has failed<br>• Fuel pump relay has failed<br>• Fuel injectors may have malfunctioned<br>• Faulty engine speed sensor |
| **DTC: P0280**<br>**2T CCM, MIL: Yes**<br>**Years: 2007, 2008**<br>**Models:** A6<br>**Engines:** All<br>**Transmissions:** All | **Cylinder 7 Injector Circuit Low Input/Short to B+ Conditions:**<br>Key on or engine running, fuses in the instrument panel and the E-box in the engine compartment must be functioning, and the ground connections between the engine ad the chassis must be well connected; and the ECM detected an unexpected voltage condition on the injector circuit<br>**Possible Causes:**<br>• Injector control circuit is open<br>• Injector power circuit (B+) is open<br>• Injector control circuit is shorted to chassis ground<br>• Injector is damaged or has failed<br>• ECM is not connected or has failed<br>• Fuel pump relay has failed<br>• Fuel injectors may have malfunctioned<br>• Faulty engine speed sensor |
| **DTC: P0282**<br>**2T CCM, MIL: Yes**<br>**Years: 2007, 2008**<br>**Models:** A6<br>**Engines:** All<br>**Transmissions:** All | **Cylinder 8 Injector Circuit Low Input/Short to Ground Conditions:**<br>Key on or engine running, fuses in the instrument panel and the E-box in the engine compartment must be functioning, and the ground connections between the engine ad the chassis must be well connected; and the ECM detected an unexpected voltage condition on the injector circuit<br>**Possible Causes:**<br>• Injector control circuit is open<br>• Injector power circuit (B+) is open<br>• Injector control circuit is shorted to chassis ground<br>• Injector is damaged or has failed<br>• ECM is not connected or has failed<br>• Fuel pump relay has failed<br>• Fuel injectors may have malfunctioned<br>• Faulty engine speed sensor |
| **DTC: P0283**<br>**2T CCM, MIL: Yes**<br>**Years: 2007, 2008**<br>**Models:** A6<br>**Engines:** All<br>**Transmissions:** All | **Cylinder 8 Injector Circuit Low Input/Short to B+ Conditions:**<br>Key on or engine running, fuses in the instrument panel and the E-box in the engine compartment must be functioning, and the ground connections between the engine ad the chassis must be well connected; and the ECM detected an unexpected voltage condition on the injector circuit<br>**Possible Causes:**<br>• Injector control circuit is open<br>• Injector power circuit (B+) is open<br>• Injector control circuit is shorted to chassis ground<br>• Injector is damaged or has failed<br>• ECM is not connected or has failed<br>• Fuel pump relay has failed<br>• Fuel injectors may have malfunctioned<br>• Faulty engine speed sensor |

| DTC | Trouble Code Title, Conditions & Possible Causes |
|---|---|
| **DTC: P0300**<br>**2T CCM, MIL: Yes**<br>**Years: 2007, 2008**<br>**Models:** A4, A6<br>**Engines:** All<br>**Transmissions:** All | **Random/Multiple Misfire Detected Conditions:**<br>Engine running under positive torque conditions, and the ECM detected a misfire or uneven engine running in two or more cylinders.<br>**Note: If the misfire is severe, the MIL will flash on/off on the first trip!**<br>**Possible Causes:**<br>• Base engine mechanical fault that affects two or more cylinders<br>• Fuel metering fault that affects two or more cylinders<br>• Fuel pressure too low or too high, fuel supply contaminated<br>• EVAP system problem or the EVAP canister is fuel saturated<br>• EGR valve is stuck open or the PCV system has a vacuum leak<br>• Ignition system fault (coil, plugs) affecting two or more cylinders<br>• MAF sensor contamination (it can cause a very lean condition)<br>• Vehicle driven while very low on fuel (less than 1/8 of a tank) |
| **DTC: P0301**<br>**2T CCM, MIL: Yes**<br>**Years: 2007, 2008**<br>**Models:** A4, A6<br>**Engines:** All<br>**Transmissions:** All | **Cylinder Number 1 Misfire Detected Conditions:**<br>Engine running under positive torque conditions, and the ECM detected a misfire or uneven engine function.<br>**Note: If the misfire is severe, the MIL will flash on/off on the 1st trip!**<br>**Possible Causes:**<br>• Air leak in the intake manifold, or in the EGR or ECM system<br>• Base engine mechanical problem<br>• Fuel delivery component problem (i.e., a contaminated, dirty or sticking fuel injector)<br>• Fuel pump relay defective<br>• Ignition coil fuses have failed<br>• Ignition system problem (dirty damaged coil or plug)<br>• Engine speed (RPM) sensor has failed<br>• Camshaft position sensors have failed<br>• Ignition coil is faulty<br>• Spark plugs are not working properly or are not gapped properly |
| **DTC: P0302**<br>**2T CCM, MIL: Yes**<br>**Years: 2007, 2008**<br>**Models:** A4, A6<br>**Engines:** All<br>**Transmissions:** All | **Cylinder Number 2 Misfire Detected Conditions:**<br>Engine running under positive torque conditions, and the ECM detected a misfire or uneven engine function.<br>**Note: If the misfire is severe, the MIL will flash on/off on the 1st trip!**<br>**Possible Causes:**<br>• Air leak in the intake manifold, or in the EGR or ECM system<br>• Base engine mechanical problem<br>• Fuel delivery component problem (i.e., a contaminated, dirty or sticking fuel injector)<br>• Fuel pump relay defective<br>• Ignition coil fuses have failed<br>• Ignition system problem (dirty damaged coil or plug)<br>• Engine speed (RPM) sensor has failed<br>• Camshaft position sensors have failed<br>• Ignition coil is faulty<br>• Spark plugs are not working properly or are not gapped properly |
| **DTC: P0303**<br>**2T CCM, MIL: Yes**<br>**Years: 2007, 2008**<br>**Models:** A4, A6<br>**Engines:** All<br>**Transmissions:** All | **Cylinder Number 3 Misfire Detected Conditions:**<br>Engine running under positive torque conditions, and the ECM detected a misfire or uneven engine function.<br>**Note: If the misfire is severe, the MIL will flash on/off on the 1st trip!**<br>**Possible Causes:**<br>• Air leak in the intake manifold, or in the EGR or ECM system<br>• Base engine mechanical problem<br>• Fuel delivery component problem (i.e., a contaminated, dirty or sticking fuel injector)<br>• Fuel pump relay defective<br>• Ignition coil fuses have failed<br>• Ignition system problem (dirty damaged coil or plug)<br>• Engine speed (RPM) sensor has failed<br>• Camshaft position sensors have failed<br>• Ignition coil is faulty<br>• Spark plugs are not working properly or are not gapped properly |

| DTC | Trouble Code Title, Conditions & Possible Causes |
|---|---|
| **DTC: P0304**<br>**2T CCM, MIL: Yes**<br>**Years:** 2007, 2008<br>**Models:** A4, A6<br>**Engines:** All<br>**Transmissions:** All | **Cylinder Number 4 Misfire Detected Conditions:**<br>Engine running under positive torque conditions, and the ECM detected a misfire or uneven engine function.<br>**Note: If the misfire is severe, the MIL will flash on/off on the 1st trip!**<br>**Possible Causes:**<br>• Air leak in the intake manifold, or in the EGR or ECM system<br>• Base engine mechanical problem<br>• Fuel delivery component problem (i.e., a contaminated, dirty or sticking fuel injector)<br>• Fuel pump relay defective<br>• Ignition coil fuses have failed<br>• Ignition system problem (dirty damaged coil or plug)<br>• Engine speed (RPM) sensor has failed<br>• Camshaft position sensors have failed<br>• Ignition coil is faulty<br>• Spark plugs are not working properly or are not gapped properly |
| **DTC: P0305**<br>**2T CCM, MIL: Yes**<br>**Years:** 2007, 2008<br>**Models:** A4, A6<br>**Engines:** All<br>**Transmissions:** All | **Cylinder Number 5 Misfire Detected Conditions:**<br>Engine running under positive torque conditions, and the ECM detected a misfire or uneven engine function.<br>**Note: If the misfire is severe, the MIL will flash on/off on the 1st trip!**<br>**Possible Causes:**<br>• Air leak in the intake manifold, or in the EGR or ECM system<br>• Base engine mechanical problem<br>• Fuel delivery component problem (i.e., a contaminated, dirty or sticking fuel injector)<br>• Fuel pump relay defective<br>• Ignition coil fuses have failed<br>• Ignition system problem (dirty damaged coil or plug)<br>• Engine speed (RPM) sensor has failed<br>• Camshaft position sensors have failed<br>• Ignition coil is faulty<br>• Spark plugs are not working properly or are not gapped properly |
| **DTC: P0306**<br>**2T CCM, MIL: Yes**<br>**Years:** 2007, 2008<br>**Models:** A4, A6<br>**Engines:** All<br>**Transmissions:** All | **Cylinder Number 6 Misfire Detected Conditions:**<br>Engine running under positive torque conditions, and the ECM detected a misfire or uneven engine function.<br>**Note: If the misfire is severe, the MIL will flash on/off on the 1st trip!**<br>**Possible Causes:**<br>• Air leak in the intake manifold, or in the EGR or ECM system<br>• Base engine mechanical problem<br>• Fuel delivery component problem (i.e., a contaminated, dirty or sticking fuel injector)<br>• Fuel pump relay defective<br>• Ignition coil fuses have failed<br>• Ignition system problem (dirty damaged coil or plug)<br>• Engine speed (RPM) sensor has failed<br>• Camshaft position sensors have failed<br>• Ignition coil is faulty<br>• Spark plugs are not working properly or are not gapped properly |
| **DTC: P0307**<br>**2T CCM, MIL: Yes**<br>**Years:** 2007, 2008<br>**Models:** A6<br>**Engines:** All<br>**Transmissions:** All | **Cylinder Number 7 Misfire Detected Conditions:**<br>Engine running under positive torque conditions, and the ECM detected a misfire or uneven engine function.<br>**Note: If the misfire is severe, the MIL will flash on/off on the 1st trip!**<br>**Possible Causes:**<br>• Air leak in the intake manifold, or in the EGR or ECM system<br>• Base engine mechanical problem<br>• Fuel delivery component problem (i.e., a contaminated, dirty or sticking fuel injector)<br>• Fuel pump relay defective<br>• Ignition coil fuses have failed<br>• Ignition system problem (dirty damaged coil or plug)<br>• Engine speed (RPM) sensor has failed<br>• Camshaft position sensors have failed<br>• Ignition coil is faulty<br>• Spark plugs are not working properly or are not gapped properly |

| DTC | Trouble Code Title, Conditions & Possible Causes |
|---|---|
| **DTC: P0308**<br>**2T CCM, MIL: Yes**<br>**Years: 2007, 2008**<br>**Models:** A6<br>**Engines:** All<br>**Transmissions:** All | **Cylinder Number 8 Misfire Detected Conditions:**<br>Engine running under positive torque conditions, and the ECM detected a misfire or uneven engine function.<br>**Note: If the misfire is severe, the MIL will flash on/off on the 1st trip!**<br>**Possible Causes:**<br>• Air leak in the intake manifold, or in the EGR or ECM system<br>• Base engine mechanical problem<br>• Fuel delivery component problem (i.e., a contaminated, dirty or sticking fuel injector)<br>• Fuel pump relay defective<br>• Ignition coil fuses have failed<br>• Ignition system problem (dirty damaged coil or plug)<br>• Engine speed (RPM) sensor has failed<br>• Camshaft position sensors have failed<br>• Ignition coil is faulty<br>• Spark plugs are not working properly or are not gapped properly |
| **DTC: P0321**<br>**1T CCM, MIL: Yes**<br>**Years: 2007, 2008**<br>**Models:** A4, A6<br>**Engines:** All<br>**Transmissions:** All | **Ignition/Distributor Engine Speed Input Circuit Range/Performance Conditions:**<br>Engine started, vehicle driven, and the ECM detected the engine speed signal was more than the calibrated value.<br>**Note: The engine will not start if there is no speed signal. If the speed signal fails when the engine is running, it will cause the engine to stall immediately.**<br>**Possible Causes:**<br>• Engine speed sensor has failed or is damaged<br>• ECM has failed<br>• Sensor wheel is damaged or doesn't fit properly<br>• Sensor wheel spacer isn't seated properly |
| **DTC: P0322**<br>**1T CCM, MIL: Yes**<br>**Years: 2007, 2008**<br>**Models:** A4, A6<br>**Engines:** All<br>**Transmissions:** All | **Ignition/Distributor Engine Input Circuit No Signal Conditions:**<br>Key on, and the ECM could not detect the engine speed signal or the signal was erratic.<br>**Note: The engine will not start if there is no speed signal. If the speed signal fails when the engine is running, it will cause the engine to stall immediately.**<br>**Possible Causes:**<br>• Engine speed sensor has failed or is damaged<br>• ECM has failed<br>• Sensor wheel is damaged or doesn't fit properly<br>• Sensor wheel spacer isn't seated properly |
| **DTC: P0324**<br>**1T CCM, MIL: Yes**<br>**Years: 2007, 2008**<br>**Models:** A4<br>**Engines:** 3.2L<br>**Transmissions:** All | **Knock Control System Error Conditions:**<br>Engine started, vehicle driven, and the ECM detected the Knock Sensor 1 (KS1) signal was too low or not recognized by the ECM<br>**Possible Causes:**<br>• Knock sensor circuit is open<br>• Knock sensor is loose (tighten to 20 Nm)<br>• Contact between the knock sensor and cylinder block is dirty, corroded or greasy<br>• Knock sensor circuit is shorted to ground, or shorted to power<br>• Knock sensor is damaged or it has failed<br>• Wrong kind of fuel used<br>• A component in the engine compartment is loose or not properly secured<br>• ECM has failed |
| **DTC: P0327**<br>**1T CCM, MIL: Yes**<br>**Years: 2007, 2008**<br>**Models:** A4, A6<br>**Engines:** All<br>**Transmissions:** All | **Knock Sensor 1 Signal Low Input Conditions:**<br>Engine started, vehicle driven, and the ECM detected the Knock Sensor 1 (KS1) signal was too low or not recognized by the ECM<br>**Possible Causes:**<br>• Knock sensor circuit is open<br>• Knock sensor is loose (tighten to 20 Nm)<br>• Contact between the knock sensor and cylinder block is dirty, corroded or greasy<br>• Knock sensor circuit is shorted to ground, or shorted to power<br>• Knock sensor is damaged or it has failed<br>• Wrong kind of fuel used<br>• A component in the engine compartment is loose or not properly secured<br>• ECM has failed |

| DTC | Trouble Code Title, Conditions & Possible Causes |
|---|---|
| **DTC: P0328**<br>**1T CCM, MIL: Yes**<br>**Years: 2007, 2008**<br>**Models:** A4, A6<br>**Engines:** All<br>**Transmissions:** All | **Knock Sensor 1 Signal High Input Conditions:**<br>Engine started, vehicle driven, and the ECM detected the Knock Sensor 1 (KS1) signal was too high<br>**Possible Causes:**<br>• Knock sensor circuit is open<br>• Knock sensor is loose (tighten to 20 Nm)<br>• Contact between the knock sensor and cylinder block is dirty, corroded or greasy<br>• Knock sensor circuit is shorted to ground, or shorted to power<br>• Knock sensor is damaged or it has failed<br>• Wrong kind of fuel used<br>• A component in the engine compartment is loose or not properly secured<br>• ECM has failed |
| **DTC: P0332**<br>**1T CCM, MIL: Yes**<br>**Years: 2007, 2008**<br>**Models:** A4, A6<br>**Engines:** All<br>**Transmissions:** All | **Knock Sensor 2 Signal Low Input Conditions:**<br>Engine started, vehicle driven, and the ECM detected the Knock Sensor 1 (KS1) signal was too low or not recognized by the ECM<br>**Possible Causes:**<br>• Knock sensor circuit is open<br>• Knock sensor is loose (tighten to 20 Nm)<br>• Contact between the knock sensor and cylinder block is dirty, corroded or greasy<br>• Knock sensor circuit is shorted to ground, or shorted to power<br>• Knock sensor is damaged or it has failed<br>• Wrong kind of fuel used<br>• A component in the engine compartment is loose or not properly secured<br>• ECM has failed |
| **DTC: P0333**<br>**1T CCM, MIL: Yes**<br>**Years: 2007, 2008**<br>**Models:** A4, A6<br>**Engines:** All<br>**Transmissions:** All | **Knock Sensor 2 Signal High Input Conditions:**<br>Engine started, vehicle driven, and the ECM detected the Knock Sensor 1 (KS1) signal was too high<br>**Possible Causes:**<br>• Knock sensor circuit is open<br>• Knock sensor is loose (tighten to 20 Nm)<br>• Contact between the knock sensor and cylinder block is dirty, corroded or greasy<br>• Knock sensor circuit is shorted to ground, or shorted to power<br>• Knock sensor is damaged or it has failed<br>• Wrong kind of fuel used<br>• A component in the engine compartment is loose or not properly secured<br>• ECM has failed |
| **DTC: P0340**<br>**1T CCM, MIL: Yes**<br>**Years: 2007, 2008**<br>**Models:** A4, A6<br>**Engines:** All<br>**Transmissions:** All | **Camshaft Position Sensor Circuit Malfunction Conditions:**<br>Engine started, battery voltage must be at least 11.5v, all electrical components must be off, parking brake must be engaged (to keep daytime driving lights off), automatic transmission selector must be in park and the ground between the engine and the chassis must be well connected. The ECM detected the CMP sensor signal was missing or it was erratic.<br>**Possible Causes:**<br>• CMP sensor circuit is open or shorted to ground<br>• CMP sensor circuit is shorted to power<br>• CMP sensor ground (return) circuit is open<br>• CMP sensor installation incorrect (Hall-effect type)<br>• CMP sensor is damaged or CMP sensor shielding damaged<br>• CMP sensor has failed<br>• ECM has failed |
| **DTC: P0341**<br>**1T CCM, MIL: Yes**<br>**Years: 2007, 2008**<br>**Models:** A4, A6<br>**Engines:** All<br>**Transmissions:** All | **Camshaft Position Sensor Circ Range/Performance Conditions:**<br>Engine started, battery voltage must be at least 11.5v, all electrical components must be off, parking brake must be engaged (to keep daytime driving lights off), automatic transmission selector must be in park and the ground between the engine and the chassis must be well connected. The ECM detected the CMP sensor signal was implausible.<br>**Possible Causes:**<br>• CMP sensor circuit is open or shorted to ground<br>• CMP sensor circuit is shorted to power<br>• CMP sensor ground (return) circuit is open<br>• CMP sensor installation incorrect (Hall-effect type)<br>• CMP sensor is damaged or CMP sensor shielding damaged<br>• ECM has failed |

| DTC | Trouble Code Title, Conditions & Possible Causes |
|---|---|
| **DTC: P0342**<br>**1T CCM, MIL: Yes**<br>**Years: 2007, 2008**<br>**Models:** A4, A6<br>**Engines:** All<br>**Transmissions:** All | **Camshaft Position Sensor "A" Circuit (Bank 1 or Single Sensor) Low Input Conditions:**<br>Engine started, battery voltage must be at least 11.5v, all electrical components must be off, parking brake must be engaged (to keep daytime driving lights off), automatic transmission selector must be in park and the ground between the engine and the chassis must be well connected. The ECM detected the CMP sensor signal exceeded the bounds of the specified maximum limit.<br>**Possible Causes:**<br>• CMP sensor circuit is open or shorted to ground<br>• CMP sensor circuit is shorted to power<br>• CMP sensor ground (return) circuit is open<br>• CMP sensor installation incorrect (Hall-effect type)<br>• CMP sensor is damaged or CMP sensor shielding damaged<br>• ECM has failed |
| **DTC: P0343**<br>**1T CCM, MIL: Yes**<br>**Years: 2007, 2008**<br>**Models:** A4, A6<br>**Engines:** All<br>**Transmissions:** All | **Camshaft Position Sensor "A" Circuit (Bank 1 or Single Sensor) High Input Conditions:**<br>Engine started, battery voltage must be at least 11.5v, all electrical components must be off, parking brake must be engaged (to keep daytime driving lights off), automatic transmission selector must be in park and the ground between the engine and the chassis must be well connected. The ECM detected the CMP sensor signal did not reach the specified minimum limit.<br>**Possible Causes:**<br>• CMP sensor circuit is open or shorted to ground<br>• CMP sensor circuit is shorted to power<br>• CMP sensor ground (return) circuit is open<br>• CMP sensor installation incorrect (Hall-effect type)<br>• CMP sensor is damaged or CMP sensor shielding damaged<br>• ECM has failed |
| **DTC: P0345**<br>**1T CCM, MIL: Yes**<br>**Years: 2007, 2008**<br>**Models:** A4, A6<br>**Engines:** All<br>**Transmissions:** All | **Camshaft Position Sensor "A" Circuit (Bank 2) Malfunction Conditions:**<br>Engine started, battery voltage must be at least 11.5v, all electrical components must be off, parking brake must be engaged (to keep daytime driving lights off), automatic transmission selector must be in park and the ground between the engine and the chassis must be well connected. The ECM detected the CMP sensor signal was missing or it was erratic.<br>**Possible Causes:**<br>• CMP sensor circuit is open or shorted to ground<br>• CMP sensor circuit is shorted to power<br>• CMP sensor ground (return) circuit is open<br>• CMP sensor installation incorrect (Hall-effect type)<br>• CMP sensor is damaged or CMP sensor shielding damaged<br>• ECM has failed |
| **DTC: P0346**<br>**1T CCM, MIL: Yes**<br>**Years: 2007, 2008**<br>**Models:** A4, A6<br>**Engines:** All<br>**Transmissions:** All | **Camshaft Position Sensor "A" Circuit (Bank 2) Range/Performance Conditions:**<br>Engine started, battery voltage must be at least 11.5v, all electrical components must be off, parking brake must be engaged (to keep daytime driving lights off), automatic transmission selector must be in park and the ground between the engine and the chassis must be well connected. The ECM detected the CMP sensor signal was implausible.<br>**Possible Causes:**<br>• CMP sensor circuit is open or shorted to ground<br>• CMP sensor circuit is shorted to power<br>• CMP sensor ground (return) circuit is open<br>• CMP sensor installation incorrect (Hall-effect type)<br>• CMP sensor is damaged or CMP sensor shielding damaged<br>• ECM has failed |
| **DTC: P0347**<br>**1T CCM, MIL: Yes**<br>**Years: 2007, 2008**<br>**Models:** A4, A6<br>**Engines:** All<br>**Transmissions:** All | **Camshaft Position Sensor "A" Circuit (Bank 2) Low Input Conditions:**<br>Engine started, battery voltage must be at least 11.5v, all electrical components must be off, parking brake must be engaged (to keep daytime driving lights off), automatic transmission selector must be in park and the ground between the engine and the chassis must be well connected. The ECM detected the CMP sensor signal exceeded the bounds of the specified maximum limit.<br>**Possible Causes:**<br>• CMP sensor circuit is open or shorted to ground<br>• CMP sensor circuit is shorted to power<br>• CMP sensor ground (return) circuit is open<br>• CMP sensor installation incorrect (Hall-effect type)<br>• CMP sensor is damaged or CMP sensor shielding damaged<br>• ECM has failed |

| DTC | Trouble Code Title, Conditions & Possible Causes |
|---|---|
| **DTC: P0348**<br>**1T CCM, MIL: Yes**<br>**Years: 2007, 2008**<br>**Models:** A4, A6<br>**Engines:** All<br>**Transmissions:** All | **Camshaft Position Sensor "A" Circuit "A" Circuit (Bank 2) High Input Conditions:**<br>Engine started, battery voltage must be at least 11.5v, all electrical components must be off, parking brake must be engaged (to keep daytime driving lights off), automatic transmission selector must be in park and the ground between the engine and the chassis must be well connected. The ECM detected the CMP sensor signal did not reach the specified minimum limit.<br>**Possible Causes:**<br>• CMP sensor circuit is open or shorted to ground<br>• CMP sensor circuit is shorted to power<br>• CMP sensor ground (return) circuit is open<br>• CMP sensor installation incorrect (Hall-effect type)<br>• CMP sensor is damaged or CMP sensor shielding damaged<br>• ECM has failed |
| **DTC: P0351**<br>**2T CCM, MIL: Yes**<br>**Years: 2007, 2008**<br>**Models:** A4, A6<br>**Engines:** All<br>**Transmissions:** All | **Ignition Coilpack A Primary/Secondary Circuit Malfunction Conditions:**<br>Engine started, battery voltage must be at least 11.5v, all electrical components must be off, parking brake must be engaged (to keep daytime driving lights off), automatic transmission selector must be in park and the ground between the engine and the chassis must be well connected. The ECM did not receive any valid pulses from the ignition module for the Ignition Coilpack A primary circuit.<br>**Note: Ignition coils and power output stages are one component and cannot be replaced individually.**<br>**Possible Causes:**<br>• Engine speed (RPM) sensor has failed<br>• Camshaft Position (CMP) sensor has failed<br>• Power Supply Relay is shorted to an open circuit<br>• There is a malfunction in voltage supply<br>• Ignition coilpack is damaged or it has failed<br>• Cylinder 1 to 4 Fuel Injector(s) have failed<br>• ECM has failed |
| **DTC: P0352**<br>**2T CCM, MIL: Yes**<br>**Years: 2007, 2008**<br>**Models:** A4, A6<br>**Engines:** All<br>**Transmissions:** All | **Ignition Coilpack B Primary/Secondary Circuit Malfunction Conditions:**<br>Engine started, battery voltage must be at least 11.5v, all electrical components must be off, parking brake must be engaged (to keep daytime driving lights off), automatic transmission selector must be in park and the ground between the engine and the chassis must be well connected. The ECM did not receive any valid pulses from the ignition module for the Ignition Coilpack B primary circuit.<br>**Note: Ignition coils and power output stages are one component and cannot be replaced individually.**<br>**Possible Causes:**<br>• Engine speed (RPM) sensor has failed<br>• Camshaft Position (CMP) sensor has failed<br>• Power Supply Relay is shorted to an open circuit<br>• There is a malfunction in voltage supply<br>• Ignition coilpack is damaged or it has failed<br>• Cylinder 1 to 4 Fuel Injector(s) have failed<br>• ECM has failed |
| **DTC: P0353**<br>**2T CCM, MIL: Yes**<br>**Years: 2007, 2008**<br>**Models:** A4, A6<br>**Engines:** All<br>**Transmissions:** All | **Ignition Coilpack C Primary/Secondary Circuit Malfunction Conditions:**<br>Engine started, battery voltage must be at least 11.5v, all electrical components must be off, parking brake must be engaged (to keep daytime driving lights off), automatic transmission selector must be in park and the ground between the engine and the chassis must be well connected. The ECM did not receive any valid pulses from the ignition module for the Ignition Coilpack C primary circuit.<br>**Note: Ignition coils and power output stages are one component and cannot be replaced individually.**<br>**Possible Causes:**<br>• Engine speed (RPM) sensor has failed<br>• Camshaft Position (CMP) sensor has failed<br>• Power Supply Relay is shorted to an open circuit<br>• There is a malfunction in voltage supply<br>• Ignition coilpack is damaged or it has failed<br>• Cylinder 1 to 4 Fuel Injector(s) have failed<br>• ECM has failed |

| DTC | Trouble Code Title, Conditions & Possible Causes |
|---|---|
| **DTC: P0354**<br>**2T CCM, MIL: Yes**<br>**Years: 2007, 2008**<br>**Models:** A4, A6<br>**Engines:** All<br>**Transmissions:** All | **Ignition Coilpack D Primary/Secondary Circuit Malfunction Conditions:**<br>Engine started, battery voltage must be at least 11.5v, all electrical components must be off, parking brake must be engaged (to keep daytime driving lights off), automatic transmission selector must be in park and the ground between the engine and the chassis must be well connected. The ECM did not receive any valid pulses from the ignition module for the Ignition Coilpack D primary circuit.<br>**Note: Ignition coils and power output stages are one component and cannot be replaced individually.**<br>**Possible Causes:**<br>&bull; Engine speed (RPM) sensor has failed<br>&bull; Camshaft Position (CMP) sensor has failed<br>&bull; Power Supply Relay is shorted to an open circuit<br>&bull; There is a malfunction in voltage supply<br>&bull; Ignition coilpack is damaged or it has failed<br>&bull; Cylinder 1 to 4 Fuel Injector(s) have failed<br>&bull; ECM has failed |
| **DTC: P0355**<br>**2T CCM, MIL: Yes**<br>**Years: 2007, 2008**<br>**Models:** A4, A6<br>**Engines:** All<br>**Transmissions:** All | **Ignition Coilpack E Primary/Secondary Circuit Malfunction Conditions:**<br>Engine started, battery voltage must be at least 11.5v, all electrical components must be off, parking brake must be engaged (to keep daytime driving lights off), automatic transmission selector must be in park and the ground between the engine and the chassis must be well connected. The ECM did not receive any valid pulses from the ignition module for the Ignition Coilpack E primary circuit.<br>**Note: Ignition coils and power output stages are one component and cannot be replaced individually.**<br>**Possible Causes:**<br>&bull; Engine speed (RPM) sensor has failed<br>&bull; Camshaft Position (CMP) sensor has failed<br>&bull; Power Supply Relay is shorted to an open circuit<br>&bull; There is a malfunction in voltage supply<br>&bull; Ignition coilpack is damaged or it has failed<br>&bull; Cylinder 1 to 4 Fuel Injector(s) have failed<br>&bull; ECM has failed |
| **DTC: P0356**<br>**2T CCM, MIL: Yes**<br>**Years: 2007, 2008**<br>**Models:** A4, A6<br>**Engines:** All<br>**Transmissions:** All | **Ignition Coilpack F Primary/Secondary Circuit Malfunction Conditions:**<br>Engine started, battery voltage must be at least 11.5v, all electrical components must be off, parking brake must be engaged (to keep daytime driving lights off), automatic transmission selector must be in park and the ground between the engine and the chassis must be well connected. The ECM did not receive any valid pulses from the ignition module for the Ignition Coilpack F primary circuit.<br>**Note: Ignition coils and power output stages are one component and cannot be replaced individually.**<br>**Possible Causes:**<br>&bull; Engine speed (RPM) sensor has failed<br>&bull; Camshaft Position (CMP) sensor has failed<br>&bull; Power Supply Relay is shorted to an open circuit<br>&bull; There is a malfunction in voltage supply<br>&bull; Ignition coilpack is damaged or it has failed<br>&bull; Cylinder 1 to 4 Fuel Injector(s) have failed<br>&bull; ECM has failed |
| **DTC: P0357**<br>**2T CCM, MIL: Yes**<br>**Years: 2007, 2008**<br>**Models:** A6<br>**Engines:** All<br>**Transmissions:** All | **Ignition Coilpack G Primary/Secondary Circuit Malfunction Conditions:**<br>Engine started, battery voltage must be at least 11.5v, all electrical components must be off, parking brake must be engaged (to keep daytime driving lights off), automatic transmission selector must be in park and the ground between the engine and the chassis must be well connected. The ECM did not receive any valid pulses from the ignition module for the Ignition Coilpack G primary circuit.<br>**Note: Ignition coils and power output stages are one component and cannot be replaced individually.**<br>**Possible Causes:**<br>&bull; Engine speed (RPM) sensor has failed<br>&bull; Camshaft Position (CMP) sensor has failed<br>&bull; Power Supply Relay is shorted to an open circuit<br>&bull; There is a malfunction in voltage supply<br>&bull; Ignition coilpack is damaged or it has failed<br>&bull; Cylinder 1 to 4 Fuel Injector(s) have failed<br>&bull; ECM has failed |

| DTC | Trouble Code Title, Conditions & Possible Causes |
|---|---|
| **DTC: P0366**<br>**2T CCM, MIL: Yes**<br>**Years:** 2007, 2008<br>**Models:** A4, A6<br>**Engines:** 3.2L, 4.2L<br>**Transmissions:** All | **Camshaft Position Sensor "B" Circuit (Bank 1) Range/Performance Conditions:**<br>Engine started, battery voltage must be at least 11.5v, all electrical components must be off, parking brake must be engaged (to keep daytime driving lights off), automatic transmission selector must be in park and the ground between the engine and the chassis must be well connected. The ECM detected the CMP sensor signal exceeded the bounds of the specified maximum limit.<br>**Possible Causes:**<br>• CMP sensor circuit is open or shorted to ground<br>• CMP sensor circuit is shorted to power<br>• CMP sensor ground (return) circuit is open<br>• CMP sensor installation incorrect (Hall-effect type)<br>• CMP sensor is damaged or CMP sensor shielding damaged<br>• ECM has failed |
| **DTC: P0367**<br>**1T CCM, MIL: Yes**<br>**Years:** 2007, 2008<br>**Models:** A4, A6<br>**Engines:** 3.2L, 4.2L<br>**Transmissions:** All | **Camshaft Position Sensor "B" Circuit (Bank 1) Low Input Conditions:**<br>Engine started, battery voltage must be at least 11.5v, all electrical components must be off, parking brake must be engaged (to keep daytime driving lights off), automatic transmission selector must be in park and the ground between the engine and the chassis must be well connected. The ECM detected the CMP sensor signal exceeded the bounds of the specified maximum limit.<br>**Possible Causes:**<br>• CMP sensor circuit is open or shorted to ground<br>• CMP sensor circuit is shorted to power<br>• CMP sensor ground (return) circuit is open<br>• CMP sensor installation incorrect (Hall-effect type)<br>• CMP sensor is damaged or CMP sensor shielding damaged<br>• ECM has failed |
| **DTC: P0368**<br>**1T CCM, MIL: Yes**<br>**Years:** 2007, 2008<br>**Models:** A4, A6<br>**Engines:** 3.2L, 4.2L<br>**Transmissions:** All | **Camshaft Position Sensor "B" Circuit (Bank 1) High Input Conditions:**<br>Engine started, battery voltage must be at least 11.5v, all electrical components must be off, parking brake must be engaged (to keep daytime driving lights off), automatic transmission selector must be in park and the ground between the engine and the chassis must be well connected. The ECM detected the CMP sensor signal did not reach the specified minimum limit.<br>**Possible Causes:**<br>• CMP sensor circuit is open or shorted to ground<br>• CMP sensor circuit is shorted to power<br>• CMP sensor ground (return) circuit is open<br>• CMP sensor installation incorrect (Hall-effect type)<br>• CMP sensor is damaged or CMP sensor shielding damaged<br>• ECM has failed |
| **DTC: P0391**<br>**1T CCM, MIL: Yes**<br>**Years:** 2007, 2008<br>**Models:** A4, A6<br>**Engines:** 3.2L, 4.2L<br>**Transmissions:** All | **Camshaft Position Sensor "B" Circuit (Bank 2) Range/Performance Conditions:**<br>Engine started, battery voltage must be at least 11.5v, all electrical components must be off, parking brake must be engaged (to keep daytime driving lights off), automatic transmission selector must be in park and the ground between the engine and the chassis must be well connected. The ECM detected the CMP sensor signal exceeded the bounds of the specified maximum limit.<br>**Possible Causes:**<br>• CMP sensor circuit is open or shorted to ground<br>• CMP sensor circuit is shorted to power<br>• CMP sensor ground (return) circuit is open<br>• CMP sensor installation incorrect (Hall-effect type)<br>• CMP sensor is damaged or CMP sensor shielding damaged<br>• ECM has failed |
| **DTC: P0392**<br>**1T CCM, MIL: Yes**<br>**Years:** 2007, 2008<br>**Models:** A4, A6<br>**Engines:** 3.2L, 4.2L<br>**Transmissions:** All | **Camshaft Position Sensor "B" Circuit (Bank 2) Low Input Conditions:**<br>Engine started, battery voltage must be at least 11.5v, all electrical components must be off, parking brake must be engaged (to keep daytime driving lights off), automatic transmission selector must be in park and the ground between the engine and the chassis must be well connected. The ECM detected the CMP sensor signal exceeded the bounds of the specified maximum limit.<br>**Possible Causes:**<br>• CMP sensor circuit is open or shorted to ground<br>• CMP sensor circuit is shorted to power<br>• CMP sensor ground (return) circuit is open<br>• CMP sensor installation incorrect (Hall-effect type)<br>• CMP sensor is damaged or CMP sensor shielding damaged<br>• ECM has failed |

| DTC | Trouble Code Title, Conditions & Possible Causes |
|---|---|
| **DTC: P0393**<br>**1T CCM, MIL: Yes**<br>**Years: 2007, 2008**<br>**Models:** A4, A6<br>**Engines:** 3.2L, 4.2L<br>**Transmissions:** All | **Camshaft Position Sensor "B" Circuit (Bank 2) High Input Conditions:**<br>Engine started, battery voltage must be at least 11.5v, all electrical components must be off, parking brake must be engaged (to keep daytime driving lights off), automatic transmission selector must be in park and the ground between the engine and the chassis must be well connected. The ECM detected the CMP sensor signal did not reach the specified minimum limit.<br>**Possible Causes:**<br>• CMP sensor circuit is open or shorted to ground<br>• CMP sensor circuit is shorted to power<br>• CMP sensor ground (return) circuit is open<br>• CMP sensor installation incorrect (Hall-effect type)<br>• CMP sensor is damaged or CMP sensor shielding damaged<br>• ECM has failed |
| **DTC: P0411**<br>**2T CCM, MIL: Yes**<br>**Years: 2007, 2008**<br>**Models:** A4<br>**Engines:** 3.2L<br>**Transmissions:** All | **Secondary Air Injection System Upstream Flow Detected Conditions:**<br>Engine started, battery voltage must be at least 11.5v, all electrical components must be off, parking brake must be engaged (to keep daytime driving lights off), automatic transmission selector must be in park and the ground between the engine and the chassis must be well connected. The ECM detected the Secondary AIR pump airflow was not diverted correctly when requested during the self-test. The pump is functioning but the quantity of air is recognized as insufficient by HO2S.<br>**Note: The solenoid valve is closed when no voltage is present.**<br>**Possible Causes:**<br>• Air pump output is blocked or restricted<br>• AIR bypass solenoid is leaking or it is restricted<br>• AIR bypass solenoid is stuck open or stuck closed<br>• Check valve (one or more) is damaged or leaking<br>• Electric air injection pump hose(s) leaking<br>• Electric air injection pump is damaged or faulty<br>• ECM has failed |
| **DTC: P0412**<br>**2T CCM, MIL: Yes**<br>**Years: 2007, 2008**<br>**Models:** A4, A6<br>**Engines:** All<br>**Transmissions:** All | **Secondary Air Injection Solenoid Circuit Malfunction Conditions:**<br>Engine started, battery voltage must be at least 11.5v, all electrical components must be off, parking brake must be engaged (to keep daytime driving lights off), automatic transmission selector must be in park and the ground between the engine and the chassis must be well connected. The ECM detected an unexpected low or high voltage condition on the AIR solenoid control circuit during testing.<br>**Possible Causes:**<br>• AIR solenoid power circuit (B+) is open (check dedicated fuse)<br>• AIR bypass solenoid control circuit is open or shorted to ground<br>• AIR diverter solenoid control circuit open or shorted to ground<br>• AIR pump control circuit is open or shorted to ground<br>• Check valve (one or more) is damaged or leaking<br>• Solid State relay is damaged or it has failed<br>• ECM has failed |
| **DTC: P0413**<br>**2T CCM, MIL: Yes**<br>**Years: 2007, 2008**<br>**Models:** A4, A6<br>**Engines:** All<br>**Transmissions:** All | **Secondary Air Injection Solenoid Circuit Open Conditions:**<br>Engine started, battery voltage must be at least 11.5v, all electrical components must be off, parking brake must be engaged (to keep daytime driving lights off), automatic transmission selector must be in park and the ground between the engine and the chassis must be well connected. The ECM detected an unexpected low or high voltage condition on the AIR solenoid control circuit during testing.<br>**Possible Causes:**<br>• AIR solenoid power circuit (B+) is open (check dedicated fuse)<br>• AIR bypass solenoid control circuit is open or shorted to ground<br>• AIR diverter solenoid control circuit open or shorted to ground<br>• AIR pump control circuit is open or shorted to ground<br>• Check valve (one or more) is damaged or leaking<br>• Solid State relay is damaged or it has failed<br>• ECM has failed |
| **DTC: P0414**<br>**2T CCM, MIL: Yes**<br>**Years: 2007, 2008**<br>**Models:** A4, A6<br>**Engines:** All<br>**Transmissions:** All | **Secondary Air Injection Solenoid Circuit Short Conditions:**<br>Engine started, battery voltage must be at least 11.5v, all electrical components must be off, parking brake must be engaged (to keep daytime driving lights off), automatic transmission selector must be in park and the ground between the engine and the chassis must be well connected. The ECM detected an unexpected low or high voltage condition on the AIR solenoid control circuit during testing.<br>**Possible Causes:**<br>• AIR solenoid power circuit (B+) is open (check dedicated fuse)<br>• AIR bypass solenoid control circuit is open or shorted to ground<br>• AIR diverter solenoid control circuit open or shorted to ground<br>• AIR pump control circuit is open or shorted to ground<br>• Check valve (one or more) is damaged or leaking<br>• Solid State relay is damaged or it has failed<br>• ECM has failed |

| DTC | Trouble Code Title, Conditions & Possible Causes |
|---|---|
| **DTC: P0418**<br>**2T CCM, MIL: Yes**<br>**Years: 2007, 2008**<br>**Models:** A4, A6<br>**Engines:** All<br>**Transmissions:** All | **Secondary Air Injection Relay (A) Circuit Malfunction Conditions:**<br>Engine started, battery voltage must be at least 11.5v, all electrical components must be off, parking brake must be engaged (to keep daytime driving lights off), automatic transmission selector must be in park and the ground between the engine and the chassis must be well connected. The ECM detected an unexpected low or high voltage condition on the AIR solenoid control circuit during testing.<br>**Possible Causes:**<br>• AIR solenoid power circuit (B+) is open (check dedicated fuse)<br>• AIR bypass solenoid control circuit is open or shorted to ground<br>• AIR diverter solenoid control circuit open or shorted to ground<br>• AIR pump control circuit is open or shorted to ground<br>• Check valve (one or more) is damaged or leaking<br>• Solid State relay is damaged or it has failed<br>• ECM has failed |
| **DTC: P0420**<br>**2T OBD/CAT1, MIL: Yes**<br>**Years: 2007, 2008**<br>**Models:** A4, A6<br>**Engines:** All<br>**Transmissions:** All | **Catalyst System Efficiency (Bank 1) Below Threshold Conditions:**<br>Engine started, battery voltage must be at least 11.5v, all electrical components must be off, parking brake must be engaged (to keep daytime driving lights off), automatic transmission selector must be in park, the exhaust system must be properly sealed between the catalytic converter and the cylinder head, coolant temperature must be at least 80 degrees Celsius and oxygen sensor heaters for oxygen sensors before the catalytic converter must be functioning properly and the ground between the engine and the chassis must be well connected. The ECM detected the switch rate of the rear HO2S-12 was close to the switch rate of front HO2S (it should be much slower).<br>**Possible Causes:**<br>• Air leaks at the exhaust manifold or in the exhaust pipes<br>• Catalytic converter is damaged, contaminated or it has failed<br>• ECT/CHT sensor has lost its calibration (the signal is incorrect)<br>• Engine cylinders misfiring, or the ignition timing is over retarded<br>• Engine oil is contaminated<br>• Front HO2S or rear HO2S is contaminated with fuel or moisture<br>• Front HO2S and/or the rear HO2S is loose in the mounting hole<br>• Front HO2S much older than the rear HO2S (HO2S-11 is lazy)<br>• Fuel system pressure is too high (check the pressure regulator)<br>• Rear HO2S wires improperly connected or the HO2S has failed |
| **DTC: P0421**<br>**2T OBD/CAT1, MIL: Yes**<br>**Years: 2007, 2008**<br>**Models:** A4<br>**Engines:** All<br>**Transmissions:** All | **Warm Up Catalyst System Efficiency (Bank 1) Below Threshold Conditions:**<br>Engine started, battery voltage must be at least 11.5v, all electrical components must be off, parking brake must be engaged (to keep daytime driving lights off), automatic transmission selector must be in park, the exhaust system must be properly sealed between the catalytic converter and the cylinder head, coolant temperature must be at least 80 degrees Celsius and oxygen sensor heaters for oxygen sensors before the catalytic converter must be functioning properly and the ground between the engine and the chassis must be well connected. The ECM detected the switch rate of the rear HO2S-12 was close to the switch rate of front HO2S (it should be much slower).<br>**Possible Causes:**<br>• Air leaks at the exhaust manifold or in the exhaust pipes<br>• Catalytic converter is damaged, contaminated or it has failed<br>• ECT/CHT sensor has lost its calibration (the signal is incorrect)<br>• Engine cylinders misfiring, or the ignition timing is over retarded<br>• Engine oil is contaminated<br>• Front HO2S or rear HO2S is contaminated with fuel or moisture<br>• Front HO2S and/or the rear HO2S is loose in the mounting hole<br>• Front HO2S much older than the rear HO2S (HO2S-11 is lazy)<br>• Fuel system pressure is too high (check the pressure regulator)<br>• Rear HO2S wires improperly connected or the HO2S has failed |

| DTC | Trouble Code Title, Conditions & Possible Causes |
|---|---|
| **DTC: P0422**<br>**2T OBD/CAT1, MIL: Yes**<br>**Years: 2007, 2008**<br>**Models:** A4, A6<br>**Engines:** All<br>**Transmissions:** All | **Main Catalyst (Bank 1) Efficiency Below Threshold Conditions:**<br>Engine started, battery voltage must be at least 11.5v, all electrical components must be off, parking brake must be engaged (to keep daytime driving lights off), automatic transmission selector must be in park, the exhaust system must be properly sealed between the catalytic converter and the cylinder head, coolant temperature must be at least 80 degrees Celsius and oxygen sensor heaters for oxygen sensors before the catalytic converter must be functioning properly and the ground between the engine and the chassis must be well connected. The ECM detected the switch rate of the rear HO2S-12 was close to the switch rate of front HO2S (it should be much slower).<br>**Possible Causes:**<br>• Air leaks at the exhaust manifold or in the exhaust pipes<br>• Catalytic converter is damaged, contaminated or it has failed<br>• ECT/CHT sensor has lost its calibration (the signal is incorrect)<br>• Engine cylinders misfiring, or the ignition timing is over retarded<br>• Engine oil is contaminated<br>• Front HO2S or rear HO2S is contaminated with fuel or moisture<br>• Front HO2S and/or the rear HO2S is loose in the mounting hole<br>• Front HO2S much older than the rear HO2S<br>• Fuel system pressure is too high (check the pressure regulator)<br>• Rear HO2S wires improperly connected or the HO2S has failed |
| **DTC: P0430**<br>**2T OBD/CAT1, MIL: Yes**<br>**Years: 2007, 2008**<br>**Models:** A4, A6<br>**Engines:** All<br>**Transmissions:** All | **Catalyst System Efficiency (Bank 2) Below Threshold Conditions:**<br>Engine started, battery voltage must be at least 11.5v, all electrical components must be off, parking brake must be engaged (to keep daytime driving lights off), automatic transmission selector must be in park, the exhaust system must be properly sealed between the catalytic converter and the cylinder head, coolant temperature must be at least 80 degrees Celsius and oxygen sensor heaters for oxygen sensors before the catalytic converter must be functioning properly and the ground between the engine and the chassis must be well connected. The ECM detected the switch rate of the rear HO2S-12 was close to the switch rate of front HO2S (it should be much slower).<br>**Possible Causes:**<br>• Air leaks at the exhaust manifold or in the exhaust pipes<br>• Catalytic converter is damaged, contaminated or it has failed<br>• ECT/CHT sensor has lost its calibration (the signal is incorrect)<br>• Engine cylinders misfiring, or the ignition timing is over retarded<br>• Engine oil is contaminated<br>• Front HO2S or rear HO2S is contaminated with fuel or moisture<br>• Front HO2S and/or the rear HO2S is loose in the mounting hole<br>• Front HO2S much older than the rear HO2S (HO2S-11 is lazy)<br>• Fuel system pressure is too high (check the pressure regulator)<br>• Rear HO2S wires improperly connected or the HO2S has failed |
| **DTC: P0431**<br>**2T OBD/CAT1, MIL: Yes**<br>**Years: 2007, 2008**<br>**Models:** A4, A6<br>**Engines:** All<br>**Transmissions:** All | **Warm Up Catalyst System Efficiency (Bank 2) Below Threshold Conditions:**<br>Engine started, battery voltage must be at least 11.5v, all electrical components must be off, parking brake must be engaged (to keep daytime driving lights off), automatic transmission selector must be in park, the exhaust system must be properly sealed between the catalytic converter and the cylinder head, coolant temperature must be at least 80 degrees Celsius and oxygen sensor heaters for oxygen sensors before the catalytic converter must be functioning properly and the ground between the engine and the chassis must be well connected. The ECM detected the switch rate of the rear HO2S-12 was close to the switch rate of front HO2S (it should be much slower).<br>**Possible Causes:**<br>• Air leaks at the exhaust manifold or in the exhaust pipes<br>• Catalytic converter is damaged, contaminated or it has failed<br>• ECT/CHT sensor has lost its calibration (the signal is incorrect)<br>• Engine cylinders misfiring, or the ignition timing is over retarded<br>• Engine oil is contaminated<br>• Front HO2S or rear HO2S is contaminated with fuel or moisture<br>• Front HO2S and/or the rear HO2S is loose in the mounting hole<br>• Front HO2S much older than the rear HO2S (HO2S-11 is lazy)<br>• Fuel system pressure is too high (check the pressure regulator)<br>• Rear HO2S wires improperly connected or the HO2S has failed |

| DTC | Trouble Code Title, Conditions & Possible Causes |
|---|---|
| **DTC: P0440**<br>**2T CCM, MIL: Yes**<br>**Years: 2007, 2008**<br>**Models:** A4, A6<br>**Engines:** All<br>**Transmissions:** All | **EVAP System Malfunction Conditions:**<br>ECT sensor is cold during startup, engine started, battery voltage must be at least 11.5v, all electrical components must be off, parking brake must be engaged (to keep daytime driving lights off), automatic transmission selector must be in park, the exhaust system must be properly sealed between the catalytic converter and the cylinder head, coolant temperature must be at least 80 degrees Celsius and oxygen sensor heaters for oxygen sensors before the catalytic converter must be functioning properly and the ground between the engine and the chassis must be well connected. The ECM detected the switch rate of the rear HO2S-12 was close to the switch rate of front HO2S (it should be much slower).<br>ECM detected a problem in the EVAP system during the EVAP System Monitor test.<br>**Possible Causes:**<br>• EVAP canister purge valve is damaged<br>• EVAP canister has an improper seal<br>• Vapor line between purge solenoid and intake manifold vacuum reservoir is damaged, or vapor line between EVAP canister purge solenoid and charcoal canister is damaged<br>• Vapor line between charcoal canister and check valve, or vapor line between check valve and fuel vapor valves is damaged<br>• ECM has failed |
| **DTC: P0441**<br>**2T CCM, MIL: Yes**<br>**Years: 2007, 2008**<br>**Models:** A4, A6<br>**Engines:** All<br>**Transmissions:** All | **EVAP Control System Incorrect Purge Flow Conditions:**<br>ECT sensor is cold during startup, engine started, battery voltage must be at least 11.5v, all electrical components must be off, parking brake must be engaged (to keep daytime driving lights off), automatic transmission selector must be in park, the exhaust system must be properly sealed between the catalytic converter and the cylinder head, coolant temperature must be at least 80 degrees Celsius and oxygen sensor heaters for oxygen sensors before the catalytic converter must be functioning properly and the ground between the engine and the chassis must be well connected. The ECM detected the switch rate of the rear HO2S-12 was close to the switch rate of front HO2S (it should be much slower).<br>ECM detected a problem in the EVAP system during the EVAP System Monitor test.<br>**Possible Causes:**<br>• EVAP canister purge valve is damaged<br>• EVAP canister has an improper seal<br>• Vapor line between purge solenoid and intake manifold vacuum reservoir is damaged, or vapor line between EVAP canister purge solenoid and charcoal canister is damaged<br>• Vapor line between charcoal canister and check valve, or vapor line between check valve and fuel vapor valves is damaged<br>• ECM has failed |
| **DTC: P0442**<br>**2T CCM, MIL: Yes**<br>**Years: 2007, 2008**<br>**Models:** A4, A6<br>**Engines:** All<br>**Transmissions:** All | **EVAP Control System Small Leak Detected Conditions:**<br>Engine started, battery voltage must be at least 11.5v, all electrical components must be off, parking brake must be engaged (to keep daytime driving lights off), automatic transmission selector must be in park, the exhaust system must be properly sealed between the catalytic converter and the cylinder head, coolant temperature must be at least 80 degrees Celsius and oxygen sensor heaters for oxygen sensors before the catalytic converter must be functioning properly and the ground between the engine and the chassis must be well connected. The ECM detected a leak in the EVAP system as small as 0.040" during the EVAP Monitor Test.<br>**Possible Causes:**<br>• Aftermarket EVAP parts that do not conform to specifications<br>• CV solenoid remains partially open when commanded to close<br>• EVAP component seals leaking (i.e., leaks in the Purge valve, fuel tank pressure sensor, canister vent solenoid, fuel vapor control valve tube assembly or fuel vapor vent valve).<br>• Fuel filler cap damaged, cross-threaded or loosely installed<br>• Loose fuel vapor hose/tube connections to EVAP components<br>• Small holes or cuts in fuel vapor hoses or EVAP canister tubes |
| **DTC: P0444**<br>**2T CCM, MIL: Yes**<br>**Years: 2007, 2008**<br>**Models:** A4, A6<br>**Engines:** All<br>**Transmissions:** All | **Evaporative Emission System Purge Control Valve Circuit Open Conditions:**<br>Engine started, battery voltage must be at least 11.5v, all electrical components must be off, parking brake must be engaged (to keep daytime driving lights off), automatic transmission selector must be in park, the exhaust system must be properly sealed between the catalytic converter and the cylinder head, coolant temperature must be at least 80 degrees Celsius and oxygen sensor heaters for oxygen sensors before the catalytic converter must be functioning properly and the ground between the engine and the chassis must be well connected. The ECM detected an unexpected voltage condition on the EVAP circuit when the device was cycled On/Off during testing.<br>**Possible Causes:**<br>• EVAP power supply circuit is open<br>• EVAP solenoid control circuit is open or shorted to ground<br>• EVAP solenoid control circuit is shorted to power (B+)<br>• EVAP solenoid valve is damaged or it has failed<br>• EVAP canister has a leak or a poor seal<br>• ECM has failed |

| DTC | Trouble Code Title, Conditions & Possible Causes |
|---|---|
| **DTC: P0445**<br>**2T CCM, MIL: Yes**<br>**Years: 2007, 2008**<br>**Models:** A4, A6<br>**Engines:** All<br>**Transmissions:** All | **Evaporative Emission System Purge Control Valve Circuit Shorted Conditions:**<br>Engine started, battery voltage must be at least 11.5v, all electrical components must be off, parking brake must be engaged (to keep daytime driving lights off), automatic transmission selector must be in park, the exhaust system must be properly sealed between the catalytic converter and the cylinder head, coolant temperature must be at least 80 degrees Celsius and oxygen sensor heaters for oxygen sensors before the catalytic converter must be functioning properly and the ground between the engine and the chassis must be well connected. The ECM detected an unexpected voltage condition on the EVAP circuit when the device was cycled On/Off during testing.<br>**Possible Causes:**<br>• EVAP power supply circuit is open<br>• EVAP solenoid control circuit is open or shorted to ground<br>• EVAP solenoid control circuit is shorted to power (B+)<br>• EVAP solenoid valve is damaged or it has failed<br>• EVAP canister has a leak or a poor seal<br>• ECM has failed |
| **DTC: P0455**<br>**2T CCM, MIL: Yes**<br>**Years: 2007, 2008**<br>**Models:** A4, A6<br>**Engines:** All<br>**Transmissions:** All | **EVAP Control System Large Leak Detected Conditions:**<br>Engine started, battery voltage must be at least 11.5v, all electrical components must be off, parking brake must be engaged (to keep daytime driving lights off), automatic transmission selector must be in park, the exhaust system must be properly sealed between the catalytic converter and the cylinder head, coolant temperature must be at least 80 degrees Celsius and oxygen sensor heaters for oxygen sensors before the catalytic converter must be functioning properly and the ground between the engine and the chassis must be well connected. The ECM detected multiple small fuel vapor leaks; or it detected a large leak in the system during the leak test.<br>**Possible Causes:**<br>• Aftermarket EVAP hardware non-conforming to specifications<br>• EVAP canister tube, EVAP canister purge outlet tube or EVAP return tube disconnected or cracked, or canister is damaged<br>• EVAP canister purge valve stuck closed, or canister damaged<br>• Fuel filler cap missing, loose (not tightened) or the wrong part<br>• Loose fuel vapor hose/tube connections to EVAP components<br>• Canister vent (CV) solenoid stuck open<br>• Fuel tank pressure (FTP) sensor has failed mechanically |
| **DTC: P0456**<br>**2T CCM, MIL: Yes**<br>**Years: 2007, 2008**<br>**Models:** A4, A6<br>**Engines:** All<br>**Transmissions:** All | **EVAP Control System Small Leak Detected Conditions:**<br>Engine started, battery voltage must be at least 11.5v, all electrical components must be off, parking brake must be engaged (to keep daytime driving lights off), automatic transmission selector must be in park, the exhaust system must be properly sealed between the catalytic converter and the cylinder head, coolant temperature must be at least 80 degrees Celsius and oxygen sensor heaters for oxygen sensors before the catalytic converter must be functioning properly and the ground between the engine and the chassis must be well connected. The ECM detected multiple small fuel vapor leaks; or it detected a large leak in the system during the leak test.<br>**Possible Causes:**<br>• Aftermarket EVAP hardware non-conforming to specifications<br>• EVAP canister tube, EVAP canister purge outlet tube or EVAP return tube disconnected or cracked, or canister is damaged<br>• EVAP canister purge valve stuck closed, or canister damaged<br>• Fuel filler cap missing, loose (not tightened) or the wrong part<br>• Loose fuel vapor hose/tube connections to EVAP components<br>• Canister vent (CV) solenoid stuck open<br>• Fuel tank pressure (FTP) sensor has failed mechanically |
| **DTC: P0458**<br>**2T CCM, MIL: Yes**<br>**Years: 2007, 2008**<br>**Models:** A6<br>**Engines:** 3.2L, 4.2L<br>**Transmissions:** All | **Evaporative Emission System Purge Control Valve Circuit Low Conditions:**<br>Engine started, battery voltage must be at least 11.5v, all electrical components must be off, parking brake must be engaged (to keep daytime driving lights off), automatic transmission selector must be in park, the exhaust system must be properly sealed between the catalytic converter and the cylinder head, coolant temperature must be at least 80 degrees Celsius and oxygen sensor heaters for oxygen sensors before the catalytic converter must be functioning properly and the ground between the engine and the chassis must be well connected. The ECM detected an unexpected voltage condition on the EVAP circuit when the device was cycled On/Off during testing.<br>**Possible Causes:**<br>• EVAP power supply circuit is open<br>• EVAP solenoid control circuit is open or shorted to ground<br>• EVAP solenoid control circuit is shorted to power (B+)<br>• EVAP solenoid valve is damaged or it has failed<br>• EVAP canister has a leak or a poor seal<br>• ECM has failed |

| DTC | Trouble Code Title, Conditions & Possible Causes |
|---|---|
| **DTC: P0459**<br>**2T CCM, MIL: Yes**<br>**Years: 2007, 2008**<br>**Models:** A6<br>**Engines:** 3.2L, 4.2L<br>**Transmissions:** All | **Evaporative Emission System Purge Control Valve Circuit High Conditions:**<br>Engine started, battery voltage must be at least 11.5v, all electrical components must be off, parking brake must be engaged (to keep daytime driving lights off), automatic transmission selector must be in park, the exhaust system must be properly sealed between the catalytic converter and the cylinder head, coolant temperature must be at least 80 degrees Celsius and oxygen sensor heaters for oxygen sensors before the catalytic converter must be functioning properly and the ground between the engine and the chassis must be well connected. The ECM detected an unexpected voltage condition on the EVAP circuit when the device was cycled On/Off during testing.<br>**Possible Causes:**<br>&bull; EVAP power supply circuit is open<br>&bull; EVAP solenoid control circuit is open or shorted to ground<br>&bull; EVAP solenoid control circuit is shorted to power (B+)<br>&bull; EVAP solenoid valve is damaged or it has failed<br>&bull; EVAP canister has a leak or a poor seal<br>&bull; ECM has failed |
| **DTC: P0491**<br>**2T CCM, MIL: Yes**<br>**Years: 2007, 2008**<br>**Models:** A4, A6<br>**Engines:** All<br>**Transmissions:** All | **Secondary Air Injection System Insufficient Flow (Bank 1) Conditions:**<br>Engine started, battery voltage must be at least 11.5v, all electrical components must be off, parking brake must be engaged (to keep daytime driving lights off), automatic transmission selector must be in park and the ground between the engine and the chassis must be well connected. The ECM detected the Secondary AIR pump airflow was not diverted correctly when requested during the self-test. The pump is functioning but the quantity of air is recognized as insufficient by HO2S<br>**Possible Causes:**<br>&bull; Air pump output is blocked or restricted<br>&bull; AIR bypass solenoid is leaking or it is restricted<br>&bull; AIR bypass solenoid is stuck open or stuck closed<br>&bull; Check valve (one or more) is damaged or leaking<br>&bull; Electric air injection pump hose(s) leaking<br>&bull; Electric air injection pump is damaged or faulty<br>&bull; ECM has failed |
| **DTC: P0492**<br>**2T CCM, MIL: Yes**<br>**Years: 2007, 2008**<br>**Models:** A4, A6<br>**Engines:** All<br>**Transmissions:** All | **Secondary Air Injection System Insufficient Flow (Bank 2) Conditions:**<br>Engine started, battery voltage must be at least 11.5v, all electrical components must be off, parking brake must be engaged (to keep daytime driving lights off), automatic transmission selector must be in park and the ground between the engine and the chassis must be well connected. The ECM detected the Secondary AIR pump airflow was not diverted correctly when requested during the self-test. The pump is functioning but the quantity of air is recognized as insufficient by HO2S<br>**Possible Causes:**<br>&bull; Air pump output is blocked or restricted<br>&bull; AIR bypass solenoid is leaking or it is restricted<br>&bull; AIR bypass solenoid is stuck open or stuck closed<br>&bull; Check valve (one or more) is damaged or leaking<br>&bull; Electric air injection pump hose(s) leaking<br>&bull; Electric air injection pump is damaged or faulty<br>&bull; ECM has failed |
| **DTC: P0501**<br>**2T CCM, MIL: Yes**<br>**Years: 2007, 2008**<br>**Models:** A4, A6<br>**Engines:** All<br>**Transmissions:** All | **Vehicle Speed Sensor or PSOM Range/Performance Conditions:**<br>Engine started; engine speed above the TCC stall speed, and the ECM detected a loss of the VSS signal over a period of time or the signal is not usable.<br>**Note: The ECM receives vehicle speed data from the VSS, TCSS, ABS module, CTM or GEM controller, depending up the application.**<br>**Possible Causes:**<br>&bull; VSS signal circuit is open or shorted to ground<br>&bull; VSS harness circuit is shorted to ground<br>&bull; VSS harness circuit is shorted to power<br>&bull; VSS circuit open between the ECM and related control module<br>&bull; VSS or wheel speed sensors circuits are damaged<br>&bull; Modules connected to VSC/VSS harness circuits are damaged<br>&bull; Mechanical drive mechanism for the VSS is damaged |

| DTC | Trouble Code Title, Conditions & Possible Causes |
|---|---|
| **DTC: P0506**<br>**2T CCM, MIL: Yes**<br>**Years: 2007, 2008**<br>**Models:** A4, A6<br>**Engines:** All<br>**Transmissions:** All | **Idle Air Control System RPM Lower Than Expected Conditions:**<br>Engine started, battery voltage must be at least 11.5v, all electrical components must be off, parking brake must be engaged (to keep daytime driving lights off), automatic transmission selector must be in park, the exhaust system must be properly sealed between the catalytic converter and the cylinder head, coolant temperature must be at least 80 degrees Celsius and oxygen sensor heaters for oxygen sensors before the catalytic converter must be functioning properly and the ground between the engine and the chassis must be well connected. The ECM detected it could not control the idle speed correctly, as it is constantly more than 100 RPM less than specification.<br>**Possible Causes:**<br>• Air inlet is plugged or the air filter element is severely clogged<br>• IAC circuit is open or shorted<br>• IAC circuit VPWR circuit is open<br>• IAC solenoid is damaged or has failed<br>• ECM has failed<br>• The VSS has failed |
| **DTC: P0507**<br>**2T CCM, MIL: Yes**<br>**Years: 2007, 2008**<br>**Models:** A4, A6<br>**Engines:** All<br>**Transmissions:** All | **Idle Air Control System RPM Higher Than Expected Conditions:**<br>Engine started, battery voltage must be at least 11.5v, all electrical components must be off, parking brake must be engaged (to keep daytime driving lights off), automatic transmission selector must be in park, the exhaust system must be properly sealed between the catalytic converter and the cylinder head, coolant temperature must be at least 80 degrees Celsius and oxygen sensor heaters for oxygen sensors before the catalytic converter must be functioning properly and the ground between the engine and the chassis must be well connected. The ECM detected it could not control the idle speed correctly, as it is constantly more than 200 RPM more than specification.<br>**Possible Causes:**<br>• Air intake leak located somewhere after the throttle body<br>• IAC control circuit is shorted to chassis ground<br>• IAC solenoid is damaged or has failed<br>• Throttle Valve Control module has failed or is clogged with carbon<br>• ECM has failed<br>• The VSS has failed |
| **DTC: P0560**<br>**2T CCM, MIL: Yes**<br>**Years: 2007, 2008**<br>**Models:** A4<br>**Engines:** All<br>**Transmissions:** All | **System Voltage Malfunction Conditions:**<br>Engine started, battery voltage must be at least 11.5v, all electrical components must be off, parking brake must be engaged (to keep daytime driving lights off), automatic transmission selector must be in park, and the ground between the engine and the chassis must be well connected. The ECM has detected a voltage value that is implausible or erratic.<br>**Possible Causes:**<br>• Alternator damaged or faulty<br>• Battery voltage low or insufficient<br>• Fuses blown or circuits open<br>• Battery connection to terminal not clean<br>• Voltage regulator has failed |
| **DTC: P0562**<br>**2T CCM, MIL: Yes**<br>**Years: 2007, 2008**<br>**Models:** A4, A6,<br>**Engines:** 3.2L, 4.2L<br>**Transmissions:** All | **System Voltage Low Conditions:**<br>Engine started, battery voltage must be at least 11.5v, all electrical components must be off, parking brake must be engaged (to keep daytime driving lights off), automatic transmission selector must be in park, and the ground between the engine and the chassis must be well connected. The ECM has detected a voltage value that is below the specified minimum limit for the system to function properly.<br>**Possible Causes:**<br>• Alternator damaged or faulty<br>• Battery voltage low or insufficient<br>• Fuses blown or circuits open<br>• Battery connection to terminal not clean<br>• Voltage regulator has failed |
| **DTC: P0563**<br>**2T CCM, MIL: Yes**<br>**Years: 2007, 2008**<br>**Models:** A4, A6,<br>**Engines:** 3.2L, 4.2L<br>**Transmissions:** All | **System Voltage High Conditions:**<br>Engine started, battery voltage must be at least 11.5v, all electrical components must be off, parking brake must be engaged (to keep daytime driving lights off), automatic transmission selector must be in park, and the ground between the engine and the chassis must be well connected. The ECM has detected a voltage value that has exceeded the specified maximum limit for the system to function properly.<br>**Possible Causes:**<br>• Alternator damaged or faulty<br>• Battery voltage low or insufficient<br>• Fuses blown or circuits open<br>• Battery connection to terminal not clean<br>• Voltage regulator has failed |

| DTC | Trouble Code Title, Conditions & Possible Causes |
|---|---|
| **DTC: P0600**<br>**1T CCM, MIL: Yes**<br>**Years:** 2007, 2008<br>**Models:** A4, A6<br>**Engines:** All<br>**Transmissions:** All | **Serial Communication Link (Data BUS) Message Missing Conditions:**<br>The Engine Control Module (ECM) communicates with all databus-capable control modules via a CAN databus. These databus-capable control modules are connected via two data bus wires which are twisted together (CAN_High and CAN_Low), and exchange information (messages). Missing information on the databus is recognized as a malfunction and stored. Trouble-free operation of the CAN-Bus requires that it have a terminal resistance. This central terminal resistor is located in the Engine Control Module (ECM).<br>**Possible Causes:**<br>   &bull; ECM has failed<br>   &bull; CAN data bus wires have short circuited to each other |
| **DTC: P0601**<br>**1T CCM, MIL: Yes**<br>**Years:** 2007, 2008<br>**Models:** A4, A6<br>**Engines:** All<br>**Transmissions:** All | **Internal Control Module Memory Check Sum Error Conditions:**<br>Key on, the ECM has detected a programming error<br>**Possible Causes:**<br>   &bull; Battery terminal corrosion, or loose battery connection<br>   &bull; Connection to the ECM interrupted, or the circuit has been opened<br>   &bull; Reprogramming error has occurred<br>   &bull; ECM has failed and needs replacement. Remember to check for Aftermarket Performance Products before replacing a ECM. |
| **DTC: P0602**<br>**1T CCM, MIL: Yes**<br>**Years:** 2007, 2008<br>**Models:** A4, A6,<br>**Engines:** All<br>**Transmissions:** All | **Control Module Programming Error Conditions:**<br>Key on, and the ECM detected a programming error in the VID block. This fault requires that the VID Block be reprogrammed, or that the EEPROM be re-flashed.<br>**Possible Causes:**<br>   &bull; During the VID reprogramming function, the Vehicle ID (VID) data block failed during reprogramming wit the Scan Tool.<br>   &bull; Battery terminal corrosion, or loose battery connection<br>   &bull; Connection to the ECM interrupted, or the circuit has been opened<br>   &bull; Reprogramming error has occurred<br>   &bull; ECM has failed and needs replacement. Remember to check for Aftermarket Performance Products before replacing a ECM. |
| **DTC: P0603**<br>**1T CCM, MIL: Yes**<br>**Years:** 2007, 2008<br>**Models:** A6<br>**Engines:** 3.2L, 4.2L<br>**Transmissions:** All | **ECM Keep Alive Memory Test Error Conditions:**<br>Key on, and the ECM detected an internal memory fault. This code will set if KAPWR to the ECM is interrupted (at the initial key on).<br>**Possible Causes:**<br>   &bull; Battery terminal corrosion, or loose battery connection<br>   &bull; KAPWR to ECM interrupted, or the circuit has been opened<br>   &bull; Reprogramming error has occurred<br>   &bull; ECM has failed and needs replacement. Remember to check for Aftermarket Performance Products before replacing a ECM. |
| **DTC: P0604**<br>**1T CCM, MIL: Yes**<br>**Years:** 2007, 2008<br>**Models:** A4, A6<br>**Engines:** All<br>**Transmissions:** All | **Internal Control Module Random Access Memory (RAM) Error Conditions:**<br>Key on, and the ECM detected an internal memory fault. This code will set if KAPWR to the ECM is interrupted (at the initial key on).<br>**Possible Causes:**<br>   &bull; Battery terminal corrosion, or loose battery connection<br>   &bull; Connection to the ECM interrupted, or the circuit has been opened<br>   &bull; Reprogramming error has occurred<br>   &bull; ECM has failed and needs replacement. Remember to check for Aftermarket Performance Products before replacing a ECM. |
| **DTC: P0605**<br>**1T CCM, MIL: Yes**<br>**Years:** 2007, 2008<br>**Models:** A4, A6<br>**Engines:** All<br>**Transmissions:** All | **ECM Read Only Memory (ROM) Test Error Conditions:**<br>Key on, and the ECM detected a ROM test error (ROM inside ECM is corrupted). The ECM is normally replaced if this code has set.<br>**Possible Causes:**<br>   &bull; An attempt was made to change the module calibration, or a module programming error may have occurred<br>   &bull; Clear the trouble codes and then check for this trouble code. If it resets, the ECM has failed and needs replacement.<br>   &bull; Aftermarket performance products may have been installed.<br>   &bull; The Transmission Control Module (TCM) has failed. |
| **DTC: P0606**<br>**1T CCM, MIL: Yes**<br>**Years:** 2007, 2008<br>**Models:** A4, A6<br>**Engines:** All<br>**Transmissions:** All | **ECM Internal Communication Error Conditions:**<br>Key on, and the ECM detected an internal communications register read back error during the initial key on check period.<br>**Possible Causes:**<br>   &bull; Clear the trouble codes and then check for this trouble code. If it resets, the ECM has failed and needs replacement.<br>   &bull; Remember to check for signs of Aftermarket Performance Products installation before replacing the ECM. |

| DTC | Trouble Code Title, Conditions & Possible Causes |
|---|---|
| **DTC: P0614**<br>**1T CCM, MIL: Yes**<br>**Years:** 2007, 2008<br>**Models:** A6<br>**Engines:** 3.2L, 4.2L<br>**Transmissions:** All | **ECM / TCM Incompatible Conditions:**<br>Key on, and the ECM detected a communication error between the Transmission control module and the ECM<br>**Possible Causes:**<br>• TCM failed<br>• ECM failed<br>• Circuit shorting between ECM and TCM<br>• Replacement control module ID doesn't match old control module ID |
| **DTC: P0627**<br>**1T CCM, MIL: Yes**<br>**Years:** 2007, 2008<br>**Models:** A6<br>**Engines:** All<br>**Transmissions:** All | **Fuel Pump "A" Control Circuit Open Conditions:**<br>Engine started, battery voltage must be at least 11.5v, all electrical components must be off, parking brake must be engaged (to keep daytime driving lights off), automatic transmission selector must be in park, and the ground between the engine and the chassis must be well connected. The ECM has detected a voltage value across the fuel pump control circuit that is out of the specified limits for the system to function properly.<br>**Possible Causes:**<br>• Fuel Pressure Regulator Valve is faulty<br>• Fuel Pressure Sensor is faulty<br>• Fuel Pump (FP) Control Module is faulty<br>• Fuel pump is faulty |
| **DTC: P0629**<br>**1T CCM, MIL: Yes**<br>**Years:** 2007, 2008<br>**Models:** A6<br>**Engines:** All<br>**Transmissions:** All | **Fuel Pump "A" Control Circuit High Conditions:**<br>Engine started, battery voltage must be at least 11.5v, all electrical components must be off, parking brake must be engaged (to keep daytime driving lights off), automatic transmission selector must be in park, and the ground between the engine and the chassis must be well connected. The ECM has detected a voltage value across the fuel pump control circuit that is above the specified limit for the system to function properly.<br>**Possible Causes:**<br>• Fuel Pressure Regulator Valve is faulty<br>• Fuel Pressure Sensor is faulty<br>• Fuel Pump (FP) Control Module is faulty<br>• Fuel pump is faulty |
| **DTC: P0638**<br>**1T CCM, MIL: Yes**<br>**Years:** 2007, 2008<br>**Models:** A4, A6<br>**Engines:** All<br>**Transmissions:** All | **Throttle Actuator Control Range/Performance Bank 1 Conditions:**<br>Engine started, battery voltage must be at least 11.5v, all electrical components must be off, parking brake must be engaged (to keep daytime driving lights off), automatic transmission selector must be in park, and the ground between the engine and the chassis must be well connected. The ECM has detected a voltage value across the throttle actuator control circuit that is out of the specified limit for the system to function properly. Both Throttle Position (TP) Sensor / Accelerator Pedal Position Sensor 2 are located at the accelerator pedal and communicate the driver's intentions to the Motronic engine control module (ECM) completely independently of each other. Both sensors are integrated into one housing.<br>**Possible Causes:**<br>• Throttle Position (TP) sensor is faulty<br>• Throttle valve control module is faulty<br>• ECM is faulty<br>• Circuit wires have short circuited to each other, to vehicle Ground (GND) or to B+.<br>• Accelerator pedal module is faulty |
| **DTC: P0641**<br>**2T CCM, MIL: Yes**<br>**Years:** 2007, 2008<br>**Models:** A6<br>**Engines:** 3.2L, 4.2L<br>**Transmissions:** All | **Sensor Reference Voltage "A" Circuit Open Conditions:**<br>Engine started, battery voltage must be at least 11.5v, all electrical components must be off, parking brake must be engaged (to keep daytime driving lights off), automatic transmission selector must be in park, and the ground between the engine and the chassis must be well connected.<br>**Possible Causes:**<br>• Circuit harness connector contacts are corroded or ingressed of water<br>• Circuit wires have shorted to each other, to battery or ground<br>• Automatic Transmission Hydraulic Pressure Sensor 1 has failed<br>• Solenoid valves in valve body are faulty<br>• Transmission Control Module (TCM) needs replacing<br>• Transmission Input Speed (RPM) Sensor has failed<br>• Transmission Output Speed (RPM) Sensor has failed |

| DTC | Trouble Code Title, Conditions & Possible Causes |
|---|---|
| **DTC: P0642**<br>**2T CCM, MIL:** Yes<br>**Years:** 2007, 2008<br>**Models:** A6<br>**Engines:** 3.2L<br>**Transmissions:** All | **Sensor Reference Voltage "A" Circuit Low Conditions:**<br>Engine started, battery voltage must be at least 11.5v, all electrical components must be off, parking brake must be engaged (to keep daytime driving lights off), automatic transmission selector must be in park, and the ground between the engine and the chassis must be well connected.<br>**Possible Causes:**<br>• Circuit harness connector contacts are corroded or ingressed of water<br>• Circuit wires have shorted to each other, to battery or ground<br>• Automatic Transmission Hydraulic Pressure Sensor 1 has failed<br>• Solenoid valves in valve body are faulty<br>• Transmission Control Module (TCM) needs replacing<br>• Transmission Input Speed (RPM) Sensor has failed<br>• Transmission Output Speed (RPM) Sensor has failed |
| **DTC: P0643**<br>**2T CCM, MIL:** Yes<br>**Years:** 2007, 2008<br>**Models:** A6<br>**Engines:** 3.2L<br>**Transmissions:** All | **Sensor Reference Voltage "A" Circuit High Conditions:**<br>Engine started, battery voltage must be at least 11.5v, all electrical components must be off, parking brake must be engaged (to keep daytime driving lights off), automatic transmission selector must be in park, and the ground between the engine and the chassis must be well connected.<br>**Possible Causes:**<br>• Circuit harness connector contacts are corroded or ingressed of water<br>• Circuit wires have shorted to each other, to battery or ground<br>• Automatic Transmission Hydraulic Pressure Sensor 1 has failed<br>• Solenoid valves in valve body are faulty<br>• Transmission Control Module (TCM) needs replacing<br>• Transmission Input Speed (RPM) Sensor has failed<br>• Transmission Output Speed (RPM) Sensor has failed |
| **DTC: P0652**<br>**2T CCM, MIL:** Yes<br>**Years:** 2007, 2008<br>**Models:** A6<br>**Engines:** 3.2L<br>**Transmissions:** All | **Sensor Reference Voltage "B" Circuit Low Conditions:**<br>Engine started, battery voltage must be at least 11.5v, all electrical components must be off, parking brake must be engaged (to keep daytime driving lights off), automatic transmission selector must be in park, and the ground between the engine and the chassis must be well connected.<br>**Possible Causes:**<br>• Circuit harness connector contacts are corroded or ingressed of water<br>• Circuit wires have shorted to each other, to battery or ground<br>• Automatic Transmission Hydraulic Pressure Sensor 1 has failed<br>• Solenoid valves in valve body are faulty<br>• Transmission Control Module (TCM) needs replacing<br>• Transmission Input Speed (RPM) Sensor has failed<br>• Transmission Output Speed (RPM) Sensor has failed |
| **DTC: P0653**<br>**2T CCM, MIL:** Yes<br>**Years:** 2007, 2008<br>**Models:** A6<br>**Engines:** 3.2L<br>**Transmissions:** All | **Sensor Reference Voltage "B" Circuit High Conditions:**<br>Engine started, battery voltage must be at least 11.5v, all electrical components must be off, parking brake must be engaged (to keep daytime driving lights off), automatic transmission selector must be in park, and the ground between the engine and the chassis must be well connected.<br>**Possible Causes:**<br>• Circuit harness connector contacts are corroded or ingressed of water<br>• Circuit wires have shorted to each other, to battery or ground<br>• Automatic Transmission Hydraulic Pressure Sensor 1 has failed<br>• Solenoid valves in valve body are faulty<br>• Transmission Control Module (TCM) needs replacing<br>• Transmission Input Speed (RPM) Sensor has failed<br>• Transmission Output Speed (RPM) Sensor has failed |
| **DTC: P0657**<br>**2T CCM, MIL:** Yes<br>**Years:** 2007, 2008<br>**Models:** A6<br>**Engines:** 3.2L<br>**Transmissions:** All | **Actuator Supply Voltage "A" Circuit Open Conditions:**<br>Engine started, battery voltage must be at least 11.5v, all electrical components must be off, parking brake must be engaged (to keep daytime driving lights off), automatic transmission selector must be in park, and the ground between the engine and the chassis must be well connected.<br>**Possible Causes:**<br>• Circuit harness connector contacts are corroded or ingressed of water<br>• Circuit wires have shorted to each other, to battery or ground<br>• Automatic Transmission Hydraulic Pressure Sensor 1 has failed<br>• Solenoid valves in valve body are faulty<br>• Transmission Control Module (TCM) needs replacing<br>• Transmission Input Speed (RPM) Sensor has failed<br>• Transmission Output Speed (RPM) Sensor has failed |

| DTC | Trouble Code Title, Conditions & Possible Causes |
|---|---|
| **DTC: P0658**<br>**2T CCM, MIL: Yes**<br>**Years: 2007, 2008**<br>**Models:** A6<br>**Engines:** 3.2L<br>**Transmissions:** All | **Actuator Supply Voltage "A" Circuit Low Conditions:**<br>Engine started, battery voltage must be at least 11.5v, all electrical components must be off, parking brake must be engaged (to keep daytime driving lights off), automatic transmission selector must be in park, and the ground between the engine and the chassis must be well connected.<br>**Possible Causes:**<br>• Circuit harness connector contacts are corroded or ingressed of water<br>• Circuit wires have shorted to each other, to battery or ground<br>• Automatic Transmission Hydraulic Pressure Sensor 1 has failed<br>• Solenoid valves in valve body are faulty<br>• Transmission Control Module (TCM) needs replacing<br>• Transmission Input Speed (RPM) Sensor has failed<br>• Transmission Output Speed (RPM) Sensor has failed |
| **DTC: P0659**<br>**2T CCM, MIL: Yes**<br>**Years: 2007, 2008**<br>**Models:** A6<br>**Engines:** 3.2L<br>**Transmissions:** All | **Actuator Supply Voltage "A" Circuit High Conditions:**<br>Engine started, battery voltage must be at least 11.5v, all electrical components must be off, parking brake must be engaged (to keep daytime driving lights off), automatic transmission selector must be in park, and the ground between the engine and the chassis must be well connected.<br>**Possible Causes:**<br>• Circuit harness connector contacts are corroded or ingressed of water<br>• Circuit wires have shorted to each other, to battery or ground<br>• Automatic Transmission Hydraulic Pressure Sensor 1 has failed<br>• Solenoid valves in valve body are faulty<br>• Transmission Control Module (TCM) needs replacing<br>• Transmission Input Speed (RPM) Sensor has failed<br>• Transmission Output Speed (RPM) Sensor has failed |
| **DTC: P0685**<br>**1T CCM, MIL: Yes**<br>**Years: 2007, 2008**<br>**Models:** A4, A6<br>**Engines:** All<br>**Transmissions:** All | **ECM Power Relay Control Circuit Open Conditions:**<br>Engine started, battery voltage must be at least 11.5v, all electrical components must be off, parking brake must be engaged (to keep daytime driving lights off), automatic transmission selector must be in park and the ground between the engine and the chassis must be well connected. The ECM detected the ECM power relay control circuit has a voltage outside requirement for proper function.<br>**Possible Causes:**<br>• Generator has failed or is damaged<br>• Fuel pump relay is faulty<br>• Circuit is grounded to power or chassis<br>• ECM has failed |
| **DTC: P0686**<br>**1T CCM, MIL: Yes**<br>**Years: 2007, 2008**<br>**Models:** A4, A6<br>**Engines:** All<br>**Transmissions:** All | **ECM/PCM Power Relay Control Circuit Low Conditions:**<br>Engine started, battery voltage must be at least 11.5v, all electrical components must be off, parking brake must be engaged (to keep daytime driving lights off), automatic transmission selector must be in park and the ground between the engine and the chassis must be well connected. The ECM detected the ECM power relay control circuit has a voltage outside requirement for proper function.<br>**Possible Causes:**<br>• Generator has failed or is damaged<br>• Fuel pump relay is faulty<br>• Circuit is grounded to power or chassis<br>• ECM has failed |
| **DTC: P0687**<br>**1T CCM, MIL: Yes**<br>**Years: 2007, 2008**<br>**Models:** A4, A6<br>**Engines:** All<br>**Transmissions:** All | **ECM/PCM Power Relay Control Circuit High Conditions:**<br>Engine started, battery voltage must be at least 11.5v, all electrical components must be off, parking brake must be engaged (to keep daytime driving lights off), automatic transmission selector must be in park and the ground between the engine and the chassis must be well connected. The ECM detected the ECM power relay control circuit has a voltage outside requirement for proper function.<br>**Possible Causes:**<br>• Generator has failed or is damaged<br>• Fuel pump relay is faulty<br>• Circuit is grounded to power or chassis<br>• ECM has failed |
| **DTC: P0688**<br>**1T CCM, MIL: Yes**<br>**Years: 2007, 2008**<br>**Models:** A4, A6<br>**Engines:** All<br>**Transmissions:** All | **ECM/PCM Power Relay Control Sense Circuit Open Conditions:**<br>Engine started, battery voltage must be at least 11.5v, all electrical components must be off, parking brake must be engaged (to keep daytime driving lights off), automatic transmission selector must be in park and the ground between the engine and the chassis must be well connected. The ECM detected the ECM power relay control circuit has a voltage outside requirement for proper function.<br>**Possible Causes:**<br>• Generator has failed or is damaged<br>• Fuel pump relay is faulty<br>• Circuit is grounded to power or chassis<br>• ECM has failed |

| DTC | Trouble Code Title, Conditions & Possible Causes |
|---|---|
| **DTC: P0700**<br>**2T CCM, MIL: Yes**<br>**Years: 2007, 2008**<br>**Models:** A6<br>**Engines:** 3.2L, 4.2L<br>**Transmissions:** A/T | **Transmission Control System Malfunction Conditions:**<br>Engine started, battery voltage must be at least 11.5v, all electrical components must be off, parking brake must be engaged (to keep daytime driving lights off), automatic transmission selector must be in park, and the ground between the engine and the chassis must be well connected. The ECM detected a malfunction int the transmission control system.<br>**Possible Causes:**<br>• Circuit harness connector contacts are corroded or ingressed of water<br>• Circuit wires have shorted to each other, to battery or ground<br>• Automatic Transmission Hydraulic Pressure Sensor 1 has failed<br>• Solenoid valves in valve body are faulty<br>• Transmission Input Speed (RPM) Sensor has failed<br>• Transmission Output Speed (RPM) Sensor has failed<br>• Engine Control Module (ECM) is faulty<br>• Voltage supply for Engine Control Module (ECM) is faulty<br>• Transmission Control Module (TCM) is faulty |
| **DTC: P0704**<br>**2T CCM, MIL: Yes**<br>**Years: 2007, 2008**<br>**Models:** A4, A6<br>**Engines:** All<br>**Transmissions:** A/T | **Clutch Switch Input Circuit Malfunction Conditions:**<br>Engine started, battery voltage must be at least 11.5v, all electrical components must be off, parking brake must be engaged (to keep daytime driving lights off), automatic transmission selector must be in park, and the ground between the engine and the chassis must be well connected. The ECM detected a voltage outside the normal performance range to allow the system to properly function.<br>**Possible Causes:**<br>• Circuit harness connector contacts are corroded or ingressed of water<br>• Circuit wires have shorted to each other, to battery or ground<br>• Automatic Transmission Hydraulic Pressure Sensor 1 has failed<br>• Solenoid valves in valve body are faulty<br>• Transmission Input Speed (RPM) Sensor has failed<br>• Transmission Output Speed (RPM) Sensor has failed<br>• Engine Control Module (ECM) is faulty<br>• Voltage supply for Engine Control Module (ECM) is faulty<br>• Transmission Control Module (TCM) is faulty |
| **DTC: P0706**<br>**2T CCM, MIL: Yes**<br>**Years: 2007, 2008**<br>**Models:** A4, A6<br>**Engines:** 3.2L, 4.2L<br>**Transmissions:** A/T | **TR Sensor Circuit Range/Performance Conditions:**<br>Engine started, battery voltage must be at least 11.5v, all electrical components must be off, parking brake must be engaged (to keep daytime driving lights off), automatic transmission selector must be in park, and the ground between the engine and the chassis must be well connected. The ECM detected a voltage outside the normal performance range to allow the system to properly function.<br>**Possible Causes:**<br>• Circuit harness connector contacts are corroded or ingressed of water<br>• Circuit wires have shorted to each other, to battery or ground<br>• Automatic Transmission Hydraulic Pressure Sensor 1 has failed<br>• Solenoid valves in valve body are faulty<br>• Transmission Input Speed (RPM) Sensor has failed<br>• Transmission Output Speed (RPM) Sensor has failed<br>• Engine Control Module (ECM) is faulty<br>• Voltage supply for Engine Control Module (ECM) is faulty<br>• Transmission Control Module (TCM) is faulty |
| **DTC: P0710**<br>**2T CCM, MIL: Yes**<br>**Years: 2007, 2008**<br>**Models:** A6<br>**Engines:** 3.2L, 4.2L<br>**Transmissions:** A/T | **Transmission Fluid Temperature Sensor Circuit Malfunction Conditions:**<br>Engine started, battery voltage must be at least 11.5v, all electrical components must be off, parking brake must be engaged (to keep daytime driving lights off), automatic transmission selector must be in park, and the ground between the engine and the chassis must be well connected. The ECM detected the Transmission fluid temperature sensor circuit was outside the normal range in the test to allow proper function.<br>**Possible Causes:**<br>• ATF is low, contaminated, dirty or burnt<br>• Circuit harness connector contacts are corroded or ingressed of water<br>• Circuit wires have shorted to each other, to battery or ground<br>• Automatic Transmission Hydraulic Pressure Sensor 1 has failed<br>• Solenoid valves in valve body are faulty<br>• Transmission Input Speed (RPM) Sensor has failed<br>• Transmission Output Speed (RPM) Sensor has failed<br>• Engine Control Module (ECM) is faulty<br>• Voltage supply for Engine Control Module (ECM) is faulty<br>• Transmission Control Module (TCM) is faulty |

| DTC | Trouble Code Title, Conditions & Possible Causes |
|---|---|
| **DTC: P0711**<br>**2T CCM, MIL: Yes**<br>**Years:** 2007, 2008<br>**Models:** A6<br>**Engines:** 3.2L, 4.2L<br>**Transmissions:** A/T | **Transmission Fluid Temperature Sensor Signal Range/Performance Conditions:**<br>Engine started, battery voltage must be at least 11.5v, all electrical components must be off, parking brake must be engaged (to keep daytime driving lights off), automatic transmission selector must be in park, and the ground between the engine and the chassis must be well connected. The ECM detected the Transmission Fluid Temperature (TFT) sensor value was not close its normal operating temperature.<br>**Possible Causes:**<br>• ATF is low, contaminated, dirty or burnt<br>• TFT sensor signal circuit has a high resistance condition<br>• TFT sensor is out-of-calibration ("skewed") or it has failed<br>• ECM has failed |
| **DTC: P0712**<br>**2T CCM, MIL: Yes**<br>**Years:** 2007, 2008<br>**Models:** A6<br>**Engines:** 3.2L, 4.2L<br>**Transmissions:** A/T | **Transmission Fluid Temperature Sensor Circuit Low Input Conditions:**<br>Engine started, battery voltage must be at least 11.5v, all electrical components must be off, parking brake must be engaged (to keep daytime driving lights off), automatic transmission selector must be in park, and the ground between the engine and the chassis must be well connected. The ECM detected the Transmission Fluid Temperature (TFT) sensor was less than its minimum self-test range in the test.<br>**Possible Causes:**<br>• TFT sensor signal circuit is shorted to chassis ground<br>• TFT sensor signal circuit is shorted to sensor ground<br>• TFT sensor is damaged, or out-of-calibration, or has failed<br>• ECM has failed |
| **DTC: P0713**<br>**2T CCM, MIL: Yes**<br>**Years:** 2007, 2008<br>**Models:** A6<br>**Engines:** 3.2L, 4.2L<br>**Transmissions:** A/T | **Transmission Fluid Temperature Sensor Circuit High Input Conditions:**<br>Engine started, battery voltage must be at least 11.5v, all electrical components must be off, parking brake must be engaged (to keep daytime driving lights off), automatic transmission selector must be in park, and the ground between the engine and the chassis must be well connected. The ECM detected the Transmission Fluid Temperature (TFT) sensor was more than its maximum self-test range in the test.<br>**Possible Causes:**<br>• TFT sensor signal circuit is open between the sensor and ECM<br>• TFT sensor ground circuit is open between sensor and ECM<br>• TFT sensor is damaged or has failed<br>• ECM has failed |
| **DTC: P0714**<br>**2T CCM, MIL: Yes**<br>**Years:** 2007, 2008<br>**Models:** A6<br>**Engines:** 3.2L, 4.2L<br>**Transmissions:** A/T | **Transmission Fluid Temperature Sensor Circuit Intermittent Conditions:**<br>Engine started, battery voltage must be at least 11.5v, all electrical components must be off, parking brake must be engaged (to keep daytime driving lights off), automatic transmission selector must be in park, and the ground between the engine and the chassis must be well connected. The ECM detected the Transmission Fluid Temperature (TFT) sensor was giving a false reading or was not reading at all.<br>**Possible Causes:**<br>• TFT sensor signal circuit is open between the sensor and ECM<br>• TFT sensor ground circuit is open between sensor and ECM<br>• TFT sensor is damaged or has failed<br>• ECM has failed |
| **DTC: P0715**<br>**1T CCM, MIL: Yes**<br>**Years:** 2007, 2008<br>**Models:** A4, A6<br>**Engines:** 3.2L, 4.2L<br>**Transmissions:** A/T | **Input/Turbine Speed Sensor Circuit Malfunction Conditions:**<br>Engine started, vehicle driven with the vehicle speed sensor indicating more than 1 mph, and the ECM detected the Transmission Vehicle Speed Sensor signals were erratic, or that they were missing for a period of time.<br>**Possible Causes:**<br>• TVSS signal circuit is open<br>• TVSS signal is shorted to chassis ground<br>• TVSS signal is shorted to sensor ground<br>• TVSS assembly is damaged or it has failed<br>• ECM has failed |
| **DTC: P0716**<br>**1T CCM, MIL: Yes**<br>**Years:** 2007, 2008<br>**Models:** A4, A6<br>**Engines:** 3.2L, 4.2L<br>**Transmissions:** A/T | **Input Turbine/Speed Sensor Circuit Range/Performance Conditions:**<br>Engine started, vehicle driven with the vehicle speed sensor indicating more than 1 mph, and the ECM detected the Transmission Vehicle Speed Sensor signals were erratic, or that they were missing for a period of time.<br>**Possible Causes:**<br>• TVSS signal circuit is open<br>• TVSS signal is shorted to chassis ground<br>• TVSS signal is shorted to sensor ground<br>• TVSS assembly is damaged or it has failed<br>• ECM has failed |

| DTC | Trouble Code Title, Conditions & Possible Causes |
|---|---|
| **DTC: P0717**<br>**1T CCM, MIL: Yes**<br>**Years: 2007, 2008**<br>**Models:** A4, A6<br>**Engines:** 3.2L, 4.2L<br>**Transmissions:** A/T | **Transmission Speed Shaft Sensor Signal Intermittent Conditions:**<br>Engine started, vehicle speed sensor indicating over 1 mph, and the ECM detected an intermittent loss of TSS signals (i.e., the TSS signals were erratic, irregular or missing).<br>**Possible Causes:**<br>• TSS connector is damaged, loose or shorted<br>• TSS signal circuit has an intermittent open condition<br>• TSS signal circuit has an intermittent short to ground condition<br>• TSS assembly is damaged or is has failed<br>• ECM has failed |
| **DTC: P0721**<br>**1T CCM, MIL: Yes**<br>**Years: 2007, 2008**<br>**Models:** A4, A6<br>**Engines:** 3.2L, 4.2L<br>**Transmissions:** A/T | **A/T Output Shaft Speed Sensor Noise Interference Conditions:**<br>Engine started, VSS signal more than 1 mph, and the ECM detected "noise" interference on the Output Shaft Speed (OSS) sensor circuit.<br>**Possible Causes:**<br>• After market add-on devices interfering with the OSS signal<br>• OSS connector is damaged, loose or shorted, or the wiring is misrouted or it is damaged<br>• OSS assembly is damaged or it has failed<br>• ECM has failed |
| **DTC: P0722**<br>**1T CCM, MIL: Yes**<br>**Years: 2007, 2008**<br>**Models:** A4, A6<br>**Engines:** 3.2L, 4.2L<br>**Transmissions:** A/T | **A/T Output Speed Sensor No Signal Conditions:**<br>Engine started, and the ECM did not detect any Vehicle Speed Sensor (VSS) sensor signals upon initial vehicle movement.<br>**Possible Causes:**<br>• After market add-on devices interfering with the VSS signal<br>• VSS sensor wiring is misrouted, damaged or shorting<br>• ECM and/or TCM has failed |
| **DTC: P0725**<br>**1T CCM, MIL: Yes**<br>**Years:** 1996<br>**Models:** A4<br>**Engines:** All<br>**Transmissions:** A/T | **Engine Speed Input Circuit Malfunction Conditions:**<br>The Transmission Control Module (TCM) does not receive a signal from the Engine Control Module (ECM).<br>**Possible Causes:**<br>• The TCM circuit is shorting to ground, B+ or is open<br>• TCM has failed<br>• ECM has failed |
| **DTC: P0727**<br>**1T CCM, MIL: Yes**<br>**Years: 2007, 2008**<br>**Models:** A4, A6<br>**Engines:** 3.2L, 4.2L<br>**Transmissions:** A/T | **Engine Speed Input Circuit No Signal Conditions:**<br>The Engine Speed (RPM) Sensor detects engine speed and reference marks. Without an engine speed signal, the engine will not start. If the engine speed signal fails while the engine is running, the engine will stop immediately.<br>**Note: There is a larger-sized gap on the sensor wheel. This gap is the reference mark and does not mean that the sensor wheel is damaged.**<br>**Possible Causes:**<br>• Engine speed sensor has failed<br>• Circuit is shorting to ground, B+ or is open<br>• Sensor wheel is damaged, run out or not properly secured<br>• ECM has failed |
| **DTC: P0729**<br>**1T CCM, MIL: Yes**<br>**Years: 2007, 2008**<br>**Models:** A6<br>**Engines:** 3.2L, 4.2L<br>**Transmissions:** A/T | **Gear 6 Incorrect Ratio Conditions:**<br>Engine started, battery voltage must be at least 11.5v, all electrical components must be off, and the ground between the engine and the chassis must be well connected. The ECM detected an incorrect ratio within the sixth gear.<br>**Possible Causes:**<br>• ATF level is low<br>• Circuit harness connector contacts are corroded or ingressed of water<br>• Circuit wires have shorted to each other, to battery or ground<br>• Automatic Transmission Hydraulic Pressure Sensor 1 has failed<br>• Solenoid valves in valve body are faulty<br>• Transmission Control Module (TCM) needs replacing<br>• Transmission Input Speed (RPM) Sensor has failed<br>• Transmission Output Speed (RPM) Sensor has failed |

| DTC | Trouble Code Title, Conditions & Possible Causes |
|---|---|
| **DTC: P0731**<br>**2T CCM, MIL: Yes**<br>**Years: 2007, 2008**<br>**Models:** A4, A6<br>**Engines:** 3.2L, 4.2L<br>**Transmissions:** A/T | **Incorrect First Gear Ratio Conditions:**<br>Engine started, vehicle operating with 1st gear commanded "on", and the ECM detected an incorrect 1st gear ratio during the test.<br>**Possible Causes:**<br>• 1st Gear solenoid harness connector not properly seated<br>• 1st Gear solenoid signal shorted to ground, or open<br>• 1st Gear solenoid wiring harness connector is damaged<br>• 1st Gear solenoid is damaged or not properly installed<br>• ATF level is low<br>• Circuit harness connector contacts are corroded or ingressed of water<br>• Circuit wires have shorted to each other, to battery or ground<br>• Automatic Transmission Hydraulic Pressure Sensor 1 has failed<br>• Transmission Control Module (TCM) needs replacing<br>• Transmission Input Speed (RPM) Sensor has failed<br>• Transmission Output Speed (RPM) Sensor has failed |
| **DTC: P0732**<br>**2T CCM, MIL: Yes**<br>**Years: 2007, 2008**<br>**Models:** A4, A6<br>**Engines:** 3.2L, 4.2L<br>**Transmissions:** A/T | **Incorrect Second Gear Ratio Conditions:**<br>Engine started, vehicle operating with 2nd Gear commanded "on", and the ECM detected an incorrect 2nd gear ratio during the test.<br>**Possible Causes:**<br>• 2nd Gear solenoid harness connector not properly seated<br>• 2nd Gear solenoid signal shorted to ground, or open<br>• 2nd Gear solenoid wring harness connector is damaged<br>• 2nd Gear solenoid is damaged or not properly installed<br>• ATF level is low<br>• Circuit harness connector contacts are corroded or ingressed of water<br>• Circuit wires have shorted to each other, to battery or ground<br>• Automatic Transmission Hydraulic Pressure Sensor 1 has failed<br>• Transmission Control Module (TCM) needs replacing<br>• Transmission Input Speed (RPM) Sensor has failed<br>• Transmission Output Speed (RPM) Sensor has failed |
| **DTC: P0733**<br>**2T CCM, MIL: Yes**<br>**Years: 2007, 2008**<br>**Models:** A4, A6<br>**Engines:** 3.2L, 4.2L<br>**Transmissions:** A/T | **Incorrect Third Gear Ratio Conditions:**<br>Engine started, vehicle operating with 3rd Gear commanded "on", and the ECM detected an incorrect 3rd gear ratio during the test.<br>**Possible Causes:**<br>• 3rd Gear solenoid harness connector not properly seated<br>• 3rd Gear solenoid signal shorted to ground, or open<br>• 3rd Gear solenoid wiring harness connector is damaged<br>• 3rd Gear solenoid is damaged or not properly installed<br>• ATF level is low<br>• Circuit harness connector contacts are corroded or ingressed of water<br>• Circuit wires have shorted to each other, to battery or ground<br>• Automatic Transmission Hydraulic Pressure Sensor 1 has failed<br>• Transmission Control Module (TCM) needs replacing<br>• Transmission Input Speed (RPM) Sensor has failed<br>• Transmission Output Speed (RPM) Sensor has failed |
| **DTC: P0734**<br>**2T CCM, MIL: Yes**<br>**Years: 2007, 2008**<br>**Models:** A4, A6<br>**Engines:** 3.2L, 4.2L<br>**Transmissions:** A/T | **Incorrect Fourth Gear Ratio Conditions:**<br>Engine started, vehicle operating with 4th Gear commanded "on", and the ECM detected an incorrect 4th gear ratio during the test.<br>**Possible Causes:**<br>• 4th Gear solenoid harness connector not properly seated<br>• 4th Gear solenoid signal shorted to ground, or open<br>• 4th Gear solenoid wiring harness connector is damaged<br>• 4th Gear solenoid is damaged or not properly installed<br>• ATF level is low<br>• Circuit harness connector contacts are corroded or ingressed of water<br>• Circuit wires have shorted to each other, to battery or ground<br>• Automatic Transmission Hydraulic Pressure Sensor 1 has failed<br>• Transmission Control Module (TCM) needs replacing<br>• Transmission Input Speed (RPM) Sensor has failed<br>• Transmission Output Speed (RPM) Sensor has failed |

| DTC | Trouble Code Title, Conditions & Possible Causes |
|---|---|
| **DTC: P0735**<br>**2T CCM, MIL: Yes**<br>**Years:** 2007, 2008<br>**Models:** A4, A6<br>**Engines:** 3.2L, 4.2L<br>**Transmissions:** A/T | **Incorrect Fifth Gear Ratio Conditions:**<br>Engine started, vehicle operating with 5th Gear commanded "on", and the ECM detected an incorrect 5th gear ratio during the test.<br>**Possible Causes:**<br>• 5th Gear solenoid harness connector not properly seated<br>• 5th Gear solenoid signal shorted to ground, or open<br>• 5th Gear solenoid wiring harness connector is damaged<br>• 5th Gear solenoid is damaged or not properly installed<br>• ATF level is low<br>• Circuit harness connector contacts are corroded or ingressed of water<br>• Circuit wires have shorted to each other, to battery or ground<br>• Automatic Transmission Hydraulic Pressure Sensor 1 has failed<br>• Transmission Control Module (TCM) needs replacing<br>• Transmission Input Speed (RPM) Sensor has failed<br>• Transmission Output Speed (RPM) Sensor has failed |
| **DTC: P0740**<br>**2T CCM, MIL: Yes**<br>**Years:** 1996<br>**Models:** A4<br>**Engines:** All<br>**Transmissions:** A/T | **TCC Solenoid Circuit Malfunction Conditions:**<br>Engine started, KOER Self-Test enabled, vehicle driven at cruise speed, and the ECM did not detect any voltage drop across the TCC solenoid circuit during the test period.<br>**Possible Causes:**<br>• TCC solenoid control circuit is open or shorted to ground<br>• TCC solenoid wiring harness connector is damaged<br>• TCC solenoid is damaged or has failed<br>• ECM has failed |
| **DTC: P0741**<br>**2T CCM, MIL: Yes**<br>**Years:** 2007, 2008<br>**Models:** A4, A6<br>**Engines:** 3.2L, 4.2L<br>**Transmissions:** A/T | **TCC Mechanical System Range/Performance Conditions:**<br>Engine started, vehicle driven in gear with VSS signals received, and the ECM detected excessive slippage while in normal operation.<br>**Possible Causes:**<br>• TCC solenoid has a mechanical failure<br>• TCC solenoid has a hydraulic failure<br>• ECM has failed |
| **DTC: P0746**<br>**1T CCM, MIL: Yes**<br>**Years:** 2007, 2008<br>**Models:** A6<br>**Engines:** 3.2L, 4.2L<br>**Transmissions:** A/T | **Pressure Control Solenoid "A" Performance or Stuck Off Conditions:**<br>Engine started, battery voltage must be at least 11.5v, all electrical components must be off, and the ground between the engine and the chassis must be well connected. The ECM detected the pressure control solenoid was in the "stuck off" position.<br>**Possible Causes:**<br>• ATF level is low<br>• Circuit harness connector contacts are corroded or ingressed of water<br>• Circuit wires have shorted to each other, to battery or ground<br>• Automatic Transmission Hydraulic Pressure Sensor 1 has failed<br>• Solenoid valves in valve body are faulty<br>• Transmission Control Module (TCM) needs replacing<br>• Transmission Input Speed (RPM) Sensor has failed<br>• Transmission Output Speed (RPM) Sensor has failed |
| **DTC: P0747**<br>**1T CCM, MIL: Yes**<br>**Years:** 2007, 2008<br>**Models:** A6<br>**Engines:** 3.2L, 4.2L<br>**Transmissions:** A/T | **Pressure Control Solenoid "A" Performance or Stuck On Conditions:**<br>Engine started, battery voltage must be at least 11.5v, all electrical components must be off, and the ground between the engine and the chassis must be well connected. The ECM detected the pressure control solenoid was in the "stuck on" position.<br>**Possible Causes:**<br>• ATF level is low<br>• Circuit harness connector contacts are corroded or ingressed of water<br>• Circuit wires have shorted to each other, to battery or ground<br>• Automatic Transmission Hydraulic Pressure Sensor 1 has failed<br>• Solenoid valves in valve body are faulty<br>• Transmission Control Module (TCM) needs replacing<br>• Transmission Input Speed (RPM) Sensor has failed<br>• Transmission Output Speed (RPM) Sensor has failed |
| **DTC: P0748**<br>**1T CCM, MIL: Yes**<br>**Years:** 2007, 2008<br>**Models:** A4, A6<br>**Engines:** 3.2L, 4.2L<br>**Transmissions:** A/T | **Pressure Control Solenoid Electrical Conditions:**<br>The valve body solenoid valve is not receiving a signal.<br>**Possible Causes:**<br>• Pressure control solenoid circuit is shorting to ground<br>• Pressure control solenoid circuit is open<br>• Valve has failed<br>• TCM has failed<br>• ECM has failed |

| DTC | Trouble Code Title, Conditions & Possible Causes |
|---|---|
| **DTC: P0751**<br>**1T CCM, MIL: Yes**<br>**Years: 2007, 2008**<br>**Models:** A4, A6<br>**Engines:** 3.2L, 4.2L<br>**Transmissions:** A/T | **Shift Solenoid "A" Performance or Stuck Off Conditions:**<br>Engine started, vehicle driven with the solenoid applied, and the ECM detected an unexpected voltage condition on the SS1/A solenoid circuit was incorrect during the test.<br>**Possible Causes:**<br>&bull; Solenoid valves in valve body are faulty<br>&bull; Solenoid circuit is shorting to ground<br>&bull; Solenoid circuit is open<br>&bull; TCM has failed or wiring is shorting<br>&bull; ECM has failed |
| **DTC: P0752**<br>**2T CCM, MIL: Yes**<br>**Years: 2007, 2008**<br>**Models:** A4, A6<br>**Engines:** 3.2L, 4.2L<br>**Transmissions:** A/T | **A/T Shift Solenoid 1/A Function Range/Performance Conditions:**<br>Engine started, vehicle driven with the solenoid applied, and the ECM detected a mechanical failure while operating the Shift Solenoid 1/A during the CCM test period.<br>**Possible Causes:**<br>&bull; SS1/A solenoid is stuck in the "on" position<br>&bull; SS1/A solenoid has a mechanical failure<br>&bull; SS1/A solenoid has a hydraulic failure<br>&bull; ECM has failed |
| **DTC: P0756**<br>**2T CCM, MIL: Yes**<br>**Years: 2007, 2008**<br>**Models:** A6<br>**Engines:** 3.2L, 4.2L<br>**Transmissions:** A/T | **A/T Shift Solenoid 2/B Function Range/Performance Conditions:**<br>Engine started, vehicle driven with the solenoid applied, and the ECM detected a mechanical failure while operating the Shift Solenoid 2/B during the CCM test period.<br>**Possible Causes:**<br>&bull; SS2/B solenoid is stuck in the "on" position<br>&bull; SS2/B solenoid has a mechanical failure<br>&bull; SS2/B solenoid has a hydraulic failure<br>&bull; ECM has failed |
| **DTC: P0776**<br>**1T CCM, MIL: Yes**<br>**Years: 2007, 2008**<br>**Models:** A6<br>**Engines:** 3.2L, 4.2L<br>**Transmissions:** A/T | **Pressure Control Solenoid "B" Performance or Stuck Off Conditions:**<br>Engine started, vehicle driven with Shift Solenoid 3/C applied, and the ECM detected a mechanical failure occurred (stuck "off") while operating Shift Solenoid 3/C during the test.<br>**Possible Causes:**<br>&bull; SS3/C solenoid may be stuck "off"<br>&bull; SS3/C solenoid has a mechanical failure<br>&bull; SS3/C solenoid has a hydraulic failure<br>&bull; ECM has failed |
| **DTC: P0777**<br>**1T CCM, MIL: Yes**<br>**Years: 2007, 2008**<br>**Models:** A6<br>**Engines:** 3.2L<br>**Transmissions:** A/T | **Pressure Control Solenoid "B" Stuck On Conditions:**<br>Engine started, vehicle driven with Shift Solenoid 3/C applied, and the ECM detected a mechanical failure occurred (stuck "on") while operating Shift Solenoid 3/C during the test.<br>**Possible Causes:**<br>&bull; SS3/C solenoid may be stuck "on"<br>&bull; SS3/C solenoid has a mechanical failure<br>&bull; SS3/C solenoid has a hydraulic failure<br>&bull; ECM has failed |
| **DTC: P0778**<br>**1T CCM, MIL: Yes**<br>**Years: 2007, 2008**<br>**Models:** A6<br>**Engines:** 3.2L, 4.2L<br>**Transmissions:** A/T | **Pressure Control Solenoid "B" Electrical Conditions:**<br>Engine started, vehicle driven with the solenoid applied, and the ECM detected an unexpected voltage condition on the SS3/C solenoid circuit was incorrect during the test..<br>**Possible Causes:**<br>&bull; Shift Solenoid connector is damaged, open or shorted<br>&bull; Shift Solenoid control circuit is open<br>&bull; Shift Solenoid control circuit is shorted to ground<br>&bull; Shift Solenoid is damaged or it has failed<br>&bull; ECM has failed |
| **DTC: P0781**<br>**2T CCM, MIL: Yes**<br>**Years: 2007, 2008**<br>**Models:** A6<br>**Engines:** 3.2L, 4.2L<br>**Transmissions:** A/T | **1-2 Shift Conditions:**<br>Engine running and vehicle driven, the ECM detected a mechanical malfunction within the transmission<br>**Possible Causes:**<br>&bull; Solenoid valves in valve body are faulty<br>&bull; Solenoid circuit is shorting to ground<br>&bull; Solenoid circuit is open<br>&bull; TCM has failed or wiring is shorting<br>&bull; ECM has failed<br>&bull; Mechanical malfunction in transmission |

| DTC | Trouble Code Title, Conditions & Possible Causes |
|---|---|
| **DTC: P0782**<br>**2T CCM, MIL: Yes**<br>**Years: 2007, 2008**<br>**Models:** A6<br>**Engines:** 3.2L, 4.2L<br>**Transmissions:** A/T | **2-3 Shift Conditions:**<br>Engine running and vehicle driven, the ECM detected a mechanical malfunction within the transmission<br>**Possible Causes:**<br>• Solenoid valves in valve body are faulty<br>• Solenoid circuit is shorting to ground<br>• Solenoid circuit is open<br>• TCM has failed or wiring is shorting<br>• ECM has failed<br>• Mechanical malfunction in transmission |
| **DTC: P0783**<br>**2T CCM, MIL: Yes**<br>**Years: 2007, 2008**<br>**Models:** A6<br>**Engines:** 3.2L, 4.2L<br>**Transmissions:** A/T | **3-4 Shift Conditions:**<br>Engine running and vehicle driven, the ECM detected a mechanical malfunction within the transmission<br>**Possible Causes:**<br>• Solenoid valves in valve body are faulty<br>• Solenoid circuit is shorting to ground<br>• Solenoid circuit is open<br>• TCM has failed or wiring is shorting<br>• ECM has failed<br>• Mechanical malfunction in transmission |
| **DTC: P0784**<br>**2T CCM, MIL: Yes**<br>**Years: 2007, 2008**<br>**Models:** A6<br>**Engines:** 3.2L, 4.2L<br>**Transmissions:** A/T | **4-5 Shift Conditions:**<br>Engine running and vehicle driven, the ECM detected a mechanical malfunction within the transmission<br>**Possible Causes:**<br>• Solenoid valves in valve body are faulty<br>• Solenoid circuit is shorting to ground<br>• Solenoid circuit is open<br>• TCM has failed or wiring is shorting<br>• ECM has failed<br>• Mechanical malfunction in transmission |
| **DTC: P0796**<br>**2T CCM, MIL: Yes**<br>**Years: 2007, 2008**<br>**Models:** A6<br>**Engines:** 3.2L, 4.2L<br>**Transmissions:** A/T | **Pressure Solenoid "C" Performance or Stuck Off Conditions:**<br>Engine started, vehicle driven with the solenoid applied, and the ECM detected an unexpected voltage condition on the SS1/C solenoid circuit was incorrect during the test.<br>**Possible Causes:**<br>• Solenoid valves in valve body are faulty<br>• Solenoid circuit is shorting to ground<br>• Solenoid circuit is open<br>• TCM has failed or wiring is shorting<br>• ECM has failed |
| **DTC: P0797**<br>**2T CCM, MIL: Yes**<br>**Years: 2007, 2008**<br>**Models:** A6<br>**Engines:** 3.2L, 4.2L<br>**Transmissions:** A/T | **Pressure Solenoid "C" Performance or Stuck On Conditions:**<br>Engine started, vehicle driven with the solenoid applied, and the ECM detected an unexpected voltage condition on the SS1/C solenoid circuit was incorrect during the test.<br>**Possible Causes:**<br>• Solenoid valves in valve body are faulty<br>• Solenoid circuit is shorting to ground<br>• Solenoid circuit is open<br>• TCM has failed or wiring is shorting<br>• ECM has failed |
| **DTC: P0798**<br>**2T CCM, MIL: Yes**<br>**Years: 2007, 2008**<br>**Models:** A6<br>**Engines:** 3.2L, 4.2L<br>**Transmissions:** A/T | **Pressure Solenoid "C" Electrical Conditions:**<br>Engine started, vehicle driven with the solenoid applied, and the ECM detected an unexpected voltage condition on the SS1/C solenoid circuit was incorrect during the test.<br>**Possible Causes:**<br>• Solenoid valves in valve body are faulty<br>• Solenoid circuit is shorting to ground<br>• Solenoid circuit is open<br>• TCM has failed or wiring is shorting<br>• ECM has failed |

| DTC | Trouble Code Title, Conditions & Possible Causes |
|---|---|
| **DTC: P0889**<br>**2T CCM, MIL: Yes**<br>**Years: 2007, 2008**<br>**Models:** A6<br>**Engines:** 3.2L, 4.2L<br>**Transmissions:** A/T | **TCM Power Relay Sense Circuit Range/Performance Conditions:**<br>The Transmission Control Module (ECM) communicates with all databus-capable control modules via a CAN databus. These databus-capable control modules are connected via two data bus wires which are twisted together (CAN_High and CAN_Low), and exchange information (messages). Missing information on the databus is recognized as a malfunction and stored. Trouble-free operation of the CAN-Bus requires that it have a terminal resistance.<br>**Possible Causes:**<br>• Solenoid valves in valve body are faulty<br>• Solenoid circuit is shorting to ground<br>• Solenoid circuit is open<br>• TCM has failed or wiring is shorting<br>• ECM has failed |
| **DTC: P0890**<br>**2T CCM, MIL: Yes**<br>**Years: 2007, 2008**<br>**Models:** A6<br>**Engines:** 3.2L, 4.2L<br>**Transmissions:** A/T | **TCM Power Relay Sense Circuit Low Conditions:**<br>The Transmission Control Module (ECM) communicates with all databus-capable control modules via a CAN databus. These databus-capable control modules are connected via two data bus wires which are twisted together (CAN_High and CAN_Low), and exchange information (messages). Missing information on the databus is recognized as a malfunction and stored. Trouble-free operation of the CAN-Bus requires that it have a terminal resistance.<br>**Possible Causes:**<br>• Solenoid valves in valve body are faulty<br>• Solenoid circuit is shorting to ground<br>• Solenoid circuit is open<br>• TCM has failed or wiring is shorting<br>• ECM has failed |
| **DTC: P0891**<br>**2T CCM, MIL: Yes**<br>**Years: 2007, 2008**<br>**Models:** A6<br>**Engines:** 3.2L, 4.2L<br>**Transmissions:** A/T | **TCM Power Relay Sense Circuit High Conditions:**<br>The Transmission Control Module (ECM) communicates with all databus-capable control modules via a CAN databus. These databus-capable control modules are connected via two data bus wires which are twisted together (CAN_High and CAN_Low), and exchange information (messages). Missing information on the databus is recognized as a malfunction and stored. Trouble-free operation of the CAN-Bus requires that it have a terminal resistance.<br>**Possible Causes:**<br>• Solenoid valves in valve body are faulty<br>• Solenoid circuit is shorting to ground<br>• Solenoid circuit is open<br>• TCM has failed or wiring is shorting<br>• ECM has failed |
| **DTC: P0892**<br>**2T CCM, MIL: Yes**<br>**Years: 2007, 2008**<br>**Models:** A6<br>**Engines:** 3.2L, 4.2L<br>**Transmissions:** A/T | **TCM Power Relay Sense Circuit Intermittent Conditions:**<br>The Transmission Control Module (ECM) communicates with all databus-capable control modules via a CAN databus. These databus-capable control modules are connected via two data bus wires which are twisted together (CAN_High and CAN_Low), and exchange information (messages). Missing information on the databus is recognized as a malfunction and stored. Trouble-free operation of the CAN-Bus requires that it have a terminal resistance.<br>**Possible Causes:**<br>• Solenoid valves in valve body are faulty<br>• Solenoid circuit is shorting to ground<br>• Solenoid circuit is open<br>• TCM has failed or wiring is shorting<br>• ECM has failed |

**Gas Engine OBD II Trouble Code List (P1xxx Codes)**

| DTC | Trouble Code Title, Conditions & Possible Causes |
|---|---|
| **DTC: P1102**<br>**1T CCM, MIL: No**<br>**Years: 2007, 2008**<br>**Models:** A4, A6<br>**Engines:** All<br>**Transmissions:** A/T | **O2 Sensor Circuit (Bank 1-Sensor 1) Short to B+ Conditions:**<br>Engine started, battery voltage must be at least 11.5v, all electrical components must be off, the ground between the engine and the chassis must be well connected, the exhaust system must be properly sealed between the catalytic converter and the cylinder head, and the oxygen sensor heater for oxygen sensor before the catalytic converter must be properly functioning. The ECM detected a voltage on the O2 sensor circuit that was outside the parameters to function properly.<br>**Note: For resistance testing of sensor heating, oxygen sensor should be cooled to ambient temperature. High temperatures at oxygen sensor may lead to inaccurate measurements.**<br>**Possible Causes:**<br>• Oxygen sensor (before catalytic converter) is faulty<br>• Oxygen sensor (behind catalytic converter) is faulty<br>• Oxygen sensor heater (before catalytic converter) is faulty<br>• Oxygen sensor heater (behind catalytic converter) is faulty<br>• Circuit wiring has a short to power or ground<br>• Engine Component Power Supply Relay is faulty<br>• E-box fuses for oxygen sensor are faulty<br>• Leaks present in the exhaust manifold or exhaust pipes<br>• HO2S signal wire and ground wire crossed in connector<br>• HO2S element is fuel contaminated or has failed<br>• ECM has failed |
| **DTC: P1105**<br>**1T CCM, MIL: No**<br>**Years: 2007, 2008**<br>**Models:** A4, A6<br>**Engines:** All<br>**Transmissions:** A/T | **O2 Sensor Circuit (Bank 1-Sensor 2) Short to B+ Conditions:**<br>Engine started, battery voltage must be at least 11.5v, all electrical components must be off, the ground between the engine and the chassis must be well connected, the exhaust system must be properly sealed between the catalytic converter and the cylinder head, and the oxygen sensor heater for oxygen sensor before the catalytic converter must be properly functioning. The ECM detected a voltage on the O2 sensor circuit that was outside the parameters to function properly.<br>**Note: For resistance testing of sensor heating, oxygen sensor should be cooled to ambient temperature. High temperatures at oxygen sensor may lead to inaccurate measurements.**<br>**Possible Causes:**<br>• Oxygen sensor (before catalytic converter) is faulty<br>• Oxygen sensor (behind catalytic converter) is faulty<br>• Oxygen sensor heater (before catalytic converter) is faulty<br>• Oxygen sensor heater (behind catalytic converter) is faulty<br>• Circuit wiring has a short to power or ground<br>• Engine Component Power Supply Relay is faulty<br>• E-box fuses for oxygen sensor are faulty<br>• Leaks present in the exhaust manifold or exhaust pipes<br>• HO2S signal wire and ground wire crossed in connector<br>• HO2S element is fuel contaminated or has failed<br>• ECM has failed |
| **DTC: P1107**<br>**1T CCM, MIL: No**<br>**Years: 2007, 2008**<br>**Models:** A4, A6<br>**Engines:** All<br>**Transmissions:** A/T | **O2 Sensor Circuit (Bank 2-Sensor 1) Voltage Too Low Conditions:**<br>Engine started, battery voltage must be at least 11.5v, all electrical components must be off, the ground between the engine and the chassis must be well connected, the exhaust system must be properly sealed between the catalytic converter and the cylinder head, and the oxygen sensor heater for oxygen sensor before the catalytic converter must be properly functioning. The ECM detected a voltage on the O2 sensor circuit that was outside the parameters to function properly.<br>**Note: For resistance testing of sensor heating, oxygen sensor should be cooled to ambient temperature. High temperatures at oxygen sensor may lead to inaccurate measurements.**<br>**Possible Causes:**<br>• Oxygen sensor (before catalytic converter) is faulty<br>• Oxygen sensor (behind catalytic converter) is faulty<br>• Oxygen sensor heater (before catalytic converter) is faulty<br>• Oxygen sensor heater (behind catalytic converter) is faulty<br>• Circuit wiring has a short to power or ground<br>• Engine Component Power Supply Relay is faulty<br>• E-box fuses for oxygen sensor are faulty<br>• Leaks present in the exhaust manifold or exhaust pipes<br>• HO2S signal wire and ground wire crossed in connector<br>• HO2S element is fuel contaminated or has failed<br>• ECM has failed |

| DTC | Trouble Code Title, Conditions & Possible Causes |
|---|---|
| **DTC: P1110**<br>**1T CCM, MIL: No**<br>**Years: 2007, 2008**<br>**Models:** A4, A6<br>**Engines:** All<br>**Transmissions:** A/T | **O2 Sensor Circuit (Bank 2-Sensor 2) Short to B+ Conditions:**<br>Engine started, battery voltage must be at least 11.5v, all electrical components must be off, the ground between the engine and the chassis must be well connected, the exhaust system must be properly sealed between the catalytic converter and the cylinder head, and the oxygen sensor heater for oxygen sensor before the catalytic converter must be properly functioning. The ECM detected a voltage on the O2 sensor circuit that was outside the parameters to function properly.<br>**Note: For resistance testing of sensor heating, oxygen sensor should be cooled to ambient temperature. High temperatures at oxygen sensor may lead to inaccurate measurements.**<br>**Possible Causes:**<br>• Oxygen sensor (before catalytic converter) is faulty<br>• Oxygen sensor (behind catalytic converter) is faulty<br>• Oxygen sensor heater (before catalytic converter) is faulty<br>• Oxygen sensor heater (behind catalytic converter) is faulty<br>• Circuit wiring has a short to power or ground<br>• Engine Component Power Supply Relay is faulty<br>• E-box fuses for oxygen sensor are faulty<br>• Leaks present in the exhaust manifold or exhaust pipes<br>• HO2S signal wire and ground wire crossed in connector<br>• HO2S element is fuel contaminated or has failed<br>• ECM has failed |
| **DTC: P1113**<br>**1T CCM, MIL: No**<br>**Years: 2007, 2008**<br>**Models:** A4, A6<br>**Engines:** All<br>**Transmissions:** All | **O2 Control (Bank 1 Sensor 1) Internal Resistance Too High Conditions:**<br>Engine started, battery voltage must be at least 11.5v, all electrical components must be off, the ground between the engine and the chassis must be well connected, the exhaust system must be properly sealed between the catalytic converter and the cylinder head, and the oxygen sensor heater for oxygen sensor before the catalytic converter must be properly functioning. The ECM detected a measurement on the O2 sensor circuit that was outside the parameters to function properly.<br>**Note: For resistance testing of sensor heating, oxygen sensor should be cooled to ambient temperature. High temperatures at oxygen sensor may lead to inaccurate measurements.**<br>**Possible Causes:**<br>• Oxygen sensor (before catalytic converter) is faulty<br>• Oxygen sensor (behind catalytic converter) is faulty<br>• Oxygen sensor heater (before catalytic converter) is faulty<br>• Oxygen sensor heater (behind catalytic converter) is faulty<br>• Circuit wiring has a short to power or ground<br>• Engine Component Power Supply Relay is faulty<br>• E-box fuses for oxygen sensor are faulty<br>• Leaks present in the exhaust manifold or exhaust pipes<br>• HO2S signal wire and ground wire crossed in connector<br>• HO2S element is fuel contaminated or has failed<br>• ECM has failed |
| **DTC: P1114**<br>**1T CCM, MIL: No**<br>**Years: 2007, 2008**<br>**Models:** A4, A6<br>**Engines:** All<br>**Transmissions:** All | **O2 Control (Bank 1 Sensor 2) Internal Resistance Too High Conditions:**<br>Engine started, battery voltage must be at least 11.5v, all electrical components must be off, the ground between the engine and the chassis must be well connected, the exhaust system must be properly sealed between the catalytic converter and the cylinder head, and the oxygen sensor heater for oxygen sensor before the catalytic converter must be properly functioning. The ECM detected a measurement on the O2 sensor circuit that was outside the parameters to function properly.<br>**Note: For resistance testing of sensor heating, oxygen sensor should be cooled to ambient temperature. High temperatures at oxygen sensor may lead to inaccurate measurements.**<br>**Possible Causes:**<br>• Oxygen sensor (before catalytic converter) is faulty<br>• Oxygen sensor (behind catalytic converter) is faulty<br>• Oxygen sensor heater (before catalytic converter) is faulty<br>• Oxygen sensor heater (behind catalytic converter) is faulty<br>• Circuit wiring has a short to power or ground<br>• Engine Component Power Supply Relay is faulty<br>• E-box fuses for oxygen sensor are faulty<br>• Leaks present in the exhaust manifold or exhaust pipes<br>• HO2S signal wire and ground wire crossed in connector<br>• HO2S element is fuel contaminated or has failed<br>• ECM has failed |

| DTC | Trouble Code Title, Conditions & Possible Causes |
|---|---|
| **DTC: P1115**<br>**1T CCM, MIL: No**<br>**Years:** 2007, 2008<br>**Models:** A4, A6<br>**Engines:** All<br>**Transmissions:** All | **O2 Control (Bank 1 Sensor 1) Short to Ground Conditions:**<br>Engine started, battery voltage must be at least 11.5v, all electrical components must be off, the ground between the engine and the chassis must be well connected, the exhaust system must be properly sealed between the catalytic converter and the cylinder head, and the oxygen sensor heater for oxygen sensor before the catalytic converter must be properly functioning. The ECM detected a measurement on the O2 sensor circuit that was outside the parameters to function properly.<br>**Note: For resistance testing of sensor heating, oxygen sensor should be cooled to ambient temperature. High temperatures at oxygen sensor may lead to inaccurate measurements.**<br>**Possible Causes:**<br>• Oxygen sensor (before catalytic converter) is faulty<br>• Oxygen sensor (behind catalytic converter) is faulty<br>• Oxygen sensor heater (before catalytic converter) is faulty<br>• Oxygen sensor heater (behind catalytic converter) is faulty<br>• Circuit wiring has a short to power or ground<br>• Engine Component Power Supply Relay is faulty<br>• E-box fuses for oxygen sensor are faulty<br>• Leaks present in the exhaust manifold or exhaust pipes<br>• HO2S signal wire and ground wire crossed in connector<br>• HO2S element is fuel contaminated or has failed<br>• ECM has failed |
| **DTC: P1116**<br>**1T CCM, MIL: No**<br>**Years:** 2007, 2008<br>**Models:** A4, A6<br>**Engines:** All<br>**Transmissions:** All | **O2 Control (Bank 1 Sensor 1) Open Conditions:**<br>Engine started, battery voltage must be at least 11.5v, all electrical components must be off, the ground between the engine and the chassis must be well connected, the exhaust system must be properly sealed between the catalytic converter and the cylinder head, and the oxygen sensor heater for oxygen sensor before the catalytic converter must be properly functioning. The ECM detected a measurement on the O2 sensor circuit that was outside the parameters to function properly.<br>**Note: For resistance testing of sensor heating, oxygen sensor should be cooled to ambient temperature. High temperatures at oxygen sensor may lead to inaccurate measurements.**<br>**Possible Causes:**<br>• Oxygen sensor (before catalytic converter) is faulty<br>• Oxygen sensor (behind catalytic converter) is faulty<br>• Oxygen sensor heater (before catalytic converter) is faulty<br>• Oxygen sensor heater (behind catalytic converter) is faulty<br>• Circuit wiring has a short to power or ground<br>• Engine Component Power Supply Relay is faulty<br>• E-box fuses for oxygen sensor are faulty<br>• Leaks present in the exhaust manifold or exhaust pipes<br>• HO2S signal wire and ground wire crossed in connector<br>• HO2S element is fuel contaminated or has failed<br>• ECM has failed |
| **DTC: P1117**<br>**1T CCM, MIL: No**<br>**Years:** 2007, 2008<br>**Models:** A4, A6<br>**Engines:** All<br>**Transmissions:** All | **O2 Control (Bank 1 Sensor 2) Open Conditions:**<br>Engine started, battery voltage must be at least 11.5v, all electrical components must be off, the ground between the engine and the chassis must be well connected, the exhaust system must be properly sealed between the catalytic converter and the cylinder head, and the oxygen sensor heater for oxygen sensor before the catalytic converter must be properly functioning. The ECM detected a measurement on the O2 sensor circuit that was outside the parameters to function properly.<br>**Note: For resistance testing of sensor heating, oxygen sensor should be cooled to ambient temperature. High temperatures at oxygen sensor may lead to inaccurate measurements.**<br>**Possible Causes:**<br>• Oxygen sensor (before catalytic converter) is faulty<br>• Oxygen sensor (behind catalytic converter) is faulty<br>• Oxygen sensor heater (before catalytic converter) is faulty<br>• Oxygen sensor heater (behind catalytic converter) is faulty<br>• Circuit wiring has a short to power or ground<br>• Engine Component Power Supply Relay is faulty<br>• E-box fuses for oxygen sensor are faulty<br>• Leaks present in the exhaust manifold or exhaust pipes<br>• HO2S signal wire and ground wire crossed in connector<br>• HO2S element is fuel contaminated or has failed<br>• ECM has failed |

| DTC | Trouble Code Title, Conditions & Possible Causes |
|---|---|
| **DTC: P1118**<br>**1T CCM, MIL: No**<br>**Years:** 2007, 2008<br>**Models:** A4, A6<br>**Engines:** All<br>**Transmissions:** All | **O2 Sensor Heater Circ. (Bank 1-Sensor2) Open Conditions:**<br>Engine started, battery voltage must be at least 11.5v, all electrical components must be off, the ground between the engine and the chassis must be well connected, the exhaust system must be properly sealed between the catalytic converter and the cylinder head, and the oxygen sensor heater for oxygen sensor before the catalytic converter must be properly functioning. The ECM detected a measurement on the O2 sensor circuit that was outside the parameters to function properly.<br>**Note: For resistance testing of sensor heating, oxygen sensor should be cooled to ambient temperature. High temperatures at oxygen sensor may lead to inaccurate measurements.**<br>**Possible Causes:**<br>• Oxygen sensor (before catalytic converter) is faulty<br>• Oxygen sensor (behind catalytic converter) is faulty<br>• Oxygen sensor heater (before catalytic converter) is faulty<br>• Oxygen sensor heater (behind catalytic converter) is faulty<br>• Circuit wiring has a short to power or ground<br>• Engine Component Power Supply Relay is faulty<br>• E-box fuses for oxygen sensor are faulty<br>• Leaks present in the exhaust manifold or exhaust pipes<br>• HO2S signal wire and ground wire crossed in connector<br>• HO2S element is fuel contaminated or has failed<br>• ECM has failed |
| **DTC: P1127**<br>**2T CCM, MIL: Yes**<br>**Years:** 2007, 2008<br>**Models:** A4, A6<br>**Engines:** All<br>**Transmissions:** All | **Long Term Fuel Trim Add. Air. Bank 1 System Too Rich Conditions:**<br>Engine started, battery voltage must be at least 11.5v, all electrical components must be off, the ground between the engine and the chassis must be well connected, the exhaust system must be properly sealed between the catalytic converter and the cylinder head, and the oxygen sensor heater for oxygen sensor before the catalytic converter must be properly functioning. The fuel mixture is so rich that the O2S control is on lean limit.<br>**Note: After exhaust system repairs, make sure exhaust system is not under stress and that it has sufficient clearance from the bodywork. If necessary, loosen double clamps and align exhaust pipe so that sufficient clearance is maintained to the bodywork and support rings carry uniform loads. Do not use any silicone sealant. Traces of silicone components which are sucked into the engine are not burned there, and they damage the oxygen sensor.**<br>**Possible Causes:**<br>• MAF sensor circuit open<br>• MAF sensor circuit shorted to ground<br>• Air leak in the manifold<br>• Secondary air injection system combi-valve stuck open<br>• Secondary air injection system electrical short<br>• Fuel pressure too high, leaks in the vacuum hose to fuel pressure regulator<br>• Fuel pressure regulator has failed<br>• Fuel injectors are dirty, faulty or do not close properly<br>• ECM has failed |
| **DTC: P1128**<br>**2T CCM, MIL: Yes**<br>**Years:** 2007, 2008<br>**Models:** A4, A6<br>**Engines:** All<br>**Transmissions:** All | **Long Term Fuel Trim Add. Air. Bank 1 System Too Lean Conditions:**<br>Engine started, battery voltage must be at least 11.5v, all electrical components must be off, the ground between the engine and the chassis must be well connected, the exhaust system must be properly sealed between the catalytic converter and the cylinder head, and the oxygen sensor heater for oxygen sensor before the catalytic converter must be properly functioning. The fuel mixture is so rich that the O2S control is on lean limit.<br>**Note: When an O2S malfunction (P0131 to P0414) is also stored with this malfunction, the O2S malfunction(s) should be repaired first.**<br>**Note: After exhaust system repairs, make sure exhaust system is not under stress and that it has sufficient clearance from the bodywork. If necessary, loosen double clamps and align exhaust pipe so that sufficient clearance is maintained to the bodywork and support rings carry uniform loads. Do not use any silicone sealant. Traces of silicone components which are sucked into the engine are not burned there, and they damage the oxygen sensor.**<br>**Possible Causes:**<br>• Fuel pressure is too low or fuel quantity supplied is too low<br>• Fuel filter faulty<br>• Transfer fuel pump has failed<br>• Fuel injector is faulty (sticking or not opening)<br>• Engine speed (RPM) sensor is faulty<br>• MAF sensor circuit open<br>• MAF sensor circuit shorted to ground<br>• Air leak in the manifold<br>• Secondary air injection system combi-valve stuck open<br>• Secondary air injection system electrical short<br>• ECM has failed |

| DTC | Trouble Code Title, Conditions & Possible Causes |
|---|---|
| **DTC: P1129**<br>**2T CCM, MIL: Yes**<br>**Years: 2007, 2008**<br>**Models:** A4, A6<br>**Engines:** All<br>**Transmissions:** A/T | **Long Term Fuel Trim at Rich Limit Conditions:**<br>Engine started, battery voltage must be at least 11.5v, all electrical components must be off, the ground between the engine and the chassis must be well connected, the exhaust system must be properly sealed between the catalytic converter and the cylinder head, and the oxygen sensor heater for oxygen sensor before the catalytic converter must be properly functioning. The ECM detected the HO2S circuit was too rich, or that it could no longer change Fuel Trim because it was at its lean limit.<br>**Possible Causes:**<br>&bull; Air intake system leaking, vacuum hoses leaking or damaged<br>&bull; Air leaks located after the MAF sensor mounting location<br>&bull; EGR valve sticking, EGR diaphragm leaking, or gasket leaking<br>&bull; EVAP vapor recovery system has failed<br>&bull; Excessive fuel pressure, leaking or contaminated fuel injectors<br>&bull; Exhaust leaks before or near the HO2S(s) mounting location<br>&bull; Fuel pressure regulator is leaking or damaged<br>&bull; HO2S circuits wet or oily, corroded, or poor terminal contact<br>&bull; HO2S is damaged or it has failed<br>&bull; HO2S signal circuit open, shorted to ground, shorted to power<br>&bull; Low fuel pressure or vehicle driven until it was out of fuel<br>&bull; Oil dipstick not seated or engine oil level too high (overfilled) |
| **DTC: P1130**<br>**2T CCM, MIL: Yes**<br>**Years: 2007, 2008**<br>**Models:** A4, A6<br>**Engines:** All<br>**Transmissions:** A/T | **Long Term Fuel Trim at Lean Limit Conditions:**<br>Engine started, battery voltage must be at least 11.5v, all electrical components must be off, the ground between the engine and the chassis must be well connected, the exhaust system must be properly sealed between the catalytic converter and the cylinder head, and the oxygen sensor heater for oxygen sensor before the catalytic converter must be properly functioning. The ECM detected the HO2S circuit was too lean, or that it could no longer change Fuel Trim because it was at its lean limit.<br>**Possible Causes:**<br>&bull; Air intake system leaking, vacuum hoses leaking or damaged<br>&bull; Air leaks located after the MAF sensor mounting location<br>&bull; EGR valve sticking, EGR diaphragm leaking, or gasket leaking<br>&bull; EVAP vapor recovery system has failed<br>&bull; Excessive fuel pressure, leaking or contaminated fuel injectors<br>&bull; Exhaust leaks before or near the HO2S(s) mounting location<br>&bull; Fuel pressure regulator is leaking or damaged<br>&bull; HO2S circuits wet or oily, corroded, or poor terminal contact<br>&bull; HO2S is damaged or it has failed<br>&bull; HO2S signal circuit open, shorted to ground, shorted to power<br>&bull; Low fuel pressure or vehicle driven until it was out of fuel<br>&bull; Oil dipstick not seated or engine oil level too high (overfilled) |
| **DTC: P1136**<br>**2T CCM, MIL: Yes**<br>**Years: 2007, 2008**<br>**Models:** A4, A6<br>**Engines:** All<br>**Transmissions:** All | **Long Term Fuel Trim Add. Fuel, Bank 1 System Too Lean Conditions:**<br>Engine started, battery voltage must be at least 11.5v, all electrical components must be off, the ground between the engine and the chassis must be well connected, the exhaust system must be properly sealed between the catalytic converter and the cylinder head, and the oxygen sensor heater for oxygen sensor before the catalytic converter must be properly functioning. The ECM detected the HO2S circuit was too lean, or that it could no longer change Fuel Trim because it was at its lean limit.<br>**Possible Causes:**<br>&bull; Air intake system leaking, vacuum hoses leaking or damaged<br>&bull; Air leaks located after the MAF sensor mounting location<br>&bull; EGR valve sticking, EGR diaphragm leaking, or gasket leaking<br>&bull; EVAP vapor recovery system has failed<br>&bull; Excessive fuel pressure, leaking or contaminated fuel injectors<br>&bull; Exhaust leaks before or near the HO2S(s) mounting location<br>&bull; Fuel pressure regulator is leaking or damaged<br>&bull; HO2S circuits wet or oily, corroded, or poor terminal contact<br>&bull; HO2S is damaged or it has failed<br>&bull; HO2S signal circuit open, shorted to ground, shorted to power<br>&bull; Low fuel pressure or vehicle driven until it was out of fuel<br>&bull; Oil dipstick not seated or engine oil level too high (overfilled) |

| DTC | Trouble Code Title, Conditions & Possible Causes |
|---|---|
| **DTC: P1137**<br>**2T CCM, MIL: Yes**<br>**Years:** 2007, 2008<br>**Models:** A4, A6<br>**Engines:** All<br>**Transmissions:** All | **Long Term Fuel Trim Add. Fuel, Bank 1 System Too Rich Conditions:**<br>Engine started, battery voltage must be at least 11.5v, all electrical components must be off, the ground between the engine and the chassis must be well connected, the exhaust system must be properly sealed between the catalytic converter and the cylinder head, and the oxygen sensor heater for oxygen sensor before the catalytic converter must be properly functioning. The ECM detected the HO2S circuit was too rich, or that it could no longer change Fuel Trim because it was at its lean limit.<br>**Possible Causes:**<br>• Air intake system leaking, vacuum hoses leaking or damaged<br>• Air leaks located after the MAF sensor mounting location<br>• EGR valve sticking, EGR diaphragm leaking, or gasket leaking<br>• EVAP vapor recovery system has failed<br>• Excessive fuel pressure, leaking or contaminated fuel injectors<br>• Exhaust leaks before or near the HO2S(s) mounting location<br>• Fuel pressure regulator is leaking or damaged<br>• HO2S circuits wet or oily, corroded, or poor terminal contact<br>• HO2S is damaged or it has failed<br>• HO2S signal circuit open, shorted to ground, shorted to power<br>• Low fuel pressure or vehicle driven until it was out of fuel<br>• Oil dipstick not seated or engine oil level too high (overfilled) |
| **DTC: P1138**<br>**2T CCM, MIL: Yes**<br>**Years:** 2007, 2008<br>**Models:** A4, A6<br>**Engines:** All<br>**Transmissions:** A/T | **Long Term Fuel Trim Add. Fuel, Bank 2 System Too Lean Conditions:**<br>Engine started, battery voltage must be at least 11.5v, all electrical components must be off, the ground between the engine and the chassis must be well connected, the exhaust system must be properly sealed between the catalytic converter and the cylinder head, and the oxygen sensor heater for oxygen sensor before the catalytic converter must be properly functioning. The ECM detected the HO2S circuit was too lean, or that it could no longer change Fuel Trim because it was at its lean limit.<br>**Possible Causes:**<br>• Air intake system leaking, vacuum hoses leaking or damaged<br>• Air leaks located after the MAF sensor mounting location<br>• EGR valve sticking, EGR diaphragm leaking, or gasket leaking<br>• EVAP vapor recovery system has failed<br>• Excessive fuel pressure, leaking or contaminated fuel injectors<br>• Exhaust leaks before or near the HO2S(s) mounting location<br>• Fuel pressure regulator is leaking or damaged<br>• HO2S circuits wet or oily, corroded, or poor terminal contact<br>• HO2S is damaged or it has failed<br>• HO2S signal circuit open, shorted to ground, shorted to power<br>• Low fuel pressure or vehicle driven until it was out of fuel<br>• Oil dipstick not seated or engine oil level too high (overfilled) |
| **DTC: P1139**<br>**2T CCM, MIL: Yes**<br>**Years:** 2007, 2008<br>**Models:** A4, A6<br>**Engines:** All<br>**Transmissions:** A/T | **Long Term Fuel Trim Add. Fuel, Bank 2 System Too Rich Conditions:**<br>Engine started, battery voltage must be at least 11.5v, all electrical components must be off, the ground between the engine and the chassis must be well connected, the exhaust system must be properly sealed between the catalytic converter and the cylinder head, and the oxygen sensor heater for oxygen sensor before the catalytic converter must be properly functioning. The ECM detected the HO2S circuit was too rich, or that it could no longer change Fuel Trim because it was at its lean limit.<br>**Possible Causes:**<br>• Air intake system leaking, vacuum hoses leaking or damaged<br>• Air leaks located after the MAF sensor mounting location<br>• EGR valve sticking, EGR diaphragm leaking, or gasket leaking<br>• EVAP vapor recovery system has failed<br>• Excessive fuel pressure, leaking or contaminated fuel injectors<br>• Exhaust leaks before or near the HO2S(s) mounting location<br>• Fuel pressure regulator is leaking or damaged<br>• HO2S circuits wet or oily, corroded, or poor terminal contact<br>• HO2S is damaged or it has failed<br>• HO2S signal circuit open, shorted to ground, shorted to power<br>• Low fuel pressure or vehicle driven until it was out of fuel<br>• Oil dipstick not seated or engine oil level too high (overfilled) |

| DTC | Trouble Code Title, Conditions & Possible Causes |
|---|---|
| **DTC: P1141**<br>**2T CCM, MIL: Yes**<br>**Years: 2007, 2008**<br>**Models:** A4, A6<br>**Engines:** All<br>**Transmissions:** All | **Load Calculation Cross Check Range/Performance Conditions:**<br>Engine started, battery voltage must be at least 11.5v, all electrical components must be off, the ground between the engine and the chassis must be well connected, the exhaust system must be properly sealed between the catalytic converter and the cylinder head, and the oxygen sensor heater for oxygen sensor before the catalytic converter must be properly functioning.<br>**Note: Vacuum in the intake system sucks in the leak detection spray with false air. Leak detection spray decreases ignition quality of the fuel mixture. This causes a drop in engine speed and changes the value produced by the Heated Oxygen Sensor.**<br>**Note: Both Throttle Position (TP) sensor and Sender 2 for accelerator pedal position are located at the accelerator pedal and communicate the driver's intentions to the ECM completely independently of each other. Both sensors are stored in one housing.**<br>**Possible Causes:**<br>• Intake system is leaking<br>• Signal is grounding<br>• ECM has failed<br>• Intake Manifold Runner Position Sensor is faulty<br>• Intake system for leaks (false air) is faulty<br>• Motor for intake flap is faulty<br>• Mass Air Flow (MAF) sensor is faulty<br>• Throttle Position (TP) sensor is faulty<br>• Throttle valve control module is faulty |
| **DTC: P1143**<br>**2T CCM, MIL: Yes**<br>**Years: 2007, 2008**<br>**Models:** A4, A6<br>**Engines:** All<br>**Transmissions:** All | **Load Calculation Cross Check Upper Limit Conditions:**<br>Engine started, battery voltage must be at least 11.5v, all electrical components must be off, the ground between the engine and the chassis must be well connected, the exhaust system must be properly sealed between the catalytic converter and the cylinder head, and the oxygen sensor heater for oxygen sensor before the catalytic converter must be properly functioning.<br>**Note: Vacuum in the intake system sucks in the leak detection spray with false air. Leak detection spray decreases ignition quality of the fuel mixture. This causes a drop in engine speed and changes the value produced by the Heated Oxygen Sensor.**<br>**Note: Both Throttle Position (TP) sensor and Sender 2 for accelerator pedal position are located at the accelerator pedal and communicate the driver's intentions to the ECM completely independently of each other. Both sensors are stored in one housing.**<br>**Possible Causes:**<br>• Intake Manifold Runner Position Sensor is faulty<br>• Intake system for leaks (false air) is faulty<br>• Motor for intake flap is faulty<br>• Mass Air Flow (MAF) sensor is faulty<br>• Throttle Position (TP) sensor is faulty<br>• Throttle valve control module is faulty<br>• Intake system is leaking<br>• Signal is grounding<br>• ECM has failed |
| **DTC: P1171**<br>**2T CCM, MIL: Yes**<br>**Years: 2007, 2008**<br>**Models:** A4, A6<br>**Engines:** All<br>**Transmissions:** All | **Throttle Actuation Potentiometer Sign.2 Range/Performance Conditions:**<br>Engine started, battery voltage must be at least 11.5v, all electrical components must be off, the ground between the engine and the chassis must be well connected, coolant temperature must be at least 80 degrees Celsius and the accelerator pedal must be properly adjusted. The ECM detected an incorrect singal from the throttle potentiometer.<br>**Note: If the complete throttle valve control module is current-less (e.g. connector disconnected) the throttle valve moves into a particular, specified mechanical position, which signals an increased idle speed with an engine at operating temperature. If only the Throttle Position (TP) actuator is current-less, the throttle valve also moves into the specified mechanical position (emergency running gap), however, since Closed Throttle Position (CTP) switch can still be recognized, an "almost normal idle RPM" is reached via the respective ignition angle retardation.**<br>**Note: Terminal assignment at throttle control module is different in vehicles with and without cruise control.**<br>**Characteristic: Steering column switch with operating module for cruise control.**<br>**Possible Causes:**<br>• Throttle valve control module has failed<br>• Throttle valve is dirty or damaged<br>• Throttle valve is not in a closed position<br>• Voltage supply of throttle valve control module is shorted or open<br>• ECM has failed |

| DTC | Trouble Code Title, Conditions & Possible Causes |
|---|---|
| **DTC: P1172**<br>**2T CCM, MIL: Yes**<br>**Years: 2007, 2008**<br>**Models:** A6<br>**Engines:** All<br>**Transmissions:** All | **Throttle Actuation Potentiometer Sign.2 Signal Too Low Conditions:**<br>Engine started, battery voltage must be at least 11.5v, all electrical components must be off, the ground between the engine and the chassis must be well connected, coolant temperature must be at least 80 degrees Celsius and the accelerator pedal must be properly adjusted. The ECM detected an incorrect singal from the throttle potentiometer.<br>**Note: If the complete throttle valve control module is current-less (e.g. connector disconnected) the throttle valve moves into a particular, specified mechanical position, which signals an increased idle speed with an engine at operating temperature. If only the Throttle Position (TP) actuator is current-less, the throttle valve also moves into the specified mechanical position (emergency running gap), however, since Closed Throttle Position (CTP) switch can still be recognized, an "almost normal idle RPM" is reached via the respective ignition angle retardation.**<br>**Note: Terminal assignment at throttle control module is different in vehicles with and without cruise control.**<br>**Characteristic: Steering column switch with operating module for cruise control.**<br>**Possible Causes:**<br>• Throttle valve control module has failed<br>• Throttle valve is dirty or damaged<br>• Throttle valve is not in a closed position<br>• Voltage supply of throttle valve control module is shorted or open<br>• ECM has failed |
| **DTC: P1173**<br>**2T CCM, MIL: Yes**<br>**Years: 2007, 2008**<br>**Models:** A6<br>**Engines:** All<br>**Transmissions:** All | **Throttle Actuation Potentiometer Sign.2 Signal Too High Conditions:**<br>Engine started, battery voltage must be at least 11.5v, all electrical components must be off, the ground between the engine and the chassis must be well connected, coolant temperature must be at least 80 degrees Celsius and the accelerator pedal must be properly adjusted. The ECM detected an incorrect singal from the throttle potentiometer.<br>**Note: If the complete throttle valve control module is current-less (e.g. connector disconnected) the throttle valve moves into a particular, specified mechanical position, which signals an increased idle speed with an engine at operating temperature. If only the Throttle Position (TP) actuator is current-less, the throttle valve also moves into the specified mechanical position (emergency running gap), however, since Closed Throttle Position (CTP) switch can still be recognized, an "almost normal idle RPM" is reached via the respective ignition angle retardation.**<br>**Note: Terminal assignment at throttle control module is different in vehicles with and without cruise control.**<br>**Characteristic: Steering column switch with operating module for cruise control.**<br>**Possible Causes:**<br>• Throttle valve control module has failed<br>• Throttle valve is dirty or damaged<br>• Throttle valve is not in a closed position<br>• Voltage supply of throttle valve control module is shorted or open<br>• ECM has failed |
| **DTC: P1176**<br>**2T CCM, MIL: Yes**<br>**Years: 2007, 2008**<br>**Models:** A4, A6<br>**Engines:** All<br>**Transmissions:** All | **O2 Correction Behind Catalyst B1 Limit Attained Conditions:**<br>Engine started, battery voltage must be at least 11.5v, all electrical components must be off, the ground between the engine and the chassis must be well connected, the exhaust system must be properly sealed between the catalytic converter and the cylinder head, the coolant temperature must be at least 80 degrees Celsius, and the oxygen sensor heater for oxygen sensor before the catalytic converter must be properly functioning. The ECM has detected a malfunction of the oxygen sensor.<br>**Note: Vacuum in the intake system sucks in the leak detection spray with false air. Leak detection spray decreases ignition quality of the fuel mixture. This causes a drop in engine speed and changes the value produced by the Heated Oxygen Sensor (HO2S).**<br>**Note: Vehicle must be raised before connector for oxygen sensor is accessible.**<br>**Note: The oxygen sensor before catalytic converter has a static regulation and can be differentiated from the oxygen sensor behind catalytic converter via a 6-pin connector.**<br>**Possible Causes:**<br>• O2 sensor circuit has shorted to ground or B+<br>• O2 sensor circuit is open<br>• ECM has failed<br>• O2 sensor has failed<br>• Intake Manifold Runner Position Sensor is faulty<br>• Intake system for leaks (false air) is faulty<br>• Motor for intake flap is faulty |

| DTC | Trouble Code Title, Conditions & Possible Causes |
|---|---|
| **DTC: P1177**<br>**2T CCM, MIL: Yes**<br>**Years: 2007, 2008**<br>**Models: A4, A6**<br>**Engines: All**<br>**Transmissions: All** | **O2 Correction Behind Catalyst B2 Limit Attained Conditions:**<br>Engine started, battery voltage must be at least 11.5v, all electrical components must be off, the ground between the engine and the chassis must be well connected, the exhaust system must be properly sealed between the catalytic converter and the cylinder head, the coolant temperature must be at least 80 degrees Celsius, and the oxygen sensor heater for oxygen sensor before the catalytic converter must be properly functioning. The ECM has detected a malfunction of the oxygen sensor.<br>**Note: Vacuum in the intake system sucks in the leak detection spray with false air. Leak detection spray decreases ignition quality of the fuel mixture. This causes a drop in engine speed and changes the value produced by the Heated Oxygen Sensor (HO2S).**<br>**Note: Vehicle must be raised before connector for oxygen sensor is accessible.**<br>**Note: The oxygen sensor before catalytic converter has a static regulation and can be differentiated from the oxygen sensor behind catalytic converter via a 6-pin connector.**<br>**Possible Causes:**<br>&bull; O2 sensor circuit has shorted to ground or B+<br>&bull; O2 sensor circuit is open<br>&bull; ECM has failed<br>&bull; O2 sensor has failed<br>&bull; Intake Manifold Runner Position Sensor is faulty<br>&bull; Intake system for leaks (false air) is faulty<br>&bull; Motor for intake flap is faulty |
| **DTC: P1196**<br>**2T CCM, MIL: Yes**<br>**Years: 2007, 2008**<br>**Models: A4, A6**<br>**Engines: All**<br>**Transmissions: All** | **O2 Sensor Heater Circuit (Bank 1-Sensor 1) Electrical Malfunction Conditions:**<br>Engine started, battery voltage must be at least 11.5v, all electrical components must be off, the ground between the engine and the chassis must be well connected, the exhaust system must be properly sealed between the catalytic converter and the cylinder head, and the oxygen sensor heater for oxygen sensor before the catalytic converter must be properly functioning.<br>**Note: For resistance testing of sensor heating, oxygen sensor should be cooled to ambient temperature. High temperatures at oxygen sensor may lead to inaccurate measurements. The ECM detected an open or shorted condition, or excessive current draw in the heater circuit.**<br>**Possible Causes:**<br>&bull; HO2S heater power circuit is open<br>&bull; HO2S heater ground circuit is open<br>&bull; HO2S signal tracking (due to oil or moisture in the connector)<br>&bull; HO2S is damaged or has failed<br>&bull; ECM has failed<br>&bull; Oxygen sensor (before catalytic converter) is faulty<br>&bull; Oxygen sensor (behind catalytic converter) is faulty<br>&bull; Oxygen sensor heater (before catalytic converter) is faulty<br>&bull; Oxygen sensor heater (behind catalytic converter) is faulty |
| **DTC: P1198**<br>**2T CCM, MIL: Yes**<br>**Years: 2007, 2008**<br>**Models: A4, A6**<br>**Engines: All**<br>**Transmissions: All** | **O2 Sensor Heater Circuit (Bank 1-Sensor 2) Electrical Malfunction Conditions:**<br>Engine started, battery voltage must be at least 11.5v, all electrical components must be off, the ground between the engine and the chassis must be well connected, the exhaust system must be properly sealed between the catalytic converter and the cylinder head, and the oxygen sensor heater for oxygen sensor before the catalytic converter must be properly functioning.<br>**Note: For resistance testing of sensor heating, oxygen sensor should be cooled to ambient temperature. High temperatures at oxygen sensor may lead to inaccurate measurements. The ECM detected an open or shorted condition, or excessive current draw in the heater circuit.**<br>**Possible Causes:**<br>&bull; HO2S heater power circuit is open<br>&bull; HO2S heater ground circuit is open<br>&bull; HO2S signal tracking (due to oil or moisture in the connector)<br>&bull; HO2S is damaged or has failed<br>&bull; ECM has failed<br>&bull; Oxygen sensor (before catalytic converter) is faulty<br>&bull; Oxygen sensor (behind catalytic converter) is faulty<br>&bull; Oxygen sensor heater (before catalytic converter) is faulty<br>&bull; Oxygen sensor heater (behind catalytic converter) is faulty |

| DTC | Trouble Code Title, Conditions & Possible Causes |
|---|---|
| **DTC: P1201**<br>**1T CCM, MIL: Yes**<br>**Years: 2007, 2008**<br>**Models:** A4, A6<br>**Engines:** All, 4.2L<br>**Transmissions:** All | **Cylinder 1 Fuel Injection Circuit Electrical Malfunction Conditions:**<br>Key on or engine running, fuses in the instrument panel and the E-box in the engine compartment must be functioning, and the ground connections between the engine ad the chassis must be well connected; and the ECM detected an unexpected voltage condition on the injector circuit<br>**Possible Causes:**<br>• Injector control circuit is open<br>• Injector power circuit (B+) is open<br>• Injector control circuit is shorted to chassis ground<br>• Injector is damaged or has failed<br>• ECM is not connected or has failed<br>• Fuel pump relay has failed<br>• Fuel injectors may have malfunctioned<br>• Faulty engine speed sensor |
| **DTC: P1202**<br>**1T CCM, MIL: Yes**<br>**Years: 2007, 2008**<br>**Models:** A4, A6<br>**Engines:** All<br>**Transmissions:** All | **Cylinder 2 Fuel Injection Circuit Electrical Malfunction Conditions:**<br>Key on or engine running, fuses in the instrument panel and the E-box in the engine compartment must be functioning, and the ground connections between the engine ad the chassis must be well connected; and the ECM detected an unexpected voltage condition on the injector circuit<br>**Possible Causes:**<br>• Injector control circuit is open<br>• Injector power circuit (B+) is open<br>• Injector control circuit is shorted to chassis ground<br>• Injector is damaged or has failed<br>• ECM is not connected or has failed<br>• Fuel pump relay has failed<br>• Fuel injectors may have malfunctioned<br>• Faulty engine speed sensor |
| **DTC: P1203**<br>**1T CCM, MIL: Yes**<br>**Years: 2007, 2008**<br>**Models:** A4, A6<br>**Engines:** All<br>**Transmissions:** All | **Cylinder 3 Fuel Injection Circuit Electrical Malfunction Conditions:**<br>Key on or engine running, fuses in the instrument panel and the E-box in the engine compartment must be functioning, and the ground connections between the engine ad the chassis must be well connected; and the ECM detected an unexpected voltage condition on the injector circuit<br>**Possible Causes:**<br>• Injector control circuit is open<br>• Injector power circuit (B+) is open<br>• Injector control circuit is shorted to chassis ground<br>• Injector is damaged or has failed<br>• ECM is not connected or has failed<br>• Fuel pump relay has failed<br>• Fuel injectors may have malfunctioned<br>• Faulty engine speed sensor |
| **DTC: P1204**<br>**1T CCM, MIL: Yes**<br>**Years: 2007, 2008**<br>**Models:** A4, A6<br>**Engines:** All<br>**Transmissions:** All | **Cylinder 4 Fuel Injection Circuit Electrical Malfunction Conditions:**<br>Key on or engine running, fuses in the instrument panel and the E-box in the engine compartment must be functioning, and the ground connections between the engine ad the chassis must be well connected; and the ECM detected an unexpected voltage condition on the injector circuit<br>**Possible Causes:**<br>• Injector control circuit is open<br>• Injector power circuit (B+) is open<br>• Injector control circuit is shorted to chassis ground<br>• Injector is damaged or has failed<br>• ECM is not connected or has failed<br>• Fuel pump relay has failed<br>• Fuel injectors may have malfunctioned<br>• Faulty engine speed sensor |

| DTC | Trouble Code Title, Conditions & Possible Causes |
|---|---|
| **DTC: P1213**<br>**1T CCM, MIL: Yes**<br>**Years: 2007, 2008**<br>**Models:** A4, A6<br>**Engines:** All<br>**Transmissions:** All | **Cylinder 1 Fuel Injection Circuit Short to B+ Conditions:**<br>Key on or engine running, fuses in the instrument panel and the E-box in the engine compartment must be functioning, and the ground connections between the engine ad the chassis must be well connected; and the ECM detected an unexpected voltage condition on the injector circuit. Wiring or fuel injector has a short circuit to positive supply.<br>**Possible Causes:**<br>• Injector control circuit is open<br>• Injector power circuit (B+) is open<br>• Injector control circuit is shorted to chassis ground<br>• Injector is damaged or has failed<br>• ECM is not connected or has failed<br>• Fuel pump relay has failed<br>• Engine speed sensor has failed |
| **DTC: P1214**<br>**1T CCM, MIL: Yes**<br>**Years: 2007, 2008**<br>**Models:** A4, A6<br>**Engines:** All<br>**Transmissions:** All | **Cylinder 2 Fuel Injection Circuit Short to B+ Conditions:**<br>Key on or engine running, fuses in the instrument panel and the E-box in the engine compartment must be functioning, and the ground connections between the engine ad the chassis must be well connected; and the ECM detected an unexpected voltage condition on the injector circuit. Wiring or fuel injector has a short circuit to positive supply.<br>**Possible Causes:**<br>• Injector control circuit is open<br>• Injector power circuit (B+) is open<br>• Injector control circuit is shorted to chassis ground<br>• Injector is damaged or has failed<br>• ECM is not connected or has failed<br>• Fuel pump relay has failed<br>• Engine speed sensor has failed |
| **DTC: P1215**<br>**1T CCM, MIL: Yes**<br>**Years: 2007, 2008**<br>**Models:** A4, A6<br>**Engines:** All<br>**Transmissions:** All | **Cylinder 3 Fuel Injection Circuit Short to B+ Conditions:**<br>Key on or engine running, fuses in the instrument panel and the E-box in the engine compartment must be functioning, and the ground connections between the engine ad the chassis must be well connected; and the ECM detected an unexpected voltage condition on the injector circuit. Wiring or fuel injector has a short circuit to positive supply.<br>**Possible Causes:**<br>• Injector control circuit is open<br>• Injector power circuit (B+) is open<br>• Injector control circuit is shorted to chassis ground<br>• Injector is damaged or has failed<br>• ECM is not connected or has failed<br>• Fuel pump relay has failed<br>• Engine speed sensor has failed |
| **DTC: P1216**<br>**1T CCM, MIL: Yes**<br>**Years: 2007, 2008**<br>**Models:** A4, A6<br>**Engines:** All<br>**Transmissions:** All | **Cylinder 4 Fuel Injection Circuit Short to B+ Conditions:**<br>Key on or engine running, fuses in the instrument panel and the E-box in the engine compartment must be functioning, and the ground connections between the engine ad the chassis must be well connected; and the ECM detected an unexpected voltage condition on the injector circuit. Wiring or fuel injector has a short circuit to positive supply.<br>**Possible Causes:**<br>• Injector control circuit is open<br>• Injector power circuit (B+) is open<br>• Injector control circuit is shorted to chassis ground<br>• Injector is damaged or has failed<br>• ECM is not connected or has failed<br>• Fuel pump relay has failed<br>• Engine speed sensor has failed |
| **DTC: P1225**<br>**1T CCM, MIL: Yes**<br>**Years: 2007, 2008**<br>**Models:** A4, A6<br>**Engines:** All<br>**Transmissions:** All | **Cylinder 1 Fuel Injection Circuit Short to Ground Conditions:**<br>Key on or engine running, fuses in the instrument panel and the E-box in the engine compartment must be functioning, and the ground connections between the engine ad the chassis must be well connected; and the ECM detected an unexpected voltage condition on the injector circuit. Wiring or fuel injector has a short circuit to ground.<br>**Possible Causes:**<br>• Injector control circuit is open<br>• Injector power circuit (B+) is open<br>• Injector control circuit is shorted to chassis ground<br>• Injector is damaged or has failed<br>• ECM is not connected or has failed<br>• Fuel pump relay has failed<br>• Engine speed sensor has failed |

| DTC | Trouble Code Title, Conditions & Possible Causes |
|---|---|
| **DTC: P1226**<br>**1T CCM, MIL: Yes**<br>**Years: 2007, 2008**<br>**Models:** A4, A6<br>**Engines:** All<br>**Transmissions:** All | **Cylinder 2 Fuel Injection Circuit Short to Ground Conditions:**<br>Key on or engine running, fuses in the instrument panel and the E-box in the engine compartment must be functioning, and the ground connections between the engine ad the chassis must be well connected; and the ECM detected an unexpected voltage condition on the injector circuit. Wiring or fuel injector has a short circuit to ground.<br>**Possible Causes:**<br>• Injector control circuit is open<br>• Injector power circuit (B+) is open<br>• Injector control circuit is shorted to chassis ground<br>• Injector is damaged or has failed<br>• ECM is not connected or has failed<br>• Fuel pump relay has failed<br>• Engine speed sensor has failed |
| **DTC: P1227**<br>**1T CCM, MIL: Yes**<br>**Years: 2007, 2008**<br>**Models:** A4, A6<br>**Engines:** All<br>**Transmissions:** All | **Cylinder 3 Fuel Injection Circuit Short to Ground Conditions:**<br>Key on or engine running, fuses in the instrument panel and the E-box in the engine compartment must be functioning, and the ground connections between the engine ad the chassis must be well connected; and the ECM detected an unexpected voltage condition on the injector circuit. Wiring or fuel injector has a short circuit to ground.<br>**Possible Causes:**<br>• Injector control circuit is open<br>• Injector power circuit (B+) is open<br>• Injector control circuit is shorted to chassis ground<br>• Injector is damaged or has failed<br>• ECM is not connected or has failed<br>• Fuel pump relay has failed<br>• Engine speed sensor has failed |
| **DTC: P1228**<br>**1T CCM, MIL: Yes**<br>**Years: 2007, 2008**<br>**Models:** A4, A6<br>**Engines:** All<br>**Transmissions:** All | **Cylinder 4 Fuel Injection Circuit Short to Ground Conditions:**<br>Key on or engine running, fuses in the instrument panel and the E-box in the engine compartment must be functioning, and the ground connections between the engine ad the chassis must be well connected; and the ECM detected an unexpected voltage condition on the injector circuit. Wiring or fuel injector has a short circuit to ground.<br>**Possible Causes:**<br>• Injector control circuit is open<br>• Injector power circuit (B+) is open<br>• Injector control circuit is shorted to chassis ground<br>• Injector is damaged or has failed<br>• ECM is not connected or has failed<br>• Fuel pump relay has failed<br>• Engine speed sensor has failed |
| **DTC: P1237**<br>**1T CCM, MIL: Yes**<br>**Years: 2007, 2008**<br>**Models:** A4, A6<br>**Engines:** All<br>**Transmissions:** All | **Cylinder 1 Fuel Injection Circuit Open Circuit Conditions:**<br>Key on or engine running, fuses in the instrument panel and the E-box in the engine compartment must be functioning, and the ground connections between the engine ad the chassis must be well connected; and the ECM detected an unexpected voltage condition on the injector circuit. Wiring or fuel injector has a short circuit that is open.<br>**Possible Causes:**<br>• Injector control circuit is open<br>• Injector power circuit (B+) is open<br>• Injector control circuit is shorted to chassis ground<br>• Injector is damaged or has failed<br>• ECM is not connected or has failed<br>• Fuel pump relay has failed<br>• Engine speed sensor has failed |
| **DTC: P1238**<br>**1T CCM, MIL: Yes**<br>**Years: 2007, 2008**<br>**Models:** A4, A6<br>**Engines:** All<br>**Transmissions:** All | **Cylinder 2 Fuel Injection Circuit Open Circuit Conditions:**<br>Key on or engine running, fuses in the instrument panel and the E-box in the engine compartment must be functioning, and the ground connections between the engine ad the chassis must be well connected; and the ECM detected an unexpected voltage condition on the injector circuit. Wiring or fuel injector has a short circuit that is open.<br>**Possible Causes:**<br>• Injector control circuit is open<br>• Injector power circuit (B+) is open<br>• Injector control circuit is shorted to chassis ground<br>• Injector is damaged or has failed<br>• ECM is not connected or has failed<br>• Fuel pump relay has failed<br>• Engine speed sensor has failed |

| DTC | Trouble Code Title, Conditions & Possible Causes |
|---|---|
| **DTC: P1239**<br>**1T CCM, MIL: Yes**<br>**Years: 2007, 2008**<br>**Models:** A4, A6<br>**Engines:** All<br>**Transmissions:** All | **Cylinder 3 Fuel Injection Circuit Open Circuit Conditions:**<br>Key on or engine running, fuses in the instrument panel and the E-box in the engine compartment must be functioning, and the ground connections between the engine ad the chassis must be well connected; and the ECM detected an unexpected voltage condition on the injector circuit. Wiring or fuel injector has a short circuit that is open.<br>**Possible Causes:**<br>• Injector control circuit is open<br>• Injector power circuit (B+) is open<br>• Injector control circuit is shorted to chassis ground<br>• Injector is damaged or has failed<br>• ECM is not connected or has failed<br>• Fuel pump relay has failed<br>• Engine speed sensor has failed |
| **DTC: P1240**<br>**1T CCM, MIL: Yes**<br>**Years: 2007, 2008**<br>**Models:** A4, A6<br>**Engines:** All<br>**Transmissions:** All | **Cylinder 4 Fuel Injection Circuit Open Circuit Conditions:**<br>Key on or engine running, fuses in the instrument panel and the E-box in the engine compartment must be functioning, and the ground connections between the engine ad the chassis must be well connected; and the ECM detected an unexpected voltage condition on the injector circuit. Wiring or fuel injector has a short circuit that is open.<br>**Possible Causes:**<br>• Injector control circuit is open<br>• Injector power circuit (B+) is open<br>• Injector control circuit is shorted to chassis ground<br>• Injector is damaged or has failed<br>• ECM is not connected or has failed<br>• Fuel pump relay has failed<br>• Engine speed sensor has failed |
| **DTC: P1250**<br>**1T CCM, MIL: Yes**<br>**Years: 2007, 2008**<br>**Models:** A4, A6<br>**Engines:** All<br>**Transmissions:** All | **Fuel Pressure Regulator Control Circuit Malfunction (Fuel Level too Low) Conditions:**<br>KOEO or KOER Self-Test enabled, and the ECM detected a lack of power (VPWR) to the Fuel Pressure Regulator Control (FPRC) solenoid circuit.<br>**Possible Causes:**<br>• FPRC solenoid valve harness circuits are open or shorted<br>• FPRC input port or output port vacuum lines are damaged<br>• FRPC solenoid is damaged<br>• Fuel level is too low<br>• ECM has failed |
| **DTC: P1296**<br>**1T CCM, MIL: Yes**<br>**Years: 2007, 2008**<br>**Models:** A4, A6<br>**Engines:** All<br>**Transmissions:** All | **Cooling System Malfunction Conditions:**<br>Key on, engine not running, the Engine Control Module (ECM) will use the intake air temperature as a replacement value for an engine start (start temperature replacement value) as soon as there is a Diagnostic Trouble Code (DTC) stored in DTC memory for the Engine Coolant Temperature (ECT) sensor. The temperature then rises according to a program stored in the ECM. When the engine has reached normal operating temperature a fixed replacement value will be displayed. This fixed value is also dependent upon the intake air temperature.<br>**Possible Causes:**<br>• Engine coolant temperature sensor has failed<br>• An open circuit or a short to B+ is present<br>• Sensor circuit is short to ground<br>• ECM has failed |
| **DTC: P1325**<br>**2T CCM, MIL: Yes**<br>**Years: 2007, 2008**<br>**Models:** A4, A6<br>**Engines:** All<br>**Transmissions:** All | **Cylinder 1-Knock Control Limit Attained Conditions:**<br>Engine started, battery voltage at least 11.5v, all electrical components off, ground connections between engine and chassis well connected, and the ECM detected the Knock Sensor signal was more than the calibrated value.<br>**Possible Causes:**<br>• Knock sensor circuit is open<br>• Knock sensor circuit is shorted to ground, or shorted to power<br>• Knock sensor is damaged or it has failed<br>• Poor fuel quality<br>• Loosen knock sensors and tighten again to 20 Nm<br>• ECM has failed |

| DTC | Trouble Code Title, Conditions & Possible Causes |
|---|---|
| **DTC: P1326**<br>**2T CCM, MIL: Yes**<br>**Years: 2007, 2008**<br>**Models:** A4, A6<br>**Engines:** All<br>**Transmissions:** All | **Cylinder 2-Knock Control Limit Attained Conditions:**<br>Engine started, battery voltage at least 11.5v, all electrical components off, ground connections between engine and chassis well connected, and the ECM detected the Knock Sensor signal was more than the calibrated value.<br>**Possible Causes:**<br>• Knock sensor circuit is open<br>• Knock sensor circuit is shorted to ground, or shorted to power<br>• Knock sensor is damaged or it has failed<br>• Poor fuel quality<br>• Loosen knock sensors and tighten again to 20 Nm<br>• ECM has failed |
| **DTC: P1327**<br>**2T CCM, MIL: Yes**<br>**Years: 2007, 2008**<br>**Models:** A4, A6<br>**Engines:** All<br>**Transmissions:** All | **Cylinder 3-Knock Control Limit Attained Conditions:**<br>Engine started, battery voltage at least 11.5v, all electrical components off, ground connections between engine and chassis well connected, and the ECM detected the Knock Sensor signal was more than the calibrated value.<br>**Possible Causes:**<br>• Knock sensor circuit is open<br>• Knock sensor circuit is shorted to ground, or shorted to power<br>• Knock sensor is damaged or it has failed<br>• Poor fuel quality<br>• Loosen knock sensors and tighten again to 20 Nm<br>• ECM has failed |
| **DTC: P1328**<br>**2T CCM, MIL: Yes**<br>**Years: 2007, 2008**<br>**Models:** A4, A6<br>**Engines:** All<br>**Transmissions:** All | **Cylinder 4-Knock Control Limit Attained Conditions:**<br>Engine started, battery voltage at least 11.5v, all electrical components off, ground connections between engine and chassis well connected, and the ECM detected the Knock Sensor signal was more than the calibrated value.<br>**Possible Causes:**<br>• Knock sensor circuit is open<br>• Knock sensor circuit is shorted to ground, or shorted to power<br>• Knock sensor is damaged or it has failed<br>• ECM has failed |
| **DTC: P1329**<br>**2T CCM, MIL: Yes**<br>**Years: 2007, 2008**<br>**Models:** A4, A6<br>**Engines:** All<br>**Transmissions:** All | **Cylinder 5-Knock Control Limit Attained Conditions:**<br>Engine started, battery voltage at least 11.5v, all electrical components off, ground connections between engine and chassis well connected, and the ECM detected the Knock Sensor signal was more than the calibrated value.<br>**Possible Causes:**<br>• Knock sensor circuit is open<br>• Knock sensor circuit is shorted to ground, or shorted to power<br>• Knock sensor is damaged or it has failed<br>• Poor fuel quality<br>• Loosen knock sensors and tighten again to 20 Nm<br>• ECM has failed |
| **DTC: P1330**<br>**2T CCM, MIL: Yes**<br>**Years: 2007, 2008**<br>**Models:** A4, A6<br>**Engines:** All<br>**Transmissions:** All | **Cylinder 6-Knock Control Limit Attained Conditions:**<br>Engine started, battery voltage at least 11.5v, all electrical components off, ground connections between engine and chassis well connected, and the ECM detected the Knock Sensor signal was more than the calibrated value.<br>**Possible Causes:**<br>• Knock sensor circuit is open<br>• Knock sensor circuit is shorted to ground, or shorted to power<br>• Knock sensor is damaged or it has failed<br>• Poor fuel quality<br>• Loosen knock sensors and tighten again to 20 Nm<br>• ECM has failed |
| **DTC: P1335**<br>**2T CCM, MIL: Yes**<br>**Years: 2007, 2008**<br>**Models:** A4, A6<br>**Engines:** All<br>**Transmissions:** All | **Engine Torque Monitoring 2 Control Limit Exceeded Conditions:**<br>Engine cold, battery voltage at least 11.5v, all electrical components off, ground connections between engine and chassis well connected, the ECM detected a signal beyond the required limit.<br>**Possible Causes:**<br>• Engine Control Module (ECM) has failed<br>• Voltage supply for Engine Control Module (ECM) is shorted<br>• Engine Coolant Temperature (ECT) sensor is faulty<br>• Intake Air Temperature (IAT) sensor is faulty<br>• Intake Manifold Runner Position Sensor is faulty<br>• Intake system for leaks (false air) is faulty<br>• Motor for intake flap is faulty<br>• Mass Air Flow (MAF) sensor is faulty |

| DTC | Trouble Code Title, Conditions & Possible Causes |
|---|---|
| **DTC: P1337**<br>**2T CCM, MIL: Yes**<br>**Years:** 2007, 2008<br>**Models:** A4, A6<br>**Engines:** All<br>**Transmissions:** All | **Camshaft Position Sensor (Bank 1) Short to Ground Conditions:**<br>Engine started, battery voltage at least 11.5v, all electrical components off, ground connections between engine and chassis well connected, and the ECM detected an unexpected low or high voltage condition on the camshaft position sensor circuit<br>**Possible Causes:**<br>• Faulty CPM sensor<br>• ECM has failed |
| **DTC: P1338**<br>**2T CCM, MIL: Yes**<br>**Years:** 2007, 2008<br>**Models:** A4, A6<br>**Engines:** All<br>**Transmissions:** All | **Camshaft Position Sensor (Bank 1) Open/Short to B+ Conditions:**<br>Engine started, battery voltage at least 11.5v, all electrical components off, ground connections between engine and chassis well connected, and the ECM detected an unexpected low or high voltage condition on the camshaft position sensor circuit<br>**Possible Causes:**<br>• Faulty CPM sensor<br>• ECM has failed |
| **DTC: P1340**<br>**2T CCM, MIL: Yes**<br>**Years:** 2007, 2008<br>**Models:** A4, A6<br>**Engines:** All<br>**Transmissions:** All | **Crankshaft Position/Camshaft Sensor Signal Out of Sequence Conditions:**<br>Engine started, battery voltage at least 11.5v, all electrical components off, ground connections between engine and chassis well connected, and the ECM detected the crankshaft position sensor and the camshaft sensor were out of sequence with each other.<br>**Note: The Engine Speed (RPM) Sensor detects engine speed and reference marks. Without an engine speed signal, the engine will not start. If the engine speed signal fails while the engine is running, the engine will stop immediately.**<br>**Possible Causes:**<br>• Engine speed sensor has failed or is contaminated (metal filings)<br>• Engine speed sensor's wheel is damaged<br>• Engine speed sensor circuit is shorted to the cable shield<br>• Engine speed sensor circuit is open<br>• ECM is faulty<br>• Canshaft position sensor is faulty |
| **DTC: P1386**<br>**2T CCM, MIL: Yes**<br>**Years:** 2007, 2008<br>**Models:** A4, A6<br>**Engines:** All<br>**Transmissions:** All | **Internal Control Module, Knock Control Circuit Error Conditions:**<br>Engine started, and the ECM detected a too high or too low voltage condition on the knock control circuits, or a miscommunication between the knock control and the ECM.<br>**Possible Causes:**<br>• ECM has failed |
| **DTC: P1387**<br>**2T CCM, MIL: Yes**<br>**Years:** 2007, 2008<br>**Models:** A4, A6<br>**Engines:** All<br>**Transmissions:** All | **Internal Control Module Altitude Sensor Error Conditions:**<br>Ignition on, the ECM detected and altitude sensor error. To achieve optimal anti-theft protection for the vehicle, an anti-theft immobilizer is instAlled. The anti-theft immobilizer is a system for enabling and locking the Engine Control Module (ECM). So that this system cannot be circumvented, it is necessary to perform adaptation of the anti-theft immobilizer using the Vehicle Diagnostic and Information System VAS 5052 in the On Board Diagnostic (OBD) function. The great availability of equipment options makes it necessary to adapt the Engine Control Module (ECM) to the vehicle (e.g. throttle valve control module or cruise control system). This "writing" function is not possible with the generic scan tool.<br>**Possible Causes:**<br>• (If ECM was replaced) ECM ID not the same as the replaced unit<br>• ECM has failed<br>• Voltage supply for Engine Control Module (ECM) has shorted |
| **DTC: P1388**<br>**2T CCM, MIL: Yes**<br>**Years:** 2007, 2008<br>**Models:** A4, A6<br>**Engines:** All<br>**Transmissions:** All | **Internal Control Module Drive By Wire Error Conditions:**<br>Ignition on, the ECM detected and drive by wire error. To achieve optimal anti-theft protection for the vehicle, an anti-theft immobilizer is instAlled. The anti-theft immobilizer is a system for enabling and locking the Engine Control Module (ECM). So that this system cannot be circumvented, it is necessary to perform adaptation of the anti-theft immobilizer using the Vehicle Diagnostic and Information System VAS 5052 in the On Board Diagnostic (OBD) function. The great availability of equipment options makes it necessary to adapt the Engine Control Module (ECM) to the vehicle (e.g. throttle valve control module or cruise control system). This "writing" function is not possible with the generic scan tool.<br>**Possible Causes:**<br>• Engine Control Module (ECM) has failed<br>• Voltage supply for Engine Control Module (ECM) has shorted |

| DTC | Trouble Code Title, Conditions & Possible Causes |
|---|---|
| **DTC: P1409**<br>**2T CCM, MIL: Yes**<br>**Years:** 2007, 2008<br>**Models:** A4, A6<br>**Engines:** All<br>**Transmissions:** All | **Tank Ventilation Valve Circuit Malfunction Conditions**<br>Key on or engine running; and the ECM detected a too high or too low voltage level in the tank ventilation valve circuit.<br>**Possible Causes:**<br>&bull; EVAP canister purge regulator valve has failed<br>&bull; Activation wire is shorting to positive<br>&bull; EVAP canister system has an improper or broken seal<br>&bull; Evaporative Emission (EVAP) canister purge regulator valve 1 is faulty<br>&bull; Leak Detection Pump (LDP) is faulty<br>&bull; Fuel filler cap is not properly closed<br>&bull; Lock ring on fuel pump not tightened<br>&bull; Hoses between EVAP canister and purge regulator valve have failed<br>&bull; ECM has failed |
| **DTC: P1410**<br>**2T CCM, MIL: Yes**<br>**Years:** 2007, 2008<br>**Models:** A4, A6<br>**Engines:** All<br>**Transmissions:** All | **Tank Ventilation Valve Circuit Short to B+:**<br>Key on or engine running; and the ECM detected a too high or too low voltage level in the tank ventilation valve circuit.<br>**Possible Causes:**<br>&bull; EVAP canister purge regulator valve has failed<br>&bull; Activation wire is shorting to positive<br>&bull; EVAP canister system has an improper or broken seal<br>&bull; Evaporative Emission (EVAP) canister purge regulator valve 1 is faulty<br>&bull; Leak Detection Pump (LDP) is faulty<br>&bull; Fuel filler cap is not properly closed<br>&bull; Lock ring on fuel pump not tightened<br>&bull; Hoses between EVAP canister and purge regulator valve have failed<br>&bull; ECM has failed |
| **DTC: P1420**<br>**2T CCM, MIL: Yes**<br>**Years:** 2007, 2008<br>**Models:** A4, A6<br>**Engines:** All<br>**Transmissions:** All | **Secondary Air Injector Valve Circuit Electrical Malfunction Conditions:**<br>The Engine Control Module activates the secondary air injection solenoid valve, but the Heated Oxygen Sensor (HO2S) does not detect secondary air injection.<br>**Note: Solenoid valve is closed when no voltage is present.**<br>**Possible Causes:**<br>&bull; Connector to the secondary air injection valve is loose or disconnected<br>&bull; Secondary air injector valve circuit short<br>&bull; Secondary air injector valve circuit is open<br>&bull; Faulty secondary air injector valve<br>&bull; ECM has failed |
| **DTC: P1421**<br>**2T CCM, MIL: Yes**<br>**Years:** 2007, 2008<br>**Models:** A4, A6<br>**Engines:** All<br>**Transmissions:** All | **Secondary Air Injector Valve Circuit Short to Ground Conditions:**<br>The Engine Control Module detects a short circuit to ground when activating the secondary air injection solenoid valve.<br>**Note: Solenoid valve is closed when no voltage is present.**<br>**Possible Causes:**<br>&bull; Connector to the secondary air injection valve is loose or disconnected<br>&bull; Secondary air injector valve circuit short<br>&bull; Secondary air injector valve circuit is open<br>&bull; Faulty secondary air injector valve<br>&bull; ECM has failed |
| **DTC: P1422**<br>**2T CCM, MIL: Yes**<br>**Years:** 2007, 2008<br>**Models:** A4, A6<br>**Engines:** All<br>**Transmissions:** All | **Secondary Air Injector Valve Circuit Short to B+ Conditions:**<br>The Engine Control Module detects a short circuit to B+ when activating the secondary air injection solenoid valve.<br>**Note: Solenoid valve is closed when no voltage is present.**<br>**Possible Causes:**<br>&bull; Connector to the secondary air injection valve is loose or disconnected<br>&bull; Secondary air injector valve circuit short<br>&bull; Secondary air injector valve circuit is open<br>&bull; Faulty secondary air injector valve<br>&bull; ECM has failed |
| **DTC: P1424**<br>**2T CCM, MIL: Yes**<br>**Years:** 2007, 2008<br>**Models:** A4, A6<br>**Engines:** All<br>**Transmissions:** All | **Secondary Air Injector System (Bank 1) Leak Detected Conditions:**<br>Ignition on or vehicle running, and the ECM detected a leak in the secondary air injector system.<br>**Possible Causes:**<br>&bull; Poor hose/pipe connections between the secondary air injector pump motor and valve<br>&bull; Faulty hoses or pipes<br>&bull; Mechanical faults in the secondary air injector system |

| DTC | Trouble Code Title, Conditions & Possible Causes |
|---|---|
| **DTC: P1425**<br>**2T CCM, MIL: Yes**<br>**Years: 2007, 2008**<br>**Models:** A4, A6<br>**Engines:** All<br>**Transmissions:** All | **Tank Ventilation Valve Short to Ground Conditions:**<br>Ignition off. The Evaporative Emission (EVAP) canister purge regulator valve in the tank venting system or activation wire has a short circuit to ground. Engine started, engine running at a steady cruise speed, canister vent solenoid enabled, and the ECM detected an unexpected voltage condition on the Canister Vent solenoid circuit.<br>**Note: Solenoid valve is closed when no voltage is present.**<br>**Possible Causes:**<br>• Activation wire has a short to ground<br>• ECM has failed<br>• EVAP canister has failed<br>• EVAP canister system has an improper or broken seal<br>• Evaporative Emission (EVAP) canister purge regulator valve is faulty<br>• Leak Detection Pump (LDP) is faulty |
| **DTC: P1426**<br>**2T CCM, MIL: Yes**<br>**Years: 2007, 2008**<br>**Models:** A4, A6<br>**Engines:** All<br>**Transmissions:** All | **Tank Ventilation Valve Open Conditions:**<br>Ignition off. The Evaporative Emission (EVAP) canister purge regulator valve in the tank venting system or activation wire has a short circuit to ground. Engine started, engine running at a steady cruise speed, canister vent solenoid enabled, and the ECM detected an unexpected voltage condition on the Canister Vent solenoid circuit.<br>**Possible Causes:**<br>• Activation wire has a short to ground<br>• ECM has failed<br>• EVAP canister has failed<br>• EVAP canister system has an improper or broken seal<br>• Evaporative Emission (EVAP) canister purge regulator valve 1 is faulty<br>• Leak Detection Pump (LDP) is faulty |
| **DTC: P1432**<br>**2T CCM, MIL: Yes**<br>**Years: 2007, 2008**<br>**Models:** A4, A6<br>**Engines:** All<br>**Transmissions:** All | **Secondary Air Injection Valve Open Conditions:**<br>The output Diagnostic Test Mode (DTM) can be activated only with the ignition switched on and the engine not running. The output DTM is interrupted if the engine is started, or if a rotary pulse from the ignition system is recognized..<br>**Possible Causes:**<br>• Fuel pump relays have failed<br>• Fuel injector has failed<br>• Hoses on the EVAP canister may be clogged<br>• EVAP canister purge regulator valve may be faulty<br>• ECM may have failed<br>• Manifold Tuning Valve (IMT) may have failed |
| **DTC: P1433**<br>**2T CCM, MIL: Yes**<br>**Years: 2007, 2008**<br>**Models:** A4, A6<br>**Engines:** All<br>**Transmissions:** All | **Secondary Air Injection System Pump Relay Circuit Open Conditions:**<br>The output Diagnostic Test Mode (DTM) can be activated only with the ignition switched on and the engine not running. The output DTM is interrupted if the engine is started, or if a rotary pulse from the ignition system is recognized..<br>**Possible Causes:**<br>• Fuel pump relays have failed<br>• Fuel injector has failed<br>• Hoses on the EVAP canister may be clogged<br>• EVAP canister purge regulator valve may be faulty<br>• ECM may have failed<br>• Manifold Tuning Valve (IMT) may have failed |
| **DTC: P1434**<br>**2T CCM, MIL: Yes**<br>**Years: 2007, 2008**<br>**Models:** A4, A6<br>**Engines:** All<br>**Transmissions:** All | **Secondary Air Injection System Pump Relay Circuit Short to B+ Conditions:**<br>The output Diagnostic Test Mode (DTM) can be activated only with the ignition switched on and the engine not running. The output DTM is interrupted if the engine is started, or if a rotary pulse from the ignition system is recognized..<br>**Possible Causes:**<br>• Fuel pump relays have failed<br>• Fuel injector has failed<br>• Hoses on the EVAP canister may be clogged<br>• EVAP canister purge regulator valve may be faulty<br>• ECM may have failed<br>• Manifold Tuning Valve (IMT) may have failed |

| DTC | Trouble Code Title, Conditions & Possible Causes |
|---|---|
| **DTC: P1435**<br>**2T CCM, MIL: Yes**<br>**Years: 2007, 2008**<br>**Models:** A4, A6<br>**Engines:** All<br>**Transmissions:** All | **Secondary Air Injection System Pump Relay Circuit Short to Ground Conditions:**<br>The output Diagnostic Test Mode (DTM) can be activated only with the ignition switched on and the engine not running. The output DTM is interrupted if the engine is started, or if a rotary pulse from the ignition system is recognized..<br>**Possible Causes:**<br>• Fuel pump relays have failed<br>• Fuel injector has failed<br>• Hoses on the EVAP canister may be clogged<br>• EVAP canister purge regulator valve may be faulty<br>• ECM may have failed<br>• Manifold Tuning Valve (IMT) may have failed |
| **DTC: P1436**<br>**2T CCM, MIL: Yes**<br>**Years: 2007, 2008**<br>**Models:** A4, A6<br>**Engines:** All<br>**Transmissions:** All | **A/C Evaporator Temperature (ACET) Circuit Low Input Conditions:**<br>Key on or engine running; and the ECM detected the ACET signal was less than the self-test minimum amount of in the self-test.<br>**Possible Causes:**<br>• ACET signal circuit shorted to sensor ground (return)<br>• ACET signal circuit shorted to chassis ground<br>• ACET sensor is damaged or has failed<br>• Check activation of Secondary Air Injection (AIR) Pump Relay<br>• ECM has failed |
| **DTC: P1471**<br>**2T CCM, MIL: Yes**<br>**Years: 2007, 2008**<br>**Models:** A4, A6<br>**Engines:** All<br>**Transmissions:** All | **EVAP Emission Control Leak Detection Pump Circuit Short to B+ Conditions:**<br>Key on, KOEO Self-Test enabled, and the ECM detected an unexpected voltage condition on the EVAP emission control leak detection pump circuit.<br>**Possible Causes:**<br>• Leak Detection Pump has failed<br>• EVAP canister system has an improper or broken seal<br>• Evaporative Emission (EVAP) canister purge regulator valve 1 is faulty<br>• Hoses between the fuel pump and the EVAP canister are faulty<br>• Fuel filler cap is loose<br>• Fuel pump seal is defective, faulty or otherwise leaking<br>• Hoses between the EVAP canister and the fuel flap unit are faulty<br>• Hoses between the EVAP canister and the evaporative emission canister purge regulator valve are faulty<br>• ECM has failed |
| **DTC: P1472**<br>**2T CCM, MIL: Yes**<br>**Years: 2007, 2008**<br>**Models:** A4, A6<br>**Engines:** All<br>**Transmissions:** All | **EVAP Emission Control Leak Detection Pump Circuit Short to Ground Conditions:**<br>Key on, KOEO Self-Test enabled, and the ECM detected an unexpected voltage condition on the EVAP emission control leak detection pump circuit.<br>**Possible Causes:**<br>• Leak Detection Pump has failed<br>• EVAP canister system has an improper or broken seal<br>• Evaporative Emission (EVAP) canister purge regulator valve 1 is faulty<br>• Hoses between the fuel pump and the EVAP canister are faulty<br>• Fuel filler cap is loose<br>• Fuel pump seal is defective, faulty or otherwise leaking<br>• Hoses between the EVAP canister and the fuel flap unit are faulty<br>• Hoses between the EVAP canister and the evaporative emission canister purge regulator valve are faulty<br>• ECM has failed |
| **DTC: P1473**<br>**2T CCM, MIL: Yes**<br>**Years: 2007, 2008**<br>**Models:** A4, A6<br>**Engines:** All<br>**Transmissions:** All | **EVAP Emission Control Leak Detection Pump Circuit Open Conditions:**<br>Key on, KOEO Self-Test enabled, and the ECM detected an unexpected voltage condition on the EVAP emission control leak detection pump circuit.<br>**Possible Causes:**<br>• Leak Detection Pump has failed<br>• EVAP canister system has an improper or broken seal<br>• Evaporative Emission (EVAP) canister purge regulator valve 1 is faulty<br>• Hoses between the fuel pump and the EVAP canister are faulty<br>• Fuel filler cap is loose<br>• Fuel pump seal is defective, faulty or otherwise leaking<br>• Hoses between the EVAP canister and the fuel flap unit are faulty<br>• Hoses between the EVAP canister and the evaporative emission canister purge regulator valve are faulty<br>• ECM has failed |

| DTC | Trouble Code Title, Conditions & Possible Causes |
|---|---|
| **DTC: P1475**<br>**2T CCM, MIL: Yes**<br>**Years: 2007, 2008**<br>**Models:** A4, A6<br>**Engines:** All<br>**Transmissions:** All | **EVAP Emission Control LDP Circuit Malfunction/Signal Circuit Open Conditions:**<br>Key on, KOEO Self-Test enabled, and the ECM detected an unexpected voltage condition on the EVAP emission control leak detection pump circuit.<br>**Possible Causes:**<br>• Leak Detection Pump has failed<br>• EVAP canister system has an improper or broken seal<br>• Evaporative Emission (EVAP) canister purge regulator valve 1 is faulty<br>• Hoses between the fuel pump and the EVAP canister are faulty<br>• Fuel filler cap is loose<br>• Fuel pump seal is defective, faulty or otherwise leaking<br>• Hoses between the EVAP canister and the fuel flap unit are faulty<br>• Hoses between the EVAP canister and the evaporative emission canister purge regulator valve are faulty<br>• ECM has failed |
| **DTC: P1476**<br>**2T CCM, MIL: Yes**<br>**Years: 2007, 2008**<br>**Models:** A4, A6<br>**Engines:** All<br>**Transmissions:** All | **EVAP Emission Control LDP Circuit Malfunction/Insufficient Vacuum Conditions:**<br>Key on, KOEO Self-Test enabled, and the ECM detected an unexpected voltage condition on the EVAP emission control leak detection pump circuit.<br>**Possible Causes:**<br>• Leak Detection Pump has failed<br>• EVAP canister system has an improper or broken seal<br>• Evaporative Emission (EVAP) canister purge regulator valve 1 is faulty<br>• Hoses between the fuel pump and the EVAP canister are faulty<br>• Fuel filler cap is loose<br>• Fuel pump seal is defective, faulty or otherwise leaking<br>• Hoses between the EVAP canister and the fuel flap unit are faulty<br>• Hoses between the EVAP canister and the evaporative emission canister purge regulator valve are faulty<br>• ECM has failed |
| **DTC: P1477**<br>**2T CCM, MIL: Yes**<br>**Years: 2007, 2008**<br>**Models:** A4, A6<br>**Engines:** All<br>**Transmissions:** All | **EVAP Emission Control LDP Circuit Malfunction Conditions:**<br>Key on, KOEO Self-Test enabled, and the ECM detected an unexpected voltage condition on the EVAP emission control leak detection pump circuit.<br>**Possible Causes:**<br>• Leak Detection Pump has failed<br>• EVAP canister system has an improper or broken seal<br>• Evaporative Emission (EVAP) canister purge regulator valve 1 is faulty<br>• Hoses between the fuel pump and the EVAP canister are faulty<br>• Fuel filler cap is loose<br>• Fuel pump seal is defective, faulty or otherwise leaking<br>• Hoses between the EVAP canister and the fuel flap unit are faulty<br>• Hoses between the EVAP canister and the evaporative emission canister purge regulator valve are faulty<br>• ECM has failed |
| **DTC: P1478**<br>**2T CCM, MIL: Yes**<br>**Years: 2007, 2008**<br>**Models:** A4, A6<br>**Engines:** All<br>**Transmissions:** All | **EVAP Emission Control LDP Circuit Clamped Tube Detected Conditions:**<br>Key on, KOEO Self-Test enabled, and the ECM detected an unexpected voltage condition on the EVAP emission control leak detection pump circuit.<br>**Possible Causes:**<br>• Leak Detection Pump has failed<br>• EVAP canister system has an improper or broken seal<br>• Evaporative Emission (EVAP) canister purge regulator valve 1 is faulty<br>• Hoses between the fuel pump and the EVAP canister are faulty<br>• Fuel filler cap is loose<br>• Fuel pump seal is defective, faulty or otherwise leaking<br>• Hoses between the EVAP canister and the fuel flap unit are faulty<br>• Hoses between the EVAP canister and the evaporative emission canister purge regulator valve are faulty<br>• ECM has failed |
| **DTC: P1500**<br>**2T CCM, MIL: Yes**<br>**Years: 2007, 2008**<br>**Models:** A4, A6<br>**Engines:** All<br>**Transmissions:** All | **Fuel Pump Relay Circuit Electrical Malfunction Conditions:**<br>Engine running the ECM detected that the fuel pump relay signal was intermittent<br>**Possible Causes:**<br>• Fuel delivery unit connector is loose or not attached<br>• Fuse 18 cause a short to the transfer fuel pump or the O2S<br>• Fuel pump has failed<br>• Fuel pump relay circuit is shorted to ground, B+ or is open<br>• Fuel Pump (FP) Relay not activated<br>• ECM has failed |

| DTC | Trouble Code Title, Conditions & Possible Causes |
|---|---|
| **DTC: P1502**<br>**2T CCM, MIL: Yes**<br>**Years: 2007, 2008**<br>**Models:** A4, A6<br>**Engines:** All<br>**Transmissions:** All | **Fuel Pump Relay Circuit Short to B+ Conditions:**<br>Engine running the ECM detected that the fuel pump relay signal was intermittent<br>**Possible Causes:**<br>• Fuel delivery unit connector is loose or not attached<br>• Fuse 18 cause a short to the transfer fuel pump or the O2S<br>• Fuel pump has failed<br>• Fuel pump relay circuit is shorted to ground, B+ or is open<br>• Fuel Pump (FP) Relay not activated<br>• ECM has failed |
| **DTC: P1512**<br>**2T CCM, MIL: Yes**<br>**Years: 2007, 2008**<br>**Models:** A4, A6<br>**Engines:** All<br>**Transmissions:** All | **Intake Manifold Changeover Valve Circuit Short to B+ Conditions:**<br>Engine started, and the ECM detected the changeover valve circuit was shorting to positive during the continuous self test.<br>**Possible Causes:**<br>• Leaky vacuum reservoir, vacuum lines loose or damaged<br>• Vacuum solenoid or vacuum actuator is damaged<br>• IMRC actuator cable/gears are seized, or the cables are improperly routed or seized<br>• IMRC housing return springs are damaged or disconnected<br>• Lever/shaft return stop may be obstructed or bent, or the lever/shaft wide open stop may be obstructed or bent, or the IMRC lever/shaft may be sticking, binding or disconnected<br>• IMRC control circuit open, shorted or the VPWR circuit is open<br>• ECM has failed |
| **DTC: P1519**<br>**2T CCM, MIL: Yes**<br>**Years: 2007, 2008**<br>**Models:** A4, A6<br>**Engines:** All<br>**Transmissions:** All | **Intake Manifold Runner Control Stuck Closed Conditions:**<br>Key on, and the ECM detected the IMRC Monitor was more than the expected calibrated range at closed throttle.<br>**Possible Causes:**<br>• IMRC monitor signal circuit shorted to power ground<br>• IMRC Monitor signal circuit shorted to signal ground (return)<br>• IMRC actuator is damaged or has failed (e.g., there may be a smAll leak in the vacuum diaphragm of the actuator)<br>• ECM has failed |
| **DTC: P1522**<br>**2T CCM, MIL: Yes**<br>**Years: 2007, 2008**<br>**Models:** A4, A6<br>**Engines:** All<br>**Transmissions:** All | **Intake Camshaft Control (Bank 2) Malfunction Conditions:**<br>Key on or engine running; and the ECM detected the intake manifold control signal for was outside of its expected calibrated range.<br>**Possible Causes:**<br>• Camshaft control circuit is open or shorted to ground<br>• Camshaft sensor is damaged or the ECM has failed<br>• Camshaft out of adjustment |
| **DTC: P1529**<br>**2T CCM, MIL: Yes**<br>**Years: 2007, 2008**<br>**Models:** A6<br>**Engines:** All<br>**Transmissions:** All | **Camshaft Control Circuit Short to B+ Conditions:**<br>Engine started and driven at an engine speed of more than 400 RPM; and the ECM detected the camshaft timing exceeded the calibrated voltage levels. The valve timing did not change from the current valve timing or it remained fixed during the testing.<br>**Note: The camshaft adjustment is load- and RPM dependant. The electrical camshaft adjustment valve 1 switches oil pressure onto camshaft adjuster (mechanical adjustment mechanism), which adjusts the camshaft.**<br>**Possible Causes:**<br>• Fuel pump has failed<br>• CPS circuit is open, shorted to ground or shorted to power<br>• ECM has failed<br>• Battery voltage below 11.5 volts<br>• Position actuator circuit may short to B+ or Ground<br>• Camshaft timing improperly set, or continuous oil flow to the VCT piston chamber<br>• Camshaft advance mechanism (the VCT unit) is sticking or binding mechanically<br>• VCT solenoid valve is stuck in open position |
| **DTC: P1530**<br>**2T CCM, MIL: Yes**<br>**Years: 2007, 2008**<br>**Models:** A6<br>**Engines:** All<br>**Transmissions:** All | **Camshaft Control Circuit Short to Ground Conditions:**<br>Engine started and driven at an engine speed of more than 400 RPM; and the ECM detected the camshaft timing exceeded the calibrated levels. The valve timing did not change from the current valve timing or it remained fixed during the testing.<br>**Note: The camshaft adjustment is load- and RPM dependant. The electrical camshaft adjustment valve 1 switches oil pressure onto camshaft adjuster (mechanical adjustment mechanism), which adjusts the camshaft.**<br>**Possible Causes:**<br>• Fuel pump has failed<br>• CPS circuit is open, shorted to ground or shorted to power<br>• ECM has failed<br>• Battery voltage below 11.5 volts<br>• Position actuator circuit may short to B+ or Ground<br>• Camshaft timing improperly set, or continuous oil flow to the VCT piston chamber<br>• Camshaft advance mechanism (the VCT unit) is sticking or binding mechanically<br>• VCT solenoid valve is stuck in open position |

| DTC | Trouble Code Title, Conditions & Possible Causes |
|---|---|
| **DTC: P1531**<br>**2T CCM, MIL:** Yes<br>**Years:** 2007, 2008<br>**Models:** A6<br>**Engines:** All<br>**Transmissions:** All | **Camshaft Control Circuit Open Conditions:**<br>Engine started and driven at an engine speed of more than 400 RPM; and the ECM detected the camshaft timing exceeded the calibrated levels. The valve timing did not change from the current valve timing or it remained fixed during the testing.<br>**Note: The camshaft adjustment is load- and RPM dependant. The electrical camshaft adjustment valve 1 switches oil pressure onto camshaft adjuster (mechanical adjustment mechanism), which adjusts the camshaft.**<br>**Possible Causes:**<br>• Fuel pump has failed<br>• CPS circuit is open, shorted to ground or shorted to power<br>• ECM has failed<br>• Battery voltage below 11.5 volts<br>• Position actuator circuit may short to B+ or Ground<br>• Camshaft timing improperly set, or continuous oil flow to the VCT piston chamber<br>• Camshaft advance mechanism (the VCT unit) is sticking or binding mechanically<br>• VCT solenoid valve is stuck in open position |
| **DTC: P1541**<br>**2T CCM, MIL:** Yes<br>**Years:** 2007, 2008<br>**Models:** A4<br>**Engines:** All<br>**Transmissions:** All | **Fuel Pump Relay Circuit Open Conditions:**<br>The ECM detected an electrical malfunction on the fuel pump relay circuit<br>**Possible Causes:**<br>• Fuel pump relay not activiated |
| **DTC: P1542**<br>**2T CCM, MIL:** Yes<br>**Years:** 2007, 2008<br>**Models:** A4, A6<br>**Engines:** All<br>**Transmissions:** All | **Throttle Actuation Potentiometer Range/Performance Conditions:**<br>Engine started, battery voltage must be at least 11.5v, all electrical components must be off, parking brake must be engaged (to keep daytime driving lights off), automatic transmission selector must be in park, the exhaust system must be properly sealed between the catalytic converter and the cylinder head, coolant temperature must be at least 80 degrees Celsius, and the ground between the engine and the chassis must be well connected. The signal from the Throttle Position Valve Module to the ECM detected was erratic, non existent or unreliable.<br>**Note: If the complete throttle valve control module is current-less (e.g. connector disconnected) the throttle valve moves into a particular, specified mechanical position, which signals an increased idle speed with an engine at operating temperature. If only the Throttle Position (TP) actuator -V60- is current-less, the throttle valve also moves into the specified mechanical position (emergency running gap), however, since Closed Throttle Position (CTP) switch -F60- can still be recognized, an "almost normal idle RPM" is reached via the respective ignition angle retardation. If the Engine Control Module (ECM) detects a malfunction at Throttle Position (TP) sensor -G69-, Throttle Position (TP) actuator -V60- is switched current-less by the Engine Control Module (ECM) and the throttle valve moves into the specified mechanical position (emergency running gap) again.**<br>**Note: Terminal assignment at throttle control module is different in vehicles with and without cruise control.**<br>**Characteristic: Steering column switch with operating module for cruise control.**<br>**Possible Causes:**<br>• Throttle valve control module is faulty<br>• Throttle valve is damaged or dirty<br>• Throttle valve must be in closed throttle position<br>• Accelerator pedal is out of adjustment (AEG engines only)<br>• Throttle position actuator is shorting to ground or power |
| **DTC: P1543**<br>**2T CCM, MIL:** Yes<br>**Years:** 2007, 2008<br>**Models:** A4, A6<br>**Engines:** All<br>**Transmissions:** All | **Throttle Actuation Potentiometer Signal Too Low Conditions:**<br>Engine started, battery voltage must be at least 11.5v, all electrical components must be off, parking brake must be engaged (to keep daytime driving lights off), automatic transmission selector must be in park, the exhaust system must be properly sealed between the catalytic converter and the cylinder head, coolant temperature must be at least 80 degrees Celsius, and the ground between the engine and the chassis must be well connected. The signal from the Throttle Position Valve Module to the ECM detected was erratic, non existent or unreliable.<br>**Note: If the complete throttle valve control module is current-less (e.g. connector disconnected) the throttle valve moves into a particular, specified mechanical position, which signals an increased idle speed with an engine at operating temperature. If only the Throttle Position (TP) actuator -V60- is current-less, the throttle valve also moves into the specified mechanical position (emergency running gap), however, since Closed Throttle Position (CTP) switch -F60- can still be recognized, an "almost normal idle RPM" is reached via the respective ignition angle retardation. If the Engine Control Module (ECM) detects a malfunction at Throttle Position (TP) sensor -G69-, Throttle Position (TP) actuator -V60- is switched current-less by the Engine Control Module (ECM) and the throttle valve moves into the specified mechanical position (emergency running gap) again.**<br>**Note: Terminal assignment at throttle control module is different in vehicles with and without cruise control.**<br>**Characteristic: Steering column switch with operating module for cruise control.**<br>**Possible Causes:**<br>• Throttle valve control module is faulty<br>• Throttle valve is damaged or dirty<br>• Throttle valve must be in closed throttle position<br>• Accelerator pedal is out of adjustment (AEG engines only)<br>• Throttle position actuator is shorting to ground or power |

| DTC | Trouble Code Title, Conditions & Possible Causes |
|---|---|
| **DTC: P1544**<br>**2T CCM, MIL: Yes**<br>**Years: 2007, 2008**<br>**Models:** A4, A6<br>**Engines:** All<br>**Transmissions:** All | **Throttle Actuation Potentiometer Signal Too High Conditions:**<br>Engine started, battery voltage must be at least 11.5v, all electrical components must be off, parking brake must be engaged (to keep daytime driving lights off), automatic transmission selector must be in park, the exhaust system must be properly sealed between the catalytic converter and the cylinder head, coolant temperature must be at least 80 degrees Celsius, and the ground between the engine and the chassis must be well connected. The signal from the Throttle Position Valve Module to the ECM detected was erratic, non existent or unreliable.<br>**Note: If the complete throttle valve control module is current-less (e.g. connector disconnected) the throttle valve moves into a particular, specified mechanical position, which signals an increased idle speed with an engine at operating temperature. If only the Throttle Position (TP) actuator –V60- is current-less, the throttle valve also moves into the specified mechanical position (emergency running gap), however, since Closed Throttle Position (CTP) switch –F60- can still be recognized, an "almost normal idle RPM" is reached via the respective ignition angle retardation. If the Engine Control Module (ECM) detects a malfunction at Throttle Position (TP) sensor –G69-, Throttle Position (TP) actuator –V60- is switched current-less by the Engine Control Module (ECM) and the throttle valve moves into the specified mechanical position (emergency running gap) again.**<br>**Note: Terminal assignment at throttle control module is different in vehicles with and without cruise control.**<br>**Characteristic: Steering column switch with operating module for cruise control.**<br>**Possible Causes:**<br>• Throttle valve control module is faulty<br>• Throttle valve is damaged or dirty<br>• Throttle valve must be in closed throttle position<br>• Accelerator pedal is out of adjustment (AEG engines only)<br>• Throttle position actuator is shorting to ground or power |
| **DTC: P1545**<br>**2T CCM, MIL: Yes**<br>**Years: 2007, 2008**<br>**Models:** A4, A6<br>**Engines:** All<br>**Transmissions:** All | **Throttle Position Control Malfunction Conditions:**<br>Engine started, battery voltage must be at least 11.5v, all electrical components must be off, parking brake must be engaged (to keep daytime driving lights off), automatic transmission selector must be in park, the exhaust system must be properly sealed between the catalytic converter and the cylinder head, coolant temperature must be at least 80 degrees Celsius, and the ground between the engine and the chassis must be well connected. The signal from the Throttle Position Valve Module to the ECM detected was erratic, non existent or unreliable.<br>**Note: If the complete throttle valve control module is current-less (e.g. connector disconnected) the throttle valve moves into a particular, specified mechanical position, which signals an increased idle speed with an engine at operating temperature. If only the Throttle Position (TP) actuator is current-less, the throttle valve also moves into the specified mechanical position (emergency running gap), however, since Closed Throttle Position (CTP) switch – can still be recognized, an "almost normal idle RPM" is reached via the respective ignition angle retardation. If the Engine Control Module (ECM) detects a malfunction at Throttle Position (TP) sensor – Throttle Position (TP) actuator is switched current-less by the Engine Control Module (ECM) and the throttle valve moves into the specified mechanical position (emergency running gap) again.**<br>**Note: Terminal assignment at throttle control module is different in vehicles with and without cruise control.**<br>**Characteristic: Steering column switch with operating module for cruise control.**<br>**Possible Causes:**<br>• Throttle valve control module is faulty<br>• Throttle valve is damaged or dirty<br>• Throttle valve must be in closed throttle position<br>• Accelerator pedal is out of adjustment (AEG engines only)<br>• Throttle position actuator is shorting to ground or power |
| **DTC: P1558**<br>**2T CCM, MIL: Yes**<br>**Years: 2007, 2008**<br>**Models:** A4, A6<br>**Engines:** All<br>**Transmissions:** All | **Throttle Actuator Electrical Malfunction Conditions:**<br>Engine started, battery voltage at least 11.5v, all electrical components off, ground connections between engine and chassis well connected, coolant temperature at least 80-degrees Celcius and the throttle valve must not be damaged or dirty; and the ECM detected the signal from the Throttle Position Valve Module to the ECM detected was erratic, non existent or unreliable (too high or too low).<br>**Possible Causes:**<br>• Throttle valve control module has failed<br>• Throttle valve control module's circuit has shorted or is open<br>• The ECM has failed |
| **DTC: P1559**<br>**2T CCM, MIL: Yes**<br>**Years: 2007, 2008**<br>**Models:** A6<br>**Engines:** All<br>**Transmissions:** All | **Idle Speed Control Throttle Position Adaptation Malfunction Conditions:**<br>Engine started, battery voltage at least 11.5v, all electrical components off, ground connections between engine and chassis well connected, coolant temperature at least 80-degrees Celcius and the throttle valve must not be damaged or dirty; and the ECM detected the signal from the Throttle Position Valve Module to the ECM detected was erratic, non existent or unreliable (too high or too low).<br>**Possible Causes:**<br>• Throttle valve control module has failed<br>• Throttle valve control module's circuit has shorted or is open<br>• The ECM has failed |

| DTC | Trouble Code Title, Conditions & Possible Causes |
|---|---|
| **DTC: P1565**<br>**2T CCM, MIL: Yes**<br>**Years: 2007, 2008**<br>**Models:** A4, A6<br>**Engines:** All<br>**Transmissions:** All | **Idle Speed Control Throttle Position Lower Limit Not Attainted Conditions:**<br>Engine started, battery voltage at least 11.5v, all electrical components off, ground connections between engine and chassis well connected, coolant temperature at least 80-degrees Celicius and the throttle valve must not be damaged or dirty; and the ECM detected the signal from the Throttle Position Valve Module to the ECM detected was erratic, non existent or unreliable (too high or too low).<br>**Possible Causes:**<br>• Alternator failed<br>• ECM failed<br>• Fuses blown or open circuits<br>• Clean Throttle Valve Control Module<br>• Accelerator cable not adjusted properly<br>• Idle speed control throttle failed<br>• Wire connections to relay carrier and ground connection of ECM may have shorted |
| **DTC: P1568**<br>**2T CCM, MIL: Yes**<br>**Years: 2007, 2008**<br>**Models:** A6<br>**Engines:** All<br>**Transmissions:** All | **Idle Speed Control Throttle Position Mechanical Malfunction Conditions:**<br>Engine started, battery voltage at least 11.5v, all electrical components off, ground connections between engine and chassis well connected, coolant temperature at least 80-degrees Celicius and the throttle valve must not be damaged or dirty; and the ECM detected the signal from the Throttle Position Valve Module to the ECM detected was erratic, non existent or unreliable (too high or too low) suggesting a mechanicl malfunction.<br>**Possible Causes:**<br>• Alternator failed<br>• ECM failed<br>• Fuses blown or open circuits<br>• Clean Throttle Valve Control Module<br>• Accelerator cable not adjusted properly<br>• Idle speed control throttle failed<br>• Wire connections to relay carrier and ground connection of ECM may have shorted |
| **DTC: P1602**<br>**2T CCM, MIL: Yes**<br>**Years: 2007, 2008**<br>**Models:** A4, A6<br>**Engines:** All<br>**Transmissions:** All | **Power Supply (B+) Terminal 15 Low Voltage Conditions:**<br>Ignition on, the ECM detected a low voltage condition on the power supply terminal (15). To achieve optimal anti-theft protection for the vehicle, an anti-theft immobilizer is instAlled. The anti-theft immobilizer is a system for enabling and locking the Engine Control Module (ECM). So that this system cannot be circumvented, it is necessary to perform adaptation of the anti-theft immobilizer using the Vehicle Diagnostic and Information System VAS 5052 in the On Board Diagnostic (OBD) function. The great availability of equipment options makes it necessary to adapt the Engine Control Module (ECM) to the vehicle (e.g. throttle valve control module or cruise control system). This "writing" function is not possible with the generic scan tool.<br>**Possible Causes:**<br>• (If ECM was replaced) ECM ID not the same as the replaced unit<br>• ECM has failed<br>• Voltage supply for Engine Control Module (ECM) has shorted |
| **DTC: P1603**<br>**2T CCM, MIL: Yes**<br>**Years: 2007, 2008**<br>**Models:** A4, A6<br>**Engines:** All<br>**Transmissions:** All | **Internal Control Module Malfunction Conditions:**<br>Ignition on, the ECM detected a control module malfunction. To achieve optimal anti-theft protection for the vehicle, an anti-theft immobilizer is instAlled. The anti-theft immobilizer is a system for enabling and locking the Engine Control Module (ECM). So that this system cannot be circumvented, it is necessary to perform adaptation of the anti-theft immobilizer using the Vehicle Diagnostic and Information System VAS 5052 in the On Board Diagnostic (OBD) function. The great availability of equipment options makes it necessary to adapt the Engine Control Module (ECM) to the vehicle (e.g. throttle valve control module or cruise control system). This "writing" function is not possible with the generic scan tool.<br>**Possible Causes:**<br>• (If ECM was replaced) ECM ID not the same as the replaced unit<br>• ECM has failed<br>• Voltage supply for Engine Control Module (ECM) has shorted |
| **DTC: P1604**<br>**2T CCM, MIL: Yes**<br>**Years: 2007, 2008**<br>**Models:** A4, A6<br>**Engines:** All<br>**Transmissions:** All | **Internal Control Module Driver Error Conditions:**<br>Ignition on, the ECM detected a control module malfunction. To achieve optimal anti-theft protection for the vehicle, an anti-theft immobilizer is instAlled. The anti-theft immobilizer is a system for enabling and locking the Engine Control Module (ECM). So that this system cannot be circumvented, it is necessary to perform adaptation of the anti-theft immobilizer using the Vehicle Diagnostic and Information System VAS 5052 in the On Board Diagnostic (OBD) function. The great availability of equipment options makes it necessary to adapt the Engine Control Module (ECM) to the vehicle (e.g. throttle valve control module or cruise control system). This "writing" function is not possible with the generic scan tool.<br>**Possible Causes:**<br>• (If ECM was replaced) ECM ID not the same as the replaced unit<br>• ECM has failed<br>• Voltage supply for Engine Control Module (ECM) has shorted |

| DTC | Trouble Code Title, Conditions & Possible Causes |
|---|---|
| **DTC: P1606**<br>**2T CCM, MIL: Yes**<br>**Years: 2007, 2008**<br>**Models:** A4, A6<br>**Engines:** All<br>**Transmissions:** All | **Rough Road Spec Engine Torque ABS-ECU Electrical Malfunction Conditions:**<br>Ignition on, the ECM detected an electrical malfunction.<br>**Possible Causes:**<br>• Check wire connection between Engine Control Module (ECM) and ABS Control Module |
| **DTC: P1612**<br>**2T CCM, MIL: Yes**<br>**Years: 2007, 2008**<br>**Models:** A4, A6<br>**Engines:** All<br>**Transmissions:** All | **Electronic Control Module Incorrect Coding Conditions:**<br>Ignition on, the ECM detected a control module malfunction. To achieve optimal anti-theft protection for the vehicle, an anti-theft immobilizer is instAlled. The anti-theft immobilizer is a system for enabling and locking the Engine Control Module (ECM). So that this system cannot be circumvented, it is necessary to perform adaptation of the anti-theft immobilizer using the Vehicle Diagnostic and Information System VAS 5052 in the On Board Diagnostic (OBD) function. The great availability of equipment options makes it necessary to adapt the Engine Control Module (ECM) to the vehicle (e.g. throttle valve control module or cruise control system). This "writing" function is not possible with the generic scan tool.<br>**Possible Causes:**<br>• (If ECM was replaced) ECM ID not the same as the replaced unit<br>• ECM has failed<br>• Voltage supply for Engine Control Module (ECM) has shorted |
| **DTC: P1630**<br>**2T CCM, MIL: Yes**<br>**Years: 2007, 2008**<br>**Models:** A4, A6<br>**Engines:** All<br>**Transmissions:** All | **Acceleration Pedal Position Sensor 1 Signal Too Low Conditions:**<br>Engine started, battery voltage at least 11.5v, all electrical components off, ground connections between engine and chassis well connected, the ECM detected that the accelerator pedal position sensor signal was too low.<br>**Note: Both the Throttle Position (TP) Sensor and Accelerator Pedal Position Sensor 2 are located at the accelerator pedal module and communicate the driver's intentions to the ECM completely independently of each other. Both sensors are stored in one housing.**<br>**Possible Causes:**<br>• Ground between engine and chassis may be broken<br>• Throttle position sensor may have failed<br>• Accelerator Pedal Position Sensor 2 has failed<br>• Throttle position sensor wiring may have shorted<br>• Faulty voltage supply<br>• ECM has failed |
| **DTC: P1631**<br>**2T CCM, MIL: Yes**<br>**Years: 2007, 2008**<br>**Models:** A4, A6<br>**Engines:** All<br>**Transmissions:** All | **Acceleration Pedal Position Sensor 1 Signal Too High Conditions:**<br>Engine started, battery voltage at least 11.5v, all electrical components off, ground connections between engine and chassis well connected, the ECM detected that the accelerator pedal position sensor signal was too high.<br>**Note: Both the Throttle Position (TP) Sensor and Accelerator Pedal Position Sensor 2 are located at the accelerator pedal module and communicate the driver's intentions to the ECM completely independently of each other. Both sensors are stored in one housing.**<br>**Possible Causes:**<br>• Ground between engine and chassis may be broken<br>• Throttle position sensor may have failed<br>• Accelerator Pedal Position Sensor 2 has failed<br>• Throttle position sensor wiring may have shorted<br>• Faulty voltage supply<br>• ECM has failed |
| **DTC: P1633**<br>**2T CCM, MIL: Yes**<br>**Years: 2007, 2008**<br>**Models:** A4, A6<br>**Engines:** All<br>**Transmissions:** All | **Acceleration Pedal Position Sensor 2 Signal Too Low Conditions:**<br>Engine started, battery voltage at least 11.5v, all electrical components off, ground connections between engine and chassis well connected, the ECM detected that the accelerator pedal position sensor signal was too low.<br>**Note: Both the Throttle Position (TP) Sensor and Accelerator Pedal Position Sensor 2 are located at the accelerator pedal module and communicate the driver's intentions to the ECM completely independently of each other. Both sensors are stored in one housing.**<br>**Possible Causes:**<br>• Ground between engine and chassis may be broken<br>• Throttle position sensor may have failed<br>• Accelerator Pedal Position Sensor 2 has failed<br>• Throttle position sensor wiring may have shorted<br>• Faulty voltage supply<br>• ECM has failed |

| DTC | Trouble Code Title, Conditions & Possible Causes |
|---|---|
| **DTC: P1634**<br>**2T CCM, MIL:** Yes<br>**Years:** 2007, 2008<br>**Models:** A4, A6<br>**Engines:** All<br>**Transmissions:** All | **Acceleration Pedal Position Sensor 2 Signal Too High Conditions:**<br>Engine started, battery voltage at least 11.5v, all electrical components off, ground connections between engine and chassis well connected, the ECM detected that the accelerator pedal position sensor signal was too high.<br>**Note: Both the Throttle Position (TP) Sensor and Accelerator Pedal Position Sensor 2 are located at the accelerator pedal module and communicate the driver's intentions to the ECM completely independently of each other. Both sensors are stored in one housing.**<br>**Possible Causes:**<br>    • Ground between engine and chassis may be broken<br>    • Throttle position sensor may have failed<br>    • Accelerator Pedal Position Sensor 2 has failed<br>    • Throttle position sensor wiring may have shorted<br>    • Faulty voltage supply<br>    • ECM has failed |
| **DTC: P1639**<br>**2T CCM, MIL:** Yes<br>**Years:** 2007, 2008<br>**Models:** A4, A6<br>**Engines:** All<br>**Transmissions:** All | **Accelerator Pedal Position Sensor 1+2 Range/Performance Conditions:**<br>Engine started, battery voltage at least 11.5v, all electrical components off, ground connections between engine and chassis well connected, the ECM detected that the accelerator pedal position sensor signal was too high.<br>**Note: Both the Throttle Position (TP) Sensor and Accelerator Pedal Position Sensor 2 are located at the accelerator pedal module and communicate the driver's intentions to the ECM completely independently of each other. Both sensors are stored in one housing.**<br>**Possible Causes:**<br>    • Ground between engine and chassis may be broken<br>    • Throttle position sensor may have failed<br>    • Accelerator Pedal Position Sensor 2 has failed<br>    • Throttle position sensor wiring may have shorted<br>    • Faulty voltage supply<br>    • ECM has failed |
| **DTC: P1640**<br>**2T CCM, MIL:** Yes<br>**Years:** 2007, 2008<br>**Models:** A4, A6<br>**Engines:** All<br>**Transmissions:** All | **Internal Control Module (EEPROM) Error Conditions:**<br>Ignition on, the ECM detected a control module malfunction (software). To achieve optimal anti-theft protection for the vehicle, an anti-theft immobilizer is instAlled. The anti-theft immobilizer is a system for enabling and locking the Engine Control Module (ECM). So that this system cannot be circumvented, it is necessary to perform adaptation of the anti-theft immobilizer using the Vehicle Diagnostic and Information System VAS 5052 in the On Board Diagnostic (OBD) function. The great availability of equipment options makes it necessary to adapt the Engine Control Module (ECM) to the vehicle (e.g. throttle valve control module or cruise control system). This "writing" function is not possible with the generic scan tool.<br>**Possible Causes:**<br>    • Engine Control Module (ECM) has failed<br>    • Voltage supply for Engine Control Module (ECM) has shorted |
| **DTC: P1648**<br>**2T CCM, MIL:** Yes<br>**Years:** 2007, 2008<br>**Models:** A4, A6<br>**Engines:** All<br>**Transmissions:** All | **Data Bus Powertrain Malfunction Conditions:**<br>Ignition on, the ECM detected a data bus malfunction (software). To achieve optimal anti-theft protection for the vehicle, an anti-theft immobilizer is instAlled. The anti-theft immobilizer is a system for enabling and locking the Engine Control Module (ECM). So that this system cannot be circumvented, it is necessary to perform adaptation of the anti-theft immobilizer using the Vehicle Diagnostic and Information System VAS 5052 in the On Board Diagnostic (OBD) function. The great availability of equipment options makes it necessary to adapt the Engine Control Module (ECM) to the vehicle (e.g. throttle valve control module or cruise control system). This "writing" function is not possible with the generic scan tool.<br>**Possible Causes:**<br>    • Ground between engine and chassis may be broken<br>    • Throttle position sensor may have failed<br>    • Accelerator Pedal Position Sensor 2 has failed<br>    • Throttle position sensor wiring may have shorted<br>    • Faulty voltage supply<br>    • ECM has failed |

| DTC | Trouble Code Title, Conditions & Possible Causes |
|---|---|
| **DTC: P1649**<br>**2T CCM, MIL:** Yes<br>**Years:** 2007, 2008<br>**Models:** A4<br>**Engines:** All<br>**Transmissions:** All | **Data Bus Powertrain Missing Message from ABS Control Module Conditions:**<br>Ignition off, the ECU is missing general Data BUS information from the central electrical control. The Engine Control Module (ECM) communicates with All databus-capable control modules via a CAN databus. These databus-capable control modules are connected via two data bus wires which are twisted together (CAN_High and CAN_Low), and exchange information (messages). Missing information on the databus is recognized as a malfunction and stored. Trouble-free operation of the CAN-bus requires that it have a terminal resistance. This central terminal resistor is located in the Engine Control Module (ECM).<br>**Possible Causes:**<br>   • Ground between engine and chassis may be broken<br>   • Throttle position sensor may have failed<br>   • Accelerator Pedal Position Sensor 2 has failed<br>   • Throttle position sensor wiring may have shorted<br>   • Faulty voltage supply<br>   • Check the Terminal resistance for CAN-bus<br>   • Data-Bus wires have short<br>   • Data-Bus components are malfunctioning<br>   • ECM has failed |
| **DTC: P1676**<br>**2T CCM, MIL:** Yes<br>**Years:** 2007, 2008<br>**Models:** A4<br>**Engines:** All<br>**Transmissions:** All | **Drive by Wire-MIL Circuit Electrical Malfunction Conditions:**<br>Key on or engine running, the ECM detected an electrical malfunction regarding the drive-by-wire circuit.<br>**Note: EPC" is an abbreviation and stands for Electronic Power Control and means "electronic engine load control". If malfunctions are recognized in the EPC system during operation of the engine, the Engine Control Module (ECM) switches on the EPC warning lamp. An entry is made in DTC memory at the same time. After a few seconds of the engine at idle, the EPC should extinguish itself.**<br>**Possible Causes:**<br>   • Circuit from the MIL to the ECM<br>   • ECM has failed<br>   • Circuit from the EPC to the ECM |
| **DTC: P1677**<br>**2T CCM, MIL:** Yes<br>**Years:** 2007, 2008<br>**Models:** A4<br>**Engines:** All<br>**Transmissions:** All | **Drive by Wire-MIL Circuit Short to B+ Conditions:**<br>Key on or engine running, the ECM detected an electrical malfunction regarding the drive-by-wire circuit.<br>**Note: EPC" is an abbreviation and stands for Electronic Power Control and means "electronic engine load control". If malfunctions are recognized in the EPC system during operation of the engine, the Engine Control Module (ECM) switches on the EPC warning lamp. An entry is made in DTC memory at the same time. After a few seconds of the engine at idle, the EPC should extinguish itself.**<br>**Possible Causes:**<br>   • Circuit from the MIL to the ECM<br>   • ECM has failed<br>   • Circuit from the EPC to the ECM |
| **DTC: P1690**<br>**2T CCM, MIL:** Yes<br>**Years:** 2007, 2008<br>**Models:** A4, A6<br>**Engines:** All<br>**Transmissions:** All | **Malfunction Indication Light Malfunction Conditions:**<br>The exhaust Malfunction Indicator Lamp (MIL) lights up when exhaust relevant malfunctions are recognized by the Engine Control Module (ECM). The Malfunction Indicator Lamp (MIL) can blink or remain lit continuously. Blinking: There is a malfunction that causes damage to the catalytic converter in this driving condition. In this case, vehicle must only be driven at reduced power! Continuously lit: There is a malfunction that causes increased emissions. Check DTC memory for Motronic control module. DTC memory must still be checked if there are driveability problems or customer complaints and the MIL is not lit, since malfunctions can be stored without causing the MIL to light immediately.<br>**Possible Causes:**<br>   • Wire from ECM to MIL is shorted or grounded<br>   • ECM has failed<br>   • MIL has failed |
| **DTC: P1691**<br>**2T CCM, MIL:** Yes<br>**Years:** 2007, 2008<br>**Models:** A4<br>**Engines:** All<br>**Transmissions:** All | **Malfunction Indication Light Open Conditions:**<br>The exhaust Malfunction Indicator Lamp (MIL) lights up when exhaust relevant malfunctions are recognized by the Engine Control Module (ECM). The Malfunction Indicator Lamp (MIL) can blink or remain lit continuously. Blinking: There is a malfunction that causes damage to the catalytic converter in this driving condition. In this case, vehicle must only be driven at reduced power! Continuously lit: There is a malfunction that causes increased emissions. Check DTC memory for Motronic control module. DTC memory must still be checked if there are driveability problems or customer complaints and the MIL is not lit, since malfunctions can be stored without causing the MIL to light immediately.<br>**Possible Causes:**<br>   • Wire from ECM to MIL is shorted or grounded<br>   • ECM has failed<br>   • MIL has failed |

| DTC | Trouble Code Title, Conditions & Possible Causes |
|-----|--------------------------------------------------|
| **DTC: P1692**<br>**2T CCM, MIL:** Yes<br>**Years:** 2007, 2008<br>**Models:** A4<br>**Engines:** All<br>**Transmissions:** All | **Malfunction Indication Light Short to Ground Conditions:**<br>The exhaust Malfunction Indicator Lamp (MIL) lights up when exhaust relevant malfunctions are recognized by the Engine Control Module (ECM). The Malfunction Indicator Lamp (MIL) can blink or remain lit continuously. Blinking: There is a malfunction that causes damage to the catalytic converter in this driving condition. In this case, vehicle must only be driven at reduced power! Continuously lit: There is a malfunction that causes increased emissions. Check DTC memory for Motronic control module. DTC memory must still be checked if there are driveability problems or customer complaints and the MIL is not lit, since malfunctions can be stored without causing the MIL to light immediately.<br>**Possible Causes:**<br>• Wire from ECM to MIL is shorted or grounded<br>• ECM has failed<br>• MIL has failed |
| **DTC: P1693**<br>**2T CCM, MIL:** Yes<br>**Years:** 2007, 2008<br>**Models:** A4, A6<br>**Engines:** All<br>**Transmissions:** All | **Malfunction Indication Light Short to B+ Conditions:**<br>The exhaust Malfunction Indicator Lamp (MIL) lights up when exhaust relevant malfunctions are recognized by the Engine Control Module (ECM). The Malfunction Indicator Lamp (MIL) can blink or remain lit continuously. Blinking: There is a malfunction that causes damage to the catalytic converter in this driving condition. In this case, vehicle must only be driven at reduced power! Continuously lit: There is a malfunction that causes increased emissions. Check DTC memory for Motronic control module. DTC memory must still be checked if there are driveability problems or customer complaints and the MIL is not lit, since malfunctions can be stored without causing the MIL to light immediately.<br>**Possible Causes:**<br>• Wire from ECM to MIL is shorted or grounded<br>• ECM has failed<br>• MIL has failed |
| **DTC: P1702**<br>**2T CCM, MIL:** Yes<br>**Years:** 2007, 2008<br>**Models:** A6<br>**Engines:** 3.2L, 4.2L<br>**Transmissions:** All | **TR Sensor Signal Intermittent Conditions:**<br>Key on or engine running; and the ECM detected the failure Trouble Code Conditions for DTC P0705 or P0708 were met intermittently.<br>**Possible Causes:**<br>• Refer to the appropriate Transmission Repair Manual or information in electronic media to perform a complete diagnosis of the automatic transmission when this code is set |

## Gas Engine OBD II Trouble Code List (P2xxx Codes)

| DTC | Trouble Code Title, Conditions & Possible Causes |
|-----|--------------------------------------------------|
| **DTC: P2004**<br>**2T CCM, MIL:** Yes<br>**Years:** 2007, 2008<br>**Models:** A6<br>**Engines:** 3.2L<br>**Transmissions:** All | **Intake Manifold Runner Control Stuck Open Bank 1 Conditions:**<br>Engine started, battery voltage must be at least 11.5v, all electrical components must be off, the ground between the engine and the chassis must be well connected. The ECM detected an unexpected voltage condition on the Intake Manifold Runner Control circuit during the CCM test period (i.e., the valve may be stuck open).<br>**Note: Intake Flap Motor and Intake Manifold Runner Position Sensor are one component and cannot be replaced individually.**<br>**Possible Causes:**<br>• Test for a sticking Accelerator or speed control cable condition: Turn the key off and disconnect accelerator and speed control cable from the throttle body. Rotate the throttle body linkage to determine if it rotates freely (the throttle body may have failed).<br>• Check the air cleaner and air inlet assembly for restrictions<br>• Check the IAC motor response (it may be damaged or sticking)<br>• Check the PCV system (valve and hoses) for leaks or plugging<br>• Check for signs of vacuum leaks in the engine or components<br>• Test TP sensor signal (due a sweep test at key on, engine off) |
| **DTC: P2008**<br>**2T CCM, MIL:** Yes<br>**Years:** 2007, 2008<br>**Models:** A6<br>**Engines:** 3.2L<br>**Transmissions:** All | **Intake Manifold Runner Control Circuit/Open Bank 1 Conditions:**<br>Engine started, battery voltage must be at least 11.5v, all electrical components must be off, the ground between the engine and the chassis must be well connected. The ECM detected an unexpected voltage condition on the Intake Manifold Runner Control circuit during the CCM test period (i.e., the valve may be stuck open).<br>**Note: Intake Flap Motor and Intake Manifold Runner Position Sensor are one component and cannot be replaced individually.**<br>**Possible Causes:**<br>• Accelerator or speed control cable sticking or binding. To test for this condition, turn the key off. Then disconnect the accelerator and speed control cable from the throttle body. Then rotate the throttle body linkage to determine if it rotates freely. If it is sticking, the throttle body may need replacement.<br>• Check the air cleaner and air inlet assembly for restrictions<br>• Check the IAC motor response (it may be damaged or sticking)<br>• Check the PCV system (valve and hoses) for leaks or plugging<br>• Check for signs of vacuum leaks in the engine or components<br>• Test TP sensor signal |

| DTC | Trouble Code Title, Conditions & Possible Causes |
|---|---|
| **DTC: P2009**<br>**2T CCM, MIL: Yes**<br>**Years: 2007, 2008**<br>**Models:** A6<br>**Engines:** 3.2L<br>**Transmissions:** All | **Intake Manifold Runner Control Circuit Low Bank 1 Conditions:**<br>Engine started, battery voltage must be at least 11.5v, all electrical components must be off, the ground between the engine and the chassis must be well connected. The ECM detected an unexpected voltage condition on the Intake Manifold Runner Control circuit during the CCM test period (i.e., the valve may be stuck open).<br>**Note: Intake Flap Motor and Intake Manifold Runner Position Sensor are one component and cannot be replaced individually.**<br>**Possible Causes:**<br>• Accelerator or speed control cable sticking or binding. To test for this condition, turn the key off. Then disconnect the accelerator and speed control cable from the throttle body. Then rotate the throttle body linkage to determine if it rotates freely. If it is sticking, the throttle body may need replacement.<br>• Check the air cleaner and air inlet assembly for restrictions<br>• Check the IAC motor response (it may be damaged or sticking)<br>• Check the PCV system (valve and hoses) for leaks or plugging<br>• Check for signs of vacuum leaks in the engine or components<br>• Test TP sensor signal |
| **DTC: P2014**<br>**2T CCM, MIL: Yes**<br>**Years: 2007, 2008**<br>**Models:** A6<br>**Engines:** 3.2L<br>**Transmissions:** All | **Intake Manifold Runner Position Sensor/Switch Circuit Bank 1 Conditions:**<br>Engine started, battery voltage must be at least 11.5v, all electrical components must be off, the ground between the engine and the chassis must be well connected. The ECM detected an unexpected voltage condition on the Intake Manifold Runner Control circuit during the CCM test period (i.e., the valve may be stuck open).<br>**Note: Intake Flap Motor and Intake Manifold Runner Position Sensor are one component and cannot be replaced individually.**<br>**Possible Causes:**<br>• Accelerator or speed control cable sticking or binding. To test for this condition, turn the key off. Then disconnect the accelerator and speed control cable from the throttle body. Then rotate the throttle body linkage to determine if it rotates freely. If it is sticking, the throttle body may need replacement.<br>• Check the air cleaner and air inlet assembly for restrictions<br>• Check the IAC motor response (it may be damaged or sticking)<br>• Check the PCV system (valve and hoses) for leaks or plugging<br>• Check for signs of vacuum leaks in the engine or components<br>• Test TP sensor signal |
| **DTC: P2015**<br>**2T CCM, MIL: Yes**<br>**Years: 2007, 2008**<br>**Models:** A6<br>**Engines:** 3.2L<br>**Transmissions::** All | **Intake Manifold Runner Position Sensor/Switch Circuit Range/Performance Bank 1 Conditions:**<br>Engine started, battery voltage must be at least 11.5v, all electrical components must be off, the ground between the engine and the chassis must be well connected. The ECM detected an unexpected voltage condition on the Intake Manifold Runner Control circuit during the CCM test period (i.e., the valve may be stuck open).<br>**Note: Intake Flap Motor and Intake Manifold Runner Position Sensor are one component and cannot be replaced individually.**<br>**Possible Causes:**<br>• Accelerator or speed control cable sticking or binding. To test for this condition, turn the key off. Then disconnect the accelerator and speed control cable from the throttle body. Then rotate the throttle body linkage to determine if it rotates freely. If it is sticking, the throttle body may need replacement.<br>• Check the air cleaner and air inlet assembly for restrictions<br>• Check the IAC motor response (it may be damaged or sticking)<br>• Check the PCV system (valve and hoses) for leaks or plugging<br>• Check for signs of vacuum leaks in the engine or components<br>• Test TP sensor signal |
| **DTC: P2017**<br>**2T CCM, MIL: Yes**<br>**Years: 2007, 2008**<br>**Models:** A6<br>**Engines:** 3.2L<br>**Transmissions:** All | **Intake Manifold Runner Position Sensor/Switch Circuit High Bank 1 Conditions:**<br>Engine started, battery voltage must be at least 11.5v, all electrical components must be off, the ground between the engine and the chassis must be well connected. The ECM detected an unexpected voltage condition on the Intake Manifold Runner Control circuit during the CCM test period (i.e., the valve may be stuck open).<br>**Note: Intake Flap Motor and Intake Manifold Runner Position Sensor are one component and cannot be replaced individually.**<br>**Possible Causes:**<br>• Accelerator or speed control cable sticking or binding. To test for this condition, turn the key off. Then disconnect the accelerator and speed control cable from the throttle body. Then rotate the throttle body linkage to determine if it rotates freely. If it is sticking, the throttle body may need replacement.<br>• Check the air cleaner and air inlet assembly for restrictions<br>• Check the IAC motor response (it may be damaged or sticking)<br>• Check the PCV system (valve and hoses) for leaks or plugging<br>• Check for signs of vacuum leaks in the engine or components<br>• Test TP sensor signal |

| DTC | Trouble Code Title, Conditions & Possible Causes |
|---|---|
| **DTC: P2088**<br>**2T CCM, MIL: Yes**<br>**Years: 2007, 2008**<br>**Models:** A6<br>**Engines:** 3.2L, 4.2L<br>**Transmissions:** All | **"A" Camshaft Position Control Circuit Low Bank 1 Conditions:**<br>Key on or engine running; and the ECM detected an unexpected voltage condition on the Camshaft Position Control circuit during the CCM test period. The relative position between the camshaft and crankshaft needs to be optimal so the engine has better torque, fuel economy and emissions.<br>**Note: camshaft adjustment is load- and RPM dependant. The electrical camshaft adjustment valve 1 switches oil pressure onto camshaft adjuster (mechanical adjustment mechanism), which adjusts the camshaft.**<br>**Possible Causes:**<br>• Camshaft position control wiring harness connector is damaged or open<br>• Camshaft adjustment valve has failed<br>• Circuit is open or grounded<br>• Assembly is damaged or it has failed (an open circuit)<br>• ECM power supply relay has failed<br>• ECM has failed |
| **DTC: P2089**<br>**2T CCM, MIL: Yes**<br>**Years: 2007, 2008**<br>**Models:** A6<br>**Engines:** 3.2L, 4.2L<br>**Transmissions:** All | **"A" Camshaft Position Control Circuit High Bank 1 Conditions:**<br>Key on or engine running; and the ECM detected an unexpected voltage condition on the Camshaft Position Control circuit during the CCM test period. The relative position between the camshaft and crankshaft needs to be optimal so the engine has better torque, fuel economy and emissions.<br>**Note: camshaft adjustment is load- and RPM dependant. The electrical camshaft adjustment valve 1 switches oil pressure onto camshaft adjuster (mechanical adjustment mechanism), which adjusts the camshaft.**<br>**Possible Causes:**<br>• Camshaft position control wiring harness connector is damaged or open<br>• Camshaft adjustment valve has failed<br>• Circuit is open or grounded<br>• Assembly is damaged or it has failed (an open circuit)<br>• ECM power supply relay has failed<br>• ECM has failed |
| **DTC: P2090**<br>**2T CCM, MIL: Yes**<br>**Years: 2007, 2008**<br>**Models:** A6<br>**Engines:** 3.2L<br>**Transmissions:** All | **"B" Camshaft Position Control Circuit Low Bank 1 Conditions:**<br>Key on or engine running; and the ECM detected an unexpected voltage condition on the Camshaft Position Control circuit during the CCM test period. The relative position between the camshaft and crankshaft needs to be optimal so the engine has better torque, fuel economy and emissions.<br>**Note: camshaft adjustment is load- and RPM dependant. The electrical camshaft adjustment valve 1 switches oil pressure onto camshaft adjuster (mechanical adjustment mechanism), which adjusts the camshaft.**<br>**Possible Causes:**<br>• Camshaft position control wiring harness connector is damaged or open<br>• Camshaft adjustment valve has failed<br>• Circuit is open or grounded<br>• Assembly is damaged or it has failed (an open circuit)<br>• ECM power supply relay has failed<br>• ECM has failed |
| **DTC: P2091**<br>**2T CCM, MIL: Yes**<br>**Years: 2007, 2008**<br>**Models:** A6<br>**Engines:** 3.2L<br>**Transmissions:** All | **"B" Camshaft Position Control Circuit High Bank 1 Conditions:**<br>Key on or engine running; and the ECM detected an unexpected voltage condition on the Camshaft Position Control circuit during the CCM test period. The relative position between the camshaft and crankshaft needs to be optimal so the engine has better torque, fuel economy and emissions.<br>**Note: camshaft adjustment is load- and RPM dependant. The electrical camshaft adjustment valve 1 switches oil pressure onto camshaft adjuster (mechanical adjustment mechanism), which adjusts the camshaft.**<br>**Possible Causes:**<br>• Camshaft position control wiring harness connector is damaged or open<br>• Camshaft adjustment valve has failed<br>• Circuit is open or grounded<br>• Assembly is damaged or it has failed (an open circuit)<br>• ECM power supply relay has failed<br>• ECM has failed |

| DTC | Trouble Code Title, Conditions & Possible Causes |
|---|---|
| **DTC: P2094**<br>**2T CCM, MIL: Yes**<br>**Years: 2007, 2008**<br>**Models:** A6<br>**Engines:** 3.2L<br>**Transmissions:** All | **"B" Camshaft Position Control Circuit Low Bank 2 Conditions:**<br>Key on or engine running; and the ECM detected an unexpected voltage condition on the Camshaft Position Control circuit during the CCM test period. The relative position between the camshaft and crankshaft needs to be optimal so the engine has better torque, fuel economy and emissions.<br>**Note: camshaft adjustment is load- and RPM dependant. The electrical camshaft adjustment valve 1 switches oil pressure onto camshaft adjuster (mechanical adjustment mechanism), which adjusts the camshaft.**<br>**Possible Causes:**<br>• Camshaft position control wiring harness connector is damaged or open<br>• Camshaft adjustment valve has failed<br>• Circuit is open or grounded<br>• Assembly is damaged or it has failed (an open circuit)<br>• ECM power supply relay has failed<br>• ECM has failed |
| **DTC: P2095**<br>**2T CCM, MIL: Yes**<br>**Years: 2007, 2008**<br>**Models:** A6<br>**Engines:** 3.2L<br>**Transmissions:** All | **"B" Camshaft Position Control Circuit High Bank 2 Conditions:**<br>Key on or engine running; and the ECM detected an unexpected voltage condition on the Camshaft Position Control circuit during the CCM test period. The relative position between the camshaft and crankshaft needs to be optimal so the engine has better torque, fuel economy and emissions.<br>**Note: camshaft adjustment is load- and RPM dependant. The electrical camshaft adjustment valve 1 switches oil pressure onto camshaft adjuster (mechanical adjustment mechanism), which adjusts the camshaft.**<br>**Possible Causes:**<br>• Camshaft position control wiring harness connector is damaged or open<br>• Camshaft adjustment valve has failed<br>• Circuit is open or grounded<br>• Assembly is damaged or it has failed (an open circuit)<br>• ECM power supply relay has failed<br>• ECM has failed |
| **DTC: P2096**<br>**2T CCM, MIL: Yes**<br>**Years: 2007, 2008**<br>**Models:** A4, A6<br>**Engines:** All<br>**Transmissions:** All | **Post Catalyst Fuel Trim System Too Lean (Bank 1) Conditions:**<br>Engine started, battery voltage must be at least 11.5v, all electrical components must be off, the ground between the engine and the chassis must be well connected, the exhaust system must be properly sealed between the catalytic converter and the cylinder head, and the oxygen sensor heater for oxygen sensor before the catalytic converter must be properly functioning. The ECM detected a problem with the fuel mixture.<br>**Note: For resistance testing of sensor heating, oxygen sensor should be cooled to ambient temperature. High temperatures at oxygen sensor may lead to inaccurate measurements.**<br>**Possible Causes:**<br>• Oxygen sensor (before catalytic converter) is faulty<br>• Oxygen sensor (behind catalytic converter) is faulty<br>• Oxygen sensor heater (before catalytic converter) is faulty<br>• Oxygen sensor heater (behind catalytic converter) is faulty<br>• Check circuits for shorts to each other, ground or power<br>• ECM has failed |
| **DTC: P2097**<br>**2T CCM, MIL: Yes**<br>**Years: 2007, 2008**<br>**Models:** A4, A6<br>**Engines:** All<br>**Transmissions:** All | **Post Catalyst Fuel Trim System Too Rich (Bank 1) Conditions:**<br>Engine started, battery voltage must be at least 11.5v, all electrical components must be off, the ground between the engine and the chassis must be well connected, the exhaust system must be properly sealed between the catalytic converter and the cylinder head, and the oxygen sensor heater for oxygen sensor before the catalytic converter must be properly functioning. The ECM detected a problem with the fuel mixture.<br>**Note: For resistance testing of sensor heating, oxygen sensor should be cooled to ambient temperature. High temperatures at oxygen sensor may lead to inaccurate measurements.**<br>**Possible Causes:**<br>• Oxygen sensor (before catalytic converter) is faulty<br>• Oxygen sensor (behind catalytic converter) is faulty<br>• Oxygen sensor heater (before catalytic converter) is faulty<br>• Oxygen sensor heater (behind catalytic converter) is faulty<br>• Check circuits for shorts to each other, ground or power<br>• ECM has failed |

| DTC | Trouble Code Title, Conditions & Possible Causes |
|---|---|
| **DTC: P2098**<br>**2T CCM, MIL: Yes**<br>**Years:** 2007, 2008<br>**Models:** A4, A6<br>**Engines:** All<br>**Transmissions:** All | **Post Catalyst Fuel Trim System Too Lean (Bank 2) Conditions:**<br>Engine started, battery voltage must be at least 11.5v, all electrical components must be off, the ground between the engine and the chassis must be well connected, the exhaust system must be properly sealed between the catalytic converter and the cylinder head, and the oxygen sensor heater for oxygen sensor before the catalytic converter must be properly functioning. The ECM detected a problem with the fuel mixture.<br>**Note: For resistance testing of sensor heating, oxygen sensor should be cooled to ambient temperature. High temperatures at oxygen sensor may lead to inaccurate measurements.**<br>**Possible Causes:**<br>• Oxygen sensor (before catalytic converter) is faulty<br>• Oxygen sensor (behind catalytic converter) is faulty<br>• Oxygen sensor heater (before catalytic converter) is faulty<br>• Oxygen sensor heater (behind catalytic converter) is faulty<br>• Check circuits for shorts to each other, ground or power<br>• ECM has failed |
| **DTC: P2099**<br>**2T CCM, MIL: Yes**<br>**Years:** 2007, 2008<br>**Models:** A4, A6<br>**Engines:** All<br>**Transmissions:** All | **Post Catalyst Fuel Trim System Too Rich (Bank 2) Conditions:**<br>Engine started, battery voltage must be at least 11.5v, all electrical components must be off, the ground between the engine and the chassis must be well connected, the exhaust system must be properly sealed between the catalytic converter and the cylinder head, and the oxygen sensor heater for oxygen sensor before the catalytic converter must be properly functioning. The ECM detected a problem with the fuel mixture.<br>**Note: For resistance testing of sensor heating, oxygen sensor should be cooled to ambient temperature. High temperatures at oxygen sensor may lead to inaccurate measurements.**<br>**Possible Causes:**<br>• Oxygen sensor (before catalytic converter) is faulty<br>• Oxygen sensor (behind catalytic converter) is faulty<br>• Oxygen sensor heater (before catalytic converter) is faulty<br>• Oxygen sensor heater (behind catalytic converter) is faulty<br>• Check circuits for shorts to each other, ground or power<br>• ECM has failed |
| **DTC: P2101**<br>**1T CCM, MIL: Yes**<br>**Years:** 2007, 2008<br>**Models:** A4, A6<br>**Engines:** All<br>**Transmissions:** All | **Throttle Actuator Control Motor Range/Performance Conditions:**<br>Engine started, battery voltage must be at least 11.5v, all electrical components must be off, parking brake must be engaged (to keep daytime driving lights off), automatic transmission selector must be in park, the exhaust system must be properly sealed between the catalytic converter and the cylinder head, coolant temperature must be at least 80 degrees Celsius. The ECM detected an unexpected low or high voltage condition on the Throttle Actuator Control Motor (TACM) circuit during the CCM test.<br>**Note: The throttle valve activation occurs via an electric motor (throttle drive) in the throttle valve control module. It is activated by the Engine Control Module (ECM) according to specifications of the two sensors, Throttle Position (TP) Sensor and Sender 2 for accelerator pedal position.**<br>**Possible Causes:**<br>• TACM wiring harness connector is damaged or open<br>• TACM wiring may be crossed in the wire harness assembly<br>• TACM (motor) circuit is open, or TACM assembly is damaged (possible open circuit)<br>• TACM or the Throttle Valve is dirty<br>• Throttle Position sensor has failed<br>• ECM has failed |
| **DTC: P2106**<br>**1T CCM, MIL: Yes**<br>**Years:** 2007, 2008<br>**Models:** A4, A6<br>**Engines:** All<br>**Transmissions:** All | **Throttle Actuator Control System – Forced Limited Power Conditions**<br>Engine started, battery voltage must be at least 11.5v, all electrical components must be off, parking brake must be engaged (to keep daytime driving lights off), automatic transmission selector must be in park, the exhaust system must be properly sealed between the catalytic converter and the cylinder head, coolant temperature must be at least 80 degrees Celsius. The ECM detected an unexpected low or high voltage condition on the Throttle Actuator Control Motor (TACM) circuit during the CCM test.<br>**Note: The throttle valve activation occurs via an electric motor (throttle drive) in the throttle valve control module. It is activated by the Engine Control Module (ECM) according to specifications of the two sensors, Throttle Position (TP) Sensor and Sender 2 for accelerator pedal position.**<br>**Possible Causes:**<br>• TACM wiring harness connector is damaged or open<br>• TACM wiring may be crossed in the wire harness assembly<br>• TACM (motor) circuit is open, or TACM assembly is damaged (possible open circuit)<br>• TACM or the Throttle Valve is dirty<br>• Throttle Position sensor has failed<br>• ECM has failed |

| DTC | Trouble Code Title, Conditions & Possible Causes |
|---|---|
| **DTC: P2122**<br>**1T CCM, MIL: Yes**<br>**Years: 2007, 2008**<br>**Models:** A4, A6<br>**Engines:** All<br>**Transmissions:** All | **Accelerator Pedal Position Sensor 'D' Circuit Low Input Conditions:**<br>Engine started, battery voltage at least 11.5v, all electrical components off, ground connections between engine and chassis well connected, the ECM detected that the accelerator pedal position sensor signal was outside the parameters to function normally.<br>**Note: Both the Throttle Position (TP) Sensor and Accelerator Pedal Position Sensor are located at the accelerator pedal module and communicate the driver's intentions to the ECM completely independently of each other. Both sensors are stored in one housing.**<br>**Possible Causes:**<br>• Ground between engine and chassis may be broken<br>• Throttle position sensor may have failed<br>• Accelerator Pedal Position Sensor has failed<br>• Throttle position sensor wiring may have shorted<br>• Throttle position sensor has failed<br>• Faulty voltage supply<br>• ECM has failed |
| **DTC: P2123**<br>**1T CCM, MIL: Yes**<br>**Years: 2007, 2008**<br>**Models:** A4, A6<br>**Engines:** All<br>**Transmissions:** All | **Accelerator Pedal Position Sensor 'D' Circuit High Input Conditions:**<br>Engine started, battery voltage at least 11.5v, all electrical components off, ground connections between engine and chassis well connected, the ECM detected that the accelerator pedal position sensor signal was outside the parameters to function normally.<br>**Note: Both the Throttle Position (TP) Sensor and Accelerator Pedal Position Sensor are located at the accelerator pedal module and communicate the driver's intentions to the ECM completely independently of each other. Both sensors are stored in one housing.**<br>**Possible Causes:**<br>• Ground between engine and chassis may be broken<br>• Throttle position sensor may have failed<br>• Accelerator Pedal Position Sensor has failed<br>• Throttle position sensor wiring may have shorted<br>• Throttle position sensor has failed<br>• Faulty voltage supply<br>• ECM has failed |
| **DTC: P2127**<br>**1T CCM, MIL: Yes**<br>**Years: 2007, 2008**<br>**Models:** A4, A6<br>**Engines:** All<br>**Transmissions:** All | **Accelerator Pedal Position Sensor 'E' Circuit Low Input Conditions:**<br>Engine started, battery voltage at least 11.5v, all electrical components off, ground connections between engine and chassis well connected, the ECM detected that the accelerator pedal position sensor signal was outside the parameters to function normally.<br>**Note: Both the Throttle Position (TP) Sensor and Accelerator Pedal Position Sensor are located at the accelerator pedal module and communicate the driver's intentions to the ECM completely independently of each other. Both sensors are stored in one housing.**<br>**Possible Causes:**<br>• Ground between engine and chassis may be broken<br>• Throttle position sensor may have failed<br>• Accelerator Pedal Position Sensor has failed<br>• Throttle position sensor wiring may have shorted<br>• Throttle position sensor has failed<br>• Faulty voltage supply<br>• ECM has failed |
| **DTC: P2128**<br>**1T CCM, MIL: Yes**<br>**Years: 2007, 2008**<br>**Models:** A4, A6<br>**Engines:** All<br>**Transmissions:** All | **Accelerator Pedal Position Sensor 'E' Circuit High Input Conditions:**<br>Engine started, battery voltage at least 11.5v, all electrical components off, ground connections between engine and chassis well connected, the ECM detected that the accelerator pedal position sensor signal was outside the parameters to function normally.<br>**Note: Both the Throttle Position (TP) Sensor and Accelerator Pedal Position Sensor are located at the accelerator pedal module and communicate the driver's intentions to the ECM completely independently of each other. Both sensors are stored in one housing.**<br>**Possible Causes:**<br>• Ground between engine and chassis may be broken<br>• Throttle position sensor may have failed<br>• Accelerator Pedal Position Sensor has failed<br>• Throttle position sensor wiring may have shorted<br>• Throttle position sensor has failed<br>• Faulty voltage supply<br>• ECM has failed |

| DTC | Trouble Code Title, Conditions & Possible Causes |
|---|---|
| **DTC: P2138**<br>**1T CCM, MIL: Yes**<br>**Years:** 2007, 2008<br>**Models:** A4, A6<br>**Engines:** All<br>**Transmissions:** All | **Throttle Position Sensor D/E Voltage Correlation Conditions:**<br>Engine started, battery voltage must be at least 11.5v, all electrical components must be off, parking brake must be engaged (to keep daytime driving lights off), automatic transmission selector must be in park; and the ECM detected the Throttle Position 'D' (TPD) and Throttle Position 'B' (TPE) sensors disagreed, or that the TPD sensor should not be in its detected position, or that the TPE sensor should not be in its detected position during testing.<br>**Note: Both the Throttle Position (TP) Sensor and Accelerator Pedal Position Sensor are located at the accelerator pedal module and communicate the driver's intentions to the ECM completely independently of each other. Both sensors are stored in one housing.**<br>**Possible Causes:**<br>• ETC TP sensor connector is damaged or shorted<br>• ETC TP sensor circuits shorted together in the wire harness<br>• ETC TP sensor signal circuit is shorted to VREF (5v)<br>• ETC TP sensor is damaged or the ECM has failed |
| **DTC: P2181**<br>**1T CCM, MIL: Yes**<br>**Years:** 2007, 2008<br>**Models:** A4, A6<br>**Engines:** All<br>**Transmissions:** All | **Cooling System Performance Malfunction Conditions:**<br>Key on, engine cold; and the Engine Coolant Temperature (ECM) detected the ECT sensor signal was more or less than the self-test limits or has failed to gain a signal. This is a thermistor-type sensor with a variable resistance that changes when exposed to different temperatures<br>**Possible Causes:**<br>• ECT sensor has failed<br>• ECT Sensor (on Radiator) has failed<br>• ECT sensor signal circuit is open (inspect wiring & connector)<br>• ECT sensor signal circuit is shorted<br>• Cooling system malfunction, or the thermostat is stuck<br>• Engine not operating at normal operating temperature<br>• EOT sensor is damaged or it has failed |
| **DTC: P2195**<br>**1T CCM, MIL: Yes**<br>**Years:** 2007, 2008<br>**Models:** A6<br>**Engines:** 3.2L, 4.2L<br>**Transmissions:** All | **O2 Sensor Signal Stuck Lean Bank 1 Sensor 1 Conditions:**<br>Engine running in closed loop, and the ECM detected the O2S indicated a lean signal, or it could no longer control Fuel Trim because it was at lean limit.<br>**Possible Causes:**<br>• Engine oil level high<br>• Camshaft timing error<br>• Cylinder compression low<br>• Exhaust leaks in front of O2S<br>• EGR valve is stuck open<br>• EGR gasket is leaking<br>• EVR diaphragm is leaking<br>• Damaged fuel pressure regulator or extremely low fuel pressure<br>• O2S circuit is open or shorted in the wiring harness<br>• Oxygen sensor (before catalytic converter) is faulty<br>• Oxygen sensor (behind catalytic converter) is faulty<br>• Oxygen sensor heater (before catalytic converter) is faulty<br>• Oxygen sensor heater (behind catalytic converter) is faulty<br>• Air leaks after the MAF sensor<br>• PCV system leaks<br>• Dip stick not seated properly |

| DTC | Trouble Code Title, Conditions & Possible Causes |
|---|---|
| **DTC: P2196**<br>**1T CCM, MIL: Yes**<br>**Years: 2007, 2008**<br>**Models:** A6<br>**Engines:** 3.2L, 4.2L<br>**Transmissions:** All | **O2 Sensor Signal Stuck Rich Bank 1 Sensor 1 Conditions:**<br>Engine running in closed loop, and the ECM detected the O2S indicated a rich signal, or it could no longer control Fuel Trim because it was at its rich limit.<br>**Possible Causes:**<br>• Engine oil level high<br>• Camshaft timing error<br>• Cylinder compression low<br>• Exhaust leaks in front of O2S<br>• EGR valve is stuck open<br>• EGR gasket is leaking<br>• EVR diaphragm is leaking<br>• Damaged fuel pressure regulator or extremely low fuel pressure<br>• O2S circuit is open or shorted in the wiring harness<br>• Oxygen sensor (before catalytic converter) is faulty<br>• Oxygen sensor (behind catalytic converter) is faulty<br>• Oxygen sensor heater (before catalytic converter) is faulty<br>• Oxygen sensor heater (behind catalytic converter) is faulty<br>• Air leaks after the MAF sensor<br>• PCV system leaks<br>• Dip stick not seated properly |
| **DTC: P2231**<br>**1T CCM, MIL: Yes**<br>**Years: 2007, 2008**<br>**Models:** A6<br>**Engines:** 3.2L, 4.2L<br>**Transmissions:** All | **O2 Sensor Signal Circuit Shorted to Heater Circuit Bank 1 Sensor 1 Conditions:**<br>Engine started, battery voltage must be at least 11.5v, all electrical components must be off, parking brake must be engaged (to keep daytime driving lights off), automatic transmission selector must be in park. The ECM detected an unexpected voltage condition, or it detected an unexpected current draw in the sensor circuit during the CCM test.<br>**Note: Vehicle must be raised before connector for oxygen sensors is accessible.**<br>**Possible Causes:**<br>• Oxygen sensor (before catalytic converter) is faulty<br>• Oxygen sensor heater (before catalytic converter) is faulty<br>• Oxygen sensor heater (before catalytic converter) is faulty<br>• Oxygen sensor heater (behind catalytic converter) is faulty<br>• O2S circuit is open or shorted in the wiring harness<br>• ECM has failed |
| **DTC: P2234**<br>**1T CCM, MIL: Yes**<br>**Years: 2007, 2008**<br>**Models:** A6<br>**Engines:** 3.2L, 4.2L<br>**Transmissions:** All | **O2 Sensor Signal Circuit Shorted to Heater Circuit Bank 2 Sensor 1 Conditions:**<br>Engine started, battery voltage must be at least 11.5v, all electrical components must be off, parking brake must be engaged (to keep daytime driving lights off), automatic transmission selector must be in park. The ECM detected an unexpected voltage condition, or it detected an unexpected current draw in the sensor circuit during the CCM test.<br>**Note: Vehicle must be raised before connector for oxygen sensors is accessible.**<br>**Possible Causes:**<br>• Oxygen sensor (before catalytic converter) is faulty<br>• Oxygen sensor heater (before catalytic converter) is faulty<br>• Oxygen sensor heater (before catalytic converter) is faulty<br>• Oxygen sensor heater (behind catalytic converter) is faulty<br>• O2S circuit is open or shorted in the wiring harness<br>• ECM has failed |
| **DTC: P2237**<br>**1T CCM, MIL: Yes**<br>**Years: 2007, 2008**<br>**Models:** A6<br>**Engines:** 3.2L, 4.2L<br>**Transmissions:** All | **O2 Sensor Positive Current Control Circuit/Open Bank 1 Sensor 1 Conditions:**<br>Engine started, battery voltage must be at least 11.5v, all electrical components must be off, parking brake must be engaged (to keep daytime driving lights off), automatic transmission selector must be in park. The ECM detected an unexpected voltage condition, or it detected an unexpected current draw in the sensor circuit during the CCM test.<br>**Note: Vehicle must be raised before connector for oxygen sensors is accessible.**<br>**Possible Causes:**<br>• Oxygen sensor (before catalytic converter) is faulty<br>• Oxygen sensor heater (before catalytic converter) is faulty<br>• Oxygen sensor heater (before catalytic converter) is faulty<br>• Oxygen sensor heater (behind catalytic converter) is faulty<br>• O2S circuit is open or shorted in the wiring harness<br>• ECM has failed |

| DTC | Trouble Code Title, Conditions & Possible Causes |
|-----|---------------------------------------------------|
| **DTC: P2240**<br>**1T CCM, MIL: Yes**<br>**Years:** 2007, 2008<br>**Models:** A6<br>**Engines:** 3.2L, 4.2L<br>**Transmissions:** All | **O2 Sensor Positive Current Control Circuit/Open Bank 2 Sensor 1 Conditions:**<br>Engine started, battery voltage must be at least 11.5v, all electrical components must be off, parking brake must be engaged (to keep daytime driving lights off), automatic transmission selector must be in park. The ECM detected an unexpected voltage condition, or it detected an unexpected current draw in the sensor circuit during the CCM test.<br>**Note: Vehicle must be raised before connector for oxygen sensors is accessible.**<br>**Possible Causes:**<br>• Oxygen sensor (before catalytic converter) is faulty<br>• Oxygen sensor heater (before catalytic converter) is faulty<br>• Oxygen sensor heater (before catalytic converter) is faulty<br>• Oxygen sensor heater (behind catalytic converter) is faulty<br>• O2S circuit is open or shorted in the wiring harness<br>• ECM has failed |
| **DTC: P2243**<br>**1T CCM, MIL: Yes**<br>**Years:** 2007, 2008<br>**Models:** A6<br>**Engines:** 3.2L, 4.2L<br>**Transmissions:** All | **O2 Sensor Reference Voltage Circuit/Open Bank 1 Sensor 1 Conditions:**<br>Engine started, battery voltage must be at least 11.5v, all electrical components must be off, parking brake must be engaged (to keep daytime driving lights off), automatic transmission selector must be in park. The ECM detected an unexpected voltage condition, or it detected an unexpected current draw in the sensor circuit during the CCM test.<br>**Note: Vehicle must be raised before connector for oxygen sensors is accessible.**<br>**Possible Causes:**<br>• Oxygen sensor (before catalytic converter) is faulty<br>• Oxygen sensor heater (before catalytic converter) is faulty<br>• Oxygen sensor heater (before catalytic converter) is faulty<br>• Oxygen sensor heater (behind catalytic converter) is faulty<br>• O2S circuit is open or shorted in the wiring harness<br>• ECM has failed |
| **DTC: P2247**<br>**1T CCM, MIL: Yes**<br>**Years:** 2007, 2008<br>**Models:** A6<br>**Engines:** 3.2L, 4.2L<br>**Transmissions:** All | **O2 Sensor Reference Voltage Circuit/Open Bank 2 Sensor 1 Conditions:**<br>Engine started, battery voltage must be at least 11.5v, all electrical components must be off, parking brake must be engaged (to keep daytime driving lights off), automatic transmission selector must be in park. The ECM detected an unexpected voltage condition, or it detected an unexpected current draw in the sensor circuit during the CCM test.<br>**Note: Vehicle must be raised before connector for oxygen sensors is accessible.**<br>**Possible Causes:**<br>• Oxygen sensor (before catalytic converter) is faulty<br>• Oxygen sensor heater (before catalytic converter) is faulty<br>• Oxygen sensor heater (before catalytic converter) is faulty<br>• Oxygen sensor heater (behind catalytic converter) is faulty<br>• O2S circuit is open or shorted in the wiring harness<br>• ECM has failed |
| **DTC: P2251**<br>**1T CCM, MIL: Yes**<br>**Years:** 2007, 2008<br>**Models:** A6<br>**Engines:** 3.2L, 4.2L<br>**Transmissions:** All | **O2 Sensor Negative Voltage Circuit/Open Bank 1 Sensor 1 Conditions:**<br>Engine started, battery voltage must be at least 11.5v, all electrical components must be off, parking brake must be engaged (to keep daytime driving lights off), automatic transmission selector must be in park. The ECM detected an unexpected voltage condition, or it detected an unexpected current draw in the sensor circuit during the CCM test.<br>**Note: Vehicle must be raised before connector for oxygen sensors is accessible.**<br>**Possible Causes:**<br>• Oxygen sensor (before catalytic converter) is faulty<br>• Oxygen sensor heater (before catalytic converter) is faulty<br>• Oxygen sensor heater (before catalytic converter) is faulty<br>• Oxygen sensor heater (behind catalytic converter) is faulty<br>• O2S circuit is open or shorted in the wiring harness<br>• ECM has failed |
| **DTC: P2254**<br>**1T CCM, MIL: Yes**<br>**Years:** 2007, 2008<br>**Models:** A6<br>**Engines:** 3.2L, 4.2L<br>**Transmissions:** All | **O2 Sensor Negative Voltage Circuit/Open Bank 2 Sensor 1 Conditions:**<br>Engine started, battery voltage must be at least 11.5v, all electrical components must be off, parking brake must be engaged (to keep daytime driving lights off), automatic transmission selector must be in park. The ECM detected an unexpected voltage condition, or it detected an unexpected current draw in the sensor circuit during the CCM test.<br>**Note: Vehicle must be raised before connector for oxygen sensors is accessible.**<br>**Possible Causes:**<br>• Oxygen sensor (before catalytic converter) is faulty<br>• Oxygen sensor heater (before catalytic converter) is faulty<br>• Oxygen sensor heater (before catalytic converter) is faulty<br>• Oxygen sensor heater (behind catalytic converter) is faulty<br>• O2S circuit is open or shorted in the wiring harness<br>• ECM has failed |

| DTC | Trouble Code Title, Conditions & Possible Causes |
|---|---|
| **DTC: P2257**<br>**1T CCM, MIL: Yes**<br>**Years: 2007, 2008**<br>**Models:** A4, A6<br>**Engines:** All<br>**Transmissions:** All | **Secondary Air Injection System Control "A" Circuit Low Conditions:**<br>Engine started, battery voltage must be at least 11.5v, all electrical components must be off, parking brake must be engaged (to keep daytime driving lights off), automatic transmission selector must be in park and the ground between the engine and the chassis must be well connected. The ECM detected an unexpected voltage condition on the AIR system control circuit during testing.<br>**Possible Causes:**<br>• AIR solenoid power circuit (B+) is open (check dedicated fuse)<br>• AIR bypass solenoid control circuit is open or shorted to ground<br>• AIR diverter solenoid control circuit open or shorted to ground<br>• AIR pump control circuit is open or shorted to ground<br>• Check valve (one or more) is damaged or leaking<br>• Solid State relay is damaged or it has failed<br>• Check activation of Secondary Air Injection (AIR) Pump Relay<br>• ECM has failed |
| **DTC: P2258**<br>**1T CCM, MIL: Yes**<br>**Years: 2007, 2008**<br>**Models:** A4, A6<br>**Engines:** All<br>**Transmissions:** All | **Secondary Air Injection System Control "A" Circuit High Conditions:**<br>Engine started, battery voltage must be at least 11.5v, all electrical components must be off, parking brake must be engaged (to keep daytime driving lights off), automatic transmission selector must be in park and the ground between the engine and the chassis must be well connected. The ECM detected an unexpected voltage condition on the AIR system control circuit during testing.<br>**Possible Causes:**<br>• AIR solenoid power circuit (B+) is open (check dedicated fuse)<br>• AIR bypass solenoid control circuit is open or shorted to ground<br>• AIR diverter solenoid control circuit open or shorted to ground<br>• AIR pump control circuit is open or shorted to ground<br>• Check valve (one or more) is damaged or leaking<br>• Solid State relay is damaged or it has failed<br>• Check activation of Secondary Air Injection (AIR) Pump Relay<br>• ECM has failed |
| **DTC: P2270**<br>**1T CCM, MIL: Yes**<br>**Years: 2007, 2008**<br>**Models:** A6<br>**Engines:** 3.2L<br>**Transmissions:** All | **O2 Sensor Signal Stuck Lean Bank 1 Sensor 2 Conditions:**<br>Engine started, battery voltage must be at least 11.5v, all electrical components must be off, parking brake must be engaged (to keep daytime driving lights off), automatic transmission selector must be in park. The ECM detected an unexpected voltage condition, or it detected an unexpected current draw in the heater circuit during the CCM test.<br>**Note: Vehicle must be raised before connector for oxygen sensors is accessible.**<br>**Possible Causes:**<br>• Oxygen sensor (before catalytic converter) is faulty<br>• Oxygen sensor heater (before catalytic converter) is faulty<br>• Oxygen sensor heater (before catalytic converter) is faulty<br>• Oxygen sensor heater (behind catalytic converter) is faulty<br>• O2S circuit is open or shorted in the wiring harness<br>• ECM has failed |
| **DTC: P2271**<br>**1T CCM, MIL: Yes**<br>**Years: 2007, 2008**<br>**Models:** A6<br>**Engines:** 3.2L<br>**Transmissions:** All | **O2 Sensor Signal Stuck Rich Bank 1 Sensor 2 Conditions:**<br>Engine started, battery voltage must be at least 11.5v, all electrical components must be off, parking brake must be engaged (to keep daytime driving lights off), automatic transmission selector must be in park. The ECM detected an unexpected voltage condition, or it detected an unexpected current draw in the heater circuit during the CCM test.<br>**Note: Vehicle must be raised before connector for oxygen sensors is accessible.**<br>**Possible Causes:**<br>• Oxygen sensor (before catalytic converter) is faulty<br>• Oxygen sensor heater (before catalytic converter) is faulty<br>• Oxygen sensor heater (before catalytic converter) is faulty<br>• Oxygen sensor heater (behind catalytic converter) is faulty<br>• O2S circuit is open or shorted in the wiring harness<br>• ECM has failed |
| **DTC: P2272**<br>**1T CCM, MIL: Yes**<br>**Years: 2007, 2008**<br>**Models:** A6<br>**Engines:** 3.2L<br>**Transmissions:** All | **O2 Sensor Signal Stuck Lean Bank 2 Sensor 2 Conditions:**<br>Engine started, battery voltage must be at least 11.5v, all electrical components must be off, parking brake must be engaged (to keep daytime driving lights off), automatic transmission selector must be in park. The ECM detected an unexpected voltage condition, or it detected an unexpected current draw in the heater circuit during the CCM test.<br>**Note: Vehicle must be raised before connector for oxygen sensors is accessible.**<br>**Possible Causes:**<br>• Oxygen sensor (before catalytic converter) is faulty<br>• Oxygen sensor heater (before catalytic converter) is faulty<br>• Oxygen sensor heater (before catalytic converter) is faulty<br>• Oxygen sensor heater (behind catalytic converter) is faulty<br>• O2S circuit is open or shorted in the wiring harness<br>• ECM has failed |

| DTC | Trouble Code Title, Conditions & Possible Causes |
|---|---|
| **DTC: P2273**<br>**1T CCM, MIL: Yes**<br>**Years: 2007, 2008**<br>**Models: A6**<br>**Engines: 3.2L**<br>**Transmissions: All** | **O2 Sensor Signal Stuck Rich Bank 2 Sensor 2 Conditions:**<br>Engine started, battery voltage must be at least 11.5v, all electrical components must be off, parking brake must be engaged (to keep daytime driving lights off), automatic transmission selector must be in park. The ECM detected an unexpected voltage condition, or it detected an unexpected current draw in the heater circuit during the CCM test.<br>**Note: Vehicle must be raised before connector for oxygen sensors is accessible.**<br>**Possible Causes:**<br>• Oxygen sensor (before catalytic converter) is faulty<br>• Oxygen sensor heater (before catalytic converter) is faulty<br>• Oxygen sensor heater (before catalytic converter) is faulty<br>• Oxygen sensor heater (behind catalytic converter) is faulty<br>• O2S circuit is open or shorted in the wiring harness<br>• ECM has failed |
| **DTC: P2294**<br>**2T CCM, MIL: Yes**<br>**Years: 2007, 2008**<br>**Models: A6**<br>**Engines: 3.2L**<br>**Transmissions: All** | **Fuel Pressure Regulator 2 Control Circuit Conditions:**<br>Engine started, battery voltage at least 11.5v, all electrical components off, ground connections between engine and chassis well connected, coolant temperature at least 80-degrees Celicius. The ECM detected a voltage condition that affected the performance of the fule pressure regulator.<br>**Possible Causes:**<br>• Fuel Pressure Regulator Valve has failed<br>• Fuel Pressure Sensor has failed<br>• Fuel Pump (FP) Control Module has failed<br>• Fuel pump has failed<br>• ECM has failed |
| **DTC: P2295**<br>**2T CCM, MIL: Yes**<br>**Years: 2007, 2008**<br>**Models: A6**<br>**Engines: 3.2L**<br>**Transmissions: All** | **Fuel Pressure Regulator 2 Control Circuit Low Conditions:**<br>Engine started, battery voltage at least 11.5v, all electrical components off, ground connections between engine and chassis well connected, coolant temperature at least 80-degrees Celicius. The ECM detected a voltage condition that affected the performance of the fule pressure regulator.<br>**Possible Causes:**<br>• Fuel Pressure Regulator Valve has failed<br>• Fuel Pressure Sensor has failed<br>• Fuel Pump (FP) Control Module has failed<br>• Fuel pump has failed<br>• ECM has failed |
| **DTC: P2296**<br>**2T CCM, MIL: Yes**<br>**Years: 2007, 2008**<br>**Models: A6**<br>**Engines: 3.2L**<br>**Transmissions: All** | **Fuel Pressure Regulator 2 Control Circuit High Conditions:**<br>Engine started, battery voltage at least 11.5v, all electrical components off, ground connections between engine and chassis well connected, coolant temperature at least 80-degrees Celicius. The ECM detected a voltage condition that affected the performance of the fule pressure regulator.<br>**Possible Causes:**<br>• Fuel Pressure Regulator Valve has failed<br>• Fuel Pressure Sensor has failed<br>• Fuel Pump (FP) Control Module has failed<br>• Fuel pump has failed<br>• ECM has failed |
| **DTC: P2400**<br>**2T CCM, MIL: Yes**<br>**Years: 2007, 2008**<br>**Models: A4, A6**<br>**Engines: All**<br>**Transmissions: All** | **EVAP Leak Detection Pump (LDP) Control Circuit Open Conditions:**<br>Engine started, battery voltage must be at least 11.5v, all electrical components must be off, parking brake must be engaged (to keep daytime driving lights off), automatic transmission selector must be in park, the exhaust system must be properly sealed between the catalytic converter and the cylinder head, coolant temperature must be at least 80 degrees Celsius and oxygen sensor heaters for oxygen sensors before the catalytic converter must be functioning properly and the ground between the engine and the chassis must be well connected. The ECM detected voltage irregularity in the leak detection pump control circuit.<br>**Possible Causes:**<br>• EVAP LDP power supply circuit is open<br>• EVAP LDP solenoid valve is damaged or it has failed<br>• EVAP LDP canister has a leak or a poor seal<br>• ECM has failed<br>• EVAP canister system has an improper seal<br>• Evaporative Emission (EVAP) canister purge regulator valve 1 has failed<br>• Leak Detection Pump (LDP) is faulty<br>• Aftermarket EVAP parts that do not conform to specifications<br>• EVAP component seals leaking (i.e., leaks in the Purge valve, fuel tank pressure sensor, canister vent solenoid, fuel vapor control valve tube assembly or fuel vapor vent valve). |

| DTC | Trouble Code Title, Conditions & Possible Causes |
|---|---|
| **DTC: P2401**<br>**2T CCM, MIL: Yes**<br>**Years:** 2007, 2008<br>**Models:** A4, A6<br>**Engines:** All<br>**Transmissions:** All | **EVAP Leak Detection Pump Control Circuit Low Conditions:**<br>Engine started, battery voltage must be at least 11.5v, all electrical components must be off, parking brake must be engaged (to keep daytime driving lights off), automatic transmission selector must be in park, the exhaust system must be properly sealed between the catalytic converter and the cylinder head, coolant temperature must be at least 80 degrees Celsius and oxygen sensor heaters for oxygen sensors before the catalytic converter must be functioning properly and the ground between the engine and the chassis must be well connected. The ECM detected voltage irregularity in the leak detection pump control circuit.<br>**Possible Causes:**<br>• EVAP LDP power supply circuit is open<br>• EVAP LDP solenoid valve is damaged or it has failed<br>• EVAP LDP canister has a leak or a poor seal<br>• ECM has failed<br>• EVAP canister system has an improper seal<br>• Evaporative Emission (EVAP) canister purge regulator valve 1 has failed<br>• Leak Detection Pump (LDP) is faulty<br>• Aftermarket EVAP parts that do not conform to specifications<br>• EVAP component seals leaking (i.e., leaks in the Purge valve, fuel tank pressure sensor, canister vent solenoid, fuel vapor control valve tube assembly or fuel vapor vent valve). |
| **DTC: P2402**<br>**2T CCM, MIL: Yes**<br>**Years:** 2007, 2008<br>**Models:** A4, A6<br>**Engines:** All<br>**Transmissions:** All | **EVAP Leak Detection Pump Control Circuit High Conditions:**<br>Engine started, battery voltage must be at least 11.5v, all electrical components must be off, parking brake must be engaged (to keep daytime driving lights off), automatic transmission selector must be in park, the exhaust system must be properly sealed between the catalytic converter and the cylinder head, coolant temperature must be at least 80 degrees Celsius and oxygen sensor heaters for oxygen sensors before the catalytic converter must be functioning properly and the ground between the engine and the chassis must be well connected. The ECM detected voltage irregularity in the leak detection pump control circuit.<br>**Possible Causes:**<br>• EVAP LDP power supply circuit is open<br>• EVAP LDP solenoid valve is damaged or it has failed<br>• EVAP LDP canister has a leak or a poor seal<br>• ECM has failed<br>• EVAP canister system has an improper seal<br>• Evaporative Emission (EVAP) canister purge regulator valve 1 has failed<br>• Leak Detection Pump (LDP) is faulty<br>• Aftermarket EVAP parts that do not conform to specifications<br>• EVAP component seals leaking (i.e., leaks in the Purge valve, fuel tank pressure sensor, canister vent solenoid, fuel vapor control valve tube assembly or fuel vapor vent valve). |
| **DTC: P2403**<br>**2T CCM, MIL: Yes**<br>**Years:** 2007, 2008<br>**Models:** A4, A6<br>**Engines:** All<br>**Transmissions:** All | **EVAP Leak Detection Pump Sense Circuit Open Conditions:**<br>Engine started, battery voltage must be at least 11.5v, all electrical components must be off, parking brake must be engaged (to keep daytime driving lights off), automatic transmission selector must be in park, the exhaust system must be properly sealed between the catalytic converter and the cylinder head, coolant temperature must be at least 80 degrees Celsius and oxygen sensor heaters for oxygen sensors before the catalytic converter must be functioning properly and the ground between the engine and the chassis must be well connected. The ECM detected voltage irregularity in the leak detection pump control circuit.<br>**Possible Causes:**<br>• EVAP LDP power supply circuit is open<br>• EVAP LDP solenoid valve is damaged or it has failed<br>• EVAP LDP canister has a leak or a poor seal<br>• ECM has failed<br>• EVAP canister system has an improper seal<br>• Evaporative Emission (EVAP) canister purge regulator valve 1 has failed<br>• Leak Detection Pump (LDP) is faulty<br>• Aftermarket EVAP parts that do not conform to specifications<br>• EVAP component seals leaking (i.e., leaks in the Purge valve, fuel tank pressure sensor, canister vent solenoid, fuel vapor control valve tube assembly or fuel vapor vent valve). |

| DTC | Trouble Code Title, Conditions & Possible Causes |
|---|---|
| **DTC: P2404**<br>**2T CCM, MIL: Yes**<br>**Years:** 2007, 2008<br>**Models:** A4, A6<br>**Engines:** All<br>**Transmissions:** All | **EVAP Leak Detection Pump Sense Circuit Range/Performance Conditions:**<br>Engine started, battery voltage must be at least 11.5v, all electrical components must be off, parking brake must be engaged (to keep daytime driving lights off), automatic transmission selector must be in park, the exhaust system must be properly sealed between the catalytic converter and the cylinder head, coolant temperature must be at least 80 degrees Celsius and oxygen sensor heaters for oxygen sensors before the catalytic converter must be functioning properly and the ground between the engine and the chassis must be well connected. The ECM detected voltage irregularity in the leak detection pump control circuit.<br>**Possible Causes:**<br>• EVAP LDP power supply circuit is open<br>• EVAP LDP solenoid valve is damaged or it has failed<br>• EVAP LDP canister has a leak or a poor seal<br>• ECM has failed<br>• EVAP canister system has an improper seal<br>• Evaporative Emission (EVAP) canister purge regulator valve 1 has failed<br>• Leak Detection Pump (LDP) is faulty<br>• Aftermarket EVAP parts that do not conform to specifications<br>• EVAP component seals leaking (i.e., leaks in the Purge valve, fuel tank pressure sensor, canister vent solenoid, fuel vapor control valve tube assembly or fuel vapor vent valve). |
| **DTC: P2414**<br>**1T CCM, MIL: Yes**<br>**Years:** 2007, 2008<br>**Models:** A6<br>**Engines:** 3.2L, 4.2L<br>**Transmissions:** All | **O2 Sensor Exhaust Sample Error Bank 1 Sensor 1 Conditions:**<br>Engine running (ground connections between the engine and the chassis must be well connected), and the ECM detected an error on the OS Sensor.<br>**Note: Intake Flap Motor and Intake Manifold Runner Position Sensor are one component and cannot be replaced individually.**<br>**Note: Vacuum in the intake system sucks in the leak detection spray with false air. Leak detection spray decreases ignition quality of the fuel mixture. This causes a drop in engine speed and changes the value produced by the Heated Oxygen Sensor (HO2S).**<br>**Possible Causes:**<br>• Intake Manifold Runner Position Sensor is damaged or has failed<br>• Intake system has leaks (false air)<br>• Motor for intake flap is faulty<br>• ECM has failed<br>• Oxygen sensor (before catalytic converter) is faulty<br>• Oxygen sensor heater (before catalytic converter) is faulty<br>• Oxygen sensor heater (before catalytic converter) is faulty<br>• Oxygen sensor heater (behind catalytic converter) is faulty<br>• O2S circuit is open or shorted in the wiring harness |
| **DTC: P2422**<br>**2T CCM, MIL: Yes**<br>**Years:** 2007, 2008<br>**Models:** A6<br>**Engines:** 3.2L, 4.2L<br>**Transmissions:** All | **Evaporative Emission System Vent Valve Stuck Closed Conditions:**<br>Engine started, battery voltage must be at least 11.5v, all electrical components must be off, parking brake must be engaged (to keep daytime driving lights off), automatic transmission selector must be in park, the exhaust system must be properly sealed between the catalytic converter and the cylinder head, coolant temperature must be at least 80 degrees Celsius and oxygen sensor heaters for oxygen sensors before the catalytic converter must be functioning properly and the ground between the engine and the chassis must be well connected. The ECM detected an unexpected EVAP malfunction.<br>**Note: Solenoid valve is closed when no voltage is present.**<br>**Possible Causes:**<br>• EVAP power supply circuit is open<br>• EVAP solenoid control circuit is open or shorted to ground<br>• EVAP solenoid control circuit is shorted to power (B+)<br>• EVAP solenoid valve is damaged or it has failed<br>• EVAP canister purge solenoid valve is faulty<br>• ECM has failed |
| **DTC: P2539**<br>**2T CCM, MIL: Yes**<br>**Years:** 2007, 2008<br>**Models:** A6<br>**Engines:** 3.2L<br>**Transmissions:** All | **Low Pressure Fuel System Sensor Circuit Conditions:**<br>Engine started, battery voltage must be at least 11.5v, all electrical components must be off, parking brake must be engaged (to keep daytime driving lights off), automatic transmission selector must be in park, the exhaust system must be properly sealed between the catalytic converter and the cylinder head, coolant temperature must be at least 80 degrees Celsius. The ECM detected an error on the fuel system sensor circuit.<br>**Note: The specified fuel pressure should be between 3000 to 5000 kPA**<br>**Possible Causes:**<br>• Fuel Pressure Regulator Valve has failed<br>• Fuel Pressure Sensor has failed<br>• Fuel Pump (FP) Control Module has failed<br>• Fuel pump has failed<br>• ECM has failed |

| DTC | Trouble Code Title, Conditions & Possible Causes |
|---|---|
| **DTC: P2540**<br>**2T CCM, MIL: Yes**<br>**Years:** 2007, 2008<br>**Models:** A6<br>**Engines:** 3.2L<br>**Transmissions:** All | **Low Pressure Fuel System Sensor Circuit Range/Performance Conditions:**<br>Engine started, battery voltage must be at least 11.5v, all electrical components must be off, parking brake must be engaged (to keep daytime driving lights off), automatic transmission selector must be in park, the exhaust system must be properly sealed between the catalytic converter and the cylinder head, coolant temperature must be at least 80 degrees Celsius. The ECM detected an error on the fuel system sensor circuit.<br>**Note: The specified fuel pressure should be between 3000 to 5000 kPA**<br>**Possible Causes:**<br>• Fuel Pressure Regulator Valve has failed<br>• Fuel Pressure Sensor has failed<br>• Fuel Pump (FP) Control Module has failed<br>• Fuel pump has failed<br>• ECM has failed |
| **DTC: P2541**<br>**2T CCM, MIL: Yes**<br>**Years:** 2007, 2008<br>**Models:** A6<br>**Engines:** 3.2L<br>**Transmissions:** All | **Low Pressure Fuel System Sensor Circuit Low Conditions:**<br>Engine started, battery voltage must be at least 11.5v, all electrical components must be off, parking brake must be engaged (to keep daytime driving lights off), automatic transmission selector must be in park, the exhaust system must be properly sealed between the catalytic converter and the cylinder head, coolant temperature must be at least 80 degrees Celsius. The ECM detected an error on the fuel system sensor circuit.<br>**Note: The specified fuel pressure should be between 3000 to 5000 kPA**<br>**Possible Causes:**<br>• Fuel Pressure Regulator Valve has failed<br>• Fuel Pressure Sensor has failed<br>• Fuel Pump (FP) Control Module has failed<br>• Fuel pump has failed<br>• ECM has failed |
| **DTC: P2626**<br>**2T CCM, MIL: Yes**<br>**Years:** 2007, 2008<br>**Models:** A6<br>**Engines:** 3.2L, 4.2L<br>**Transmissions:** All | **O2 Sensor Pumping Current Trim Circuit/Open Bank 1 Sensor 1 Conditions:**<br>Engine started, battery voltage must be at least 11.5v, all electrical components must be off, parking brake must be engaged (to keep daytime driving lights off), automatic transmission selector must be in park, the exhaust system must be properly sealed between the catalytic converter and the cylinder head, coolant temperature must be at least 80 degrees Celsius and oxygen sensor heaters for oxygen sensors before the catalytic converter must be functioning properly and the ground between the engine and the chassis must be well connected. The ECM detected a voltage value that doesn't fAll within the desired parameters for a properly functioning O2 system.<br>**Possible Causes:**<br>• Check activation of Recirculation Pump Relay<br>• Oxygen sensor (before catalytic converter) is faulty<br>• Oxygen sensor (behind catalytic converter) is faulty<br>• Oxygen sensor heater (before catalytic converter) is faulty<br>• Oxygen sensor heater (behind catalytic converter) is faulty |
| **DTC: P2629**<br>**1T CCM, MIL: Yes**<br>**Years:** 2007, 2008<br>**Models:** A6<br>**Engines:** 3.2L, 4.2L<br>**Transmissions:** All | **O2 Sensor Pumping Current Trim Circuit/Open Bank 2 Sensor 1 Conditions:**<br>Engine started, battery voltage must be at least 11.5v, all electrical components must be off, parking brake must be engaged (to keep daytime driving lights off), automatic transmission selector must be in park, the exhaust system must be properly sealed between the catalytic converter and the cylinder head, coolant temperature must be at least 80 degrees Celsius and oxygen sensor heaters for oxygen sensors before the catalytic converter must be functioning properly and the ground between the engine and the chassis must be well connected. The ECM detected a voltage value that doesn't fAll within the desired parameters for a properly functioning O2 system.<br>**Possible Causes:**<br>• Check activation of Recirculation Pump Relay<br>• Oxygen sensor (before catalytic converter) is faulty<br>• Oxygen sensor (behind catalytic converter) is faulty<br>• Oxygen sensor heater (before catalytic converter) is faulty<br>• Oxygen sensor heater (behind catalytic converter) is faulty |

| DTC | Trouble Code Title, Conditions & Possible Causes |
|---|---|
| **DTC: P2637**<br>**1T CCM, MIL: Yes**<br>**Years: 2007, 2008**<br>**Models:** A6<br>**Engines:** 3.2L, 4.2L<br>**Transmissions:** All | **Torque Management Feedback Signal "A" Conditions:**<br>Engine started, battery voltage must be at least 11.5v, all electrical components must be off, parking brake must be engaged (to keep daytime driving lights off), automatic transmission selector must be in park, the exhaust system must be properly sealed between the catalytic converter and the cylinder head, coolant temperature must be at least 80 degrees Celsius and oxygen sensor heaters for oxygen sensors before the catalytic converter must be functioning properly and the ground between the engine and the chassis must be well connected. The ECM detected a voltage value on the torque management circuits that doesn't fAll within the desired parameters<br>**Possible Causes:**<br>• Engine Control Module (ECM) has failed<br>• Voltage supply for Engine Control Module (ECM) is damaged<br>• Engine Coolant Temperature (ECT) sensor has failed<br>• Intake Air Temperature (IAT) sensor has failed<br>• Intake Manifold Runner Position Sensor has failed<br>• Intake system has leaks (false air)<br>• Motor for intake flap has failed<br>• Mass Air Flow (MAF) sensor has failed |
| **DTC: P2714**<br>**2T CCM, MIL: Yes**<br>**Years: 2007, 2008**<br>**Models:** A6<br>**Engines:** 3.2L, 4.2L<br>**Transmissions:** All | **Pressure Control Solenoid "D" Performance or Stuck Off Conditions:**<br>Engine started, battery voltage must be at least 11.5v, all electrical components must be off, and the ground between the engine and the chassis must be well connected. The ECM detected the pressure control solenoid was in the "stuck off" position.<br>**Possible Causes:**<br>• ATF level is low<br>• Circuit harness connector contacts are corroded or ingressed of water<br>• Circuit wires have shorted to each other, to battery or ground<br>• Automatic Transmission Hydraulic Pressure Sensor 1 has failed<br>• Solenoid valves in valve body are faulty<br>• Transmission Control Module (TCM) needs replacing<br>• Transmission Input Speed (RPM) Sensor has failed<br>• Transmission Output Speed (RPM) Sensor has failed |
| **DTC: P2715**<br>**2T CCM, MIL: Yes**<br>**Years: 2007, 2008**<br>**Models:** A6<br>**Engines:** 3.2L, 4.2L<br>**Transmissions:** All | **Pressure Control Solenoid "D" Performance or Stuck On Conditions:**<br>Engine started, battery voltage must be at least 11.5v, all electrical components must be off, and the ground between the engine and the chassis must be well connected. The ECM detected the pressure control solenoid was in the "stuck on" position.<br>**Possible Causes:**<br>• ATF level is low<br>• Circuit harness connector contacts are corroded or ingressed of water<br>• Circuit wires have shorted to each other, to battery or ground<br>• Automatic Transmission Hydraulic Pressure Sensor 1 has failed<br>• Solenoid valves in valve body are faulty<br>• Transmission Control Module (TCM) needs replacing<br>• Transmission Input Speed (RPM) Sensor has failed<br>• Transmission Output Speed (RPM) Sensor has failed |
| **DTC: P2716**<br>**2T CCM, MIL: Yes**<br>**Years: 2007, 2008**<br>**Models:** A6<br>**Engines:** 3.2L, 4.2L<br>**Transmissions:** All | **Pressure Control Solenoid "D" Electrical Malfunction Conditions:**<br>Engine started, battery voltage must be at least 11.5v, all electrical components must be off, and the ground between the engine and the chassis must be well connected. The ECM detected the pressure control solenoid was experiencing electrical malfunctions.<br>**Possible Causes:**<br>• ATF level is low<br>• Circuit harness connector contacts are corroded or ingressed of water<br>• Circuit wires have shorted to each other, to battery or ground<br>• Automatic Transmission Hydraulic Pressure Sensor 1 has failed<br>• Solenoid valves in valve body are faulty<br>• Transmission Control Module (TCM) needs replacing<br>• Transmission Input Speed (RPM) Sensor has failed<br>• Transmission Output Speed (RPM) Sensor has failed |

| DTC | Trouble Code Title, Conditions & Possible Causes |
|---|---|
| **DTC: P2723**<br>**2T CCM, MIL: Yes**<br>**Years:** 2007, 2008<br>**Models:** A6<br>**Engines:** 3.2L<br>**Transmissions:** All | **Pressure Control Solenoid "E" Performance or Stuck Off Conditions:**<br>Engine started, battery voltage must be at least 11.5v, all electrical components must be off, and the ground between the engine and the chassis must be well connected. The ECM detected the pressure control solenoid was in the "stuck off" position.<br>**Possible Causes:**<br>• ATF level is low<br>• Circuit harness connector contacts are corroded or ingressed of water<br>• Circuit wires have shorted to each other, to battery or ground<br>• Automatic Transmission Hydraulic Pressure Sensor 1 has failed<br>• Solenoid valves in valve body are faulty<br>• Transmission Control Module (TCM) needs replacing<br>• Transmission Input Speed (RPM) Sensor has failed<br>• Transmission Output Speed (RPM) Sensor has failed |
| **DTC: P2724**<br>**2T CCM, MIL: Yes**<br>**Years:** 2007, 2008<br>**Models:** A6<br>**Engines:** 3.2L, 4.2L<br>**Transmissions:** All | **Pressure Control Solenoid "E" Performance or Stuck On Conditions:**<br>Engine started, battery voltage must be at least 11.5v, all electrical components must be off, and the ground between the engine and the chassis must be well connected. The ECM detected the pressure control solenoid was in the "stuck on" position.<br>**Possible Causes:**<br>• ATF level is low<br>• Circuit harness connector contacts are corroded or ingressed of water<br>• Circuit wires have shorted to each other, to battery or ground<br>• Automatic Transmission Hydraulic Pressure Sensor 1 has failed<br>• Solenoid valves in valve body are faulty<br>• Transmission Control Module (TCM) needs replacing<br>• Transmission Input Speed (RPM) Sensor has failed<br>• Transmission Output Speed (RPM) Sensor has failed |
| **DTC: P2732**<br>**2T CCM, MIL: Yes**<br>**Years:** 2007, 2008<br>**Models:** A6<br>**Engines:** 3.2L, 4.2L<br>**Transmissions:** All | **Pressure Control Solenoid "F" Performance or Stuck Off Conditions:**<br>Engine started, battery voltage must be at least 11.5v, all electrical components must be off, and the ground between the engine and the chassis must be well connected. The ECM detected the pressure control solenoid was in the "stuck off" position.<br>**Possible Causes:**<br>• ATF level is low<br>• Circuit harness connector contacts are corroded or ingressed of water<br>• Circuit wires have shorted to each other, to battery or ground<br>• Automatic Transmission Hydraulic Pressure Sensor 1 has failed<br>• Solenoid valves in valve body are faulty<br>• Transmission Control Module (TCM) needs replacing<br>• Transmission Input Speed (RPM) Sensor has failed<br>• Transmission Output Speed (RPM) Sensor has failed |
| **DTC: P2733**<br>**2T CCM, MIL: Yes**<br>**Years:** 2007, 2008<br>**Models:** A6<br>**Engines:** 3.2L, 4.2L<br>**Transmissions:** All | **Pressure Control Solenoid "F" Performance or Stuck On Conditions:**<br>Engine started, battery voltage must be at least 11.5v, all electrical components must be off, and the ground between the engine and the chassis must be well connected. The ECM detected the pressure control solenoid was in the "stuck on" position.<br>**Possible Causes:**<br>• ATF level is low<br>• Circuit harness connector contacts are corroded or ingressed of water<br>• Circuit wires have shorted to each other, to battery or ground<br>• Automatic Transmission Hydraulic Pressure Sensor 1 has failed<br>• Solenoid valves in valve body are faulty<br>• Transmission Control Module (TCM) needs replacing<br>• Transmission Input Speed (RPM) Sensor has failed<br>• Transmission Output Speed (RPM) Sensor has failed |

## Gas Engine OBD II Trouble Code List (P3xxx Codes)

| DTC | Trouble Code Title, Conditions & Possible Causes |
|---|---|
| **DTC: P3081**<br>**Years:** 2007, 2008<br>**Models:** A4, A6<br>**Engines:** All<br>**Transmissions:** All | **Engine Temperature Too Low Conditions:**<br>Engine running and the ECM has detected that the engine temperature is too low.<br>**Possible Causes:**<br>• Engine hasn't completely warmed up<br>• Radiator malfunction<br>• Thermostat malfunction<br>• ECM failure |

## Gas Engine OBD II Trouble Code List (U1xxx Codes)

| DTC | Trouble Code Title, Conditions & Possible Causes |
|---|---|
| **DTC: U0001**<br>**1T CCM, MIL: No**<br>**Years:** 2007, 2008<br>**Models:** A4, A6<br>**Engines:** 3.2L, 4.2L<br>**Transmissions:** All | **High Speed CAN Communication Bus Conditions:**<br>The Engine Control Module (ECM) communicates with All databus-capable control modules via a CAN databus. These databus-capable control modules are connected via two data bus wires which are twisted together (CAN_High and CAN_Low), and exchange information (messages). Missing information on the databus is recognized as a malfunction and stored. Trouble-free operation of the CAN-Bus requires that it have a terminal resistance. This central terminal resistor is located in the Engine Control Module (ECM).<br>**Possible Causes:**<br>• ECM has failed<br>• CAN data bus wires have short circuited to each other |
| **DTC: U0101**<br>**2T CCM, MIL: Yes**<br>**Years:** 2007, 2008<br>**Models:** A4, A6<br>**Engines:** 3.2L, 4.2L<br>**Transmissions:** All | **Lost Communication With TCM Conditions:**<br>Key on, and the ECM detected that it has lost communication with the Transmission Control Module (TCM) during its initial startup. The Engine Control Module (ECM) communicates with All databus-capable control modules via a CAN databus. These databus-capable control modules are connected via two data bus wires which are twisted together (CAN_High and CAN_Low), and exchange information (messages). Missing information on the databus is recognized as a malfunction and stored. Trouble-free operation of the CAN-Bus requires that it have a terminal resistance.<br>**Possible Causes:**<br>• ECM has failed<br>• Terminal resistance for CAN-bus are faulty<br>• Can data bus wires have short circuited to each other<br>• TCM has failed |
| **DTC: U0155**<br>**1T CCM, MIL: No**<br>**Years:** 2007, 2008<br>**Models:** A4, A6<br>**Engines:** 3.2L, 4.2L<br>**Transmissions:** All | **Lost Communication With Instrument Cluster Conditions:**<br>Key on, and the ECM detected that it has lost communication with the Instrument Cluster Panel (I/P) during its initial startup. The Engine Control Module (ECM) communicates with All databus-capable control modules via a CAN databus. These databus-capable control modules are connected via two data bus wires which are twisted together (CAN_High and CAN_Low), and exchange information (messages). Missing information on the databus is recognized as a malfunction and stored. Trouble-free operation of the CAN-Bus requires that it have a terminal resistance.<br>**Possible Causes:**<br>• ECM has failed<br>• Terminal resistance for CAN-bus are faulty<br>• Can data bus wires have short circuited to each other |
| **DTC: U0302**<br>**1T CCM, MIL: No**<br>**Years:** 2007, 2008<br>**Models:** A6<br>**Engines:** 3.2L, 4.2L<br>**Transmissions:** All | **Software Incompatibility with Transmission Control Module Conditions:**<br>Key on, and the ECM detected a software incompatibility condition with the Transmission Control Module during its initial startup. The Engine Control Module (ECM) communicates with All databus-capable control modules via a CAN databus. These databus-capable control modules are connected via two data bus wires which are twisted together (CAN_High and CAN_Low), and exchange information (messages). Missing information on the databus is recognized as a malfunction and stored. Trouble-free operation of the CAN-Bus requires that it have a terminal resistance.<br>**Possible Causes:**<br>• ECM or TCM has failed or is not properly coded<br>• Terminal resistance for CAN-bus are faulty<br>• Can data bus wires have short circuited to each other |
| **DTC: U0402**<br>**1T CCM, MIL: No**<br>**Years:** 2007, 2008<br>**Models:** A6<br>**Engines:** 3.2L, 4.2L<br>**Transmissions:** All | **Invalid Data Received From Transmission Control Module Conditions:**<br>Key on, and the ECM detected a software invalid data from the Cruise Control Module during its initial startup. The Engine Control Module (ECM) communicates with All databus-capable control modules via a CAN databus. These databus-capable control modules are connected via two data bus wires which are twisted together (CAN_High and CAN_Low), and exchange information (messages). Missing information on the databus is recognized as a malfunction and stored. Trouble-free operation of the CAN-Bus requires that it have a terminal resistance.<br>**Possible Causes:**<br>• ECM or TCM has failed<br>• Terminal resistance for CAN-bus are faulty<br>• Can data bus wires have short circuited to each other |

# BMW

## 3 • 5 • 7 Series

## SPECIFICATIONS AND MAINTENANCE CHARTS

### ENGINE AND VEHICLE IDENTIFICATION

| | | | Engine | | | | | Model Year | |
|---|---|---|---|---|---|---|---|---|---|
| Code | Liters (cc) | Cu. In. | Cyl. | Fuel Sys. | Engine Type | Eng. Mfg. | | Code ① | Year |
| N52B30 | 3.0 (2996) | 183 | 6 | ②③ | DOHC | BMW | | 7 | 2007 |
| N54B30 | 3.0 (2979) | 181 | 6 | ④ | SOHC | BMW | | 8 | 2008 |
| N62B48 | 4.8 (4841) | 293 | 8 | ⑤ | DOHC | BMW | | | |

DOHC: Double Overhead Camshaft

SOHC: Single Overhead Camshaft

① 10th digit of the Vehicle Identification Number (VIN)

② Siemens MSV80 3-Series

③ Siemens MSV70 5-Series

④ Siemens MSD80

⑤ Bosch ME9

22205_BMWC_C0001

## GENERAL ENGINE SPECIFICATIONS

| Year | Body Type | Model | Engine Displ. Liters (cc) | Engine ID/VIN | Fuel System Type | Net Horsepower @ rpm | Net Torque @ rpm (ft. lbs.) | Bore x Stroke (in.) | Compression Ratio | Oil Pressure @ rpm |
|---|---|---|---|---|---|---|---|---|---|---|
| 2007 | E93 | 328i Convertible | 3.0 (2996) | N52B30 | ① | 230@6500 | 200@2750 | 3.35x3.46 | 10.7:1 | 7.3@idle |
| | E92 | 328i Coupe | 3.0 (2996) | N52B30 | ① | 230@6500 | 200@2750 | 3.35x3.46 | 10.7:1 | 7.3@idle |
| | E90 | 328i Sedan | 3.0 (2996) | N52B30 | ① | 230@6500 | 200@2750 | 3.35x3.46 | 10.5:1 | 7.3@idle |
| | E91 | 328i Wagon | 3.0 (2996) | N52B30 | ① | 230@6500 | 200@2750 | 3.35x3.46 | 10.5:1 | 7.3@idle |
| | E92 | 328xi Coupe | 3.0 (2996) | N52B30 | ① | 230@6500 | 200@2750 | 3.35x3.46 | 10.7:1 | 7.3@idle |
| | E90 | 328xi Sedan | 3.0 (2996) | N52B30 | ① | 230@6500 | 200@2750 | 3.35x3.46 | 10.7:1 | 7.3@idle |
| | E91 | 328xi Wagon | 3.0 (2996) | N52B30 | ① | 230@6500 | 200@2750 | 3.35x3.46 | 10.7:1 | 7.3@idle |
| | E93 | 335i Convertible | 3.0 (2979) | N54B30 | ② | 300@5800 | 300@1400 | 3.31x3.53 | 10.2:1 | 21.8@idle |
| | E92 | 335i Coupe | 3.0 (2979) | N54B30 | ② | 300@5800 | 300@1400 | 3.31x3.53 | 10.2:1 | 21.8@idle |
| | E90 | 335i Sedan | 3.0 (2979) | N54B30 | ② | 300@5800 | 300@1400 | 3.31x3.53 | 10.2:1 | 21.8@idle |
| | E90 | 335xi Sedan | 3.0 (2979) | N54B30 | ② | 300@5800 | 300@1400 | 3.31x3.53 | 10.2:1 | 21.8@idle |
| | E60 | 525i Sedan | 3.0 (2996) | N52B30 | ③ | 215@6248 | 185@2748 | 3.35x3.46 | 10.7:1 | 7.3@idle |
| | E60 | 525xi Sedan | 3.0 (2996) | N52B30 | ③ | 215@6248 | 185@2748 | 3.35x3.46 | 10.7:1 | 7.3@idle |
| | E60 | 530i Sedan | 3.0 (2996) | N52B30 | ③ | 255@6650 | 220@2750 | 3.35x3.46 | 10.7:1 | 7.3@idle |
| | E60 | 530xi Sedan | 3.0 (2996) | N52B30 | ③ | 255@6650 | 220@2750 | 3.35x3.46 | 10.7:1 | 7.3@idle |
| | E61 | 530xi Sport Wagon | 3.0 (2996) | N52B30 | ③ | 255@6650 | 220@2750 | 3.35x3.46 | 10.7:1 | 7.3@idle |
| | E60 | 550i Sedan | 4.8 (4841) | N62B48 | ④ | 360@6300 | 360@3400 | 3.66x3.48 | 10.5:1 | 14.5@idle |
| | E60 | 750i Sedan | 4.8 (4841) | N62B48 | ④ | 360@6300 | 360@3400 | 3.66x3.48 | 10.5:1 | 14.5@idle |
| | E60 | 750Li Sedan | 4.8 (4841) | N62B48 | ④ | 360@6300 | 360@3400 | 3.66x3.48 | 10.5:1 | 14.5@idle |
| 2008 | E93 | 328i Convertible | 3.0 (2996) | N52B30 | ① | 230@6500 | 200@2750 | 3.35x3.46 | 10.7:1 | 7.3@idle |
| | E92 | 328i Coupe | 3.0 (2996) | N52B30 | ① | 230@6500 | 200@2750 | 3.35x3.46 | 10.7:1 | 7.3@idle |
| | E90 | 328i Sedan | 3.0 (2996) | N52B30 | ① | 230@6500 | 200@2750 | 3.35x3.46 | 10.5:1 | 7.3@idle |
| | E91 | 328i Wagon | 3.0 (2996) | N52B30 | ① | 230@6500 | 200@2750 | 3.35x3.46 | 10.5:1 | 7.3@idle |
| | E92 | 328xi Coupe | 3.0 (2996) | N52B30 | ① | 230@6500 | 200@2750 | 3.35x3.46 | 10.7:1 | 7.3@idle |
| | E90 | 328xi Sedan | 3.0 (2996) | N52B30 | ① | 230@6500 | 200@2750 | 3.35x3.46 | 10.7:1 | 7.3@idle |
| | E91 | 328xi Wagon | 3.0 (2996) | N52B30 | ① | 230@6500 | 200@2750 | 3.35x3.46 | 10.7:1 | 7.3@idle |
| | E93 | 335i Convertible | 3.0 (2979) | N54B30 | ② | 300@5800 | 300@1400 | 3.31x3.53 | 10.2:1 | 21.8@idle |
| | E92 | 335i Coupe | 3.0 (2979) | N54B30 | ② | 300@5800 | 300@1400 | 3.31x3.53 | 10.2:1 | 21.8@idle |
| | E90 | 335i Sedan | 3.0 (2979) | N54B30 | ② | 300@5800 | 300@1400 | 3.31x3.53 | 10.2:1 | 21.8@idle |
| | E90 | 335xi Sedan | 3.0 (2979) | N54B30 | ② | 300@5800 | 300@1400 | 3.31x3.53 | 10.2:1 | 21.8@idle |
| | E60 | 528i Sedan | 3.0 (2996) | N52B30 | ① | 230@6500 | 200@2750 | 3.35x3.46 | 10.7:1 | 7.3@idle |
| | E60 | 528xi Sedan | 3.0 (2996) | N52B30 | ① | 230@6500 | 200@2750 | 3.35x3.46 | 10.7:1 | 7.3@idle |
| | E60 | 535i Sedan | 3.0 (2996) | N52B30 | ② | 300@5800 | 300@1400 | 3.35x3.46 | 10.2:1 | 7.3@idle |
| | E60 | 535xi Sedan | 3.0 (2996) | N52B30 | ② | 300@5800 | 300@1400 | 3.35x3.46 | 10.2:1 | 7.3@idle |
| | E61 | 535xi Sport Wagon | 3.0 (2996) | N52B30 | ② | 255@6650 | 220@2750 | 3.35x3.46 | 10.2:1 | 7.3@idle |
| | E60 | 550i Sedan | 4.8 (4841) | N62B48 | ④ | 360@6300 | 360@3400 | 3.66x3.48 | 10.5:1 | 14.5@idle |
| | E60 | 750i Sedan | 4.8 (4841) | N62B48 | ④ | 360@6300 | 360@3400 | 3.66x3.48 | 10.5:1 | 14.5@idle |
| | E60 | 750Li Sedan | 4.8 (4841) | N62B48 | ④ | 360@6300 | 360@3400 | 3.66x3.48 | 10.5:1 | 14.5@idle |

① Siemens MSV80

② Siemens MSD80

③ Siemens MSV70

④ Bosch ME9

## ENGINE TUNE-UP SPECIFICATIONS

| Year | Engine Displacement Liters | Engine ID/VIN | Spark Plug Gap (in.) | Ignition Timing (deg.) | | Fuel Pump (psi) | Idle Speed (rpm) | | Valve Clearance | |
|------|------|------|------|------|------|------|------|------|------|------|
| | | | | MT | AT | | MT | AT | In. | Ex. |
| **2007** | 3.0 | N52B30 | ① | ② | ② | 70-76 | ② | ② | HYD | HYD |
| | 3.0 | N54B30 | ① | ② | ② | 70-76 | ② | ② | HYD | HYD |
| | 4.8 | N62B48 | ① | ② | ② | 48-54 | ② | ② | HYD | HYD |
| **2008** | 3.0 | N52B30 | ① | ② | ② | 70-76 | ② | ② | HYD | HYD |
| | 3.0 | N54B30 | ① | ② | ② | 70-76 | ② | ② | HYD | HYD |
| | 4.8 | N62B48 | ① | ② | ② | 48-54 | ② | ② | HYD | HYD |

NOTE: The Vehicle Emission Control Information label reflects specification changes during production and must be used if they differ from this chart.

B: Before Top Dead Center

HYD: Hydraulic

① Three mass and four-mass electrodes cannot be adjusted

   Dual mass electrodes: 0.035-0.039 inches

   All others: 0.028-0.031 inches

② Controlled by the Engine Control Module (ECM) and cannot be adjusted

③ At idle, pressure measured at injectors

22205_BMWC_C0003

## CAPACITIES

| Year | Body Type | Model | Engine Displacement Liters (cc) | Engine ID/VIN | Engine Oil with Filter (qts.) | Transmission (pts.) 5-Spd | Transmission (pts.) 6-Spd | Transmission (pts.) Auto. | Drive Axle Front (pts.) | Drive Axle Rear (pts.) | Fuel Tank (gal.) | Cooling System (qts.) |
|------|-----------|-------|-------------------------------|---------------|-------------------------------|------|------|------|------|------|------|------|
| 2007 | E93 | 328i Convertible | 3.0 (2996) | N52B30 | 6.9 | — | ① | ② | 1.3 | 2.5 | 16.4 | ③ |
| | E92 | 328i Coupe | 3.0 (2996) | N52B30 | 6.9 | — | ① | ② | 1.3 | 2.5 | 16.4 | ③ |
| | E90 | 328i Sedan | 3.0 (2996) | N52B30 | 6.9 | — | ① | ② | 1.3 | 2.5 | 16.4 | ③ |
| | E91 | 328i Wagon | 3.0 (2996) | N52B30 | 6.9 | — | ① | ② | 1.3 | 2.5 | 16.4 | ③ |
| | E92 | 328xi Coupe | 3.0 (2996) | N52B30 | 6.9 | — | ① | ② | 1.3 | 2.5 | 16.4 | ③ |
| | E90 | 328xi Sedan | 3.0 (2996) | N52B30 | 6.9 | — | ① | ② | 1.3 | 2.5 | 16.4 | ③ |
| | E91 | 328xi Wagon | 3.0 (2996) | N52B30 | 6.9 | — | ① | ② | 1.3 | 2.5 | 16.4 | ③ |
| | E93 | 335i Convertible | 3.0 (2979) | N54B30 | 6.9 | — | ① | ② | 1.3 | 2.5 | 16.4 | ③ |
| | E92 | 335i Coupe | 3.0 (2979) | N54B30 | 6.9 | — | ① | ② | 1.3 | 2.5 | 16.4 | ③ |
| | E90 | 335i Sedan | 3.0 (2979) | N54B30 | 6.9 | — | ① | ② | 1.3 | 2.5 | 16.4 | ③ |
| | E90 | 335xi Sedan | 3.0 (2979) | N54B30 | 6.9 | | ① | ② | 1.3 | 2.5 | 13.5 | ③ |
| | E60 | 525i Sedan | 3.0 (2996) | N52B30 | 6.9 | — | ① | ② | 1.3 | 2.5 | 18.5 | ④ |
| | E60 | 525xi Sedan | 3.0 (2996) | N52B30 | 6.9 | — | ① | ② | 1.3 | 2.5 | 18.5 | ④ |
| | E60 | 530i Sedan | 3.0 (2996) | N52B30 | 6.9 | — | ① | ② | 1.3 | 2.5 | 18.5 | ④ |
| | E60 | 530xi Sedan | 3.0 (2996) | N52B30 | 6.9 | | ① | ② | 1.3 | 2.5 | 18.5 | ④ |
| | E61 | 530xi Sport Wagon | 3.0 (2996) | N52B30 | 6.9 | | ① | ② | 1.3 | 2.5 | 18.5 | ④ |
| | E60 | 550i Sedan | 4.8 (4841) | N62B48 | 6.9 | — | ① | ② | 1.3 | 2.5 | 23.3 | ④ |
| | E60 | 750i Sedan | 4.8 (4841) | N62B48 | 6.9 | | ① | ② | 1.3 | 2.5 | 23.3 | ④ |
| | E60 | 750Li Sedan | 4.8 (4841) | N62B48 | 6.9 | — | ① | ② | 1.3 | 2.5 | 23.3 | ④ |
| 2008 | E93 | 328i Convertible | 3.0 (2996) | N52B30 | 6.9 | — | ① | ② | 1.3 | 2.5 | 16.4 | ③ |
| | E92 | 328i Coupe | 3.0 (2996) | N52B30 | 6.9 | — | ① | ② | 1.3 | 2.5 | 16.4 | ③ |
| | E90 | 328i Sedan | 3.0 (2996) | N52B30 | 6.9 | — | ① | ② | 1.3 | 2.5 | 16.4 | ③ |
| | E91 | 328i Wagon | 3.0 (2996) | N52B30 | 6.9 | — | ① | ② | 1.3 | 2.5 | 16.4 | ③ |
| | E92 | 328xi Coupe | 3.0 (2996) | N52B30 | 6.9 | — | ① | ② | 1.3 | 2.5 | 16.4 | ③ |
| | E90 | 328xi Sedan | 3.0 (2996) | N52B30 | 6.9 | — | ① | ② | 1.3 | 2.5 | 16.4 | ③ |
| | E91 | 328xi Wagon | 3.0 (2996) | N52B30 | 6.9 | — | ① | ② | 1.3 | 2.5 | 16.4 | ③ |
| | E93 | 335i Convertible | 3.0 (2979) | N54B30 | 6.9 | — | ① | ② | 1.3 | 2.5 | 16.4 | ③ |
| | E92 | 335i Coupe | 3.0 (2979) | N54B30 | 6.9 | — | ① | ② | 1.3 | 2.5 | 16.4 | ③ |
| | E90 | 335i Sedan | 3.0 (2979) | N54B30 | 6.9 | — | ① | ② | 1.3 | 2.5 | 16.4 | ③ |
| | E90 | 335xi Sedan | 3.0 (2979) | N54B30 | 6.9 | | ① | ② | 1.3 | 2.5 | 13.5 | ③ |
| | E60 | 528i Sedan | 3.0 (2996) | N52B30 | 6.9 | — | ① | ② | 1.3 | 2.5 | 18.5 | ④ |
| | E60 | 528xi Sedan | 3.0 (2996) | N52B30 | 6.9 | — | ① | ② | 1.3 | 2.5 | 18.5 | ④ |
| | E60 | 535i Sedan | 3.0 (2996) | N52B30 | 6.9 | — | ① | ② | 1.3 | 2.5 | 18.5 | ④ |
| | E60 | 535xi Sedan | 3.0 (2996) | N52B30 | 6.9 | — | ① | ② | 1.3 | 2.5 | 18.5 | ④ |
| | E61 | 535xi Sport Wagon | 3.0 (2996) | N52B30 | 6.9 | — | ① | ② | 1.3 | 2.5 | 18.5 | ④ |
| | E60 | 550i Sedan | 4.8 (4841) | N62B48 | 6.9 | — | ① | ② | 1.3 | 2.5 | 23.3 | ④ |
| | E60 | 750i Sedan | 4.8 (4841) | N62B48 | 6.9 | | ① | ② | 1.3 | 2.5 | 23.3 | ④ |
| | E60 | 750Li Sedan | 4.8 (4841) | N62B48 | 6.9 | | ① | ② | 1.3 | 2.5 | 23.3 | ④ |

NOTE: All capacities are approximate. Add fluid gradually and ensure a proper fluid level is obtained.

NOTE: Capacities given are service, not overhaul capacities

① GS6-37BZ/DZ / MECH Transmission: 3.18 pts.

  E53 GS6-53DZ/BZ / MECH Transmission 3.18 pts.

  GS6-17BG/DG / MECH Transmission 2.76 pts.

② GA6HP19Z Transmission: 19-21.2 pts.

  GA6L45R Transmission: 19.0 pts.

③ M/T: 8.7 qts. A/T 8.9 qts.

④ M/T: 10.6 qts. A/T 11.2 qts.

## FLUID SPECIFICATIONS

| Year | Model | Engine Size Liters | Engine Oil | Man. Trans. | Auto. Trans. | Drive Axle Front | Drive Axle Rear | Transfer Case | Power Steering Fluid | Brake Master Cylinder | Cooling System |
|------|-------|-------------------|-----------|-------------|-------------|-------|------|-------|------------------|------------------|---------|
| 2007 | 328i Convertible | 3.0 (2996) | 5W-30 | ① | NA | — | ② | — | Dexron III | DOT 4 | ③ |
| | 328i Coupe | 3.0 (2996) | 5W-30 | ① | NA | — | ② | — | Dexron III | DOT 4 | ③ |
| | 328i Sedan | 3.0 (2996) | 5W-30 | ① | NA | — | ② | — | Dexron III | DOT 4 | ③ |
| | 328i Wagon | 3.0 (2996) | 5W-30 | ① | NA | — | ② | — | Dexron III | DOT 4 | ③ |
| | 328xi Coupe | 3.0 (2996) | 5W-30 | ① | NA | — | ② | — | Dexron III | DOT 4 | ③ |
| | 328xi Sedan | 3.0 (2996) | 5W-30 | ① | NA | — | ② | — | Dexron III | DOT 4 | ③ |
| | 328xi Wagon | 3.0 (2996) | 5W-30 | ① | NA | — | ② | — | Dexron III | DOT 4 | ③ |
| | 335i Convertible | 3.0 (2979) | 5W-30 | ① | NA | — | ② | — | Dexron III | DOT 4 | ③ |
| | 335i Coupe | 3.0 (2979) | 5W-30 | ① | NA | — | ② | — | Dexron III | DOT 4 | ③ |
| | 335i Sedan | 3.0 (2979) | 5W-30 | ① | NA | — | ② | — | Dexron III | DOT 4 | ③ |
| | 335xi Sedan | 3.0 (2979) | 5W-30 | ① | NA | — | ② | — | Dexron III | DOT 4 | ③ |
| | 525i Sedan | 3.0 (2996) | 5W-30 | ① | M-1375-4 | — | ② | — | Dexron III | DOT 4 | ③ |
| | 525xi Sedan | 3.0 (2996) | 5W-30 | ① | M-1375-4 | — | ② | — | Dexron III | DOT 4 | ③ |
| | 530i Sedan | 3.0 (2996) | 5W-30 | ① | M-1375-4 | — | ② | — | Dexron III | DOT 4 | ③ |
| | 530xi Sedan | 3.0 (2996) | 5W-30 | ① | M-1375-4 | — | ② | — | Dexron III | DOT 4 | ③ |
| | 530xi Sport Wagon | 3.0 (2996) | 5W-30 | ① | M-1375-4 | — | ② | — | Dexron III | DOT 4 | ③ |
| | 550i Sedan | 4.8 (4841) | 5W-30 | ① | M-1375-4 | — | ② | — | Dexron III | DOT 4 | ③ |
| | 750i Sedan | 4.8 (4841) | 5W-30 | ① | M-1375-4 | — | ② | — | Dexron III | DOT 4 | ③ |
| | 750Li Sedan | 4.8 (4841) | 5W-30 | ① | M-1375-4 | — | ② | — | Dexron III | DOT 4 | ③ |
| 2008 | 328i Convertible | 3.0 (2996) | 5W-30 | ① | NA | — | ② | — | Dexron III | DOT 4 | ③ |
| | 328i Coupe | 3.0 (2996) | 5W-30 | ① | NA | — | ② | — | Dexron III | DOT 4 | ③ |
| | 328i Sedan | 3.0 (2996) | 5W-30 | ① | NA | — | ② | — | Dexron III | DOT 4 | ③ |
| | 328i Wagon | 3.0 (2996) | 5W-30 | ① | NA | — | ② | — | Dexron III | DOT 4 | ③ |
| | 328xi Coupe | 3.0 (2996) | 5W-30 | ① | NA | — | ② | — | Dexron III | DOT 4 | ③ |
| | 328xi Sedan | 3.0 (2996) | 5W-30 | ① | NA | — | ② | — | Dexron III | DOT 4 | ③ |
| | 328xi Wagon | 3.0 (2996) | 5W-30 | ① | NA | — | ② | — | Dexron III | DOT 4 | ③ |
| | 335i Convertible | 3.0 (2979) | 5W-30 | ① | NA | — | ② | — | Dexron III | DOT 4 | ③ |
| | 335i Coupe | 3.0 (2979) | 5W-30 | ① | NA | — | ② | — | Dexron III | DOT 4 | ③ |
| | 335i Sedan | 3.0 (2979) | 5W-30 | ① | NA | — | ② | — | Dexron III | DOT 4 | ③ |
| | 335xi Sedan | 3.0 (2979) | 5W-30 | ① | NA | — | ② | — | Dexron III | DOT 4 | ③ |
| | 528i Sedan | 3.0 (2996) | 5W-30 | ① | M-1375-4 | — | ② | — | Dexron III | DOT 4 | ③ |
| | 528xi Sedan | 3.0 (2996) | 5W-30 | ① | M-1375-4 | — | ② | — | Dexron III | DOT 4 | ③ |
| | 535i Sedan | 3.0 (2996) | 5W-30 | ① | M-1375-4 | — | ② | — | Dexron III | DOT 4 | ③ |
| | 535xi Sedan | 3.0 (2996) | 5W-30 | ① | M-1375-4 | — | ② | — | Dexron III | DOT 4 | ③ |
| | 535xi Sport Wagon | 3.0 (2996) | 5W-30 | ① | M-1375-4 | — | ② | — | Dexron III | DOT 4 | ③ |
| | 550i Sedan | 4.8 (4841) | 5W-30 | ① | M-1375-4 | — | ② | — | Dexron III | DOT 4 | ③ |
| | 750i Sedan | 4.8 (4841) | 5W-30 | ① | M-1375-4 | — | ② | — | Dexron III | DOT 4 | ③ |
| | 750Li Sedan | 4.8 (4841) | 5W-30 | ① | M-1375-4 | | ② | — | Dexron III | DOT 4 | ③ |

NA: Not Available

DOT: Department Of Transpotation

① Orange Label: Dexron AFT

  Green Label: Synthetic Mobil SCH630

  Yellow Label: MTF-LT-1 or MTF-LT-2

  6-Speed: MTF-LT-3

  No label: MIL-L-2105

② Limited slip: SAF-XJ. Non-limited slip: SAF-XO

③ BMW Lonmg Life Coolant

## VALVE SPECIFICATIONS

| Year | Engine Displacement Liters | Engine ID/VIN | Seat Angle (deg.) | Face Angle (deg.) | Spring Test Pressure (lbs. @ in.) | Spring Installed Height (in.) | Stem-to-Guide Clearance (in.) | | Stem Diameter (in.) | |
|------|------|------|------|------|------|------|------|------|------|------|
| | | | | | | | Intake | Exhaust | Intake | Exhaust |
| 2007 | 3.0 | N52B30 | ① | 45 | NA | NA | ② 0.0197 | ② 0.0197 | 0.0234-0.0235 | 0.0234-0.0235 |
| | 3.0 | N54B30 | ① | 45 | NA | NA | ② 0.0197 | ② 0.0197 | 0.0234-0.0235 | 0.0234-0.0235 |
| | 4.8 | N62B48 | ① | 45 | NA | NA | ② 0.0197 | ② 0.0197 | 0.0234-0.0235 | 0.0234-0.0235 |
| 2008 | 3.0 | N52B30 | ① | 45 | NA | NA | ② 0.0197 | ② 0.0197 | 0.0234-0.0235 | 0.0234-0.0235 |
| | 3.0 | N54B30 | ① | 45 | NA | NA | ② 0.0197 | ② 0.0197 | 0.0234-0.0235 | 0.0234-0.0235 |
| | 4.8 | N62B48 | ① | 45 | NA | NA | ② 0.0197 | ② 0.0197 | 0.0234-0.0235 | 0.0234-0.0235 |

NA: Not Available

① Valve seat angle: 45 degrees

Correction angle outside: 15 degrees

Correction angle inside: 60 degrees

② To measure: Insert a new valve into guide

with end of valve flush with end of guide.

Use a dial indicator to measure axial valve head movement.

## CAMSHAFT SPECIFICATIONS
### All measurements in inches unless noted

| Year | Engine Displacement Liters | Engine Code/ID | Journal Dia. | Brg. Oil Clearance | Shaft End-play | Circle Runout | Lobe Height | |
|------|------|------|------|------|------|------|------|------|
| | | | | | | | Intake | Exhaust |
| 2007 | 3.0 | N52B30 | NA | NA | 0.0007-0.0063 | 0.0021-0.0038 | NA | NA |
| | 3.0 | N54B30 | NA | NA | 0.0007-0.0063 | 0.0021-0.0038 | NA | NA |
| | 4.8 | N62B48 | NA | NA | 0.0025-0.0058 | 0.0001-0.0032 | NA | NA |
| 2008 | 3.0 | N52B30 | NA | NA | 0.0007-0.0063 | 0.0021-0.0038 | NA | NA |
| | 3.0 | N54B30 | NA | NA | 0.0007-0.0063 | 0.0021-0.0038 | NA | NA |
| | 4.8 | N62B48 | NA | NA | 0.0025-0.0058 | 0.0001-0.0032 | NA | NA |

NA: Not Available

## CRANKSHAFT AND CONNECTING ROD SPECIFICATIONS

All measurements are given in inches.

| Year | Engine Displacement Liters | Engine ID/VIN | Crankshaft | | | | Connecting Rod | | |
| --- | --- | --- | --- | --- | --- | --- | --- | --- | --- |
| | | | Main Brg. Journal Dia. | Main Brg. Oil Clearance | Shaft End-play | Thrust on No. | Journal Diameter | Oil Clearance | Side Clearance |
| 2007 | 3.0 | N52B30 | ① | 0.0007-0.0023 | 0.0031-0.0064 | NA | 1.7706-1.7712 | 0.0007-0.0022 | NA |
| | 3.0 | N54B30 | ② | 0.0007-0.0017 | 0.0024-0.0098 | NA | 1.9682-2.0742 | 0.0007-0.0023 | NA |
| | 4.8 | N62B48 | ③ | 0.0009-0.0020 | 0.0031-0.0096 | NA | 2.2677-2.8740 | 0.0011-0.0027 | NA |
| 2008 | 3.0 | N52B30 | ① | 0.0007-0.0023 | 0.0031-0.0064 | NA | 1.7706-1.7712 | 0.0007-0.0022 | NA |
| | 3.0 | N54B30 | ② | 0.0007-0.0017 | 0.0024-0.0098 | NA | 1.9682-2.0742 | 0.0007-0.0023 | NA |
| | 4.8 | N62B48 | ③ | 0.0009-0.0020 | 0.0031-0.0096 | NA | 2.2677-2.8740 | 0.0011-0.0027 | NA |

① Standard yellow 2.3615-2.3618 inches
  Standard green: 2.3613-2.3615 inches
  Standard white: 2.3611-2.3613 inches

② Designation S1: 2.1837-2.1839 inches
  Designation S2: 2.1834-2.1836 inches
  Designation S1: 2.1832-2.1834 inches

③ Standard yellow 2.7548-2.7551 inches for bearing 1, 2.7552-2.7555 inches for other bearing positions
  Standard green: 2.7545-2.7548 inches for bearing 1, 2.7549-2.7552 inches for other bearing positions
  Standard violet: 2.75436-2.7545 inches for bearing 1, 2.7547-2.7549 inches for other bearing positions

22205_BMWC_C0006

## PISTON AND RING SPECIFICATIONS

All measurements are given in inches

| Year | Engine Size Liters | Engine ID/VIN | Piston Clearance | Ring Gap | | | Ring Side Clearance | | |
| --- | --- | --- | --- | --- | --- | --- | --- | --- | --- |
| | | | | Top Compression | Bottom Compression | Oil Control | Top Compression | Bottom Compression | Oil Control |
| 2007 | 3.0 | N52B30 | 0.0004-0.0016 | 0.0078-0.0157 | 0.0078-0.0157 | 0.0078-0.0177 | 0.0008-0.0024 | 0.0008-0.0024 | 0.0005-0.0023 |
| | 3.0 | N54B30 | ① | 0.0071-0.0130 | 0.0118-0.0197 | — | 0.0008-0.0024 | 0.0008-0.0024 | — |
| | 4.8 | N62B48 | 0.0001-0.0014 | 0.0059-0.0098 | 0.0078-0.0136 | 0.0078-0.0157 | 0.0005-0.0023 | 0.0007-0.0026 | 0.0007-0.0026 |
| 2008 | 3.0 | N52B30 | 0.0004-0.0016 | 0.0078-0.0157 | 0.0078-0.0157 | 0.0078-0.0177 | 0.0008-0.0024 | 0.0008-0.0024 | 0.0005-0.0023 |
| | 3.0 | N54B30 | ① | 0.0071-0.0130 | 0.0118-0.0197 | — | 0.0008-0.0024 | 0.0008-0.0024 | — |
| | 4.8 | N62B48 | 0.0001-0.0014 | 0.0059-0.0098 | 0.0078-0.0136 | 0.0078-0.0157 | 0.0005-0.0023 | 0.0007-0.0026 | 0.0007-0.0026 |

① New Piston: 0.000-0.001
  Used Piston: 0.0007.0.0018

22205_BMWC_C0007

## TORQUE SPECIFICATIONS
All readings in ft. lbs.

| Year | Engine Displacement Liters | Engine ID/VIN | Cylinder Head Bolts | Main Bearing Bolts | Rod Bearing Bolts | Crankshaft Damper Bolts | Flywheel Bolts | Manifold | | Spark Plugs | Lug Nut |
|------|------|------|------|------|------|------|------|------|------|------|------|
| | | | | | | | | Intake | Exhaust | | |
| 2007 | 3.0 | N52B30 | ① | ② | ③ | 302 | ④ | ⑤ | ⑥ | ⑦ | 80 |
| | 3.0 | N54B30 | ⑧ | ⑨ | ③ | 18.5 | ⑩ | ⑤ | ⑥ | ⑦ | 80 |
| | 4.8 | N62B48 | ① | ⑪ | ③ | 302 | ④ | ⑤ | ⑥ | ⑦ | 88.5 |
| 2008 | 3.0 | N52B30 | ① | ② | ③ | 302 | ④ | ⑤ | ⑥ | ⑦ | 80 |
| | 3.0 | N54B30 | ⑧ | ⑨ | ③ | 18.5 | ⑩ | ⑤ | ⑥ | ⑦ | 80 |
| | 4.8 | N62B48 | ① | ⑪ | ③ | 302 | ④ | ⑤ | ⑥ | ⑦ | 88.5 |

① Cast iron block. Replace, wash and oil bolts
   Step 1: 22 ft. lbs.
   Step 2: 90 degrees
   Step 3: 90 degrees

② Cast iron block. Replace, wash and oil bolts
   Step 1: 14.8 ft. lbs.
   Step 2: 50 degrees

③ Replace, wash and oil connecting rod bolts
   Step 1: 14.8 ft. lbs.
   Step 2: 70 degrees

④ New micro-encapsulated screws:
   Automatic transmission: 88 ft. lbs.
   Manual transmission: 77.4 ft. lbs.

⑤ All M6 fasteners: 88 inch lbs.
   All M7 fasteners: 11 ft. lbs.
   All M8 fasteners: 16 ft. lbs.

⑥ Coat with Molykkote HSC compound or equivalent
   All M6 fasteners: 88 inch lbs.
   All M7 fasteners: 14.8 ft. lbs.

⑦ M12x1.25: 14.8-19.1 ft. lbs.
   M14x1.25: 21.4-24.3 ft. lbs.

⑧ Always use new bolts.
   Step 1: Tighten bolts 1-10 (M11) to 22 ft. lbs.
   Step 2: Tighten bolts 11-14 (M9) to 22 ft. lbs.
   Step 3: Tighten bolts 1-14 90 degrees
   Step 4: Tighten bolts 1-10 (M11) 90 degrees
   Step 5: Tighten bolts 1-14  45 degrees

⑨ Always use new bolts.
   Step 1: 14.8 ft. lbs.
   Step 2: 70 degrees

⑩ Always use new bolts.
   Automatic transmission:
   Step 1: 22.2 ft. lbs.
   Step 2: 92.5 ft. lbs.
   Manual transmission: 88.8 ft. lbs.

⑪ Always use new bolts. Do not remove coating
   Step 1: 14.75 ft. lbs.
   Step 2: 100 degrees

22205_BMWC_C0008

## WHEEL ALIGNMENT

| Year | Model | | Caster Range (+/-Deg.) | Caster Preferred Setting (Deg.) | Camber Range (+/-Deg.) | Camber Preferred Setting (Deg.) | Toe-in (Deg.) | Steering Axis Inclination (Deg.) |
|------|-------|---|------|------|------|------|------|------|
| 2007 | 3 Series ① | F | 0.15 | 0 | 0.25 | -0.18 | 0.14 +/- 0.10 | — |
| | | R | — | — | 0.15 | -1.30 | 0.18 +/- 0.06 | — |
| | 3 Series ② | F | 0.15 | 0 | 0.20 | -0.33 | 0.14 +/- 0.10 | — |
| | | R | — | — | 0.15 | -1.50 | 0.18 +/- 0.06 | — |
| | 3 Series ③ | F | 0.15 | 0 | 0.20 | 3 | 0.14 +/- 0.10 | — |
| | | R | — | — | 0.15 | -1.20 | 0.18 +/- 0.06 | — |
| | 5 Series | F | 0.15 | 0 | 20 | -0.12 | ④ | — |
| | | R | — | — | 0.20 | -2 | 0.18+/- 0.10 | — |
| | 7 Series ① | F | 0.33 | 7.65 | 0.33 | -0.1 | 0.16 +/- 0.13 | — |
| | | R | — | — | 0.33 | 0.5 | 0.16 +/- 0.16 | — |
| | 7 Series ② | F | 0.50 | +0.20 | 0.50 | -0.60 | 0.12 +/- 0.08 | — |
| | | R | — | — | 0.08 | +0.45 | 0.15 +/- 0.08 | — |
| 2008 | 3 Series ① | F | 0.15 | 0 | 0.25 | -0.18 | 0.14 +/- 0.10 | — |
| | | R | — | — | 0.15 | -1.30 | 0.18 +/- 0.06 | — |
| | 3 Series ② | F | 0.15 | 0 | 0.20 | -0.33 | 0.14 +/- 0.10 | — |
| | | R | — | — | 0.15 | -1.50 | 0.18 +/- 0.06 | — |
| | 3 Series ③ | F | 0.15 | 0 | 0.20 | 3 | 0.14 +/- 0.10 | — |
| | | R | — | — | 0.15 | -1.20 | 0.18 +/- 0.06 | — |
| | 5 Series | F | 0.15 | 0 | 20 | -0.12 | ④ | — |
| | | R | — | — | 0.20 | -2 | 0.18+/- 0.10 | — |
| | 7 Series ① | F | 0.33 | 7.65 | 0.33 | -0.1 | 0.16 +/- 0.13 | — |
| | | R | — | — | 0.33 | 0.5 | 0.16 +/- 0.16 | — |
| | 7 Series ② | F | 0.50 | +0.20 | 0.50 | -0.60 | 0.12 +/- 0.08 | — |
| | | R | — | — | 0.08 | +0.45 | 0.15 +/- 0.08 | — |

NOTE: Load vehicle with 150 lbs. on the front seats, 150 lbs on the rear seats and 46 lbs in the trunk with a full fuel tank.

① Standard suspension

② Sport suspension

③ Rough road package

④ 535xi Sport Wagon: 0.18+/- 0.10.

Except 535xi Sport Wagon: 0.08+/- 0.10.

## TIRE, WHEEL AND BALL JOINT SPECIFICATIONS

| Year | Model | OEM Tires | | Tire Pressures (psi) | | Wheel Size | Ball Joint Inspection | Lug Nut Torque (ft. lbs.) |
|------|-------|-----------|----------|-------|------|------------|----------------------|---------------------------|
| | | Standard | Optional | Front | Rear | | | |
| 2007 | 3-Series (Except 335xi) | P205/55R16 | ① | ② | ② | NA | NA | 89 |
| | 335xi | P225/45R17 | ① | ② | ② | NA | NA | 89 |
| | 5-Series | P225/50R17 | ③ | ② | ② | NA | NA | 89 |
| | 7-Series | P245/50R18 | ④ | ② | ② | NA | NA | 89 |
| 2008 | 3-Series (Except 335xi) | P205/55R16 | ① | ② | ② | NA | NA | 89 |
| | 335xi | P225/45R17 | ① | ② | ② | NA | NA | 89 |
| | 5-Series | P225/50R17 | ③ | ② | ② | NA | NA | 89 |
| | 7-Series | P245/50R18 | ④ | ② | ② | NA | NA | 89 |

① P225/45R17, P25/40R18, P255/40R17, P255/35R18

② See door sticker

③ P245/40R18, P245/35R19, P275/35R18, P275/30R19

④ P245/45R19, P245/40R19, P245/50R18, P245/40R20, P275/35R20

22205_BMWC_C0014

## BRAKE SPECIFICATIONS

All measurements in inches unless noted

| Year | Body Type | Model | | Brake Disc Original Thickness | Brake Disc Minimum Thickness | Maximum Runout | Minimum Pad Lining Thickness Front | Minimum Pad Lining Thickness Rear | Brake Caliper Bracket Bolts (ft. lbs.) | Brake Caliper Mounting Bolts (ft. lbs.) |
|---|---|---|---|---|---|---|---|---|---|---|
| 2007 | E93 | 328i Convertible | F | ① | ② | NA | 0.118 | — | 81 | 22 |
| | | | R | ③ | ② | NA | — | ④ | 48 | 22 |
| | E92 | 328i Coupe | F | ① | ② | NA | 0.118 | — | 81 | 22 |
| | | | R | ③ | ② | NA | — | ④ | 48 | 22 |
| | E90 | 328i Sedan | F | ① | ② | NA | 0.118 | — | 81 | 26 |
| | | | R | ③ | ② | NA | — | ④ | 48 | 26 |
| | E91 | 328i Wagon | F | ① | ② | NA | 0.118 | — | 81 | 22 |
| | | | R | ③ | ② | NA | — | ④ | 48 | 22 |
| | E92 | 328xi Coupe | F | ① | ② | NA | 0.118 | — | 81 | 22 |
| | | | R | ③ | ② | NA | — | ④ | 48 | 22 |
| | E90 | 328xi Sedan | F | ① | ② | NA | 0.118 | — | 81 | 26 |
| | | | R | ③ | ② | NA | — | ④ | 48 | 26 |
| | E91 | 328xi Wagon | F | ① | ② | NA | 0.118 | — | 81 | 22 |
| | | | R | ③ | ② | NA | — | ④ | 48 | 22 |
| | E93 | 335i Convertible | F | ① | ② | NA | 0.118 | — | 81 | 22 |
| | | | R | ③ | ② | NA | — | ④ | 48 | 22 |
| | E92 | 335i Coupe | F | ① | ② | NA | 0.118 | — | 81 | 22 |
| | | | R | ③ | ② | NA | — | ④ | 48 | 22 |
| | E90 | 335i Sedan | F | ① | ② | NA | 0.118 | — | 81 | 26 |
| | | | R | ③ | ② | NA | — | ④ | 48 | 26 |
| | E90 | 335xi Sedan | F | ① | ② | NA | 0.118 | — | 81 | 26 |
| | | | R | ③ | ② | NA | — | ④ | 48 | 26 |
| | E60 | 525i Sedan | F | ⑤ | ② | NA | 0.118 | — | 81 | 22 |
| | | | R | ⑥ | ② | NA | — | ⑤ | 48 | 22 |
| | E60 | 525xi Sedan | F | ⑤ | ② | NA | 0.118 | — | 81 | 22 |
| | | | R | ⑥ | ② | NA | — | ⑤ | 48 | 22 |
| | E60 | 530i Sedan | F | ⑤ | ② | NA | 0.118 | — | 81 | 22 |
| | | | R | ⑥ | ② | NA | — | ⑤ | 48 | 22 |
| | E60 | 530xi Sedan | F | ⑤ | ② | NA | 0.118 | — | 81 | 22 |
| | | | R | ⑥ | ② | NA | — | ⑤ | 48 | 22 |
| | E61 | 530xi Sport Wagon | F | ⑤ | ② | NA | 0.118 | — | 81 | 22 |
| | | | R | ⑥ | ② | NA | — | ⑤ | 48 | 22 |
| | E60 | 550i Sedan | F | ⑤ | ② | NA | 0.118 | — | 81 | 22 |
| | | | R | ⑥ | ② | NA | — | ⑤ | 48 | 22 |
| | E65 | 750i | F | ⑤ | ② | NA | 0.118 | — | 81 | 22 |
| | | | R | ⑥ | ② | NA | — | ⑤ | 48 | 22 |
| | E65 | 750Li | F | ⑤ | ② | NA | 0.118 | — | 81 | 22 |
| | | | R | ⑥ | ② | NA | — | ⑤ | 48 | 22 |

## BRAKE SPECIFICATIONS
All measurements in inches unless noted

| Year | Body Type | Model | | Brake Disc Original Thickness | Brake Disc Minimum Thickness | Brake Disc Maximum Runout | Minimum Pad Lining Thickness Front | Minimum Pad Lining Thickness Rear | Brake Caliper Bracket Bolts (ft. lbs.) | Brake Caliper Mounting Bolts (ft. lbs.) |
|------|-----------|-------|---|---|---|---|---|---|---|---|
| 2008 | E93 | 328i Convertible | F | ① | ② | NA | 0.118 | — | 81 | 22 |
| | | | R | ③ | ② | NA | — | ④ | 48 | 22 |
| | E92 | 328i Coupe | F | ① | ② | NA | 0.118 | — | 81 | 22 |
| | | | R | ③ | ② | NA | — | ④ | 48 | 22 |
| | E90 | 328i Sedan | F | ① | ② | NA | 0.118 | — | 81 | 26 |
| | | | R | ③ | ② | NA | — | ④ | 48 | 26 |
| | E91 | 328i Wagon | F | ① | ② | NA | 0.118 | — | 81 | 22 |
| | | | R | ③ | ② | NA | — | ④ | 48 | 22 |
| | E92 | 328xi Coupe | F | ① | ② | NA | 0.118 | — | 81 | 22 |
| | | | R | ③ | ② | NA | — | ④ | 48 | 22 |
| | E90 | 328xi Sedan | F | ① | ② | NA | 0.118 | — | 81 | 26 |
| | | | R | ③ | ② | NA | — | ④ | 48 | 26 |
| | E91 | 328xi Wagon | F | ① | ② | NA | 0.118 | — | 81 | 22 |
| | | | R | ③ | ② | NA | — | ④ | 48 | 22 |
| | E93 | 335i Convertible | F | ① | ② | NA | 0.118 | — | 81 | 22 |
| | | | R | ③ | ② | NA | — | ④ | 48 | 22 |
| | E92 | 335i Coupe | F | ① | ② | NA | 0.118 | — | 81 | 22 |
| | | | R | ③ | ② | NA | — | ④ | 48 | 22 |
| | E90 | 335i Sedan | F | ① | ② | NA | 0.118 | — | 81 | 26 |
| | | | R | ③ | ② | NA | — | ④ | 48 | 26 |
| | E90 | 335xi Sedan | F | ① | ② | NA | 0.118 | — | 81 | 26 |
| | | | R | ③ | ② | NA | — | ④ | 48 | 26 |
| | E60 | 528i Sedan | F | ⑤ | ② | NA | 0.118 | — | 81 | 22 |
| | | | R | ⑥ | ② | NA | — | ⑤ | 48 | 22 |
| | E60 | 528xi Sedan | F | ⑤ | ② | NA | 0.118 | — | 81 | 22 |
| | | | R | ⑥ | ② | NA | — | ⑤ | 48 | 22 |
| | E60 | 535i Sedan | F | ⑤ | ② | NA | 0.118 | — | 81 | 22 |
| | | | R | ⑥ | ② | NA | — | ⑤ | 48 | 22 |
| | E60 | 535xi Sedan | F | ⑤ | ② | NA | 0.118 | — | 81 | 22 |
| | | | R | ⑥ | ② | NA | — | ⑤ | 48 | 22 |
| | E61 | 535xi Sport Wagon | F | ⑤ | ② | NA | 0.118 | — | 81 | 22 |
| | | | R | ⑥ | ② | NA | — | ⑤ | 48 | 22 |
| | E60 | 550i Sedan | F | ⑤ | ② | NA | 0.118 | — | 81 | 22 |
| | | | R | ⑥ | ② | NA | — | ⑤ | 48 | 22 |
| | E65 | 750i | F | ⑤ | ② | NA | 0.118 | — | 81 | 22 |
| | | | R | ⑥ | ② | NA | — | ⑤ | 48 | 22 |
| | E65 | 750Li | F | ⑤ | ② | NA | 0.118 | — | 81 | 22 |
| | | | R | ⑥ | ② | NA | — | ⑤ | 48 | 22 |

NA: Not Available

F: Front

R: Rear

① Depending on brake package: .866in., .945 in. or 1.18 in.

② Minimum thickness is stamped in the brake disk shell
   Maximum machining limit per side: 0.315 inches

③ Depending on brake package: .394 in., .787 in. or .945 in.

④ Rear brake pad wear limit: 0.118 inches
   Parking brake shoe wear limit: 0.059 inches

⑤ Depending on brake package: .945 in., 1.18 in. or 1.41in.

⑥ Depending on brake package: .787 in. or .945 in.

## SCHEDULED MAINTENANCE INTERVALS
### BMW—3 SERIES, 5 SERIES and 7 SERIES

| TO BE SERVICED | TYPE OF SERVICE | SERVICE INTERVALS | | | |
|---|---|---|---|---|---|
| | | INITIAL 1200 MILES | OIL SERVICE | INSPECTION I | INSPECTION II |
| Oil level | S/I | ✓ | | | |
| Engine oil | R | ① | | | |
| Engine oil & filter | R ② | | ✓ | ✓ | ✓ |
| Engine air cleaner element | R ③ | | | | ✓ |
| Spark plugs | R | | | | ✓ |
| Fuel filter | R ④ | | | | ✓ |
| Fuel, vapor lines & fuel cap | S/I | ✓ | | ✓ | ✓ |
| Cooling system | S/I | ✓ | | ✓ | ✓ |
| Exhaust pipe & muffler | S/I | ✓ | | ✓ | ✓ |
| Catalytic converter & shielding | S/I | ✓ | | ✓ | ✓ |
| Throttle linkage | S/I | | | ✓ | ✓ |
| Engine (check for leakage) | S/I | ✓ | | | |
| Engine drive belts | S/I | | | | ✓ |
| Maintenance Indicators | RE | | ⑤ | ✓ | ✓ |
| Engine coolant | R | | | ⑥ | ⑥ |
| Oxygen sensor | R ⑦ | | | | |
| Intake air dust separators | S/I ⑧ | | | | ✓ |
| Brake & clutch fluids ⑥ | S/I | | | ✓ | ✓ |
| Brake pads & discs | S/I | | | ✓ | ✓ |
| Parking brake system | S/I | | | ✓ | ✓ |
| Power steering system | S/I | | | ✓ | ✓ |
| Rear axle fluid | S/I | | | ✓ | ✓ |
| Steering play, suspension track rods, front axle joints, steering linkage & joint disc | S/I | | | ✓ | ✓ |
| Transmission fluid/oil | S/I | | | ✓ | ⑨ |
| Wheel centering hubs | S/I | | | ✓ | ✓ |
| Rear axle fluid ⑩ | R | | ✓ | | ✓ |
| OBD system for codes | S/I | ✓ | | ✓ | ✓ |

R: Replace          S/I: Service or Inspect          RE: Reset

Note: BMW does not rely solely on vehicle mileage to determine service intervals. An on-oboard diagnostic center, monitors engine operating conditions, along with mileage, to determine the most effective maintenance intervals. The information is then conveyed to the driver through the service indicator lights, located in the center of the instrument p[anel.

① Service is not required for 328 models.

② On vehicles operated less than 6200 miles per year, more frequent service may be required.

③ Replace more frequently if vehicle is operated in dusty conditions.

④ Recommended service for California models, required for all other models.

⑤ Reset the oil service indicator lights only.

⑥ Replace every 2 years with inspection service.

⑦ Replace every 100,000 miles on all models.

⑧ Required service for manual transmission models only.

⑨ Change fluid (A/T) or oil (M/T) at inspection.

⑩ At first oil service, then at each inspection.

## FREQUENT OPERATION MAINTENANCE (SEVERE SERVICE)

**If a vehicle is operated under any of the following conditions it is considered severe service**

- Extremely dusty areas.

- 50% or more of the vehicle operation is in 32°C (90°F) or higher temperatures, or constant operation in temperatures below 0°C (32°F).

- Prolonged idling (vehicle operation in stop and go traffic).

- Frequent short running periods (engine does not warm to normal operating temperatures).

- Police, taxi, delivery usage or trailer towing usage.

22205_BMWC_C0012

## PRECAUTIONS

Before servicing any vehicle, please be sure to read all of the following precautions, which deal with personal safety, prevention of component damage, and important points to take into consideration when servicing a motor vehicle:

• Never open, service or drain the radiator or cooling system when the engine is hot; serious burns can occur from the steam and hot coolant.

• Observe all applicable safety precautions when working around fuel. Whenever servicing the fuel system, always work in a well-ventilated area. Do not allow fuel spray or vapors to come in contact with a spark, open flame, or excessive heat (a hot drop light, for example). Keep a dry chemical fire extinguisher near the work area. Always keep fuel in a container specifically designed for fuel storage; also, always properly seal fuel containers to avoid the possibility of fire or explosion. Refer to the additional fuel system precautions later in this section.

• Fuel injection systems often remain pressurized, even after the engine has been turned **OFF**. The fuel system pressure must be relieved before disconnecting any fuel lines. Failure to do so may result in fire and/or personal injury.

• Brake fluid often contains polyglycol ethers and polyglycols. Avoid contact with the eyes and wash your hands thoroughly after handling brake fluid. If you do get brake fluid in your eyes, flush your eyes with clean, running water for 15 minutes. If eye irritation persists, or if you have taken brake fluid internally, IMMEDIATELY seek medical assistance.

• The EPA warns that prolonged contact with used engine oil may cause a number of skin disorders, including cancer. You should make every effort to minimize your exposure to used engine oil. Protective gloves should be worn when changing oil. Wash your hands and any other exposed skin areas as soon as possible after exposure to used engine oil. Soap and water, or waterless hand cleaner should be used.

• All new vehicles are now equipped with an air bag system, often referred to as a Supplemental Restraint System (SRS) or Supplemental Inflatable Restraint (SIR) system. The system must be disabled before performing service on or around system components, steering column, instrument panel components, wiring and sensors. Failure to follow safety and disabling procedures could result in accidental air bag deployment, possible personal injury and unnecessary system repairs.

• Always wear safety goggles when working with, or around, the air bag system. When carrying a non-deployed air bag, be sure the bag and trim cover are pointed away from your body. When placing a non-deployed air bag on a work surface, always face the bag and trim cover upward, away from the surface. This will reduce the motion of the module if it is accidentally deployed. Refer to the additional air bag system precautions later in this section.

• Clean, high quality brake fluid from a sealed container is essential to the safe and proper operation of the brake system. You should always buy the correct type of brake fluid for your vehicle. If the brake fluid becomes contaminated, completely flush the system with new fluid. Never reuse any brake fluid. Any brake fluid that is removed from the system should be discarded. Also, do not allow any brake fluid to come in contact with a painted surface; it will damage the paint.

• Never operate the engine without the proper amount and type of engine oil; doing so WILL result in severe engine damage.

• Timing belt maintenance is extremely important. Many models utilize an interference-type, non-freewheeling engine. If the timing belt breaks, the valves in the cylinder head may strike the pistons, causing potentially serious (also time-consuming and expensive) engine damage. Refer to the maintenance interval charts for the recommended replacement interval for the timing belt, and to the timing belt section for belt replacement and inspection.

• Disconnecting the negative battery cable on some vehicles may interfere with the functions of the on-board computer system(s) and may require the computer to undergo a relearning process once the negative battery cable is reconnected.

• When servicing drum brakes, only disassemble and assemble one side at a time, leaving the remaining side intact for reference.

• Only an MVAC-trained, EPA-certified automotive technician should service the air conditioning system or its components.

## BRAKES

### GENERAL INFORMATION

*PRECAUTIONS*

• Certain components within the ABS system are not intended to be serviced or repaired individually.

• Do not use rubber hoses or other parts not specifically specified for and ABS system. When using repair kits, replace all parts included in the kit. Partial or incorrect repair may lead to functional problems and require the replacement of components.

• Lubricate rubber parts with clean, fresh brake fluid to ease assembly. Do not use shop air to clean parts; damage to rubber components may result.

• Use only DOT 3 brake fluid from an unopened container.

• If any hydraulic component or line is removed or replaced, it may be necessary to bleed the entire system.

• A clean repair area is essential. Always clean the reservoir and cap thoroughly before removing the cap. The slightest amount of dirt in the fluid may plug an orifice and impair the system function. Perform repairs after components have been thoroughly cleaned; use only denatured alcohol

## ANTI-LOCK BRAKE SYSTEM (ABS)

to clean components. Do not allow ABS components to come into contact with any substance containing mineral oil; this includes used shop rags.

• The Anti-Lock control unit is a microprocessor similar to other computer units in the vehicle. Ensure that the ignition switch is **OFF** before removing or installing controller harnesses. Avoid static electricity discharge at or near the controller.

• If any arc welding is to be done on the vehicle, the control unit should be unplugged before welding operations begin.

## BRAKES                                    BLEEDING THE BRAKE SYSTEM

### BLEEDING PROCEDURE

*BLEEDING PROCEDURE*

1. Connect a brake bleeding tool with a maximum 29 PSI (2 Bar) filling pressure.

➡**A second person is needed to help carry out this work.**

2. Disconnect retaining tabs and remove the master cylinder cover.
3. Flushing brake system completely
4. Connect bleeder hose with collecting tray to bleeder valve on rear right brake caliper.
    a. Open bleeder valve and purge until clear, bubble-free brake fluid emerges.

b. Close bleed valve.
5. Follow same procedure on rear left, front right and front left wheel brake.
6. Bleeding rear-axle brake circuit
    a. Connect bleeder hose with collecting tray to bleeder valve on rear right brake caliper.
    b. Bleed right brake caliper. After completing routine, press brake pedal 5 times to floor; clear and bubble-free brake fluid must flow out.
    c. Close bleed valve.
7. Repeat procedure at rear left.
8. Bleeding front-axle brake circuit
    a. Connect bleeder hose with collect-

ing tray to bleeder valve on front right brake caliper.
    b. Bleed right front brake caliper. After completing routine, press brake pedal 5 times to floor, clear and bubble-free brake fluid must flow out.
    c. Close bleed valve.
9. Repeat procedure at front left.
10. Disconnect the brake bleeding tool and remove from the master cylinder.
11. Check and adjust the brake fluid level.
12. Inspect the rubber seal in the brake fluid master cylinder cap. Replace as necessary.

## BRAKES                                              FRONT DISC BRAKES

### ❊❊ CAUTION

**Dust and dirt accumulating on brake parts during normal use may contain asbestos fibers from production or aftermarket brake linings. Breathing excessive concentrations of asbestos fibers can cause serious bodily harm. Exercise care when servicing brake parts. Do not sand or grind brake lining unless equipment used is designed to contain the dust residue. Do not clean brake parts with compressed air or by dry brushing. Cleaning should be done by dampening the brake components with a fine mist of water, then wiping the brake components clean with a dampened cloth. Dispose of cloth and all residue containing asbestos fibers in an impermeable container with the appropriate label. Follow practices prescribed by the Occupational Safety and Health Administration (OSHA) and the Environmental Protection Agency (EPA) for the handling, processing, and disposing of dust or debris that may contain asbestos fibers.**

### BRAKE CALIPER

*REMOVAL & INSTALLATION*

**TRW Caliper**

*See Figures 1 and 2.*

1. Before servicing the vehicle, refer to the precautions.
2. Remove wheels.
3. Press clutch pedal down to floor and secure with pedal support.

### ❊❊ WARNING

**The pedal support may only be removed when the brake lines are reconnected. This prevents brake fluid from emerging from the expansion tank and air from entering the system when the brake lines are opened.**

4. Disconnect brake hose from holder.

### ❊❊ WARNING

**Grip brake hose at square head to prevent connecting piece from turning in retaining bracket.**

5. Disconnect brake hose from brake line.
6. Disconnect brake hose from brake caliper.

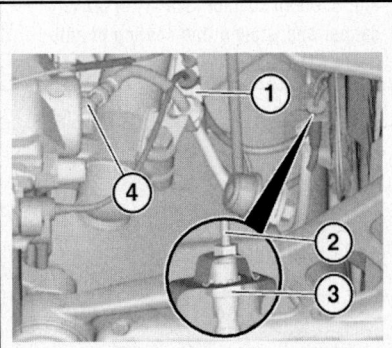

22205_BMWC_G0003

**Fig. 1 Pull the brake hose out of the holder (1) by gripping the brake hose at the square head (3), disconnect the hose from the line (2), and disconnect the hose from the caliper (4)**

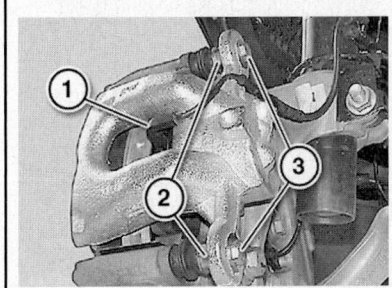

22205_BMWC_G0004

**Fig. 2 Pull brake pad wear sensor (1) towards rear out of lining, unscrew guide bolts (3) and if necessary, grip at hexagon head (2) to prevent rotation**

7. Pull brake pad wear sensor towards rear out of lining (left side only).
8. Unscrew guide bolts. If necessary, grip at hexagon head to prevent rotation.
9. Remove brake caliper by pulling upwards.

*To install:*

10. Install brake caliper. Replace the guide bolts and tighten to 26 ft. lbs. (35 Nm) on E90 and 22 ft. lbs. (30 Nm) on all others.
11. If necessary, replace brake pad wear sensor.

### ❊❊ WARNING

**Never twist brake hose when installing it and avoid all contact with parts attached rigidly to the body.**

12. Connect brake hose to brake caliper. Tighten to 18 ft. lbs. (24 Nm).
13. Connect brake hose to brake line. Tighten to 8 ft. lbs. (12 Nm) on E90 series and 12 ft. lbs. (17 Nm) on E60 series.

14. Connect brake hose to holder.
15. Remove clutch pedal.
16. Check and fill brake fluid to proper level.
17. Bleed the brake system.
18. Install the wheels.

### Teves Caliper

*See Figures 3 through 6.*

1. Before servicing the vehicle, refer to the precautions.
2. Remove wheels
3. Remove brake pad wear sensor

> **✳✳ WARNING**
>
> **The brake pad wear sensor must be replaced once it has been removed (brake pad wear sensor loses its retention capability in the brake pad).**

4. Disconnect brake hose to brake caliper.
5. Disconnect brake hose to brake line.
6. Disconnect brake hose to holder.
7. Lift out retaining spring.
8. Remove plastic plugs.
9. Remove guide bolts and lift out brake caliper towards rear of vehicle.

*To install:*

10. Clean contact face of brake piston and apply a thin coating of anti-squeak compound.
11. Clean contact faces of brake pad hammer heads/brake caliper housing and coat with anti-squeak compound.
12. Clean contact face of brake caliper and apply a thin coating of anti-squeak compound.
13. Clean brake carrier at hammerhead guides and apply a thin coating of anti-squeak compound.
14. Clean guide bolts only; do not grease. Check threads.

➡ **Replace all guide bolts which are not in perfect condition.**

15. Install brake caliper. Replace the guide bolts and tighten to 26 ft. lbs. (35 Nm) on E90 and 22 ft. lbs. (30 Nm) on all others.

> **✳✳ WARNING**
>
> **Never twist brake hose when installing it and avoid all contact with parts attached rigidly to the body.**

16. Connect brake hose to brake caliper. Tighten to 18 ft. lbs. (24 Nm).
17. Connect brake hose to brake line.

Tighten to 8 ft. lbs. (12 Nm) on E90 series and 12 ft. lbs. (17 Nm) on E60 series.
18. Connect brake hose to holder.
19. Install plastic plugs.
20. Install the retaining spring.
21. Install the brake pad wear sensor.
22. Check and fill brake fluid to proper level.
23. Bleed the brake system.
24. Install the wheels.

## DISC BRAKE PADS

### REMOVAL & INSTALLATION

*See Figures 7 through 10.*

1. Before servicing the vehicle, refer to the precautions.
2. Remove the caliper and suspend it out of the way using wire.

➡ **Do not allow brake caliper to hang from the brake hose.**

3. Remove the brake pads.

*To install:*

4. Press brake piston fully back with special tool 34 1 050.

> **✳✳ WARNING**
>
> **When pressing piston back, note brake fluid level in expansion tank. Overflowing brake fluid will damage the paintwork.**

5. Check dust sleeve for damage and replace if necessary.
6. Clean contact face of brake piston and apply a thin coating of anti-squeak compound.

> **✳✳ WARNING**
>
> **Dust sleeve must not come into contact with anti-squeak compound as this may cause the dust sleeve to swell.**

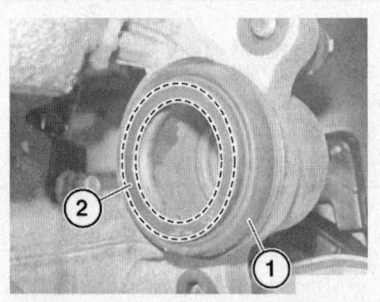

22205_BMWC_G0007

**Fig. 3 Clean contact face (2) of brake piston and apply a thin coating of anti-squeak compound**

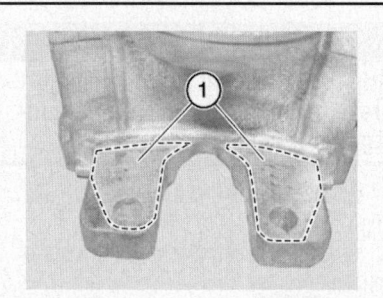

22205_BMWC_G0009

**Fig. 5 Clean contact face (1) of brake caliper and apply a thin coating of anti-squeak compound**

22205_BMWC_G0008

**Fig. 4 Clean contact faces (1) of brake pad hammer heads/brake caliper housing and coat with anti-squeak compound**

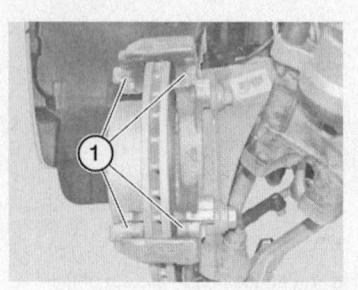

22205_BMWC_G0010

**Fig. 6 Clean brake carrier at hammerhead guides (1) and apply a thin coating of anti-squeak compound.**

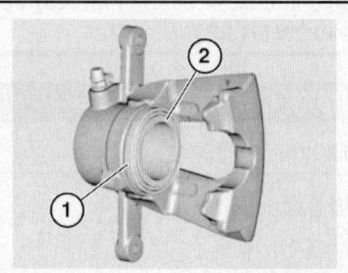

22205_BMWC_G0013

**Fig. 7 Check dust sleeve (1) for damage and replace if necessary. Clean contact face (2) of brake piston and apply a thin coating of anti-squeak compound**

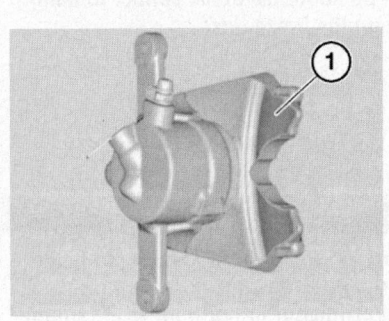

Fig. 8 Check dust sleeve (1) for damage and replace if necessary. Clean contact face (2) of brake piston and apply a thin coating of anti-squeak compound

Fig. 9 Remove lining retaining springs (1), as necessary

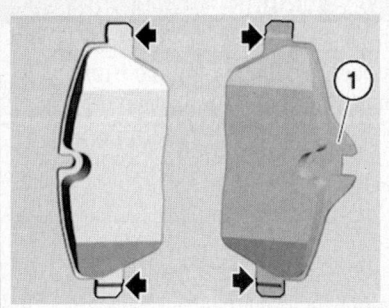

Fig. 10 Brake pad with indentation (1) is intended for accommodating the brake pad wear sensor and must be fitted on the piston side

7. Clean contact face of brake caliper and apply a thin coating of anti-squeak compound.
8. Clean hammerhead guides and apply a thin coating of anti-squeak compound.
9. Replace the lining retaining springs, as necessary.
10. Clean contact face of brake carrier and apply a thin coating of anti-squeak compound.
11. Install the new brake pads.

➡**Brake pad with indentation is intended for accommodating the brake pad wear sensor and must be fitted on the piston side.**

12. Install the caliper.
13. Check and fill brake fluid to proper level.

14. Fully depress brake pedal several times so that brake linings contact brake discs.

➡**When replacing pads, reset CBS display in accordance with factory specification.**

15. Install the wheels.

**BRAKES**

**REAR DISC BRAKES**

❊❊ **CAUTION**

Dust and dirt accumulating on brake parts during normal use may contain asbestos fibers from production or aftermarket brake linings. Breathing excessive concentrations of asbestos fibers can cause serious bodily harm. Exercise care when servicing brake parts. Do not sand or grind brake lining unless equipment used is designed to contain the dust residue. Do not clean brake parts with compressed air or by dry brushing. Cleaning should be done by dampening the brake components with a fine mist of water, then wiping the brake components clean with a dampened cloth. Dispose of cloth and all residue containing asbestos fibers in an impermeable container with the appropriate label. Follow practices prescribed by the Occupational Safety and Health Administration (OSHA) and the Environmental Protection Agency (EPA) for the handling, processing, and disposing of dust or debris that may contain asbestos fibers.

**BRAKE CALIPER**

*REMOVAL & INSTALLATION*

**TRW Caliper**

*See Figure 11.*

1. Before servicing the vehicle, refer to the precautions.
2. Remove wheels.
3. Press clutch pedal down to floor and secure with pedal support.

❊❊ **WARNING**

The pedal support may only be removed when the brake lines are reconnected. This prevents brake fluid from emerging from the expansion tank and air from entering the system when the brake lines are opened.

4. Disconnect brake hose from holder.

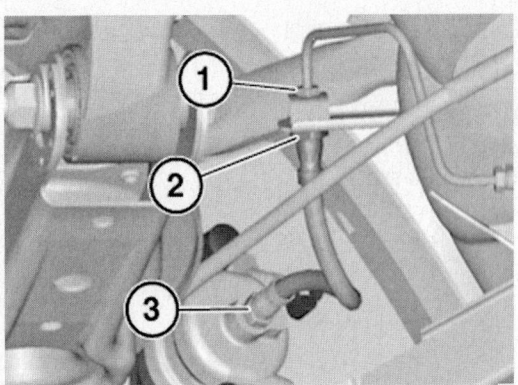

Fig. 11 Grip the brake hose at the square head (2), disconnect the hose from the line (1) and caliper (3)

> **※ WARNING**
>
> **Grip brake hose at square head to prevent connecting piece from turning in retaining bracket.**

5. Disconnect brake hose from brake line.
6. Disconnect brake hose from brake caliper.
7. Pull brake pad wear sensor towards rear out of lining (right side only).
8. Unscrew guide bolts. If necessary, grip at hexagon head to prevent rotation.
9. Remove brake caliper by pulling upwards.

*To install:*
10. Install brake caliper. Replace the guide bolts and tighten to 26 ft. lbs. (35 Nm) on E90 and 22 ft. lbs. (30 Nm) on all others.
11. If necessary, replace brake pad wear sensor.

> **※ WARNING**
>
> **Never twist brake hose when installing it and avoid all contact with parts attached rigidly to the body.**

12. Connect brake hose to brake caliper. Tighten to 18 ft. lbs. (24 Nm).
13. Connect brake hose to brake line. Tighten to 8 ft. lbs. (12 Nm) on E90 series and 12 ft. lbs. (17 Nm) on E60 series.
14. Connect brake hose to holder.
15. Remove clutch pedal.
16. Check and fill brake fluid to proper level.
17. Bleed the brake system.
18. Install the wheels.

**Teves Caliper**

1. Before servicing the vehicle, refer to the precautions.
2. Remove wheels.
3. Remove brake pad wear sensor (right side only).

> **※ WARNING**
>
> **The brake pad wear sensor must be replaced once it has been removed**

(brake pad wear sensor loses its retention capability in the brake pad).

4. Disconnect brake hose to brake caliper.
5. Disconnect brake hose to brake line.
6. Disconnect brake hose to holder.
7. Lift out retaining spring.
8. Remove plastic plugs.
9. Remove guide bolts and lift out brake caliper towards rear of vehicle.

*To install:*
10. Clean guide bolts only; do not grease. Check threads.

➡ **Replace all guide bolts which are not in perfect condition.**

11. Install brake caliper. Replace the guide bolts and tighten to 26 ft. lbs. (35 Nm) on E90 and 22 ft. lbs. (30 Nm) on all others.

> **※ WARNING**
>
> **Never twist brake hose when installing it and avoid all contact with parts attached rigidly to the body.**

12. Connect brake hose to brake caliper. Tighten to 18 ft. lbs. (24 Nm).
13. Connect brake hose to brake line. Tighten to 8 ft. lbs. (12 Nm) on E90 series and 12 ft. lbs. (17 Nm) on E60 series.
14. Install plastic plugs.
15. Install the retaining spring.
16. Install the brake pad wear sensor.
17. Check and fill brake fluid to proper level.
18. Bleed the brake system.
19. Install the wheels.

## DISC BRAKE PADS

*REMOVAL AND INSTALLATION*

1. Before servicing the vehicle, refer to the precautions.
2. Remove the caliper and suspend it out of the way using wire.

➡ **Do not allow brake caliper to hang from the brake hose.**

3. Remove the brake pads.

*To install:*
4. Press brake piston fully back with special tool 34 1 050.

> **※ WARNING**
>
> **When pressing piston back, note brake fluid level in expansion tank. Overflowing brake fluid will damage the paintwork.**

5. Check dust sleeve for damage and replace if necessary.

> **※ WARNING**
>
> **Dust sleeve must not come into contact with anti-squeak compound as this may cause the dust sleeve to swell.**

6. Clean contact face of brake caliper and apply a thin coating of anti-squeak compound.
7. Clean hammerhead guides and apply a thin coating of anti-squeak compound.
8. Replace the lining retaining springs, as necessary.
9. Clean contact face of brake carrier and apply a thin coating of anti-squeak compound.
10. Install the new brake pads.

➡ **Brake pad with indentation is intended for accommodating the brake pad wear sensor and must be fitted on the piston side.**

11. Install the caliper.
12. Check and fill brake fluid to proper level.
13. Fully depress brake pedal several times so that brake linings contact brake discs.

➡ **When replacing pads, reset CBS display in accordance with factory specification.**

14. Install the wheels.

## BRAKES                                      PARKING BRAKE

### PARKING BRAKE SHOES

*REMOVAL & INSTALLATION*

*See Figures 12 through 15.*

1. Before servicing the vehicle, refer to the precautions.

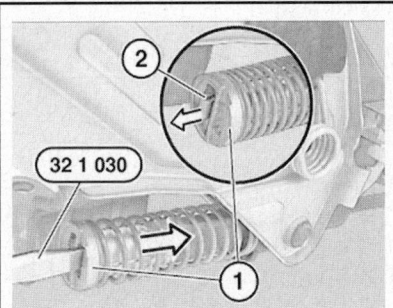

Fig. 12 Actuate parking brake lever. Screw in special tool 32 1 030 partially. Press stop (1) of adjusting spring back to such an extent that retaining hook (2) engages in stop (1)

2. Remove rear brake disc.
3. Actuate parking brake lever. Screw in special tool 32 1 030 partially. Press stop of adjusting spring back to such an extent that retaining hook engages in stop.
4. Disconnect return upper spring with brake spring pliers.
5. Disconnect return lower spring with brake spring pliers.

Fig. 13 Disconnect return upper spring (1) with brake spring pliers

6. Turn clamping pins with special tool 34 4 000 through 90° and disconnect. Remove brake shoes.

*To install:*

7. Install brake shoes.
8. Turn clamping pins with special tool 34 4 000 through 90° to lock.
9. Check and if necessary replace return springs.
10. Connect return lower spring with brake spring pliers.
11. Connect return upper spring with brake spring pliers.

➡ **Pay attention to installation position of adjustment screw.**

12. Apply a thin coat of grease to bush and screw threads.
13. Lever out restraining hook with a suitable screwdriver.

Fig. 14 Disconnect return lower spring (1) with brake spring pliers

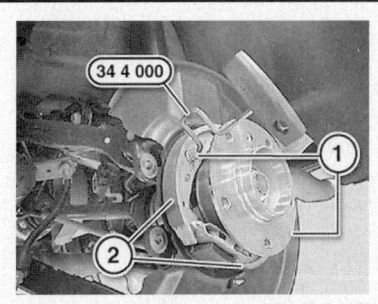

Fig. 15 Turn clamping pins (1) with special tool 34 4 000 through 90° and disconnect. Remove break shores (2)

14. Restraining hook must detach from stop of adjusting spring.
15. Install rear brake disc.
16. Adjusting parking brake.

## CHASSIS ELECTRICAL     AIR BAG (SUPPLEMENTAL RESTRAINT SYSTEM)

### GENERAL INFORMATION

#### ✳✳ CAUTION

**These vehicles are equipped with an air bag system. The system must be disarmed before performing service on, or around, system components, the steering column, instrument panel components, wiring and sensors. Failure to follow the safety precautions and the disarming procedure could result in accidental air bag deployment, possible injury and unnecessary system repairs.**

*SERVICE PRECAUTIONS*

Disconnect and isolate the battery negative cable before beginning any airbag system component diagnosis, testing, removal, or installation procedures. Allow system capacitor to discharge for two minutes before beginning any component service. This will disable the airbag system. Failure to disable the airbag system may result in accidental airbag deployment, personal injury, or death.

Do not place an intact undeployed airbag face down on a solid surface. The airbag will propel into the air if accidentally deployed and may result in personal injury or death.

When carrying or handling an undeployed airbag, the trim side (face) of the airbag should be pointing towards the body to minimize possibility of injury if accidental deployment occurs. Failure to do this may result in personal injury or death.

Replace airbag system components with OEM replacement parts. Substitute parts may appear interchangeable, but internal differences may result in inferior occupant protection. Failure to do so may result in occupant personal injury or death.

Wear safety glasses, rubber gloves, and long sleeved clothing when cleaning powder residue from vehicle after an airbag deployment. Powder residue emitted from a deployed airbag can cause skin irritation. Flush affected area with cool water if irritation is experienced. If nasal or throat irritation is experienced, exit the vehicle for fresh air until the irritation ceases. If irritation continues, see a physician.

Do not use a replacement airbag that is not in the original packaging. This may result in improper deployment, personal injury, or death.

The factory installed fasteners, bolts and bolts used to fasten airbag components have a special coating and are specifically designed for the airbag system. Do not use substitute fasteners. Use only original equipment fasteners listed in the parts catalog when fastener replacement is required.

During, and following, any child restraint anchor service, due to impact event or vehicle repair, carefully inspect all mounting hardware, tether straps, and anchors for proper installation, operation, or damage. If a child restraint anchor is found damaged in any way, the anchor must be replaced. Failure to do this may result in personal injury or death.

Deployed and non-deployed airbags may or may not have live pyrotechnic material within the airbag inflator.

Do not dispose of driver/passenger/curtain airbags or seat belt tensioners unless you are sure of complete deployment. Refer to the Hazardous Substance Control System for proper disposal.

Dispose of deployed airbags and tensioners consistent with state, provincial, local, and federal regulations.

After any airbag component testing or service, do not connect the battery negative cable. Personal injury or death may result if the system test is not performed first.

If the vehicle is equipped with the Occupant Classification System (OCS), do not connect the battery negative cable before performing the OCS Verification Test using the scan tool and the appropriate diagnostic information. Personal injury or death may result if the system test is not performed properly.

Never replace both the Occupant Restraint Controller (ORC) and the Occupant Classification Module (OCM) at the same time. If both require replacement, replace one, then perform the Airbag System test before replacing the other.

Both the ORC and the OCM store Occupant Classification System (OCS) calibration data, which they transfer to one another when one of them is replaced. If both are replaced at the same time, an irreversible fault will be set in both modules and the OCS may malfunction and cause personal injury or death.

If equipped with OCS, the Seat Weight Sensor is a sensitive, calibrated unit and must be handled carefully. Do not drop or handle roughly. If dropped or damaged, replace with another sensor. Failure to do so may result in occupant injury or death.

If equipped with OCS, the front passenger seat must be handled carefully as well. When removing the seat, be careful when setting on floor not to drop. If dropped, the sensor may be inoperative, could result in occupant injury, or possibly death.

If equipped with OCS, when the passenger front seat is on the floor, no one should sit in the front passenger seat. This uneven force may damage the sensing ability of the seat weight sensors. If sat on and damaged, the sensor may be inoperative, could result in occupant injury, or possibly death.

### DISARMING THE SYSTEM

1. Before servicing the vehicle, refer to the precautions.
2. Place the ignition switch in the **OFF** position.
3. Disconnect the negative battery terminal and cover the battery terminal to prevent accidental contact.
4. Once the battery has been disconnected, wait for a period of approximately 10 minutes allowing the capacitor in the control unit to discharge.

### ARMING THE SYSTEM

When repairs are completed, connect the negative battery cable.

## DRIVETRAIN

### AUTOMATIC TRANSMISSION ASSEMBLY

*REMOVAL & INSTALLATION*

#### GA6L45R Transmission
*See Figures 16 through 23.*

**✳✳ WARNING**

**Aluminum-magnesium materials. No steel bolts/bolts may be used due to the threat of electrochemical corrosion. A magnesium crankcase requires aluminum bolts/bolts exclusively. Aluminum bolts/bolts must be replaced each time they are removed. The end faces of these bolts/bolts are painted blue for the purposes of reliable identification. Jointing torque and angle of rotation must be observed without fail (risk of damage).**

1. Before servicing the vehicle, refer to the precautions.
2. Disconnect the negative battery cable.
3. Remove underbody protection with bracket at front and rear.

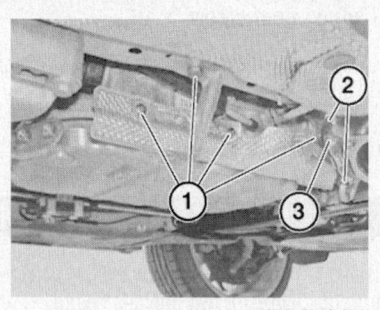

Fig. 16 Remove bolts (1), remove bracket and heat shield, Disconnect lines (2) and remove bracket (3)

22205_BMWC_G0022

4. Remove complete exhaust system.
5. Remove heat shields.
6. Support engine with lifter when removing transmission.
7. Remove bolts, remove bracket and heat shield, Disconnect lines and remove bracket.
8. Disconnect plug connector, remove bolts and remove retaining plate.
9. Remove aluminum bolts on right next to cable retaining plate with special tool 00 9 010.

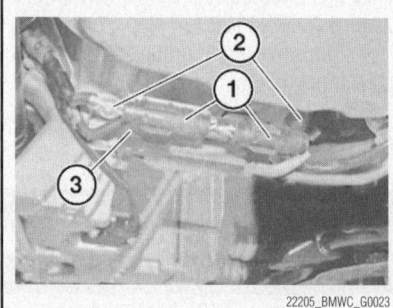

Fig. 17 Disconnect plug connector (1), remove bolts (2) and remove retaining plate (3)

22205_BMWC_G0023

**✳✳ WARNING**

**Blue aluminum bolts/bolts must be replaced.**

10. Grip clamping sleeve, loosen nut, detach retainer downward using a screwdriver and pull cable out of holder.
11. Remove screw and disconnect hydraulic lines to transmission fluid cooler.
12. Remove nut and bracket from transmission oil lines on oil pan.
13. Support transmission with special tools 23 4 050, 00 2 030.

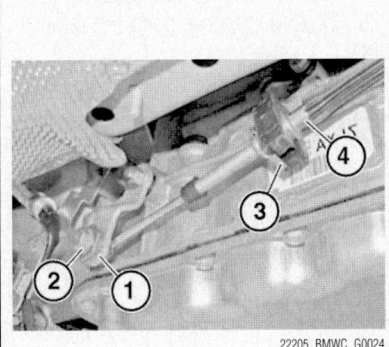

**Fig. 18 Grip clamping sleeve (1), loosen nut (2), detach retainer (3) downward using a screwdriver and pull cable (4) out of holder**

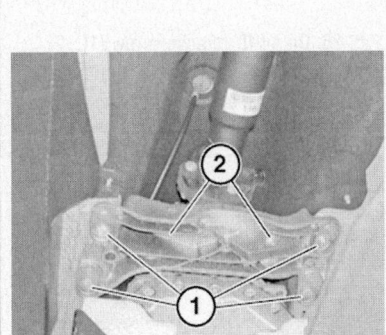

**Fig. 19 Remove bolts (1) and nuts (2) and remove transmission cross-member**

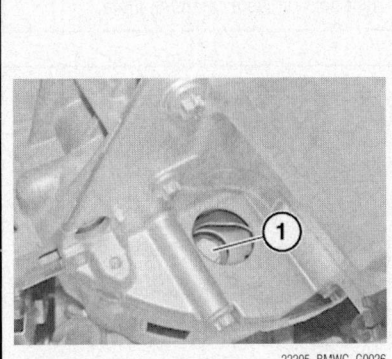

**Fig. 20 Crank engine at vibration damper in direction of rotation until screw (1) is visible in opening**

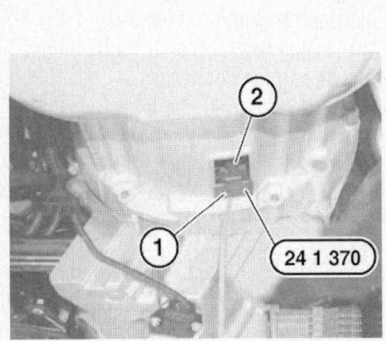

**Fig. 21 Check installation of special tool 21 1 370 (1). Lug (2) must point in opposite direction to direction of travel**

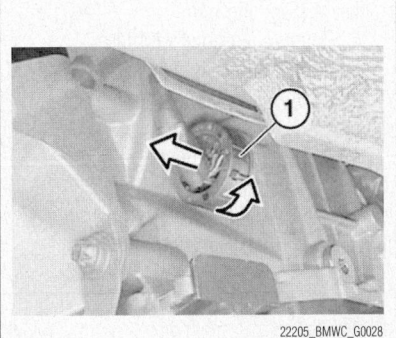

**Fig. 22 Unlock and disconnect plug (1) by turning**

14. Remove bolts (1) and nuts (2) and remove transmission cross-member.

15. Remove driveshaft from transmission.

16. Remove center bearing.

17. Tie driveshaft to one side.

### ✳✳ WARNING

**Do not allow driveshaft to hang from fixed ball joint (risk of damage).**

18. Crank engine at vibration damper in direction of rotation until screw is visible in opening.

19. Remove all bolts of torque converter with special tool 24 1 110.

20. Crank engine further and remove remaining 5 bolts.

21. Insert special tool 24 1 370 into opening of transmission housing and secure with screw.

➡**Check installation location. Lug must point in opposite direction to direction of travel.**

22. Unlock and disconnect plug by turning. Remove cable from retainers. Insert special tool 24 2 390 in sealing sleeve.

### ✳✳ WARNING

**Do not touch pins.**

23. Remove transmission-to-engine fasteners.

24. Remove the transmission.

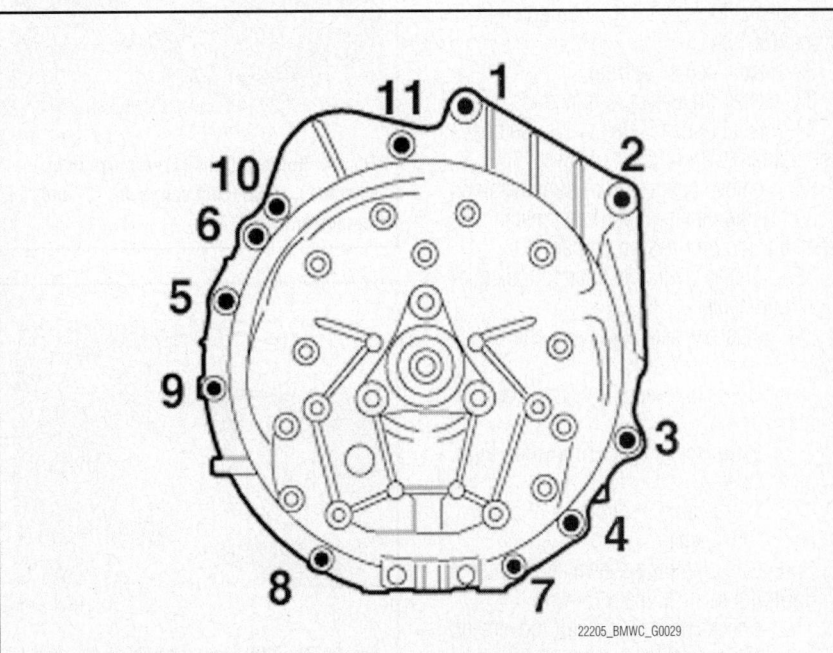

**Fig. 23 Transmission to engine fastener torque sequence—GA6L45R transmission**

*To install:*

25. Turn the driveplate so the hole is accessible from opening on engine oil pan. Rotate torque converter until hole in torque converter is flush with hole in driveplate.

26. Check that dowel sleeves are correctly seated.

➡**Replace damaged dowel sleeves.**

27. Install the transmission and tighten fasteners in sequence shown to the specified torque.

### ✳✳ WARNING

**Blue aluminum fasteners must be replaced with new fasteners.**

- Steel Torx® bolts M8: 14 ft. lbs. (19 Nm).
- Steel Torx® bolts M12: 49 ft. lbs. (66 Nm).
- Aluminum M10: 15 ft. lbs. (20 Nm) plus 90–110° of rotation.
- Aluminum M12: 19 ft. lbs. (25 Nm) plus 130° of rotation.
- Cover plate: 7 ft. lbs. (9 Nm).

28. Remove special tool 24 2 390. Connect cable and secure with retainers. Connect and lock the plug.

### ✳✳ WARNING

**Do not touch pins.**

29. Remove special tool 24 1 370.

30. Crank engine at vibration damper in direction of rotation until screw is visible in opening.

31. Install all bolts of torque converter with special tool 24 1 110 and tighten to 42 ft. lbs. (56 Nm).

32. Install center bearing.

33. Install driveshaft to transmission.

34. Install transmission cross-member and tighten to 14 ft. lbs. (19 Nm).

35. Remove support from transmission.

36. Using new sealing rings, install transmission oil lines on oil pan.

37. Connect hydraulic lines to transmission fluid cooler.

38. Adjust the selector lever cable as follows:

  a. Grip clamping sleeve and slacken nut.

  b. Press selector lever forwards into park position.

  c. Press cable in direction of arrow and remove again.

  d. Grip clamping sleeve and tighten down nut to 11 ft. lbs. (15 Nm).

  e. Check cable for ease of movement.

  f. Move selector lever to "P" position.

  g. Check whether parking gear is engaged by turning driveshaft.

39. Install retaining plate, tighten bolts and connect plug connector.

40. Install bracket, clip lines, install bracket and heat shield and tighten bolts.

41. Install heat shields and tighten to 6 ft. lbs. (8 Nm).

42. Install complete exhaust system.

43. Install underbody protection with bracket at front and rear.

44. Connect the negative battery cable.

45. Check and fill transmission fluid to proper level.

### GA6HP19Z Transmission

*See Figures 24 through 35.*

1. Before servicing the vehicle, refer to the precautions.

2. Disconnect the negative battery cable.

3. Remove underbody protection with bracket at front and rear.

4. Remove fan cowl with fan.

5. Remove complete exhaust system.

6. On AWD, remove the reinforcement plate.

7. Remove heat shields.

8. Support engine with lifter.

9. Remove bolts, remove heat shield, disconnect connector and remove bracket.

10. Remove cables from holder.

11. On 2WD, remove bolts and remove holder.

12. On AWD, remove screw, remove clamp and remove shift cable head using a screwdriver from ball head of shift cable lever.

13. Disconnect plug connector (1), remove bolts (2) and remove cover plate (3).

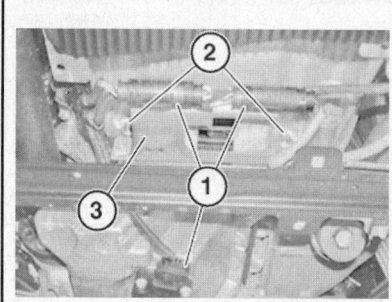
22205_BMWC_G0032

**Fig. 26 On AWD, remove screw (1), remove clamp (2) and remove shift cable head (3) using a screwdriver from ball head of shift cable lever**

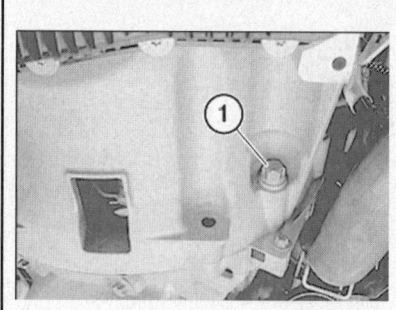
22205_BMWC_G0033

**Fig. 27 Remove aluminum bolt (1) on right next to cable retaining plate**

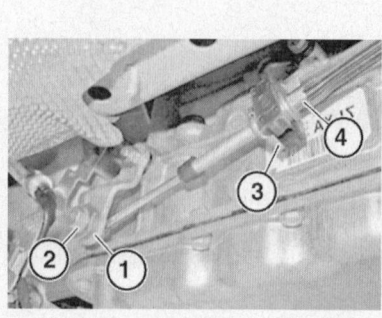
22205_BMWC_G0024

**Fig. 28 Grip clamping sleeve (1), loosen nut (2), detach retainer (3) downwards using a screwdriver and pull cable (4) out of holder—selector cable version 1**

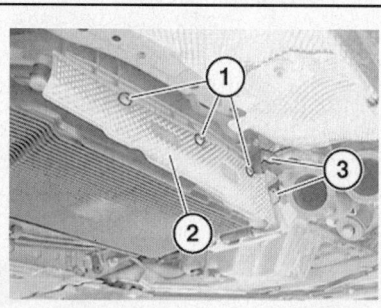
22205_BMWC_G0030

**Fig. 24 Remove bolts (1), remove heat shield (2), disconnect connector (3) and remove bracket**

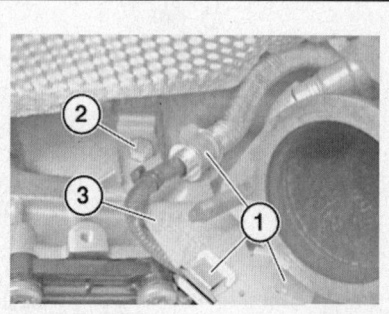
22205_BMWC_G0031

**Fig. 25 On 2WD, remove cables (1) from holder (3) and then remove bolts (2)**

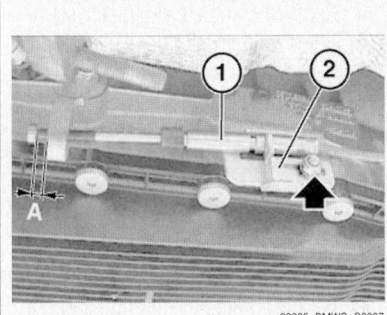

**Fig. 29 Remove the nut and disengage the cable—selector cable version 2**

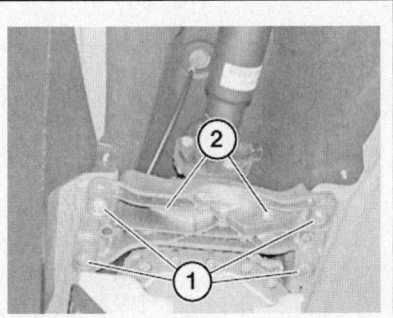

**Fig. 30 Remove bolts (1) and nuts (2) and remove transmission cross-member**

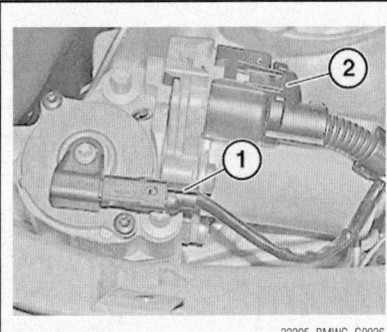

**Fig. 31 On AWD, disconnect plugs (1) from the servo motor (2)**

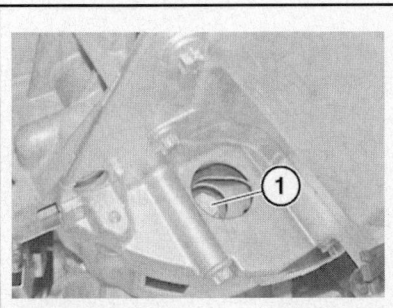

**Fig. 32 Crank engine at vibration damper in direction of rotation until screw (1) is visible in opening**

14. Remove aluminum bolt (1) on right next to cable retaining plate.

### ✳✳ WARNING

**Blue aluminum bolts/bolts must be replaced.**

15. Remove selector cable as follows:
   a. For version 1, grip clamping sleeve, loosen nut, detach retainer downwards using a screwdriver and pull cable out of holder.
   b. For version 2, remove the nut and disengage the cable.

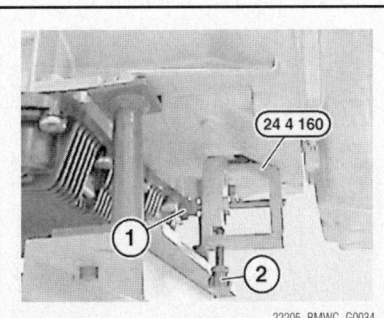

**Fig. 33 Crank engine at vibration damper in direction of rotation until screw (1) is visible in opening**

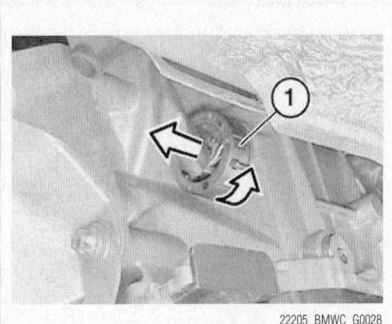

**Fig. 34 Unlock and disconnect plug (1) by turning**

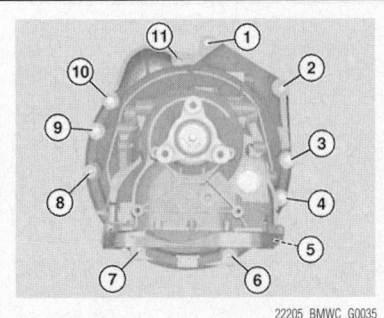

**Fig. 35 Transmission to engine fastener torque sequence—GA6HP19Z transmission**

16. Remove screw and disconnect hydraulic lines to transmission fluid cooler.
17. Remove nut and bracket from transmission oil lines on oil pan.
18. Support transmission with special tools 23 4 050, 00 2 030.
19. Remove bolts (1) and nuts (2) and remove transmission cross-member.
20. On AWD, disconnect plugs from the servo motor.
21. Remove driveshaft from transmission. On AWD, remove front driveshaft.
22. Crank engine at vibration damper in direction of rotation until screw is visible in opening.
23. Insert special tool 24 4 160 into opening of transmission housing and clamp gently with screw (1). Raise by turning screw (2) and clamp down. Then tighten down screw (1).
24. Unlock and disconnect plug by turning. Remove cable from retainers. Insert special tool 24 2 390 in sealing sleeve.

### ✳✳ WARNING

**Do not touch pins.**

25. Remove transmission-to-engine fasteners.
26. Remove the transmission.

### *To install:*

27. Turn the driveplate so the hole is accessible from opening on engine oil pan. Rotate torque converter until hole in torque converter is flush with hole in driveplate.
28. Check that dowel sleeves are correctly seated.

➡ **Replace damaged dowel sleeves.**

29. Install the transmission and tighten fasteners in sequence shown to the specified torque.

### ✳✳ WARNING

**Blue aluminum fasteners must be replaced with new fasteners.**

- Steel hex bolts M8: 18 ft. lbs. (24 Nm).
- Steel hex bolts M10: 33 ft. lbs. (45 Nm).
- Steel hex bolts M12: 61 ft. lbs. (82 Nm).
- Steel Torx® bolts M6: 7 ft. lbs. (9 Nm).
- Steel Torx® bolts M8: 16 ft. lbs. (21 Nm).
- Steel Torx® bolts M10: 31 ft. lbs. (42 Nm).
- Steel Torx® bolts M12: 53 ft. lbs. (72 Nm).

- Steel Torx® bolts M8: 14 ft. lbs. (19 Nm).
- Steel Torx® bolts M12: 49 ft. lbs. (66 Nm).
- Aluminum M10x30: 15 ft. lbs. (20 Nm) plus 90–110°of rotation.
- Aluminum M10x85: 15 ft. lbs. (20 Nm) plus 180–200°of rotation.
- Aluminum M12: 19 ft. lbs. (25 Nm) plus 130°of rotation.
- Cover plate: 7 ft. lbs. (9 Nm).

30. Remove special tool 24 2 390. Connect cable and secure with retainers. Connect and lock the plug.

### ❊❊ WARNING

**Do not touch pins.**

31. Remove special tool 24 1 370.
32. Crank engine at vibration damper in direction of rotation until screw is visible in opening.
33. Install all bolts of torque converter with special tool 24 1 110 and tighten as to specification.
- M8: 19 ft. lbs. (26 Nm)
- M10 8.8: 33 ft. lbs. (45 Nm)
- M10 10.9: 42 ft. lbs. (56 Nm)

34. Install driveshaft to transmission.
35. Install transmission cross-member and tighten to 14 ft. lbs. (19 Nm).
36. Remove support from transmission.
37. Using new sealing rings, install transmission oil lines on oil pan.
38. Connect hydraulic lines to transmission fluid cooler.
39. Adjust the selector lever cable version 1 as follows:
   a. Grip clamping sleeve and slacken nut.
   b. Press selector lever forwards into park position.
   c. Press cable in direction of arrow and remove again.
   d. Grip clamping sleeve and tighten down nut to 11 ft. lbs. (15 Nm).
   e. Check cable for ease of movement.
   f. Move selector lever to "P" position.
   g. Check whether parking gear is engaged by turning driveshaft.
40. Adjust the selector lever cable version 2 as follows:
   a. Adjust cable by means of holder until spacing A = 1 mm is obtained.
   b. Tighten the nut.
41. Install retaining plate, tighten bolts and connect plug connector.
42. Install bracket, clip lines, install bracket and heat shield and tighten bolts.
43. Install heat shields and tighten to 6 ft. lbs. (8 Nm).
44. Install complete exhaust system.

45. Install underbody protection with bracket at front and rear.
46. Connect the negative battery cable.
47. Check and fill transmission fluid to proper level.

### GA6HP26Z Transmission

*See Figures 36 through 45.*

1. Before servicing the vehicle, refer to the precautions.
2. Disconnect the negative battery cable.
3. Remove underbody protection with bracket at front and rear.

4. Remove complete exhaust system.
5. Remove the reinforcement plate.
6. Remove the oxygen sensors.
7. Remove the heat shields.
8. Disconnect plug from speed sensor.
9. Remove screw and remove speed sensor.
10. Disconnect plug from oil level sensor and Disconnect cable from fixtures.
11. Disconnect plug and Disconnect cable duct from transmission housing. Tie wiring harness to one side.
12. Remove the locking nut and disengage the selector cable.

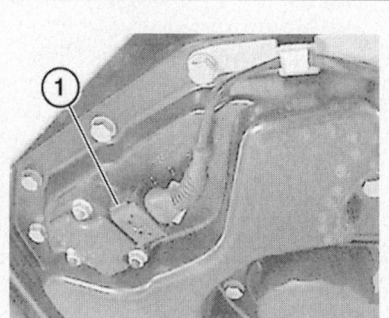

**Fig. 36 Disconnect plug (1) from oil level sensor and Disconnect cable from fixtures**

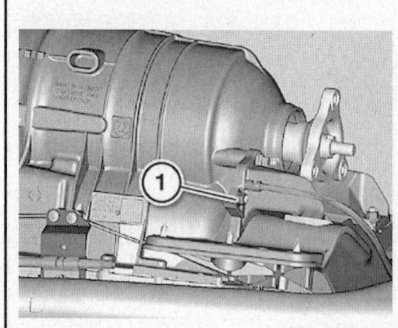

**Fig. 39 Remove grounding strap (1)**

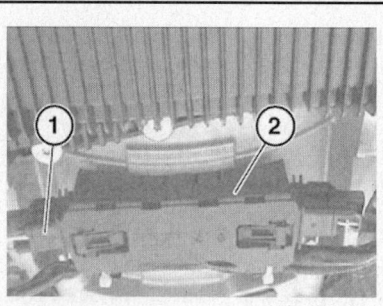

**Fig. 37 Disconnect plug (1) and Disconnect cable duct (2) from transmission housing**

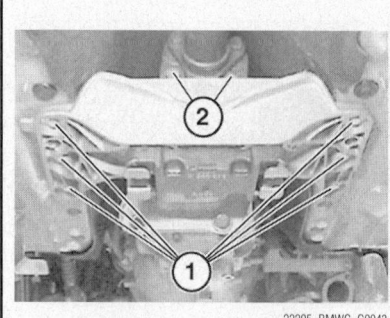

**Fig. 40 Remove bolts (1) and nuts (2) and remove transmission cross-member**

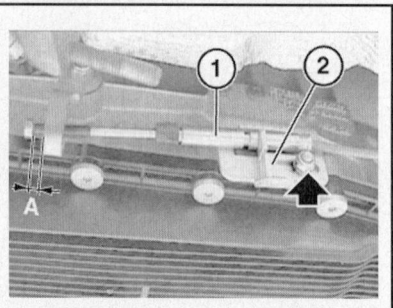

**Fig. 38 Remove the nut and disengage the cable (1). Use the holder (2) to adjust the cable during installation**

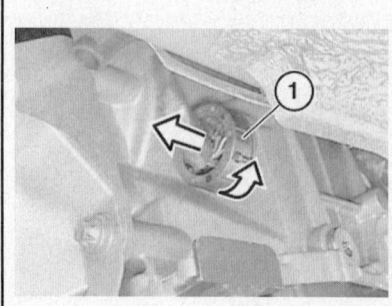

**Fig. 41 Unlock and disconnect plug (1) by turning**

13. Place the transmission on a jack.

14. Remove grounding strap.

15. Remove bolts and nuts and remove transmission cross-member

16. Remove driveshaft from transmission.

17. Remove center bearing.

18. Tie driveshaft to one side.

19. Unlock and disconnect plug by turning. Remove cable from retainers. Insert special tool 24 2 390 in sealing sleeve.

**✳✳ WARNING**

**Do not touch pins.**

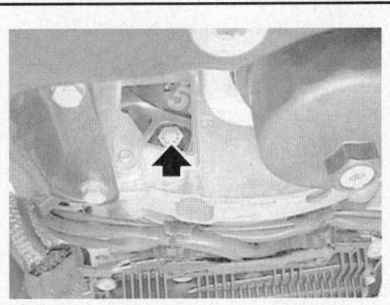

22205_BMWC_G0044

**Fig. 42 Crank engine at vibration damper in direction of rotation until screw (1) is visible in opening**

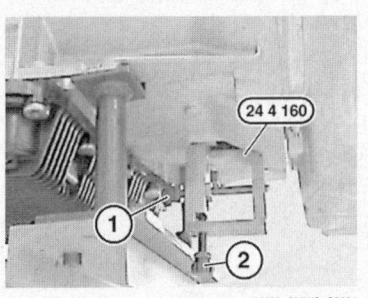

22205_BMWC_G0034

**Fig. 43 Crank engine at vibration damper in direction of rotation until screw (1) is visible in opening**

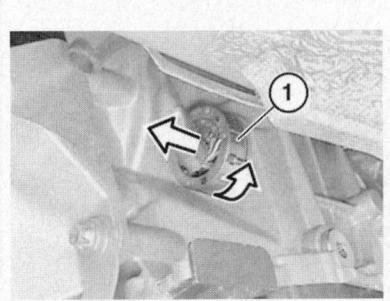

22205_BMWC_G0028

**Fig. 44 Unlock and disconnect plug (1) by turning**

20. Remove screw and disconnect hydraulic lines to transmission fluid cooler.

21. Crank engine at vibration damper in direction of rotation until screw is visible in opening.

22. Insert special tool 24 4 160 into opening of transmission housing and clamp gently with screw. Raise by turning screw and clamp down. Then tighten down screw.

23. Unlock and disconnect plug by turning. Remove cable from retainers. Insert special tool 24 2 390 in sealing sleeve.

**✳✳ WARNING**

**Do not touch pins.**

24. Remove transmission-to-engine fasteners.

25. Remove the transmission.

**To install:**

26. Turn the driveplate so the hole is accessible from opening on engine oil pan. Rotate torque converter until hole in torque converter is flush with hole in driveplate.

27. Check that dowel sleeves are correctly seated.

➡**Replace damaged dowel sleeves.**

28. Install the transmission and tighten fasteners in sequence shown to the specified torque.

**✳✳ WARNING**

**Blue aluminum fasteners must be replaced with new fasteners.**

- Steel hex bolts M8: 18 ft. lbs. (24 Nm).
- Steel hex bolts M10: 33 ft. lbs. (45 Nm).
- Steel hex bolts M12: 61 ft. lbs. (82 Nm).
- Steel Torx® bolts M6: 7 ft. lbs. (9 Nm).
- Steel Torx® bolts M8: 16 ft. lbs. (21 Nm).
- Steel Torx® bolts M10: 31 ft. lbs. (42 Nm).
- Steel Torx® bolts M12: 53 ft. lbs. (72 Nm).
- Steel Torx® bolts M8: 14 ft. lbs. (19 Nm).
- Steel Torx® bolts M12: 49 ft. lbs. (66 Nm).
- Aluminum M10x30: 15 ft. lbs. (20 Nm) plus 90–110° of rotation.
- Aluminum M10x85: 15 ft. lbs. (20 Nm) plus 180–200° of rotation.
- Aluminum M12: 19 ft. lbs. (25 Nm) plus 130° of rotation.
- Cover plate: 7 ft. lbs. (9 Nm).

29. Remove special tool 24 2 390. Connect cable and secure with retainers. Connect and lock the plug.

**✳✳ WARNING**

**Do not touch pins.**

30. Remove special tool 24 1 370.

31. Crank engine at vibration damper in direction of rotation until screw is visible in opening.

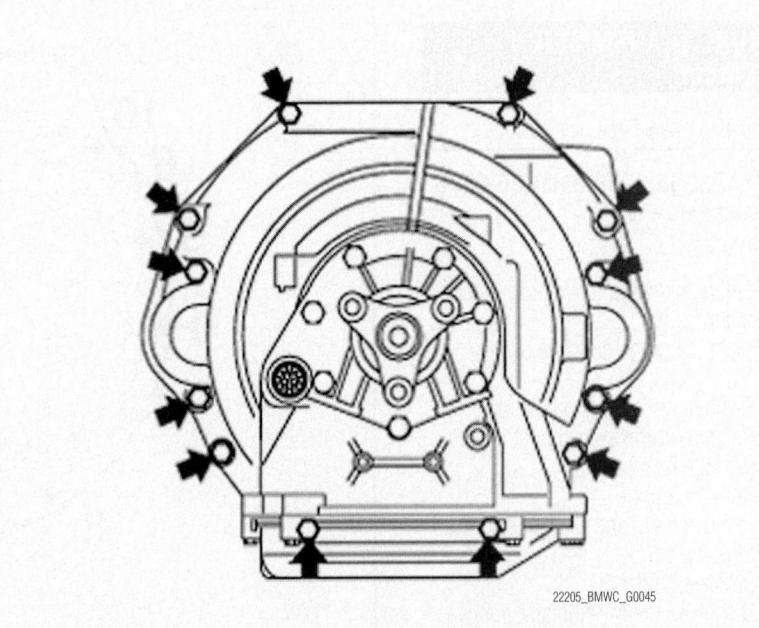

22205_BMWC_G0045

**Fig. 45 Transmission to engine fastener locations—GA6HP26Z transmission**

32. Install all bolts of torque converter with special tool 24 1 110 and tighten as to specification.
- M8: 19 ft. lbs. (26 Nm)
- M10 8.8: 33 ft. lbs. (45 Nm)
- M10 10.9: 42 ft. lbs. (56 Nm)

33. Connect hydraulic lines to transmission fluid cooler.

> **✳✳ WARNING**
> **Do not touch pins. sealing sleeve.**

34. Connect plug by turning. Secure cable to retainers.

35. Install center bearing.

36. Install driveshaft to transmission.

37. Install transmission cross-member. Tighten to 22 ft. lbs. (30 Nm).

38. Install grounding strap.

39. Remove transmission on a jack.

40. Adjust cable by means of holder until spacing A = 1 mm is obtained. Tighten the nut.

41. Connect plug and clip cable duct to transmission housing.

42. Connect oil level sensor and clip cable to fixtures.

43. Install speed sensor.

44. Connect plug to speed sensor.

45. Install the heat shields.

46. Install the oxygen sensors.

47. Install the reinforcement plate.

48. Install complete exhaust system.

49. Install underbody protection with bracket at front and rear.

50. Connect the negative battery cable.

51. Check and fill transmission fluid to proper level.

## MANUAL TRANSMISSION ASSEMBLY

### REMOVAL & INSTALLATION

#### GS6-17BZ, GS6-17BG and GS6-53BZ Transmissions

*See Figure 46.*

1. Before servicing the vehicle, refer to the precautions.

2. Disconnect the negative battery cable.

3. Remove underbody protection.

4. Remove complete exhaust system.

5. Remove heat shields.

6. Support engine.

7. Remove bolts and remove exhaust system bracket.

8. Remove holder.

9. Disconnect connector, and remove the cable by releasing the bolts and removing the bracket.

10. Disconnect cable from holder. Remove screw and remove holder.

11. Disconnect plug from reverse gear switch.

12. Disconnect cables from fixtures on transmission.

13. Secure transmission to transmission jack with tensioning strap.

14. Remove cross-member.

15. Remove driveshaft from transmission.

16. Remove center bearing.

17. Tie driveshaft to one side.

> **✳✳ WARNING**
> **Do not disconnect the pressure line of clutch slave cylinder.**

18. Relieve tension on clutch slave cylinder slowly; otherwise air will be drawn in through sealing sleeve.

19. Remove nuts and remove clutch slave cylinder.

20. Disconnect locking clip and pull shift rod out of shift rod joint). When installing, grease shift rod pin.

21. Unlock bearing pin in direction of arrow as shown and remove. Lift out shift arm.

> **✳✳ WARNING**
> **Aluminum bolts/bolts must be replaced.**

22. Remove transmission-to-engine bolts.

> **✳✳ WARNING**
> **Do not suspend the transmission from the transmission input shaft during removal and installation as this will deform the clutch plate.**

23. Pull out transmission towards rear and remove.

*To install:*

24. Remove and clean remove bearing and remove lever (do not grease).

25. Push on grease scraper ring 21 2 221 as far as it will go.

26. Grease splines (1) of input shaft with a brush.

27. Disconnect grease scraper ring.

28. Check lubrication of transmission input shaft for sticky consistency. If grease is sticky, clean input shaft and replace clutch plate.

29. Check clutch plate for friction rust in splines and replace if necessary.

30. Remove any grease and lining abrasion from splines of clutch plate by mechanical means (with a cloth).

31. Check dowel sleeves for correct seating. Replace damaged sleeves.

32. Install transmission and tighten transmission-to-engine bolts in sequence shown to specification.
- Steel hex bolts M8: 18 ft. lbs. (24 Nm)

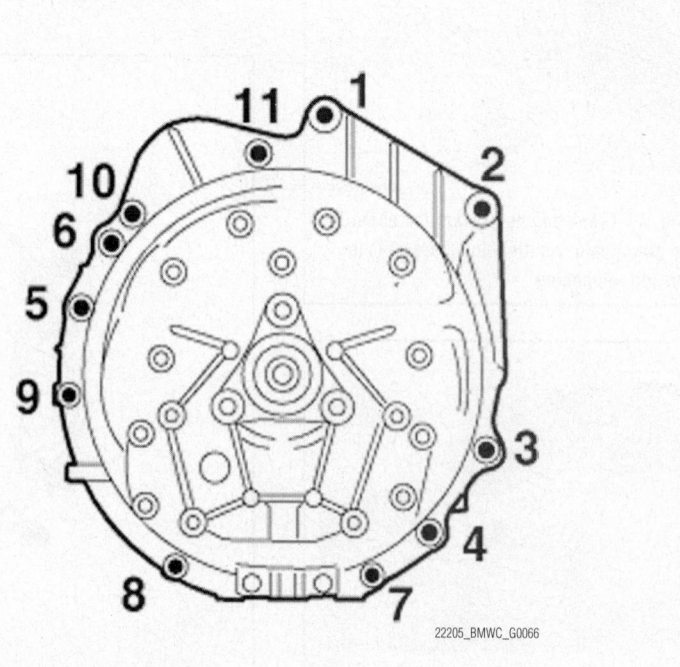

22205_BMWC_G0066

**Fig. 46 Transmission-to-engine bolt tightening sequence**

- Steel hex bolts M10: 33 ft. lbs. (45 Nm)
- Steel hex bolts M12: 61 ft. lbs. (82 Nm)
- Steel Torx® bolts M6: 7 ft. lbs. (9 Nm)
- Steel Torx® bolts M8: 16 ft. lbs. (21 Nm)
- Steel Torx® bolts M10: 31 ft. lbs. (42 Nm)
- Steel Torx® bolts M12: 53 ft. lbs. (72 Nm)
- Steel Torx® bolts M8: 14 ft. lbs. (19 Nm)
- Steel Torx® bolts M12: 49 ft. lbs. (66 Nm)
- Aluminum M10x30: 15 ft. lbs. (20 Nm) plus 90–110°of rotation
- Aluminum M10x85: 15 ft. lbs. (20 Nm) plus 180–200°of rotation
- Aluminum M12: 19 ft. lbs. (25 Nm) plus 130°of rotation

### ✻✻ WARNING

**Aluminum bolts/bolts must be replaced.**

33. Ensure correct position of cover plate. Tighten to 6 ft. lbs. (8 Nm).
34. Insert the shift arm and lock in place with the bearing pin.
35. Grease shift rod pin, install and lock using clip joint.
36. Install clutch slave cylinder.
37. Install center bearing.
38. Install driveshaft on transmission.
39. Install cross-member. Tighten bolts to specification.
   - M10 bolts: 31 ft. lbs. (42 Nm)
   - M12 bolts: 53 ft. lbs. (72 Nm)
40. Remove transmission jack.
41. Connect cables to fixtures on transmission.
42. Connect plug from reverse gear switch.
43. Install holder and clip cable to holder.
44. Install the cable by securing the bolts and installing the bracket. Connect the connector.
45. Install holder.
46. Install exhaust system bracket.
47. Remove the support from the engine.
48. Install heat shields.
49. Install complete exhaust system.
50. Install underbody protection.
51. Connect the negative battery cable.
52. Check and fill transmission fluid to proper level.

### GS6-37BZ Transmission

### ✻✻ WARNING

**Aluminum-magnesium materials. No steel bolts/bolts may be used due to the threat of electrochemical corrosion. A magnesium crankcase requires aluminum bolts/bolts exclusively. Aluminum bolts/bolts must be replaced each time they are removed. The end faces of these bolts/bolts are painted blue for the purposes of reliable identification. Jointing torque and angle of rotation must be observed without fail (risk of damage).**

1. Before servicing the vehicle, refer to the precautions.
2. Disconnect the negative battery cable.
3. Remove underbody protection and bracket at front and rear
4. Remove reinforcement plate.
5. Remove complete exhaust system
6. Support engine with lifter when removing transmission
7. Remove heat shields.
8. Disconnect plug from oil level sensor.
9. Disconnect connector, remove bolts, remove cover plate and unscrew holder.
10. Remove aluminum screw on right next to cable retaining plate with special tool 00 9 010.
11. Disconnect plug from reversing light switch. Remove cable from retainers.
12. Disconnect plugs and from servomotor.
13. Remove front driveshaft at output flange of transfer case and tie to one side.
14. Remove driveshaft from transmission.
15. Remove centre bearing.
16. Tie driveshaft to one side.
17. Support transmission with special tools 23 4 050, 00 2 030.
18. Remove transmission cross-member.
19. Lift off retainer and disconnect shift rod.
20. Unlock bearing pin in direction of arrow and remove. Lift out shift arm.

➡**Do not disconnect the pressure line of clutch slave cylinder.**

### ✻✻ WARNING

**Slowly relieve tension on clutch slave cylinder otherwise air is drawn in through sealing sleeve.**

21. Remove nuts and remove clutch slave cylinder.

### ✻✻ WARNING

**Do not allow the transmission to hang off from the transmission input shaft as this will deform the clutch plate.**

22. Remove the transmission-to-engine bolts.
23. Pull out transmission towards rear and remove.

#### To install:

24. Grease splines (1) of input shaft with a brush.
25. Disconnect grease scraper ring.
26. Check lubrication of transmission input shaft for sticky consistency. If grease is sticky, clean input shaft and replace clutch plate.
27. Check clutch plate for friction rust in splines and replace if necessary.
28. Remove any grease and lining abrasion from splines of clutch plate by mechanical means (with a cloth).
29. Check dowel sleeves for correct seating. Replace damaged sleeves.
30. Install transmission and tighten transmission-to-engine bolts to specification.
   - Steel hex bolts M8: 18 ft. lbs. (24 Nm)
   - Steel hex bolts M10: 33 ft. lbs. (45 Nm)
   - Steel hex bolts M12: 61 ft. lbs. (82 Nm)
   - Steel Torx® bolts M6: 7 ft. lbs. (9 Nm)
   - Steel Torx® bolts M8: 16 ft. lbs. (21 Nm)
   - Steel Torx® bolts M10: 31 ft. lbs. (42 Nm)
   - Steel Torx® bolts M12: 53 ft. lbs. (72 Nm)
   - Steel Torx® bolts M8: 14 ft. lbs. (19 Nm)
   - Steel Torx® bolts M12: 49 ft. lbs. (66 Nm)
   - Aluminum M10x30: 15 ft. lbs. (20 Nm) plus 90–110° of rotation
   - Aluminum M10x85: 15 ft. lbs. (20 Nm) plus 180–200° of rotation
   - Aluminum M12: 19 ft. lbs. (25 Nm) plus 130° of rotation

### ✻✻ WARNING

**Aluminum bolts/bolts must be replaced.**

31. Install clutch slave cylinder.
32. Connect shift arm and lock bearing pin.

33. Connect shift rod and secure with retainer.

34. Install cross-member. Tighten bolts to specification.
- M10 bolts: 31 ft. lbs. (42 Nm)
- M12 bolts: 53 ft. lbs. (72 Nm)

35. Remove transmission support.

36. Install centre bearing.

37. Install driveshaft to transmission.

38. Install front driveshaft at output flange of transfer case.

39. Connect plugs and to servomotor.

40. Connect plug to reversing light switch and secure cable to retainers.

41. Install aluminum screw on right next to cable retaining plate and tighten using specifications above.

42. Ensure correct position of cover plate. Tighten to 6 ft. lbs. (8 Nm).

43. Connect plug to oil level sensor.

44. Install heat shields.

45. Remove engine support.

46. Install complete exhaust system

47. Install reinforcement plate.

48. Install underbody protection and bracket at front and rear

49. Connect the negative battery cable.

50. Check and fill transmission fluid to proper level.

## CLUTCH

### REMOVAL & INSTALLATION

A self adjusting clutch can be identified by 3 openings each with a pressure piece of adjustment ring and a pressure spring.

1. Before servicing the vehicle, refer to the precautions.

2. Remove transmission.

3. Block flywheel with special tool 11 9 260.

4. Remove bolts and remove SAC clutch from flywheel.

5. Remove clutch plate from flywheel.

6. Clean flywheel, check for wear and damage. Replace damaged flywheel.

7. Check grooved ball bearing in crankshaft for ease of movement and tightness and if necessary replace grooved ball bearing.

### ❋❋ WARNING

**Always replace clutch plates fouled e.g. by oil, cleaning agent.**

8. Check clutch plate for damage and friction rust in hub profile and replace if necessary.

9. Check clutch disk for wear:
   a. Measure lining protrusion at lining rivets head.

b. Replace clutch plate if lining protrusion at head is less than 1 mm.

10. Push clutch disk onto the cleaned transmission input shaft and make sure it slides smoothly.

### ❋❋ WARNING

**Do not clean SAC clutch with high-pressure cleaner or washing machine as this may impair function of adjustment unit. Clean friction surface only.**

11. Check SAC clutch for wear and damage and replace if necessary.

### *To install:*

➡**Bolts must be replaced.**

12. Set SAC clutch (1) down on a clean surface.

13. Install special tool 21 2 180 in SAC clutch.

➡**Locking hooks of special tool 21 2 180 must engage in openings in SAC clutch.**

14. Press special tool 21 2 180 together at handles as far as it will go and grip firmly.

15. At same time tighten down knurled bolts.

16. Adjustment ring of SAC clutch is now secured in its original position (wear position). Do not reset adjustment ring to new position.

17. Insert special tool 21 2 170 only in area of bores for dowel pins.

18. Fit special tool 21 2 170 and tighten down at knurled screw.

19. If installing a new clutch plate, screw in spindle until adjustment ring of SAC clutch can be turned with special tool 21 2 180 at handles.

20. If installing a new clutch plate, press special tool 21 2 180 together at handles as far as it will go and grip firmly. At same time tighten down knurled bolts. Adjustment ring of SAC clutch is now secured in new position.

21. Screw in spindle until diaphragm spring is tensioned up to stop.

### ❋❋ WARNING

**Handle clutch disk carefully, do not touch surfaces of friction pads.**

22. Install clutch plate in correct position. Note designation for "engine side" / "transmission side".

23. Centre clutch plate with special tool.

24. Fit SAC clutch on flywheel paying attention to dowel pins. The SAC clutch must be secured by way of dowel pins.

25. Insert bolts and tighten to 18 ft. lbs. (25 Nm) for M8 8.8 bolts and 27 ft. lbs. (37 Nm) for M8 10.9 bolts. If ZNS bolts are used (shiny zinc coating), replace bolts with new ones and tighten to 11 ft. lbs. (15 Nm), then rotate and additional 85–95°.

26. Remove spindle until tension is completely removed from diaphragm spring.

27. Remove knurled screw and remove special tool 21 2 170 from SAC clutch.

28. Remove knurled bolts and remove special tool 21 2 180 from SAC clutch.

29. Withdraw special tool from clutch plate with aid of accompanying screw.

30. Carefully unscrew locking piece clockwise or counterclockwise with a hexagon socket wrench.

➡**A slight snapping of the plate spring while unscrewing is possible.**

31. Withdraw special tool from clutch plate with aid of accompanying screw.

### BLEEDING

1. Before servicing the vehicle, refer to the precautions.

2. Remove underbody protection from transmission.

3. Remove micro filter housing as necessary.

4. If the vehicle uses a plastic slave cylinder, perform the following procedure prior to bleeding the clutch hydraulic system.

   a. Connect bleeder unit to brake fluid expansion tank.

### ❋❋ WARNING

**Charging pressure should not exceed 29 PSI (2 bar).**

   b. Connect bleeder hose to bleed valve.

   c. Open bleed valve and flush until clear brake fluid emerges without air bubbles.

   d. Close bleed valve.

   e. Switch off bleeder unit and remove from brake fluid expansion tank.

   f. Correct brake fluid level in expansion tank.

5. Remove nuts and remove clutch slave cylinder (pressure line remains connected).

6. Fit special tool 21 5 030 on clutch slave cylinder.

7. Press piston rod with aid of spindle completely into clutch slave cylinder.

8. Connect bleeder unit to brake fluid expansion tank.

**Charging pressure should not exceed 29 PSI (2 bar).**

9. Connect bleeder hose to bleed valve.

10. Hold clutch slave cylinder in illustrated position as shown with special tool 21 5 030.

11. Open bleeder valve.

12. If bubble-free brake fluid emerges, retract piston rod of clutch slave cylinder with aid of spindle a little and press in again.

13. If no air bubbles escape, close bleeder valve, otherwise repeat procedure.

**Do not under any circumstances remove special tool 21 5 030 from clutch slave cylinder when brake system is pressurized. Piston with push rod can jump out of clutch slave cylinder.**

14. Switch off bleeder unit or remove from brake fluid expansion tank.

15. Slowly retract piston rod of clutch slave cylinder with special tool 21 5 030.

16. Remove special tool 21 5 030 from clutch slave cylinder.

17. Install clutch slave cylinder to transmission. Replace self-locking nuts and tighten to 16 ft. lbs. (22 Nm).

18. Correct brake fluid level in expansion tank.

## TRANSFER CASE ASSEMBLY

### REMOVAL & INSTALLATION

1. Before servicing the vehicle, refer to the precautions.

2. Remove underbody protection.

3. Remove complete exhaust system.

4. Remove heat shields.

5. Remove front driveshaft at output flange of transfer case and tie to one side. Discard the bolts.

6. Remove driveshaft from transmission.

7. Remove centre bearing.

8. Tie driveshaft to one side.

9. Support transmission with special tools 00 2 030 / 23 4 050.

10. Remove bolts and remove metal plate.

11. Disconnect plugs from servomotor.

12. Remove transmission cross-member.

13. Remove the transfer case bolts.

14. Remove transfer case.

### To install:

15. Pay attention to dowel pin alignment. Grease dowel pin with Weicon Anti-Seize.

16. Install transfer case. Tighten bolts to specification.

**ZNS bolts and nuts have a shiny zinc coating and must be replaced each time they are removed. Torque and angle of rotation must be observed.**

- M10 10.9 with ribbed teeth: 30 ft. lbs. (40 Nm) plus an additional 45° rotation.
- M10: 15 ft. lbs. (20 Nm) plus an additional 90° rotation.
- M12x55 10.9: 41 ft. lbs. (55 Nm) plus an additional 90° rotation.

17. Install cross-member. Tighten bolts to specification.

- M10 bolts: 31 ft. lbs. (42 Nm)
- M12 bolts: 53 ft. lbs. (72 Nm)

18. Connect plugs to servomotor.

19. Install metal plate.

20. Remove transmission support.

21. Install front driveshaft at output flange of transfer case and tighten new M10 bolt to 15 ft. lbs. (20 Nm) plus an additional 90° rotation.

22. Install heat shields.

23. Install complete exhaust system.

24. Install underbody protection.

25. Check gear oil level and top up if necessary.

## FRONT HALFSHAFT

### REMOVAL & INSTALLATION

*See Figure 47.*

1. Before servicing the vehicle, refer to the precautions.

2. Remove front wheel.

3. Expand turning lock sufficiently to avoid damaging thread when releasing collar nut.

4. Remove collar nut; to do so, press brake pedal to floor.

5. Remove reinforcement plate.

6. Remove tie rod end from ball joint.

7. On the right side only, remove guide joint from ball joint.

8. Remove control arm from ball joint.

9. Remove stabilizer link from spring strut.

10. Turn ball joint to one side.

11. Press halfshaft out of wheel hub and tie up.

12. Press halfshaft with special tool 31 5 110 out of front differential and remove.

13. Press halfshaft with special tool 31 5 110 out of bearing block and remove.

### To install:

14. Replace shaft seal in front differential and coat sealing lips with front differential oil.

15. Replace shaft seal in bearing block and coat sealing lips with front differential oil.

16. Replace retaining ring and slide in halfshaft over resistance of retaining ring. Halfshaft must snap audibly into place.

**High installation forces indicate that the spline teeth on the halfshaft/rear differential side gear are damaged/deformed. Check spline teeth, replaced damaged parts.**

17. Replace collar nut, oil collar nut/wheel bearing contact surface only and tighten to 311 ft. lbs. (420 Nm).

**Do not oil thread of shaft journal or collar nut.**

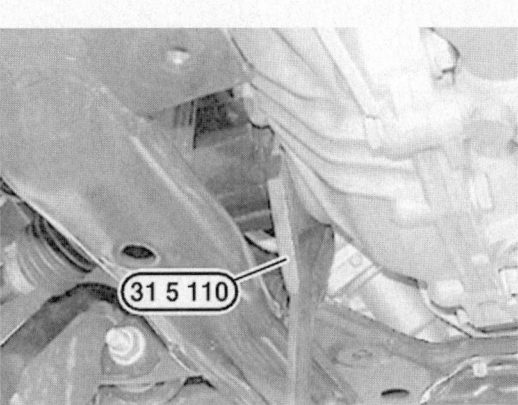

22205_BMWC_G0079

**Fig. 47 Press halfshaft with special tool 31 5 110**

18. Secure collar nut by positive peening on flat areas of halfshaft.

19. Install front wheel.

20. Check front differential oil level, correct if necessary

## REAR HALFSHAFT

### REMOVAL & INSTALLATION

*See Figure 48.*

1. Before servicing the vehicle, refer to the precautions.

2. Remove rear wheel.

### ✳ WARNING

**Expand turning lock sufficiently to avoid damaging thread when releasing collar nut.**

3. Remove collar nut, activate handbrake for this purpose.

4. On the left side, remove strut if necessary.

5. On the left side, lower exhaust system in rear area.

6. Remove bolts and remove with washers.

7. Press output shaft off drive flange using a suitable tool; if necessary, raise wheel carrier with workshop jack approximately 20 mm.

**To install:**

8. Before installing output shaft, make sure that drive flange is fully engaged in rear differential.

9. Replace bolts and washers and tighten to specification.
- M8: 38 ft. lbs. (52 Nm).
- M10 ZNS3: 52 ft. lbs. (70 Nm).

10. Replace collar nut, oil collar nut/wheel bearing contact surface only and tighten to specification.
- M27: 311 ft. lbs. (420 Nm).
- M24: 185 ft. lbs. (250 Nm).

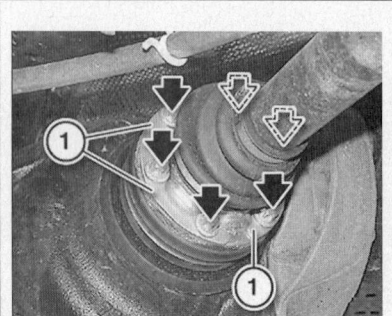

**Fig. 48 Remove bolts and remove with washers (1)**

### ✳ WARNING

**No oil permitted on thread of shaft journal or collar nut.**

11. Secure collar nut by positive peening on flat areas of output shaft.

12. Check that output shaft is correctly seated in rear differential.

13. On the left side, raise exhaust system in rear area.

14. On the left side, install strut if necessary.

15. Install rear wheel.

## REAR PINION SEAL

### REMOVAL & INSTALLATION

*See Figures 49 through 54.*

1. Before servicing the vehicle, refer to the precautions.

2. Remove propeller shaft from rear differential and tie back

3. Removing drive flange:

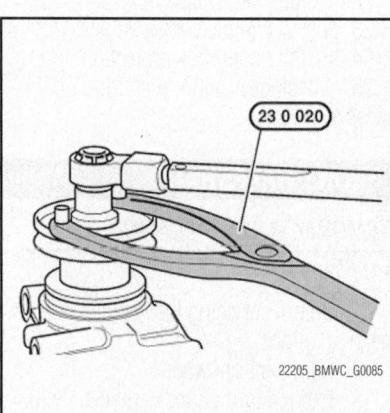

**Fig. 49 Brace drive flange with special tool 23 0 020 and Remove collar nut**

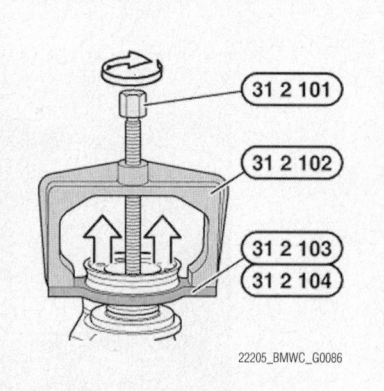

**Fig. 50 On automatic transmission equipped vehicles, remove drive flange with special tools 31 2 101, 31 2 102, 31 2 103 / 31 2 104**

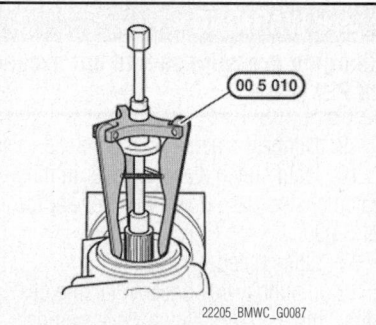

**Fig. 51 On manual transmission equipped vehicles, remove drive flange with special tool 33 1 150**

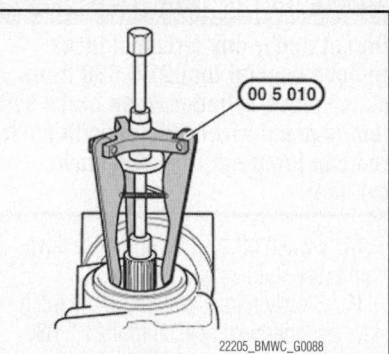

**Fig. 52 Withdraw shaft seal with special tool 00 5 010**

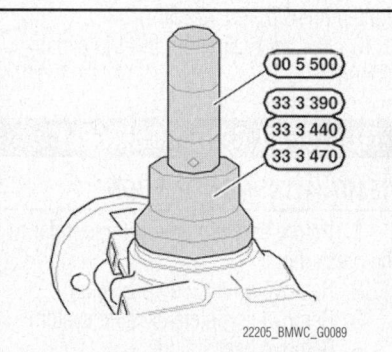

**Fig. 53 Shaft seal installation on automatic transmission equipped vehicles**

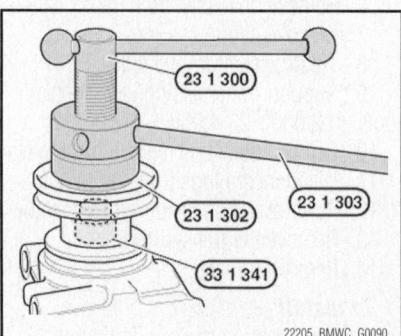

**Fig. 54 Shaft seal installation on automatic transmission equipped vehicles**

4. Lift out retaining plate.

5. Mark position of collar nut on drive shaft with peening tool.

6. Brace drive flange with special tool 23 0 020 and Remove collar nut.

7. On automatic transmission equipped vehicles, remove drive flange with special tools 31 2 101, 31 2 102, 31 2 103 / 31 2 104.

8. On manual transmission equipped vehicles, remove drive flange with special tool 33 1 150.

9. Withdraw shaft seal with special tool 00 5 010.

10. Drive in new shaft seal with following special tools (depending on rear differential) as far as it will go.

11. Coat sealing lips of shaft seal and sealing surface of drive flange with differential oil.

12. Fit drive flange.

13. Press on drive flange with special tools 23 1 300, 33 1 341 and 23 1 303, if necessary 23 1 302 until collar nut can be screwed on.

> **⁑ WARNING**
>
> **Do not under any circumstances tighten down collar nut beyond marker points in order to avoid damaging the clamping sleeve.**

14. Tighten down collar nut to point where marker points are aligned.

15. Drive in new retaining plate as far as it will go.

16. Correct rear differential fluid level

## ENGINE COOLING

### THERMOSTAT

*REMOVAL & INSTALLATION*

#### N52 and N54 Engines

*See Figure 55.*

1. Before servicing the vehicle, refer to the precautions.

2. Drain coolant.

3. Remove hose clips.

4. Remove coolant hoses.

5. Unlock and detach coolant hoses.

6. Disconnect plug connection.

7. Remove bolts.

8. Remove coolant thermostat.

*To install:*

9. Install coolant thermostat.

10. Install bolts.

11. Connect plug connection.

12. Connect and lock coolant hoses.

13. Install coolant hoses.

14. Fasten hose clips.

15. Vent cooling system and check for leaks.

➡**Before filling, turn ignition ON. Set blower to low level. Seat heating controller to maximum temperature. This ensures that the heater valves are fully** opened and the auxiliary water pump starts up. Important: The auxiliary water pump must deliver coolant in order to ensure fully venting.

a. Pour coolant into expansion tank up to **MAX** mark. Perform filling operation slowly.

b. Start engine and run at idle speed for approx. one minute (cap open). Then adjust coolant level to **MAX**.

c. Close cap and run engine up to operating temperature until main thermostat opens. Check cooling circuit and drain plug for leaks.

> **⁑ WARNING**
>
> **The engine must be cooled down before the coolant level is checked. Coolant temperature must not exceed 30°C. If ambient temperature is above 30°C, allow engine to cool down to ambient temperature at least.**

d. Check coolant level and adjust to **MAX**.

➡**Do not fill coolant expansion tank overMAXlevel as overfilling will cause the coolant to overflow.**

#### N62 Engine

*See Figures 56 and 57.*

1. Before servicing the vehicle, refer to the precautions.

2. Remove front underbody protection.

3. Remove coolant drain plug on radiator. Drain and dispose of coolant.

4. Unlock plug and remove.

5. Unlock and detach water hose.

6. Remove nut.

7. Remove bolts and remove coolant thermostat.

➡**Coolant thermostat is integrated in cover and can only be replaced as a complete unit.**

*To install:*

8. Clean sealing surfaces.

9. Replace sealing ring.

10. Vent cooling system and check for leaks.

➡**Before filling, turn ignition ON. Set blower to low level. Seat heating controller to maximum temperature. This ensures that the heater valves are fully opened and the auxiliary water pump starts up.**

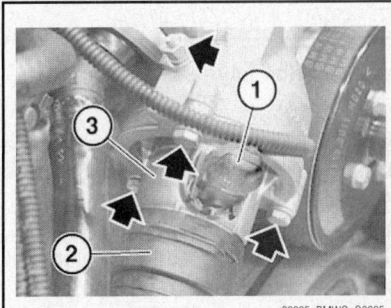

22205_BMWC_G0095

**Fig. 56 Unlock plug (1), unlock and detach water hose (2) and remove coolant thermostat (3)**

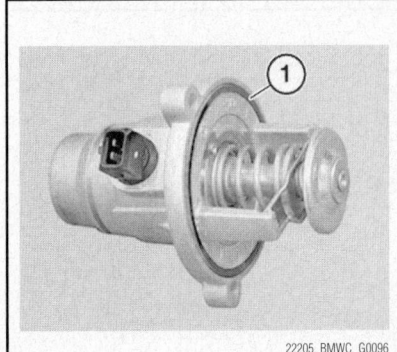

22205_BMWC_G0096

**Fig. 57 Coolant thermostat sealing ring**

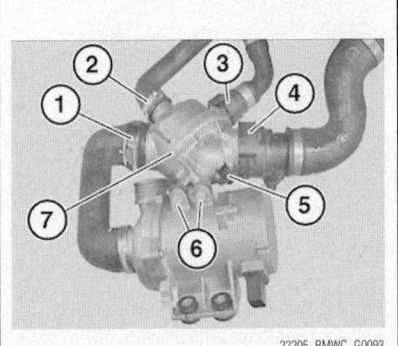

22205_BMWC_G0093

**Fig. 55 Thermostat assembly**

**Important:** The auxiliary water pump must deliver coolant in order to ensure fully venting.

a. Pour coolant into expansion tank up to **MAX** mark. Perform filling operation slowly.

b. Start engine and run at idle speed for approx. one minute (cap open). Then adjust coolant level to **MAX**.

c. Close cap and run engine up to operating temperature until main thermostat opens. Check cooling circuit and drain plug for leaks.

### ⁂ WARNING

**The engine must be cooled down before the coolant level is checked. Coolant temperature must not exceed 30°C. If ambient temperature is above 30°C, allow engine to cool down to ambient temperature at least.**

d. Check coolant level and adjust to **MAX**.

➡ Do not fill coolant expansion tank over MAX level as overfilling will cause the coolant to overflow.

## WATER PUMP

*REMOVAL & INSTALLATION*

### N52 and N54 Engines
*See Figure 58.*

### ⁂ WARNING

**Aluminum-magnesium material. No steel fasteners may be used due to the threat of electrochemical corrosion. A magnesium crankcase requires aluminum fasteners exclusively. Aluminum fasteners must be replaced each time they are removed. The end faces of aluminum**

fasteners are painted blue for purposes of identification. Torque specifications and torque angles must be observed for risk of damage.

### ⁂ WARNING

**If a water pump that has already been operated is reused, it must be filled with coolant immediately after removal.**

1. Before servicing the vehicle, refer to the precautions.
2. Remove coolant thermostat.
3. Remove hose clamps.
4. Remove coolant hoses.
5. Disconnect plug connection.
6. Remove bolts and discard aluminum bolts.
7. Remove electric water pump.

*To install:*

### ⁂ WARNING

**If the electric water pump is reused, it must be rotated one turn due to the breakaway torque at the blade wheels.**

8. Install electric water pump.
9. Install new bolts and tighten to 7 ft. lbs. (10 Nm) plus 90° additional rotation.
10. Connect plug connection.
11. Install coolant hoses.
12. Fasten hose clamps.
13. Install coolant thermostat.

### N62 Engines
*See Figures 59 through 62.*

1. Before servicing the vehicle, refer to the precautions.
2. Drain and dispose of coolant from radiator.
3. Remove fan cowl.
4. Remove electric fan.
5. Remove alternator drive belt.

6. Unlock and detach plug connections.
7. Remove screw and remove vacuum line holder.
8. Unlock and detach all coolant hoses on water pump.
9. Remove bolts and remove belt pulley.
10. Remove vibration damper.

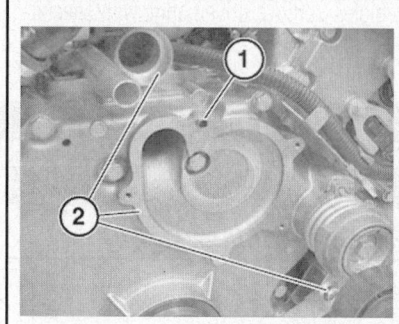

Fig. 60 Check adapter sleeve for correct seating and damage; replace if necessary

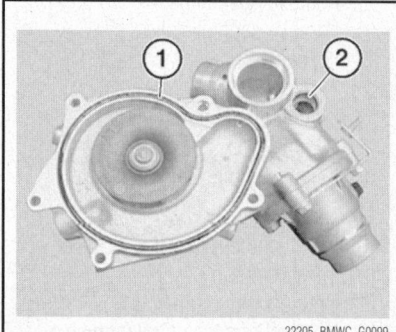

Fig. 61 Clean sealing surfaces and replace sealing rings

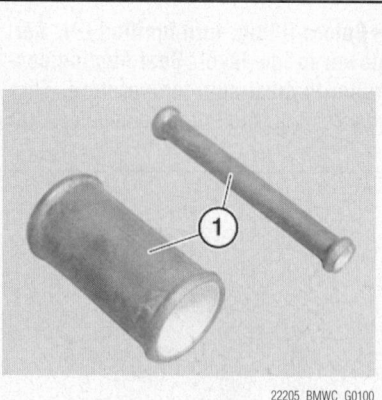

Fig. 62 Replace both coolant pipes (1) and coat sealing surfaces with anti-friction rubber coating. Thin coolant pipe only with water-cooled alternator

Fig. 58 Water pump assembly

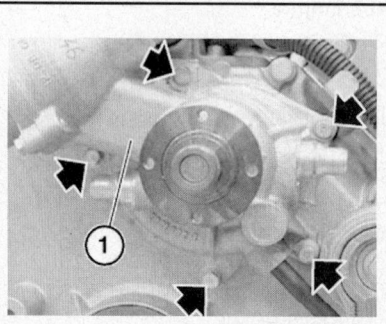

Fig. 59 Water pump (1) mounting bolt locations

11. Remove bolts and remove water pump.

12. Remove coolant thermostat and coolant sensor of faulty water pump.

13. Check coolant thermostat and coolant sensor for damage and replace if necessary.

**To install:**

14. Install coolant thermostat and coolant sensor in new water pump.

15. Check adapter sleeve for correct seating and damage; replace if necessary.

16. Clean sealing surfaces.

17. Replace sealing rings.

18. Replace both coolant pipes and coat sealing surfaces with anti-friction rubber coating.

➡**Thin coolant pipe only with water-cooled alternator.**

19. Install water pump.

20. Install vibration damper.

21. Install belt pulley.

22. Connect and lock all coolant hoses on water pump.

23. Install vacuum line holder.

24. Connect and lock plug connections.

25. Install alternator drive belt.

26. Install electric fan.

27. Install fan cowl.

28. Vent cooling system and check for leaks.

➡**Before filling, turn ignition ON. Set blower to low level. Seat heating controller to maximum temperature. This ensures that the heater valves are fully opened and the auxiliary water pump starts up. Important: The auxiliary water pump must deliver coolant in order to ensure fully venting.**

   a. Pour coolant into expansion tank up to **MAX** mark. Perform filling operation slowly.

   b. Start engine and run at idle speed for approx. one minute (cap open). Then adjust coolant level to **MAX**.

   c. Close cap and run engine up to operating temperature until main thermostat opens. Check cooling circuit and drain plug for leaks.

✳✳ **WARNING**

**The engine must be cooled down before the coolant level is checked. Coolant temperature must not exceed 30°C. If ambient temperature is above 30°C, allow engine to cool down to ambient temperature at least.**

   d. Check coolant level and adjust to **MAX**.

➡**Do not fill coolant expansion tank over MAX level as overfilling will cause the coolant to overflow.**

---

## ENGINE ELECTRICAL             CHARGING SYSTEM

### ALTERNATOR

*REMOVAL & INSTALLATION*

**N52 Engine**

*See Figure 63.*

✳✳ **WARNING**

**Aluminum-magnesium materials. No steel bolts/bolts may be used due to the threat of electrochemical corrosion. A magnesium crankcase requires aluminum bolts/bolts exclusively. Aluminum bolts/bolts must be replaced each time they are Removed. The end faces of aluminum bolts/bolts are painted blue for the purposes of reliable identification. Jointing torque and angle of rotation must be observed without fail (risk of damage).**

1. Before servicing the vehicle, refer to the precautions.

2. Disconnect the negative battery cable.

3. Remove intake filter housing .

4. Remove alternator drive belt.

5. Unlock plug and remove.

6. Remove nut and disconnect B+ wire.

7. Remove aluminum bolts and discard

8. Remove alternator.

**To install:**

9. Install alternator.

10. Install new aluminum bolts tighten to

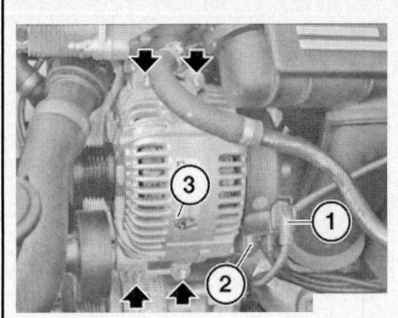

Fig. 63 Alternator (3), mounting bolts (arrows), B+ wire (1) and plug (2)

11. Connect B+ wire and tighten nut to 14 ft. lbs. (19 Nm).

12. Connect and lock plug.

13. Install alternator drive belt.

14. Install intake filter housing .

15. Connect the negative battery cable.

**N54 Engine**

*See Figures 64 and 65.*

1. Before servicing the vehicle, refer to the precautions.

2. Disconnect the negative battery cable.

3. Remove charge air duct.

4. Remove alternator drive belt.

5. Remove the A/C compressor and place to one side.

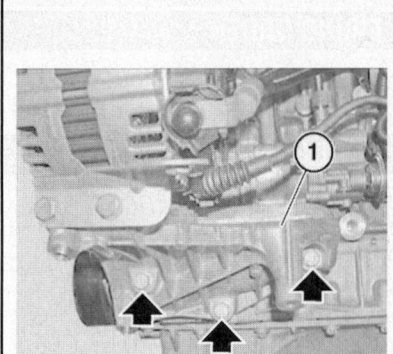

Fig. 64 Remove bolts a few turns until bracket (1) is loose. Do not remove bracket

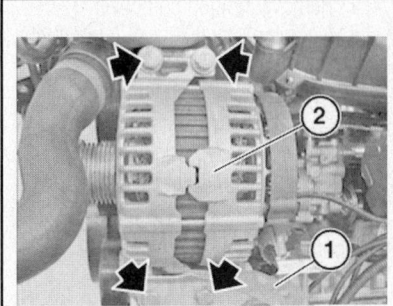

Fig. 65 Remove bracket (1) and alternator (2)

➡**Do not disconnect refrigerant lines from compressor.**

6. Remove bolts a few turns until bracket is loose. Do not remove bracket.

7. Unlock plug and remove.

8. Remove protective cap and Remove nut underneath.

9. Disconnect B+ wire.

10. Remove bracket and alternator.

### To install:

11. Install bracket and alternator.

12. Connect B+ wire.

13. Install protective cap and Remove nut underneath.

14. Install and lock plug.

➡**Do not disconnect refrigerant lines from compressor.**

15. Install the A/C compressor.

16. Install alternator drive belt

17. Install charge air duct.

18. Connect the negative battery cable.

### N62 Engine

*See Figure 66.*

1. Before servicing the vehicle, refer to the precautions.

2. Disconnect the negative battery cable.

3. Remove suction filter housing.

4. Remove electric fan with cowl.

5. Remove alternator drive belt.

6. Disconnect plug from alternator.

7. Unscrew nut and place B+ line to one side.

8. Remove alternator.

### To install:

9. Install alternator and tighten bolts to 15 ft. lbs. (21 Nm).

10. Connect B+ line and tighten to 115 inch lbs. (13 Nm).

11. Connect plug to alternator.

12. Install alternator drive belt.

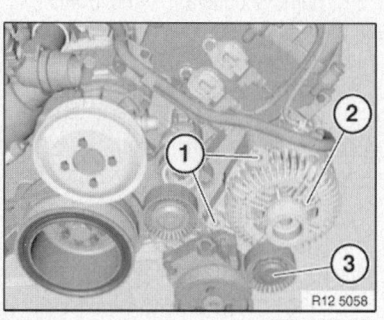

22205_BMWC_G0101

**Fig. 66 Alternator (2), mounting bolts (1) and guide pulley (3)**

13. Install electric fan with cowl.

14. Install suction filter housing.

15. Connect the negative battery cable.

16. Check alternator for correct operation.

---

## ENGINE ELECTRICAL

## IGNITION SYSTEM

### FIRING ORDER

*See Figures 67 and 68.*

### IGNITION COIL

*REMOVAL & INSTALLATION*

#### N52 and N54 Engines

*See Figures 69 and 70.*

1. Before servicing the vehicle, refer to the precautions.

2. Read out fault memory of DME control unit.

3. If equipped, remove engine cover.

4. On N62 engine, remove the intake filter housing.

5. Remove ignition coil cover.

6. Unlock plug retainer of ignition coil and disconnect plug.

7. Pull ignition coil up and out.

### To install:

8. Check that rubber seal of ignition coil is correctly seated.

9. Push plug with plug retainer open onto ignition coil.

10. Carefully close plug retainer.

11. Install ignition coil cover.

12. If equipped, install engine cover.

13. On N62 engine, install the intake filter housing.

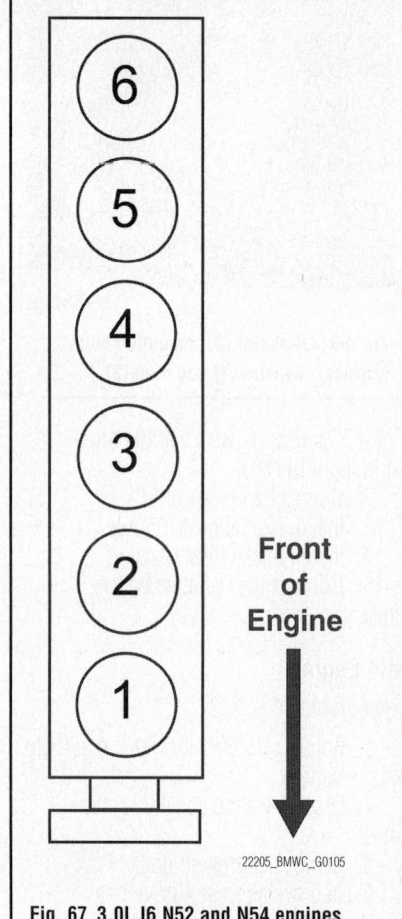

22205_BMWC_G0105

**Fig. 67 3.0L I6 N52 and N54 engines Firing order: 1–5–3–6–2–4 Distributorless ignition system**

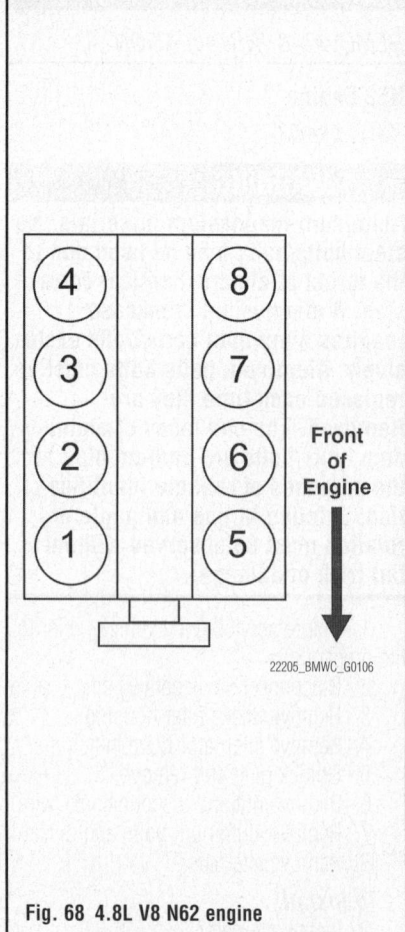

22205_BMWC_G0106

**Fig. 68 4.8L V8 N62 engine Firing order: 1–5–4–8–6–3–7–2 Distributorless ignition system**

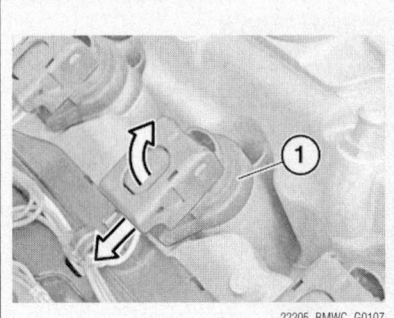

**Fig. 69 Unlock plug retainer of ignition coil (1) and disconnect plug**

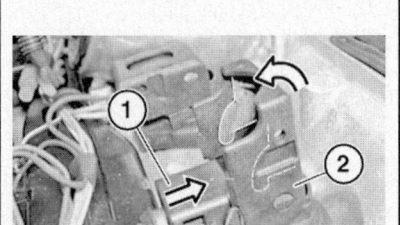

**Fig. 70 Push plug (1) with plug retainer (2) open onto ignition coil**

➡**The plug retainer must snap into place without great effort.**

14. Clear the fault memory.

### N62 Engine

*See Figures 71 and 72.*

1. Before servicing the vehicle, refer to the precautions.

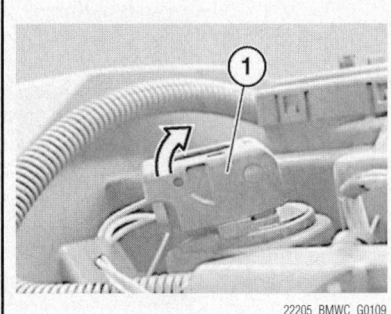

**Fig. 71 Unlock plug fastener (1) of ignition coil**

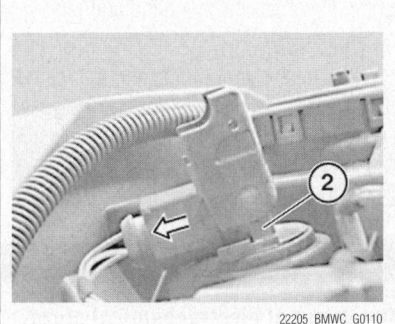

**Fig. 72 Disconnect ignition coil (2) connector in direction of arrow**

2. Read out fault memory of DME control unit.
3. Remove the intake filter housing.
4. Remove holder.
5. Pull ignition coil covers out of rubber grommets
6. Unlock plug fastener of ignition coil.
7. Disconnect ignition coil connector.
8. Pull ignition coil up and out.

*To install:*

9. Check that rubber seal of ignition coil is correctly seated.
10. Connect ignition coil connector.
11. Lock plug fastener of ignition coil.
12. Push ignition coil covers into rubber grommets.
13. Install holder.
14. Install the intake filter housing.
15. Clear the fault memory.
16. Check for proper ignition system operation.

### IGNITION TIMING

*ADJUSTMENT*

The ignition timing is controlled by the Digital Motor Electronics (DME). No adjustments are necessary.

### SPARK PLUGS

*REMOVAL & INSTALLATION*

1. Before servicing the vehicle, refer to the precautions.
2. Disconnect the negative battery cable.
3. If equipped, removal engine cover.
4. Remove ignition coils.
5. Unscrew spark plugs with special tool 12 1 171 (N52 and N62) or 12 1 220 (N54).

*To install:*

6. Screw spark plugs in special tool.
7. Tighten spark plugs to 18 ft. lbs. (24 Nm).
8. Install ignition coils.
9. If equipped, install engine cover.
10. Connect the negative battery cable.

## ENGINE ELECTRICAL

### STARTER

*REMOVAL & INSTALLATION*

### N52 and N54 Engines

*See Figure 73.*

**✳✳ WARNING**

**Aluminum-magnesium materials. No steel bolts/bolts may be used due to the threat of electrochemical corrosion. A magnesium crankcase requires aluminum bolts/bolts exclusively. Aluminum bolts/bolts must be replaced each time they are Removed. The end faces of aluminum bolts/bolts are painted blue for**

**the purposes of reliable identification. Jointing torque and angle of rotation must be observed without fail (risk of damage).**

1. Disconnect the negative battery cable.
2. Remove intake air manifold
3. Unlock plug and remove.
4. Remove the B+ cable from the starter.
5. Remove and discard the starter mounting bolts.
6. Remove starter motor.

*To install:*

7. Check starter pinion and ring gear for damage, replace damaged parts if necessary.

## STARTING SYSTEM

**Fig. 73 Starter mounting bolt locations**

8. Install starter motor.

9. Install the starter mounting bolts and tighten to specification.
- M10x85: 15 ft. lbs. (20 Nm) plus an additional 180° rotation
- M10x30: 15 ft. lbs. (20 Nm) plus an additional 90° rotation

10. Install the B+ cable from the starter and tighten nut to 10 ft. lbs. (13 Nm).

11. Unlock plug and Install.

12. Install intake air manifold

13. Connect the negative battery cable.

### N62 Engine

*See Figure 74.*

1. Before servicing the vehicle, refer to the precautions.

2. Disconnect the negative battery cable.

3. Remove reinforcement plate.

4. On E60 models, remove right exhaust manifold.

5. Remove heat shield.

6. Remove the B+ cable from the starter.

**Fig. 74 Starter mounting bolt locations**

22205_BMWC_G0111

7. Unlock plug and remove.

8. Remove bolts and pull starter motor out of transmission mounting and remove.

#### *To install:*

9. Check starter pinion and ring gear for damage, replace damaged parts if necessary.

10. Install starter motor in transmission mounting and tighten bolts to 33 ft. lbs. (45 Nm).

11. Connect the B+ cable to the starter and tighten the nut.

12. Install and lock plug.

13. Install heat shield.

14. On E60 models, install right exhaust manifold.

15. Install reinforcement plate.

16. Connect the negative battery cable.

## ENGINE MECHANICAL

➥Disconnecting the negative battery cable may interfere with the functions of the on board computer systems and may require the computer to undergo a relearning process, once the negative battery cable is reconnected.

### ACCESSORY DRIVE BELTS

#### *ACCESSORY BELT ROUTING*

*See Figures 75 through 77.*

#### *INSPECTION*

Inspect the drive belt for signs of glazing or cracking. A glazed belt will be perfectly smooth from slippage, while a good belt will have a slight texture of fabric visible. Cracks will usually start at the inner edge of the belt and run outward. All worn or damaged drive belts should be replaced immediately.

#### *ADJUSTMENT*

The belt tension is maintained by an automatic tensioner. No adjustment is possible.

#### *REMOVAL & INSTALLATION*

### N52 Engine

*See Figure 78.*

#### ✸✸ WARNING

**Aluminum-magnesium materials. No steel bolts/bolts may be used due to**

the threat of electrochemical corrosion. A magnesium crankcase requires aluminum bolts/bolts exclusively. Aluminum bolts/bolts must be replaced each time they are Removed. The end faces of aluminum bolts/bolts are painted blue for the purposes of reliable identification. Jointing torque and angle of rotation must be observed without fail (risk of damage).

1. Before servicing the vehicle, refer to the precautions.

2. Remove fan cowl with electric fan.

3. Mark the direction of rotation of the drive belt if it is to be reused.

4. Turn belt tensioner clockwise until bore is flush on housing.

5. By holding belt tensioner under tension, the load is removed from tensioning pulley.

6. Secure belt tensioner with special tool 11 3 340.

7. Remove drive belt upwards.

8. Check drive belt for correct installation position and, if reusing, observe direction of rotation.

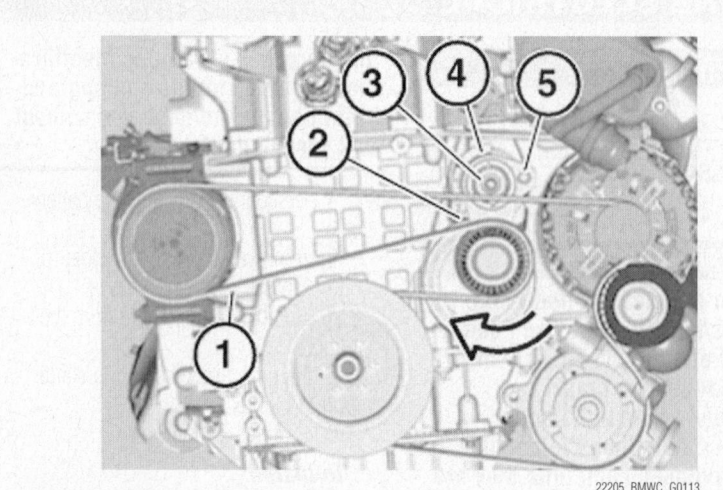

**Fig. 75 Accessory belt routing—N52 engine**

22205_BMWC_G0113

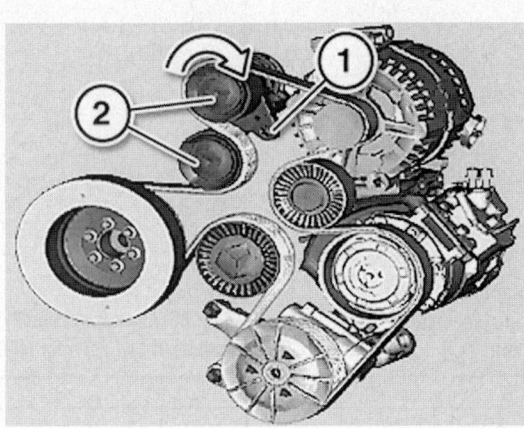

22205_BMWC_G0186

**Fig. 76 Accessory belt routing—N54 engine**

22205_BMWC_G0191

**Fig. 77 Accessory belt routing—N62 engine**

22205_BMWC_G0114

**Fig. 78 Secure belt tensioner with special tool 11 3 340**

## N54 Engine

*See Figure 79.*

➡**Mark the direction of rotation of the drive belt if it is to be reused. Depending on the build date (version), the idler pulleys can be fitted with and without grooves.**

1. Before servicing the vehicle, refer to the precautions.
2. Remove fan cowl.
3. Remove hose clip.
4. Remove quick-connect fastener 90∞ on boost pressure pipe in direction of arrow.
5. Pull off air hose.
6. Disconnect line from holder in direction of arrow.
7. Remove coolant hose from holder.
8. Remove screw.
9. Fold air duct down.

➡**Do not remove air duct.**

10. Turn belt tensioner in direction of arrow until bore is flush on housing.
11. Secure belt tensioner in place with special tool 11 3 340.
12. Remove drive belt (1).

*To install:*

13. Install belt in the previously marked direction if reusing the old belt.
14. Pretension tensioning pulley in direction of arrow.
15. Remove special tool 11 3 340.
16. On boost pressure pipe, bring lock back 90° into installation position.
17. Recirculated air hose must audibly snap into place.
18. Check that drive belt for is in correct installation position.

22205_BMWC_G0185

**Fig. 79 Turn belt tensioner (1) in direction of arrow until bore is flush on housing**

## N62 Engine

➡Mark the direction of rotation of the drive belt if it is to be reused. Depending on the build date (version), the idler pulleys can be fitted with and without grooves.

1. Before servicing the vehicle, refer to the precautions.
2. Turn belt tensioner in direction of arrow until bore is flush on housing.
3. Secure belt tensioner in place with special tool 11 3 340.
4. Remove drive belt (1).

### To install:

5. Install belt in the previously marked direction if reusing the old belt.
6. Pretension tensioning pulley in direction of arrow.
7. Remove special tool 11 3 340.
8. Check that drive belt for is in correct installation position.

## CAMSHAFT AND VALVE LIFTERS

### REMOVAL & INSTALLATION

### N52 and N54 Engines

*See Figures 80 through 108.*

### ✳✳ WARNING

Aluminum bolts/bolts must be replaced each time they are Removed. The end faces of aluminum bolts/bolts are painted blue for the purposes of reliable identification. Jointing torque and angle of rotation must be observed without fail (risk of damage).

1. Before servicing the vehicle, refer to the precautions.
2. Remove cylinder head cover.
3. Remove the underbody protection.

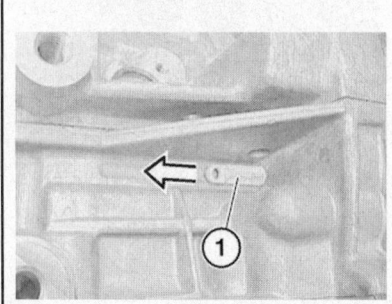

Fig. 80 Remove fastener (1) in direction of arrow

Fig. 81 Slide special tool 11 0 300 in direction of arrow into special tool bore and secure crankshaft

Fig. 82 With 1st cylinder in firing TDC position, cams of inlet camshaft (1) at 1st cylinder point upwards at an angle

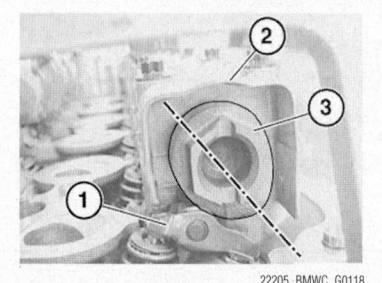

Fig. 83 With 1st cylinder in firing TDC position, cams of exhaust camshaft (3) at 6th cylinder point downwards at an angle

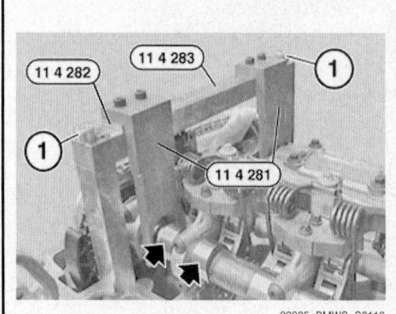

Fig. 84 Secure the special tools to cylinder head with bolts (1)

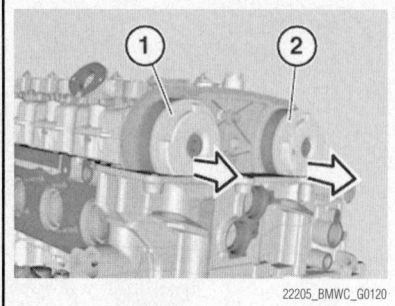

Fig. 85 Adjustment unit (1) from exhaust camshaft and (2) from inlet camshaft

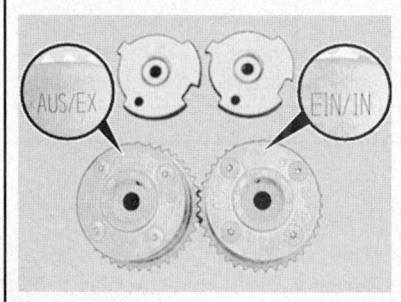

Fig. 86 Inlet and exhaust adjustment units are different. VANOS is marked with AUS/EX for the exhaust camshaft and EIN/IN for the inlet camshaft

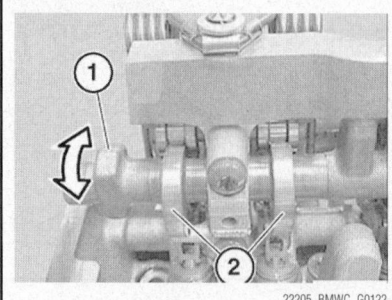

Fig. 87 If necessary, move eccentric shaft (1) on twin surface to minimum lift (2)

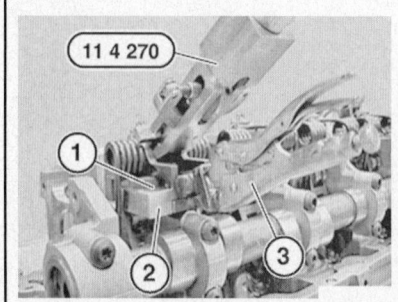

Fig. 88 Secure special tool 11 4 270 (1) with gripping pliers (3) to guide block (2)

4. Check the camshaft timing.

a. Remove fastener in direction of arrow.

b. Rotate crankshaft at central bolt into TDC position.

c. Slide special tool 11 0 300 in direction of arrow into special tool bore and secure crankshaft.

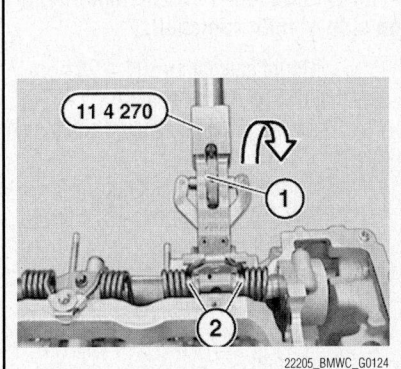

Fig. 89 Secure both bearing pins (2) in torsion springs with knurled screw (1) of special tool 11 4 270. Press special tool 11 4 270 in direction of arrow as far as it will go

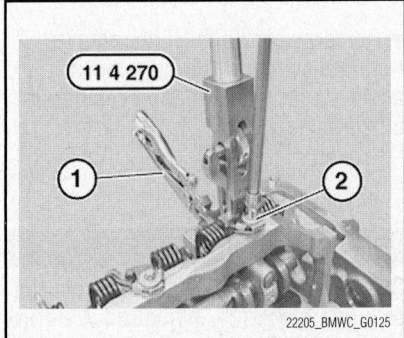

Fig. 90 Remove screw (2) of torsion spring

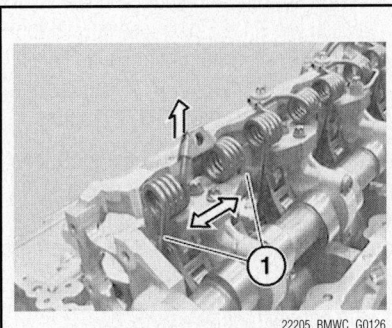

Fig. 91 Press torsion spring apart at positions (1)

---

**✵✵ WARNING**

**On vehicles with optional extra SA205 (automatic transmission), there is a large bore for the TDC position shortly before the special tool bore. This bore can be confused with the special tool bore.**

d. If the flywheel is secured in the correct special tool bore with special tool 11 0 300, the engine can no longer be moved at the central bolt.

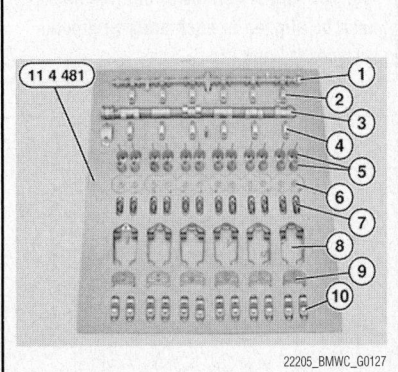

Fig. 92 Place all components in clean and neat order in special tool 11 4 481

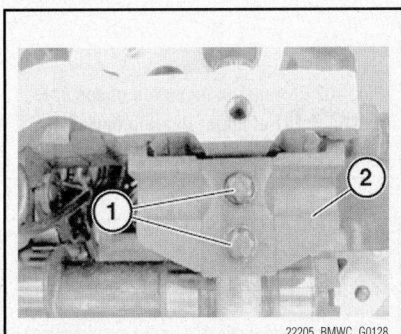

Fig. 93 Remove bolts (1) on guide block (2)

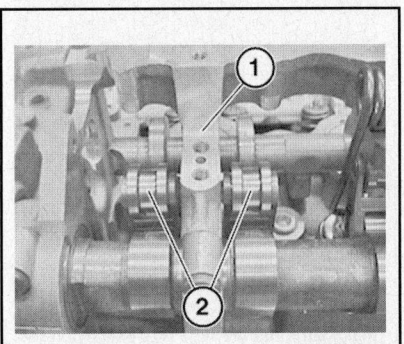

Fig. 94 Lift out intermediate levers (2)

---

e. With 1st cylinder in firing TDC position, cams of inlet camshaft at 1st cylinder point upwards at an angle.

f. The timings are correct when the part numbers on the inlet and exhaust camshafts point upwards.

g. With 1st cylinder in firing TDC position, cams of exhaust camshaft at 6th cylinder point downwards at an angle.

Fig. 95 All intermediate levers are classified and must be reinstalled in the same positions in an engine which has already been in use

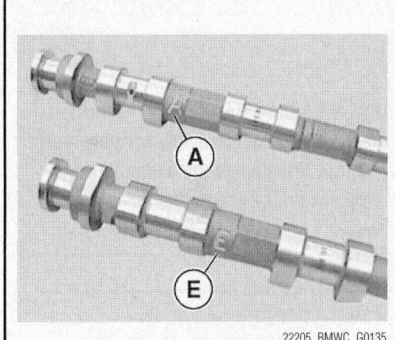

Fig. 96 Markings of inlet and exhaust camshafts are different. Mixing up the inlet (E) and exhaust (A) camshaft will result in engine damage

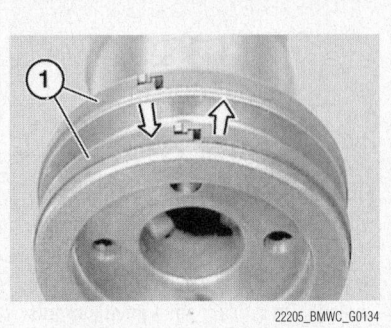

Fig. 97 Check plain compression rings (1) for damage and replace if necessary

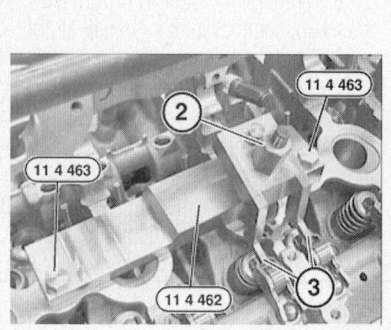

Fig. 98 Press down cam followers (3) on cylinder no. 2 with spindle nut (2) of special tool 11 4 462

Fig. 99 Before mounting the exhaust camshaft on the correct cam follower seat (1), pay attention to the hydraulic valve clearance adjustment element and the valve

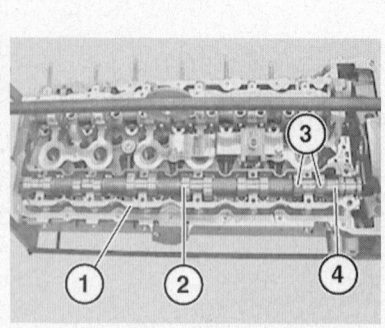

Fig. 100 Position lower bearing bank (1) with exhaust camshaft (2) cam followers. Align exhaust camshaft (2) so that cylinder nos. 2 and 4 are at valve overlap and the cams (3) on cylinder no. 1 point upwards at an angle. Part number (4) on twin surface of exhaust camshaft (2) points upwards

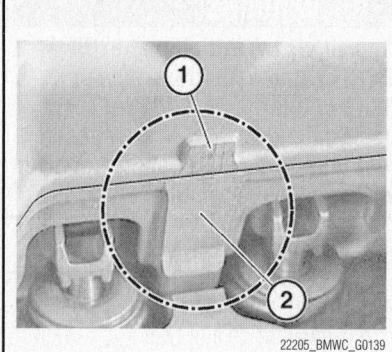

Fig. 101 Upper and lower bearing banks must be aligned to each other at ground surfaces (1) and (2)

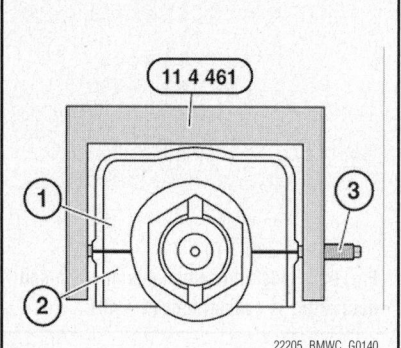

Fig. 102 Schematic depiction of special tool 11 4 461 at upper bearing bank (1) and lower bearing bank (2)

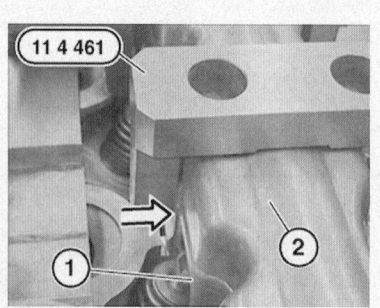

Fig. 103 Position special tool 11 4 461 over screw connection of bearing banks. Make sure that the legs rest exactly on the ground surfaces of the upper bearing bank (2) and lower bearing bank (1)

h. Cam follower (1) is not actuated.

➡When the engine is installed, the position of the exhaust camshaft (3) for the timing can only be checked with a mirror.

i. Secure special tool 11 4 283 to cylinder head with bolts.

➡Fit special tool 11 4 282 underneath on side of inlet camshaft.

j. Mount special tool 11 4 281 on inlet and exhaust camshafts.
5. Remove inlet and exhaust adjustment units.
a. Remove chain tensioner.
b. Remove central bolts on inlet and exhaust adjustment units.
c. Remove exhaust adjustment unit from exhaust camshaft.
d. Remove inlet adjustment unit from inlet camshaft.

### ✳✳ WARNING

Inlet and exhaust adjustment units are different. VANOS is marked with AUS/EX for the exhaust camshaft and EIN/IN for the inlet camshaft.

e. If necessary, move eccentric shaft on twin surface to minimum lift.

➡Oil spray nozzle must be removed from 3rd cylinder. During removal make a note of installation position of oil spray nozzle.

f. Secure special tool 11 4 270 with gripping pliers to guide block.

### ✳✳ WARNING

Special tool 11 4 270 is only secured to guide block (2). Adjusting the gripping pliers (3) on special tool 11 4 270 is not permitted.

g. Secure both bearing pins (2) in torsion springs with knurled screw (1) of special tool 11 4 270. Press special tool 11 4 270 in direction of arrow as far as it will go.
h. Remove screw of torsion spring.

➡To avoid jamming of screw with torsion spring, it is necessary when releasing screw to relieve the pretension on special tool 11 4 270 uniformly.

i. Relieve tension on torsion spring with special tool 11 4 270.

➡Metal lug cannot be disassembled and must not be removed.

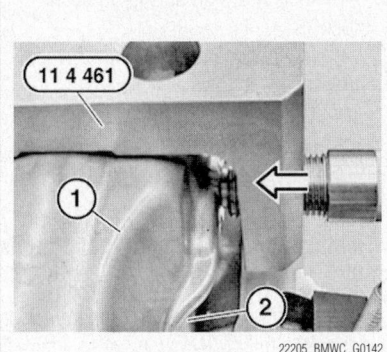

**Fig. 104 Initially tighten screw of special tool 11 4 461 to ground surfaces of upper bearing bank (1) and lower bearing bank (2)**

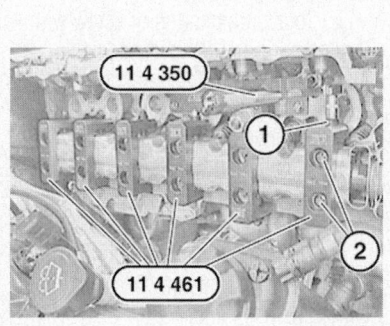

**Fig. 105 Mount special tools 11 4 461 with screw (1) to inside of cylinder head. Position special tools 11 4 461 so that screw connections (2) of bearing bank are easily accessible**

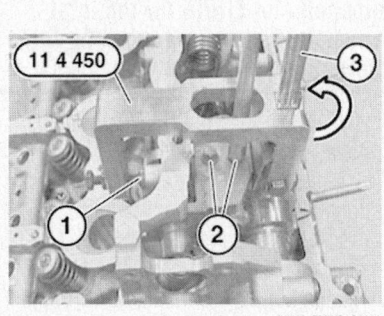

**Fig. 106 Secure special tool 11 4 450 to bolt connection (1) of eccentric shaft. Turn eccentric lever (3) on special tool 11 4 450 in direction of arrow. Guide block is now pretensioned. Insert bolts (2) of guide blocks**

    j. Press torsion spring apart at positions. Remove torsion spring towards top.
    k. Place all components in clean and neat order in special tool 11 4 481.

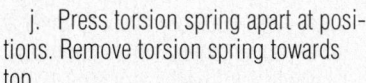

> ※※ **WARNING**
>
> **All components must be reinstalled in the same positions in an engine which has already been in use.**

    l. Remove bolts on guide block.
    m. Place all guide blocks in neat order in special tool 11 4 481.
    n. Lift out intermediate levers (2).
    o. Place all intermediate levers in neat order in special tool 11 4 481.

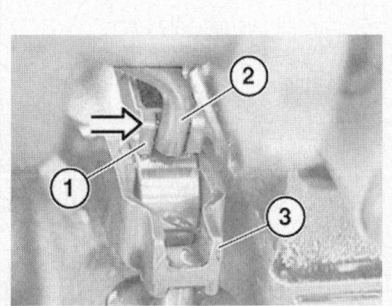

**Fig. 107 Insert torsion spring (2) in intermediate lever (1) (arrow) and check that cam follower (3) is in correct installation position**

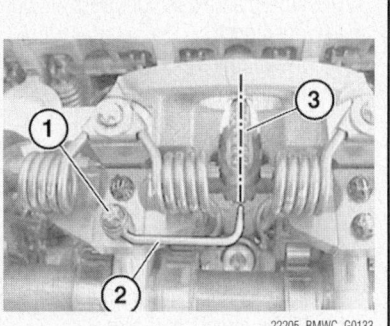

**Fig. 108 At cylinder no. 3, adjust oil spray nozzle (2) so that oil spray points precisely towards spline teeth (3). Insert screw (1) with oil spray nozzle (2) (external)**

> ※※ **WARNING**
>
> **All intermediate levers are classified and must be reinstalled in the same positions in an engine which has already been in use.**

    6. Remove inlet camshaft.

➡ **All bearing caps are marked with numbers from 1 to 6. Bearing cap is a thrust bearing.**

    a. Remove bolts on bearing caps 1 to 6.
    b. Set all bearing caps down in special tool 11 4 481 in a tidy and orderly fashion.
    c. Remove inlet camshaft towards top.

> ※※ **WARNING**
>
> **Markings of inlet and exhaust camshafts are different. Mixing up the inlet and exhaust camshaft will result in engine damage.**

    7. Remove exhaust camshaft.
    a. The screw connection of the bearing banks must be Removed from the outside inwards.
    b. Lift out upper and lower bearing banks with exhaust camshaft.
    c. Remove upper bearing bank.
    d. Remove exhaust camshaft from lower bearing bank.

***To install:***
    8. Check plain compression rings for damage and replace if necessary.
    a. Plain compression rings are engaged at joint.
    b. Press plain compression rings apart upwards and downwards and removed towards front.

> ※※ **WARNING**
>
> **Plain compression rings can easily break.**

    9. Install exhaust camshaft.
    a. Mounting bearing bank. Pre-install special tool 11 4 462 on cylinder no. 2.
    b. Insert special tool 11 4 463 in screw connection of cylinder head cover.

➡ **Special tool 11 4 463 is a special screw.**

    c. Press down cam followers on cylinder no. 2 with spindle nut of special tool 11 4 462.
    d. Before mounting the exhaust camshaft on the correct cam follower seat, pay attention to the hydraulic valve clearance adjustment element and the valve.

e. Position lower bearing bank with exhaust camshaft cam followers.

f. Align exhaust camshaft so that cylinder nos. 2 and 4 are at valve overlap and the cams on cylinder no. 1 point upwards at an angle. Part number on twin surface of exhaust camshaft points upwards.

### ✳✳ WARNING

**There must be no adhesive residues in the cylinder head tapped holes.**

g. Clean tapped holes.

h. Fit upper bearing bank. Insert bolts dry and tension down upper bearing bank with exhaust camshaft at bearing points 3 and 5 through a ½ bolt turn.

i. Join exhaust camshaft to upper and lower bearing banks with torque wrench from inside outwards to 6 ft. lbs. (8 Nm).

j. Remove all bolts of upper bearing bank from outside inwards by 90°.

k. Upper and lower bearing banks must be aligned to each other at ground surfaces (1 and 2).

Make sure that the thrust piece and the legs of special tools 11 4 461 rest on the milled surfaces.

➡**Schematic depiction of special tool 11 4 461 at upper bearing bank and lower bearing bank.**

**Pretension all special tools 11 4 461 with special tool 11 4 350 only.**

l. Tighten screw on thrust piece to 13 inch lbs. (2 Nm).

m. Position special tool 11 4 461 over screw connection of bearing banks.

n. Make sure that the legs rest exactly on the ground surfaces of the upper bearing bank and lower bearing bank.

o. Initially tighten screw of special tool 11 4 461 to ground surfaces of upper bearing bank and lower bearing bank.

p. Tighten bolts on thrust piece to 13 inch lbs. (2 Nm).

### ✳✳ WARNING

**Set special tool 11 4 350 to 13 inch lbs. (2 Nm).**

q. Pretension all special tools 11 4 461 with special tool 11 4 350 only.

r. Mount special tools 11 4 461 with screw to inside of cylinder head.

s. Mount special tool 11 4 461 with screw facing outwards on cylinder no. 2.

t. Position special tools 11 4 461 so that screw connections of bearing bank are easily accessible.

u. Insert bolts dry. Tighten upper and lower bearing banks bolts from inside outwards with special tool 00 9 120 to 6 ft. lbs. (8Nm) plus an additional 60° of rotation.

### ✳✳ WARNING

**Remove special tool 11 4 461 only when exhaust camshaft screw connection is completed.**

10. Install inlet camshaft.

a. Clean all bearing points and lubricate with oil.

b. Insert inlet camshaft so that part number on twin surface points upwards.

c. Position inlet camshaft so that cams point upwards at an angle.

d. Connect special tool 11 4 281 to twin surface.

e. Tighten bearing bolts to 7 ft. lbs. (9 Nm).

11. Install intermediate levers.

a. Mixing up the guide blocks will cause the engine to suffer idle-speed fluctuations. This will result in maladjustment of uniform distribution.

b. All contact surfaces of guide block must be clean and free from oil and grease. If necessary, clean contact surfaces.

c. Mixing up the intermediate levers will cause the engine to suffer idle-speed fluctuations.

Installation:

d. All contact surfaces must be clean and free from oil and grease. If necessary, clean contact surfaces.

e. All intermediate levers are classified and must be reinstalled in the same positions in an engine which has already been in use.

f. Before installing intermediate levers, make sure cam followers are correctly positioned.

g. Install intermediate levers.

h. Fit guide block cleanly into opening. Tighten bolts hand-tight.

i. Check that intermediate levers are in correct installation position.

j. Remove bolts by a ¼ turn.

k. Secure special tool 11 4 450 to bolt connection of eccentric shaft.

l. Turn eccentric lever on special tool 11 4 450 in direction of arrow.

m. Guide block is now pretensioned. Insert bolts of guide blocks and tighten to 7 ft. lbs. (10 Nm).

n. At cylinder no. 3, the guide block can be pre-installed with one screw (internal) only.

o. Oil spray nozzle is fitted only after torsion spring has been installed.

p. Install torsion spring on guide block.

q. Insert torsion spring in intermediate lever.

r. Check that cam follower is in correct installation position.

s. Secure special tool 11 4 270 with gripping pliers to guide block.

➡**Replace torsion spring if metal lug is faulty.**

t. Secure both bearing pins in torsion springs with knurled screw of special tool 11 4 270.

u. Check torsion spring on intermediate lever to ensure correct installation position.

v. Press special tool 11 4 270 in direction of arrow as far as it will go.

w. Insert screw of torsion spring and tighten to 7 ft. lbs. (10 Nm).

x. To avoid jamming of screw with torsion spring, it is necessary when inserting screw to increase pretension on special tool 11 4 270 uniformly.

y. Remove special tool 11 4 270.

z. At cylinder no. 3, adjust oil spray nozzle so that oil spray points precisely towards spline teeth.

aa. Insert screw with oil spray nozzle (external) and tighten to 7 ft. lbs. (10 Nm).

12. Install inlet and exhaust adjustment units.

a. To facilitate installation of the inlet and exhaust adjustment units, turn the sensor gears at the opening downwards.

### ✳✳ WARNING

**Do not mixing up the inlet and exhaust adjustment units. VANOS is marked with AUS/EX for the exhaust camshaft and EIN/IN for the inlet camshaft. Sensor gears can be fitted alternatively.**

b. Position inlet and exhaust adjustment units on camshafts.

c. Insert new central bolts but do not tighten until after valve timing is checked. Grip inlet and exhaust camshafts at dihedron when tightening. Tighten bolts to 15 ft. lbs. (20 Nm) plus an additional 180° rotation.

13. Check valve timing.

a. Install special tool 11 4 280 to secure the central bolts on the inlet and exhaust adjustment units and camshafts.

b. Press clamping rail by hand against guide rail and make sure timing chain is guided in clamping rail.

c. Rotate crankshaft at central bolt into TDC position.

d. Slide special tool 11 0 300 in direction of arrow into special tool bore and secure crankshaft.

> ✸✸ **WARNING**
>
> **On vehicles with optional extra SA205 (automatic transmission), there is a large bore for the TDC position shortly before the special tool bore. This bore can be confused with the special tool bore.**

e. If the flywheel is secured in the correct special tool bore with special tool 11 0 300, the engine can no longer be moved at the central bolt.

f. With 1st cylinder in firing TDC position, cams of inlet camshaft at 1st cylinder point upwards at an angle.

g. The timings are correct when the part numbers on the inlet and exhaust camshafts point upwards.

h. With 1st cylinder in firing TDC position, cams of exhaust camshaft at 6th cylinder point downwards at an angle.

i. Cam follower is not actuated.

➡ **When the engine is installed, the position of the exhaust camshaft for the timing can only be checked with a mirror.**

j. Secure special tool 11 4 283 to cylinder head with bolts.

➡ **Fit special tool 11 4 282 underneath on side of inlet camshaft.**

k. Mount special tool 11 4 281 on inlet and exhaust camshafts.

l. Make sure the chain tensioner is Removed.

m. Turn sensor gears (2) in direction of arrow until locating pins (1) on special tool 11 4 290 match up.

n. Slide on special tool 11 4 290 in direction of arrow.

o. Secure special tool 11 4 290 with bolts (1).

p. Screw special tool 11 9 340 into cylinder head.

q. Pretension timing chain with special tool 00 9 250 to 0.6 Nm.

r. Secure both central bolts of inlet and exhaust adjustment units with special tool 00 9 120 to inlet and exhaust camshafts. Tighten bolts to 15 ft. lbs. (20 Nm) plus an additional 180°rotation.

14. Install the chain tensioner.

15. Install fastener with bore facing outwards.

## N62 Engine

### Intake Camshafts

*See Figures 109 through 130.*

1. Before servicing the vehicle, refer to the precautions.

2. Read and document the fault memory.

3. Remove servomotor for eccentric shaft.

4. Remove ignition coils.

5. Remove cylinder head cover.

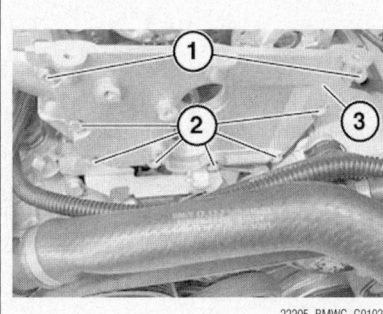

**Fig. 109 Timing cover (3) and mounting bolts (1 and 2)**

6. Remove spark plugs.

7. Remove timing case cover.

a. Remove both solenoid valves on.

b. Remove alternator (air-cooled).

c. Remove screw, remove holder.

d. Remove line from holder.

e. Remove bolts.

f. Remove timing case cover.

8. Remove intake and exhaust adjustment units as follows:

9. If necessary, remove fan cowl.

> ✸✸ **WARNING**
>
> **Screw is a special screw and must not be replaced with a normal M8 screw.**

10. Remove screw.

11. Disconnect oil line from retainers and remove.

12. Crank engine at central bolt in direction of rotation to firing TDC position of 1st cylinder.

13. When removing the left intake cam, with 1st cylinder in firing TDC position, cams of intake and exhaust camshafts at 5th cylinder point upwards at an angle as shown.

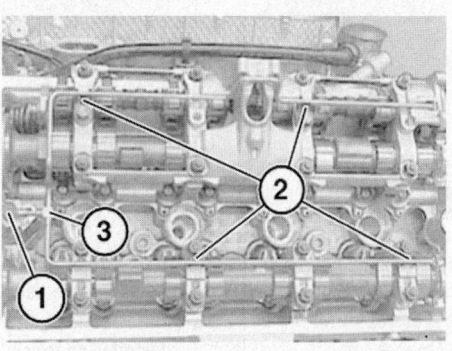

**Fig. 110 Screw (1) is a special screw and must not be replaced with a normal M8 screw. Disconnect oil line (3) from retainers (2) and remove**

**Fig. 111 When removing the left intake cam, with 1st cylinder in firing TDC position, cams of intake and exhaust camshafts at 5th cylinder point upwards at an angle as shown**

Fig. 112  In firing TDC position, cam of exhaust camshaft at 1st cylinder points upwards at an angle. Cam of intake camshaft points downwards at an angle

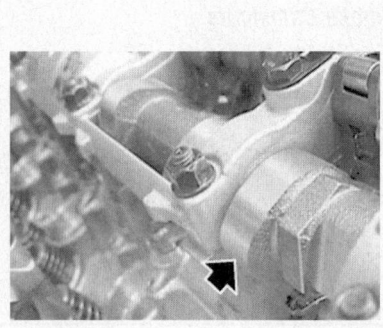

Fig. 115  On the right intake camshaft, rotate intake camshaft in direction of rotation until cam on 1st cylinder is positioned horizontally as shown

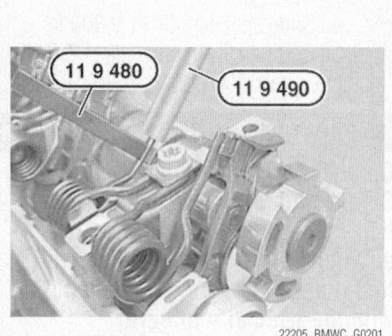

Fig. 118  Raise one end of torsion spring (1) with special tool 11 9 480. Support end of torsion spring protected with special tool 11 9490 on intake camshaft

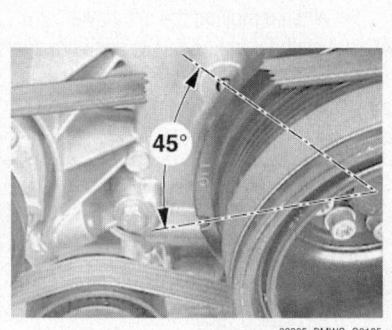

Fig. 113  Crank engine at central bolt and secure vibration damper with special tool 11 9 190 in firing TDC position of 1st cylinder

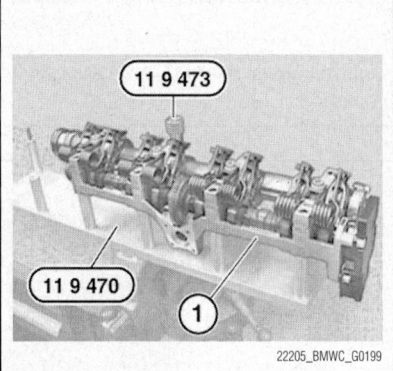

Fig. 116  Place bearing bracket (1) with intake camshaft and eccentric shaft as illustrated on special tool 11 9 470

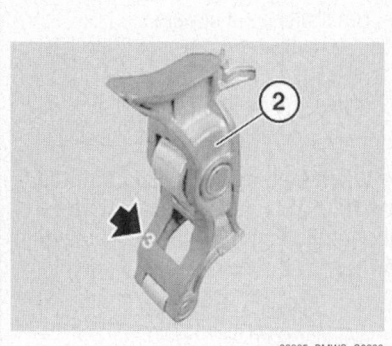

Fig. 119  Intermediate levers (2) are divided into individual tolerance classes. Only intermediate levers of the same tolerance class may be fitted in a single cylinder head

Fig. 114  Vigorously press back tensioner rail (3) several times to remove oil in chain tensioner. Feed out exhaust adjustment unit (2)

Fig. 117  Remove the eccentric shaft sensor (2) only if necessary for service

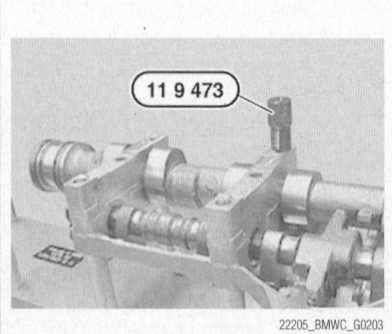

Fig. 120  Unscrew nut (special tool 11 9 473)

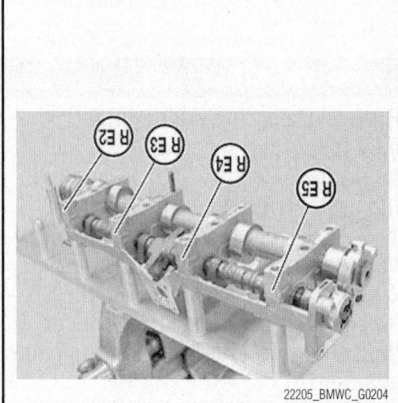

Fig. 121 Camshaft bearing caps of cylinders 1–4 and 5–8 must not be mixed up

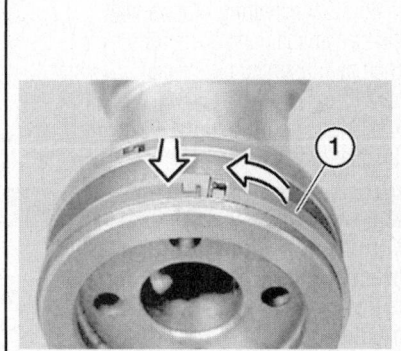

Fig. 122 Press compression ring (1) on one side into groove, pull up on other side and remove catch

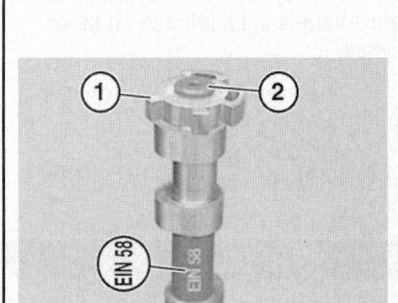

Fig. 123 If necessary, Remove screw (2) and remove sensor gear (1)

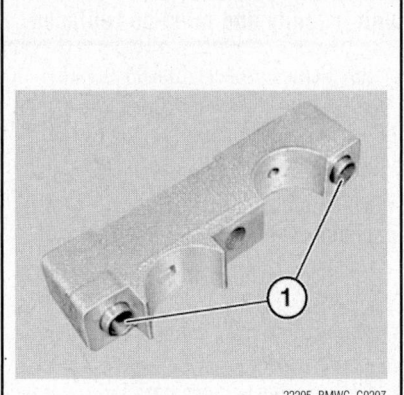

Fig. 124 Check dowel sleeves (1) for damage and correct installation position.

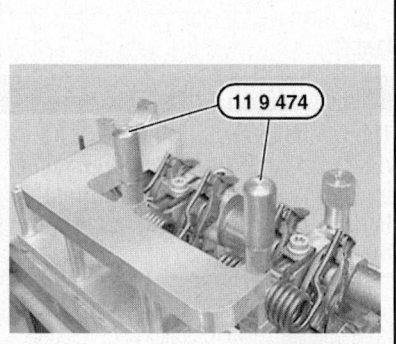

Fig. 125 Remove special tool 11 9 474

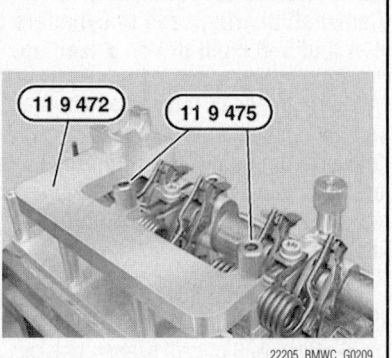

Fig. 126 Remove special tool 11 9 472 and special tool 11 9 475

Fig. 127 Ends of compression rings point upward

Fig. 128 The marking (1) on the hexagon drive of the intake camshaft faces upward

Fig. 129 Intake adjustment unit (1) is marked with EIN and IN. Exhaust adjustment unit (2) is marked with AUS and EX

14. When removing the right intake cam, crank engine at central bolt in direction of rotation to firing TDC position of 1st cylinder.

15. In firing TDC position, cam of exhaust camshaft at 1st cylinder points upwards at an angle. Cam of intake camshaft points downwards at an angle.

➡**Alignment hole for TDC position is at front on timing case cover.**

16. Crank engine at central bolt and secure vibration damper with special tool 11 9 190 in firing TDC position of 1st cylinder.

Fig. 130 Fit special tool 11 9 463, secure screw (1) in thread for oil line and tighten down by hand

**※※ WARNING**

**When the engine is shut down, the intake and exhaust adjustment unit is normally locked in its initial position. The situation may arise in some individual cases where this initial position is not reached and the camshaft can continue to be rotated in the adjustment range of the adjustment unit. In order to avoid incorrect timing adjustment, it is essential to check the locking of the adjustment unit and if necessary perform locking by rotating the camshafts.**

17. Checking locking of intake adjustment unit in initial position by engaging the hexagon head of intake camshaft and attempt to rotate intake camshaft carefully against direction of rotation.

18. If there is no fixed connection between intake camshaft and intake adjustment unit, rotate intake camshaft against direction of rotation as far as it will go.

19. The intake adjustment unit is locked in the initial position when the intake camshaft is non-positively connected to the intake adjustment unit.

20. Checking locking of exhaust adjustment unit in initial position by engaging hexagon head of exhaust camshaft and attempt to rotate exhaust camshaft carefully in direction of rotation.

21. If there is no fixed connection between exhaust camshaft and exhaust adjustment unit, rotate exhaust camshaft in direction of rotation as far as it will go.

22. The exhaust adjustment unit is locked in the initial position when the exhaust camshaft is non-positively connected to the exhaust adjustment unit.

**※※ WARNING**

**If the intake or exhaust adjustment unit of the camshafts cannot be**

locked as described, the adjustment unit is faulty and must be replaced.

23. Remove special tool 11 9 190.

24. Crank engine at central bolt against direction of rotation to 45∞ before TDC position.

➡**When slackening bolts, grip camshafts at hexagon head.**

25. Remove bolts of exhaust and intake adjustment unit.

26. Remove screw on exhaust adjustment unit.

27. Vigorously press back tensioner rail several times to remove oil in chain tensioner.

28. Feed out exhaust adjustment unit.

29. Remove screw on intake adjustment unit.

30. Feed out intake adjustment unit.

31. Secure timing chain to prevent it from sliding down.

32. Remove the intake camshaft as follows:

**※※ WARNING**

**The intake camshaft must first be rotated so that when the bearing bracket is removed the intermediate levers do not slip out and damage the camshaft.**

33. On the left intake camshaft rotate the camshaft against direction of rotation until lettering on 8th cylinder points upwards in cylinder axis and cam is horizontal.

34. On the right intake camshaft, rotate intake camshaft in direction of rotation until cam on 1st cylinder is positioned horizontally as shown.

**※※ WARNING**

**Camshaft bearing caps of cylinders 1–4 and 5–8 must not be mixed up.**

➡**Bearing caps of intake camshaft are marked on cylinder bank 5–8 with R E1 to R E5 from intake side. Bearing caps of intake camshaft are marked on cylinder bank 1–4 with L E1 to L E5 from intake side.**

35. Remove nuts and remove bearing cap E1.

36. Remove 8 nuts of bearing bracket from outside to inside.

➡**Rocker arms are freely accessible after bearing bracket has been removed.**

37. Do **NOT** remove rocker arm on intake side.

**※※ WARNING**

**Rocker arms are divided into individual tolerance classes. The tolerance classes are designated as illustrated with the numbers from 1–4. Used rocker arms may only be reused in the same position. When replacing rocker arms on intake side: install rocker arms of the same tolerance class in the same position.**

38. Clamp special tool 11 9 470 in a vice as shown.

**※※ WARNING**

**Do not tilt bearing bracket.**

39. Carefully lift out bearing bracket.

40. Place bearing bracket with intake camshaft and eccentric shaft as illustrated on special tool 11 9 470.

41. Secure bearing bracket with a nut (special tool 11 9 473).

**※※ WARNING**

**The lower section of the bearing bracket is machined with the cylinder head and must not be mixed up.**

➡**Lower section of bearing bracket remains on cylinder head.**

42. Remove the eccentric shaft sensor only if necessary for service.

➡**Removal of the intermediate levers and torsion springs is described on the 8th cylinder. The same procedure is applicable to all other cylinders.**

43. Raise one end of torsion spring with special tool 11 9 480.

44. Lift out intermediate lever and set down in an orderly fashion.

**※※ WARNING**

**Keep holding torsion spring with special tool 11 9 480.**

45. Connect special tool 11 9490 to end of torsion spring.

46. Support end of torsion spring protected with special tool 11 9490 on intake camshaft.

❊❊ **WARNING**

**Intermediate levers are divided into individual tolerance classes. Only intermediate levers of the same tolerance class may be fitted in a single cylinder head. The tolerance classes are designated as illustrated with the numbers from 1 to 5. Used intermediate levers may only be reused in the same position.**

47. Raise second end of torsion spring with special tool 11 9 480.

48. Lift out intermediate lever and set down in an orderly fashion.

❊❊ **WARNING**

**Keep holding torsion spring with special tool 11 9 480.**

49. Connect special tool 11 9490 to second end of torsion spring.

50. Support end of torsion spring protected with special tool 11 9 490 on intake camshaft.

51. Remove screw.

52. Remove torsion spring and special tool 11 9 490.

53. Remove intermediate levers and torsion springs of cylinders 5 to 7 according to the same procedure and set down in an orderly fashion.

54. Unscrew nut (special tool 11 9 473).

❊❊ **WARNING**

**Camshaft bearing caps of cylinders 1–4 and 5–8 must not be mixed up.**

➡Bearing caps are marked in graphic with cylinders 1–4 designated as L E2 to L E5and 5–8 designated as R E2 to R E5.

55. Remove bearing caps E2 to E5 and place to one side in order.

56. Lift out intake camshaft.

*To install:*

57. Install the intake camshaft as follows:

58. If necessary, replace plain compression rings.

❊❊ **WARNING**

**Plain compression rings can easily break.**

59. Plain compression rings are engaged at joint.

60. Press compression ring on one side into groove, pull up on other side and remove catch.

61. Carefully pull compression ring apart and remove towards front.

➡Intake camshaft of cylinder bank 5–8 is marked with "EIN 58". Intake camshaft of cylinder bank 1 to 4 is marked with "EIN 14".

62. If necessary, replace sensor gear.

63. Remove screw and remove sensor gear.

64. Clean all bearings and cams of intake camshaft and lubricate with engine oil.

➡Camshaft has a groove and sensor gear has a lug for fastening purposes.

65. Fit sensor gear and align to groove in camshaft.

66. Insert screw and tighten to 22 ft. lbs. (30 Nm).

67. Installing plain compression rings:

❊❊ **WARNING**

**Plain compression rings can easily break.**

68. Carefully pull compression ring apart and install from front.

69. Press compression ring on one side into groove, install catch on other side.

70. Insert intake camshaft so that cams point upwards at cylinder 5 (for the left intake cam) and cylinder 4 (for the right).

71. Make sure bearing shells of eccentric shaft are engaged in bearing bracket.

➡Bearing shell is guided in a groove in bearing cap.

72. Check dowel sleeves for damage and correct installation position.

➡Bearing caps are marked in graphic with R E2 to R E5 for the right side and L E2 to L E5 for the right side.

73. Fit bearing caps E2 to E5.

➡Initially tighten special tool 11 9 473 without play only.

74. Secure bearing bracket and bearing cap with nut (special tool 11 9 473).

➡The mounting of the bearing bracket described later can only be performed if the first bearing of the intake camshaft is aligned with special tool 11 9 472.

75. Fit special tool 11 9 472 as shown in graphic and align intake camshaft.

76. Fit centering sleeves (special tool 11 9 475) and align special tool 11 9 472 to bearing caps 2 and 3.

77. Insert special tool 11 9 474 and initially tighten without play.

➡Special tools 11 9 472, 11 9 474 and 11 9 475 remain fitted until all torsion springs and intermediate levers have been installed.

➡On the left side, installation of the intermediate levers and torsion springs is described on the 8th cylinder. The same procedure is applicable to cylinders 5 to 7. On the right side, installation of the intermediate levers and torsion springs is described on the 4th cylinder. The same procedure is applicable to cylinders 1 to 3.

78. Connect special tool 11 9 490 to ends of torsion spring.

79. Install torsion spring.

80. Insert screw and tighten down.

❊❊ **WARNING**

**Intermediate levers are divided into individual tolerance classes.**

81. Only intermediate levers of the same tolerance class may be fitted in a single engine.

82. The tolerance classes are designated as illustrated with the numbers from 1 to 5.

83. Used intermediate levers may only be reused in the same position.

84. Lubricate all sliding surfaces on intermediate lever with engine oil.

85. Raise torsion spring with special tool 11 9 480.

86. Remove special tool 11 9 490.

87. Hold torsion spring with special tool 11 9 480.

88. Install intermediate lever from above.

89. Insert end of torsion spring into guide on intermediate lever.

90. Raise second end of torsion spring with special tool 11 9 480.

91. Remove special tool 11 9 490.

92. Hold torsion spring with special tool 11 9 480.

93. Install intermediate lever from above.

94. Insert end of torsion spring into guide on intermediate lever.

95. Install eccentric shaft sensor.

96. Insert bolts and tighten down eccentric shaft sensor.

97. Remove special tool 11 9 474.

98. Remove special tool 11 9 472 and special tool 11 9 475.

99. Ends of compression rings point upward.

100. Make sure compression rings are engaged at ends.

**⁂ WARNING**

**Rocker arms slip slightly when bearing bracket is fitted.**

101. Make sure rocker arms are secured as illustrated on hydraulic valve clearance compensating elements and on valves.
102. Align rockers straight.
103. Remove special tool 11 9 473.
104. Remove bearing bracket from special tool 11 9 470.

**✿ WARNING**

**Do not tilt bearing bracket.**

105. Lower bearing bracket from above and carefully bring into contact with cylinder head.
106. Insert nuts and tighten by hand without play.

**⁂ WARNING**

**Make sure none of the intermediate levers or rocker arms have slipped out.**

107. Tighten nuts from inside to outside to specification.
  • M6: 7 ft. lbs. (10 Nm )
  • M7: 10 ft. lbs. (14 Nm)
  • M7- 8.8: 11 ft. lbs. (15 Nm)

**✿ WARNING**

**Camshaft bearing caps of cylinders 1–4 and 5–8 must not be mixed up.**

108. Fit bearing cap R E1 in such a way that marking is legible from intake side.
109. Install nuts and tighten to specification.
  • M6: 7 ft. lbs. (10 Nm )
  • M7: 10 ft. lbs. (14 Nm)
  • M7- 8.8: 11 ft. lbs. (15 Nm)
110. Rotate intake camshaft in direction of rotation until cam on 5th cylinder points upwards at an angle as shown in illustration.

➡ **The marking (1) on the hexagon drive of the intake camshaft faces upward.**

111. Install intake and exhaust adjustment units as follows:

**⁂ WARNING**

**Intake and exhaust adjustment units are different. Mixing up the intake and exhaust adjustment units will cause damage to the engine. Intake adjustment unit is marked with EIN and IN. Exhaust adjustment unit is marked with AUS and EX.**

➡ **Position of intake adjustment unit to timing chain can be freely selected.**

112. Pull timing chain up.
113. Feed intake adjustment unit into timing chain and fit onto intake camshaft.
114. Replace screw.
115. Install screw on intake adjustment unit.
116. Tighten screw without play and then slacken off again by half a turn.

➡ **Position of exhaust adjustment unit to timing chain can be freely selected.**

117. Pull timing chain up.
118. Feed exhaust adjustment unit into timing chain.
119. Press tensioner rail back and fit exhaust adjustment unit onto exhaust camshaft.

➡ **If tensioner rail cannot be pressed back far enough to enable fitting of exhaust adjustment unit:**

120. Remove chain tensioner.
121. Place chain tensioner on a level surface and compress slowly and carefully.
122. Repeat this procedure twice.
123. Replace sealing ring.
124. Install chain tensioner and tighten to 48 ft. lbs. (65 Nm).
125. Fit exhaust adjustment unit.
126. Replace screw.
127. Install screw on exhaust adjustment unit.
128. Tighten screw without play and then slacken off again by half a turn.
129. Get special tool kit 11 9 460 ready for securing camshafts.

➡ **Special tool 11 9 461 for securing intake camshaft. Special tool 11 9 462 for securing exhaust camshaft. Special tool 11 9 463 (holder with screw).**

130. Place special tool 11 9 461 on intake camshaft and align intake camshaft so that special tool 11 9 461 rests without a gap on cylinder head.
131. Fit special tool 11 9 463, secure screw in thread for oil line and tighten down by hand.

**⁂ WARNING**

**Make sure bolts of intake and exhaust adjustment units have been slackened off by a half turn. Crank engine at central bolt from 45∞ before TDC position in direction of rotation to firing TDC position. Secure vibration damper with special**

tool 11 9 190 in firing TDC position of 1st cylinder.

➡ **When tightening down screw, grip camshaft at hexagon head.**

132. Tighten down screw of intake adjustment unit to 59 ft. lbs. (80 Nm).
133. Remove screw and remove special tools 11 9 463 / 11 9 461 from intake camshaft.
134. Place special tool 11 9 462 on exhaust camshaft and align exhaust camshaft so that special tool 11 9 462 rests without a gap on cylinder head.
135. Fit special tool 11 9 463, secure screw in thread for oil line and tighten down by hand.

➡ **When tightening down screw, grip camshaft at hexagon head. Tighten down screw of exhaust adjustment unit to 59 ft. lbs. (80 Nm).**

136. Remove screw and remove special tools 11 9 463 / 11 9 462 from exhaust camshaft.
137. Remove special tool 11 9 190.
138. Crank engine at central bolt twice in direction of rotation until engine returns to firing TDC position of 1st cylinder.
139. Secure vibration damper with special tool 11 9 190 in firing TDC position of 1st cylinder.
140. Place special tool 11 9 461 on intake camshaft and check timing.

➡ **The timing is correctly adjusted when special tool 11 9 461 rests flat on the cylinder head or protrudes by up to 0.5 mm to the exhaust side.**

141. Remove special tool 11 9 461 from intake camshaft.
142. Place special tool 11 9 462 on exhaust camshaft and check timing.

➡ **The timing is correctly adjusted when special tool 11 9 462 rests flat on the cylinder head or protrudes by up to 0.5 mm to the exhaust side.**

143. Remove all special tools.

**⁂ WARNING**

**Screw is a special screw and must not be replaced with a normal M8 screw.**

144. Clip oil line into retainers.
145. Insert screw and tighten down.
146. Install timing case cover.
  a. Free sealing face on timing case cover of seal debris and clean.

b. Check adapter sleeves for damage and correct installation position; replace if necessary.

c. Free sealing face of seal debris and clean.

d. Replace gasket and check for correct installation position.

147. Install spark plugs.

148. Install cylinder head cover.

149. Install ignition coils.

### Exhaust Camshafts

*See Figures 131 through 152.*

1. Before servicing the vehicle, refer to the precautions.

2. Read and document the fault memory.

3. Remove servomotor for eccentric shaft.

4. Remove ignition coils.

5. Remove cylinder head cover.

6. Remove spark plugs.

7. Remove timing case cover.

   a. Remove both solenoid valves.

   b. Remove alternator (air-cooled).

   c. Remove screw, remove holder.

   d. Remove line from holder.

   e. Remove bolts.

   f. Remove timing case cover.

8. Remove exhaust and exhaust adjustment units as follows:

9. If necessary, remove fan cowl.

**✳✳ WARNING**

**Screw is a special screw and must not be replaced with a normal M8 screw.**

10. Remove screw.

11. Disconnect oil line from retainers and remove.

12. Crank engine at central bolt in direction of rotation to firing TDC position of 1st cylinder.

13. When removing the left exhaust cam, with 1st cylinder in firing TDC position, cams of exhaust and exhaust camshafts at 5th cylinder point upwards at an angle as shown.

14. When removing the right exhaust cam, crank engine at central bolt in direction of rotation to firing TDC position of 1st cylinder.

15. In firing TDC position, cam of exhaust camshaft at 1st cylinder points upwards at an angle. Cam of exhaust camshaft points downwards at an angle.

➡**Alignment hole for TDC position is at front on timing case cover.**

16. Crank engine at central bolt and

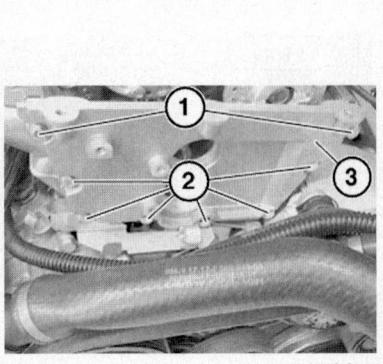

22205_BMWC_G0192

**Fig. 131  Timing cover (3) and mounting bolts (1 and 2)**

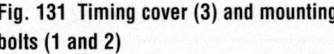

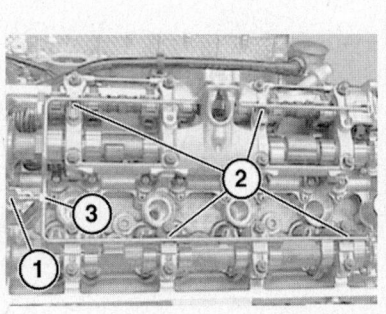

22205_BMWC_G0193

**Fig. 132  Screw (1) is a special screw and must not be replaced with a normal M8 screw. Disconnect oil line (3) from retainers (2) and remove**

22205_BMWC_G0194

**Fig. 133  When removing the left exhaust cam, with 1st cylinder in firing TDC position, cams of exhaust and exhaust camshafts at 5th cylinder point upwards at an angle as shown**

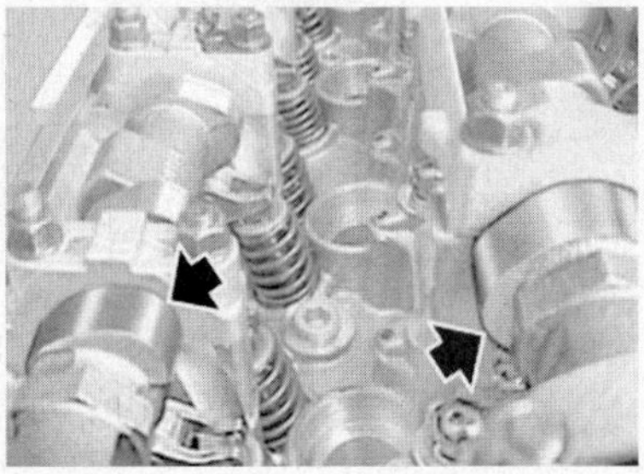

22205_BMWC_G0212

**Fig. 134  In firing TDC position, cam of exhaust camshaft at 1st cylinder points upwards at an angle. Cam of exhaust camshaft points downwards at an angle**

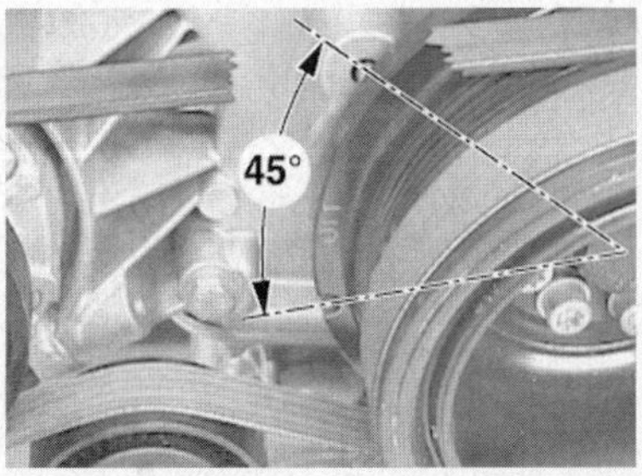

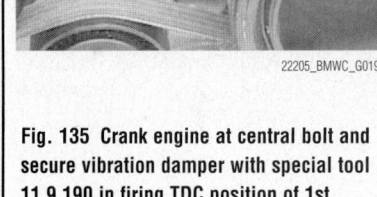

22205_BMWC_G0195

**Fig. 135  Crank engine at central bolt and secure vibration damper with special tool 11 9 190 in firing TDC position of 1st cylinder**

22205_BMWC_G0196

**Fig. 136  Vigorously press back tensioner rail (3) several times to remove oil in chain tensioner. Feed out exhaust adjustment unit (2)**

22205_BMWC_G0213

**Fig. 137 On the right exhaust camshaft, rotate exhaust camshaft in direction of rotation until cam on 1st cylinder is positioned horizontally as shown**

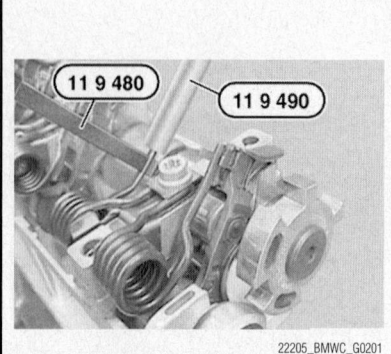

22205_BMWC_G0201

**Fig. 140 Raise one end of torsion spring (1) with special tool 11 9 480. Support end of torsion spring protected with special tool 11 9490 on exhaust camshaft**

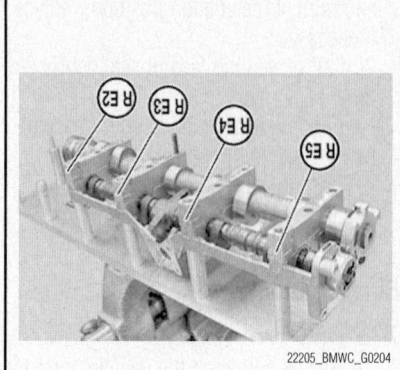

22205_BMWC_G0204

**Fig. 143 Camshaft bearing caps of cylinders 1–4 and 5–8 must not be mixed up**

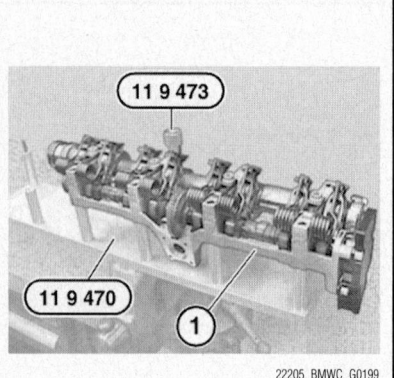

22205_BMWC_G0199

**Fig. 138 Place bearing bracket (1) with exhaust camshaft and eccentric shaft as illustrated on special tool 11 9 470**

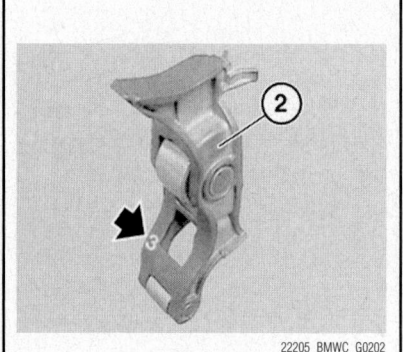

22205_BMWC_G0202

**Fig. 141 Intermediate levers (2) are divided into individual tolerance classes. Only intermediate levers of the same tolerance class may be fitted in a single cylinder head**

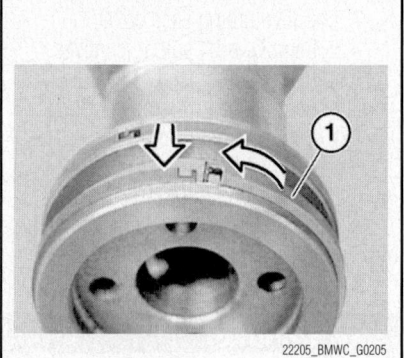

22205_BMWC_G0205

**Fig. 144 Press compression ring (1) on one side into groove, pull up on other side and remove catch**

22205_BMWC_G0200

**Fig. 139 Remove the eccentric shaft sensor (2) only if necessary for service**

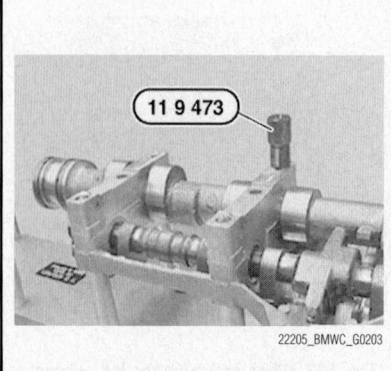

22205_BMWC_G0203

**Fig. 142 Unscrew nut (special tool 11 9 473)**

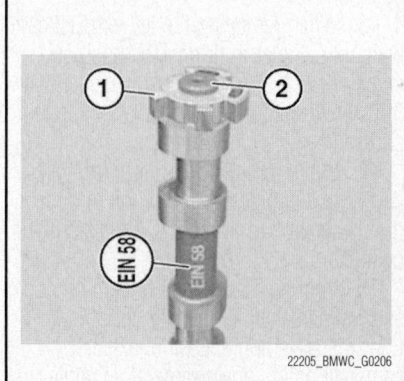

22205_BMWC_G0206

**Fig. 145 If necessary, Remove screw (2) and remove sensor gear (1)**

secure vibration damper with special tool 11 9 190 in firing TDC position of 1st cylinder.

**When the engine is shut down, the exhaust and exhaust adjustment unit is normally locked in its initial position. The situation may arise in some individual cases where this initial position is not reached and the camshaft can continue to be rotated in the adjustment range of the adjustment unit. In order to avoid incorrect timing adjustment, it is essential to check the locking of the adjustment unit and if necessary perform locking by rotating the camshafts.**

17. Checking locking of exhaust adjustment unit in initial position by engaging the hexagon head of exhaust camshaft and attempt to rotate exhaust camshaft carefully against direction of rotation.

18. If there is no fixed connection between exhaust camshaft and exhaust adjustment unit, rotate exhaust camshaft against direction of rotation as far as it will go.

19. The exhaust adjustment unit is locked in the initial position when the exhaust camshaft is non-positively connected to the exhaust adjustment unit.

20. Checking locking of exhaust adjustment unit in initial position by engaging hexagon head of exhaust camshaft and attempt to rotate exhaust camshaft carefully in direction of rotation.

21. If there is no fixed connection between exhaust camshaft and exhaust adjustment unit, rotate exhaust camshaft in direction of rotation as far as it will go.

22. The exhaust adjustment unit is locked in the initial position when the exhaust camshaft is non-positively connected to the exhaust adjustment unit.

**If the exhaust or exhaust adjustment unit of the camshafts cannot be locked as described, the adjustment unit is faulty and must be replaced.**

23. Remove special tool 11 9 190.

24. Crank engine at central bolt against direction of rotation to 45° before TDC position.

➡**When slackening bolts, grip camshafts at hexagon head.**

25. Remove bolts of exhaust and exhaust adjustment unit.

26. Remove screw on exhaust adjustment unit.

27. Vigorously press back tensioner rail several times to remove oil in chain tensioner.

28. Feed out exhaust adjustment unit.

29. Remove screw on exhaust adjustment unit.

30. Feed out exhaust adjustment unit.

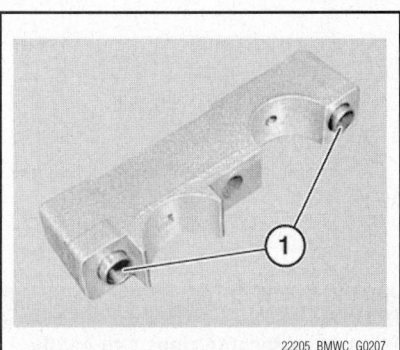

**Fig. 146  Check dowel sleeves (1) for damage and correct installation position.**

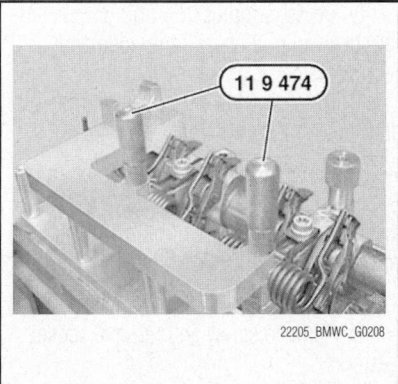

**Fig. 147  Remove special tool 11 9 474**

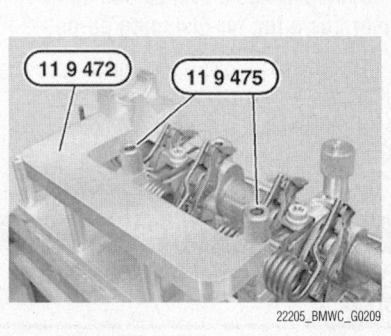

**Fig. 148  Remove special tool 11 9 472 and special tool 11 9 475**

**Fig. 149  Ends of compression rings point upwards**

**Fig. 150  The marking (1) on the hexagon drive of the exhaust camshaft faces upwards**

**Fig. 151  Exhaust adjustment unit (1) is marked with EIN and IN. Exhaust adjustment unit (2) is marked with AUS and EX**

**Fig. 152  Fit special tool 11 9 463, secure screw (1) in thread for oil line and tighten down by hand**

31. Secure timing chain to prevent it from sliding down.

32. Remove the exhaust camshaft as follows:

### ✪ WARNING

**The exhaust camshaft must first be rotated so that when the bearing bracket is removed the intermediate levers do not slip out and damage the camshaft.**

33. On the left exhaust camshaft rotate the camshaft against direction of rotation until lettering on 8th cylinder points upwards in cylinder axis and cam is horizontal.

34. On the right exhaust camshaft, rotate exhaust camshaft in direction of rotation until cam on 1st cylinder is positioned horizontally as shown.

### ✪ WARNING

**Camshaft bearing caps of cylinders 1–4 and 5–8 must not be mixed up.**

➡Bearing caps of exhaust camshaft are marked on cylinder bank 5–8 with R E1 to R E5 from exhaust side. Bearing caps of exhaust camshaft are marked on cylinder bank 1–4 with L E1 to L E5 from exhaust side.

35. Remove nuts and remove bearing cap E1.

36. Remove 8 nuts of bearing bracket from outside to inside.

➡Rocker arms are freely accessible after bearing bracket has been removed.

37. Do **NOT** remove rocker arm on exhaust side.

### ❋ WARNING

**Rocker arms are divided into individual tolerance classes. The tolerance classes are designated as illustrated with the numbers from 1–4. Used rocker arms may only be reused in the same position. When replacing rocker arms on exhaust side: install rocker arms of the same tolerance class in the same position.**

38. Clamp special tool 11 9 470 in a vice as shown.

### ❋ WARNING

**Do not tilt bearing bracket.**

39. Carefully lift out bearing bracket.

40. Place bearing bracket with exhaust camshaft and eccentric shaft as illustrated on special tool 11 9 470.

41. Secure bearing bracket with a nut (special tool 11 9 473).

### ❋ WARNING

**The lower section of the bearing bracket is machined with the cylinder head and must not be mixed up.**

➡Lower section of bearing bracket remains on cylinder head.

42. Remove the eccentric shaft sensor only if necessary for service.

➡Removal of the intermediate levers and torsion springs is described on the 8th cylinder. The same procedure is applicable to all other cylinders.

43. Raise one end of torsion spring with special tool 11 9 480.

44. Lift out intermediate lever and set down in an orderly fashion.

### ❋ WARNING

**Keep holding torsion spring with special tool 11 9 480.**

45. Connect special tool 11 9490 to end of torsion spring.

46. Support end of torsion spring protected with special tool 11 9490 on exhaust camshaft.

### ❋ WARNING

**Intermediate levers are divided into individual tolerance classes. Only intermediate levers of the same tolerance class may be fitted in a single cylinder head. The tolerance classes are designated as illustrated with the numbers from 1 to 5. Used intermediate levers may only be reused in the same position.**

47. Raise second end of torsion spring with special tool 11 9 480.

48. Lift out intermediate lever and set down in an orderly fashion.

### ❋ WARNING

**Keep holding torsion spring with special tool 11 9 480.**

49. Connect special tool 11 9490 to second end of torsion spring.

50. Support end of torsion spring protected with special tool 11 9 490 on exhaust camshaft.

51. Remove screw.

52. Remove torsion spring and special tool 11 9 490.

53. Remove intermediate levers and torsion springs of cylinders 5 to 7 according to the same procedure and set down in an orderly fashion.

54. Unscrew nut (special tool 11 9 473).

### ❋ WARNING

**Camshaft bearing caps of cylinders 1–4 and 5–8 must not be mixed up.**

➡Bearing caps are marked in graphic with cylinders 1–4 designated as L E2 to L E5and 5–8 designated as R E2 to R E5.

55. Remove bearing caps E2 to E5 and place to one side in order.

56. Lift out exhaust camshaft.

*To install:*

57. Install the exhaust camshaft as follows:

58. If necessary, replace plain compression rings.

### ❋ WARNING

**Plain compression rings can easily break.**

59. Plain compression rings are engaged at joint.

60. Press compression ring on one side into groove, pull up on other side and remove catch.

61. Carefully pull compression ring apart and remove towards front.

➡Exhaust camshaft of cylinder bank 5–8 is marked with "EIN 58". Exhaust camshaft of cylinder bank 1 to 4 is marked with "EIN 14".

62. If necessary, replace sensor gear.

63. Remove screw and remove sensor gear.

64. Clean all bearings and cams of exhaust camshaft and lubricate with engine oil.

➡Camshaft has a groove and sensor gear has a lug for fastening purposes.

65. Fit sensor gear and align to groove in camshaft.

66. Insert screw and tighten to 22 ft. lbs. (30 Nm).

67. Installing plain compression rings:

### ❋ WARNING

**Plain compression rings can easily break.**

68. Carefully pull compression ring apart and install from front.

69. Press compression ring on one side into groove, install catch on other side.

70. Insert exhaust camshaft so that cams point upwards at cylinder 5 (for the left exhaust cam) and cylinder 4 (for the right).

71. Make sure bearing shells of eccentric shaft are engaged in bearing bracket.

**➡Bearing shell is guided in a groove in bearing cap.**

72. Check dowel sleeves for damage and correct installation position.

**➡Bearing caps are marked in graphic with R E2 to R E5 for the right side and L E2 to L E5 for the right side.**

73. Fit bearing caps E2 to E5.

**➡Initially tighten special tool 11 9 473 without play only.**

74. Secure bearing bracket and bearing cap with nut (special tool 11 9 473).

**➡The mounting of the bearing bracket described later can only be performed if the first bearing of the exhaust camshaft is aligned with special tool 11 9 472.**

75. Fit special tool 11 9 472 as shown in graphic and align exhaust camshaft.

76. Fit centering sleeves (special tool 11 9 475) and align special tool 11 9 472 to bearing caps 2 and 3.

77. Insert special tool 11 9 474 and initially tighten without play.

**➡Special tools 11 9 472, 11 9 474 and 11 9 475 remain fitted until all torsion springs and intermediate levers have been installed.**

**➡On the left side, installation of the intermediate levers and torsion springs is described on the 8th cylinder. The same procedure is applicable to cylinders 5 to 7. On the right side, installation of the intermediate levers and torsion springs is described on the 4th cylinder. The same procedure is applicable to cylinders 1 to 3.**

78. Connect special tool 11 9 490 to ends of torsion spring.

79. Install torsion spring.

80. Insert screw and tighten down.

**⁂ WARNING**

**Intermediate levers are divided into individual tolerance classes.**

81. Only intermediate levers of the same tolerance class may be fitted in a single engine.

82. The tolerance classes are designated as illustrated with the numbers from 1 to 5.

83. Used intermediate levers may only be reused in the same position.

84. Lubricate all sliding surfaces on intermediate lever with engine oil.

85. Raise torsion spring with special tool 11 9 480.

86. Remove special tool 11 9 490.

87. Hold torsion spring with special tool 11 9 480.

88. Install intermediate lever from above.

89. Insert end of torsion spring into guide on intermediate lever.

90. Raise second end of torsion spring with special tool 11 9 480.

91. Remove special tool 11 9 490.

92. Hold torsion spring with special tool 11 9 480.

93. Install intermediate lever from above.

94. Insert end of torsion spring into guide on intermediate lever.

95. Install eccentric shaft sensor.

96. Insert bolts and tighten down eccentric shaft sensor.

97. Remove special tool 11 9 474.

98. Remove special tool 11 9 472 and special tool 11 9 475.

99. Ends of compression rings point upwards.

100. Make sure compression rings are engaged at ends.

**⁂ WARNING**

**Rocker arms slip slightly when bearing bracket is fitted.**

101. Make sure rocker arms are secured as illustrated on hydraulic valve clearance compensating elements and on valves.

102. Align rockers straight.

103. Remove special tool 11 9 473.

104. Remove bearing bracket from special tool 11 9 470.

**⁂ WARNING**

**Do not tilt bearing bracket.**

105. Lower bearing bracket from above and carefully bring into contact with cylinder head.

106. Insert nuts and tighten by hand without play.

**⁂ WARNING**

**Make sure none of the intermediate levers or rocker arms have slipped out.**

107. Tighten nuts from inside to outside to specification.

- M6: 7 ft. lbs. (10 Nm )
- M7: 10 ft. lbs. (14 Nm)
- M7- 8.8: 11 ft. lbs. (15 Nm)

**⁂ WARNING**

**Camshaft bearing caps of cylinders 1–4 and 5–8 must not be mixed up.**

108. Fit bearing cap R E1 in such a way that marking is legible from exhaust side.

109. Install nuts and tighten to specification.

- M6: 7 ft. lbs. (10 Nm )
- M7: 10 ft. lbs. (14 Nm)
- M7- 8.8: 11 ft. lbs. (15 Nm)

110. Rotate exhaust camshaft in direction of rotation until cam on 5th cylinder points upwards at an angle as shown in illustration.

**➡The marking (1) on the hexagon drive of the exhaust camshaft faces upwards.**

111. Install exhaust and exhaust adjustment units as follows:

**⁂ WARNING**

**Exhaust and exhaust adjustment units are different. Mixing up the exhaust and exhaust adjustment units will cause damage to the engine. Exhaust adjustment unit is marked with EIN and IN. Exhaust adjustment unit is marked with AUS and EX.**

**➡Position of exhaust adjustment unit to timing chain can be freely selected.**

112. Pull timing chain up.

113. Feed exhaust adjustment unit into timing chain and fit onto exhaust camshaft.

114. Replace screw.

115. Install screw on exhaust adjustment unit.

116. Tighten screw without play and then slacken off again by half a turn.

**➡Position of exhaust adjustment unit to timing chain can be freely selected.**

117. Pull timing chain up.

118. Feed exhaust adjustment unit into timing chain.

119. Press tensioner rail back and fit exhaust adjustment unit onto exhaust camshaft.

**➡If tensioner rail cannot be pressed back far enough to enable fitting of exhaust adjustment unit:**

120. Remove chain tensioner.

121. Place chain tensioner on a level surface and compress slowly and carefully.

122. Repeat this procedure twice.

123. Replace sealing ring.

124. Install chain tensioner and tighten to 48 ft. lbs. (65 Nm).

125. Fit exhaust adjustment unit.

126. Replace screw.

127. Install screw on exhaust adjustment unit.

128. Tighten screw without play and then slacken off again by half a turn.

129. Get special tool kit 11 9 460 ready for securing camshafts.

➡**Special tool 11 9 461 for securing exhaust camshaft. Special tool 11 9 462 for securing exhaust camshaft. Special tool 11 9 463 (holder with screw).**

130. Place special tool 11 9 461 on exhaust camshaft and align exhaust camshaft so that special tool 11 9 461 rests without a gap on cylinder head.

131. Fit special tool 11 9 463, secure screw in thread for oil line and tighten down by hand.

### ❊❊ WARNING

**Make sure bolts of exhaust and exhaust adjustment units have been slackened off by a half turn. Crank engine at central bolt from 45° before TDC position in direction of rotation to firing TDC position. Secure vibration damper with special tool 11 9 190 in firing TDC position of 1st cylinder.**

➡**When tightening down screw, grip camshaft at hexagon head.**

132. Tighten down screw of exhaust adjustment unit to 59 ft. lbs. (80 Nm).

133. Remove screw and remove special tools 11 9 463 / 11 9 461 from exhaust camshaft.

134. Place special tool 11 9 462 on exhaust camshaft and align exhaust camshaft so that special tool 11 9 462 rests without a gap on cylinder head.

135. Fit special tool 11 9 463, secure screw in thread for oil line and tighten down by hand.

➡**When tightening down screw, grip camshaft at hexagon head. Tighten down screw of exhaust adjustment unit to 59 ft. lbs. (80 Nm).**

136. Remove screw and remove special tools 11 9 463 / 11 9 462 from exhaust camshaft.

137. Remove special tool 11 9 190.

138. Crank engine at central bolt twice in direction of rotation until engine returns to firing TDC position of 1st cylinder.

139. Secure vibration damper with special tool 11 9 190 in firing TDC position of 1st cylinder.

140. Place special tool 11 9 461 on exhaust camshaft and check timing.

➡**The timing is correctly adjusted when special tool 11 9 461 rests flat on the cylinder head or protrudes by up to 0.5 mm to the exhaust side.**

141. Remove special tool 11 9 461 from exhaust camshaft.

142. Place special tool 11 9 462 on exhaust camshaft and check timing.

➡**The timing is correctly adjusted when special tool 11 9 462 rests flat on the cylinder head or protrudes by up to 0.5 mm to the exhaust side.**

143. Remove all special tools.

### ❊❊ WARNING

**Screw is a special screw and must not be replaced with a normal M8 screw.**

144. Clip oil line into retainers.

145. Insert screw and tighten down.

146. Install timing case cover.

　a. Free sealing face on timing case cover of seal debris and clean.

　b. Check adapter sleeves for damage and correct installation position; replace if necessary.

　c. Free sealing face of seal debris and clean.

　d. Replace gasket and check for correct installation position.

147. Install spark plugs.

148. Install cylinder head cover.

149. Install ignition coils.

## CRANKSHAFT FRONT SEAL

### REMOVAL & INSTALLATION

#### N52 and N54 Engine

1. Before servicing the vehicle, refer to the precautions.

2. Remove vibration damper

### ❊❊ WARNING

**Do not Remove central bolt. If the central bolt is Removed, the sprocket wheels of the timing chain and the oil pump will no longer be non-positively connected to the crankshaft. Inlet and exhaust camshafts can turn in relation to crankshaft.**

3. Turn back special tool 11 9 222.

4. Push special tool 11 9 221 onto crankshaft.

➡**When bolts are tightened down (special tool 11 9 224), crankshaft seal is pressed inwards approximately 1 mm and thus slackened for subsequent removal.**

5. Insert bolts (special tool 11 9 224) and tighten down to approximately 15 ft. lbs. (20 Nm).

6. Screw special tool 11 0 371 to 59 ft. lbs. (80 Nm) into crankshaft seal.

7. Screw in spindle 11 0 372.

8. Remove crankshaft seal from housing.

➡**Repeat the operation several times if necessary.**

9. Carefully saw open crankshaft seal at cutting line.

10. Remove crankshaft seal from special tool 11 0 371.

### To install:

### ❊❊ WARNING

**The following text describes installation and sealing between the engine block and crankshaft seal. The engine block will not be leak proof at the outside of the crankshaft seal if you fail to comply with the individual work steps and the work sequence.**

11. Clean sealing surface and degrease thoroughly in area of housing partition.

12. Apply a light coat of oil to running surface of crankshaft seal.

13. Screw special tool 11 9 232 with bolts (special tool 11 9 234) to crankshaft.

➡**Support sleeve is supplied with crankshaft seal.**

14. When crankshaft seal is installed, only support sleeve may be used as a slip sleeve.

15. Crankshaft seal has a groove on both left and right sides.

### ❊❊ WARNING

**After installation, the grooves must be filled with sealing compound.**

➡**The required parts are available from the BMW Parts Service (Electronic Parts Catalogue ETK).**

16. Remove screw caps from injector.

17. Screw on metering needle.

18. Insert piston for pressing out. Injector contains the sealing compound Loctite, manufacturer's number 128357. Bottle contains the primer Loctite, manufacturer's number 171000.

19. Push support sleeve with crankshaft seal onto special tool 11 9 232.

### ⁕⁕ WARNING

**Support sleeve remains on special tool 11 9 232, until crankshaft seal is drawn in.**

20. Align groove centrally to housing partition.

21. Coat both grooves on crankshaft seal with Loctite primer, manufacturer's number 171000, and expose to air for approx. one minute.

22. Draw in crankshaft seal with special tool 11 9 231 in conjunction with special tool 11 9 233 until flush.

23. Before filling with sealing compound, moisten brush with Loctite primer, manufacturer's number 171000. Insert brush as far as possible into grooves on crankshaft seal in order to coat housing partition on engine block.

24. Using injector, fill both grooves flush with Loctite sealing compound, manufacturer's number 128357.

➡**Loctite primer, manufacturer's number 171000, binds the Loctite sealing compound, manufacturer's number 128357, and prevents leakage.**

25. Coat surface of sealing compound in both grooves with Loctite primer, manufacturer's number 171000.

26. Install vibration damper.

### N62 Engine
*See Figures 153 through 155.*

1. Before servicing the vehicle, refer to the precautions.

2. Remove vibration damper

3. Position lever for removing radial shaft seal horizontally and install special tool 11 9 410.

4. Turn lever so that it grips behind radial shaft seal.

5. Turn bolt on special tool 11 9 410 to remove radial shaft seal.

➡**Radial shaft seal may only be supported with a "support sleeve".**

6. If the radial shaft seal is stored for longer than six months without the support sleeve, its operational reliability can no longer be guaranteed and it must not be reused.

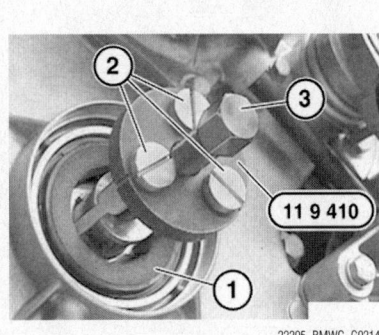

22205_BMWC_G0214

**Fig. 153 Special tool 11 9 410 to remove radial shaft seal (1) by positioning the lever (2) behind the radial seal and turning bolt (1) to remove**

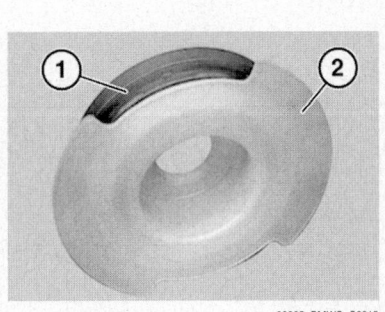

22205_BMWC_G0215

**Fig. 154 If the radial shaft seal (1) is stored for longer than six months without the support sleeve (2), its operational reliability can no longer be guaranteed and it must not be reused**

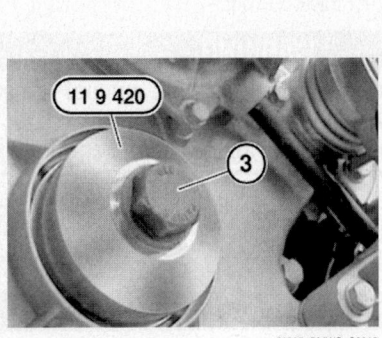

22205_BMWC_G0216

**Fig. 155 Using special tool 11 9 420 and central bolt, install radial shaft seal flush with timing case cover**

### ⁕⁕ WARNING

**The sealing lip of the radial shaft seal is highly sensitive and must not be kinked under any circumstances.**

7. Do not touch the sealing lip with your fingers.

8. Remove support sleeve from radial shaft seal.

*To install:*

9. Fit radial shaft seal on timing case cover.

10. Using special tool 11 9 420 and central bolt, install radial shaft seal flush with timing case cover.

11. Install vibration damper

### CYLINDER HEAD

*REMOVAL & INSTALLATION*

### N52 and N54 Engines
*See Figures 156 through 162.*

### ⁕⁕ WARNING

**Aluminum-magnesium material. No steel fasteners may be used due to the threat of electrochemical corrosion. A magnesium crankcase requires aluminum fasteners exclusively. Aluminum fasteners must be replaced each time they are removed. The end faces of aluminum fasteners are painted blue for purposes of identification. Torque specifications and torque angles must be observed for risk of damage.**

1. Before servicing the vehicle, refer to the precautions.

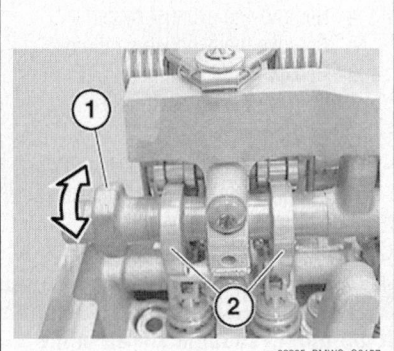

22205_BMWC_G0157

**Fig. 156 Pretension eccentric shaft (1) upwards in direction of arrow. Remove stop screw between 1st and 2nd cylinders (2)**

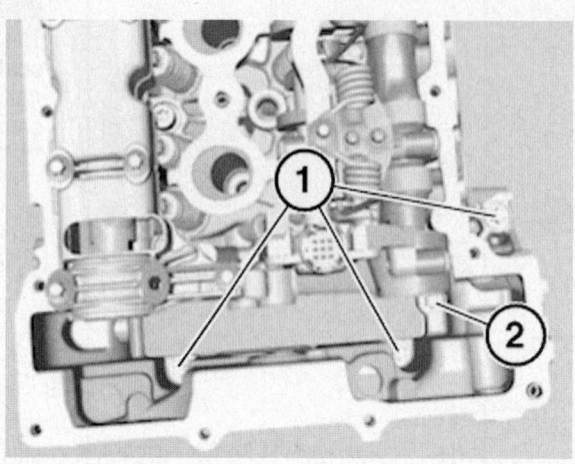

Fig. 157 Secure bolt (2) with a gripper against falling down. Remove and discard bolts (1) and (2)

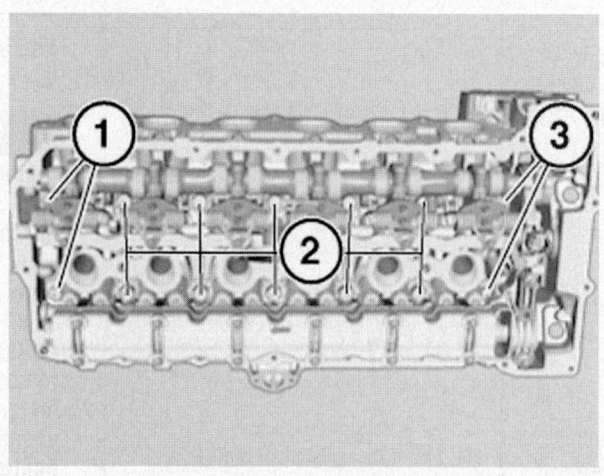

Fig. 158 Use the correct special tool for each bolt size M10 (2) and M9 (1) and (3)

2. Drain the cooling system and engine oil.

3. Relieve the fuel system pressure.

4. Remove the exhaust system.

5. Remove the exhaust manifolds.

6. Remove the intake manifold.

7. Remove the coolant hoses from cylinder head.

8. Remove the cylinder head cover.

9. Remove the inlet and exhaust adjustment unit. See "Engine Mechanical, Camshaft, Removal & Installation."

### ✸✸ WARNING
**If the timing chain is stored in the gear case, the crankshaft must not be rotated. Only during assembly can the timing chain be lifted out.**

10. Remove the timing chain module and fasteners.

11. Remove the eccentric shaft sensor and fasteners towards front.

12. Remove the magnet wheel and fastener towards front.

### ✸✸ WARNING
**After removing, secure magnet wheel in a plastic bag. Magnet wheel must be protected against metal chips.**

13. Pretension eccentric shaft upwards. Remove mini stop screw between first and second cylinders

### ✸✸ WARNING
**Secure bolt (2) with a gripper against falling down.**

14. Remove and discard bolt (2).

➡ Bolt (2) can only be Removed when

the timing chain module is pressed forward slightly.

15. Remove and discard bolts (1).

16. Remove and discard M10 cylinder head bolts (1) with special tool 11 8 580.

17. Remove and discard M9 cylinder head bolts (2) with special tool 11 4 420.

18. Remove and discard M9 cylinder head bolts (1 and 3) with special tool 11 4 420.

19. Remove and discard M10 cylinder head bolts (2) with special tool 11 8 580 from outside inwards.

### ✸✸ WARNING
**All cylinder head bolts must be replaced.**

20. Secure special tool 11 0 320 with existing cylinder head cover bolts (1). Tighten to 7 ft. lbs. (9 Nm).

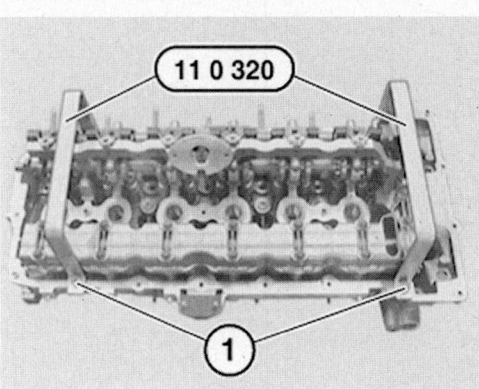

Fig. 159 Secure special tool 11 0 320 with existing cylinder head cover bolts (1)

Fig. 160 Insert special tool 11 4 430 into bores.

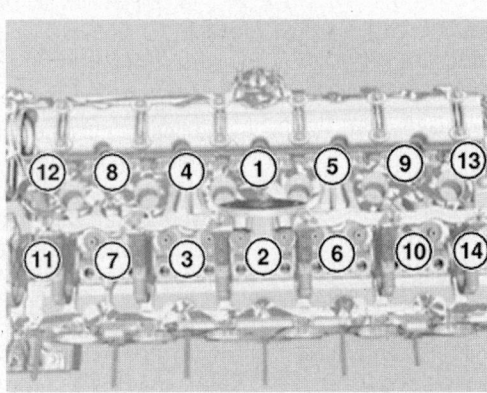

Fig. 161 Cylinder head bolt torque sequence—N52 engine

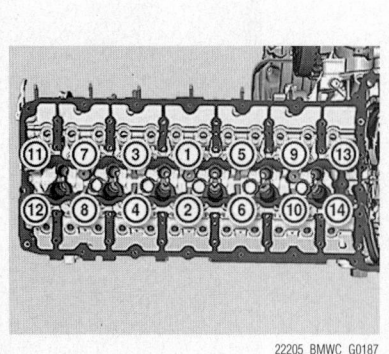

Fig. 162 Cylinder head bolt torque sequence—N52 engine

**❊❊ WARNING**

**Do not rest cylinder head on sealing surface.**

21. Insert special tool 11 4 430 into bores.

### To install:

22. Install bolts 1-10 with special tool No. 115190. Install bolts 11-14 with special tool No. 114420.

23. Torque following the tightening sequence as follows:
- Step 1: Tighten bolts 1-14 to 22 ft. lbs. (30 Nm)
- Step 2: Tighten bolts 1-14 an additional 90 degrees
- Step 3: Tighten bolts 1-10 an additional 90 degrees
- Step 4: Tighten bolts 1-14 an additional 45 degrees

24. The balance of installation is the reverse of the removal procedure.

25. Fill the cooling system.

26. Start the engine and check for leaks.

### N62 Engines

#### Left Side

*See Figures 163 and 164.*

1. Before servicing the vehicle, refer to the precautions.

2. Drain the cooling system.

3. Relieve the fuel system pressure.

4. Disconnect the negative battery cable.

5. Remove the left exhaust manifold.

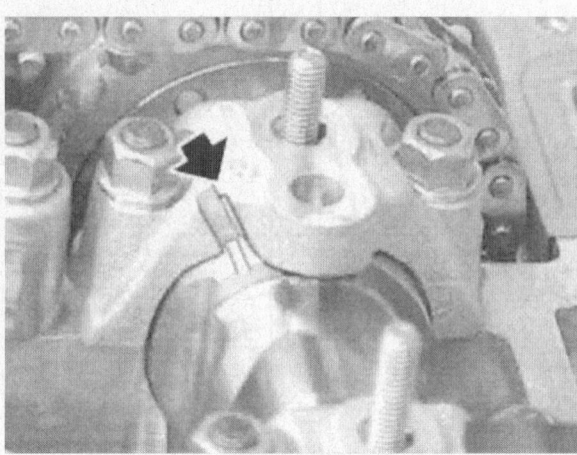

Fig. 163 Rotate the eccentric shaft (1), pull on the torsion spring and rotate out from roller (3)

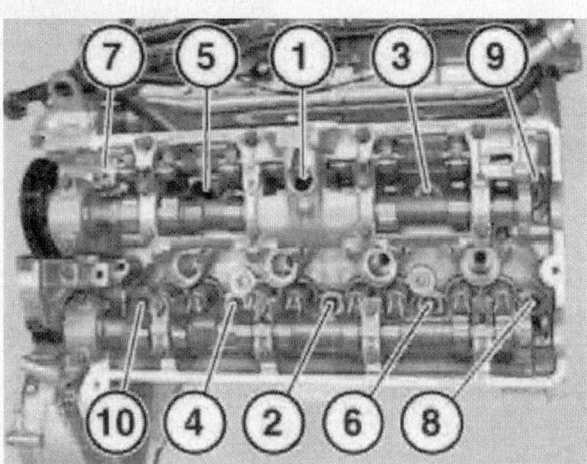

Fig. 164 Cylinder head bolt sequence—N62 engine

➡To remove the exhaust manifolds, the engine must be secured in the installation position with the special tool and then the front axle bracket lowered. Reinstall front axle bracket provisionally and remove special tool for securing engine in installation position.

6. Remove the ignition coils.
7. Remove the eccentric shaft servo motor.
8. Remove the cylinder head cover.
9. Remove the intake air manifold.
10. Remove the timing case cover.
11. Remove the engine wiring harness and secure aside.
12. Remove the spark plugs.
13. Remove the vent hose from cylinder head.
14. Remove the hose and check valve.

➡Do not remove the guide tube for the dipstick.

15. Rotate the eccentric shaft at the dihedral head to reduce tension on the torsion spring. Pull on the torsion spring with a strap and rotate it out from the roller.
16. Rotate the dihedral head eccentric shaft back to the minimum stroke position.
17. Remove the mounting screw and spring mount from locating pin.
18. Remove the eccentric shaft sensor.
19. Remove the guide rail from cylinder head.
20. Remove the bolts between cylinder head and timing case cover.
21. Remove the cylinder head bolts in sequence one turn from 10 to 1.
22. Remove the rotate the eccentric shaft and remove bolt 7.

23. Remove the rotate the eccentric shaft back and remove the remaining cylinder head bolts.
24. Remove the cylinder head.

*To install:*

25. Thoroughly clean all mounting surfaces and check the head for warpage. Take care not to drop any pieces of gasket or debris into the oil or coolant passages. Check the condition of the head locating dowel sleeves and clean out the bolt threads with a tap.
26. Coat the joint between the engine block and timing case cover with Drei Bond 1209.
27. Mount the cylinder head on the block and use new bolts.

**✳✳ WARNING**

**Do not remove the coating on the head bolts.**

28. Apply oil to the threads and washer contact area. Rotate the eccentric shaft to insert bolt 7.
29. Torque bolts in sequence to specification.
- Step 1: Tighten to 22 ft. lbs. (30 Nm).
- Step 2: Tighten an additional 90° rotation
- Step 3: Tighten an additional 90° rotation
30. Installation is the reverse of removal.
31. Fill the cooling system.
32. Start the engine and check for leaks.

*Right Side*

1. Before servicing the vehicle, refer to the precautions.
2. Drain the cooling system.
3. Relieve the fuel system pressure.
4. Disconnect the negative battery cable.
5. Remove the right exhaust manifold.

➡To remove the exhaust manifolds, the engine must be secured in the installation position with the special tool and then the front axle bracket lowered. Reinstall front axle bracket provisionally and remove special tool for securing engine in installation position.

6. Remove the ignition coils.
7. Remove the eccentric shaft servo motor.
8. Remove the cylinder head cover.
9. Remove the intake manifold.
10. Remove the timing case cover.

11. Remove the engine wiring harness and secure aside.

12. Remove the spark plugs.

13. Remove the vent hose from cylinder head.

14. Remove the hose and check valve

15. Remove chain tensioning piston by one turn. Rotate the eccentric shaft at the dihedral head to reduce tension on the torsion spring. Pull on the torsion spring with a strap and rotate it out from the roller.

16. Rotate the dihedral head eccentric shaft back to the minimum stroke position.

17. Remove the mounting screw and spring mount from locating pin.

18. Remove the eccentric shaft sensor.

19. Remove the guide rail from cylinder head.

20. Remove the bolts between cylinder head and timing case cover.

21. Remove the cylinder head bolts in sequence one turn from 10 to 1.

22. Remove the rotate the eccentric shaft and remove bolt 7.

23. Remove the rotate the eccentric shaft back and remove the remaining cylinder head bolts.

24. Remove the cylinder head.

25. Remove the chain tensioning piston.

*To install:*

26. Place piston on a level surface and compress slowly. Repeat twice.

27. Thoroughly clean all mounting surfaces and check the head for warpage. Take care not to drop any pieces of gasket or debris into the oil or coolant passages. Check the condition of the head locating dowel sleeves and clean out the bolt threads with a tap.

28. Coat the joint between the engine block and timing case cover with Drei Bond 1209.

29. Replace sealing ring on piston. Install piston and tighten unit there is no play.

30. Mount the cylinder head on the block and use new bolts.

### ❉❉ WARNING

**Do not remove the coating on the head bolts.**

31. Apply oil to the threads and washer contact area. Rotate the eccentric shaft to insert bolt 7.

32. Torque bolts in sequence to specification.

- Step 1: Tighten to 22 ft. lbs. (30 Nm)
- Step 2: Tighten an additional 90° rotation

- Step 3: Tighten an additional 90° rotation

33. Installation is the reverse of removal.

34. Fill the cooling system.

35. Start the engine and check for leaks.

## ENGINE ASSEMBLY

*REMOVAL & INSTALLATION*

### N52 and N54 Engines

### ❉❉ WARNING

**Aluminum-magnesium material. No steel fasteners may be used due to the threat of electrochemical corrosion. A magnesium crankcase requires aluminum fasteners exclusively. Aluminum fasteners must be replaced each time they are removed. The end faces of aluminum fasteners are painted blue for purposes of identification. Torque specifications and torque angles must be observed for risk of damage.**

1. Before servicing the vehicle, refer to the precautions.

2. Drain the cooling system and engine oil.

3. Relieve the fuel system pressure.

4. Raise the hood all the way up into the service position. This usually requires disconnecting the gas struts and securing the hood hinges with bolts.

5. Secure the engine with a holding tool to prevent it from tilting.

6. Disconnect the negative battery cable.

7. Remove the exhaust system.

8. Remove the transmission assembly.

9. Remove the intake filter housing.

10. Remove the fan cowl with electric fan.

11. Remove the radiator and hoses.

12. Remove the water pump.

13. Remove the thermostat.

14. Remove the all coolant hoses from engine.

15. Remove the intake manifold.

16. Remove the vacuum line and hose from brake booster.

17. Disconnect the ignition wiring harness, secure and set aside.

18. Remove the engine wiring harness, secure and set aside.

19. Remove the fuel injector rail, secure and set aside.

➡**Do not disconnect coolant pipe from crankcase.**

20. Remove the A/C compressor, set down and secure on front axle carrier.

➡**Do not disconnect hydraulic lines.**

21. Remove the power steering pump, set down and secure on front axle carrier.

22. Verify that all remaining fluid lines or electrical leads have been disconnected and properly placed aside..

23. Remove the engine mount nuts and bolts and carefully lift the engine from the vehicle.

*To install:*

24. Installation is the reverse of removal.

25. Replace all aluminum bolts and tighten to specification.

- M6: 61 inch lbs. (7 Nm)
- M7: 10 ft. lbs. (13 Nm)
- M8: 16 ft. lbs. (22 Nm)
- M10: 31 ft. lbs. (42 Nm)

26. Fill engine oil and cooling system.

27. Start engine and check for leaks.

28. Check DME function.

### N62 Engines

1. Before servicing the vehicle, refer to the precautions.

2. Drain the cooling system.

3. Relieve the fuel system pressure.

4. Raise the hood all the way up into the service position. This usually requires disconnecting the gas struts and securing the hood hinges with bolts.

5. Secure the engine with a holding tool to prevent it from tilting.

6. Disconnect the negative battery cable.

7. Remove the transmission assembly.

8. Remove the radiator and hoses.

9. Remove the expansion tank.

10. Remove the intake manifold.

11. Remove the heater hoses.

12. Remove the engine wiring harness from the control unit.

13. Remove the spark plug wiring harness.

14. Remove the negative lead from spring strut dome.

15. Remove the vacuum line and hose from vacuum pump.

16. Remove the line on tank venting valve.

17. Remove the both oxygen monitor and control sensors.

18. Remove the negative lead from engine support arm.

19. Remove the vibration damper and tensioning pulley fro A/C compressor drive belt.

20. Remove the fasteners from power steering pump, secure pump aside.

21. Remove the fasteners from a/c compressors, secure compressor aside.

22. Remove the fasteners from abs control unit, secure unit aside.

23. Remove the lower universal joint from the steering spindle.

24. Remove the if equipped with dynamic drive, unlock and detach the hydraulic line.

25. Remove the verify that all remaining fluid lines or electrical leads have been disconnected and properly placed aside.

26. Remove the engine mount nuts and bolts and carefully lift the engine from the vehicle.

*To install:*

27. Installation is the reverse of the removal procedure.

28. Observe the following torque specifications.

- M6: 61 inch lbs. (7 Nm)
- M7: 10 ft. lbs. (13 Nm)
- M8: 16 ft. lbs. (22 Nm)
- M10: 31 ft. lbs. (42 Nm)

## EXHAUST MANIFOLD

### REMOVAL & INSTALLATION

#### N52 Engine

1. Before servicing the vehicle, refer to the precautions.

2. Remove the ignition coil cover.

3. Remove the coolant expansion tank.

4. Remove the underbody protection.

5. Remove the complete exhaust system.

6. On AWD vehicles, remove the reinforcement plate.

7. Remove the oxygen sensor plug from cylinders number 4 and 6 and remove the exhaust assembly.

8. Remove the manifold for cylinders number 1 and 3 downwards.

9. Remove the manifold for cylinders number 4 and 6 downwards.

*To install:*

10. Remove the old gasket from the cylinder head and exhaust manifold and replace the gasket. The gasket beads face the exhaust manifolds.

11. Installation is the reverse of removal.

12. Coat screw connections with CRC copper paste.

13. Install new nuts and tighten the exhaust manifolds to 15 ft. lbs. (20 Nm).

#### N54 Engine

The N54 exhaust manifold is an integral part of the turbocharger assembly and is not serviced separately.

#### N62 Engine

1. Before servicing the vehicle, refer to the precautions.

2. Remove oxygen sensors.

3. Remove complete exhaust system

4. Secure engine in installation position with special tool

5. Lower front axle support.

6. Remove lower universal joint of steering spindle.

7. Remove nuts on exhaust manifold from below.

8. Remove exhaust manifold towards bottom and pull out towards rear.

*To install:*

9. Replace seals.

10. Seal beads faces the exhaust manifold.

11. Coat threads with copper paste.

12. Install new nuts and tighten the exhaust manifolds to 15 ft. lbs. (20 Nm).

## INTAKE MANIFOLD

### REMOVAL & INSTALLATION

#### N52 and N54 Engines

*See Figure 165.*

1. Before servicing the vehicle, refer to the precautions.

2. Disconnect the negative battery cable.

3. If necessary, remove the strut.

4. Remove the intake filter housing and mass air flow sensor.

5. Remove the ignition coil cover.

6. Remove the vent hose and cable holder.

7. Remove the electrical connectors and engine wiring harness, set aside and secure.

8. Remove the oil pressure switch plug connector.

➡**Do not detach fuel line.**

9. Remove the fuel rail, set aside and secure.

10. Remove the intake manifold fasteners.

11. Remove the tank vent line behind throttle valve assembly.

12. Remove the intake manifold.

*To install:*

13. Replace all seals and gaskets.

14. Installation is the reverse of the removal procedure.

15. Tighten intake manifold bolts to 11 ft. lbs. (15 Nm)

#### N62 Engines

1. Before servicing the vehicle, refer to the precautions.

2. Disconnect the negative battery cable.

3. Remove the engine covers.

4. Remove the intake hose.

5. Remove the injection pipe.

6. Remove the differential pressure sensor and servomotor connectors.

7. Remove the vent hose and cable holder.

8. Remove the intake manifold fasteners and the intake manifold.

06041_BMWC_G0002

**Fig. 165 Intake manifold fastener locations (1, 3) and fuel rail (2)**

*To install:*

9. Clean sealing faces on intake air manifold and engine.

10. Replace gasket.

11. Installation is the reverse of the removal.

12. Observe the following torque specifications.

- M6: 88 inch lbs. (10 Nm)
- M7: 11 ft. lbs. (15 Nm)
- M8: 16 ft. lbs. (22 Nm)

## OIL PAN

### REMOVAL & INSTALLATION

#### N52 and N54 Engines

*See Figure 166.*

> ❋❋ **WARNING**
>
> **Aluminum-magnesium material. No steel fasteners may be used due to the threat of electrochemical corrosion. A magnesium crankcase requires aluminum fasteners exclusively. Aluminum fasteners must be replaced each time they are removed. The end faces of aluminum fasteners are painted blue for purposes of identification. Torque specifications and torque angles must be observed for risk of damage.**

1. Before servicing the vehicle, refer to the precautions.

2. Install engine support tool or equivalent.

3. Disconnect the negative battery cable.

4. Remove the lower front axle.

5. Remove the left drive shaft.

6. Remove the right drive shaft.

7. Remove the front axle differential.

8. Remove the engine oil.

9. On vehicles equipped with automatic transmission, oil lines must be detached from the engine oil pan.

10. If necessary, remove vane pump and set it aside.

11. Remove the two bolts securing oil pan to transmission.

12. Remove the oil return hose.

13. Remove oil pan bolts and remove the oil pan.

14. If necessary, remove oil level sensor bolts and oil level sensor.

*To install:*

15. Clean the mounting surfaces and install a new gasket and all seals.

> ❋❋ **WARNING**
>
> **There must be no adhesive residues in the oil pan retaining threads. Clean retaining threads.**

16. Install oil pan.

17. Replace all aluminum fasteners and tighten 70 inch lbs (8 Nm) plus an additional 90° rotation

18. Installation is the reverse of removal.

19. Fill the engine with oil.

20. Start the engine and check for leaks.

#### N62 Engine

##### Lower

*See Figure 167.*

1. Before servicing the vehicle, refer to the precautions.

2. Remove reinforcement plate

3. Remove oil drain plug and drain engine oil.

4. Unlock plug connection (1) on oil level sensor and disconnect.

5. Remove bolts and remove oil pan bottom section (2).

*To install:*

6. Clean sealing surfaces.

7. Install a new gasket and tighten new bolts to 9 ft. lbs. (12 Nm).

8. Connect the oil level sensor.

9. Fill the engine with engine oil.

10. Install reinforcement plate.

11. Start engine and check for leaks.

##### Upper

*See Figures 168 and 169.*

1. Remove design cover.

2. Remove center front panel.

3. Remove electric fan.

4. Remove alternator drive belt.

5. Remove A/C compressor drive belt.

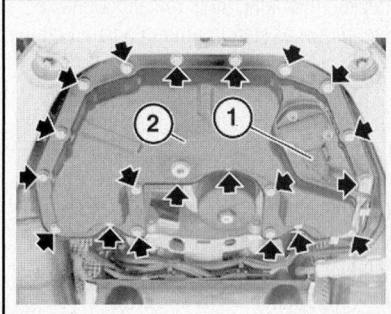

22205_BMWC_G0219

**Fig. 167  Oil level sensor connection (1) and oil pan bottom section (2)**

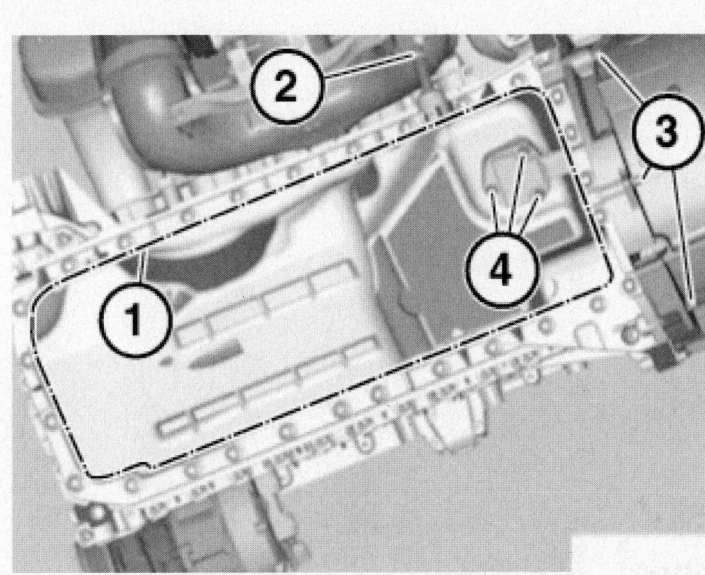

06041_BMWC_G0003

**Fig. 166  Oil pan bolts (1), oil return hose (2), transmission bolts (3) and oil level sensor (4)**

22205_BMWC_G0220

**Fig. 168  Oil pan mounting bolt locations—1 of 2**

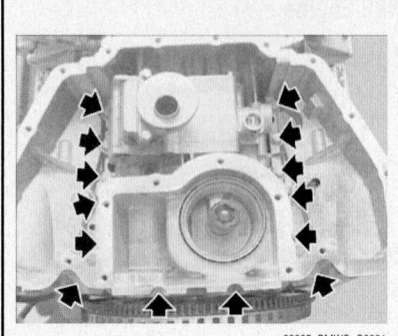

22205_BMWC_G0221

**Fig. 169 Oil pan mounting bolt locations—2 of 2**

6. Remove acoustic cover.

7. Secure engine in installation position with special tool.

8. Remove nuts on left and right engine mounts at top.

9. Remove reinforcement plate.

10. Drain off engine oil.

11. Remove lower oil pan section.

12. Lower front axle support.

➡**To remove the oil pan, you must lower the front axle support. There is no need to perform a front axle alignment check.**

13. Remove power steering pump on bracket, hydraulic lines remain connected. Secure power steering pump against falling down.

14. Unlock and disconnect hydraulic lines from heat exchanger with special tool 17 0 030.

15. Remove hydraulic lines on automatic transmission.

16. Remove holder for hydraulic lines. Remove hydraulic lines. Installation location: on radiator at bottom left.

17. Transmission fluid emerges when hydraulic lines are removed.

18. Catch and dispose of escaping transmission fluid.

19. Disconnect cable duct of oxygen sensors from automatic transmission.

20. Remove bolts and remove oil pan top section.

### To install:

21. Replace sealing ring on dipstick guide tube.

22. Clean sealing faces of seal debris and clean.

23. Replace gasket.

24. Install oil pan.

25. Insert all bolts on transmission end.

26. Tighten all bolts on oil pan to specification.

• M6: 8 ft. lbs. (10 Nm)
• M8: 15 ft. lbs. (20 Nm)

27. Check fluid level in automatic transmission and adjust if necessary.

## OIL PUMP

### REMOVAL & INSTALLATION

#### N52 and N54 Engines

*See Figure 170.*

1. Before servicing the vehicle, refer to the precautions.

2. Disconnect the negative battery cable.

3. Remove the oil pan.

4. Remove the oil pump intake pipe fasteners and oil pump intake pipe, pull towards transmission.

5. Remove the oil pump pulley bolt.

6. Remove the oil pump mounting bolts.

➡**Timing chain of triangular drive is pressed upwards by chain tensioner.**

7. Do not remove pulley from assembly.

8. Remove the oil pump pulley, pull towards front of engine bay.

9. Remove the oil pump.

### To install:

10. Check the seals on the oil pipes and replace it if necessary. Lubricate the seals with oil and the oil pipes.

11. Check the seal in the oil pump and replace it if necessary.

12. Align twin surface on oil pump to sprocket wheel.

13. Replace all aluminum fasteners and tighten to specification.

• Oil pump to bedplate: 25 ft. lbs. (34 Nm) plus an additional 180° rotation
• Chain module to crankcase and oil pump 35 inch lbs. (4 Nm) plus an additional 45° rotation
• Pulley to oil pump: 15 ft. lbs. (20 Nm) plus an additional 45° rotation
• Intake pipe to bedplate: 35 inch lbs. (4 Nm) plus an additional 100° rotation

14. Installation is the reverse of removal.

15. Fill the engine with oil.

16. Start the engine and check for leaks.

#### N62 Engine

1. Before servicing the vehicle, refer to the precautions.

2. Drain engine oil.

3. Remove upper oil sump section.

4. Remove reinforcement plate.

5. Remove oil pump sprocket wheel.

6. Remove screws, remove oil pump with aid of a second person.

### To install:

➡**Use new, lightly oiled sealing rings during installation.**

7. Apply light coat of oil and replace sealing ring.

8. Insert oil line in crankcase.

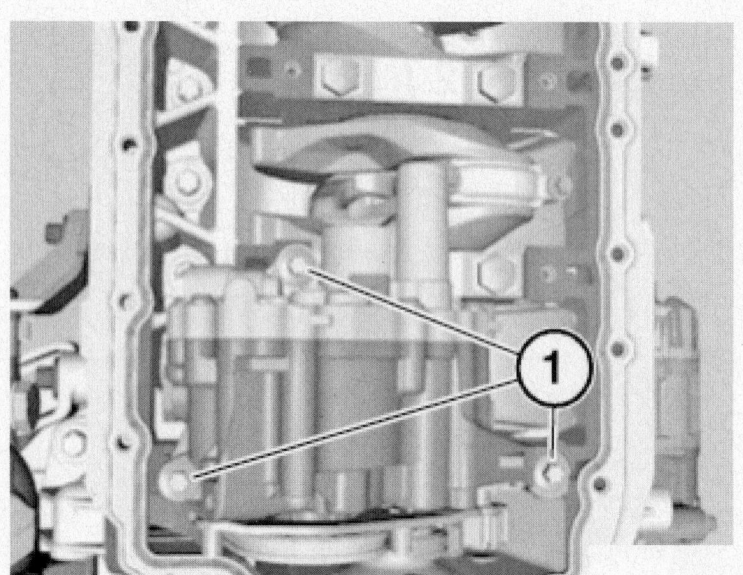

06041_BMWC_G0004

**Fig. 170 Oil pump mounting bolts (1)**

※※ **WARNING**

**Do not damage sealing ring.**

9. Insert oil line in bore hole in oil pump and make sure oil lines are correctly seated.

10. Install oil pump with aid of a second person. Tighten bolts to 25 ft. lbs. (34 Nm).

11. Install oil pump sprocket wheel and tighten to 22 ft. lbs. (30 Nm).

12. Install the reinforcement plate.

13. Install oil sump.

14. Fill engine with oil, start engine and check for leaks.

## PISTON AND RING

*POSITIONING*

*See Figures 171 through 173.*

## REAR MAIN SEAL

*REMOVAL & INSTALLATION*

**N52 and N54 Engines**

*See Figures 174 through 179.*

1. Before servicing the vehicle, refer to the precautions.

2. Remove the transmission.

3. Remove the flywheel assembly.

➡Crankshaft radial seal has six removal openings for removal with special tool 11 9 200. If necessary, remove rubber coating on top side of crankshaft radial seal and expose a removal opening.

4. Fit special tool 11 9 200. Insert sheet metal bolts into removal opening of crank-shaft radial seal and fasten without play (do not overtighten sheet metal bolts).

5. Screw in spindle (1) slowly and carefully and detach crankshaft radial seal.

**To install:**

※※ **WARNING**

**The following text describes installa-tion and sealing between the engine block and crankshaft radial seal. The engine block will not be leak proof at the outside of the crankshaft radial seal if you fail to comply with the individual work steps and the work sequence.**

6. Clean sealing surface and degrease thoroughly in area of housing partition.

7. Apply a light coat of oil to running surface of crankshaft radial seal.

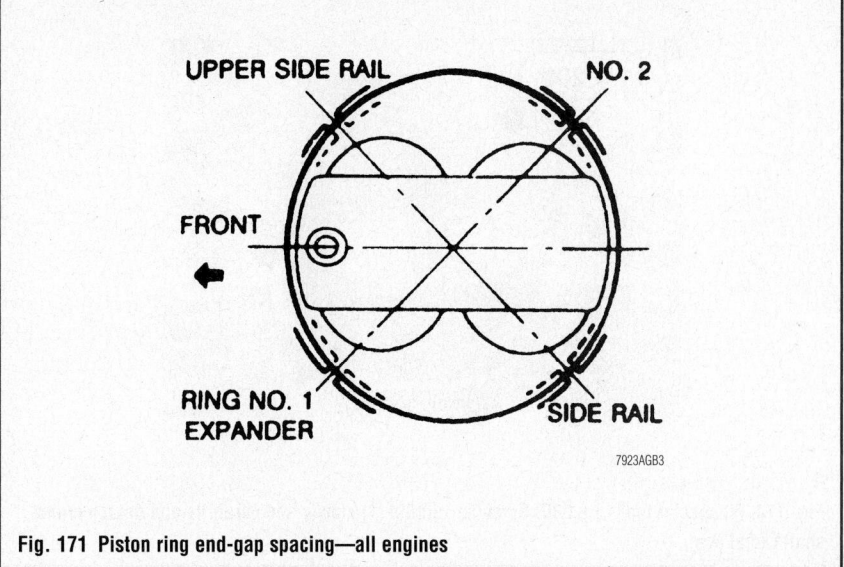

Fig. 171 Piston ring end-gap spacing—all engines

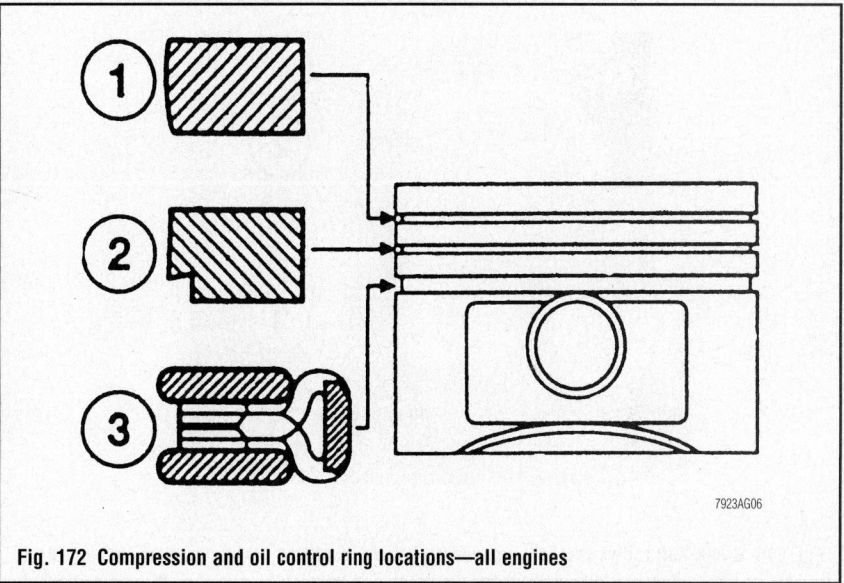

Fig. 172 Compression and oil control ring locations—all engines

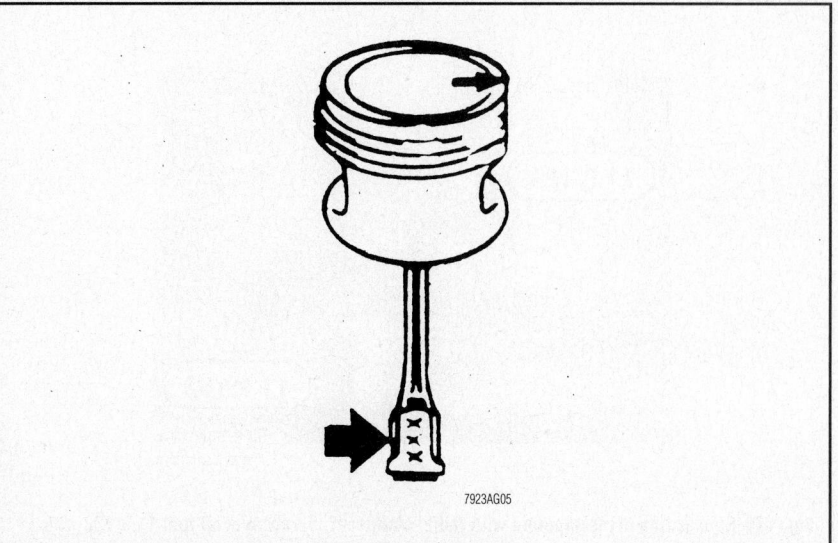

Fig. 173 Connecting rod-to-piston positioning—all engines

Fig. 174  Fit special tool 11 9 200. Screw in spindle (1) slowly and carefully and detach crankshaft radial seal

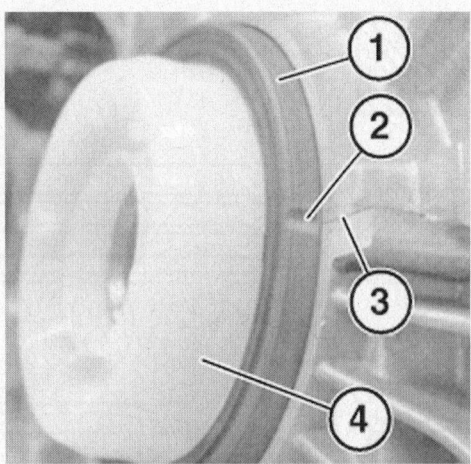

Fig. 175  Crankshaft radial seal (1), grooves (2) housing partition (3) and support bushing (4)

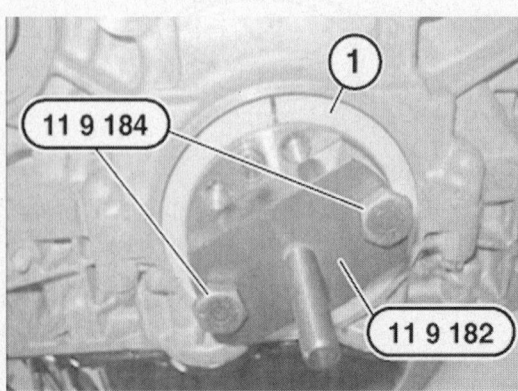

Fig. 176  Spacer ring (1) is supplied with radial shaft seal. Screw special tool 11 9 182 with bolts (special tool 11 9 184) to crankshaft

➡Support bushing is contained in scope of delivery of crankshaft radial seal.

8.  When crankshaft radial seal is installed, only support bushing may be used as a slip bushing.

9.  Crankshaft radial seal has a groove on both left and right sides.

### ❊❊ WARNING

**After installation, grooves must be filled with sealing compound.**

➡The required parts are available from the BMW Parts Service (ETK).

10.  Remove screw caps from injector.
11.  Screw on metering needle.
12.  Insert piston for pressing out.
13.  Injector contains the sealing compound Loctite, manufacturer's number 128357.
14.  Bottle contains the primer Loctite, manufacturer's number 171000.
15.  Fit support bushing with crankshaft radial seal on crankshaft.
16.  Align groove centrally to housing partition.
17.  Coat both grooves on crankshaft radial seal with Loctite primer, manufacturer's number 171000, and expose to air for approx. one minute.
18.  Push crankshaft radial seal by hand as far as possible onto running surface.
19.  Carefully remove support sleeve.

➡Spacer ring is supplied with radial shaft seal.

20.  Screw special tool 11 9 182 with bolts (special tool 11 9 184) to crankshaft.
21.  Fit spacer ring on preassembled radial shaft seal.
22.  Draw in radial shaft seal and spacer ring with special tool 11 9 181 in conjunction with special tool 11 9 183.
23.  Then remove spacer ring again.
24.  Before filling with sealing compound, moisten brush with Loctite primer, manufacturer's number 171000. Insert brush as far as possible into grooves on crankshaft radial seal in order to coat housing partition on engine block.
25.  Using injector, fill both grooves flush with Loctite sealing compound, manufacturer's number 128357.

➡Loctite primer, manufacturer's number 171000, binds the Loctite sealing compound, manufacturer's number 128357, and prevents leakage.

26.  Coat surface of sealing compound in both grooves with Loctite primer, manufacturer's number 171000.

**Fig. 177** Draw in radial shaft seal and spacer ring with special tool 11 9 181 in conjunction with special tool 11 9 183

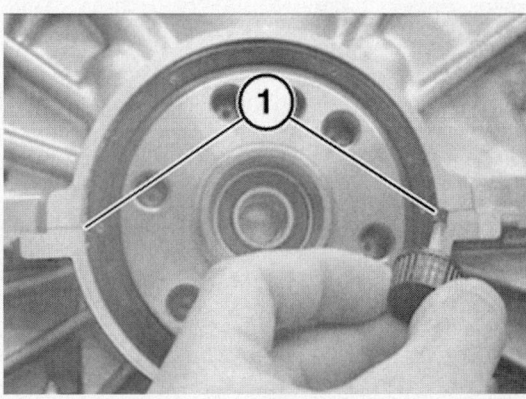

**Fig. 178** Insert brush as far as possible into grooves (1) on crankshaft radial seal in order to coat housing partition on engine block

**Fig. 179** Insert brush as far as possible into grooves (1) on crankshaft radial seal in order to coat housing partition on engine block

## N62 Engine

*See Figure 180.*

1. Before servicing the vehicle, refer to the precautions.
2. Remove transmission.
3. Drain engine oil.
4. Remove flywheel.
5. Remove the upper and lower mounting bolts, then remove end cover.

### ✳✳ WARNING

**The sealing lip of the radial shaft seal is highly sensitive and must not be kinked. Do not touch the sealing lip with your fingers. If the radial shaft seal is stored for longer than six months without the support sleeve, its operational reliability can no longer be guaranteed and it must not be reused.**

➡The support sleeve remains in the radial shaft seal and is used as a slip sleeve in the installation described later.

➡The radial shaft seal can only be replaced completely with the end cover. The gasket is an integral part of the end cover and cannot be replaced individually.

### To install:

6. Check dowel sleeves for damage and correct installation position.
7. Keep sealing faces clean and free of oil.
8. Coat contact points on joint along oil sump with Drei Bond 1209 (refer to BMW Parts Service).
9. Lightly oil running surface of crankshaft.
10. Fit end cover with support sleeve on crankshaft and push on carefully.

**Fig. 180** Fit end cover (1) with support sleeve (2) on crankshaft and push on carefully

11. Insert screws and initially tighten without play.

12. Tighten bolts securely.

13. Install flywheel.

14. Install transmission.

15. Fill engine with oil.

16. Start the engine and check for leaks.

## TIMING CHAIN, SPROCKETS, AND FRONT COVER

### REMOVAL & INSTALLATION

#### N52 and N54 Engines

*See Figures 181 through 192.*

1. Before servicing the vehicle, refer to the precautions.

2. Remove cylinder head cover.

3. Remove all spark plugs.

4. Remove chain tensioner.
   a. Remove chain tensioner.

➡ Have a cleaning cloth ready. A small quantity of engine oil will emerge after the screw connection has been Removed.

#### ❄ WARNING

**Make sure no engine oil runs onto belt drive.**

5. Remove crankshaft front radial seal.

6. Remove accessory drive belt and tensioner.

7. Remove vibration damper.

8. Remove fastener in direction of arrow.

9. Rotate crankshaft at central bolt into TDC position.

10. Slide special tool 11 0 300 in direction of arrow into special tool bore and secure crankshaft.

#### ❄ WARNING

**On vehicles with optional extra SA205 (automatic transmission), there is a large bore for the TDC position shortly before the special tool bore. This bore can be confused with the special tool bore. If the flywheel is secured in the correct special tool bore with special tool 11 0 300, the engine can no longer be moved at the central bolt.**

11. Do not remove special tool 11 0 300 to Remove central bolt. Employ a second person for gripping when releasing central bolt.

12. Screw special tool 11 9 280 onto hub of vibration damper.

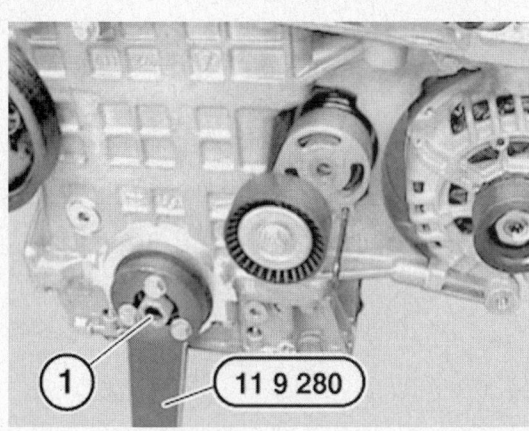

22205_BMWC_G0175

**Fig. 181  Do not remove special tool 11 0 300 to Remove central bolt (1). Screw special tool 11 9 280 onto hub of vibration damper**

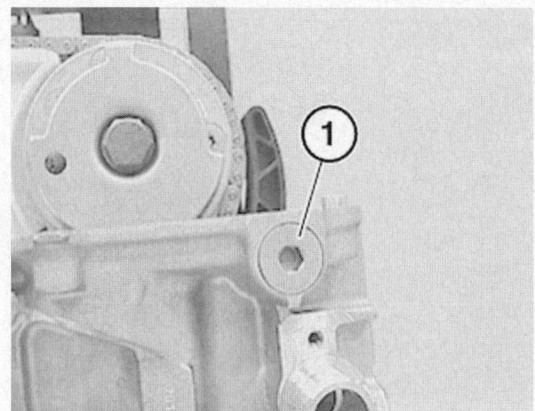

22205_BMWC_G0176

**Fig. 182  Open plug at top of cylinder head**

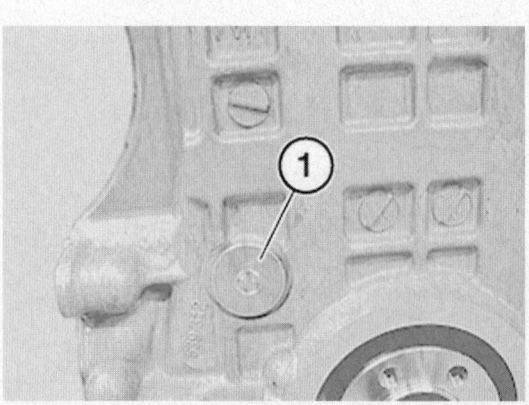

22205_BMWC_G0177

**Fig. 183  Open plug at lower left of engine block**

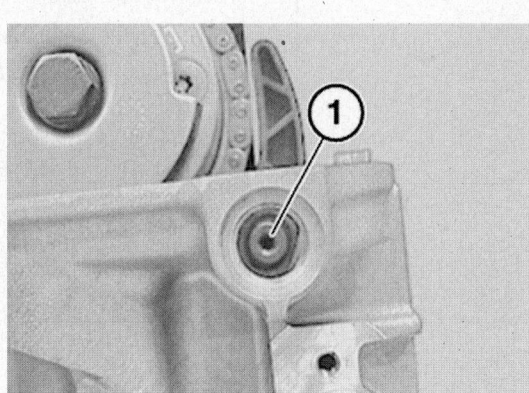

**Fig. 184 Remove bearing pin (1) from timing chain module on cylinder head**

**Fig. 185 Remove bearing pin (1) from timing chain module on crankcase**

**Fig. 186 Collar (see arrow) on sprocket wheel (2) points to engine**

13. Remove central bolt.

14. Remove hub towards front.

15. Open plug at top of cylinder head.

16. Open plug at lower left of engine block.

17. Remove bearing pin from timing chain module on cylinder head.

18. Remove bearing pin from timing chain module on crankcase.

✳✳ **WARNING**

**Install special tool 11 4 280 to Remove the central bolts on the inlet and exhaust adjustment units.**

19. Secure special tool 11 4 283 to cylinder head with bolts.

➡**Fit special tool 11 4 282 underneath on side of inlet camshaft.**

20. Mount special tool 11 4 281 on inlet and exhaust camshafts.

21. Do not remove special tool 11 4 280.

22. Remove inlet and exhaust adjustment unit. Refer to "Engine Mechanical, Camshaft and Lifters, Removal & Installation."

23. Remove bolts (1) from timing chain module on cylinder head.

24. Remove chain module with timing chain and sprocket wheel upwards in direction of arrow.

*To install:*

✳✳ **WARNING**

**Note installation direction of sprocket wheel. Collar on sprocket wheel points to engine. Incorrect assembly will result in engine damage.**

25. Pull timing chain upwards until sprocket wheel engages chain guide.

26. On N54 engines, special friction plates are required between the friction surfaces.

✳✳ **WARNING**

**The engine will incur damage if the plates are damaged or are not fitted.**

　a. Friction plates are clipped into place on sprocket wheel/oil pump module.

　b. Make sure friction plate is in correct installation position.

　c. Push on friction plate without retainers.

　d. Insert chain module from above and secure with bolt.

　e. Make sure gear wheels are in correct installation position.

　f. Insert hub with friction plate.

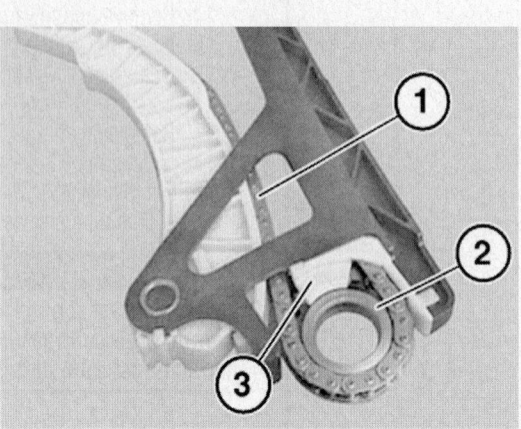

22205_BMWC_G0182

**Fig. 187 Pull timing chain (1) upwards until sprocket wheel (2) engages chain guide (3)**

22205_BMWC_G0188

**Fig. 188 On N54 engines, special friction plates (1 and 2) are required between the friction surfaces**

22205_BMWC_G0189

**Fig. 189 Make sure friction plate (3) is in correct installation position**

27. Install timing chain and sprocket wheel in this position.

➡**Always hold timing chain under tension. Timing chain may jam on chain guide.**

28. Install hub with central bolt.

29. Tighten down special tool 11 5 200 with bolts to hub.

30. Install bolts from timing chain module on cylinder head and tighten to 7 ft. lbs. (9 Nm).

31. Install bearing pin from timing chain module on crankcase and tighten to 15 ft. lbs. (20 Nm).

32. Install bearing pin from timing chain module on cylinder head and tighten to10 ft. lbs. (14 Nm).

33. Install plug and tighten to 19 ft. lbs. (25 Nm).

34. If not previously done, remove tensioner for drive belt.

35. Screw in special tool 11 4 362 from special tool kit 11 4 360.

36. Mount special tool 11 9 280 on 11 5 200.

37. Support special tool 11 9 280 on special tool 11 4 362.

38. Special tool 11 0 300 secures crankshaft.

39. Tighten central bolt to 74 ft. lbs. (100 Nm) plus an additional 360° of rotation.

40. Install inlet and exhaust adjustment units.

41. Install chain tensioner.

42. Crank engine twice.

43. Check and if necessary adjust valve timing. Refer to "Engine Mechanical, Camshaft and Lifters, Removal & Installation."

44. Install vibration damper.

45. Install accessory drive belt and tensioner. Tighten tensioner mounting bolt to 19 ft. lbs. (25 Nm) plus an additional 90° rotation.

46. Install crankshaft front radial seal.

47. Install chain tensioner.

➡**No sealing ring is fitted during series-production assembly. A sealing ring must be fitted by service personnel when the chain tensioner is fitted.**

48. If the chain tensioner is reused, its oil chamber must be drained. Place chain tensioner on a level working surface and slowly compress.

49. Repeat procedure twice.

50. Install fastener with bore facing outwards.

51. Install tensioner and tighten to 41 ft. lbs. (55 Nm).

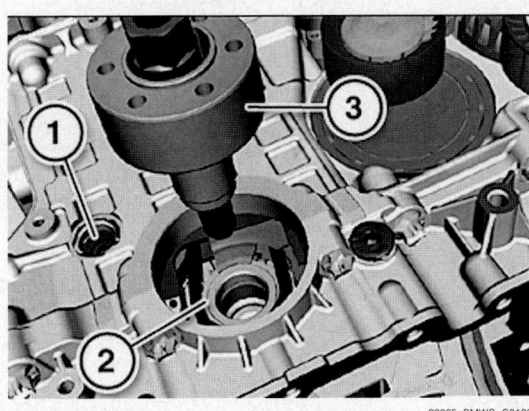

**Fig. 190 Insert chain module from above and secure with bolt (1). Make sure gear wheels (2) are in correct installation position, then insert hub (3) with friction plate**

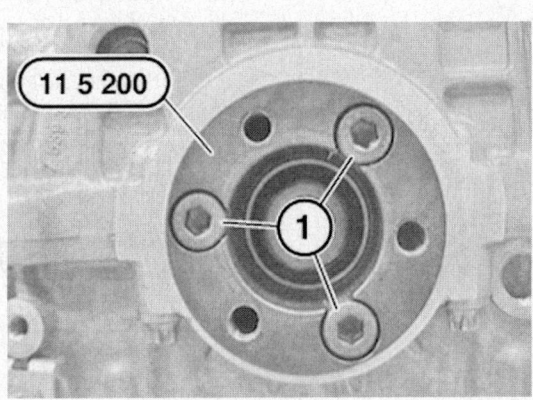

**Fig. 191 Tighten down special tool 11 5 200 with bolts (1) to hub**

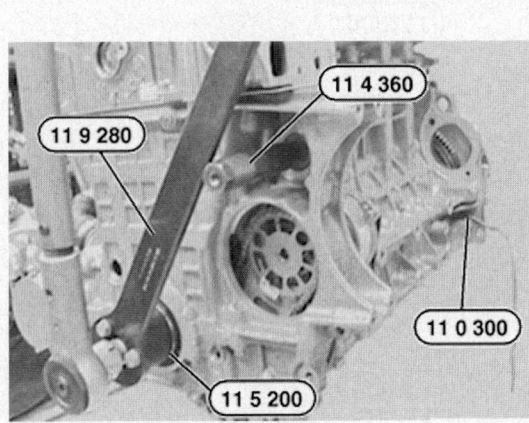

**Fig. 192 Screw in special tool 11 4 362 from special tool kit 11 4 360**

52. Install all spark plugs.
53. Install cylinder head cover.

### N62 Engine

*See Figures 193 through 195.*

➡The official BMW procedure calls for timing chain removal and installation to be performed with the engine removed from the vehicle. It may be possible to perform this procedure with the engine still installed in the vehicle.

1. Before servicing the vehicle, refer to the precautions.
2. Remove lower timing case cover.
  a. Remove engine.
  b. Remove both cylinder heads.
  c. Remove lower oil sump section.
  d. Remove upper oil sump section.
  e. Remove vibration damper.
  f. Remove water pump.
  g. Remove tensioner from poly-V-belt.
  h. Remove alternator with housing.
  i. Remove hub for vibration damper.
  j. Remove flywheel.
  k. Remove rear coolant cap.
  l. Drive out water pipe with special tool 23 1 040 at cutout towards front.
3. Remove screws along lines.
4. Remove timing case cover towards front. Remove timing chain.
5. Remove timing chain.
6. Remove screws.
7. Remove guide rails.
8. Remove tensioning rails.

*To install:*

**❉❉ WARNING**

**Maintain tension of timing chains when installing timing case cover. Observe sparking protection on timing case cover.**

9. Make sure timing chain is correctly installed when placing it in guide rail.
10. Connect special tools 11 2 001 and 11 2 002 to crankshaft.
11. Insert special tool 11 2 007 and remove sprocket wheel with special tool 11 2 003.
12. Check sprocket wheels for wear, replace if necessary.
13. Install lower timing case cover.
  a. Replace water pipe and sealing ring.

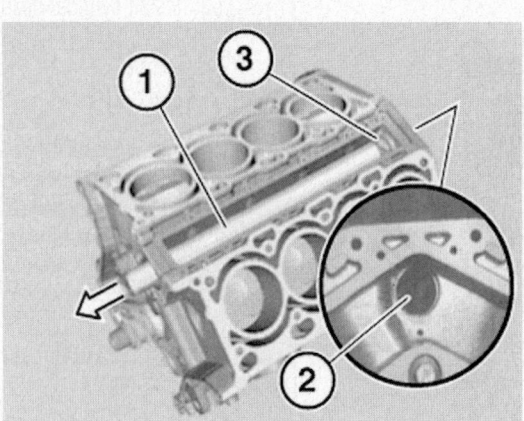

**Fig. 193  Drive out water pipe (1) with special tool 23 1 040 at cutout (2) towards front and replace water pipe and sealing ring (3)**

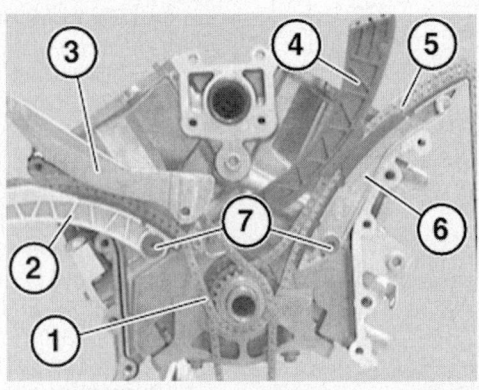

**Fig. 194  Timing chain (1 and 5), tensioning rails (2 and 4), guide rails (3 and 6) and bolts (7)**

b. Clean sealing surfaces and replace seal.

c. Install timing case cover, insert all screws and initially tighten to approximately 4 ft. lbs. (5 Nm).

### ❊❊ WARNING

**Once all screws have been tightened down, retighten them in a second operation.**

d. Fully tighten all screws in alternate sequence.
- M6: 8 ft. lbs. (10 Nm)
- M7: 11 ft. lbs. (15 Nm)
- M8: 16 ft. lbs. (22 Nm)

e. Replace radial seal in timing case at bottom.

f. Install rear coolant cap.

g. Install flywheel.

h. Install hub for vibration damper.

i. Install alternator with housing.

j. Install tensioner from poly-V-belt.

k. Install water pump.

l. Install vibration damper.

m. Install upper oil sump section.

n. Install lower oil sump section.

o. Install both cylinder heads.

p. Install engine.

### VALVE LASH

*ADJUSTMENT*

All engines are equipped with hydraulic valve lash adjusters. No adjustments are possible.

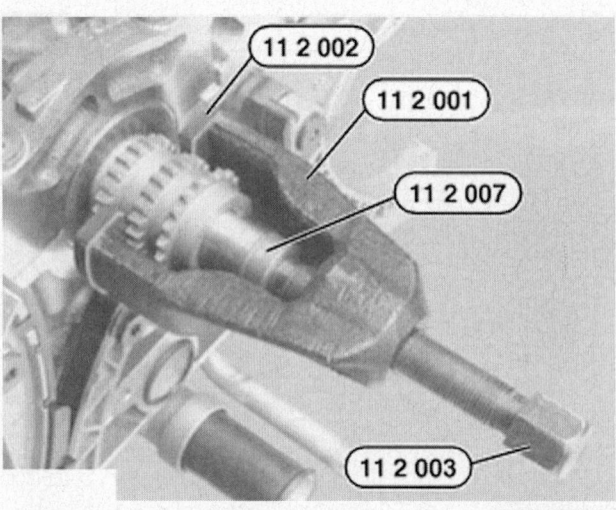

**Fig. 195  Connect special tools 11 2 001 and 11 2 002 to crankshaft. Insert special tool 11 2 007 and remove sprocket wheel with special tool 11 2 003**

## ENGINE PERFORMANCE & EMISSION CONTROL

### CAMSHAFT POSITION (CMP) SENSOR

#### LOCATION

*See Figures 196 through 198.*

#### REMOVAL & INSTALLATION

#### N52 and N54 Engines

1. Before servicing the vehicle, refer to the precautions.
2. Read out fault memory of DME control unit; if necessary, work through test schedules
3. Switch ignition **OFF**.
4. Remove radiator cover.
5. Remove plug connector and pull off.
6. Remove screw.
7. Remove pulse generator.

***To install:***

8. Installation is the reverse of removal.
9. If equipped, replace sealing ring and coat with anti-seize agent.

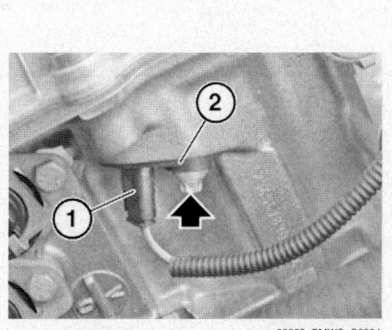

**Fig. 196 Intake CMP sensor (2) and connector (1) location—N52 and N54 engines**

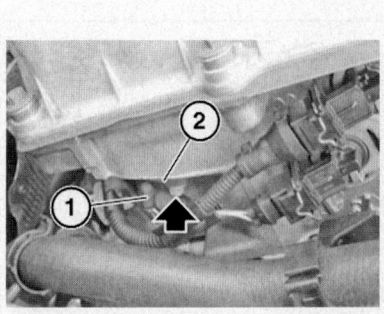

**Fig. 197 Exhaust CMP sensor (2) and connector (1) location—N52 and N54 engines**

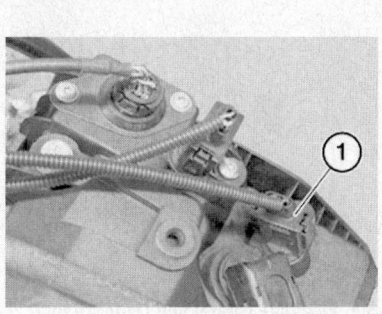

**Fig. 198 CMP sensor (1) location—N62 engine left side shown, right side similar**

10. Tighten sensor to 7 ft. lbs. (9 Nm).
11. Check for stored fault messages, rectify faults and clear the fault memory.

#### N62 Engine

1. Before servicing the vehicle, refer to the precautions.
2. Read out fault memory of DME control unit; if necessary, work through test schedules
3. Switch ignition **OFF**.
4. Remove acoustic cover.
5. Remove left ignition coil cover
6. Remove plug connector and pull off.
7. Remove screw.
8. Remove pulse generator.

***To install:***

9. Installation is the reverse of removal.
10. Replace sealing ring and coat with anti-seize agent.
11. Check for stored fault messages, rectify faults and clear the fault memory.

### CRANKSHAFT POSITION (CKP) SENSOR

#### LOCATION

*See Figures 199 and 200.*

#### REMOVAL & INSTALLATION

#### N52 and N54 Engines

#### ❊❊ WARNING

**Aluminum-magnesium materials. No steel screws/bolts may be used due to the threat of electrochemical**

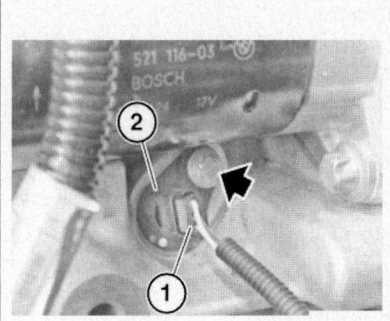

**Fig. 199 CKP sensor (2) and connector (1) location—N52 and N54 engines**

**corrosion. A magnesium crankcase requires aluminum screws/bolts exclusively. Aluminum screws/bolts must be replaced each time they are released. The end faces of aluminum screws/bolts are painted blue for the purposes of reliable identification. Jointing torque and angle of rotation must be observed without fail (risk of damage).**

1. Before servicing the vehicle, refer to the precautions.
2. Read out fault memory of DME control unit; if necessary, work through test schedules
3. Switch ignition **OFF**.
4. Remove intake air manifold.
5. Disconnect plug
6. Disconnect the crankshaft pulse generator.
7. Remove and discard the aluminum bolts.
8. Remove pulse generator from crankcase.

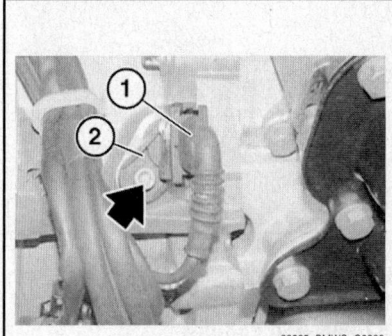

**Fig. 200 CKP sensor (2) and connector (1) location—N62 engine**

*To install:*

9. Replace sealing ring.
10. Install pulse generator to crankcase.
11. Using new aluminum bolts, tighten to 2 ft. lbs (3 Nm) plus an additional 45° of rotation.
12. Connect the crankshaft pulse generator.
13. Install intake air manifold.

### N62 Engine

1. Before servicing the vehicle, refer to the precautions.
2. Read out fault memory of DME control unit; if necessary, work through test schedules
3. Switch ignition **OFF**.
4. If necessary, remove underbody protection.
5. Unlock plug and remove.
6. Remove screw.
7. Withdraw pulse generator from transmission housing.

*To install:*

8. Replace sealing ring.
9. Install pulse generator to transmission housing.
10. Replace screw.
11. Check for stored fault messages, rectify faults and clear the fault memory.

## DIGITAL MOTOR ELECTRONIC (DME) CONTROL UNIT

### LOCATION

*See Figures 201 and 202.*

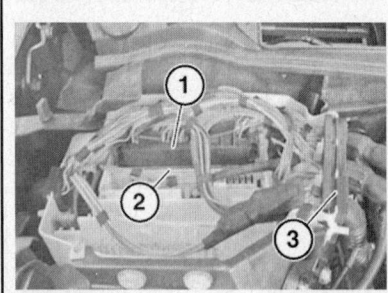

**Fig. 201 DME control unit (2), connector (1) and cover (3) location—N52 and N54 engines**

### REMOVAL & INSTALLATION

#### N52 and N54 Engines

*See Figure 203.*

1. Before servicing the vehicle, refer to the precautions.

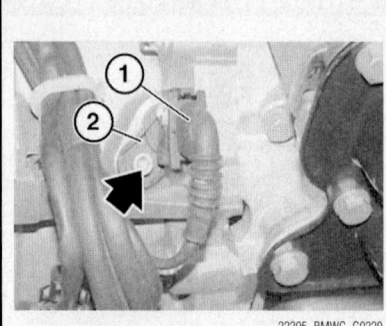

**Fig. 202 DME control unit (1) location—N62 engine**

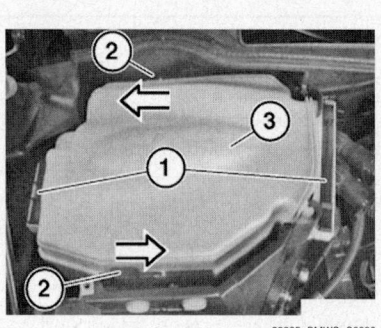

**Fig. 203 Unlock fasteners (1) from below and slide upwards approximately 10 mm. Unlock locks (2) in direction of arrow and remove cover (3)**

2. Read out fault memory of DME control unit; if necessary, work through test schedules
3. Switch ignition **OFF**.
4. Disconnect the negative battery cable.
5. Remove the air filter housing.
6. Unlock fasteners from below and slide upwards approximately 10 mm.
7. Unlock locks in direction of arrow.
8. Remove cover.
9. Unlock plug and remove.
10. Unlock control unit and remove towards top.

*To install:*

11. Installation is the reverse of removal.
12. Note device identification number and coding.
13. As necessary, code and program the new control unit.
14. Check for stored fault messages, rectify faults and clear the fault memory.

### N62 Engine

*See Figures 204 and 205.*

1. Before servicing the vehicle, refer to the precautions.
2. Read out fault memory of DME control unit; if necessary, work through test schedules
3. Switch ignition **OFF**.
4. Disconnect the negative battery cable.
5. Remove the right fresh-air duct.
6. There is a sliding "stopper" cover that needs to first be removed. Do so by sliding it the direction of the white arrow in the photo.
7. Unscrew the retaining bolts to the cover and lift it out. Some models are equipped with an armor plating that needs to be unscrewed as well.
8. Unlock (via small tabs) and detach all of the plug connections on the DME control unit.
9. Unlock the DME control unit itself and remove out towards top.

*To install:*

10. Slide the DME control unit back into the box.
11. Lock the small tabs and reconnect all of the plug connectors

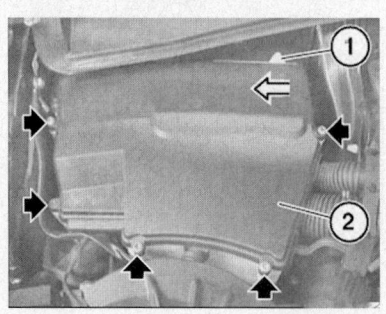

**Fig. 204 Slide the locking stopper the direction of the white arrow**

**Fig. 205 Unlock the DME control unit (1) itself and remove out towards top**

12. Screw on the retaining bolts of the cover
13. Slide the "stopper" cover back over the DME box
14. Replace the fresh air duct.
15. Check for stored fault messages, rectify faults and clear the fault memory.

## ENGINE COOLANT TEMPERATURE (ECT) SENSOR

### LOCATION

*See Figures 206 through 208.*

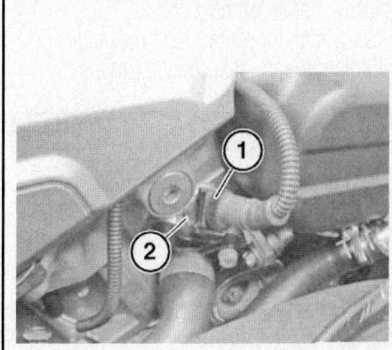

**Fig. 206 ECT location—N52 engine**

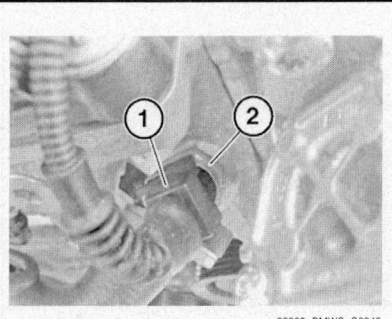

**Fig. 207 ECT sensor (2) and connector (1) location—N54 engine**

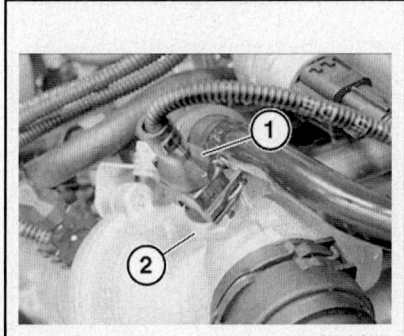

**Fig. 208 ECT location—N62 engine**

### REMOVAL & INSTALLATION

#### N52 Engine

*See Figure 209.*

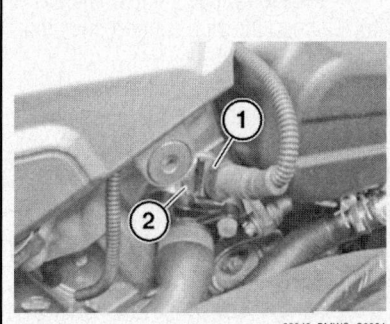

**Fig. 209 Unlock the plug on the sensor (1) and remove the temperature sensor (2)**

1. Before servicing the vehicle, refer to the precautions.
2. Read out fault memory of DME control unit; if necessary, work through test schedules
3. Switch ignition **OFF**.
4. Though you can avoid it, for easier access, remove the intake duct.
5. Unlock the plug on the sensor (1) and remove it.
6. Remove the temperature sensor (2).

**To install:**
7. Reposition the sensor and lock the plug.
8. After returning the temperature sensor back to its position, make sure to vent the cooling system and check for leaks.
9. If necessary, top off the coolant in the reservoir.
10. Check for stored fault messages, rectify faults and clear the fault memory.

#### N54 Engine

*See Figure 210.*

**Fig. 210 Disconnect the oil line (1) to gain access to the sensor**

1. Before servicing the vehicle, refer to the precautions.
2. Read out fault memory of DME control unit; if necessary, work through test schedules
3. Switch ignition **OFF**.
4. Remove the acoustic cover.
5. Remove the fan cowl.
6. Disconnect the oil line to gain access to the sensor.
7. Unlock the plug on the sensor and remove it.
8. Remove the temperature sensor.

**To install:**
9. Install the sensor and tighten to 10 ft. lbs. (14 Nm). Lock the plug.
10. Connect the oil line using new O-rings and tighten to 14 ft. lbs. (19 Nm).
11. Install the fan cowl.
12. Install the acoustic cover.
13. After returning the temperature sensor back to its position, make sure to vent the cooling system and check for leaks.
14. If necessary, top off the coolant in the reservoir.
15. Check for stored fault messages, rectify faults and clear the fault memory.

#### N62 Engine

*See Figure 211.*

1. Before servicing the vehicle, refer to the precautions.
2. Read out fault memory of DME control unit; if necessary, work through test schedules
3. Switch ignition **OFF**.

➡ **Catch and recycle any escaping coolant.**

4. Unlock the plug connector and remove it.
5. Remove the temperature sensor from its position.

**To install:**
6. After returning the temperature sensor back to its position, make sure to vent the cooling system and check for leaks.

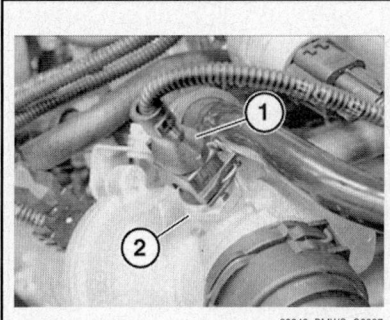

**Fig. 211 Unlock the plug on the sensor (1) and remove the temperature sensor (2)**

7. If necessary, top off the coolant in the reservoir.

8. Check for stored fault messages, rectify faults and clear the fault memory.

## HEATED OXYGEN (HO2S) SENSOR

### LOCATION

See Figure 212.

### REMOVAL & INSTALLATION

#### N52 and N54 Engines

1. Before servicing the vehicle, refer to the precautions.

2. Read out fault memory of DME control unit; if necessary, work through test schedules

3. Switch ignition **OFF**.

4. Remove the exhaust system.

5. Disconnect the plug connection from the sensors.

6. Disconnect the oxygen sensor from cylinders 4 to 6.

7. The oxygen sensor at cylinders 1 to 3 is accessible from above without the exhaust system having to be removed.

#### To install:

➡ The threads of a new oxygen sensors are already coated with an anti-seize compound. If an oxygen sensor is to be used again, apply a thin and even coat of an anti-seize compound to the thread only.

➡ Do not clean the oxygen sensor section which protrudes into the exhaust line and ensure that it avoids all contact with any lubricants.

➡ Observe cable routing of the oxygen sensor so it doesn't interfere with any other system or the exhaust pipes.

8. Tighten sensor to 37 ft. lbs. (50 Nm).

9. Connect the oxygen sensors and their respective cables.

- The cable color for the sensor that leads to cylinders 1 to 3 is black.
- The cable color for the sensor that leads to cylinders 4 to 6 is gray.

10. Connect the plug connection to the sensors.

11. Check for stored fault messages, rectify faults and clear the fault memory.

#### N62 Engines

See Figures 213 and 214.

1. Before servicing the vehicle, refer to the precautions.

2. Read out fault memory of DME control unit; if necessary, work through test schedules

3. Switch ignition **OFF**.

4. Remove the reinforcement plate.

5. Unlock and detach the plug connection.

6. Disconnect the control sensor cable.

7. Remove the sensor.

#### To install:

➡ The threads of a new oxygen sensors are already coated with an anti-seize compound. If an oxygen sensor is to be used again, apply a thin and even coat

of an anti-seize compound to the thread only.

➡ Do not clean the oxygen sensor section which protrudes into the exhaust line and ensure that it avoids all contact with any lubricants.

➡ Observe cable routing of the oxygen sensor so it doesn't interfere with any other system or the exhaust pipes.

8. Replace the sensor, the plug connection and the sensor cable.

9. For the right sensor, to prevent fraying, make sure that the wiring harness is correctly placed on heat shield.

10. Check for stored fault messages, rectify faults and clear the fault memory.

## KNOCK SENSOR (KS)

### LOCATION

See Figures 215 and 216.

### REMOVAL & INSTALLATION

#### N52 and N54 Engines

➡ No steel bolts or bolts may be used due to the threat of electrochemical corrosion. A magnesium crankcase

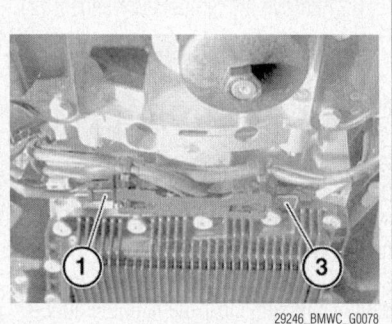

Fig. 213 The location of the plug connection (3) and the sensor cable (1)

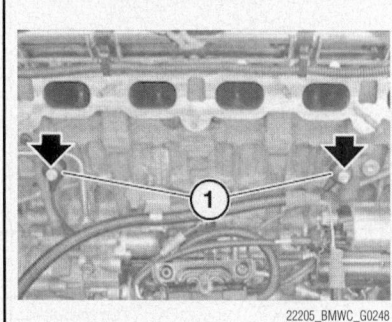

Fig. 215 KS sensor location—N52 and N54 engine

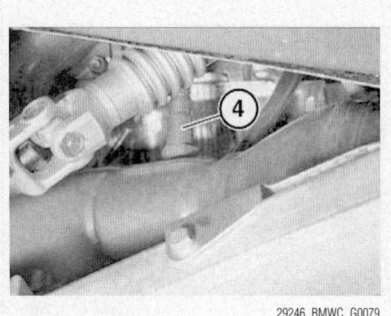

Fig. 212 HO2S sensor (4) location—N62 engine

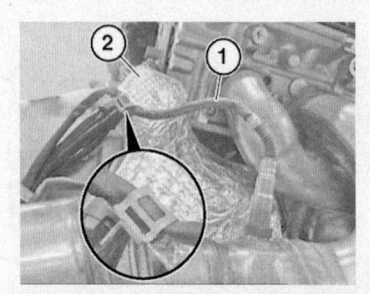

Fig. 214 For the right sensor, to prevent fraying, make sure that the wiring harness (1) is correctly placed on heat shield (2)

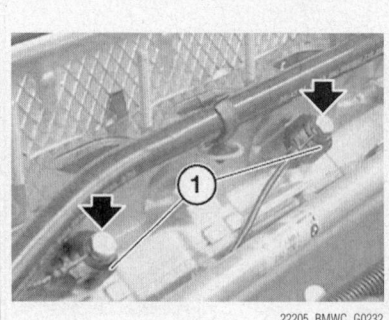

Fig. 216 KS sensor location—N62 engine right side shown, left side similar

**requires aluminum bolts and bolts exclusively. The end of the aluminum bolts and bolts are painted blue for the purposes of reliable identification.**

1. Before servicing the vehicle, refer to the precautions.
2. Read out fault memory of DME control unit; if necessary, work through test schedules
3. Switch ignition **OFF**.
4. Disconnect the battery.
5. Remove the air intake manifold.
6. Unlock the plug connection and remove it.
7. Unscrew the bolts on both knock sensors and remove the sensors.

*To install:*
8. Clean the support face of the knock sensors on engine block.
9. Install the bolts on both knock sensors to replace the sensors.
10. Check for stored fault messages, rectify faults and clear the fault memory.

### N62 Engines

1. Before servicing the vehicle, refer to the precautions.
2. Read out fault memory of DME control unit; if necessary, work through test schedules
3. Switch ignition **OFF**.
4. Disconnect the negative battery cable.
5. Remove the air intake manifold.

### ❋❋ WARNING

**Fitting the knock sensors incorrectly during the assembly will cause engine damage.**

6. Disconnect the plug-in wires to the sensors
7. Remove the bolts that retain the sensors.
8. Remove the sensors

*To install:*
9. Replace the sensors
10. Tighten the bolts and attach the plug-in connections.
11. Check for stored fault messages, rectify faults and clear the fault memory.

### MASS AIR FLOW (MAF) SENSOR

*LOCATION*

See Figures 217 and 218.

*REMOVAL & INSTALLATION*

### N52 and N54 Engines

1. Before servicing the vehicle, refer to the precautions.

2. Read out fault memory of DME control unit; if necessary, work through test schedules
3. Switch ignition **OFF**.
4. Remove the bolts that hold the sensor in place.
5. Unlock the plug and remove it.
6. Pull the mass airflow sensor out of the upper section of the intake filter housing.

*To install:*
7. Replace the sensor, screw it down and connect the plug-in connector.
8. Check stored fault messages.
9. Rectify faults.
10. Check for stored fault messages, rectify faults and clear the fault memory.

### N62 Engine

1. Before servicing the vehicle, refer to the precautions.
2. Read out fault memory of DME control unit; if necessary, work through test schedules
3. Switch ignition **OFF**.
4. Open the clips to the top section of the intake filter housing.

5. Remove the connector by unlocking it and pulling it aside.
6. Unscrew the two bolts holding the mass airflow sensor to the upper section of the intake filter housing.

*To install:*
7. Install the sensor and tighten down the two bolts.
8. Replace the connector.
9. Replace the cover to the intake filter housing and fasten the clips.
10. Check for stored fault messages, rectify faults and clear the fault memory.

### THROTTLE POSITION SENSOR (TPS)

*LOCATION*

See Figures 219 and 220.

*REMOVAL & INSTALLATION*

1. Before servicing the vehicle, refer to the precautions.

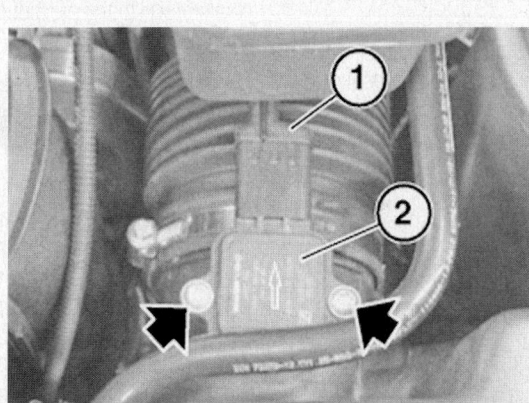

**Fig. 217  MAF/IAT sensor location–N52 and N54 engines**

29246_BMWC_G0065

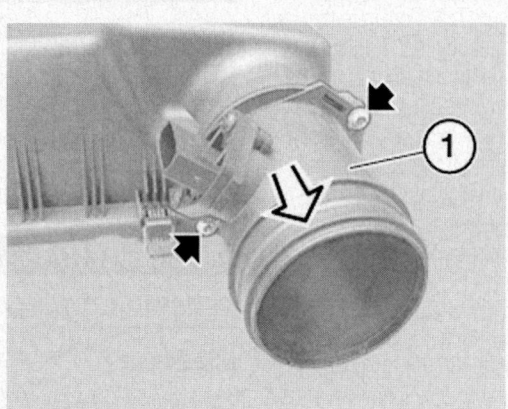

**Fig. 218  MAF/IAT sensor location–N62 engine**

22205_BMWC_G0231

**Fig. 219 TPS sensor (2) and connector (1) location–N52 and N54 engines**

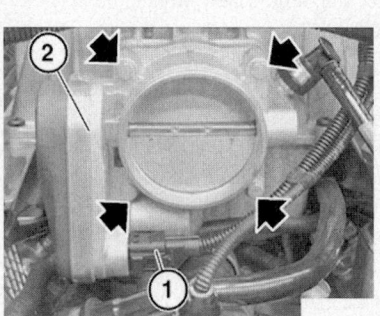

**Fig. 220 TPS sensor (2) and connector (1) location–N62 engine**

2. Read out fault memory of DME control unit; if necessary, work through test schedules
3. Switch ignition **OFF**.
4. Remove air intake hose.
5. Unlock plug and remove.
6. Remove throttle assembly.

### To install:

7. Replace sealing ring of throttle assembly.
8. Install throttle assembly. Tighten bolts to 6 ft. lbs. (9 Nm).
9. Install air intake hose.
10. Check for stored fault messages, rectify faults and clear the fault memory.

## FUEL                    GASOLINE FUEL INJECTION SYSTEM

### FUEL SYSTEM SERVICE PRECAUTIONS

Safety is the most important factor when performing not only fuel system maintenance but any type of maintenance. Failure to conduct maintenance and repairs in a safe manner may result in serious personal injury or death. Maintenance and testing of the vehicle's fuel system components can be accomplished safely and effectively by adhering to the following rules and guidelines.

• To avoid the possibility of fire and personal injury, always disconnect the negative battery cable unless the repair or test procedure requires that battery voltage be applied.

• Always relieve the fuel system pressure prior to disconnecting any fuel system component (injector, fuel rail, pressure regulator, etc.), fitting or fuel line connection. Exercise extreme caution whenever relieving fuel system pressure to avoid exposing skin, face and eyes to fuel spray. Please be advised that fuel under pressure may penetrate the skin or any part of the body that it contacts.

• Always place a shop towel or cloth around the fitting or connection prior to loosening to absorb any excess fuel due to spillage. Ensure that all fuel spillage (should it occur) is quickly removed from engine surfaces. Ensure that all fuel soaked cloths or towels are deposited into a suitable waste container.

• Always keep a dry chemical (Class B) fire extinguisher near the work area.

• Do not allow fuel spray or fuel vapors to come into contact with a spark or open flame.

• Always use a back-up wrench when loosening and tightening fuel line connection fittings. This will prevent unnecessary stress and torsion to fuel line piping.

• Always replace worn fuel fitting O-rings with new Do not substitute fuel hose or equivalent where fuel pipe is installed.

Before servicing the vehicle, make sure to also refer to the precautions in the beginning of this section as well.

### RELIEVING FUEL SYSTEM PRESSURE

To relieve the pressure in the system, locate fuel pump relay located on the cowl. The relay can sometimes be distinguished by the orange color of the housing. Unplug and remove the relay, and place it in a safe location. With the fuel pump relay removed, start the engine and operate it until it stalls. Crank the engine for 10 seconds after it stalls to remove any residual pressure.

### FUEL FILTER

#### REMOVAL & INSTALLATION

The fuel filter is an integral component of the fuel level/fuel pump assembly in the fuel tank and is not normally serviced. Refer to "Gasoline Fuel Injection System, Fuel Pump, Removal & Installation."

### FUEL INJECTORS

#### REMOVAL & INSTALLATION

#### N52 Engine

*See Figures 221 through 223.*

1. Before servicing the vehicle, refer to the precautions.

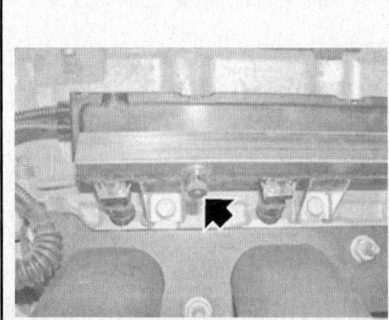

**Fig. 221 Connect compressed air line to compressed air valve. Blow fuel back into tank with a short blast of compressed air maximum of 43 PSI (3 bar)**

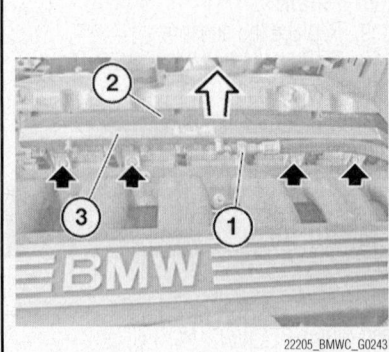

**Fig. 222 Unlock and detach fuel line (1), detach connector strip (2) in direction of arrow, remove injection pipe (3).**

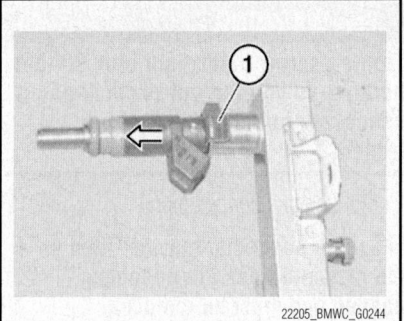

**Fig. 223 Pry out retainers (1) and pull fuel injectors out of injection pipe**

2. Read out fault memory of DME control unit; if necessary, work through test schedules

3. Switch ignition **OFF**.

4. Remove clean air pipe.

5. Remove ignition coil cover.

6. If necessary, unclip plug connection from holder and disconnect. Unclip wiring harnesses from holder and connector strip. Disconnect holder from injection pipe.

7. Remove protective cap from compressed air valve. Connect compressed air line to compressed air valve. Blow fuel back into tank with a short blast of compressed air maximum of 43 PSI (3 bar).

8. Unlock and detach fuel line.

9. Disconnect connector strip in direction of arrow.

10. Remove injection pipe.

11. Seal fuel hose with special tool 13 5 281.

12. Pry out retainers and pull fuel injectors out of injection pipe.

### To install:

13. Replace sealing rings on fuel injectors and coat with anti-friction rubber coating.

14. Install injectors and capture with retainers.

15. Remove special tool 13 5 281.

16. Install injection pipe.

17. Connect connector strip in reverse direction of arrow.

18. Connect fuel line and lock.

19. Install ignition coil cover.

20. Install clean air pipe.

21. Check for stored fault messages, rectify faults and clear the fault memory.

### N54 Engine

*See Figures 224 through 230.*

1. Before servicing the vehicle, refer to the precautions.

2. Read out fault memory of DME control unit; if necessary, work through test schedules

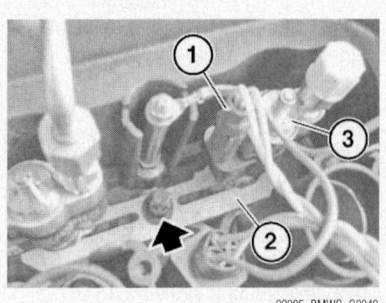

**Fig. 224 Unlock plug (1), remove holding-down element (2) and remove injector (3)**

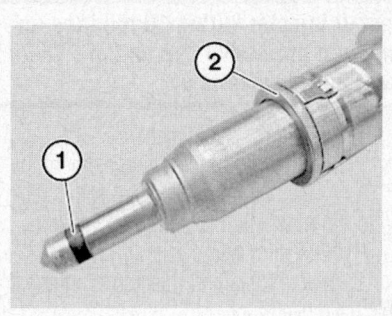

**Fig. 225 Uncoupling element (2) and the PTFE sealing ring (1)**

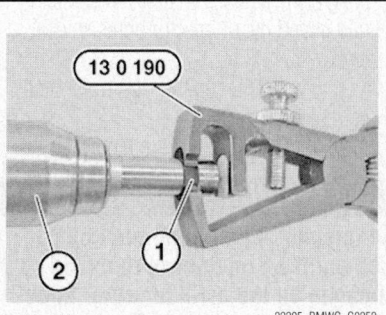

**Fig. 226 Remove PTFE sealing ring (1) with special tool 13 0 191 from injector (2)**

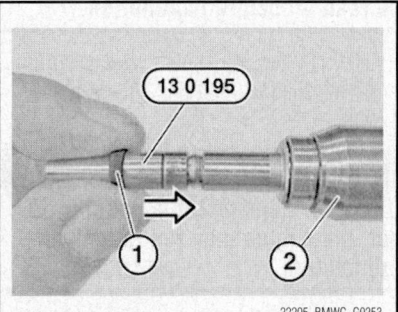

**Fig. 227 Use fingers and mounting taper 13 0 195 to slide PTFE sealing ring (1) onto injector (2)**

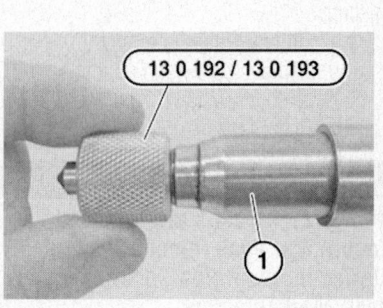

**Fig. 228 To bring the expanded PTFE sealing ring to its installation dimension, slide three mounting sleeves with decreasing diameters onto the injector**

**Fig. 229 The adjustment value is printed in two blocks of three digits on the injector. The adjustment value must be read off before installation**

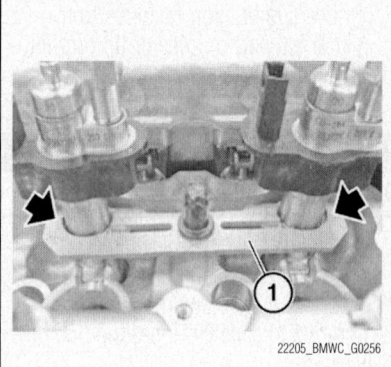

**Fig. 230 Make sure holding-down element (1) is correctly seated**

3. Switch ignition **OFF**.

4. Remove the pressure line at the injector.

   a. Unlock plug and remove.

   b. Remove screw and remove holding-down element.

   c. Remove injector.

### ✴✴ WARNING

**If several injectors are removed, ensure that each injector is reinstalled in its original location (cylinder). Mark injectors.**

5. If the injector is stuck in its bore, perform the following procedure to free it:

   a. Mount special tool 13 0 180 on injector.

   b. Mount special tool 13 5 250.

   c. Carefully knock out injector with special tools 13 0 180 and 13 5 250.

   d. After removing, fit protective caps to injector tip and fuel line connection.

### To install:

6. Installing a used fuel injector:

   a. Replace uncoupling element and the PTFE sealing ring.

### ✴ WARNING

**A PTFE seal which has been heated once by engine operation must be replaced before the fuel injector is reinstalled.**

   b. Before replacing PTFE sealing ring, make sure hands and work surface are clean and free of oil. Avoid mechanical contact with injector tip.

   c. Remove PTFE sealing ring with special tool 13 0 191 from injector.

   d. Use a lint-free cloth only to remove combustion residues from cylindrical part of injector tip (do not use ultrasound or other tools/agents).

### ✴ WARNING

**Do not clean injector tip.**

   e. Slide new PTFE sealing ring onto mounting taper 13 0 195.

   f. Use fingers and mounting taper 13 0 195 to slide PTFE sealing ring onto injector.

➡**Do not use fingernails to slide PTFE sealing ring on. Do not use any lubricating agents. The sealing ring is expanded when slid on.**

   g. To bring the expanded PTFE sealing ring to its installation dimension, slide three mounting sleeves with decreasing diameters onto the injector.

   h. Slide mounting sleeve with large opening first onto injector. Do not use any lubricating agents.

   i. First slide mounting sleeve 13 0 192 (large diameter) onto injector. Then slide mounting sleeve 13 0 193 (medium diameter) onto injector. Finally, press injector into mounting sleeve 13 0 194 (small diameter).

### ✴✴ WARNING

**Install injector within 10 minutes or slide on protective cap as the PTFE sealing ring swells up.**

7. Installing a new injector:

   a. Use a new uncoupling element.

   b. Remove protective cap from injector tip max. 10 min. before installation (PTFE sealing ring swells up).

8. Before installing injector in engines that have been run:

   a. Clean contact surfaces of uncoupling elements in cylinder head.

   b. Clean injector bore: To do so, preferably slide injector without uncoupling element but with new PTFE sealing ring in and out of injector bores several times.

   c. The PTFE sealing ring must then be replaced.

   d. Replace uncoupling element.

### ✴✴ WARNING

**An injector adjustment must be carried out if an injector is replaced or changed on the cylinder side. Injector adjustment is carried out with the aid of a so-called adjustment value. The adjustment value is printed in two blocks of three digits on the injector. The adjustment value must be read off before installation.**

9. Enter the adjustment value according to the installation position (cylinder) of the injector.

### ✴✴ WARNING

**If injector adjustment is not carried out, the engine may run roughly or fail to start.**

10. Install injector and holding-down element. Make sure holding-down element is correctly seated.

### ✴✴ WARNING

**Tighten screw hand-tight only so that holding-down element is slack and if necessary injector can still be turned.**

11. Connect contact plug.

➡**Copper seals that may be fitted on the pressure lines are no longer needed and must be removed.**

12. Connect pressure line, tightening nuts hand-tight only in the process.

### ✴✴ WARNING

**Connect pressure line without tension only.**

13. To ensure distortion-free installation of the pressure line and to avoid damaging the thread, it must be possible for both nuts to be screwed on easily by hand.

14. If the nuts cannot be screwed on easily by hand, the injector must if necessary be turned a little.

15. Turn injector if necessary in direction of arrow until nuts on pressure line can be easily screwed on by hand.

16. Tighten nuts on pressure line hand-tight.

17. Then tighten down screw for holding-down element to 10 ft. lbs. (13 Nm).

18. Adhere to tightening sequence:
- First tighten down nut on injector
- Then tighten down nut on high-pressure rail

19. When tightening nut on injector, grip hexagon head of injector with wrench.

20. Coat screw connection with transmission oil. Tighten down nuts with special tool 37 1 151 to 19 ft. lbs. (25 Nm).

21. Using a BMW DIS Tester perform the injector adjustment:

22. Check fuel system for leaks.

23. Check for stored fault messages, rectify faults and clear the fault memory.

### N62 Engine

*See Figures 231 through 233.*

1. Before servicing the vehicle, refer to the precautions.

2. Read out fault memory of DME control unit; if necessary, work through test schedules

3. Switch ignition **OFF**.

4. Remove the acoustic cover.

5. Unlock plugs on servomotors and disconnect.

6. Disconnect plug connections of knock sensors from cable strips.

7. Unclip cable.

**Fig. 231 Unlock plugs (1) on servomotors and disconnect. Disconnect plug connections (2) of knock sensors from cable strips. Unclip cable (3)**

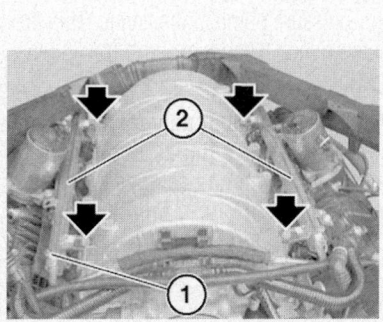

**Fig. 232 Remove protective cap from compressed air connection (1), and remove both injection pipes (2) towards top**

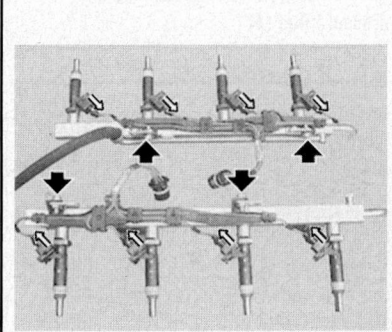

**Fig. 233 Injector rail and injector assembly**

8. Unlock and disconnect following plug connections.

- Fuel rail, left and right, on cable duct
- Throttle-valve assembly
- All solenoid valves
- Both temperature sensors
- Oil-pressure switch
- Tank venting valve
- Alternator
- Differential pressure sensor
- Servomotor on intake air manifold
- Pulse generator on left exhaust and inlet camshafts
- Left eccentric shaft sensor

9. Remove brackets, place cable ducts to one side.

10. Remove protective cap from compressed air connection. Blow fuel back into tank with a short blast of compressed air 44 PSI (3 bar) at compressed air connection.

11. Remove both injection pipes towards top.

12. Unlock and disconnect fuel hose. Seal fuel hose with special tool 13 5 281.

13. Unlock connector and remove.

14. Lever out retainers.

15. Pull fuel injectors out of injection pipe.

### To install:

16. Replace sealing rings on fuel injectors and coat with anti-friction agent.

17. Install fuel injectors in injection pipe and push in retainers.

18. Install the connector and lock it.

19. Connect fuel hose and lock it.

20. Install both injection pipes.

21. Replace cable ducts and install brackets.

22. Connect and lock following plug connections.

- Left eccentric shaft sensor
- Pulse generator on left exhaust and inlet camshafts
- Servomotor on intake air manifold
- Differential pressure sensor
- Alternator
- Tank venting valve
- Oil-pressure switch
- Both temperature sensors
- All solenoid valves
- Throttle-valve assembly
- Fuel rail, left and right, on cable duct

23. Install cable in clip.

24. Connect knock sensors to cable strips.

25. Connect and lock plugs on servomotors.

26. Install the acoustic cover.

27. Check for stored fault messages, rectify faults and clear the fault memory.

## FUEL PUMP

### REMOVAL & INSTALLATION

#### N52 and N54 Engine

*See Figure 234.*

The fuel pump is mounted through the top of the fuel tank along with the fuel level sending unit. The fuel tank should not be filled more than ⅓ of the total fuel tank capacity to prevent fuel leakage during fuel pump removal. If the fuel tank is filled beyond this level, the fuel level must be reduced using an approved fuel removal device.

1. Before servicing the vehicle, refer to the precautions.

2. Relive the fuel system pressure and disconnect the negative battery cable.

3. Drain the fuel, if filled beyond ⅓ of the capacity of the fuel tank. Drain the fuel tank enough to prevent spillage when removing the pump using an approved fuel removal device.

➡**The fuel pump must be removed through the top of the fuel tank, thus the location of the fuel tank determines whether the fuel pump is accessed by removal of the rear seat, or removal of the trim panels in the trunk.**

4. Remove the rear seat, or the trim panels in the trunk, depending on fuel tank location to access the top of the fuel tank

➡**On models which require removal of the rear seat, the insulation mat under the seat must be cut in a "U" shape to allow the insulation to be folded up to access the top of the fuel tank.**

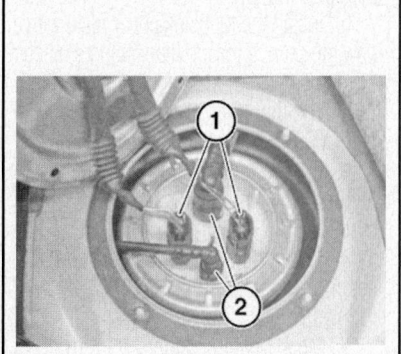

**Fig. 234 Disconnect plug connection (1), unlock and detach vent lines (2)**

5. Remove the fasteners securing the metal cover located above the fuel tank, and remove the cover.

6. Remove the electrical connector at the top of the combination fuel pump and fuel level sending unit assembly.

7. Remove the fuel feed and return lines.

8. Match mark the combination fuel pump and fuel level sending unit assembly to the fuel tank to ensure proper installation during reassembly.

9. Remove the fasteners or fastener securing the combination fuel pump and fuel level sending unit assembly to the fuel tank. The fasteners are one of 2 types.

10. If the fuel pump assembly is fastened to the fuel tank with a series of 6 mm nuts:

   a. Loosen the nuts evenly using a crisscross sequence and carefully lift the cover and place aside.

   b. Compress the large plastic tongue to remove the fuel pump, and lift the pump along with the fuel sending unit out of the fuel tank.

11. If the fuel pump assembly is fastened to the fuel tank with a large sealing ring:

   a. Use tool No. 16-1-020 to loosen the sealing ring in a counterclockwise direction.

   b. With the seal ring removed, lift the fuel pump assembly out of the fuel tank.

**To install:**

➡ **Always use a new seal or gasket when installing the fuel pump or fuel level gauge sending unit assembly.**

12. Install the fuel pump into the fuel tank taking care not to bend or damage the fuel sending unit assembly.

13. If the fuel pump is held in place by a plastic bracket in the fuel tank perform the following:

   a. Make sure the fuel pump is fully snapped in place.

   b. Install the fuel tank cover plate with a new gasket and torque the fasteners using a crisscross pattern to 57 inch lbs. (6.5 Nm).

14. If the fuel pump is held in place with a sealing ring perform the following:

   a. Ensure the pump is properly aligned with the fuel tank matchmarks made during disassembly.

   b. Install a new seal and torque the sealing ring using tool No. 16-1-020 as follows:

   • Metal sealing rings: 26 ft. lbs. (35 Nm)

   • Plastic sealing rings: 41 ft. lbs. (55 Nm)

15. The balance of the assembly is in reverse order of disassembly.

16. Connect the negative battery cable.

17. Once the vehicle is started, check for leaks. If a strong fuel odor is present, or any fuel leakage is noted, stop the engine immediately and repair as necessary.

### N54 Engine (High Pressure Pump)

*See Figures 235 through 238.*

1. Before servicing the vehicle, refer to the precautions.

2. Disconnect battery negative terminal.

⁂ **WARNING**

**Electric fuel pump starts up automatically each time door is opened.**

3. Remove intake air manifold.

4. Unlock plug and remove.

5. Unlock and detach fuel line.

6. Catch and dispose of escaping fuel.

7. Seal fuel lines with special tools 13 5 281 and 13 5 282.

8. Remove nut, remove bolt and remove feed line.

9. Seal feed line with matching plug from special tool kit 32 1 270.

⁂ **WARNING**

**Wear full face guard and protective gloves.**

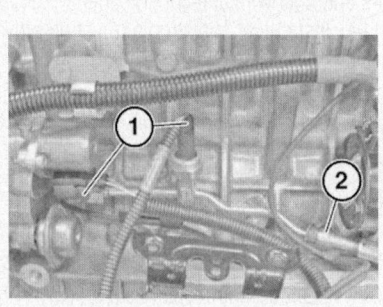

**Fig. 235  Unlock plug (1) and remove. Unlock and detach fuel line (2)**

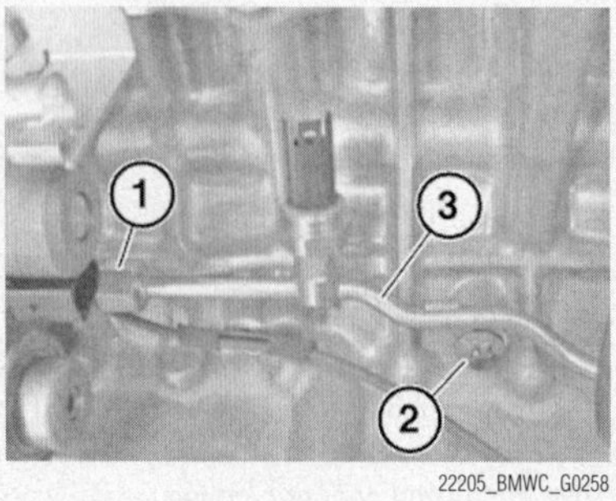

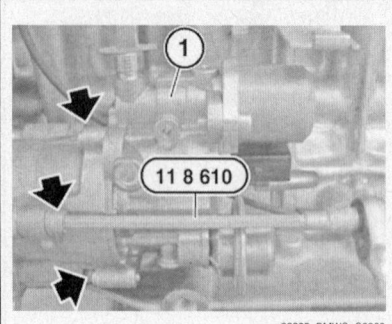

**Fig. 236  Remove nut (1), remove bolt (2) and remove feed line (3)**

10. Remove nut, slacken nut, unscrew bolt and disconnect high-pressure line.

11. Seal connections of high-pressure line with matching plugs from special tool kit 32 1 270.

12. Seal fuel line connections of high-pressure pump with matching plugs from special tool kit 32 1 270.

13. Remove screws with special tool 11 8 610.

14. Disconnect high-pressure pump and remove.

➡ **Engine oil can escape when pump is detached; have a cleaning cloth ready.**

*To install:*

15. Replace sealing ring and clean contact faces.

16. When installing, turn high-pressure pump until bores for screws are flush.

17. Tighten screws of high-pressure pump with special tool 11 8 610 hand-tight only. It must still be possible to turn the high-pressure pump at the flange. This prevents twisting when high-pressure line is tightened.

➡ **Copper seals that may be fitted on the high-pressure line are no longer needed and must be removed. If reusing pressure line, lightly grease**

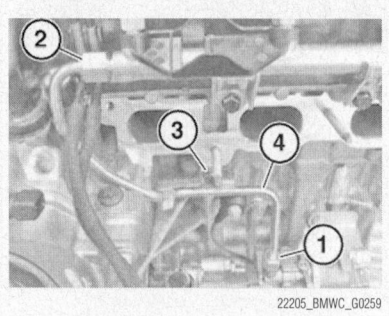

**Fig. 237  Remove nut (1), slacken nut (2), unscrew bolt (3) and disconnect high-pressure line (4)**

**Fig. 238  Remove screws with special tool 11 8 610**

threads of pressure connections. Threads of new pressure lines are already coated.

18. Pre-install high-pressure line. Tighten screw connections hand-tight only. You must still be able to move the high-pressure line at the holder.

19. Only when high-pressure pump has if necessary turned into position and is thus pre-installed without twisting with high pressure-line should screws of high-pressure pump be tightened down with special tool 11 8 610. Tighten to 7 ft. lbs. (9 Nm).

20. Follow sequence of screw connections:
  • Coat with transmission oil. Tighten down nut with special tool 13 5 020 to 22 ft. lbs. (30 Nm).
  • Coat with transmission oil. Tighten nut to 22 ft. lbs. (30 Nm).
  • Tighten down screw for holder to 10 ft. lbs. (13 Nm).

➡Feed lines with soldered holder must be replaced by new version with Elastomer clamp holder. Observe installation position of fuel low-pressure sensor. Lines for solenoid switches are laid behind fuel line.

21. Install feed line. In so doing, tighten screw connection hand-tight only

22. Follow sequence of screw connections:
  • Coat with transmission oil and tighten down nut with special tool 13 5 020 to 22 ft. lbs. (30 Nm).
  • Replace aluminum screw, coat with transmission oil and tighten to 8 ft. lbs. (10 Nm) plus an additional 90° of rotation.

23. Once the vehicle is started, check for leaks. If a strong fuel odor is present, or any fuel leakage is noted, stop the engine immediately and repair as necessary.

### N62 Engine

*See Figure 239.*

1. Before servicing the vehicle, refer to the precautions.

2. Remove left underbody paneling.

3. Seal off water hoses at fuel pump with special tool 13 3 010.

4. Unfasten plug connection and disconnect.

5. Remove hose clamps and detach fuel hoses.

6. Unscrew bolt.

7. Remove fuel pump in direction of arrow.

8. Disconnect fuel pump from fuel pump holder.

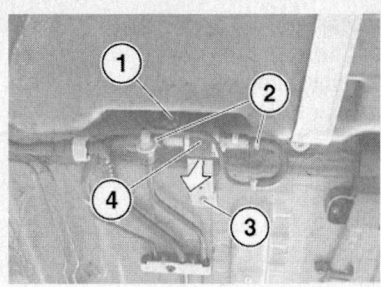

22205_BMWC_G0264

**Fig. 239 Plug connection (1), hose clamps (2), bolt (3) and fuel pump (4)**

*To install:*

9. Installation is the reverse of removal.

10. Observe direction of flow of fuel pump when installing.

## FUEL TANK

*REMOVAL & INSTALLATION*

### 3-Series

1. Before servicing the vehicle, refer to the precautions.

2. Relive the fuel system pressure and disconnect the negative battery cable.

3. Drain the fuel tank enough to prevent spillage when removing the tank.

4. Remove rear seat bench.

5. Remove guide tube for handbrake cables.

6. Remove right wheel arch trim.

7. Remove rear left and right underbody paneling.

8. Remove left and right underbody cover.

9. Remove screws and remove cover from left and right sides of fuel tank.

10. Remove the fuel sending unit.

11. Remove hose clamp and detach fuel filler hose from fuel filler pipe.

12. Unlock quick-release fastener and detach vent line.

13. Disconnect vent line from holder.

14. Support the fuel tank using a transmission jack.

15. Unfasten and discard the fuel tank-to-body self-locking nut.

16. Remove screws for tightening straps on left and right and remove tightening straps.

17. Carefully lower fuel tank.

**✳✳ WARNING**

**Carefully feed the vent line through the body when lowering the fuel tank.**

*To install:*

➡Note rubber mount with spacer bush. Wide collar on spacer bush points to screw head.

18. Carefully raise fuel tank.

**✳✳ WARNING**

**Carefully feed the vent line through the body when raising the fuel tank.**

19. Install tightening straps and tighten screws to 14 ft. lbs. (19 Nm).

20. Install a new fuel tank-to-body self-locking nut and tighten to 14 ft. lbs. (19 Nm).

21. Connect vent line to holder.

22. Connect vent line and lock quick-release fastener.

23. Connect fuel filler hose to fuel filler pipe and secure hose clamp.

24. Install the fuel sending unit.

25. Install cover from left and right sides of fuel tank.

26. Install left and right underbody cover.

27. Install rear left and right underbody paneling.

28. Install right wheel arch trim.

29. Install guide tube for handbrake cables.

30. Install rear seat bench.

### 5 and 7-Series

1. Before servicing the vehicle, refer to the precautions.

2. Relive the fuel system pressure and disconnect the negative battery cable.

3. Drain the fuel tank enough to prevent spillage when removing the tank.

4. Remove rear right wheel arch trim.

5. Remove complete propeller shaft.

6. Remove handbrake Bowden cables from wheel carrier and unclip from fuel tank.

7. Remove right strut for rear axle.

8. Unfasten hose clip.

9. Disconnect filler vent line from fuel filler pipe.

10. Disconnect service vent line from carbon canister.

11. Unclip both lines and remove retaining clip.

12. Unlock and disconnect plug connections of feed line.

13. If necessary, disconnect line for independent heating.

14. Remove screw.

15. Support the fuel tank with a transmission jack.

16. Remove screws for tightening straps on left and right and remove tightening straps.

17. Lower tank until plug on top side of right tank half is accessible.

18. Unlock plug and detach from delivery unit.

➡ **Get a second person to feed out vent lines to wheel arch while removing tank towards bottom.**

19. Feed vent lines and through body.

### ✸✸ WARNING
**Do not kink lines.**

*To install:*

20. Installation is the reverse of removal.

➡ **Note rubber mount with spacer bush. Wide collar on spacer bush points to screw head.**

21. Carefully raise fuel tank.

### ✸✸ WARNING
**Carefully feed the vent line through the body when raising the fuel tank.**

22. Tightening strap screws to 14 ft. lbs. (19 Nm).

23. Install a new fuel tank-to-body self-locking nut and tighten to 14 ft. lbs. (19 Nm).

24. Tighten the fuel filler-to-body nut to 2 ft. lbs. (3 Nm). Seal nut with under seal.

### IDLE SPEED

#### ADJUSTMENT

Idle speed is maintained by the by the Digital Motor Electronics (DME). No adjustment is necessary or possible.

### THROTTLE BODY

#### REMOVAL & INSTALLATION

1. Before servicing the vehicle, refer to the precautions.

2. Read out fault memory of DME control unit; if necessary, work through test schedules

3. Switch ignition **OFF**.

4. Remove air intake hose.

5. Unlock plug and remove.

6. Remove throttle assembly.

*To install:*

7. Replace sealing ring of throttle assembly.

8. Install throttle assembly. Tighten bolts to 6 ft. lbs. (9 Nm).

9. Install air intake hose.

10. Check for stored fault messages, rectify faults and clear the fault memory.

## HEATING & AIR CONDITIONING SYSTEM

### BLOWER MOTOR

#### REMOVAL & INSTALLATION

**3-Series**

*See Figure 240.*

1. Before servicing the vehicle, refer to the precautions.

2. Remove trim for instrument panel, bottom right

3. Partially release air duct at side

4. Disconnect plug connection.

5. Carefully raise lug and feed out housing by turning clockwise.

*To install:*

6. Install housing into vehicle.

➡ **Lug must not be damaged and must snap audibly into place.**

7. Connect plug.

8. Connect air duct.

9. Install instrument panel trim.

**5 and 7-Series**

*See Figures 241 through 245.*

1. Before servicing the vehicle, refer to the precautions.

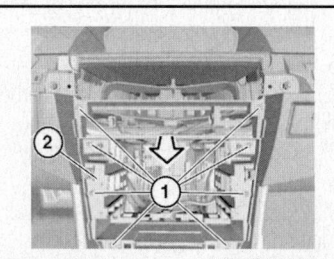

22205_BMWC_G0265

**Fig. 241 Remove screws (1) and remove middle function carrier (2) in direction of arrow**

2. Remove center console.

3. Remove audio system controller / Car Communication Computer.

4. Remove trim panel for pedal assembly.

5. Remove right glove box with housing.

6. Disconnect rear compartment air duct by moving it toward the rear of the vehicle.

7. Remove screws.

8. Remove middle function carrier in direction of arrow.

9. If necessary, remove temperature sensor for cold-air distributor.

10. Unfasten plug connection and disconnect.

11. Remove clips.

12. Unfasten plug connection and disconnect.

13. Remove rear compartment air duct in direction of arrow.

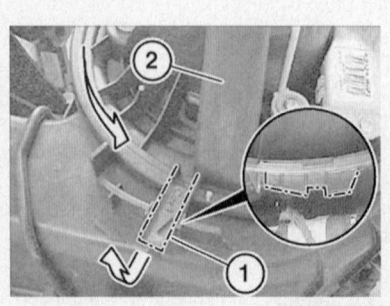

22205_BMWC_G0279

**Fig. 240 Carefully raise lug (1) and feed out housing (2) by turning clockwise**

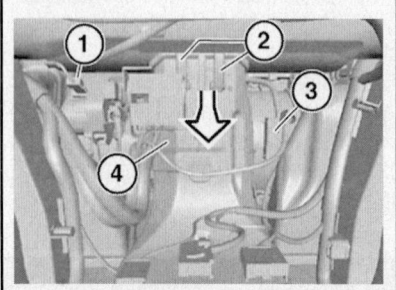

22205_BMWC_G0266

**Fig. 242 Plug connection (1), clips (2), plug connection (3) and rear compartment air duct (4)**

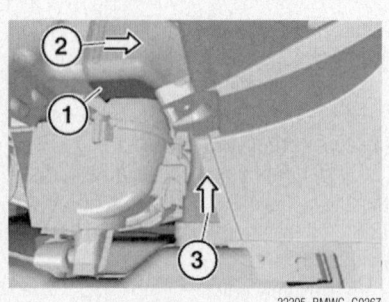

22205_BMWC_G0267

**Fig. 243 Disconnect air ducts (1) in direction of arrow (2) and (3)—left shown, right similar**

14. Disconnect air duct in direction of arrows.

15. Unlock plug (1) and remove.

16. Remove screws (2).

17. Remove cover (3) in direction of arrow.

18. Remove screws (1).

19. Remove fan for heater - A/C unit (2) in direction of arrow.

### To install:

20. Installation is the reverse of removal noting the following:

21. Make sure heater - A/C unit fan is correctly seated.

22. Connect and lock the plug connections

23. Ensure correct cable routing.

24. Make sure seal is correctly seated in cover.

25. Install cover in reverse direction of arrow.

26. Connect air ducts in reverse direction of arrow.

27. Make sure right air duct is correctly seated.

28. Install rear compartment air duct in reverse direction of arrow.

29. Make sure rear compartment air duct is correctly seated.

30. Remove middle function carrier in reverse direction of arrow.

31. Make sure middle function carrier is correctly seated.

32. Ensure correct cable routing.

## HEATER CORE

### REMOVAL & INSTALLATION

#### 3-Series

*See Figures 246 through 249.*

1. Before servicing the vehicle, refer to the precautions.

➡**The heater case assembly must be removed to remove the heater core on all 3-Series vehicles.**

2. Disconnect the negative battery cable.

3. Drain the cooling system into a clean container for reuse.

4. Discharge and recover the air conditioning system refrigerant.

5. Remove trim panel for pedal assembly.

6. Remove heater/air conditioner assembly.

7. Remove lower section of micro filter housing.

8. Remove expansion valve.

9. Remove instrument panel trim.

10. Remove complete steering column.

11. Disconnect plug connection for control unit for Car Access System.

12. Remove cover cap from bulkhead.

13. Disconnect left and right foot well heating duct with adapter from heater/air conditioner and remove downward from right rear compartment heating duct.

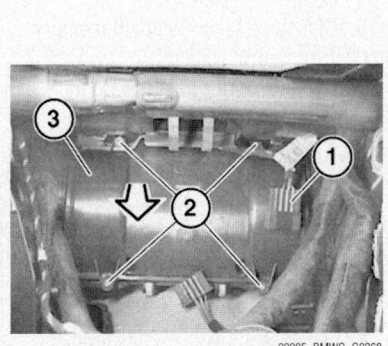

Fig. 244 Plug (1), screws (2) and cover (3)

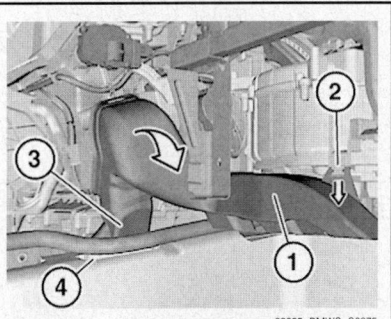

Fig. 246 Disconnect foot well heating duct (1) in area (2) in direction of arrow. Disconnect foot well heating duct (1) with adapter (3) from heater/air conditioner and remove in direction of arrow from rear compartment heating duct (4)

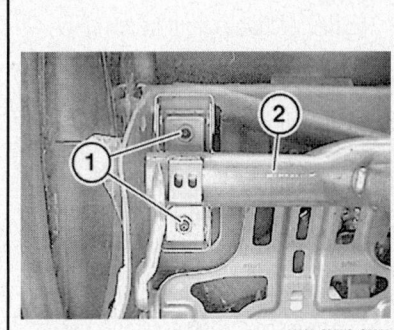

Fig. 248 Remove the support tube fasteners—left side shown, right side similar

Fig. 245 Remove screws (1) and remove fan for heater - A/C unit (2) in direction of arrow

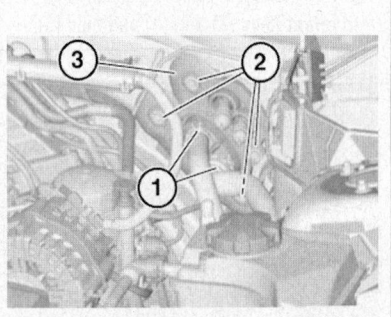

Fig. 247 Unclip refrigerant lines (1), unscrew nuts (2) and feed out media passage sealing plate (3) and associated rubber grommet

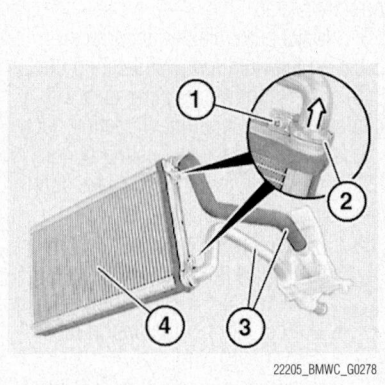

Fig. 249 Heater core assembly as removed from heater case

14. Feed out media passage sealing plate and associated rubber grommet.

### ✳✻ WARNING

**Carefully blow through aluminum double pipe to remove remaining coolant from heat exchanger for heating system.**

15. Disconnect all necessary cable ties on support tube.

16. Remove wiring harnesses/cable ducts and lay to one side.

17. If necessary, remove control units/disconnect plug connections.

18. Remove the support tube fasteners.

19. Disconnect support tube and remove from vehicle with assistance of a second person.

20. Remove heater assembly.

21. Slide rubber grommet with foam seal forward slightly. Carefully feed pipe holder with pipes past rubber grommet.

22. If necessary, remove coolant hose.

23. Remove heater core from heater/air conditioner.

24. Remove pipes from heater core.

### To install:

25. Installation is the reverse of removal, noting the following:

26. When installing the heater core, make sure that the return-flow connection (larger opening) marked with a black dot is positioned at the top. Inflow connection accordingly may only be positioned at the bottom.

27. Make sure heater assembly is correctly seated.

28. Make sure condensate drain is correctly seated in grommet.

29. Make sure wiring harnesses/cable ducts are correctly routed and secured.

30. Makes sure media passage sealing plate and associated rubber grommet are correctly seated.

31. Make sure retainer is correctly seated on left/right foot well heating duct.

32. Insert foot well heating duct with adapter in rear compartment heating duct.

33. Connect foot well heating duct in area to heater/air conditioner. Make sure assembly is locked exactly at point.

34. Tighten fasteners to specification:
- Bulkhead adapter: 9 inch lbs. (1 Nm)
- Air conditioner to support tube: 7 ft. lbs. (9 Nm)
- Heater to support tube: 3 ft. lbs. (4 Nm)
- Support tube to body: 16 ft. lbs. (21 Nm)

- Support tube to support, steering column upper section: 14 ft. lbs. (19 Nm)

### 5 and 7-Series

*See Figures 250 through 254.*

1. Before servicing the vehicle, refer to the precautions.

➡**The heater case assembly must be removed to remove the heater core on all 5 and 7-Series vehicles.**

2. Disconnect the negative battery cable.

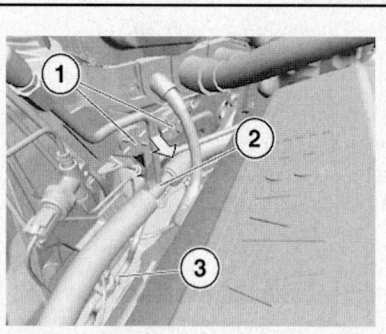

22205_BMWC_G0270

**Fig. 250 Unscrew nuts (1), detach refrigerant line (2) in direction of arrow and unclip refrigerant line (3)**

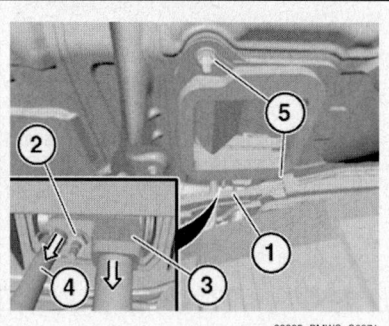

22205_BMWC_G0271

**Fig. 251 Rubber grommet (1), nut (2), refrigerant lines (3) and (4) and nuts (5)**

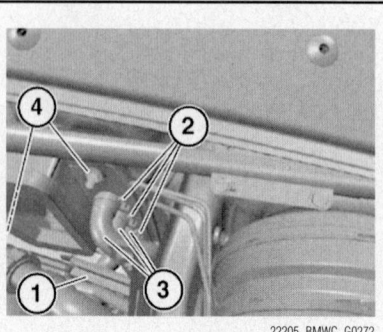

22205_BMWC_G0272

**Fig. 252 Rubber grommet (1), spring clamps (2), coolant hoses (3) and nuts (4)**

3. Drain the cooling system into a clean container for reuse.

4. Discharge and recover the air conditioning system refrigerant.

5. Remove fresh air duct.

6. Remove trim for instrument panel.

7. Remove temperature sensor for cold-air distributor.

8. Disconnect and unclip the refrigerant line.

9. Feed out rubber grommet.

10. Remove nut and detach refrigerant lines and in direction of arrow.

11. Remove nuts on right air duct.

12. Feed out rubber grommet.

13. Remove spring clamps and detach coolant hoses.

### ✳✻ WARNING

**Carefully blow through aluminum triple pipe to remove remaining coolant from heat exchanger for heating system.**

14. Remove nuts on left air duct.

15. Disconnect rear compartment air duct toward rear of vehicle.

16. Remove left air duct toward the left of the vehicle.

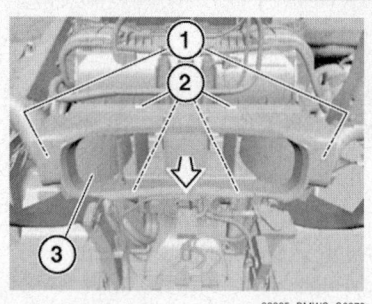

22205_BMWC_G0273

**Fig. 253 Carefully lever out locks (1) and (2) and detach air distributor (3) in direction of arrow**

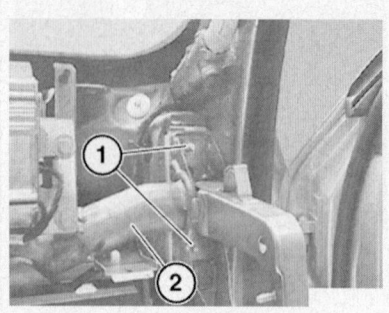

22205_BMWC_G0274

**Fig. 254 Remove bolts (1) and detach support tube (2) and remove from vehicle**

17. Remove right air duct toward the right and upward.

18. Carefully lever out locks and.

19. Disconnect air distributor in direction of arrow.

20. If necessary, remove CD changer.

21. Set fuse-carrier (3) down in foot well.

22. Disconnect support tube (2) and remove from vehicle with assistance of a second person.

23. If necessary, unlock and disconnect plug connections on heater. Remove heater toward the rear of the vehicle.

***To install:***

24. Installation is the reverse of removal, noting the following:

25. When installing heater, make sure left and right air ducts are correctly seated.

26. Check left and right air ducts for leaks.

27. If necessary, reseal left and right air ducts with butylene tape (sourcing reference: BMW Parts Service).

28. If left and right air ducts are resealed, make sure flaps can move freely.

29. Tighten fasteners to specification:
- Support tube-to-body: 16 ft. lbs. (21 Nm)
- Steering column-to-instrument panel / support tube: 16 ft. lbs. (21 Nm)
- Air conditioner-to-bulkhead: 5 ft. lbs. (6 Nm)
- Expansion valve-to- A/C heater: 5 ft. lbs. (6 Nm)
- Refrigerant line-to-assembly compartment partition reinforcement: 6 ft. lbs. (8 Nm)

# STEERING

## POWER STEERING GEAR

### REMOVAL & INSTALLATION

#### 3-Series

*See Figures 255 through 256.*

1. Before servicing the vehicle, refer to the precautions.

2. Draw off and dispose of hydraulic fluid from fluid reservoir

3. Remove front underbody protection.

4. Remove both tie rod ends from ball joint.

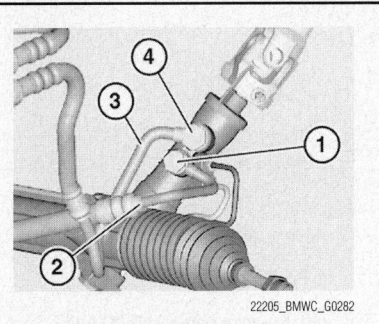

**Fig. 255 Remove banjo bolts (1 and 4), then disconnect pressure line (2) and return line (3) from power steering gear**

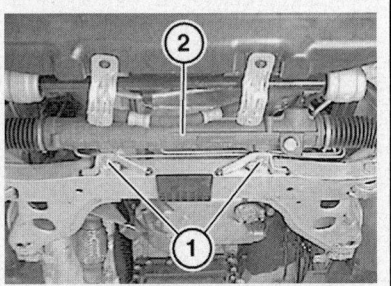

**Fig. 256 Remove nuts and remove screws (1) towards bottom, then remove power steering gear (2) towards front**

5. Remove lower steering spindle from power steering gear

6. If necessary, remove heat shield from power steering gear.

7. Remove banjo bolts and disconnect pressure line and return line from power steering gear.

8. If necessary, remove hydraulic lines with bracket from power steering gear.

9. Remove nuts and remove screws towards bottom.

10. Remove power steering gear towards front.

***To install:***

11. Install power steering gear. Using new bolts and nuts, tighten to 41 ft. lbs. (56 Nm) plus 90° of additional rotation.

12. Install hydraulic lines with bracket to power steering gear.

13. Replace all sealing rings.

14. Connect pressure line and return line from power steering gear. Tighten pressure line to 22 ft. lbs. (30 Nm) and return line to 26 ft. lbs. (35 Nm).

### ✲✲ WARNING

**Make sure hydraulic lines are laid without tension and with sufficient spacing to adjoining components.**

15. Install heat shield on power steering gear.

16. Install lower steering spindle on power steering gear

17. Install both tie rod ends on ball joint.

18. Install front underbody protection.

19. Fill and bleed hydraulic system.

20. Check pipe connections for leaks.

21. Perform chassis alignment check.

22. Carry out steering angle sensor adjustment.

#### 5 and 7-Series

*See Figure 257.*

1. Before servicing the vehicle, refer to the precautions.

2. Disconnect the negative battery cable.

3. Draw off and dispose of hydraulic fluid from fluid reservoir.

4. Secure engine in installation position.

5. Remove both tie rod ends from ball joint.

6. Lower front axle support.

7. Remove left mounting bracket or engine mount from front axle carrier.

8. If necessary, remove right engine mount from front axle carrier.

9. If necessary, remove heat shield from power steering gear.

10. Disconnect pressure line and return line from power steering gear.

11. Remove screw. Remove nuts and remove screws towards bottom.

12. On vehicles with active steering, disconnect plug connections.

13. Raise power steering gear and unclip wiring harness.

14. Move power steering gear to right and remove towards rear.

***To install:***

15. Install power steering gear.

16. Replace screws and self-locking nuts and tighten to specification.

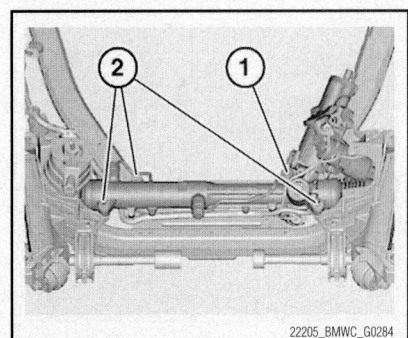

**Fig. 257 Remove screw (1), release nuts (2) and remove screws towards bottom**

- 2WD: 42 ft. lbs. (56 Nm) plus an additional 90° rotation
- AWD: 89 ft. lbs. (120 Nm)

17. Replace all sealing rings.

18. Make sure hydraulic lines are laid without tension and with sufficient spacing to adjoining components. Tighten to 26 ft. lbs. (35 Nm).

➡ **When replacing the power steering gear, the Banjo bolt for pressure line (expansion hose) must be replaced by a banjo bolt with non-return valve.**

19. Fill and bleed hydraulic system.
20. Check pipe connections for leaks.
21. Perform chassis alignment check.
22. Carry out steering angle sensor adjustment/adjustment for active front steering.
23. If necessary, carry out initial Dynamic Drive operation.

### POWER STEERING PUMP

*REMOVAL & INSTALLATION*

#### 3-Series

*See Figures 258 and 259.*

### ✷✷ WARNING

**Aluminum-magnesium materials. No steel screws/bolts may be used due to the threat of electrochemical corrosion. A magnesium crankcase requires aluminum screws/bolts exclusively. Aluminum screws/bolts must be replaced each time they are released. The end faces of aluminum screws/bolts are painted blue for the purposes of reliable identification. Jointing torque and angle of rotation must be observed without fail (risk of damage).**

1. Before servicing the vehicle, refer to the precautions.
2. Draw off and dispose of hydraulic fluid from fluid reservoir.
3. Remove intake filter housing.
4. Remove belt pulley.
5. Remove front underbody protection.
6. Remove radiator seal.
7. Remove hose clamp and detach suction line from vane pump.
8. Remove bolt and remove bracket with refrigerant line.
9. Remove screws.
10. Remove banjo bolt and disconnect pressure line.
11. Remove bolts and remove vane pump toward bottom.

#### To install:

12. Markings on suction line and vane pump connection must match up.
13. Replace aluminum screws.
14. Replace all sealing rings.
15. Make sure pressure line and return lines are laid without tension and with sufficient spacing to adjoining components. Tighten pressure line to 22 ft. lbs. (30 Nm) and return line to 26 ft. lbs. (35 Nm).
16. Tighten pump fasteners to specification as follows:

- Secure vane pump with screws
- Tighten side screws to 18 inch lbs. (2 Nm)
- Tighten front screws to 18 inch lbs. (2 Nm)
- Tighten down front screws to 15 ft. lbs. (20 Nm) plus an additional 90° rotation
- Remove screws at side and check screw fastening points for gap freedom
- Tighten down side screws to 15 ft. lbs. (20 Nm) plus an additional 90° rotation

17. Fill and bleed hydraulic system
18. Check pipe connections for leaks

#### 5 and 7-Series

*See Figures 260 and 261.*

### ✷✷ WARNING

**Adhere to the utmost cleanliness. Do not allow any dirt to enter the hydraulic system. Close off pipe connections with plugs.**

1. Draw off and dispose of hydraulic fluid from fluid reservoir.
2. Remove belt pulley.
3. Remove front underbody protection.
4. Remove banjo bolt and detach pressure line from vane pump.
5. Remove hose clamp and detach suction line from vane pump.
6. Remove bolt and nut.
7. Remove vane pump downward.

#### To install:

8. Install vane pump and tighten bolt and nut to 16 ft. lbs. (21 Nm).
9. Connect suction line to vane pump and tighten hose clamp.

➡ **Markings on suction line and vane pump connection must match up.**

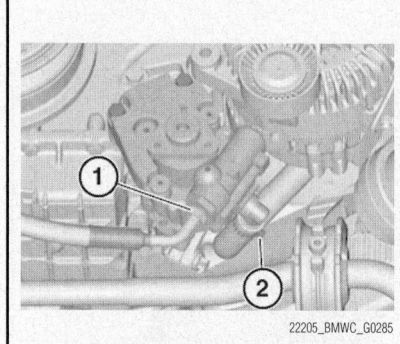

22205_BMWC_G0285

**Fig. 260 Remove banjo bolt (1) and detach pressure line from vane pump. Remove hose clamp (2) and detach suction line from vane pump**

22205_BMWC_G0280

**Fig. 258 Hose clamp (1) and suction line (2)**

22205_BMWC_G0281

**Fig. 259 Pressure line banjo bolt (1) and pump mounting bolts**

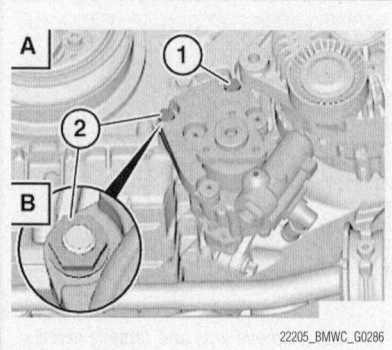

22205_BMWC_G0286

**Fig. 261 Pump mounting bolt locations**

10. Connect pressure line to vane pump. Replace all sealing rings and tighten banjo bolt to 24 ft. lbs. (33 Nm).

➡ **Make sure pressure line is laid without tension and with sufficient spacing to adjoining components.**

11. Fill and bleed hydraulic system
12. Check pipe connections for leaks

### BLEEDING

1. Before servicing the vehicle, refer to the precautions.

### ❋❋ WARNING

**Adhere to the utmost cleanliness. Do not allow any dirt to enter the hydraulic system. Using contaminated equipment to add fluid may introduce dirt particles into the fluid**

reservoir and significantly reduce the service life of the power steering system. Do not use any filler funnels or similar.

➡ **The fill level may only be checked or adjusted when the engine is stopped. The fluid temperature should be approximately 68°F (20°C) here. Ensure that the cap is fully screwed in prior to the fill level check.**

2. Thoroughly clean fluid reservoir and its immediate surroundings
3. Check and correct fill level

➡ **The fill level can come to rest above the MAX mark when the engine is at normal operating temperature. This is dictated by the design in that the marking on the dipstick is referred to a fluid**

temperature of 68°F (20°C). With the engine at normal operating temperature (approximately 122–140°F (50–60°C) fluid temperature adjust a fill height .39 in. (10 mm) above the MAX mark. Do not under any circumstances draw off the fluid to the MAX mark when the engine is at normal operating temperature.

4. Start engine
5. Turn steering wheel left and right twice in each case up to full lock; if necessary, top up hydraulic fluid (e.g. if hydraulic system is completely drained)
6. Move steering wheel to straight-ahead position and turn off engine
7. Check and correct fill level with engine stopped
8. Check hydraulic system for leaks

---

## SUSPENSION

## FRONT SUSPENSION

### COIL SPRING

#### REMOVAL & INSTALLATION
*See Figures 262 through 266.*

### ❋❋ CAUTION

**This procedure calls for the spring to be compressed. A compressed spring has high potential energy and if removed suddenly can cause severe damage and personal injury.**

1. Before servicing the vehicle, refer to the precautions.
2. Remove front spring strut.
3. Remove nut, remove holder and remove screw.
4. Expand ball joint with special tool 31 2 230 and detach from spring strut.
5. Clamp special tool 31 3 341 in vice.
6. Install insert 31 3 358 in special tool 31 3 354.

7. Position special tools 31 3 355 and 31 3 354 with insert 31 3 358 from above on special tool 31 3 341 until locking pins can be felt and heard to snap into place.

8. Check seating of special tools 31 3 355 and 31 3 354 with insert 31 3 358, correct if necessary.
9. Clean coil spring to remove coarse dirt and take up with special tools 31 3 355 and 31 3 354 with insert 31 3 358.

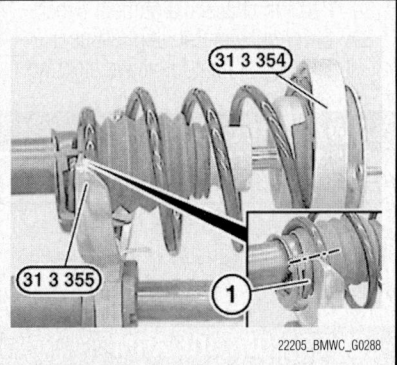
Fig. 263 Twist spring strut until end of coil spring (1)

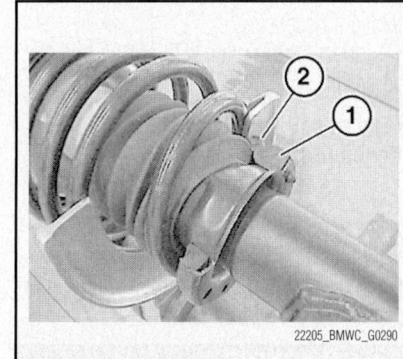
Fig. 265 Lower end of coil spring (2) must rest on stop of spring pad (1)

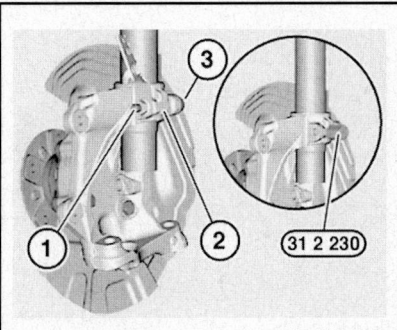
Fig. 262 Remove nut (1), remove holder (2) and remove screw (3)

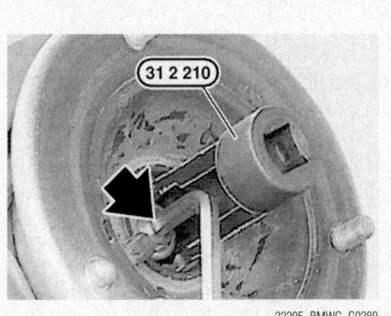
Fig. 264 Remove nut with special tool 31 2 210 (grip piston rod in the process)

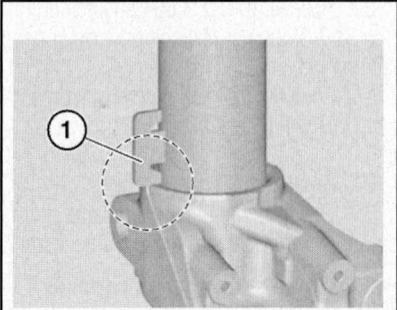
Fig. 266 Make sure ball joint contacts stop correctly

10. Twist spring strut until end of coil spring is flush with end of special tool 31 3 355.

> ☀☀ **WARNING**
>
> **Special tool 31 3 354 and centering ring 31 3 358 must rest correctly on upper spring plate.**

11. Lower coil of coil spring must rest completely in recess of special tool 31 3 355.

12. Compress coil spring until stress on piston rod is relieved.

13. Remove cap.

14. Remove nut with special tool 31 2 210 (grip piston rod in the process).

15. Remove support bearing, dust sleeve and supporting ring.

16. Remove shock absorber with auxiliary damper, gaiter and lower spring pad sideways from tensioned coil spring.

17. If necessary, remove auxiliary damper, gaiter and spring pad from shock absorber.

18. Relieve tension on coil spring.

19. Remove coil spring with spring plate from special tools 31 3 355, 31 3 358 and 31 3 354.

### To install:

20. Check spring pad for damage, replace if necessary.

Connect spring plate with spring pad to coil spring.

➡ **End of coil spring must be positively aligned to spring pad.**

21. Accommodate coil spring with spring plate with special tools 31 3 355, 31 3 358 and 31 3 354. Twist coil spring until lower end of coil spring is flush with end of special tool 31 3 355.

> ☀☀ **WARNING**
>
> **Do not compress coil spring to full extent.**

22. Special tool 31 3 354 and centering ring 31 3 358 must rest correctly on upper spring plate.

23. Lower coil of coil spring must rest completely in recess of special tool 31 3 355.

24. Tension coil spring.

25. Check auxiliary damper, gaiter and spring pad for damage, replace if necessary.

➡ **Make sure spring pad is correctly seated on shock absorber.**

26. Insert shock absorber in tensioned coil spring.

27. Connect thrust washer to piston rod.

28. Check support bearing for damage, replace if necessary.

29. Connect dust sleeve and support bearing to piston rod.

30. Replace nut and tighten down with special tool 31 2 210 (grip piston rod in the process) to 47 ft. lbs. (64 Nm).

31. Fit cover cap.

> ☀☀ **WARNING**
>
> **Spring pad must rest positively on spring plate. Upper end of coil spring must rest on stop of spring pad. Lower end of coil spring must rest on stop of spring pad.**

32. Relieve tension on coil spring.

33. Check installation position of gaiter, correct fold if necessary.

34. Make sure ball joint contacts stop correctly.

35. Expand ball joint with special tool 31 2 230, align by way of gap to positioning pins on back of spring strut and press together up to stop.

36. Keep press fit of ball joint and spring strut in lower area clean and free from oil and grease. Screw head must point in direction of travel.

37. Replace self-locking nut and tighten to 60 ft. lbs. (81 Nm).

38. Perform chassis alignment check.

39. Carry out steering angle sensor adjustment/adjustment for active steering.

## LOWER BALL JOINT

### REMOVAL & INSTALLATION

The lower ball joint is an integral part of the lower control arm and is not serviced separately.

## LOWER CONTROL ARM

### REMOVAL & INSTALLATION

### 3-Series

*See Figure 267.*

1. Before servicing the vehicle, refer to the precautions.

2. Remove front wheel.

3. Remove front underbody protection.

4. If necessary, remove jointed rod of ride-height sensor from control arm.

5. Remove lower ball joint nut; if necessary, grip at Torx® socket (T40).

6. Remove lower control arm nut. Remove screw towards front.

7. If necessary remove bracket with ride-height sensor.

8. Remove control arm.

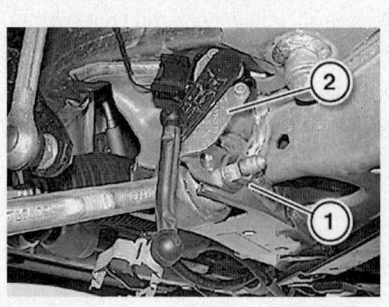

22205_BMWC_G0292

Fig. 267 Remove nut (1), then remove screw towards front. If necessary remove bracket (2) with ride-height sensor

### To install:

9. Install control arm.

➡ **Screw head must point in direction of travel. Replace self-locking nut.**

10. Temporarily tighten down bolt and nut. Once the vehicle is lowered to normal ride height, finish tightening to 50 ft. lbs. (68 Nm) plus an additional 90° rotation.

11. Keep control arm to ball joint connection clean and free from oil and grease.

12. Replace self-locking nut and tighten to 122 ft. lbs. (165 Nm).

13. Perform chassis alignment check.

14. Carry out steering angle sensor adjustment/adjustment for active front steering.

### 5 and 7-Series

*See Figures 268 and 269.*

1. Before servicing the vehicle, refer to the precautions.

2. If necessary, remove steering gear cover at side.

3. Right side only on version with xenon headlight:

   a. Release nut that attaches the rod to control arm.

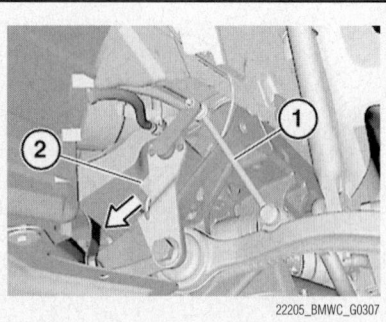

22205_BMWC_G0307

Fig. 268 Remove jointed rod (1) from control arm and disconnect plug connection (2) on ride-height sensor

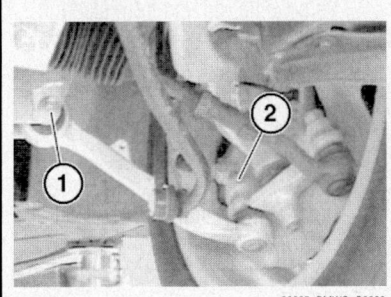

**Fig. 269 Unscrew nut (1) and remove screw towards rear. Release nut (2) and if necessary, grip at Torx®socket**

   b. Remove jointed rod from control arm.

   c. Disconnect plug connection on ride-height sensor.

   4. Unscrew nut and remove screw toward rear.

   5. Only on right side: If necessary, remove holder with ride-height sensor.

   6. Release nut; if necessary, grip at Torx® socket.

   7. Remove control arm.

### To install:

   8. Install control arm.

   9. Replace self-locking nut.

   10. Temporarily tighten down bolt and nut. Once the vehicle is lowered to normal ride height, finish tightening to 74 ft. lbs. (100 Nm) plus an additional 90° rotation.

   11. Keep control arm to steering knuckle connection clean and free from oil and grease.

   12. Replace self-locking nut and tighten to122 ft. lbs. (165 Nm) on 2WD and 59 ft. lbs. (80 Nm) on AWD.

   13. Perform chassis alignment check.

   14. Carry out steering angle sensor adjustment/adjustment for active front steering.

### CONTROL ARM BUSHING REPLACEMENT

   The bushings must be pressed out of the housing bores. BMW bushings are notoriously hard to press out of the housings. Use a high capacity hydraulic press, penetrating lubricant and the proper sized mandrels for the press. Do not use sockets to try to replace the bushings. Mark the relationship of the bushing to the bore for correct replacement positioning.

## MACPHERSON STRUT

### REMOVAL & INSTALLATION

#### 3-Series

*See Figure 270.*

   1. Before servicing the vehicle, refer to the precautions.

   2. Remove front wheel.

   3. Remove stabilizer link from spring strut.

   4. Disconnect plug connection for pulse generator and expose line up to holder on spring strut.

   5. Disconnect plug connection for brake pad sensor and expose line up to brake caliper.

   6. Remove front brake disc.

   7. If necessary, remove jointed rod of ride-height sensor from control arm.

   8. Slacken control arm bolt connection on front axle carrier in order to prevent control arm rubber mount from being damaged.

   9. Remove control arm from steering knuckle.

   10. Remove tie rod end from steering knuckle.

   11. Remove tension strut from steering knuckle.

### ✳✳ WARNING

**Secure spring strut against falling out.**

   12. Remove tension strut (on spring strut dome) and mark position of threaded pins to wheel arch.

   13. Release nuts and remove spring strut with steering knuckle towards bottom.

### To install:

   14. If necessary, replace faulty sealing washer.

   15. Clean contact surface in spring strut dome.

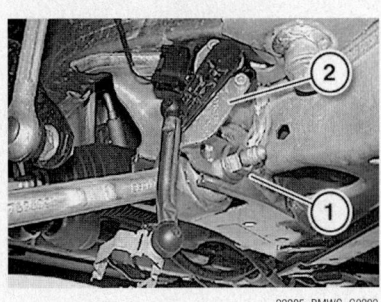

**Fig. 270 Sealing washer (1) and plate insert (2)**

   16. Align spring strut using centering pin to bore in wheel arch or studs to wheel arch and push upwards.

   17. Replace self-locking nuts and tighten to 25 ft. lbs. (34 Nm).

   18. The remainder of the installation is the reverse of removal.

   19. Perform chassis alignment check.

   20. Carry out steering angle sensor adjustment/adjustment for active front steering.

#### 5 and 7-Series

   1. Before servicing the vehicle, refer to the precautions.

   2. Remove steering knuckle

   3. If necessary, release expander rivet.

   4. If necessary, unclip spring strut dome cover from front cross-strut.

   5. Make position of studs in relation to wheel arch.

   6. Secure spring strut against falling out.

   7. Unscrew nuts.

   8. Remove spring strut downwards out of wheel arch.

### To install:

   9. Align spring strut using centering pin to bore in wheel arch or studs to wheel arch and push upwards.

   10. Replace self-locking nuts and tighten to 25 ft. lbs (34 Nm).

   11. Carry out wheel alignment check if a spring strut with support bearing was or has been installed without centering pin.

## STABILIZER BAR

### REMOVAL & INSTALLATION

#### 3-Series

*See Figures 271 and 272.*

   1. Before servicing the vehicle, refer to the precautions.

   2. Remove front underbody protection.

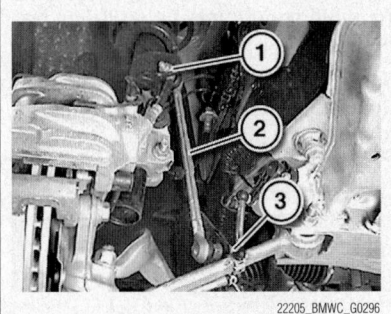

**Fig. 271 Release nut (1) and remove bracket for brake hose. Release nut (3) and remove stabilizer link (2)**

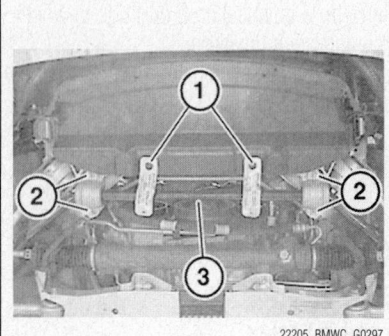

**Fig. 272 Remove bolts (1) and remove brackets. Release nuts (2); if necessary, grip screws after repair. Remove stabilizer (3)**

3. Remove stabilizer link on both sides from stabilizer.

4. Remove bolts and remove brackets.

5. Release nuts; if necessary, grip screws after repair.

6. Remove stabilizer; if necessary, press off front axle carrier with a suitable tool.

### To install:

7. If studs are damaged, repair as follows:

   a. Remove bolt.

   b. Raise locking nut in area of bore and detach from front axle carrier.

   c. Drive out studs in upwards direction and remove/feed out through an opening in front axle carrier.

   d. Insert new screws from above.

8. Check both rubber mounts for damage, replace if necessary.

9. Install stabilizer and secure with brackets.

10. Replace locking nut and tighten temporarily. Once vehicle is at normal ride height, tighten to 50 ft. lbs. (68 Nm) plus an additional 90° rotation.

### ✳✳ WARNING

**To avoid complaints being made by the customer about noise (e.g. grating), set the car on its wheels and tighten down the stabilizer mounting to specified torque.**

11. Connect stabilizer link on both sides from stabilizer. Tighten nuts to 43 ft. lbs. (58 Nm).

12. Install front underbody protection.

### 5 and 7-Series

1. Before servicing the vehicle, refer to the precautions.

2. Remove front assembly underside protection.

3. If necessary, remove wheel suspension cover on both sides.

4. Remove stabilizer link on both sides from stabilizer.

5. If necessary, remove holder for hydraulic lines from retaining bracket.

6. Release nut and screw at both ends.

7. Remove stabilizer from front axle carrier.

### To install:

8. Check both rubber mounts for damage, replace if necessary.

9. Replace self-locking nuts and tighten stabilizer bar mounts to 22 ft. lbs. (30 Nm).

10. Replace self-locking nuts and tighten stabilizer link to 48 ft. lbs. (65 Nm).

11. If necessary, install wheel suspension cover on both sides.

12. Install front assembly underside protection.

## STEERING KNUCKLE

### REMOVAL & INSTALLATION

### 3-Series
*See Figure 273.*

1. Before servicing the vehicle, refer to the precautions.

2. Remove front brake disk.

3. Remove front wheel speed sensor.

4. If necessary, remove jointed rod of ride-height sensor from control arm.

5. Slacken control arm bolt connection on front axle carrier in order to prevent control arm rubber mount from being elongated.

6. Remove control arm from steering knuckle.

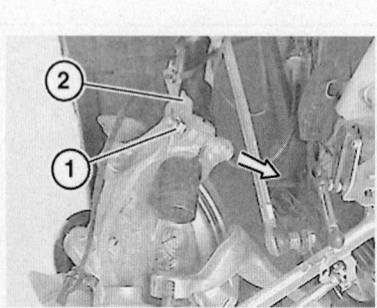

**Fig. 273 Release nut (1), remove holder (2) and remove screw**

7. Remove tie rod end from steering knuckle.

8. Remove strut from steering knuckle.

9. Support steering knuckle with workshop jack and a suitable mounting.

10. Release nut, remove holder and remove screw.

11. Spread steering knuckle with special tool 31 2 230.

12. Lower workshop jack and remove steering knuckle.

### To install:

13. Keep press fit of swivel gearing and spring strut in lower area clean and free from oil and grease.

14. Expand steering knuckle with special tool 31 2 230, align by way of gap to positioning pins on back of spring strut and raise up to stop.

15. Make sure steering knuckle contacts stop correctly.

16. Replace self-locking nut.

17. Install the nut, holder and screw. Tighten to 60 ft. lbs. (81 Nm).

➡**Screw head must point in direction of travel.**

### ✳✳ WARNING

**Check sensor head and line from pulse generator prior to installation for external damage, replacing if necessary.**

18. The remainder of the installation is the reverse of removal.

19. Perform chassis alignment check.

20. Carry out steering angle sensor adjustment/adjustment for active front steering.

### 5 and 7-Series

1. Before servicing the vehicle, refer to the precautions.

2. Remove front brake disc.

3. Remove front pulse generator from steering knuckle.

4. Remove control arm from steering knuckle.

5. Remove tie rod end from steering knuckle.

6. Remove bracket for stabilizer link from steering knuckle.

7. Remove tension strut from steering knuckle.

8. Spread steering knuckle with special tool 31 2 230.

9. Lower steering knuckle with workshop jack.

*To install:*

10. Keep press fit of swivel gearing and spring strut in lower area clean and free from oil and grease.

11. Spread steering knuckle with special tool 31 2 230, align by means of gap to web on back of spring strut and raise as far as it will go.

12. Replace self-locking nut.

13. Install the nut, holder and screw. Tighten to 60 ft. lbs. (81 Nm).

14. Install tension strut from steering knuckle.

15. Install bracket for stabilizer link from steering knuckle.

16. Install tie rod end from steering knuckle.

17. Install control arm from steering knuckle.

18. Install front pulse generator from steering knuckle.

19. Install front brake disc.

## WHEEL BEARINGS

### REMOVAL & INSTALLATION

*See Figure 274.*

1. Before servicing the vehicle, refer to the precautions.

2. Remove brake disk.

3. Remove bolts.

4. Press wheel bearing off steering knuckle with a suitable tool.

**To install:**

5. When reusing the wheel bearing, recut all the threads in the wheel bearing.

6. Keep contact surface of steering knuckle and wheel bearing clean and free from oil and grease.

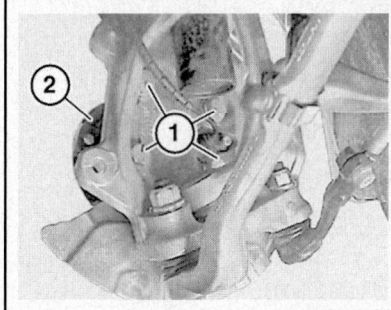

Fig. 274 Remove bolts (1) and press wheel bearing (2) off steering knuckle with a suitable tool

7. Replace microencapsulated screws and tighten to 81 ft. lbs. (110 Nm).

### ADJUSTMENT

Wheel bearings cannot be adjusted. The front wheel bearings are pressed into the hub and are not available separately. If a front wheel bearing is in need of replacement, it is replaces as a unit with the hub.

## SUSPENSION

### COIL SPRING

**REMOVAL & INSTALLATION**

#### 3-Series

*See Figures 275 and 276.*

1. Before servicing the vehicle, refer to the precautions.

2. Remove rear wheel.

3. Insert lower spring plate 33 5 012 centrally into coil spring and turn to lowest coil.

Guide spindles 33 5 013, 33 5 014, 33 5 015 from below through camber arm and lower spring plate 33 5 012.

4. Insert upper spring plate 33 5 011 sideways into coil spring and turn to uppermost coil.

## REAR SUSPENSION

5. Align special tools 33 5 011, 33 5 012, 33 5 013, 33 5 014, 33 5 015 centrally to obtain the biggest possible contact surface on the coil spring.

6. Check installation position of special tools 33 5 011, 33 5 012 and 33 5 013, 33 5 014, 33 5 015, correct if necessary.

7. Tension coil spring using special tools 33 5 016 and 33 5 020, gripping spindle of spring tensioner with special tool 33 5 017 in the process.

8. Remove coil spring upwards.

**To install:**

9. Bottom end of coil spring must be flush with opening of spring plate 33 5 012.

10. Check spring mounts for damage, replace if necessary.

11. Remove spring plate and position with upper spring pad on coil spring.

➡**Upper spring pad must come into contact with end of coil spring.**

12. Lower spring pad must be positively seated in the designated receptacle in the camber arm

a. Lower spring pad must come into contact with end of coil spring.

b. Lower spring pad must rest flush on last coil.

13. Align coil spring by way of spring plate to opening in side member and relieve tension.

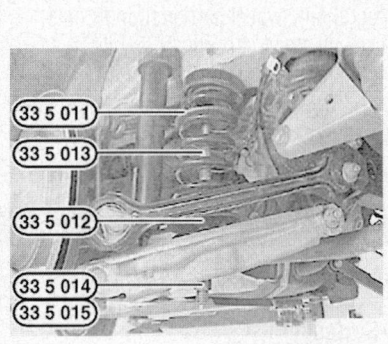

Fig. 275 Align special tools 33 5 011, 33 5 012, 33 5 013, 33 5 014, 33 5 015 centrally to obtain the biggest possible contact surface on the coil spring

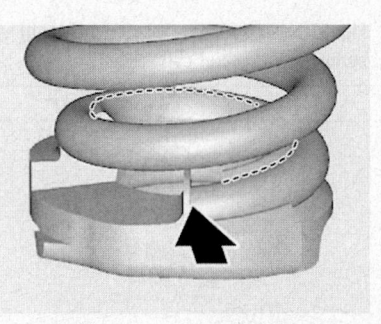

Fig. 276 Lower spring pad must be positively seated in the designated receptacle in the camber arm. The spring pad must come into contact with end of coil spring (arrow) and must rest flush on last coil (broken line)

14. Remove special tools 33 5 011, 33 5 012 and 33 5 013, 33 5 014, 33 5 015.

15. Check headlight adjustment, correct if necessary

### 5 and 7-Series

*See Figures 277 through 279.*

1. Before servicing the vehicle, refer to the precautions.

2. Remove rear wheel.

3. Remove rear spring strut shock absorber

4. Clamp special tool 31 3 341 in vice.

5. Fit special tools 31 3 357 from above on special tool 31 3 341 until locking pins (1) can be felt and heard to snap into place.

6. Check seating of special tools 31 3 357, correct if necessary.

7. Clean coil spring to remove all coarse dirt and mount on special tools 31 3 357.

### ❈❈ WARNING

**Coils of coil spring must be located completely in recesses of special tools 31 3 357!**

8. Compress coil spring until stress on piston rod is relieved.

9. Release nut (gripping piston rod in the process).

10. Remove plate and support bearing with spring pad.

11. Remove shock absorber with support pot, auxiliary damper, protective tube and spring pad sideways from tensioned coil spring.

12. If necessary, remove support pot, auxiliary damper (1) with protective tube and spring pad (2) from shock absorber.

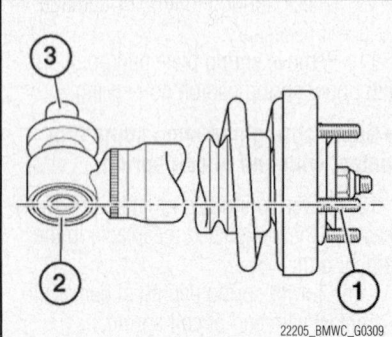

**Fig. 277 One threaded pin on the thrust bearing must be on one plane with the bore in the rubber mount and the opposite side of the bush**

22205_BMWC_G0309

13. Relieve tension on coil spring and remove from special tools 31 3 357.

### *To install:*

14. Coils of coil spring must be located completely in recesses of special tools 31 3 357!

15. Tension coil spring.

16. Check auxiliary damper (1) with protective tube and spring pad (2) for damage, replace if necessary.

Connect support pot to piston rod.

17. Insert shock absorber in tensioned coil spring.

18. Check support bearing for damage, replace if necessary.

19. Connect support bearing and plate to piston rod.

20. Replace nut and tighten down (gripping piston rod in so doing) to 20 ft. lbs. (27 Nm).

### ❈❈ WARNING

**One threaded pin on the thrust bearing must be on one plane with the bore in the rubber mount and the opposite side of the bush.**

21. Align support bearing by way of spring pad to end of coil spring.

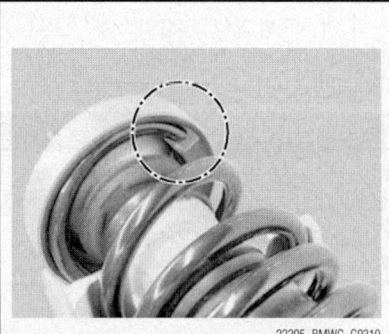

**Fig. 278 Align support bearing by way of spring pad to end of coil spring**

22205_BMWC_G0310

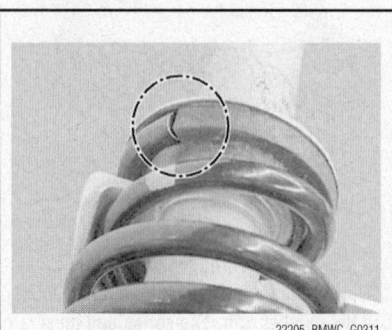

**Fig. 279 Lower end of coil spring must rest on stop of spring pad**

22205_BMWC_G0311

22. Lower end of coil spring must rest on stop of spring pad.

23. Relieve tension on coil spring.

24. Install rear spring strut shock absorber.

### LOWER CONTROL ARM

#### REMOVAL & INSTALLATION

### 5 and 7-Series

1. Before servicing the vehicle, refer to the precautions.

2. Remove rear wheel.

3. Remove jointed rod of ride-height sensor from lower control arm.

4. Remove stabilizer link from lower control arm.

5. Mark position of eccentric screw to rear axle carrier.

6. Support wheel carrier with workshop jack.

7. Release nut and remove eccentric washer.

8. Remove eccentric screw towards front.

9. Release nut and pull out bolt towards rear.

10. Unfasten nut.

11. Remove screw towards rear and remove lower control arm.

### *To install:*

12. Install lower control arm

13. Note insertion direction of eccentric screw. Align eccentric screw by means of marking to rear axle carrier. Refit eccentric washer.

14. Replace self-locking nuts. Temporarily tighten swinging arm to rear axle support nuts. Once vehicle is at normal ride height, tighten to 74 ft. lbs. (100 Nm)

15. Replace self-locking nuts. Temporarily tighten bolt connection, swinging arm, integral link, wheel carrier support nuts. Once vehicle is at normal ride height, tighten to 178 ft. lbs. (240 Nm)

16. Check that output shaft is correctly seated in rear differential.

17. Perform chassis alignment check

### STABILIZER BAR

#### REMOVAL & INSTALLATION

### 3-Series

*See Figures 280 and 281.*

1. Before servicing the vehicle, refer to the precautions.

2. Remove both rear coil springs.

3. Remove left control arm on wheel carrier.

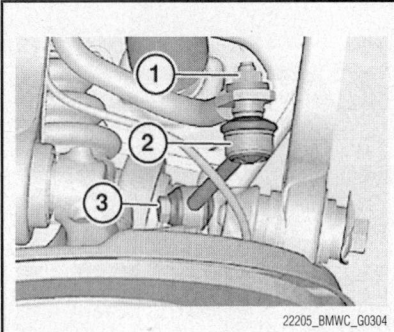

**Fig. 280 Stabilizer link (2) and mounting nut (1) and bolt (3)**

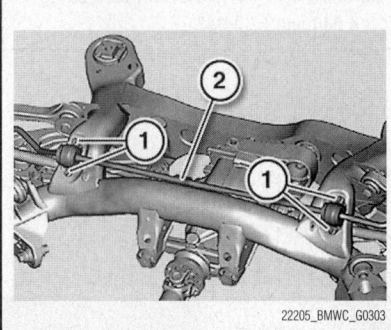

**Fig. 281 Stabilizer bar (2) and mounting bolts (1)**

4. Remove both stabilizer links from stabilizer.

5. Lower rear axle carrier.

6. Remove bolts and remove both rubber mounts from stabilizer.

7. Turn stabilizer and remove sideways.

*To install:*

8. Check both rubber mounts for damage, replace if necessary.

9. Tightening fasteners to specification.

- Retaining bracket, stabilizer bar to rear axle support: 16 ft. lbs. (21 Nm)
- Stabilizer link to stabilizer: 43 ft. lbs. (58 Nm)
- Stabilizer link to wheel carrier: 16 ft. lbs. (21 Nm)

10. Raise rear axle carrier.

11. Install both rear coil springs.

12. Install left control arm on wheel carrier.

13. Bleed braking system.

### 5 and 7-Series

1. Before servicing the vehicle, refer to the precautions.

2. Remove stabilizer link on both sides from stabilizer

3. If necessary, remove vibration damper from right retaining bracket of stabilizer.

4. Release nuts at both ends.

5. Remove stabilizer sideways.

*To install:*

6. Check rubber mount on both sides, replace if necessary.

7. Replace self-locking nuts and tighten stabilizer bar mounting bolts to 16 ft. lbs. (22 Nm)..

8. Replace self-locking nuts and tighten stabilizer link nuts to 16 ft. lbs. (22 Nm).

## TRAILING ARM

### REMOVAL & INSTALLATION

### 3-Series

*See Figure 282.*

1. Before servicing the vehicle, refer to the precautions.

2. Remove rear wheel.

3. Release nuts.

4. Unscrew bolts.

5. Remove trailing arm downwards.

6. Note insertion direction of bolts.

7. Replace self-locking nuts and tighten temporarily.

8. Lower vehicle to normal ride height and tighten bolts to 74 ft. lbs. (100 Nm).

9. Perform chassis alignment check.

## UPPER CONTROL ARM

### REMOVAL & INSTALLATION

### 3-Series

*See Figures 283 and 284.*

1. Before servicing the vehicle, refer to the precautions.

2. Remove rear wheel.

3. Remove bolt.

4. Release nut and remove screw towards front.

5. Unclip line holder for pulse generator or brake pad sensor on control arm.

6. Remove control arm.

*To install:*

7. Install control arm.

### ❊❊ WARNING

**Make sure during installation that in the connection area to the wheel carrier the tapered end of the bearing bushing (of the rubber mount) points in the direction of the wheel carrier.**

8. Install control arm to rear axle carrier bolt and point head in direction of travel.

9. Replace self-locking nut and tighten to 74 ft. lbs. (100 Nm).

10. If necessary, raise brake disk in drive flange area with workshop jack.

11. Replace control arm to wheel carrier bolt and tighten temporarily. Fully tighten the bolt once the vehicle is at ride

**Fig. 283 Make sure during installation that in the connection area to the wheel carrier the tapered end of the bearing bushing (of the rubber mount) points in the direction of the wheel carrier**

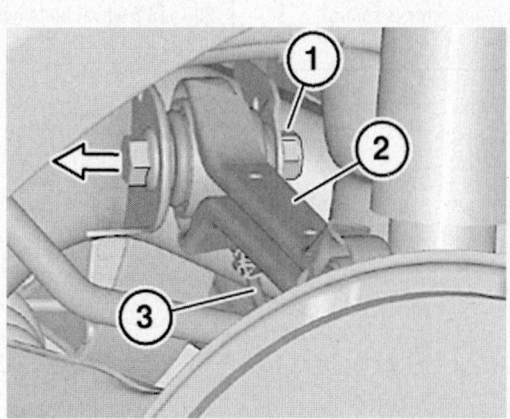

**Fig. 282 Release nut (1), remove control arm (2) and line holder (3) for pulse generator**

height to 74 ft. lbs. (100 Nm) plus 90° additional rotation.

12. Perform chassis alignment check.

### 5 and 7-Series

*See Figure 284.*

1. Before servicing the vehicle, refer to the precautions.

2. Remove rear wheel.

3. Unscrew nut and remove control arm at top from wheel carrier.

4. Unscrew nut and pull out bolt and remove control arm.

#### *To install:*

5. Keep control arm to wheel carrier connection clean and free from oil and grease.

6. Replace self-locking nut. Tighten upper control arm to wheel carrier nut to 130 ft. lbs. (175 Nm).

7. Note insertion direction of screw. Replace self-locking nut. Temporarily tighten nut. When vehicle is at normal ride height, tighten to 74 ft. lbs. (100 Nm).

8. Perform chassis alignment check

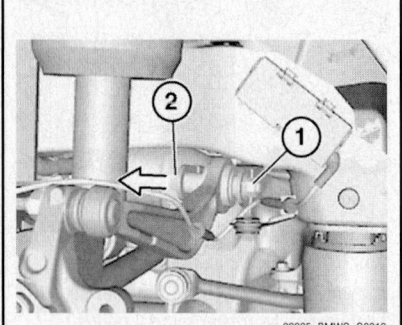

**Fig. 284 Unscrew nut and remove control arm at top from wheel carrier. Unscrew nut and pull out bolt and remove control arm**

## WHEEL BEARINGS

### *REMOVAL & INSTALLATION*

*See Figures 285 and 286.*

1. Before servicing the vehicle, refer to the precautions.

2. Remove drive flange of rear axle shaft.

3. Remove output shaft.

4. Remove brake disk.

5. Remove pulse generator.

### ✳✳ WARNING

**Check sensor head and line from pulse generator prior to installation for external damage, replacing if necessary.**

6. Force drive flange with special tools 33 2 116 / 33 2 201, 33 2 160, 33 4 200 and 5 wheel bolts out of wheel bearing.

➡**Rounded inside edge of special tool 33 2 160 must point to drive flange.**

7. Press retaining ring together using pliers and remove.

8. Pull out wheel bearing with special tools 33 4 041, 33 4 042, 33 4 031, 33 4 048 and 33 4 043.

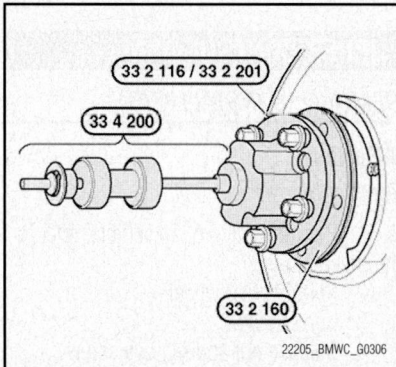

**Fig. 285 Force drive flange with special tools 33 2 116 / 33 2 201, 33 2 160, 33 4 200 and 5 wheel bolts out of wheel bearing**

**Fig. 286 Pull out wheel bearing with special tools 33 4 041, 33 4 042, 33 4 031, 33 4 048 and 33 4 043**

#### *To install:*

9. Install retaining ring.

10. Check seating of retaining ring, correct if necessary.

11. Draw in new wheel bearing with special tools 33 4 041, 33 4 042, 33 4 049, 33 4 047 and 33 4 043.

### ✳✳ WARNING

**Do not reuse old wheel bearing! The wheel bearing is destroyed when the drive flange is removed and cannot be reused.**

12. Replace wheel bearing.

13. Install output shaft.

14. Oil drive flange lightly and attach to splines of output shaft.

15. Draw drive flange into wheel bearing.

16. Install brake disk.

17. Install pulse generator.

18. Adjust handbrake.

### *ADJUSTMENT*

Wheel bearings cannot be adjusted. The rear bearings must be replaced as a unit and never be reused once removed.

## SPECIFICATIONS AND MAINTENANCE CHARTS

### ENGINE AND VEHICLE IDENTIFICATION

| | | | Engine | | | | | | Model Year | |
|---|---|---|---|---|---|---|---|---|---|---|
| Code | Liters (cc) | Cu. In. | Cyl. | Fuel Sys. | Engine Type | Eng. Mfg. | | Code ① | | Year |
| M54 | 3.0 (2979) | 182 | 6 | SMPI | DOHC | BMW | | 6 | | 2006 |
| N52K | 3.0 (2979) | 182 | 6 | SMPI | DOHC | BMW | | 7 | | 2007 |
| N52K | 3.0 (2979) | 182 | 6 | SMPI | DOHC | BMW | | 8 | | 2008 |

SMPI: Sequencial Multi-port Fuel Injection

DOHC: Double Overhead Camshaft

① 10th position of the VIN

22205_BMX3_C0001

### GENERAL ENGINE SPECIFICATIONS

| Year | Body Type | Model | Engine Displacement Liters (cc) | Engine ID/VIN | Fuel System Type | Net Horsepower @ rpm | Net Torque @ rpm (ft. lbs.) | Bore x Stroke (in.) | Com- pression Ratio | Oil Pressure @ rpm |
|---|---|---|---|---|---|---|---|---|---|---|
| 2006 | E83 | X3 | 3.0 (2979) | M54 | SMPI | 225@5900 | 214@3500 | 3.31x3.53 | 10.2:1 | 7.4@700 |
| 2007 | E83 | X3 | 3.0 (2979) | N52K | SMPI | 260@6600 | 225@2700 | 3.31x3.53 | 10.2:1 | 7.4@700 |
| 2008 | E83 | X3 | 3.0 (2979) | N52K | SMPI | 260@6600 | 225@2750 | 3.31x3.53 | 10.2:1 | 7.4@700 |

SMPI: Sequencial Multi-port Fuel Injection

22205_BMX3_C0002

### ENGINE TUNE-UP SPECIFICATIONS

| Year | Engine Displacement Liters | Engine ID/VIN | Spark Plug Gap (in.) | Ignition Timing (deg.) MT | Ignition Timing (deg.) AT | Fuel Pump (psi) | Idle Speed (rpm) MT | Idle Speed (rpm) AT | Valve Clearance In. | Valve Clearance Ex. |
|---|---|---|---|---|---|---|---|---|---|---|
| 2006 | 3.0 | M54 | 0.024-0.028 | ① | ① | 72 | ② | ② | HYD | HYD |
| 2007 | 3.0 | N52K | 0.024-0.028 | ① | ① | 72 | ② | ② | HYD | HYD |
| 2008 | 3.0 | N52K | 0.024-0.028 | ① | ① | 72 | ② | ② | HYD | HYD |

NOTE: The Vehicle Emission Control Information label reflects specification changes during production and must be used if they differ from this chart.

HYD: Hydraulic

① Ignition timing is regulated by the Engine Control Module (ECM) and cannot be adjusted

② Idle speed is controlled by the Engine Control Module (ECM) and cannot be adjusted

22205_BMX3_C0003

## CAPACITIES

| Year | Model | Body Type | Engine Displacement Liters | Engine ID/VIN | Engine Oil with Filter (qts.) | Transmission (pts.) | | | Drive Axle | | Fuel Tank (gal.) | Cooling System (qts.) |
|------|-------|-----------|----------------------------|---------------|-------------------------------|---------|--------|-------|-------|------|------------------|------------------------|
| | | | | | | 5-Spd | 6-Spd | Auto. | Front (pts.) | Rear (pts.) | | |
| 2006 | X3 | E83 | 3.0 | M54 | 8 | — | 3.2 | 7 | — | 3.4 | 17.7 | 11.1 |
| 2007 | X3 | E83 | 3.0 | N52K | 8 | — | 3.2 | 7 | — | 3.4 | 17.7 | 11.1 |
| 2008 | X3 | E83 | 3.0 | N52K | 8 | — | 3.2 | 7 | — | 3.4 | 17.7 | 11.1 |

NOTE: All capacities are approximate. Add fluid gradually and ensure a proper fluid level is obtained.

NOTE: Capacities given are service, not overhaul capacities

22205_BMX3_C0014

## FLUID SPECIFICATIONS

| Year | Model | Engine Displ. Liters (VIN) | Engine Oil | Man. Trans. | Auto. Trans. | Drive Axle Front | Drive Axle Rear | Transfer Case | Power Steering Fluid | Brake Master Cylinder | Cooling System |
|------|-------|----------------------------|------------|-------------|--------------|------------------|------------------|---------------|----------------------|------------------------|----------------|
| 2006 | X3 | 3.0 (M54) | ① | ② | ③ | ④ | ④ | ⑤ | Dexron® III | DOT 4 | Ethylene glycol |
| 2007 | X3 | 3.0 (N52K) | ① | ② | ③ | ④ | ④ | ⑤ | Dexron® III | DOT 4 | Ethylene glycol |
| 2008 | X3 | 3.0 (N52K) | ① | ② | ③ | ④ | ④ | ⑤ | Dexron® III | DOT 4 | Ethylene glycol |

DOT: Department Of Transpotation

① SAE, API 5W-50

② Exxon MTF-LT1

③ ESSO LT 711 41, BMW Part No. 83 22 9 407 807

④ BMW Synthetic Final Drive Oil

⑤ BMW Part No. 83 22 0 397 244 or TF0870

22205_BMX3_C0004

## VALVE SPECIFICATIONS

| Year | Engine Displacement Liters | Engine ID/VIN | Seat Angle (deg.) | Face Angle (deg.) | Spring Test Pressure (lbs. @ in.) | Spring Installed Height (in.) | Stem-to-Guide Clearance (in.) Intake | Stem-to-Guide Clearance (in.) Exhaust | Stem Diameter (in.) Intake | Stem Diameter (in.) Exhaust |
|------|----------------------------|---------------|-------------------|-------------------|-----------------------------------|-------------------------------|--------|---------|--------|---------|
| 2006 | 3.0 | M54 | ① | 45 | NA | NA | NA | 0.0197 | 0.025-0.0400 | 0.040-0.0550 |
| 2007 | 3.0 | N52K | ① | 45 | NA | NA | NA | NA | NA | NA |
| 2008 | 3.0 | N52K | ① | 45 | NA | NA | NA | NA | NA | NA |

NA: Not Available

① Valve seat angle: 45 degrees

Correction angle outside: 15 degrees

Correction angle inside: 60 degrees

22205_BMX3_C0005

## CAMSHAFT AND BEARING SPECIFICATIONS

All measurements are given in inches.

| Year | Engine Displacement Liters | Engine VIN | Journal Diameter | Brg. Oil Clearance | Shaft End-play | Runout | Journal Bore | Lobe Lift Intake | Lobe Lift Exhaust |
|------|------|------|------|------|------|------|------|------|------|
| 2006 | 3.0 | M54 | 2.0505-2.0515 | 0.0010-0.0030 | 0.0010-0.0060 | 0.002 | ① | 0.2449 | 0.2587 |
| 2007 | 3.0 | N52K | 1.1260-1.1270 | 0.0010-0.0030 | 0.0035-0.0075 | 0.001 | 1.1280-1.1290 | 0.2173 | 0.2168 |
| 2008 | 3.0 | N52K | 1.1260-1.1270 | 0.0009-0.0029 | 0.0013-0.0019 | 0.004 | 1.1280-1.1290 | 0.2156 | 0.2173 |

NA: Information not available

① Intake: 1.8532-1.8542 in.

Exhaust: 1.5635-1.5645 in.

22205_BMX3_C0006

## CRANKSHAFT AND CONNECTING ROD SPECIFICATIONS

All measurements are given in inches.

| Year | Engine Displacement Liters | Engine ID/VIN | Crankshaft Main Brg. Journal Dia. | Main Brg. Oil Clearance | Shaft End-play | Thrust on No. | Connecting Rod Journal Diameter | Oil Clearance | Side Clearance |
|------|------|------|------|------|------|------|------|------|------|
| 2006 | 3.0 | M54 | ① | 0.0007-0.0023 | 0.0031-0.0064 | 5 | 1.7720-1.7706 | 0.0007-0.0022 | 0.0060-0.016 |
| 2007 | 3.0 | N52K | ① | 0.016 | NA | 5 | | 0.0007-0.0022 | NA |
| 2008 | 3.0 | N52K | ① | 0.016 | NA | 5 | 1.7720- | 0.0007-0.0022 | NA |

① Standard yellow 2.3615-2.3618 inches

Standard green: 2.3613-2.3615 inches

Standard white: 2.3611-2.3613 inches

22205_BMX3_C0007

## PISTON AND RING SPECIFICATIONS

All measurements are given in inches

| Year | Engine Displacement Liters | Engine ID/VIN | Piston Clearance | Ring Gap Top Compression | Ring Gap Bottom Compression | Ring Gap Oil Control | Ring Side Clearance Top Compression | Ring Side Clearance Bottom Compression | Ring Side Clearance Oil Control |
|------|------|------|------|------|------|------|------|------|------|
| 2006 | 3.0 | M54 | 0.0004-0.0015 | 0.0078-0.0057 | 0.0078-0.0157 | 0.0079-0.0157 | NA | NA | NA |
| 2007 | 3.0 | N52K | 0.0059 | 0.0078-0.0129 | 0.011-0.0196 | NA | NA | NA | NA |
| 2008 | 3.0 | N52K | 0.0059 | 0.0078-0.0129 | 0.011-0.0196 | NA | NA | NA | NA |

22205_BMX3_C0008

## TORQUE SPECIFICATIONS
All readings in ft. lbs.

| Year | Engine Displacement Liters | Engine ID/VIN | Cylinder Head Bolts | Main Bearing Bolts | Rod Bearing Bolts | Crankshaft Damper Bolts | Flywheel Bolts | Manifold Intake | Manifold Exhaust | Spark Plugs | Lug Nut |
|------|------|------|------|------|------|------|------|------|------|------|------|
| 2006 | 3.0 | M54 | ① | ② | ③ | 240 | ④ | ⑤ | ⑥ | ⑦ | 103 |
| 2007 | 3.0 | N52K | ① | ② | ③ | 302 | ④ | ⑤ | ⑥ | ⑦ | 103 |
| 2008 | 3.0 | N52K | ① | ② | ③ | 302 | ④ | ⑤ | ⑥ | ⑦ | 103 |

① Aluminum block. Replace, wash and oil bolts

Step 1: 30 ft. lbs.

Step 2: 90 degrees

Step 3: 90 degrees

② Aluminum block. Replace, wash and oil bolts

Step 1: 14.8 ft. lbs.

Step 2: 70 degrees

③ Tigthen to 78 ft. lbs.

④ New micro-encapsulated screws:

Automatic transmission: 88 ft. lbs.

Manual transmission: 77.4 ft. lbs.

⑤ Replace nuts and torque to 11 ft. lbs.

⑥ Coat with Molykkote HSC compound or equivalent

All M6 fasteners: 88 inch lbs.

All M7 fasteners: 14.8 ft. lbs.

⑦ M12x1.25:14.8-19.1 ft. lbs.

M14x1.25: 21.4-24.3 ft. lbs.

22205_BMX3_C0009

## WHEEL ALIGNMENT

| Year | Model | | Caster Range (+/-Deg.) | Caster Preferred Setting (Deg.) | Camber Range (+/-Deg.) | Camber Preferred Setting (Deg.) | Toe-in (Deg.) | Steering Axis Inclination (Deg.) |
|------|------|------|------|------|------|------|------|------|
| 2006 | X3 | F | NA | 0 | NA | 2 | 0 | — |
|      |    | R | — | — | NA | -2 | 0 | 0.06 |
| 2007 | X3 | F | NA | 0 | NA | 2 | 0 | — |
|      |    | R | — | — | NA | -2 | 0 | 0.06 |
| 2008 | X3 | F | NA | 0 | NA | 2 | 0 | — |
|      |    | R | — | — | NA | -2 | 0 | 0.06 |

① Standard suspension

22205_BMX3_C0010

## TIRE, WHEEL AND BALL JOINT SPECIFICATIONS

| Year | Model | OEM Tires | | Tire Pressures (psi) | | Wheel Size | Ball Joint Inspection | Lug Nut Torque (ft. lbs.) |
|------|-------|-----------|-----------|-------|------|------------|----------------------|------------------|
| | | Standard | Optional | Front | Rear | | | |
| 2006 | X3 | 235/50HR17 | 235/50HR18 | ③ | ③ | NA | NA | 103 |
| 2007 | X3 | 235/55HR18 | ① ② | ③ | ③ | ④ ⑤ | NA | 103 |
| 2008 | X3 | 235/55HR18 | ① ② | ③ | ③ | ④ ⑤ | NA | 103 |

OEM: Original Equipment Manufacturer

PSI: Pounds Per Square Inch

NA: Information not available

① Optional wheel size 235/50/R18

② Optional wheel size 255/45/R19

③ See placard on vehicle

④ Optional wheel with 18x8

⑤ Optional wheel size 19x9.5

22205_BMX3_C0011

## BRAKE SPECIFICATIONS
All measurements in inches unless noted

| Year | Model | | Brake Disc | | | Brake Drum Diameter | | | Minimum Lining Thickness | | Brake Caliper | |
|------|-------|---|-----------|-----------|-----------|-----------------|-----------|-----------------|-------|------|--------------------|-------------------|
| | | | Original Thickness | Minimum Thickness | Maximum Runout | Original Inside Diameter | Max. Wear Limit | Maximum Machine Diameter | Front | Rear | Bracket Bolts (ft. lbs.) | Mounting Bolts (ft. lbs.) |
| 2006 | X3 | F | — | 0.921① | 0.003 ② | — | — | — | 0.118 | — | 81 | 23 |
| | | R | — | 0.803 ① | 0.003 ② | 7.280 | — | 0.031 | 0.118 | — | 49 | 21 |
| 2007 | X3 | F | — | 0.921 ① | 0.003 ② | — | — | — | 0.118 | — | 81 | 23 |
| | | R | — | 0.803 ① | 0.003 ② | 7.280 | — | 0.031 | 0.118 | — | 49 | 21 |
| 2008 | X3 | F | — | 0.921 ① | 0.003 ② | — | — | — | 0.118 | — | 81 | 23 |
| | | R | — | 0.803 ① | 0.003 ② | 7.280 | — | 0.031 | 0.118 | — | 49 | 21 |

F: Front

R: Rear

① Minimum thickness is stamped in the brake disk shell

② Only with precision turned brake disc

22205_BMX3_C0012

## SCHEDULED MAINTENANCE INTERVALS
### BMW—X3

| TO BE SERVICED | TYPE OF SERVICE | SERVICE INTERVALS | | | |
| --- | --- | --- | --- | --- | --- |
| | | INITIAL 1200 MILES | OIL SERVICE | INSPECTION I | INSPECTION II |
| Oil level | S/I | ✓ | | | |
| Engine oil | R | ① | | | |
| Engine oil & filter | R ② | | ✓ | ✓ | ✓ |
| Engine air cleaner element | R ③ | | | | ✓ |
| Spark plugs | R | | | | ✓ |
| Fuel filter | R ④ | | | | ✓ |
| Fuel, vapor lines & fuel cap | S/I | ✓ | | ✓ | ✓ |
| Cooling system | S/I | ✓ | | ✓ | ✓ |
| Exhaust pipe & muffler | S/I | ✓ | | ✓ | ✓ |
| Catalytic converter & shielding | S/I | ✓ | | ✓ | ✓ |
| Throttle linkage | S/I | | | ✓ | ✓ |
| Engine (check for leakage) | S/I | ✓ | | | |
| Engine drive belts | S/I | | | | ✓ |
| Maintenance Indicators | RE | | ⑤ | ✓ | ✓ |
| Engine coolant | R | | | ⑥ | ⑥ |
| Oxygen sensor | R ⑦ | | | | |
| Intake air dust separators | S/I ⑧ | | | | ✓ |
| Brake & clutch fluids ⑥ | S/I | | | ✓ | ✓ |
| Brake pads & discs | S/I | | | ✓ | ✓ |
| Parking brake system | S/I | | | ✓ | ✓ |
| Power steering system | S/I | | | ✓ | ✓ |
| Rear axle fluid | S/I | | | ✓ | ✓ |
| Steering play, suspension track rods, front axle joints, steering linkage & joint | S/I | | | ✓ | ✓ |
| Transmission fluid/oil | S/I | | | ✓ | ⑨ |
| Wheel centering hubs | S/I | | | ✓ | ✓ |
| Rear axle fluid ⑩ | R | | ✓ | | ✓ |
| OBD system for codes | S/I | ✓ | | ✓ | ✓ |

R: Replace　　　S/I: Service or Inspect　　　RE: Reset

Note: BMW does not rely solely on vehicle mileage to determine service intervals. An on-oboard diagnostic center, monitors engine operating conditions, along with mileage, to determine the most effective maintenance intervals. The information is then conveyed to the driver through the service indicator lights, located in the center of the instrument p[anel.

① Service is not required for 325 models.

② On vehicles operated less than 6200 miles per year, more frequent service may be required.

③ Replace more frequently if vehicle is operated in dusty conditions.

④ Recommended service for California models, required for all other models.

⑤ Reset the oil service indicator lights only.

⑥ Replace every 2 years with inspection service.

⑦ Replace every 100,000 miles on all models.

⑧ Required service for manual transmission models only.

⑨ Change fluid (A/T) or oil (M/T) at inspection.

⑩ At first oil service, then at each inspection.

## FREQUENT OPERATION MAINTENANCE (SEVERE SERVICE)

If a vehicle is operated under any of the following conditions it is considered severe service

- Extremely dusty areas.

- 50% or more of the vehicle operation is in 32°C (90°F) or higher temperatures, or constant operation in temperatures below 0°C (32°F).

- Prolonged idling (vehicle operation in stop and go traffic).

- Frequent short running periods (engine does not warm to normal operating temperatures).

- Police, taxi, delivery usage or trailer towing usage.

## PRECAUTIONS

Before servicing any vehicle, please be sure to read all of the following precautions, which deal with personal safety, prevention of component damage, and important points to take into consideration when servicing a motor vehicle:

• Never open, service or drain the radiator or cooling system when the engine is hot; serious burns can occur from the steam and hot coolant.

• Observe all applicable safety precautions when working around fuel. Whenever servicing the fuel system, always work in a well-ventilated area. Do not allow fuel spray or vapors to come in contact with a spark, open flame, or excessive heat (a hot drop light, for example). Keep a dry chemical fire extinguisher near the work area. Always keep fuel in a container specifically designed for fuel storage; also, always properly seal fuel containers to avoid the possibility of fire or explosion. Refer to the additional fuel system precautions later in this section.

• Fuel injection systems often remain pressurized, even after the engine has been turned **OFF**. The fuel system pressure must be relieved before disconnecting any fuel lines. Failure to do so may result in fire and/or personal injury.

• Brake fluid often contains polyglycol ethers and polyglycols. Avoid contact with the eyes and wash your hands thoroughly after handling brake fluid. If you do get brake fluid in your eyes, flush your eyes with clean, running water for 15 minutes. If eye irritation persists, or if you have taken

brake fluid internally, IMMEDIATELY seek medical assistance.

• The EPA warns that prolonged contact with used engine oil may cause a number of skin disorders, including cancer. You should make every effort to minimize your exposure to used engine oil. Protective gloves should be worn when changing oil. Wash your hands and any other exposed skin areas as soon as possible after exposure to used engine oil. Soap and water, or waterless hand cleaner should be used.

• All new vehicles are now equipped with an air bag system, often referred to as a Supplemental Restraint System (SRS) or Supplemental Inflatable Restraint (SIR) system. The system must be disabled before performing service on or around system components, steering column, instrument panel components, wiring and sensors. Failure to follow safety and disabling procedures could result in accidental air bag deployment, possible personal injury and unnecessary system repairs.

• Always wear safety goggles when working with, or around, the air bag system. When carrying a non-deployed air bag, be sure the bag and trim cover are pointed away from your body. When placing a non-deployed air bag on a work surface, always face the bag and trim cover upward, away from the surface. This will reduce the motion of the module if it is accidentally deployed. Refer to the additional air bag system precautions later in this section.

• Clean, high quality brake fluid from a sealed container is essential to the safe and

proper operation of the brake system. You should always buy the correct type of brake fluid for your vehicle. If the brake fluid becomes contaminated, completely flush the system with new fluid. Never reuse any brake fluid. Any brake fluid that is removed from the system should be discarded. Also, do not allow any brake fluid to come in contact with a painted surface; it will damage the paint.

• Never operate the engine without the proper amount and type of engine oil; doing so WILL result in severe engine damage.

• Timing belt maintenance is extremely important. Many models utilize an interference-type, non-freewheeling engine. If the timing belt breaks, the valves in the cylinder head may strike the pistons, causing potentially serious (also time-consuming and expensive) engine damage. Refer to the maintenance interval charts for the recommended replacement interval for the timing belt, and to the timing belt section for belt replacement and inspection.

• Disconnecting the negative battery cable on some vehicles may interfere with the functions of the on-board computer system(s) and may require the computer to undergo a relearning process once the negative battery cable is reconnected.

• When servicing drum brakes, only disassemble and assemble one side at a time, leaving the remaining side intact for reference.

• Only an MVAC-trained, EPA-certified automotive technician should service the air conditioning system or its components.

## BRAKES

## ANTI-LOCK BRAKE SYSTEM (ABS)

### GENERAL INFORMATION

*PRECAUTIONS*

• Certain components within the ABS system are not intended to be serviced or repaired individually.

• Do not use rubber hoses or other parts not specifically specified for and ABS system. When using repair kits, replace all parts included in the kit. Partial or incorrect repair may lead to functional problems and require the replacement of components.

• Lubricate rubber parts with clean, fresh brake fluid to ease assembly. Do not

use shop air to clean parts; damage to rubber components may result.

• Use only DOT 3 brake fluid from an unopened container.

• If any hydraulic component or line is removed or replaced, it may be necessary to bleed the entire system.

• A clean repair area is essential. Always clean the reservoir and cap thoroughly before removing the cap. The slightest amount of dirt in the fluid may plug an orifice and impair the system function. Perform repairs after components have been thoroughly cleaned; use only denatured alcohol

to clean components. Do not allow ABS components to come into contact with any substance containing mineral oil; this includes used shop rags.

• The Anti-Lock control unit is a microprocessor similar to other computer units in the vehicle. Ensure that the ignition switch is **OFF** before removing or installing controller harnesses. Avoid static electricity discharge at or near the controller.

• If any arc welding is to be done on the vehicle, the control unit should be unplugged before welding operations begin.

**BRAKES**                                      **BLEEDING THE BRAKE SYSTEM**

## BLEEDING PROCEDURE

*BLEEDING PROCEDURE*

### ✳ WARNING

Clean, high quality brake fluid is essential to the safe and proper operation of the brake system. You should always buy the highest quality brake fluid that is available. If the brake fluid becomes contaminated, drain and flush the system, then refill the master cylinder with new fluid. Never reuse any brake fluid. Any brake fluid that is removed from the system should be discarded. Also, do not allow any brake fluid to come in contact with a painted surface; it will damage the paint.

### ❖ CAUTION

Brake fluid contains polyglycol ethers and polyglycols. Avoid contact with the eyes and wash your hands thoroughly after handling brake fluid. If you do get brake fluid in your eyes, flush your eyes with clean, running water for 15 minutes. If eye irritation persists, or if you have taken brake fluid internally, IMMEDIATELY seek medical assistance.

1. Remove the reservoir cap and fill the brake reservoir with brake fluid.
2. Connect a vinyl tube to the wheel cylinder bleeder screw and insert the other end of the tube in a clear container.
3. Slowly depress the brake pedal several times.
4. While depressing the brake pedal fully, loosen the bleeder screw until fluid runs out. Then close the bleeder screw and release the brake pedal.
5. Repeat these steps until there are no more bubbles in the fluid escaping to the clear container.
6. Tighten the bleeder screw to 80 inch lbs. (9 Nm).

7. Repeat the above procedure for each wheel.

*BLEEDING THE ABS SYSTEM*
*See Figure 1.*

### ✳✳ CAUTION

Brake fluid contains polyglycol ethers and polyglycols. Avoid contact with the eyes and wash your hands thoroughly after handling brake fluid. If you do get brake fluid in your eyes, flush your eyes with clean, running water for 15 minutes. If eye irritation persists, or if you have taken brake fluid internally, IMMEDIATELY seek medical assistance.

### ✳✳ WARNING

Clean, high quality brake fluid is essential to the safe and proper operation of the brake system. You should always buy the highest quality brake fluid that is available. If the brake fluid becomes contaminated, drain and flush the system, then refill the master cylinder with new fluid. Never reuse any brake fluid. Any brake fluid that is removed from the system should be discarded. Also, do not

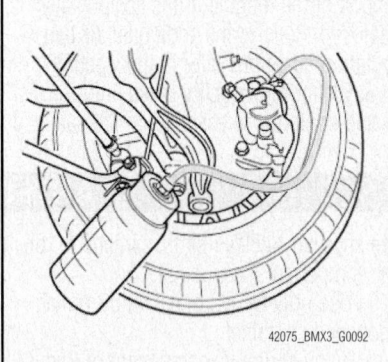

42075_BMX3_G0092

**Fig. 1 Using a clear plastic tube and bottle at bleeder screw to observe fluid bubbles**

allow any brake fluid to come in contact with a painted surface; it will damage the paint.

1. Remove the reservoir cap and fill the brake reservoir with brake fluid.
2. Connect a clear plastic tube to the wheel cylinder bleeder plug and insert the other end of the tube into a clear plastic bottle.
3. Activate pre—charging pump with Diagnosis and Information System (DIS).
4. Select and operate according to the instructions on the DIS screen:
   - Select: Service Functions
   - Select: Bleeding ABS/DSC3 Hydraulics

### ✳✳ CAUTION

You must obey the maximum operating time of the ABS motor to prevent the motor pump from burning.

5. Connect bleeder unit to expansion tank and switch on.
6. Pump the brake pedal several times, and then loosen the bleeder screw until fluid starts to run out without bubbles. Then close the bleeder screw.
7. Bleed fluid from each wheel until there are no more bubbles exiting with the brake fluid. Run bleeding routine with DIS and bleeder valve open.
8. Press brake pedal 5 times to the floor; clear and bubble-free brake fluid must flow out.
9. Close bleeder valve.
10. Repeat bleeding procedure on each wheel.
11. After completion of the repair or correction of the problem, erase any stored fault codes.
12. Disconnect the DIS system.
13. Fill the brake reservoir with the proper amount of brake fluid.
14. Check for fluid leaks at all connections.

## BRAKES                                                    FRONT DISC BRAKES

### ⁑ CAUTION

Dust and dirt accumulating on brake parts during normal use may contain asbestos fibers from production or aftermarket brake linings. Breathing excessive concentrations of asbestos fibers can cause serious bodily harm. Exercise care when servicing brake parts. Do not sand or grind brake lining unless equipment used is designed to contain the dust residue. Do not clean brake parts with compressed air or by dry brushing. Cleaning should be done by dampening the brake components with a fine mist of water, then wiping the brake components clean with a dampened cloth. Dispose of cloth and all residue containing asbestos fibers in an impermeable container with the appropriate label. Follow practices prescribed by the Occupational Safety and Health Administration (OSHA) and the Environmental Protection Agency (EPA) for the handling, processing, and disposing of dust or debris that may contain asbestos fibers.

### BRAKE CALIPER

#### REMOVAL & INSTALLATION

1. Before servicing the vehicle, refer to the precautions in the beginning of this section.
2. Remove or disconnect the following:
   - Negative battery cable
   - Wheel assembly
3. Apply the brake pedal slightly with a brake clamp.
   - Brake pipe from the connection with the brake hose
   - Connector for the wear indicator on the left side
   - Caliper guide bolts
   - Brake caliper

#### To install:

4. Install or connect the following:
   - Caliper and torque the guide bolts to 22 ft. lbs. (30 Nm)
   - Brake hose to the brake pipe to 13 ft. lbs. (18 Nm)
5. Set the wheel in a straight-ahead position.
6. Be sure brake hose is positively attached to the mounting fixture.
   - Wear indicator on the left side
   - Brake clamp
   - Front wheels
7. Bleed the brakes.

### DISC BRAKE PADS

#### REMOVAL & INSTALLATION

1. Before servicing the vehicle, refer to the precautions in the beginning of this section.
2. Remove the front wheels.
3. Remove the disk pad retaining spring from the caliper.
4. Remove the calipers from the disk.
5. Use a special tool, 34–1–050, to force piston back into caliper.
6. Remove the outer brake pad (the inner pad is held in place with a spring in the piston).

#### To install:

7. Be sure the pads marked "L" and "R" are inserted properly on left and right sides, respectively.
8. Apply anti-squeak compound to all mounting surfaces.
9. Install the calipers.
10. Reposition the retaining spring.
11. Install the front wheels.
12. Fully depress brake pedal several times to set proper contact of pads with rotor.
13. Hold ignition key for at least 30 seconds in position "1" without starting engine. This clear any fault codes stored in system and prevent the wear indicator light from coming on.
14. Bleed brake system. If necessary.

## BRAKES                                                     REAR DISC BRAKES

### ⁑ CAUTION

Dust and dirt accumulating on brake parts during normal use may contain asbestos fibers from production or aftermarket brake linings. Breathing excessive concentrations of asbestos fibers can cause serious bodily harm. Exercise care when servicing brake parts. Do not sand or grind brake lining unless equipment used is designed to contain the dust residue. Do not clean brake parts with compressed air or by dry brushing. Cleaning should be done by dampening the brake components with a fine mist of water, then wiping the brake components clean with a dampened cloth. Dispose of cloth and all residue containing asbestos fibers in an impermeable container with the appropriate label. Follow practices prescribed by the Occupational Safety and Health Administration (OSHA) and the Environmental Protection Agency (EPA) for the handling, processing, and disposing of dust or debris that may contain asbestos fibers.

### BRAKE CALIPER

#### REMOVAL & INSTALLATION

1. Before servicing the vehicle, refer to the precautions in the beginning of this section.
2. Remove or disconnect the following:
   - Negative battery cable
   - Wheel assembly
3. Apply the brake pedal slightly with a brake clamp.
   - Slacken union nut
   - Brake hose from the caliper fitting
   - Connector for the wear indicator on the right side
   - Caliper guide bolts and remove the brake caliper

#### To install:

4. Install or connect the following:
   - Caliper and torque the guide bolts to 21 ft. lbs. (28 Nm)
   - Brake hose to the brake pipe and torque to 14 ft. lbs. (19 Nm)
   - Wear indicator
5. Remove the brake clamp.
6. Install the rear wheels.
7. Bleed the brakes.

### DISC BRAKE PADS

#### REMOVAL & INSTALLATION

1. Before servicing the vehicle, refer to the precautions in the beginning of this section.
2. Remove the rear wheels.
3. Remove the plastic plugs from the inside of the caliper.
4. Disconnect the plug connection for the wear indicator.

5. Remove the calipers from the disk.

6. Lift out the pad retaining spring from the caliper.

7. Use a special tool, 34–1–050, to force piston back into caliper.

8. Remove the outer brake pad (the inner pad is held in place with a spring in the piston).

### To install:

9. Apply anti-squeak compound to all mounting surfaces.

10. Reposition retaining spring.

11. Install calipers.

12. Install rear wheels.

13. Fully depress brake pedal several times to set proper contact of pads with rotor.

14. Hold ignition key for at least 30 seconds in position "1" without starting engine. This clear any fault codes stored in system and prevent the wear indicator light from coming on.

15. Bleed brake system. If necessary.

---

## BRAKES                                PARKING BRAKE

### PARKING BRAKE SHOES

#### REMOVAL & INSTALLATION

*See Figures 2 through 4.*

1. Before servicing the vehicle, refer to the Precautions Section.

2. Remove rear disc brake rotor. Refer to Rear Disc Brakes, Rotor removal and installation.

3. Release parking brake lever.

4. Lock adjuster unit (ASZE):

a. Press stop (1) of adjusting spring back.

b. Move retaining hook (2) so that it engages in stop (1). Use special tool 32 1 030.

5. Disconnect upper return spring (1) with brake spring pliers.

6. Disconnect lower return spring with brake spring pliers.

➡**Check and if necessary replace return spring (1). Pay attention to installation position of adjustment screw (2).**

7. Apply a thin coat of grease to bushing and screw threads.

8. Turn clamping pins (1) with special tool 34 4 000 through 90° and disconnect.

9. Remove brake shoes (2).

### To install:

10. Installation is the reverse of the removal procedure.

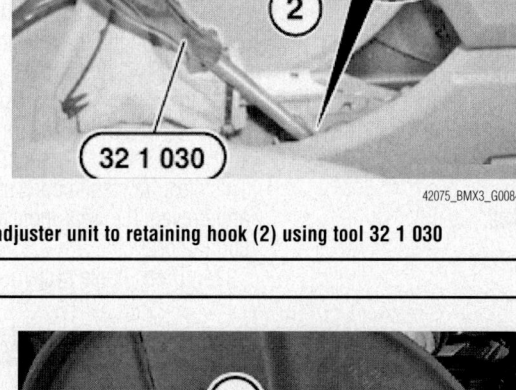

Fig. 2 Locking adjuster unit to retaining hook (2) using tool 32 1 030

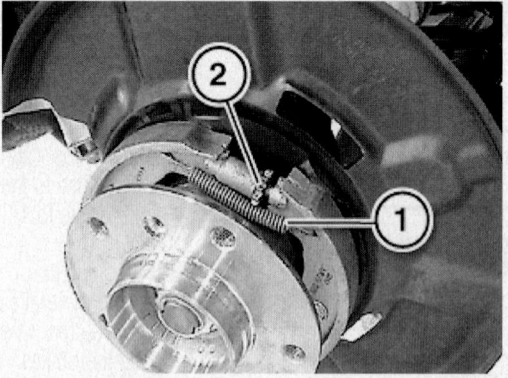

Fig. 3 Disconnecting return springs on parking brake

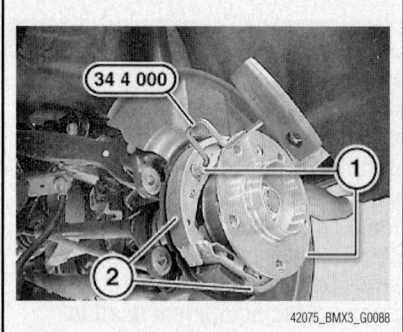

Fig. 4 Using special tool 34 4 000 to remove parking brake shoes

## CHASSIS ELECTRICAL — AIR BAG (SUPPLEMENTAL RESTRAINT SYSTEM)

### GENERAL INFORMATION

#### ✳✳ CAUTION

**These vehicles are equipped with an air bag system. The system must be disarmed before performing service on, or around, system components, the steering column, instrument panel components, wiring and sensors. Failure to follow the safety precautions and the disarming procedure could result in accidental air bag deployment, possible injury and unnecessary system repairs.**

#### *SERVICE PRECAUTIONS*

Disconnect and isolate the battery negative cable before beginning any airbag system component diagnosis, testing, removal, or installation procedures. Allow system capacitor to discharge for two minutes before beginning any component service. This will disable the airbag system. Failure to disable the airbag system may result in accidental airbag deployment, personal injury, or death.

Do not place an intact undeployed airbag face down on a solid surface. The airbag will propel into the air if accidentally deployed and may result in personal injury or death.

When carrying or handling an undeployed airbag, the trim side (face) of the airbag should be pointing towards the body to minimize possibility of injury if accidental deployment occurs. Failure to do this may result in personal injury or death.

Replace airbag system components with OEM replacement parts. Substitute parts may appear interchangeable, but internal differences may result in inferior occupant protection. Failure to do so may result in occupant personal injury or death.

Wear safety glasses, rubber gloves, and long sleeved clothing when cleaning powder residue from vehicle after an airbag deployment. Powder residue emitted from a deployed airbag can cause skin irritation. Flush affected area with cool water if irritation is experienced. If nasal or throat irritation is experienced, exit the vehicle for fresh air until the irritation ceases. If irritation continues, see a physician.

Do not use a replacement airbag that is not in the original packaging. This may result in improper deployment, personal injury, or death.

The factory installed fasteners, screws and bolts used to fasten airbag components have a special coating and are specifically designed for the airbag system. Do not use substitute fasteners. Use only original equipment fasteners listed in the parts catalog when fastener replacement is required.

During, and following, any child restraint anchor service, due to impact event or vehicle repair, carefully inspect all mounting hardware, tether straps, and anchors for proper installation, operation, or damage. If a child restraint anchor is found damaged in any way, the anchor must be replaced. Failure to do this may result in personal injury or death.

Deployed and non-deployed airbags may or may not have live pyrotechnic material within the airbag inflator.

Do not dispose of driver/passenger/curtain airbags or seat belt tensioners unless you are sure of complete deployment. Refer to the Hazardous Substance Control System for proper disposal.

Dispose of deployed airbags and tensioners consistent with state, provincial, local, and federal regulations.

After any airbag component testing or service, do not connect the battery negative cable. Personal injury or death may result if the system test is not performed first.

If the vehicle is equipped with the Occupant Classification System (OCS), do not connect the battery negative cable before performing the OCS Verification Test using the scan tool and the appropriate diagnostic information. Personal injury or death may result if the system test is not performed properly.

Never replace both the Occupant Restraint Controller (ORC) and the Occupant Classification Module (OCM) at the same time. If both require replacement, replace one, then perform the Airbag System test before replacing the other.

Both the ORC and the OCM store Occupant Classification System (OCS) calibration data, which they transfer to one another when one of them is replaced. If both are replaced at the same time, an irreversible fault will be set in both modules and the OCS may malfunction and cause personal injury or death.

If equipped with OCS, the Seat Weight Sensor is a sensitive, calibrated unit and must be handled carefully. Do not drop or handle roughly. If dropped or damaged, replace with another sensor. Failure to do so may result in occupant injury or death.

If equipped with OCS, the front passenger seat must be handled carefully as well. When removing the seat, be careful when setting on floor not to drop. If dropped, the sensor may be inoperative, could result in occupant injury, or possibly death.

If equipped with OCS, when the passenger front seat is on the floor, no one should sit in the front passenger seat. This uneven force may damage the sensing ability of the seat weight sensors. If sat on and damaged, the sensor may be inoperative, could result in occupant injury, or possibly death.

#### *DISARMING THE SYSTEM*

1. Before servicing the vehicle, refer to the precautions in the beginning of this section.

2. Place the ignition switch in the **OFF** position.

3. Disconnect the negative battery terminal and cover the battery terminal to prevent accidental contact.

4. Once the battery has been disconnected, wait for a period of approximately 3 minutes allowing the capacitor in the control unit to discharge. Once the capacitor is discharged, a trigger pulse cannot be generated inadvertently.

#### *ARMING THE SYSTEM*

1. Before servicing the vehicle, refer to the precautions in the beginning of this section.

2. Place the ignition switch in the **OFF** position.

3. Attach the sensors, the steering column connector and the seat belt tensioner connectors.

4. Connect the negative battery terminal.

5. Place the ignition switch in the **ON** position. Check that the SRS light illuminates for 6 seconds and then turns off. If it illuminates in any other pattern, check the components and their connections for proper operation and recheck operation of the warning light.

# DRIVETRAIN

## AUTOMATIC TRANSMISSION ASSEMBLY

### REMOVAL & INSTALLATION

*See Figures 5 and 6.*

1. Before servicing the vehicle, refer to the precautions in the beginning of this section.
2. Remove or disconnect the following:
   - Negative battery cable
   - Engine under guard at front
   - Reinforcement plate from undercarriage
   - Exhaust system
   - Cables for oxygen sensors, mark the locations for installation purposes
   - Front propeller shaft
   - Vent line from transmission at bracket
   - Exhaust bracket
   - Nut and bushing from shift lever
   - Shift control cable from body bracket
   - Electrical connection and cable from transmission
   - Hydraulic cooler lines from transmission
3. Position the proper transmission cradle, 00–2–030, 24–5–301/305.
4. Support the transmission and transfer case with transmission cradle.
5. Remove or disconnect the following:
   - Wiring cables from transmission
   - Crossmember
   - Propeller shaft flange nuts
   - Center bearing support and support the propeller shaft
   - Propeller shaft from transmission and tie aside
   - Hex bolt near protective cap on transmission flange
   - Protective cap
6. Protect torque converter from slipping out. Insert special tools 24–4–131/137 in opening in transmission housing. Clamp torque converter.
7. Remove or disconnect the following:
   - Bolts from torque converter through protective cap opening using special tool 24–1–110
   - Transmission housing–to–engine bolts
8. Pull the transmission, with transfer case, out toward rear.

   *To install:*

➡ **The sheet metal flywheel is equipped with three recesses for the torque converter retaining tabs. When guiding the engine and transmission together, the three mounting tabs on the torque converter must be aligned with the three indentations in the sheet metal flywheel. Turning on the torque converter, or engine, is no longer possible after guiding the engine and transmission together and can lead to damage.**

9. Note the position and condition of dowel sleeves on flywheel housing. Replace dowels if damaged.
10. Remove special tools 24–4–131/137 before install transmission to engine.
11. Position the transmission to the engine and install the housing bolts. Torque the bolts as follows:
    a. M8 hex bolts: 18 ft. lbs. (24 Nm).
    b. M10 hex bolts: 33 ft. lbs. (45 Nm).
    c. M12 hex bolts: 60 ft. lbs. (82 Nm).
    d. M8 Torx® bolts: 15 ft. lbs. (21 Nm).
    e. M10 Torx® bolts: 31 ft. lbs. (42 Nm).
    f. M12 Torx® bolts: 53 ft. lbs. (72 Nm).
12. Install and torque the torque converter–to–flywheel bolts to the following:
    a. M8 hex bolts: 19 ft. lbs. (26 Nm).
    b. M10 hex bolts: 33 ft. lbs. (45 Nm).
13. Install or connect the following:
    - Protective cap over flywheel bolt opening
    - Hex bolt near protective cap
    - Propeller shaft into center support; torque mount–to–body bolts to 15 ft. lbs. (21 Nm).
    - Propeller shaft flange new nuts; torque to 74 ft. lbs. (100 Nm)
14. Install the transmission crossmember. Torque the bolts as follows:
    a. To rubber mounts: 55 ft. lbs. (74 Nm).
    b. To body: 30 ft. lbs. (41 Nm).
15. Install or connect the following:
    - Transmission cables
    - Transmission cooler lines with new sealing ring
    - Wiring cable to transmission
    - Shift control cable to body bracket
    - Shift selector lever bushing and nut
    - Exhaust hanger bracket
    - Vent line to bracket
    - Engine guard
    - Reinforcement plate
    - Oxygen sensor connectors
    - Heat shield
    - Exhaust system
    - Negative battery cable
16. Refill transmission and check operation, then recheck fluid level.

## TRANSFER CASE ASSEMBLY

### REMOVAL & INSTALLATION

1. Before servicing the vehicle, refer to the precautions in the beginning of this section.
2. Remove or disconnect the following:
   - Negative battery cable
   - Exhaust system
   - Heat shield
   - Front propeller shaft set aside and secure
   - Propeller shaft nuts, bend shaft downward at center bearing
   - Propeller shaft from transmission and tie aside

➡ **Automatic transmissions: Use special tools 00–2–030, 24–5–301, 24–5–305. Manual transmission: Support with hydraulic lifter.**

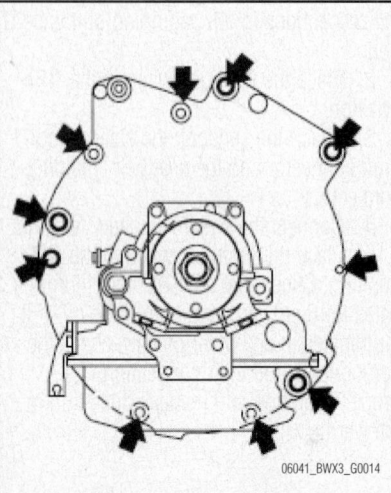

06041_BWX3_G0014

**Fig. 5 Showing transmission housing bolts—Automatic transmission**

06041_BWX3_G0015

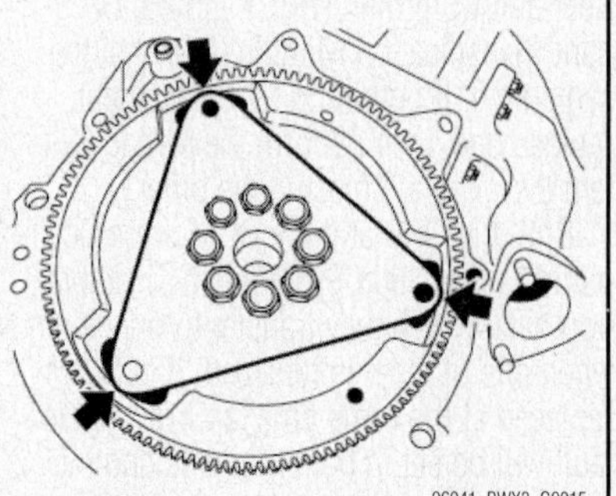

**Fig. 6 Showing three recesses for the torque converter retaining tabs—Automatic transmission**

3. Support the transmission
- Crossmember
- Electrical connectors to servomotor

4. Support the transfer case. Remove the retaining bolts and remove the transfer case.

***To install:***

5. Check the condition of the dowel pins and replace if necessary.

6. Be sure the mating surfaces are clean.

7. Apply a thin coat of anti-seize grease to transfer case splines.

8. Replace the sealing ring of the drive-shaft of the transfer case.

9. Install the transfer case and torque the retaining bolts to 32 ft. lbs. (43 Nm).

10. Reposition the propeller shaft in the center support bearing and to the flange. Torque the center support bearing nuts to 15 ft. lbs. (21 Nm) and the flange nuts to 24 ft. lbs. (32 Nm).

11. Install or connect the following:
- Cable to transfer case

12. Install the transmission crossmember. Torque the bolts as follows:

    a. To rubber mounts: 55 ft. lbs. (74 Nm).

    b. To body: 30 ft. lbs. (41 Nm).

13. Install or connect the following:
- Electrical connectors to servomotor
- Propeller shaft and center bearing
- Reinforcement plate
- Heat shields
- Exhaust system

14. Refill the transfer case.

## FRONT DRIVESHAFT

### *REMOVAL & INSTALLATION*

1. Before servicing the vehicle, refer to the precautions in the beginning of this section.

2. Remove or disconnect the following:
- Negative battery cable
- Bolts on rear flange (4)
- Bolts on front flange (4)
- Remove Shaft

***To install:***

3. Install or connect the following:
- Shaft
- Bolts on front flange (4)
- Bolts on rear flange (4)
- Negative battery cable

➡**Torque shaft to specifications below:**

- Compression nut M10—50 ft. lbs. (64 Nm)
- Torx® bolt universal joint M10—62 ft. lbs. (85 Nm)
- Torx® bolt constant velocity joint M10—52 ft. lbs. (70 Nm)

- Compression nut constant velocity joint M8—24 ft. lbs. (32 Nm)
- Finned nut M8—32 ft. lbs. (43 Nm)
- Finned nut M10—52 ft. lbs. (70 Nm)
- ZNS Screws and nuts, shiny zinc coating, all versions with universal or constant velocity joint

➡ **For ZNS hardware, replace bolts and nuts. Jointing torque and angle of rotation must be observed without fail.**

- Universal or constant velocity joint M10—M 10.9 screws with ribbed teeth 29 ft. lbs. (40 Nm) plus 45°
- M10—M 10.9 joining torque 14.5 ft. lbs. (20 Nm) Plus 90°
- M12 x 55—M 10.9 replace screws and shims 52 ft. lbs. (70 Nm)
- Pivot to center propeller shaft (version without slide) 16 ft. lbs. (21 Nm)
- Center mount to body 16 ft. lbs. (21 Nm)
- Universal joint bolts M10—10.9 to 29 ft. lbs. (40 Nm) plus 45°

## FRONT HALFSHAFT

### *REMOVAL & INSTALLATION*

1. Before servicing the vehicle, refer to the precautions in the beginning of this section.

2. Remove or disconnect the following:
- Negative battery cable
- Front wheel
- Reinforcement plate
- Front splash guard
- ABS pulse generator
- Brake caliper from disc and tie out of the way with the hose still connected
- Steering gear tie rod from swivel bearing
- Tension strut, with guide joint, from swivel bearing
- Control arm from swivel bearing
- Collar nut on halfshaft, press half-shaft out of drive flange
- Halfshaft

***To install:***

3. Push in output shaft over the resistance of the retaining ring until it snaps in place.

➡**Use a new snap ring on halfshaft spline**

4. Install or connect the following:
- Collar nut
- Control arm to swivel bearing
- Tension strut to swivel bearing

- Tie rod to swivel bearing
- Brake caliper to disc
- ABS pulse generator
- Front wheel
- Splash guard
- Reinforcement plate

5. Check the front axle differential fluid level.

### *CV-BOOTS INSPECTION*

1. Before servicing the vehicle, refer to the Precautions Section.

2. Check the driveshaft boots for damage and deterioration.
- Raise front of vehicle
- Rotate axle and inspect for cracked or ripped CV boot material on inner and outer CV joints on both sides of vehicle

3. Replace boot if damaged or deteriorated.

## FRONT PINION SEAL

### *REMOVAL & INSTALLATION*

*See Figures 7 through 13.*

1. Before servicing the vehicle, refer to the Precautions Section.

2. Remove or disconnect the following:
- The exhaust system
- The driveshaft

3. Remove the drive flange.

4. Remove nuts.

5. Before removing collar nut, mark drive flange (1) and collar nut (2) to drive-shaft (3) with center punch or color marker pen.

6. Remover the Lever shaft seal (1) with a screwdriver (2) out of bearing block (3).

***To install:***

7. Protective sleeve (2) serves to protect the sealing lips of shaft seal (1) from being damaged when the output shaft is inserted into the bearing block.

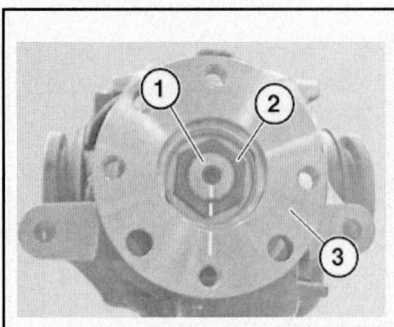

42075_BMX3_G0093

**Fig. 7 Marking components before removal—drive flange (1), collar nut (2), driveshaft (3)**

➡ **Check that no seal residue is left in the final drive.**

8. Clean end face of drive flange and apply a thin coating of grease.

9. Coat sealing lips of shaft seal and sealing surface of drive flange with differential oil.

10. Drive in shaft seal with special tool 31 5 130 as far as it will go.

11. Attach drive flange according to the markings made earlier.

12. Press on drive flange with special tools 23 1 300, 33 1 341 and 23 1 303, if necessary 23 1 302, until collar nut can be screwed on.

➡ **Do not tighten down collar nut beyond marker points, made before removal, in order to avoid damaging the clamping sleeve.**

13. Tighten down collar nut to point where marker points are aligned.

14. Drive in new retaining plate with special tools 00 5 500 and 33 3 480.

15. Install new self-locking nuts on compression strut. Tighten to 57 ft. lbs. (77 Nm).

16. Install the drive flange. Tighten to 47 ft. lbs. (64 Nm).

17. Install the driveshaft.

18. Refill the oil in the final drive.

## REAR AXLE HOUSING

### REMOVAL & INSTALLATION

*See Figures 14 through 16.*

1. Before servicing the vehicle, refer to the Precautions Section.

2. Remove or disconnect the following:
   - The exhaust system
   - The driveshaft from the rear differential

➡ **Match mark the shaft and flange**
   - Output shafts
   - Compression strut

➡ **Support the rear differential with a jack.**

3. Remove release screws

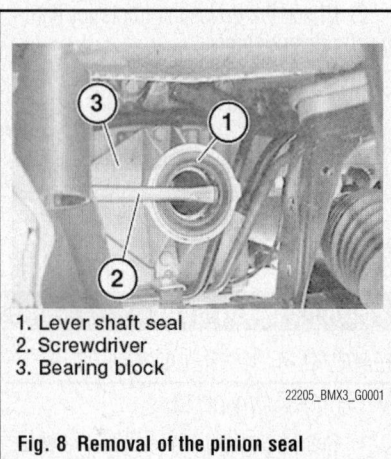

1. Lever shaft seal
2. Screwdriver
3. Bearing block

22205_BMX3_G0001

**Fig. 8  Removal of the pinion seal**

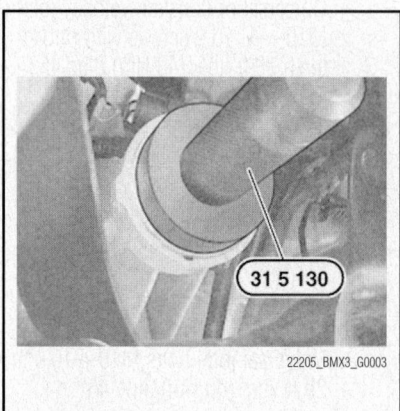

31 5 130

22205_BMX3_G0003

**Fig. 11  Installation of pinion seal**

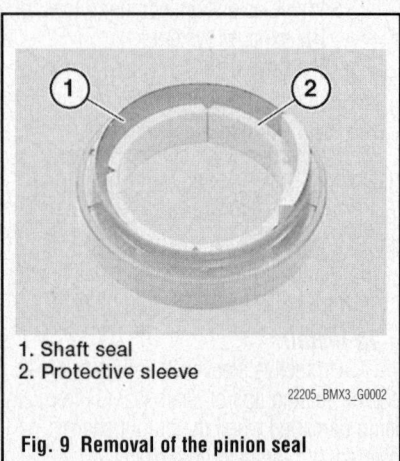

1. Shaft seal
2. Protective sleeve

22205_BMX3_G0002

**Fig. 9  Removal of the pinion seal**

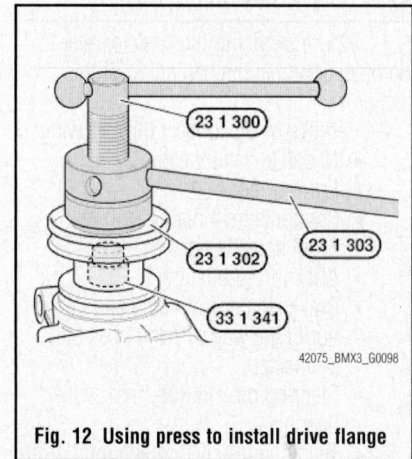

23 1 300
23 1 302
23 1 303
33 1 341

42075_BMX3_G0098

**Fig. 12  Using press to install drive flange**

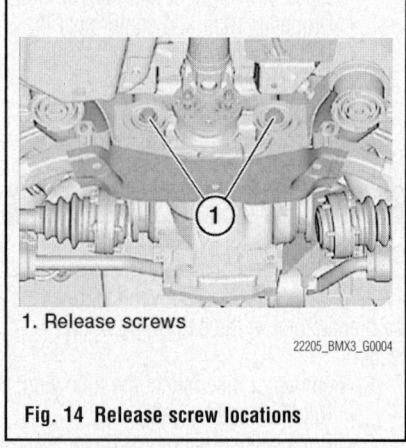

1. Release screws

22205_BMX3_G0004

**Fig. 14  Release screw locations**

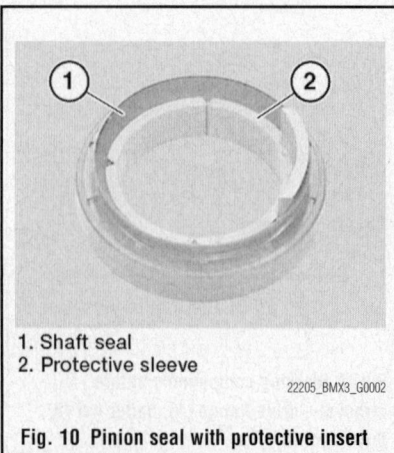

1. Shaft seal
2. Protective sleeve

22205_BMX3_G0002

**Fig. 10  Pinion seal with protective insert**

00 5 500
33 3 480
33 3 490

42075_BMX3_G0099

**Fig. 13  Driving in new retaining plate with tools 00 5 500 and 33 3 480**

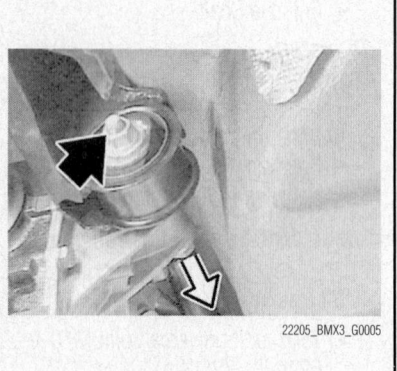

22205_BMX3_G0005

**Fig. 15  Carrier bolt in center of vehicle**

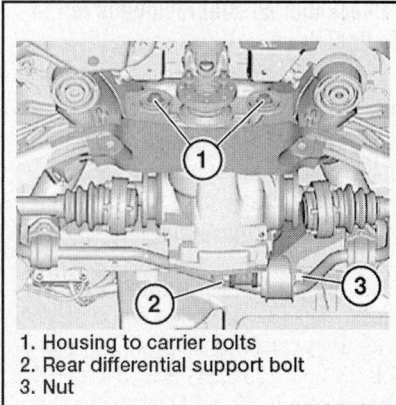

1. Housing to carrier bolts
2. Rear differential support bolt
3. Nut

**Fig. 16 Housing bolts**

22205_BMX3_G0006

42075_BMX3_G0093

**Fig. 17 Marking components before removal—drive flange (1), collar nut (2), driveshaft (3)**

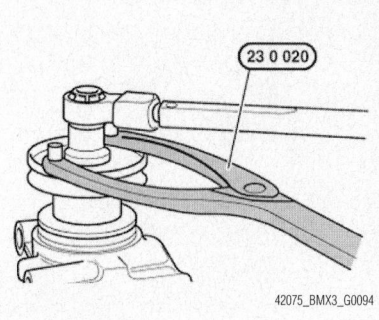

42075_BMX3_G0094

**Fig. 18 Using special tool 23 0 020 to brace drive flange**

42075_BMX3_G0095

**Fig. 19 Using special tool 33 1 150 to remove drive flange**

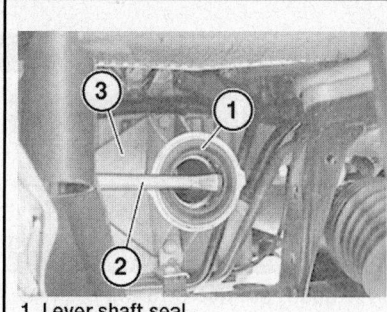

1. Lever shaft seal
2. Screwdriver
3. Bearing block

22205_BMX3_G0001

**Fig. 20 Removal of the pinion seal**

4. Remove bolt in the center of the vehicle, if necessary, remove spacer.

5. Slowly lower the jack and tip the rear differential out toward the rear of the vehicle.

### To install:

6. Install the rear differential the jack into the vehicle.

7. Insert the bolts (1), do not tighten.

8. Insert the bolts and nut (2 and 3), do not tighten.

➡**If necessary, fit spacer.**

9. Remove jack.

10. Install compression nut.

11. Tighten screws to 78 ft. lbs. 100 (Nm).

12. Tighten nut to 120 ft. lbs. 165 (Nm).

➡**Check differential oil**

13. Install the driveshaft to the rear differential using the match marks made at disassembly.
   output shafts

14. Install the exhaust system

## REAR DRIVESHAFT

### REMOVAL & INSTALLATION

*See Figure 17.*

1. Before servicing the vehicle, refer to the Precautions Section.

2. Remove the exhaust system

3. Before removing collar nut, mark drive flange to driveshaft with center punch or color marker pen.

4. Remove the driveshaft

### To install:

5. Install new self-locking nuts on compression strut. Tighten to 57 ft. lbs. (77 Nm).

6. Install the drive flange. Tighten to 47 ft. lbs. (64 Nm).

7. Install the exhaust system.

## REAR HALFSHAFT

### REMOVAL & INSTALLATION

1. Before servicing the vehicle, refer to the precautions in the beginning of this section.

2. Remove or disconnect the following:
   • Rear wheel
   • Rear muffler from center muffler
   • Flange bolts to differential
   • Depressurize ride control system, if equipped

3. Raise the rear wheel carrier with a proper jack and support.

4. Press the halfshaft from the wheel carrier.

5. Remove the halfshaft.

### To install:

6. Position the halfshaft and press into the wheel carrier.

7. Fit the shims, then install the bolts on the inner halfshaft flange to 61 ft. lbs. (83 Nm).

8. Attach the rear muffler to the center muffler.

9. Install the rear wheel.

## REAR PINION SEAL

### REMOVAL & INSTALLATION

*See Figures 17 through 24.*

1. Before servicing the vehicle, refer to the Precautions Section.

2. Remove or disconnect the following:
   • The exhaust system
   • The driveshaft

3. Remove the drive flange:
   a. Support rear axle carrier at front middle with workshop jack.
   b. Remove nuts and remove compression strut.

4. Before removing collar nut, mark drive flange (1) and collar nut (2) to driveshaft (3) with center punch or color marker pen.

5. Brace drive flange with special tool 23 0 020 and release collar nut.

6. Remove drive flange with special tool 33 1 150.

7. Remover the Lever shaft seal (1) with a screwdriver (2) out of bearing block (3).

### To install:

8. Protective sleeve (2) serves to protect the sealing lips of shaft seal (1) from being damaged when the output shaft is inserted into the bearing block.

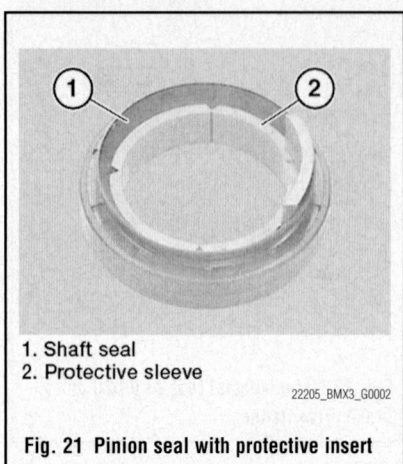

1. Shaft seal
2. Protective sleeve

22205_BMX3_G0002

**Fig. 21  Pinion seal with protective insert**

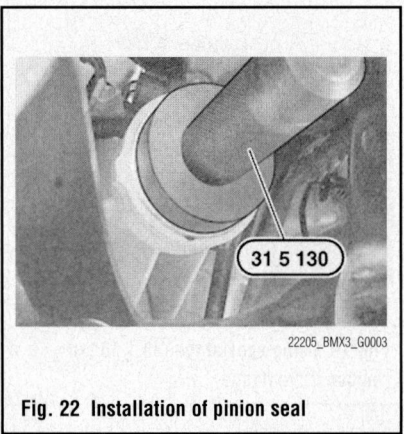

22205_BMX3_G0003

**Fig. 22  Installation of pinion seal**

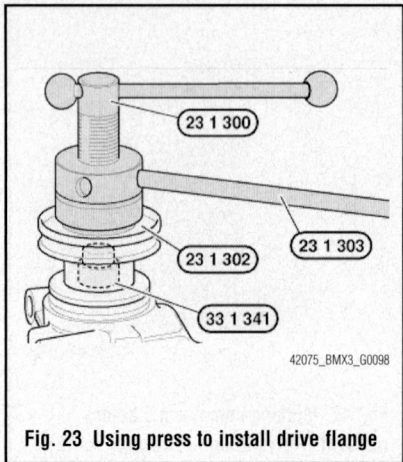

42075_BMX3_G0098

**Fig. 23  Using press to install drive flange**

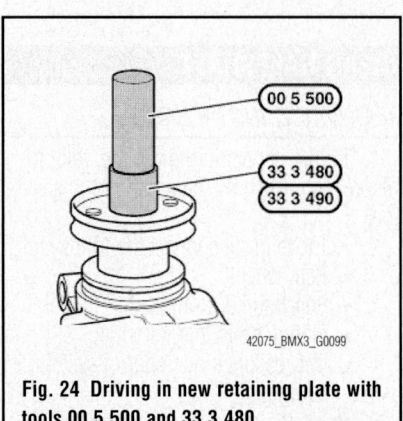

42075_BMX3_G0099

**Fig. 24  Driving in new retaining plate with tools 00 5 500 and 33 3 480**

➡**Check that no seal residue is left in the final drive.**

9. Drive in shaft seal with special tool 31 5 130 as far as it will go.
10. Clean end face of drive flange and apply a thin coating of grease.
11. Coat sealing lips of shaft seal and sealing surface of drive flange with differential oil.
12. Attach drive flange according to the markings made earlier.
13. Press on drive flange with special tools 23 1 300, 33 1 341 and 23 1 303, if necessary 23 1 302, until collar nut can be screwed on.

➡**Do not tighten down collar nut beyond marker points, made before removal, in order to avoid damaging the clamping sleeve.**

14. Tighten down collar nut to point where marker points are aligned.
15. Drive in new retaining plate with special tools 00 5 500 and 33 3 480.
16. Install new self-locking nuts on compression strut. Tighten to 57 ft. lbs. (77 Nm).
17. Install the drive flange. Tighten to 47 ft. lbs. (64 Nm).
18. The driveshaft.
19. The exhaust system.
20. Refill the oil in the final drive.

# ENGINE COOLING

## THERMOSTAT

*REMOVAL & INSTALLATION*
*See Figures 25 through 27.*

### ✷✷ CAUTION

**Never open, service or drain the radiator or cooling system when hot; serious burns can occur from the steam and hot coolant. Also, when draining engine coolant, keep in mind that cats and dogs are attracted to ethylene glycol antifreeze and could drink any that is left in an uncovered container or in puddles on the ground. This will prove fatal in sufficient quantities. Always drain coolant into a sealable container. Coolant should be reused unless it is contaminated or is several years old.**

1. Before servicing the vehicle, refer to the Precautions Section.
2. Remove or disconnect the following:
   • The engine fan. Refer to Engine Fan removal and installation

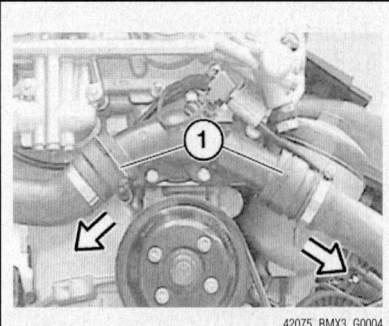

42075_BMX3_G0004

**Fig. 25  Location and removal of coolant hoses from thermostat housing**

• Engine coolant. Close drain plug after coolant is removed. Tighten plug to 19 ft. lbs. (25 Nm)

➡**The drain plug is located on the exhaust side on cylinder 2 in the engine block.**

• Coolant hose clamps and coolant hoses
• Bolts on thermostat housing
• Thermostat housing

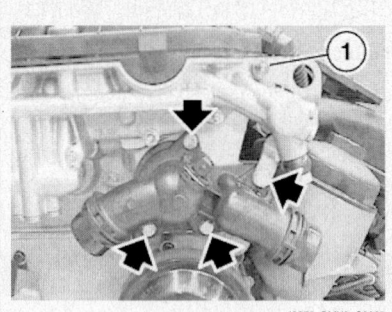

42075_BMX3_G0005

**Fig. 26  Bolt removal from thermostat housing**

*To install:*
Keep sealing faces clean and free of oil. The coolant thermostat is integrated in the coolant thermostat housing and can only be replaced as a single unit.

3. Install or connect the following:
   • New seal for thermostat housing
   • Thermostat housing
   • Coolant hoses to thermostat housing and tighten clamps
   • Engine fan

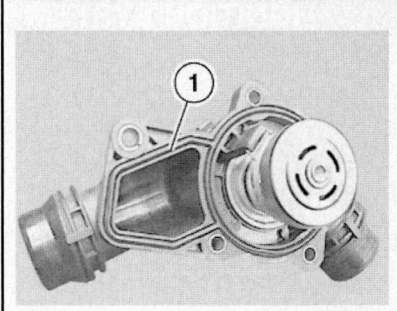

**Fig. 27 Interior view of thermostat housing**

4. Refill with proper engine coolant.

5. Test drive the engine until normal operating temperature is achieved.

6. Check for leaks and proper function. Fill fluid as necessary.

## WATER PUMP

### REMOVAL & INSTALLATION

*See Figure 28.*

1. Before servicing the vehicle, refer to the precautions in the beginning of this section.

2. Remove the alternator drive belt.

3. Drain the cooling system. Drain plug is located on exhaust side of block, next to cylinder number 2.

4. Remove the water pump pulley.

5. Remove the 4 water pump retaining bolts. Use 2 M6 bolts in holes next to mounting bolt holes and screw in until water pump releases from timing cover.

### To install:

6. When installing, use a new O-ring.

7. Tighten mounting bolts as follows:
   a. M6 bolts: 88 inch lbs. (10 Nm).

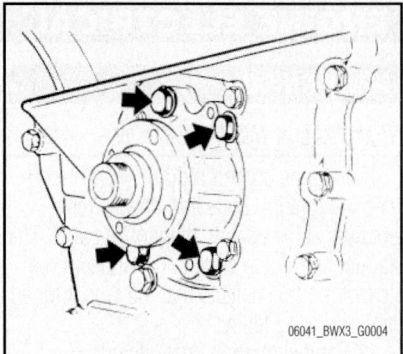

**Fig. 28 Showing water pump bolts—3.0L engine**

   b. M7 bolts: 132 inch lbs. (15 Nm).
   c. M8 bolts: 16 ft. lbs. (22 Nm).

8. Install water pump pulley.

9. Refill cooling system.

10. Install the alternator drive belt.

11. Start the vehicle, check for leaks and repair as necessary.

## ENGINE ELECTRICAL

### ALTERNATOR

### REMOVAL & INSTALLATION

*See Figure 29.*

➡When the battery is disconnected the radio code, on-board computer and clock settings will be lost. The radio code should be obtained before disconnecting the battery or radio. Once the battery has been reconnected, the radio will not function unless the code is keyed in.

1. Read out stored fault codes (if applicable).

2. Switch off ignition.

3. Disconnect negative battery cable.

4. Remove or disconnect the following:
   - Suction filter housing
   - Fan clutch
   - Alternator drive belt
   - Power steering pump from mounting (move aside)
   - Alternator air hose (if equipped)
   - Electrical connections from alternator
   - Alternator mounting bolts (versions without idler pulley)
   - Idler pulley cover and idler pulley bolt (versions with idler pulley)
   - Alternator

### To install:

5. To install, reverse removal procedure.

➡Replace mounting hardware.

6. Torque bolts to 8 ft. lbs. (10 Nm) plus 180°.

7. If equipped with idler pulley, turn lock of tensioning roller to engage alternator groove.

8. With scan tool, clear fault code memory.

## CHARGING SYSTEM

**Fig. 29 Showing idler pulley bolt and alternator bolt (3.0L engines)**

## IGNITION COIL

### REMOVAL & INSTALLATION

The engine control module (ECM) regulates the ignition coils and checks their function via a separate diagnostic cable. The diagnostic trouble code is generated if the control module detects that the signal for an ignition coil is faulty.

1. Fault symptoms may include:
   - Malfunction indicator lamp (MIL) lit
   - The engine is not firing on all cylinders

The diagnostic trouble code can be diagnosed when the engine is running and will indicate which ignition coil is involved.

## IGNITION TIMING

### ADJUSTMENT

The Digital Motor Electronics (DME) control unit carries out the operation of all ignition and fuel injection functions. Ignition timing is fully electronically controlled; there is no vacuum advance or manual adjustment. Ignition functions are calculated from internal maps and from the same sensors used for the fuel injection system. On vehicles with an automatic transmission, the control unit will retard ignition timing briefly when the transmission is about to shift up or down. For this reason, there is a data link between the DME control unit and the transmission control unit.

Since the ignition timing is controlled by the DME, checking and adjusting the timing is impossible. There is no method of setting dynamic or static timing.

## SPARK PLUGS

### REMOVAL & INSTALLATION

1. Before servicing the vehicle, refer to the Precautions Section.

2. Remove ignition coils. Refer to Ignition Coil Pack removal and installation.

3. Remove spark plug connector from the spark plug.

4. Clean loose debris away from area of spark plug to keep contaminants from entering engine when spark plug is removed.

5. Remove the spark plug using a spark plug socket and wrench.

### To install:

6. Be sure the spark plug gap is set to 0.028 in. (0.712mm).

7. Carefully install the spark plug and tighten to 21–24 ft. lbs. (29–33 Nm).

8. Install ignition coils. Refer to Ignition Coil Pack removal and installation.

➡ Disconnecting the negative battery cable may interfere with the functions of the on board computer systems and may require the computer to undergo a relearning process, once the negative battery cable is reconnected.

### ❋ WARNING

The engine is constructed with Aluminum-magnesium materials. No steel screws/bolts may be used due to the threat of electrochemical corrosion. A magnesium crankcase requires aluminum screws/bolts exclusively. Aluminum screws/bolts must be replaced each time they are released. The end faces of aluminum screws/bolts are painted blue for the purposes of reliable identification. Jointing torque and angle of rotation must be observed without fail or risk of damage.

## ACCESSORY DRIVE BELTS

### ACCESSORY BELT ROUTING
See Figure 30.

### INSPECTION

Inspect the drive belt for signs of glazing or cracking. A glazed belt will be perfectly smooth from slippage, while a good belt will

**Fig. 30 Accessory drive belt routing**

42075_BMX3_G0045

have a slight texture of fabric visible. Cracks will usually start at the inner edge of the belt and run outward. All worn or damaged drive belts should be replaced immediately.

### ADJUSTMENT

The accessory drive belt adjustment is maintained by an automatic tensioner.

### REMOVAL & INSTALLATION
See Figures 31 through 33.

1. Before servicing the vehicle, refer to the Precautions Section.

➡ If the drive belt is to be reused, mark direction of travel and reinstall drive belt in same direction of travel.

**Fig. 31 Removing A/C compressor drive belt**

42075_BMX3_G0046

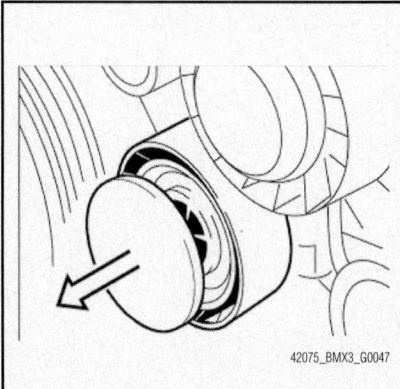

**Fig. 32 Removing dust cap for drive belt tensioner**

42075_BMX3_G0047

**Fig. 33 Releasing tension off drive belt for removal/installation**

2. Remove A/C compressor drive belt:

a. Release tension from drive belt by attaching hexagon head wrench or Torx® socket to belt tensioner.

b. Remove A/C drive belt.

3. Remove alternator/accessory drive belt:

a. Remove dust cap from accessory drive belt tensioner.

b. Relieve tension on drive belt.

c. Remove accessory drive belt.

**To install:**

➡**Check drive belts for coolant and oil residues, replace if necessary.**

4. Installation is the reverse of the removal procedure.

5. Route drive belts around pulleys in same order of removal.

## CAMSHAFT AND VALVE LIFTERS

### INSPECTION

1. Check the camshaft journals for wear. If the journals are badly worn, replace the camshaft.

2. Check the cam lobes for damage. If the lobe is damaged or excessively worn, replace the camshaft.

3. Check the cam surface for abnormal wear or damage, and replace if necessary.

4. Check each bearing for damage. If the bearing surface is excessively damaged, replace the cylinder head assembly or camshaft-bearing cap, as necessary.

### REMOVAL & INSTALLATION

See Figures 34 through 39.

1. Before servicing the vehicle, refer to the precautions in the beginning of this section.

2. Disconnect the negative battery cable.

3. Remove the camshaft (Double VANOS) adjustment unit as follows:

a. Remove the intake filter housing with Mass Air Flow (MAF) sensor.

b. Remove the fan impeller with fan clutch and cowl.

c. Remove the cylinder head covers.

d. Remove the spark plugs.

e. Remove the plastic cover for intake camshaft.

f. Remove the oil pressure pipe.

g. Install a special tool, 11–3–450, with banjo bolt on oil pressure connection unit.

h. Cover the camshaft adjustment unit.

i. Connect a compressed air hose to the special tool fitting on the oil line connection.

j. Rotate the engine, a least 2 turns in direction of rotation, to return the camshafts to TDC position (front cam lobes pointing to 10 o'clock and 2 o'clock positions).

k. Install a special locking tool, 11–2–300, into hole to lock flywheel in TDC position.

l. Remove the 2 studs from the rear of the cylinder head outer edge. Secure the camshafts with locking tools, 11–2–240.

m. Disconnect the compressed air hose from the special tool.

n. Using care so no oil drips onto belt drive, remove the screw plugs from the camshaft adjustment unit cover corresponding to the forward ends of the camshafts. Have a container to catch oil that runs out.

o. Using short, flat nose pliers, remove sealing caps through screw plug openings, then remove the fitting screws (left-hand thread) from both camshaft forward ends.

p. Detach the plug connection from the camshaft sensor and solenoid valves on both camshafts.

q. Remove the engine lifting eye from the camshaft adjustment unit cover.

r. Remove nuts and remove the camshaft adjustment unit

### ❋❋ WARNING

**Once camshaft adjust unit is removed, DO NOT crank engine. The toothed shaft on the intake side camshaft may slip out of the spline teeth and valves could rest of the piston.**

4. Remove the cylinder for the chain tensioning piston. Use caution as piston is under spring pressure.

5. Press down on the secondary chain tensioner at the top and lock the chain in place with a special tool, 11–3–292.

**Fig. 34 Attaching compressed air hose to oil line connector—3.0L engines**

**Fig. 35 Showing location of retaining nuts on camshaft adjustment unit (Double VANOS)—3.0L engines**

6. Remove or disconnect the following:

- Sensor gear from exhaust camshaft
- Spring plate from behind sensor gear
- Nuts on intake camshaft and remove corrugated washer
- Screws from chain gear on exhaust camshaft
- Toothed shaft with sleeve
- Secondary (upper) chain tensioner
- Release screw-in pins from exhaust camshaft chain gear
- Exhaust camshaft chain gear (leave chain on end of camshaft)
- Thrust washer on intake camshaft
- Intake camshaft sensor gear

### ❋❋ CAUTION

**DO NOT release screws from front ends of either camshaft.**

- Studs from between both camshafts

7. Release the special locking tool so flywheel is free.

8. Lift the timing chain off from end of exhaust camshaft and hold it under tension by lifting it straight up.

9. Rotate the engine against the normal direction of rotation about 30°, using the crankshaft bolt.

**⁂ CAUTION**

To prevent the intake camshaft from moving the bearing inserts, remove the camshaft no. 1 bearing cap nuts and remove the bearing cap.

10. Install a special tool, 11-2-260, to the cylinder head and install retaining bolts into tool and into spark plug holes for cylinders 1 and 4.

11. Turn the eccentric shaft of the special tool to pre-tension the bearing caps. Remove the nuts on all bearing caps.

12. Relieve the tension from the eccentric shaft of the special tool and remove the tool.

13. Remove the bearing caps and keep in order as removed.

14. Remove the camshaft. Repeat the process for the other camshaft.

➡ If cylinder head is to be removed, remove the complete bearing strip with the bucket tappets. Keep in order of original locations.

### To install:

15. Check bearing points of bucket tappets for signs of scoring.

➡ Bearing strips are marked "A" for exhaust side and "E" for intake side.

16. Be sure centering dowels are in place on retaining pins at bearing points 2 and 7.

17. Install the bearing strips.

18. Oil all teeth, camshafts, bearings, bearing caps, and friction washers before installation.

**⁂ CAUTION**

Bucket tappets expand when not subjected to load by the camshaft and therefore require some time before they can be pushed back down.

During a rapid assembly sequence, the "closed" valves may still be open and therefore be in contact with the piston. After assembly, wait at least 30 minutes before cranking the engine back to the TDC position.

19. Pull up the timing chain and feed in the exhaust camshaft. Position the timing chain onto the end of the exhaust camshaft.

**⁂ CAUTION**

To prevent damaging valves when fitting camshafts, no pistons should be in TDC position.

20. Install the camshafts so the cam tips on the intake and exhaust valves on no. 1 cylinder face each other at about the 10 o'clock and 2 o'clock positions.

21. Install the bearing caps (caps are marked from the exhaust side, A1 through A7 for the exhaust side, and E1 through E7 for the intake side).

22. Install the special tool, 11-3-260, onto the cylinder head as during removal.

23. Turn the eccentric shaft to pre-tension the bearing caps. Tighten the bearing cap bolts as follows:

    a. M6 bolts: 88 inch lbs. (10 Nm).
    b. M7 bolts: 120 inch lbs. (14 Nm).
    c. M8 bolts: 15 ft. lbs. (20 Nm).

24. Remove the special tool.

25. Install camshaft locking tools, 11-3-240 on back end of camshafts. Use an open-end wrench to align camshafts (if necessary, machine wrench head so it does not contact cylinder head) so there is no gap at the locking tools.

26. Install the middle locking tool, 11-3-244 so it adjoins other tools and is bolted into spark plug hole.

27. Lift the timing chain straight up off the exhaust camshaft and hold it under tension.

28. Rotate the engine from 30° BTDC to TDC, in normal direction of rotation.

29. Using special tool, 11-2-300, lock flywheel in this position.

30. Install sensor gear onto intake camshaft. Install the thrust washer over the sensor gear and torque the gear screw-in pins to 15 ft. lbs. (20 Nm).

31. Feed the chain wheel onto the timing chain so the arrow on the chain wheel faces the upper edge of the cylinder head.

32. Install a special tool, 11-4-220 into the cylinder head in the chain tensioning piston bore and bring tool adjustment screw into contact with the timing chain tensioning rail, but no further at this time.

33. Check the arrow mark on the chain wheel and adjust if needed so it still points to upper edge of the cylinder head.

34. Install and tighten the retaining screw-in pins in the exhaust chain wheel to 15 ft. lbs. (20 Nm).

35. Install the secondary (upper) chain tensioner.

36. Position the toothed sleeve into the exhaust camshaft to the toothed gaps are opposed. Secure the toothed shaft. Insert the pin of the toothed shaft into the tooth gaps of the splines on the camshaft and toothed sleeve.

37. Push in the toothed shaft on the exhaust camshaft until the elongated holes in the tooth sleeve wheel are centered over the bolt holes.

38. Place forward chain wheels onto special tool, 116-6-180, and position tooth gap on intake chain wheel as shown as feed on the timing chain.

**⁂ CAUTION**

DO NOT alter position of chain wheels and chain when removing the special tool.

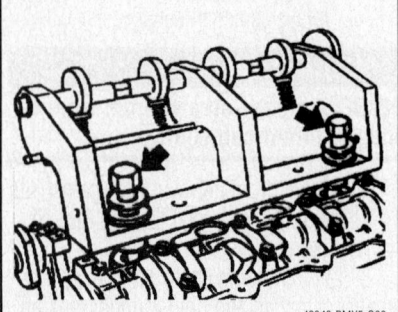

42348-BMX5-G08

**Fig. 36 Installing special tool to pretension camshaft bearing caps for removal—3.0L engines**

42348-BMX5-G09

**Fig. 37 Setting timing chain and chain wheel mark in position during installation of chain—3.0L engines**

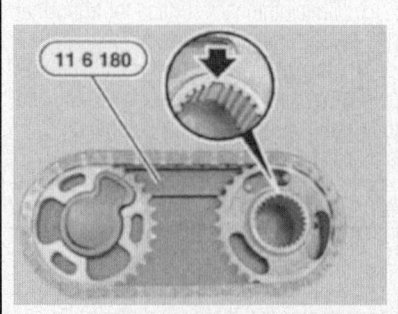

11 6 180

42348-BMX5-G10

**Fig. 38 Aligning forward timing chain on chain wheels with special tool—3.0L engines**

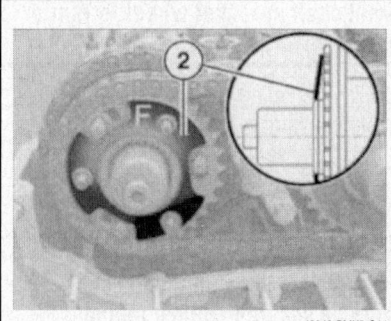

**Fig. 39 Installing the cup spring on the exhaust camshaft chain wheel—3.0L engines**

39. Remove the special tool and position chain wheels on camshafts, so that tooth spaces oppose each other on intake camshaft.

40. Align chain with sprocket wheels so tooth spaces are positioned exactly over each other on intake side.

41. Install and secure the toothed shaft into the tooth gaps of the splines of the camshaft and chain wheel. Push in the toothed shaft until about 0.4 inch (1mm) of splines can still be seen.

42. Note the installation direction of the corrugated washed so "FRONT" is visible. Install the washer and tighten retaining nuts snug only at this time.

43. Install 4 bolts on exhaust side to retain chain wheel. Initially tighten to about 36 inch lbs. (5 Nm), then slack off one-half turn.

44. Install the thrust washer over the exhaust side chain wheel.

45. Note the installation direction of the cup spring (2) so the "F" stamp is forward. If "F" is no longer visible on used engine, install the cup spring so the small locating diameter of the spring points to the sensor gear. Install the cup spring.

46. Position the sensor gear over the cup spring so the arrow on the sensor gear is in line with the upper edge of the cylinder head. Install retaining nuts but do not fully tighten.

47. Pull out the exhaust side toothed shaft to the stop.

48. Press down the secondary chain tensioner at the top and remove the special locking tool.

49. Use a special tool, 11–4–200, to preload timing chain tensioning rail by rotating the adjusting screws of the tool with a small wrench.

50. Preload the exhaust side cup spring slightly be pressing on the sensor gear and hand-tighten the gear retaining nuts (do not fully tighten yet).

51. Remove any remaining gasket material from mating face on front of cylinder head.

52. Check dowel sleeves for damage and for correct installation position.

53. Make sure sealing face is clean and free of oil.

54. Fit special tool, 11–6–150 without gasket over both toothed shafts. Tighten retaining nuts to hold tool in place. Hand-tighten only until special tool is uniformly in contact with cylinder head.

55. Insert bolts on exhaust side chain wheel and tighten to about 36 inch lbs. (5 Nm).

56. Initially tighten nuts on both exhaust and intake chain wheels to about 36 inch lbs. (5 Nm).

57. Torque bolts on exhaust chain wheel to 15 ft. lbs. (20 Nm).

58. Torque nuts on exhaust and intake chain wheels screw-in pins to 88 inch lbs. (10 Nm).

59. Pull back the special tool from the flywheel.

60. Remove the special tools from the rear of the camshafts.

61. Crank the engine twice in the direction of rotation until the cam lobe tips on the front of the camshafts point inward at about the 10 o'clock and 2 o'clock positions.

62. Lock the flywheel with the special tool to the crankshaft is at Top Dead Center (TDC) position.

### ✳✳ WARNING

**DO NOT turn the engine against normal rotation.**

63. Install the special camshaft locking tool, 11–3–240, onto rear end of camshafts to hold its position.

➡**The camshaft timing is correctly set if the special tool rests on the camshafts without a gap on the cylinder head or any protrusion up to 0.04 inch (1mm) on the intake side.**

64. Remove the special tool from the camshaft forward ends.

65. Install the camshaft adjustment (Double VANOS) unit as follows:

   a. Ensure dowel sleeves on forward mating face of cylinder head are not damaged and are correctly installed.

   b. Keep sealing faces clean and free of oil.

   c. Apply a thin and even coat of sealing compound, Drei Bond 1209, to contact surface edges of the separating face between the cylinder head and the camshaft adjustment unit.

   d. Install or connect the following:
   • New seal on mating face
   • Camshaft adjustment unit
   • Screws for hydraulic pistons on toothed shaft end on both camshafts (through front openings); torque to 88 inch lbs. (10 Nm). Screws have left-hand threads.
   • Sealing caps into camshaft adjustment unit openings
   • Screw plugs, with new seal rings; torque to 37 ft. lbs. (50 Nm)
   • Engine lifting eye

66. Set camshaft timing, if necessary.

67. Complete installation in reverse of removal procedure.

### CRANKSHAFT FRONT SEAL

*REMOVAL & INSTALLATION*

1. Before servicing the vehicle, refer to the Precautions Section.

2. Remove or disconnect the following:
   • Splash guard
   • Engine fan. Refer to Engine Fan removal and installation
   • Accessory drive belts. Refer to Accessory Drive Belts removal and installation
   • Crankshaft damper. Refer to Crankshaft Damper removal and installation

3. Fit special tools 11 2 383 and 11 2 385 to crankshaft.

4. Align groove of special tool 11 2 385 to keyway of crankshaft.

5. Screw in special tool 11 2 380 until it has made firm contact with radial seal.

6. Remove radial seal by tightening bolt on special tool 11 2 380.

*To install:*

7. Coat sealing lips of new radial sealing ring with oil.

8. Using special tool 11 3 280 and centering bolt, install new radial seal flush with timing case cover.

9. Install or connect the following:
   • Crankshaft damper
   • Accessory drive belts
   • Engine fan
   • Splash guard

### CYLINDER HEAD

*REMOVAL & INSTALLATION*

*See Figures 40 through 48.*

1. Before servicing the vehicle, refer to the precautions in the beginning of this section.

2. Properly relieve the fuel system pressure.

3. Remove or disconnect the following:

- Negative battery cable
- Ignition coils
- Exhaust system
- Cylinder head cover
- Spark plugs
- Intake manifold
- Drain cooling system
- Thermostat housing
- Coolant pipe from side of block
- coolant hoses from cylinder head
- inlet and exhaust adjustment unit

### ✳✳ WARNING

**Fit new cylinder head screws.**

### ✳✳ WARNING

**Do not wash off bolt coating.**

➡ **There must be no coolant, water or engine oil in the pocket holes.**

4. Release screws (1).

5. Unclip timing chain module (2) at junction (3) and remove towards top.

6. Set down timing chain.

### ✳✳ WARNING

**If the timing chain is stowed in the gearcase, the crankshaft must no longer be rotated. This would cause the timing chain on the**

1. Screw
2. Camshaft magnet wheel

22205_BMX3_G0009

**Fig. 42  Camshaft magnet wheel**

**crankshaft sprocket wheel to jam or jump.**

7. The timing chain is lifted out with a hook only during assembly.

8. Release bolts (2) for eccentric shaft sensor (1).

9. Remove eccentric shaft sensor (1) towards front

➡ **Screw (1) is not magnetic and must be secured against falling down.**

10. Release screw (1).

11. Remove magnet wheel (2) toward front.

➡ **Magnet wheel is highly magnetic and must be protected against metal filings/borings. After removing, place magnet wheel in a plastic bag with a seal.**

12. Pretension eccentric shaft (1) upward in direction of arrow.

13. Remove stop screw between 1st and 2nd cylinders.

1. Release screws
2. Timing chain module
3. Junction

22205_BMX3_G0007

**Fig. 40  Timing chain removal**

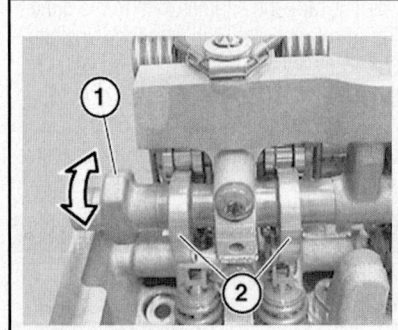

1. Eccentric shaft
2. Stop screw between 1st and 2nd cylinders

22205_BMX3_G0010

**Fig. 43  Eccentric shaft**

1. M10 cylinder head bolts
2. M9 cylinder head bolts

22205_BMX3_G0012

**Fig. 45  Cylinder head bolts**

1. Eccentric shaft sensor
2. Release bolts

22205_BMX3_G0008

**Fig. 41  Timing chain eccentric shaft sensor removal**

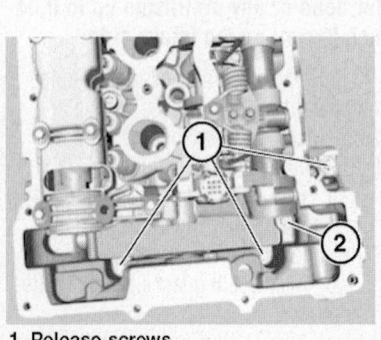

1. Release screws
2. Securing bolt

22205_BMX3_G0011

**Fig. 44  Release screws**

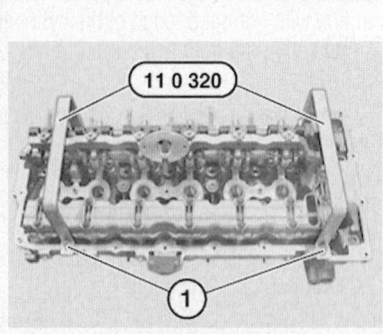

1 and 3. M9 cylinder head bolts
2. M10 g0013 cylinder head bolts

22205_BMX3_G0013

**Fig. 46  Different cylinder head bolts**

➡ **Bolt (2) can only be released when the timing chain module is pressed forward slightly.**

➡**Secure bolt (2) with a gripper against falling down.**

14. Release screw (2).
15. Release screws (1).

➡**Observe different bolt heads.**

16. Release M10 cylinder head bolts (1) with special tool 11 8 580.
17. Release M9 cylinder head bolts (2) with special tool 11 4 420.

➡**Picture shows inlet and exhaust camshafts removed.**

➡**Observe different M9 bolt lengths (1 and 3).**

18. Release M9 cylinder head bolts (1 and 3) with special tool 11 4 420.
19. Release M10 cylinder head bolts (2) with special tool 11 8 580 from outside inwards.

➡**All cylinder head bolts (1, 2 and 3) must be replaced.**

➡**Jointing torque and angle of rotation must be observed without fail.**

1. Special tool 11 0 320
   with existing cylinder head cover bolts

22205_BMX3_G0014

**Fig. 47 Secure special tool 11 0 320**

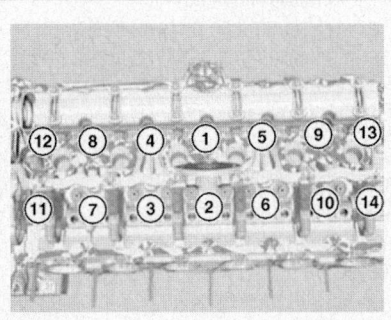

22205_BMX3_G0015

**Fig. 48 sequence for tightening cylinder head bolts**

20. Secure special tool 11 0 320 with existing cylinder head cover bolts (1).

➡**Removing and install cylinder head with a second person helping.**

21. Weight of cylinder head with add-on parts is approx. 40 kg.

❋❋ **WARNING**

**Do not rest cylinder head on sealing surface. Risk of damage to valves**

22. Insert special tool 11 4 430 into bores
23. Remove coarse residues on sealing faces

❋❋ **WARNING**

**Do not use any metal-cutting tools.**

❋❋ **WARNING**

**There must be no coolant, water or engine oil in the pocket holes.**

➡**Observe sequence for tightening cylinder head bolts without fail.**

24. Fit new cylinder head screws.
25. Insert cylinder head bolts (1 to 10) with special tool 11 8 580.
26. Observe sequence for tightening cylinder head bolts without fail.
27. Tighten cylinder head bolts, following the sequence shown, in 3 steps, to the following:
   a. Step 1: 22 ft. lbs. (30 Nm).
   b. Step 2: Additional 90°.
   c. Step 3: Additional 90°.
   d. Step 3: Additional 45°.
28. Install or connect the following:
   • Timing chain guide
   • Timing case cover–to–cylinder head bolts
   • Camshafts
   • Remove exhaust system.
   • Drain coolant
   • Remove both exhaust manifolds
   • Remove intake air manifold

**ENGINE ASSEMBLY**

*REMOVAL & INSTALLATION*

*See Figures 49 and 50.*

1. Before servicing the vehicle, refer to the precautions in the beginning of this section.
2. Disconnect the negative battery cable.
3. Remove or disconnect the following:
4. Remove the heater bulkhead by pulling off the sealing strips from the

firewall, turning the toggle retainers about 90°, lifting out the cover, and removing the heater bulkhead.
   • Engine cover
   • Both oxygen sensors
   • Hood to full open (assembly) position
   • Front splash guard
   • Reinforcement plate from undercarriage
   • Vacuum line from vacuum accumulator

➡**On automatic transmissions, remove transmission fluid line bolt on engine oil sump.**

   • Transmission
   • A/C compressor from mounting (move aside, with lines connected)
5. Drain the cooling system (drain plug is located on exhaust side of cylinder number 2 in engine block). Reinstall drain plug with new seal ring.
6. Pull the locks and disconnect the coolant hoses from the water pump.
7. Remove or disconnect the following:
   • Radiator
   • Water hoses from pipe and from heating valve
   • Tension strut

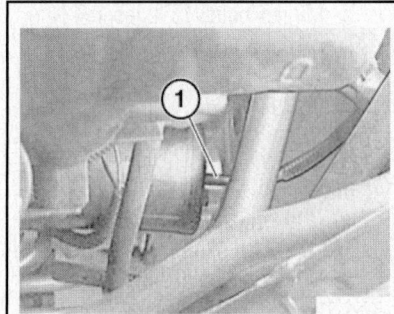

06041_BWX3_G0002

**Fig. 49 Showing vacuum line and vacuum accumulator—3.0L engine**

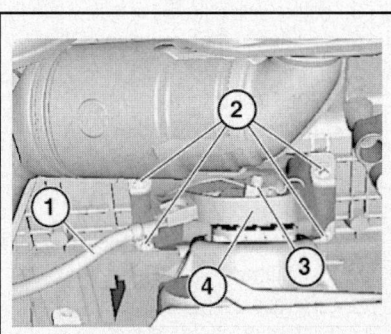

06041_BWX3_G0003

**Fig. 50 Showing right ground wire and nut on right engine mount—3.0L engine**

- Wiring harness section for engine (lay aside)

8. Siphon off some fluid from power steering reservoir. Unbolt reservoir and tie out of the way, leaving lines connected.

9. Remove or disconnect the following:
- Fan clutch with water pump impeller
- Alternator drive belt
- Vane pump for power steering (tie out of way, leaving lines connected)

10. Install an engine lift and chains, attaching to eye hooks.

11. Disconnect the right ground wire.

12. Unbolt the left and right engine mounts.

13. Carefully lift out the engine assembly.

### To install:

14. Install in reverse of removal procedure.

15. Tighten components to the following torque settings:
  a. Radiator bolts: 72–88 inch lbs. (9–10 Nm)
  b. Power steering pump bracket bolts: 16 ft. lbs. (22 Nm)

16. Tighten the automatic transmission–to–engine bolts to the following specifications:
  a. M8 hex bolts: 18 ft. lbs. (24 Nm).
  b. M10 hex bolts: 33 ft. lbs. (45 Nm).
  c. M12 hex bolts: 60 ft. lbs. (82 Nm).
  d. M8 Torx® bolts: 15 ft. lbs. (21 Nm).
  e. M10 Torx® bolts: 31 ft. lbs. (42 Nm).
  f. M12 Torx® bolts: 53 ft. lbs. (72 Nm).

### EXHAUST MANIFOLD

#### REMOVAL & INSTALLATION

1. Before servicing the vehicle, refer to the precautions in the beginning of this section.

2. Remove or disconnect the following:
- Negative battery cable
- Intermediate muffler
- Reinforcement plate from undercarriage
- Bracket, rubber and both stabilizers
- Oxygen control sensors from intake manifold, make sure to mark locations for reinstallation purposes
- Fuel injector cover

- Plug connections from intake manifold, make sure to mark all connection locations for reinstallation purposes
- Oxygen (O2s) sensors
- Lower front axle
- Front exhaust manifold with catalytic converter
- Rear exhaust manifold with catalytic converter

### To install:

3. Clean all mating faces.

4. Install new gaskets.

5. Position the rear exhaust manifold and front exhaust manifold and install new retaining nuts as follows:
  a. M6 nuts: 88 inch lbs. (10 Nm).
  b. M7 nuts: 15 ft. lbs. (20 Nm).

6. Install or connect the following:
- O2s monitor sensors on exhaust pipes; torque to 37 ft. lbs. (50 Nm)
- Connections on intake manifold and components
- Fuel injector cover
- Oxygen control sensors on intake manifold; torque to 37 ft. lbs. (50 Nm)
- Bracket, rubber and both stabilizers from struts; torque nuts to 74 ft. lbs. (100 Nm)
- Reinforcement plate from undercarriage; torque bolts to 74 ft. lbs. (100 Nm)
- Negative battery cable

### INTAKE MANIFOLD

#### REMOVAL & INSTALLATION

*See Figures 51 and 52.*

1. Before servicing the vehicle, refer to the precautions in the beginning of this section.

2. Properly relieve the fuel system pressure.

3. Remove or disconnect the following:
- Negative battery cable
- Engine cover
- Oxygen (O2s) sensor connectors (mark connector locations before removing)
- Wiring from clips on intake manifold
- Battery positive lead from intake manifold
- Battery positive lead retainer
- Vent hose from cylinder head cover
- Retainer from top of engine and connection for intake air temperature sensor

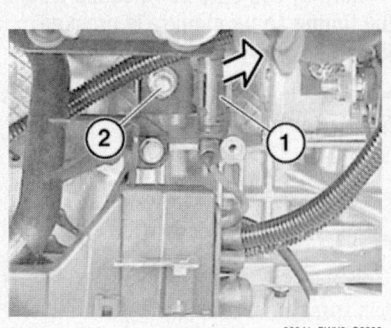

**Fig. 51 Showing knock sensor connector and nut on manifold support—3.0L engine**

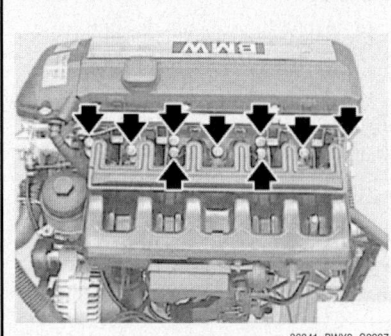

**Fig. 52 Showing intake manifold nuts— 3.0L engine**

- Fuel injector terminal strip and place aside
- Tank venting valve located near the dipstick
- Fuel line from clip and connection to pipe located near the water pump
- Dipstick tube guide
- Return line from dipstick tube
- Throttle assembly
- Knock sensor connector and nut on manifold support
- Intake manifold nuts
- Vacuum line on reverse side of intake manifold, if equipped
- Intake manifold

### To install:

4. Check all intake manifold gaskets and replace if necessary.

5. Inspect the rubber dampers at manifold connection and replace if needed.

6. Install intake manifold and torque bolts as follows:
  a. M6 bolts: 88 inch lbs. (10 Nm).
  b. M7 bolts: 132 inch lbs. (15 Nm).
  c. M8 bolts: 16 ft. lbs. (22 Nm).

7. Install or connect the following:
- Knock sensor connector and manifold support nut
- Throttle assembly
- Return line to dipstick tube
- Dipstick tube guide
- Fuel line to clip and connection to pipe located near the water pump
- Tank venting valve located near the dipstick
- Fuel injector terminal strip
- Retainer to top of engine and connection for intake air temperature sensor
- Vent hose to cylinder head cover
- Battery positive lead retainer
- Battery positive lead to intake manifold
- Wiring to clips on intake manifold
- O$_2$s sensor connectors
- Engine cover
- Negative battery cable

## OIL PAN

### REMOVAL & INSTALLATION

*See Figure 53.*

1. Before servicing the vehicle, refer to the precautions in the beginning of this section.
2. Drain the engine oil.
3. Remove or disconnect the following:
4. Attach an engine lift and hoist to engine.
5. Remove top left and right nuts on engine mounts
6. Raise the engine about 0.15 inch (5mm).
7. Remove and disconnect the following:
- Front splash guard
- Left and right swivel bearings
- Output shafts
- Propeller shaft at front

- Alternator drive belt from belt pulley for vane pump
- Power steering pump and move aside with the lines still connected
- Return hose from oil separator
- Dipstick tube
- Steering spindle from steering gear tie rods

8. Release the screws on the oil pan.
9. Move the pan backward when removing.
10. Remove transmission bolts

#### To install:
11. Ensure mating surfaces are clean.
12. Install a new gasket and position the oil pan in place.
13. Install and tighten the oil pan bolts, but do not fully tighten the transmission end bolts.
14. Tighten the engine end bolts, then tighten the transmission end bolts.
15. Restore the front axle assembly to position.
16. Install or connect the following:
- Dipstick tube
- Return hose from oil separator
- Power steering pump
- Alternator drive belt
- Propeller shaft
- Output shafts
- Swivel bearings
- Splash guard

17. Reposition the engine onto the mount and torque the upper nuts.
18. Remove the engine hoist
19. Reconnect the negative battery cable.
20. Refill the engine oil.

## OIL PUMP

### REMOVAL & INSTALLATION

*See Figure 54.*

1. Before servicing the vehicle, refer to the precautions in the beginning of this section.

2. Drain the engine oil.
3. Remove or disconnect the following:
- Negative battery cable
- Oil pan
- Oil pump sprocket wheel and chain
- If the oil pump is integrated with the oil scraper, remove oil pump with the scraper
- If the oil pump and scraper are two parts, remove oil pump intake pipe
- Oil pump

#### To install:
4. Check dowel sleeves for damage and correct installation position.
5. Check the seals on the oil pipes and replace it if necessary. Lubricate the seals with oil.
6. Check the seal in the oil pump and replace it if necessary.
7. If the oil pump is integrated with the oil scraper, install oil pump with scraper
8. If the oil pump and scraper are two parts, install oil pump intake pipe
9. Install or connect the following:
- Oil pump. Torque the bolts to 17 ft. lbs. (22 Nm).
- Oil pump sprocket wheel and chain. Torque the nut to 35 ft. lbs. (47 Nm).
- Oil pan
- Negative battery cable

10. Fill the engine with clean oil.
11. Start the vehicle and check for leaks, repair if necessary.

### INSPECTION

Check the oil pressure produced by the engine. The engine should be at normal operating temperature. If the oil quality and type or oil filter condition cannot be determined, replace the oil and filter before testing the oil pump and oil pressure.
1. Start the engine and check the oil pressure at different engine speeds (RPM).
2. The oil pump is capable of producing 7.4 psi at 700 RPM.
3. If the oil pressure is low, remove oil pump and inspect:
- Oil pump case for cracks or damage or oil seepage
- Relief plunger for smooth operation
- Relief spring for deformation or a break

4. Inspect the surface that mates with the oil filter for any damage

## PISTON AND RING

### POSITIONING

*See Figures 55 through 57.*

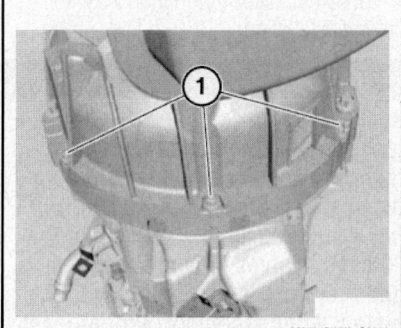

06041_BWX3_G0009

**Fig. 53 Showing transmission bolts**

06041_BWX3_G0010

**Fig. 54 Showing oil pump mounting bolts**

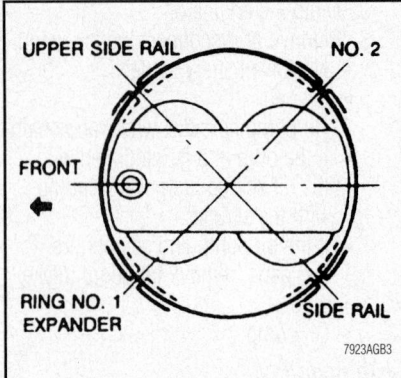

**Fig. 55 Piston ring end-gap spacing**

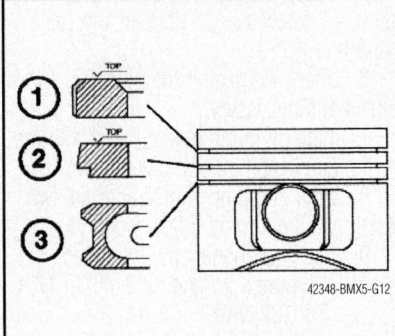

**Fig. 56 Piston ring locations and identification—3.0L engines**

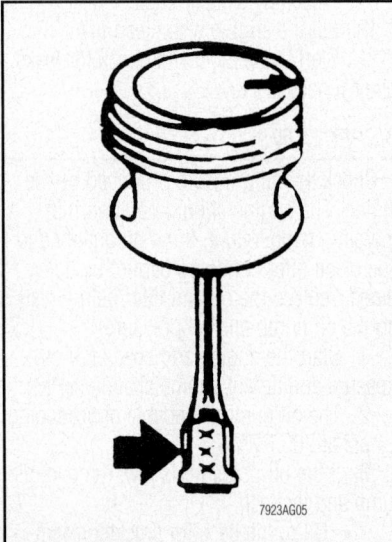

**Fig. 57 Showing connecting rod-to-piston positioning**

## REAR MAIN SEAL

### REMOVAL & INSTALLATION

The rear main bearing oil seal can be replaced after the transmission and flywheel has been removed from the engine.

1. Before servicing the vehicle, refer to the precautions in the beginning of this section.
2. Drain the transmission fluid.
3. Remove or disconnect the following:
   - Negative battery cable
   - Transmission
   - Flywheel assembly
   - Oil seal, using a suitable seal removal tool

*To install:*

4. Coat the sealing lips of the new seal with oil.
5. Install or connect the following:
   - New seal into the end cover housing with a suitable seal installation tool
   - Flywheel
   - Transmission
   - Negative battery cable
6. Fill the transmission with new fluid.
7. Start the engine and check that oil pressure is present; if the oil pressure lamp does not extinguish within 5–7 seconds of starting the engine, turn the engine **OFF**.
8. Check and top off all fluid levels.

## TIMING CHAIN, SPROCKETS, FRONT COVER AND SEAL

### REMOVAL & INSTALLATION

*See Figures 58 through 60.*

1. Before servicing the vehicle, refer to the Precautions Section.
2. Disconnect the negative battery cable.
3. Remove the camshaft (Double VANOS) adjustment unit as follows:
   a. Remove the intake filter housing with Mass Air Flow (MAF) sensor.
   b. Remove the fan impeller with fan clutch and cowl.
   c. Remove the cylinder head covers.
   d. Remove the spark plugs.
   e. Remove the plastic cover for intake camshaft.
   f. Remove the oil pressure pipe.
   g. Install a special tool, 11–3–450, with banjo bolt on oil pressure connection unit.
   h. Cover the camshaft adjustment unit.
   i. Connect a compressed air hose to the special tool fitting on the oil line connection.
   j. Rotate the engine, a least 2 turns in direction of rotation, to return the camshafts to TDC position (front cam lobes pointing to 10 o'clock and 2 o'clock positions).
   k. Install a special locking tool, 11–2–300, into hole to lock flywheel in TDC position.

l. Remove the 2 studs from the rear of the cylinder head outer edge. Secure the camshafts with locking tools, 11–2–240.
   m. Disconnect the compressed air hose from the special tool.

**✳✳ WARNING**

**Once camshaft adjust unit is removed, DO NOT crank engine. The toothed shaft on the intake side camshaft may slip out of the spline teeth and valves could rest on the piston.**

4. Remove the cylinder for the chain tensioning piston. Use caution as piston is under spring pressure.
5. Press down on the secondary chain tensioner at the top and lock the chain in place with a special tool, 11–3–292.
6. Remove or disconnect the following:
   - Nuts on intake camshaft and remove corrugated washer
   - Screws from chain gear on exhaust camshaft
   - Toothed shaft with sleeve
   - Secondary (upper) chain tensioner
   - Release screw-in pins from exhaust camshaft chain gear
   - Exhaust camshaft chain gear (leave chain on end of camshaft)
   - Thrust washer on intake camshaft
   - Intake camshaft sensor gear

**✳✳ CAUTION**

**DO NOT release screws from front ends of either camshaft.**

7. Lift the timing chain off from end of exhaust camshaft and hold it under tension by lifting it straight up.
8. Remove bolts holding sprockets on end of camshafts.
9. Remove sprockets and timing chain.

*To install:*

10. Oil all teeth and friction washers before installation.

**✳✳ CAUTION**

**Bucket tappets expand when not subjected to load by the camshaft and therefore require some time before they can be pushed back down. During a rapid assembly sequence, the "closed" valves may still be open and therefore be in contact with the piston. After assembly, wait at least 30 minutes before cranking the engine back to the TDC position.**

11. Install new timing chain. Position the timing chain onto the end of the exhaust camshaft.

### ❊❊ CAUTION

**To prevent damaging valves when adjusting camshafts, no pistons should be in TDC position.**

12. Position the camshafts so the cam tips on the intake and exhaust valves on no. 1 cylinder face each other at about the 10 o'clock and 2 o'clock positions.

13. Install camshaft locking tools, 11–3–240 on back end of camshafts. Use an open-end wrench to align camshafts (if necessary, machine wrench head so it does not contact cylinder head) so there is no gap at the locking tools.

14. Install the middle locking tool, 11–3–244 so it adjoins other tools and is bolted into spark plug hole.

15. Lift the timing chain straight up off the exhaust camshaft and hold it under tension.

16. Rotate the engine from 30° BTDC to TDC, in normal direction of rotation.

17. Using special tool, 11–2–300, lock flywheel in this position.

18. Install sensor gear onto intake camshaft. Install the thrust washer over the sensor gear and torque the gear screw-in pins to 15 ft. lbs. (20 Nm).

19. Feed the chain wheel onto the timing chain so the arrow on the chain wheel faces the upper edge of the cylinder head.

20. Install a special tool, 11–4–220 into the cylinder head in the chain tensioning piston bore and bring tool adjustment screw into contact with the timing chain tensioning rail, but no further at this time.

21. Check the arrow mark on the chain wheel and adjust if needed so it still points to upper edge of the cylinder head.

22. Install and tighten the retaining screw-in pins in the exhaust chain wheel to 15 ft. lbs. (20 Nm).

23. Install the secondary (upper) chain tensioner.

24. Position the toothed sleeve into the exhaust camshaft to the toothed gaps are opposed. Secure the toothed shaft. Insert the pin of the toothed shaft into the tooth gaps of the splines on the camshaft and toothed sleeve.

25. Push in the toothed shaft on the exhaust camshaft until the elongated holes in the tooth sleeve wheel are centered over the bolt holes.

26. Place forward chain wheels onto special tool, 116–6–180, and position tooth gap on intake chain wheel as shown as feed on the timing chain.

### ❊❊ CAUTION

**DO NOT alter position of chain wheels and chain when removing the special tool.**

27. Remove the special tool and position chain wheels on camshafts, so that tooth spaces oppose each other on intake camshaft.

28. Align chain with sprocket wheels so tooth spaces are positioned exactly over each other on intake side.

29. Install and secure the toothed shaft into the tooth gaps of the splines of the camshaft and chain wheel. Push in the toothed shaft until about 0.4 inch (1mm) of splines can still be seen.

30. Note the installation direction of the corrugated washed so "FRONT" is visible. Install the washer and tighten retaining nuts snug only at this time.

31. Install 4 bolts on exhaust side to retain chain wheel. Initially tighten to about 36 inch lbs. (5 Nm), then slack off one-half turn.

32. Install the thrust washer over the exhaust side chain wheel.

33. Note the installation direction of the cup spring (2) so the "F" stamp is forward. If "F" is no longer visible on used engine, install the cup spring so the small locating diameter of the spring points to the sensor gear. Install the cup spring.

34. Position the sensor gear over the cup spring so the arrow on the sensor gear is in line with the upper edge of the cylinder head. Install retaining nuts but do not fully tighten.

35. Pull out the exhaust side toothed shaft to the stop.

36. Press down the secondary chain tensioner at the top and remove the special locking tool.

37. Use a special tool, 11–4–200, to preload timing chain tensioning rail by rotating the adjusting screws of the tool with a small wrench.

38. Preload the exhaust side cup spring slightly be pressing on the sensor gear and hand-tighten the gear retaining nuts (do not fully tighten yet).

39. Remove any remaining gasket material from mating face on front of cylinder head.

40. Check dowel sleeves for damage and for correct installation position.

41. Make sure sealing face is clean and free of oil.

42. Fit special tool, 11–6–150 without gasket over both toothed shafts. Tighten retaining nuts to hold tool in place. Hand-tighten only until special tool is uniformly in contact with cylinder head.

43. Insert bolts on exhaust side chain wheel and tighten to about 36 inch lbs. (5 Nm).

44. Initially tighten nuts on both exhaust and intake chain wheels to about 36 inch lbs. (5 Nm).

42348-BMX5-G09

**Fig. 58 Setting timing chain and chain wheel mark in position during installation of chain—3.0L engines**

42348-BMX5-G10

**Fig.59 Aligning forward timing chain on chain wheels with special tool—3.0L engines**

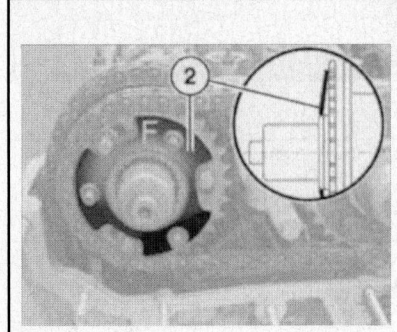

42348-BMX5-G11

**Fig. 60 Installing the cup spring on the exhaust camshaft chain wheel—3.0L engines**

45. Torque bolts on exhaust chain wheel to 15 ft. lbs. (20 Nm).

46. Torque nuts on exhaust and intake chain wheels screw-in pins to 88 inch lbs. (10 Nm).

47. Pull back the special tool from the flywheel.

48. Remove the special tools from the rear of the camshafts.

49. Crank the engine twice in the direction of rotation until the cam lobe tips on the front of the camshafts point inward at about the 10 o'clock and 2 o'clock positions.

50. Lock the flywheel with the special tool to the crankshaft is at Top Dead Center (TDC) position.

### ✳ WARNING

**DO NOT turn the engine against normal rotation.**

51. Install the special camshaft locking tool, 11–3–240, onto rear end of camshafts to hold its position.

➡**The camshaft timing is correctly set if the special tool rests on the camshafts without a gap on the cylinder head or any protrusion up to 0.04 inch (1mm) on the intake side.**

52. Remove the special tool from the camshaft forward ends.

53. Install the camshaft adjustment (Double VANOS) unit as follows:

  a. Ensure dowel sleeves on forward mating face of cylinder head are not damaged and are correctly installed.

  b. Keep sealing faces clean and free of oil.

  c. Apply a thin and even coat of sealing compound, Drei Bond 1209, to contact surface edges of the separating face between the cylinder head and the camshaft adjustment unit.

  d. Install or connect the following:

    • New seal on mating face

    • Camshaft adjustment unit

    • Screws for hydraulic pistons on toothed shaft end on both camshafts (through front openings); torque to 88 inch lbs. (10 Nm). Screws have left-hand threads.

    • Sealing caps into camshaft adjustment unit openings

    • Screw plugs, with new seal rings; torque to 37 ft. lbs. (50 Nm)

    • Engine lifting eye

54. Set camshaft timing, if necessary.

55. Complete installation in reverse of removal procedure.

### VALVE LASH

#### *ADJUSTMENT*

These engines are equipped with hydraulic valve lash adjusters. This design does not permit adjustments nor are adjustments possible.

## ENGINE PERFORMANCE & EMISSION CONTROL

### ✳ WARNING

**The engine is constructed with Aluminum-magnesium materials. No steel screws/bolts may be used due to the threat of electrochemical corrosion. A magnesium crankcase requires aluminum screws/bolts exclusively. Aluminum screws/bolts must be replaced each time they are released. The end faces of aluminum screws/bolts are painted blue for the purposes of reliable identification. Jointing torque and angle of rotation must be observed without fail or risk of damage.**

### CAMSHAFT POSITION (CMP) SENSOR/ CAMSHAFT PULSE GENERATOR

#### *LOCATION*

See Figures 61 and 62.

#### *REMOVAL & INSTALLATION*

1. Before servicing the vehicle, refer to the Precautions Section.

2. Prior to servicing the Camshaft Pulse Generator read out fault memory of the DME control unit.

3. Switch off ignition.

➡**If necessary remove the radiator cover.**

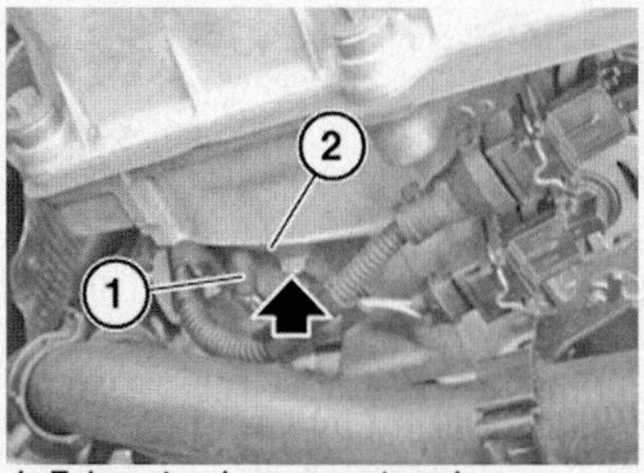

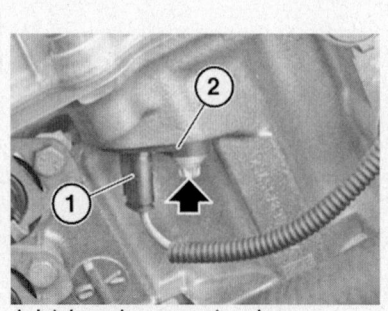

1. Exhaust pulse generator plug
2. Exhaust pulse generator

22205_BMX3_G0018

**Fig. 61 Camshaft Position Sensor/Pulse Generator Exhaust**

1. Intake pulse generator plug
2. Intake pulse generator

22205_BMX3_G0019

**Fig. 62 Camshaft Position Sensor/Pulse Generator Intake**

4. Unlock the plug and remove the release bolt.

#### *To install:*

5. Replace O-ring.

6. Install the Camshaft Pulse Generator.

7. Install the new screws and torque to 6.5 ft. lbs. (9 Nm).

8. Attach the connector.

9. Clear the fault memory.

### CRANKSHAFT POSITION (CKP) SENSOR/PULSE GENERATOR

#### *LOCATION*

See Figure 63.

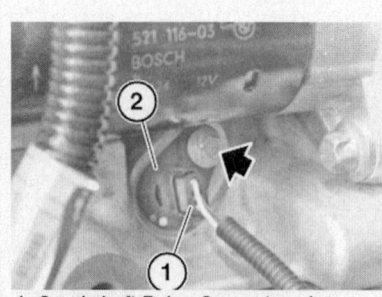

1. Crankshaft Pulse Generator plug
2. Crankshaft Pulse Generator

22205_BMX3_G0017

**Fig. 63 Crankshaft position sensor/Pulse Generator**

The Crankshaft Position (CKP)sensor/ Pulse Generator is underneath the starter motor.

### REMOVAL & INSTALLATION

1. Before servicing the vehicle, refer to the Precautions Section.

2. Prior to servicing the Crankshaft Position (CKP)sensor /Pulse Generator read out fault memory of the DME control unit.

3. Switch off ignition.

4. Remove the intake air manifold.

5. Detach the plug.

6. Remove the screw.

7. Remove the Crankshaft Position (CKP)sensor /Pulse Generator.

**To install:**

8. Replace O-ring.

9. Install the Crankshaft Position (CKP)sensor /Pulse Generator.

10. Install the new screws and torque to 2.7inch lbs. (3 Nm) plus 45°.

11. Attach the connector.

12. Clear the fault memory.

## ELECTRONIC CONTROL MODULE (DME)

### LOCATION

*See Figure 64.*

### REMOVAL & INSTALLATION

*See Figures 65 and 66.*

1. Before servicing the vehicle, refer to the Precautions Section.

2. Code the control unit.

### ✷✷ WARNING

**It is essential to read the fault memory with the MoDiC or the BMW DIS and to create a fault memory printout.**

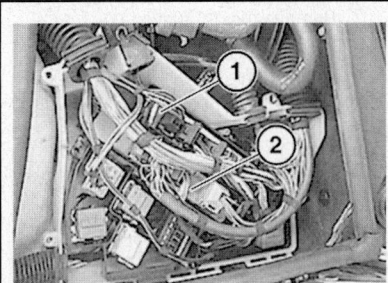

1. Electronic Control Module (DME)
2. Control Module for the Automatic Transmission

22205_BMX3_G0020

**Fig. 64 Electronic Control Module (DME) and Automatic Transmission Control Module**

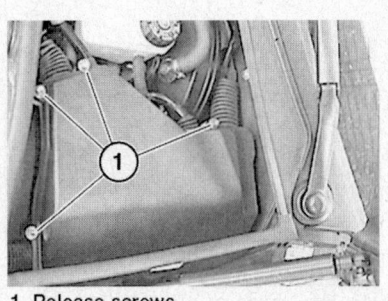

1. Release screws

22205_BMX3_G0021

**Fig. 65 Cover screws removal and location**

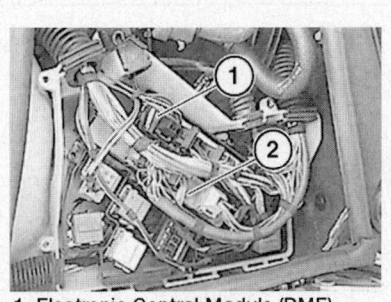

1. Electronic Control Module (DME)
2. Control Module for the Automatic Transmission

22205_BMX3_G0020

**Fig. 66 Electronic Control Module (DME)**

3. Turn off ignition.

4. Remove screws from Control Module cover.

5. Remove cover.

6. Remove (DME ) connectors and unit (1).

**To install:**

➡ Observe unit number and coding.

7. Install the (DME) unit and cover.

8. Check memory of the (DME).

9. Rectify all faults.

10. Clear memory.

## ENGINE COOLANT TEMPERATURE (ECT) SENSOR

### LOCATION

*See Figure 67.*

The Engine Coolant Temperature (ECT) Sensor is located on the cylinder head at the font.

### REMOVAL & INSTALLATION

*See Figure 67.*

1. Before servicing the vehicle, refer to the Precautions Section.

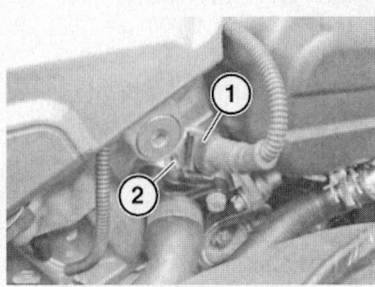

1. Engine Coolant Temperature (ECT) Sensor plug
2. Engine Coolant Temperature (ECT) Sensor

22205_BMX3_G0026

**Fig. 67 Engine Coolant Temperature (ECT) Sensor**

### ✷✷ CAUTION

**Only perform service on the cooling system after it has cooled down.**

2. Prior to servicing the Engine Coolant Temperature (ECT) Sensor read out fault memory of the DME control unit.

3. Disconnect the negative battery cable.

4. Remove the intake duct.

5. Unlock and remove the plug.

6. Remove the temperature sensor.

**To install:**

7. Install a new seal onto the sensor.

8. Install the Engine Coolant Temperature (ECT) Sensor,

9. Torque to 13 ft. lbs. (18 Nm).

10. Check memory of the (DME).

11. Rectify all faults.

12. Clear memory.

## HEATED OXYGEN (HO2S) SENSOR

### LOCATION

The Heated Oxygen (HO2S) Sensors are located in the exhaust manifolds.

### REMOVAL & INSTALLATION

1. Before servicing the vehicle, refer to the Precautions Section.

### ✷✷ CAUTION

**Only perform service on the cooling system after it has cooled down.**

2. Prior to servicing the Heated Oxygen (HO2S)Sensors read out fault memory of the DME control unit.

3. Disconnect the negative battery cable.

4. Remove the intake duct.

5. Unlock and remove the plug.
6. Remove the sensor.

### To install:

7. Install a new seal onto the sensor.
8. Install the Heated Oxygen (HO2S) Sensor,
9. Torque to 26 ft. lbs. (35 Nm).
10. Check memory of the (DME).
11. Rectify all faults.
12. Clear memory.

## KNOCK SENSOR (KS)

### LOCATION

See Figure 68.

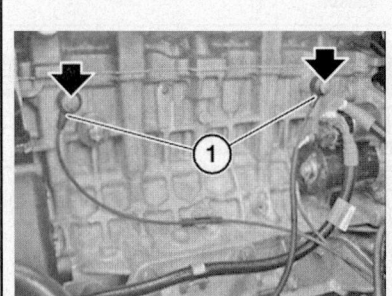

1. Knock Sensors

22205_BMX3_G0023

**Fig. 68 Knock Sensor locations**

### REMOVAL & INSTALLATION

See Figures 69 and 70.

1. Before servicing the vehicle, refer to the Precautions Section.
2. Prior to servicing the Camshaft Pulse Generator read out fault memory of the DME control unit.
3. Switch off ignition.

1. Knock sensor connector

22205_BMX3_G0022

**Fig. 69 Knock Sensor connector**

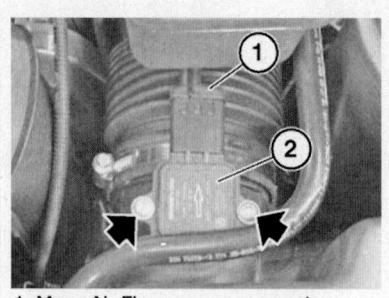

1. Mass Air Flow sensor connector
2. Mass Air Flow sensor

22205_BMX3_G0025

**Fig. 70 Mass Air Flow (MAF) sensor**

4. Disconnect the negative battery cable.
5. Remove the intake manifold.
6. Unlock and remove the plug.
7. Remove the screws from both knock sensors.
8. Remove both knock sensors.
9. Clean knock sensor area on block and sensor face.
10. Replace aluminum screws and tighten to 7.5 ft. lbs. (10Nm) plus 90°.
11. Check fault messages and clear memory.

## MASS AIR FLOW (MAF) SENSOR

### LOCATION

See Figure 70.

### OPERATION

The Mass Air Flow (MAF)sensor is placed in the stream of intake air. It measures the intake flow rate by measuring a part of the entire intake flow. The mass air flow sensor controls the temperature of the hot wire to a certain amount. The heat generated by the hot wire is reduced as the intake air flows around it. The more air, the greater the heat loss. Therefore, the electric current supplied to hot wire is changed to maintain the temperature of the hot wire as air flow increases. The ECM detects the air flow by means of this current change.

### REMOVAL & INSTALLATION

See Figure 70.

1. Before servicing the vehicle, refer to the Precautions Section.

2. Prior to servicing the Mass Air Flow (MAF) sensor read out fault memory of the DME control unit.
3. Switch off ignition.
4. Remove the screws shown by the black arrows.
5. Pull Mass Air Flow (MAF) sensor out of the upper section of the intake filter housing.

### To install:

6. Insert the Mass Air Flow (MAF) sensor into the upper section of the intake filter housing.
7. Tighten the new screws to 27 inch lbs. (3 Nm).
8. Check fault messages and clear memory.

## THROTTLE POSITION SENSOR (TPS)

### LOCATION

See Figure 71.

1. Throttle Position Sensor (TPS)

22205_BMX3_G0024

**Fig. 71 Throttle Position Sensor (TPS) location**

### REMOVAL & INSTALLATION

1. Switch ignition off.
2. Remove the air intake hose.
3. Unlock plug and remove.
4. Remove the screws.
5. Remove the Throttle Position Sensor (TPS) .

### To install:

6. Install a new O-ring
7. Install the Throttle Position Sensor (TPS) .
8. Tighter new screws to 6.5 ft. lbs. (9 Nm).
9. Check for stored faults.
10. Clear all fault memory.

**FUEL**                **GASOLINE FUEL INJECTION SYSTEM**

## FUEL SYSTEM SERVICE PRECAUTIONS

Safety is the most important factor when performing not only fuel system maintenance but any type of maintenance. Failure to conduct maintenance and repairs in a safe manner may result in serious personal injury or death. Maintenance and testing of the vehicle's fuel system components can be accomplished safely and effectively by adhering to the following rules and guidelines.

• To avoid the possibility of fire and personal injury, always disconnect the negative battery cable unless the repair or test procedure requires that battery voltage be applied.

• Always relieve the fuel system pressure prior to disconnecting any fuel system component (injector, fuel rail, pressure regulator, etc.), fitting or fuel line connection. Exercise extreme caution whenever relieving fuel system pressure to avoid exposing skin, face and eyes to fuel spray. Please be advised that fuel under pressure may penetrate the skin or any part of the body that it contacts.

• Always place a shop towel or cloth around the fitting or connection prior to loosening to absorb any excess fuel due to spillage. Ensure that all fuel spillage (should it occur) is quickly removed from engine surfaces. Ensure that all fuel soaked cloths or towels are deposited into a suitable waste container.

• Always keep a dry chemical (Class B) fire extinguisher near the work area.

• Do not allow fuel spray or fuel vapors to come into contact with a spark or open flame.

• Always use a back-up wrench when loosening and tightening fuel line connection fittings. This will prevent unnecessary stress and torsion to fuel line piping.

• Always replace worn fuel fitting O-rings with new Do not substitute fuel hose or equivalent where fuel pipe is installed.

Before servicing the vehicle, make sure to also refer to the precautions in the beginning of this section as well.

## RELIEVING FUEL SYSTEM PRESSURE

### ✳✳ WARNING

**Fuel lines are under about 44–73 psi (3–5 bar) of pressure. Manufacturer**

does not provide a specific pressure relieving procedure, but instructs that any fuel system disconnect will spill fuel, so be prepared to catch and clean up any spilled fuel.

A safe way to relieve the pressure in the system is to locate the fuel pump relay located on the cowl. Unplug and remove the relay, and place it in a safe location. With the fuel pump relay removed, start the engine and operate it until it stalls. Crank the engine for 10 seconds after it stalls to remove any residual pressure.

## FUEL FILTER

### *REMOVAL & INSTALLATION*

1. Before servicing the vehicle, refer to the precautions in the beginning of this section.
2. Properly relieve the fuel system pressure.
3. Remove or disconnect the following:
   • Negative battery cable
   • Fuel pressure regulator and seal the fuel line before and after the filter with special tool 16–1–020
   • Clips and fuel line from the filter
   • Fuel filter

**To install:**
   • New fuel filter
   • New seal
   • Fuel lines onto the correct fittings. Tighten the fuel line clamps until tight, but not to the point where the fuel lines become excessively pinched or damaged, then tighten the mounting bracket until snug.
   • Negative battery cable and cycle the ignition **ON** and **OFF** several times to build fuel pressure
4. Start the vehicle and check for leaks, repair if necessary.

## FUEL INJECTORS

### *REMOVAL & INSTALLATION*

➡**This procedure involves removing the complete fuel rail with injectors.**

1. Before servicing the vehicle, refer to the precautions in the beginning of this section.
2. Properly relieve the fuel system pressure.
3. Mark the locations of each oxygen sensor connector for proper reinstallation.

4. Remove or disconnect the following:
   • Negative battery cable
   • Connectors from clips on fuel rail
   • Intake Air Temperature (IAT) sensor connector
   • Plug connection on solenoid valve for camshaft adjustment (VANOS) unit
   • Terminal strip for fuel injectors from fuel rail
5. Unclip the fuel line from its holder, then unlock the fuel line from the cylinder head at the rear (be prepared to catch any fuel).
6. Remove the retaining bolts and remove the fuel rail with the fuel injectors.

**To install:**
7. Coat the sealing rings of the injectors with anti-seize compound prior to fitting.
8. Install the fuel injectors and fuel rail to the engine.
9. Install or connect the following:
   • Fuel line
   • Terminal strip to fuel injectors
   • Connectors as removed
   • Negative battery cable

## FUEL PUMP

### *REMOVAL & INSTALLATION*

*See Figure 72.*

1. Before servicing the vehicle, refer to the precautions in the beginning of this section.
2. Drain the fuel tank.
3. Properly relieve the fuel system pressure.
4. Remove or disconnect the following:
   • Negative battery cable
   • Rear seat bench

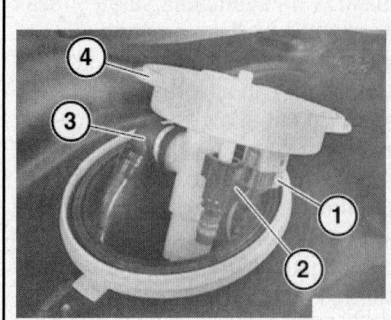

06041_BWX3_G0012

**Fig. 72 Showing service cap, fuel line, vent line and electrical connector on fuel pump**

- Rubber plug above the sender unit and fold the rubber mat back
- Metal cover
- Fuel gauge level sending unit electrical connector
- Fuel lines and service vent line
- Rotary connection with special tool 16–1–020

5. Raise the fuel level sensor and expose the spiral hose.

6. Remove the fuel level sensor and fuel pump from the tank.

### To install:

→**Always use a new seal or gasket when installing the fuel pump or fuel level gauge sending unit assembly.**

7. Install the fuel pump into the fuel tank, taking care not to bend or damage the fuel sending unit assembly.

8. Install a new seal and torque the sealing ring using tool No. 16-1-020 as follows:

   a. Metal sealing rings: 26 ft. lbs. (35 Nm).

   b. Plastic sealing rings: 41 ft. lbs. (55 Nm).

9. Install or connect the following:
- Fuel lines and service vent line
- Fuel gauge level sending unit electrical connector
- Metal cover
- Rubber plug above the sender unit
- Rear seat bench
- Negative battery cable

10. Start the vehicle and check for leaks, repair if necessary.

## FUEL TANK

### REMOVAL & INSTALLATION

*See Figures 73 through 75.*

### ❋❋ CAUTION

**Observe all applicable safety precautions when working around fuel. Whenever servicing the fuel system, always work in a well ventilated area. Do not allow fuel spray or vapors to come in contact with a spark or open flame. Keep a dry chemical fire extinguisher near the work area. Always keep fuel in a container specifically designed for fuel storage; also, always properly seal fuel containers to avoid the possibility of fire or explosion.**

1. Before servicing the vehicle, refer to the Precautions Section.

2. Relieve the fuel system pressure.

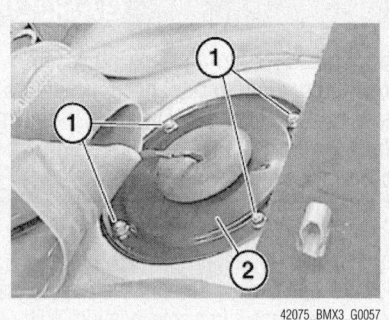

**Fig. 73 Removing cover (2) and bolts (1) for fuel tank connection access**

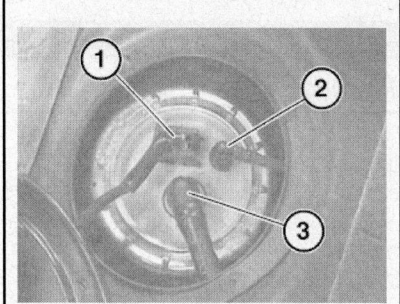

**Fig. 74 Removing connections (1), (2), and (3) from the fuel pump assembly**

3. Drain the fuel tank.

4. Remove or disconnect the following:
- Rear seat bench
- Underbody paneling
- Handbrake Bowden cables
- Complete driveshaft

5. Remove bolts (1) and remove cover (2) from right side of fuel tank.

6. Disconnect plug (1) and service vent line (3) from fuel pump assembly along with hose (2).

7. Remove hose clamps and detach fuel filler hose from fuel tank.

8. Unclip fuel vent line from clips on fuel tank.

9. Disconnect fuel feed line and fuel return line at the fittings joining the fuel lines to the front of the vehicle.

10. Remove bolt and unclip lines from fuel line holder.

→**Do not kink fuel lines when moving them aside. Seal the lines with special tools 13 5 281 and 13 5 282.**

11. Support the fuel tank.

12. Remove bolts for fuel tank mounting strap on the left and right sides.

13. Lower fuel tank until line (1) is accessible.

14. Unclip line (1) from the holders.

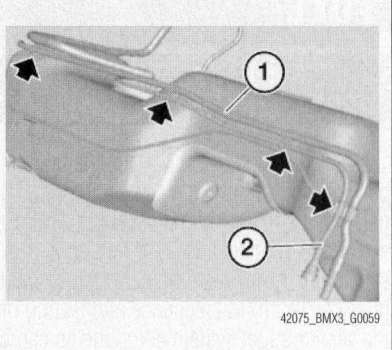

**Fig. 75 Accessing fuel line (1) after lowering fuel tank**

15. Move fuel tank downwards.

16. Feed out fuel line (2).

→**Do not kink fuel line.**

17. Remove fuel tank.

### To install:

18. Raise fuel tank to a height that will allow the installation of the vehicle's fuel line through the clips on the tank.

19. Clip the fuel lines to the tank.

20. Attach the fuel tank mounting straps around the fuel tank. Tighten the bolts in the straps to 14 ft. lbs. (19 Nm).

21. Install or connect the following:
- The fuel lines beneath the vehicle
- The fuel lines under the back seat area
- The fuel fill hose and fuel fill vent hose
- Clamps holding fuel supply and return lines
- Electrical connection at fuel pump assembly
- Driveshaft
- Handbrake Bowden cables
- Underbody paneling
- Rear seat bench

22. Fill tank with at least 5 gallons of fuel and test function of fuel pump.

23. Check for fuel leakage in fuel delivery system.

## IDLE SPEED

### ADJUSTMENT

Idle speed is maintained by the Powertrain Control Module (PCM). No adjustment is necessary or possible.

## THROTTLE BODY

### REMOVAL & INSTALLATION

*See Figures 76 and 77.*

1. Before servicing the vehicle, refer to the Precautions Section.

**Fig. 76 Location of resonance flap and idle-speed control valve**

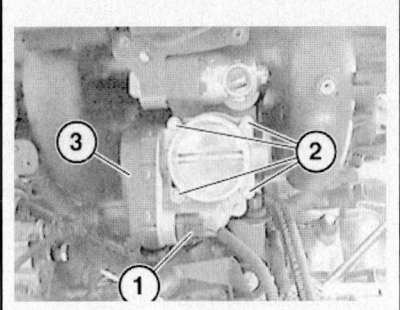

**Fig. 77 Removing throttle body**

- Disconnect plug from tank venting valve
- Remove screws on cable duct mounting
- Remove screw and remove oil dipstick guide tube
- Unlock plug (1) and detach from throttle body (3)
- Remove screws (2) and remove throttle body (3)

**To install:**

3. Replace sealing ring between throttle body and intake manifold.

➡**A faulty throttle adjusts the adaptation values stored in the DME control unit. These adaptation values must be reset after the throttle assembly has been replaced.**

4. The installation continues in the reverse of the removal procedure.

2. Remove or disconnect the following:
- The negative battery terminal
- Intake filter housing
- Intake hose (between intake filter housing and throttle body)
- Disconnect the resonance flap (1)

- Disconnect the idle—speed control valve (2)
- Disconnect the oil pressure switch
- Disconnect the oil temperature switch

## HEATING & AIR CONDITIONING ⬛ SYSTEM

### BLOWER MOTOR

#### REMOVAL & INSTALLATION

*See Figures 78 through 84.*

1. Before servicing the vehicle, refer to the Precautions Section.
2. Remove the fuel injector cover:
   a. Unclip cover caps from bolts holding fuel injector cover.
   b. Remove bolts.
   c. Remove fuel injector cover.
3. Remove the ignition coil cover:
   a. Unclip cover caps from bolts holding ignition coil cover.
   b. Remove bolts.
   c. Remove sealing cap from cylinder head cover (1).
   d. Remove ignition coil cover.
   e. Replace cover cap on cylinder head to prevent foreign material entry.

4. Detach hose (1) in direction of arrow from fitting (2).
5. Remove cover strip (3) in direction of arrow from guide (4).

6. Pull lock (1) in direction of arrow and feed out flap (2).
7. Unclip retainers (1).
8. Remove bolt (2).

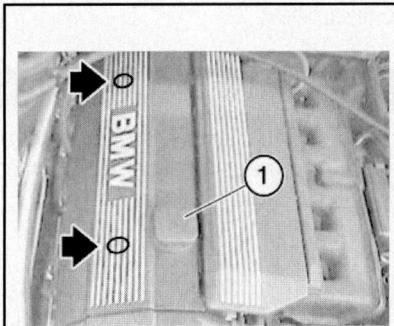

**Fig. 79 Location of cylinder head cover cap and mounting cover caps for fuel injectors**

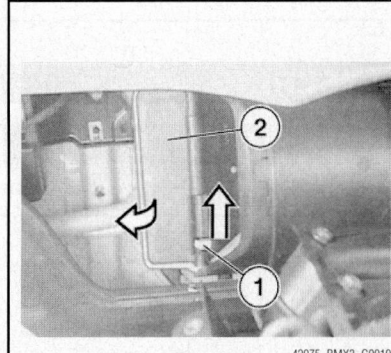

**Fig. 81 Removing cover shroud of blower motor**

**Fig. 78 Location of cover caps for removal of fuel injector cover**

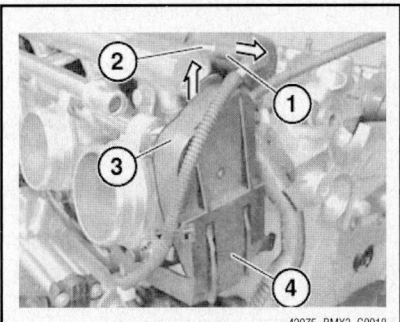

**Fig. 80 Removing hose and fitting for blower motor access**

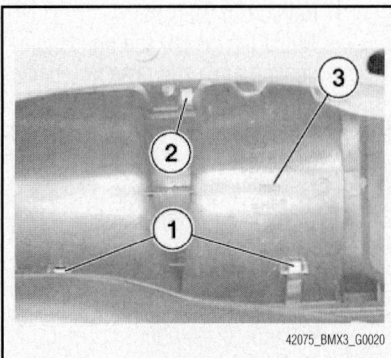

**Fig. 82 Removing blower motor fan cover**

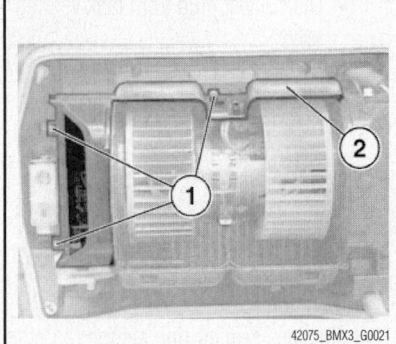

Fig. 83 Removing blower motor fan baffle

Fig. 84 Disconnecting the blower motor from retainer

9. Pull out fan cover (3).
10. Remove bolts (1).
11. Pull out fan baffle (2).
12. Disconnect the electrical connection to fan motor.
13. Release fan motor retainer with special tool 51 3 057.
14. Remove blower motor and squirrel fan.

### To install:

15. Place blower motor and fan into position.
16. Make sure blower motor is seated correctly. Snap into retainer.
17. Connect electrical connections.
18. Installation continues in the reverse of the removal procedure.

## HEATER CORE

### REMOVAL & INSTALLATION

1. Before servicing the vehicle, refer to the precautions in the beginning of this section.
2. Disconnect the negative battery cable.
3. On 3.0L engines, remove the heater bulkhead as follows:
    a. Pull off the weatherstrips on top of the bulkhead.
    b. Release the cowl retaining toggle clips.
    c. Lift out the cover.
    d. Remove the screws and remove the heater bulkhead.
4. Remove or disconnect the following:
    • Heater hoses from firewall connections
    • Air duct for left footwell from mountings
    • Left footwell vent
    • Heater pipes from under dash
    • Heater duct for right footwell
    • Air duct for rear footwell on right
    • Cover from heater housing
    • Evaporator temperature sensor (pull out to one side)
    • Heater core

### To install:

5. Install or connect the following:
    • Heater core
    • Evaporator temperature sensor
    • Heater housing cover
    • Air ducts in footwells
    • Heater pipes under dash
    • Heater hoses at firewall connections
    • Heater bulkhead
6. Reconnect the negative battery cable
7. Check coolant level

## STEERING

### POWER STEERING GEAR

#### REMOVAL & INSTALLATION

1. Before servicing the vehicle, refer to the precautions in the beginning of this section.
2. Set the steering gear in the straight ahead position by aligning the marks on the steering gear and spindle.
3. Drain the power steering fluid.
4. Remove or disconnect the following:
    • Negative battery cable
    • Front wheels
    • Reinforcement plate
    • Nuts from the left and right engine support arms and raise the engine slightly
    • 2 bolts accessible through holes in frame
    • Lower clamping screw
    • Steering gear clamps (after locking wheels in straight-ahead position)
    • Tie rod by pressing it off with special tool 32–3–090
    • Stabilizer end-links
    • Heat shield from power steering gear
    • Hydraulic lines, set aside and secure
    • Electrical connector on EH converter
    • If necessary, pressure line from front axle carrier
    • Self-locking nuts and brace the front axle support
    • Banjo bolts and slide the steering gear out through the left wheel opening

#### To install:

5. Install the steering gear through the left side wheel opening
6. Install new sealing rings and banjo bolts. Torque the bolts as follows:
    a. M10: 108 inch lbs. (12 Nm).
    b. M14: 25 ft. lbs. (35 Nm).
    c. M16: 29 ft. lbs. (40 Nm).
    d. M18: 34 ft. lbs. (45 Nm).
7. Install or connect the following:
    • Front axle support screws. Torque the screws to 74 ft. lbs. (100 Nm).
    • Self-locking nuts. Torque the nuts to 74 ft. lbs. (100 Nm).
    • Tie rod. Torque the castle nut to 58 ft. lbs. (80 Nm).
    • Steering gear to the spindle. Torque the fastener to 18 ft. lbs. (24 Nm).
    • Steering gear clamp
    • Engine support arm nuts. Torque the nuts to 60 ft. lbs. (85 Nm).
    • Reinforcement plate
    • Splash guard
    • Both front wheels
    • Negative battery cable
8. Fill and bleed the power steering system.
9. Start the vehicle and check for leaks, repair if necessary.

### POWER STEERING PUMP

#### REMOVAL & INSTALLATION

*See Figures 85 through 87.*

1. Before servicing the vehicle, refer to the Precautions Section.
2. Remove accessory drive belts. Refer to Accessory Drive Belts removal and installation.
3. Suction off hydraulic fluid from power steering reservoir.
4. Remove front underbody protection
5. Remove hose clamp (1) with special tool 32 1 260.
6. Detach suction line (2) from power steering pump (3).

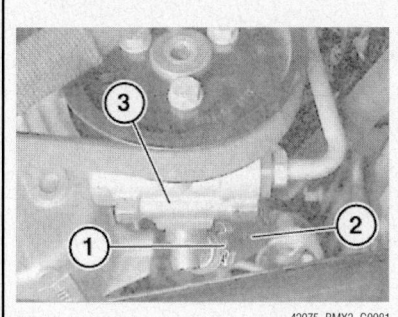

**Fig. 85 Using tool 32 1 260 to detach suction line on power steering pump**

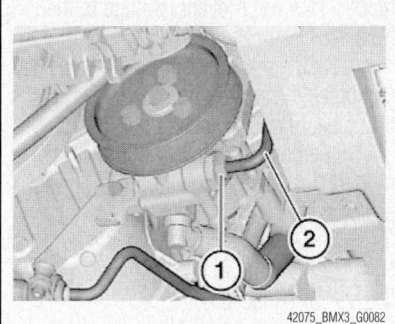

**Fig. 86 Removing banjo blot (1) and pressure line (2) from power steering pump**

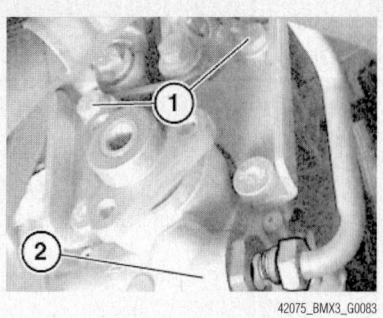

**Fig. 87 Pulley and bolt (1) removal from power steering pump**

7. Release banjo bolt (1) and detach pressure line (2) from power steering pump.

8. Remove pulley from power steering pump.

9. Remove power steering bracket bolts (1) from engine crankcase.

10. Remove power steering pump (2) with bracket through wheel arch.

***To install:***

11. Install or connect the following:
- Power steering pump. Tighten bracket bolts to 16 ft. lbs. (21 Nm)
- Pulley on power steering pump. Tighten bolts to 21 ft. lbs. (28 Nm)
- Replace sealing ring on pressure line. Tighten banjo bolt to 27 ft. lbs. (36 Nm)
- Suction line on power steering pump

- Hose clamp with special tool 32 1 260.
- Front underbody protection

12. Fill power steering reservoir with hydraulic fluid.

13. Install accessory drive belts. Refer to Accessory Drive Belts removal and installation.

14. Bleed hydraulic system. Refer to Power Steering Pump bleeding.

15. Check pipe connections for leaks.

### BLEEDING

1. Before servicing the vehicle, refer to the Precautions Section.

2. With engine off, turn the steering wheel fully to the right and left several times.

➡ **Do not allow the fluid level in the reservoir tank to go below the MIN level line. Check and add fluid as needed.**

3. Run the engine at idle speed. Turn the steering wheel fully to the right and then fully to the left. Hold for about three seconds. Check for fluid leakage.

4. Repeat the above step several times at three second intervals.

➡ **Do not hold the steering wheel in the locked position for more than ten seconds.**

5. Check for air bubbles or cloudy fluid. If found, repeat the bleeding procedure.

6. Stop the engine and check the fluid level. Fill as required.

## SUSPENSION

## FRONT SUSPENSION

### LOWER BALL JOINT

*REMOVAL & INSTALLATION*

1. Before servicing the vehicle, refer to the precautions in the beginning of this section.

2. Remove or disconnect the following:
- Negative battery cable
- Coil spring
- Upper control arm from trailing arm
- Shock absorber from the swinging arm

3. Using special tool 33–3–331/2/3 and pull the ball joint out of the steering knuckle.

***To install:***

4. Install or connect the following:
- Ball joint into the steering knuckle with special tools 33–3–332/3/4
- Shock absorber to the swinging arm
- Upper control arm to trailing arm
- Coil spring
- Negative battery cable

### LOWER CONTROL ARM

*REMOVAL & INSTALLATION*

1. Before servicing the vehicle, refer to the precautions in the beginning of this section.

2. Remove or disconnect the following:
- Negative battery cable
- Front wheel
- Control arm from the front axle support and loosen the nut from the control arm to swivel bearing
- Control arm from the swivel bearing by pressing it off with special tool 31–2–240

***To install:***

3. Install or connect the following:
- Lower control arm to the swivel bearing. Torque the nut to 58 ft. lbs. (80 Nm).
- Lower control arm to the front axle support. Torque the nut to 74 ft.

lbs. (100 Nm) plus an additional 90°
- Front wheel
- Negative battery cable

4. Check and adjust the front end alignment as needed.

### BUSHING REPLACEMENT

1. Before servicing the vehicle, refer to the precautions in the beginning of this section.

2. Remove or disconnect the following:
- Negative battery cable
- Control arm and tie it back to prevent damage to the ball joint
- Control arm bushing by installing special tools 31–1–331/2/3

***To install:***

3. Install or connect the following:
- Control arm bushing by installing special tools 31–1–331/2/3
- Control arm
- Negative battery cable

## MACPHERSON STRUT

### REMOVAL & INSTALLATION

1. Before servicing the vehicle, refer to the precautions in the beginning of this section.

2. Mark the position of the threaded pin to the wheel arch to retain the camber setting when installed.

3. Remove or disconnect the following:
- Negative battery cable
- Tire and wheel assembly

➡ **Carefully mark position of threaded pin to strut tower to ensure original camber is retained during installation.**

- Two of the nuts on the spring strut support bearing
- Center strut bracket nut
- Speed sensor/brake wear cable and disconnect the plug housing
- Swivel bearing and tie it aside
- Remaining nut on the spring strut support bearing
- Strut assembly

### To install:

4. Install or connect the following:
- Strut assembly
- One nut to the spring strut support bearing and hand tighten at this time
- Swivel bearing. Torque the new self-locking nut to 176 ft. lbs. (250 Nm).
- Speed sensor/brake wear cable and connect the housing plug
- Center strut bracket. Torque the nut to 74 ft. lbs. (100 Nm).

5. Align the three upper spring strut support bearing nuts and match the threaded pin with the mark made during the removal procedure. When aligned properly torque the nuts to 25 ft. lbs. (34 Nm).

6. Install the tire and wheel assembly.

7. Reconnect the negative battery cable.

### OVERHAUL

### ✷✷ CAUTION

**This procedure calls for the spring to be compressed. A compressed spring has high potential energy and if released suddenly can cause severe damage and personal injury.**

1. Before servicing the vehicle, refer to the precautions in the beginning of this section.

2. Disconnect the negative battery cable.

3. Remove the strut assembly from the vehicle and mount in a vise using a strut holder. This will prevent damage to the strut tube

4. Using a proper spring compressor, compress the spring until the stress on the thrust bearing is released.

5. Remove the top nut of the strut mount. Counterhold the strut rod during removal.

6. Pull the strut mount off the strut rod. Note the positioning of the spacers and washer for replacement.

7. Pull the spring off the strut and place aside in a safe area.

8. Slowly release the compression of the spring.

### To install:

9. Install or connect the following:
- Spring in the compressor and compress
- Spring and strut mount with all the spacers and washers in their original positions. Torque the new strut rod nut: 47 ft. lbs. (65 Nm).

10. Release the spring slowly and check that it seats in the spring holders. Install the strut in the vehicle.

11. Connect the negative battery cable.

## STABILIZER BAR

### REMOVAL & INSTALLATION

*See Figures 88 through 90.*

1. Before servicing the vehicle, refer to the Precautions Section.

2. Remove or disconnect the following:
- Underside protection
- Reinforcement plate
- Partially detach front wheel arch trim on both sides toward rear of front fenders
- Double pivot of lower steering spindle from power steering gear
- Stabilizer links on both sides from stabilizer bar

### ✷✷ WARNING

**When supporting components, make sure that the vehicle can no longer be raised or lowered and that the vehicle does not lift off the locating plates on the lifting platform.**

3. If necessary, position special tool 00 2 040 with a 2nd person helping on a workshop jack.

4. Insert special tools 31 4 051 and 31 4 052 into corresponding mountings of special tool 00 2 040.

5. Align special tool 00 2 040 to front axle carrier.

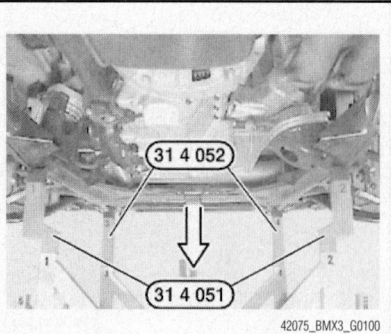

**Fig. 88 Safely support the vehicle with special tools 31 4 051 and 31 4 052**

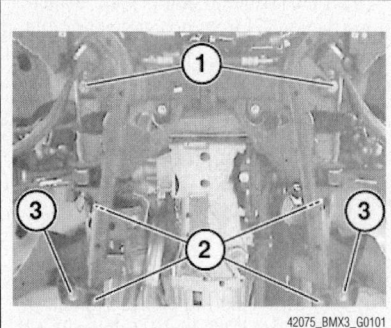

**Fig. 89 Location of bolts (1), (2), and (3) to lower front axle carrier**

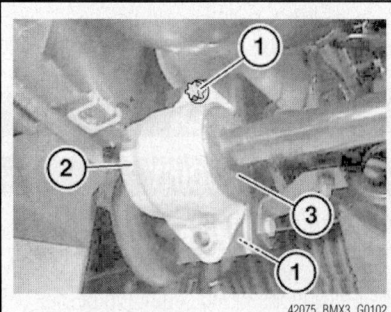

**Fig. 90 Removing stabilizer bar bracket bolts (1), retaining bracket (2), and rubber mount (3)**

6. If necessary, lower special tool 33 3 274.

7. Support front axle carrier by raising special tool 00 2 040.

➡ **Pay attention to power steering hoses and lines when lowering and raising. Hoses and lines must not be kinked, tensioned, or bent with force.**

8. Remove bolts (1) from frame.

9. Remove bolts (2) and (3) to release front axle carrier.

10. If necessary, disconnect pressure line for power steering from front axle carrier.

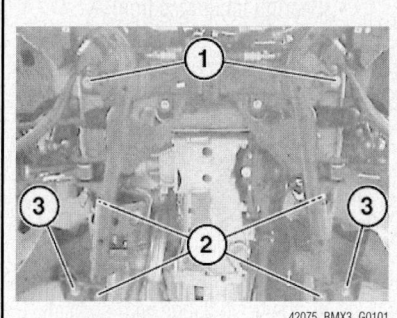

**Fig. 91 Location of bolts (1), (2), and (3) holding front axle carrier**

**Fig. 92 Steering knuckle (1) and bolts (2) with expander tool 31 2 230 installed for removal**

**Fig. 93 Crimping halfshaft nut to axle during installation**

11. Carefully lower front axle carrier with workshop jack.

12. Release bolts (1) and twist stabilizer bar towards the front of vehicle.

13. Remove retaining bracket (2) and rubber mount (3) on both sides.

14. Remove stabilizer bar sideways through wheel arch.

**To install:**

15. Check rubber mount (3) on both sides. Replace if damaged.

➡**Keep retaining bracket (2), rubber mount (3), and stabilizer bar clean and free from oil and grease.**

16. Place stabilizer bar into position with rubber mountings and brackets.

17. Install stabilizer bar into place with new bolts. Tighten to 42 ft. lbs. (56 Nm).

18. Install front axle carrier into place with new bolts. Tighten in order (1), (2), then (3) bolts.

    a. Tighten (1) M12 bolts to 84 ft. lbs. (113 Nm).

    b. Tighten (2) M12 bolts to 74 ft. lbs. (100 Nm).

    c. Tighten (3) M10 bolts to 37 ft. lbs. (50 Nm).

➡**Pay attention to power steering hoses and lines when raising and lowering axle carrier.**

19. Installation continues in reverse of the removal procedure.

20. Install stabilizer to stabilizer link. Tighten to 48 ft. lbs. (65 Nm).

21. Replace M10 bolts and install reinforcement plate to front axle support/engine carrier. Tighten to 55 ft. lbs. (74 Nm).

## STEERING KNUCKLE

### REMOVAL & INSTALLATION

*See Figures 92 and 93.*

1. Before servicing the vehicle, refer to the Precautions Section.

2. Raise and support the vehicle safely.

3. Remove front wheel.

4. Remove or disconnect the following:

- The stabilizer link from the spring strut. Use a fixed wrench as a counterhold
- The ABS sensor. Hang up the sensor using a piece of wire
- The ABS sensor cable from the spring strut
- The brake caliper mounting bolts. Hang the caliper up using a piece of wire
- The halfshaft bolt. Use a screwdriver as a counterhold on the brake disc
- The brake disc. Detach the end of the halfshaft in the hub by knocking the drive shaft into the hub approximately ⅓–½ inch (10–15mm). Use a rubber or copper mallet
- The tie rod ends from the steering arm

5. Measure the position of the steering knuckle and spring strut for installation purposes.

6. Support steering knuckle (1) with workshop jack and a suitable mounting.

7. Remove the bolts (2) retaining the spring strut and the steering knuckle.

8. Expand steering knuckle (1) with special tool 31 2 230.

9. Lower workshop jack.

10. Suspend the drive shaft from a hook.

➡**Take care not to damage the halfshaft boot.**

11. Disconnect the ball joint pinion from the control arm. Refer to Lower Ball Joint removal and installation.

12. Remove steering knuckle.

**To install:**

13. Install the steering knuckle in approximately the same position from which it was removed.

14. Install a new nut on the ball joint. Tighten to 59 ft. lbs. (80 Nm).

➡**Make sure that the mating surfaces on the ball joint and link are clean.**

15. Clean the splines on the halfshaft.

16. Turn the steering knuckle and bring the halfshaft into the hub.

17. Install halfshaft bolt and finger-tighten. Lubricate the bolt.

18. Install bolts retaining the spring strut in the steering knuckle. Use new bolts and nuts. Tighten to 74 ft. lbs. (100 Nm).

19. Install or connect the following:

- The stabilizer link to the spring strut. Use a new nut. Counterhold using a wrench so that the boot is not damaged. Tighten to 48 ft. lbs. (65 Nm)
- The ABS sensor cable
- The ABS sensor onto the steering knuckle. Tighten the ABS sensor to 82 inch lbs. (8 Nm)

➡**Ensure that the sensor seat in the steering knuckle is absolutely clean. Clean the ABS sensor with a soft brush.**

- The tie rod end onto the steering knuckle. Use a new nut. Tighten to 59 ft. lbs. (80 Nm)
- The brake disc.

➡**Ensure that the brake disc and wheel rim hub mating surfaces are clean.**

20. Install new halfshaft nut. Tighten to 310 ft. lbs. (420 Nm). Crimp nut to axle.

21. Install the brake caliper. Use new bolts.

22. Install the wheel. Tighten to 103 ft. lbs. (140 Nm).

## UPPER CONTROL ARM

### REMOVAL & INSTALLATION

1. Before servicing the vehicle, refer to the precautions in the beginning of this section.

2. Remove or disconnect the following:
- Negative battery cable
- Wheel assembly
- Coil spring
- Upper control arm from the stabilizer link from upper control arm
- If necessary, jointed rod of ride-height sensor from upper control arm

3. Support rear differential and release bearing bolts on rear differential. Halfshafts do not have to be removed.

4. Remove or disconnect the following:
- Upper control arm through bolts towards rear
- Upper control arm

**To install:**

5. Installation is reverse of removal.

6. Install and torque the following to correct torque specifications:
- Upper control arm. Torque the bolt to 78 ft. lbs. (106 Nm).
- Upper control arm to the steering knuckle. Torque the bolt to 74 ft. lbs. (100 Nm).
- Wheel assembly
- Negative battery cable

## WHEEL BEARINGS

### REMOVAL & INSTALLATION

➡The wheel bearings are only removed if they are worn. They cannot be removed without destroying them (due to side thrust created by the bearing puller). They cannot be disassembled, repacked or adjusted.

1. Before servicing the vehicle, refer to the precautions in the beginning of this section.

2. Remove or disconnect the following:
- Negative battery cable

### ✻✻ CAUTION

To avoid damaging the dust sleeve, use special tool tools 33–2–160 to drive out and draw in the drive flange after removing the brake disk. Rounded inside edge of special tool must point to drive flange.

- Front brake disc
- Drive flange by installing special tools 33–2–116, 160 and 33–4–200

- Bearing inner race from the flange
- Swivel bearing
- Bearing by installing special tools 31–2–113, 33–4–266, 33–3–266 and 33–4–266/261

**To install:**

3. Install or connect the following:
- Wheel bearing with the wider chamfer facing the swivel bearing to the drive flange with special tools 31–2–113, 33–4–266, 33–3–266 and 33–4–266/261
- Snap ring and circlip
- Bearing inner race to the drive flange
- Drive flange to the swivel bearing by using special tool 33–2–116, 160 and 33–4–200
- Swivel bearing
- Front brake disc
- Negative battery cable

### ADJUSTMENT

Wheel bearings cannot be adjusted and must be replaced as a unit and never be reused once removed.

## SUSPENSION

### COIL SPRING

#### REMOVAL & INSTALLATION

*See Figure 94.*

1. Before servicing the vehicle, refer to the Precautions Section.

2. Remove or disconnect the following:
- Rear wheel
- Output shaft from rear differential and tie back
- Stabilizer link on both sides from stabilizer
- Jointed rod from sensor lever of ride—height sensor, if necessary

➡Brake hose must not be exposed to tensile loads.

- Bolt and bracket with brake hose

3. Support trailing arm with a workshop jack.

4. Remove lower bolt of shock absorber.

5. Lower workshop jack.

6. Press trailing arm downwards.

7. Pull out coil spring.

**To install:**

8. Position coil spring with upper spring pad and lower spring pad mounted in control arm and align to take-up locator/frame side member.

9. Raise trailing arm using a workshop jack.

➡Spring pads must be positively seated in the take-up locator/frame side member or control arm to prevent the coil spring from springing out of the centering mount.

10. Installation continues in the reverse of the removal procedure.

11. Tighten shock absorber to trailing arm to 74 ft. lbs. (100 Nm).

12. Tighten wheel bolts to 103 ft. lbs. (140 Nm).

13. Check headlight adjustment, correct if necessary.

## REAR SUSPENSION

### CONTROL ARMS/LINKS

#### REMOVAL & INSTALLATION

**Upper Control Arm**

*See Figure 95.*

1. Before servicing the vehicle, refer to the Precautions Section.

2. Raise the vehicle.

3. Remove or disconnect the following:
- The wheel
- The brake caliper. Hang the brake caliper on a hook in the sub-frame.

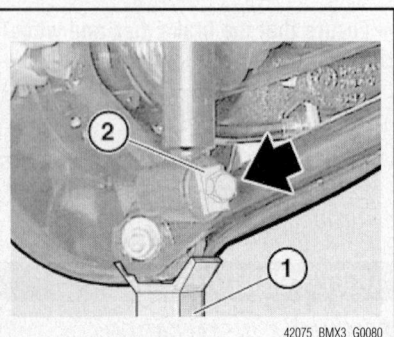

**Fig. 94 Using workshop jack to support trailing arm in removal of rear coil spring**

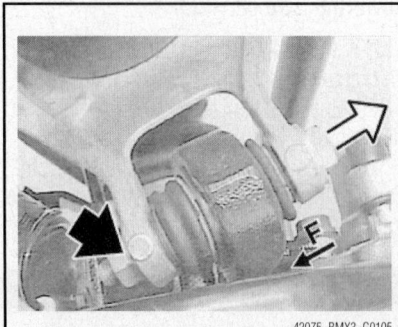

**Fig. 95 Location of nut and bolt for removal of upper control arm**

- The coil spring. Refer to Rear Suspension, Coil Spring removal and installation
- The stabilizer link from upper control arm

4. Support rear differential

5. Remove nut and bolt from control arm as illustrated.

➡**Observe bolt insertion direction (F is direction of travel).**

6. Remove or disconnect the following:
- The bolt for the inner rear control arm mounting
- The two bolts for the inner front control arm mounting
- The control arm

### To install:

7. Install or connect the following:
- The bolts for the inner rear control arm mounting
- The bolts for the front inner control arm mounting
- The bolts for the outer control arm mounting

➡**Replace self-locking nuts and bolts.**

8. Tighten the M12 bolts for the inner rear control arm mounting to 57 ft. lbs. (77 Nm).

9. Tighten the M12 bolts for the front inner control arm mounting to 57 ft. lbs. (77 Nm).

10. Tighten the M12 bolts for the upper control arm to the trailing arm to 78 ft. lbs. (106 Nm).

➡**Tighten the bolts when the rear suspension is in the normal position.**

11. Install the wheel and tighten bolts to 103 ft. lbs. (140 Nm).

### LOWER CONTROL ARM

#### REMOVAL & INSTALLATION

*See Figures 96 through 98.*

1. Before servicing the vehicle, refer to the Precautions Section.
2. Raise the vehicle.
3. Remove or disconnect the following:
- The wheel
- The stabilizer from rear axle carrier
- The coil spring. Refer to Rear Suspension, Coil Spring removal and installation
4. Support rear differential

➡**To avoid damaging the output shaft, it will be necessary to support the trailing arm with the workshop jack.**

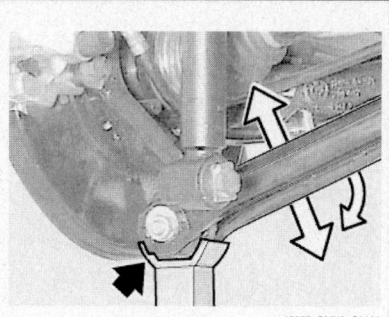

**Fig. 96 Using workshop jack to support the trailing arm**

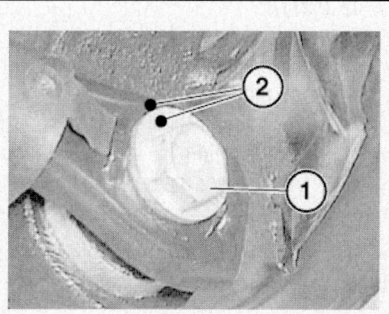

**Fig. 97 Marking eccentric bolt (1) to lower control arm center marks (2)**

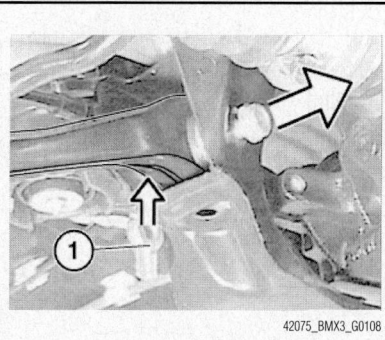

**Fig. 98 Removing thread plate (1) to remove lower control arm**

5. Support trailing arm from underneath using a workshop jack.

6. Mark position of eccentric bolt (1) to lower control arm with center marks (2).

7. Remove nut and eccentric washer.

8. Remove eccentric bolt towards front.

9. Remove bolt and remove thread plate (1) downward.

10. Press rear differential to side and remove bolt.

11. Remove lower control arm at side.

### To install:

➡**Tighten all joints and rubber bushings with the rear suspension in the**

normal position. Weld seam of control arm must point upwards.

12. Install or connect the following:
- Thread plate into opening in rear axle carrier. Tighten bolt to 57 ft. lbs. (77 Nm).
- Eccentric bolt. Align by means of marking to lower control arm. Reinstall eccentric washer. Replace self-locking nut. Tighten bolt to 74 ft. lbs. (100 Nm).
- The coil spring. Refer to Rear Suspension, Coil Spring removal and installation
- The stabilizer to rear axle carrier
- The wheel. Tighten bolts to 103 ft. lbs. (140 Nm).

### SHOCK ABSORBER

#### REMOVAL & INSTALLATION

1. Before servicing the vehicle, refer to the precautions in the beginning of this section.
2. Remove or disconnect the following:
- Negative battery cable
- Wheel
- Brake caliper
- Stabilizer link from spring/strut
- Tie rod end from swivel bearing
- Line for pulse generator from spring/strut
- If necessary, brake pad wear sensor
- Swivel bearing from spring/strut, set aside and secure

➡**If the centering pin missing, mark the position of studs in relation to wheel arch.**

3. Secure spring so that it doesn't fall out. Remove mounting nuts.
Remove shock absorber downwards out of wheel arch.

### To install:

4. Installation is reverse of removal procedure.

5. Replace all bolts and self-locking nuts

6. Tighten all following fasteners to correct torque specifications:
- Spring/strut support: 25 ft. lbs. (30 Nm)
- Shock absorber piston rod on thrust bearing: 47 ft. lbs. (64 Nm)
- Spring strut shock absorber to pivot mount: 73 ft. lbs. (100 Nm)

#### TESTING

1. Check the rubber parts for damage or deterioration.

2. Check the spring for correct height, deformation, deterioration, or damage.

3. Check the shock absorber for abnormal resistance or unusual sounds.

4. Check for oil seepage around seals.

5. Replace as needed.

## UPPER CONTROL ARM

### REMOVAL & INSTALLATION

*See Figure 99.*

1. Before servicing the vehicle, refer to the Precautions Section.

2. Raise the vehicle.

3. Remove or disconnect the following:
- The wheel
- The brake caliper. Hang the brake caliper on a hook in the sub-frame.
- The coil spring. Refer to Rear Suspension, Coil Spring removal and installation
- The stabilizer link from upper control arm

4. Support rear differential

5. Remove nut and bolt from control arm as illustrated.

➡**Observe bolt insertion direction (F is direction of travel).**

6. Remove or disconnect the following:
- The bolt for the inner rear control arm mounting
- The two bolts for the inner front control arm mounting
- The control arm

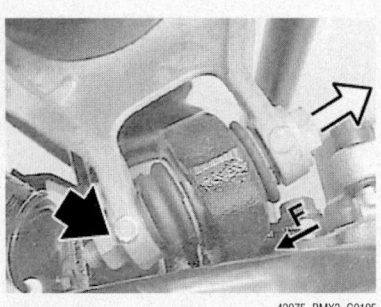

42075_BMX3_G0105

**Fig. 99 Location of nut and bolt for removal of upper control arm**

### To install:

7. Install or connect the following:
- The bolts for the inner rear control arm mounting
- The bolts for the front inner control arm mounting
- The bolts for the outer control arm mounting

➡**Replace self-locking nuts and bolts.**

8. Tighten the M12 bolts for the inner rear control arm mounting to 57 ft. lbs. (77 Nm).

9. Tighten the M12 bolts for the front inner control arm mounting to 57 ft. lbs. (77 Nm).

10. Tighten the M12 bolts for the upper control arm to the trailing arm to 78 ft. lbs. (106 Nm).

➡**Tighten the bolts when the rear suspension is in the normal position.**

11. Install the wheel and tighten bolts to 103 ft. lbs. (140 Nm).

## WHEEL HUB AND BEARING

### REMOVAL & INSTALLATION

1. Before servicing the vehicle, refer to the precautions in the beginning of this section.

2. Remove or disconnect the following:
- Negative battery cable
- Wheel assembly
- Collar nut
- Brake disc
- Rear axle shaft drive flange
- Wheel bearing inner race from the drive flange
- Retaining ring
- Wheel bearing with special tools 33–3–261/2/3

### To install:

3. Install or connect the following:
- New wheel bearing with special tools 33–3–261/4/5
- Retaining ring
- Wheel bearing inner race to the drive flange.
- Drive flange to the axle shaft
- Brake disc
- Collar nut
- Wheel assembly
- Negative battery cable

### ADJUSTMENT

Wheel bearings cannot be adjusted and must be replaced as a unit and never be reused once removed

# BMW

X5

## SPECIFICATIONS AND MAINTENANCE CHARTS

### ENGINE AND VEHICLE IDENTIFICATION

| | Engine | | | | | | | Model Year | |
|---|---|---|---|---|---|---|---|---|---|
| Code | Liters (cc) | Cu. In. | Cyl. | Fuel Sys. | Engine Type | Eng. Mfg. | | Code ① | Year |
| M54 | 3.0 (2979) | 182 | 6 | SMPI | DOHC | BMW | | 6 | 2006 |
| N62 | 4.4 (4398) | 268 | 8 | SMPI | DOHC | BMW | | 7 | 2007 |
| N62 | 4.8 (4799) | 292 | 8 | SMPI | DOHC | BMW | | 8 | 2008 |
| N52K | 3.0 (2979) | 182 | 6 | SMPI | DOHC | BMW | | | |
| N62TU | 4.8 (4799) | 292 | 8 | SMPI | DOHC | BMW | | | |
| N52K | 3.0 (2979) | 182 | 6 | SMPI | DOHC | BMW | | | |
| N62TU | 4.8 (4799) | 292 | 8 | SMPI | DOHC | BMW | | | |

DOHC: Double Overhead Camshaft

SMPI: Sequential Multi-Port Injection

① 10th position of VIN

22205_BMX5_C0001

### GENERAL ENGINE SPECIFICATIONS

| Year | Model | Engine Displacement Liters (ID/VIN) | Net Horsepower @ rpm | Net Torque @ rpm (ft. lbs.) | Bore x Stroke (in.) | Com-pression Ratio | Oil Pressure @ rpm |
|---|---|---|---|---|---|---|---|
| 2006 | X5 3.0i | 3.0 (M54) | 225@5900 | 214@3500 | 3.31x3.53 | 10.2:1 | 7.4@700 |
| | X5 4.4i | 4.4 (M62) | 315@5400 | 324@3600 | 3.62x3.26 | 10.0:1 | 7.4@580 |
| | X5 4.8is | 4.6 (N62) | 355@6200 | 396@3500 | 3.66x3.35 | 10.5:1 | 7.4@580 |
| 2007 | X5 3.0si | 3.0 (N52K) | 260@6650 | 225@2750 | 3.31x3.53 | 10.2:1 | 7.4@700 |
| | X5 4.8i | 4.8 (N62TU) | 260@6650 | 350@3400 | 3.66x3.48 | 10.2:1 | NA |
| 2008 | X5 3.0si | 3.0 (N52K) | 260@6650 | 225@2750 | 3.31x3.53 | 10.2:1 | 7.4@700 |
| | X5 4.8i | 4.8 (N62TU) | 350@6300 | 350@3400 | 3.66x3.48 | 10.2:1 | NA |

NA: Not Available

22205_BMX5_C0002

## ENGINE TUNE-UP SPECIFICATIONS

| Year | Engine Displacement Liters (ID) | Spark Plug Gap (in.) | Ignition Timing (deg.) | Fuel Pump (psi) | Idle Speed (rpm) | Valve Clearance | |
|------|------|------|------|------|------|------|------|
| | | | | | | In. | Ex. |
| 2006 | 3.0 (M54) | 0.024-0.028 | ① | 48-54 | ② | HYD | HYD |
| | 4.4 (N62) | 0.024-0.028 | ① | 48-54 | ② | HYD | HYD |
| | 4.8 (N62) | 0.024-0.028 | ① | 48-54 | ② | HYD | HYD |
| 2007 | 3.0 (N52K) | 0.024-0.028 | ① | 48-54 | ② | HYD | HYD |
| | 4.8 (N62TU) | NA | ① | 48-54 | ② | HYD | HYD |
| 2008 | 3.0 (N52K) | 0.024-0.028 | ① | 48-54 | ② | HYD | HYD |
| | 4.8 (N62TU) | NA | ① | 48-54 | ② | HYD | HYD |

NOTE: The Vehicle Emission Control Information label often reflects specification changes made during production.

The label figures must be used if they differ from those in this chart.

HYD: Hydraulic

① Ignition timing is regulated by the Electronic Control Module (ECM), and cannot be adjusted.

② Idle speed is controled by the Electronic Control Module (ECM), and cannot be adjusted.

NA: Not Available

22205_BMX5_C0003

## CAPACITIES

| Year | Model | Engine Displacement Liters (VIN) | Engine Oil with Filter (qts.) | Automatic Trans. (qts.) | Manual Trans. (qts.) | Rear Drive Axle (pts.) | Fuel Tank (gal.) | Cooling System (qts.) |
|------|------|------|------|------|------|------|------|------|
| 2006 | X5 3.0i | 3.0 (M54) | 8.0 | 7.0 | 3.2 | 3.4 | 24.6 | 11.1 |
| | X5 4.4i | 4.4 (N62) | 8.5 | 11.7 | NA | 3.4 | 24.6 | 13.7 |
| | X5 4.8is | 4.8 (N62) | 8.5 | 11.7 | NA | 3.4 | 24.6 | 13.7 |
| 2007 | X5 3.0si | 3.0 (N52K) | 8.0 | 7.0 | 3.2 | 3.4 | 24.6 | 11.1 |
| | X5 4.8i | 4.8 (N62TU) | 8.5 | 11.7 | NA | 3.0 | 24.6 | 13.7 |
| 2008 | X5 3.0si | 3.0 (N52K) | 8.0 | 7.0 | 3.2 | 3.4 | 24.6 | 11.1 |
| | X5 4.8i | 4.8 (N62TU) | 8.5 | 11.7 | NA | 3.0 | 24.6 | 13.7 |

NOTE: All capacities are approximate. Add fluid gradually and check to be sure a proper fluid level is obtained.

NA: Not available.

22205_BMX5_C0005

## VALVE SPECIFICATIONS

| Year | Engine Displacement Liters (VIN) | Seat Angle (deg.) | Face Angle (deg.) | Spring Test Pressure (lbs. @ in.) | Spring Installed Height (in.) | Stem-to-Guide Clearance (in.) | | Stem Diameter (in.) | |
|---|---|---|---|---|---|---|---|---|---|
| | | | | | | Intake | Exhaust | Intake | Exhaust |
| 2006 | 3.0 (M54) | 45 | 45 | NA | NA | 0.0197 | 0.0197 | 0.2372-0.2340 | 0.2378-0.2384 |
| | 4.4 (N62) | 45 | 45 | NA | NA | 0.0197 | 0.0197 | 0.2156-0.2159 | 0.2146-0.2150 |
| | 4.8 (N62) | 45 | 45 | NA | NA | 0.0016 | 0.0016 | 0.1998-0.2378 | 0.1992-0.2384 |
| 2007 | 3.0 (N52K) | 45 | 45 | NA | NA | 0.0197 | 0.0197 | 0.2372-0.2340 | 0.2378-0.2384 |
| | 4.8 (N62TU) | 45 | 45 | NA | NA | 0.0016 | 0.0016 | 0.1998-0.2378 | 0.1992-0.2384 |
| 2008 | 3.0 (N52K) | 45 | 45 | NA | NA | 0.0197 | 0.0197 | 0.2372-0.2340 | 0.2378-0.2384 |
| | 4.8 (N62TU) | 45 | 45 | NA | NA | 0.0016 | 0.0016 | 0.1998-0.2378 | 0.1992-0.2384 |

NA: Not available

22205_BMX5_C0006

## CAMSHAFT AND BEARING SPECIFICATIONS

All measurements are given in inches.

| Year | Engine Displacement Liters | Engine VIN | Journal Diameter | Brg. Oil Clearance | Shaft End-play | Runout | Journal Bore | Lobe Lift | |
|---|---|---|---|---|---|---|---|---|---|
| | | | | | | | | Intake | Exhaust |
| 2006 | 3.0 | M54 | 2.0505-2.0515 | 0.0010-0.0030 | 0.0010-0.0060 | 0.002 | ① | 0.2449 | 0.2587 |
| | 4.4 | N62 | 1.0605-1.0615 | 0.0010-0.0030 | 0.0035-0.0075 | 0.002 | 1.0625-1.0635 | 0.2560 | 0.2560 |
| | 4.8 | N62 | 1.1260-1.1270 | 0.0010-0.0030 | 0.0035-0.0075 | 0.001 | 1.1280-1.1290 | 0.2173 | 0.2168 |
| 2007 | 3.0 | N52K | 1.1260-1.1270 | 0.0010-0.0030 | 0.0035-0.0075 | 0.001 | 1.1280-1.1290 | 0.2173 | 0.2168 |
| | 4.8 | N62TU | 2.4400-2.4410 | 0.0015-0.0060 | 0.0020-0.0080 | NA | 2.4430-2.4460 | 0.2261 | 0.2296 |
| 2008 | 3.0 | N52K | 1.1260-1.1270 | 0.0009-0.0029 | 0.0013-0.0019 | 0.004 | 1.1280-1.1290 | 0.2156 | 0.2173 |
| | 4.8 | N62TU | 2.4400-2.4410 | 0.0015-0.0060 | 0.0020-0.0080 | NA | 2.4430-2.4460 | 0.2261 | 0.2296 |

NA: Information not available

① Intake: 1.8532-1.8542 in.
Exhaust: 1.5635-1.5645 in.

22205_BMX5_C0007

## CRANKSHAFT AND CONNECTING ROD SPECIFICATIONS

All measurements are given in inches.

| Year | Engine Displacement Liters (VIN) | Crankshaft | | | | Connecting Rod | | |
|---|---|---|---|---|---|---|---|---|
| | | Main Brg. Journal Dia. | Main Brg. Oil Clearance | Shaft End-play | Thrust on No. | Journal Diameter | Oil Clearance | Side Clearance |
| 2006 | 3.0 (M54) | ① | 0.0007-0.0029 | 0.0031-0.0064 | 5 | 1.7720-1.7706 | 0.0007-0.0022 | 0.0060-0.0160 |
| | 4.4 (N62) | ① | 0.0007-0.0018 | 0.0033-0.0101 | 3 | 1.8901-1.8887 | 0.0007-0.0022 | 0.0060-0.0196 |
| | 4.8 (N62) | ① | 0.0007-0.0018 | 0.0033-0.0101 | 3 | 1.8901-1.8904 | 0.0007-0.0022 | 0.0060-0.0196 |
| 2007 | 3.0 (N52K) | ① | 0.0007-0.0029 | 0.0031-0.0064 | 5 | 1.7720-1.7706 | 0.0007-0.0022 | 0.0060-0.0160 |
| | 4.8 (N62TU) | ① | 0.0009-0.0020 | 0.0032-0.0092 | 3 | NA | 0.0011-0.0028 | NA |
| 2008 | 3.0 (N52K) | ① | 0.0007-0.0029 | 0.0031-0.0064 | 5 | 1.7720-1.7706 | 0.0007-0.0022 | 0.0060-0.0160 |
| | 4.8 (N62TU) | ① | 0.0009-0.0020 | 0.0032-0.0092 | 3 | NA | 0.0011-0.0028 | NA |

NA: Not Available

① Standard green: 2.3613-2.3615 inches

   Standard white: 2.3611-2.3613 inches

22205_BMX5_C0008

## PISTON AND RING SPECIFICATIONS

All measurements are given in inches.

| Year | Engine Displacement Liters (ID) | Piston Clearance | Ring Gap | | | Ring Side Clearance | | |
|---|---|---|---|---|---|---|---|---|
| | | | Top Compression | Bottom Compression | Oil Control | Top Compression | Bottom Compression | Oil Control |
| 2006 | 3.0 (M54) | 0.0004-0.0016 | 0.0039-0.0118 | 0.0078-0.0157 | 0.0098-0.0197 | 0.0008-0.0024 | 0.0012-0.0026 | 0.0007-0.0024 |
| | 4.4 (N62) | 0.0002-0.0015 | 0.0039-0.0118 | 0.0078-0.0157 | 0.0078-0.0354 | 0.0008-0.0024 | 0.0008-0.0024 | ① |
| | 4.8 (N62) | 0.0002-0.0015 | 0.0039-0.0118 | 0.0078-0.0157 | 0.0078-0.0354 | 0.0008-0.0022 | 0.0008-0.0022 | ① |
| 2007 | 3.0 (N52K) | 0.0004-0.0016 | 0.0039-0.0118 | 0.0078-0.0157 | 0.0098-0.0197 | 0.0008-0.0024 | 0.0012-0.0026 | 0.0007-0.0024 |
| | 4.8 (N62TU) | 0.0001-0.0017 | 0.0039-0.0118 | 0.0078-0.0157 | 0.0078-0.0354 | 0.0008-0.0022 | 0.0008-0.0022 | ① |
| 2008 | 3.0 (N52K) | 0.0004-0.0016 | 0.0039-0.0118 | 0.0078-0.0157 | 0.0098-0.0197 | 0.0008-0.0024 | 0.0012-0.0026 | 0.0007-0.0024 |
| | 4.8 (N62TU) | 0.0001-0.0017 | 0.0039-0.0118 | 0.0078-0.0157 | 0.0078-0.0354 | 0.0008-0.0022 | 0.0008-0.0022 | ① |

① Does not require measurement.

22205_BMX5_C0009

# TORQUE SPECIFICATIONS

All readings in ft. lbs.

| Year | Engine Displacement Liters (ID) | Cylinder Head Bolts | Main Bearing Bolts | Rod Bearing Bolts | Crankshaft Damper Bolts | Flywheel Bolts | Manifold | | Spark Plugs | Oil Pan Drain Plug |
|------|----|----|----|----|----|----|----|----|----|----|
| | | | | | | | Intake | Exhaust | | |
| 2006 | 3.0 (M54) | ① | ② | ③ | ④ | 77 | ⑤ | 15 | ⑥ | ⑦ |
| | 4.4 (N62) | ① | NS | ⑧ | 25 | 77 | ⑤ | 7 | ⑥ | ⑦ |
| | 4.8 (N62) | ① | NS | ⑧ | 25 | 77 | ⑤ | 7 | ⑥ | ⑦ |
| 2007 | 3.0 (N52K) | ① | ② | ③ | ④ | 77 | ⑤ | 15 | ⑥ | ⑦ |
| | 4.8 (N62TU) | ① | NS | ④ | 25 | 77 | ⑤ | 7 | ⑥ | ⑦ |
| 2008 | 3.0 (N52K) | ① | ② | ③ | ④ | 77 | ⑤ | 15 | ⑥ | ⑦ |
| | 4.8 (N62TU) | ① | NS | ⑧ | 25 | 77 | ⑤ | 7 | ⑥ | ⑦ |

① See repair procedure for torque information.

② Step 1: 15 ft. lbs.

Step 2: Additional 70 degrees.

③ Step 1: 4 ft. lbs.

Step 2: 15 ft. lbs.

Step 3: Additional 70 degrees.

④ Two-piece: 25 ft. lbs.

One-piece: 302 ft. lbs.

⑤ M12 X 1.5 bolts: 18 ft. lbs.

M18 X 1.5 bolts: 22 ft. lbs.

M22 X 1.5 bolts: 44 ft. lbs.

⑥ With thread M12x1.25: 15-19 ft. lbs.

With thread M14x1.25: 20-24 ft. lbs.

⑦ M6 bolts: 7 ft. lbs.

M7 bolts: 11 ft. lbs.

M8 bolts: 16 ft. lbs.

⑧ Step 1: 4 ft. lbs.

Step 2: 15 ft. lbs.

Step 3: Additional 80 degrees.

22205_BMX5_C0010

# WHEEL ALIGNMENT

| Year | Model | | Caster | | Camber | | Toe-in (in.) |
|------|-------|---|--------|--------|--------|--------|------|
| | | | Range (+/-Deg.) | Preferred Setting (Deg.) | Range (+/-Deg.) | Preferred Setting (Deg.) | |
| 2006 | X5 | F | 0.50 | +0.83 | 0.42 | -0.20 | 0.30+/-0.13 |
| | | R | — | — | 0.33 | +0.83 | 0.30+/-0.13 |
| 2007 | X5 | F | 0.50 | +0.83 | 0.42 | -0.20 | 0.30+/-0.13 |
| | | R | — | — | 0.33 | +0.83 | 0.30+/-0.13 |
| 2008 | X5 | F | 0.50 | +0.83 | 0.42 | -0.20 | 0.30+/-0.13 |
| | | R | — | — | 0.33 | +0.83 | 0.30+/-0.13 |

22205_BMX5_C0012

## TIRE, WHEEL AND BALL JOINT SPECIFICATIONS

| Year | Model | Engine | OEM Tires Standard | OEM Tires Optional | Tire Pressures (psi) Front | Tire Pressures (psi) Rear | Wheel Size | Ball Joint Inspection | Lug Nut (ft. lbs.) |
|------|-------|--------|----------|----------|-------|------|-----------|------------|-----------|
| 2006 | X5 3.0i | 3.0(M54) | 235/65R17 | 255/55R18 | 32 | 32 | 7.5J/8.5J | NA | 81-96 |
| | X5 4.4i | 4.4(N62) | 255/55R18 | 255/50R19 ① | 32 ① | 32 ① | 9J ① | NA | 81-96 |
| | | | | 285/45R19 ② | 32 ② | 32 ② | 10J ② | NA | |
| | X5 4.8is | 4.8(N62) | 275/40R20 ① | NA | 32 ① | 32 ① | 9J ① | NA | 81-96 |
| | | | 315/35R20 ② | NA | 32 ② | 32 ② | 10J ② | NA | |
| 2007 | X5 3.0si | 3.0 (N52K) | 235/65R17 | 255/55R18 | 32 | 32 | 7.5J/8.5J | NA | 81-96 |
| | X5 4.8i | 4.8 (N62TU) | 255/55R18 | NA | 32 ① | 32 ① | 9J ① | NA | 81-96 |
| | | | 255/55R18 | NA | 32 ② | 32 ② | 10J ② | NA | |
| 2008 | X5 3.0si | 3.0 (N52K) | 255/55R18 | NA | 32 | 32 | 7.5J | NA | 81-96 |
| | X5 4.8i | 4.8 (N62TU) | 255/55R18 | NA | 32 | 32 | 8.5J | NA | 81-96 |

OEM: Original Equipment Manufacturer

PSI: Pounds Per Square Inch

NA: Not Available

① Front

② Rear

22205_BMX5_C0011

## BRAKE SPECIFICATIONS

All measurements in inches unless noted

| Year | Model | | Brake Disc Original Thickness | Brake Disc Minimum Thickness | Brake Disc Maximum Run-out | Brake Drum Diameter Original Inside Diameter | Brake Drum Diameter Max. Wear Limit | Brake Drum Diameter Maximum Machine Diameter | Min. Lining Thickness | Brake Caliper Bracket Bolts (ft. lbs.) | Brake Caliper Mounting Bolts (ft. lbs.) |
|------|-------|---|----------|----------|----------|----------|------|----------|------|---------|---------|
| 2006 | M54 | F | 1.118 | ① | 0.005 | NA | NA | NA | 0.118 | 81 | 23 |
| | | R | 0.470 | ① | 0.005 | NA | NA | NA | ① | 49 | 21 |
| | N62 | F | 1.118 | ① | 0.007 | NA | NA | NA | 0.118 | 81 | 23 |
| | | R | 0.470 ② | ① | 0.007 | NA | NA | NA | ① | 49 | 21 |
| 2007 | N52K | F | 0.803 | ① | 0.005 | NA | NA | NA | 0.118 | 81 | 23 |
| | | R | 0.470 | ① | 0.005 | NA | NA | NA | ① | 49 | 21 |
| | N62TU | F | 1.118 | ① | 0.007 | NA | NA | NA | 0.118 | 81 | 23 |
| | | R | 0.470 ② | ① | 0.007 | NA | NA | NA | ① | 49 | 21 |
| 2008 | N52K | F | 0.803 | ① | 0.005 | NA | NA | NA | 0.118 | 81 | 23 |
| | | R | 0.470 | ① | 0.005 | NA | NA | NA | ① | 49 | 21 |
| | N62TU | F | 1.118 | ① | 0.007 | NA | NA | NA | 0.118 | 81 | 23 |
| | | R | 0.470 ② | ① | 0.007 | NA | NA | NA | ① | 49 | 21 |

NA: Not Available

① Minimum thickness is stamped in the brake disc shell

22205_BMX5_C0013

## SCHEDULED MAINTENANCE INTERVALS
### BMW—X5

| TO BE SERVICED | TYPE OF SERVICE | SERVICE INTERVALS | | | |
|---|---|---|---|---|---|
| | | INITIAL 1200 MILES | OIL SERVICE | INSPECTION I | INSPECTION II |
| Oil level | S/I | ✓ | | | |
| Engine oil | R | ✓ | ✓ | ✓ | ✓ |
| Engine oil & filter | R ① | ✓ | ✓ | ✓ | ✓ |
| Cabin Filter | R | ✓ | ✓ | ✓ | ✓ |
| Engine air cleaner element | R ② | | | | ✓ |
| Spark plugs | R | | | | ✓ |
| Fuel filter | R ③ | | | | ✓ |
| Fuel, vapor lines & fuel cap | S/I | ✓ | ✓ | ✓ | ✓ |
| Cooling system | S/I | ✓ | ✓ | ✓ | ✓ |
| Exhaust pipe & muffler | S/I | ✓ | ✓ | ✓ | ✓ |
| Catalytic converter & shielding | S/I | ✓ | ✓ | ✓ | ✓ |
| Throttle linkage | S/I | | | ✓ | ✓ |
| Engine (check for leakage) | S/I | ✓ | | | |
| Engine drive belts | S/I | | | | ✓ |
| Maintenance Indicators | RE | | ④ | ✓ | ✓ |
| Engine coolant | R | | | ⑤ | ⑤ |
| Oxygen sensor | R ⑥ | | | | |
| Brake & clutch fluids ⑥ | S/I | ✓ | ✓ | ✓ | ✓ |
| Brake pads & discs | S/I | ✓ | ✓ | ✓ | ✓ |
| Parking brake system | S/I | ✓ | ✓ | ✓ | ✓ |
| Power steering system | S/I | ✓ | ✓ | ✓ | ✓ |
| Rear axle fluid | S/I | ✓ | ✓ | ✓ | ✓ |
| Steering play, suspension track rods, front axle joints, steering linkage & joint disc | S/I | ✓ | ✓ | ✓ | ✓ |
| Transmission fluid/oil | S/I | ✓ | ✓ | ✓ | ✓ |
| Wheel centering hubs | S/I | | | ✓ | ✓ |
| Rear axle fluid | R | ✓ | ✓ | | ✓ |
| OBD system for codes | S/I | ✓ | | ✓ | ✓ |

R: Replace        S/I: Service or Inspect        RE: Reset

Note: BMW does not rely solely on vehicle mileage to determine service intervals. An on-oboard diagnostic center, monitors engine operating conditions, along with mileage, to determine the most effective maintenance intervals. The information is then conveyed to the driver through the service indicator lights, located in the center of the instrument panel.

Note: Maintenance and most wear items are covered by the manufacturer. Refer to the operator's manual for additional information.

① On vehicles operated less than 6200 miles per year, more frequent service may be required.

② Replace more frequently if vehicle is operated in dusty conditions.

③ Recommended service for California models, required for all other models.

④ Reset the oil service indicator lights only.

⑤ Replace every 2 years with inspection service.

⑥ Replace every 100,000 miles on all models.

## FREQUENT OPERATION MAINTENANCE (SEVERE SERVICE)

If a vehicle is operated under any of the following conditions it is considered severe service

- Extremely dusty areas.

- 50% or more of the vehicle operation is in 32°C (90°F) or higher temperatures, or constant operation in
   in temperatures below 0°C (32°F).

- Prolonged idling (vehicle operation in stop and go traffic).

- Frequent short running periods (engine does not warm to normal operating temperatures).

- Police, taxi, delivery usage or trailer towing usage.

## PRECAUTIONS

Before servicing any vehicle, please be sure to read all of the following precautions, which deal with personal safety, prevention of component damage, and important points to take into consideration when servicing a motor vehicle:

• Never open, service or drain the radiator or cooling system when the engine is hot; serious burns can occur from the steam and hot coolant.

• Observe all applicable safety precautions when working around fuel. Whenever servicing the fuel system, always work in a well-ventilated area. Do not allow fuel spray or vapors to come in contact with a spark, open flame, or excessive heat (a hot drop light, for example). Keep a dry chemical fire extinguisher near the work area. Always keep fuel in a container specifically designed for fuel storage; also, always properly seal fuel containers to avoid the possibility of fire or explosion. Refer to the additional fuel system precautions later in this section.

• Fuel injection systems often remain pressurized, even after the engine has been turned **OFF**. The fuel system pressure must be relieved before disconnecting any fuel lines. Failure to do so may result in fire and/or personal injury.

• Brake fluid often contains polyglycol ethers and polyglycols. Avoid contact with the eyes and wash your hands thoroughly after handling brake fluid. If you do get brake fluid in your eyes, flush your eyes with clean, running water for 15 minutes. If eye irritation persists, or if you have taken

brake fluid internally, IMMEDIATELY seek medical assistance.

• The EPA warns that prolonged contact with used engine oil may cause a number of skin disorders, including cancer. You should make every effort to minimize your exposure to used engine oil. Protective gloves should be worn when changing oil. Wash your hands and any other exposed skin areas as soon as possible after exposure to used engine oil. Soap and water, or waterless hand cleaner should be used.

• All new vehicles are now equipped with an air bag system, often referred to as a Supplemental Restraint System (SRS) or Supplemental Inflatable Restraint (SIR) system. The system must be disabled before performing service on or around system components, steering column, instrument panel components, wiring and sensors. Failure to follow safety and disabling procedures could result in accidental air bag deployment, possible personal injury and unnecessary system repairs.

• Always wear safety goggles when working with, or around, the air bag system. When carrying a non-deployed air bag, be sure the bag and trim cover are pointed away from your body. When placing a non-deployed air bag on a work surface, always face the bag and trim cover upward, away from the surface. This will reduce the motion of the module if it is accidentally deployed. Refer to the additional air bag system precautions later in this section.

• Clean, high quality brake fluid from a sealed container is essential to the safe and

proper operation of the brake system. You should always buy the correct type of brake fluid for your vehicle. If the brake fluid becomes contaminated, completely flush the system with new fluid. Never reuse any brake fluid. Any brake fluid that is removed from the system should be discarded. Also, do not allow any brake fluid to come in contact with a painted surface; it will damage the paint.

• Never operate the engine without the proper amount and type of engine oil; doing so WILL result in severe engine damage.

• Timing belt maintenance is extremely important. Many models utilize an interference-type, non-freewheeling engine. If the timing belt breaks, the valves in the cylinder head may strike the pistons, causing potentially serious (also time-consuming and expensive) engine damage. Refer to the maintenance interval charts for the recommended replacement interval for the timing belt, and to the timing belt section for belt replacement and inspection.

• Disconnecting the negative battery cable on some vehicles may interfere with the functions of the on-board computer system(s) and may require the computer to undergo a relearning process once the negative battery cable is reconnected.

• When servicing drum brakes, only disassemble and assemble one side at a time, leaving the remaining side intact for reference.

• Only an MVAC-trained, EPA-certified automotive technician should service the air conditioning system or its components.

## BRAKES

### GENERAL INFORMATION

#### PRECAUTIONS

• Certain components within the ABS system are not intended to be serviced or repaired individually.

• Do not use rubber hoses or other parts not specifically specified for and ABS system. When using repair kits, replace all parts included in the kit. Partial or incorrect repair may lead to functional problems and require the replacement of components.

• Lubricate rubber parts with clean, fresh brake fluid to ease assembly. Do not

use shop air to clean parts; damage to rubber components may result.

• Use only DOT 3 brake fluid from an unopened container.

• If any hydraulic component or line is removed or replaced, it may be necessary to bleed the entire system.

• A clean repair area is essential. Always clean the reservoir and cap thoroughly before removing the cap. The slightest amount of dirt in the fluid may plug an orifice and impair the system function. Perform repairs after components have been thoroughly cleaned; use

## ANTI-LOCK BRAKE SYSTEM (ABS)

only denatured alcohol to clean components. Do not allow ABS components to come into contact with any substance containing mineral oil; this includes used shop rags.

• The Anti-Lock control unit is a microprocessor similar to other computer units in the vehicle. Ensure that the ignition switch is **OFF** before removing or installing controller harnesses. Avoid static electricity discharge at or near the controller.

• If any arc welding is to be done on the vehicle, the control unit should be unplugged before welding operations begin.

## BRAKES                                                    BLEEDING THE BRAKE SYSTEM

### BLEEDING PROCEDURE

*BLEEDING PROCEDURE*

*See Figures 1 through 3.*

The brake bleeding should be completed with the use of a brake fluid changer. Connect the bleeder unit to the brake fluid reservoir and turn the unit on.

1. Remove the heater bulkhead to access the brake fluid reservoir.
2. Connect a bleeder hose with a suitable container to the bleeder valve on the right rear caliper. Open the bleeder valve and purge the fluid until clear, bubble-free fluid appears.
3. Close the bleeder valve.

4. Repeat this procedure at each wheel in the following order:
   a. Right rear
   b. Left rear
   c. Right front
   d. Left front
5. Switch off the bleeder unit and remove from the brake fluid reservoir.

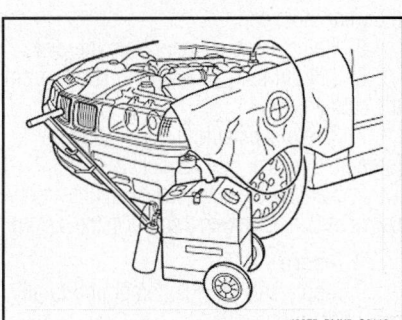

42075_BMX5_G0116

Fig. 1 A Brake Fluid Changer (bleeder unit) should be used to bleed the brake system.

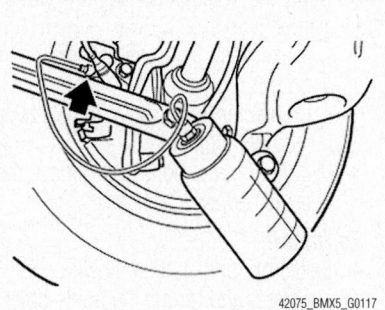

42075_BMX5_G0117

Fig. 2 Connect a bleeder hose with a suitable container to catch the fluid as you bleed each wheel.

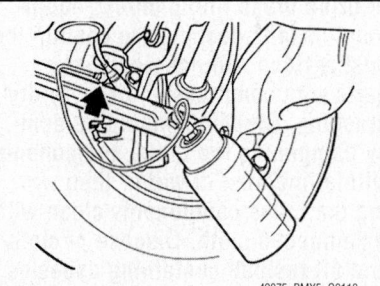

42075_BMX5_G0118

Fig. 3 When you disconnect the bleeder unit from the reservoir, check that gasket seal is in good condition before installing the cap.

## BRAKES                                                    FRONT DISC BRAKES

### ❋❋ CAUTION

**Dust and dirt accumulating on brake parts during normal use may contain asbestos fibers from production or aftermarket brake linings. Breathing excessive concentrations of asbestos fibers can cause serious bodily harm. Exercise care when servicing brake parts. Do not sand or grind brake lining unless equipment used is designed to contain the dust residue. Do not clean brake parts with compressed air or by dry brushing. Cleaning should be done by dampening the brake components with a fine mist of water, then wiping the brake components clean with a dampened cloth. Dispose of cloth and all residue containing asbestos fibers in an impermeable container with the appropriate label. Follow practices prescribed by the Occupational Safety and Health Administration (OSHA) and the Environmental Protection Agency (EPA) for the handling, processing, and disposing of dust or debris that may contain asbestos fibers.**

### BRAKE CALIPER

*REMOVAL & INSTALLATION*

1. Before servicing the vehicle, refer to the precautions section.
2. Remove or disconnect the following:
   • Negative battery cable
   • Wheel assembly
3. Apply the brake pedal slightly with a brake clamp.
   • Brake pipe from the connection with the brake hose
   • Connector for the wear indicator on the left side
   • Caliper guide bolts
   • Brake caliper

*To install:*

4. Install or connect the following:
   • Caliper and torque the guide bolts to 81 ft. lbs. (110 Nm)
   • Brake hose to the brake pipe to 13 ft. lbs. (18 Nm)
5. Set the wheel in a straight-ahead position.
6. Be sure brake hose is positively attached to the mounting fixture.
   • Wear indicator on the left side
   • Brake clamp
   • Front wheels
7. Bleed the brakes.

### DISC BRAKE PADS

*REMOVAL & INSTALLATION*

1. Before servicing the vehicle, refer to the precautions section.
2. Remove the front wheels.
3. Remove the disk pad retaining spring from the caliper.
4. Remove the calipers from the disk.
5. Use a special tool, 34–1–050, to force piston back into caliper.
6. Remove the outer brake pad (the inner pad is held in place with a spring in the piston).

*To install:*

7. Be sure the pads marked "L" and "R" are inserted properly on left and right sides, respectively.
8. Apply anti-squeak compound to all mounting surfaces.
9. Install the calipers.
10. Reposition the retaining spring.
11. Install the front wheels.
12. Fully depress brake pedal several times to set proper contact of pads with rotor.
13. Hold ignition key for at least 30 seconds in position "1" without starting engine. This clear any fault codes stored in system and prevent the wear indicator light from coming on.
14. Bleed brake system. If necessary.

### ✳✳ CAUTION

Dust and dirt accumulating on brake parts during normal use may contain asbestos fibers from production or aftermarket brake linings. Breathing excessive concentrations of asbestos fibers can cause serious bodily harm. Exercise care when servicing brake parts. Do not sand or grind brake lining unless equipment used is designed to contain the dust residue. Do not clean brake parts with compressed air or by dry brushing. Cleaning should be done by dampening the brake components with a fine mist of water, then wiping the brake components clean with a dampened cloth. Dispose of cloth and all residue containing asbestos fibers in an impermeable container with the appropriate label. Follow practices prescribed by the Occupational Safety and Health Administration (OSHA) and the Environmental Protection Agency (EPA) for the handling, processing, and disposing of dust or debris that may contain asbestos fibers.

### BRAKE CALIPER

#### REMOVAL & INSTALLATION

1. Before servicing the vehicle, refer to the precautions section.
2. Remove or disconnect the following:
   - Negative battery cable
   - Wheel assembly
3. Apply the brake pedal slightly with a brake clamp.
   - Brake hose from the caliper fitting
   - Connector for the wear indicator on the right side
   - Caliper guide bolts and remove the brake caliper

#### To install:
4. Install or connect the following:
   - Caliper and torque the guide bolts to 21 ft. lbs. (28 Nm)
   - Brake hose to the brake pipe and torque to 14 ft. lbs. (19 Nm)
   - Wear indicator
5. Remove the brake clamp.
6. Install the rear wheels.
7. Bleed the brakes.

### DISC BRAKE PADS

#### REMOVAL & INSTALLATION

1. Before servicing the vehicle, refer to the precautions section.
2. Remove the rear wheels.
3. Remove the plastic plugs from the inside of the caliper.
4. Disconnect the plug connection for the wear indicator.
5. Remove the calipers from the disk.
6. Lift out the pad retaining spring from the caliper.
7. Use a special tool, 34–1–050, to force piston back into caliper.
8. Remove the outer brake pad (the inner pad is held in place with a spring in the piston).

#### To install:
9. Apply anti-squeak compound to all mounting surfaces.
10. Reposition retaining spring.
11. Install calipers.
12. Install rear wheels.
13. Fully depress brake pedal several times to set proper contact of pads with rotor.
14. Hold ignition key for at least 30 seconds in position "1" without starting engine. This clear any fault codes stored in system and prevent the wear indicator light from coming on.
15. Bleed brake system. If necessary.

### PARKING BRAKE SHOES

#### REMOVAL & INSTALLATION

*See Figures 4 and 5.*

1. Remove the rear brake rotors.
2. Disconnect the return springs with brake spring pliers.
3. Twist the clamping pins with Special Tool 34-4-000 90° and disconnect. Remove the parking brake shoes.

#### To install:
4. Installation is the reverse order of removal. Inspect the brake components and replace if necessary.
5. After installation, the parking brake shoes must be adjusted.

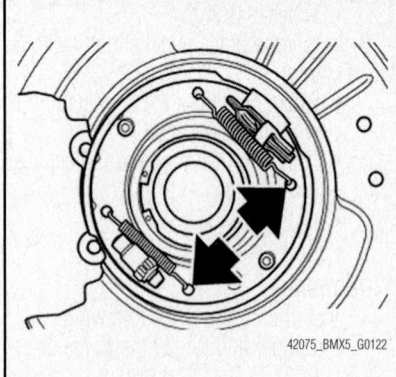

42075_BMX5_G0122

**Fig. 4 Disconnect the brake return springs . . .**

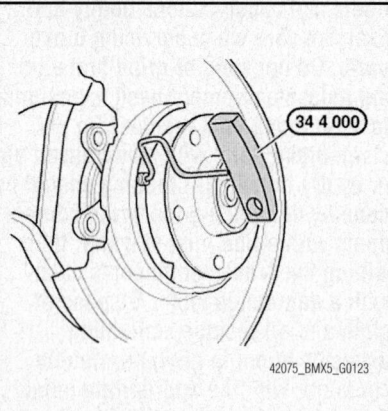

42075_BMX5_G0123

**Fig. 5 . . . and clamping pins to remove the parking brake shoes**

## CHASSIS ELECTRICAL          AIR BAG (SUPPLEMENTAL RESTRAINT SYSTEM)

### GENERAL INFORMATION

### ✷✷ CAUTION

**These vehicles are equipped with an air bag system. The system must be disarmed before performing service on, or around, system components, the steering column, instrument panel components, wiring and sensors. Failure to follow the safety precautions and the disarming procedure could result in accidental air bag deployment, possible injury and unnecessary system repairs.**

### SERVICE PRECAUTIONS

Disconnect and isolate the battery negative cable before beginning any airbag system component diagnosis, testing, removal, or installation procedures. Allow system capacitor to discharge for two minutes before beginning any component service. This will disable the airbag system. Failure to disable the airbag system may result in accidental airbag deployment, personal injury, or death.

Do not place an intact undeployed airbag face down on a solid surface. The airbag will propel into the air if accidentally deployed and may result in personal injury or death.

When carrying or handling an undeployed airbag, the trim side (face) of the airbag should be pointing towards the body to minimize possibility of injury if accidental deployment occurs. Failure to do this may result in personal injury or death.

Replace airbag system components with OEM replacement parts. Substitute parts may appear interchangeable, but internal differences may result in inferior occupant protection. Failure to do so may result in occupant personal injury or death.

Wear safety glasses, rubber gloves, and long sleeved clothing when cleaning powder residue from vehicle after an airbag deployment. Powder residue emitted from a deployed airbag can cause skin irritation. Flush affected area with cool water if irritation is experienced. If nasal or throat irritation is experienced, exit the vehicle for fresh air until the irritation ceases. If irritation continues, see a physician.

Do not use a replacement airbag that is not in the original packaging. This may result in improper deployment, personal injury, or death.

The factory installed fasteners, screws and bolts used to fasten airbag components have a special coating and are specifically designed for the airbag system. Do not use substitute fasteners. Use only original equipment fasteners listed in the parts catalog when fastener replacement is required.

During, and following, any child restraint anchor service, due to impact event or vehicle repair, carefully inspect all mounting hardware, tether straps, and anchors for proper installation, operation, or damage. If a child restraint anchor is found damaged in any way, the anchor must be replaced. Failure to do this may result in personal injury or death.

Deployed and non-deployed airbags may or may not have live pyrotechnic material within the airbag inflator.

Do not dispose of driver/passenger/curtain airbags or seat belt tensioners unless you are sure of complete deployment. Refer to the Hazardous Substance Control System for proper disposal.

Dispose of deployed airbags and tensioners consistent with state, provincial, local, and federal regulations.

After any airbag component testing or service, do not connect the battery negative cable. Personal injury or death may result if the system test is not performed first.

If the vehicle is equipped with the Occupant Classification System (OCS), do not connect the battery negative cable before performing the OCS Verification Test using the scan tool and the appropriate diagnostic information. Personal injury or death may result if the system test is not performed properly.

Never replace both the Occupant Restraint Controller (ORC) and the Occupant Classification Module (OCM) at the same time. If both require replacement, replace one, then perform the Airbag System test before replacing the other.

Both the ORC and the OCM store Occupant Classification System (OCS) calibration data, which they transfer to one another when one of them is replaced. If both are replaced at the same time, an irreversible fault will be set in both modules and the OCS may malfunction and cause personal injury or death.

If equipped with OCS, the Seat Weight Sensor is a sensitive, calibrated unit and must be handled carefully. Do not drop or handle roughly. If dropped or damaged, replace with another sensor. Failure to do so may result in occupant injury or death.

If equipped with OCS, the front passenger seat must be handled carefully as well. When removing the seat, be careful when setting on floor not to drop. If dropped, the sensor may be inoperative, could result in occupant injury, or possibly death.

If equipped with OCS, when the passenger front seat is on the floor, no one should sit in the front passenger seat. This uneven force may damage the sensing ability of the seat weight sensors. If sat on and damaged, the sensor may be inoperative, could result in occupant injury, or possibly death.

### DISARMING THE SYSTEM

1. Before servicing the vehicle, refer to the precautions section.
2. Place the ignition switch in the **OFF** position.
3. Disconnect the negative battery terminal and cover the battery terminal to prevent accidental contact.
4. Once the battery has been disconnected, wait for a period of approximately 5 seconds allowing the capacitor in the control unit to discharge. Once the capacitor is discharged, a trigger pulse cannot be generated inadvertently.

### ARMING THE SYSTEM

1. Before servicing the vehicle, refer to the precautions section.
2. Place the ignition switch in the **OFF** position.
3. Attach the sensors, the steering column connector and the seat belt tensioner connectors.
4. Connect the negative battery terminal.
5. Place the ignition switch in the **ON** position. Check that the SRS light illuminates for 6 seconds and extinguishes. If it illuminates in any other pattern, check the components and their connections for proper operation and recheck operation of the warning light.

# DRIVETRAIN

## AUTOMATIC TRANSMISSION ASSEMBLY

### REMOVAL & INSTALLATION

#### 3.0L Engines

1. Before servicing the vehicle, refer to the precautions section.
2. Remove or disconnect the following:
   - Negative battery cable
   - Exhaust system
   - Cables for oxygen sensors, mark the locations for installation purposes
   - Reinforcement plate from undercarriage
   - Engine under guard at front
   - Stabilizer bar and slide forward
   - Front propeller shaft
   - Vent line from transmission at bracket
   - Exhaust bracket
   - Nut and bushing from shift lever
   - Shift control cable from body bracket
   - Electrical connection and cable from transmission
   - Hydraulic cooler lines from transmission
3. Position the proper transmission cradle, 00–2–030, 24–5–301/305.
4. Support the transmission and transfer case with transmission cradle.
5. Remove or disconnect the following:
   - Oxygen (O2s) sensor wiring cable at transmission crossmember
   - Wiring cable from transmission
   - Crossmember
   - Propeller shaft flange nuts
   - Center bearing support and support the propeller shaft
   - Propeller shaft from transmission and tie aside
   - Hex bolt near protective cap on transmission flange
   - Protective cap
   - Bolts from torque converter through protective cap opening
   - Transmission housing–to–engine bolts
6. Pull the transmission, with transfer case, out toward rear.

#### To install:

7. Note the position and condition of dowel sleeves on flywheel housing. Replace dowels if damaged.
8. Position the transmission to the engine and install the housing bolts. Torque the bolts as follows:

   a. M8 hex bolts: 18 ft. lbs. (24 Nm).
   b. M10 hex bolts: 33 ft. lbs. (45 Nm).
   c. M12 hex bolts: 60 ft. lbs. (82 Nm).
   d. M8 Torx bolts: 15 ft. lbs. (21 Nm).
   e. M10 Torx bolts: 31 ft. lbs. (42 Nm).
   f. M12 Torx bolts: 53 ft. lbs. (72 Nm).
9. Install and torque the torque converter–to–flywheel bolts to the following:
   a. M8 hex bolts: 19 ft. lbs. (26 Nm).
   b. M10 hex bolts: 33 ft. lbs. (45 Nm).
10. Install or connect the following:
    - Protective cap over flywheel bolt opening
    - Hex bolt near protective cap
    - Propeller shaft into center support; torque mount–to–body bolts to 15 ft. lbs. (21 Nm)
    - Propeller shaft flange new nuts; torque to 74 ft. lbs. (100 Nm)
11. Install the transmission crossmember. Torque the bolts as follows:
    a. To rubber mounts: 55 ft. lbs. (74 Nm).
    b. To body: 30 ft. lbs. (41 Nm).
12. Install or connect the following:
    - Transmission cable
    - O2s sensor wiring cables
    - Transmission cooler lines with new sealing ring
    - Wiring cable to transmission
    - Shift control cable to body bracket
    - Shift selector lever bushing and nut
    - Exhaust hanger bracket
    - Vent line to bracket

- Stabilizer bar
- Engine guard
- Reinforcement plate
- Oxygen sensor connectors
- Heat shield
- Exhaust system
- Negative battery cable

13. Refill transmission and check operation, then recheck fluid level.

#### 4.4L (N62) & 4.8L Engines

*See Figures 6 and 7.*

1. Drain the transmission fluid.
2. Disconnect the negative battery cable.
3. Remove the engine skid plates.
4. Remove the reinforcement plate.
5. Remove the center muffler.
6. Remove the front and rear heat shields.
7. Disconnect the right and left stabilizer links.
8. Remove the driveshaft from the front differential. Tie the driveshaft up to one side.
9. Disconnect the speed sensor electrical connector and remove the speed sensor.
10. Press in the retaining lugs with a screwdriver, remove the cable duct and tie off to one side.
11. Disconnect the remaining electrical connectors.
12. Grip the clamping sleeve, slacken off the nut, detach the retainer towards the bottom using a suitable pry tool and pull the selector cable out of the holder.
13. Disconnect the oil cooler lines.

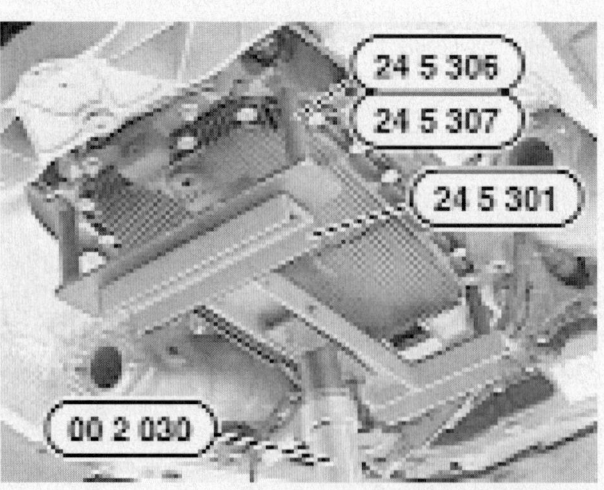

42075_BMX5_G0072

**Fig. 6 Assembly the Special Tools to a transmission jack to secure the transmission assembly—4.4L (N62) & 4.8L Engines**

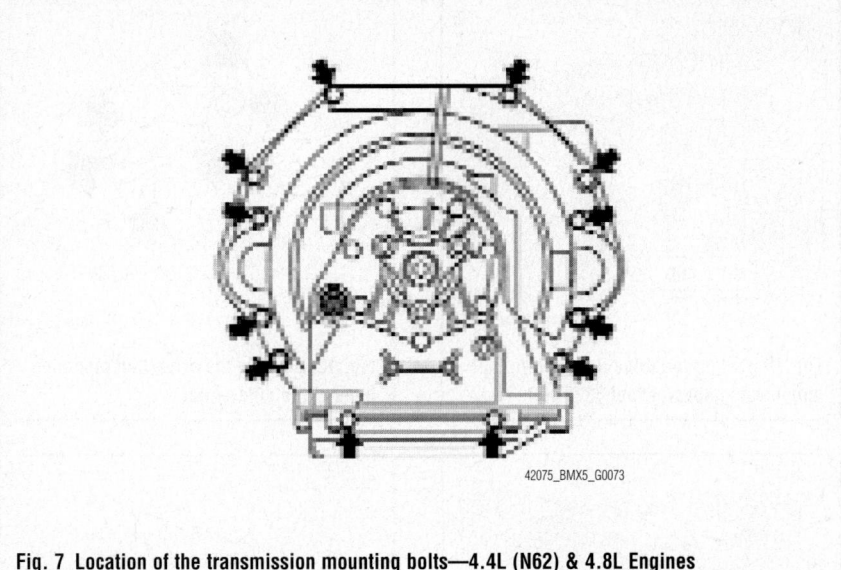

**Fig. 7 Location of the transmission mounting bolts—4.4L (N62) & 4.8L Engines**

14. Attach Special Tools 24-5-307, 310 and 00-2-030 combined with 24-5-307 to a suitable transmission jack.

15. Remove the transfer case from the transmission.

16. Turning the crankshaft at the crankshaft damper clockwise to access and remove the torque converter bolts.

17. Remove the transmission mounting bolts and lower the transmission assembly.

18. Installation is the reverse order of removal.

19. Note the following torques for the transmission mounting bolts:

    a. M8 Hex screws: 18 ft. lbs. (24 Nm).

    b. M10 Hex screws: 33 ft. lbs. (45 Nm).

    c. M12 Hex screws: 60 ft. lbs. (82 Nm).

    d. M8 Torx bolts: 16 ft. lbs. (21 Nm).

    e. M10 Torx bolts: 31 ft. lbs. (42 Nm).

    f. M12 Torx bolts: 54 ft. lbs. (72 Nm).

20. Refill the transmission with fluid to the correct level.

## TRANSFER CASE ASSEMBLY

### REMOVAL & INSTALLATION

#### 3.0L Engines

1. Before servicing the vehicle, refer to the precautions section.

2. Remove or disconnect the following:
- Negative battery cable
- Exhaust system
- Heat shield at right front
- Heat shield at rear
- Reinforcement plate from undercarriage
- Stabilizer bar
- Front propeller shaft

3. Support the transmission
- Crossmember
- Vent line from transfer case
- Driveshaft–to–transmission flange nuts and discard the nuts
- Propeller shaft nuts, bend shaft downward at center bearing
- Propeller shaft from transmission and tie aside

4. Support the transfer case. Remove the retaining bolts and remove the transfer case.

**To install:**

5. Check the condition of the dowel pins and replace if necessary.

6. Be sure the mating surfaces are clean.

7. Apply a thin coat of grease to transfer case splines

8. Replace the sealing ring of the driveshaft of the transfer case.

9. Install the transfer case and torque the retaining bolts to 32 ft. lbs. (43 Nm).

10. Reposition the propeller shaft in the center support bearing and to the flange. Torque the center support bearing nuts to 15 ft. lbs. (21 Nm) and the flange nuts to 24 ft. lbs. (32 Nm).

11. Install or connect the following:
- Cable to transfer case
- Vent line

12. Install the transmission crossmember. Torque the bolts as follows:

    a. To rubber mounts: 55 ft. lbs. (74 Nm).

    b. To body: 30 ft. lbs. (41 Nm).

13. Install or connect the following:
- Propeller shaft
- Front stabilizer
- Reinforcement plate
- Heat shields
- Exhaust system

14. Refill the transfer case.

#### 4.4L (N62) & 4.8L Engines

*See Figure 8.*

1. Drain the transmission fluid.

2. Remove the center muffler and catalytic converter.

3. Remove the front and rear heat shields.

4. Grip the clamping sleeve, slacken off the nut, detach the retainer towards the bottom using a suitable pry tool and pull the selector cable out of the holder.

5. Remove the driveshaft from the transmission. Tie the driveshaft up to one side.

6. Support the transmission with a suitable jack.

7. Remove the mounting bolts and remove transmission cross-member.

8. Disconnect the electrical connectors from the servomotor.

9. Remove the transfer case mounting bolts.

10. Remove the transfer case.

11. Installation is the reverse order of removal. Tighten the transfer case mounting bolts to 32 ft. lbs. (43 Nm).

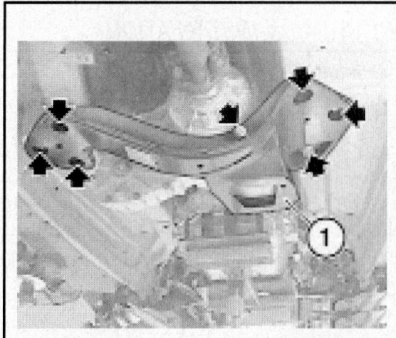

**Fig. 8 Location of the mounting bolts to remove the transmission crossmember (1)—4.4L (N62) & 4.8L Engines**

## FRONT HALFSHAFT

### REMOVAL & INSTALLATION

1. Before servicing the vehicle, refer to the precautions section.

2. Remove or disconnect the following:
- Negative battery cable
- Front wheel

- Reinforcement plate
- Front splash guard
- ABS pulse generator
- Brake caliper from disc and tie out of the way with the hose still connected
- Steering gear tie rod from swivel bearing
- Tension strut, with guide joint, from swivel bearing
- Control arm from swivel bearing
- Collar nut on halfshaft, press halfshaft out of drive flange
- Halfshaft

### To install:

3. Push in output shaft over the resistance of the retaining ring until it snaps in place.

➡ **Use a new snap ring on halfshaft spline**

4. Install or connect the following:
- Collar nut
- Control arm to swivel bearing
- Tension strut to swivel bearing
- Tie rod to swivel bearing
- Brake caliper to disc
- ABS pulse generator
- Front wheel
- Splash guard
- Reinforcement plate

5. Check the front axle differential fluid level.

## FRONT PINION SEAL

### REMOVAL & INSTALLATION

*See Figures 9 through 11.*

1. Remove the driveshaft.
2. Matchmark the pinion nut in relation to the drive flange on the pinion.
3. Secure the pinion flange with Special Tool 23-0-020 and remove the pinion nut.
4. Using Special Tool 33-1-150, pull the pinion flange from the pinion.

**Fig. 9 Secure the drive flange with the Special Tool to remove the pinion nut.**

42075_BMX5_G0078

**Fig. 10 Pull off the drive flange from the pinion using Special Tool 33-1-150**

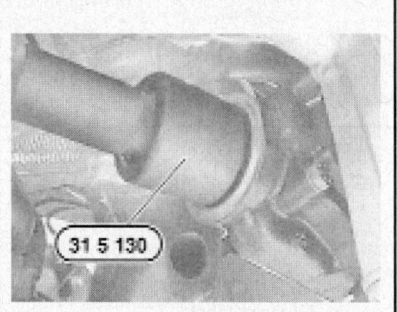
42075_BMX5_G0079

**Fig. 11 Drive in a new pinion seal with Special Tool 31-5-130**

5. Using a suitable pry tool, pry out the pinion seal from the front differential.

### To install:

6. Drive in a new pinion seal with Special Tool 31-5-130 as far as it will go. Coat the seal with differential oil.
7. Install the pinion flange and tighten the pinion nut until the matchmark is aligned.
8. Install the driveshaft.
9. Fill the front differential to the correct level.

## REAR DRIVESHAFT

### REMOVAL & INSTALLATION

*See Figures 12 and 13.*

➡ **Do not move the vehicle by driving it once the driveshaft has been removed.**

1. Remove the reinforcement plate.
2. Remove the driveshaft mounting bolts.
3. Push the driveshaft back and remove the flexible disc with the center flange.
4. Disconnect the driveshaft from the transfer case.
5. Installation is the reverse order of removal. Any self-locking nuts must be replaced.

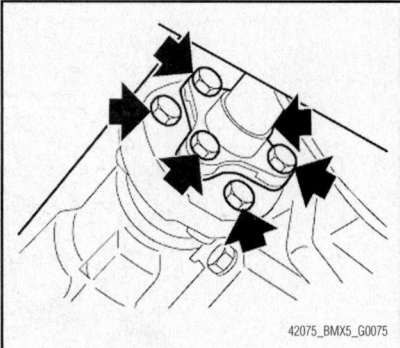
42075_BMX5_G0075

**Fig. 12 Remove the driveshaft mounting bolts at the differential**

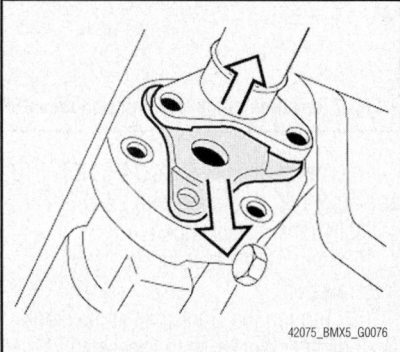
42075_BMX5_G0076

**Fig. 13 Slide the driveshaft and remove the center flange**

## REAR HALFSHAFT

### REMOVAL & INSTALLATION

*See Figures 14 and 15.*

1. Remove the exhaust system.
2. Remove the rear heat shield.
3. Remove the driveshaft mounting nuts at the transmission side.
4. Remove the mounting nuts at the pinion side.
5. With a suitable pry tool, press the CV joint off the input flange of the differential.
6. Support the driveshaft center mount, and loosen the center bearing mounting nuts.

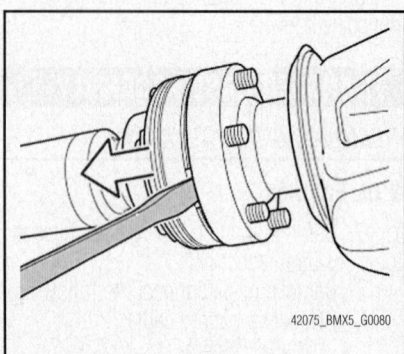
42075_BMX5_G0080

**Fig. 14 Pry the CV-joint off the input flange of the differential**

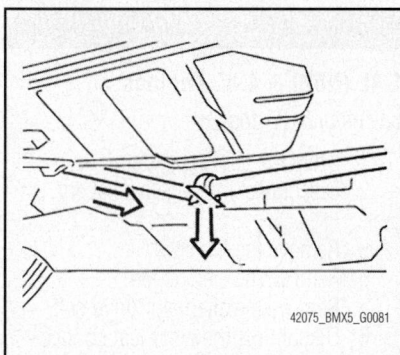

Fig. 15 Bend the driveshaft down on the center bearing to remove

7. Bend the driveshaft downwards on the center bearing. Slide the driveshaft off the output flange off the transmission, then remove the CV-joint from the differential.

8. Installation is reverse order of removal. Replace all self-locking nuts with new ones.

## REAR PINION SEAL

### REMOVAL & INSTALLATION

*See Figures 16 through 22.*

1. Pry the rear driveshaft from the rear differential, or remove the rear driveshaft if necessary.

2. Lift out the retaining plate.

3. Matchmark the position of the pinion nut on the pinion shaft with a peening tool.

4. Secure the pinion flange with Special Tool 23-0-020 and remove the pinion nut.

5. Using Special Tool 33-1-150, remove the pinion flange.

6. Remove the pinion seal with Special Tool 00-5-010.

### To install:

7. Drive in a new pinion seal using Special Tool 00-5-500 and 33-3-470.

8. Coat the sealing surfaces of the pinion seal and pinion flange with differential oil.

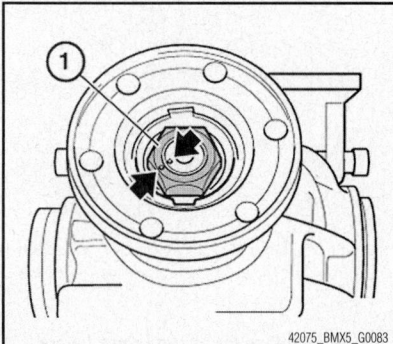

Fig. 16 Matchmark the pinion nut to the pinion shaft

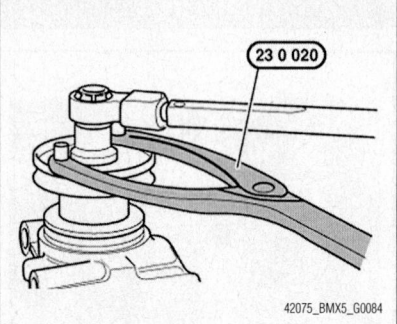

Fig. 17 Secure the pinion flange to remove the pinion nut

Fig. 18 Remove the pinion flange using Special Tool 33-1-150

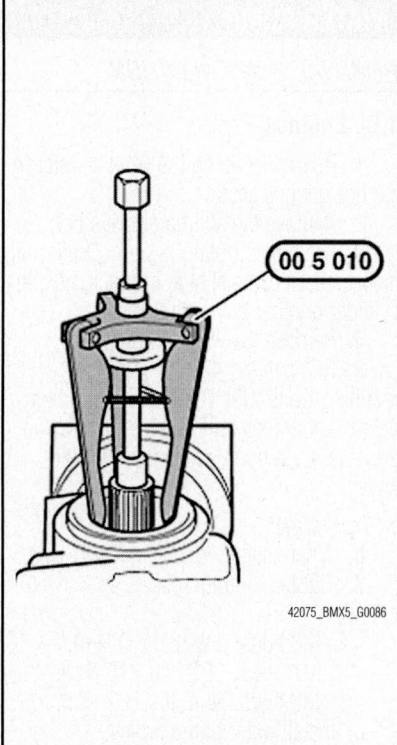

Fig. 19 Remove the pinion seal with the Special Tool

9. Fit the pinion flange and drive it on with Special Tools 23-1-300, 302, 303 and 33-1-341.

10. Tighten the pinion nut until the matchmarks are aligned.

11. Drive in a new retaining plate using Special Tools 00-5-500 and 33-3-480.

12. Install the rear driveshaft.

13. Drain and refill the differential oil to the correct level.

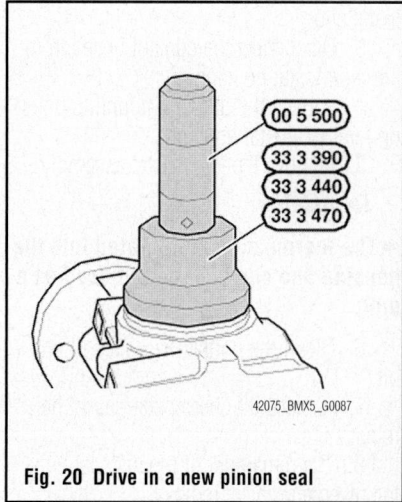

Fig. 20 Drive in a new pinion seal

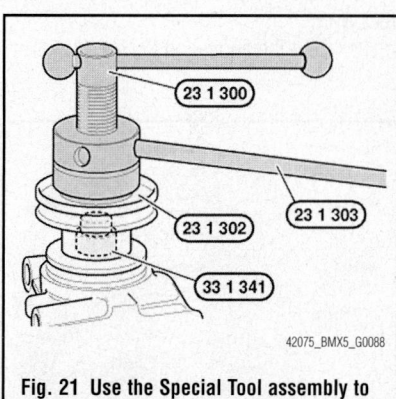

Fig. 21 Use the Special Tool assembly to press on the pinion flange

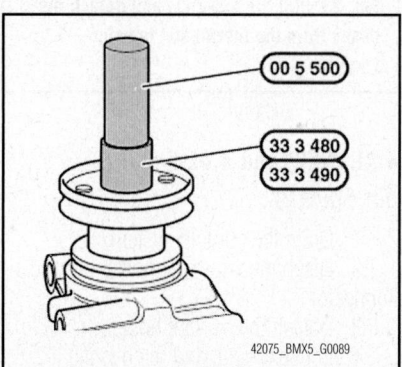

Fig. 22 Drive on the retaining plate as far as it will go

# ENGINE COOLING

## THERMOSTAT

### REMOVAL & INSTALLATION

#### 3.0L Engines

*See Figure 23.*

1. Drain the cooling system.
2. Remove the engine fan. For additional information, please refer to the following topic: "Engine Fan, Removal & Installation".
3. Disconnect the coolant hoses from the thermostat housing.
4. Loosen the nut and mounting bolts on the thermostat housing.
5. Remove the thermostat assembly.

#### To install:

➡ **The thermostat is integrated into the housing and can only be replaced as a unit.**

6. Clean the contact surfaces of any oil.
7. Replace the thermostat gasket before reinstalling.
8. The remainder of the installation is the reverse order of removal.
9. Refill the cooling system to the correct level.
10. Start the engine and check for leaks.

**Fig. 23 Pull the locks (1) and detach the hoses from the thermostat housing—3.0L Engines**

#### 4.4L (N62) and 4.8L Engines

*See Figure 24.*

1. Drain the cooling system.
2. Disconnect the thermostat electrical connector.
3. Detach the coolant hose.
4. Loose the nut and remove the mounting bolts.
5. Remove the thermostat.

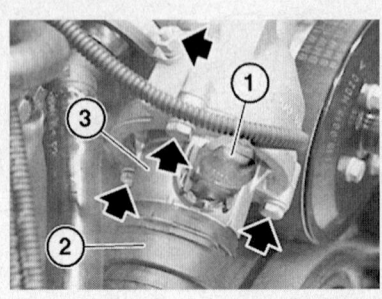

**Fig. 24 Disconnect the electrical connector (1) and disconnect the coolant hose (2) to remove the thermostat (3)—4.4L (N62) & 4.8L Engines shown**

#### To install:

➡ **The thermostat is integrated into the cover and can only be replaced as a unit.**

6. Installation is the reverse order of removal.
7. Refill the cooling system to the correct level.
8. Start the engine and check for leaks.

## WATER PUMP

### REMOVAL & INSTALLATION

#### 3.0L Engines

1. Before servicing the vehicle, refer to the precautions section.
2. Remove the alternator drive belt.
3. Drain the cooling system. Drain plug is located on exhaust side of block, next to cylinder number 2.
4. Remove the water pump pulley.
5. Remove the 4 water pump retaining bolts. Use 2 M6 bolts in holes next to mounting bolt holes and screw in until water pump releases from timing cover.

#### To install:

6. When installing, use a new O-ring.
7. Tighten mounting bolts as follows:
   a. M6 bolts: 7 ft. lbs. (10 Nm).
   b. M7 bolts: 11 ft. lbs. (15 Nm).
   c. M8 bolts: 16 ft. lbs. (22 Nm).
8. Install water pump pulley.
9. Refill cooling system.
10. Install the alternator drive belt.
11. Start the vehicle, check for leaks and repair as necessary.

#### 4.4L (N62) & 4.8L Engines

*See Figures 25 and 26.*

1. Drain the cooling system.
2. Disconnect the negative battery cable.
3. Remove the fan cowl.
4. Remove the electric fan.
5. Remove the alternator drive belt.
6. Disconnect the electrical connectors.
7. Remove the vacuum line holder.
8. Disconnect all of the coolant hoses on the water pump.
9. Remove the water pump pulley bolts and remove the water pump pulley.
10. Remove the crankshaft damper. For additional information, please refer to the following topic: "Crankshaft Damper, Removal & Installation".

**Fig. 25 Remove the mounting bolts to remove the water pump (1)—4.4L (N62) & 4.8L Engines**

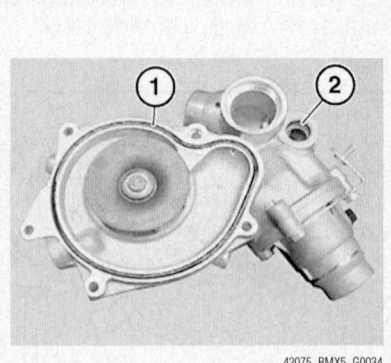

**Fig. 26 Replace the sealing rings before installing the water pump—4.4L (N62) & 4.8L Engines**

11. Remove the mounting bolts and remove the water pump.

*To install:*

12. Clean the water pump and contact surface of any gasket material.

13. Replace the gasket and sealing rings on the water pump.

14. Install the water pump.

15. The remainder of the installation is the reverse order of removal.

16. Refill the cooling system to the correct level.

17. Start the engine and check for leaks.

## ENGINE ELECTRICAL

### ALTERNATOR

➡**When the battery is disconnected the radio code, on-board computer and clock settings will be lost. The radio code should be obtained before disconnecting the battery or radio. Once the battery has been reconnected, the radio will not function unless the code is keyed in.**

### REMOVAL & INSTALLATION

#### 3.0L Engines

1. Before servicing the vehicle, refer to the precautions section.

2. Read out stored fault codes (if applicable).

3. Switch off ignition.

4. Disconnect negative battery cable.

5. Remove or disconnect the following:
- Suction filter housing
- Fan clutch
- Alternator drive belt
- Power steering pump from mounting (move aside)
- Alternator air hose (if equipped)
- Electrical connections from alternator
- Alternator mounting bolts (versions without idler pulley)
- Idler pulley cover and idler pulley bolt (versions with idler pulley)
- Alternator

**To install:**

6. To install, reverse removal procedure.

7. If equipped with idler pulley, turn lock of tensioning roller to engage alternator groove.

## CHARGING SYSTEM

8. With scan tool, clear fault code memory.

#### 4.4L and 4.8L Engines

1. Switch off the ignition.

2. Disconnect negative battery cable.

3. Remove or disconnect the following:
- Alternator drive belt
- Fan cowl with the electric fan
- Idler pulley
- Alternator mounting bolts and pull the alternator forward to reveal the electrical connections.
- Electrical connections from alternator
- Alternator

4. Installation is the reverse order of removal.

## ENGINE ELECTRICAL

### IGNITION COIL

### REMOVAL & INSTALLATION

#### 3.0L Engines

1. Disconnect the negative battery cable.

2. Remove the fuel injector cover.

3. Remove the ignition coil cover.

4. Disconnect the electrical connector from the ignition coils.

5. Remove the bolts and pull out the ignition coils.

➡**Note the location of the grounding straps before removal.**

*To install:*

6. Install the ignition coil and two bolts.

7. Connect the electrical connector to the ignition coil.

8. Install the ignition coil and fuel injector covers.

9. Connect the negative battery cable.

#### 4.4L and 4.8L Engines

1. Disconnect the negative battery cable.

2. Remove the engine acoustic cover by turning the fasteners 90 degrees counterclockwise.

3. Remove the ignition coil covers.

4. Disconnect the electrical connectors for the ignition coils.

5. Remove the bolts and remove the ignition coils.

*To install:*

6. Install the ignition coil, ground strap and two bolts.

7. Connect the electrical connector to the ignition coil.

8. Install the ignition coil and acoustic covers.

9. Connect the negative battery cable.

### IGNITION TIMING

The Digital Motor Electronics (DME) control unit controls all ignition and fuel injection functions. Ignition timing is fully electronically controlled; there is no vacuum advance or manual adjustment. Ignition functions are calculated from internal maps and from the same sensors used for the fuel injection system. On vehicles with an automatic transmission, the control unit will retard ignition timing briefly when the transmission is about to shift up or down. For this reason, there is a data link between the DME control unit and the transmission control unit.

## IGNITION SYSTEM

Since the ignition timing is controlled by the DME, checking and adjusting the timing is impossible. There is no method of setting dynamic or static timing.

### ADJUSTMENT

The ignition system is controlled by the Digital Motor Electronics control unit. Adjustment is not possible.

### SPARK PLUGS

### REMOVAL & INSTALLATION

1. Disconnect the negative battery cable.

2. Remove the fuel injector cover.

3. Remove the ignition coil cover.

4. Disconnect the electrical connector from the ignition coils.

5. Remove the bolts and pull out the ignition coil.

6. Using a spark plug socket with an extension, remove the spark plug.

*To install:*

7. Using a spark plug socket with an extension, install the spark plug and tighten as follows:

a. Plugs with M12 x 1.25 threads: to 17 ft. lbs. (23 Nm).

b. Plugs with M14 x 1.25 threads: to 22 ft. lbs. (30 Nm).

8. Install the ignition coil, ground strap and two bolts.

9. Connect the electrical connector to the ignition coil.

10. Install the ignition coil and fuel injector covers.

11. Connect the negative battery cable.

## ENGINE ELECTRICAL

### STARTER

#### REMOVAL & INSTALLATION

➡When the battery is disconnected, the radio code, on-board computer and clock settings will be lost. The radio code should be obtained before disconnecting the battery or radio. Once the battery has been reconnected, the radio will not function unless the code is keyed in.

1. Before servicing the vehicle, refer to the precautions section.

2. If needed, read the stored fault memories from the control module.

3. Relieve the fuel system pressure.

4. Set the ignition switch to the **OFF** position.

5. Remove or disconnect the following:
- Negative battery cable
- Reinforcement plate
- Positive battery cable from the starter
- Starter electrical connectors
- Heat shield
- Starter from the transmission mount

## STARTING SYSTEM

- Control leads
- Starter

### To install:

6. Install or connect the following:
- Starter to transmission mount
- Control leads
- Heat shield torque bolts to 38 ft. lbs. (47 Nm).
- Starter electrical connectors
- Positive battery cable to the starter
- Reinforcement plate
- Negative battery cable

## ENGINE MECHANICAL

➡Disconnecting the negative battery cable may interfere with the functions of the on board computer systems and may require the computer to undergo a relearning process, once the negative battery cable is reconnected.

### ACCESSORY DRIVE BELTS

#### ACCESSORY BELT ROUTING

See Figures 27 and 28.

#### INSPECTION

Inspect the drive belt for signs of glazing or cracking. A glazed belt will be perfectly smooth from slippage, while a good belt will have a slight texture of fabric visible. Cracks will usually start at the inner edge of the belt and run outward. All worn or damaged drive belts should be replaced immediately.

#### ADJUSTMENT

All models use an auto tensioner for the accessory drive belts. If the belt tension is not at the proper specification, the tensioner must be replaced.

#### REMOVAL & INSTALLATION

#### 3.0L Engines

See Figure 29.

1. Disconnect the negative battery cable.

➡If the belt is going to be reused, mark the direction of travel for proper installation.

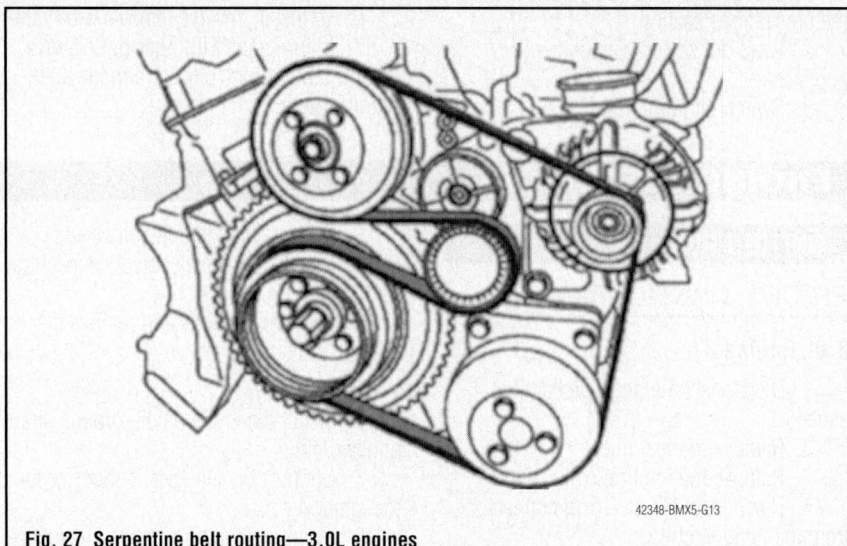

Fig. 27 Serpentine belt routing—3.0L engines

42348-BMX5-G13

Fig. 28 Alternator belt routing—4.4L and 4.8L engines

42348-BMX5-G14

**Fig. 29 Releasing the tension to remove the alternator belt—3.0L Engines**

2. Remove the fan clutch.

3. Remove the splash guard for clearance.

4. If equipped with an auto tensioner, push back the belt tensioner at the bolt connection of the guide pulley and remove the A/C compressor belt.

5. If equipped with mechanical drive belt, push back the belt tensioner at the hexagonal head and remove the A/C compressor belt.

6. Remove the cap from the tensioner pulley.

7. Release the tension using a socket wrench and remove the alternator belt.

8. Installation is the reverse order of removal.

### 4.8L Engines

1. Disconnect the negative battery cable.

➡**If the belt is going to be reused, mark the direction of travel for proper installation.**

2. Remove the alternator belt as follows:

a. Using a Torx® socket, release the tension on the belt tensioner and secure it with Special Tool 11-3-340.

b. Remove the alternator belt.

3. Remove A/C compressor belt as follows:

a. Using a Torx® socket, release the tension on the belt tensioner and secure it with Special Tool 11-3-340.

b. Remove the A/C compressor belt.

➡**If you get power steering fluid on the belts, they must be replaced.**

*To install:*

4. Install the A/C compressor and ensure it is correctly positioned on the pulleys.

5. Remove Special Tool 11-3-340.

6. Install the alternator belt and ensure it is correctly position on the pulleys.

7. Remove Special Tool 11-3-340.

## CAMSHAFT AND VALVE LIFTERS

### REMOVAL & INSTALLATION

#### 3.0L Engines

*See Figures 30 through 35.*

1. Before servicing the vehicle, refer to the precautions section.

2. Disconnect the negative battery cable.

3. Remove the camshaft (Double VANOS) adjustment unit as follows:

a. Remove the intake filter housing with Mass Air Flow (MAF) sensor.

b. Remove the fan impeller with fan clutch and cowl.

c. Remove the cylinder head covers.

d. Remove the spark plugs.

e. Remove the plastic cover for intake camshaft.

f. Remove the oil pressure pipe.

g. Install a special tool, 11–3–450, with banjo bolt on oil pressure connection unit.

h. Cover the camshaft adjustment unit.

i. Connect a compressed air hose to the special tool fitting on the oil line connection.

**Fig. 30 Attaching compressed air hose to oil line connector—3.0L engines**

**Fig. 31 Showing location of retaining nuts on camshaft adjustment unit (Double VANOS)—3.0L engines**

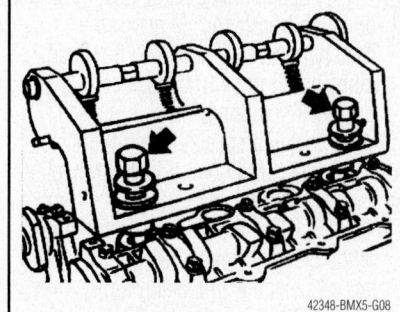

**Fig. 32 Installing special tool to pre-tension camshaft bearing caps for removal—3.0L engines**

**Fig. 33 Setting timing chain and chain wheel mark in position during installation of chain—3.0L engines**

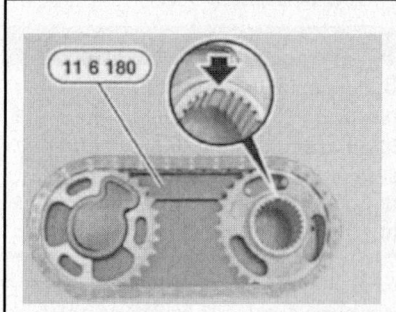

**Fig. 34 Aligning forward timing chain on chain wheels with special tool—3.0L engines**

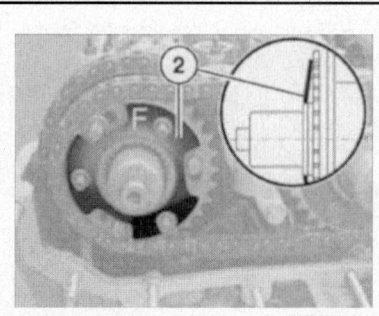

**Fig. 35 Installing the cup spring on the exhaust camshaft chain wheel—3.0L engines**

j. Rotate the engine, a least 2 turns in direction of rotation, to return the camshafts to TDC position (front cam lobes pointing to 10 o'clock and 2 o'clock positions).

k. Install a special locking tool, 11–2–300, into hole to lock flywheel in TDC position.

l. Remove the 2 studs from the rear of the cylinder head outer edge. Secure the camshafts with locking tools, 11–2–240.

m. Disconnect the compressed air hose from the special tool.

n. Using care so no oil drips onto belt drive, remove the screw plugs from the camshaft adjustment unit cover corresponding to the forward ends of the camshafts. Have a container to catch oil that runs out.

o. Using short, flat nose pliers, remove sealing caps through screw plug openings, then remove the fitting screws (left-hand thread) from both camshaft forward ends.

p. Detach the plug connection from the camshaft sensor and solenoid valves on both camshafts.

q. Remove the engine lifting eye from the camshaft adjustment unit cover.

r. Remove nuts and remove the camshaft adjustment unit

### ✳✳ WARNING

**Once camshaft adjust unit is removed, DO NOT crank engine. The toothed shaft on the intake side camshaft may slip out of the spline teeth and valves could rest of the piston.**

4. Remove the cylinder for the chain tensioning piston. Use caution as piston is under spring pressure.

5. Press down on the secondary chain tensioner at the top and lock the chain in place with a special tool, 11–3–292.

6. Remove or disconnect the following:

- Sensor gear from exhaust camshaft
- Spring plate from behind sensor gear
- Nuts on intake camshaft and remove corrugated washer
- Screws from chain gear on exhaust camshaft
- Toothed shaft with sleeve
- Secondary (upper) chain tensioner
- Release screw-in pins from exhaust camshaft chain gear

- Exhaust camshaft chain gear (leave chain on end of camshaft)
- Thrust washer on intake camshaft
- Intake camshaft sensor gear

### ✳✳ CAUTION

**DO NOT release screws from front ends of either camshaft.**

- Studs from between both camshafts

7. Release the special locking tool so flywheel is free.

8. Lift the timing chain off from end of exhaust camshaft and hold it under tension by lifting it straight up.

9. Rotate the engine against the normal direction of rotation about 30 degrees, using the crankshaft bolt.

### ✳✳ CAUTION

**To prevent the intake camshaft from moving the bearing inserts, remove the camshaft no. 1 bearing cap nuts and remove the bearing cap.**

10. Install a special tool, 11–2–260, to the cylinder head and install retaining bolts into tool and into spark plug holes for cylinders no.1 and no. 4.

11. Turn the eccentric shaft of the special tool to pre-tension the bearing caps. Remove the nuts on all bearing caps.

12. Relieve the tension from the eccentric shaft of the special tool and remove the tool.

13. Remove the bearing caps and keep in order as removed.

14. Remove the camshaft. Repeat the process for the other camshaft.

➡ If cylinder head is to be removed, remove the complete bearing strip with the bucket tappets. Keep in order of original locations.

*To install:*

15. Check bearing points of bucket tappets for signs of scoring.

➡ Bearing strips are marked "A" for exhaust side and "E" for intake side.

16. Be sure centering dowels are in place on retaining pins at bearing points 2 and 7.

17. Install the bearing strips.

18. Oil all teeth, camshafts, bearings, bearing caps, and friction washers before installation.

### ✳✳ CAUTION

**Bucket tappets expand when not subjected to load by the camshaft and therefore require some time before**

they can be pushed back down. During a rapid assembly sequence, the "closed" valves may still be open and therefore be in contact with the piston. After assembly, wait at least 30 minutes before cranking the engine back to the TDC position.

19. Pull up the timing chain and feed in the exhaust camshaft. Position the timing chain onto the end of the exhaust camshaft.

### ✳✳ CAUTION

**To prevent damaging valves when fitting camshafts, no pistons should be in TDC position.**

20. Install the camshafts so the cam tips on the intake and exhaust valves on no. 1 cylinder face each other at about the 10 o'clock and 2 o'clock positions.

21. Install the bearing caps (caps are marked from the exhaust side, A1 through A7 for the exhaust side, and E1 through E7 for the intake side).

22. Install the special tool, 11–3–260, onto the cylinder head as during removal.

23. Turn the eccentric shaft to pre-tension the bearing caps. Tighten the bearing cap bolts as follows:

a. M6 bolts: 7 ft. lbs. (10 Nm).
b. M7 bolts: 10 ft. lbs. (14 Nm).
c. M8 bolts: 15 ft. lbs. (20 Nm).

24. Remove the special tool.

25. Install camshaft locking tools, 11–3–240 on back end of camshafts. Use an open-end wrench to align camshafts (if necessary, machine wrench head so it does not contact cylinder head) so there is no gap at the locking tools.

26. Install the middle locking tool, 11–3–244 so it adjoins other tools and is bolted into spark plug hole.

27. Lift the timing chain straight up off the exhaust camshaft and hold it under tension.

28. Rotate the engine from 30 degrees BTDC to TDC, in normal direction of rotation.

29. Using special tool, 11–2–300, lock flywheel in this position.

30. Install sensor gear onto intake camshaft. Install the thrust washer over the sensor gear and torque the gear screw-in pins to 15 ft. lbs. (20 Nm).

31. Feed the chain wheel onto the timing chain so the arrow on the chain wheel faces the upper edge of the cylinder head.

32. Install a special tool, 11–4–220 into the cylinder head in the chain ten-

sioning piston bore and bring tool adjustment screw into contact with the timing chain tensioning rail, but no further at this time.

33. Check the arrow mark on the chain wheel and adjust if needed so it still points to upper edge of the cylinder head.

34. Install and tighten the retaining screw-in pins in the exhaust chain wheel to 15 ft. lbs. (20 Nm).

35. Install the secondary (upper) chain tensioner.

36. Position the toothed sleeve into the exhaust camshaft to the toothed gaps are opposed. Secure the toothed shaft. Insert the pin of the toothed shaft into the tooth gaps of the splines on the camshaft and toothed sleeve.

37. Push in the toothed shaft on the exhaust camshaft until the elongated holes in the tooth sleeve wheel are centered over the bolt holes.

38. Place forward chain wheels onto special tool, 116–6–180, and position tooth gap on intake chain wheel as shown as feed on the timing chain.

### ✳✳ CAUTION

**DO NOT alter position of chain wheels and chain when removing the special tool.**

39. Remove the special tool and position chain wheels on camshafts, so that tooth spaces oppose each other on intake camshaft.

40. Align chain with sprocket wheels so tooth spaces are positioned exactly over each other on intake side.

41. Install and secure the toothed shaft into the tooth gaps of the splines of the camshaft and chain wheel. Push in the toothed shaft until about 0.4 inch (1 mm) of splines can still be seen.

42. Note the installation direction of the corrugated washed so "FRONT" is visible. Install the washer and tighten retaining nuts snug only at this time.

43. Install 4 bolts on exhaust side to retain chain wheel. Initially tighten to about 3 ft. lbs. (5 Nm), then slack off one-half turn.

44. Install the thrust washer over the exhaust side chain wheel.

45. Note the installation direction of the cup spring (2) so the "F" stamp is forward. If "F" is no longer visible on used engine, install the cup spring so the small locating diameter of the spring points to the sensor gear. Install the cup spring.

46. Position the sensor gear over the cup spring so the arrow on the sensor gear is in line with the upper edge of the cylinder head. Install retaining nuts but do not fully tighten.

47. Pull out the exhaust side toothed shaft to the stop.

48. Press down the secondary chain tensioner at the top and remove the special locking tool.

49. Use a special tool, 11–4–200, to preload timing chain tensioning rail by rotating the adjusting screws of the tool with a small wrench.

50. Preload the exhaust side cup spring slightly be pressing on the sensor gear and hand-tighten the gear retaining nuts (do not fully tighten yet).

51. Remove any remaining gasket material from mating face on front of cylinder head.

52. Check dowel sleeves for damage and for correct installation position.

53. Make sure sealing face is clean and free of oil.

54. Fit special tool, 11–6–150 without gasket over both toothed shafts. Tighten retaining nuts to hold tool in place. Hand-tighten only until special tool is uniformly in contact with cylinder head.

55. Insert bolts on exhaust side chain wheel and tighten to about 3 ft. lbs. (5 Nm).

56. Initially tighten nuts on both exhaust and intake chain wheels to about 3 ft. lbs. (5 Nm).

57. Torque bolts on exhaust chain wheel to 15 ft. lbs. (20 Nm).

58. Torque nuts on exhaust and intake chain wheels screw-in pins to 7 ft. lbs. (10 Nm).

59. Pull back the special tool from the flywheel.

60. Remove the special tools from the rear of the camshafts.

61. Crank the engine twice in the direction of rotation until the cam lobe tips on the front of the camshafts point inward at about the 10 o'clock and 2 o'clock positions.

62. Lock the flywheel with the special tool to the crankshaft is at Top Dead Center (TDC) position.

### ✳✳ WARNING

**DO NOT turn the engine against normal rotation.**

63. Install the special camshaft locking tool, 11–3–240, onto rear end of camshafts to hold its position.

➡**The camshaft timing is correctly set if the special tool rests on the camshafts without a gap on the cylinder head or any protrusion up to 0.04 inch (1 mm) on the intake side.**

64. Remove the special tool from the camshaft forward ends.

65. Install the camshaft adjustment (Double VANOS) unit as follows:

a. Ensure dowel sleeves on forward mating face of cylinder head are not damaged and are correctly installed.

b. Keep sealing faces clean and free of oil.

c. Apply a thin and even coat of sealing compound, Drei Bond 1209, to contact surface edges of the separating face between the cylinder head and the camshaft adjustment unit.

d. Install or connect the following:
- New seal on mating face
- Camshaft adjustment unit
- Screws for hydraulic pistons on toothed shaft end on both camshafts (through front openings); torque to 7 ft. lbs. (10 Nm). Screws have left-hand threads.
- Sealing caps into camshaft adjustment unit openings
- Screw plugs, with new seal rings; torque to 37 ft. lbs. (50 Nm)
- Engine lifting eye

66. Set camshaft timing, if necessary.

67. Complete installation in reverse of removal procedure.

### 4.4L (N62) and 4.8L Engines

#### *Left Intake*

1. Disconnect the negative battery cable.

2. Remove the acoustic cover and ignition coil cover.

3. Disconnect the electrical connector on the servomotor.

4. Unbolt and remove the servomotor.

5. Remove the ignition coils. For additional information, please refer to the following topic: "Ignition Coils, Removal & Installation".

6. Remove the spark plugs.

7. Remove the cylinder head cover. For additional information, please refer to the following topic: "Cylinder Head Covers, Removal & Installation".

8. Remove the upper timing chain cover.

9. Remove the intake and exhaust adjustment unit.

10. Rotate the intake camshaft against the direction of rotation until the lettering on the 8th cylinder points upwards and cam is horizontal.

11. Remove the nuts and remove bearing cap RE1.

12. Remove the eight nuts of the bearing bracket from the outside to the inside

13. Clamp Special Tool 11-9-470 in a vice to accept the bearing bracket.

14. Carefully lift out the bearing bracket and place on Special Tool 11-9-470 and secure with a nut.

15. Remove the intermediate levers and torsion springs on each cylinder as follows:

   a. Raise one end of the torsion spring with Special Tool 11-9-480.

   b. Left out the intermediate lever.

➡ **Keep all valvetrain components in order for reinstallation is the same position.**

   c. Attach Special Tool 11-9-490 to the end of the torsion spring and support it on the intake camshaft.

   d. Raise the second end of the torsion spring with Special Tool 11-9-480. Remove the intermediate lever.

   e. Attach Special Tool 11-9-490 to the second end of the torsion spring and support it on the intake camshaft.

   f. Remove the center bolt securing the torsion spring and remove the torsion spring assembly.

   g. Repeat this procedure for the remaining cylinders.

16. Remove the bearing cap nuts and remove the bearing caps.

➡ **Keep the bearing caps in order for reinstallation.**

17. Remove the intake camshaft.

### To install:

18. Install the camshaft into the bearing bracket so the 5th cylinder cam lobes point upward.

19. Ensure the bearing shells of the eccentric shaft are engaged in the bearing bracket.

20. Install the bearing caps in the correct order.

21. Hand tighten all of the bearing cap nuts, then tighten the nuts using Special Tool 11-9-473.

22. Install Special Tool 11-9-472 to the bearing bracket and align the intake camshaft.

23. Fit the centering sleeves and align

Special Tool 11-9-472 to bearing caps 2 and 3.

24. Install Special Tool 11-9-474 and tighten so there is no play.

➡ **These alignment tools stay in place while the torsion springs and intermediate levers are being installed.**

25. Install the torsion springs and intermediate levers in the reverse order they were removed, using Special Tools 11-9-480 and 11-9-490.

26. Remove the camshaft alignment tools.

27. Remove the bearing bracket from Special Tool 11-9-470.

28. Without tilting the bearing bracket assembly, slowly lower the bearing bracket and carefully bring it into contact with the cylinder head. Install the bearing bracket nuts and hand tighten. Then tighten the bearing bracket nuts from the inside to the outside.

29. Install bearing cap RE1 and tighten the nuts.

30. Rotate the camshaft in the direction of rotation until the marking on the hex portion of the camshaft faces upwards.

31. The remainder of the installation is the reverse order of removal.

### Left Exhaust

1. Disconnect the negative battery cable.

2. Remove the acoustic cover and ignition coil cover.

3. Disconnect the electrical connector on the servomotor.

4. Unbolt and remove the servomotor.

5. Remove the ignition coils. For additional information, please refer to the following topic: "Ignition Coils, Removal & Installation".

6. Remove the spark plugs.

7. Remove the cylinder head cover. For additional information, please refer to the following topic: "Cylinder Head Covers, Removal & Installation".

8. Remove the upper timing chain cover.

9. Remove the intake and exhaust adjustment unit.

10. Remove the slide rail.

11. Rotate the camshaft at the hexagonal head until the cam lobe on the 6th cylinder is facing upwards.

12. Remove the nuts and remove bearing cap RA1.

13. Remove the nuts and remove the bearing caps RA2 through RA5.

14. Remove the exhaust camshaft.

### To install:

15. Lubricate the bearing caps and exhaust camshaft journals with clean engine oil.

16. Install the exhaust camshaft. Rotate until the exhaust cam on the 6th cylinder is facing upwards.

17. Install bearing caps RA2 through RA5. Make sure the retaining clips for the oil line are fitted on the bearing caps 3 and 5. Hand tighten the caps at first and then tighten in ½ turn increments evenly from the inside to the outside.

18. Install bearing cap RA1.

19. Install the slide rail with a new gasket.

20. Rotate the camshaft against the direction of rotation until the cam lobe on the 5th cylinder points upwards.

21. The remainder of the installation is the reverse order of removal.

### Right Intake

1. Disconnect the negative battery cable.

2. Remove the acoustic cover and ignition coil cover.

3. Disconnect the electrical connector on the servomotor.

4. Unbolt and remove the servomotor.

5. Remove the ignition coils. For additional information, please refer to the following topic: "Ignition Coils, Removal & Installation".

6. Remove the spark plugs.

7. Remove the cylinder head cover. For additional information, please refer to the following topic: "Cylinder Head Covers, Removal & Installation".

8. Remove the upper timing chain cover.

9. Remove the intake and exhaust adjustment unit.

10. Rotate the intake camshaft against the direction of rotation until the cam love on the 1st cylinder is positioned horizontally.

11. Remove the nuts and remove bearing cap LE1.

12. Remove the eight nuts of the bearing bracket from the outside to the inside

13. Clamp Special Tool 11-9-470 in a vice to accept the bearing bracket.

14. Carefully lift out the bearing bracket and place on Special Tool 11-9-470 and secure with a nut.

15. Remove the intermediate levers and torsion springs on each cylinder as follows:

   a. Raise one end of the torsion spring with Special Tool 11-9-480.

   b. Left out the intermediate lever.

➡️**Keep all valvetrain components in order for reinstallation is the same position.**

   c. Attach Special Tool 11-9-490 to the end of the torsion spring and support it on the intake camshaft.

   d. Raise the second end of the torsion spring with Special Tool 11-9-480. Remove the intermediate lever.

   e. Attach Special Tool 11-9-490 to the second end of the torsion spring and support it on the intake camshaft.

   f. Remove the center bolt securing the torsion spring and remove the torsion spring assembly.

   g. Repeat this procedure for the remaining cylinders.

16. Remove the bearing cap nuts and remove the bearing caps.

➡️**Keep the bearing caps in order for reinstallation.**

17. Remove the intake camshaft.

### To install:

18. Install the camshaft into the bearing bracket so the 4th cylinder cam lobes point upward.

19. Ensure the bearing shells of the eccentric shaft are engaged in the bearing bracket.

20. Install the bearing caps in the correct order.

21. Hand tighten all of the bearing cap nuts, then tighten the nuts using Special Tool 11-9-473.

22. Install Special Tool 11-9-472 to the bearing bracket and align the intake camshaft.

23. Fit the centering sleeves and align Special Tool 11-9-472 to bearing caps 2 and 3.

24. Install Special Tool 11-9-474 and tighten so there is no play.

➡️**These alignment tools stay in place while the torsion springs and intermediate levers are being installed.**

25. Install the torsion springs and intermediate levers in the reverse order they were removed, using Special Tools 11-9-480 and 11-9-490.

26. Remove the camshaft alignment tools.

27. Remove the bearing bracket from Special Tool 11-9-470.

28. Without tilting the bearing bracket assembly, slowly lower the bearing bracket and carefully bring it into contact with the cylinder head. Install the bearing bracket nuts and hand tighten. Then tighten the

bearing bracket nuts from the inside to the outside.

29. Install bearing cap LE1 and tighten the nuts.

30. Rotate the camshaft against the direction of rotation until the markings on the 1st cylinder hexagonal portion face upwards.

31. The remainder of the installation is the reverse order of removal.

### Right Exhaust

1. Disconnect the negative battery cable.

2. Remove the acoustic cover and ignition coil cover.

3. Disconnect the electrical connector on the servomotor.

4. Unbolt and remove the servomotor.

5. Remove the ignition coils. For additional information, please refer to the following topic: "Ignition Coils, Removal & Installation".

6. Remove the spark plugs.

7. Remove the cylinder head cover. For additional information, please refer to the following topic: "Cylinder Head Covers, Removal & Installation".

8. Remove the upper timing chain cover.

9. Remove the intake and exhaust adjustment unit.

10. Remove the slide rail.

11. Rotate the camshaft at the hexagonal head until the cam lobe on the 2nd cylinder is facing upwards.

12. Remove the nuts and remove bearing cap LA1.

13. Remove the nuts and remove the bearing caps LA2 through LA5.

14. Remove the exhaust camshaft.

### To install:

15. Lubricate the bearing caps and exhaust camshaft journals with clean engine oil.

16. Install the exhaust camshaft. Rotate until the exhaust cam lobe on the 2nd cylinder is facing upwards.

17. Install bearing caps LA2 through LA5. Make sure the retaining clips for the oil line are fitted on the bearing caps 3 and 5. Hand tighten the caps at first and then tighten in ½ turn increments evenly from the inside to the outside.

18. Install bearing cap LA1.

19. Install the slide rail with a new gasket.

20. Rotate the camshaft against the direction of rotation until the cam lobe on the 1st cylinder points upwards.

21. The remainder of the installation is the reverse order of removal.

## CRANKSHAFT FRONT SEAL

### REMOVAL & INSTALLATION

➡️**BMW refers to the crankshaft seals as 'radial' seals.**

### 3.0L Engines

*See Figures 36 and 37.*

1. Remove the crankshaft damper. For additional information, please refer to the following topic: "Crankshaft Damper, Removal & Installation".

2. Install Special Tool 11-2-383 to the crankshaft.

3. Fit Special Tool 11-2-385 to the crankshaft and align the groove with the featherkey of the crankshaft.

4. Screw on Special Tool 11-2-380 until it is securely connected to the crankshaft seal. Remove the seal by tightening the bolt.

### To install:

5. Coat the sealing lips of the new crankshaft front seal with clean engine oil.

6. Use Special Tool 11-1-280 to install the seal until it is flush with the timing chain cover.

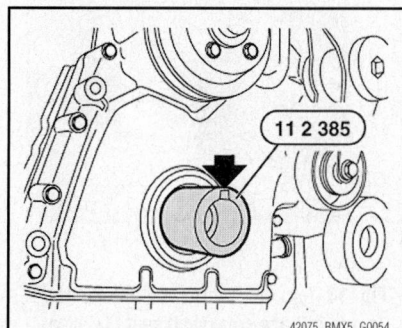

**Fig. 36 Align the groove of Special Tool 11-2-385 with the featherkey of the crankshaft—3.0L Engines**

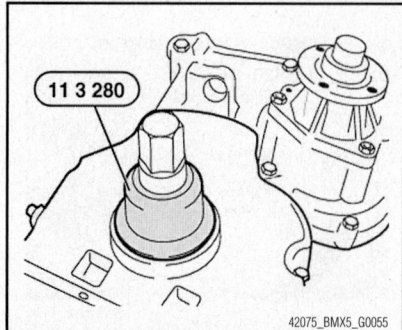

**Fig. 37 Turn the center bolt of Special Tool 11-1-280 to install the new front seal—3.0L Engines**

7. Install the crankshaft damper.

8. Start the engine and check for leaks.

### 4.4L (N62) and 4.8L Engines

*See Figures 38 and 39.*

1. Remove the crankshaft damper. For additional information, please refer to the following topic: "Crankshaft Damper, Removal & Installation".

2. Install Special Tool 11-9-410 to the crankshaft seal and turn the lever so it grips behind the seal.

3. Turn the center bolt of the seal removal tool to pop out the seal.

*To install:*

### ✱✱ CAUTION

**The sealing lip of the crankshaft seal is highly sensitive and must not be kinked or touched with your fingers at any time.**

4. Fit a new crankshaft seal on the timing chain cover.

5. Using Special Tool 11-9-420, install

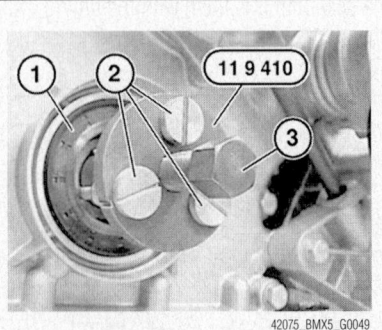

42075_BMX5_G0049

**Fig. 38 Turn the levers (2) of the Special Tool to grip the crankshaft seal (1), then turn the center bolt (3) to remove—4.4L (N62) & 4.8L Engines**

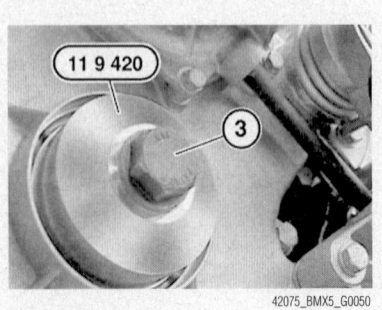

42075_BMX5_G0050

**Fig. 39 Turn the center bolt (3) to install a new crankshaft front seal.**

the seal so it is flush with the timing chain cover.

6. Install the crankshaft damper.

7. Start the engine and check for leaks.

### CYLINDER HEAD

*REMOVAL & INSTALLATION*

#### 3.0L Engines

*See Figure 40.*

1. Before servicing the vehicle, refer to the precautions section.

2. Properly relieve the fuel system pressure.

3. Remove or disconnect the following:
- Negative battery cable
- Left side exhaust manifold
- Cylinder head cover
- Spark plugs
- Intake manifold
- Drain cooling system
- Thermostat housing
- Coolant pipe from side of block
- Double VANOS adjustment unit and camshafts with bearings
- Timing case cover–to–cylinder head bolts
- Timing chain guide

4. Remove the cylinder head bolts in

reverse of sequence shown and remove the cylinder head.

*To install:*

5. Clean all sealing material from mating faces

### ✱✱ CAUTION

**There must be no oil in the cylinder head bolt holes in the block and timing case cover or there is a possibility of cracking and distorting torque values.**

6. Be sure that dowel sleeve are correctly positioned in block.

7. Apply permanently elastic sealing compound, Drei Bond 1209, at joints to timing case cover.

8. Install a new cylinder head gasket.

9. Position the cylinder head onto the block and install new cylinder head bolts, applying a light coat of oil to the washer contact area and the threads of the new bolts.

10. Tighten cylinder head bolts, following the sequence shown, in 3 steps, to the following:
   a. Step 1: 30 ft. lbs. (40 Nm).
   b. Step 2: Additional 90 degrees.
   c. Step 3: Additional 90 degrees.

11. Install or connect the following:
- Timing chain guide

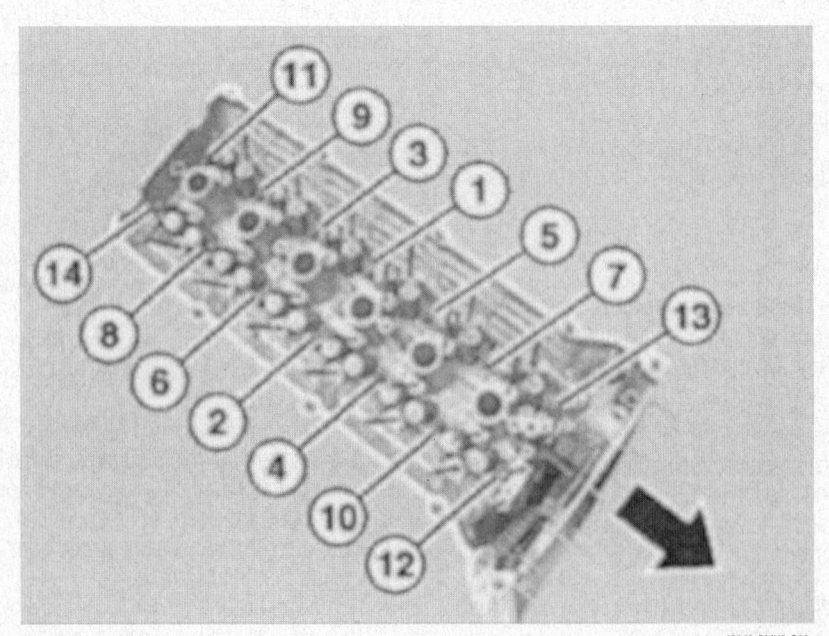

42348-BMX5-G02

**Fig. 40 Cylinder head bolt tightening sequence—3.0L engines**

- Timing case cover–to–cylinder head bolts
- Camshafts
- Double VANOS adjustment unit

12. Complete installation in reverse of the removal procedure.

### 4.4L (N62) and 4.8L Engines

#### Left Side

*See Figure 41.*

1. Disconnect the negative battery cable.

2. Drain the cooling system.

3. Remove the left exhaust manifold.

4. Remove the ignition coils. For additional information, please refer to the following topic: "Ignition Coils, Removal & Installation".

5. Remove the servomotor for the left eccentric shaft.

6. Remove the intake manifold. For additional information, please refer to the following topic: "Intake Manifold, Removal & Installation".

7. Remove the left cylinder head cover. For additional information, please refer to the following topic: "Cylinder Head Cover, Removal & Installation".

8. Remove the left timing case cover. For additional information, please refer to the following topic: "Timing Chain Cover & Seal, Removal & Installation".

9. Disconnect the wiring harness and lay to one side.

10. Remove the spark plugs.

11. Remove the vent hose from the cylinder head.

12. Detach the hose from the check valve.

13. Remove the left side camshaft adjustment unit.

14. Remove the eccentric shaft sensor.

15. Detach the guide rail from the cylinder head.

16. Remove the bolts between the cylinder head and timing chain cover.

17. Using a socket wrench with a short extension, remove the cylinder head mounting bolts in the reverse order of the tightening sequence.

18. Lift off the cylinder head and gasket.

#### To install:

19. Clean the contact surfaces on the cylinder head and engine block of any old gasket material.

20. Coat any joints between the engine block and timing chain cover with liquid gasket material.

21. Install a new cylinder head gasket.

22. Install the cylinder head. Install new bolts and tighten in the sequence shown as follows:

   a. Step 1: 22 ft. lbs. (30 Nm)
   b. Step 2: Plus 80 degrees
   c. Step 3: Plus another 80 degrees

23. The remainder of the installation is the reverse order of removal.

24. Refill the cooling system to the correct level.

25. Start the engine and check for leaks.

#### Right Side

*See Figure 42.*

1. Disconnect the negative battery cable.

2. Drain the cooling system.

3. Remove the right exhaust manifold.

4. Remove the ignition coils. For additional information, please refer to the following topic: "Ignition Coils, Removal & Installation".

5. Remove the servomotor for the right eccentric shaft.

6. Remove the intake manifold. For additional information, please refer to the following topic: "Intake Manifold, Removal & Installation".

7. Remove the right cylinder head cover. For additional information, please refer to the following topic: "Cylinder Head Cover, Removal & Installation".

8. Remove the right timing case cover. For additional information, please refer to the following topic: "Timing Chain Cover & Seal, Removal & Installation".

9. Disconnect the wiring harness and lay to one side.

10. Remove the spark plugs.

11. Remove the vent hose from the cylinder head.

12. Detach the hose from the check valve.

13. Remove the right side camshaft adjustment unit.

14. Slacken off the chain tensioner piston by one turn.

15. Remove the right side eccentric shaft sensor.

16. Detach the guide rail from the cylinder head.

17. Remove the bolts between the cylinder head and timing chain cover.

18. Using a socket wrench with a short extension, remove the cylinder head mounting bolts in the reverse order of the tightening sequence.

19. Lift off the cylinder head and gasket.

#### To install:

20. Clean the contact surfaces on the cylinder head and engine block of any old gasket material.

21. Coat any joints between the engine block and timing chain cover with liquid gasket material.

22. Install a new cylinder head gasket.

23. Install the cylinder head. Install new bolts and tighten in the sequence shown as follows:

   a. Step 1: 22 ft. lbs. (30 Nm)
   b. Step 2: Plus 80 degrees
   c. Step 3: Plus another 80 degrees

24. The remainder of the installation is the reverse order of removal.

25. Refill the cooling system to the correct level.

26. Start the engine and check for leaks.

### ENGINE ASSEMBLY

#### REMOVAL & INSTALLATION

#### 3.0L Engines

1. Before servicing the vehicle, refer to the precautions section.

2. Disconnect the negative battery cable.

3. Remove or disconnect the following:

4. Remove the heater bulkhead by pulling off the sealing strips from the firewall, turning the toggle retainers about 90 degrees, lifting out the cover, and removing the heater bulkhead.

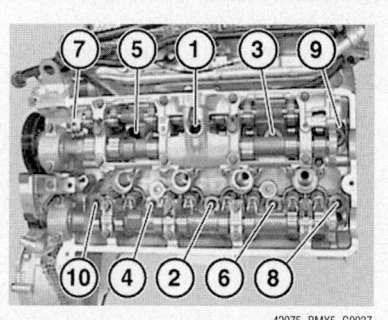
42075_BMX5_G0037

**Fig. 41 Left cylinder head tightening sequence—4.4L (N62) & 4.8L Engines**

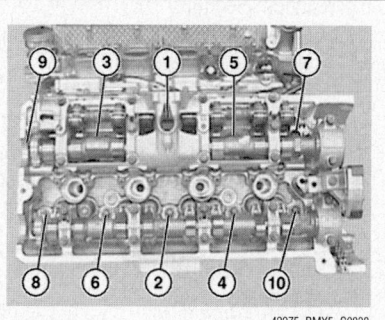
42075_BMX5_G0038

**Fig. 42 Right cylinder head tightening sequence—4.4L (N62) & 4.8L Engines**

- Engine cover
- Both oxygen sensors
- Hood to full open (assembly) position
- Front splash guard
- Reinforcement plate from undercarriage
- Transmission
- A/C compressor from mounting (move aside, with lines connected)

5. Drain the cooling system (drain plug is located on exhaust side of cylinder number 2 in engine block). Reinstall drain plug with new seal ring.

6. Pull the locks and disconnect the coolant hoses from the water pump.

7. Remove or disconnect the following:
- Radiator
- Water hoses from pipe and from heating valve
- Wiring harness section for engine (lay aside)

8. Siphon off some fluid from power steering reservoir. Unbolt reservoir and tie out of the way, leaving lines connected.

9. Remove or disconnect the following:
- Fan clutch with water pump impeller
- Alternator drive belt
- Vane pump for power steering (tie out of way, leaving lines connected)

10. Install an engine lift and chains, attaching to eye hooks.

11. Disconnect the right ground wire.

12. Unbolt the left and right engine mounts.

13. Carefully lift out the engine assembly.

***To install:***

14. Install in reverse of removal procedure.

15. Tighten components to the following torque settings:
   a. Radiator bolts: 6–7 ft. lbs. (9–10 Nm)
   b. Power steering pump bracket bolts: 16 ft. lbs. (22 Nm)

16. Tighten the automatic transmission–to–engine bolts to the following specifications:
   a. M8 hex bolts: 18 ft. lbs. (24 Nm).
   b. M10 hex bolts: 33 ft. lbs. (45 Nm).
   c. M12 hex bolts: 60 ft. lbs. (82 Nm).
   d. M8 Torx bolts: 15 ft. lbs. (21 Nm).
   e. M10 Torx bolts: 31 ft. lbs. (42 Nm).
   f. M12 Torx bolts: 53 ft. lbs. (72 Nm).

### 4.4L (N62) and 4.8L Engines

1. Before servicing the vehicle, refer to the precautions section.

2. Disconnect negative battery cable.

3. Fully open the hood and properly secure it in place.

4. Drain the cooling system.

5. Drain the engine oil.

6. Remove or disconnect the following:
- Front splash guard and reinforcement plate
- Exhaust system
- Transmission
- Driveshafts
- Accessory drive belts
- A/C compressor and hang securely without disconnect the lines.
- Air intake assembly
- Transmission oil cooler
- Radiator
- All engine oil pipes
- Windshield washer reservoir
- Heater end plate
- Fuel lines
- Heater hoses
- Vacuum hoses
- Wiring harness from control box unit
- Partition wall from the engine compartment

7. Using Special Tool 11-0-020 engine leveling unit attached to a suitable engine crane, safely support the engine on the suspension lugs.

8. Carefully lift out the engine.

9. Installation is the reverse order of removal.

10. Refill the cooling system to the correct level.

11. Refill the engine with oil to the correct level.

12. Start the engine and check for leaks.

## EXHAUST MANIFOLD

### REMOVAL & INSTALLATION

#### 3.0L Engines

1. Before servicing the vehicle, refer to the precautions section.

2. Remove or disconnect the following:
- Negative battery cable
- Intermediate muffler
- Reinforcement plate from undercarriage
- Bracket, rubber and both stabilizers
- Oxygen control sensors from intake manifold, make sure to mark locations for reinstallation purposes
- Fuel injector cover
- Plug connections from intake manifold, make sure to mark all connec-

tion locations for reinstallation purposes
- Oxygen ($O_2$s) sensors
- Front exhaust manifold with catalytic converter
- Rear exhaust manifold with catalytic converter

***To install:***

3. Clean all mating faces.

4. Install new gaskets.

5. Position the rear exhaust manifold and front exhaust manifold and install new retaining nuts as follows:
   a. M6 nuts: 7 ft. lbs. (10 Nm).
   b. M7 nuts: 15 ft. lbs. (20 Nm).

6. Install or connect the following:
- $O_2$s monitor sensors on exhaust pipes; torque to 37 ft. lbs. (50 Nm)
- Connections on intake manifold and components
- Fuel injector cover
- Oxygen control sensors on intake manifold; torque to 37 ft. lbs. (50 Nm)
- Bracket, rubber and both stabilizers from struts; torque nuts to 74 ft. lbs. (100 Nm)
- Reinforcement plate from undercarriage; torque bolts to 74 ft. lbs. (100 Nm)
- Negative battery cable

### 4.4L (N62) and 4.8L Engines

1. Before servicing the vehicle, refer to the precautions section.

2. Remove or disconnect the following:
- Negative battery cable
- Exhaust system
- Reinforcement plate
- Propeller shaft, left side manifold only
- Screw connection at the exhaust manifold
- Exhaust manifold downward and discard the gaskets

***To install:***

3. Remove the old gasket off of the cylinder head and exhaust manifold and replace the gasket. The gasket beads face the exhaust manifolds.

4. Install or connect the following:
- Exhaust manifold with new gaskets. Torque the bolts to 10 ft. lbs. (15 Nm).
- Screw connection at the exhaust manifold
- Propeller shaft, left side manifold only
- Reinforcement plate
- Exhaust system
- Negative battery cable

## INTAKE MANIFOLD

### REMOVAL & INSTALLATION

#### 3.0L Engines

1. Before servicing the vehicle, refer to the precautions section.
2. Properly relieve the fuel system pressure.
3. Remove or disconnect the following:
   - Negative battery cable
   - Engine cover
   - Oxygen ($O_2s$) sensor connectors (mark connector locations before removing)
   - Wiring from clips on intake manifold
   - Battery positive lead from intake manifold
   - Battery positive lead retainer
   - Vent hose from cylinder head cover
   - Retainer from top of engine and connection for intake air temperature sensor
   - Fuel injector terminal strip and place aside
   - Tank venting valve located near the dipstick
   - Fuel line from clip and connection to pipe located near the water pump
   - Dipstick tube guide
   - Return line from dipstick tube
   - Throttle assembly
   - Knock sensor connector and nut on manifold support
   - Intake manifold nuts
   - Vacuum line on reverse side of intake manifold, if equipped
   - Intake manifold

**To install:**

4. Check all intake manifold gaskets and replace if necessary.
5. Inspect the rubber dampers at manifold connection and replace if needed.
6. Install intake manifold and torque bolts as follows:
   a. M6 bolts: 7 ft. lbs. (10 Nm).
   b. M7 bolts: 11 ft. lbs. (15 Nm).
   c. M8 bolts: 16 Nm (22 Nm).
7. Install or connect the following:
   - Knock sensor connector and manifold support nut
   - Throttle assembly
   - Return line to dipstick tube
   - Dipstick tube guide
   - Fuel line to clip and connection to pipe located near the water pump
   - Tank venting valve located near the dipstick
   - Fuel injector terminal strip
   - Retainer to top of engine and connection for intake air temperature sensor

- Vent hose to cylinder head cover
- Battery positive lead retainer
- Battery positive lead to intake manifold
- Wiring to clips on intake manifold
- $O_2s$ sensor connectors
- Engine cover
- Negative battery cable

#### 4.4L (N62) and 4.8L Engines

*See Figure 43.*

1. Disconnect the negative battery cable.
2. Remove the engine appearance cover and acoustic cover.
3. Remove the center heater bulkhead.
4. Remove the air intake assembly.
5. Remove the fuel rail. For additional information, please refer to the following topic: "Fuel Rail, Removal & Installation".
6. Disconnect the electrical connector for the differential pressure sensor and servomotor.
7. Detach the vent hose. Remove the cable from the holder.
8. Remove the intake manifold mounting bolts.
9. With the help of an assistant, lift intake manifold upwards and remove.

**To install:**

10. Clean the contact surfaces of the manifold of any old gasket material.
11. Install the intake manifold with new gaskets.
12. The remainder of the installation is the reverse order of removal.

## OIL PAN

### REMOVAL & INSTALLATION

#### 3.0L Engines

1. Before servicing the vehicle, refer to the precautions section.
2. Drain the engine oil.
3. Remove or disconnect the following:

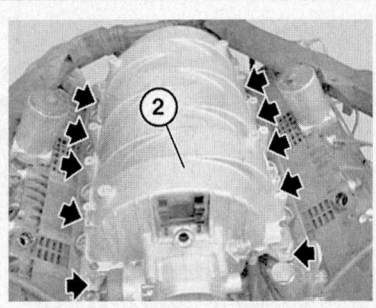

**Fig. 43 Remove all of the mounting bolts to remove the intake manifold—4.4L (N62) & 4.8L Engines**

42075_BMX5_G0029

- Bulkhead heater
- Cover over fuel injectors
- Upper section of suction filter housing with Mass Air Flow (MAF) sensor

4. Attach an engine lift and hoist to engine.
5. Remove top left and right nuts on engine mounts
6. Raise the engine about 0.15 inch (5mm).
7. Remove and disconnect the following:
   - Front splash guard
   - Reinforcement plate
   - Left and right swivel bearings
   - Output shafts
   - Propeller shaft at front
   - Alternator drive belt from belt pulley for vane pump
   - Power steering pump and move aside with the lines still connected
   - Return hose from oil separator
   - Dipstick tube
   - Steering spindle from steering gear tie rods
8. Temporarily reinstall the left and right swivel bearings and connect with a bolt to the spring strut.
9. Support the entire front axle assembly.

➡**The steering gear will remain bolted to the front axle support. Return hose must not be over-stretched when front axle is lowered.**

10. Detach front axle support from engine carrier and lower about 3–4 inches (90-100mm).
11. Remove the front axle differential.
12. Release the screws on the oil pan.
13. Move the pan backward when removing.

**To install:**

14. Ensure mating surfaces are clean.
15. Install a new gasket and position the oil pan in place.
16. Install and tighten the oil pan bolts, but do not fully tighten the transmission end bolts.
17. Tighten the engine end bolts, then tighten the transmission end bolts.
18. Restore the front axle assembly to position.
19. Reconnect the steering gear tie rods.
20. Install or connect the following:
   - Dipstick tube
   - Return hose from oil separator
   - Power steering pump
   - Alternator drive belt
   - Propeller shaft

- Output shafts
- Swivel bearings
- Reinforcement plate
- Splash guard

21. Reposition the engine onto the mount and torque the upper nuts.

22. Remove the engine hoist

23. Install the upper section of suction filter housing with MAF sensor

24. Install the cover of the fuel injectors.

25. Install the bulkhead heater.

26. Reconnect the negative battery cable.

27. Refill the engine oil.

### 4.4L (N62) and 4.8L Engines

1. Before servicing the vehicle, refer to the precautions section.

2. Drain the engine oil.

3. Remove or disconnect the following:

- Negative battery cable
- Reinforcement plate
- Oil level switch plug
- Cable guide clips
- Lower oil pan section

➡ **To remove the upper section of the oil pan, proceed with the following steps.**

4. Remove or disconnect the following:

- Upper nuts on the left and right engine mounts
- Front splash guard
- Positive battery cable from the starter
- Left and right swivel bearings
- Output shafts
- Bearing pedestal from the right output shaft
- Propeller shaft
- Steering spindle from the steering gear and support the front axle
- Front axle support from the engine carrier and slightly lower the axle support
- Drive belt
- Vane pump from the oil pan
- Adjustable plate from the oil pan after releasing the tension from the A/C compressor belt
- Guide tube for the oil dipstick
- Oil return line from the oil separator to the oil pan
- Oil pump snorkel
- Cable guide for the positive lead
- Upper oil pan section towards the rear of the vehicle

**To install:**

5. Clean the mounting surfaces.

6. Check the seals on the oil pipes and replace it if necessary. Lubricate the seals with oil.

7. Install or connect the following:

- Install upper oil pan. Torque the bolts to 89 inch lbs. (10 Nm) and lower the engine.
- Cable guide for the positive lead
- Banjo bolt for the oil return pipe from the oil filter at the oil pan
- Drive belt
- Left and right engine mounts Torque the bottom bolts to 32 ft. lbs. (43 Nm).
- Lower oil pan with a new gasket. Torque the bolts, beginning in the middle and working to the outside to 89 inch lbs. (10 Nm).
- Plug for the level switch, making sure to replace the O-ring
- Steering spindle to the steering gear
- Propeller shaft
- Bearing pedestal
- Output shafts
- Positive battery cable
- Engine splash guards
- Oil dipstick guide tube, making sure to replace the O-ring
- Reinforcement plate
- Negative battery cable

8. Fill the engine with clean oil.

9. Start the vehicle and check for leaks, repair if necessary.

### OIL PUMP

*REMOVAL & INSTALLATION*

1. Drain the engine oil.

2. Remove or disconnect the following:

- Negative battery cable
- Oil pan
- Oil pump sprocket wheel and chain
- Oil pump

**To install**

3. Check the seals on the oil pipes and replace it if necessary. Lubricate the seals with oil.

4. Check the seal in the oil pump and replace it if necessary. Screw the hexagon adapter back into the oil pump until it stops.

5. Install or connect the following:

- Oil pump. Torque the bolts to 17 ft. lbs. (22 Nm).
- Oil pump sprocket wheel and chain. Torque the nut to 35 ft. lbs. (47 Nm).
- Oil pan
- Negative battery cable

6. Fill the engine with clean oil.

7. Start the vehicle and check for leaks, repair if necessary.

### PISTON AND RING

*POSITIONING*

*See Figures 44 through 47.*

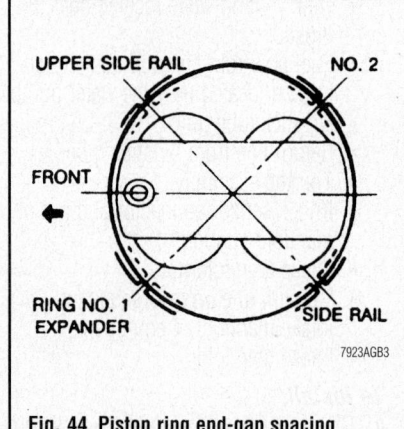

**Fig. 44 Piston ring end-gap spacing**

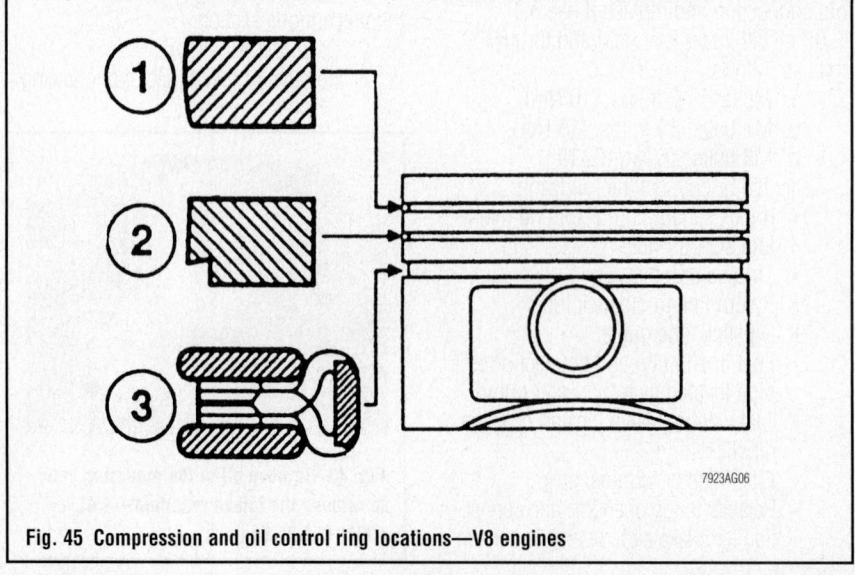

**Fig. 45 Compression and oil control ring locations—V8 engines**

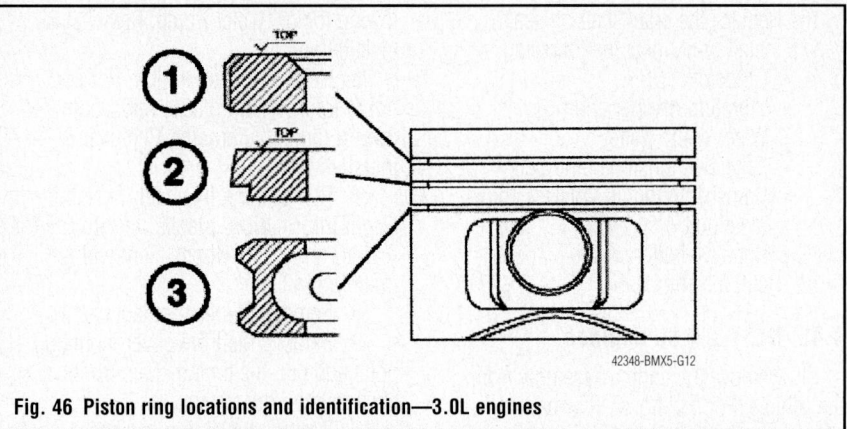

Fig. 46 Piston ring locations and identification—3.0L engines

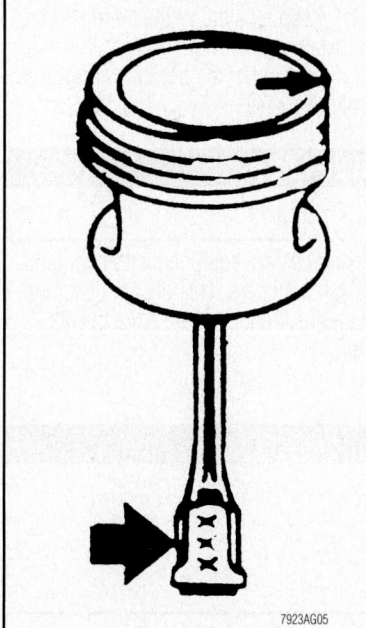

Fig. 47 Showing connecting rod-to-piston positioning

## REAR MAIN SEAL

### REMOVAL & INSTALLATION

#### 3.0L Engines

The rear main bearing oil seal can be replaced after the transmission and flywheel has been removed from the engine.

1. Before servicing the vehicle, refer to the precautions section.
2. Drain the transmission fluid.
3. Remove or disconnect the following:
   - Negative battery cable
   - Transmission
   - Flywheel assembly
   - Oil seal, using a suitable seal removal tool

##### *To install:*
4. Coat the sealing lips of the new seal with oil.

5. Install or connect the following:
   - New seal into the end cover housing with a suitable seal installation tool
   - Flywheel
   - Transmission
   - Negative battery cable
6. Fill the transmission with new fluid.
7. Start the engine and check that oil pressure is present; if the oil pressure lamp does not extinguish within 5–7 seconds of starting the engine, turn the engine **OFF**.
8. Check and top off all fluid levels.

#### 4.4L (N62) and 4.8L Engines

*See Figures 48 and 49.*

1. Drain the engine oil.
2. Remove the transmission assembly from the vehicle.
3. Remove the flywheel assembly.
4. Remove the mounting bolts and remove the end cover with rear main seal.

➡**The rear main seal is an integral part of the end cover and they must be replaced as a unit.**

##### *To install:*
5. Lightly coat the running surface of the crankshaft with clean engine oil.

Fig. 48 Remove the mounting bolts (1 & 2) to remove the end cover—4.4L (N62) & 4.8L Engines

Fig. 49 Fit the end cover (1) with the support sleeve (2) onto the crankshaft, then push on carefully—4.4L (N62) & 4.8L Engines

6. Fit a new end cover with the support sleeve and carefully push the assembly onto the crankshaft.
7. Install the end cover mounting bolts.
8. Install the flywheel assembly.
9. Install the transmission assembly into the vehicle.
10. Fill the engine with oil to the correct level.
11. Start the engine and check for leaks.

## TIMING CHAIN, SPROCKETS, FRONT COVER AND SEAL

### REMOVAL & INSTALLATION

#### 3.0L Engines

➡**This procedure covers lower timing case cover removal only. Procedure for removal of timing chain is covered under "Camshaft" for this engine.**

1. Before servicing the vehicle, refer to the precautions section.
2. Drain the engine oil.
3. Remove or disconnect the following:
   - Negative battery cable
   - Camshaft adjustment unit (Double VANOS), as described under "Camshaft."
   - Alternator drive belt tensioner and belt
   - Water pump pulley
   - Vibration damper and hub, if equipped
   - Oil pan
4. Drive out the dowel pins from the timing case cover toward the rear.
5. Remove the lower timing case cover bolts from cylinder head.
6. Remove the timing case cover bolts and cover.

**To install:**

7. Check cylinder head and gasket for possible damage.

8. Clean all mating surfaces.

9. Drive dowel pins into timing case cover until they protrude about 0.08-0.12 inch (2-3mm).

10. Use grease to hold timing case cover in place during installation.

11. Apply sealing compound, Drei Bond 1209, to joints at cylinder head gasket left and right, and thinly and evenly to entire timing case cover sealing surface adjoining the cylinder head.

12. Install all timing cover retaining bolts to about 3 ft. lbs. (5 Nm).

13. Drive in the dowel pins until flush.

14. Torque all timing cover retaining screws, in an alternating sequence, to the following:

    a. M6 bolts: 7 ft. lbs. (10 Nm).

    b. M7 bolts: 11 ft. lbs. (15 Nm).

    c. M8 bolts: 16 ft. lbs. (22 Nm).

    d. M10 bolts: 35 ft. lbs. (47 Nm).

    e. Repeat the torque sequence a second time.

15. Install and tighten the bolts connecting the timing case cover to the cylinder head.

16. Replace the crankshaft oil seal.

17. Install or connect the following:

- Oil pan
- Vibration damper and hub
- Water pump pulley
- Drive belt tensioner and belt
- Camshaft (Double VANOS) adjustment unit
- Negative battery cable.

18. Refill the engine oil.

### 4.4L (N62) & 4.8L Engines

1. Remove the engine assembly from the vehicle. For additional information, please refer to the following topic: "Engine Assembly, Removal & Installation".

2. Remove the cylinder heads. For additional information, please refer to the following topic: "Cylinder Heads, Removal & Installation".

3. Remove the oil pan. For additional information, please refer to the following topic: "Oil Pan, Removal & Installation".

4. Remove the crankshaft damper. For additional information, please refer to the following topic: "Crankshaft Damper, Removal & Installation".

5. Remove the water pump. For additional information, please refer to the fol-

lowing topic: "Water Pump, Removal & Installation".

6. Remove the alternator. For additional information, please refer to the following topic: "Alternator, Removal & Installation".

7. Remove the flywheel. For additional information, please refer to the following topic: "Flywheel, Removal & Installation".

8. Remove the rear coolant cap.

9. Remove the front cover mounting bolts and pull the timing chain front cover towards the front to remove.

10. Remove the timing chains.

11. Installation is the reverse order of removal. Keep tension on the timing chains when installing the front cover.

12. Fully tighten the front cover bolts in alternate sequence.

## VALVE LASH

### ADJUSTMENT

All engines are equipped with hydraulic valve lash adjusters. This design does not permit adjustments nor are adjustments possible.

## ENGINE PERFORMANCE & EMISSION CONTROL

## COMPONENT LOCATIONS

*See Figure 50.*

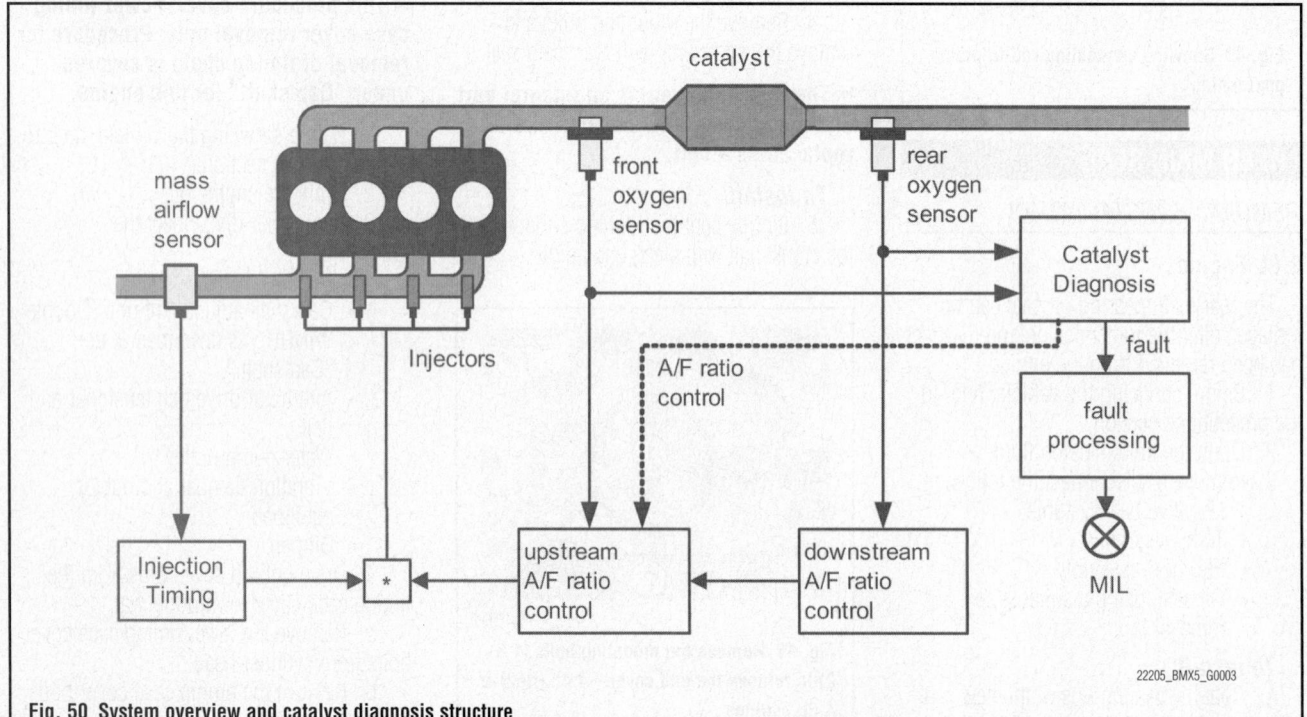

Fig. 50 System overview and catalyst diagnosis structure

22205_BMX5_G0003

**FUEL** GASOLINE FUEL INJECTION SYSTEM

## FUEL SYSTEM SERVICE PRECAUTIONS

Safety is the most important factor when performing not only fuel system maintenance but any type of maintenance. Failure to conduct maintenance and repairs in a safe manner may result in serious personal injury or death. Maintenance and testing of the vehicle's fuel system components can be accomplished safely and effectively by adhering to the following rules and guidelines.

- To avoid the possibility of fire and personal injury, always disconnect the negative battery cable unless the repair or test procedure requires that battery voltage be applied.
- Always relieve the fuel system pressure prior to disconnecting any fuel system component (injector, fuel rail, pressure regulator, etc.), fitting or fuel line connection. Exercise extreme caution whenever relieving fuel system pressure to avoid exposing skin, face and eyes to fuel spray. Please be advised that fuel under pressure may penetrate the skin or any part of the body that it contacts.
- Always place a shop towel or cloth around the fitting or connection prior to loosening to absorb any excess fuel due to spillage. Ensure that all fuel spillage (should it occur) is quickly removed from engine surfaces. Ensure that all fuel soaked cloths or towels are deposited into a suitable waste container.
- Always keep a dry chemical (Class B) fire extinguisher near the work area.
- Do not allow fuel spray or fuel vapors to come into contact with a spark or open flame.
- Always use a back-up wrench when loosening and tightening fuel line connection fittings. This will prevent unnecessary stress and torsion to fuel line piping.
- Always replace worn fuel fitting O-rings with new Do not substitute fuel hose or equivalent where fuel pipe is installed.

Before servicing the vehicle, make sure to also refer to the precautions in the beginning of this section as well.

## RELIEVING FUEL SYSTEM PRESSURE

### ✳✳ WARNING

**Fuel lines are under about 43.5–72.5 psi (3–5 bar) of pressure. Manufacturer does not provide a specific pressure relieving procedure, but instructs that any fuel system disconnect will spill fuel, so be prepared to catch and clean up any spilled fuel.**

A safe way to relieve the pressure in the system is to locate the fuel pump relay located on the cowl. Unplug and remove the relay, and place it in a safe location. With the fuel pump relay removed, start the engine and operate it until it stalls. Crank the engine for 10 seconds after it stalls to remove any residual pressure.

## FUEL FILTER

### REMOVAL & INSTALLATION

1. Before servicing the vehicle, refer to the precautions section.
2. Properly relieve the fuel system pressure.
3. Remove or disconnect the following:
   - Negative battery cable
   - Fuel pressure regulator and seal the fuel line before and after the filter with special tool 13–3–010
   - Clips and fuel line from the filter
   - Fuel filter

### To install:
   - New fuel filter
   - Fuel lines onto the correct fittings. Tighten the fuel line clamps until tight, but not to the point where the fuel lines become excessively pinched or damaged, then tighten the mounting bracket until snug.
   - Negative battery cable and cycle the ignition **ON** and **OFF** several times to build fuel pressure
4. Start the vehicle and check for leaks, repair if necessary.

## FUEL PUMP

### REMOVAL & INSTALLATION

1. Before servicing the vehicle, refer to the precautions section.
2. Drain the fuel tank.
3. Properly relieve the fuel system pressure.
4. Remove or disconnect the following:
   - Negative battery cable
   - Rear seat bench
   - Rubber plug above the sender unit and fold the rubber mat back
   - Metal cover
   - Fuel gauge level sending unit electrical connector

   - Fuel lines
   - Rotary connection with special tool 16–1–020
5. Raise the fuel level sensor and expose the spiral hose.
6. Remove the fuel level sensor and fuel pump from the tank.

### To install:

➡**Always use a new seal or gasket when installing the fuel pump or fuel level gauge sending unit assembly.**

7. Install the fuel pump into the fuel tank, taking care not to bend or damage the fuel sending unit assembly.
8. Install a new seal and torque the sealing ring using tool No. 16-1-020 as follows:
   a. Metal sealing rings: 26 ft. lbs. (35 Nm).
   b. Plastic sealing rings: 41 ft. lbs. (55 Nm).
9. Install or connect the following:
   - Fuel lines
   - Fuel gauge level sending unit electrical connector
   - Metal cover
   - Rubber plug above the sender unit
   - Rear seat bench
   - Negative battery cable
10. Start the vehicle and check for leaks, repair if necessary.

## FUEL TANK

### REMOVAL & INSTALLATION

1. Drain the any remaining fuel from the tank through the filler hose.
2. Remove the rear bench seat.
3. Remove the rubber plugs covering the access holes.
4. Fold back the rubber mat, remove the screws and fold back the metal covers.
5. On both sides, disconnect the electrical connectors from the fuel level sensors and detach the tank vent lines.
6. Raise and safely secure the vehicle.
7. Remove the driveshaft.
8. Remove the fuel tank underbody protection.
9. Remove the fuel filter.
10. Remove the rear right wheel, then remove the wheel well trim.
11. Matchmark all of the fill, vent and overflow hoses of the fuel tank, and then disconnect.
12. Support the fuel tank with a suitable jack.

13. Remove the mounting bolts for the tank straps. Begin with the right, left and then remove the center support last.

14. Slowly lower the fuel tank assembly. Remove the line bundle from the wheel well and remove the tank assembly from the vehicle.

15. Installation is the reverse order of removal.

## DLE SPEED

### ADJUSTMENT

Idle speed is maintained by the Digital Motor Electronics control unit. No adjustment is necessary or possible.

## THROTTLE BODY

### REMOVAL & INSTALLATION

#### 3.0L Engines

*See Figure 51.*

1. Disconnect the negative battery cable.
2. Remove the air intake assembly.
3. Disconnect the resonance flap and idle speed control valve.
4. Disconnect the oil pressure and oil temperature switches.

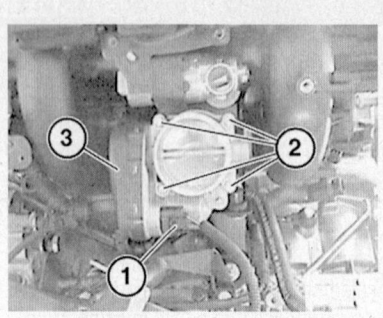

**Fig. 51 Remove the electrical connector (1) and mounting bolts (2) to remove the throttle body (3)—3.0L Engines**

5. Disconnect the tank venting valve.
6. Remove the mounting screws on the cable duct mounting.
7. Detach the throttle body electrical connector.
8. Remove the mounting bolts and remove the throttle body.

#### To install:

9. Replace the throttle body gasket and install the throttle body assembly.
10. The remainder of the installation is the reverse order of assembly.

#### 4.4L (N62) & 4.8L Engines

*See Figure 52.*

1. Disconnect the negative battery cable.
2. Remove the air intake assembly.
3. Detach the throttle body electrical connector.
4. Remove the mounting bolts and remove the throttle body.
5. Installation is the reverse order of removal.

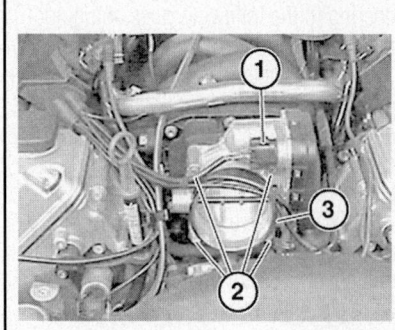

**Fig. 52 Remove the electrical connector (1) and mounting bolts (2) to remove the throttle body (3)**

## HEATING & AIR CONDITIONING SYSTEM

### BLOWER MOTOR

#### REMOVAL & INSTALLATION

*See Figure 53.*

1. Remove the instrument panel. For additional information, refer to the following topic: "Instrument Panel, Removal & Installation".

2. Unclip the retainers and remove the cover for the blower motors.

3. Disconnect the electrical connector, loose the mounting bolts and remove the blower motor.

4. Installation is the reverse order of removal.

### HEATER CORE

#### REMOVAL & INSTALLATION

1. Before servicing the vehicle, refer to the precautions section.

2. Disconnect the negative battery cable.

3. On 3.0L engines, remove the heater bulkhead as follows:

   a. Pull off the weatherstripping on top of the bulkhead.

   b. Release the cowl retaining toggle clips.

   c. Lift out the cover.

   d. Remove the screws and remove the heater bulkhead.

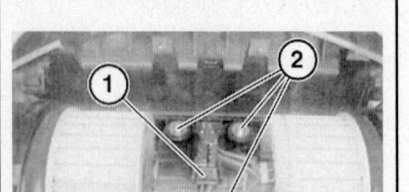

**Fig. 53 Disconnect the wiring harness (1), remove the mounting bolts (2) and pull the blower motor (3) in the direction shown to remove.**

4. Remove or disconnect the following:
- Heater hoses from firewall connections
- Air duct for left foot well from mountings
- Left foot well vent
- Heater pipes from under dash
- Heater duct for right foot well
- Air duct for rear foot well on right
- Cover from heater housing
- Evaporator temperature sensor (pull out to one side)

#### To install:

5. Install or connect the following:
- Evaporator temperature sensor
- Heater housing cover
- Air ducts in foot wells
- Heater pipes under dash
- Heater hoses at firewall connections
- Heater bulkhead (3.0L)

6. Reconnect the negative battery cables

7. Check coolant level

# STEERING

## POWER STEERING GEAR

### REMOVAL & INSTALLATION

1. Before servicing the vehicle, refer to the precautions section.

2. Set the steering gear in the straight ahead position by aligning the marks on the steering gear and spindle.

3. Drain the power steering fluid.

4. Remove or disconnect the following:
   - Negative battery cable
   - Front wheels
   - Reinforcement plate
   - Nuts from the left and right engine support arms and raise the engine slightly
   - 2 bolts accessible through holes in frame
   - Lower clamping screw
   - Steering gear clamps (after locking wheels in straight-ahead position)
   - Tie rod by pressing it off with special tool 32–3–090
   - Self-locking nuts and brace the front axle support
   - Banjo bolts and slide the steering gear out through the left wheel opening

#### To install:

5. Install the steering gear through the left side wheel opening

6. Install new sealing rings and banjo bolts. Torque the bolts as follows:
   a. M10: 7 ft. lbs. (12 Nm).
   b. M14: 25 ft. lbs. (35 Nm).
   c. M16: 29 ft. lbs. (40 Nm).
   d. M18: 34 ft. lbs. (45 Nm).

7. Install or connect the following:
   - Front axle support screws. Torque the screws to 74 ft. lbs. (100 Nm).
   - Self-locking nuts. Torque the nuts to 74 ft. lbs. (100 Nm).
   - Tie rod. Torque the castle nut to 58 ft. lbs. (80 Nm).

- Steering gear to the spindle. Torque the fastener to 18 ft. lbs. (24 Nm).
- Steering gear clamp
- Engine support arm nuts. Torque the nuts to 60 ft. lbs. (85 Nm).
- Reinforcement plate
- Splash guard
- Both front wheels
- Negative battery cable

8. Fill and bleed the power steering system.

9. Start the vehicle and check for leaks, repair if necessary.

## POWER STEERING PUMP

### REMOVAL & INSTALLATION

See Figures 54 and 55.

1. Disconnect the negative battery cable.

2. Drain the power steering reservoir.

3. Remove the engine splash guard and skid plates, if equipped.

4. Remove the accessory drive belt. For additional information, refer to the following topic: "Accessory Drive Belts, Removal & Installation".

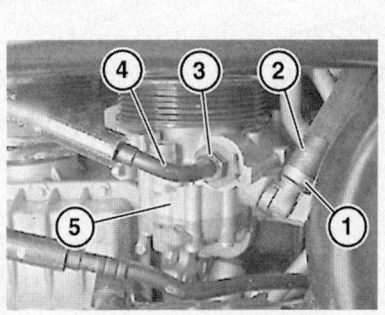

Fig. 54 Disconnect the hose clamp (1), intake line (2), union nut (3) and pressure line (4) from the power steering pump (5).

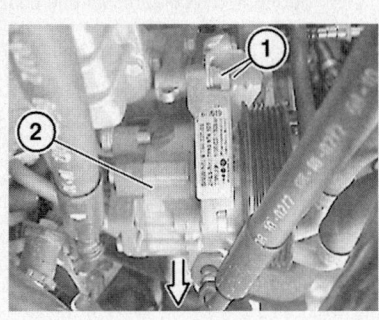

Fig. 55 Remove the mounting bolts (1) to remove the power steering pump (2).

5. Remove the hose camp with Special Tool 32-1-260.

6. Disconnect the intake line from the power steering pump.

7. Release the union nut.

8. Detach the pressure line from the power steering pump and lay to one side.

9. Remove the mounting bolts and remove the power steering pump from the vehicle.

10. Installation is the reverse order of removal. Replace the o-rings on the power steering lines before installing them.

11. Bleed the power steering system.

### BLEEDING

1. Fill the power steering reservoir to the correct level.

2. Start the engine.

3. Turn the steering wheel left and right twice to full lock. If necessary, top off the power steering fluid during this procedure (usually necessary if the system was completely drained).

4. Move the steering wheel to the straight-ahead position and turn off the engine.

5. Check and fill the power steering reservoir to the correct level.

6. Check the system for any leaks.

# SUSPENSION

# FRONT SUSPENSION

## COIL SPRING

### REMOVAL & INSTALLATION

### ☀ CAUTION

**This procedure calls for the spring to be compressed. A compressed spring has high potential energy and if released suddenly can cause severe damage and personal injury.**

1. Before servicing the vehicle, refer to the precautions section.

2. Disconnect the negative battery cable.

3. Remove the strut from the vehicle and mount in a vise using a strut holder. This will prevent damage to the strut tube

4. Using a proper spring compressor, compress the spring until the stress on the thrust bearing is released.

5. Remove the top nut of the strut mount. Counter-hold the strut rod during removal.

6. Pull the strut mount off the strut rod. Note the positioning of the spacers and washer for replacement.

7. Pull the spring off the strut and place aside in a safe area.

8. Slowly release the compression of the spring.

**To install:**

9. Install or connect the following:
- Spring in the compressor and compress
- Spring and strut mount with all the spacers and washers in their original positions. Torque the

new strut rod nut: 47 ft. lbs. (65 Nm).

10. Release the spring slowly and check that it seats in the spring holders. Install the strut in the vehicle.

11. Connect the negative battery cable.

## LOWER BALL JOINT

### REMOVAL & INSTALLATION

*See Figure 56.*

1. Before servicing the vehicle, refer to the precautions section.

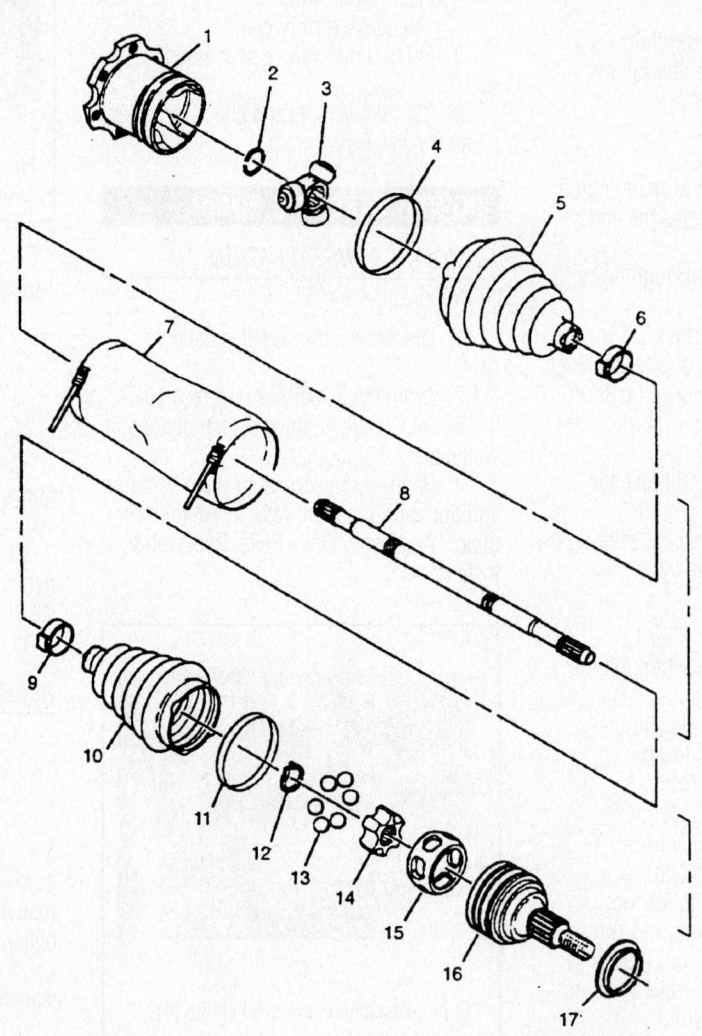

## Legend

(1) Tripot Housing Assembly
(2) Spacer Ring
(3) Tripot Joint Spider Assembly
(4) Swage Ring
(5) Tripot Joint Seal
(6) Small Seal Retaining Clamp
(7) Drive Axle Seal Cover (Optional)
(8) Drive Axle Shaft

(9) CV Joint Seal
(10) Race Retaining Ring
(11) Ball
(12) CV Joint Inner Race
(13) CV Joint Cage
(14) CV Joint Outer Race
(15) Deflector Ring

9308KG10

**Fig. 56 Remove the lower ball joint from the steering knuckle**

2. Remove or disconnect the following:
- Negative battery cable
- Push rod/integral link assembly and properly support the wheel carrier
- Shock absorber from the swinging arm
- Circlip

3. Using special tool 33–4–191, 192, 193 and 33–3–333 pull the ball joint out of the steering knuckle.

### To install:
4. Install or connect the following:
- Ball joint into the steering knuckle with special tools 33–4–191, 192, 194 and 33–3–333
- New circlip
- Shock absorber and remove the support from the wheel carrier
- Push rod/integral link
- Negative battery cable

## LOWER CONTROL ARM

### REMOVAL & INSTALLATION

1. Before servicing the vehicle, refer to the precautions section.
2. Remove or disconnect the following:

- Negative battery cable
- Front wheel
- Control arm from the front axle support and loosen the nut from the control arm to swivel bearing
- Control arm from the swivel bearing by pressing it off with special tool 31–2–240

### To install:
3. Install or connect the following:
- Lower control arm to the swivel bearing. Torque the nut to 58 ft. lbs. (80 Nm).
- Lower control arm to the front axle support. Torque the nut to 74 ft. lbs. (100 Nm) plus an additional 90 degrees.
- Front wheel
- Negative battery cable
4. Check and adjust the front end alignment as needed.

## MACPHERSON STRUT

### REMOVAL & INSTALLATION

See removal of Strut & Spring Assembly in this section.

## STABILIZER BAR

### REMOVAL & INSTALLATION

#### 3.0L Engines
*See Figure 57.*

1. Lower the front axle support as follows:
   a. Remove the front wheels.
   b. Remove the skid plates, if equipped.
   c. Remove the reinforcement plate.
   d. Remove the lower steering spindle from the power steering gearbox.
   e. Disconnect the tie-rod end from the wheel knuckle.
   f. Disconnect the upper control arm.
   g. Disconnect the lower control arm.
   h. Disconnect the engine mount from the front axle carrier.
   i. Loosen the bolts and lower the front axle carrier.
2. Remove the rubber mounts and remove the stabilizer bar.
3. Installation is the reverse order of removal.

#### 4.4L (N62) and 4.8L Engines

1. Disconnect the stabilizer links on both sides of the stabilizer.
2. Remove the left steering gear cover and remove from the front axle carrier on the right.
3. Lower the front axle carrier.
4. Remove the stabilizer bar out towards the rear.
5. Installation is the reverse order of removal.

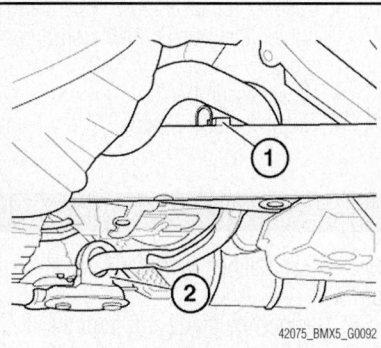

Fig. 57 Remove the rubber mounts and remove the stabilizer bar—3.0L Engines

## STEERING KNUCKLE

### REMOVAL & INSTALLATION

*See Figures 58 through 60.*

1. Raise and safely support the vehicle.
2. Remove the front wheel.

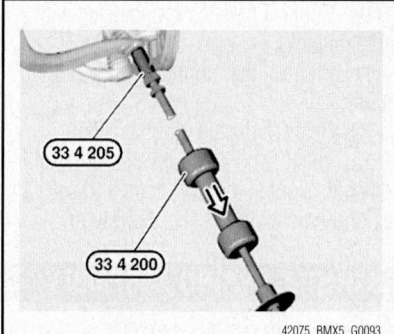

Fig. 58 Drive out the lower ball joint from the steering knuckle

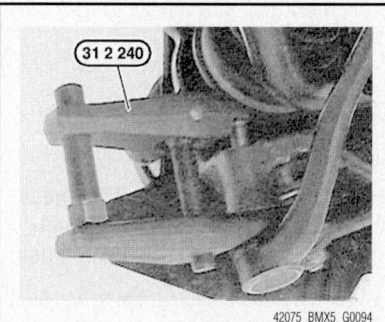

Fig. 59 Use Special Tool 31-2-240 to disconnect the control arm from the steering knuckle.

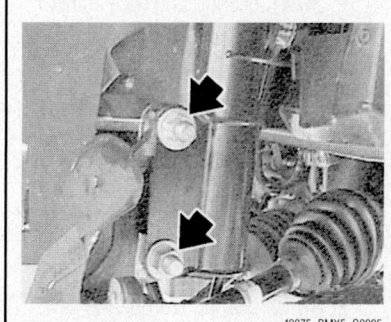

Fig. 60 Remove the mounting bolts to remove the steering knuckle.

3. Have an assistant depress the brake pedal, and remove the hub nut.
4. Remove the front brake rotor.
5. Remove the front pulse generator.
6. Disconnect the tie-rod ends from the steering knuckle.
7. Drive out the lower ball joint using Special Tool 33-4-200.
8. Remove the lower control arm from the steering knuckle using Special Tool 31-2-240.
9. Remove the axle shaft. For additional information, refer to the following topic: "Axle Shaft, Bearing & Seal, Removal & Installation".

10. Support the steering knuckle with a suitable jack.

11. Remove the mounting nuts and bolts.

12. Remove the steering knuckle.

13. Installation is the reverse order of removal. Tighten the new bolts and self-locking nuts to 176 ft. lbs. (250 Nm).

## STRUT & SPRING ASSEMBLY

### REMOVAL & INSTALLATION

1. Before servicing the vehicle, refer to the precautions section.

2. Mark the position of the threaded pin to the wheel arch to retain the camber setting when installed.

3. Remove or disconnect the following:

- Negative battery cable
- Tire and wheel assembly

➡**Carefully mark position of threaded pin to strut tower to ensure original camber is retained during installation.**

- Two of the nuts on the spring strut support bearing
- Center strut bracket nut
- Speed sensor/brake wear cable and disconnect the plug housing
- Swivel bearing and tie it aside
- Remaining nut on the spring strut support bearing
- Strut assembly

### To install:

4. Install or connect the following:

- Strut assembly
- One nut to the spring strut support bearing and hand tighten at this time
- Swivel bearing. Torque the new self-locking nut to 176 ft. lbs. (250 Nm).

- Speed sensor/brake wear cable and connect the housing plug
- Center strut bracket. Torque the nut to 74 ft. lbs. (100 Nm).

5. Align the three upper spring strut support bearing nuts and match the threaded pin with the mark made during the removal procedure. When aligned properly torque the nuts to 25 ft. lbs. (34 Nm).

6. Install the tire and wheel assembly.

7. Reconnect the negative battery cable.

## UPPER CONTROL ARM

### REMOVAL & INSTALLATION

1. Before servicing the vehicle, refer to the precautions section.

2. Remove or disconnect the following:

- Negative battery cable
- Wheel assembly
- Fuse for the air supply system, if equipped with air suspension and loosen the pipes on the distributor block
- Upper control arm from the steering knuckle
- Plastic shim and unhook the lines
- Upper control arm

### To install:

3. Install or connect the following:

- Upper control arm. Torque the bolt to 74 ft. lbs. (100 Nm).
- Plastic shim and connect the lines
- Upper control arm to the steering knuckle. Torque the bolt to 122 ft. lbs. (165 Nm).
- Fuse for the air supply system and tighten the pipes on the distributor block
- Wheel assembly
- Negative battery cable

## WHEEL BEARINGS

### REMOVAL & INSTALLATION

➡**The wheel bearings are only removed if they are worn. They cannot be removed without destroying them (due to side thrust created by the bearing puller). They cannot be disassembled, repacked or adjusted.**

1. Before servicing the vehicle, refer to the precautions section.

2. Remove or disconnect the following:

- Negative battery cable
- Swivel bearing and clamp it in a vise
- Drive flange by installing special tools 33–2–116, 150 and 33–4–200
- Bearing inner race from the flange
- Circlip
- Snap ring
- Bearing by installing special tools 31–2–113, 33–3–261, 262 and 266

### To install:

3. Install or connect the following:

- Wheel bearing with the wider chamfer facing the steering knuckle to the drive flange with special tools 33–2–261, 264, 268 and 31–2–113
- Snap ring and circlip
- inner race to the drive flange
- Drive flange to the swivel bearing by using special tool 33–3–261, 266, 268 and 31–2–113
- Swivel bearing
- Negative battery cable

### ADJUSTMENT

Wheel bearings cannot be adjusted and must be replaced as a unit and never be reused once removed.

## SUSPENSION

## REAR SUSPENSION

### COIL SPRING

#### REMOVAL & INSTALLATION

1. Raise and safely support the vehicle.

2. Remove the rear wheel.

3. Remove the rear pulse generator from the wheel carrier.

4. Remove the shock absorber. For additional information, refer to the following topic: "Shock Absorber, Removal & Installation".

5. Remove the brake caliper. For additional information, refer to the following

topic: "Brake Caliper, Removal & Installation".

6. Remove the guide arm from the wheel carrier.

7. Remove the control arm.

8. Disconnect the stabilizer link.

9. Disconnect the integral and swinging arm from the axle carrier.

10. Lower the wheel carrier, pull out the coil spring to the side.

11. Installation is the reverse order of removal.

12. After installation, the headlight adjustment must be checked.

### AIR SPRING

#### REMOVAL & INSTALLATION

See Figure 61.

1. Before servicing the vehicle, drain and deactivate the air suspension system as follows:

a. Pull the fuse for the air spring system control unit.

b. Loosen the banjo bolts of the valve unit (located beneath the right side underbody paneling) until the sound of escaping air can be

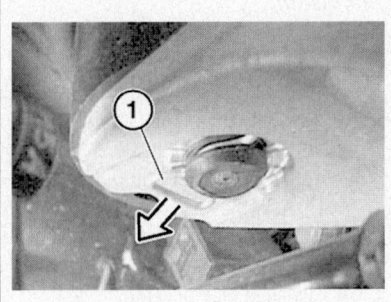

**Fig. 61 Detach the locking clip from the swinging arm and remove the air spring**

heard. Retighten the banjo bolts.

2. Raise and safely support the vehicle.

3. Remove the rear wheel.

4. Remove the luggage compartment floor trim.

5. Disconnect the air line to the springs by pressing the coupling downwards, pressing the retaining ring together and detaching the air line. Remove the top clips.

6. Detach the locking clip from the swinging arm to remove the air spring.

7. Installation is the reverse order of removal.

## CONTROL LINKS

### REMOVAL & INSTALLATION

*See Figures 62 and 63.*

1. Raise and safely support the vehicle.

2. Remove the rear wheel.

3. Remove the brake caliper assembly.

4. Remove the ABS pulse generator.

5. Support the wheel carrier with a suitable jack stand.

6. Remove the shock absorber from the swinging arm.

7. Remove the mounting nut.

8. Remove the upper mounting bolt.

9. Remove the control arm.

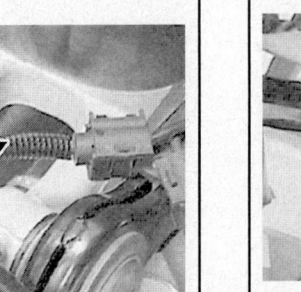

**Fig. 62 Remove the mounting nut at the wheel carrier**

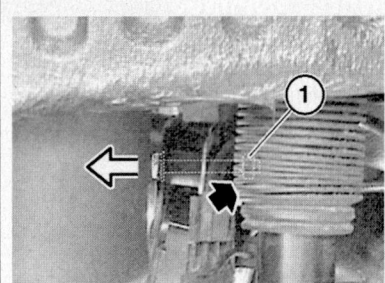

**Fig. 63 Remove the upper mounting bolt (1) to remove the rear control arm**

10. Installation is the reverse order of removal. Hand tighten the control arm nuts until the vehicle is resting on the ground. Then tighten the nuts to 74 ft. lbs. (100 Nm).

11. Check and adjust the alignment as necessary.

## PUSH ROD/INTEGRAL LINK

### REMOVAL & INSTALLATION

*See Figure 64.*

1. Raise and safely support the vehicle.

2. Remove the rear wheel.

3. Remove the brake caliper assembly.

4. Remove the ABS pulse generator.

5. Remove the stabilizer link from the swinging arm.

6. Remove the guide arm from the wheel carrier.

7. Disconnect the control arm from the wheel carrier.

8. Support the wheel carrier with a suitable jack stand.

9. Remove the lower nut and bolt. Remove the upper bolt. Press the swinging arm downward and remove the integral link.

10. Installation is the reverse order of removal. Hand tighten the bolts until the vehicle is resting at ride height on the ground. Then tighten the bolts to 184 ft. lbs. (250 Nm).

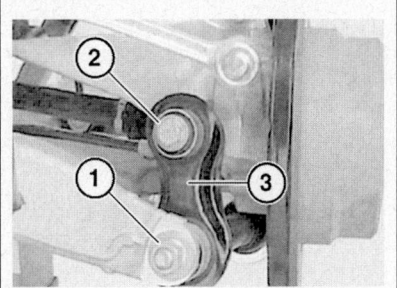

**Fig. 64 Remove the lower bolts (1), upper bolts (2) and the remove the integral link (3)**

11. Check and adjust the alignment as necessary.

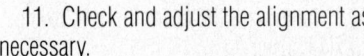

## SWINGING ARM

### REMOVAL & INSTALLATION

*See Figures 65 through 67.*

1. Raise and safely support the vehicle.

2. Remove the rear wheel.

3. Disconnect the stabilizer link from the swinging arm.

4. Remove the shock absorber from the swinging arm.

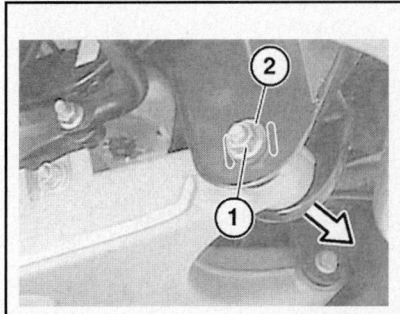

**Fig. 65 Remove the eccentric bolt and nut.**

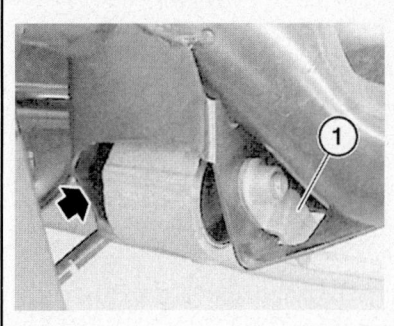

**Fig. 66 Remove the bolt and self-locking nut**

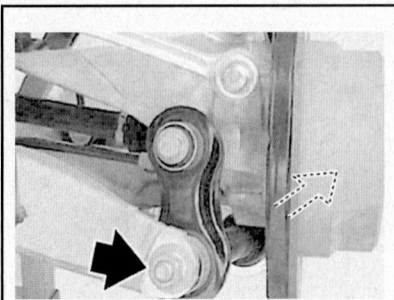

**Fig. 67 Remove the bolt that attaches to the integral link**

5. Matchmark the eccentric bolt to the rear axle carrier.

6. Support the wheel carrier with a suitable jack stand.

7. Remove the eccentric bolt and nut.

8. Remove the bolt and self-locking nut.

9. Remove the bolt that attaches to the integral link.

10. Remove the swinging arm.

11. Installation is the reverse order of removal. Hand tighten the bolts until the vehicle is resting at ride height on the ground. Then tighten the bolts to 184 ft. lbs. (250 Nm).

12. Check and adjust the alignment as necessary.

## WHEEL BEARINGS

### REMOVAL & INSTALLATION

1. Before servicing the vehicle, refer to the precautions section.

2. Remove or disconnect the following:
- Negative battery cable
- Wheel assembly
- Collar nut
- Brake disc
- Drive flange by installing special tool 33–2–116, 33–4–201, 202 and 203
- Inner race from the drive flange
- Wheel bearing

### To install:

3. Install or connect the following:
- Wheel bearing
- Inner race to the drive flange. Torque the bolts to 74 ft. lbs. (100 Nm).
- Drive flange to the axle shaft. Torque the collar nut to 310 ft. lbs. (420 Nm).
- Brake disc
- Wheel assembly
- Negative battery cable

### ADJUSTMENT

Wheel bearings cannot be adjusted and must be replaced as a unit and never be reused once removed.

# BMW

## Diagnostic Trouble Codes

---

## DIAGNOSTIC TROUBLE CODES

### OBD II VEHICLE APPLICATIONS

*BMW*

**328i, 328Ci, 328xi**
2007–2008
- E93, E92, E91 and E90 . . . . N52K, N51

**335i, 335Ci, 335xi**
2007–2008
- E93, E92 and E90 . . . . . . . . . . . . N54

**525i, 525xi**
2007–2008
- E60 . . . . . . . . . . . . . . . . . . . . . . . .M52

**530i, 530xi**
2007–2008
- E60 and E61 . . . . . . . . . . . . . . . . N52

**550i**
2007–2008
- E60 . . . . . . . . . . . . . . . . . . . . . . . N62

**650i**
2007–2008
- E63 and E64 . . . . . . . . . . . . . . . N62

**750i, 750Li**
2007–2008
- E65 and E66 . . . . . . . . . . . . N62, N73

**760Li**
2007–2008
- E66 . . . . . . . . . . . . . . . . . . . . . . . N73

**X3**
2007–2008
- E83 and E86 . . . . . . . . . . . . . . . N52K

**X5**
2007–2008
- 3.0L . . . . . . . . . . . . . . . . . . . . . N52K
- 4.8L . . . . . . . . . . . . . . . . . . . N62Tu

**Z4 3.0i, Z4 3.0si, Z4 M**
2007–2008
- E85 and E86 . . . . . . . . . . . . N52, S54

## Gas Engine OBD II Trouble Code List (P0xxx Codes)

| DTC | Trouble Code Title, Conditions & Possible Causes |
|---|---|
| **DTC: P0010**<br>**2T MIL: Yes**<br>**Years:** 2007, 2008<br>**Models:** 335i, 335Ci, 335xi, 525i, 525xi, 550i, 650i, 750i, 750Li, 760Li, X3, X5, Z4 3.0i, Z4 3.0Si, Z4 M<br>**Engines:** All<br>**Transmissions:** All | **"A" Camshaft Position Actuator Circuit (Bank 1) Conditions:**<br>Key on or engine running; and the DME detected an unexpected high voltage or low voltage condition on the camshaft position sensor. The relative position between the camshaft and crankshaft needs to be optimal so the engine has better torque, fuel economy and emissions.<br>**Note: The camshaft adjustment is load- and RPM-dependant. The electrical camshaft adjustment valve 1 switches oil pressure onto camshaft adjuster (mechanical adjustment mechanism), which adjusts the camshaft.**<br>**Possible Causes:**<br>• Fuel pump has failed<br>• Actuator circuit is open<br>• Battery voltage below 11.5 volts<br>• Position actuator circuit may short to B+ or Ground |
| **DTC: P0011**<br>**2T MIL: Yes**<br>**Years:** 2007, 2008<br>**Models:** 335i, 335Ci, 335xi, 525i, 525xi, 550i, 650i, 750i, 750Li, 760Li, X3, X5, Z4 3.0i, Z4 3.0Si, Z4 M<br>**Engines:** All<br>**Transmissions:** All | **"A" Camshaft Position Timing Over-Advanced (Bank 1) Conditions:**<br>Engine started and driven at an engine speed of more than 400rpm; and the DME detected the camshaft timing exceeded the maximum calibrated advance value, or the camshaft remained in an advanced position during the CCM test. The valve timing did not change from the current valve timing or it remained fixed during the testing.<br>**Note: The camshaft adjustment is load- and RPM-dependant. The electrical camshaft adjustment valve 1 switches oil pressure onto camshaft adjuster (mechanical adjustment mechanism), which adjusts the camshaft.**<br>**Possible Causes:**<br>• Fuel pump has failed<br>• CPS circuit is open, shorted to ground or shorted to power<br>• Battery voltage below 11.5 volts<br>• Position actuator circuit may short to B+ or Ground<br>• Camshaft timing improperly set, or continuous oil flow to the VCT piston chamber<br>• Camshaft advance mechanism (the VCT unit) is sticking or binding mechanically<br>• VCT solenoid valve is stuck in open position |
| **DTC: P0012**<br>**2T MIL: Yes**<br>**Years:** 2007, 2008<br>**Models:** 335i, 335Ci, 335xi, 525i, 525xi, 550i, 650i, 750i, 750Li, 760Li, X3, X5, Z4 3.0i, Z4 3.0Si, Z4 M<br>**Engines:** All<br>**Transmissions:** All | **"A" Camshaft Position Over-Retarded (Bank 1) Conditions:**<br>Engine started and driven at an engine speed of more than 400rpm; and the DME detected the camshaft timing exceeded the minimum calibrated retarded value, or the camshaft remained in an retarded position during the CCM test. The valve timing did not change from the current valve timing or it remained fixed during the testing.<br>**Note: The camshaft adjustment is load- and RPM dependant. The electrical camshaft adjustment valve 1 switches oil pressure onto camshaft adjuster (mechanical adjustment mechanism), which adjusts the camshaft.**<br>**Possible Causes:**<br>• Fuel pump has failed<br>• CPS circuit is open, shorted to ground or shorted to power<br>• Battery voltage below 11.5 volts<br>• Position actuator circuit may short to B+ or Ground<br>• Camshaft timing improperly set, or continuous oil flow to the VCT piston chamber<br>• Camshaft advance mechanism (the VCT unit) is sticking or binding mechanically<br>• VCT solenoid valve is stuck in open position |
| **DTC: P0013**<br>**2T MIL: Yes**<br>**Years:** 2007, 2008<br>**Models:** 335i, 335Ci, 335xi, 525i, 525xi, 550i, 650i, 750i, 750Li, 760Li, X3, X5, Z4 3.0i, Z4 3.0Si, Z4 M<br>**Engines:** All<br>**Transmissions:** All | **"B" Camshaft Position Actuator Circuit (Bank 1) Conditions:**<br>Key on or engine running; and the DME detected an unexpected high voltage or low voltage condition on the camshaft position sensor. The relative position between the camshaft and crankshaft needs to be optimal so the engine has better torque, fuel economy and emissions.<br>**Note: The camshaft adjustment is load- and RPM dependant. The electrical camshaft adjustment valve 1 switches oil pressure onto camshaft adjuster (mechanical adjustment mechanism), which adjusts the camshaft.**<br>**Possible Causes:**<br>• Fuel pump has failed<br>• Battery voltage below 11.5 volts<br>• Position actuator circuit may short to B+ or Ground |

| DTC | Trouble Code Title, Conditions & Possible Causes |
|-----|--------------------------------------------------|
| **DTC: P0014**<br>**2T MIL: Yes**<br>**Years:** 2007, 2008<br>**Models:** 335i, 335Ci, 335xi, 525i, 525xi, 550i, 650i, 750i, 750Li, 760Li, X3, X5, Z4 3.0i, Z4 3.0Si, Z4 M<br>**Engines:** All<br>**Transmissions:** All | **"B" Camshaft Position Timing Over-Advanced (Bank 1) Conditions:**<br>Engine started and driven at an engine speed of more than 400rpm; and the DME detected the camshaft timing exceeded the maximum calibrated advance value, or the camshaft remained in an advanced position during the CCM test. The valve timing did not change from the current valve timing or it remained fixed during the testing. The VANOS is in the end position.<br>**Note: The camshaft adjustment is load- and RPM dependant. The electrical camshaft adjustment valve 1 switches oil pressure onto camshaft adjuster (mechanical adjustment mechanism), which adjusts the camshaft.**<br>**Possible Causes:**<br>• Fuel pump has failed<br>• CPS circuit is open, shorted to ground or shorted to power<br>• Battery voltage below 11.5 volts<br>• Position actuator circuit may short to B+ or Ground<br>• Camshaft timing improperly set, or continuous oil flow to the VCT piston chamber<br>• Camshaft advance mechanism (the VCT unit) is sticking or binding mechanically<br>• VCT solenoid valve is stuck in open position |
| **DTC: P0015**<br>**2T MIL: Yes**<br>**Years:** 2007, 2008<br>**Models:** 335i, 335Ci, 335xi, 525i, 525xi, 550i, 650i, 750i, 750Li, 760Li, X3, X5, Z4 3.0i, Z4 3.0Si, Z4 M<br>**Engines:** All<br>**Transmissions:** All | **"B" Camshaft Position Over-Retarded (Bank 1) Conditions:**<br>Engine started and driven at an engine speed of more than 400rpm; and the DME detected the camshaft timing exceeded the minimum calibrated retarded value, or the camshaft remained in an retarded position during the CCM test. The valve timing did not change from the current valve timing or it remained fixed during the testing.<br>**Note: The camshaft adjustment is load- and RPM dependant. The electrical camshaft adjustment valve 1 switches oil pressure onto camshaft adjuster (mechanical adjustment mechanism), which adjusts the camshaft.**<br>**Possible Causes:**<br>• Fuel pump has failed<br>• CPS circuit is open, shorted to ground or shorted to power<br>• Battery voltage below 11.5 volts<br>• Position actuator circuit may short to B+ or Ground<br>• Camshaft timing improperly set, or continuous oil flow to the VCT piston chamber<br>• Camshaft advance mechanism (the VCT unit) is sticking or binding mechanically<br>• VCT solenoid valve is stuck in open position |
| **DTC: P0016**<br>**2T MIL: Yes**<br>**Years:** 2007, 2008<br>**Models:** 335i, 335Ci, 335xi, 525i, 525xi, 530i, 530xi, 550i, 650i, 750i, 750Li, 760Li<br>**Engines:** All<br>**Transmissions:** All | **Crankshaft Position - Camshaft Position Correlation Bank 1 Sensor A Conditions:**<br>Engine started, engine running, and the DME detected a deviation between the crankshaft position sensor signal and the camshaft position sensor. A rationality error has been detected for camshaft position out of phase with crankshaft.<br>**Possible Causes:**<br>• Camshaft Position (CMP) sensor is faulty<br>• CMP circuit short to ground, power or open<br>• Engine Speed (RPM) sensor is faulty |
| **DTC: P0017**<br>**2T MIL: Yes**<br>**Years:** 2007, 2008<br>**Models:** 335i, 335Ci, 335xi, 525i, 525xi, 530i, 530xi, 550i, 650i, 750i, 750Li, 760Li, M5, M6, X3, X5, Z4 3.0i, Z4 3.0Si, Z4 M<br>**Engines:** All<br>**Transmissions:** All | **Crankshaft Position - Camshaft Position Correlation Bank 1 Sensor B Conditions:**<br>Engine started, engine running, and the DME detected a deviation between the crankshaft position sensor signal and the camshaft position sensor. A rationality error has been detected for camshaft position out of phase with crankshaft.<br>**Possible Causes:**<br>• Camshaft Position (CMP) sensor is faulty<br>• CMP circuit short to ground, power or open<br>• Engine Speed (RPM) sensor is faulty |
| **DTC: P0018**<br>**2T MIL: Yes**<br>**Years:** 2007, 2008<br>**Models:** 335i, 335Ci, 335xi, 525i, 525xi, 530i, 530xi, 550i, 650i, 750i, 750Li, 760Li<br>**Engines:** All<br>**Transmissions:** All | **Crankshaft Position - Camshaft Position Correlation Bank 2 Sensor A Conditions:**<br>Engine started, engine running, and the DME detected a deviation between the crankshaft position sensor signal and the camshaft position sensor. A rationality error has been detected for camshaft position out of phase with crankshaft.<br>**Possible Causes:**<br>• Camshaft Position (CMP) sensor is faulty<br>• CMP circuit short to ground, power or open<br>• Engine Speed (RPM) sensor is faulty |
| **DTC: P0019**<br>**2T MIL: Yes**<br>**Years:** 2007, 2008<br>**Models:** 335i, 335Ci, 335xi, 525i, 525xi, 530i, 530xi, 550i, 650i, 750i, 750Li, 760Li<br>**Engines:** All<br>**Transmissions:** All | **Crankshaft Position - Camshaft Position Correlation Bank 2 Sensor B Conditions:**<br>Engine started, engine running, and the DME detected a deviation between the crankshaft position sensor signal and the camshaft position sensor. A rationality error has been detected for camshaft position out of phase with crankshaft.<br>**Possible Causes:**<br>• Camshaft Position (CMP) sensor is faulty<br>• CMP circuit short to ground, power or open<br>• Engine Speed (RPM) sensor is faulty |

| DTC | Trouble Code Title, Conditions & Possible Causes |
|---|---|
| **DTC: P0020**<br>**2T MIL:** Yes<br>**Years:** 2007, 2008<br>**Models:** 335i, 335Ci, 335xi, 525i, 525xi, 550i, 650i, 750i, 750Li, 760Li, M5, M6, X3, X5, Z4 3.0i, Z4 3.0Si, Z4 M<br>**Engines:** All<br>**Transmissions:** All | **"A" Camshaft Position Timing Over-Advanced (Bank 2) Conditions:**<br>Engine started and driven at an engine speed of more than 400rpm; and the DME detected the camshaft timing exceeded the maximum calibrated advance value, or the camshaft remained in an advanced position during the CCM test. The valve timing did not change from the current valve timing or it remained fixed during the testing.<br>**Possible Causes:**<br>• Fuel pump has failed<br>• CPS circuit is open, shorted to ground or shorted to power<br>• Battery voltage below 11.5 volts<br>• Position actuator circuit may short to B+ or Ground<br>• Camshaft timing improperly set, or continuous oil flow to the VCT piston chamber<br>• Camshaft advance mechanism (the VCT unit) is sticking or binding mechanically<br>• VCT solenoid valve is stuck in open position |
| **DTC: P0021**<br>**2T MIL:** Yes<br>**Years:** 2007, 2008<br>**Models:** 335i, 335Ci, 335xi, 525i, 525xi, 550i, 650i, 750i, 750Li, 760Li, X3, X5, Z4 3.0i, Z4 3.0Si, Z4 M<br>**Engines:** All<br>**Transmissions:** All | **"A" Camshaft Position Actuator Circuit (Bank 2) Conditions:**<br>Key on or engine running; and the DME detected an unexpected high voltage or low voltage condition on the camshaft position sensor. The relative position between the camshaft and crankshaft needs to be optimal so the engine has better torque, fuel economy and emissions.<br>**Possible Causes:**<br>• Fuel pump has failed<br>• Actuator circuit is open, shorted to ground or shorted to power<br>• Battery voltage below 11.5 volts<br>• Position actuator circuit may short to B+ or Ground |
| **DTC: P0022**<br>**2T MIL:** Yes<br>**Years:** 2007, 2008<br>**Models:** 335i, 335Ci, 335xi, 525i, 525xi, 550i, 650i, 750i, 750Li, 760Li, X3, X5, Z4 3.0i, Z4 3.0Si, Z4 M<br>**Engines:** All<br>**Transmissions:** All | **"A" Camshaft Position Over-Retarded (Bank 2) Conditions:**<br>Engine started and driven at an engine speed of more than 400rpm; and the DME detected the camshaft timing exceeded the minimum calibrated retarded value, or the camshaft remained in an retarded position during the CCM test. The valve timing did not change from the current valve timing or it remained fixed during the testing.<br>**Possible Causes:**<br>• Fuel pump has failed<br>• CPS circuit is open, shorted to ground or shorted to power<br>• Battery voltage below 11.5 volts<br>• Position actuator circuit may short to B+ or Ground<br>• Camshaft timing improperly set, or continuous oil flow to the VCT piston chamber<br>• Camshaft advance mechanism (the VCT unit) is sticking or binding mechanically<br>• VCT solenoid valve is stuck in open position |
| **DTC: P0023**<br>**2T MIL:** Yes<br>**Years:** 2007, 2008<br>**Models:** 335i, 335Ci, 335xi, 525i, 525xi, 530i, 530xi, 550i, 650i, 750i, 750Li, 760Li<br>**Engines:** All<br>**Transmissions:** All | **"B" Camshaft Position Actuator Circuit (Bank 2) Conditions:**<br>Key on or engine running; and the DME detected an unexpected high voltage or low voltage condition on the camshaft position sensor. The relative position between the camshaft and crankshaft needs to be optimal so the engine has better torque, fuel economy and emissions.<br>**Possible Causes:**<br>• Fuel pump has failed<br>• Actuator circuit is open, shorted to ground or shorted to power<br>• Battery voltage below 11.5 volts<br>• Position actuator circuit may short to B+ or Ground |
| **DTC: P0024**<br>**2T MIL:** Yes<br>**Years:** 2007, 2008<br>**Models:** 335i, 335Ci, 335xi, 525i, 525xi, 550i, 650i, 750i, 750Li, 760Li, M5, M6, X3, X5, Z4 3.0i, Z4 3.0Si, Z4 M<br>**Engines:** All<br>**Transmissions:** All | **"B" Camshaft Position Timing Over-Advanced (Bank 2) Conditions:**<br>Engine started and driven at an engine speed of more than 400rpm; and the DME detected the camshaft timing exceeded the maximum calibrated advance value, or the camshaft remained in an advanced position during the CCM test. The valve timing did not change from the current valve timing or it remained fixed during the testing. The engine speed should be more than 500rpm and the VANOS is in the end position.<br>**Possible Causes:**<br>• Fuel pump has failed<br>• CPS circuit is open, shorted to ground or shorted to power<br>• Battery voltage below 11.5 volts<br>• Position actuator circuit may short to B+ or Ground<br>• Camshaft timing improperly set, or continuous oil flow to the VCT piston chamber<br>• Camshaft advance mechanism (the VCT unit) is sticking or binding mechanically<br>• VCT solenoid valve is stuck in open position |

| DTC | Trouble Code Title, Conditions & Possible Causes |
|---|---|
| **DTC: P0025**<br>**2T MIL: Yes**<br>**Years:** 2007, 2008<br>**Models:** 335i, 335Ci, 335xi, 525i, 525xi, 550i, 650i, 750i, 750Li, 760Li, M5, M6, X3, X5, Z4 3.0i, Z4 3.0Si, Z4 M<br>**Engines:** All<br>**Transmissions:** All | **"B" Camshaft Position Over-Retarded (Bank 2) Conditions:**<br>Engine started and driven at an engine speed of more than 600rpm; and the DME detected the camshaft timing exceeded the minimum calibrated retarded value, or the camshaft remained in an retarded position during the CCM test. The valve timing did not change from the current valve timing or it remained fixed during the testing.<br>**Possible Causes:**<br>• Fuel pump has failed<br>• CPS circuit is open, shorted to ground or shorted to power<br>• Battery voltage below 11.5 volts<br>• Position actuator circuit may short to B+ or Ground<br>• Camshaft timing improperly set, or continuous oil flow to the VCT piston chamber<br>• Camshaft advance mechanism (the VCT unit) is sticking or binding mechanically<br>• VCT solenoid valve is stuck in open position |
| **DTC: P0030**<br>**2T MIL: Yes**<br>**Years:** 2007, 2008<br>**Models:** 335i, 335Ci, 335xi, 525i, 525xi, 550i, 650i, 750i, 750Li, 760Li, M5, M6, X3, X5, Z4 3.0i, Z4 3.0Si, Z4 M<br>**Engines:** All<br>**Transmissions:** All | **HO2S Heater (Bank 1 Sensor 1) Control Circuit Malfunction Conditions:**<br>Engine started, battery voltage must be at least 11.5v, all electrical components must be off, the ground between the engine and the chassis must be well connected, the exhaust system must be properly sealed between the catalytic converter and the cylinder head, the coolant temperature must be 80 degrees Celsius, and the oxygen sensor heater for oxygen sensor before the catalytic converter must be properly functioning. The DME detected the HO2S signal was in a negative voltage range referred to as "character shift downward". This code sets when the HO2S signal remains in a low state (usually less than 156 mv). In effect, it does not switch properly between 0.1v and 1.1v in closed loop operation.<br>**Possible Causes:**<br>• HO2S is contaminated (due to presence of silicone in fuel)<br>• HO2S signal and ground circuit wires crossed in wiring harness<br>• HO2S signal circuit is shorted to sensor or chassis ground<br>• HO2S element has failed (internal short condition) |
| **DTC: P0031**<br>**2T MIL: Yes**<br>**Years:** 2007, 2008<br>**Models:** 335i, 335Ci, 335xi, 525i, 525xi, 550i, 650i, 750i, 750Li, 760Li, M5, M6, X3, X5, Z4 3.0i, Z4 3.0Si, Z4 M<br>**Engines:** All<br>**Transmissions:** All | **HO2S Heater (Bank 1 Sensor 1) Circuit Low Input Conditions:**<br>Engine started, battery voltage must be at least 11.5v, all electrical components must be off, the ground between the engine and the chassis must be well connected, the exhaust system must be properly sealed between the catalytic converter and the cylinder head, the coolant temperature must be 80 degrees Celsius, and the oxygen sensor heater for oxygen sensor before the catalytic converter must be properly functioning. The DME detected the HO2S signal was in a negative voltage range referred to as "character shift downward". This code sets when the HO2S signal remains in a low state. In effect, it does not switch properly in the closed loop operation. The HO2S (before the three-way catalytic converter) has a short circuit to ground that has lasted longer than 200 seconds.<br>**Possible Causes:**<br>• HO2S is contaminated (due to presence of silicone in fuel)<br>• HO2S signal and ground circuit wires crossed in wiring harness<br>• HO2S signal circuit is shorted to sensor or chassis ground<br>• HO2S element has failed (internal short condition) |
| **DTC: P0032**<br>**2T MIL: Yes**<br>**Years:** 2007, 2008<br>**Models:** 335i, 335Ci, 335xi, 525i, 525xi, 550i, 650i, 750i, 750Li, 760Li, M5, M6, X3, X5, Z4 3.0i, Z4 3.0Si, Z4 M<br>**Engines:** All<br>**Transmissions:** All | **HO2S Heater (Bank 1 Sensor 1) Circuit High Input Conditions:**<br>Engine started, battery voltage must be at least 11.5v, all electrical components must be off, the ground between the engine and the chassis must be well connected, the exhaust system must be properly sealed between the catalytic converter and the cylinder head, the coolant temperature must be 80 degrees Celsius, and the oxygen sensor heater for oxygen sensor before the catalytic converter must be properly functioning. The DME detected the HO2S signal remained in a high state.<br>**Note: The HO2S signal circuit may be shorted to the heater power circuit due to tracking inside of the HO2S connector. Remove the connector and visually inspect the connector for signs of oil or water.**<br>**Possible Causes:**<br>• HO2S signal shorted to heater power circuit inside connector<br>• HO2S signal circuit shorted to ground or to system voltage |
| **DTC: P0036**<br>**2T MIL: Yes**<br>**Years:** 2007, 2008<br>**Models:** 335i, 335Ci, 335xi, 525i, 525xi, 530i, 530xi, 550i, 650i, 750i, 750Li, 760Li, M5, M6, X3, X5, Z4 3.0i, Z4 3.0Si, Z4 M<br>**Engines:** All<br>**Transmissions:** All | **HO2S Heater (Bank 1 Sensor 2) Control Circuit Malfunction Conditions:**<br>Engine started, battery voltage must be at least 11.5v, all electrical components must be off, the ground between the engine and the chassis must be well connected, the exhaust system must be properly sealed between the catalytic converter and the cylinder head, the coolant temperature must be 80 degrees Celsius, and the oxygen sensor heater for oxygen sensor before the catalytic converter must be properly functioning. The DME detected the HO2S signal was in a negative voltage range referred to as "character shift downward". This code sets when the HO2S signal remains in a low state.<br>**Possible Causes:**<br>• HO2S is contaminated (due to presence of silicone in fuel)<br>• HO2S signal and ground circuit wires crossed in wiring harness<br>• HO2S signal circuit is shorted to sensor or chassis ground<br>• HO2S element has failed (internal short condition) |

| DTC | Trouble Code Title, Conditions & Possible Causes |
|---|---|
| **DTC: P0037**<br>**2T MIL:** Yes<br>**Years:** 2007, 2008<br>**Models:** 335i, 335Ci, 335xi, 525i, 525xi, 550i, 650i, 750i, 750Li, 760Li, M5, M6, X3, X5, Z4 3.0i, Z4 3.0Si, Z4 M<br>**Engines:** All<br>**Transmissions:** All | **HO2S Heater (Bank 1 Sensor 2) Circuit Low Input Conditions:**<br>Engine started, battery voltage must be at least 11.5v, all electrical components must be off, the ground between the engine and the chassis must be well connected, the exhaust system must be properly sealed between the catalytic converter and the cylinder head, the coolant temperature must be 80 degrees Celsius, and the oxygen sensor heater for oxygen sensor before the catalytic converter must be properly functioning. The DME detected the HO2S signal was in a negative voltage range referred to as "character shift downward". This code sets when the HO2S signal remains in a low state. In effect, it does not switch properly in the closed loop operation. The HO2S (before the three-way catalytic converter) has a short circuit to ground that has lasted longer than 200 seconds.<br>**Possible Causes:**<br>• HO2S is contaminated (due to presence of silicone in fuel)<br>• HO2S signal and ground circuit wires crossed in wiring harness<br>• HO2S signal circuit is shorted to sensor or chassis ground<br>• HO2S element has failed (internal short condition) |
| **DTC: P0038**<br>**2T MIL:** Yes<br>**Years:** 2007, 2008<br>**Models:** 335i, 335Ci, 335xi, 525i, 525xi, 550i, 650i, 750i, 750Li, 760Li, M5, M6, X3, X5, Z4 3.0i, Z4 3.0Si, Z4 M<br>**Engines:** All<br>**Transmissions:** All | **HO2S Heater (Bank 1 Sensor 2) Circuit High Input Conditions:**<br>Engine started, battery voltage must be at least 11.5v, all electrical components must be off, the ground between the engine and the chassis must be well connected, the exhaust system must be properly sealed between the catalytic converter and the cylinder head, the coolant temperature must be 80 degrees Celsius, and the oxygen sensor heater for oxygen sensor before the catalytic converter must be properly functioning. The DME detected the HO2S signal remained in a high state.<br>**Note: The HO2S signal circuit may be shorted to the heater power circuit due to tracking inside of the HO2S connector. Remove the connector and visually inspect the connector for signs of oil or water.**<br>**Possible Causes:**<br>• HO2S signal shorted to heater power circuit inside connector<br>• HO2S signal circuit shorted to ground or to system voltage |
| **DTC: P0040**<br>**2T MIL:** Yes<br>**Years:** 2007, 2008<br>**Models:** 335i, 335Ci, 335xi, 525i, 525xi, 530i, 530xi, 550i, 650i, 750i, 750Li, 760Li<br>**Engines:** All<br>**Transmissions:** All | **O2 Sensor Signals Swapped (Bank 1 Sensor 1/Bank 2 Sensor 1) Conditions:**<br>Engine started, battery voltage must be at least 11.5v, all electrical components must be off, the ground between the engine and the chassis must be well connected, the exhaust system must be properly sealed between the catalytic converter and the cylinder head, and the coolant temperature must be 80 degrees Celsius. The DME detected the O2 signals were mixed and reading implausible results from both. The Lambda controllers for the two cylinder banks display mutual displays exceeding 20 percent. The conditions for monitoring faults must be present longer than 10 seconds from the point at which the lambda control assumes active operation.<br>**Possible Causes:**<br>• HO2S-11 and HO2S-21 harness connectors are swapped<br>• HO2S-11 and HO2S-21 wiring is crossed inside the harness<br>• HO2S-11 and HO2S-21 wires are crossed at 104-pin connector<br>• Connector coding and color mixed with correct catalytic converter |
| **DTC: P0050**<br>**2T MIL:** Yes<br>**Years:** 2007, 2008<br>**Models:** 335i, 335Ci, 335xi, 525i, 525xi, 530i, 530xi, 550i, 650i, 750i, 750Li, 760Li, M5, M6, X3, X5, Z4 3.0i, Z4 3.0Si, Z4 M<br>**Engines:** All<br>**Transmissions:** All | **HO2S Heater (Bank 2 Sensor 1) Control Circuit Malfunction Conditions:**<br>Engine started, battery voltage must be at least 11.5v, all electrical components must be off, the ground between the engine and the chassis must be well connected, the exhaust system must be properly sealed between the catalytic converter and the cylinder head, and the coolant temperature must be 80 degrees Celsius. The DME detected the HO2S signal was in a negative voltage range referred to as "character shift downward".<br>**Possible Causes:**<br>• HO2S is contaminated (due to presence of silicone in fuel)<br>• HO2S signal and ground circuit wires crossed in wiring harness<br>• HO2S signal circuit is shorted to sensor or chassis ground<br>• HO2S element has failed (internal short condition) |
| **DTC: P0051**<br>**2T MIL:** Yes<br>**Years:** 2007, 2008<br>**Models:** 335i, 335Ci, 335xi, 525i, 525xi, 530i, 530xi, 550i, 650i, 750i, 750Li, 760Li, M5, M6, X3, X5, Z4 3.0i, Z4 3.0Si, Z4 M<br>**Engines:** All<br>**Transmissions:** All | **HO2S Heater (Bank 2 Sensor 1) Circuit Low Input Conditions:**<br>Engine started, battery voltage must be at least 11.5v, all electrical components must be off, the ground between the engine and the chassis must be well connected, the exhaust system must be properly sealed between the catalytic converter and the cylinder head, and the coolant temperature must be 80 degrees Celsius. The DME detected the HO2S signal was in a negative voltage range referred to as "character shift downward". This code sets when the HO2S signal remains in a low state. In effect, it does not switch properly in the closed loop operation. The HO2S (before the three-way catalytic converter) has a short circuit to ground that has lasted longer than a specified time.<br>**Possible Causes:**<br>• HO2S is contaminated (due to presence of silicone in fuel)<br>• HO2S signal and ground circuit wires crossed in wiring harness<br>• HO2S signal circuit is shorted to sensor or chassis ground<br>• HO2S element has failed (internal short condition) |

| DTC | Trouble Code Title, Conditions & Possible Causes |
|---|---|
| **DTC: P0052**<br>**2T MIL: Yes**<br>**Years:** 2007, 2008<br>**Models:** 335i, 335Ci, 335xi, 525i, 525xi, 530i, 530xi, 550i, 650i, 750i, 750Li, 760Li, M5, M6, X3, X5, Z4 3.0i, Z4 3.0Si, Z4 M<br>**Engines:** All<br>**Transmissions:** All | **HO2S Heater (Bank 2 Sensor 1) Circuit High Input Conditions:**<br>Engine started, battery voltage must be at least 11.5v, all electrical components must be off, the ground between the engine and the chassis must be well connected, the exhaust system must be properly sealed between the catalytic converter and the cylinder head, and the coolant temperature must be 80 degrees Celsius. The DME detected the HO2S signal was in a negative voltage range referred to as "character shift downward". This code sets when the HO2S signal remains in a low state. In effect, it does not switch properly in the closed loop operation. The HO2S (before the three-way catalytic converter) has a short circuit to ground that has lasted longer than a specified time.<br>**Possible Causes:**<br>• HO2S is contaminated (due to presence of silicone in fuel)<br>• HO2S signal and ground circuit wires crossed in wiring harness<br>• HO2S signal circuit is shorted to sensor or chassis ground<br>• HO2S element has failed (internal short condition) |
| **DTC: P0056**<br>**2T MIL: Yes**<br>**Years:** 2007, 2008<br>**Models:** 335i, 335Ci, 335xi, 525i, 525xi, 530i, 530xi, 550i, 650i, 750i, 750Li, 760Li, M5, M6, X3, X5, Z4 3.0i, Z4 3.0Si, Z4 M<br>**Engines:** All<br>**Transmissions:** All | **HO2S Heater (Bank 2 Sensor 2) Circuit High Input Conditions:**<br>Engine started, battery voltage must be at least 11.5v, all electrical components must be off, the ground between the engine and the chassis must be well connected, the exhaust system must be properly sealed between the catalytic converter and the cylinder head, and the coolant temperature must be 80 degrees Celsius. The DME detected the HO2S signal remained in a high state.<br>**Note: The HO2S signal circuit may be shorted to the heater power circuit due to tracking inside of the HO2S connector. Remove the connector and visually inspect the connector for signs of oil or water.**<br>**Possible Causes:**<br>• HO2S signal shorted to heater power circuit inside connector<br>• HO2S signal circuit shorted to ground or to system voltage |
| **DTC: P0057**<br>**2T MIL: Yes**<br>**Years:** 2007, 2008<br>**Models:** 335i, 335Ci, 335xi, 525i, 525xi, 530i, 530xi, 550i, 650i, 750i, 750Li, 760Li, M5, M6, X3, X5, Z4 3.0i, Z4 3.0Si, Z4 M<br>**Engines:** All<br>**Transmissions:** All | **HO2S Heater (Bank 2 Sensor 2) Control Circuit Malfunction Conditions:**<br>Engine started, battery voltage must be at least 11.5v, all electrical components must be off, the ground between the engine and the chassis must be well connected, the exhaust system must be properly sealed between the catalytic converter and the cylinder head, and the coolant temperature must be 80 degrees Celsius. The DME detected the HO2S signal was in a negative voltage range referred to as "character shift downward".<br>**Possible Causes:**<br>• HO2S is contaminated (due to presence of silicone in fuel)<br>• HO2S signal and ground circuit wires crossed in wiring harness<br>• HO2S signal circuit is shorted to sensor or chassis ground<br>• HO2S element has failed (internal short condition) |
| **DTC: P0058**<br>**2T MIL: Yes**<br>**Years:** 2007, 2008<br>**Models:** 335i, 335Ci, 335xi, 525i, 525xi, 530i, 530xi, 550i, 650i, 750i, 750Li, 760Li, M5, M6, X3, X5, Z4 3.0i, Z4 3.0Si, Z4 M<br>**Engines:** All<br>**Transmissions:** All | **HO2S Heater (Bank 2 Sensor 2) Circuit Low Input Conditions:**<br>Engine started, battery voltage must be at least 11.5v, all electrical components must be off, the ground between the engine and the chassis must be well connected, the exhaust system must be properly sealed between the catalytic converter and the cylinder head, and the coolant temperature must be 80 degrees Celsius. The DME detected the HO2S signal was in a negative voltage range referred to as "character shift downward". This code sets when the HO2S signal remains in a low state. In effect, it does not switch properly in the closed loop operation. The HO2S (before the three-way catalytic converter) has a short circuit to ground that has lasted longer than a specified time. The difference between the outside and coolant temperature is greater than 3 degrees Celsius.<br>**Possible Causes:**<br>• HO2S is contaminated (due to presence of silicone in fuel)<br>• HO2S signal and ground circuit wires crossed in wiring harness<br>• HO2S signal circuit is shorted to sensor or chassis ground<br>• HO2S element has failed (internal short condition) |
| **DTC: P0071**<br>**2T MIL: Yes**<br>**Years:** 2007, 2008<br>**Models:** 335i, 335Ci, 335xi, 525i, 525xi, 530i, 530xi, 550i, 650i, 750i, 750Li, 760Li,<br>**Engines:** All<br>**Transmissions:** All | **Ambient Air Temperature Sensor Range/Performance Conditions:**<br>Key on or engine running (at over 800rpm), the vehicle velocity is over 25mph, the ambient temperature is 20 degrees above or below the model figure for four seconds. This is a thermistor-type sensor with a variable resistance that changes when exposed to different temperatures. This means: the higher the temperature, the lower the resistance value.<br>**Possible Causes:**<br>• IAT sensor signal circuit is grounded (check wiring & connector)<br>• Resistance value between sockets 33 and 36 out of range<br>• IAT sensor has an open circuit<br>• IAT sensor is damaged or it has failed<br>• Ambient temperature sensor at the cluster is defective |

| DTC | Trouble Code Title, Conditions & Possible Causes |
|---|---|
| **DTC: P0100**<br>**2T MIL:** Yes<br>**Years:** 2007, 2008<br>**Models:** 328i, 328Ci, 328xi, 335i, 335Ci, 335xi, 525i, 525xi, 530i, 530xi, 550i, 650i, 750i, 750Li, 760Li, M5, M6, X3, X5, Z4 3.0i, Z4 3.0Si, Z4 M<br>**Engines:** All<br>**Transmissions:** All | **Mass or Volume Air Flow Circuit "A" Conditions**<br>Engine running, with the system voltage more than 11.0v, and the temperature must be at least 185-degrees (F) and all electrical equipment (A/C, lights, etc) must be off. The DME has detected that the MAF signal was out of a calculated range with the engine (or undetectable) for a certain period of time. The engine speed is greater than 200rpm.<br>**Possible Causes:**<br>• Mass air flow (MAF) sensor has failed or is damaged<br>• Signal and ground wires of Mass Air Flow (MAF) sensor has short circuited |
| **DTC: P0101**<br>**2T MIL:** Yes<br>**Years:** 2007, 2008<br>**Models:** 328i, 328Ci, 328xi, 335i, 335Ci, 335xi, 525i, 525xi, 530i, 530xi, 550i, 650i, 750i, 750Li, 760Li, M5, M6, X3, X5, Z4 3.0i, Z4 3.0Si, Z4 M<br>**Engines:** All<br>**Transmissions:** All | **Mass or Volume Air Flow Circuit Range/Performance Conditions**<br>Engine running, with the system voltage more than 11.0v, and the temperature must be at least 185-degrees (F) and all electrical equipment (A/C, lights, etc) must be off. The DME has detected that the MAF signal was out of a calculated range with the engine (or undetectable) for a certain period of time.<br>**Possible Causes:**<br>• Mass air flow (MAF) sensor has failed or is damaged<br>• Signal and ground wires of Mass Air Flow (MAF) sensor has short circuited |
| **DTC: P0102**<br>**2T MIL:** Yes<br>**Years:** 2007, 2008<br>**Models:** 335i, 335Ci, 335xi, 525i, 525xi, 550i, 650i, 750i, 750Li, 760Li, M5, M6<br>**Engines:** All<br>**Transmissions:** All | **MAF Sensor Circuit Low Input Conditions:**<br>Key on, engine started, and the DME detected the MAF sensor signal was less than the minimum calibrated value. The engine temperature must beat least 185-degrees (F) and all electrical equipment (A/C, lights, etc) must be off. The DME has detected that the MAF signal was less than the required minimum. The engine speed is greater than 150rpm and the battery voltage is greater than 6 volts.<br>**Possible Causes:**<br>• Check for leaks between MAF sensor and throttle valve control module<br>• Voltage supply faulty.<br>• Sensor power circuit open from fuel pump relay to MAF sensor<br>• Sensor signal circuit open (may be disconnected) from DME and MAF<br>• Faulty ground cable resistance between connector terminal 1 and Ground<br>• MAF Sensor malfunction |
| **DTC: P0103**<br>**2T MIL:** Yes<br>**Years:** 2007, 2008<br>**Models:** 335i, 335Ci, 335xi, 525i, 525xi, 550i, 650i, 750i, 750Li, 760Li, M5, M6<br>**Engines:** All<br>**Transmissions:** All | **MAF Sensor Circuit High Input Conditions:**<br>Key on, engine started, and the DME detected the MAF sensor signal was more than the minimum calibrated value. The engine temperature must beat least 185-degrees (F) and all electrical equipment (A/C, lights, etc) must be off. The DME has detected that the MAF signal was more than the required minimum. The engine speed is greater than 150rpm and the battery voltage is greater than 6 volts.<br>**Possible Causes:**<br>• Check for leaks between MAF sensor and throttle valve control module<br>• Voltage supply faulty.<br>• Sensor power circuit open from fuel pump relay to MAF sensor<br>• Sensor signal circuit open (may be disconnected) from DME and MAF<br>• Faulty ground cable resistance between connector terminal 1 and Ground<br>• MAF Sensor malfunction |
| **DTC: P0111**<br>**2T MIL:** Yes<br>**Years:** 2007, 2008<br>**Models:** 328i, 328Ci, 328xi, 335i, 335Ci, 335xi, 525i, 525xi, 530i, 530xi, 550i, 650i, 750i, 750Li, 760Li, M5, M6, X3, X5, Z4 3.0i, Z4 3.0Si, Z4 M<br>**Engines:** All<br>**Transmissions:** All | **Intake Air Temperature Sensor Circuit Low Input Conditions:**<br>Key on or engine running, the temperature must beat least 185-degrees (F) and all electrical equipment (A/C, lights, etc) must be off; and the DME detected the IAT sensor signal was less than the self-test minimum. This is a thermistor-type sensor with a variable resistance that changes when exposed to different temperatures. This means: the higher the temperature, the lower the resistance value.<br>**Possible Causes:**<br>• IAT sensor signal circuit is grounded (check wiring & connector)<br>• Resistance value between sockets 33 and 36 out of range<br>• IAT sensor has an open circuit<br>• IAT sensor is damaged or it has failed |

| DTC | Trouble Code Title, Conditions & Possible Causes |
|---|---|
| **DTC: P0112**<br>**2T MIL: Yes**<br>**Years:** 2007, 2008<br>**Models:** 335i, 335Ci, 335xi, 525i, 525xi, 550i, 650i, 750i, 750Li, 760Li, M5, M6,<br>**Engines:** All<br>**Transmissions:** All | **Intake Air Temperature Sensor Circuit Low Input Conditions:**<br>Key on or Engine running, the temperature must beat least 185-degrees (F) and all electrical equipment (A/C, lights, etc) must be off; and the DME detected the IAT sensor signal was less than the self-test minimum. This is a thermistor-type sensor with a variable resistance that changes when exposed to different temperatures. This means: the higher the temperature, the lower the resistance value.<br>**Possible Causes:**<br>  • IAT sensor signal circuit is grounded (check wiring & connector)<br>  • Resistance value between sockets 33 and 36 out of range<br>  • IAT sensor has an open circuit<br>  • IAT sensor is damaged or it has failed |
| **DTC: P0113**<br>**2T MIL: Yes**<br>**Years:** 2007, 2008<br>**Models:** 335i, 335Ci, 335xi, 525i, 525xi, 550i, 650i, 750i, 750Li, 760Li, M5, M6, X3, X5, Z4 3.0i, Z4 3.0Si, Z4 M<br>**Engines:** All<br>**Transmissions:** All | **Intake Air Temperature Sensor Circuit High Input Conditions:**<br>Key on or engine running, the temperature must beat least 185-degrees (F) and all electrical equipment (A/C, lights, etc) must be off; and the DME detected the IAT sensor signal was more than the self-test maximum. This is a thermistor-type sensor with a variable resistance that changes when exposed to different temperatures. This means: the higher the temperature, the lower the resistance value.<br>**Possible Causes:**<br>  • IAT sensor signal circuit is open (inspect wiring & connector)<br>  • IAT sensor signal circuit is shorted<br>  • Resistance value between sockets 33 and 36 out of range<br>  • IAT sensor is damaged or it has failed |
| **DTC: P0116**<br>**2T MIL: Yes**<br>**Years:** 2007, 2008<br>**Models:** 328i, 328Ci, 328xi, 335i, 335Ci, 335xi, 525i, 525xi, 530i, 530xi, 550i, 650i, 750i, 750Li, 760Li, M5, M6, X3, X5, Z4 3.0i, Z4 3.0Si, Z4 M<br>**Engines:** All<br>**Transmissions:** All | **ECT Sensor Signal Range/Performance Conditions:**<br>Engine started (cold), battery voltage must be 11.5, and all equipment must be off. The DME detected the ECT sensor exceeded the required calibrated value, or the engine is at idle and doesn't reach operating temperature quickly enough; the Catalyst, Fuel System, HO2S and Misfire Monitor did not complete, or the timer expired. Testing completion of procedure, the engine's temperature must rise uniformly during idle.<br>**Possible Causes:**<br>  • Check for low coolant level or incorrect coolant mixture<br>  • DME detects a short circuit wiring in the ECT<br>  • CHT sensor is out-of-calibration or it has failed<br>  • ECT sensor is out-of-calibration or it has failed |
| **DTC: P0117**<br>**2T MIL: Yes**<br>**Years:** 2007, 2008<br>**Models:** 335i, 335Ci, 335xi, 525i, 525xi, 550i, 650i, 750i, 750Li, 760Li, M5, M6, X3, X5, Z4 3.0i, Z4 3.0Si, Z4 M<br>**Engines:** All<br>**Transmissions:** All | **ECT Sensor Circuit Low Input Conditions:**<br>Engine started (cold) for 10 seconds, battery voltage must be 11.5, and all equipment must be off. The DME detected the ECT sensor signal was less than the self-test minimum. This is a thermistor-type sensor with a variable resistance that changes when exposed to different temperatures<br>**Possible Causes:**<br>  • ECT sensor signal circuit is grounded in the wiring harness<br>  • ECT sensor doesn't react to changes in temperature<br>  • ECT sensor is damaged or the DME has failed |
| **DTC: P0118**<br>**2T MIL: Yes**<br>**Years:** 2007, 2008<br>**Models:** 335i, 335Ci, 335xi, 525i, 525xi, 550i, 650i, 750i, 750Li, 760Li, M5, M6, X3, X5, Z4 3.0i, Z4 3.0Si, Z4 M<br>**Engines:** All<br>**Transmissions:** All | **ECT Sensor Circuit High Input Conditions:**<br>Engine started (cold) for 10 seconds, battery voltage must be 11.5, and all equipment must be off. The DME detected the ECT sensor signal was more than the self-test maximum. This is a thermistor-type sensor with a variable resistance that changes when exposed to different temperatures<br>**Possible Causes:**<br>  • ECT sensor signal circuit is open (inspect wiring & connector)<br>  • ECT sensor signal circuit is shorted to ground<br>  • ECT sensor is damaged or it has failed |
| **DTC: P0120**<br>**2T MIL: Yes**<br>**Years:** 2007, 2008<br>**Models:** 328i, 328Ci, 328xi, 335i, 335Ci, 335xi, 525i, 525xi, 530i, 530xi, 550i, 650i, 750i, 750Li, 760Li, M5, M6, X3, X5, Z4 3.0i, Z4 3.0Si, Z4 M<br>**Engines:** All<br>**Transmissions:** All | **Throttle/Pedal Position Sensor (A) Circuit Malfunction Conditions:**<br>Engine started, at idle (to 1320rpm), the temperature must be 80 degrees Celsius. The throttle position sensor supplies implausible signal to the DME. The throttle valve activation occurs via an electric motor (throttle drive) in the throttle valve control module. It is activated by the Engine Control Module (DME) according to specifications of the two sensors, Throttle Position (TP) Sensor and Accelerator Pedal Position Sensor 2.<br>**Possible Causes:**<br>  • TP sensor signal circuit is open (inspect wiring & connector)<br>  • TP sensor signal circuit is shorted to ground<br>  • TP sensor or module is damaged or it has failed<br>  • Throttle valve is damaged or dirty<br>  • Throttle valve control module is faulty |

| DTC | Trouble Code Title, Conditions & Possible Causes |
|---|---|
| **DTC: P0121**<br>**2T MIL:** Yes<br>**Years:** 2007, 2008<br>**Models:** 328i, 328Ci, 328xi, 335i, 335Ci, 335xi, 525i, 525xi, 530i, 530xi, 550i, 650i, 750i, 750Li, 760Li, M5, M6, X3, X5, Z4 3.0i, Z4 3.0Si, Z4 M<br>**Engines:** All<br>**Transmissions:** All | **Throttle/Pedal Position Sensor Signal Range/Performance Conditions:**<br>Engine started; then immediately following a condition where the engine was running under at off-idle, the DME detected the TP sensor signal indicated the throttle did not return to its previous closed position during the Rationality test. The engine speed is greater than 1320rpm.<br>**Possible Causes:**<br>• Throttle plate is binding, dirty or sticking<br>• Throttle valve is damaged or dirty<br>• Throttle valve control module is faulty<br>• TP sensor signal circuit open (inspect wiring & connector)<br>• TP sensor ground circuit open (inspect wiring & connector)<br>• TP sensor and/or control module is damaged or has failed<br>• MAF sensor signal is damaged, has failed or a short is present |
| **DTC: P0122**<br>**2T MIL:** Yes<br>**Years:** 2007, 2008<br>**Models:** 328i, 328Ci, 328xi, 335i, 335Ci, 335xi, 525i, 525xi, 530i, 530xi, 550i, 650i, 750i, 750Li, 760Li, M5, M6, X3, X5, Z4 3.0i, Z4 3.0Si, Z4 M<br>**Engines:** All<br>**Transmissions:** All | **Throttle/Pedal Position Sensor Circuit Low Input Conditions:**<br>Engine started, at idle, the temperature must be at least 80 degrees Celsius. The throttle position sensor supplies implausible signal to the DME.<br>**Possible Causes:**<br>• TP sensor signal circuit open (inspect wiring & connector)<br>• TP sensor signal shorted to ground (inspect wiring & connector)<br>• TP sensor is damaged or has failed<br>• Throttle control module's voltage supply is shorted or open |
| **DTC: P0123**<br>**2T MIL:** Yes<br>**Years:** 2007, 2008<br>**Models:** 328i, 328Ci, 328xi, 335i, 335Ci, 335xi, 525i, 525xi, 530i, 530xi, 550i, 650i, 750i, 750Li, 760Li, M5, M6, X3, X5, Z4 3.0i, Z4 3.0Si, Z4 M<br>**Engines:** All<br>**Transmissions:** All | **TP Sensor Circuit High Input Conditions:**<br>Engine started, at idle, the temperature must be at least 80 degrees Celsius. The DME detected the TP sensor signal was more than the self-test maximum during testing.<br>**Possible Causes:**<br>• TP sensor not seated correctly in housing (may be damaged)<br>• TP sensor signal is circuit shorted to ground or system voltage<br>• TP sensor ground circuit is open (check the wiring harness)<br>• TP sensor and/or DME has failed |
| **DTC: P0125**<br>**2T MIL:** Yes<br>**Years:** 2007, 2008<br>**Models:** 328i, 328Ci, 328xi, 335i, 335Ci, 335xi, 525i, 525xi, 530i, 530xi, 550i, 650i, 750i, 750Li, 760Li, M5, M6, X3, X5, Z4 3.0i, Z4 3.0Si, Z4 M<br>**Engines:** All<br>**Transmissions:** All | **ECT Sensor Insufficient for Closed Loop Fuel Control Conditions:**<br>Engine started (cold), battery voltage must be 11.5, and all equipment must be off. The DME detected the ECT sensor exceeded the required calibrated value, or the engine is at idle and doesn't reach operating temperature quickly enough; the Catalyst, Fuel System, HO2S and Misfire Monitor did not complete, or the timer expired. Testing completion of procedure, the engine's temperature must rise uniformly during idle.<br>**Possible Causes:**<br>• Check for low coolant level or incorrect coolant mixture<br>• DME detects a short circuit wiring in the ECT<br>• CHT sensor is out-of-calibration or it has failed<br>• ECT sensor is out-of-calibration or it has failed |
| **DTC: P0130**<br>**2T MIL:** Yes<br>**Years:** 2007, 2008<br>**Models:** 328i, 328Ci, 328xi, 335i, 335Ci, 335xi, 525i, 525xi, 530i, 530xi, 550i, 650i, 750i, 750Li, 760Li, M5, M6, X3, X5, Z4 3.0i, Z4 3.0Si, Z4 M<br>**Engines:** All<br>**Transmissions:** All | **O2 Sensor Circuit Bank 1 Sensor 1 Conditions:**<br>Engine running, battery voltage 11.5, all electrical components off, ground between engine and chassis well connected and the exhaust system must be properly sealed between catalytic converter and the cylinder head. The DME detected the HO2S signal was implausible or not detected. The response rate for the sensor signal period is greater than 3.8/second. The engine speed is 1280 to 2400rpm, the catalyst temperature is greater than 300 degrees Celsius and the heater has been on for less than 90 seconds.<br>**Possible Causes:**<br>• Oxygen sensor heater for oxygen sensor (HO2S) before catalytic converter is faulty<br>• HO2S is contaminated (due to presence of silicone in fuel)<br>• HO2S signal and ground circuit wires crossed in wiring harness<br>• HO2S signal circuit is shorted to sensor or chassis ground<br>• HO2S element before the catalytic converter has failed (internal short condition)<br>• Leaks present in the exhaust manifold or exhaust pipes |

| DTC | Trouble Code Title, Conditions & Possible Causes |
|---|---|
| **DTC: P0131**<br>**2T MIL: Yes**<br>**Years:** 2007, 2008<br>**Models:** 328i, 328Ci, 328xi, 335i, 335Ci, 335xi, 525i, 525xi, 530i, 530xi, 550i, 650i, 750i, 750Li, 760Li, M5, M6, X3, X5, Z4 3.0i, Z4 3.0Si, Z4 M<br>**Engines:** All<br>**Transmissions:** All | **HO2S (Bank 1 Sensor 1) Circuit Low Input Conditions:**<br>Engine running, battery voltage 11.5, all electrical components off, ground between engine and chassis well connected and the exhaust system must be properly sealed between catalytic converter and the cylinder head. The DME detected the HO2S signal was in a negative voltage range referred to as "character shift downward". This code sets when the HO2S signal remains in a low state for a measured period of time. In effect, it does not switch properly in the closed loop operation.<br>**Possible Causes:**<br>• HO2S is contaminated (due to presence of silicone in fuel)<br>• HO2S signal and ground circuit wires crossed in wiring harness<br>• HO2S signal circuit is shorted to sensor or chassis ground<br>• HO2S element has failed (internal short condition)<br>• Leaks present in the exhaust manifold or exhaust pipes |
| **DTC: P0132**<br>**2T MIL: Yes**<br>**Years:** 2007, 2008<br>**Models:** 328i, 328Ci, 328xi, 335i, 335Ci, 335xi, 525i, 525xi, 530i, 530xi, 550i, 650i, 750i, 750Li, 760Li, M5, M6, X3, X5, Z4 3.0i, Z4 3.0Si, Z4 M<br>**Engines:** All<br>**Transmissions:** All | **HO2S (Bank 1 Sensor 1) Circuit High Input Conditions:**<br>Engine running, battery voltage 11.5, all electrical components off, ground between engine and chassis well connected and the exhaust system must be properly sealed between catalytic converter and the cylinder head. The DME detected the HO2S signal was in a high state. This code sets when the HO2S signal remains in a high state for a measured period of time. In effect, it does not switch properly in the closed loop operation.<br>**Note: The HO2S signal circuit may be shorted to the heater power circuit due to tracking inside of the HO2S connector. Remove the connector and visually inspect the connector for signs of oil or water.**<br>**Possible Causes:**<br>• HO2S is contaminated (due to presence of silicone in fuel)<br>• HO2S signal and ground circuit wires crossed in wiring harness<br>• HO2S signal circuit is shorted to sensor or chassis ground<br>• HO2S element has failed (internal short condition)<br>• Leaks present in the exhaust manifold or exhaust pipes |
| **DTC: P0133**<br>**2T MIL: Yes**<br>**Years:** 2007, 2008<br>**Models:** 328i, 328Ci, 328xi, 335i, 335Ci, 335xi, 525i, 525xi, 530i, 530xi, 550i, 650i, 750i, 750Li, 760Li, M5, M6, X3, X5, Z4 3.0i, Z4 3.0Si, Z4 M<br>**Engines:** All<br>**Transmissions:** All | **HO2S (Bank 1 Sensor 1) Circuit Slow Response Conditions:**<br>Engine running, battery voltage 11.5, all electrical components off, ground between engine and chassis well connected and the exhaust system must be properly sealed between catalytic converter and the cylinder head. The DME detected the HO2S amplitude and frequency were out of the normal range (e.g., the HO2S rich to lean switch) during the HO2S Monitor test. The response rate for the sensor signal period is greater than 3.8/second. The engine speed is 1280 to 2400rpm, the catalyst temperature is greater than 300 degrees Celsius and the heater has been on for less than 90 seconds.<br>**Possible Causes:**<br>• HO2S before the three-way catalytic converter is contaminated (due to presence of silicone in fuel); Run the engine for three minutes at 3500rpm as a self-cleaning effect<br>• HO2S signal circuit open<br>• Leaks present in the exhaust manifold or exhaust pipes<br>• HO2S is damaged or has failed |
| **DTC: P0134**<br>**2T MIL: Yes**<br>**Years:** 2007, 2008<br>**Models:** 328i, 328Ci, 328xi, 335i, 335Ci, 335xi, 525i, 525xi, 530i, 530xi, 550i, 650i, 750i, 750Li, 760Li, M5, M6, X3, X5, Z4 3.0i, Z4 3.0Si, Z4 M<br>**Engines:** All<br>**Transmissions:** All | **HO2S (Bank 1 Sensor 1) Circuit No Activity Conditions:**<br>Engine running, battery voltage 11.5, all electrical components off, ground between engine and chassis well connected and the exhaust system must be properly sealed between catalytic converter and the cylinder head. The DME detected the HO2S signal failed to meet the maximum or minimum voltage levels (i.e., it failed the voltage range check).<br>**Possible Causes:**<br>• Leaks present in the exhaust manifold or exhaust pipes<br>• HO2S signal wire and ground wire crossed in connector (voltage jumps)<br>• HO2S element is fuel contaminated or has failed |
| **DTC: P0135**<br>**2T MIL: Yes**<br>**Years:** 2007, 2008<br>**Models:** 328i, 328Ci, 328xi, 335i, 335Ci, 335xi, 525i, 525xi, 530i, 530xi, 550i, 650i, 750i, 750Li, 760Li, M5, M6, X3, X5, Z4 3.0i, Z4 3.0Si, Z4 M<br>**Engines:** All<br>**Transmissions:** All | **HO2S (Bank 1 Sensor 1) Heater Circuit Malfunction Conditions:**<br>Engine running, battery voltage 11.5, all electrical components off, ground between engine and chassis well connected and the exhaust system must be properly sealed between catalytic converter and the cylinder head. The DME detected an unexpected voltage condition, or it detected excessive current draw in the heater circuit during the CCM test. The response rate for the sensor signal period is greater than 3.8/second. The engine speed is 1280 to 2400rpm, the catalyst temperature is greater than 300 degrees Celsius and the heater has been on for less than 90 seconds.<br>**Possible Causes:**<br>• HO2S heater power circuit is open or heater ground circuit open<br>• HO2S signal tracking (due to oil or moisture in the connector)<br>• HO2S is damaged or has failed |

| DTC | Trouble Code Title, Conditions & Possible Causes |
|---|---|
| **DTC: P0136**<br>**2T MIL: Yes**<br>**Years:** 2007, 2008<br>**Models:** 328i, 328Ci, 328xi, 335i, 335Ci, 335xi, 525i, 525xi, 530i, 530xi, 550i, 650i, 750i, 750Li, 760Li, M5, M6, X3, X5, Z4 3.0i, Z4 3.0Si, Z4 M<br>**Engines:** All<br>**Transmissions:** All | **HO2S (Bank 1 Sensor 2) Circuit Malfunction Conditions:**<br>Engine running, battery voltage 11.5, all electrical components off, ground between engine and chassis well connected and the exhaust system must be properly sealed between catalytic converter and the cylinder head. The DME detected the HO2S signal failed to meet the maximum or minimum voltage levels (i.e., it failed the voltage range check). The heater has been on for less than 90 seconds, the fuel system status is in fuel cut-off, the output voltage is between 400mV and 500mV and it is 120 seconds after engine start up.<br>**Possible Causes:**<br>• Leaks present in the exhaust manifold or exhaust pipes<br>• HO2S signal wire and ground wire crossed in connector<br>• HO2S element is fuel contaminated or has failed |
| **DTC: P0137**<br>**2T MIL: Yes**<br>**Years:** 2007, 2008<br>**Models:** 328i, 328Ci, 328xi, 335i, 335Ci, 335xi, 525i, 525xi, 530i, 530xi, 550i, 650i, 750i, 750Li, 760Li, M5, M6, X3, X5, Z4 3.0i, Z4 3.0Si, Z4 M<br>**Engines:** All<br>**Transmissions:** All | **HO2S (Bank 1 Sensor 2) Circuit Low Input Conditions:**<br>Engine running, battery voltage 11.5, all electrical components off, ground between engine and chassis well connected and the exhaust system must be properly sealed between catalytic converter and the cylinder head. The DME detected the HO2S signal remained in a high state.<br>**Note: The HO2S signal circuit may be shorted to the heater power circuit due to "tracking inside of the HO2S connector. Remove the connector and visually inspect the connector for signs of oil or water.**<br>**Possible Causes:**<br>• HO2S signal shorted to heater power circuit in the connector<br>• HO2S signal circuit shorted to ground (for more than 200 seconds) or to system voltage |
| **DTC: P0138**<br>**2T MIL: Yes**<br>**Years:** 2007, 2008<br>**Models:** 328i, 328Ci, 328xi, 335i, 335Ci, 335xi, 525i, 525xi, 530i, 530xi, 550i, 650i, 750i, 750Li, 760Li, M5, M6, X3, X5, Z4 3.0i, Z4 3.0Si, Z4 M<br>**Engines:** All<br>**Transmissions:** All | **HO2S (Bank 1 Sensor 2) Circuit High Input Conditions:**<br>Engine running, battery voltage 11.5, all electrical components off, ground between engine and chassis well connected and the exhaust system must be properly sealed between catalytic converter and the cylinder head. The DME detected the HO2S signal remained in a high state.<br>**Note: The HO2S signal circuit may be shorted to the heater power circuit due to "tracking inside of the HO2S connector. Remove the connector and visually inspect the connector for signs of oil or water.**<br>**Possible Causes:**<br>• HO2S signal shorted to heater power circuit in the positive connector<br>• HO2S signal circuit shorted to ground or to system voltage<br>• HO2S has failed |
| **DTC: P0139**<br>**2T MIL: Yes**<br>**Years:** 2007, 2008<br>**Models:** 328i, 328Ci, 328xi, 335i, 335Ci, 335xi, 525i, 525xi, 530i, 530xi, 550i, 650i, 750i, 750Li, 760Li, M5, M6<br>**Engines:** All<br>**Transmissions:** All | **HO2S (Bank 1 Sensor 2) Slow Response Conditions:**<br>Engine running, battery voltage 11.5, all electrical components off, ground between engine and chassis well connected and the exhaust system must be properly sealed between catalytic converter and the cylinder head. The DME detected the HO2S amplitude and frequency were out of the normal range during the HO2S Monitor test. The heater has been on for less than 90 seconds, the fuel system status is in fuel cut-off, the output voltage is between 400mV and 500mV and it is 120 seconds after engine start up.<br>**Possible Causes:**<br>• HO2S signal shorted to heater power circuit in the connector<br>• HO2S signal circuit shorted to VREF or to system voltage |
| **DTC: P0140**<br>**2T MIL: Yes**<br>**Years:** 2007, 2008<br>**Models:** 328i, 328Ci, 328xi, 335i, 335Ci, 335xi, 525i, 525xi, 530i, 530xi, 550i, 650i, 750i, 750Li, 760Li, M5, M6, X3, X5, Z4 3.0i, Z4 3.0Si, Z4 M<br>**Engines:** All<br>**Transmissions:** All | **HO2S (Bank 1 Sensor 2) No Activity Conditions:**<br>Engine running, battery voltage 11.5, all electrical components off, ground between engine and chassis well connected and the exhaust system must be properly sealed between catalytic converter and the cylinder head. The DME detected the HO2S signal failed to meet the maximum or minimum voltage levels (i.e., it failed the voltage range check).<br>**Possible Causes:**<br>• HO2S before the three-way catalytic converter is contaminated (due to presence of silicone in fuel); Run the engine for three minutes at 3500rpm as a self-cleaning effect<br>• Leaks present in the exhaust manifold or exhaust pipes<br>• HO2S signal wire and ground wire crossed in connector (voltage jumps)<br>• HO2S element is contaminated or has failed |
| **DTC: P0141**<br>**2T MIL: Yes**<br>**Years:** 2007, 2008<br>**Models:** 328i, 328Ci, 328xi, 335i, 335Ci, 335xi, 525i, 525xi, 530i, 530xi, 550i, 650i, 750i, 750Li, 760Li, M5, M6, X3, X5, Z4 3.0i, Z4 3.0Si, Z4 M<br>**Engines:** All<br>**Transmissions:** All | **HO2S (Bank 1 Sensor 2) Malfunction Conditions:**<br>Engine running, battery voltage 11.5, all electrical components off, ground between engine and chassis well connected and the exhaust system must be properly sealed between catalytic converter and the cylinder head. The DME detected the HO2S signal failed to meet the maximum or minimum voltage levels (i.e., it failed the voltage range check). The engine speed is greater than 40rpm, the battery voltage must be between 10.7 and 15.5 volts, and the fault occurs 200 seconds after engine start up.<br>**Possible Causes:**<br>• Leaks present in the exhaust manifold or exhaust pipes<br>• HO2S signal wire and ground wire crossed in connector<br>• HO2S element is fuel contaminated or has failed |

| DTC | Trouble Code Title, Conditions & Possible Causes |
|---|---|
| **DTC: P0150**<br>**2T MIL:** Yes<br>**Years:** 2007, 2008<br>**Models:** 328i, 328Ci, 328xi, 335i, 335Ci, 335xi, 525i, 525xi, 530i, 530xi, 550i, 650i, 750i, 750Li, 760Li, M5, M6, X3, X5, Z4 3.0i, Z4 3.0Si, Z4 M<br>**Engines:** All<br>**Transmissions:** All | **HO2S (Bank 2 Sensor 1) Circuit Malfunction Conditions:**<br>Engine running, battery voltage 11.5, all electrical components off, ground between engine and chassis well connected and the exhaust system must be properly sealed between catalytic converter and the cylinder head. The DME detected the HO2S signal failed to meet the maximum or minimum voltage levels (i.e., it failed the voltage range check). The response rate for the sensor signal period is greater than 3.8/second. The engine speed is 1280 to 2400rpm, the catalyst temperature is greater than 300 degrees Celsius and the heater has been on for less than 90 seconds.<br>**Possible Causes:**<br>• Leaks present in the exhaust manifold or exhaust pipes<br>• HO2S signal wire and ground wire crossed in connector<br>• HO2S element is fuel contaminated or has failed |
| **DTC: P0151**<br>**2T MIL:** Yes<br>**Years:** 2007, 2008<br>**Models:** 328i, 328Ci, 328xi, 335i, 335Ci, 335xi, 525i, 525xi, 530i, 530xi, 550i, 650i, 750i, 750Li, 760Li, M5, M6, X3, X5, Z4 3.0i, Z4 3.0Si, Z4 M<br>**Engines:** All<br>**Transmissions:** All | **HO2S (Bank 2 Sensor 1) Low Input Conditions:**<br>Engine running, battery voltage 11.5, all electrical components off, ground between engine and chassis well connected and the exhaust system must be properly sealed between catalytic converter and the cylinder head. The DME detected the HO2S signal remained in a high state.<br>**Note: The HO2S signal circuit may be shorted to the heater power circuit due to "tracking inside of the HO2S connector. Remove the connector and visually inspect the connector for signs of oil or water.**<br>**Possible Causes:**<br>• HO2S is contaminated (due to presence of silicone in fuel)<br>• HO2S signal tracking (due to oil or moisture in the connector)<br>• HO2S signal circuit is open or shorted to VREF |
| **DTC: P0152**<br>**2T MIL:** Yes<br>**Years:** 2007, 2008<br>**Models:** 328i, 328Ci, 328xi, 335i, 335Ci, 335xi, 525i, 525xi, 530i, 530xi, 550i, 650i, 750i, 750Li, 760Li, M5, M6, X3, X5, Z4 3.0i, Z4 3.0Si, Z4 M<br>**Engines:** All<br>**Transmissions:** All | **HO2S (Bank 2 Sensor 1) Circuit High Input Conditions:**<br>Engine running, battery voltage 11.5, all electrical components off, ground between engine and chassis well connected and the exhaust system must be properly sealed between catalytic converter and the cylinder head. The DME detected the HO2S signal remained in a high state (more than 1.5v).<br>**Note: The HO2S signal circuit may be shorted to the heater power circuit due to "tracking inside of the HO2S connector. Remove the connector and visually inspect the connector for signs of oil or water.**<br>**Possible Causes:**<br>• HO2S is contaminated (due to presence of silicone in fuel)<br>• HO2S signal tracking (due to oil or moisture in the connector)<br>• HO2S signal circuit is open or shorted to VREF |
| **DTC: P0153**<br>**2T MIL:** Yes<br>**Years:** 2007, 2008<br>**Models:** 328i, 328Ci, 328xi, 335i, 335Ci, 335xi, 525i, 525xi, 530i, 530xi, 550i, 650i, 750i, 750Li, 760Li, M5, M6, X3, X5, Z4 3.0i, Z4 3.0Si, Z4 M<br>**Engines:** All<br>**Transmissions:** All | **HO2S (Bank 2 Sensor 1) Circuit Slow Response Conditions:**<br>Engine running, battery voltage 11.5, all electrical components off, ground between engine and chassis well connected and the exhaust system must be properly sealed between catalytic converter and the cylinder head. The DME detected the HO2S amplitude and frequency were out of the normal range during the HO2S Monitor test.<br>**Possible Causes:**<br>• HO2S is contaminated (due to presence of silicone in fuel)<br>• Leaks present in the exhaust manifold or exhaust pipes<br>• HO2S is damaged or has failed |
| **DTC: P0154**<br>**2T MIL:** Yes<br>**Years:** 2007, 2008<br>**Models:** 328i, 328Ci, 328xi, 335i, 335Ci, 335xi, 525i, 525xi, 530i, 530xi, 550i, 650i, 750i, 750Li, 760Li, M5, M6, X3, X5, Z4 3.0i, Z4 3.0Si, Z4 M<br>**Engines:** All<br>**Transmissions:** All | **HO2S (Bank 2 Sensor 1) Circuit No Activity Conditions:**<br>Engine running, battery voltage 11.5, all electrical components off, ground between engine and chassis well connected and the exhaust system must be properly sealed between catalytic converter and the cylinder head. The DME detected the HO2S signal failed to meet the maximum or minimum voltage (i.e., it failed the voltage check).<br>**Possible Causes:**<br>• Leaks present in the exhaust manifold or exhaust pipes<br>• HO2S signal wire and ground wire crossed in connector<br>• HO2S element is fuel contaminated or has failed |
| **DTC: P0155**<br>**2T MIL:** Yes<br>**Years:** 2007, 2008<br>**Models:** 328i, 328Ci, 328xi, 335i, 335Ci, 335xi, 525i, 525xi, 530i, 530xi, 550i, 650i, 750i, 750Li, 760Li, M5, M6, X3, X5, Z4 3.0i, Z4 3.0Si, Z4 M<br>**Engines:** All<br>**Transmissions:** All | **HO2S (Bank 2 Sensor 1) Heater Circuit Malfunction Conditions:**<br>Engine running, battery voltage 11.5, all electrical components off, ground between engine and chassis well connected and the exhaust system must be properly sealed between catalytic converter and the cylinder head. The DME detected an open or shorted condition, or excessive current draw in the heater circuit. The response rate for the sensor signal period is greater than 3.8/second. The engine speed is 1280 to 2400rpm, the catalyst temperature is greater than 300 degrees Celsius and the heater has been on for less than 90 seconds.<br>**Possible Causes:**<br>• HO2S heater power circuit is open<br>• HO2S heater ground circuit is open<br>• HO2S signal tracking (due to oil or moisture in the connector)<br>• HO2S is damaged or has failed |

| DTC | Trouble Code Title, Conditions & Possible Causes |
|---|---|
| **DTC: P0156**<br>**2T MIL: Yes**<br>**Years:** 2007, 2008<br>**Models:** 328i, 328Ci, 328xi, 335i, 335Ci, 335xi, 525i, 525xi, 530i, 530xi, 550i, 650i, 750i, 750Li, 760Li, M5, M6, X3, X5, Z4 3.0i, Z4 3.0Si, Z4 M<br>**Engines:** All<br>**Transmissions:** All | **HO2S (Bank 2 Sensor 2) Circuit No Activity Conditions:**<br>Engine running, battery voltage 11.5, all electrical components off, ground between engine and chassis well connected and the exhaust system must be properly sealed between catalytic converter and the cylinder head. The DME detected the HO2S signal failed to meet the maximum or minimum voltage (i.e., it failed the voltage check). The heater has been on for less than 90 seconds, the fuel system status is in fuel cut-off, the output voltage is between 400mV and 500mV and it is 120 seconds after engine start up.<br>**Possible Causes:**<br>&bull; Leaks present in the exhaust manifold or exhaust pipes<br>&bull; HO2S signal wire and ground wire crossed in connector<br>&bull; HO2S element is fuel contaminated or has failed |
| **DTC: P0157**<br>**2T MIL: Yes**<br>**Years:** 2007, 2008<br>**Models:** 328i, 328Ci, 328xi, 335i, 335Ci, 335xi, 525i, 525xi, 530i, 530xi, 550i, 650i, 750i, 750Li, 760Li, M5, M6, X3, X5, Z4 3.0i, Z4 3.0Si, Z4 M<br>**Engines:** All<br>**Transmissions:** All | **HO2S (Bank 2 Sensor 2) Circuit Low Voltage Conditions:**<br>Engine running, battery voltage 11.5, all electrical components off, ground between engine and chassis well connected and the exhaust system must be properly sealed between catalytic converter and the cylinder head. The DME detected the HO2S signal remained in a high state.<br>**Note: The HO2S signal circuit may be shorted to the heater power circuit due to "tracking inside of the HO2S connector. Remove the connector and visually inspect the connector for signs of oil or water.**<br>**Possible Causes:**<br>&bull; HO2S is contaminated (due to presence of silicone in fuel)<br>&bull; HO2S signal tracking (due to oil or moisture in the connector)<br>&bull; HO2S signal circuit is open or shorted to VREF |
| **DTC: P0158**<br>**2T MIL: Yes**<br>**Years:** 2007, 2008<br>**Models:** 328i, 328Ci, 328xi, 335i, 335Ci, 335xi, 525i, 525xi, 530i, 530xi, 550i, 650i, 750i, 750Li, 760Li, M5, M6, X3, X5, Z4 3.0i, Z4 3.0Si, Z4 M<br>**Engines:** All<br>**Transmissions:** All | **HO2S (Bank 2 Sensor 2) Circuit High Input Conditions:**<br>Engine running, battery voltage 11.5, all electrical components off, ground between engine and chassis well connected and the exhaust system must be properly sealed between catalytic converter and the cylinder head. The DME detected the HO2S signal remained in a high state (i.e., more than 1.5v).<br>**Note: The HO2S signal circuit may be shorted to the heater power circuit due to "tracking inside of the HO2S connector. Remove the connector and visually inspect the connector for signs of oil or water.**<br>**Possible Causes:**<br>&bull; HO2S signal shorted to the heater power circuit (due to oil or moisture in the connector)<br>&bull; HO2S signal circuit shorted to VREF or to system voltage |
| **DTC: P0159**<br>**2T MIL: Yes**<br>**Years:** 2007, 2008<br>**Models:** 328i, 328Ci, 328xi, 335i, 335Ci, 335xi, 525i, 525xi, 530i, 530xi, 550i, 650i, 750i, 750Li, 760Li, M5, M6<br>**Engines:** All<br>**Transmissions:** All | **HO2S (Bank 2 Sensor 2) Circuit Slow Response Conditions:**<br>Engine running, battery voltage 11.5, all electrical components off, ground between engine and chassis well connected and the exhaust system must be properly sealed between catalytic converter and the cylinder head. The DME detected the HO2S amplitude and frequency were out of the normal range during the HO2S Monitor test. The heater has been on for less than 90 seconds, the fuel system status is in fuel cut-off, the output voltage is between 400mV and 500mV and it is 120 seconds after engine start up.<br>**Possible Causes:**<br>&bull; HO2S is contaminated (due to presence of silicone in fuel)<br>&bull; Leaks present in the exhaust manifold or exhaust pipes<br>&bull; HO2S is damaged or has failed |
| **DTC: P0160**<br>**2T MIL: Yes**<br>**Years:** 2007, 2008<br>**Models:** 328i, 328Ci, 328xi, 335i, 335Ci, 335xi, 525i, 525xi, 530i, 530xi, 550i, 650i, 750i, 750Li, 760Li, M5, M6<br>**Engines:** All<br>**Transmissions:** All | **HO2S (Bank 2 Sensor 2) Circuit No Activity Detected Conditions:**<br>Engine running, battery voltage 11.5, all electrical components off, ground between engine and chassis well connected and the exhaust system must be properly sealed between catalytic converter and the cylinder head. The DME detected the HO2S signal failed to meet the maximum or minimum voltage (i.e., it failed the voltage check).<br>**Possible Causes:**<br>&bull; Leaks present in the exhaust manifold or exhaust pipes<br>&bull; HO2S signal wire and ground wire crossed in connector<br>&bull; HO2S element is fuel contaminated or has failed |
| **DTC: P0161**<br>**2T MIL: Yes**<br>**Years:** 2007, 2008<br>**Models:** 328i, 328Ci, 328xi, 335i, 335Ci, 335xi, 525i, 525xi, 530i, 530xi, 550i, 650i, 750i, 750Li, 760Li, M5, M6, X3, X5, Z4 3.0i, Z4 3.0Si, Z4 M<br>**Engines:** All<br>**Transmissions:** All | **HO2S (Bank 2 Sensor 2) Heater Circuit Malfunction Conditions:**<br>Engine running, battery voltage 11.5, all electrical components off, ground between engine and chassis well connected and the exhaust system must be properly sealed between catalytic converter and the cylinder head. The DME detected an open or shorted condition, or excessive current draw in the heater circuit. The engine speed is greater than 40rpm, the battery voltage must be between 10.7 and 15.5 volts, and the fault occurs 200 seconds after engine start up.<br>**Possible Causes:**<br>&bull; HO2S heater power circuit or the heater ground circuit is open<br>&bull; HO2S signal tracking (due to oil or moisture in the connector)<br>&bull; HO2S has failed, or the DME has failed |

| DTC | Trouble Code Title, Conditions & Possible Causes |
|---|---|
| **DTC: P0171**<br>**2T MIL: Yes**<br>**Years:** 2007, 2008<br>**Models:** 328i, 328Ci, 328xi, 335i, 335Ci, 335xi, 525i, 525xi, 530i, 530xi, 550i, 650i, 750i, 750Li, 760Li, M5, M6, X3, X5, Z4 3.0i, Z4 3.0Si, Z4 M<br>**Engines:** All<br>**Transmissions:** All | **Fuel System Too Lean (Cylinder Bank 1) Conditions:**<br>Key on or engine running, all electrical components off and coolant temperature at least 80 degrees Celsius; and the DME detected the Bank 1 Adaptive Fuel Control System reached its rich correction limit (a lean A/F condition). The fuel status is in a closed loop pattern, the coolant temperature is between 69 and 100 degrees Celsius, and the engine speed is between 800 and 6000rpm.<br>**Possible Causes:**<br>• Air leaks after the MAF sensor, or leaks in the PCV system<br>• Exhaust leaks before or near where the HO2S is mounted<br>• Fuel injector(s) restricted or not supplying enough fuel<br>• Fuel pump not supplying enough fuel during high fuel demand conditions<br>• Leaking EGR gasket, or leaking EGR valve diaphragm<br>• MAF sensor dirty (causes DME to underestimate airflow)<br>• Vehicle running out of fuel or engine oil dip stick not seated |
| **DTC: P0172**<br>**2T MIL: Yes**<br>**Years:** 2007, 2008<br>**Models:** 328i, 328Ci, 328xi, 335i, 335Ci, 335xi, 525i, 525xi, 530i, 530xi, 550i, 650i, 750i, 750Li, 760Li, M5, M6, X3, X5, Z4 3.0i, Z4 3.0Si, Z4 M<br>**Engines:** All<br>**Transmissions:** All | **Fuel System Too Rich (Cylinder Bank 1) Conditions:**<br>Key on or engine running, all electrical components off and coolant temperature at least 80 degrees Celsius; and the DME detected the Bank 1 Adaptive Fuel Control System reached its rich correction limit (a rich A/F condition). The fuel status is in a closed loop pattern, the coolant temperature is between 69 and 100 degrees Celsius, and the engine speed is between 800 and 6000rpm.<br>**Possible Causes:**<br>• Camshaft timing is incorrect, or the engine has an oil overfill condition<br>• EVAP vapor recovery system failure (may be pulling vacuum)<br>• Fuel pressure regulator is damaged or leaking<br>• HO2S element is contaminated with alcohol or water<br>• MAF or MAP sensor values are incorrect or out-of-range<br>• One of more fuel injectors is leaking |
| **DTC: P0174**<br>**2T MIL: Yes**<br>**Years:** 2007, 2008<br>**Models:** 328i, 328Ci, 328xi, 335i, 335Ci, 335xi, 525i, 525xi, 530i, 530xi, 550i, 650i, 750i, 750Li, 760Li, M5, M6, X3, X5, Z4 3.0i, Z4 3.0Si, Z4 M<br>**Engines:** All<br>**Transmissions:** All | **Fuel System Too Lean (Cylinder Bank 2) Conditions:**<br>Key on or engine running, all electrical components off and coolant temperature at least 80 degrees Celsius; and the DME detected the Bank 2 Fuel Control System reached its lean correction limit. The fuel status is in a closed loop pattern, the coolant temperature is between 69 and 100 degrees Celsius, and the engine speed is between 800 and 6000rpm.<br>**Possible Causes:**<br>• Air leaks after the MAF sensor, or leaks in the PCV system<br>• Exhaust leaks before or near where the HO2S is mounted<br>• Fuel injector(s) restricted or not supplying enough fuel<br>• Fuel pump not supplying enough fuel during high fuel demand conditions<br>• Leaking EGR gasket, or leaking EGR valve diaphragm<br>• MAF sensor dirty (causes DME to underestimate airflow)<br>• Vehicle running out of fuel or engine oil dip stick not seated |
| **DTC: P0175**<br>**2T MIL: Yes**<br>**Years:** 2007, 2008<br>**Models:** 328i, 328Ci, 328xi, 335i, 335Ci, 335xi, 525i, 525xi, 530i, 530xi, 550i, 650i, M5, M6, Z4 3.0i, Z4 3.0Si, Z4 M<br>**Engines:** All<br>**Transmissions:** All | **Fuel System Too Rich (Cylinder Bank 2) Conditions:**<br>Key on or engine running, all electrical components off and coolant temperature at least 80 degrees Celsius; and the DME detected the Bank 2 Adaptive Fuel Control System reached its rich correction limit (a rich A/F condition). The fuel status is in a closed loop pattern, the coolant temperature is between 69 and 100 degrees Celsius, and the engine speed is between 800 and 6000rpm.<br>**Possible Causes:**<br>• Air leaks after the MAF sensor, or leaks in the PCV system<br>• Exhaust leaks before or near where the HO2S is mounted<br>• Fuel injector(s) restricted or not supplying enough fuel<br>• Fuel pump not supplying enough fuel during high fuel demand conditions<br>• Leaking EGR gasket, or leaking EGR valve diaphragm<br>• MAF sensor dirty (causes DME to underestimate airflow)<br>• Vehicle running out of fuel or engine oil dip stick not seated |
| **DTC: P0201**<br>**2T MIL: Yes**<br>**Years:** 2007, 2008<br>**Models:** 328i, 328Ci, 328xi, 335i, 335Ci, 335xi, 525i, 525xi, 530i, 530xi, 550i, 650i, 750i, 750Li, 760Li, M5, M6, X3, X5, Z4 3.0i, Z4 3.0Si, Z4 M<br>**Engines:** All<br>**Transmissions:** All | **Cylinder 1 Injector Circuit Malfunction Conditions:**<br>Engine started, and the DME detected the fuel injector "1" control circuit was in a high state when it should have been low, or in a low state when it should have been high (wiring harness & injector okay). The battery voltage should be between 9.5 and 17 volts while the engine speed is less than 40rpm.<br>**Possible Causes:**<br>• Injector 1 connector is damaged, open or shorted<br>• Injector 1 control circuit is open, shorted to ground or to power (the injector driver circuit may be damaged) |

| DTC | Trouble Code Title, Conditions & Possible Causes |
|---|---|
| **DTC: P0202**<br>**2T MIL: Yes**<br>**Years:** 2007, 2008<br>**Models:** 328i, 328Ci, 328xi, 335i, 335Ci, 335xi, 525i, 525xi, 530i, 530xi, 550i, 650i, 750i, 750Li, 760Li, M5, M6, X3, X5, Z4 3.0i, Z4 3.0Si, Z4 M<br>**Engines:** All<br>**Transmissions:** All | **Cylinder 2 Injector Circuit Malfunction Conditions:**<br>Engine started, and the DME detected the fuel injector "2" control circuit was in a high state when it should have been low, or in a low state when it should have been high (wiring harness & injector okay). The battery voltage should be between 9.5 and 17 volts while the engine speed is less than 40rpm.<br>**Possible Causes:**<br>• Injector 2 connector is damaged, open or shorted<br>• Injector 2 control circuit is open, shorted to ground or to power (the injector driver circuit may be damaged) |
| **DTC: P0203**<br>**2T MIL: Yes**<br>**Years:** 2007, 2008<br>**Models:** 328i, 328Ci, 328xi, 335i, 335Ci, 335xi, 525i, 525xi, 530i, 530xi, 550i, 650i, 750i, 750Li, 760Li, M5, M6, X3, X5, Z4 3.0i, Z4 3.0Si, Z4 M<br>**Engines:** All<br>**Transmissions:** All | **Cylinder 3 Injector Circuit Malfunction Conditions:**<br>Engine started, and the DME detected the fuel injector "3" control circuit was in a high state when it should have been low, or in a low state when it should have been high (wiring harness & injector okay). The battery voltage should be between 9.5 and 17 volts while the engine speed is less than 40rpm.<br>**Possible Causes:**<br>• Injector 3 connector is damaged, open or shorted<br>• Injector 3 control circuit is open, shorted to ground or to power (the injector driver circuit may be damaged) |
| **DTC: P0204**<br>**2T MIL: Yes**<br>**Years:** 2007, 2008<br>**Models:** 328i, 328Ci, 328xi, 335i, 335Ci, 335xi, 525i, 525xi, 530i, 530xi, 550i, 650i, 750i, 750Li, 760Li, M5, M6, X3, X5, Z4 3.0i, Z4 3.0Si, Z4 M<br>**Engines:** All<br>**Transmissions:** All | **Cylinder 4 Injector Circuit Malfunction Conditions:**<br>Engine started, and the DME detected the fuel injector "4" control circuit was in a high state when it should have been low, or in a low state when it should have been high (wiring harness & injector okay). The battery voltage should be between 9.5 and 17 volts while the engine speed is less than 40rpm.<br>**Possible Causes:**<br>• Injector 4 connector is damaged, open or shorted<br>• Injector 4 control circuit is open, shorted to ground or to power (the injector driver circuit may be damaged) |
| **DTC: P0205**<br>**2T MIL: Yes**<br>**Years:** 2007, 2008<br>**Models:** 328i, 328Ci, 328xi, 335i, 335Ci, 335xi, 525i, 525xi, 530i, 530xi, 550i, 650i, 750i, 750Li, 760Li, M5, M6, X3, X5, Z4 3.0i, Z4 3.0Si, Z4 M<br>**Engines:** All<br>**Transmissions:** All | **Cylinder 5 Injector Circuit Malfunction Conditions:**<br>Engine started, and the DME detected the fuel injector "5" control circuit was in a high state when it should have been low, or in a low state when it should have been high (wiring harness & injector okay). The battery voltage should be between 9.5 and 17 volts while the engine speed is less than 40rpm.<br>**Possible Causes:**<br>• Injector 5 connector is damaged, open or shorted<br>• Injector 5 control circuit is open, shorted to ground or to power (the injector driver circuit may be damaged) |
| **DTC: P0206**<br>**2T MIL: Yes**<br>**Years:** 2007, 2008<br>**Models:** 328i, 328Ci, 328xi, 335i, 335Ci, 335xi, 525i, 525xi, 530i, 530xi, 550i, 650i, 750i, 750Li, 760Li, M5, M6, X3, X5, Z4 3.0i, Z4 3.0Si, Z4 M<br>**Engines:** All<br>**Transmissions:** All | **Cylinder 6 Injector Circuit Malfunction Conditions:**<br>Engine started, and the DME detected the fuel injector "6" control circuit was in a high state when it should have been low, or in a low state when it should have been high (wiring harness & injector okay). The battery voltage should be between 9.5 and 17 volts while the engine speed is less than 40rpm.<br>**Possible Causes:**<br>• Injector 6 connector is damaged, open or shorted<br>• Injector 6 control circuit is open, shorted to ground or to power (the injector driver circuit may be damaged) |
| **DTC: P0207**<br>**2T MIL: Yes**<br>**Years:** 2007, 2008<br>**Models:** 328i, 328Ci, 328xi, 335i, 335Ci, 335xi, 525i, 525xi, 530i, 530xi, 550i, 650i, M5, M6, Z4 3.0i, Z4 3.0Si, Z4 M<br>**Engines:** All<br>**Transmissions:** All | **Cylinder 7 Injector Circuit Malfunction Conditions:**<br>Engine started, and the DME detected the fuel injector "7" control circuit was in a high state when it should have been low, or in a low state when it should have been high (wiring harness & injector okay). The battery voltage should be between 9.5 and 17 volts while the engine speed is less than 40rpm.<br>**Note: Monitor the INJIF PID Fault "flags" with the Scan Tool. The appropriate INJF PID "flag" will read Yes when this code is set.**<br>**Possible Causes:**<br>• Injector 7 connector is damaged, open or shorted<br>• Injector 7 control circuit is open, shorted to ground or to power (the injector driver circuit may be damaged) |

| DTC | Trouble Code Title, Conditions & Possible Causes |
|---|---|
| **DTC: P0208**<br>**2T MIL: Yes**<br>**Years:** 2007, 2008<br>**Models:** 328i, 328Ci, 328xi, 335i, 335Ci, 335xi, 525i, 525xi, 530i, 530xi, 550i, 650i, 750i, 750Li, 760Li, M5, M6, X3, X5, Z4 3.0i, Z4 3.0Si, Z4 M<br>**Engines:** All<br>**Transmissions:** All | **Cylinder 8 Injector Circuit Malfunction Conditions:**<br>Engine started, and the DME detected the fuel injector "8" control circuit was in a high state when it should have been low, or in a low state when it should have been high (wiring harness & injector okay). The battery voltage should be between 9.5 and 17 volts while the engine speed is less than 40rpm.<br>**Note: Monitor the INJIF PID Fault "flags" with the Scan Tool. The appropriate INJF PID "flag" will read Yes when this code is set.**<br>**Possible Causes:**<br>• Injector 8 connector is damaged, open or shorted<br>• Injector 8 control circuit is open, shorted to ground or to power (the injector driver circuit may be damaged) |
| **DTC: P0209**<br>**2T MIL: Yes**<br>**Years:** 2007, 2008<br>**Models:** 750i, 750Li, 760Li<br>**Engines:** All<br>**Transmissions:** All | **Cylinder 9 Injector Circuit Malfunction Conditions:**<br>Engine started, and the DME detected the fuel injector "9" control circuit was in a high state when it should have been low, or in a low state when it should have been high (wiring harness & injector okay).<br>**Note: Monitor the INJIF PID Fault "flags" with the Scan Tool. The appropriate INJF PID "flag" will read Yes when this code is set.**<br>**Possible Causes:**<br>• Injector 8 connector is damaged, open or shorted<br>• Injector 8 control circuit is open, shorted to ground or to power (the injector driver circuit may be damaged) |
| **DTC: P0210**<br>**2T MIL: Yes**<br>**Years:** 2007, 2008<br>**Models:** 750i, 750Li, 760Li<br>**Engines:** All<br>**Transmissions:** All | **Cylinder 10 Injector Circuit Malfunction Conditions:**<br>Engine started, and the DME detected the fuel injector "10" control circuit was in a high state when it should have been low, or in a low state when it should have been high (wiring harness & injector okay).<br>**Note: Monitor the INJIF PID Fault "flags" with the Scan Tool. The appropriate INJF PID "flag" will read Yes when this code is set.**<br>**Possible Causes:**<br>• Injector 8 connector is damaged, open or shorted<br>• Injector 8 control circuit is open, shorted to ground or to power (the injector driver circuit may be damaged) |
| **DTC: P0211**<br>**2T MIL: Yes**<br>**Years:** 2007, 2008<br>**Models:** 750i, 750Li, 760Li<br>**Engines:** All<br>**Transmissions:** All | **Cylinder 11 Injector Circuit Malfunction Conditions:**<br>Engine started, and the DME detected the fuel injector "11" control circuit was in a high state when it should have been low, or in a low state when it should have been high (wiring harness & injector okay).<br>**Note: Monitor the INJIF PID Fault "flags" with the Scan Tool. The appropriate INJF PID "flag" will read Yes when this code is set.**<br>**Possible Causes:**<br>• Injector 8 connector is damaged, open or shorted<br>• Injector 8 control circuit is open, shorted to ground or to power (the injector driver circuit may be damaged) |
| **DTC: P0212**<br>**2T MIL: Yes**<br>**Years:** 2007, 2008<br>**Models:** 750i, 750Li, 760Li<br>**Engines:** All<br>**Transmissions:** All | **Cylinder 12 Injector Circuit Malfunction Conditions:**<br>Engine started, and the DME detected the fuel injector "12" control circuit was in a high state when it should have been low, or in a low state when it should have been high (wiring harness & injector okay).<br>**Note: Monitor the INJIF PID Fault "flags" with the Scan Tool. The appropriate INJF PID "flag" will read Yes when this code is set.**<br>**Possible Causes:**<br>• Injector 8 connector is damaged, open or shorted<br>• Injector 8 control circuit is open, shorted to ground or to power (the injector driver circuit may be damaged) |
| **DTC: P0221**<br>**2T MIL: Yes**<br>**Years:** 2007, 2008<br>**Models:** 328i, 328Ci, 328xi, 335i, 335Ci, 335xi, 525i, 525xi, 530i, 530xi, 550i, 650i, 750i, 750Li, 760Li, M5, M6, X3, X5, Z4 3.0i, Z4 3.0Si, Z4 M<br>**Engines:** All<br>**Transmissions:** All | **Throttle Position Sensor 'B' Signal Performance Conditions:**<br>Engine started, battery voltage at least 11.5v, all electrical components off, ground connections between engine and chassis well connected, coolant temperature at least 80-degrees Celsius and the throttle valve must not be damaged or dirty; and the DME detected the TP Sensor 'B' circuit was out of its normal operating range during a condition with the throttle wide open, or with it completely closed. The throttle valve activation occurs via an electric motor (throttle drive) in the throttle valve control module. It is activated by the DME according to specifications of the two sensors, Throttle Position Sensor and Accelerator Pedal Position Sensor 2. Slowly depress accelerator pedal up to Wide Open Throttle (WOT) stop while observing the percentage display on the PID data function of the scan tool. The percentage display must increase uniformly. The engine speed is greater than 1320rpm.<br>**Possible Causes:**<br>• Throttle body is damaged<br>• Throttle linkage is binding or sticking<br>• ETC TP Sensor 'B' signal circuit to the DME is open<br>• ETC TP Sensor 'B' ground circuit is open<br>• ETC TP Sensor 'B' is damaged or it has failed |

| DTC | Trouble Code Title, Conditions & Possible Causes |
|---|---|
| **DTC: P0222**<br>**2T MIL: Yes**<br>**Years:** 2007, 2008<br>**Models:** 328i, 328Ci, 328xi, 335i, 335Ci, 335xi, 525i, 525xi, 530i, 530xi, 550i, 650i, 750i, 750Li, 760Li, M5, M6, X3, X5, Z4 3.0i, Z4 3.0Si, Z4 M<br>**Engines:** All<br>**Transmissions:** All | **Throttle Position Sensor 'B' Circuit Low Input Conditions:**<br>Engine started, battery voltage at least 11.5v, all electrical components off, ground connections between engine and chassis well connected, coolant temperature at least 80-degrees Celsius and the throttle valve must not be damaged or dirty; and the DME detected the TP Sensor 'B' circuit was out of its normal operating range during a condition with the throttle wide open, or with it completely closed. The throttle valve activation occurs via an electric motor (throttle drive) in the throttle valve control module. It is activated by the DME according to specifications of the two sensors, Throttle Position Sensor and Accelerator Pedal Position Sensor 2. Slowly depress accelerator pedal up to Wide Open Throttle (WOT) stop while observing the percentage display on the PID data function of the scan tool. The percentage display must increase uniformly.<br>**Possible Causes:**<br>• ETC TP Sensor 'B' connector is damaged or shorted<br>• ETC TP Sensor 'B' signal circuit is shorted to ground<br>• ETC TP Sensor 'B' is damaged or it has failed |
| **DTC: P0223**<br>**2T MIL: Yes**<br>**Years:** 2007, 2008<br>**Models:** 328i, 328Ci, 328xi, 335i, 335Ci, 335xi, 525i, 525xi, 530i, 530xi, 550i, 650i, 750i, 750Li, 760Li, M5, M6, X3, X5, Z4 3.0i, Z4 3.0Si, Z4 M<br>**Engines:** All<br>**Transmissions:** All | **Throttle Position Sensor 'B' Circuit High Input Conditions:**<br>Engine started, battery voltage at least 11.5v, all electrical components off, ground connections between engine and chassis well connected, coolant temperature at least 80-degrees Celsius and the throttle valve must not be damaged or dirty; and the DME detected the TP Sensor 'B' circuit was out of its normal operating range during a condition with the throttle wide open, or with it completely closed. The throttle valve activation occurs via an electric motor (throttle drive) in the throttle valve control module. It is activated by the DME according to specifications of the two sensors, Throttle Position Sensor and Accelerator Pedal Position Sensor 2. Slowly depress accelerator pedal up to Wide Open Throttle (WOT) stop while observing the percentage display on the PID data function of the scan tool. The percentage display must increase uniformly.<br>**Possible Causes:**<br>• ETC TP Sensor 'B' connector is damaged or open<br>• ETC TP Sensor 'B' signal circuit is open<br>• ETC TP Sensor 'B' signal circuit is shorted to VREF (5v)<br>• ETC TP Sensor 'B' is damaged or it has failed |
| **DTC: P0261**<br>**2T MIL: Yes**<br>**Years:** 2007, 2008<br>**Models:** 328i, 328Ci, 328xi, 335i, 335Ci, 335xi, 525i, 525xi, 530i, 530xi, 550i, 650i, 750i, 750Li, 760Li, M5, M6, X3, X5, Z4 3.0i, Z4 3.0Si, Z4 M<br>**Engines:** All<br>**Transmissions:** All | **Cylinder 1 Injector Circuit Low Input/Short to Ground Conditions:**<br>Key on or engine running, fuses in the instrument panel and the E-box in the engine compartment must be functioning, and the ground connections between the engine ad the chassis must be well connected; and the DME detected an unexpected voltage condition on the injector circuit.<br>**Possible Causes:**<br>• Injector 1 control circuit is open<br>• Injector 1 power circuit (B+) is open<br>• Injector 1 control circuit is shorted to chassis ground<br>• Injector 1 is damaged or has failed<br>• DME is not connected or has failed |
| **DTC: P0262**<br>**2T MIL: Yes**<br>**Years:** 2007, 2008<br>**Models:** 328i, 328Ci, 328xi, 335i, 335Ci, 335xi, 525i, 525xi, 530i, 530xi, 550i, 650i, 750i, 750Li, 760Li, M5, M6, X3, X5, Z4 3.0i, Z4 3.0Si, Z4 M<br>**Engines:** All<br>**Transmissions:** All | **Cylinder 1 Injector Circuit Low Input/Short to B+ Conditions:**<br>Key on or engine running, fuses in the instrument panel and the E-box in the engine compartment must be functioning, and the ground connections between the engine ad the chassis must be well connected; and the DME detected an unexpected voltage condition on the injector circuit.<br>**Possible Causes:**<br>• Injector control circuit is open<br>• Injector power circuit (B+) is open<br>• Injector control circuit is shorted to chassis ground<br>• Injector is damaged or has failed<br>• DME is not connected or has failed<br>• Fuel pump relay has failed<br>• Fuel injectors may have malfunctioned<br>• Faulty engine speed sensor |
| **DTC: P0264**<br>**2T MIL: Yes**<br>**Years:** 2007, 2008<br>**Models:** 328i, 328Ci, 328xi, 335i, 335Ci, 335xi, 525i, 525xi, 530i, 530xi, 550i, 650i, 750i, 750Li, 760Li, M5, M6, X3, X5, Z4 3.0i, Z4 3.0Si, Z4 M<br>**Engines:** All<br>**Transmissions:** All | **Cylinder 2 Injector Circuit Low Input/Short to Ground Conditions:**<br>Key on or engine running, fuses in the instrument panel and the E-box in the engine compartment must be functioning, and the ground connections between the engine ad the chassis must be well connected; and the DME detected an unexpected voltage condition on the injector circuit.<br>**Possible Causes:**<br>• Injector control circuit is open<br>• Injector power circuit (B+) is open<br>• Injector control circuit is shorted to chassis ground<br>• Injector is damaged or has failed<br>• DME is not connected or has failed<br>• Fuel pump relay has failed<br>• Fuel injectors may have malfunctioned<br>• Faulty engine speed sensor |

| DTC | Trouble Code Title, Conditions & Possible Causes |
|---|---|
| **DTC: P0265**<br>**2T MIL: Yes**<br>**Years:** 2007, 2008<br>**Models:** 328i, 328Ci, 328xi, 335i, 335Ci, 335xi, 525i, 525xi, 530i, 530xi, 550i, 650i, 750i, 750Li, 760Li, M5, M6, X3, X5, Z4 3.0i, Z4 3.0Si, Z4 M<br>**Engines:** All<br>**Transmissions:** All | **Cylinder 2 Injector Circuit Low Input/Short to B+ Conditions:**<br>Key on or engine running, fuses in the instrument panel and the E-box in the engine compartment must be functioning, and the ground connections between the engine ad the chassis must be well connected; and the DME detected an unexpected voltage condition on the injector circuit.<br>**Possible Causes:**<br>• Injector control circuit is open<br>• Injector power circuit (B+) is open<br>• Injector control circuit is shorted to chassis ground<br>• Injector is damaged or has failed<br>• DME is not connected or has failed<br>• Fuel pump relay has failed<br>• Fuel injectors may have malfunctioned<br>• Faulty engine speed sensor |
| **DTC: P0267**<br>**2T MIL: Yes**<br>**Years:** 2007, 2008<br>**Models:** 328i, 328Ci, 328xi, 335i, 335Ci, 335xi, 525i, 525xi, 530i, 530xi, 550i, 650i, 750i, 750Li, 760Li, M5, M6, X3, X5, Z4 3.0i, Z4 3.0Si, Z4 M<br>**Engines:** All<br>**Transmissions:** All | **Cylinder 3 Injector Circuit Low Input/Short to Ground Conditions:**<br>Key on or engine running, fuses in the instrument panel and the E-box in the engine compartment must be functioning, and the ground connections between the engine ad the chassis must be well connected; and the DME detected an unexpected voltage condition on the injector circuit.<br>**Possible Causes:**<br>• Injector control circuit is open<br>• Injector power circuit (B+) is open<br>• Injector control circuit is shorted to chassis ground<br>• Injector is damaged or has failed<br>• DME is not connected or has failed<br>• Fuel pump relay has failed<br>• Fuel injectors may have malfunctioned<br>• Faulty engine speed sensor |
| **DTC: P0268**<br>**2T MIL: Yes**<br>**Years:** 2007, 2008<br>**Models:** 328i, 328Ci, 328xi, 335i, 335Ci, 335xi, 525i, 525xi, 530i, 530xi, 550i, 650i, 750i, 750Li, 760Li, M5, M6, X3, X5, Z4 3.0i, Z4 3.0Si, Z4 M<br>**Engines:** All<br>**Transmissions:** All | **Cylinder 3 Injector Circuit Low Input/Short to B+ Conditions:**<br>Key on or engine running, fuses in the instrument panel and the E-box in the engine compartment must be functioning, and the ground connections between the engine ad the chassis must be well connected; and the DME detected an unexpected voltage condition on the injector circuit.<br>**Possible Causes:**<br>• Injector control circuit is open<br>• Injector power circuit (B+) is open<br>• Injector control circuit is shorted to chassis ground<br>• Injector is damaged or has failed<br>• DME is not connected or has failed<br>• Fuel pump relay has failed<br>• Fuel injectors may have malfunctioned<br>• Faulty engine speed sensor |
| **DTC: P0270**<br>**2T MIL: Yes**<br>**Years:** 2007, 2008<br>**Models:** 328i, 328Ci, 328xi, 335i, 335Ci, 335xi, 525i, 525xi, 530i, 530xi, 550i, 650i, 750i, 750Li, 760Li, M5, M6, X3, X5, Z4 3.0i, Z4 3.0Si, Z4 M<br>**Engines:** All<br>**Transmissions:** All | **Cylinder 4 Injector Circuit Low Input/Short to Ground Conditions:**<br>Key on or engine running, fuses in the instrument panel and the E-box in the engine compartment must be functioning, and the ground connections between the engine ad the chassis must be well connected; and the DME detected an unexpected voltage condition on the injector circuit.<br>**Possible Causes:**<br>• Injector control circuit is open<br>• Injector power circuit (B+) is open<br>• Injector control circuit is shorted to chassis ground<br>• Injector is damaged or has failed<br>• DME is not connected or has failed<br>• Fuel pump relay has failed<br>• Fuel injectors may have malfunctioned<br>• Faulty engine speed sensor |

| DTC | Trouble Code Title, Conditions & Possible Causes |
|---|---|
| **DTC: P0271**<br>**2T MIL: Yes**<br>**Years:** 2007, 2008<br>**Models:** 328i, 328Ci, 328xi, 335i, 335Ci, 335xi, 525i, 525xi, 530i, 530xi, 550i, 650i, 750i, 750Li, 760Li, M5, M6, X3, X5, Z4 3.0i, Z4 3.0Si, Z4 M<br>**Engines:** All<br>**Transmissions:** All | **Cylinder 4 Injector Circuit Low Input/Short to B+ Conditions:**<br>Key on or engine running, fuses in the instrument panel and the E-box in the engine compartment must be functioning, and the ground connections between the engine ad the chassis must be well connected; and the DME detected an unexpected voltage condition on the injector circuit.<br>**Possible Causes:**<br>• Injector control circuit is open<br>• Injector power circuit (B+) is open<br>• Injector control circuit is shorted to chassis ground<br>• Injector is damaged or has failed<br>• DME is not connected or has failed<br>• Fuel pump relay has failed<br>• Fuel injectors may have malfunctioned<br>• Faulty engine speed sensor |
| **DTC: P0273**<br>**2T MIL: Yes**<br>**Years:** 2007, 2008<br>**Models:** 328i, 328Ci, 328xi, 335i, 335Ci, 335xi, 525i, 525xi, 530i, 530xi, 550i, 650i, 750i, 750Li, 760Li, M5, M6, X3, X5, Z4 3.0i, Z4 3.0Si, Z4 M<br>**Engines:** All<br>**Transmissions:** All | **Cylinder 5 Injector Circuit Low Input/Short to Ground Conditions:**<br>Key on or engine running, fuses in the instrument panel and the E-box in the engine compartment must be functioning, and the ground connections between the engine ad the chassis must be well connected; and the DME detected an unexpected voltage condition on the injector circuit.<br>**Possible Causes:**<br>• Injector control circuit is open<br>• Injector power circuit (B+) is open<br>• Injector control circuit is shorted to chassis ground<br>• Injector is damaged or has failed<br>• DME is not connected or has failed<br>• Fuel pump relay has failed<br>• Fuel injectors may have malfunctioned<br>• Faulty engine speed sensor |
| **DTC: P0274**<br>**2T MIL: Yes**<br>**Years:** 2007, 2008<br>**Models:** 328i, 328Ci, 328xi, 335i, 335Ci, 335xi, 525i, 525xi, 530i, 530xi, 550i, 650i, 750i, 750Li, 760Li, M5, M6, X3, X5, Z4 3.0i, Z4 3.0Si, Z4 M<br>**Engines:** All<br>**Transmissions:** All | **Cylinder 5 Injector Circuit Low Input/Short to B+ Conditions:**<br>Key on or engine running, fuses in the instrument panel and the E-box in the engine compartment must be functioning, and the ground connections between the engine ad the chassis must be well connected; and the DME detected an unexpected voltage condition on the injector circuit.<br>**Possible Causes:**<br>• Injector control circuit is open<br>• Injector power circuit (B+) is open<br>• Injector control circuit is shorted to chassis ground<br>• Injector is damaged or has failed<br>• DME is not connected or has failed<br>• Fuel pump relay has failed<br>• Fuel injectors may have malfunctioned<br>• Faulty engine speed sensor |
| **DTC: P0276**<br>**2T MIL: Yes**<br>**Years:** 2007, 2008<br>**Models:** 328i, 328Ci, 328xi, 335i, 335Ci, 335xi, 525i, 525xi, 530i, 530xi, 550i, 650i, 750i, 750Li, 760Li, M5, M6, X3, X5, Z4 3.0i, Z4 3.0Si, Z4 M<br>**Engines:** All<br>**Transmissions:** All | **Cylinder 6 Injector Circuit Low Input/Short to Ground Conditions:**<br>Key on or engine running, fuses in the instrument panel and the E-box in the engine compartment must be functioning, and the ground connections between the engine ad the chassis must be well connected; and the DME detected an unexpected voltage condition on the injector circuit.<br>**Possible Causes:**<br>• Injector control circuit is open<br>• Injector power circuit (B+) is open<br>• Injector control circuit is shorted to chassis ground<br>• Injector is damaged or has failed<br>• DME is not connected or has failed<br>• Fuel pump relay has failed<br>• Fuel injectors may have malfunctioned<br>• Faulty engine speed sensor |

| DTC | Trouble Code Title, Conditions & Possible Causes |
|---|---|
| **DTC: P0277**<br>**2T MIL: Yes**<br>**Years:** 2007, 2008<br>**Models:** 328i, 328Ci, 328xi, 335i, 335Ci, 335xi, 525i, 525xi, 530i, 530xi, 550i, 650i, 750i, 750Li, 760Li, M5, M6, X3, X5, Z4 3.0i, Z4 3.0Si, Z4 M<br>**Engines:** All<br>**Transmissions:** All | **Cylinder 6 Injector Circuit Low Input/Short to B+ Conditions:**<br>Key on or engine running, fuses in the instrument panel and the E-box in the engine compartment must be functioning, and the ground connections between the engine ad the chassis must be well connected; and the DME detected an unexpected voltage condition on the injector circuit.<br>**Possible Causes:**<br>• Injector control circuit is open<br>• Injector power circuit (B+) is open<br>• Injector control circuit is shorted to chassis ground<br>• Injector is damaged or has failed<br>• DME is not connected or has failed<br>• Fuel pump relay has failed<br>• Fuel injectors may have malfunctioned<br>• Faulty engine speed sensor |
| **DTC: P0279**<br>**2T MIL: Yes**<br>**Years:** 2007, 2008<br>**Models:** 335i, 335Ci, 335xi, 525i, 525xi, 530i, 530xi, 750i, 750Li, 760Li, M5, M6, Z4 3.0i, Z4 3.0Si, Z4 M<br>**Engines:** All<br>**Transmissions:** All | **Cylinder 7 Injector Circuit Low Input/Short to Ground Conditions:**<br>Key on or engine running, fuses in the instrument panel and the E-box in the engine compartment must be functioning, and the ground connections between the engine ad the chassis must be well connected; and the DME detected an unexpected voltage condition on the injector circuit.<br>**Possible Causes:**<br>• Injector control circuit is open<br>• Injector power circuit (B+) is open<br>• Injector control circuit is shorted to chassis ground<br>• Injector is damaged or has failed<br>• DME is not connected or has failed<br>• Fuel pump relay has failed<br>• Fuel injectors may have malfunctioned<br>• Faulty engine speed sensor |
| **DTC: P0280**<br>**2T MIL: Yes**<br>**Years:** 2007, 2008<br>**Models:** 335i, 335Ci, 335xi, 525i, 525xi, 530i, 530xi, 750i, 750Li, 760Li, M5, M6, Z4 3.0i, Z4 3.0Si, Z4 M<br>**Engines:** All<br>**Transmissions:** All | **Cylinder 7 Injector Circuit Low Input/Short to B+ Conditions:**<br>Key on or engine running, fuses in the instrument panel and the E-box in the engine compartment must be functioning, and the ground connections between the engine ad the chassis must be well connected; and the DME detected an unexpected voltage condition on the injector circuit.<br>**Possible Causes:**<br>• Injector control circuit is open<br>• Injector power circuit (B+) is open<br>• Injector control circuit is shorted to chassis ground<br>• Injector is damaged or has failed<br>• DME is not connected or has failed<br>• Fuel pump relay has failed<br>• Fuel injectors may have malfunctioned<br>• Faulty engine speed sensor |
| **DTC: P0282**<br>**2T MIL: Yes**<br>**Years:** 2007, 2008<br>**Models:** 335i, 335Ci, 335xi, 525i, 525xi, 530i, 530xi, 750i, 750Li, 760Li, M5, M6, Z4 3.0i, Z4 3.0Si, Z4 M<br>**Engines:** All<br>**Transmissions:** All | **Cylinder 8 Injector Circuit Low Input/Short to Ground Conditions:**<br>Key on or engine running, fuses in the instrument panel and the E-box in the engine compartment must be functioning, and the ground connections between the engine ad the chassis must be well connected; and the DME detected an unexpected voltage condition on the injector circuit.<br>**Possible Causes:**<br>• Injector control circuit is open<br>• Injector power circuit (B+) is open<br>• Injector control circuit is shorted to chassis ground<br>• Injector is damaged or has failed<br>• DME is not connected or has failed<br>• Fuel pump relay has failed<br>• Fuel injectors may have malfunctioned<br>• Faulty engine speed sensor |

| DTC | Trouble Code Title, Conditions & Possible Causes |
|---|---|
| **DTC: P0283**<br>**2T MIL:** Yes<br>**Years:** 2007, 2008<br>**Models:** 335i, 335Ci, 335xi, 525i, 525xi, 530i, 530xi, 750i, 750Li, 760Li, M5, M6, Z4 3.0i, Z4 3.0Si, Z4 M<br>**Engines:** All<br>**Transmissions:** All | **Cylinder 8 Injector Circuit Low Input/Short to B+ Conditions:**<br>Key on or engine running, fuses in the instrument panel and the E-box in the engine compartment must be functioning, and the ground connections between the engine ad the chassis must be well connected; and the DME detected an unexpected voltage condition on the injector circuit.<br>**Possible Causes:**<br>• Injector control circuit is open<br>• Injector power circuit (B+) is open<br>• Injector control circuit is shorted to chassis ground<br>• Injector is damaged or has failed<br>• DME is not connected or has failed<br>• Fuel pump relay has failed<br>• Fuel injectors may have malfunctioned<br>• Faulty engine speed sensor |
| **DTC: P0285**<br>**2T MIL:** Yes<br>**Years:** 2007, 2008<br>**Models:** 750i, 750Li, 760Li<br>**Engines:** All<br>**Transmissions:** All | **Cylinder 9 Injector Circuit Low Input/Short to Ground Conditions:**<br>Key on or engine running, fuses in the instrument panel and the E-box in the engine compartment must be functioning, and the ground connections between the engine ad the chassis must be well connected; and the DME detected an unexpected voltage condition on the injector circuit.<br>**Possible Causes:**<br>• Injector control circuit is open<br>• Injector power circuit (B+) is open<br>• Injector control circuit is shorted to chassis ground<br>• Injector is damaged or has failed<br>• DME is not connected or has failed<br>• Fuel pump relay has failed<br>• Fuel injectors may have malfunctioned<br>• Faulty engine speed sensor |
| **DTC: P0286**<br>**2T MIL:** Yes<br>**Years:** 2007, 2008<br>**Models:** 750i, 750Li, 760Li<br>**Engines:** All<br>**Transmissions:** All | **Cylinder 9 Injector Circuit Low Input/Short to B+ Conditions:**<br>Key on or engine running, fuses in the instrument panel and the E-box in the engine compartment must be functioning, and the ground connections between the engine ad the chassis must be well connected; and the DME detected an unexpected voltage condition on the injector circuit.<br>**Possible Causes:**<br>• Injector control circuit is open<br>• Injector power circuit (B+) is open<br>• Injector control circuit is shorted to chassis ground<br>• Injector is damaged or has failed<br>• DME is not connected or has failed<br>• Fuel pump relay has failed<br>• Fuel injectors may have malfunctioned<br>• Faulty engine speed sensor |
| **DTC: P0288**<br>**2T MIL:** Yes<br>**Years:** 2007, 2008<br>**Models:** 750i, 750Li, 760Li<br>**Engines:** All<br>**Transmissions:** All | **Cylinder 10 Injector Circuit Low Input/Short to Ground Conditions:**<br>Key on or engine running, fuses in the instrument panel and the E-box in the engine compartment must be functioning, and the ground connections between the engine ad the chassis must be well connected; and the DME detected an unexpected voltage condition on the injector circuit.<br>**Possible Causes:**<br>• Injector control circuit is open<br>• Injector power circuit (B+) is open<br>• Injector control circuit is shorted to chassis ground<br>• Injector is damaged or has failed<br>• DME is not connected or has failed<br>• Fuel pump relay has failed<br>• Fuel injectors may have malfunctioned<br>• Faulty engine speed sensor |

| DTC | Trouble Code Title, Conditions & Possible Causes |
|---|---|
| **DTC: P0289**<br>**2T MIL: Yes**<br>**Years:** 2007, 2008<br>**Models:** 750i, 750Li, 760Li<br>**Engines:** All<br>**Transmissions:** All | **Cylinder 10 Injector Circuit Low Input/Short to B+ Conditions:**<br>Key on or engine running, fuses in the instrument panel and the E-box in the engine compartment must be functioning, and the ground connections between the engine ad the chassis must be well connected; and the DME detected an unexpected voltage condition on the injector circuit.<br>**Possible Causes:**<br>• Injector control circuit is open<br>• Injector power circuit (B+) is open<br>• Injector control circuit is shorted to chassis ground<br>• Injector is damaged or has failed<br>• DME is not connected or has failed<br>• Fuel pump relay has failed<br>• Fuel injectors may have malfunctioned<br>• Faulty engine speed sensor |
| **DTC: P0291**<br>**2T MIL: Yes**<br>**Years:** 2007, 2008<br>**Models:** 750i, 750Li, 760Li<br>**Engines:** All<br>**Transmissions:** All | **Cylinder 11 Injector Circuit Low Input/Short to Ground Conditions:**<br>Key on or engine running, fuses in the instrument panel and the E-box in the engine compartment must be functioning, and the ground connections between the engine ad the chassis must be well connected; and the DME detected an unexpected voltage condition on the injector circuit.<br>**Possible Causes:**<br>• Injector control circuit is open<br>• Injector power circuit (B+) is open<br>• Injector control circuit is shorted to chassis ground<br>• Injector is damaged or has failed<br>• DME is not connected or has failed<br>• Fuel pump relay has failed<br>• Fuel injectors may have malfunctioned<br>• Faulty engine speed sensor |
| **DTC: P0292**<br>**2T MIL: Yes**<br>**Years:** 2007, 2008<br>**Models:** 750i, 750Li, 760Li<br>**Engines:** All<br>**Transmissions:** All | **Cylinder 11 Injector Circuit Low Input/Short to B+ Conditions:**<br>Key on or engine running, fuses in the instrument panel and the E-box in the engine compartment must be functioning, and the ground connections between the engine ad the chassis must be well connected; and the DME detected an unexpected voltage condition on the injector circuit.<br>**Possible Causes:**<br>• Injector control circuit is open<br>• Injector power circuit (B+) is open<br>• Injector control circuit is shorted to chassis ground<br>• Injector is damaged or has failed<br>• DME is not connected or has failed<br>• Fuel pump relay has failed<br>• Fuel injectors may have malfunctioned<br>• Faulty engine speed sensor |
| **DTC: P0294**<br>**2T MIL: Yes**<br>**Years:** 2007, 2008<br>**Models:** 750i, 750Li, 760Li<br>**Engines:** All<br>**Transmissions:** All | **Cylinder 12 Injector Circuit Low Input/Short to Ground Conditions:**<br>Key on or engine running, fuses in the instrument panel and the E-box in the engine compartment must be functioning, and the ground connections between the engine ad the chassis must be well connected; and the DME detected an unexpected voltage condition on the injector circuit.<br>**Possible Causes:**<br>• Injector control circuit is open<br>• Injector power circuit (B+) is open<br>• Injector control circuit is shorted to chassis ground<br>• Injector is damaged or has failed<br>• DME is not connected or has failed<br>• Fuel pump relay has failed<br>• Fuel injectors may have malfunctioned<br>• Faulty engine speed sensor |

| DTC | Trouble Code Title, Conditions & Possible Causes |
|---|---|
| **DTC: P0295**<br>**2T MIL:** Yes<br>**Years:** 2007, 2008<br>**Models:** 750i, 750Li, 760Li<br>**Engines:** All<br>**Transmissions:** All | **Cylinder 12 Injector Circuit Low Input/Short to B+ Conditions:**<br>Key on or engine running, fuses in the instrument panel and the E-box in the engine compartment must be functioning, and the ground connections between the engine ad the chassis must be well connected; and the DME detected an unexpected voltage condition on the injector circuit.<br>**Possible Causes:**<br>• Injector control circuit is open<br>• Injector power circuit (B+) is open<br>• Injector control circuit is shorted to chassis ground<br>• Injector is damaged or has failed<br>• DME is not connected or has failed<br>• Fuel pump relay has failed<br>• Fuel injectors may have malfunctioned<br>• Faulty engine speed sensor |
| **DTC: P0298**<br>**Years:** 2007, 2008<br>**Models:** 335i, 335Ci, 335xi, 525i, 525xi, 530i, 530xi, 550i, 650i, 750i, 750Li, 760Li, Z4 3.0i, Z4 3.0Si, Z4 M<br>**Engines:** All<br>**Transmissions:** All | **Engine Oil Over Temperature Conditions:**<br>The oil temperature difference of greater than 100 degrees within one second. The ignition must be on. The DME detected an error in the Engine Oil Temperature sensor. This occurs during attempted start value calibration.<br>**Possible Causes:**<br>• Replace the oil temperature sensor<br>• Engine Oil temperature is too high<br>• Engine coolant temperature is too high<br>• Highest possible gear engaged in transmission<br>• Check coolant |
| **DTC: P0300**<br>**1T MISFIRE, MIL:** Yes<br>**Years:** 2007, 2008<br>**Models:** 328i, 328Ci, 328xi, 335i, 335Ci, 335xi, 525i, 525xi, 530i, 530xi, 550i, 650i, 750i, 750Li, 760Li, M5, M6, X3, X5, Z4 3.0i, Z4 3.0Si, Z4 M<br>**Engines:** All<br>**Transmissions:** All | **Random/Multiple Misfire Detected Conditions:**<br>Engine running at an RPM greater than 400 but less than 6400 the DME detected a misfire or uneven engine running in two or more cylinders within 200 crankshaft rotations. Engine speed is between 480 and 4500rpm, load change is 0.4ms at ignition with a speed change of 2800rpms and the ASC is not active.<br>**Note: If the misfire is severe, the MIL will flash on/off on the first trip!**<br>**Possible Causes:**<br>• Fuel metering fault that affects two or more cylinders<br>• Fuel pressure too low or too high, fuel supply contaminated<br>• EVAP system problem or the EVAP canister is fuel saturated<br>• EGR valve is stuck open or the PCV system has a vacuum leak<br>• Ignition system fault (coil, plugs) affecting two or more cylinders<br>• MAF sensor contamination (it can cause a very lean condition)<br>• Vehicle driven while very low on fuel (less than 1/8 of a tank) |
| **DTC: P0301**<br>**1T MISFIRE, MIL:** Yes<br>**Years:** 2007, 2008<br>**Models:** 328i, 328Ci, 328xi, 335i, 335Ci, 335xi, 525i, 525xi, 530i, 530xi, 550i, 650i, 750i, 750Li, 760Li, M5, M6, X3, X5, Z4 3.0i, Z4 3.0Si, Z4 M<br>**Engines:** All<br>**Transmissions:** All | **Cylinder Number 1 Misfire Detected Conditions:**<br>Engine running at an RPM greater than 400 but less than 6400 the DME detected a misfire or uneven engine running in two or more cylinders within 200 crankshaft rotations. Engine speed is between 480 and 4500rpm, load change is 0.4ms at ignition with a speed change of 2800rpms and the ASC is not active.<br>**Note: If the misfire is severe, the MIL will flash on/off on the first trip!**<br>**Possible Causes:**<br>• Air leak in the intake manifold, or in the EGR or DME system<br>• Base engine mechanical problem<br>• Fuel delivery component problem (i.e., a contaminated, dirty or sticking fuel injector)<br>• Fuel pump relay defective<br>• Ignition coil fuses have failed<br>• Ignition system problem (dirty damaged coil or plug)<br>• Engine speed (RPM) sensor has failed<br>• Camshaft position sensors have failed<br>• Ignition coil is faulty<br>• Spark plugs are not working properly or are not gapped properly |

| DTC | Trouble Code Title, Conditions & Possible Causes |
|---|---|
| **DTC: P0302**<br>**1T MISFIRE, MIL: Yes**<br>**Years:** 2007, 2008<br>**Models:** 328i, 328Ci, 328xi, 335i, 335Ci, 335xi, 525i, 525xi, 530i, 530xi, 550i, 650i, 750i, 750Li, 760Li, M5, M6, X3, X5, Z4 3.0i, Z4 3.0Si, Z4 M<br>**Engines:** All<br>**Transmissions:** All | **Cylinder Number 2 Misfire Detected Conditions:**<br>Engine running at an RPM greater than 400 but less than 6400 the DME detected a misfire or uneven engine running in two or more cylinders within 200 crankshaft rotations. Engine speed is between 480 and 4500rpm, load change is 0.4ms at ignition with a speed change of 2800rpms and the ASC is not active.<br>**Note: If the misfire is severe, the MIL will flash on/off on the 1st trip!**<br>**Possible Causes:**<br>&bull; Air leak in the intake manifold, or in the EGR or DME system<br>&bull; Base engine mechanical problem<br>&bull; Fuel delivery component problem (i.e., a contaminated, dirty or sticking fuel injector)<br>&bull; Fuel pump relay defective<br>&bull; Ignition coil fuses have failed<br>&bull; Ignition system problem (dirty damaged coil or plug)<br>&bull; Engine speed (RPM) sensor has failed<br>&bull; Camshaft position sensors have failed<br>&bull; Ignition coil is faulty<br>&bull; Spark plugs are not working properly or are not gapped properly |
| **DTC: P0303**<br>**1T MISFIRE, MIL: Yes**<br>**Years:** 2007, 2008<br>**Models:** 328i, 328Ci, 328xi, 335i, 335Ci, 335xi, 525i, 525xi, 530i, 530xi, 550i, 650i, 750i, 750Li, 760Li, M5, M6, X3, X5, Z4 3.0i, Z4 3.0Si, Z4 M<br>**Engines:** All<br>**Transmissions:** All | **Cylinder Number 3 Misfire Detected Conditions:**<br>Engine running at an RPM greater than 400 but less than 6400 the DME detected a misfire or uneven engine running in two or more cylinders within 200 crankshaft rotations. Engine speed is between 480 and 4500rpm, load change is 0.4ms at ignition with a speed change of 2800rpms and the ASC is not active.<br>**Note: If the misfire is severe, the MIL will flash on/off on the 1st trip!**<br>**Possible Causes:**<br>&bull; Air leak in the intake manifold, or in the EGR or DME system<br>&bull; Base engine mechanical problem<br>&bull; Fuel delivery component problem (i.e., a contaminated, dirty or sticking fuel injector)<br>&bull; Fuel pump relay defective<br>&bull; Ignition coil fuses have failed<br>&bull; Ignition system problem (dirty damaged coil or plug)<br>&bull; Engine speed (RPM) sensor has failed<br>&bull; Camshaft position sensors have failed<br>&bull; Ignition coil is faulty<br>&bull; Spark plugs are not working properly or are not gapped properly |
| **DTC: P0304**<br>**1T MISFIRE, MIL: Yes**<br>**Years:** 2007, 2008<br>**Models:** 328i, 328Ci, 328xi, 335i, 335Ci, 335xi, 525i, 525xi, 530i, 530xi, 550i, 650i, 750i, 750Li, 760Li, M5, M6, X3, X5, Z4 3.0i, Z4 3.0Si, Z4 M<br>**Engines:** All<br>**Transmissions:** All | **Cylinder Number 4 Misfire Detected Conditions:**<br>Engine running at an RPM greater than 400 but less than 6400 the DME detected a misfire or uneven engine running in two or more cylinders within 200 crankshaft rotations. Engine speed is between 480 and 4500rpm, load change is 0.4ms at ignition with a speed change of 2800rpms and the ASC is not active.<br>**Note: If the misfire is severe, the MIL will flash on/off on the 1st trip!**<br>**Possible Causes:**<br>&bull; Air leak in the intake manifold, or in the EGR or DME system<br>&bull; Base engine mechanical problem<br>&bull; Fuel delivery component problem (i.e., a contaminated, dirty or sticking fuel injector)<br>&bull; Fuel pump relay defective<br>&bull; Ignition coil fuses have failed<br>&bull; Ignition system problem (dirty damaged coil or plug)<br>&bull; Engine speed (RPM) sensor has failed<br>&bull; Camshaft position sensors have failed<br>&bull; Ignition coil is faulty<br>&bull; Spark plugs are not working properly or are not gapped properly |

| DTC | Trouble Code Title, Conditions & Possible Causes |
|---|---|
| **DTC: P0305**<br>**1T MISFIRE, MIL: Yes**<br>**Years:** 2007, 2008<br>**Models:** 328i, 328Ci, 328xi, 335i, 335Ci, 335xi, 525i, 525xi, 530i, 530xi, 550i, 650i, 750i, 750Li, 760Li, M5, M6, X3, X5, Z4 3.0i, Z4 3.0Si, Z4 M<br>**Engines:** All<br>**Transmissions:** All | **Cylinder Number 5 Misfire Detected Conditions:**<br>Engine running under positive torque conditions, and the DME detected a misfire or uneven engine function. Engine speed is between 480 and 4500rpm, load change is 0.4ms at ignition with a speed change of 2800rpms and the ASC is not active.<br>**Note: If the misfire is severe, the MIL will flash on/off on the 1st trip!**<br>**Possible Causes:**<br>• Air leak in the intake manifold, or in the EGR or DME system<br>• Base engine mechanical problem<br>• Fuel delivery component problem (i.e., a contaminated, dirty or sticking fuel injector)<br>• Fuel pump relay defective<br>• Ignition coil fuses have failed<br>• Ignition system problem (dirty damaged coil or plug)<br>• Engine speed (RPM) sensor has failed<br>• Camshaft position sensors have failed<br>• Ignition coil is faulty<br>• Spark plugs are not working properly or are not gapped properly |
| **DTC: P0306**<br>**1T MISFIRE, MIL: Yes**<br>**Years:** 2007, 2008<br>**Models:** 328i, 328Ci, 328xi, 335i, 335Ci, 335xi, 525i, 525xi, 530i, 530xi, 550i, 650i, 750i, 750Li, 760Li, M5, M6, X3, X5, Z4 3.0i, Z4 3.0Si, Z4 M<br>**Engines:** All<br>**Transmissions:** All | **Cylinder Number 6 Misfire Detected Conditions:**<br>Engine running under positive torque conditions, and the DME detected a misfire or uneven engine function. Engine speed is between 480 and 4500rpm, load change is 0.4ms at ignition with a speed change of 2800rpms and the ASC is not active.<br>**Note: If the misfire is severe, the MIL will flash on/off on the 1st trip!**<br>**Possible Causes:**<br>• Air leak in the intake manifold, or in the EGR or DME system<br>• Base engine mechanical problem<br>• Fuel delivery component problem (i.e., a contaminated, dirty or sticking fuel injector)<br>• Fuel pump relay defective<br>• Ignition coil fuses have failed<br>• Ignition system problem (dirty damaged coil or plug)<br>• Engine speed (RPM) sensor has failed<br>• Camshaft position sensors have failed<br>• Ignition coil is faulty<br>• Spark plugs are not working properly or are not gapped properly |
| **DTC: P0307**<br>**1T MISFIRE, MIL: Yes**<br>**Years:** 2007, 2008<br>**Models:** 328i, 328Ci, 328xi, 335i, 335Ci, 335xi, 525i, 525xi, 530i, 530xi, 550i, 650i, 750i, 750Li, 760Li, M5, M6, X3, X5, Z4 3.0i, Z4 3.0Si, Z4 M<br>**Engines:** All<br>**Transmissions:** All | **Cylinder Number 7 Misfire Detected Conditions:**<br>Engine running under positive torque conditions, and the DME detected a misfire or uneven engine function. Engine speed is between 480 and 4500rpm, load change is 0.4ms at ignition with a speed change of 2800rpms and the ASC is not active.<br>**Note: If the misfire is severe, the MIL will flash on/off on the 1st trip!**<br>**Possible Causes:**<br>• Air leak in the intake manifold, or in the EGR or DME system<br>• Base engine mechanical problem<br>• Fuel delivery component problem (i.e., a contaminated, dirty or sticking fuel injector)<br>• Fuel pump relay defective<br>• Ignition coil fuses have failed<br>• Ignition system problem (dirty damaged coil or plug)<br>• Engine speed (RPM) sensor has failed<br>• Camshaft position sensors have failed<br>• Ignition coil is faulty<br>• Spark plugs are not working properly or are not gapped properly |
| **DTC: P0308**<br>**1T MISFIRE, MIL: Yes**<br>**Years:** 2007, 2008<br>**Models:** 328i, 328Ci, 328xi, 335i, 335Ci, 335xi, 525i, 525xi, 530i, 530xi, 550i, 650i, 750i, 750Li, 760Li, M5, M6, X3, X5, Z4 3.0i, Z4 3.0Si, Z4 M<br>**Engines:** All<br>**Transmissions:** All | **Cylinder Number 8 Misfire Detected Conditions:**<br>Engine running under positive torque conditions, and the DME detected a misfire or uneven engine function. Engine speed is between 480 and 4500rpm, load change is 0.4ms at ignition with a speed change of 2800rpms and the ASC is not active.<br>**Note: If the misfire is severe, the MIL will flash on/off on the 1st trip!**<br>**Possible Causes:**<br>• Air leak in the intake manifold, or in the EGR or DME system<br>• Base engine mechanical problem<br>• Fuel delivery component problem (i.e., a contaminated, dirty or sticking fuel injector)<br>• Fuel pump relay defective<br>• Ignition coil fuses have failed<br>• Ignition system problem (dirty damaged coil or plug)<br>• Engine speed (RPM) sensor has failed<br>• Camshaft position sensors have failed<br>• Ignition coil is faulty<br>• Spark plugs are not working properly or are not gapped properly |

| DTC | Trouble Code Title, Conditions & Possible Causes |
|---|---|
| **DTC: P0313**<br>**1T MISFIRE, MIL: Yes**<br>**Years:** 2007, 2008<br>**Models:** 352i, 530i, 530xi, 750i, 750Li, 760Li, Z4 3.0i, Z4 3.0Si, Z4 M<br>**Engines:** All<br>**Transmissions:** All | **Misfire Detected with Low Fuel Conditions:**<br>Engine running under positive torque conditions, and the DME detected a misfire or uneven engine function.<br>**Note: If the misfire is severe, the MIL will flash on/off on the 1st trip!**<br>**Possible Causes:**<br>• Air leak in the intake manifold, or in the EGR or DME system<br>• Base engine mechanical problem<br>• Fuel delivery component problem (i.e., a contaminated, dirty or sticking fuel injector)<br>• Fuel pump relay defective<br>• Ignition coil fuses have failed<br>• Ignition system problem (dirty damaged coil or plug)<br>• Engine speed (RPM) sensor has failed<br>• Camshaft position sensors have failed<br>• Ignition coil is faulty<br>• Spark plugs are not working properly or are not gapped properly |
| **DTC: P0325**<br>**2T MIL: Yes**<br>**Years:** 2007, 2008<br>**Models:** 328i, 328Ci, 328xi, 335i, 335Ci, 335xi, 525i, 525xi, 530i, 530xi, 550i, 650i, 750i, 750Li, 760Li, M5, M6, X3, X5, Z4 3.0i, Z4 3.0Si, Z4 M<br>**Engines:** All<br>**Transmissions:** All | **Knock Sensor 1 Circuit Malfunction Conditions:**<br>Engine started, vehicle driven at 1520rpm for 3 seconds or to a temperature of 40 degrees Celsius, and the DME detected the Knock Sensor 1 (KS1) signal was not recognized. The engine speed is greater than 2080rpm but less than 6000rpm and the coolant temperature is greater than 40.5 degrees Celsius.<br>**Possible Causes:**<br>• Knock sensor circuit is open<br>• Knock sensor is loose (tighten to 20 NM)<br>• Contact between the knock sensor and cylinder block is dirty, corroded or greasy<br>• Knock sensor circuit is shorted to ground, or shorted to power<br>• Knock sensor is damaged or it has failed<br>• Wrong kind of fuel used<br>• A component in the engine compartment is loose or not properly secured |
| **DTC: P0327**<br>**2T MIL: Yes**<br>**Years:** 2007, 2008<br>**Models:** 328i, 328Ci, 328xi, 335i, 335Ci, 335xi, 525i, 525xi, 530i, 530xi, 550i, 650i, 750i, 750Li, 760Li, M5, M6, X3, X5, Z4 3.0i, Z4 3.0Si, Z4 M<br>**Engines:** All<br>**Transmissions:** All | **Knock Sensor 1 Signal Low Input Conditions:**<br>Engine started, vehicle driven at 2000rpm for 3 seconds or to a temperature of 40 degrees Celsius, and the DME detected the Knock Sensor 1 (KS1) signal was too low or not recognized by the DME<br>**Possible Causes:**<br>• Knock sensor circuit is open<br>• Knock sensor is loose (tighten to 20 NM)<br>• Contact between the knock sensor and cylinder block is dirty, corroded or greasy<br>• Knock sensor circuit is shorted to ground, or shorted to power<br>• Knock sensor is damaged or it has failed<br>• Wrong kind of fuel used<br>• A component in the engine compartment is loose or not properly secured |
| **DTC: P0328**<br>**2T MIL: Yes**<br>**Years:** 2007, 2008<br>**Models:** 335i, 335Ci, 335xi, 525i, 525xi, 530i, 530xi, 550i, 650i, 750i, 750Li, 760Li, Z4 3.0i, Z4 3.0Si, Z4 M<br>**Engines:** All<br>**Transmissions:** All | **Knock Sensor 1 Signal High Input Conditions:**<br>Engine started, vehicle driven at 1600rpm for 3 seconds or to a temperature of 40 degrees Celsius, and the DME detected the Knock Sensor 1 (KS1) signal was too high<br>**Possible Causes:**<br>• Knock sensor circuit is open<br>• Knock sensor is loose (tighten to 20 NM)<br>• Contact between the knock sensor and cylinder block is dirty, corroded or greasy<br>• Knock sensor circuit is shorted to ground, or shorted to power<br>• Knock sensor is damaged or it has failed<br>• Wrong kind of fuel used<br>• A component in the engine compartment is loose or not properly secured |
| **DTC: P0330**<br>**2T MIL: Yes**<br>**Years:** 2007, 2008<br>**Models:** 328i, 328Ci, 328xi, 335i, 335Ci, 335xi, 525i, 525xi, 530i, 530xi, 550i, 650i, 750i, 750Li, 760Li, M5, M6, X3, X5, Z4 3.0i, Z4 3.0Si, Z4 M<br>**Engines:** All<br>**Transmissions:** All | **Knock Sensor 1 Circuit Malfunction Conditions:**<br>Engine started, vehicle driven at 1520rpm for 3 seconds or to a temperature of 40 degrees Celsius, and the DME detected the Knock Sensor 1 (KS1) signal was not recognized. The engine speed is greater than 2080rpm but less than 6000rpm and the coolant temperature is greater than 40.5 degrees Celsius.<br>**Possible Causes:**<br>• Knock sensor circuit is open<br>• Knock sensor is loose (tighten to 20 NM)<br>• Contact between the knock sensor and cylinder block is dirty, corroded or greasy<br>• Knock sensor circuit is shorted to ground, or shorted to power<br>• Knock sensor is damaged or it has failed<br>• Wrong kind of fuel used<br>• A component in the engine compartment is loose or not properly secured |

| DTC | Trouble Code Title, Conditions & Possible Causes |
|---|---|
| **DTC: P0332**<br>**2T MIL: Yes**<br>**Years:** 2007, 2008<br>**Models:** 328i, 328Ci, 328xi, 335i, 335Ci, 335xi, 525i, 525xi, 530i, 530xi, 550i, 650i, 750i, 750Li, 760Li, M5, M6, X3, X5, Z4 3.0i, Z4 3.0Si, Z4 M<br>**Engines:** All<br>**Transmissions:** All | **Knock Sensor 2 Signal Low Input Conditions:**<br>Engine started, vehicle driven, and the DME detected the Knock Sensor 1 (KS1) signal was too low or not recognized by the DME<br>**Possible Causes:**<br>• Knock sensor circuit is open<br>• Knock sensor is loose (tighten to 20 NM)<br>• Contact between the knock sensor and cylinder block is dirty, corroded or greasy<br>• Knock sensor circuit is shorted to ground, or shorted to power<br>• Knock sensor is damaged or it has failed<br>• Wrong kind of fuel used<br>• A component in the engine compartment is loose or not properly secured |
| **DTC: P0333**<br>**2T MIL: Yes**<br>**Years:** 2007, 2008<br>**Models:** 750i, 750Li, 760Li<br>**Engines:** All<br>**Transmissions:** All | **Knock Sensor 2 Signal High Input Conditions:**<br>Engine started, vehicle driven, and the DME detected the Knock Sensor 1 (KS1) signal was too high<br>**Possible Causes:**<br>• Knock sensor circuit is open<br>• Knock sensor is loose (tighten to 20 NM)<br>• Contact between the knock sensor and cylinder block is dirty, corroded or greasy<br>• Knock sensor circuit is shorted to ground, or shorted to power<br>• Knock sensor is damaged or it has failed<br>• Wrong kind of fuel used<br>• A component in the engine compartment is loose or not properly secured |
| **DTC: P0335**<br>**2T MIL: Yes**<br>**Years:** 2007, 2008<br>**Models:** 328i, 328Ci, 328xi, 335i, 335Ci, 335xi, 525i, 525xi, 530i, 530xi, 550i, 650i, 750i, 750Li, 760Li, M5, M6, X3, X5, Z4 3.0i, Z4 3.0Si, Z4 M<br>**Engines:** All<br>**Transmissions:** All | **Camshaft Position Sensor "A" Circ Malfunction Conditions:**<br>Engine started, battery voltage must be at least 11.5v, all electrical components must be off, parking brake must be engaged (to keep daytime driving lights off), automatic transmission selector must be in park and the ground between the engine and the chassis must be well connected. The DME detected the CMP sensor signal was implausible. Engine speed is greater than 500rpm, and the fault is tolerable as long as there are no misfired occurring at the same time.<br>**Possible Causes:**<br>• CMP sensor circuit is open or shorted to ground<br>• CMP sensor circuit is shorted to power<br>• CMP sensor ground (return) circuit is open<br>• CMP sensor installation incorrect (Hall-effect type)<br>• CMP sensor is damaged or CMP sensor shielding damaged |
| **DTC: P0336**<br>**2T MIL: Yes**<br>**Years:** 2007, 2008<br>**Models:** 335i, 335Ci, 335xi, 525i, 525xi, 530i, 530xi, 550i, 650i, 750i, 750Li, 760Li, Z4 3.0i, Z4 3.0Si, Z4 M<br>**Engines:** All<br>**Transmissions:** All | **Camshaft Position Sensor "A" Circ Range/Performance Conditions:**<br>Engine started (and engine speed is less than 25rpm), battery voltage must be at least 11.5v, all electrical components must be off, parking brake must be engaged (to keep daytime driving lights off), automatic transmission selector must be in park and the ground between the engine and the chassis must be well connected. The DME detected the CMP sensor signal was implausible.<br>**Possible Causes:**<br>• CMP sensor circuit is open or shorted to ground<br>• CMP sensor circuit is shorted to power<br>• CMP sensor ground (return) circuit is open<br>• CMP sensor installation incorrect (Hall-effect type)<br>• CMP sensor is damaged or CMP sensor shielding damaged |
| **DTC: P0339**<br>**2T MIL: Yes**<br>**Years:** 2007, 2008<br>**Models:** 335i, 335Ci, 335xi, 525i, 525xi, 530i, 530xi, 750i, 750Li, 760Li, Z4 3.0i, Z4 3.0Si, Z4 M<br>**Engines:** All<br>**Transmissions:** All | **Camshaft Position Sensor Circuit Malfunction Conditions:**<br>Engine started, battery voltage must be at least 11.5v, all electrical components must be off, parking brake must be engaged (to keep daytime driving lights off), automatic transmission selector must be in park and the ground between the engine and the chassis must be well connected. The DME detected the CMP sensor signal was missing or it was erratic. There is no signal or an invalid one, and the engine speed is greater than 200rpm for two cycles.<br>**Possible Causes:**<br>• CMP sensor circuit is open or shorted to ground<br>• CMP sensor circuit is shorted to power<br>• CMP sensor ground (return) circuit is open<br>• CMP sensor installation incorrect (Hall-effect type)<br>• CMP sensor is damaged or CMP sensor shielding damaged<br>• CMP sensor has failed |

| DTC | Trouble Code Title, Conditions & Possible Causes |
|---|---|
| **DTC: P0340**<br>**2T MIL: Yes**<br>**Years:** 2007, 2008<br>**Models:** 328i, 328Ci, 328xi, 335i, 335Ci, 335xi, 525i, 525xi, 530i, 530xi, 550i, 650i, 750i, 750Li, 760Li, M5, M6, X3, X5, Z4 3.0i, Z4 3.0Si, Z4 M<br>**Engines:** All<br>**Transmissions:** All | **Camshaft Position Sensor Circuit Malfunction Conditions:**<br>Engine started, battery voltage must be at least 11.5v, all electrical components must be off, parking brake must be engaged (to keep daytime driving lights off), automatic transmission selector must be in park and the ground between the engine and the chassis must be well connected. The DME detected the CMP sensor signal was missing or it was erratic. There is no signal or an invalid one, and the engine speed is greater than 200rpm for two cycles.<br>**Possible Causes:**<br>• CMP sensor circuit is open or shorted to ground<br>• CMP sensor circuit is shorted to power<br>• CMP sensor ground (return) circuit is open<br>• CMP sensor installation incorrect (Hall-effect type)<br>• CMP sensor is damaged or CMP sensor shielding damaged<br>• CMP sensor has failed |
| **DTC: P0341**<br>**2T MIL: Yes**<br>**Years:** 2007, 2008<br>**Models:** 335i, 335Ci, 335xi, 525i, 525xi, 530i, 530xi, 550i, 650i, 750i, 750Li, 760Li, Z4 3.0i, Z4 3.0Si, Z4 M<br>**Engines:** All<br>**Transmissions:** All | **Camshaft Position Sensor Circ Range/Performance Conditions:**<br>Engine started, battery voltage must be at least 11.5v, all electrical components must be off, parking brake must be engaged (to keep daytime driving lights off), automatic transmission selector must be in park and the ground between the engine and the chassis must be well connected. The DME detected the CMP sensor signal was implausible.<br>**Possible Causes:**<br>• CMP sensor circuit is open or shorted to ground<br>• CMP sensor circuit is shorted to power<br>• CMP sensor ground (return) circuit is open<br>• CMP sensor installation incorrect (Hall-effect type)<br>• CMP sensor is damaged or CMP sensor shielding damaged |
| **DTC: P0342**<br>**2T MIL: Yes**<br>**Years:** 2007, 2008<br>**Models:** 335i, 335Ci, 335xi, 525i, 525xi, 530i, 530xi, 550i, 650i, 750i, 750Li, 760Li, Z4 3.0i, Z4 3.0Si, Z4 M<br>**Engines:** All<br>**Transmissions:** All | **Camshaft Position Sensor "A" Circuit (Bank 1 or Single Sensor) Low Input Conditions:**<br>Engine started, battery voltage must be at least 11.5v, all electrical components must be off, parking brake must be engaged (to keep daytime driving lights off), automatic transmission selector must be in park and the ground between the engine and the chassis must be well connected. The DME detected the CMP sensor signal exceeded the bounds of the specified maximum limit.<br>**Possible Causes:**<br>• CMP sensor circuit is open or shorted to ground<br>• CMP sensor circuit is shorted to power<br>• CMP sensor ground (return) circuit is open<br>• CMP sensor installation incorrect (Hall-effect type)<br>• CMP sensor is damaged or CMP sensor shielding damaged |
| **DTC: P0343**<br>**2T MIL: Yes**<br>**Years:** 2007, 2008<br>**Models:** 335i, 335Ci, 335xi, 525i, 525xi, 530i, 530xi, 550i, 650i, 750i, 750Li, 760Li, Z4 3.0i, Z4 3.0Si, Z4 M<br>**Engines:** All<br>**Transmissions:** All | **Camshaft Position Sensor "A" Circuit (Bank 1 or Single Sensor) High Input Conditions:**<br>Engine started, battery voltage must be at least 11.5v, all electrical components must be off, parking brake must be engaged (to keep daytime driving lights off), automatic transmission selector must be in park and the ground between the engine and the chassis must be well connected. The DME detected the CMP sensor signal did not reach the specified minimum limit.<br>**Possible Causes:**<br>• CMP sensor circuit is open or shorted to ground<br>• CMP sensor circuit is shorted to power<br>• CMP sensor ground (return) circuit is open<br>• CMP sensor installation incorrect (Hall-effect type)<br>• CMP sensor is damaged or CMP sensor shielding damaged |
| **DTC: P0344**<br>**2T MIL: Yes**<br>**Years:** 2007, 2008<br>**Models:** 335i, 335Ci, 335xi, 525i, 525xi, 530i, 530xi, 750i, 750Li, 760Li, Z4 3.0i, Z4 3.0Si, Z4 M<br>**Engines:** All<br>**Transmissions:** All | **Camshaft Position Sensor Circuit Malfunction Conditions:**<br>Engine started, battery voltage must be at least 11.5v, all electrical components must be off, parking brake must be engaged (to keep daytime driving lights off), automatic transmission selector must be in park and the ground between the engine and the chassis must be well connected. The DME detected the CMP sensor signal was missing or it was erratic. There is no signal or an invalid one, and the engine speed is greater than 200rpm for two cycles.<br>**Possible Causes:**<br>• CMP sensor circuit is open or shorted to ground<br>• CMP sensor circuit is shorted to power<br>• CMP sensor ground (return) circuit is open<br>• CMP sensor installation incorrect (Hall-effect type)<br>• CMP sensor is damaged or CMP sensor shielding damaged<br>• CMP sensor has failed |

| DTC | Trouble Code Title, Conditions & Possible Causes |
|---|---|
| **DTC: P0345**<br>**2T MIL: Yes**<br>**Years:** 2007, 2008<br>**Models:** 335i, 335Ci, 335xi, 525i, 525xi, 530i, 530xi, 550i, 650i, 750i, 750Li, 760Li, M5, M6, X3, X5, Z4 3.0i, Z4 3.0Si, Z4 M<br>**Engines:** All<br>**Transmissions:** All | **Camshaft Position Sensor "A" Circuit (Bank 2) Conditions:**<br>Engine started, battery voltage must be at least 11.5v, all electrical components must be off, parking brake must be engaged (to keep daytime driving lights off), automatic transmission selector must be in park and the ground between the engine and the chassis must be well connected. The DME detected the CMP sensor signal was missing or it was erratic.<br>**Possible Causes:**<br>• CMP sensor circuit is open or shorted to ground<br>• CMP sensor circuit is shorted to power<br>• CMP sensor ground (return) circuit is open<br>• CMP sensor installation incorrect (Hall-effect type)<br>• CMP sensor is damaged or CMP sensor shielding damaged |
| **DTC: P0346**<br>**2T MIL: Yes**<br>**Years:** 2007, 2008<br>**Models:** 335i, 335Ci, 335xi, 525i, 525xi, 530i, 530xi, 550i, 650i, 750i, 750Li, 760Li, Z4 3.0i, Z4 3.0Si, Z4 M<br>**Engines:** All<br>**Transmissions:** All | **Camshaft Position Sensor "A" Circuit (Bank 2) Range/Performance Conditions:**<br>Engine started, battery voltage must be at least 11.5v, all electrical components must be off, parking brake must be engaged (to keep daytime driving lights off), automatic transmission selector must be in park and the ground between the engine and the chassis must be well connected. The DME detected the CMP sensor signal was implausible.<br>**Possible Causes:**<br>• CMP sensor circuit is open or shorted to ground<br>• CMP sensor circuit is shorted to power<br>• CMP sensor ground (return) circuit is open<br>• CMP sensor installation incorrect (Hall-effect type)<br>• CMP sensor is damaged or CMP sensor shielding damaged |
| **DTC: P0347**<br>**2T MIL: Yes**<br>**Years:** 2007, 2008<br>**Models:** 335i, 335Ci, 335xi, 525i, 525xi, 530i, 530xi, 550i, 650i, 750i, 750Li, 760Li, Z4 3.0i, Z4 3.0Si, Z4 M<br>**Engines:** All<br>**Transmissions:** All | **Camshaft Position Sensor "A" Circuit (Bank 2) Low Input Conditions:**<br>Engine started, battery voltage must be at least 11.5v, all electrical components must be off, parking brake must be engaged (to keep daytime driving lights off), automatic transmission selector must be in park and the ground between the engine and the chassis must be well connected. The DME detected the CMP sensor signal exceeded the bounds of the specified maximum limit.<br>**Possible Causes:**<br>• CMP sensor circuit is open or shorted to ground<br>• CMP sensor circuit is shorted to power<br>• CMP sensor ground (return) circuit is open<br>• CMP sensor installation incorrect (Hall-effect type)<br>• CMP sensor is damaged or CMP sensor shielding damaged |
| **DTC: P0348**<br>**2T MIL: Yes**<br>**Years:** 2007, 2008<br>**Models:** 335i, 335Ci, 335xi, 525i, 525xi, 530i, 530xi, 550i, 650i, 750i, 750Li, 760Li, Z4 3.0i, Z4 3.0Si, Z4 M<br>**Engines:** All<br>**Transmissions:** All | **Camshaft Position Sensor "A" Circuit (Bank 2) High Input Conditions:**<br>Engine started, battery voltage must be at least 11.5v, all electrical components must be off, parking brake must be engaged (to keep daytime driving lights off), automatic transmission selector must be in park and the ground between the engine and the chassis must be well connected. The DME detected the CMP sensor signal did not reach the specified minimum limit.<br>**Possible Causes:**<br>• CMP sensor circuit is open or shorted to ground<br>• CMP sensor circuit is shorted to power<br>• CMP sensor ground (return) circuit is open<br>• CMP sensor installation incorrect (Hall-effect type)<br>• CMP sensor is damaged or CMP sensor shielding damaged |
| **DTC: P0351**<br>**2T MIL: Yes**<br>**Years:** 2007, 2008<br>**Models:** 335i, 335Ci, 335xi, 525i, 525xi, 530i, 530xi, 550i, 650i, 750i, 750Li, 760Li, Z4 3.0i, Z4 3.0Si, Z4 M<br>**Engines:** All<br>**Transmissions:** All | **Ignition Coilpack A Primary/Secondary Circuit Malfunction Conditions:**<br>Engine started, battery voltage must be at least 11.5v, all electrical components must be off, parking brake must be engaged (to keep daytime driving lights off), automatic transmission selector must be in park and the ground between the engine and the chassis must be well connected. The DME did not receive any valid pulses from the ignition module for the Ignition Coilpack A primary circuit.<br>**Note: Ignition coils and power output stages are one component and cannot be replaced individually.**<br>**Possible Causes:**<br>• Engine speed (RPM) sensor has failed<br>• Camshaft Position (CMP) sensor has failed<br>• Power Supply Relay is shorted to an open circuit<br>• There is a malfunction in voltage supply<br>• Ignition coilpack is damaged or it has failed<br>• Cylinder 1 to 4 Fuel Injector(s) have failed |

| DTC | Trouble Code Title, Conditions & Possible Causes |
|---|---|
| **DTC: P0353**<br>**2T MIL: Yes**<br>**Years:** 2007, 2008<br>**Models:** 335i, 335Ci, 335xi, 525i, 525xi, 530i, 530xi, 550i, 650i, 750i, 750Li, 760Li, Z4 3.0i, Z4 3.0Si, Z4 M<br>**Engines:** All<br>**Transmissions:** All | **Ignition Coilpack C Primary/Secondary Circuit Malfunction Conditions:**<br>Engine started, battery voltage must be between 9 and 17 volts. The DME did not receive any valid pulses from the ignition module for the Ignition Coilpack C primary circuit. Voltage supplied and ground must be connected for ignition system spark plugs and coils. Check wiring harness, ground connection and plug-in contacts. Visual inspection of spark plug, ignition coil (replace if damaged). After excluding all of these faults, replace the control module. The injection is deactivated with a combustion miss and supplementary recognition of a rough running diagnosis.<br>**Note: Ignition coils and power output stages are one component and cannot be replaced individually.**<br>**Possible Causes:**<br>    • Engine speed (RPM) sensor has failed<br>    • Camshaft Position (CMP) sensor has failed<br>    • Power Supply Relay is shorted to an open circuit<br>    • There is a malfunction in voltage supply<br>    • Ignition coilpack is damaged or it has failed<br>    • Cylinder 1 to 4 Fuel Injector(s) have failed |
| **DTC: P0354**<br>**2T MIL: Yes**<br>**Years:** 2007, 2008<br>**Models:** 335i, 335Ci, 335xi, 525i, 525xi, 530i, 530xi, 550i, 650i, 750i, 750Li, 760Li, Z4 3.0i, Z4 3.0Si, Z4 M<br>**Engines:** All<br>**Transmissions:** All | **Ignition Coilpack D Primary/Secondary Circuit Malfunction Conditions:**<br>Engine started, battery voltage must be between 9 and 17 volts. The DME did not receive any valid pulses from the ignition module for the Ignition Coilpack C primary circuit. Voltage supplied and ground must be connected for ignition system spark plugs and coils. Check wiring harness, ground connection and plug-in contacts. Visual inspection of spark plug, ignition coil (replace if damaged). After excluding all of these faults, replace the control module. The injection is deactivated with a combustion miss and supplementary recognition of a rough running diagnosis.<br>**Note: Ignition coils and power output stages are one component and cannot be replaced individually.**<br>**Possible Causes:**<br>    • Engine speed (RPM) sensor has failed<br>    • Camshaft Position (CMP) sensor has failed<br>    • Power Supply Relay is shorted to an open circuit<br>    • There is a malfunction in voltage supply<br>    • Ignition coilpack is damaged or it has failed<br>    • Cylinder 1 to 4 Fuel Injector(s) have failed |
| **DTC: P0355**<br>**2T MIL: Yes**<br>**Years:** 2007, 2008<br>**Models:** 335i, 335Ci, 335xi, 525i, 525xi, 530i, 530xi, 550i, 650i, 750i, 750Li, 760Li, Z4 3.0i, Z4 3.0Si, Z4 M<br>**Engines:** All<br>**Transmissions:** All | **Ignition Coilpack E Primary/Secondary Circuit Malfunction Conditions:**<br>Engine started, battery voltage must be between 9 and 17 volts. The DME did not receive any valid pulses from the ignition module for the Ignition Coilpack C primary circuit. Voltage supplied and ground must be connected for ignition system spark plugs and coils. Check wiring harness, ground connection and plug-in contacts. Visual inspection of spark plug, ignition coil (replace if damaged). After excluding all of these faults, replace the control module. The injection is deactivated with a combustion miss and supplementary recognition of a rough running diagnosis.<br>**Note: Ignition coils and power output stages are one component and cannot be replaced individually.**<br>**Possible Causes:**<br>    • Engine speed (RPM) sensor has failed<br>    • Camshaft Position (CMP) sensor has failed<br>    • Power Supply Relay is shorted to an open circuit<br>    • There is a malfunction in voltage supply<br>    • Ignition coilpack is damaged or it has failed<br>    • Cylinder 1 to 4 Fuel Injector(s) have failed |
| **DTC: P0356**<br>**2T MIL: Yes**<br>**Years:** 2007, 2008<br>**Models:** 335i, 335Ci, 335xi, 525i, 525xi, 530i, 530xi, 550i, 650i, 750i, 750Li, 760Li, Z4 3.0i, Z4 3.0Si, Z4 M<br>**Engines:** All<br>**Transmissions:** All | **Ignition Coilpack F Primary/Secondary Circuit Malfunction Conditions:**<br>Engine started, battery voltage must be between 9 and 17 volts. The DME did not receive any valid pulses from the ignition module for the Ignition Coilpack C primary circuit. Voltage supplied and ground must be connected for ignition system spark plugs and coils. Check wiring harness, ground connection and plug-in contacts. Visual inspection of spark plug, ignition coil (replace if damaged). After excluding all of these faults, replace the control module. The injection is deactivated with a combustion miss and supplementary recognition of a rough running diagnosis.<br>**Note: Ignition coils and power output stages are one component and cannot be replaced individually.**<br>**Possible Causes:**<br>    • Engine speed (RPM) sensor has failed<br>    • Camshaft Position (CMP) sensor has failed<br>    • Power Supply Relay is shorted to an open circuit<br>    • There is a malfunction in voltage supply<br>    • Ignition coilpack is damaged or it has failed<br>    • Cylinder 1 to 4 Fuel Injector(s) have failed |

| DTC | Trouble Code Title, Conditions & Possible Causes |
|---|---|
| **DTC: P0357**<br>**2T MIL: Yes**<br>**Years:** 2007, 2008<br>**Models:** 335i, 335Ci, 335xi, 525i, 525xi, 530i, 530xi, 550i, 650i, 750i, 750Li, 760Li, Z4 3.0i, Z4 3.0Si, Z4 M<br>**Engines:** All<br>**Transmissions:** All | **Ignition Coilpack G Primary/Secondary Circuit Malfunction Conditions:**<br>Engine started, battery voltage must be between 9 and 17 volts. The DME did not receive any valid pulses from the ignition module for the Ignition Coilpack C primary circuit. Voltage supplied and ground must be connected for ignition system spark plugs and coils. Check wiring harness, ground connection and plug-in contacts. Visual inspection of spark plug, ignition coil (replace if damaged). After excluding all of these faults, replace the control module. The injection is deactivated with a combustion miss and supplementary recognition of a rough running diagnosis.<br>**Note: Ignition coils and power output stages are one component and cannot be replaced individually.**<br>**Possible Causes:**<br>    • Engine speed (RPM) sensor has failed<br>    • Camshaft Position (CMP) sensor has failed<br>    • Power Supply Relay is shorted to an open circuit<br>    • There is a malfunction in voltage supply<br>    • Ignition coilpack is damaged or it has failed<br>    • Cylinder 1 to 4 Fuel Injector(s) have failed |
| **DTC: P0358**<br>**2T MIL: Yes**<br>**Years:** 2007, 2008<br>**Models:** 335i, 335Ci, 335xi, 525i, 525xi, 530i, 530xi, 550i, 650i, 750i, 750Li, 760Li, Z4 3.0i, Z4 3.0Si, Z4 M<br>**Engines:** All<br>**Transmissions:** All | **Ignition Coilpack H Primary/Secondary Circuit Malfunction Conditions:**<br>Engine started, battery voltage must be between 9 and 17 volts. The DME did not receive any valid pulses from the ignition module for the Ignition Coilpack C primary circuit. Voltage supplied and ground must be connected for ignition system spark plugs and coils. Check wiring harness, ground connection and plug-in contacts. Visual inspection of spark plug, ignition coil (replace if damaged). After excluding all of these faults, replace the control module. The injection is deactivated with a combustion miss and supplementary recognition of a rough running diagnosis.<br>**Note: Ignition coils and power output stages are one component and cannot be replaced individually.**<br>**Possible Causes:**<br>    • Engine speed (RPM) sensor has failed<br>    • Camshaft Position (CMP) sensor has failed<br>    • Power Supply Relay is shorted to an open circuit<br>    • There is a malfunction in voltage supply<br>    • Ignition coilpack is damaged or it has failed<br>    • Cylinder 1 to 4 Fuel Injector(s) have failed |
| **DTC: P0365**<br>**2T MIL: Yes**<br>**Years:** 2007, 2008<br>**Models:** All<br>**Engines:** All<br>**Transmissions:** All | **Camshaft Position Sensor "B" Circuit (Bank 1) Conditions:**<br>Engine started, battery voltage must be at least 11.5v, all electrical components must be off, parking brake must be engaged (to keep daytime driving lights off), automatic transmission selector must be in park and the ground between the engine and the chassis must be well connected. The DME detected the CMP sensor signal exceeded the bounds of the specified maximum limit. Flank number within three camshaft revolutions not 0, 1, 11, 12, 13. The number of phase flanks per cycle is implausible.<br>**Possible Causes:**<br>    • CMP sensor circuit is open or shorted to ground<br>    • CMP sensor circuit is shorted to power<br>    • CMP sensor ground (return) circuit is open<br>    • CMP sensor installation incorrect (Hall-effect type)<br>    • CMP sensor is damaged or CMP sensor shielding damaged |
| **DTC: P0366**<br>**2T MIL: Yes**<br>**Years:** 2007, 2008<br>**Models:** 335i, 335Ci, 335xi, 525i, 525xi, 530i, 530xi, 550i, 650i, 750i, 750Li, 760Li, Z4 3.0i, Z4 3.0Si, Z4 M<br>**Engines:** All<br>**Transmissions:** All | **Camshaft Position Sensor "B" Circuit (Bank 1) Range/Performance Conditions:**<br>Engine started, battery voltage must be at least 11.5v, all electrical components must be off, parking brake must be engaged (to keep daytime driving lights off), automatic transmission selector must be in park and the ground between the engine and the chassis must be well connected. The DME detected the CMP sensor signal exceeded the bounds of the specified maximum limit.<br>**Possible Causes:**<br>    • CMP sensor circuit is open or shorted to ground<br>    • CMP sensor circuit is shorted to power<br>    • CMP sensor ground (return) circuit is open<br>    • CMP sensor installation incorrect (Hall-effect type)<br>    • CMP sensor is damaged or CMP sensor shielding damaged |
| **DTC: P0367**<br>**2T MIL: Yes**<br>**Years:** 2007, 2008<br>**Models:** 335i, 335Ci, 335xi, 525i, 525xi, 530i, 530xi, 550i, 650i, 750i, 750Li, 760Li, Z4 3.0i, Z4 3.0Si, Z4 M<br>**Engines:** All<br>**Transmissions:** All | **Camshaft Position Sensor "B" Circuit (Bank 1) Low Input Conditions:**<br>Engine started, battery voltage must be at least 11.5v, all electrical components must be off, parking brake must be engaged (to keep daytime driving lights off), automatic transmission selector must be in park and the ground between the engine and the chassis must be well connected. The DME detected the CMP sensor signal exceeded the bounds of the specified maximum limit.<br>**Possible Causes:**<br>    • CMP sensor circuit is open or shorted to ground<br>    • CMP sensor circuit is shorted to power<br>    • CMP sensor ground (return) circuit is open<br>    • CMP sensor installation incorrect (Hall-effect type)<br>    • CMP sensor is damaged or CMP sensor shielding damaged |

| DTC | Trouble Code Title, Conditions & Possible Causes |
|---|---|
| **DTC: P0368**<br>**2T MIL: Yes**<br>**Years:** 2007, 2008<br>**Models:** 335i, 335Ci, 335xi, 525i, 525xi, 530i, 530xi, 550i, 650i, 750i, 750Li, 760Li, Z4 3.0i, Z4 3.0Si, Z4 M<br>**Engines:** All<br>**Transmissions:** All | **Camshaft Position Sensor "B" Circuit (Bank 1) High Input Conditions:**<br>Engine turning over for at least nine faults, battery voltage must be at least 11.5v, there must be multiple reference points lost, signal faults or intermittent contact on KWG signal wire. VVT emergency default mode (max stroke) active, VANOS emergency default mode (spec. 120 degrees) active, RPM sensor emergency default mode active.<br>**Possible Causes:**<br>• CMP sensor circuit is open or shorted to ground<br>• CMP sensor circuit is shorted to power<br>• CMP sensor ground (return) circuit is open<br>• CMP sensor installation incorrect (Hall-effect type)<br>• Defective KWG<br>• Excessive gap between KWG and KW (or deformed KW) |
| **DTC: P0369**<br>**2T MIL: Yes**<br>**Years:** 2007, 2008<br>**Models:** 335i, 335Ci, 335xi, 525i, 525xi, 530i, 530xi, Z4 3.0i, Z4 3.0Si, Z4 M<br>**Engines:** All<br>**Transmissions:** All | **Crankshaft Position Sensor Rationality Check Conditions:**<br>Engine started, battery voltage must be at least 11.5v, all electrical components must be off, parking brake must be engaged (to keep daytime driving lights off), automatic transmission selector must be in park and the ground between the engine and the chassis must be well connected. The DME detected the CMP sensor signal did not reach the specified minimum or maximum limit, or the difference between the actual and target position was incorrectly reported. This fault occurs 120 seconds after start up.<br>**Possible Causes:**<br>• CMP sensor circuit is open or shorted to ground<br>• CMP sensor circuit is shorted to power<br>• CMP sensor ground (return) circuit is open<br>• CMP sensor installation incorrect (Hall-effect type)<br>• CMP sensor is damaged or CMP sensor shielding damaged |
| **DTC: P0370**<br>**2T MIL: Yes**<br>**Years:** 2007, 2008<br>**Models:** 335i, 335Ci, 335xi, 525i, 525xi, 530i, 530xi, 550i, 650i, 750i, 750Li, 760Li, Z4 3.0i, Z4 3.0Si, Z4 M<br>**Engines:** All<br>**Transmissions:** All | **Crankshaft Position Sensor Timing Reference High Conditions:**<br>Engine started, battery voltage must be at least 11.5v, all electrical components must be off, parking brake must be engaged (to keep daytime driving lights off), automatic transmission selector must be in park and the ground between the engine and the chassis must be well connected. The DME detected the CMP sensor signal did not reach the specified minimum limit.<br>**Possible Causes:**<br>• CMP sensor circuit is open or shorted to ground<br>• CMP sensor circuit is shorted to power<br>• CMP sensor ground (return) circuit is open<br>• CMP sensor installation incorrect (Hall-effect type)<br>• CMP sensor is damaged or CMP sensor shielding damaged |
| **DTC: P0372**<br>**2T MIL: Yes**<br>**Years:** 2007, 2008<br>**Models:** 335i, 335Ci, 335xi, 525i, 525xi, 530i, 530xi, 550i, 650i, 750i, 750Li, 760Li, Z4 3.0i, Z4 3.0Si, Z4 M<br>**Engines:** All<br>**Transmissions:** All | **Crankshaft Position Sensor Timing Reference High Resolution Signal "A" Too Few Pulses Conditions:**<br>Engine started, battery voltage must be at least 11.5v, all electrical components must be off, parking brake must be engaged (to keep daytime driving lights off), automatic transmission selector must be in park and the ground between the engine and the chassis must be well connected. The DME detected the CMP sensor signal did not reach the specified minimum limit.<br>**Possible Causes:**<br>• CMP sensor circuit is open or shorted to ground<br>• CMP sensor circuit is shorted to power<br>• CMP sensor ground (return) circuit is open<br>• CMP sensor installation incorrect (Hall-effect type)<br>• CMP sensor is damaged or CMP sensor shielding damaged |
| **DTC: P0373**<br>**2T MIL: Yes**<br>**Years:** 2007, 2008<br>**Models:** 335i, 335Ci, 335xi, 525i, 525xi, 530i, 530xi, 550i, 650i, 750i, 750Li, 760Li, Z4 3.0i, Z4 3.0Si, Z4 M<br>**Engines:** All<br>**Transmissions:** All | **Crankshaft Position Sensor Timing Reference High Resolution Signal "A" Intermittent/Erratic Pulses Conditions:**<br>Engine started, battery voltage must be at least 11.5v, all electrical components must be off, parking brake must be engaged (to keep daytime driving lights off), automatic transmission selector must be in park and the ground between the engine and the chassis must be well connected. The DME detected the CMP sensor signal did not reach the specified minimum limit.<br>**Possible Causes:**<br>• CMP sensor circuit is open or shorted to ground<br>• CMP sensor circuit is shorted to power<br>• CMP sensor ground (return) circuit is open<br>• CMP sensor installation incorrect (Hall-effect type)<br>• CMP sensor is damaged or CMP sensor shielding damaged |

| DTC | Trouble Code Title, Conditions & Possible Causes |
|---|---|
| **DTC: P0390**<br>**2T MIL: Yes**<br>**Years:** 2007, 2008<br>**Models:** 335i, 335Ci, 335xi, 525i, 525xi, 530i, 530xi, 550i, 650i, 750i, 750Li, 760Li, M5, M6, X3, X5, Z4 3.0i, Z4 3.0Si, Z4 M<br>**Engines:** All<br>**Transmissions:** All | **Camshaft Position Sensor "B" Circuit (Bank 2) Conditions:**<br>Engine started, battery voltage must be at least 11.5v, all electrical components must be off, parking brake must be engaged (to keep daytime driving lights off), automatic transmission selector must be in park and the ground between the engine and the chassis must be well connected. The DME detected the CMP sensor signal was missing or it was erratic.<br>**Possible Causes:**<br>• CMP sensor circuit is open or shorted to ground<br>• CMP sensor circuit is shorted to power<br>• CMP sensor ground (return) circuit is open<br>• CMP sensor installation incorrect (Hall-effect type)<br>• CMP sensor is damaged or CMP sensor shielding damaged |
| **DTC: P0391**<br>**2T MIL: Yes**<br>**Years:** 2007, 2008<br>**Models:** 335i, 335Ci, 335xi, 525i, 525xi, 530i, 530xi, 550i, 650i, 750i, 750Li, 760Li, Z4 3.0i, Z4 3.0Si, Z4 M<br>**Engines:** All<br>**Transmissions:** All | **Camshaft Position Sensor "B" Circuit (Bank 2) Range/Performance Conditions:**<br>Engine started, battery voltage must be at least 11.5v, all electrical components must be off, parking brake must be engaged (to keep daytime driving lights off), automatic transmission selector must be in park and the ground between the engine and the chassis must be well connected. The DME detected the CMP sensor signal exceeded the bounds of the specified maximum limit.<br>**Possible Causes:**<br>• CMP sensor circuit is open or shorted to ground<br>• CMP sensor circuit is shorted to power<br>• CMP sensor ground (return) circuit is open<br>• CMP sensor installation incorrect (Hall-effect type)<br>• CMP sensor is damaged or CMP sensor shielding damaged |
| **DTC: P0392**<br>**2T MIL: Yes**<br>**Years:** 2007, 2008<br>**Models:** 335i, 335Ci, 335xi, 525i, 525xi, 530i, 530xi, 550i, 650i, 750i, 750Li, 760Li, Z4 3.0i, Z4 3.0Si, Z4 M<br>**Engines:** All<br>**Transmissions:** All | **Camshaft Position Sensor "B" Circuit (Bank 2) Low Input Conditions:**<br>Engine started, battery voltage must be at least 11.5v, all electrical components must be off, parking brake must be engaged (to keep daytime driving lights off), automatic transmission selector must be in park and the ground between the engine and the chassis must be well connected. The DME detected the CMP sensor signal exceeded the bounds of the specified maximum limit.<br>**Possible Causes:**<br>• CMP sensor circuit is open or shorted to ground<br>• CMP sensor circuit is shorted to power<br>• CMP sensor ground (return) circuit is open<br>• CMP sensor installation incorrect (Hall-effect type)<br>• CMP sensor is damaged or CMP sensor shielding damaged |
| **DTC: P0393**<br>**2T MIL: Yes**<br>**Years:** 2007, 2008<br>**Models:** 335i, 335Ci, 335xi, 525i, 525xi, 530i, 530xi, 550i, 650i, 750i, 750Li, 760Li, Z4 3.0i, Z4 3.0Si, Z4 M<br>**Engines:** All<br>**Transmissions:** All | **Camshaft Position Sensor "B" Circuit (Bank 2) High Input Conditions:**<br>Engine started, battery voltage must be at least 11.5v, all electrical components must be off, parking brake must be engaged (to keep daytime driving lights off), automatic transmission selector must be in park and the ground between the engine and the chassis must be well connected. The DME detected the CMP sensor signal did not reach the specified minimum limit.<br>**Possible Causes:**<br>• CMP sensor circuit is open or shorted to ground<br>• CMP sensor circuit is shorted to power<br>• CMP sensor ground (return) circuit is open<br>• CMP sensor installation incorrect (Hall-effect type)<br>• CMP sensor is damaged or CMP sensor shielding damaged |
| **DTC: P0394**<br>**2T MIL: Yes**<br>**Years:** 2007, 2008<br>**Models:** M5, M6<br>**Engines:** All<br>**Transmissions:** All | **Camshaft Position Sensor "B" Circuit (Bank 2) Conditions:**<br>Engine started, battery voltage must be at least 11.5v, all electrical components must be off, parking brake must be engaged (to keep daytime driving lights off), automatic transmission selector must be in park and the ground between the engine and the chassis must be well connected. The DME detected the CMP sensor signal was missing or it was erratic.<br>**Possible Causes:**<br>• CMP sensor circuit is open or shorted to ground<br>• CMP sensor circuit is shorted to power<br>• CMP sensor ground (return) circuit is open<br>• CMP sensor installation incorrect (Hall-effect type)<br>• CMP sensor is damaged or CMP sensor shielding damaged |

| DTC | Trouble Code Title, Conditions & Possible Causes |
|---|---|
| **DTC: P0411**<br>**2T MIL: Yes**<br>**Years:** 2007, 2008<br>**Models:** 335i, 335Ci, 335xi, 525i, 525i, 530i, 530xi, 550i, 650i, 750i, 750i, 760Li, M5, M6, X3, X5, Z4 3.0i, Z4 3.0Si, Z4 M<br>**Engines:** All<br>**Transmissions:** All | **Secondary Air Injection System Upstream Flow Detected Conditions:**<br>Engine started, battery voltage must be at least 11.5v, all electrical components must be off, parking brake must be engaged (to keep daytime driving lights off), automatic transmission selector must be in park and the ground between the engine and the chassis must be well connected. The DME detected the Secondary AIR pump airflow was not diverted correctly when requested during the self-test. The pump is functioning but the quantity of air is recognized as insufficient by HO2S.<br>**Note: The solenoid valve is closed when no voltage is present.**<br>**Possible Causes:**<br>• Air pump output is blocked or restricted<br>• AIR bypass solenoid is leaking or it is restricted<br>• AIR bypass solenoid is stuck open or stuck closed<br>• Check valve (one or more) is damaged or leaking<br>• Electric air injection pump hose(s) leaking<br>• Electric air injection pump is damaged or faulty |
| **DTC: P0412**<br>**2T MIL: Yes**<br>**Years:** 2007, 2008<br>**Models:** 328i, 328Ci, 328xi, 335i, 335Ci, 335xi, 525i, 525xi, 530i, 530xi, 550i, 650i, 750i, 750Li, 760Li, M5, M6, X3, X5, Z4 3.0i, Z4 3.0Si, Z4 M<br>**Engines:** All<br>**Transmissions:** All | **Secondary Air Injection Solenoid Circuit Malfunction Conditions:**<br>Engine started, battery voltage must be at least 11.5v, all electrical components must be off, parking brake must be engaged (to keep daytime driving lights off), automatic transmission selector must be in park and the ground between the engine and the chassis must be well connected. The DME detected an unexpected low or high voltage condition on the AIR solenoid control circuit during testing.<br>**Possible Causes:**<br>• AIR solenoid power circuit (B+) is open (check dedicated fuse)<br>• AIR bypass solenoid control circuit is open or shorted to ground<br>• AIR diverter solenoid control circuit open or shorted to ground<br>• AIR pump control circuit is open or shorted to ground<br>• Check valve (one or more) is damaged or leaking<br>• Solid State relay is damaged or it has failed |
| **DTC: P0413**<br>**2T MIL: Yes**<br>**Years:** 2007, 2008<br>**Models:** 335i, 335Ci, 335xi, 525i, 525xi, 530i, 530xi, 550i, 650i, 750i, 750Li, 760Li, M5, M6, X3, X5, Z4 3.0i, Z4 3.0Si, Z4 M<br>**Engines:** All<br>**Transmissions:** All | **Secondary Air Injection Solenoid Circuit Open Conditions:**<br>Engine started, battery voltage must be at least 11.5v, all electrical components must be off, parking brake must be engaged (to keep daytime driving lights off), automatic transmission selector must be in park and the ground between the engine and the chassis must be well connected. The DME detected an unexpected low or high voltage condition on the AIR solenoid control circuit during testing.<br>**Possible Causes:**<br>• AIR solenoid power circuit (B+) is open (check dedicated fuse)<br>• AIR bypass solenoid control circuit is open or shorted to ground<br>• AIR diverter solenoid control circuit open or shorted to ground<br>• AIR pump control circuit is open or shorted to ground<br>• Check valve (one or more) is damaged or leaking<br>• Solid State relay is damaged or it has failed |
| **DTC: P0414**<br>**2T MIL: Yes**<br>**Years:** 2007, 2008<br>**Models:** 335i, 335Ci, 335xi, 525i, 525xi, 530i, 530xi, 550i, 650i, 750i, 750Li, 760Li, M5, M6, X3, X5, Z4 3.0i, Z4 3.0Si, Z4 M<br>**Engines:** All<br>**Transmissions:** All | **Secondary Air Injection Solenoid Circuit Short Conditions:**<br>Engine started, battery voltage must be at least 11.5v, all electrical components must be off, parking brake must be engaged (to keep daytime driving lights off), automatic transmission selector must be in park and the ground between the engine and the chassis must be well connected. The DME detected an unexpected low or high voltage condition on the AIR solenoid control circuit during testing.<br>**Possible Causes:**<br>• AIR solenoid power circuit (B+) is open (check dedicated fuse)<br>• AIR bypass solenoid control circuit is open or shorted to ground<br>• AIR diverter solenoid control circuit open or shorted to ground<br>• AIR pump control circuit is open or shorted to ground<br>• Check valve (one or more) is damaged or leaking<br>• Solid State relay is damaged or it has failed |
| **DTC: P0418**<br>**2T MIL: Yes**<br>**Years:** 2007, 2008<br>**Models:** 335i, 335Ci, 335xi, 525i, 525xi, 530i, 530xi, 550i, 650i, 750i, 750Li, 760Li, M5, M6, X3, X5, Z4 3.0i, Z4 3.0Si, Z4 M<br>**Engines:** All<br>**Transmissions:** All | **Secondary Air Injection Relay (A) Circuit Malfunction Conditions:**<br>Engine started, battery voltage must be at least 11.5v, all electrical components must be off, parking brake must be engaged (to keep daytime driving lights off), automatic transmission selector must be in park and the ground between the engine and the chassis must be well connected. The DME detected an unexpected low or high voltage condition on the AIR solenoid control circuit during testing. The fuel status is in a closed loop pattern, the coolant temperature is between 69 and 100 degrees Celsius, and the engine speed is between 800 and 6000rpm.<br>**Possible Causes:**<br>• AIR solenoid power circuit (B+) is open (check dedicated fuse)<br>• AIR bypass solenoid control circuit is open or shorted to ground<br>• AIR diverter solenoid control circuit open or shorted to ground<br>• AIR pump control circuit is open or shorted to ground<br>• Check valve (one or more) is damaged or leaking<br>• Solid State relay is damaged or it has failed |

| DTC | Trouble Code Title, Conditions & Possible Causes |
|---|---|
| **DTC: P0420**<br>**MIL: Yes**<br>**Years:** 2007, 2008<br>**Models:** 328i, 328Ci, 328xi, 335i, 335Ci, 335xi, 525i, 525xi, 530i, 530xi, 550i, 650i, 750i, 750Li, 760Li, M5, M6, X3, X5, Z4 3.0i, Z4 3.0Si, Z4 M<br>**Engines:** All<br>**Transmissions:** All | **Catalyst System Efficiency (Bank 1) Below Threshold Conditions:**<br>Engine started, battery voltage must be at least 11.5v, all electrical components must be off, parking brake must be engaged (to keep daytime driving lights off), automatic transmission selector must be in park, the exhaust system must be properly sealed between the catalytic converter and the cylinder head, coolant temperature must be at least 80 degrees Celsius and oxygen sensor heaters for oxygen sensors before the catalytic converter must be functioning properly and the ground between the engine and the chassis must be well connected. The DME detected the switch rate of the rear HO2S-12 was close to the switch rate of front HO2S (it should be much slower). The exhaust-gas mass airflow is less than 22g/sec. The engine speed is between 980 and 1920rpm, the catalyst temperature is greater than 300 degrees Celsius, the fuel system status is in a closed loop and the purge vapor factor is less than 3.5.<br>**Possible Causes:**<br>• Air leaks at the exhaust manifold or in the exhaust pipes<br>• Catalytic converter is damaged, contaminated or it has failed<br>• ECT/CHT sensor has lost its calibration (the signal is incorrect)<br>• Engine cylinders misfiring, or the ignition timing is over retarded<br>• Engine oil is contaminated<br>• Front HO2S or rear HO2S is contaminated with fuel or moisture<br>• Front HO2S and/or the rear HO2S is loose in the mounting hole<br>• Front HO2S much older than the rear HO2S (HO2S-11 is lazy)<br>• Fuel system pressure is too high (check the pressure regulator)<br>• Rear HO2S wires improperly connected or the HO2S has failed |
| **DTC: P0423**<br>**2T MIL: Yes**<br>**Years:** 2007, 2008<br>**Models:** 750i, 750Li, 760Li<br>**Engines:** All<br>**Transmissions:** All | **Heated Catalyst System Efficiency Below Threshold Conditions:**<br>Engine started, battery voltage must be at least 11.5v, all electrical components must be off, parking brake must be engaged (to keep daytime driving lights off), automatic transmission selector must be in park, the exhaust system must be properly sealed between the catalytic converter and the cylinder head, coolant temperature must be at least 80 degrees Celsius and oxygen sensor heaters for oxygen sensors before the catalytic converter must be functioning properly and the ground between the engine and the chassis must be well connected. The coolant temperature is less than 90 degrees Celsius, catalyst temperature is less than 300 degrees Celsius, engine crank time less than five seconds, vehicle speed is less than three mph and the engine speed is less than 200rpm.<br>**Possible Causes:**<br>• Air leaks at the exhaust manifold or in the exhaust pipes<br>• Catalytic converter is damaged, contaminated or it has failed<br>• ECT/CHT sensor has lost its calibration (the signal is incorrect)<br>• Engine cylinders misfiring, or the ignition timing is over retarded<br>• Engine oil is contaminated<br>• Front HO2S or rear HO2S is contaminated with fuel or moisture<br>• Front HO2S and/or the rear HO2S is loose in the mounting hole<br>• Front HO2S much older than the rear HO2S (HO2S-11 is lazy)<br>• Fuel system pressure is too high (check the pressure regulator)<br>• Rear HO2S wires improperly connected or the HO2S has failed |
| **DTC: P0430**<br>**MIL: Yes**<br>**Years:** 2007, 2008<br>**Models:** 328i, 328Ci, 328xi, 335i, 335Ci, 335xi, 525i, 525xi, 530i, 530xi, 550i, 650i, 750i, 750Li, 760Li, M5, M6, X3, X5, Z4 3.0i, Z4 3.0Si, Z4 M<br>**Engines:** All<br>**Transmissions:** All | **Catalyst System Efficiency (Bank 2) Below Threshold Conditions:**<br>Engine started, battery voltage must be at least 11.5v, all electrical components must be off, parking brake must be engaged (to keep daytime driving lights off), automatic transmission selector must be in park, the exhaust system must be properly sealed between the catalytic converter and the cylinder head, coolant temperature must be at least 80 degrees Celsius and oxygen sensor heaters for oxygen sensors before the catalytic converter must be functioning properly and the ground between the engine and the chassis must be well connected. The DME detected the switch rate of the rear HO2S-12 was close to the switch rate of front HO2S (it should be much slower). The engine speed is between 980 and 1920rpm, the catalyst temperature is greater than 300 degrees Celsius, the fuel system status is in a closed loop and the purge vapor factor is less than 3.5.<br>**Possible Causes:**<br>• Air leaks at the exhaust manifold or in the exhaust pipes<br>• Catalytic converter is damaged, contaminated or it has failed<br>• ECT/CHT sensor has lost its calibration (the signal is incorrect)<br>• Engine cylinders misfiring, or the ignition timing is over retarded<br>• Engine oil is contaminated<br>• Front HO2S or rear HO2S is contaminated with fuel or moisture<br>• Front HO2S and/or the rear HO2S is loose in the mounting hole<br>• Front HO2S much older than the rear HO2S (HO2S-11 is lazy)<br>• Fuel system pressure is too high (check the pressure regulator)<br>• Rear HO2S wires improperly connected or the HO2S has failed |

| DTC | Trouble Code Title, Conditions & Possible Causes |
|---|---|
| **DTC: P0433**<br>**2T MIL:** Yes<br>**Years:** 2007, 2008<br>**Models:** 750i, 750Li, 760Li<br>**Engines:** All<br>**Transmissions:** All | **Heated Catalyst System Efficiency Below Threshold Conditions:**<br>Engine started, battery voltage must be at least 11.5v, all electrical components must be off, parking brake must be engaged (to keep daytime driving lights off), automatic transmission selector must be in park, the exhaust system must be properly sealed between the catalytic converter and the cylinder head, coolant temperature must be at least 80 degrees Celsius and oxygen sensor heaters for oxygen sensors before the catalytic converter must be functioning properly and the ground between the engine and the chassis must be well connected. The coolant temperature is less than 90 degrees Celsius, catalyst temperature is less than 300 degrees Celsius, engine crank time less than five seconds, vehicle speed is less than three mph and the engine speed is less than 200rpm.<br>**Possible Causes:**<br>• Air leaks at the exhaust manifold or in the exhaust pipes<br>• Catalytic converter is damaged, contaminated or it has failed<br>• ECT/CHT sensor has lost its calibration (the signal is incorrect)<br>• Engine cylinders misfiring, or the ignition timing is over retarded<br>• Engine oil is contaminated<br>• Front HO2S or rear HO2S is contaminated with fuel or moisture<br>• Front HO2S and/or the rear HO2S is loose in the mounting hole<br>• Front HO2S much older than the rear HO2S (HO2S-11 is lazy)<br>• Fuel system pressure is too high (check the pressure regulator)<br>• Rear HO2S wires improperly connected or the HO2S has failed |
| **DTC: P0440**<br>**2T MIL:** Yes<br>**Years:** 2007, 2008<br>**Models:** 328i, 328Ci, 328xi, 335i, 335Ci, 335xi, 525i, 525xi, 530i, 530xi, 550i, 650i, 750i, 750Li, 760Li, M5, M6, X3, X5, Z4 3.0i, Z4 3.0Si, Z4 M<br>**Engines:** All<br>**Transmissions:** All | **EVAP System Malfunction Conditions:**<br>ECT sensor is cold during startup, engine started, battery voltage must be at least 11.5v, all electrical components must be off, parking brake must be engaged (to keep daytime driving lights off), automatic transmission selector must be in park, the exhaust system must be properly sealed between the catalytic converter and the cylinder head, coolant temperature must be at least 80 degrees Celsius and oxygen sensor heaters for oxygen sensors before the catalytic converter must be functioning properly and the ground between the engine and the chassis must be well connected. The DME detected the switch rate of the rear HO2S-12 was close to the switch rate of front HO2S (it should be much slower). DME detected a problem in the EVAP system during the EVAP System Monitor test. The fuel system adaptation has finished, the coolant temperature is greater than 60 degrees Celsius, normal purge is on, vehicle speed is zero, and engine is at idle.<br>**Possible Causes:**<br>• EVAP canister purge valve is damaged<br>• EVAP canister has an improper seal<br>• Vapor line between purge solenoid and intake manifold vacuum reservoir is damaged, or vapor line between EVAP canister purge solenoid and charcoal canister is damaged<br>• Vapor line between charcoal canister and check valve, or vapor line between check valve and fuel vapor valves is damaged |
| **DTC: P0441**<br>**2T MIL:** Yes<br>**Years:** 2007, 2008<br>**Models:** 328i, 328Ci, 328xi, 335i, 335Ci, 335xi, 525i, 525xi, 530i, 530xi, 550i, 650i, 750i, 750Li, 760Li, M5, M6, X3, X5, Z4 3.0i, Z4 3.0Si, Z4 M<br>**Engines:** All<br>**Transmissions:** All | **EVAP Control System Incorrect Purge Flow Conditions:**<br>ECT sensor is cold during startup, engine started, battery voltage must be at least 11.5v, all electrical components must be off, parking brake must be engaged (to keep daytime driving lights off), automatic transmission selector must be in park, the exhaust system must be properly sealed between the catalytic converter and the cylinder head, coolant temperature must be at least 80 degrees Celsius and oxygen sensor heaters for oxygen sensors before the catalytic converter must be functioning properly and the ground between the engine and the chassis must be well connected. The DME detected the switch rate of the rear HO2S-12 was close to the switch rate of front HO2S (it should be much slower). DME detected a problem in the EVAP system during the EVAP System Monitor test.<br>**Possible Causes:**<br>• EVAP canister purge valve is damaged<br>• EVAP canister has an improper seal<br>• Vapor line between purge solenoid and intake manifold vacuum reservoir is damaged, or vapor line between EVAP canister purge solenoid and charcoal canister is damaged<br>• Vapor line between charcoal canister and check valve, or vapor line between check valve and fuel vapor valves is damaged |

| DTC | Trouble Code Title, Conditions & Possible Causes |
|---|---|
| **DTC: P0442**<br>**2T MIL:** Yes<br>**Years:** 2007, 2008<br>**Models:** 328i, 328Ci, 328xi, 335i, 335Ci, 335xi, 525i, 525xi, 530i, 530xi, 550i, 650i, 750i, 750Li, 760Li, M5, M6, X3, X5, Z4 3.0i, Z4 3.0Si, Z4 M<br>**Engines:** All<br>**Transmissions:** All | **EVAP Control System Small Leak Detected Conditions:**<br>Engine started, battery voltage must be at least 11.5v, all electrical components must be off, parking brake must be engaged (to keep daytime driving lights off), automatic transmission selector must be in park, the exhaust system must be properly sealed between the catalytic converter and the cylinder head, coolant temperature must be at least 80 degrees Celsius and oxygen sensor heaters for oxygen sensors before the catalytic converter must be functioning properly and the ground between the engine and the chassis must be well connected. The DME detected a leak in the EVAP system as small as 0.040 inches during the EVAP Monitor Test. The fuel system adaptation has finished, the coolant temperature is greater than 60 degrees Celsius, normal purge is on, vehicle speed is zero, and engine is at idle. Engine start temperature must be greater than 2 degrees Celsius and the last driving cycle greater than 20 minutes.<br>**Possible Causes:**<br>• Aftermarket EVAP parts that do not conform to specifications<br>• CV solenoid remains partially open when commanded to close<br>• EVAP component seals leaking (i.e., leaks in the Purge valve, fuel tank pressure sensor, canister vent solenoid, fuel vapor control valve tube assembly or fuel vapor vent valve).<br>• Fuel filler cap damaged, cross-threaded or loosely installed<br>• Loose fuel vapor hose/tube connections to EVAP components<br>• Small holes or cuts in fuel vapor hoses or EVAP canister tubes |
| **DTC: P0443**<br>**2T MIL:** Yes<br>**Years:** 2007, 2008<br>**Models:** 328i, 328Ci, 328xi, 335i, 335Ci, 335xi, 525i, 525xi, 530i, 530xi, 550i, 650i, 750i, 750Li, 760Li, M5, M6, X3, X5, Z4 3.0i, Z4 3.0Si, Z4 M<br>**Engines:** All<br>**Transmissions:** All | **EVAP Vapor Management Valve Circuit Malfunction Conditions:**<br>Engine started, battery voltage must be at least 11.5v, all electrical components must be off, parking brake must be engaged (to keep daytime driving lights off), automatic transmission selector must be in park, the exhaust system must be properly sealed between the catalytic converter and the cylinder head, coolant temperature must be at least 80 degrees Celsius and oxygen sensor heaters for oxygen sensors before the catalytic converter must be functioning properly and the ground between the engine and the chassis must be well connected. The DME detected an unexpected high or low voltage condition on the Vapor Management Valve (VMV) circuit when the device was cycled On/Off during testing.<br>**Possible Causes:**<br>• EVAP power supply circuit is open<br>• EVAP solenoid control circuit is open or shorted to ground<br>• EVAP solenoid control circuit is shorted to power (B+)<br>• EVAP solenoid valve is damaged or it has failed |
| **DTC: P0444**<br>**2T MIL:** Yes<br>**Years:** 2007, 2008<br>**Models:** All<br>**Engines:** All<br>**Transmissions:** All | **Evaporative Emission System Purge Control Valve Circuit Open Conditions:**<br>Engine started, battery voltage must be at least 11.5v, all electrical components must be off, parking brake must be engaged (to keep daytime driving lights off), automatic transmission selector must be in park, the exhaust system must be properly sealed between the catalytic converter and the cylinder head, coolant temperature must be at least 80 degrees Celsius and oxygen sensor heaters for oxygen sensors before the catalytic converter must be functioning properly and the ground between the engine and the chassis must be well connected. The DME detected an unexpected voltage condition on the EVAP circuit when the device was cycled On/Off during testing.<br>**Possible Causes:**<br>• EVAP power supply circuit is open<br>• EVAP solenoid control circuit is open or shorted to ground<br>• EVAP solenoid control circuit is shorted to power (B+)<br>• EVAP solenoid valve is damaged or it has failed<br>• EVAP canister has a leak or a poor seal |
| **DTC: P0445**<br>**2T MIL:** Yes<br>**Years:** 2007, 2008<br>**Models:** All<br>**Engines:** All<br>**Transmissions:** All | **Evaporative Emission System Purge Control Valve Circuit Shorted Conditions:**<br>Engine started, battery voltage must be at least 11.5v, all electrical components must be off, parking brake must be engaged (to keep daytime driving lights off), automatic transmission selector must be in park, the exhaust system must be properly sealed between the catalytic converter and the cylinder head, coolant temperature must be at least 80 degrees Celsius and oxygen sensor heaters for oxygen sensors before the catalytic converter must be functioning properly and the ground between the engine and the chassis must be well connected. The DME detected an unexpected voltage condition on the EVAP circuit when the device was cycled On/Off during testing.<br>**Possible Causes:**<br>• EVAP power supply circuit is open<br>• EVAP solenoid control circuit is open or shorted to ground<br>• EVAP solenoid control circuit is shorted to power (B+)<br>• EVAP solenoid valve is damaged or it has failed<br>• EVAP canister has a leak or a poor seal |

| DTC | Trouble Code Title, Conditions & Possible Causes |
|---|---|
| **DTC: P0446**<br>**2T MIL: Yes**<br>**Years:** 2007, 2008<br>**Models:** 328i, 328Ci, 328xi, 335i, 335Ci, 335i, 525i, 525xi,<br>**Engines:** All<br>**Transmissions:** All | **EVAP Control System Large Leak Detected Conditions:**<br>Engine started, battery voltage must be at least 11.5v, all electrical components must be off, parking brake must be engaged (to keep daytime driving lights off), automatic transmission selector must be in park, the exhaust system must be properly sealed between the catalytic converter and the cylinder head, coolant temperature must be at least 80 degrees Celsius and oxygen sensor heaters for oxygen sensors before the catalytic converter must be functioning properly and the ground between the engine and the chassis must be well connected. The DME detected multiple small fuel vapor leaks; or it detected a large leak in the system during the leak test.<br>**Possible Causes:**<br>• Aftermarket EVAP hardware non-conforming to specifications<br>• EVAP canister tube, EVAP canister purge outlet tube or EVAP return tube disconnected or cracked, or canister is damaged<br>• EVAP canister purge valve stuck closed, or canister damaged<br>• Fuel filler cap missing, loose (not tightened) or the wrong part<br>• Loose fuel vapor hose/tube connections to EVAP components<br>• Canister vent (CV) solenoid stuck open<br>• Fuel tank pressure (FTP) sensor has failed mechanically |
| **DTC: P0455**<br>**2T MIL: Yes**<br>**Years:** 2007, 2008<br>**Models:** 328i, 328Ci, 328xi, 335i, 335Ci, 335xi, 525i, 525xi, 530i, 530xi, 550i, 650i, 750i, 750Li, 760Li, M5, M6, X3, X5, Z4 3.0i, Z4 3.0Si, Z4 M<br>**Engines:** All<br>**Transmissions:** All | **EVAP Control System Large Leak Detected Conditions:**<br>Engine started, battery voltage must be at least 11.5v, all electrical components must be off, parking brake must be engaged (to keep daytime driving lights off), automatic transmission selector must be in park, the exhaust system must be properly sealed between the catalytic converter and the cylinder head, coolant temperature must be at least 80 degrees Celsius and oxygen sensor heaters for oxygen sensors before the catalytic converter must be functioning properly and the ground between the engine and the chassis must be well connected. The DME detected multiple small fuel vapor leaks; or it detected a large leak in the system during the leak test.<br>**Possible Causes:**<br>• Aftermarket EVAP hardware non-conforming to specifications<br>• EVAP canister tube, EVAP canister purge outlet tube or EVAP return tube disconnected or cracked, or canister is damaged<br>• EVAP canister purge valve stuck closed, or canister damaged<br>• Fuel filler cap missing, loose (not tightened) or the wrong part<br>• Loose fuel vapor hose/tube connections to EVAP components<br>• Canister vent (CV) solenoid stuck open<br>• Fuel tank pressure (FTP) sensor has failed mechanically |
| **DTC: P0456**<br>**2T MIL: Yes**<br>**Years:** 2007, 2008<br>**Models:** All<br>**Engines:** All<br>**Transmissions:** All | **EVAP Control System Small Leak Detected Conditions:**<br>Engine started, battery voltage must be at least 11.5v, all electrical components must be off, parking brake must be engaged (to keep daytime driving lights off), automatic transmission selector must be in park, the exhaust system must be properly sealed between the catalytic converter and the cylinder head, coolant temperature must be at least 80 degrees Celsius and oxygen sensor heaters for oxygen sensors before the catalytic converter must be functioning properly and the ground between the engine and the chassis must be well connected. The DME detected multiple small fuel vapor leaks; or it detected a large leak in the system during the leak test.<br>**Possible Causes:**<br>• Aftermarket EVAP hardware non-conforming to specifications<br>• EVAP canister tube, EVAP canister purge outlet tube or EVAP return tube disconnected or cracked, or canister is damaged<br>• EVAP canister purge valve stuck closed, or canister damaged<br>• Fuel filler cap missing, loose (not tightened) or the wrong part<br>• Loose fuel vapor hose/tube connections to EVAP components<br>• Canister vent (CV) solenoid stuck open<br>• Fuel tank pressure (FTP) sensor has failed mechanically |
| **DTC: P0458**<br>**2T MIL: Yes**<br>**Years:** 2007, 2008<br>**Models:** 335i, 335Ci, 335xi, 525i, 525xi, 530i, 530xi, 550i, 650i, 750i, 750Li, 760Li, Z4 3.0i, Z4 3.0Si, Z4 M<br>**Engines:** All<br>**Transmissions:** All | **Evaporative Emission System Purge Control Valve Circuit Low Conditions:**<br>Engine started, battery voltage must be at least 11.5v, all electrical components must be off, parking brake must be engaged (to keep daytime driving lights off), automatic transmission selector must be in park, the exhaust system must be properly sealed between the catalytic converter and the cylinder head, coolant temperature must be at least 80 degrees Celsius and oxygen sensor heaters for oxygen sensors before the catalytic converter must be functioning properly and the ground between the engine and the chassis must be well connected. The DME detected an unexpected voltage condition on the EVAP circuit when the device was cycled On/Off during testing.<br>**Possible Causes:**<br>• EVAP power supply circuit is open<br>• EVAP solenoid control circuit is open or shorted to ground<br>• EVAP solenoid control circuit is shorted to power (B+)<br>• EVAP solenoid valve is damaged or it has failed<br>• EVAP canister has a leak or a poor seal |

| DTC | Trouble Code Title, Conditions & Possible Causes |
|---|---|
| **DTC: P0459**<br>**2T MIL: Yes**<br>**Years:** 2007, 2008<br>**Models:** 335i, 335Ci, 335xi, 525i, 525xi, 530i, 530xi, 550i, 650i, 750i, 750Li, 760Li, Z4 3.0i, Z4 3.0Si, Z4 M<br>**Engines:** All<br>**Transmissions:** All | **Evaporative Emission System Purge Control Valve Circuit High Conditions:**<br>Engine started, battery voltage must be at least 11.5v, all electrical components must be off, parking brake must be engaged (to keep daytime driving lights off), automatic transmission selector must be in park, the exhaust system must be properly sealed between the catalytic converter and the cylinder head, coolant temperature must be at least 80 degrees Celsius and oxygen sensor heaters for oxygen sensors before the catalytic converter must be functioning properly and the ground between the engine and the chassis must be well connected. The DME detected an unexpected voltage condition on the EVAP circuit when the device was cycled On/Off during testing.<br>**Possible Causes:**<br>• EVAP power supply circuit is open<br>• EVAP solenoid control circuit is open or shorted to ground<br>• EVAP solenoid control circuit is shorted to power (B+)<br>• EVAP solenoid valve is damaged or it has failed<br>• EVAP canister has a leak or a poor seal |
| **DTC: P0460**<br>**Years:** 2007, 2008<br>**Models:** 335i, 335Ci, 335xi, 525i, 525xi, 530i, 530xi, 550i, 650i, 750i, 750Li, 760Li, Z4 3.0i, Z4 3.0Si, Z4 M<br>**Engines:** All<br>**Transmissions:** All | **Fuel Level Sensor "A" Circuit Malfunction Conditions:**<br>KOEO or KOER Self-Test enabled, and the DME detected a lack of power (VPWR) to the Fuel Pressure Regulator Control (FPRC) solenoid circuit. Cluster received incorrect fuel level from CAN or no message at all, calculated consumption does not correspond to transmitted fuel quantity.<br>**Possible Causes:**<br>• FPRC solenoid valve harness circuits are open or shorted<br>• FPRC input port or output port vacuum lines are damaged<br>• FRPC solenoid is damaged<br>• Fuel level is too low<br>• Check fuel level sensor |
| **DTC: P0461**<br>**Years:** 2007, 2008<br>**Models:** 335i, 335Ci, 335xi, 525i, 525xi, 530i, 530xi, 550i, 650i, 750i, 750Li, 760Li, Z4 3.0i, Z4 3.0Si, Z4 M<br>**Engines:** All<br>**Transmissions:** All | **Fuel Level Sensor "A" Circuit Range/Performance Conditions:**<br>KOEO or KOER Self-Test enabled, and the DME detected a lack of power (VPWR) to the Fuel Pressure Regulator Control (FPRC) solenoid circuit. Cluster received incorrect fuel level from CAN, calculated consumption does not correspond to transmitted fuel quantity. There is a stuck fuel level sensor, and the fault is recorded after driving roughly 50 miles (or 2.6 gallons of gas).<br>**Possible Causes:**<br>• FPRC solenoid valve harness circuits are open or shorted<br>• FPRC input port or output port vacuum lines are damaged<br>• FRPC solenoid is damaged<br>• Fuel level is too low<br>• Check fuel level sensor |
| **DTC: P0477**<br>**Years:** 2007, 2008<br>**Models:** 335i, 335Ci, 335xi, 525i, 525xi, 530i, 530xi, 550i, 650i, 750i, 750Li, 760Li, Z4 3.0i, Z4 3.0Si, Z4 M<br>**Engines:** All<br>**Transmissions:** All | **Exhaust Pressure Control Valve Low Conditions:**<br>Engine started, battery voltage must be at least 11.5v, all electrical components must be off, parking brake must be engaged (to keep daytime driving lights off), automatic transmission selector must be in park, the exhaust system must be properly sealed between the catalytic converter and the cylinder head, coolant temperature must be at least 80 degrees Celsius and oxygen sensor heaters for oxygen sensors before the catalytic converter must be functioning properly and the ground between the engine and the chassis must be well connected. The DME detected an unexpected voltage condition on the EVAP circuit when the device was cycled On/Off during testing. The driver circuit has detected a short to ground.<br>**Possible Causes:**<br>• EVAP power supply circuit is open<br>• EVAP solenoid control circuit is open or shorted to ground<br>• EVAP solenoid control circuit is shorted to power (B+)<br>• EVAP solenoid valve is damaged or it has failed<br>• EVAP canister has a leak or a poor seal<br>• Check wiring harness, otherwise replace DME |
| **DTC: P0478**<br>**Years:** 2007, 2008<br>**Models:** 335i, 335Ci, 335xi, 525i, 525xi, 530i, 530xi, 550i, 650i, 750i, 750Li, 760Li, Z4 3.0i, Z4 3.0Si, Z4 M<br>**Engines:** All<br>**Transmissions:** All | **Exhaust Pressure Control Valve High Conditions:**<br>Engine started, battery voltage must be at least 11.5v, all electrical components must be off, parking brake must be engaged (to keep daytime driving lights off), automatic transmission selector must be in park, the exhaust system must be properly sealed between the catalytic converter and the cylinder head, coolant temperature must be at least 80 degrees Celsius and oxygen sensor heaters for oxygen sensors before the catalytic converter must be functioning properly and the ground between the engine and the chassis must be well connected. The DME detected an unexpected voltage condition on the EVAP circuit when the device was cycled On/Off during testing. The driver circuit has detected a short to battery voltage<br>**Possible Causes:**<br>• EVAP power supply circuit is open<br>• EVAP solenoid control circuit is open or shorted to ground<br>• EVAP solenoid control circuit is shorted to power (B+)<br>• EVAP solenoid valve is damaged or it has failed<br>• EVAP canister has a leak or a poor seal<br>• Check wiring harness, otherwise replace DME |

| DTC | Trouble Code Title, Conditions & Possible Causes |
|---|---|
| **DTC: P0479**<br>**Years:** 2007, 2008<br>**Models:** 335i, 335Ci, 335xi, 525i, 525xi, 530i, 530xi, 550i, 650i, 750i, 750Li, 760Li, Z4 3.0i, Z4 3.0Si, Z4 M<br>**Engines:** All<br>**Transmissions:** All | **Exhaust Pressure Control Valve Intermittent Conditions:**<br>Engine started, battery voltage must be at least 11.5v, all electrical components must be off, parking brake must be engaged (to keep daytime driving lights off), automatic transmission selector must be in park, the exhaust system must be properly sealed between the catalytic converter and the cylinder head, coolant temperature must be at least 80 degrees Celsius and oxygen sensor heaters for oxygen sensors before the catalytic converter must be functioning properly and the ground between the engine and the chassis must be well connected. The DME detected an unexpected voltage condition on the EVAP circuit when the device was cycled On/Off during testing. The driver circuit has detected a implausible signal.<br>**Possible Causes:**<br>• EVAP power supply circuit is open<br>• EVAP solenoid control circuit is open or shorted to ground<br>• EVAP solenoid control circuit is shorted to power (B+)<br>• EVAP solenoid valve is damaged or it has failed<br>• EVAP canister has a leak or a poor seal<br>• Check wiring harness, otherwise replace DME |
| **DTC: P0491**<br>**2T MIL:** Yes<br>**Years:** 2007, 2008<br>**Models:** 328i, 328Ci, 328xi, 335i, 335Ci, 335xi, 525i, 525xi, 530i, 530xi, 550i, 650i, 750i, 750Li, 760Li, M5, M6, X3, X5, Z4 3.0i, Z4 3.0Si, Z4 M<br>**Engines:** All<br>**Transmissions:** All | **Secondary Air Injection System Insufficient Flow (Bank 1) Conditions:**<br>Engine started, battery voltage must be at least 11.5v, all electrical components must be off, parking brake must be engaged (to keep daytime driving lights off), automatic transmission selector must be in park and the ground between the engine and the chassis must be well connected. The DME detected the Secondary AIR pump airflow was not diverted correctly when requested during the self-test. The pump is functioning but the quantity of air is recognized as insufficient by HO2S. The secondary air pump is on, the oxygen sensor is heated up the cold start enrichment is activated and the coolant temperature is between negative 12 and 30 degrees Celsius.<br>**Possible Causes:**<br>• Air pump output is blocked or restricted<br>• AIR bypass solenoid is leaking or it is restricted<br>• AIR bypass solenoid is stuck open or stuck closed<br>• Check valve (one or more) is damaged or leaking<br>• Electric air injection pump hose(s) leaking<br>• Electric air injection pump is damaged or faulty |
| **DTC: P0492**<br>**2T MIL:** Yes<br>**Years:** 2007, 2008<br>**Models:** 328i, 328Ci, 328xi, 335i, 335Ci, 335xi, 525i, 525xi, 530i, 530xi, 550i, 650i, 750i, 750Li, 760Li, M5, M6, X3, X5, Z4 3.0i, Z4 3.0Si, Z4 M<br>**Engines:** All<br>**Transmissions:** All | **Secondary Air Injection System Insufficient Flow (Bank 2) Conditions:**<br>Engine started, battery voltage must be at least 11.5v, all electrical components must be off, parking brake must be engaged (to keep daytime driving lights off), automatic transmission selector must be in park and the ground between the engine and the chassis must be well connected. The DME detected the Secondary AIR pump airflow was not diverted correctly when requested during the self-test. The pump is functioning but the quantity of air is recognized as insufficient by HO2S. The secondary air pump is on, the oxygen sensor is heated up the cold start enrichment is activated and the coolant temperature is between negative 12 and 30 degrees Celsius.<br>**Possible Causes:**<br>• Air pump output is blocked or restricted<br>• AIR bypass solenoid is leaking or it is restricted<br>• AIR bypass solenoid is stuck open or stuck closed<br>• Check valve (one or more) is damaged or leaking<br>• Electric air injection pump hose(s) leaking<br>• Electric air injection pump is damaged or faulty |
| **DTC: P0500**<br>**MIL:** Yes<br>**Years:** 2007, 2008<br>**Models:** 328i, 328Ci, 328xi, 335i, 335Ci, 335xi, 525i, 525xi, 530i, 530xi, 550i, 650i, 750i, 750Li, 760Li, M5, M6, X3, X5, Z4 3.0i, Z4 3.0Si, Z4 M<br>**Engines:** All<br>**Transmissions:** All | **Vehicle Speed Sensor "A" Malfunction Conditions:**<br>Engine started; engine speed above the TCC stall speed, and the DME detected a loss of the VSS signal over a period of time or the signal is not usable.<br>**Note: The DME receives vehicle speed data from the VSS, TCSS, ABS module, CTM or GEM controller, depending up the application. Speed Signal from DSC too high because of possible tampering. Check DSC and wires. The engine speed is greater than 2000rpm, engine load greater than 3.5msec/rev., and the vehicle speed is more than 55mph. The fuel system status is in fuel cut-off mode.**<br>**Possible Causes:**<br>• VSS signal circuit is open or shorted to ground<br>• VSS harness circuit is shorted to ground<br>• VSS harness circuit is shorted to power<br>• VSS circuit open between the DME and related control module<br>• VSS or wheel speed sensors circuits are damaged<br>• Modules connected to VSC/VSS harness circuits are damaged<br>• Mechanical drive mechanism for the VSS is damaged |

| DTC | Trouble Code Title, Conditions & Possible Causes |
|---|---|
| **DTC: P0501**<br>**MIL: Yes**<br>**Years:** 2007, 2008<br>**Models:** 335i, 335Ci, 335xi, 525i, 525xi, 530i, 530xi, 550i, 650i, 750i, 750Li, 760Li, Z4 3.0i, Z4 3.0Si, Z4 M<br>**Engines:** All<br>**Transmissions:** All | **Vehicle Speed Sensor or PSOM Range/Performance Conditions:**<br>Engine started; engine speed above the TCC stall speed, and the DME detected a loss of the VSS signal over a period of time or the signal is not usable.<br>**Note: The DME receives vehicle speed data from the VSS, TCSS, ABS module, CTM or GEM controller, depending up the application. The engine speed is between 1000 and 1320rpm. The fuel system status is in the fuel cut-off mode. The coolant temperature is greater than 60 degrees Celsius. The vehicle speed is zero.**<br>**Possible Causes:**<br>• VSS signal circuit is open or shorted to ground<br>• VSS harness circuit is shorted to ground<br>• VSS harness circuit is shorted to power<br>• VSS circuit open between the DME and related control module<br>• VSS or wheel speed sensors circuits are damaged<br>• Modules connected to VSC/VSS harness circuits are damaged<br>• Mechanical drive mechanism for the VSS is damaged |
| **DTC: P0503**<br>**MIL: Yes**<br>**Years:** 2007, 2008<br>**Models:** 335i, 335Ci, 335xi, 525i, 525xi, 530i, 530xi, 550i, 650i, 750i, 750Li, 760Li, Z4 3.0i, Z4 3.0Si, Z4 M<br>**Engines:** All<br>**Transmissions:** All | **Vehicle Speed Sensor "A" Intermittent/Erratic/High Conditions:**<br>Engine started; engine speed above the TCC stall speed, and the DME detected a loss of the VSS signal over a period of time or the signal is not usable.<br>**Note: The DME receives vehicle speed data from the VSS, TCSS, ABS module, CTM or GEM controller, depending up the application. Speed Signal from DSC too high because of possible tampering. Check DSC and wires**<br>**Possible Causes:**<br>• VSS signal circuit is open or shorted to ground<br>• VSS harness circuit is shorted to ground<br>• VSS harness circuit is shorted to power<br>• VSS circuit open between the DME and related control module<br>• VSS or wheel speed sensors circuits are damaged<br>• Modules connected to VSC/VSS harness circuits are damaged<br>• Mechanical drive mechanism for the VSS is damaged |
| **DTC: P0505**<br>**2T MIL: Yes**<br>**Years:** 2007, 2008<br>**Models:** 328i, 328Ci, 328xi, 335i, 335Ci, 335xi, 525i, 525xi, 530i, 530xi, 550i, 650i, 750i, 750Li, 760Li, M5, M6, X3, X5, Z4 3.0i, Z4 3.0Si, Z4 M<br>**Engines:** All<br>**Transmissions:** All | **Idle Air Control Valve Malfunction Conditions:**<br>Engine started, battery voltage at least 11.5v, all electrical components off, ground connections between engine and chassis well connected, coolant temperature at least 80-degrees Celsius. The DME detected deviation from the normal operating parameters of the Idle Air Control Valve. The vehicle speed can be zero mph and the engine load must be less than 1.5ms.<br>**Possible Causes:**<br>• Charge air system leaks<br>• Recirculating valve for turbocharger is faulty<br>• Turbocharging system is damaged<br>• Vacuum diaphragm for turbocharger needs adjusting<br>• Wastegate bypass regulator valve is faulty |
| **DTC: P0506**<br>**2T MIL: Yes**<br>**Years:** 2007, 2008<br>**Models:** All<br>**Engines:** All<br>**Transmissions:** All | **Idle Air Control System RPM Lower Than Expected Conditions:**<br>Engine started, battery voltage must be at least 11.5v, all electrical components must be off, parking brake must be engaged (to keep daytime driving lights off), automatic transmission selector must be in park, the exhaust system must be properly sealed between the catalytic converter and the cylinder head, coolant temperature must be at least 80 degrees Celsius and oxygen sensor heaters for oxygen sensors before the catalytic converter must be functioning properly and the ground between the engine and the chassis must be well connected. The DME detected it could not control the idle speed correctly, as it is constantly more than 100 rpm less than specification.<br>**Possible Causes:**<br>• Air inlet is plugged or the air filter element is severely clogged<br>• IAC circuit is open or shorted<br>• IAC circuit VPWR circuit is open<br>• IAC solenoid is damaged or has failed<br>• The VSS has failed |

| DTC | Trouble Code Title, Conditions & Possible Causes |
|---|---|
| **DTC: P0507**<br>**2T MIL: Yes**<br>**Years:** 2007, 2008<br>**Models:** 328i, 328Ci, 328xi, 335i, 335Ci, 335xi, 525i, 525xi, 530i, 530xi, 550i, 650i, 750i, 750Li, 760Li, Z4 3.0i, Z4 3.0Si, Z4 M<br>**Engines:** All<br>**Transmissions:** All | **Idle Air Control System RPM Higher Than Expected Conditions:**<br>Engine started, battery voltage must be at least 11.5v, all electrical components must be off, parking brake must be engaged (to keep daytime driving lights off), automatic transmission selector must be in park, the exhaust system must be properly sealed between the catalytic converter and the cylinder head, coolant temperature must be at least 80 degrees Celsius and oxygen sensor heaters for oxygen sensors before the catalytic converter must be functioning properly and the ground between the engine and the chassis must be well connected. The DME detected it could not control the idle speed correctly, as it is constantly more than 200 rpm more than specification.<br>**Possible Causes:**<br>• Air intake leak located somewhere after the throttle body<br>• IAC control circuit is shorted to chassis ground<br>• IAC solenoid is damaged or has failed<br>• Throttle Valve Control module has failed or is clogged with carbon<br>• The VSS has failed |
| **DTC: P0512**<br>**Years:** 2007, 2008<br>**Models:** 335i, 335Ci, 335xi, 525i, 525xi, 530i, 530xi, 550i, 650i, 750i, 750Li, 760Li, Z4 3.0i, Z4 3.0Si, Z4 M<br>**Engines:** All<br>**Transmissions:** All | **Starter Request Circuit Malfunction Conditions:**<br>The engine is on for more than one second and the injection and ignition have not yet released. Engine rpm present before DME triggers the starter, or the starter relay sticks in IVM, crankshaft sensor intermittent contact, the starter is grounded.<br>**Possible Causes:**<br>• Check the starter and starter relay.<br>• The DME has failed |
| **DTC: P0520**<br>**Years:** 2007, 2008<br>**Models:** 335i, 335Ci, 335xi, 525i, 525xi, 530i, 530xi, 550i, 650i, 750i, 750Li, 760Li, Z4 3.0i, Z4 3.0Si, Z4 M<br>**Engines:** All<br>**Transmissions:** All | **Engine Oil Pressure Sensor/Switch Circuit Malfunction Conditions:**<br>The ignition must be on. The DME detected an error in the Engine Oil Pressure sensor. There is a short to ground. The plug has fallen off of the oil pressure switch. There is an open circuit in the harness. The pressure switch is defective.<br>**Possible Causes:**<br>• Replace the oil pressure sensor<br>• Check coolant |
| **DTC: P0530**<br>**Years:** 2007, 2008<br>**Models:** 335i, 335Ci, 335xi, 525i, 525xi, 530i, 530xi, 550i, 650i, 750i, 750Li, 760Li, Z4 3.0i, Z4 3.0Si, Z4 M<br>**Engines:** All<br>**Transmissions:** All | **A/C Refrigerant Pressure Sensor "A" Circuit Malfunction Conditions:**<br>The DME detected an implausible condition on the sensor.<br>**Possible Causes:**<br>• The A/C Refrigerant Pressure Sensor "A" has failed.<br>• The DME has failed |
| **DTC: P0532**<br>**Years:** 2007, 2008<br>**Models:** 335i, 335Ci, 335xi, 525i, 525xi, 530i, 530xi, 550i, 650i, 750i, 750Li, 760Li, Z4 3.0i, Z4 3.0Si, Z4 M<br>**Engines:** All<br>**Transmissions:** All | **A/C Refrigerant Pressure Sensor "A" Circuit Low Conditions:**<br>The DME detected a low condition on the sensor.<br>**Possible Causes:**<br>• The A/C Refrigerant Pressure Sensor "A" has failed.<br>• The DME has failed |
| **DTC: P0533**<br>**Years:** 2007, 2008<br>**Models:** 335i, 335Ci, 335xi, 525i, 525xi, 530i, 530xi, 550i, 650i, 750i, 750Li, 760Li, Z4 3.0i, Z4 3.0Si, Z4 M<br>**Engines:** All<br>**Transmissions:** All | **A/C Refrigerant Pressure Sensor "A" Circuit High Conditions:**<br>The DME detected a high condition on the sensor.<br>**Possible Causes:**<br>• The A/C Refrigerant Pressure Sensor "A" has failed.<br>• The DME has failed |

| DTC | Trouble Code Title, Conditions & Possible Causes |
|---|---|
| **DTC: P0560**<br>**Years:** 2007, 2008<br>**Models:** 335i, 335Ci, 335xi, 525i, 525xi, 530i, 530xi, 550i, 650i, 750i, 750Li, 760Li, Z4 3.0i, Z4 3.0Si, Z4 M<br>**Engines:** All<br>**Transmissions:** All | **System Voltage Malfunction Conditions:**<br>Engine started, battery voltage must be at least 11.5v, all electrical components must be off, parking brake must be engaged (to keep daytime driving lights off), automatic transmission selector must be in park, and the ground between the engine and the chassis must be well connected. The DME has detected a voltage value that is implausible or erratic. Engine speed must be greater than 1400rpm.<br>**Possible Causes:**<br>• Alternator damaged or faulty<br>• Battery voltage low or insufficient<br>• Fuses blown or circuits open<br>• Battery connection to terminal not clean<br>• Voltage regulator has failed |
| **DTC: P0561**<br>**Years:** 2007, 2008<br>**Models:** 335i, 335Ci, 335xi, 525i, 525xi, 530i, 530xi, 550i, 650i, 750i, 750Li, 760Li, Z4 3.0i, Z4 3.0Si, Z4 M<br>**Engines:** All<br>**Transmissions:** All | **System Voltage Unstable Conditions:**<br>Engine started, battery voltage must be at least 11.5v, all electrical components must be off, parking brake must be engaged (to keep daytime driving lights off), automatic transmission selector must be in park, and the ground between the engine and the chassis must be well connected. The DME has detected a voltage value that is too erratic for the system to function properly.<br>**Possible Causes:**<br>• Alternator damaged or faulty<br>• Battery voltage low or insufficient<br>• Fuses blown or circuits open<br>• Battery connection to terminal not clean<br>• Voltage regulator has failed |
| **DTC: P0562**<br>**Years:** 2007, 2008<br>**Models:** 335i, 335Ci, 335xi, 525i, 525xi, 530i, 530xi, 550i, 650i, 750i, 750Li, 760Li, Z4 3.0i, Z4 3.0Si, Z4 M<br>**Engines:** All<br>**Transmissions:** All | **System Voltage Low Conditions:**<br>Engine started, battery voltage must be at least 11.5v, all electrical components must be off, parking brake must be engaged (to keep daytime driving lights off), automatic transmission selector must be in park, and the ground between the engine and the chassis must be well connected. The DME has detected a voltage value that is below the specified minimum limit for the system to function properly.<br>**Possible Causes:**<br>• Alternator damaged or faulty<br>• Battery voltage low or insufficient<br>• Fuses blown or circuits open<br>• Battery connection to terminal not clean<br>• Voltage regulator has failed |
| **DTC: P0563**<br>**Years:** 2007, 2008<br>**Models:** 335i, 335Ci, 335xi, 525i, 525xi, 530i, 530xi, 550i, 650i, 750i, 750Li, 760Li, Z4 3.0i, Z4 3.0Si, Z4 M<br>**Engines:** All<br>**Transmissions:** All | **System Voltage High Conditions:**<br>Engine started for 18 seconds, battery voltage must be at least 11.5v, all electrical components must be off, parking brake must be engaged (to keep daytime driving lights off), automatic transmission selector must be in park, and the ground between the engine and the chassis must be well connected. The DME has detected a voltage value that has exceeded the specified maximum limit for the system to function properly. The vehicle was connected to 24 volts for too long after a jump start. ADC in ECU is defective. Delete stored fault codes from log. If fault reoccurs replace the ECU.<br>**Possible Causes:**<br>• Alternator damaged or faulty<br>• Battery voltage low or insufficient<br>• Fuses blown or circuits open<br>• Battery connection to terminal not clean<br>• Voltage regulator has failed |
| **DTC: P0571**<br>**Years:** 2007, 2008<br>**Models:** 335i, 335Ci, 335xi, 525i, 525xi, 530i, 530xi, 550i, 650i, 750i, 750Li, 760Li, Z4 3.0i, Z4 3.0Si, Z4 M<br>**Engines:** All<br>**Transmissions:** All | **Cruise/Brake Switch (A) Circuit Malfunction Conditions:**<br>Engine started, battery voltage must be at least 11.5v, all electrical components must be off, parking brake must be engaged (to keep daytime driving lights off), automatic transmission selector must be in park, and the ground between the engine and the chassis must be well connected. The DME has detected a voltage value that is implausible or erratic.<br>**Possible Causes:**<br>• Brake light switch is faulty<br>• Control circuit is shorted to chassis ground |

| DTC | Trouble Code Title, Conditions & Possible Causes |
|---|---|
| **DTC: P0597**<br>**2T MIL: Yes (U.S. only)**<br>**Years:** 2007, 2008<br>**Models:** 335i, 335Ci, 335xi, 525i, 525xi, 530i, 530xi, 550i, 650i, 750i, 750Li, 760Li, Z4 3.0i, Z4 3.0Si, Z4 M<br>**Engines:** All<br>**Transmissions:** All | **Thermostat Heater Control Circuit Open Conditions:**<br>The engine's warm up performance is monitored by comparing measured coolant temperature with the modeled coolant temperature to detect a defective coolant thermostat that is reading false. The engine temperature must be less than 65 degrees Celsius, engine speed greater than 800rpm (with the vehicle speed greater than 10 but less than 90km/h) and the ambient temperature greater than −8 degrees Celsius. The thermostat should be wide open when cold, but is in error if it opens below desired control temperature<br>**Possible Causes:**<br>• Check for low coolant level or incorrect coolant mixture<br>• DME detects a short circuit wiring in the ECT<br>• CHT sensor is out-of-calibration or it has failed<br>• ECT sensor is out-of-calibration or it has failed<br>• Replace the thermostat |
| **DTC: P0598**<br>**2T MIL: Yes (U.S. only)**<br>**Years:** 2007, 2008<br>**Models:** 335i, 335Ci, 335xi, 525i, 525xi, 530i, 530xi, 550i, 650i, 750i, 750Li, 760Li, Z4 3.0i, Z4 3.0Si, Z4 M<br>**Engines:** All<br>**Transmissions:** All | **Thermostat Heater Control Circuit Low Conditions:**<br>The engine's warm up performance is monitored by comparing measured coolant temperature with the modeled coolant temperature to detect a defective coolant thermostat that is reading false. The engine temperature must be less than 65 degrees Celsius, engine speed greater than 800rpm (with the vehicle speed greater than 10 but less than 90km/h) and the ambient temperature greater than −8 degrees Celsius. The thermostat should be wide open when cold, but is in error if it opens below desired control temperature<br>**Possible Causes:**<br>• Check for low coolant level or incorrect coolant mixture<br>• DME detects a short circuit wiring in the ECT<br>• CHT sensor is out-of-calibration or it has failed<br>• ECT sensor is out-of-calibration or it has failed<br>• Replace the thermostat |
| **DTC: P0599**<br>**2T MIL: Yes (U.S. only)**<br>**Years:** 2007, 2008<br>**Models:** 335i, 335Ci, 335xi, 525i, 525xi, 530i, 530xi, 550i, 650i, 750i, 750Li, 760Li, Z4 3.0i, Z4 3.0Si, Z4 M<br>**Engines:** All<br>**Transmissions:** All | **Thermostat Heater Control Circuit High Conditions:**<br>The engine's warm up performance is monitored by comparing measured coolant temperature with the modeled coolant temperature to detect a defective coolant thermostat that is reading false. The engine temperature must be less than 65 degrees Celsius, engine speed greater than 800rpm (with the vehicle speed greater than 10 but less than 90 km/h) and the ambient temperature greater than −8 degrees Celsius. The thermostat should be wide open when cold, but is in error if it opens below desired control temperature<br>**Possible Causes:**<br>• Check for low coolant level or incorrect coolant mixture<br>• DME detects a short circuit wiring in the ECT<br>• CHT sensor is out-of-calibration or it has failed<br>• ECT sensor is out-of-calibration or it has failed<br>• Replace the thermostat |
| **DTC: P0600**<br>**2T MIL: Yes**<br>**Years:** 2007, 2008<br>**Models:** 328i, 328Ci, 328xi, 335i, 335Ci, 335xi, 525i, 525xi, 530i, 530xi, 550i, 650i, 750i, 750Li, 760Li, M5, M6, X3, X5, Z4 3.0i, Z4 3.0Si, Z4 M<br>**Engines:** All<br>**Transmissions:** All | **Serial Communication Link (Data BUS) Message Missing Conditions:**<br>The Engine Control Module (DME) communicates with all databus-capable control modules via a CAN databus. These databus-capable control modules are connected via two data bus wires which are twisted together (CAN_High and CAN_Low), and exchange information (messages). Missing information on the databus is recognized as a malfunction and stored. Trouble-free operation of the CAN-Bus requires that it have a terminal resistance. This central terminal resistor is located in the Engine Control Module (DME).<br>**Possible Causes:**<br>• CAN data bus wires have short circuited to each other |
| **DTC: P0601**<br>**2T MIL: Yes**<br>**Years:** 2007, 2008<br>**Models:** 328i, 328Ci, 328xi, 335i, 335Ci, 335xi, 525i, 525xi, 530i, 530xi, 550i, 650i, 750i, 750Li, 760Li, M5, M6, X3, X5, Z4 3.0i, Z4 3.0Si, Z4 M<br>**Engines:** All<br>**Transmissions:** All | **Internal Control Module Memory Check Sum Error Conditions:**<br>Key on, the DME has detected a programming error. The RAM and ROM check displays an invalid check-sum at power up/down.<br>**Possible Causes:**<br>• Battery terminal corrosion, or loose battery connection<br>• Connection to the DME interrupted, or the circuit has been opened<br>• Reprogramming error has occurred and needs replacement. Remember to check for Aftermarket Performance Products before replacing a DME. |

| DTC | Trouble Code Title, Conditions & Possible Causes |
|---|---|
| **DTC: P0604**<br>**2T MIL: Yes**<br>**Years:** 2007, 2008<br>**Models:** 328i, 328Ci, 328xi, 335i, 335Ci, 335xi, 525i, 525xi, 530i, 530xi, 550i, 650i, 750i, 750Li, 760Li, M5, M6, X3, X5, Z4 3.0i, Z4 3.0Si, Z4 M<br>**Engines:** All<br>**Transmissions:** All | **Internal Control Module Random Access Memory (RAM) Error Conditions:**<br>Key on, and the DME detected an internal memory fault. This code will set if KAPWR to the DME is interrupted (at the initial key on). Watchdog on.<br>**Possible Causes:**<br>• Battery terminal corrosion, or loose battery connection<br>• Connection to the DME interrupted, or the circuit has been opened<br>• Reprogramming error has occurred and needs replacement. Remember to check for Aftermarket Performance Products before replacing a DME. |
| **DTC: P0605**<br>**Years:** 2007, 2008<br>**Models:** 335i, 335Ci, 335xi, 525i, 525xi, 550i, 650i, M5, M6, Z4 3.0i, Z4 3.0Si, Z4 M<br>**Engines:** All<br>**Transmissions:** All | **DME Read Only Memory (ROM) Test Error Conditions:**<br>Key on, and the DME detected a ROM test error (ROM inside DME is corrupted). The DME is normally replaced if this code has set.<br>**Possible Causes:**<br>• An attempt was made to change the module calibration, or a module programming error may have occurred<br>• Clear the trouble codes and then check for this trouble code. If it resets, the DME has failed and needs replacement.<br>• Aftermarket performance products may have been installed.<br>• The Transmission Control Module (TCM) has failed. |
| **DTC: P0606**<br>**2T MIL: Yes**<br>**Years:** 2007, 2008<br>**Models:** 335i, 335Ci, 335xi, 525i, 525xi, 550i, 650i, M5, M6, Z4 3.0i, Z4 3.0Si, Z4 M<br>**Engines:** All<br>**Transmissions:** All | **DME Internal Communication Error Conditions:**<br>Key on, and the DME detected an internal communications register read back error during the initial key on check period.<br>**Possible Causes:**<br>• Clear the trouble codes and then check for this trouble code. If it resets, the DME has failed and needs replacement.<br>• Remember to check for signs of Aftermarket Performance Products installation before replacing the DME. |
| **DTC: P0620**<br>**Years:** 2007, 2008<br>**Models:** 335i, 335Ci, 335xi, 525i, 525xi, 530i, 530xi, 550i, 650i, 750i, 750Li, 760Li, Z4 3.0i, Z4 3.0Si, Z4 M<br>**Engines:** All<br>**Transmissions:** All | **Generator Control Circuit Error Conditions:**<br>The engine is running for at least 25 seconds, and there is no communication faults at the BSD Interface<br>**Possible Causes:**<br>• Wires have short circuited to each other<br>• ECU has failed<br>• Generator failure |
| **DTC: P0704**<br>**Years:** 2007, 2008<br>**Models:** 335i, 335Ci, 335xi, 525i, 525xi, 530i, 530xi, 550i, 650i, 750i, 750Li, 760Li, Z4 3.0i, Z4 3.0Si, Z4 M<br>**Engines:** All<br>**Transmissions:** All | **Clutch Switch Input Circuit Malfunction Conditions:**<br>Engine started, battery voltage must be at least 11.5v, all electrical components must be off, parking brake must be engaged (to keep daytime driving lights off), automatic transmission selector must be in park, and the ground between the engine and the chassis must be well connected. The DME detected a voltage outside the normal performance range to allow the system to properly function.<br>**Possible Causes:**<br>• Circuit harness connector contacts are corroded or ingresses of water<br>• Circuit wires have shorted to each other, to battery or ground<br>• Automatic Transmission Hydraulic Pressure Sensor 1 has failed<br>• Solenoid valves in valve body are faulty<br>• Transmission Input Speed (RPM) Sensor has failed<br>• Transmission Output Speed (RPM) Sensor has failed<br>• Engine Control Module (DME) is faulty<br>• Voltage supply for Engine Control Module (DME) is faulty<br>• Transmission Control Module (TCM) is faulty |

| DTC | Trouble Code Title, Conditions & Possible Causes |
|---|---|
| **DTC: P0705**<br>**2T MIL: Yes**<br>**Years:** 2007, 2008<br>**Models:** 328i, 328Ci, 328xi, 335i, 335Ci, 335xi, 525i, 525xi, 530i, 530xi, 550i, 650i, 750i, 750Li, 760Li, M5, M6, X3, X5, Z4 3.0i, Z4 3.0Si, Z4 M<br>**Engines:** All<br>**Transmissions:** A/T | **TR Sensor Circuit Malfunction Conditions:**<br>Engine started, battery voltage must be at least 11.5v, all electrical components must be off, parking brake must be engaged (to keep daytime driving lights off), automatic transmission selector must be in park, and the ground between the engine and the chassis must be well connected. The DME detected a voltage or signal outside the normal performance range to allow the system to properly function. The engine speed is between 200 and 440rpm.<br>**Possible Causes:**<br>• Circuit harness connector contacts are corroded or ingresses of water<br>• Circuit wires have shorted to each other, to battery or ground<br>• Automatic Transmission Hydraulic Pressure Sensor 1 has failed<br>• Solenoid valves in valve body are faulty<br>• Transmission Input Speed (RPM) Sensor has failed<br>• Transmission Output Speed (RPM) Sensor has failed<br>• Engine Control Module (DME) is faulty<br>• Voltage supply for Engine Control Module (DME) is faulty<br>• Transmission Control Module (TCM) is faulty |
| **DTC: P0710**<br>**Years:** 2007, 2008<br>**Models:** 335i, 335Ci, 335xi, 525i, 525xi, 530i, 530xi, Z4 3.0i, Z4 3.0Si, Z4 M<br>**Engines:** All<br>**Transmissions:** A/T | **Transmission Fluid Temperature Sensor Circuit Malfunction Conditions:**<br>Engine started, battery voltage must be at least 11.5v, all electrical components must be off, parking brake must be engaged (to keep daytime driving lights off), automatic transmission selector must be in park, and the ground between the engine and the chassis must be well connected. The DME detected the Transmission fluid temperature sensor circuit was outside the normal range in the test to allow proper function.<br>**Possible Causes:**<br>• ATF is low, contaminated, dirty or burnt<br>• Circuit harness connector contacts are corroded or ingresses of water<br>• Circuit wires have shorted to each other, to battery or ground<br>• Automatic Transmission Hydraulic Pressure Sensor 1 has failed<br>• Solenoid valves in valve body are faulty<br>• Transmission Input Speed (RPM) Sensor has failed<br>• Transmission Output Speed (RPM) Sensor has failed<br>• Engine Control Module (DME) is faulty<br>• Voltage supply for Engine Control Module (DME) is faulty<br>• Transmission Control Module (TCM) is faulty |
| **DTC: P0711**<br>**Years:** 2007, 2008<br>**Models:** 335i, 335Ci, 335xi, 525i, 525xi, 530i, 530xi, Z4 3.0i, Z4 3.0Si, Z4 M<br>**Engines:** All<br>**Transmissions:** A/T | **Transmission Fluid Temperature Sensor Signal Range/Performance Conditions:**<br>Engine started, battery voltage must be at least 11.5v, all electrical components must be off, parking brake must be engaged (to keep daytime driving lights off), automatic transmission selector must be in park, and the ground between the engine and the chassis must be well connected. The DME detected the Transmission Fluid Temperature (TFT) sensor value was not close its normal operating temperature.<br>**Possible Causes:**<br>• ATF is low, contaminated, dirty or burnt<br>• TFT sensor signal circuit has a high resistance condition<br>• TFT sensor is out-of-calibration ("skewed") or it has failed |
| **DTC: P0712**<br>**Years:** 2007, 2008<br>**Models:** 335i, 335Ci, 335xi, 525i, 525xi, 530i, 530xi, Z4 3.0i, Z4 3.0Si, Z4 M<br>**Engines:** All<br>**Transmissions:** A/T | **Transmission Fluid Temperature Sensor Circuit Low Input Conditions:**<br>Engine started, battery voltage must be at least 11.5v, all electrical components must be off, parking brake must be engaged (to keep daytime driving lights off), automatic transmission selector must be in park, and the ground between the engine and the chassis must be well connected. The DME detected the Transmission Fluid Temperature (TFT) sensor was less than its minimum self-test range in the test.<br>**Possible Causes:**<br>• TFT sensor signal circuit is shorted to chassis ground<br>• TFT sensor signal circuit is shorted to sensor ground<br>• TFT sensor is damaged, or out-of-calibration, or has failed |
| **DTC: P0713**<br>**Years:** 2007, 2008<br>**Models:** 335i, 335Ci, 335xi, 525i, 525xi, 530i, 530xi, Z4 3.0i, Z4 3.0Si, Z4 M<br>**Engines:** All<br>**Transmissions:** A/T | **Transmission Fluid Temperature Sensor Circuit High Input Conditions:**<br>Engine started, battery voltage must be at least 11.5v, all electrical components must be off, parking brake must be engaged (to keep daytime driving lights off), automatic transmission selector must be in park, and the ground between the engine and the chassis must be well connected. The DME detected the Transmission Fluid Temperature (TFT) sensor was more than its maximum self-test range in the test.<br>**Possible Causes:**<br>• TFT sensor signal circuit is open between the sensor and DME<br>• TFT sensor ground circuit is open between sensor and DME<br>• TFT sensor is damaged or has failed |

| DTC | Trouble Code Title, Conditions & Possible Causes |
|---|---|
| **DTC: P0714**<br>**Years:** 2007, 2008<br>**Models:** 335i, 335Ci, 335xi, 525i, 525xi, 530i, 530xi, Z4 3.0i, Z4 3.0Si, Z4 M<br>**Engines:** All<br>**Transmissions:** A/T | **Transmission Fluid Temperature Sensor Circuit Intermittent Conditions:**<br>Engine started, battery voltage must be at least 11.5v, all electrical components must be off, parking brake must be engaged (to keep daytime driving lights off), automatic transmission selector must be in park, and the ground between the engine and the chassis must be well connected. The DME detected the Transmission Fluid Temperature (TFT) sensor was giving a false reading or was not reading at all.<br>**Possible Causes:**<br>• TFT sensor signal circuit is open between the sensor and DME<br>• TFT sensor ground circuit is open between sensor and DME<br>• TFT sensor is damaged or has failed |
| **DTC: P0715**<br>**2T MIL: Yes**<br>**Years:** 2007, 2008<br>**Models:** 328i, 328Ci, 328xi, 335i, 335Ci, 335xi, 525i, 525xi, 530i, 530xi, 550i, 650i, 750i, 750Li, 760Li, M5, M6, X3, X5, Z4 3.0i, Z4 3.0Si, Z4 M<br>**Engines:** All<br>**Transmissions:** A/T | **Input/Turbine Speed Sensor Circuit Malfunction Conditions:**<br>Engine started, vehicle driven with the vehicle speed sensor indicating more than 1 mph, and the DME detected the Transmission Vehicle Speed Sensor signals were erratic, or that they were missing for a period of time. The engine speed is greater than 600rpm. Any gear can be selected, output speed must be greater than 600rpm, and wheel speed greater than 400rpm.<br>**Possible Causes:**<br>• TVSS signal circuit is open<br>• TVSS signal is shorted to chassis ground<br>• TVSS signal is shorted to sensor ground<br>• TVSS assembly is damaged or it has failed |
| **DTC: P0720**<br>**2T MIL: Yes**<br>**Years:** 2007, 2008<br>**Models:** 328i, 328Ci, 328xi, 335i, 335Ci, 335xi, 525i, 525xi, 530i, 530xi, 550i, 650i, 750i, 750Li, 760Li, M5, M6, X3, X5, Z4 3.0i, Z4 3.0Si, Z4 M<br>**Engines:** All<br>**Transmissions:** A/T | **Output/Turbine Speed Sensor Circuit Malfunction Conditions:**<br>Engine started, vehicle driven with the vehicle speed sensor indicating more than 1 mph, and the DME detected the Transmission Vehicle Speed Sensor signals were erratic, or that they were missing for a period of time. The engine speed is greater than 600rpm. Any gear can be selected, output speed must be greater than 600rpm, and wheel speed greater than 400rpm.<br>**Possible Causes:**<br>• TVSS signal circuit is open<br>• TVSS signal is shorted to chassis ground<br>• TVSS signal is shorted to sensor ground<br>• TVSS assembly is damaged or it has failed |
| **DTC: P0721**<br>**Years:** 2007, 2008<br>**Models:** 335i, 335Ci, 335xi, 525i, 525xi, 530i, 530xi, Z4 3.0i, Z4 3.0Si, Z4 M<br>**Engines:** All<br>**Transmissions:** A/T | **A/T Output Shaft Speed Sensor Noise Interference Conditions:**<br>Engine started, VSS signal more than 1 mph, and the DME detected "noise" interference on the Output Shaft Speed (OSS) sensor circuit.<br>**Possible Causes:**<br>• After market add-on devices interfering with the OSS signal<br>• OSS connector is damaged, loose or shorted, or the wiring is misrouted or it is damaged<br>• OSS assembly is damaged or it has failed |
| **DTC: P0722**<br>**2T MIL: Yes**<br>**Years:** 2007, 2008<br>**Models:** 335i, 335Ci, 335xi, 525i, 525xi, 530i, 530xi, Z4 3.0i, Z4 3.0Si, Z4 M<br>**Engines:** All<br>**Transmissions:** A/T | **A/T Output Speed Sensor No Signal Conditions:**<br>Engine started, and the DME did not detect any Vehicle Speed Sensor (VSS) sensor signals upon initial vehicle movement.<br>**Possible Causes:**<br>• After market add-on devices interfering with the VSS signal<br>• VSS sensor wiring is misrouted, damaged or shorting<br>• DME and/or TCM has failed |
| **DTC: P0727**<br>**1T MIL: Yes**<br>**Years:** 2007, 2008<br>**Models:** 335i, 335Ci, 335xi, 525i, 525xi, 530i, 530xi,<br>**Engines:** All<br>**Transmissions:** A/T | **Engine Speed Input Circuit No Signal Conditions:**<br>The Engine Speed (RPM) Sensor detects engine speed and reference marks. Without an engine speed signal, the engine will not start. If the engine speed signal fails while the engine is running, the engine will stop immediately.<br>**Note: There is a larger-sized gap on the sensor wheel. This gap is the reference mark and does not mean that the sensor wheel is damaged.**<br>**Possible Causes:**<br>• Engine speed sensor has failed<br>• Circuit is shorting to ground, B+ or is open<br>• Sensor wheel is damaged, run out or not properly secured |

| DTC | Trouble Code Title, Conditions & Possible Causes |
|---|---|
| **DTC: P0731**<br>**2T MIL: Yes**<br>**Years:** 2007, 2008<br>**Models:** 328i, 328Ci, 328xi, 335i, 335Ci, 335xi, 525i, 525xi, 530i, 530xi, 550i, 650i, 750i, 750Li, 760Li, M5, M6, X3, X5, Z4 3.0i, Z4 3.0Si, Z4 M<br>**Engines:** All<br>**Transmissions:** A/T | **Incorrect First Gear Ratio Conditions:**<br>Engine started, vehicle operating with 1st gear commanded "on", and the DME detected an incorrect 1st gear ratio during the test.<br>**Possible Causes:**<br>• 1st Gear solenoid harness connector not properly seated<br>• 1st Gear solenoid signal shorted to ground, or open<br>• 1st Gear solenoid wiring harness connector is damaged<br>• 1st Gear solenoid is damaged or not properly installed<br>• ATF level is low<br>• Circuit harness connector contacts are corroded or ingresses of water<br>• Circuit wires have shorted to each other, to battery or ground<br>• Automatic Transmission Hydraulic Pressure Sensor 1 has failed<br>• Transmission Control Module (TCM) needs replacing<br>• Transmission Input Speed (RPM) Sensor has failed<br>• Transmission Output Speed (RPM) Sensor has failed |
| **DTC: P0732**<br>**2T MIL: Yes**<br>**Years:** 2007, 2008<br>**Models:** 328i, 328Ci, 328xi, 335i, 335Ci, 335xi, 525i, 525xi, 530i, 530xi, 550i, 650i, 750i, 750Li, 760Li, M5, M6, X3, X5, Z4 3.0i, Z4 3.0Si, Z4 M<br>**Engines:** All<br>**Transmissions:** A/T | **Incorrect Second Gear Ratio Conditions:**<br>Engine started, vehicle operating with 2nd Gear commanded "on", and the DME detected an incorrect 2nd gear ratio during the test. Input speed must be greater than 400rpm, and output speed must be greater than 250rpm for 10ms of continuous time.<br>**Possible Causes:**<br>• 2nd Gear solenoid harness connector not properly seated<br>• 2nd Gear solenoid signal shorted to ground, or open<br>• 2nd Gear solenoid wring harness connector is damaged<br>• 2nd Gear solenoid is damaged or not properly installed<br>• ATF level is low<br>• Circuit harness connector contacts are corroded or ingresses of water<br>• Circuit wires have shorted to each other, to battery or ground<br>• Automatic Transmission Hydraulic Pressure Sensor 1 has failed<br>• Transmission Control Module (TCM) needs replacing<br>• Transmission Input Speed (RPM) Sensor has failed<br>• Transmission Output Speed (RPM) Sensor has failed |
| **DTC: P0733**<br>**2T MIL: Yes**<br>**Years:** 2007, 2008<br>**Models:** 328i, 328Ci, 328xi, 335i, 335Ci, 335xi, 525i, 525xi, 530i, 530xi, 550i, 650i, 750i, 750Li, 760Li, M5, M6, X3, X5, Z4 3.0i, Z4 3.0Si, Z4 M<br>**Engines:** All<br>**Transmissions:** A/T | **Incorrect Third Gear Ratio Conditions:**<br>Engine started, vehicle operating with 3rd Gear commanded "on", and the DME detected an incorrect 3rd gear ratio during the test. Input speed must be greater than 400rpm, and output speed must be greater than 250rpm for 10ms of continuous time.<br>**Possible Causes:**<br>• 3rd Gear solenoid harness connector not properly seated<br>• 3rd Gear solenoid signal shorted to ground, or open<br>• 3rd Gear solenoid wiring harness connector is damaged<br>• 3rd Gear solenoid is damaged or not properly installed<br>• ATF level is low<br>• Circuit harness connector contacts are corroded or ingresses of water<br>• Circuit wires have shorted to each other, to battery or ground<br>• Automatic Transmission Hydraulic Pressure Sensor 1 has failed<br>• Transmission Control Module (TCM) needs replacing<br>• Transmission Input Speed (RPM) Sensor has failed<br>• Transmission Output Speed (RPM) Sensor has failed |
| **DTC: P0734**<br>**2T MIL: Yes**<br>**Years:** 2007, 2008<br>**Models:** 328i, 328Ci, 328xi, 335i, 335Ci, 335xi, 525i, 525xi, 530i, 530xi, 550i, 650i, 750i, 750Li, 760Li, M5, M6, X3, X5, Z4 3.0i, Z4 3.0Si, Z4 M<br>**Engines:** All<br>**Transmissions:** A/T | **Incorrect Fourth Gear Ratio Conditions:**<br>Engine started, vehicle operating with 4th Gear commanded "on", and the DME detected an incorrect 4th gear ratio during the test. Input speed must be greater than 400rpm, and output speed must be greater than 250rpm for 10ms of continuous time.<br>**Possible Causes:**<br>• 4th Gear solenoid harness connector not properly seated<br>• 4th Gear solenoid signal shorted to ground, or open<br>• 4th Gear solenoid wiring harness connector is damaged<br>• 4th Gear solenoid is damaged or not properly installed<br>• ATF level is low<br>• Circuit harness connector contacts are corroded or ingresses of water<br>• Circuit wires have shorted to each other, to battery or ground<br>• Automatic Transmission Hydraulic Pressure Sensor 1 has failed<br>• Transmission Control Module (TCM) needs replacing<br>• Transmission Input Speed (RPM) Sensor has failed<br>• Transmission Output Speed (RPM) Sensor has failed |

| DTC | Trouble Code Title, Conditions & Possible Causes |
|---|---|
| **DTC: P0735**<br>**2T MIL: Yes**<br>**Years:** 2007, 2008<br>**Models:** 328i, 328Ci, 328xi, 335i, 335Ci, 335xi, 525i, 525xi, 530i, 530xi, 550i, 650i, 750i, 750Li, 760Li, M5, M6, X3, X5, Z4 3.0i, Z4 3.0Si, Z4 M<br>**Engines:** All<br>**Transmissions:** A/T | **Incorrect Fifth Gear Ratio Conditions:**<br>Engine started, vehicle operating with 5th Gear commanded "on", and the DME detected an incorrect 5th gear ratio during the test. Input speed must be greater than 400rpm, and output speed must be greater than 250rpm for 10ms of continuous time.<br>**Possible Causes:**<br>• 5th Gear solenoid harness connector not properly seated<br>• 5th Gear solenoid signal shorted to ground, or open<br>• 5th Gear solenoid wiring harness connector is damaged<br>• 5th Gear solenoid is damaged or not properly installed<br>• ATF level is low<br>• Circuit harness connector contacts are corroded or ingresses of water<br>• Circuit wires have shorted to each other, to battery or ground<br>• Automatic Transmission Hydraulic Pressure Sensor 1 has failed<br>• Transmission Control Module (TCM) needs replacing<br>• Transmission Input Speed (RPM) Sensor has failed<br>• Transmission Output Speed (RPM) Sensor has failed |
| **DTC: P0740**<br>**2T MIL: Yes**<br>**Years:** 2007, 2008<br>**Models:** 328i, 328Ci, 328xi, 335i, 335Ci, 335xi, 525i, 525xi, 530i, 530xi, 550i, 650i, 750i, 750Li, 760Li, M5, M6, X3, X5, Z4 3.0i, Z4 3.0Si, Z4 M<br>**Engines:** All<br>**Transmissions:** A/T | **TCC Solenoid Circuit Malfunction Conditions:**<br>Engine started, KOER Self-Test enabled, vehicle driven at cruise speed, and the DME did not detect any voltage drop across the TCC solenoid circuit during the test period.<br>**Possible Causes:**<br>• TCC solenoid control circuit is open or shorted to ground<br>• TCC solenoid wiring harness connector is damaged<br>• TCC solenoid is damaged or has failcd |
| **DTC: P0741**<br>**2T MIL: Yes**<br>**Years:** 2007, 2008<br>**Models:** All<br>**Engines:** All<br>**Transmissions:** A/T | **TCC Mechanical System Range/Performance Conditions:**<br>Engine started, vehicle driven in gear with VSS signals received, and the DME detected excessive slippage while in normal operation. The TCC is stuck off.<br>**Possible Causes:**<br>• TCC solenoid has a mechanical failure<br>• TCC solenoid has a hydraulic failure |
| **DTC: P0743**<br>**2T MIL: Yes**<br>**Years:** 2007, 2008<br>**Models:** 328i, 328Ci, 328xi, 335i, 335Ci, 335xi, 525i, 525xi, 530i, 530xi, 550i, 650i, 750i, 750Li, 760Li, M5, M6, X3, X5, Z4 3.0i, Z4 3.0Si, Z4 M<br>**Engines:** All<br>**Transmissions:** A/T | **TCC Solenoid Circuit Malfunction Conditions:**<br>Engine started, KOER Self-Test enabled, vehicle driven at cruise speed, and the DME did not detect any voltage drop across the TCC solenoid circuit during the test period.<br>**Possible Causes:**<br>• TCC solenoid control circuit is open or shorted to ground<br>• TCC solenoid wiring harness connector is damaged<br>• TCC solenoid is damaged or has failed |
| **DTC: P0745**<br>**2T MIL: Yes**<br>**Years:** 2007, 2008<br>**Models:** 335i, 335Ci, 335xi, 525i, 525xi, Z4 3.0i, Z4 3.0Si, Z4 M<br>**Engines:** All<br>**Transmissions:** A/T | **Pressure Regulator Valve 1 Plausibility Conditions:**<br>The current to/from the pressure regulator valve is either higher or lower than the threshold value.<br>**Possible Causes:**<br>• Pressure control solenoid circuit is shorting to ground<br>• Pressure control solenoid circuit is open<br>• Valve has failed<br>• TCM has failed |
| **DTC: P0748**<br>**2T MIL: Yes**<br>**Years:** 2007, 2008<br>**Models:** 328i, 328Ci, 328xi, 335i, 335Ci, 335xi, 525i, 525xi, 530i, 530xi, 550i, 650i, 750i, 750Li, 760Li, M5, M6, X3, X5, Z4 3.0i, Z4 3.0Si, Z4 M<br>**Engines:** All<br>**Transmissions:** A/T | **Pressure Regulator Valve 2 Upper Threshold Conditions:**<br>The signal to/from the pressure regulator valve has been interrupted.<br>**Possible Causes:**<br>• Pressure control solenoid circuit is shorting to ground<br>• Pressure control solenoid circuit is open<br>• Valve has failed<br>• TCM has failed |

| DTC | Trouble Code Title, Conditions & Possible Causes |
|---|---|
| **DTC: P0753**<br>**2T MIL:** Yes<br>**Years:** 2007, 2008<br>**Models:** 328i, 328Ci, 328xi, 335i, 335Ci, 335xi, 525i, 525xi, 530i, 530xi, 550i, 650i, 750i, 750Li, 760Li, M5, M6, X3, X5, Z4 3.0i, Z4 3.0Si, Z4 M<br>**Engines:** All<br>**Transmissions:** A/T | **Solenoid Valve 1 Upper Threshold Conditions:**<br>The signal to/from the pressure regulator valve is interrupted or short circuited to supply.<br>**Possible Causes:**<br>• Pressure control solenoid circuit is shorting to ground<br>• Pressure control solenoid circuit is open<br>• Valve has failed<br>• TCM has failed |
| **DTC: P0755**<br>**2T MIL:** Yes<br>**Years:** 2007, 2008<br>**Models:** 335i, 335Ci, 335xi, 525i, 525xi, Z4 3.0i, Z4 3.0Si, Z4 M<br>**Engines:** All<br>**Transmissions:** A/T | **Shift Solenoid "B" Circuit Continuity Short to Battery Conditions:**<br>Engine started, vehicle driven with the solenoid applied, and the DME detected an unexpected voltage condition on the SS1/B solenoid circuit was incorrect during the test.<br>**Possible Causes:**<br>• Solenoid valves in valve body are faulty<br>• Solenoid circuit is shorting to ground<br>• Solenoid circuit is open<br>• TCM has failed or wiring is shorting |
| **DTC: P0758**<br>**2T MIL:** Yes<br>**Years:** 2007, 2008<br>**Models:** 328i, 328Ci, 328xi, 335i, 335Ci, 335xi, 525i, 525xi, 530i, 530xi, 550i, 650i, 750i, 750Li, 760Li, M5, M6, X3, X5, Z4 3.0i, Z4 3.0Si, Z4 M<br>**Engines:** All<br>**Transmissions:** A/T | **Solenoid Valve 2 Upper Threshold Conditions:**<br>The signal to/from the pressure regulator valve is interrupted or short circuited to supply.<br>**Possible Causes:**<br>• Pressure control solenoid circuit is shorting to ground<br>• Pressure control solenoid circuit is open<br>• Valve has failed<br>• TCM has failed |
| **DTC: P0760**<br>**2T MIL:** Yes<br>**Years:** 2007, 2008<br>**Models:** 335i, 335Ci, 335xi, 525i, 525xi, Z4 3.0i, Z4 3.0Si, Z4 M<br>**Engines:** All<br>**Transmissions:** A/T | **Shift Solenoid "C" Circuit Continuity Short to Battery Conditions:**<br>Engine started, vehicle driven with the solenoid applied, and the DME detected an unexpected voltage condition on the SS1/C solenoid circuit was incorrect during the test.<br>**Possible Causes:**<br>• Solenoid valves in valve body are faulty<br>• Solenoid circuit is shorting to ground<br>• Solenoid circuit is open<br>• TCM has failed or wiring is shorting |
| **DTC: P0761**<br>**2T MIL:** Yes<br>**Years:** 2007, 2008<br>**Models:** 335i, 335Ci, 335xi, 525i, 525xi, Z4 3.0i, Z4 3.0Si, Z4 M<br>**Engines:** All<br>**Transmissions:** A/T | **Solenoid Valve 3 Plausibility Conditions:**<br>The signal to/from the pressure regulator valve is interrupted or does not exist.<br>**Possible Causes:**<br>• Pressure control solenoid circuit is shorting to ground<br>• Pressure control solenoid circuit is open<br>• Valve has failed<br>• TCM has failed |
| **DTC: P0762**<br>**2T MIL:** Yes<br>**Years:** 2007, 2008<br>**Models:** 335i, 335Ci, 335xi, 525i, 525xi, Z4 3.0i, Z4 3.0Si, Z4 M<br>**Engines:** All<br>**Transmissions:** A/T | **A/T Shift Solenoid 3/C Function Range/Performance Conditions:**<br>Engine started, vehicle driven with Shift Solenoid 3/C applied, and the DME detected a mechanical failure occurred (stuck "on") while operating Shift Solenoid 3/C during the test.<br>**Possible Causes:**<br>• SS3/C solenoid may be stuck "on"<br>• SS3/C solenoid has a mechanical failure<br>• SS3/C solenoid has a hydraulic failure |
| **DTC: P0763**<br>**2T MIL:** Yes<br>**Years:** 2007, 2008<br>**Models:** 328i, 328Ci, 328xi, 335i, 335Ci, 335xi, 525i, 525xi, 530i, 530xi, 550i, 650i, 750i, 750Li, 760Li, M5, M6, X3, X5, Z4 3.0i, Z4 3.0Si, Z4 M<br>**Engines:** All<br>**Transmissions:** A/T | **Solenoid Valve 3 Upper Threshold Conditions:**<br>The signal to/from the pressure regulator valve is interrupted or short circuited to supply.<br>**Possible Causes:**<br>• Pressure control solenoid circuit is shorting to ground<br>• Pressure control solenoid circuit is open<br>• Valve has failed<br>• TCM has failed |

| DTC | Trouble Code Title, Conditions & Possible Causes |
|---|---|
| **DTC: P0775**<br>**2T MIL:** Yes<br>**Years:** 2007, 2008<br>**Models:** 335i, 335Ci, 335xi, 525i, 525xi, Z4 3.0i, Z4 3.0Si, Z4 M<br>**Engines:** All<br>**Transmissions:** A/T | **Pressure Regulator Valve 2 Plausibility Conditions:**<br>The current to/from the pressure regulator valve is either higher or lower than the threshold value.<br>**Possible Causes:**<br>• Pressure control solenoid circuit is shorting to ground<br>• Pressure control solenoid circuit is open<br>• Valve has failed<br>• TCM has failed |
| **DTC: P0778**<br>**2T MIL:** Yes<br>**Years:** 2007, 2008<br>**Models:** 335i, 335Ci, 335xi, 525i, 525xi, 530i, 530xi, 550i, 650i, 750i, 750Li, 760Li, Z4 3.0i, Z4 3.0Si, Z4 M<br>**Engines:** All<br>**Transmissions:** A/T | **Pressure Control Solenoid "B" Electrical Conditions:**<br>Engine started, vehicle driven with the solenoid applied, and the DME detected an unexpected voltage condition on the SS3/C solenoid circuit was incorrect during the test..<br>**Possible Causes:**<br>• Shift Solenoid connector is damaged, open or shorted<br>• Shift Solenoid control circuit is open<br>• Shift Solenoid control circuit is shorted to ground<br>• Shift Solenoid is damaged or it has failed |
| **DTC: P0781**<br>**2T MIL:** Yes<br>**Years:** 2007, 2008<br>**Models:** 335i, 335Ci, 335xi, 525i, 525xi, 530i, 530xi, Z4 3.0i, Z4 3.0Si, Z4 M<br>**Engines:** All<br>**Transmissions:** A/T | **1-2 Shift Range Monitoring Conditions:**<br>Engine running and vehicle driven, the DME detected a mechanical malfunction within the transmission. The output speed is greater than 300rpm, the transmission oil temperature is greater than 0 degrees Celsius, the engine speed is greater or equal to 600rpm and the range position is P, R, or N.<br>**Possible Causes:**<br>• Solenoid valves in valve body are faulty<br>• Solenoid circuit is shorting to ground<br>• Solenoid circuit is open<br>• TCM has failed or wiring is shorting<br>• Mechanical malfunction in transmission |
| **DTC: P0782**<br>**2T MIL:** Yes<br>**Years:** 2007, 2008<br>**Models:** 335i, 335Ci, 335xi, 525i, 525xi, Z4 3.0i, Z4 3.0Si, Z4 M<br>**Engines:** All<br>**Transmissions:** A/T | **2-3 Shift Range Monitoring Conditions:**<br>Engine running and vehicle driven, the DME detected a mechanical malfunction within the transmission. The output speed is greater than 300rpm, the transmission oil temperature is greater than 0 degrees Celsius, the engine speed is greater or equal to 600rpm and the range position is P, R, or N.<br>**Possible Causes:**<br>• Solenoid valves in valve body are faulty<br>• Solenoid circuit is shorting to ground<br>• Solenoid circuit is open<br>• TCM has failed or wiring is shorting<br>• Mechanical malfunction in transmission |
| **DTC: P0783**<br>**2T MIL:** Yes<br>**Years:** 2007, 2008<br>**Models:** 335i, 335Ci, 335xi, 525i, 525xi, Z4 3.0i, Z4 3.0Si, Z4 M<br>**Engines:** All<br>**Transmissions:** A/T | **3-4 Shift Range Monitoring Conditions:**<br>Engine running and vehicle driven, the DME detected a mechanical malfunction within the transmission. The output speed is greater than 300rpm, the transmission oil temperature is greater than 0 degrees Celsius, the engine speed is greater or equal to 600rpm and the range position is P, R, or N.<br>**Possible Causes:**<br>• Solenoid valves in valve body are faulty<br>• Solenoid circuit is shorting to ground<br>• Solenoid circuit is open<br>• TCM has failed or wiring is shorting<br>• Mechanical malfunction in transmission |
| **DTC: P0784**<br>**2T MIL:** Yes<br>**Years:** 2007, 2008<br>**Models:** 335i, 335Ci, 335xi, 525i, 525xi, 530i, 530xi, Z4 3.0i, Z4 3.0Si, Z4 M<br>**Engines:** All<br>**Transmissions:** A/T | **4-5 Shift Range Monitoring Conditions:**<br>Engine running and vehicle driven, the DME detected a mechanical malfunction within the transmission. The output speed is greater than 300rpm, the transmission oil temperature is greater than 0 degrees Celsius, the engine speed is greater or equal to 600rpm and the range position is P, R, or N.<br>**Possible Causes:**<br>• Solenoid valves in valve body are faulty<br>• Solenoid circuit is shorting to ground<br>• Solenoid circuit is open<br>• TCM has failed or wiring is shorting<br>• Mechanical malfunction in transmission |

| DTC | Trouble Code Title, Conditions & Possible Causes |
|---|---|
| **DTC: P0798**<br>**2T MIL: Yes**<br>**Years:** 2007, 2008<br>**Models:** 335i, 335Ci, 335xi, 525i, 525xi, 530i, 530xi, 750i, 750Li, 760Li, Z4 3.0i, Z4 3.0Si, Z4 M<br>**Engines:** All<br>**Transmissions:** A/T | **Pressure Regulator Valve 2 Upper Threshold Conditions:**<br>The signal to/from the pressure regulator valve has been interrupted.<br>**Possible Causes:**<br>• Pressure control solenoid circuit is shorting to ground<br>• Pressure control solenoid circuit is open<br>• Valve has failed<br>• TCM has failed |
| **DTC: P0970**<br>**2T MIL: Yes**<br>**Years:** 2007, 2008<br>**Models:** 335i, 335Ci, 335xi, 525i, 525xi, 530i, 530xi, Z4 3.0i, Z4 3.0Si, Z4 M<br>**Engines:** All<br>**Transmissions:** A/T | **Pressure Regulator Valve 3 Lower Threshold Conditions:**<br>The signal to/from the pressure regulator valve is interrupted or short circuited to ground.<br>**Possible Causes:**<br>• Pressure control solenoid circuit is shorting to ground<br>• Pressure control solenoid circuit is open<br>• Valve has failed<br>• TCM has failed |
| **DTC: P0971**<br>**2T MIL: Yes**<br>**Years:** 2007, 2008<br>**Models:** 335i, 335Ci, 335xi, 525i, 525xi, 530i, 530xi, Z4 3.0i, Z4 3.0Si, Z4 M<br>**Engines:** All<br>**Transmissions:** A/T | **Pressure Regulator Valve 3 Upper Threshold Conditions:**<br>The signal to/from the pressure regulator valve is interrupted or short circuited to supply.<br>**Possible Causes:**<br>• Pressure control solenoid circuit is shorting to ground<br>• Pressure control solenoid circuit is open<br>• Valve has failed<br>• TCM has failed |
| **DTC: P0973**<br>**2T MIL: Yes**<br>**Years:** 2007, 2008<br>**Models:** 335i, 335Ci, 335xi, 525i, 525xi, 530i, 530xi, Z4 3.0i, Z4 3.0Si, Z4 M<br>**Engines:** All<br>**Transmissions:** A/T | **Solenoid Valve 1 Lower Threshold Conditions:**<br>The signal to/from the pressure regulator valve is interrupted or short circuited to ground.<br>**Possible Causes:**<br>• Pressure control solenoid circuit is shorting to ground<br>• Pressure control solenoid circuit is open<br>• Valve has failed<br>• TCM has failed |

## Gas Engine OBD II Trouble Code List (P1xxx Codes)

| DTC | Trouble Code Title, Conditions & Possible Causes |
|---|---|
| **DTC: P1000**<br>**MIL: No**<br>**Years:** 2007, 2008<br>**Models:** 335i, 335Ci, 335xi, 525i, 525xi, 530i, 530xi, 550i, 650i, 750i, 750Li, 760Li, Z4 3.0i, Z4 3.0Si, Z4 M<br>**Engines:** All<br>**Transmissions:** A/T | **Valvetronic (VVT) System Minimum Stroke Adaptation Number of Stops Exceeded Conditions:**<br>After the ignition is on for 500ms, the minimum number of stroke adaptations was exceeded. It is a distribution balance issue, and raising the minimum stroke at idle doesn't produce the desired results.<br>**Possible Causes:**<br>• Mechanical components have worn |
| **DTC: P1001**<br>**Years:** 2007, 2008<br>**Models:** 335i, 335Ci, 335xi, 525i, 525xi, 530i, 530xi, 550i, 650i, 750i, 750Li, 760Li, Z4 3.0i, Z4 3.0Si, Z4 M<br>**Engines:** All<br>**Transmissions:** A/T | **Valvetronic (VVT) Limp Home Request High Input Conditions:**<br>After 500ms the there is detected a short to battery voltage. If there are simultaneous CAN faults detected, Terminal 15 is probably open at VVT-SG, otherwise check Kb-B.<br>**Possible Causes:**<br>• Resolve the primary fault. |
| **DTC: P1002**<br>**Years:** 2007, 2008<br>**Models:** 335i, 335Ci, 335xi, 525i, 525xi, 530i, 530xi, 550i, 650i, 750i, 750Li, 760Li, Z4 3.0i, Z4 3.0Si, Z4 M<br>**Engines:** All<br>**Transmissions:** A/T | **Valvetronic (VVT) Limp Home Request Low Input Conditions:**<br>After 500ms the there is detected a short to ground. If there are simultaneous CAN faults detected, Terminal 15 is probably open at VVT-SG, otherwise check Kb-B.<br>**Possible Causes:**<br>• Resolve the primary fault. |

| DTC | Trouble Code Title, Conditions & Possible Causes |
|---|---|
| **DTC: P1003**<br>**Years:** 2007, 2008<br>**Models:** 335i, 335Ci, 335xi, 525i, 525xi, 530i, 530xi, 550i, 650i, 750i, 750Li, 760Li, Z4 3.0i, Z4 3.0Si, Z4 M<br>**Engines:** All<br>**Transmissions:** A/T | **Valvetronic (VVT) Limp Home Request Open Circuit Conditions:**<br>After 500ms the there is detected a short. If there are simultaneous CAN faults detected, Terminal 15 is probably open at VVT-SG, otherwise check Kb-B.<br>**Possible Causes:**<br>• Resolve the primary fault. |
| **DTC: P1004**<br>**Years:** 2007, 2008<br>**Models:** 335i, 335Ci, 335xi, 525i, 525xi, 530i, 530xi, 550i, 650i, 750i, 750Li, 760Li, Z4 3.0i, Z4 3.0Si, Z4 M<br>**Engines:** All<br>**Transmissions:** A/T | **Valvetronic (VVT) Guiding Sensor Solenoid Loss (Bank 1) Conditions:**<br>After 3ms and the DME and VVT are active, it is detected that the sensor is missing a magnet.<br>**Possible Causes:**<br>• Defective sensor. |
| **DTC: P1005**<br>**Years:** 2007, 2008<br>**Models:** 335i, 335Ci, 335xi, 525i, 525xi, 530i, 530xi, 550i, 650i, 750i, 750Li, 760Li, Z4 3.0i, Z4 3.0Si, Z4 M<br>**Engines:** All<br>**Transmissions:** A/T | **Valvetronic (VVT) Guiding Sensor Reset Error (Bank 1) Conditions:**<br>After 1.5ms and the DME and VVT are active, it is detected that the sensor is not properly resetting.<br>**Possible Causes:**<br>• Plug contact problem at Pin 6 on sensor<br>• Oil in plug at sensor<br>• Replace sensor<br>• Repair gaskets |
| **DTC: P1006**<br>**Years:** 2007, 2008<br>**Models:** 335i, 335Ci, 335xi, 525i, 525xi, 530i, 530xi, 550i, 650i, 750i, 750Li, 760Li, Z4 3.0i, Z4 3.0Si, Z4 M<br>**Engines:** All<br>**Transmissions:** A/T | **Valvetronic (VVT) Guiding Sensor Parity Error (Bank 1) Conditions:**<br>After 12ms and the DME and VVT are active, it is detected that the sensor is not properly communicating. The plug is defective on the sensor or there is an open circuit<br>**Possible Causes:**<br>• Plug contact problem at Pin 6 on sensor<br>• Oil in plug at sensor<br>• Replace sensor<br>• Repair gaskets |
| **DTC: P1007**<br>**Years:** 2007, 2008<br>**Models:** 335i, 335Ci, 335xi, 525i, 525xi, 530i, 530xi, 550i, 650i, 750i, 750Li, 760Li, Z4 3.0i, Z4 3.0Si, Z4 M<br>**Engines:** All<br>**Transmissions:** A/T | **Valvetronic (VVT) Guiding Sensor Gradient Error (Bank 1) Conditions:**<br>After 9ms and the DME and VVT are active, it is detected that the sensor has a gradient violation/identity, causing a reading of implausible sensor data. The plug is defective on the sensor or there is an open circuit<br>**Possible Causes:**<br>• Replace sensor |
| **DTC: P1008**<br>**Years:** 2007, 2008<br>**Models:** 335i, 335Ci, 335xi, 525i, 525xi, 530i, 530xi, 550i, 650i, 750i, 750Li, 760Li, Z4 3.0i, Z4 3.0Si, Z4 M<br>**Engines:** All<br>**Transmissions:** A/T | **Valvetronic (VVT) Guiding Sensor Solenoid Loss (Bank 2) Conditions:**<br>After 3ms and the DME and VVT are active, it is detected that the sensor is missing a magnet.<br>**Possible Causes:**<br>• Defective sensor. |
| **DTC: P1009**<br>**Years:** 2007, 2008<br>**Models:** 335i, 335Ci, 335xi, 525i, 525xi, 530i, 530xi, 550i, 650i, 750i, 750Li, 760Li, Z4 3.0i, Z4 3.0Si, Z4 M<br>**Engines:** All<br>**Transmissions:** A/T | **Valvetronic (VVT) Guiding Sensor Reset Error (Bank 2) Conditions:**<br>After 1.5ms and the DME and VVT are active, it is detected that the sensor is not properly resetting.<br>**Possible Causes:**<br>• Plug contact problem at Pin 6 on sensor<br>• Oil in plug at sensor<br>• Replace sensor<br>• Repair gaskets |

| DTC | Trouble Code Title, Conditions & Possible Causes |
|---|---|
| **DTC: P1010**<br>**Years:** 2007, 2008<br>**Models:** 335i, 335Ci, 335xi, 525i, 525xi, 530i, 530xi, 550i, 650i, 750i, 750Li, 760Li, Z4 3.0i, Z4 3.0Si, Z4 M<br>**Engines:** All<br>**Transmissions:** A/T | **Valvetronic (VVT) Guiding Sensor Parity Error (Bank 2) Conditions:**<br>After 12ms and the DME and VVT active, it is detected that the sensor is not properly communicating. The plug is defective on the sensor or there is an open circuit<br>**Possible Causes:**<br>• Plug contact problem at Pin 6 on sensor<br>• Oil in plug at sensor<br>• Replace sensor<br>• Repair gaskets |
| **DTC: P1011**<br>**Years:** 2007, 2008<br>**Models:** 335i, 335Ci, 335xi, 525i, 525xi, 530i, 530xi, 550i, 650i, 750i, 750Li, 760Li, Z4 3.0i, Z4 3.0Si, Z4 M<br>**Engines:** All<br>**Transmissions:** A/T | **Valvetronic (VVT) Guiding Sensor Gradient Error (Bank 2) Conditions:**<br>After 9ms and the DME and VVT are active, it is detected that the sensor has a gradient violation/identity, causing a reading of implausible sensor data. The plug is defective on the sensor or there is an open circuit<br>**Possible Causes:**<br>• Replace sensor |
| **DTC: P1012**<br>**Years:** 2007, 2008<br>**Models:** 335i, 335Ci, 335xi, 525i, 525xi, 530i, 530xi, 550i, 650i, 750i, 750Li, 760Li, Z4 3.0i, Z4 3.0Si, Z4 M<br>**Engines:** All<br>**Transmissions:** A/T | **Valvetronic (VVT) Reference Sensor Solenoid Loss (Bank 1) Conditions:**<br>After 3ms and the DME and VVT are active, it is detected that the sensor is missing a magnet.<br>**Possible Causes:**<br>• Defective sensor. |
| **DTC: P1013**<br>**Years:** 2007, 2008<br>**Models:** 335i, 335Ci, 335xi, 525i, 525xi, 530i, 530xi, 550i, 650i, 750i, 750Li, 760Li, Z4 3.0i, Z4 3.0Si, Z4 M<br>**Engines:** All<br>**Transmissions:** A/T | **Valvetronic (VVT) Reference Sensor Reset Error (Bank 1) Conditions:**<br>After 1.5ms and the DME and VVT are active, it is detected that the sensor is not properly resetting.<br>**Possible Causes:**<br>• Plug contact problem at Pin 6 on sensor<br>• Oil in plug at sensor<br>• Replace sensor<br>• Repair gaskets |
| **DTC: P1014**<br>**Years:** 2007, 2008<br>**Models:** 335i, 335Ci, 335xi, 525i, 525xi, 530i, 530xi, 550i, 650i, 750i, 750Li, 760Li, Z4 3.0i, Z4 3.0Si, Z4 M<br>**Engines:** All<br>**Transmissions:** A/T | **Valvetronic (VVT) Reference Sensor Parity Error (Bank 1) Conditions:**<br>After 12ms and the DME and VVT are active, it is detected that the sensor is not properly communicating. The plug is defective on the sensor or there is an open circuit<br>**Possible Causes:**<br>• Plug contact problem at Pin 6 on sensor<br>• Oil in plug at sensor<br>• Replace sensor<br>• Repair gaskets |
| **DTC: P1015**<br>**Years:** 2007, 2008<br>**Models:** 335i, 335Ci, 335xi, 525i, 525xi, 530i, 530xi, 550i, 650i, 750i, 750Li, 760Li, Z4 3.0i, Z4 3.0Si, Z4 M<br>**Engines:** All<br>**Transmissions:** A/T | **Valvetronic (VVT) Reference Sensor Gradient Error (Bank 1) Conditions:**<br>After 9ms and the DME and VVT are active, it is detected that the sensor has a gradient violation/identity, causing a reading of implausible sensor data. The plug is defective on the sensor or there is an open circuit<br>**Possible Causes:**<br>• Replace sensor |
| **DTC: P1022**<br>**MIL:** No<br>**Years:** 2007, 2008<br>**Models:** 335i, 335Ci, 335xi, 525i, 525xi, 530i, 530xi, 550i, 650i, 750i, 750Li, 760Li, Z4 3.0i, Z4 3.0Si, Z4 M<br>**Engines:** All<br>**Transmissions:** A/T | **Valvetronic (VVT), Eccentric Shaft Sensor 2 Circuit Low Input Conditions:**<br>With the engine running, the fault is a low voltage supply to the sensor after 3ms. The DME, VVT are active at 4.5 to 5.5 volts<br>**Possible Causes:**<br>• Short to ground in wiring harness<br>• VVT-SG defective<br>• Short circuit within the sensor<br>• Check plug and wiring harness for sensor defect |

| DTC | Trouble Code Title, Conditions & Possible Causes |
|---|---|
| **DTC: P1023**<br>**MIL:** No<br>**Years:** 2007, 2008<br>**Models:** 335i, 335Ci, 335xi, 525i, 525xi, 530i, 530xi, 550i, 650i, 750i, 750Li, 760Li, Z4 3.0i, Z4 3.0Si, Z4 M<br>**Engines:** All<br>**Transmissions:** A/T | **Valvetronic (VVT) Self-Learning Function Faulty Adjustment Range (Bank 1) Conditions:**<br>With the engine running, the fault is an out of range adjustment for the self-learning function of the VVT. Check the balance spring installation and the mechanical components for wear.<br>**Possible Causes:**<br>• Stuck at upper travel limit<br>• Travel limit worn or deformed<br>• Wear in idler lever<br>• Torque compensation spring missing or not connected |
| **DTC: P1024**<br>**MIL:** No<br>**Years:** 2007, 2008<br>**Models:** 335i, 335Ci, 335xi, 525i, 525xi, 530i, 530xi, 550i, 650i, 750i, 750Li, 760Li, Z4 3.0i, Z4 3.0Si, Z4 M<br>**Engines:** All<br>**Transmissions:** A/T | **Valvetronic (VVT) Self-Learning Function Faulty Lower Learning Range (Bank 1) Conditions:**<br>With the engine running, the fault is an out of range adjustment for the self-learning function of the VVT at the lower range. Check the installation and the mechanical components for wear.<br>**Possible Causes:**<br>• Stuck at upper travel limit<br>• Defective sensor magnet missing)<br>• Travel limit worn or deformed<br>• Wear in idler lever<br>• Torque compensation spring missing or not connected |
| **DTC: P1025**<br>**MIL:** No<br>**Years:** 2007, 2008<br>**Models:** 335i, 335Ci, 335xi, 525i, 525xi, 530i, 530xi, 550i, 650i, 750i, 750Li, 760Li, Z4 3.0i, Z4 3.0Si, Z4 M<br>**Engines:** All<br>**Transmissions:** A/T | **Valvetronic (VVT) Self-Learning Function No Positions Stored (Bank 1) Conditions:**<br>With the engine running, no travel limit has been initialized. Check the installation and the mechanical components for wear. This fault is because the system is operated for the first time with a new VVT-SG as there is no automatic limit initialization.<br>**Possible Causes:**<br>• Conduct travel limit initialization. |
| **DTC: P1026**<br>**MIL:** No<br>**Years:** 2007, 2008<br>**Models:** 335i, 335Ci, 335xi, 525i, 525xi, 530i, 530xi, 550i, 650i, 750i, 750Li, 760Li, Z4 3.0i, Z4 3.0Si, Z4 M<br>**Engines:** All<br>**Transmissions:** A/T | **Valvetronic (VVT) Self-Learning Function Faulty Adjustment Range (Bank 2) Conditions:**<br>With the engine running, the fault is an out of range adjustment for the self-learning function of the VVT. Check the balance spring installation and the mechanical components for wear.<br>**Possible Causes:**<br>• Stuck at upper travel limit<br>• Travel limit worn or deformed<br>• Wear in idler lever<br>• Torque compensation spring missing or not connected |
| **DTC: P1027**<br>**MIL:** No<br>**Years:** 2007, 2008<br>**Models:** 335i, 335Ci, 335xi, 525i, 525xi, 530i, 530xi, 550i, 650i, 750i, 750Li, 760Li, Z4 3.0i, Z4 3.0Si, Z4 M<br>**Engines:** All<br>**Transmissions:** A/T | **Valvetronic (VVT) Self-Learning Function Faulty Lower Learning Range (Bank 2) Conditions:**<br>With the engine running, the fault is an out of range adjustment for the self-learning function of the VVT at the lower range. Check the installation and the mechanical components for wear.<br>**Possible Causes:**<br>• Stuck at upper travel limit<br>• Defective sensor (magnet missing)<br>• Travel limit worn or deformed<br>• Wear in idler lever<br>• Torque compensation spring missing or not connected |
| **DTC: P1028**<br>**MIL:** No<br>**Years:** 2007, 2008<br>**Models:** 335i, 335Ci, 335xi, 525i, 525xi, 530i, 530xi, 550i, 650i, 750i, 750Li, 760Li, Z4 3.0i, Z4 3.0Si, Z4 M<br>**Engines:** All<br>**Transmissions:** A/T | **Valvetronic (VVT) Self-Learning Function No Positions Stored (Bank 2) Conditions:**<br>With the engine running, no travel limit has been initialized. Check the sensor installation and the mechanical components for wear. This fault is because the system is operated for the first time with a new VVT-SG as there is no automatic limit initialization.<br>**Possible Causes:**<br>• Conduct travel limit initialization. |

| DTC | Trouble Code Title, Conditions & Possible Causes |
|---|---|
| **DTC: P1030**<br>**MIL:** No<br>**Years:** 2007, 2008<br>**Models:** 335i, 335Ci, 335xi, 525i, 525xi, 530i, 530xi, 550i, 650i, 750i, 750Li, 760Li, Z4 3.0i, Z4 3.0Si, Z4 M<br>**Engines:** All<br>**Transmissions:** A/T | **Valvetronic (VVT) Actuator Monitoring Position Control, Control Deviation (Bank 1) Conditions:**<br>The engine is on and running for 45ms at 9.6 to 15.5 volts and a sluggish monitoring movement, direction or rotation was detected. This function monitors the VVT system for resistance to motion and is always active when the driver circuits are released for operation, there's no control with pulse-duty factor, no relay/enable fault, no under voltage, the travel limits are initialized, and there is no reference sensor faults after sensor switching. Check the wiring harness for shorts, the sensor installation, and for wear and mechanical sticking.<br>**Possible Causes:**<br>• Low battery charge<br>• Open motor control circuit<br>• Motor shorted to ground<br>• Loose sensor or it is operating at the limit |
| **DTC: P1031**<br>**MIL:** No<br>**Years:** 2007, 2008<br>**Models:** 335i, 335Ci, 335xi, 525i, 525xi, 530i, 530xi, 550i, 650i, 750i, 750Li, 760Li, Z4 3.0i, Z4 3.0Si, Z4 M<br>**Engines:** All<br>**Transmissions:** A/T | **Valvetronic (VVT) Actuator Monitoring Recognition of Direction of Rotation Plausibility (Bank 1) Conditions:**<br>The engine is on and running for 63 to 498ms at 9.6 to 15.5 volts. Check to determine whether adjustment can be approved. Check VVT system once before each power application within one driving cycle for correct servo motor polarity and sticking in the system.<br>**Possible Causes:**<br>• Low battery charge<br>• Open motor control circuit<br>• Motor shorted to ground<br>• Loose sensor or it is operating at the limit |
| **DTC: P1033**<br>**MIL:** No<br>**Years:** 2007, 2008<br>**Models:** 335i, 335Ci, 335xi, 525i, 525xi, 530i, 530xi, 550i, 650i, 750i, 750Li, 760Li, Z4 3.0i, Z4 3.0Si, Z4 M<br>**Engines:** All<br>**Transmissions:** A/T | **Valvetronic (VVT) Actuator Monitoring Position Control, Control Deviation (Bank 2) Conditions:**<br>The engine is on and running for 45ms at 9.6 to 15.5 volts and a sluggish monitoring movement, direction or rotation was detected. This function monitors the VVT system for resistance to motion and is always active when the driver circuits are released for operation, there's no control with pulse-duty factor, no relay/enable fault, no under voltage, the travel limits are initialized, and there is no reference sensor faults after sensor switching. Check the wiring harness for shorts, the sensor installation, and for wear and mechanical sticking.<br>**Possible Causes:**<br>• Low battery charge<br>• Open motor control circuit<br>• Motor shorted to ground<br>• Loose sensor or it is operating at the limit |
| **DTC: P1034**<br>**MIL:** No<br>**Years:** 2007, 2008<br>**Models:** 335i, 335Ci, 335xi, 525i, 525xi, 530i, 530xi, 550i, 650i, 750i, 750Li, 760Li, Z4 3.0i, Z4 3.0Si, Z4 M<br>**Engines:** All<br>**Transmissions:** A/T | **Valvetronic (VVT) Actuator Monitoring Recognition of Direction of Rotation Plausibility (Bank 1) Conditions:**<br>The engine is on and running for 63 to 498ms at 9.6 to 15.5 volts. Check to determine whether adjustment can be approved. Check VVT system once before each power application within one driving cycle for correct servo motor polarity and sticking in the system.<br>**Possible Causes:**<br>• Low battery charge<br>• Open motor control circuit<br>• Motor shorted to ground<br>• Loose sensor or it is operating at the limit |
| **DTC: P1035**<br>**MIL:** No<br>**Years:** 2007, 2008<br>**Models:** 335i, 335Ci, 335xi, 525i, 525xi, 530i, 530xi, 550i, 650i, 750i, 750Li, 760Li, Z4 3.0i, Z4 3.0Si, Z4 M<br>**Engines:** All<br>**Transmissions:** A/T | **Valvetronic (VVT) CAN Message Monitoring Faulty Desired Message (Bank 1) Conditions:**<br>After the DME is active or the engine running for 500ms, with a voltage of 7 on Terminal 87, the monitoring system displayed a faulty message.<br>**Possible Causes:**<br>• ECU failure (or SZL or ZGM)<br>• Bus system failure<br>• Defective bus controller (SZL or ZGM)<br>• Short circuit in CAN wire or open circuit.<br>• Defective DME<br>• Defective VVT-SG |
| **DTC: P1036**<br>**MIL:** No<br>**Years:** 2007, 2008<br>**Models:** 335i, 335Ci, 335xi, 525i, 525xi, 530i, 530xi, 550i, 650i, 750i, 750Li, 760Li, Z4 3.0i, Z4 3.0Si, Z4 M<br>**Engines:** All<br>**Transmissions:** A/T | **Valvetronic (VVT) CAN Timeout VVT-Desired Message (Bank 1) Conditions:**<br>After the key has been on for 800ms (within two messages) or the engine running for 400ms, with a battery voltage at 10 volts, the difference between the deactivation and the starting positions exceeds specification. No suspension of BUS activity.<br>**Possible Causes:**<br>• ECU failure (or SZL or ZGM)<br>• Bus system failure<br>• Defective bus controller (SZL or ZGM)<br>• Short circuit in CAN wire or open circuit.<br>• Defective DME<br>• Defective VVT-SG |

| DTC | Trouble Code Title, Conditions & Possible Causes |
|---|---|
| **DTC: P1037**<br>**MIL:** No<br>**Years:** 2007, 2008<br>**Models:** 335i, 335Ci, 335xi, 525i, 525xi, 530i, 530xi, 550i, 650i, 750i, 750Li, 760Li, Z4 3.0i, Z4 3.0Si, Z4 M<br>**Engines:** All<br>**Transmissions:** A/T | **Valvetronic (VVT) CAN Timeout Message (Bank 1) Conditions:**<br>After the key has been on for 800ms (within two messages) or the engine running for 400ms, with a battery voltage at 10 volts, the difference between the deactivation and the starting positions exceeds specification. No suspension of BUS activity.<br>**Possible Causes:**<br>• ECU failure (or SZL or ZGM)<br>• Bus system failure<br>• Defective bus controller (SZL or ZGM)<br>• Short circuit in CAN wire or open circuit.<br>• Defective DME<br>• Defective VVT-SG |
| **DTC: P1038**<br>**MIL:** No<br>**Years:** 2007, 2008<br>**Models:** 335i, 335Ci, 335xi, 525i, 525xi, 530i, 530xi, 550i, 650i, 750i, 750Li, 760Li, Z4 3.0i, Z4 3.0Si, Z4 M<br>**Engines:** All<br>**Transmissions:** A/T | **Valvetronic (VVT) CAN Message Monitoring Faulty Desired Message (Bank 2) Conditions:**<br>After the DME is active or the engine running for 500ms, with a voltage of 7 on Terminal 87, the monitoring system displayed a faulty message.<br>**Possible Causes:**<br>• ECU failure (or SZL or ZGM)<br>• Bus system failure<br>• Defective bus controller (SZL or ZGM)<br>• Short circuit in CAN wire or open circuit.<br>• Defective DME<br>• Defective VVT-SG |
| **DTC: P1039**<br>**MIL:** No<br>**Years:** 2007, 2008<br>**Models:** 335i, 335Ci, 335xi, 525i, 525xi, 530i, 530xi, 550i, 650i, 750i, 750Li, 760Li, Z4 3.0i, Z4 3.0Si, Z4 M<br>**Engines:** All<br>**Transmissions:** A/T | **Valvetronic (VVT) CAN Timeout VVT-Desired Message (Bank 2) Conditions:**<br>After the key has been on for 800ms (within two messages) or the engine running for 400ms, with a battery voltage at 10 volts, the difference between the deactivation and the starting positions exceeds specification. No suspension of BUS activity.<br>**Possible Causes:**<br>• ECU failure (or SZL or ZGM)<br>• Bus system failure<br>• Defective bus controller (SZL or ZGM)<br>• Short circuit in CAN wire or open circuit.<br>• Defective DME<br>• Defective VVT-SG |
| **DTC: P1040**<br>**MIL:** No<br>**Years:** 2007, 2008<br>**Models:** 335i, 335Ci, 335xi, 525i, 525xi, 530i, 530xi, 550i, 650i, 750i, 750Li, 760Li, Z4 3.0i, Z4 3.0Si, Z4 M<br>**Engines:** All<br>**Transmissions:** A/T | **Valvetronic (VVT) CAN Timeout Message (Bank 2) Conditions:**<br>After the key has been on for 800ms (within two messages) or the engine running for 400ms, with a battery voltage at 10 volts, the difference between the deactivation and the starting positions exceeds specification. No suspension of BUS activity.<br>**Possible Causes:**<br>• ECU failure (or SZL or ZGM)<br>• Bus system failure<br>• Defective bus controller (SZL or ZGM)<br>• Short circuit in CAN wire or open circuit.<br>• Defective DME<br>• Defective VVT-SG |
| **DTC: P1041**<br>**2T MIL:** Yes<br>**Years:** 2007, 2008<br>**Models:** 335i, 335Ci, 335xi, 525i, 525xi, 550i, 650i, 750i, 750Li, 760Li, X3, X5, Z4 3.0i, Z4 3.0Si, Z4 M<br>**Engines:** All<br>**Transmissions:** A/T | **Valvetronic (VVT) Actuator Control Module EEPROM Error (Bank 1) Conditions:**<br>Ignition on for 50ms, the DME detected a control module malfunction (software). To achieve optimal anti-theft protection for the vehicle, an anti-theft immobilizer is installed. The anti-theft immobilizer is a system for enabling and locking the Engine Control Module (DME). So that this system cannot be circumvented, it is necessary to perform adaptation of the anti-theft immobilizer using the Vehicle Diagnostic and Information System VAS 5052 in the On Board Diagnostic (OBD) function. The great availability of equipment options makes it necessary to adapt the Engine Control Module (DME) to the vehicle (e.g. throttle valve control module or cruise control system). This "writing" function is not possible with the generic scan tool.<br>**Possible Causes:**<br>• Engine Control Module (DME) has failed<br>• Voltage supply for Engine Control Module (DME) has shorted |
| **DTC: P1042**<br>**2T MIL:** Yes<br>**Years:** 2007, 2008<br>**Models:** 335i, 335Ci, 335xi, 525i, 525xi, 550i, 650i, 750i, 750Li, 760Li, X3, X5, Z4 3.0i, Z4 3.0Si, Z4 M<br>**Engines:** All<br>**Transmissions:** A/T | **Valvetronic (VVT) Actuator Control Module Random Access Memory Error (Bank 1) Conditions:**<br>Key on for 50ms, and the DME detected an internal memory fault. This code will set if KAPWR to the DME is interrupted (at the initial key on).<br>**Possible Causes:**<br>• Battery terminal corrosion, or loose battery connection<br>• Connection to the DME interrupted, or the circuit has been opened<br>• Reprogramming error has occurred and needs replacement. Remember to check for Aftermarket Performance Products before replacing a DME. |

| DTC | Trouble Code Title, Conditions & Possible Causes |
|---|---|
| **DTC: P1043**<br>**2T MIL: Yes**<br>**Years:** 2007, 2008<br>**Models:** 335i, 335Ci, 335xi, 525i, 525xi, 550i, 650i, 750i, 750Li, 760Li, X3, X5, Z4 3.0i, Z4 3.0Si, Z4 M<br>**Engines:** All<br>**Transmissions:** A/T | **Valvetronic (VVT) Actuator Control Module Read Only Memory Error (Bank 1) Conditions:**<br>Key on for 50ms, and the DME detected an internal memory fault. This code will set if KAPWR to the DME is interrupted (at the initial key on).<br>**Possible Causes:**<br>• Battery terminal corrosion, or loose battery connection<br>• Connection to the DME interrupted, or the circuit has been opened<br>• Reprogramming error has occurred and needs replacement. Remember to check for Aftermarket Performance Products before replacing a DME. |
| **DTC: P1044**<br>**2T MIL: Yes**<br>**Years:** 2007, 2008<br>**Models:** 335i, 335Ci, 335xi, 525i, 525xi, 550i, 650i, 750i, 750Li, 760Li, X3, X5, Z4 3.0i, Z4 3.0Si, Z4 M<br>**Engines:** All<br>**Transmissions:** A/T | **Valvetronic (VVT) Actuator Control Module EEPROM Error (Bank 2) Conditions:**<br>Ignition on for 50ms, the DME detected a control module malfunction (software). To achieve optimal anti-theft protection for the vehicle, an anti-theft immobilizer is installed. The anti-theft immobilizer is a system for enabling and locking the Engine Control Module (DME). So that this system cannot be circumvented, it is necessary to perform adaptation of the anti-theft immobilizer using the Vehicle Diagnostic and Information System VAS 5052 in the On Board Diagnostic (OBD) function. The great availability of equipment options makes it necessary to adapt the Engine Control Module (DME) to the vehicle (e.g. throttle valve control module or cruise control system). This "writing" function is not possible with the generic scan tool.<br>**Possible Causes:**<br>• Engine Control Module (DME) has failed<br>• Voltage supply for Engine Control Module (DME) has shorted |
| **DTC: P1045**<br>**2T MIL: Yes**<br>**Years:** 2007, 2008<br>**Models:** 335i, 335Ci, 335xi, 525i, 525xi, 550i, 650i, 750i, 750Li, 760Li, X3, X5, Z4 3.0i, Z4 3.0Si, Z4 M<br>**Engines:** All<br>**Transmissions:** A/T | **Valvetronic (VVT) Actuator Control Module Random Access Memory Error (Bank 2) Conditions:**<br>Key on for 50ms, and the DME detected an internal memory fault. This code will set if KAPWR to the DME is interrupted (at the initial key on).<br>**Possible Causes:**<br>• Battery terminal corrosion, or loose battery connection<br>• Connection to the DME interrupted, or the circuit has been opened<br>• Reprogramming error has occurred and needs replacement. Remember to check for Aftermarket Performance Products before replacing a DME. |
| **DTC: P1046**<br>**2T MIL: Yes**<br>**Years:** 2007, 2008<br>**Models:** 335i, 335Ci, 335xi, 525i, 525xi, 550i, 650i, 750i, 750Li, 760Li, X3, X5, Z4 3.0i, Z4 3.0Si, Z4 M<br>**Engines:** All<br>**Transmissions:** A/T | **Valvetronic (VVT) Actuator Control Module Read Only Memory Error (Bank 2) Conditions:**<br>Key on for 50ms, and the DME detected an internal memory fault. This code will set if KAPWR to the DME is interrupted (at the initial key on).<br>**Possible Causes:**<br>• Battery terminal corrosion, or loose battery connection<br>• Connection to the DME interrupted, or the circuit has been opened<br>• Reprogramming error has occurred and needs replacement. Remember to check for Aftermarket Performance Products before replacing a DME. |
| **DTC: P1047**<br>**2T MIL: Yes**<br>**Years:** 2007, 2008<br>**Models:** 335i, 335Ci, 335xi, 525i, 525xi, 550i, 650i, 750i, 750Li, 760Li, X3, X5, Z4 3.0i, Z4 3.0Si, Z4 M<br>**Engines:** All<br>**Transmissions:** A/T | **Valvetronic (VVT) Actuator Control Circuit High Input (Bank 1) Conditions:**<br>Key on for 3ms, and the DME detected a short to positive.<br>**Possible Causes:**<br>• Short to battery voltage in Kb-B<br>• KS ground in Kb-B<br>• KS sensor motor to ground<br>• KS motor winding |
| **DTC: P1048**<br>**2T MIL: Yes**<br>**Years:** 2007, 2008<br>**Models:** 335i, 335Ci, 335xi, 525i, 525xi, 550i, 650i, 750i, 750Li, 760Li, X3, X5, Z4 3.0i, Z4 3.0Si, Z4 M<br>**Engines:** All<br>**Transmissions:** A/T | **Valvetronic (VVT) Actuator Control Circuit Low Input (Bank 1) Conditions:**<br>Key on for 3ms, and the DME detected a short to ground.<br>**Possible Causes:**<br>• Short to ground in Kb-B<br>• KS ground in Kb-B<br>• KS motor to ground<br>• KS motor winding |

| DTC | Trouble Code Title, Conditions & Possible Causes |
|---|---|
| **DTC: P1050**<br>**2T MIL: Yes**<br>**Years:** 2007, 2008<br>**Models:** 335i, 335Ci, 335xi, 525i, 525xi, 550i, 650i, 750i, 750Li, 760Li, X3, X5, Z4 3.0i, Z4 3.0Si, Z4 M<br>**Engines:** All<br>**Transmissions:** A/T | **Valvetronic (VVT) Control Circuit (Bank 1) Conditions:**<br>Key on for 3ms, and the DME detected that the Control Circuit triggered a general fault.<br>**Possible Causes:**<br>• This fault is usually overwritten by three other faults before the user recognizes it. |
| **DTC: P1051**<br>**2T MIL: Yes**<br>**Years:** 2007, 2008<br>**Models:** 335i, 335Ci, 335xi, 525i, 525xi, 550i, 650i, 750i, 750Li, 760Li, X3, X5, Z4 3.0i, Z4 3.0Si, Z4 M<br>**Engines:** All<br>**Transmissions:** A/T | **Valvetronic (VVT) Control Circuit High Input (Bank 1) Conditions:**<br>Key on for 3ms, and the DME detected a short to positive. Check wiring harness or otherwise replace the servo.<br>**Possible Causes:**<br>• Short to battery voltage in Kb-B<br>• KS ground in Kb-B<br>• KS motor to ground<br>• KS motor winding |
| **DTC: P1052**<br>**2T MIL: Yes**<br>**Years:** 2007, 2008<br>**Models:** 335i, 335Ci, 335xl, 525l, 525xi, 550i, 650i, 750i, 750Li, 760Li, X3, X5, Z4 3.0i, Z4 3.0Si, Z4 M<br>**Engines:** All<br>**Transmissions:** A/T | **Valvetronic (VVT) Control Circuit Low Input (Bank 1) Conditions:**<br>Key on for 3ms, and the DME detected a short to ground. Check wiring harness or otherwise replace the servo.<br>**Possible Causes:**<br>• Short to ground in Kb-B<br>• KS ground in Kb-B<br>• KS motor to ground<br>• KS motor winding |
| **DTC: P1054**<br>**2T MIL: Yes**<br>**Years:** 2007, 2008<br>**Models:** 335i, 335Ci, 335xi, 525i, 525xi, 550i, 650i, 750i, 750Li, 760Li, X3, X5, Z4 3.0i, Z4 3.0Si, Z4 M<br>**Engines:** All<br>**Transmissions:** A/T | **Valvetronic (VVT) Control Circuit (Bank 2) Conditions:**<br>Key on for 3ms, and the DME detected that the Control Circuit triggered a general fault.<br>**Possible Causes:**<br>• This fault is usually overwritten by three other faults before the user recognizes it. |
| **DTC: P1055**<br>**2T MIL: Yes**<br>**Years:** 2007, 2008<br>**Models:** 335i, 335Ci, 335xi, 525i, 525xi, 550i, 650i, 750i, 750Li, 760Li, X3, X5, Z4 3.0i, Z4 3.0Si, Z4 M<br>**Engines:** All<br>**Transmissions:** A/T | **Valvetronic (VVT) Supply Voltage Control Motor High Input (Bank 1) Conditions:**<br>Key on for 200ms, the DME detected that the supply voltage was too high (more than 17 volts). This is a jump-start detection that throws a fault if there is 24 volts of power for longer than 21 seconds.<br>**Possible Causes:**<br>• Check electrical system for faults. |
| **DTC: P1056**<br>**2T MIL: Yes**<br>**Years:** 2007, 2008<br>**Models:** 335i, 335Ci, 335xi, 525i, 525xi, 550i, 650i, 750i, 750Li, 760Li, X3, X5, Z4 3.0i, Z4 3.0Si, Z4 M<br>**Engines:** All<br>**Transmissions:** A/T | **Valvetronic (VVT) Supply Voltage Control Motor Low Input (Bank 1) Conditions:**<br>Key on for 200ms, the DME detected that the supply voltage was too low (less than 5 volts).<br>**Possible Causes:**<br>• VVT fuse is faulty<br>• Defective load reduction relay<br>• Plug/Kb-B open circuit in power supply circuit<br>• Plug/Kb-B open circuit in relay supply circuit |

| DTC | Trouble Code Title, Conditions & Possible Causes |
|---|---|
| **DTC: P1057**<br>**2T MIL: Yes**<br>**Years:** 2007, 2008<br>**Models:** 335i, 335Ci, 335xi, 525i, 525xi, 550i, 650i, 750i, 750Li, 760Li, X3, X5, Z4 3.0i, Z4 3.0Si, Z4 M<br>**Engines:** All<br>**Transmissions:** A/T | **Valvetronic (VVT) Supply Voltage Control Motor Electrical (Bank 1) Conditions:**<br>Key on for 50ms, the DME detected that the supply voltage was irregular.<br>**Possible Causes:**<br>• Short to ground at power input or defective ECU (capacitor preload). |
| **DTC: P1058**<br>**2T MIL: Yes**<br>**Years:** 2007, 2008<br>**Models:** 335i, 335Ci, 335xi, 525i, 525xi, 550i, 650i, 750i, 750Li, 760Li, X3, X5, Z4 3.0i, Z4 3.0Si, Z4 M<br>**Engines:** All<br>**Transmissions:** A/T | **Valvetronic (VVT) Supply Voltage Control Motor High Input (Bank 2) Conditions:**<br>Key on for 200ms, the DME detected that the supply voltage was too high (more than 17 volts). This is a jump-start detection that throws a fault if there is 24 volts of power for longer than 21 seconds.<br>**Possible Causes:**<br>• Check electrical system for faults. |
| **DTC: P1059**<br>**2T MIL: Yes**<br>**Years:** 2007, 2008<br>**Models:** 335i, 335Ci, 335xi, 525i, 525xi, 550i, 650i, 750i, 750Li, 760Li, X3, X5, Z4 3.0i, Z4 3.0Si, Z4 M<br>**Engines:** All<br>**Transmissions:** A/T | **Valvetronic (VVT) Supply Voltage Control Motor Low Input (Bank 2) Conditions:**<br>Key on for 200ms, the DME detected that the supply voltage was too low (less than 5 volts).<br>**Possible Causes:**<br>• VVT fuse is faulty<br>• Defective load reduction relay<br>• Plug/Kb-B open circuit in power supply circuit<br>• Plug/Kb-B open circuit in relay supply circuit |
| **DTC: P1060**<br>**2T MIL: Yes**<br>**Years:** 2007, 2008<br>**Models:** 335i, 335Ci, 335xi, 525i, 525xi, 550i, 650i, 750i, 750Li, 760Li, X3, X5, Z4 3.0i, Z4 3.0Si, Z4 M<br>**Engines:** All<br>**Transmissions:** A/T | **Valvetronic (VVT) Supply Voltage Control Motor Electrical (Bank 2) Conditions:**<br>Key on for 50ms, the DME detected that the supply voltage was irregular.<br>**Possible Causes:**<br>• Short to ground at power input or defective ECU (capacitor preload). |
| **DTC: P1061**<br>**2T MIL: Yes**<br>**Years:** 2007, 2008<br>**Models:** 335i, 335Ci, 335xi, 525i, 525xi, 550i, 650i, 750i, 750Li, 760Li, X3, X5, Z4 3.0i, Z4 3.0Si, Z4 M<br>**Engines:** All<br>**Transmissions:** A/T | **Valvetronic (VVT) Limp Home Request RPM and Charge Limitation (Bank 1) Conditions:**<br>After 3000ms the charge difference between the two banks (caused by other VVT faults) lead to an rpm charge limit.<br>**Possible Causes:**<br>• Resolve the primary fault. |
| **DTC: P1062**<br>**2T MIL: Yes**<br>**Years:** 2007, 2008<br>**Models:** 335i, 335Ci, 335xi, 525i, 525xi, 550i, 650i, 750i, 750Li, 760Li, X3, X5, Z4 3.0i, Z4 3.0Si, Z4 M<br>**Engines:** All<br>**Transmissions:** A/T | **Valvetronic (VVT) Limp Home Request Full Stroke Position Not Reached (Bank 1) Conditions:**<br>After 3000ms the eccentric angle fails to close at the full stroke position. Other VVT fault issues maximum stroke command but the position is not reached.<br>**Possible Causes:**<br>• Resolve the primary fault. |

| DTC | Trouble Code Title, Conditions & Possible Causes |
|---|---|
| **DTC: P1063**<br>**2T MIL:** Yes<br>**Years:** 2007, 2008<br>**Models:** 335i, 335Ci, 335xi, 525i, 525xi, 530i, 530xi, 550i, 650i, 750i, 750Li, 760Li, Z4 3.0i, Z4 3.0Si, Z4 M<br>**Engines:** All<br>**Transmissions:** A/T | **Valvetronic (VVT) Limp Home Request Air Mass Plausibility (Bank 1) Conditions:**<br>After 4000ms the eccentric overload angle detected a fault with the mass airflow plausibility. Other VVT fault issues maximum stroke command but the position is not reached.<br>**Possible Causes:**<br>• Resolve the primary fault. |
| **DTC: P1064**<br>**MIL:** No<br>**Years:** 2007, 2008<br>**Models:** 335i, 335Ci, 335xi, 525i, 525xi, 530i, 530xi, 550i, 650i, 750i, 750Li, 760Li, Z4 3.0i, Z4 3.0Si, Z4 M<br>**Engines:** All<br>**Transmissions:** A/T | **Valvetronic (VVT) Value Comparison Starting Position/Parking Position Plausibility (Bank 1) Conditions:**<br>After 500ms the difference between the deactivation and the starting positions exceeds specification. Usually occurs after repairs.<br>**Possible Causes:**<br>• Turn off ignition and wait until the HR releases. Turn on the ignition and the VVT will initialize automatically. |
| **DTC: P1065**<br>**MIL:** No<br>**Years:** 2007, 2008<br>**Models:** 335i, 335Ci, 335xi, 525i, 525xi, 550i, 650i, 750i, 750Li, 760Li, X3, X5, Z4 3.0i, Z4 3.0Si, Z4 M<br>**Engines:** All<br>**Transmissions:** A/T | **Valvetronic (VVT) CAN Timeout No Signal Conditions:**<br>After the key has been on for 800ms (within two messages) or the engine running for 400ms, with a battery voltage at 10 volts, the difference between the deactivation and the starting positions exceeds specification. No suspension of BUS activity. CAN signal is missing, therefore considered in a time out.<br>**Possible Causes:**<br>• ECU failure (or SZL or ZGM)<br>• Bus system failure<br>• Defective bus controller (SZL or ZGM)<br>• Short circuit in CAN wire or open circuit. |
| **DTC: P1066**<br>**MIL:** No<br>**Years:** 2007, 2008<br>**Models:** 335i, 335Ci, 335xi, 525i, 525xi, 550i, 650i, 750i, 750Li, 760Li, X3, X5, Z4 3.0i, Z4 3.0Si, Z4 M<br>**Engines:** All<br>**Transmissions:** A/T | **Valvetronic (VVT) CAN Message Monitoring Faulty Actual Message Conditions:**<br>After the DME is active or the engine running for 500ms, with a voltage of 7 on Terminal 87, the monitoring system displayed a faulty message.<br>**Possible Causes:**<br>• ECU failure (or SZL or ZGM)<br>• Bus system failure<br>• Defective bus controller (SZL or ZGM)<br>• Short circuit in CAN wire or open circuit.<br>• Defective DME<br>• Defective VVT-SG |
| **DTC: P1067**<br>**Years:** 2007, 2008<br>**Models:** 335i, 335Ci, 335xi, 525i, 525xi, 530i, 530xi, 550i, 650i, 750i, 750Li, 760Li, Z4 3.0i, Z4 3.0Si, Z4 M<br>**Engines:** All<br>**Transmissions:** A/T | **Valvetronic (VVT) Reference Sensor Solenoid Loss (Bank 2) Conditions:**<br>After 3ms and the DME and VVT are active, it is detected that the sensor is missing a magnet.<br>**Possible Causes:**<br>• Defective sensor. |
| **DTC: P1068**<br>**Years:** 2007, 2008<br>**Models:** 335i, 335Ci, 335xi, 525i, 525xi, 530i, 530xi, 550i, 650i, 750i, 750Li, 760Li, Z4 3.0i, Z4 3.0Si, Z4 M<br>**Engines:** All<br>**Transmissions:** A/T | **Valvetronic (VVT) Reference Sensor Reset Error (Bank 2) Conditions:**<br>After 1.5ms and the DME and VVT are active, it is detected that the sensor is not properly resetting.<br>**Possible Causes:**<br>• Plug contact problem at Pin 6 on sensor<br>• Oil in plug at sensor<br>• Replace sensor<br>• Repair gaskets |
| **DTC: P1069**<br>**Years:** 2007, 2008<br>**Models:** 335i, 335Ci, 335xi, 525i, 525xi, 530i, 530xi, 550i, 650i, 750i, 750Li, 760Li, Z4 3.0i, Z4 3.0Si, Z4 M<br>**Engines:** All<br>**Transmissions:** A/T | **Valvetronic (VVT) Reference Sensor Parity Error (Bank 2) Conditions:**<br>After 12ms and the DME and VVT are active, it is detected that the sensor is not properly communicating. The plug is defective on the sensor or there is an open circuit.<br>**Possible Causes:**<br>• Plug contact problem at Pin 6 on sensor<br>• Oil in plug at sensor<br>• Replace sensor<br>• Repair gaskets |

| DTC | Trouble Code Title, Conditions & Possible Causes |
|---|---|
| **DTC: P1070**<br>**Years:** 2007, 2008<br>**Models:** 335i, 335Ci, 335xi, 525i, 525xi, 530i, 530xi, 550i, 650i, 750i, 750Li, 760Li, Z4 3.0i, Z4 3.0Si, Z4 M<br>**Engines:** All<br>**Transmissions:** A/T | **Valvetronic (VVT) Reference Sensor Gradient Error (Bank 2) Conditions:**<br>After 9ms and the DME and VVT are active, it is detected that the sensor has a gradient violation/identity, causing a reading of implausible sensor data. The plug is defective on the sensor or there is an open circuit.<br>**Possible Causes:**<br>• Replace sensor |
| **DTC: P1071**<br>**2T MIL:** Yes<br>**Years:** 2007, 2008<br>**Models:** 335i, 335Ci, 335xi, 525i, 525xi, 530i, 530xi, 550i, 650i, 750i, 750Li, 760Li, Z4 3.0i, Z4 3.0Si, Z4 M<br>**Engines:** All<br>**Transmissions:** A/T | **Valvetronic (VVT) Control Module Watchdog or Temperature Sensor Error (Bank 1) Conditions:**<br>Key on for 10ms, and the DME detected an internal fault relating to an internal temperature sensor. Ignore single isolated appearances and only respond to repeated occurrences by replacing the VVT-SG.<br>**Possible Causes:**<br>• Battery terminal corrosion, or loose battery connection<br>• Connection to the DME interrupted, or the circuit has been opened<br>• Reprogramming error has occurred and needs replacement. Remember to check for Aftermarket Performance Products before replacing a DME. |
| **DTC: P1072**<br>**2T MIL:** Yes<br>**Years:** 2007, 2008<br>**Models:** 335i, 335Ci, 335xi, 525i, 525xi, 530i, 530xi, 550i, 650i, 750i, 750Li, 760Li, Z4 3.0i, Z4 3.0Si, Z4 M<br>**Engines:** All<br>**Transmissions:** A/T | **Valvetronic (VVT) Control Module Watchdog or Temperature Sensor Error (Bank 2) Conditions:**<br>Key on for 10ms, and the DME detected an internal fault relating to an internal temperature sensor. Ignore single isolated appearances and only respond to repeated occurrences by replacing the VVT-SG.<br>**Possible Causes:**<br>• Battery terminal corrosion, or loose battery connection<br>• Connection to the DME interrupted, or the circuit has been opened<br>• Reprogramming error has occurred and needs replacement. Remember to check for Aftermarket Performance Products before replacing a DME. |
| **DTC: P1075**<br>**MIL:** No<br>**Years:** 2007, 2008<br>**Models:** 335i, 335Ci, 335xi, 525i, 525xi, 530i, 530xi, 550i, 650i, 750i, 750Li, 760Li, Z4 3.0i, Z4 3.0Si, Z4 M<br>**Engines:** All<br>**Transmissions:** A/T | **Valvetronic (VVT) Overload Protection (Bank 1) Conditions:**<br>After the ignition is on for 45ms, the temperature was recorded as too high.<br>**Possible Causes:**<br>• Sticking in VVT mechanicals, pinion gear, etc.<br>• Loose sensor<br>• Sensor servo motor has short circuit |
| **DTC: P1076**<br>**MIL:** No<br>**Years:** 2007, 2008<br>**Models:** 335i, 335Ci, 335xi, 525i, 525xi, 530i, 530xi, 550i, 650i, 750i, 750Li, 760Li, Z4 3.0i, Z4 3.0Si, Z4 M<br>**Engines:** All<br>**Transmissions:** A/T | **Valvetronic (VVT) Overload Protection ECU Temperature High Input (Bank 1) Conditions:**<br>After the ignition is on for 45ms, the ECU temperature was recorded as too high.<br>**Possible Causes:**<br>• Sticking in VVT mechanicals, pinion gear, etc.<br>• Loose sensor<br>• Sensor servo motor has short circuit |
| **DTC: P1077**<br>**MIL:** No<br>**Years:** 2007, 2008<br>**Models:** 335i, 335Ci, 335xi, 525i, 525xi, 530i, 530xi, 550i, 650i, 750i, 750Li, 760Li, Z4 3.0i, Z4 3.0Si, Z4 M<br>**Engines:** All<br>**Transmissions:** A/T | **Valvetronic (VVT) Overload Protection Control Motor Temperature High Input (Bank 1) Conditions:**<br>After the ignition is on for 45ms, the E motor temperature was recorded as too high.<br>**Possible Causes:**<br>• Sticking in VVT mechanicals, pinion gear, etc.<br>• Loose sensor<br>• Sensor servo motor has short circuit |

| DTC | Trouble Code Title, Conditions & Possible Causes |
|---|---|
| **DTC: P1078**<br>**MIL:** No<br>**Years:** 2007, 2008<br>**Models:** 335i, 335Ci, 335xi, 525i, 525xi, 530i, 530xi, 550i, 650i, 750i, 750Li, 760Li, Z4 3.0i, Z4 3.0Si, Z4 M<br>**Engines:** All<br>**Transmissions:** A/T | **Valvetronic (VVT) Overload Protection Control Motor Current High Input (Bank 1) Conditions:**<br>After the ignition is on for 45ms, the E motor activation current is too high.<br>**Possible Causes:**<br>• Sticking in VVT mechanicals, pinion gear, etc.<br>• Loose sensor<br>• Sensor servo motor has short circuit |
| **DTC: P1079**<br>**MIL:** No<br>**Years:** 2007, 2008<br>**Models:** 335i, 335Ci, 335xi, 525i, 525xi, 530i, 530xi, 550i, 650i, 750i, 750Li, 760Li, Z4 3.0i, Z4 3.0Si, Z4 M<br>**Engines:** All<br>**Transmissions:** A/T | **Valvetronic (VVT) Overload Protection (Bank 1) Conditions:**<br>After the ignition is on for 45ms, the temperature was recorded as too high.<br>**Possible Causes:**<br>• Sticking in VVT mechanicals, pinion gear, etc.<br>• Loose sensor<br>• Sensor servo motor has short circuit |
| **DTC: P1080**<br>**MIL:** No<br>**Years:** 2007, 2008<br>**Models:** 335i, 335Ci, 335xi, 525i, 525xi, 530i, 530xi, 550i, 650i, 750i, 750Li, 760Li, Z4 3.0i, Z4 3.0Si, Z4 M<br>**Engines:** All<br>**Transmissions:** A/T | **Valvetronic (VVT) Overload Protection ECU Temperature High Input (Bank 2) Conditions:**<br>After the ignition is on for 45ms, the ECU temperature was recorded as too high.<br>**Possible Causes:**<br>• Sticking in VVT mechanicals, pinion gear, etc.<br>• Loose sensor<br>• Sensor servo motor has short circuit |
| **DTC: P1081**<br>**MIL:** No<br>**Years:** 2007, 2008<br>**Models:** 335i, 335Ci, 335xi, 525i, 525xi, 530i, 530xi, 550i, 650i, 750i, 750Li, 760Li, Z4 3.0i, Z4 3.0Si, Z4 M<br>**Engines:** All<br>**Transmissions:** A/T | **Valvetronic (VVT) Overload Protection Control Motor Temperature High Input (Bank 2) Conditions:**<br>After the ignition is on for 45ms, the E motor temperature was recorded as too high.<br>**Possible Causes:**<br>• Sticking in VVT mechanicals, pinion gear, etc.<br>• Loose sensor<br>• Sensor servo motor has short circuit |
| **DTC: P1082**<br>**MIL:** No<br>**Years:** 2007, 2008<br>**Models:** 335i, 335Ci, 335xi, 525i, 525xi, 530i, 530xi, 550i, 650i, 750i, 750Li, 760Li, Z4 3.0i, Z4 3.0Si, Z4 M<br>**Engines:** All<br>**Transmissions:** A/T | **Valvetronic (VVT) Overload Protection Control Motor Current High Input (Bank 2) Conditions:**<br>After the ignition is on for 45ms, the E motor activation current is too high.<br>**Possible Causes:**<br>• Sticking in VVT mechanicals, pinion gear, etc.<br>• Loose sensor<br>• Sensor servo motor has short circuit |
| **DTC: P1084**<br>**2T MIL:** Yes<br>**Years:** 2007, 2008<br>**Models:** 335i, 335Ci, 335xi, 525i, 525xi, 550i, 650i, 750i, 750Li, 760Li, X3, X5, Z4 3.0i, Z4 3.0Si, Z4 M<br>**Engines:** All<br>**Transmissions:** A/T | **Fuel System Too Lean Conditions:**<br>Key on or engine running, all electrical components off and coolant temperature at least 80 degrees Celsius; and the DME detected the Bank 1 Adaptive Fuel Control System reached its rich correction limit (a lean A/F condition). The fuel status is in a closed loop pattern, the coolant temperature is between 69 and 100 degrees Celsius, and the engine speed is between 800 and 6000rpm.<br>**Possible Causes:**<br>• Air leaks after the MAF sensor, or leaks in the PCV system<br>• Exhaust leaks before or near where the HO2S is mounted<br>• Fuel injector(s) restricted or not supplying enough fuel<br>• Fuel pump not supplying enough fuel during high fuel demand conditions<br>• Leaking EGR gasket, or leaking EGR valve diaphragm<br>• MAF sensor dirty (causes DME to underestimate airflow)<br>• Vehicle running out of fuel or engine oil dip stick not seated |

| DTC | Trouble Code Title, Conditions & Possible Causes |
|---|---|
| **DTC: P1085**<br>**2T MIL: Yes**<br>**Years:** 2007, 2008<br>**Models:** 335i, 335Ci, 335xi, 525i, 525xi, 550i, 650i, 750i, 750Li, 760Li, X3, X5, Z4 3.0i, Z4 3.0Si, Z4 M<br>**Engines:** All<br>**Transmissions:** A/T | **Fuel System Too Rich Conditions:**<br>Key on or engine running, all electrical components off and coolant temperature at least 80 degrees Celsius; and the DME detected the Bank 2 Adaptive Fuel Control System reached its rich correction limit (a rich A/F condition). The fuel status is in a closed loop pattern, the coolant temperature is between 69 and 100 degrees Celsius, and the engine speed is between 800 and 6000rpm.<br>**Possible Causes:**<br>• Air leaks after the MAF sensor, or leaks in the PCV system<br>• Exhaust leaks before or near where the HO2S is mounted<br>• Fuel injector(s) restricted or not supplying enough fuel<br>• Fuel pump not supplying enough fuel during high fuel demand conditions<br>• Leaking EGR gasket, or leaking EGR valve diaphragm<br>• MAF sensor dirty (causes DME to underestimate airflow)<br>• Vehicle running out of fuel or engine oil dip stick not seated |
| **DTC: P1086**<br>**2T MIL: Yes**<br>**Years:** 2007, 2008<br>**Models:** 335i, 335Ci, 335xi, 525i, 525xi, 550i, 650i, 750i, 750Li, 760Li, X3, X5, Z4 3.0i, Z4 3.0Si, Z4 M<br>**Engines:** All<br>**Transmissions:** A/T | **Fuel System Too Lean Conditions:**<br>Key on or engine running, all electrical components off and coolant temperature at least 80 degrees Celsius; and the DME detected the Bank 2 Fuel Control System reached its lean correction limit. The fuel status is in a closed loop pattern, the coolant temperature is between 69 and 100 degrees Celsius, and the engine speed is between 800 and 6000rpm.<br>**Possible Causes:**<br>• Air leaks after the MAF sensor, or leaks in the PCV system<br>• Exhaust leaks before or near where the HO2S is mounted<br>• Fuel injector(s) restricted or not supplying enough fuel<br>• Fuel pump not supplying enough fuel during high fuel demand conditions<br>• Leaking EGR gasket, or leaking EGR valve diaphragm<br>• MAF sensor dirty (causes DME to underestimate airflow)<br>• Vehicle running out of fuel or engine oil dip stick not seated |
| **DTC: P1104**<br>**Years:** 2007, 2008<br>**Models:** 335i, 335Ci, 335xi, 525i, 525xi, 530i, 530xi, 550i, 650i, 750i, 750Li, 760Li, Z4 3.0i, Z4 3.0Si, Z4 M<br>**Engines:** All<br>**Transmissions:** A/T | **Differential Pressure Sensor Intake Manifold Pressure Too Low (Bank 1) Conditions:**<br>Engine started, battery voltage must be at least 11v, and the differential pressure sensor detected a control deviation at the minimum limit. The closed loop control of the differential pressure in the intake manifold is suspended and replaced by a direct specification.<br>**Possible Causes:**<br>• Sensor's voltage supply on Terminal 87<br>• Sensor's ground connection faulty<br>• Signal wire to DME faulty<br>• Replace sensor |
| **DTC: P1105**<br>**Years:** 2007, 2008<br>**Models:** 335i, 335Ci, 335xi, 525i, 525xi, 530i, 530xi, 550i, 650i, 750i, 750Li, 760Li, Z4 3.0i, Z4 3.0Si, Z4 M<br>**Engines:** All<br>**Transmissions:** A/T | **Differential Pressure Sensor Intake Manifold Pressure Too High (Bank 1) Conditions:**<br>Engine started, battery voltage must be at least 11v, and the differential pressure sensor detected a control deviation at the maximum limit. The closed loop control of the differential pressure in the intake manifold is suspended and replaced by a direct specification.<br>**Possible Causes:**<br>• Sensor's voltage supply on Terminal 87<br>• Sensor's ground connection faulty<br>• Signal wire to DME faulty<br>• Replace sensor |
| **DTC: P1111**<br>**Years:** 2007, 2008<br>**Models:** 335i, 335Ci, 335xi, 525i, 525xi, 530i, 530xi, 550i, 650i, 750i, 750Li, 760Li, Z4 3.0i, Z4 3.0Si, Z4 M<br>**Engines:** All<br>**Transmissions:** All | **O2 Control (Bank 1) System Too Lean Conditions:**<br>Engine started, battery voltage must be at least 11.5v, all electrical components must be off, the ground between the engine and the chassis must be well connected, the exhaust system must be properly sealed between the catalytic converter and the cylinder head, and the oxygen sensor heater for oxygen sensor before the catalytic converter must be properly functioning. The DME detected a measurement on the O2 sensor circuit that was outside the parameters to function properly.<br>**Note: For resistance testing of sensor heating, oxygen sensor should be cooled to ambient temperature. High temperatures at oxygen sensor may lead to inaccurate measurements.**<br>**Note: When an O2S malfunction (P0131 to P0414) is also stored with this malfunction, the O2S malfunction(s) should be repaired first.**<br>**Possible Causes:**<br>• Oxygen sensor (before catalytic converter) is faulty<br>• Oxygen sensor (behind catalytic converter) is faulty<br>• Oxygen sensor heater (before catalytic converter) is faulty<br>• Oxygen sensor heater (behind catalytic converter) is faulty<br>• Circuit wiring has a short to power or ground<br>• Engine Component Power Supply Relay is faulty<br>• E-box fuses for oxygen sensor are faulty<br>• Leaks present in the exhaust manifold or exhaust pipes<br>• HO2S signal wire and ground wire crossed in connector<br>• HO2S element is fuel contaminated or has failed |

| DTC | Trouble Code Title, Conditions & Possible Causes |
|---|---|
| **DTC: P1112**<br>**Years:** 2007, 2008<br>**Models:** 335i, 335Ci, 335xi, 525i, 525xi, 530i, 530xi, 550i, 650i, 750i, 750Li, 760Li, Z4 3.0i, Z4 3.0Si, Z4 M<br>**Engines:** All<br>**Transmissions:** All | **O2 Control (Bank 1) System Too Rich Conditions:**<br>Engine started, battery voltage must be at least 11.5v, all electrical components must be off, the ground between the engine and the chassis must be well connected, the exhaust system must be properly sealed between the catalytic converter and the cylinder head, and the oxygen sensor heater for oxygen sensor before the catalytic converter must be properly functioning. The DME detected a measurement on the O2 sensor circuit that was outside the parameters to function properly.<br>**Note: For resistance testing of sensor heating, oxygen sensor should be cooled to ambient temperature. High temperatures at oxygen sensor may lead to inaccurate measurements.**<br>**Note: When an O2S malfunction (P0131 to P0414) is also stored with this malfunction, the O2S malfunction(s) should be repaired first.**<br>**Possible Causes:**<br>• Oxygen sensor (before catalytic converter) is faulty<br>• Oxygen sensor (behind catalytic converter) is faulty<br>• Oxygen sensor heater (before catalytic converter) is faulty<br>• Oxygen sensor heater (behind catalytic converter) is faulty<br>• Circuit wiring has a short to power or ground<br>• Engine Component Power Supply Relay is faulty<br>• E-box fuses for oxygen sensor are faulty<br>• Leaks present in the exhaust manifold or exhaust pipes<br>• HO2S signal wire and ground wire crossed in connector<br>• HO2S element is fuel contaminated or has failed |
| **DTC: P1129**<br>**MIL: No (Oil warning lamp)**<br>**Years:** 2007, 2008<br>**Models:** 335i, 335Ci, 335xi, 525i, 525xi, 530i, 530xi, 550i, 650i, 750i, 750Li, 760Li, Z4 3.0i, Z4 3.0Si, Z4 M<br>**Engines:** All<br>**Transmissions:** A/T | **Engine Oil Level Sensor Signal Oil Level Too Low Conditions:**<br>Engine started, and the oil sensor has detected that the level is too low.<br>**Possible Causes:**<br>• Top off the oil |
| **DTC: P1130**<br>**2T MIL: Yes**<br>**Years:** 2007, 2008<br>**Models:** 335i, 335Ci, 335xi, 525i, 525xi, 530i, 530xi, 550i, 650i, 750i, 750Li, 760Li, Z4 3.0i, Z4 3.0Si, Z4 M<br>**Engines:** All<br>**Transmissions:** A/T | **Long Term Fuel Trim at Lean Limit Conditions:**<br>Engine started, battery voltage must be at least 11.5v, all electrical components must be off, the ground between the engine and the chassis must be well connected, the exhaust system must be properly sealed between the catalytic converter and the cylinder head, and the oxygen sensor heater for oxygen sensor before the catalytic converter must be properly functioning. The DME detected the HO2S circuit was too lean, or that it could no longer change Fuel Trim because it was at its lean limit.<br>**Possible Causes:**<br>• Air intake system leaking, vacuum hoses leaking or damaged<br>• Air leaks located after the MAF sensor mounting location<br>• EGR valve sticking, EGR diaphragm leaking, or gasket leaking<br>• EVAP vapor recovery system has failed<br>• Excessive fuel pressure, leaking or contaminated fuel injectors<br>• Exhaust leaks before or near the HO2S(s) mounting location<br>• Fuel pressure regulator is leaking or damaged<br>• HO2S circuits wet or oily, corroded, or poor terminal contact<br>• HO2S is damaged or it has failed<br>• HO2S signal circuit open, shorted to ground, shorted to power<br>• Low fuel pressure or vehicle driven until it was out of fuel<br>• Oil dipstick not seated or engine oil level too high (overfilled) |
| **DTC: P1134**<br>**2T MIL: Yes**<br>**Years:** 2007, 2008<br>**Models:** All<br>**Engines:** All<br>**Transmissions:** All | **HO2S Heater Circuit Current Malfunction Conditions:**<br>Engine running, battery voltage 11.5, all electrical components off, ground between engine and chassis well connected and the exhaust system must be properly sealed between catalytic converter and the cylinder head. The DME detected an open or shorted condition, or excessive current draw in the heater circuit. The response rate for the sensor signal period is greater than 3.8/second. The engine speed is 1280 to 2400rpm, the catalyst temperature is greater than 300 degrees Celsius and the heater has been on for less than 90 seconds.<br>**Possible Causes:**<br>• HO2S heater power circuit is open<br>• HO2S heater ground circuit is open<br>• HO2S signal tracking (due to oil or moisture in the connector)<br>• HO2S is damaged or has failed |

| DTC | Trouble Code Title, Conditions & Possible Causes |
|---|---|
| **DTC: P1135**<br>**2T MIL:** Yes<br>**Years:** 2007, 2008<br>**Models:** All<br>**Engines:** All<br>**Transmissions:** All | **HO2S Heater Circuit Current Malfunction Conditions:**<br>Engine running, battery voltage 11.5, all electrical components off, ground between engine and chassis well connected and the exhaust system must be properly sealed between catalytic converter and the cylinder head. The DME detected an open or shorted condition, or excessive current draw in the heater circuit. The response rate for the sensor signal period is greater than 3.8/second. The engine speed is 1280 to 2400rpm, the catalyst temperature is greater than 300 degrees Celsius and the heater has been on for less than 90 seconds.<br>**Possible Causes:**<br>• HO2S heater power circuit is open<br>• HO2S heater ground circuit is open<br>• HO2S signal tracking (due to oil or moisture in the connector)<br>• HO2S is damaged or has failed |
| **DTC: P1136**<br>**2T MIL:** Yes<br>**Years:** 2007, 2008<br>**Models:** All<br>**Engines:** All<br>**Transmissions:** All | **HO2S Heater Circuit Heater Resistance Conditions:**<br>Engine running, battery voltage 11.5, all electrical components off, ground between engine and chassis well connected and the exhaust system must be properly sealed between catalytic converter and the cylinder head. The DME detected an open or shorted condition, or excessive current draw in the heater circuit. The response rate for the sensor signal period is greater than 3.8/second. The engine speed is 1280 to 2400rpm, the catalyst temperature is greater than 300 degrees Celsius and the heater has been on for less than 90 seconds.<br>**Possible Causes:**<br>• HO2S heater power circuit is open<br>• HO2S heater ground circuit is open<br>• HO2S signal tracking (due to oil or moisture in the connector)<br>• HO2S is damaged or has failed |
| **DTC: P1137**<br>**2T MIL:** Yes<br>**Years:** 2007, 2008<br>**Models:** 335i, 335Ci, 335xi, 525i, 525xi, 530i, 530xi, M5, M6<br>**Engines:** All<br>**Transmissions:** All | **Long Term Fuel Trim Add. Fuel, Bank 1 System Too Rich Conditions:**<br>Engine started, battery voltage must be at least 11.5v, all electrical components must be off, the ground between the engine and the chassis must be well connected, the exhaust system must be properly sealed between the catalytic converter and the cylinder head, and the oxygen sensor heater for oxygen sensor before the catalytic converter must be properly functioning. The DME detected the HO2S circuit was too rich, or that it could no longer change Fuel Trim because it was at its lean limit.<br>**Possible Causes:**<br>• Air intake system leaking, vacuum hoses leaking or damaged<br>• Air leaks located after the MAF sensor mounting location<br>• EGR valve sticking, EGR diaphragm leaking, or gasket leaking<br>• EVAP vapor recovery system has failed<br>• Excessive fuel pressure, leaking or contaminated fuel injectors<br>• Exhaust leaks before or near the HO2S(s) mounting location<br>• Fuel pressure regulator is leaking or damaged<br>• HO2S circuits wet or oily, corroded, or poor terminal contact<br>• HO2S is damaged or it has failed<br>• HO2S signal circuit open, shorted to ground, shorted to power<br>• Low fuel pressure or vehicle driven until it was out of fuel<br>• Oil dipstick not seated or engine oil level too high (overfilled) |
| **DTC: P1138**<br>**2T MIL:** Yes<br>**Years:** 2007, 2008<br>**Models:** 335i, 335Ci, 335xi, 525i, 525xi, 530i, 530xi, 750i, 750Li, 760Li,<br>**Engines:** All<br>**Transmissions:** All | **HO2S Circuit Malfunction Conditions:**<br>Engine running, battery voltage 11.5, all electrical components off, ground between engine and chassis well connected and the exhaust system must be properly sealed between catalytic converter and the cylinder head. The DME detected the HO2S signal failed to meet the maximum or minimum voltage levels (i.e., it failed the voltage range check). The heater has been on for less than 90 seconds, the fuel system status is in fuel cut-off, the output voltage is between 400mV and 500mV and it is 120 seconds after engine start up.<br>**Possible Causes:**<br>• Leaks present in the exhaust manifold or exhaust pipes<br>• HO2S signal wire and ground wire crossed in connector<br>• HO2S element is fuel contaminated or has failed |
| **DTC: P1139**<br>**2T MIL:** Yes<br>**Years:** 2007, 2008<br>**Models:** 335i, 335Ci, 335xi, 525i, 525xi, 530i, 530xi, 750i, 750Li, 760Li, M5, M6<br>**Engines:** All<br>**Transmissions:** All | **HO2S Circuit Malfunction Conditions:**<br>Engine running, battery voltage 11.5, all electrical components off, ground between engine and chassis well connected and the exhaust system must be properly sealed between catalytic converter and the cylinder head. The DME detected the HO2S signal failed to meet the maximum or minimum voltage levels (i.e., it failed the voltage range check). The heater has been on for less than 90 seconds, the fuel system status is in fuel cut-off, the output voltage is between 400mV and 500mV and it is 120 seconds after engine start up.<br>**Possible Causes:**<br>• Leaks present in the exhaust manifold or exhaust pipes<br>• HO2S signal wire and ground wire crossed in connector<br>• HO2S element is fuel contaminated or has failed |

| DTC | Trouble Code Title, Conditions & Possible Causes |
|---|---|
| **DTC: P1151**<br>**2T MIL: Yes**<br>**Years:** 2007, 2008<br>**Models:** M5, M6<br>**Engines:** All<br>**Transmissions:** All | **HO2S Heater Circuit Current Malfunction Conditions:**<br>Engine running, battery voltage 11.5, all electrical components off, ground between engine and chassis well connected and the exhaust system must be properly sealed between catalytic converter and the cylinder head. The DME detected an open or shorted condition, or excessive current draw in the heater circuit. The response rate for the sensor signal period is greater than 3.8/second. The engine speed is 1280 to 2400rpm, the catalyst temperature is greater than 300 degrees Celsius and the heater has been on for less than 90 seconds.<br>**Possible Causes:**<br>• HO2S heater power circuit is open<br>• HO2S heater ground circuit is open<br>• HO2S signal tracking (due to oil or moisture in the connector)<br>• HO2S is damaged or has failed |
| **DTC: P1152**<br>**2T MIL: Yes**<br>**Years:** 2007, 2008<br>**Models:** M5, M6<br>**Engines:** All<br>**Transmissions:** All | **HO2S Heater Circuit Current Malfunction Conditions:**<br>Engine running, battery voltage 11.5, all electrical components off, ground between engine and chassis well connected and the exhaust system must be properly sealed between catalytic converter and the cylinder head. The DME detected an open or shorted condition, or excessive current draw in the heater circuit. The response rate for the sensor signal period is greater than 3.8/second. The engine speed is 1280 to 2400rpm, the catalyst temperature is greater than 300 degrees Celsius and the heater has been on for less than 90 seconds.<br>**Possible Causes:**<br>• HO2S heater power circuit is open<br>• HO2S heater ground circuit is open<br>• HO2S signal tracking (due to oil or moisture in the connector)<br>• HO2S is damaged or has failed |
| **DTC: P1153**<br>**2T MIL: Yes**<br>**Years:** 2007, 2008<br>**Models:** M5, M6<br>**Engines:** All<br>**Transmissions:** All | **HO2S Heater Circuit Current Malfunction Conditions:**<br>Engine running, battery voltage 11.5, all electrical components off, ground between engine and chassis well connected and the exhaust system must be properly sealed between catalytic converter and the cylinder head. The DME detected an open or shorted condition, or excessive current draw in the heater circuit. The response rate for the sensor signal period is greater than 3.8/second. The engine speed is 1280 to 2400rpm, the catalyst temperature is greater than 300 degrees Celsius and the heater has been on for less than 90 seconds.<br>**Possible Causes:**<br>• HO2S heater power circuit is open<br>• HO2S heater ground circuit is open<br>• HO2S signal tracking (due to oil or moisture in the connector)<br>• HO2S is damaged or has failed |
| **DTC: P1155**<br>**2T MIL: Yes**<br>**Years:** 2007, 2008<br>**Models:** 335i, 335Ci, 335xi, 525i, 525xi, 530i, 530xi, 750i, 750Li, 760Li, Z4 3.0i, Z4 3.0Si, Z4 M<br>**Engines:** All<br>**Transmissions:** All | **HO2S Heater Circuit Current Malfunction Conditions:**<br>Engine running, battery voltage 11.5, all electrical components off, ground between engine and chassis well connected and the exhaust system must be properly sealed between catalytic converter and the cylinder head. The DME detected an open or shorted condition, or excessive current draw in the heater circuit. The response rate for the sensor signal period is greater than 3.8/second. The engine speed is 1280 to 2400rpm, the catalyst temperature is greater than 300 degrees Celsius and the heater has been on for less than 90 seconds.<br>**Possible Causes:**<br>• HO2S heater power circuit is open<br>• HO2S heater ground circuit is open<br>• HO2S signal tracking (due to oil or moisture in the connector)<br>• HO2S is damaged or has failed |
| **DTC: P1156**<br>**2T MIL: Yes**<br>**Years:** 2007, 2008<br>**Models:** M5, M6<br>**Engines:** All<br>**Transmissions:** All | **HO2S Heater Circuit Current Malfunction-Circuit Continuity Conditions:**<br>Engine running, battery voltage 11.5, all electrical components off, ground between engine and chassis well connected and the exhaust system must be properly sealed between catalytic converter and the cylinder head. The DME detected an open or shorted condition, or excessive current draw in the heater circuit. The response rate for the sensor signal period is greater than 3.8/second. The engine speed is 1280 to 2400rpm, the catalyst temperature is greater than 300 degrees Celsius and the heater has been on for less than 90 seconds.<br>**Possible Causes:**<br>• HO2S heater power circuit is open<br>• HO2S heater ground circuit is open<br>• HO2S signal tracking (due to oil or moisture in the connector)<br>• HO2S is damaged or has failed |

| DTC | Trouble Code Title, Conditions & Possible Causes |
|---|---|
| **DTC: P1157**<br>**2T MIL: Yes**<br>**Years:** 2007, 2008<br>**Models:** 335i, 335Ci, 335xi, 525i, 525xi, 530i, 530xi, 750i, 750Li, 760Li, Z4 3.0i, Z4 3.0Si, Z4 M<br>**Engines:** All<br>**Transmissions:** All | **HO2S Heater Circuit Current Malfunction-Heater Resistance Conditions:**<br>Engine running, battery voltage 11.5, all electrical components off, ground between engine and chassis well connected and the exhaust system must be properly sealed between catalytic converter and the cylinder head. The DME detected an open or shorted condition, or excessive current draw in the heater circuit. The response rate for the sensor signal period is greater than 3.8/second. The engine speed is 1280 to 2400rpm, the catalyst temperature is greater than 300 degrees Celsius and the heater has been on for less than 90 seconds.<br>**Possible Causes:**<br>• HO2S heater power circuit is open<br>• HO2S heater ground circuit is open<br>• HO2S signal tracking (due to oil or moisture in the connector)<br>• HO2S is damaged or has failed |
| **DTC: P1178**<br>**2T MIL: Yes**<br>**Years:** 2007, 2008<br>**Models:** 328i, 328Ci, 328xi, M5, M6<br>**Engines:** All<br>**Transmissions:** All | **O2 Sensor Switching Time Conditions:**<br>Engine running, battery voltage 11.5, all electrical components off, ground between engine and chassis well connected and the exhaust system must be properly sealed between catalytic converter and the cylinder head. The DME detected the O2S signal was implausible or not detected, the switching time range from lean to rich and vies versa was too slow. The exhaust temperature is greater than 380 degrees Celsius, the fuel system is in a closed loop, and the engine speed is between 2000 and 3200rpm.<br>**Possible Causes:**<br>• Oxygen sensor heater for oxygen sensor (HO2S) before catalytic converter is faulty<br>• O2S is contaminated (due to presence of silicone in fuel)<br>• O2S signal and ground circuit wires crossed in wiring harness<br>• O2S signal circuit is shorted to sensor or chassis ground<br>• O2S element before the catalytic converter has failed (internal short condition)<br>• Leaks present in the exhaust manifold or exhaust pipes |
| **DTC: P1179**<br>**2T MIL: Yes**<br>**Years:** 2007, 2008<br>**Models:** 328i, 328Ci, 328xi, M5, M6<br>**Engines:** All<br>**Transmissions:** All | **O2 Sensor Switching Time Conditions:**<br>Engine running, battery voltage 11.5, all electrical components off, ground between engine and chassis well connected and the exhaust system must be properly sealed between catalytic converter and the cylinder head. The DME detected the O2S signal was implausible or not detected, the switching time range from lean to rich and vies versa was too slow. The exhaust temperature is greater than 380 degrees Celsius, the fuel system is in a closed loop, and the engine speed is between 2000 and 3200rpm.<br>**Possible Causes:**<br>• Oxygen sensor heater for oxygen sensor (HO2S) before catalytic converter is faulty<br>• O2S is contaminated (due to presence of silicone in fuel)<br>• O2S signal and ground circuit wires crossed in wiring harness<br>• O2S signal circuit is shorted to sensor or chassis ground<br>• O2S element before the catalytic converter has failed (internal short condition)<br>• Leaks present in the exhaust manifold or exhaust pipes |
| **DTC: P1197**<br>**Years:** 2007, 2008<br>**Models:** 335i, 335Ci, 335xi, 525i, 525xi, 530i, 530xi, 550i, 650i, 750i, 750Li, 760Li, Z4 3.0i, Z4 3.0Si, Z4 M<br>**Engines:** All<br>**Transmissions:** All | **Differential Pressure Sensor Intake Manifold High Input (Bank 1) Conditions:**<br>Engine started, battery voltage must be at least 11v, and the differential pressure sensor wiring shorted to battery voltage. The closed loop control of the differential pressure in the intake manifold is suspended and replaced by a direct specification.<br>**Possible Causes:**<br>• Sensor's voltage supply on Terminal 87<br>• Sensor's ground connection faulty<br>• Signal wire to DME faulty<br>• Replace sensor |
| **DTC: P1198**<br>**Years:** 2007, 2008<br>**Models:** 335i, 335Ci, 335xi, 525i, 525xi, 530i, 530xi, 550i, 650i, 750i, 750Li, 760Li, Z4 3.0i, Z4 3.0Si, Z4 M<br>**Engines:** All<br>**Transmissions:** All | **Differential Pressure Sensor Intake Manifold Low Input (Bank 1) Conditions:**<br>Engine started, battery voltage must be at least 11v, and the differential pressure sensor wiring shorted to ground. The closed loop control of the differential pressure in the intake manifold is suspended and replaced by a direct specification.<br>**Possible Causes:**<br>• Sensor's voltage supply on Terminal 87<br>• Sensor's ground connection faulty<br>• Signal wire to DME faulty<br>• Replace sensor |
| **DTC: P1199**<br>**Years:** 2007, 2008<br>**Models:** 335i, 335Ci, 335xi, 525i, 525xi, 530i, 530xi, 550i, 650i, 750i, 750Li, 760Li, Z4 3.0i, Z4 3.0Si, Z4 M<br>**Engines:** All<br>**Transmissions:** All | **Differential Pressure Sensor Intake Manifold Pressure Plausibility (Bank 1) Conditions:**<br>Engine started, battery voltage must be at least 11v, and the differential pressure sensor signal is malfunction or is not present. The closed loop control of the differential pressure in the intake manifold is suspended and replaced by a direct specification.<br>**Possible Causes:**<br>• Sensor's voltage supply on Terminal 87<br>• Sensor's ground connection faulty<br>• Signal wire to DME faulty<br>• Replace sensor |

| DTC | Trouble Code Title, Conditions & Possible Causes |
|---|---|
| **DTC: P1327**<br>**MIL:** Yes<br>**Years:** 2007, 2008<br>**Models:** 335i, 335Ci, 335xi, 525i, 525xi, 530i, 530xi, 550i, 650i, 750i, 750Li, 760Li, Z4 3.0i, Z4 3.0Si, Z4 M<br>**Engines:** All<br>**Transmissions:** All | **Knock Sensor 2 Signal Low Input Conditions:**<br>Engine started, vehicle driven at 2000rpm for 3 seconds or to a temperature of 40 degrees Celsius, and the DME detected the Knock Sensor 1 (KS1) signal was too low or not recognized by the DME.<br>**Possible Causes:**<br>• Knock sensor circuit is open<br>• Knock sensor is loose (tighten to 20 NM)<br>• Contact between the knock sensor and cylinder block is dirty, corroded or greasy<br>• Knock sensor circuit is shorted to ground, or shorted to power<br>• Knock sensor is damaged or it has failed<br>• Wrong kind of fuel used<br>• A component in the engine compartment is loose or not properly secured |
| **DTC: P1328**<br>**MIL:** Yes<br>**Years:** 2007, 2008<br>**Models:** 335i, 335Ci, 335xi, 525i, 525xi, 530i, 530xi, 550i, 650i, 750i, 750Li, 760Li, Z4 3.0i, Z4 3.0Si, Z4 M<br>**Engines:** All<br>**Transmissions:** All | **Knock Sensor 2 Signal High Input Conditions:**<br>Engine started, vehicle driven at 1600rpm for 3 seconds or to a temperature of 40 degrees Celsius, and the DME detected the Knock Sensor 1 (KS1) signal was too high.<br>**Possible Causes:**<br>• Knock sensor circuit is open<br>• Knock sensor is loose (tighten to 20 NM)<br>• Contact between the knock sensor and cylinder block is dirty, corroded or greasy<br>• Knock sensor circuit is shorted to ground, or shorted to power<br>• Knock sensor is damaged or it has failed<br>• Wrong kind of fuel used<br>• A component in the engine compartment is loose or not properly secured |
| **DTC: P1329**<br>**2T MIL:** Yes<br>**Years:** 2007, 2008<br>**Models:** 335i, 335Ci, 335xi, 525i, 525xi, 530i, 530xi, 550i, 650i, 750i, 750Li, 760Li, Z4 3.0i, Z4 3.0Si, Z4 M<br>**Engines:** All<br>**Transmissions:** All | **Knock Sensor 3 Signal Low Input Conditions:**<br>Engine started, vehicle driven at 2000rpm for 3 seconds or to a temperature of 40 degrees Celsius, and the DME detected the Knock Sensor 1 (KS1) signal was too low or not recognized by the DME.<br>**Possible Causes:**<br>• Knock sensor circuit is open<br>• Knock sensor is loose (tighten to 20 NM)<br>• Contact between the knock sensor and cylinder block is dirty, corroded or greasy<br>• Knock sensor circuit is shorted to ground, or shorted to power<br>• Knock sensor is damaged or it has failed<br>• Wrong kind of fuel used<br>• A component in the engine compartment is loose or not properly secured |
| **DTC: P1330**<br>**MIL:** Yes<br>**Years:** 2007, 2008<br>**Models:** 335i, 335Ci, 335xi, 525i, 525xi, 530i, 530xi, 550i, 650i, 750i, 750Li, 760Li, Z4 3.0i, Z4 3.0Si, Z4 M<br>**Engines:** All<br>**Transmissions:** All | **Knock Sensor 3 Signal High Input Conditions:**<br>Engine started, vehicle driven at 1600rpm for 3 seconds or to a temperature of 40 degrees Celsius, and the DME detected the Knock Sensor 1 (KS1) signal was too high.<br>**Possible Causes:**<br>• Knock sensor circuit is open<br>• Knock sensor is loose (tighten to 20 NM)<br>• Contact between the knock sensor and cylinder block is dirty, corroded or greasy<br>• Knock sensor circuit is shorted to ground, or shorted to power<br>• Knock sensor is damaged or it has failed<br>• Wrong kind of fuel used<br>• A component in the engine compartment is loose or not properly secured |
| **DTC: P1332**<br>**2T MIL:** Yes<br>**Years:** 2007, 2008<br>**Models:** 335i, 335Ci, 335xi, 525i, 525xi, 530i, 530xi, 550i, 650i, 750i, 750Li, 760Li, Z4 3.0i, Z4 3.0Si, Z4 M<br>**Engines:** All<br>**Transmissions:** All | **Knock Sensor 4 Signal Low Input Conditions:**<br>Engine started, vehicle driven at 2000rpm for 3 seconds or to a temperature of 40 degrees Celsius, and the DME detected the Knock Sensor 1 (KS1) signal was too low or not recognized by the DME.<br>**Possible Causes:**<br>• Knock sensor circuit is open<br>• Knock sensor is loose (tighten to 20 NM)<br>• Contact between the knock sensor and cylinder block is dirty, corroded or greasy<br>• Knock sensor circuit is shorted to ground, or shorted to power<br>• Knock sensor is damaged or it has failed<br>• Wrong kind of fuel used<br>• A component in the engine compartment is loose or not properly secured |

| DTC | Trouble Code Title, Conditions & Possible Causes |
|---|---|
| **DTC: P1333**<br>**2T MIL: Yes**<br>**Years:** 2007, 2008<br>**Models:** 335i, 335Ci, 335xi, 525i, 525xi, 530i, 530xi, 550i, 650i, 750i, 750Li, 760Li, Z4 3.0i, Z4 3.0Si, Z4 M<br>**Engines:** All<br>**Transmissions:** All | **Knock Sensor 4 Signal High Input Conditions:**<br>Engine started, vehicle driven at 1600rpm for 3 seconds or to a temperature of 40 degrees Celsius, and the DME detected the Knock Sensor 1 (KS1) signal was too high.<br>**Possible Causes:**<br>• Knock sensor circuit is open<br>• Knock sensor is loose (tighten to 20 NM)<br>• Contact between the knock sensor and cylinder block is dirty, corroded or greasy<br>• Knock sensor circuit is shorted to ground, or shorted to power<br>• Knock sensor is damaged or it has failed<br>• Wrong kind of fuel used<br>• A component in the engine compartment is loose or not properly secured |
| **DTC: P1340**<br>**2T MIL: Yes**<br>**Years:** 2007, 2008<br>**Models:** 335i, 335Ci, 335xi, 525i, 525xi, 530i, 530xi, 550i, 650i, M5, M6<br>**Engines:** All<br>**Transmissions:** All | **Crankshaft Position/Camshaft Sensor Signal Out of Sequence Conditions:**<br>Engine started, battery voltage at least 11.5v, all electrical components off, ground connections between engine and chassis well connected, and the DME detected the crankshaft position sensor and the camshaft sensor were out of sequence with each other.<br>**Note: The Engine Speed (RPM) Sensor detects engine speed and reference marks. Without an engine speed signal, the engine will not start. If the engine speed signal fails while the engine is running, the engine will stop immediately.**<br>**Possible Causes:**<br>• Engine speed sensor has failed or is contaminated (metal filings)<br>• Engine speed sensor's wheel is damaged<br>• Engine speed sensor circuit is shorted to the cable shield<br>• Engine speed sensor circuit is open<br>• DME is faulty<br>• Camshaft position sensor is faulty |
| **DTC: P1342**<br>**1T MISFIRE, MIL: Yes**<br>**Years:** 2007, 2008<br>**Models:** 335i, 335Ci, 335xi, 525i, 525xi, 530i, 530xi, 550i, 650i, 750i, 750Li, 760Li, Z4 3.0i, Z4 3.0Si, Z4 M<br>**Engines:** All<br>**Transmissions:** All | **Random/Multiple Misfire Detected Conditions:**<br>Engine running at an RPM greater than 400 but less than 6400 the DME detected a misfire or uneven engine running in two or more cylinders within 200 crankshaft rotations. Engine speed is between 480 and 4500rpm, load change is 0.4ms at ignition with a speed change of 2800rpms and the ASC is not active.<br>**Note: If the misfire is severe, the MIL will flash on/off on the first trip!**<br>**Possible Causes:**<br>• Fuel metering fault that affects two or more cylinders<br>• Fuel pressure too low or too high, fuel supply contaminated<br>• EVAP system problem or the EVAP canister is fuel saturated<br>• EGR valve is stuck open or the PCV system has a vacuum leak<br>• Ignition system fault (coil, plugs) affecting two or more cylinders<br>• MAF sensor contamination (it can cause a very lean condition)<br>• Vehicle driven while very low on fuel (less than 1/8 of a tank) |
| **DTC: P1344**<br>**1T MISFIRE, MIL: Yes**<br>**Years:** 2007, 2008<br>**Models:** 335i, 335Ci, 335xi, 525i, 525xi, 530i, 530xi, 550i, 650i, 750i, 750Li, 760Li, Z4 3.0i, Z4 3.0Si, Z4 M<br>**Engines:** All<br>**Transmissions:** All | **Random/Multiple Misfire Detected Conditions:**<br>Engine running at an RPM greater than 400 but less than 6400 the DME detected a misfire or uneven engine running in two or more cylinders within 200 crankshaft rotations. Engine speed is between 480 and 4500rpm, load change is 0.4ms at ignition with a speed change of 2800rpms and the ASC is not active.<br>**Note: If the misfire is severe, the MIL will flash on/off on the first trip!**<br>**Possible Causes:**<br>• Fuel metering fault that affects two or more cylinders<br>• Fuel pressure too low or too high, fuel supply contaminated<br>• EVAP system problem or the EVAP canister is fuel saturated<br>• EGR valve is stuck open or the PCV system has a vacuum leak<br>• Ignition system fault (coil, plugs) affecting two or more cylinders<br>• MAF sensor contamination (it can cause a very lean condition)<br>• Vehicle driven while very low on fuel (less than 1/8 of a tank) |
| **DTC: P1346**<br>**1T MISFIRE, MIL: Yes**<br>**Years:** 2007, 2008<br>**Models:** 335i, 335Ci, 335xi, 525i, 525xi, 530i, 530xi, 550i, 650i, 750i, 750Li, 760Li, Z4 3.0i, Z4 3.0Si, Z4 M<br>**Engines:** All<br>**Transmissions:** All | **Random/Multiple Misfire Detected Conditions:**<br>Engine running at an RPM greater than 400 but less than 6400 the DME detected a misfire or uneven engine running in two or more cylinders within 200 crankshaft rotations. Engine speed is between 480 and 4500rpm, load change is 0.4ms at ignition with a speed change of 2800rpms and the ASC is not active.<br>**Note: If the misfire is severe, the MIL will flash on/off on the first trip!**<br>**Possible Causes:**<br>• Fuel metering fault that affects two or more cylinders<br>• Fuel pressure too low or too high, fuel supply contaminated<br>• EVAP system problem or the EVAP canister is fuel saturated<br>• EGR valve is stuck open or the PCV system has a vacuum leak<br>• Ignition system fault (coil, plugs) affecting two or more cylinders<br>• MAF sensor contamination (it can cause a very lean condition)<br>• Vehicle driven while very low on fuel (less than 1/8 of a tank) |

| DTC | Trouble Code Title, Conditions & Possible Causes |
|---|---|
| **DTC: P1348**<br>**1T MISFIRE, MIL: Yes**<br>**Years:** 2007, 2008<br>**Models:** 335i, 335Ci, 335xi, 525i, 525xi, 530i, 530xi, 550i, 650i, 750i, 750Li, 760Li, Z4 3.0i, Z4 3.0Si, Z4 M<br>**Engines:** All<br>**Transmissions:** All | **Random/Multiple Misfire Detected Conditions:**<br>Engine running at an RPM greater than 400 but less than 6400 the DME detected a misfire or uneven engine running in two or more cylinders within 200 crankshaft rotations. Engine speed is between 480 and 4500rpm, load change is 0.4ms at ignition with a speed change of 2800rpms and the ASC is not active.<br>**Note: If the misfire is severe, the MIL will flash on/off on the first trip!**<br>**Possible Causes:**<br>• Fuel metering fault that affects two or more cylinders<br>• Fuel pressure too low or too high, fuel supply contaminated<br>• EVAP system problem or the EVAP canister is fuel saturated<br>• EGR valve is stuck open or the PCV system has a vacuum leak<br>• Ignition system fault (coil, plugs) affecting two or more cylinders<br>• MAF sensor contamination (it can cause a very lean condition)<br>• Vehicle driven while very low on fuel (less than 1/8 of a tank) |
| **DTC: P1350**<br>**1T MISFIRE, MIL: Yes**<br>**Years:** 2007, 2008<br>**Models:** 335i, 335Ci, 335xi, 525i, 525xi, 530i, 530xi, 550i, 650i, 750i, 750Li, 760Li, Z4 3.0i, Z4 3.0Si, Z4 M<br>**Engines:** All<br>**Transmissions:** All | **Random/Multiple Misfire Detected Conditions:**<br>Engine running at an RPM greater than 400 but less than 6400 the DME detected a misfire or uneven engine running in two or more cylinders within 200 crankshaft rotations. Engine speed is between 480 and 4500rpm, load change is 0.4ms at ignition with a speed change of 2800rpms and the ASC is not active.<br>**Note: If the misfire is severe, the MIL will flash on/off on the first trip!**<br>**Possible Causes:**<br>• Fuel metering fault that affects two or more cylinders<br>• Fuel pressure too low or too high, fuel supply contaminated<br>• EVAP system problem or the EVAP canister is fuel saturated<br>• EGR valve is stuck open or the PCV system has a vacuum leak<br>• Ignition system fault (coil, plugs) affecting two or more cylinders<br>• MAF sensor contamination (it can cause a very lean condition)<br>• Vehicle driven while very low on fuel (less than 1/8 of a tank) |
| **DTC: P1352**<br>**1T MISFIRE, MIL: Yes**<br>**Years:** 2007, 2008<br>**Models:** All<br>**Engines:** All<br>**Transmissions:** All | **Random/Multiple Misfire Detected Conditions:**<br>Engine running at an RPM greater than 400 but less than 6400 the DME detected a misfire or uneven engine running in two or more cylinders within 200 crankshaft rotations. Engine speed is between 480 and 4500rpm, load change is 0.4ms at ignition with a speed change of 2800rpms and the ASC is not active.<br>**Note: If the misfire is severe, the MIL will flash on/off on the first trip!**<br>**Possible Causes:**<br>• Fuel metering fault that affects two or more cylinders<br>• Fuel pressure too low or too high, fuel supply contaminated<br>• EVAP system problem or the EVAP canister is fuel saturated<br>• EGR valve is stuck open or the PCV system has a vacuum leak<br>• Ignition system fault (coil, plugs) affecting two or more cylinders<br>• MAF sensor contamination (it can cause a very lean condition)<br>• Vehicle driven while very low on fuel (less than 1/8 of a tank) |
| **DTC: P1354**<br>**1T MISFIRE, MIL: Yes**<br>**Years:** 2007, 2008<br>**Models:** 550i, 650i, M5, M6<br>**Engines:** All<br>**Transmissions:** All | **Random/Multiple Misfire Detected Conditions:**<br>Engine running at an RPM greater than 400 but less than 6400 the DME detected a misfire or uneven engine running in two or more cylinders within 200 crankshaft rotations. Engine speed is between 480 and 4500rpm, load change is 0.4ms at ignition with a speed change of 2800rpms and the ASC is not active.<br>**Note: If the misfire is severe, the MIL will flash on/off on the first trip!**<br>**Possible Causes:**<br>• Fuel metering fault that affects two or more cylinders<br>• Fuel pressure too low or too high, fuel supply contaminated<br>• EVAP system problem or the EVAP canister is fuel saturated<br>• EGR valve is stuck open or the PCV system has a vacuum leak<br>• Ignition system fault (coil, plugs) affecting two or more cylinders<br>• MAF sensor contamination (it can cause a very lean condition)<br>• Vehicle driven while very low on fuel (less than 1/8 of a tank) |

| DTC | Trouble Code Title, Conditions & Possible Causes |
|---|---|
| **DTC: P1355**<br>**1T MISFIRE, MIL: Yes**<br>**Years:** 2007, 2008<br>**Models:** 550i, 650i, M5, M6<br>**Engines:** All<br>**Transmissions:** All | **Random/Multiple Misfire Detected Conditions:**<br>Engine running at an RPM greater than 400 but less than 6400 the DME detected a misfire or uneven engine running in two or more cylinders within 200 crankshaft rotations. Engine speed is between 480 and 4500rpm, load change is 0.4ms at ignition with a speed change of 2800rpms and the ASC is not active.<br>**Note: If the misfire is severe, the MIL will flash on/off on the first trip!**<br>**Possible Causes:**<br>&bull; Fuel metering fault that affects two or more cylinders<br>&bull; Fuel pressure too low or too high, fuel supply contaminated<br>&bull; EVAP system problem or the EVAP canister is fuel saturated<br>&bull; EGR valve is stuck open or the PCV system has a vacuum leak<br>&bull; Ignition system fault (coil, plugs) affecting two or more cylinders<br>&bull; MAF sensor contamination (it can cause a very lean condition)<br>&bull; Vehicle driven while very low on fuel (less than 1/8 of a tank) |
| **DTC: P1356**<br>**1T MISFIRE, MIL: Yes**<br>**Years:** 2007, 2008<br>**Models:** 550i, 650i, M5, M6<br>**Engines:** All<br>**Transmissions:** All | **Random/Multiple Misfire Detected Conditions:**<br>Engine running at an RPM greater than 400 but less than 6400 the DME detected a misfire or uneven engine running in two or more cylinders within 200 crankshaft rotations. Engine speed is between 480 and 4500rpm, load change is 0.4ms at ignition with a speed change of 2800rpms and the ASC is not active.<br>**Note: If the misfire is severe, the MIL will flash on/off on the first trip!**<br>**Possible Causes:**<br>&bull; Fuel metering fault that affects two or more cylinders<br>&bull; Fuel pressure too low or too high, fuel supply contaminated<br>&bull; EVAP system problem or the EVAP canister is fuel saturated<br>&bull; EGR valve is stuck open or the PCV system has a vacuum leak<br>&bull; Ignition system fault (coil, plugs) affecting two or more cylinders<br>&bull; MAF sensor contamination (it can cause a very lean condition)<br>&bull; Vehicle driven while very low on fuel (less than 1/8 of a tank) |
| **DTC: P1357**<br>**1T MISFIRE, MIL: Yes**<br>**Years:** 2007, 2008<br>**Models:** 550i, 650i, M5, M6<br>**Engines:** All<br>**Transmissions:** All | **Random/Multiple Misfire Detected Conditions:**<br>Engine running at an RPM greater than 400 but less than 6400 the DME detected a misfire or uneven engine running in two or more cylinders within 200 crankshaft rotations. Engine speed is between 480 and 4500rpm, load change is 0.4ms at ignition with a speed change of 2800rpms and the ASC is not active.<br>**Note: If the misfire is severe, the MIL will flash on/off on the first trip!**<br>**Possible Causes:**<br>&bull; Fuel metering fault that affects two or more cylinders<br>&bull; Fuel pressure too low or too high, fuel supply contaminated<br>&bull; EVAP system problem or the EVAP canister is fuel saturated<br>&bull; EGR valve is stuck open or the PCV system has a vacuum leak<br>&bull; Ignition system fault (coil, plugs) affecting two or more cylinders<br>&bull; MAF sensor contamination (it can cause a very lean condition)<br>&bull; Vehicle driven while very low on fuel (less than 1/8 of a tank) |
| **DTC: P1377**<br>**MIL: Yes**<br>**Years:** 2007, 2008<br>**Models:** 335i, 335Ci, 335xi, 525i, 525xi, 530i, 530xi, 550i, 650i, 750i, 750Li, 760Li, Z4 3.0i, Z4 3.0Si, Z4 M<br>**Engines:** All<br>**Transmissions:** All | **Camshaft Position Sensor Master Camshaft Not Defined Conditions:**<br>Engine started, battery voltage must be at least 11.5v, after five camshaft revolutions, the DME detected the CMP sensor signal was implausible. Perhaps a defect in the power supply. Reduced power occurs and once the engine is turned off, it is impossible to restart it.<br>**Possible Causes:**<br>&bull; CMP sensor circuit is open or shorted to ground<br>&bull; CMP sensor circuit is shorted to power<br>&bull; CMP sensor ground (return) circuit is open<br>&bull; CMP sensor installation incorrect (Hall-effect type)<br>&bull; CMP sensor is damaged or CMP sensor shielding damaged |
| **DTC: P1381**<br>**MIL: Yes**<br>**Years:** 2007, 2008<br>**Models:** 530i, 530xi, 550i, 650i, Z4 3.0i, Z4 3.0Si, Z4 M<br>**Engines:** All<br>**Transmissions:** All | **Control Module Self-Test, Knock Control Offset (Bank 1) Conditions:**<br>Engine started, vehicle driven at 1520rpm for 10 seconds or to a temperature of 40 degrees Celsius, and the DME detected the Knock chip is defective.<br>**Possible Causes:**<br>&bull; Knock sensor circuit is open<br>&bull; Knock sensor is loose (tighten to 20 NM)<br>&bull; Contact between the knock sensor and cylinder block is dirty, corroded or greasy<br>&bull; Knock sensor circuit is shorted to ground, or shorted to power<br>&bull; Knock sensor is damaged or it has failed<br>&bull; Wrong kind of fuel used<br>&bull; A component in the engine compartment is loose or not properly secured |

| DTC | Trouble Code Title, Conditions & Possible Causes |
|---|---|
| **DTC: P1382**<br>**MIL: Yes**<br>**Years:** 2007, 2008<br>**Models:** 530i, 530xi, 550i, 650i, Z4 3.0i, Z4 3.0Si, Z4 M<br>**Engines:** All<br>**Transmissions:** All | **Control Module Self-Test, Knock Control Test Pulse (Bank 1) Conditions:**<br>Engine started, vehicle driven at 1520rpm for 10 seconds or to a temperature of 40 degrees Celsius, and the DME detected the Knock chip is defective.<br>**Possible Causes:**<br>• Knock sensor circuit is open<br>• Knock sensor is loose (tighten to 20 NM)<br>• Contact between the knock sensor and cylinder block is dirty, corroded or greasy<br>• Knock sensor circuit is shorted to ground, or shorted to power<br>• Knock sensor is damaged or it has failed<br>• Wrong kind of fuel used<br>• A component in the engine compartment is loose or not properly secured |
| **DTC: P1384**<br>**2T MIL: Yes**<br>**Years:** 2007, 2008<br>**Models:** 335i, 335Ci, 335xi, 525i, 525xi, 530i, 530xi, 550i, 650i, 750i, 750Li, 760Li, Z4 3.0i, Z4 3.0Si, Z4 M<br>**Engines:** All<br>**Transmissions:** All | **Knock Sensor 3 Circuit Malfunction Conditions:**<br>Engine started, vehicle driven at 1520rpm for 3 seconds or to a temperature of 40 degrees Celsius, and the DME detected the Knock Sensor 1 (KS1) signal was not recognized. The engine speed is greater than 2080rpm but less than 6000rpm and the coolant temperature is greater than 40.5 degrees Celsius.<br>**Possible Causes:**<br>• Knock sensor circuit is open<br>• Knock sensor is loose (tighten to 20 NM)<br>• Contact between the knock sensor and cylinder block is dirty, corroded or greasy<br>• Knock sensor circuit is shorted to ground, or shorted to power<br>• Knock sensor is damaged or it has failed<br>• Wrong kind of fuel used<br>• A component in the engine compartment is loose or not properly secured |
| **DTC: P1385**<br>**2T MIL: Yes**<br>**Years:** 2007, 2008<br>**Models:** 335i, 335Ci, 335xi, 525i, 525xi, 530i, 530xi, 550i, 650i, 750i, 750Li, 760Li, Z4 3.0i, Z4 3.0Si, Z4 M<br>**Engines:** All<br>**Transmissions:** All | **Knock Sensor 4 Circuit Malfunction Conditions:**<br>Engine started, vehicle driven at 1520rpm for 3 seconds or to a temperature of 40 degrees Celsius, and the DME detected the Knock Sensor 1 (KS1) signal was not recognized. The engine speed is greater than 2080rpm but less than 6000rpm and the coolant temperature is greater than 40.5 degrees Celsius.<br>**Possible Causes:**<br>• Knock sensor circuit is open<br>• Knock sensor is loose (tighten to 20 NM)<br>• Contact between the knock sensor and cylinder block is dirty, corroded or greasy<br>• Knock sensor circuit is shorted to ground, or shorted to power<br>• Knock sensor is damaged or it has failed<br>• Wrong kind of fuel used<br>• A component in the engine compartment is loose or not properly secured |
| **DTC: P1386**<br>**MIL: Yes**<br>**Years:** 2007, 2008<br>**Models:** 335i, 335Ci, 335xi, 525i, 525xi, 530i, 530xi, 550i, 650i, 750i, 750Li, 760Li, Z4 3.0i, Z4 3.0Si, Z4 M<br>**Engines:** All<br>**Transmissions:** All | **Control Module Self-Test, Knock Control Circuit Baseline Test (Bank 1) Conditions:**<br>Engine started, vehicle driven at 1520rpm for 10 seconds or to a temperature of 40 degrees Celsius, and the DME detected the Knock chip is defective. The engine speed is greater than 2080rpm but less than 6000rpm and the coolant temperature is greater than 40.5 degrees Celsius.<br>**Possible Causes:**<br>• Knock sensor circuit is open<br>• Knock sensor is loose (tighten to 20 NM)<br>• Contact between the knock sensor and cylinder block is dirty, corroded or greasy<br>• Knock sensor circuit is shorted to ground, or shorted to power<br>• Knock sensor is damaged or it has failed<br>• Wrong kind of fuel used<br>• A component in the engine compartment is loose or not properly secured |
| **DTC: P1396**<br>**2T MIL: Yes**<br>**Years:** 2007, 2008<br>**Models:** All<br>**Engines:** All<br>**Transmissions:** All | **Camshaft Position Sensor "A" Circ Malfunction Conditions:**<br>Engine started, battery voltage must be at least 11.5v, all electrical components must be off, parking brake must be engaged (to keep daytime driving lights off), automatic transmission selector must be in park and the ground between the engine and the chassis must be well connected. The DME detected the CMP sensor signal was implausible. Engine speed is greater than 500rpm, and the fault is tolerable as long as there are no misfired occurring at the same time.<br>**Possible Causes:**<br>• CMP sensor circuit is open or shorted to ground<br>• CMP sensor circuit is shorted to power<br>• CMP sensor ground (return) circuit is open<br>• CMP sensor installation incorrect (Hall-effect type)<br>• CMP sensor is damaged or CMP sensor shielding damaged |

| DTC | Trouble Code Title, Conditions & Possible Causes |
|---|---|
| **DTC: P1400**<br>**2T MIL: Yes**<br>**Years:** 2007, 2008<br>**Models:** 750i, 750Li, 760Li<br>**Engines:** All<br>**Transmissions:** All | **Heated Catalyst System Minimum Battery Voltage Conditions:**<br>Engine started, battery voltage must be at least 11.5v, all electrical components must be off, parking brake must be engaged (to keep daytime driving lights off), automatic transmission selector must be in park, the exhaust system must be properly sealed between the catalytic converter and the cylinder head, coolant temperature must be at least 80 degrees Celsius and oxygen sensor heaters for oxygen sensors before the catalytic converter must be functioning properly and the ground between the engine and the chassis must be well connected. The coolant temperature is less than 90 degrees Celsius, catalyst temperature is less than 300 degrees Celsius, engine crank time less than five seconds, vehicle speed is less than three mph and the engine speed is less than 200rpm.<br>**Possible Causes:**<br>• Air leaks at the exhaust manifold or in the exhaust pipes<br>• Catalytic converter is damaged, contaminated or it has failed<br>• ECT/CHT sensor has lost its calibration (the signal is incorrect)<br>• Engine cylinders misfiring, or the ignition timing is over retarded<br>• Engine oil is contaminated<br>• Front HO2S or rear HO2S is contaminated with fuel or moisture<br>• Front HO2S and/or the rear HO2S is loose in the mounting hole<br>• Front HO2S much older than the rear HO2S (HO2S-11 is lazy)<br>• Fuel system pressure is too high (check the pressure regulator)<br>• Rear HO2S wires improperly connected or the HO2S has failed |
| **DTC: P1401**<br>**2T MIL: Yes**<br>**Years:** 2007, 2008<br>**Models:** 750i, 750Li, 760Li<br>**Engines:** All<br>**Transmissions:** All | **Heated Catalyst System Minimum Battery Voltage Conditions:**<br>Engine started, battery voltage must be at least 11.5v, all electrical components must be off, parking brake must be engaged (to keep daytime driving lights off), automatic transmission selector must be in park, the exhaust system must be properly sealed between the catalytic converter and the cylinder head, coolant temperature must be at least 80 degrees Celsius and oxygen sensor heaters for oxygen sensors before the catalytic converter must be functioning properly and the ground between the engine and the chassis must be well connected. The coolant temperature is less than 90 degrees Celsius, catalyst temperature is less than 300 degrees Celsius, engine crank time less than five seconds, vehicle speed is less than three mph and the engine speed is less than 200rpm.<br>**Possible Causes:**<br>• Air leaks at the exhaust manifold or in the exhaust pipes<br>• Catalytic converter is damaged, contaminated or it has failed<br>• ECT/CHT sensor has lost its calibration (the signal is incorrect)<br>• Engine cylinders misfiring, or the ignition timing is over retarded<br>• Engine oil is contaminated<br>• Front HO2S or rear HO2S is contaminated with fuel or moisture<br>• Front HO2S and/or the rear HO2S is loose in the mounting hole<br>• Front HO2S much older than the rear HO2S (HO2S-11 is lazy)<br>• Fuel system pressure is too high (check the pressure regulator)<br>• Rear HO2S wires improperly connected or the HO2S has failed |
| **DTC: P1403**<br>**2T MIL: Yes**<br>**Years:** 2007, 2008<br>**Models:** 750i, 750Li, 760Li<br>**Engines:** All<br>**Transmissions:** All | **Heated Catalyst System Minimum Battery Voltage Conditions:**<br>Engine started, battery voltage must be at least 11.5v, all electrical components must be off, parking brake must be engaged (to keep daytime driving lights off), automatic transmission selector must be in park, the exhaust system must be properly sealed between the catalytic converter and the cylinder head, coolant temperature must be at least 80 degrees Celsius and oxygen sensor heaters for oxygen sensors before the catalytic converter must be functioning properly and the ground between the engine and the chassis must be well connected. The coolant temperature is less than 90 degrees Celsius, catalyst temperature is less than 300 degrees Celsius, engine crank time less than five seconds, vehicle speed is less than three mph and the engine speed is less than 200rpm.<br>**Possible Causes:**<br>• Air leaks at the exhaust manifold or in the exhaust pipes<br>• Catalytic converter is damaged, contaminated or it has failed<br>• ECT/CHT sensor has lost its calibration (the signal is incorrect)<br>• Engine cylinders misfiring, or the ignition timing is over retarded<br>• Engine oil is contaminated<br>• Front HO2S or rear HO2S is contaminated with fuel or moisture<br>• Front HO2S and/or the rear HO2S is loose in the mounting hole<br>• Front HO2S much older than the rear HO2S (HO2S-11 is lazy)<br>• Fuel system pressure is too high (check the pressure regulator)<br>• Rear HO2S wires improperly connected or the HO2S has failed |

| DTC | Trouble Code Title, Conditions & Possible Causes |
|---|---|
| **DTC: P1403**<br>**2T MIL: Yes**<br>**Years:** 2007, 2008<br>**Models:** 328i, 328Ci, 328xi,<br>**Engines:** All<br>**Transmissions:** All | **Evaporative Emission System Shut Off Valve Conditions:**<br>Engine started, battery voltage must be at least 11.5v, all electrical components must be off, parking brake must be engaged (to keep daytime driving lights off), automatic transmission selector must be in park, the exhaust system must be properly sealed between the catalytic converter and the cylinder head, coolant temperature must be at least 80 degrees Celsius and oxygen sensor heaters for oxygen sensors before the catalytic converter must be functioning properly and the ground between the engine and the chassis must be well connected. The DME detected an unexpected condition on the EVAP shut off valve when the device was cycled On/Off during testing.<br>**Possible Causes:**<br>• EVAP power supply circuit is open<br>• EVAP solenoid control circuit is open or shorted to ground<br>• EVAP solenoid control circuit is shorted to power (B+)<br>• EVAP solenoid valve is damaged or it has failed<br>• EVAP canister has a leak or a poor seal |
| **DTC: P1404**<br>**2T MIL: Yes**<br>**Years:** 2007, 2008<br>**Models:** 750i, 750Li, 760Li<br>**Engines:** All<br>**Transmissions:** All | **Heated Catalyst System Minimum Battery Voltage Conditions:**<br>Engine started, battery voltage must be at least 11.5v, all electrical components must be off, parking brake must be engaged (to keep daytime driving lights off), automatic transmission selector must be in park, the exhaust system must be properly sealed between the catalytic converter and the cylinder head, coolant temperature must be at least 80 degrees Celsius and oxygen sensor heaters for oxygen sensors before the catalytic converter must be functioning properly and the ground between the engine and the chassis must be well connected. The coolant temperature is less than 90 degrees Celsius, catalyst temperature is less than 300 degrees Celsius, engine crank time less than five seconds, vehicle speed is less than three mph and the engine speed is less than 200rpm.<br>**Possible Causes:**<br>• Air leaks at the exhaust manifold or in the exhaust pipes<br>• Catalytic converter is damaged, contaminated or it has failed<br>• ECT/CHT sensor has lost its calibration (the signal is incorrect)<br>• Engine cylinders misfiring, or the ignition timing is over retarded<br>• Engine oil is contaminated<br>• Front HO2S or rear HO2S is contaminated with fuel or moisture<br>• Front HO2S and/or the rear HO2S is loose in the mounting hole<br>• Front HO2S much older than the rear HO2S (HO2S-11 is lazy)<br>• Fuel system pressure is too high (check the pressure regulator)<br>• Rear HO2S wires improperly connected or the HO2S has failed |
| **DTC: P1411**<br>**Years:** 2007, 2008<br>**Models:** 335i, 335Ci, 335xi, 525i,<br>525xi, 530i, 530xi, 550i, 650i, 750i,<br>750Li, 760Li, Z4 3.0i, Z4 3.0Si,<br>Z4 M<br>**Engines:** All<br>**Transmissions:** All | **Secondary Air Pump Valve Plausibility Conditions:**<br>The Engine Control Module detects an implausible signal when activating the secondary air injection solenoid valve.<br>**Note: Solenoid valve is closed when no voltage is present.**<br>**Possible Causes:**<br>• Connector to the secondary air injection valve is loose or disconnected<br>• Secondary air injector valve circuit short<br>• Secondary air injector valve circuit is open<br>• Faulty secondary air injector valve |
| **DTC: P1412**<br>**Years:** 2007, 2008<br>**Models:** 335i, 335Ci, 335xi, 525i,<br>525xi, 530i, 530xi,<br>**Engines:** All<br>**Transmissions:** All | **Secondary Air Pump Valve Plausibility Conditions:**<br>The Engine Control Module detects an implausible signal when activating the secondary air injection solenoid valve. The max flow limit check detected leakage between the air pump and the valve.<br>**Note: Solenoid valve is closed when no voltage is present.**<br>**Possible Causes:**<br>• Connector to the secondary air injection valve is loose or disconnected<br>• Secondary air injector valve circuit short<br>• Secondary air injector valve circuit is open<br>• Faulty secondary air injector valve |
| **DTC: P1413**<br>**Years:** 2007, 2008<br>**Models:** 335i, 335Ci, 335xi, 525i,<br>525xi, 530i, 530xi, 550i, 650i, 750i,<br>750Li, 760Li, Z4 3.0i, Z4 3.0Si,<br>Z4 M<br>**Engines:** All<br>**Transmissions:** All | **Secondary Air Injector Pump Relay Control Circuit Signal Low Conditions:**<br>The Engine Control Module detects a short circuit when activating the secondary air injection solenoid valve.<br>**Note: Solenoid valve is closed when no voltage is present.**<br>**Possible Causes:**<br>• Connector to the secondary air injection valve is loose or disconnected<br>• Secondary air injector valve circuit short<br>• Secondary air injector valve circuit is open<br>• Faulty secondary air injector valve |

| DTC | Trouble Code Title, Conditions & Possible Causes |
|---|---|
| **DTC: P1414**<br>**Years:** 2007, 2008<br>**Models:** 335i, 335Ci, 335xi, 525i, 525xi, 530i, 530xi, 550i, 650i, 750i, 750Li, 760Li, Z4 3.0i, Z4 3.0Si, Z4 M<br>**Engines:** All<br>**Transmissions:** All | **Secondary Air Injector Pump Relay Control Circuit Signal High Conditions:**<br>The Engine Control Module detects a short circuit when activating the secondary air injection solenoid valve.<br>**Note: Solenoid valve is closed when no voltage is present.**<br>**Possible Causes:**<br>• Connector to the secondary air injection valve is loose or disconnected<br>• Secondary air injector valve circuit short<br>• Secondary air injector valve circuit is open<br>• Faulty secondary air injector valve |
| **DTC: P1418**<br>**Years:** 2007, 2008<br>**Models:** 335i, 335Ci, 335xi, 525i, 525xi, 530i, 530xi,<br>**Engines:** All<br>**Transmissions:** All | **Secondary Air Pump Valve Plausibility Conditions:**<br>The Engine Control Module detects an implausible signal when activating the secondary air injection solenoid valve. The secondary air valve or tube is blocked.<br>**Note: Solenoid valve is closed when no voltage is present.**<br>**Possible Causes:**<br>• Connector to the secondary air injection valve is loose or disconnected<br>• Secondary air injector valve circuit short<br>• Secondary air injector valve circuit is open<br>• Faulty secondary air injector valve |
| **DTC: P1454**<br>**2T MIL: Yes**<br>**Years:** 2007, 2008<br>**Models:** 750i, 750Li, 760Li<br>**Engines:** All<br>**Transmissions:** All | **Secondary Air Injector Valve Circuit Electrical Malfunction (Open) Conditions:**<br>The Engine Control Module activates the secondary air injection solenoid valve, but the Heated Oxygen Sensor (HO2S) does not detect secondary air injection.<br>**Note: Solenoid valve is closed when no voltage is present.**<br>**Possible Causes:**<br>• Connector to the secondary air injection valve is loose or disconnected<br>• Secondary air injector valve circuit short<br>• Secondary air injector valve circuit is open<br>• Faulty secondary air injector valve |
| **DTC: P1456**<br>**2T MIL: Yes**<br>**Years:** 2007, 2008<br>**Models:** 750i, 750Li, 760Li<br>**Engines:** All<br>**Transmissions:** All | **Heated Catalyst System Efficiency Above Threshold Conditions:**<br>Engine started, battery voltage must be at least 11.5v, all electrical components must be off, parking brake must be engaged (to keep daytime driving lights off), automatic transmission selector must be in park, the exhaust system must be properly sealed between the catalytic converter and the cylinder head, coolant temperature must be at least 80 degrees Celsius and oxygen sensor heaters for oxygen sensors before the catalytic converter must be functioning properly and the ground between the engine and the chassis must be well connected. The coolant temperature is less than 90 degrees Celsius, catalyst temperature is less than 300 degrees Celsius, engine crank time less than five seconds, vehicle speed is less than three mph and the engine speed is less than 200rpm.<br>**Possible Causes:**<br>• Air leaks at the exhaust manifold or in the exhaust pipes<br>• Catalytic converter is damaged, contaminated or it has failed<br>• ECT/CHT sensor has lost its calibration (the signal is incorrect)<br>• Engine cylinders misfiring, or the ignition timing is over retarded<br>• Engine oil is contaminated<br>• Front HO2S or rear HO2S is contaminated with fuel or moisture<br>• Front HO2S and/or the rear HO2S is loose in the mounting hole<br>• Front HO2S much older than the rear HO2S (HO2S-11 is lazy)<br>• Fuel system pressure is too high (check the pressure regulator)<br>• Rear HO2S wires improperly connected or the HO2S has failed |

| DTC | Trouble Code Title, Conditions & Possible Causes |
|---|---|
| **DTC: P1459**<br>**2T MIL:** Yes<br>**Years:** 2007, 2008<br>**Models:** 750i, 750Li, 760Li<br>**Engines:** All<br>**Transmissions:** All | **Heated Catalyst System Efficiency Below Threshold Conditions:**<br>Engine started, battery voltage must be at least 11.5v, all electrical components must be off, parking brake must be engaged (to keep daytime driving lights off), automatic transmission selector must be in park, the exhaust system must be properly sealed between the catalytic converter and the cylinder head, coolant temperature must be at least 80 degrees Celsius and oxygen sensor heaters for oxygen sensors before the catalytic converter must be functioning properly and the ground between the engine and the chassis must be well connected. The coolant temperature is less than 90 degrees Celsius, catalyst temperature is less than 300 degrees Celsius, engine crank time less than five seconds, vehicle speed is less than three mph and the engine speed is less than 200rpm.<br>**Possible Causes:**<br>• Air leaks at the exhaust manifold or in the exhaust pipes<br>• Catalytic converter is damaged, contaminated or it has failed<br>• ECT/CHT sensor has lost its calibration (the signal is incorrect)<br>• Engine cylinders misfiring, or the ignition timing is over retarded<br>• Engine oil is contaminated<br>• Front HO2S or rear HO2S is contaminated with fuel or moisture<br>• Front HO2S and/or the rear HO2S is loose in the mounting hole<br>• Front HO2S much older than the rear HO2S (HO2S-11 is lazy)<br>• Fuel system pressure is too high (check the pressure regulator)<br>• Rear HO2S wires improperly connected or the HO2S has failed |
| **DTC: P1463**<br>**2T MIL:** Yes<br>**Years:** 2007, 2008<br>**Models:** 750i, 750Li, 760Li<br>**Engines:** All<br>**Transmissions:** All | **Heated Catalyst System Rationality Check Conditions:**<br>Engine started, battery voltage must be at least 11.5v, all electrical components must be off, parking brake must be engaged (to keep daytime driving lights off), automatic transmission selector must be in park, the exhaust system must be properly sealed between the catalytic converter and the cylinder head, coolant temperature must be at least 80 degrees Celsius and oxygen sensor heaters for oxygen sensors before the catalytic converter must be functioning properly and the ground between the engine and the chassis must be well connected. The coolant temperature is less than 90 degrees Celsius, catalyst temperature is less than 300 degrees Celsius, engine crank time less than five seconds, vehicle speed is less than three mph and the engine speed is less than 200rpm.<br>**Possible Causes:**<br>• Air leaks at the exhaust manifold or in the exhaust pipes<br>• Catalytic converter is damaged, contaminated or it has failed<br>• ECT/CHT sensor has lost its calibration (the signal is incorrect)<br>• Engine cylinders misfiring, or the ignition timing is over retarded<br>• Engine oil is contaminated<br>• Front HO2S or rear HO2S is contaminated with fuel or moisture<br>• Front HO2S and/or the rear HO2S is loose in the mounting hole<br>• Front HO2S much older than the rear HO2S (HO2S-11 is lazy)<br>• Fuel system pressure is too high (check the pressure regulator)<br>• Rear HO2S wires improperly connected or the HO2S has failed |
| **DTC: P1464**<br>**2T MIL:** Yes<br>**Years:** 2007, 2008<br>**Models:** 750i, 750Li, 760Li<br>**Engines:** All<br>**Transmissions:** All | **Heated Catalyst System Rationality Check Conditions:**<br>Engine started, battery voltage must be at least 11.5v, all electrical components must be off, parking brake must be engaged (to keep daytime driving lights off), automatic transmission selector must be in park, the exhaust system must be properly sealed between the catalytic converter and the cylinder head, coolant temperature must be at least 80 degrees Celsius and oxygen sensor heaters for oxygen sensors before the catalytic converter must be functioning properly and the ground between the engine and the chassis must be well connected. The coolant temperature is less than 90 degrees Celsius, catalyst temperature is less than 300 degrees Celsius, engine crank time less than five seconds, vehicle speed is less than three mph and the engine speed is less than 200rpm.<br>**Possible Causes:**<br>• Air leaks at the exhaust manifold or in the exhaust pipes<br>• Catalytic converter is damaged, contaminated or it has failed<br>• ECT/CHT sensor has lost its calibration (the signal is incorrect)<br>• Engine cylinders misfiring, or the ignition timing is over retarded<br>• Engine oil is contaminated<br>• Front HO2S or rear HO2S is contaminated with fuel or moisture<br>• Front HO2S and/or the rear HO2S is loose in the mounting hole<br>• Front HO2S much older than the rear HO2S (HO2S-11 is lazy)<br>• Fuel system pressure is too high (check the pressure regulator)<br>• Rear HO2S wires improperly connected or the HO2S has failed |

| DTC | Trouble Code Title, Conditions & Possible Causes |
|---|---|
| **DTC: P1466**<br>**2T MIL:** Yes<br>**Years:** 2007, 2008<br>**Models:** 750i, 750Li, 760Li<br>**Engines:** All<br>**Transmissions:** All | **Heated Catalyst System Rationality Check Conditions:**<br>Engine started, battery voltage must be at least 11.5v, all electrical components must be off, parking brake must be engaged (to keep daytime driving lights off), automatic transmission selector must be in park, the exhaust system must be properly sealed between the catalytic converter and the cylinder head, coolant temperature must be at least 80 degrees Celsius and oxygen sensor heaters for oxygen sensors before the catalytic converter must be functioning properly and the ground between the engine and the chassis must be well connected. The coolant temperature is less than 90 degrees Celsius, catalyst temperature is less than 300 degrees Celsius, engine crank time less than five seconds, vehicle speed is less than three mph and the engine speed is less than 200rpm.<br>**Possible Causes:**<br>• Air leaks at the exhaust manifold or in the exhaust pipes<br>• Catalytic converter is damaged, contaminated or it has failed<br>• ECT/CHT sensor has lost its calibration (the signal is incorrect)<br>• Engine cylinders misfiring, or the ignition timing is over retarded<br>• Engine oil is contaminated<br>• Front HO2S or rear HO2S is contaminated with fuel or moisture<br>• Front HO2S and/or the rear HO2S is loose in the mounting hole<br>• Front HO2S much older than the rear HO2S (HO2S-11 is lazy)<br>• Fuel system pressure is too high (check the pressure regulator)<br>• Rear HO2S wires improperly connected or the HO2S has failed |
| **DTC: P1467**<br>**2T MIL:** Yes<br>**Years:** 2007, 2008<br>**Models:** 750i, 750Li, 760Li<br>**Engines:** All<br>**Transmissions:** All | **Heated Catalyst System Rationality Check Conditions:**<br>Engine started, battery voltage must be at least 11.5v, all electrical components must be off, parking brake must be engaged (to keep daytime driving lights off), automatic transmission selector must be in park, the exhaust system must be properly sealed between the catalytic converter and the cylinder head, coolant temperature must be at least 80 degrees Celsius and oxygen sensor heaters for oxygen sensors before the catalytic converter must be functioning properly and the ground between the engine and the chassis must be well connected. The coolant temperature is less than 90 degrees Celsius, catalyst temperature is less than 300 degrees Celsius, engine crank time less than five seconds, vehicle speed is less than three mph and the engine speed is less than 200rpm.<br>**Possible Causes:**<br>• Air leaks at the exhaust manifold or in the exhaust pipes<br>• Catalytic converter is damaged, contaminated or it has failed<br>• ECT/CHT sensor has lost its calibration (the signal is incorrect)<br>• Engine cylinders misfiring, or the ignition timing is over retarded<br>• Engine oil is contaminated<br>• Front HO2S or rear HO2S is contaminated with fuel or moisture<br>• Front HO2S and/or the rear HO2S is loose in the mounting hole<br>• Front HO2S much older than the rear HO2S (HO2S-11 is lazy)<br>• Fuel system pressure is too high (check the pressure regulator)<br>• Rear HO2S wires improperly connected or the HO2S has failed |
| **DTC: P1500**<br>**2T MIL:** Yes<br>**Years:** 2007, 2008<br>**Models:** 335i, 335Ci, 335xi, 525i, 525xi, 530i, 530xi, M5, M6, Z4 3.0i, Z4 3.0Si, Z4 M<br>**Engines:** All<br>**Transmissions:** All | **Idle Air Control Valve Malfunction Conditions:**<br>Engine started, battery voltage at least 11.5v, all electrical components off, ground connections between engine and chassis well connected, coolant temperature at least 80-degrees Celsius. The DME detected deviation from the normal operating parameters of the Idle Air Control Valve. The vehicle speed can be zero mph and the engine load must be less than 1.5ms.<br>**Possible Causes:**<br>• Charge air system leaks<br>• Recirculating valve for turbocharger is faulty<br>• Turbocharging system is damaged<br>• Vacuum diaphragm for turbocharger needs adjusting<br>• Wastegate bypass regulator valve is faulty |
| **DTC: P1501**<br>**2T MIL:** Yes<br>**Years:** 2007, 2008<br>**Models:** 335i, 335Ci, 335xi, 525i, 525xi, 530i, 530xi, M5, M6, Z4 3.0i, Z4 3.0Si, Z4 M<br>**Engines:** All<br>**Transmissions:** All | **Idle Air Control Valve Malfunction Conditions:**<br>Engine started, battery voltage at least 11.5v, all electrical components off, ground connections between engine and chassis well connected, coolant temperature at least 80-degrees Celsius. The DME detected deviation from the normal operating parameters of the Idle Air Control Valve. The vehicle speed can be zero mph and the engine load must be less than 1.5ms.<br>**Possible Causes:**<br>• Charge air system leaks<br>• Recirculating valve for turbocharger is faulty<br>• Turbocharging system is damaged<br>• Vacuum diaphragm for turbocharger needs adjusting<br>• Wastegate bypass regulator valve is faulty |

| DTC | Trouble Code Title, Conditions & Possible Causes |
|---|---|
| **DTC: P1502**<br>**2T MIL: Yes**<br>**Years:** 2007, 2008<br>**Models:** 335i, 335Ci, 335xi, 525i, 525xi, 530i, 530xi, M5, M6, Z4 3.0i, Z4 3.0Si, Z4 M<br>**Engines:** All<br>**Transmissions:** All | **Idle Air Control Valve Circuit Short to B+ Conditions:**<br>Engine running the DME detected that the idle air control valve was intermittent.<br>**Possible Causes:**<br>• Fuel delivery unit connector is loose or not attached<br>• Fuse 18 cause a short to the transfer fuel pump or the O2S<br>• Fuel pump has failed<br>• Fuel pump relay circuit is shorted to ground, B+ or is open<br>• Fuel Pump (FP) Relay not activated |
| **DTC: P1503**<br>**2T MIL: Yes**<br>**Years:** 2007, 2008<br>**Models:** 335i, 335Ci, 335xi, 525i, 525xi, 530i, 530xi, M5, M6, Z4 3.0i, Z4 3.0Si, Z4 M<br>**Engines:** All<br>**Transmissions:** All | **Idle Air Control Valve Circuit Short to Ground Conditions:**<br>Engine running the DME detected that the idle air control valve was intermittent.<br>**Possible Causes:**<br>• Fuel delivery unit connector is loose or not attached<br>• Fuse 18 cause a short to the transfer fuel pump or the O2S<br>• Fuel pump has failed<br>• Fuel pump relay circuit is shorted to ground, B+ or is open<br>• Fuel Pump (FP) Relay not activated |
| **DTC: P1504**<br>**2T MIL: Yes**<br>**Years:** 2007, 2008<br>**Models:** 335i, 335Ci, 335xi, 525i, 525xi, 530i, 530xi, M5, M6, Z4 3.0i, Z4 3.0Si, Z4 M<br>**Engines:** All<br>**Transmissions:** All | **Idle Air Control Valve Circuit Continuity-Open Load Conditions:**<br>Engine running the DME detected that the idle air control valve signal was intermittent.<br>**Possible Causes:**<br>• Fuel delivery unit connector is loose or not attached<br>• Fuse 18 cause a short to the transfer fuel pump or the O2S<br>• Fuel pump has failed<br>• Fuel pump relay circuit is shorted to ground, B+ or is open<br>• Fuel Pump (FP) Relay not activated |
| **DTC: P1506**<br>**2T MIL: Yes**<br>**Years:** 2007, 2008<br>**Models:** 335i, 335Ci, 335xi, 525i, 525xi, 530i, 530xi, M5, M6, Z4 3.0i, Z4 3.0Si, Z4 M<br>**Engines:** All<br>**Transmissions:** All | **Idle Air Control Valve Circuit Short to B+ Conditions:**<br>Engine running the DME detected that the idle air control valve signal was intermittent.<br>**Possible Causes:**<br>• Fuel delivery unit connector is loose or not attached<br>• Fuse 18 cause a short to the transfer fuel pump or the O2S<br>• Fuel pump has failed<br>• Fuel pump relay circuit is shorted to ground, B+ or is open<br>• Fuel Pump (FP) Relay not activated |
| **DTC: P1507**<br>**2T MIL: Yes**<br>**Years:** 2007, 2008<br>**Models:** 335i, 335Ci, 335xi, 525i, 525xi, 530i, 530xi, M5, M6, Z4 3.0i, Z4 3.0Si, Z4 M<br>**Engines:** All<br>**Transmissions:** All | **Idle Air Control Valve Circuit Short to Ground Conditions:**<br>Engine running the DME detected that the idle air control valve signal was intermittent.<br>**Possible Causes:**<br>• Fuel delivery unit connector is loose or not attached<br>• Fuse 18 cause a short to the transfer fuel pump or the O2S<br>• Fuel pump has failed<br>• Fuel pump relay circuit is shorted to ground, B+ or is open<br>• Fuel Pump (FP) Relay not activated |
| **DTC: P1508**<br>**2T MIL: Yes**<br>**Years:** 2007, 2008<br>**Models:** 335i, 335Ci, 335xi, 525i, 525xi, 530i, 530xi, M5, M6, Z4 3.0i, Z4 3.0Si, Z4 M<br>**Engines:** All<br>**Transmissions:** All | **Idle Air Control Valve Circuit Continuity-Open Load Conditions:**<br>Engine running the DME detected that the idle air control valve signal was intermittent.<br>**Possible Causes:**<br>• Fuel delivery unit connector is loose or not attached<br>• Fuse 18 cause a short to the transfer fuel pump or the O2S<br>• Fuel pump has failed<br>• Fuel pump relay circuit is shorted to ground, B+ or is open<br>• Fuel Pump (FP) Relay not activated |
| **DTC: P1511**<br>**2T MIL: Yes**<br>**Years:** 2007, 2008<br>**Models:** 328i, 328Ci, 328xi,<br>**Engines:** All<br>**Transmissions:** All | **Differentiated Intake Manifold Control Circuit Electrical Conditions:**<br>Engine started, and the DME detected the changeover valve circuit was faulting during the continuous self test.<br>**Possible Causes:**<br>• Leaky vacuum reservoir, vacuum lines loose or damaged<br>• Vacuum solenoid or vacuum actuator is damaged<br>• IMRC actuator cable/gears are seized, or the cables are improperly routed or seized<br>• IMRC housing return springs are damaged or disconnected<br>• Lever/shaft return stop may be obstructed or bent, or the lever/shaft wide open stop may be obstructed or bent, or the IMRC lever/shaft may be sticking, binding or disconnected<br>• IMRC control circuit open, shorted or the VPWR circuit is open |

| DTC | Trouble Code Title, Conditions & Possible Causes |
|---|---|
| **DTC: P1512**<br>**2T MIL:** Yes<br>**Years:** 2007, 2008<br>**Models:** 335i, 335Ci, 335xi, 525i, 525xi, 530i, 530xi, 550i, 650i, 750i, 750Li, 760Li, Z4 3.0i, Z4 3.0Si, Z4 M<br>**Engines:** All<br>**Transmissions:** All | **Differentiated Intake Manifold Control Circuit Signal Low Conditions:**<br>Engine started, and the DME detected the changeover valve circuit was shorting to negative during the continuous self test.<br>**Possible Causes:**<br>• Leaky vacuum reservoir, vacuum lines loose or damaged<br>• Vacuum solenoid or vacuum actuator is damaged<br>• IMRC actuator cable/gears are seized, or the cables are improperly routed or seized<br>• IMRC housing return springs are damaged or disconnected<br>• Lever/shaft return stop may be obstructed or bent, or the lever/shaft wide open stop may be obstructed or bent, or the IMRC lever/shaft may be sticking, binding or disconnected<br>• IMRC control circuit open, shorted or the VPWR circuit is open |
| **DTC: P1513**<br>**2T MIL:** Yes<br>**Years:** 2007, 2008<br>**Models:** 335i, 335Ci, 335xi, 525i, 525xi, 530i, 530xi, 550i, 650i, 750i, 750Li, 760Li, Z4 3.0i, Z4 3.0Si, Z4 M<br>**Engines:** All<br>**Transmissions:** All | **Differentiated Intake Manifold Control Circuit Signal High Conditions:**<br>Engine started, and the DME detected the changeover valve circuit was shorting to positive during the continuous self test.<br>**Possible Causes:**<br>• Leaky vacuum reservoir, vacuum lines loose or damaged<br>• Vacuum solenoid or vacuum actuator is damaged<br>• IMRC actuator cable/gears are seized, or the cables are improperly routed or seized<br>• IMRC housing return springs are damaged or disconnected<br>• Lever/shaft return stop may be obstructed or bent, or the lever/shaft wide open stop may be obstructed or bent, or the IMRC lever/shaft may be sticking, binding or disconnected<br>• IMRC control circuit open, shorted or the VPWR circuit is open |
| **DTC: P1515**<br>**MIL:** Yes<br>**Years:** 2007, 2008<br>**Models:** 335i, 335Ci, 335xi, 525i, 525xi, 530i, 530xi, 550i, 650i, 750i, 750Li, 760Li, Z4 3.0i, Z4 3.0Si, Z4 M<br>**Engines:** All<br>**Transmissions:** All | **Engine Off Timer Plausibility Conditions:**<br>The DME detected an implausible instrument cluster and/or power module signal. CAN signal failure while the DME was operating at a range between 6 and 16 volts. CAN Bus lost communications. The system time (time pulse) is implausible relative to the DME's internal counter.<br>**Possible Causes:**<br>• Check the CAN bus<br>• Check instrument cluster/power module/Can Signal if no other CAN faults are detected |
| **DTC: P1517**<br>**Years:** 2007, 2008<br>**Models:** 335i, 335Ci, 335xi, 525i, 525xi, 530i, 530xi, 550i, 650i, 750i, 750Li, 760Li, Z4 3.0i, Z4 3.0Si, Z4 M<br>**Engines:** All<br>**Transmissions:** All | **Rough Road Detection, No Wheel Speed Signal Conditions:**<br>The DME detected an electrical malfunction on the main relay circuit<br>**Possible Causes:**<br>• Possible failure of wheel speed sensor on drive axle |
| **DTC: P1518**<br>**Years:** 2007, 2008<br>**Models:** 335i, 335Ci, 335xi, 525i, 525xi, 530i, 530xi, 550i, 650i, 750i, 750Li, 760Li, Z4 3.0i, Z4 3.0Si, Z4 M<br>**Engines:** All<br>**Transmissions:** All | **Rough Road Detection, Wheel Speed Too High Conditions:**<br>The DME detected an electrical malfunction on the main relay circuit<br>**Possible Causes:**<br>• Possible failure of wheel speed sensor on drive axle |
| **DTC: P1520**<br>**Years:** 2007, 2008<br>**Models:** 335i, 335Ci, 335xi, 525i, 525xi, 530i, 530xi, 550i, 650i, 750i, 750Li, 760Li, Z4 3.0i, Z4 3.0Si, Z4 M<br>**Engines:** All<br>**Transmissions:** All | **Engine Oil Quality Sensor Level Measurement Error Conditions:**<br>After 60 seconds with the ignition on, a miscommunication between the oil sensor and the DME was detected.<br>**Possible Causes:**<br>• Replace the oil temperature sensor<br>• Check BSD wire, voltage supply and oil level sensor ground for an open circuit. If okay, replace the oil level sensor. |
| **DTC: P1521**<br>**Years:** 2007, 2008<br>**Models:** 335i, 335Ci, 335xi, 525i, 525xi, 530i, 530xi, 550i, 650i, 750i, 750Li, 760Li, Z4 3.0i, Z4 3.0Si, Z4 M<br>**Engines:** All<br>**Transmissions:** All | **Engine Oil Quality Sensor Communication Error Conditions:**<br>After 60 seconds with the ignition on, a miscommunication between the oil sensor and the DME was detected.<br>**Possible Causes:**<br>• Replace the oil temperature sensor<br>• Check BSD wire, voltage supply and oil level sensor ground for an open circuit. If okay, replace the oil level sensor. |

| DTC | Trouble Code Title, Conditions & Possible Causes |
|---|---|
| **DTC: P1526**<br>**2T MIL:** Yes<br>**Years:** 2007, 2008<br>**Models:** M5, M6<br>**Engines:** All<br>**Transmissions:** All | **Camshaft Control Circuit Ground Conditions:**<br>Engine started and driven at an engine speed of more than 400rpm; and the DME detected the camshaft timing exceeded the calibrated levels. The valve timing did not change from the current valve timing or it remained fixed during the testing.<br>**Note: The camshaft adjustment is load- and RPM dependant. The electrical camshaft adjustment valve 1 switches oil pressure onto camshaft adjuster (mechanical adjustment mechanism), which adjusts the camshaft.**<br>**Possible Causes:**<br>• Fuel pump has failed<br>• CPS circuit is open, shorted to ground or shorted to power<br>• Battery voltage below 11.5 volts<br>• Position actuator circuit may short to B+ or Ground<br>• Camshaft timing improperly set, or continuous oil flow to the VCT piston chamber<br>• Camshaft advance mechanism (the VCT unit) is sticking or binding mechanically<br>• VCT solenoid valve is stuck in open position |
| **DTC: P1529**<br>**2T MIL:** Yes<br>**Years:** 2007, 2008<br>**Models:** 328i, 328Ci, 328xi, 335i, 335Ci, 335xi, 525i, 525xi, 530i, 530xi, M5, M6<br>**Engines:** All<br>**Transmissions:** All | **Camshaft Control Circuit Short to B+ Conditions:**<br>Engine started and driven at an engine speed of more than 400rpm; and the DME detected the camshaft timing exceeded the calibrated voltage levels. The valve timing did not change from the current valve timing or it remained fixed during the testing.<br>**Note: The camshaft adjustment is load- and RPM dependant. The electrical camshaft adjustment valve 1 switches oil pressure onto camshaft adjuster (mechanical adjustment mechanism), which adjusts the camshaft.**<br>**Possible Causes:**<br>• Fuel pump has failed<br>• CPS circuit is open, shorted to ground or shorted to power<br>• Battcry voltage below 11.5 volts<br>• Position actuator circuit may short to B+ or Ground<br>• Camshaft timing improperly set, or continuous oil flow to the VCT piston chamber<br>• Camshaft advance mechanism (the VCT unit) is sticking or binding mechanically<br>• VCT solenoid valve is stuck in open position |
| **DTC: P1530**<br>**2T MIL:** Yes<br>**Years:** 2007, 2008<br>**Models:** 328i, 328Ci, 328xi, 335i, 335Ci, 335xi, 525i, 525xi, 530i, 530xi, M5, M6<br>**Engines:** All<br>**Transmissions:** All | **Camshaft Control Circuit Short to Ground Conditions:**<br>Engine started and driven at an engine speed of more than 400rpm; and the DME detected the camshaft timing exceeded the calibrated levels. The valve timing did not change from the current valve timing or it remained fixed during the testing.<br>**Note: The camshaft adjustment is load- and RPM dependant. The electrical camshaft adjustment valve 1 switches oil pressure onto camshaft adjuster (mechanical adjustment mechanism), which adjusts the camshaft.**<br>**Possible Causes:**<br>• Fuel pump has failed<br>• CPS circuit is open, shorted to ground or shorted to power<br>• Battery voltage below 11.5 volts<br>• Position actuator circuit may short to B+ or Ground<br>• Camshaft timing improperly set, or continuous oil flow to the VCT piston chamber<br>• Camshaft advance mechanism (the VCT unit) is sticking or binding mechanically<br>• VCT solenoid valve is stuck in open position |
| **DTC: P1532**<br>**2T MIL:** Yes<br>**Years:** 2007, 2008<br>**Models:** 328i, 328Ci, 328xi, 335i, 335Ci, 335xi, 525i, 525xi, 530i, 530xi, M5, M6<br>**Engines:** All<br>**Transmissions:** All | **Camshaft Control Circuit Open Conditions:**<br>Engine started and driven at an engine speed of more than 400rpm; and the DME detected the camshaft timing exceeded the calibrated levels. The valve timing did not change from the current valve timing or it remained fixed during the testing.<br>**Note: The camshaft adjustment is load- and RPM dependant. The electrical camshaft adjustment valve 1 switches oil pressure onto camshaft adjuster (mechanical adjustment mechanism), which adjusts the camshaft.**<br>**Possible Causes:**<br>• Fuel pump has failed<br>• CPS circuit is open, shorted to ground or shorted to power<br>• Battery voltage below 11.5 volts<br>• Position actuator circuit may short to B+ or Ground<br>• Camshaft timing improperly set, or continuous oil flow to the VCT piston chamber<br>• Camshaft advance mechanism (the VCT unit) is sticking or binding mechanically<br>• VCT solenoid valve is stuck in open position |
| **DTC: P1535**<br>**2T MIL:** No<br>**Years:** 2007, 2008<br>**Models:** 335i, 335Ci, 335xi, 525i, 525xi, 530i, 530xi, 550i, 650i, 750i, 750Li, 760Li, Z4 3.0i, Z4 3.0Si, Z4 M<br>**Engines:** All<br>**Transmissions:** All | **Differentiated Intake Manifold Coil Temperature Limit Value Exceeded Conditions:**<br>This fault stems from the All-N62 equipped with an infinitely adjustable control of induction system. The temperature of the coil has exceeded the predetermined limits.<br>**Possible Causes:**<br>• Faulty controller<br>• Faulty coil |

| DTC | Trouble Code Title, Conditions & Possible Causes |
|---|---|
| **DTC: P1536**<br>**2T MIL: No**<br>**Years:** 2007, 2008<br>**Models:** 335i, 335Ci, 335xi, 525i, 525xi, 530i, 530xi, 550i, 650i, 750i, 750Li, 760Li, Z4 3.0i, Z4 3.0Si, Z4 M<br>**Engines:** All<br>**Transmissions:** All | **Differentiated Intake Manifold Controller Monitoring, Control Deviation Conditions:**<br>This fault stems from the All-N62 equipped with an infinitely adjustable control of induction system.<br>**Possible Causes:**<br>• Faulty controller |
| **DTC: P1537**<br>**2T MIL: No**<br>**Years:** 2007, 2008<br>**Models:** 335i, 335Ci, 335xi, 525i, 525xi, 530i, 530xi, 550i, 650i, 750i, 750Li, 760Li, Z4 3.0i, Z4 3.0Si, Z4 M<br>**Engines:** All<br>**Transmissions:** All | **Differentiated Intake Manifold Potentiometer Voltage in Lower Diagnosis Range Conditions:**<br>This fault stems from the All-N62 equipped with an infinitely adjustable control of induction system.<br>**Possible Causes:**<br>• Faulty potentiometer |
| **DTC: P1538**<br>**2T MIL: No**<br>**Years:** 2007, 2008<br>**Models:** 335i, 335Ci, 335xi, 525i, 525xi, 530i, 530xi, 550i, 650i, 750i, 750Li, 760Li, Z4 3.0i, Z4 3.0Si, Z4 M<br>**Engines:** All<br>**Transmissions:** All | **Differentiated Intake Manifold Potentiometer Voltage in Upper Diagnosis Range Conditions:**<br>This fault stems from the All-N62 equipped with an infinitely adjustable control of induction system.<br>**Possible Causes:**<br>• Faulty potentiometer |
| **DTC: P1539**<br>**2T MIL: No**<br>**Years:** 2007, 2008<br>**Models:** 335i, 335Ci, 335xi, 525i, 525xi, 530i, 530xi, 550i, 650i, 750i, 750Li, 760Li, Z4 3.0i, Z4 3.0Si, Z4 M<br>**Engines:** All<br>**Transmissions:** All | **Differentiated Intake Manifold Coil Temperature Threshold Exceeded Conditions:**<br>This fault stems from the All-N62 equipped with an infinitely adjustable control of induction system. The temperature of the coil has exceeded the predetermined limits.<br>**Possible Causes:**<br>• Faulty controller<br>• Faulty coil |
| **DTC: P1551**<br>**MIL: Yes**<br>**Years:** 2007, 2008<br>**Models:** 335i, 335Ci, 335xi, 525i, 525xi, 530i, 530xi, 550i, 650i, 750i, 750Li, 760Li, Z4 3.0i, Z4 3.0Si, Z4 M<br>**Engines:** All<br>**Transmissions:** All | **Engine Off Timer Timeout Conditions:**<br>The DME detected a CAN signal failure while the DME was operating at a range between 6 and 16 volts. CAN Bus lost communications.<br>**Possible Causes:**<br>• Check the CAN bus<br>• Check instrument cluster/power module/Can Signal |
| **DTC: P1603**<br>**2T MIL: Yes**<br>**Years:** 2007, 2008<br>**Models:** 750i, 750Li, 760Li, M5, M6<br>**Engines:** All<br>**Transmissions:** All | **Control Module Self Test, Torque Monitoring Conditions:**<br>Ignition on, the DME detected a control module malfunction. The torque monitoring feature compares the torque demand (from accelerator pedal, FGR, electrical equipment and transmission) with the torque provided (calculated from HFH, injector valves, ignition angle, throttle valve angle, differential pressure and lambda). Deviations trigger fuel-supply safety shutdown to prevent vehicle from autonomous acceleration.<br>**Possible Causes:**<br>• Check HFM for contamination and replace if necessary<br>• Replace the DME |

| DTC | Trouble Code Title, Conditions & Possible Causes |
|---|---|
| **DTC: P1614**<br>**MIL:** No<br>**Years:** 2007, 2008<br>**Models:** 335i, 335Ci, 335xi, 525i, 525xi, 530i, 530xi, 550i, 650i, 750i, 750Li, 760Li, Z4 3.0i, Z4 3.0Si, Z4 M<br>**Engines:** All<br>**Transmissions:** All | **Serial Communication Link ACC Malfunction Conditions:**<br>Key on after 800ms or engine running for 100ms, the DME detected an electrical malfunction regarding the Adaptive Cruise Control circuit. Check the ACC fuse, the ECU or CAN for fault, measure the resistance on the BUS, check the wiring harness. If this fault occurs when the ACC is not installed, the DME has been incorrectly initialized for the ACC version.<br>**Possible Causes:**<br>• Circuit from the MIL to the DME<br>• ECU Failure, BUS system failure<br>• Circuit from the EPC to the DME |
| **DTC: P1619**<br>**Years:** 2007, 2008<br>**Models:** 750i, 750Li, 760Li<br>**Engines:** All<br>**Transmissions:** All | **Thermostat Control Circuit Ground Conditions:**<br>The engine's warm up performance is monitored by comparing measured coolant temperature with the modeled coolant temperature to detect a defective coolant thermostat that is reading false. The engine temperature must be less than 65 degrees Celsius, engine speed greater than 800rpm (with the vehicle speed greater than 10 but less than 90km/h) and the ambient temperature greater than −8 degrees Celsius. The thermostat should be wide open when cold, but is in error if it opens below desired control temperature. There is a difference between the coolant temperature at the engine and the radiator outlet of less than 5 degrees Celsius.<br>**Possible Causes:**<br>• Check for low coolant level or incorrect coolant mixture<br>• DME detects a short circuit wiring in the ECT<br>• CHT sensor is out-of-calibration or it has failed<br>• ECT sensor is out-of-calibration or it has failed<br>• Replace the thermostat |
| **DTC: P1620**<br>**Years:** 2007, 2008<br>**Models:** 750i, 750Li, 760Li<br>**Engines:** All<br>**Transmissions:** All | **Thermostat Control Circuit Open Conditions:**<br>The engine's warm up performance is monitored by comparing measured coolant temperature with the modeled coolant temperature to detect a defective coolant thermostat that is reading false. The engine temperature must be less than 65 degrees Celsius, engine speed greater than 800rpm (with the vehicle speed greater than 10 but less than 90km/h) and the ambient temperature greater than −8 degrees Celsius. The thermostat should be wide open when cold, but is in error if it opens below desired control temperature. There is a difference between the coolant temperature at the engine and the radiator outlet of less than 5 degrees Celsius.<br>**Possible Causes:**<br>• Check for low coolant level or incorrect coolant mixture<br>• DME detects a short circuit wiring in the ECT<br>• CHT sensor is out-of-calibration or it has failed<br>• ECT sensor is out-of-calibration or it has failed<br>• Replace the thermostat |
| **DTC: P1628**<br>**2T MIL:** Yes<br>**Years:** 2007, 2008<br>**Models:** 335i, 335Ci, 335xi, 525i, 525xi, 530i, 530xi, 550i, 650i, 750i, 750Li, 760Li, Z4 3.0i, Z4 3.0Si, Z4 M<br>**Engines:** All<br>**Transmissions:** All | **Throttle Valve Adaptation Spring Test Malfunction During Opening (Bank 1) Conditions:**<br>Engine started, and the battery voltage is greater than 7 volts, this fault is only diagnosed during the throttle valve's adaptation phase. DVE fails to close from emergency air position. Vehicle speed is zero, throttle pedal is less than 14.9 percent, coolant temperature is between 5.3 and 100.5 degrees Celsius and the intake air temperature is 5.3 degrees Celsius.<br>**Possible Causes:**<br>• Plate is sticking<br>• Defective return spring<br>• DVE electrical activation defective<br>• DVE motor wiring may have shorted<br>• Throttle valve is contaminated with foreign objects |
| **DTC: P1629**<br>**2T MIL:** Yes<br>**Years:** 2007, 2008<br>**Models:** 335i, 335Ci, 335xi, 525i, 525xi, 530i, 530xi, 550i, 650i, 750i, 750Li, 760Li, Z4 3.0i, Z4 3.0Si, Z4 M<br>**Engines:** All<br>**Transmissions:** All | **Throttle Valve Adaptation Spring Test Stop, Spring Does Not Open (Bank 1) Conditions:**<br>Engine started, and the battery voltage is greater than 7 volts, this fault is only diagnosed during the throttle valve's adaptation phase. Emergency air position not achieved from closed valve. Vehicle speed is zero, throttle pedal is less than 14.9 percent, coolant temperature is between 5.3 and 100.5 degrees Celsius and the intake air temperature is 5.3 degrees Celsius.<br>**Possible Causes:**<br>• Plate is sticking<br>• Defective return spring<br>• DVE electrical activation defective.<br>• DVE motor wiring may have shorted<br>• Throttle valve is contaminated with foreign objects |

| DTC | Trouble Code Title, Conditions & Possible Causes |
|---|---|
| **DTC: P1631**<br>**2T MIL:** Yes<br>**Years:** 2007, 2008<br>**Models:** 335i, 335Ci, 335xi, 525i, 525xi, 530i, 530xi, 550i, 650i, 750i, 750Li, 760Li, Z4 3.0i, Z4 3.0Si, Z4 M<br>**Engines:** All<br>**Transmissions:** All | **Throttle Valve Adaptation Spring Test (Bank 1) Conditions:**<br>Engine started, and the battery voltage is greater than 7 volts, this fault is only diagnosed during the throttle valve's adaptation phase. It is not possible to move DVE from emergency air position. Vehicle speed is zero, throttle pedal is less than 14.9 percent, coolant temperature is between 5.3 and 100.5 degrees Celsius and the intake air temperature is 5.3 degrees Celsius.<br>**Possible Causes:**<br>• Plate is sticking<br>• Defective return spring<br>• DVE electrical activation defective.<br>• Throttle position sensor wiring may have shorted<br>• Throttle valve is contaminated with foreign objects |
| **DTC: P1633**<br>**2T MIL:** Yes<br>**Years:** 2007, 2008<br>**Models:** 335i, 335Ci, 335xi, 525i, 525xi, 530i, 530xi, 550i, 650i, 750i, 750Li, 760Li, Z4 3.0i, Z4 3.0Si, Z4 M<br>**Engines:** All<br>**Transmissions:** All | **Throttle Valve Adaptation Limp-Home Position Unknown Conditions:**<br>Engine started, and the battery voltage is greater than 7 volts, this fault is only diagnosed during the throttle valve's adaptation phase. Check on throttle valve's emergency default position as determined during throttle valve adaptation. There is a failure to reach emergency air position with DVE switched off. Vehicle speed is zero, throttle pedal is less than 14.9 percent, coolant temperature is between 5.3 and 100.5 degrees Celsius and the intake air temperature is 5.3 degrees Celsius.<br>**Possible Causes:**<br>• Plate is sticking<br>• Defective return spring<br>• DVE electrical activation defective.<br>• DVE motor wiring may have shorted<br>• Throttle valve is contaminated with foreign objects |
| **DTC: P1634**<br>**2T MIL:** Yes<br>**Years:** 2007, 2008<br>**Models:** 335i, 335Ci, 335xi, 525i, 525xi, 530i, 530xi, 550i, 650i, 750i, 750Li, 760Li, M5, M6, X3, X5, Z4 3.0i, Z4 3.0Si, Z4 M<br>**Engines:** All<br>**Transmissions:** All | **Throttle Valve Adaptation Spring Test Failed (Bank 1) Conditions:**<br>Engine started, and the battery voltage is greater than 7 volts, this fault is only diagnosed during the throttle valve's adaptation phase. Vehicle speed is zero, throttle pedal is less than 14.9 percent, coolant temperature is between 5.3 and 100.5 degrees Celsius and the intake air temperature is 5.3 degrees Celsius.<br>**Possible Causes:**<br>• Plate is sticking<br>• Defective return spring<br>• DVE electrical activation defective.<br>• Throttle position sensor wiring may have shorted<br>• Throttle valve is contaminated with foreign objects |
| **DTC: P1635**<br>**2T MIL:** Yes<br>**Years:** 2007, 2008<br>**Models:** 335i, 335Ci, 335xi, 525i, 525xi, 530i, 530xi, 550i, 650i, 750i, 750Li, 760Li, Z4 3.0i, Z4 3.0Si, Z4 M<br>**Engines:** All<br>**Transmissions:** All | **Throttle Valve Adaptation Lower Mechanical Stop Not Adapted (Bank 1) Conditions:**<br>Engine started, and the battery voltage is greater than 7 volts, there was no throttle valve adaptation conducted yet and the fuel-supply safety was shutdown until adaptation is successful. There was a failure to reach the lower mechanical travel limit<br>**Possible Causes:**<br>• Travel stop contaminated.<br>• Defected DVE<br>• Throttle valve contaminated |
| **DTC: P1636**<br>**2T MIL:** Yes<br>**Years:** 2007, 2008<br>**Models:** 335i, 335Ci, 335xi, 525i, 525xi, 530i, 530xi, 550i, 650i, 750i, 750Li, 760Li, M5, M6, X3, X5, Z4 3.0i, Z4 3.0Si, Z4 M<br>**Engines:** All<br>**Transmissions:** All | **Throttle Valve Position Control, Range Check (Bank 1) Conditions:**<br>Engine started, and the battery voltage is greater than 7 volts, and the comparison of the throttle valve's actual angle to its specified angle is great than 0.2 seconds and less than 0.5 seconds greater than a valve calculated from rpm and temperature readings. The throttle valve activation occurs via an electric motor (throttle drive) in the throttle valve control module. It is activated by the Engine Control Module (DME) according to specifications of the two sensors, Throttle Position (TP) Sensor and Accelerator Pedal Position Sensor 2.<br>**Possible Causes:**<br>• TP sensor signal circuit is open (inspect wiring & connector)<br>• TP sensor signal circuit is shorted to ground<br>• TP sensor or module is damaged or it has failed<br>• Throttle valve is damaged or dirty<br>• Throttle valve control module is faulty |

| DTC | Trouble Code Title, Conditions & Possible Causes |
|---|---|
| **DTC: P1637**<br>**2T MIL:** Yes<br>**Years:** 2007, 2008<br>**Models:** 335i, 335Ci, 335xi, 525i, 525xi, 530i, 530xi, 550i, 650i, 750i, 750Li, 760Li, M5, M6, X3, X5, Z4 3.0i, Z4 3.0Si, Z4 M<br>**Engines:** All<br>**Transmissions:** All | **Throttle Valve Position Control, Control Deviation (Bank 1) Conditions:**<br>Engine started, and the battery voltage is greater than 7 volts, and the comparison of the throttle valve's actual angle to its specified angle is great than 0.2 seconds and less than 0.5 seconds greater than a valve calculated from rpm and temperature readings. The throttle valve activation occurs via an electric motor (throttle drive) in the throttle valve control module. It is activated by the Engine Control Module (DME) according to specifications of the two sensors, Throttle Position (TP) Sensor and Accelerator Pedal Position Sensor 2.<br>**Possible Causes:**<br>• TP sensor signal circuit is open (inspect wiring & connector)<br>• TP sensor signal circuit is shorted to ground<br>• TP sensor or module is damaged or it has failed<br>• Throttle valve is damaged or dirty<br>• Throttle valve control module is faulty |
| **DTC: P1638**<br>**2T MIL:** Yes<br>**Years:** 2007, 2008<br>**Models:** 335i, 335Ci, 335xi, 525i, 525xi, 530i, 530xi, 550i, 650i, 750i, 750Li, 760Li, M5, M6, X3, X5, Z4 3.0i, Z4 3.0Si, Z4 M<br>**Engines:** All<br>**Transmissions:** All | **Throttle Valve Position Control Throttle Stuck Temporarily (Bank 1) Conditions:**<br>Engine started, and the battery voltage is greater than 7 volts, and despite control signal to throttle valve, no position change was detected in 0.6 seconds. The throttle valve activation occurs via an electric motor (throttle drive) in the throttle valve control module. It is activated by the Engine Control Module (DME) according to specifications of the two sensors, Throttle Position (TP) Sensor and Accelerator Pedal Position Sensor 2.<br>**Possible Causes:**<br>• TP sensor signal circuit is open (inspect wiring & connector)<br>• TP sensor signal circuit is shorted to ground<br>• TP sensor or module is damaged or it has failed<br>• Throttle valve is damaged or dirty<br>• Throttle valve control module is faulty |
| **DTC: P1639**<br>**2T MIL:** Yes<br>**Years:** 2007, 2008<br>**Models:** 335i, 335Ci, 335xi, 525i, 525xi, 530i, 530xi, 550i, 650i, 750i, 750Li, 760Li, M5, M6, X3, X5, Z4 3.0i, Z4 3.0Si, Z4 M<br>**Engines:** All<br>**Transmissions:** All | **Accelerator Pedal Position Sensor 1+2 Range/Performance Conditions:**<br>Engine started, battery voltage at least 11.5v, all electrical components off, ground connections between engine and chassis well connected, the DME detected that the accelerator pedal position sensor signal was too high.<br>**Note: Both the Throttle Position (TP) Sensor and Accelerator Pedal Position Sensor 2 are located at the accelerator pedal module and communicate the driver's intentions to the DME completely independently of each other. Both sensors are stored in one housing.**<br>**Possible Causes:**<br>• Ground between engine and chassis may be broken<br>• Throttle position sensor may have failed<br>• Accelerator Pedal Position Sensor 2 has failed<br>• Throttle position sensor wiring may have shorted<br>• Faulty voltage supply |
| **DTC: P1640**<br>**2T MIL:** Yes<br>**Years:** 2007, 2008<br>**Models:** 750i, 750Li, 760Li, M5, M6<br>**Engines:** All<br>**Transmissions:** All | **Internal Control Module (EEPROM) Error Conditions:**<br>Ignition on, the DME detected a control module malfunction (software). To achieve optimal anti-theft protection for the vehicle, an anti-theft immobilizer is installed. The anti-theft immobilizer is a system for enabling and locking the Engine Control Module (DME). So that this system cannot be circumvented, it is necessary to perform adaptation of the anti-theft immobilizer using the Vehicle Diagnostic and Information System VAS 5052 in the On Board Diagnostic (OBD) function. The great availability of equipment options makes it necessary to adapt the Engine Control Module (DME) to the vehicle (e.g. throttle valve control module or cruise control system). This "writing" function is not possible with the generic scan tool.<br>**Possible Causes:**<br>• Engine Control Module (DME) has failed<br>• Voltage supply for Engine Control Module (DME) has shorted |
| **DTC: P1641**<br>**Years:** 2007, 2008<br>**Models:** 335i, 335Ci, 335xi, 525i, 525xi, 530i, 530xi, 550i, 650i, 750i, 750Li, 760Li, Z4 3.0i, Z4 3.0Si, Z4 M<br>**Engines:** All<br>**Transmissions:** All | **Throttle Valve Adaptation Stop Due to Environmental Conditions Conditions:**<br>Engine started, and the battery voltage is greater than 7 volts, this fault is only diagnosed during the throttle valve's adaptation phase. Environmental conditions for throttle valve adaptation were not present and adaptation aborted and failed to satisfy conditions. If the previous adaptation was valid, the fault code entry is for information only.<br>**Possible Causes:**<br>• There is no action required |
| **DTC: P1642**<br>**Years:** 2007, 2008<br>**Models:** 335i, 335Ci, 335xi, 525i, 525xi, 530i, 530xi, 550i, 650i, 750i, 750Li, 760Li, Z4 3.0i, Z4 3.0Si, Z4 M<br>**Engines:** All<br>**Transmissions:** All | **Throttle Valve Adaptation Stop Due to Environmental Values Conditions:**<br>Engine started, and the battery voltage is greater than 7 volts, this fault is only diagnosed during the throttle valve's adaptation phase. Environmental conditions for throttle valve adaptation were not present and adaptation aborted and failed to satisfy conditions. If the previous adaptation was valid, the fault code entry is for information only.<br>**Possible Causes:**<br>• There is no action required |

| DTC | Trouble Code Title, Conditions & Possible Causes |
|---|---|
| **DTC: P1643**<br>**Years:** 2007, 2008<br>**Models:** 335i, 335Ci, 335xi, 525i, 525xi, 530i, 530xi, 550i, 650i, 750i, 750Li, 760Li, Z4 3.0i, Z4 3.0Si, Z4 M<br>**Engines:** All<br>**Transmissions:** All | **Throttle Valve Actuator Start Test Amplifier Balancing Plausibility Conditions:**<br>Engine started, and the battery voltage is greater than 7 volts, this fault is only diagnosed during the throttle valve's adaptation phase. The fault during amplifier calibration leads to operation with unamplified signal from potentiometer 1. The throttle valve activation occurs via an electric motor (throttle drive) in the throttle valve control module. It is activated by the Engine Control Module (DME) according to specifications of the two sensors, Throttle Position (TP) Sensor and Accelerator Pedal Position Sensor 2.<br>**Possible Causes:**<br>• TP sensor signal circuit is open (inspect wiring & connector)<br>• TP sensor signal circuit is shorted to ground<br>• TP sensor or module is damaged or it has failed<br>• Throttle valve is damaged or dirty<br>• Throttle valve control module is faulty |
| **DTC: P1644**<br>**Years:** 2007, 2008<br>**Models:** 335i, 335Ci, 335xi, 525i, 525xi, 530i, 530xi, 550i, 650i, 750i, 750Li, 760Li, Z4 3.0i, Z4 3.0Si, Z4 M<br>**Engines:** All<br>**Transmissions:** All | **Throttle Valve Adaptation Stop Relearning Lower Mechanical Stop Conditions:**<br>Engine started, and the battery voltage is greater than 7 volts, this fault is only diagnosed during the throttle valve's adaptation phase. There was no throttle valve adaptation conducted yet and the fuel-supply safety was shutdown until adaptation is successful. The fault was triggered during an attempt to repeat the initialization and there was a failure to reach the lower mechanical travel limit.<br>**Possible Causes:**<br>• Travel stop contaminated.<br>• Defected DVE<br>• Throttle valve contaminated |
| **DTC: P1645**<br>**Years:** 2007, 2008<br>**Models:** 335i, 335Ci, 335xi, 525i, 525xi, 530i, 530xi, 550i, 650i, 750i, 750Li, 760Li, Z4 3.0i, Z4 3.0Si, Z4 M<br>**Engines:** All<br>**Transmissions:** All | **Internal Control Module Random Access Memory (RAM) Reading Error Conditions:**<br>Key on, and the DME detected an internal memory fault. This code will set if KAPWR to the DME is interrupted (at the initial key on).<br>**Possible Causes:**<br>• Battery terminal corrosion, or loose battery connection<br>• Connection to the DME interrupted, or the circuit has been opened<br>• Reprogramming error has occurred and needs replacement. Remember to check for Aftermarket Performance Products before replacing a DME. |
| **DTC: P1649**<br>**Years:** 2007, 2008<br>**Models:** 335i, 335Ci, 335xi, 525i, 525xi, 530i, 530xi, 550i, 650i, 750i, 750Li, 760Li, Z4 3.0i, Z4 3.0Si, Z4 M<br>**Engines:** All<br>**Transmissions:** All | **Internal Control Module Random Access Memory (RAM) Writing Error Conditions:**<br>Key on, and the DME detected an internal memory fault. This code will set if KAPWR to the DME is interrupted (at the initial key on).<br>**Possible Causes:**<br>• Battery terminal corrosion, or loose battery connection<br>• Connection to the DME interrupted, or the circuit has been opened<br>• Reprogramming error has occurred and needs replacement. Remember to check for Aftermarket Performance Products before replacing a DME. |
| **DTC: P1650**<br>**Years:** 2007, 2008<br>**Models:** 335i, 335Ci, 335xi, 525i, 525xi, 530i, 530xi, 550i, 650i, 750i, 750Li, 760Li, Z4 3.0i, Z4 3.0Si, Z4 M<br>**Engines:** All<br>**Transmissions:** All | **Start While Engine is Running Conditions:**<br>Engine speed must be at least 1200rpm, the starter is engaged with the engine running and the DME detected an internal memory fault. This code will set if KAPWR to the DME is interrupted (at the initial key on). Check the starter and the starter relay and inspect CAS-SG.<br>**Possible Causes:**<br>• Battery terminal corrosion, or loose battery connection<br>• Connection to the DME interrupted, or the circuit has been opened<br>• Reprogramming error has occurred and needs replacement. Remember to check for Aftermarket Performance Products before replacing a DME. |
| **DTC: P1660**<br>**Years:** 2007, 2008<br>**Models:** 335i, 335Ci, 335xi, 525i, 525xi, 530i, 530xi, 550i, 650i, 750i, 750Li, 760Li, Z4 3.0i, Z4 3.0Si, Z4 M<br>**Engines:** All<br>**Transmissions:** All | **EWS (Electronic Immobilizer) Telegram Error Conditions:**<br>Key on the DME detected an electrical malfunction regarding the EWS. Check the EWS fuse, the ECU or CAN for fault, turn the ignition off and then on to repeat start calibration.<br>**Possible Causes:**<br>• Circuit from the MIL to the DME<br>• ECU Failure, BUS system failure<br>• Circuit from the EPC to the DME<br>• Defective CAS |
| **DTC: P1661**<br>**Years:** 2007, 2008<br>**Models:** 335i, 335Ci, 335xi, 525i, 525xi, 530i, 530xi, 550i, 650i, 750i, 750Li, 760Li, Z4 3.0i, Z4 3.0Si, Z4 M<br>**Engines:** All<br>**Transmissions:** All | **Timeout EWS (Electronic Immobilizer) Telegram Conditions:**<br>Key on the DME detected an electrical malfunction regarding the EWS. Check the EWS fuse, the ECU or CAN for fault, turn the ignition off and then on to repeat start calibration.<br>**Possible Causes:**<br>• Circuit from the MIL to the DME<br>• ECU Failure, BUS system failure<br>• Circuit from the EPC to the DME<br>• Defective CAS |

| DTC | Trouble Code Title, Conditions & Possible Causes |
|---|---|
| **DTC: P1662**<br>**Years:** 2007, 2008<br>**Models:** 335i, 335Ci, 335xi, 525i, 525xi, 530i, 530xi, 550i, 650i, 750i, 750Li, 760Li, Z4 3.0i, Z4 3.0Si, Z4 M<br>**Engines:** All<br>**Transmissions:** All | **EWS (Electronic Immobilizer) Telegram Parity Error Conditions:**<br>Key on the DME detected an electrical malfunction regarding the EWS. Check the EWS fuse, the ECU or CAN for fault, turn the ignition off and then on to repeat start calibration.<br>**Possible Causes:**<br>• Circuit from the MIL to the DME<br>• ECU Failure, BUS system failure<br>• Circuit from the EPC to the DME<br>• Defective CAS |
| **DTC: P1663**<br>**Years:** 2007, 2008<br>**Models:** 335i, 335Ci, 335xi, 525i, 525xi, 530i, 530xi, 550i, 650i, 750i, 750Li, 760Li, Z4 3.0i, Z4 3.0Si, Z4 M<br>**Engines:** All<br>**Transmissions:** All | **EWS (Electronic Immobilizer) Rolling Code Faulty Storage in EEPROM Conditions:**<br>Key on the DME detected an electrical malfunction regarding the EWS. Check the EWS fuse, the ECU or CAN for fault, turn the ignition off and then on to repeat start calibration.<br>**Possible Causes:**<br>• Circuit from the MIL to the DME<br>• ECU Failure, BUS system failure<br>• Circuit from the EPC to the DME<br>• Defective CAS |
| **DTC: P1664**<br>**Years:** 2007, 2008<br>**Models:** 335i, 335Ci, 335xi, 525i, 525xi, 530i, 530xi, 550i, 650i, 750i, 750Li, 760Li, Z4 3.0i, Z4 3.0Si, Z4 M<br>**Engines:** All<br>**Transmissions:** All | **EWS (Electronic Immobilizer) Writing/Reading Error in EEPROM Conditions:**<br>Key on the DME detected an electrical malfunction regarding the EWS. Check the EWS fuse, the ECU or CAN for fault, turn the ignition off and then on to repeat start calibration.<br>**Possible Causes:**<br>• Circuit from the MIL to the DME<br>• ECU Failure, BUS system failure<br>• Circuit from the EPC to the DME<br>• Defective CAS |
| **DTC: P1665**<br>**Years:** 2007, 2008<br>**Models:** 335i, 335Ci, 335xi, 525i, 525xi, 530i, 530xi, 550i, 650i, 750i, 750Li, 760Li, Z4 3.0i, Z4 3.0Si, Z4 M<br>**Engines:** All<br>**Transmissions:** All | **EWS (Electronic Immobilizer) Tampering Via Rolling Code Conditions:**<br>Key on the DME detected an electrical malfunction regarding the EWS. Check the EWS fuse, the ECU or CAN for fault, turn the ignition off and then on to repeat start calibration.<br>**Possible Causes:**<br>• Circuit from the MIL to the DME<br>• ECU Failure, BUS system failure<br>• Circuit from the EPC to the DME<br>• Defective CAS |
| **DTC: P1666**<br>**Years:** 2007, 2008<br>**Models:** 335i, 335Ci, 335xi, 525i, 525xi, 530i, 530xi, 550i, 650i, 750i, 750Li, 760Li, Z4 3.0i, Z4 3.0Si, Z4 M<br>**Engines:** All<br>**Transmissions:** All | **EWS (Electronic Immobilizer) Tampering/Start Value Not Yet Programmed Conditions:**<br>Key on the DME detected an electrical malfunction regarding the EWS. Check the EWS fuse, the ECU or CAN for fault, turn the ignition off and then on to repeat start calibration.<br>**Possible Causes:**<br>• Circuit from the MIL to the DME<br>• ECU Failure, BUS system failure<br>• Circuit from the EPC to the DME<br>• Defective CAS |
| **DTC: P1667**<br>**Years:** 2007, 2008<br>**Models:** 335i, 335Ci, 335xi, 525i, 525xi, 530i, 530xi, 550i, 650i, 750i, 750Li, 760Li, Z4 3.0i, Z4 3.0Si, Z4 M<br>**Engines:** All<br>**Transmissions:** All | **EWS (Electronic Immobilizer) Start Value Not Yet Programmed Conditions:**<br>Key on the DME detected an electrical malfunction regarding the EWS. Check the EWS fuse, the ECU or CAN for fault, turn the ignition off and then on to repeat start calibration.<br>**Possible Causes:**<br>• Circuit from the MIL to the DME<br>• ECU Failure, BUS system failure<br>• Circuit from the EPC to the DME<br>• Defective CAS |
| **DTC: P1668**<br>**Years:** 2007, 2008<br>**Models:** 335i, 335Ci, 335xi, 525i, 525xi, 530i, 530xi, 550i, 650i, 750i, 750Li, 760Li, Z4 3.0i, Z4 3.0Si, Z4 M<br>**Engines:** All<br>**Transmissions:** All | **EWS (Electronic Immobilizer) Start Value Destroyed Conditions:**<br>Key on the DME detected an electrical malfunction regarding the EWS. Check the EWS fuse, the ECU or CAN for fault, turn the ignition off and then on to repeat start calibration.<br>**Possible Causes:**<br>• Circuit from the MIL to the DME<br>• ECU Failure, BUS system failure<br>• Circuit from the EPC to the DME |

| DTC | Trouble Code Title, Conditions & Possible Causes |
|---|---|
| **DTC: P1677**<br>**MIL: No**<br>**Years:** 2007, 2008<br>**Models:** 335i, 335Ci, 335xi, 525i, 525xi, 530i, 530xi, 550i, 650i, 750i, 750Li, 760Li, Z4 3.0i, Z4 3.0Si, Z4 M<br>**Engines:** All<br>**Transmissions:** All | **Adaptive Cruise Control No Activity Detected Conditions:**<br>Key on after 800ms or engine running for 100ms, the DME detected an electrical malfunction regarding the Adaptive Cruise Control circuit. Check the ACC fuse, the ECU or CAN for fault, measure the resistance on the BUS, check the wiring harness. If this fault occurs when the ACC is not installed, the DME has been incorrectly initialized for the ACC version.<br>**Possible Causes:**<br>• Circuit from the MIL to the DME<br>• ECU Failure, BUS system failure<br>• Circuit from the EPC to the DME |
| **DTC: P1680**<br>**MIL: No**<br>**Years:** 2007, 2008<br>**Models:** 335i, 335Ci, 335xi, 525i, 525xi, 530i, 530xi, 550i, 650i, 750i, 750Li, 760Li, Z4 3.0i, Z4 3.0Si, Z4 M<br>**Engines:** All<br>**Transmissions:** All | **Electronic Throttle Control Monitor Level 2/3 ADC Processor Fault Conditions:**<br>Key on, engine running to at least 1200rpm, the DME has detected an internal fault in the computer or internal fault in the control modules. The Torque monitoring feature compares the torque demand (from accelerator pedal, FGR, electrical equipment, transmission) with the torque provided (calculated from HFM, injector valves, ignition angle, throttle valve angle, differential pressure, lambda). Deviations trigger fuel-supply safety shutdown to prevent vehicle from autonomous acceleration. Internal fault in computer or in electronic control modules (check whether all ADC channels have been converted). If additional fault codes are entered in the DME, resolve these issues. If fault remains, replace the DME.<br>**Possible Causes:**<br>• Battery terminal corrosion, loose battery connection, or faulty<br>• Connection to the DME interrupted, or the circuit has been opened<br>• Reprogramming error has occurred and needs replacement.<br>• Voltage supply for Engine Control Module (DME) is faulty |
| **DTC: P1719**<br>**MIL: Yes** (U.S. only)<br>**Years:** 2007, 2008<br>**Models:** 335i, 335Ci, 335xi, 525i, 525xi, 530i, 530xi, 550i, 650i, 750i, 750Li, 760Li, M5, M6, X3, X5, Z4 3.0i, Z4 3.0Si, Z4 M<br>**Engines:** All<br>**Transmissions:** All | **CAN Level Wrong Value Conditions:**<br>The Engine Control Module (DME) communicates with all databus-capable control modules via a CAN databus. These databus-capable control modules are connected via two data bus wires which are twisted together (CAN_High and CAN_Low), and exchange information (messages). Missing information on the databus is recognized as a malfunction and stored. Trouble-free operation of the CAN-Bus requires that it have a terminal resistance. This central terminal resistor is located in the Engine Control Module (DME). The ignition is on for 800ms or the engine running for 15 seconds, the voltage is greater than 10 volts. This applies to vehicles with MIL only and does not affect bus activity.<br>**Possible Causes:**<br>• CAN data bus wires have short circuited to each other<br>• ECU has failed<br>• BUS system failure<br>• Defective bus controller (EGS) |
| **DTC: P1720**<br>**MIL: Yes** (U.S. only)<br>**Years:** 2007, 2008<br>**Models:** 335i, 335Ci, 335xi, 525i, 525xi, 530i, 530xi, 550i, 650i, 750i, 750Li, 760Li, M5, M6, X3, X5, Z4 3.0i, Z4 3.0Si, Z4 M<br>**Engines:** All<br>**Transmissions:** All | **CAN Message Timeout Conditions:**<br>The Engine Control Module (DME) communicates with all databus-capable control modules via a CAN databus. These databus-capable control modules are connected via two data bus wires which are twisted together (CAN_High and CAN_Low), and exchange information (messages). Missing information on the databus is recognized as a malfunction and stored. Trouble-free operation of the CAN-Bus requires that it have a terminal resistance. This central terminal resistor is located in the Engine Control Module (DME). The ignition is on for 800ms or the engine running for 15 seconds, the voltage is greater than 10 volts. This applies to vehicles with MIL only and does not affect bus activity.<br>**Possible Causes:**<br>• CAN data bus wires have short circuited to each other<br>• ECU has failed<br>• BUS system failure<br>• Defective bus controller (EGS) |
| **DTC: P1721**<br>**MIL: Yes** (U.S. only)<br>**Years:** 2007, 2008<br>**Models:** 335i, 335Ci, 335xi, 525i, 525xi, 530i, 530xi, 550i, 650i, 750i, 750Li, 760Li, M5, M6, X3, X5, Z4 3.0i, Z4 3.0Si, Z4 M<br>**Engines:** All<br>**Transmissions:** A/T | **CAN Timeout ASC/DSC Conditions:**<br>The Engine Control Module (DME) communicates with all databus-capable control modules via a CAN databus. These databus-capable control modules are connected via two data bus wires which are twisted together (CAN_High and CAN_Low), and exchange information (messages). Missing information on the databus is recognized as a malfunction and stored. Trouble-free operation of the CAN-Bus requires that it have a terminal resistance. This central terminal resistor is located in the Engine Control Module (DME). The ignition is on for 800ms or the engine running for 15 seconds, the voltage is greater than 10 volts. This applies to vehicles with MIL only and does not affect bus activity.<br>**Possible Causes:**<br>• CAN data bus wires have short circuited to each other<br>• ECU has failed<br>• BUS system failure<br>• Defective bus controller (EGS) |

| DTC | Trouble Code Title, Conditions & Possible Causes |
|---|---|
| **DTC: P1727**<br>**2T MIL: Yes**<br>**Years:** 2007, 2008<br>**Models:** 335i, 335Ci, 335xi, 525i, 525xi, 530i, 530xi, 550i, 650i, 750i, 750Li, 760Li, M5, M6, X3, X5, Z4 3.0i, Z4 3.0Si, Z4 M<br>**Engines:** All<br>**Transmissions:** A/T | **Engine Speed Signal Plausibility Conditions:**<br>The CAN Message signal error flag alive counter or check sum is sending/receiving an incorrect signal. There is no alteration of the alive counter and no wrong check sum. The DME-CAN Connection is okay. The CAN-Bus is okay and the ignition is on.<br>**Possible Causes:**<br>• Pressure control solenoid circuit is shorting to ground<br>• Pressure control solenoid circuit is open<br>• Valve has failed<br>• TCM has failed |
| **DTC: P1728**<br>**2T MIL: Yes**<br>**Years:** 2007, 2008<br>**Models:** 335i, 335Ci, 335xi, 525i, 525xi, 530i, 530xi, 550i, 650i, 750i, 750Li, 760Li, M5, M6, X3, X5, Z4 3.0i, Z4 3.0Si, Z4 M<br>**Engines:** All<br>**Transmissions:** A/T | **Engine Overspeed Plausibility Conditions:**<br>The engine speed is over 10,000rpm. The range position is D, and the park-lock sensor is okay.<br>**Possible Causes:**<br>• Pressure control solenoid circuit is shorting to ground<br>• Pressure control solenoid circuit is open<br>• Valve has failed<br>• TCM has failed |
| **DTC: P1747**<br>**MIL: Yes (U.S. only)**<br>**Years:** 2007, 2008<br>**Models:** 335i, 335Ci, 335xi, 525i, 525xi, 530i, 530xi, 550i, 650i, 750i, 750Li, 760Li, M5, M6, X3, X5, Z4 3.0i, Z4 3.0Si, Z4 M<br>**Engines:** All<br>**Transmissions:** All | **CAN-Bus Plausibility-Disabled Conditions:**<br>The Engine Control Module (DME) communicates with all databus-capable control modules via a CAN databus. These databus-capable control modules are connected via two data bus wires which are twisted together (CAN_High and CAN_Low), and exchange information (messages). Missing information on the databus is recognized as a malfunction and stored. Trouble-free operation of the CAN-Bus requires that it have a terminal resistance. This central terminal resistor is located in the Engine Control Module (DME). The ignition is on for 800ms or the engine running for 15 seconds, the voltage is greater than 10 volts. This applies to vehicles with MIL only and does not affect bus activity.<br>**Possible Causes:**<br>• CAN data bus wires have short circuited to each other<br>• ECU has failed<br>• BUS system failure<br>• Defective bus controller (EGS) |
| **DTC: P1753**<br>**2T MIL: Yes**<br>**Years:** 2007, 2008<br>**Models:** 328i, 328Ci, 328xi, 750i, 750Li, 760Li,<br>**Engines:** All<br>**Transmissions:** A/T | **Pressure Regulator Valve 4 Upper Threshold Conditions:**<br>The signal to/from the pressure regulator valve has been interrupted.<br>**Possible Causes:**<br>• Pressure control solenoid circuit is shorting to ground<br>• Pressure control solenoid circuit is open<br>• Valve has failed<br>• TCM has failed |
| **DTC: P1762**<br>**2T MIL: Yes**<br>**Years:** 2007, 2008<br>**Models:** 335i, 335Ci, 335xi, 525i, 525xi, 530i, 530xi, 550i, 650i, 750i, 750Li, 760Li, M5, M6, X3, X5, Z4 3.0i, Z4 3.0Si, Z4 M<br>**Engines:** All<br>**Transmissions:** A/T | **Shift Solenoid C Short to Power Conditions:**<br>The shift solenoid valve is shorting to power after 50ms of operation.<br>**Possible Causes:**<br>• Pressure control solenoid circuit is shorting to ground<br>• Pressure control solenoid circuit is open<br>• Valve has failed<br>• TCM has failed |
| **DTC: P1763**<br>**2T MIL: Yes**<br>**Years:** 2007, 2008<br>**Models:** All<br>**Engines:** All<br>**Transmissions:** A/T | **Shift Solenoid C Short to Ground Conditions:**<br>The shift solenoid valve is shorting to ground after 50ms of operation.<br>**Possible Causes:**<br>• Pressure control solenoid circuit is shorting to ground<br>• Pressure control solenoid circuit is open<br>• Valve has failed<br>• TCM has failed |

| DTC | Trouble Code Title, Conditions & Possible Causes |
|---|---|
| **DTC: P1764**<br>**2T MIL: Yes**<br>**Years:** 2007, 2008<br>**Models:** 335i, 335Ci, 335xi, 525i, 525xi, 530i, 530xi, 550i, 650i, 750i, 750Li, 760Li, M5, M6, X3, X5, Z4 3.0i, Z4 3.0Si, Z4 M<br>**Engines:** All<br>**Transmissions:** A/T | **Shift Solenoid C Short Circuit Continuity-Disconnection Conditions:**<br>The shift solenoid valve has a disconnection continuity after 50ms of operation.<br>**Possible Causes:**<br>• Pressure control solenoid circuit is shorting to ground<br>• Pressure control solenoid circuit is open<br>• Valve has failed<br>• TCM has failed |
| **DTC: P1771**<br>**2T MIL: Yes**<br>**Years:** 2007, 2008<br>**Models:** 335i, 335Ci, 335xi, 525i, 525xi, 530i, 530xi, 550i, 650i, 750i, 750Li, 760Li, M5, M6, X3, X5, Z4 3.0i, Z4 3.0Si, Z4 M<br>**Engines:** All<br>**Transmissions:** A/T | **Engine Torque Plausibility Conditions:**<br>The CAN Message signal error flag alive counter or check sum is sending/receiving an incorrect signal. There is no alteration of the alive counter and no wrong check sum. The DME-CAN Connection is okay. The CAN-Bus is okay and the ignition is on.<br>**Possible Causes:**<br>• Throttle solenoid circuit is shorting to ground<br>• Throttle valve sensor has failed<br>• Valve has failed<br>• DME/TCM has failed |
| **DTC: P1810**<br>**2T MIL: Yes**<br>**Years:** 2007, 2008<br>**Models:** 335i, 335Ci, 335xi, 525i, 525xi, 530i, 530xi, Z4 3.0i, Z4 3.0Si, Z4 M<br>**Engines:** All<br>**Transmissions:** A/T | **Input/Turbine Speed Sensor Circuit Malfunction Upper Threshold Conditions:**<br>Engine started, vehicle driven with the vehicle speed sensor indicating more than 1 mph, and the DME detected the Transmission Vehicle Speed Sensor signals were erratic, or that they were missing for a period of time. The engine speed is greater than 600rpm. Any gear can be selected, output speed must be greater than 600rpm, and wheel speed greater than 400rpm.<br>**Possible Causes:**<br>• TVSS signal circuit is open<br>• TVSS signal is shorted to chassis ground<br>• TVSS signal is shorted to sensor ground<br>• TVSS assembly is damaged or it has failed |
| **DTC: P1811**<br>**2T MIL: Yes**<br>**Years:** 2007, 2008<br>**Models:** 335i, 335Ci, 335xi, 525i, 525xi, 530i, 530xi, Z4 3.0i, Z4 3.0Si, Z4 M<br>**Engines:** All<br>**Transmissions:** A/T | **Input/Turbine Speed Sensor Circuit Malfunction Lower Threshold Conditions:**<br>Engine started, vehicle driven with the vehicle speed sensor indicating more than 1 mph, and the DME detected the Transmission Vehicle Speed Sensor signals were erratic, or that they were missing for a period of time. The engine speed is greater than 600rpm. Any gear can be selected, output speed must be greater than 600rpm, and wheel speed greater than 400rpm.<br>**Possible Causes:**<br>• TVSS signal circuit is open<br>• TVSS signal is shorted to chassis ground<br>• TVSS signal is shorted to sensor ground<br>• TVSS assembly is damaged or it has failed |
| **DTC: P1812**<br>**2T MIL: Yes**<br>**Years:** 2007, 2008<br>**Models:** 335i, 335Ci, 335xi, 525i, 525xi, 530i, 530xi, Z4 3.0i, Z4 3.0Si, Z4 M<br>**Engines:** All<br>**Transmissions:** A/T | **Output Speed Sensor Circuit Malfunction Upper Threshold Conditions:**<br>Engine started, vehicle driven with the vehicle speed sensor indicating more than 1 mph, and the DME detected the Transmission Vehicle Speed Sensor signals were erratic, or that they were missing for a period of time. The engine speed is greater than 600rpm. Any gear can be selected, output speed must be greater than 600rpm, and wheel speed greater than 400rpm. The sensor supply status is okay.<br>**Possible Causes:**<br>• TVSS signal circuit is open<br>• TVSS signal is shorted to chassis ground<br>• TVSS signal is shorted to sensor ground<br>• TVSS assembly is damaged or it has failed |
| **DTC: P1813**<br>**2T MIL: Yes**<br>**Years:** 2007, 2008<br>**Models:** 335i, 335Ci, 335xi, 525i, 525xi, 530i, 530xi, Z4 3.0i, Z4 3.0Si, Z4 M<br>**Engines:** All<br>**Transmissions:** A/T | **Output Speed Sensor Circuit Malfunction Lower Threshold Conditions:**<br>Engine started, vehicle driven with the vehicle speed sensor indicating more than 1 mph, and the DME detected the Transmission Vehicle Speed Sensor signals were erratic, or that they were missing for a period of time. The engine speed is greater than 600rpm. Any gear can be selected, output speed must be greater than 600rpm, and wheel speed greater than 400rpm. The sensor supply status is okay.<br>**Possible Causes:**<br>• TVSS signal circuit is open<br>• TVSS signal is shorted to chassis ground<br>• TVSS signal is shorted to sensor ground<br>• TVSS assembly is damaged or it has failed |

| DTC | Trouble Code Title, Conditions & Possible Causes |
|---|---|
| **DTC: P1814**<br>**2T MIL: Yes**<br>**Years:** 2007, 2008<br>**Models:** 335i, 335Ci, 335xi, 525i, 525xi, 530i, 530xi, Z4 3.0i, Z4 3.0Si, Z4 M<br>**Engines:** All<br>**Transmissions:** A/T | **Output Speed Sensor Circuit Malfunction Plausibility Conditions:**<br>Engine started, vehicle driven with the vehicle speed sensor indicating more than 1 mph, and the DME detected the Transmission Vehicle Speed Sensor signals were erratic, or that they were missing for a period of time. There was a negative gradient of the signal which was greater than the threshold at 1000rpm or 10 seconds after start. The engine speed is greater than 600rpm. Any gear can be selected, output speed must be greater than 600rpm, and wheel speed greater than 400rpm. The sensor supply status is okay.<br>**Possible Causes:**<br>• TVSS signal circuit is open<br>• TVSS signal is shorted to chassis ground<br>• TVSS signal is shorted to sensor ground<br>• TVSS assembly is damaged or it has failed |
| **DTC: P1881**<br>**2T MIL: Yes**<br>**Years:** 2007, 2008<br>**Models:** 335i, 335Ci, 335xi, 525i, 525xi, 530i, 530xi, Z4 3.0i, Z4 3.0Si, Z4 M<br>**Engines:** All<br>**Transmissions:** A/T | **1-2 Shift Range Monitoring Upper Threshold Conditions:**<br>Engine running and vehicle driven, the DME detected a mechanical malfunction within the transmission. The output speed is greater than 300rpm, the transmission oil temperature is greater than 0 degrees Celsius, the engine speed is greater or equal to 600rpm and the range position is P, R, or N.<br>**Possible Causes:**<br>• Solenoid valves in valve body are faulty<br>• Solenoid circuit is shorting to ground<br>• Solenoid circuit is open<br>• TCM has failed or wiring is shorting<br>• Mechanical malfunction in transmission |
| **DTC: P1882**<br>**2T MIL: Yes**<br>**Years:** 2007, 2008<br>**Models:** 335i, 335Ci, 335xi, 525i, 525xi, 530i, 530xi, Z4 3.0i, Z4 3.0Si, Z4 M<br>**Engines:** All<br>**Transmissions:** A/T | **2-3 Shift Range Monitoring Upper Threshold Conditions:**<br>Engine running and vehicle driven, the DME detected a mechanical malfunction within the transmission. The output speed is greater than 300rpm, the transmission oil temperature is greater than 0 degrees Celsius, the engine speed is greater or equal to 600rpm and the range position is P, R, or N.<br>**Possible Causes:**<br>• Solenoid valves in valve body are faulty<br>• Solenoid circuit is shorting to ground<br>• Solenoid circuit is open<br>• TCM has failed or wiring is shorting<br>• Mechanical malfunction in transmission |
| **DTC: P1883**<br>**2T MIL: Yes**<br>**Years:** 2007, 2008<br>**Models:** 335i, 335Ci, 335xi, 525i, 525xi, 530i, 530xi, Z4 3.0i, Z4 3.0Si, Z4 M<br>**Engines:** All<br>**Transmissions:** A/T | **3-4 Shift Upper Threshold Conditions:**<br>Engine running and vehicle driven, the DME detected a mechanical malfunction within the transmission. The ratio of the input speed minus the output speed is greater than the threshold value. The output speed is greater than 300rpm, the transmission oil temperature is greater than 0 degrees Celsius, the engine speed is greater or equal to 600rpm and the range position is P, R, or N.<br>**Possible Causes:**<br>• Solenoid valves in valve body are faulty<br>• Solenoid circuit is shorting to ground<br>• Solenoid circuit is open<br>• TCM has failed or wiring is shorting<br>• Mechanical malfunction in transmission |
| **DTC: P1884**<br>**2T MIL: Yes**<br>**Years:** 2007, 2008<br>**Models:** 335i, 335Ci, 335xi, 525i, 525xi, 530i, 530xi, Z4 3.0i, Z4 3.0Si, Z4 M<br>**Engines:** All<br>**Transmissions:** A/T | **4-5 Shift Upper Threshold Conditions:**<br>Engine running and vehicle driven, the DME detected a mechanical malfunction within the transmission. The ratio of the input speed minus the output speed is greater than the threshold value. The output speed is greater than 300rpm, the transmission oil temperature is greater than 0 degrees Celsius, the engine speed is greater or equal to 600rpm and the range position is P, R, or N.<br>**Possible Causes:**<br>• Solenoid valves in valve body are faulty<br>• Solenoid circuit is shorting to ground<br>• Solenoid circuit is open<br>• TCM has failed or wiring is shorting<br>• Mechanical malfunction in transmission |
| **DTC: P1885**<br>**2T MIL: Yes**<br>**Years:** 2007, 2008<br>**Models:** 335i, 335Ci, 335xi, 525i, 525xi, 530i, 530xi, Z4 3.0i, Z4 3.0Si, Z4 M<br>**Engines:** All<br>**Transmissions:** A/T | **5-6 Shift Upper Threshold Conditions:**<br>Engine running and vehicle driven, the DME detected a mechanical malfunction within the transmission. The ratio of the input speed minus the output speed is greater than the threshold value. The output speed is greater than 300rpm, the transmission oil temperature is greater than 0 degrees Celsius, the engine speed is greater or equal to 600rpm and the range position is P, R, or N.<br>**Possible Causes:**<br>• Solenoid valves in valve body are faulty<br>• Solenoid circuit is shorting to ground<br>• Solenoid circuit is open<br>• TCM has failed or wiring is shorting<br>• Mechanical malfunction in transmission |

| DTC | Trouble Code Title, Conditions & Possible Causes |
|---|---|
| **DTC: P1889**<br>**2T MIL:** Yes<br>**Years:** 2007, 2008<br>**Models:** 335i, 335Ci, 335xi, 525i, 525xi, 530i, 530xi, Z4 3.0i, Z4 3.0Si, Z4 M<br>**Engines:** All<br>**Transmissions:** A/T | **System Power Supply (B+) Terminal 15 Malfunction Conditions:**<br>Ignition on, the DME detected a low voltage condition on the power supply terminal (15). To achieve optimal anti-theft protection for the vehicle, an anti-theft immobilizer is installed. The anti-theft immobilizer is a system for enabling and locking the Engine Control Module (DME). So that this system cannot be circumvented, it is necessary to perform adaptation of the anti-theft immobilizer using the Vehicle Diagnostic and Information System VAS 5052 in the On Board Diagnostic (OBD) function. The great availability of equipment options makes it necessary to adapt the Engine Control Module (DME) to the vehicle (e.g. throttle valve control module or cruise control system). This "writing" function is not possible with the generic scan tool.<br>**Possible Causes:**<br>   • (If DME was replaced) DME ID not the same as the replaced unit<br>   • Voltage supply for Engine Control Module (DME) has shorted |
| **DTC: P1890**<br>**2T MIL:** Yes<br>**Years:** 2007, 2008<br>**Models:** 335i, 335Ci, 335xi, 525i, 525xi, 530i, 530xi, Z4 3.0i, Z4 3.0Si, Z4 M<br>**Engines:** All<br>**Transmissions:** A/T | **TCC Power Supply Upper Threshold Conditions:**<br>Engine started to 400rpm, FET enabled, vehicle driven in gear with VSS signals received, and the DME detected a the battery voltage was less than the threshold value (less than 9 volts) while in normal operation.<br>**Possible Causes:**<br>   • TCC solenoid has a mechanical failure<br>   • TCC solenoid has a hydraulic failure |
| **DTC: P1891**<br>**2T MIL:** Yes<br>**Years:** 2007, 2008<br>**Models:** 335i, 335Ci, 335xi, 525i, 525xi, 530i, 530xi, Z4 3.0i, Z4 3.0Si, Z4 M<br>**Engines:** All<br>**Transmissions:** A/T | **TCC Power Supply Upper Threshold Conditions:**<br>Engine started to 400rpm, FET enabled, vehicle driven in gear with VSS signals received, and the DME detected a the battery voltage was greater than the threshold value (greater than 16 volts) while in normal operation.<br>**Possible Causes:**<br>   • TCC solenoid has a mechanical failure<br>   • TCC solenoid has a hydraulic failure |
| **DTC: P1892**<br>**2T MIL:** Yes<br>**Years:** 2007, 2008<br>**Models:** 335i, 335Ci, 335xi, 525i, 525xi, 530i, 530xi, Z4 3.0i, Z4 3.0Si, Z4 M<br>**Engines:** All<br>**Transmissions:** A/T | **TCC Power Supply Lower Threshold Conditions:**<br>Engine started to 400rpm, FET enabled, vehicle driven in gear with VSS signals received, and the DME detected a the battery voltage was less than the threshold value (less than 7 volts) while in normal operation.<br>**Possible Causes:**<br>   • TCC solenoid has a mechanical failure<br>   • TCC solenoid has a hydraulic failure |
| **DTC: P1893**<br>**2T MIL:** Yes<br>**Years:** 2007, 2008<br>**Models:** 335i, 335Ci, 335xi, 525i, 525xi, 530i, 530xi, Z4 3.0i, Z4 3.0Si, Z4 M<br>**Engines:** All<br>**Transmissions:** A/T | **TCC Power Supply Circuit Continuity Power Short Conditions:**<br>Engine started to 400rpm, FET enabled, vehicle driven in gear with VSS signals received, and the DME detected a the battery voltage shorting to power.<br>**Possible Causes:**<br>   • TCC solenoid has a mechanical failure<br>   • TCC solenoid has a hydraulic failure |
| **DTC: P1894**<br>**2T MIL:** Yes<br>**Years:** 2007, 2008<br>**Models:** 335i, 335Ci, 335xi, 525i, 525xi, 530i, 530xi, Z4 3.0i, Z4 3.0Si, Z4 M<br>**Engines:** All<br>**Transmissions:** A/T | **TCC Power Supply Circuit Continuity Ground Short Conditions:**<br>Engine started to 400rpm, FET enabled, vehicle driven in gear with VSS signals received, and the DME detected a the battery voltage shorting to ground.<br>**Possible Causes:**<br>   • TCC solenoid has a mechanical failure<br>   • TCC solenoid has a hydraulic failure |
| **DTC: P1895**<br>**2T MIL:** Yes<br>**Years:** 2007, 2008<br>**Models:** 335i, 335Ci, 335xi, 525i, 525xi, 530i, 530xi, Z4 3.0i, Z4 3.0Si, Z4 M<br>**Engines:** All<br>**Transmissions:** A/T | **TCC Power Supply Circuit Continuity Disconnection Conditions:**<br>Engine started to 400rpm, FET enabled, vehicle driven in gear with VSS signals received, and the DME detected a the battery voltage seems to be disconnected.<br>**Possible Causes:**<br>   • TCC solenoid has a mechanical failure<br>   • TCC solenoid has a hydraulic failure |

**Gas Engine OBD II Trouble Code List (P2xxx Codes)**

| DTC | Trouble Code Title, Conditions & Possible Causes |
|---|---|
| **DTC: P2088**<br>**2T MIL: Yes**<br>**Years:** 2007, 2008<br>**Models:** 335i, 335Ci, 335xi, 525i, 525xi, 530i, 530xi, 550i, 650i, 750i, 750Li, 760Li, Z4 3.0i, Z4 3.0Si, Z4 M<br>**Engines:** All<br>**Transmissions:** All | **Inlet "A" Camshaft Position Control Circuit Low Bank 1 Conditions:**<br>Key on or engine running; and the DME detected an unexpected voltage condition on the Camshaft Position Control circuit during the CCM test period. The relative position between the camshaft and crankshaft needs to be optimal so the engine has better torque, fuel economy and emissions.<br>**Note: camshaft adjustment is load- and RPM dependant. The electrical camshaft adjustment valve 1 switches oil pressure onto camshaft adjuster (mechanical adjustment mechanism), which adjusts the camshaft.**<br>**Possible Causes:**<br>• Camshaft position control wiring harness connector is damaged or open<br>• Camshaft adjustment valve has failed<br>• Circuit is open or grounded<br>• Assembly is damaged or it has failed (an open circuit)<br>• DME power supply relay has failed |
| **DTC: P2089**<br>**2T MIL: Yes**<br>**Years:** 2007, 2008<br>**Models:** 335i, 335Ci, 335xi, 525i, 525xi, 530i, 530xi, 550i, 650i, 750i, 750Li, 760Li, Z4 3.0i, Z4 3.0Si, Z4 M<br>**Engines:** All<br>**Transmissions:** All | **Inlet "A" Camshaft Position Control Circuit High Bank 1 Conditions:**<br>Key on or engine running; and the DME detected an unexpected voltage condition on the Camshaft Position Control circuit during the CCM test period. The relative position between the camshaft and crankshaft needs to be optimal so the engine has better torque, fuel economy and emissions.<br>**Note: camshaft adjustment is load- and RPM dependant. The electrical camshaft adjustment valve 1 switches oil pressure onto camshaft adjuster (mechanical adjustment mechanism), which adjusts the camshaft.**<br>**Possible Causes:**<br>• Camshaft position control wiring harness connector is damaged or open<br>• Camshaft adjustment valve has failed<br>• Circuit is open or grounded<br>• Assembly is damaged or it has failed (an open circuit)<br>• DME power supply relay has failed |
| **DTC: P2090**<br>**2T MIL: Yes**<br>**Years:** 2007, 2008<br>**Models:** 335i, 335Ci, 335xi, 525i, 525xi, 530i, 530xi, 550i, 650i, 750i, 750Li, 760Li, Z4 3.0i, Z4 3.0Si, Z4 M<br>**Engines:** All<br>**Transmissions:** All | **Outlet "B" Camshaft Position Control Circuit Low Bank 1 Conditions:**<br>Key on or engine running; and the DME detected an unexpected voltage condition on the Camshaft Position Control circuit during the CCM test period. The relative position between the camshaft and crankshaft needs to be optimal so the engine has better torque, fuel economy and emissions.<br>**Note: camshaft adjustment is load- and RPM dependant. The electrical camshaft adjustment valve 1 switches oil pressure onto camshaft adjuster (mechanical adjustment mechanism), which adjusts the camshaft.**<br>**Possible Causes:**<br>• Camshaft position control wiring harness connector is damaged or open<br>• Camshaft adjustment valve has failed<br>• Circuit is open or grounded<br>• Assembly is damaged or it has failed (an open circuit)<br>• DME power supply relay has failed |
| **DTC: P2091**<br>**2T MIL: Yes**<br>**Years:** 2007, 2008<br>**Models:** 335i, 335Ci, 335xi, 525i, 525xi, 530i, 530xi, 550i, 650i, 750i, 750Li, 760Li, Z4 3.0i, Z4 3.0Si, Z4 M<br>**Engines:** All<br>**Transmissions:** All | **Outlet "B" Camshaft Position Control Circuit High Bank 1 Conditions:**<br>Key on or engine running; and the DME detected an unexpected voltage condition on the Camshaft Position Control circuit during the CCM test period. The relative position between the camshaft and crankshaft needs to be optimal so the engine has better torque, fuel economy and emissions.<br>**Note: camshaft adjustment is load- and RPM dependant. The electrical camshaft adjustment valve 1 switches oil pressure onto camshaft adjuster (mechanical adjustment mechanism), which adjusts the camshaft.**<br>**Possible Causes:**<br>• Camshaft position control wiring harness connector is damaged or open<br>• Camshaft adjustment valve has failed<br>• Circuit is open or grounded<br>• Assembly is damaged or it has failed (an open circuit)<br>• DME power supply relay has failed |
| **DTC: P2092**<br>**2T MIL: Yes**<br>**Years:** 2007, 2008<br>**Models:** 335i, 335Ci, 335xi, 525i, 525xi, 530i, 530xi, 550i, 650i, 750i, 750Li, 760Li, Z4 3.0i, Z4 3.0Si, Z4 M<br>**Engines:** All<br>**Transmissions:** All | **Inlet "A" Camshaft Position Control Circuit Low Bank 1 Conditions:**<br>Key on or engine running; and the DME detected an unexpected voltage condition on the Camshaft Position Control circuit during the CCM test period. The relative position between the camshaft and crankshaft needs to be optimal so the engine has better torque, fuel economy and emissions.<br>**Note: camshaft adjustment is load- and RPM dependant. The electrical camshaft adjustment valve 1 switches oil pressure onto camshaft adjuster (mechanical adjustment mechanism), which adjusts the camshaft.**<br>**Possible Causes:**<br>• Camshaft position control wiring harness connector is damaged or open<br>• Camshaft adjustment valve has failed<br>• Circuit is open or grounded<br>• Assembly is damaged or it has failed (an open circuit)<br>• DME power supply relay has failed |

| DTC | Trouble Code Title, Conditions & Possible Causes |
|---|---|
| **DTC: P2093**<br>**2T MIL: Yes**<br>**Years:** 2007, 2008<br>**Models:** 335i, 335Ci, 335xi, 525i, 525xi, 530i, 530xi, 550i, 650i, 750i, 750Li, 760Li<br>**Engines:** All<br>**Transmissions:** All | **Inlet "A" Camshaft Position Control Circuit Low Bank 2 Conditions:**<br>Key on or engine running; and the DME detected an unexpected voltage condition on the Camshaft Position Control circuit during the CCM test period. The relative position between the camshaft and crankshaft needs to be optimal so the engine has better torque, fuel economy and emissions.<br>**Note: camshaft adjustment is load- and RPM dependant. The electrical camshaft adjustment valve 1 switches oil pressure onto camshaft adjuster (mechanical adjustment mechanism), which adjusts the camshaft.**<br>**Possible Causes:**<br>&bull; Camshaft position control wiring harness connector is damaged or open<br>&bull; Camshaft adjustment valve has failed<br>&bull; Circuit is open or grounded<br>&bull; Assembly is damaged or it has failed (an open circuit)<br>&bull; DME power supply relay has failed |
| **DTC: P2094**<br>**2T MIL: Yes**<br>**Years:** 2007, 2008<br>**Models:** 335i, 335Ci, 335xi, 525i, 525xi, 530i, 530xi, 550i, 650i, 750i, 750Li, 760Li<br>**Engines:** All<br>**Transmissions:** All | **Outlet "B" Camshaft Position Control Circuit Low Bank 2 Conditions:**<br>Key on or engine running; and the DME detected an unexpected voltage condition on the Camshaft Position Control circuit during the CCM test period. The relative position between the camshaft and crankshaft needs to be optimal so the engine has better torque, fuel economy and emissions.<br>**Note: camshaft adjustment is load- and RPM dependant. The electrical camshaft adjustment valve 1 switches oil pressure onto camshaft adjuster (mechanical adjustment mechanism), which adjusts the camshaft.**<br>**Possible Causes:**<br>&bull; Camshaft position control wiring harness connector is damaged or open<br>&bull; Camshaft adjustment valve has failed<br>&bull; Circuit is open or grounded<br>&bull; Assembly is damaged or it has failed (an open circuit)<br>&bull; DME power supply relay has failed |
| **DTC: P2095**<br>**2T MIL: Yes**<br>**Years:** 2007, 2008<br>**Models:** 335i, 335Ci, 335xi, 525i, 525xi, 530i, 530xi, 550i, 650i, 750i, 750Li, 760Li<br>**Engines:** All<br>**Transmissions:** All | **Outlet "B" Camshaft Position Control Circuit High Bank 2 Conditions:**<br>Key on or engine running; and the DME detected an unexpected voltage condition on the Camshaft Position Control circuit during the CCM test period. The relative position between the camshaft and crankshaft needs to be optimal so the engine has better torque, fuel economy and emissions.<br>**Note: camshaft adjustment is load- and RPM dependant. The electrical camshaft adjustment valve 1 switches oil pressure onto camshaft adjuster (mechanical adjustment mechanism), which adjusts the camshaft.**<br>**Possible Causes:**<br>&bull; Camshaft position control wiring harness connector is damaged or open<br>&bull; Camshaft adjustment valve has failed<br>&bull; Circuit is open or grounded<br>&bull; Assembly is damaged or it has failed (an open circuit)<br>&bull; DME power supply relay has failed |
| **DTC: P2096**<br>**2T MIL: Yes**<br>**Years:** 2007, 2008<br>**Models:** 335i, 335Ci, 335xi, 525i, 525xi, 530i, 530xi, 550i, 650i, 750i, 750Li, 760Li, Z4 3.0i, Z4 3.0Si, Z4 M<br>**Engines:** All<br>**Transmissions:** All | **Post Catalyst Fuel Trim System Too Lean (Bank 1) Conditions:**<br>Engine started, battery voltage must be at least 11.5v, all electrical components must be off, the ground between the engine and the chassis must be well connected, the exhaust system must be properly sealed between the catalytic converter and the cylinder head, and the oxygen sensor heater for oxygen sensor before the catalytic converter must be properly functioning. The DME detected a problem with the fuel mixture. Trim control 1 segment (precision controller with oxygen sensor behind cat.) below delta lambda threshold of less than −0.03. Fault monitoring criterion must remain present for over one second. The engine speed must be between 1060 and 3000 rpm and the catalytic converter temperature must be 280 degrees Celsius.<br>**Note: For resistance testing of sensor heating, oxygen sensor should be cooled to ambient temperature. High temperatures at oxygen sensor may lead to inaccurate measurements.**<br>**Possible Causes:**<br>&bull; Oxygen sensor (before catalytic converter) is faulty<br>&bull; Oxygen sensor (behind catalytic converter) is faulty<br>&bull; Oxygen sensor heater (before catalytic converter) is faulty<br>&bull; Oxygen sensor heater (behind catalytic converter) is faulty<br>&bull; Check circuits for shorts to each other, ground or power |

| DTC | Trouble Code Title, Conditions & Possible Causes |
|---|---|
| **DTC: P2097**<br>**2T MIL:** Yes<br>**Years:** 2007, 2008<br>**Models:** 335i, 335Ci, 335xi, 525i, 525xi, 530i, 530xi, 550i, 650i, 750i, 750Li, 760Li, Z4 3.0i, Z4 3.0Si, Z4 M<br>**Engines:** All<br>**Transmissions:** All | **Post Catalyst Fuel Trim System Too Rich (Bank 1) Conditions:**<br>Engine started, battery voltage must be at least 11.5v, all electrical components must be off, the ground between the engine and the chassis must be well connected, the exhaust system must be properly sealed between the catalytic converter and the cylinder head, and the oxygen sensor heater for oxygen sensor before the catalytic converter must be properly functioning. The DME detected a problem with the fuel mixture. Trim control 1 segment (precision controller with oxygen sensor behind cat.) below delta lambda threshold of less than −0.03. Fault monitoring criterion must remain present for over one second. The engine speed must be between 1060 and 3000 rpm and the catalytic converter temperature must be 280 degrees Celsius.<br>**Note: For resistance testing of sensor heating, oxygen sensor should be cooled to ambient temperature. High temperatures at oxygen sensor may lead to inaccurate measurements.**<br>**Possible Causes:**<br>• Oxygen sensor (before catalytic converter) is faulty<br>• Oxygen sensor (behind catalytic converter) is faulty<br>• Oxygen sensor heater (before catalytic converter) is faulty<br>• Oxygen sensor heater (behind catalytic converter) is faulty<br>• Check circuits for shorts to each other, ground or power |
| **DTC: P2098**<br>**2T MIL:** Yes<br>**Years:** 2007, 2008<br>**Models:** 335i, 335Ci, 335xi, 525i, 525xi, 530i, 530xi, 550i, 650i, 750i, 750Li, 760Li, Z4 3.0i, Z4 3.0Si, Z4 M<br>**Engines:** All<br>**Transmissions:** All | **Post Catalyst Fuel Trim System Too Lean (Bank 2) Conditions:**<br>Engine started, battery voltage must be at least 11.5v, all electrical components must be off, the ground between the engine and the chassis must be well connected, the exhaust system must be properly sealed between the catalytic converter and the cylinder head, and the oxygen sensor heater for oxygen sensor before the catalytic converter must be properly functioning. The DME detected a problem with the fuel mixture. Trim control 1 segment (precision controller with oxygen sensor behind cat.) below delta lambda threshold of less than −0.03. Fault monitoring criterion must remain present for over one second. The engine speed must be between 1060 and 3000 rpm and the catalytic converter temperature must be 280 degrees Celsius.<br>**Note: For resistance testing of sensor heating, oxygen sensor should be cooled to ambient temperature. High temperatures at oxygen sensor may lead to inaccurate measurements.**<br>**Possible Causes:**<br>• Oxygen sensor (before catalytic converter) is faulty<br>• Oxygen sensor (behind catalytic converter) is faulty<br>• Oxygen sensor heater (before catalytic converter) is faulty<br>• Oxygen sensor heater (behind catalytic converter) is faulty<br>• Check circuits for shorts to each other, ground or power |
| **DTC: P2099**<br>**2T MIL:** Yes<br>**Years:** 2007, 2008<br>**Models:** 335i, 335Ci, 335xi, 525i, 525xi, 530i, 530xi, 550i, 650i, 750i, 750Li, 760Li, Z4 3.0i, Z4 3.0Si, Z4 M<br>**Engines:** All<br>**Transmissions:** All | **Post Catalyst Fuel Trim System Too Rich (Bank 2) Conditions:**<br>Engine started, battery voltage must be at least 11.5v, all electrical components must be off, the ground between the engine and the chassis must be well connected, the exhaust system must be properly sealed between the catalytic converter and the cylinder head, and the oxygen sensor heater for oxygen sensor before the catalytic converter must be properly functioning. The DME detected a problem with the fuel mixture. Trim control 1 segment (precision controller with oxygen sensor behind cat.) below delta lambda threshold of less than −0.03. Fault monitoring criterion must remain present for over one second. The engine speed must be between 1060 and 3000 rpm and the catalytic converter temperature must be 400 degrees Celsius.<br>**Note: For resistance testing of sensor heating, oxygen sensor should be cooled to ambient temperature. High temperatures at oxygen sensor may lead to inaccurate measurements.**<br>**Possible Causes:**<br>• Oxygen sensor (before catalytic converter) is faulty<br>• Oxygen sensor (behind catalytic converter) is faulty<br>• Oxygen sensor heater (before catalytic converter) is faulty<br>• Oxygen sensor heater (behind catalytic converter) is faulty<br>• Check circuits for shorts to each other, ground or power |
| **DTC: P2100**<br>**2T MIL:** Yes<br>**Years:** 2007, 2008<br>**Models:** 335i, 335Ci, 335xi, 525i, 525xi, 530i, 530xi, 550i, 650i, 750i, 750Li, 760Li, Z4 3.0i, Z4 3.0Si, Z4 M<br>**Engines:** All<br>**Transmissions:** All | **Throttle Actuator "A" Control Motor Circuit Open Conditions:**<br>Engine started, battery voltage must be at least 7v, coolant temperature must be at least 80 degrees Celsius. The DME detected an unexpected low or high voltage condition on the Throttle Actuator Control Motor (TACM) circuit during the CCM test.<br>**Note: The throttle valve activation occurs via an electric motor (throttle drive) in the throttle valve control module. It is activated by the Engine Control Module (DME) according to specifications of the two sensors, Throttle Position (TP) Sensor and Sender 2 for accelerator pedal position.**<br>**Possible Causes:**<br>• TACM wiring harness connector is damaged or open<br>• TACM wiring may be crossed in the wire harness assembly<br>• TACM (motor) circuit is open, or TACM assembly is damaged (possible open circuit)<br>• TACM or the Throttle Valve is dirty<br>• Throttle Position sensor has failed<br>• Heater defective |

| DTC | Trouble Code Title, Conditions & Possible Causes |
|-----|--------------------------------------------------|
| **DTC: P2102**<br>**2T MIL:** Yes<br>**Years:** 2007, 2008<br>**Models:** 335i, 335Ci, 335xi, 525i, 525xi, 530i, 530xi, 550i, 650i, 750i, 750Li, 760Li, Z4 3.0i, Z4 3.0Si, Z4 M<br>**Engines:** All<br>**Transmissions:** All | **Throttle Actuator "A" Control Motor Circuit Low Conditions:**<br>Engine started, battery voltage must be at least 7v, coolant temperature must be at least 80 degrees Celsius. The DME detected an unexpected low or high voltage condition on the Throttle Actuator Control Motor (TACM) circuit during the CCM test.<br>**Note: The throttle valve activation occurs via an electric motor (throttle drive) in the throttle valve control module. It is activated by the Engine Control Module (DME) according to specifications of the two sensors, Throttle Position (TP) Sensor and Sender 2 for accelerator pedal position.**<br>**Possible Causes:**<br>• TACM wiring harness connector is damaged or open<br>• TACM wiring may be crossed in the wire harness assembly<br>• TACM (motor) circuit is open, or TACM assembly is damaged (possible open circuit)<br>• TACM or the Throttle Valve is dirty<br>• Throttle Position sensor has failed<br>• Heater defective |
| **DTC: P2103**<br>**2T MIL:** Yes<br>**Years:** 2007, 2008<br>**Models:** 335i, 335Ci, 335xi, 525i, 525xi, 530i, 530xi, 550i, 650i, 750i, 750Li, 760Li, Z4 3.0i, Z4 3.0Si, Z4 M<br>**Engines:** All<br>**Transmissions:** All | **Throttle Actuator "A" Control Motor Circuit High Conditions:**<br>Engine started, battery voltage must be at least 7v, coolant temperature must be at least 80 degrees Celsius. The DME detected an unexpected low or high voltage condition on the Throttle Actuator Control Motor (TACM) circuit during the CCM test.<br>**Note: The throttle valve activation occurs via an electric motor (throttle drive) in the throttle valve control module. It is activated by the Engine Control Module (DME) according to specifications of the two sensors, Throttle Position (TP) Sensor and Sender 2 for accelerator pedal position.**<br>**Possible Causes:**<br>• TACM wiring harness connector is damaged or open<br>• TACM wiring may be crossed in the wire harness assembly<br>• TACM (motor) circuit is open, or TACM assembly is damaged (possible open circuit)<br>• TACM or the Throttle Valve is dirty<br>• Throttle Position sensor has failed<br>• Heater defective |
| **DTC: P2122**<br>**2T MIL:** Yes<br>**Years:** 2007, 2008<br>**Models:** 335i, 335Ci, 335xi, 525i, 525xi, 530i, 530xi, 550i, 650i, 750i, 750Li, 760Li, Z4 3.0i, Z4 3.0Si, Z4 M<br>**Engines:** All<br>**Transmissions:** All | **Accelerator Pedal Position Sensor 'D' Circuit Low Input Conditions:**<br>Engine started, battery voltage at least 11.5v, all electrical components off, ground connections between engine and chassis well connected, the DME detected that the accelerator pedal position sensor signal was outside the parameters to function normally.<br>**Note: Both the Throttle Position (TP) Sensor and Accelerator Pedal Position Sensor are located at the accelerator pedal module and communicate the driver's intentions to the DME completely independently of each other. Both sensors are stored in one housing.**<br>**Possible Causes:**<br>• Ground between engine and chassis may be broken<br>• Throttle position sensor may have failed<br>• Accelerator Pedal Position Sensor has failed<br>• Throttle position sensor wiring may have shorted<br>• Throttle position sensor has failed<br>• Faulty voltage supply |
| **DTC: P2123**<br>**2T MIL:** Yes<br>**Years:** 2007, 2008<br>**Models:** 335i, 335Ci, 335xi, 525i, 525xi, 530i, 530xi, 550i, 650i, 750i, 750Li, 760Li, Z4 3.0i, Z4 3.0Si, Z4 M<br>**Engines:** All<br>**Transmissions:** All | **Accelerator Pedal Position Sensor 'D' Circuit High Input Conditions:**<br>Engine started, battery voltage at least 11.5v, all electrical components off, ground connections between engine and chassis well connected, the DME detected that the accelerator pedal position sensor signal was outside the parameters to function normally.<br>**Note: Both the Throttle Position (TP) Sensor and Accelerator Pedal Position Sensor are located at the accelerator pedal module and communicate the driver's intentions to the DME completely independently of each other. Both sensors are stored in one housing.**<br>**Possible Causes:**<br>• Ground between engine and chassis may be broken<br>• Throttle position sensor may have failed<br>• Accelerator Pedal Position Sensor has failed<br>• Throttle position sensor wiring may have shorted<br>• Throttle position sensor has failed<br>• Faulty voltage supply |

| DTC | Trouble Code Title, Conditions & Possible Causes |
|---|---|
| **DTC: P2127**<br>**2T MIL: Yes**<br>**Years:** 2007, 2008<br>**Models:** 335i, 335Ci, 335xi, 525i, 525xi, 530i, 530xi, 550i, 650i, 750i, 750Li, 760Li, Z4 3.0i, Z4 3.0Si, Z4 M<br>**Engines:** All<br>**Transmissions:** All | **Accelerator Pedal Position Sensor 'E' Circuit Low Input Conditions:**<br>Engine started, battery voltage at least 11.5v, all electrical components off, ground connections between engine and chassis well connected, the DME detected that the accelerator pedal position sensor signal was outside the parameters to function normally.<br>**Note: Both the Throttle Position (TP) Sensor and Accelerator Pedal Position Sensor are located at the accelerator pedal module and communicate the driver's intentions to the DME completely independently of each other. Both sensors are stored in one housing.**<br>**Possible Causes:**<br>• Ground between engine and chassis may be broken<br>• Throttle position sensor may have failed<br>• Accelerator Pedal Position Sensor has failed<br>• Throttle position sensor wiring may have shorted<br>• Throttle position sensor has failed<br>• Faulty voltage supply |
| **DTC: P2128**<br>**2T MIL: Yes**<br>**Years:** 2007, 2008<br>**Models:** 335i, 335Ci, 335xi, 525i, 525xi, 530i, 530xi, 550i, 650i, 750i, 750Li, 760Li, Z4 3.0i, Z4 3.0Si, Z4 M<br>**Engines:** All<br>**Transmissions:** All | **Accelerator Pedal Position Sensor 'E' Circuit High Input Conditions:**<br>Engine started, battery voltage at least 11.5v, all electrical components off, ground connections between engine and chassis well connected, the DME detected that the accelerator pedal position sensor signal was outside the parameters to function normally.<br>**Note: Both the Throttle Position (TP) Sensor and Accelerator Pedal Position Sensor are located at the accelerator pedal module and communicate the driver's intentions to the DME completely independently of each other. Both sensors are stored in one housing.**<br>**Possible Causes:**<br>• Ground between engine and chassis may be broken<br>• Throttle position sensor may have failed<br>• Accelerator Pedal Position Sensor has failed<br>• Throttle position sensor wiring may have shorted<br>• Throttle position sensor has failed<br>• Faulty voltage supply |
| **DTC: P2138**<br>**2T MIL: Yes**<br>**Years:** 2007, 2008<br>**Models:** 335i, 335Ci, 335xi, 525i, 525xi, 530i, 530xi, 550i, 650i, 750i, 750Li, 760Li, Z4 3.0i, Z4 3.0Si, Z4 M<br>**Engines:** All<br>**Transmissions:** All | **Throttle Position Sensor D/E Voltage Correlation Conditions:**<br>Engine started, battery voltage must be at least 11.5v, all electrical components must be off, parking brake must be engaged (to keep daytime driving lights off), automatic transmission selector must be in park; and the DME detected the Throttle Position 'D' (TPD) and Throttle Position 'B' (TPE) sensors disagreed, or that the TPD sensor should not be in its detected position, or that the TPE sensor should not be in its detected position during testing.<br>**Note: Both the Throttle Position (TP) Sensor and Accelerator Pedal Position Sensor are located at the accelerator pedal module and communicate the driver's intentions to the DME completely independently of each other. Both sensors are stored in one housing.**<br>**Possible Causes:**<br>• ETC TP sensor connector is damaged or shorted<br>• ETC TP sensor circuits shorted together in the wire harness<br>• ETC TP sensor signal circuit is shorted to VREF (5v)<br>• ETC TP sensor is damaged or the DME has failed |
| **DTC: P2177**<br>**2T MIL: Yes**<br>**Years:** 2007, 2008<br>**Models:** 335i, 335Ci, 335xi, 525i, 525xi, 530i, 530xi, 550i, 650i, 750i, 750Li, 760Li, Z4 3.0i, Z4 3.0Si, Z4 M<br>**Engines:** All<br>**Transmissions:** All | **System Too Lean Off Idle Bank 1 Conditions:**<br>Engine started, battery voltage must be at least 11.5v, all electrical components must be off, the ground between the engine and the chassis must be well connected, the exhaust system must be properly sealed between the catalytic converter and the cylinder head, and the oxygen sensor heater for oxygen sensor before the catalytic converter must be properly functioning. The DME detected the system indicated a lean signal, or it could no longer control bank 1 because it was at its lean limit.<br>**Possible Causes:**<br>• Intake Manifold Runner Position Sensor has failed<br>• Intake system has leaks (false air)<br>• Motor for intake flap is faulty<br>• Oxygen sensor (before catalytic converter) is faulty<br>• Oxygen sensor (behind catalytic converter) is faulty<br>• Oxygen sensor heater (before catalytic converter) is faulty<br>• Oxygen sensor heater (behind catalytic converter) is faulty<br>• Check circuits for shorts to each other, ground or power<br>• Fuel Injector(s) may have failed |

| DTC | Trouble Code Title, Conditions & Possible Causes |
|---|---|
| **DTC: P2178**<br>**2T MIL:** Yes<br>**Years:** 2007, 2008<br>**Models:** 335i, 335Ci, 335xi, 525i, 525xi, 530i, 530xi, 550i, 650i, 750i, 750Li, 760Li, Z4 3.0i, Z4 3.0Si, Z4 M<br>**Engines:** All<br>**Transmissions:** All | **System Too Rich Off Idle Bank 1 Conditions:**<br>Engine started, battery voltage must be at least 11.5v, all electrical components must be off, the ground between the engine and the chassis must be well connected, the exhaust system must be properly sealed between the catalytic converter and the cylinder head, and the oxygen sensor heater for oxygen sensor before the catalytic converter must be properly functioning. The DME detected the system indicated a rich signal, or it could no longer control bank 1 because it was at its rich limit.<br>**Possible Causes:**<br>• Intake Manifold Runner Position Sensor has failed<br>• Intake system has leaks (false air)<br>• Motor for intake flap is faulty<br>• Oxygen sensor (before catalytic converter) is faulty<br>• Oxygen sensor (behind catalytic converter) is faulty<br>• Oxygen sensor heater (before catalytic converter) is faulty<br>• Oxygen sensor heater (behind catalytic converter) is faulty<br>• Check circuits for shorts to each other, ground or power<br>• Fuel Injector(s) may have failed |
| **DTC: P2179**<br>**2T MIL:** Yes<br>**Years:** 2007, 2008<br>**Models:** 335i, 335Ci, 335xi, 525i, 525xi, 530i, 530xi, 550i, 650i, 750i, 750Li, 760Li, Z4 3.0i, Z4 3.0Si, Z4 M<br>**Engines:** All<br>**Transmissions:** All | **System Too Lean Off Idle Bank 2 Conditions:**<br>Engine started, battery voltage must be at least 11.5v, all electrical components must be off, the ground between the engine and the chassis must be well connected, the exhaust system must be properly sealed between the catalytic converter and the cylinder head, and the oxygen sensor heater for oxygen sensor before the catalytic converter must be properly functioning. The DME detected the system indicated a lean signal, or it could no longer control bank 2 because it was at its lean limit.<br>**Possible Causes:**<br>• Intake Manifold Runner Position Sensor has failed<br>• Intake system has leaks (false air)<br>• Motor for intake flap is faulty<br>• Oxygen sensor (before catalytic converter) is faulty<br>• Oxygen sensor (behind catalytic converter) is faulty<br>• Oxygen sensor heater (before catalytic converter) is faulty<br>• Oxygen sensor heater (behind catalytic converter) is faulty<br>• Check circuits for shorts to each other, ground or power<br>• Fuel Injector(s) may have failed |
| **DTC: P2180**<br>**2T MIL:** Yes<br>**Years:** 2007, 2008<br>**Models:** 335i, 335Ci, 335xi, 525i, 525xi, 530i, 530xi, 550i, 650i, 750i, 750Li, 760Li, Z4 3.0i, Z4 3.0Si, Z4 M<br>**Engines:** All<br>**Transmissions:** All | **System Too Rich Off Idle Bank 2 Conditions:**<br>Engine started, battery voltage must be at least 11.5v, all electrical components must be off, the ground between the engine and the chassis must be well connected, the exhaust system must be properly sealed between the catalytic converter and the cylinder head, and the oxygen sensor heater for oxygen sensor before the catalytic converter must be properly functioning. The DME detected the system indicated a rich signal, or it could no longer control bank 2 because it was at its rich limit.<br>**Possible Causes:**<br>• Intake Manifold Runner Position Sensor has failed<br>• Intake system has leaks (false air)<br>• Motor for intake flap is faulty<br>• Oxygen sensor (before catalytic converter) is faulty<br>• Oxygen sensor (behind catalytic converter) is faulty<br>• Oxygen sensor heater (before catalytic converter) is faulty<br>• Oxygen sensor heater (behind catalytic converter) is faulty<br>• Check circuits for shorts to each other, ground or power<br>• Fuel Injector(s) may have failed |
| **DTC: P2183**<br>**2T MIL:** Yes<br>**Years:** 2007, 2008<br>**Models:** 335i, 335Ci, 335xi, 525i, 525xi, 530i, 530xi, 550i, 650i, 750i, 750Li, 760Li, Z4 3.0i, Z4 3.0Si, Z4 M<br>**Engines:** All<br>**Transmissions:** All | **ECT Sensor Signal Range/Performance Rationality Conditions:**<br>Engine started (cold), battery voltage must be 11.5, and all equipment must be off. The DME detected the ECT sensor exceeded the required calibrated value, or the engine is at idle and doesn't reach operating temperature quickly enough; the Catalyst, Fuel System, HO2S and Misfire Monitor did not complete, or the timer expired. Testing completion of procedure, the engine's temperature must rise uniformly during idle. The ECT is greater than 101.3 degrees Celsius for more than 60 seconds. The engine speed is greater than 1100rpm. The ambient temperature is greater than negative 7 degrees Celsius. The vehicle speed less than 62.5mph.<br>**Possible Causes:**<br>• Check for low coolant level or incorrect coolant mixture<br>• DME detects a short circuit wiring in the ECT<br>• CHT sensor is out-of-calibration or it has failed<br>• ECT sensor is out-of-calibration or it has failed |

| DTC | Trouble Code Title, Conditions & Possible Causes |
|---|---|
| **DTC: P2184**<br>**2T MIL: Yes**<br>**Years:** 2007, 2008<br>**Models:** 335i, 335Ci, 335xi, 525i, 525xi, 530i, 530xi, 550i, 650i, 750i, 750Li, 760Li, Z4 3.0i, Z4 3.0Si, Z4 M<br>**Engines:** All<br>**Transmissions:** All | **ECT Sensor 2 Circuit Range Check (Minimum) Conditions:**<br>Engine started (cold) for 10 seconds, battery voltage must be 11.5, and all equipment must be off. The DME detected the ECT sensor signal was less than the self-test minimum. This is a thermistor-type sensor with a variable resistance that changes when exposed to different temperatures<br>**Possible Causes:**<br>• ECT sensor signal circuit is grounded in the wiring harness<br>• ECT sensor doesn't react to changes in temperature<br>• ECT sensor is damaged or the DME has failed |
| **DTC: P2185**<br>**2T MIL: Yes**<br>**Years:** 2007, 2008<br>**Models:** 335i, 335Ci, 335xi, 525i, 525xi, 530i, 530xi, 550i, 650i, 750i, 750Li, 760Li, Z4 3.0i, Z4 3.0Si, Z4 M<br>**Engines:** All<br>**Transmissions:** All | **ECT Sensor 2 Circuit Range Check (Maximum) Conditions:**<br>Engine started (cold) for 10 seconds, battery voltage must be 11.5, and all equipment must be off. The DME detected the ECT sensor signal was less than the self-test minimum. This is a thermistor-type sensor with a variable resistance that changes when exposed to different temperatures<br>**Possible Causes:**<br>• ECT sensor signal circuit is grounded in the wiring harness<br>• ECT sensor doesn't react to changes in temperature<br>• ECT sensor is damaged or the DME has failed |
| **DTC: P2186**<br>**2T MIL: Yes**<br>**Years:** 2007, 2008<br>**Models:** 335i, 335Ci, 335xi, 525i, 525xi, 530i, 530xi, 550i, 650i, 750i, 750Li, 760Li, Z4 3.0i, Z4 3.0Si, Z4 M<br>**Engines:** All<br>**Transmissions:** All | **ECT Sensor 2 Circuit High Input Conditions:**<br>Engine started (cold) for 10 seconds, battery voltage must be 11.5, and all equipment must be off. The DME detected the ECT sensor signal was more than the self-test maximum. This is a thermistor-type sensor with a variable resistance that changes when exposed to different temperatures<br>**Possible Causes:**<br>• ECT sensor signal circuit is open (inspect wiring & connector)<br>• ECT sensor signal circuit is shorted to ground<br>• ECT sensor is damaged or it has failed |
| **DTC: P2187**<br>**2T MIL: Yes**<br>**Years:** 2007, 2008<br>**Models:** 335i, 335Ci, 335xi, 525i, 525xi, 530i, 530xi, 550i, 650i, 750i, 750Li, 760Li, Z4 3.0i, Z4 3.0Si, Z4 M<br>**Engines:** All<br>**Transmissions:** All | **System Too Lean at Idle Bank 1 Conditions:**<br>Engine started, battery voltage must be at least 11v, all electrical components must be off, the ground between the engine and the chassis must be well connected, the exhaust system must be properly sealed between the catalytic converter and the cylinder head, and the oxygen sensor heater for oxygen sensor before the catalytic converter must be properly functioning. The engine temperature must be greater than 63 degrees Celsius for approximately 10 to 20 minutes. The air intake temperature must be less than or equal to 80 degrees Celsius, and the engine speed must be less than or equal to 800rpm.<br>**Possible Causes:**<br>• Evaporative Emission (EVAP) canister purge regulator valve is faulty<br>• Exhaust system components are damaged<br>• Fuel injectors are faulty<br>• Fuel pressure regulator and residual pressure have failed<br>• Fuel Pump (FP) in fuel tank is faulty<br>• Intake system has leaks (false air)<br>• Secondary Air Injection (AIR) system has an improper seal<br>• Intake Manifold Runner Position Sensor has failed<br>• Motor for intake flap is faulty<br>• Oxygen sensor (before catalytic converter) is faulty<br>• Oxygen sensor (behind catalytic converter) is faulty<br>• Oxygen sensor heater (before catalytic converter) is faulty<br>• Oxygen sensor heater (behind catalytic converter) is faulty<br>• Check circuits for shorts to each other, ground or power |

| DTC | Trouble Code Title, Conditions & Possible Causes |
|---|---|
| **DTC: P2188**<br>**2T MIL: Yes**<br>**Years:** 2007, 2008<br>**Models:** 335i, 335Ci, 335xi, 525i, 525xi, 530i, 530xi, 550i, 650i, 750i, 750Li, 760Li, Z4 3.0i, Z4 3.0Si, Z4 M<br>**Engines:** All<br>**Transmissions:** All | **System Too Rich at Idle Bank 1 Conditions:**<br>Engine started, battery voltage must be at least 11v, all electrical components must be off, the ground between the engine and the chassis must be well connected, the exhaust system must be properly sealed between the catalytic converter and the cylinder head, and the oxygen sensor heater for oxygen sensor before the catalytic converter must be properly functioning. he engine temperature must be greater than 63 degrees Celsius for approximately 10 to 20 minutes. The air intake temperature must be less than or equal to 80 degrees Celsius, and the engine speed must be less than or equal to 800rpm.<br>**Possible Causes:**<br>• Evaporative Emission (EVAP) canister purge regulator valve is faulty<br>• Exhaust system components are damaged<br>• Fuel injectors are faulty<br>• Fuel pressure regulator and residual pressure have failed<br>• Fuel Pump (FP) in fuel tank is faulty<br>• Intake system has leaks (false air)<br>• Secondary Air Injection (AIR) system has an improper seal<br>• Intake Manifold Runner Position Sensor has failed<br>• Motor for intake flap is faulty<br>• Oxygen sensor (before catalytic converter) is faulty<br>• Oxygen sensor (behind catalytic converter) is faulty<br>• Oxygen sensor heater (before catalytic converter) is faulty<br>• Oxygen sensor heater (behind catalytic converter) is faulty<br>• Check circuits for shorts to each other, ground or power |
| **DTC: P2189**<br>**2T MIL: Yes**<br>**Years:** 2007, 2008<br>**Models:** 335i, 335Ci, 335xi, 525i, 525xi, 530i, 530xi, 550i, 650i, 750i, 750Li, 760Li, Z4 3.0i, Z4 3.0Si, Z4 M<br>**Engines:** All<br>**Transmissions:** All | **System Too Lean at Idle Bank 2 Conditions:**<br>Engine started, battery voltage must be at least 11v, all electrical components must be off, the ground between the engine and the chassis must be well connected, the exhaust system must be properly sealed between the catalytic converter and the cylinder head, and the oxygen sensor heater for oxygen sensor before the catalytic converter must be properly functioning. he engine temperature must be greater than 63 degrees Celsius for approximately 10 to 20 minutes. The air intake temperature must be less than or equal to 80 degrees Celsius, and the engine speed must be less than or equal to 800rpm.<br>**Possible Causes:**<br>• Evaporative Emission (EVAP) canister purge regulator valve is faulty<br>• Exhaust system components are damaged<br>• Fuel injectors are faulty<br>• Fuel pressure regulator and residual pressure have failed<br>• Fuel Pump (FP) in fuel tank is faulty<br>• Intake system has leaks (false air)<br>• Secondary Air Injection (AIR) system has an improper seal<br>• Intake Manifold Runner Position Sensor has failed<br>• Motor for intake flap is faulty<br>• Oxygen sensor (before catalytic converter) is faulty<br>• Oxygen sensor (behind catalytic converter) is faulty<br>• Oxygen sensor heater (before catalytic converter) is faulty<br>• Oxygen sensor heater (behind catalytic converter) is faulty<br>• Check circuits for shorts to each other, ground or power |
| **DTC: P2190**<br>**2T MIL: Yes**<br>**Years:** 2007, 2008<br>**Models:** 335i, 335Ci, 335xi, 525i, 525xi, 530i, 530xi, 550i, 650i, 750i, 750Li, 760Li, Z4 3.0i, Z4 3.0Si, Z4 M<br>**Engines:** All<br>**Transmissions:** All | **System Too Rich at Idle Bank 2 Conditions:**<br>Engine started, battery voltage must be at least 11v, all electrical components must be off, the ground between the engine and the chassis must be well connected, the exhaust system must be properly sealed between the catalytic converter and the cylinder head, and the oxygen sensor heater for oxygen sensor before the catalytic converter must be properly functioning. he engine temperature must be greater than 63 degrees Celsius for approximately 10 to 20 minutes. The air intake temperature must be less than or equal to 80 degrees Celsius, and the engine speed must be less than or equal to 800rpm.<br>**Possible Causes:**<br>• Evaporative Emission (EVAP) canister purge regulator valve is faulty<br>• Exhaust system components are damaged<br>• Fuel injectors are faulty<br>• Fuel pressure regulator and residual pressure have failed<br>• Fuel Pump (FP) in fuel tank is faulty<br>• Intake system has leaks (false air)<br>• Secondary Air Injection (AIR) system has an improper seal<br>• Intake Manifold Runner Position Sensor has failed<br>• Motor for intake flap is faulty<br>• Oxygen sensor (before catalytic converter) is faulty<br>• Oxygen sensor (behind catalytic converter) is faulty<br>• Oxygen sensor heater (before catalytic converter) is faulty<br>• Oxygen sensor heater (behind catalytic converter) is faulty<br>• Check circuits for shorts to each other, ground or power |

| DTC | Trouble Code Title, Conditions & Possible Causes |
|---|---|
| **DTC: P2191**<br>**2T MIL: Yes**<br>**Years:** 2007, 2008<br>**Models:** 335i, 335Ci, 335xi, 525i, 525xi, 530i, 530xi, 550i, 650i, 750i, 750Li, 760Li, Z4 3.0i, Z4 3.0Si, Z4 M<br>**Engines:** All<br>**Transmissions:** All | **System Too Lean at Higher Load Bank 1 Conditions:**<br>Engine started, battery voltage must be at least 11v, all electrical components must be off, the ground between the engine and the chassis must be well connected, the exhaust system must be properly sealed between the catalytic converter and the cylinder head, and the oxygen sensor heater for oxygen sensor before the catalytic converter must be properly functioning. The engine temperature must be greater than 63 degrees Celsius for approximately 10 to 20 minutes. The air intake temperature must be less than or equal to 80 degrees Celsius.<br>**Possible Causes:**<br>• Evaporative Emission (EVAP) canister purge regulator valve is faulty<br>• Exhaust system components are damaged<br>• Fuel injectors are faulty<br>• Fuel pressure regulator and residual pressure have failed<br>• Fuel Pump (FP) in fuel tank is faulty<br>• Intake system has leaks (false air)<br>• Secondary Air Injection (AIR) system has an improper seal<br>• Intake Manifold Runner Position Sensor has failed<br>• Motor for intake flap is faulty<br>• Oxygen sensor (before catalytic converter) is faulty<br>• Oxygen sensor (behind catalytic converter) is faulty<br>• Oxygen sensor heater (before catalytic converter) is faulty<br>• Oxygen sensor heater (behind catalytic converter) is faulty<br>• Check circuits for shorts to each other, ground or power |
| **DTC: P2192**<br>**2T MIL: Yes**<br>**Years:** 2007, 2008<br>**Models:** 335i, 335Ci, 335xi, 525i, 525xi, 530i, 530xi, 550i, 650i, 750i, 750Li, 760Li, Z4 3.0i, Z4 3.0Si, Z4 M<br>**Engines:** All<br>**Transmissions:** All | **System Too Rich at Higher Load Bank 1 Conditions:**<br>Engine started, battery voltage must be at least 11v, all electrical components must be off, the ground between the engine and the chassis must be well connected, the exhaust system must be properly sealed between the catalytic converter and the cylinder head, and the oxygen sensor heater for oxygen sensor before the catalytic converter must be properly functioning. The engine temperature must be greater than 63 degrees Celsius for approximately 10 to 20 minutes. The air intake temperature must be less than or equal to 80 degrees Celsius.<br>**Possible Causes:**<br>• Evaporative Emission (EVAP) canister purge regulator valve is faulty<br>• Exhaust system components are damaged<br>• Fuel injectors are faulty<br>• Fuel pressure regulator and residual pressure have failed<br>• Fuel Pump (FP) in fuel tank is faulty<br>• Intake system has leaks (false air)<br>• Secondary Air Injection (AIR) system has an improper seal<br>• Intake Manifold Runner Position Sensor has failed<br>• Motor for intake flap is faulty<br>• Oxygen sensor (before catalytic converter) is faulty<br>• Oxygen sensor (behind catalytic converter) is faulty<br>• Oxygen sensor heater (before catalytic converter) is faulty<br>• Oxygen sensor heater (behind catalytic converter) is faulty<br>• Check circuits for shorts to each other, ground or power |

| DTC | Trouble Code Title, Conditions & Possible Causes |
|---|---|
| **DTC: P2193**<br>**2T MIL: Yes**<br>**Years:** 2007, 2008<br>**Models:** 335i, 335Ci, 335xi, 525i, 525xi, 530i, 530xi, 550i, 650i, 750i, 750Li, 760Li, Z4 3.0i, Z4 3.0Si, Z4 M<br>**Engines:** All<br>**Transmissions:** All | **System Too Lean at Higher Load Bank 2 Conditions:**<br>Engine started, battery voltage must be at least 11.5v, all electrical components must be off, the ground between the engine and the chassis must be well connected, the exhaust system must be properly sealed between the catalytic converter and the cylinder head, and the oxygen sensor heater for oxygen sensor before the catalytic converter must be properly functioning. DME detected the system indicated a lean signal, or it could no longer control bank 2 because it was at its lean limit. The engine temperature must be greater than 63 degrees Celsius for approximately 10 to 20 minutes. The air intake temperature must be less than or equal to 80 degrees Celsius.<br>**Possible Causes:**<br>• Evaporative Emission (EVAP) canister purge regulator valve is faulty<br>• Exhaust system components are damaged<br>• Fuel injectors are faulty<br>• Fuel pressure regulator and residual pressure have failed<br>• Fuel Pump (FP) in fuel tank is faulty<br>• Intake system has leaks (false air)<br>• Secondary Air Injection (AIR) system has an improper seal<br>• Intake Manifold Runner Position Sensor has failed<br>• Motor for intake flap is faulty<br>• Oxygen sensor (before catalytic converter) is faulty<br>• Oxygen sensor (behind catalytic converter) is faulty<br>• Oxygen sensor heater (before catalytic converter) is faulty<br>• Oxygen sensor heater (behind catalytic converter) is faulty<br>• Check circuits for shorts to each other, ground or power |
| **DTC: P2194**<br>**2T MIL: Yes**<br>**Years:** 2007, 2008<br>**Models:** 335i, 335Ci, 335xi, 525i, 525xi, 530i, 530xi, 550i, 650i, 750i, 750Li, 760Li, Z4 3.0i, Z4 3.0Si, Z4 M<br>**Engines:** All<br>**Transmissions:** All | **System Too Rich at Higher Load Bank 2 Conditions:**<br>Engine started, battery voltage must be at least 11v, all electrical components must be off, the ground between the engine and the chassis must be well connected, the exhaust system must be properly sealed between the catalytic converter and the cylinder head, and the oxygen sensor heater for oxygen sensor before the catalytic converter must be properly functioning. The engine temperature must be greater than 63 degrees Celsius for approximately 10 to 20 minutes. The air intake temperature must be less than or equal to 80 degrees Celsius.<br>**Possible Causes:**<br>• Evaporative Emission (EVAP) canister purge regulator valve is faulty<br>• Exhaust system components are damaged<br>• Fuel injectors are faulty<br>• Fuel pressure regulator and residual pressure have failed<br>• Fuel Pump (FP) in fuel tank is faulty<br>• Intake system has leaks (false air)<br>• Secondary Air Injection (AIR) system has an improper seal<br>• Intake Manifold Runner Position Sensor has failed<br>• Motor for intake flap is faulty<br>• Oxygen sensor (before catalytic converter) is faulty<br>• Oxygen sensor (behind catalytic converter) is faulty<br>• Oxygen sensor heater (before catalytic converter) is faulty<br>• Oxygen sensor heater (behind catalytic converter) is faulty<br>• Check circuits for shorts to each other, ground or power |
| **DTC: P2195**<br>**2T MIL: Yes**<br>**Years:** 2007, 2008<br>**Models:** 335i, 335Ci, 335xi, 525i, 525xi, 530i, 530xi, 550i, 650i, 750i, 750Li, 760Li, Z4 3.0i, Z4 3.0Si, Z4 M<br>**Engines:** All<br>**Transmissions:** All | **O2 Sensor Signal Stuck Lean Bank 1 Sensor 1 Conditions:**<br>Engine running in closed loop, and the DME detected the O2S indicated a lean signal, or it could no longer control Fuel Trim because it was at lean limit.<br>**Possible Causes:**<br>• Engine oil level high<br>• Camshaft timing error<br>• Cylinder compression low<br>• Exhaust leaks in front of O2S<br>• EGR valve is stuck open<br>• EGR gasket is leaking<br>• EVR diaphragm is leaking<br>• Damaged fuel pressure regulator or extremely low fuel pressure<br>• O2S circuit is open or shorted in the wiring harness<br>• Oxygen sensor (before catalytic converter) is faulty<br>• Oxygen sensor (behind catalytic converter) is faulty<br>• Oxygen sensor heater (before catalytic converter) is faulty<br>• Oxygen sensor heater (behind catalytic converter) is faulty<br>• Air leaks after the MAF sensor<br>• PCV system leaks<br>• Dip stick not seated properly |

| DTC | Trouble Code Title, Conditions & Possible Causes |
|---|---|
| **DTC: P2196**<br>**2T MIL: Yes**<br>**Years:** 2007, 2008<br>**Models:** 335i, 335Ci, 335xi, 525i, 525xi, 530i, 530xi, 550i, 650i, 750i, 750Li, 760Li, Z4 3.0i, Z4 3.0Si, Z4 M<br>**Engines:** All<br>**Transmissions:** All | **O2 Sensor Signal Stuck Rich Bank 1 Sensor 1 Conditions:**<br>Engine running in closed loop, and the DME detected the O2S indicated a rich signal, or it could no longer control Fuel Trim because it was at its rich limit. The sensor temperature is heated up. The relative engine load change is less than or equal to 3 percent per camshaft revolution.<br>**Possible Causes:**<br>• Engine oil level high<br>• Camshaft timing error<br>• Cylinder compression low<br>• Exhaust leaks in front of O2S<br>• EGR valve is stuck open<br>• EGR gasket is leaking<br>• EVR diaphragm is leaking<br>• Damaged fuel pressure regulator or extremely low fuel pressure<br>• O2S circuit is open or shorted in the wiring harness<br>• Oxygen sensor (before catalytic converter) is faulty<br>• Oxygen sensor (behind catalytic converter) is faulty<br>• Oxygen sensor heater (before catalytic converter) is faulty<br>• Oxygen sensor heater (behind catalytic converter) is faulty<br>• Air leaks after the MAF sensor<br>• PCV system leaks<br>• Dip stick not seated properly |
| **DTC: P2197**<br>**2T MIL: Yes**<br>**Years:** 2007, 2008<br>**Models:** 335i, 335Ci, 335xi, 525i, 525xi, 530i, 530xi, 550i, 650i, 750i, 750Li, 760Li, Z4 3.0i, Z4 3.0Si, Z4 M<br>**Engines:** All<br>**Transmissions:** All | **O2 Sensor Signal Stuck Lean Bank 2 Sensor 1 Conditions:**<br>Engine running in closed loop, and the DME detected the O2S indicated a lean signal, or it could no longer control Fuel Trim because it was at lean limit. The sensor temperature is heated up. The relative engine load change is less than or equal to 3 percent per camshaft revolution.<br>**Possible Causes:**<br>• Engine oil level high<br>• Camshaft timing error<br>• Cylinder compression low<br>• Exhaust leaks in front of O2S<br>• EGR valve is stuck open<br>• EGR gasket is leaking<br>• EVR diaphragm is leaking<br>• Damaged fuel pressure regulator or extremely low fuel pressure<br>• O2S circuit is open or shorted in the wiring harness<br>• Oxygen sensor (before catalytic converter) is faulty<br>• Oxygen sensor (behind catalytic converter) is faulty<br>• Oxygen sensor heater (before catalytic converter) is faulty<br>• Oxygen sensor heater (behind catalytic converter) is faulty<br>• Air leaks after the MAF sensor<br>• PCV system leaks<br>• Dip stick not seated properly |
| **DTC: P2198**<br>**2T MIL: Yes**<br>**Years:** 2007, 2008<br>**Models:** 335i, 335Ci, 335xi, 525i, 525xi, 530i, 530xi, 550i, 650i, 750i, 750Li, 760Li, Z4 3.0i, Z4 3.0Si, Z4 M<br>**Engines:** All<br>**Transmissions:** All | **O2 Sensor Signal Stuck Rich Bank 2 Sensor 1 Conditions:**<br>Engine running in closed loop, and the DME detected the O2S indicated a rich signal, or it could no longer control Fuel Trim because it was at its rich limit. The sensor temperature is heated up. The relative engine load change is less than or equal to 3 percent per camshaft revolution.<br>**Possible Causes:**<br>• Engine oil level high<br>• Camshaft timing error<br>• Cylinder compression low<br>• Exhaust leaks in front of O2S<br>• EGR valve is stuck open<br>• EGR gasket is leaking<br>• EVR diaphragm is leaking<br>• Damaged fuel pressure regulator or extremely low fuel pressure<br>• O2S circuit is open or shorted in the wiring harness<br>• Oxygen sensor (before catalytic converter) is faulty<br>• Oxygen sensor (behind catalytic converter) is faulty<br>• Oxygen sensor heater (before catalytic converter) is faulty<br>• Oxygen sensor heater (behind catalytic converter) is faulty<br>• Air leaks after the MAF sensor<br>• PCV system leaks<br>• Dip stick not seated properly |

| DTC | Trouble Code Title, Conditions & Possible Causes |
|---|---|
| **DTC: P2228**<br>**2T MIL:** Yes<br>**Years:** 2007, 2008<br>**Models:** 335i, 335Ci, 335xi, 525i, 525xi, 530i, 530xi, 550i, 650i, 750i, 750Li, 760Li, Z4 3.0i, Z4 3.0Si, Z4 M<br>**Engines:** All<br>**Transmissions:** All | **Barometric Circuit Low Conditions:**<br>Engine started, the temperature must beat least 185-degrees (F) and all electrical equipment (A/C, lights, etc) must be off. The DME detected the BARO sensor was out of range during the CCM test. The BARO sensor signal should be in 4.5v. The BARO sensor is a variable capacitance unit used to detect altitude. There is a short to ground and the internal voltage measurement in ambient pressure sensor is greater than 4.7998 and the ambient pressure is greater than 1150 hPa.<br>**Possible Causes:**<br>• Sensor has deteriorated (response time too slow) or has failed<br>• MAP sensor signal circuit is shorted to ground<br>• MAP sensor circuit (5v) is open<br>• MAP sensor is damaged or it has failed<br>• BARO sensor signal circuit is shorted to ground<br>• BARO sensor circuit (5v) is open<br>• BARO sensor is damaged or it has failed<br>• Replace the DME |
| **DTC: P2229**<br>**2T MIL:** Yes<br>**Years:** 2007, 2008<br>**Models:** 335i, 335Ci, 335xi, 525i, 525xi, 530i, 530xi, 550i, 650i, 750i, 750Li, 760Li, Z4 3.0i, Z4 3.0Si, Z4 M<br>**Engines:** All<br>**Transmissions:** All | **Barometric Circuit High Conditions:**<br>Engine started, the temperature must beat least 185-degrees (F) and all electrical equipment (A/C, lights, etc) must be off. The DME detected the BARO sensor was out of range during the CCM test. The BARO sensor signal should be in 4.5v. The BARO sensor is a variable capacitance unit used to detect altitude. There is a short to battery voltage and the internal voltage measurement in ambient pressure sensor is greater than 4.7998 and the ambient pressure is greater than 1150 hPa.<br>**Possible Causes:**<br>• Sensor has deteriorated (response time too slow) or has failed<br>• MAP sensor signal circuit is shorted to ground<br>• MAP sensor circuit (5v) is open<br>• MAP sensor is damaged or it has failed<br>• BARO sensor signal circuit is shorted to ground<br>• BARO sensor circuit (5v) is open<br>• BARO sensor is damaged or it has failed<br>• Replace the DME |
| **DTC: P2231**<br>**2T MIL:** Yes<br>**Years:** 2007, 2008<br>**Models:** 335i, 335Ci, 335xi, 525i, 525xi, 530i, 530xi, 550i, 650i, 750i, 750Li, 760Li, Z4 3.0i, Z4 3.0Si, Z4 M<br>**Engines:** All<br>**Transmissions:** All | **O2 Sensor Signal Circuit Shorted to Heater Circuit Bank 1 Sensor 1 Conditions:**<br>Engine started, battery voltage must be at least 11.5v, all electrical components must be off, parking brake must be engaged (to keep daytime driving lights off), automatic transmission selector must be in park. The DME detected an unexpected voltage condition, or it detected an unexpected current draw in the sensor circuit during the CCM test.<br>**Note: Vehicle must be raised before connector for oxygen sensors is accessible.**<br>**Possible Causes:**<br>• Oxygen sensor (before catalytic converter) is faulty<br>• Oxygen sensor heater (before catalytic converter) is faulty<br>• Oxygen sensor heater (before catalytic converter) is faulty<br>• Oxygen sensor heater (behind catalytic converter) is faulty<br>• O2S circuit is open or shorted in the wiring harness |
| **DTC: P2234**<br>**2T MIL:** Yes<br>**Years:** 2007, 2008<br>**Models:** 335i, 335Ci, 335xi, 525i, 525xi, 530i, 530xi, 550i, 650i, 750i, 750Li, 760Li, Z4 3.0i, Z4 3.0Si, Z4 M<br>**Engines:** All<br>**Transmissions:** All | **O2 Sensor Signal Circuit Shorted to Heater Circuit Bank 2 Sensor 1 Conditions:**<br>Engine started, battery voltage must be at least 11.5v, all electrical components must be off, parking brake must be engaged (to keep daytime driving lights off), automatic transmission selector must be in park. The DME detected an unexpected voltage condition, or it detected an unexpected current draw in the sensor circuit during the CCM test.<br>**Note: Vehicle must be raised before connector for oxygen sensors is accessible.**<br>**Possible Causes:**<br>• Oxygen sensor (before catalytic converter) is faulty<br>• Oxygen sensor heater (before catalytic converter) is faulty<br>• Oxygen sensor heater (before catalytic converter) is faulty<br>• Oxygen sensor heater (behind catalytic converter) is faulty<br>• O2S circuit is open or shorted in the wiring harness |
| **DTC: P2237**<br>**2T MIL:** Yes<br>**Years:** 2007, 2008<br>**Models:** 335i, 335Ci, 335xi, 525i, 525xi, 530i, 530xi, 550i, 650i, 750i, 750Li, 760Li, Z4 3.0i, Z4 3.0Si, Z4 M<br>**Engines:** All<br>**Transmissions:** All | **O2S Sensor Positive Current Control Circuit/Open Circuit (Bank 1 Sensor 1) Conditions:**<br>Engine started, the fault criterion must remain present for over 2 seconds, the Lambda specification is outside three percent relative to Lambda = 1, and the closed loop lambda control is active.<br>**Possible Causes:**<br>• Check plugs<br>• Check wiring harness<br>• Replace sensor |

| DTC | Trouble Code Title, Conditions & Possible Causes |
|---|---|
| **DTC: P2240**<br>**2T MIL: Yes**<br>**Years:** 2007, 2008<br>**Models:** 335i, 335Ci, 335xi, 525i, 525xi, 530i, 530xi, 550i, 650i, 750i, 750Li, 760Li, Z4 3.0i, Z4 3.0Si, Z4 M<br>**Engines:** All<br>**Transmissions:** All | **O2S Sensor Positive Current Control Circuit/Open Circuit (Bank 2 Sensor 1) Conditions:**<br>Engine started, the fault criterion must remain present for over 2 seconds, the Lambda specification is outside three percent relative to Lambda = 1, and the closed loop lambda control is active.<br>**Possible Causes:**<br>• Check plugs<br>• Check wiring harness<br>• Replace sensor |
| **DTC: P2243**<br>**2T MIL: Yes**<br>**Years:** 2007, 2008<br>**Models:** 335i, 335Ci, 335xi, 525i, 525xi, 530i, 530xi, 550i, 650i, 750i, 750Li, 760Li, Z4 3.0i, Z4 3.0Si, Z4 M<br>**Engines:** All<br>**Transmissions:** All | **O2 Sensor Reference Voltage Circuit/Open Bank 1 Sensor 1 Conditions:**<br>Engine started, battery voltage must be at least 11.5v, all electrical components must be off, parking brake must be engaged (to keep daytime driving lights off), automatic transmission selector must be in park. The DME detected an unexpected voltage condition, or it detected an unexpected current draw in the sensor circuit during the CCM test. The voltage is out of range. The sensor temperature is heated up and the battery voltage is between 11 and 16 volts.<br>**Note: Vehicle must be raised before connector for oxygen sensors is accessible.**<br>**Possible Causes:**<br>• Oxygen sensor (before catalytic converter) is faulty<br>• Oxygen sensor heater (before catalytic converter) is faulty<br>• Oxygen sensor heater (before catalytic converter) is faulty<br>• Oxygen sensor heater (behind catalytic converter) is faulty<br>• O2S circuit is open or shorted in the wiring harness |
| **DTC: P2247**<br>**2T MIL: Yes**<br>**Years:** 2007, 2008<br>**Models:** 335i, 335Ci, 335xi, 525i, 525xi, 530i, 530xi, 550i, 650i, 750i, 750Li, 760Li, Z4 3.0i, Z4 3.0Si, Z4 M<br>**Engines:** All<br>**Transmissions:** All | **O2 Sensor Reference Voltage Circuit/Open Bank 2 Sensor 1 Conditions:**<br>Engine started, battery voltage must be at least 11.5v, all electrical components must be off, parking brake must be engaged (to keep daytime driving lights off), automatic transmission selector must be in park. The DME detected an unexpected voltage condition, or it detected an unexpected current draw in the sensor circuit during the CCM test. The voltage is out of range. The sensor temperature is heated up and the battery voltage is between 11 and 16 volts.<br>**Note: Vehicle must be raised before connector for oxygen sensors is accessible.**<br>**Possible Causes:**<br>• Oxygen sensor (before catalytic converter) is faulty<br>• Oxygen sensor heater (before catalytic converter) is faulty<br>• Oxygen sensor heater (before catalytic converter) is faulty<br>• Oxygen sensor heater (behind catalytic converter) is faulty<br>• O2S circuit is open or shorted in the wiring harness |
| **DTC: P2251**<br>**2T MIL: Yes**<br>**Years:** 2007, 2008<br>**Models:** 335i, 335Ci, 335xi, 525i, 525xi, 530i, 530xi, 550i, 650i, 750i, 750Li, 760Li, Z4 3.0i, Z4 3.0Si, Z4 M<br>**Engines:** All<br>**Transmissions:** All | **O2 Sensor Negative Voltage Circuit/Open Bank 1 Sensor 1 Conditions:**<br>Engine started, battery voltage must be at least 11.5v, all electrical components must be off, parking brake must be engaged (to keep daytime driving lights off), automatic transmission selector must be in park. The DME detected an unexpected voltage condition, or it detected an unexpected current draw in the sensor circuit during the CCM test. Fault monitoring criterion must remain present for over five seconds. The voltage is within critical range. The sensor temperature is heated up and the battery voltage is between 11 and 16 volts.<br>**Note: Vehicle must be raised before connector for oxygen sensors is accessible.**<br>**Possible Causes:**<br>• Oxygen sensor (before catalytic converter) is faulty<br>• Oxygen sensor heater (before catalytic converter) is faulty<br>• Oxygen sensor heater (before catalytic converter) is faulty<br>• Oxygen sensor heater (behind catalytic converter) is faulty<br>• O2S circuit is open or shorted in the wiring harness |
| **DTC: P2254**<br>**2T MIL: Yes**<br>**Years:** 2007, 2008<br>**Models:** 335i, 335Ci, 335xi, 525i, 525xi, 530i, 530xi, 550i, 650i, 750i, 750Li, 760Li, Z4 3.0i, Z4 3.0Si, Z4 M<br>**Engines:** All<br>**Transmissions:** All | **O2 Sensor Negative Voltage Circuit/Open Bank 2 Sensor 1 Conditions:**<br>Engine started, battery voltage must be at least 11.5v, all electrical components must be off, parking brake must be engaged (to keep daytime driving lights off), automatic transmission selector must be in park. The DME detected an unexpected voltage condition, or it detected an unexpected current draw in the sensor circuit during the CCM test. Fault monitoring criterion must remain present for over five seconds. The voltage is within critical range. The sensor temperature is heated up and the battery voltage is between 11 and 16 volts.<br>**Note: Vehicle must be raised before connector for oxygen sensors is accessible.**<br>**Possible Causes:**<br>• Oxygen sensor (before catalytic converter) is faulty<br>• Oxygen sensor heater (before catalytic converter) is faulty<br>• Oxygen sensor heater (before catalytic converter) is faulty<br>• Oxygen sensor heater (behind catalytic converter) is faulty<br>• O2S circuit is open or shorted in the wiring harness |

| DTC | Trouble Code Title, Conditions & Possible Causes |
|---|---|
| **DTC: P2270**<br>**2T MIL: Yes**<br>**Years:** 2007, 2008<br>**Models:** 335i, 335Ci, 335xi, 525i, 525xi, 530i, 530xi, 550i, 650i, 750i, 750Li, 760Li, Z4 3.0i, Z4 3.0Si, Z4 M<br>**Engines:** All<br>**Transmissions:** All | **O2 Sensor Signal Stuck Lean Bank 1 Sensor 2 Conditions:**<br>Engine started, battery voltage must be at least 11.5v, all electrical components must be off, parking brake must be engaged (to keep daytime driving lights off), automatic transmission selector must be in park. The DME detected an unexpected voltage condition, or it detected an unexpected current draw in the heater circuit during the CCM test.<br>**Note: Vehicle must be raised before connector for oxygen sensors is accessible.**<br>**Possible Causes:**<br>• Oxygen sensor (before catalytic converter) is faulty<br>• Oxygen sensor heater (before catalytic converter) is faulty<br>• Oxygen sensor heater (before catalytic converter) is faulty<br>• Oxygen sensor heater (behind catalytic converter) is faulty<br>• O2S circuit is open or shorted in the wiring harness |
| **DTC: P2271**<br>**2T MIL: Yes**<br>**Years:** 2007, 2008<br>**Models:** 335i, 335Ci, 335xi, 525i, 525xi, 530i, 530xi, 550i, 650i, 750i, 750Li, 760Li, Z4 3.0i, Z4 3.0Si, Z4 M<br>**Engines:** All<br>**Transmissions:** All | **O2 Sensor Signal Stuck Rich Bank 1 Sensor 2 Conditions:**<br>Engine started, battery voltage must be at least 11.5v, all electrical components must be off, parking brake must be engaged (to keep daytime driving lights off), automatic transmission selector must be in park. The DME detected an unexpected voltage condition, or it detected an unexpected current draw in the heater circuit during the CCM test.<br>**Note: Vehicle must be raised before connector for oxygen sensors is accessible.**<br>**Possible Causes:**<br>• Oxygen sensor (before catalytic converter) is faulty<br>• Oxygen sensor heater (before catalytic converter) is faulty<br>• Oxygen sensor heater (before catalytic converter) is faulty<br>• Oxygen sensor heater (behind catalytic converter) is faulty<br>• O2S circuit is open or shorted in the wiring harness |
| **DTC: P2272**<br>**2T MIL: Yes**<br>**Years:** 2007, 2008<br>**Models:** 335i, 335Ci, 335xi, 525i, 525xi, 530i, 530xi, 550i, 650i, 750i, 750Li, 760Li, Z4 3.0i, Z4 3.0Si, Z4 M<br>**Engines:** All<br>**Transmissions:** All | **O2 Sensor Signal Stuck Lean Bank 2 Sensor 2 Conditions:**<br>Engine started, battery voltage must be at least 11.5v, all electrical components must be off, parking brake must be engaged (to keep daytime driving lights off), automatic transmission selector must be in park. The DME detected an unexpected voltage condition, or it detected an unexpected current draw in the heater circuit during the CCM test.<br>**Note: Vehicle must be raised before connector for oxygen sensors is accessible.**<br>**Possible Causes:**<br>• Oxygen sensor (before catalytic converter) is faulty<br>• Oxygen sensor heater (before catalytic converter) is faulty<br>• Oxygen sensor heater (before catalytic converter) is faulty<br>• Oxygen sensor heater (behind catalytic converter) is faulty<br>• O2S circuit is open or shorted in the wiring harness |
| **DTC: P2273**<br>**2T MIL: Yes**<br>**Years:** 2007, 2008<br>**Models:** 335i, 335Ci, 335xi, 525i, 525xi, 530i, 530xi, 550i, 650i, 750i, 750Li, 760Li, Z4 3.0i, Z4 3.0Si, Z4 M<br>**Engines:** All<br>**Transmissions:** All | **O2 Sensor Signal Stuck Rich Bank 2 Sensor 2 Conditions:**<br>Engine started, battery voltage must be at least 11.5v, all electrical components must be off, parking brake must be engaged (to keep daytime driving lights off), automatic transmission selector must be in park. The DME detected an unexpected voltage condition, or it detected an unexpected current draw in the heater circuit during the CCM test.<br>**Note: Vehicle must be raised before connector for oxygen sensors is accessible.**<br>**Possible Causes:**<br>• Oxygen sensor (before catalytic converter) is faulty<br>• Oxygen sensor heater (before catalytic converter) is faulty<br>• Oxygen sensor heater (before catalytic converter) is faulty<br>• Oxygen sensor heater (behind catalytic converter) is faulty<br>• O2S circuit is open or shorted in the wiring harness |
| **DTC: P2400**<br>**2T MIL: Yes**<br>**Years:** 2007, 2008<br>**Models:** 335i, 335Ci, 335xi, 525i, 525xi, 530i, 530xi, 550i, 650i, 750i, 750Li, 760Li, Z4 3.0i, Z4 3.0Si, Z4 M<br>**Engines:** All<br>**Transmissions:** All | **EVAP Leak Detection Pump (LDP) Control Circuit Open Conditions:**<br>Engine started, battery voltage must be at least 11.5v, all electrical components must be off, parking brake must be engaged (to keep daytime driving lights off), automatic transmission selector must be in park, the exhaust system must be properly sealed between the catalytic converter and the cylinder head, coolant temperature must be at least 80 degrees Celsius and oxygen sensor heaters for oxygen sensors before the catalytic converter must be functioning properly and the ground between the engine and the chassis must be well connected. The DME detected voltage irregularity in the leak detection pump control circuit.<br>**Possible Causes:**<br>• EVAP LDP power supply circuit is open<br>• EVAP LDP solenoid valve is damaged or it has failed<br>• EVAP LDP canister has a leak or a poor seal<br>• EVAP canister system has an improper seal<br>• Evaporative Emission (EVAP) canister purge regulator valve 1 has failed<br>• Leak Detection Pump (LDP) is faulty<br>• Aftermarket EVAP parts that do not conform to specifications<br>• EVAP component seals leaking (i.e., leaks in the Purge valve, fuel tank pressure sensor, canister vent solenoid, fuel vapor control valve tube assembly or fuel vapor vent valve). |

| DTC | Trouble Code Title, Conditions & Possible Causes |
|---|---|
| **DTC: P2401**<br>**2T MIL: Yes**<br>**Years:** 2007, 2008<br>**Models:** 335i, 335Ci, 335xi, 525i, 525xi, 530i, 530xi, 550i, 650i, 750i, 750Li, 760Li, Z4 3.0i, Z4 3.0Si, Z4 M<br>**Engines:** All<br>**Transmissions:** All | **EVAP Leak Detection Pump Control Circuit Low Conditions:**<br>Engine started, battery voltage must be at least 11.5v, all electrical components must be off, parking brake must be engaged (to keep daytime driving lights off), automatic transmission selector must be in park, the exhaust system must be properly sealed between the catalytic converter and the cylinder head, coolant temperature must be at least 80 degrees Celsius and oxygen sensor heaters for oxygen sensors before the catalytic converter must be functioning properly and the ground between the engine and the chassis must be well connected. The DME detected voltage irregularity in the leak detection pump control circuit.<br>**Possible Causes:**<br>• EVAP LDP power supply circuit is open<br>• EVAP LDP solenoid valve is damaged or it has failed<br>• EVAP LDP canister has a leak or a poor seal<br>• EVAP canister system has an improper seal<br>• Evaporative Emission (EVAP) canister purge regulator valve 1 has failed<br>• Leak Detection Pump (LDP) is faulty<br>• Aftermarket EVAP parts that do not conform to specifications<br>• EVAP component seals leaking (i.e., leaks in the Purge valve, fuel tank pressure sensor, canister vent solenoid, fuel vapor control valve tube assembly or fuel vapor vent valve). |
| **DTC: P2402**<br>**2T MIL: Yes**<br>**Years:** 2007, 2008<br>**Models:** 335i, 335Ci, 335xi, 525i, 525xi, 530i, 530xi, 550i, 650i, 750i, 750Li, 760Li, Z4 3.0i, Z4 3.0Si, Z4 M<br>**Engines:** All<br>**Transmissions:** All | **EVAP Leak Detection Pump Control Circuit High Conditions:**<br>Engine started, battery voltage must be at least 11.5v, all electrical components must be off, parking brake must be engaged (to keep daytime driving lights off), automatic transmission selector must be in park, the exhaust system must be properly sealed between the catalytic converter and the cylinder head, coolant temperature must be at least 80 degrees Celsius and oxygen sensor heaters for oxygen sensors before the catalytic converter must be functioning properly and the ground between the engine and the chassis must be well connected. The DME detected voltage irregularity in the leak detection pump control circuit.<br>**Possible Causes:**<br>• EVAP LDP power supply circuit is open<br>• EVAP LDP solenoid valve is damaged or it has failed<br>• EVAP LDP canister has a leak or a poor seal<br>• EVAP canister system has an improper seal<br>• Evaporative Emission (EVAP) canister purge regulator valve 1 has failed<br>• Leak Detection Pump (LDP) is faulty<br>• Aftermarket EVAP parts that do not conform to specifications<br>• EVAP component seals leaking (i.e., leaks in the Purge valve, fuel tank pressure sensor, canister vent solenoid, fuel vapor control valve tube assembly or fuel vapor vent valve). |
| **DTC: P2414**<br>**2T MIL: Yes**<br>**Years:** 2007, 2008<br>**Models:** 335i, 335Ci, 335xi, 525i, 525xi, 530i, 530xi, 550i, 650i, 750i, 750Li, 760Li, Z4 3.0i, Z4 3.0Si, Z4 M<br>**Engines:** All<br>**Transmissions:** All | **O2 Sensor Exhaust Sample Error Bank 1 Sensor 1 Conditions:**<br>Engine running (ground connections between the engine and the chassis must be well connected), and the DME detected an error on the OS Sensor.<br>**Note: Intake Flap Motor and Intake Manifold Runner Position Sensor are one component and cannot be replaced individually.**<br>**Note: Vacuum in the intake system sucks in the leak detection spray with false air. Leak detection spray decreases ignition quality of the fuel mixture. This causes a drop in engine speed and changes the value produced by the Heated Oxygen Sensor (HO2S). The voltage is out of range. The sensor temperature is heated up and the battery voltage is between 11 and 16 volts.**<br>**Possible Causes:**<br>• Intake Manifold Runner Position Sensor is damaged or has failed<br>• Intake system has leaks (false air)<br>• Motor for intake flap is faulty<br>• Oxygen sensor (before catalytic converter) is faulty<br>• Oxygen sensor heater (before catalytic converter) is faulty<br>• Oxygen sensor heater (before catalytic converter) is faulty<br>• Oxygen sensor heater (behind catalytic converter) is faulty<br>• O2S circuit is open or shorted in the wiring harness |

| DTC | Trouble Code Title, Conditions & Possible Causes |
|---|---|
| **DTC: P2415**<br>**2T MIL: Yes**<br>**Years:** 2007, 2008<br>**Models:** 335i, 335Ci, 335xi, 525i, 525xi, 530i, 530xi, 550i, 650i, 750i, 750Li, 760Li, Z4 3.0i, Z4 3.0Si, Z4 M<br>**Engines:** All<br>**Transmissions:** All | **O2 Sensor Exhaust Sample Error Bank 2 Sensor 1 Conditions:**<br>Engine running (ground connections between the engine and the chassis must be well connected), and the DME detected an error on the OS Sensor.<br>**Note: Intake Flap Motor and Intake Manifold Runner Position Sensor are one component and cannot be replaced individually. The voltage is out of range. The sensor temperature is heated up and the battery voltage is between 11 and 16 volts.**<br>**Note: Vacuum in the intake system sucks in the leak detection spray with false air. Leak detection spray decreases ignition quality of the fuel mixture. This causes a drop in engine speed and changes the value produced by the Heated Oxygen Sensor (HO2S).**<br>**Possible Causes:**<br>• Intake Manifold Runner Position Sensor is damaged or has failed<br>• Intake system has leaks (false air)<br>• Motor for intake flap is faulty<br>• Oxygen sensor (before catalytic converter) is faulty<br>• Oxygen sensor heater (before catalytic converter) is faulty<br>• Oxygen sensor heater (before catalytic converter) is faulty<br>• Oxygen sensor heater (behind catalytic converter) is faulty<br>• O2S circuit is open or shorted in the wiring harness |
| **DTC: P2418**<br>**2T MIL: Yes**<br>**Years:** 2007, 2008<br>**Models:** 335i, 335Ci, 335xi, 525i, 525xi, 530i, 530xi, 550i, 650i, 750i, 750Li, 760Li, Z4 3.0i, Z4 3.0Si, Z4 M<br>**Engines:** All<br>**Transmissions:** All | **Evaporative Emission System Switching Valve Control Circuit Open Conditions:**<br>Engine started, battery voltage must be at least 11.5v, all electrical components must be off, parking brake must be engaged (to keep daytime driving lights off), automatic transmission selector must be in park, the exhaust system must be properly sealed between the catalytic converter and the cylinder head, coolant temperature must be at least 80 degrees Celsius and oxygen sensor heaters for oxygen sensors before the catalytic converter must be functioning properly and the ground between the engine and the chassis must be well connected. The DME detected an unexpected EVAP malfunction.<br>**Note: Solenoid valve is closed when no voltage is present.**<br>**Possible Causes:**<br>• EVAP power supply circuit is open<br>• EVAP solenoid control circuit is open or shorted to ground<br>• EVAP solenoid control circuit is shorted to power (B+)<br>• EVAP solenoid valve is damaged or it has failed<br>• EVAP canister purge solenoid valve is faulty |
| **DTC: P2419**<br>**2T MIL: Yes**<br>**Years:** 2007, 2008<br>**Models:** 335i, 335Ci, 335xi, 525i, 525xi, 530i, 530xi, 550i, 650i, 750i, 750Li, 760Li, Z4 3.0i, Z4 3.0Si, Z4 M<br>**Engines:** All<br>**Transmissions:** All | **Evaporative Emission System Switching Valve Control Circuit Low Conditions:**<br>Engine started, battery voltage must be at least 11.5v, all electrical components must be off, parking brake must be engaged (to keep daytime driving lights off), automatic transmission selector must be in park, the exhaust system must be properly sealed between the catalytic converter and the cylinder head, coolant temperature must be at least 80 degrees Celsius and oxygen sensor heaters for oxygen sensors before the catalytic converter must be functioning properly and the ground between the engine and the chassis must be well connected. The DME detected an unexpected EVAP malfunction.<br>**Note: Solenoid valve is closed when no voltage is present.**<br>**Possible Causes:**<br>• EVAP power supply circuit is open<br>• EVAP solenoid control circuit is open or shorted to ground<br>• EVAP solenoid control circuit is shorted to power (B+)<br>• EVAP solenoid valve is damaged or it has failed<br>• EVAP canister purge solenoid valve is faulty |
| **DTC: P2420**<br>**2T MIL: Yes**<br>**Years:** 2007, 2008<br>**Models:** 335i, 335Ci, 335xi, 525i, 525xi, 530i, 530xi, 550i, 650i, 750i, 750Li, 760Li, Z4 3.0i, Z4 3.0Si, Z4 M<br>**Engines:** All<br>**Transmissions:** All | **Evaporative Emission System Switching Valve Control Circuit High Conditions:**<br>Engine started, battery voltage must be at least 11.5v, all electrical components must be off, parking brake must be engaged (to keep daytime driving lights off), automatic transmission selector must be in park, the exhaust system must be properly sealed between the catalytic converter and the cylinder head, coolant temperature must be at least 80 degrees Celsius and oxygen sensor heaters for oxygen sensors before the catalytic converter must be functioning properly and the ground between the engine and the chassis must be well connected. The DME detected an unexpected EVAP malfunction.<br>**Note: Solenoid valve is closed when no voltage is present.**<br>**Possible Causes:**<br>• EVAP power supply circuit is open<br>• EVAP solenoid control circuit is open or shorted to ground<br>• EVAP solenoid control circuit is shorted to power (B+)<br>• EVAP solenoid valve is damaged or it has failed<br>• EVAP canister purge solenoid valve is faulty |

| DTC | Trouble Code Title, Conditions & Possible Causes |
|---|---|
| **DTC: P2430**<br>**Years:** 2007, 2008<br>**Models:** 335i, 335Ci, 335xi, 525i, 525xi, 530i, 530xi, 550i, 650i, 750i, 750Li, 760Li, Z4 3.0i, Z4 3.0Si, Z4 M<br>**Engines:** All<br>**Transmissions:** All | **Secondary Air Mass Flow Sensor Rationality Check Conditions:**<br>Engine started, battery voltage must be at least 11.5v, all electrical components must be off, parking brake must be engaged (to keep daytime driving lights off), automatic transmission selector must be in park, the exhaust system must be properly sealed between the catalytic converter and the cylinder head, coolant temperature must be at least 80 degrees Celsius and oxygen sensor heaters for oxygen sensors before the catalytic converter must be functioning properly and the ground between the engine and the chassis must be well connected. The DME detected an unexpected secondary air system malfunction. It is disconnected or stuck.<br>**Note: Solenoid valve is closed when no voltage is present.**<br>**Possible Causes:**<br>• EVAP power supply circuit is open<br>• EVAP solenoid control circuit is open or shorted to ground<br>• EVAP solenoid control circuit is shorted to power (B+)<br>• EVAP solenoid valve is damaged or it has failed<br>• EVAP canister purge solenoid valve is faulty |
| **DTC: P2540**<br>**Years:** 2007, 2008<br>**Models:** 335i, 335Ci, 335xi, 525i, 525xi, 530i, 530xi, 550i, 650i, 750i, 750Li, 760Li, Z4 3.0i, Z4 3.0Si, Z4 M<br>**Engines:** All<br>**Transmissions:** All | **Secondary Air System Vent Valve Stuck Closed Conditions:**<br>Engine started, battery voltage must be at least 11.5v, all electrical components must be off, parking brake must be engaged (to keep daytime driving lights off), automatic transmission selector must be in park, the exhaust system must be properly sealed between the catalytic converter and the cylinder head, coolant temperature must be at least 80 degrees Celsius and oxygen sensor heaters for oxygen sensors before the catalytic converter must be functioning properly and the ground between the engine and the chassis must be well connected. The DME detected an unexpected secondary air system malfunction.<br>**Note: Solenoid valve is closed when no voltage is present.**<br>**Possible Causes:**<br>• EVAP power supply circuit is open<br>• EVAP solenoid control circuit is open or shorted to ground<br>• EVAP solenoid control circuit is shorted to power (B+)<br>• EVAP solenoid valve is damaged or it has failed<br>• EVAP canister purge solenoid valve is faulty |
| **DTC: P2626**<br>**2T MIL: Yes**<br>**Years:** 2007, 2008<br>**Models:** 335i, 335Ci, 335xi, 525i, 525xi, 530i, 530xi, 550i, 650i, 750i, 750Li, 760Li, Z4 3.0i, Z4 3.0Si, Z4 M<br>**Engines:** All<br>**Transmissions:** All | **O2 Sensor Pumping Current Trim Circuit/Open Bank 1 Sensor 1 Conditions:**<br>Engine started and the fault entry trips after the criterion remains present for 1 second, battery voltage must be at least 11v, the trailing throttle overrun abort is present for at least two seconds, the O2 sensor is heated to an adequate temperature and the exhaust gas temperature before the catalytic converter is less than 750 degrees Celsius.<br>**Possible Causes:**<br>• Check activation of Recirculation Pump Relay<br>• Oxygen sensor (before catalytic converter) is faulty<br>• Oxygen sensor (behind catalytic converter) is faulty<br>• Oxygen sensor heater (before catalytic converter) is faulty<br>• Oxygen sensor heater (behind catalytic converter) is faulty |
| **DTC: P2629**<br>**2T MIL: Yes**<br>**Years:** 2007, 2008<br>**Models:** 335i, 335Ci, 335xi, 525i, 525xi, 530i, 530xi, 550i, 650i, 750i, 750Li, 760Li, Z4 3.0i, Z4 3.0Si, Z4 M<br>**Engines:** All<br>**Transmissions:** All | **O2 Sensor Pumping Current Trim Circuit/Open Bank 2 Sensor 1 Conditions:**<br>Engine started, battery voltage must be at least 11.5v, all electrical components must be off, parking brake must be engaged (to keep daytime driving lights off), automatic transmission selector must be in park, the exhaust system must be properly sealed between the catalytic converter and the cylinder head, coolant temperature must be at least 80 degrees Celsius and oxygen sensor heaters for oxygen sensors before the catalytic converter must be functioning properly and the ground between the engine and the chassis must be well connected. The DME detected a voltage value that doesn't fall within the desired parameters for a properly functioning O2 system.<br>**Possible Causes:**<br>• Check activation of Recirculation Pump Relay<br>• Oxygen sensor (before catalytic converter) is faulty<br>• Oxygen sensor (behind catalytic converter) is faulty<br>• Oxygen sensor heater (before catalytic converter) is faulty<br>• Oxygen sensor heater (behind catalytic converter) is faulty |
| **DTC: P2713**<br>**2T MIL: Yes**<br>**Years:** 2007, 2008<br>**Models:** 335i, 335Ci, 335xi, 525i, 525xi, 530i, 530xi, 750i, 750Li, 760Li, Z4 3.0i, Z4 3.0Si, Z4 M<br>**Engines:** All<br>**Transmissions:** A/T | **Pressure Regulator Valve 4 Plausibility Conditions:**<br>The current to/from the pressure regulator valve is either higher or lower than the threshold value.<br>**Possible Causes:**<br>• Pressure control solenoid circuit is shorting to ground<br>• Pressure control solenoid circuit is open<br>• Valve has failed<br>• TCM has failed |

| DTC | Trouble Code Title, Conditions & Possible Causes |
|---|---|
| **DTC: P2716**<br>**2T MIL: Yes**<br>**Years:** 2007, 2008<br>**Models:** 335i, 335Ci, 335xi, 525i, 525xi, 530i, 530xi, 750i, 750Li, 760Li, Z4 3.0i, Z4 3.0Si, Z4 M<br>**Engines:** All<br>**Transmissions:** A/T | **Pressure Control Solenoid 4 Electrical Malfunction Conditions:**<br>Engine started, battery voltage must be at least 11.5v, all electrical components must be off, and the ground between the engine and the chassis must be well connected. The DME detected the pressure control solenoid was experiencing electrical malfunctions.<br>**Possible Causes:**<br>• ATF level is low<br>• Circuit harness connector contacts are corroded or ingresses of water<br>• Circuit wires have shorted to each other, to battery or ground<br>• Automatic Transmission Hydraulic Pressure Sensor 1 has failed<br>• Solenoid valves in valve body are faulty<br>• Transmission Control Module (TCM) needs replacing<br>• Transmission Input Speed (RPM) Sensor has failed<br>• Transmission Output Speed (RPM) Sensor has failed |
| **DTC: P2720**<br>**2T MIL: Yes**<br>**Years:** 2007, 2008<br>**Models:** 335i, 335Ci, 335xi, 525i, 525xi, 530i, 530xi, 750i, 750Li, 760Li, Z4 3.0i, Z4 3.0Si, Z4 M<br>**Engines:** All<br>**Transmissions:** A/T | **Pressure Regulator Valve 4 Lower Threshold Conditions:**<br>The signal to/from the pressure regulator valve is interrupted or short circuited to ground.<br>**Possible Causes:**<br>• Pressure control solenoid circuit is shorting to ground<br>• Pressure control solenoid circuit is open<br>• Valve has failed<br>• TCM has failed |
| **DTC: P2721**<br>2T MIL: Yes<br>**Years:** 2007, 2008<br>**Models:** 335i, 335Ci, 335xi, 525i, 525xi, 530i, 530xi, 750i, 750Li, 760Li, Z4 3.0i, Z4 3.0Si, Z4 M<br>**Engines:** All<br>**Transmissions:** A/T | **Pressure Regulator Valve 4 Upper Threshold Conditions:**<br>The signal to/from the pressure regulator valve is interrupted or short circuited to supply.<br>**Possible Causes:**<br>• Pressure control solenoid circuit is shorting to ground<br>• Pressure control solenoid circuit is open<br>• Valve has failed<br>• TCM has failed |
| **DTC: P2722**<br>**2T MIL: Yes**<br>**Years:** 2007, 2008<br>**Models:** 335i, 335Ci, 335xi, 525i, 525xi, 530i, 530xi, 750i, 750Li, 760Li, Z4 3.0i, Z4 3.0Si, Z4 M<br>**Engines:** All<br>**Transmissions:** A/T | **Pressure Regulator Valve 5 Plausibility Conditions:**<br>The current to/from the pressure regulator valve is either higher or lower than the threshold value.<br>**Possible Causes:**<br>• Pressure control solenoid circuit is shorting to ground<br>• Pressure control solenoid circuit is open<br>• Valve has failed<br>• TCM has failed |
| **DTC: P2725**<br>**2T MIL: Yes**<br>**Years:** 2007, 2008<br>**Models:** 335i, 335Ci, 335xi, 525i, 525xi, 530i, 530xi, 750i, 750Li, 760Li, Z4 3.0i, Z4 3.0Si, Z4 M<br>**Engines:** All<br>**Transmissions:** A/T | **Pressure Regulator Valve 5 No Signal Conditions:**<br>The signal to/from the pressure regulator valve is interrupted and there is no signal.<br>**Possible Causes:**<br>• Pressure control solenoid circuit is shorting to ground<br>• Pressure control solenoid circuit is open<br>• Valve has failed<br>• TCM has failed |
| **DTC: P2729**<br>**2T MIL: Yes**<br>**Years:** 2007, 2008<br>**Models:** 335i, 335Ci, 335xi, 525i, 525xi, 530i, 530xi, 750i, 750Li, 760Li, Z4 3.0i, Z4 3.0Si, Z4 M<br>**Engines:** All<br>**Transmissions:** A/T | **Pressure Regulator Valve 5 Lower Threshold Conditions:**<br>The signal to/from the pressure regulator valve is interrupted or short circuited to ground.<br>**Possible Causes:**<br>• Pressure control solenoid circuit is shorting to ground<br>• Pressure control solenoid circuit is open<br>• Valve has failed<br>• TCM has failed |
| **DTC: P2730**<br>**2T MIL: Yes**<br>**Years:** 2007, 2008<br>**Models:** 335i, 335Ci, 335xi, 525i, 525xi, 530i, 530xi, 750i, 750Li, 760Li, Z4 3.0i, Z4 3.0Si, Z4 M<br>**Engines:** All<br>**Transmissions:** A/T | **Pressure Regulator Valve 5 Upper Threshold Conditions:**<br>The signal to/from the pressure regulator valve is interrupted or short circuited to supply.<br>**Possible Causes:**<br>• Pressure control solenoid circuit is shorting to ground<br>• Pressure control solenoid circuit is open<br>• Valve has failed<br>• TCM has failed |

| DTC | Trouble Code Title, Conditions & Possible Causes |
|---|---|
| **DTC: P2761**<br>**2T MIL:** Yes<br>**Years:** 2007, 2008<br>**Models:** 335i, 335Ci, 335xi, 525i, 525xi, 530i, 530xi, 750i, 750Li, 760Li, Z4 3.0i, Z4 3.0Si, Z4 M<br>**Engines:** All<br>**Transmissions:** A/T | **Pressure Regulator Valve 4 Plausibility Conditions:**<br>The current to/from the pressure regulator valve is either higher or lower than the threshold value.<br>**Possible Causes:**<br>• Pressure control solenoid circuit is shorting to ground<br>• Pressure control solenoid circuit is open<br>• Valve has failed<br>• TCM has failed |
| **DTC: P2763**<br>**2T MIL:** Yes<br>**Years:** 2007, 2008<br>**Models:** 335i, 335Ci, 335xi, 525i, 525xi, 530i, 530xi, 750i, 750Li, 760Li, Z4 3.0i, Z4 3.0Si, Z4 M<br>**Engines:** All<br>**Transmissions:** A/T | **Pressure Regulator Valve 4 Upper Threshold Conditions:**<br>The signal to/from the pressure regulator valve is interrupted or short circuited to power.<br>**Possible Causes:**<br>• Pressure control solenoid circuit is shorting to ground<br>• Pressure control solenoid circuit is open<br>• Valve has failed<br>• TCM has failed |
| **DTC: P2764**<br>**2T MIL:** Yes<br>**Years:** 2007, 2008<br>**Models:** 335i, 335Ci, 335xi, 525i, 525xi, 530i, 530xi, 750i, 750Li, 760Li, Z4 3.0i, Z4 3.0Si, Z4 M<br>**Engines:** All<br>**Transmissions:** A/T | **Pressure Regulator Valve 4 Lower Threshold Conditions:**<br>The signal to/from the pressure regulator valve is interrupted or short circuited to ground.<br>**Possible Causes:**<br>• Pressure control solenoid circuit is shorting to ground<br>• Pressure control solenoid circuit is open<br>• Valve has failed<br>• TCM has failed |

## Gas Engine OBD II Trouble Code List (P3xxx Codes)

| DTC | Trouble Code Title, Conditions & Possible Causes |
|---|---|
| **DTC: P3012**<br>**2T MIL:** Yes<br>**Years:** 2007, 2008<br>**Models:** 335i, 335Ci, 335xi, 525i, 525xi, 530i, 530xi, 550i, 650i, 750i, 750Li, 760Li, Z4 3.0i, Z4 3.0Si, Z4 M<br>**Engines:** All<br>**Transmissions:** All | **O2S Sensor Circuit Adaptation Value to High (Bank 1 Sensor 1) Conditions:**<br>Engine started and the fault entry trips after 10 seconds, after the calibration runs after engine is in idle phases, battery voltage must be at least 11v, and an offset correction value was found to be above the maximum approved threshold value. The battery voltage must be between 11 and 16 volts.<br>**Possible Causes:**<br>• O2S signal shorted to power circuit inside connector<br>• O2S signal circuit shorted to ground or to system voltage<br>• Check wiring harness<br>• Replace control module |
| **DTC: P3013**<br>**2T MIL:** Yes<br>**Years:** 2007, 2008<br>**Models:** 335i, 335Ci, 335xi, 525i, 525xi, 530i, 530xi, 550i, 650i, 750i, 750Li, 760Li, Z4 3.0i, Z4 3.0Si, Z4 M<br>**Engines:** All<br>**Transmissions:** All | **O2S Sensor Circuit Adaptation Value to High (Bank 2 Sensor 1) Conditions:**<br>Engine started and the fault entry trips after 10 seconds, after the calibration runs after engine is in idle phases, battery voltage must be at least 11v, and an offset correction value was found to be above the maximum approved threshold value. The battery voltage must be between 11 and 16 volts.<br>**Possible Causes:**<br>• O2S signal shorted to power circuit inside connector<br>• O2S signal circuit shorted to ground or to system voltage<br>• Check wiring harness<br>• Replace control module |
| **DTC: P3014**<br>**2T MIL:** Yes<br>**Years:** 2007, 2008<br>**Models:** 335i, 335Ci, 335xi, 525i, 525xi, 530i, 530xi, 550i, 650i, 750i, 750Li, 760Li, Z4 3.0i, Z4 3.0Si, Z4 M<br>**Engines:** All<br>**Transmissions:** All | **O2S Sensor WRAF-IC Supply Voltage Too Low (Bank 1 Sensor 1) Conditions:**<br>Engine started and the fault entry trips after the criterion remains present for 2 seconds, battery voltage must be at least 11v, and an the monitor for the CJ125 chip power supply was found that the voltage was too low. The battery voltage must be between 11 and 16 volts.<br>**Possible Causes:**<br>• Internal HW fault in control module<br>• Replace control module |

| DTC | Trouble Code Title, Conditions & Possible Causes |
|---|---|
| **DTC: P3015**<br>**2T MIL:** Yes<br>**Years:** 2007, 2008<br>**Models:** 335i, 335Ci, 335xi, 525i, 525xi, 530i, 530xi, 550i, 650i, 750i, 750Li, 760Li, Z4 3.0i, Z4 3.0Si, Z4 M<br>**Engines:** All<br>**Transmissions:** All | **O2S Sensor WRAF-IC Supply Voltage Too Low (Bank 2 Sensor 1) Conditions:**<br>Engine started and the fault entry trips after the criterion remains present for 2 seconds, battery voltage must be at least 11v, and an the monitor for the CJ125 chip power supply was found that the voltage was too low. The battery voltage must be between 11 and 16 volts.<br>**Possible Causes:**<br>• Internal HW fault in control module<br>• Replace control module |
| **DTC: P3016**<br>**2T MIL:** Yes<br>**Years:** 2007, 2008<br>**Models:** 335i, 335Ci, 335xi, 525i, 525xi, 530i, 530xi, 550i, 650i, 750i, 750Li, 760Li, Z4 3.0i, Z4 3.0Si, Z4 M<br>**Engines:** All<br>**Transmissions:** All | **O2 Sensor Calibration Resistance at WRAF-IC Plausibility Bank 1 Sensor 1 Conditions:**<br>Engine running for 25 seconds, battery voltage 11.5, ground between engine and chassis well connected and the exhaust system must be properly sealed between catalytic converter and the cylinder head. The DME detected the O2S signal was implausible or not detected. There is no overrun cutoff and the exhaust gas temperature is less than 400 degrees Celsius. The battery voltage must be between 11 and 16 volts.<br>**Possible Causes:**<br>• Oxygen sensor for oxygen sensor (O2S) before catalytic converter is faulty<br>• O2S is contaminated (due to presence of silicone in fuel)<br>• O2S signal and ground circuit wires crossed in wiring harness<br>• O2S signal circuit is shorted to sensor or chassis ground<br>• O2S element before the catalytic converter has failed (internal short condition)<br>• Leaks present in the exhaust manifold or exhaust pipes |
| **DTC: P3017**<br>**2T MIL:** Yes<br>**Years:** 2007, 2008<br>**Models:** 335i, 335Ci, 335xi, 525i, 525xi, 530i, 530xi, 550i, 650i, 750i, 750Li, 760Li, Z4 3.0i, Z4 3.0Si, Z4 M<br>**Engines:** All<br>**Transmissions:** All | **O2 Sensor Calibration Resistance at WRAF-IC Plausibility Bank 2 Sensor 1 Conditions:**<br>Engine running for 25 seconds, battery voltage 11.5, ground between engine and chassis well connected and the exhaust system must be properly sealed between catalytic converter and the cylinder head. The DME detected the O2S signal was implausible or not detected. There is no overrun cutoff and the exhaust gas temperature is less than 400 degrees Celsius. The battery voltage must be between 11 and 16 volts.<br>**Possible Causes:**<br>• Oxygen sensor for oxygen sensor (O2S) before catalytic converter is faulty<br>• O2S is contaminated (due to presence of silicone in fuel)<br>• O2S signal and ground circuit wires crossed in wiring harness<br>• O2S signal circuit is shorted to sensor or chassis ground<br>• O2S element before the catalytic converter has failed (internal short condition)<br>• Leaks present in the exhaust manifold or exhaust pipes |
| **DTC: P3018**<br>**2T MIL:** Yes<br>**Years:** 2007, 2008<br>**Models:** 335i, 335Ci, 335xi, 525i, 525xi, 530i, 530xi, 550i, 650i, 750i, 750Li, 760Li, Z4 3.0i, Z4 3.0Si, Z4 M<br>**Engines:** All<br>**Transmissions:** All | **O2S Sensor Lambda Controller Value Above Threshold due to Open Pumping Current Circuit (Bank 1 Sensor 1) Conditions:**<br>Engine started and the fault entry trips after the criterion remains present for 2 seconds, battery voltage must be at least 11v, and there is a communications fault between the ECU and the CJ125 (SPI bus) chip. The sensor temperature status is heated up and the fuel system status is closed loop.<br>**Possible Causes:**<br>• Internal HW fault in control module<br>• Replace control module |
| **DTC: P3019**<br>**2T MIL:** Yes<br>**Years:** 2007, 2008<br>**Models:** 335i, 335Ci, 335xi, 525i, 525xi, 530i, 530xi, 550i, 650i, 750i, 750Li, 760Li, Z4 3.0i, Z4 3.0Si, Z4 M<br>**Engines:** All<br>**Transmissions:** All | **O2S Sensor Lambda Controller Value Above Threshold due to Open Pumping Current Circuit (Bank 2 Sensor 1) Conditions:**<br>Engine started and the fault entry trips after the criterion remains present for 2 seconds, battery voltage must be at least 11v, and there is a communications fault between the ECU and the CJ125 (SPI bus) chip. The sensor temperature status is heated up and the fuel system status is closed loop.<br>**Possible Causes:**<br>• Internal HW fault in control module<br>• Replace control module |
| **DTC: P3020**<br>**2T MIL:** Yes<br>**Years:** 2007, 2008<br>**Models:** 335i, 335Ci, 335xi, 525i, 525xi, 530i, 530xi, 550i, 650i, 750i, 750Li, 760Li, Z4 3.0i, Z4 3.0Si, Z4 M<br>**Engines:** All<br>**Transmissions:** All | **O2S Sensor Signal Voltage Too Low During Coast Down Fuel Cut-Off Due to Open Pumping Current Circuit (Bank 1 Sensor 1) Conditions:**<br>Engine started and the fault monitoring criterion must remain present for over 2 seconds, battery voltage must be at least 11v, and the voltage at CJ125 is at 1.5v so the sensor is heated adequately, there is an open pump current wire.<br>**Possible Causes:**<br>• Check plugs<br>• Check wiring harness<br>• Replace sensor |

| DTC | Trouble Code Title, Conditions & Possible Causes |
|---|---|
| **DTC: P3021**<br>**2T MIL: Yes**<br>**Years:** 2007, 2008<br>**Models:** 335i, 335Ci, 335xi, 525i, 525xi, 530i, 530xi, 550i, 650i, 750i, 750Li, 760Li, Z4 3.0i, Z4 3.0Si, Z4 M<br>**Engines:** All<br>**Transmissions:** All | **O2S Sensor Signal Voltage Too Low During Coast Down Fuel Cut-Off Due to Open Pumping Current Circuit (Bank 2 Sensor 1) Conditions:**<br>Engine started and the fault monitoring criterion must remain present for over 2 seconds, battery voltage must be at least 11v, and the voltage at CJ125 is at 1.5v so the sensor is heated adequately, there is an open pump current wire.<br>**Possible Causes:**<br>• Check plugs<br>• Check wiring harness<br>• Replace sensor |
| **DTC: P3022**<br>**2T MIL: Yes**<br>**Years:** 2007, 2008<br>**Models:** 335i, 335Ci, 335xi, 525i, 525xi, 530i, 530xi, 550i, 650i, 750i, 750Li, 760Li, Z4 3.0i, Z4 3.0Si, Z4 M<br>**Engines:** All<br>**Transmissions:** All | **O2S Sensor Disturbed SPI Communication to WRAF-IC (Bank 1 Sensor 1) Conditions:**<br>Engine started and the fault entry trips after the criterion remains present for 2 seconds, battery voltage must be at least 11v, and there is a communications fault between the ECU and the CJ125 (SPI bus) chip.<br>**Possible Causes:**<br>• Internal HW fault in control module<br>• Replace control module |
| **DTC: P3023**<br>**2T MIL: Yes**<br>**Years:** 2007, 2008<br>**Models:** 335i, 335Ci, 335xi, 525i, 525xi, 530i, 530xi, 550i, 650i, 750i, 750Li, 760Li, Z4 3.0i, Z4 3.0Si, Z4 M<br>**Engines:** All<br>**Transmissions:** All | **O2S Sensor Disturbed SPI Communication to WRAF-IC (Bank 2 Sensor 1) Conditions:**<br>Engine started and the fault entry trips after the criterion remains present for 2 seconds, battery voltage must be at least 11v, and there is a communications fault between the ECU and the CJ125 (SPI bus) chip.<br>**Possible Causes:**<br>• Internal HW fault in control module<br>• Replace control module |
| **DTC: P3024**<br>**2T MIL: Yes**<br>**Years:** 2007, 2008<br>**Models:** 335i, 335Ci, 335xi, 525i, 525xi, 530i, 530xi, 550i, 650i, 750i, 750Li, 760Li, Z4 3.0i, Z4 3.0Si, Z4 M<br>**Engines:** All<br>**Transmissions:** All | **O2S Sensor Initialization Error WRAF-IC (Bank 1 Sensor 1) Conditions:**<br>Engine started and the fault entry trips after the criterion remains present for 2 seconds, battery voltage must be at least 11v, and there is a communications fault between the ECU and the CJ125 (SPI bus) chip.<br>**Possible Causes:**<br>• Internal HW fault in control module<br>• Replace control module |
| **DTC: P3025**<br>**2T MIL: Yes**<br>**Years:** 2007, 2008<br>**Models:** 335i, 335Ci, 335xi, 525i, 525xi, 530i, 530xi, 550i, 650i, 750i, 750Li, 760Li, Z4 3.0i, Z4 3.0Si, Z4 M<br>**Engines:** All<br>**Transmissions:** All | **O2S Sensor Initialization Error WRAF-IC (Bank 2 Sensor 1) Conditions:**<br>Engine started and the fault entry trips after the criterion remains present for 2 seconds, battery voltage must be at least 11v, and there is a communications fault between the ECU and the CJ125 (SPI bus) chip.<br>**Possible Causes:**<br>• Internal HW fault in control module<br>• Replace control module |
| **DTC: P3026**<br>**2T MIL: Yes**<br>**Years:** 2007, 2008<br>**Models:** 335i, 335Ci, 335xi, 525i, 525xi, 530i, 530xi, 550i, 650i, 750i, 750Li, 760Li, Z4 3.0i, Z4 3.0Si, Z4 M<br>**Engines:** All<br>**Transmissions:** All | **O2 Sensor Operating Temperature Not Reached Bank 1 Sensor 1 Conditions:**<br>Engine running for 40 seconds, battery voltage 11.5, ground between engine and chassis well connected and the exhaust system must be properly sealed between catalytic converter and the cylinder head. The DME detected the O2S signal was implausible or not detected. Monitoring of the sensor's ceramic temperature and control single pulse-duty factor detects reduced heater performance at the oxygen sensor before the catalytic converter. The heater controller value stays above the threshold (97 percent). There is no overrun cutoff and the exhaust gas temperature is less than 400 degrees Celsius.<br>**Possible Causes:**<br>• Oxygen sensor heater for oxygen sensor (HO2S) before catalytic converter is faulty<br>• O2S is contaminated (due to presence of silicone in fuel)<br>• O2S signal and ground circuit wires crossed in wiring harness<br>• O2S signal circuit is shorted to sensor or chassis ground<br>• O2S element before the catalytic converter has failed (internal short condition)<br>• Leaks present in the exhaust manifold or exhaust pipes |

| DTC | Trouble Code Title, Conditions & Possible Causes |
|---|---|
| **DTC: P3027**<br>**2T MIL:** Yes<br>**Years:** 2007, 2008<br>**Models:** 335i, 335Ci, 335xi, 525i, 525xi, 530i, 530xi, 550i, 650i, 750i, 750Li, 760Li, Z4 3.0i, Z4 3.0Si, Z4 M<br>**Engines:** All<br>**Transmissions:** All | **O2 Sensor Operating Temperature Not Reached Bank 2 Sensor 1 Conditions:**<br>Engine running for 40 seconds, battery voltage 11.5, ground between engine and chassis well connected and the exhaust system must be properly sealed between catalytic converter and the cylinder head. The DME detected the O2S signal was implausible or not detected. Monitoring of the sensor's ceramic temperature and control single pulse-duty factor detects reduced heater performance at the oxygen sensor before the catalytic converter. The heater controller value stays above the threshold (97 percent). There is no overrun cutoff and the exhaust gas temperature is less than 400 degrees Celsius.<br>**Possible Causes:**<br>• Oxygen sensor heater for oxygen sensor (HO2S) before catalytic converter is faulty<br>• O2S is contaminated (due to presence of silicone in fuel)<br>• O2S signal and ground circuit wires crossed in wiring harness<br>• O2S signal circuit is shorted to sensor or chassis ground<br>• O2S element before the catalytic converter has failed (internal short condition)<br>• Leaks present in the exhaust manifold or exhaust pipes |
| **DTC: P3028**<br>**2T MIL:** Yes<br>**Years:** 2007, 2008<br>**Models:** 335i, 335Ci, 335xi, 525i, 525xi, 530i, 530xi, 550i, 650i, 750i, 750Li, 760Li, Z4 3.0i, Z4 3.0Si, Z4 M<br>**Engines:** All<br>**Transmissions:** All | **O2 Sensor Heater Control No Activity Detected Bank 1 Sensor 1 Conditions:**<br>Engine running for 40 seconds, battery voltage 11.5, ground between engine and chassis well connected and the exhaust system must be properly sealed between catalytic converter and the cylinder head. The DME detected the O2S signal was implausible or not detected. Monitoring of the sensor's ceramic temperature and control single pulse-duty factor detects reduced heater performance at the oxygen sensor before the catalytic converter. There is no overrun cutoff and the exhaust gas temperature is less than 400 degrees Celsius.<br>**Possible Causes:**<br>• Oxygen sensor heater for oxygen sensor (HO2S) before catalytic converter is faulty<br>• O2S is contaminated (due to presence of silicone in fuel)<br>• O2S signal and ground circuit wires crossed in wiring harness<br>• O2S signal circuit is shorted to sensor or chassis ground<br>• O2S element before the catalytic converter has failed (internal short condition)<br>• Leaks present in the exhaust manifold or exhaust pipes |
| **DTC: P3029**<br>**2T MIL:** Yes<br>**Years:** 2007, 2008<br>**Models:** 335i, 335Ci, 335xi, 525i, 525xi, 530i, 530xi, 550i, 650i, 750i, 750Li, 760Li, Z4 3.0i, Z4 3.0Si, Z4 M<br>**Engines:** All<br>**Transmissions:** All | **O2 Sensor Heater Control No Activity Detected Bank 2 Sensor 1 Conditions:**<br>Engine running for 40 seconds, battery voltage 11.5, ground between engine and chassis well connected and the exhaust system must be properly sealed between catalytic converter and the cylinder head. The DME detected the O2S signal was implausible or not detected. Monitoring of the sensor's ceramic temperature and control single pulse-duty factor detects reduced heater performance at the oxygen sensor before the catalytic converter. There is no overrun cutoff and the exhaust gas temperature is less than 400 degrees Celsius.<br>**Possible Causes:**<br>• Oxygen sensor heater for oxygen sensor (HO2S) before catalytic converter is faulty<br>• O2S is contaminated (due to presence of silicone in fuel)<br>• O2S signal and ground circuit wires crossed in wiring harness<br>• O2S signal circuit is shorted to sensor or chassis ground<br>• O2S element before the catalytic converter has failed (internal short condition)<br>• Leaks present in the exhaust manifold or exhaust pipes |
| **DTC: P3037**<br>**2T MIL:** Yes<br>**Years:** 2007, 2008<br>**Models:** 335i, 335Ci, 335xi, 525i, 525xi, 530i, 530xi, 550i, 650i, 750i, 750Li, 760Li, Z4 3.0i, Z4 3.0Si, Z4 M<br>**Engines:** All<br>**Transmissions:** All | **O2S Sensor Positive Current Control Circuit/Open Circuit (Bank 1 Sensor 1) Conditions:**<br>Engine started, the fault criterion must remain present for over 2 seconds, the Lambda specification is outside three percent relative to Lambda = 1, and the closed loop lambda control is active.<br>**Possible Causes:**<br>• Check plugs<br>• Check wiring harness<br>• Replace sensor |
| **DTC: P3200**<br>**2T MIL:** Yes<br>**Years:** 2007, 2008<br>**Models:** 335i, 335Ci, 335xi, 525i, 525xi, 530i, 530xi, 550i, 650i, 750i, 750Li, 760Li, Z4 3.0i, Z4 3.0Si, Z4 M<br>**Engines:** All<br>**Transmissions:** All | **Power CAN, CAN Chip Defective Conditions:**<br>The Engine Control Module (DME) communicates with all databus-capable control modules via a CAN databus. These databus-capable control modules are connected via two data bus wires which are twisted together (CAN_High and CAN_Low), and exchange information (messages). Missing information on the databus is recognized as a malfunction and stored. Trouble-free operation of the CAN-Bus requires that it have a terminal resistance. This central terminal resistor is located in the Engine Control Module (DME). The ignition is on for 800ms or the engine running for 360ms, the voltage is greater than 10 volts. This applies to vehicles with ARS only and does not affect bus activity.<br>**Possible Causes:**<br>• CAN data bus wires have short circuited to each other<br>• ECU has failed<br>• BUS system failure<br>• Defective bus controller (EGS) |

| DTC | Trouble Code Title, Conditions & Possible Causes |
|---|---|
| **DTC: P3201**<br>**2T MIL: Yes**<br>**Years:** 2007, 2008<br>**Models:** 335i, 335Ci, 335xi, 525i, 525xi, 530i, 530xi, 550i, 650i, 750i, 750Li, 760Li, Z4 3.0i, Z4 3.0Si, Z4 M<br>**Engines:** All<br>**Transmissions:** All | **Power CAN, DPRAM-CAN Chip Defective Conditions:**<br>The Engine Control Module (DME) communicates with all databus-capable control modules via a CAN databus. These databus-capable control modules are connected via two data bus wires which are twisted together (CAN_High and CAN_Low), and exchange information (messages). Missing information on the databus is recognized as a malfunction and stored. Trouble-free operation of the CAN-Bus requires that it have a terminal resistance. This central terminal resistor is located in the Engine Control Module (DME). The ignition is on for 800ms or the engine running for 360ms, the voltage is greater than 10 volts. This applies to vehicles with ARS only and does not affect bus activity.<br>**Possible Causes:**<br>• CAN data bus wires have short circuited to each other<br>• ECU has failed<br>• BUS system failure<br>• Defective bus controller (EGS) |
| **DTC: P3202**<br>**2T MIL: Yes**<br>**Years:** 2007, 2008<br>**Models:** 335i, 335Ci, 335xi, 525i, 525xi, 530i, 530xi, 550i, 650i, 750i, 750Li, 760Li, Z4 3.0i, Z4 3.0Si, Z4 M<br>**Engines:** All<br>**Transmissions:** All | **Powertrain CAN, CAN Chip Cut-Off Conditions:**<br>The Engine Control Module (DME) communicates with all databus-capable control modules via a CAN databus. These databus-capable control modules are connected via two data bus wires which are twisted together (CAN_High and CAN_Low), and exchange information (messages). Missing information on the databus is recognized as a malfunction and stored. Trouble-free operation of the CAN-Bus requires that it have a terminal resistance. This central terminal resistor is located in the Engine Control Module (DME). The ignition is on for 800ms or the engine running for 360ms, the voltage is greater than 10 volts. This applies to vehicles with ARS only and does not affect bus activity.<br>**Possible Causes:**<br>• CAN data bus wires have short circuited to each other<br>• ECU has failed<br>• BUS system failure<br>• Defective bus controller (EGS) |
| **DTC: P3203**<br>**2T MIL: Yes**<br>**Years:** 2007, 2008<br>**Models:** 335i, 335Ci, 335xi, 525i, 525xi, 530i, 530xi, 550i, 650i, 750i, 750Li, 760Li, Z4 3.0i, Z4 3.0Si, Z4 M<br>**Engines:** All<br>**Transmissions:** All | **Local CAN, LoCAN Chip Defective Conditions:**<br>The Engine Control Module (DME) communicates with all databus-capable control modules via a CAN databus. These databus-capable control modules are connected via two data bus wires which are twisted together (CAN_High and CAN_Low), and exchange information (messages). Missing information on the databus is recognized as a malfunction and stored. Trouble-free operation of the CAN-Bus requires that it have a terminal resistance. This central terminal resistor is located in the Engine Control Module (DME). The ignition is on for 800ms or the engine running for 360ms, the voltage is greater than 10 volts. This applies to vehicles with ARS only and does not affect bus activity.<br>**Possible Causes:**<br>• CAN data bus wires have short circuited to each other<br>• ECU has failed<br>• BUS system failure<br>• Defective bus controller (EGS) |
| **DTC: P3204**<br>**2T MIL: Yes**<br>**Years:** 2007, 2008<br>**Models:** 335i, 335Ci, 335xi, 525i, 525xi, 530i, 530xi, 550i, 650i, 750i, 750Li, 760Li, Z4 3.0i, Z4 3.0Si, Z4 M<br>**Engines:** All<br>**Transmissions:** All | **Local CAN, DPRAM-LoCAN Chip Defective Conditions:**<br>The Engine Control Module (DME) communicates with all databus-capable control modules via a CAN databus. These databus-capable control modules are connected via two data bus wires which are twisted together (CAN_High and CAN_Low), and exchange information (messages). Missing information on the databus is recognized as a malfunction and stored. Trouble-free operation of the CAN-Bus requires that it have a terminal resistance. This central terminal resistor is located in the Engine Control Module (DME). The ignition is on for 800ms or the engine running for 360ms, the voltage is greater than 10 volts. This applies to vehicles with ARS only and does not affect bus activity.<br>**Possible Causes:**<br>• CAN data bus wires have short circuited to each other<br>• ECU has failed<br>• BUS system failure<br>• Defective bus controller (EGS) |
| **DTC: P3205**<br>**2T MIL: Yes**<br>**Years:** 2007, 2008<br>**Models:** 335i, 335Ci, 335xi, 525i, 525xi, 530i, 530xi, 550i, 650i, 750i, 750Li, 760Li, Z4 3.0i, Z4 3.0Si, Z4 M<br>**Engines:** All<br>**Transmissions:** All | **Local CAN, DPRAM-LoCAN Chip Cut-Off Conditions:**<br>The Engine Control Module (DME) communicates with all databus-capable control modules via a CAN databus. These databus-capable control modules are connected via two data bus wires which are twisted together (CAN_High and CAN_Low), and exchange information (messages). Missing information on the databus is recognized as a malfunction and stored. Trouble-free operation of the CAN-Bus requires that it have a terminal resistance. This central terminal resistor is located in the Engine Control Module (DME). The ignition is on for 800ms or the engine running for 360ms, the voltage is greater than 10 volts. This applies to vehicles with ARS only and does not affect bus activity.<br>**Possible Causes:**<br>• CAN data bus wires have short circuited to each other<br>• ECU has failed<br>• BUS system failure<br>• Defective bus controller (EGS) |

| DTC | Trouble Code Title, Conditions & Possible Causes |
|---|---|
| **DTC: P3206**<br>**MIL: Yes (U.S. only)**<br>**Years:** 2007, 2008<br>**Models:** 335i, 335Ci, 335xi, 525i, 525xi, 530i, 530xi, 550i, 650i, 750i, 750Li, 760Li, Z4 3.0i, Z4 3.0Si, Z4 M<br>**Engines:** All<br>**Transmissions:** All | **CAN Timeout ARS Conditions:**<br>The Engine Control Module (DME) communicates with all databus-capable control modules via a CAN databus. These databus-capable control modules are connected via two data bus wires which are twisted together (CAN_High and CAN_Low), and exchange information (messages). Missing information on the databus is recognized as a malfunction and stored. Trouble-free operation of the CAN-Bus requires that it have a terminal resistance. This central terminal resistor is located in the Engine Control Module (DME). The ignition is on for 800ms or the engine running for 360ms, the voltage is greater than 10 volts. This applies to vehicles with ARS only and does not affect bus activity.<br>**Possible Causes:**<br>• CAN data bus wires have short circuited to each other<br>• ECU has failed<br>• BUS system failure<br>• Defective bus controller (EGS) |
| **DTC: P3207**<br>**MIL: Yes (U.S. only)**<br>**Years:** 2007, 2008<br>**Models:** 335i, 335Ci, 335xi, 525i, 525xi, 530i, 530xi, 550i, 650i, 750i, 750Li, 760Li, Z4 3.0i, Z4 3.0Si, Z4 M<br>**Engines:** All<br>**Transmissions:** All | **CAN Message Monitoring ARS No Signal Conditions:**<br>The Engine Control Module (DME) communicates with all databus-capable control modules via a CAN databus. These databus-capable control modules are connected via two data bus wires which are twisted together (CAN_High and CAN_Low), and exchange information (messages). Missing information on the databus is recognized as a malfunction and stored. Trouble-free operation of the CAN-Bus requires that it have a terminal resistance. This central terminal resistor is located in the Engine Control Module (DME). The ignition is on for 800ms or the engine running for 360ms, the voltage is greater than 10 volts. This applies to vehicles with ARS only and does not affect bus activity. After 40ms, no signal was received.<br>**Possible Causes:**<br>• CAN data bus wires have short circuited to each other<br>• ECU has failed<br>• BUS system failure<br>• Defective bus controller (EGS) |
| **DTC: P3208**<br>**MIL: Yes (U.S. only)**<br>**Years:** 2007, 2008<br>**Models:** 335i, 335Ci, 335xi, 525i, 525xi, 530i, 530xi, 550i, 650i, 750i, 750Li, 760Li, Z4 3.0i, Z4 3.0Si, Z4 M<br>**Engines:** All<br>**Transmissions:** All | **CAN Message Monitoring ARS Plausibility Conditions:**<br>The Engine Control Module (DME) communicates with all databus-capable control modules via a CAN databus. These databus-capable control modules are connected via two data bus wires which are twisted together (CAN_High and CAN_Low), and exchange information (messages). Missing information on the databus is recognized as a malfunction and stored. Trouble-free operation of the CAN-Bus requires that it have a terminal resistance. This central terminal resistor is located in the Engine Control Module (DME). The ignition is on for 800ms or the engine running for 360ms, the voltage is greater than 10 volts. This applies to vehicles with ARS only and does not affect bus activity.<br>**Possible Causes:**<br>• CAN data bus wires have short circuited to each other<br>• ECU has failed<br>• BUS system failure<br>• Defective bus controller (EGS) |
| **DTC: P3209**<br>**MIL: Yes (U.S. only)**<br>**Years:** 2007, 2008<br>**Models:** 335i, 335Ci, 335xi, 525i, 525xi, 530i, 530xi, 550i, 650i, 750i, 750Li, 760Li, Z4 3.0i, Z4 3.0Si, Z4 M<br>**Engines:** All<br>**Transmissions:** All | **CAN Message Monitoring ASC/DSC Alive Check Malfunction Conditions:**<br>The Engine Control Module (DME) communicates with all databus-capable control modules via a CAN databus. These databus-capable control modules are connected via two data bus wires which are twisted together (CAN_High and CAN_Low), and exchange information (messages). Missing information on the databus is recognized as a malfunction and stored. Trouble-free operation of the CAN-Bus requires that it have a terminal resistance. This central terminal resistor is located in the Engine Control Module (DME). The ignition is on for 800ms or the engine running for 20ms, the voltage is greater than 10 volts<br>**Possible Causes:**<br>• CAN data bus wires have short circuited to each other<br>• ECU has failed<br>• BUS system failure<br>• Defective bus controller (EGS) |
| **DTC: P3210**<br>**MIL: Yes (U.S. only)**<br>**Years:** 2007, 2008<br>**Models:** 335i, 335Ci, 335xi, 525i, 525xi, 530i, 530xi, 550i, 650i, 750i, 750Li, 760Li, Z4 3.0i, Z4 3.0Si, Z4 M<br>**Engines:** All<br>**Transmissions:** All | **CAN Message Monitoring ASC/DSC Plausibility Conditions:**<br>The Engine Control Module (DME) communicates with all databus-capable control modules via a CAN databus. These databus-capable control modules are connected via two data bus wires which are twisted together (CAN_High and CAN_Low), and exchange information (messages). Missing information on the databus is recognized as a malfunction and stored. Trouble-free operation of the CAN-Bus requires that it have a terminal resistance. This central terminal resistor is located in the Engine Control Module (DME). The ignition is on for 800ms or the engine running for 20ms, the voltage is greater than 10 volts<br>**Possible Causes:**<br>• CAN data bus wires have short circuited to each other<br>• ECU has failed<br>• BUS system failure<br>• Defective bus controller (EGS) |

| DTC | Trouble Code Title, Conditions & Possible Causes |
|---|---|
| **DTC: P3211**<br>**MIL: Yes (U.S. only)**<br>**Years:** 2007, 2008<br>**Models:** 335i, 335Ci, 335xi, 525i, 525xi, 530i, 530xi, 550i, 650i, 750i, 750Li, 760Li, Z4 3.0i, Z4 3.0Si, Z4 M<br>**Engines:** All<br>**Transmissions:** All | **CAN Message Monitoring CAS No Signal Conditions:**<br>The Engine Control Module (DME) communicates with all databus-capable control modules via a CAN databus. These databus-capable control modules are connected via two data bus wires which are twisted together (CAN_High and CAN_Low), and exchange information (messages). Missing information on the databus is recognized as a malfunction and stored. Trouble-free operation of the CAN-Bus requires that it have a terminal resistance. This central terminal resistor is located in the Engine Control Module (DME). The ignition is on for 800ms or the engine running for 400ms, the voltage is greater than 10 volts. This applies to vehicles with CAS only and does not affect bus activity. After 40ms, no signal was received.<br>**Possible Causes:**<br>• CAN data bus wires have short circuited to each other<br>• ECU has failed<br>• BUS system failure<br>• Defective bus controller (EGS) |
| **DTC: P3212**<br>**MIL: Yes (U.S. only)**<br>**Years:** 2007, 2008<br>**Models:** 335i, 335Ci, 335xi, 525i, 525xi, 530i, 530xi, 550i, 650i, 750i, 750Li, 760Li, Z4 3.0i, Z4 3.0Si, Z4 M<br>**Engines:** All<br>**Transmissions:** All | **CAN Message Monitoring CAS Plausibility Conditions:**<br>The Engine Control Module (DME) communicates with all databus-capable control modules via a CAN databus. These databus-capable control modules are connected via two data bus wires which are twisted together (CAN_High and CAN_Low), and exchange information (messages). Missing information on the databus is recognized as a malfunction and stored. Trouble-free operation of the CAN-Bus requires that it have a terminal resistance. This central terminal resistor is located in the Engine Control Module (DME). The ignition is on for 800ms or the engine running for 400ms, the voltage is greater than 10 volts. This applies to vehicles with CAS only and does not affect bus activity.<br>**Possible Causes:**<br>• CAN data bus wires have short circuited to each other<br>• ECU has failed<br>• BUS system failure<br>• Defective bus controller (EGS) |
| **DTC: P3213**<br>**MIL: Yes (U.S. only)**<br>**Years:** 2007, 2008<br>**Models:** 335i, 335Ci, 335xi, 525i, 525xi, 530i, 530xi, 550i, 650i, 750i, 750Li, 760Li, Z4 3.0i, Z4 3.0Si, Z4 M<br>**Engines:** All<br>**Transmissions:** All | **CAN Message Monitoring ETC Alive Check Malfunction Conditions:**<br>The Engine Control Module (DME) communicates with all databus-capable control modules via a CAN databus. These databus-capable control modules are connected via two data bus wires which are twisted together (CAN_High and CAN_Low), and exchange information (messages). Missing information on the databus is recognized as a malfunction and stored. Trouble-free operation of the CAN-Bus requires that it have a terminal resistance. This central terminal resistor is located in the Engine Control Module (DME). The ignition is on for 800ms or the engine running for 20ms, the voltage is greater than 10 volts<br>**Possible Causes:**<br>• CAN data bus wires have short circuited to each other<br>• ECU has failed<br>• BUS system failure<br>• Defective bus controller (EGS) |
| **DTC: P3214**<br>**MIL: Yes (U.S. only)**<br>**Years:** 2007, 2008<br>**Models:** 335i, 335Ci, 335xi, 525i, 525xi, 530i, 530xi, 550i, 650i, 750i, 750Li, 760Li, Z4 3.0i, Z4 3.0Si, Z4 M<br>**Engines:** All<br>**Transmissions:** All | **CAN Message Monitoring ETC Plausibility Conditions:**<br>The Engine Control Module (DME) communicates with all databus-capable control modules via a CAN databus. These databus-capable control modules are connected via two data bus wires which are twisted together (CAN_High and CAN_Low), and exchange information (messages). Missing information on the databus is recognized as a malfunction and stored. Trouble-free operation of the CAN-Bus requires that it have a terminal resistance. This central terminal resistor is located in the Engine Control Module (DME). The ignition is on for 800ms or the engine running for 20ms, the voltage is greater than 10 volts<br>**Possible Causes:**<br>• CAN data bus wires have short circuited to each other<br>• ECU has failed<br>• BUS system failure<br>• Defective bus controller (EGS) |
| **DTC: P3215**<br>**MIL: Yes (U.S. only)**<br>**Years:** 2007, 2008<br>**Models:** 335i, 335Ci, 335xi, 525i, 525xi, 530i, 530xi, 550i, 650i, 750i, 750Li, 760Li, Z4 3.0i, Z4 3.0Si, Z4 M<br>**Engines:** All<br>**Transmissions:** All | **CAN Message Monitoring IHKA (Automatic Heating and Air Conditioning) No Signal Conditions:**<br>The Engine Control Module (DME) communicates with all databus-capable control modules via a CAN databus. These databus-capable control modules are connected via two data bus wires which are twisted together (CAN_High and CAN_Low), and exchange information (messages). Missing information on the databus is recognized as a malfunction and stored. Trouble-free operation of the CAN-Bus requires that it have a terminal resistance. This central terminal resistor is located in the Engine Control Module (DME). The ignition is on for 800ms or the engine running for 360ms, the voltage is greater than 10 volts. This applies to vehicles with ARS only and does not affect bus activity. After 40ms, no signal was received.<br>**Possible Causes:**<br>• CAN data bus wires have short circuited to each other<br>• ECU has failed<br>• BUS system failure<br>• Defective bus controller (EGS) |

| DTC | Trouble Code Title, Conditions & Possible Causes |
|---|---|
| **DTC: P3216**<br>**MIL: Yes (U.S. only)**<br>**Years:** 2007, 2008<br>**Models:** 335i, 335Ci, 335xi, 525i, 525xi, 530i, 530xi, 550i, 650i, 750i, 750Li, 760Li, Z4 3.0i, Z4 3.0Si, Z4 M<br>**Engines:** All<br>**Transmissions:** All | **CAN Timeout Instrument Pack Conditions:**<br>The Engine Control Module (DME) communicates with all databus-capable control modules via a CAN databus. These databus-capable control modules are connected via two data bus wires which are twisted together (CAN_High and CAN_Low), and exchange information (messages). Missing information on the databus is recognized as a malfunction and stored. Trouble-free operation of the CAN-Bus requires that it have a terminal resistance. This central terminal resistor is located in the Engine Control Module (DME). The ignition is on for 800ms or the engine running for 15 seconds, the voltage is greater than 10 volts. This applies to vehicles with MIL only and does not affect bus activity.<br>**Possible Causes:**<br>  • CAN data bus wires have short circuited to each other<br>  • ECU has failed<br>  • BUS system failure<br>  • Defective bus controller (EGS) |
| **DTC: P3217**<br>**MIL: Yes (U.S. only)**<br>**Years:** 2007, 2008<br>**Models:** 335i, 335Ci, 335xi, 525i, 525xi, 530i, 530xi, 550i, 650i, 750i, 750Li, 760Li, Z4 3.0i, Z4 3.0Si, Z4 M<br>**Engines:** All<br>**Transmissions:** All | **CAN Message Monitoring Instrument Pack Plausibility Conditions:**<br>The Engine Control Module (DME) communicates with all databus-capable control modules via a CAN databus. These databus-capable control modules are connected via two data bus wires which are twisted together (CAN_High and CAN_Low), and exchange information (messages). Missing information on the databus is recognized as a malfunction and stored. Trouble-free operation of the CAN-Bus requires that it have a terminal resistance. This central terminal resistor is located in the Engine Control Module (DME). The ignition is on for 800ms or the engine running for 15 seconds, the voltage is greater than 10 volts. This applies to vehicles with MIL only and does not affect bus activity.<br>**Possible Causes:**<br>  • CAN data bus wires have short circuited to each other<br>  • ECU has failed<br>  • BUS system failure<br>  • Defective bus controller (EGS) |
| **DTC: P3219**<br>**MIL: Yes (U.S. only)**<br>**Years:** 2007, 2008<br>**Models:** 335i, 335Ci, 335xi, 525i, 525xi, 530i, 530xi, 550i, 650i, 750i, 750Li, 760Li, Z4 3.0i, Z4 3.0Si, Z4 M<br>**Engines:** All<br>**Transmissions:** All | **CAN Message Monitoring SZL (Switch Cluster Steering Column) Alive Check Malfunction Conditions:**<br>The Engine Control Module (DME) communicates with all databus-capable control modules via a CAN databus. These databus-capable control modules are connected via two data bus wires which are twisted together (CAN_High and CAN_Low), and exchange information (messages). Missing information on the databus is recognized as a malfunction and stored. Trouble-free operation of the CAN-Bus requires that it have a terminal resistance. This central terminal resistor is located in the Engine Control Module (DME). The ignition is on for 800ms or the engine running for 20ms, the voltage is greater than 10 volts<br>**Possible Causes:**<br>  • CAN data bus wires have short circuited to each other<br>  • ECU has failed<br>  • BUS system failure<br>  • Defective bus controller (EGS) |
| **DTC: P3220**<br>**MIL: Yes (U.S. only)**<br>**Years:** 2007, 2008<br>**Models:** 335i, 335Ci, 335xi, 525i, 525xi, 530i, 530xi, 550i, 650i, 750i, 750Li, 760Li, Z4 3.0i, Z4 3.0Si, Z4 M<br>**Engines:** All<br>**Transmissions:** All | **CAN Message Monitoring SZL (Switch Cluster Steering Column) No Signal Conditions:**<br>The Engine Control Module (DME) communicates with all databus-capable control modules via a CAN databus. These databus-capable control modules are connected via two data bus wires which are twisted together (CAN_High and CAN_Low), and exchange information (messages). Missing information on the databus is recognized as a malfunction and stored. Trouble-free operation of the CAN-Bus requires that it have a terminal resistance. This central terminal resistor is located in the Engine Control Module (DME). The ignition is on for 800ms or the engine running for 360ms, the voltage is greater than 10 volts. This applies to vehicles with SZL only and does not affect bus activity. After 40ms, no signal was received.<br>**Possible Causes:**<br>  • CAN data bus wires have short circuited to each other<br>  • ECU has failed<br>  • BUS system failure<br>  • Defective bus controller (EGS) |
| **DTC: P3221**<br>**MIL: Yes (U.S. only)**<br>**Years:** 2007, 2008<br>**Models:** 335i, 335Ci, 335xi, 525i, 525xi, 530i, 530xi, 550i, 650i, 750i, 750Li, 760Li, Z4 3.0i, Z4 3.0Si, Z4 M<br>**Engines:** All<br>**Transmissions:** All | **CAN Message Monitoring SZL (Switch Cluster Steering Column) Plausibility Conditions:**<br>The Engine Control Module (DME) communicates with all databus-capable control modules via a CAN databus. These databus-capable control modules are connected via two data bus wires which are twisted together (CAN_High and CAN_Low), and exchange information (messages). Missing information on the databus is recognized as a malfunction and stored. Trouble-free operation of the CAN-Bus requires that it have a terminal resistance. This central terminal resistor is located in the Engine Control Module (DME). The ignition is on for 800ms or the engine running for 15 seconds, the voltage is greater than 10 volts. This applies to vehicles with SZL only and does not affect bus activity.<br>**Possible Causes:**<br>  • CAN data bus wires have short circuited to each other<br>  • ECU has failed<br>  • BUS system failure<br>  • Defective bus controller (EGS) |

| DTC | Trouble Code Title, Conditions & Possible Causes |
|---|---|
| **DTC: P3223**<br>**Years:** 2007, 2008<br>**Models:** 335i, 335Ci, 335xi, 525i, 525xi, 530i, 530xi, 550i, 650i, 750i, 750Li, 760Li, Z4 3.0i, Z4 3.0Si, Z4 M<br>**Engines:** All<br>**Transmissions:** All | **Generator Mechanical Error Conditions:**<br>The engine is running for at least 25 seconds, and there is no communication faults at the BSD Interface. Mechanical problems exist within the generator. An alternator is not approved for E60, E65 and E66 models specifically.<br>**Possible Causes:**<br>  • Wires have short circuited to each other<br>  • ECU has failed<br>  • Generator failure or installed incorrectly |
| **DTC: P3225**<br>**Years:** 2007, 2008<br>**Models:** 335i, 335Ci, 335xi, 525i, 525xi, 530i, 530xi, 550i, 650i, 750i, 750Li, 760Li, Z4 3.0i, Z4 3.0Si, Z4 M<br>**Engines:** All<br>**Transmissions:** All | **Generator Communication Error Conditions:**<br>The engine is running for at least 25 seconds, and there is no communication faults at the BSD Interface<br>**Possible Causes:**<br>  • Wires have short circuited to each other<br>  • ECU has failed<br>  • Generator failure |
| **DTC: P3226**<br>**Years:** 2007, 2008<br>**Models:** 335i, 335Ci, 335xi, 525i, 525xi, 530i, 530xi, 550i, 650i, 750i, 750Li, 760Li, Z4 3.0i, Z4 3.0Si, Z4 M<br>**Engines:** All<br>**Transmissions:** All | **E-Box Control Circuit Fan High Conditions:**<br>The engine is running for at least 800ms and the voltage is between 9.5 and 17.8 volts, and there is a short to the battery voltage.<br>**Possible Causes:**<br>  • Wires have short circuited to each other<br>  • ECU has failed<br>  • Generator failure |
| **DTC: P3227**<br>**Years:** 2007, 2008<br>**Models:** 335i, 335Ci, 335xi, 525i, 525xi, 530i, 530xi, 550i, 650i, 750i, 750Li, 760Li, Z4 3.0i, Z4 3.0Si, Z4 M<br>**Engines:** All<br>**Transmissions:** All | **E-Box Control Circuit Fan Low Conditions:**<br>The engine is running for at least 800ms and the voltage is between 9.5 and 17.8 volts, and there is a short to the ground.<br>**Possible Causes:**<br>  • Wires have short circuited to each other<br>  • ECU has failed<br>  • Generator failure |
| **DTC: P3228**<br>**Years:** 2007, 2008<br>**Models:** 335i, 335Ci, 335xi, 525i, 525xi, 530i, 530xi, 550i, 650i, 750i, 750Li, 760Li, Z4 3.0i, Z4 3.0Si, Z4 M<br>**Engines:** All<br>**Transmissions:** All | **E-Box Control Circuit Open Circuit Conditions:**<br>The engine is running for at least 800ms and the voltage is between 9.5 and 17.8 volts, and there is a short.<br>**Possible Causes:**<br>  • Wires have short circuited to each other<br>  • ECU has failed<br>  • Generator failure |
| **DTC: P3231**<br>**Years:** 2007, 2008<br>**Models:** 335i, 335Ci, 335xi, 525i, 525xi, 530i, 530xi, 550i, 650i, 750i, 750Li, 760Li, Z4 3.0i, Z4 3.0Si, Z4 M<br>**Engines:** All<br>**Transmissions:** All | **Control Module Monitoring Error Response Plausibility Conditions:**<br>Key on, engine running to at least 1200rpm, the DME has detected an internal fault in the computer or internal fault in the control modules. The fault is deactivated and cannot be entered. If additional fault codes are entered in the DME, resolve these issues. If fault remains, replace the DME.<br>**Possible Causes:**<br>  • Battery terminal corrosion, loose battery connection, or faulty<br>  • Connection to the DME interrupted, or the circuit has been opened<br>  • Reprogramming error has occurred and needs replacement.<br>  • Voltage supply for Engine Control Module (DME) is faulty |

| DTC | Trouble Code Title, Conditions & Possible Causes |
|---|---|
| **DTC: P3232**<br>**Years:** 2007, 2008<br>**Models:** 335i, 335Ci, 335xi, 525i, 525xi, 530i, 530xi, 550i, 650i, 750i, 750Li, 760Li, Z4 3.0i, Z4 3.0Si, Z4 M<br>**Engines:** All<br>**Transmissions:** All | **Control Module Monitoring Ignition Timing Plausibility Conditions:**<br>Key on, engine running to at least 1200rpm, the DME has detected an internal fault in the computer or internal fault in the control modules. The Torque monitoring feature compares the torque demand (from accelerator pedal, FGR, electrical equipment, transmission) with the torque provided (calculated from HFM, injector valves, ignition angle, throttle valve angle, differential pressure, lambda). Deviations trigger fuel-supply safety shutdown to prevent vehicle from autonomous acceleration. Internal fault in computer or in electronic control modules (check whether all ADC channels have been converted). If additional fault codes are entered in the DME, resolve these issues. If fault remains, replace the DME.<br>**Possible Causes:**<br>• Battery terminal corrosion, loose battery connection, or faulty<br>• Connection to the DME interrupted, or the circuit has been opened<br>• Reprogramming error has occurred and needs replacement.<br>• Voltage supply for Engine Control Module (DME) is faulty |
| **DTC: P3233**<br>**Years:** 2007, 2008<br>**Models:** 335i, 335Ci, 335xi, 525i, 525xi, 530i, 530xi, 550i, 650i, 750i, 750Li, 760Li, Z4 3.0i, Z4 3.0Si, Z4 M<br>**Engines:** All<br>**Transmissions:** All | **Control Module Monitoring Relative Charge Plausibility Conditions:**<br>Key on, engine running to at least 1200rpm, the DME has detected an internal fault in the computer or internal fault in the control modules. The Torque monitoring feature compares the torque demand (from accelerator pedal, FGR, electrical equipment, transmission) with the torque provided (calculated from HFM, injector valves, ignition angle, throttle valve angle, differential pressure, lambda). Deviations trigger fuel-supply safety shutdown to prevent vehicle from autonomous acceleration. Internal fault in computer or in electronic control modules (check whether all ADC channels have been converted). If additional fault codes are entered in the DME, resolve these issues. If fault remains, replace the DME.<br>**Possible Causes:**<br>• Battery terminal corrosion, loose battery connection, or faulty<br>• Connection to the DME interrupted, or the circuit has been opened<br>• Reprogramming error has occurred and needs replacement.<br>• Voltage supply for Engine Control Module (DME) is faulty |
| **DTC: P3236**<br>**Years:** 2007, 2008<br>**Models:** 335i, 335Ci, 335xi, 525i, 525xi, 530i, 530xi, 550i, 650i, 750i, 750Li, 760Li, Z4 3.0i, Z4 3.0Si, Z4 M<br>**Engines:** All<br>**Transmissions:** All | **Control Module Monitoring Injection Time Relative Fuel Quantity Plausibility Conditions:**<br>Key on, engine running to at least 1200rpm for 4.1 seconds, the DME has detected an internal fault in the computer or internal fault in the control modules. The Torque monitoring feature compares the torque demand (from accelerator pedal, FGR, electrical equipment, transmission) with the torque provided (calculated from HFM, injector valves, ignition angle, throttle valve angle, differential pressure, lambda). Deviations trigger fuel-supply safety shutdown to prevent vehicle from autonomous acceleration. Internal fault in computer or in electronic control modules (check whether all ADC channels have been converted). If additional fault codes are entered in the DME, resolve these issues. If fault remains, replace the DME.<br>**Possible Causes:**<br>• Battery terminal corrosion, loose battery connection, or faulty<br>• Connection to the DME interrupted, or the circuit has been opened<br>• Reprogramming error has occurred and needs replacement.<br>• Voltage supply for Engine Control Module (DME) is faulty |
| **DTC: P3237**<br>**Years:** 2007, 2008<br>**Models:** 335i, 335Ci, 335xi, 525i, 525xi, 530i, 530xi, 550i, 650i, 750i, 750Li, 760Li, Z4 3.0i, Z4 3.0Si, Z4 M<br>**Engines:** All<br>**Transmissions:** All | **Control Module Monitoring Fuel Correction Error Conditions:**<br>Key on, engine running to at least 1200rpm, the DME has detected an internal fault in the computer or internal fault in the control modules. The Torque monitoring feature compares the torque demand (from accelerator pedal, FGR, electrical equipment, transmission) with the torque provided (calculated from HFM, injector valves, ignition angle, throttle valve angle, differential pressure, lambda). Deviations trigger fuel-supply safety shutdown to prevent vehicle from autonomous acceleration. Internal fault in computer or in electronic control modules (check whether all ADC channels have been converted). If additional fault codes are entered in the DME, resolve these issues. If fault remains, replace the DME.<br>**Possible Causes:**<br>• Battery terminal corrosion, loose battery connection, or faulty<br>• Connection to the DME interrupted, or the circuit has been opened<br>• Reprogramming error has occurred and needs replacement.<br>• Voltage supply for Engine Control Module (DME) is faulty |

| DTC | Trouble Code Title, Conditions & Possible Causes |
|---|---|
| **DTC: P3238**<br>**Years:** 2007, 2008<br>**Models:** 335i, 335Ci, 335xi, 525i, 525xi, 530i, 530xi, 550i, 650i, 750i, 750Li, 760Li, Z4 3.0i, Z4 3.0Si, Z4 M<br>**Engines:** All<br>**Transmissions:** All | **Control Module Monitoring TPU Chip Defective Conditions:**<br>Key on, engine running to at least 1200rpm, the DME has detected a programming error. The Torque monitoring feature compares the torque demand (from accelerator pedal, FGR, electrical equipment, transmission) with the torque provided (calculated from HFM, injector valves, ignition angle, throttle valve angle, differential pressure, lambda). Deviations trigger fuel-supply safety shutdown to prevent vehicle from autonomous acceleration. Internal fault in computer or in electronic control modules (check whether all ADC channels have been converted). If additional fault codes are entered in the DME, resolve these issues. If fault remains, replace the DME.<br>**Possible Causes:**<br>• Battery terminal corrosion, loose battery connection, or faulty<br>• Connection to the DME interrupted, or the circuit has been opened<br>• Reprogramming error has occurred and needs replacement.<br>• Voltage supply for Engine Control Module (DME) is faulty |
| **DTC: P3247**<br>**Years:** 2007, 2008<br>**Models:** 335i, 335Ci, 335xi, 525i, 525xi, 530i, 530xi, 550i, 650i, 750i, 750Li, 760Li, Z4 3.0i, Z4 3.0Si, Z4 M<br>**Engines:** All<br>**Transmissions:** All | **Internal Control Module NVRAM Backup Error Conditions:**<br>Key on, the DME has detected a programming error in that the counters in NVRAM and the RAM backup not identical. The RAM test is initiated each time the DME starts.<br>**Possible Causes:**<br>• Battery terminal corrosion, loose battery connection, or faulty<br>• Connection to the DME interrupted, or the circuit has been opened<br>• Reprogramming error has occurred and needs replacement.<br>• Voltage supply for Engine Control Module (DME) is faulty |
| **DTC: P3300**<br>**Years:** 2007, 2008<br>**Models:** 335i, 335Ci, 335xi, 525i, 525xi, 530i, 530xi, 550i, 650i, 750i, 750Li, 760Li, Z4 3.0i, Z4 3.0Si, Z4 M<br>**Engines:** All<br>**Transmissions:** All | **Ignition Coil Cylinder 1 High Input or None-Impedance Conditions:**<br>Engine started, battery voltage must be at least 11.5v, all electrical components must be off, parking brake must be engaged (to keep daytime driving lights off), automatic transmission selector must be in park and the ground between the engine and the chassis must be well connected. The DME received high pulses from the ignition module for the Ignition Coil 1 primary circuit.<br>**Possible Causes:**<br>• Engine speed (RPM) sensor has failed<br>• Camshaft Position (CMP) sensor has failed<br>• Power Supply Relay is shorted to an open circuit<br>• There is a malfunction in voltage supply<br>• Ignition coilpack is damaged or it has failed<br>• Cylinder 1 to 4 Fuel Injector(s) have failed |
| **DTC: P3301**<br>**Years:** 2007, 2008<br>**Models:** 335i, 335Ci, 335xi, 525i, 525xi, 530i, 530xi, 550i, 650i, 750i, 750Li, 760Li, Z4 3.0i, Z4 3.0Si, Z4 M<br>**Engines:** All<br>**Transmissions:** All | **Ignition Coil Cylinder 1 Contact Resistance or High-Impedance Conditions:**<br>Engine started, battery voltage must be at least 11.5v, all electrical components must be off, parking brake must be engaged (to keep daytime driving lights off), automatic transmission selector must be in park and the ground between the engine and the chassis must be well connected. The DME received high pulses from the ignition module for the Ignition Coil 1 primary circuit.<br>**Possible Causes:**<br>• Engine speed (RPM) sensor has failed<br>• Camshaft Position (CMP) sensor has failed<br>• Power Supply Relay is shorted to an open circuit<br>• There is a malfunction in voltage supply<br>• Ignition coilpack is damaged or it has failed<br>• Cylinder 1 to 4 Fuel Injector(s) have failed |
| **DTC: P3302**<br>**Years:** 2007, 2008<br>**Models:** 335i, 335Ci, 335xi, 525i, 525xi, 530i, 530xi, 550i, 650i, 750i, 750Li, 760Li, Z4 3.0i, Z4 3.0Si, Z4 M<br>**Engines:** All<br>**Transmissions:** All | **Ignition Coil Cylinder 1 Cut-Off Due to Over-temperature Condition or No Signal Conditions:**<br>Engine started, battery voltage must be at least 11.5v, all electrical components must be off, parking brake must be engaged (to keep daytime driving lights off), automatic transmission selector must be in park and the ground between the engine and the chassis must be well connected. The DME received no pulses from the ignition module for the Ignition Coil 1 primary circuit due to an elevated temperature reading.<br>**Possible Causes:**<br>• Engine speed (RPM) sensor has failed<br>• Camshaft Position (CMP) sensor has failed<br>• Power Supply Relay is shorted to an open circuit<br>• There is a malfunction in voltage supply<br>• Ignition coilpack is damaged or it has failed<br>• Cylinder 1 to 4 Fuel Injector(s) have failed |

| DTC | Trouble Code Title, Conditions & Possible Causes |
|---|---|
| **DTC: P3303**<br>**Years:** 2007, 2008<br>**Models:** 335i, 335Ci, 335xi, 525i, 525xi, 530i, 530xi, 550i, 650i, 750i, 750Li, 760Li, Z4 3.0i, Z4 3.0Si, Z4 M<br>**Engines:** All<br>**Transmissions:** All | **Ignition Coil Cylinder 5 High Input or None-Impedance Conditions:**<br>Engine started, battery voltage must be at least 11.5v, all electrical components must be off, parking brake must be engaged (to keep daytime driving lights off), automatic transmission selector must be in park and the ground between the engine and the chassis must be well connected. The DME received high pulses from the ignition module for the Ignition Coil 1 primary circuit.<br>**Possible Causes:**<br>&bull; Engine speed (RPM) sensor has failed<br>&bull; Camshaft Position (CMP) sensor has failed<br>&bull; Power Supply Relay is shorted to an open circuit<br>&bull; There is a malfunction in voltage supply<br>&bull; Ignition coilpack is damaged or it has failed<br>&bull; Cylinder 1 to 4 Fuel Injector(s) have failed |
| **DTC: P3304**<br>**Years:** 2007, 2008<br>**Models:** 335i, 335Ci, 335xi, 525i, 525xi, 530i, 530xi, 550i, 650i, 750i, 750Li, 760Li, Z4 3.0i, Z4 3.0Si, Z4 M<br>**Engines:** All<br>**Transmissions:** All | **Ignition Coil Cylinder 5 Contact Resistance or High-Impedance Conditions:**<br>Engine started, battery voltage must be at least 11.5v, all electrical components must be off, parking brake must be engaged (to keep daytime driving lights off), automatic transmission selector must be in park and the ground between the engine and the chassis must be well connected. The DME received high pulses from the ignition module for the Ignition Coil 1 primary circuit.<br>**Possible Causes:**<br>&bull; Engine speed (RPM) sensor has failed<br>&bull; Camshaft Position (CMP) sensor has failed<br>&bull; Power Supply Relay is shorted to an open circuit<br>&bull; There is a malfunction in voltage supply<br>&bull; Ignition coilpack is damaged or it has failed<br>&bull; Cylinder 1 to 4 Fuel Injector(s) have failed |
| **DTC: P3305**<br>**Years:** 2007, 2008<br>**Models:** 335i, 335Ci, 335xi, 525i, 525xi, 530i, 530xi, 550i, 650i, 750i, 750Li, 760Li, Z4 3.0i, Z4 3.0Si, Z4 M<br>**Engines:** All<br>**Transmissions:** All | **Ignition Coil Cylinder 5 Cut-Off Due to Over-temperature Condition or No Signal Conditions:**<br>Engine started, battery voltage must be at least 11.5v, all electrical components must be off, parking brake must be engaged (to keep daytime driving lights off), automatic transmission selector must be in park and the ground between the engine and the chassis must be well connected. The DME received no pulses from the ignition module for the Ignition Coil 1 primary circuit due to an elevated temperature reading.<br>**Possible Causes:**<br>&bull; Engine speed (RPM) sensor has failed<br>&bull; Camshaft Position (CMP) sensor has failed<br>&bull; Power Supply Relay is shorted to an open circuit<br>&bull; There is a malfunction in voltage supply<br>&bull; Ignition coilpack is damaged or it has failed<br>&bull; Cylinder 1 to 4 Fuel Injector(s) have failed |
| **DTC: P3306**<br>**Years:** 2007, 2008<br>**Models:** 335i, 335Ci, 335xi, 525i, 525xi, 530i, 530xi, 550i, 650i, 750i, 750Li, 760Li, Z4 3.0i, Z4 3.0Si, Z4 M<br>**Engines:** All<br>**Transmissions:** All | **Ignition Coil Cylinder 4 High Input or None-Impedance Conditions:**<br>Engine started, battery voltage must be at least 11.5v, all electrical components must be off, parking brake must be engaged (to keep daytime driving lights off), automatic transmission selector must be in park and the ground between the engine and the chassis must be well connected. The DME received high pulses from the ignition module for the Ignition Coil 1 primary circuit.<br>**Possible Causes:**<br>&bull; Engine speed (RPM) sensor has failed<br>&bull; Camshaft Position (CMP) sensor has failed<br>&bull; Power Supply Relay is shorted to an open circuit<br>&bull; There is a malfunction in voltage supply<br>&bull; Ignition coilpack is damaged or it has failed<br>&bull; Cylinder 1 to 4 Fuel Injector(s) have failed |
| **DTC: P3307**<br>**Years:** 2007, 2008<br>**Models:** 335i, 335Ci, 335xi, 525i, 525xi, 530i, 530xi, 550i, 650i, 750i, 750Li, 760Li, Z4 3.0i, Z4 3.0Si, Z4 M<br>**Engines:** All<br>**Transmissions:** All | **Ignition Coil Cylinder 4 Contact Resistance or High-Impedance Conditions:**<br>Engine started, battery voltage must be at least 11.5v, all electrical components must be off, parking brake must be engaged (to keep daytime driving lights off), automatic transmission selector must be in park and the ground between the engine and the chassis must be well connected. The DME received high pulses from the ignition module for the Ignition Coil 1 primary circuit.<br>**Possible Causes:**<br>&bull; Engine speed (RPM) sensor has failed<br>&bull; Camshaft Position (CMP) sensor has failed<br>&bull; Power Supply Relay is shorted to an open circuit<br>&bull; There is a malfunction in voltage supply<br>&bull; Ignition coilpack is damaged or it has failed<br>&bull; Cylinder 1 to 4 Fuel Injector(s) have failed |

| DTC | Trouble Code Title, Conditions & Possible Causes |
|---|---|
| **DTC: P3308**<br>**Years:** 2007, 2008<br>**Models:** 335i, 335Ci, 335xi, 525i, 525xi, 530i, 530xi, 550i, 650i, 750i, 750Li, 760Li, Z4 3.0i, Z4 3.0Si, Z4 M<br>**Engines:** All<br>**Transmissions:** All | **Ignition Coil Cylinder 4 Cut-Off Due to Over-temperature Condition or No Signal Conditions:**<br>Engine started, battery voltage must be at least 11.5v, all electrical components must be off, parking brake must be engaged (to keep daytime driving lights off), automatic transmission selector must be in park and the ground between the engine and the chassis must be well connected. The DME received no pulses from the ignition module for the Ignition Coil 1 primary circuit due to an elevated temperature reading.<br>**Possible Causes:**<br>• Engine speed (RPM) sensor has failed<br>• Camshaft Position (CMP) sensor has failed<br>• Power Supply Relay is shorted to an open circuit<br>• There is a malfunction in voltage supply<br>• Ignition coilpack is damaged or it has failed<br>• Cylinder 1 to 4 Fuel Injector(s) have failed |
| **DTC: P3309**<br>**Years:** 2007, 2008<br>**Models:** 335i, 335Ci, 335xi, 525i, 525xi, 530i, 530xi, 550i, 650i, 750i, 750Li, 760Li, Z4 3.0i, Z4 3.0Si, Z4 M<br>**Engines:** All<br>**Transmissions:** All | **Ignition Coil Cylinder 8 High Input or None-Impedance Conditions:**<br>Engine started, battery voltage must be at least 11.5v, all electrical components must be off, parking brake must be engaged (to keep daytime driving lights off), automatic transmission selector must be in park and the ground between the engine and the chassis must be well connected. The DME received high pulses from the ignition module for the Ignition Coil 1 primary circuit.<br>**Possible Causes:**<br>• Engine speed (RPM) sensor has failed<br>• Camshaft Position (CMP) sensor has failed<br>• Power Supply Relay is shorted to an open circuit<br>• There is a malfunction in voltage supply<br>• Ignition coilpack is damaged or it has failed<br>• Cylinder 1 to 4 Fuel Injector(s) have failed |
| **DTC: P3310**<br>**Years:** 2007, 2008<br>**Models:** 335i, 335Ci, 335xi, 525i, 525xi, 530i, 530xi, 550i, 650i, 750i, 750Li, 760Li, Z4 3.0i, Z4 3.0Si, Z4 M<br>**Engines:** All<br>**Transmissions:** All | **Ignition Coil Cylinder 8 Contact Resistance or High-Impedance Conditions:**<br>Engine started, battery voltage must be at least 11.5v, all electrical components must be off, parking brake must be engaged (to keep daytime driving lights off), automatic transmission selector must be in park and the ground between the engine and the chassis must be well connected. The DME received high pulses from the ignition module for the Ignition Coil 1 primary circuit.<br>**Possible Causes:**<br>• Engine speed (RPM) sensor has failed<br>• Camshaft Position (CMP) sensor has failed<br>• Power Supply Relay is shorted to an open circuit<br>• There is a malfunction in voltage supply<br>• Ignition coilpack is damaged or it has failed<br>• Cylinder 1 to 4 Fuel Injector(s) have failed |
| **DTC: P3311**<br>**Years:** 2007, 2008<br>**Models:** 335i, 335Ci, 335xi, 525i, 525xi, 530i, 530xi, 550i, 650i, 750i, 750Li, 760Li, Z4 3.0i, Z4 3.0Si, Z4 M<br>**Engines:** All<br>**Transmissions:** All | **Ignition Coil Cylinder 8 Cut-Off Due to Over-temperature Condition or No Signal Conditions:**<br>Engine started, battery voltage must be at least 11.5v, all electrical components must be off, parking brake must be engaged (to keep daytime driving lights off), automatic transmission selector must be in park and the ground between the engine and the chassis must be well connected. The DME received no pulses from the ignition module for the Ignition Coil 1 primary circuit due to an elevated temperature reading.<br>**Possible Causes:**<br>• Engine speed (RPM) sensor has failed<br>• Camshaft Position (CMP) sensor has failed<br>• Power Supply Relay is shorted to an open circuit<br>• There is a malfunction in voltage supply<br>• Ignition coilpack is damaged or it has failed<br>• Cylinder 1 to 4 Fuel Injector(s) have failed |
| **DTC: P3312**<br>**Years:** 2007, 2008<br>**Models:** 335i, 335Ci, 335xi, 525i, 525xi, 530i, 530xi, 550i, 650i, 750i, 750Li, 760Li, Z4 3.0i, Z4 3.0Si, Z4 M<br>**Engines:** All<br>**Transmissions:** All | **Ignition Coil Cylinder 6 High Input or None-Impedance Conditions:**<br>Engine started, battery voltage must be at least 11.5v, all electrical components must be off, parking brake must be engaged (to keep daytime driving lights off), automatic transmission selector must be in park and the ground between the engine and the chassis must be well connected. The DME received high pulses from the ignition module for the Ignition Coil 1 primary circuit.<br>**Possible Causes:**<br>• Engine speed (RPM) sensor has failed<br>• Camshaft Position (CMP) sensor has failed<br>• Power Supply Relay is shorted to an open circuit<br>• There is a malfunction in voltage supply<br>• Ignition coilpack is damaged or it has failed<br>• Cylinder 1 to 4 Fuel Injector(s) have failed |

| DTC | Trouble Code Title, Conditions & Possible Causes |
|---|---|
| **DTC: P3313**<br>**Years:** 2007, 2008<br>**Models:** 335i, 335Ci, 335xi, 525i, 525xi, 530i, 530xi, 550i, 650i, 750i, 750Li, 760Li, Z4 3.0i, Z4 3.0Si, Z4 M<br>**Engines:** All<br>**Transmissions:** All | **Ignition Coil Cylinder 6 Contact Resistance or High-Impedance Conditions:**<br>Engine started, battery voltage must be at least 11.5v, all electrical components must be off, parking brake must be engaged (to keep daytime driving lights off), automatic transmission selector must be in park and the ground between the engine and the chassis must be well connected. The DME received high pulses from the ignition module for the Ignition Coil 1 primary circuit.<br>**Possible Causes:**<br>• Engine speed (RPM) sensor has failed<br>• Camshaft Position (CMP) sensor has failed<br>• Power Supply Relay is shorted to an open circuit<br>• There is a malfunction in voltage supply<br>• Ignition coilpack is damaged or it has failed<br>• Cylinder 1 to 4 Fuel Injector(s) have failed |
| **DTC: P3314**<br>**Years:** 2007, 2008<br>**Models:** 335i, 335Ci, 335xi, 525i, 525xi, 530i, 530xi, 550i, 650i, 750i, 750Li, 760Li, Z4 3.0i, Z4 3.0Si, Z4 M<br>**Engines:** All<br>**Transmissions:** All | **Ignition Coil Cylinder 6 Cut-Off Due to Over-temperature Condition or No Signal Conditions:**<br>Engine started, battery voltage must be at least 11.5v, all electrical components must be off, parking brake must be engaged (to keep daytime driving lights off), automatic transmission selector must be in park and the ground between the engine and the chassis must be well connected. The DME received no pulses from the ignition module for the Ignition Coil 1 primary circuit due to an elevated temperature reading.<br>**Possible Causes:**<br>• Engine speed (RPM) sensor has failed<br>• Camshaft Position (CMP) sensor has failed<br>• Power Supply Relay is shorted to an open circuit<br>• There is a malfunction in voltage supply<br>• Ignition coilpack is damaged or it has failed<br>• Cylinder 1 to 4 Fuel Injector(s) have failed |
| **DTC: P3315**<br>**Years:** 2007, 2008<br>**Models:** 335i, 335Ci, 335xi, 525i, 525xi, 530i, 530xi, 550i, 650i, 750i, 750Li, 760Li, Z4 3.0i, Z4 3.0Si, Z4 M<br>**Engines:** All<br>**Transmissions:** All | **Ignition Coil Cylinder 31 High Input or None-Impedance Conditions:**<br>Engine started, battery voltage must be at least 11.5v, all electrical components must be off, parking brake must be engaged (to keep daytime driving lights off), automatic transmission selector must be in park and the ground between the engine and the chassis must be well connected. The DME received high pulses from the ignition module for the Ignition Coil 3 primary circuit.<br>**Possible Causes:**<br>• Engine speed (RPM) sensor has failed<br>• Camshaft Position (CMP) sensor has failed<br>• Power Supply Relay is shorted to an open circuit<br>• There is a malfunction in voltage supply<br>• Ignition coilpack is damaged or it has failed<br>• Cylinder 1 to 4 Fuel Injector(s) have failed |
| **DTC: P3316**<br>**Years:** 2007, 2008<br>**Models:** 335i, 335Ci, 335xi, 525i, 525xi, 530i, 530xi, 550i, 650i, 750i, 750Li, 760Li, Z4 3.0i, Z4 3.0Si, Z4 M<br>**Engines:** All<br>**Transmissions:** All | **Ignition Coil Cylinder 3 Contact Resistance or High-Impedance Conditions:**<br>Engine started, battery voltage must be at least 11.5v, all electrical components must be off, parking brake must be engaged (to keep daytime driving lights off), automatic transmission selector must be in park and the ground between the engine and the chassis must be well connected. The DME received high pulses from the ignition module for the Ignition Coil 3 primary circuit.<br>**Possible Causes:**<br>• Engine speed (RPM) sensor has failed<br>• Camshaft Position (CMP) sensor has failed<br>• Power Supply Relay is shorted to an open circuit<br>• There is a malfunction in voltage supply<br>• Ignition coilpack is damaged or it has failed<br>• Cylinder 1 to 4 Fuel Injector(s) have failed |
| **DTC: P3317**<br>**Years:** 2007, 2008<br>**Models:** 335i, 335Ci, 335xi, 525i, 525xi, 530i, 530xi, 550i, 650i, 750i, 750Li, 760Li, Z4 3.0i, Z4 3.0Si, Z4 M<br>**Engines:** All<br>**Transmissions:** All | **Ignition Coil Cylinder 3 Cut-Off Due to Over-temperature Condition or No Signal Conditions:**<br>Engine started, battery voltage must be at least 11.5v, all electrical components must be off, parking brake must be engaged (to keep daytime driving lights off), automatic transmission selector must be in park and the ground between the engine and the chassis must be well connected. The DME received no pulses from the ignition module for the Ignition Coil 3 primary circuit due to an elevated temperature reading.<br>**Possible Causes:**<br>• Engine speed (RPM) sensor has failed<br>• Camshaft Position (CMP) sensor has failed<br>• Power Supply Relay is shorted to an open circuit<br>• There is a malfunction in voltage supply<br>• Ignition coilpack is damaged or it has failed<br>• Cylinder 1 to 4 Fuel Injector(s) have failed |

| DTC | Trouble Code Title, Conditions & Possible Causes |
|---|---|
| **DTC: P3318**<br>**Years:** 2007, 2008<br>**Models:** 335i, 335Ci, 335xi, 525i, 525xi, 530i, 530xi, 550i, 650i, 750i, 750Li, 760Li, Z4 3.0i, Z4 3.0Si, Z4 M<br>**Engines:** All<br>**Transmissions:** All | **Ignition Coil Cylinder 7 High Input or None-Impedance Conditions:**<br>Engine started, battery voltage must be at least 11.5v, all electrical components must be off, parking brake must be engaged (to keep daytime driving lights off), automatic transmission selector must be in park and the ground between the engine and the chassis must be well connected. The DME received high pulses from the ignition module for the Ignition Coil 1 primary circuit.<br>**Possible Causes:**<br>• Engine speed (RPM) sensor has failed<br>• Camshaft Position (CMP) sensor has failed<br>• Power Supply Relay is shorted to an open circuit<br>• There is a malfunction in voltage supply<br>• Ignition coilpack is damaged or it has failed<br>• Cylinder 1 to 4 Fuel Injector(s) have failed |
| **DTC: P3319**<br>**Years:** 2007, 2008<br>**Models:** 335i, 335Ci, 335xi, 525i, 525xi, 530i, 530xi, 550i, 650i, 750i, 750Li, 760Li, Z4 3.0i, Z4 3.0Si, Z4 M<br>**Engines:** All<br>**Transmissions:** All | **Ignition Coil Cylinder 7 Contact Resistance or High-Impedance Conditions:**<br>Engine started, battery voltage must be at least 11.5v, all electrical components must be off, parking brake must be engaged (to keep daytime driving lights off), automatic transmission selector must be in park and the ground between the engine and the chassis must be well connected. The DME received high pulses from the ignition module for the Ignition Coil 1 primary circuit.<br>**Possible Causes:**<br>• Engine speed (RPM) sensor has failed<br>• Camshaft Position (CMP) sensor has failed<br>• Power Supply Relay is shorted to an open circuit<br>• There is a malfunction in voltage supply<br>• Ignition coilpack is damaged or it has failed<br>• Cylinder 1 to 4 Fuel Injector(s) have failed |
| **DTC: P3320**<br>**Years:** 2007, 2008<br>**Models:** 335i, 335Ci, 335xi, 525i, 525xi, 530i, 530xi, 550i, 650i, 750i, 750Li, 760Li, Z4 3.0i, Z4 3.0Si, Z4 M<br>**Engines:** All<br>**Transmissions:** All | **Ignition Coil Cylinder 7 Cut-Off Due to Over-temperature Condition or No Signal Conditions:**<br>Engine started, battery voltage must be at least 11.5v, all electrical components must be off, parking brake must be engaged (to keep daytime driving lights off), automatic transmission selector must be in park and the ground between the engine and the chassis must be well connected. The DME received no pulses from the ignition module for the Ignition Coil 1 primary circuit due to an elevated temperature reading.<br>**Possible Causes:**<br>• Engine speed (RPM) sensor has failed<br>• Camshaft Position (CMP) sensor has failed<br>• Power Supply Relay is shorted to an open circuit<br>• There is a malfunction in voltage supply<br>• Ignition coilpack is damaged or it has failed<br>• Cylinder 1 to 4 Fuel Injector(s) have failed |

## Gas Engine OBD II Trouble Code List (U1xxx Codes)

| DTC | Trouble Code Title, Conditions & Possible Causes |
|---|---|
| **DTC: U1115**<br>**Years:** 2007, 2008<br>**Models:** 335i, 335Ci, 335xi, 525i, 525xi, 530i, 530xi, 550i, 650i, 750i, 750Li, 760Li, Z4 3.0i, Z4 3.0Si, Z4 M<br>**Engines:** All<br>**Transmissions:** All | **Lost Communication With Vehicle Mode Status Conditions:**<br>Key on for 800ms, and the DME detected that it has lost communication with the Vehicle Mode Status during its initial startup for 40ms (2 message cycles). The Engine Control Module (DME) communicates with all databus-capable control modules via a CAN databus. These databus-capable control modules are connected via two data bus wires which are twisted together (CAN_High and CAN_Low), and exchange information (messages). Missing information on the databus is recognized as a malfunction and stored. Trouble-free operation of the CAN-Bus requires that it have a terminal resistance.<br>**Possible Causes:**<br>• DME has failed<br>• Terminal resistance for CAN-bus are faulty<br>• Can data bus wires have short circuited to each other |
| **DTC: U1116**<br>**Years:** 2007, 2008<br>**Models:** 335i, 335Ci, 335xi, 525i, 525xi, 530i, 530xi, 550i, 650i, 750i, 750Li, 760Li, Z4 3.0i, Z4 3.0Si, Z4 M<br>**Engines:** All<br>**Transmissions:** All | **Lost Communication With Vehicle Mode Status Check Sum Error Conditions:**<br>Key on for 800ms, and the DME detected that it has lost communication with the Vehicle Mode Status during its initial startup for 40ms (2 message cycles). The Engine Control Module (DME) communicates with all databus-capable control modules via a CAN databus. These databus-capable control modules are connected via two data bus wires which are twisted together (CAN_High and CAN_Low), and exchange information (messages). Missing information on the databus is recognized as a malfunction and stored. Trouble-free operation of the CAN-Bus requires that it have a terminal resistance.<br>**Possible Causes:**<br>• DME has failed<br>• Terminal resistance for CAN-bus are faulty<br>• Can data bus wires have short circuited to each other |

| DTC | Trouble Code Title, Conditions & Possible Causes |
|---|---|
| **DTC: U1120**<br>**Years:** 2007, 2008<br>**Models:** 335i, 335Ci, 335xi, 525i, 525xi, 530i, 530xi, 550i, 650i, 750i, 750Li, 760Li, Z4 3.0i, Z4 3.0Si, Z4 M<br>**Engines:** All<br>**Transmissions:** All | **Lost Communication With Steering Angle Sensor Module Conditions:**<br>Key on for 800ms, and the DME detected that it has lost communication with the Steering Angle Sensor during its initial startup for 40ms (2 message cycles). The Engine Control Module (DME) communicates with all databus-capable control modules via a CAN databus. These databus-capable control modules are connected via two data bus wires which are twisted together (CAN_High and CAN_Low), and exchange information (messages). Missing information on the databus is recognized as a malfunction and stored. Trouble-free operation of the CAN-Bus requires that it have a terminal resistance.<br>**Possible Causes:**<br>• DME has failed<br>• Terminal resistance for CAN-bus are faulty<br>• Can data bus wires have short circuited to each other<br>• The steering angle sensor module is faulty |
| **DTC: U1121**<br>**Years:** 2007, 2008<br>**Models:** 335i, 335Ci, 335xi, 525i, 525xi, 530i, 530xi, 550i, 650i, 750i, 750Li, 760Li, Z4 3.0i, Z4 3.0Si, Z4 M<br>**Engines:** All<br>**Transmissions:** All | **Lost Communication With Power Management Battery Voltage Conditions:**<br>Key on for 800ms, and the DME detected that it has lost communication with the power management battery voltage during its initial startup for 40ms (2 message cycles). The Engine Control Module (DME) communicates with all databus-capable control modules via a CAN databus. These databus-capable control modules are connected via two data bus wires which are twisted together (CAN_High and CAN_Low), and exchange information (messages). Missing information on the databus is recognized as a malfunction and stored. Trouble-free operation of the CAN-Bus requires that it have a terminal resistance.<br>**Possible Causes:**<br>• DME has failed<br>• Terminal resistance for CAN-bus are faulty<br>• Can data bus wires have short circuited to each other |
| **DTC: U1121**<br>**Years:** 2007, 2008<br>**Models:** 335i, 335Ci, 335xi, 525i, 525xi, 530i, 530xi, 550i, 650i, 750i, 750Li, 760Li, Z4 3.0i, Z4 3.0Si, Z4 M<br>**Engines:** All<br>**Transmissions:** All | **Lost Communication With Power Management Charge Voltage Conditions:**<br>Key on for 800ms, and the DME detected that it has lost communication with the power management charge voltage during its initial startup for 40ms (2 message cycles). The Engine Control Module (DME) communicates with all databus-capable control modules via a CAN databus. These databus-capable control modules are connected via two data bus wires which are twisted together (CAN_High and CAN_Low), and exchange information (messages). Missing information on the databus is recognized as a malfunction and stored. Trouble-free operation of the CAN-Bus requires that it have a terminal resistance.<br>**Possible Causes:**<br>• DME has failed<br>• Terminal resistance for CAN-bus are faulty<br>• Can data bus wires have short circuited to each other |
| **DTC: U1129**<br>**Years:** 2007, 2008<br>**Models:** 335i, 335Ci, 335xi, 525i, 525xi, 530i, 530xi, 550i, 650i, 750i, 750Li, 760Li, Z4 3.0i, Z4 3.0Si, Z4 M<br>**Engines:** All<br>**Transmissions:** All | **Lost Communication With Reverse Status Conditions:**<br>Key on for 800ms, and the DME detected that it has lost communication with the power management charge voltage during its initial startup for 40ms (2 message cycles). The Engine Control Module (DME) communicates with all databus-capable control modules via a CAN databus. These databus-capable control modules are connected via two data bus wires which are twisted together (CAN_High and CAN_Low), and exchange information (messages). Missing information on the databus is recognized as a malfunction and stored. Trouble-free operation of the CAN-Bus requires that it have a terminal resistance.<br>**Possible Causes:**<br>• DME has failed<br>• Terminal resistance for CAN-bus are faulty<br>• Can data bus wires have short circuited to each other |
| **DTC: U1134**<br>**Years:** 2007, 2008<br>**Models:** 335i, 335Ci, 335xi, 525i, 525xi, 530i, 530xi, 550i, 650i, 750i, 750Li, 760Li, Z4 3.0i, Z4 3.0Si, Z4 M<br>**Engines:** All<br>**Transmissions:** All | **Lost Communication With Lamp Status Conditions:**<br>Key on for 800ms, and the DME detected that it has lost communication with the lamp status during its initial startup for 40ms (2 message cycles). The Engine Control Module (DME) communicates with all databus-capable control modules via a CAN databus. These databus-capable control modules are connected via two data bus wires which are twisted together (CAN_High and CAN_Low), and exchange information (messages). Missing information on the databus is recognized as a malfunction and stored. Trouble-free operation of the CAN-Bus requires that it have a terminal resistance.<br>**Possible Causes:**<br>• DME has failed<br>• Terminal resistance for CAN-bus are faulty<br>• Can data bus wires have short circuited to each other |

| DTC | Trouble Code Title, Conditions & Possible Causes |
|---|---|
| **DTC: U1135**<br>**Years:** 2007, 2008<br>**Models:** 335i, 335Ci, 335xi, 525i, 525xi, 530i, 530xi, 550i, 650i, 750i, 750Li, 760Li, Z4 3.0i, Z4 3.0Si, Z4 M<br>**Engines:** All<br>**Transmissions:** All | **Lost Communication With Status Water Valve Conditions:**<br>Key on for 800ms, and the DME detected that it has lost communication with the status water valve during its initial startup for 40ms (2 message cycles). The Engine Control Module (DME) communicates with all databus-capable control modules via a CAN databus. These databus-capable control modules are connected via two data bus wires which are twisted together (CAN_High and CAN_Low), and exchange information (messages). Missing information on the databus is recognized as a malfunction and stored. Trouble-free operation of the CAN-Bus requires that it have a terminal resistance.<br>**Possible Causes:**<br>• DME has failed<br>• Terminal resistance for CAN-bus are faulty<br>• Can data bus wires have short circuited to each other |

# MERCEDES-BENZ

C • E • S Class

## SPECIFICATIONS AND MAINTENANCE CHARTS

### ENGINE AND VEHICLE IDENTIFICATION CHART

| Engine | | | | | | | Model Year | |
|---|---|---|---|---|---|---|---|---|
| Code ① | Liters (cc) | Cu. In. | Cyl. | Fuel Sys. | Type | Eng. Mfg. | Code ② | Year |
| M272.920 | 2.5 (2469) | 151 | V6 | ME9.7 MFI | DOHC | MB | 6 | 2006 |
| M272.940 | 3.0 (2996) | 183 | V6 | ME9.7 MFI | DOHC | MB | 7 | 2007 |
| M272.941 | 3.0 (2996) | 183 | V6 | ME9.7 MFI | DOHC | MB | 8 | 2008 |
| M272.947 | 3.0 (2996) | 183 | V6 | ME9.7 MFI | DOHC | MB | | |
| M272.948 | 3.0 (2996) | 183 | V6 | ME9.7 MFI | DOHC | MB | | |
| M272.960 | 3.5 (3498) | 213 | V6 | ME9.7 MFI | DOHC | MB | | |
| M272.961 | 3.5 (3498) | 213 | V6 | ME9.7 MFI | DOHC | MB | | |
| M272.964 | 3.5 (3498) | 213 | V6 | ME9.7 MFI | DOHC | MB | | |
| M272.970 | 3.5 (3498) | 213 | V6 | ME9.7 MFI | DOHC | MB | | |
| M272.972 | 3.5 (3498) | 213 | V6 | ME9.7 MFI | DOHC | MB | | |
| M112.972 | 3.7 (3724 ) | 223 | V6 | ME2.8 SFI | SOHC | MB | | |
| M113.941 | 4.3 (4266) | 260 | V8 | ME2.8 SFI | SOHC | MB | | |
| M113.948 | 4.3 (4266) | 260 | V8 | ME2.8 SFI | SOHC | MB | | |
| M113.960 | 5.0 (4966) | 303 | V8 | ME2.8 SFI | SOHC | MB | | |
| M113.967 | 5.0 (4966) | 303 | V8 | ME2.8 SFI | SOHC | MB | | |
| M113.969 | 5.0 (4966) | 303 | V8 | ME2.8 SFI | SOHC | MB | | |
| M273.960 | 5.5 (5461) | 333 | V8 | ME9.7 MFI | DOHC | MB | | |
| M273.961 | 5.5 (5461) | 333 | V8 | ME9.7 MFI | DOHC | MB | | |
| M273.968 | 5.5 (5461) | 333 | V8 | ME9.7 MFI | DOHC | MB | | |

MFI: Multiport Fuel Injection

SFI: Sequential Fuel Injection

SOHC: Singe overhead camshaft

DOHC: Double overhead camshafts

① Stamped on engine block

② 11th digit of the VIN

22205_MBCA_C0001

## GENERAL ENGINE SPECIFICATIONS

| Year | Engine Displacement Liters | Engine ID/VIN | Net Horsepower @ rpm | Net Torque@rpm (ft. lbs.) | Bore x Stroke (in.) | Compression Ratio | Oil Pressure @ rpm |
|------|------|------|------|------|------|------|------|
| 2006 | 2.5 | M272.920 | 201@6200 | 181@2700 | 3.46x2.69 | 11.2:1 | NA |
|      | 3.0 | M272.940 | 228@6000 | 221@2700 | 3.46x3.23 | 11.0:1 | NA |
|      | 3.0 | M272.941 | 228@6000 | 221@2700 | 3.46x3.23 | 11.0:1 | NA |
|      | 3.5 | M272.960 | 268@6000 | 258@2400 | 3.66x3.39 | 10.7:1 | NA |
|      | 3.5 | M272.964 | 268@6000 | 258@2400 | 3.54x3.31 | 10.7:1 | NA |
|      | 3.5 | M272.970 | 268@6000 | 258@2400 | 3.66x3.39 | 10.7:1 | NA |
|      | 3.5 | M272.972 | 268@6000 | 258@2400 | 3.54x3.31 | 10.7:1 | NA |
|      | 3.7 | M112.972 | 241@5800 | 258@3000 | 3.28x3.78 | 10.0:1 | 43.5@3000 |
|      | 4.3 | M113.941 | 275@5750 | 295@3000 | 3.54x3.31 | 10.0:1 | 43.5@3000 |
|      | 4.3 | M113.948 | 275@5750 | 295@3000 | 3.54x3.31 | 10.0:1 | 43.5@3000 |
|      | 5.0 | M113.960 | 302@5600 | 339@2700 | 3.82x3.31 | 10.0:1 | 43.5@3000 |
|      | 5.0 | M113.967 | 302@5600 | 339@2700 | 3.82x3.31 | 10.0:1 | 43.5@3000 |
|      | 5.0 | M113.969 | 302@5600 | 339@2700 | 3.82x3.31 | 10.0:1 | 43.5@3000 |
| 2007 | 2.5 | M272.920 | 201@6200 | 181@2700 | 3.46x2.69 | 11.2:1 | NA |
|      | 3.0 | M272.940 | 228@6000 | 221@2700 | 3.46x3.23 | 11.0:1 | NA |
|      | 3.0 | M272.941 | 228@6000 | 221@2700 | 3.46x3.23 | 11.0:1 | NA |
|      | 3.5 | M272.960 | 268@6000 | 258@2400 | 3.66x3.39 | 10.7:1 | NA |
|      | 3.5 | M272.964 | 268@6000 | 258@2400 | 3.54x3.31 | 10.7:1 | NA |
|      | 3.5 | M272.970 | 268@6000 | 258@2400 | 3.66x3.39 | 10.7:1 | NA |
|      | 3.5 | M272.972 | 268@6000 | 258@2400 | 3.54x3.31 | 10.7:1 | NA |
|      | 5.5 | M273.960 | 382@6000 | 391@2800 | 3.86x3.56 | 10.7:1 | NA |
|      | 5.5 | M273.961 | 382@6000 | 391@2800 | 3.86x3.56 | 10.7:1 | NA |
|      | 5.5 | M273.968 | 382@6000 | 391@2800 | 3.86x3.56 | 10.7:1 | NA |
| 2008 | 3.0 | M272.947 | 228@6000 | 221@2700 | 3.46x3.23 | 11.1:1 | NA |
|      | 3.0 | M272.948 | 228@6000 | 221@2700 | 3.46x3.23 | 11.1:1 | NA |
|      | 3.5 | M272.961 | 268@6000 | 258@2400 | 3.66x3.39 | 10.7:1 | NA |
|      | 3.5 | M272.964 | 268@6000 | 258@2400 | 3.54x3.31 | 10.7:1 | NA |
|      | 3.5 | M272.972 | 268@6000 | 258@2400 | 3.54x3.31 | 10.7:1 | NA |
|      | 5.5 | M273.960 | 382@6000 | 391@2800 | 3.86x3.56 | 10.7:1 | NA |
|      | 5.5 | M273.961 | 382@6000 | 391@2800 | 3.86x3.56 | 10.7:1 | NA |
|      | 5.5 | M273.968 | 382@6000 | 391@2800 | 3.86x3.56 | 10.7:1 | NA |

NA - Not Available

## GASOLINE ENGINE TUNE-UP SPECIFICATIONS

| Year | Engine Displacement Liters | Engine ID/VIN | Spark Plug Gap (in.) | Ignition Timing (deg.) | | Fuel Pump (psi) | Idle Speed (rpm) | | Valve Clearance | |
|---|---|---|---|---|---|---|---|---|---|---|
| | | | | MT | AT | | MT | AT | In. | Ex. |
| 2006 | 2.5 | M272.920 | ① | ② | ② | NA | ② | ② | HYD. | HYD. |
| | 3.0 | M272.940 | ① | - | ② | NA | - | ② | HYD. | HYD. |
| | 3.0 | M272.941 | ① | - | ② | NA | - | ② | HYD. | HYD. |
| | 3.5 | M272.960 | ① | ② | ② | NA | ② | ② | HYD. | HYD. |
| | 3.5 | M272.964 | ① | ② | ② | NA | ② | ② | HYD. | HYD. |
| | 3.5 | M272.970 | ① | - | ② | NA | - | ② | HYD. | HYD. |
| | 3.5 | M272.972 | ① | - | ② | NA | - | ② | HYD. | HYD. |
| | 3.7 | M112.972 | ① | - | ② | NA | - | ② | HYD. | HYD. |
| | 4.3 | M113.941 | ① | - | ② | NA | - | ② | HYD. | HYD. |
| | 4.3 | M113.948 | ① | - | ② | NA | - | ② | HYD. | HYD. |
| | 5.0 | M113.960 | ① | - | ② | NA | - | ② | HYD. | HYD. |
| | 5.0 | M113.967 | ① | - | ② | NA | - | ② | HYD. | HYD. |
| | 5.0 | M113.969 | ① | - | ② | NA | - | ② | HYD. | HYD. |
| 2007 | 2.5 | M272.920 | ① | ② | ② | NA | ② | ② | HYD. | HYD. |
| | 3.0 | M272.940 | ① | - | ② | NA | - | ② | HYD. | HYD. |
| | 3.0 | M272.941 | ① | - | ② | NA | - | ② | HYD. | HYD. |
| | 3.5 | M272.960 | ① | ② | ② | NA | ② | ② | HYD. | HYD. |
| | 3.5 | M272.964 | ① | ② | ② | NA | ② | ② | HYD. | HYD. |
| | 3.5 | M272.970 | ① | - | ② | NA | - | ② | HYD. | HYD. |
| | 3.5 | M272.972 | ① | - | ② | NA | - | ② | HYD. | HYD. |
| | 5.5 | M273.960 | ① | - | ② | NA | - | ② | HYD. | HYD. |
| | 5.5 | M273.961 | ① | - | ② | NA | - | ② | HYD. | HYD. |
| | 5.5 | M273.968 | ① | - | ② | NA | - | ② | HYD. | HYD. |
| 2008 | 3.0 | M272.947 | ① | ② | ② | NA | ② | ② | HYD. | HYD. |
| | 3.0 | M272.948 | ① | - | ② | NA | - | ② | HYD. | HYD. |
| | 3.5 | M272.961 | ① | - | ② | NA | - | ② | HYD. | HYD. |
| | 3.5 | M272.964 | ① | - | ② | NA | - | ② | HYD. | HYD. |
| | 3.5 | M272.972 | ① | - | ② | NA | - | ② | HYD. | HYD. |
| | 5.5 | M273.960 | ① | - | ② | NA | - | ② | HYD. | HYD. |
| | 5.5 | M273.961 | ① | - | ② | NA | - | ② | HYD. | HYD. |
| | 5.5 | M273.968 | ① | - | ② | NA | - | ② | HYD. | HYD. |

① Spark plugs using Iridium or other precious metals should not be gapped.

Other park plugs should be gapped at 0.031 in.

② Ignition timing and idle speed are controlled by the ME engine controller and not adjustable.

22205_MBCA_C0003

## CAPACITIES

| Year | Model | Engine Displacement Liters | Engine ID/VIN | Engine Oil with Filter (qts.) | Transmission (pts.) Man. | Transmission (pts.) Auto. | Drive Axle Front (pts.) | Drive Axle Rear (pts.) | Fuel Tank (gal.) | Cooling System (qts.) |
|------|-------|------|------|------|------|------|------|------|------|------|
| 2006 | C230 | 2.5 | 272.920 | 8.5 | 2.5 | 19.1 | - | ① | 16.4 | 7.5 |
| | C280 | 3.0 | 272.940 | 8.5 | - | 19.1 | - | ① | 16.4 | 7.5 |
| | C280 4MATIC | 3.0 | 272.941 | 8.5 | - | 19.1 | 0.97 | ① | 16.4 | 7.5 |
| | C350 | 3.5 | 272.960 | 8.5 | 2.5 | 19.1 | - | ① | 16.4 | 7.5 |
| | E350 | 3.5 | 272.964 | 8.5 | - | 19.1 | - | ① | 21.1 | 10.7 |
| | E350 4MATIC | 3.5 | 272.972 | 8.5 | - | 19.1 | 1.27 | ① | 21.1 | 10.7 |
| | E500 | 5.0 | 113.967 | 8.0 | - | 19.1 | - | ① | 21.1 | 11.9 |
| | E500 4MATIC | 5.0 | 113.969 | 8.0 | - | 19.1 | 1.27 | ① | 21.1 | 11.9 |
| | S350 | 3.7 | 112.972 | 8.5 | - | 19.1 | - | ① | 23.2 | 12.1 |
| | S430 | 4.3 | 113.941 | 8.5 | - | 19.1 | - | ① | 23.2 | 12.1 |
| | S430 4MATIC | 4.3 | 113.948 | 8.5 | - | 18-18.9 | 1.27 | ① | 23.2 | 12.1 |
| | S500 | 5.0 | 113.960 | 8.5 | - | 19.1 | - | ① | 23.2 | 12.1 |
| | S500 4MATIC | 5.0 | 113.960 | 8.5 | - | 19.1 | 1.27 | ① | 23.2 | 12.1 |
| 2007 | C230 | 2.5 | 272.920 | 8.5 | 2.5 | 19.1 | - | ① | 16.4 | 7.5 |
| | C280 | 3.0 | 272.940 | 8.5 | - | 19.1 | - | ① | 16.4 | 7.5 |
| | C280 4MATIC | 3.0 | 272.941 | 8.5 | - | 19.1 | 0.97 | ① | 16.4 | 7.5 |
| | C350 | 3.5 | 272.960 | 8.5 | 2.5 | 19.1 | - | ① | 16.4 | 7.5 |
| | E350 | 3.5 | 272.964 | 8.5 | - | 19.1 | - | ① | 21.1 | 10.7 |
| | E350 4MATIC | 3.5 | 272.972 | 8.5 | - | 19.1 | 1.27 | ① | 21.1 | 10.7 |
| | E550 | 5.5 | 273.960 | 8.0 | - | 18-18.9 | - | ① | 21.1 | 11.9 |
| | E550 4MATIC | 5.5 | 273.960 | 8.0 | - | 18-18.9 | 1.27 | ① | 21.1 | 11.9 |
| | S550 | 5.5 | 273.961 | 8.0 | - | 19.1 | - | ① | 21.1 | 11.9 |
| | S550 4MATIC | 5.5 | 273.968 | 8.0 | - | 19.1 | 1.27 | ① | 21.1 | 11.9 |
| 2008 | C300 | 3.0 | 272.947 | NA | 2.5 | 19.1 | - | ① | 17.4 | NA |
| | C300 4MATIC | 3.0 | 272.948 | NA | - | 19.1 | 0.97 | ① | 17.4 | NA |
| | C350 | 3.5 | 272.961 | NA | - | 19.1 | - | ① | 17.4 | NA |
| | E350 | 3.5 | 272.960 | 8.5 | - | 19.1 | - | ① | 21.1 | 10.7 |
| | E350 4MATIC | 3.5 | 272.960 | 8.5 | - | 19.1 | 1.27 | ① | 21.1 | 10.7 |
| | E550 | 5.5 | 272.960 | 8.0 | - | 19.1 | - | ① | 21.1 | 11.9 |
| | E550 4MATIC | 5.5 | 272.960 | 8.0 | - | 18-18.9 | 1.27 | ① | 21.1 | 11.9 |
| | S550 | 5.5 | 272.961 | 8.0 | - | 19.1 | - | ① | 21.1 | 11.9 |
| | S550 4MATIC | 5.5 | 272.968 | 8.0 | - | 19.1 | 1.27 | ① | 21.1 | 11.9 |

NOTE: All capacities are approximate. Add fluid gradually and check to be sure a proper fluid level is obtained.

NA: Not Available

① 240 mm Diameter Rear Differential: 4.0 pts.

215 mm Diameter Rear Differential: 2.5 pts.

210 mm Diameter Rear Differential: 3.4 pts.

198 mm Diameter Rear Differential: 2.3 pts.

187 mm Diameter Rear Differential: 2.1 pts.

22205_MBCA_C0004

## FLUID SPECIFICATIONS

| Year | Model | Engine Size Liters | Engine ID/VIN | Engine Oil | Man. Trans. | Auto. Trans. | Drive Axle Front | Drive Axle Rear | Transfer Case | Power Steering Fluid | Brake Master Cylinder | Cooling System |
|------|-------|------|---------|---|-------|----------|---|---|-------|--------|------|-----|
| 2006 | C230 | 2.5 | 272.920 | ① | MB317 | ATF 3353 |  | ② | - | CHF11S | DOT4 | LLC |
|  | C280 | 3.0 | 272.940 | ① | - | ATF 3353 |  | ② | - | CHF11S | DOT4 | LLC |
|  | C280 4MATIC | 3.0 | 272.941 | ① | - | ATF 3353 | ② | ② | 85W-90 | CHF11S | DOT4 | LLC |
|  | C350 | 3.5 | 272.960 | ① | MB317 | ATF 3353 |  | ② | - | CHF11S | DOT4 | LLC |
|  | E350 | 3.5 | 272.964 | ① | - | ATF 3353 |  | ② | - | CHF11S | DOT4 | LLC |
|  | E350 4MATIC | 3.5 | 272.972 | ① | - | ATF 3353 | ② | ② | 85W-90 | CHF11S | DOT4 | LLC |
|  | E500 | 5.0 | 113.967 | ① | - | ATF 3353 |  | ② | - | CHF11S | DOT4 | LLC |
|  | E500 4MATIC | 5.0 | 113.969 | ① | - | ATF 3353 | ② | ② | 85W-90 | CHF11S | DOT4 | LLC |
|  | S350 | 3.7 | 112.972 | ① | - | ATF 3353 |  | ② | - | CHF11S | DOT4 | LLC |
|  | S430 | 4.3 | 113.941 | ① | - | ATF 3353 |  | ② | - | CHF11S | DOT4 | LLC |
|  | S430 4MATIC | 4.3 | 113.948 | ① | - | ATF 3353 | ② | ② | 85W-90 | CHF11S | DOT4 | LLC |
|  | S500 | 5.0 | 113.960 | ① | - | ATF 3353 |  | ② | - | CHF11S | DOT4 | LLC |
|  | S500 4MATIC | 5.0 | 113.960 | ① | - | ATF 3353 | ② | ② | 85W-90 | CHF11S | DOT4 | LLC |
| 2007 | C230 | 2.5 | 272.920 | ① | MB317 | ATF 3353 |  | ② | - | CHF11S | DOT4 | LLC |
|  | C280 | 3.0 | 272.940 | ① | - | ATF 3353 |  | ② | - | CHF11S | DOT4 | LLC |
|  | C280 4MATIC | 3.0 | 272.941 | ① | - | ATF 3353 | ② | ② | 85W-90 | CHF11S | DOT4 | LLC |
|  | C350 | 3.5 | 272.960 | ① | MB317 | ATF 3353 |  | ② | - | CHF11S | DOT4 | LLC |
|  | E350 | 3.5 | 272.964 | ① | - | ATF 3353 |  | ② | - | CHF11S | DOT4 | LLC |
|  | E350 4MATIC | 3.5 | 272.972 | ① | - | ATF 3353 | ② | ② | 85W-90 | CHF11S | DOT4 | LLC |
|  | E550 | 5.5 | 273.960 | ① | - | ATF 3353 |  | ② | - | CHF11S | DOT4 | LLC |
|  | E550 4MATIC | 5.5 | 273.960 | ① | - | ATF 3353 | ② | ② | 85W-90 | CHF11S | DOT4 | LLC |
|  | S550 | 5.5 | 273.961 | ① | - | ATF 3353 |  | ② | - | CHF11S | DOT4 | LLC |
|  | S550 4MATIC | 5.5 | 273.968 | ① | - | ATF 3353 | ② | ② | 85W-90 | CHF11S | DOT4 | LLC |
| 2008 | C300 | 3.0 | 272.947 | ① | MB317 | ATF 3353 |  | ② | - | CHF11S | DOT4 | LLC |
|  | C300 4MATIC | 3.0 | 272.948 | ① | - | ATF 3353 | ② | ② | 85W-90 | CHF11S | DOT4 | LLC |
|  | C350 | 3.5 | 272.961 | ① | - | ATF 3353 |  | ② | 85W-90 | CHF11S | DOT4 | LLC |
|  | E350 | 3.5 | 272.960 | ① | - | ATF 3353 |  | ② | - | CHF11S | DOT4 | LLC |
|  | E350 4MATIC | 3.5 | 272.960 | ① | - | ATF 3353 | ② | ② | 85W-90 | CHF11S | DOT4 | LLC |
|  | E550 | 5.5 | 272.960 | ① | - | ATF 3353 |  | ② | - | CHF11S | DOT4 | LLC |
|  | E550 4MATIC | 5.5 | 272.960 | ① | - | ATF 3353 | ② | ② | 85W-90 | CHF11S | DOT4 | LLC |
|  | S550 | 5.5 | 272.960 | ① | - | ATF 3353 |  | ② | - | CHF11S | DOT4 | LLC |
|  | S550 4MATIC | 5.5 | 272.960 | ① | - | ATF 3353 | ② | ② | 85W-90 | CHF11S | DOT4 | LLC |

DOT: Department Of Transpotation

LLC: Long Life Coolant

① In cold climate: 5W-30 (Semi-Synthetic)

In moderate climate: 10w-30 (Semi-Synthetic)

In hot climate: 20w-40 (Semi-Synthetic)

② 75W-90 or 85W-90

22205_MBCA_C0013

## CRANKSHAFT AND CONNECTING ROD SPECIFICATIONS

All measurements are given in inches.

| Year | Engine Displacement Liters | Engine ID/VIN | Main Brg. Oil Clearance | Shaft End-play | Thrust on No. | Connecting Rod | | |
|---|---|---|---|---|---|---|---|---|
| | | | | | | Journal Diameter | Oil Clearance | Side Clearance |
| 2006 | 2.5 | M272.920 | 0.0008-0.0018 | 0.0039-0.0104 | 3 | NA | 0.0008-0.0018 | NA |
| | 3.0 | M272.940 M272.941 | 0.0008-0.0018 | 0.0039-0.0104 | 3 | NA | 0.0008-0.0018 | NA |
| | 3.5 | M272.960 M272964 | 0.0008-0.0018 | 0.0039-0.0104 | 3 | NA | 0.0008-0.0018 | NA |
| | 3.5 | M272.970 M272.972 | 0.0008-0.0018 | 0.0039-0.0104 | 3 | NA | 0.0008-0.0018 | NA |
| | 3.7 | M112.972 | 0.0010-0.0019 | 0.0039-0.0098 | 3 | NA | 0.0008-0.0026 | NA |
| | 4.3 | M113.941 M113.948 | 0.0010-0.0019 | 0.0039-0.0098 | 3 | NA | 0.0008-0.0026 | NA |
| | 5.0 | M113.960 | 0.0010-0.0019 | 0.0039-0.0098 | 3 | NA | 0.0008-0.0026 | NA |
| | 5.0 | M113.967 M113.969 | 0.0010-0.0019 | 0.0039-0.0098 | 3 | NA | 0.0008-0.0026 | NA |
| 2007 | 2.5 | M272.920 | 0.0008-0.0018 | 0.0039-0.0104 | 3 | NA | 0.0008-0.0018 | NA |
| | 3.0 | M272.940 M272.941 | 0.0008-0.0018 | 0.0039-0.0104 | 3 | NA | 0.0008-0.0018 | NA |
| | 3.5 | M272.960 M272964 | 0.0008-0.0018 | 0.0039-0.0104 | 3 | NA | 0.0008-0.0018 | NA |
| | 3.5 | M272.970 M272.972 | 0.0008-0.0018 | 0.0039-0.0104 | 3 | NA | 0.0008-0.0018 | NA |
| | 5.5 | M273.960 | 0.0008-0.0018 | 0.0039-0.0104 | 3 | NA | 0.0008-0.0018 | NA |
| | 5.5 | M273.961 | 0.0008-0.0018 | 0.0039-0.0104 | 3 | NA | 0.0008-0.0018 | NA |
| | 5.5 | M273.968 | 0.0008-0.0018 | 0.0039-0.0104 | 3 | NA | 0.0008-0.0018 | NA |
| 2008 | 3.0 | M272.947 | 0.0008-0.0018 | 0.0039-0.0104 | 3 | NA | 0.0008-0.0018 | NA |
| | 3.0 | M272.948 | 0.0008-0.0018 | 0.0039-0.0104 | 3 | NA | 0.0008-0.0018 | NA |
| | 3.5 | M272.961 | 0.0008-0.0018 | 0.0039-0.0104 | 3 | NA | 0.0008-0.0018 | NA |
| | 3.5 | M272.964 | 0.0008-0.0018 | 0.0039-0.0104 | 3 | NA | 0.0008-0.0018 | NA |
| | 3.5 | M272.972 | 0.0008-0.0018 | 0.0039-0.0104 | 3 | NA | 0.0008-0.0018 | NA |
| | 5.5 | M273.960 | 0.0008-0.0018 | 0.0039-0.0104 | 3 | NA | 0.0008-0.0018 | NA |
| | 5.5 | M273.961 | 0.0008-0.0018 | 0.0039-0.0104 | 3 | NA | 0.0008-0.0018 | NA |
| | 5.5 | M273.968 | 0.0008-0.0018 | 0.0039-0.0104 | 3 | NA | 0.0008-0.0018 | NA |

NA - Not Available

22205_MBCA_C0006

# PISTON AND RING SPECIFICATIONS

All measurements are given in inches.

| Year | Engine Size Liters | Engine ID/VIN | Piston Clearance | Ring Gap | | | Ring Side Clearance | | |
|------|------|------|------|------|------|------|------|------|------|
| | | | | Top Compression | Bottom Compression | Oil Control | Top Compression | Bottom Compression | Oil Control |
| 2006 | 2.5 | M272.920 | 0.0001-0.0016 | 0.0078-0.0136 | 0.0117-0.0195 | 0.0078-0.0351 | 0.0012-0.0031 | 0.0008-0.0023 | NA |
| | 3.0 | M272.940 M272.941 | 0.0001-0.0016 | 0.0078-0.0136 | 0.0117-0.0195 | 0.0078-0.0351 | 0.0012-0.0031 | 0.0008-0.0023 | NA |
| | 3.5 | M272.960 M272.964 | 0.0001-0.0020 | 0.0078-0.0136 | 0.0117-0.0195 | 0.0078-0.0351 | 0.0008-0.0027 | 0.0006-0.0019 | NA |
| | 3.5 | M272.970 M272.972 | 0.0001-0.0020 | 0.0078-0.0136 | 0.0117-0.0195 | 0.0078-0.0351 | 0.0008-0.0027 | 0.0006-0.0019 | NA |
| | 3.7 | M112.972 | 0.0017 | 0.0078-0.0137 | 0.0078-0.0157 | NA | 0.0004-0.0023 | 0.0003-0.0011 | NA |
| | 4.3 | M113.941 M113.948 | 0.0017 | 0.0078-0.0137 | 0.0078-0.0157 | NA | 0.0004-0.0023 | 0.0003-0.0011 | NA |
| | 5.0 | M113.960 M113.967 | 0.0017 | 0.0078-0.0137 | 0.0078-0.0157 | NA | 0.0004-0.0023 | 0.0003-0.0011 | NA |
| | 5.0 | M113.969 | 0.0017 | 0.0078-0.0137 | 0.0078-0.0157 | NA | 0.0004-0.0023 | 0.0003-0.0011 | NA |
| 2007 | 2.5 | M272.920 | 0.0001-0.0016 | 0.0078-0.0136 | 0.0117-0.0195 | 0.0078-0.0351 | 0.0012-0.0031 | 0.0008-0.0023 | NA |
| | 3.0 | M272.940 | 0.0001-0.0016 | 0.0078-0.0136 | 0.0117-0.0195 | 0.0078-0.0351 | 0.0012-0.0031 | 0.0008-0.0023 | NA |
| | 3.0 | M272.941 | 0.0001-0.0016 | 0.0078-0.0136 | 0.0117-0.0195 | 0.0078-0.0351 | 0.0012-0.0031 | 0.0008-0.0023 | NA |
| | 3.5 | M272.960 | 0.0001-0.0020 | 0.0078-0.0136 | 0.0117-0.0195 | 0.0078-0.0351 | 0.0008-0.0027 | 0.0006-0.0019 | NA |
| | 3.5 | M272.964 | 0.0001-0.0020 | 0.0078-0.0136 | 0.0117-0.0195 | 0.0078-0.0351 | 0.0008-0.0027 | 0.0006-0.0019 | NA |
| | 3.5 | M272.970 | 0.0001-0.0020 | 0.0078-0.0136 | 0.0117-0.0195 | 0.0078-0.0351 | 0.0008-0.0027 | 0.0006-0.0019 | NA |
| | 3.5 | M272.972 | 0.0001-0.0020 | 0.0078-0.0136 | 0.0117-0.0195 | 0.0078-0.0351 | 0.0008-0.0027 | 0.0006-0.0019 | NA |
| | 5.5 | M273.960 | 0.0002-0.0016 | 0.0078-0.0136 | 0.0117-0.0195 | 0.0078-0.0351 | 0.0008-0.0027 | 0.0000-0.0023 | NA |
| | 5.5 | M273.961 | 0.0002-0.0016 | 0.0078-0.0136 | 0.0117-0.0195 | 0.0078-0.0351 | 0.0008-0.0027 | 0.0000-0.0023 | NA |
| | 5.5 | M273.968 | 0.0002-0.0016 | 0.0078-0.0136 | 0.0117-0.0195 | 0.0078-0.0351 | 0.0008-0.0027 | 0.0000-0.0023 | NA |

22205_MBCA_C0007

## PISTON AND RING SPECIFICATIONS

All measurements are given in inches.

| Year | Engine Size Liters | Engine ID/VIN | Piston Clearance | Ring Gap | | | Ring Side Clearance | | |
|------|------|------|------|------|------|------|------|------|------|
| | | | | Top Compression | Bottom Compression | Oil Control | Top Compression | Bottom Compression | Oil Control |
| 2008 | 3.0 | M272.947 | 0.0001-0.0016 | 0.0078-0.0136 | 0.0117-0.0195 | 0.0078-0.0351 | 0.0012-0.0031 | 0.0008-0.0023 | NA |
| | 3.0 | M272.948 | 0.0001-0.0016 | 0.0078-0.0136 | 0.0117-0.0195 | 0.0078-0.0351 | 0.0012-0.0031 | 0.0008-0.0023 | NA |
| | 3.5 | M272.961 | 0.0001-0.0020 | 0.0078-0.0136 | 0.0117-0.0195 | 0.0078-0.0351 | 0.0008-0.0027 | 0.0006-0.0019 | NA |
| | 3.5 | M272.964 | 0.0001-0.0020 | 0.0078-0.0136 | 0.0117-0.0195 | 0.0078-0.0351 | 0.0008-0.0027 | 0.0006-0.0019 | NA |
| | 3.5 | M272.972 | 0.0001-0.0020 | 0.0078-0.0136 | 0.0117-0.0195 | 0.0078-0.0351 | 0.0008-0.0027 | 0.0006-0.0019 | NA |
| | 5.5 | M273.960 | 0.0002-0.0016 | 0.0078-0.0136 | 0.0117-0.0195 | 0.0078-0.0351 | 0.0008-0.0027 | 0.0000-0.0023 | NA |
| | 5.5 | M273.961 | 0.0002-0.0016 | 0.0078-0.0136 | 0.0117-0.0195 | 0.0078-0.0351 | 0.0008-0.0027 | 0.0000-0.0023 | NA |
| | 5.5 | M273.968 | 0.0002-0.0016 | 0.0078-0.0136 | 0.0117-0.0195 | 0.0078-0.0351 | 0.0008-0.0027 | 0.0000-0.0023 | NA |

NA: Not Available

22205_MBCA_C0009

# TORQUE SPECIFICATIONS
All readings in ft. lbs.

| Year | Engine Displacement Liters | Engine ID/VIN | Cylinder Head Bolts | Main Bearing Bolts | Rod Bearing Bolts | Crankshaft Damper Bolts | Flywheel Bolts | Manifold Intake | Manifold Exhaust | Spark Plugs | Lug Nut |
|------|------|------|------|------|------|------|------|------|------|------|------|
| 2006 | 2.5 | M272.920 | ① | ② | ③ | ④ | ⑤ | 7 | 12 | 17 | ⑥ |
|      | 3.0 | M272.940 | ① | ② | ③ | ④ | ⑤ | 7 | 12 | 17 | ⑥ |
|      | 3.0 | M272.941 | ① | ② | ③ | ④ | ⑤ | 7 | 12 | 17 | ⑥ |
|      | 3.5 | M272.960 | ① | ② | ③ | ④ | ⑤ | 7 | 12 | 17 | ⑥ |
|      | 3.5 | M272.964 | ① | ② | ③ | ④ | ⑤ | 7 | 12 | 17 | ⑥ |
|      | 3.5 | M272.970 | ① | ② | ③ | ④ | ⑤ | 7 | 12 | 17 | ⑥ |
|      | 3.5 | M272.972 | ① | ② | ③ | ④ | ⑤ | 7 | 12 | 17 | ⑥ |
|      | 3.7 | M112.972 | ③ | ⑧ | ⑨ | ⑦ | ⑩ | 15 | 12 | 21 | ⑥ |
|      | 4.3 | M113.941 | ③ | ⑧ | ⑨ | ⑦ | ⑩ | 15 | 12 | 21 | ⑥ |
|      | 4.3 | M113.948 | ③ | ⑧ | ⑨ | ⑦ | ⑩ | 15 | 12 | 21 | ⑥ |
|      | 5.0 | M113.960 | ③ | ⑧ | ⑨ | ⑦ | ⑩ | 15 | 12 | 21 | ⑥ |
|      | 5.0 | M113.967 | ③ | ⑧ | ⑨ | ⑦ | ⑩ | 15 | 12 | 21 | ⑥ |
|      | 5.0 | M113.969 | ③ | ⑧ | ⑨ | ⑦ | ⑩ | 15 | 12 | 21 | ⑥ |
| 2007 | 2.5 | M272.920 | ① | ② | ③ | ④ | ⑤ | 7 | 12 | 17 | ⑥ |
|      | 3.0 | M272.940 | ① | ② | ③ | ④ | ⑤ | 7 | 12 | 17 | ⑥ |
|      | 3.0 | M272.941 | ① | ② | ③ | ④ | ⑤ | 7 | 12 | 17 | ⑥ |
|      | 3.5 | M272.960 | ① | ② | ③ | ④ | ⑤ | 7 | 12 | 17 | ⑥ |
|      | 3.5 | M272.964 | ① | ② | ③ | ④ | ⑤ | 7 | 12 | 17 | ⑥ |
|      | 3.5 | M272.970 | ① | ② | ③ | ④ | ⑤ | 7 | 12 | 17 | ⑥ |
|      | 3.5 | M272.972 | ① | ② | ③ | ④ | ⑤ | 7 | 12 | 17 | ⑥ |
|      | 5.5 | M273.960 | ① | ② | ⑪ | ④ | ⑤ | 7 | 12 | 17 | ⑥ |
|      | 5.5 | M273.961 | ① | ② | ⑪ | ⑦ | ⑩ | 7 | 12 | 17 | ⑫ |
|      | 5.5 | M273.968 | ① | ② | ⑪ | ⑦ | ⑩ | 7 | 12 | 17 | ⑫ |
| 2008 | 3.0 | M272.947 | ① | ② | ③ | ④ | ⑤ | 7 | 12 | 17 | ⑥ |
|      | 3.0 | M272.948 | ① | ② | ③ | ④ | ⑤ | 7 | 12 | 17 | ⑥ |
|      | 3.5 | M272.961 | ① | ② | ③ | ④ | ⑤ | 7 | 12 | 17 | ⑥ |
|      | 3.5 | M272.964 | ① | ② | ③ | ④ | ⑤ | 7 | 12 | 17 | ⑥ |
|      | 3.5 | M272.972 | ① | ② | ③ | ④ | ⑤ | 7 | 12 | 17 | ⑥ |
|      | 5.5 | M273.960 | ① | ② | ⑪ | ⑦ | ⑩ | 7 | 12 | 17 | ⑫ |
|      | 5.5 | M273.961 | ① | ② | ⑪ | ⑦ | ⑩ | 7 | 12 | 17 | ⑫ |
|      | 5.5 | M273.968 | ① | ② | ⑪ | ⑦ | ⑩ | 7 | 12 | 17 | ⑫ |

① M8 Cylinder head to timing case
Step 1: 11 ft. lbs.
Step 2: plus 90 degrees
M11 Cylinder head to crankcase
Step 1: 15 ft. lbs.
Step 2: 30 ft. lbs.
Step 3: plus 90 degrees
Step 4: plus 90 degrees

② M8 Side crankshaft bearing cap
Step 1: 22 ft. lbs.
M8 Crankshaft bearing cap
Step 1: 15 ft. lbs.
Step 2: plus 90 degrees
M10 Cranksahft bearing cap
Step 1: 4 ft. lbs.
Step 2: 22 ft. lbs.
Step 3: plus 90 degrees

③ Step 1: 4 ft. lbs.
Step 2: 15 ft. lbs.
Step 3: plus 90 degrees

④ Step 1: 148 ft. lbs.
Step 2: 95 degrees

⑤ Step 1: 15 ft. lbs.
Step 2: 33 ft. lbs.
Step 3: plus 90 degrees

⑥ C-Class: 81 ft. lbs.
E-Class: 96 ft. lbs.
S-Class: 111 ft. lbs.

⑦ Step 1: 7 ft. lbs.
Step 2: 22 ft. lbs.
Step 3: plus 90 degrees
Step 4: plus 90 degrees

⑧ M8 vertical bolts
Step 1: 15 ft. lbs.
Step 2: plus 90 degrees
M8 side bolts: 22 ft. lbs.
M10 bolts
Step 1: 44 inch lbs.
Step 2: 22 ft. lbs.
Step 3: plus 90 degrees

⑨ Step 1: 44 inch lbs.
Step 2: 18 ft. lbs
Step 3. 90 degrees

⑩ Step 1: 33 ft. lbs.
Step 2: 90 degrees

⑪ Step 1: 4 ft. lbs.
Step 2: 19 ft. lbs.
Step 3: plus 90 degrees

## TIRE, WHEEL AND BALL JOINT SPECIFICATIONS

| Year | Model | OEM Tires | | Tire Pressures (psi) | | Wheel Size | Ball Joint Inspection |
|------|-------|-----------|----------|-------|------|------------|------------------------|
| | | Standard | Optional | Front | Rear | | |
| 2006 | C230 Sport | F: 225/45ZR17 | None | ① | ① | 7.5J | NS |
| | | R: 245/40ZR17 | | | | 8.5J | |
| | C280 | 205/55R16 | None | ① | ① | 7-J | NS |
| | C350 | 205/55R16 | None | ① | ① | 7-J | NS |
| | C350 Sport | F: 225/45ZR17 | None | ① | ① | 7.5J | NS |
| | | R: 245/40ZR17 | | | | 8.5J | |
| | E350 | 245/45R17 | 245/40ZR18 | ① | ① | 8.0J | NS |
| | E500 | 245/45R17 | 245/40ZR18 | ① | ① | 8.0J | NS |
| | S350 | 225/55R17 | 245/45ZR18 | ① | ① | 7.5-J | NS |
| | S430 | 225/55R17 | 245/45ZR18 | ① | ① | 7.5-J | NS |
| | S500 | 225/55R17 | 245/45ZR18 | ① | ① | 7.5-J | NS |
| 2007 | C230 Sport | F: 225/45ZR17 | None | ① | ① | 7.5J | NS |
| | | R: 245/40ZR17 | | | | 8.5J | |
| | C280 | 205/55R16 | None | ① | ① | 7-J | NS |
| | C350 | 205/55R16 | None | ① | ① | 7-J | NS |
| | C350 Sport | F: 225/45ZR17 | None | ① | ① | 7.5J | NS |
| | | R: 245/40ZR17 | | | | 8.5J | |
| | E350 | 245/45R17 | 245/40ZR18 | ① | ① | 8.0J | NS |
| | E550 | 245/45R17 | 245/40ZR18 | ① | ① | 8.0J | NS |
| | S550 | 245/45R17 | 245/40ZR18 | ① | ① | 8.0J | NS |
| 2008 | C300 | 245/45R17 | 245/40ZR18 | ① | ① | 8.0J | NS |
| | C350 | 245/45R17 | 245/40ZR18 | ① | ① | 8.0J | NS |
| | C350 Sport | F: 225/45ZR17 | None | ① | ① | 7.5J | NS |
| | | R: 245/40ZR17 | | | | 8.5J | |
| | E350 | 245/45R17 | 245/40ZR18 | ① | ① | 8.0J | NS |
| | E550 | 245/45R17 | 245/40ZR18 | ① | ① | 8.0J | NS |
| | S550 | 245/45R17 | 245/40ZR18 | ① | ① | 8.0J | NS |

Note: Always follow the specifications provided on the vehicle sticker.

OEM: Original Equipment Manufacturer

PSI: Pounds Per Square Inch

STD: Standard

OPT: Optional

F: Front

R: Rear

NS: Not specified by manufacturer

① Refer to pressure on door sticker.

22205_MBCA_C0010

# BRAKE SPECIFICATIONS

All measurements in inches unless noted

| Year | Model | Front Brake Disc | | | Rear Brake Disc | | | Minimum Lining Thickness | | Brake Caliper | |
|---|---|---|---|---|---|---|---|---|---|---|---|
| | | Original Thickness | Minimum Thickness | Max. Runout | Original Thickness | Minimum Thickness | Max. Runout | Front | Rear | Bracket Bolts (ft. lbs.) | Mounting Bolts (ft. lbs.) |
| 2006 | C230 | 1.092 | 1.014 | NA | 0.390 | 0.324 | NA | 0.468 | 0.429 | NA | ① |
| | C280 | 1.092 | 1.014 | NA | 0.390 | 0.324 | NA | 0.468 | 0.429 | NA | ① |
| | C280 4MATIC | 1.092 | 1.034 | NA | 0.390 | 0.324 | NA | 0.554 | 0.429 | NA | ① |
| | C350 | 1.092 | 1.034 | NA | 0.390 | 0.324 | NA | 0.468 | 0.429 | NA | ① |
| | E350 | 1.248 | 1.170 | NA | 0.858 | 0.780 | NA | 0.117 | 0.117 | NA | 85 |
| | E350 4MATIC | 1.248 | 1.170 | NA | 0.858 | 0.780 | NA | 0.117 | 0.117 | NA | 85 |
| | E500 | 1.248 | 1.170 | NA | 0.858 | 0.780 | NA | 0.117 | 0.117 | NA | 85 |
| | E500 4MATIC | 1.248 | 1.170 | NA | 0.858 | 0.780 | NA | 0.117 | 0.117 | NA | 85 |
| | S350 | 1.248 | 1.189 | NA | 0.936 | 0.858 | NA | 0.117 | 0.117 | NA | 85 |
| | S430 | 1.248 | 1.189 | NA | 0.936 | 0.858 | NA | 0.117 | 0.117 | NA | 85 |
| | S430 4MATIC | 1.248 | 1.189 | NA | 0.936 | 0.858 | NA | 0.117 | 0.117 | NA | 85 |
| | S500 | 1.248 | 1.189 | NA | 0.936 | 0.858 | NA | 0.117 | 0.117 | NA | 85 |
| | S500 4MATIC | 1.248 | 1.189 | NA | 0.936 | 0.858 | NA | 0.117 | 0.117 | NA | 85 |
| 2007 | C230 | 1.092 | 1.014 | NA | 0.390 | 0.324 | NA | 0.468 | 0.429 | NA | ① |
| | C280 | 1.092 | 1.014 | NA | 0.390 | 0.324 | NA | 0.468 | 0.429 | NA | ① |
| | C280 4MATIC | 1.092 | 1.034 | NA | 0.390 | 0.324 | NA | 0.554 | 0.429 | NA | ① |
| | C350 | 1.092 | 1.034 | NA | 0.390 | 0.324 | NA | 0.468 | 0.429 | NA | ① |
| | E350 | 1.248 | 1.170 | NA | 0.858 | 0.780 | NA | 0.117 | 0.117 | NA | 85 |
| | E350 4MATIC | 1.248 | 1.170 | NA | 0.858 | 0.780 | NA | 0.117 | 0.117 | NA | 85 |
| | E550 | 1.248 | 1.170 | NA | 0.858 | 0.780 | NA | 0.117 | 0.117 | NA | 85 |
| | E550 4MATIC | 1.248 | 1.170 | NA | 0.858 | 0.780 | NA | 0.117 | 0.117 | NA | 85 |
| | S550 | 1.248 | 1.189 | NA | 0.936 | 0.858 | NA | 0.117 | 0.117 | NA | 85 |
| | S550 4MATIC | 1.248 | 1.189 | NA | 0.936 | 0.858 | NA | 0.117 | 0.117 | NA | 85 |
| 2008 | C300 | 1.248 | 1.170 | NA | 0.858 | 0.780 | NA | 0.117 | 0.117 | NA | 85 |
| | C300 4MATIC | 1.248 | 1.170 | NA | 0.858 | 0.780 | NA | 0.117 | 0.117 | NA | 85 |
| | C350 | 1.248 | 1.170 | NA | 0.858 | 0.780 | NA | 0.117 | 0.117 | NA | 85 |
| | E350 | 1.248 | 1.170 | NA | 0.858 | 0.780 | NA | 0.117 | 0.117 | NA | 85 |
| | E350 4MATIC | 1.248 | 1.170 | NA | 0.858 | 0.780 | NA | 0.117 | 0.117 | NA | 85 |
| | E550 | 1.248 | 1.170 | NA | 0.858 | 0.780 | NA | 0.117 | 0.117 | NA | 85 |
| | E550 4MATIC | 1.248 | 1.170 | NA | 0.858 | 0.780 | NA | 0.117 | 0.117 | NA | 85 |
| | S550 | 1.248 | 1.189 | NA | 0.936 | 0.858 | NA | 0.117 | 0.117 | NA | 85 |
| | S550 4MATIC | 1.248 | 1.189 | NA | 0.936 | 0.858 | NA | 0.117 | 0.117 | NA | 85 |

NA: Not available from the manufacturer

① Front: 85 ft. lbs.

Rear: 41 ft. lbs.

22205_MBCA_C0011

## SCHEDULED MAINTENANCE INTERVALS
### MERCEDES

**Mercedes-Benz vehicles for this model year follow a maintenance schedule as programmed into the vehicle. Maintenance services are performed as they are requested by vehicle.**

| TO BE SERVICED | TYPE OF SERVICE | SERVICE INTERVALS | | |
|---|---|---|---|---|
| | | A | B | Every 60,000 MILES |
| Air cleaner filter | Replace | | ✓ | |
| Brake system | Inspect/Service | ✓ | ✓ | |
| Brake fluid | Flush | | | ✓ |
| Cabin air filter | Replace | | ✓ | |
| Door hinges & latches | Lubricate | | ✓ | |
| Engine coolant | Flush | | | ✓ |
| Engine oil & filter | Replace | ✓ | ✓ | |
| Exterior lamps | Inspect/Service | ✓ | ✓ | |
| Fault codes | Read/Clear | ✓ | ✓ | |
| Fluid levels | Top off | ✓ | ✓ | |
| Fuel filter | Replace | | | ✓ |
| Hood locks | Lubricate | | ✓ | |
| Interior switches | Inspect/Service | | ✓ | |
| Parking brake | Inspect/Adjust | | ✓ | |
| Power steering filter | Replace | | | ✓ |
| Service indicator light | Reset | ✓ | ✓ | |
| Spark plugs | Replace | | | ✓ |
| Steering & suspension | Inspect/Service | ✓ | ✓ | |
| Sunroof glides | Lubricate | | ✓ | |
| Tires (air pressure) | Inspect/Adjust | ✓ | ✓ | |
| Throttle Linkage | Adjust/Lubricate | ✓ | ✓ | |
| Transmission fluid & filter | Replace | | | ✓ |
| Windshield wipers | Replace | ✓ | ✓ | |

22205_MBCA_C0012

## PRECAUTIONS

Before servicing any vehicle, please be sure to read all of the following precautions, which deal with personal safety, prevention of component damage, and important points to take into consideration when servicing a motor vehicle:

• Never open, service or drain the radiator or cooling system when the engine is hot; serious burns can occur from the steam and hot coolant.

• Observe all applicable safety precautions when working around fuel. Whenever servicing the fuel system, always work in a well-ventilated area. Do not allow fuel spray or vapors to come in contact with a spark, open flame, or excessive heat (a hot drop light, for example). Keep a dry chemical fire extinguisher near the work area. Always keep fuel in a container specifically designed for fuel storage; also, always properly seal fuel containers to avoid the possibility of fire or explosion. Refer to the additional fuel system precautions later in this section.

• Fuel injection systems often remain pressurized, even after the engine has been turned **OFF**. The fuel system pressure must be relieved before disconnecting any fuel lines. Failure to do so may result in fire and/or personal injury.

• Brake fluid often contains polyglycol ethers and polyglycols. Avoid contact with the eyes and wash your hands thoroughly after handling brake fluid. If you do get brake fluid in your eyes, flush your eyes with clean, running water for 15 minutes. If eye irritation persists, or if you have taken brake fluid internally, IMMEDIATELY seek medical assistance.

• The EPA warns that prolonged contact with used engine oil may cause a number of skin disorders, including cancer. You should make every effort to minimize your exposure to used engine oil. Protective gloves should be worn when changing oil. Wash your hands and any other exposed skin areas as soon as possible after exposure to used engine oil. Soap and water, or waterless hand cleaner should be used.

• All new vehicles are now equipped with an air bag system, often referred to as a Supplemental Restraint System (SRS) or Supplemental Inflatable Restraint (SIR) system. The system must be disabled before performing service on or around system components, steering column, instrument panel components, wiring and sensors. Failure to follow safety and disabling procedures could result in accidental air bag deployment, possible personal injury and unnecessary system repairs.

• Always wear safety goggles when working with, or around, the air bag system. When carrying a non-deployed air bag, be sure the bag and trim cover are pointed away from your body. When placing a non-deployed air bag on a work surface, always face the bag and trim cover upward, away from the surface. This will reduce the motion of the module if it is accidentally deployed. Refer to the additional air bag system precautions later in this section.

• Clean, high quality brake fluid from a sealed container is essential to the safe and proper operation of the brake system. You should always buy the correct type of brake fluid for your vehicle. If the brake fluid becomes contaminated, completely flush the system with new fluid. Never reuse any brake fluid. Any brake fluid that is removed from the system should be discarded. Also, do not allow any brake fluid to come in contact with a painted surface; it will damage the paint.

• Never operate the engine without the proper amount and type of engine oil; doing so WILL result in severe engine damage.

• Timing belt maintenance is extremely important. Many models utilize an interference-type, non-freewheeling engine. If the timing belt breaks, the valves in the cylinder head may strike the pistons, causing potentially serious (also time-consuming and expensive) engine damage. Refer to the maintenance interval charts for the recommended replacement interval for the timing belt, and to the timing belt section for belt replacement and inspection.

• Disconnecting the negative battery cable on some vehicles may interfere with the functions of the on-board computer system(s) and may require the computer to undergo a relearning process once the negative battery cable is reconnected.

• When servicing drum brakes, only disassemble and assemble one side at a time, leaving the remaining side intact for reference.

• Only an MVAC-trained, EPA-certified automotive technician should service the air conditioning system or its components.

## BRAKES

## ANTI-LOCK BRAKE SYSTEM (ABS)

### GENERAL INFORMATION

#### PRECAUTIONS

• Certain components within the ABS system are not intended to be serviced or repaired individually.

• Do not use rubber hoses or other parts not specifically specified for and ABS system. When using repair kits, replace all parts included in the kit. Partial or incorrect repair may lead to functional problems and require the replacement of components.

• Lubricate rubber parts with clean, fresh brake fluid to ease assembly. Do not use shop air to clean parts; damage to rubber components may result.

• Use only DOT 3 brake fluid from an unopened container.

• If any hydraulic component or line is removed or replaced, it may be necessary to bleed the entire system.

• A clean repair area is essential. Always clean the reservoir and cap thoroughly before removing the cap. The slightest amount of dirt in the fluid may plug an orifice and impair the system function. Perform repairs after components have been thoroughly cleaned; use only denatured alcohol to clean components.

Do not allow ABS components to come into contact with any substance containing mineral oil; this includes used shop rags.

• The Anti-Lock control unit is a microprocessor similar to other computer units in the vehicle. Ensure that the ignition switch is **OFF** before removing or installing controller harnesses. Avoid static electricity discharge at or near the controller.

• If any arc welding is to be done on the vehicle, the control unit should be unplugged before welding operations begin.

## BLEEDING PROCEDURE

### BLEEDING PROCEDURE

*See Figure 1.*

1. Before servicing the vehicle, refer to the precautions.

2. Attach a pressure bleeding unit and adjust the pressure to 29 PSI (2.0 bar).

3. Remove dust caps of bleed screws (1, 2, 3, 4, 5, 6) on the brake calipers.

4. Open bleed screw (1) and allow approx. 80 cm 3 of brake fluid bleed.

➡**New brake fluid must flow out free of bubbles.**

5. Close bleed screw (1) and pull off bleed hose.

6. Repeat process at the bleed screws (2, 3, 4, 5, 6).

➡**The numbering of the bleed screws (1, 2, 3, 4, 5, 6) represents the opening sequence.**

7. On Vehicles with manual transmission, remove cover from bell housing and repeat process at bleed screw of the central clutch release bearing.

8. Attach dust caps of bleed screws (1, 2, 3, 4, 5, 6) and cover to bell housing.

9. Tighten bleeding screws to 5 ft. lbs. (7 Nm).

10. Refill the fluid level in the master cylinder and check the system for leaks.

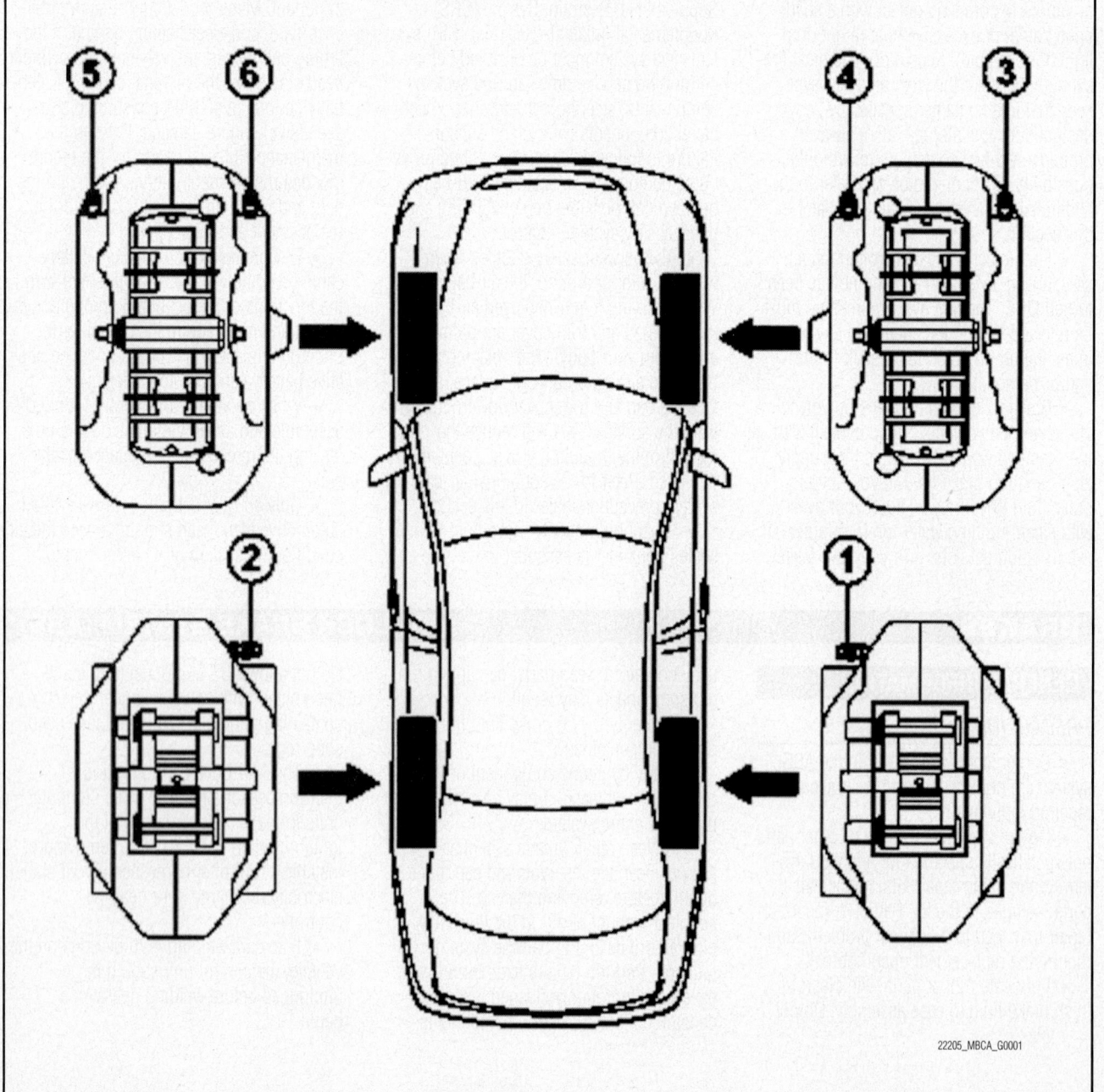

22205_MBCA_G0001

**Fig. 1 The numbering of the bleed screws (1, 2, 3, 4, 5, 6) represents the opening sequence**

## BRAKES

## FRONT DISC BRAKES

### ✳✳ CAUTION

Dust and dirt accumulating on brake parts during normal use may contain asbestos fibers from production or aftermarket brake linings. Breathing excessive concentrations of asbestos fibers can cause serious bodily harm. Exercise care when servicing brake parts. Do not sand or grind brake lining unless equipment used is designed to contain the dust residue. Do not clean brake parts with compressed air or by dry brushing. Cleaning should be done by dampening the brake components with a fine mist of water, then wiping the brake components clean with a dampened cloth. Dispose of cloth and all residue containing asbestos fibers in an impermeable container with the appropriate label. Follow practices prescribed by the Occupational Safety and Health Administration

(OSHA) and the Environmental Protection Agency (EPA) for the handling, processing, and disposing of dust or debris that may contain asbestos fibers.

### BRAKE CALIPER

*REMOVAL & INSTALLATION*

#### C-Class

*See Figures 2 through 4.*

1. Before servicing the vehicle, refer to the precautions.
2. Remove front wheels.
3. Disconnect brake hose from brake line

### ✳✳ WARNING

Seal off line connections immediately with plugs. Do not allow brake fluid supply reservoir to run completely empty.

4. Unclip brake hose from bracket.
5. Disconnect brake hose from brake caliper.
6. Disconnect brake pad contact sensor from the pad contact sensor connector.
7. Unscrew bolts brake pad contact sensor.
8. Unscrew bolts of top and bottom of brake caliper and remove from steering knuckle.
9. Remove the brake pads.

➡On floating calipers the brake pads are removed after the caliper is removed. On fixed calipers, the brake pads are removed prior to caliper removal.

#### *To install:*

10. Installation is the reverse of removal.
11. Observe installation position of inner brake lining. The brake lining with printed-on arrow must be mounted on the side of the piston. The printed-on arrow must point in the direction of rotation of the wheel.

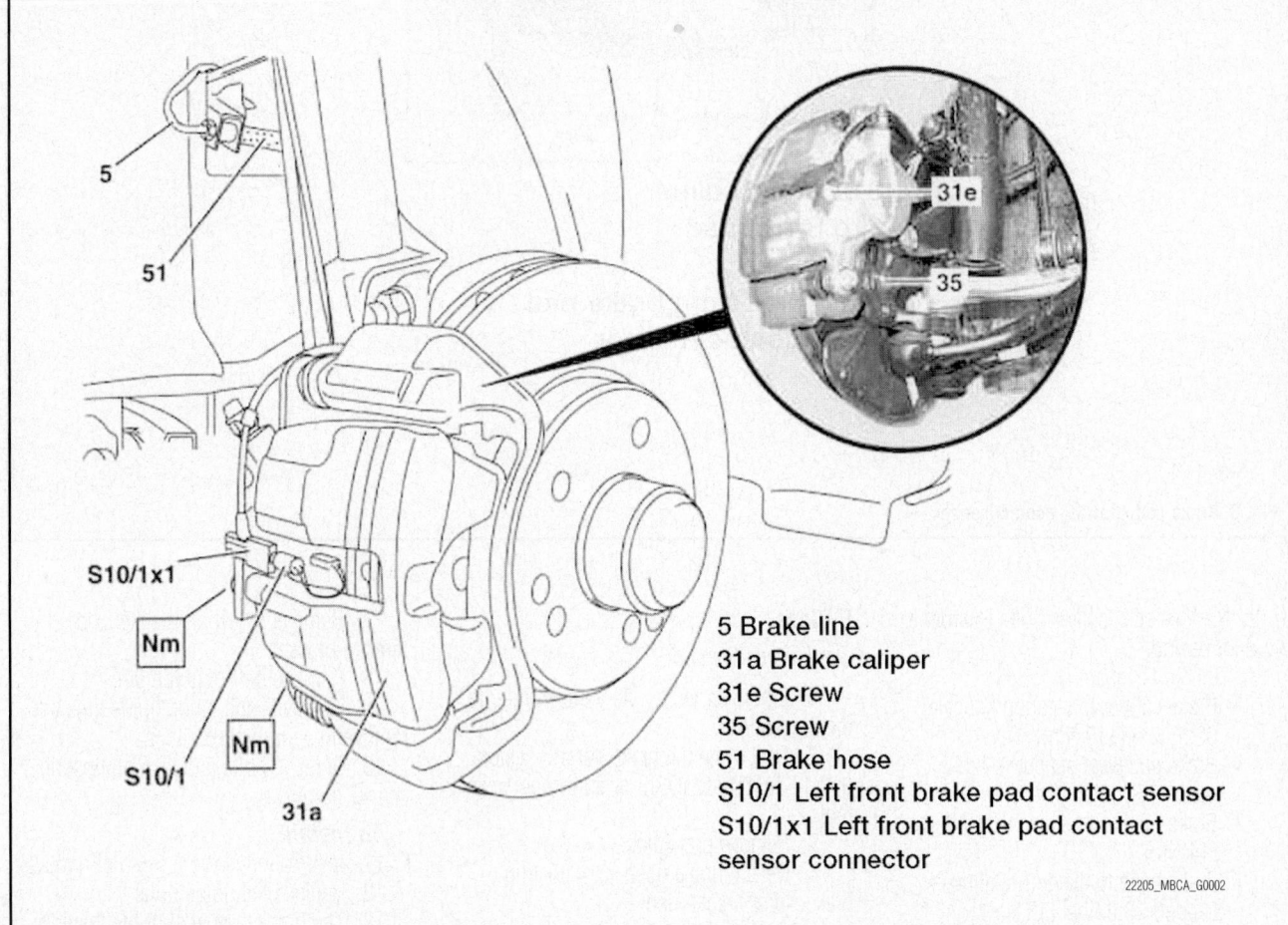

5 Brake line
31a Brake caliper
31e Screw
35 Screw
51 Brake hose
S10/1 Left front brake pad contact sensor
S10/1x1 Left front brake pad contact sensor connector

22205_MBCA_G0002

Fig. 2 Floating caliper used on luxury models

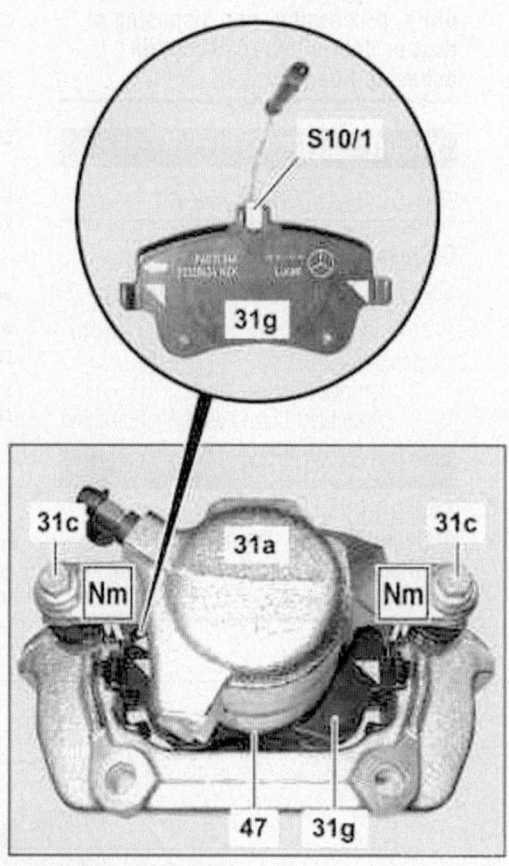

31a Brake caliper
31g Brake pad
47 Pistons
S10/1 Left front brake pad
   contact sensor

22205_MBCA_G0003

Fig. 3 Brake pad showing contact sensor

12. Replace self-locking bolts in order to avoid loosening.

13. Tighten fasteners as follows:
- Brake caliper to steering knuckle: 85 ft. lbs. (115 Nm).
- Brake pad wear sensor to brake caliper: 6 ft. lbs. (8 Nm).
- Brake pipe to brake hose: 10 ft. lbs. (14 Nm).
- Brake hose to fixed or floating caliper: 13 ft. lbs. (18 Nm).

14. Actuate the brake pedal several times to seat the brake pads against the rotor.

### E-Class

*See Figures 5 and 6.*

1. Before servicing the vehicle, refer to the precautions.

2. Deactivate the brake control system using STAR DIAGNOSIS or equivalent scan tool.

3. Remove brake hose.

4. Detach brake hose from floating brake caliper brake line.

5. Remove brake pads.

6. Detach electric feed line of brake pad contact sensor from floating brake caliper.

7. Remove screw on bracket and remove bracket.

8. Remove bolt of guide pin.

9. Remove caps, undo guide pins and remove floating brake caliper.

10. Remove bolts and take off floating caliper mount.

### To install:

11. Installation is the reverse of removal.

12. Replace the guide bolts.

13. The bracket must abut the floating brake caliper.

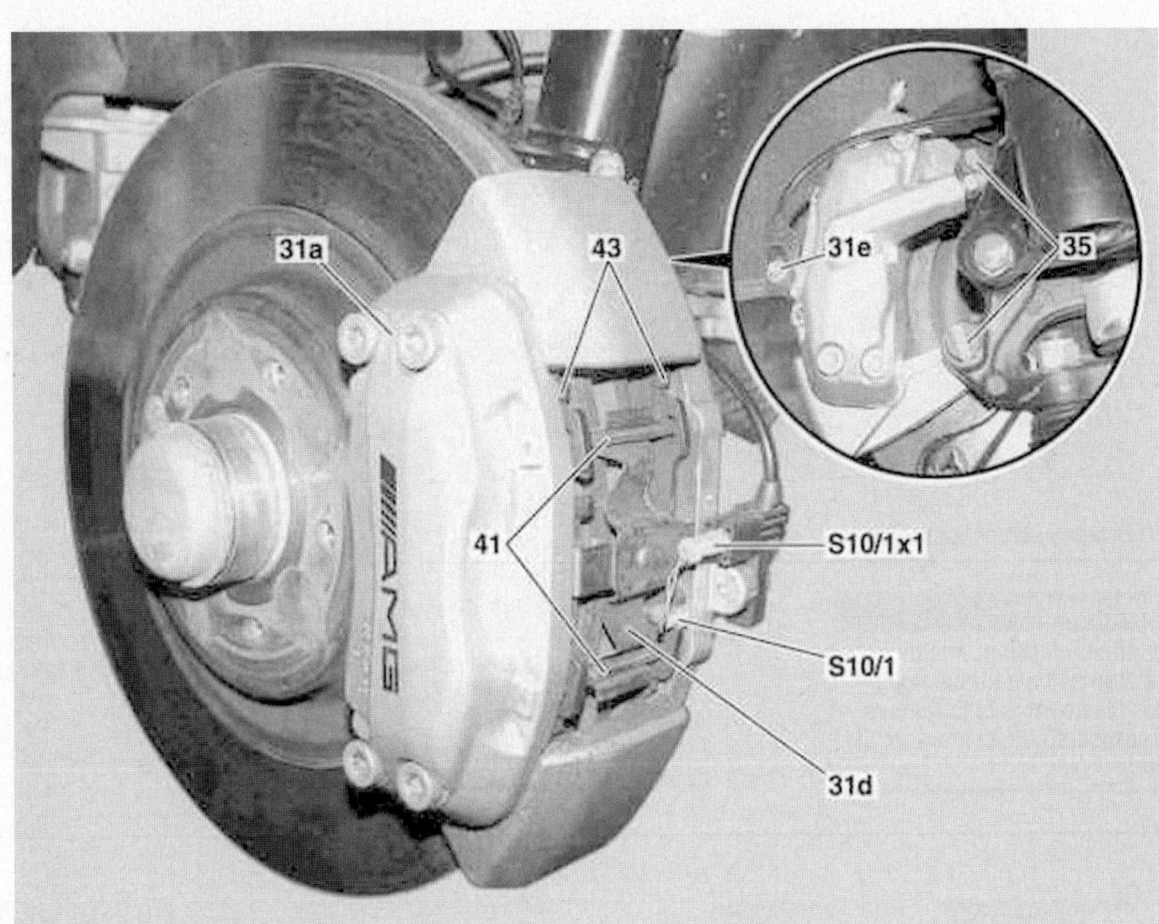

31a Fixed caliper
31d Pad retaining plate
31e Bolt for contact sensor connector
35 Bolts
41 Retaining pins

43 Brake linings
S10/1 Left front brake pad contact sensor
S10/1x1 Left front brake pad contact sensor connector

22205_MBCA_G0004

**Fig. 4 Fixed caliper used on sport models**

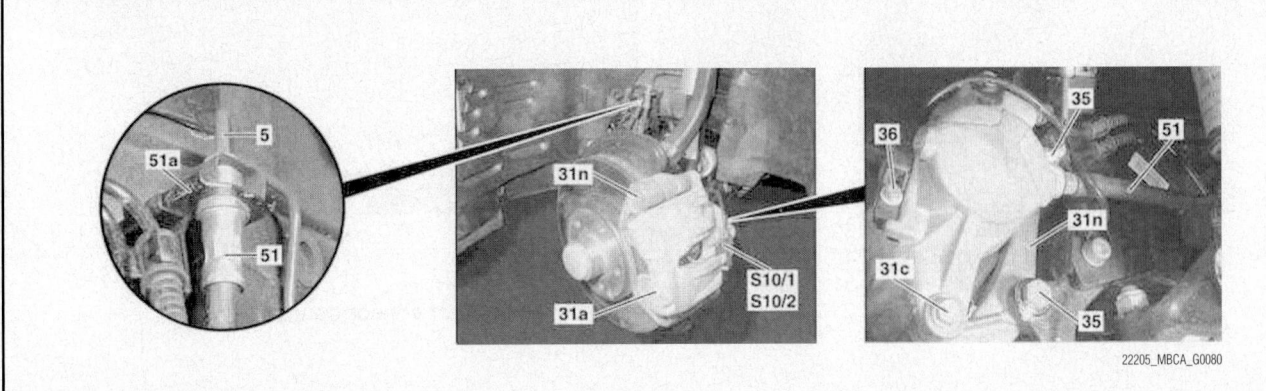

22205_MBCA_G0080

**Fig. 5 Removing the front brake caliper**

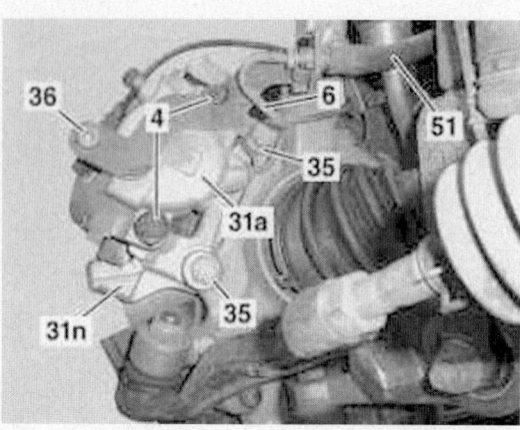

Fig. 6 Front brake caliper mounting bolts

**⁑ WARNING**

Replace all self locking, micro-encapsulated and aluminum bolts and nuts. Chase threads to remove micro-encapsulated residue from old bolts/nuts.

14. Tighten bolts/nuts to specification as follows:
- Self-locking bolt, brake caliper to steering knuckle M12: 85 ft. lbs. (115 Nm)
- Self-locking bolt, brake caliper to steering knuckle M14: 59 ft. lbs.

(80 Nm) plus an additional 45° of rotation
- Bolt, brake pad contact sensor to brake caliper: 6 ft. lbs. (8 Nm)
- Bolt, brake hose clip to floating brake caliper: 15 ft. lbs. (20Nm)

15. Tighten guide pin, brake caliper to brake caliper support screw to specification:
- Innentorx T45: 20 ft. lbs. (27 Nm)
- HexagonSW13: 30 ft. lbs. (40 Nm)

16. Tighten guide pin to specification:
- Inbus size 7: 20 ft. lbs. (27 Nm)
- Inbus size 9: 41 ft. lbs (55 Nm)

17. Bleed brake system.

## S-Class

*See Figures 7 through 9.*

1. Before servicing the vehicle, refer to the precautions.

2. Remove brake pads from brake caliper.

3. Unscrew bolt and remove right front brake pad contact sensor from brake caliper.

4. Remove brake hose.

5. Detach brake line from brake caliper.

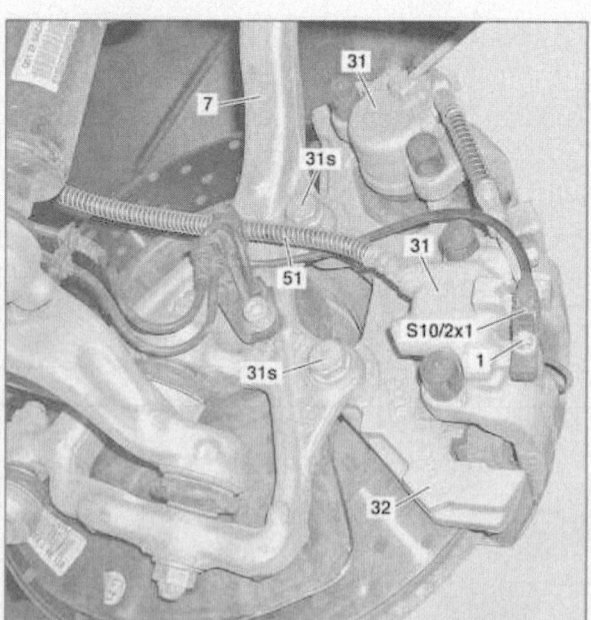
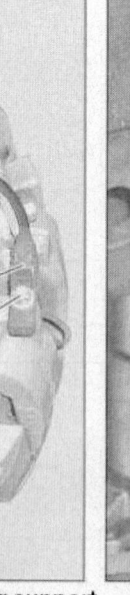

1. Screw
7. Steering knuckle
31. Brake caliper
31s. Screw
32. Caliper support
43. Brake pads
51. Brake hose
S10/2x1. Right front brake pad contact sensor connector

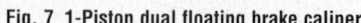

Fig. 7 1-Piston dual floating brake caliper

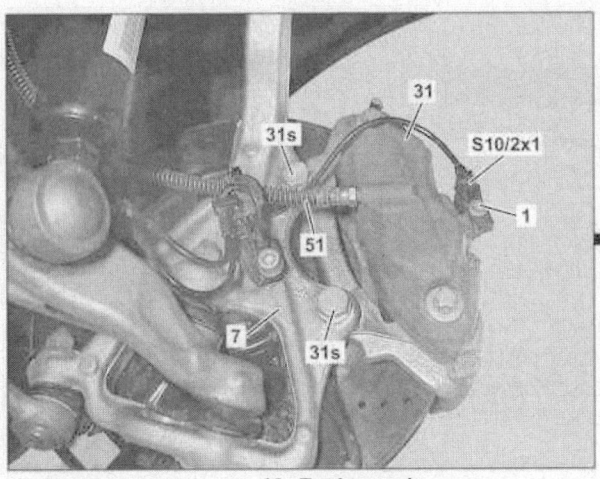

1. Screw
7. Steering knuckle
31. Brake caliper
31s. Screw
43. Brake pads
51. Brake hose
S10/2x1. Right front brake pad contact sensor connector

22205_MBCA_G0157

**Fig. 8  4-Piston brake caliper**

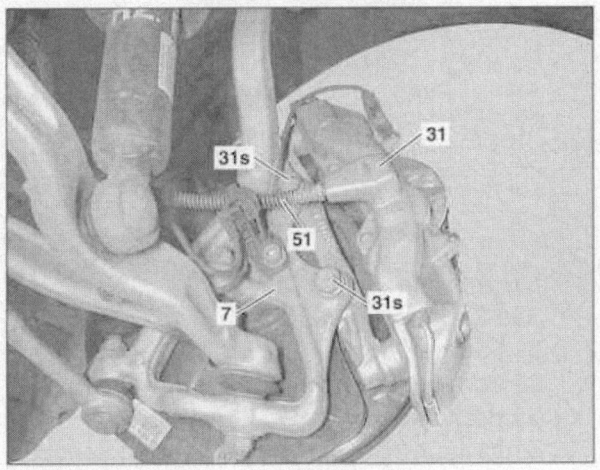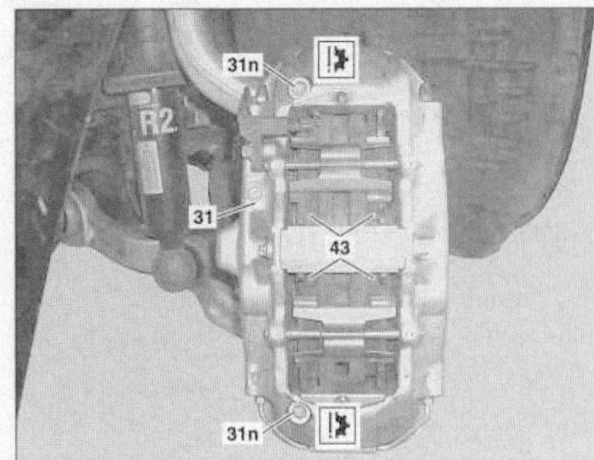

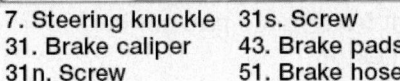

7. Steering knuckle
31. Brake caliper
31n. Screw
31s. Screw
43. Brake pads
51. Brake hose

22205_MBCA_G0158

**Fig. 9  8-Piston brake caliper**

➡Seal off both line ends immediately with stop plugs to prevent dirt from entering. The brake fluid expansion reservoir must not run empty.

6. Unscrew bolts and remove brake caliper or brake caliper support from steering knuckle.

### ❋❋ WARNING

**Under no circumstances loosen the bolts (31n) on the 8-piston-brake calipers otherwise brake system malfunctions can occur**

#### To install:

7. Installation is the reverse of removal.

8. Tighten bolts/nuts to specification as follows:

- Bolt, brake pad contact sensor to brake caliper: 6 ft. lbs. (8 Nm)
- Self-locking bolt, caliper support to steering knuckle: 133 ft. lbs. (180 Nm)

## DISC BRAKE PADS

### REMOVAL & INSTALLATION

#### C-Class

*See Figures 10 through 13.*

1. Before servicing the vehicle, refer to the precautions.

2. Remove front wheels.

3. Disconnect brake pad contact sensor from the pad contact sensor connector.

4. Unscrew bolts brake pad contact sensor.

5. Remove bolts and detach floating brake caliper upwards.

6. Hang brake caliper out of the way using a piece of wire.

7. On 4MATIC models, drive out retaining pins using a punch and remove retaining plates

8. Remove the brake pads.

➡On floating calipers the brake pads are removed after the caliper is removed. On fixed calipers, the brake pads are removed prior to caliper removal.

#### To install:

9. Installation is the reverse of removal.

10. Pay attention to installation position of inner brake pads (31g). The brake pad

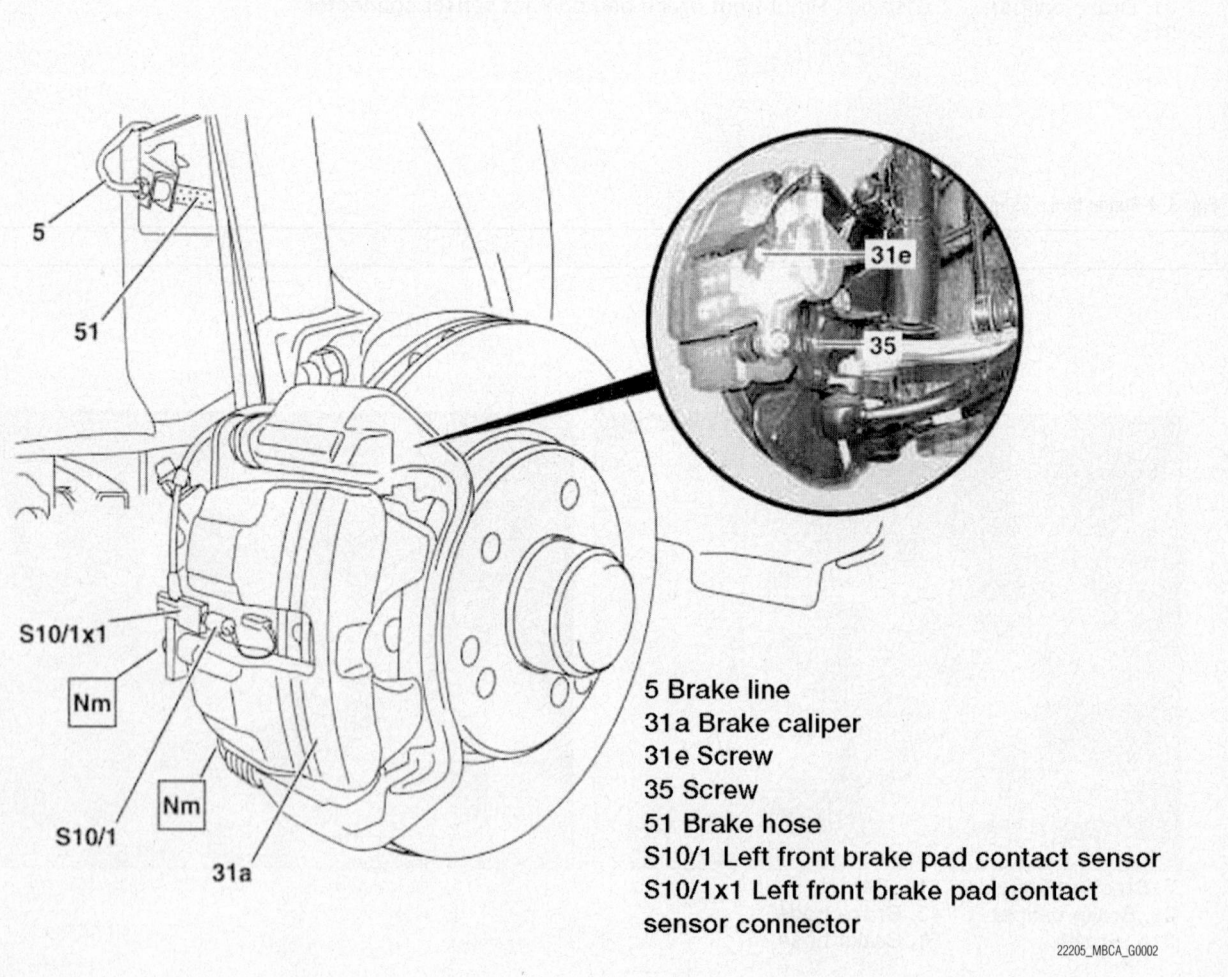

5 Brake line
31a Brake caliper
31e Screw
35 Screw
51 Brake hose
S10/1 Left front brake pad contact sensor
S10/1x1 Left front brake pad contact sensor connector

22205_MBCA_G0002

**Fig. 10 Front floating caliper used on luxury models**

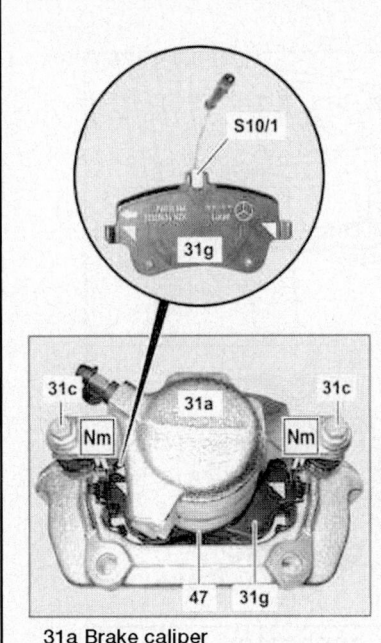

31a Brake caliper
31g Brake pad
47 Pistons
S10/1 Left front brake pad
       contact sensor

22205_MBCA_G0003

**Fig. 11 Brake pad showing contact sensor**

with the printed-on arrow must be assembled on the side of the brake piston. The arrow must point in the direction of rotation of the wheel during forward travel.

11. Press back brake piston using resetting device

12. Clean contact surfaces of brake pads in brake carrier and floating brake caliper.

13. Replace self-locking bolts in order to avoid loosening.

14. Tighten fasteners as follows:
- Brake caliper to steering knuckle: 85 ft. lbs. (115 Nm).
- Brake pad wear sensor to brake caliper: 6 ft. lbs. (8 Nm).
- Brake pipe to brake hose: 10 ft. lbs. (14 Nm).
- Brake hose to fixed or floating caliper: 13 ft. lbs. (18 Nm).

15. Actuate the brake pedal several times to seat the brake pads against the rotor.

### E-Class

*See Figures 14 and 15.*

1. Before servicing the vehicle, refer to the precautions.

2. Deactivate the brake control system using STAR DIAGNOSIS or equivalent scan tool.

3. Lower brake fluid level in brake fluid reservoir down to **MIN** mark.

4. Remove the front wheels.

5. Check brake pad thickness and brake discs.

➡**Replace brake pads and brake discs in complete sets if necessary.**

6. Check brake lining thickness.

7. Inspect condition of brake discs.

8. Separate brake pad contact sensor connector.

9. Unclip electrical feed line of brake pad contact sensor from retaining clips (arrow) on steering knuckle.

10. Unclip electrical feed line of rpm sensor from retaining clips (arrow) on steering knuckle

11. Unclip electrical feed line of brake pad contact sensor from retaining clips (arrow) on longitudinal member.

12. Remove clips and protective caps (arrows).

13. Remove bolts from guide pins.

14. Fold down floating caliper.

### ✳✳ WARNING

**Do not kink or tension brake hose otherwise it will be damaged.**

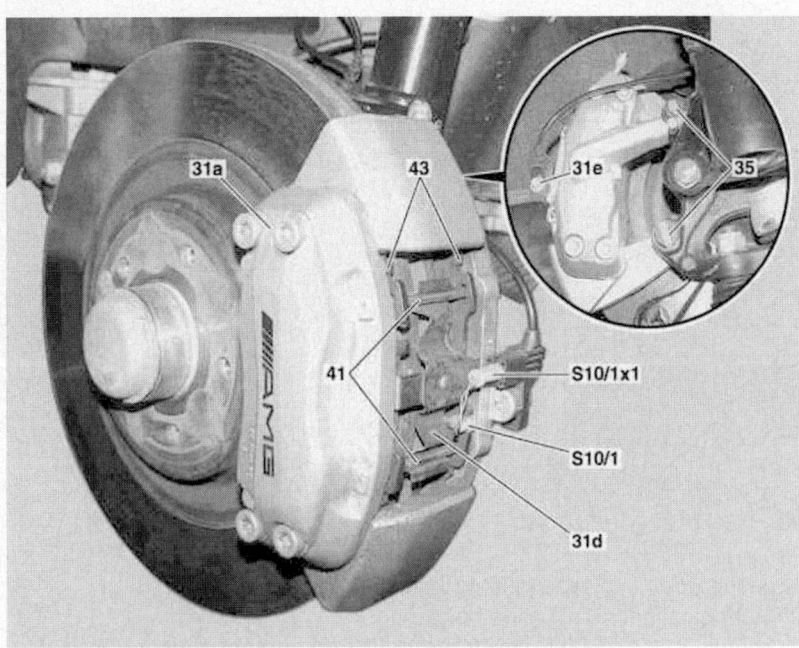

31a Fixed caliper
31d Pad retaining plate
31e Bolt for contact sensor connector
35 Bolts
41 Retaining pins

43 Brake linings
S10/1 Left front brake pad contact sensor
S10/1x1 Left front brake pad contact
          sensor connector

22205_MBCA_G0004

**Fig. 12 Front fixed caliper used on sport models**

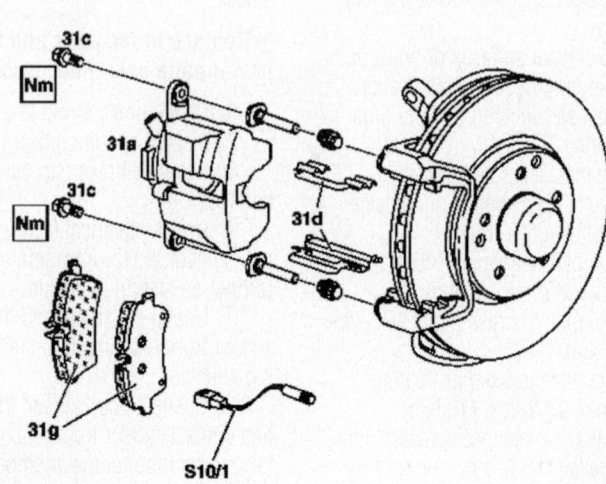

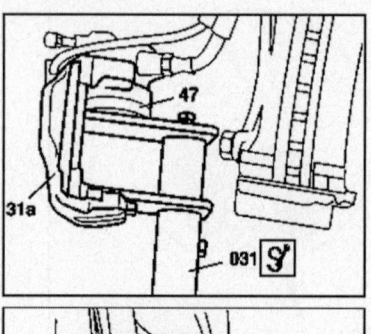

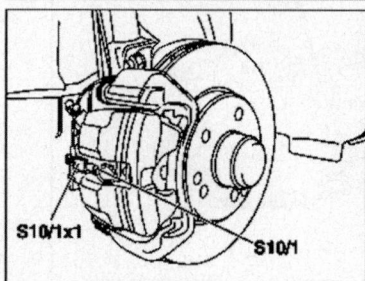

31c Screw
S10/1 Left front brake pad contact
31d Retaining plate sensor
031 Pusher tool
31g Brake pad
S10/1x1 Left front brake pad contact
31a Floating caliper

47 Dust boot sensor connector
31a Floating caliper
31c Screw
31g Brake pad
47 Dust boot
S10/1 Left front brake pad contact sensor

22205_MBCA_G0006

Fig. 13 Exploded view of floating caliper brake pad components

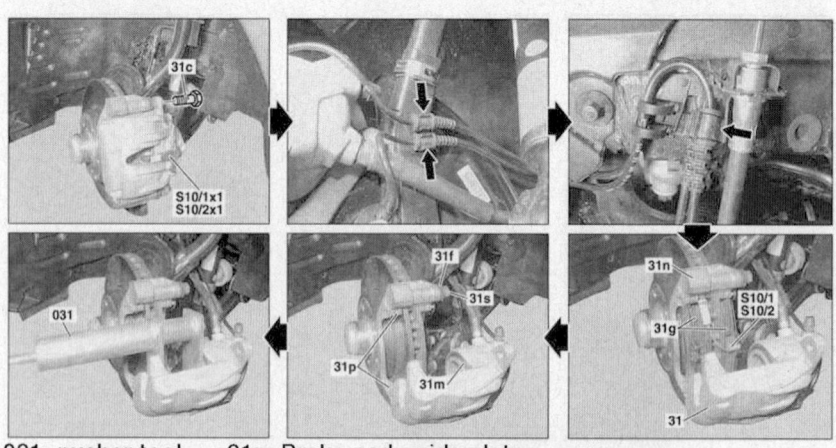

031. pusher tool
31. Floating caliper
31c. Screw
31f. Bellows
31g. Brake pad
31m. Dust boot
31n. Caliper support

31p. Brake pad guide plates
31s. Guide bolts
S10/1. Left front brake pad contact sensor
S10/1x1. Left front brake pad contact sensor connector
S10/2. Right front brake pad contact sensor
S10/2x1. Right front brake pad contact sensor connector

22205_MBCA_G0082

Fig. 14 Removing the front brake pads—without 4MATIC

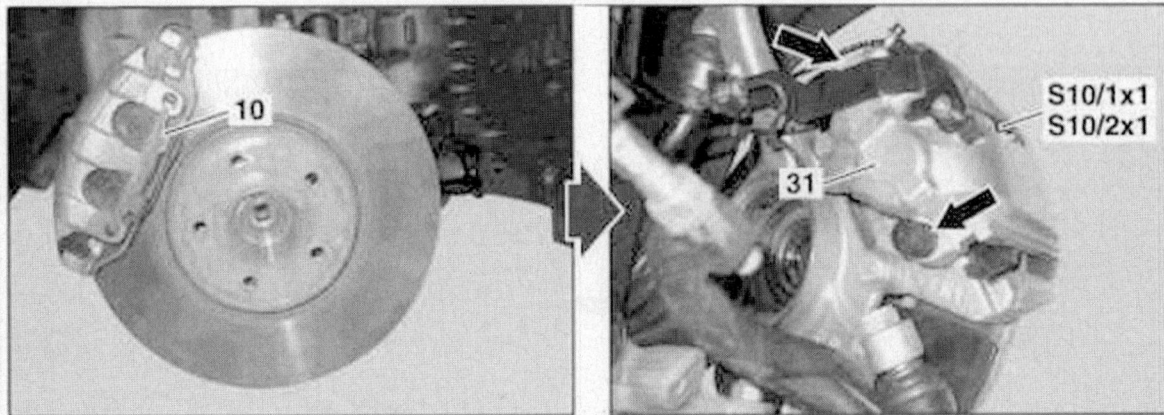

10. Clamp           S10/1x1. Left front brake pad contact sensor connector
31. Floating caliper S10/2x1. Right front brake pad contact sensor connector

22205_MBCA_G0083

**Fig. 15 Removing the front brake pads—with 4MATIC**

15. Remove floating brake caliper with brake pads and fasten to the vehicle free of tension.

16. Remove brake pads.

17. Replace brake pad guide plates.

18. Check boots for damage and correct seat as well as guide pins for ease of movement.

19. Check floating brake caliper for leaks and dust boot for damage and correct seat. If floating brake caliper is leaky or dust boot is damaged or not seated correctly, replace floating brake caliper.

20. Pull brake pad contact sensor out of pad backing plate of brake pads.

21. Notes on installing brake pad wear sensor contact sensors.

22. Press back brake piston using pusher tool. Secure remaining brake pistons of floating brake caliper with wedges to prevent them from falling out.

23. If the brake piston is difficult to move, replace floating brake caliper.

24. On vehicle using Bosch floating brake caliper, if after pressing back the brake piston the dust boot projects beyond the brake piston, its seat should be correctly. Check for folds forming in the dust boots.

25. Clean contact surfaces of brake pads at floating caliper.

26. Grease contact surfaces of brake pads using Molykote Cu-7439 brake pad paste.

*To install:*

27. Installation is the reverse of removal.

28. Install brake pads without grease. The anti-squeal shims enclosed with the repair kit must be fitted on the piston side.

⁜⁜ **WARNING**

**Replace all self locking, micro-encapsulated and aluminum bolts and nuts. Chase threads to remove micro-encapsulated residue from old bolts/nuts.**

29. Tighten guide pin, brake caliper to brake caliper support screw to specification:
  - Innentorx T45: 20 ft. lbs. (27 Nm)
  - HexagonSW13: 30 ft. lbs. (40 Nm)
30. Tighten guide pin to specification:
  - Inbus size 7: 20 ft. lbs. (27 Nm)
  - Inbus size 9: 41 ft. lbs (55 Nm)
31. Activate the brake control system using STAR DIAGNOSIS or equivalent scan tool.

32. Operate the brake pedal several times until the brake pads contact the brake discs.

33. Check brake fluid level, correct if necessary.

**S-Class**

*Floating Caliper*
*See Figure 16.*

1. Before servicing the vehicle, refer to the precautions.

2. Unscrew the cap on the brake fluid expansion reservoir and suction off some brake fluid.

3. Remove wheels Remove/install wheels.

4. Unscrew bolts and remove cover panel.

5. Separate brake pad contact sensor from electrical feed line.

6. Unclip the electrical feed line from the cable retainer on the wheel carrier.

7. Pry off spring steel sheet and remove.

8. Pry off and remove protective caps from boots.

9. Remove guide pins.

10. Remove brake caliper.

⁜⁜ **WARNING**

**Do not subject brake hoses to tensile loads or kink. Attach brake caliper without tension to vehicle to relieve load on brake hose, otherwise the brake hose will be damaged.**

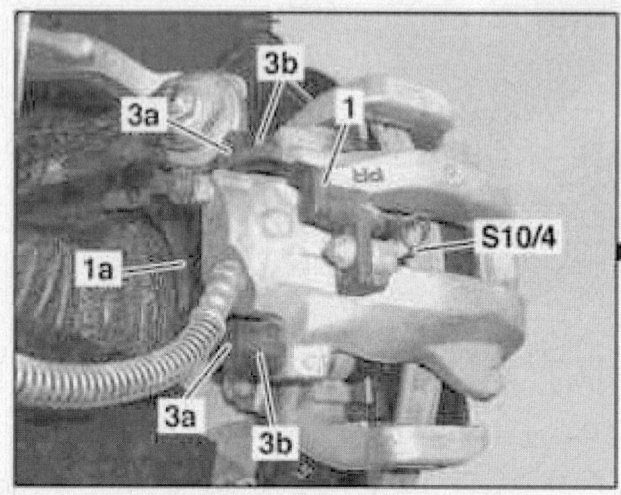

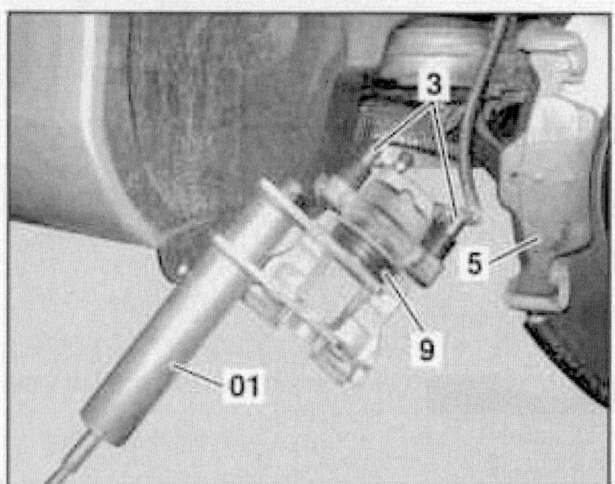
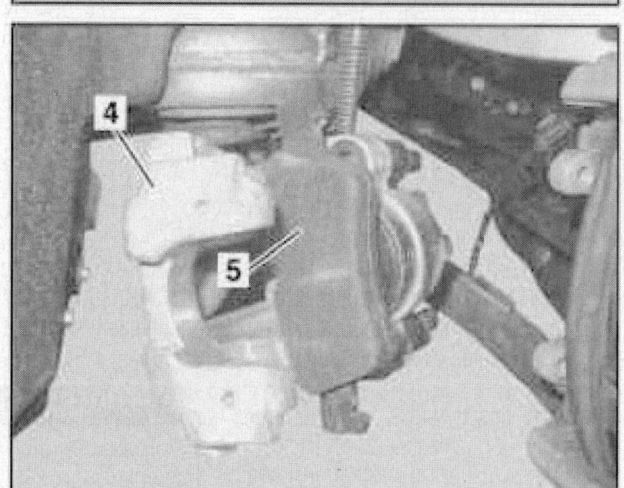

01. Pusher tool
1. Electrical feed line
1a. Cable retainer
2. Spring steel sheet
3. Guide bolts
3a. Protective cap
3b. Boot
4. Brake caliper

5. Brake pad
6. Brake hose
7. Caliper support
8. Brake disk
9. Dust boot
S10/4. Right rear brake pad contact sensor

22205_MBCA_G0159

Fig. 16 Floating caliper

➡**Do not detach brake hoses.**

**Take out brake pads.**

11.  Pull brake pad contact sensor out of the pad backing plate of the brake pad.

12.  Check brake pads and brake discs.

### ✳✳ WARNING

**Replace brake pads and brake discs in complete sets if necessary. Check brake lining thickness.**

13.  Inspect condition of brake discs.

14.  Check brake caliper for leaks.

15.  Check dust boot on brake piston for damage.

16.  Push back brake piston using pusher tool.

17.  Clean contact surfaces of brake pads on brake caliper and brake caliper support using LU cleaner and a rag.

### ✳✳ WARNING

**Do not damage dust boot in the process. Do not use sharp-edged or pointed tools, otherwise the brake caliper or brake caliper support will be damaged.**

18.  Grease contact surfaces of brake pads lightly with Molykote Cu 7439 brake pad paste.

### *To install:*

19.  Installation is the reverse of removal.

20.  Replace bolts.

21.  Coat ATE brake cylinder paste lightly on the contact surfaces of the bolt heads. Do not grease the threads. Wipe off excess brake cylinder paste after tightening the bolts.

22.  Insert inner brake pad with riveted-on spring in brake piston. Insert outer brake pad in brake caliper support and mount brake caliper.

23.  Tighten bolts/nuts to specification as follows:
- Guide pin, brake caliper to brake caliper support: 21 ft. lbs. (28 Nm)
- Guide pin to caliper support: 21 ft. lbs. (28 Nm)

24.  Press brake pedal several times until the brake pads touch the brake discs.

25.  Inspect fluid level in expansion reservoir, adjust to correct level if necessary.

26.  Check function and effectiveness of brake system.

### *Fixed Caliper*

*See Figures 17 and 18.*

1.  Before servicing the vehicle, refer to the precautions.

2.  Unscrew the cap on the brake fluid expansion reservoir and suction off some brake fluid.

3.  Remove wheels.

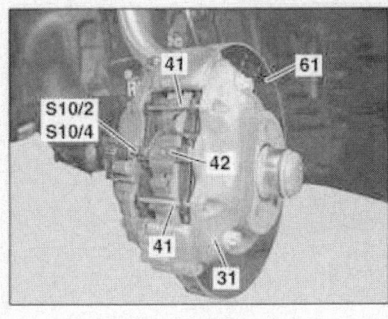
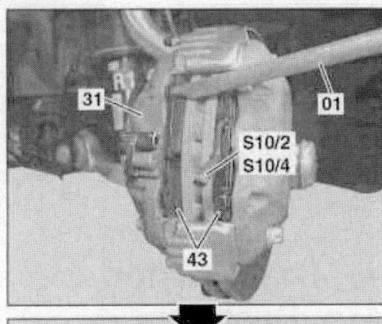
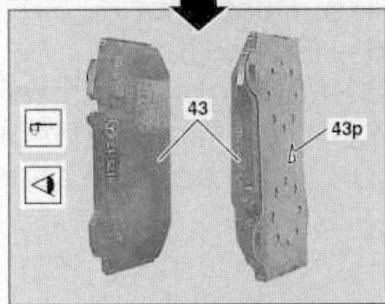
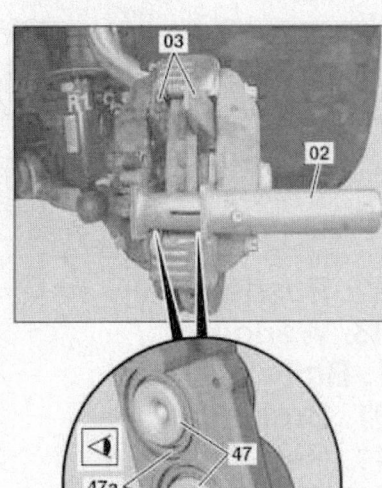

01. Lever
02. Pusher tool
03. Wedge
31. Brake caliper
41. Retaining pin
42. Spring steel sheet 8-piston brake caliper
43. Brake pad

43p. Arrow
47. Brake piston
47a. Dust boot
61. Brake disk
S10/2. Right front brake pad contact sensor
S10/4. Right rear brake pad contact sensor

22205_MBCA_G0160

**Fig. 17  4-Piston fixed caliper**

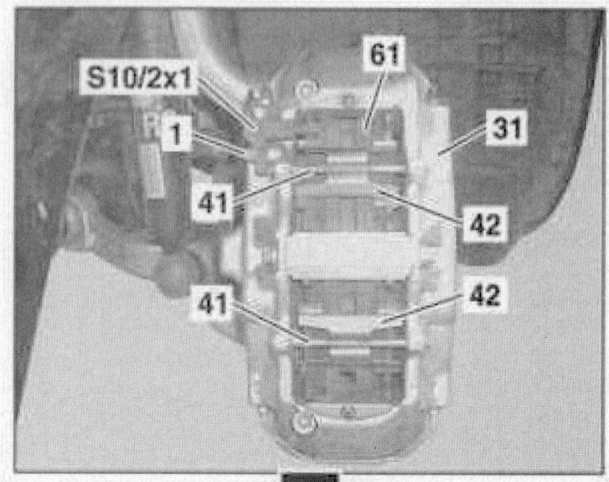
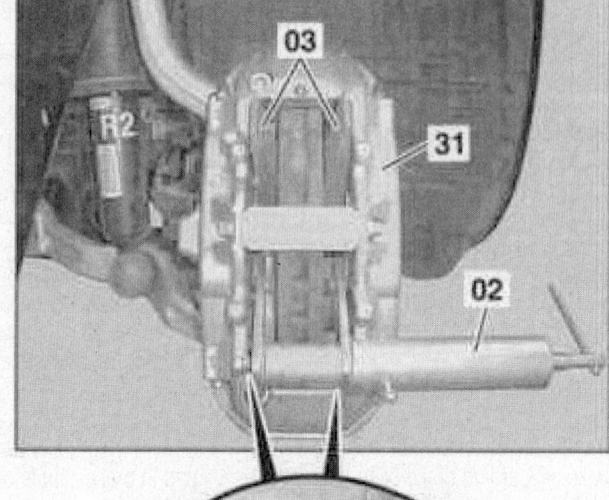
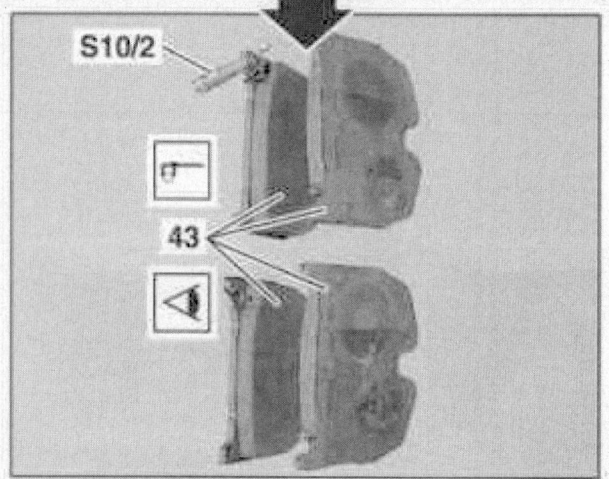
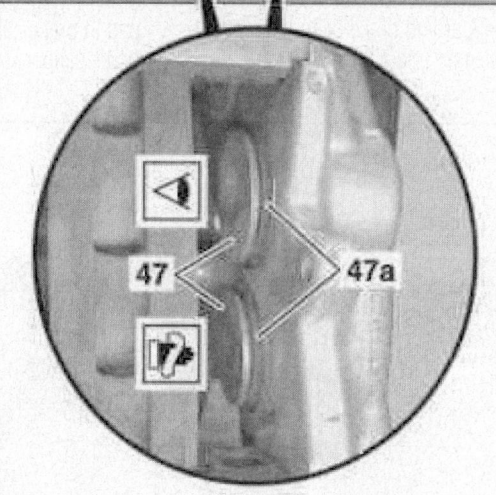

02. Pusher tool
03. Wedge
1. Bolt
31. Brake caliper
41. Retaining pin
42. Spring steel sheet
43. Brake pad

47. Brake piston
47a. Dust boot
61. Brake disk
S10/2. Right front brake pad contact sensor
S10/2x1. Right front brake pad
            contact sensor connector

22205_MBCA_G0161

**Fig. 18 8-Piston fixed caliper**

4. Check brake pads and brake discs.
5. Check brake lining thickness.
6. Inspect condition of brake discs.
7. Separate electrical connector of front right contact sensor or rear right contact sensor.
8. Pull right front brake pad contact sensor out of bracket on spring steel sheet.

9. Remove bolt and right front brake pad contact sensor connector from brake caliper.
10. Hammer out retaining pins using a drift.
11. Remove spring steel sheet.
12. Pull brake pads out of brake caliper.

13. Pull right front or right rear brake pad contact sensor out of the brake pad.
14. Check brake caliper for leaks.
15. Check brake piston dust boots.
16. Press back brake piston using pusher tool.
17. If necessary, clean the perforation in the brake discs.

18. Clean contact surfaces of brake pads on brake caliper and brake caliper support using LU cleaner and a rag.

### ✷✷ WARNING

**Do not damage dust boot in the process. Do not use sharp-edged or pointed tools, otherwise the brake caliper or brake caliper support will be damaged.**

19. Grease contact surfaces of brake pads lightly with Molykote Cu 7439 brake pad paste.

*To install:*

20. Installation is the reverse of removal.
21. Replace brake pads and brake discs in complete sets if necessary.
22. Mount all brake pads so that the arrow on the lining caliper points in the direction of rotation of the wheel when driving forwards. In the event of noise complaints only fit running direction-bound brake pads.
23. Press the brake pedal several times until the brake pads contact the brake discs.
24. Inspect fluid level in expansion reservoir, adjust to correct level if necessary.
25. Check function and effect of brake system.

**BRAKES**                                          **REAR DISC BRAKES**

### ✷✷ CAUTION

Dust and dirt accumulating on brake parts during normal use may contain asbestos fibers from production or aftermarket brake linings. Breathing excessive concentrations of asbestos fibers can cause serious bodily harm. Exercise care when servicing brake parts. Do not sand or grind brake lining unless equipment used is designed to contain the dust residue. Do not clean brake parts with compressed air or by dry brushing. Cleaning should be done by dampening the brake components with a fine mist of water, then wiping the brake components clean with a dampened cloth. Dispose of cloth and all residue containing asbestos fibers in an impermeable container with the appropriate label. Follow practices prescribed by the Occupational Safety and Health Administration (OSHA) and the Environmental Protection Agency (EPA) for the handling, processing, and disposing of dust or debris that may contain asbestos fibers.

### BRAKE CALIPER

*REMOVAL & INSTALLATION*

**C-Class**
*See Figure 19.*

1. Before servicing the vehicle, refer to the precautions.

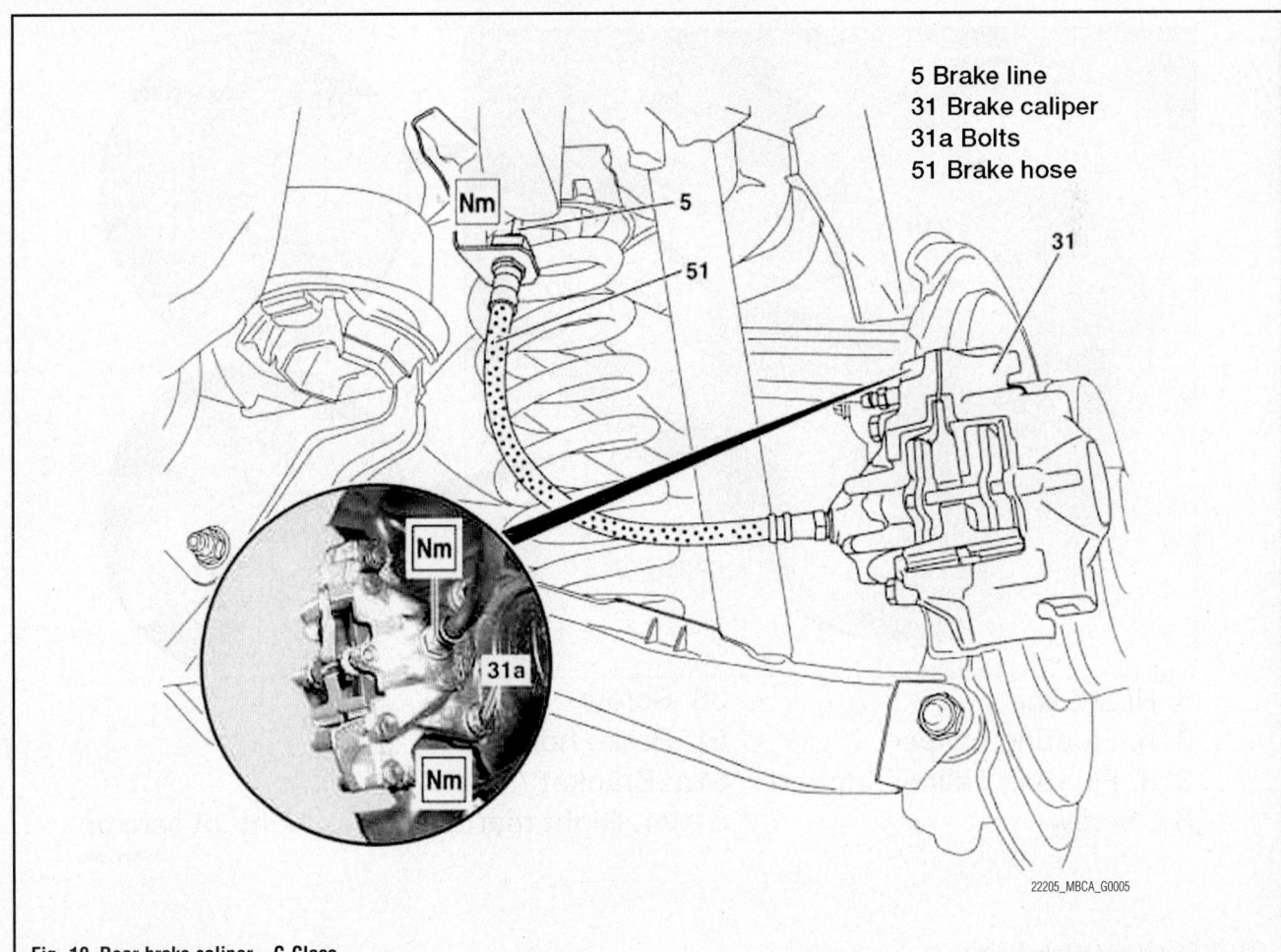

5 Brake line
31 Brake caliper
31a Bolts
51 Brake hose

22205_MBCA_G0005

**Fig. 19 Rear brake caliper—C-Class**

2. Remove rear wheels.

3. Disconnect brake hose from brake line

### ✲✲ WARNING

**Seal off line connections immediately with plugs. Do not allow brake fluid supply reservoir to run completely empty.**

4. Disconnect brake hose from brake caliper.

5. Unscrew bolts of brake caliper and remove from wheel carrier.

➡**On floating calipers the brake pads are removed after the caliper is removed. On fixed calipers, the brake pads are removed prior to caliper removal.**

**To install:**

6. Installation is the reverse of removal.

7. Install new micro-encapsulated bolts.

8. Tighten fasteners as follows:
- Brake caliper to steering knuckle: 41 ft. lbs. (55 Nm).

- Brake pipe to brake hose: 10 ft. lbs. (14 Nm).
- Brake hose to fixed or floating caliper: 13 ft. lbs. (18 Nm).

9. Actuate the brake pedal several times to seat the brake pads against the rotor.

### E-Class

*See Figure 20.*

1. Before servicing the vehicle, refer to the precautions.

2. Deactivate the brake control system using STAR DIAGNOSIS or equivalent scan tool.

3. Remove brake hose.

4. Remove bolt and remove electrical feed line of brake pad contact sensor from floating brake caliper.

5. Remove brake pads.

6. Detach floating caliper carrier from wheel carrier.

**To install:**

7. Installation is the reverse of removal.

### ✲✲ WARNING

**Replace all self locking, micro-encapsulated and aluminum bolts and nuts. Chase threads to remove micro-encapsulated residue from old bolts/nuts.**

8. Tighten bolts/nuts to specification as follows:
- Bolt, brake pad contact sensor to brake caliper: 6 ft. lbs. (8Nm)
- Self-locking bolt, brake caliper to wheel carrier: 85 ft. lbs. (115 Nm)

9. Bleed brake system.

### S-Class

*See Figures 21 and 22.*

1. Before servicing the vehicle, refer to the precautions.

2. Remove brake pads on rear axle.

3. Remove bolt and remove right rear brake pad contact sensor connector from brake caliper.

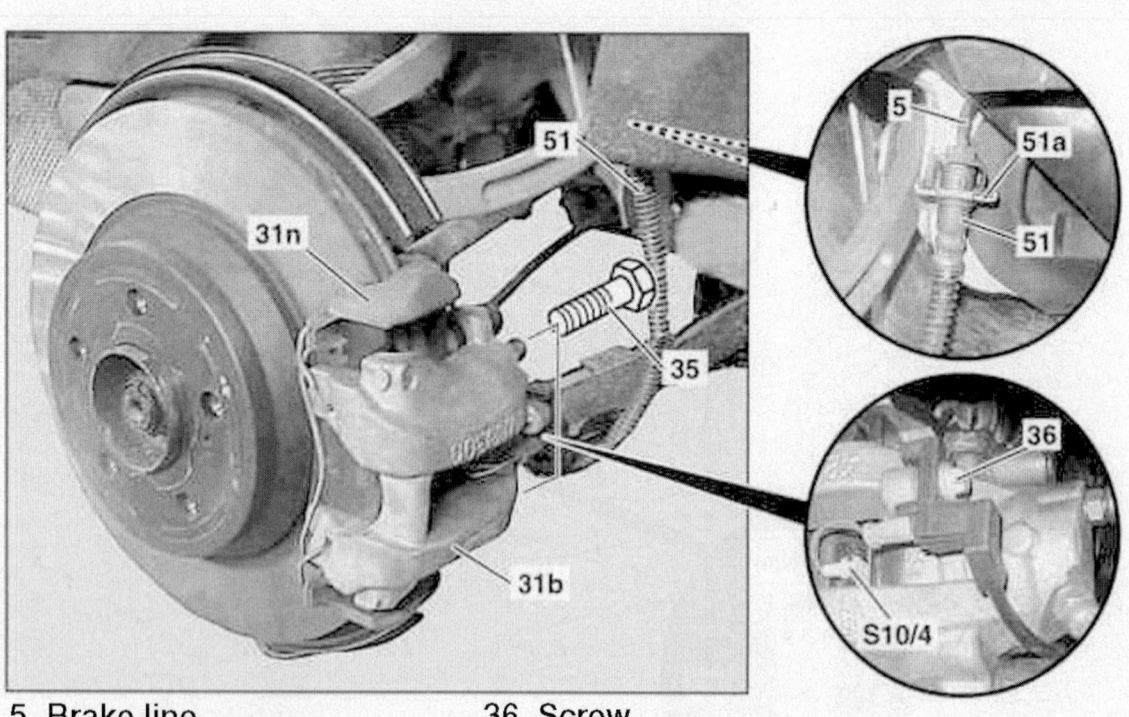

5. Brake line
31b. Floating caliper
31n. Floating caliper support
35. Screw

36. Screw
51. Brake hose
51a. Bracket
S10/4. Right rear brake pad contact sensor

22205_MBCA_G0084

**Fig. 20 Rear brake caliper—E-Class**

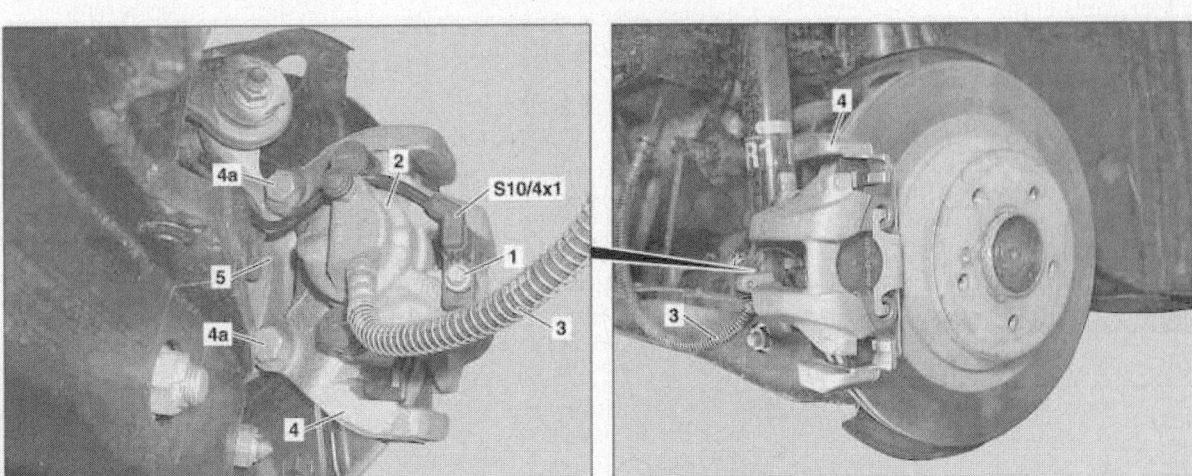

1. Screw
2. Brake caliper
3. Brake hose
4. Caliper support
4a. Screw
5. Wheel carrier
S10/4x1. Right rear brake pad contact sensor connector

22205_MBCA_G0165

**Fig. 21  1-Piston brake caliper**

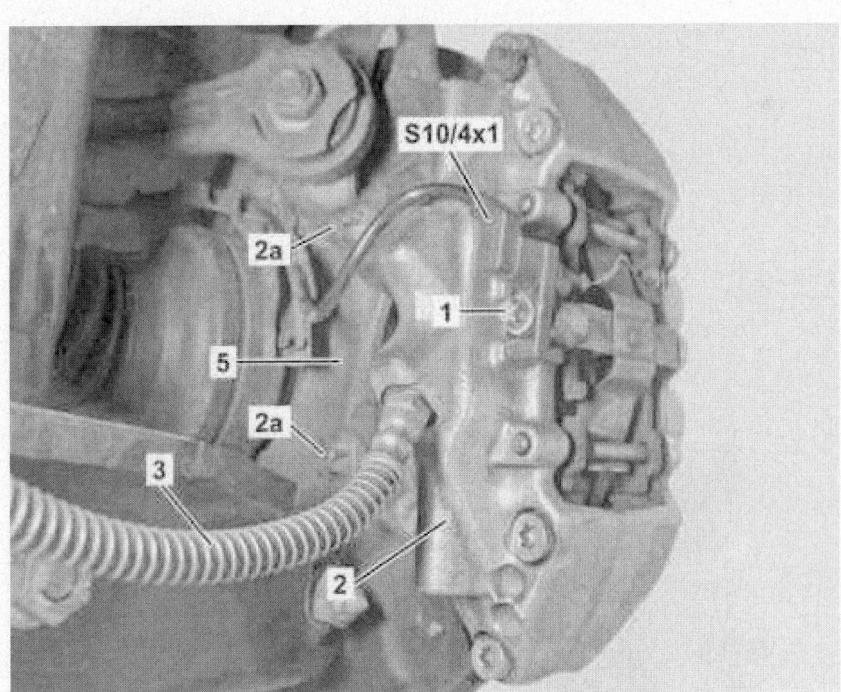

1. Screw
2. Brake caliper
2a. Screw
3. Brake hose
5. Wheel carrier
S10/4x1. Right rear brake pad contact sensor connector

22205_MBCA_G0166

**Fig. 22  4-Piston brake caliper**

4. Remove brake hose.

5. Remove bolts and detach brake caliper from wheel carrier.

### To install:

6. Installation is the reverse of removal.

7. Tighten bolts/nuts to specification as follows:

- Bolt, brake pad contact sensor to brake caliper: 6 ft. lbs. (8 Nm)
- Self-locking bolt, caliper support to wheel carrier: 81 ft. lbs. (110 Nm)

## DISC BRAKE PADS

### REMOVAL & INSTALLATION

#### C-Class

*See Figures 23 and 24.*

1. Before servicing the vehicle, refer to the precautions.

2. Remove front wheels.

3. Disconnect brake pad contact sensor from the pad contact sensor connector.

4. Unscrew bolts brake pad contact sensor.

5. Drive out retaining pins using a punch and remove retaining springs.

6. Remove the brake pads.

### To install:

7. Installation is the reverse of removal.

8. Install new micro-encapsulated bolts.

9. Tighten fasteners as follows:

- Brake caliper to steering knuckle: 41 ft. lbs. (55 Nm).
- Brake pipe to brake hose: 10 ft. lbs. (14 Nm).
- Brake hose to fixed or floating caliper: 13 ft. lbs. (18 Nm).

10. Actuate the brake pedal several times to seat the brake pads against the rotor.

#### E-Class

*See Figure 25.*

1. Before servicing the vehicle, refer to the precautions.

2. Deactivate the brake control system using STAR DIAGNOSIS or equivalent scan tool.

3. Unscrew the cap on the brake fluid expansion reservoir and suction off some brake fluid.

4. Remove rear wheels.

5. Remove brake pad contact sensor.

6. Detach clip.

7. Remove protective caps.

8. Remove guide pin.

9. Remove floating brake caliper with inner brake pad.

10. Remove brake pads.

11. Check brake pad thickness and brake discs.

12. Check brake lining thickness.

13. Inspect condition of brake discs.

14. Check floating brake caliper for leaks and dust boot for damage and correct seat.

15. Push back brake piston using pusher tool. If brake pads have been removed from several brake calipers, secure their brake pistons beforehand with wedges to prevent them from falling out.

16. If brake pistons are sluggish, replace floating brake caliper.

17. Clean contact surfaces of brake pads at floating caliper and at brake caliper support.

### To install:

18. Installation is the reverse of removal.

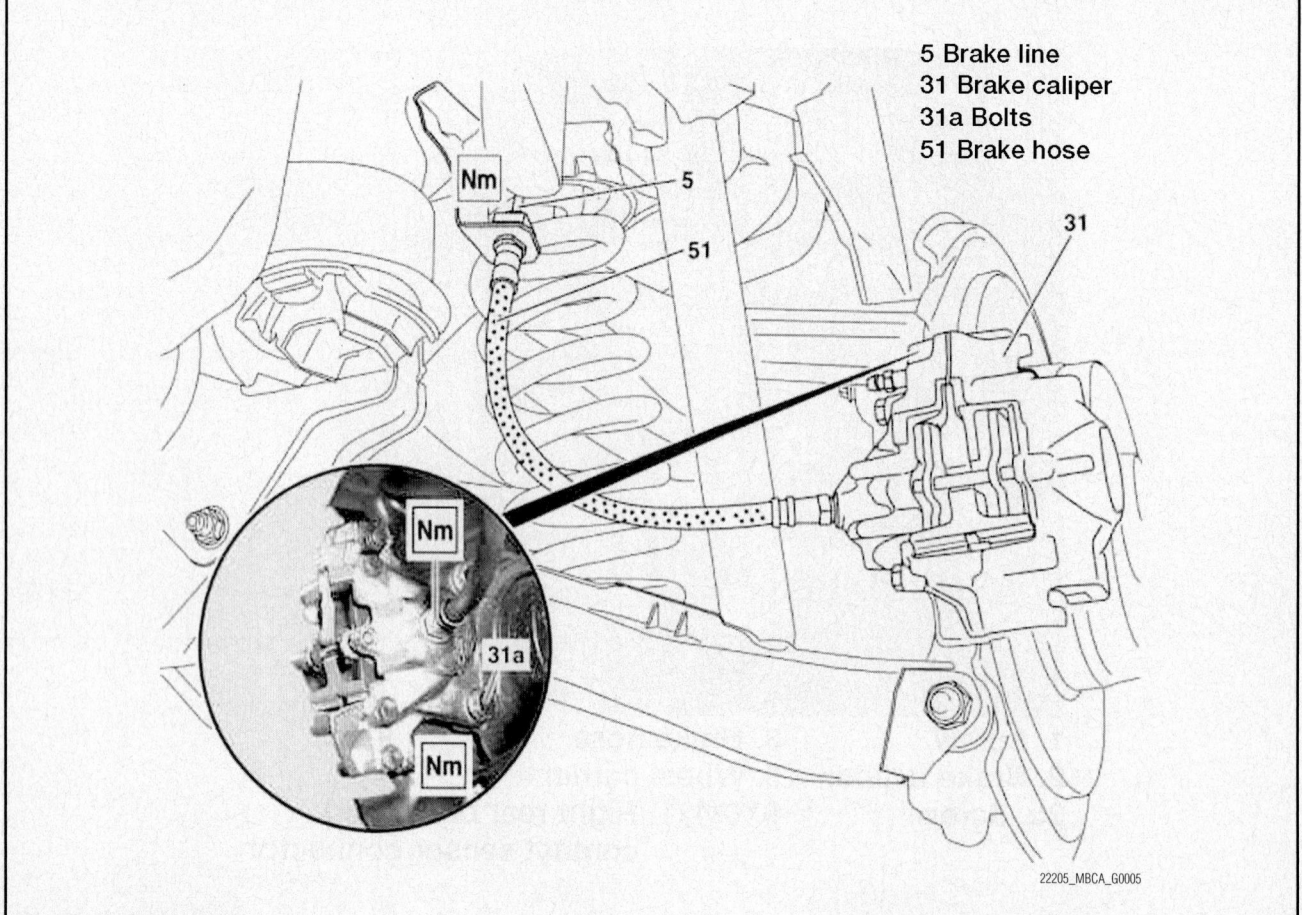

5 Brake line
31 Brake caliper
31a Bolts
51 Brake hose

22205_MBCA_G0005

**Fig. 23 Rear brake caliper**

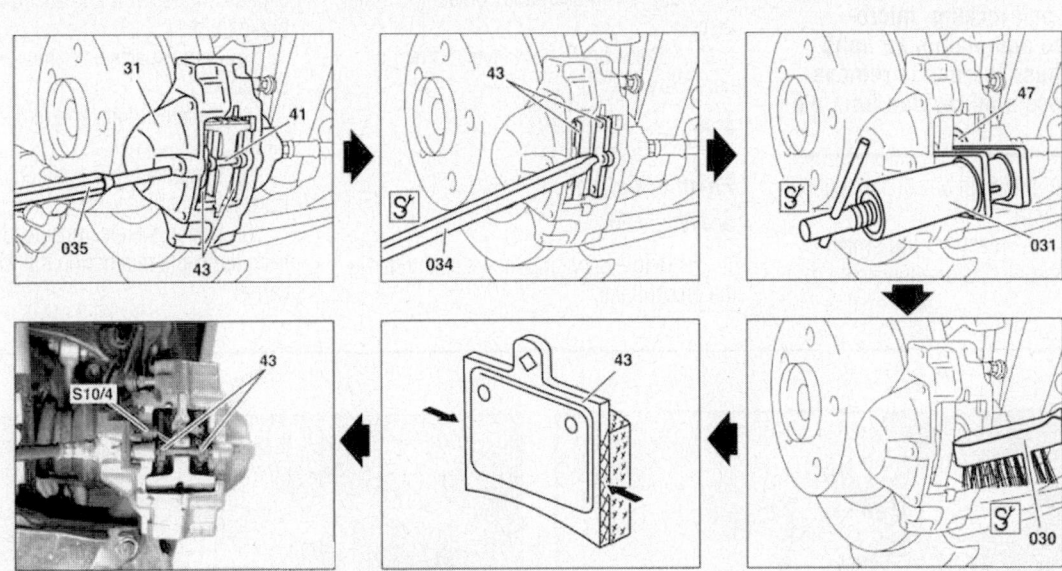

030 Brake caliper brush
035 Drift
43 Brake pad
031 Pusher tool

31 Brake caliper
47 Brake piston
034 Lever
41 Retaining pin
S10/4 Right rear brake pad contact sensor

22205_MBCA_G0007

**Fig. 24  Procedures to service rear brake pads**

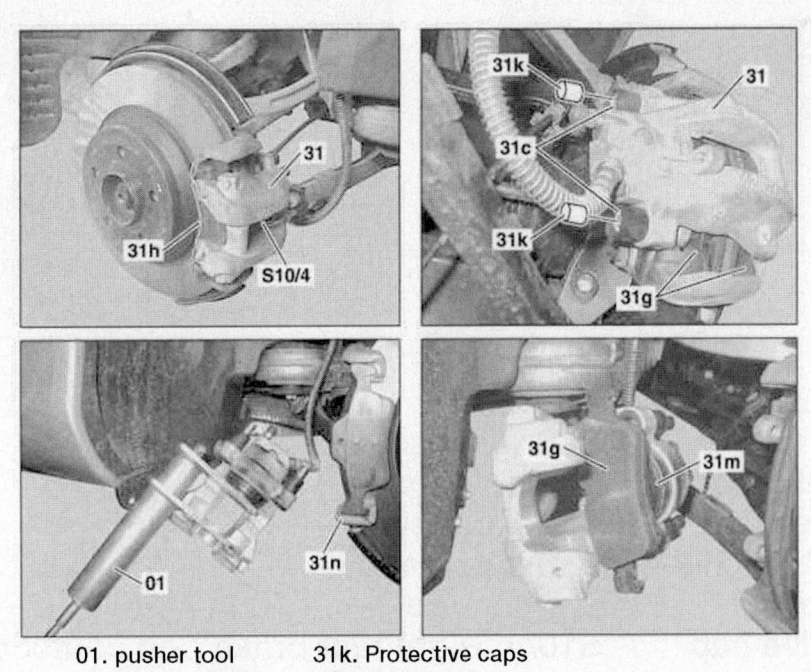

01. pusher tool
31. Floating caliper
31C. Guide bolts
31g. Brake pad
31h. Clamp

31k. Protective caps
31M. Dust boot
31n. Caliper support
S10/4. Right rear brake pad contact sensor

22205_MBCA_G0085

**Fig. 25  Rear brake pads**

**⁕⁕ WARNING**

Replace all self locking, micro-encapsulated and aluminum bolts and nuts. Chase threads to remove micro-encapsulated residue from old bolts/nuts.

19. Tighten guide pin on caliper support: 21 ft. lbs. (28 Nm).

20. Activate the brake control system using STAR DIAGNOSIS or equivalent scan tool.

21. Operate the brake pedal several times until the brake pads contact the brake discs.

22. Check brake fluid level, correct if necessary.

### S-Class

#### *Floating Caliper*

*See Figure 26.*

1. Before servicing the vehicle, refer to the precautions.

2. Unscrew the cap on the brake fluid expansion reservoir and suction off some brake fluid.

3. Remove wheels Remove/install wheels.

4. Unscrew bolts and remove cover panel.

5. Separate brake pad contact sensor from electrical feed line.

6. Unclip the electrical feed line from the cable retainer on the wheel carrier.

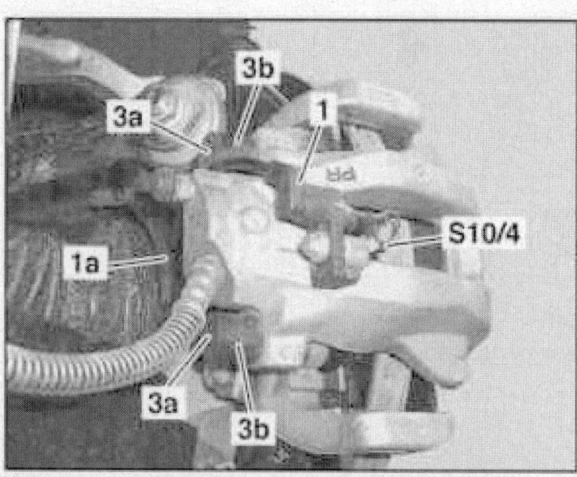

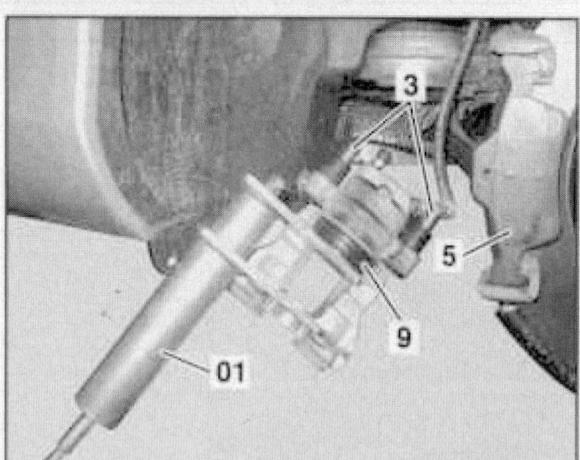
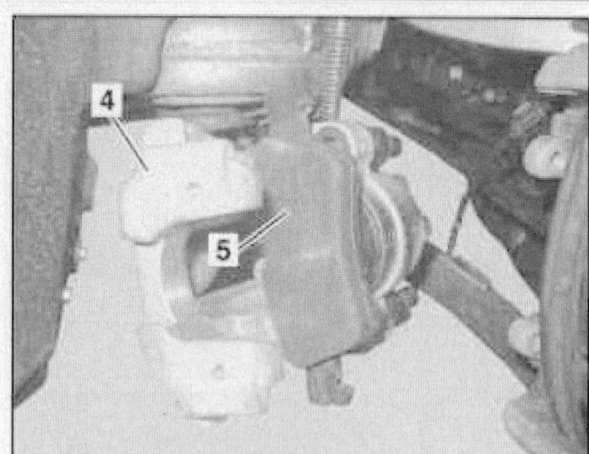

01. Pusher tool
1. Electrical feed line
1a. Cable retainer
2. Spring steel sheet
3. Guide bolts
3a. Protective cap
3b. Boot
4. Brake caliper
5. Brake pad
6. Brake hose
7. Caliper support
8. Brake disk
9. Dust boot
S10/4. Right rear brake pad contact sensor

22205_MBCA_G0159

**Fig. 26  Floating caliper**

7. Pry off spring steel sheet and remove.

8. Pry off and remove protective caps from boots.

9. Remove guide pins.

10. Remove brake caliper.

### ✳✳ WARNING

**Do not subject brake hoses to tensile loads or kink. Attach brake caliper without tension to vehicle to relieve load on brake hose, otherwise the brake hose will be damaged.**

➡**Do not detach brake hoses.**

**Take out brake pads.**

11. Pull brake pad contact sensor out of the pad backing plate of the brake pad.

12. Check brake pads and brake discs.

### ✳✳ WARNING

**Replace brake pads and brake discs in complete sets if necessary. Check brake lining thickness.**

13. Inspect condition of brake discs.

14. Check brake caliper for leaks.

15. Check dust boot on brake piston for damage.

16. Push back brake piston using pusher tool.

17. Clean contact surfaces of brake pads on brake caliper and brake caliper support using LU cleaner and a rag.

### ✳✳ WARNING

**Do not damage dust boot in the process. Do not use sharp-edged or pointed tools, otherwise the brake caliper or brake caliper support will be damaged.**

18. Grease contact surfaces of brake pads lightly with Molykote Cu 7439 brake pad paste.

**To install:**

19. Installation is the reverse of removal.

20. Replace bolts.

21. Coat ATE brake cylinder paste lightly on the contact surfaces of the bolt heads. Do not grease the threads. Wipe off excess brake cylinder paste after tightening the bolts.

22. Insert inner brake pad with riveted-on spring in brake piston. Insert outer brake pad in brake caliper support and mount brake caliper.

23. Tighten bolts/nuts to specification as follows:

- Guide pin, brake caliper to brake caliper support: 21 ft. lbs. (28 Nm)
- Guide pin to caliper support: 21 ft. lbs. (28 Nm)

24. Press brake pedal several times until the brake pads touch the brake discs.

25. Inspect fluid level in expansion reservoir, adjust to correct level if necessary.

26. Check function and effectiveness of brake system.

### Fixed Caliper

*See Figures 27 and 28.*

1. Before servicing the vehicle, refer to the precautions.

2. Unscrew the cap on the brake fluid expansion reservoir and suction off some brake fluid.

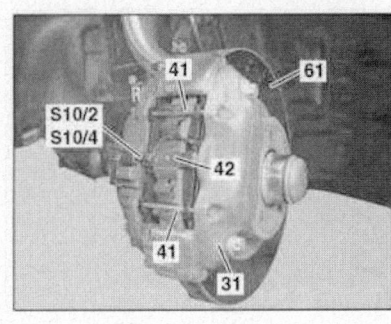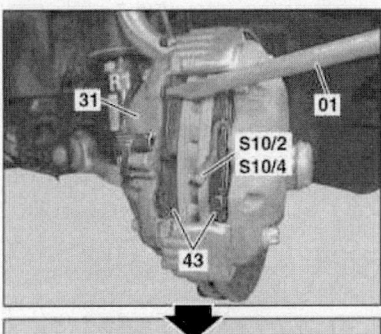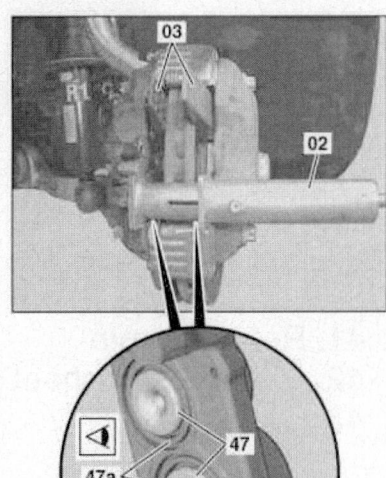

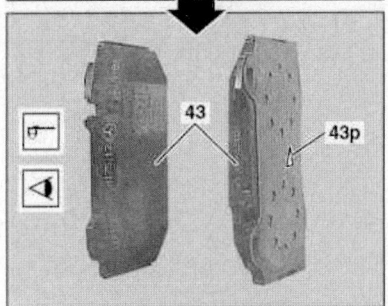

01. Lever
02. Pusher tool
03. Wedge
31. Brake caliper
41. Retaining pin
42. Spring steel sheet 8-piston brake caliper
43. Brake pad

43p. Arrow
47. Brake piston
47a. Dust boot
61. Brake disk
S10/2. Right front brake pad contact sensor
S10/4. Right rear brake pad contact sensor

22205_MBCA_G0160

**Fig. 27 4-Piston fixed caliper**

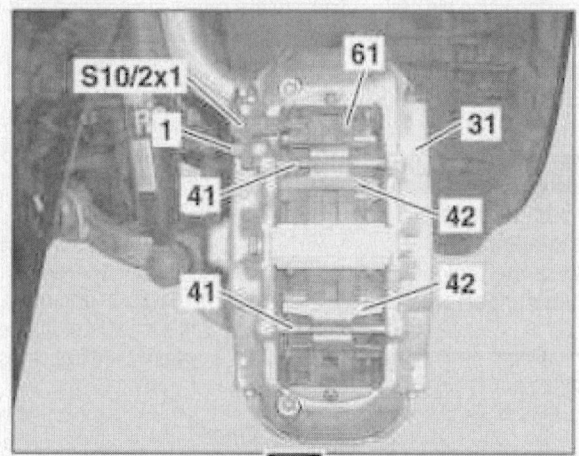
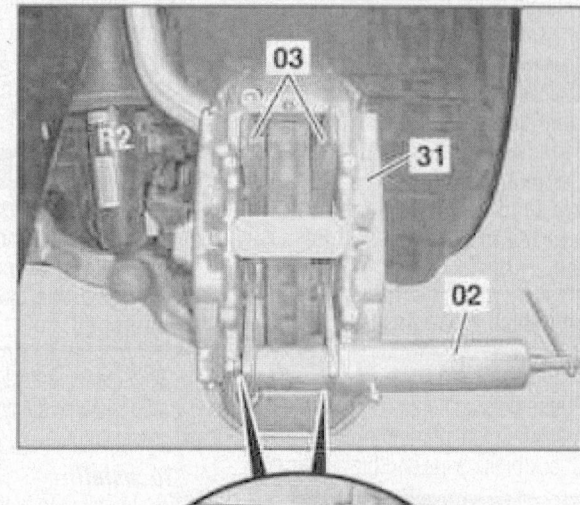
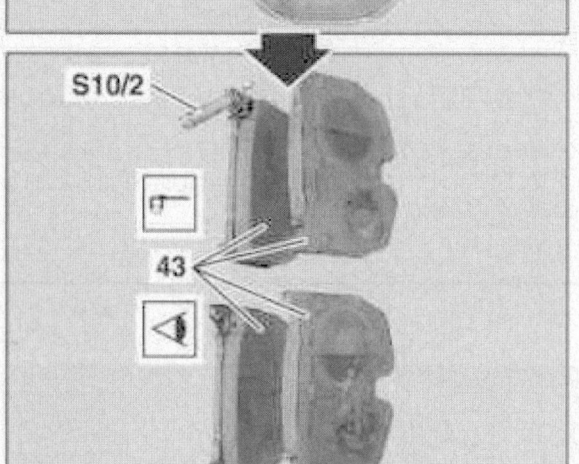
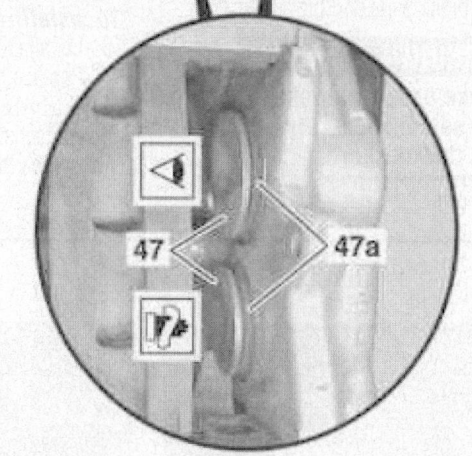

02. Pusher tool
03. Wedge
1. Bolt
31. Brake caliper
41. Retaining pin
42. Spring steel sheet
43. Brake pad

47. Brake piston
47a. Dust boot
61. Brake disk
S10/2. Right front brake pad contact sensor
S10/2x1. Right front brake pad
        contact sensor connector

22205_MBCA_G0161

**Fig. 28 8-Piston fixed caliper**

3. Remove wheels.
4. Check brake pads and brake discs.
5. Check brake lining thickness.
6. Inspect condition of brake discs.
7. Separate electrical connector of front right contact sensor or rear right contact sensor.
8. Pull right front brake pad contact sensor out of bracket on spring steel sheet.
9. Remove bolt and right front brake pad contact sensor connector from brake caliper.

10. Hammer out retaining pins using a drift.
11. Remove spring steel sheet.
12. Pull brake pads out of brake caliper.
13. Pull right front or right rear brake pad contact sensor out of the brake pad.
14. Check brake caliper for leaks.
15. Check brake piston dust boots.
16. Press back brake piston using pusher tool.
17. If necessary, clean the perforation in the brake discs.

18. Clean contact surfaces of brake pads on brake caliper and brake caliper support using LU cleaner and a rag.

**✳✳ WARNING**

**Do not damage dust boot in the process. Do not use sharp-edged or pointed tools, otherwise the brake caliper or brake caliper support will be damaged.**

19. Grease contact surfaces of brake pads lightly with Molykote Cu 7439 brake pad paste.

**To install:**

20. Installation is the reverse of removal.

21. Replace brake pads and brake discs in complete sets if necessary.

22. Mount all brake pads so that the arrow on the lining caliper points in the direction of rotation of the wheel when driving forwards. In the event of noise complaints only fit running direction-bound brake pads.

23. Press the brake pedal several times until the brake pads contact the brake discs.

24. Inspect fluid level in expansion reservoir, adjust to correct level if necessary.

25. Check function and effect of brake system.

## BRAKES

## PARKING BRAKE

### PARKING BRAKE SHOES

*REMOVAL & INSTALLATION*

#### C-Class

*See Figure 29.*

1. Before servicing the vehicle, refer to the precautions.
2. Remove the wheels.
3. Remove the brake disc.
4. Release the adjustable cable slack adjuster.
5. Carefully unhook the retracting and return springs from the brake shoes.
6. Remove the shoes.
7. Remove the shoe adjuster.

➡ **If the brake shoes are burnt, the shaft flange brake shoes, retracting springs and retaining springs must be replaced. On vehicles with internally ventilated brake discs, the brake discs must also be replaced.**

**To install:**

8. Installation is the reverse of removal.
9. Coat all bearing and slide surfaces at the expansion lock with MB long-term grease.
10. The retracting spring must be underneath the bolt. Make sure that the retracting spring is seated correctly.
11. Turn back the adjusting mechanism and insert into both brake shoes so that the

flat area on the adjusting screw is pointing upward.

➡ **The adjusting screw must point in the direction of travel on the left-hand side and toward the rear on the right-hand side.**

12. Make sure the spring are seated correctly.
13. Adjust the cable slack adjuster.

#### E-Class and S-Class

1. Before servicing the vehicle, refer to the precautions.
2. Deactivate the brake control system using STAR DIAGNOSIS or equivalent scan tool.

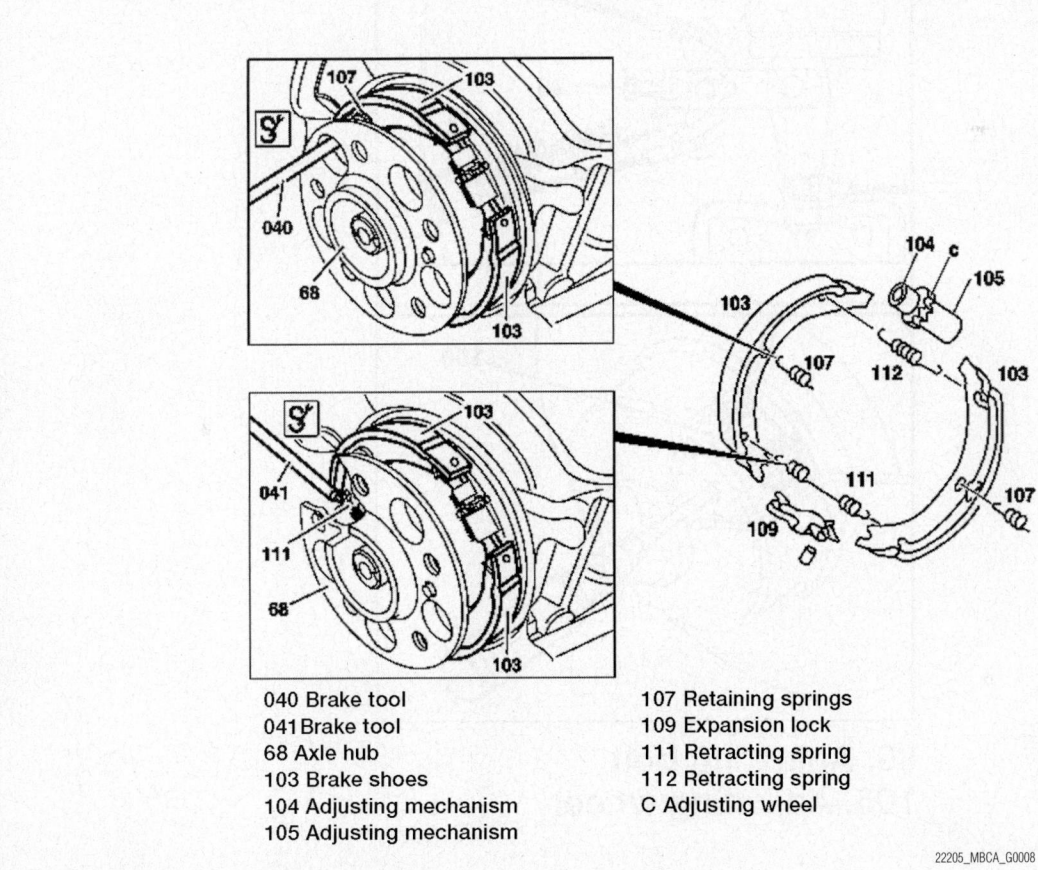

040 Brake tool
041 Brake tool
68 Axle hub
103 Brake shoes
104 Adjusting mechanism
105 Adjusting mechanism

107 Retaining springs
109 Expansion lock
111 Retracting spring
112 Retracting spring
C Adjusting wheel

22205_MBCA_G0008

**Fig. 29 Exploded view of the parking brake assembly**

3. Remove brake disc from the rear axle.

4. Detach front retracting spring using removal and installation tool.

5. Detach rear retracting spring using removal and installation tool.

6. Remove upper retaining spring using installation tool together. Squeeze the upper retaining spring together and turn 90° counterclockwise or clockwise.

7. Take off top brake shoe off adjusting mechanism. If the brake shoes are burnt or brake disc surfaces are damaged, the brake shoes, retracting springs, retaining springs and brake discs must be replaced.

8. Remove lower retaining spring using installation tool and remove lower brake shoes clockwise.

9. Remove expansion lock. Fold open expansion lock and press pin out of expansion lock and rear brake cable.

10. Inspect cap.

### To install:

11. Installation is the reverse of removal.

12. Check rear brake cable. Check for ease of movement and chafing marks, replace if necessary, install new rear brake cable.

13. Lightly grease the pin as well as the bearing and slide surfaces on the expansion lock with MB long-life grease.

14. Apply light coating of MB long-life grease to thread of thrust piece and inner cylinder of adjusting mechanism. Turn back the adjusting mechanism and place on the lower brake shoe in such a way that the adjustment wheel points upward.

15. Adjust parking brake.

### ADJUSTMENT

### C-Class

*See Figure 30.*

1. Before servicing the vehicle, refer to the precautions.

2. Operate parking brake and check pedal travel of parking brake pedal.

3. Raise and safely support the vehicle securely on jackstands.

4. Slacken adjusting screw.

5. Undo a wheel bolt at the left and right rear wheel.

6. Remove wheels.

7. Use screwdriver to rotate adjusting wheel until the brake shoes abut the parking brake drum and the rear wheel or brake disk can no longer be turned by hand.

8. Turn back adjusting wheel until the wheel or brake disk is able to rotate completely freely by hand.

9. Turn in adjusting screw until the brake cables no longer sag.

10. Operate parking brake pedal several times and check pedal travel of the parking brake pedal, then release the parking brake.

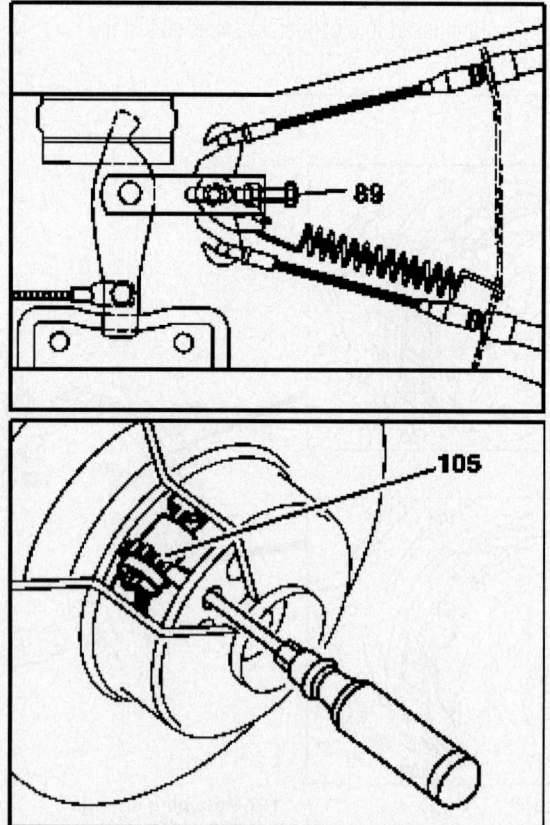

**89. Adjusting bolt**
**105. Adjusting wheel**

22205_MBCA_G0170

**Fig. 30 Adjusting the parking brake**

11. Operate parking brake lever several times, check parking brake lever travel, and then release the parking brake.

12. Check for unobstructed movement of rear wheels or brake disks.

13. Install wheels.

### E-Class

*See Figure 31.*

1. Before servicing the vehicle, refer to the precautions.

2. Operate parking brake pedal and check pedal travel of parking brake pedal.

3. Remove wheels.

4. Rotate adjusting wheel by means of a screwdriver until the brake shoes abut the parking brake drum and the rear wheel or brake disk can no longer be turned by hand.

5. Turn the left adjusting wheel in the direction of travel and the right adjusting wheel opposite to the direction of travel.

6. Turn back adjusting wheel by 10 teeth. Back off the left and right adjusting wheel by the same number of teeth. It must be possible to rotate the rear wheel or brake disk completely freely by hand.

7. Release automatic slack adjuster.

8. Operate parking brake pedal several times and check pedal travel of the parking brake pedal, then release the parking brake.

9. Check for unobstructed movement of the rear wheels or brake disks.

10. Fit wheel bolts or light alloy wheels.

11. Install wheels.

### S-Class

*See Figure 32.*

1. Before servicing the vehicle, refer to the precautions.

2. Release parking brake.

3. Hoist vehicle with lifting platform until wheels are free.

4. Remove rear wheels.

5. Turn right and left adjusting wheel using screwdriver until the parking brake shoes contact the parking brake drums.

6. Check brake disks for locking by turning manually.

7. Loosen off left and right adjusting wheel.

8. Check brake disks for unobstructed movement.

9. Operate parking brake.

10. Check brake disks for locking by turning manually.

11. Release parking brake.

12. Check brake disks for unobstructed movement.

13. Install rear wheels.

14. Lower vehicle.

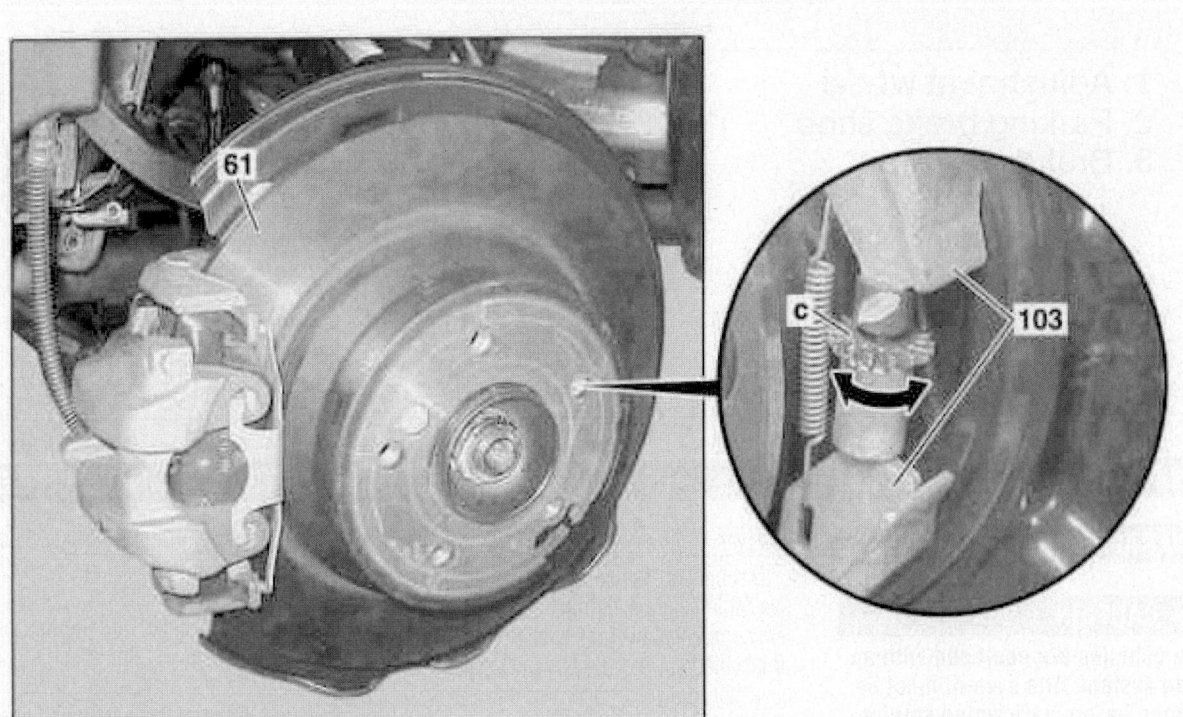

**61. Brake disk**
**103. Brake shoe**
**c. Thumbwheel**

22205_MBCA_G0169

**Fig. 31 Adjusting the parking brake**

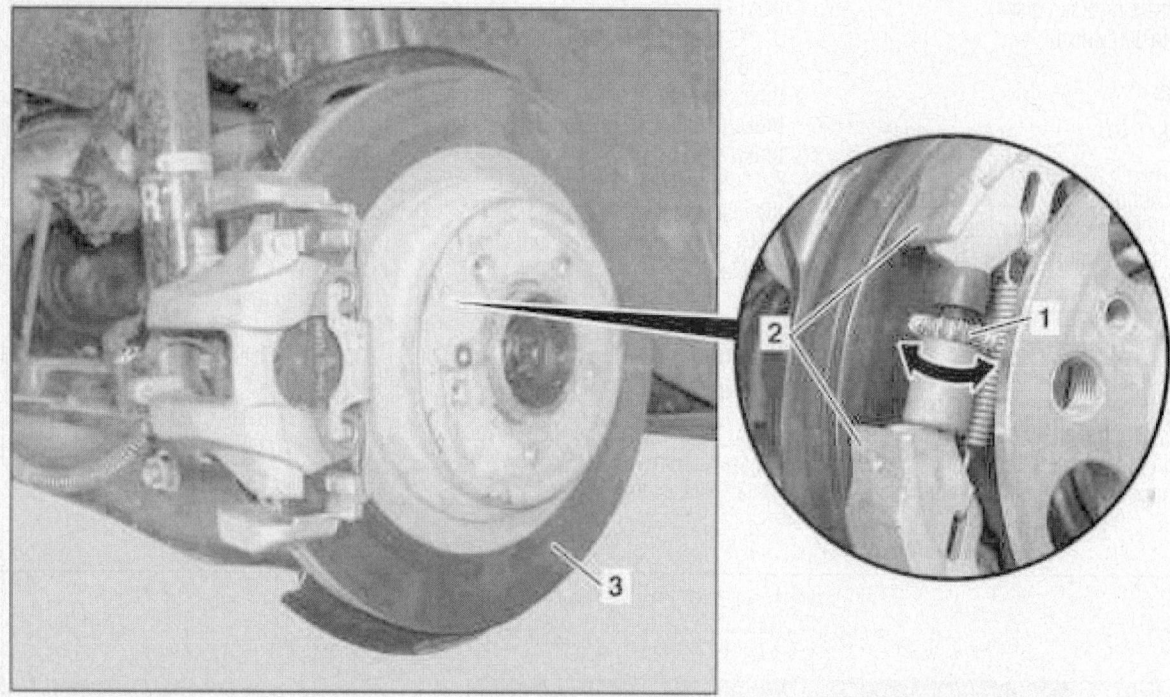

1. Adjustment wheel
2. Parking brake shoe
3. Brake disk

22205_MBCA_G0168

Fig. 32 Adjusting the parking brake

## CHASSIS ELECTRICAL  AIR BAG (SUPPLEMENTAL RESTRAINT SYSTEM)

### GENERAL INFORMATION

#### ☀ CAUTION

**These vehicles are equipped with an air bag system. The system must be disarmed before performing service on, or around, system components, the steering column, instrument panel components, wiring and sensors. Failure to follow the safety precautions and the disarming procedure could result in accidental air bag deployment, possible injury and unnecessary system repairs.**

### SERVICE PRECAUTIONS

Disconnect and isolate the battery negative cable before beginning any airbag system component diagnosis, testing, removal, or installation procedures. Allow system capacitor to discharge for two minutes before beginning any component service. This will disable the airbag system. Failure to disable the airbag system may result in accidental airbag deployment, personal injury, or death.

Do not place an intact undeployed airbag face down on a solid surface. The airbag will propel into the air if accidentally deployed and may result in personal injury or death.

When carrying or handling an undeployed airbag, the trim side (face) of the airbag should be pointing towards the body to minimize possibility of injury if accidental deployment occurs. Failure to do this may result in personal injury or death.

Replace airbag system components with OEM replacement parts. Substitute parts may appear interchangeable, but internal differences may result in inferior occupant protection. Failure to do so may result in occupant personal injury or death.

Wear safety glasses, rubber gloves, and long sleeved clothing when cleaning powder residue from vehicle after an airbag deployment. Powder residue emitted from a deployed airbag can cause skin irritation. Flush affected area with cool water if irritation is experienced. If nasal or throat irritation is experienced, exit the vehicle for fresh air until the irritation ceases. If irritation continues, see a physician.

Do not use a replacement airbag that is not in the original packaging. This may result in improper deployment, personal injury, or death.

The factory installed fasteners, screws and bolts used to fasten airbag components have a special coating and are specifically designed for the airbag system. Do not use substitute fasteners. Use only original equipment fasteners listed in the parts catalog when fastener replacement is required.

During, and following, any child restraint anchor service, due to impact event or vehicle repair, carefully inspect all mounting hardware, tether straps, and anchors for proper installation, operation, or damage. If a child restraint anchor is found damaged in

any way, the anchor must be replaced. Failure to do this may result in personal injury or death.

Deployed and non-deployed airbags may or may not have live pyrotechnic material within the airbag inflator.

Do not dispose of driver/passenger/curtain airbags or seat belt tensioners unless you are sure of complete deployment. Refer to the Hazardous Substance Control System for proper disposal.

Dispose of deployed airbags and tensioners consistent with state, provincial, local, and federal regulations.

After any airbag component testing or service, do not connect the battery negative cable. Personal injury or death may result if the system test is not performed first.

### DISARMING THE SYSTEM

To avoid personal injury when working on vehicles equipped with an air bag, the negative battery cable must be disconnected

and insulated before working on the system. Failure to do so may result in accidental deployment of the air bag.

### ARMING THE SYSTEM

To rearm the air bag system, reattach the battery cable(s).

### CLOCKSPRING CENTERING

#### C-Class

*See Figure 33.*

1. If the clock spring contact was twisted, bring the clock spring contact into the center position before reinstalling.

2. Turn the fanfare horns and airbag clock spring contact counterclockwise as far as the limit stop (turn gently against the limit stop, risk of damage), then turn approximately 2.5 turns clockwise until the screws are visible.

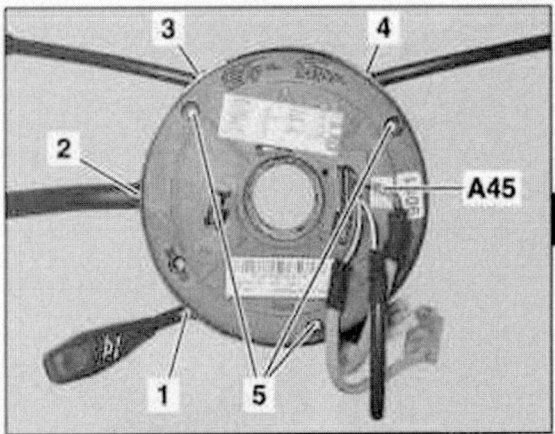
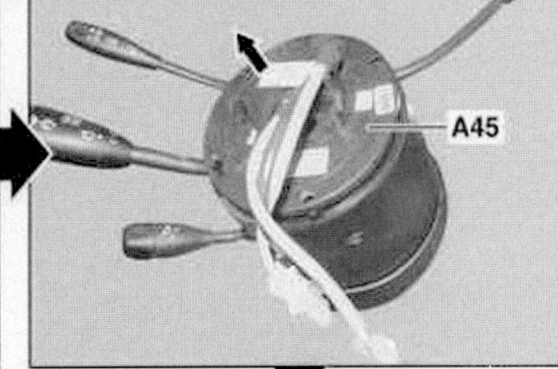
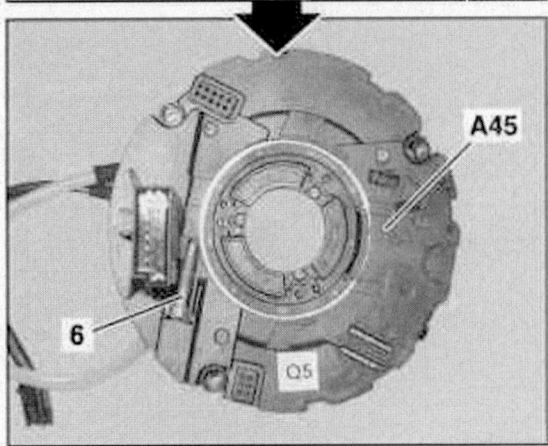

2. Switch trim
3. Switch trim
4. Switch trim
5. Screw
6. Clamping bolt
A45. Fanfare horns and airbag
     clock spring contact

22205_MBCA_G0061

**Fig. 33 Fanfare horns and airbag clock spring**

## E-Class

*See Figure 34.*

1. If the clock spring contact was twisted, bring the clock spring contact into the center position before reinstalling:

2. Turn clock spring contact clockwise up to the limit stop (turn gently against the limit stop, risk of damage), then turn counterclockwise until the arrows are opposite one another and the cable loop with black points is visible in the window.

## S-Class

*See Figure 35.*

1. If the clock spring contact was twisted, bring the clock spring contact into the center position before reinstalling.

2. Turn clock spring contact clockwise up to the limit stop (turn gently against the limit stop, risk of damage), then turn counterclockwise until the arrows are opposite one another and the cable loop with black spots is visible in the window.

3. Check plug contact is correctly seated on steering angle sensor.

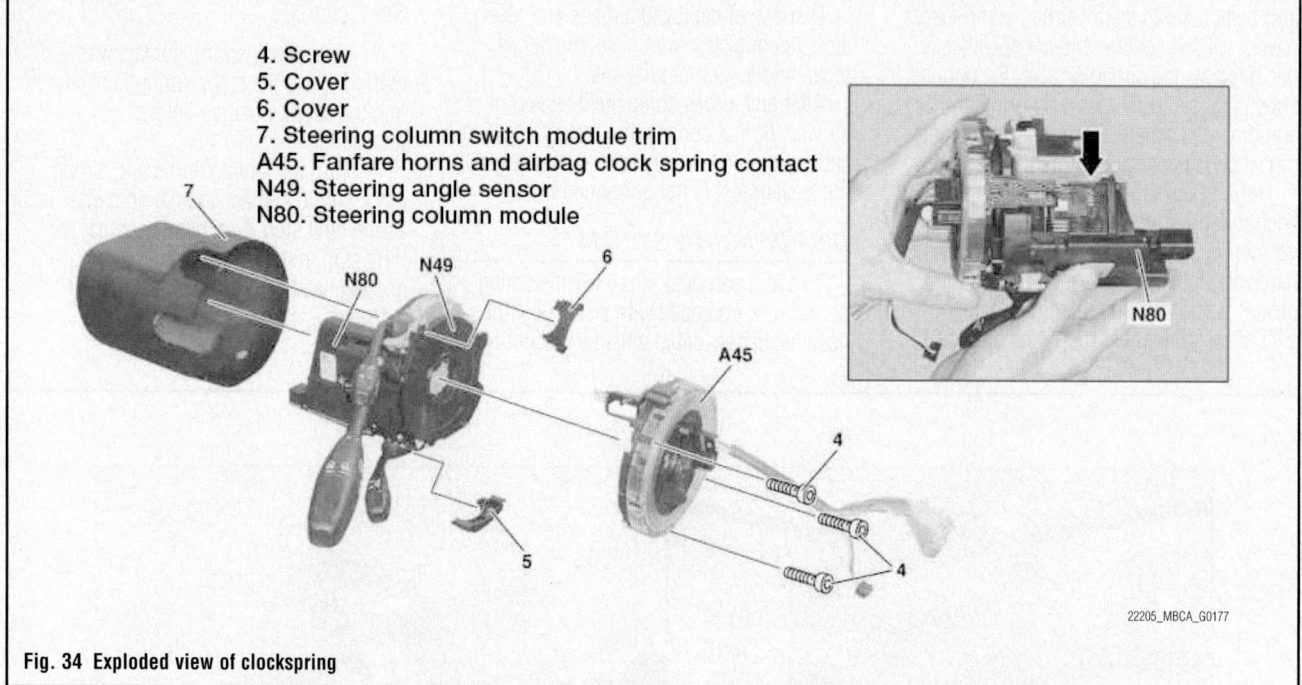

4. Screw
5. Cover
6. Cover
7. Steering column switch module trim
A45. Fanfare horns and airbag clock spring contact
N49. Steering angle sensor
N80. Steering column module

22205_MBCA_G0177

**Fig. 34 Exploded view of clockspring**

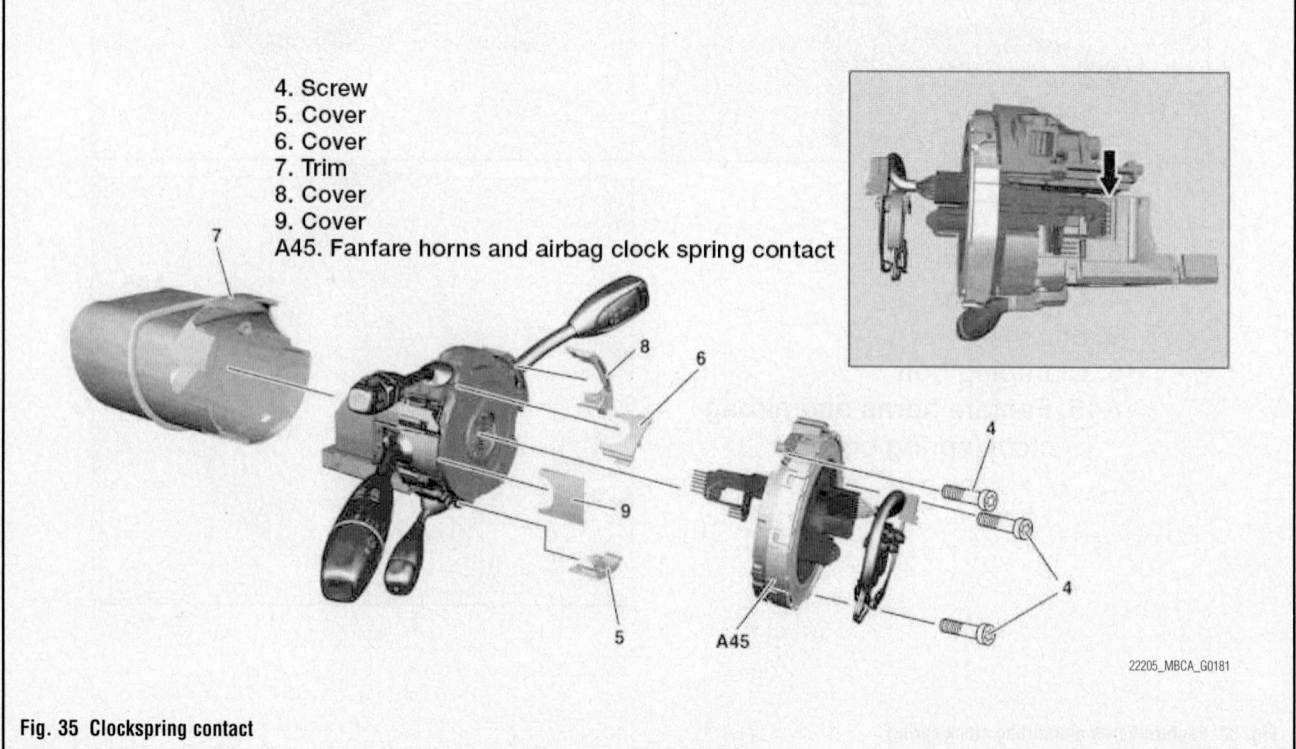

4. Screw
5. Cover
6. Cover
7. Trim
8. Cover
9. Cover
A45. Fanfare horns and airbag clock spring contact

22205_MBCA_G0181

**Fig. 35 Clockspring contact**

# DRIVETRAIN

## AUTOMATIC TRANSMISSION ASSEMBLY

### REMOVAL & INSTALLATION

*See Figure 36.*

1. Before servicing the vehicle, refer to the precautions.

➡ **Thoroughly clean the area around the cooling lines. The smallest dirt particles introduced into the hydraulic system can lead to malfunctions.**

2. Remove air filter housing.
3. Remove lower engine compartment paneling.
4. Drain transmission oil.
5. Remove complete exhaust system.

6. Support transmission with transmission jack.
7. Remove rear engine crossmember with engine crossmember.

➡ **Do not detach engine mount from rear engine mount.**

8. Disconnect front propeller shaft section from transmission. The flexible coupling remains on the front propeller shaft.
9. Undo fitted sleeves of flexible coupling and slide back front propeller shaft.

➡ **If the fitted sleeves are jammed, prepare drift for loosening fitted sleeves in flexible couplings.**

10. Press off retainer using a suitable tool.

11. Unhook shift rod from range selector lever on transmission.
12. Disconnect the transmission electrical connectors.
13. Unclip cover and remove bolts.
14. Detach pipe from engine oil pan.
15. Detach oil cooling line to oil cooler from transmission.
16. Remove the bolt and turn lead pipe down out of the way.
17. Remove transmission downwards at an angle using transmission jack and transmission. Secure torque converter to prevent it from falling out.

### To install:

18. Installation is the reverse of removal.

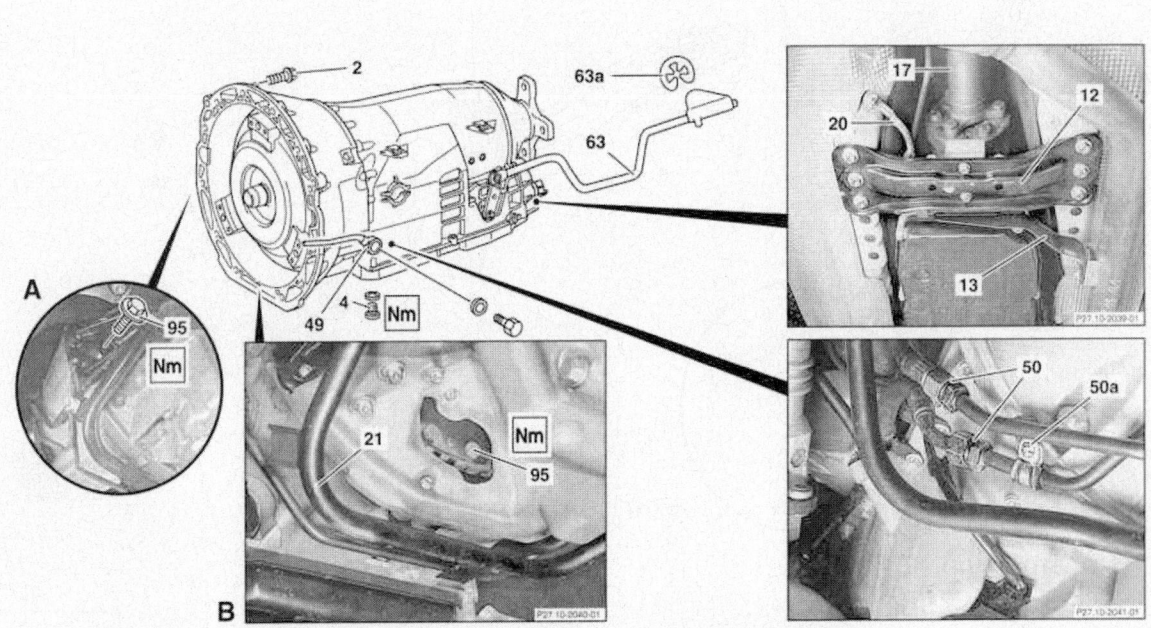

2 Bolts
4 Oil drain screw, transmission oil pan
12 Rear engine crossmember
13 Exhaust bracket
17 Front propeller shaft
20 Ground cable
21 Lead pipe
49 Oil cooling line

63 Shift rod
63a Fuse
95 Screw
50 Oil cooling line
50a Screw
A Remove this letter
B Except engine 111.955

22205_MBCA_G0009

**Fig. 36 Important steps of transmission removal**

※ **WARNING**

**Replace all self locking, micro-encapsulated and aluminum bolts and nuts. Chase threads to remove micro-encapsulated residue from old bolts/nuts.**

19. Tighten fasteners to the following specifications:
- Oil line bracket to oil pan: 7 ft. lbs. (9 Nm).
- Automatic transmission to crankcase: 29 ft. lbs. (39 Nm).
- Automatic transmission to engine oil pan: 29 ft. lbs. (39 Nm).
- Heat shield to transmission housing: 8 ft. lbs. (11 Nm).

- Self-locking nut, flexible coupling to transmission or front propeller shaft: M10 - 30 ft. lbs. (40 Nm) and M12 - 45 ft. lbs. (60 Nm).
- Torque converter to drive plate connection: Straight threaded 31 ft. lbs. (42 Nm)
- Torque converter to drive plate connection: Helical cut:
- Stage 1: 3 ft. lbs (4 Nm)
- Stage 2: 22 ft. lbs. (30 Nm)
- Stage 3: Tighten an additional 90° rotation
20. Fill transmission with gear oil.
21. Check oil level in automatic transmission, correct if necessary.
22. Perform basic programming.

23. Read out fault memory with STAR DIAGNOSIS or equivalent scan tool and erase if necessary.
24. Perform transmission adaptation after the transmission is changed or repaired.
25. Check transmission for proper function and leaks.

## MANUAL TRANSMISSION ASSEMBLY

### REMOVAL & INSTALLATION

*See Figure 37.*

1. Before servicing the vehicle, refer to the precautions.
2. Remove air filter housing.

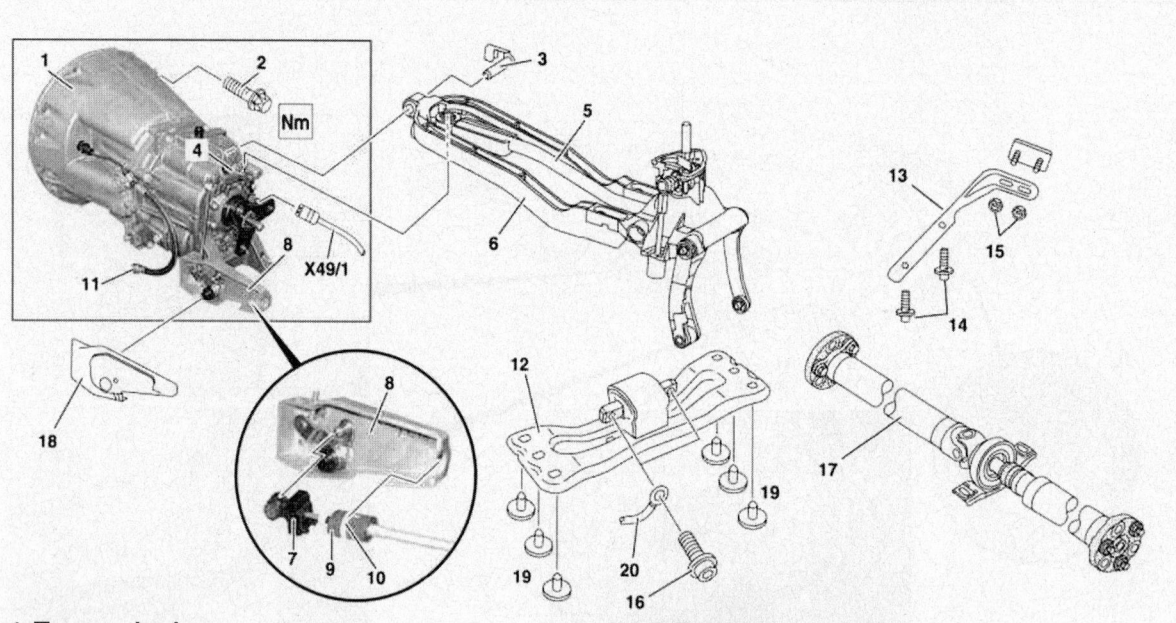

1 Transmission
2 Bolt
3 Safety bolt
4 Connecting piece
5 Shift rod
6 Connection for center-mounted shift mechanism / transmission
7 Selector cable
8 Bracket
9 Nut
10 Spacer ring
11 Hydraulic line

12 Engine support
13 Exhaust bracket
14 Bolts
15 Nuts
16 Bolt
17 Front propeller shaft
18 Cap
19 Bolts
20 Ground cable
X49/1 Backup lamp switch connector

22205_MBCA_G0010

Fig. 37 Exploded view of the manual transmission shift lever assembly—C-Class

3. Remove lower engine compartment paneling.

4. Remove complete exhaust system.

5. Detach front propeller shaft from transmission.

6. Undo fitted sleeves of flexible coupling and slide back front propeller shaft.

➡ **If the fitted sleeves are jammed, prepare drift for loosening fitted sleeves in flexible couplings.**

7. Separate plug connection of reversing light switch.

8. Disconnect clutch hydraulic line.

9. Support transmission with transmission jack.

10. Remove engine support with engine to do so, remove bolts and support. detach ground line.

➡ **The engine mount remains on the engine support.**

11. On transmissions with a selector cable perform the following:
   a. Release locking pin and pull out.
   b. Detach shift rod from ball head on transmission.

   c. Detach cap from bracket.
   d. Detach selector cable from bracket.
   e. Secure selector cable on inside of transmission tunnel.

12. On transmissions with a single rod gearshift perform the following:
   a. Detach pipe from engine oil pan.
   b. Unscrew bolt.
   c. Turn lead pipe down out of the way.

13. Remove transmission downwards at an angle.

14. Check thrust bearing and deep-groove ball bearing of crankshaft.

### To install:

15. Installation is the reverse of removal.

### ❊❊ WARNING

**Replace all self locking, micro-encapsulated and aluminum bolts and nuts. Chase threads to remove micro-encapsulated residue from old bolts/nuts.**

16. Tighten bolts/nuts to specification as follows:

- Oil pan to crankcase, timing case cover or rear cover: 7 ft. lbs. (9 Nm)
- Transmission to engine: 30 ft. lbs. (40 Nm)

17. Check transmission oil level; correct as necessary.

18. Bleed clutch operating system

19. Check shift mechanism for ease of movement. On vehicles with selector cable shift mechanism check if selector cable shift mechanism is sluggish. If so adjust selector cable.

20. Perform basic programming.

21. Read out fault memory with STAR DIAGNOSIS or equivalent scan tool and erase if necessary.

### CLUTCH

#### REMOVAL & INSTALLATION

**C-Class**

*See Figure 38.*

1. Before servicing the vehicle, refer to the precautions.

2. Remove transmission.

3. Unscrew hexagon socket bolts.

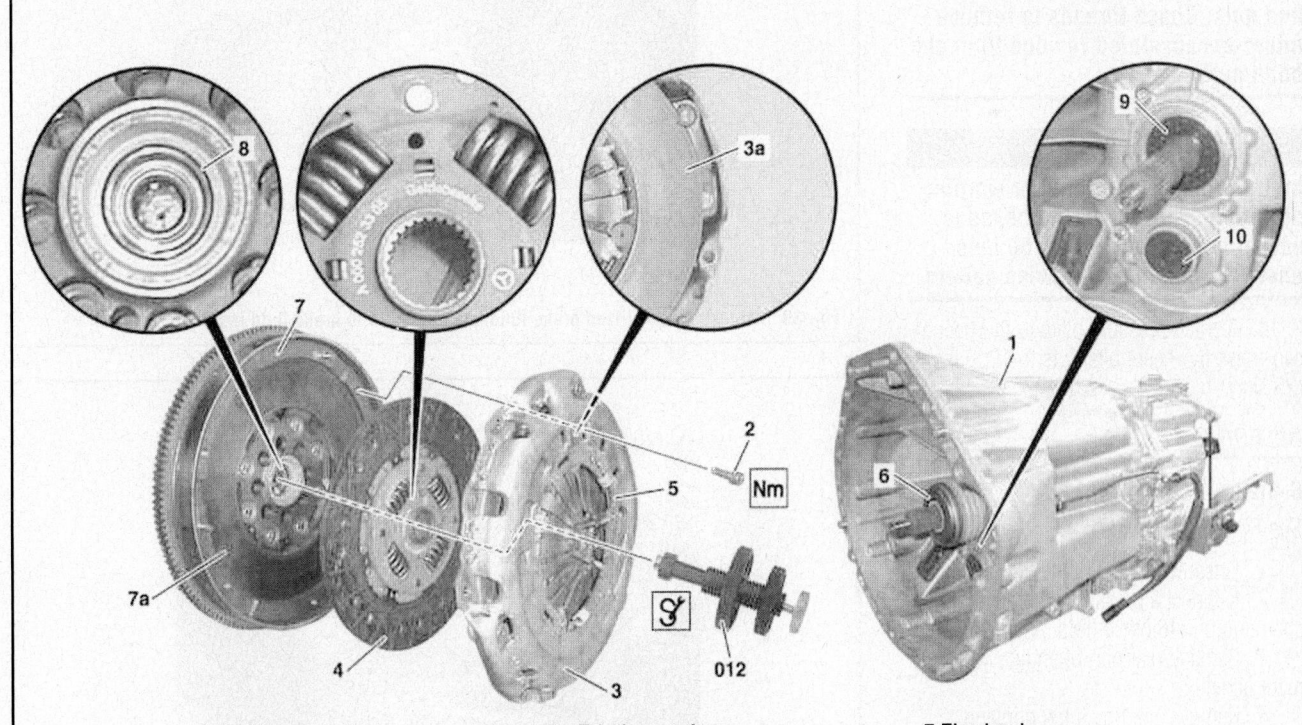

012 Alignment tool
1 Transmission
2 Hexagon socket head screw
3 Pressure plate

3a Friction surface
4 Clutch plate
5 Adjusting ring
6 Central clutch release bearing

7 Flywheel
7a Friction surface
8 Guide bearing
9 Radial shaft seal

22205_MBCA_G0011

**Fig. 38 Exploded view of the clutch assembly**

**✳✳ WARNING**

**In order to prevent tilting or warping of the pressure plate, the hexagon socket head screws must be loosened in steps in a crosswise pattern.**

4. Remove pressure plate and clutch disc.

5. Check pressure plate and clutch plate

6. Inspect central clutch release bearing

7. Inspect flywheel.

8. Check guide bearing for wear.

9. Check radial shaft sealing ring and cover for leaks.

### To install:

10. Installation is the reverse of removal.

11. Clean friction surface of pressure plate, degrease with LU Cleaner (BR00.45-Z-1028-04A) and dress with coarse abrasive cloth.

12. Fit pressure plate with clutch plate on flywheel and tighten.

13. Using centering tool, attach clutch plate to pressure plate centrally.

**✳✳ WARNING**

**Replace all self locking, micro-encapsulated and aluminum bolts and nuts. Chase threads to remove micro-encapsulated residue from old bolts/nuts.**

**✳✳ WARNING**

**In order to prevent tilting or warping of the pressure plate, the hexagon socket head screws must be loosened in steps in a crosswise pattern.**

14. Tighten pressure plate to flywheel or two-mass flywheel bolts/nuts to 19 ft. lbs. (25 Nm).

### BLEEDING

#### C-Class

*See Figures 39 and 40.*

1. Unscrew cap on brake fluid reservoir.

2. Connect a pressurized brake fluid changing unit to brake fluid reservoir.

3. Remove rear part of engine compartment paneling.

4. Remove cap on clutch housing.

5. Open bleed screw on central clutch operator.

6. Allow brake fluid to flow out. Brake fluid must emerge clean and free of bubbles.

7. Close bleed screw on central clutch operator. Tighten to 7 ft. lbs. (9 Nm).

8. Correct brake fluid level in reservoir.

9. Screw cap onto brake fluid reservoir.

10. Check function of clutch operator.

11. Check system for leakage.

### TRANSFER CASE ASSEMBLY

#### REMOVAL & INSTALLATION

#### C-Class

*See Figure 41.*

On 4MATIC models, a "transfer case" assembly is located on the end of the transmission and is used to transmit torque to the front drive axles.

1. Before servicing the vehicle, refer to the precautions.

2. Raise and safely support the vehicle securely.

3. Detach exhaust shielding plate for rear engine mount from reinforcement brace

4. Remove exhaust bracket.

5. Detach transmission ground from transfer case.

6. Mark propeller shaft of front axle gear relative to flange of transfer case.

7. Detach front axle gear propeller shaft from transfer case.

8. Support automatic transmission with transmission jack.

9. Remove rear engine crossmember.

10. Detach front propeller shaft from transfer case.

➡**The flexible coupling remains on the propeller shaft.**

11. Disconnect front propeller shaft section from transmission. The flexible coupling remains on the front propeller shaft.

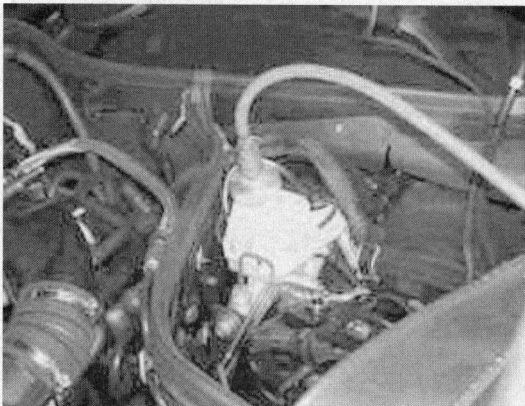

22205_MBCA_G0012

**Fig. 39 Connect a pressurized brake fluid changing unit to brake fluid reservoir**

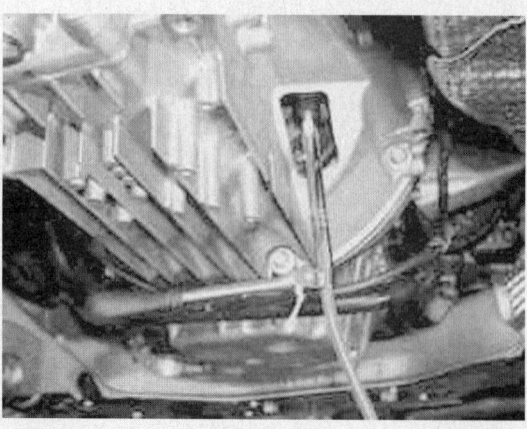

22205_MBCA_G0013

**Fig. 40 Allow brake fluid to flow out. Brake fluid must emerge clean and free of bubbles**

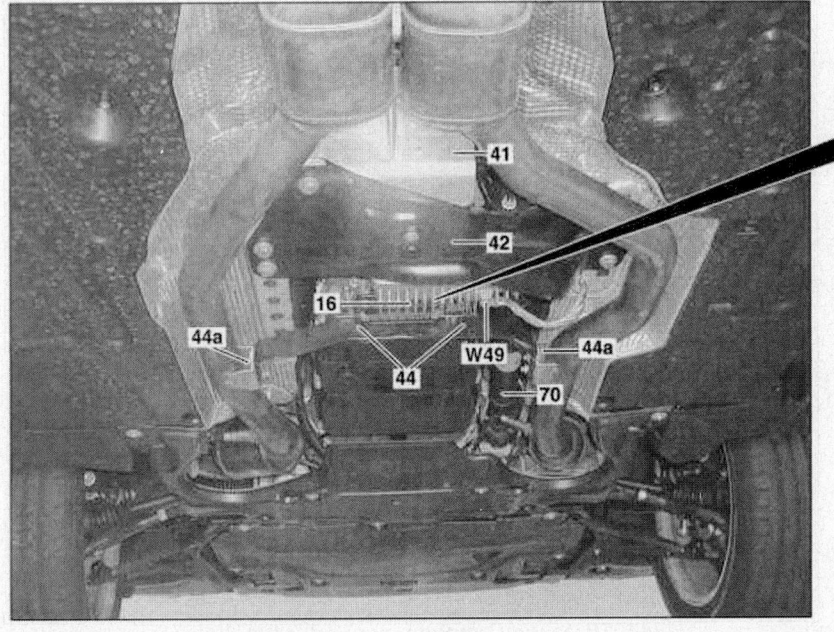
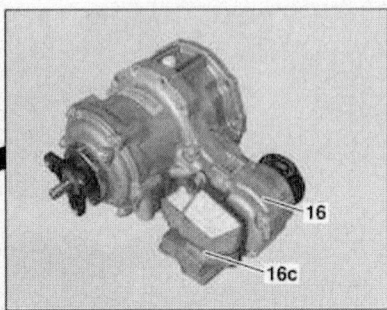

16 Transfer case
16c Vibration damper
41 Exhaust shielding plate
42 Rear engine crossmember
44 Exhaust bracket
44a Threaded plates
70 Front axle gear propeller shaft
W49 Ground (transmission)

22205_MBCA_G0014

**Fig. 41 Transfer case mounted to the rear of the transmission**

12. Undo fitted sleeves of flexible coupling and slide back front propeller shaft.

➡ **If the fitted sleeves are jammed, prepare drift for loosening fitted sleeves in flexible couplings.**

13. Slightly lower automatic transmission.
14. Detach transfer case from automatic transmission.

### To install:

15. Installation is the reverse of removal.
16. Coat plug-type splines in transfer case with grease Klueber Microlube GNY 202.
17. Mount transfer case to automatic transmission.

➡ **Both centering sleeves (arrows) have to be firmly seated in the automatic transmission (17), if necessary replace centering sleeves.**

18. Check oil level in transfer case, correct if necessary

### ❊❊ WARNING

**Replace all self locking, micro-encapsulated and aluminum bolts and nuts. Chase threads to remove micro-encapsulated residue from old bolts/nuts.**

19. Tighten bolts/nuts to specification as follows:

- Nut and bolt, transfer case to automatic transmission: 19 ft. lbs. (25 Nm)
- Oil filler opening screw plug: 22 ft. lbs. (30 Nm)
- Inspection hole screw plug: 22 ft. lbs. (30 Nm)
- Bolt of front axle gear propeller shaft to transfer case flange:
- Step 1: 11 ft. lbs. (15 Nm)
- Step 2: Tighten an additional 60° rotation.

### E-Class

*See Figure 42.*

On 4MATIC models, a "transfer case" assembly is located on the end of the transmission and is used to transmit torque to the front drive axles.

### ❊❊ WARNING

**Before starting work the area around the separation points on the automatic transmission and transfer case must be cleaned thoroughly. Even the smallest dirt particles in the hydraulic components can lead to malfunctions and a total failure of the automatic transmission and transfer case.**

1. Before servicing the vehicle, refer to the precautions.
2. Remove engine trim panel.
3. Remove hot film mass air flow sensor.
4. Remove rear section of lower engine compartment paneling.
5. Remove rear part section of bottom section of soundproofing.
6. Detach exhaust shielding plate for propeller shaft behind transfer case.
7. Detach three way catalytic converter bracket.
8. Detach propeller shaft of front axle gear from flange of transfer case. Tie back the propeller shaft otherwise the oxygen sensor may be damaged.
9. Detach rear propeller shaft from flexible coupling.
10. Loosen fitted sleeves of flexible coupling and push back propeller shaft.
11. Support automatic transmission with transmission jack.
12. Remove rear engine crossmember.
13. Remove flexible coupling from flange on rear axle gear.
14. Unhook shift rod at automatic transmission. Push the retainer upwards until the pin of the shift rod can be removed.
15. Remove crossmember.
16. Unhook rear exhaust system from the exhaust rubber mounts, lower slightly and tie securely.

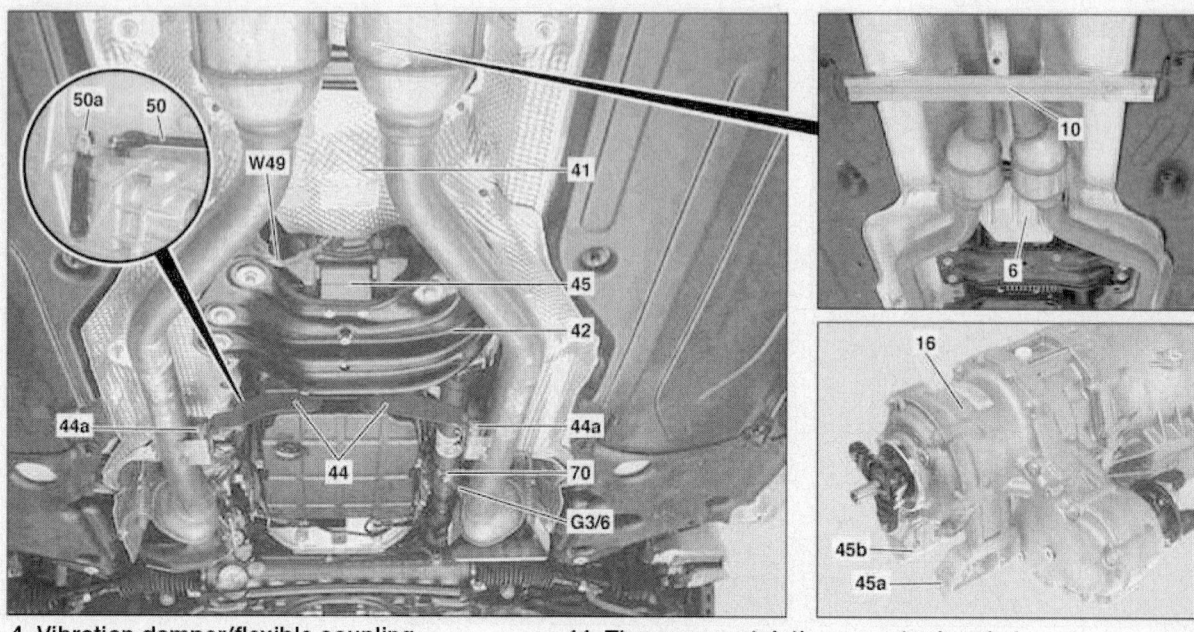

4. Vibration damper/flexible coupling
5. Reinforcement behind engine crossmember
6. Exhaust shielding plate
10. Crossmember
16. Transfer case
42. Rear engine crossmember

44. Three way catalytic converter bracket
44a. Threaded plates
50. Shift rod
50a. Fuse
70. Front axle gear propeller shaft

22205_MBCA_G0193

**Fig. 42  Transfer case mounted to the rear of the transmission**

---

**✳✳ WARNING**

**Lower the exhaust system carefully so that the rear bumper is not damaged by the rear muffler.**

17. Remove banjo bolt from left and right oil line at automatic transmission.
18. Slightly lower automatic transmission.
19. Remove reinforcement of rear engine crossmember.
20. Remove vibration damper/flexible coupling.
21. Detach transfer case from automatic transmission and remove Install.

***To install:***
22. Coat plug-type splines in transfer case with grease.
23. Mount transfer case on automatic transmission. Both centering sleeves (arrows) must seat firmly in the automatic transmission, replace centering sleeves if necessary.
24. Install reinforcement of rear engine crossmember. Do not tighten the bolt yet.

25. Attach flexible coupling to rear axle gear output flange.
26. Attach rear propeller shaft to flexible coupling.
27. Install rear engine crossmember.
28. Tighten bolt for reinforcement of rear engine crossmember.
29. Remove transmission jack and transmission platform.
30. Attach rear exhaust system to exhaust rubber mounts.
31. Check oil level in transfer case and correct if necessary.
32. Install vibration damper/flexible coupling.
33. Install the cross member.
34. Attach exhaust shielding plate for propeller shaft behind transfer case.
35. Install catalytic converter bracket.
36. Screw banjo bolt of left and right oil line into automatic transmission.
37. Attach propeller shaft of front axle gear to flange of transfer case.

38. Replace retainer of shift rod in order to avoid a malfunction due to a broken retainer.
39. Attach shift rod to automatic transmission and mount retainer.
40. Install rear part piece of engine compartment paneling.
41. Install rear part section of soundproofing bottom section.
42. Install hot film mass air flow sensor.
43. Reinstall engine cover.
44. Lower vehicle with lifting platform.
45. Tighten bolts/nuts to specification as follows:

- Banjo bolt or union nut, oil cooler line to torque converter and transmission housing: Tighten to 4 ft. lbs. (5 Nm) plus an additional 90° rotation
- Self-locking nut, flexible coupling to rear propeller shaft or rear axle differential M10: 30 ft. lbs. (40 Nm)
- Self-locking nut, flexible coupling to rear propeller shaft or rear axle differential M12: 44 ft. lbs. (60 Nm)

- Bolt, crossmember to body: 19 ft. lbs. (25 Nm)
- Nut and bolt, transfer case to automatic transmission: 19 ft. lbs. (25 Nm)
- Bolt, vibration damper to transfer case: 15 ft. lbs. (20 Nm)

### S-Class

S-Class transfer case information was not available at time of publication.

## DIFFERENTIAL ASSEMBLY

### REMOVAL & INSTALLATION

### C-Class

*See Figure 43.*

1. Before servicing the vehicle, refer to the precautions.
2. Open engine hood.
3. Remove left and right intake air duct upstream of air filter.
4. Detach air filter housing from above vertically from the cylinder head covers.
5. Remove air filter housing.

6. Loosen bolt on exhaust shielding plate for propeller shaft of front axle gear.
7. Mount engine support bracket and raise engine slightly.
8. Detach return line of rack-and-pinion steering from the top of the front axle carrier. Nuts between the engine and radiator and nut in front of the right-hand engine mount.
9. Raise vehicle.
10. Detach front and rear lower engine compartment paneling.
11. Detach return line of rack-and-pinion steering from the bottom of the front axle carrier. The nut is located in front of the steering coupling.
12. Remove right and left front axle shaft.
13. Remove intermediate shaft.
14. Detach stabilizer bar on left and right of the link rods.
15. Detach the exhaust shielding plate above the right rubber boot of the rack-and-pinion steering.
16. Remove retaining panel of rack-and-pinion steering.
17. Pull rack-and-pinion steering slightly backwards and tie in place.

18. Remove right O2 sensor downstream of TWC.
19. Unclip electrical feed line of O2 sensor from retaining clamps.
20. Undo bolt and nut and remove exhaust shielding plate.
21. Mark front axle gear propeller shaft in relation to front axle gear flange.
22. Detach front axle gear propeller shaft from front axle gear flange.
23. Tie front axle gear propeller shaft to one side.
24. Detach the left and right engine mount at the bottom of the front axle carrier.
25. Remove headlight range control's linkage rod from lower control arm.
26. Support front subframe.
27. Remove bolts of front axle carrier.
28. Secure front axle carrier against falling and lower front subframe.
29. Detach exhaust shielding plate for front axle gear propeller shaft.
30. Detach mounting support of front axle gear.
31. Detach high-pressure line of rack-and-pinion steering from the oil pan.

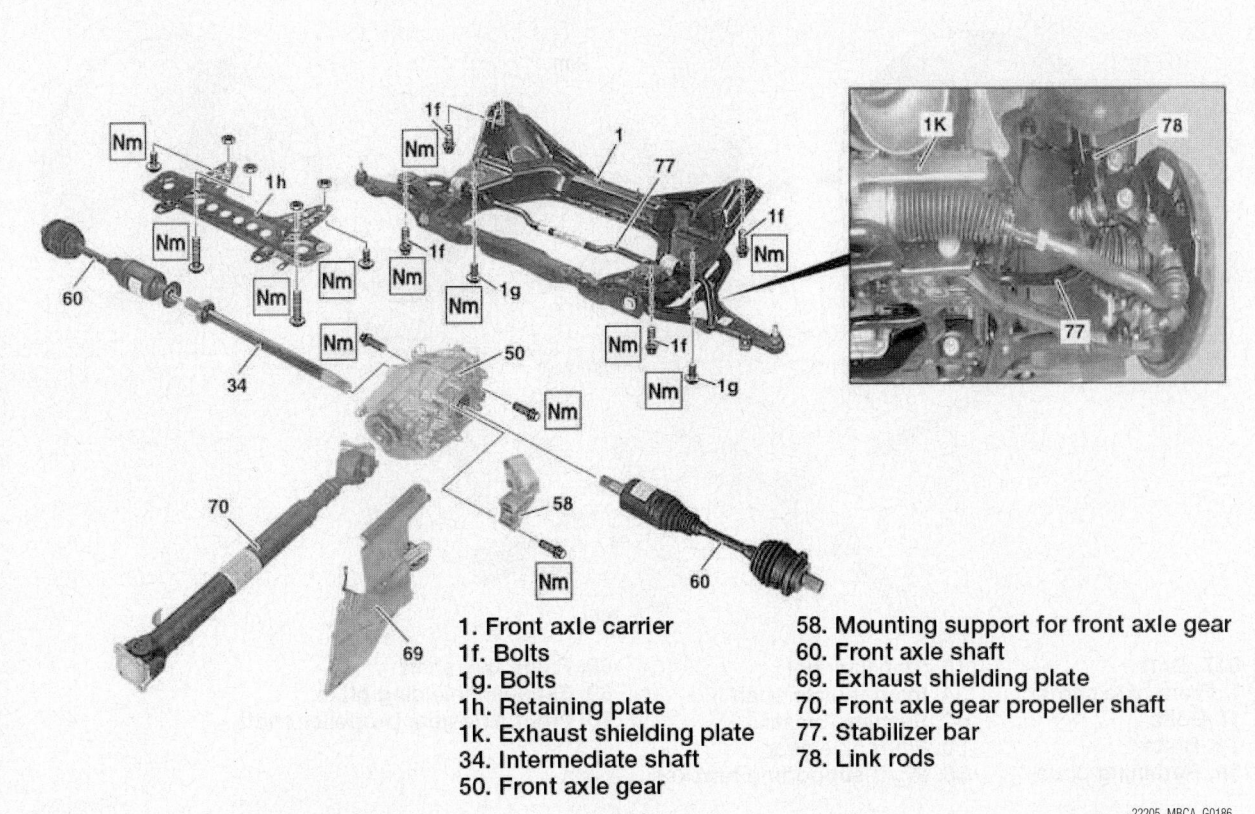

1. Front axle carrier
1f. Bolts
1g. Bolts
1h. Retaining plate
1k. Exhaust shielding plate
34. Intermediate shaft
50. Front axle gear
58. Mounting support for front axle gear
60. Front axle shaft
69. Exhaust shielding plate
70. Front axle gear propeller shaft
77. Stabilizer bar
78. Link rods

22205_MBCA_G0186

**Fig. 43 Exploded view of front axle assembly**

32. Insert drift into radial sealing ring on outside right of front axle gear.

33. Detach front axle gear and remove.

### To install:

34. Installation is the reverse of removal.

35. The bolts of front axle carrier are self-tapping, so the following points must be observed:

- Never re-tap the thread in the side member
- Replace screws
- Use only original bolts
- Do not crossthread bolts
- Do not use an impact wrench

36. Tighten bolts/nuts to specification as follows:

- Bolt, front engine mount to front axle carrier: 26 ft. lbs. (35 Nm)
- Self-locking nut of connecting rod to stabilizer bar: 41 ft. lbs. (55 Nm)
- Bolt, front axle carrier to front end: 74 ft. lbs. (100 Nm)
- Bolt, front axle gear propeller shaft to front axle gear flange: 11 ft. lbs.

(15 Nm) plus an additional 60° rotation

- Collar bolt connecting mounting support to front axle final drive and engine mount carrier: 16 ft. lbs. (22 Nm)
- Collar bolt, front axle gear to engine oil pan: 48 ft. lbs. (65 Nm)
- Oil filler opening screw plug: 22 ft. lbs. (30 Nm)
- Bolted connection, rack-and-pinion steering to front axle carrier: 37 ft. lbs. (50 Nm) plus an additional 90° rotation
- Bolted connection, supporting plate to front axle carrier: 37 ft. lbs. (50 Nm) plus an additional 60° rotation

37. When installing air filter housing, apply antifriction agent to the rubber retainers on the cylinder head covers and the sealing ring of the hot film mass air flow sensor in the air filter housing.

38. Check front axle gear oil level, refill if necessary

39. Read out fault memory with STAR DIAGNOSIS and erase if necessary.

### E-Class

*See Figure 44.*

1. Before servicing the vehicle, refer to the precautions.

2. Remove the left and right air guide duct.

3. Remove air filter housing.

4. Mount engine lifting device and raise engine.

5. Lift vehicle on lifting platform.

6. Remove catalytic converter.

7. Remove right and left front axle shaft.

8. Remove intermediate shaft.

9. Remove shield for right bellows on steering gear.

10. Release engine mount at bottom.

11. Mark propeller shaft of front axle gear relative to flange on front axle gear.

12. Detach propeller shaft of front axle gear from flange of front axle gear.

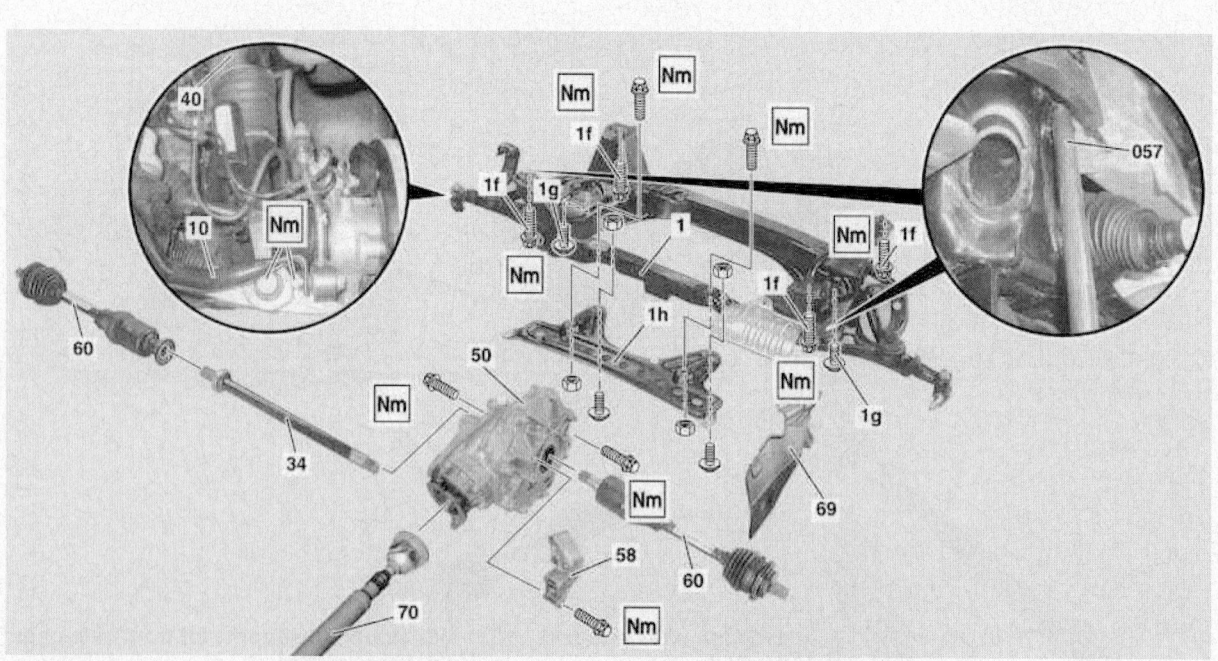

| | | |
|---|---|---|
| 057. Drift | 10. Stabilizer bar | 60. Front axle shaft |
| 1. Front axle carrier | 34. Intermediate shaft | 69. Exhaust shielding plate |
| 1f. Bolts | 40. Suspension strut | 70. Front axle gear propeller shaft |
| 1g. Bolts | 50. Front axle gear | |
| 1h. Retaining plate | 58. Right supporting bracket | |

22205_MBCA_G0187

**Fig. 44 Exploded view of front axle assembly**

13. Tie front axle gear propeller shaft to one side.

14. Detach high pressure line for power steering pump and electrical line for speed-sensitive power steering on front axle carrier.

15. Remove retaining panel of rack-and-pinion steering.

16. Pull rack-and-pinion steering slightly backwards and tie in place.

17. Support the front axle carrier with a transmission jack.

18. Remove bolts from front axle carrier.

19. Lower front subframe and secure front axle carrier to prevent it from falling out.

20. Remove exhaust shielding plate.

21. Detach right supporting bracket. Lower the engine slightly to provide better access to the bolts.

22. Insert drift into radial sealing ring on outside right of front axle gear.

23. Detach and lift out front axle gear.

*To install:*

24. Installation is the reverse of removal.

25. The bolts of front axle carrier are self- tapping, so the following points must be observed:

- Never re-tap the thread in the side member
- Replace screws
- Use only original bolts
- Do not crossthread bolts
- Do not use an impact wrench

26. Tighten bolts/nuts to specification as follows:

- Bolt, front axle carrier to longitudinal member: ft. lbs. (120 Nm)
- Bolt, front axle gear propeller shaft to front axle gear flange M8: ft. lbs. (15 Nm) plus an additional 60° of rotation
- Bolt, front axle gear propeller shaft to front axle gear flange M10: ft. lbs. (40 Nm) plus an additional 60° of rotation
- Collar bolt of front axle gear to engine oil pan: ft. lbs. (65 Nm)
- Collar bolt of right support clamp: ft. lbs. (22 Nm)
- Oil filler opening screw plug: ft. lbs. (30 Nm)
- Bolted connection, rack-and-pinion steering to front axle carrier: ft. lbs. (50 Nm) plus an additional 60° of rotation

- Bolted connection, supporting plate to front axle carrier: ft. lbs. (50 Nm) plus an additional 30° of rotation

27. When installing air filter housing, apply antifriction agent to the rubber retainers on the cylinder head covers and the sealing ring of the hot film mass air flow sensor in the air filter housing.

28. Check front axle gear oil level and correct if necessary.

29. Connect STAR DIAGNOSIS, read out and erase fault memory.

30. Perform a wheel alignment check.

**S-Class**

*See Figure 45.*

**❊❊ WARNING**

**Absolute cleanliness is essential when working on the front axle. Even minute dirt particles can prevent the front axle transmission from functioning correctly.**

1. Before servicing the vehicle, refer to the precautions.

2. Detach propeller shaft from front axle transmission.

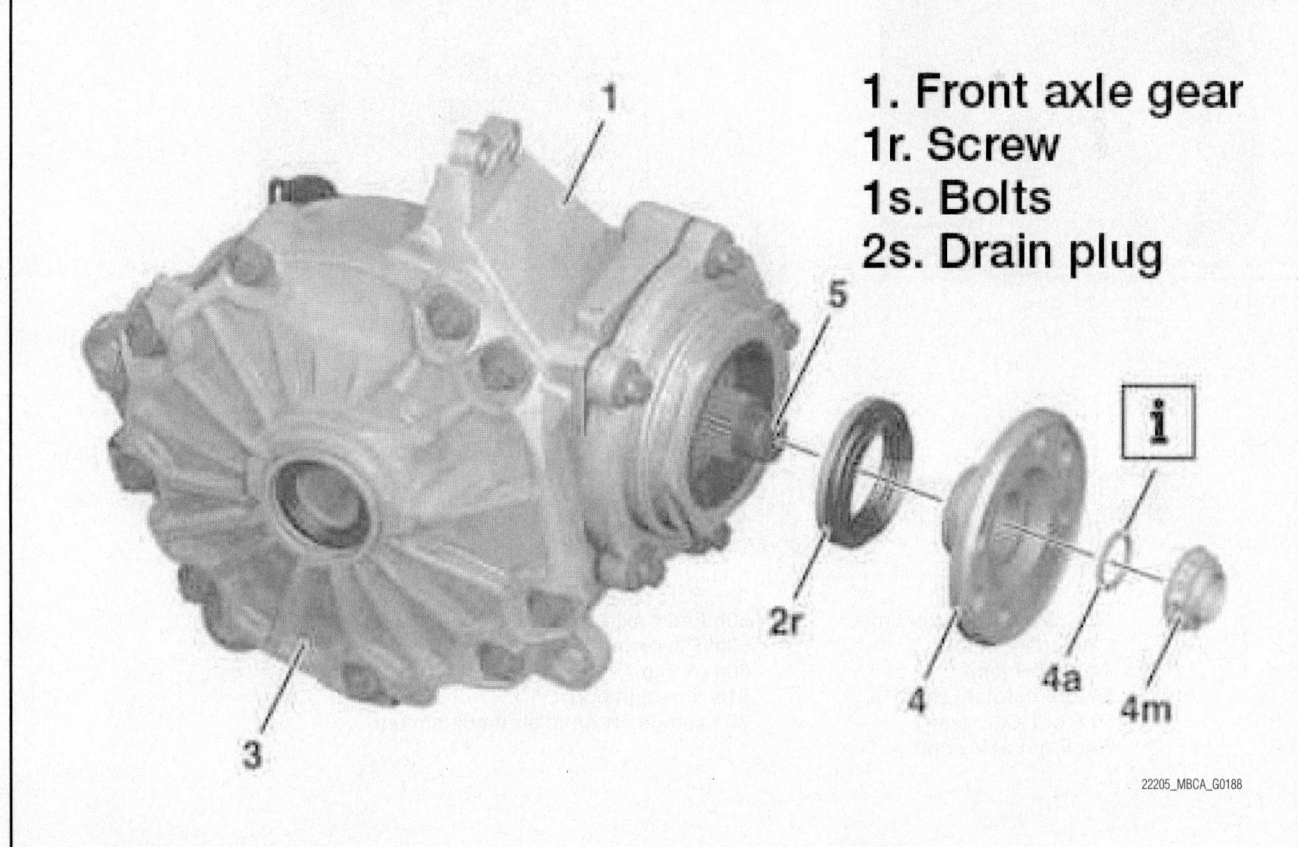

1. Front axle gear
1r. Screw
1s. Bolts
2s. Drain plug

22205_MBCA_G0188

**Fig. 45 Exploded view of front axle assembly**

3. Remove front axle carrier.

4. Remove intermediate shaft.

5. Remove bolts and remove front axle differential.

### To install:

6. Tighten bolts/nuts to specification as follows:

- Collar bolt, front axle gear to support brace: 48 ft. lbs. (65 Nm)
- Collar bolt, front axle transmission to engine support
- Step 1: 4 ft. lbs. (5 Nm)
- Step 2: 37 ft. lbs. (50 Nm)
- Step 3: tighten an additional 95° rotation
- Oil drain screw, front axle gear: 15 ft. lbs. (20 Nm)

7. Check that guide sleeve is seated correctly in oil pan.

8. Attach front axle transmission to front axle.

9. Remove drain screw and drain oil from front axle transmission.

10. Fill front axle with oil.

11. Install intermediate shaft.

12. Install front axle carrier.

13. Attach propeller shaft to front axle transmission.

14. Perform engine test run and check front axle for leaks and that it is functioning correctly.

### FRONT HALFSHAFT

#### REMOVAL & INSTALLATION

*See Figure 46.*

1. Before servicing the vehicle, refer to the precautions.

2. Remove the front wheels.

3. Unscrew hexagon collar bolt from front axle shaft.

4. Take off bottom engine compartment panel at front.

5. Detach linkage for headlamp range adjustment on lower transverse control arm.

6. Pry supporting joint off steering knuckle.

7. Turn the steering knuckle outwards at the front and then pull outwards, in the process move the front axle shaft out of the front axle shaft flange.

8. If necessary to loosen the front axle shaft or screw in the hexagon-collar bolt a few threads again and hammer out front axle shaft slightly with a rubber hammer.

9. Press out and remove front axle shaft.

### ✳✳ WARNING

**Do not damage the plastic guard ring, rubber boots and rpm sensor rotor. To avoid damage to the wheel bearings the vehicle must no longer be moved on its wheels when the front axle shaft is removed.**

10. Press out and remove left front axle shaft.

11. Check joints and boots for leaks and damage.

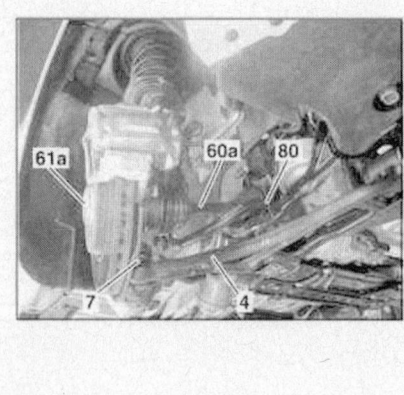

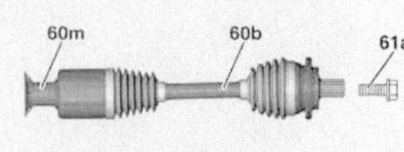
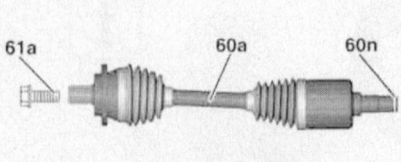
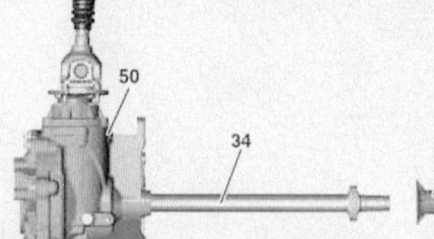

4 Transverse control arms
5 Steering knuckle
7 Support joint
34 Intermediate shaft
50 Front axle gear
60a Front axle shaft

60b Front axle shaft
60m Plastic protective ring
60n Circlip
61a Hexagon collar
80 Linkage for headlamp adjustment

22205_MBCA_G0015

**Fig. 46 Exploded view of the front halfshaft assembly**

12. Check sealing ring in the front axle gear of the right-hand front axle shaft for damage.

***To install:***

13. Installation is the reverse of removal.

### ❋❋ WARNING

**Replace all self locking, micro-encapsulated and aluminum bolts and nuts. Chase threads to remove micro-encapsulated residue from old bolts/nuts.**

14. Tighten the self-locking nut, supporting joint to steering knuckle bolts/nuts to specification as follows:
- Stage 1: 21 ft. lbs. (28 Nm).
- Stage 2: Tighten 110° additional rotation

15. Tighten the collar bolt, front axle shaft to front axle shaft flange to specification as follows:
- Stage 1: 82 ft. lbs. (110 Nm).
- Stage 2: Tighten 60° additional rotation

### DIFFERENTIAL ASSEMBLY

*REMOVAL & INSTALLATION*

#### C-Class

*See Figure 47.*

1. Before servicing the vehicle, refer to the precautions.
2. Drain oil from rear axle differential.
3. Oil filler/drain screw on rear axle differential modified.
4. Remove right rear wheel.

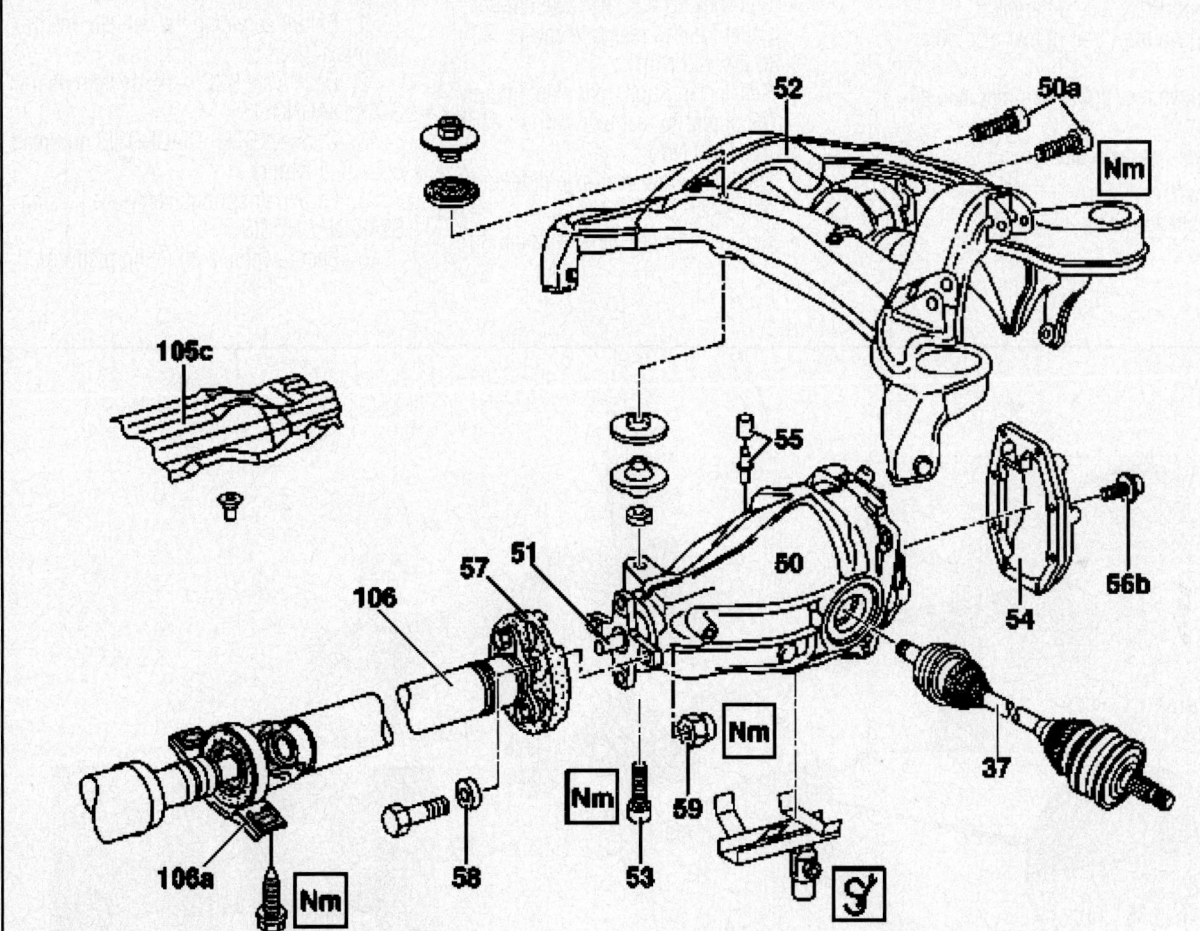

37. Rear axle shaft
50. Rear axle center assembly
50a. Rear bolts of rear axle differential
51. Input shaft
52. Rear axle carrier
53. Front bolt of rear axle differential
54. End cover
55. Breather
56b. Bolts of end cover
57. Flex disk
58. Fitting sleeve
59. Self-locking nut
105c. Heat shield
106. Rear propeller shaft
106a. Propeller shaft intermediate bearing

22205_MBCA_G0190

**Fig. 47 Rear differential assembly**

5. Remove rear axle shaft on right.

6. Remove heat shield.

7. Remove rear propeller shaft from rear axle differential.

8. Drift for loosening fitting sleeves in flex disks.

9. Mount rear propeller shaft to rear axle center assembly.

10. Loosen fitting sleeves of flexible coupling from flange.

11. Drift for loosening fitting sleeves in flex disks.

12. Loosen bolts on propeller shaft intermediate bearing.

13. Support rear axle differential.

14. Remove front bolt of rear axle differential.

15. Remove rear bolts from rear axle differential.

16. Lower rear axle differential.

### To install:

17. Installation is the reverse of removal.

18. Replace breather. Drive in new breather evenly and slowly into the rear axle differential to avoid deformation and damage.

### ✳✳ WARNING

**Replace all self locking, micro-encapsulated and aluminum bolts and nuts. Chase threads to remove micro-encapsulated residue from old bolts/nuts.**

19. Tighten bolts/nuts to specification as follows:

- Self-locking nut, rear axle differential at front to rear axle carrier: 33 ft. lbs. (45 Nm)
- Self-locking bolt, rear axle differential at rear to rear axle carrier: 81 ft. lbs. (110 Nm)
- Oil filler screw, rear axle differential: 37 ft. lbs. (50 Nm)
- Self-locking nut, flexible coupling to rear propeller shaft or rear axle differential M10: 30 ft. lbs. (40 Nm)
- Self-locking nut, flexible coupling to rear propeller shaft or rear axle differential M12: 44 ft. lbs. (60 Nm)
- Bolt, propeller shaft intermediate bearing to frame floor assembly: 30 ft. lbs. (40 Nm)

20. Check oil level in rear axle differential and correct if necessary.

21. Check rear axle differential for leaks.

### E-Class

*See Figure 48.*

1. Before servicing the vehicle, refer to the precautions.

2. Deactivate SBC brake system using STAR DIAGNOSIS.

3. Connect STAR DIAGNOSIS and read out fault memory.

4. Empty air springs completely using STAR DIAGNOSIS.

5. Secure vehicle on lifting platform.

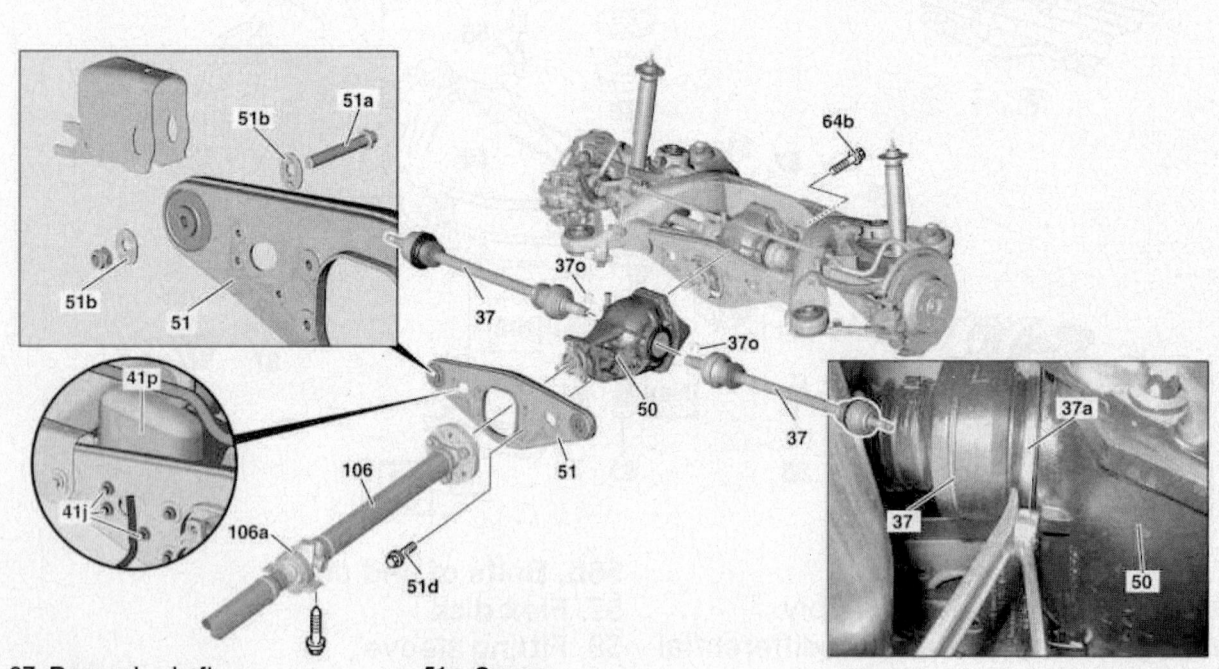

37. Rear axle shaft
37a. Protective ring
37o. Circlip
41j. Bolts
41p. Additional volume reservoir
50. Rear axle center assembly
51. Front center assembly mount
51a. Screw
51b. Shims
51d. Bolts
64b. Screw
106. Propeller shaft
106a. Propeller shaft intermediate bearing

22205_MBCA_G0191

**Fig. 48 Rear differential assembly**

6. Remove rear muffler.

7. Remove exhaust system at flange connection.

8. Detach complete exhaust system.

9. Remove exhaust system from pipe connection.

10. Unhook brake cables from automatic slack adjuster of parking brake.

11. Detach propeller shaft from flange on rear axle center assembly.

12. Undo fitting sleeves of flexible coupling on flange of rear axle center assembly.

13. Remove bolts from propeller shaft intermediate bearing and remove rear part of propeller shaft.

14. Remove right rear axle shaft.

15. Remove bolts and remove left and right additional volume reservoirs from front center assembly mounting.

16. Support rear axle center assembly with transmission jack and transmission platform and lash down securely.

17. Remove the screw for the rear center assembly mounting on the rear axle carrier.

18. Remove both bolts of the rear differential mount on the rear axle carrier.

19. Remove left and right bolts of front center assembly mounting.

20. Lower the rear axle center assembly slightly.

21. Lift out the left rear axle shaft from the rear axle center assembly.

22. Lower rear axle center assembly further.

23. Remove bolts and remove front center assembly mounting.

### To install:

24. Installation is the reverse of removal.

### ✳✳ WARNING

**Replace all self locking, micro-encapsulated and aluminum bolts and nuts. Chase threads to remove micro-encapsulated residue from old bolts/nuts.**

25. Tighten bolts/nuts to specification as follows:
- Oil filler screw, rear axle differential: 37 ft. lbs. (50 Nm)
- Self-locking nut, flexible coupling to rear propeller shaft or rear axle differential M10: 30 ft. lbs. (40 Nm)
- Self-locking nut, flexible coupling to rear propeller shaft or rear axle differential M12: 44 ft. lbs. (60 Nm)
- Bolt, Self-tapping screw propeller shaft intermediate bearing to frame floor assembly: 19 ft. lbs. (25 Nm)

- Self-safety bolt, rear axle center assembly at rear at rear axle cross-member rear M14: 59 ft. lbs. (80 Nm) plus an additional 60° rotation
- Self-safety bolt, rear axle center assembly at rear at rear axle cross-member front M10: 48 ft. lbs. (65 Nm)
- Self-locking bolt, front crossmember to rear axle carrier: 37 ft. lbs (50 Nm) plus an additional 90° rotation

26. Check oil level in rear axle center assembly and correct if necessary.

27. Activate SBC brake system using STAR DIAGNOSIS.

28. Connect STAR DIAGNOSIS and read out fault memory.

29. Perform a wheel alignment check.

### S-Class

*See Figure 49.*

### ✳✳ WARNING

**While working on the rear axle care must be taken to ensure that the surface of all aluminum components do not get any scratches, cracks and notches. Otherwise the service life of the parts will be affected.**

1. Before servicing the vehicle, refer to the precautions.

2. Move parking brake into the assembly position using STAR DIAGNOSIS.

3. Secure vehicle on vehicle lift.

4. Remove exhaust system at flange connection.

5. Detach propeller shaft from joint flange on rear axle differential.

6. Loosen fitting sleeves of flexible coupling on joint flange of rear axle differential.

7. Remove right rear axle shaft.

8. Unscrew nuts and remove controller unit from bracket. Do not disconnect electrical connectors.

9. Loosen bracket of controller unit on front differential mount.

10. Support rear axle center assembly with transmission jack and transmission platform and lash down securely.

11. Mark installation position of sleeve.

12. Remove bolt from rear axle differential mount and remove bolt with sleeve.

13. Remove bolts and remove right exhaust rubber mount.

14. Remove left and right bolts of front center assembly mount.

15. Lower the rear axle center assembly slightly.

16. Lift out the left rear axle shaft from the rear axle center assembly.

17. Lower rear axle center assembly further.

18. Remove bolts and remove front center assembly mount from rear axle center assembly.

19. Remove bolts and adapter from rear axle differential.

20. Detach extension bracket from rear axle differential.

### To install:

21. Installation is the reverse of removal.

22. Replace locking ring on inner spline profile of rear axle shaft.

### ✳✳ WARNING

**Replace all self locking, micro-encapsulated and aluminum bolts and nuts. Chase threads to remove micro-encapsulated residue from old bolts/nuts.**

23. Tighten bolts/nuts to specification as follows:
- Self-locking bolt, front crossmember to rear axle carrier: 37 ft. lbs. (50 Nm) plus an additional 90° of rotation
- Self-locking nut securing rear extension bracket to rear axle carrier: 74 ft. lbs. (100 Nm) plus an additional 90° of rotation
- Adapter to rear axle differential: 74 ft. lbs. (100 Nm) plus an additional 90° of rotation
- Adapter to front transverse bridge: 37 ft. lbs. (50 Nm) plus an additional 90° of rotation
- Self-locking nut, flex disk to rear propeller shaft or rear axle center assembly: 43 ft. lbs. (58 Nm)
- Nut, electric parking brake controller unit to bracket: 5 ft. lbs. (7 Nm)
- Bolt, bracket of electric parking brake control unit to front crossmember: 11 ft. lbs. (15 Nm)
- Bolt, exhaust rubber mount to rear axle carrier: 15 ft. lbs. (20 Nm)
- Oil filler screw, rear axle differential: 37 ft. lbs. (50 Nm)

24. Check oil level in rear axle center assembly and correct if necessary.

25. Perform a wheel alignment check.

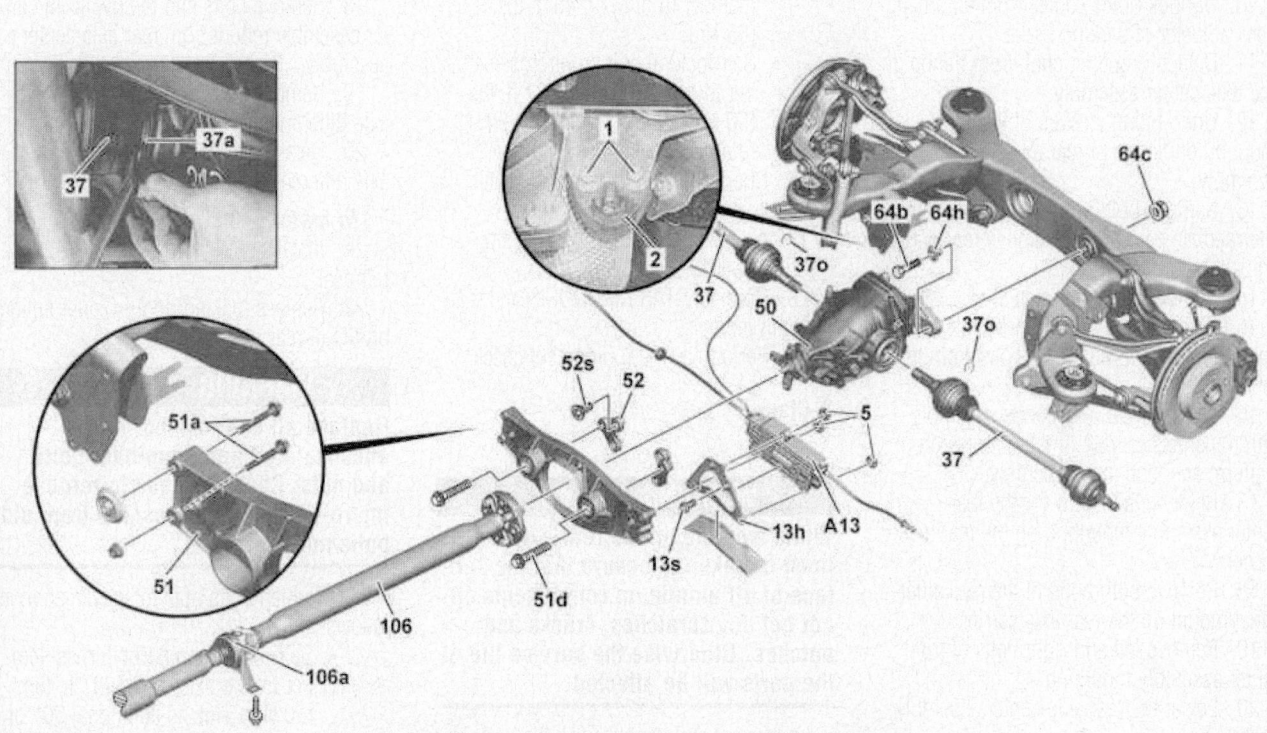

1. Bolts
2. Exhaust rubber mount
5. Nuts
13h. Bracket
13s. Screw
37. Rear axle shaft
37a. Protective ring

37o. Circlip
50. Rear axle center assembly
51. Front center assembly mount
51a. Screw
51d. Screw
52. Adapter
52s. Screw

64b. Screw
64h. Sleeve
106. Propeller shaft
106a. Propeller shaft intermediate bearing
A13. Electric parking brake controller unit

22205_MBCA_G0192

**Fig. 49 Rear differential assembly**

## REAR HALFSHAFT

### REMOVAL & INSTALLATION

*See Figure 50.*

1. Before servicing the vehicle, refer to the precautions.

2. Raise and safely support the vehicle securely.

3. Remove twelve-point collar nut from rear axle shaft.

4. Remove each wheel Remove/install wheels, rotate if necessary.

5. Remove fixed brake caliper from the rear axle.

6. Remove stabilizer bar link rod.

7. Remove left rear rpm sensor or right rear rpm sensor from wheel carrier.

8. Remove torque strut from wheel carrier.

9. Detach camber strut from wheel carrier.

10. Detach tie rod from wheel carrier.

11. Remove thrust arm from wheel carrier.

12. Fold wheel carrier outwards.

13. Pull rear axle shaft out of rear axle differential.

14. Check condition of pinion seal in rear axle differential, replace if necessary.

### To install:

15. Installation is the reverse of removal.

16. Check oil level in rear axle differential and correct if necessary

17. Check rear axle differential for leaks

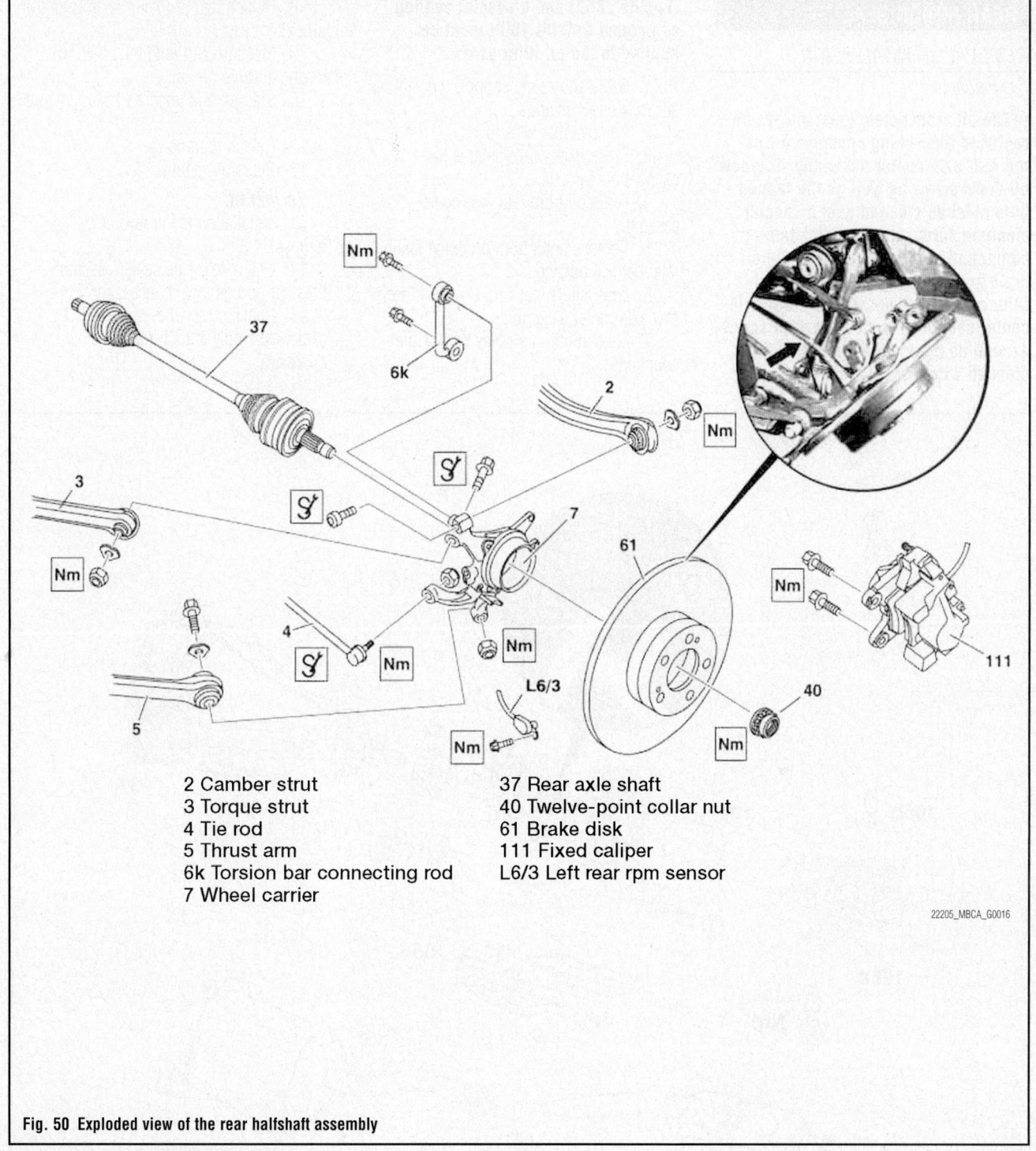

2 Camber strut
3 Torque strut
4 Tie rod
5 Thrust arm
6k Torsion bar connecting rod
7 Wheel carrier

37 Rear axle shaft
40 Twelve-point collar nut
61 Brake disk
111 Fixed caliper
L6/3 Left rear rpm sensor

22205_MBCA_G0016

**Fig. 50 Exploded view of the rear halfshaft assembly**

**✷✷ WARNING**

**Replace all self locking, micro-encapsulated and aluminum bolts and nuts. Chase threads to remove micro-encapsulated residue from old bolts/nuts.**

18. Tighten fasteners to specification:
- Nut, link rod to stabilizer bar: 30 ft. lbs. (40 Nm)

- Bolt, link rod to wheel carrier: 30 ft. lbs. (40 Nm)
- Self-locking nut, camber strut to wheel carrier: 52 ft. lbs. (70 Nm)
- Self-locking nut, torque strut to wheel carrier: 52 ft. lbs. (70 Nm)
- Self-locking nut, thrust arm to wheel: 52 ft. lbs. (70 Nm)

- Collar nut, rear axle shaft to rear axle shaft flange: 163 ft. lbs. (220 Nm)
- Self-locking nut, tie rod at wheel carrier:
- Stage 1: 15 ft. lbs. (20 Nm)
- Stage 2: Tighten an additional 45° rotation

## REAR PINION SEAL

### REMOVAL & INSTALLATION

*See Figure 51.*

➡The oil drain screw must always be replaced after being unscrewed from the rear axle center assembly. The new oil drain screw as well as the tapped hole must be cleaned with a special cleaning spray and a special sealing compound must be applied to the oil drain screw. Always install new oil filler screw after unscrewing rear axle center section. The new oil filler screw as well as the tapped hole must be cleaned with a special cleaning spray

(Loctite 7063) and a special sealing compound Omnifit 100H must be applied to the oil filler screw.

1. Raise and safely support the vehicle securely on jackstands.
2. Remove rear wheels.
3. Press back brake pads at fixed calipers of rear axle.
4. Drain oil from rear axle center assembly.
5. Remove bolts from propeller shaft intermediate bearing.
6. Disconnect rear propeller shaft from rear axle center assembly.
7. Loosen fitting sleeves on flexible couplings.

8. Detach rear propeller shaft from front propeller shaft .
9. Measure and note total friction torque of drive pinion.
10. Slacken and unscrew hexagon collar nut.
11. Detach joint flange.
12. Press out pinion seal.

### *To install:*

13. Installation is the reverse of removal.
14. Check rear axle center assembly breather and replace if necessary.
15. Check oil level in rear axle center assembly and correct if necessary.

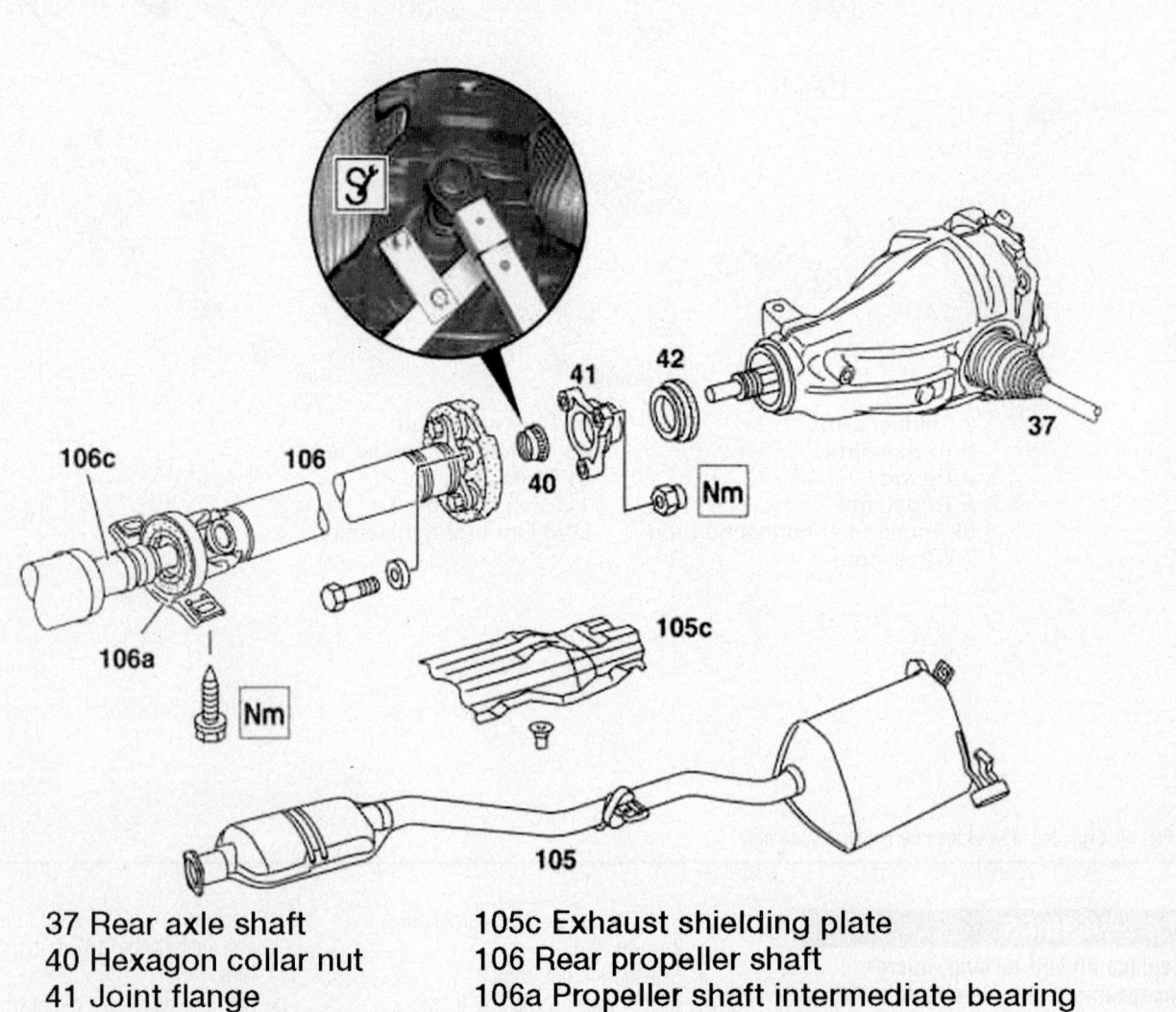

37 Rear axle shaft
40 Hexagon collar nut
41 Joint flange
42 Radial sealing ring
105 Exhaust system

105c Exhaust shielding plate
106 Rear propeller shaft
106a Propeller shaft intermediate bearing
106c Front propeller shaft

22205_MBCA_G0017

**Fig. 51 Exploded view of the differential carrier**

## ENGINE COOLING

### ENGINE FAN

#### REMOVAL & INSTALLATION

##### C-Class

*See Figures 52 and 53.*

1. Before servicing the vehicle, refer to the precautions.

2. Remove left and right intake air duct.

3. Remove front engine cover. Do not tilt front engine cover, but lift upwards vertically otherwise the retaining lugs can break off.

4. Disconnect connector of the electric suction fan engine and AC with integrated control.

5. Remove fanfare horns.

6. Remove bolts from steering oil cooler on front end crossmember.

7. Remove hydraulic line for power steering from front end crossmember.

8. Remove bolt for windshield washer reservoir filler pipe.

9. Remove front end crossmember from crossmember at the front by removing bolts.

10. Remove bolts at front end crossmember on right and left.

11. Detach hood cable.

12. Remove retaining clips for air ducting.

13. Push coolant line out of fixture.

14. Remove cable ties at front end crossmember.

15. Disconnect plug connection of hood lock on right.

16. Remove lock carrier with the hood release cables.

➡ **If the rubber mounts are firmly attached in the radiator, do not pull front end crossmember upwards with force, but unscrew retaining pin on right and left. Press together retaining bracket of radiator on right and left, then unhook front end crossmember and remove.**

17. Drain coolant at radiator.

18. Notes on coolant.

19. Disconnect coolant hose from thermostat housing to engine radiator, at engine radiator and place to one side.

20. Press clamp on right and left. At the same time take out electric suction fan engine and AC with integrated control toward top.

21. Take out engine and air conditioning electric suction fan with integrated control toward top.

#### To install:

22. Installation is the reverse of removal.

23. Ensure that plates of electric suction fan engine and AC with integrated control are seated in supports of radiator on right and left.

24. Carry out operational check of hood lock.

25. Check gap dimension of the engine hood.

26. Connect STAR DIAGNOSIS and read out fault memory.

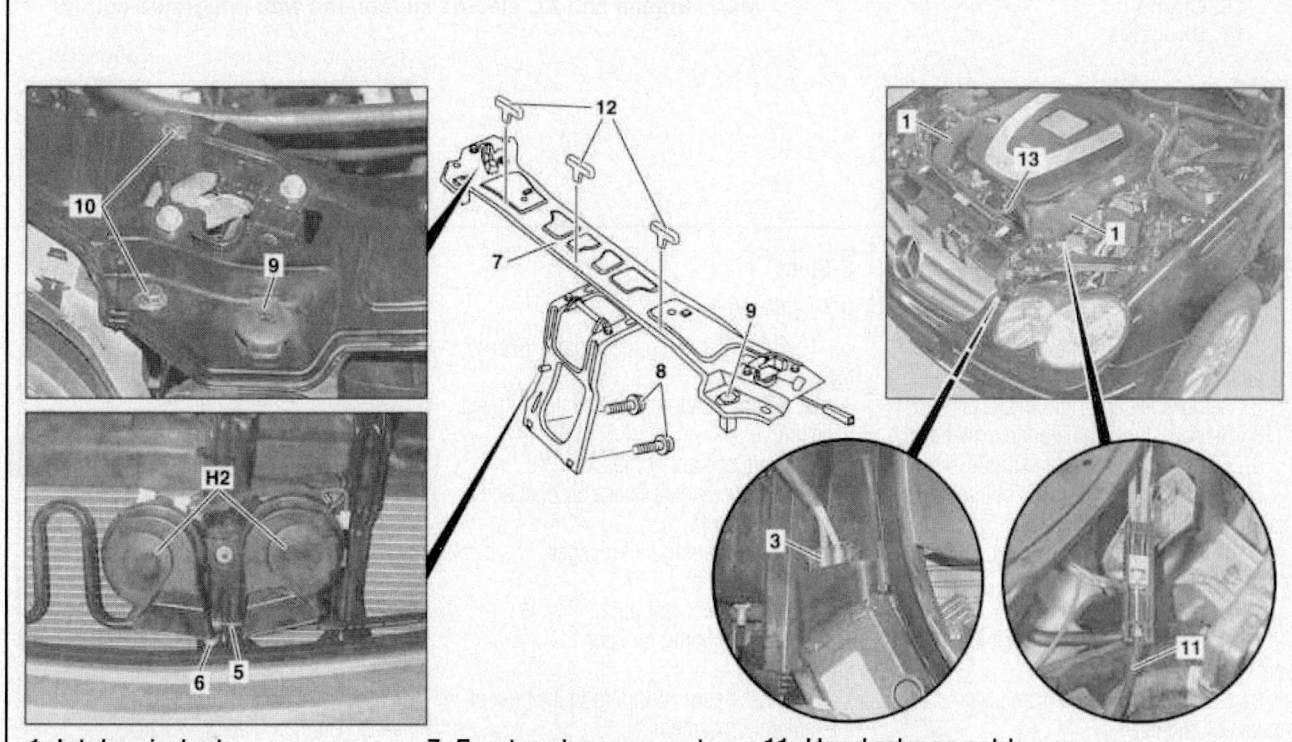

1. Intake air duct
3. Plug connection
5. Power steering hydraulic line
6. Screw
7. Front end crossmember
8. Bolts
9. Retaining pin
10. Bolts
11. Hood release cable
12. Retaining clips
13. Coolant line
H2. Fanfare horns

22205_MBCA_G0194

**Fig. 52 Removing the front crossmember**

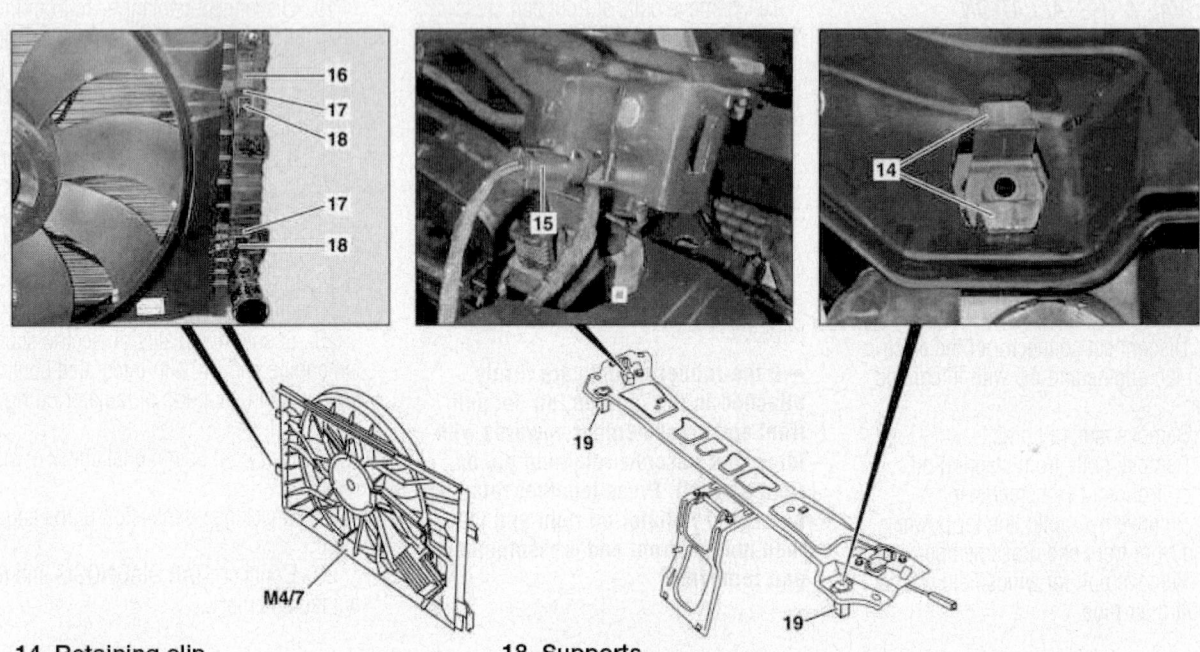

14. Retaining clip
15. Plug connection for right hood lock
16. Clamp
17. Shackles

18. Supports
19. Rubber mounts
M4/7. Engine and AC electric suction fan with integrated control

22205_MBCA_G0195

**Fig. 53 Engine fan assembly**

### E-Class

*See Figures 54 and 55.*

1. Before servicing the vehicle, refer to the precautions.
2. Remove left and right intake air duct.
3. Remove upper radiator crossmember.
4. Remove plug from fan shroud.
5. Detach expansion rivets of upper mounts of radiator.
6. Tilt radiator at top against direction of travel (arrow).
7. Detach clamping springs.
8. Unclip coolant lines on left and right from brackets at fan shroud.
9. Remove fan shroud from lower mounts of radiator.
10. While removing the fan shroud ensure that coolant hose does not buckle.

*To install:*

11. Installation is the reverse of removal.
12. Ensure correct position of retaining tabs of fan shroud in lower mounts of radiator.

### S-Class

*See Figure 56.*

1. Before servicing the vehicle, refer to the precautions.
2. Remove bottom engine compartment paneling.
3. Drain coolant at radiator.
4. Remove engine intake air duct and front engine cover.
5. Remove engine oil dipstick.
6. Remove hose.
7. Remove coolant pipe.
8. Remove upper radiator crossmember.
9. Detach edge stripping in the area of the radiator crossmember.
10. Detach hose from radiator and temperature control switch housing and place to the side with bracket.
11. Disconnect electrical connector on electrical suction fan.
12. Remove feed line for electrical suction fan from fan shroud (arrow).

13. Remove clamps on the left and right on fan shroud.
14. Take out fan shroud towards top.

*To install:*

15. Installation is the reverse of removal.

## RADIATOR

*REMOVAL & INSTALLATION*

### C-Class

*See Figure 57.*

1. Before servicing the vehicle, refer to the precautions.
2. Remove electric fan.
3. Drain coolant at radiator.
4. Take off coolant line from radiator to coolant pump at radiator. Use screwdriver to push back clamp until it locks in place.
5. Take off coolant line from radiator to thermostat housing on radiator. Use screwdriver to push back clamp until it locks in place.

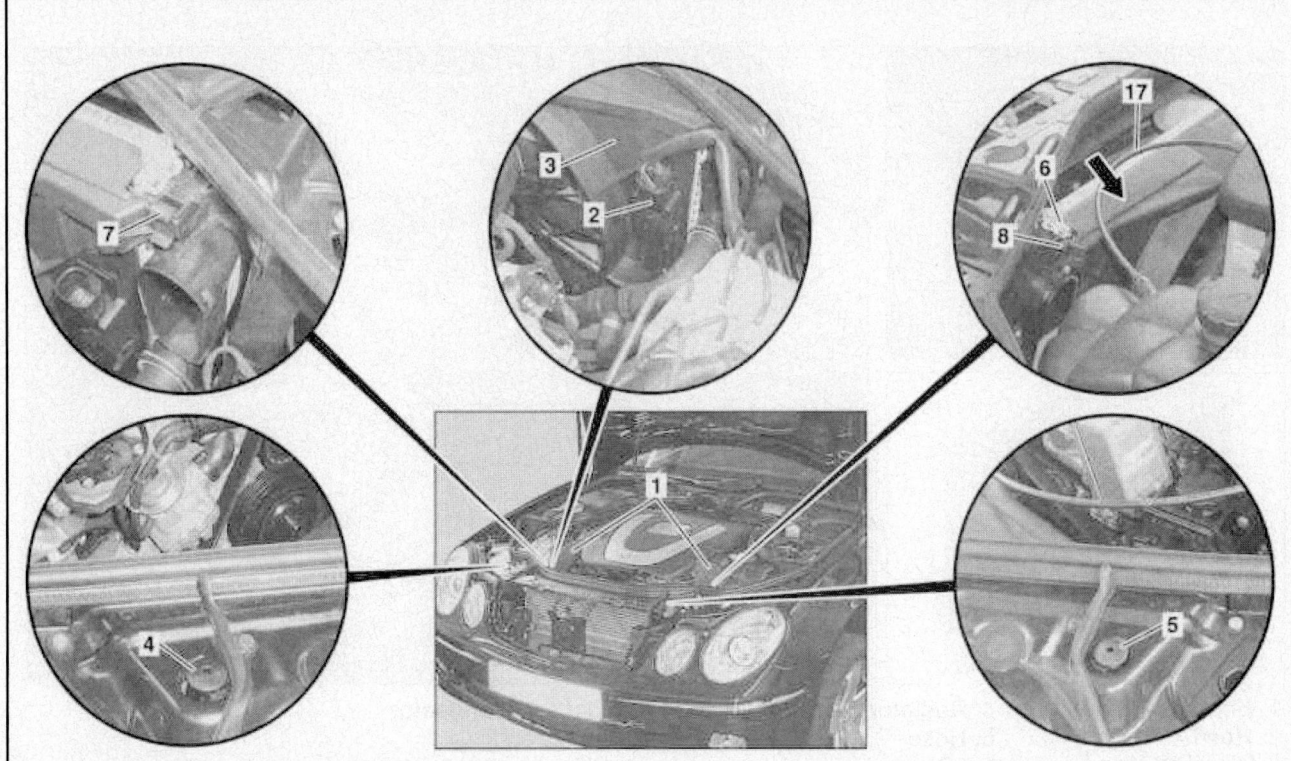

1. Intake air duct
2. Connector
3. Fan shroud
4. Expanding rivet
5. Expanding rivet
6. Radiator
7. Clamping spring
8. Clamping spring
17. Coolant hose

22205_MBCA_G0196

**Fig. 54  Removing the front crossmember**

3. Fan shroud
9. Coolant line
10. Bracket
3. Fan shroud
11. Bracket
12. Bracket
13. Lower mount at the radiator
14. Lower mount at the radiator
15. Retaining tab of the fan shroud
16. Retaining tab of the fan shroud

22205_MBCA_G0197

**Fig. 55  Engine fan assembly**

1. Engine oil dipstick
2. Hose
3. Weather strip
4. Radiator crossmember
5. Hose
5a. Bracket
6. Electrical connector
7. Fan shroud
7a. Clamp

22205_MBCA_G0198

**Fig. 56 Engine fan assembly**

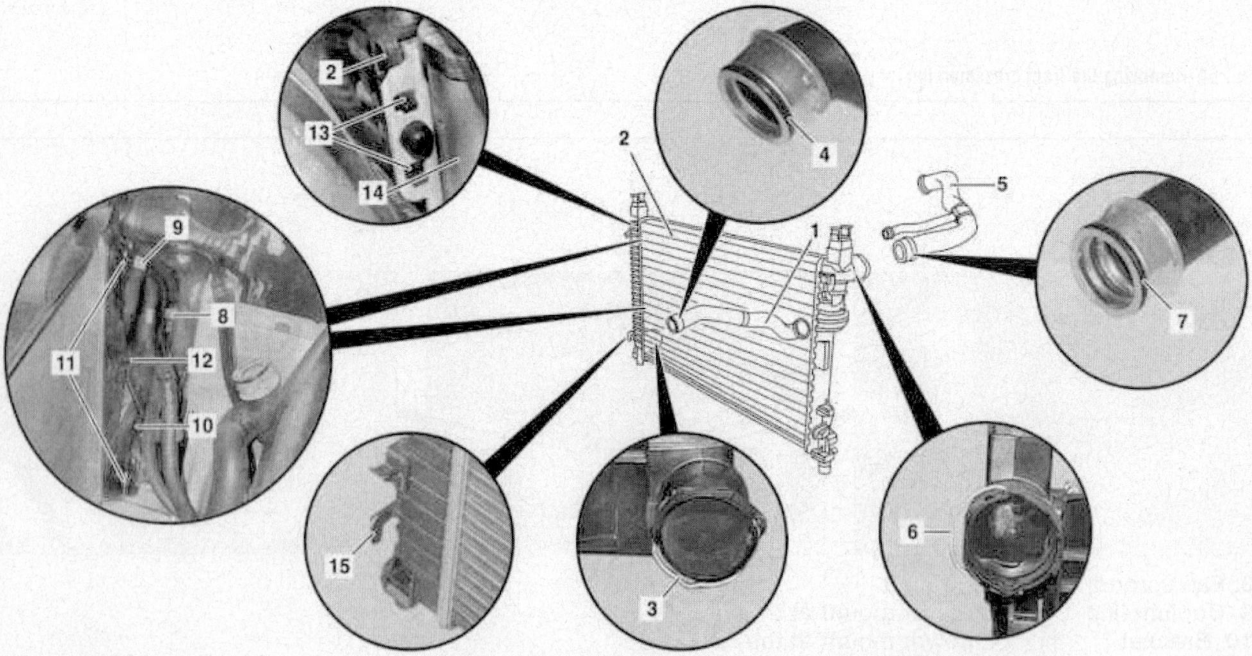

1. Coolant line on radiator to coolant pump
2. Radiator
3. Clamp
4. O-ring
5. Coolant line on radiator to thermostat housing
6. Clamp
7. O-ring
8. Bracket
9. ATF line
10. ATF line
11. Clamps
12. Bracket
13. Clamps
14. Capacitor
15. Bracket

22205_MBCA_G0199

**Fig. 57 Radiator assembly**

6. Take off bracket of ATF lines.

7. Detach ATF lines on radiator.

8. Push ATF line out of bracket.

9. Push back clamps on radiator on right and left. Pull condenser forward at the top for this step and detach at the bottom.

10. Lift radiator up and out.

11. Install in the reverse order.

12. Inspecting cooling system for leaks.

13. Check for leaks with engine running.

### To install:

14. Installation is the reverse of removal.

15. Ensure that radiators lugs at the bottom sit in the rubber supports.

16. Ensure that condenser is located at the bottom in the brackets on the right and left; at the same time the clamps must be locked in place.

17. Check clips, replace if necessary Replace O-rings of ATF lines.

18. Replace O-ring of coolant line for radiator to thermostat housing.

19. Replace O-ring for coolant line on radiator to coolant pump.

### E-Class

*See Figures 58 through 60.*

1. Before servicing the vehicle, refer to the precautions.

2. Pull out front engine cover from mounts.

3. Remove left and right intake air duct.

4. Remove fan shroud.

5. Remove front and center sections of lower engine compartment paneling.

6. Unhook low temperature radiator with attached lines.

7. Drain coolant at radiator.

8. Remove upper radiator crossmember.

9. Detach the coolant hose from the connection fitting at the coolant hose.

10. Detach coolant hoses from radiator.

11. Remove oil lines for automatic transmission from radiator at top.

12. Unclip power steering cooler from mount on cooler.

13. Unclip the top bracket of the condenser from the mount at the radiator and pry off the radiator from the condenser.

14. Unclip bottom bracket for air ducts from mounts on radiator.

15. Lift out radiator from bottom radiator mounts, bottom condenser brackets and top brackets of air ducts.

### To install:

16. Installation is the reverse of removal.

17. Check oil level in automatic transmission, adjust to correct level if necessary.

18. Inspect cooling system for leaks.

### S-Class

*See Figures 61 and 62.*

1. Before servicing the vehicle, refer to the precautions.

2. Remove bottom engine compartment paneling.

3. Drain coolant from radiator.

4. Discharge air conditioning system.

5. Remove oil lines at radiator.

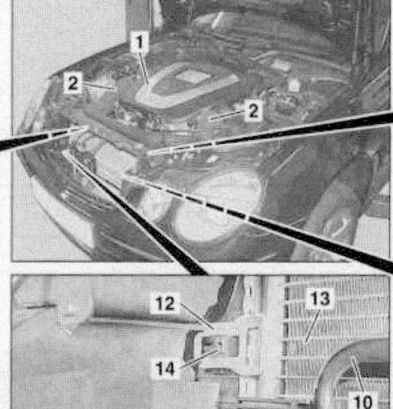

1. Front engine cover
2. Intake air duct
3. Radiator
4. Coolant hose
5. Coolant hose
6. Coolant hose
7. Coolant hose
8. ATF line for automatic transmission
9. ATF line for automatic transmission
10. Power steering radiator
11. Mount on radiator
12. Condenser, top bracket
13. Capacitor
14. Mount on radiator

22205_MBCA_G0200

**Fig. 58 Coolant hose connections**

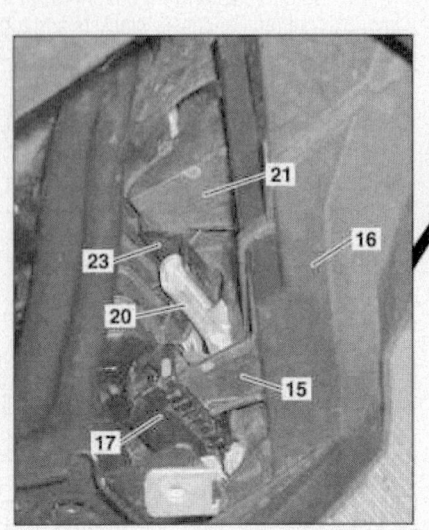
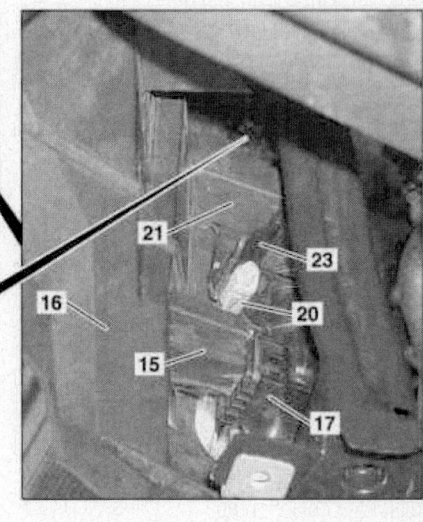

11. Mount on radiator
15. Bottom air ducting bracket
16. Air ducting
17. Supports on cooler

20. Bottom condenser bracket
21. Top air ducting bracket
23. Supports on cooler

22205_MBCA_G0201

**Fig. 59  Radiator supports**

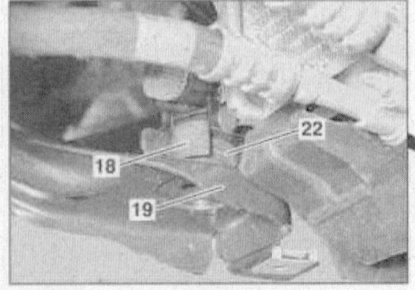
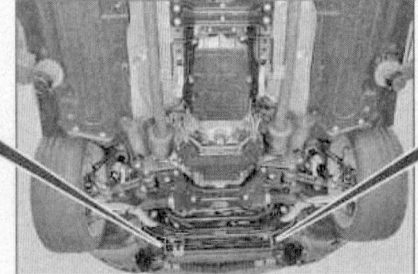

18. Bottom radiator mounts
19. Bottom radiator mounts
22. Rubber bushing

22205_MBCA_G0202

**Fig. 60  Radiator mounting**

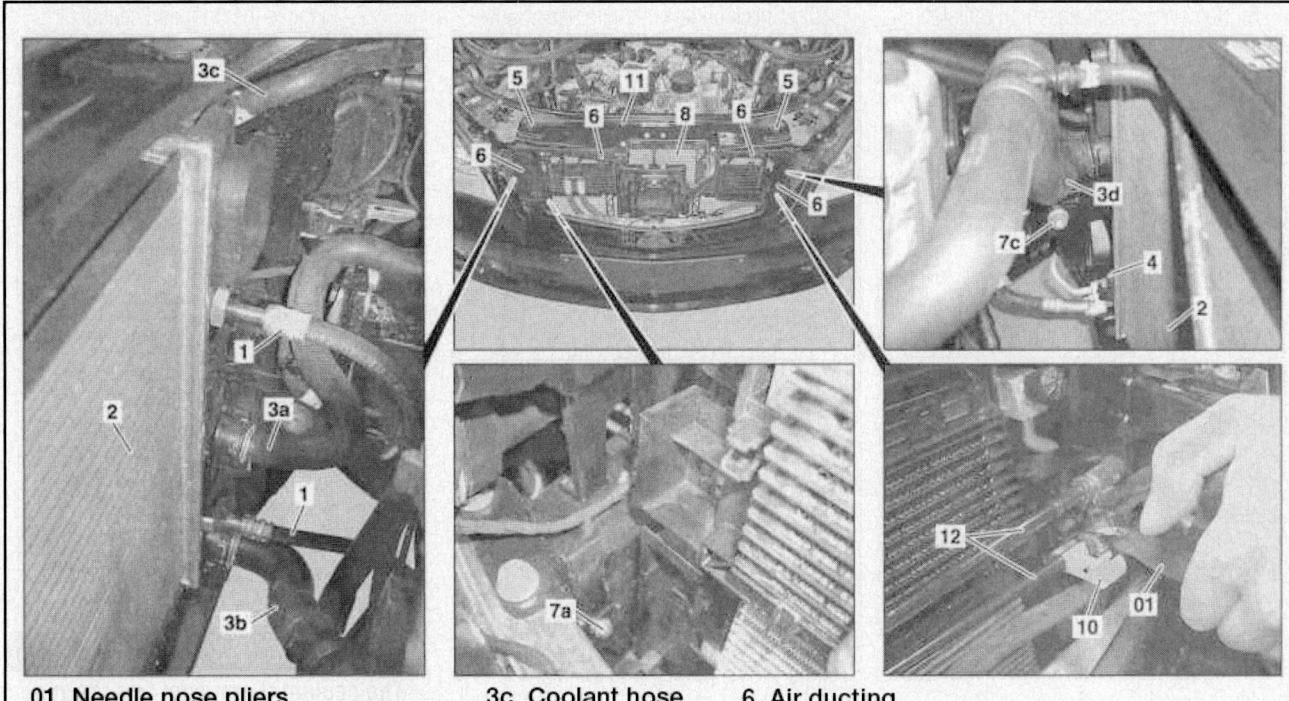

01. Needle nose pliers
1. ATF line of automatic transmission
2. Radiator
3a, 3b. Coolant hose
3c. Coolant hose
3d. Coolant hose
4. Bolt
5. Bracket
6. Air ducting
7a. Bolt
7c. Bolt
8. Capacitor

22205_MBCA_G0203

**Fig. 61 Radiator mounting**

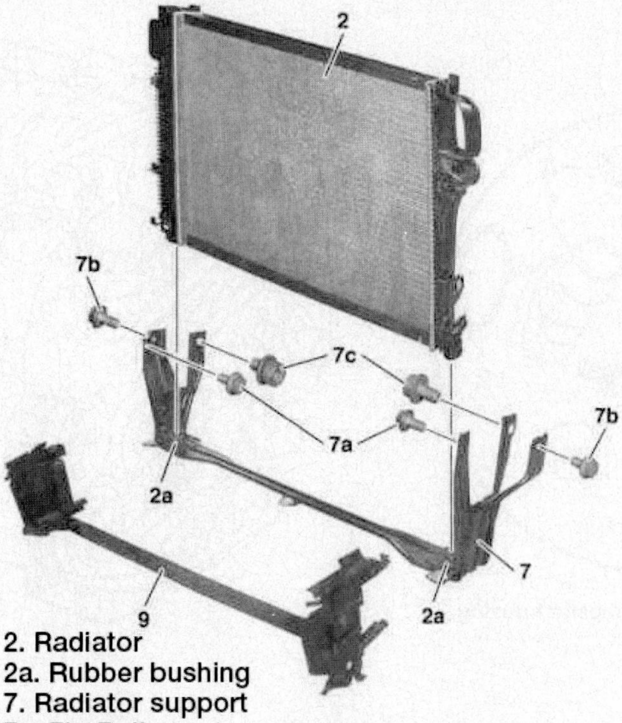

2. Radiator
2a. Rubber bushing
7. Radiator support
7a, 7b. Bolt
7c. Bolt
9. Air ducting

22205_MBCA_G0204

**Fig. 62 Radiator mounting**

6. Remove coolant hoses on radiator.

7. Remove coolant hose on radiator.

8. Take off coolant hose at radiator.

9. Remove fan shroud.

10. Unscrew bolt.

11. Remove radiator bracket.

12. Remove upper radiator crossmember.

13. Remove air ducts.

14. Remove bolts.

15. Disconnect AC line on condenser.

16. Unclip condenser using needle nose pliers on the left and right from the radiator.

17. Remove bolts and radiator support downwards.

18. Remove bottom air ducting.

19. Unhook condenser below out of radiator.

20. Hook guard plate onto radiator and remove radiator upwards.

*To install:*

21. Installation is the reverse of removal.

22. Tighten bolts/nuts to specification as follows:

- Bolt, radiator support to longitudinal member M6: 8 ft. lbs. (10 Nm)
- Bolt, radiator support to longitudinal member M12: 89 ft. lbs. (120Nm)

23. Check oil level in automatic transmission and correct

## THERMOSTAT

### REMOVAL & INSTALLATION

#### M112 and M113 Engines

*See Figure 63.*

1. Before servicing the vehicle, refer to the precautions.

2. Drain the engine coolant.

3. Detach the cooling hose from the thermostat housing.

4. Remove the bolts securing the thermostat housing, then remove it and the thermostat from the engine.

*To install:*

5. Installation is the reverse of removal.

6. Tighten the thermostat housing bolts to 10 ft. lbs. (14 Nm) on M112 engines and 7 ft. lbs. (10 Nm) on M113 engines.

#### M272 and M273 Engines

*See Figures 64 and 65.*

1. Before servicing the vehicle, refer to the precautions.

2. Remove the front and central sections lower engine compartment paneling.

3. Drain coolant from radiator.

4. Remove the left intake air duct and the front engine cover.

5. Dismount hose from electric air pump and air shutoff valve.

6. Dismount the top coolant hose from the coolant thermostat housing and place it on the side.

7. Dismount coolant hoses unscrew bolts and remove coolant pipe.

8. Remove poly-V-belt.

9. Remove guide pulley of poly-V-belt.

10. Unlatch electrical connector at three-disk thermostat valve and disconnect.

11. Remove bolts.

12. Pull out coolant thermostat housing with coolant thermostat from timing case cover.

### ✷✷ WARNING

**The coolant thermostat and thermostat housing are a single coolant thermostat must not be coolant thermostat housing damaged in the process.**

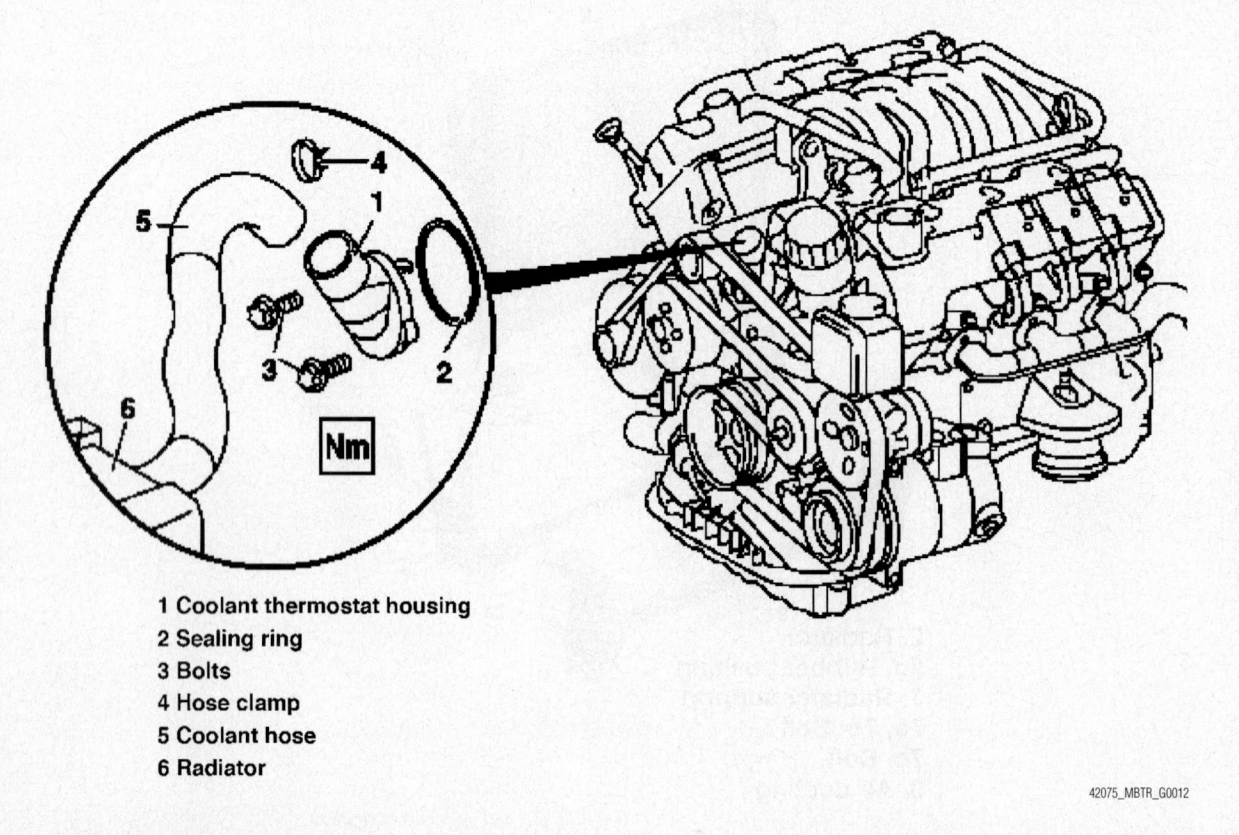

1 Coolant thermostat housing
2 Sealing ring
3 Bolts
4 Hose clamp
5 Coolant hose
6 Radiator

42075_MBTR_G0012

**Fig. 63 Exploded view of the thermostat housing**

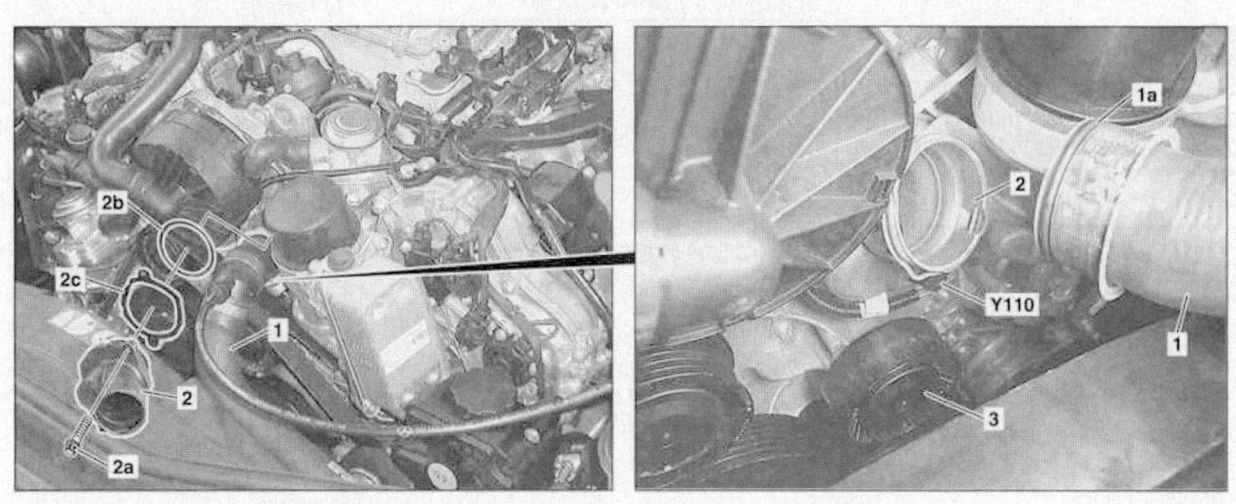

22205_MBCA_G0018

**Fig. 64  Coolant thermostat and housing—272 and 273 engine**

| | |
|---|---|
| 1 Upper coolant hose | 6 Coolant pipe |
| 1a O-ring | 6a Coolant hoses |
| 2 Coolant thermostat housing | 6b Bolts |
| 2a Bolt | 7 Hose |
| 2b O-ring | 8 Air shutoff valve |
| 2c Gasket | M33 Electric air pump |
| 3 Pulley | Y110 Three-disk thermostat valve |

22205_MBCA_G0019

**Fig. 65  Components that must be removed to access the coolant thermostat—272 and 273 engine**

*To install:*

13. Installation is the reverse of removal.
14. Replace coolant O-rings and gaskets.

### ✳✳ WARNING

**Replace all self locking, micro-encapsulated and aluminum bolts and nuts. Chase threads to remove micro-encapsulated residue from old bolts/nuts.**

15. Tighten bolts/nuts to specification as follows:

- Coolant thermostat housing to timing case cover: 19 ft. lbs. (25 Nm)
- Poly-V-belt guide pulley to coolant pump: 26 ft. lbs. (35 Nm)
- M273 Engine coolant pipe to coolant thermostat housing: 7 ft. lbs. (9 Nm)
- M273 Engine coolant pipe to cylinder head front cover: 7 ft. lbs. (9 Nm)

16. Check and fill engine coolant to proper level.
17. Start engine and check for leaks.

### WATER PUMP

#### REMOVAL & INSTALLATION

**M112 and M113 Engines**

*See Figure 66.*

1. Before servicing the vehicle, refer to the precautions.
2. Drain the cooling system.
3. Remove fan clutch, fan and shroud.
4. Remove engine cover if equipped.
5. Remove accessory drive belt and tensioner.
6. Remove secondary air injection switchover valve if equipped.
7. Remove power steering pump and position the pump aside, leaving the hoses attached.
8. Remove water pump coolant hoses.
9. Remove oil-to-water heat exchanger coolant hoses, if equipped.
10. Remove belt pulley.
11. Remove shock absorber from coolant pump if equipped.
12. Remove water pump mounting bolts.
13. Remove water pump.

14. Clean all gasket material from the sealing surfaces.

*To install:*

15. Install the water pump and gasket. Torque the bolts as follows:

- Engines with pre-tapped holes: M6 bolts to 88 inch lbs. (10 Nm), M8 bolts to 18 ft. lbs. (25 Nm)
- Engines without pre-tapped holes: M6 bolts to 10 ft. lbs. (14 Nm), M8 bolts to 26 ft. lbs. (35 Nm)

16. Install belt pulley and torque the mounting bolts to 88 inch lbs. (10 Nm).
17. Install oil-to-water heat exchanger coolant hoses.
18. Install water pump coolant hoses.
19. Install power steering pump.
20. Install secondary air injection switchover valve.
21. Install accessory drive belt and tensioner.
22. Install fan clutch, fan and shroud.
23. Install engine cover.
24. Fill and check coolant level.
25. Start engine and check for leaks.

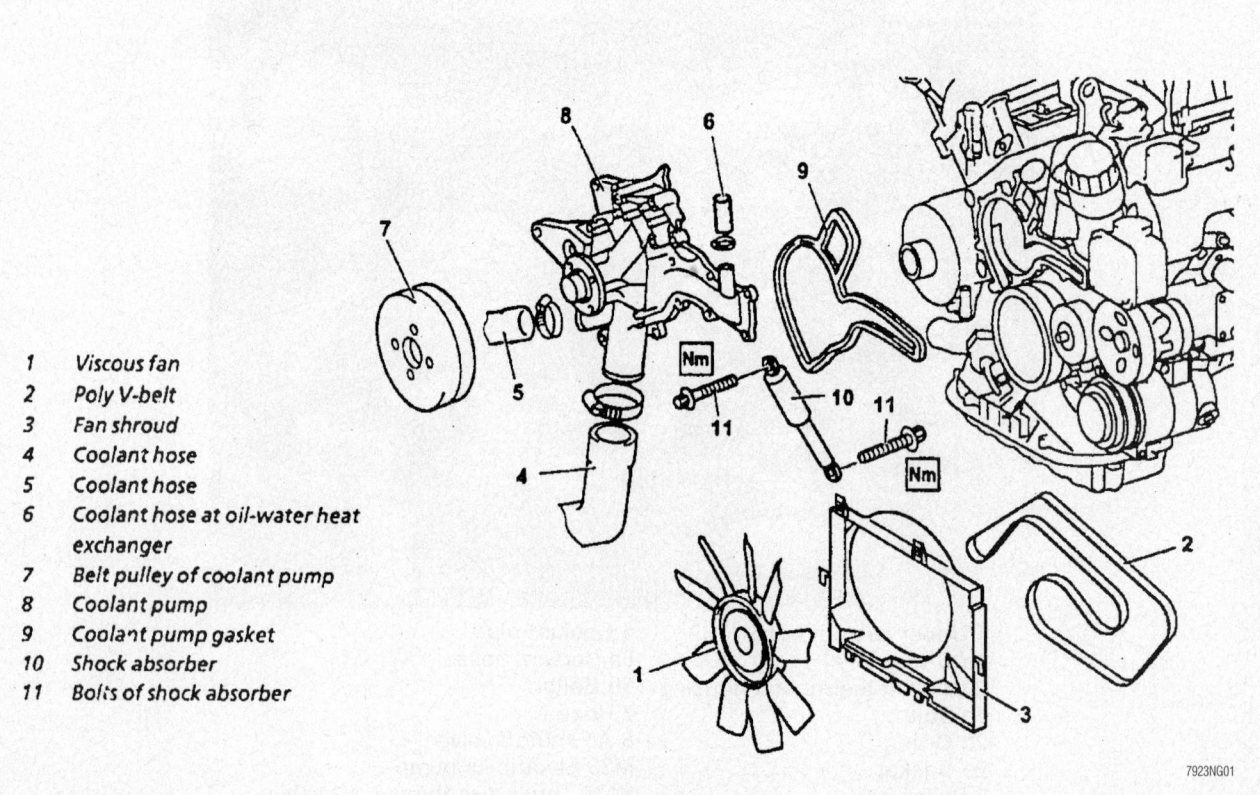

| | |
|---|---|
| 1 | Viscous fan |
| 2 | Poly V-belt |
| 3 | Fan shroud |
| 4 | Coolant hose |
| 5 | Coolant hose |
| 6 | Coolant hose at oil-water heat exchanger |
| 7 | Belt pulley of coolant pump |
| 8 | Coolant pump |
| 9 | Coolant pump gasket |
| 10 | Shock absorber |
| 11 | Bolts of shock absorber |

7923NG01

**Fig. 66 Exploded view of the water pump mounting**

**M272 and M273 Engines**

*See Figure 67.*

1. Before servicing the vehicle, refer to the precautions.
2. Drain coolant from radiator.
3. Remove fan shroud.
4. Remove poly-V-belt.
5. Remove guide pulleys .
6. Pry off the retaining clamp of the electrical feed line from the coolant pump.
7. Dismount coolant hose from coolant pump.
8. Unscrew bolts of coolant pump.
9. Remove coolant pump from timing case cover.

10. Clean sealing surfaces on coolant pump and timing case cover with a rag.
11. Installation is the reverse of removal.
12. Inspect cooling system for leaks.
13. Replace the O-ring on the coolant hose.

### ✴✴ WARNING

**Replace all self locking, micro-encapsulated and aluminum bolts and nuts. Chase threads to remove micro-encapsulated residue from old bolts/nuts.**

14. Tighten bolts in correct sequence; M7 bolts to 18 ft. lbs. (25 Nm), M8 bolts to 15 ft. lbs. (20 Nm).

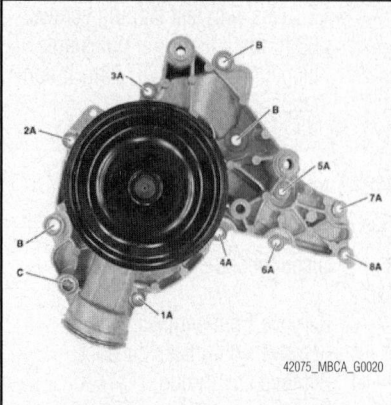

42075_MBCA_G0020

**Fig. 67 Tighten the M7 (A bolts) in the proper sequence**

## ENGINE ELECTRICAL

## CHARGING SYSTEM

### ALTERNATOR

*REMOVAL & INSTALLATION*

**M112 and M113 Engines**

*See Figure 68.*

1. Before servicing the vehicle, refer to the precautions.

2. Disconnect the negative battery cable.
3. To gain easier access to the alternator, remove or place aside any components in the way such as the air intake tube, engine fan and cooling hoses.
4. Remove the accessory drive belt.
5. Remove the engine under covers.
6. Detach the electrical connections from the alternator.

7. Remove the bolt(s) securing the alternator.
8. Remove the alternator out from the bottom except on S-Class; remove it from the top.

*To install:*

9. Installation is the reverse of removal.
10. Tighten the fasteners to specification:

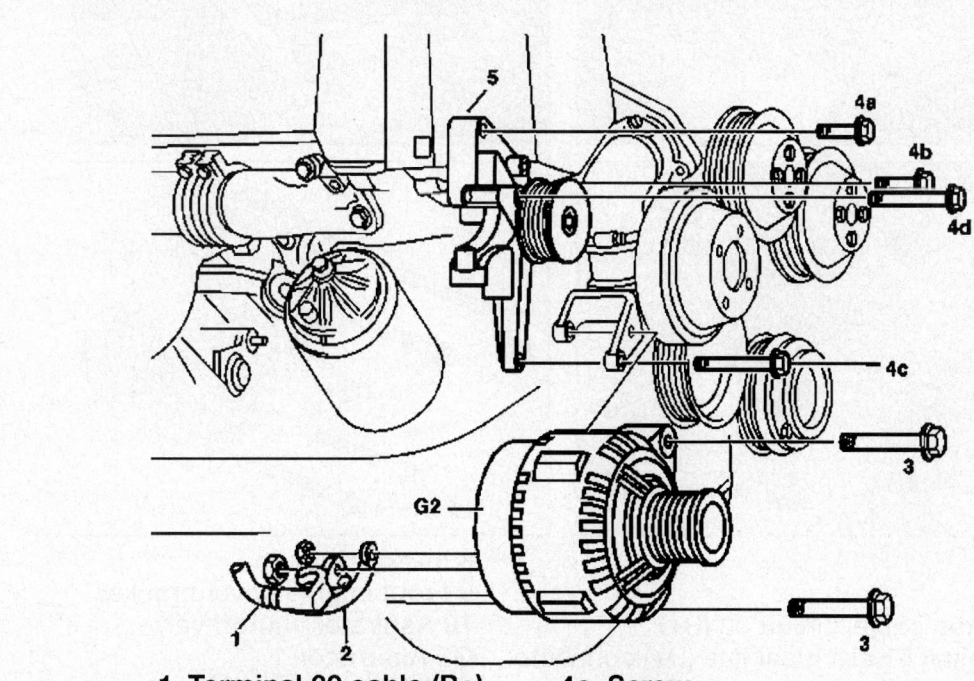

1. Terminal 30 cable (B+)
2. Terminal 61 cable (D+)
3. Bolts
4a. Screw
4b. Screw
4c. Screw
4d. Screw
5. Bracket
G2. Generator

42075_MBCA_G0003

**Fig. 68 Exploded view of the alternator mounting—E Class, others similar**

- E Class with self tapping screws: 33 ft. lbs. (45 Nm) if holes are tapped, 42 ft. lbs. (57 Nm) if non-tapped.
- All others: 31 ft. lbs. (42 Nm).

### M272 and M273 Engines

*See Figure 69.*

1. Disconnect the negative battery cable.
2. Remove front engine cover.
3. Remove right intake air duct.
4. Remove fan shroud.
5. Remove poly-V-belt.
6. Remove bracket for front engine cover and place with air pump switchover valve aside.

➡Do not detach lines at switchover valve.

7. Remove electric air pump, as necessary.
8. Remove aspirator shutoff valve at right cylinder head.
9. Raise and safely support the vehicle securely on jackstands.
10. Dismount lower engine compartment paneling.
11. Remove cap and disconnect electrical line (B+) from alternator by unscrewing nut.
12. Disconnect plug for electrical line from alternator.
13. Remove lower bolts from alternator.
14. Lower vehicle with lifting platform.
15. Remove upper bolts from alternator.
16. Take alternator down and out.

### To install:

17. Installation is the reverse of removal.

### ✳✳ WARNING

**Replace all self locking, micro-encapsulated and aluminum bolts and nuts. Chase threads to remove micro-encapsulated residue from old bolts/nuts.**

18. Tighten bolts/nuts to specification as follows:

- Alternator to timing case: 15 ft. lbs. (20 Nm)
- Nut of circuit B+ to alternator: 11 ft. lbs. (15 Nm)
- Switchover valve bracket to cylinder head: 7 ft. lbs. (9 Nm).

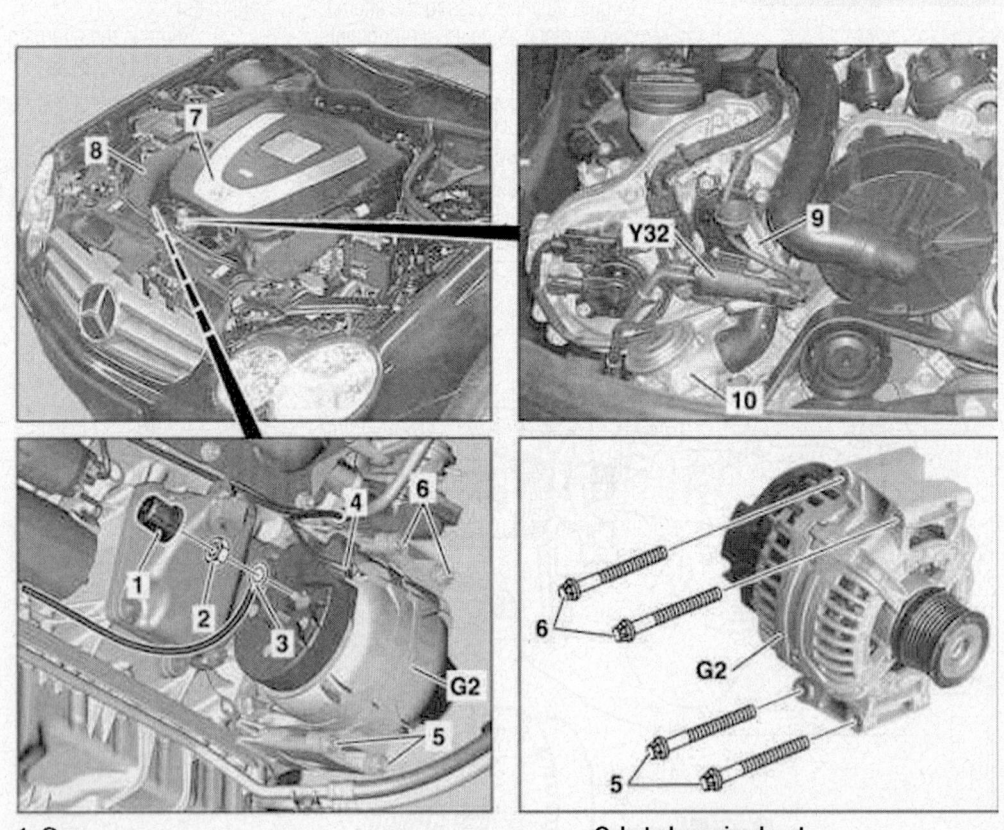

1 Cap
2 Nut
3 Electric cable, circuit 30 (B+)
4 Terminal 61 electrical line (D+) connector
5 Lower screws
6 Upper bolts
7 Front engine cover

8 Intake air duct
9 Front engine cover bracket
10 Aspirator shutoff valve
G2 Alternator
Y32 Air pump switchover valve

22205_MBCA_G0021

**Fig. 69 Exploded view of alternator assembly—M272 and M273 engines**

**ENGINE ELECTRICAL**
**IGNITION SYSTEM**

## FIRING ORDER

*See Figures 70 and 71.*

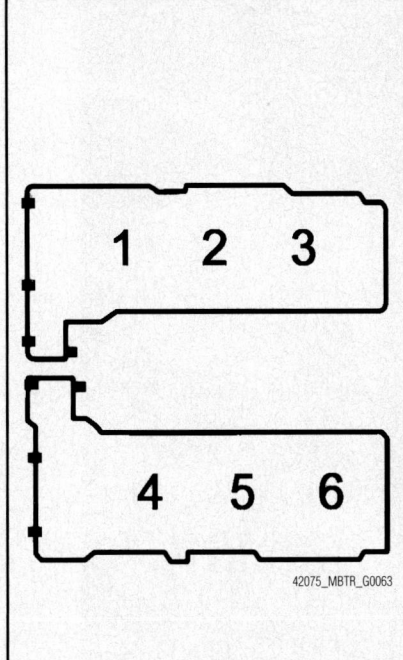

**Fig. 70 M112 and M272 engine firing order 1-4-3-6-2-5 Distributorless ignition**

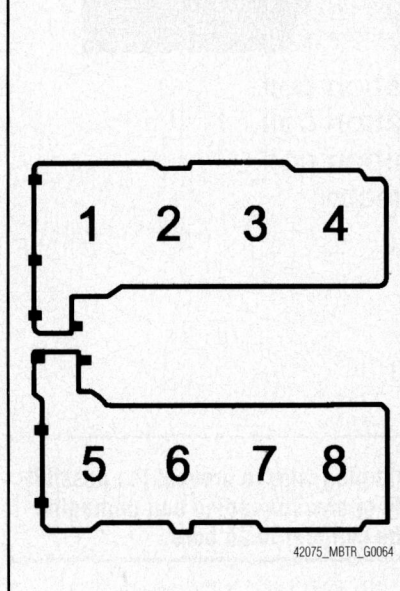

**Fig. 71 M113 and M273 engine firing order 1-5-4-2-6-3-7-8 Distributorless ignition**

## IGNITION COIL

### REMOVAL & INSTALLATION

#### M112 and M113 Engines

1. Before servicing the vehicle, refer to the precautions.
2. Remove the upper part of the air cleaner to access the coil packs.
3. Detach the electrical connector and spark plug wires from the coil pack.
4. Remove the mounting screw, then remove the coil pack.

*To install:*

5. Installation is the reverse of removal.
6. Tighten the mounting screw until just snug.
7. Be sure to connect the correct plug wire to the correct coil terminal.
8. Check the terminal a and b identification stamps on the coil, match that to the **A** and **B** identification stamps on the cylinder head between the spark plugs.

#### M272 and M273 Engines

*See Figure 72.*

1. Before servicing the vehicle, refer to the precautions.
2. Turn key in ignition switch **OFF**.
3. Remove air filter housing
4. Disconnect electrical connectors on ignition coils
5. Unscrew bolts (1) for ignition coils
6. Detach ignition coils for cylinders upwards using a spark plug connector (a) and remove

*To install:*

7. Installation is the reverse of removal.

### ❈❈ WARNING

**Replace all self locking, micro-encapsulated and aluminum bolts and nuts. Chase threads to remove micro-encapsulated residue from old bolts/nuts.**

8. Tighten ignition coil to cylinder head cover to 7 ft. lbs. (9 Nm).
9. Connect STAR DIAGNOSIS or equivalent scan tool and read out fault memory.

## IGNITION TIMING

### ADJUSTMENT

All engines are equipped with a Distributorless Ignition System (DIS). No adjustment is necessary.

## SPARK PLUGS

### REMOVAL & INSTALLATION

*See Figure 73.*

1. Before servicing the vehicle, refer to the precautions.
2. Disconnect the negative battery cable, and if the vehicle has been run recently, allow the engine to thoroughly cool.
3. Remove the engine cover; if necessary, remove the air filter housing to gain access to it.
4. Remove the ignition coils.
5. If equipped, twist, then pull the spark plug wires from the plugs. Be sure to label them first to prevent mixing them up during installation.
6. Using compressed air, blow any water or debris from the spark plug well to assure that no harmful contaminants are allowed to enter the combustion chamber when the spark plug is removed. If compressed air is not available, use a rag or a brush to clean the area.

➡**Remove the spark plugs when the engine is cold, if possible, to prevent damage to the threads. If removal of the plugs is difficult, apply a few drops of penetrating oil or silicone spray to the area around the base of the plug, and allow it a few minutes to work.**

7. Using a spark plug socket that is equipped with a rubber insert to properly hold the plug, turn the spark plug counterclockwise to loosen and remove the spark plug from the bore.

### ❈❈ WARNING

**Be sure not to use a flexible extension on the socket. Use of a flexible extension may allow a shear force to be applied to the plug. A shear force could break the plug off in the cylinder head, leading to costly and frustrating repairs.**

*To install:*

8. Using a wire feeler gauge, check and adjust the spark plug gap. When using a gauge, the proper size should pass between the electrodes with a slight drag. The next larger size should not be able to pass while the next smaller size should pass freely.

1  Bolts
T1/1 Cylinder 1 ignition coil
T1/2 Cylinder 2 ignition coil
T1/3 Cylinder 3 ignition coil
T1/4 Cylinder 4 ignition coil
T1/5 Cylinder 5 ignition coil
T1/6 Cylinder 6 ignition coil
a  Spark plug connector

22205_MBCA_G0022

**Fig. 72 Ignition coil—M272 engine**

9.  Carefully thread the plug into the bore by hand. If resistance is felt before the plug is almost completely threaded, back the plug out and begin threading again. In small, hard to reach areas, an old spark plug wire and boot could be used as a threading tool. The boot will hold the plug while you twist the end of the wire and the wire is supple enough to twist before it would allow the plug to crossthread.

**❊❊ WARNING**

**Do not use the spark plug socket to thread the plugs. Always carefully thread the plug by hand or using an old plug wire to prevent the possibility of crossthreading and damaging the cylinder head bore.**

10.  Installation is the reverse of removal.
11.  Carefully tighten the spark plug to 17 ft. lbs. (23 Nm).

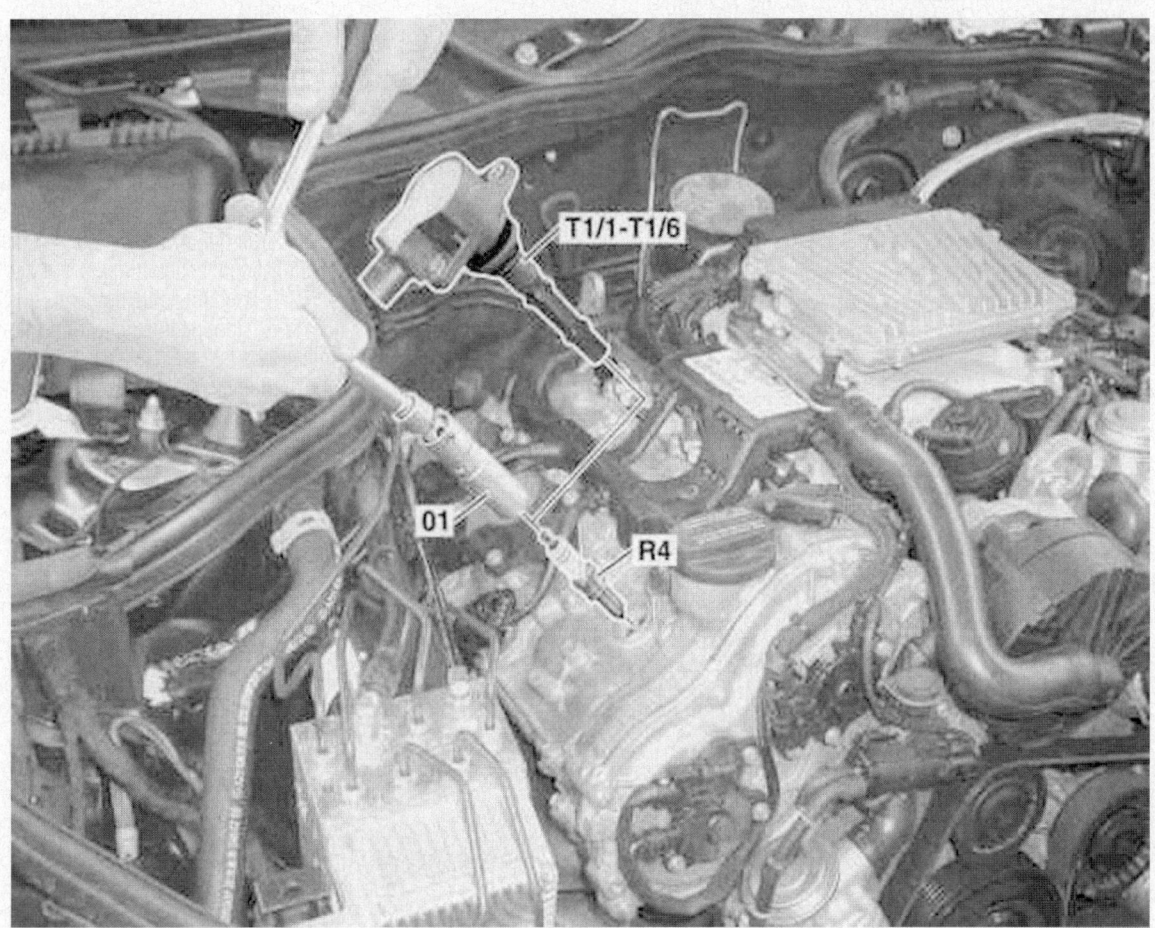

01 Spark plug wrench
R4 Spark plugs
T1/1 Cylinder 1 ignition coil
T1/2 Cylinder 2 ignition coil

T1/3 Cylinder 3 ignition coil
T1/4 Ignition coil cylinder 4
T1/5 Ignition coil cylinder 5
T1/6 Cylinder 6 ignition coil

22205_MBCA_G0024

**Fig. 73** Removing the spark plugs

## STARTER

### REMOVAL & INSTALLATION

#### M112 and M113 Engines

1. Before servicing the vehicle, refer to the precautions.
2. Turn the steering wheel to full left position to gain access if necessary.
3. Disconnect the negative battery cable.
4. Remove engine under cover
5. Remove exhaust system and engine mount, if necessary for additional clearance
6. Remove starter electrical connections
7. Remove starter mounting bolts
8. Remove starter assembly

**To install:**

9. Install starter assembly.
10. Tighten the mounting bolts to 31 ft. lbs. (42 Nm).
11. If necessary, install exhaust system and engine mount.
12. Connect starter electrical connections
13. Connect the negative battery cable.
14. Start the engine and check for proper operation.

#### M272 and M273 Engines

*See Figure 74.*

1. Before servicing the vehicle, refer to the precautions.

2. Raise and safely support the vehicle securely on jackstands.
3. Dismount lower engine compartment paneling.
4. Detach the right catalytic converter from exhaust manifold and the catalytic converter.
   bracket.
5. Remove complete exhaust system.
6. Remove exhaust shielding plate on the right for front axle gear propeller shaft.
7. Pull out protective cap and dismantle electrical connection circuit 30 at starter.
8. Dismount cable holder from starter).
9. Dismount electrical connection terminal 50 at starter.
10. Remove bolts and starter from the bottom.

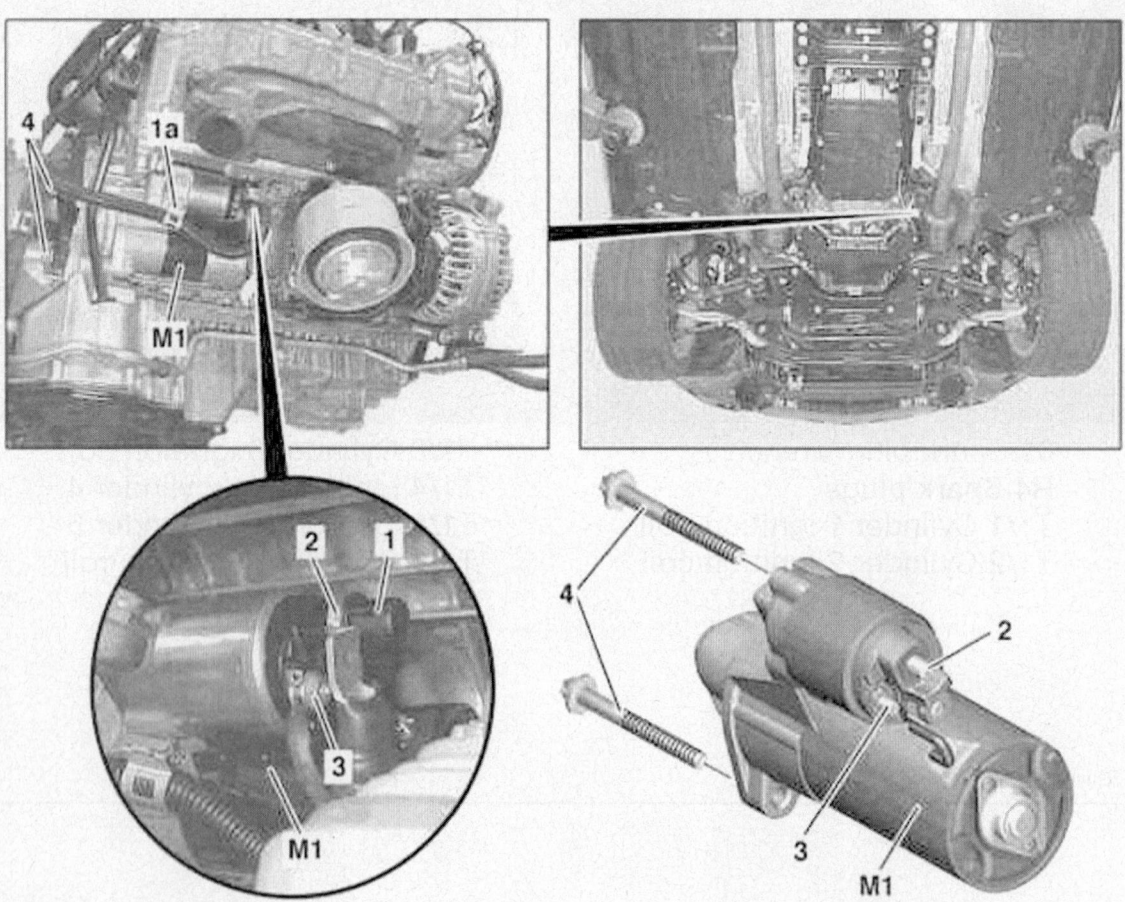

1 Protective cap
1a Cable holder
2 Electrical connection circuit 30
3 Electrical connection circuit 50
4 Bolts
M1 Starter

22205_MBCA_G0025

**Fig. 74 Exploded view of starter assembly—M272 and M273 engines**

11. Crank engine at center bolt of crankshaft in the direction of rotation of the engine and check the ring gear at the flywheel/drive plate for wear and damage.

**To install:**

12. Installation is the reverse of removal.

13. Tighten bolts/nuts to specification as follows:

- Nut for connection of circuit 30: 11 ft. lbs. (15 Nm)
- Nut for connection of circuit 50: 6 ft. lbs. (8 Nm)
- Bolt connecting starter to crankcase: 30 ft. lbs. (40 Nm)

## ENGINE MECHANICAL

➡**Disconnecting the negative battery cable may interfere with the functions of the on board computer systems and may require the computer to undergo a relearning process, once the negative battery cable is reconnected.**

## ACCESSORY DRIVE BELTS

### ACCESSORY BELT ROUTING

*See Figures 75 and 76.*

### INSPECTION

Inspect the drive belt for signs of glazing or cracking. A glazed belt will be perfectly smooth from slippage, while a good belt will have a slight texture of fabric visible.

Cracks will usually start at the inner edge of the belt and run outward. All worn or damaged drive belts should be replaced immediately.

### ADJUSTMENT

The belt is tensioned automatically. If the belt is slipping, either the belt has stretched beyond its serviceable limit or the tensioner is faulty.

### REMOVAL & INSTALLATION

1. Before servicing the vehicle, refer to the precautions.
2. If additional access is needed, remove the cooling fan and fan shroud.
3. Using a breaker bar and appropriate sized Torx socket (or

appropriate drive belt removal tool), relieve the belt tension by rotating the tensioner.

4. Remove the belt from the pulleys.

**To install:**

5. Installation is the reverse of removal.
6. Be sure to follow the belt routing diagram.

## CAMSHAFT AND VALVE LIFTERS

### INSPECTION

1. Before servicing the vehicle, refer to the precautions.
2. Remove the camshaft from the engine.
3. Check the camshaft bearing journals for damage and binding.
4. If the journals are binding, check the cylinder head for damage.
5. Check the cylinder head for clogged oil holes.
6. Check the camshaft surface for abnormal wear and damage. Replace the camshaft, as required.

### REMOVAL & INSTALLATION

**M112 and M113 Engines**

*See Figure 77.*

1. Before servicing the vehicle, refer to the precautions.
2. Disconnect the negative battery cable and remove the cylinder head cover. Rotate the engine clockwise to position the crankshaft 40° After Top Dead Center (ATDC).

3. Remove generator.
4. Remove timing chain tensioner.
5. Remove Camshaft Position Sensor (CMP).

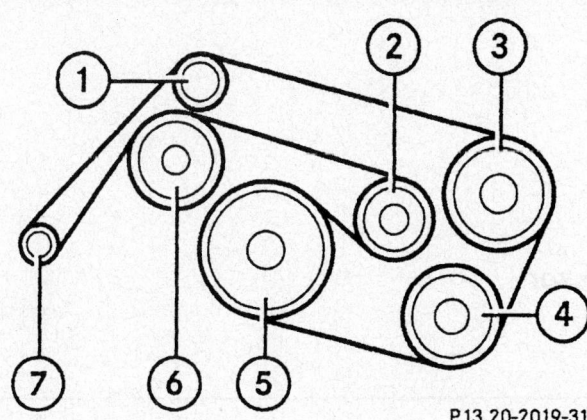

P13.20-2019-31

**1** Idler pulley
**2** Automatic belt tensioner
**3** Power steering pump
**4** Air conditioner compressor
**5** Crankshaft
**6** Coolant pump, fan
**7** Generator (alternator)

42348-BENZ-G02

**Fig. 75 Accessory drive belt routing–M112 and M113 engines**

1. **Belt pulley/vibration damper**
2. **Tensioning pulley**
3. **Pulley**
4. **Coolant pump belt pulley**
5. **Generator belt pulley**
6. **Pulley**
7. **Power steering pump belt pulley**
8. **Belt pulley on refrigerant compressor**

42075_MBCA_G0009

**Fig. 76 Accessory drive belt routing—M272 and M273 engines**

6. Remove cable-tie the timing chain to the camshaft sprocket. Lock the camshafts using camshaft locking tools.

7. Remove camshaft gears.

8. Remove camshaft bearing bridge.

9. Remove camshafts.

**To install:**

10. Apply clean engine oil to the camshaft contact surfaces.

11. Install camshafts.

12. Install camshaft bearing bridge.

➡ **The camshafts can be rotated 40° ATDC of the No. 1 cylinder without the valves touching the pistons.**

13. Position the camshaft so that the groove points towards the contact surface of the cylinder head cover, then attach the camshaft fixing plate. Repeat this step for the other camshaft.

14. Install camshaft sprockets. Torque the bolt to 37 ft. lbs. (50 Nm) plus 90–100°.

15. Install timing chain.

16. Install CMP sensor. Torque the mounting bolt to 70 inch lbs. (8 Nm).

17. Install timing chain tensioner with new gasket. Torque to 59 ft. lbs. (80 Nm).

18. Install generator.

19. Install cylinder head cover.

20. Connect the negative battery cable, start the engine and check operation.

### M272 and M273 Engines

*See Figures 78 through 82.*

1. Before servicing the vehicle, refer to the precautions.

2. Remove the air filter housing assembly.

3. Remove the covers from the front of the cylinder heads.

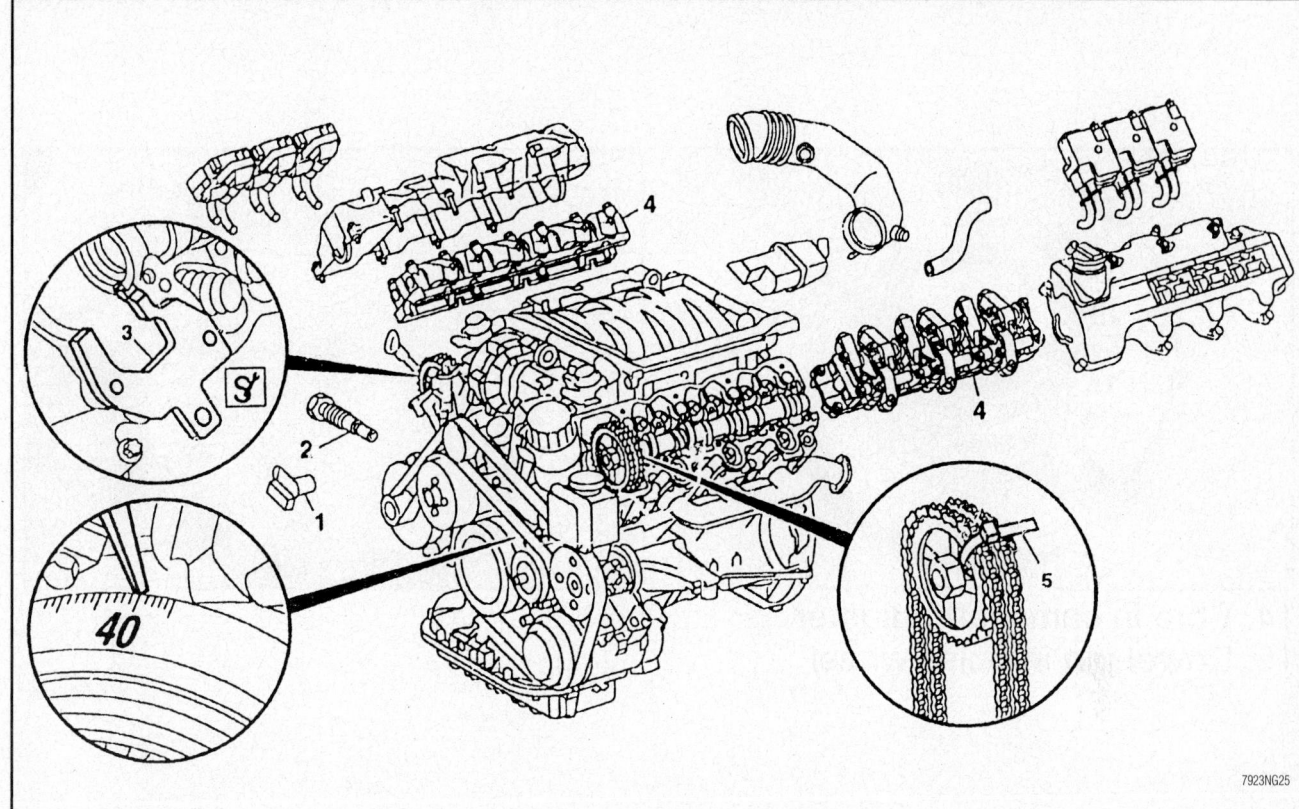

Fig. 77 Exploded view of the camshaft mounting, showing related components

Fig. 78 Lock the compensators as shown—M272 and M273 engines

Fig. 79 Loosen the center valves in the direction of the arrows

4. Detach the crankcase ventilation hose from the centrifuge housing. Remove the bolts securing the housing, then remove the bolt securing the centrifuge on the end of the camshaft.

5. Rotate the engine in the direction of rotation so that it is at 40° After Top Dead Center (ATDC) of cylinder 1.

6. Lock the gear backlash compensators using a drift pin (refer to the illustration). This must be done to avoid damaging the camshaft adjusters.

7. Loosen, but do not remove, the center valves in the direction of the arrows (refer to the illustration). Use a Torx wrench on the rears of the camshafts to counterhold them.

8. Remove the cylinder head covers.

9. Remove the center valves on the exhaust camshafts and remove the pulse wheels. Discard the pulse wheels as they must be replaced with new ones during installation.

➡When removing the pulse wheels, check if the dowel pin is still intact. If not, remove the pin from the adjusters.

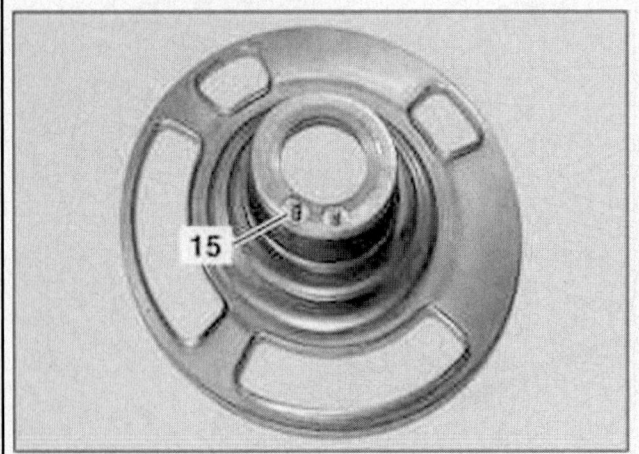

## 14. Bore in camshaft adjuster
## 15. Dowel pin in pulse wheel

22205_MBCA_G0027

**Fig. 80 Camshaft adjuster as shown on exhaust camshaft**

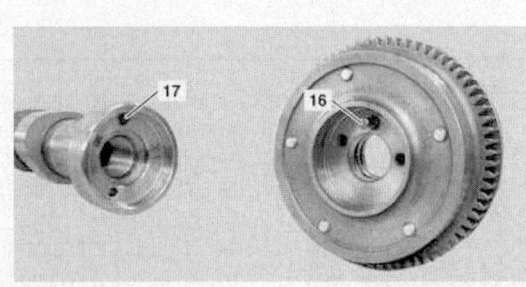

22205_MBCA_G0028

**Fig. 81 Camshaft adjuster as it fits into the exhaust camshaft**

10. Counterhold the exhaust camshafts and pull the camshaft adjuster forward off of the exhaust camshafts.

11. Remove the bearing caps on the exhaust camshaft, then remove the camshaft.

12. Make sure the marks on the timing chain still align, then remove the chain tensioner.

13. Remove the tensioning rail bolt. Pull the rail upward and out of the engine.

14. Remove the center valves on the intake camshafts and remove the pulse wheels. Discard the pulse wheels as they must be replaced with new ones during installation.

15. Counterhold the intake camshafts and pull the camshaft adjuster forward off of the camshafts.

16. Remove the bearing caps on the intake camshaft, then remove the camshaft.

### To install:

17. Installation is the reverse of removal.

18. Lubricate the contact surfaces of the adjusters and valves with oil during reassembly.

19. Align the adjusters with the marks on the timing chain and align the dowel pins of the new pulse wheels with the bores in the adjusters.

20. Tighten fasteners to specification:
- Bearing caps: 71 inch lbs. (8 Nm)
- Center valves: 107 ft. lbs. (145 Nm).

## CRANKSHAFT FRONT SEAL

### REMOVAL & INSTALLATION

#### M112 and M113 Engines

1. Before servicing the vehicle, refer to the precautions.

2. Remove engine cooling fan and clutch

3. Remove fan shroud

➡**The fan clutch is equipped with right-hand thread.**

4. Remove accessory drive belt

5. Remove crankshaft damper

6. Remove crankshaft seal, using a seal pick

### To install:

7. Apply clean engine oil to the end of the crankshaft to ease installation.

8. Install crankshaft seal, using a suitable seal driver

9. Install crankshaft damper

10. Install accessory drive belt

5. Center valve of exhaust camshaft
6. Center valve of intake camshaft
7. Pulse wheel of exhaust camshaft
8. Pulse wheel for intake camshaft
9. Camshaft adjuster of exhaust camshaft
10. Camshaft adjuster for intake camshaft
11. Auxiliary bearing cap of exhaust camshaft
12. Auxiliary bearing cap of intake camshaft
13. Color coding

22205_MBCA_G0026

**Fig. 82 Camshaft assembly on cylinder head**

11. Install fan shroud, engine cooling fan, and fan clutch

### M272 and M273 Engines

*See Figure 83.*

1. Before servicing the vehicle, refer to the precautions.
2. Remove front engine cover.
3. Remove intake air ducts from air filter.
4. Remove poly-V-belt.
5. Install retaining lock for crankshaft/starter ring gear.
6. Remove belt pulley / vibration damper.

7. Press out crankshaft radial sealing ring from mounting bore with a screwdriver.

**✳✳ WARNING**

**Do not damage crankshaft and mounting bore. Use a clean rag as a base while pressing out; otherwise there can be leakage.**

8. Check tread of crankshaft radial sealing ring at hub of belt pulley/vibration damper.
9. Check for scoring and galling, if necessary, replace belt pulley/vibration damper.

***To install:***

10. Installation is the reverse of removal.

**✳✳ WARNING**

**Replace all self locking, micro-encapsulated and aluminum bolts and nuts. Chase threads to remove micro-encapsulated residue from old bolts/nuts.**

11. Tighten bolts to specification as follows:
• Step 1: Tighten bolt to 148 ft. lbs. (200 Nm)

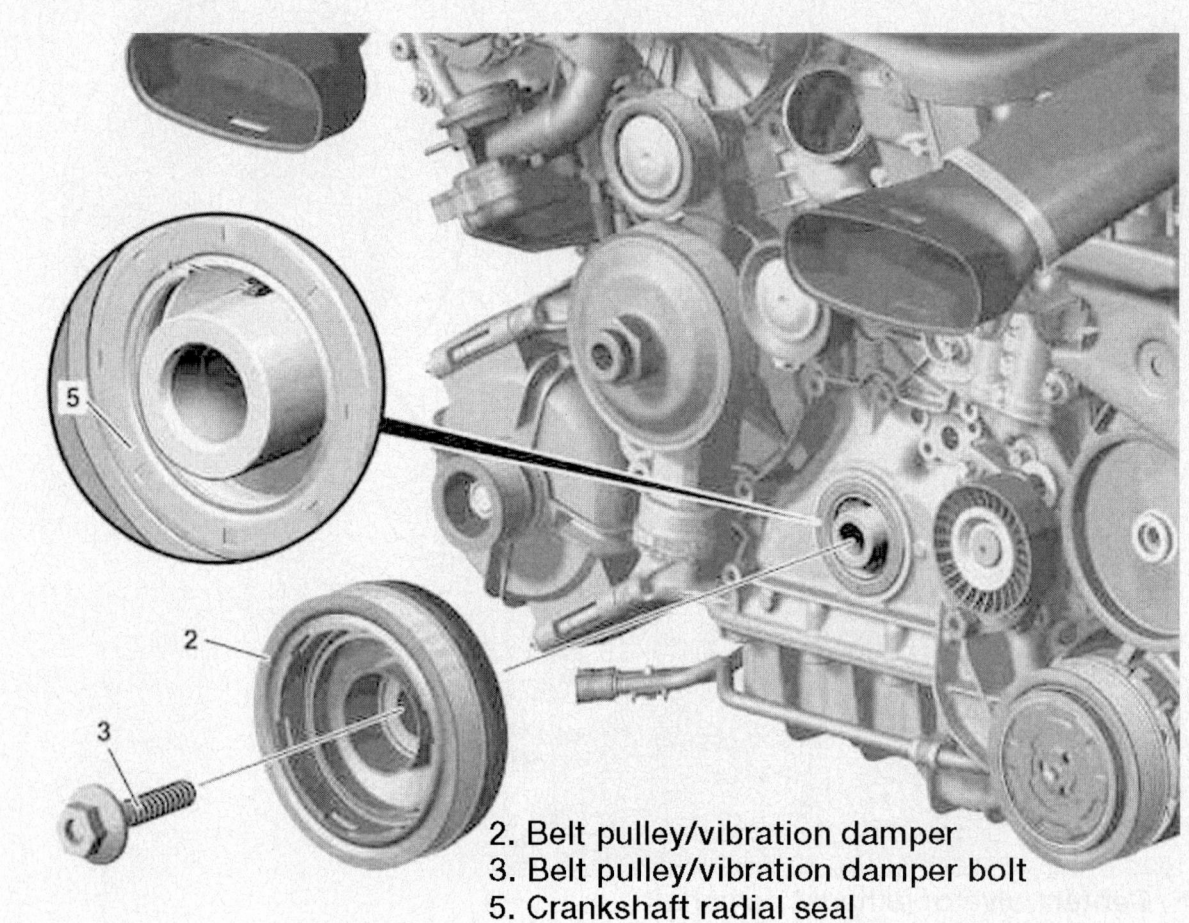

2. Belt pulley/vibration damper
3. Belt pulley/vibration damper bolt
5. Crankshaft radial seal

22205_MBCA_G0029

**Fig. 83 Exploded view of the crankshaft damper and seal**

- Step 2: Tighten bolt an additional 90° rotation
12. Start engine and check for leaks.

## CYLINDER HEAD

### REMOVAL & INSTALLATION

#### M112 and M113 Engines

*See Figures 84 through 87.*

1. Before servicing the vehicle, refer to the precautions.

2. Properly relieve the fuel system pressure and drain the engine coolant. Place a guard plate behind the radiator/condenser to protect it from damage.

3. Disconnect the negative battery cable.

4. Remove fan clutch, fan and shroud.

5. Remove engine cover.

6. Remove air cleaner housing, resonance pipe and body.

7. Remove fuel line.

8. Remove ignition coils.

9. Remove cylinder head covers.

➡The intake manifold system must not be disassembled.

10. Remove intake manifold.

11. Remove vacuum switchover valve.

12. Remove camshaft Position (CMP) sensor.

13. Lock the Automatic belt tensioner by rotating the tensioner counterclockwise until a 5mm drift or pin fits through the tensioner, and then remove the serpentine belt.

14. Remove power steering pump and position it aside leaving the hoses attached.

15. Remove serpentine belt.

16. Remove heater hose at the firewall.

17. Remove exhaust system.

18. Rotate the engine clockwise to position the crankshaft 40° After Top Dead Center (ATDC).

**✳✳ WARNING**

**The engine must not be rotated backwards.**

19. Lock the camshafts using camshaft locking tools.

20. Remove generator.

21. Remove timing chain tensioner.

22. Remove camshaft gears. Attach to the chain with a cable tie.

23. Remove camshaft bearing bridges.

24. Remove camshafts.

25. Remove timing case-to-cylinder head bolts.

26. Remove cylinder head bolts in the reverse order of the illustrated tightening sequence.

**✳✳ CAUTION**

**Never use a prybar between the head and block.**

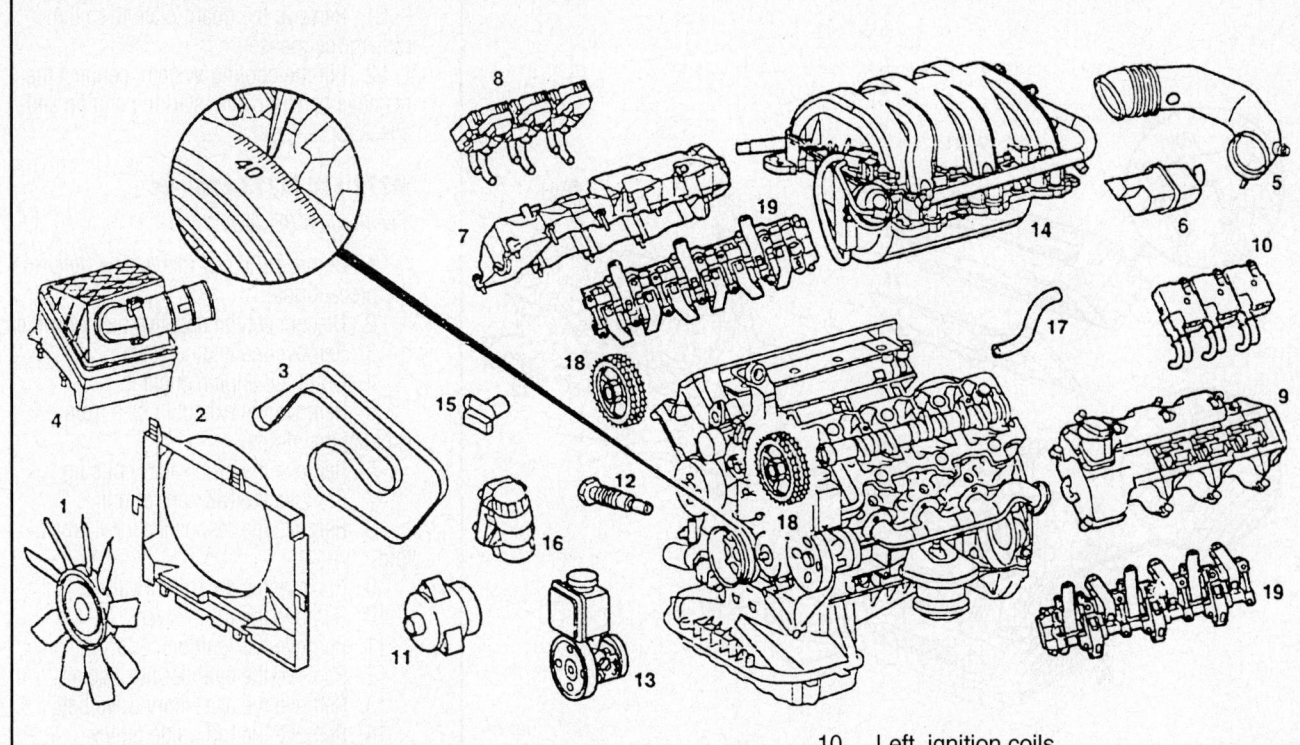

1    Viscous fan
2    Fan shroud
3    Poly V-belt
4    Air cleaner housing with HFM-SFI
5    Resonance pipe
6    Resonance body
7    Right cylinder head cover
8    Right ignition coils
9    Left cylinder head cover

10    Left ignition coils
11    Generator
12    Chain tensioner
13    Power steering pump with reservoir
14    Intake manifold
15    Camshaft position sensor
16    Oil filter housing
17    Heating hose
18    Camshaft gears
19    Camshaft bearing bridges

7923NG45

**Fig. 84 Exploded view of the cylinder head accessory components—M112 shown, M113 engines similar.**

27. Remove the cylinder head and clean all gasket material from the sealing surfaces. Be sure the cylinder head locating dowels are positioned in the engine block.

28. Inspect length of the cylinder head bolt shaft. New bolt length is 141.5mm and the maximum permissible length is 144.5mm. Replace bolts that measure greater than the maximum permissible length.

*To install:*

### ✳✳ WARNING

**Replace all self locking, micro-encapsulated and aluminum bolts and nuts. Chase threads to remove micro-encapsulated residue from old bolts/nuts.**

29. Clean the head bolt threads and apply clean engine oil to the thread and head contact surfaces.

30. Install the cylinder head gasket and head. Torque the head bolts to specification in sequence shown:
    a. Step 1: 7 ft. lbs. (10 Nm)
    b. Step 2: 22 ft. lbs. (30 Nm)
    c. Step 3: Additional 90° rotation
    d. Step 4: Additional 90° rotation

31. Install timing case-to-cylinder head bolts and torque to 15 ft. lbs. (20 Nm).

32. Install camshafts.

33. Install camshaft bearing bridges.

34. Install camshaft gear. Torque the mounting bolt to 37 ft. lbs. (50 Nm) plus an additional 90°.

35. Install timing chain tensioner with new gasket. Torque to 59 ft. lbs. (80 Nm).

36. Install generator.

37. Remove the camshaft locking plates.

38. Install exhaust system. Torque mounting nuts to 15 ft. lbs. (20 Nm).

39. Install heater hose.

40. Install power steering pump.

41. Install serpentine belt.

42. Install CMP sensor.

43. Install vacuum switchover valve.

44. Install intake manifold.

45. Install cylinder head covers.

46. Install ignition coils.

47. Install fuel line.

48. Install air cleaner housing, resonance pipe and body.

49. Install engine cover.

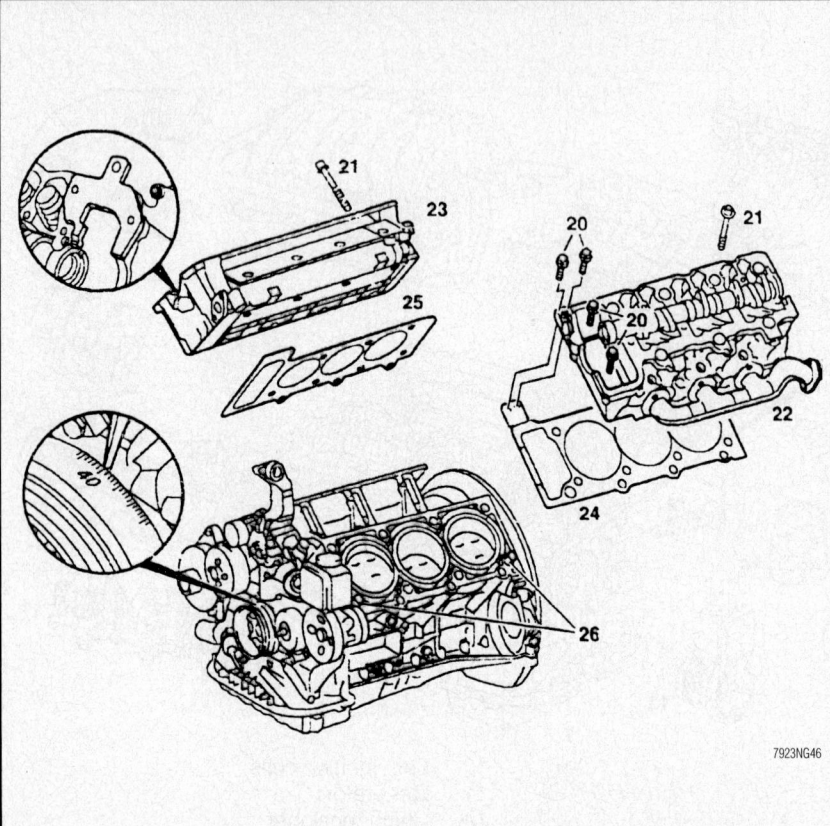

**Fig. 85 Exploded view of the cylinder head removal—M112 shown, M113 engines similar.**

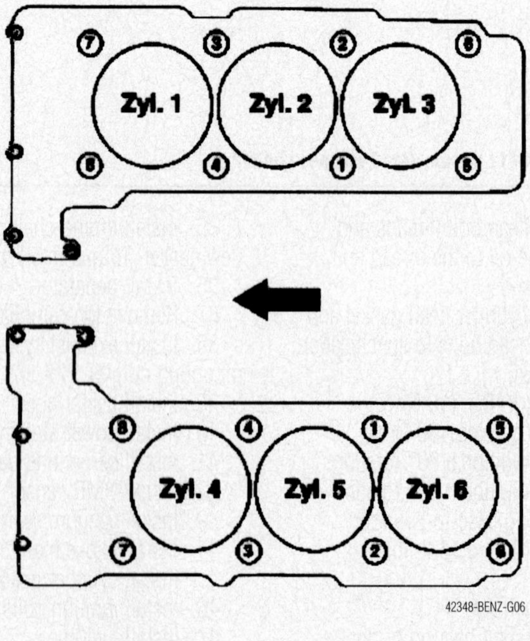

Zyl. 1  Zyl. 2  Zyl. 3

Zyl. 4  Zyl. 5  Zyl. 6

42348-BENZ-G06

**Fig. 86 Cylinder head bolt tightening sequence—M112 engine.**

50. Install fan clutch, fan and shroud.

51. Remove the guard plate from the radiator/condenser.

52. Fill the cooling system, connect the negative battery cable, start the engine and check for leaks.

### M272 and M273 Engines

*See Figures 88 through 91.*

1. Before servicing the vehicle, refer to the precautions.

2. Disconnect the negative battery cable.

3. Remove the engine cover.

4. Drain the engine coolant.

5. Remove the exhaust pipes from exhaust manifolds.

6. Remove the air cleaner housing.

7. Remove the ME control unit.

8. Remove the resonance intake manifold.

9. Remove the air shutoff valve.

10. Remove the front cover.

11. Remove the ignition coils.

12. Remove the cylinder head cover.

13. Remove the accessory drive belt.

14. Remove the top guide pulley.

15. Remove the alternator.

16. Rotate the engine clockwise to position the crankshaft 40° After Top Dead Center (ATDC).

**✴✴ WARNING**

**The engine must not be rotated backwards.**

17. Remove the timing chain tensioner.

18. Remove the camshaft adjusters.

19. Remove the guide rail pin on the right cylinder head.

20. Remove the air pump and aspirator valve.

21. Remove the oil dipstick guide tube bolt.

22. Remove the guide bolts for slide rail on the left cylinder head.

23. Remove the cylinder head bolts in the reverse order of the tightening sequence shown.

**✴✴ CAUTION**

**Never use a prybar between the head and block.**

➡The rear left bolt on the right cylinder head cannot be removed until the head is removed from the engine. Pull the bolt up slightly and use a cable tie to keep it from sliding back down.

24. Remove the cylinder head and clean all gasket material from the sealing surfaces.

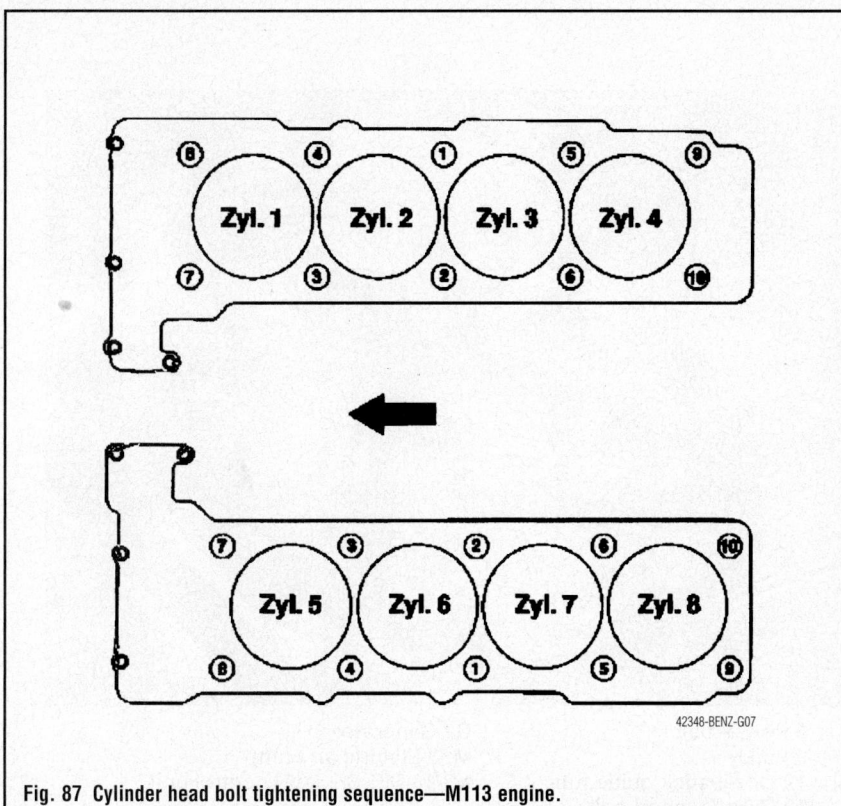

Fig. 87 Cylinder head bolt tightening sequence—M113 engine.

*To install:*

25. Installation is the reverse of removal.

26. Use a new cylinder head gasket and bolts.

27. Tighten the M8 cylinder head to timing case bolts to 11 ft. lbs. (15 Nm) plus an additional 90° rotation.

28. Tighten the M11 cylinder head to crankcase bolts as follows:

- Step 1: Tighten bolts to 15 ft. lbs. (20 Nm)
- Step 2: Tighten bolts to 30 ft. lbs. (40 Nm)
- Step 3: Tighten an additional 90° rotation
- Step 4: Tighten an additional 90° rotation

29. Tighten the M272 engine cylinder head bolts in the sequence shown.

- Step 1: Tighten cylinder head bolts (1-8) to stage 1
- Step 2: Tighten cylinder head bolts (1-8) to stage 2
- Step 3: Tighten cylinder head bolts (1-8) to stage 2 (follow-up tightening)
- Step 4: Tighten cylinder head bolts (1-8) to stage 3

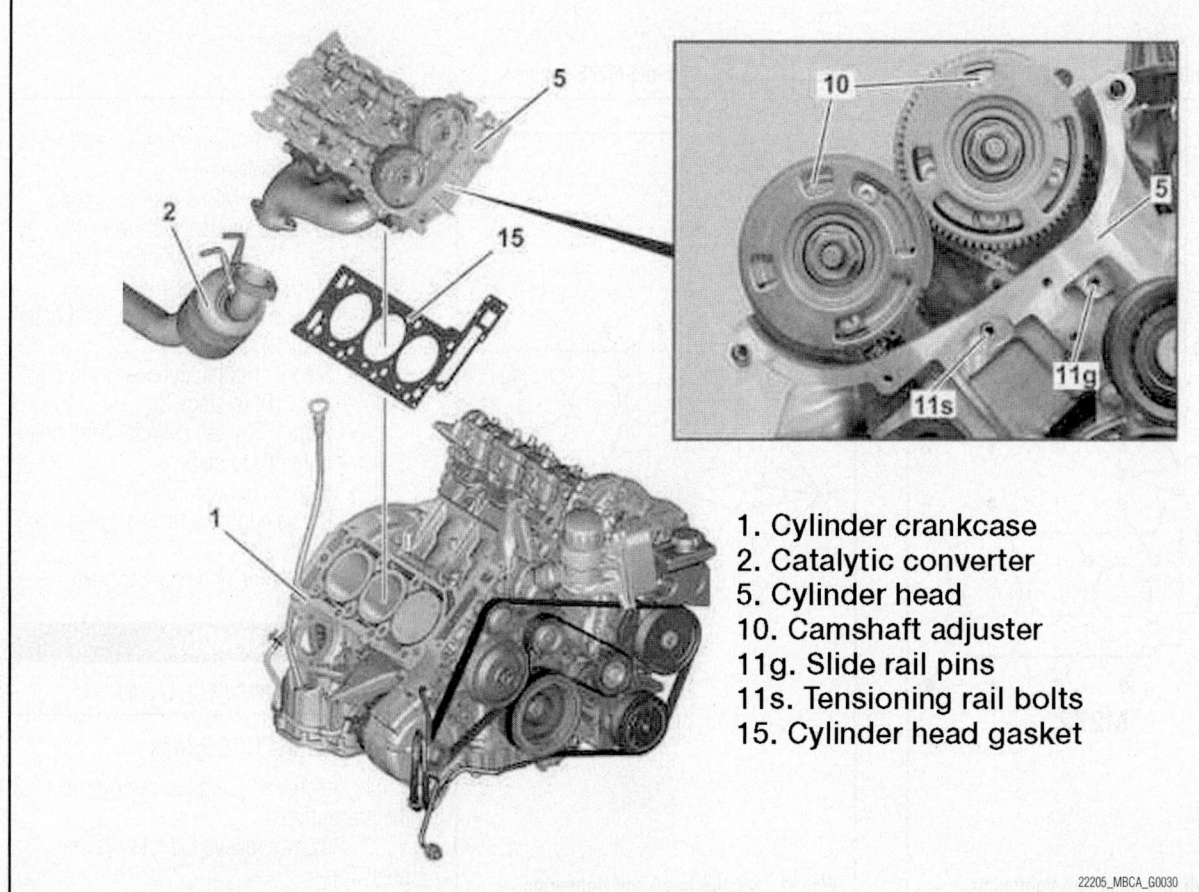

1. Cylinder crankcase
2. Catalytic converter
5. Cylinder head
10. Camshaft adjuster
11g. Slide rail pins
11s. Tensioning rail bolts
15. Cylinder head gasket

Fig. 88 Cylinder head and components—M272 and M273 engines

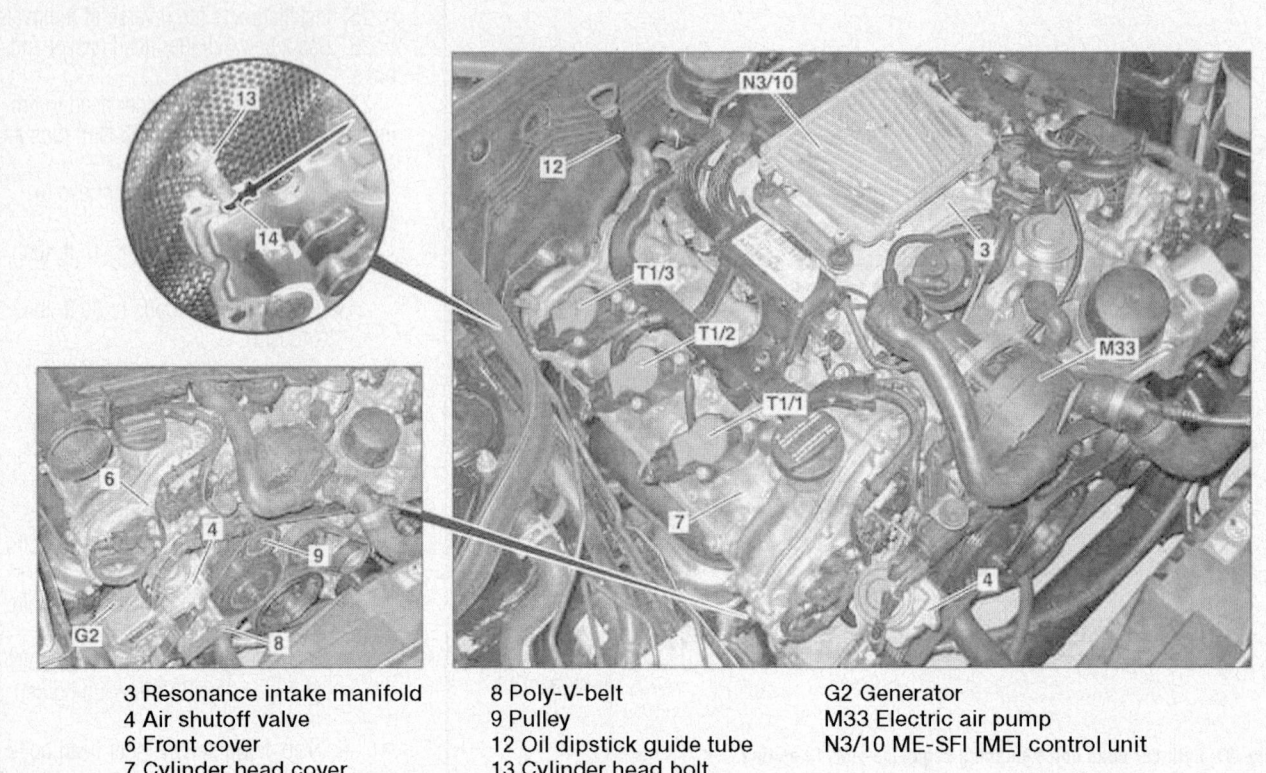

3 Resonance intake manifold
4 Air shutoff valve
6 Front cover
7 Cylinder head cover

8 Poly-V-belt
9 Pulley
12 Oil dipstick guide tube
13 Cylinder head bolt
14 Cable tie

G2 Generator
M33 Electric air pump
N3/10 ME-SFI [ME] control unit

22205_MBCA_G0031

**Fig. 89 Removing components necessary to access cylinder head—M272 and M273 engines**

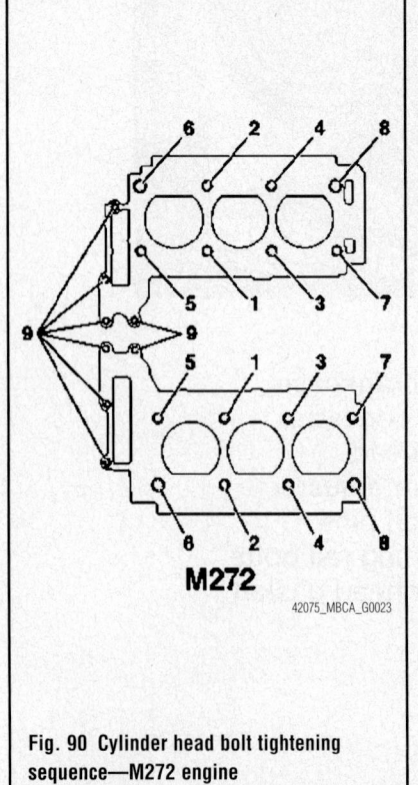

**Fig. 90 Cylinder head bolt tightening sequence—M272 engine**

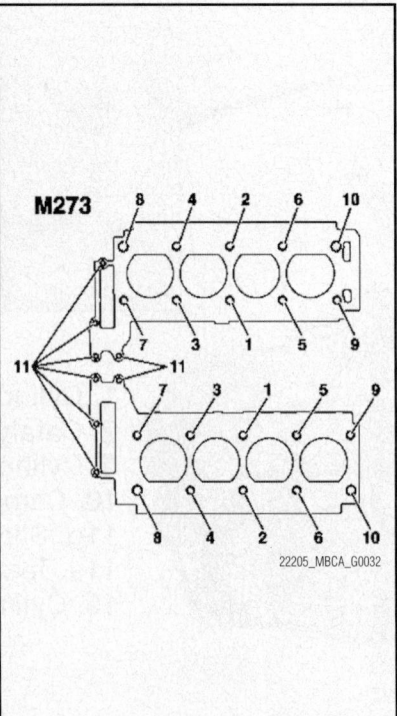

22205_MBCA_G0032

**Fig. 91 Cylinder head bolt tightening sequence—M273 engines**

- Step 5: Tighten cylinder head bolts (1-8) to stage 4
30. Tighten the M273 engine cylinder head bolts in the sequence shown.
- Step 1: Tighten cylinder head bolts (1 to 10) to stage 1
- Step 2: Tighten cylinder head bolts (1 to 10) to stage 2
- Step 3: Tighten cylinder head bolts (1 to 10) to stage 3
- Step 4: Tighten cylinder head bolts (1 to 10) to stage 4
- Step 5: Tighten bolts (11)
31. Change engine oil and oil filter element.
32. Start engine and check for leaks.

## ENGINE ASSEMBLY

*REMOVAL & INSTALLATION*

### M112 and M113 Engines

1. Before servicing the vehicle, refer to the precautions.
2. Properly relieve the fuel system pressure.
3. Drain the engine coolant and engine oil.

4. Drain the transmission fluid on automatic transmissions.

5. Disconnect the negative battery cable.

6. Remove the engine under cover.

7. Remove the accessory drive belt.

8. Remove the fan shroud, clutch and fan assembly.

9. Remove the coolant pipes and hoses.

10. Remove the air cleaner housing.

11. Remove the resonance pipe and body.

12. Remove the mass air flow (MAF) sensor.

13. Remove the radiator.

14. Drain the power steering fluid and detach the lines from the pump and plug the lines.

15. Remove the fuel lines.

16. Remove the wiring, vacuum and cable connections on the engine and transmission.

17. Cover the air conditioning condenser using a piece of sheet metal, plywood or plastic.

18. Remove the air conditioning compressor and position it aside with the hoses attached.

19. Remove the exhaust system and bracket.

20. Remove the clutch slave cylinder, if equipped.

21. Remove the lines from the side of the engine and transmission.

22. Remove the driveshaft vibration damper .

23. Remove the driveshaft.

24. On 4MATIC models, remove the front halfshafts, shields and front driveshaft.

25. Remove the transmission linkage.

26. Remove the rear engine crossmember.

27. Remove the bolts from front engine mounts.

28. Make sure all lines, hoses and electrical connections are detached. Support the transmission assembly and lift the engine out of the vehicle.

*To install:*

29. Install the engine assembly.

30. Install the front engine mounts. Torque the bolts to 26 ft. lbs. (35 Nm).

31. Install the rear engine crossmember to body. Torque the bolts to 37 ft. lbs. (50 Nm).

32. Install the rear engine mount to crossmember. Torque the bolts to 30 ft. lbs. (40 Nm).

33. Install the transmission linkage.

34. Install the on 4MATIC models, front halfshafts, shields and front driveshaft. Torque the M10 bolts to 30 ft. lbs. (40 Nm), and M12 bolts to 44 ft. lbs. (60 Nm).

35. Install the driveshaft. Torque the M10 bolts to 30 ft. lbs. (40 Nm), and M12 bolts to 44 ft. lbs. (60 Nm).

36. Install the driveshaft vibration damper.

37. Install the lines from the side of the engine and transmission.

38. Install the clutch slave cylinder, if equipped.

39. Install the exhaust system and bracket. Torque the bolts to 15 ft. lbs. (20 Nm).

40. Install the air conditioning compressor. Torque the bolts to 15 ft. lbs. (20 Nm).

41. Install the wiring, vacuum and cable connections on the engine and transmission.

42. Install the fuel lines. Torque the bolts to 28 ft. lbs. (38 Nm).

43. Install the power steering pump line. Torque to 30 ft. lbs. (40 Nm).

44. Install the radiator.

45. Install the MAF sensor.

46. Install the resonance pipe and body.

47. Install the air cleaner housing.

48. Install the coolant pipes and hoses.

49. Install the fan shroud, clutch and fan assembly.

50. Install the accessory drive belt.

51. Install the engine under cover.

52. Install the negative battery cable.

53. Install the coolant.

54. Install the engine oil.

55. Install the transmission fluid.

### M272 and M273 Engines

1. Before servicing the vehicle, refer to the precautions.

2. Properly relieve the fuel system pressure.

3. Drain the engine coolant and engine oil.

4. Drain the transmission fluid on automatic transmissions.

5. Disconnect the negative battery cable.

6. Remove the engine under cover.

7. Remove the accessory drive belt.

8. Remove the fan shroud, clutch and fan assembly.

9. Remove the coolant pipes and hoses.

10. Remove the air cleaner housing.

11. Remove the resonance pipe and body.

12. Remove the mass air flow (MAF) sensor.

13. Remove the radiator.

14. Drain the power steering fluid and detach the lines from the pump and plug the lines.

15. Remove the fuel lines.

16. Remove the wiring, vacuum and cable connections on the engine and transmission.

17. Cover the air conditioning condenser using a piece of sheet metal, plywood or plastic.

18. Remove the air conditioning compressor and position it aside with the hoses attached.

19. Remove the exhaust system and bracket.

20. Remove the clutch slave cylinder, if equipped.

21. Remove the lines from the side of the engine and transmission.

22. Remove the driveshaft vibration damper .

23. Remove the driveshaft.

24. Remove the transmission linkage.

25. Remove the rear engine crossmember.

26. Remove the bolts from front engine mounts.

27. Make sure all lines, hoses and electrical connections are detached. Support the transmission assembly and lift the engine out of the vehicle.

*To install:*

28. Install the engine assembly.

29. Install the front engine mounts. Torque the bolts to 26 ft. lbs. (35 Nm).

30. Install the rear engine mount. Torque the bolts to 21 ft. lbs. (28 Nm).

31. Install the rear engine mount to transmission. Torque the bolts to 37 ft. lbs. (50 Nm).

32. Install the transmission linkage.

33. Install the driveshaft. Torque the M10 bolts to 30 ft. lbs. (40 Nm), and M12 bolts to 44 ft. lbs. (60 Nm).

34. Install the driveshaft vibration damper.

35. Install the lines from the side of the engine and transmission.

36. Install the clutch slave cylinder, if equipped.

37. Install the exhaust system and bracket. Torque the bolts to 15 ft. lbs. (20 Nm).

38. Install the air conditioning compressor. Torque the bolts to 15 ft. lbs. (20 Nm).

39. Install the wiring, vacuum and cable connections on the engine and transmission.

40. Install the fuel lines. Torque the bolts to 15 ft. lbs. (20 Nm).

41. Install the power steering pump line. Torque to 30 ft. lbs. (40 Nm).
42. Install the radiator.
43. Install the MAF sensor.
44. Install the resonance pipe and body.
45. Install the air cleaner housing.
46. Install the coolant pipes and hoses.
47. Install the fan shroud, clutch and fan assembly.
48. Install the accessory drive belt.
49. Install the engine under cover.
50. Install the negative battery cable.
51. Install the coolant.
52. Install the engine oil.
53. Install the transmission fluid.

## EXHAUST MANIFOLD

### REMOVAL & INSTALLATION

#### M112 and M113 Engines

*See Figure 92.*

1. Before servicing the vehicle, refer to the precautions.
2. Properly relieve the fuel system pressure.
3. Disconnect the negative battery cable.
4. Remove the engine undercover.
5. Remove the exhaust pipes.
6. Remove the air filter assembly.
7. Remove the ignition coils for additional clearance, if necessary.
8. Remove the heat shields on 4matic models.
9. Remove the auxiliary coolant pump for additional clearance, if necessary.
10. Remove the exhaust manifold.

➡Inspect the rivet nuts in the manifold and replace as needed.

#### To install:

11. Installation is the reverse of removal.
12. Clean sealing surfaces at exhaust manifold and at cylinder head.
13. Use new gaskets.

### ✳✳ WARNING

Replace all self locking, micro-encapsulated and aluminum bolts and nuts. Chase threads to remove micro-encapsulated residue from old bolts/nuts.

14. Torque the manifold mounting bolts to 12 ft. lbs. (16 Nm) and the exhaust system to the manifolds to 15 ft. lbs. (20 Nm).

#### M272 and M273 Engines

*See Figure 93.*

1. Before servicing the vehicle, refer to the precautions.

2. Properly relieve the fuel system pressure.
3. Disconnect the negative battery cable.
4. Remove the engine undercover.
5. Remove the air filter housing assembly.
6. Remove the upper exhaust manifold nuts.
7. Remove the catalytic converter and exhaust pipe.
8. Remove the lower exhaust manifold nuts and manifold.

#### To install:

9. Installation is the reverse of removal.
10. Clean sealing surfaces at exhaust manifold and at cylinder head.
11. Use new gaskets.

### ✳✳ WARNING

Replace all self locking, micro-encapsulated and aluminum bolts and nuts. Chase threads to remove micro-encapsulated residue from old bolts/nuts.

12. Torque the manifold mounting bolts to 12 ft. lbs. (16 Nm) and the exhaust system to the manifolds to 15 ft. lbs. (20 Nm).

## INTAKE MANIFOLD

### REMOVAL & INSTALLATION

#### M112 and M113 Engines

*See Figure 94.*

1. Before servicing the vehicle, refer to the precautions.
2. Properly relieve the fuel system pressure.
3. Disconnect the negative battery cable.
4. Remove the cylinder head cover.
5. Remove the mass air flow (MAF) sensor with intake pipe.
6. Remove the fuel rail with injectors.
7. Remove the vacuum, electrical and cable connectors.
8. Remove the exhaust gas recirculation (EGR) valve.

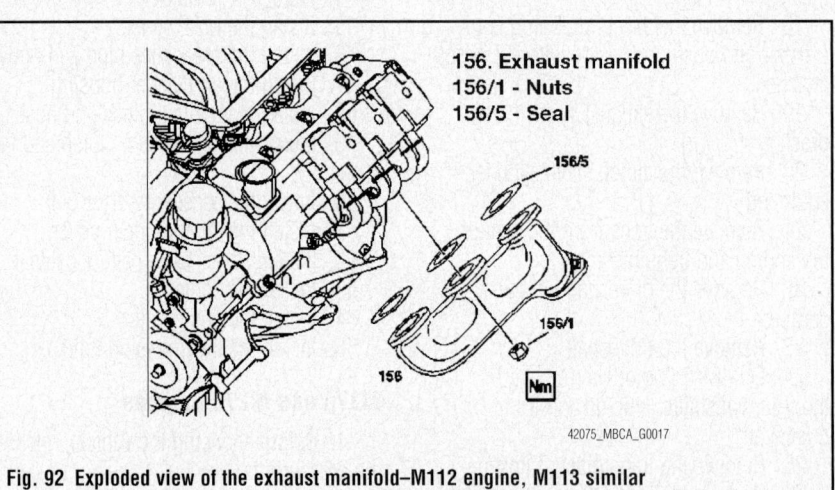

156. Exhaust manifold
156/1 - Nuts
156/5 - Seal

42075_MBCA_G0017

**Fig. 92 Exploded view of the exhaust manifold—M112 engine, M113 similar**

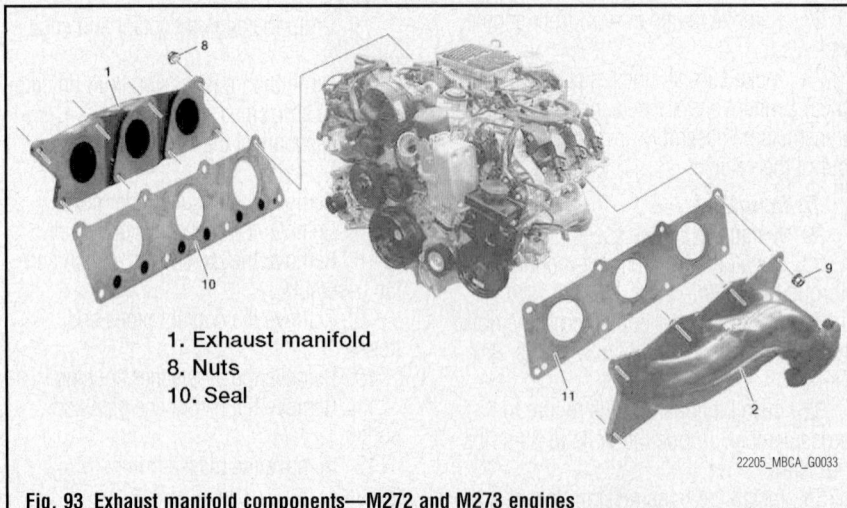

1. Exhaust manifold
8. Nuts
10. Seal

22205_MBCA_G0033

**Fig. 93 Exhaust manifold components—M272 and M273 engines**

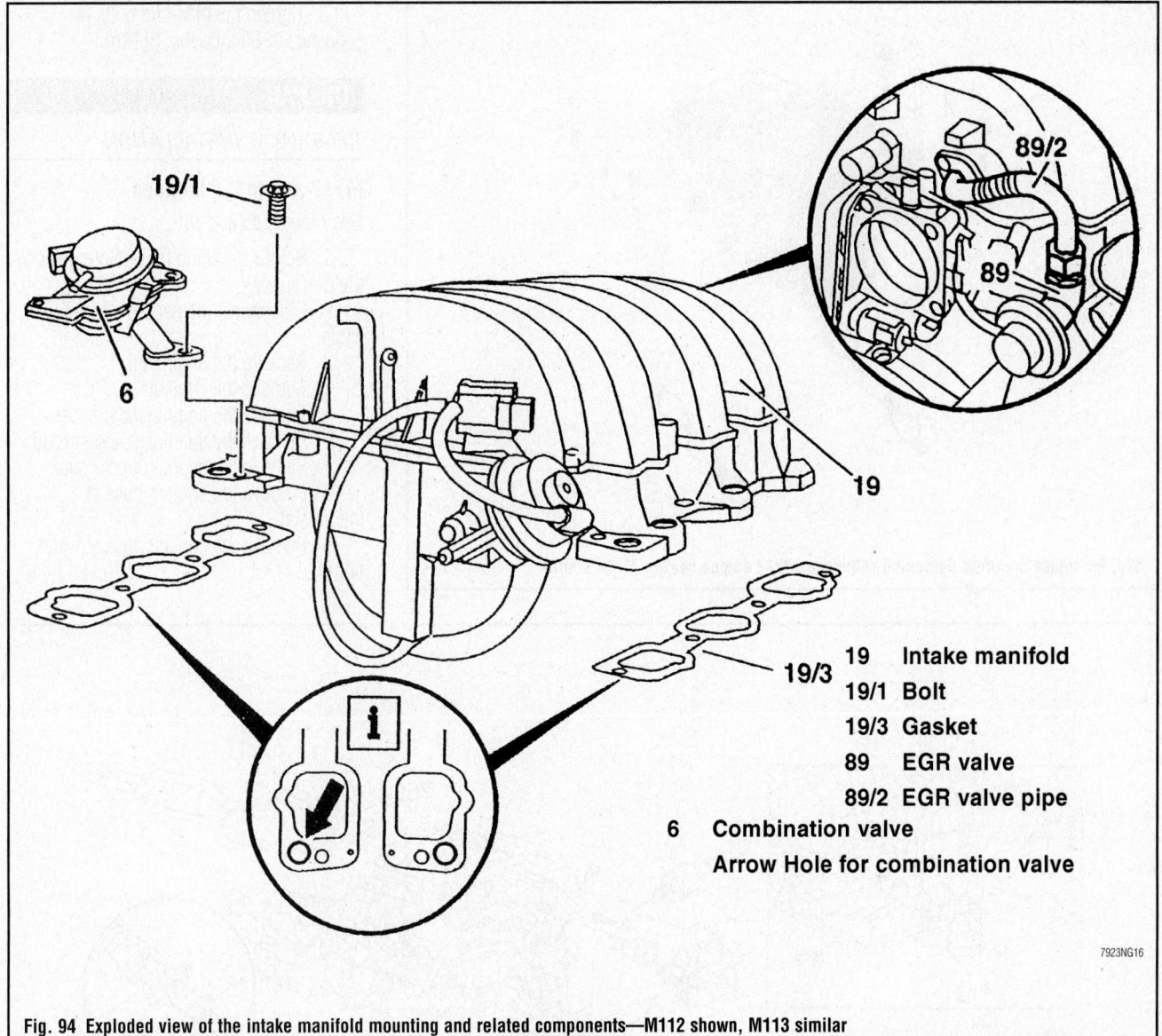

19 Intake manifold
19/1 Bolt
19/3 Gasket
89 EGR valve
89/2 EGR valve pipe
6 Combination valve
Arrow Hole for combination valve

7923NG16

**Fig. 94 Exploded view of the intake manifold mounting and related components—M112 shown, M113 similar**

9. Remove the combination valve.
10. Remove the manifold mounting bolts.
11. Remove the intake manifold.
12. Place clean shop rags into the intake passages to prevent dirt from entering.

***To install:***
13. Clean all gasket material from the sealing surfaces.
14. Install the new intake manifold gasket. Verify the secondary air injection passage opening in the gasket and install the intake manifold. Torque the mounting bolts to 15 ft. lbs. (20 Nm).
15. Install the combination valve. Torque the bolts to 15 ft. lbs. (20 Nm).
16. Install the EGR valve.
17. Install the vacuum, electrical and cable connectors.
18. Install the fuel rail with new injector seals.

19. Install the MAF sensor with the air intake pipe.
20. Install the cylinder head cover. Torque the bolts to 88 inch lbs. (10 Nm).
21. Connect the negative battery cable.
22. Start the vehicle and check for leaks.

**M272 and M273 Engines**
*See Figure 95.*

1. Before servicing the vehicle, refer to the precautions.
2. Properly relieve the fuel system pressure.
3. Remove the engine cover, the remove the bolts securing the engine control module and place aside.
4. Remove the Mass Air Flow (MAF) sensor with intake pipe.
5. Disconnect the fuel feed line.

6. Label and detach all vacuum, electrical and cable connectors from the manifold.
7. Remove the manifold mounting bolts in the reverse order of tightening sequence.
8. Carefully lift the intake manifold, then detach the hoses from it.
9. Place clean shop rags into the intake passages to prevent dirt from entering.

***To install:***
10. Installation is the reverse of removal.
11. Use new gaskets.

**※ WARNING**

**Replace all self locking, micro-encapsulated and aluminum bolts and nuts. Chase threads to remove micro-encapsulated residue from old bolts/nuts.**

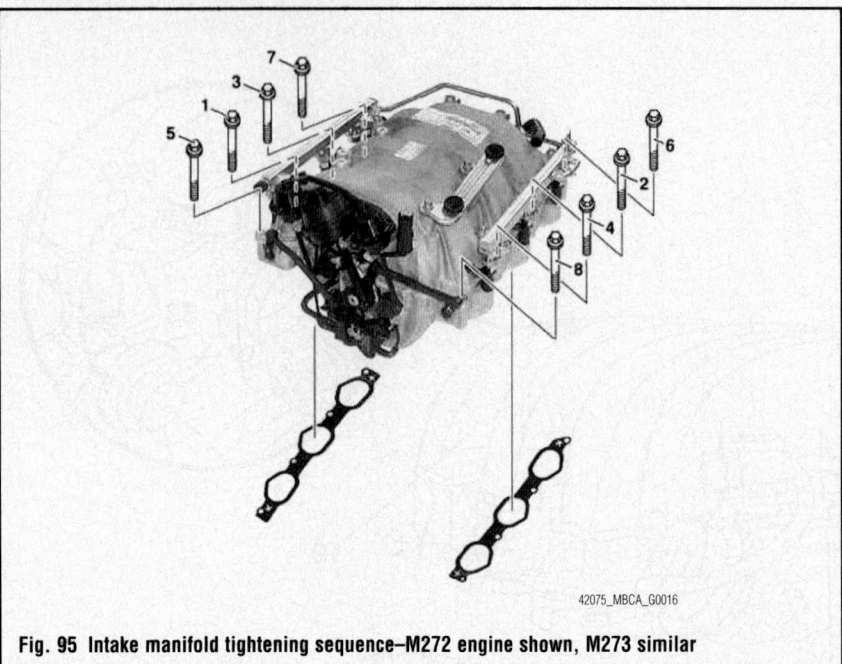

**Fig. 95 Intake manifold tightening sequence—M272 engine shown, M273 similar**

42075_MBCA_G0016

12. Tighten the manifold bolts in sequence to 80 inch lbs. (9 Nm).

## OIL PAN

### REMOVAL & INSTALLATION

#### M112 and M113 Engines

*See Figures 96 and 97.*

1. Before servicing the vehicle, refer to the precautions.
2. Disconnect the negative battery cable.
3. Remove the engine oil.
4. Remove the coolant.
5. Remove the engine under cover.
6. Remove the fan clutch and shroud.
7. Remove the engine upper cover.
8. Remove the coolant hose at thermostat.
9. Remove the coolant hose at water pump.

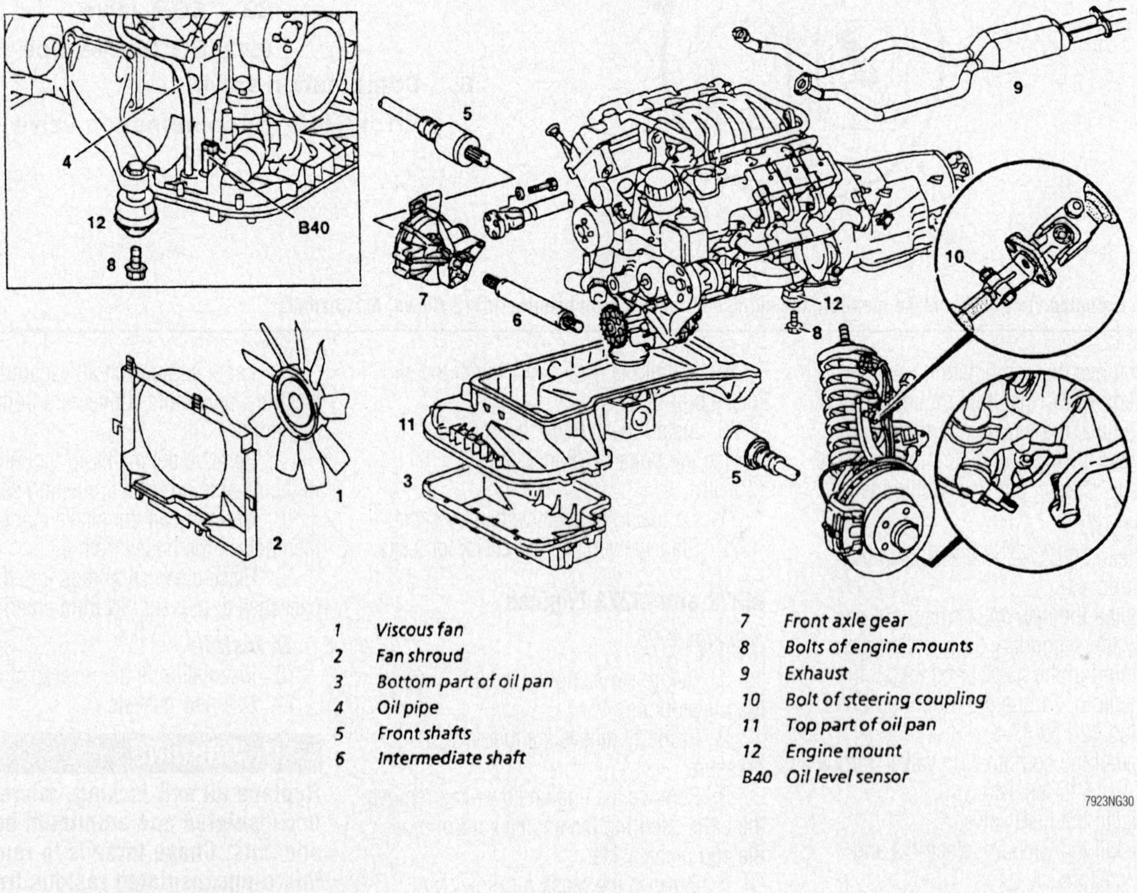

| | | | |
|---|---|---|---|
| 1 | Viscous fan | 7 | Front axle gear |
| 2 | Fan shroud | 8 | Bolts of engine mounts |
| 3 | Bottom part of oil pan | 9 | Exhaust |
| 4 | Oil pipe | 10 | Bolt of steering coupling |
| 5 | Front shafts | 11 | Top part of oil pan |
| 6 | Intermediate shaft | 12 | Engine mount |
| | | B40 | Oil level sensor |

7923NG30

**Fig. 96 Exploded view of the mounting of the upper oil pan and related components**

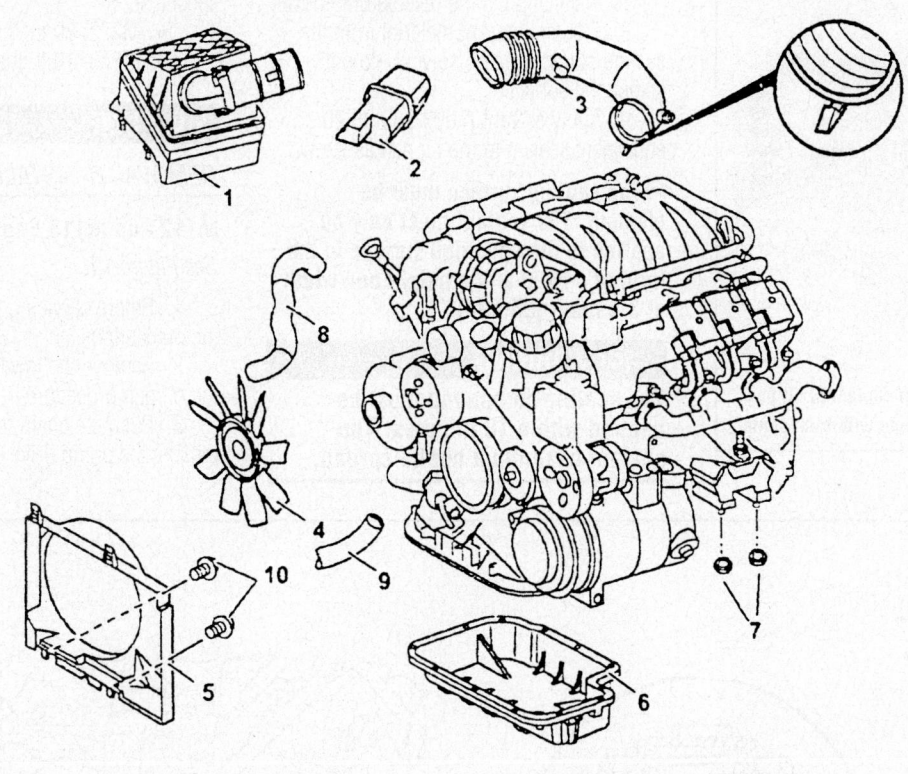

| | | | |
|---|---|---|---|
| 1 | Air cleaner housing | 6 | Bottom part of oil pan |
| 2 | Resonance body | 7 | Nuts |
| 3 | Resonance pipe | 8 | Coolant pipe |
| 4 | Viscous fan | 9 | Coolant pipe |
| 5 | Fan shroud | 10 | Bolts of fan shroud |

7923NG31

**Fig. 97 Exploded view of the mounting of the lower oil pan and related components**

10. Remove the mounting bolts and the oil pan.

➡It may be necessary to tap on the oil pan with a rubber mallet to dislodge it from the engine block.

11. Clean all gasket material from the sealing surfaces.

**To install:**

12. Install the oil pan with a new gasket. Torque the oil pan bolts to 84 inch lbs. (10 Nm).

13. Install the coolant hose at water pump.

14. Install the coolant hose at thermostat.

15. Install the fan clutch and shroud.

16. Install the engine upper cover.

17. Install the coolant.

18. Fill the engine with oil.

19. Connect the negative battery cable.

### M272 and M273 Engines

*See Figures 98 through 100.*

1. Before servicing the vehicle, refer to the precautions.

2. Disconnect the negative battery cable.

3. Remove the engine under cover.

4. Remove the engine oil.

5. Remove the bolts from the oil pan.

➡It may be necessary to tap on the oil pan with a rubber mallet to dislodge it from the engine block.

6. Remove the oil pan.

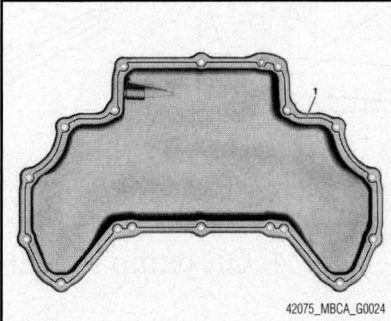

42075_MBCA_G0024

**Fig. 98 Application of sealant to oil pan bottom section for engine with rear sump**

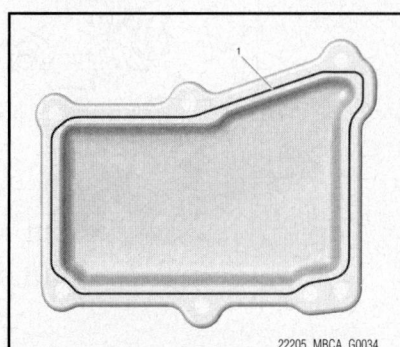

22205_MBCA_G0034

**Fig. 99 Application of sealant to oil pan bottom section for engine with middle sump**

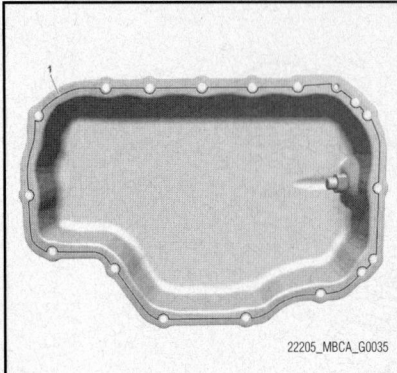

**Fig. 100 Application of sealant to oil pan bottom section for engine with front sump**

22205_MBCA_G0035

### To install:

7. Installation is the reverse of removal.

8. Clean all gasket material from the sealing surfaces using solvent, do not scrape the surface.

9. Apply sealant A 003 989 98 20 or equivalent sealant to the oil pan as shown.

➡The sealing surface must be cleaned. The sealant must only be applied to the specified surface in the form of a bead and a height and width of 2.0 mm ( 0.5 mm).

### ✳✳ WARNING

**The sealing compound must be applied within 10 minutes. The sealant bead must not be spread.**

10. Tighten the pan mounting bolts to specification:
- M272: 80 inch lbs. (9 Nm)
- M273: 10 ft. lbs. (14 Nm)

### OIL PUMP

#### REMOVAL & INSTALLATION

#### M112 and M113 Engines

*See Figure 101.*

1. Before servicing the vehicle, refer to the precautions.

2. Remove the lower oil pan. Refer to the oil pan procedure in this section.

3. Push the chain tensioner back and remove the pump drive chain.

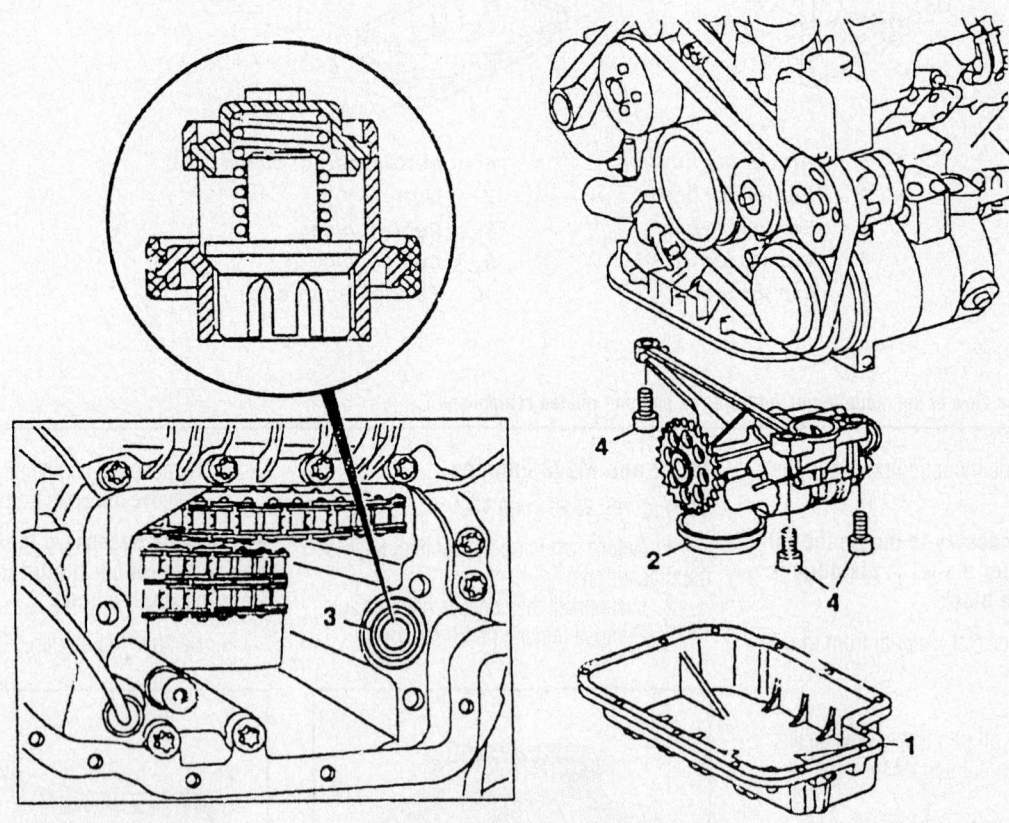

1. Oil pan
2. Oil pump
3. Oil return check valve
4. Oil pump mounting bolts

7923NG80

**Fig. 101 Exploded view of the oil pump mounting and related component—3.2L (112) and 4.3L (113) engines**

4. Unbolt and remove the oil pump.

*To install:*

5. Fill the oil pump with clean engine oil and install it to the engine.

6. Tighten the bolts to 15 ft. lbs. (20 Nm).

7. Install the pump drive chain.

8. Install the oil pan.

### M272 and M273 Engines

*See Figure 102.*

1. Before servicing the vehicle, refer to the precautions.

2. Remove the lower oil pan.

3. Remove the bolts from the oil suction tube

4. Push the chain tensioner back and remove the pump drive chain.

5. Unbolt and remove the oil pump together with the suction tube.

*To install:*

6. Fill the oil pump with clean engine oil and install it to the engine.

7. Tighten the fasteners to specification

- Oil pump to crankcase:
- Step1: 44 inch lbs. (5 Nm)
- Step2: 15 ft. lbs. (20 Nm)
- Suction tube bolts: 80 inch lbs. (9 Nm)

8. Install the pump drive chain.

9. Install the oil pan.

## PISTON AND RING

### POSITIONING

*See Figures 103 and 104.*

## REAR MAIN SEAL

### REMOVAL & INSTALLATION

### M112 and M113 Engines

*See Figure 105.*

1. Before servicing the vehicle, refer to the precautions.

2. Remove the flywheel.

3. Place a hook type seal removal tool behind the sealing lip of the oil seal. Pry out the oil seal.

*To install:*

4. Place the seal onto the crankshaft flange. Do not lubricate the seal, install it dry.

5. Use tool 111 589 08 43 00 or an equivalent seal installation tool to push the seal evenly into the sealing flange.

6. The seal should be pushed to a depth so that it is 1mm below the sealing flange.

7. Install the flywheel.

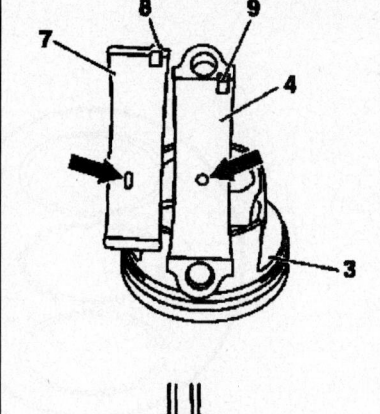

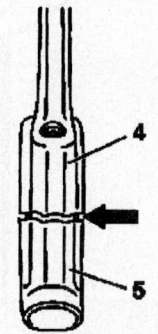

42348-MCLS-G09

**Fig. 103 Install pistons so the code next to the piston pin and group number are pointing in the direction of travel, and the anti-twist lock (8) and the groove (9) are matched**

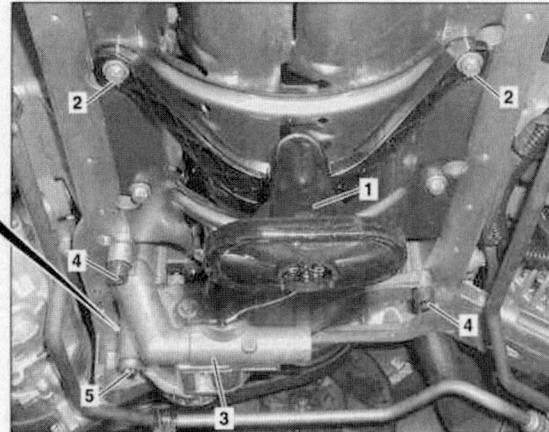

1 Oil suction pipe
2 Bolt 4 Bolt
3 Oil pump
5 Bolt

6 Chain tensioner
7 Oil pump chain
8 Oil pump gear

22205_MBCA_G0036

**Fig. 102 Oil pump and chain**

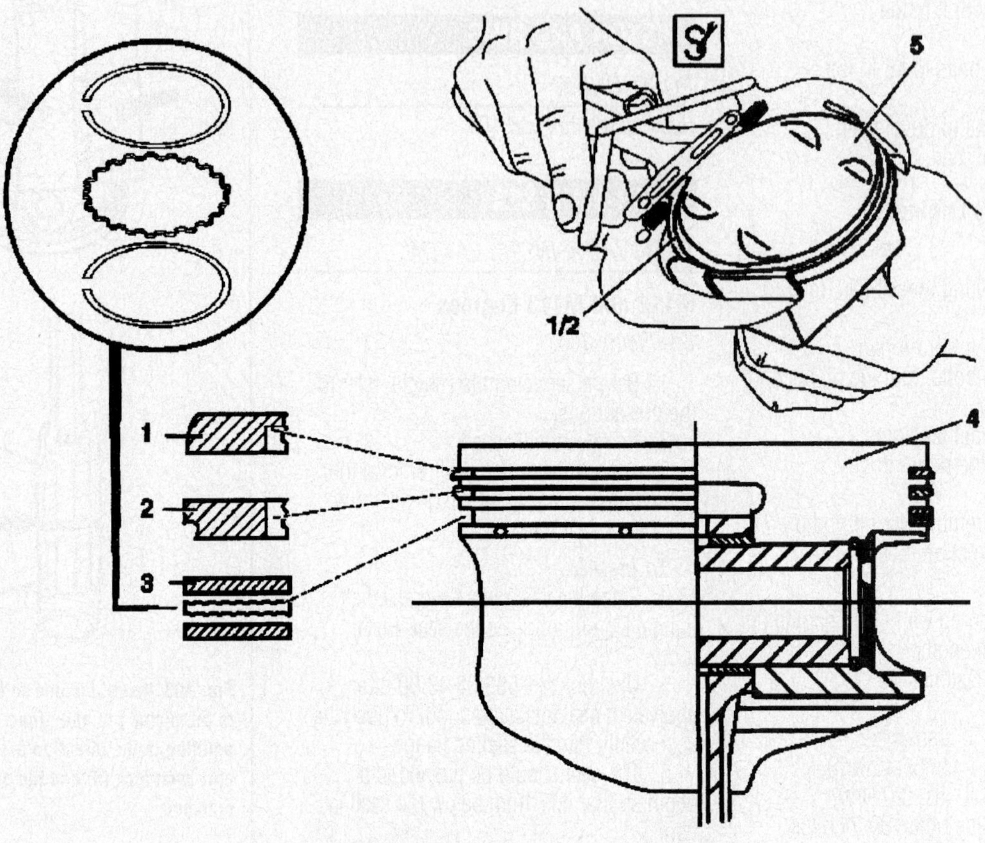

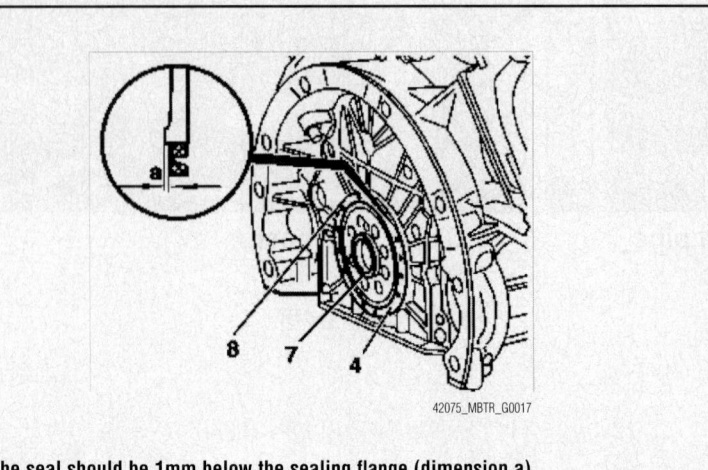

1    Compression ring (plain compression ring)

2    Compression ring ( taper-faced hook
     scraper ring)

3    Oil scraper ring

4    Piston

5    Piston crown

42348-MCLS-G10

**Fig. 104  Ring positioning**

**Fig. 105  The seal should be 1mm below the sealing flange (dimension a)**

42075_MBTR_G0017

## M272 and M273 Engines

*See Figures 106 through 111.*

1. Before servicing the vehicle, refer to the precautions.

2. Remove transmission.

3. Remove flywheel/driven plate.

4. Drain the engine oil.

5. Remove oil pan bottom section.

6. Clean crankcase and oil pan in area of end cover.

7. Unscrew bolts.

8. Pry off end cover with crankshaft radial sealing ring from crankcase.

9. Remove end cover.

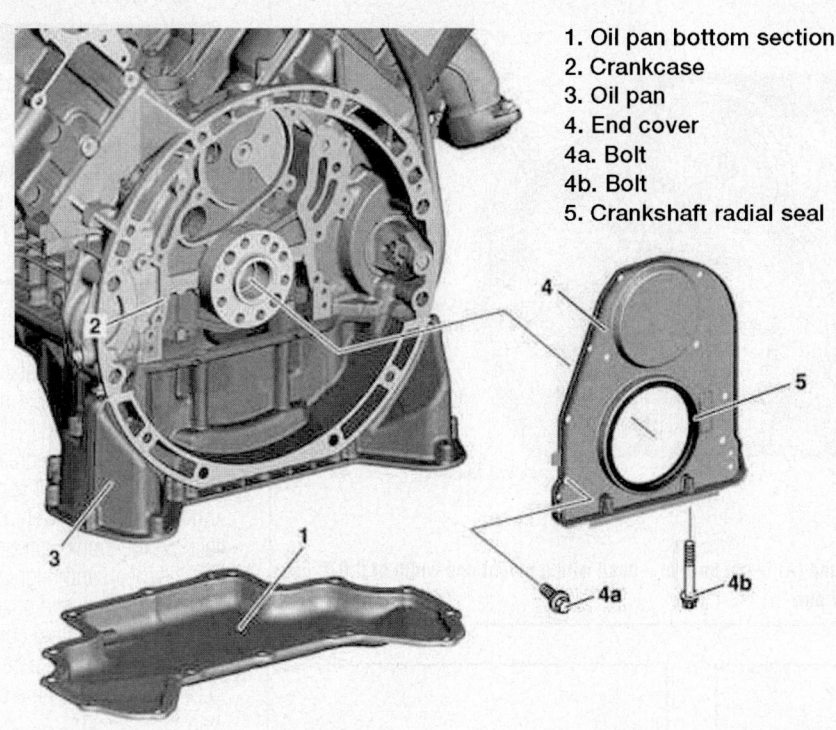

1. Oil pan bottom section
2. Crankcase
3. Oil pan
4. End cover
4a. Bolt
4b. Bolt
5. Crankshaft radial seal

22205_MBCA_G0037

Fig. 106 The crankshaft radial sealing ring (1) is vulcanized in the end cover (2) and it can therefore only be replaced together with the end cover (2). Do not reuse the end cover (2) with the crankshaft radial sealing ring (1) after removal

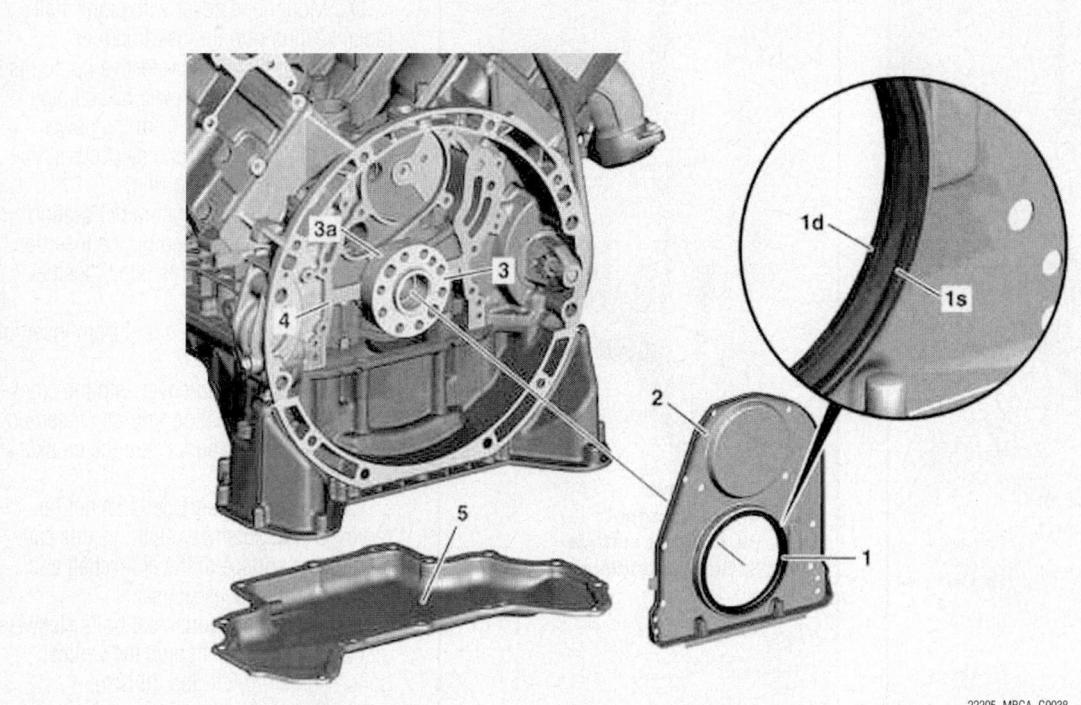

22205_MBCA_G0038

Fig. 107 Rear main seal and crankshaft end cover

**Fig. 108** Apply sealing compound (A) in the form of a bead with a height and width of 0.078 in. (2.0 mm) the crankcase and oil pan

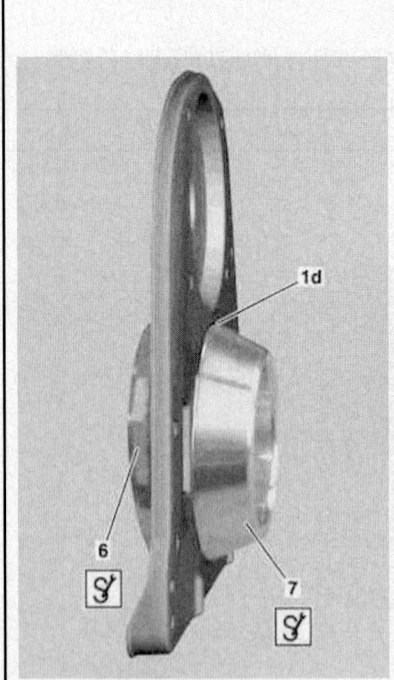

Insertion tool (6)
Insertion tool (7).
Lip (1d)

**Fig. 109** Insertion tool is fitted together with the insertion tool. Ensure that the lip points away from the transmission side

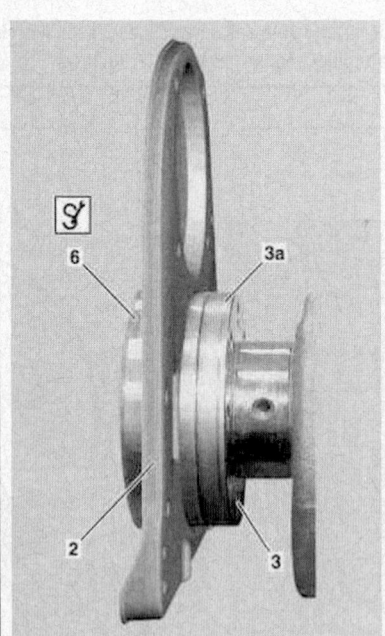

2. End cover
6. Insertion tool
3a. Contact surface
3. Crankshaft flange

**Fig. 110** Mount the end cover with the crankshaft radial shaft sealing ring and insertion tool on the contact surface on the crankshaft flange

**To install:**

> ❋❋ **WARNING**
>
> **The crankshaft radial sealing ring is vulcanized in the end cover and it can therefore only be replaced together with the end cover. Do not reuse the end cover with the crankshaft radial sealing ring after removal.**

10. The crankshaft radial sealing ring has both a lip and a protective lip. The lip, protective lip and contact surface on the crankshaft flange must not be lubricated for assembly purposes!

11. Clean and degrease crankcase and oil pan in the area of application of the sealing compound as well as the contact surface on the crankshaft flange.

12. Clean and degrease the sealing surface on the new end cover.

13. Apply sealing compound in the form of a bead with a height and width of 0.078 in. (2.0 mm) the crankcase and oil pan.

14. Use exclusively approved sealing compounds. Do not spread bead.

15. Mount end cover within 5 minutes after applying the sealing compound.

16. Mount end cover with crankshaft radial sealing ring on insertion tool.

17. Mount end cover with crankshaft radial sealing ring on insertion tool.

18. Ensure that the protective lip points in the direction of the power takeoff side and the lip points away from the power takeoff side or else leaks may occur at the crankshaft radial sealing ring.

19. The rear crankshaft radial sealing ring can only be mounted on the insertion tool if the insertion tool is fitted together with the insertion tool.

20. Separate insertion tool from insertion tool.

21. Mount the end cover with the crankshaft radial shaft sealing ring and insertion tool on the contact surface on the crankshaft flange.

22. The contact surface must not be removed with an emery cloth, as this can destroy the surface of the crankshaft and render the crankshaft unusable.

23. Tighten the horizontal bolts stepwise diagonally and then tighten the vertical bolts. Tighten to 6 ft. lbs. (8 Nm).

24. Detach the insertion tool from crankshaft flange

25. Lift protective lip carefully and check for correct installation position of the lip.

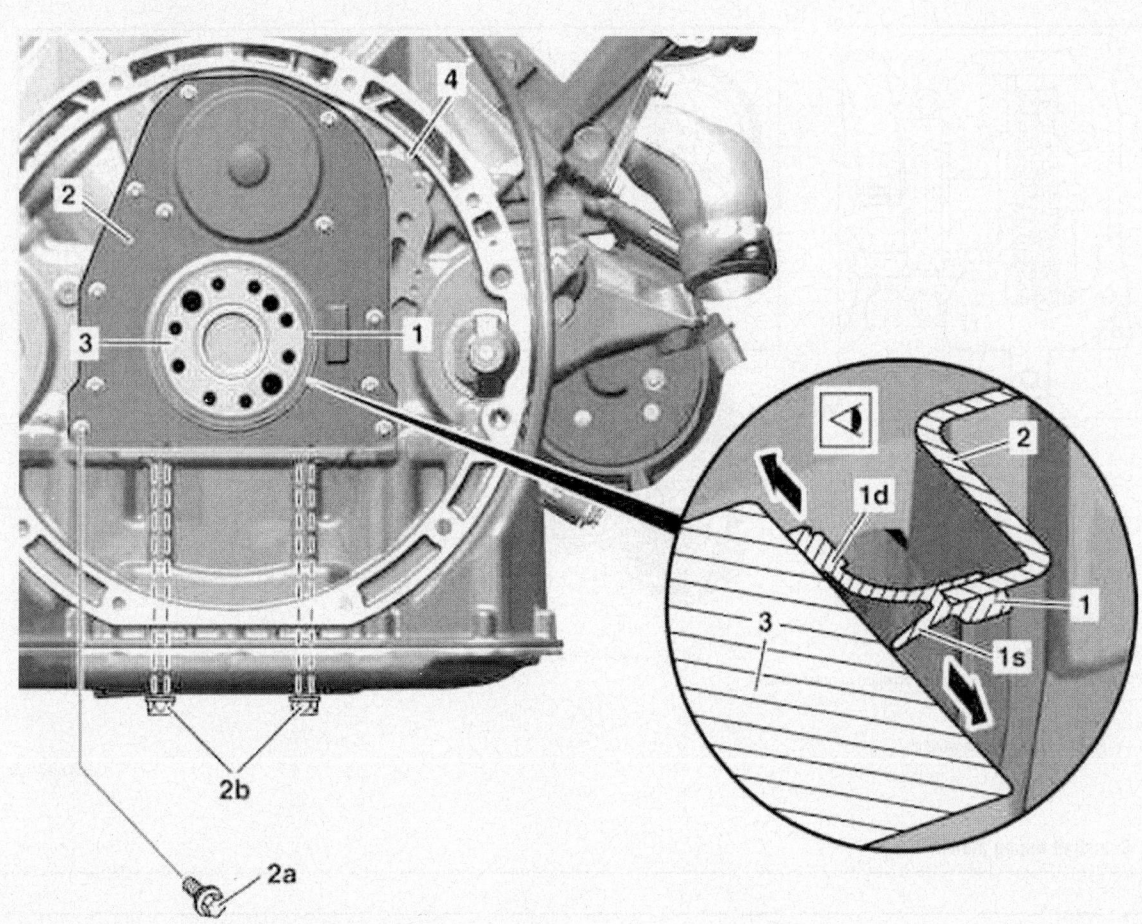

1 Crankshaft radial shaft sealing ring
1d Extruded edge seal
1s Protective lip
2 End cover

2a Bolts
2b Bolts
3 Crankshaft flange
4 Crankcase

22205_MBCA_G0042

**Fig. 111 Properly securing the end cover**

## ✳✳ WARNING

**A wrong installation position leads to leakage.**

26. The lip must point over it whole extent towards the crankcase.
27. Check for correct installation position of the protective lip

## ✳✳ WARNING

**An incorrect installation position leads to premature wear of the crankshaft radial sealing ring.**

28. The protective lip must point over it whole extent towards the crankcase.
29. Fill engine with oil, start engine and check for leaks.

## TIMING CHAIN, SPROCKETS, FRONT COVER AND SEAL

### REMOVAL & INSTALLATION

#### M112 and M113 Engines

*See Figures 112 through 114.*

1. Before servicing the vehicle, refer to the precautions.
2. Disconnect the negative battery cable.
3. Drain the cooling system and engine oil.
4. Remove engine cover.
5. Remove engine cooling fan, fan clutch, and fan shroud.

→The fan clutch is equipped with right-hand thread.

6. Place a guard plate behind the radiator/condenser to protect it from damage during removal and installation.
7. Remove accessory drive belt and tensioner.
8. Remove cylinder head covers.
9. Remove coolant hoses.
10. Remove A/C compressor and wire aside.
11. Remove oil pan.
12. Remove crankshaft pulley.
13. Remove timing chain tensioner.
14. Remove oil filter housing and heat exchanger.
15. Remove water pump pulley.
16. Remove timing chain cover.

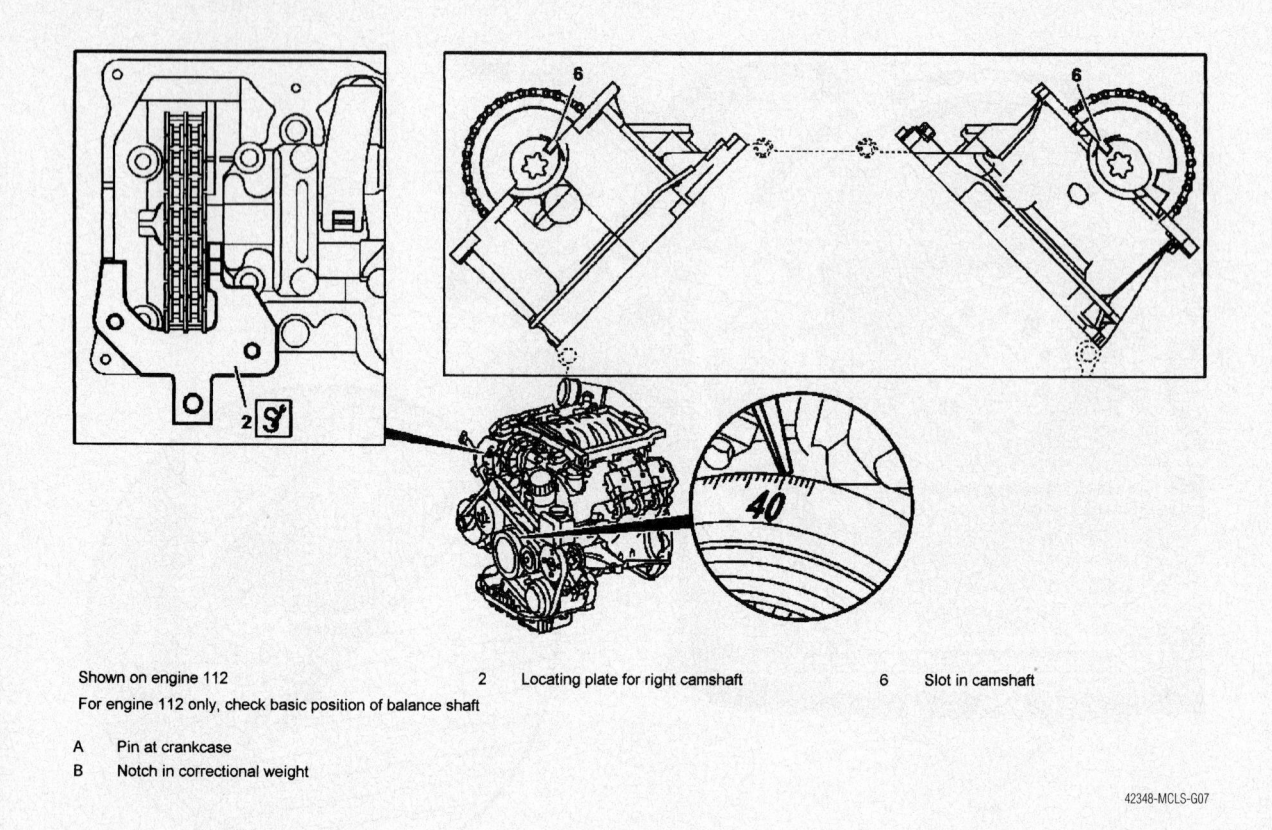

Shown on engine 112

For engine 112 only, check basic position of balance shaft

2    Locating plate for right camshaft          6    Slot in camshaft

A    Pin at crankcase
B    Notch in correctional weight

42348-MCLS-G07

**Fig. 112  Camshaft timing marks**

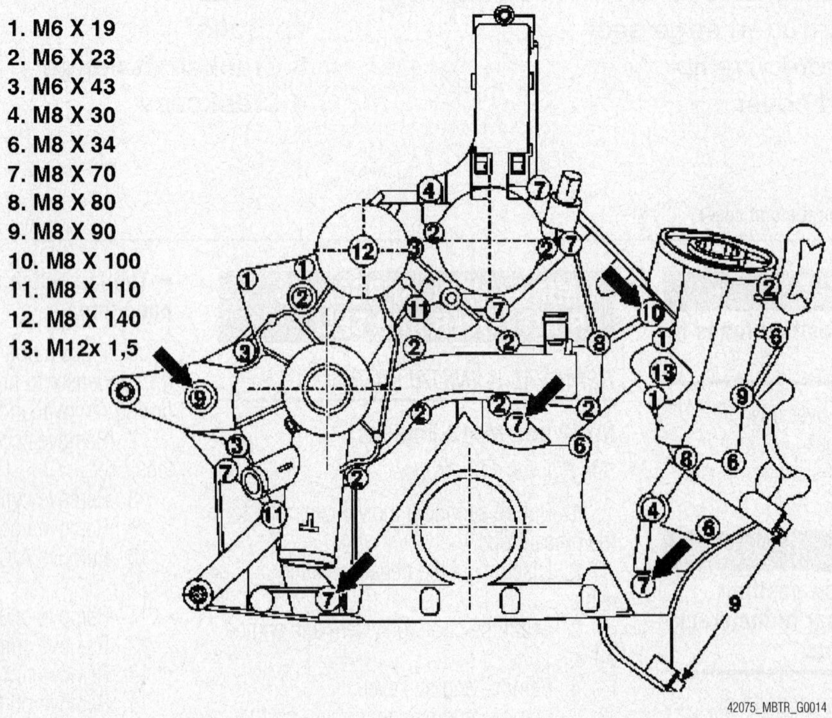

1. M6 X 19
2. M6 X 23
3. M6 X 43
4. M8 X 30
6. M8 X 34
7. M8 X 70
8. M8 X 80
9. M8 X 90
10. M8 X 100
11. M8 X 110
12. M8 X 140
13. M12x 1,5

42075_MBTR_G0014

**Fig. 113  Apply sealant to the bolts marked with arrows**

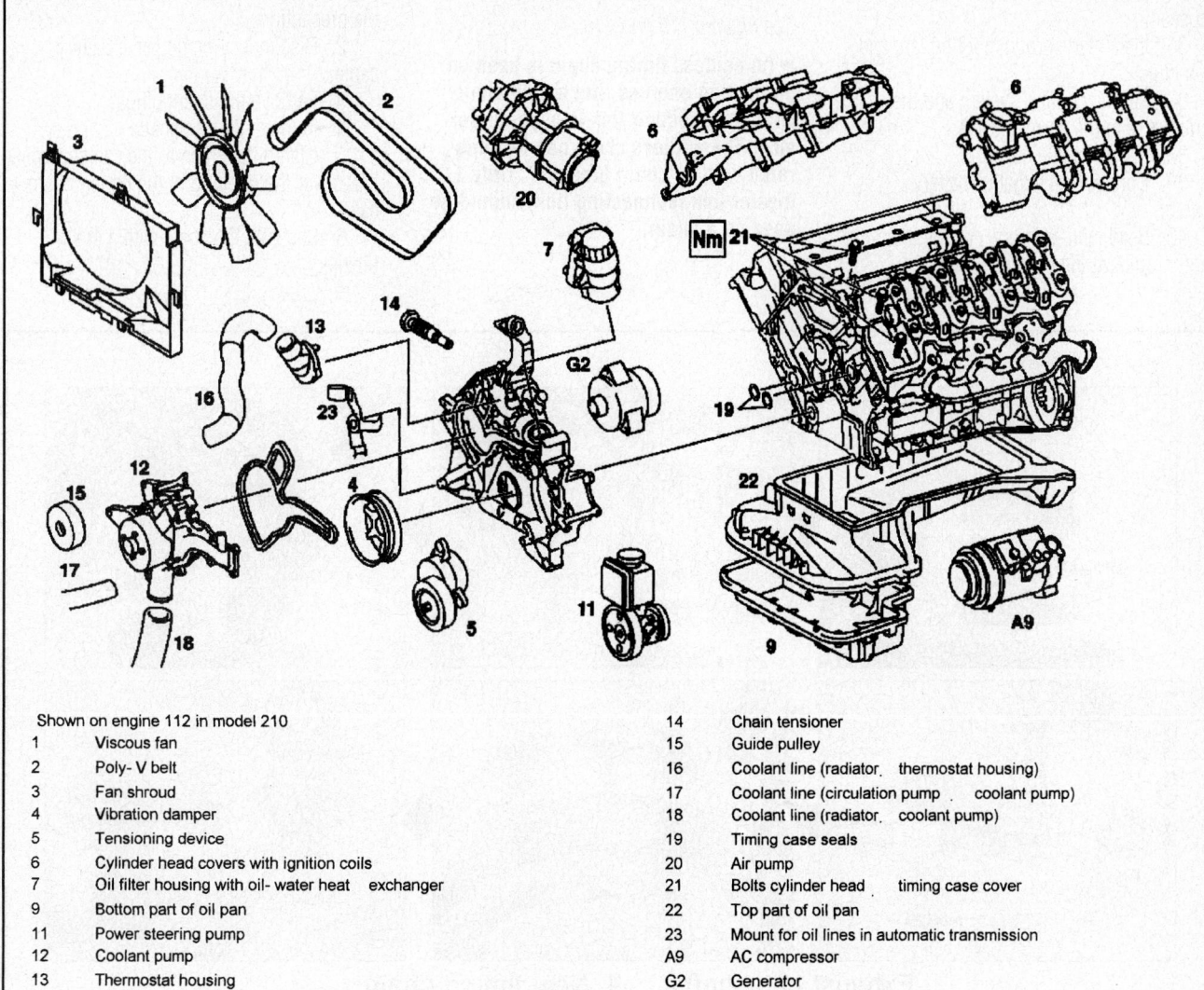

Shown on engine 112 in model 210
| | |
|---|---|
| 1 | Viscous fan |
| 2 | Poly- V belt |
| 3 | Fan shroud |
| 4 | Vibration damper |
| 5 | Tensioning device |
| 6 | Cylinder head covers with ignition coils |
| 7 | Oil filter housing with oil- water heat exchanger |
| 9 | Bottom part of oil pan |
| 11 | Power steering pump |
| 12 | Coolant pump |
| 13 | Thermostat housing |
| 14 | Chain tensioner |
| 15 | Guide pulley |
| 16 | Coolant line (radiator . thermostat housing) |
| 17 | Coolant line (circulation pump . coolant pump) |
| 18 | Coolant line (radiator . coolant pump) |
| 19 | Timing case seals |
| 20 | Air pump |
| 21 | Bolts cylinder head . timing case cover |
| 22 | Top part of oil pan |
| 23 | Mount for oil lines in automatic transmission |
| A9 | AC compressor |
| G2 | Generator |

42348-MCLS-G08

**Fig. 114 Exploded view of timing cover components**

➡The timing chain cover bolts are different lengths and diameters. Note their locations for installation.

**✴✴ WARNING**

**The engine must not be rotated backward.**

17. Rotate the crankshaft clockwise to align the crankshaft timing marks at 40° After Top Dead Center (ATDC). Insure that the grooves in the camshafts (6 in the illustration) align with the cylinder head cover mating surface on the intake side of the cylinder heads.

18. Lock the camshafts using the Camshaft Locking tools 112 589 00 32 00 and 112 589 01 32 00.

19. Remove camshaft sprockets bolts.
20. Remove timing chain.

**To install:**

21. Align the copper plated links of the timing chain with the marks on the camshaft sprockets, the crankshaft sprocket, and the balancer sprocket for the M112 engines. The M113 engines use an idler sprocket in place of the balance shaft.

22. Install camshaft sprockets along with the timing chain. Tighten the attaching bolts to 37 ft. lbs. (50 Nm) plus 90–100°.

23. Pull the crankshaft seal from the cover. Insert a new one using a suitable seal driver.

24. Clean the sealing surfaces and apply a bead of silicone sealant to the timing chain cover.

➡The cover must be installed within ten minutes after the sealant is applied.

25. Install timing cover. Apply sealant to the bolts indicated in the illustration. Tighten the bolts to 15 ft. lbs. (20 Nm)
26. Install CT sensor.
27. Install timing chain tensioner.
28. Install generator.
29. Install crankshaft damper.
30. Install air pump, if equipped.
31. Install air conditioning compressor and the power steering pump.
32. Install cylinder head covers.
33. Install oil pan.
34. Install accessory drive belt and tensioner.

35. Install upper and lower coolant hoses.

36. Install fan shroud, cooling fan, and fan clutch.

37. Fill the cooling system and the crankcase to the correct levels.

38. Install engine cover.

39. Connect the negative battery cable.

40. Read fault memory, encode the radio and normalize the power windows.

### M272 and M273 Engines

*See Figures 115 and 116.*

➡An endless timing chain is used on production engines, but a split chain with a connecting link is used for service. The endless chain can be separated with a "chain breaker". Only 1 master link (connecting link) should be used on a chain.

1. Before servicing the vehicle, refer to the precautions.

2. Disconnect the negative battery cable.

3. Remove the spark plugs.

4. Remove the camshafts.

5. Clamp the chain to the camshaft gear and cover the opening of the timing chain case.

6. Separate the chain with a chain breaker.

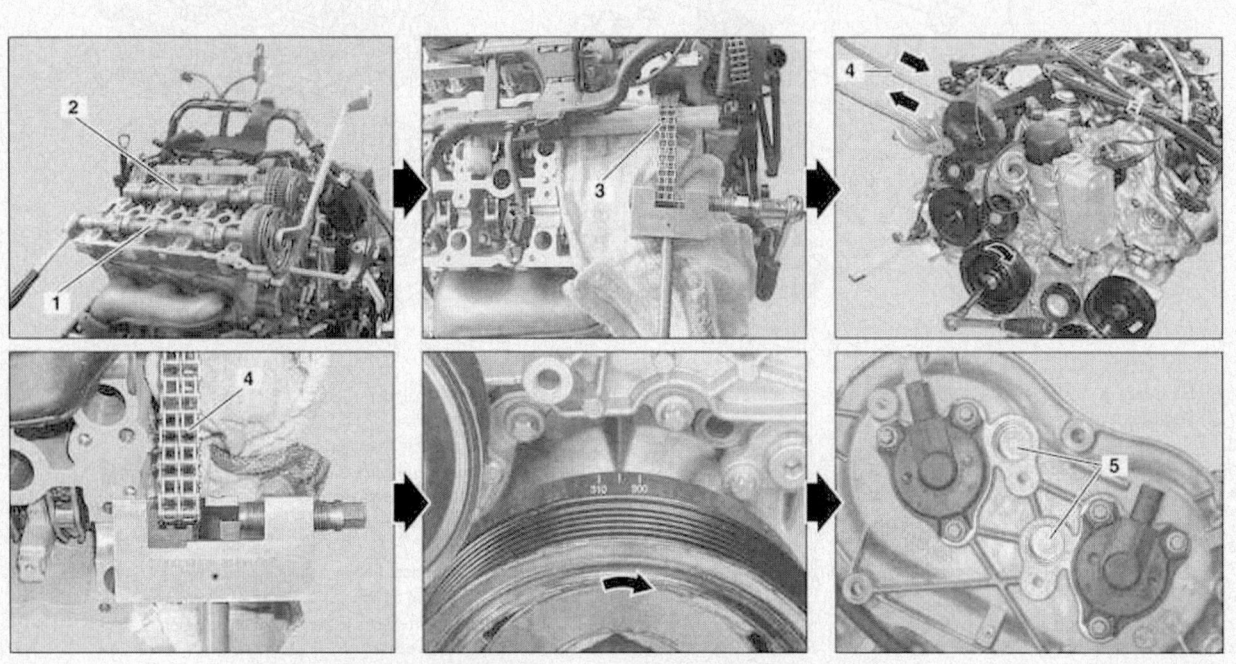

1. Exhaust camshaft
2. Intake camshaft
3. Old timing chain
4. New timing chain
5. Markings on pulse wheels

22205_MBCA_G0043

**Fig. 115 Illustrating the timing chain removal procedure**

6. Markings on camshaft adjuster
7. Markings on camshaft adjuster

22205_MBCA_G0044

**Fig. 116 Markings on the crankshaft damper and camshaft adjusters**

**To install:**

7. Attach a new timing chain to the old chain with a master link, center plate and end plate. Using a socket wrench on the crankshaft, slowly rotate the engine in the direction of normal rotation. Simultaneously, pull the old chain through until the master link is uppermost on the camshaft sprocket. Be sure to keep tension on the chain throughout this procedure.

8. Disconnect the old timing chain and connect the ends of the new chain with the master link. Insert the new connecting link from the rear so the lockwashers can be seen from the front.

9. Rotate the engine in the direction of rotation until the timing marks align at 55° Before Top Dead Center (BTDC).

10. The markings on the pulse wheels of the intake and exhaust camshafts at the left cylinder head must be located centrally in the bores of the camshaft sensors. If not, rotate the engine another revolution.

11. Rotate the engine in the direction of rotation 95° so that it is at 40° After Top Dead Center (ATDC) of cylinder 1.

12. Install the camshafts on the right cylinder head. The marks should point upwards and match with the markings on the cylinder head covers.

13. Rotate the engine in the direction of rotation until the timing marks align at 55° BTDC again. The markings on the pulse wheels of the intake and exhaust camshafts at the left cylinder head must be located centrally in the bores of the camshaft sensors.

14. Install the spark plugs.

15. Connect the negative battery cable.

16. Start the engine and check operation.

## VALVE LASH

### ADJUSTMENT

Mercedes-Benz engines use hydraulic valve lifters. There is no provision for valve clearance adjustments.

# ENGINE PERFORMANCE & EMISSION CONTROL

## CAMSHAFT POSITION (CMP) SENSOR

### LOCATION
See Figure 117.

### REMOVAL & INSTALLATION

1. Before servicing the vehicle, refer to the precautions.

2. Remove engine trim panel.

3. Detach the electrical connector at the intake camshaft Hall sensor.

4. Remove bolt.

5. Remove intake camshaft Hall sensor.

**To install:**

6. Installation is the reverse of removal.

7. Replace sealing rings.

8. Tighten camshaft position sensor to 6 ft. lbs. (8 Nm).

## CRANKSHAFT POSITION (CKP) SENSOR

### LOCATION
See Figure 118.

### REMOVAL & INSTALLATION

1. Before servicing the vehicle, refer to the precautions.

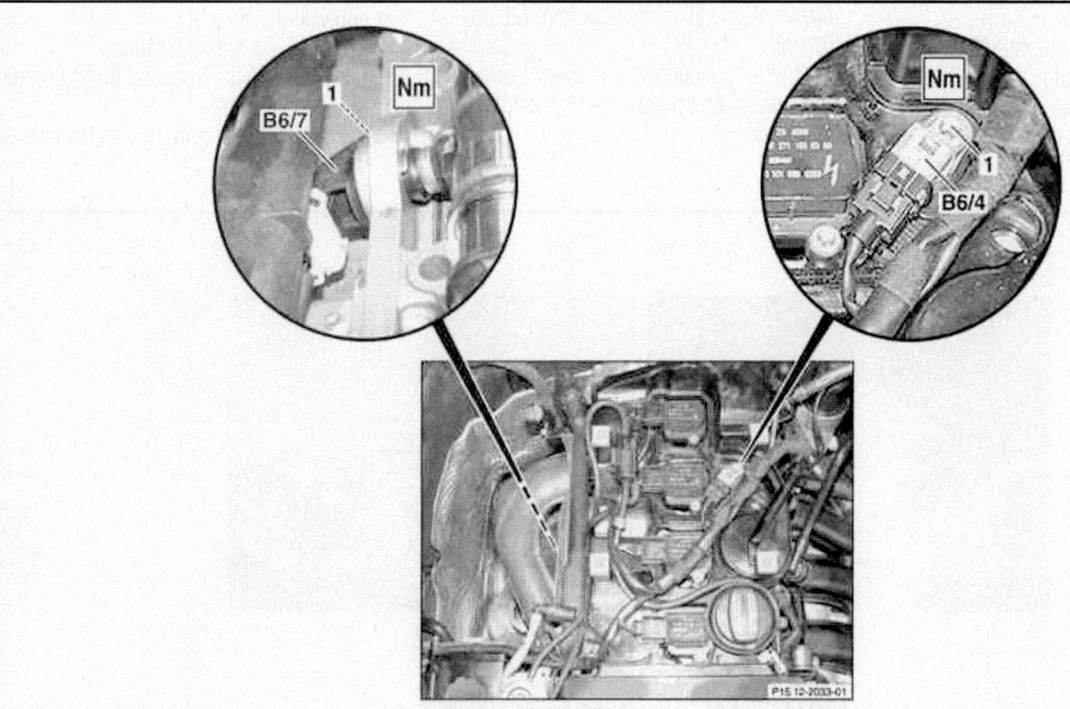

1 Bolt
B6/4 Left intake camshaft Hall sensor
B6/7 Right exhaust camshaft Hall sensor

22205_MBCA_G0050

**Fig. 117 CMP sensor location**

**1. Crankshaft position sensor**
**B70 electrical connector**

22205_MBCA_G0045

**Fig. 118  CKP sensor location**

2. Remove air filter housing.
3. Remove hot film mass air flow sensor.
4. Remove air duct housing.
5. Release and disconnect electrical connector at crankshaft Hall sensor.
6. Remove bolt.
7. Pull crankshaft Hall sensor out of crankcase.

***To install:***
8. Installation is the reverse of removal.
9. Tighten crankshaft Hall sensor to crankcase bolt to 6 ft. lbs. (8 Nm).
10. Carry out first initialization and then sensor rotor adaptation using STAR DIAGNOSIS

## ELECTRONIC CONTROL MODULE (ECM)

### LOCATION
*See Figure 119.*

### REMOVAL & INSTALLATION
1. Before servicing the vehicle, refer to the precautions.
2. Remove air filter housing.
3. Move shift interlocks for electrical connectors on control unit outward and remove electrical connectors from ME control unit.
4. Pull out control unit upwards out of right control unit bracket and left control unit bracket.

5. Detach ball-type support braces from control unit.
6. Detach right control unit bracket and left control unit bracket as required from resonance intake manifold.

***To install:***
7. Installation is the reverse of removal.
8. Tighten control unit holder to resonance intake manifold bolt to 6 ft. lbs. (8 Nm).
9. Tighten ball head support to control unit to 3 ft. lbs. (4 Nm).
10. When replacing the control unit, perform the following using STAR DIAGNOSIS or equivalent scan tool.
- Sensor rotor adaptation
- Release the transport protection on control unit
- Personalize and activate the control unit

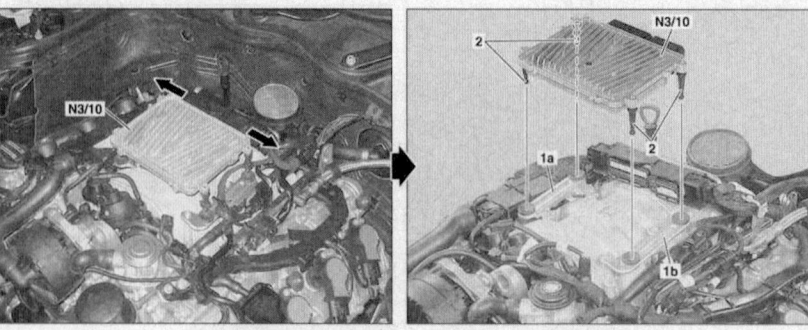

**1a. Right control unit bracket**
**2. Ball head support**
**N3/10. ME control unit**
**1b. Left control unit bracket**

22205_MBCA_G0047

**Fig. 119  ECM location**

## ENGINE COOLANT TEMPERATURE (ECT) SENSOR

### LOCATION

*See Figure 120.*

### REMOVAL & INSTALLATION

1. Before servicing the vehicle, refer to the precautions.
2. Remove rear section of lower engine compartment paneling.
3. Open engine hood and raise to vertical position.
4. Reduce pressure in cooling system, to do this, slowly open the cap at expansion reservoir and reclose.
5. Remove air filter housing.
6. Release and disconnect electrical connector at crankshaft Hall sensor.
7. Unlatch the electrical connector at the coolant temperature sensor and disconnect.
8. Unscrew bolt and remove clamping plate.
9. Fit new O-ring to new coolant temperature sensor.
10. Pull coolant temperature sensor out of cylinder head and immediately insert new coolant temperature sensor.

*To install:*

11. Installation is the reverse of removal.
12. Tighten clamping plate for coolant temperature sensor to cylinder head to 6 ft. lbs. (8 Nm).
13. Pour in coolant and bleed cooling system.
14. Read out fault memory and erase if required.

## HEATED OXYGEN (HO2S) SENSOR

### LOCATION

*See Figure 121.*

### REMOVAL & INSTALLATION

1. Before servicing the vehicle, refer to the precautions.
2. Raise and safely support the vehicle securely on jackstands.
3. Remove rear section of lower engine compartment paneling.
4. Unclip electrical connector and electrical feed line for corresponding O2 sensor out of retaining clamps on transmission.
5. Remove heat shield for steering below left catalytic converter.
6. Disconnect electrical connector for corresponding O2 sensor.
7. Remove complete exhaust system.
8. Unscrew corresponding O2 sensor out of catalytic converter completely.

*To install:*

9. Installation is the reverse of removal.
10. Apply heat-resistant lubricant to thread of O2 sensor.
11. Tighten O2 sensor to 37 ft. lbs. (50 Nm).
12. Connect STAR DIAGNOSIS and read out fault memory

## INTAKE AIR TEMPERATURE (IAT) SENSOR

### LOCATION

The IAT sensor is an integral part of the MAF sensor and cannot be serviced separately.

## KNOCK SENSOR (KS)

### LOCATION

*See Figure 122.*

### REMOVAL & INSTALLATION

1. Before servicing the vehicle, refer to the precautions.
2. Remove air filter housing.

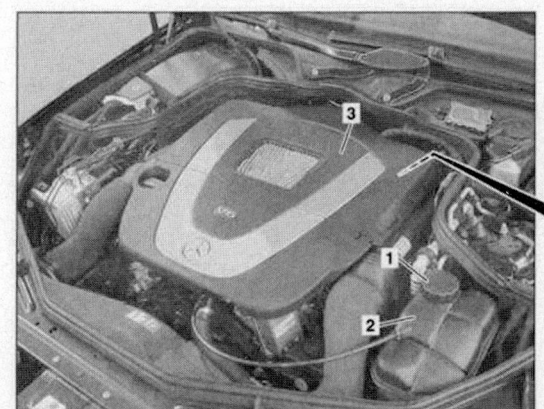
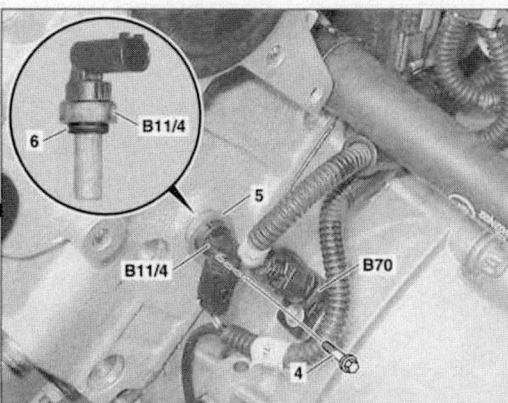

1. Cap
2. Coolant expansion reservoir
3. Air filter housing
4. Bolt

5. Clamping plate
6. O-ring
B11/4. Coolant temperature sensor
B70. Crankshaft Hall sensor

22205_MBCA_G0048

**Fig. 120 ECT sensor location**

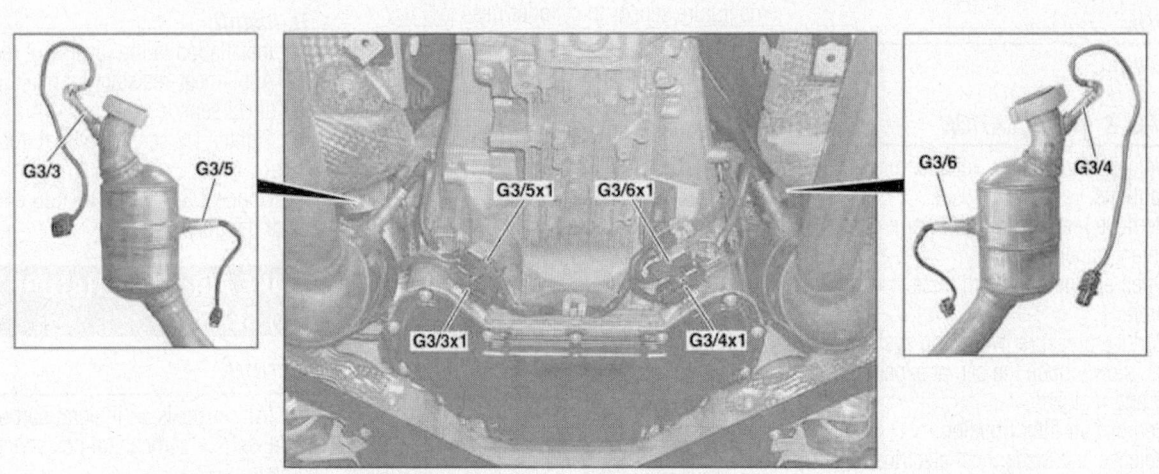

G3/3 Left O2 sensor upstream of TWC

G3/3x1 Connector for left O2 sensor upstream of TWC

G3/4 Right O2 sensor upstream of TWC

G3/4x1 Connector for right O2 sensor upstream of TWC

G3/5 Left O2 sensor downstream of TWC

G3/5x1 Connector for left O2 sensor downstream of TWC

G3/6 Right O2 sensor downstream of TWC

G3/6x1 Connector for right O2 sensor downstream of TWC

22205_MBCA_G0049

**Fig. 121 HO2S sensor location**

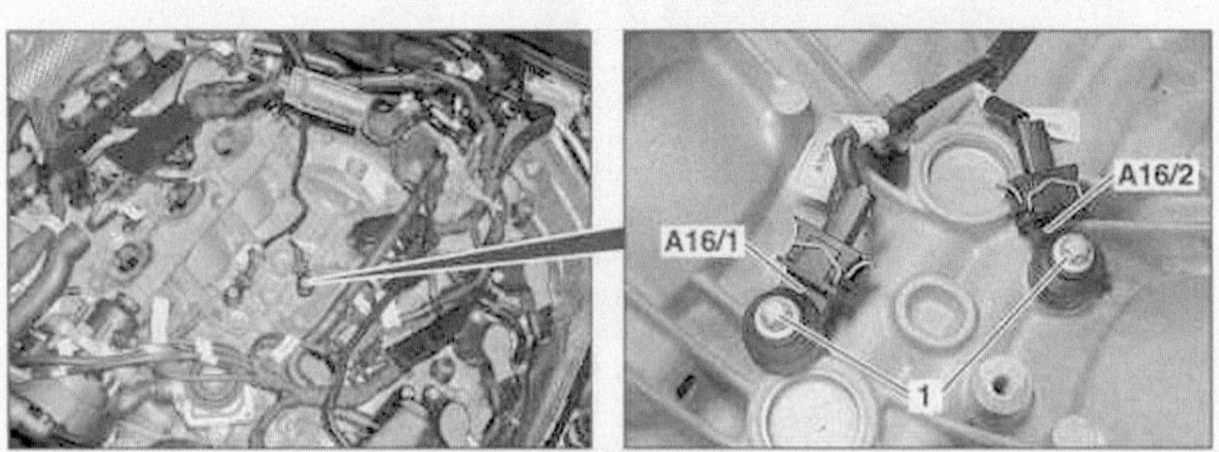

## A16/1 Knock sensor 1, right
## A16/2 Left knock sensor 2

22205_MBCA_G0046

**Fig. 122 KS sensor location**

3. Remove resonance intake manifold.

4. Disconnect electrical connector at right knock sensor or left knock sensor.

5. Remove bolt and take out right knock sensor or left knock sensor.

**To install:**

6. Installation is the reverse of removal.

7. Tighten knock sensor to crankcase bolt to 15 ft. lbs. (20 Nm).

8. Connect STAR DIAGNOSIS and read out fault memory.

## MASS AIR FLOW (MAF) SENSOR

### LOCATION

See Figure 123.

### REMOVAL & INSTALLATION

1. Before servicing the vehicle, refer to the precautions.

2. Remove air filter housing.

3. Remove the electrical connector at the hot film mass air flow sensor.

4. Disengage the right notch of the hot film mass air flow sensor at the resonance intake manifold.

5. Unlatch left notch for hot film MAF sensor on resonance intake manifold and tip hot film MAF sensor somewhat to the rear.

6. Disengage the spring steel sheet and remove the hot film mass air flow sensor from the air duct housing.

7. Check seal between air duct housing and hot film MAF sensor for damage; replace if necessary.

8. Check right notch, left notch and spring steel sheet for damage; replace if necessary.

**To install:**

9. Installation is the reverse of removal.

10. Connect up STAR DIAGNOSIS and read out fault memory.

## THROTTLE POSITION SENSOR (TPS)

### LOCATION

See Figure 124.

### REMOVAL & INSTALLATION

1. Before servicing the vehicle, refer to the precautions.

2. Remove air filter housing.

3. Remove hot film mass air flow sensor.

4. Remove air guide housing.

5. Release and disconnect electrical connector at throttle valve actuator.

6. Pull bleed hose out of throttle valve actuator.

7. Remove bolts.

8. Remove throttle valve actuator.

9. Remove gasket.

**To install:**

10. Installation is the reverse of removal.

11. Tighten bolt connecting throttle valve actuator to resonance intake manifold to 7 ft. lbs. (9 Nm).

12. Connect STAR DIAGNOSIS and read out fault memory.

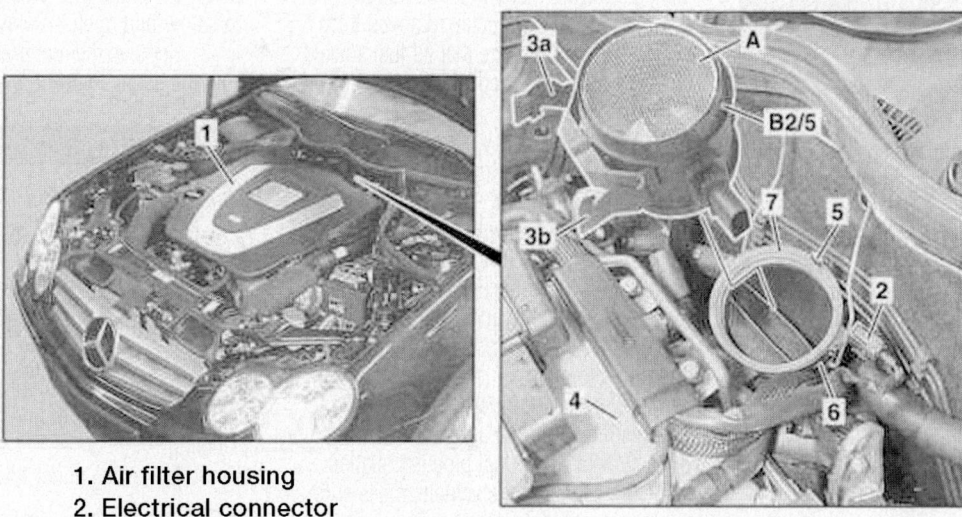

1. Air filter housing
2. Electrical connector
3a. Right notch
3b. Left notch
4. Resonance intake pipe
5. Spring steel sheet
6. Air duct housing
7. Seal
A. Grid
B2/5. Hot film MAF sensor

22205_MBCA_G0051

**Fig. 123 MAF sensor location**

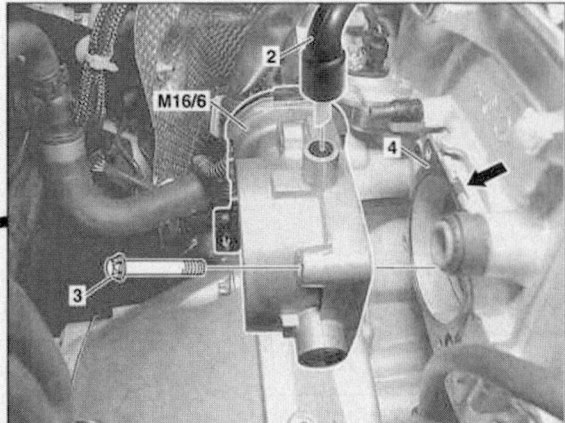

1. Air duct housing
2. Venting hose
3. Bolt
4. Seal
B2/5. Hot film mass air flow sensor
M16/6. Throttle valve actuator

22205_MBCA_G0052

**Fig. 124 TPS sensor location**

## FUEL

## GASOLINE FUEL INJECTION SYSTEM

### FUEL SYSTEM SERVICE PRECAUTIONS

Safety is the most important factor when performing not only fuel system maintenance but any type of maintenance. Failure to conduct maintenance and repairs in a safe manner may result in serious personal injury or death. Maintenance and testing of the vehicle's fuel system components can be accomplished safely and effectively by adhering to the following rules and guidelines.

• To avoid the possibility of fire and personal injury, always disconnect the negative battery cable unless the repair or test procedure requires that battery voltage be applied.

• Always relieve the fuel system pressure prior to disconnecting any fuel system component (injector, fuel rail, pressure regulator, etc.), fitting or fuel line connection. Exercise extreme caution whenever relieving fuel system pressure to avoid exposing skin, face and eyes to fuel spray. Please be advised that fuel under pressure may penetrate the skin or any part of the body that it contacts.

• Always place a shop towel or cloth around the fitting or connection prior to loosening to absorb any excess fuel due to spillage. Ensure that all fuel spillage (should it occur) is quickly removed from engine surfaces. Ensure that all fuel soaked cloths or towels are deposited into a suitable waste container.

• Always keep a dry chemical (Class B) fire extinguisher near the work area.

• Do not allow fuel spray or fuel vapors to come into contact with a spark or open flame.

• Always use a back-up wrench when loosening and tightening fuel line connection fittings. This will prevent unnecessary stress and torsion to fuel line piping.

• Always replace worn fuel fitting O-rings with new Do not substitute fuel hose or equivalent where fuel pipe is installed.

Before servicing the vehicle, make sure to also refer to the precautions in the beginning of this section as well.

### RELIEVING FUEL SYSTEM PRESSURE

1. Before servicing the vehicle, refer to the precautions.
2. Disconnect the negative battery cable.
3. Connect a fuel pressure gauge with a pressure release valve to the service port on the fuel supply rail.

4. Place the fuel release tube into a container and open the valve.
5. Remove the fuel pressure gauge from the service port on the fuel supply rail.

### FUEL FILTER

*REMOVAL & INSTALLATION*

**C-Class**

*See Figure 125.*

1. Raise and safely support the vehicle securely on jackstands.
2. Loosen under floor paneling by removing rear nuts of under floor paneling.
3. Lower the under floor paneling.

➡**Only press downwards until assembly clearance at the fuel filter is guaranteed.**

4. Remove protective cap.
5. Loosen hose clamps.
6. Detach fuel lines.
7. Remove bolt.
8. Remove fuel filter.

*To install:*

9. Installation is the reverse of removal.
10. Replace hose clamps.
11. Start the engine and check for leaks.

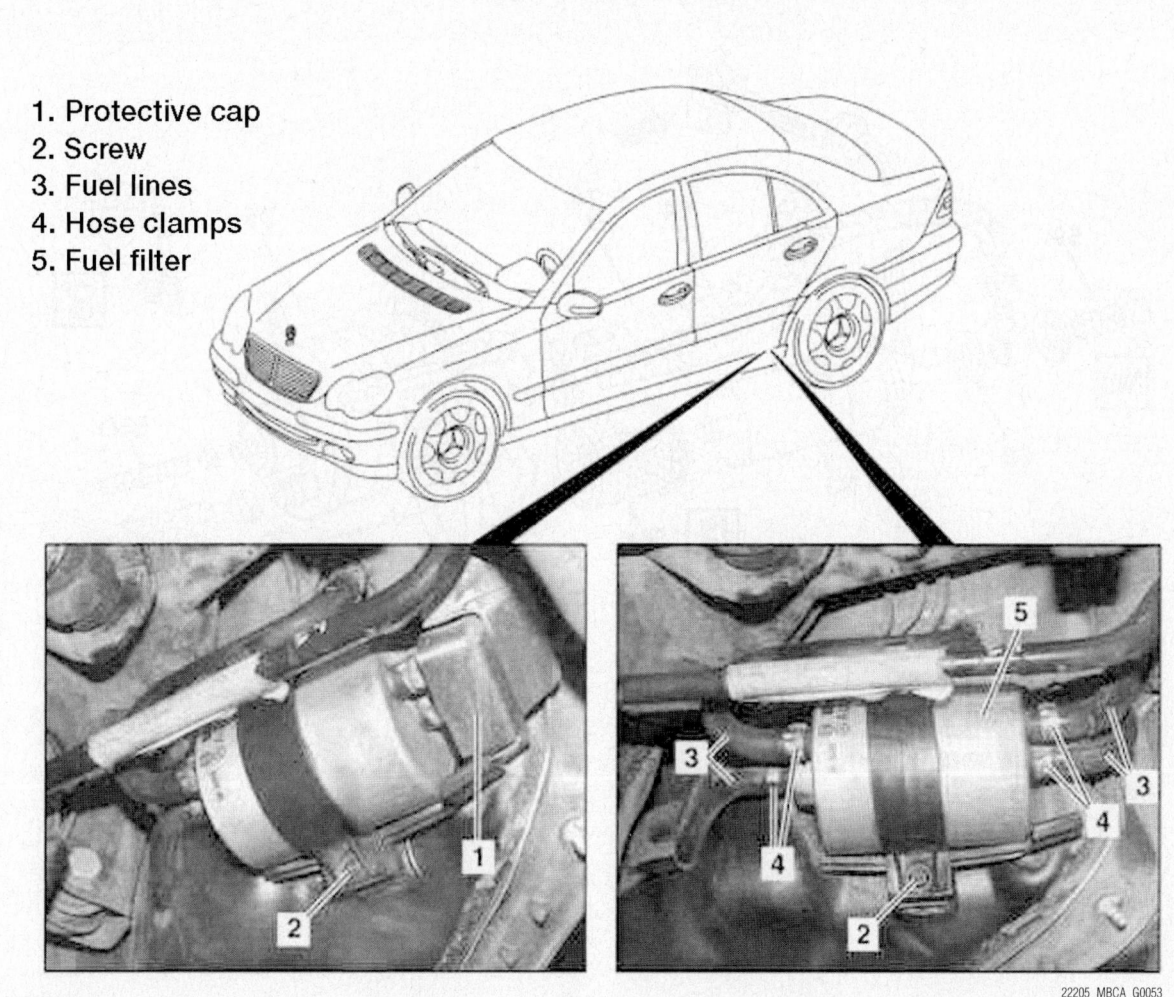

1. Protective cap
2. Screw
3. Fuel lines
4. Hose clamps
5. Fuel filter

22205_MBCA_G0053

**Fig. 125 Fuel filter location**

## E-Class

1. Before servicing the vehicle, refer to the precautions in the beginning of this section.
2. Relieve the fuel system pressure.
3. Disconnect the negative battery cable.
4. Empty the fuel tank into a suitable container.
5. Remove the rear seat cushion.
6. Lift the carpeting to gain access to the fuel pump and filter covers.

➡ **The fuel filter assembly is located under the driver's side cover.**

7. Remove fuel filter cover.
8. Remove fuel filter electrical connector.

➡ **Be sure not to kink the fuel pipes.**

9. Remove supply and return fuel pipes clips and the pipes.

10. Remove union nut mounting the fuel pump-to-the tank.
11. Remove fuel filter from the tank.

**To install:**

➡ **Lightly oil the fuel pump sealing O-ring to simplify the installation.**

12. Install fuel filter into the tank using a new union nut and O-ring. Tighten the union nut to 59 ft. lbs. (80 Nm).
13. Install supply and return lines.
14. Install fuel filter electrical connector.
15. Install fuel filter access cover and the rear seat.
16. Fill the fuel tank.
17. Connect the negative battery cable.
18. Read fault memory, encode the radio and normalize the power windows.
19. Start the vehicle and check for leaks.

## S-Class

*See Figure 126.*

1. Before servicing the vehicle, refer to the precautions in the beginning of this section.
2. Properly relieve the fuel system pressure.
3. Remove cover box.
4. Remove gas cap.
5. Remove fuel pump cover.
6. Remove pressure hoses.
7. Remove fuel filter.
8. Remove the connecting plug from the old filter and install it on a new filter using a new gasket.

**To install:**

9. Install fuel filter.
10. Install attaching screws.
11. Install pressure hoses.
12. Install fuel pump cover.

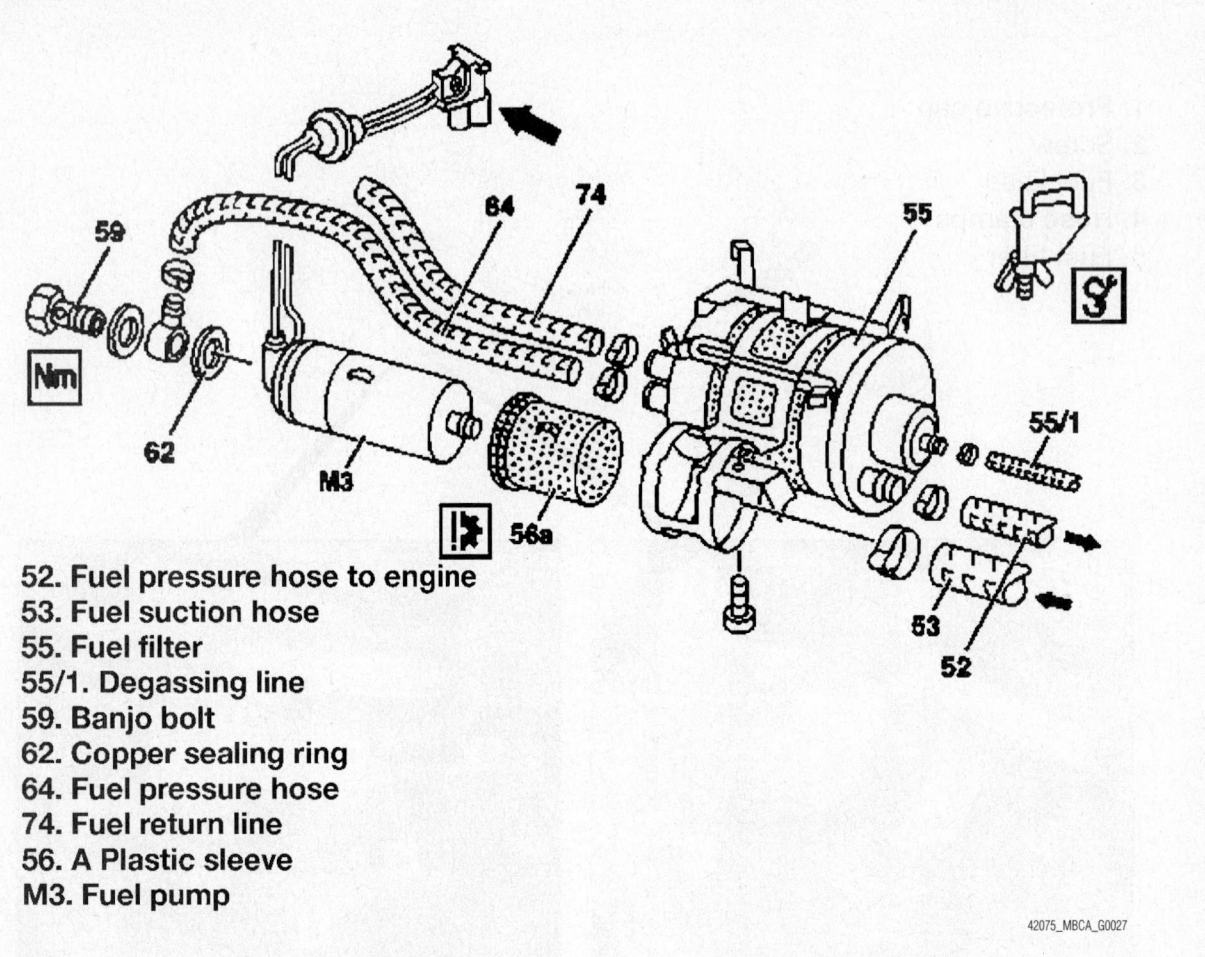

52. Fuel pressure hose to engine
53. Fuel suction hose
55. Fuel filter
55/1. Degassing line
59. Banjo bolt
62. Copper sealing ring
64. Fuel pressure hose
74. Fuel return line
56. A Plastic sleeve
M3. Fuel pump

42075_MBCA_G0027

**Fig. 126 Fuel pump and filter assembly—S-Class**

13. Install gas cap.
14. Install cover box and check for proper sealing.

## FUEL INJECTORS

### REMOVAL & INSTALLATION

#### M112 and M113 Engines

*See Figures 127 and 128.*

1. Before servicing the vehicle, refer to the precautions.
2. Relieve the fuel system pressure.
3. Remove the air intake hose and air filter assembly.
4. Remove the screw securing the fuel rail plastic cover, then remove the cover.
5. If equipped, release any residual pressure in the fuel rail using the service valve on the end of the rail.
6. Detach the fuel feed line and separate the electrical connections from the injectors.

7. Remove the bolt(s) securing the fuel rail, then carefully pull it up and out of the intake manifold.
8. If necessary to remove the injectors, remove the retaining lock then pull the injector out of the fuel rail.

#### To install:

9. Installation is the reverse of removal.
10. Use new O-rings for the injectors and coat them lightly with oil.

#### M272 and M273 Engines

*See Figure 129.*

1. Remove air filter housing.
2. Remove fuel distributor.
3. Lever out safety clips for fuel injection valves.
4. Pull out fuel injection valves from fuel distributor.

#### To install:

5. Replace O-rings.
6. Slide securing clamps on fuel injection valves.

➡ **Mount safety clips in such a way that the cone of the safety clips encloses the molding at the fuel injection valves.**

7. Apply liquid lubricant to O-rings.
8. Insert fuel injection valves into fuel distributor in the correct position.

➡ **The shape of safety clips and connection fittings determines the installation position of the fuel injection valves. The safety clips must latch in audibly.**

9. Install fuel distributor.
10. Install air filter housing.

## FUEL PUMP

### REMOVAL & INSTALLATION

#### C-Class

*See Figure 130.*

1. Before servicing the vehicle, refer to the precautions.

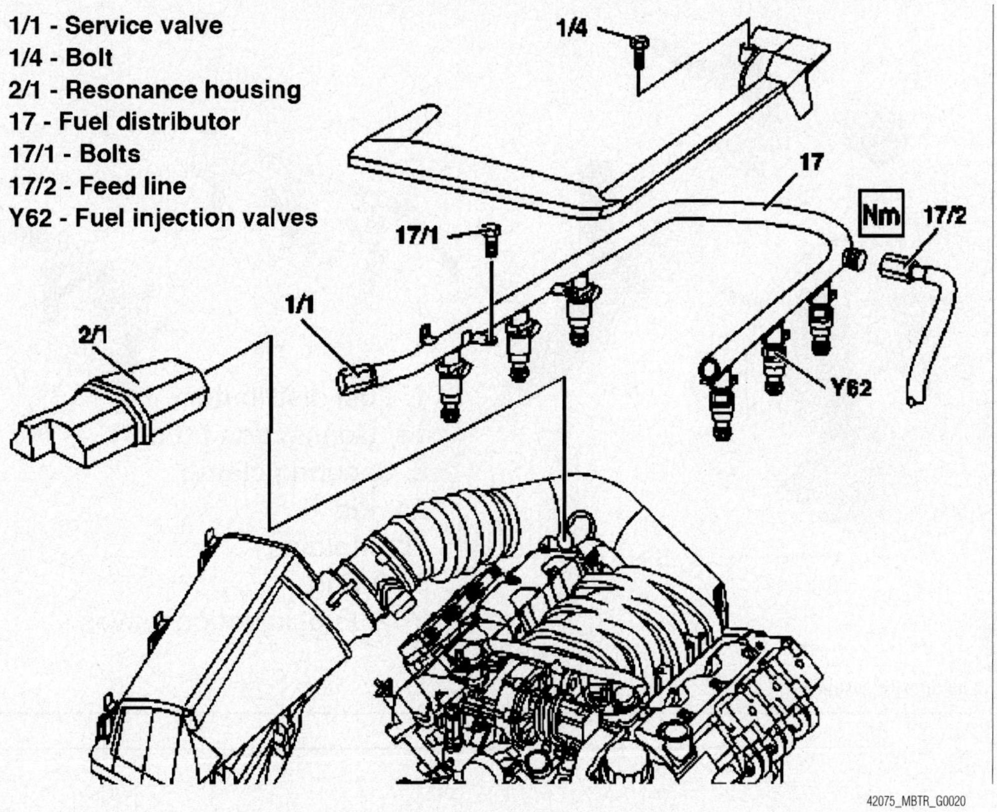

1/1 - Service valve
1/4 - Bolt
2/1 - Resonance housing
17 - Fuel distributor
17/1 - Bolts
17/2 - Feed line
Y62 - Fuel injection valves
1/4
17
Nm
17/2
17/1
1/1
2/1
Y62
42075_MBTR_G0020

Fig. 127 Exploded view of the fuel rail

17
16/7 - Anti-twist lock
17. Fuel distributor
Y62 - Fuel injection valves
Arrow: Square catch
Y62
16/7
42075_MBTR_G0021

Fig. 128 Be sure to engage the square catch (arrow) in the retaining lock

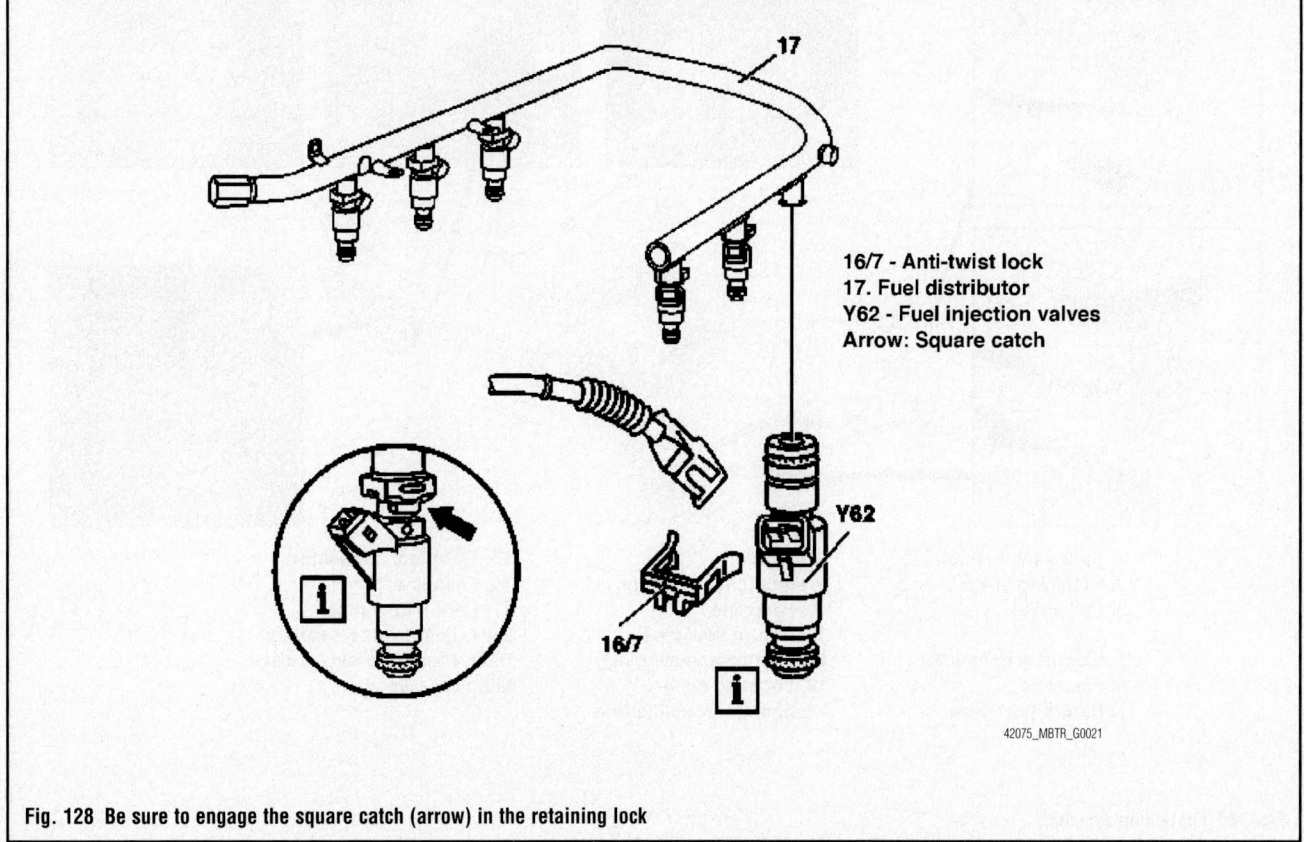

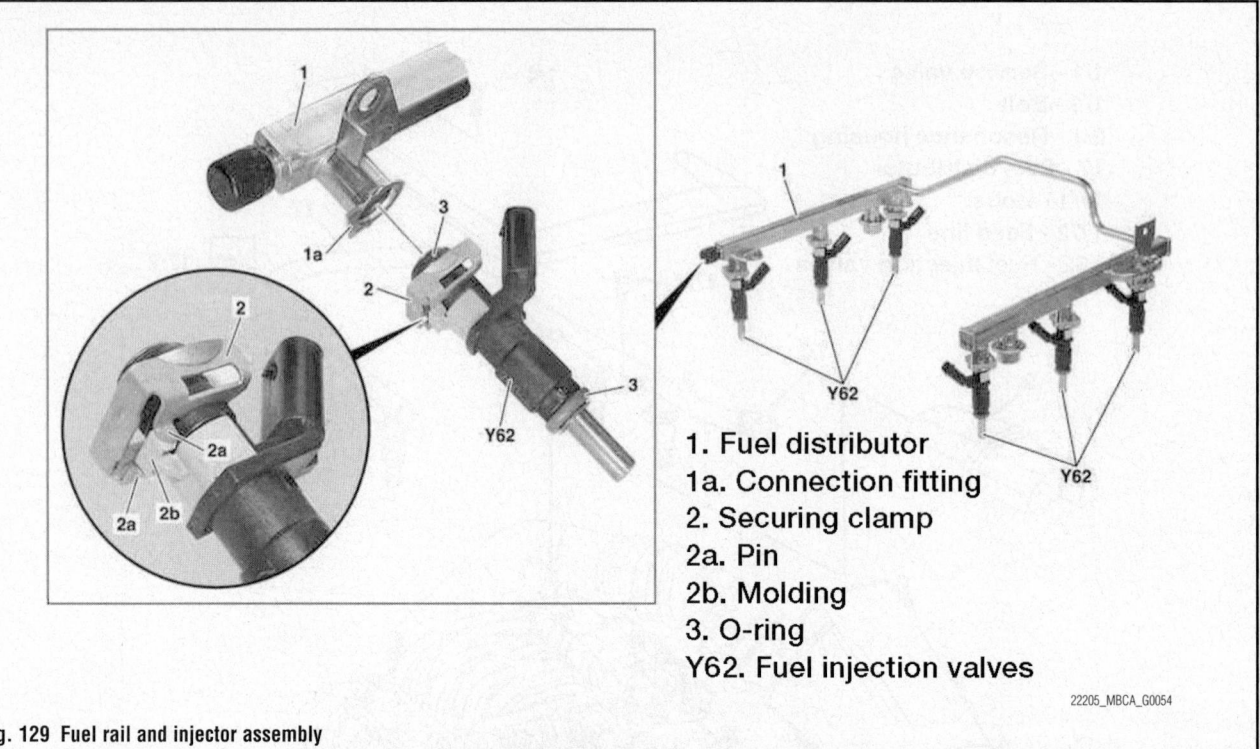

1. Fuel distributor
1a. Connection fitting
2. Securing clamp
2a. Pin
2b. Molding
3. O-ring
Y62. Fuel injection valves

22205_MBCA_G0054

**Fig. 129 Fuel rail and injector assembly**

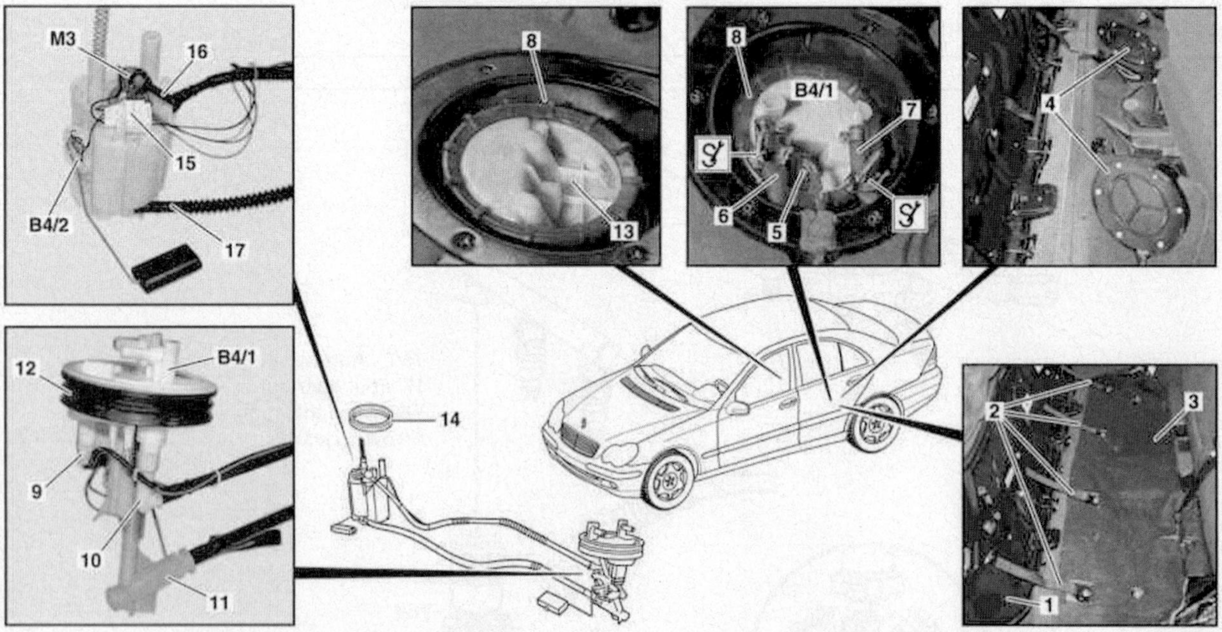

| | | |
|---|---|---|
| 1. Rear seat cushion | 8. Ring nut | 15. Electrical connector |
| 2. Retaining strap | 9. Electrical connector | 16. Feed line |
| 3. Felt cover | 10. Feed line | 17. Return flow pipe |
| 4. Cap | 11. Return flow pipe | B4/1. Left fuel level sensor |
| 5. Electrical connector | 12. Sectional sealing ring | B4/2. Right fuel level sensor |
| 6. Feed line | 13. Locating cover | M3. Fuel pump |
| 7. Return flow pipe | 14. Sectional sealing ring | |

22205_MBCA_G0055

**Fig. 130 Fuel pump assembly**

2. Pump out fuel.

3. Fold rear seat cushion forwards.

4. Remove rear seat cushion.

5. Unscrew bolts of retaining strap for rear seat cushion.

6. Lift felt cover and place to one side.

7. Remove right cap.

8. Unscrew ring bolt on locating cover.

9. Open plug at fuel tank.

10. Remove locating cover together with sectional sealing ring.

11. Disconnect electrical connector at fuel level sensor.

12. Remove feed line at fuel pump.

13. Pull fuel level sensor outwards slightly and detach return line at fuel level sensor.

14. Remove fuel level sensor with fuel pump.

### To install:

15. Installation is the reverse of removal.

16. Tighten ring nut to fuel tank to 63 ft. lbs. (85 Nm).

### E-Class

1. Before servicing the vehicle, refer to the precautions.

2. Relieve the fuel system pressure.

3. Disconnect the negative battery cable.

4. Empty the fuel tank into a suitable container.

5. Remove the rear seat cushion.

6. Lift the carpeting to gain access to the fuel pump cover.

7. Remove fuel filter.

8. Remove fuel pump cover.

9. Remove fuel pump electrical connector.

➡ **Be sure not to kink the fuel pipes.**

10. Remove supply and return fuel pipes clips and the pipes.

11. Remove union nut mounting the fuel pump-to-the tank.

12. Remove fuel pump from the tank.

### To install:

➡ **Lightly oil the fuel pump sealing O-ring to simplify the installation.**

13. Install fuel pump into the tank using a new union nut and O-ring.

14. Tighten the union nut to 59 ft. lbs. (80 Nm).

15. Install supply and return lines to the fuel pump.

16. Install fuel pump electrical connector.

17. Install fuel filter.

18. Install fuel pump access cover and the rear seat.

19. Fill the fuel tank.

20. Connect the negative battery cable.

21. Read fault memory, encode the radio and normalize the power windows.

22. Start the vehicle and check for leaks.

### S-Class

*See Figures 131 and 132.*

1. Before servicing the vehicle, refer to the precautions.

2. Drain fuel tank.

3. To remove the left fuel level sensor:

a. Remove fuel filter.

b. Unscrew nuts.

c. Detach ground line at left tank half fuel level sensor.

d. Disconnect left tank half fuel level sensor connector.

e. Remove fuel gage sensor, left half of tank.

### ❋❋ WARNING

**Handle fuel level sensor carefully in order to avoid deforming the float linkage.**

4. To remove the right fuel level sensor:

a. Remove fuel pump.

b. Press notches and remove right fuel level sensor out of bracket on swirl pot at the side.

### To install:

5. Installation is the reverse of removal.

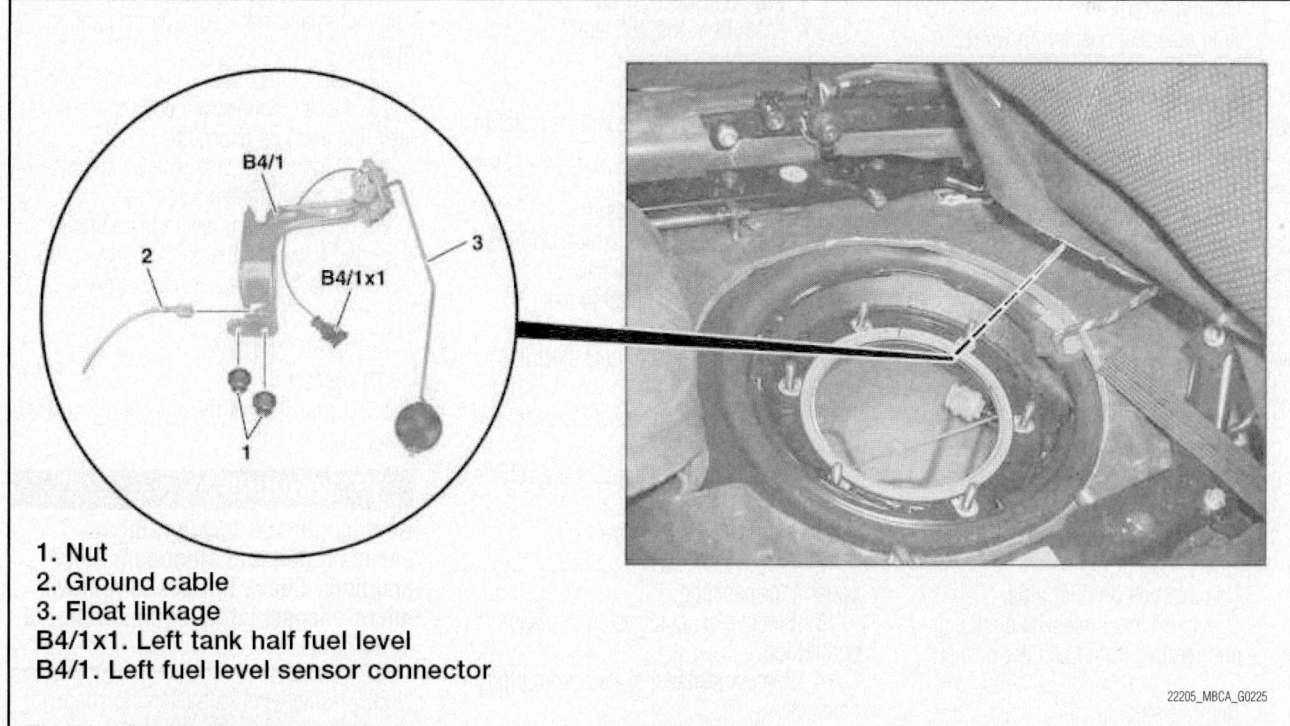

1. Nut
2. Ground cable
3. Float linkage
B4/1x1. Left tank half fuel level
B4/1. Left fuel level sensor connector

22205_MBCA_G0225

**Fig. 131 Left fuel level sensor**

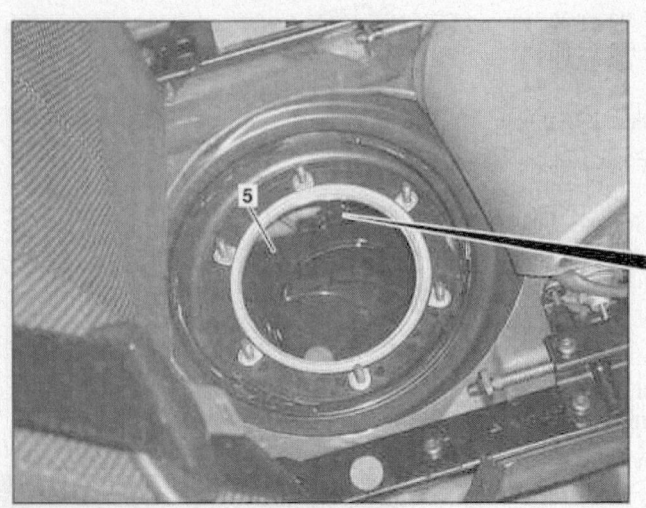 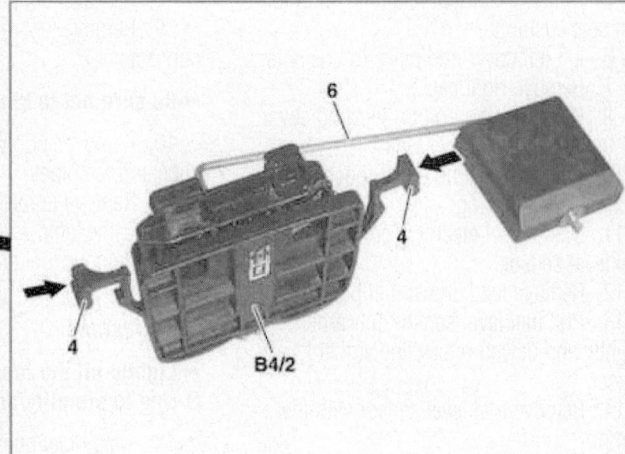

4. Detent
5. Splash bowl
6. Float linkage
B4/2. Right fuel level sensor

22205_MBCA_G0226

**Fig. 132 Right fuel level sensor**

## FUEL TANK

### REMOVAL & INSTALLATION

#### C-Class

*See Figures 133 and 134.*

1. Before servicing the vehicle, refer to the precautions.
2. Drain fuel tank.
3. Fold rear seat cushion forwards.
4. Remove rear seat cushion.
5. Remove left and right trim in the rear section.
6. Remove cover on the left and right service opening.
7. Unlock electrical connectors on fuel filter, fixing cover and tank pressure sensor, disconnect and set aside.
8. Remove battery.
9. Unhook parking brake cable on relay lever.
10. Pull jacket for parking brake cable.
11. from bracket.
12. Remove under floor paneling.
13. Remove rear axle.
14. Receive right rear fender liner.
15. Detach hose clamps.
16. Detach fuel hoses.
17. Remove bolt on filler pipe.
18. Clamp off fuel hose with clamping device and remove from fuel filter or fuel line.
19. Remove shield.
20. Support fuel tank.

21. Unscrew nuts for retaining straps on underbody.
22. Remove fuel tank.

**To install:**

23. Installation is the reverse of removal.
24. Tighten fuel tank retaining straps to specification:
   - M8 15 ft. lbs. (20 Nm)
   - M18 44 ft. lbs. (60 Nm)

#### E-Class

1. Before servicing the vehicle, refer to the precautions.
2. Drain fuel tank.
3. Remove rear seat cushion.
4. Position rear seat cushion upright.
5. Enable access to left cap.
6. Fold back the insulation mat.
7. Remove the left cap.
8. Release and disconnect electrical connector.
9. Unscrew nut from grub screw of the fuel tank.
10. Raise and safely support the vehicle securely.
11. Remove crossmember.
12. Remove both rear sections of the under floor paneling.
13. Remove exhaust system at flange connection.
14. Remove exhaust system from pipe connection.
15. Remove rear muffler.

16. Remove complete exhaust system.
17. Remove heat shields.
18. Detach propeller shaft from rear axle center assembly.
19. Unscrew bolts from propeller shaft center support bearing and remove rear part of the propeller shaft.
20. Remove heat shield.
21. Unhook rear brake cables from automatic slack adjuster on parking brake.
22. Disconnect filler hose at fuel tank.
23. Detach connector hose from fuel feed line and fuel return line.
24. Support fuel tank with a transmission jack and a transmission plate.
25. Detach retaining straps on under floor and lower the fuel tank slightly.
26. Detach connector hoses from vent lines.
27. Lower fuel tank.

**To install:**

28. Installation is the reverse of removal.

### ✴✴ WARNING

**Replace all self locking, micro-encapsulated and aluminum bolts and nuts. Chase threads to remove micro-encapsulated residue from old bolts/nuts.**

29. Tighten fuel tank retaining straps to underbody to 15 ft. lbs. (20 Nm).

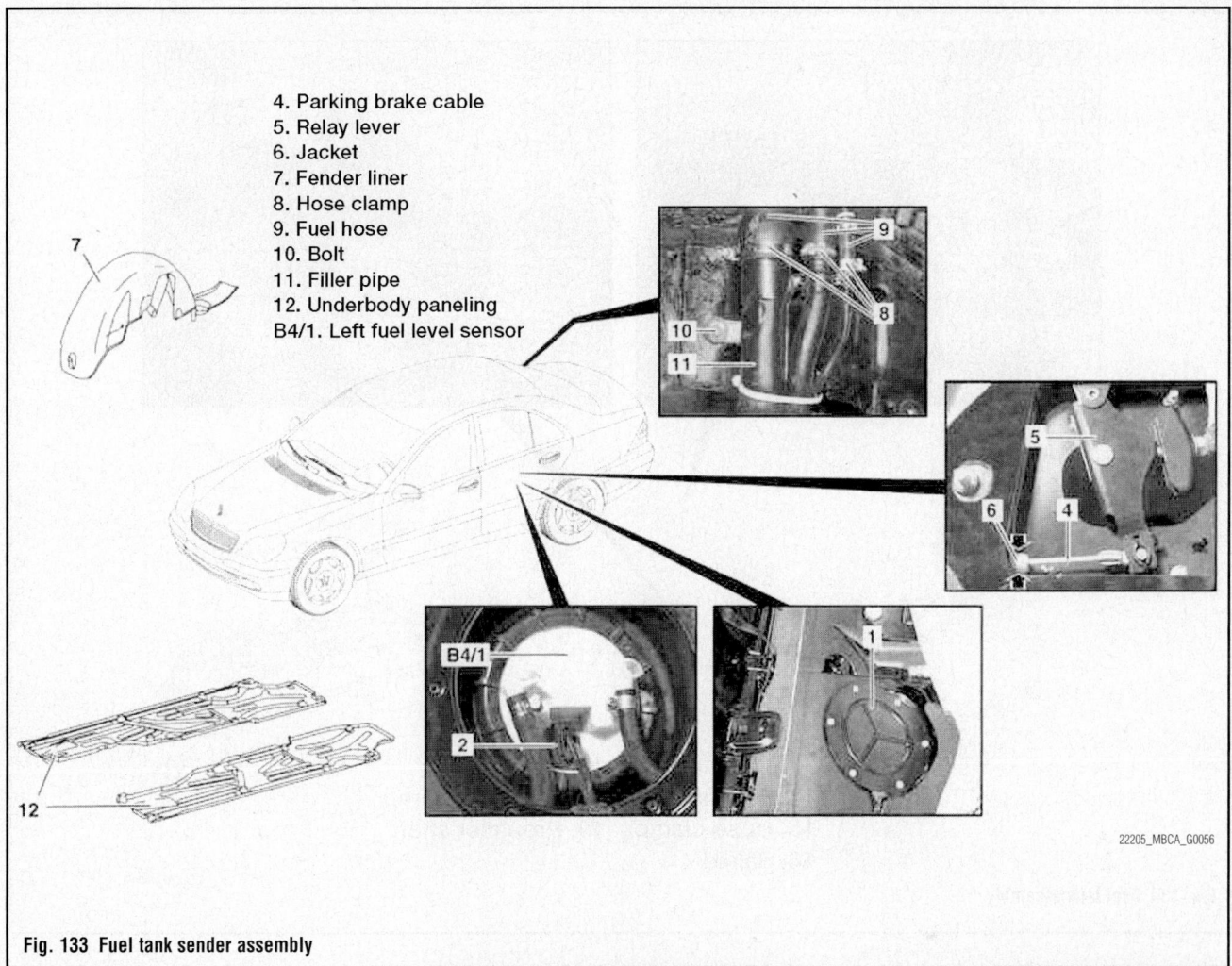

4. Parking brake cable
5. Relay lever
6. Jacket
7. Fender liner
8. Hose clamp
9. Fuel hose
10. Bolt
11. Filler pipe
12. Underbody paneling
B4/1. Left fuel level sensor

22205_MBCA_G0056

**Fig. 133 Fuel tank sender assembly**

## S-Class

*See Figure 135.*

1. Before servicing the vehicle, refer to the precautions.

2. Drain fuel tank.

3. Detach rear SAM control unit with fuse and relay module (N10/2) and place into luggage compartment with the lines connected.

4. Remove hose from fuel tank.

5. Disconnect electrical connector.

6. Remove center rear seat cushions or continuous rear seat cushions and unscrew nuts.

7. Remove under floor paneling.

8. Remove center skid plates.

9. Remove complete rear axle.

10. Remove heat shield.

11. Remove fuel hose from fuel tank.

12. Remove fuel return hose from fuel tank.

13. Remove fuel filler flap with filler neck recess.

14. Unscrew bolts.

15. Remove bolts.

16. Take out fuel tank.

### ✳✳ WARNING

**Never handle fuel tank at the filler neck or vent line since these can be damaged.**

#### To install:

17. Installation is the reverse of removal.

18. Tighten bolts/nuts to specification as follows:

- Nut or bolt, fuel tank to vehicle floor: 11 ft. lbs. (15 Nm)
- Bolt for filler neck on body: 11 ft. lbs. (15 Nm)

➡ **Filler neck and vent line cannot be replaced separately.**

### IDLE SPEED

#### ADJUSTMENT

Ignition timing and idle speed are controlled by the ME engine controller and not adjustable.

### THROTTLE BODY

#### REMOVAL & INSTALLATION

*See Figure 136.*

1. Before servicing the vehicle, refer to the precautions.

2. Remove air filter housing.

3. Remove hot film mass air flow sensor.

4. Remove air guide housing.

5. Release and disconnect electrical connector at throttle valve actuator.

6. Pull bleed hose out of throttle valve actuator.

7. Remove bolts.

8. Remove throttle valve actuator.

9. Remove gasket.

#### To install:

10. Installation is the reverse of removal.

11. Tighten bolt connecting throttle valve actuator to resonance intake manifold to 7 ft. lbs. (9 Nm).

12. Connect STAR DIAGNOSIS and read out fault memory.

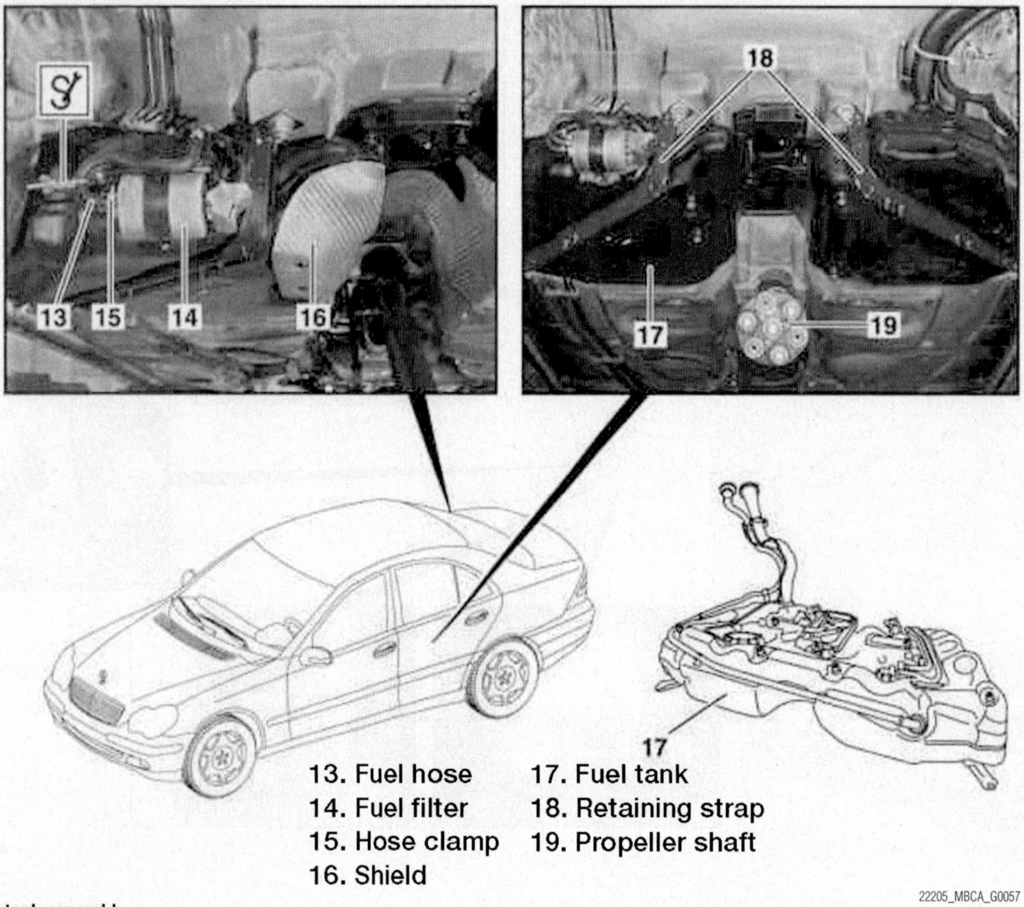

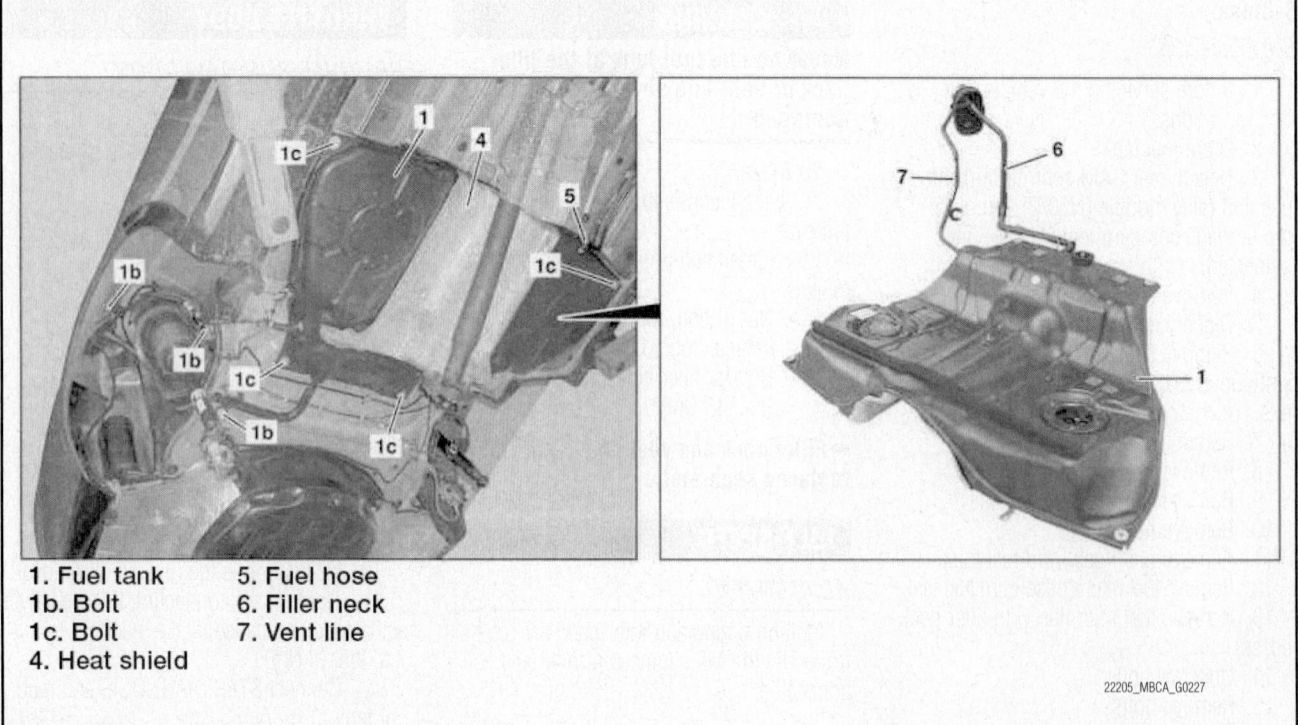

13. Fuel hose
14. Fuel filter
15. Hose clamp
16. Shield

17. Fuel tank
18. Retaining strap
19. Propeller shaft

17

22205_MBCA_G0057

**Fig. 134 Fuel tank assembly**

1. Fuel tank
1b. Bolt
1c. Bolt
4. Heat shield

5. Fuel hose
6. Filler neck
7. Vent line

22205_MBCA_G0227

**Fig. 135 Fuel tank view from under vehicle**

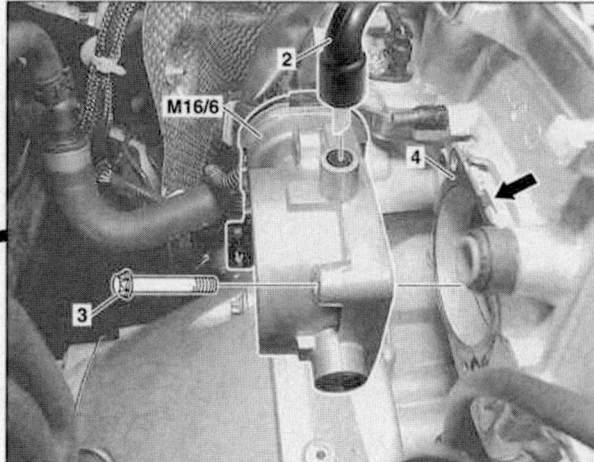

1. Air duct housing    4. Seal
2. Venting hose    B2/5. Hot film mass air flow sensor
3. Bolt    M16/6. Throttle valve actuator

22205_MBCA_G0052

**Fig. 136 Exploded view of throttle body**

## HEATING & AIR CONDITIONING SYSTEM

### BLOWER MOTOR

*REMOVAL & INSTALLATION*

#### C-Class

*See Figure 137.*

1. Before servicing the vehicle, refer to the precautions.
2. Remove cover below instrument panel (right).
3. Disconnect coupling from electronic blower control unit.
4. Expose lead at blower motor.
5. Unscrew bolts Torx bit set.
6. Remove blower motor with blower.

#### ✳✳ WARNING
**Do not lay blower motor on the control unit fan wheel.**

**To install:**
7. Installation is the reverse of removal.

#### E-Class

*See Figure 138.*

1. Before servicing the vehicle, refer to the precautions.

2. Remove cover below right instrument panel.
3. Disconnect electrical connector.
4. Unscrew screws from blower base.
5. Remove blower base.
6. Unscrew screws and remove blower motor with blower regulator.
7. Remove electronic blower regulator from blower motor.

#### ✳✳ WARNING
**Do not place blower motor on fan wheel, since the blower motor will otherwise be unbalanced.**

**To install:**
8. Installation is the reverse of removal.
9. Perform a function test.

#### S-Class

*See Figures 139 and 140.*

1. Before servicing the vehicle, refer to the precautions.
2. Switch ignition **OFF** and withdraw transmitter key.
3. Remove cover below instrument panel on right.
4. Expose electrical wiring harness on

housing of blower motor and detach electrical connector.
5. Remove bolts.
6. Remove housing with blower motor from air conditioner housing.
7. Unscrew bolt.
8. Remove blower regulator from housing and detach electrical connector.

**To install:**
9. Installation is the reverse of removal.
10. Perform a function test.

### HEATER CORE

*REMOVAL & INSTALLATION*

#### C-Class

*See Figures 141 and 142.*

1. Before servicing the vehicle, refer to the precautions.
2. Disconnect the negative battery cable.
3. Detach lower engine compartment paneling.
4. Remove bottom section of soundproofing.
5. Drain coolant from radiator.
6. Remove steering wheel.

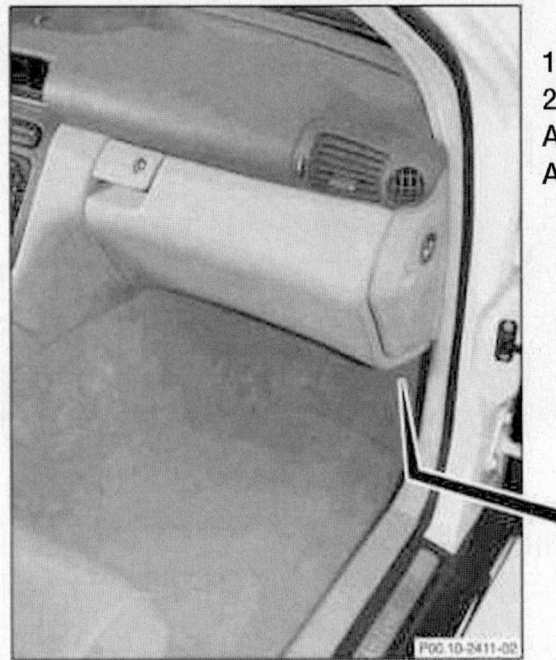

1. Bolts
2. Coupling connection at electronic blower
A32m1. Blower motor
A32n1. Electronic blower control unit

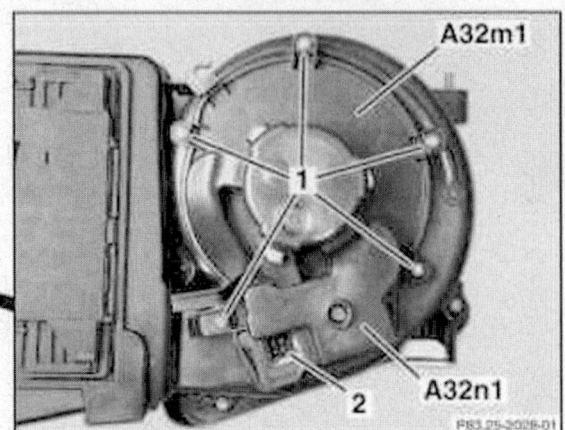

Fig. 137 Blower motor location

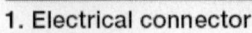
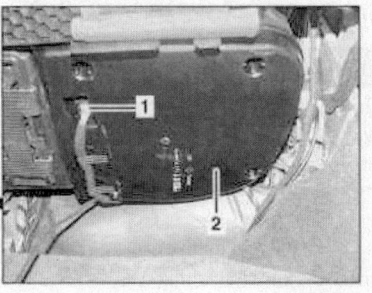
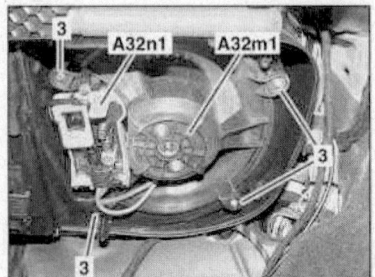

1. Electrical connector
2. Blower base
3. Bolts
A32m1. Blower motor
A32n1. Blower regulator

Fig. 138 Blower motor location

1. Wiring harness
2. Housing
3. Electrical connector
4. Bolts

22205_MBCA_G0247

**Fig. 139 Blower motor location**

---

※※ **WARNING**

**Store airbag units with deployment side facing up; do not expose to temperatures greater than 100 °C. When working on these units, disconnect the power supply.**

7. Remove cover below instrument panel (on left).
8. Remove bottom section of instrument panel on driver's side.

9. Remove instrument cluster.
10. Remove accelerator pedal.
11. Fold back floor covering at front left.

➡**Air duct to left rear compartment must be accessible.**

12. Remove air duct to left rear compartment.
13. Remove jacket tube from instrument panel carrier and fold down into foot well.

※※ **WARNING**

**Do not remove bottom steering shaft from steering coupling.**

14. Remove three tensioning springs.
15. Remove fitting.
16. Pull out socket yokes and discard.
17. Pull heating water pipes out of heat exchanger.
18. Plug heating water pipes and openings on heat exchanger.

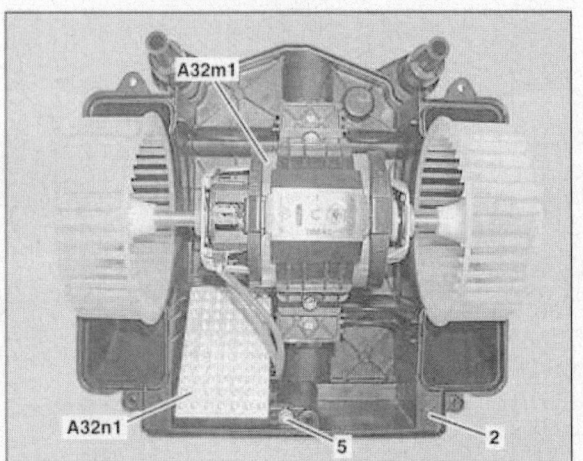

2. Housing
5. Bolt
6. Electrical connector

A32n1. Blower regulator
A32m1. Blower motor

22205_MBCA_G0248

**Fig. 140  Disassembling the housing to access the blower motor**

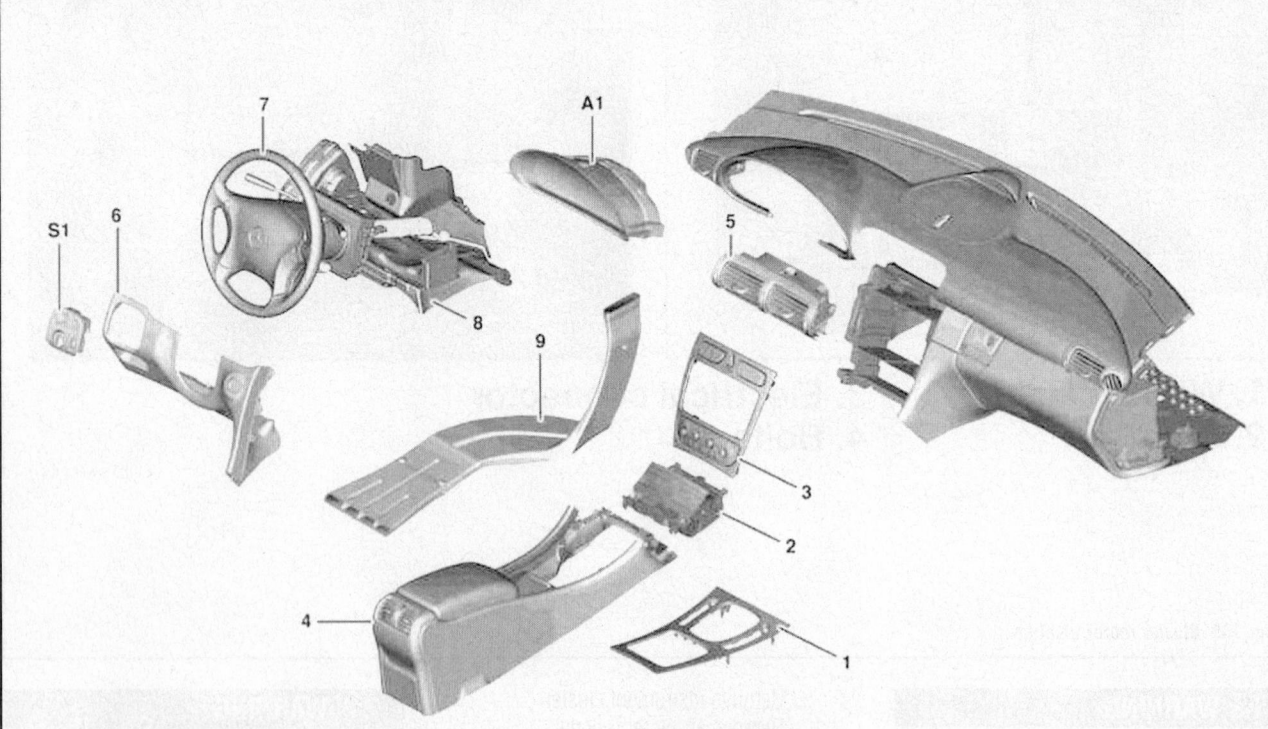

1. Cover on shift lever
2. Ashtray housing
3. Cover on center console
4. Center console
5. Center air nozzle
6. Instrument panel bottom section on driver's side
7. Steering wheel
8. Cover below instrument panel
9. Air duct to left rear compartment
A1. Instrument cluster
S1. Light switch module

22205_MBCA_G0059

**Fig. 141  Instrument and console panels**

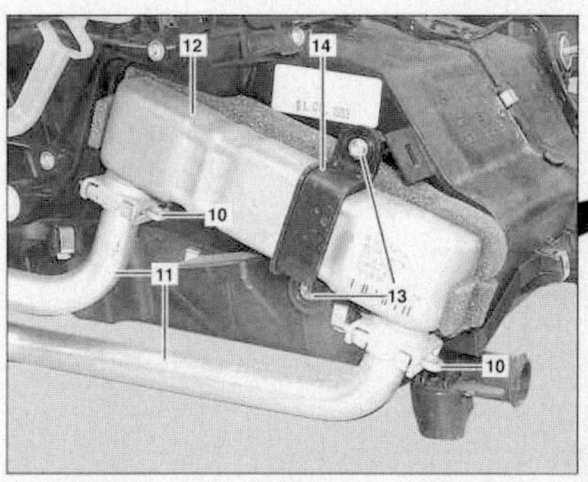

9. Fitting  
9a. Tensioning springs  
10. Socket yokes  
11. Heating water pipes  
12. Heat exchanger  
13. Bolts  
14. Bracket

22205_MBCA_G0060.

**Fig. 142 Heater core mounted in A/C unit**

19. Remove bolts.
20. Remove bracket.
21. Pull out heat exchanger.
22. Clean AC/heater housing from inside, if necessary remove coolant running out.

**To install:**
23. Installation is the reverse of removal.
24. Install new sealing rings, as old sealing rings can result in leaks.

### ❊❊ WARNING
**Replace all self locking, micro-encapsulated and aluminum bolts and nuts. Chase threads to remove micro-encapsulated residue from old bolts/nuts.**

25. Carry out function check.

### E-Class
*See Figures 143 and 144.*

1. Before servicing the vehicle, refer to the precautions.

2. Remove AC housing.
3. Remove right center air outlet air flap actuator motor.
4. Unscrew bolts and detach center air outlet air flap actuator motor.
5. Unscrew bolts and detach actuator motor.
6. for left center air outlet air flap.
7. Disconnect electrical connector.
8. Expose positive lead and ground lead of the heater booster at the AC housing.
9. Detach seal.
10. Remove bracket and bracket from A/C housing.
11. Unscrew bolt.
12. Open bracket of clamping bracket in direction of arrow (arrow A) and slide out clamping bracket in direction of arrow (arrow B).
13. Unclip clips al round.
14. Detach AC housing.
15. Remove air distributor housing and put to side.
16. Remove heat exchanger.

**To install:**
17. Installation is the reverse of removal.
18. Replace seals.

### ❊❊ WARNING
**Do not damage sealing tape on heat exchanger.**

### S-Class
*See Figure 145.*

1. Before servicing the vehicle, refer to the precautions.
2. Remove evaporator.
3. Remove all actuator motors, temperature sensors and the electrical wiring harness from the heat exchanger housing.

**To install:**
4. Installation is the reverse of removal.
5. Synchronize the actuator motors.
6. Connect STAR DIAGNOSIS tool and read out fault memory.

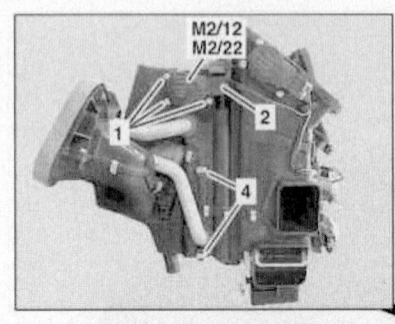
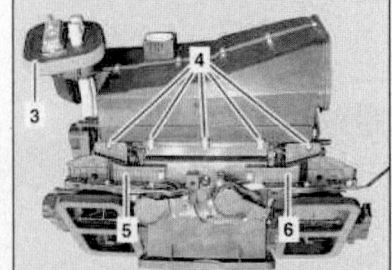
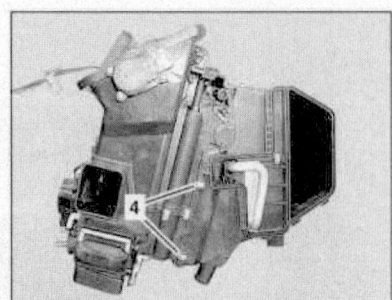
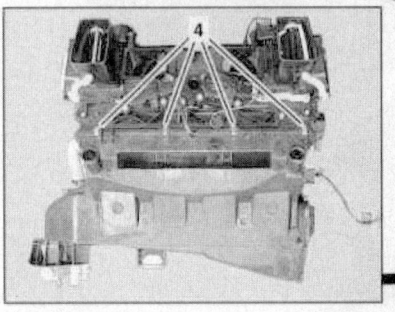

1. Bolts     5. Bracket
2. Electrical connector     6. Bracket
3. Seal     M2/12. Left center air outlet air flap actuator motor
4. Clamps     M2/22. Center air outlet air flap actuator motor

22205_MBCA_G0237

**Fig. 143 Disassembling the A/C unit**

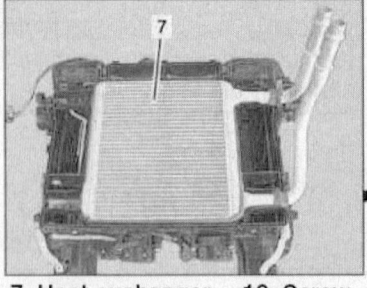
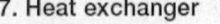
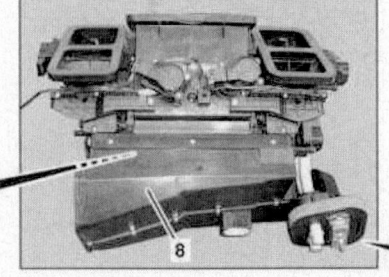
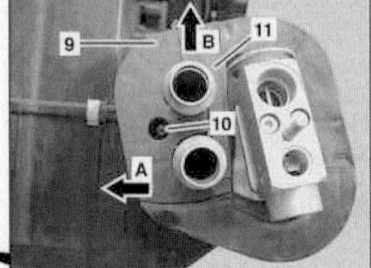

7. Heat exchanger     10. Screw
8. AC housing     11. Clamp fixture
9. Bow

22205_MBCA_G0238

**Fig. 144 Heater core mounted in A/C unit**

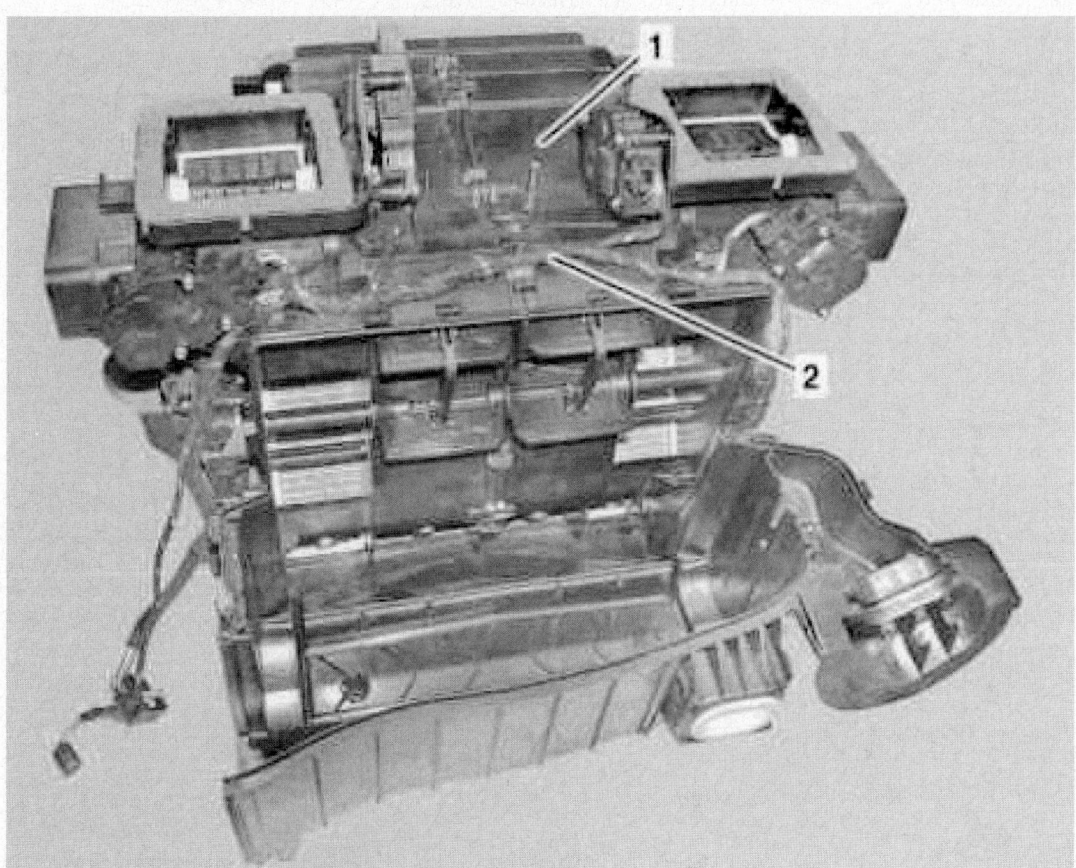

1. Heat exchanger housing
2. Wiring harness

22205_MBCA_G0254

**Fig. 145 Heater core mounted in A/C unit**

## STEERING

### POWER STEERING GEAR

#### REMOVAL & INSTALLATION

#### C-Class

*See Figure 146.*

1. Before servicing the vehicle, refer to the precautions.

2. Remove front wheels Remove/install wheels, rotate if necessary.

3. Extract fluid out of the power steering pump's fluid reservoir.

4. Extend or pull out adjustable steering column fully.

5. Deactivate easy entry/exit function.

6. Turn steering wheel to central position (front wheels in straight ahead position) and secure with the retaining device.

#### ✳✳ WARNING

**On a disengaged steering coupling the steering wheel should not be turned because the spiral contact coil will be destroyed.**

7. Remove lower engine compartment paneling.

8. Remove bottom sections of soundproofing.

9. Press tie rod joints off steering knuckles.

10. Remove exhaust shielding plate to front plate over steering coupling.

11. Remove the bolt of the steering coupling to steering shaft.

12. Pull the lower steering shaft upward to extract it from the steering coupling.

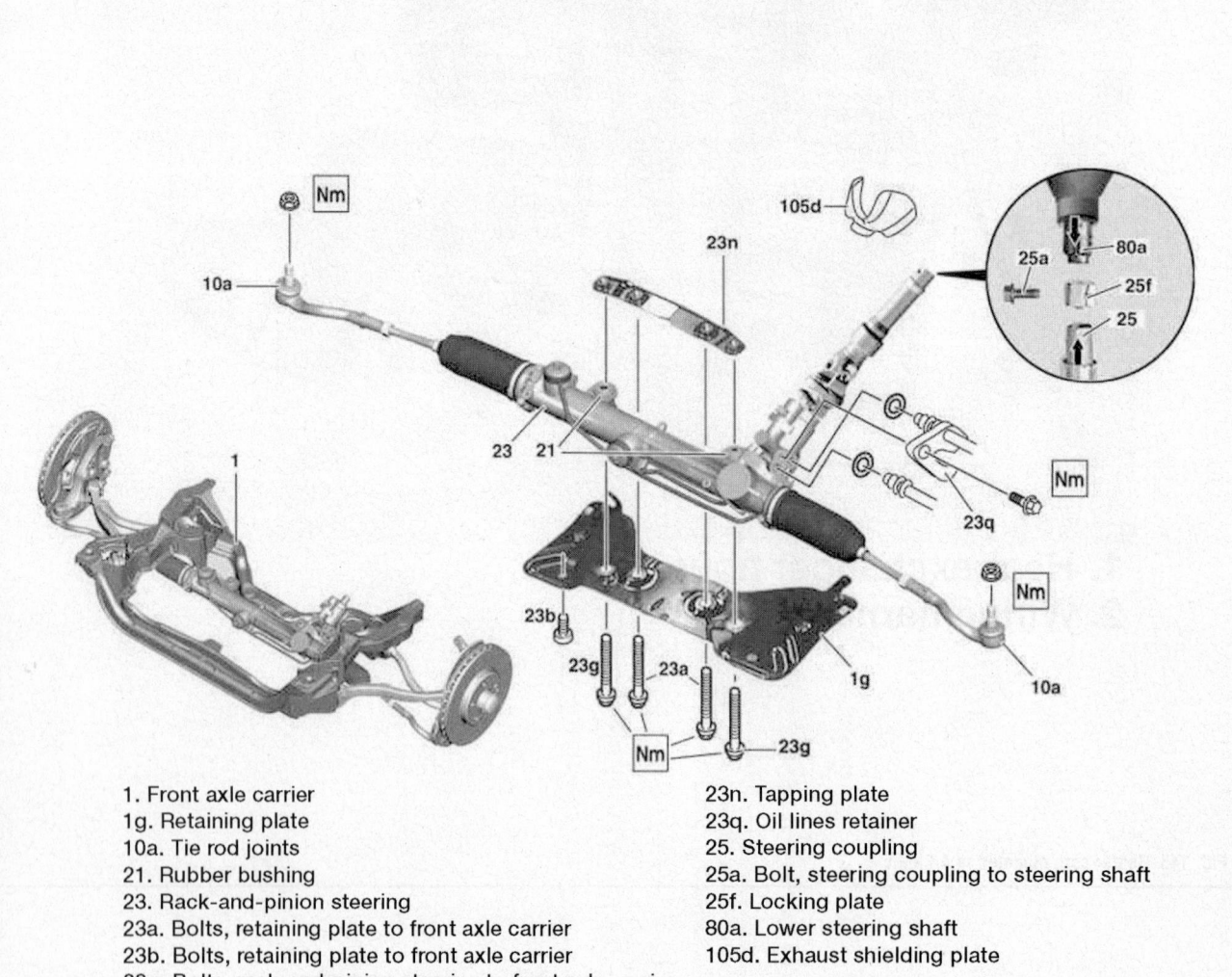

1. Front axle carrier
1g. Retaining plate
10a. Tie rod joints
21. Rubber bushing
23. Rack-and-pinion steering
23a. Bolts, retaining plate to front axle carrier
23b. Bolts, retaining plate to front axle carrier
23g. Bolts, rack-and-pinion steering to front axle carrier

23n. Tapping plate
23q. Oil lines retainer
25. Steering coupling
25a. Bolt, steering coupling to steering shaft
25f. Locking plate
80a. Lower steering shaft
105d. Exhaust shielding plate

22205_MBCA_G0062

**Fig. 146 Exploded view of the power rack and steering assembly**

## ❊❊ WARNING

**Do not apply excessive force (hammer blows, excessive lever action) because the lower steering shaft is sensitive to lateral loads and could be damaged.**

13. Detach oil lines to rack-and-pinion steering.

14. Remove bolts of retaining plate on front axle carrier and remove support bracket from front axle carrier.

15. Detach electrical connector to rack-and-pinion steering.

16. Remove bolts of rack-and-pinion steering at front axle carrier.

17. Detach oil line retaining plate from rack-and-pinion steering and pull out oil lines. Seal off line connections with plugs.

18. Remove threaded hole plate of rack-and-pinion steering.

19. Remove rack-and-pinion steering.

20. Check rubber mount of rack-and-pinion steering mounting for signs of damage.

21. Check protective cap on input shaft of rack-and-pinion steering for damage. Replace as necessary.

### To install:

22. Replace threaded hole plate of rack-and-pinion steering.

23. Insert rack-and-pinion steering and fasten to front axle carrier using bolts for rack-and-pinion steering to front axle carrier.

24. Mount support bracket using bolts for support bracket to front axle carrier.

➡ **The bolts for the support bracket to the front axle carrier, threaded hole plate and the area around the bolted connections must be free of oil and grease. Under exposure to heat the lubricants can liquefy, allowing bolts and nuts to loosen.**

25. Insert oil lines and mount retaining plate for oil lines to rack-and-pinion steering. Replace O-rings to prevent any escape of oil.

26. Mount oil lines onto rack-and-pinion steering.

27. Fit steering coupling by turning the rack-and-pinion steering to center position.

28. Slide on new locking plate until the lug latches into the steering coupling.

29. Insert lower steering shaft as far as the end stop in the steering coupling.

## ❊❊ WARNING

**Do not apply excessive force (hammer blows, excessive lever action) because the lower steering shaft is sensitive to lateral loads and could be damaged. When doing so observe the steering coupling journal and the groove of the lower steering shaft (arrows).**

30. Turn in bolt for steering coupling to steering shaft.

31. Mount exhaust shielding plate to front plate under steering coupling.

32. Mount tie rod joints onto steering knuckles.

33. Fill power steering pump and bleed.

34. Check steering for function and leaks.

➡ **There is a grease cushion under the cover cap. A slight escape of grease is to be regarded as normal and should not be confused with leakage.**

35. Install front wheels.

36. Check front axle toe and adjust if necessary.

37. Install lower engine compartment paneling.

### E-Class

*See Figure 147.*

1. Before servicing the vehicle, refer to the precautions.

2. Fully extend or pull out adjustable steering column.

3. Deactivating entry/exit aid.

4. Turn steering wheel to center position (front wheels in straight-ahead position) and fix in place with clamping device.

## ❊❊ WARNING

**With a disengaged steering coupling the steering wheel should not be turned because the spiral contact coil will be irreparably damaged.**

5. Open engine hood.

6. Remove front engine cover.

7. Remove engine trim panel.

8. Remove air filter housing.

9. Remove trim panels of cylinder head cover.

10. Remove intake air duct leading to left air filter.

11. Remove intake air ducts upstream of air filter.

12. Suction off oil from expansion reservoir of power steering.

13. Remove front wheels.

14. Remove bottom section of sound-proofing.

15. Remove engine compartment paneling.

16. Press tie rod joints off steering knuckles.

17. Detach right heat shield from front axle carrier.

18. Detach left heat shield from front axle carrier.

19. Detach steering coupling from lower steering shaft.

20. Pull the lower steering shaft upward to extract it from the steering coupling.

21. Separate electrical connector from SPS solenoid valve.

22. Remove the bracket from the rack-and-pinion steering.

23. Remove the bracket for the oil lines on the rack-and-pinion steering and then pull out the oil lines.

24. Support rack-and-pinion-steering using transmission jack and platform.

25. Detach rack-and-pinion steering from front axle carrier.

26. Carefully lower rack-and-pinion steering and remove.

27. Inspect rack-and-pinion steering.

28. Evaluating steering rack when repairing accident vehicles.

29. Check rubber mount of rack-and-pinion steering mounting for damage.

30. Check protective cap on input shaft of rack-and-pinion steering for damage.

### To install:

## ❊❊ WARNING

**Replace all self locking, micro-encapsulated and aluminum bolts and nuts. Chase threads to remove micro-encapsulated residue from old bolts/nuts.**

31. Installation is the reverse of removal.

32. Tighten bolts/nuts to specification as follows:

- Bolt, steering coupling to steering shaft: 22 ft. lbs. (30 Nm)
- Bolt, oil line retaining plate to rack-and-pinion steering: 13 ft. lbs. (18 Nm)
- Bolted connection, rack-and-pinion steering to front axle carrier: 52 ft. lbs. (70 Nm) plus an additional 90° of rotation
- Self-locking nut, tie rod to steering: 37 ft. lbs. (50 Nm) plus an additional 60° rotation

33. Fill power steering pump and bleed.

34. Check front axle toe and adjust if necessary.

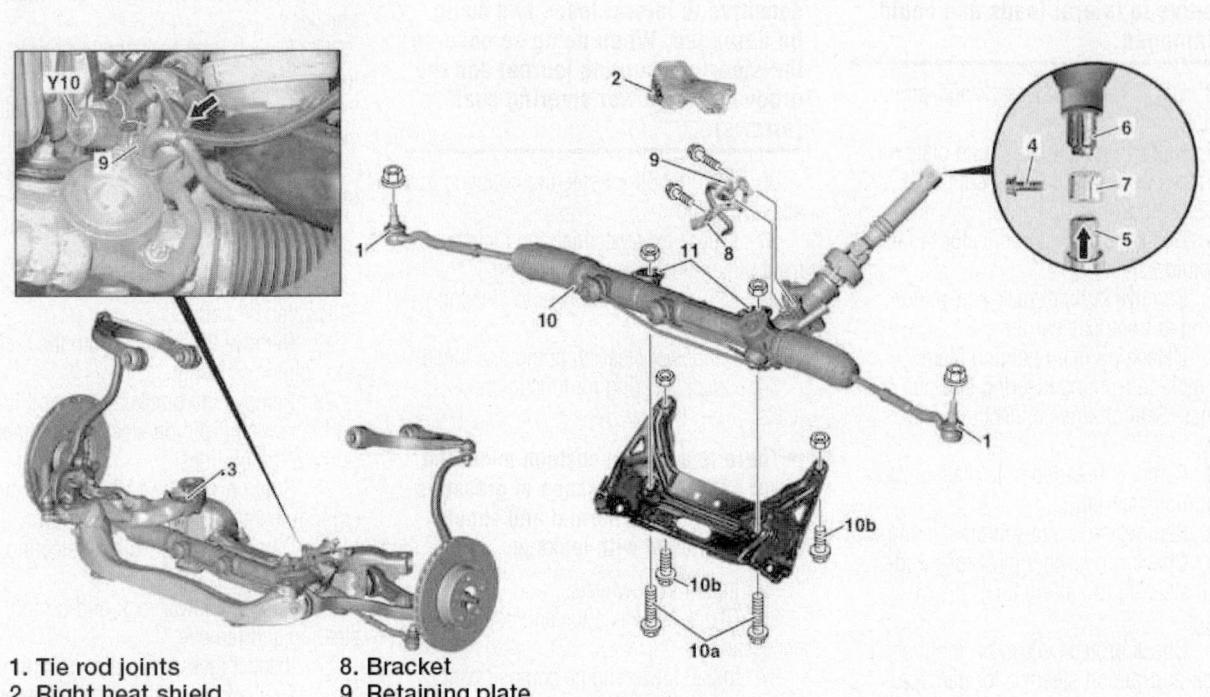

1. Tie rod joints
2. Right heat shield
3. Front axle carrier
4. Bolt
5. Steering coupling
6. Lower steering shaft
7. Locking plate
8. Bracket
9. Retaining plate
10. Rack-and-pinion steering
10a. Bolts
10b. Bolt
11. Rubber bushing
Y10. SPS [PML] solenoid valve

22205_MBCA_G0256

**Fig. 147 Exploded view of the power rack and steering assembly**

### S-Class

*See Figure 148.*

1. Before servicing the vehicle, refer to the precautions.

**✷✷ WARNING**

**Before opening the hydraulic system, thoroughly clean the area surrounding the separation point. Even the smallest dirt particles, introduced into the hydraulic components, can lead to malfunctions and a total failure of the hydraulic system.**

2. Deactivate easy entry/exit function.
3. Turn steering wheel to center position (front wheels in straight-ahead position) and fix in place with holding device.

**✷✷ WARNING**

**On a disengaged steering coupling the steering wheel should not be turned because the spiral contact coil will be destroyed.**

4. Remove the left engine intake air duct.
5. Remove engine trim panel.
6. Extract oil from power steering expansion reservoir.
7. Remove front wheels.
8. Remove bottom engine compartment paneling.
9. Detach tie rods from wheel carriers.
10. Detach steering coupling from lower steering shaft; to do so, remove bolt from locking plate.

11. Pull lower steering shaft upward out of steering coupling.
12. Disconnect electrical connector from solenoid valve.
13. Detach clamp from rack-and-pinion steering.
14. Detach clamping plate from rack-and-pinion steering and pull out oil lines.
15. Support rack-and-pinion steering using transmission jack.
16. Detach rack-and-pinion steering from front axle carrier by unscrewing bolts.
17. Remove retaining plate.
18. Remove rack-and-pinion steering.

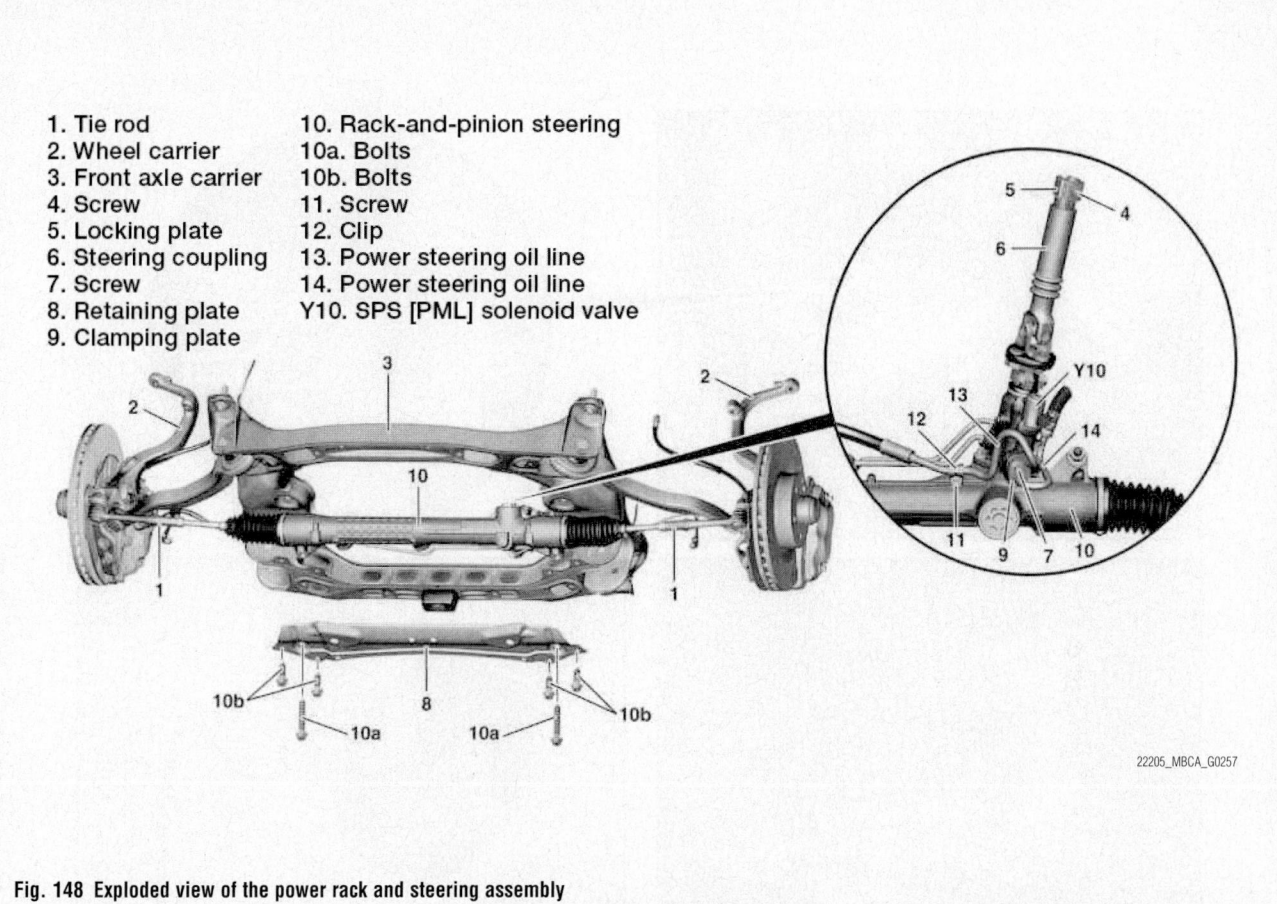

1. Tie rod
2. Wheel carrier
3. Front axle carrier
4. Screw
5. Locking plate
6. Steering coupling
7. Screw
8. Retaining plate
9. Clamping plate
10. Rack-and-pinion steering
10a. Bolts
10b. Bolts
11. Screw
12. Clip
13. Power steering oil line
14. Power steering oil line
Y10. SPS [PML] solenoid valve

22205_MBCA_G0257

**Fig. 148 Exploded view of the power rack and steering assembly**

*To install:*

> ※※ **WARNING**
>
> **Replace all self locking, micro-encapsulated and aluminum bolts and nuts. Chase threads to remove micro-encapsulated residue from old bolts/nuts.**

19. Installation is the reverse of removal.
20. Tighten bolts/nuts to specification as follows:

- Bolted connection, rack-and-pinion steering to front axle carrier: 52 ft. lbs. (70 Nm) plus an additional 90° of rotation
- Bolt, retaining plate of rack-and-pinion steering to front axle carrier: 74 ft. lbs. (100 Nm)
- Bolt for clamp plate, high-pressure line / return line to rack-and-pinion steering: 15 ft. lbs. (20 Nm)
- Bolt, steering coupling to steering shaft: 22 ft. lbs. (30 Nm)
- Self-locking nut, tie rod to steering: 37 ft. lbs. (50 Nm) plus an additional 60° rotation

21. Fill power steering pump and bleed.
22. Check front axle toe and adjust if necessary.

## POWER STEERING PUMP

*REMOVAL & INSTALLATION*

### C-Class

*See Figures 149 and 150.*

1. Before servicing the vehicle, refer to the precautions.

> ※※ **WARNING**
>
> **Before opening the hydraulic system of the power steering, thoroughly clean the area surrounding the separation point. Even minute particles of dirt, introduced into the hydraulic components, can result in malfunctions or damage to the power steering.**

2. Remove air filter housing.
3. Remove poly-V belt.
4. Remove power steering expansion reservoir.
5. Detach feed line of expansion reservoir on power steering pump.
6. Detach high pressure expansion hose from power steering pump, to do this remove banjo bolt. Use a rag to clear up any escaping oil and seal the line connections using stop plugs.
7. Detach plug from power steering pump pressure regulator valve.
8. Remove coolant compressor from timing case cover and engine support, to do this unscrew the bolts.
9. Separate refrigerant compressor connector and put down refrigerant compressor with pipe system connected towards the radiator.
10. Remove screw.
11. Detach ground line from longitudinal member.
12. Detach power steering pump from timing case cover by unscrewing the bolts.
13. Remove power steering pump.

*To install:*

14. Installation is the reverse of removal.
15. Replace sealing rings.
16. Tighten fasteners to specification:

- Bolt, power steering pump to timing case cover M8 x 30: 15 ft. lbs. (20 Nm)
- Bolt, power steering pump to timing case cover M8 x 44: 26 ft. lbs. (35 Nm)

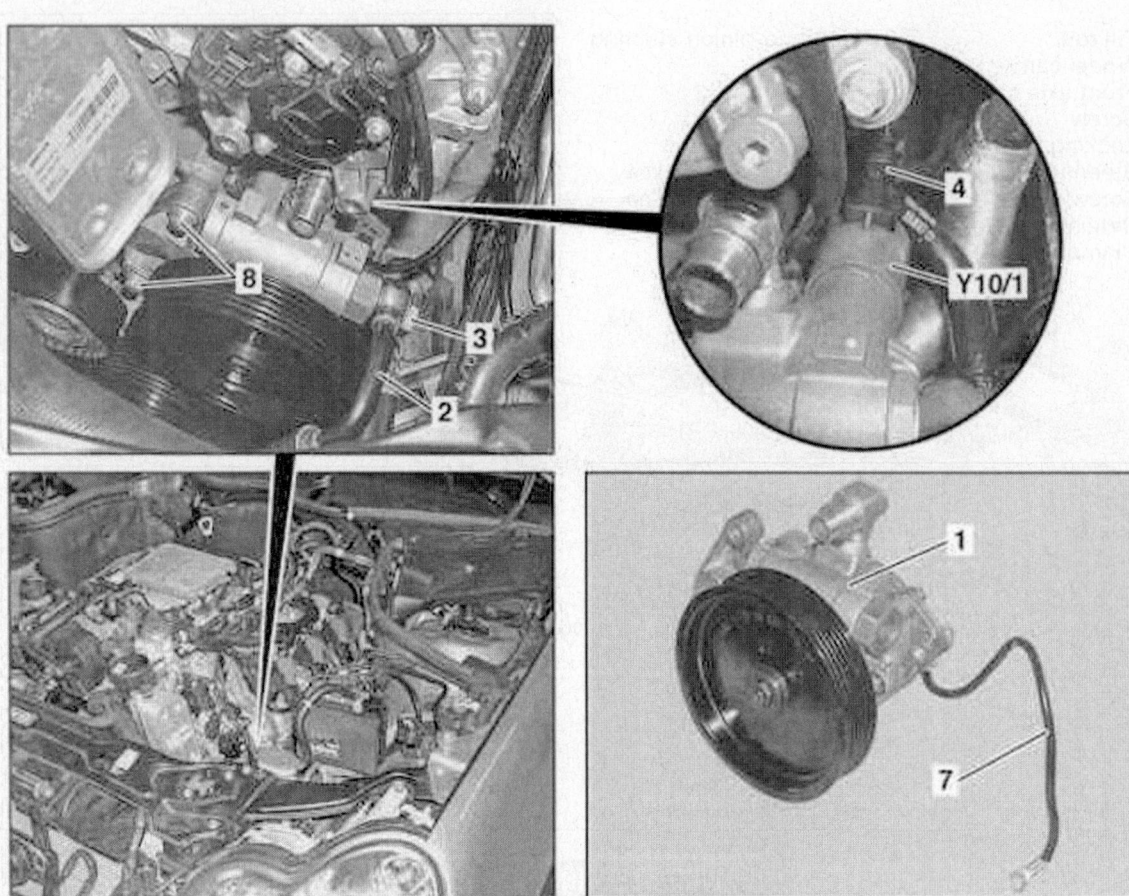

1. Power steering pump
2. High pressure expansion hose
3. Banjo bolt
4. Connector
7. Ground cable
8. Bolts
Y10/1. Power steering pump pressure
       regulator valve

22205_MBCA_G0063

**Fig. 149 Power steering reservoir and high pressure hose**

- Banjo bolt, high pressure expansion hose: 33 ft. lbs. (45 Nm)
- Bolt, refrigerant compressor to timing case cover or engine support: 15 ft. lbs. (20 Nm)

17. Fill and bleed power steering pump.

### E-Class

*See Figures 151 and 152.*

1. Before servicing the vehicle, refer to the precautions.

### ✳✳ WARNING

**Before opening the hydraulic system of the power steering, thoroughly clean the area surrounding the separation point. Even minute particles of dirt, introduced into the hydraulic components, can result in malfunctions or damage to the power steering.**

2. Open engine hood and raise to vertical position.
3. Remove poly-V belt.
4. Remove front engine cover.

5. Remove intake air duct upstream of air filter.
6. Suction off oil from expansion reservoir.
7. Detach return line from expansion reservoir.
8. Undo screws for front cover on cylinder head.
9. Release clip.
10. Detach expansion reservoir from power steering pump.
11. Detach pressure line from power steering pump by undoing banjo bolt.
12. Detach plug from power steering pump pressure regulator valve.

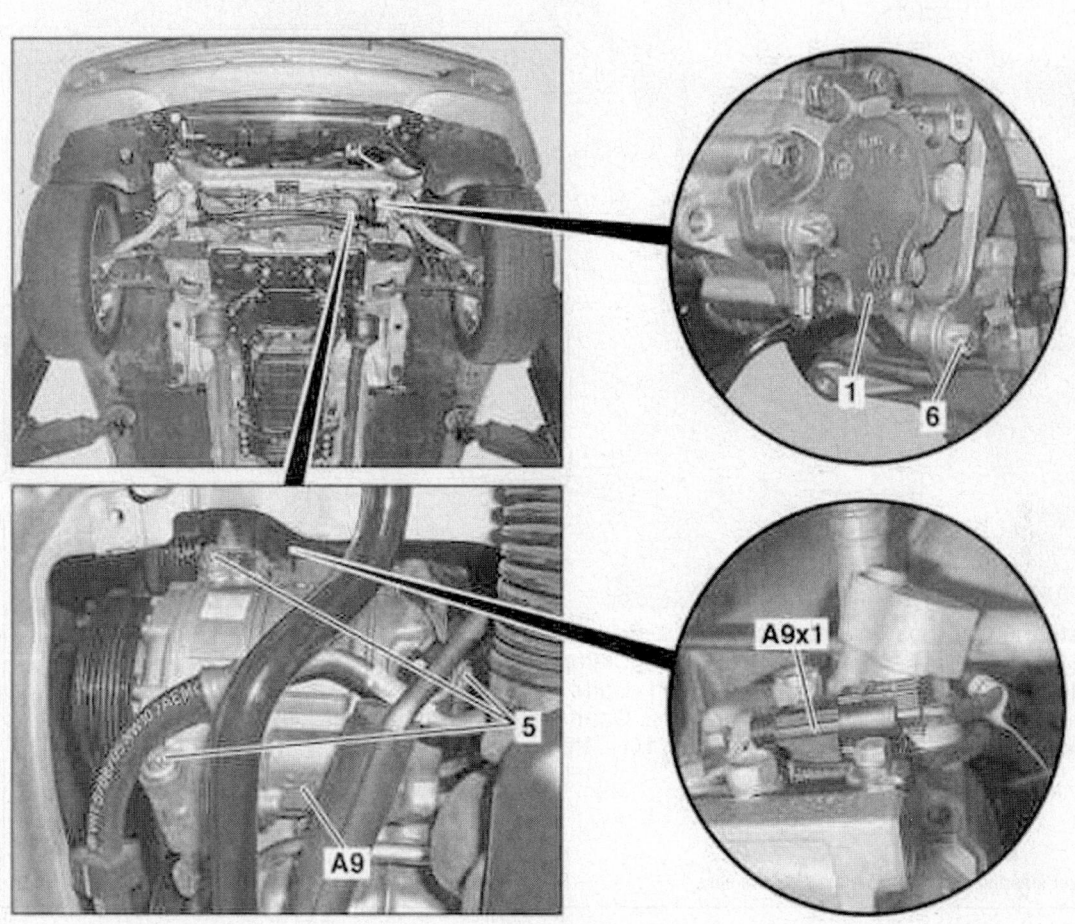

1. Power steering pump    A9. Refrigerant compressor
5. Bolts                    A9x1. Refrigerant compressor connector
6. Screw

22205_MBCA_G0064

**Fig. 150 Power steering pump mounting**

13. Detach power steering pump from timing case cover; to do so, remove bolts.

14. Raise and safely support the vehicle.

15. Remove bolts.

16. Detach refrigerant compressor connector and remove refrigerant compressor from timing case cover and engine support. Place refrigerant compressor with pipe system connected aside.

17. Detach ground line from power steering pump by unscrewing bolt.

18. Remove screw.

19. Lower vehicle.

20. Remove power steering pump.

**To install:**

21. Installation is the reverse of removal.

22. Tighten bolts/nuts to specification as follows:

- Bolt, power steering pump to timing case cover: 26 ft. lbs. (35 Nm)
- Hose clamp, hydraulic line to power steering pump expansion reservoir: 2 ft. lbs. (3 Nm)
- Bolt, ground line to power steering pump: 15 ft. lbs. (20 Nm)
- Banjo bolt, power steering pressure line to power steering pump: 33 ft. lbs. (45 Nm)
- Bolt, power steering expansion reservoir to front cover at cylinder head: 7 ft. lbs. (9 Nm)
- Bolt, refrigerant compressor to timing case cover or engine support: 15 ft. lbs. (20 Nm)

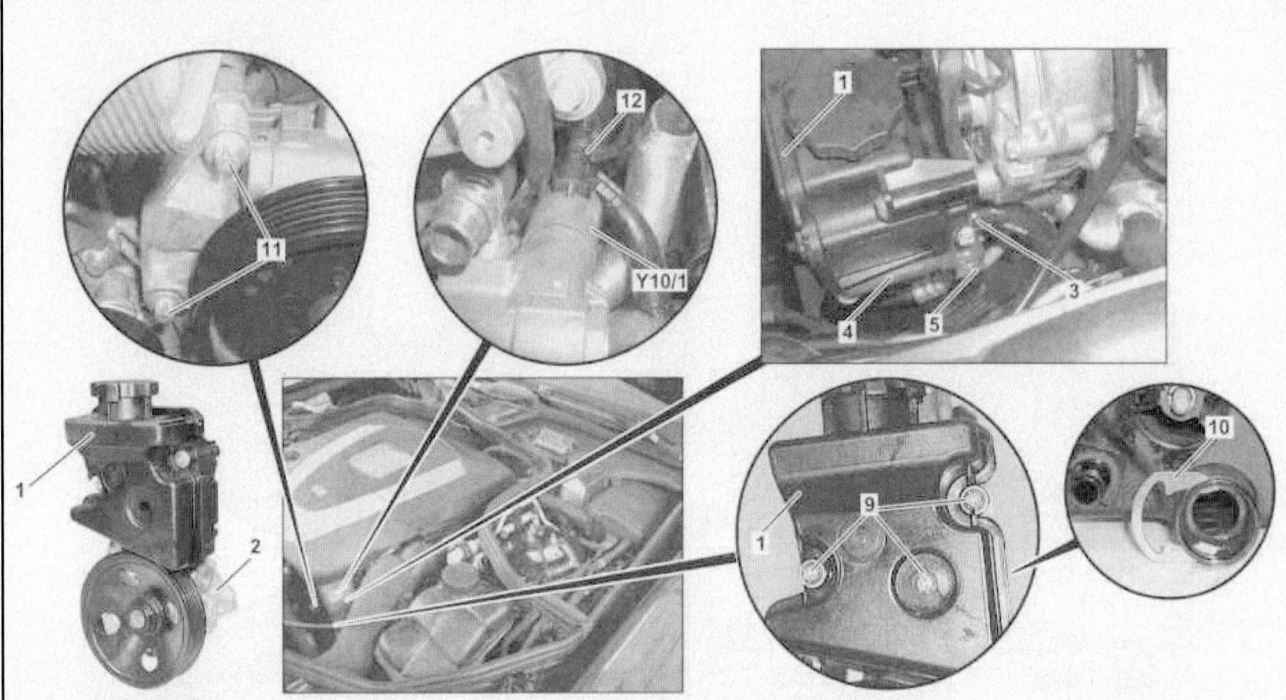

1. Power steering expansion reservoir
2. Power steering pump
3. Return flow pipe
4. Power steering pressure line
5. Banjo bolt
9. Bolts
10. Clamp
11. Bolts
12. Connector
Y10/1. Power steering pump pressure regulator valve

22205_MBCA_G0258

**Fig. 151 Power steering reservoir and high pressure hose**

23. Fill and bleed power steering pump.
24. Check function of power steering and for leaks.

### S-Class

*See Figure 153.*

1. Before servicing the vehicle, refer to the precautions.

> **⁕⁕ WARNING**
>
> **Before opening the hydraulic system of the power steering, thoroughly clean the area surrounding the separation point. Even minute particles of dirt, introduced into the hydraulic components, can result in malfunctions or damage to the power steering.**

2. Remove air filter housing.
3. Remove left-hand air filter housing.
4. Remove poly-V belt.
5. Remove expansion reservoir .

6. Remove windshield washer fluid reservoir.
7. Detach pressure line from the power steering pump .
8. Disconnect the electrical connector at the pressure regulator valve.
9. Detach power steering pump from timing case cover by unscrewing the bolts.
10. Detach ground line from power steering pump.
11. Remove screw.
12. Remove power steering pump.

**To install:**

13. Installation is the reverse of removal.
14. Tighten bolts/nuts to specification as follows:

- Bolt, power steering pump to timing case cover M8 x 30: 15 ft. lbs. (20 Nm)
- Bolt, power steering pump to timing case cover M8 x 34: 26 ft. lbs. (35 Nm)

- Bolt, power steering pump to timing case cover M8 x 44: 26 ft. lbs. (35 Nm)
- Bolt, ground line to power steering pump: 15 ft. lbs. (20 Nm)
- Banjo bolt, power steering pressure line to power steering pump: 30 ft. lbs. (40 Nm)

15. Fill and bleed power steering pump.
16. Check function of power steering and for leaks.

### BLEEDING

1. Before servicing the vehicle, refer to the precautions.

> **⁕⁕ WARNING**
>
> **Do not start the engine! If the engine is started there is a risk of air entering the hydraulic steering system, which is then extremely difficult to remove.**

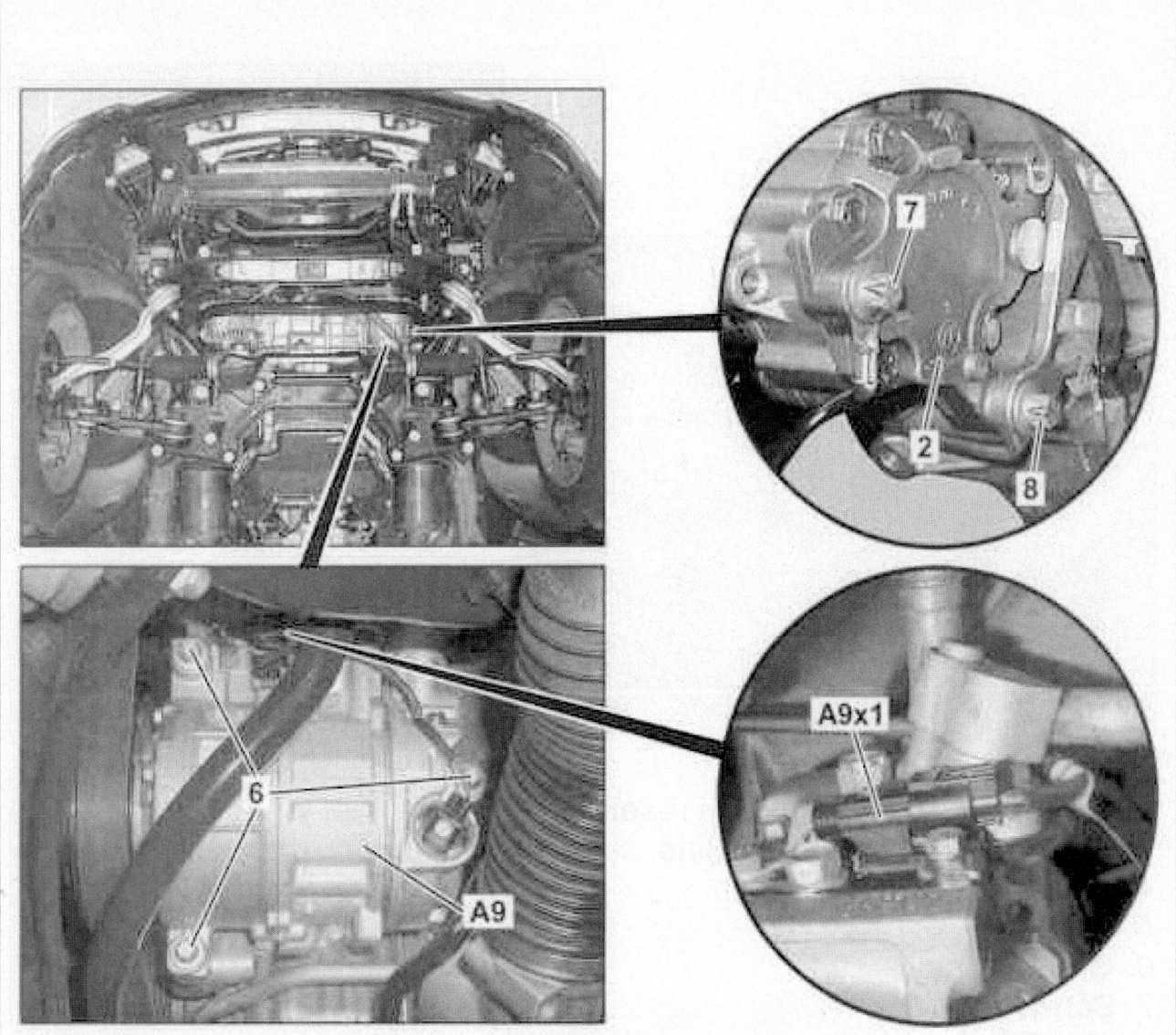

2. Power steering pump
6. Bolts
7. Screw

8. Screw
A9. Refrigerant compressor
A9x1. Refrigerant compressor connector

22205_MBCA_G0259

**Fig. 152 Power steering pump mounting**

2. Fill the expansion reservoir with MB steering gear oil up to approximately 10 mm below the top edge of the reservoir. Add MB steering gear oil continuously, until the fluid level remains constant.

3. Slowly turn steering wheel from steering stop to steering stop until no more bubbles are ascertained in the expansion reservoir. Repeat turning the steering from stop to stop several times (up to 30 times). During this process

MB steering gear oil must be added to the expansion reservoir by an assistant.

4. Start the engine and allow engine to run for approximately 1 min at idle speed. During this process an assistant must observe the steering gear oil level in the expansion reservoir and add MB steering gear oil if necessary.

5. When the engine is running, turn the steering wheel slowly several times from

steering stop to steering stop, in the meantime refilling the expansion reservoir with MB steering gear oil.

6. Repeat process until the steering gear oil level remains constant in the expansion reservoir and there are no more bubbles.

7. Check oil level in expansion reservoir and fill to maximum fill level.

8. Check steering system for leaks.

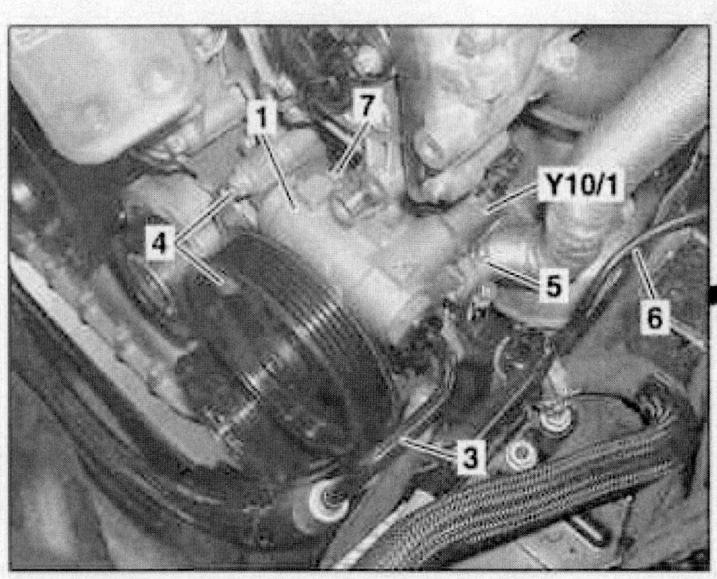

1. Power steering pump
2. Power steering expansion reservoir
3. Power steering pressure line
4. Bolts
5. Screw
6. Ground cable
7. Screw
Y10/1. Power steering pump pressure regulator valve

22205_MBCA_G0260

Fig. 153 Power steering pump mounting

## COIL SPRING

### REMOVAL & INSTALLATION

#### C-Class and E-Class

*See Figure 154.*

1. Remove front spring strut.
2. Clamp spring compressor in a vise and install clamping plates.
3. Install suspension strut in spring compressor.
4. Clamp spring until the load is taken off the upper spring cup.
5. Unscrew slotted nut.
6. Remove ball bearing.
7. Remove suspension strut.
8. Release spring compressor and remove spring and upper spring cup.
9. Check suspension strut.

**To assemble:**

10. Lay the spring in the tensioning device.
11. Tension spring.
12. Insert suspension strut with the lower rubber mount and stop buffer.
13. Mount the upper rubber mount, ball bearing and upper support bearing.
14. Attach slotted nut adequately and tighten nut to piston rod to 15 ft. lbs. (20 Nm).
15. Release spring and remove spring strut from spring compressor Install.
16. Install front strut assembly.
17. Perform chassis alignment.

## LOWER BALL JOINT

### REMOVAL & INSTALLATION

The lower ball joint is an integral part of the lower control arm and cannot be serviced separately.

## LOWER CONTROL ARM

### REMOVAL AND & INSTALLATION

#### C-Class

*Except 4MATIC*

*See Figure 155.*

These vehicles use a cross strut (lower control arm) and a torque strut. This procedure covers the service of both struts.

1. Before servicing the vehicle, refer to the precautions.
2. Remove front wheel.
3. Unclip brake hose bracket from suspension strut.

4. Unplug wheel speed sensor connector in wheelhouse.
5. Remove left or right front brake pad contact sensor on brake caliper.
6. Remove brake caliper and tie to one side.

➡**Do not detach brake line.**

7. Detach stabilizer bar connecting rod from suspension strut. Use Allen wrench to

apply counter pressure to the ball joint as required.
8. Detach cross strut from the front axle carrier.
9. Detach the stabilizer bar from the front axle carrier.
10. Detach torque strut from front axle carrier.
11. Remove track rod from steering knuckle assembly.

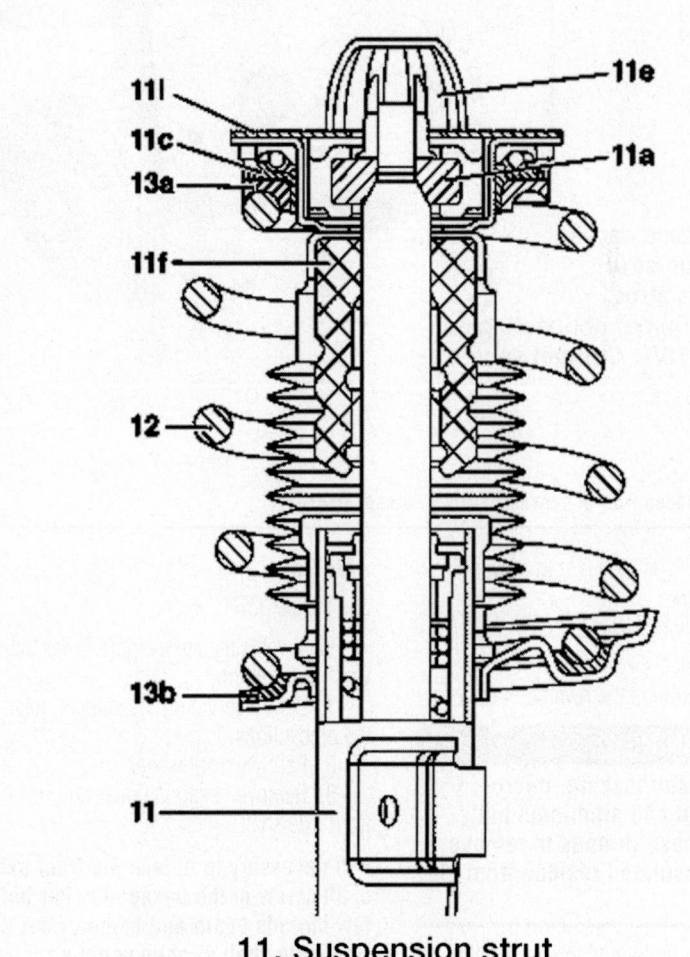

11. Suspension strut
11a. Grooved nut
11c. Ball bearing
11e. Cap
11f. Stop buffer
11i. Support bearing
12. Spring
13a. Upper rubber mount
13b. Lower rubber mount

22205_MBCA_G0065

**Fig. 154 Exploded view of the MacPherson strut assembly**

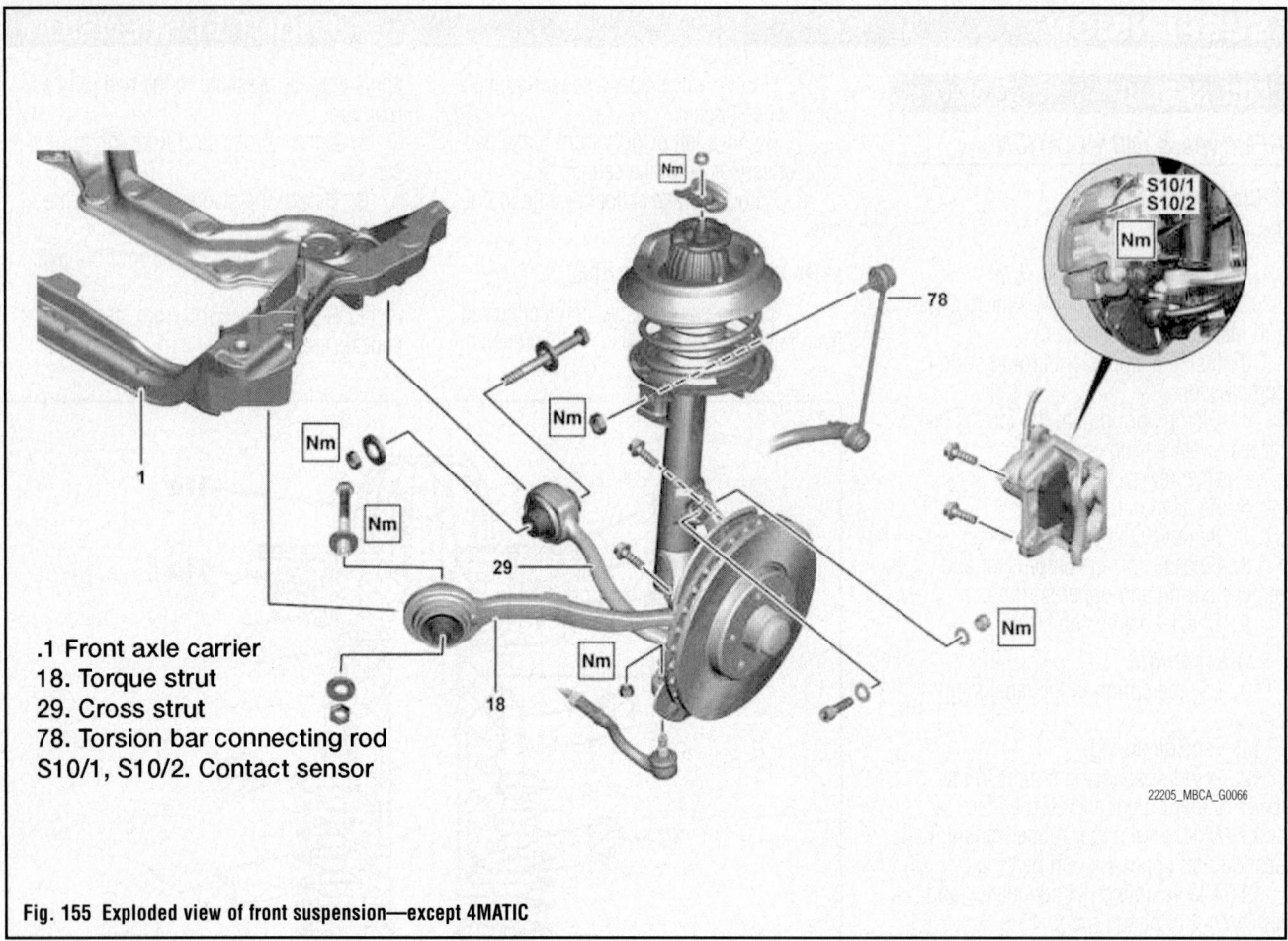

.1 Front axle carrier
18. Torque strut
29. Cross strut
78. Torsion bar connecting rod
S10/1, S10/2. Contact sensor

22205_MBCA_G0066

**Fig. 155 Exploded view of front suspension—except 4MATIC**

12. Secure front-axle half to prevent it from falling down.

13. Unscrew rebound stop.

***To install:***

14. Installation is the reverse of removal.

### ❋❋ WARNING

**Replace all self locking, micro-encapsulated and aluminum bolts and nuts. Chase threads to remove micro-encapsulated residue from old bolts/nuts.**

15. Tighten bolts/nuts to specification as follows:

- Self-locking nut, connecting rod to spring strut: 44 ft. lbs. (60 Nm)
- Nut of tie rod to steering knuckle: 37 ft. lbs. (50 Nm) plus an additional 60° rotation
- Self-locking nut for cross strut on front axle carrier: 59 ft. lbs. (80 Nm) plus an additional 120° rotation
- Self-locking bolt, torque strut to front axle carrier: 59 ft. lbs. (80 Nm) plus an additional 120° rotation

16. Perform a wheel alignment check.

***4MATIC***

*See Figure 156.*

These vehicles use a single lower transverse control arm.

1. Before servicing the vehicle, refer to the precautions.

2. Remove front wheel.

3. Remove hexagon collar bolt from front axle shaft.

➡**If necessary to detach the front axle shaft screw in the hexagon collar bolt a few threads again and hammer out the front axle shaft slightly using a rubber hammer.**

4. Detach front and rear lower engine compartment paneling.

5. Detach linkage for headlamp range adjustment on lower transverse control arm.

6. Guide the brake line from the bracket on the strut assembly.

7. Disconnect the connectors for the brake pad contact sensor wear indicator and the rpm sensor in the wheelhouse and unclip the lines.

➡**To detach, press in the spring (arrow). The plug for the brake pad**

wear sensor is present on the right-hand side only.

8. Detach stabilizer bar link rod from suspension strut. If necessary, counterhold ball head bolt using Allen wrench.

9. Press the tie rod joint from the stub axle unit.

10. Detach brake caliper and tie up to one side with the brake line connected.

11. Installing protective washers to protect brake disc dust shields.

12. Detach lower control arm from front axle carrier. Swivel the track rod upward to install and remove the rear bolt. During this process the steering must be turned all the way to the left for access at the right lower control arm, and all the way to the right for the left lower control arm.

13. Observe installation position if repair bolts have already been installed. Install repair bolts and washers in the same position again.

14. Pull the steering knuckle with the lower transverse control arm outwards from the bottom and move the front axle shaft out of the front axle shaft flange.

15. Support front-axle half.

16. Detach top of strut from strut dome.

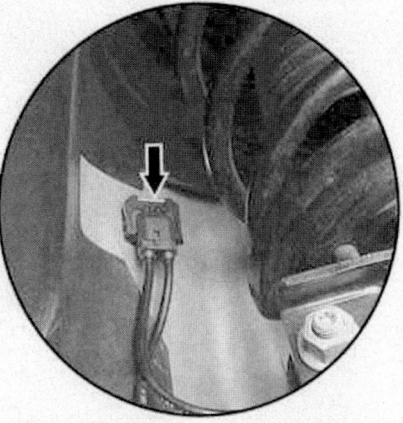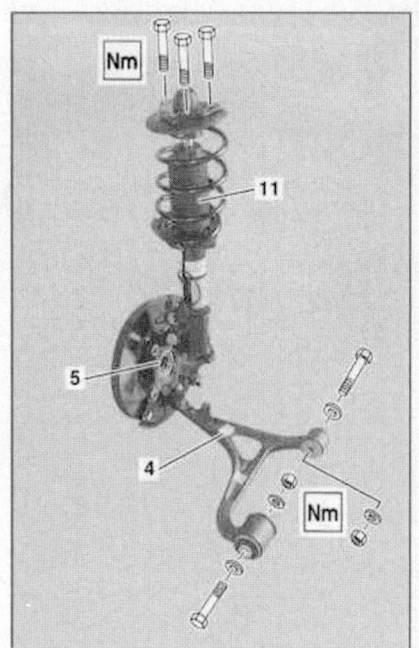

4. Transverse control arms
5. Steering knuckle
10a. Tie rod joint
11. Suspension strut
23. Stabilizer bar link rod
33. Brake caliper
60. Front axle shaft
61. A Hexagon collar bolt
Arrow: Spring

22205_MBCA_G0067

**Fig. 156 Exploded view of front suspension—4MATIC**

17. Remove front-axle half Checking.

18. Check lower transverse control arm and steering knuckle for damage or deformation and replace if necessary.

19. Check suspension strut for leaks and damage and replace if necessary.

*To install:*

20. Installation is the reverse of removal.

### ✳ WARNING

**Replace all self locking, micro-encapsulated and aluminum bolts and nuts. Chase threads to remove micro-encapsulated residue from old bolts/nuts.**

21. Tighten bolts/nuts to specification as follows:
- Self-locking nut, connecting rod to spring strut: 44 ft. lbs. (60 Nm)
- Self-locking nut, transverse control arm to frame: 81 ft. lbs. (110 Nm)
- Collar bolt, front axle shaft to front axle shaft: 81 ft. lbs. (110 Nm) plus an additional 60° rotation
- Self-locking nut, tie rod to steering knuckle: 52 ft. lbs. (70 Nm)

➡**Do not final tighten nuts until vehicle is at normal ride height.**

22. Perform a wheel alignment check

### E-Class

#### *Except 4MATIC*

*See Figure 157.*

1. Before servicing the vehicle, refer to the precautions.

### ✳ WARNING

**When working on the front end assembly, be careful not to scratch, mark or notch the surfaces of aluminum parts. Otherwise the service life of the parts will be affected.**

2. Remove front wheel.

3. Remove rear part section of bottom part of noise encapsulation.

4. Remove rear section of lower engine compartment paneling.

5. Remove left front rpm sensor or right front rpm sensor from steering knuckle and hang up to one side.

6. Detach suspension strut from spring control arm.

7. Remove link rod of stabilizer bar from stabilizer bar pivot bushing.

8. Detach spring control arm from front-axle carrier.

9. Press track control arm from stub axle assembly.

10. Check rubber bushing of spring control arm for signs of damage.

11. Check stabilizer bar pivot bushing in spring control arm for signs of damage.

12. Check supporting joint in steering knuckle for signs of damage.

*To install:*

13. Installation is the reverse of removal.

➡**Tighten nuts and bolts of chassis components only when vehicle is at ride height.**

14. Tighten bolts/nuts to specification as follows:
- Bolt and nut, front suspension strut to spring control arm: 122 ft. lbs. (165 Nm)
- Self-locking nut, spring control arm to front axle carrier: 59 ft. lbs. (80 Nm) plus an additional 120° rotation

15. Perform chassis alignment.

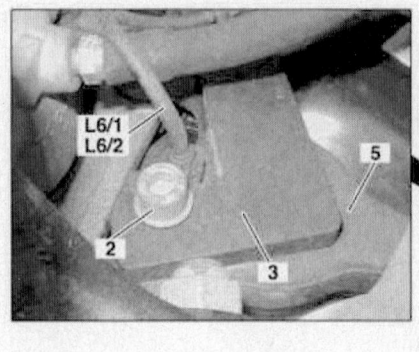
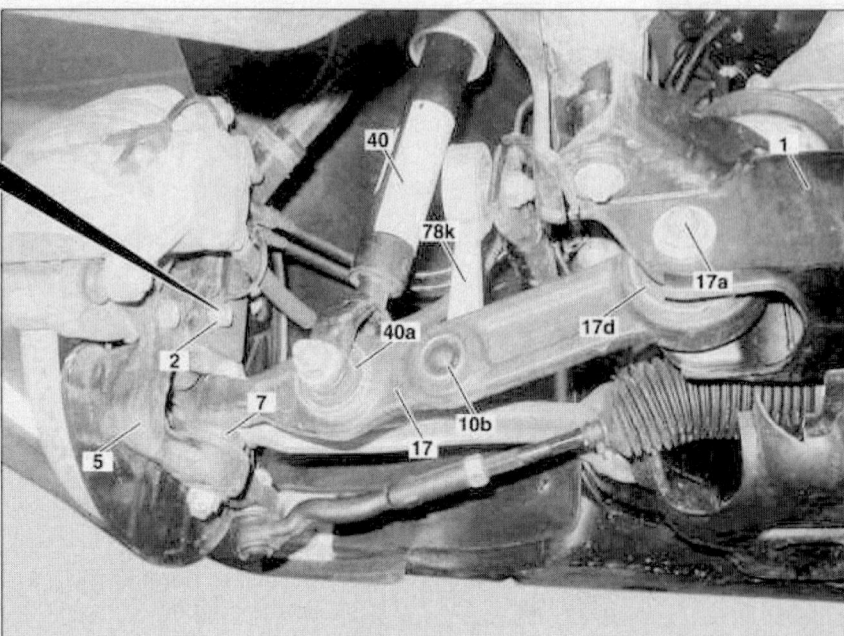

1. Front axle carrier
2. Bolt
3. Scuff protection
5. Steering knuckle
7. Support joint
10b. Stabilizer bar pivot bushing
17. Spring control arm
17a. Bolt
17d. Rubber bushing
40. Suspension strut
40a. Rubber bushing
78k. Link rod
L6/1. Left front rpm sensor
L6/2. Right front rpm sensor

22205_MBCA_G0264

**Fig. 157 Lower control arm—except 4MATIC**

16. Check headlamp setting and correct if necessary.

### 4MATIC

*See Figure 158.*

1. Before servicing the vehicle, refer to the precautions.

> ✳✳ **WARNING**
>
> When working on the front end assembly, be careful not to scratch, mark or notch the surfaces of aluminum parts. Otherwise the service life of the parts will be affected.

2. Remove air filter housing with air ducts.

3. Remove front wheel .

4. Detach hexagon collar bolt of front axle shaft.

5. Detach front axle brake caliper from steering knuckle and hang brake caliper to one side.

➡ The brake hose does not have to be unscrewed from the brake caliper.

6. Pry supporting joint off steering knuckle.

7. Press track rod joint out of steering knuckle.

8. Pry follower joint of upper transverse control arm off steering knuckle.

> ✳✳ **WARNING**
>
> The follower joint must be removed, otherwise the follower joint is damaged by hard pulling or excessive angles.

9. Detach link rod from stabilizer bar.

10. Move steering knuckle out of supporting joint.

11. Remove steering knuckle and move the right front axle shaft or left front axle shaft.

12. From the front axle shaft flange.

➡ To loosen the front axle shaft, if necessary screw in the hexagon collar bolt a few threads again and hammer out the front axle shaft slightly using a rubber hammer.

13. Detach lower engine compartment paneling.

14. Remove stabilizer bar on front axle.

15. Loosen front shock-absorber strut on lower transverse control arm and secure.

16. Detach lower transverse control arm from front axle carrier.

> ✳✳ **WARNING**
>
> The bolt must not be twisted. It is essential to counterhold the bolt when loosening the nut. Otherwise, when the repair kit is fitted, the fixing lugs for camber adjustment and caster adjustment in the rubber mount are sheared off.

*To install:*

17. Installation is the reverse of removal.

> ✳✳ **WARNING**
>
> Replace all self locking, micro-encapsulated and aluminum bolts and nuts. Chase threads to remove micro-encapsulated residue from old bolts/nuts.

 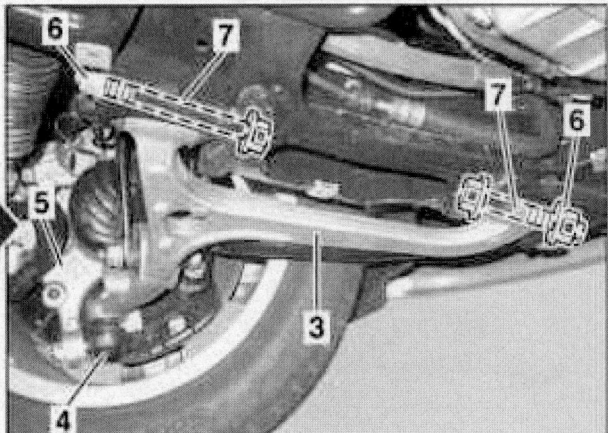

1. Stabilizer bar
2. Shock-absorber strut
3. Lower transverse control arm
4. Support joint

5. Steering knuckle
6. Nut
7. Bolt

22205_MBCA_G0265

Fig. 158 Lower control arm—except 4MATIC

➡**Tighten nuts and bolts of chassis components only when vehicle is at ride height.**

18. Tighten bolts/nuts to specification as follows:

- Collar bolt, front axle shaft to front axle shaft: 52 ft. lbs. (70 Nm) plus an additional 90° rotation
- Self-locking nut, lower wishbone to front axle carrier: 81 ft. lbs. (110 Nm)
- Self-locking nut, supporting joint to steering: 52 ft. lbs. (70 Nm) plus an additional 50° rotation
- Self-locking nut, upper wishbone follower joint to steering knuckle: 15 ft. lbs. (20 Nm) plus an additional 90° rotation

19. Perform chassis alignment.
20. Check headlamp setting and correct if necessary.

**S-Class**

**Except 4MATIC**

*See Figure 159.*

1. Before servicing the vehicle, refer to the precautions.

**✲✲ WARNING**

**When working on the front end assembly, be careful not to scratch, mark or notch the surfaces of aluminum parts. Otherwise the service life of the parts will be affected.**

2. Remove front wheel.
3. Remove engine compartment paneling.
4. Loosen nut. Do not twist bolt. Always counterhold the bolt when slackening the nut, otherwise the lock tabs for camber adjustment in the rubber mount will be sheared off.
5. Open retaining plate and push upwards.
6. Remove bolts from suspension strut.
7. Using a suitable lever, loosen spring strut at ball joint.
8. Detach spring control arm from steering knuckle.

**✲✲ WARNING**

**Support the steering knuckle using the transmission jack. The upper ball joint on the upper transverse control arm will otherwise be damaged.**

9. Pull bolt out of front axle carrier.
10. Remove spring control arm downwards Checking.
11. Check rubber mount in spring control arm for damage.
12. Check rubber boots and ball joints in spring control arm for damage.

*To install:*

13. Installation is the reverse of removal.
14. Screw nut of bolt all the way on until it abuts, but do not tighten yet so that the rubber mount can twist when the vehicle is rocked to settle the suspension. Do not tighten nut until in ready-to-drive condition.

**✲✲ WARNING**

**Replace all self locking, micro-encapsulated and aluminum bolts and nuts. Chase threads to remove micro-encapsulated residue from old bolts/nuts.**

➡**Tighten nuts and bolts of chassis components only when vehicle is at ride height.**

15. Tighten bolts/nuts to specification as follows:

- Self-locking nut, spring control arm to front axle carrier Hexagon: 59 ft.

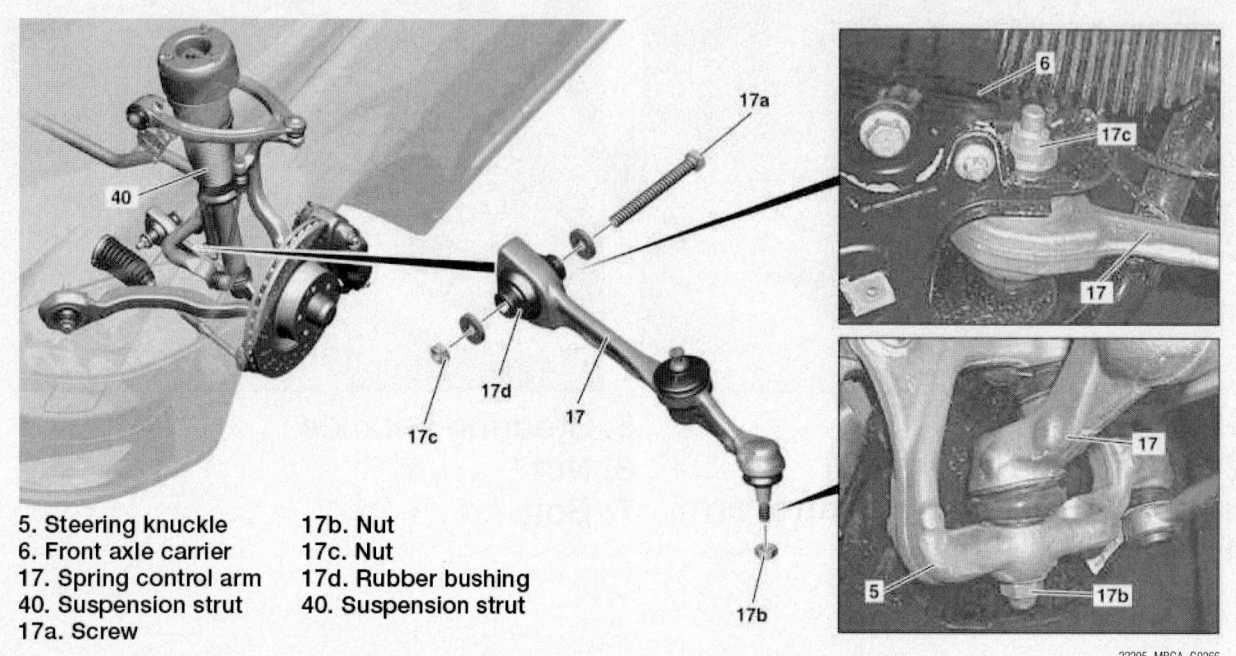

**5. Steering knuckle**
**6. Front axle carrier**
**17. Spring control arm**
**40. Suspension strut**
**17a. Screw**

**17b. Nut**
**17c. Nut**
**17d. Rubber bushing**
**40. Suspension strut**

22205_MBCA_G0266

**Fig. 159 Exploded view of the lower control arm—except 4MATIC**

---

lbs. (80 Nm) plus an additional 150° rotation
- Self-locking nut, spring control arm to front axle carrier Torx: 59 ft. lbs. (80 Nm) plus an additional 180° rotation
- Bolt, front suspension strut to spring control arm/ball joint: 15 ft. lbs. (20 Nm)

16. Perform a wheel alignment check.

### 4MATIC

*See Figure 160.*

1. Before servicing the vehicle, refer to the precautions.

### ✳✳ WARNING

**When working on the front end assembly, be careful not to scratch, mark or notch the surfaces of aluminum parts. Otherwise the service life of the parts will be affected.**

2. Remove wheels.
3. Remove engine compartment paneling.

4. Unscrew nut and press suspension strut out of spring control arm.
5. Loosen nut.

### ✳✳ WARNING

**Do not twist bolt. Always counterhold the bolt when slackening the nut, otherwise the lock tabs for camber adjustment in the rubber mount will be sheared off.**

6. Detach spring control arm from steering knuckle and remove downward.
7. Check rubber mount in spring control arm for damage.
8. Check rubber boot and ball joint in spring control arm for damage.

#### To install:

9. Installation is the reverse of removal.
10. Screw nut of bolt all the way on until it abuts, but do not tighten yet so that the rubber mount can twist when the vehicle is rocked to settle the suspension. Do not tighten nut until in ready-to-drive condition.

### ✳✳ WARNING

**Replace all self locking, micro-encapsulated and aluminum bolts and nuts. Chase threads to remove micro-encapsulated residue from old bolts/nuts.**

➡ Tighten nuts and bolts of chassis components only when vehicle is at ride height.

11. Tighten bolts/nuts to specification as follows:
- Self-locking nut, spring control arm to front axle carrier Hexagon: 59 ft. lbs. (80 Nm) plus an additional 150° rotation
- Self-locking nut, spring control arm to front axle carrier Torx: 59 ft. lbs. (80 Nm) plus an additional 180° rotation
- Self-locking nut, track control arm to stub axle: 37 ft. lbs. (50 Nm) plus an additional 90° rotation

12. Perform a wheel alignment check.

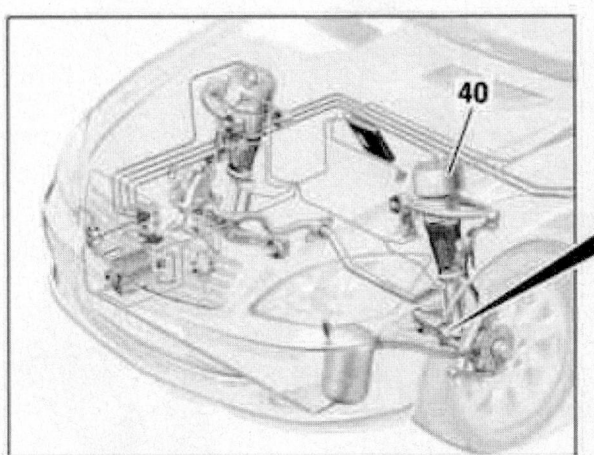

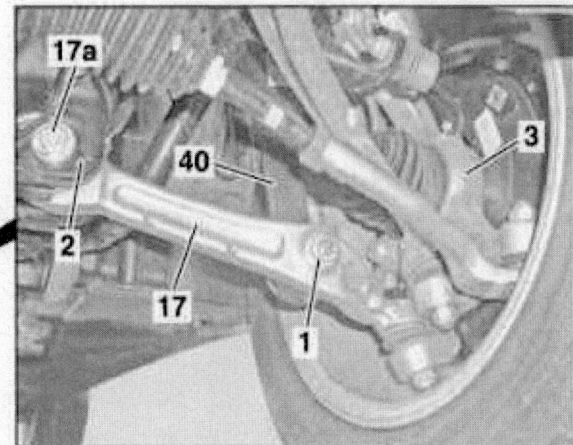

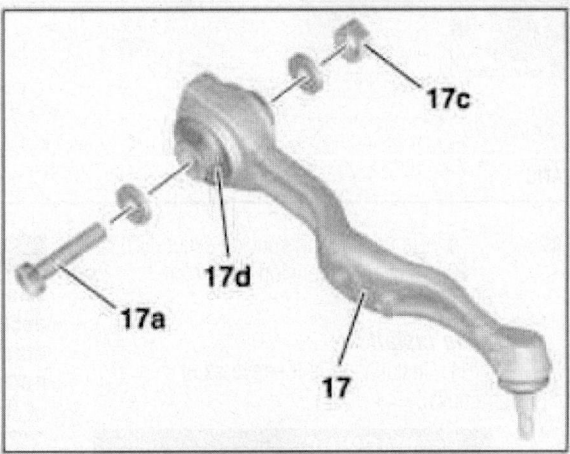

1. Nut
2. Front axle carrier
3. Steering knuckle
17. Spring control arm

17a. Screw
17c. Nut
17d. Rubber bushing
40. Suspension strut

22205_MBCA_G0267

**Fig. 160 Exploded view of the lower control arm—4MATIC**

### MACPHERSON STRUT

*REMOVAL & INSTALLATION*

#### C-Class

##### Except 4MATIC

*See Figure 161.*

1. Before servicing the vehicle, refer to the precautions.
2. Remove the front wheels.

3. Unclip bracket for brake hose (arrow) on suspension strut.
4. Remove cable ties. Push bracket (arrow) on brake hose along as far as possible upwards and fasten.
5. Detach stabilizer bar link rod from suspension strut.
6. Detach suspension strut from wheel carrier.
7. Unscrew nut for rebound stop on suspension strut tower.

8. Remove suspension strut.

*To install:*

9. Installation is the reverse of removal.

### ✴✴ WARNING

**Replace all self locking, micro-encapsulated and aluminum bolts and nuts. Chase threads to remove micro-encapsulated residue from old bolts/nuts.**

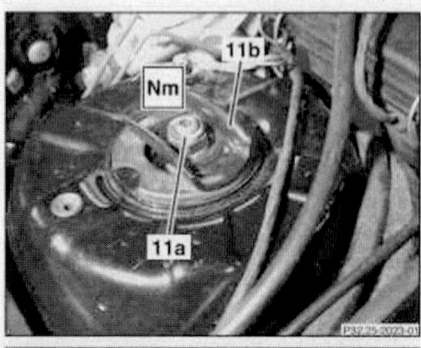

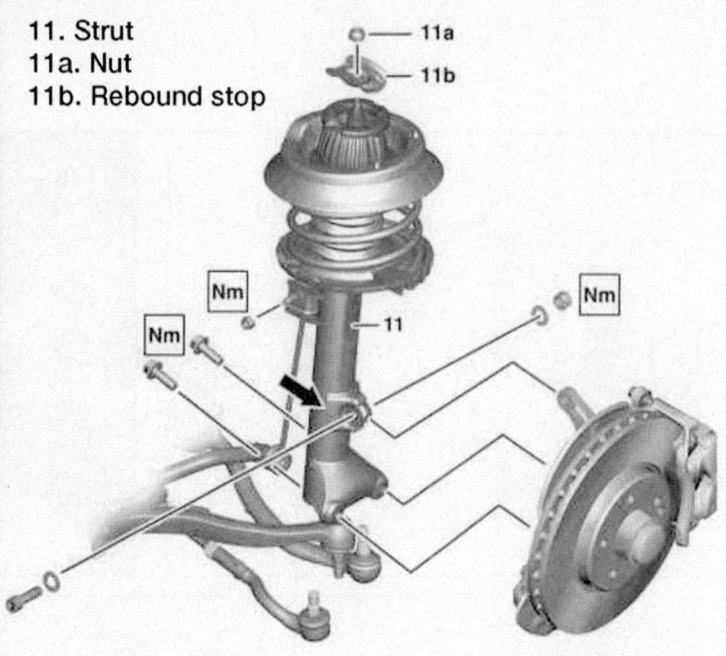

11. Strut
11a. Nut
11b. Rebound stop

Fig. 161 Front strut assembly—except 4MATIC

10. Tighten bolts/nuts to specification as follows:

- Nut, rebound stop to piston rod: 44 ft. lbs. (60 Nm)
- Self-locking bolt, suspension strut to steering knuckle at bottom: 81 ft. lbs. (110 Nm)
- Self-locking nut, suspension strut to steering knuckle at top: 74 ft. lbs. (100 Nm) plus an additional 90° rotation

11. Perform a wheel alignment check

### 4MATIC

*See Figure 162.*

1. Before servicing the vehicle, refer to the precautions.
2. Remove the front wheels, rotate if necessary.
3. Unclip rpm sensor wire from suspension strut.
4. Detach bracket from line for rpm sensor/brake pad contact sensor/wear indicator from suspension strut.
5. Guide the brake line from the bracket on the strut assembly.
6. Detach brake caliper.
7. Detach stabilizer bar link rod from suspension strut.
8. Detach suspension strut from steering knuckle.

9. Tie up steering knuckle securely.
10. Detach suspension strut from dome.

#### To install:

11. Installation is the reverse of removal.

### ❊❊ WARNING

**Replace all self locking, micro-encapsulated and aluminum bolts and nuts. Chase threads to remove micro-encapsulated residue from old bolts/nuts.**

12. Tighten bolts/nuts to specification as follows:

- Self-locking bolt, suspension strut to steering knuckle at bottom: 81 ft. lbs. (110 Nm)
- Self-locking nut, suspension strut to steering knuckle at top: 74 ft. lbs. (100 Nm) plus an additional 90° rotation

13. Perform a wheel alignment check.

### E-Class

#### Except 4MATIC

*See Figure 163.*

1. Before servicing the vehicle, refer to the precautions.

### ❊❊ WARNING

**When working on the front end assembly, be careful not to scratch, mark or notch the surfaces of aluminum parts. Otherwise the service life of the parts will be affected.**

2. Position wheels straight ahead.
3. Remove front wheels.
4. Unscrew nuts.
5. Detach suspension strut from spring control arm.
6. Press spring control arm downwards and remove spring strut towards the rear.

#### To install:

7. Installation is the reverse of removal.

### ❊❊ WARNING

**Replace all self locking, micro-encapsulated and aluminum bolts and nuts. Chase threads to remove micro-encapsulated residue from old bolts/nuts.**

➡Tighten nuts and bolts of chassis components only when vehicle is at ride height.

8. Tighten bolts/nuts to specification as follows:

- Nut, front suspension strut to front end: 22 ft. lbs. (30 Nm)

11. Suspension strut
23. Stabilizer bar link rod
33. Brake caliper

22205_MBCA_G0069

Fig. 162 Front strut assembly—4MATIC

7. Strut
17. Lower control arm

22205_MBCA_G0268

Fig. 163 Front strut assembly—except 4MATIC

- Bolt, front suspension strut to spring control arm: 122 ft. lbs. (165 Nm)

9. Check headlamp setting and correct if necessary.

### 4MATIC
*See Figure 164.*

1. Before servicing the vehicle, refer to the precautions.

> ※※ **WARNING**
>
> **When working on the front end assembly, be careful not to scratch, mark or notch the surfaces of aluminum parts. Otherwise the service life of the parts will be affected.**

2. Remove front axle shaft.
3. Unscrew nuts.
4. Press transverse control arm downwards and remove and remove suspension strut.

**To install:**
5. Installation is the reverse of removal.

> ※※ **WARNING**
>
> **Replace all self locking, micro-encapsulated and aluminum bolts and nuts. Chase threads to remove micro-encapsulated residue from old bolts/nuts.**

➡️ **Tighten nuts and bolts of chassis components only when vehicle is at ride height.**

6. Tighten bolts/nuts to specification as follows:
- Nut, front suspension strut to front end: 22 ft. lbs. (30 Nm)

7. Check headlamp setting and correct if necessary.

### S-Class

#### Except 4MATIC
*See Figure 165.*

1. Before servicing the vehicle, refer to the precautions.

> ※※ **WARNING**
>
> **When working on the front end assembly, be careful not to scratch, mark or notch the surfaces of aluminum parts. Otherwise the service life of the parts will be affected.**

2. Place front wheels in straight-ahead position.
3. Loosen nuts on front end.

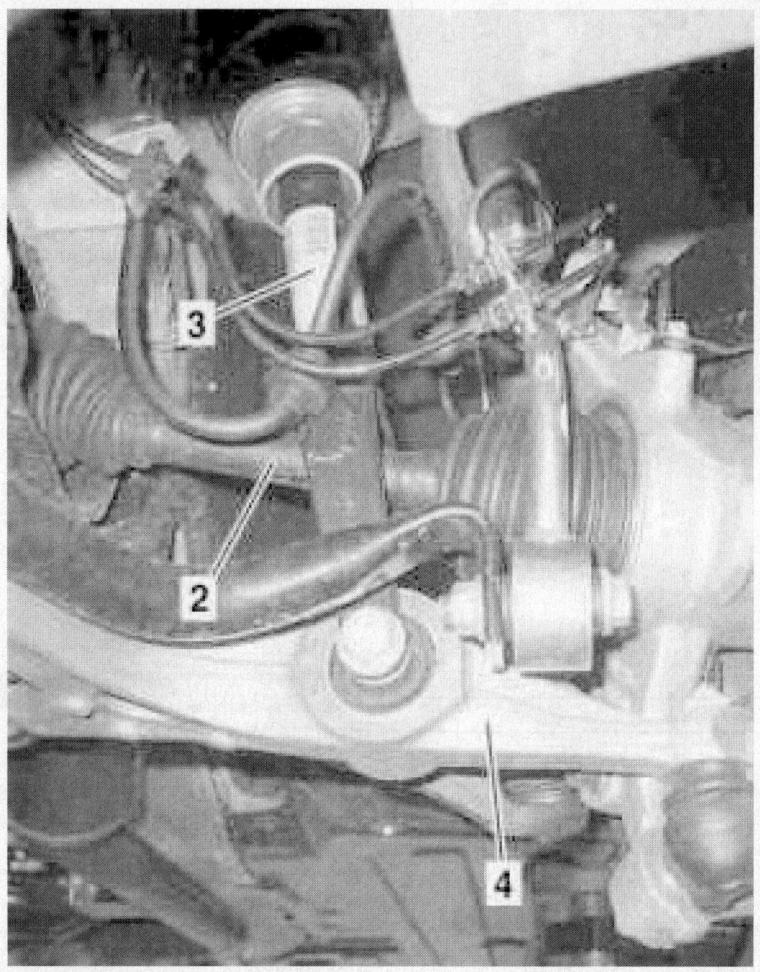

1. Nuts
2. Front axle shaft
3. Suspension strut
4. Transverse control arm

22205_MBCA_G0269

Fig. 164 Front strut assembly—4MATIC

4. Disconnect electrical connector in wheel well.

5. Unscrew nut of stabilizer bar link rod from wheel carrier and press out stabilizer bar link rod.

6. Detach line bracket from suspension strut.

7. Remove bolts from spring strut.

8. Loosen suspension strut at ball joint of spring control arm using a suitable lever.

9. Unscrew nuts, press wheel carrier downwards and remove suspension strut downwards to the side.

10. Check suspension strut.

**To install:**

11. Installation is the reverse of removal.

**⁑ WARNING**

**Replace all self locking, micro-encapsulated and aluminum bolts and nuts. Chase threads to remove micro-encapsulated residue from old bolts/nuts.**

➡ **Tighten nuts and bolts of chassis components only when vehicle is at ride height.**

12. Tighten bolts/nuts to specification as follows:
- Nut, front suspension strut to front end: 22 ft. lbs. (30 Nm)
- Bolt, front suspension strut to spring control arm/ball joint: 15 ft. lbs. (20 Nm)

### 4MATIC

*See Figure 166.*

1. Before servicing the vehicle, refer to the precautions.

**⁑ WARNING**

**When working on the front end assembly, be careful not to scratch, mark or notch the surfaces of aluminum parts. Otherwise the service life of the parts will be affected.**

2. Unscrew nuts.

3. Remove front wheel.

4. Disconnect electrical line of front axle damping valve unit from front axle electrical distributor connector.

5. Unclip brake hose and electrical line from bracket.

6. Unscrew nut and press suspension strut out of spring control arm.

7. Remove suspension strut downward toward rear.

8. Check suspension strut.

**To install:**

9. Installation is the reverse of removal.

**⁑ WARNING**

**Replace all self locking, micro-encapsulated and aluminum bolts and nuts. Chase threads to remove micro-encapsulated residue from old bolts/nuts.**

➡ **Tighten nuts and bolts of chassis components only when vehicle is at ride height.**

10. Tighten bolts/nuts to specification as follows:
- Nut, front suspension strut to front end: 26 ft. lbs. (35 Nm)
- Nut, front suspension strut to spring control arm: 81 ft. lbs. (110 Nm)

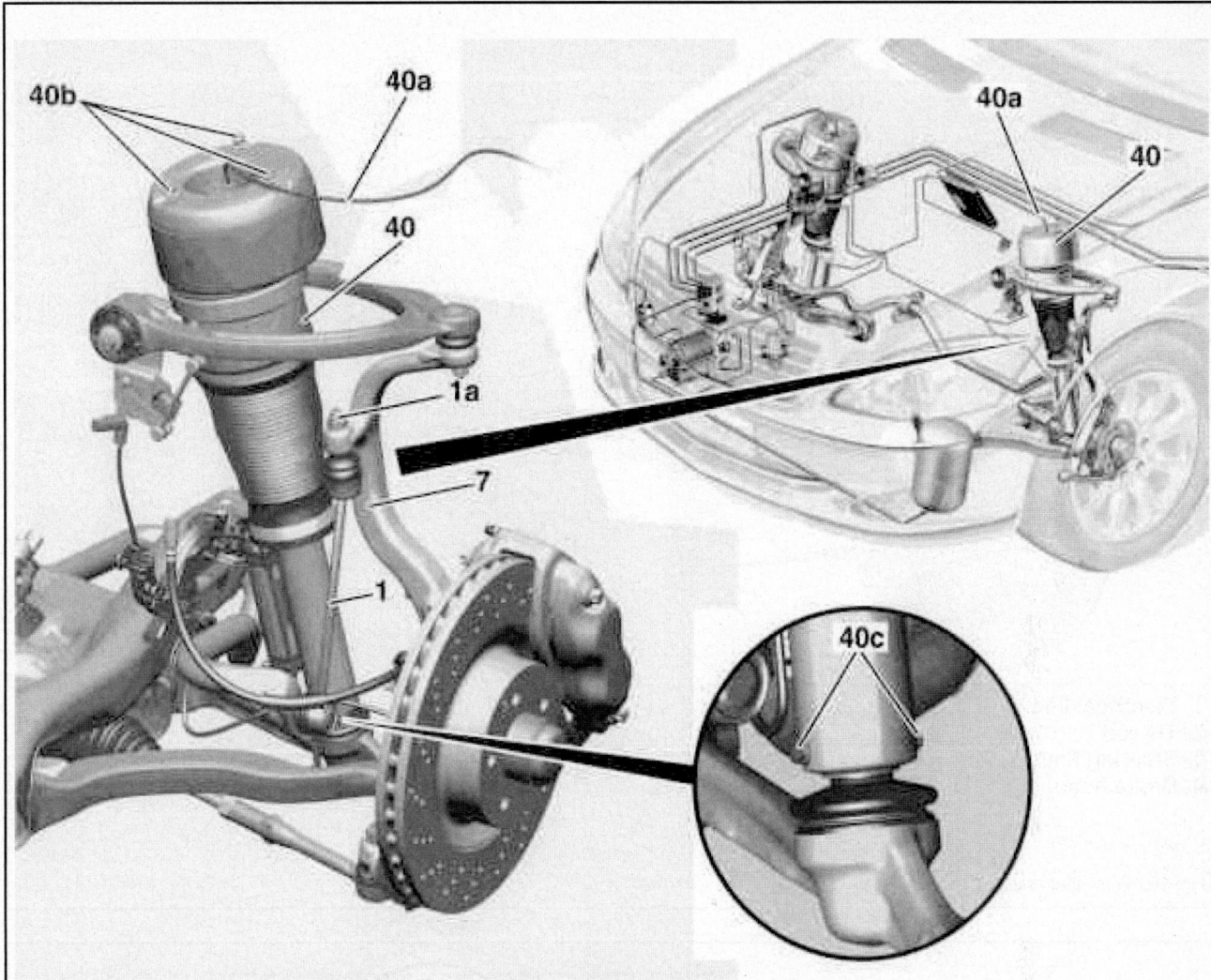

1. Stabilizer bar link rod
1a. Nut
7. Wheel carrier
40. Suspension strut

40a. High pressure line
40b. Nut
40c. Screw

22205_MBCA_G0270

**Fig. 165 Front strut assembly—except 4MATIC**

## STABILIZER BAR

*REMOVAL & INSTALLATION*

### C-Class

*Except 4MATIC*

*See Figures 167 and 168.*

1. Before servicing the vehicle, refer to the precautions.
2. Remove under floor panels.
3. Remove diagonal struts from front axle carrier.
4. Remove stiffening plate.
5. Detach stabilizer bar from the link rods.

6. Detach stabilizer bar from front axle carrier.
7. Detach rubber mount of stabilizer bar and check.

*To install:*

8. Installation is the reverse of removal.

### ❋❋ WARNING

**Replace all self locking, micro-encapsulated and aluminum bolts and nuts. Chase threads to remove micro-encapsulated residue from old bolts/nuts.**

9. Tighten bolts/nuts to specification as follows:

- Self-locking nut of connecting rod to stabilizer bar: 44 ft. lbs. (60 Nm)
- Bolt, stabilizer bar retaining bracket to front axle carrier: 30 ft. lbs. (40 Nm)

*4MATIC*

*See Figure 169.*

1. Before servicing the vehicle, refer to the precautions.
2. Remove the front wheels.
3. Take off bottom engine compartment panel at front.

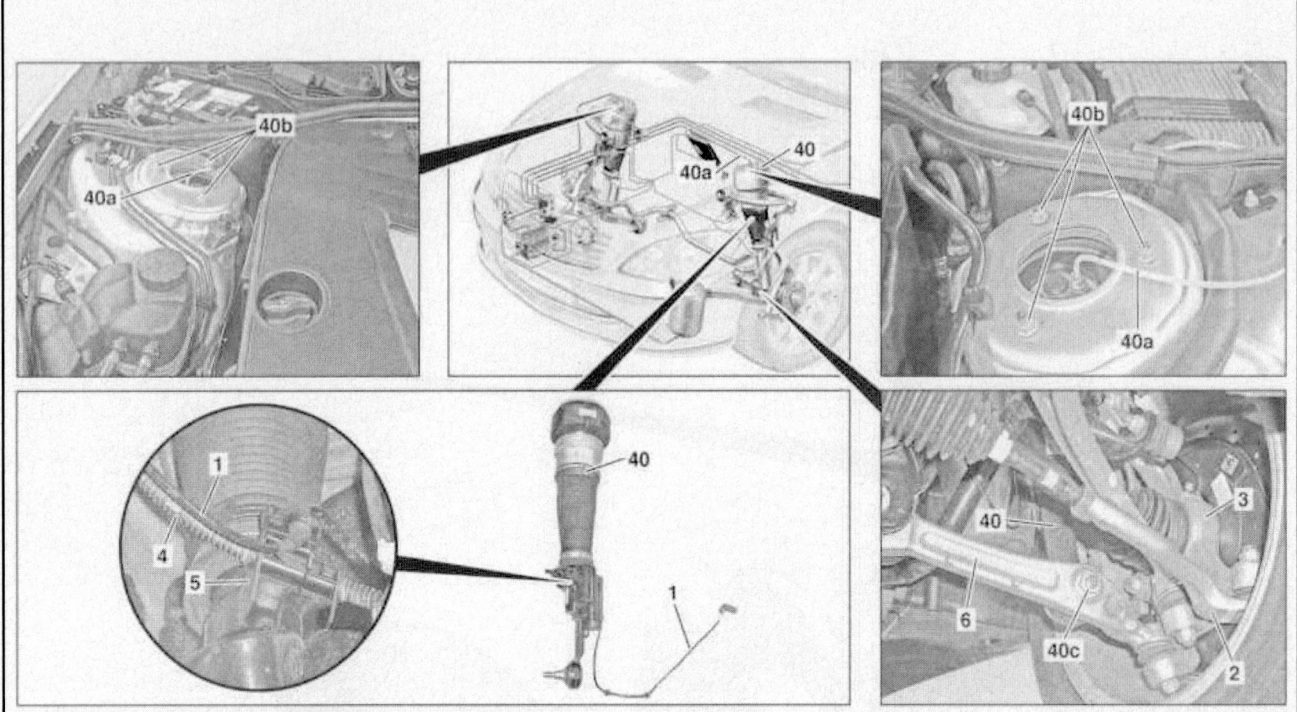

1. Electrical line
2. Tie rod
3. Steering knuckle
4. Brake hose
5. Bracket
6. Spring control arm
40. Suspension strut
40a. High pressure line
40b. Nuts
40c. Nut

22205_MBCA_G0271

**Fig. 166 Front strut assembly—4MATIC**

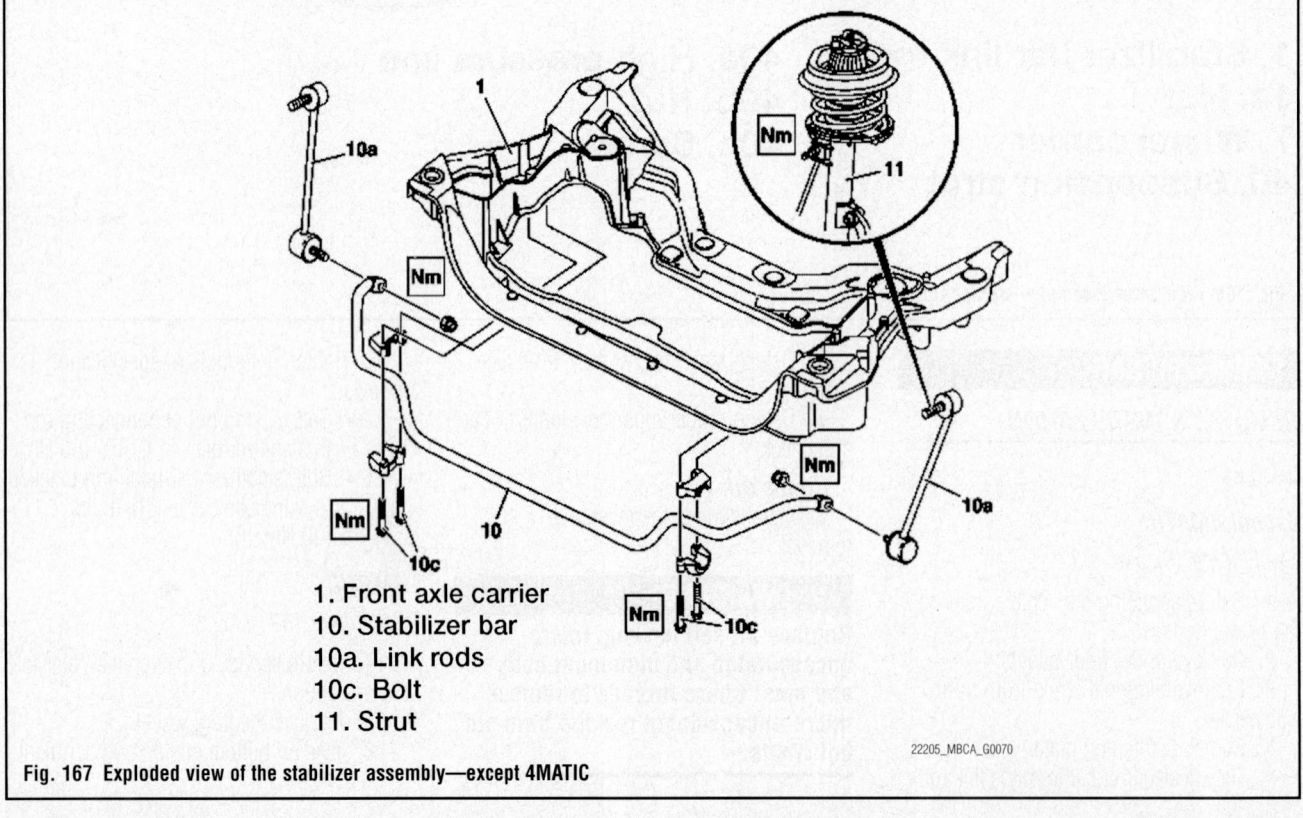

1. Front axle carrier
10. Stabilizer bar
10a. Link rods
10c. Bolt
11. Strut

22205_MBCA_G0070

**Fig. 167 Exploded view of the stabilizer assembly—except 4MATIC**

2. Bolt
3. Stiffening plate
4. Bolt

5. Bolt
6. Struts

22205_MBCA_G0071

**Fig. 168  Diagonal struts on the front axle carrier**

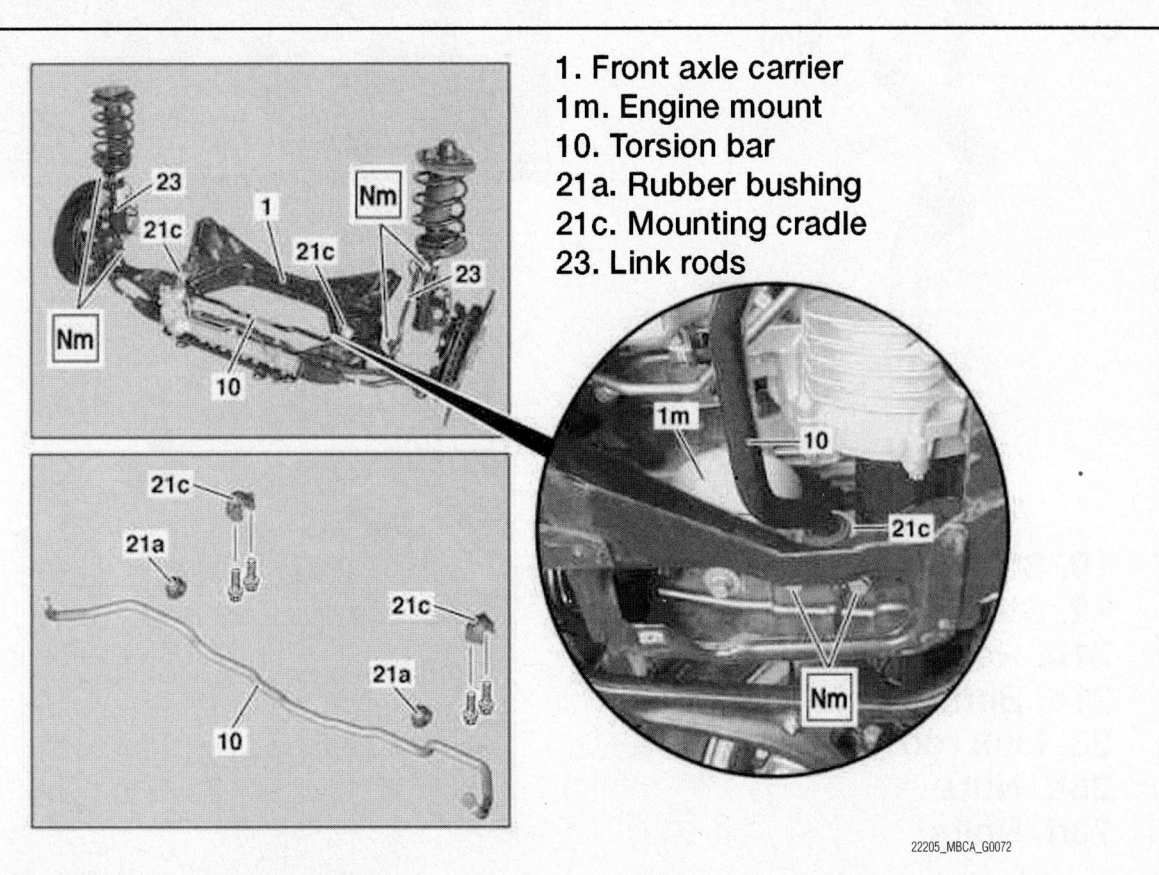

1. Front axle carrier
1m. Engine mount
10. Torsion bar
21a. Rubber bushing
21c. Mounting cradle
23. Link rods

22205_MBCA_G0072

**Fig. 169  Exploded view of the stabilizer assembly—4MATIC**

4. Detach linkage for front axle sensor (headlamp range adjustment) on right lower transverse control arm.

5. Detach front axle sensor (headlamp range adjustment) on longitudinal member and attach securely to the body.

6. Remove right front axle shaft.

7. Remove left front axle shaft.

8. Remove left and right engine mount.

9. Unclip left brake hose from brake hose clip on left suspension strut.

10. Raise engine with the engine support bracket.

11. Remove the stabilizer bar from the link rod.

12. Remove stabilizer bar from front axle carrier.

13. Remove stabilizer bar.

14. Detach fastening bracket from the.

15. Rubber mounts.

16. Detach rubber mounts.

17. Checking.

18. Check rubber mounts and replace if necessary.

### To install:

19. Installation is the reverse of removal.

### ❊❊ WARNING

**Replace all self locking, micro-encapsulated and aluminum bolts and nuts. Chase threads to remove micro-encapsulated residue from old bolts/nuts.**

20. Tighten bolts/nuts to specification as follows:

- Self-locking nut of connecting rod to stabilizer bar: 41 ft. lbs. (55 Nm)
- Bolt of stabilizer bar retaining bracket to front axle carrier: 19 ft. lbs. (25 Nm)

### E-Class

### Except 4MATIC

*See Figure 170.*

1. Before servicing the vehicle, refer to the precautions.

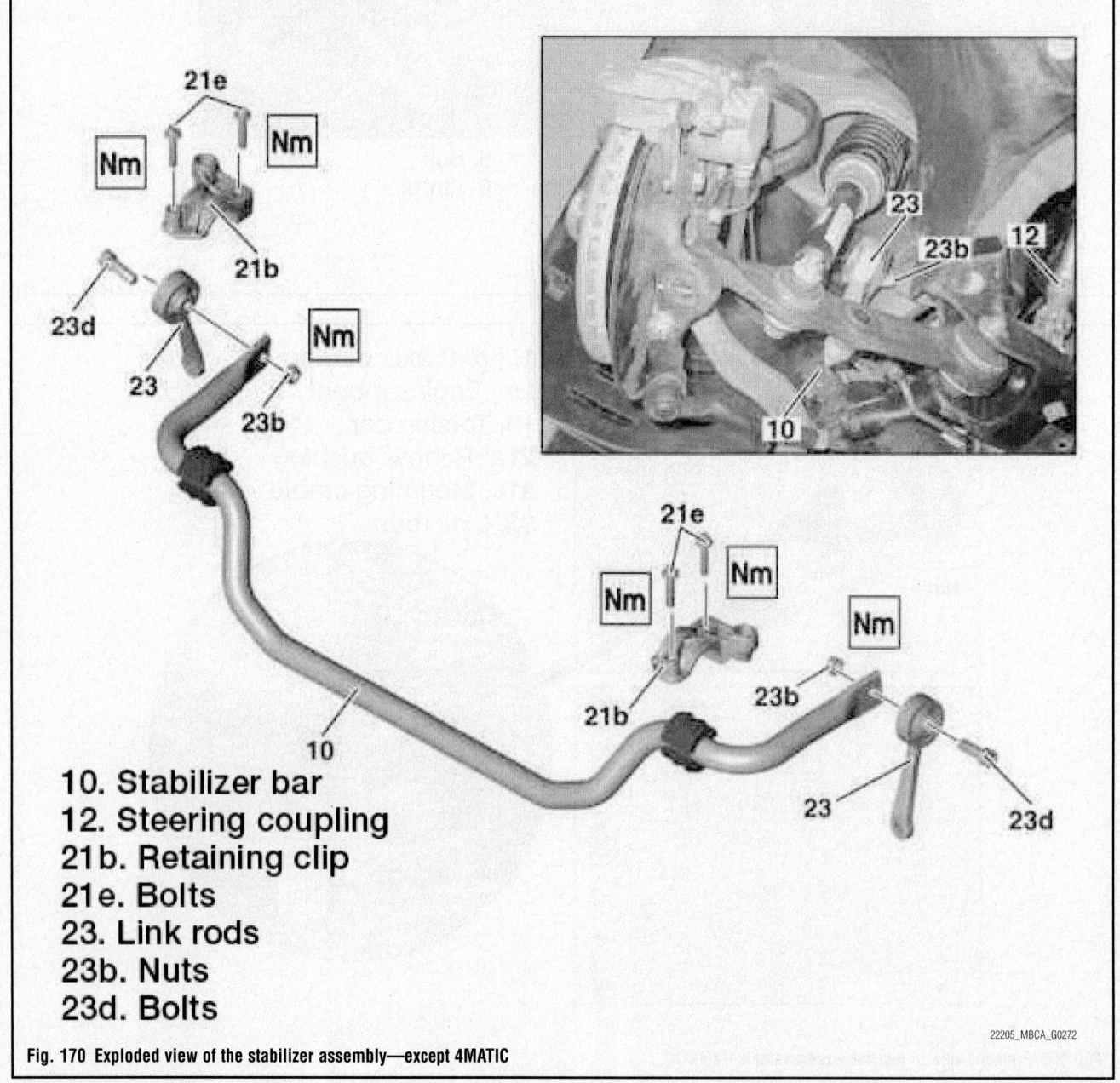

10. Stabilizer bar
12. Steering coupling
21b. Retaining clip
21e. Bolts
23. Link rods
23b. Nuts
23d. Bolts

22205_MBCA_G0272

**Fig. 170 Exploded view of the stabilizer assembly—except 4MATIC**

**⁕⁕ WARNING**

When working on the front-end assembly, be careful not to scratch, mark or notch the surfaces of aluminum parts. Otherwise the service life of the parts will be affected.

2. Remove trim panels of the cylinder head cover and of the charge air manifold.

3. Remove air filter housing with air ducts.

4. Remove engine trim panel.

5. Attach engine support frame.

6. Remove bottom section of soundproofing.

7. Remove engine compartment paneling.

8. Remove bolt of left and right front engine mount from front axle carrier.

9. Raise engine.

10. Unscrew nuts and remove link rods from stabilizer bar.

11. Support front axle carrier with transmission jack and transmission platform.

12. Remove bolts from front axle carrier.

13. Lower front axle carrier.

14. Remove bolts and remove retaining bracket.

15. Remove stabilizer bar.

**To install:**

16. Installation is the reverse of removal.

17. The retaining brackets on the stabilizer bar (10) must be replaced.

**⁕⁕ WARNING**

Replace all self locking, micro-encapsulated and aluminum bolts and nuts. Chase threads to remove micro-encapsulated residue from old bolts/nuts.

➡Only tighten nuts and bolts of chassis components when vehicle is at normal ride height.

18. Tighten bolts/nuts to specification as follows:

- Self-locking nut of connecting rod to stabilizer bar: 96 ft. lbs. (130 Nm)
- Bolt, retaining bracket to longitudinal member: 37 ft. lbs. (50 Nm)
- Bolt, retaining bracket to front end: 37 ft. lbs. (50 Nm)
- Bolt, front axle carrier to longitudinal member: 74 ft. lbs. (100 Nm)

19. Perform a wheel alignment check

### 4MATIC

See Figure 171.

1. Before servicing the vehicle, refer to the precautions.

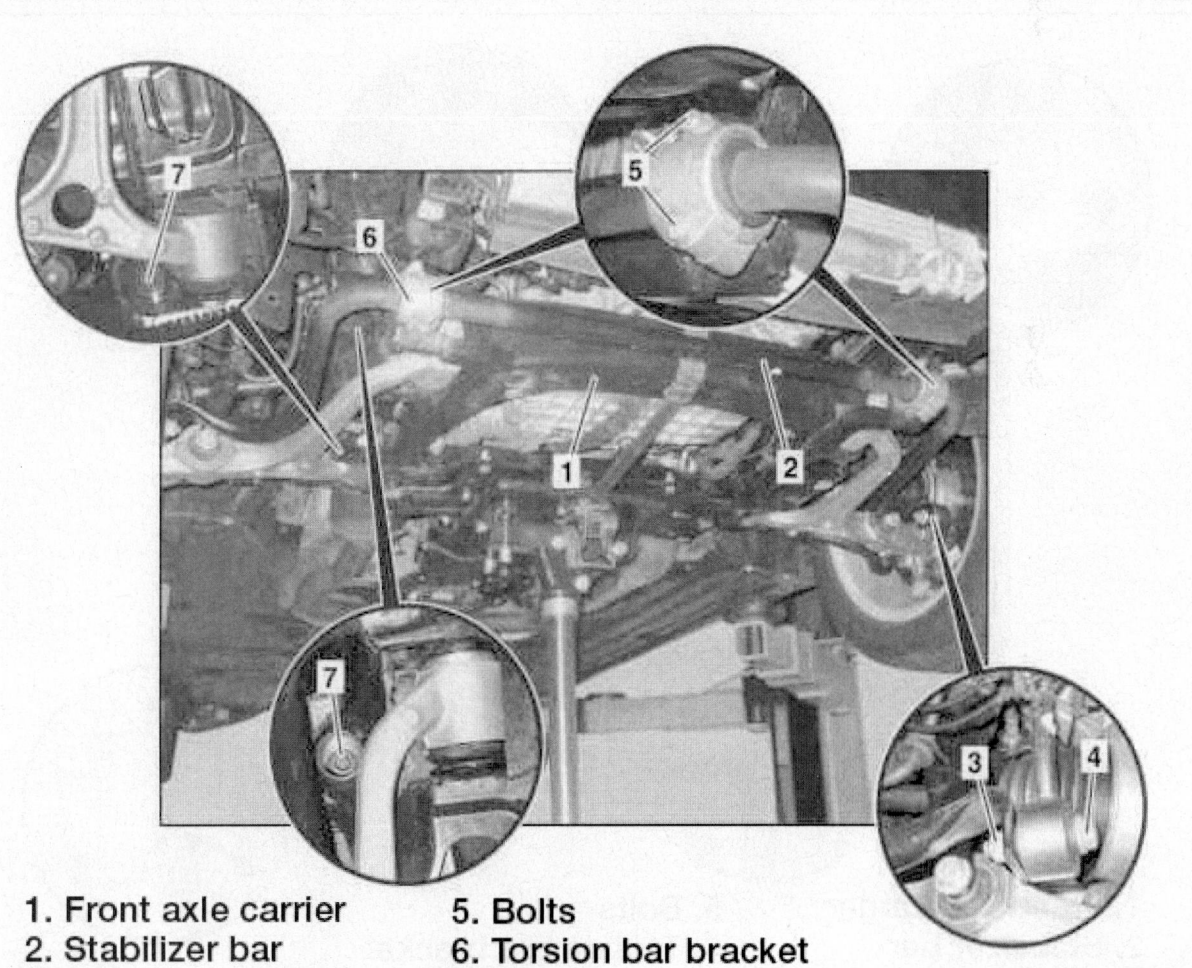

1. Front axle carrier
2. Stabilizer bar
3. Nut
4. Screw
5. Bolts
6. Torsion bar bracket
7. Bolts

22205_MBCA_G0273

Fig. 171 Exploded view of the stabilizer assembly—4MATIC

**❋❋ WARNING**

**When working on the front-end assembly, be careful not to scratch, mark or notch the surfaces of aluminum parts. Otherwise the service life of the parts will be affected.**

2. Remove air filter housing with air ducts.

3. Remove cover over fuse box on left suspension strut.

4. Mount engine hoist.

5. Remove lower engine compartment paneling.

6. Remove soundproofing.

7. Remove bolts of left and right front engine mounts from the front axle carrier.

8. Raise engine.

9. Remove nut and remove screw.

10. Support front subframe.

11. Remove bolts from front axle carrier.

12. Lower the front axle carrier until the bolts are accessible.

13. Remove bolts and remove stabilizer bar holder.

14. Remove stabilizer bar.

*To install:*

15. Installation is the reverse of removal.

**❋❋ WARNING**

**Replace all self locking, micro-encapsulated and aluminum bolts and nuts. Chase threads to remove micro-encapsulated residue from old bolts/nuts.**

➡ **Only tighten nuts and bolts of chassis components when vehicle is at normal ride height.**

16. Tighten bolts/nuts to specification as follows:
- Self-locking nut of connecting rod to torsion bar: 70 ft. lbs. (95 Nm)
- Bolt, torsion bar to torsion bar bracket: 16 ft. lbs. (22 Nm)
- Bolt, front axle carrier to longitudinal member: 89 ft. lbs. (120 Nm)

**S-Class**

*See Figure 172.*

1. Before servicing the vehicle, refer to the precautions.

**❋❋ WARNING**

**When working on the front-end assembly, be careful not to scratch, mark or notch the surfaces of aluminum parts. Otherwise the service life of the parts will be affected.**

2. Remove wheels.

3. Remove bottom engine compartment paneling.

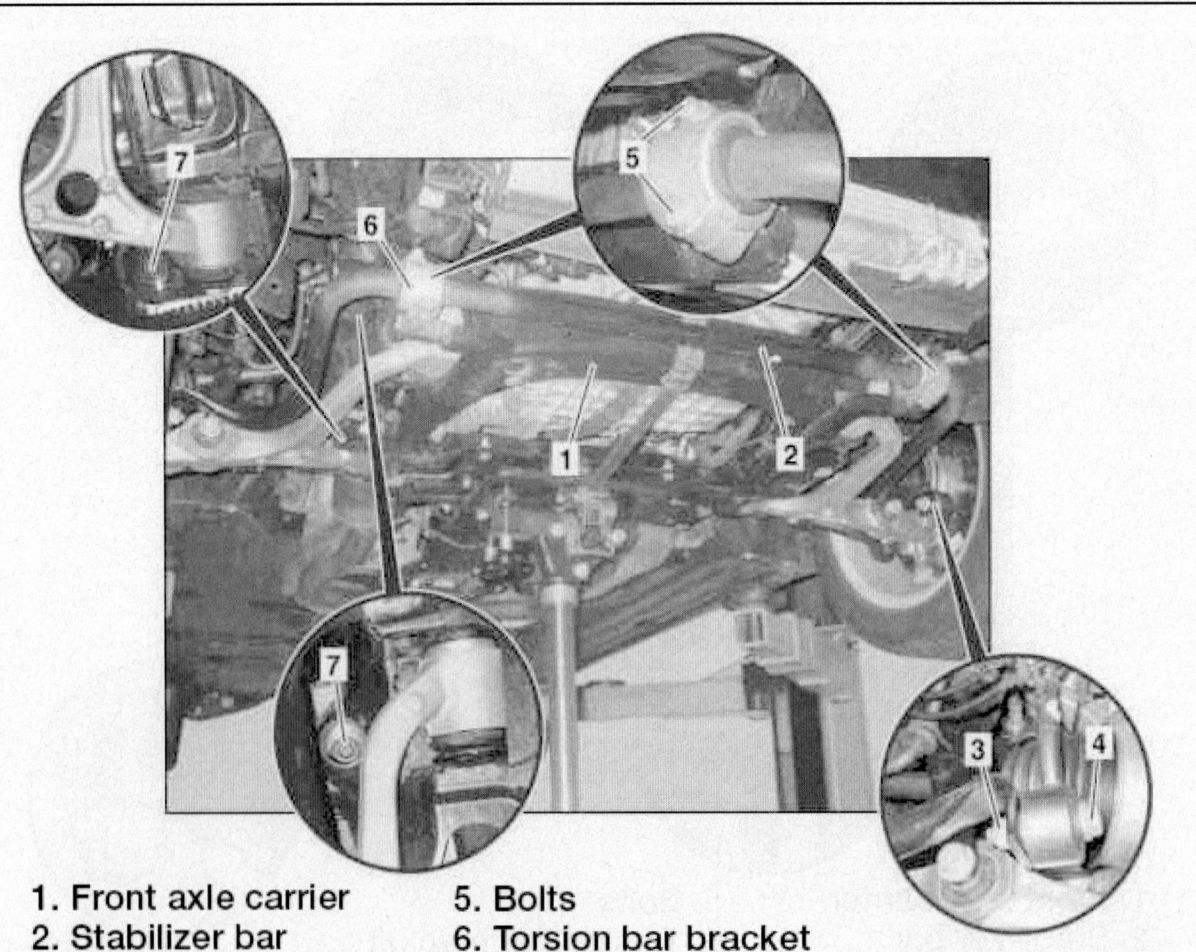

1. Front axle carrier
2. Stabilizer bar
3. Nut
4. Screw
5. Bolts
6. Torsion bar bracket
7. Bolts

22205_MBCA_G0273

**Fig. 172 Exploded view of the stabilizer assembly**

4. Unscrew nuts of link rods to stabilizer bar and press link rods out of stabilizer bar.

5. Remove bolts and remove retaining bracket.

6. Unclip lines from brackets.

7. Detach left or right suspension strut from spring control arm.

8. Detach left or right spring control arm from front axle carrier.

9. Remove stabilizer bar toward rear.

*To install:*

10. Installation is the reverse of removal.

### ✳✳ WARNING

**Replace all self locking, micro-encapsulated and aluminum bolts and nuts. Chase threads to remove micro-encapsulated residue from old bolts/nuts.**

➡**Only tighten nuts and bolts of chassis components when vehicle is at normal ride height.**

11. Tighten bolts/nuts to specification as follows:

- Nut, link rod to stabilizer bar: 67 ft. lbs. (90 Nm)

- Bolt, torsion bar to front axle carrier: 37 ft. lbs. (50 Nm)

### STEERING KNUCKLE

*REMOVAL & INSTALLATION*

#### C-Class

##### *Except 4MATIC*

*See Figure 173.*

1. Before servicing the vehicle, refer to the precautions.

2. Remove front wheel.

3. Remove brake disc.

4. Remove front wheel hub.

5. Detach brake cover plate.

6. Remove rpm sensor.

7. Remove track rod from steering knuckle assembly.

8. Detach cross strut from steering knuckle.

9. Detach torque strut from steering knuckle.

10. Detach steering knuckle from suspension strut.

11. Remove steering knuckle assembly.

*To install:*

12. Installation is the reverse of removal.

### ✳✳ WARNING

**Replace all self locking, micro-encapsulated and aluminum bolts and nuts. Chase threads to remove micro-encapsulated residue from old bolts/nuts.**

13. Tighten bolts/nuts to specification as follows:

- Nut of tie rod to steering knuckle: 37 ft. lbs. (50 Nm) plus 60°additional rotation

- Self-locking nut, radius rod to steering knuckle: 37 ft. lbs. (50 Nm) plus 60°additional rotation

- Self-locking nut, transverse link to steering knuckle assembly: 37 ft. lbs. (50 Nm) plus 60°additional rotation

##### *4MATIC*

*See Figure 174.*

1. Before servicing the vehicle, refer to the precautions.

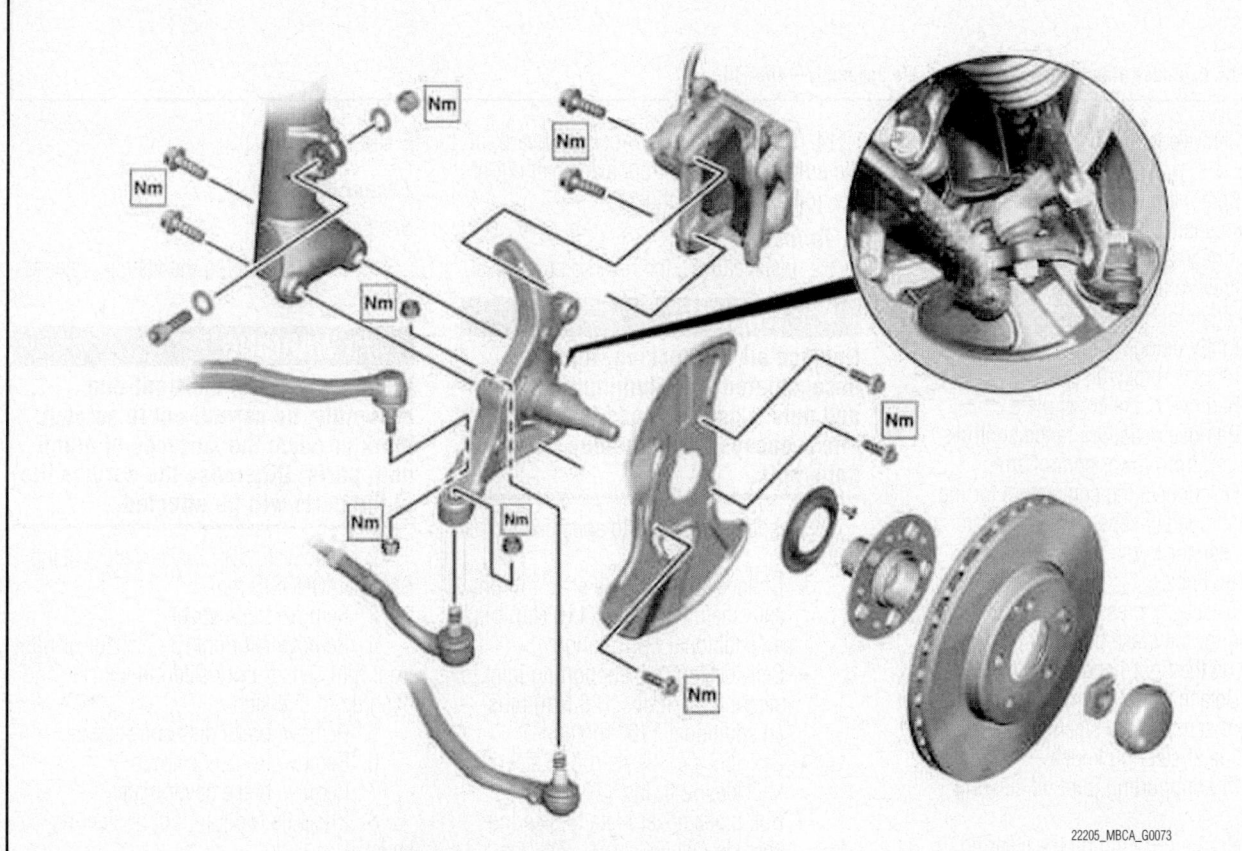

**Fig. 173 Exploded view of the steering knuckle assembly—except 4MATIC**

22205_MBCA_G0073

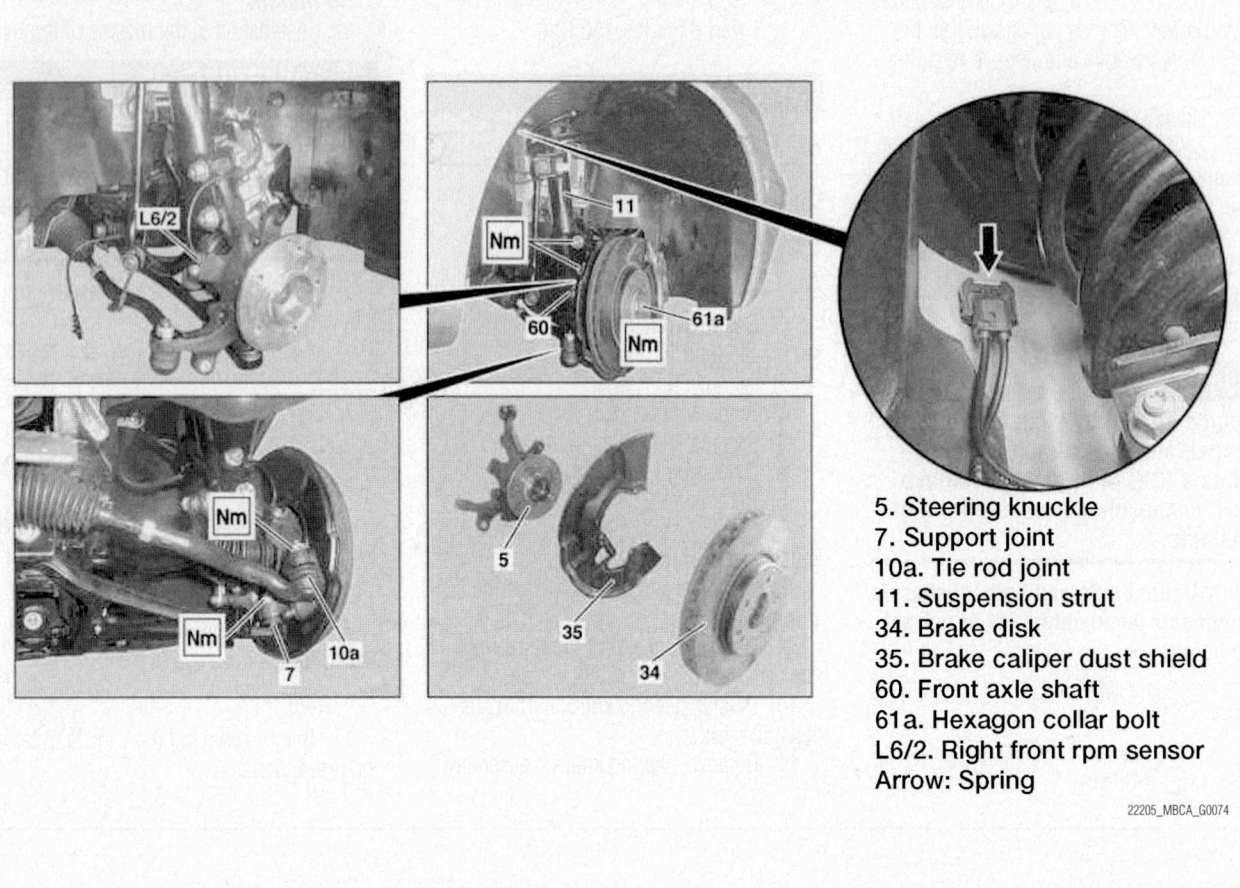

5. Steering knuckle
7. Support joint
10a. Tie rod joint
11. Suspension strut
34. Brake disk
35. Brake caliper dust shield
60. Front axle shaft
61a. Hexagon collar bolt
L6/2. Right front rpm sensor
Arrow: Spring

22205_MBCA_G0074

**Fig. 174 Exploded view of the steering knuckle assembly—4MATIC**

2. Remove front wheel, rotate if necessary.

3. Remove hexagon collar bolt from front axle shaft.

4. If necessary, to loosen the front axle shaft, screw in the hexagon collar bolt a few threads again and hammer out the front axle shaft slightly using a rubber hammer.

5. Remove brake disc.

6. Remove brake cover plate.

7. Remove headlight range control's linkage rod from lower control arm.

8. Disconnect the connectors for the brake pad contact sensor wear indicator and the rpm sensor in the wheelhouse and unclip the lines.

9. To detach, press in the spring (arrow). The plug for the brake pad wear sensor is present on the right-hand side only.

10. Detach right front rpm sensor or left front rpm sensor from steering knuckle and unclip line at steering knuckle.

11. Pry supporting joint off steering knuckle.

12. Pry tie rod joint off the steering knuckle.

13. Remove strut from steering knuckle assembly.

14. Remove steering knuckle forward, in the process move the front axle shaft out of the front axle shaft flange.

**To install:**

15. Installation is the reverse of removal.

> ❋❋ **WARNING**
>
> **Replace all self locking, micro-encapsulated and aluminum bolts and nuts. Chase threads to remove micro-encapsulated residue from old bolts/nuts.**

16. Tighten bolts/nuts to specification as follows:

- Collar bolt, front axle shaft to front axle shaft: 81 ft. lbs. (110 Nm) plus an additional 60° rotation
- Self-locking nut, supporting joint to steering: 21 ft. lbs. (28 Nm) plus an additional 110° rotation
- Self-locking nut, tie rod to steering knuckle: 52 ft. lbs. (70 Nm)
- Bolt brake cover plate to steering knuckle / wheel carrier: 7 ft. lbs. (10 Nm)

17. Perform a wheel alignment check.

### E-Class

#### Except 4MATIC

*See Figure 175.*

1. Before servicing the vehicle, refer to the precautions.

> ❋❋ **WARNING**
>
> **When working on the front-end assembly, be careful not to scratch, mark or notch the surfaces of aluminum parts. Otherwise the service life of the parts will be affected.**

2. Deactivate SBC brake system using STAR DIAGNOSIS tool.

3. Remove front wheel.

4. Remove left front rpm sensor or right front rpm sensor from steering knuckle and hang up to one side.

5. Remove brake disk at front axle.

6. Remove front wheel hub.

7. Remove brake cover plate.

8. Press tie rod joint out of steering knuckle.

9. Press radius rod out of steering knuckle.

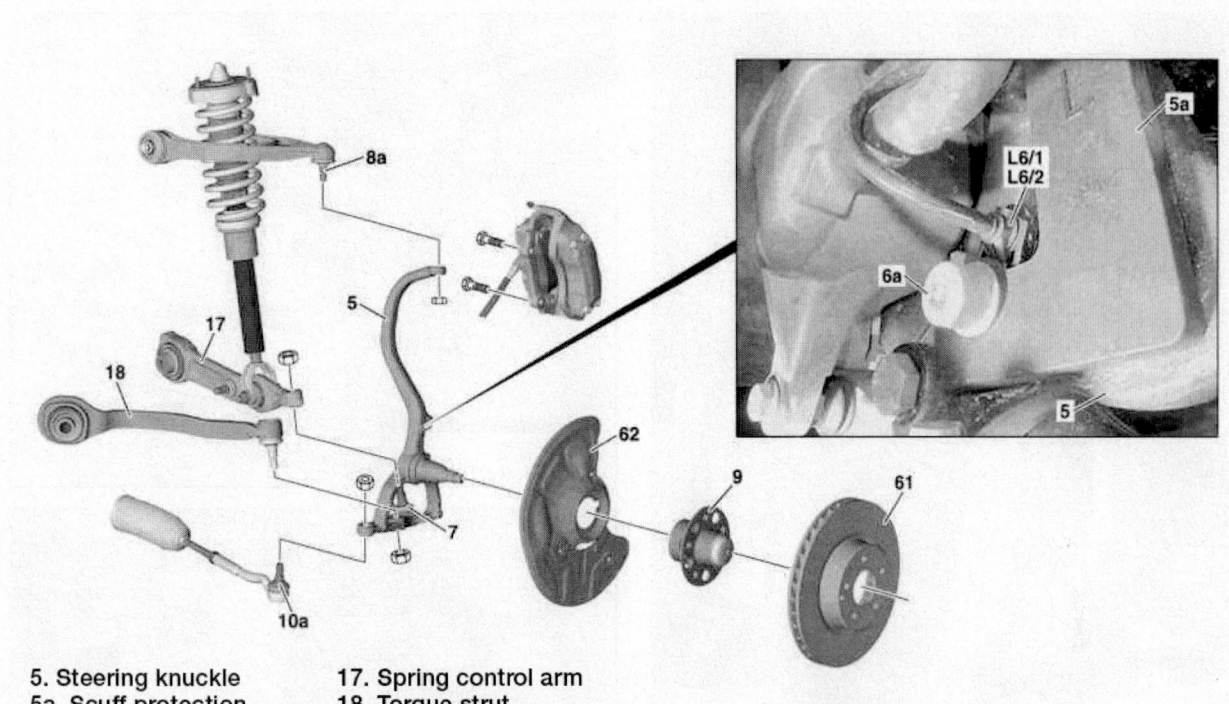

5. Steering knuckle
5a. Scuff protection
6a. Bolt
7. Support joint
8a. Follower joint
9. Front wheel hub
10a. Tie rod joint

17. Spring control arm
18. Torque strut
61. Brake disk
62. Brake caliper dust shield
L6/1. Left front rpm sensor
L6/2. Right front rpm sensor

22205_MBCA_G0275

**Fig. 175 Exploded view of the steering knuckle assembly—except 4MATIC**

10. Press spring control arm out of steering knuckle.

11. Press follower joint out of steering knuckle.

12. Remove steering knuckle.

13. Check supporting joint.

**To install:**

14. Installation is the reverse of removal.

15. Observe correct installed position; the scuff protection must be bolted to the front rpm sensor as shown. The scuff protection must not be mounted between the rpm sensor and steering knuckle.

### ✳✳ WARNING

**Replace all self locking, micro-encapsulated and aluminum bolts and nuts. Chase threads to remove micro-encapsulated residue from old bolts/nuts.**

➡**Only tighten nuts and bolts of chassis components when vehicle is at normal ride height.**

16. Tighten bolts/nuts to specification as follows:

- Self-locking nut, supporting joint to steering knuckle: 52 ft. lbs. (70 Nm) plus an additional 50° rotation
- Self-locking nut, upper wishbone follower joint to steering knuckle: 15 ft. lbs. (20 Nm) plus an additional 90° rotation
- Self-locking nut, link rod to steering knuckle: 70 ft. lbs. (95 Nm)
- Collar bolt, front axle shaft to front axle shaft: 52 ft. lbs. (70 Nm) plus an additional 50° rotation

17. Adjust wheel bearing play.

18. Activate SBC brake system using STAR DIAGNOSIS tool.

### 4MATIC

*See Figure 176.*

1. Before servicing the vehicle, refer to the precautions.

### ✳✳ WARNING

**When working on the front-end assembly, be careful not to scratch, mark or notch the surfaces of aluminum parts. Otherwise the service life of the parts will be affected.**

2. Remove front wheel.

3. Detach hexagon collar bolt from front axle shaft.

➡**If necessary to loosen the front axle shaft, screw in the hexagon collar bolt a few threads again and hammer out the front axle shaft slightly using a rubber hammer.**

4. Deactivate SBC brake system using STAR DIAGNOSIS tool.

5. Remove brake disk.

6. Remove brake caliper support.

7. Detach brake cover plate.

8. Detach connector on left front or right front of steering knuckle and unclip line from steering knuckle.

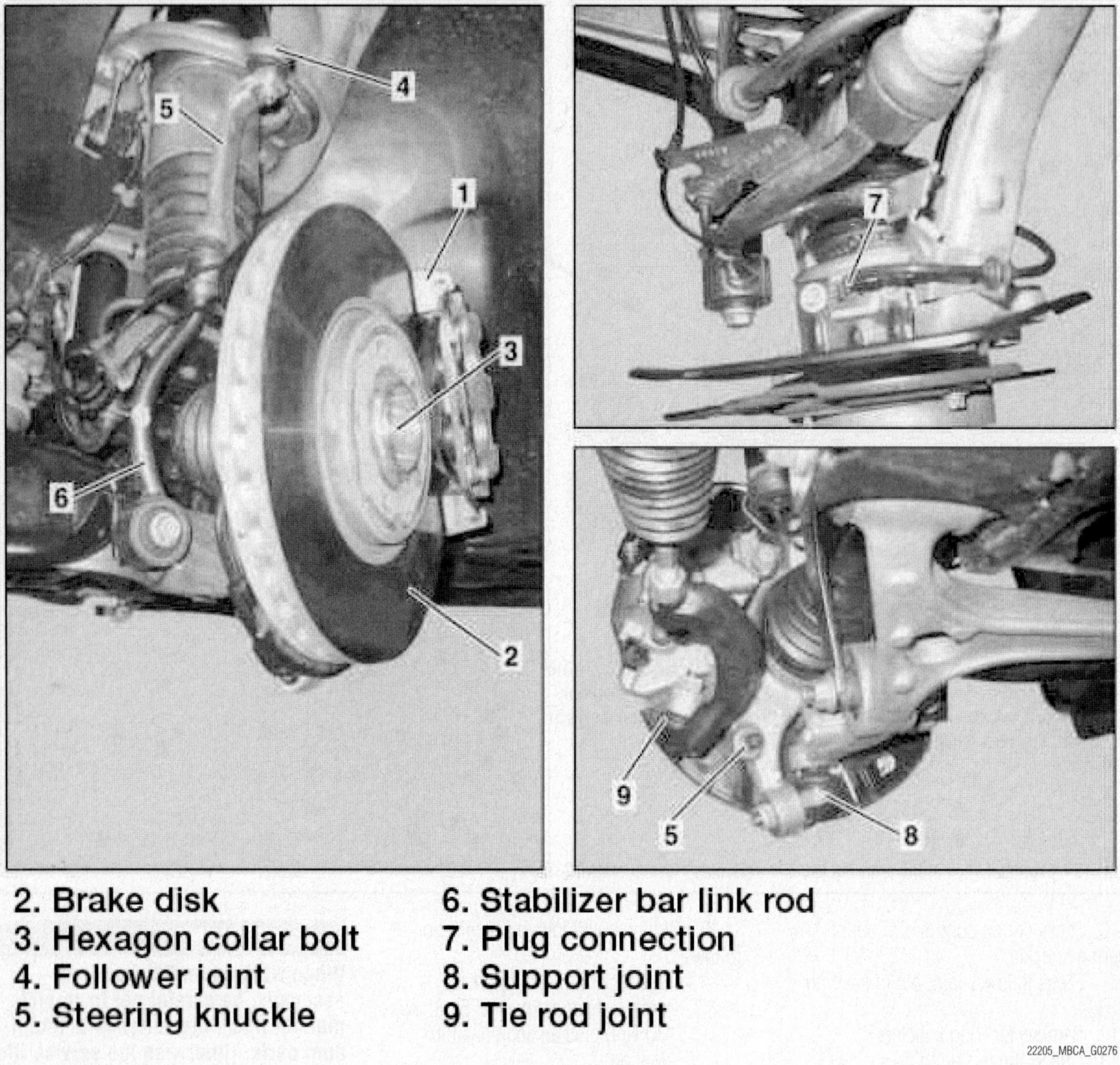
1
2
3
4
5
6
7
8
9
5
2. Brake disk
3. Hexagon collar bolt
4. Follower joint
5. Steering knuckle
6. Stabilizer bar link rod
7. Plug connection
8. Support joint
9. Tie rod joint
22205_MBCA_G0276

**Fig. 176 Steering knuckle assembly—4MATIC**

9. Pry tie rod joint off steering knuckle.
10. Press supporting joint from steering knuckle.
11. Pry follower joint off steering knuckle.
12. Detach torsion bar link rod from steering knuckle and pull out.
13. Remove steering knuckle, in the process move the front axle shaft out of the front axle shaft flange.

***To install:***

14. Installation is the reverse of removal.
15. Observe correct installed position; the scuff protection must be bolted to the front rpm sensor as shown. The scuff protection must not be mounted between the rpm sensor and steering knuckle.

**WARNING**

**Replace all self locking, micro-encapsulated and aluminum bolts and nuts. Chase threads to remove micro-encapsulated residue from old bolts/nuts.**

**➡Only tighten nuts and bolts of chassis components when vehicle is at normal ride height.**

16. Tighten bolts/nuts to specification as follows:
- Self-locking nut, supporting joint to steering knuckle: 52 ft. lbs. (70 Nm) plus an additional 50° rotation
- Self-locking nut, upper wishbone follower joint to steering knuckle: 15 ft. lbs. (20 Nm) plus an additional 90° rotation
- Self-locking nut, link rod to steering knuckle: 70 ft. lbs. (95 Nm)
- Collar bolt, front axle shaft to front axle shaft: 52 ft. lbs. (70 Nm) plus an additional 50° rotation

17. Adjust wheel bearing play.
18. Activate SBC brake system using STAR DIAGNOSIS tool.

## S-Class

### *Except 4MATIC*

*See Figure 177.*

1. Before servicing the vehicle, refer to the precautions.

> ※※ **WARNING**
>
> **When working on the front-end assembly, be careful not to scratch, mark or notch the surfaces of aluminum parts. Otherwise the service life of the parts will be affected.**

2. Remove front wheel.
3. Detach rpm sensor from steering knuckle.
4. Remove front wheel hub.
5. Remove brake cover plate.
6. Remove link rod from steering knuckle.
7. Remove tie rod joint from steering knuckle.
8. Remove torque strut from front axle carrier.
9. Detach spring control arm from steering knuckle.
10. Detach follower joint from steering knuckle.
11. Remove steering knuckle.

12. Remove torque strut from steering knuckle.
13. Check rubber boot and supporting ball joint of spring control arm for damage.
14. Check rubber boot and ball joint of torque strut for damage.

### To install:

15. Installation is the reverse of removal.
16. Observe correct installed position; the scuff protection must be bolted to the front rpm sensor as shown. The scuff protection must not be mounted between the rpm sensor and steering knuckle.

> ※※ **WARNING**
>
> **Replace all self locking, micro-encapsulated and aluminum bolts and nuts. Chase threads to remove micro-encapsulated residue from old bolts/nuts.**

➡ **Only tighten nuts and bolts of chassis components when vehicle is at normal ride height.**

17. Tighten bolts/nuts to specification as follows:
- Bolt brake cover plate to steering knuckle or wheel carrier M6: 10 ft. lbs. (14 Nm)

- Bolt brake cover plate to steering knuckle or wheel carrier M8: 13 ft. lbs. (18 Nm)
- Nut, link rod to steering knuckle: 67 ft. lbs. (90 Nm)
- Self-locking nut, track control arm to stub axle: 30 ft. lbs. (40 Nm) plus an additional 150° rotation
- Self-locking nut, radius rod to steering: 30 ft. lbs. (40 Nm) plus an additional 150° rotation
- Self-locking nut, upper wishbone follower joint to steering knuckle: 15 ft. lbs. (20 Nm) plus an additional 90° rotation

### *4MATIC*

*See Figure 178.*

1. Before servicing the vehicle, refer to the precautions.

> ※※ **WARNING**
>
> **When working on the front-end assembly, be careful not to scratch, mark or notch the surfaces of aluminum parts. Otherwise the service life of the parts will be affected.**

2. Remove front wheel.
3. Remove bolt from front axle shaft.

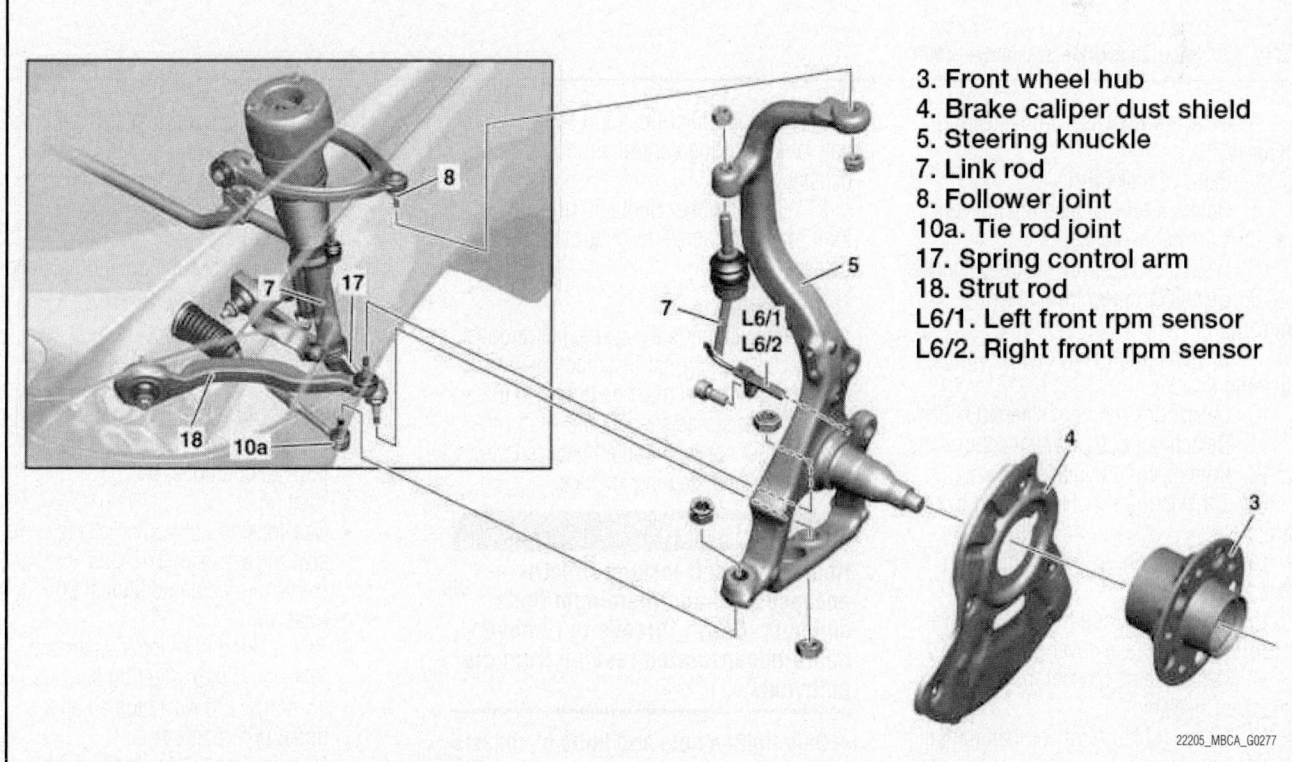

3. Front wheel hub
4. Brake caliper dust shield
5. Steering knuckle
7. Link rod
8. Follower joint
10a. Tie rod joint
17. Spring control arm
18. Strut rod
L6/1. Left front rpm sensor
L6/2. Right front rpm sensor

22205_MBCA_G0277

**Fig. 177 Steering knuckle assembly—except 4MATIC**

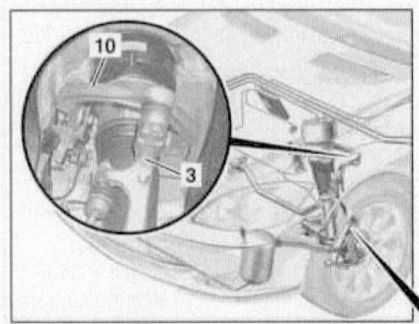

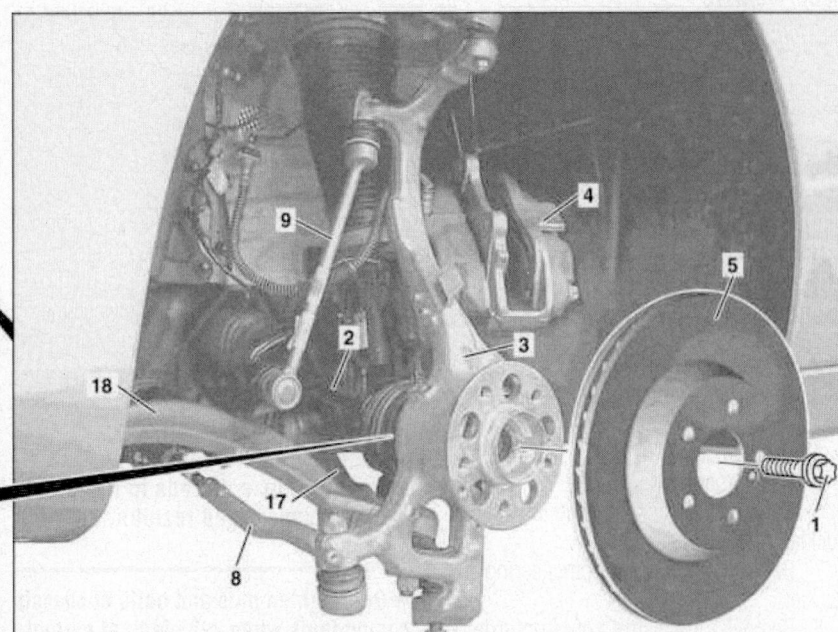

1. Screw
2. Front axle shaft
3. Steering knuckle
4. Brake caliper
5. Brake disk
6. Brake caliper dust shield
7. Cable holder
8. Tie rod
9. Link rod
10. Upper transverse control arm
17. Spring control arm
18. Torque strut
L6/1. Left front rpm sensor
L6/2. Right front rpm sensor

22205_MBCA_G0278

**Fig. 178 Steering knuckle assembly—4MATIC**

4. Detach rpm sensor from steering knuckle.

5. Remove brake disc.

6. Unclip electrical feed line for rpm sensor from cable bracket.

7. Detach tie rod from steering knuckle.

8. Detach strut rod from steering knuckle.

9. Detach spring control arm from steering knuckle.

10. Detach link rod from steering knuckle.

11. Detach upper transverse control arm.

12. from steering knuckle.

13. Tilt steering knuckle outward to side while.

14. moving front axle shaft out of front axle.

15. Shaft flange, and lower to remove.

16. Remove brake cover plate.

17. Detach cable bracket from steering knuckle Checking.

18. Check rubber boot and ball joint of tie rod for wear and damage.

19. Check rubber boot and ball joint of strut rod for wear and damage.

20. Check rubber boot and supporting ball joint of spring control arm for damage.

21. Check rubber boot and upper ball joint of upper transverse control arm for damage.

### To install:

22. Installation is the reverse of removal.

23. Observe correct installed position; the scuff protection must be bolted to the front rpm sensor as shown. The scuff protection must not be mounted between the rpm sensor and steering knuckle.

### ☀❋ WARNING

**Replace all self locking, micro-encapsulated and aluminum bolts and nuts. Chase threads to remove micro-encapsulated residue from old bolts/nuts.**

➡**Only tighten nuts and bolts of chassis components when vehicle is at normal ride height.**

24. Tighten bolts/nuts to specification as follows:

- Bolt, front axle shaft to wheel hub:
- Step 1: Tighten to 118 ft. lbs. (160 Nm)
- Step 2: Release the bolt completely
- Step 3: Tighten to 89 ft. lbs. (120 Nm)
- Step 4: Tighten an additional 90° rotation
- Self-locking nut, track control arm to stub: 37 ft. lbs. (50 Nm) plus an additional 90° rotation
- Self-locking nut, radius rod to steering knuckle: 37 ft. lbs. (50 Nm) plus an additional 90° rotation
- Self-locking nut, upper wishbone follower joint to steering knuckle: 15 ft. lbs. (20 Nm) plus an additional 90° rotation
- Nut, link rod to steering knuckle: 67 ft. lbs. (90 Nm)

## UPPER BALL JOINT

### REMOVAL & INSTALLATION

The upper ball joint is an integral part of the upper control arm. Control arm replacement is necessary if the ball joint becomes worn or damaged.

## UPPER CONTROL ARM

### REMOVAL & INSTALLATION

#### E-Class

**Except 4MATIC**

*See Figure 179.*

1. Before servicing the vehicle, refer to the precautions.

> ✳✳ **WARNING**
>
> **When working on the front-end assembly, be careful not to scratch, mark or notch the surfaces of aluminum parts. Otherwise the service life of the parts will be affected.**

2. Remove front wheel.
3. Press follower joint out of steering knuckle.
4. Left upper transverse control arm.

5. Remove left front suspension strut.
6. Remove fuse box.
7. Detach upper wishbone from front end.
8. Upper transverse control arm modified.
9. Inspect follower joint.
10. Remove right front suspension strut.
11. Remove air filter housing .
12. Remove combination filter housing.
13. Detach upper wishbone from front end.
14. Upper transverse control arm modified.
15. Inspect follower joint.

**To install:**

16. Installation is the reverse of removal.
17. Observe correct installed position; the scuff protection must be bolted to the front rpm sensor as shown. The scuff protection must not be mounted between the rpm sensor and steering knuckle.

> ✳✳ **WARNING**
>
> **Replace all self locking, micro-encapsulated and aluminum bolts and nuts. Chase threads to remove micro-encapsulated residue from old bolts/nuts.**

➡ **Only tighten nuts and bolts of chassis components when vehicle is at normal ride height.**

18. Tighten bolts/nuts to specification as follows:
- Self-locking nut of upper wishbone on front end: 37 ft. lbs. (50 Nm)

#### 4MATIC

*See Figure 180.*

1. Before servicing the vehicle, refer to the precautions.

> ✳✳ **WARNING**
>
> **When working on the front-end assembly, be careful not to scratch, mark or notch the surfaces of aluminum parts. Otherwise the service life of the parts will be affected.**

2. Remove front wheel.
3. Press follower joint out of steering knuckle.
4. Remove Left upper transverse control arm as follows:
   a. Remove nuts.
   b. Loosen suspension strut on lower transverse control arm.

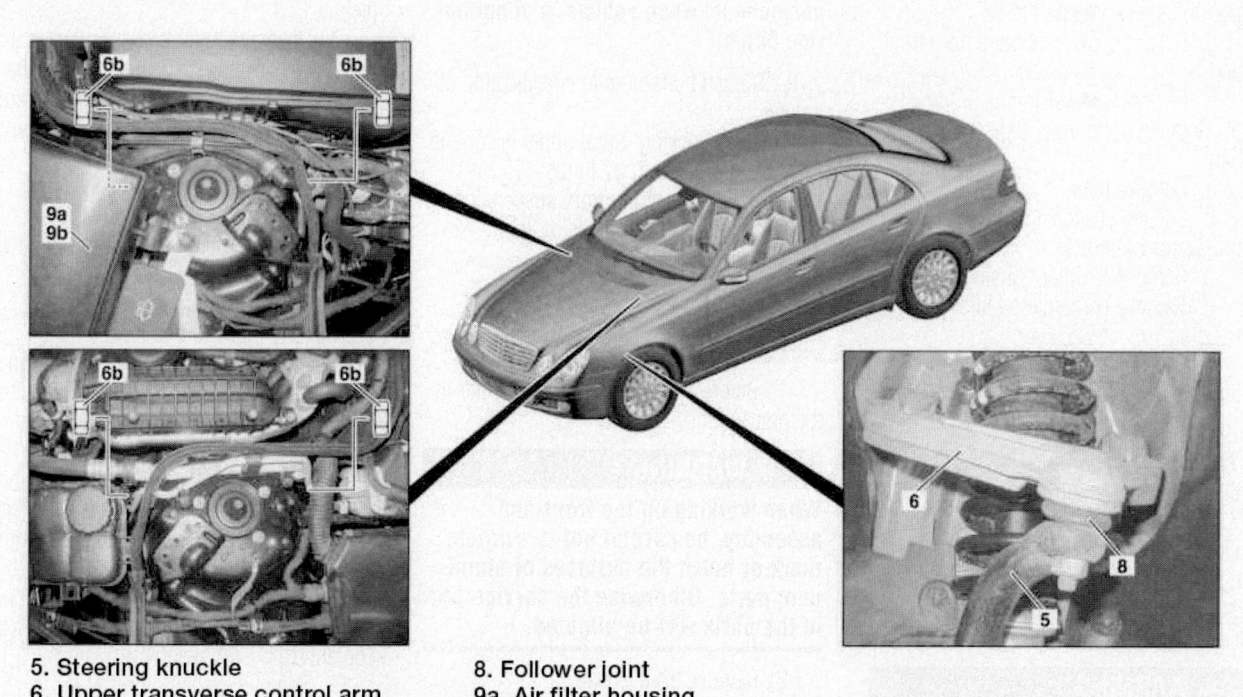

5. Steering knuckle
6. Upper transverse control arm
6b. Nuts
8. Follower joint
9a. Air filter housing
9b. Combination filter housing

22205_MBCA_G0280

**Fig. 179 Upper control arm assembly**

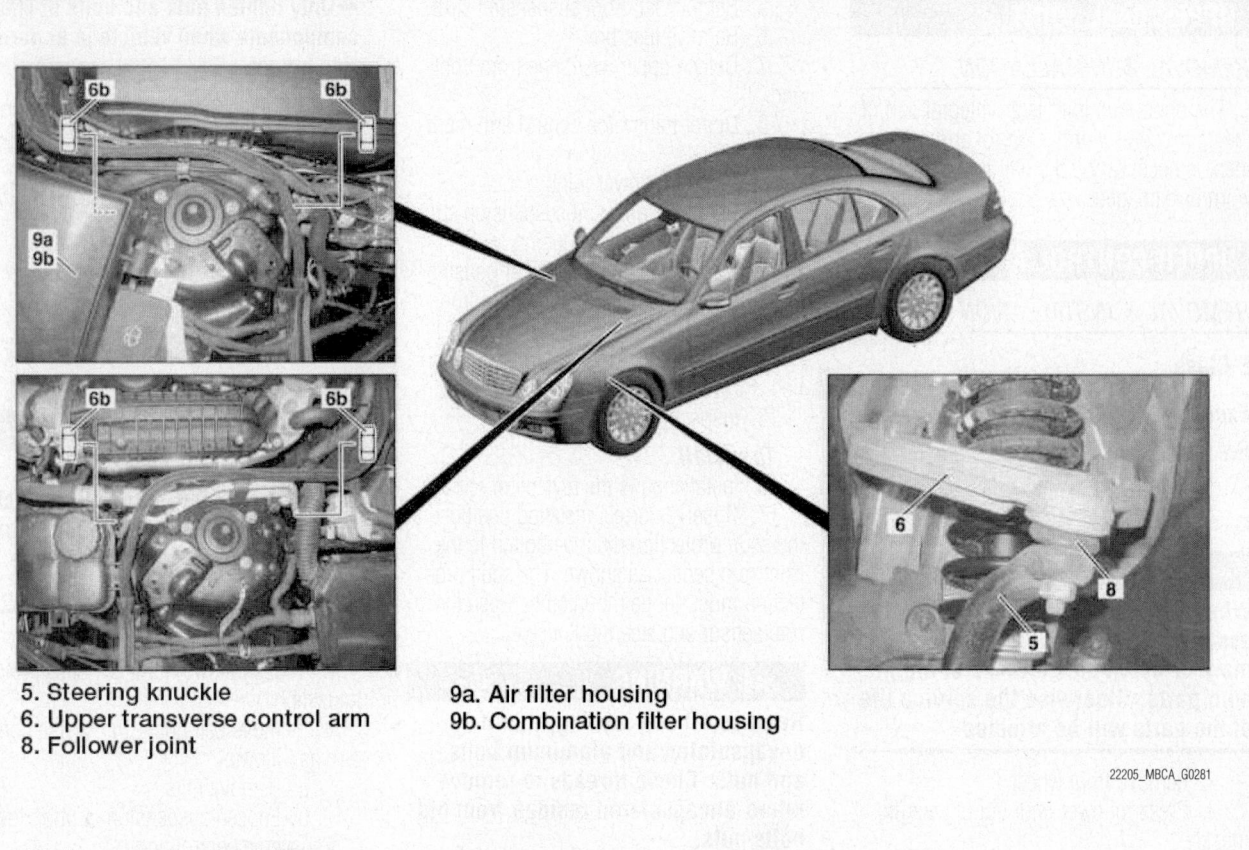

5. Steering knuckle
6. Upper transverse control arm
8. Follower joint

9a. Air filter housing
9b. Combination filter housing

22205_MBCA_G0281

**Fig. 180 Upper control arm assembly**

c. Remove Fuse and relay box in left front of engine compartment.

d. Detach upper wishbone from front end.

e. Inspect follower joint.

5. Remove right upper transverse control arm:

a. Remove nuts.

b. Loosen suspension strut on lower transverse control arm.

c. Remove air filter housing.

d. Remove combination filter housing.

e. Detach upper wishbone from front end.

f. Inspect follower joint.

### To install:

6. Installation is the reverse of removal.

7. Observe correct installed position; the scuff protection must be bolted to the front rpm sensor as shown. The scuff protection must not be mounted between the rpm sensor and steering knuckle.

### ✱✱ WARNING

**Replace all self locking, micro-encapsulated and aluminum bolts and nuts. Chase threads to remove micro-encapsulated residue from old bolts/nuts.**

➡**Only tighten nuts and bolts of chassis components when vehicle is at normal ride height.**

8. Tighten bolts/nuts to specification as follows:

- Self-locking nut of upper wishbone on front end: 37 ft. lbs. (50 Nm)
- Nut connecting front suspension strut to front end: 22 ft. lbs. (30 Nm)

### S-Class

*See Figure 181.*

1. Before servicing the vehicle, refer to the precautions.

### ✱✱ WARNING

**When working on the front-end assembly, be careful not to scratch, mark or notch the surfaces of aluminum parts. Otherwise the service life of the parts will be affected.**

2. Remove front wheel.

3. Unscrew compressed air terminal from suspension strut.

4. Detach level sensor from upper transverse control arm.

5. Detach follower joint from steering knuckle.

6. Support suspension strut using floor jack.

7. Unscrew nuts.

8. Lower floor jack carefully until the threaded studs of the suspension strut are moved out of the front end left upper transverse control arm.

9. Remove the left engine intake air duct .

10. Unclip expansion reservoir of power steering from bracket.

11. Detach vacuum line from brake booster and place to one side.

12. Undo nuts and remove bolts from upper transverse control arm.

13. Remove upper transverse control arm.

14. Check follower joint and rubber boot for signs of damage.

15. Right upper transverse control arm.

16. Remove starter battery.

17. Detach battery base from body.

18. Disconnect electrical connector from coolant expansion reservoir.

19. Remove expansion reservoir for coolant and place it along with the connected coolant lines to one side.

20. Undo nuts and remove bolts from upper transverse control arm.

21. Remove upper transverse control arm.

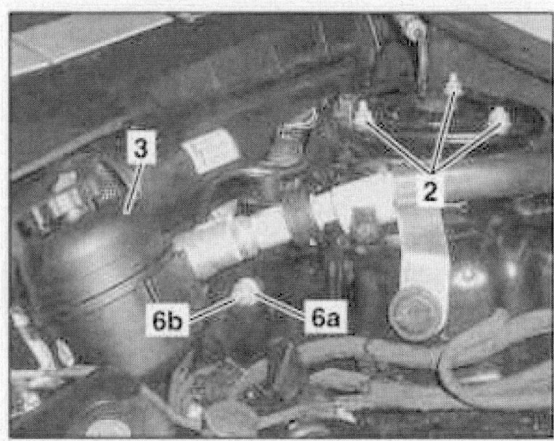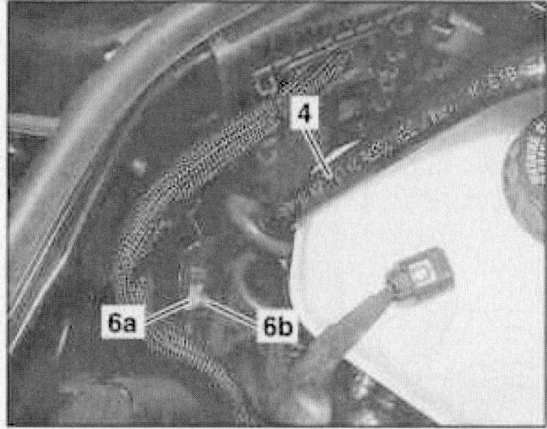

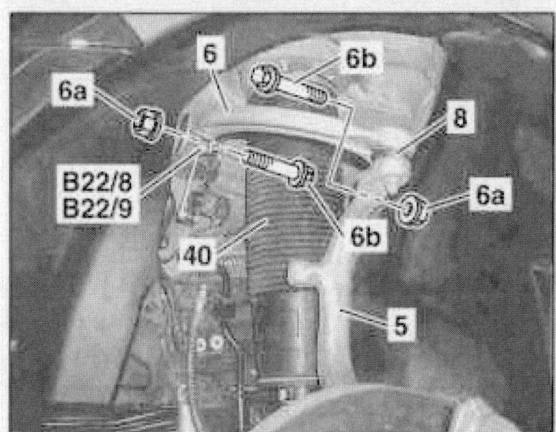

2. Nuts
3. Power steering expansion reservoir
4. Vacuum line
5. Steering knuckle
6. Upper transverse control arm
6a. Nuts

6b. Bolts
8. Follower joint
9. Cable tie
40. Suspension strut
B22/8. Left front level sensor
B22/9. Right front level sensor

22205_MBCA_G0279

**Fig. 181 Upper control arm assembly**

22. Check follower joint and rubber boot for signs of damage.

***To install:***

23. Installation is the reverse of removal.

24. Observe correct installed position; the scuff protection must be bolted to the front rpm sensor as shown. The scuff protection must not be mounted between the rpm sensor and steering knuckle.

### ❊❊ WARNING

**Replace all self locking, micro-encapsulated and aluminum bolts and nuts. Chase threads to remove micro-encapsulated residue from old bolts/nuts.**

➡**Only tighten nuts and bolts of chassis components when vehicle is at normal ride height.**

25. Tighten bolts/nuts to specification as follows:

- Nut, front suspension strut to front end: 26 ft. lbs. (35 Nm)
- Air suspension pressure line to suspension strut: 4 ft. lbs. (5 Nm)
- Self-locking nut of upper wishbone on front end: 37 ft. lbs. (50 Nm)
- Self-locking nut, upper wishbone follower joint to steering knuckle: 15 ft. lbs. (20 Nm) plus an additional 90° rotation

### WHEEL BEARINGS

*REMOVAL & INSTALLATION*

#### C-Class and E-Class

***Except 4MATIC***

*See Figure 182.*

1. Before servicing the vehicle, refer to the precautions.
2. Remove front wheel.
3. Remove brake disc.
4. Pull off hub cap.
5. Remove outer tapered roller bearing.
6. Undo hexagon socket head screw, remove clamping nut and remove outer tapered roller bearing.

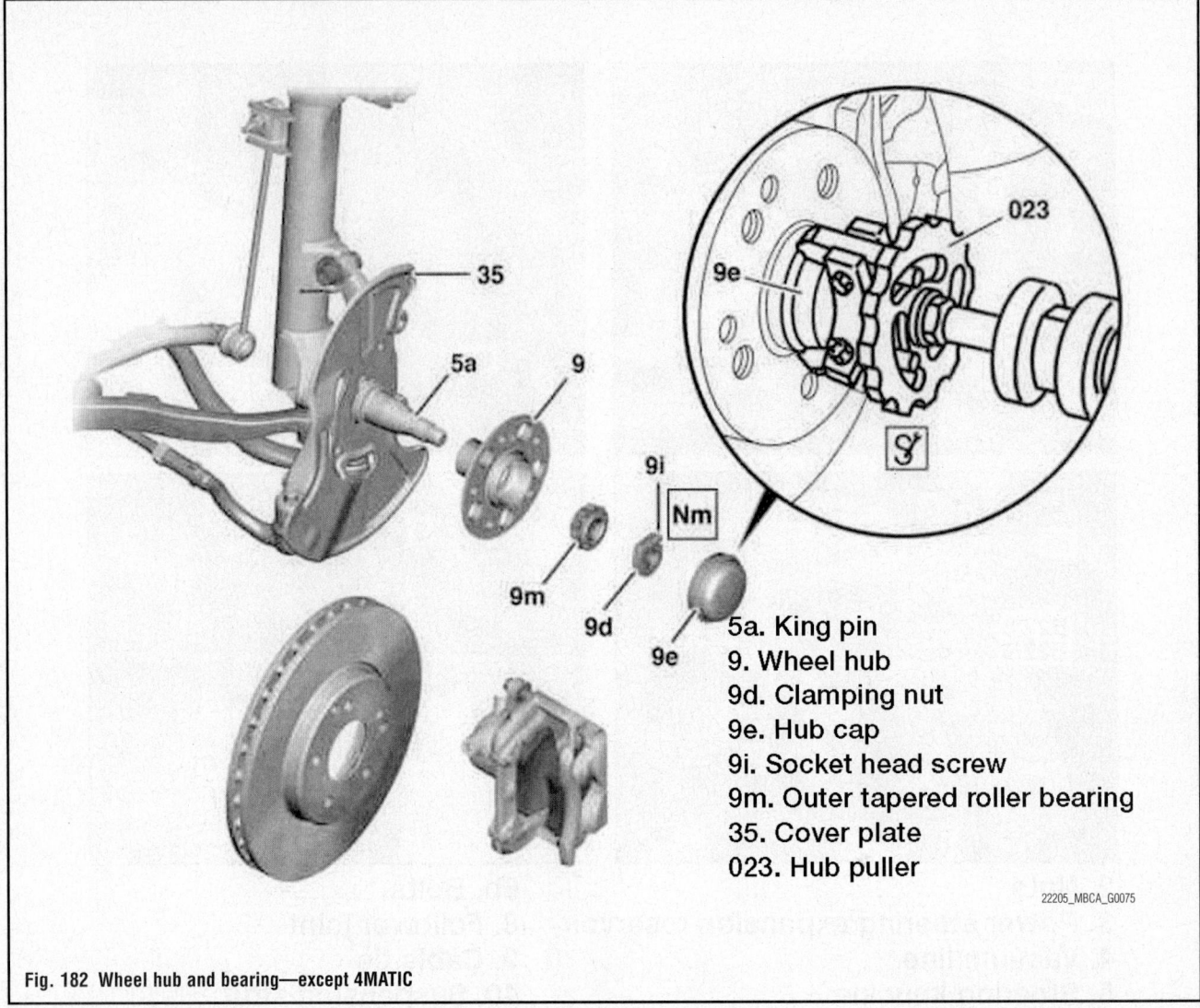

5a. King pin
9. Wheel hub
9d. Clamping nut
9e. Hub cap
9i. Socket head screw
9m. Outer tapered roller bearing
35. Cover plate
023. Hub puller

22205_MBCA_G0075

**Fig. 182 Wheel hub and bearing—except 4MATIC**

7. Pull off front wheel hub cover.

8. The pulse ring for the rpm recording is integrated in the radial sealing ring , therefore do not damage the radial sealing ring.

9. Check brake cover plate.

10. Check kingpin for discoloration and wear on bearing seats or contact surface of radial sealing ring.

11. Discoloration can occur, for example, due to heat formation as a result of incorrectly adjusted bearing play. If necessary, replace stub axle.

12. Check front wheel hub inspecting tapered roller bearing and bearing races.

**To install:**

13. Installation is the reverse of removal.

14. Pack the front wheel hub with grease and coat or pack radial sealing ring with grease as well as the space between the sealing lip and tapered roller bearings. Push the front wheel hub with inner tapered roller bearing and radial sealing ring onto the steering knuckle spindle.

15. Adjust wheel bearing play as follows:

a. Turn-in wheel bolt opposite from brake rotor attachment bolt to immobilize brake rotor.

b. Loosen hexagon socket bolt and tighten clamp nut lightly while turning the front wheel hub.

c. Mount dial indicator with dial indicator holder on the brake disc.

d. Adjust wheel bearing play by turning (arrow) the clamping nut in stages while pushing and pulling the brake disc firmly back and forth. Wheel bearing play at front axle should be 0.00039–0.00078 in. (0.01–0.02 mm).

➡Do not twist or tilt the wheel hub during the measurement. Correct measurement is only possible by pulling and pushing parallel to the wheel axle. Incorrect measurement will result in the set wheel bearing play being too small. This can then lead to wheel bearing damage.

e. Tighten the Allen bolt, check the wheel bearing end float again.

f. Remove the dial indicator with dial indicator holder.

**❊❊ WARNING**

**Replace all self locking, micro-encapsulated and aluminum bolts and nuts. Chase threads to remove micro-encapsulated residue from old bolts/nuts.**

16. Tighten screw for clamp nut to adjust front wheel bearing end float to 8 ft. lbs. (11 Nm)

**4MATIC**

*See Figure 183.*

1. Before servicing the vehicle, refer to the precautions.

2. Remove steering knuckle.

3. Press out front axle shaft flange.

4. Remove circlip.

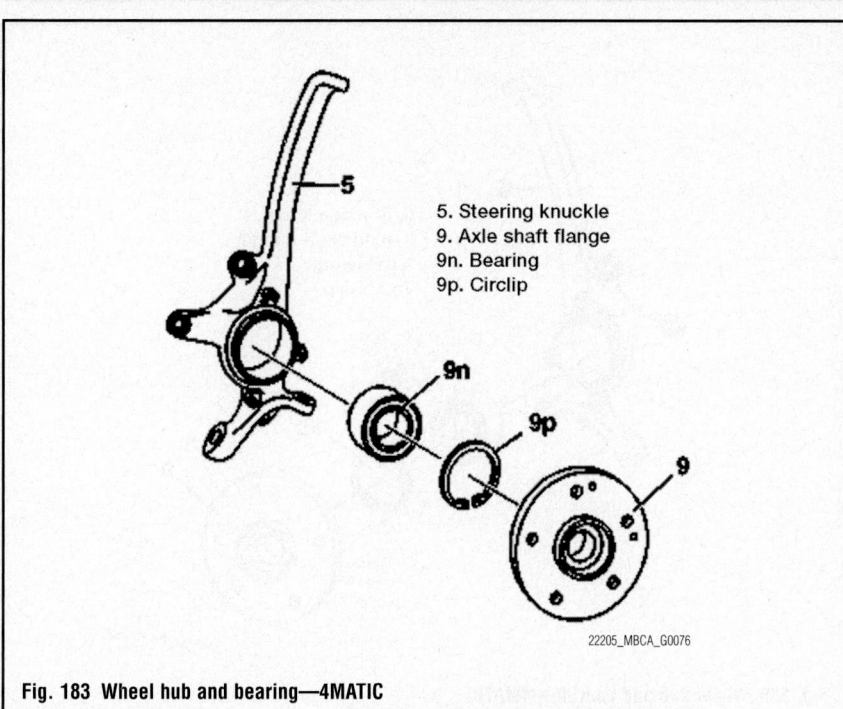

5. Steering knuckle
9. Axle shaft flange
9n. Bearing
9p. Circlip

22205_MBCA_G0076

**Fig. 183 Wheel hub and bearing—4MATIC**

5. Pull out double-row angular ball bearing.

6. If necessary detach the remaining bearing inner race from the front axle shaft flange.

*To install:*

7. Installation is the reverse of removal.

8. This bearing unit is sealed and does not require greasing or adjustment.

### S-Class

#### Except 4MATIC

*See Figure 184.*

1. Before servicing the vehicle, refer to the precautions.

2. Remove front wheel.

3. Remove brake disk.

4. Pull off hub cap.

5. Loosen hexagon socket head screw and remove clamping nut.

6. Remove outer tapered roller bearing.

7. Detach front wheel hub from steering knuckle spindle.

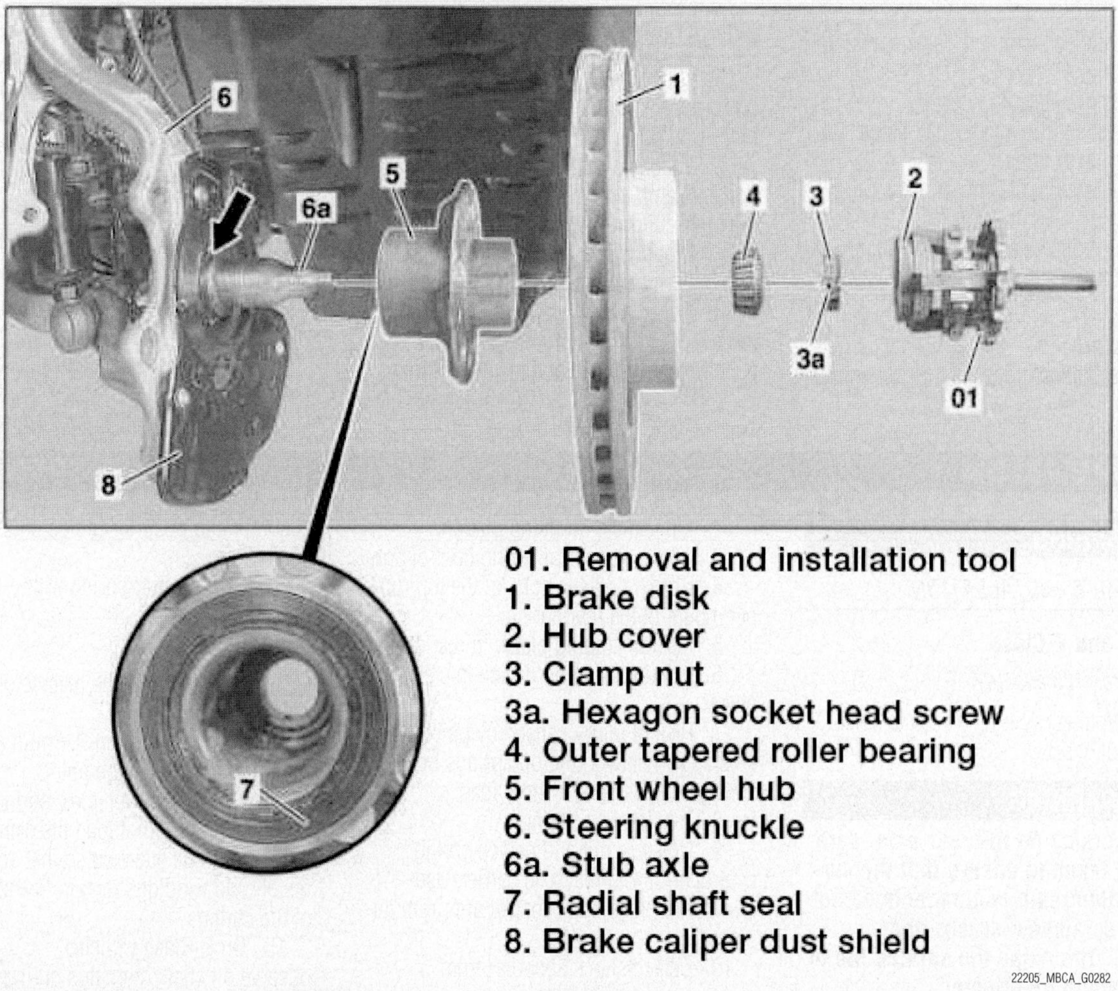

01. **Removal and installation tool**
1. **Brake disk**
2. **Hub cover**
3. **Clamp nut**
3a. **Hexagon socket head screw**
4. **Outer tapered roller bearing**
5. **Front wheel hub**
6. **Steering knuckle**
6a. **Stub axle**
7. **Radial shaft seal**
8. **Brake caliper dust shield**

22205_MBCA_G0282

**Fig. 184 Wheel hub and bearing—except 4MATIC**

8. Replace radial shaft seal.

9. Check brake cover plate for damage.

10. Check steering knuckle spindle for discoloration and wear on bearing seats.

11. Check surface (arrow) of radial shaft sealing ring on steering knuckle spindle.

12. Check threaded holes of wheel bolts in front wheel hub.

13. Check inner and outer tapered roller bearing and bearing races.

### To install:

14. Installation is the reverse of removal.

15. Tighten bolts/nuts to specification as follows:

- Bolt for clamp nut used to adjust front-axle wheel bearing play: 8 ft. lbs. (11 Nm)
- Bolt brake cover plate to steering knuckle or wheel carrier M6: 10 ft. lbs. (14 Nm)
- Bolt brake cover plate to steering knuckle or wheel carrier M8: 13 ft. lbs. (18 Nm)

16. Adjust wheel bearing play.

### 4MATIC

*See Figure 185.*

1. Before servicing the vehicle, refer to the precautions.

2. Remove steering knuckle.

3. Press out front axle shaft flange.

4. Remove circlip.

5. Pull out double-row angular ball bearing.

6. If necessary detach the remaining bearing inner race from the front axle shaft flange.

### To install:

7. Installation is the reverse of removal.

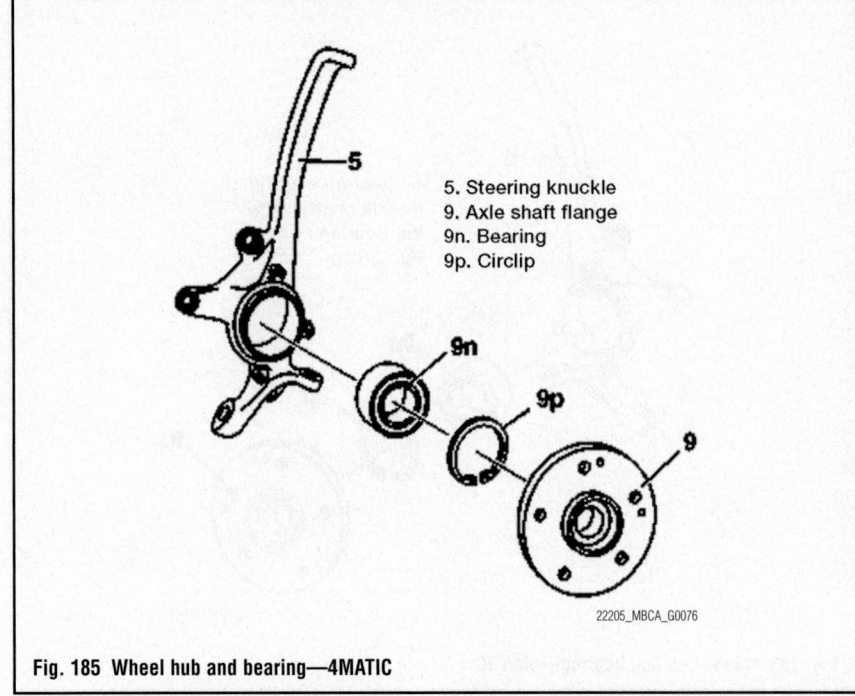

5. Steering knuckle
9. Axle shaft flange
9n. Bearing
9p. Circlip

22205_MBCA_G0076

**Fig. 185 Wheel hub and bearing—4MATIC**

8. This bearing unit is sealed and does not require greasing or adjustment.

### ADJUSTMENT

1. Turn-in wheel bolt opposite from brake rotor attachment bolt to immobilize brake rotor.

2. Loosen hexagon socket bolt and tighten clamp nut lightly while turning the front wheel hub.

3. Mount dial indicator with dial indicator holder on the brake disc.

4. Adjust wheel bearing play by turning (arrow) the clamping nut in stages while pushing and pulling the brake disc firmly back and forth. Wheel bearing play at front axle should be 0.00039–0.00078 in. (0.01–0.02 mm).

➡Do not twist or tilt the wheel hub during the measurement. Correct measurement is only possible by pulling and pushing parallel to the wheel axle. Incorrect measurement will result in the set wheel bearing play being too small. This can then lead to wheel bearing damage.

5. Tighten the Allen bolt, check the wheel bearing end float again.

6. Remove the dial indicator with dial indicator holder.

## SUSPENSION

### COIL SPRING

#### REMOVAL & INSTALLATION

**C-Class and E-Class**

*See Figures 186 and 187.*

1. Before servicing the vehicle, refer to the precautions.

#### ✳✳ WARNING

**While working on the rear axle, care must be taken to ensure that the surface of aluminum components do not get any scratches, cracks and notches. Otherwise the service life of the parts will be affected.**

2. Lift vehicle on lifting platform.

3. Remove cover of spring control arm.

4. Remove plastic cone in the top rubber mount of the rear spring.

5. Insert clamping plates in rear spring.

6. Insert tensioning device in clamping plates.

7. Risk of injury caused by pinching or crushing when working on springs or spring bodies that are under load.

8. Tension rear spring.

9. Unscrew nut and detach spring control arm from rear axle carrier. Use crowbar to press track control arm from its mount.

10. Detach shock absorber from spring control arm. Push together shock

## REAR SUSPENSION

absorber and position at side of spring control arm.

11. Remove rear spring with rubber bushing.

### To install:

12. Installation is the reverse of removal.

13. Clean track control arm in area surrounding the anchorage point.

14. On C-Class vehicles, springs which have been relieved of load the clamping plates are to be attached so that approximately two windings remain free at the ends of the spring.

15. On E-Class vehicles, observe oil chalk markings during installation.

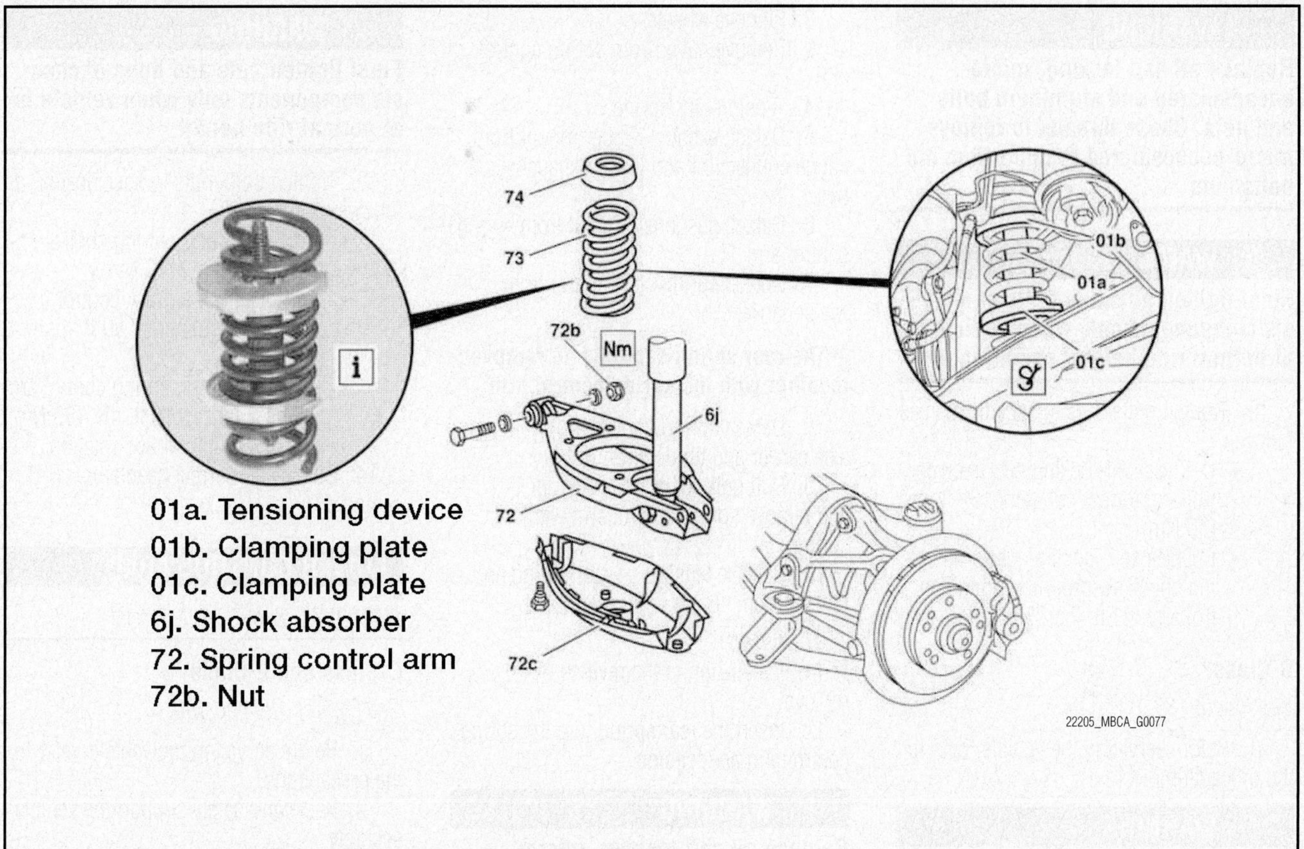
74
73
72b
Nm
6j
72
72c
01b
01a
01c
i
01a. Tensioning device
01b. Clamping plate
01c. Clamping plate
6j. Shock absorber
72. Spring control arm
72b. Nut
22205_MBCA_G0077

**Fig. 186 Typical rear coil spring mounting—models without air suspension**

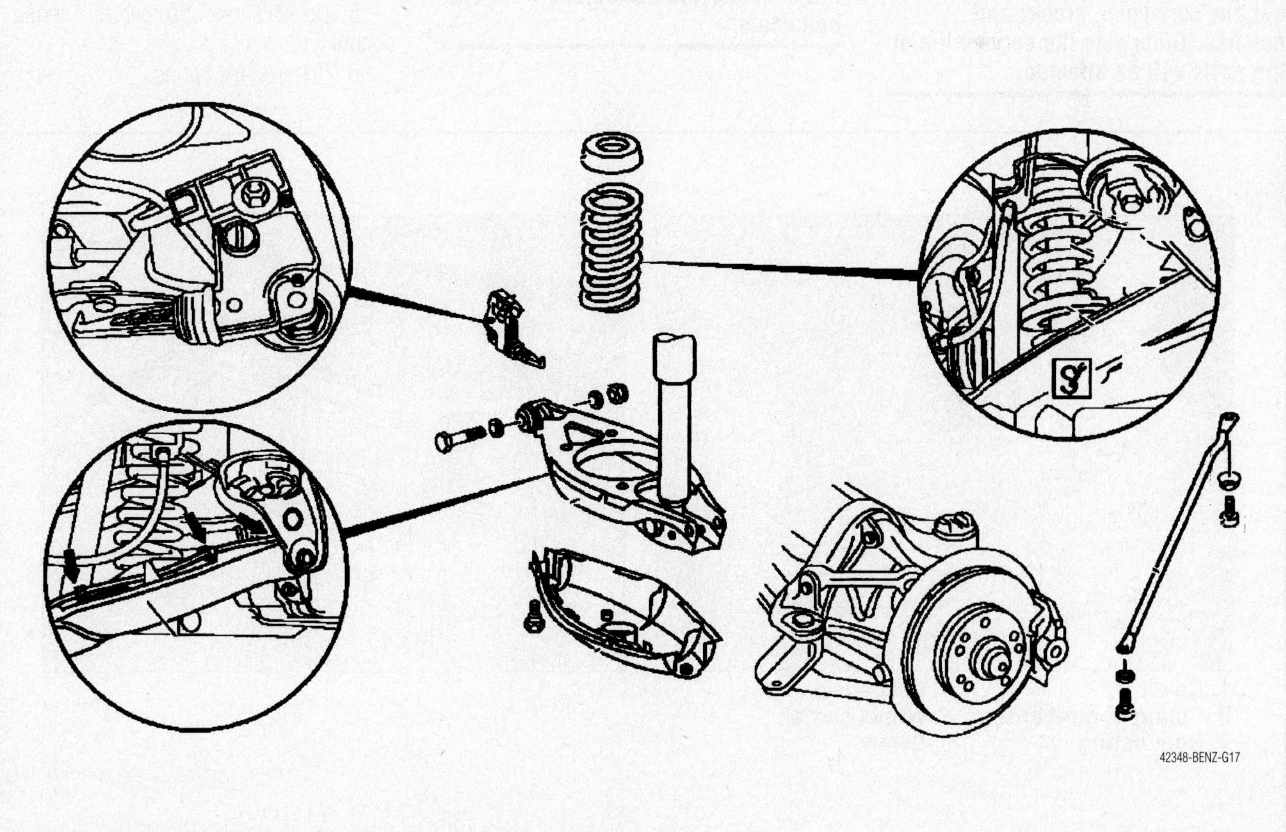
42348-BENZ-G17

**Fig. 187 Typical rear coil spring mounting—models with air suspension**

**✳✳ WARNING**

Replace all self locking, micro-encapsulated and aluminum bolts and nuts. Chase threads to remove micro-encapsulated residue from old bolts/nuts.

**✳ WARNING**

Final tighten nuts and bolts of chassis components only when vehicle is at normal ride height. condition.

16. Tighten bolts/nuts to specification as follows:
- C-Class self-locking nut, track control arm to rear subframe: 52 ft. lbs. (70 Nm)
- E-Class self-locking nut for attaching shock absorber to spring control arm: 41 ft. lbs. (55 Nm)

### S-Class

*See Figure 188.*

1. Before servicing the vehicle, refer to the precautions.

**✳✳ WARNING**

While working on the rear axle, care must be taken to ensure that the surface of aluminum components do not get any scratches, cracks and notches. Otherwise the service life of the parts will be affected.

2. Remove wheels.
3. Remove cover from spring control arm.
4. Tension rear spring.
5. Detach exhaust rubber mount from flange connection and lower exhaust system.
6. Detach suspension strut from spring control arm.
7. Detach spring control arm from wheel carrier.

➡ **The rear spring can only be removed together with the spring control arm.**

8. Detach spring control arm from rear axle carrier and tilt downward.
9. Pull bolt out of wheel carrier and remove spring control arm with rear spring.
10. Release tension on rear spring and remove from spring control arm.

### To install:

11. Installation is the reverse of removal.
12. Insert the rear spring into the spring control arm and tension.

**✳✳ WARNING**

Replace all self locking, micro-encapsulated and aluminum bolts and nuts. Chase threads to remove micro-encapsulated residue from old bolts/nuts.

**✳✳ WARNING**

Final tighten nuts and bolts of chassis components only when vehicle is at normal ride height.

13. Tighten bolts/nuts to specification as follows:
- Nut, rear strut to spring control arm: 81 ft. lbs. (110 Nm)
- Self-locking nut, track control arm to rear subframe: 89 ft. lbs. (120 Nm)
- Self-locking nut, spring control arm to wheel carrier: 59 ft. lbs. (80 Nm) plus an additional 90° rotation
14. Check and correct headlamp setting.

## LOWER CONTROL ARM

### REMOVAL & INSTALLATION

#### C-Class and E-Class

*See Figures 189 and 190.*

1. Before servicing the vehicle, refer to the precautions.
2. Raise and safely support the vehicle securely.
3. Remove rear wheels.
4. Detach cover from spring control arm.
5. Detach shock absorber from spring control arm.
6. Remove rear spring.

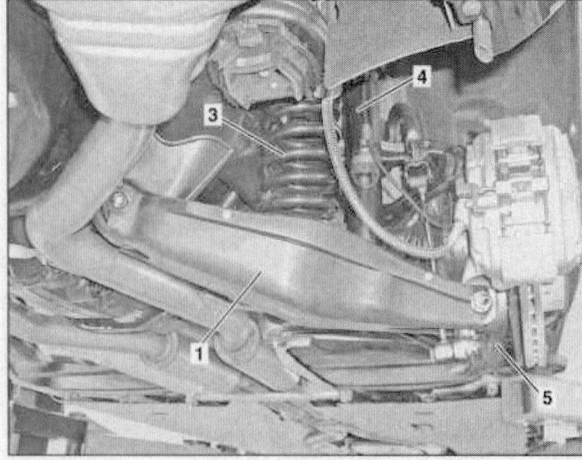
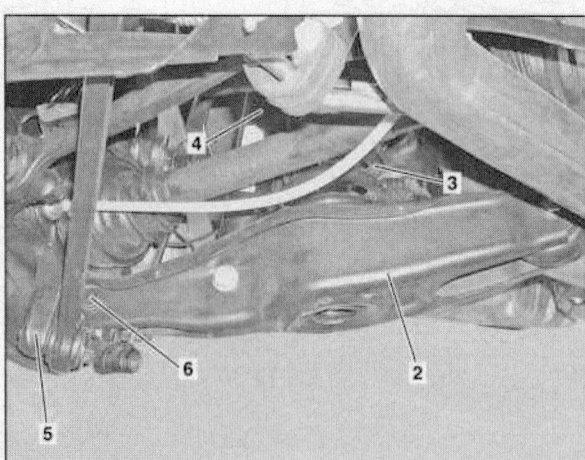

1. Cover
2. Spring control arm
3. Rear spring
4. Suspension strut
5. Wheel carrier
6. Screw

22205_MBCA_G0283

**Fig. 188 Rear coil spring mounting**

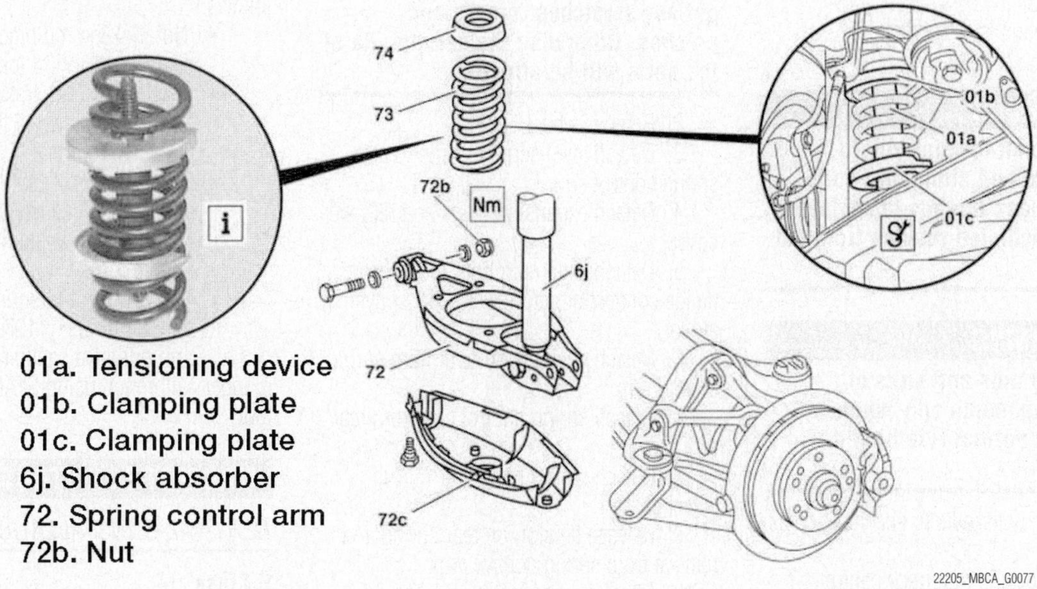
74
73
i
72b
Nm
6j
72
72c
01b
01a
01c
01a. Tensioning device
01b. Clamping plate
01c. Clamping plate
6j. Shock absorber
72. Spring control arm
72b. Nut
22205_MBCA_G0077

Fig. 189 Typical rear coil spring mounting—models without air suspension

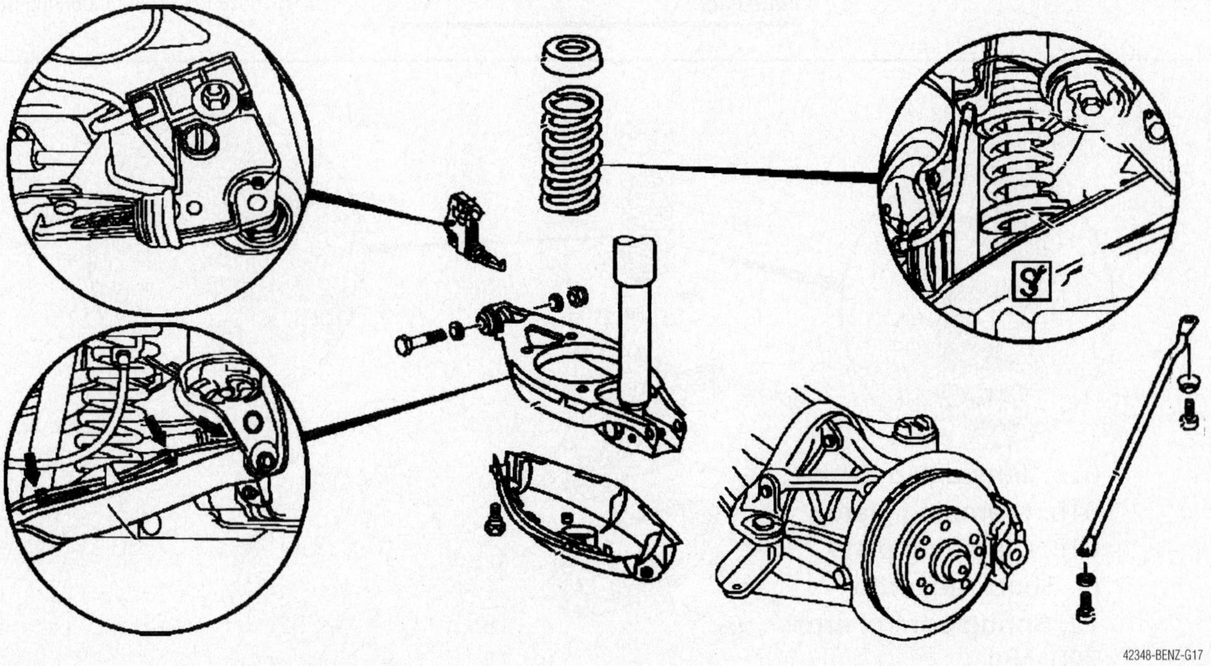
42348-BENZ-G17

Fig. 190 Typical rear coil spring mounting—models with air suspension

7. Detach spring control arm from rear axle carrier and wheel carrier and remove.

8. Check rubber mount in spring control arm.

**To install:**

9. Installation is the reverse of removal.

### ✳✳ WARNING

Replace all self locking, micro-encapsulated and aluminum bolts and nuts. Chase threads to remove micro-encapsulated residue from old bolts/nuts.

### ✳✳ WARNING

Final tighten nuts and bolts of chassis components only when vehicle is at normal ride height. condition.

10. Tighten bolts/nuts to specification as follows:

- Self-locking nut, track control arm to rear subframe: 52 ft. lbs. (70 Nm)
- Self-locking nut, spring control arm to wheel carrier 89 ft. lbs. (120 Nm)

### S-Class

*See Figure 191.*

1. Before servicing the vehicle, refer to the precautions.

### ✳✳ WARNING

While working on the rear axle care must be taken to ensure that the surface of aluminum components do not get any scratches, cracks and notches. Otherwise the service life of the parts will be affected.

2. Remove wheel.

3. Detach level sensor linkage from spring control arm.

4. Detach expansion rivets and remove cover .

5. Detach exhaust rubber mount from flange connection and lower exhaust system at rear.

6. Detach suspension strut from spring control arm.

7. Detach spring control arm from rear axle carrier.

8. Detach spring control arm from wheel carrier.

9. Release tension on rear spring and remove from spring control arm.

10. Check rubber mount in rear axle carrier.

**To install:**

11. Installation is the reverse of removal.

### ✳✳ WARNING

Replace all self locking, micro-encapsulated and aluminum bolts and nuts. Chase threads to remove micro-encapsulated residue from old bolts/nuts.

➡**Final tighten nuts and bolts of chassis components only when vehicle is at normal ride height.**

12. Tighten bolts/nuts to specification as follows:

- Nut, linkage rod bracket to spring control arm: 16 ft. lbs. (22 Nm)
- Self-locking nut, track control arm to rear subframe: 44 ft. lbs. (60 Nm) plus an additional 90° rotation
- Self-locking nut, spring control arm to wheel carrier: 59 ft. lbs. (80 Nm) plus an additional 90° rotation
- Nut, rear strut to spring control arm: 81 ft. lbs. (110Nm)

13. Carry out level calibration and plunger calibration using STAR DIAGNOSIS tool.

## MACPHERSON STRUTS

### REMOVAL & INSTALLATION

### S Class

*See Figure 192.*

1. Before servicing the vehicle, refer to the precautions.

2. Relieve the pressure in the strut if equipped with Active Body Control.

3. Unscrew the pressure line connection.

4. Remove the track control arm shield.

5. Remove upper strut mounting bolts.

6. Remove lower strut mounting bolt.

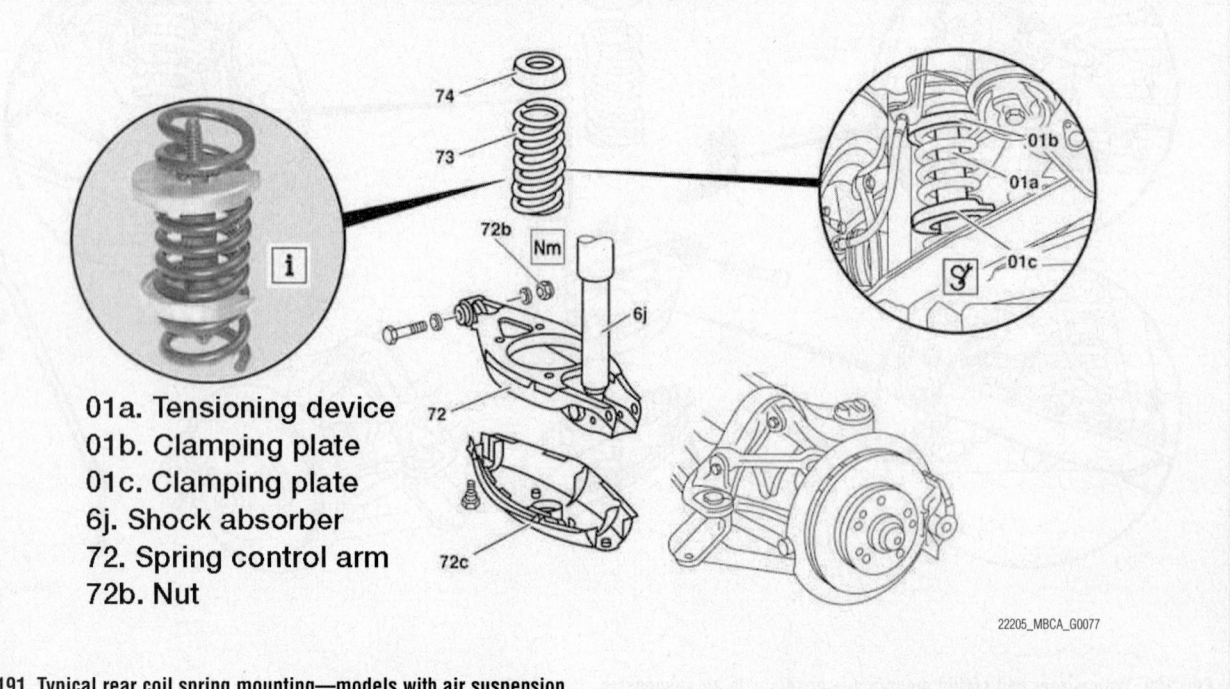

01a. Tensioning device
01b. Clamping plate
01c. Clamping plate
6j. Shock absorber
72. Spring control arm
72b. Nut

22205_MBCA_G0077

**Fig. 191 Typical rear coil spring mounting—models with air suspension**

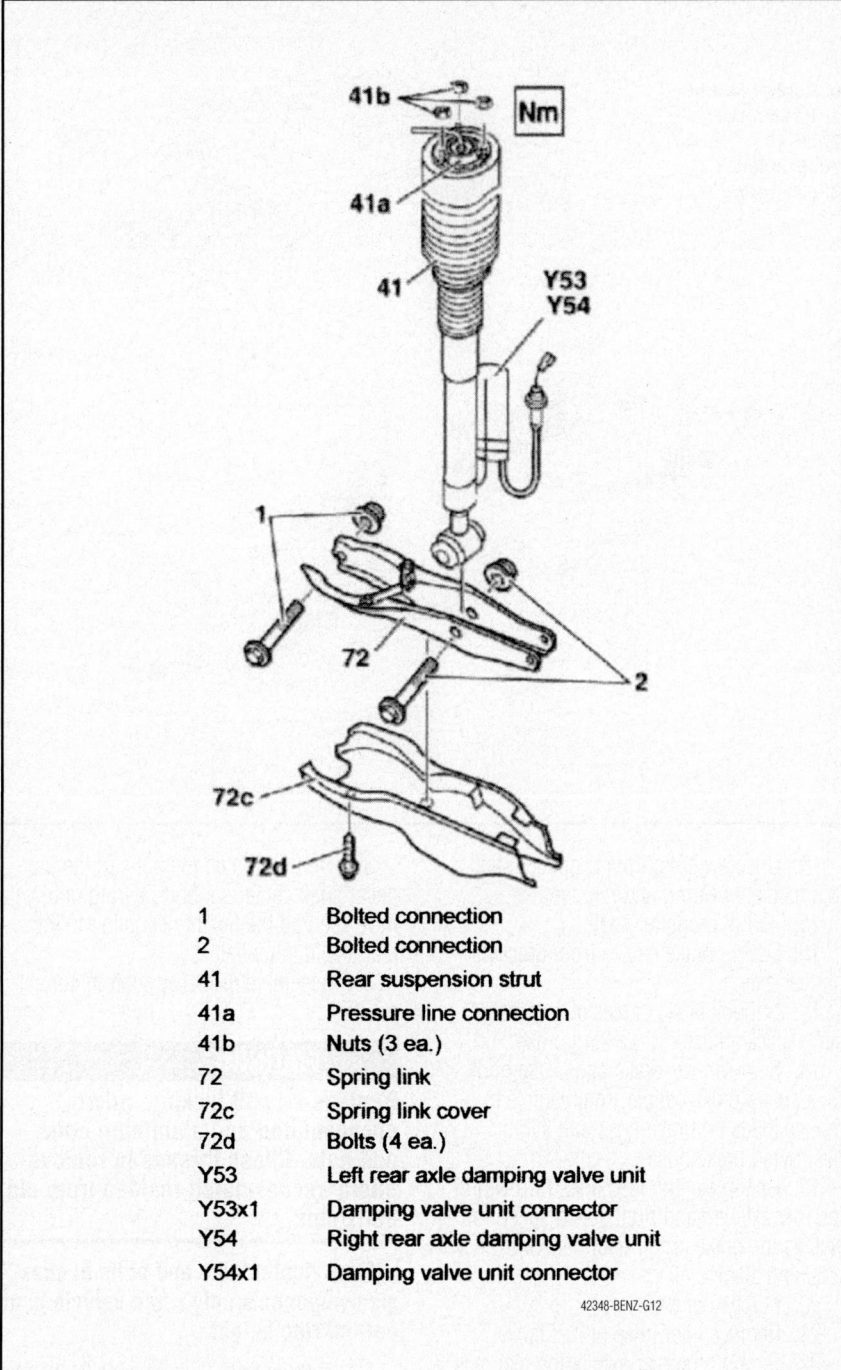

| | |
|---|---|
| 1 | Bolted connection |
| 2 | Bolted connection |
| 41 | Rear suspension strut |
| 41a | Pressure line connection |
| 41b | Nuts (3 ea.) |
| 72 | Spring link |
| 72c | Spring link cover |
| 72d | Bolts (4 ea.) |
| Y53 | Left rear axle damping valve unit |
| Y53x1 | Damping valve unit connector |
| Y54 | Right rear axle damping valve unit |
| Y54x1 | Damping valve unit connector |

42348-BENZ-G12

**Fig. 192 Rear strut assembly—S Class**

7. Remove upper thrust arm bolt.
8. Remove strut assembly.

**To install:**
9. Install strut assembly.
10. Torque the lower bolts to 81 ft. lbs. (110 Nm), torque the upper nuts 15 ft. lbs. (20 Nm).
11. Install upper thrust arm bolt. Torque to 37 ft. lbs. (50 Nm) plus 90 °.
12. Install track arm control shield.
13. If equipped with Active Body Control, pressurize the strut.

### SHOCK ABSORBER

*REMOVAL & INSTALLATION*

#### C-Class and E-Class

1. Before servicing the vehicle, refer to the precautions.
2. Detach upper shock absorber mount. The vehicle must be on its wheels for removing the upper shock absorber mount.
3. Remove plate and rubber mount.

4. Raise and safely support the vehicle securely.
5. Remove rear axle switch for roll bar.
6. Remove track control arm shield.
7. Detach shock absorber from spring control arm.
8. Remove mounting parts for shock absorber.

**To install:**
9. Installation is the reverse of removal.

### ❉❉ WARNING
**Replace all self locking, micro-encapsulated and aluminum bolts and nuts. Chase threads to remove micro-encapsulated residue from old bolts/nuts.**

10. Tighten bolts/nuts to specification as follows:
- Nut for attaching shock absorber to frame floor 20 ft. lbs. (30 Nm)
- Self-locking nut for attaching shock absorber to spring control arm 45 ft. lbs. (55 Nm)

### STABILIZER BAR

*REMOVAL & INSTALLATION*

#### C-Class

*See Figure 193.*

1. Before servicing the vehicle, refer to the precautions.
2. Remove rear wheels.
3. Unhook the exhaust system from the rear rubber retainers (6 pieces), lower and secure to prevent it from falling down.
4. Detach torsion bar on link rods.
5. Unscrew retaining bracket from bracket of torsion bar on rear axle carrier.
6. Take off torsion bar.
7. Check link rods for wear and damage.

**To install:**
8. Installation is the reverse of removal.

### ❉❉ WARNING
**Replace all self locking, micro-encapsulated and aluminum bolts and nuts. Chase threads to remove micro-encapsulated residue from old bolts/nuts.**

9. Tighten bolts/nuts to specification as follows:
- Nut, link rod to stabilizer bar: 30 ft. lbs. (40 Nm)
- Bolt, link rod to wheel carrier: 30 ft. lbs. (40 Nm)
- Bolt, bracket to stabilizer bar bracket: 22 ft. lbs. (30 Nm)

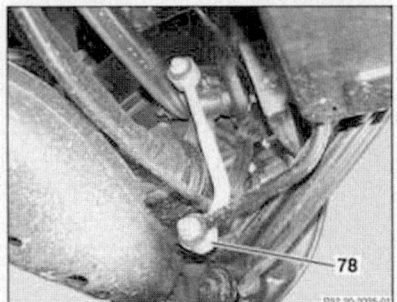

2c. Rubber Mount
77. Torsion bar
77b. Bolt
77d. Bracket
78. Link rods

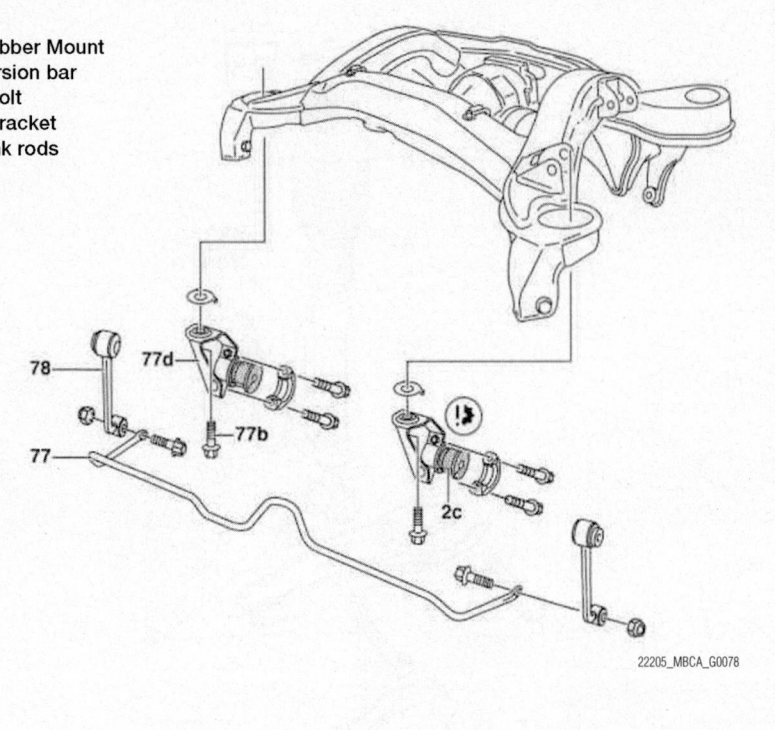

**Fig. 193 Rear stabilizer bar assembly**

## E-Class

*See Figures 194 and 195.*

1. Before servicing the vehicle, refer to the precautions.

### ✳✳ WARNING

**While working on the rear axle, care must be taken to ensure that the surface of aluminum components do not get any scratches, cracks and notches. Otherwise the service life of the parts will be affected.**

2. Deactivate SBC brake system using STAR DIAGNOSIS tool.

3. Secure vehicle on the lifting platform.

4. Remove rear wheels.

5. Remove crossmember.

6. Remove rear muffler.

7. Remove exhaust system as of flange connection.

8. Remove complete exhaust system.

9. Remove exhaust system from pipe connection.

10. Remove rear section components on under floor paneling.

11. Detach and remove heat shields.

12. Detach propeller shaft from flange on rear axle center assembly.

13. Undo fitting sleeves of flexible coupling on flange of rear axle center assembly.

14. Unscrew bolts from propeller shaft intermediate bearing and remove rear.

15. part of propeller shaft.

16. Detach brake cables from brackets on rear axle.

17. Unhook brake cables at automatic cable slack adjuster of parking brake.

18. Remove rear brake pads. Use hook to suspend brake caliper from vehicle to reduce loads on brake hose and avoid destroying brake hose.

19. Unclip electric feed lines (arrows) of the rear axle left and right speed sensor as well as the brake lining wear sensor from all retaining clamps.

20. Empty fuel tank.

21. Remove filler hose of fuel tank.

22. Support the rear axle at the rear axle center assembly with a transmission jack.

23. Detach rear axle carrier from front of frame; to do this, unscrew bolts.

24. Unclip linkage of rear axle level sensor from stabilizer bar.

25. Detach stabilizer bar from left and right link rod.

26. Detach left and right mount of stabilizer bar from rear axle carrier.

27. Remove stabilizer bar to the right.

28. Check condition of air springs.

### To install:

29. Installation is the reverse of removal.

30. Do not tighten the bolts of the propeller shaft center support bearing until you have secured the flexible coupling to the rear axle differential.

31. Assemble propeller shaft in correct position.

### ✳✳ WARNING

**Replace all self locking, micro-encapsulated and aluminum bolts and nuts. Chase threads to remove micro-encapsulated residue from old bolts/nuts.**

➡ **Final tighten nuts and bolts of chassis components only when vehicle is at normal ride height.**

32. Tighten bolts/nuts to specification as follows:

- Bolt, crossmember to body: 19 ft. lbs. (25 Nm)
- Self-locking nut, flexible coupling to rear propeller shaft or rear axle differential M10: 30 ft. lbs. (40 Nm)
- Self-locking nut, flexible coupling to rear propeller shaft or rear axle differential M12: 44 ft. lbs. (60 Nm)
- Self-locking bolt, front rear axle carrier rubber mount to frame floor: 67 ft. lbs. (90 Nm) plus an additional 60° rotation

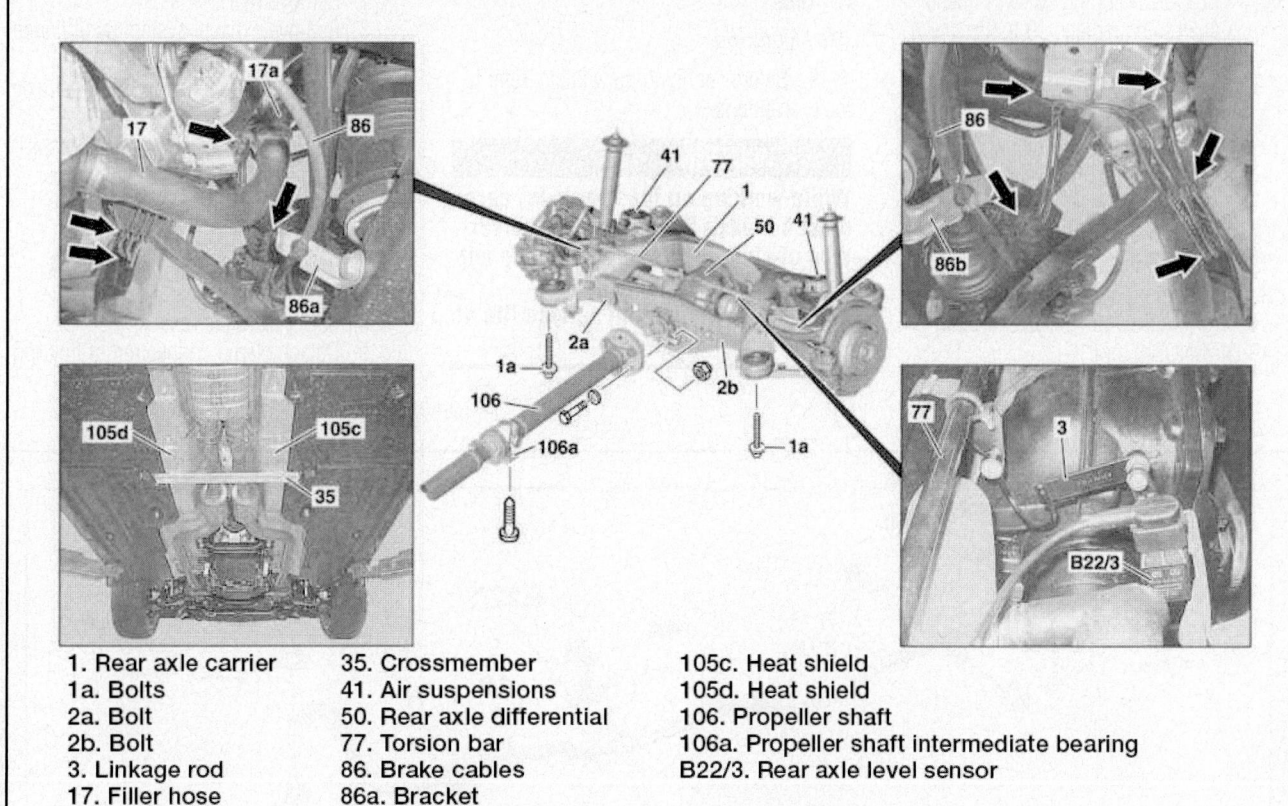

1. Rear axle carrier
1a. Bolts
2a. Bolt
2b. Bolt
3. Linkage rod
17. Filler hose
17a. Filler neck

35. Crossmember
41. Air suspensions
50. Rear axle differential
77. Torsion bar
86. Brake cables
86a. Bracket
86b. Bracket

105c. Heat shield
105d. Heat shield
106. Propeller shaft
106a. Propeller shaft intermediate bearing
B22/3. Rear axle level sensor

22205_MBCA_G0285

**Fig. 194 Rear axle assembly**

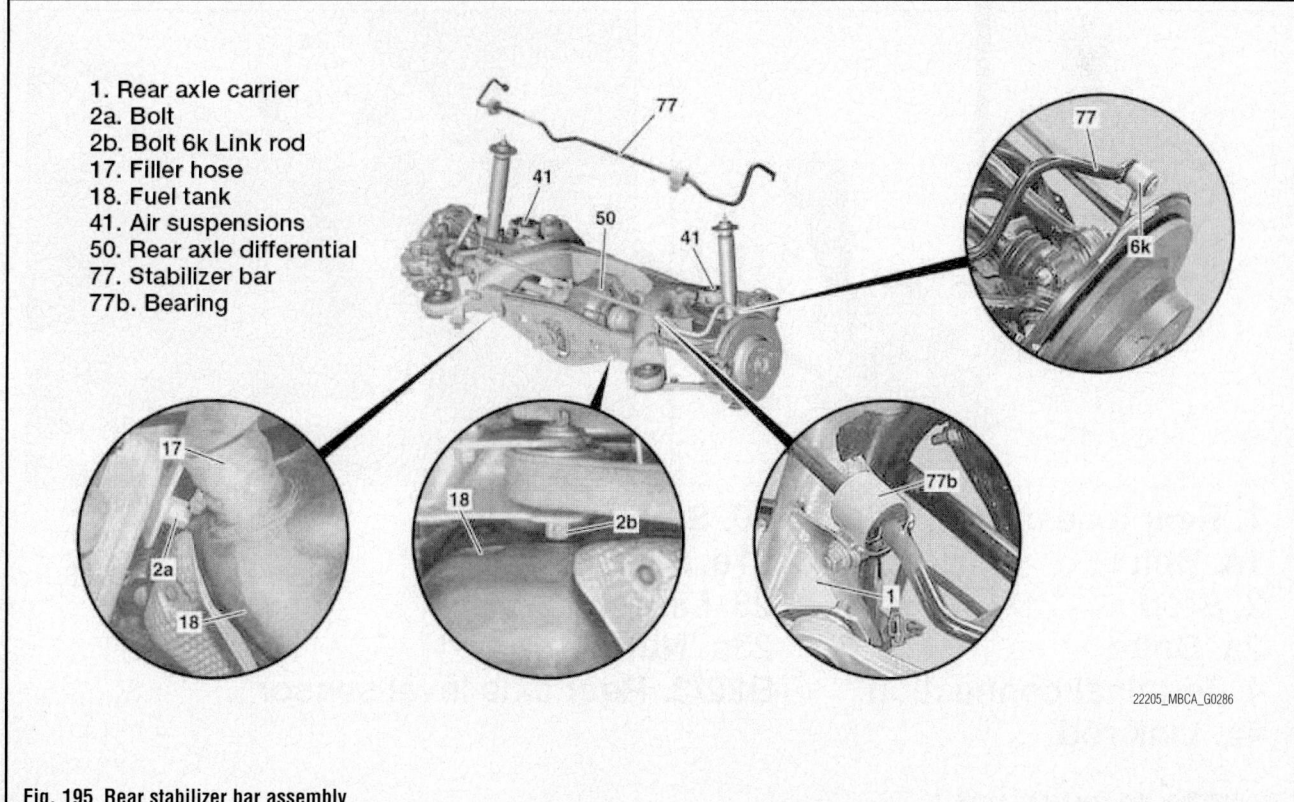

1. Rear axle carrier
2a. Bolt
2b. Bolt 6k Link rod
17. Filler hose
18. Fuel tank
41. Air suspensions
50. Rear axle differential
77. Stabilizer bar
77b. Bearing

22205_MBCA_G0286

**Fig. 195 Rear stabilizer bar assembly**

- Bolt, Self-tapping screw propeller shaft intermediate: 19 ft. lbs. (25 Nm)
- Nut, link rod to stabilizer bar: 37 ft. lbs. (50 Nm)
- Bolt, torsion bar to rear subframe: 52 ft. lbs. (70 Nm)
- Nut, rear air spring at spring control arm: 111 ft. lbs. (150 Nm)

33. Check oil level in rear axle center assembly and correct if necessary.

34. Activate SBC brake system using STAR DIAGNOSIS tool.

### S-Class

*See Figure 196.*

1. Before servicing the vehicle, refer to the precautions.

### ✳✳ WARNING

**While working on the rear axle, care must be taken to ensure that the surface of aluminum components do not get any scratches, cracks and notches. Otherwise the service life of the parts will be affected.**

2. Remove rear wheels.

3. Remove exhaust system as of flange connection.

4. Detach propeller shaft from rear axle differential.

5. Support rear axle carrier with transmission jack and transmission platform.

6. Remove bolts from left and right stop.

7. Remove bolts from left and right front rubber mount of rear axle carrier.

8. Lower front rear axle carrier.

9. Detach clamp connection of link rod of level sensor from stabilizer bar.

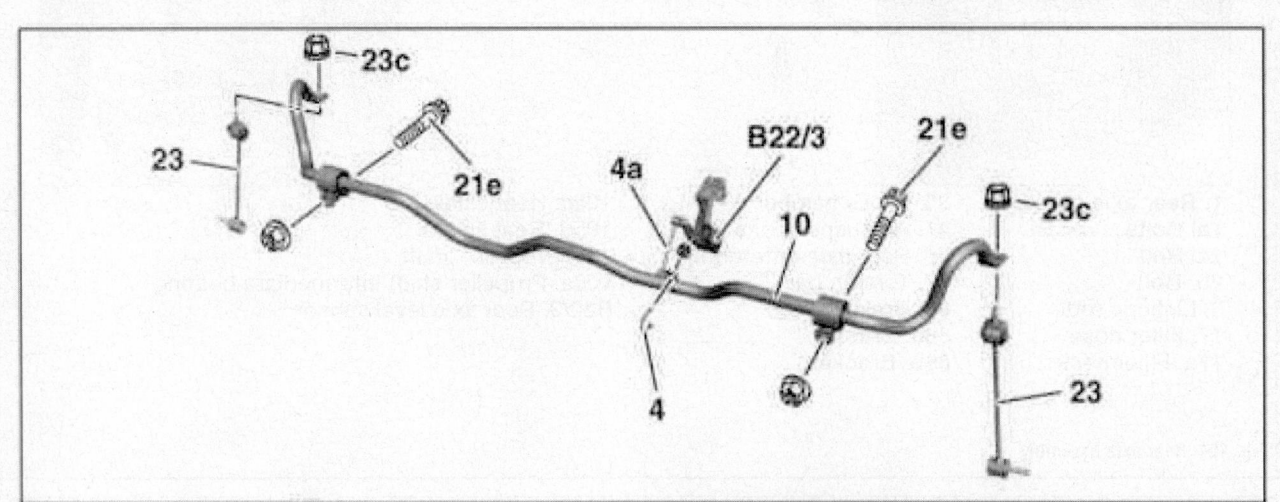

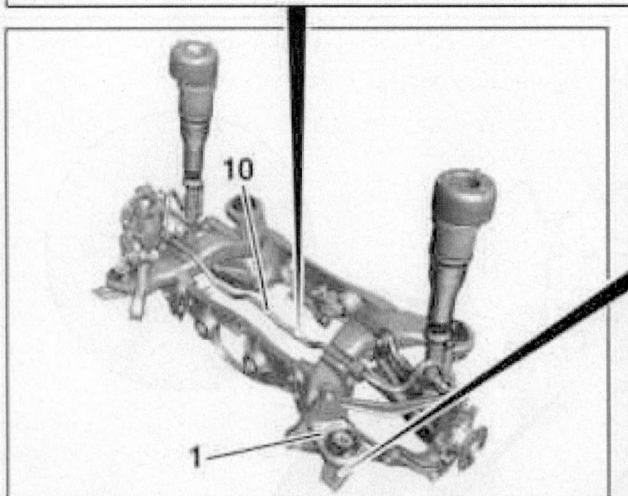

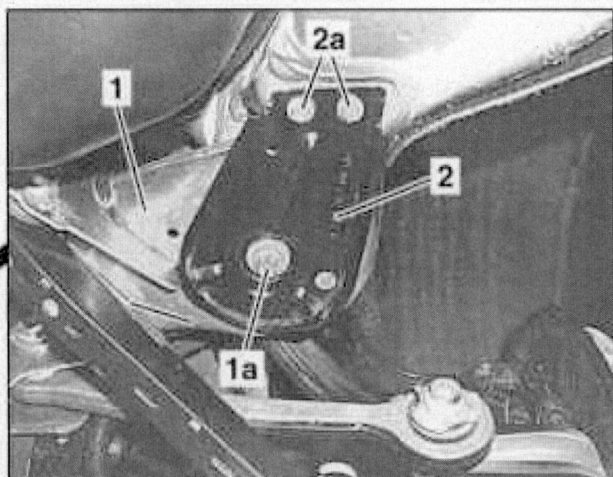

1. **Rear axle carrier**
1a. **Bolt**
2. **Stop**
2a. **Bolts**
4. **Terminal connection**
4a. **Link rod**

10. **Stabilizer bar**
21e. **Bolts**
23. **Link rod**
23c. **Nut**
B22/3. **Rear axle level sensor**

**Fig. 196 Rear stabilizer bar assembly**

22205_MBCA_G0287

10. Unscrew nuts of link rods to stabilizer bar and press link rods out of stabilizer bar.

11. Unscrew bolts.

12. Remove stabilizer bar to left side of vehicle.

**To install:**

13. Installation is the reverse of removal.

14. Do not tighten the bolts of the propeller shaft center support bearing until you have secured the flexible coupling to the rear axle differential.

15. Assemble propeller shaft in correct position.

### ❊❊ WARNING

**Replace all self locking, micro-encapsulated and aluminum bolts and nuts. Chase threads to remove micro-encapsulated residue from old bolts/nuts.**

➡**Final tighten nuts and bolts of chassis components only when vehicle is at normal ride height.**

16. Tighten bolts/nuts to specification as follows:

- Self-locking bolt, front rear axle carrier rubber mount to frame floor: 59 ft. lbs. (80 Nm) plus an additional 90° rotation
- Bolt, stop to longitudinal member: 26 ft. lbs. (35 Nm)
- Nut of retaining shell of torsion bar intermediate lever to torsion bar: 4 ft. lbs. (5 Nm)
- Nut, link rod to stabilizer bar: 37 ft. lbs. (50 Nm)
- Bolt, stabilizer bar to rear axle carrier: 52 ft. lbs. (70 Nm)

### WHEEL BEARINGS

*REMOVAL & INSTALLATION*

#### C-Class

*See Figure 197.*

1. Before servicing the vehicle, refer to the precautions.

2. Detach twelve-point nut of rear axle shaft.

3. Secure vehicle on the lifting platform.

4. Remove brake shoes of parking brake.

5. Remove link rod of stabilizer bar.

6. Detach brake cable from wheel carrier.

7. Detach rear rpm sensor from wheel carrier.

8. Detach camber strut from wheel carrier.

9. Remove torque strut from wheel carrier.

10. Remove thrust arm from wheel carrier.

11. Press rear axle shaft out of wheel carrier.

12. Remove bolt securing spring control arm to wheel carrier and pull wheel carrier upward out of spring control arm.

**To install:**

13. Installation is the reverse of removal.

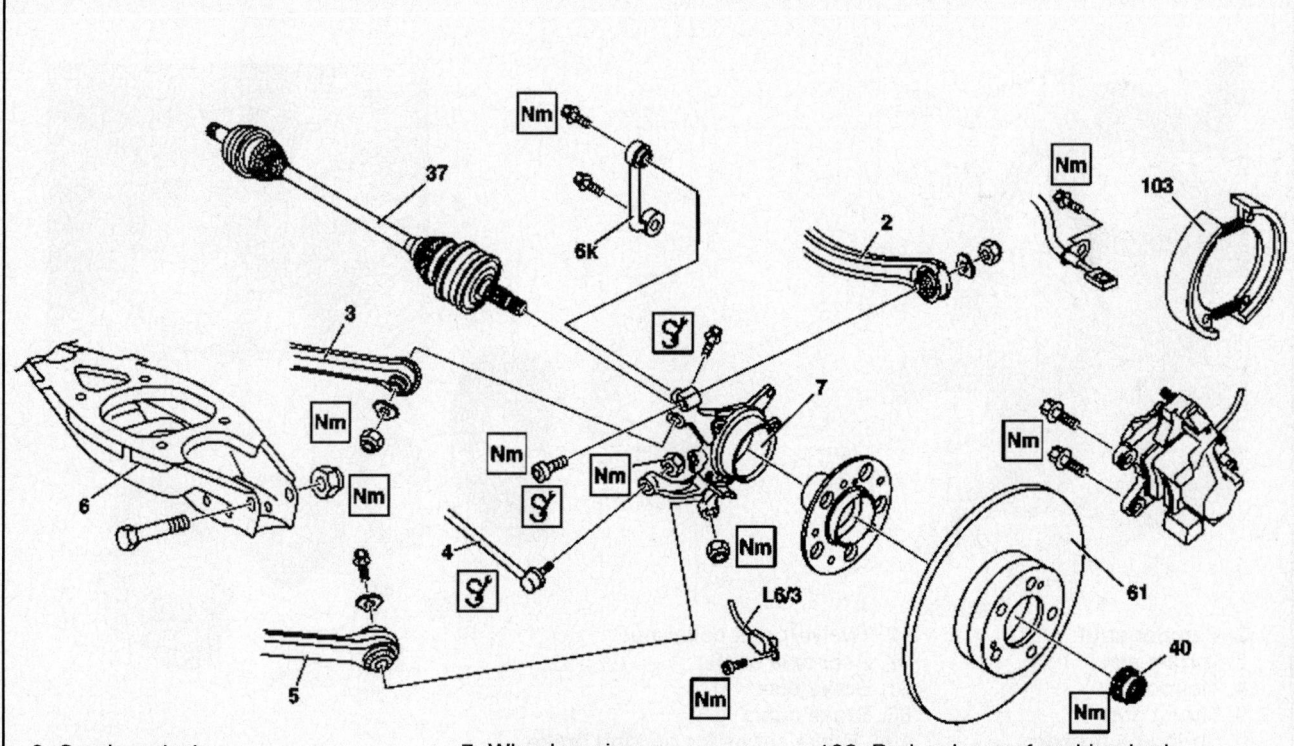

2. Camber strut
3. Torque strut
4. Tie rod
5. Thrust arm 6 Spring control arm
6k. Torsion bar connecting rod

7. Wheel carrier
37. Rear axle shaft
40. Twelve-point collar nut
61. Brake disk

103. Brake shoes of parking brake
L6/3. Left rear rpm sensor

22205_MBCA_G0079

**Fig. 197 Rear wheel carrier assembly**

**✳✳ WARNING**

Do not damage the supporting joint boot when moving in the wheel carrier.

**✳✳ WARNING**

Replace all self locking, micro-encapsulated and aluminum bolts and nuts. Chase threads to remove micro-encapsulated residue from old bolts/nuts.

**✳✳ WARNING**

Final tighten nuts and bolts of chassis components only when vehicle is at normal ride height. condition.

14. Tighten bolts/nuts to specification as follows:
- Collar nut, rear axle shaft to rear axle shaft flange: 163 ft. lbs. (220 Nm)

- Self-locking bolt, brake cable at rear wheel carrier:15 ft. lbs. (20 Nm)
- Nut, link rod to stabilizer bar: 30 ft. lbs. (40 Nm)
- Bolt, link rod to wheel carrier: 30 ft. lbs. (40 Nm)

15. Adjust parking brake.

### E-Class

*See Figure 198.*

1. Before servicing the vehicle, refer to the precautions.

**✳✳ WARNING**

While working on the rear axle care must be taken to ensure that the surface of all aluminum components do not get any scratches, cracks and notches. Otherwise the service life of the parts will be affected.

2. Deactivate SBC brake system using STAR DIAGNOSIS tool.

3. Remove twelve-point collar nut from rear axle shaft.

4. Remove rear wheel.

5. Remove rear brake pads.

6. Detach brake caliper at rear axle. Secure rear brake caliper to vehicle to relieve load on brake hose to prevent brake hose from being damaged during repair work.

7. Remove brake disk.

8. Remove brake shoes of parking brake.

9. Pull brake cable out of wheel carrier.

10. Detach left rear rpm sensor or right rear rpm sensor from wheel carrier.

11. Detach stabilizer bar link rod from wheel carrier.

12. Raise wheel carrier with transmission jack and transmission platform until the rear axle shaft is approximately horizontal.

13. Detach torque strut from wheel carrier.

14. Detach camber strut from wheel carrier.

15. Detach tie rod from wheel carrier.

16. Detach thrust arm from wheel carrier.

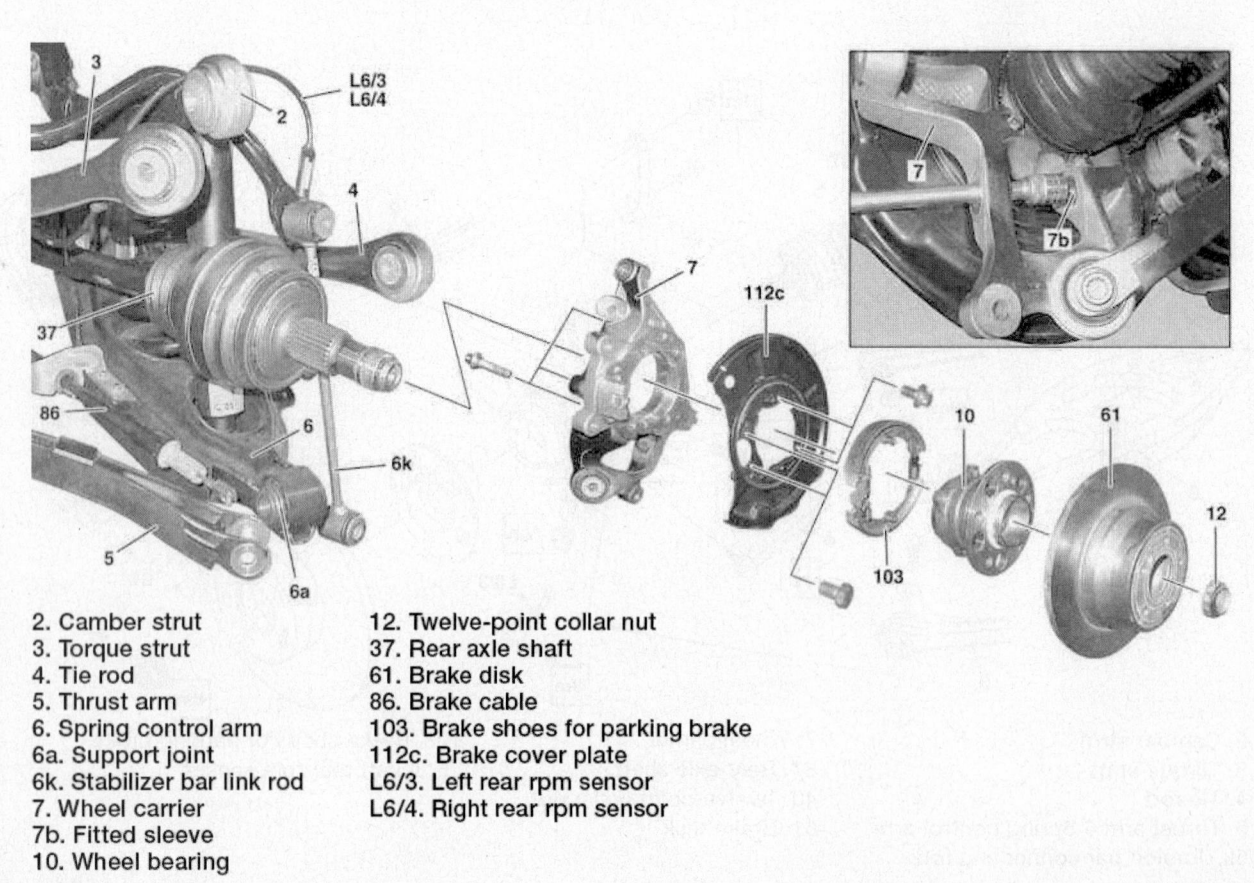

2. Camber strut
3. Torque strut
4. Tie rod
5. Thrust arm
6. Spring control arm
6a. Support joint
6k. Stabilizer bar link rod
7. Wheel carrier
7b. Fitted sleeve
10. Wheel bearing

12. Twelve-point collar nut
37. Rear axle shaft
61. Brake disk
86. Brake cable
103. Brake shoes for parking brake
112c. Brake cover plate
L6/3. Left rear rpm sensor
L6/4. Right rear rpm sensor

**Fig. 198 Rear wheel carrier assembly**

22205_MBCA_G0288

17. Lower transmission jack.

18. Unscrew bolt holding spring control arm to wheel carrier.

19. Check that rear axle shaft is firmly seated in rear axle shaft flange.

20. Pry off wheel carrier supporting joint and remove wheel carrier.

➡**Suspend rear axle shaft on vehicle to prevent damage to the rubber boot on the rear axle shaft.**

**To install:**

21. Installation is the reverse of removal.

### ❋❋ WARNING

**Do not damage the supporting joint boot when moving in the wheel carrier.**

### ❋❋ WARNING

**Replace all self locking, micro-encapsulated and aluminum bolts and nuts. Chase threads to remove micro-encapsulated residue from old bolts/nuts.**

### ❋❋ WARNING

**Final tighten nuts and bolts of chassis components only when vehicle is at normal ride height. condition.**

22. Tighten rear axle shaft to rear axle shaft flange collar nut to 237 ft. lbs. (320 Nm)

### S-Class

*See Figure 199.*

1. Before servicing the vehicle, refer to the precautions.

### ❋❋ WARNING

**While working on the rear axle care must be taken to ensure that the surface of all aluminum components do not get any scratches, cracks and notches. Otherwise the service life of the parts will be affected.**

2. Remove/install wheels.

3. Remove twelve-point collar nut from rear axle shaft.

4. Move electronic parking brake into assembly position using STAR DIAGNOSIS tool.

5. Detach brake caliper at rear axle. Attach brake caliper to vehicle without tension to relieve brake hose so that it is not damaged during repair.

6. Remove brake disc.

7. Remove brake shoes.

8. Detach brake cover plate from wheel carrier.

➡**The brake cover plate cannot be removed, but it must be twisted in order to be able to remove the bolts of the struts.**

9. Detach rpm sensor from wheel carrier.

10. Pull brake cable out of wheel carrier.

11. Detach link rod from wheel carrier.

12. Raise wheel carrier using transmission jack until rear axle shaft is roughly horizontal.

13. Detach torque strut from wheel carrier.

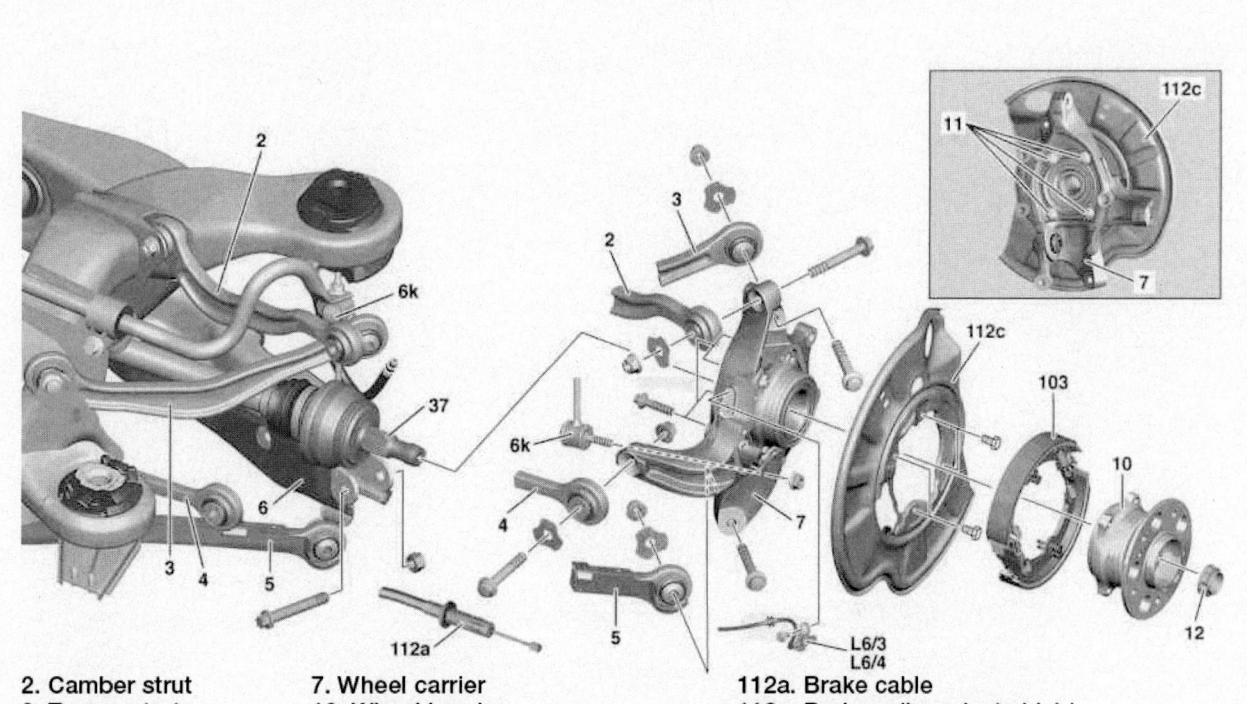

| | | |
|---|---|---|
| 2. Camber strut | 7. Wheel carrier | 112a. Brake cable |
| 3. Torque strut | 10. Wheel bearing | 112c. Brake caliper dust shield |
| 4. Tie rod | 11. Bolts | L6/3. Left rear rpm sensor |
| 5. Thrust arm | 12. Twelve-point collar nut | L6/4. Right rear rpm sensor |
| 6. Spring control arm | 37. Rear axle shaft | |
| 6k. Link rod | 103. Brake shoes for parking brake | |

22205_MBCA_G0289

**Fig. 199 Rear wheel carrier assembly**

14. Detach camber strut from wheel carrier.

15. Detach tie rod from wheel carrier.

16. Detach thrust arm from wheel carrier.

17. Check that rear axle shaft is firmly seated in rear axle shaft flange.

18. Detach spring control arm from wheel carrier.

19. Remove wheel carrier with brake cover plate and wheel bearing from rear axle shaft.

20. Pull rear axle shaft flange out of angular contact ball bearing.

21. Detach wheel bearing from wheel carrier.

22. Press rear axle shaft flange off wheel carrier.

**To install:**

23. Installation is the reverse of removal.

---

❋❋ **WARNING**

**Do not damage the supporting joint boot when moving in the wheel carrier.**

---

❋❋ **WARNING**

**Replace all self locking, micro-encapsulated and aluminum bolts**

and nuts. Chase threads to remove micro-encapsulated residue from old bolts/nuts.

---

❋❋ **WARNING**

**Final tighten nuts and bolts of chassis components only when vehicle is at normal ride height condition.**

---

24. Tighten rear axle shaft to rear axle shaft flange collar nut to 237 ft. lbs. (320 Nm)

25. Adjust parking brake.

26. Perform a wheel alignment check.

# MERCEDES-BENZ

## ML350 • ML500 • ML550

# 8

## SPECIFICATIONS AND MAINTENANCE CHARTS

### ENGINE AND VEHICLE IDENTIFICATION CHART

| | | Engine Code | | | | | Model Year | |
|---|---|---|---|---|---|---|---|---|
| Code | Liters (cc) | Cu. In. | Cyl. | Fuel Sys. | Eng. Mfg. | | Code ① | Year |
| M272.967 | 3.5 (3498) | 213 | 6 | SFI | MB | | 6 | 2006 |
| M113.964 | 5.0 (4966) | 303 | 8 | SFI | MB | | 7 | 2007 |
| M273.963 | 5.5 (5461) | 333 | 8 | SFI | MB | | 8 | 2008 |

SFI: Sequential Fuel Injection

MB: Mercedes-Benz

① 10th digit of the VIN

22205_MBML_C0001

### GENERAL ENGINE SPECIFICATIONS

| Year | Model | Engine Displacement Liters | Engine ID | Net Horsepower @ rpm | Net Torque @ rpm (ft. lbs.) | Bore x Stroke (in.) | Compression Ratio | Oil Pressure @ rpm |
|---|---|---|---|---|---|---|---|---|
| 2006 | ML350 | 3.5 | M272 | 268@5750 | 258@2400 | 3.66x3.39 | 10.0:1 | 43.5@3000 |
| | ML500 | 5.0 | M113 | 302@5600 | 339@2700 | 3.82x3.31 | 10.7:1 | 43.5@3000 |
| 2007 | ML350 | 3.5 | M272 | 268@5750 | 258@2400 | 3.66x3.39 | 10.7:1 | 43.5@3000 |
| | ML500 | 5.0 | M113 | 302@5600 | 339@2700 | 3.82x3.31 | 10.0:1 | 43.5@3000 |
| 2008 | ML350 | 3.5 | M272 | 268@5750 | 258@2400 | 3.66x3.39 | 10.7:1 | 43.5@3000 |
| | ML550 | 5.5 | M273 | 382@5600 | 391@2800 | 3.66x3.39 | 10.7:1 | 43.5@3000 |

22205_MBML_C0002

### GASOLINE ENGINE TUNE-UP SPECIFICATIONS

| Year | Engine Displacement Liters | Engine ID | Spark Plug Gap (in.) | Ignition Timing (deg.) | Fuel Pump (psi) | Idle Speed (rpm) | Valve Clearance In. | Valve Clearance Ex. |
|---|---|---|---|---|---|---|---|---|
| 2006 | 3.5 | M272 | 0.039 | ① | 55 | 700 | ② | ② |
| | 5.0 | M113 | 0.039 | ① | 55 | 700 | ② | ② |
| 2007 | 3.5 | M272 | 0.039 | ① | 55 | 700 | ② | ② |
| | 5.0 | M113 | 0.039 | ① | 55 | 700 | ② | ② |
| 2008 | 3.5 | M272 | 0.039 | ① | 55 | 700 | ② | ② |
| | 5.5 | M273 | 0.039 | ① | 55 | 700 | ② | ② |

① ECM controlled

② Hydraulic lash adjusters

22205_MBML_C0004

## CAPACITIES

| Year | Model | Engine Displacement Liters | Engine ID | Engine Oil with Filter (qts.) | Transmission (pts.) | Transfer Case (pts.) | Drive Axle | | Fuel Tank (gal.) | Cooling System (qts.) |
|------|-------|------|------|------|------|------|------|------|------|------|
| | | | | | | | Front (pts.) | Rear (pts.) | | |
| 2006 | ML350 | 3.5 | M272 | 8.5 | 18 | 3.2 | 2.5 | 3.5 | 25.1 | 12.7 |
| | ML500 | 5.0 | M113 | 8.5 | 18 | 3.2 | 2.5 | 3.5 | 25.1 | 12.7 |
| 2007 | ML350 | 3.5 | M272 | 8.5 | 18 | 3.2 | 2.5 | 3.5 | 25.1 | 12.7 |
| | ML500 | 5.0 | M113 | 8.5 | 18 | 3.2 | 2.5 | 3.5 | 25.1 | 12.7 |
| 2008 | ML350 | 3.5 | M272 | 8.5 | 18 | 3.2 | 2.5 | 3.5 | 25.1 | 12.7 |
| | ML550 | 5.5 | M273 | 8.5 | 18 | 3.2 | 2.5 | 3.5 | 25.1 | 12.7 |

Note: All capacities are approximate. Add fluid gradually and check to be sure a proper fluid level is obtained.

22205_MBML_C0003

## VALVE SPECIFICATIONS

| Year | Engine Displacement Liters | Engine ID | Seat Angle (deg.) | Face Angle (deg.) | Spring Test Pressure (lbs. @ in.) | Spring Installed Height (in.) | Stem-to-Guide Clearance (in.) | | Stem Diameter (in.) | |
|------|------|------|------|------|------|------|------|------|------|------|
| | | | | | | | Intake | Exhaust | Intake | Exhaust |
| 2006 | 3.7 | M272 | 45 | NA | NA | NA | NA | NA | 0.275 | 0.274 |
| | 5.0 | M113 | 45 | NA | NA | NA | NA | NA | 0.275 | 0.274 |
| 2007 | 3.7 | M272 | 45 | NA | NA | NA | NA | NA | 0.275 | 0.274 |
| | 5.0 | M113 | 45 | NA | NA | NA | NA | NA | 0.275 | 0.274 |
| 2008 | 3.7 | M272 | 45 | NA | NA | NA | NA | NA | 0.275 | 0.274 |
| | 5.5 | M273 | 45 | NA | NA | NA | NA | NA | 0.275 | 0.274 |

NA: Information not available

22205_MBML_C0005

## CRANKSHAFT AND CONNECTING ROD SPECIFICATIONS

All measurements are given in inches.

| Year | Engine Displacement Liters | Engine ID/VIN | Crankshaft | | | | Connecting Rod | | |
|------|---|---|---|---|---|---|---|---|---|
| | | | Main Brg. Journal Dia. | Main Brg. Oil Clearance | Shaft End-play | Thrust on No. | Journal Diameter | Oil Clearance | Side Clearance |
| 2006 | 3.5 | M272 | NA | 0.0008-0.0018 | 0.0039-0.0104 | 3 | NA | 0.0008-0.0018 | NA |
| | 5.0 | M113 | NA | 0.0010-0.0019 | 0.0039-0.0104 | 3 | NA | 0.0010-0.0019 | NA |
| 2007 | 3.5 | M272 | NA | 0.0008-0.0018 | 0.0039-0.0104 | 3 | NA | 0.0008-0.0018 | NA |
| | 5.0 | M113 | NA | 0.0010-0.0019 | 0.0039-0.0104 | 3 | NA | 0.0010-0.0019 | NA |
| 2008 | 3.5 | M272 | NA | 0.0008-0.0018 | 0.0039-0.0104 | 3 | NA | 0.0008-0.0018 | NA |
| | 5.5 | M273 | NA | 0.0008-0.0018 | 0.0039-0.0104 | 3 | NA | 0.0008-0.0018 | NA |

NA: Not Available

22205_MBML_C0006

## PISTON AND RING SPECIFICATIONS

All measurements are given in inches.

| Year | Engine Displacement Liters | Engine ID/VIN | Piston Clearance | Ring Gap | | | Ring Side Clearance | | |
|------|---|---|---|---|---|---|---|---|---|
| | | | | Top Compression | Bottom Compression | Oil Control | Top Compression | Bottom Compression | Oil Control |
| 2006 | 3.5 | M272 | 0.0001 | NA | NA | NA | NA | NA | NA |
| | 5.0 | M113 | 0.0001 | NA | NA | NA | NA | NA | NA |
| 2007 | 3.5 | M272 | 0.0001 | NA | NA | NA | NA | NA | NA |
| | 5.0 | M113 | 0.0001 | NA | NA | NA | NA | NA | NA |
| 2008 | 3.5 | M272 | 0.0001 | NA | NA | NA | NA | NA | NA |
| | 5.5 | M273 | 0.0001 | NA | NA | NA | NA | NA | NA |

NA: Not Available

22205_MBML_C0007

## TORQUE SPECIFICATIONS
All readings in ft. lbs.

| Year | Engine Displacement Liters | Engine ID | Cylinder Head Bolts | Main Bearing Bolts | Rod Bearing Bolts | Crankshaft Damper Bolts | Flywheel Bolts | Manifold Intake | Manifold Exhaust | Spark Plugs | Oil Pan Drain Plug |
|------|------|------|------|------|------|------|------|------|------|------|------|
| 2006 | 3.5 | M272 | ① | ② | ③ | ④ | ⑤ | 15 | 12 | 21 | 22 |
|      | 5.0 | M113 | ① | ② | ③ | ⑥ | ⑤ | 15 | 12 | 21 | 22 |
| 2007 | 3.5 | M272 | ① | ② | ③ | ④ | ⑤ | 15 | 12 | 21 | 22 |
|      | 5.0 | M113 | ① | ② | ③ | ⑥ | ⑤ | 15 | 12 | 21 | 22 |
| 2008 | 3.5 | M272 | ① | ② | ③ | ④ | ⑤ | 15 | 12 | 21 | 22 |
|      | 5.5 | M273 | ① | ② | ③ | ④ | ⑤ | 15 | 12 | 21 | 22 |

① Step 1: 8 ft. lbs.
  Step 2: 22 ft. lbs.
  Step 3: plus 90 degrees
  Step 4: plus 90 degrees

② M8 vertical bolts:
  Step 1: 15 ft. lbs.
  Step 2: plus 90 degrees
  M8 side bolts: 22 ft. lbs.
  M10 bolts:
  Step 1: 44 inch lbs.
  Step 2: 22 ft. lbs.
  Step 3: plus 90 degrees

③ Step 1: 44 inch lbs.
  Step 2: 18 ft. lbs
  Step 3. 90 degrees

④ Step 1: 148 ft. lbs.
  Step 2: 95 degrees

⑤ Step 1: 33 ft. lbs.
  Step 2: 90 degrees

22205_MBML_C0008

## WHEEL ALIGNMENT

| Year | Model | | Caster Range (+/-Deg.) | Caster Preferred Setting (Deg.) ① | Camber Range (+/-Deg.) | Camber Preferred Setting (Deg.) | Toe-in (Deg.) |
|------|-------|---|------|------|------|------|------|
| 2006 | ML Series | F | 0.008 | 6.01 | 0.005 | 0.01 | 0.005 +/-0.002 |
|      |           | R | 0.5 | 8.24 | 0.005 | 0.01 | 0.005 +/-0.001 |
| 2007 | ML Series | F | 0.008 | 6.01 | 0.005 | 0.01 | 0.005 +/-0.002 |
|      |           | R | 0.5 | 8.24 | 0.005 | 0.01 | 0.005 +/-0.001 |
| 2008 | ML Series | F | 0.008 | 6.01 | 0.005 | 0.01 | 0.005 +/-0.002 |
|      |           | R | 0.5 | 8.24 | 0.005 | 0.01 | 0.005 +/-0.001 |

① Check with wheel turned 20 degrees.

22205_MBML_C0009

## TIRE, WHEEL AND BALL JOINT SPECIFICATIONS

| Year | Model | OEM Tires Standard | OEM Tires Optional | Tire Pressures (psi) Front | Tire Pressures (psi) Rear | Wheel Size | Ball Joint Inspection | Lug Nut Torque (ft. lbs.) |
|------|-------|----------|----------|-------|------|------------|-----------------|---------------------------|
| 2006 | ML350 | P235/65R17 | None | ① | ① | 8J | NS | 111 |
|      | ML500 | P255/55R18 | None | ① | ① | 8.5J | NS | 111 |
| 2007 | ML350 | P235/65R17 | None | ① | ① | 8J | NS | 111 |
|      | ML500 | P255/55R18 | None | ① | ① | 8.5J | NS | 111 |
| 2008 | ML350 | P255/50R19 | None | ① | ① | 8J | NS | 111 |
|      | ML550 | P255/50R19 | None | ① | ① | 8J | NS | 111 |

NS: Not Specified

① See placard on vehicle

22205_MBML_C0010

## BRAKE SPECIFICATIONS

All measurements in inches unless noted

| Year | Model | | Brake Disc Original Thickness | Brake Disc Minimum Thickness | Brake Disc Maximum Runout | Minimum Lining Thickness | Brake Caliper Bracket Bolts (ft. lbs.) | Brake Caliper Mounting Bolts (ft. lbs.) |
|------|-------|---|----------|----------|---------|----------|----------|----------|
| 2006 | ML350 | F | 1.26 | 1.16 | NA | NA | ① | ② |
|      |       | R | 0.86 | 0.76 | NA | NA | ③ | ④ |
|      | ML500 | F | 1.26 | 1.16 | NA | NA | ① | ② |
|      |       | R | 0.86 | 0.76 | NA | NA | ③ | ④ |
| 2007 | ML350 | F | 1.26 | 1.16 | NA | NA | ① | ② |
|      |       | R | 0.86 | 0.76 | NA | NA | ③ | ④ |
|      | ML500 | F | 1.26 | 1.16 | NA | NA | ① | ② |
|      |       | R | 0.86 | 0.76 | NA | NA | ③ | ④ |
| 2008 | ML350 | F | 1.26 | 1.16 | NA | NA | ① | ② |
|      |       | R | 0.86 | 0.76 | NA | NA | ③ | ④ |
|      | ML550 | F | 1.26 | 1.16 | NA | NA | ① | ② |
|      |       | R | 0.55 | 0.45 | NA | NA | ③ | ④ |

NA: Information not available

① Step 1. 59 ft. lbs.

  Step 2. Plus 45 degrees

② 41 ft. lbs.

③ Step 1. 44 ft. lbs.

  Step 2. Plus 45 degrees

④ 26 ft. lbs.

22205_MBML_C0011

## PRECAUTIONS

Before servicing any vehicle, please be sure to read all of the following precautions, which deal with personal safety, prevention of component damage, and important points to take into consideration when servicing a motor vehicle:

• Never open, service or drain the radiator or cooling system when the engine is hot; serious burns can occur from the steam and hot coolant.

• Observe all applicable safety precautions when working around fuel. Whenever servicing the fuel system, always work in a well-ventilated area. Do not allow fuel spray or vapors to come in contact with a spark, open flame, or excessive heat (a hot drop light, for example). Keep a dry chemical fire extinguisher near the work area. Always keep fuel in a container specifically designed for fuel storage; also, always properly seal fuel containers to avoid the possibility of fire or explosion. Refer to the additional fuel system precautions later in this section.

• Fuel injection systems often remain pressurized, even after the engine has been turned **OFF**. The fuel system pressure must be relieved before disconnecting any fuel lines. Failure to do so may result in fire and/or personal injury.

• Brake fluid often contains polyglycol ethers and polyglycols. Avoid contact with the eyes and wash your hands thoroughly after handling brake fluid. If you do get brake fluid in your eyes, flush your eyes with clean, running water for 15 minutes. If eye irritation persists, or if you have taken

brake fluid internally, IMMEDIATELY seek medical assistance.

• The EPA warns that prolonged contact with used engine oil may cause a number of skin disorders, including cancer. You should make every effort to minimize your exposure to used engine oil. Protective gloves should be worn when changing oil. Wash your hands and any other exposed skin areas as soon as possible after exposure to used engine oil. Soap and water, or waterless hand cleaner should be used.

• All new vehicles are now equipped with an air bag system, often referred to as a Supplemental Restraint System (SRS) or Supplemental Inflatable Restraint (SIR) system. The system must be disabled before performing service on or around system components, steering column, instrument panel components, wiring and sensors. Failure to follow safety and disabling procedures could result in accidental air bag deployment, possible personal injury and unnecessary system repairs.

• Always wear safety goggles when working with, or around, the air bag system. When carrying a non-deployed air bag, be sure the bag and trim cover are pointed away from your body. When placing a non-deployed air bag on a work surface, always face the bag and trim cover upward, away from the surface. This will reduce the motion of the module if it is accidentally deployed. Refer to the additional air bag system precautions later in this section.

• Clean, high quality brake fluid from a sealed container is essential to the safe and

proper operation of the brake system. You should always buy the correct type of brake fluid for your vehicle. If the brake fluid becomes contaminated, completely flush the system with new fluid. Never reuse any brake fluid. Any brake fluid that is removed from the system should be discarded. Also, do not allow any brake fluid to come in contact with a painted surface; it will damage the paint.

• Never operate the engine without the proper amount and type of engine oil; doing so WILL result in severe engine damage.

• Timing belt maintenance is extremely important. Many models utilize an interference-type, non-freewheeling engine. If the timing belt breaks, the valves in the cylinder head may strike the pistons, causing potentially serious (also time-consuming and expensive) engine damage. Refer to the maintenance interval charts for the recommended replacement interval for the timing belt, and to the timing belt section for belt replacement and inspection.

• Disconnecting the negative battery cable on some vehicles may interfere with the functions of the on-board computer system(s) and may require the computer to undergo a relearning process once the negative battery cable is reconnected.

• When servicing drum brakes, only disassemble and assemble one side at a time, leaving the remaining side intact for reference.

• Only an MVAC-trained, EPA-certified automotive technician should service the air conditioning system or its components.

## BRAKES                                    ANTI-LOCK BRAKE SYSTEM (ABS)

### GENERAL INFORMATION

*PRECAUTIONS*

• Certain components within the ABS system are not intended to be serviced or repaired individually.

• Do not use rubber hoses or other parts not specifically specified for and ABS system. When using repair kits, replace all parts included in the kit. Partial or incorrect repair may lead to functional problems and require the replacement of components.

• Lubricate rubber parts with clean, fresh brake fluid to ease assembly. Do not

use shop air to clean parts; damage to rubber components may result.

• Use only DOT 3 brake fluid from an unopened container.

• If any hydraulic component or line is removed or replaced, it may be necessary to bleed the entire system.

• A clean repair area is essential. Always clean the reservoir and cap thoroughly before removing the cap. The slightest amount of dirt in the fluid may plug an orifice and impair the system function. Perform repairs after components have been thoroughly cleaned; use only denatured alcohol

to clean components. Do not allow ABS components to come into contact with any substance containing mineral oil; this includes used shop rags.

• The Anti-Lock control unit is a microprocessor similar to other computer units in the vehicle. Ensure that the ignition switch is **OFF** before removing or installing controller harnesses. Avoid static electricity discharge at or near the controller.

• If any arc welding is to be done on the vehicle, the control unit should be unplugged before welding operations begin.

## BRAKES

### BLEEDING PROCEDURE

*BLEEDING PROCEDURE*

When any part of the hydraulic system has been disconnected for repair or replacement, air may get into the lines and cause spongy pedal action (because air can be compressed and brake fluid cannot). To correct this condition, it is necessary to bleed the hydraulic system so to be sure all air is purged.

When bleeding the brake system, bleed one brake cylinder at a time, beginning at the cylinder with the longest hydraulic line (farthest from the master cylinder) first. The sequence for calipers with single bleeder screws is: right rear, left rear, right front then left front. For calipers with 2 bleeder screws, follow the illustration (note the difference in sequence for right and left hand drive vehicles). ALWAYS keep the master cylinder reservoir filled with brake fluid during the bleeding operation. Never use brake fluid that has been drained from the hydraulic system, no matter how clean it is.

The primary and secondary hydraulic brake systems are separate and are bled independently. During the bleeding operation, do not allow the reservoir to run dry. Keep the master cylinder reservoir filled with brake fluid.

1. Clean all dirt from around the master cylinder fill cap, remove the cap and fill the master cylinder with brake fluid until the level is within ¼ in. (6mm) of the top edge of the reservoir.

2. Clean the bleeder screws at all 4 wheels. The bleeder screws are located on the top of the brake calipers.

3. Attach a length of rubber hose over the bleeder screw and place the other end of the hose in a glass jar, submerged in brake fluid.

4. Open the bleeder screw ½–¾ turn. Have an assistant slowly depress the brake pedal.

### ❊❊ CAUTION

**Brake fluid contains polyglycol ethers and polyglycols. Avoid contact with the eyes and wash your hands thoroughly after handling brake fluid. If you do get brake fluid in your eyes, flush your eyes with clean, running water for 15 minutes.**

**If eye irritation persists, or if you have taken brake fluid internally, IMMEDIATELY seek medical assistance.**

5. Close the bleeder screw and tell your assistant to allow the brake pedal to return slowly. Continue this process to purge all air from the system.

6. When bubbles cease to appear at the end of the bleeder hose, close the bleeder screw and remove the hose. Tighten the bleeder screw until just snug.

7. Check the master cylinder fluid level and add fluid accordingly. Do this after bleeding each wheel.

8. Repeat the bleeding operation at the remaining 3 wheels.

9. Fill the master cylinder reservoir to the proper level.

10. After the bleeding procedure, make sure you have a hard pedal that does not sink. Perform a test drive and engage the ABS system at least once. Be sure to do this safely, for example in an empty parking lot. Make sure the pedal remains hard after the ABS engagement. If not, repeat the bleeding procedure.

## BRAKES

### ❊❊ CAUTION

**Dust and dirt accumulating on brake parts during normal use may contain asbestos fibers from production or aftermarket brake linings. Breathing excessive concentrations of asbestos fibers can cause serious bodily harm. Exercise care when servicing brake parts. Do not sand or grind brake lining unless equipment used is designed to contain the dust residue. Do not clean brake parts with compressed air or by dry brushing. Cleaning should be done by dampening the brake components with a fine mist of water, then wiping the brake components clean with a dampened cloth. Dispose of cloth and all residue containing asbestos fibers in an impermeable container with the appropriate label. Follow practices prescribed by the Occupational Safety and Health Administration (OSHA) and the Environmental Protection Agency (EPA) for the handling, processing, and disposing of dust or debris that may contain asbestos fibers.**

### BRAKE CALIPER

*REMOVAL & INSTALLATION*

See Figure 1.

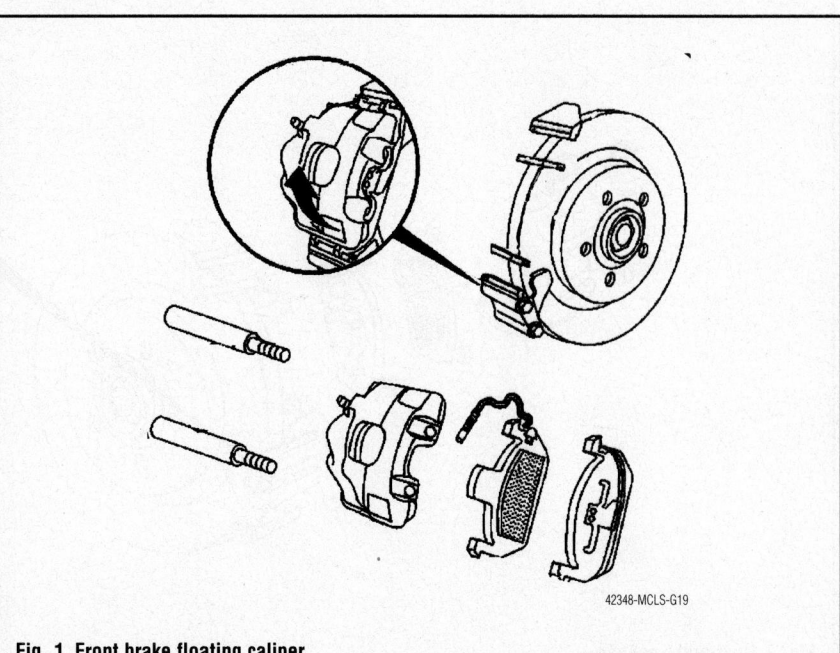

42348-MCLS-G19

**Fig. 1 Front brake floating caliper**

1. Before servicing the vehicle, refer to the precautions in the beginning of this section.

2. Remove the wheels.

3. Loosen the hydraulic line at the caliper, then remove the caliper from the carrier. Be sure to hold the pin with a back-up wrench when removing the caliper bolts.

4. Remove the caliper from the hydraulic line.

**To install:**

5. Thread the caliper onto the hydraulic line and hand-tighten it. Fit the caliper into place on the carrier.

6. Torque the bolts to 22 ft. lbs. (30 Nm).

7. Using new washers, tighten the hydraulic line bolt to 24 ft. lbs. (33 Nm), and bleed the brakes.

## DISC BRAKE PADS

*REMOVAL & INSTALLATION*

1. Before servicing the vehicle, refer to the precautions in the beginning of this section.

2. Remove the front wheels.

3. Hold the lower guide pin with an open wrench and remove the bolt securing the caliper to the guide pin.

4. Pivot the caliper up on the upper guide pin and slide the pads straight out to remove them.

**To install:**

5. Compress the caliper piston into the bore.

6. Fit the new pads into the carrier and pivot the caliper into place.

7. The original bolts are micro-encapsulated with a thread locking compound. Install a new bolt or clean the old bolt and apply a thread-locking compound.

8. When tightening the bolt, be sure to use a back-up wrench to hold the guide pin. Torque the bolt to 22 ft. lbs. (30 Nm).

9. Install the wheels.

## BRAKE DISC (ROTOR)

*REMOVAL & INSTALLATION*

1. Before servicing the vehicle, refer to the precautions in the beginning of this section.

2. Raise and support the vehicle, then remove the wheels.

3. Remove the brake caliper.

4. Remove the screw securing the disc to the wheel hub, then remove the disc.

5. Installation is the reverse of removal. Tighten the disc securing screw to 17 ft. lbs. (23 Nm).

## BRAKES

### ❋❋ CAUTION

Dust and dirt accumulating on brake parts during normal use may contain asbestos fibers from production or aftermarket brake linings. Breathing excessive concentrations of asbestos fibers can cause serious bodily harm. Exercise care when servicing brake parts. Do not sand or grind brake lining unless equipment used is designed to contain the dust residue. Do not clean brake parts with compressed air or by dry brushing. Cleaning should be done by dampening the brake components with a fine mist of water, then wiping the brake components clean with a dampened cloth. Dispose of cloth and all residue containing asbestos fibers in an impermeable container with the appropriate label. Follow practices prescribed by the Occupational Safety and

## REAR DISC BRAKES

Health Administration (OSHA) and the Environmental Protection Agency (EPA) for the handling, processing, and disposing of dust or debris that may contain asbestos fibers.

### BRAKE CALIPER

*REMOVAL & INSTALLATION*
*See Figure 2.*

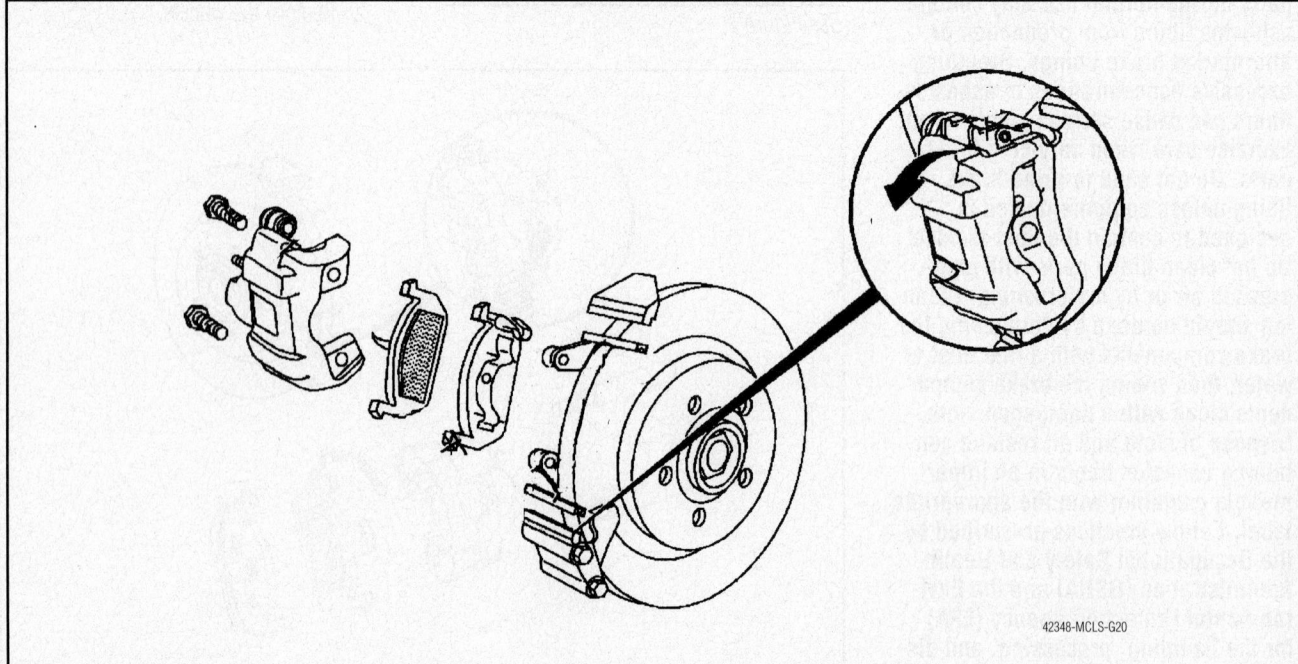

42348-MCLS-G20

**Fig. 2 Rear brake caliper**

1. Before servicing the vehicle, refer to the precautions in the beginning of this section.

2. If equipped with ABS, make sure the ignition switch stays **OFF** and pump the brake pedal 25–35 times to relieve the system pressure.

3. Remove the wheels.

4. Disconnect the parking brake cable.

5. Loosen the hydraulic line.

6. Use a back-up wrench to hold the guide pins and remove the caliper bolts.

7. Lift the caliper off the carrier and unscrew it from the hydraulic line. Discard the washers.

### To install:

8. Thread the caliper onto the hydraulic line and hand-tighten it. Fit the caliper into place on the carrier. Torque the bolts to 17 ft. lbs. (24 Nm).

9. Connect the brake line. Torque to 13 ft. lbs. (18 Nm).

10. Bleed the brakes.

11. Install the wheels.

## DISC BRAKE PADS

### REMOVAL & INSTALLATION

1. Before servicing the vehicle, refer to the precautions in the beginning of this section.

2. Remove the rear wheels.

3. Remove the parking brake cable clip from the caliper. Disconnect the parking brake cable.

4. Hold the guide pin with a back-up wrench and remove the upper mounting bolt from the brake caliper.

5. Swing the caliper downward and remove the brake pads.

### To install:

6. Retract the piston into the housing by rotating the piston clockwise.

7. Install the new brake pads onto the pad carrier.

8. Install the caliper to the pad carrier using a new self locking bolt or a thread locking compound and torque to 17 ft. lbs. (24 Nm).

9. Attach the hand brake cable to the caliper.

10. Check the parking brake operation and adjust the cable if necessary.

11. Install the wheels.

## ROTOR

### REMOVAL & INSTALLATION

1. Before servicing the vehicle, refer to the precautions in the beginning of this section.

2. Raise and support the vehicle, then remove the wheels. Make sure the wheels are blocked and the transmission is in Park, then disengage the parking brake.

3. Remove the brake caliper.

4. Back off parking brake shoes.

5. Remove the screw securing the disc to the wheel hub, then remove the disc.

6. Installation is the reverse of removal. Tighten the disc securing screw to 17 ft. lbs. (23 Nm).

## BRAKES | PARKING BRAKE

## PARKING BRAKE SHOES

### REMOVAL & INSTALLATION

*See Figure 3.*

1. Before servicing the vehicle, refer to the precautions in the beginning of this section.

2. Raise and support the vehicle, then remove the wheels. Make sure the wheels are blocked and the transmission is in Park, then disengage the parking brake.

3. Remove the brake disc.

4. Carefully unhook the return springs from the brake shoes.

5. Remove the retaining springs. Remove the shoes.

6. Remove the shoe adjuster.

7. Installation is the reverse of removal.

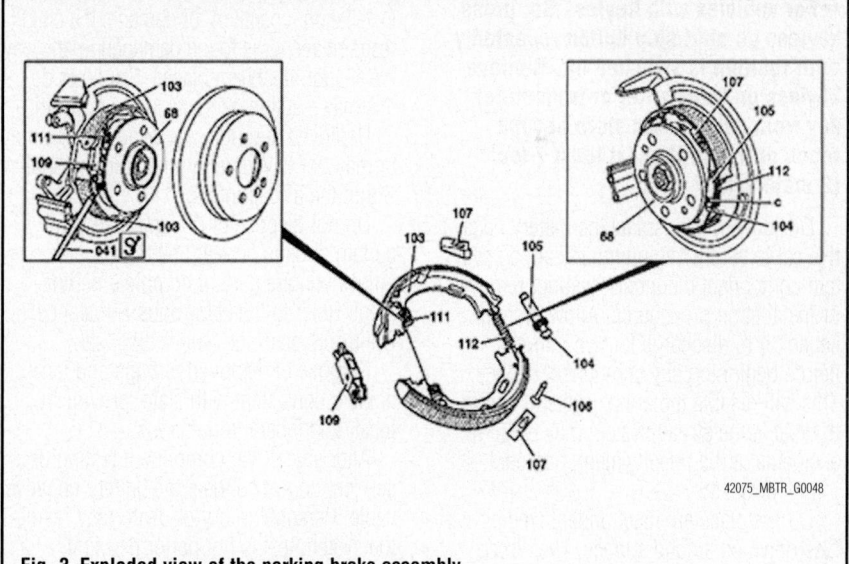

42075_MBTR_G0048

**Fig. 3 Exploded view of the parking brake assembly**

## CHASSIS ELECTRICAL | AIR BAG (SUPPLEMENTAL RESTRAINT SYSTEM)

### GENERAL INFORMATION

**✳✳ CAUTION**

**These vehicles are equipped with an air bag system. The system must be disarmed before performing service on, or around, system components, the steering column, instrument panel components, wiring and sensors. Failure to follow the safety precautions and the disarming procedure could result in accidental air bag deployment, possible injury and unnecessary system repairs.**

### SERVICE PRECAUTIONS

➡**For vehicles with memory option (driver seat, steering column and mirrors), the easy entry/exit function must be disabled. Otherwise, damage can occur when battery is reconnected. The system is deactivated in the Comfort menu of multifunction display using buttons on steering wheel.**

➡**For vehicles with Keyless Go, press keyless go start/stop button repeatedly until ignition is switched off. Remove keyless go transmitter or transmitter key from vehicle and store beyond reach of transmitter (at least 7 feet (2 meters)).**

Disconnect and isolate the battery negative cable before beginning any airbag system component diagnosis, testing, removal, or installation procedures. Allow system capacitor to discharge for two minutes before beginning any component service. This will disable the airbag system. Failure to disable the airbag system may result in accidental airbag deployment, personal injury, or death.

Do not place an intact undeployed airbag face down on a solid surface. The airbag will propel into the air if accidentally deployed and may result in personal injury or death.

When carrying or handling an undeployed airbag, the trim side (face) of the airbag should be pointing towards the body to minimize possibility of injury if accidental deployment occurs. Failure to do this may result in personal injury or death.

Replace airbag system components with OEM replacement parts. Substitute parts may appear interchangeable, but internal differences may result in inferior occupant

protection. Failure to do so may result in occupant personal injury or death.

Wear safety glasses, rubber gloves, and long sleeved clothing when cleaning powder residue from vehicle after an airbag deployment. Powder residue emitted from a deployed airbag can cause skin irritation. Flush affected area with cool water if irritation is experienced. If nasal or throat irritation is experienced, exit the vehicle for fresh air until the irritation ceases. If irritation continues, see a physician.

Do not use a replacement airbag that is not in the original packaging. This may result in improper deployment, personal injury, or death.

The factory installed fasteners, screws and bolts used to fasten airbag components have a special coating and are specifically designed for the airbag system. Do not use substitute fasteners. Use only original equipment fasteners listed in the parts catalog when fastener replacement is required.

During, and following, any child restraint anchor service, due to impact event or vehicle repair, carefully inspect all mounting hardware, tether straps, and anchors for proper installation, operation, or damage. If a child restraint anchor is found damaged in any way, the anchor must be replaced. Failure to do this may result in personal injury or death.

Deployed and non-deployed airbags may or may not have live pyrotechnic material within the airbag inflator.

Do not dispose of driver/passenger/ curtain airbags or seat belt tensioners unless you are sure of complete deployment. Refer to the Hazardous Substance Control System for proper disposal.

Dispose of deployed airbags and tensioners consistent with state, provincial, local, and federal regulations.

After any airbag component testing or service, do not connect the battery negative cable. Personal injury or death may result if the system test is not performed first.

If the vehicle is equipped with the Occupant Classification System (OCS), do not connect the battery negative cable before performing the OCS Verification Test using the scan tool and the appropriate diagnostic information. Personal injury or death may result if the system test is not performed properly.

Never replace both the Occupant Restraint Controller (ORC) and the Occupant Classification Module (OCM) at the same time. If both require replacement, replace one, then perform the Airbag System test before replacing the other.

Both the ORC and the OCM store Occupant Classification System (OCS) calibration data, which they transfer to one another when one of them is replaced. If both are replaced at the same time, an irreversible fault will be set in both modules and the OCS may malfunction and cause personal injury or death.

If equipped with OCS, the Seat Weight Sensor is a sensitive, calibrated unit and must be handled carefully. Do not drop or handle roughly. If dropped or damaged, replace with another sensor. Failure to do so may result in occupant injury or death.

If equipped with OCS, the front passenger seat must be handled carefully as well. When removing the seat, be careful when setting on floor not to drop. If dropped, the sensor may be inoperative, could result in occupant injury, or possibly death.

If equipped with OCS, when the passenger front seat is on the floor, no one should sit in the front passenger seat. This uneven force may damage the sensing ability of the seat weight sensors. If sat on and damaged, the sensor may be inoperative, could result in occupant injury, or possibly death.

### DISARMING THE SYSTEM

To avoid personal injury when working on vehicles equipped with an air bag, the negative battery cable must be disconnected and insulated before working on the system. Failure to do so may result in accidental deployment of the air bag.

### ARMING THE SYSTEM

To rearm the air bag system, reattach the battery cable(s).

### CLOCKSPRING CENTERING

See Figure 4.

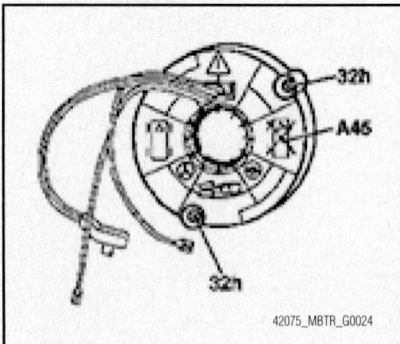

42075_MBTR_G0024

**Fig. 4 Clockspring securing screws (32h) and contact (A45)**

1. Before servicing the vehicle, refer to the precautions in the beginning of this section.

2. Tighten the mounting screws (32h) completely into the horn and airbag clockspring contact (A45).

3. Rotate the clockspring contact (A45) counterclockwise until a slight resistance is noticeable (clockspring contact is rolled up completely).

4. Rotate the clockspring contact approximately 3 to 3.5 revolutions clockwise, until the mounting screws (32h) can be unscrewed again through the openings and fix the clockspring contact (A45) in position.

➡**The entire rotation range of the clockspring contact is approximately 6 to 7 revolutions.**

## DRIVETRAIN

➡**For vehicles with memory option (driver seat, steering column and mirrors), the easy entry/exit function must be disabled. Otherwise, damage can occur when battery is reconnected. The system is deactivated in the Comfort menu of multifunction display using buttons on steering wheel.**

➡**For vehicles with Keyless Go, press keyless go start/stop button repeatedly until ignition is switched off. Remove keyless go transmitter or transmitter key from vehicle and store beyond reach of transmitter (at least 7 feet (2 meters).**

### AUTOMATIC TRANSMISSION ASSEMBLY

*REMOVAL & INSTALLATION*

See Figure 5.

1. Before servicing the vehicle, refer to the precautions in the beginning of this section.
2. Drain the transmission fluid.
3. Remove or disconnect the following:
   - Negative battery cable
   - Transfer case
   - Dipstick tube
   - Electrical connector shield
   - Electrical connector
   - Shift rod
   - Park interlock cable
   - Oxygen sensors
   - Transmission cooler lines
   - Torque converter cover
   - Torque converter-to-flexplate bolts
   - Rear transmission mount
   - Ground strap
   - Transmission-to-engine mounting bolts
   - Transmission

**To install:**

4. Position the transmission on the engine and install the mounting bolts.

5. Install or connect the following:
   - Ground strap
   - Rear transmission mount
   - Torque converter-to-flexplate bolts. Tighten to 31 ft. lbs. (42 Nm).
   - Torque converter cover
   - Transmission cooler lines
   - Oxygen sensors
   - Park interlock cable
   - Shift rod
   - Electrical connector and shield
   - Transmission dipstick tube
   - Transfer case
   - Negative battery cable
   - Transmission fluid

### TRANSFER CASE ASSEMBLY

*REMOVAL & INSTALLATION*

See Figure 6.

1. Before servicing the vehicle, refer to the precautions in the beginning of this section.

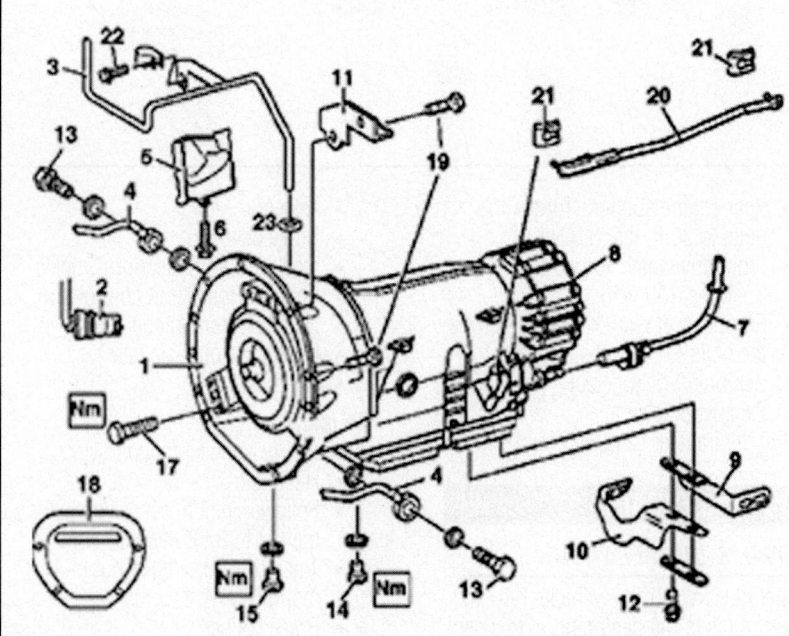

1 Torque converter housing
2 13-pin plug connector
3 Oil filler pipe
4 Oil cooling lines
5 Shield
6 Shield fixing bolt
7 Cable for park lock interlock
8 Adapter housing for transfer case
9 Left exhaust bracket
10 Right exhaust bracket
11 Retaining plate
12 Exhaust bracket bolt
13 Bolts
14 Oil drain screw
15 Oil drain screw for torque converter
17 Torque converter bolt
18 Cover
19 Bolts
20 Shift rod
21 Securing clip
22 Bolt, oil filler pipe
23 O-ring

42075_MBTR_G0027

**Fig. 5 Exploded view of the transmission mounting components**

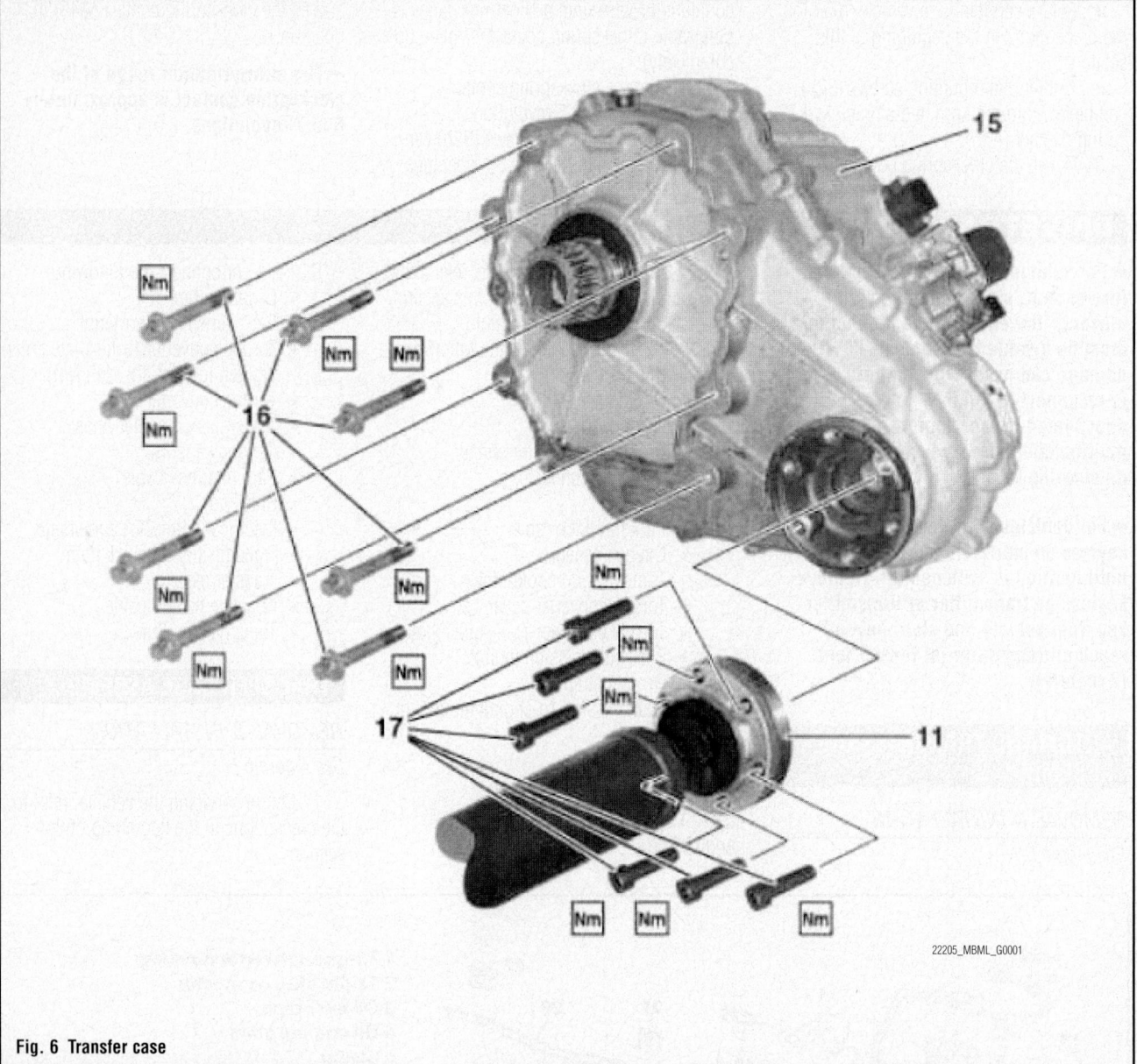

22205_MBML_G0001

**Fig. 6 Transfer case**

2. Drain the fluid and remove the exhaust brackets. Suspend the exhaust with a piece of wire.

3. Remove or disconnect the following:
- Front half of rear driveshaft
- Front driveshaft
- Servomotor connector

4. Support the transmission and remove the engine support along with the rear engine mount. Lower the assembly slightly.

5. Remove or disconnect the following:
- Transfer case bolts

6. Remove the transfer case and drain the residual oil.

*To install:*

7. Install or connect the following:
- Transfer case with new gasket. Torque the bolts to 15 ft. lbs. (20 Nm).

- Rear engine support. Torque the bolts to 30 ft. lbs. (40 Nm).
- Front driveshaft. Torque the bolts to 37 ft. lbs. (50 Nm).
- Front half of rear driveshaft. Torque the bolts to 30 ft. lbs. (40 Nm).
- Servomotor connector
- Exhaust brackets

8. Fill the transfer case with fluid.

## FRONT HALFSHAFT

*REMOVAL & INSTALLATION*

1. Before servicing the vehicle, refer to the precautions in the beginning of this section.

2. Remove or disconnect the following:
- Negative battery cable
- Front wheel

- Axle nut
- Brake caliper
- Track rod from steering knuckle
- Follower joint from control arm
- Halfshaft from flange
- Halfshaft from axle gear
- Halfshaft

*To install:*

3. Install or connect the following:
- Halfshaft
- Follower joint to control arm. Torque the nut to 37 ft. lbs. (50 Nm).
- Track rod. Torque the nut to 41 ft. lbs. (55 Nm).
- Brake caliper
- Axle nut. Torque the nut to 361 ft. lbs. (490 Nm).
- Front wheel
- Negative battery cable

## REAR HALFSHAFT

### REMOVAL & INSTALLATION

1. Before servicing the vehicle, refer to the precautions in the beginning of this section.
2. Remove or disconnect the following:
   • Wheel
   • Center axle hold-down bolt (in hub)
   • Wheel speed sensor
3. Push the follower joint out of carrier.
4. Press the track rod joint out of carrier.
5. Pull the axle shaft out of axle shaft flange.
6. Press the axle shaft out of axle carrier.

### To install:

7. Install or connect the following:
   • Axle shaft. Torque the nut to 184 ft. lbs. then an additional 45 degrees.
   • Axle lock ring
   • Torque Strut. Torque the nut to 37 ft. lbs. then an additional 90 degrees.
   • Camber Strut. Torque the nut to 41 ft. lbs. then an additional 90 degrees.
   • Wheel speed sensor
   • Center axle hold down bolt. Torque the nut to 361 ft. lbs. (490 Nm).
   • Wheel

## REAR PINION SEAL

### REMOVAL & INSTALLATION

See Figures 7 and 8.

1. Before servicing the vehicle, refer to the precautions in the beginning of this section.
2. Remove or disconnect the following:

➡ **Always matchmark the position of the driveshaft flange before removing it.**

   • Driveshaft
3. Using a torque wrench on the flange nut, measure the rotation torque reading. When reinstalling the flange, the nut must be tightened until the same rotation torque reading is reached.
   • Drive flange
   • Pinion seal

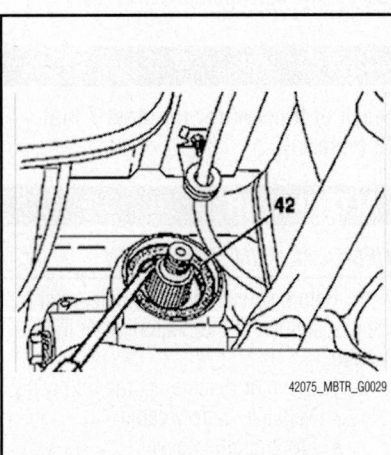

**Fig. 7  Removing the rear pinion seal**

### To install:

4. Coat the pinion seal lip clean hypoid gear oil.
5. Install or connect the following using new hardware:
   • Pinion seal
   • Drive flange

➡ **New self-locking hexagon collar nut. The nut must be tightened until the same rotation torque reading measured during removal is reached.**

   • Driveshaft

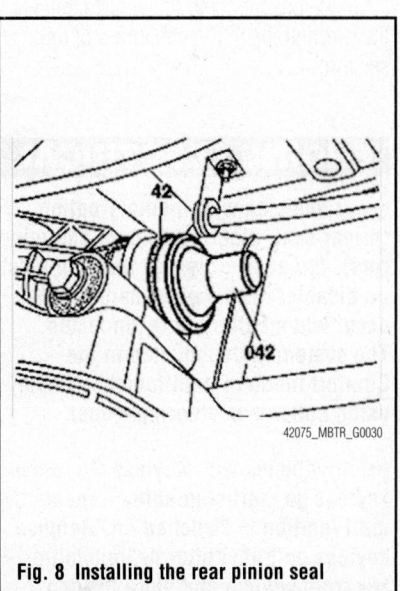

**Fig. 8  Installing the rear pinion seal**

## ENGINE COOLING

➡ For vehicles with memory option (driver seat, steering column and mirrors), the easy entry/exit function must be disabled. Otherwise, damage can occur when battery is reconnected. The system is deactivated in the Comfort menu of multifunction display using buttons on steering wheel.

➡ For vehicles with Keyless Go, press keyless go start/stop button repeatedly until ignition is switched off. Remove keyless go transmitter or transmitter key from vehicle and store beyond reach of transmitter (at least 7 feet (2 meters)).

## THERMOSTAT

### REMOVAL & INSTALLATION

See Figure 9.

1. Before servicing the vehicle, refer to the precautions in the beginning of this section.

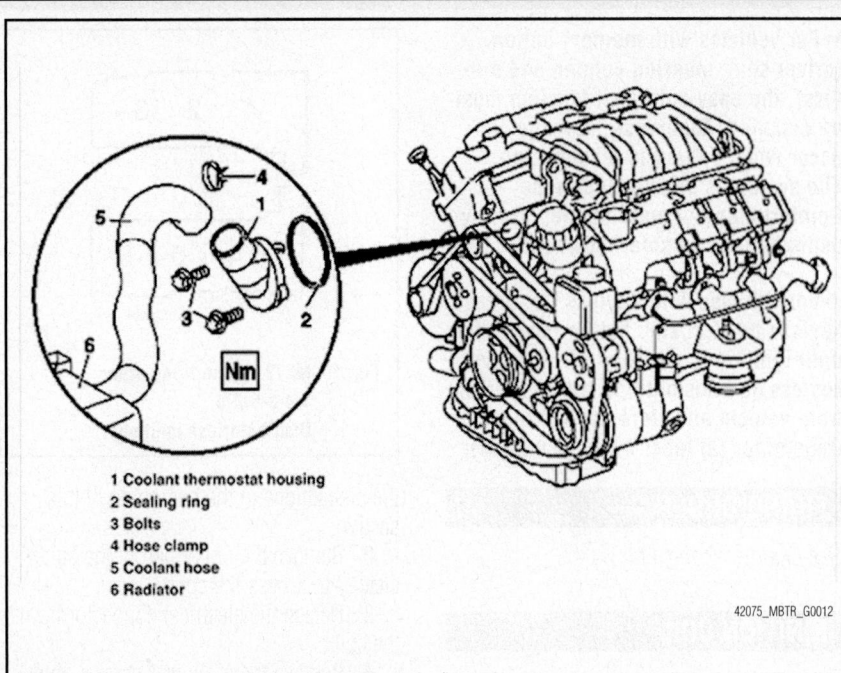

1 Coolant thermostat housing
2 Sealing ring
3 Bolts
4 Hose clamp
5 Coolant hose
6 Radiator

**Fig. 9  Exploded view of the thermostat housing**

2. Drain the engine coolant.

3. Detach the cooling hose from the thermostat housing.

4. Remove the bolts securing the thermostat housing, then remove it and the thermostat from the engine.

5. Installation is the reverse of removal. Tighten the thermostat housing bolts to 18 ft. lbs. (25 Nm).

## WATER PUMP

### REMOVAL & INSTALLATION

1. Before servicing the vehicle, refer to the precautions in the beginning of this section.

2. Remove or disconnect the following:
- Negative battery cable
- Cooling fan
- Drain the engine coolant
- Engine cover

3. Lock the automatic belt tensioner by rotating the tensioner counterclockwise until a 5mm drift or pin fits through the tensioner.
- Serpentine belt
- Pry off wiring retaining clamp from water pump
- Coolant hoses from the water pump
- Belt pulley
- Water pump mounting bolts
- Water pump

4. Clean and dry the gasket mating surface for the water pump.

**To install:**

5. Install or connect the following:
- Water pump and gasket. Tighten the self-tapping bolt to timing case cover bolts to 18 ft. lbs. (25 Nm) and the pump to crankcase bolts to 15 ft. lbs. (15 Nm).
- Coolant hoses to the water pump
- Serpentine belt and remove the locking pin
- Engine cover
- Fan

6. Fill the engine with coolant.

7. Connect the negative battery cable.

8. Read fault memory, encode the radio and normalize the power windows.

9. Start the vehicle and check for leaks.

## ENGINE ELECTRICAL

➡ For vehicles with memory option (driver seat, steering column and mirrors), the easy entry/exit function must be disabled. Otherwise, damage can occur when battery is reconnected. The system is deactivated in the Comfort menu of multifunction display using buttons on steering wheel.

➡ For vehicles with Keyless Go, press keyless go start/stop button repeatedly until ignition is switched off. Remove keyless go transmitter or transmitter key from vehicle and store beyond reach of transmitter (at least 7 feet (2 meters)).

## ALTERNATOR

### REMOVAL & INSTALLATION

1. Before servicing the vehicle, refer to the precautions in the beginning of this section.

2. Remove or disconnect the following:
- Negative battery cable
- Front engine cover
- Left and right engine intake air ducts

## CHARGING SYSTEM

- Accessory drive belt
- Alternator electrical wires
- Alternator bolts
- Alternator

**To install:**

3. Install the alternator assembly. Torque the mounting bolts to 20 ft. lbs. (15 Nm).

4. Install or connect the following:
- Alternator electrical wires
- Right inner fender liner
- Accessory drive belt
- Connect the negative battery cable

5. Start the engine and check for proper operation.

## ENGINE ELECTRICAL

➡ For vehicles with memory option (driver seat, steering column and mirrors), the easy entry/exit function must be disabled. Otherwise, damage can occur when battery is reconnected. The system is deactivated in the Comfort menu of multifunction display using buttons on steering wheel.

➡ For vehicles with Keyless Go, press keyless go start/stop button repeatedly until ignition is switched off. Remove keyless go transmitter or transmitter key from vehicle and store beyond reach of transmitter (at least 7 feet (2 meters)).

## FIRING ORDER

*See Figures 10 and 11.*

## IGNITION COIL

### REMOVAL & INSTALLATION

1. Before servicing the vehicle, refer to

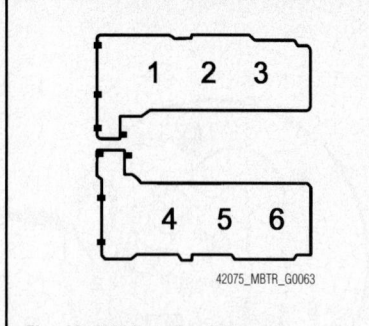

42075_MBTR_G0063

**Fig. 10 M272 engine firing order
1-4-3-6-2-5
Distributorless ignition**

the precautions in the beginning of this section.

2. Remove the upper part of the air cleaner to access the coils.

3. Detach the electrical connector from the coil.

4. Remove the mounting screws, then remove the coil.

## IGNITION SYSTEM

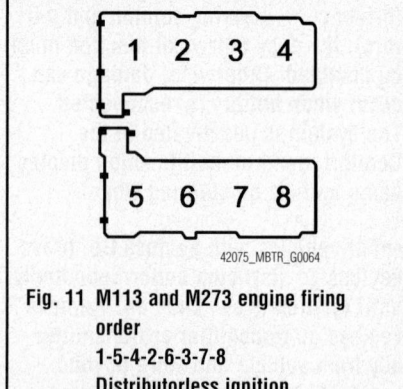

42075_MBTR_G0064

**Fig. 11 M113 and M273 engine firing order
1-5-4-2-6-3-7-8
Distributorless ignition**

5. Installation is the reverse of removal. Tighten the mounting screws to 7 ft. lbs. (9 Nm).

## IGNITION TIMING

### ADJUSTMENT

The ignition timing is controlled by the

Powertrain Control Module (PCM). No adjustment is necessary or possible.

## SPARK PLUGS

### REMOVAL & INSTALLATION

1. Before servicing the vehicle, refer to the Precautions Section.
2. Disconnect the negative battery cable, and if the vehicle has been run recently, allow the engine to thoroughly cool.
3. Remove ignition coils.

➡ **Remove the spark plugs when the engine is cold, if possible, to prevent damage to the threads. If removal of the plugs is difficult, apply a few drops of penetrating oil or silicone spray to** the area around the base of the plug, and allow it a few minutes to work.

4. Using a spark plug socket that is equipped with a rubber insert to properly hold the plug, turn the spark plug counterclockwise to loosen and remove the spark plug from the bore.

### To install:

5. Using a wire feeler gauge, check and adjust the spark plug gap. When using a gauge, the proper size should pass between the electrodes with a slight drag. The next larger size should not be able to pass while the next smaller size should pass freely.
6. Carefully thread the plug into the bore by hand. If resistance is felt before the plug is almost completely threaded, back the plug out and begin threading again. In small, hard to reach areas, an old spark plug wire and boot could be used as a threading tool. The boot will hold the plug while you twist the end of the wire and the wire is supple enough to twist before it would allow the plug to cross thread.

### ✳✳ WARNING

**Do not use the spark plug socket to thread the plugs. Always carefully thread the plug by hand or using an old plug wire to prevent the possibility of cross threading and damaging the cylinder head bore.**

7. Carefully tighten the spark plug to 17 ft. lbs. (23 Nm).

## ENGINE ELECTRICAL

➡ **For vehicles with memory option (driver seat, steering column and mirrors), the easy entry/exit function must be disabled. Otherwise, damage can occur when battery is reconnected. The system is deactivated in the Comfort menu of multifunction display using buttons on steering wheel.**

➡ **For vehicles with Keyless Go, press keyless go start/stop button repeatedly until ignition is switched off. Remove keyless go transmitter or transmitter key from vehicle and store beyond reach of transmitter (at least 7 feet (2 meters)).**

## STARTER

### REMOVAL & INSTALLATION

See Figure 12.

1. Before servicing the vehicle, refer to the precautions in the beginning of this section.
2. Remove or disconnect the following:
   - Negative battery cable
   - Bottom engine compartment panel
   - Right catalytic converter
   - Heat shield
   - Pry cable holder from transmission bracket
   - Starter bolts
   - Protective cap and starter electrical connectors

## STARTING SYSTEM

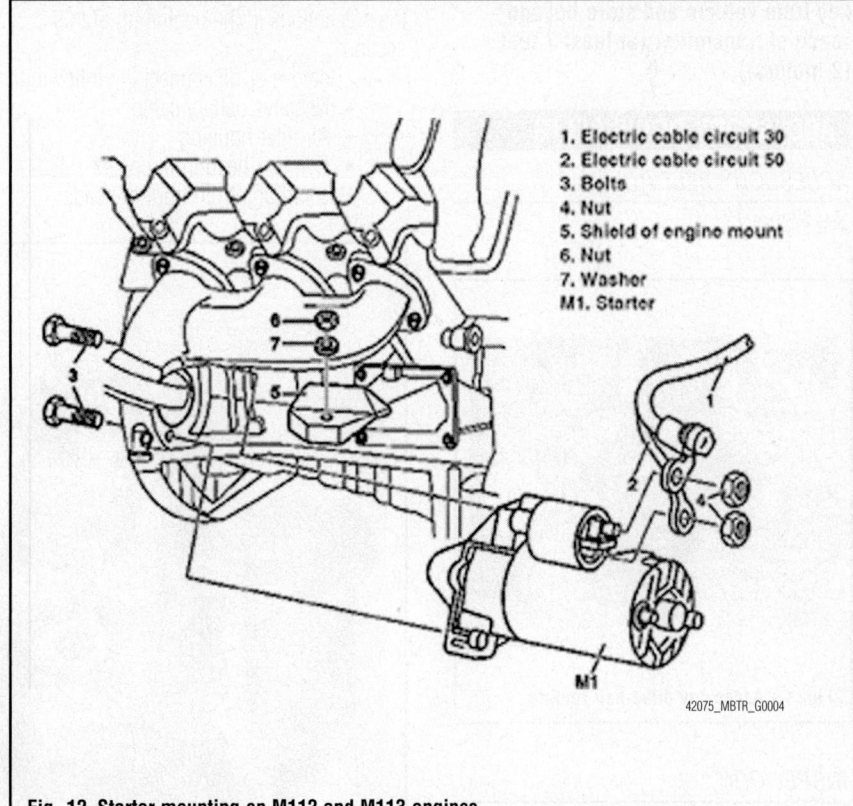

1. Electric cable circuit 30
2. Electric cable circuit 50
3. Bolts
4. Nut
5. Shield of engine mount
6. Nut
7. Washer
M1. Starter

42075_MBTR_G0004

**Fig. 12 Starter mounting on M112 and M113 engines**

### To install:
3. Install or connect the following:
   - Starter motor. Torque the mounting bolts to 30 ft. lbs. (40 Nm).
   - Starter electrical connectors
   - Heat shield
   - Right catalytic converter
   - Bottom engine compartment panel
   - Negative battery cable

## ENGINE MECHANICAL

➡Disconnecting the negative battery cable may interfere with the functions of the on board computer systems and may require the computer to undergo a relearning process, once the negative battery cable is reconnected.

➡For vehicles with memory option (driver seat, steering column and mirrors), the easy entry/exit function must be disabled. Otherwise, damage can occur when battery is reconnected. The system is deactivated in the Comfort menu of multifunction display using buttons on steering wheel.

➡For vehicles with Keyless Go, press keyless go start/stop button repeatedly until ignition is switched off. Remove keyless go transmitter or transmitter key from vehicle and store beyond reach of transmitter (at least 7 feet (2 meters)).

## ACCESSORY DRIVE BELTS

### ACCESSORY BELT ROUTING

See Figure 13.

**Fig. 13 Accessory drive belt routing**

### INSPECTION

Inspect the drive belt for signs of glazing or cracking. A glazed belt will be perfectly smooth from slippage, while a good belt will have a slight texture of fabric visible. Cracks will usually start at the inner edge of the belt and run outward. All worn or damaged drive belts should be replaced immediately.

### ADJUSTMENT

The belt is tensioned automatically. If the belt is slipping, either the belt has stretched beyond its serviceable limit or the tensioner is faulty.

### REMOVAL & INSTALLATION

1. Before servicing the vehicle, refer to the Precautions Section.
2. Remove engine front cover.
3. Using a breaker bar and appropriate sized Torx socket (or appropriate drive belt removal tool), relieve the belt tension by rotating the tensioner.
4. Remove the belt from the pulleys.
5. Installation is the reverse of removal. Be sure to follow the belt routing diagram.

## CAMSHAFT AND VALVE LIFTERS

### REMOVAL & INSTALLATION

See Figure 14.

1. Before servicing the vehicle, refer to the precautions in the beginning of this section.
2. Remove or disconnect the following:
   - Negative battery cable
   - Air filter housing
   - Cylinder head front covers
   - centrifuge from front of heads

## ❋❋ WARNING

**If the gears are not held in place, the camshaft adjuster will be damaged beyond repair.**

3. Lock the exhaust camshaft gear backlash compensation. Insert a 3mm drift into hole in front cover of camshaft adjuster to lock the toothed gears together. The gears may need to be aligned to allow the drift to be inserted.
4. Rotate the engine clockwise to position the crankshaft 40 degrees After Top Dead Center (ATDC).

## ❋❋ WARNING

**The engine must not be rotated backward.**

5. Lock the camshafts using the Camshaft Locking tools 112 589 00 32 00 and 112 589 01 32 00.
6. Loosen but do not remove intake and exhaust cam to sprocket bolts.

5. Center valve of exhaust camshaft
6. Center valve of intake camshaft
7. Pulse wheel of exhaust camshaft
8. Pulse wheel of intake camshaft
9. Camshaft adjuster of exhaust camshaft
10. Camshaft adjuster of intake camshaft
11. Auxiliary bearing cap of intake camshaft
13. Color coding

22205_MBML_G0003

**Fig. 14 Camshaft components**

## ❊❊ WARNING

**Hold camshafts from turning on opposite end.**

7. Remove or disconnect the following:
- Cylinder head cover
- Intake and exhaust cam to sprocket bolts

➡**Check for sheared or broken components.**

8. Hold rear of exhaust camshaft and remove camshaft adjuster

9. Remove exhaust camshaft bearing caps and camshaft.

10. Mark timing chain and intake camshaft sprocket for installation.

11. Remove or disconnect the following:
- Chain tensioner
- Tensioning rail bolt
- Pull tensioning rail up out of front of engine

12. Hold rear of intake camshaft and remove camshaft adjuster

13. Remove intake camshaft bearing caps and camshaft.

*To install:*

➡**Be sure to install the correct camshaft for the corresponding cylinder head.**

14. Apply clean engine oil to the camshaft contact surfaces.

15. Install or connect the following:
- Intake camshaft
- Camshaft bearing caps

16. Tighten the mounting bolt to 70 inch lbs. (8 Nm)

17. Position the camshafts, then attach the camshaft fixing plate.

18. Align marks on intake camshaft adjuster and timing chain.

19. Install intake camshaft adjuster into camshaft.

20. Install new pulse wheel.

21. Install camshaft center valve and bolt.

➡**There is a risk of shearing the camshaft adjuster alignment pin.**

22. Apply sealant to tensioning rail bolt. Install tensioning rail and bolt.

## ❊❊ WARNING

**On 5.5L engine, replace chain tensioner with M24X1.5 thread with new tensioner. Reusing old tensioner will overstretch the chain.**

23. Install chain tensioner. Ensure timing marks still align and chain is tensioned.

24. Install exhaust camshaft and bearing caps

25. Tighten the mounting bolt to 70 inch lbs. (8 Nm)

26. Install camshaft adjuster, new pulse wheel and center valve, aligning dowel pin and gear.

➡**There is a risk of shearing the camshaft adjuster alignment pin.**

27. Tighten the attaching bolt to 107 ft. lbs. (145 Nm)

28. Remove drift from exhaust camshaft adjuster.

29. Remove the camshaft locking tools.

30. Install centrifuge.
- Cylinder head cover
- Air filter housing
- Negative battery cable

## CRANKSHAFT FRONT SEAL

### REMOVAL & INSTALLATION

1. Before servicing the vehicle, refer to the precautions in the beginning of this section.

2. Position engine at TDC.

3. Remove drive belt.

4. Install retaining lock for crankshaft/starter ring. Raise vehicle. Remove or disconnect the following:
- Left catalytic converter
- Front driveshaft heat shield
- Cover plate at left front of bell housing.
- Install retaining lock 112 589 03 40 00 or equivalent in opening

5. Remove pulley /vibration damper bolt.

6. Remove pulley /vibration damper.

7. Remove crankshaft seal.

8. Inspect sealing surface on pulley /vibration damper.

*To install:*

9. Apply clean engine oil to the end of the crankshaft to ease installation.

10. Install or connect the following:
- Crankshaft seal, using a suitable seal driver
- Crankshaft pulley and tighten the bolt to 148 ft. lbs. (200 Nm), then tighten another 90 degrees
- Remove retaining lock
- Drive belt

11. Start engine and check for leaks. Check oil level.

## CYLINDER HEAD

### REMOVAL & INSTALLATION

**Right Cylinder Head**

*See Figures 15 through 19.*

1. Before servicing the vehicle, refer to the precautions in the beginning of this section.

2. Turn ignition off.

3. Remove catalytic converter from manifold.

4. Drain and recycle the engine coolant.

5. Remove or disconnect the following:
- Negative battery cable
- Engine cover
- Air cleaner housing, resonance pipe and body
- Brake booster vacuum line at intake manifold
- engine wiring harness
- Relieve the fuel system pressure at service valve
- Fuel line

➡**The intake manifold system must not be disassembled.**

- Intake manifold. See intake manifold.
- Dipstick tube bolt at cylinder head
- Drive belt
- Aspirator valve shutoff valve at head
- Air pump
- Air pump switchover valve
- Ignition coils
- Centrifuge
- Valve cover

6. Rotate the engine clockwise to position the crankshaft 40 degrees after top dead center.

## ❊❊ WARNING

**The engine must not be rotated backward.**

7. Lock the camshafts using the Camshaft Locking tools 112 589 00 32 00 and 112 589 01 32 00.

8. Remove or disconnect the following:
- Generator, then the timing chain tensioner
- Camshaft adjusters
- Camshaft gears and attach them to the chain with a cable tie
- Side rail bolts and tensioning rail bolts

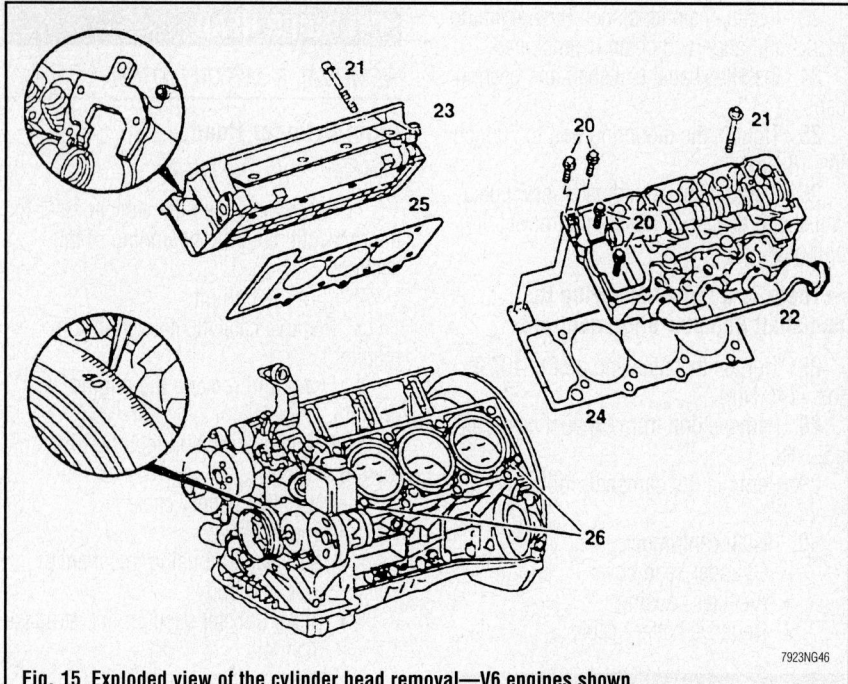

Fig. 15 Exploded view of the cylinder head removal—V6 engines shown

10. Measure the length of the cylinder head bolt shafts, new bolt length is 6.69 inches (170mm) and the maximum permissible length is 6.77 inches (172mm). Replace bolts that measure greater than the maximum permissible length.

### To install:

➡ **Due to tolerances, the cylinder head and block cannot be surfaced.**

11. Clean the head bolt threads, then apply clean engine oil to the thread and head contact surfaces.

12. Install the cylinder head to the engine block and tighten the head bolts according to sequence as follows:

    a. Step 1. 15 ft. lbs. (20 Nm).
    b. Step 2. 30 ft. lbs. (40 Nm).

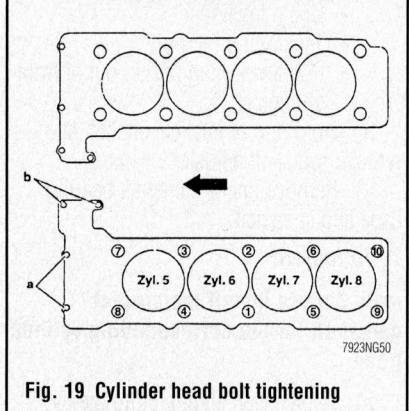

Fig. 19 Cylinder head bolt tightening sequence—V8 engines

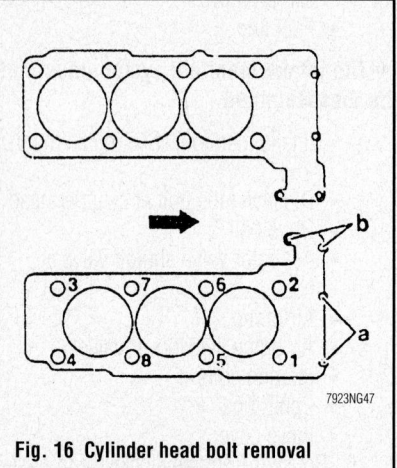

Fig. 16 Cylinder head bolt removal sequence—V6 engines

- Cylinder head off the engine block
- All gasket material from the sealing surfaces of the cylinder head and engine block. Be careful not to gouge or scratch the surface of the aluminum head. Be sure the cylinder head locating dowels are positioned in the engine block. Clean and dry the head bolt holes using compressed air.

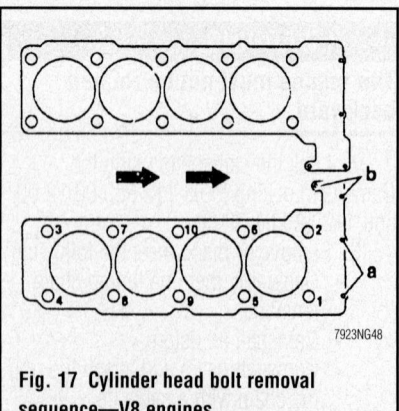

Fig. 17 Cylinder head bolt removal sequence—V8 engines

9. Loosen and remove the cylinder head bolts in stages following the illustrated sequence.

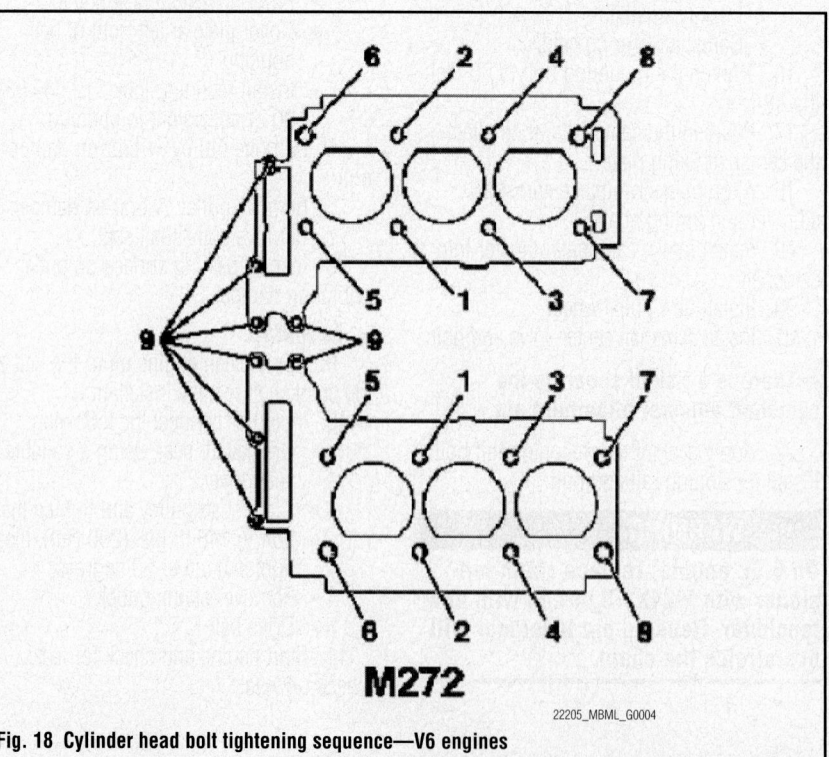

**M272**

Fig. 18 Cylinder head bolt tightening sequence—V6 engines

c. Step 3. 90 degrees.
d. Step 4. additional 90 degrees.
13. Install or connect the following:
- Cylinder head-to-timing case bolt. Tighten to 15 ft. lbs. (20 Nm), then an additional 90 degrees
- Front cover-to-cylinder head bolts. Tighten to 6 ft. lbs. (9 Nm)
- Side rail bolts and tensioning rail bolts. Apply sealant to bolts.
- Camshaft gears
- Camshaft adjusters

### ※※ WARNING

**On 5.5L engine, replace chain tensioner with M24X1.5 thread with new tensioner. Reusing old tensioner will overstretch the chain.**

- Timing chain tensioner with a new gasket and tighten to 59 ft. lbs. (80 Nm).
- Generator
14. Remove the camshaft locking plates.
- Cylinder head covers and tighten the bolts to 89 inch lbs. (10 Nm)
- Centrifuge
- Ignition coils
- Air pump switchover valve
- Air pump
- Aspirator valve shutoff valve at head
- Drive belt
- Dipstick tube bolt at cylinder head
- Intake manifold
- Fuel line
- engine wiring harness
- Brake booster vacuum line at intake manifold
- Air cleaner housing, resonance pipe and body
- Engine cover
15. Fill the engine with coolant.
16. Connect the negative battery cable.
17. Clear fault memory.
18. Start the vehicle and check for leaks.

### Left Cylinder Head

1. Before servicing the vehicle, refer to the precautions in the beginning of this section.
2. Turn ignition off.
3. Remove catalytic converter from manifold.
4. Drain and recycle the engine coolant.
5. Remove or disconnect the following:
- Negative battery cable
- Engine cover
- Air cleaner housing, resonance pipe and body
- SFI control unit
- Mass air flow sensor

- Purge control valve line from crankcase ventilation vacuum line
- Brake booster vacuum line at intake manifold
- engine wiring harness
- Engine coolant sensor connector
- Relieve the fuel system pressure at service valve
- Fuel line

➡ **The intake manifold system must not be disassembled.**

- Intake manifold
- Fluid from power steering reservoir
- Power steering reservoir return line
- Power steering reservoir
- Drive belt, tensioner and guide pulley
- Oil filter housing and heat exchanger
- Aspirator valve shutoff valve at head
- Coolant hose at heat exchanger
- Coolant hose at thermostat housing
- Thermostat housing
- Front cover
- Oil separator
- Ignition coils
- Valve cover
6. Rotate the engine clockwise to position the crankshaft 40 degrees after top dead center.

### ※※ WARNING

**The engine must not be rotated backward.**

7. Lock the camshafts using the Camshaft Locking tools 112 589 00 32 00 and 112 589 01 32 00.
8. Remove or disconnect the following:
- Generator, then the timing chain tensioner
- Camshaft adjusters
- Camshaft gears and attach them to the chain with a cable tie
- Slide rail pins
9. Loosen and remove the cylinder head bolts in stages following the illustrated sequence.
10. Remove cylinder head off the engine block.
11. Remove all gasket material from the sealing surfaces of the cylinder head and engine block. Be careful not to gouge or scratch the surface of the aluminum head. Be sure the cylinder head locating dowels are positioned in the engine block. Clean and dry the head bolt holes using compressed air.
12. Measure the length of the cylinder head bolt shafts, new bolt length is

6.69 inches (170mm) and the maximum permissible length is 6.77 inches (172mm). Replace bolts that measure greater than the maximum permissible length.

**To install:**

➡ **Due to tolerances, the cylinder head and block cannot be surfaced.**

13. Clean the head bolt threads, then apply clean engine oil to the thread and head contact surfaces.
14. Install the cylinder head to the engine block and tighten the head bolts according to sequence as follows:
a. Step 1. 15 ft. lbs. (20 Nm).
b. Step 2. 30 ft. lbs. (40 Nm).
c. Step 3. 90 degrees.
d. Step 4. additional 90 degrees.
15. Install or connect the following:
- Cylinder head-to-timing case bolt. Tighten to 15 ft. lbs. (20 Nm), then an additional 90 degrees
- Front cover-to-cylinder head bolts. Tighten to 6 ft. lbs. (9 Nm)
- Slide rail pins. Apply sealant to pins.
- Camshaft gears
- Camshaft adjusters

### ※※ WARNING

**On 5.5L engine, replace chain tensioner with M24X1.5 thread with new tensioner. Reusing old tensioner will overstretch the chain.**

- Timing chain tensioner with a new gasket and tighten to 59 ft. lbs. (80 Nm).
- Generator
16. Remove the camshaft locking plates.
- Cylinder head covers and tighten the bolts to 89 inch lbs. (10 Nm)
- Ignition coils
- Oil separator
- Front cover
- Thermostat housing
- Coolant hose at thermostat housing
- Coolant hose at heat exchanger
- Aspirator valve shutoff valve at head
- Oil filter housing and heat exchanger
- Drive belt, tensioner and guide pulley
- Power steering reservoir
- Power steering reservoir return line
- Fluid from power steering reservoir
- Intake manifold
- Fuel line
- engine wiring harness
- Brake booster vacuum line at intake manifold

- Purge control valve line from crankcase ventilation vacuum line
- Mass air flow sensor
- SFI control unit
- Air cleaner housing, resonance pipe and body
- Engine cover

17. Fill the engine with coolant.
18. Replace oil filter and oil
19. Connect the negative battery cable.
20. Clear fault memory.
21. Start the vehicle and check for leaks.

## ENGINE ASSEMBLY

### REMOVAL & INSTALLATION

#### 3.5L Engine—Removed From Above

1. Before servicing the vehicle, refer to the precautions in the beginning of this section.
2. Verify that the rear engine lifting eyes are correct. The left lifting eye is marked with a star and code number 04, and the right lifting eye is marked with a star and code number 02.
3. Leave shifter in neutral, parking brake on and steering wheel in straight ahead position.
4. Remove or disconnect the following:
- Negative battery cable
- front engine cover
- Air filter housing and ducts
- MAF sensor
- Purge line from purge switchover valve and crankcase system
- Brake booster vacuum hose from the rear of the intake manifold
- Engine wiring harness
- Relieve the fuel system pressure
- Fuel feed line
- Drain the power steering fluid from the pump
- Hoses from the power steering pump, then plug the openings
- Transmission lines at radiator
- Coolant
- Heater hose at heater
- Coolant hoses from the water pump and thermostat housing
- Electric fan
- Oil dipstick from tube
- Serpentine belt

5. Lock the automatic belt tensioner by rotating the tensioner counterclockwise until a 5mm drift or pin fits through the tensioner.
6. Secure vehicle to lifting platform.
- Lower engine paneling
- Power steering line bracket on oil pan

- O2 sensor connectors and brackets at transmission
- Left downstream O2 sensor
- exhaust system
- Air conditioning compressor and position it aside, leaving the hoses attached
- Transmission cooler lines at transmission
- Transmission cooler line bracket on oil pan
- Support engine at front and remove transmission.
- Engine mount bolts

7. Attach engine hoist to engine and lift engine from vehicle.

#### To install:

8. Lower engine into vehicle.
9. Install or connect the following:
- engine mount bolts
- transmission
- Transmission cooler line bracket on oil pan
- Transmission cooler lines at transmission
- Air conditioning compressor
- exhaust system
- Left downstream O2 sensor
- O2 sensor connectors and brackets at transmission
- Power steering line bracket on oil pan
- Lower engine paneling
- Serpentine belt
- Oil dipstick from tube
- Electric fan
- Coolant hoses from the water pump and thermostat housing
- Heater hose at heater
- Transmission lines at radiator
- Hoses to the power steering pump
- Fuel feed line
- Engine wiring harness
- Brake booster vacuum hose to the rear of the intake manifold
- Purge line to purge switchover valve and crankcase system
- MAF sensor
- Fill the power steering reservoir
- Coolant
- Air filter housing and ducts
- front engine cover
- Negative battery cable

#### 3.5L, 5.0L & 5.5L Engines— Removed From Below

1. Before servicing the vehicle, refer to the precautions in the beginning of this section.
2. Verify that the rear engine lifting eyes are correct. The left lifting eye is marked

with a star and code number 04, and the right lifting eye is marked with a star and code number 02.
3. Leave shifter in neutral, parking brake on and steering wheel in straight ahead position.
4. Secure vehicle to lifting platform.
5. Remove or disconnect the following:
- Negative battery cable
- front engine cover
- Air filter housing and ducts
- Suction refrigerant from A/C system
- Engine wiring harness
- left downstream O2 sensor
- exhaust system
- Purge line from purge switchover valve and crankcase system
- Brake booster vacuum hose from the rear of the intake manifold
- Drain the power steering fluid from the pump
- Hoses from the power steering pump, then plug the openings
- Transmission lines at radiator

6. Lock the automatic belt tensioner by rotating the tensioner counterclockwise until a 5mm drift or pin fits through the tensioner.
- Serpentine belt
- Air conditioning compressor and position it aside, leaving the hoses attached
- Relieve the fuel system pressure
- Fuel feed line
- Coolant
- Heater hose at heater
- Coolant hoses from the water pump and thermostat housing
- Stiffening bridge under driveshaft
- Rear driveshaft from transfer case
- Front wheels
- Front inner fender liners
- Under floor panels
- Brackets from steering knuckles
- Front wheel speed sensors
- Front guide pins and brake calipers
- Link rods from strut assembly
- Strut assembly from lower control arm
- Upper control arm from knuckle

7. Support front axle carrier.
- Rear engine cross member
- Stiffening plate bolts and longitudinal member bolts

8. Lower engine, transmission and front axle carrier assembly from vehicle.

#### To install:

9. Raise engine, transmission and front axle carrier assembly into vehicle.

10. Install or connect the following:
- Rear engine cross member
- Stiffening plate bolts and longitudinal member bolts
- Upper control arm to knuckle
- Strut assembly to lower control arm
- Link rods to strut assembly
- Front guide pins and brake calipers
- Front wheel speed sensors
- Strap brackets to steering knuckles
- Under floor panels
- Front inner fender liners
- Front wheels
- Rear driveshaft to transfer case
- Stiffening bridge under driveshaft
- Coolant hoses to the water pump and thermostat housing
- Heater hose at heater
- Fuel feed line
- Air conditioning compressor
- Serpentine belt
- Transmission lines at radiator
- Hoses to the power steering pump
- Brake booster vacuum hose to the intake manifold
- Purge line from purge switchover valve and crankcase system
- exhaust system
- left downstream O2 sensor
- Engine wiring harness
- Air filter housing and ducts
- Negative battery cable
11. Fill the engine with coolant and oil.
12. Fill the power steering reservoir.
13. Connect the negative battery cable.
14. Perform wheel alignment
15. Charge A/C system

### EXHAUST MANIFOLD

#### REMOVAL & INSTALLATION

1. Before servicing the vehicle, refer to the precautions in the beginning of this section.
2. Remove or disconnect the following:
- front engine cover
- Intake air ducts
- Negative battery cable
- Accessible exhaust manifold nuts
3. Raise vehicle.
4. Remove or disconnect the following:
- Right and left catalytic converters
- Starter heat shield
- Remaining exhaust manifold nuts
- Exhaust manifold
5. Clean the gasket mating surfaces.

**To install:**

6. Install or connect the following:
- Exhaust manifold using new gaskets and nuts. Tighten the nuts to 12 ft. lbs. (16 Nm).

- Starter heat shield
- Right and left catalytic converters
- Negative battery cable
- Intake air ducts and front cover
7. Start the vehicle and check for leaks.

### INTAKE MANIFOLD

#### REMOVAL & INSTALLATION

*See Figure 20.*

1. Before servicing the vehicle, refer to the precautions in the beginning of this section.
2. Remove or disconnect the following:
- Air cleaner and ducts
- Right SFI control unit connector
- Three electrical connectors and cable ties at left rear of intake manifold
- SFI control unit from left bracket and set aside. Leave left connector installed.
- MAF sensor
- Air duct housing
- Crankcase ventilation system hoses
- Right SFI control unit holder
- Feed duct (wiring harness holder) from rear of intake manifold
- Feed duct (wiring harness holder) from right side of intake manifold

- Left rear engine lifting eye
- Feed duct (wiring harness holder) from left cylinder head
- Injector harness connectors
- Left and right intake tumble flap position sensor connectors at rear of intake manifold
- Electrical connector at left front of intake manifold
- Air pump electrical connector
- Relieve fuel system pressure at the service valve
- Fuel line at injector rail
- Intake manifold bolts
3. Raise intake manifold and remove pneumatic lines from rear of manifold.
4. Remove the intake manifold from vehicle.
5. Place clean shop rags into the intake passages to prevent dirt from entering. Clean the gasket mating surfaces.
6. If replacing the intake manifold, remove the following components and transfer to new manifold:
- Left SFI control unit bracket
- Fuel rail and injectors
- throttle valve actuator
- Left and right intake tumble flap position sensors

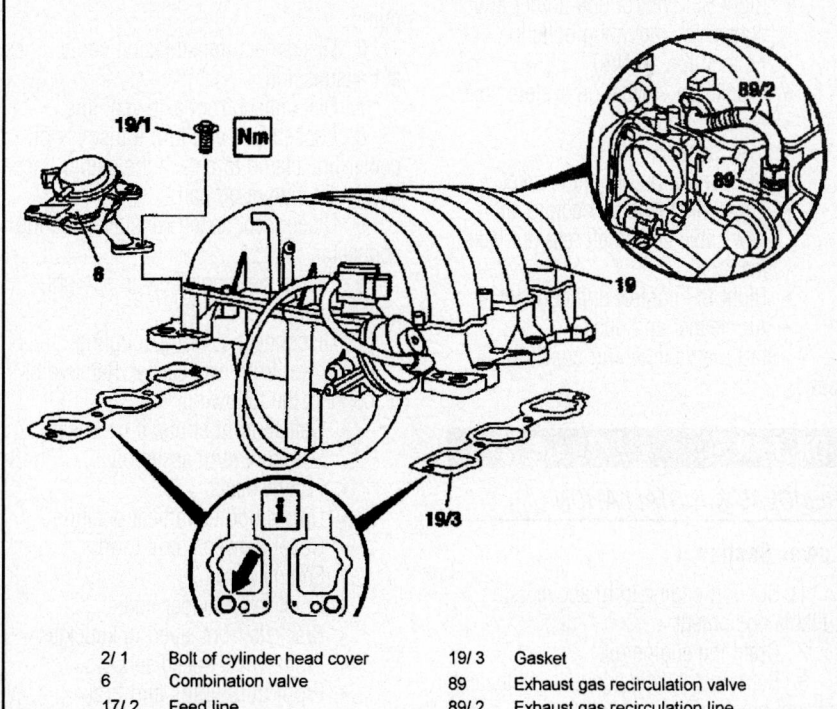

| 2/ 1 | Bolt of cylinder head cover | 19/ 3 | Gasket |
| 6 | Combination valve | 89 | Exhaust gas recirculation valve |
| 17/ 2 | Feed line | 89/ 2 | Exhaust gas recirculation line |
| 19 | Intake manifold | Arrow | Hole for combination valve |
| 19/ 1 | Bolt | | |

42348-MCLS-G06

**Fig. 20 Exploded view of the intake manifold and related components—Typical**

**To install:**

7. Remove the shop rags from the intake passages.

8. Install or connect the following:
- Intake manifold to the engine and tighten the mounting bolts to 80 inch lbs. (9 Nm)
- Fuel line at injector rail and tighten the mounting bolts to 80 inch lbs. (9 Nm)
- Air pump electrical connector
- Electrical connector at left front of intake manifold
- Left and right intake tumble flap position sensor connectors at rear of intake manifold
- Injector harness connectors
- Feed duct (wiring harness holder) to left cylinder head and tighten the mounting bolts to 71 inch lbs. (8 Nm)
- Left rear engine lifting eye and tighten the bolt to 15 ft. lbs. (20 Nm)
- Feed duct (wiring harness holder) to right side of intake manifold and tighten the mounting bolts to 71 inch lbs. (8 Nm)
- Feed duct (wiring harness holder) to rear of intake manifold and tighten the mounting bolts to 71 inch lbs. (8 Nm)
- Right SFI control unit holder and tighten the mounting bolts to 71 inch lbs. (8 Nm)
- Crankcase ventilation system hoses
- Air duct housing
- MAF sensor
- SFI control unit to left bracket.
- Three electrical connectors and new cable ties at left rear of intake manifold
- Right SFI control unit connector
- Air cleaner and ducts

9. Start the vehicle and check for leaks.

## OIL PAN

### REMOVAL & INSTALLATION

#### Lower Section

1. Support engine from above using suitable equipment.

2. Drain the engine oil.

3. Remove catalytic converter bolts and exhaust brackets.

4. Remove engine mount bolts.

5. Raise the engine.

6. Remove the power steering high pressure line bracket.

7. Remove the lower oil pan bolts.

8. Remove lower oil pan. Use a heat gun and a wedge to separate, if necessary.

9. Clean lower and upper oil pan sealing surfaces.

**To install:**

10. Apply a 2 mm bead of sealant along inside of the bolt holes. Do not spread. Parts must be assembled within 10 minutes.

11. Install lower pan and bolts. Tighten to 7 ft. lbs. (9 Nm).

12. Install power steering high pressure line bracket. Tighten to 7 ft. lbs. (10 Nm).

13. Install motor mount bolt. Tighten to 39 ft. lbs. (53 Nm).

14. Install catalytic converter bolts and exhaust bracket bolts. Tighten to 15 ft. lbs. (20 Nm).

15. Remove engine support.

16. Fill the engine oil.

#### Upper Section

1. Before servicing the vehicle, refer to the precautions in the beginning of this section.

2. Removing oil dipstick guide tube.

3. Remove lower oil pan.

4. Remove oil level check switch.

5. Remove oil pickup tube from oil pump.

6. Disconnect transmission cooler lines at transmission.

7. Disconnect front axle vent line.

8. Loosen bolt and turn transmission cooler line clamp to detach the cooler lines from right side of oil pan.

9. Disconnect speed sensitive steering connector.

10. Clamp steering wheel in straight ahead position.

11. Disconnect steering coupling.

12. Lower front axle carrier. Remove or disconnect the following:
- Deflate front struts, if necessary. See front strut assembly.
- Front wheels
- Open hood to vertical position
- Cover between strut towers
- Engine cover
- Front inner fender liners
- Brackets from steering knuckles
- Front wheel speed sensors
- Front guide pins and brake calipers
- Link rods from strut assembly
- Strut assembly from lower control arm
- Upper control arm from knuckle

13. Support front axle carrier. Ensure engine is supported as described in lower oil pan procedure.
- Rear engine cross member

14. Lower front axle carrier assembly.

15. Remove upper oil pan.

➡ **The upper pan bolts are different lengths and diameters. Note their locations for installation.**

16. Remove upper pan Use a heat gun and a wedge to separate, if necessary.

**To install:**

17. Clean the sealing surfaces and apply a bead of silicone sealant.

➡ **The pan must be installed within ten minutes after the sealant is applied.**

18. Install or connect the following:
- Upper pan. Tighten the 6mm bolts to 80 inch lbs. (9 Nm), and the 8mm bolts to 15 ft. lbs. (20 Nm).
- Front axle carrier
- Steering coupling. Tighten to 24 ft. lbs. (33 Nm).
- Speed sensitive steering connector
- Transmission cooler lines and clamp. Tighten bracket bolt to 9 ft. lbs. (12 Nm). Tighten banjo fitting to 44 inch lbs. (5 Nm), then an additional 90 degrees. Tighten hexalobular bolt to 9 ft. lbs. (12 Nm).
- Front axle vent line
- Oil pickup tube to oil pump
- Oil level check switch
- Lower oil pan Tighten the bolts to 90 inch lbs. (10 Nm).
- Engine mount bolts. Tighten to 39 ft. lbs. (53 Nm).
- Oil dipstick guide tube

19. Fill the crankcase to the correct level.

20. Connect the negative battery cable.

## OIL PUMP

### REMOVAL & INSTALLATION

1. Before servicing the vehicle, refer to the precautions in the beginning of this section.

2. Remove the lower oil pan. Refer to the oil pan procedure in this section.

3. Remove oil pickup tube and oil sensor.

4. Unbolt the oil pump.

5. Push the chain tensioner back and remove the pump drive chain from drive gear.

6. Remove the oil pump.

## To install:

7. Fill the oil pump with clean engine oil and install it to the engine. Tighten the bolts to 15 ft. lbs. (20 Nm).

8. Install the pump drive chain.

9. Install oil pickup tube and oil sensor. Tighten pickup tube bolts to 7 ft. lbs. (9 Nm).

10. Install the lower oil pan. Refer to the oil pan procedure in this section.

## PISTON AND RING

### POSITIONING

*See Figures 21 and 22.*

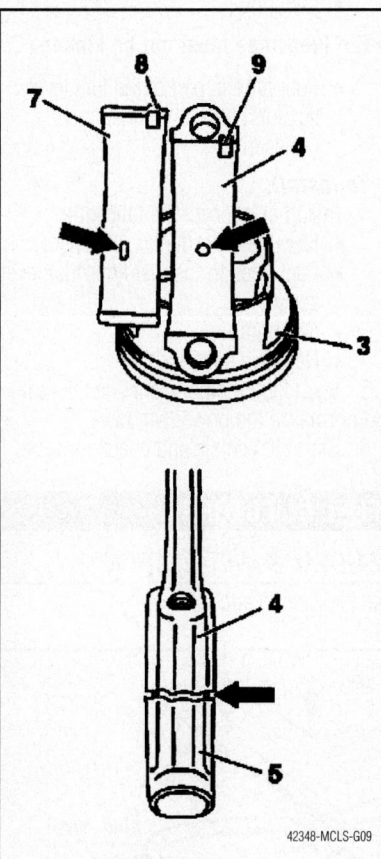

**Fig. 21 Install pistons so the code next to the piston pin and group number are pointing in the direction of travel, and the anti-twist lock (8) and the groove (9) are matched**

## REAR MAIN SEAL

### REMOVAL & INSTALLATION

1. Before servicing the vehicle, refer to the precautions in the beginning of this section.

2. Remove the flywheel.

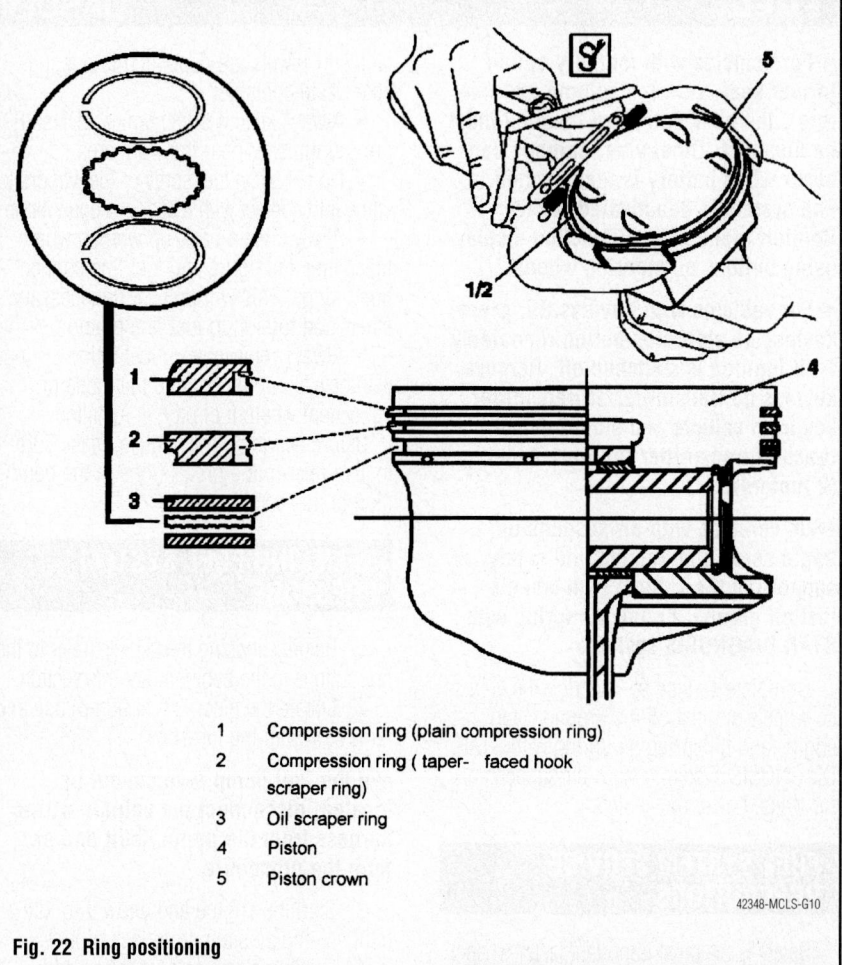

1  Compression ring (plain compression ring)
2  Compression ring ( taper-  faced hook scraper ring)
3  Oil scraper ring
4  Piston
5  Piston crown

**Fig. 22 Ring positioning**

3. Remove rear cover bolts and cover. The oil seal is part of the rear cover and cannot be replaced separately.

## To install:

4. Do not lubricate the seal, install it dry.

5. Clean and degrease new rear cover, crankshaft flange and sealing surfaces on block and oil pan.

6. Apply sealing compound to block and oil pan.

➡ **Rear cover must be installed within 5 minutes of applying sealant.**

7. Mount rear cover with seal on tool 271 589 00 43 00 or an equivalent. Part of the tool is used to mount seal onto another part of the tool for installation. Ensure the outer seal lip is facing out and the inner lip is facing in.

8. Install rear cover with seal using tool.

9. Install and tighten rear cover bolts to 6 ft. lbs. (8 Nm).

10. Remove tool. Check seal lip for proper position.

11. Install the flywheel.

## TIMING CHAIN COVER AND SEAL

### REMOVAL & INSTALLATION

➡**This procedure requires special Mercedes tools and components that are only available to an authorized Mercedes Benz service center.**

## TIMING CHAIN AND SPROCKETS

### REMOVAL & INSTALLATION

➡**This procedure requires special Mercedes tools and components that are only available to an authorized Mercedes Benz service center.**

## VALVE LASH

### ADJUSTMENT

These vehicles are equipped with Hydraulic Lash Adjusters (HLA's) which do not require periodic adjustment.

➡For vehicles with memory option (driver seat, steering column and mirrors), the easy entry/exit function must be disabled. Otherwise, damage can occur when battery is reconnected. The system is deactivated in the Comfort menu of multifunction display using buttons on steering wheel.

➡For vehicles with Keyless Go, press keyless go start/stop button repeatedly until ignition is switched off. Remove keyless go transmitter or transmitter key from vehicle and store beyond reach of transmitter (at least 7 feet (2 meters)).

➡On vehicles with air suspension, begin service by raising and safely supporting the vehicle with wheels just off ground. Empty air spring with STAR DIAGNOSIS system.

Complete service by lowering the vehicle so wheels are just off the ground. Start engine and fill spring air spring with STAR DIAGNOSIS system. Check air suspension for leaks. Lower the vehicle.

## FUEL SYSTEM SERVICE PRECAUTIONS

Safety is the most important factor when performing not only fuel system maintenance but any type of maintenance. Failure to conduct maintenance and repairs in a safe manner may result in serious personal injury or death. Maintenance and testing of the vehicle's fuel system components can be accomplished safely and effectively by adhering to the following rules and guidelines.

• To avoid the possibility of fire and personal injury, always disconnect the negative battery cable unless the repair or test procedure requires that battery voltage be applied.

• Always relieve the fuel system pressure prior to disconnecting any fuel system component (injector, fuel rail, pressure regulator, etc.), fitting or fuel line connection. Exercise extreme caution whenever relieving fuel system pressure to avoid exposing skin, face and eyes to fuel spray. Please be advised that fuel under pressure may penetrate the skin or any part of the body that it contacts.

• Always place a shop towel or cloth around the fitting or connection prior to loosening to absorb any excess fuel due to spillage. Ensure that all fuel spillage (should it occur) is quickly removed from engine surfaces. Ensure that all fuel soaked

cloths or towels are deposited into a suitable waste container.

• Always keep a dry chemical (Class B) fire extinguisher near the work area.

• Do not allow fuel spray or fuel vapors to come into contact with a spark or open flame.

• Always use a back-up wrench when loosening and tightening fuel line connection fittings. This will prevent unnecessary stress and torsion to fuel line piping.

• Always replace worn fuel fitting O-rings with new Do not substitute fuel hose or equivalent where fuel pipe is installed.

Before servicing the vehicle, make sure to also refer to the precautions in the beginning of this section as well.

## RELIEVING FUEL SYSTEM PRESSURE

1. Before servicing the vehicle, refer to the precautions in the beginning of this section.
2. Locate the electric fuel pump fuse and remove it from the fuse box.

➡If the fuel pump fuse cannot be located, disconnect the vehicle wiring harness from the pump itself and perform the procedure.

3. Start the engine and allow it to idle until the engine stalls from lack of fuel.
4. Crank the engine over for an additional 15–20 seconds.
5. Reinstall the pump fuse when repairs are completed.

## FUEL FILTER

### REMOVAL & INSTALLATION

1. Before servicing the vehicle, refer to the precautions in the beginning of this section.

➡Fuel pump is located in right side, fuel filter/level sensor is in left side.

2. Relieve the pressure in the fuel tank by opening, then tightening the filler cap.
3. Remove or disconnect the following:
• Negative battery cable
• Fuel pump.

➡The fuel lines must not be kinked.

• Fuel filter is part of the fuel level sensor. Entire unit must be replaced.

### To install:
4. Install or connect the following:
• New fuel filter/fuel level sensor
• Fuel lines to the fuel filter/fuel level sensor
• fuel pump
• Negative battery cable
5. Read fault memory, encode the radio and normalize the power windows.
6. Start the vehicle and check for leaks.

## FUEL INJECTORS

### REMOVAL & INSTALLATION
*See Figures 23 and 24.*

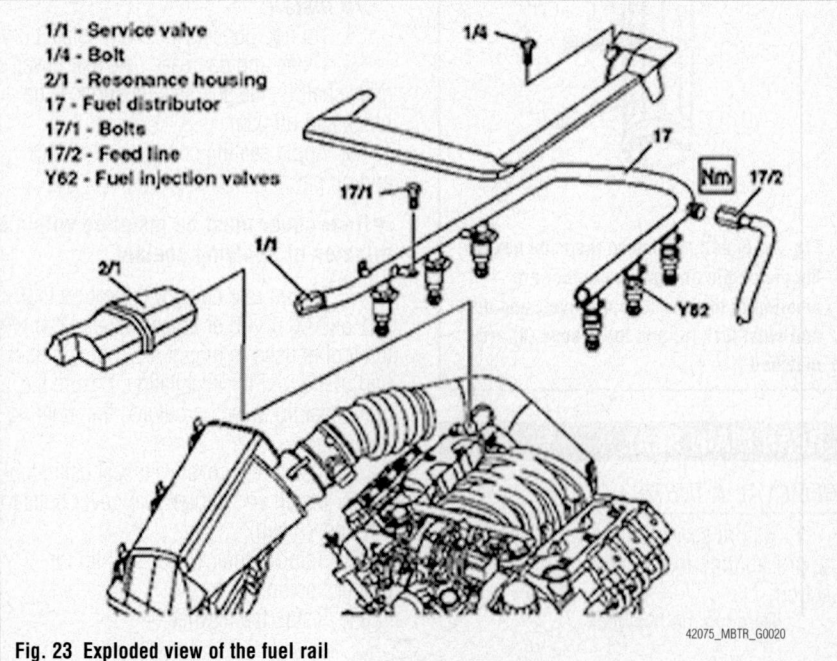

1/1 - Service valve
1/4 - Bolt
2/1 - Resonance housing
17 - Fuel distributor
17/1 - Bolts
17/2 - Feed line
Y62 - Fuel injection valves

**Fig. 23 Exploded view of the fuel rail**

42075_MBTR_G0020

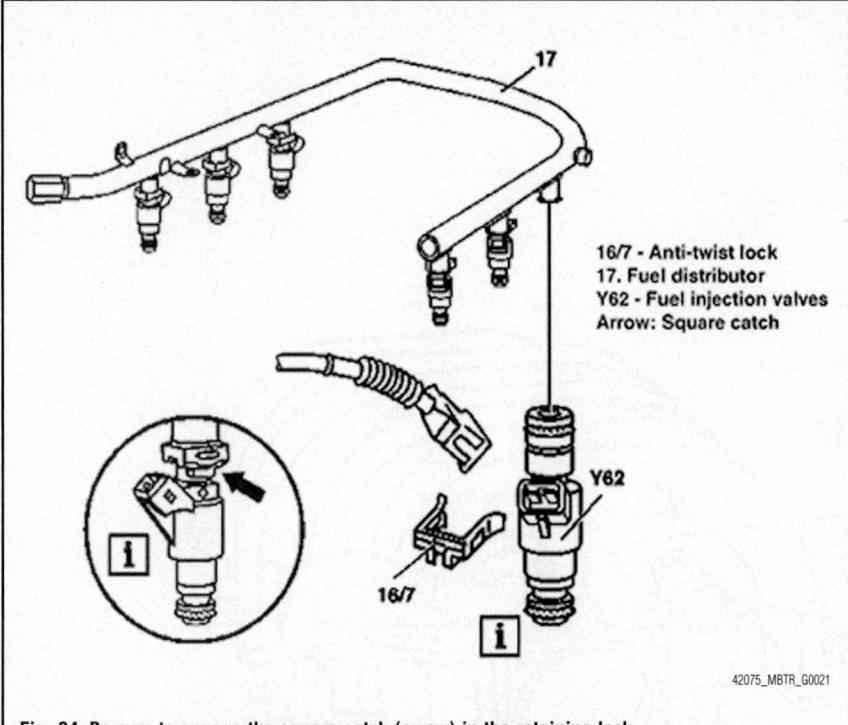

16/7 - Anti-twist lock
17 - Fuel distributor
Y62 - Fuel injection valves
Arrow: Square catch

**Fig. 24 Be sure to engage the square catch (arrow) in the retaining lock**

1. Before servicing the vehicle, refer to the precautions in the beginning of this section.
2. Relieve the fuel system pressure.
3. Remove the air intake hose and air filter assembly.
4. Remove the screw securing the fuel rail plastic cover, then remove the cover.
5. Release any residual pressure in the fuel rail using the service valve on the end of the rail.
6. Detach the fuel feed line and separate the electrical connections from the injectors.
7. Remove the bolt securing the fuel rail, then carefully pull it up and out of the intake manifold.
8. If necessary to remove the injectors, remove the retaining lock then pull the injector out of the fuel rail.
9. Installation is the reverse of removal. Use new O-rings for the injectors and coat them lightly with oil.

## FUEL PUMP

*REMOVAL & INSTALLATION*

1. Before servicing the vehicle, refer to the precautions in the beginning of this section.
2. Relieve the fuel system pressure.
3. Disconnect the negative battery cable.
4. Empty the fuel tank into a suitable container.
5. Raise the rear seat approximately 20 inches (50 cm).

6. Lift the carpeting to gain access to the fuel pump cover.

➡**Fuel pump is located in right side, fuel filter/level sensor is in left side.**

7. Remove or disconnect the following:
   • Fuel pump cover
   • Fuel pump electrical connector

➡**Be sure not to kink the fuel pipes.**

   • Supply and return fuel pipes clips and the pipes
   • Union nut mounting the fuel pump-to-the tank
   • Fuel pump from the tank

*To install:*

➡**Lightly oil the fuel pump sealing O-ring to simplify the installation.**

8. Install or connect the following:
   • Fuel pump into the tank using a new union nut and O-ring. Tighten the union nut to 50 ft. lbs. (65 Nm).
   • Supply and return lines to the fuel pump
   • Fuel pump electrical connector
   • Fuel pump access cover and the rear seat
9. Fill the fuel tank.
   • Negative battery cable
10. Read fault memory, encode the radio and normalize the power windows.
11. Start the vehicle and check for leaks.

## FUEL TANK

*REMOVAL & INSTALLATION*

1. Before servicing the vehicle, refer to the precautions in the beginning of this section.
2. Relieve the fuel system pressure and drain the fuel tank.
3. Remove underfloor panel to access fuel tank.
4. Remove exhaust system.
5. Matchmark the position of the driveshaft on the rear axle flange. Remove the bolts, then secure the driveshaft out of the way. Discard the bolts, as new ones must be used during installation.
6. Remove filler neck from fuel tank.
7. Disconnect feed line.
8. Support the fuel tank with a transmission jack, then remove the fuel tank tensioning straps.
9. Remove fuel tank bolts.
10. Remove fuel tank self-locking nut.
11. Drain fuel tank.
12. Lower the fuel tank, then detach the lines and electrical connections from the fuel tank. Remove the tank from the vehicle.
13. Installation is the reverse of removal. Tighten the fuel tank fasteners to 20 ft. lbs. (23 Nm). Refer to the appropriate procedures in this section for the tightening specifications of other components.

## IDLE SPEED

*ADJUSTMENT*

Idle speed is maintained by the Powertrain Control Module (PCM). No adjustment is necessary or possible.

## THROTTLE BODY

*REMOVAL & INSTALLATION*

See Figure 25.

1. Before servicing the vehicle, refer to the precautions in the beginning of this section.
2. Remove the air intake hose assembly from the throttle body assembly.
3. Detach the crankcase ventilation hose and the electrical connection from the throttle body.
4. Remove the bolts securing the throttle body to the intake manifold, then remove it. It may need a slight tap to break it free from the gasket.
5. Installation is the reverse of removal. Be sure to use a new gasket. Tighten the mounting bolts to 80 inch lbs. (9 Nm).

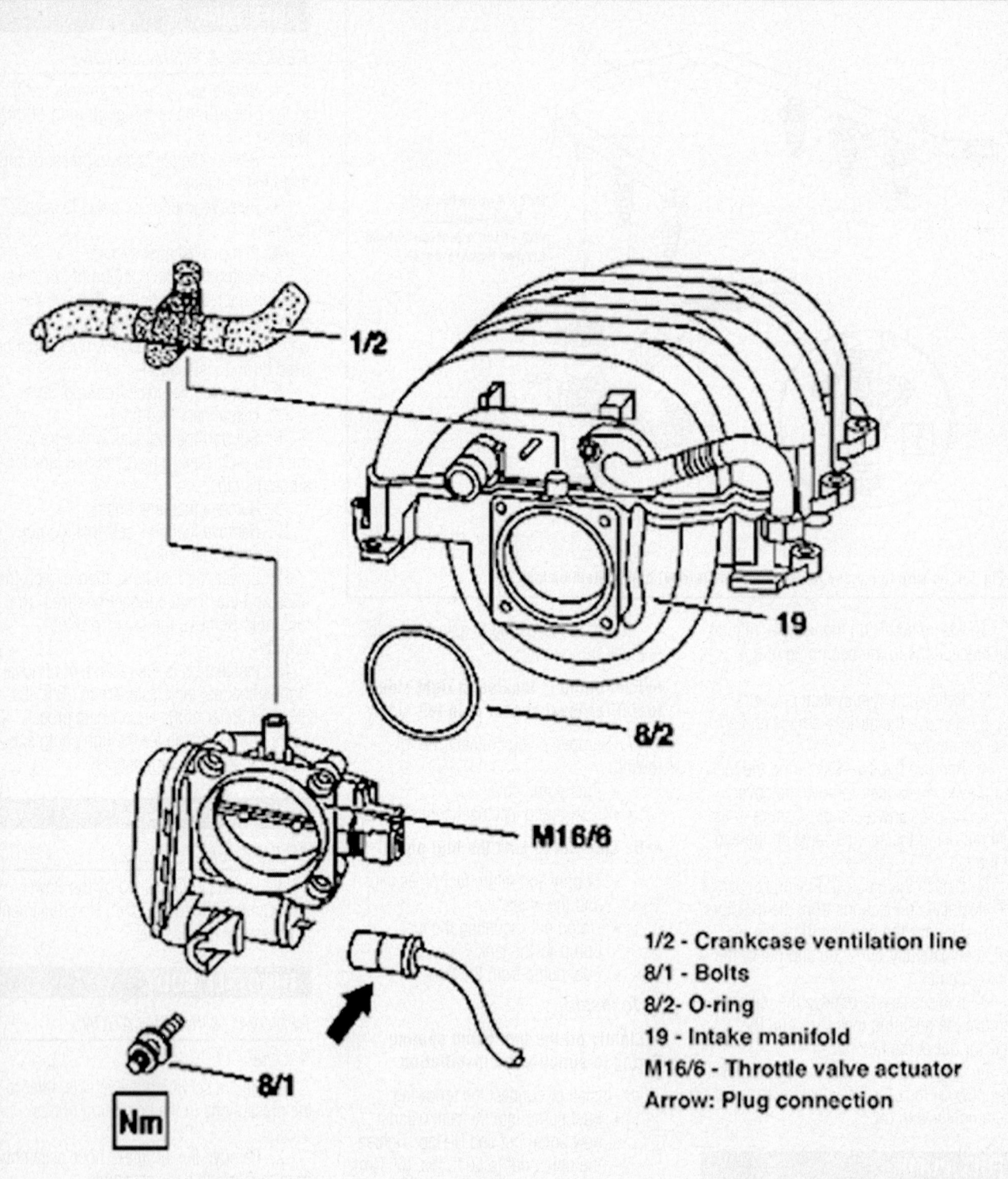

1/2 - Crankcase ventilation line
8/1 - Bolts
8/2- O-ring
19 - Intake manifold
M16/6 - Throttle valve actuator
Arrow: Plug connection

42075_MBTR_G0019

Fig. 25 Exploded view of the throttle body assembly

# HEATING & AIR CONDITIONING SYSTEM

## PRECAUTIONS

Before servicing the air conditioning system on any vehicle, please be sure to read all of the following precautions, which deal with personal safety, prevention of component damage, and important points to take into consideration when servicing the air conditioning system:

• When removing refrigerant components from a vehicle, immediately cap (seal) the component to minimize the entry of moisture from the atmosphere.

• When installing refrigerant components to a vehicle, do not remove the caps (unseal) until just before connecting the components. Connect all refrigerant loop components as quickly as possible to minimize the entry of moisture into system.

• Only use the specified lubricant from a sealed container. Immediately reseal containers of lubricant. Without proper sealing, lubricant will become moisture saturated and should not be used.

• Avoid breathing A/C refrigerant and lubricant vapor or mist. Exposure may irritate eyes, nose and throat. Remove HFC-134a (R-134a) from the A/C system, using certified service equipment meeting requirements of SAE J2210 HFC-134a (R-134a) recycling equipment, or J2209 HFC-134a (R-134a) recovery equipment. If accidental system discharge occurs, ventilate work area before resuming service. Additional health and safety information may be obtained from refrigerant and lubricant manufacturers.

• Do not allow lubricant to come in contact with Styrofoam parts. Damage may result.

## BLOWER MOTOR

### REMOVAL & INSTALLATION

*See Figure 26.*

1. Before servicing the vehicle, refer to the precautions in the beginning of this section.
2. Move passenger seat back.

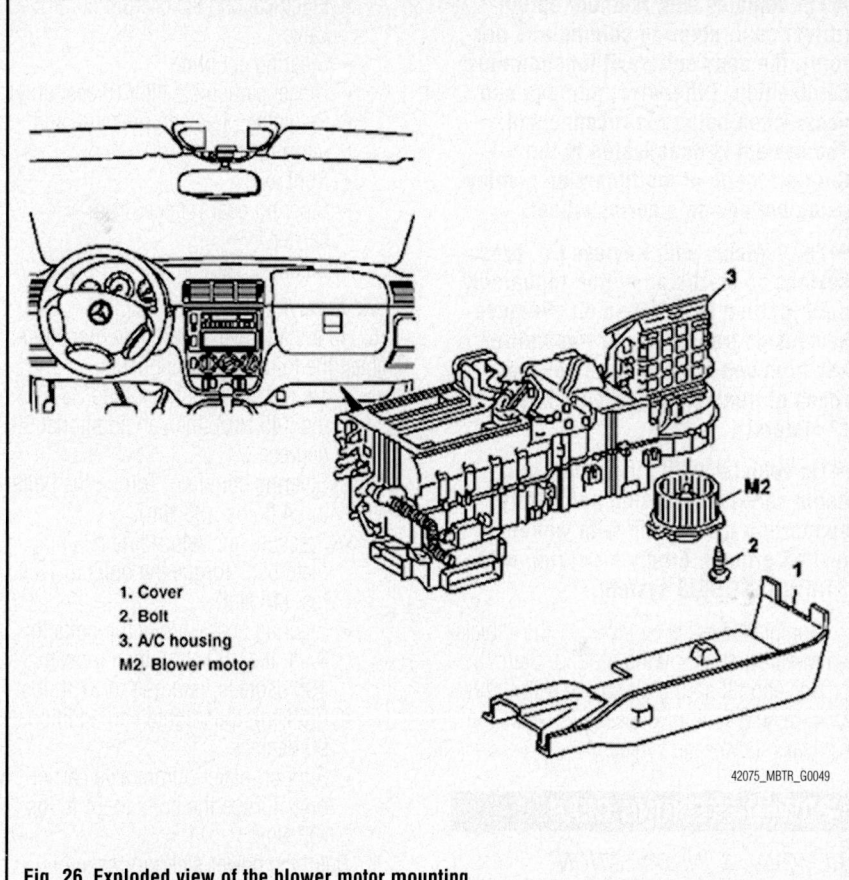

1. Cover
2. Bolt
3. A/C housing
M2. Blower motor

42075_MBTR_G0049

**Fig. 26 Exploded view of the blower motor mounting**

3. Remove the screws and lower the locking lever securing the passenger's side lower footwell panel. Lower the panel, unplug the electrical connection, then remove the panel.

4. Remove the screws securing the blower motor assembly, then detach the connector and remove the motor from the vehicle.

5. Installation is the reverse of removal.

## HEATER CORE

### REMOVAL & INSTALLATION

1. Before servicing the vehicle, refer to the precautions in the beginning of this section.

➡Cover interior area before disconnecting hoses to prevent coolant spills. Plug hoses and heater core tubes.

2. Drain cooling system.
3. Remove or disconnect the following:
   • Negative battery cable
   • Instrument panel
   • Left footwell air pipe (do not disconnect wiring)
   • Heater hoses
   • Heater core
4. To install, reverse the removal procedure.
5. Connect the negative battery cable.
6. Fill the engine with coolant and check for proper heater operation.

## STEERING

➡For vehicles with memory option (driver seat, steering column and mirrors), the easy entry/exit function must be disabled. Otherwise, damage can occur when battery is reconnected. The system is deactivated in the Comfort menu of multifunction display using buttons on steering wheel.

➡For vehicles with Keyless Go, press keyless go start/stop button repeatedly until ignition is switched off. Remove keyless go transmitter or transmitter key from vehicle and store beyond reach of transmitter (at least 7 feet (2 meters)).

➡On vehicles with air suspension, begin service by raising and safely supporting the vehicle with wheels just off ground. Empty air spring with STAR DIAGNOSIS system.

Complete service by lowering the vehicle so wheels are just off the ground. Start engine and fill spring air spring with STAR DIAGNOSIS system. Check air suspension for leaks. Lower the vehicle.

### POWER STEERING GEAR

#### REMOVAL & INSTALLATION

1. Before servicing the vehicle, refer to the precautions in the beginning of this section.
2. Disconnect the negative battery cable.
3. Center the steering wheel and turn the key to the lock position. Remove the key from the ignition switch.
4. Remove engine trim panel and drain the power steering reservoir.
5. Remove or disconnect the following:
   - Lower engine compartment panel
   - Outer tie rod end from knuckle

   - Electrical connector from control valve
   - Steering coupling
   - Diesel particulate filter (diesel only)
   - Retaining plate and pressure and return lines
   - front axle
   - Steering gear retainer from axle carrier
   - Steering gear

#### To install:

6. To install, reverse removal procedure. Tighten the following components:
   - Tie rod. Torque the nuts to 33 ft. lbs. (45 Nm), then an additional 90 degrees.
   - Steering coupling. Torque the bolts to 24 ft. lbs. (33 Nm).
   - Pressure and return line retaining plate bolt. Torque the bolts to 13 ft. lbs. (18 Nm).
   - Steering gear. Torque the bolts to 37 ft. lbs. (50 Nm), then loosen 180 degrees, retorque to 37 ft. lbs. (50 Nm), then tighten an additional 90 degrees.
   - Support plate to front axle carrier bolt. Torque the bolts to 24 ft. lbs. (32 Nm).

7. Refill the power steering reservoir with fresh fluid. Start the engine and turn the steering wheel from lock-to-lock several times. The system will bleed automatically. Check for leaks. Check alignment.

### POWER STEERING PUMP

#### REMOVAL & INSTALLATION

1. Before servicing the vehicle, refer to the precautions in the beginning of this section.
2. Disconnect the negative battery cable.
3. Drain the power steering fluid.

4. Remove or disconnect the following:
   - Engine cover and air cleaner
   - Reservoir
5. Discharge A/C system and remove suction line from compressor.
6. Remove or disconnect the following:
   - Accessory drive belt
   - Power steering lines at pump
   - Power steering pump

#### To install:

7. Install or connect the following:
   - Power steering pump. Torque the mounting bolts to 18 ft. lbs. (25 Nm)
   - Power steering lines
   - A/C suction line at compressor
   - Accessory drive belt
   - Reservoir
   - Engine cover and air cleaner
8. Refill the reservoir with fresh fluid. Start the engine and turn the steering wheel from lock-to-lock several times. The system will bleed automatically. Check for leaks.
9. Recharge A/C system.

#### BLEEDING

1. Do not start engine.
2. Inspect the power steering fluid level and top up as needed.
3. Raise the vehicle until the front wheels are off the ground.
4. Turn the steering wheel from stop to stop about 30 times. During this process continue to top off the fluid as needed.
5. Start the engine and check the fluid level, top off if necessary. Allow to idle for 1 minute.
6. Turn the steering wheel slowly from stop to stop several times at idle speed.
7. The system has been bled when air bubbles no longer rise to the surface in the fluid reservoir and the fluid level no longer drops.

**SUSPENSION** | **FRONT SUSPENSION**

→For vehicles with memory option (driver seat, steering column and mirrors), the easy entry/exit function must be disabled. Otherwise, damage can occur when battery is reconnected. The system is deactivated in the Comfort menu of multifunction display using buttons on steering wheel.

→For vehicles with Keyless Go, press keyless go start/stop button repeatedly until ignition is switched off. Remove keyless go transmitter or transmitter key from vehicle and store beyond reach of transmitter (at least 7 feet (2 meters)).

→On vehicles with air suspension, begin service by raising and safely supporting the vehicle with wheels just off ground. Empty air spring with STAR DIAGNOSIS system.

Complete service by lowering the vehicle so wheels are just off the ground. Start engine and fill spring air spring with STAR DIAGNOSIS system. Check air suspension for leaks. Lower the vehicle.

## COIL SPRING

### REMOVAL & INSTALLATION

1. Remove strut and spring assembly.
2. Install assembly into strut compressor tool.
3. Compress spring until top plate tension is released.
4. Remove upper nut and remove shock absorber.
5. Remove bellows and damper.
6. Release tension on spring and remove spring from tool.

**To install:**
7. Installation is the reverse of removal. Use new hardware.
8. Tighten upper nut to 22 ft. lbs. (30Nm).

## LOWER BALL JOINT

### REMOVAL & INSTALLATION

1. Remove the lower control arm.
2. Remove slotted lock nut from bottom of knuckle using 385 589 00 07 00 or equivalent.
3. Press lower ball joint from knuckle.

**To install:**
4. Press new ball joint into knuckle.
5. Install slotted lock nut and tighten to 221 ft. lbs. (300Nm).
6. Install lower control arm.

## LOWER CONTROL ARM

### REMOVAL & INSTALLATION

The ball joint and bushings on the upper control arm are not replaceable. If necessary, the entire control arm must be replaced.

1. Before servicing the vehicle, refer to the precautions in the beginning of this section.
2. Raise and safely support the vehicle with wheels just off ground.
3. Empty air spring with STAR DIAGNOSIS system.
4. Turn ignition off.
5. Raise vehicle to working height.
6. Remove the wheel.
7. Remove the lower engine panel.
8. Remove the nuts and swivel the stabilizer bar down.
9. Mark installation position of lower control arm to knuckle. Note location of spacer washer at front inner pivot.
10. Remove the bolts and nut from the front of the control arm, then remove the clamp from the rear of the arm.
11. Remove lower shock absorber nut and bolt.
12. Remove lower ball joint nut and separate ball joint from knuckle.
13. Remove the arm from the vehicle.

**To install:**
14. Installation is the reverse of removal. Use new hardware.
15. When installing the control arm, tighten the nuts and bolts until just seated. When the vehicle is lowered, tighten fully.
16. Front frame bolt to 199 ft. lbs. (270 Nm).
17. Shock absorber nut to 195 ft. lbs. (265 Nm).
18. Stabilizer bar nuts to 37 ft. lbs. (50 Nm).
19. Inside rear clamp bolts to 162 ft. lbs. (220 Nm).
20. Outside rear clamp bolts to 133 ft. lbs. (180 Nm).
21. Start engine and fill spring air spring with STAR DIAGNOSIS system.
22. Check air suspension for leaks. Repair as necessary.
23. Turn off engine and lower vehicle.
24. Perform a wheel alignment.

## STABILIZER BAR

### REMOVAL & INSTALLATION

1. Before servicing the vehicle, refer to the precautions in the beginning of this section.

2. Remove or disconnect the following:
- Engine undercover
- Link rod from strut assembly
- Mounting brackets
- Stabilizer bar

**To install:**
3. Install or connect the following:
- Stabilizer bar
- Torque the fasteners as follows:
- Mounting bracket. Tighten to 30 ft. lbs. (40 Nm)
- Stabilizer link rod to front shock. Tighten the nut to 74 ft. lbs. (100 Nm), then loosen 180 degrees, and then tighten to 148 ft. lbs. (200 Nm).
- Stabilizer link rod to bar. Tighten the nut to 74 ft. lbs. (100 Nm), then loosen 180 degrees, then tighten to 33 ft. lbs. (45 Nm) then tighten an additional 60 degrees.
- Engine undercover

## STEERING KNUCKLE

### REMOVAL & INSTALLATION

1. Before servicing the vehicle, refer to the precautions in the beginning of this section.
2. Raise and safely support the vehicle with wheels just off ground.
3. Empty air spring with STAR DIAGNOSIS system.
4. Turn ignition off.
5. Raise vehicle to working height.
6. Remove the wheel.
7. Remove front disc.
8. Remove wheel speed sensor.
9. Cut tie and remove bracket.
10. Remove tie rod end.
11. Remove axle shaft nut and press axle from axle shaft flange.
12. Remove upper ball joint from knuckle.
13. Remove lower ball joint from knuckle.
14. Remove knuckle.

**To install:**
→Always replace any self-locking nuts after one use. When installing suspension components, tighten the fasteners until just seated. When the vehicle is lowered to normal ride height, fully tighten fasteners.

15. Install or connect the following:
- Knuckle
- Upper control arm nut. Tighten the nut to 15 ft. lbs. (20 Nm), then an additional 90 degrees.

- Lower control arm nut. Tighten the nut to 192 ft. lbs. (260 Nm), then an additional 45 degrees.
- Axle shaft and nut. Tighten the nut to 169 ft. lbs. (260 Nm).
- Tie rod end. Tighten the nut to 33 ft. lbs. (45 Nm), then an additional 90 degrees.
- Bracket
- Wheel speed sensor. Tighten the bolt to 6 ft. lbs. (8 Nm)
- Disc brake
- Wheel

16. Start engine and fill spring air spring with STAR DIAGNOSIS system.
17. Check air suspension for leaks. Repair as necessary.
18. Turn off engine and lower vehicle.

## STRUT & SPRING ASSEMBLY

### REMOVAL & INSTALLATION

1. Before servicing the vehicle, refer to the precautions in the beginning of this section.
2. Raise hood to vertical position. Remove strut tower brace.
3. Raise and safely support the vehicle with wheels just off ground.
4. Empty air spring with STAR DIAGNOSIS system.
5. Turn ignition off.
6. Raise vehicle to working height.
7. Remove wheel.
8. Remove the upper strut nuts.
9. On vehicles with air suspension, remove the air line. Remove front inner fender liner.
10. Remove cable ties and brake hose from bracket.
11. Disconnect active damping wiring from ABS sensor bracket.
12. Disconnect active damping wiring connector.
13. Remove tie rod end from knuckle.
14. Remove stabilizer link rod from shock absorber.
15. Remove upper ball joint from knuckle.
16. Remove front axle shaft.
17. Remove lower strut assembly bolt and nut.
18. Press lower control arm down and remove strut assembly.

**To install:**

➡️Always replace any self-locking nuts after one use. When installing suspension components, tighten the fasteners until just seated. When the vehicle is lowered to normal ride height, fully tighten fasteners.

Install or connect the following:
- Strut assembly
- Lower strut assembly bolt. Tighten to 195 ft. lbs. (265 Nm).
- Axle shaft and nut. Tighten the nut to 169 ft. lbs. (260 Nm).
- Upper control arm nut. Tighten the nut to 15 ft. lbs. (20 Nm), then an additional 90 degrees.
- Stabilizer link rod to front shock. Tighten the nut to 74 ft. lbs. (100 Nm), then loosen 180 degrees, and then tighten to 148 ft. lbs. (200 Nm).
- Tie rod end. Tighten the nut to 33 ft. lbs. (45 Nm), then an additional 90 degrees.
- Wiring
- Brake hose
- Inner fender liner
- Wheel

19. Start engine and fill spring air spring with STAR DIAGNOSIS system.
20. Check air suspension for leaks. Repair as necessary.
21. Turn off engine and lower vehicle.
22. Perform a wheel alignment.

## UPPER BALL JOINT

### REMOVAL & INSTALLATION

The ball joints are not replaceable. If necessary, the entire control arm must be replaced.

## UPPER CONTROL ARM

### REMOVAL & INSTALLATION

The ball joint and bushings on the upper control arm are not replaceable. If necessary, the entire control arm must be replaced.

1. Before servicing the vehicle, refer to the precautions in the beginning of this section.
2. Remove air cleaner.
3. Raise and safely support the vehicle with wheels just off ground.
4. Empty air spring with STAR DIAGNOSIS system.
5. Turn ignition off.
6. Raise vehicle to working height.
7. Remove the wheel.
8. Remove the nut from the upper control arm, then use a puller to separate the ball joint from the knuckle. Secure the knuckle to the shock absorber to prevent it from tilting down too far.
9. Remove the bolt securing the front level sensor to the control arm.
10. Remove the nuts in engine area securing the control arm to the frame.

An assistant must hold the bolts in the wheel well area. Remove the bolts.
11. Remove the upper control arm.
12. Installation is the reverse of removal. Note the following:
- When installing the control arm, tighten the nuts until just seated. When the vehicle is lowered, tighten the frame nuts to 45 ft. lbs. (61 Nm) and the knuckle nut to 15 ft. lbs. (20 Nm), then an additional 90 degrees.

13. Start engine and fill spring air spring with STAR DIAGNOSIS system.
14. Check air suspension for leaks. Repair as necessary.
15. Turn off engine and lower vehicle.
16. Perform a wheel alignment

## WHEEL HUB AND BEARING

### REMOVAL & INSTALLATION

1. Before servicing the vehicle, refer to the precautions in the beginning of this section.
2. Raise and safely support the vehicle with wheels just off ground.
3. Empty air spring with STAR DIAGNOSIS system.
4. Turn ignition off.
5. Raise vehicle to working height.
6. Remove the wheel.
7. Remove front disc.
8. Remove wheel speed sensor.
9. Cut tie and remove bracket.
10. Remove tie rod end.
11. Remove axle shaft nut and press axle from axle shaft flange.
12. Remove upper ball joint from knuckle.
13. Remove lower ball joint from knuckle.
14. Remove knuckle.
15. Press out front axle shaft flange.
16. Remove circlip.
17. Press wheel bearing out of knuckle.

**To install:**

➡️Always replace any self-locking nuts after one use. When installing suspension components, tighten the fasteners until just seated. When the vehicle is lowered to normal ride height, fully tighten fasteners.

18. Align new bearing with knuckle. Dots, line or Black inner race must face toward knuckle. Magnetic side of wheel bearing must face toward knuckle.
19. Press wheel bearing into knuckle.
20. Install circlip, making sure it is seated.
21. Press in axle shaft flange.

Install or connect the following:

- Knuckle
- Upper control arm nut. Tighten the nut to 15 ft. lbs. (20 Nm), then an additional 90 degrees.
- Lower control arm nut. Tighten the nut to 192 ft. lbs. (260 Nm), then an additional 45 degrees.

- Axle shaft and nut. Tighten the nut to 169 ft. lbs. (260 Nm).
- Tie rod end. Tighten the nut to 33 ft. lbs. (45 Nm), then an additional 90 degrees.
- Bracket
- wheel speed sensor. Tighten the bolt to 6 ft. lbs. (8 Nm)

- disc brake
- Wheel

22. Start engine and fill spring air spring with STAR DIAGNOSIS system.

23. Check air suspension for leaks. Repair as necessary.

24. Turn off engine and lower vehicle.

## SUSPENSION

➥For vehicles with memory option (driver seat, steering column and mirrors), the easy entry/exit function must be disabled. Otherwise, damage can occur when battery is reconnected. The system is deactivated in the Comfort menu of multifunction display using buttons on steering wheel.

➥For vehicles with Keyless Go, press keyless go start/stop button repeatedly until ignition is switched off. Remove keyless go transmitter or transmitter key from vehicle and store beyond reach of transmitter (at least 7 feet (2 meters)).

➥On vehicles with air suspension, begin service by raising and safely supporting the vehicle with wheels just off ground. Empty air spring with STAR DIAGNOSIS system.

Complete service by lowering the vehicle so wheels are just off the ground. Start engine and fill spring air spring with STAR DIAGNOSIS system. Check air suspension for leaks. Lower the vehicle.

### AIR SPRING

#### REMOVAL & INSTALLATION

➥This applies to vehicles with air suspension.

1. Before servicing the vehicle, refer to the precautions in the beginning of this section.

2. Raise and safely support the vehicle with wheels just off ground.

3. Empty air spring with STAR DIAGNOSIS system.

4. Turn ignition off.

5. Raise vehicle to working height.

6. Remove the wheel.

7. Remove cover.

8. Remove air line from air spring and cap openings.

9. Mark all spring related parts and components to aid in installation. Do not twist spring during removal.

10. Pull spring down until it snaps loose

from the retaining clip. The clip must be replaced.

11. Compress and remove spring.

**To install:**

➥Always replace any self-locking nuts after one use. When installing suspension components, tighten the fasteners until just seated. When the vehicle is lowered to normal ride height, fully tighten fasteners.

12. Compress spring and install cover. Make sure the curved portion of the cover clears the air line fitting.

13. Remove remains of old clip and install new clip into body.

14. Install spring, making sure clip latches and lower part of spring is centered properly in lower arm.

15. Install air line.

16. Install cover.

17. Install wheel and lower vehicle so tire is just off ground.

18. Start engine and fill spring air spring with STAR DIAGNOSIS system.

19. Check air suspension for leaks. Repair as necessary.

Turn off engine and lower vehicle.

### COIL SPRING

#### REMOVAL & INSTALLATION

➥This applies to vehicles without air suspension.

1. Before servicing the vehicle, refer to the precautions in the beginning of this section.

2. Raise and safely support the vehicle.

3. Remove the wheel.

4. Mark all spring related parts and components to aid in installation.

5. Snap out and smooth edges of hole on lower control arms without access hole.

6. Install spring tensioning Tool 202 589 02 31 00 or equivalent.

7. Compress and remove spring.

**To install:**

➥Always replace any self-locking nuts after one use. When installing

suspension components, tighten the fasteners until just seated. When the vehicle is lowered to normal ride height, fully tighten fasteners.

8. Compress spring and install with boot, plastic ring and spacer washer.

9. Remove tool.

10. Install wheel and lower vehicle.

11. If rear springs were replaced, check vehicle level and adjust if necessary.

### LOWER CONTROL ARM

#### REMOVAL & INSTALLATION

1. Before servicing the vehicle, refer to the precautions in the beginning of this section.

2. Raise and safely support the vehicle with wheels just off ground.

3. Empty air spring with STAR DIAGNOSIS system.

4. Turn ignition off.

5. Raise vehicle to working height.

6. Remove the wheel.

7. Remove rear spring.

8. Remove exhaust system.

9. Remove bolt and emergency brake cable bracket.

10. Remove stabilizer bar link.

11. Remove lower shock absorber bolt.

12. Remove bolts at body-to-front of rear differential carrier.

13. Mark then remove rear driveshaft from rear differential.

14. Slightly lower front of rear differential carrier (off-road package vehicles only).

15. Remove inner bolts from lower control arm and lower control arm.

16. Remove nut and separate lower ball joint from hub.

17. Remove lower control arm.

**To install:**

➥Always replace any self-locking nuts after one use. When installing suspension components, tighten the fasteners until just seated. When the vehicle is lowered to normal ride height, fully tighten fasteners.

18. Install components in reverse order.
19. Install wheel and lower vehicle so tire is just off ground.
20. Start engine and fill spring air spring with STAR DIAGNOSIS system.
21. Check air suspension for leaks. Repair as necessary. Turn off engine and lower vehicle.
22. Tighten fasteners.
- Link rod nuts to 133 ft. lbs. (180 Nm).
- Lower shock absorber .mounting bolt. Torque to 74 ft. lbs. (100 Nm) then an additional 120 degrees.
- Driveshaft bolts to 47 ft. lbs. (64 Nm).
- Inner exhaust bracket bolts to 9 ft. lbs. (12 Nm).
- Outer exhaust bracket bolts to 15 ft. lbs. (20 Nm).
- Inner exhaust bracket bolts to 15 ft. lbs. (20 Nm).
- body-to-front of rear differential carrier 74 ft. lbs. (100 Nm) then back off 180 degrees, retorque to 74 ft. lbs. (100 Nm) then tighten an additional 90 degrees.

## SHOCK ABSORBER

### REMOVAL & INSTALLATION

1. Before servicing the vehicle, refer to the precautions in the beginning of this section.
2. Remove interior panel over rear wheel arch.
3. Carefully cut damping mat and fold up to access upper mounting nuts. Do not tear damping mat or damage ADS sensor wiring on right side.
4. Partially raise vehicle.
5. Remove rear wheel.
6. Remove fender liner.
7. Disconnect wiring.
8. using a jack, support suspension to prevent drooping and damage to air spring.
9. Remove sensor (right side) and top shock mounting nuts.
10. Remove lower shock mounting bolt.
11. Compress and remove the shock absorber. Make sure to remove top cover plate and sealing rings.

### To install:

12. Install or connect the following:
- cover plate and new sealing rings
- Shock absorber
- Upper shock nuts, do not tighten at this time
- Lower shock bolt, do not tighten at this time

13. Using jack, raise lower suspension arm to normal ride height.
14. Tighten fasteners.
- Top shock absorber mounting nuts. Torque to 26 ft. lbs. (35 Nm)
- Lower shock absorber .mounting bolt. Torque to 74 ft. lbs. (100 Nm) then an additional 120 degrees.
15. Lower and remove jack.
16. Install or connect the following:
- Wiring connector
- Sensor on right side
- Fender liner

➡ **If new shock absorber has a protective tube, the inner fender liner must be cut for clearance.**

- Rear wheel
- Seal damping mat
- Interior panel

17. Install wheel and lower vehicle so tire is just off ground.
18. Start engine and fill spring air spring with STAR DIAGNOSIS system.
19. Check air suspension for leaks. Repair as necessary.
Turn off engine and lower vehicle.

## STABILIZER BAR

### REMOVAL & INSTALLATION

See Figure 27.

1. Before servicing the vehicle, refer to the precautions in the beginning of this section.
2. Remove or disconnect the following:
- Link rod from control arm to torsion bar

- Stabilizer bar clamps
- Stabilizer bar

### To install:

3. Install or connect the following:
- Stabilizer bar and brackets with new bushings. Torque the bracket nuts to 81 ft. lbs. (110 Nm).
- Link rod. Torque the link rod nuts to 133 ft. lbs. (180 Nm).

## WHEEL BEARINGS

### REMOVAL & INSTALLATION

1. Before servicing the vehicle, refer to the precautions in the beginning of this section.
2. Raise and safely support the vehicle.
3. Remove the rear axle shaft.
4. Remove the wheel.
5. Remove the brake disc.
6. Disconnect the parking brake cable and remove the parking brake shoes.

➡ **Removal of the hub and bearing requires the use of the following extraction and installation tool 210 589 03 43 00 or equivalent.**

7. Pull the hub from the knuckle using the hub extraction tool.
8. Remove the circlip from in front of the bearing.
9. Pull the bearing out of the hub using the bearing extraction tool.
10. Pull the inner bearing race from the hub.
11. Installation is the reverse of removal. Be sure the circlip is fully seated in its groove.

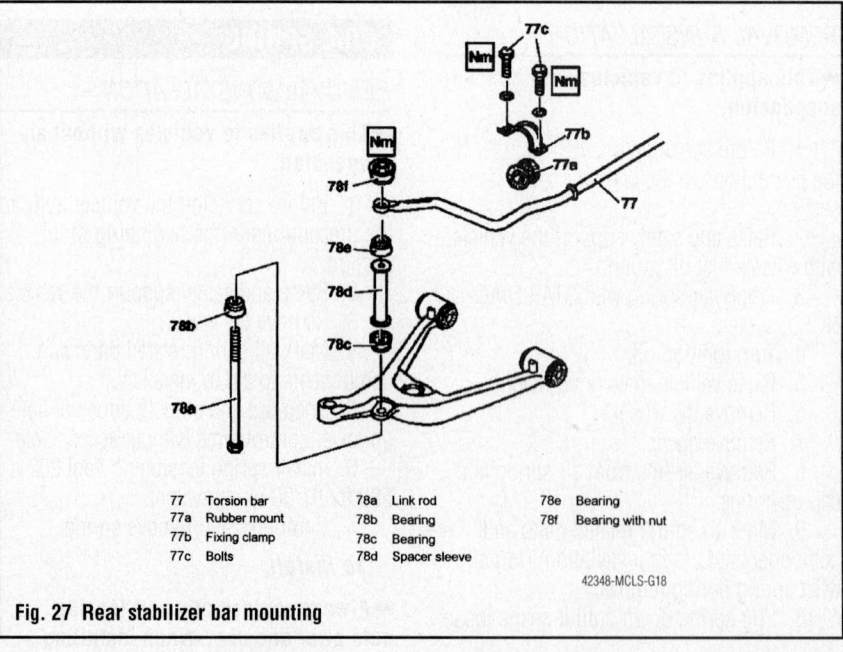

| 77 | Torsion bar | 78a | Link rod | 78e | Bearing |
| 77a | Rubber mount | 78b | Bearing | 78f | Bearing with nut |
| 77b | Fixing clamp | 78c | Bearing | | |
| 77c | Bolts | 78d | Spacer sleeve | | |

42348-MCLS-G18

**Fig. 27 Rear stabilizer bar mounting**

# MINI

## Cooper • Cooper S

**9**

## SPECIFICATIONS AND MAINTENANCE CHARTS

### ENGINE AND VEHICLE IDENTIFICATION

| Code ① | Liters (cc) | Cu. In. | Cyl. | Fuel Sys. | Engine Type | Eng. Mfg. | Code ② | Year |
|---|---|---|---|---|---|---|---|---|
| | | **Engine** | | | | | **Model Year** | |
| W10 | 1.6 (1598) | 97.5 | 4 | NA | DOHC | BMW | 6 | 2006 |
| W11 | 1.6 (1598) | 97.5 | 4 | NA | DOHC | BMW | 7 | 2007 |
| N12 | 1.6 (1598) | 97.5 | 4 | NA | DOHC | BMW | 8 | 2008 |
| N14 | 1.6 (1598) | 97.5 | 4 | NA | DOHC | BMW | | |

NA: Not available

SOHC: Single Overhead Camshaft

① 8th position of VIN

② 10th position of VIN

22205_MINI_C0001

### GENERAL ENGINE SPECIFICATIONS

| Year | Model | Series | Engine Displacement Liters (VIN) | Net Horsepower @ rpm | Net Torque @ rpm (ft. lbs.) | Bore x Stroke (in.) | Compression Ratio | Oil Pressure @ rpm |
|---|---|---|---|---|---|---|---|---|
| **2006** | Cooper | R50 | 1.6 (W10) | 115@6000 | 111@4500 | 3.03x3.37 | 10.6:1 | 25-80@3000 |
| | Cooper S | R53 | 1.6 (W11) | 168@6000 | 162@4000 | 3.03x3.37 | 8.3:1 | 25-80@3000 |
| **2007** | Cooper | R56 | 1.6 (N12) | 118@6000 | 114@4250 | 3.03x3.38 | 11.0:1 | 25-80@3000 |
| | Cooper S | R56 | 1.6 (N14) | 172@5500 | 177@1600 | 2.87x3.38 | 10.5:1 | 25-80@3000 |
| | Cooper Convertible | R52 | 1.6 (W10) | 115@6000 | 111@4500 | 3.03x3.37 | 10.6:1 | 25-80@3000 |
| | Cooper S Convertible | R52 | 1.6 (W11) | 168@6000 | 162@4000 | 3.03x3.37 | 8.3:1 | 25-80@3000 |
| **2008** | Cooper | R56 | 1.6 (N12) | 118@6000 | 114@4250 | 3.03x3.38 | 11.0:1 | 25-80@3000 |
| | Cooper S | R56 | 1.6 (N14) | 172@5500 | 177@1600 | 2.87x3.38 | 10.5:1 | 25-80@3000 |
| | Cooper Convertible | R52 | 1.6 (W10) | 115@6000 | 111@4500 | 3.03x3.37 | 10.6:1 | 25-80@3000 |
| | Cooper S Convertible | R52 | 1.6 (W11) | 168@6000 | 162@4000 | 3.03x3.37 | 8.3:1 | 25-80@3000 |

22205_MINI_C0002

## ENGINE TUNE-UP SPECIFICATIONS

| Year | Model | Series | Engine Displacement Liters (VIN) | Spark Plug Gap (in.) | Ignition Timing (deg.) | Fuel Pump (psi) | Idle Speed (rpm) | Valve Clearance In. | Valve Clearance Ex. |
|------|-------|--------|------|------|------|------|------|------|------|
| 2006 | Cooper | R50 | 1.6 (W10) | NA | ① | NA | ② | HYD | HYD |
|  | Cooper S | R53 | 1.6 (W11) | NA | ① | NA | ② | HYD | HYD |
| 2007 | Cooper | R56 | 1.6 (N12) | NA | ① | NA | ② | HYD | HYD |
|  | Cooper S | R56 | 1.6 (N14) | NA | ① | NA | ② | HYD | HYD |
|  | Cooper Convertible | R52 | 1.6 (W10) | NA | ① | NA | ② | HYD | HYD |
|  | Cooper S Convertible | R52 | 1.6 (W11) | NA | ① | NA | ② | HYD | HYD |
| 2008 | Cooper | R56 | 1.6 (N12) | NA | ① | NA | ② | HYD | HYD |
|  | Cooper S | R56 | 1.6 (N14) | NA | ① | NA | ② | HYD | HYD |
|  | Cooper Convertible | R52 | 1.6 (W10) | NA | ① | NA | ② | HYD | HYD |
|  | Cooper S Convertible | R52 | 1.6 (W11) | NA | ① | NA | ② | HYD | HYD |

NOTE: The Vehicle Emission Control Information label often reflects specification changes made during production. The label figures must be used if they differ from those in this chart.

NA: Not available

HYD: Hydraulic

① Ignition timing is regulated by the Electronic Control Module (ECM), and cannot be adjusted.

② Idle speed is controled by the Electronic Control Module (ECM), and cannot be adjusted.

22205_MINI_C0003

## CAPACITIES

| Year | Model | Series | Engine Displacement Liters (VIN) | Engine Oil with Filter (qts.) | Automatic Transaxle (qts.) | Manual Transaxle (qts.) | Fuel Tank (gal.) | Cooling System (qts.) |
|------|-------|--------|------|------|------|------|------|------|
| 2006 | Cooper | R50 | 1.6 (W10) | 4.8 | 6.3 | 1.8 | 13.2 | 7 |
|  | Cooper S | R53 | 1.6 (W11) | 5.1 | 5.3 | 1.8 | 13.2 | 7 |
| 2007 | Cooper | R56 | 1.6 (N12) | 4.4 | 5.3 | 1.8 | 10.6 | 5.5 |
|  | Cooper S | R56 | 1.6 (N14) | 4.4 | 5.3 | 1.6 | 13.2 | 5.5 |
|  | Cooper Convertible | R52 | 1.6 (W10) | 4.8 | 5.3 | 1.8 | 13.2 | 7 |
|  | Cooper S Convertible | R52 | 1.6 (W11) | 5.1 | 5.3 | 1.8 | 13.2 | 7 |
| 2008 | Cooper | R56 | 1.6 (N12) | 4.4 | 5.3 | 1.8 | 10.6 | 5.5 |
|  | Cooper S | R56 | 1.6 (N14) | 4.4 | 5.3 | 1.6 | 13.2 | 5.5 |
|  | Cooper Convertible | R52 | 1.6 (W10) | 4.8 | 5.3 | 1.8 | 13.2 | 7 |
|  | Cooper S Convertible | R52 | 1.6 (W11) | 5.1 | 5.3 | 1.8 | 13.2 | 7 |

22205_MINI_C0004

# TORQUE SPECIFICATIONS
All readings in ft. lbs.

| Year | Model | Engine Displacement Liters (VIN) | Cylinder Head Bolts | Main Bearing Bolts | Rod Bearing Bolts | Crankshaft Damper Bolts | Flywheel Bolts | Manifold Intake | Manifold Exhaust | Spark Plugs | Oil Pan Drain Plug |
|------|-------|------|------|------|------|------|------|------|------|------|------|
| 2006 | Cooper | 1.6 (W10) | ① | ② | 15 | NA | 59 | 19 | 18 | 20 | 18 |
| | Cooper S | 1.6 (W11) | ① | ② | 15 | NA | 66 | 19 | 18 | 20 | 18 |
| 2007 | Cooper | 1.6 (N12) | ③ | ④ | ⑤ | NA | ⑥ | 11 | 18 | 20 | 22 |
| | Cooper S | 1.6 (N14) | ③ | ④ | ⑤ | NA | ⑤ | ⑦ | 18 | 20 | 18 |
| | Convertible | 1.6 (W10) | ① | ② | 15 | NA | 59 | 19 | 18 | 20 | 18 |
| | S Convertible | 1.6 (W11) | ① | ② | 15 | NA | 66 | 19 | 18 | 20 | 18 |
| 2008 | Cooper | 1.6 (N12) | ③ | ④ | ⑤ | NA | ⑥ | ⑦ | 18 | 20 | 22 |
| | Cooper S | 1.6 (N14) | ③ | ④ | ⑤ | NA | ⑤ | 19 | 18 | 20 | 18 |
| | Convertible | 1.6 (W10) | ① | ② | 15 | NA | 59 | 19 | 18 | 20 | 18 |
| | S Convertible | 1.6 (W11) | ① | ② | 15 | NA | 66 | 19 | 18 | 20 | 18 |

NA: Not available

① Step 1: Bolts 1 through 10: 30 ft. lbs.
   Step: 2 Bolts 1 through 10 tighten an additional 90 degrees
   Step: 3 Bolts 11 through 12 21 ft. lbs.

② Inner (M10) bolts: 44 ft. lbs.
   Outer (M8) bolts: 26 ft. lbs.

③ Step 1: Bolts 1 through 10: 22 ft. lbs.
   Step 2: Bolts 1 through 10 tighten an additional 90 degrees
   Step 3: Bolts 1 through 10 tighten an additional 90 degrees
   Step 4: Bolts 11 through 12: 11 ft. lbs.
   Step 5: Bolts 11 through 12 tighten an additional 90 degrees
   Step 6: Bolts 11 through 12 tighten an additional 90 degrees
   Step 7: Screw replaced and tightened to 22 ft. lbs.

④ Inner (M9) bolts: 22 ft. lbs. plus 150 degrees
   Outer (M6) bolts: 7 ft. lbs.

⑤ Step 1: 44 inch lbs.
   Step 2: 132 inch lbs.
   Step 3: Plus 90 degrees

⑥ Step 1: 72 inch lbs.
   Step 2: 22 ft.lbs.
   Step 3: Plus 90 degrees

⑦ M8 bolts: 11 ft.lbs.
   Hexagon nut: 15 ft. lbs.

22205_MINI_C0005

## WHEEL ALIGNMENT

| Year | Model | | Caster Range (+/-Deg.) | Caster Preferred Setting (Deg.) | Camber Range (+/-Deg.) | Camber Preferred Setting (Deg.) | Toe-in (Deg.) |
|---|---|---|---|---|---|---|---|
| 2006 | Cooper | F | NA | ① | 0.42 | -0.50 | 0.3+/-0.08 |
| | (R50) | R | — | — | 0.33 | -1.75 | 0.4+/-0.13 |
| | Cooper S | F | NA | ① | 0.42 | -0.50 | 0.3+/-0.08 |
| | (R53) | R | — | — | 0.33 | -1.75 | 0.4+/-0.13 |
| 2007 | Cooper | F | NA | ① | 0.42 | -0.50 | 0.2+/-0.17 |
| | (R56) | R | — | — | 0.33 | -1.75 | 0.4+/-0.13 |
| | Cooper S | F | NA | ① | 0.42 | -0.50 | 0.2+/-0.17 |
| | (R56) | R | — | — | 0.33 | -1.75 | 0.4+/-0.13 |
| | Cooper Convertible | F | NA | ① | 0.42 | -0.50 | 0.3+/-0.08 |
| | (R52) | R | — | — | 0.33 | -1.75 | 0.4+/-0.13 |
| | Cooper S Convertible | F | NA | ① | 0.42 | -0.50 | 0.3+/-0.08 |
| | (R52) | R | — | — | 0.33 | -1.75 | 0.4+/-0.13 |
| 2008 | Cooper | F | NA | ① | 0.42 | -0.50 | 0.2+/-0.17 |
| | (R56) | R | — | — | 0.33 | -1.75 | 0.4+/-0.13 |
| | Cooper S | F | NA | ① | 0.42 | -0.50 | 0.2+/-0.17 |
| | (R56) | R | — | — | 0.33 | -1.75 | 0.4+/-0.13 |
| | Cooper Convertible | F | NA | ① | 0.42 | -0.50 | 0.3+/-0.08 |
| | (R52) | R | — | — | 0.33 | -1.75 | 0.4+/-0.13 |
| | Cooper S Convertible | F | NA | ① | 0.42 | -0.50 | 0.3+/-0.08 |
| | (R52) | R | — | — | 0.33 | -1.75 | 0.4+/-0.13 |

① Difference between left/right max. 0.5 degrees

22205_MINI_C0006

## TIRE, WHEEL AND BALL JOINT SPECIFICATIONS

| Year | Model | OEM Tires Standard | OEM Tires Optional | Tire Pressures (psi) Front | Tire Pressures (psi) Rear | Wheel Size | Ball Joint | Lug Nut (ft. lbs) |
|---|---|---|---|---|---|---|---|---|
| 2006 | Cooper | 175/65R15 | 195/55R16 | ① | ① | 15 X 5.5J | NA | 103 |
| | Cooper S | 195/55R16 | 205/45R17 | ① | ① | 16 X 6.5J | NA | 103 |
| 2007 | Cooper | 175/65R15 | 195/55R16 | ① | ① | 15 X 5.5J | N/A | 103 |
| | Cooper S | 195/55R16 | 205/45R17 | ① | ① | 16 X 6.5J | NA | 103 |
| | Cooper Convertible | 175/65R15 | 195/55R16 | ① | ① | 15 X 5.5J | NA | 103 |
| | Cooper S Convertible | 195/55R16 | 205/45R17 | ① | ① | 16 X 6.5J | NA | 103 |
| 2008 | Cooper | 175/65R15 | 195/55R16 | ① | ① | 15 X 5.5J | NA | 103 |
| | Cooper S | 195/55R16 | 205/45R17 | ① | ① | 16 X 6.5J | NA | 103 |
| | Cooper Convertible | 175/65R15 | 195/55R16 | ① | ① | 15 X 5.5J | NA | 103 |
| | Cooper S Convertible | 195/55R16 | 205/45R17 | ① | ① | 16 X 6.5J | NA | 103 |

OEM: Original Equipment Manufacturer

PSI: Pounds Per Square Inch

① See specification in owners manual

22205_MINI_C0007

# BRAKE SPECIFICATIONS
All measurements in inches unless noted

| Year | Model | | Brake Disc | | | Brake Drum Diameter | | | Min. Lining Thickness | Brake Caliper | |
|---|---|---|---|---|---|---|---|---|---|---|---|
| | | | Original Thickness | Minimum Thickness | Maximum Run-out | Original Inside Diameter | Max. Wear Limit | Maximum Machine Diameter | | Bracket Bolts (ft. lbs.) | Mounting Bolts (ft. lbs.) |
| 2006 | R50 | F | NA | ① | NA | NA | NA | NA | NA | 81 | 22-26 |
| | | R | NA | ① | NA | NA | NA | NA | NA | 48 | 22-26 |
| | R52 | F | NA | ① | NA | NA | NA | NA | NA | 81 | 22-26 |
| | | R | NA | ① | NA | NA | NA | NA | NA | 48 | 22-26 |
| | R53 | F | NA | ① | NA | NA | NA | NA | NA | 81 | 22-26 |
| | | R | NA | ① | NA | NA | NA | NA | NA | 48 | 22-26 |
| 2007 | R52 | F | NA | ① | NA | NA | NA | NA | NA | 81 | 22-26 |
| | | R | NA | ① | NA | NA | NA | NA | NA | 48 | 22-26 |
| | R55 | F | NA | ① | NA | NA | NA | NA | NA | 81 | 26 |
| | | R | NA | ① | NA | NA | NA | NA | NA | 48 | 26 |
| | R56 | F | NA | ① | NA | NA | NA | NA | NA | 81 | 26 |
| | | R | NA | ① | NA | NA | NA | NA | NA | 48 | 26 |
| 2008 | R52 | F | NA | ① | NA | NA | NA | NA | NA | 81 | 22-26 |
| | | R | NA | ① | NA | NA | NA | NA | NA | 48 | 22-26 |
| | R55 | F | NA | ① | NA | NA | NA | NA | NA | 81 | 26 |
| | | R | NA | ① | NA | NA | NA | NA | NA | 48 | 26 |
| | R56 | F | NA | ① | NA | NA | NA | NA | NA | 81 | 26 |
| | | R | NA | ① | NA | NA | NA | NA | NA | 48 | 26 |

F: Front

R: Rear

NA: Not available

① Minimum thickness is stamped in the brake disc shell

22205_MINI_C0008

## PRECAUTIONS

Before servicing any vehicle, please be sure to read all of the following precautions, which deal with personal safety, prevention of component damage, and important points to take into consideration when servicing a motor vehicle:

• Never open, service or drain the radiator or cooling system when the engine is hot; serious burns can occur from the steam and hot coolant.

• Observe all applicable safety precautions when working around fuel. Whenever servicing the fuel system, always work in a well-ventilated area. Do not allow fuel spray or vapors to come in contact with a spark, open flame, or excessive heat (a hot drop light, for example). Keep a dry chemical fire extinguisher near the work area. Always keep fuel in a container specifically designed for fuel storage; also, always properly seal fuel containers to avoid the possibility of fire or explosion. Refer to the additional fuel system precautions later in this section.

• Fuel injection systems often remain pressurized, even after the engine has been turned OFF. The fuel system pressure must be relieved before disconnecting any fuel lines. Failure to do so may result in fire and/or personal injury.

• Brake fluid often contains polyglycol ethers and polyglycols. Avoid contact with the eyes and wash your hands thoroughly after handling brake fluid. If you do get brake fluid in your eyes, flush your eyes with clean, running water for 15 minutes. If eye irritation persists, or if you have taken

brake fluid internally, IMMEDIATELY seek medical assistance.

• The EPA warns that prolonged contact with used engine oil may cause a number of skin disorders, including cancer. You should make every effort to minimize your exposure to used engine oil. Protective gloves should be worn when changing oil. Wash your hands and any other exposed skin areas as soon as possible after exposure to used engine oil. Soap and water, or waterless hand cleaner should be used.

• All new vehicles are now equipped with an air bag system, often referred to as a Supplemental Restraint System (SRS) or Supplemental Inflatable Restraint (SIR) system. The system must be disabled before performing service on or around system components, steering column, instrument panel components, wiring and sensors. Failure to follow safety and disabling procedures could result in accidental air bag deployment, possible personal injury and unnecessary system repairs.

• Always wear safety goggles when working with, or around, the air bag system. When carrying a non-deployed air bag, be sure the bag and trim cover are pointed away from your body. When placing a non-deployed air bag on a work surface, always face the bag and trim cover upward, away from the surface. This will reduce the motion of the module if it is accidentally deployed. Refer to the additional air bag system precautions later in this section.

• Clean, high quality brake fluid from a sealed container is essential to the safe and

proper operation of the brake system. You should always buy the correct type of brake fluid for your vehicle. If the brake fluid becomes contaminated, completely flush the system with new fluid. Never reuse any brake fluid. Any brake fluid that is removed from the system should be discarded. Also, do not allow any brake fluid to come in contact with a painted surface; it will damage the paint.

• Never operate the engine without the proper amount and type of engine oil; doing so WILL result in severe engine damage.

• Timing belt maintenance is extremely important. Many models utilize an interference-type, non-freewheeling engine. If the timing belt breaks, the valves in the cylinder head may strike the pistons, causing potentially serious (also time-consuming and expensive) engine damage. Refer to the maintenance interval charts for the recommended replacement interval for the timing belt, and to the timing belt section for belt replacement and inspection.

• Disconnecting the negative battery cable on some vehicles may interfere with the functions of the on-board computer system(s) and may require the computer to undergo a relearning process once the negative battery cable is reconnected.

• When servicing drum brakes, only disassemble and assemble one side at a time, leaving the remaining side intact for reference.

• Only an MVAC-trained, EPA-certified automotive technician should service the air conditioning system or its components.

## BRAKES
## ANTI-LOCK BRAKE SYSTEM (ABS)

### GENERAL INFORMATION

*PRECAUTIONS*

• Certain components within the ABS system are not intended to be serviced or repaired individually.

• Do not use rubber hoses or other parts not specifically specified for and ABS system. When using repair kits, replace all parts included in the kit. Partial or incorrect repair may lead to functional problems and require the replacement of components.

• Lubricate rubber parts with clean, fresh brake fluid to ease assembly. Do not

use shop air to clean parts; damage to rubber components may result.

• Use only DOT 3 brake fluid from an unopened container.

• If any hydraulic component or line is removed or replaced, it may be necessary to bleed the entire system.

• A clean repair area is essential. Always clean the reservoir and cap thoroughly before removing the cap. The slightest amount of dirt in the fluid may plug an orifice and impair the system function. Perform repairs after components have been thoroughly cleaned; use only denatured alcohol

to clean components. Do not allow ABS components to come into contact with any substance containing mineral oil; this includes used shop rags.

• The Anti-Lock control unit is a microprocessor similar to other computer units in the vehicle. Ensure that the ignition switch is OFF before removing or installing controller harnesses. Avoid static electricity discharge at or near the controller.

• If any arc welding is to be done on the vehicle, the control unit should be unplugged before welding operations begin.

# BRAKES                                BLEEDING THE BRAKE SYSTEM

## BLEEDING PROCEDURE

*BLEEDING PROCEDURE*

**With ABS/ASC+T**

1. Before servicing the vehicle, refer to the Precautions Section.
2. Connect pressurized brake bleeder to the reservoir.

### ※ CAUTION

**Charging pressure should not exceed 2 bar (29 psi.)**

*Rear Brake Circuit*

1. Connect bleeder hose and collecting container to the right rear brake.
2. Open the bleeder valve and flush until clear brake fluid emerges with no air bubbles.
3. Close the bleed valve.
4. Repeat for the left rear brake.

*Front Brake Circuit*

1. Connect bleeder hose and collecting container to the right front brake.
2. Open the bleeder valve.
3. Fully depress brake pedal at least 12

times until brake fluid emerges clear and without air bubbles.
4. Hold the brake pedal down.
5. Close the bleeder valve
6. Repeat for the left front brake.
7. Remove the pressurized brake bleeder.
8. Test for proper brake operation.

**With DSC**

➡**This procedure requires the use of a factory or equivalent scan tool. Refer to scan tool documentation.**

1. Before servicing the vehicle, refer to the precautions in the beginning of this section.
2. Connect the scan tool and set for service function 'Bleeding ABS/DSC Hydraulics'.
3. Connect pressurized brake bleeder to the reservoir.

### ※ CAUTION

**Charging pressure should not exceed 2 bar (29 psi.)**

*Flushing The Brake System*

1. Connect bleeder hose and collecting container to the right rear brake.

2. Open the bleeder valve and flush until clear brake fluid emerges with no air bubbles.
3. Close the bleed valve.
4. Repeat for the left rear, right front, and left front brakes.

*Bleeding The Rear Brake Circuit*

1. Connect bleeder hose and collecting container to the right rear brake.
2. Open the bleeder valve.
3. Run the scan tool bleeding routine.
4. Press the brake pedal 5 times. Clear and bubble-free fluid must flow out.
5. Close the bleed valve.
6. Repeat for left rear brake.

*Bleeding The Front Brake Circuit*

1. Connect bleeder hose and collecting container to the right front brake.
2. Open the bleeder valve.
3. Run the scan tool bleeding routine.
4. Press the brake pedal 5 times. Clear and bubble-free fluid must flow out.
5. Close the bleed valve.
6. Repeat for left front brake.
7. Remove the pressurized brake bleeder.
8. Test for proper brake operation.

# BRAKES                                        FRONT DISC BRAKES

### ※ CAUTION

**Dust and dirt accumulating on brake parts during normal use may contain asbestos fibers from production or aftermarket brake linings. Breathing excessive concentrations of asbestos fibers can cause serious bodily harm. Exercise care when servicing brake parts. Do not sand or grind brake lining unless equipment used is designed to contain the dust residue. Do not clean brake parts with compressed air or by dry brushing. Cleaning should be done by dampening the brake components with a fine mist of water, then wiping the brake components clean with a dampened cloth. Dispose of cloth and all residue containing asbestos fibers in an impermeable container with the appropriate label. Follow practices prescribed by the Occupational Safety and Health Administration (OSHA) and the Environmental Protection Agency (EPA) for the handling, processing, and disposing of**

**dust or debris that may contain asbestos fibers.**

## BRAKE CALIPER

*REMOVAL & INSTALLATION*

1. Before servicing the vehicle, refer to the precautions section.
2. Apply the brake pedal slightly with a brake clamp.
3. Remove or disconnect the following:
   - Wheel assembly
   - Brake hose from caliper
   - Retaining spring across caliper
   - Plastic plugs over guide pin bolts
   - Guide pin bolts
   - Caliper from rotor

*To install:*
4. Position the caliper into place.
5. Clean but do not grease guide pin bolts
6. Install the torque the guide pin bolts to 23 ft. lbs. (31 Nm).
7. Replace the plastic plugs over the guide pin bolts.
8. Install the retainer spring.

9. Install the brake pipe and banjo bolt to the caliper; torque to 30 ft. lbs. (40 Nm).
10. Remove the brake clamp.
11. Install the front wheels.
12. Bleed the brakes.

## DISC BRAKE PADS

*REMOVAL & INSTALLATION*

1. Before servicing the vehicle, refer to the precautions section.
2. Remove the front wheels.
3. Remove the disc pad retaining spring from the caliper, from bottom and then from the top.
4. Remove the plastic plugs over the caliper guide pin bolts.
5. Remove the guide pin bolts and the calipers from the rotor.
6. Press the piston back into caliper.

### ※ CAUTION

**Watch the brake fluid level in reservoir during this procedure.**

7. Remove the outer brake pad (inner pad is held in place with a spring in the piston).

**To install:**

8. Check piston dust sleeves for damage; replace if needed.

9. Clean all mating surfaces.

10. Apply anti-squeak compound to all mounting surfaces.

## BRAKES

### BRAKE CALIPER

*REMOVAL & INSTALLATION*

1. Before servicing the vehicle, refer to the precautions section.

2. Apply the brake pedal slightly with a brake clamp.

3. Remove or disconnect the following:
   - Wheel assembly
   - Retaining spring from bottom, then top, if applicable
   - Plastic plugs from guide pin bolts
   - Handbrake cable from handbrake lever and at rear caliper
   - Brake hose from caliper
   - Caliper guide bolts
   - Rear caliper; remove toward rear

**To install:**

4. Position the caliper into place.

5. Install the torque the guide bolts to 21 ft. lbs. (28 Nm).

6. Attach brake hose to caliper at torque

11. Install calipers; torque bolts to 21 ft. lbs. (31 Nm). Install the plastic plugs.

12. Reposition retaining spring at the top, then at the bottom.

13. Install front wheels.

banjo bolt, with new seals, to 33 ft. lbs. (45 Nm).

7. Install the handbrake cable to caliper and to handbrake.

8. Install the plastic plugs over the guide pin bolts.

9. Install the retaining spring at the top, then at the bottom, if applicable.

10. Remove the brake clamp.

11. Install the rear wheels.

12. Adjust the parking brake.

13. Bleed the brakes.

### DISC BRAKE PADS

*REMOVAL & INSTALLATION*

#### 2006 Cooper Coupe and Cooper S Coupe

*See Figure 1.*

1. Before servicing the vehicle, refer to the precautions section.

2. Remove the rear wheels.

3. Remove the retaining spring from the top and then from the bottom, if applicable.

4. Remove the plastic plugs from the inside of the caliper.

5. Remove the caliper guide pin bolts and remove the caliper from the rotor.

6. Use special tools, 34 6 301, 34 6 306/7/8, force piston back into caliper, as shown.

7. Remove the disc pads.

**To install:**

8. Check condition of dust sleeve on piston; replace if needed.

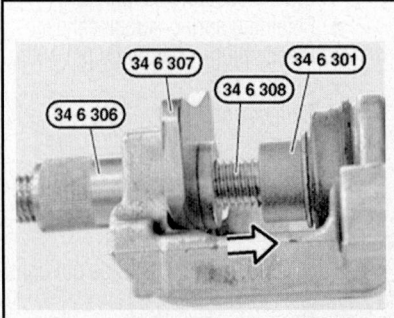

**Fig. 1 Push piston into caliper for removal of disc pads**

22205_MINI_G0001

## REAR DISC BRAKES

9. Clean all contact surfaces.

10. Apply anti-squeak compound to all mounting surfaces.

11. Install the new disc pads evenly in their mounted positions.

12. Clean caliper guide pin bolts; do not apply grease.

13. Install calipers and guide pin bolts. Torque bolts to 21 ft. lbs. (28 Nm).

14. Install the plastic plugs.

15. Install the retaining spring at the bottom and then at the top, if applicable.

16. Install rear wheels.

17. Fully depress brake pedal several times to set proper contact of pads with rotor.

18. Bleed brake system, if necessary.

#### 2007–08 Cooper and Cooper S

1. Before servicing the vehicle, refer to the precautions section.

2. Remove the rear wheels.

3. Pull brake pad wear sensor towards front out of pad (left side only).

4. Release guide screw.

5. If necessary, grip at hexagon head.

6. Feed brake hose out of holder.

7. Tilt brake caliper upwards.

8. Remove brake pads in direction of arrow from brake console.

➡ **Mark any worn brake pads. In the event of one-sided brake pad wear, do not change brake pads round.**

9. Observe minimum thickness of brake pads.

10. Clean brake pads.

➡ **Do not apply grease to brake pad backplate.**

11. Check minimum brake disc thickness: Position special tool No. 34 1 280 at three measuring points in area and measure. Compare measurement result and lowest value with setpoint value. New brake pads may only be installed if the brake disc thickness is greater than or equal to the minimum brake disc thickness.

➡ **The minimum thickness of the brake disc is designed so that it holds over the service life of a further set of brake**

14. Fully depress brake pedal several times to set proper contact of pads with rotor.

15. Check fluid level and bleed brake system, if necessary.

**pads if it is greater than or equal to the minimum brake disc thickness.**

12. Remove pad retaining springs.

*To install:*

13. Press brake piston fully back with special tool No. 34 1 050.

14. Check dust sleeve for damage and replace if necessary.

15. Clean contact face of brake piston and apply a thin coating of anti-squeak compound. Dust sleeve must not come into contact with anti-squeak com-

pound as this may cause the dust sleeve to swell.

16. Clean contact face of brake caliper and apply a thin coating of anti-squeak compound.

17. Clean hammerhead guides and apply a thin coating of anti-squeak compound.

18. Install pad retaining springs. Clean contact face of brake carrier and apply a thin coating of anti-squeak compound.

➡**Brake pad with indentation is intended for accommodating the brake pad wear sensor and must be fitted on the piston side.**

19. The remainder of installation is the reverse of removal, noting the following:

- Fully depress brake pedal several times so that brake pads contact brake discs
- When installing new brake pads at front and rear axles, brake fluid level must be brought up to "MAX" marking
- If necessary, when replacing pads, reset CBS display in accordance with factory specification

## CHASSIS ELECTRICAL

### GENERAL INFORMATION

### ✳✳ CAUTION

**These vehicles are equipped with an air bag system. The system must be disarmed before performing service on, or around, system components, the steering column, instrument panel components, wiring and sensors. Failure to follow the safety precautions and the disarming procedure could result in accidental air bag deployment, possible injury and unnecessary system repairs.**

#### SERVICE PRECAUTIONS

Disconnect and isolate the battery negative cable before beginning any airbag system component diagnosis, testing, removal, or installation procedures. Allow system capacitor to discharge for two minutes before beginning any component service. This will disable the airbag system. Failure to disable the airbag system may result in accidental airbag deployment, personal injury, or death.

Do not place an intact undeployed airbag face down on a solid surface. The airbag will propel into the air if accidentally deployed and may result in personal injury or death.

When carrying or handling an undeployed airbag, the trim side (face) of the airbag should be pointing towards the body to minimize possibility of injury if accidental deployment occurs. Failure to do this may result in personal injury or death.

Replace airbag system components with OEM replacement parts. Substitute parts may appear interchangeable, but internal differences may result in inferior occupant protection. Failure to do so may result in occupant personal injury or death.

## AIR BAG (SUPPLEMENTAL RESTRAINT SYSTEM)

Wear safety glasses, rubber gloves, and long sleeved clothing when cleaning powder residue from vehicle after an airbag deployment. Powder residue emitted from a deployed airbag can cause skin irritation. Flush affected area with cool water if irritation is experienced. If nasal or throat irritation is experienced, exit the vehicle for fresh air until the irritation ceases. If irritation continues, see a physician.

Do not use a replacement airbag that is not in the original packaging. This may result in improper deployment, personal injury, or death.

The factory installed fasteners, screws and bolts used to fasten airbag components have a special coating and are specifically designed for the airbag system. Do not use substitute fasteners. Use only original equipment fasteners listed in the parts catalog when fastener replacement is required.

During, and following, any child restraint anchor service, due to impact event or vehicle repair, carefully inspect all mounting hardware, tether straps, and anchors for proper installation, operation, or damage. If a child restraint anchor is found damaged in any way, the anchor must be replaced. Failure to do this may result in personal injury or death.

Deployed and non-deployed airbags may or may not have live pyrotechnic material within the airbag inflator.

Do not dispose of driver/passenger/curtain airbags or seat belt tensioners unless you are sure of complete deployment. Refer to the Hazardous Substance Control System for proper disposal.

Dispose of deployed airbags and tensioners consistent with state, provincial, local, and federal regulations.

After any airbag component testing or service, do not connect the battery negative cable. Personal injury or death may result if the system test is not performed first.

If the vehicle is equipped with the Occupant Classification System (OCS), do not connect the battery negative cable before performing the OCS Verification Test using the scan tool and the appropriate diagnostic information. Personal injury or death may result if the system test is not performed properly.

Never replace both the Occupant Restraint Controller (ORC) and the Occupant Classification Module (OCM) at the same time. If both require replacement, replace one, then perform the Airbag System test before replacing the other.

Both the ORC and the OCM store Occupant Classification System (OCS) calibration data, which they transfer to one another when one of them is replaced. If both are replaced at the same time, an irreversible fault will be set in both modules and the OCS may malfunction and cause personal injury or death.

If equipped with OCS, the Seat Weight Sensor is a sensitive, calibrated unit and must be handled carefully. Do not drop or handle roughly. If dropped or damaged, replace with another sensor. Failure to do so may result in occupant injury or death.

If equipped with OCS, the front passenger seat must be handled carefully as well. When removing the seat, be careful when setting on floor not to drop. If dropped, the sensor may be inoperative, could result in occupant injury, or possibly death.

If equipped with OCS, when the passenger front seat is on the floor, no one should sit in the front passenger seat. This uneven force may damage the sensing ability of the seat weight sensors. If sat on and damaged, the sensor may be inoperative, could result in occupant injury, or possibly death.

#### DISARMING THE SYSTEM

1. Before servicing the vehicle, refer to the precautions section.

2. Place the ignition switch in the **OFF** position.

3. Disconnect the negative battery terminal and cover the battery terminal to prevent accidental contact.

4. Once the battery has been disconnected, wait for a short period of time to allow the capacitor in the control unit to discharge. Once the capacitor is discharged, a trigger pulse cannot be generated inadvertently.

*ARMING THE SYSTEM*

1. Before servicing the vehicle, refer to the precautions section.

2. Place the ignition switch in the **OFF** position.

3. Attach the sensors, the steering column connector and the seat belt tensioner connectors.

4. Connect the negative battery terminal.

5. Place the ignition switch in the **ON** position. Check that the SRS light illuminates for 6 seconds and extinguishes. If it

illuminates in any other pattern, check the components and their connections for proper operation and recheck operation of the warning light.

*CLOCKSPRING CENTERING*

1. Turn spring counterclockwise as far as it will go.

2. Turn spring clockwise as far as it will go.

3. Turn spring back to center position and secure so that centering pin is at bottom position.

## DRIVETRAIN

### AUTOMATIC TRANSAXLE ASSEMBLY

*REMOVAL & INSTALLATION*

See Figures 2 through 5.

1. Before servicing the vehicle, refer to the precautions section.

#### ✳✳ WARNING

**Use only the approved automatic transmission fluid in this automatic transmission. Failure to comply with this requirement will result in serious damage to the automatic transmission.**

➡**An incorrectly adjusted gearshift mechanism can result in gear teeth noises being transmitted to the passenger compartment. Adjust selector lever.**

2. Switch off ignition.
3. Disconnect the negative battery cable.
4. Remove air intake filter housing.
5. Remove intake filter housing gaiter.
6. Drain coolant.
7. Secure engine in installation position.
8. Lower front axle support.

9. Remove starter.
10. Remove rubber mounts for transmission mounting.
11. Remove left and right output shafts.
12. For N14 engines, release screws and release holder from engine.
13. Unfasten hose clips.
14. Disconnect coolant hoses from oil cooler.
15. Disconnect multiple connectors (1/2) of EGS control unit.
16. Remove wiring harness.
17. Pay attention to routing of wiring harness.
18. Release cable lock nut.
19. Disconnect plug from gear position switch.
20. Remove hose from holder.
21. Slide cable locking sleeve in direction of arrow.
22. Remove cable upwards from holder.
23. Support transmission with special tools No. 23 4 150 and No. 00 2 030.
24. Secure transmission with tensioning strap.
25. Release nut through opening for starter.

26. Crank engine further and release remaining 5 nuts.
27. Release bolts and remove transmission.

#### ✳✳ WARNING

**Transmission mounting bolts differ in length. Note installation position. Installing the wrong bolts may cause serious damage.**

*To install:*
28. Installation is the reverse of removal, noting the following torque specifications:
- Cable lock nut: 9 ft. lbs. (12 Nm).

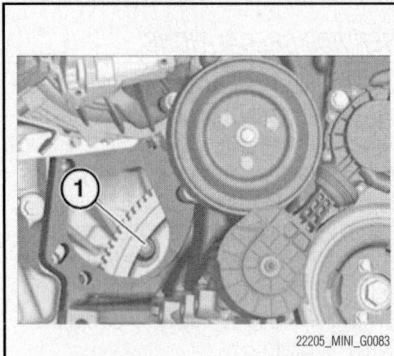

22205_MINI_G0083

**Fig. 4 Release nuts**

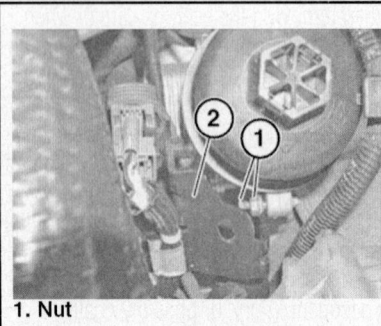

1. Nut
2. Holder

22205_MINI_G0081

**Fig. 2 Release screws and holder from engine—2007–08 Cooper S models**

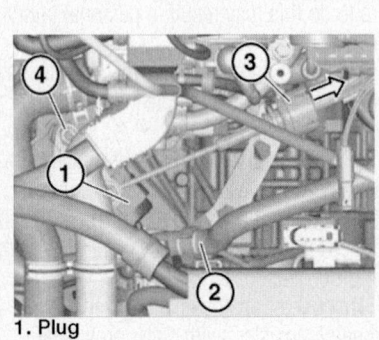

1. Plug
2. Hose
3. Cable locking sleeve

22205_MINI_G0082

**Fig. 3 Disconnect plug, remove hose, and remove cable**

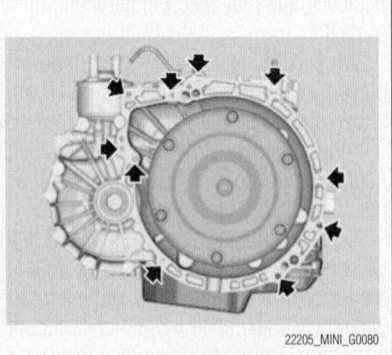

22205_MINI_G0080

**Fig. 5 Automatic transaxle bolts**

- Converter bolt: 42 ft. lbs. (57 Nm).
- Transmission to engine bolts: 28 ft. lbs. (38 Nm).

29. Check that dowel sleeves are correctly seated. Replace damaged dowel sleeves.

30. Check transmission fluid level.

## MANUAL TRANSAXLE ASSEMBLY

### REMOVAL & INSTALLATION

#### Cooper Coupe

*See Figures 6 through 8.*

1. Before servicing the vehicle, refer to the precautions section.

2. Remove or disconnect the following:
- Negative battery cable
- Battery and battery box
- Manifold heat shield
- Engine stabilizer (upper)
- Fuel and vent pipes from bracket near stabilizer
- Drain transaxle
- Front left wheel well liner
- Driveshafts with steering knuckle carrier
- Lower stabilizer

**Fig. 6 Installing engine lifting eye bracket—Cooper Coupe**

**Fig. 7 Identifying the location of the transaxle mount—Cooper Coupe**

- Front subframe
- Gearshift cables from ball joint attachment
- Gearshift cable bracket
- Clutch slave cylinder from transaxle
- Reverse lamp connector from transaxle
- Brake booster pipe from manifold (push down circular ring to release)
- Coolant pressure cap from fill tower
- Oxygen sensor bracket, coolant hose clamp, and bolt

3. Install a engine lifting eye bracket, 11–8–260 as shown.

4. Support the engine with lifting equipment.

5. Raise equipment enough to take weight of engine and transaxle.

6. Remove the upper bracket retaining bolt (1), mount to transaxle bolts (2) and remove the transaxle mount.

7. Lower the engine about 1.5 inches (40mm).

### ✳✳ CAUTION

**DO NOT lower engine too much or exhaust system could be damaged. Also watch A/C pipe to compressor when lowering engine.**

8. Remove or disconnect the following:
- Starter heat shield
- Oxygen sensor wiring from clip
- Starter
- Closure plate bracket around inner driveshaft opening

9. Support transaxle with suitable jack.

10. Remove the transaxle retaining bolts.

11. Remove the transaxle.

➡**Shorter 2 bolts are located into oil pan.**

#### To install:

12. Clean all mating surfaces.

13. Position the transaxle into the vehicle.

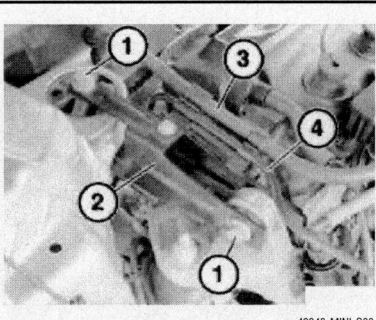

**Fig. 8 Showing the upper stabilizer bracket and bolt locations, plus the location of the fuel and vent pipes—Cooper Coupe**

14. Install and torque the transaxle–to–engine housing bolts to 63 ft. lbs. (85 Nm).

➡**2 shorter bolts go directly into oil pan.**

15. Remove jack.

16. Install and torque closure plate bracket bolts to 7 ft. lbs. (9 Nm).

17. Install or connect the following:
- Starter; torque bolts to 63 ft. lbs. (85 Nm).
- Starter heat shield; torque bolts to 7 ft. lbs. (9 Nm)
- Starter electrical connections
- Oxygen sensor to clip near starter heat shield

18. Raise the engine back into normal position and install the transaxle mount. Torque the bolts as follows:
- Mount bracket–to<TRANSAXLE>: 28 ft. lbs. (38 Nm)
- Mount–to–upper bracket: 49 ft. lbs. (66 Nm)

19. Slowly release engine tension from lift equipment. Remove the equipment.

20. Remove the engine lifting eye bracket.

21. Install or connect the following:
- Coolant hose, clamp and bolt
- Oxygen sensor bracket (near coolant hose)
- Brake booster pipe to manifold
- Reverse light switch connector
- Clutch slave cylinder; torque bolts to 18 ft. lbs. (24 Nm)
- Gearshift cable and bracket
- Front subframe
- Lower stabilizer bracket (2); torque bolts (1) to 74 ft. lbs. (100 Nm)
- Driveshafts and steering knuckle carrier
- Left wheel well liner

22. Refill the transaxle with proper oil.

23. Install the upper stabilizer bolts. Torque the bolts to 74 ft. lbs. (100 Nm).

24. Attach the fuel and vent pipes to the upper stabilizer.

25. Install the manifold heat shield.

26. Install and connect the battery.

27. Start the engine and check transaxle operation.

#### Cooper S Coupe

*See Figures 9 and 10.*

1. Before servicing the vehicle, refer to the precautions section.

2. Remove or disconnect the following:
- Negative battery cable
- Battery and battery box
- Intake filter housing

- Manifold heat shield
- Engine stabilizer (upper)
- Fuel and vent pipes from bracket near stabilizer
- Drain transaxle
- Front left wheel well liner
- Driveshafts with steering knuckle carrier
- Lower stabilizer
- Crush tubes
- Front subframe
- Coolant expansion tank cap
- Oxygen sensor bracket, coolant hose clamp, and bolt

3. Install an engine lifting eye bracket, 11–8–260 as shown.

4. Remove the gearshift cables from ball joint attachment, with special tool 23–4<010, then remove the gearshift cable mounting bracket.

5. Remove the clutch slave cylinder from the transaxle.

6. Disconnect the reverse light switch connector from the transaxle.

7. Open the hood to the full upright position and install strut extensions, 51–2–160, to hold the hood in this position.

8. Support the engine with lifting equipment.

9. Raise equipment enough to take weight of engine and transaxle.

10. Remove or disconnect the following:
- Throttle housing
- Supercharger intake hose
- Detach other pipes by quick-fit couplings
- Slave cylinder hose from transaxle and move aside
- Closure plate around inner drive-shaft opening
- Starter heat shield
- Oxygen sensor from clip on heat shield
- Starter connections and move wiring harness aside

- Starter
- Transaxle mount

11. Lower the engine about 5 inches (135 mm).

12. Support transaxle with suitable jack.

13. Remove the transaxle retaining bolts.

14. Remove the transaxle.

### To install:

15. Clean all mating surfaces.

16. Position the transaxle into the vehicle.

17. Install and torque the transaxle–to–engine housing bolts to 63 ft. lbs. (85 Nm).

18. Remove jack.

19. Raise the engine to normal position.

20. Install or connect the following:
- Transaxle mount; torque bolts to 49 ft. lbs. (66 Nm)
- Starter; torque bolts to 63 ft. lbs. (85 Nm)
- Starter electrical connections and wiring harness
- Oxygen sensor to clip near starter heat shield
- Starter heat shield; torque bolts to 7 ft. lbs. (9 Nm)
- Closure plate bolts to 7 ft. lbs. (9 Nm).

21. Install the lower support bracket and torque the bolts as follows:
- Mount bracket–to<TRANSAXLE>: 28 ft. lbs. (38 Nm)
- Mount–to–upper bracket: 49 ft. lbs. (66 Nm)

22. Install or connect the following:
- MAP sensor
- Slave cylinder hose to transaxle
- Supercharger, pipes and hoses
- Reverse light switch connector
- Slave cylinder to transaxle; torque bolts to 18 ft. lbs. (24 Nm)
- Gearshift cables and bracket

23. Slowly release engine tension from lift equipment. Remove the equipment.

24. Remove the engine lifting eye bracket.

25. Install or connect the following:
- Coolant hose, clamp and bolt
- Oxygen sensor bracket (near coolant hose)

26. Install the front subframe; torque bolts to 74 ft. lbs. (100 Nm)

27. Reinstall the MFE to its normal position. Torque the bolts as follows:
- M8x30 bolts: 17 ft. lbs. (22 Nm)
- M6x16 bolts: 3 ft. lbs. (5 Nm)

28. Install or connect the following:
- Crush member–to–subframe: 74 ft. lbs. (100 Nm)
- Lower stabilizer bracket (2); torque bolts (1) to 74 ft. lbs. (100 Nm)
- Driveshafts and steering knuckle carrier
- Left wheel well liner

29. Refill the transaxle with proper oil.

30. Install the upper stabilizer bolts. Torque the bolts to 74 ft. lbs. (100 Nm).

31. Attach the fuel and vent pipes to the upper stabilizer.

32. Install the manifold heat shield.

33. Install and connect the battery.

34. Start the engine and check transaxle operation.

## CLUTCH

### REMOVAL & INSTALLATION

1. Before servicing the vehicle, refer to the precautions section.

2. Remove or disconnect the following:

3. Remove the transmission.

4. Using a holding tool to restrain or lock the crankshaft pulley in place (keep it from turning).

5. Slacken the pressure plate bolts evenly, in an alternating sequence, then remove all bolts.

6. Remove the pressure plate and disc.

### To install:

7. Position the clutch disc onto the transmission input shaft and check for free movement.

8. Install the pressure plate and clutch disc onto the flywheel, using a special tool, 21–6–100 (Cooper Coupe) or 21–2–210 (Cooper S Coupe).

9. Install new pressure plate retaining bolts. Tighten them gradually and evenly, in an alternating pattern. Final torque setting is 15 ft. lbs. (20 Nm) for Cooper Coupe or to 17 ft. lbs. (23 Nm) for Cooper S Coupe.

➡During the tightening process, rotate the special holding tool. This will help to centralize the clutch disc.

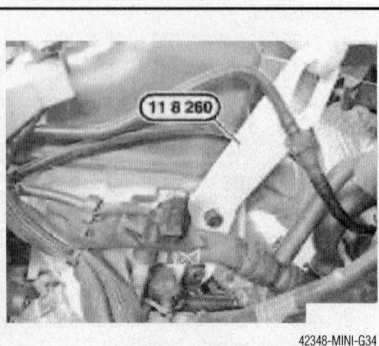

42348-MINI-G34

**Fig. 9 Installing engine lifting eye bracket—Cooper S Coupe**

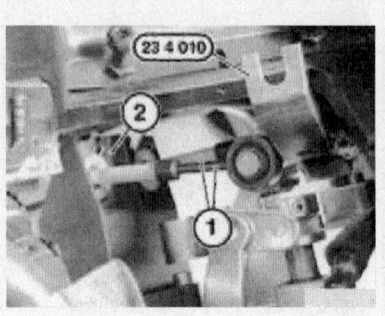

42348-MINI-G35

**Fig. 10 Disconnect the gearshift cables from the ball joints—Cooper S Coupe**

10. Remove the special tool from the clutch.

11. Install the transmission.

12. Remove the holding tool from the crankshaft pulley.

### BLEEDING

1. Before servicing the vehicle, refer to the precautions in the beginning of this section.

2. Remove the battery compartment.

3. Remove the dust cap from the bleeder screw.

4. Attach Special Service Tool 21 5 030 to the clutch slave cylinder.

5. Open the bleed screw.

➡**Operating pressure must not exceed 1 bar (14.5 psi.). Refer to equipment operating instructions.**

6. Close bleed screw when no further bubbles appear.

7. Fill the brake fluid to the correct level.

## FRONT HALFSHAFT

### REMOVAL & INSTALLATION

*See Figures 11 and 12.*

1. Before servicing the vehicle, refer to the precautions section.

2. Remove or disconnect the following:
   • Front wheel
   • Front wheel hub nut
   • Drain transaxle
   • Brake caliper from disc (tie out of way; hose connected)
   • Tie rod ball joint from steering knuckle
   • ABS sensor from steering knuckle
   • Control arm from steering knuckle

3. On right side driveshaft only, remove bolts holding the intermediate shaft housing to the bracket.

4. Pull the driveshaft from transaxle (discard snap ring)

5. Remove the bolt holding the steering knuckle to the McPherson strut, then lift the steering knuckle out with the driveshaft.

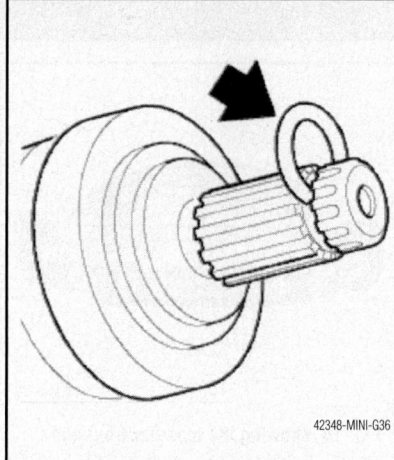

42348-MINI-G36

**Fig. 11 Installing a new snap ring on driveshaft inner spline**

42348-MINI-G37

**Fig. 12 Showing special seal protector tool installed in transaxle**

### To install:

6. Install a new snap ring on the end of the driveshaft inner spline.

7. Install a special seal protector tool, 24–8–120, into side of transaxle.

8. Position the driveshaft to the transaxle and insert into to seal. Pull on the special tool handle to remove once the driveshaft is in position.

9. Push in output shaft over the resistance of the retaining ring until it snaps in place.

10. Install the steering knuckle to the McPherson strut. Torque the retaining bolt to 60 ft. lbs. (81 Nm).

11. Install the intermediate shaft housing to the bracket. Torque the retaining bolts to 18 ft. lbs. (25 Nm).

12. Install or connect the following:
   • Control arm to steering knuckle; torque new nut to 41 ft. lbs. (56 Nm)
   • ABS sensor to steering knuckle; torque to 6 ft. lbs. (8 Nm)
   • Tie rod to steering knuckle; torque new ball joint nut to 38 ft. lbs. (52 Nm)
   • Brake caliper to disc; torque caliper guide bolts to 23 ft. lbs. (31 Nm)
   • Front wheel hub nut; torque new nut to 134 ft. lbs. (182 Nm)
   • Front wheel

13. Refill the transaxle.

### CV-JOINTS OVERHAUL

1. Before servicing the vehicle, refer to the precautions section.

2. Remove the driveshaft.

3. Remove the bellows clamps.

4. Slide the bellows away from the inner CV joint.

5. Hold the shaft firmly and drive the inner CV joint off the shaft.

6. Remove the bellows.

### To install:

7. Install a new bellows and seal onto the shaft.

8. Generously pack new joint with grease. Be sure the joint rests on the new snap ring on the shaft.

9. Press the snap ring into the shaft groove, then drive the CV joint onto the shaft.

10. Slide the bellows onto the joint and shaft and make sure the seal bearing of the bellows fits into the grooves on the shaft on one end and the grooves on the CV joint on the other end.

11. Install the bellows clamps.

12. Install the driveshaft.

# ENGINE COOLING

## THERMOSTAT

### REMOVAL & INSTALLATION

#### 2007–08 Cooper and Cooper S

*See Figure 13.*

1. Drain coolant.
2. For N14 engine, remove the intake air manifold.
3. Release lock on coolant pipe in direction of arrow.
4. Disconnect the thermostat plug connection.
5. Using Special Tool No. 17 2 050, detach all coolant hoses from thermostat.
6. Disconnect the coolant temperature sensor plug connection.
7. Loosen the nut and remove the screws.
8. Remove the seal.

**To install:**

9. Installation is the reverse of removal.

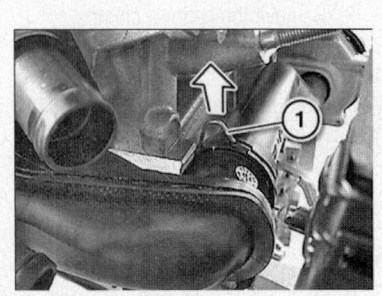

1. Coolant pipe lock

22205_MINI_G0002

**Fig. 13 Release lock on coolant pipe**

## WATER PUMP

### REMOVAL & INSTALLATION

#### 2006 Cooper Coupe

*See Figure 14.*

1. Before servicing the vehicle, refer to the precautions section.
2. Disconnect the battery.
3. Drain the cooling system.
4. Remove the alternator drive belt and alternator.
5. Drain the cooling system. Drain plug is located on exhaust side of block, next to cylinder number 2.
6. Remove the lower modular front end (MFE) as follows:
   a. Discharge the A/C system.
   b. Remove the engine compartment under tray.

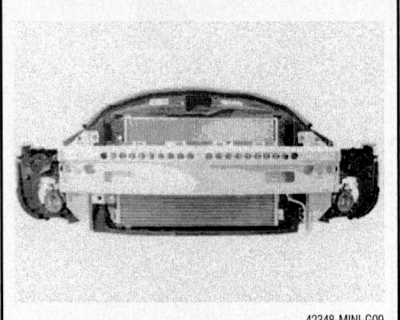

42348-MINI-G09

**Fig. 14 Showing the modular front end (MFE) assembly—Cooper Coupe**

   c. Remove the front bumper assembly.
   d. Remove both front wheel well liners.
   e. Remove or disconnect the following:
   - Cooling fan connectors
   - A/C pipe from condenser (plug openings)
   - A/C pipe from A/C hose (plug openings)
   - ABS speed sensor connector from clip (both sides)
   - Front fog lamp connector (both sides)
   - Horn connector (both sides)
   - Feed harness through MFE into wheel housing (both sides)
   - Subframe crash tube bolts (both sides)
   - Upper radiator hose
   - Nuts securing MFE and bumper carrier assembly
   - MFE

7. Insert bracing tools, 11–8–401/2, to provide access room.
8. Detach the hoses from the water pump.
9. Remove the bolts and remove the water pump.

**To install:**

10. Install or connect the following:
   - Water pump mounting bolts; torque to (30 Nm)
   - Hoses to water pump
   - Impact tubes; torque bolts to 74 ft. lbs. (100 Nm)
   - MFE
   - Alternator
   - Battery

11. Refill the cooling system. Start the engine and check for leaks.

#### 2006 Cooper S Coupe

*See Figures 15 and 16.*

1. Before servicing the vehicle, refer to the precautions section.
2. Drain the cooling system.

3. Remove the negative battery cable.
4. Remove the supercharger
5. Remove the water pump bolts.
6. Remove the water pump.

**To install:**

7. Clean and remove any residual debris or material from the mounting surfaces for the water pump.
8. Align the water pump drive with the supercharger drive.
9. Install the water pump to the supercharger. Torque the bolts to 18 ft. lbs. (25 Nm).
10. Install a new sealing ring to the water pump and lubricate the seal.
11. Install the supercharger.
12. Reconnect the negative battery cable.
13. Fill and bleed the cooling system.
14. Start the vehicle, check for leaks and repair as necessary.

#### 2007–08 Cooper and Cooper S

1. Remove drive belt tensioner.
2. Remove screws.
3. Remove seal.

**To install:**

4. Clean sealing surfaces.
5. Installation is the reverse of removal.

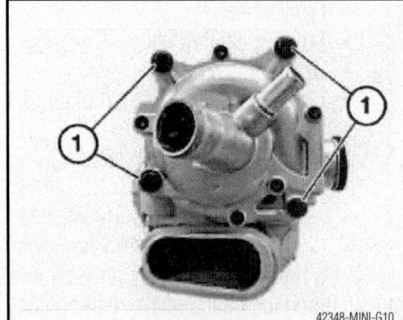

42348-MINI-G10

**Fig. 15 Showing the location of the water pump mounting bolts—Cooper S Coupe**

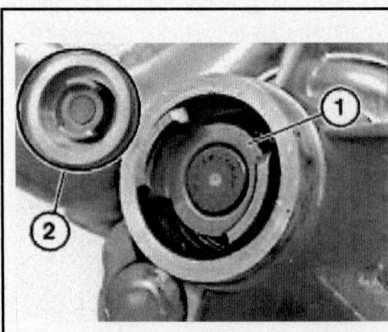

42348-MINI-G11

**Fig. 16 Aligning water pump drive (1) to supercharger drive (2)—Cooper S Coupe**

**ENGINE ELECTRICAL**                                    **CHARGING SYSTEM**

### ALTERNATOR

*REMOVAL & INSTALLATION*

➡When the battery is disconnected the radio code, on-board computer and clock settings will be lost. The radio code should be obtained before disconnecting the battery or radio. Once the battery has been reconnected, the radio will not function unless the code is keyed in.

#### 2006 Cooper Coupe

1. Before servicing the vehicle, refer to the precautions section.
2. Check for stored fault codes, then erase code memory.
3. Switch off ignition.
4. Disconnect negative battery cable.
5. Remove or disconnect the following:
   • Alternator drive belt
   • Electrical connections from alternator
   • Alternator mounting bolts
   • Alternator

**To install:**
6. To install, reverse removal procedure.
7. Torque alternator mounting bolts to 18 ft. lbs. (25 Nm) and the power lead to the alternator stud to 7 ft. lbs. (10 Nm).
8. Check for any stored fault codes.
9. Clear fault code memory.

#### 2006 Cooper S Coupe

*See Figure 17.*

1. Before servicing the vehicle, refer to the precautions section.
2. Check for stored fault codes, then erase fault code memory.
3. Switch off ignition.

**Fig. 17 Installing special tools to move modular front end for alternator access—Cooper S Coupe models**

4. Disconnect the battery.
5. Remove or disconnect the following:
   • Alternator drive belt
   • Front bumper cover
6. Loosen the retainers for the modular front end and insert bracing tools, 11–8–401/2, to provide access room.
7. Detach the electrical connections from the alternator.
8. Remove the mounting bolts and remove the alternator.

**To install:**
9. To install, reverse removal procedure.
10. Torque alternator mounting bolts to 18 ft. lbs. (25 Nm) and the power lead to the alternator stud to 7 ft. lbs. (10 Nm).
11. Check for any stored fault codes.
12. Clear fault code memory.

#### 2007–08 Cooper and Cooper S

*See Figure 18.*

1. Before servicing the vehicle, refer to the precautions section.
2. Check for stored fault codes, then erase fault code memory.
3. Switch off ignition.

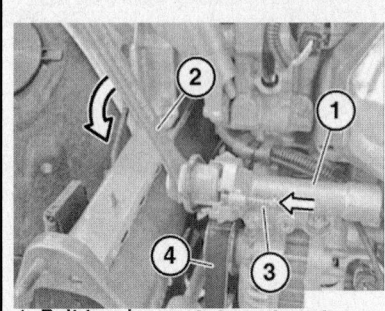

1. Belt tensioner   3. Locating pin
2. Wrench           4. Drive belt

**Fig. 18 Drive belt removal**

4. Disconnect the battery.
5. Move the front panel into assembly position.
6. For N14 engines, remove the bolt, then remove the bracket and place to one side.
7. Bring belt tensioner with wrench into assembly position and hold.
8. Secure assembly position of belt tensioner by sliding locating pin in direction of arrow.

**✷✷ CAUTION**

**Remove wrench again from belt tensioner.**

9. Unlock connector and remove.
10. Release nut and remove battery positive lead.
11. Release screws and remove the belt tensioner.
12. Release screw with joint extension.
13. Remove alternator.

**To install:**
14. Installation is the reverse of removal.

**ENGINE ELECTRICAL**                                    **IGNITION SYSTEM**

### FIRING ORDER

The firing order for these engines is: 1–3–4–2

### IGNITION COIL

*REMOVAL & INSTALLATION*

#### 2006 Cooper and Cooper S

*See Figure 19.*

1. Before servicing the vehicle, refer to the precautions in the beginning of this section.
2. Disconnect the negative battery cable.
3. Disconnect the ignition coil wiring harness.
4. Remove the spark plug wires.
5. Remove the three screws and the ignition coil.

**To install:**
6. Install the ignition coil and tighten the three screws to 12 Nm (106 inch lbs.).
7. Install the spark plug wires.

8. Connect the ignition coil wiring harness.

#### 2007–08 Cooper and Cooper S

1. Before servicing the vehicle, refer to the precautions section.
2. Check for stored fault codes.
3. Turn ignition off.
4. Remove upper engine cover.
5. Unlock the plug retainer of ignition coil and disconnect the plug.
6. Pull the ignition coil up and out.

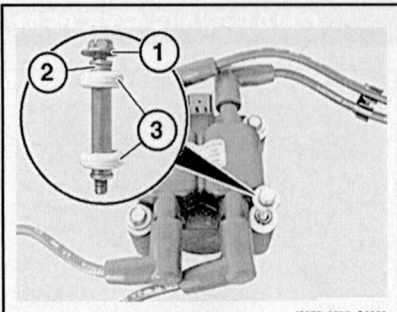

**Fig. 19 Screw (1) and spacer tube (2) are only available as a single set
Check insulating rings (3), replace if necessary**

***To install:***
7. Installation is the reverse of removal.

## IGNITION TIMING

### ADJUSTMENT

The ignition timing is controlled by the Powertrain Control Module (PCM). No adjustment is necessary or possible.

## ENGINE ELECTRICAL

### STARTER

### REMOVAL & INSTALLATION

#### 2006 Cooper and Cooper S

1. Before servicing the vehicle, refer to the precautions section.
2. Remove or disconnect the following:
   • Battery
   • Exhaust system from manifold
   • Exhaust manifold
   • Heat shield from starter
   • Oxygen sensor cable from wire clip
   • Alternator connectors
   • Starter solenoid connectors
   • Starter

***To install:***
3. Install or connect the following:
   • Starter to transmission; torque mounting bolts to 63 ft. lbs. (85 Nm)
   • Alternator connector on starter; torque to 10 ft. lbs. (14 Nm)
   • Oxygen sensor cable to wire clip
   • Heat shield for starter; torque bolts to 7 ft. lbs. (9 Nm)
   • Exhaust manifold
   • Exhaust pipes to manifold
   • Battery

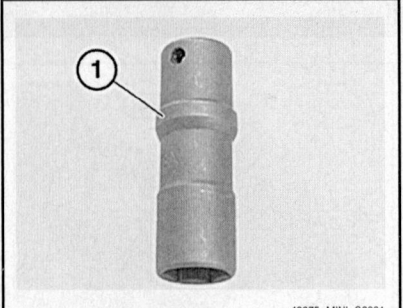

**Fig. 20 Special Service Tool 12 1 170—
(1) This collar must be ground off for proper fit**

## SPARK PLUGS

### REMOVAL & INSTALLATION

#### 2006 Cooper and Cooper S
*See Figure 20.*

1. Remove the spark plug wire connectors.
2. Using Special Service Tool 12 1 170, remove the spark plugs.

#### 2007–08 Cooper and Cooper S
*See Figures 21 and 22.*

1. Turn ignition off.
2. Disconnect the negative battery cable.
3. Remove intake filter housing.
4. For N12 engine, remove the tank venting valve.
5. Remove the right wheel.
6. For N14 engine, remove the bolts and lay the vacuum tank to one side.

**Fig. 21 Remove starter screw**

3. Clean the spark plugs.
4. If the electrode has traces of wet carbon, allow it to dry and then clean with a spark plug cleaner.
5. Check the spark plug for thread damage and insulator damage. If abnormal, replace the spark plug.
6. Using Special Service Tool 12 1 170, install the spark plugs and tighten to 27 Nm (20 ft. lbs.)
7. Reinstall the ignition coils.

#### 2007–08 Cooper and Cooper S

1. Turn ignition off.
2. Remove the ignition coils.
3. Unscrew and remove spark plugs with Special Tool 12 1 220.

***To install:***
4. Tighten spark plugs, using Special Tool 12 1 220 and Special Tool 12 1 172. If special tool 12 1 172 is not used, torque to 17 ft. lbs. (23 Nm).

## STARTING SYSTEM

7. Release screw.
8. Unlock plug and remove.
9. Release nut and remove battery positive lead.
10. Release screws.
11. Remove bracket and starter motor.

***To install:***
12. Install starter and fit screws.
13. Press starter in direction of arrow and tighten down.
14. The remainder of installation is the reverse of removal.

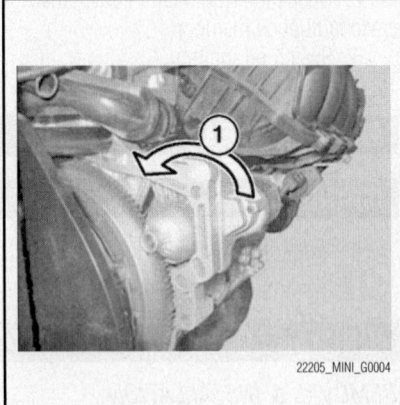

**Fig. 22 Install starter**

## ENGINE MECHANICAL

➡Disconnecting the negative battery cable may interfere with the functions of the on board computer systems and may require the computer to undergo a relearning process, once the negative battery cable is reconnected.

### ACCESSORY DRIVE BELTS

#### ACCESSORY BELT ROUTING

See Figure 23.

#### INSPECTION

Inspect the drive belt for signs of glazing or cracking. A glazed belt will be perfectly smooth from slippage, while a good belt will have a slight texture of fabric visible. Cracks will usually start at the inner edge of the belt and run outward. All worn or damaged drive belts should be replaced immediately.

#### ADJUSTMENT

No adjustment is possible.

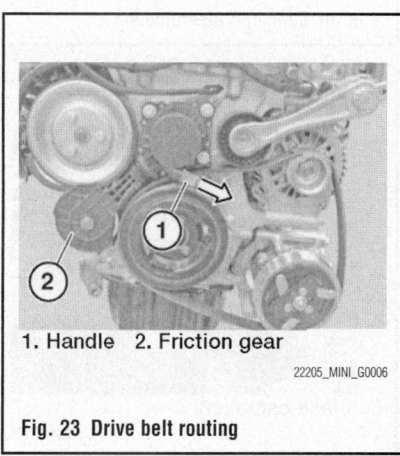

1. Handle  2. Friction gear

22205_MINI_G0006

**Fig. 23  Drive belt routing**

#### REMOVAL & INSTALLATION

##### 2006 Cooper and Cooper S

1. Before servicing the vehicle, refer to the precautions.
2. Disconnect the negative battery cable.
3. Remove the engine undercover.
4. Remove the right front wheel.
5. Remove the right front inner fender.
6. Using Special Service Tools 11 8 410, relieve the tension on the drive belt. Block the tensioner with Special Service Tool 11 8 470.
7. Remove the accessory drive belt.
8. Installation is the reverse of the removal procedure.

##### 2007–08 Cooper and Cooper S

See Figures 24 and 25.

1. Remove right wheel arch cover.
2. Remove right headlight.
3. Remove lock bridge.
4. Bring belt tensioner with wrench into assembly position.
5. Secure assembly position of belt tensioner by sliding locating pin in direction of arrow.

### ✳✳ CAUTION
**Remove wrench again from belt tensioner.**

6. Remove drive belt from alternator.
7. Move friction wheel into servicing position.
8. In order to release the frictional connection between crankshaft and coolant pump, it is necessary to move the friction gear into the servicing position.
9. Firmly pull the handle in direction of arrow until friction gear is separated from belt pulley.
10. To secure friction gear in servicing position, suspend pull cable on housing.

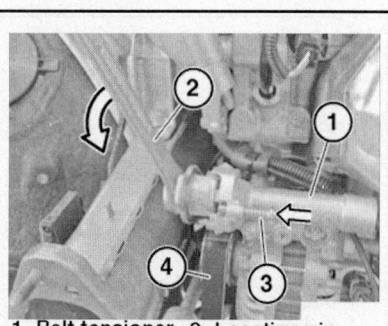

1. Belt tensioner  3. Locating pin
2. Wrench       4. Drive belt

22205_MINI_G0003

**Fig. 24  Drive belt removal**

1. Handle  2. Friction gear

22205_MINI_G0006

**Fig. 25  Separate friction gear from belt pulley**

#### To install:

11. Installation is the reverse of the removal procedure.

### CAMSHAFT AND VALVE LIFTERS

#### REMOVAL & INSTALLATION

##### 2006 Cooper Coupe and Cooper S Coupe

See Figure 26.

1. Before servicing the vehicle, refer to the precautions section.
   a. Remove or disconnect the following:
   • Battery
   • Spark plugs
   • Wheel well liners
   • Cylinder head cover
   • Left engine mount
   • Hydraulic chain tensioner
2. Remove bolts from rocker arm shafts in sequence shown.
3. Install a special engine holding tool, 11–8–370, onto cylinder block and fixture of engine mount.
4. Remove the camshaft (CMP) sensor connector and then the sensor.
5. Rotate the engine until the triangular adjustment mark on the camshaft gear is at the 12 o'clock position. Apply a paint mark across the adjustment mark and timing chain for reassembly reference. Also mark the vibration damper and timing case cover with a paint reference mark.

➡Brass color timing chain links are of no importance to chain timing.

6. Install a special locking tool, 11–8–250, onto camshaft gear and loosen, but do not remove, the camshaft gear center bolt. Remove the special tool.

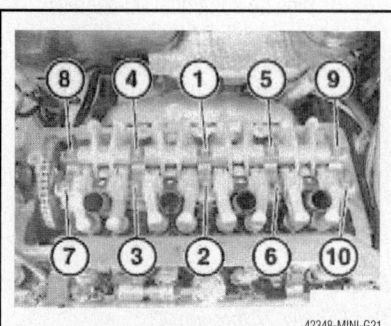

42348-MINI-G21

**Fig. 26  Showing rocker arm shaft bolt removal and tightening sequence—Cooper Coupe**

7. Make sure that paint reference mark on camshaft gear and timing chain are aligned, then remove the camshaft gear center bolt.

8. Remove all the camshaft bearing caps and the camshaft. Be sure to keep bearing caps in same order and orientation as removed.

**To install:**

9. Check components for signs of wear or damage. Replace components as necessary.

➡ **If camshaft is replaced with a new unit, rocker arms must also be replaced.**

**✳✳ CAUTION**

**Install bearing caps in same positions as removed.**

10. Lubricate camshaft bearing journals and rocker arm rolling areas with clean engine oil.

11. Install timing chain to the camshaft gear.

12. Ensure that the timing reference paint marks are aligned.

13. Install the camshaft gear center bolt.

14. Install the camshaft gear locking tool, 11–8–250, then torque the center bolt to 75 ft. lbs. (102 Nm). Remove the locking tool.

15. Apply a thin coat of engine oil to the camshaft seal.

16. Remove the engine holding tool, 11–8–370.

17. Install or connect the following:
- CMP sensor and connector
- Rocker arm shafts; torque bolts first evenly by hand, then to 22 ft. lbs. (30 Nm) in sequence shown.
- Hydraulic chain tensioner

18. Install the left engine (hydra) mount. Torque the bolts as follows:
- M10x110 bolts: 41 ft. lbs. (56 Nm), then an additional 90 degrees
- Other bolt: 74 ft. lbs. (100 Nm)

19. Install or connect the following:
- Cylinder head cover
- Wheel well liners
- Spark plugs
- Battery

## 2007–08 Cooper

### Intake

*See Figures 27 through 33.*

1. Before servicing the vehicle, refer to the precautions section.

2. Remove cylinder head cover.

3. Remove adjusting unit for intake camshaft.

4. Remove intermediate lever.

5. Remove exhaust camshaft.

**✳✳ WARNING**

**The screws of the bearing bridge must not be opened. Releasing the bearing bridge will result in damage to the cylinder head.**

➡ **The bearing cap marked 5 is a thrust bearing.**

6. Release screws of bearing caps 1 to 5.

7. Set all bearing caps down in special tool No. 11 4 481.

8. Remove camshaft.

**To install:**

9. Clean all bearing points and lubricate with oil.

10. Check plain compression rings for damage and replace if necessary.

11. The plain compression rings have catches at the joint. Press plain compression rings apart upwards and downwards and remove towards front, being careful as they can break easily. Make sure they can move freely.

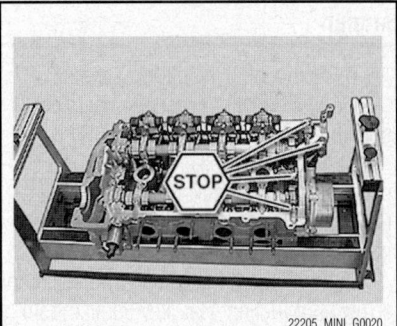

22205_MINI_G0020

**Fig. 27 DO NOT open bearing bridge screws**

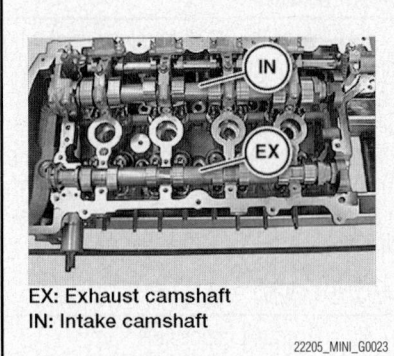

EX: Exhaust camshaft
IN: Intake camshaft

22205_MINI_G0023

**Fig. 30 Camshaft identification**

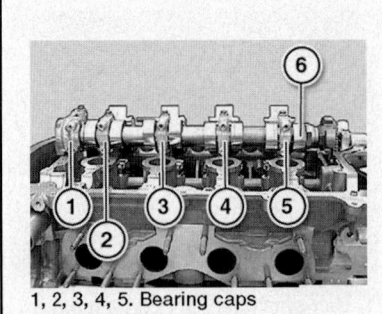

1, 2, 3, 4, 5. Bearing caps
6. Camshaft

22205_MINI_G0021

**Fig. 28 Remove camshaft**

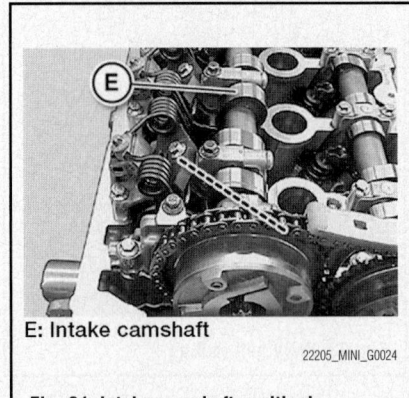

E: Intake camshaft

22205_MINI_G0024

**Fig. 31 Intake camshaft positioning**

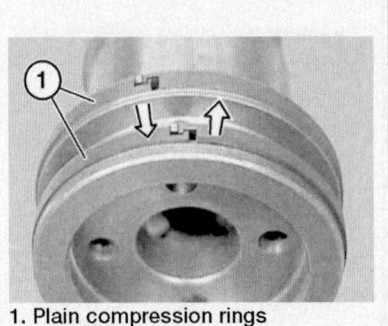

1. Plain compression rings

22205_MINI_G0022

**Fig. 29 Plain compression rings**

1. Pairing letters

22205_MINI_G0025

**Fig. 32 Pay attention to pairing letters**

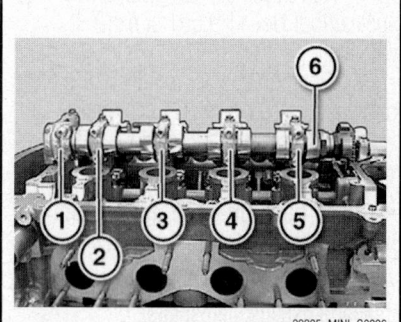

**Fig. 33 Bearing cap tightening sequence**

5. Release central bolt of intake adjustment unit.

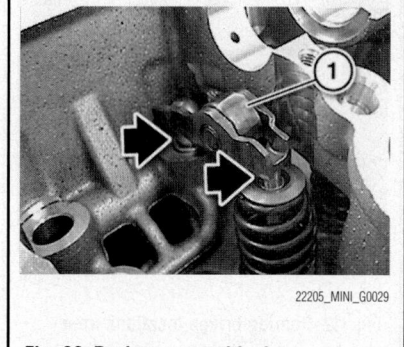

**Fig. 38 Rocker arm positioning**

12. Insert camshaft so that "IN" marking points upwards.

13. Position inlet camshaft so that cams point upwards at an angle.

14. Attach special tool No. 11 9 551 to twin surface.

15. Make sure plain compression rings cable can move freely.

16. All bearing caps are identified from 5 to 10.

17. Tighten bearing caps from inside outwards. Tighten to 7 ft. lbs. (10 Nm).

18. Adjust valve timing.

19. The remainder of installation is the reverse of removal.

### Exhaust

*See Figures 34 through 43.*

1. Before servicing the vehicle, refer to the precautions section.

2. Remove cylinder head cover.

3. Remove vacuum pump.

4. Remove exhaust adjusting unit for exhaust camshaft.

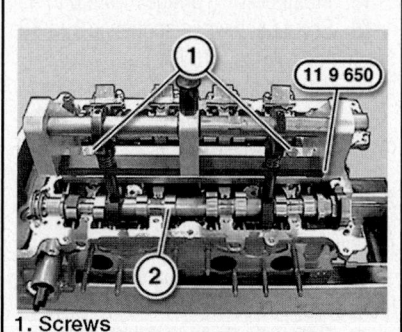

1. Screws

**Fig. 35 Install special tool**

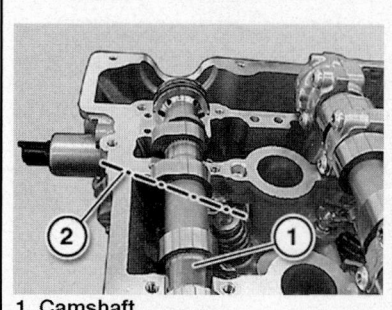

1. Camshaft
2. Installation position

**Fig. 39 Camshaft positioning**

**Fig. 36 Release bearing caps (shown without special tool for clarity)**

EX: Exhaust camshaft
IN: Intake camshaft

**Fig. 40 Camshaft identification**

**Fig. 34 DO NOT open bearing bridge screws**

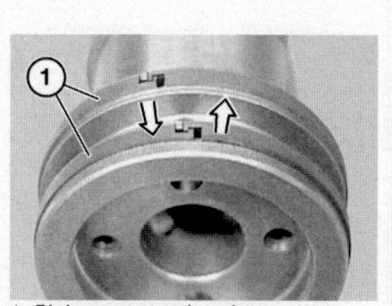

1. Plain compression rings

**Fig. 37 Plain compression rings**

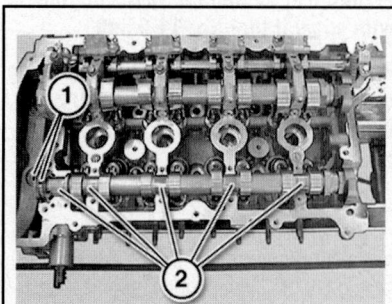

1. Plain compression rings
2. Bearing points

**Fig. 41 Positioning plain compression rings**

**Fig. 42 Bearing bridge locations from 0 to 4**

**Fig. 43 Bearing bridge screw tightening sequence (shown without special tool for clarity)**

6. Secure special tool No. 11 9 650 on cylinder head with screws in spark plug holes.

7. With special tool No. 11 9 650 installed, release bearing caps from 10 to 1.

8. Set all bearing caps down in special tool No. 11 4 480.

9. Check plain compression rings for damage and replace if necessary.

10. The plain compression rings have catches at the joint. Press plain compression rings apart upwards and downwards and remove towards front, being careful as they can break easily. Make sure they can move freely.

➡ **Removal on engine: Block engine with special tool No. 11 9 590.**

➡ **Removed cylinder head: When using special tool No. 11 9 000, it will be necessary to remove the aluminum profile insert.**

**To install:**

11. Before installing exhaust camshaft, make sure roller rocker arm is correctly seated HVCA element and valve.

12. Lubricate all bearing points with engine oil.

13. Insert camshaft, paying close attention to installation position.

**✹✹ WARNING**

**Both camshafts have different identifications. Mixing up the two camshafts will result in engine damage.**

14. Make sure plain compression rings can move freely.

15. Align plain compressing rings in downward direction.

16. Lubricate all bearing points with engine oil.

17. Secure special tool No. 11 9 650 on cylinder head with screws in spark plug holes.

18. Fit all bearing bridges from 0 to 4.

19. Secure screws in sequence 1 to 10. Tighten to 7 ft. lbs. (10 Nm).

20. Adjust valve timing.

21. The remainder of installation is the reverse of removal.

### 2007–08 Cooper S

*See Figures 44 through 56.*

1. Before servicing the vehicle, refer to the precautions section.

2. Remove cylinder head cover.

3. Check timing.

4. Remove chain tensioner.

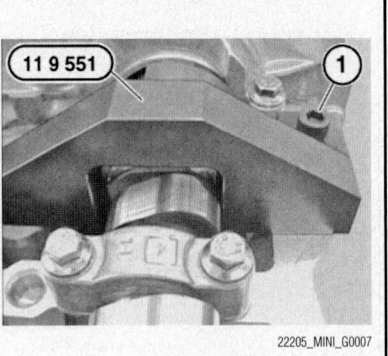

**Fig. 44 Position camshaft tool**

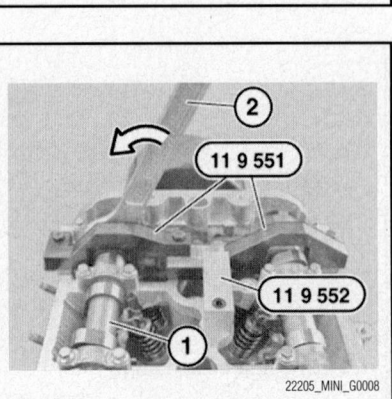

**Fig. 45 Locking camshafts**

5. To release central bolts, always use special tool No. 11 9 551 of exhaust camshaft.

6. Position special tool No. 11 9 551 on twin surface of exhaust camshaft.

7. Secure special tool No. 11 9 551 with a screw.

➡ **Check function of adjustment unit locking by rotating camshaft.**

8. Mount special tool No. 11 9 551 on inlet and exhaust camshafts.

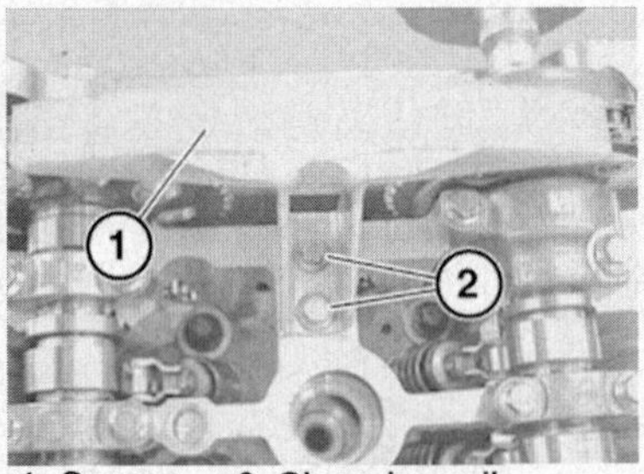

1. Screws    2. Clamping rail

**Fig. 46 Release screws and remove rail**

**Fig. 47 Release screw**

1. Timing chain
2. Sprocket wheel
3. Central bolt

**Fig. 48 Removing camshaft sprocket**

Fig. 49 Set down VANOS adjustment

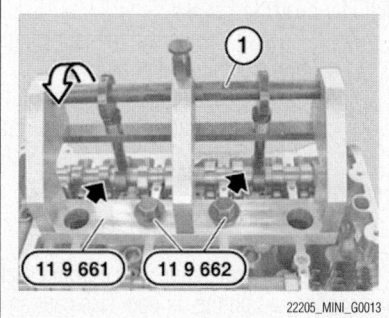

Fig. 50 Screw special tool into spark plug holes and turn intake camshaft eccentric shaft

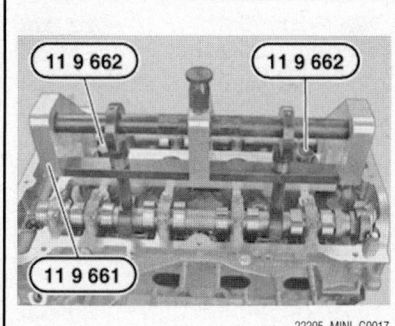

Fig. 51 Screw special tool into exhaust camshaft spark plug holes

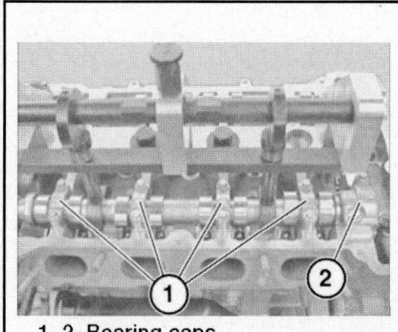

1, 2. Bearing caps

Fig. 52 Intake camshaft bearing caps

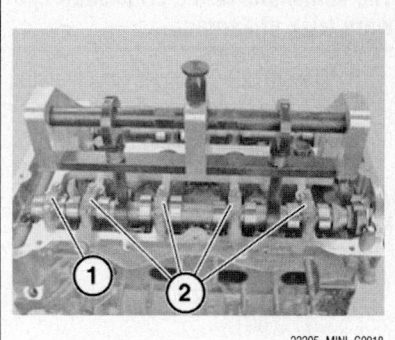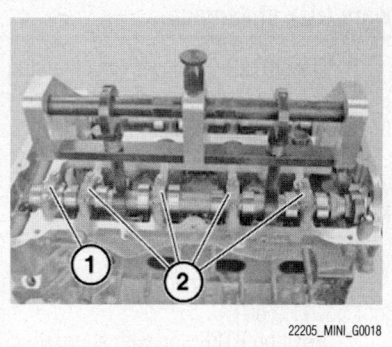

Fig. 53 Exhaust camshaft bearing caps

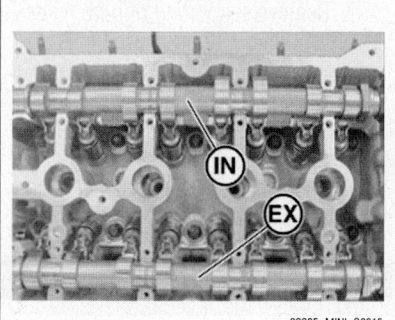

Fig. 54 Identifying intake and exhaust camshafts

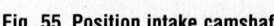

1. Exhaust camshaft

Fig. 55 Position intake camshaft

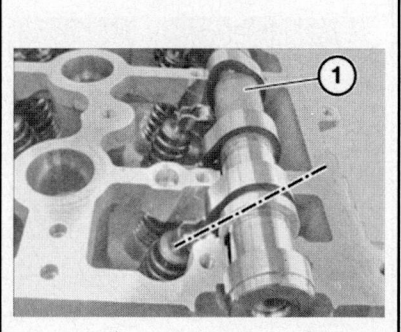

Fig. 56 Position exhaust camshaft

9. Screw in special tool No. 11 9 552 on cylinder head with a screw.

10. To release central bolts, always use special tool No. 11 9 551.

11. Release screws.

12. Remove clamping rail.

13. Release screw.

14. Release central bolt.

15. Feed out sprocket wheel from timing chain towards front.

16. Release central bolt.

17. For exhaust camshaft, do not remove VANOS unit.

18. For intake camshaft, set down VANOS adjustment unit on special tool No. 11 4 480.

➡**With the cylinder head removed, it will be necessary to remove the aluminum profile insert when using special tool No. 11 9 000.**

19. Screw special tool No. 11 9 661 with special tool No. 11 9 662 into spark plug holes.

20. For intake camshaft, turn eccentric shaft in direction of ring and lock.

21. Release all screws on bearing caps.

22. Bearing cap No. 1 is a thrust bearing and has the number 0.

23. Bearing cap No. 2 is a thrust bearing and has the number 5.

24. All intake bearing caps are identified with numbers from 6 to 9.

25. All exhaust bearing caps are identified with numbers from 1 to 4.

26. Intake camshaft is identified with designation (IN), and exhaust camshaft is identified with the designation (EX).

27. Insert camshafts so that designations (IN and EX) can be read from above.

***To install:***

28. Position intake camshaft so that cam of the intake camshaft points upward at an angle.

29. Position exhaust camshaft so that cam of exhaust camshaft points inward at an angle.

30. The remainder of installation is the reverse of removal. Tighten to bearing caps to 7 ft. lbs. (10 Nm).

31. Adjust valve timing.

### CRANKSHAFT FRONT SEAL

*REMOVAL & INSTALLATION*

#### 2007–08 Cooper and Cooper S

*See Figures 57 through 59.*

1. Before servicing the vehicle, refer to the precautions section.

2. Remove A/C line from compressor.

3. Remove vibration damper.

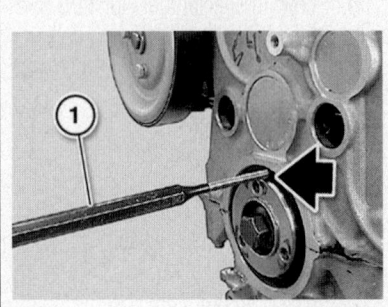

**Fig. 57 Push PTFE ring in until it tilts out at the bottom**

**Fig. 58 Install special tool on crankshaft**

**Fig. 59 Push PTFE ring over supporting ring**

> ※※ **WARNING**
>
> PTFE ring is supplied with a supporting ring. Supporting ring is required as an installation tool. Do not touch inner sealing face of PTFE ring with fingers (risk of damage).

> ※※ **WARNING**
>
> Do not release central bolt. If the central bolt is released, the sprocket wheels of the timing chain and the oil pump will no longer be non-positively connected to the crankshaft.

**The camshafts to the crankshaft can warp (risk of damage).**

4. Drive PTFE ring inwards with a drift until PTFE ring tilts outwards at bottom. Do not allow PTFE ring to slip inward.

5. Secure special tool No. 11 9 601 with screws to crankshaft and tighten to 11 ft. lbs. (15 Nm).

### To install:

6. Apply a light coating of oil to special tool No. 11 9 601.

7. Position PTFE ring with supporting ring on special tool No. 11 9 601.

8. Push PTFE ring over supporting ring in direction of arrow up to crankcase.

9. Remove supporting ring from special tool No. 11 9 601. Supporting ring is no longer needed.

10. Draw in PTFE ring with special tool No. 11 9 602 in conjunction with special tool No. 11 9 603 until flush.

11. The remainder of installation is the reverse of removal.

## CYLINDER HEAD

### REMOVAL & INSTALLATION

#### 2006 Cooper Coupe

*See Figures 60 through 65.*

1. Before servicing the vehicle, refer to the precautions section.

2. Drain the cooling system.

3. Remove or disconnect the following:
- Wheel well liners
- Battery and battery container
- Vent hose and engine control DME connector from cylinder head
- Fuel rail cover
- Fuel injector wiring harness (move aside)
- Heater hoses from cylinder head
- Top hose from thermostat housing
- Exhaust manifold from block
- Spark plugs
- Fuel line from fuel rail (plug openings)
- Lines from stabilizer bar bracket
- Vacuum line to brake booster from intake manifold
- CMP sensor connector
- Dipstick

4. Remove the intake manifold bolts in reverse of the order as shown (start with bolt 5), then lift the manifold over the dipstick tube.

5. Disconnect the line from the filler neck to the expansion tank, then the engine wiring harness can be moved around the thermostat housing.

6. Tie back the intake manifold from the cylinder head.

7. Remove or disconnect the following:
- Engine stabilizer bar bracket
- Oxygen sensor plug connector
- Holder for oxygen sensor plug from cylinder head
- Coolant distributor pipe screw
- Coolant temperature sensor connector

8. Support the engine with a trolley jack and rubber pad on the oil pan. Use caution so oil pan is not damaged.

9. Remove the engine carrier bolts and engine mount nut.

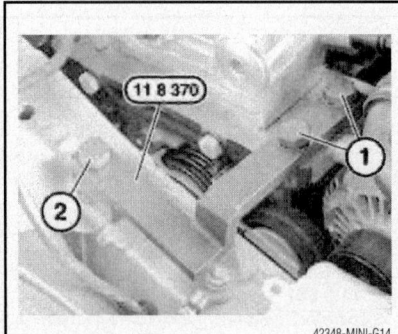

**Fig. 61 Mounting special holding tool on engine—Cooper Coupe**

**Fig. 60 Showing intake manifold bolt tightening sequence—Cooper Coupe**

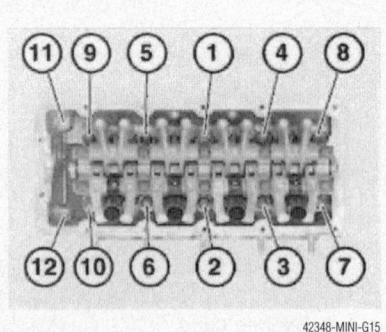

**Fig. 62 Cylinder head bolt tightening sequence—Cooper Coupe**

10. Remove the engine carrier.

11. Use a special tool, 11–8–200, and remove the hydraulic engine mount.

12. Mount a special engine holding tool, 11–8–370, on the cylinder block and body fixtures, as shown.

13. Remove the camshaft sensor.

14. Remove the plugs from the cover on each side of the camshaft.

15. Rotate the crankshaft until the triangular adjustment mark on the camshaft gear is at 12 o'clock. Apply a paint reference

**Fig. 63 Aligning camshaft and timing chain reference marks—Cooper Coupe**

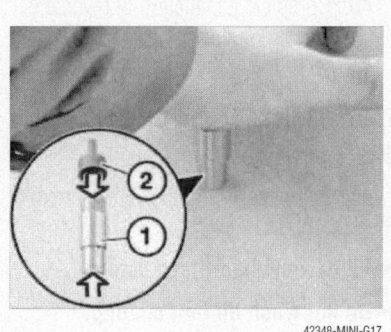

**Fig. 64 Reassembling timing chain tensioner clamping fixture—Cooper Coupe**

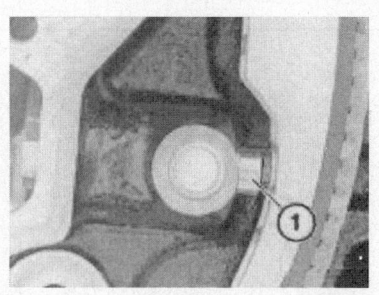

**Fig. 65 Showing the released position of the timing chain clamping fixture (with tension on the timing chain)—Cooper Coupe**

mark from camshaft and across timing chain for reassembly reference.

➡**The brass-colored chain links are of no importance to the timing.**

16. Install a special clamping fixture tool, 11–8–250, to camshaft gear. Slacken, but do not remove the center bolt from the camshaft gear.

17. Remove the wiring harness holder, timing chain tensioner and clamping fixture tool.

18. Remove center bolt from camshaft gear.

19. Remove the camshaft gear from the timing chain and secure the chain to prevent it from falling.

20. Remove the bolts from timing chain guides (through the plug openings).

21. Remove the clamping rail and timing chain guides.

➡**The timing chain cover is designed so that the timing chain can remain on the crankshaft gear without any gear teeth being skipped.**

**✳✳ CAUTION**

**DO NOT rotate crankshaft.**

22. Remove the cylinder head retainers 11 and 12 first, then remove the cylinder head bolts in reverse of the order shown.

23. Remove the cylinder head.

*To install:*

24. Clean all sealing material from mating faces

**✳✳ CAUTION**

**There must be no oil in the cylinder head bolt holes in the block and timing case cover or there is a possibility of cracking and distorting torque values.**

25. Install a new cylinder head gasket.

26. Position the cylinder head onto the block and install new cylinder head bolts (do not clean compound applied to new bolts).

27. Tighten cylinder head bolts, following the sequence shown for bolts 1 through 10, in 2 steps, to the following:
    a. Step 1: 30 ft. lbs. (40 Nm)
    b. Step 2: Additional 90 degrees

28. Tighten cylinder head retainers number 11 and 12 to 21 ft. lbs. (28 Nm).

29. Install or connect the following:
- Clamping rail and timing chain guides
- Chain guide bolts; torque to 21 ft. lbs. (28 Nm)

- Timing chain onto camshaft gear
- Center bolt in camshaft gear

30. Align the camshaft and timing chain paint marks made during removal.

31. Install camshaft gear holding special tool, 11–8–250, then torque center camshaft gear bolt to 75 ft. lbs. (102 Nm).

32. Move the timing chain tensioner into transition position. Place the timing chain tensioner clamping fixture (1) on a level surface and remove the cap (2).

33. Place palm of hand against the clamping fixture and exert continuous pressure until fixture is completely compressed. Replace clamping fixture cap. Position the clamping fixture in place.

34. Install the timing chain tensioner and torque screw plug to 46 ft. lbs. (63 Nm). Install the cable holder.

**✳✳ CAUTION**

**Timing chain tensioner is in the transition position. Ensure timing chain is correctly arranged inside the channel of the timing chain guides.**

35. Use a prybar to lever the clamping rail until the timing chain tensioner applies tension to the timing chain (do not lever directly on the timing chain).

36. Exam the released position of the clamping fixture as shown.

37. Complete installation in reverse of the removal procedure.

38. Refill the cooling system.

**2006 Cooper S Coupe**

*See Figure 66.*

1. Before servicing the vehicle, refer to the precautions section.

2. Disconnect or remove the following:
- Battery
- Intercooler
- ECU connectors
- Intake filter housing

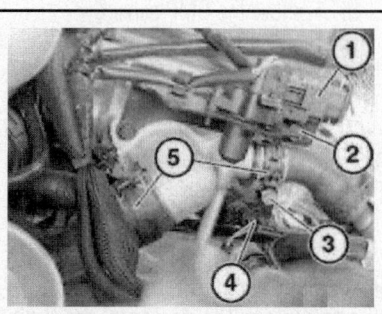

**Fig. 66 Showing components to remove from coolant housing area of engine—Cooper S Coupe**

3. Slacken the module front end (MFE) and install the extension tools, 11–8–401/2 to keep MFE extended for access.

4. Disconnect or remove the following:
- Throttle assembly
- Cylinder head cover
- Fuel tank venting valve
- Fuel line from fuel rail (quick-disconnect fitting); plug openings

> ⁂ **CAUTION**
>
> **Fuel system may be under pressure; be prepared to open line cautiously and ready to catch spilling fuel.**

5. Release both pipes from the engine stabilizer bracket and move to one side.

6. Disconnect or remove the following:
- Intake manifold
- Supercharger outlet pipe
- Engine stabilizer support bracket
- Cap from coolant reservoir
- Drain the cooling system
- Oxygen sensor connector (1)
- Oxygen sensor connector bracket from cylinder head (2)
- Coolant rail support bolt (3)
- Coolant sensor connector (4)
- Coolant hoses (5)

7. Disconnect the camshaft (CMP) sensor connector.

8. Remove the dipstick.

9. Remove the exhaust heat shield and exhaust manifold bolts from cylinder head.

10. Remove the spark plugs.

11. Support the engine with a suitable jack.

12. Remove the engine mount support bracket.

13. Remove the engine hydra-mount, with special tool, 11–8–200.

14. Install a special engine retainer brace, 11–8–370, to cylinder block and engine mount chassis location.

15. Remove the CMP sensor.

16. Remove both plugs from the front of the cylinder head.

17. Remove both fender well liners.

18. Rotate the engine until the camshaft gear triangular timing mark is at the 12 o'clock position. Make a paint mark across the timing mark and timing chain for reassembly reference.

➡**The copper colored link has no relation to timing. The design of the timing chain cover will allow the chain to stay on the crankshaft gear without skipping any teeth.**

> ⁂ **CAUTION**
>
> **DO NOT rotate the engine with timing chain disconnected.**

19. Install camshaft gear holding tool, 11–8–250, then slacken, but do not remove, the camshaft gear center bolt.

20. Remove the wiring harness holder, timing chain tensioner and clamping fixture tool.

21. Remove center bolt from camshaft gear.

22. Remove the camshaft gear from the timing chain and secure the chain to prevent it from falling.

23. Remove the bolts from timing chain guides (through the plug openings).

24. Remove the clamping rail and timing chain guides.

➡**The timing chain cover is designed so that the timing chain can remain on the crankshaft gear without any gear teeth being skipped.**

> ⁂ **CAUTION**
>
> **DO NOT rotate crankshaft.**

25. Remove the cylinder head retainers 11 and 12 first, then remove the cylinder head bolts in reverse of the order shown.

26. Remove the cylinder head.

**To install:**

27. Clean all sealing material from mating faces.

> ⁂ **CAUTION**
>
> **There must be no oil in the cylinder head bolt holes in the block and timing case cover or there is a possibility of cracking and distorting torque values.**

28. Install a new cylinder head gasket.

29. Position the cylinder head onto the block and install new cylinder head bolts (do not clean compound applied to new bolts).

30. Tighten cylinder head bolts, following the sequence shown for bolts 1 through 10, in 2 steps, to the following:
   a. Step 1: 30 ft. lbs. (40 Nm)
   b. Step 2: Additional 90 degrees

31. Tighten cylinder head retainers number 11 and 12 to 21 ft. lbs. (28 Nm).

32. Install or connect the following:
- Clamping rail and timing chain guides
- Chain guide bolts; torque to 21 ft. lbs. (28 Nm)
- Timing chain onto camshaft gear
- Center bolt in camshaft gear

33. Align the camshaft and timing chain paint marks made during removal.

34. Install camshaft gear holding special tool, 11–8–250, then torque center camshaft gear bolt to 75 ft. lbs. (102 Nm).

35. Move the timing chain tensioner into transition position. Place the timing chain tensioner clamping fixture (1) on a level surface and remove the cap (2).

36. Place palm of hand against the clamping fixture and exert continuous pressure until fixture is completely compressed. Replace clamping fixture cap. Position the clamping fixture in place.

37. Install the timing chain tensioner and torque screw plug to 46 ft. lbs. (63 Nm). Install the cable holder.

> ⁂ **CAUTION**
>
> **Timing chain tensioner is in the transition position. Ensure timing chain is correctly arranged inside the channel of the timing chain guides.**

38. Use a prybar to lever the clamping rail until the timing chain tensioner applies tension to the timing chain (do not lever directly on the timing chain).

39. Exam the released position of the clamping fixture as shown.

40. Complete installation in reverse of the removal procedure.

41. Refill the cooling system.

### 2007–08 Cooper and Cooper S

*See Figures 67 through 73.*

1. Before servicing the vehicle, refer to the precautions section.

➡**Fit new cylinder head screws.**

➡**Do not wash off bolt coating.**

➡**There must not be any coolant, water or oil present in the pocket holes (risk of corrosion and cracking).**

2. Remove exhaust system.

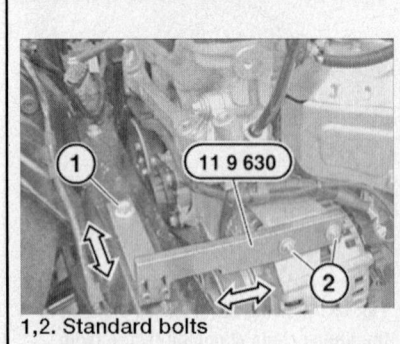

1,2. Standard bolts

22205_MINI_G0033

**Fig. 67 Secure special tool**

3. Drain coolant.
4. Drain engine oil.
5. Remove exhaust manifold.
6. Remove intake air manifold.
7. Remove oil dipstick.
8. Detach coolant hoses from cylinder head.
9. Remove cylinder head cover.
10. Remove inlet and exhaust adjustment unit.
11. Secure crankshaft with special tool No. 11 9 590.

➡**Remove and install cylinder head in installed state.**

12. Suspend engine with engine crane.

➡**Remove and install cylinder head in installed state.**

13. Move front panel into assembly position.
14. Release upper alternator screws, do not remove alternator.
15. Remove right engine mount.
16. Secure special tool No. 11 9 630 with standard bolts.
17. Release bolts.

➡**If the timing chain is stowed in the gearcase, the crankshaft must no**

longer be rotated. The timing chain may jam on the crankshaft gear.

18. Release screw.
19. Release cylinder head bolts with special tool No. 11 2 250.
20. Release cylinder head bolts from outside inward (10 to 1).

➡**Remove shims with a magnet.**

➡ **Do not use any metal-cutting tools for gasket removal.**

21. Use special tool No. 11 4 471 to remove coarse gasket remnants from sealing faces of cylinder head and crankcase.
22. Remove fine gasket remnants with special tool No. 11 4 472.

➡**There must not be any coolant, water or oil present in the pocket holes (risk of corrosion and cracking).**

23. Clean all pocket holes.

*To install:*
24. Replace cylinder head gasket.

➡**Fit new cylinder head screws. Do not wash off bolt coating. Attach shims to cylinder head bolts.**

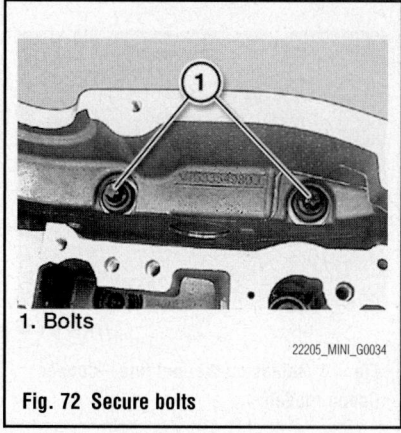

1. Bolts

22205_MINI_G0034

**Fig. 72  Secure bolts**

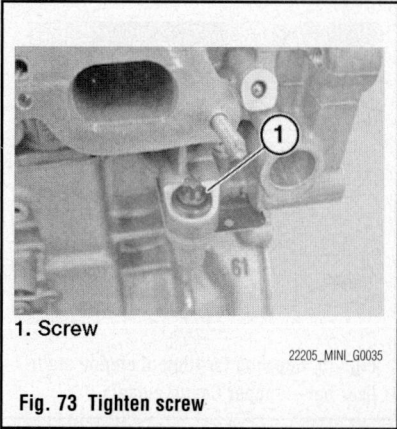

1. Screw

22205_MINI_G0035

**Fig. 73  Tighten screw**

✳✳ **WARNING**

**Do not allow shims to drop into engine.**

25. Secure cylinder head bolts from inside outward (1 to 10), using the following sequence:
 • Step 1: Tighten to 22 ft. lbs. (30 Nm)
 • Step 2: Turn angle 90°
 • Step 3: Turn angle 90°
26. Secure bolts, using the following sequence:
 • Step 1: Tighten to 11 ft. lbs. (15 Nm)
 • Step 2: Turn angle 90°
 • Step 3: Turn angle 90°
27. Tighten the screw to 22 ft. lbs. (30 Nm).
28. The remainder of installation is reverse of removal.

**ENGINE ASSEMBLY**

*REMOVAL & INSTALLATION*

**2006 Cooper Coupe**
*See Figures 74 through 77.*

1. Before servicing the vehicle, refer to the precautions section.
2. Disconnect the battery and battery container.

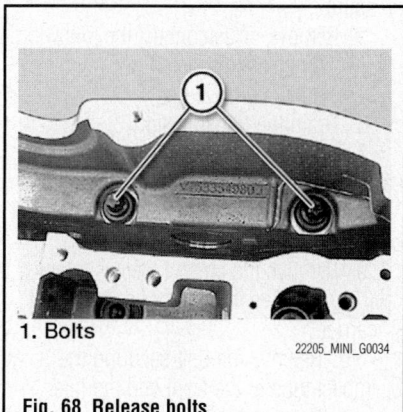

1. Bolts

22205_MINI_G0034

**Fig. 68  Release bolts**

11 2 250

22205_MINI_G0036

**Fig. 70  Release cylinder head bolts**

1. Screw

22205_MINI_G0035

**Fig. 69  Release screw**

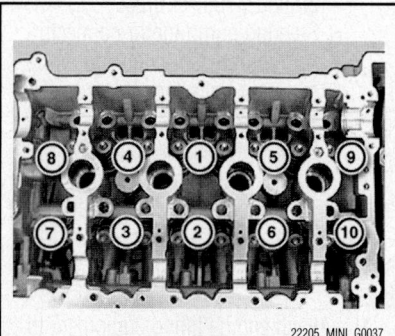

22205_MINI_G0037

**Fig. 71  Cylinder head bolts (illustration shows camshafts removed)**

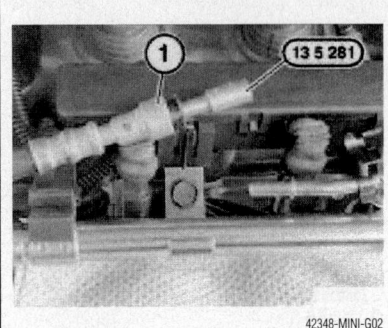

**Fig. 74 Releasing the fuel line—Cooper Coupe models**

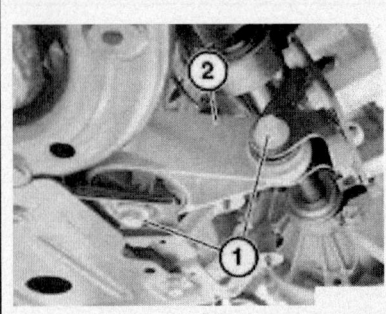

**Fig. 75 Showing location of engine stabilizer bar—Cooper Coupe models**

**Fig. 76 Showing left transmission bracket—Cooper Coupe models**

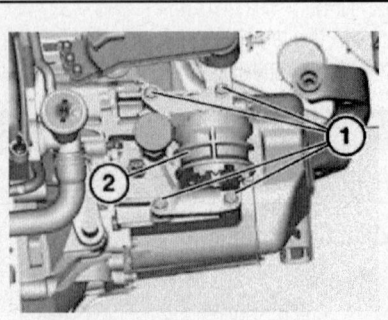

**Fig. 77 Showing right transmission bracket—Cooper Coupe models**

3. Properly relieve the fuel system pressure.

4. Remove or disconnect the following:
- Front bumper cover
- Bumper bracket
- Air cleaner housing
- Clutch cylinder
- Top stabilizer bar bracket
- Exhaust manifold with catalytic converter
- Both driveshafts
- Auxiliary drive belts

5. Drain the cooling system.

6. Remove the upper engine hood support and install a special support tool, 51–2–160.

7. Loosen the retainers for the modular front end and insert bracing tools, 11–8–401/2, to provide access room.

8. Remove or disconnect the following:
- Coolant hoses
- Heater hoses
- Overflow hose

9. Release the fuel line (1), then plug the line opening.

### ✳✳ CAUTION

**Fuel may still be under pressure. Use caution when releasing any fuel lines. Prepare to catch fuel spillage.**

10. Remove the starter heat shield.

11. Release the brake booster line connection at the booster by pressing downward on clip ring and pulling line upward.

12. Remove or disconnect the following:
- Plug connector from fuse box
- Fuse box (position aside)
- Grounding cable from left spring strut
- Circular connector (twist upper and lower halves in opposite directions)
- Transmission shift cables from retaining clips and ball connections
- Shift control housing
- Engine stabilizer brace
- Steering pump motor connector
- A/C compressor connector and 2 retaining bolts (position compressor aside)
- Both transmission brackets

13. Install a special holding tool, 11–8–352, on transmission as a lifting eye.

14. Install a special lifting tool, 11–8–351, on front of engine as a lifting eye.

15. Attach engine lifting equipment to lifting eyes.

16. Remove engine mount retaining nut and remove the engine.

### To install:

17. Install in reverse of removal procedure.

18. Torque retainers to the following:
- Engine mount retaining nut: 50 ft. lbs. (68 Nm)
- Mount bracket to transmission bolts: 49 ft. lbs. (66 Nm)
- A/C compressor mounting bolts: 18 ft. lbs. (25 Nm)
- Engine stabilizer bar bolts: 74 ft. lbs. (100 Nm)
- Shift cable bracket bolt: 16 ft. lbs. (22 Nm)
- Ground cable at left strut: 7 ft. lbs. (9 Nm)

19. When engine is fully installed and assembled, refill transmission and cooling systems.

### 2006 Cooper S Coupe

*See Figures 78 through 80.*

1. Before servicing the vehicle, refer to the precautions section.

2. Install the intercooler protector, 11–8–480 to ensure intercooler is not damaged during engine removal and installation.

3. Disconnect the battery and battery container.

4. Properly relieve the fuel system pressure.

5. Remove or disconnect the following:
- Air cleaner housing
- Auxiliary drive belt

6. Drain the transmission.

7. Remove the lower engine stabilizer bar bracket bolts and remove the bracket.

8. Disconnect the steering pump motor connector.

9. Remove the crush tubes as follows:
   a. Remove the bumper and bumper carrier.
   b. Remove the bolt securing the impact tube to the front end module.
   c. Remove the bolts retaining the impact tube to the chassis subframe.
   d. Remove the impact (crush) tube.

10. Loosen the retainers for the modular front end and insert bracing tools, 11–8–401/2, to provide access room.

11. Remove or disconnect the following:
- Subframe
- Driveshafts
- Exhaust manifold
- Starter motor heat shield
- Starter motor connections
- Oil pressure connector
- Wiring harness from brackets on starter motor and cooling pipe (note routing for installation; move wiring harness aside)

**Fig. 78 Showing installed position of the intercooler protector—Cooper S Coupe models**

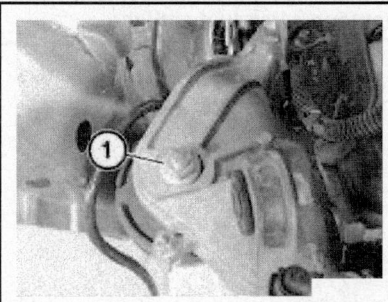

**Fig. 79 Identifying location of engine mount retaining nut—Cooper S Coupe models**

**Fig. 80 Identifying location of transaxle–to–upper mount retaining bolt—Cooper S Coupe models**

- Reverse lamp connector from transaxle
- Intercooler
- Throttle valve
- Tank venting valve lower pipe from stabilizer bracket
- Fuel line from fuel rail (quick-disconnect fitting); plug openings
- Fusebox cover and connector from fusebox
- Grounding cable from left spring strut
- Circular connector (twist upper and lower halves in opposite directions)

- Drain cooling system
- Pipes from heater matrix
- Expansion tank pipes at heater pipe junction
- Upper radiator hose
- Top engine stabilizer bracket
- Clutch slave cylinder
- Transmission shift cables from retaining clips and ball connections and bracket
- A/C compressor connector and retaining bolts (position compressor aside)
- MAP sensor connector (near water housing)
- Coolant hose from water housing

12. Install a special lifting tools, 11–8–351 and 11–8–351, as a lifting eye.

13. Remove the supercharger inlet retaining bolt and release the hose clip.

14. Remove the supercharger inlet pipe to access and disconnect additional hose fittings.

15. Remove the brake booster pipe from the engine and the slave cylinder pipe from the transaxle (move aside).

16. Attach engine lifting equipment to lifting eyes.

17. Remove engine mount retaining nut and remove the engine.

18. Remove the transaxle–to–upper mount retaining bolt.

19. Lower and remove the engine.

### To install:

20. Install in reverse of removal procedure.

21. Torque retainers to the following:
- Engine mount retaining nut: 50 ft. lbs. (68 Nm)
- Mount bracket to transmission bolts: 49 ft. lbs. (66 Nm)
- Coolant pipe to cylinder head: 18 ft. lbs. (25 Nm)
- A/C compressor mounting bolts: 18 ft. lbs. (25 Nm)
- Ground cable at left strut: 7 ft. lbs. (9 Nm)
- Starter motor to transaxle bolts: 63 ft. lbs. (85 Nm)
- Alternator connection on starter: 10 ft. lbs. (14 Nm)
- Starter motor heat shield bolts: 7 ft. lbs. (9 Nm)
- Crush tube to Subframe bolts: (165 Nm)
- Engine stabilizer bar bolts: 74 ft. lbs. (100 Nm)

22. When engine is fully installed and assembled, refill transmission and cooling systems.

### 2007–08 Cooper and Cooper S

*See Figures 81 through 83.*

1. Before servicing the vehicle, refer to the precautions section.
2. Remove exhaust system.
3. Drain engine oil.
4. Disconnect negative battery lead.
5. Remove both drive shafts.
6. Remove air cleaner housing.
7. Remove fan cowl with electric fan.
8. Detach all coolant hoses from engine.
9. Detach vacuum line from brake booster.

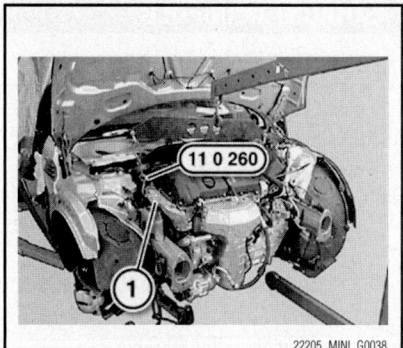

**Fig. 81 Attach lifting tool to engine**

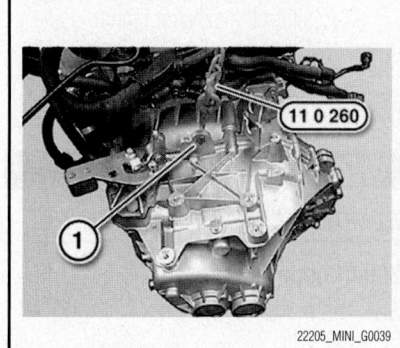

**Fig. 82 Attach lifting tool to engine**

**Fig. 83 Remove engine**

10. Unfasten engine wiring harness and lay to one side.

11. Remove complete front panel.

12. Attach special tool No. 11 0 260 to lifting eye at engine end.

13. Attach special tool No. 11 0 260 to lifting eye at transmission end.

14. Release transmission and engine mounts.

15. Remove engine with special tool No. 11 0 260 towards front.

*To install:*

16. Installation is the reverse of removal.

17. Check function of DME.

## EXHAUST MANIFOLD

### REMOVAL & INSTALLATION

1. Before servicing the vehicle, refer to the precautions section.

2. Move front end into assembly position.

3. Remove or disconnect the following:
- Exhaust system from manifold
- Both oxygen sensor connectors
- Heat shield
- Exhaust manifold bolts
- Exhaust manifold

*To install:*

4. Clean all mating faces.

5. Install new gaskets.

6. Position the exhaust manifold and torque retaining bolts to 18 ft. lbs. (24 Nm).

7. The remainder of installation is the reverse of removal.

## INTAKE MANIFOLD

### REMOVAL & INSTALLATION

#### 2006 Cooper Coupe

*See Figure 84.*

1. Before servicing the vehicle, refer to the precautions section.

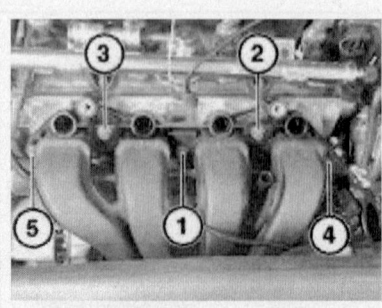

**Fig. 84 Showing intake manifold bolt removal and tightening sequence—Cooper Coupe**

42348-MINI-G20

2. Properly relieve the fuel system pressure.

> **⁂ CAUTION**
>
> **Fuel system may still be under pressure; use caution when disconnecting any fuel system components. Be prepared to catch fuel spillage.**

3. Remove or disconnect the following:
- Battery
- Air cleaner housing
- Throttle assembly
- Cover from fuel rail
- Brake booster line from intake manifold (push attaching ring down to expose line)
- Crankcase vent valve from inspection hole cover
- Fuel line from fuel rail (quick-disconnect attachment)
- Plug connector from intake air temperature/manifold air pressure sensor (TMAP)
- Knock sensor plug from fuel rail wiring harness
- Tank vent line and unclip at fuel rail
- Vacuum line from intake manifold
- Retaining screws of fuel rail
- Injectors and fuel rail (plug all openings)
- Support or wire fuel rail assembly out of the way
- Dipstick
- Coolant line below intake manifold

4. Remove the intake manifold bolts, starting from the center and working outward in an alternating pattern.

5. Remove the intake manifold

*To install:*

6. Check all intake manifold gaskets and replace if necessary.

7. Install intake manifold and torque bolts to 19 ft. lbs. (26 Nm), in the sequence shown.

8. Install or connect the following:
- Coolant line below intake manifold
- Dipstick
- Fuel rail and injectors
- Intake manifold vacuum line
- Tank venting line on fuel rail
- Knock sensor plug to fuel rail wiring harness
- Connector to intake air temperature/manifold air pressure sensor (TMAP)
- Fuel line to fuel rail
- Crankcase vent valve to inspection hole cover
- Brake booster line to intake manifold

- Fuel rail cover
- Throttle assembly
- Air cleaner housing
- Battery

#### 2006 Cooper S Coupe

1. Before servicing the vehicle, refer to the precautions section.

2. Drain the cooling system

3. Properly relieve the fuel system pressure.

> **⁂ CAUTION**
>
> **Fuel system may still be under pressure; use caution when disconnecting any fuel system components. Be prepared to catch fuel spillage.**

4. Remove or disconnect the following:
- Battery
- Air cleaner housing
- Throttle assembly
- Fuel injector rail
- Engine vent control valve
- Coolant hose from upper connection at intake manifold
- Knock sensor plug from manifold
- Air intake sensor connector at manifold
- Top tank vent valve line
- TMAP sensor connector

5. Remove the intake manifold nuts, starting from the center and working outward in an alternating pattern.

6. Remove the intake manifold

*To install:*

7. Position the intake manifold and install the nuts. Torque the nuts in an alternating pattern working outward from the center to 19 ft. lbs. (26 Nm).

8. Install or connect the following:
- TMAP sensor connector
- Top tank vent valve line
- Air intake sensor connector at manifold
- Knock sensor plug from manifold
- Coolant hose from upper connection at intake manifold
- Engine vent control valve
- Fuel injector rail
- Throttle assembly
- Air cleaner housing
- Battery

9. Refill the cooling system.

#### 2007–08 Cooper

*See Figures 85 through 90.*

1. Before servicing the vehicle, refer to the precautions section.

2. Remove suction filter housing.

3. Remove engine cover.

1. Engine wiring harness
2. Plug connection
3. Plug connection on tank valve.

22205_MINI_G0041

**Fig. 85 Disconnect wiring harness and plug connections**

1. Plug connection
2. Tank vent valve

22205_MINI_G0042

**Fig. 86 Disconnect plug connection and tank vent valve**

1. Plug connection on solenoid valve
2. Nut

22205_MINI_G0043

**Fig. 87 Disconnect plug connection on solenoid valve**

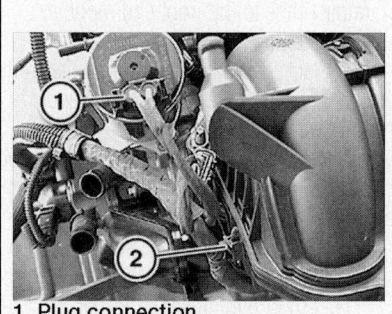

1. Plug connection
2. Engine wiring harness

22205_MINI_G0044

**Fig. 88 Unfasten engine wiring harness**

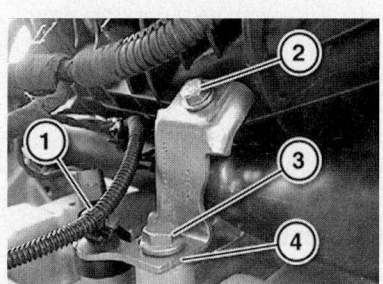

1. Cable
2, 3. Screws
4. Holder

22205_MINI_G0045

**Fig. 89 Release cable, screws, and take off holder**

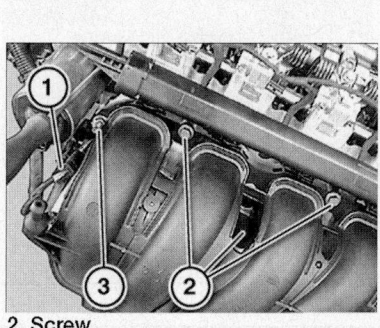

2. Screw
3. Nuts

22205_MINI_G0046

**Fig. 90 Release screw and nuts**

16. Take off holder.
17. Release screw.
18. Unscrew nuts.

***To install:***

19. Replace all seals.
20. The remainder of installation is the reverse of removal. Torque the intake manifold to cylinder head to 11 ft. lbs. (15 Nm).

**2007–08 Cooper S**

*See Figures 91 through 93.*

1. Before servicing the vehicle, refer to the precautions section.
2. Disconnect the negative battery cable.

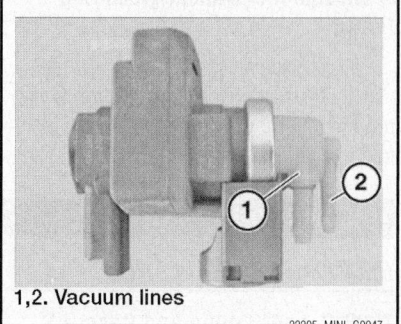

1,2. Vacuum lines

22205_MINI_G0047

**Fig. 91 Disconnect plug connection at EPPC (picture shows EPPC removed)**

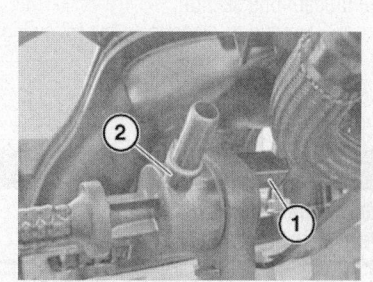

1. Plug connection
2. Tank vent valve

22205_MINI_G0048

**Fig. 92 Detach hose from tank vent valve**

1. Nuts

22205_MINI_G0049

**Fig. 93 Unscrew nuts**

4. Unfasten engine wiring harness on intake manifold.
5. Disconnect plug connection.
6. Disconnect plug connection on tank vent valve.
7. Disconnect plug connection.
8. Release tank vent valve.
9. Disconnect plug connection on solenoid valve.

10. Release engine breathers and hold to one side.
11. Loosen nut.
12. Disconnect plug connection.
13. Unfasten engine wiring harness on intake manifold.
14. Release cable at intake manifold holder.
15. Release screws.

3. Remove suction filter housing.

4. Disconnect vacuum lines on vacuum connection.

5. Disconnect plug connection at EPPC.

6. Disconnect plug connection on tank vent valve.

7. Detach hose from tank vent valve.

8. Unscrew nuts.

**To install:**

➡ **OUT connector on EPPC is identified with a green ring.**

➡ **Vacuum line is fitted with a green ring (OUT).**

➡ **Vacuum line without green ring (VAC).**

9. Replace all seals.

10. The remainder of installation is the reverse of removal. Torque the intake manifold to cylinder head to 11 ft. lbs. (15 Nm).

## OIL PAN

### REMOVAL & INSTALLATION

#### 2006 Cooper Coupe and Cooper S Coupe

*See Figures 94 through 96.*

1. Before servicing the vehicle, refer to the precautions section.

2. Remove or disconnect the following:
- Battery
- Alternator drive belt
- Drain engine oil

#### ✳ CAUTION

**Cover the alternator to prevent oil from dripping on it.**

3. Remove the impact (crush) tube as follows:

a. Remove the bumper and bumper carrier.

b. Remove the bolt securing the impact tube to the front end module.

c. Remove the impact tube.

4. Slacken retainers from modular front end (MFE) and push MFE outward and restrain with special tools 11–8–401/2.

5. Detach the A/C compressor connector.

6. Remove the compressor and lower out of the way. Secure it to MFE.

7. Unclip high-pressure A/C hose.

8. Remove or disconnect the following:
- Lower engine stabilizer bar
- Bracket from oil pan
- 2 forward transaxle–to–oil pan bolts and upper bolts (1)

9. Remove the oil pan bolts in the specified sequence.

**To install:**

10. Ensure mating surfaces are clean.

11. Install a new gasket and position the oil pan in place.

12. Install and tighten the oil pan bolts to 23 ft. lbs. (31 Nm).

13. Install and tighten oil pan–to–transaxle bolts to 23 ft. lbs. (31 Nm).

➡ **Shorter bolts go in lower locations.**

14. Install the lower stabilizer–to–oil pan holder and torque bolts to 74 ft. lbs. (100 Nm).

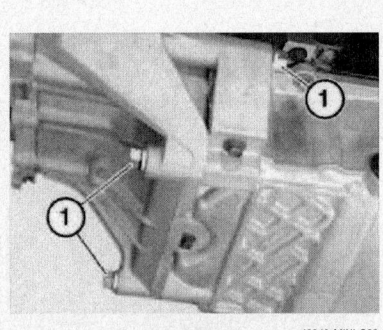

**Fig. 95 Showing location of the transaxle–to–oil pan bolts to remove**

15. Install the lower engine stabilizer bar to oil pan and torque the bolts to 33 ft. lbs. (45 Nm).

16. Install the A/C compressor and torque the mounting bolts to 18 ft. lbs. (25 Nm).

17. Reconnect the A/C compressor connector.

18. Restore the MFE to normal position. Torque the bolts as follows:
- M8x30 bolts: 16 ft. lbs. (22 Nm)
- m6x16 bolts: 3 ft. lbs. (5 Nm)

19. Install the impact tube and torque the bolts to 74 ft. lbs. (100 Nm).

20. Install the bumper and bumper carrier. Torque nuts and bolts to 16 ft. lbs. (22 Nm).

21. Install or connect the following:
- Splash guard
- Alternator drive belt
- Battery

22. Refill the engine oil. Start the engine and wait until the oil indicator lamp goes out. Switch the engine off and wait about 5 minutes, then recheck the oil level.

#### 2007–08 Cooper and Cooper S

*See Figure 97.*

1. Before servicing the vehicle, refer to the precautions section.

2. Drain engine oil.

3. Release oil pan bolts in area of line.

4. Release screw over exhaust manifold with special tools No. 11 9 582 and No. 11 9 581.

5. Clean sealing face with special tool No. 11 4 470.

6. Remove protruding or surplus sealing beads with a suitable tool.

**To install:**

➡ **Do not use adhesive sealing bead.**

➡ **Do not use liquid seal.**

➡ **A metal substrate gasket is available for repairs. See manufacturer's part's service.**

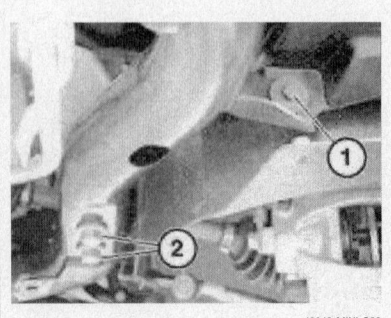

**Fig. 94 Showing location of impact tube retaining bolts**

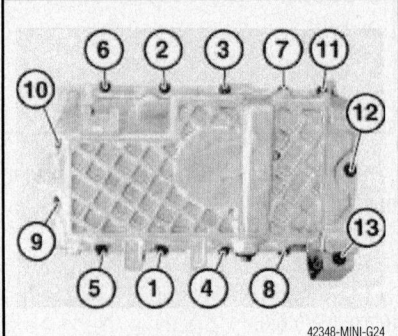

**Fig. 96 Oil pan bolt removal and installation sequence**

**Fig. 97 Oil pan bolts**

7. Installation is the reverse of removal. Tighten oil pan bolts to 9 ft. lbs. (11 Nm).

## OIL PUMP

### REMOVAL & INSTALLATION

#### 2006 Cooper Coupe and Cooper S Coupe

*See Figure 98.*

1. Before servicing the vehicle, refer to the precautions section.
2. Drain the engine oil.
3. Remove or disconnect the following:
   - Negative battery cable
   - Timing chain cover
   - Oil pump cover in reverse of order shown
   - Oil pump

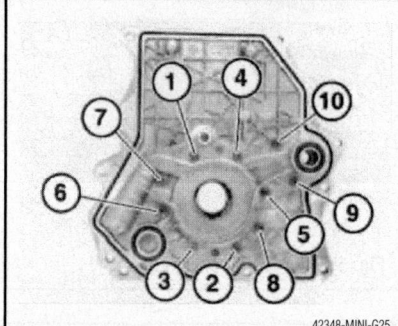

**Fig. 98 Showing oil pump cover bolt tightening sequence**

**To install:**

4. Check oil pump gears, pressure relief valve and housing for signs of wear or damage.
5. Install or connect the following:
6. Fill the rotor cavity with clean engine oil before installing the oil pump.
   - Oil pump
   - Oil pump cover; torque bolts to 13 ft. lbs. (18 Nm)
   - Timing chain cover
   - Negative battery cable
7. Fill the engine with clean oil.
8. Start the vehicle and check for leaks, repair if necessary.

#### 2007–08 Cooper and Cooper S

*See Figures 99 and 100.*

1. Before servicing the vehicle, refer to the precautions section.
2. Drain the engine oil.
3. Remove oil pan.
4. Pull off cover in direction of arrow.
5. Grip crankshaft central bolt to release central bolt.

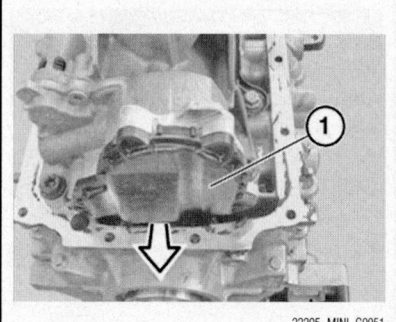

**Fig. 99 Cover removal**

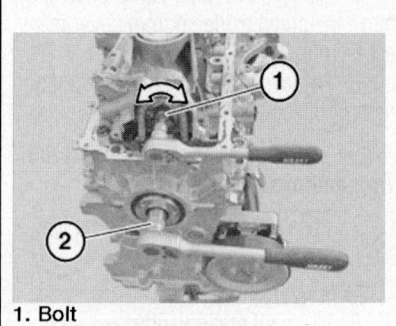

1. Bolt
2. Crankshaft central bolt

**Fig. 100 Release central bolt**

6. Release bolts and fuel pump.

**To install:**

7. Installation is the reverse of removal. Tighten oil pump bolts to 18 ft. lbs. (25 Nm).

## PISTON AND RING

### POSITIONING

*See Figures 101 through 103.*

➡**Offset position of ring end gaps by 120° from each other, but not above piston pin boss.**

## REAR MAIN SEAL

### REMOVAL & INSTALLATION

#### 2006 Cooper Coupe and Cooper S Coupe

The rear main bearing oil seal can be replaced after the transmission.

1. Before servicing the vehicle, refer to the precautions section.
2. Drain the transmission fluid.
3. Remove or disconnect the following:
   - Negative battery cable
   - Transmission

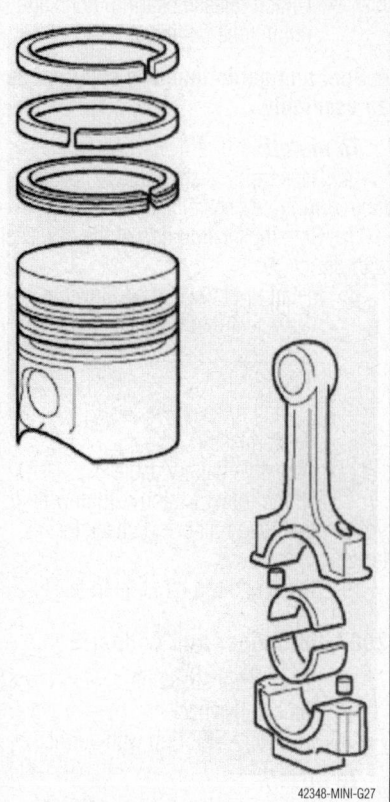

**Fig. 101 Showing orientation of piston, rings and connecting rod**

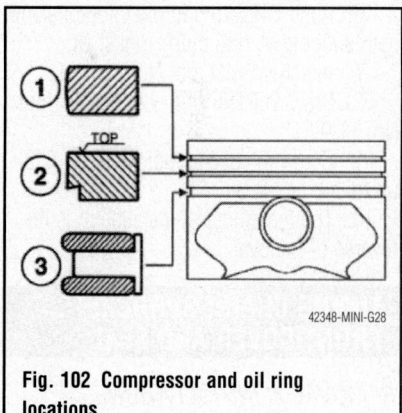

**Fig. 102 Compressor and oil ring locations**

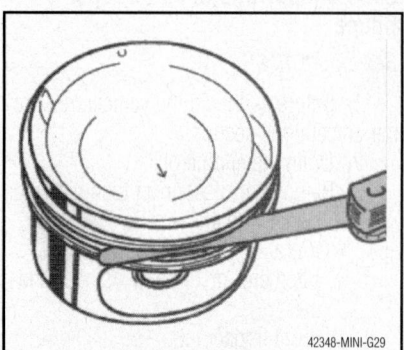

**Fig. 103 Showing piston positioning arrow pointing toward front of block**

- Clutch release bearing, bolts and guide tube (with seal)

➡ **Seal and guide tube are supplied as an assembly.**

### To install:

4. Rape input shaft splines to protect them during seal installation.

5. Coat the sealing lips of the new seal with oil.

6. Install or connect the following:
- New seal into transaxle housing; torque guide sleeve bolts to 4 ft. lbs. (6 Nm)
- Clutch release bearing
- Transmission
- Negative battery cable

7. Fill the transmission with new fluid.

8. Start the engine and check for oil leaks.

9. Check and top off all fluid levels.

### 2007–08 Cooper and Cooper S

1. Remove transmission.
2. Remove flywheel.
3. Break off PTFE ring with a drift.

### To install:

4. Secure special tool No. 11 9 611 with supplied screws to crankshaft.

5. Position PTFE ring with supporting ring on special tool No. 11 9 611.

6. Push PTFE ring in direction of arrow over supporting ring onto crankshaft.

7. Attach special tool No. 11 9 612.

8. Draw in PTFE ring with special tool No. 11 9 613.

9. Screw in special tool No. 11 9 612 up to engine block.

10. The remainder of installation is the reverse of removal.

## TIMING CHAIN, SPROCKETS, FRONT COVER AND SEAL

### REMOVAL & INSTALLATION

### 2006 Cooper Coupe and Cooper S Coupe

*See Figure 104.*

1. Before servicing the vehicle, refer to the precautions section.
2. Drain the engine oil.
3. Remove or disconnect the following:
- Negative battery cable
- Vibration damper
- Alternator drive belt tensioner and belt
- Front impact tube

4. Slacken the modular front end (MFE) retainers and restrain MFE outward with special tools 11–8–401/2.

**Fig. 104  Timing cover bolt tightening sequence**

5. Remove the water pump bolts and move the pump aside so front cover is accessible.

6. Remove timing chain cover bolts in reverse of sequence shown.

➡ **Pay attention to the location of the Torx and oval-head bolts.**

7. Remove O-ring seals and housing seal.

### To install:

8. Clean all mating surfaces.
9. Install new timing chain cover seals.

➡ **If oil pump was removed, fill the rotor cavity with clean engine oil before installing the oil pump.**

10. Install timing chain cover. Install and torque the bolts, in the sequence shown, as follows:
- Torx bolts: 9 ft. lbs. (12 Nm)
- Oval head bolts: 13 ft. lbs. (18 Nm)
- M6 bolts: 9 ft. lbs. (12 Nm)

11. Install water pump and torque the mounting bolts to 41 ft. lbs. (56 Nm) plus an additional 90 degrees.

12. Restore MFE to its normal position.

13. Install the impact tube.

14. Install the belt tensioner, then install the auxiliary drive belt.

15. Install the vibration damper.

16. Connect the negative battery cable.

17. Refill the engine oil.

### 2007–08 Cooper and Cooper S

*See Figures 105 through 114.*

➡ **Modified procedure for timing adjustment.**

➡ **The timing is not determined at firing TDC of cylinder No. 1.**

➡ **All pistons are in the 90° position.**

1. Before servicing the vehicle, refer to the precautions section.
2. Remove cylinder head cover.

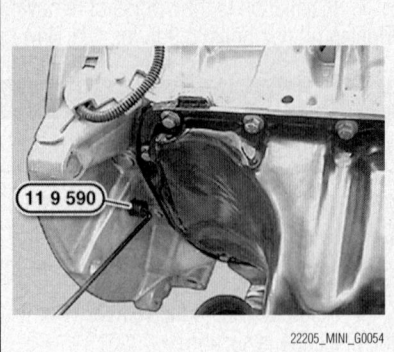

**Fig. 105  Special tool No. 11 9 590**

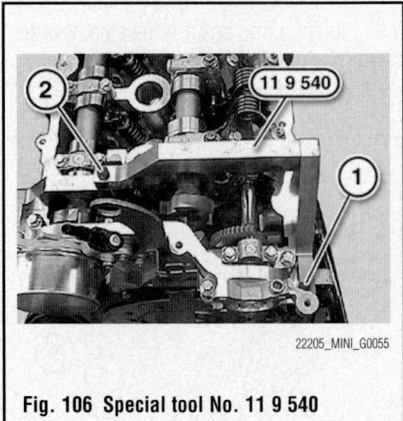

**Fig. 106  Special tool No. 11 9 540**

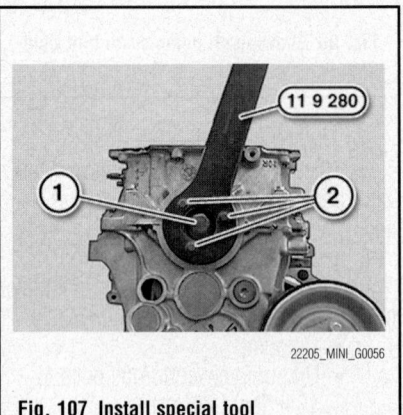

**Fig. 107  Install special tool**

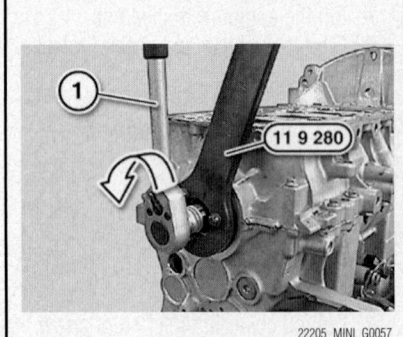

**Fig. 108  Release central bolt**

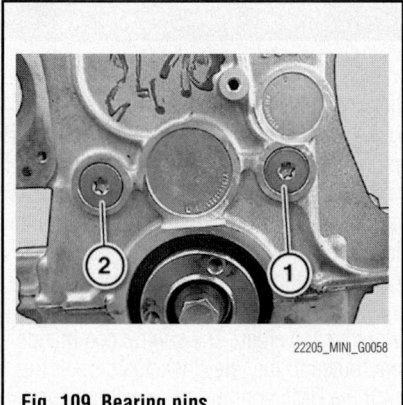

**Fig. 109 Bearing pins**

1. Timing chain sprocket wheel
2. Oil pump sprocket wheel

**Fig. 111 Sprocket wheels**

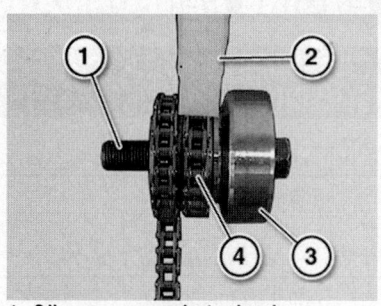

1. Oil pump sprocket wheel
2. Timing chain guide rail
3. Hub on crankshaft
4. Timing chain sprocket wheel

**Fig. 113 Install position of both sprocket wheels**

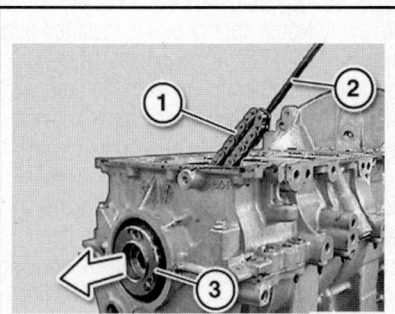

1. Oil pump chain
2. Hook
3. Hub

**Fig. 110 Remove hub and chain module with timing chain**

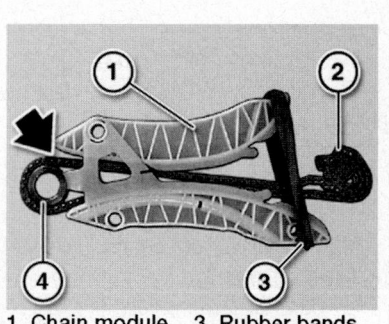

1. Chain module     3. Rubber bands
2. Timing chain     4. Sprocket wheel

**Fig. 112 Chain module assembly**

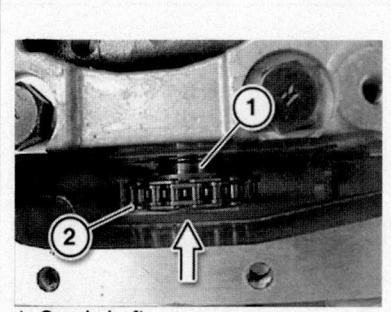

1. Crankshaft
2. Oil pump sprocket wheel

**Fig. 114 Attach oil pump sprocket wheel**

3. Remove all spark plugs.
4. Remove vibration damper.
5. Remove chain tensioner.
6. Remove both VANOS adjustment units.
7. Remove PTFE ring at front.
8. Remove belt tensioner.
9. Position crankshaft with special tool No. 11 9 590. Do not remove special tool No. 11 9 590 during repair work. Do not remove special tool No. 11 9 540.
10. Fit special tool No. 11 9 280 on hub for vibration damper with screws.

➡**You will need another person for gripping when releasing the central bolt.**

11. Release central bolt in direction of arrow.
12. Release bearing pins.
13. Remove hub toward front.
14. Remove chain module with timing chain.

15. Using a hook, pull oil pump chain upwards.

*To install:*

16. Secure chain module with rubber bands to facilitate assembly.
17. Pull timing chain upwards until sprocket wheel rests against chain guide.
18. Install timing chain and sprocket wheel in this position.

➡**Always keep timing chain tensioned; it is possible for timing chain to jam on chain module.**

19. Attach oil pump sprocket wheel in direction of arrow to crankshaft.
20. Insert chain module with timing chain and secure.
21. Attach crankshaft hub.
22. Screw in central bolt. Central bolt torque: 37 ft. lbs. (50 Nm), plus torque angle: 100°.

23. Remove special tool No. 11 9 280 from hub.
24. Secure central bolt with special tool No. 00 9 120.
25. Install VANOS adjustment units.
26. Install sprocket wheel for exhaust camshaft.
27. Crank engine twice.
28. Check timing.
29. Install PTFE ring.
30. Assemble engine.

## VALVE LASH

### ADJUSTMENT

All engines are equipped with hydraulic valve lash adjusters. This design does not permit adjustments nor are adjustments possible.

# ENGINE PERFORMANCE & EMISSION CONTROL

## CAMSHAFT POSITION (CMP) SENSOR

### LOCATION

See Figure 115.

### REMOVAL & INSTALLATION

1. Switch off ignition.
2. Read out fault memory of DME control unit.
3. Check stored fault messages.
4. For 2006 vehicles, remove the right engine support arm.
5. Unlock and remove plug.
6. Release screw and camshaft sensor.

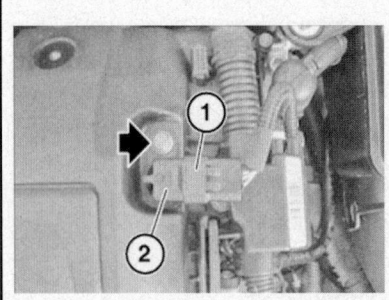

**Fig. 115 Camshaft Position Sensor (Pulse Generator)**

### To install:

7. Replace sealing ring and coat with anti-seize agent.
8. The remainder of installation is the reverse of removal.
9. Clear the fault memory.

## CRANKSHAFT POSITION (CKP) SENSOR

### LOCATION

See Figure 116.

### REMOVAL & INSTALLATION

1. Before servicing the vehicle, refer to the precautions section.
2. Switch off ignition.
3. Read out fault memory of DME control unit.
4. For 2006 vehicles, remove intake air manifold.

➡**For purposes of clarity, the following work step is shown on the engine after it has been removed.**

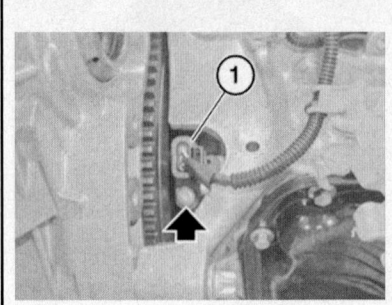

**Fig. 116 Crankshaft Position Sensor (Pulse Generator)**

5. Remove cover.
6. Unlock plug and remove.
7. Release screw and remove crankshaft position sensor.
8. Check stored fault messages.

### To install:

9. Replace sealing ring.
10. The remainder of installation is the reverse of removal.
11. Clear the fault memory.

## ELECTRONIC CONTROL MODULE (ECM)

### LOCATION

See Figure 117.

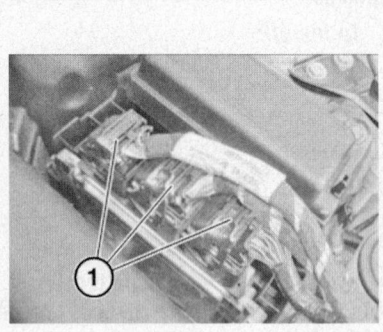

**Fig. 117 DME location**

### REMOVAL & INSTALLATION

See Figures 117 through 119.

When replacing the DME/DDE control unit, observe the following: In each case read out the hardware/software status of the relevant control unit using the BMW diagnosis system. Comply with the instructions of the DIS diagnosis system on the steps pertaining to coding and programming. On

vehicles with electronic vehicle immobilization, comply with the instructions of the BMW diagnosis system. Each control unit is programmed with certain basic values, which serve as mean values. The control unit receives different input values, depending on engine condition, which are compared with the stored values. The adaptive system compares the input values with the stored map values. The control commands are routed to the relevant actuators.

If the DME control unit is without current for a long time (more than one hour), its adaptive system loses the stored values. When a cleared control unit is restarted or a new control unit is installed, the adaptive system must read in and store the input values of the associated engine as new basic values itself.

This procedure could lead to erratic idling and disturbed overrunning of the engine after starting. Depending on the engine it could require some time before all values are adapted to the engine condition.

Therefore observe the following procedure before replacing or reinstalling a DME/DDE control unit:

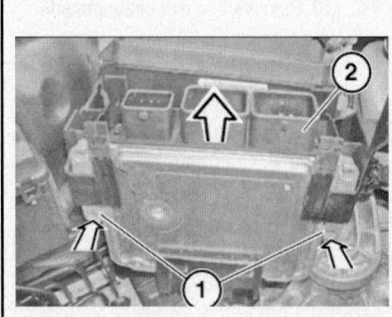

**Fig. 118 Remove DME—2007–08 models**

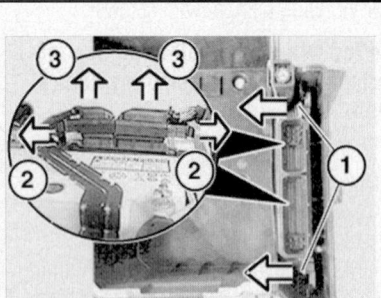

1. Clips
2. Direction to unlock plug connection
3. Direction to detach

**Fig. 119 Remove DME—2006 models**

If possible before exchanging control unit, run engine up to operating temperature. Exchange control units and run the vehicle at alternating engine speeds.

1. Before servicing the vehicle, refer to the precautions section.

> ✳✳ **WARNING**
>
> **Before beginning, always make sure to communicate with the fault memory with a BMW DIS (or equivalent OBD-II scan tool) for existing faults. It may be helpful to print out the results. Once the installation is complete, rerun the scan and correct the remaining faults.**

> ✳✳ **WARNING**
>
> **Make sure that all electrical accessories are off and the ignition is switched off.**

2. Disconnect the negative battery cable.

> ✳✳ **WARNING**
>
> **Take precautions against electrostatic damage.**

3. Connect diagnosis system.
4. Read fault memory.
5. Check stored fault messages.
6. Rectify faults.
7. Clear fault memory.
8. For 2007–08 models, remove the DME as follows:
   - Unlock and remove cover
   - Unlock plug and remove
   - Press locks in direction of arrow and remove control unit towards top (Locks are accessible through bores)
9. For 2006 models, remove the DME as follows:
   - Release clips and lift out DME control unit
   - Unlock plug connections outwards and detach in direction of arrow.
   - Remove control unit

**To install:**
10. Installation is the reverse of removal.
11. Check stored fault messages.
12. Clear fault memory.

## ENGINE COOLANT TEMPERATURE (ECT) SENSOR

### LOCATION
*See Figure 120.*

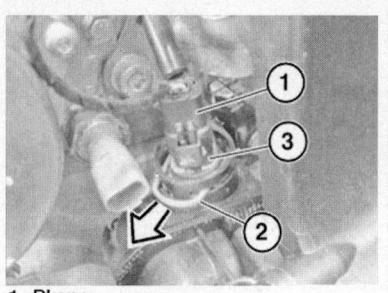

1. Plug
2. Lock
3. Coolant temperature sensor

22205_MINI_G0068

**Fig. 120 Coolant Temperature Sensor—2007 models**

### *REMOVAL & INSTALLATION*

#### 2006 Cooper Coupe and Cooper S Coupe
*See Figure 121.*

> ✳✳ **WARNING**
>
> **There is a danger of scalding so only perform this task on an engine that has completely cooled down.**

> ✳✳ **CAUTION**
>
> **Before beginning, always make sure to communicate with the fault memory with a BMW DIS (or equivalent OBD-II scan tool) for existing faults. It may be helpful to print out the results. Once the installation is complete, rerun the scan and correct the remaining faults.**

> ✳✳ **CAUTION**
>
> **Make sure that all electrical accessories are off and the ignition is switched off.**

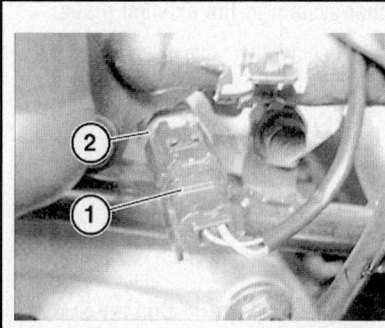

29246_MINI_G0005

**Fig. 121 The plug connection and the coolant temperature sensor**

1. Remove the intake filter housing.
2. Drain the coolant down to below the height of the thermostat housing.
3. For the R50 (W10 Cooper) model only: Remove the battery housing.
4. Unlock the plug connection and remove it.
5. Remove the coolant temperature sensor.

**To install:**
6. Replace the coolant temperature sensor.
7. Install the plug connection and lock down the tab.
8. Refill and vent the cooling system.
9. Clear the fault memory.

#### 2007–08 Cooper and Cooper S
*See Figure 122.*

> ✳✳ **WARNING**
>
> **There is a danger of scalding so only perform this task on an engine that has completely cooled down.**

1. Before servicing the vehicle, refer to the precautions section.
2. Read out fault memory of DME control unit.
3. Check stored fault messages.
4. Switch off ignition.

➡**Coolant can escape when temperature sensor is being replaced. Catch and dispose of coolant.**

5. Release screw or clamps, as applicable.
6. Pull intake muffler towards top and detach clean air pipe.
7. Detach intake muffler from air filter housing and remove.
8. Remove clean air pipe.
9. Unlock and detach plugs.
10. Unlock and disconnect line.

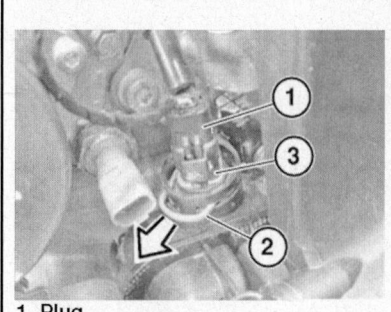

1. Plug
2. Lock
3. Coolant temperature sensor

22205_MINI_G0068

**Fig. 122 Coolant temperature sensor**

11. Carefully pull cable duct upwards slightly.

12. Unlock plug and remove.

13. Detach lock and remove temperature sensor.

*To install:*

14. Replace sealing ring.

15. If necessary, add coolant.

16. Check cooling system for leaks.

17. Clear the fault memory.

18. The remainder of installation is the reverse of removal.

## HEATED OXYGEN (HO2S) SENSOR

### LOCATION

See Figure 123.

**Fig. 123 Oxygen sensor**

### REMOVAL & INSTALLATION

#### 2006 Cooper Coupe and Cooper S Coupe

See Figures 124 and 125.

### ✳✳ CAUTION

**Before beginning, always make sure to communicate with the fault memory with a BMW DIS (or equivalent OBD-II scan tool) for existing faults.**

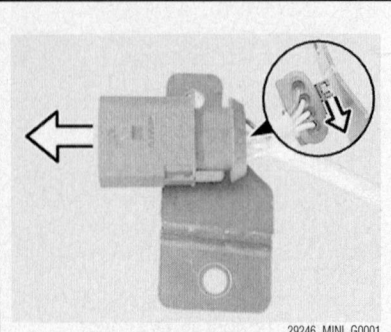

**Fig. 124 The plug connectors and the retaining clips**

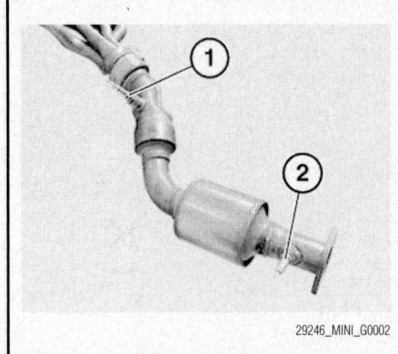

**Fig. 125 The upstream oxygen sensor (1) and the downstream oxygen sensor (2)**

It may be helpful to print out the results. Once the installation is complete, rerun the scan and correct the remaining faults.

### ✳✳ CAUTION

**Make sure that all electrical accessories are off and the ignition is switched off.**

1. Remove the heat shield.

2. Detach the connector from the retaining clips.

3. Remove the oxygen sensors from their respective positions upstream and downstream.

*To install:*

➡**The threads of a new oxygen sensors are already coated with an anti-seize compound. If an oxygen sensor is to be used again, apply a thin and even coat of an anti-seize compound to the thread only.**

➡**Do not clean the oxygen sensor section which protrudes into the exhaust line and ensure that it avoids all contact with any lubricants.**

➡**Observe cable routing of the oxygen sensor so it doesn't interfere with any other system or the exhaust pipes.**

4. Replace the oxygen sensors in their respective positions upstream and downstream.

5. Attach the connector to the retaining clips.

6. Replace the heat shield.

7. Check for any remaining faults. Rectify them and clear the fault memory.

#### 2007–08 Cooper and Cooper S

See Figure 126.

1. Before servicing the vehicle, refer to the precautions section.

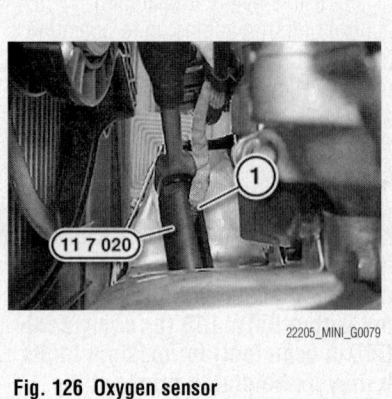

**Fig. 126 Oxygen sensor**

2. Read fault memory.

➡**If an oxygen sensor is to be reused, only apply a thin and uniform coat of Never Seez Compound (refer to BMW Parts Service) to thread.**

➡**The part of the oxygen control sensor which projects into the exhaust system branch (sensor ceramic) must not be cleaned or come into contact with lubricant.**

3. Disconnect plug connection for lambda control sensor.

4. Release oxygen sensor with special tool No. 11 7 020.

*To install:*

5. Installation is reverse of removal.

6. Check function of DME.

## INTAKE AIR TEMPERATURE (IAT) SENSOR

### LOCATION

IAT sensor is located behind the front bumper.

### REMOVAL & INSTALLATION

1. Before servicing the vehicle, refer to the precautions section.

2. Remove front grill or bumper trim, as applicable.

3. Disconnect connector, and remove sensor.

*To install:*

4. Installation is the reverse of removal.

## KNOCK SENSOR (KS)

### LOCATION

The knock sensor is located under the intake manifold next to the starter motor.

### REMOVAL & INSTALLATION

See Figure 127.

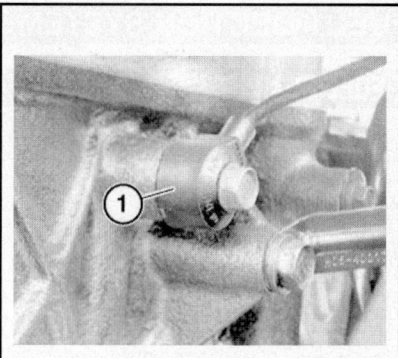

Fig. 127 The knock sensor (1)

**✳✳ CAUTION**

Before beginning, always make sure to communicate with the fault memory with a BMW DIS (or equivalent OBD-II scan tool) for existing faults. It may be helpful to print out the results. Once the installation is complete, rerun the scan and correct the remaining faults.

**✳✳ CAUTION**

Make sure that all electrical accessories are off and the ignition is switched off.

1. For the R50 (W10 Cooper) only: Remove the intake air manifold.
2. For the R53 (W11 Cooper S) only: Remove the exhaust turbocharger.
3. Unlock and disconnect the knock sensor plug connection.
4. Unscrew the knock sensor screw and remove knock sensor.

*To install:*
5. Clean the surface of knock sensors where they contact the engine block.
6. Observe the position of the knock sensor in relation to the engine block. It should be positioned at an angle of 20 degrees to the perpendicular of the engine block.
7. Replace the knock sensor and knock sensor screw.
8. Connect and lock the knock sensor plug connection.
9. Clear the fault memory.

## MANIFOLD ABSOLUTE PRESSURE (MAP) SENSOR

### LOCATION

*See Figure 128.*

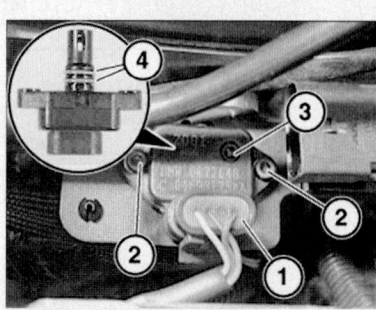

1. MAP sensor connector
2. MAP sensor retaining screws
3. MAP sensor
4. O-ring

22205_MINI_G0070

Fig. 128 MAP Sensor—2006 Cooper S

### REMOVAL & INSTALLATION

#### 2006 Cooper Coupe and Cooper S Coupe

*See Figure 129.*

**✳✳ CAUTION**

Before beginning, always make sure to communicate with the fault memory with a BMW DIS (or equivalent OBD-II scan tool) for existing faults. It may be helpful to print out the results. Once the installation is complete, rerun the scan and correct the remaining faults.

1. Disconnect oxygen sensor connector.
2. Disconnect MAP sensor connector.
3. Remove MAP sensor retaining screws and remove sensor.
4. Replace O-ring

*To install:*
5. Installation is the reverse of removal.

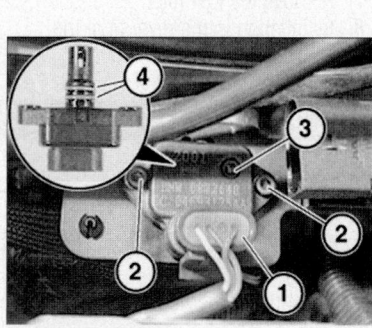

1. MAP sensor connector
2. MAP sensor retaining screws
3. MAP sensor
4. O-ring

22205_MINI_G0070

Fig. 129 MAP Sensor

## MASS AIR FLOW (MAF) SENSOR

### LOCATION

*See Figure 130.*

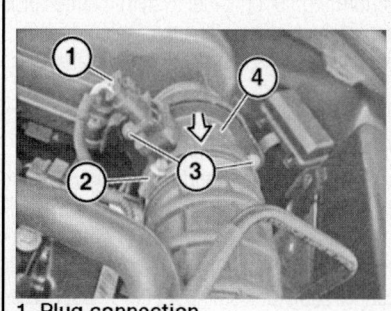

1. Plug connection
2. Clamp
3. Screws

22205_MINI_G0069

Fig. 130 Mass Air Flow Sensor

### REMOVAL & INSTALLATION

#### 2007–08 Cooper and Cooper S

*See Figure 131.*

1. Before servicing the vehicle, refer to the precautions section.
2. Read out fault memory of DME control unit.
3. Check stored fault messages.
4. Switch off ignition.
5. Disconnect plug connection.
6. Release clamp.
7. Release screws.
8. Remove air-mass sensor in direction of arrow.

*To install:*
9. Installation is the reverse of removal.
10. Check stored fault messages.
11. Clear the fault memory.

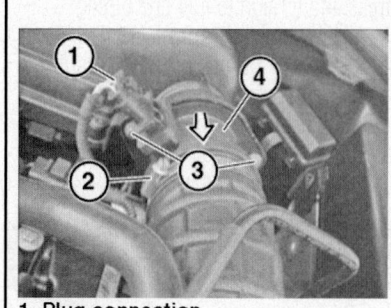

1. Plug connection
2. Clamp
3. Screws

22205_MINI_G0069

Fig. 131 Remove mass air flow sensor

**FUEL**                                    **GASOLINE FUEL INJECTION SYSTEM**

## FUEL SYSTEM SERVICE PRECAUTIONS

Safety is the most important factor when performing not only fuel system maintenance but any type of maintenance. Failure to conduct maintenance and repairs in a safe manner may result in serious personal injury or death. Maintenance and testing of the vehicle's fuel system components can be accomplished safely and effectively by adhering to the following rules and guidelines.

• To avoid the possibility of fire and personal injury, always disconnect the negative battery cable unless the repair or test procedure requires that battery voltage be applied.

• Always relieve the fuel system pressure prior to disconnecting any fuel system component (injector, fuel rail, pressure regulator, etc.), fitting or fuel line connection. Exercise extreme caution whenever relieving fuel system pressure to avoid exposing skin, face and eyes to fuel spray. Please be advised that fuel under pressure may penetrate the skin or any part of the body that it contacts.

• Always place a shop towel or cloth around the fitting or connection prior to loosening to absorb any excess fuel due to spillage. Ensure that all fuel spillage (should it occur) is quickly removed from engine surfaces. Ensure that all fuel soaked cloths or towels are deposited into a suitable waste container.

• Always keep a dry chemical (Class B) fire extinguisher near the work area.

• Do not allow fuel spray or fuel vapors to come into contact with a spark or open flame.

• Always use a back-up wrench when loosening and tightening fuel line connection fittings. This will prevent unnecessary stress and torsion to fuel line piping.

• Always replace worn fuel fitting O-rings with new Do not substitute fuel hose or equivalent where fuel pipe is installed.

Before servicing the vehicle, make sure to also refer to the precautions in the beginning of this section as well.

## RELIEVING FUEL SYSTEM PRESSURE

*See Figure 132.*

1. Install special tool, 13–5–220.
2. Fit a suitable length of hose onto the special tool and route the hose into a fuel container.

**Fig. 132 Using special tool to relieve fuel rail pressure**

42348-MINI-G30

3. Screw in check valve (1) of the special tool to release the fuel pressure from the injector rail.
4. Hold an absorbent cloth around the special tool and remove the hose and tool.

### ✳✳ WARNING

**Other parts of the fuel system may have some residual pressure. Always open fittings slowly and be prepared to catch any fuel.**

## FUEL FILTER

### REMOVAL & INSTALLATION

1. Before servicing the vehicle, refer to the precautions in the beginning of this section.
2. Disconnect the negative battery cable.
3. Remove the rear seat.
4. Remove the access cover.
5. Disconnect the fuel line.
6. Remove the locking ring.
7. Remove the fuel filter.
8. Installation is the reverse of the removal procedure.

## FUEL INJECTORS

### REMOVAL & INSTALLATION

1. Before servicing the vehicle, refer to the precautions in the beginning of this section.
2. Disconnect the negative battery cable.
3. Relieve fuel system pressure.
4. Disconnect the fuel line.
5. Remove the fuel rail with injectors attached.
6. Installation is the reverse of the removal procedure.

## FUEL PUMP

### REMOVAL & INSTALLATION

1. Before servicing the vehicle, refer to the precautions section.
2. Drain the fuel tank.
3. Remove or disconnect the following:
   • Negative battery cable
   • Rear seat
   • Fuel pump access plate (under trim panel)
   • Electrical connector
4. Unscrew the outer ring with special tool 16–1–020.
5. Lift the cap and detach the fuel line and electrical connector from the fuel level sensor.
6. Remove the fuel level sensor and fuel pump from the tank.

### *To install:*

➡**Always use a new seal or gasket when installing the fuel pump or fuel level gauge sending unit assembly.**

7. Install or connect the following:
   • Fuel pump into the fuel tank taking care not to bend or damage the fuel sending unit assembly
   • New seal and torque the sealing ring, using tool No. 16-1-020, to 26 ft. lbs. (35 Nm).
   • Fuel gauge level sending unit electrical connector
   • Metal cover
   • Rear seat bench
   • Negative battery cable
8. Refill fuel tank.
9. Start the vehicle and check for leaks, repair if necessary.

## FUEL TANK

### REMOVAL & INSTALLATION

1. Before servicing the vehicle, refer to the precautions in the beginning of this section.
2. Disconnect the negative battery cable.
3. Drain the fuel tank.
4. Remove the muffler.
5. Remove the exhaust heat shield.
6. Remove the rear storage compartment bracket.
7. Remove the parking brake cables.
8. Remove the rear seat.
9. Remove trim panels.
10. Remove the left wheel arch trim.
11. Disconnect the fuel level sensor and wiring harness.

12. Remove the fuel tank filler hose.

13. Remove the vent hose from the fuel filler pipe.

14. Support the fuel tank and remove the mounting bolts.

15. Lower the tank and disconnect the right vent line.

16. Remove the fuel tank.

### To install:

17. Raise the fuel tank into position and attach the right vent line.

18. Install the mounting bolts and tighten them to 22–28 Nm (16–20 ft. lbs.).

19. Install the vent hose to the fuel filler pipe.

20. Install the fuel tank filler hose.

21. Connect the fuel level sensor and wiring harness.

22. Install the wheel arch trim.

23. Install the trim panels.

24. Install the rear seat.

25. Install and adjust the parking brake cables.

26. Install the rear storage compartment bracket.

27. Install the exhaust heat shield.

28. Install the muffler.

29. Add fuel.

30. Connect the negative battery cable.

## IDLE SPEED

### ADJUSTMENT

Idle speed is maintained by the Powertrain Control Module (PCM). No adjustment is necessary or possible.

## THROTTLE BODY

### REMOVAL & INSTALLATION

#### 2006 Cooper Coupe

*See Figure 133.*

1. Before servicing the vehicle, refer to the precautions section.

2. Read out fault memory of DME control unit.

3. Check stored fault messages.

4. Switch off ignition.

5. Remove intake filter housing.

6. Remove intake fitting.

7. Unlock line of tank venting valve on throttle valve assembly and disconnect.

8. Unlock plug and remove.

9. Loosen screws.

10. Remove throttle valve assembly.

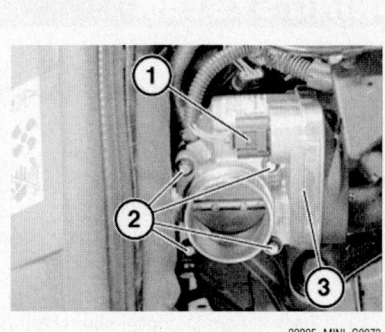

22205_MINI_G0073

**Fig. 133 Throttle assembly—2006 Cooper**

### To install:

11. Replace sealing ring.

12. The remainder of installation is the reverse of removal.

➡A faulty throttle valve causes the adaptation values stored in the DME control unit to be modified. These adaptation values must be reset after the throttle valve has been replaced.

➡Replacement only: Delete adaptation values and reset. Connect diagnosis system: Diagnosis, model, DME engine, control unit function, DME 200, component activation, reset adaptation values, electronic throttle.

13. Clear the fault memory.

#### 2007–08 Cooper

*See Figure 134.*

1. Before servicing the vehicle, refer to the precautions section.

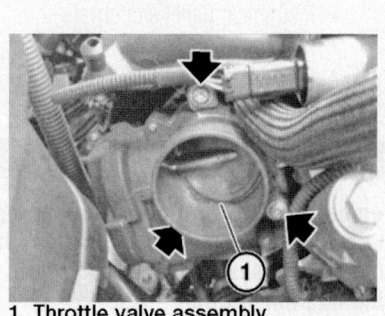

1. Throttle valve assembly

22205_MINI_G0071

**Fig. 134 Throttle assembly—2007–08 Cooper**

2. Read out fault memory of DME control unit.

3. Check stored fault messages.

4. Switch off ignition.

5. Unfasten screws.

6. Carefully feed out throttle valve assembly towards top until plug for cable connection is accessible.

7. Unlock connector and remove.

8. Remove throttle valve assembly.

### To install:

9. Replace sealing ring.

10. The remainder of installation is the reverse of removal.

11. Clear the fault memory.

#### 2007–08 Cooper S

*See Figure 135.*

1. Before servicing the vehicle, refer to the precautions section.

2. Read out fault memory of DME control unit.

3. Check stored fault messages.

4. Switch off ignition.

5. Remove sound generator.

6. Release clamps and detach air intake hose.

7. Release screws and carefully feed out throttle valve assembly towards top until plug is accessible.

8. Unlock plug and disconnect.

9. Detach cable ties.

10. Remove throttle valve assembly.

### To install:

11. Replace sealing ring.

12. The remainder of installation is the reverse of removal.

13. Clear the fault memory.

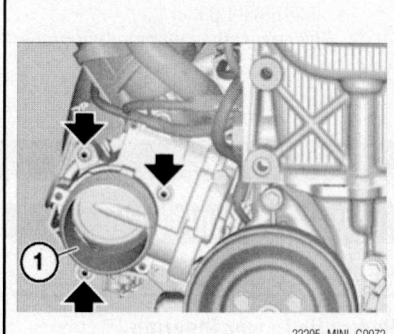

22205_MINI_G0072

**Fig. 135 Throttle assembly—2007–08 Cooper S**

# HEATING & AIR CONDITIONING SYSTEM

## BLOWER MOTOR

### REMOVAL & INSTALLATION

1. Before servicing the vehicle, refer to the precautions section.
2. Disconnect the negative battery cable.
3. Drain the cooling system.
4. Discharge and recover the A/C refrigerant.
5. Remove the instrument panel by removing or disconnecting the following:
   - Battery box
   - Intake filter housing
   - Heater hoses
   - A/C pipe from firewall fittings (plug openings)
   - Heater locating stud nut
   - Steering column from steering gear
   - Steering column from rubber bellows
   - Left and right A-pillar trim
   - Radio
   - Cover from beneath center controls
   - Left and right kick panels
   - Connector behind left kick panel
   - Upper connector from blower control
   - Instrument panel end covers
   - Bolts behind end covers
   - Lower section of steering column trim panel
   - Electrical harness from instrument panel trim
   - Connectors on steering column
   - Heater connector
   - Lower bolts on instrument panel support (next to steering column lower end)
   - Instrument panel
6. Disconnect the fan motor wiring harness.
7. Remove the fan motor.

### To install:

8. Install the fan motor and connect the wiring harness.
9. Install the instrument panel in reverse of the removal procedure.
10. Refill the cooling system.
11. Reconnect the negative battery cable.
12. Evacuate and recharge the A/C system.
13. Check the coolant level after starting the engine and running for several minutes.

## HEATER CORE

### REMOVAL & INSTALLATION

See Figure 136.

1. Before servicing the vehicle, refer to the precautions section.
2. Disconnect the negative battery cable.
3. Drain the cooling system.
4. Discharge and recover the A/C refrigerant.
5. Remove the instrument panel by removing or disconnecting the following:
   - Battery box
   - Intake filter housing
   - Heater hoses
   - A/C pipe from firewall fittings (plug openings)
   - Heater locating stud nut
   - Steering column from steering gear
   - Steering column from rubber bellows
   - Left and right A-pillar trim
   - Radio
   - Cover from beneath center controls
   - Left and right kick panels
   - Connector behind left kick panel
   - Upper connector from blower control
   - Instrument panel end covers
   - Bolts behind end covers
   - Lower section of steering column trim panel

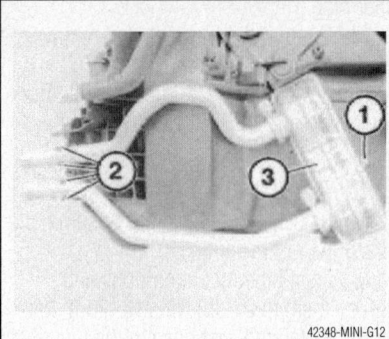

42348-MINI-G12

**Fig. 136 Removing heater pipes from heater housing**

   - Electrical harness from instrument panel trim
   - Connectors on steering column
   - Heater connector
   - Lower bolts on instrument panel support (next to steering column lower end)
   - Instrument panel
6. Remove the cover panel from the heater core.
7. Remove the screw and pipes from the connector on the side of the heater housing.
8. Remove the heater core.

### To install:

9. Install the heater core into the housing.
10. Connect the heater pipes on the side of the housing.
11. Install the heater housing side cover.
12. Install the instrument panel in reverse of the removal procedure.
13. Refill the cooling system.
14. Reconnect the negative battery cable.
15. Evacuate and recharge the A/C system.
16. Check the coolant level after starting the engine and running for several minutes.

# STEERING

## POWER STEERING GEAR

### REMOVAL & INSTALLATION

**Hydraulic Power Steering**

See Figure 137.

1. Before servicing the vehicle, refer to the precautions section.

### ✳✳ CAUTION

**It is essential to maintain cleanliness of components, especially when removing hoses or otherwise opening the hydraulic system. Always plug all openings to seal against debris getting into the system.**

2. Remove the front wheels.
3. Apply a reference mark on the steering tie rod with paint, for reassembly reference.
4. Remove the nuts on the left and right tie rod ends and separate the tie rod ends from the steering knuckle.
5. Release the nuts (1) at the bottom of the stabilizer bar.

6. Drain the steering fluid from the reservoir.
7. Remove the nut on the clamp at the lower end of the steering column (near the firewall).
8. Detach the high pressure line and low pressure line from the steering gear (banjo bolts). Plug the openings.
9. Remove the heat shield from the steering gear.
10. Detach the line bracket from the ends of the steering gear housing, then remove

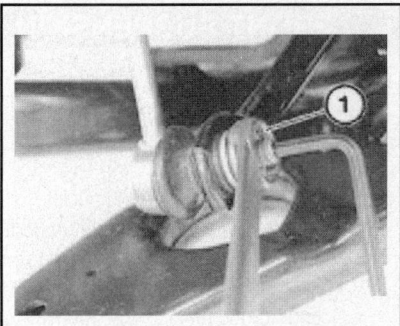

**Fig. 137 Releasing nuts on bottom of the stabilizer bar**

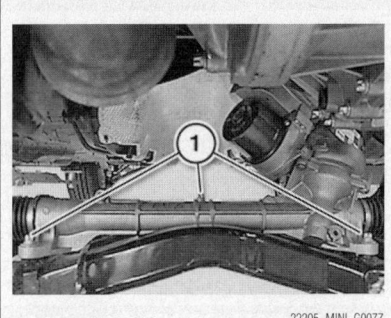

**Fig. 138 Electronic power steering**

the 4 bolts holding the steering gear to the chassis.

11. Remove the steering gear.

*To install:*

12. Position the steering gear to the chassis mountings. Torque the retaining bolts to 41 ft. lbs. (56 Nm).

13. Install or connect the following:
- Line bracket on end of steering gear
- Heat shield; torque bolts to 14 ft. lbs. (19 Nm)
- Power steering pipes to steering gear; torque high pressure pipe bolt to 25 ft. lbs. (34 Nm) and low pressure pipe bolt to 30 ft. lbs. (40 Nm)
- Pinch bolt at steering gear to steering column; torque new nut to 16 ft. lbs. (22Nm)
- New self-locking nuts of bottom of stabilizer bar; torque to 41 ft. lbs. (56 Nm)
- Tie rod end ball joint nut to steering knuckle; torque new nut to 38 ft. lbs. (52 Nm)

14. Reset tie rod to steering gear by screwing connection in until paint marks align (this is an initial setting).

15. Bleed the power steering system.

16. Check and adjust alignment.

**Electronic Power Steering**

*See Figure 138.*

1. Before servicing the vehicle, refer to the precautions section.

### ✳✳ WARNING

**Steering gear: Check connection of steering gear for corrosion,**
clean contacts if necessary. The steering gear must be replaced if the corrosion is too far advanced.

### ✳✳ WARNING

**Connecting cable: In the event of moisture/corrosion inside the two plug connections, check the insulation of the connecting cable. If the insulation reveals any noticeable/striking features, it will be necessary to replace the part. Otherwise it will be sufficient to replace the contacts or plug housing.**

2. Disconnect the negative battery cable.

3. Remove both tie rod ends from swivel bearing.

4. Replacement: Remove both tie rod ends from steering gear.

5. Lower front axle support.

6. Release screws.

### ✳✳ WARNING

**Do not allow the control head of the steering gear to strike other components. This may result in damage to the steering gear.**

7. First swing steering gear in direction of travel to right and then remove towards front.

*To install:*

➡For replacement: On cars with 18" tires, it will be necessary to replace steering stop limiters.

➡Deformation elements must point to steering gear.

8. The remainder of installation is the reverse of removal.

9. After installation, check alignment.

10. For replacement: Carry out programming/coding.

11. For models with Dynamic Stability Control (DSC): Carry out steering angle sensor adjustment.

➡Only cars with DSC are fitted with a steering angle sensor (integrated in the steering column switch cluster).

### POWER STEERING PUMP

*REMOVAL & INSTALLATION*

1. Before servicing the vehicle, refer to the precautions in the beginning of this section.

2. Disconnect the negative battery cable.

3. Remove the power steering cooling fan.

4. Drain the power steering fluid.

5. Disconnect the wiring harness connectors.

6. Disconnect the pressure and return lines.

7. Remove the power steering pump and bracket.

8. Installation is the reverse of the removal procedure.

9. Bleed the power steering hydraulic system.

*BLEEDING*

### ✳✳ CAUTION

**Thoroughly clean the reservoir and parts in immediate working area before removing the oil reservoir cap. No dirt must enter the system.**

1. Check power steering fluid level and top up as needed.

2. Start the engine and turn the steering wheel 2 times to the left and right.

3. Stop the engine and check the fluid level. Adjust if needed.

4. Repeat this process if the fluid level went down significantly or if presence of bubbles is still noted.

## SUSPENSION

## FRONT SUSPENSION

### COIL SPRING

*REMOVAL & INSTALLATION*

### ✳ CAUTION

**This procedure calls for the spring to be compressed. A compressed spring has high potential energy and if released suddenly can cause severe damage and personal injury.**

➡**Springs with identical color code must be used in pairs (color code is on the end of the spring coil).**

1. Before servicing the vehicle, refer to the precautions section.
2. Remove the strut from the vehicle and mount in a vise using a strut holder. This will prevent damage to the strut tube
3. Using a proper spring compressor, 31–3–341 with 31–3–355, compress the spring until the lock pins on the coil spring holding tools are heard and felt to lock in place.
4. Only tighten the coil springs until the stress on the thrust bearing is relieved.
5. Remove the retaining nut and the coil spring and strut assembly.
6. Slowly release the compression of the spring.

**To install:**

7. Check the condition of the spring pad and replace it, if necessary.
8. Be sure the spring pad fits over the tongue on the lower spring seat.
9. Check the protective sleeve on top of the upper spring seat. During installation, make sure the tabs of the sleeve fit correctly over the trim.
10. Be sure spring seats are fit from the chamfered side of the special tools and that the lock pins are heard and felt to lock in position.
11. Recheck the fit of the spring seats.
12. There must be 3 spring coils between the spring retainers when properly positioned. The end of the spring must be located under end of spring retainer. Coil spring must lie completely in the recess when tensioned in the spring retainer.
13. Insert the spring strut into the coil spring. Mount the protective sleeve, auxiliary spring, and upper spring seat.
14. Screw the self-locking nut onto the piston rod.
15. Release the coil spring until it is fully resting on the lower spring plate.

➡**The end of the coil spring must be aligned correctly at the rubber seal of the spring seat.**

16. Fully tighten the new self-locking nut to 47 ft. lbs. (64 Nm).
17. Fully release the coil spring and remove the spring compressor tool.
18. Install the strut assembly.

### LOWER BALL JOINT

*REMOVAL & INSTALLATION*

➡**Vehicle uses ball joints on lower control arm and on tie rod ends. Lower ball joints can be separated from steering knuckle after raising front end of vehicle. Complete removal of lower control arm requires removal of subframe.**

### LOWER CONTROL ARM

*REMOVAL & INSTALLATION*

See Subframe.

### MACPHERSON STRUT

*REMOVAL & INSTALLATION*

1. Before servicing the vehicle, refer to the precautions section.
2. Remove or disconnect the following:
   • Tire and wheel assembly
   • ABS cable and brake hose from retainers on strut bracket
   • Brake caliper from disc and tie out of way
   • Nut from stabilizer end.
   • Tie rod end ball joint from steering knuckle
   • Transverse link ball joint from steering knuckle
3. With steering knuckle supported, remove the clamping screw from the lower end of the McPherson strut and detach the steering knuckle from the strut.
4. Remove the 3 upper strut retaining nuts (on top of strut tower).
5. Remove the strut assembly.

**To install:**

6. Install or connect the following:
   • Strut assembly
   • 3 new nuts on strut tower; torque to 25 ft. lbs. (34 Nm)
7. Fit the support bracket into the gap and press the steering knuckle upward until the bolt fits in the bracket hole (lower end of strut). Torque the bolt to 60 ft. lbs. (81 Nm).
8. Remove the steering knuckle support.

9. Install or connect the following:
   • Transverse link to steering knuckle; torque new self-locking nut to 41 ft. lbs. (56 Nm)
   • Tie rod end to steering knuckle; torque new self-locking nut to 38 ft. lbs. (52 Nm)
   • New self-locking nut on stabilizer end; torque to 41 ft. lbs. (56 Nm)
   • Brake caliper; torque guide pin bolts to 23 ft. lbs. (31 Nm)
   • ABS cable and brake hose to strut retainers
   • Front wheel

### STABILIZER BAR

*REMOVAL & INSTALLATION*

See Subframe.

### STEERING KNUCKLE

*REMOVAL & INSTALLATION*

See Wheel Bearings.

### SUBFRAME (AXLE CARRIER)

*REMOVAL & INSTALLATION*

*See Figure 139.*

1. Before servicing the vehicle, refer to the precautions section.
2. Remove or disconnect the following:
   • Front wheels
   • Front bumper and bumper carrier
   • Impact tube
   • Bulkhead nuts and clamp nut holding reservoir tank to bulkhead
   • Tie rod ends from steering knuckle
   • Left and right stabilizer rods from stabilizer bar
   • Lower control arm ball joints from steering knuckle

42348-MINI-G40

**Fig. 139 Illustrating how the power steering reservoir should be held after the subframe is removed**

- Power steering pump electrical plug
- Steering column shaft from steering gear
- Lower engine stabilizer bracket

3. Position a jack at the jacking point on chassis subframe.

4. Remove or disconnect the following:
- Center bolts for left and right side of subframe at vehicle body
- Bolts holding left and right retaining bushing housing on body
- All bolts from rear end of subframe to body

5. Lower the subframe and power steering reservoir through the engine compartment.

6. Remove the cable assembly between the subframe and steering gear (if equipped).

7. Use a 0.16 inch (4mm) rod to hold the power steering reservoir vertically, as shown.

8. Remove or disconnect the following:
- Stabilizer bar
- Lower control arms from subframe
- Retaining clips for lower covering from subframe
- Steering gear heat shield
- Power steering line and power steering gear from subframe
- Power steering pump from subframe
- Stone guard

**To install:**

9. Install or connect the following:
- New retaining clips for ABS wiring harness to subframe
- Stone guard
- Power steering pump to subframe; torque mounting bracket bolts to 14 ft. lbs. (19 Nm)
- Steering gear to subframe; torque mounting bolts to 41 ft. lbs. (56 Nm)

- Power steering line bracket
- Steering gear heat shield
- Bottom covering to subframe clips
- Lower control arms to subframe; torque mounting bolts 74 ft. lbs. (100 Nm)
- Stabilizer bar to subframe; torque mounting bolts to 122 ft. lbs. (165 Nm)

10. Position the subframe and power steering reservoir into position. Install the rear subframe bolts. Torque bolts to 71 ft. lbs. (100 Nm).

11. Install, but do not tighten, bolts holding left and right retaining bushing housing to body.

12. Install the center bolts for the front left and right side subframe mountings to body. Torque bolts to 74 ft. lbs. (100 Nm).

13. Fully tighten the left and right bushing housing bolt to 44 ft. lbs. (59 Nm), plus an additional 90 degrees, then an additional 15 degrees.

14. Remove the jack from the subframe jacking point.

15. Install or connect the following:
- Lower engine stabilizer; torque bolts to 74 ft. lbs. (100 Nm)
- Steering gear to lower end of steering column; tighten new pinch bolt nut to 16 ft. lbs. (22 Nm)
- High voltage connector to power steering pump
- Lower control arm ball joints to steering knuckle; torque new nuts to 41 ft. lbs. (56 Nm)
- Left and right stabilizer rods to stabilizer bar; torque new nuts to 41 ft. lbs. (56 Nm)
- Tie rod end ball joints to steering knuckle; torque new nuts to 38 ft. lbs. (52 Nm)

- Power steering reservoir clamp bolt and mounting bolts to bulkhead; torque nuts to 14 ft. lbs. (19 Nm)
- Impact tube
- Bumper carrier and bumper
- Front wheels

16. Check and adjust wheel alignment.

## WHEEL BEARINGS

### REMOVAL & INSTALLATION

1. Before servicing the vehicle, refer to the precautions section.

2. Remove or disconnect the following:
- Front wheels
- Hub nut
- Brake caliper
- Rotor
- ABS sensor
- Wheel hub from steering knuckle and driveshaft
- Wheel bearings

**To install:**

3. Install or connect the following:
- Wheel bearing into hub
- Wheel hub to steering knuckle and driveshaft; torque bolts to 41 ft. lbs. (56 Nm)
- ABS sensor; torque to 6 ft. lbs. (8 Nm)
- Brake rotor and caliper
- New flanged hub nut; torque to 134 ft. lbs. (182 Nm)

4. Stake flanged nut into groove of thread on driveshaft.

### ADJUSTMENT

Wheel bearings cannot be adjusted and must be replaced as a unit and never be reused once removed.

## SUSPENSION

### COIL SPRING

#### REMOVAL & INSTALLATION

#### ※※ CAUTION

**This procedure calls for the spring to be compressed. A compressed spring has high potential energy and if released suddenly can cause severe damage and personal injury.**

➡**Springs with identical color code must be used in pairs (color code is on the end of the spring coil).**

1. Before servicing the vehicle, refer to the precautions section.

2. Remove the strut from the vehicle and mount in a vise using a strut holder. This will prevent damage to the strut tube.

3. Using a proper spring compressor, 31–3–341 with 31–3–355, compress the spring until the lock pins on the coil spring holding tools are heard and felt to lock in place.

4. Only tighten the coil springs until the stress on the thrust bearing is relieved.

5. Remove the retaining nut and the coil spring and strut assembly.

6. Slowly release the compression of the spring.

## REAR SUSPENSION

**To install:**

7. Check the condition of the spring pad and replace it, if necessary.

8. Be sure the spring pad fits over the tongue on the lower spring seat.

9. Check the protective sleeve on top of the upper spring seat. During installation, make sure the tabs of the sleeve fit correctly over the trim.

10. Be sure spring seats are fit from the chamfered side of the special tools and that the lock pins are heard and felt to lock in position.

11. Recheck the fit of the spring seats.

12. There must be 3 spring coils between the spring retainers when properly

positioned. The end of the spring must be located under end of spring retainer. Coil spring must lie completely in the recess when tensioned in the spring retainer.

13. Insert the spring strut into the coil spring. Mount the protective sleeve, auxiliary spring, and upper spring seat.

14. Screw the self-locking nut onto the piston rod.

15. Release the coil spring until it is fully resting on the lower spring plate.

➡**The end of the coil spring must be aligned correctly at the rubber seal of the spring seat.**

16. Fully tighten the new self-locking nut to 47 ft. lbs. (64 Nm).

17. Fully release the coil spring and remove the spring compressor tool.

18. Install the strut assembly.

## LOWER CONTROL ARM

### REMOVAL & INSTALLATION

*See Figure 140.*

1. Before servicing the vehicle, refer to the precautions section.

2. Remove rear wheel. If necessary, remove spare wheel.

3. Remove underbody paneling.

4. Only on right side: Remove complete exhaust system and heat shield.

➡**To obtain balanced vehicle handling, it is recommended that the control arms be replaced in pairs.**

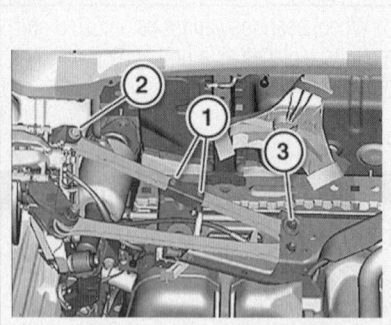

22205_MINI_G0078

**Fig. 140 Upper and lower control arms**

5. Mark position of eccentric adjustment washer to trailing arm to simplify subsequent adjustment of rear axle.

6. Release screw.

7. Loosen nut.

8. Remove screw and then remove control arm.

### *To install:*

9. Installation is the reverse of removal. Tighten the bolt and the self-locking nuts to 74 ft. lbs. (100 Nm).

10. Check alignment.

## STRUT

### REMOVAL & INSTALLATION

1. Before servicing the vehicle, refer to the precautions section.

2. Place a jack or other support to relieve the load on the trailing arm.

3. Remove the lower strut mounting bolt.

4. Release the ABS sensor and brake hose from the retainers on the strut.

5. Remove the 2 upper retaining bolts for the strut.

6. Support the strut and remove it from the vehicle.

### *To install:*

7. Position the strut and install the upper mounting bolts. Torque the bolts to 41 ft. lbs. (56 Nm).

8. Be sure that the rubber grommets for the ABS sensor and brake hose are correctly installed.

9. Install the ABS sensor and brake hose to the strut retainers.

10. Install the lower strut bolt and torque to 103 ft. lbs. (140 Nm).

11. Remove the jack from under the trailing arm.

## UPPER CONTROL ARM

### REMOVAL & INSTALLATION

*See Figure 140.*

1. Before servicing the vehicle, refer to the precautions section.

2. Remove rear wheel. If necessary, remove spare wheel.

3. Remove underbody paneling.

4. Only on right side: Remove heat shield (on COOPER S on left and right).

➡**To obtain balanced vehicle handling, it is recommended that the control arms be replaced in pairs.**

5. Release screw.

6. Release nut and remove screw.

7. Unscrew nut.

8. Remove screw and then remove control arm.

### *To install:*

9. Installation is the reverse of removal. Tighten the self-locking nuts to 74 ft. lbs. (100 Nm).

10. Check alignment.

## WHEEL BEARINGS

### REMOVAL & INSTALLATION

1. Before servicing the vehicle, refer to the precautions section.

2. Remove or disconnect the following:
   - Wheel assembly
   - Brake caliper (tie to body without strain on hose)
   - ABS sensor from trailing arm
   - Stabilizer bar, if equipped
   - Brake rotor
   - Wheel hub from trailing arm
   - Wheel bearing from hub

### *To install:*

3. Install or connect the following:
   - Wheel bearing
   - Wheel hub to trailing arm; torque bolts to 41 ft. lbs. (56 Nm)
   - Brake rotor; torque bolts to 20 ft. lbs. (27 Nm)
   - Stabilizer bar, if equipped; torque new nuts to 41 ft. lbs. (56 Nm)
   - ABS sensor
   - Brake caliper; torque guide pin bolts to 48 ft. lbs. (65 Nm)
   - Wheel assembly

### ADJUSTMENT

Wheel bearings cannot be adjusted and must be replaced as a unit and never be reused once removed.

# MINI

## Diagnostic Trouble Codes

# 10

## DIAGNOSTIC TROUBLE CODES

### OBD II VEHICLE APPLICATIONS

*MINI*

**Cooper**
2007–2008
- R56 Series, N12 Engine ........ RC33

**Cooper**
2007–2008
- R56 Series, N14 Engine ........ RE33

## Gas Engine OBD II Trouble Code List (P0xxx Codes)

| DTC | Trouble Code Title, Conditions & Possible Causes |
|---|---|
| **DTC: P0030**<br>**2T CCM, MIL: Yes**<br>**Years:** 2007, 2008<br>**Models:** Cooper<br>**Engines:** 1.6L, 1.6L Turbo<br>**Transmissions:** All | **HO2S Heater (Bank 1 Sensor 1) Control Circuit Malfunction Conditions:**<br>Engine started, battery voltage must be at least 11.5v, all electrical components must be off, the ground between the engine and the chassis must be well connected, the exhaust system must be properly sealed between the catalytic converter and the cylinder head, the coolant temperature must be 80 degrees Celsius, and the oxygen sensor heater for oxygen sensor before the catalytic converter must be properly functioning. The DME detected the HO2S signal was in a negative voltage range referred to as "character shift downward". This code sets when the HO2S signal remains in a low state (usually less than 156 mv). In effect, it does not switch properly between 0.1v and 1.1v in closed loop operation.<br>**Possible Causes:**<br>• HO2S is contaminated (due to presence of silicone in fuel)<br>• HO2S signal and ground circuit wires crossed in wiring harness<br>• HO2S signal circuit is shorted to sensor or chassis ground<br>• HO2S element has failed (internal short condition) |
| **DTC: P0031**<br>**2T CCM, MIL: Yes**<br>**Years:** 2007, 2008<br>**Models:** Cooper<br>**Engines:** 1.6L, 1.6L Turbo<br>**Transmissions:** All | **HO2S Heater (Bank 1 Sensor 1) Circuit Low Input Conditions:**<br>Engine started, battery voltage must be at least 11.5v, all electrical components must be off, the ground between the engine and the chassis must be well connected, the exhaust system must be properly sealed between the catalytic converter and the cylinder head, the coolant temperature must be 80 degrees Celsius, and the oxygen sensor heater for oxygen sensor before the catalytic converter must be properly functioning. The DME detected the HO2S signal was in a negative voltage range referred to as "character shift downward". This code sets when the HO2S signal remains in a low state. In effect, it does not switch properly in the closed loop operation. The HO2S (before the three-way catalytic converter) has a short circuit to ground that has lasted longer than 200 seconds.<br>**Possible Causes:**<br>• HO2S is contaminated (due to presence of silicone in fuel)<br>• HO2S signal and ground circuit wires crossed in wiring harness<br>• HO2S signal circuit is shorted to sensor or chassis ground<br>• HO2S element has failed (internal short condition) |
| **DTC: P0032**<br>**2T CCM, MIL: Yes**<br>**Years:** 2007, 2008<br>**Models:** Cooper<br>**Engines:** 1.6L, 1.6L Turbo<br>**Transmissions:** All | **HO2S Heater (Bank 1 Sensor 1) Circuit High Input Conditions:**<br>Engine started, battery voltage must be at least 11.5v, all electrical components must be off, the ground between the engine and the chassis must be well connected, the exhaust system must be properly sealed between the catalytic converter and the cylinder head, the coolant temperature must be 80 degrees Celsius, and the oxygen sensor heater for oxygen sensor before the catalytic converter must be properly functioning. The DME detected the HO2S signal remained in a high state.<br>**Note: The HO2S signal circuit may be shorted to the heater power circuit due to tracking inside of the HO2S connector. Remove the connector and visually inspect the connector for signs of oil or water.**<br>**Possible Causes:**<br>• HO2S signal shorted to heater power circuit inside connector<br>• HO2S signal circuit shorted to ground or to system voltage |
| **DTC: P0036**<br>**2T CCM, MIL: Yes**<br>**Years:** 2007, 2008<br>**Models:** Cooper<br>**Engines:** 1.6L, 1.6L Turbo<br>**Transmissions:** All | **HO2S Heater (Bank 1 Sensor 2) Control Circuit Malfunction Conditions:**<br>Engine started, battery voltage must be at least 11.5v, all electrical components must be off, the ground between the engine and the chassis must be well connected, the exhaust system must be properly sealed between the catalytic converter and the cylinder head, the coolant temperature must be 80 degrees Celsius, and the oxygen sensor heater for oxygen sensor before the catalytic converter must be properly functioning. The DME detected the HO2S signal was in a negative voltage range referred to as "character shift downward". This code sets when the HO2S signal remains in a low state.<br>**Possible Causes:**<br>• HO2S is contaminated (due to presence of silicone in fuel)<br>• HO2S signal and ground circuit wires crossed in wiring harness<br>• HO2S signal circuit is shorted to sensor or chassis ground<br>• HO2S element has failed (internal short condition) |
| **DTC: P0037**<br>**2T CCM, MIL: Yes**<br>**Years:** 2007, 2008<br>**Models:** Cooper<br>**Engines:** 1.6L, 1.6L Turbo<br>**Transmissions:** All | **HO2S Heater (Bank 1 Sensor 2) Circuit Low Input Conditions:**<br>Engine started, battery voltage must be at least 11.5v, all electrical components must be off, the ground between the engine and the chassis must be well connected, the exhaust system must be properly sealed between the catalytic converter and the cylinder head, the coolant temperature must be 80 degrees Celsius, and the oxygen sensor heater for oxygen sensor before the catalytic converter must be properly functioning. The DME detected the HO2S signal was in a negative voltage range referred to as "character shift downward". This code sets when the HO2S signal remains in a low state. In effect, it does not switch properly in the closed loop operation. The HO2S (before the three-way catalytic converter) has a short circuit to ground that has lasted longer than 200 seconds.<br>**Possible Causes:**<br>• HO2S is contaminated (due to presence of silicone in fuel)<br>• HO2S signal and ground circuit wires crossed in wiring harness<br>• HO2S signal circuit is shorted to sensor or chassis ground<br>• HO2S element has failed (internal short condition) |

| DTC | Trouble Code Title, Conditions & Possible Causes |
|---|---|
| **DTC: P0038**<br>**2T CCM, MIL: Yes**<br>**Years:** 2007, 2008<br>**Models:** Cooper<br>**Engines:** 1.6L, 1.6L Turbo<br>**Transmissions:** 1.6L, 1.6L Turbo | **HO2S Heater (Bank 1 Sensor 2) Circuit High Input Conditions:**<br>Engine started, battery voltage must be at least 11.5v, all electrical components must be off, the ground between the engine and the chassis must be well connected, the exhaust system must be properly sealed between the catalytic converter and the cylinder head, the coolant temperature must be 80 degrees Celsius, and the oxygen sensor heater for oxygen sensor before the catalytic converter must be properly functioning. The DME detected the HO2S signal remained in a high state.<br>**Note: The HO2S signal circuit may be shorted to the heater power circuit due to tracking inside of the HO2S connector. Remove the connector and visually inspect the connector for signs of oil or water.**<br>**Possible Causes:**<br>&bull; HO2S signal shorted to heater power circuit inside connector<br>&bull; HO2S signal circuit shorted to ground or to system voltage |
| **DTC: P0053**<br>**2T CCM, MIL: Yes**<br>**Years:** 2007, 2008<br>**Models:** Cooper<br>**Engines:** 1.6L, 1.6L Turbo<br>**Transmissions:** All | **HO2S Heater (Bank 1 Sensor 1) Control Circuit Malfunction Conditions:**<br>Engine started, battery voltage must be at least 10.96v, all electrical components must be off, the ground between the engine and the chassis must be well connected, the exhaust system must be properly sealed between the catalytic converter and the cylinder head, and the coolant temperature must be 80 degrees Celsius. The DME detected the HO2S signal was in a negative voltage range referred to as "character shift downward". The resistance is out of limits. The engine speed is less than 7008rpm (6208 for A/T) and the exhaust temperatures are between 350.006 and 649.995 degrees Celsius.<br>**Possible Causes:**<br>&bull; HO2S is contaminated (due to presence of silicone in fuel)<br>&bull; HO2S signal and ground circuit wires crossed in wiring harness<br>&bull; HO2S signal circuit is shorted to sensor or chassis ground<br>&bull; HO2S element has failed (internal short condition) |
| **DTC: P0054**<br>**2T CCM, MIL: Yes**<br>**Years:** 2007, 2008<br>**Models:** Cooper<br>**Engines:** 1.6L, 1.6L Turbo<br>**Transmissions:** All | **HO2S Heater (Bank 1 Sensor 2) Circuit High Input Conditions:**<br>Engine started, battery voltage must be at least 11.5v, all electrical components must be off, the ground between the engine and the chassis must be well connected, the exhaust system must be properly sealed between the catalytic converter and the cylinder head, the coolant temperature must be 80 degrees Celsius, and the oxygen sensor heater for oxygen sensor before the catalytic converter must be properly functioning. The DME detected the HO2S signal remained in a high state. The resistance is out of limits. The engine speed is less than 7008rpm (6208 for A/T) and the exhaust temperatures are between 350.006 and 649.995 degrees Celsius.<br>**Note: The HO2S signal circuit may be shorted to the heater power circuit due to tracking inside of the HO2S connector. Remove the connector and visually inspect the connector for signs of oil or water.**<br>**Possible Causes:**<br>&bull; HO2S signal shorted to heater power circuit inside connector<br>&bull; HO2S signal circuit shorted to ground or to system voltage |
| **DTC: P0070**<br>**2T CCM, MIL: Yes**<br>**Years:** 2007, 2008<br>**Models:** Cooper<br>**Engines:** 1.6L, 1.6L Turbo<br>**Transmissions:** All | **Ambient Air Temperature Sensor Malfunction Conditions:**<br>Key on or engine running (at over 800rpm), the vehicle velocity is over 25mph for 26 seconds, the ambient temperature is 20 degrees above or below the model figure for four seconds. This is a thermistor-type sensor with a variable resistance that changes when exposed to different temperatures. This means: the higher the temperature, the lower the resistance value.<br>**Possible Causes:**<br>&bull; IAT sensor signal circuit is grounded (check wiring & connector)<br>&bull; Resistance value between sockets 33 and 36 out of range<br>&bull; IAT sensor has an open circuit<br>&bull; IAT sensor is damaged or it has failed<br>&bull; Ambient temperature sensor at the cluster is defective |
| **DTC: P0106**<br>**2T CCM, MIL: Yes**<br>**Years:** 2007, 2008<br>**Models:** Cooper<br>**Engines:** 1.6L, 1.6L Turbo<br>**Transmissions:** All | **Manifold Pressure Sensor Circuit High Conditions:**<br>Engine started, battery voltage must be at least 11v, and the differential pressure sensor detected a control deviation at the minimum limit. The closed loop control of the differential pressure in the intake manifold is suspended and replaced by a direct specification.<br>**Possible Causes:**<br>&bull; Sensor's voltage supply on Terminal 87<br>&bull; Sensor's ground connection faulty<br>&bull; Signal wire to DME faulty<br>&bull; Replace sensor |
| **DTC: P0107**<br>**2T CCM, MIL: Yes**<br>**Years:** 2007, 2008<br>**Models:** Cooper<br>**Engines:** 1.6L, 1.6L Turbo<br>**Transmissions:** All | **Manifold Pressure Sensor Circuit Low Conditions:**<br>Engine started, battery voltage must be at least 11v, and the differential pressure sensor detected a control deviation at the minimum limit. The closed loop control of the differential pressure in the intake manifold is suspended and replaced by a direct specification. The MAP was too low (less than 105.0016kPa); engine stopped.<br>**Possible Causes:**<br>&bull; Sensor's voltage supply on Terminal 87<br>&bull; Sensor's ground connection faulty<br>&bull; Signal wire to DME faulty<br>&bull; Replace sensor |

| DTC | Trouble Code Title, Conditions & Possible Causes |
|---|---|
| **DTC: P0108**<br>**2T CCM, MIL: Yes**<br>**Years:** 2007, 2008<br>**Models:** Cooper<br>**Engines:** 1.6L, 1.6L Turbo<br>**Transmissions:** All | **Manifold Pressure Sensor Circuit Short to Battery Conditions:**<br>Engine started, battery voltage must be at least 11v, and the differential pressure sensor detected a control deviation at the minimum limit. The closed loop control of the differential pressure in the intake manifold is suspended and replaced by a direct specification. The MAP was too low (less than 105.0016kPa); engine stopped.<br>**Possible Causes:**<br>• Sensor's voltage supply on Terminal 87<br>• Sensor's ground connection faulty<br>• Signal wire to DME faulty<br>• Replace sensor |
| **DTC: P0112**<br>**2T CCM, MIL: Yes**<br>**Years:** 2007, 2008<br>**Models:** Cooper<br>**Engines:** 1.6L, 1.6L Turbo<br>**Transmissions:** All | **Intake Air Temperature Sensor Circuit Low Input Conditions:**<br>Key on or Engine running, the temperature must beat least 185-degrees (F) and all electrical equipment (A/C, lights, etc) must be off; and the DME detected the IAT sensor signal was less than the self-test minimum. This is a thermistor-type sensor with a variable resistance that changes when exposed to different temperatures. This means: the higher the temperature, the lower the resistance value.<br>**Possible Causes:**<br>• IAT sensor signal circuit is grounded (check wiring & connector)<br>• Resistance value between sockets 33 and 36 out of range<br>• IAT sensor has an open circuit<br>• IAT sensor is damaged or it has failed |
| **DTC: P0113**<br>**2T CCM, MIL: Yes**<br>**Years:** 2007, 2008<br>**Models:** Cooper<br>**Engines:** 1.6L, 1.6L Turbo<br>**Transmissions:** All | **Intake Air Temperature Sensor Circuit High Input Conditions:**<br>Key on or engine running, the temperature must beat least 185-degrees (F) and all electrical equipment (A/C, lights, etc) must be off; and the DME detected the IAT sensor signal was more than the self-test maximum. This is a thermistor-type sensor with a variable resistance that changes when exposed to different temperatures. This means: the higher the temperature, the lower the resistance value.<br>**Possible Causes:**<br>• IAT sensor signal circuit is open (inspect wiring & connector)<br>• IAT sensor signal circuit is shorted<br>• Resistance value between sockets 33 and 36 out of range<br>• IAT sensor is damaged or it has failed |
| **DTC: P0114**<br>**2T CCM, MIL: Yes**<br>**Years:** 2007, 2008<br>**Models:** Cooper<br>**Engines:** 1.6L, 1.6L Turbo<br>**Transmissions:** All | **Intake Air Temperature Sensor Circuit Intermittent Failure Conditions:**<br>Key on or engine running, the temperature must beat least 185-degrees (F) and all electrical equipment (A/C, lights, etc) must be off; and the DME detected the IAT sensor signal was more than the self-test maximum. This is a thermistor-type sensor with a variable resistance that changes when exposed to different temperatures. This means: the higher the temperature, the lower the resistance value. The gradient between filtered and current intake air sensor values exceeds 9.75 degrees Celsius.<br>**Possible Causes:**<br>• IAT sensor signal circuit is open (inspect wiring & connector)<br>• IAT sensor signal circuit is shorted<br>• Resistance value between sockets 33 and 36 out of range<br>• IAT sensor is damaged or it has failed |
| **DTC: P0116**<br>**2T CCM, MIL: Yes**<br>**Years:** 2007, 2008<br>**Models:** Cooper<br>**Engines:** 1.6L, 1.6L Turbo<br>**Transmissions:** All | **ECT Sensor Circuit Range/Performance:**<br>Engine started (cold) for 10 seconds, battery voltage must be 11.5, and all equipment must be off. The DME detected the ECT sensor signal was less than the self-test minimum. This is a thermistor-type sensor with a variable resistance that changes when exposed to different temperatures<br>**Possible Causes:**<br>• ECT sensor signal circuit is grounded in the wiring harness<br>• ECT sensor doesn't react to changes in temperature<br>• ECT sensor is damaged or the DME has failed |
| **DTC: P0117**<br>**2T CCM, MIL: Yes**<br>**Years:** 2007, 2008<br>**Models:** Cooper<br>**Engines:** 1.6L, 1.6L Turbo<br>**Transmissions:** All | **ECT Sensor Circuit Low Input Conditions:**<br>Engine started (cold) for 10 seconds, battery voltage must be 11.5, and all equipment must be off. The DME detected the ECT sensor signal was less than the self-test minimum. This is a thermistor-type sensor with a variable resistance that changes when exposed to different temperatures<br>**Possible Causes:**<br>• ECT sensor signal circuit is grounded in the wiring harness<br>• ECT sensor doesn't react to changes in temperature<br>• ECT sensor is damaged or the DME has failed |

| DTC | Trouble Code Title, Conditions & Possible Causes |
|---|---|
| **DTC: P0118**<br>**2T CCM, MIL: Yes**<br>**Years:** 2007, 2008<br>**Models:** Cooper<br>**Engines:** 1.6L, 1.6L Turbo<br>**Transmissions:** All | **ECT Sensor Circuit High Input Conditions:**<br>Engine started (cold) for 10 seconds, battery voltage must be 11.5, and all equipment must be off. The DME detected the ECT sensor signal was more than the self-test maximum. This is a thermistor-type sensor with a variable resistance that changes when exposed to different temperatures<br>**Possible Causes:**<br>   &bull; ECT sensor signal circuit is open (inspect wiring & connector)<br>   &bull; ECT sensor signal circuit is shorted to ground<br>   &bull; ECT sensor is damaged or it has failed |
| **DTC: P0119**<br>**2T CCM, MIL: Yes**<br>**Years:** 2007, 2008<br>**Models:** Cooper<br>**Engines:** 1.6L, 1.6L Turbo<br>**Transmissions:** All | **ECT Sensor Circuit Continuity Conditions:**<br>Engine started (cold) for 10 seconds, battery voltage must be 11.5, and all equipment must be off. The DME detected the ECT sensor signal was out of the specified range. This is a thermistor-type sensor with a variable resistance that changes when exposed to different temperatures<br>**Possible Causes:**<br>   &bull; ECT sensor signal circuit is open (inspect wiring & connector)<br>   &bull; ECT sensor signal circuit is shorted to ground<br>   &bull; ECT sensor is damaged or it has failed |
| **DTC: P0122**<br>**2T CCM, MIL: Yes**<br>**Years:** 2007, 2008<br>**Models:** Cooper<br>**Engines:** 1.6L, 1.6L Turbo<br>**Transmissions:** All | **Throttle/Pedal Position Sensor Circuit Low Input Conditions:**<br>Engine started, at idle, the temperature must be at least 80 degrees Celsius. The throttle position sensor supplies implausible signal to the DME.<br>**Possible Causes:**<br>   &bull; TP sensor signal circuit open (inspect wiring & connector)<br>   &bull; TP sensor signal shorted to ground (inspect wiring & connector)<br>   &bull; TP sensor is damaged or has failed<br>   &bull; Throttle control module's voltage supply is shorted or open |
| **DTC: P0123**<br>**2T CCM, MIL: Yes**<br>**Years:** 2007, 2008<br>**Models:** Cooper<br>**Engines:** 1.6L, 1.6L Turbo<br>**Transmissions:** All | **TP Sensor Circuit High Input Conditions:**<br>Engine started, at idle, the temperature must be at least 80 degrees Celsius. The DME detected the TP sensor signal was more than the self-test maximum during testing.<br>**Possible Causes:**<br>   &bull; TP sensor not seated correctly in housing (may be damaged)<br>   &bull; TP sensor signal is circuit shorted to ground or system voltage<br>   &bull; TP sensor ground circuit is open (check the wiring harness)<br>   &bull; TP sensor and/or DME has failed |
| **DTC: P0125**<br>**2T CCM, MIL: Yes**<br>**Years:** 2007, 2008<br>**Models:** Cooper<br>**Engines:** 1.6L, 1.6L Turbo<br>**Transmissions:** All | **ECT Sensor Insufficient for Closed Loop Fuel Control Conditions:**<br>Engine started (cold), battery voltage must be 11.5, and all equipment must be off. The DME detected the ECT sensor exceeded the required calibrated value, or the engine is at idle and doesn't reach operating temperature quickly enough; the Catalyst, Fuel System, HO2S and Misfire Monitor did not complete, or the timer expired. Testing completion of procedure, the engine's temperature must rise uniformly during idle.<br>**Possible Causes:**<br>   &bull; Check for low coolant level or incorrect coolant mixture<br>   &bull; DME detects a short circuit wiring in the ECT<br>   &bull; CHT sensor is out-of-calibration or it has failed<br>   &bull; ECT sensor is out-of-calibration or it has failed |
| **DTC: P0128**<br>**2T CCM, MIL: Yes**<br>**Years:** 2007, 2008<br>**Models:** Cooper<br>**Engines:** 1.6L, 1.6L Turbo<br>**Transmissions:** All | **Coolant Thermostat (Coolant Temperature Below Thermostat Regulating Temperature) Conditions:**<br>The engine's warm up performance is monitored by comparing measured coolant temperature with the modeled coolant temperature to detect a defective coolant thermostat. The engine temperature must be less than 65 degrees Celsius, engine speed greater than 800rpm (with the vehicle speed greater than 10 but less than 90km/h) and the ambient temperature greater than −8 degrees Celsius. The thermostat should be wide open when cold, but is in error if it opens below desired control temperature.<br>**Possible Causes:**<br>   &bull; Check for low coolant level or incorrect coolant mixture<br>   &bull; DME detects a short circuit wiring in the ECT<br>   &bull; CHT sensor is out-of-calibration or it has failed<br>   &bull; ECT sensor is out-of-calibration or it has failed<br>   &bull; Replace the thermostat |

| DTC | Trouble Code Title, Conditions & Possible Causes |
|---|---|
| **DTC: P0130**<br>**2T CCM, MIL: Yes**<br>**Years:** 2007, 2008<br>**Models:** Cooper<br>**Engines:** 1.6L, 1.6L Turbo<br>**Transmissions:** All | **O2 Sensor Circuit Bank 1 Sensor 1 Conditions:**<br>Engine running, battery voltage 11.5, all electrical components off, ground between engine and chassis well connected and the exhaust system must be properly sealed between catalytic converter and the cylinder head. The DME detected the HO2S signal was implausible or not detected. The engine speed is less than 8000 rpm.<br>**Possible Causes:**<br>• Oxygen sensor heater for oxygen sensor (HO2S) before catalytic converter is faulty<br>• HO2S is contaminated (due to presence of silicone in fuel)<br>• HO2S signal and ground circuit wires crossed in wiring harness<br>• HO2S signal circuit is shorted to sensor or chassis ground<br>• HO2S element before the catalytic converter has failed (internal short condition)<br>• Leaks present in the exhaust manifold or exhaust pipes |
| **DTC: P0131**<br>**2T CCM, MIL: Yes**<br>**Years:** 2007, 2008<br>**Models:** Cooper<br>**Engines:** 1.6L, 1.6L Turbo<br>**Transmissions:** All | **HO2S (Bank 1 Sensor 1) Circuit Low Input Conditions:**<br>Engine running, battery voltage 11.5, all electrical components off, ground between engine and chassis well connected and the exhaust system must be properly sealed between catalytic converter and the cylinder head. The DME detected the HO2S signal was in a negative voltage range referred to as "character shift downward". This code sets when the HO2S signal remains in a low state for a measured period of time. In effect, it does not switch properly in the closed loop operation. Engine speed is less than 8000rpm.<br>**Possible Causes:**<br>• HO2S is contaminated (due to presence of silicone in fuel)<br>• HO2S signal and ground circuit wires crossed in wiring harness<br>• HO2S signal circuit is shorted to sensor or chassis ground<br>• HO2S element has failed (internal short condition)<br>• Leaks present in the exhaust manifold or exhaust pipes |
| **DTC: P0132**<br>**2T CCM, MIL: Yes**<br>**Years:** 2007, 2008<br>**Models:** Cooper<br>**Engines:** 1.6L, 1.6L Turbo<br>**Transmissions:** All | **HO2S (Bank 1 Sensor 1) Circuit High Input Conditions:**<br>Engine running, battery voltage 11.5, all electrical components off, ground between engine and chassis well connected and the exhaust system must be properly sealed between catalytic converter and the cylinder head. The DME detected the HO2S signal was in a high state. This code sets when the HO2S signal remains in a high state for a measured period of time. In effect, it does not switch properly in the closed loop operation.<br>**Note: The HO2S signal circuit may be shorted to the heater power circuit due to tracking inside of the HO2S connector. Remove the connector and visually inspect the connector for signs of oil or water.**<br>**Possible Causes:**<br>• HO2S is contaminated (due to presence of silicone in fuel)<br>• HO2S signal and ground circuit wires crossed in wiring harness<br>• HO2S signal circuit is shorted to sensor or chassis ground<br>• HO2S element has failed (internal short condition)<br>• Leaks present in the exhaust manifold or exhaust pipes |
| **DTC: P0133**<br>**2T CCM, MIL: Yes**<br>**Years:** 2007, 2008<br>**Models:** Cooper<br>**Engines:** 1.6L, 1.6L Turbo<br>**Transmissions:** All | **HO2S (Bank 1 Sensor 1) Circuit Slow Response Conditions:**<br>Engine running, battery voltage 11.5, all electrical components off, ground between engine and chassis well connected and the exhaust system must be properly sealed between catalytic converter and the cylinder head. The DME detected the HO2S amplitude and frequency were out of the normal range (e.g., the HO2S rich to lean switch) during the HO2S Monitor test. The engine speed is 1984 to 3488rpm (1888 to 3296 for A/T), the coolant temperature is greater than 80.25 degrees Celsius and the vehicle speed is between 24.85 and 68.35mph. The ambient pressure is greater than 75.00114kPa.<br>**Possible Causes:**<br>• HO2S before the three-way catalytic converter is contaminated (due to presence of silicone in fuel); Run the engine for three minutes at 3500rpm as a self-cleaning effect<br>• HO2S signal circuit open<br>• Leaks present in the exhaust manifold or exhaust pipes<br>• HO2S is damaged or has failed |
| **DTC: P0135**<br>**2T CCM, MIL: Yes**<br>**Years:** 2007, 2008<br>**Models:** Cooper<br>**Engines:** All<br>**Transmissions:** All | **HO2S (Bank 1 Sensor 1) Heater Circuit Malfunction Conditions:**<br>Engine running, battery voltage is between 11 and 16 volts, all electrical components off, ground between engine and chassis well connected and the exhaust system must be properly sealed between catalytic converter and the cylinder head. The DME detected an unexpected voltage condition, or it detected excessive current draw in the heater circuit during the CCM test. The engine load is 25 to 160kg/h. The exhaust gas temperature is between 450 and 700 degrees Celsius.<br>**Possible Causes:**<br>• HO2S heater power circuit is open or heater ground circuit open<br>• HO2S signal tracking (due to oil or moisture in the connector)<br>• HO2S is damaged or has failed |

| DTC | Trouble Code Title, Conditions & Possible Causes |
|---|---|
| **DTC: P0136**<br>**2T CCM, MIL: Yes**<br>**Years:** 2007, 2008<br>**Models:** Cooper<br>**Engines:** 1.6L, 1.6L Turbo<br>**Transmissions:** All | **HO2S (Bank 1 Sensor 2) Circuit Malfunction Conditions:**<br>Engine running, battery voltage 11.5, all electrical components off, ground between engine and chassis well connected and the exhaust system must be properly sealed between catalytic converter and the cylinder head. The DME detected the HO2S signal failed to meet the maximum or minimum voltage levels (i.e., it failed the voltage range check). The heater has been on for less than 90 seconds, the fuel system status is in fuel cut-off, the output voltage is between 400mV and 500mV and it is 120 seconds after engine start up. The engine speed is less than 8000rpm.<br>**Possible Causes:**<br>• Leaks present in the exhaust manifold or exhaust pipes<br>• HO2S signal wire and ground wire crossed in connector<br>• HO2S element is fuel contaminated or has failed |
| **DTC: P0137**<br>**2T CCM, MIL: Yes**<br>**Years:** 2007, 2008<br>**Models:** Cooper<br>**Engines:** 1.6L, 1.6L Turbo<br>**Transmissions:** All | **HO2S (Bank 1 Sensor 2) Circuit Low Input Conditions:**<br>Engine running, battery voltage 11.5, all electrical components off, ground between engine and chassis well connected and the exhaust system must be properly sealed between catalytic converter and the cylinder head. The DME detected the HO2S signal remained in a high state.<br>**Note: The HO2S signal circuit may be shorted to the heater power circuit due to "tracking inside of the HO2S connector. Remove the connector and visually inspect the connector for signs of oil or water.**<br>**Possible Causes:**<br>• HO2S signal shorted to heater power circuit in the connector<br>• HO2S signal circuit shorted to ground (for more than 200 seconds) or to system voltage |
| **DTC: P0138**<br>**2T CCM, MIL: Yes**<br>**Years:** 2007, 2008<br>**Models:** Cooper<br>**Engines:** 1.6L, 1.6L Turbo<br>**Transmissions:** All | **HO2S (Bank 1 Sensor 2) Circuit High Input Conditions:**<br>Engine running, battery voltage 11.5, all electrical components off, ground between engine and chassis well connected and the exhaust system must be properly sealed between catalytic converter and the cylinder head. The DME detected the HO2S signal remained in a high state.<br>**Note: The HO2S signal circuit may be shorted to the heater power circuit due to "tracking inside of the HO2S connector. Remove the connector and visually inspect the connector for signs of oil or water.**<br>**Possible Causes:**<br>• HO2S signal shorted to heater power circuit in the positive connector<br>• HO2S signal circuit shorted to ground or to system voltage<br>• HO2S has failed |
| **DTC: P0141**<br>**2T CCM, MIL: Yes**<br>**Years:** 2007, 2008<br>**Models:** Cooper<br>**Engines:** 1.6L, 1.6L Turbo<br>**Transmissions:** All | **HO2S (Bank 1 Sensor 2) Malfunction Conditions:**<br>Engine running, battery voltage 11.5, all electrical components off, ground between engine and chassis well connected and the exhaust system must be properly sealed between catalytic converter and the cylinder head. The DME detected the HO2S signal failed to meet the maximum or minimum voltage levels (i.e., it failed the voltage range check). The engine speed is greater than 40rpm, the battery voltage must be between 10.7 and 15.5 volts, and the fault occurs 200 seconds after engine start up.<br>**Possible Causes:**<br>• Leaks present in the exhaust manifold or exhaust pipes<br>• HO2S signal wire and ground wire crossed in connector<br>• HO2S element is fuel contaminated or has failed |
| **DTC: P0153**<br>**2T CCM, MIL: Yes**<br>**Years:** 2007, 2008<br>**Models:** Cooper<br>**Engines:** 1.6L, 1.6L Turbo<br>**Transmissions:** All | **HO2S (Bank 2 Sensor 1) Circuit Slow Response Conditions:**<br>Engine running, battery voltage 11.5, all electrical components off, ground between engine and chassis well connected and the exhaust system must be properly sealed between catalytic converter and the cylinder head. The DME detected the HO2S amplitude and frequency were out of the normal range during the HO2S Monitor test. For the 1999 M62: The idle speed variation is between 1400 and 2600rpm, the engine load variation is between 20 and 54 while the catalyst temperature should be greater than 360 degrees Celsius.<br>**Possible Causes:**<br>• HO2S is contaminated (due to presence of silicone in fuel)<br>• Leaks present in the exhaust manifold or exhaust pipes<br>• HO2S is damaged or has failed |
| **DTC: P0154**<br>**2T CCM, MIL: Yes**<br>**Years:** 2007, 2008<br>**Models:** Cooper<br>**Engines:** 1.6L, 1.6L Turbo<br>**Transmissions:** All | **HO2S (Bank 2 Sensor 1) Circuit No Activity Conditions:**<br>Engine running, battery voltage 11.5, all electrical components off, ground between engine and chassis well connected and the exhaust system must be properly sealed between catalytic converter and the cylinder head. The DME detected the HO2S signal failed to meet the maximum or minimum voltage (i.e., it failed the voltage check).<br>**Possible Causes:**<br>• Leaks present in the exhaust manifold or exhaust pipes<br>• HO2S signal wire and ground wire crossed in connector<br>• HO2S element is fuel contaminated or has failed |

| DTC | Trouble Code Title, Conditions & Possible Causes |
|---|---|
| **DTC: P0171**<br>**2T CCM, MIL: Yes**<br>**Years:** 2007, 2008<br>**Models:** Cooper<br>**Engines:** 1.6L, 1.6L Turbo<br>**Transmissions:** All | **Fuel System Too Lean (Cylinder Bank 1) Conditions:**<br>Key on or engine running, all electrical components off and coolant temperature at least 80 degrees Celsius; and the DME detected the Bank 1 Adaptive Fuel Control System reached its rich correction limit (a lean A/F condition). The fuel status is in a closed loop pattern, the coolant temperature is greater than 7 degrees Celsius, and the engine speed is less than 1400rpm.<br>**Possible Causes:**<br>• Air leaks after the MAF sensor, or leaks in the PCV system<br>• Exhaust leaks before or near where the HO2S is mounted<br>• Fuel injector(s) restricted or not supplying enough fuel<br>• Fuel pump not supplying enough fuel during high fuel demand conditions<br>• Leaking EGR gasket, or leaking EGR valve diaphragm<br>• MAF sensor dirty (causes DME to underestimate airflow)<br>• Vehicle running out of fuel or engine oil dip stick not seated |
| **DTC: P0172**<br>**2T CCM, MIL: Yes**<br>**Years:** 2007, 2008<br>**Models:** Cooper<br>**Engines:** 1.6L, 1.6L Turbo<br>**Transmissions:** All | **Fuel System Too Rich (Cylinder Bank 1) Conditions:**<br>Key on or engine running, all electrical components off and coolant temperature at least 80 degrees Celsius; and the DME detected the Bank 1 Adaptive Fuel Control System reached its rich correction limit (a rich A/F condition). The fuel status is in a closed loop pattern, the coolant temperature is greater than 7 degrees Celsius, and the engine speed is less than 1400rpm.<br>**Possible Causes:**<br>• Camshaft timing is incorrect, or the engine has an oil overfill condition<br>• EVAP vapor recovery system failure (may be pulling vacuum)<br>• Fuel pressure regulator is damaged or leaking<br>• HO2S element is contaminated with alcohol or water<br>• MAF or MAP sensor values are incorrect or out-of-range<br>• One of more fuel injectors is leaking |
| **DTC: P0201**<br>**2T CCM, MIL: Yes**<br>**Years:** 2007, 2008<br>**Models:** Cooper<br>**Engines:** 1.6L, 1.6L Turbo<br>**Transmissions:** All | **Cylinder 1 Injector Circuit Malfunction Conditions:**<br>Engine started, and the DME detected the fuel injector "1" control circuit was in a high state when it should have been low, or in a low state when it should have been high (wiring harness & injector okay). The battery voltage should be between 9.5 and 17 volts while the engine speed is less than 40rpm.<br>**Possible Causes:**<br>• Injector 1 connector is damaged, open or shorted<br>• Injector 1 control circuit is open, shorted to ground or to power<br>(the injector driver circuit may be damaged) |
| **DTC: P0202**<br>**2T CCM, MIL: Yes**<br>**Years:** 2007, 2008<br>**Models:** Cooper<br>**Engines:** 1.6L, 1.6L Turbo<br>**Transmissions:** All | **Cylinder 2 Injector Circuit Malfunction Conditions:**<br>Engine started, and the DME detected the fuel injector "2" control circuit was in a high state when it should have been low, or in a low state when it should have been high (wiring harness & injector okay). The battery voltage should be between 9.5 and 17 volts while the engine speed is less than 40rpm.<br>**Possible Causes:**<br>• Injector 2 connector is damaged, open or shorted<br>• Injector 2 control circuit is open, shorted to ground or to power<br>(the injector driver circuit may be damaged) |
| **DTC: P0203**<br>**2T CCM, MIL: Yes**<br>**Years:** 2007, 2008<br>**Models:** Cooper<br>**Engines:** 1.6L, 1.6L Turbo<br>**Transmissions:** All | **Cylinder 3 Injector Circuit Malfunction Conditions:**<br>Engine started, and the DME detected the fuel injector "3" control circuit was in a high state when it should have been low, or in a low state when it should have been high (wiring harness & injector okay). The battery voltage should be between 9.5 and 17 volts while the engine speed is less than 40rpm.<br>**Possible Causes:**<br>• Injector 3 connector is damaged, open or shorted<br>• Injector 3 control circuit is open, shorted to ground or to power<br>(the injector driver circuit may be damaged) |
| **DTC: P0204**<br>**2T CCM, MIL: Yes**<br>**Years:** 2007, 2008<br>**Models:** Cooper<br>**Engines:** 1.6L, 1.6L Turbo<br>**Transmissions:** All | **Cylinder 4 Injector Circuit Malfunction Conditions:**<br>Engine started, and the DME detected the fuel injector "4" control circuit was in a high state when it should have been low, or in a low state when it should have been high (wiring harness & injector okay). The battery voltage should be between 9.5 and 17 volts while the engine speed is less than 40rpm.<br>**Possible Causes:**<br>• Injector 4 connector is damaged, open or shorted<br>• Injector 4 control circuit is open, shorted to ground or to power (the injector driver circuit may be damaged) |

| DTC | Trouble Code Title, Conditions & Possible Causes |
|---|---|
| **DTC: P0218**<br>**MIL: No**<br>**Years:** 2007, 2008<br>**Models:** Cooper<br>**Engines:** 1.6L, 1.6L Turbo<br>**Transmissions:** All | **Engine Oil Over Temperature Conditions:**<br>The oil temperature difference of greater than 100 degrees within one second. The ignition must be on. The DME detected an error in the Engine Oil Temperature sensor. This occurs during attempted start value calibration.<br>**Possible Causes:**<br>• Replace the oil temperature sensor<br>• Engine Oil temperature is too high<br>• Engine coolant temperature is too high<br>• Highest possible gear engaged in transmission<br>• Check coolant |
| **DTC: P0222**<br>**2T CCM, MIL: Yes**<br>**Years:** 2007, 2008<br>**Models:** Cooper<br>**Engines:** 1.6L, 1.6L Turbo<br>**Transmissions:** All | **Throttle Position Sensor 'B' Circuit Low Input Conditions:**<br>Engine started, battery voltage at least 11.5v, all electrical components off, ground connections between engine and chassis well connected, coolant temperature at least 80-degrees Celsius and the throttle valve must not be damaged or dirty; and the DME detected the TP Sensor 'B' circuit was out of its normal operating range during a condition with the throttle wide open, or with it completely closed. The throttle valve activation occurs via an electric motor (throttle drive) in the throttle valve control module. It is activated by the DME according to specifications of the two sensors, Throttle Position Sensor and Accelerator Pedal Position Sensor 2. Slowly depress accelerator pedal up to Wide Open Throttle (WOT) stop while observing the percentage display on the PID data function of the scan tool. The percentage display must increase uniformly.<br>**Possible Causes:**<br>• ETC TP Sensor 'B' connector is damaged or shorted<br>• ETC TP Sensor 'B' signal circuit is shorted to ground<br>• ETC TP Sensor 'B' is damaged or it has failed |
| **DTC: P0223**<br>**2T CCM, MIL: Yes**<br>**Years:** 2007, 2008<br>**Models:** Cooper<br>**Engines:** 1.6L, 1.6L Turbo<br>**Transmissions:** All | **Throttle Position Sensor 'B' Circuit High Input Conditions:**<br>Engine started, battery voltage at least 11.5v, all electrical components off, ground connections between engine and chassis well connected, coolant temperature at least 80-degrees Celsius and the throttle valve must not be damaged or dirty; and the DME detected the TP Sensor 'B' circuit was out of its normal operating range during a condition with the throttle wide open, or with it completely closed. The throttle valve activation occurs via an electric motor (throttle drive) in the throttle valve control module. It is activated by the DME according to specifications of the two sensors, Throttle Position Sensor and Accelerator Pedal Position Sensor 2. Slowly depress accelerator pedal up to Wide Open Throttle (WOT) stop while observing the percentage display on the PID data function of the scan tool. The percentage display must increase uniformly.<br>**Possible Causes:**<br>• ETC TP Sensor 'B' connector is damaged or open<br>• ETC TP Sensor 'B' signal circuit is open<br>• ETC TP Sensor 'B' signal circuit is shorted to VREF (5v)<br>• ETC TP Sensor 'B' is damaged or it has failed |
| **DTC: P0261**<br>**2T CCM, MIL: Yes**<br>**Years:** 2007, 2008<br>**Models:** Cooper<br>**Engines:** 1.6L, 1.6L Turbo<br>**Transmissions:** All | **Cylinder 1 Injector Circuit Low Input/Short to Ground Conditions:**<br>Key on or engine running, fuses in the instrument panel and the E-box in the engine compartment must be functioning, and the ground connections between the engine ad the chassis must be well connected; and the DME detected an unexpected voltage condition on the injector circuit.<br>**Possible Causes:**<br>• Injector 1 control circuit is open<br>• Injector 1 power circuit (B+) is open<br>• Injector 1 control circuit is shorted to chassis ground<br>• Injector 1 is damaged or has failed<br>• DME is not connected or has failed |
| **DTC: P0262**<br>**2T CCM, MIL: Yes**<br>**Years:** 2007, 2008<br>**Models:** Cooper<br>**Engines:** 1.6L, 1.6L Turbo<br>**Transmissions:** All | **Cylinder 1 Injector Circuit Low Input/Short to B+ Conditions:**<br>Key on or engine running, fuses in the instrument panel and the E-box in the engine compartment must be functioning, and the ground connections between the engine ad the chassis must be well connected; and the DME detected an unexpected voltage condition on the injector circuit.<br>**Possible Causes:**<br>• Injector control circuit is open<br>• Injector power circuit (B+) is open<br>• Injector control circuit is shorted to chassis ground<br>• Injector is damaged or has failed<br>• DME is not connected or has failed<br>• Fuel pump relay has failed<br>• Fuel injectors may have malfunctioned<br>• Faulty engine speed sensor |

| DTC | Trouble Code Title, Conditions & Possible Causes |
|---|---|
| **DTC: P0264**<br>**2T CCM, MIL: Yes**<br>**Years:** 2007, 2008<br>**Models:** Cooper<br>**Engines:** 1.6L, 1.6L Turbo<br>**Transmissions:** All | **Cylinder 2 Injector Circuit Low Input/Short to Ground Conditions:**<br>Key on or engine running, fuses in the instrument panel and the E-box in the engine compartment must be functioning, and the ground connections between the engine ad the chassis must be well connected; and the DME detected an unexpected voltage condition on the injector circuit.<br>**Possible Causes:**<br>• Injector control circuit is open<br>• Injector power circuit (B+) is open<br>• Injector control circuit is shorted to chassis ground<br>• Injector is damaged or has failed<br>• DME is not connected or has failed<br>• Fuel pump relay has failed<br>• Fuel injectors may have malfunctioned<br>• Faulty engine speed sensor |
| **DTC: P0265**<br>**2T CCM, MIL: Yes**<br>**Years:** 2007, 2008<br>**Models:** Cooper<br>**Engines:** 1.6L, 1.6L Turbo<br>**Transmissions:** All | **Cylinder 2 Injector Circuit Low Input/Short to B+ Conditions:**<br>Key on or engine running, fuses in the instrument panel and the E-box in the engine compartment must be functioning, and the ground connections between the engine ad the chassis must be well connected; and the DME detected an unexpected voltage condition on the injector circuit.<br>**Possible Causes:**<br>• Injector control circuit is open<br>• Injector power circuit (B+) is open<br>• Injector control circuit is shorted to chassis ground<br>• Injector is damaged or has failed<br>• DME is not connected or has failed<br>• Fuel pump relay has failed<br>• Fuel injectors may have malfunctioned<br>• Faulty engine speed sensor |
| **DTC: P0267**<br>**2T CCM, MIL: Yes**<br>**Years:** 2007, 2008<br>**Models:** Cooper<br>**Engines:** 1.6L, 1.6L Turbo<br>**Transmissions:** All | **Cylinder 3 Injector Circuit Low Input/Short to Ground Conditions:**<br>Key on or engine running, fuses in the instrument panel and the E-box in the engine compartment must be functioning, and the ground connections between the engine ad the chassis must be well connected; and the DME detected an unexpected voltage condition on the injector circuit.<br>**Possible Causes:**<br>• Injector control circuit is open<br>• Injector power circuit (B+) is open<br>• Injector control circuit is shorted to chassis ground<br>• Injector is damaged or has failed<br>• DME is not connected or has failed<br>• Fuel pump relay has failed<br>• Fuel injectors may have malfunctioned<br>• Faulty engine speed sensor |
| **DTC: P0268**<br>**2T CCM, MIL: Yes**<br>**Years:** 2007, 2008<br>**Models:** Cooper<br>**Engines:** 1.6L, 1.6L Turbo<br>**Transmissions:** All | **Cylinder 3 Injector Circuit Low Input/Short to B+ Conditions:**<br>Key on or engine running, fuses in the instrument panel and the E-box in the engine compartment must be functioning, and the ground connections between the engine ad the chassis must be well connected; and the DME detected an unexpected voltage condition on the injector circuit.<br>**Possible Causes:**<br>• Injector control circuit is open<br>• Injector power circuit (B+) is open<br>• Injector control circuit is shorted to chassis ground<br>• Injector is damaged or has failed<br>• DME is not connected or has failed<br>• Fuel pump relay has failed<br>• Fuel injectors may have malfunctioned<br>• Faulty engine speed sensor |

| DTC | Trouble Code Title, Conditions & Possible Causes |
|---|---|
| **DTC: P0270**<br>**2T CCM, MIL: Yes**<br>**Years:** 2007, 2008<br>**Models:** Cooper<br>**Engines:** 1.6L, 1.6L Turbo<br>**Transmissions:** All | **Cylinder 4 Injector Circuit Low Input/Short to Ground Conditions:**<br>Key on or engine running, fuses in the instrument panel and the E-box in the engine compartment must be functioning, and the ground connections between the engine ad the chassis must be well connected; and the DME detected an unexpected voltage condition on the injector circuit.<br>**Possible Causes:**<br>• Injector control circuit is open<br>• Injector power circuit (B+) is open<br>• Injector control circuit is shorted to chassis ground<br>• Injector is damaged or has failed<br>• DME is not connected or has failed<br>• Fuel pump relay has failed<br>• Fuel injectors may have malfunctioned<br>• Faulty engine speed sensor |
| **DTC: P0271**<br>**2T CCM, MIL: Yes**<br>**Years:** 2007, 2008<br>**Models:** Cooper<br>**Engines:** 1.6L, 1.6L Turbo<br>**Transmissions:** All | **Cylinder 4 Injector Circuit Low Input/Short to B+ Conditions:**<br>Key on or engine running, fuses in the instrument panel and the E-box in the engine compartment must be functioning, and the ground connections between the engine ad the chassis must be well connected; and the DME detected an unexpected voltage condition on the injector circuit.<br>**Possible Causes:**<br>• Injector control circuit is open<br>• Injector power circuit (B+) is open<br>• Injector control circuit is shorted to chassis ground<br>• Injector is damaged or has failed<br>• DME is not connected or has failed<br>• Fuel pump relay has failed<br>• Fuel injectors may have malfunctioned<br>• Faulty engine speed sensor |
| **DTC: P0300**<br>**2T MISFIRE, MIL: Yes**<br>**Years:** 2007, 2008<br>**Models:** Cooper<br>**Engines:** 1.6L, 1.6L Turbo<br>**Transmissions:** All | **Random/Multiple Misfire Detected Conditions:**<br>Engine running at an RPM greater than 600 but less than 7000 the DME detected a misfire or uneven engine running in two or more cylinders within 1000 engine revolutions. The sum of misfires caused an increase in emissions for the first 1000 revolutions after start up, or the sum of misfires caused catalyst damage after the first 200 engine revolutions. Time after start less than one second.<br>**Note: If the misfire is severe, the MIL will flash on/off on the first trip!**<br>**Possible Causes:**<br>• Fuel metering fault that affects two or more cylinders<br>• Fuel pressure too low or too high, fuel supply contaminated<br>• EVAP system problem or the EVAP canister is fuel saturated<br>• EGR valve is stuck open or the PCV system has a vacuum leak<br>• Ignition system fault (coil, plugs) affecting two or more cylinders<br>• MAF sensor contamination (it can cause a very lean condition)<br>• Vehicle driven while very low on fuel (less than 1/8 of a tank) |
| **DTC: P0301**<br>**2T MISFIRE, MIL: Yes**<br>**Years:** 2007, 2008<br>**Models:** Cooper<br>**Engines:** 1.6L, 1.6L Turbo<br>**Transmissions:** All | **Cylinder Number 1 Misfire Detected Conditions:**<br>Engine running at an RPM greater than 600 but less than 7000 the DME detected a misfire or uneven engine running in two or more cylinders within 1000 engine revolutions. The sum of misfires caused an increase in emissions for the first 1000 revolutions after start up, or the sum of misfires caused catalyst damage after the first 200 engine revolutions. Time after start less than one second.<br>**Note: If the misfire is severe, the MIL will flash on/off on the first trip!**<br>**Possible Causes:**<br>• Air leak in the intake manifold, or in the EGR or DME system<br>• Base engine mechanical problem<br>• Fuel delivery component problem (i.e., a contaminated, dirty or sticking fuel injector)<br>• Fuel pump relay defective<br>• Ignition coil fuses have failed<br>• Ignition system problem (dirty damaged coil or plug)<br>• Engine speed (RPM) sensor has failed<br>• Camshaft position sensors have failed<br>• Ignition coil is faulty<br>• Spark plugs are not working properly or are not gapped properly |

| DTC | Trouble Code Title, Conditions & Possible Causes |
|---|---|
| **DTC: P0302**<br>**2T MISFIRE, MIL: Yes**<br>**Years:** 2007, 2008<br>**Models:** Cooper<br>**Engines:** 1.6L, 1.6L Turbo<br>**Transmissions:** All | **Cylinder Number 2 Misfire Detected Conditions:**<br>Engine running at an RPM greater than 600 but less than 7000 the DME detected a misfire or uneven engine running in two or more cylinders within 1000 engine revolutions. The sum of misfires caused an increase in emissions for the first 1000 revolutions after start up, or the sum of misfires caused catalyst damage after the first 200 engine revolutions. Time after start less than one second.<br>**Note: If the misfire is severe, the MIL will flash on/off on the 1st trip!**<br>**Possible Causes:**<br>• Air leak in the intake manifold, or in the EGR or DME system<br>• Base engine mechanical problem<br>• Fuel delivery component problem (i.e., a contaminated, dirty or sticking fuel injector)<br>• Fuel pump relay defective<br>• Ignition coil fuses have failed<br>• Ignition system problem (dirty damaged coil or plug)<br>• Engine speed (RPM) sensor has failed<br>• Camshaft position sensors have failed<br>• Ignition coil is faulty<br>• Spark plugs are not working properly or are not gapped properly |
| **DTC: P0303**<br>**2T MISFIRE, MIL: Yes**<br>**Years:** 2007, 2008<br>**Models:** Cooper<br>**Engines:** 1.6L, 1.6L Turbo<br>**Transmissions:** All | **Cylinder Number 3 Misfire Detected Conditions:**<br>Engine running at an RPM greater than 600 but less than 7000 the DME detected a misfire or uneven engine running in two or more cylinders within 1000 engine revolutions. The sum of misfires caused an increase in emissions for the first 1000 revolutions after start up, or the sum of misfires caused catalyst damage after the first 200 engine revolutions. Time after start less than one second.<br>**Note: If the misfire is severe, the MIL will flash on/off on the 1st trip!**<br>**Possible Causes:**<br>• Air leak in the intake manifold, or in the EGR or DME system<br>• Base engine mechanical problem<br>• Fuel delivery component problem (i.e., a contaminated, dirty or sticking fuel injector)<br>• Fuel pump relay defective<br>• Ignition coil fuses have failed<br>• Ignition system problem (dirty damaged coil or plug)<br>• Engine speed (RPM) sensor has failed<br>• Camshaft position sensors have failed<br>• Ignition coil is faulty<br>• Spark plugs are not working properly or are not gapped properly |
| **DTC: P0304**<br>**2T MISFIRE, MIL: Yes**<br>**Years:** 2007, 2008<br>**Models:** Cooper<br>**Engines:** 1.6L, 1.6L Turbo<br>**Transmissions:** All | **Cylinder Number 4 Misfire Detected Conditions:**<br>Engine running at an RPM greater than 600 but less than 7000 the DME detected a misfire or uneven engine running in two or more cylinders within 1000 engine revolutions. The sum of misfires caused an increase in emissions for the first 1000 revolutions after start up, or the sum of misfires caused catalyst damage after the first 200 engine revolutions. Time after start less than one second.<br>**Note: If the misfire is severe, the MIL will flash on/off on the 1st trip!**<br>**Possible Causes:**<br>• Air leak in the intake manifold, or in the EGR or DME system<br>• Base engine mechanical problem<br>• Fuel delivery component problem (i.e., a contaminated, dirty or sticking fuel injector)<br>• Fuel pump relay defective<br>• Ignition coil fuses have failed<br>• Ignition system problem (dirty damaged coil or plug)<br>• Engine speed (RPM) sensor has failed<br>• Camshaft position sensors have failed<br>• Ignition coil is faulty<br>• Spark plugs are not working properly or are not gapped properly |

| DTC | Trouble Code Title, Conditions & Possible Causes |
|---|---|
| **DTC: P0313**<br>**2T MISFIRE, MIL: Yes**<br>**Years:** 2007, 2008<br>**Models:** Cooper<br>**Engines:** 1.6L, 1.6L Turbo<br>**Transmissions:** All | **Misfire Detected with Low Fuel Conditions:**<br>Engine running under positive torque conditions, and the DME detected a misfire or uneven engine function as well as an indication of low fuel level when another misfire was detected.<br>**Note: If the misfire is severe, the MIL will flash on/off on the 1st trip!**<br>**Possible Causes:**<br>• Air leak in the intake manifold, or in the EGR or DME system<br>• Base engine mechanical problem<br>• Fuel delivery component problem (i.e., a contaminated, dirty or sticking fuel injector)<br>• Fuel pump relay defective<br>• Ignition coil fuses have failed<br>• Ignition system problem (dirty damaged coil or plug)<br>• Engine speed (RPM) sensor has failed<br>• Camshaft position sensors have failed<br>• Ignition coil is faulty<br>• Spark plugs are not working properly or are not gapped properly |
| **DTC: P0324**<br>**2T CCM, MIL: Yes**<br>**Years:** 2007, 2008<br>**Models:** Cooper<br>**Engines:** 1.6L, 1.6L Turbo<br>**Transmissions:** All | **Knock Control System Error Conditions:**<br>Engine started, vehicle driven, and the DME detected the Knock Sensor 1 (KS1) signal was too low or not recognized by the DME.<br>**Possible Causes:**<br>• Knock sensor circuit is open<br>• Knock sensor is loose (tighten to 20 NM)<br>• Contact between the knock sensor and cylinder block is dirty, corroded or greasy<br>• Knock sensor circuit is shorted to ground, or shorted to power<br>• Knock sensor is damaged or it has failed<br>• Wrong kind of fuel used<br>• A component in the engine compartment is loose or not properly secured |
| **DTC: P0326**<br>**2T CCM, MIL: Yes**<br>**Years:** 2007, 2008<br>**Models:** Cooper<br>**Engines:** 1.6L, 1.6L Turbo<br>**Transmissions:** All | **Knock Sensor Circuit Malfunction Conditions:**<br>Engine started, vehicle driven at 1520rpm for 3 seconds or to a temperature of 40 degrees Celsius, and the DME detected the Knock Sensor 1 (KS1) signal was not recognized. The engine speed is greater than 2016rpm and the coolant temperature is greater than 50.25 degrees Celsius. The difference between raw and filtered knock sensor signal is less than 0.0499 to 0.0698 volts.<br>**Possible Causes:**<br>• Knock sensor circuit is open<br>• Knock sensor is loose (tighten to 20 NM)<br>• Contact between the knock sensor and cylinder block is dirty, corroded or greasy<br>• Knock sensor circuit is shorted to ground, or shorted to power<br>• Knock sensor is damaged or it has failed<br>• Wrong kind of fuel used<br>• A component in the engine compartment is loose or not properly secured |
| **DTC: P0335**<br>**2T CCM, MIL: Yes**<br>**Years:** 2007, 2008<br>**Models:** Cooper<br>**Engines:** 1.6L, 1.6L Turbo<br>**Transmissions:** All | **Camshaft Position Sensor "A" Circ Malfunction Conditions:**<br>Engine started, battery voltage must be at least 11.5v, all electrical components must be off, parking brake must be engaged (to keep daytime driving lights off), automatic transmission selector must be in park and the ground between the engine and the chassis must be well connected. The DME detected the CMP sensor signal was implausible or missing. Engine speed is greater than 500rpm, and the fault is tolerable as long as there are no misfired occurring at the same time.<br>**Possible Causes:**<br>• CMP sensor circuit is open or shorted to ground<br>• CMP sensor circuit is shorted to power<br>• CMP sensor ground (return) circuit is open<br>• CMP sensor installation incorrect (Hall-effect type)<br>• CMP sensor is damaged or CMP sensor shielding damaged |
| **DTC: P0336**<br>**2T CCM, MIL: Yes**<br>**Years:** 2007, 2008<br>**Models:** Cooper<br>**Engines:** 1.6L, 1.6L Turbo<br>**Transmissions:** All | **Camshaft Position Sensor "A" Circ Range/Performance Conditions:**<br>Engine started (and engine speed is less than 25rpm), battery voltage must be at least 11.5v, all electrical components must be off, parking brake must be engaged (to keep daytime driving lights off), automatic transmission selector must be in park and the ground between the engine and the chassis must be well connected. The DME detected the CMP sensor signal was implausible.<br>**Possible Causes:**<br>• CMP sensor circuit is open or shorted to ground<br>• CMP sensor circuit is shorted to power<br>• CMP sensor ground (return) circuit is open<br>• CMP sensor installation incorrect (Hall-effect type)<br>• CMP sensor is damaged or CMP sensor shielding damaged |

| DTC | Trouble Code Title, Conditions & Possible Causes |
|---|---|
| **DTC: P0340**<br>**2T CCM, MIL: Yes**<br>**Years:** 2007, 2008<br>**Models:** Cooper<br>**Engines:** 1.6L, 1.6L Turbo<br>**Transmissions:** All | **Camshaft Position Sensor Circuit Malfunction Conditions:**<br>Engine started, battery voltage must be at least 11.5v, all electrical components must be off, parking brake must be engaged (to keep daytime driving lights off), automatic transmission selector must be in park and the ground between the engine and the chassis must be well connected. The DME detected the CMP sensor signal was missing or it was erratic. There is no signal or an invalid one, and the engine speed is greater than 200rpm for two cycles.<br>**Possible Causes:**<br>• CMP sensor circuit is open or shorted to ground<br>• CMP sensor circuit is shorted to power<br>• CMP sensor ground (return) circuit is open<br>• CMP sensor installation incorrect (Hall-effect type)<br>• CMP sensor is damaged or CMP sensor shielding damaged<br>• CMP sensor has failed |
| **DTC: P0341**<br>**2T CCM, MIL: Yes**<br>**Years:** 2007, 2008<br>**Models:** Cooper<br>**Engines:** 1.6L, 1.6L Turbo<br>**Transmissions:** All | **Camshaft Position Sensor Circ Range/Performance Conditions:**<br>Engine started, battery voltage must be at least 11.5v, all electrical components must be off, parking brake must be engaged (to keep daytime driving lights off), automatic transmission selector must be in park and the ground between the engine and the chassis must be well connected. The DME detected the CMP sensor signal was implausible.<br>**Possible Causes:**<br>• CMP sensor circuit is open or shorted to ground<br>• CMP sensor circuit is shorted to power<br>• CMP sensor ground (return) circuit is open<br>• CMP sensor installation incorrect (Hall-effect type)<br>• CMP sensor is damaged or CMP sensor shielding damaged |
| **DTC: P0351**<br>**2T CCM, MIL: Yes**<br>**Years:** 2007, 2008<br>**Models:** Cooper<br>**Engines:** All<br>**Transmissions:** All | **Ignition Coilpack A Primary/Secondary Circuit Malfunction Conditions:**<br>Engine started, battery voltage must be at least 11.5v, all electrical components must be off, parking brake must be engaged (to keep daytime driving lights off), automatic transmission selector must be in park and the ground between the engine and the chassis must be well connected. The DME did not receive any valid pulses from the ignition module for the Ignition Coilpack A primary circuit.<br>**Note: Ignition coils and power output stages are one component and cannot be replaced individually.**<br>**Possible Causes:**<br>• Engine speed (RPM) sensor has failed<br>• Camshaft Position (CMP) sensor has failed<br>• Power Supply Relay is shorted to an open circuit<br>• There is a malfunction in voltage supply<br>• Ignition coilpack is damaged or it has failed<br>• Cylinder 1 to 4 Fuel Injector(s) have failed |
| **DTC: P0352**<br>**2T CCM, MIL: Yes**<br>**Years:** 2007, 2008<br>**Models:** Cooper<br>**Engines:** All<br>**Transmissions:** All | **Ignition Coilpack A Primary/Secondary Circuit Malfunction Conditions:**<br>Engine started, battery voltage must be at least 11.5v, all electrical components must be off, parking brake must be engaged (to keep daytime driving lights off), automatic transmission selector must be in park and the ground between the engine and the chassis must be well connected. The DME did not receive any valid pulses from the ignition module for the Ignition Coilpack A primary circuit.<br>**Note: Ignition coils and power output stages are one component and cannot be replaced individually.**<br>**Possible Causes:**<br>• Engine speed (RPM) sensor has failed<br>• Camshaft Position (CMP) sensor has failed<br>• Power Supply Relay is shorted to an open circuit<br>• There is a malfunction in voltage supply<br>• Ignition coilpack is damaged or it has failed<br>• Cylinder 1 to 4 Fuel Injector(s) have failed |
| **DTC: P0353**<br>**2T CCM, MIL: Yes**<br>**Years:** 2007, 2008<br>**Models:** Cooper<br>**Engines:** All<br>**Transmissions:** All | **Ignition Coilpack A Primary/Secondary Circuit Malfunction Conditions:**<br>Engine started, battery voltage must be at least 11.5v, all electrical components must be off, parking brake must be engaged (to keep daytime driving lights off), automatic transmission selector must be in park and the ground between the engine and the chassis must be well connected. The DME did not receive any valid pulses from the ignition module for the Ignition Coilpack A primary circuit.<br>**Note: Ignition coils and power output stages are one component and cannot be replaced individually.**<br>**Possible Causes:**<br>• Engine speed (RPM) sensor has failed<br>• Camshaft Position (CMP) sensor has failed<br>• Power Supply Relay is shorted to an open circuit<br>• There is a malfunction in voltage supply<br>• Ignition coilpack is damaged or it has failed<br>• Cylinder 1 to 4 Fuel Injector(s) have failed |

| DTC | Trouble Code Title, Conditions & Possible Causes |
|---|---|
| **DTC: P0354**<br>**2T CCM, MIL: Yes**<br>**Years:** 2007, 2008<br>**Models:** Cooper<br>**Engines:** All<br>**Transmissions:** All | **Ignition Coilpack A Primary/Secondary Circuit Malfunction Conditions:**<br>Engine started, battery voltage must be at least 11.5v, all electrical components must be off, parking brake must be engaged (to keep daytime driving lights off), automatic transmission selector must be in park and the ground between the engine and the chassis must be well connected. The DME did not receive any valid pulses from the ignition module for the Ignition Coilpack A primary circuit.<br>**Note: Ignition coils and power output stages are one component and cannot be replaced individually.**<br>**Possible Causes:**<br>• Engine speed (RPM) sensor has failed<br>• Camshaft Position (CMP) sensor has failed<br>• Power Supply Relay is shorted to an open circuit<br>• There is a malfunction in voltage supply<br>• Ignition coilpack is damaged or it has failed<br>• Cylinder 1 to 4 Fuel Injector(s) have failed |
| **DTC: P0420**<br>**MIL: Yes**<br>**Years:** 2007, 2008<br>**Models:** Cooper<br>**Engines:** 1.6L, 1.6L Turbo<br>**Transmissions:** All | **Catalyst System Efficiency (Bank 1) Below Threshold Conditions:**<br>Engine started for longer than one second, battery voltage must be at least 11.5v, all electrical components must be off, parking brake must be engaged (to keep daytime driving lights off), automatic transmission selector must be in park, the exhaust system must be properly sealed between the catalytic converter and the cylinder head, coolant temperature must be at least 80 degrees Celsius and oxygen sensor heaters for oxygen sensors before the catalytic converter must be functioning properly and the ground between the engine and the chassis must be well connected. The DME detected the switch rate of the rear HO2S-12 was close to the switch rate of front HO2S (it should be much slower). The coolant temperature is greater than 80.25 degrees Celsius. The fuel system is in closed loop. The vehicle speed is between 28 and 80.8mph. The engine speed is between 1984 and 3648rpm. Exhaust gas temperature is between 450 and 700 degrees Celsius. Ambient pressure is 75.001kPa.<br>**Possible Causes:**<br>• Air leaks at the exhaust manifold or in the exhaust pipes<br>• Catalytic converter is damaged, contaminated or it has failed<br>• ECT/CHT sensor has lost its calibration (the signal is incorrect)<br>• Engine cylinders misfiring, or the ignition timing is over retarded<br>• Engine oil is contaminated<br>• Front HO2S or rear HO2S is contaminated with fuel or moisture<br>• Front HO2S and/or the rear HO2S is loose in the mounting hole<br>• Front HO2S much older than the rear HO2S (HO2S-11 is lazy)<br>• Fuel system pressure is too high (check the pressure regulator)<br>• Rear HO2S wires improperly connected or the HO2S has failed |
| **DTC: P0441**<br>**2T CCM, MIL: Yes**<br>**Years:** 2007, 2008<br>**Models:** Cooper<br>**Engines:** 1.6L, 1.6L Turbo<br>**Transmissions:** All | **EVAP Emission System Incorrect Purge Flow Conditions:**<br>ECT sensor is cold during startup, engine started, battery voltage must be at least 11.5v, all electrical components must be off. The coolant temperature is less than 60 degrees Celsius, and the ambient pressure is greater than 76.2994kPa. The air intake temperature at start is between 9.04 and 16.04 degrees Celsius. The change in barometric pressure since engine start is less than 0.9998kPa. The vehicle speed is less than 74.56mph, and the purge valve has opened enough on previous driving cycle. The DME detected the switch rate of the rear HO2S-12 was close to the switch rate of front HO2S (it should be much slower). DME detected a problem in the EVAP system during the EVAP System Monitor test.<br>**Possible Causes:**<br>• EVAP canister purge valve is damaged<br>• EVAP canister has an improper seal<br>• Vapor line between purge solenoid and intake manifold vacuum reservoir is damaged, or vapor line between EVAP canister purge solenoid and charcoal canister is damaged<br>• Vapor line between charcoal canister and check valve, or vapor line between check valve and fuel vapor valves is damaged |
| **DTC: P0442**<br>**2T CCM, MIL: Yes**<br>**Years:** 2007, 2008<br>**Models:** Cooper<br>**Engines:** 1.6L, 1.6L Turbo<br>**Transmissions:** All | **EVAP Emission System Small Leak Detected Conditions:**<br>Engine started, battery voltage must be at least 11.5v, all electrical components must be off. The DME detected a leak in the EVAP system as small as 0.040 inches during the EVAP Monitor Test. The coolant temperature is less than 60 degrees Celsius, and the ambient pressure is greater than 76.2994kPa. The air intake temperature at start is between 9.04 and 16.04 degrees Celsius. The change in barometric pressure since engine start is less than 0.9998kPa. The vehicle speed is less than 74.56mph, and the purge valve has opened enough on previous driving cycle.<br>**Possible Causes:**<br>• Aftermarket EVAP parts that do not conform to specifications<br>• CV solenoid remains partially open when commanded to close<br>• EVAP component seals leaking (i.e., leaks in the Purge valve, fuel tank pressure sensor, canister vent solenoid, fuel vapor control valve tube assembly or fuel vapor vent valve).<br>• Fuel filler cap damaged, cross-threaded or loosely installed<br>• Loose fuel vapor hose/tube connections to EVAP components<br>• Small holes or cuts in fuel vapor hoses or EVAP canister tubes |

| DTC | Trouble Code Title, Conditions & Possible Causes |
|---|---|
| **DTC: P0443**<br>**2T CCM, MIL: Yes**<br>**Years:** 2007, 2008<br>**Models:** Cooper<br>**Engines:** 1.6L, 1.6L Turbo<br>**Transmissions:** All | **EVAP Vapor Management Valve Circuit Malfunction Conditions:**<br>Engine started, battery voltage must be at least 11.5v, all electrical components must be off, parking brake must be engaged (to keep daytime driving lights off), automatic transmission selector must be in park, the exhaust system must be properly sealed between the catalytic converter and the cylinder head, coolant temperature must be at least 80 degrees Celsius and oxygen sensor heaters for oxygen sensors before the catalytic converter must be functioning properly and the ground between the engine and the chassis must be well connected. The DME detected an unexpected high or low voltage condition on the Vapor Management Valve (VMV) circuit when the device was cycled On/Off during testing.<br>**Possible Causes:**<br>• EVAP power supply circuit is open<br>• EVAP solenoid control circuit is open or shorted to ground<br>• EVAP solenoid control circuit is shorted to power (B+)<br>• EVAP solenoid valve is damaged or it has failed |
| **DTC: P0444**<br>**2T CCM, MIL: Yes**<br>**Years:** 2007, 2008<br>**Models:** Cooper<br>**Engines:** 1.6L, 1.6L Turbo<br>**Transmissions:** All | **Evaporative Emission System Purge Control Valve Circuit Open Conditions:**<br>Engine started, battery voltage must be at least 11.5v, all electrical components must be off, parking brake must be engaged (to keep daytime driving lights off), automatic transmission selector must be in park, the exhaust system must be properly sealed between the catalytic converter and the cylinder head, coolant temperature must be at least 80 degrees Celsius and oxygen sensor heaters for oxygen sensors before the catalytic converter must be functioning properly and the ground between the engine and the chassis must be well connected. The DME detected an unexpected voltage condition on the EVAP circuit when the device was cycled On/Off during testing.<br>**Possible Causes:**<br>• EVAP power supply circuit is open<br>• EVAP solenoid control circuit is open or shorted to ground<br>• EVAP solenoid control circuit is shorted to power (B+)<br>• EVAP solenoid valve is damaged or it has failed<br>• EVAP canister has a leak or a poor seal |
| **DTC: P0445**<br>**2T CCM, MIL: Yes**<br>**Years:** 2007, 2008<br>**Models:** Cooper<br>**Engines:** 1.6L, 1.6L Turbo<br>**Transmissions:** All | **Evaporative Emission System Purge Control Valve Circuit Shorted Conditions:**<br>Engine started, battery voltage must be at least 11.5v, all electrical components must be off, parking brake must be engaged (to keep daytime driving lights off), automatic transmission selector must be in park, the exhaust system must be properly sealed between the catalytic converter and the cylinder head, coolant temperature must be at least 80 degrees Celsius and oxygen sensor heaters for oxygen sensors before the catalytic converter must be functioning properly and the ground between the engine and the chassis must be well connected. The DME detected an unexpected voltage condition on the EVAP circuit when the device was cycled On/Off during testing.<br>**Possible Causes:**<br>• EVAP power supply circuit is open<br>• EVAP solenoid control circuit is open or shorted to ground<br>• EVAP solenoid control circuit is shorted to power (B+)<br>• EVAP solenoid valve is damaged or it has failed<br>• EVAP canister has a leak or a poor seal |
| **DTC: P0455**<br>**2T CCM, MIL: Yes**<br>**Years:** 2007, 2008<br>**Models:** Cooper<br>**Engines:** 1.6L, 1.6L Turbo<br>**Transmissions:** All | **EVAP Control System Large Leak Detected Conditions:**<br>Engine started, battery voltage must be at least 11.5v, all electrical components must be off. The coolant temperature is less than 60 degrees Celsius, and the ambient pressure is greater than 76.2994kPa. The air intake temperature at start is between 9.04 and 16.04 degrees Celsius. The change in barometric pressure since engine start is less than 0.9998kPa. The vehicle speed is less than 74.56mph, and the purge valve has opened enough on previous driving cycle. The DME detected multiple small fuel vapor leaks; or it detected a large leak in the system during the leak test.<br>**Possible Causes:**<br>• Aftermarket EVAP hardware non-conforming to specifications<br>• EVAP canister tube, EVAP canister purge outlet tube or EVAP return tube disconnected<br>  or cracked, or canister is damaged<br>• EVAP canister purge valve stuck closed, or canister damaged<br>• Fuel filler cap missing, loose (not tightened) or the wrong part<br>• Loose fuel vapor hose/tube connections to EVAP components<br>• Canister vent (CV) solenoid stuck open<br>• Fuel tank pressure (FTP) sensor has failed mechanically |

| DTC | Trouble Code Title, Conditions & Possible Causes |
|---|---|
| **DTC: P0456**<br>**2T CCM, MIL: Yes**<br>**Years:** 2007, 2008<br>**Models:** Cooper<br>**Engines:** 1.6L, 1.6L Turbo<br>**Transmissions:** All | **EVAP Control System Small Leak Detected Conditions:**<br>Engine started, battery voltage must be at least 11.5v, all electrical components must be off. The coolant temperature is less than 60 degrees Celsius, and the ambient pressure is greater than 76.2994kPa. The air intake temperature at start is between 9.04 and 16.04 degrees Celsius. The change in barometric pressure since engine start is less than 0.9998kPa. The vehicle speed is less than 74.56mph, and the purge valve has opened enough on previous driving cycle. The DME detected multiple small fuel vapor leaks; or it detected a large leak in the system during the leak test.<br>**Possible Causes:**<br>• Aftermarket EVAP hardware non-conforming to specifications<br>• EVAP canister tube, EVAP canister purge outlet tube or EVAP return tube disconnected or cracked, or canister is damaged<br>• EVAP canister purge valve stuck closed, or canister damaged<br>• Fuel filler cap missing, loose (not tightened) or the wrong part<br>• Loose fuel vapor hose/tube connections to EVAP components<br>• Canister vent (CV) solenoid stuck open<br>• Fuel tank pressure (FTP) sensor has failed mechanically |
| **DTC: P0458**<br>**2T CCM, MIL: Yes**<br>**Years:** 2007, 2008<br>**Models:** Cooper<br>**Engines:** 1.6L, 1.6L Turbo<br>**Transmissions:** All | **Evaporative Emission System Purge Control Valve Circuit High:**<br>Engine started, battery voltage must be at least 11.5v, all electrical components must be off, parking brake must be engaged (to keep daytime driving lights off), automatic transmission selector must be in park, the exhaust system must be properly sealed between the catalytic converter and the cylinder head, coolant temperature must be at least 80 degrees Celsius and oxygen sensor heaters for oxygen sensors before the catalytic converter must be functioning properly and the ground between the engine and the chassis must be well connected. The DME detected an unexpected voltage condition on the EVAP circuit when the device was cycled On/Off during testing.<br>**Possible Causes:**<br>• EVAP power supply circuit is open<br>• EVAP solenoid control circuit is open or shorted to ground<br>• EVAP solenoid control circuit is shorted to power (B+)<br>• EVAP solenoid valve is damaged or it has failed<br>• EVAP canister has a leak or a poor seal |
| **DTC: P0459**<br>**2T CCM, MIL: Yes**<br>**Years:** 2007, 2008<br>**Models:** Cooper<br>**Engines:** 1.6L, 1.6L Turbo<br>**Transmissions:** All | **Evaporative Emission System Purge Control Valve Circuit Low:**<br>Engine started, battery voltage must be at least 11.5v, all electrical components must be off, parking brake must be engaged (to keep daytime driving lights off), automatic transmission selector must be in park, the exhaust system must be properly sealed between the catalytic converter and the cylinder head, coolant temperature must be at least 80 degrees Celsius and oxygen sensor heaters for oxygen sensors before the catalytic converter must be functioning properly and the ground between the engine and the chassis must be well connected. The DME detected an unexpected voltage condition on the EVAP circuit when the device was cycled On/Off during testing.<br>**Possible Causes:**<br>• EVAP power supply circuit is open<br>• EVAP solenoid control circuit is open or shorted to ground<br>• EVAP solenoid control circuit is shorted to power (B+)<br>• EVAP solenoid valve is damaged or it has failed<br>• EVAP canister has a leak or a poor seal |
| **DTC: P0460**<br>**MIL: No**<br>**Years:** 2007, 2008<br>**Models:** Cooper<br>**Engines:** 1.6L, 1.6L Turbo<br>**Transmissions:** All | **Fuel Level Signal 1:**<br>The DME detected a high condition on the pump.<br>**Possible Causes:**<br>• The fuel pump has failed.<br>• The DME has failed<br>• Check wiring |
| **DTC: P0461**<br>**MIL: No**<br>**Years:** 2007, 2008<br>**Models:** Cooper<br>**Engines:** 1.6L, 1.6L Turbo<br>**Transmissions:** All | **Fuel Level Signal 1:**<br>The DME detected a high condition on the pump.<br>**Possible Causes:**<br>• The fuel pump has failed.<br>• The DME has failed<br>• Check wiring |
| **DTC: P0462**<br>**MIL: No**<br>**Years:** 2007, 2008<br>**Models:** Cooper<br>**Engines:** 1.6L, 1.6L Turbo<br>**Transmissions:** All | **Fuel Level Signal 1:**<br>The DME detected a high condition on the pump.<br>**Possible Causes:**<br>• The fuel pump has failed.<br>• The DME has failed<br>• Check wiring |

| DTC | Trouble Code Title, Conditions & Possible Causes |
|---|---|
| **DTC: P0463**<br>**MIL:** No<br>**Years:** 2007, 2008<br>**Models:** Cooper<br>**Engines:** 1.6L, 1.6L Turbo<br>**Transmissions:** All | **Fuel Level Signal 1:**<br>The DME detected a high condition on the pump.<br>**Possible Causes:**<br>• The fuel pump has failed.<br>• The DME has failed<br>• Check wiring |
| **DTC: P0500**<br>**2T MIL:** Yes<br>**Years:** 2007, 2008<br>**Models:** Cooper<br>**Engines:** 1.6L, 1.6L Turbo<br>**Transmissions:** All | **Vehicle Speed Sensor "A" Malfunction Conditions:**<br>Engine started; engine speed above the TCC stall speed, and the DME detected a loss of the VSS signal over a period of time or the signal is not usable.<br>**Note: The DME receives vehicle speed data from the VSS, TCSS, ABS module, CTM or GEM controller, depending up the application. Speed Signal from DSC too high because of possible tampering. Check DSC and wires.**<br>**Possible Causes:**<br>• VSS signal circuit is open or shorted to ground<br>• VSS harness circuit is shorted to ground<br>• VSS harness circuit is shorted to power<br>• VSS circuit open between the DME and related control module<br>• VSS or wheel speed sensors circuits are damaged<br>• Modules connected to VSC/VSS harness circuits are damaged<br>• Mechanical drive mechanism for the VSS is damaged |
| **DTC: P0506**<br>**2T CCM, MIL:** Yes<br>**Years:** 2007, 2008<br>**Models:** Cooper<br>**Engines:** 1.6L, 1.6L Turbo<br>**Transmissions:** All | **Idle Air Control System RPM Lower Than Expected Conditions:**<br>Engine started, battery voltage must be at least 10.96v, all electrical components must be off, parking brake must be engaged (to keep daytime driving lights off), automatic transmission selector must be in park, the exhaust system must be properly sealed between the catalytic converter and the cylinder head, coolant temperature must be between 80.25 and 110.25 degrees Celsius and oxygen sensor heaters for oxygen sensors before the catalytic converter must be functioning properly and the ground between the engine and the chassis must be well connected. The DME detected it could not control the idle speed correctly, as it is constantly more than 100 rpm less than specification.<br>**Possible Causes:**<br>• Air inlet is plugged or the air filter element is severely clogged<br>• IAC circuit is open or shorted<br>• IAC circuit VPWR circuit is open<br>• IAC solenoid is damaged or has failed<br>• The VSS has failed |
| **DTC: P0507**<br>**2T CCM, MIL:** Yes<br>**Years:** 2007, 2008<br>**Models:** Cooper<br>**Engines:** 1.6L, 1.6L Turbo<br>**Transmissions:** All | **Idle Air Control System RPM Higher Than Expected Conditions:**<br>Engine started, battery voltage must be at least 10.96v, all electrical components must be off, parking brake must be engaged (to keep daytime driving lights off), automatic transmission selector must be in park, the exhaust system must be properly sealed between the catalytic converter and the cylinder head, coolant temperature must be between 80.25 and 110.25 degrees Celsius and oxygen sensor heaters for oxygen sensors before the catalytic converter must be functioning properly and the ground between the engine and the chassis must be well connected. The DME detected it could not control the idle speed correctly, as it is constantly more than 200 rpm more than specification.<br>**Possible Causes:**<br>• Air intake leak located somewhere after the throttle body<br>• IAC control circuit is shorted to chassis ground<br>• IAC solenoid is damaged or has failed<br>• Throttle Valve Control module has failed or is clogged with carbon<br>• The VSS has failed |
| **DTC: P0532**<br>**MIL:** No<br>**Years: Years:** 2007, 2008<br>**Models:** Cooper<br>**Engines:** 1.6L, 1.6L Turbo<br>**Transmissions:** All | **A/C Refrigerant Pressure Sensor "A" Circuit Low Conditions:**<br>The DME detected a low condition on the sensor.<br>**Possible Causes:**<br>• The A/C Refrigerant Pressure Sensor "A" has failed.<br>• The DME has failed |
| **DTC: P0533**<br>**MIL:** No<br>**Years:** 2007, 2008<br>**Models:** Cooper<br>**Engines:** 1.6L, 1.6L Turbo<br>**Transmissions:** All | **A/C Refrigerant Pressure Sensor "A" Circuit High Conditions:**<br>The DME detected a high condition on the sensor.<br>**Possible Causes:**<br>• The A/C Refrigerant Pressure Sensor "A" has failed.<br>• The DME has failed |

| DTC | Trouble Code Title, Conditions & Possible Causes |
|---|---|
| **DTC: P0562**<br>**MIL: No**<br>**Years:** 2007, 2008<br>**Models:** Cooper<br>**Engines:** 1.6L, 1.6L Turbo<br>**Transmissions:** All | **System Voltage Low Conditions:**<br>Engine started, battery voltage must be at least 11.5v, all electrical components must be off, parking brake must be engaged (to keep daytime driving lights off), automatic transmission selector must be in park, and the ground between the engine and the chassis must be well connected. The DME has detected a voltage value that is below the specified minimum limit for the system to function properly.<br>**Possible Causes:**<br>• Alternator damaged or faulty<br>• Battery voltage low or insufficient<br>• Fuses blown or circuits open<br>• Battery connection to terminal not clean<br>• Voltage regulator has failed |
| **DTC: P0563**<br>**MIL: No**<br>**Years:** 2007, 2008<br>**Models:** Cooper<br>**Engines:** 1.6L, 1.6L Turbo<br>**Transmissions:** All | **System Voltage High Conditions:**<br>Engine started for 18 seconds, battery voltage must be at least 11.5v, all electrical components must be off, parking brake must be engaged (to keep daytime driving lights off), automatic transmission selector must be in park, and the ground between the engine and the chassis must be well connected. The DME has detected a voltage value that has exceeded the specified maximum limit for the system to function properly. The vehicle was connected to 24 volts for too long after a jump start. ADC in ECU is defective. Delete stored fault codes from log. If fault reoccurs replace the ECU.<br>**Possible Causes:**<br>• Alternator damaged or faulty<br>• Battery voltage low or insufficient<br>• Fuses blown or circuits open<br>• Battery connection to terminal not clean<br>• Voltage regulator has failed |
| **DTC: P0571**<br>**MIL: No**<br>**Years:** 2007, 2008<br>**Models:** Cooper<br>**Engines:** 1.6L, 1.6L Turbo<br>**Transmissions:** All | **Cruise/Brake Switch (A) Circuit Malfunction Conditions:**<br>Engine started, battery voltage must be at least 11.5v, all electrical components must be off, parking brake must be engaged (to keep daytime driving lights off), automatic transmission selector must be in park, and the ground between the engine and the chassis must be well connected. The DME has detected a voltage value that is implausible or erratic.<br>**Possible Causes:**<br>• Brake light switch is faulty<br>• Control circuit is shorted to chassis ground |
| **DTC: P0600**<br>**2T CCM, MIL: Yes**<br>**Years:** 2007, 2008<br>**Models:** Cooper<br>**Engines:** 1.6L, 1.6L Turbo<br>**Transmissions:** All | **Serial Communication Link (Data BUS) Message Missing Conditions:**<br>The Engine Control Module (DME) communicates with all databus-capable control modules via a CAN databus. These databus-capable control modules are connected via two data bus wires which are twisted together (CAN_High and CAN_Low), and exchange information (messages). Missing information on the databus is recognized as a malfunction and stored. Trouble-free operation of the CAN-Bus requires that it have a terminal resistance. This central terminal resistor is located in the Engine Control Module (DME).<br>**Possible Causes:**<br>• CAN data bus wires have short circuited to each other |
| **DTC: P0601**<br>**2T CCM, MIL: Yes**<br>**Years:** 2007, 2008<br>**Models:** Cooper<br>**Engines:** 1.6L, 1.6L Turbo<br>**Transmissions:** All | **Internal Control Module Memory Check Sum Error Conditions:**<br>Key on, the DME has detected a programming error. The RAM and ROM check displays an invalid check-sum at power up/down.<br>**Possible Causes:**<br>• Battery terminal corrosion, or loose battery connection<br>• Connection to the DME interrupted, or the circuit has been opened<br>• Reprogramming error has occurred and needs replacement. Remember to check for Aftermarket Performance Products before replacing a DME. |
| **DTC: P0603**<br>**2T CCM, MIL: Yes**<br>**Years:** 2007, 2008<br>**Models:** Cooper<br>**Engines:** 1.6L, 1.6L Turbo<br>**Transmissions:** All | **DME Keep Alive Memory Test Error Conditions:**<br>Key on, and the DME detected an internal memory fault. This code will set if KAPWR to the DME is interrupted (at the initial key on). Watchdog on.<br>**Possible Causes:**<br>• Battery terminal corrosion, or loose battery connection<br>• KAPWR to DME interrupted, or the circuit has been opened<br>• Reprogramming error has occurred and needs replacement. Remember to check for Aftermarket Performance Products before replacing a DME. |

| DTC | Trouble Code Title, Conditions & Possible Causes |
|---|---|
| **DTC: P0604**<br>**2T CCM, MIL: Yes**<br>**Years:** 2007, 2008<br>**Models:** Cooper<br>**Engines:** 1.6L, 1.6L Turbo<br>**Transmissions:** All | **Internal Control Module Random Access Memory (RAM) Error Conditions:**<br>Key on, and the DME detected an internal memory fault. This code will set if KAPWR to the DME is interrupted (at the initial key on). Watchdog on.<br>**Possible Causes:**<br>• Battery terminal corrosion, or loose battery connection<br>• Connection to the DME interrupted, or the circuit has been opened<br>• Reprogramming error has occurred and needs replacement. Remember to check for Aftermarket Performance Products before replacing a DME. |
| **DTC: P0627**<br>**MIL: No**<br>**Years:** 2007, 2008<br>**Models:** Cooper<br>**Engines:** 1.6L, 1.6L Turbo<br>**Transmissions:** All | **Fuel Pump "A" Control Circuit/Open:**<br>The DME detected a high condition on the pump.<br>**Possible Causes:**<br>• The fuel pump has failed.<br>• The DME has failed<br>• Check wiring |
| **DTC: P0628**<br>**MIL: No**<br>**Years:** 2007, 2008<br>**Models:** Cooper<br>**Engines:** 1.6L, 1.6L Turbo<br>**Transmissions:** All | **Fuel Pump "A" Control Circuit/Low:**<br>The DME detected a high condition on the pump.<br>**Possible Causes:**<br>• The fuel pump has failed.<br>• The DME has failed<br>• Check wiring |
| **DTC: P0629**<br>**MIL: No**<br>**Years:** 2007, 2008<br>**Models:** Cooper<br>**Engines:** 1.6L, 1.6L Turbo<br>**Transmissions:** All | **Fuel Pump "A" Control Circuit/High:**<br>The DME detected a high condition on the pump.<br>**Possible Causes:**<br>• The fuel pump has failed.<br>• The DME has failed<br>• Check wiring |
| **DTC: P0646**<br>**MIL: No**<br>**Years:** 2007, 2008<br>**Models:** Cooper<br>**Engines:** 1.6L, 1.6L Turbo<br>**Transmissions:** All | **A/C Compressor Circuit High Conditions:**<br>The DME detected a high condition on the sensor.<br>**Possible Causes:**<br>• The A/C Compressor has failed.<br>• The DME has failed<br>• Check wiring |
| **DTC: P0647**<br>**MIL: No**<br>**Years:** 2007, 2008<br>**Models:** Cooper<br>**Engines:** 1.6L, 1.6L Turbo<br>**Transmissions:** All | **A/C Compressor Circuit Low Conditions:**<br>The DME detected a low condition on the sensor.<br>**Possible Causes:**<br>• The A/C Compressor has failed.<br>• The DME has failed<br>• Check wiring |
| **DTC: P0704**<br>**MIL: No**<br>**Years:** 2007, 2008<br>**Models:** Cooper<br>**Engines:** 1.6L, 1.6L Turbo<br>**Transmissions:** A/T | **Clutch Switch Input Circuit Malfunction Conditions:**<br>Engine started, battery voltage must be at least 11.5v, all electrical components must be off, parking brake must be engaged (to keep daytime driving lights off), automatic transmission selector must be in park, and the ground between the engine and the chassis must be well connected. The DME detected a voltage outside the normal performance range to allow the system to properly function.<br>**Possible Causes:**<br>• Circuit harness connector contacts are corroded or ingresses of water<br>• Circuit wires have shorted to each other, to battery or ground<br>• Automatic Transmission Hydraulic Pressure Sensor 1 has failed<br>• Solenoid valves in valve body are faulty<br>• Transmission Input Speed (RPM) Sensor has failed<br>• Transmission Output Speed (RPM) Sensor has failed<br>• Engine Control Module (DME) is faulty<br>• Voltage supply for Engine Control Module (DME) is faulty<br>• Transmission Control Module (TCM) is faulty |

| DTC | Trouble Code Title, Conditions & Possible Causes |
|---|---|
| **DTC: P0705**<br>**2T CCM, MIL: Yes**<br>**Years:** 2007, 2008<br>**Models:** Cooper<br>**Engines:** 1.6L, 1.6L Turbo<br>**Transmissions:** A/T | **TR Sensor Circuit Malfunction Conditions:**<br>Engine started, battery voltage must be at least 11.5v, all electrical components must be off, parking brake must be engaged (to keep daytime driving lights off), automatic transmission selector must be in park, and the ground between the engine and the chassis must be well connected. The DME detected a voltage or signal outside the normal performance range to allow the system to properly function. The engine speed is between 200 and 440rpm.<br>**Possible Causes:**<br>• Circuit harness connector contacts are corroded or ingresses of water<br>• Circuit wires have shorted to each other, to battery or ground<br>• Automatic Transmission Hydraulic Pressure Sensor 1 has failed<br>• Solenoid valves in valve body are faulty<br>• Transmission Input Speed (RPM) Sensor has failed<br>• Transmission Output Speed (RPM) Sensor has failed<br>• Engine Control Module (DME) is faulty<br>• Voltage supply for Engine Control Module (DME) is faulty<br>• Transmission Control Module (TCM) is faulty |
| **DTC: P0712**<br>**MIL: No**<br>**Years:** 2007, 2008<br>**Models:** Cooper<br>**Engines:** 1.6L, 1.6L Turbo<br>**Transmissions:** A/T | **Oil Temperature Sensor Circuit Low Input Conditions:**<br>Engine started, battery voltage must be at least 11.5v, all electrical components must be off, parking brake must be engaged (to keep daytime driving lights off), automatic transmission selector must be in park, and the ground between the engine and the chassis must be well connected. The DME detected the oil temperature sensor was less than its minimum self-test range in the test.<br>**Possible Causes:**<br>• Sensor signal circuit is open between the sensor and DME<br>• Sensor ground circuit is open between sensor and DME<br>• Sensor is damaged or has failed |
| **DTC: P0713**<br>**MIL: No**<br>**Years:** 2007, 2008<br>**Models:** Cooper<br>**Engines:** 1.6L, 1.6L Turbo<br>**Transmissions:** A/T | **Oil Temperature Sensor Circuit High Input Conditions:**<br>Engine started, battery voltage must be at least 11.5v, all electrical components must be off, parking brake must be engaged (to keep daytime driving lights off), automatic transmission selector must be in park, and the ground between the engine and the chassis must be well connected. The DME detected the oil temperature sensor was more than its maximum self-test range in the test.<br>**Possible Causes:**<br>• Sensor signal circuit is open between the sensor and DME<br>• Sensor ground circuit is open between sensor and DME<br>• Sensor is damaged or has failed |
| **DTC: P0721**<br>**MIL: No**<br>**Years:** 2007, 2008<br>**Models:** Cooper<br>**Engines:** 1.6L, 1.6L Turbo<br>**Transmissions:** A/T | **A/T Output Shaft Speed Sensor Noise Interference Conditions:**<br>Engine started, VSS signal more than 1 mph, and the DME detected "noise" interference on the Output Shaft Speed (OSS) sensor circuit. The calculation of the road speed impossible, as the indicated speed is less than the minimum road speed value and the timer expired.<br>**Possible Causes:**<br>• After market add-on devices interfering with the OSS signal<br>• OSS connector is damaged, loose or shorted, or the wiring is misrouted or it is damaged<br>• OSS assembly is damaged or it has failed<br>• Failure of the ABS CAN vehicle speed sensor |

## Gas Engine OBD II Trouble Code List (P1xxx Codes)

| DTC | Trouble Code Title, Conditions & Possible Causes |
|---|---|
| **DTC: P1104**<br>**2T CCM, MIL: Yes**<br>**Years:** 2007, 2008<br>**Models:** Cooper<br>**Engines:** 1.6L, 1.6L Turbo<br>**Transmissions:** All | **Manifold Pressure Sensor Plausibility Conditions:**<br>Engine started, battery voltage must be at least 11v, and the differential pressure sensor detected a control deviation at the minimum limit. The closed loop control of the pressure in the intake manifold is suspended and replaced by a direct specification. The MAP was too low (less than 105.0016kPa); engine stopped.<br>**Possible Causes:**<br>• Sensor's voltage supply on Terminal 87<br>• Sensor's ground connection faulty<br>• Signal wire to DME faulty<br>• Replace sensor |

| DTC | Trouble Code Title, Conditions & Possible Causes |
|---|---|
| **DTC: P1106**<br>**2T CCM, MIL: Yes**<br>**Years:** 2007, 2008<br>**Models:** Cooper<br>**Engines:** 1.6L, 1.6L Turbo<br>**Transmissions:** All | **Manifold Pressure Too Low at Full Load for Low Engine Speed Conditions:**<br>Engine started, battery voltage must be at least 11v, and the differential pressure sensor detected a control deviation at the minimum limit. The closed loop control of the differential pressure in the intake manifold is suspended and replaced by a direct specification. The engine speed is less than 4000rpm. The manifold pressure is less than 600hPa.<br>**Possible Causes:**<br>• Sensor's voltage supply on Terminal 87<br>• Sensor's ground connection faulty<br>• Signal wire to DME faulty<br>• Replace sensor |
| **DTC: P1107**<br>**2T CCM, MIL: Yes**<br>**Years:** 2007, 2008<br>**Models:** Cooper<br>**Engines:** 1.6L, 1.6L Turbo<br>**Transmissions:** All | **Manifold Pressure Too Low at Idle Conditions:**<br>Engine started, battery voltage must be at least 11v, and the differential pressure sensor detected a control deviation at the minimum limit. The closed loop control of the differential pressure in the intake manifold is suspended and replaced by a direct specification. The engine speed is less than 1504rpm. The manifold pressure is less than 120hPa.<br>**Possible Causes:**<br>• Sensor's voltage supply on Terminal 87<br>• Sensor's ground connection faulty<br>• Signal wire to DME faulty<br>• Replace sensor |
| **DTC: P1108**<br>**2T CCM, MIL: Yes**<br>**Years:** 2007, 2008<br>**Models:** Cooper<br>**Engines:** 1.6L, 1.6L Turbo<br>**Transmissions:** All | **Manifold Pressure Too Low at Stable and in Full Load for Low Engine Speed Conditions:**<br>Engine started, battery voltage must be at least 11v, and the differential pressure sensor detected a control deviation at the minimum limit. The closed loop control of the differential pressure in the intake manifold is suspended and replaced by a direct specification. The engine speed is less than 4000rpm. The manifold pressure is less than 600hPa.<br>**Possible Causes:**<br>• Sensor's voltage supply on Terminal 87<br>• Sensor's ground connection faulty<br>• Signal wire to DME faulty<br>• Replace sensor |
| **DTC: P1109**<br>**2T CCM, MIL: Yes**<br>**Years:** 2007, 2008<br>**Models:** Cooper<br>**Engines:** 1.6L, 1.6L Turbo<br>**Transmissions:** All | **Manifold Pressure Too High During Deceleration Conditions:**<br>Engine started, battery voltage must be at least 11v, and the differential pressure sensor detected a control deviation at the minimum limit. The closed loop control of the differential pressure in the intake manifold is suspended and replaced by a direct specification. The engine speed is greater than 1696rpm. The manifold pressure is greater than 600hPa.<br>**Possible Causes:**<br>• Sensor's voltage supply on Terminal 87<br>• Sensor's ground connection faulty<br>• Signal wire to DME faulty<br>• Replace sensor |
| **DTC: P1122**<br>**2T CCM, MIL: Yes**<br>**Years:** 2007, 2008<br>**Models:** Cooper<br>**Engines:** 1.6L, 1.6L Turbo<br>**Transmissions:** All | **Accelerator Pedal Position Sensor 'D' Circuit Low Input Conditions:**<br>Engine started, battery voltage at least 11.5v, all electrical components off, ground connections between engine and chassis well connected, the DME detected that the accelerator pedal position sensor signal was outside the parameters to function normally.<br>**Note: Both the Throttle Position (TP) Sensor and Accelerator Pedal Position Sensor are located at the accelerator pedal module and communicate the driver's intentions to the DME completely independently of each other. Both sensors are stored in one housing.**<br>**Possible Causes:**<br>• Ground between engine and chassis may be broken<br>• Throttle position sensor may have failed<br>• Accelerator Pedal Position Sensor has failed<br>• Throttle position sensor wiring may have shorted<br>• Throttle position sensor has failed<br>• Faulty voltage supply |

| DTC | Trouble Code Title, Conditions & Possible Causes |
|---|---|
| **DTC: P1123**<br>**2T CCM, MIL:** Yes<br>**Years:** 2007, 2008<br>**Models:** Cooper<br>**Engines:** 1.6L, 1.6L Turbo<br>**Transmissions:** All | **Accelerator Pedal Position Sensor 'D' Circuit High Input Conditions:**<br>Engine started, battery voltage at least 11.5v, all electrical components off, ground connections between engine and chassis well connected, the DME detected that the accelerator pedal position sensor signal was outside the parameters to function normally.<br>**Note: Both the Throttle Position (TP) Sensor and Accelerator Pedal Position Sensor are located at the accelerator pedal module and communicate the driver's intentions to the DME completely independently of each other. Both sensors are stored in one housing.**<br>**Possible Causes:**<br>• Ground between engine and chassis may be broken<br>• Throttle position sensor may have failed<br>• Accelerator Pedal Position Sensor has failed<br>• Throttle position sensor wiring may have shorted<br>• Throttle position sensor has failed<br>• Faulty voltage supply |
| **DTC: P1125**<br>**2T CCM, MIL:** Yes<br>**Years:** 2007, 2008<br>**Models:** Cooper<br>**Engines:** 1.6L, 1.6L Turbo<br>**Transmissions:** All | **Throttle/Pedal Position Sensor Circuit Small Plausibility Error Conditions:**<br>Engine started, at idle, the temperature must be at least 80 degrees Celsius. The throttle position sensor supplies implausible signal to the DME. The difference between the TPS1 and the TPS2 is greater than five percent.<br>**Possible Causes:**<br>• TP sensor signal circuit open (inspect wiring & connector)<br>• TP sensor signal shorted to ground (inspect wiring & connector)<br>• TP sensor is damaged or has failed<br>• Throttle control module's voltage supply is shorted or open |
| **DTC: P1126**<br>**2T CCM, MIL:** Yes<br>**Years:** 2007, 2008<br>**Models:** Cooper<br>**Engines:** 1.6L, 1.6L Turbo<br>**Transmissions:** All | **Throttle/Pedal Position Sensor Circuit Large Plausibility Error Conditions:**<br>Engine started, at idle, the temperature must be at least 80 degrees Celsius. The throttle position sensor supplies implausible signal to the DME. The difference between the TPS1 and the TPS2 is greater than five percent.<br>**Possible Causes:**<br>• TP sensor signal circuit open (inspect wiring & connector)<br>• TP sensor signal shorted to ground (inspect wiring & connector)<br>• TP sensor is damaged or has failed<br>• Throttle control module's voltage supply is shorted or open |
| **DTC: P1143**<br>**2T CCM, MIL:** Yes<br>**Years:** 2007, 2008<br>**Models:** Cooper<br>**Engines:** 1.6L, 1.6L Turbo<br>**Transmissions:** All | **O2 Sensor Signal Stuck Lean Bank 1 Sensor 2 Conditions:**<br>Engine started, battery voltage must be at least 11.5v, all electrical components must be off, parking brake must be engaged (to keep daytime driving lights off), automatic transmission selector must be in park. The DME detected an unexpected voltage condition, or it detected an unexpected current draw in the heater circuit during the CCM test. Coolant temperature must been at least 80.25 degrees Celsius. The vehicle speed is greater than 27.96 and less than 80.76. The engine speed is between 1984 and 3647rpm. Ambient pressure is greater than 75.001kPa and the engine stability load is 6.94g/s.<br>**Note: Vehicle must be raised before connector for oxygen sensors is accessible.**<br>**Possible Causes:**<br>• Oxygen sensor (before catalytic converter) is faulty<br>• Oxygen sensor heater (before catalytic converter) is faulty<br>• Oxygen sensor heater (before catalytic converter) is faulty<br>• Oxygen sensor heater (behind catalytic converter) is faulty<br>• O2S circuit is open or shorted in the wiring harness |
| **DTC: P1144**<br>**2T CCM, MIL:** Yes<br>**Years:** 2007, 2008<br>**Models:** Cooper<br>**Engines:** 1.6L, 1.6L Turbo<br>**Transmissions:** All | **O2 Sensor Signal Stuck Rich Bank 1 Sensor 2 Conditions:**<br>Engine started, battery voltage must be at least 11.5v, all electrical components must be off, parking brake must be engaged (to keep daytime driving lights off), automatic transmission selector must be in park. The DME detected an unexpected voltage condition, or it detected an unexpected current draw in the heater circuit during the CCM test. Coolant temperature must been at least 80.25 degrees Celsius. The vehicle speed is greater than 27.96 and less than 80.76. The engine speed is between 1984 and 3647rpm. Ambient pressure is greater than 75.001kPa and the engine stability load is 6.94g/s.<br>**Note: Vehicle must be raised before connector for oxygen sensors is accessible.**<br>**Possible Causes:**<br>• Oxygen sensor (before catalytic converter) is faulty<br>• Oxygen sensor heater (before catalytic converter) is faulty<br>• Oxygen sensor heater (before catalytic converter) is faulty<br>• Oxygen sensor heater (behind catalytic converter) is faulty<br>• O2S circuit is open or shorted in the wiring harness |

| DTC | Trouble Code Title, Conditions & Possible Causes |
|---|---|
| **DTC: P1222**<br>**2T CCM, MIL: Yes**<br>**Years:** 2007, 2008<br>**Models:** Cooper<br>**Engines:** 1.6L, 1.6L Turbo<br>**Transmissions:** All | **Accelerator Pedal Position Sensor 'E' Circuit Low Input Conditions:**<br>Engine started, battery voltage at least 11.5v, all electrical components off, ground connections between engine and chassis well connected, the DME detected that the accelerator pedal position sensor signal was outside the parameters to function normally.<br>**Note: Both the Throttle Position (TP) Sensor and Accelerator Pedal Position Sensor are located at the accelerator pedal module and communicate the driver's intentions to the DME completely independently of each other. Both sensors are stored in one housing.**<br>**Possible Causes:**<br>    • Ground between engine and chassis may be broken<br>    • Throttle position sensor may have failed<br>    • Accelerator Pedal Position Sensor has failed<br>    • Throttle position sensor wiring may have shorted<br>    • Throttle position sensor has failed<br>    • Faulty voltage supply |
| **DTC: P1223**<br>**2T CCM, MIL: Yes**<br>**Years:** 2007, 2008<br>**Models:** Cooper<br>**Engines:** 1.6L, 1.6L Turbo<br>**Transmissions:** All | **Accelerator Pedal Position Sensor 'E' Circuit High Input Conditions:**<br>Engine started, battery voltage at least 11.5v, all electrical components off, ground connections between engine and chassis well connected, the DME detected that the accelerator pedal position sensor signal was outside the parameters to function normally.<br>**Note: Both the Throttle Position (TP) Sensor and Accelerator Pedal Position Sensor are located at the accelerator pedal module and communicate the driver's intentions to the DME completely independently of each other. Both sensors are stored in one housing.**<br>**Possible Causes:**<br>    • Ground between engine and chassis may be broken<br>    • Throttle position sensor may have failed<br>    • Accelerator Pedal Position Sensor has failed<br>    • Throttle position sensor wiring may have shorted<br>    • Throttle position sensor has failed<br>    • Faulty voltage supply |
| **DTC: P1224**<br>**2T CCM, MIL: Yes**<br>**Years:** 2007, 2008<br>**Models:** Cooper<br>**Engines:** 1.6L, 1.6L Turbo<br>**Transmissions:** All | **Throttle Position Sensor D/E Voltage Correlation Conditions:**<br>Engine started, battery voltage must be at least 11.5v, all electrical components must be off, parking brake must be engaged (to keep daytime driving lights off), automatic transmission selector must be in park; and the DME detected the Throttle Position 'D' (TPD) and Throttle Position 'B' (TPE) sensors disagreed, or that the TPD sensor should not be in its detected position, or that the TPE sensor should not be in its detected position during testing.<br>**Note: Both the Throttle Position (TP) Sensor and Accelerator Pedal Position Sensor are located at the accelerator pedal module and communicate the driver's intentions to the DME completely independently of each other. Both sensors are stored in one housing.**<br>**Possible Causes:**<br>    • ETC TP sensor connector is damaged or shorted<br>    • ETC TP sensor circuits shorted together in the wire harness<br>    • ETC TP sensor signal circuit is shorted to VREF (5v)<br>    • ETC TP sensor is damaged or the DME has failed |
| **DTC: P1229**<br>**2T CCM, MIL: Yes**<br>**Years:** 2007, 2008<br>**Models:** Cooper<br>**Engines:** 1.6L, 1.6L Turbo<br>**Transmissions:** All | **Throttle/Pedal Position Sensor Adaptation Outside Tolerance Conditions:**<br>Engine started, at idle, the temperature must be at least 80 degrees Celsius. The throttle position sensor supplies implausible signal to the DME and is outside the specified tolerance. The measured max/min TPS values within the limits is greater than 0.0244 volts.<br>**Possible Causes:**<br>    • TP sensor signal circuit open (inspect wiring & connector)<br>    • TP sensor signal shorted to ground (inspect wiring & connector)<br>    • TP sensor is damaged or has failed<br>    • Throttle control module's voltage supply is shorted or open |
| **DTC: P1234**<br>**MIL: No**<br>**Years:** 2007, 2008<br>**Models:** Cooper<br>**Engines:** 1.6L, 1.6L Turbo<br>**Transmissions:** All | **Electrical Fuel Pump Circuit Short to Battery Conditions:**<br>The DME detected a low condition on the pump.<br>**Possible Causes:**<br>    • The fuel pump has failed.<br>    • The DME has failed<br>    • Check wiring |
| **DTC: P1236**<br>**MIL: No**<br>**Years:** 2007, 2008<br>**Models:** Cooper<br>**Engines:** 1.6L, 1.6L Turbo<br>**Transmissions:** All | **Electrical Fuel Pump Short to Ground or Open Circuit Conditions:**<br>The DME detected a high condition on the pump.<br>**Possible Causes:**<br>    • The fuel pump has failed.<br>    • The DME has failed<br>    • Check wiring |

| DTC | Trouble Code Title, Conditions & Possible Causes |
|---|---|
| **DTC: P1320**<br>**2T MISFIRE, MIL: Yes**<br>**Years:** 2007, 2008<br>**Models:** Cooper<br>**Engines:** 1.6L, 1.6L Turbo<br>**Transmissions:** All | **Misfire Detected Crankshaft Segment Adaptation Conditions:**<br>Engine running under positive torque conditions, and the DME detected a misfire or uneven engine function as well as the crankshaft adaptation at its limit.<br>**Note: If the misfire is severe, the MIL will flash on/off on the 1st trip!**<br>**Possible Causes:**<br>• Air leak in the intake manifold, or in the EGR or DME system<br>• Base engine mechanical problem<br>• Fuel delivery component problem (i.e., a contaminated, dirty or sticking fuel injector)<br>• Fuel pump relay defective<br>• Ignition coil fuses have failed<br>• Ignition system problem (dirty damaged coil or plug)<br>• Engine speed (RPM) sensor has failed<br>• Camshaft position sensors have failed<br>• Ignition coil is faulty<br>• Spark plugs are not working properly or are not gapped properly |
| **DTC: P1321**<br>**2T MISFIRE, MIL: Yes**<br>**Years:** 2007, 2008<br>**Models:** Cooper<br>**Engines:** 1.6L, 1.6L Turbo<br>**Transmissions:** All | **Misfire Crank Wheel Tooth Count Conditions:**<br>Engine running under positive torque conditions, and the DME detected a misfire or uneven engine function as well as a tooth error of plus or minus one or two teeth during the count.<br>**Note: If the misfire is severe, the MIL will flash on/off on the 1st trip!**<br>**Possible Causes:**<br>• Air leak in the intake manifold, or in the EGR or DME system<br>• Base engine mechanical problem<br>• Fuel delivery component problem (i.e., a contaminated, dirty or sticking fuel injector)<br>• Fuel pump relay defective<br>• Ignition coil fuses have failed<br>• Ignition system problem (dirty damaged coil or plug)<br>• Engine speed (RPM) sensor has failed<br>• Camshaft position sensors have failed<br>• Ignition coil is faulty<br>• Spark plugs are not working properly or are not gapped properly |
| **DTC: P1366**<br>**2T CCM, MIL: Yes**<br>**Years:** 2007, 2008<br>**Models:** Cooper<br>**Engines:** 1.6L, 1.6L Turbo<br>**Transmissions:** All | **Ignition Coilpack A Primary/Secondary Circuit Malfunction Open Circuit/Short to Ground Conditions:**<br>Engine started, battery voltage must be at least 11.5v, all electrical components must be off, parking brake must be engaged (to keep daytime driving lights off), automatic transmission selector must be in park and the ground between the engine and the chassis must be well connected. The DME did not receive any valid pulses from the ignition module for the Ignition Coilpack A primary circuit.<br>**Note: Ignition coils and power output stages are one component and cannot be replaced individually.**<br>**Possible Causes:**<br>• Engine speed (RPM) sensor has failed<br>• Camshaft Position (CMP) sensor has failed<br>• Power Supply Relay is shorted to an open circuit<br>• There is a malfunction in voltage supply<br>• Ignition coilpack is damaged or it has failed<br>• Cylinder 1 to 4 Fuel Injector(s) have failed |
| **DTC: P1367**<br>**2T CCM, MIL: Yes**<br>**Years:** 2007, 2008<br>**Models:** Cooper<br>**Engines:** 1.6L, 1.6L Turbo<br>**Transmissions:** All | **Ignition Coilpack A Primary/Secondary Circuit Malfunction Open Circuit/Short to Ground Conditions:**<br>Engine started, battery voltage must be at least 11.5v, all electrical components must be off, parking brake must be engaged (to keep daytime driving lights off), automatic transmission selector must be in park and the ground between the engine and the chassis must be well connected. The DME did not receive any valid pulses from the ignition module for the Ignition Coilpack A primary circuit.<br>**Note: Ignition coils and power output stages are one component and cannot be replaced individually.**<br>**Possible Causes:**<br>• Engine speed (RPM) sensor has failed<br>• Camshaft Position (CMP) sensor has failed<br>• Power Supply Relay is shorted to an open circuit<br>• There is a malfunction in voltage supply<br>• Ignition coilpack is damaged or it has failed<br>• Cylinder 1 to 4 Fuel Injector(s) have failed |
| **DTC: P1407**<br>**MIL: No**<br>**Years:** 2007, 2008<br>**Models:** Cooper<br>**Engines:** 1.6L, 1.6L Turbo<br>**Transmissions:** All | **Fuel Level Signal 1:**<br>The DME detected a high condition on the pump.<br>**Possible Causes:**<br>• The fuel pump has failed.<br>• The DME has failed<br>• Check wiring |

| DTC | Trouble Code Title, Conditions & Possible Causes |
|---|---|
| **DTC: P1409**<br>**MIL:** No<br>**Years:** 2007, 2008<br>**Models:** Cooper<br>**Engines:** 1.6L, 1.6L Turbo<br>**Transmissions:** All | **Fuel Level 1 CAN Error:**<br>The DME detected a high condition on the pump.<br>**Possible Causes:**<br>   • The fuel pump has failed.<br>   • The DME has failed<br>   • Check wiring |
| **DTC: P1433**<br>**MIL:** No<br>**Years:** 2007, 2008<br>**Models:** Cooper<br>**Engines:** 1.6L, 1.6L Turbo<br>**Transmissions:** All | **Fuel Level Sensor 2 (FSTEsig) CAN Signal Check:**<br>The DME detected a high condition on the pump.<br>**Possible Causes:**<br>   • The fuel pump has failed.<br>   • The DME has failed<br>   • Check wiring |
| **DTC: P1434**<br>**2T CCM, MIL:** Yes<br>**Years:** 2007, 2008<br>**Models:** Cooper<br>**Engines:** 1.6L, 1.6L Turbo<br>**Transmissions:** All | **Diagnostic Module Tank Leakage (DM-TL):**<br>Engine started, battery voltage must be at least 11.5v, all electrical components must be off. The coolant temperature is less than 60 degrees Celsius, and the ambient pressure is greater than 76.2994kPa. The air intake temperature at start is between 9.04 and 16.04 degrees Celsius. The change in barometric pressure since engine start is less than 0.9998kPa. The vehicle speed is less than 74.56mph, and the purge valve has opened enough on previous driving cycle. The DME detected multiple small fuel vapor leaks; or it detected a large leak in the system during the leak test.<br>**Possible Causes:**<br>   • Aftermarket EVAP hardware non-conforming to specifications<br>   • EVAP canister tube, EVAP canister purge outlet tube or EVAP return tube disconnected or cracked, or canister is damaged<br>   • EVAP canister purge valve stuck closed, or canister damaged<br>   • Fuel filler cap missing, loose (not tightened) or the wrong part<br>   • Loose fuel vapor hose/tube connections to EVAP components<br>   • Canister vent (CV) solenoid stuck open<br>   • Fuel tank pressure (FTP) sensor has failed mechanically |
| **DTC: P1436**<br>**2T CCM, MIL:** Yes<br>**Years:** 2007, 2008<br>**Models:** Cooper<br>**Engines:** 1.6L, 1.6L Turbo<br>**Transmissions:** All | **EVAP Leak Detection Pump (LDP) Control Circuit Open Conditions:**<br>Engine started, battery voltage must be at least 11.5v, all electrical components must be off, parking brake must be engaged (to keep daytime driving lights off), automatic transmission selector must be in park, the exhaust system must be properly sealed between the catalytic converter and the cylinder head, coolant temperature must be at least 80 degrees Celsius and oxygen sensor heaters for oxygen sensors before the catalytic converter must be functioning properly and the ground between the engine and the chassis must be well connected. The DME detected voltage irregularity in the leak detection pump control circuit.<br>**Possible Causes:**<br>   • EVAP LDP power supply circuit is open<br>   • EVAP LDP solenoid valve is damaged or it has failed<br>   • EVAP LDP canister has a leak or a poor seal<br>   • EVAP canister system has an improper seal<br>   • Evaporative Emission (EVAP) canister purge regulator valve 1 has failed<br>   • Leak Detection Pump (LDP) is faulty<br>   • Aftermarket EVAP parts that do not conform to specifications<br>   • EVAP component seals leaking (i.e., leaks in the Purge valve, fuel tank pressure sensor, canister vent solenoid, fuel vapor control valve tube assembly or fuel vapor vent valve). |
| **DTC: P1437**<br>**2T CCM, MIL:** Yes<br>**Years:** 2007, 2008<br>**Models:** Cooper<br>**Engines:** 1.6L, 1.6L Turbo<br>**Transmissions:** All | **EVAP Emission Control LDP Circuit Malfunction Pump Problem Conditions:**<br>Key on, KOEO Self-Test enabled, and the DME detected an unexpected voltage condition on the EVAP emission control leak detection pump circuit. The reed switch level stays low after activation of solenoids within the time threshold of more than 1 second. The coolant temperature is less than 60 degrees Celsius, and the ambient pressure is greater than 76.2994kPa. The air intake temperature at start is between 9.04 and 16.04 degrees Celsius. The change in barometric pressure since engine start is less than 0.9998kPa. The vehicle speed is less than 74.56mph, and the purge valve has opened enough on previous driving cycle.<br>**Possible Causes:**<br>   • Leak Detection Pump has failed<br>   • EVAP canister system has an improper or broken seal<br>   • Evaporative Emission (EVAP) canister purge regulator valve 1 is faulty<br>   • Hoses between the fuel pump and the EVAP canister are faulty<br>   • Fuel filler cap is loose<br>   • Fuel pump seal is defective, faulty or otherwise leaking<br>   • Hoses between the EVAP canister and the fuel flap unit are faulty<br>   • Hoses between the EVAP canister and the evaporative emission canister purge regulator valve are faulty |

| DTC | Trouble Code Title, Conditions & Possible Causes |
|---|---|
| **DTC: P1442**<br>**2T CCM, MIL: Yes**<br>**Years:** 2007, 2008<br>**Models:** Cooper<br>**Engines:** 1.6L, 1.6L Turbo<br>**Transmissions:** All | **EVAP Leak Detection Pump Control Circuit Low Conditions:**<br>Engine started, battery voltage must be at least 11.5v, all electrical components must be off, parking brake must be engaged (to keep daytime driving lights off), automatic transmission selector must be in park, the exhaust system must be properly sealed between the catalytic converter and the cylinder head, coolant temperature must be at least 80 degrees Celsius and oxygen sensor heaters for oxygen sensors before the catalytic converter must be functioning properly and the ground between the engine and the chassis must be well connected. The DME detected voltage irregularity in the leak detection pump control circuit.<br>**Possible Causes:**<br>• EVAP LDP power supply circuit is open<br>• EVAP LDP solenoid valve is damaged or it has failed<br>• EVAP LDP canister has a leak or a poor seal<br>• EVAP canister system has an improper seal<br>• Evaporative Emission (EVAP) canister purge regulator valve 1 has failed<br>• Leak Detection Pump (LDP) is faulty<br>• Aftermarket EVAP parts that do not conform to specifications<br>• EVAP component seals leaking (i.e., leaks in the Purge valve, fuel tank pressure sensor, canister vent solenoid, fuel vapor control valve tube assembly or fuel vapor vent valve). |
| **DTC: P1443**<br>**2T CCM, MIL: Yes**<br>**Years:** 2007, 2008<br>**Models:** Cooper<br>**Engines:** 1.6L, 1.6L Turbo<br>**Transmissions:** All | **EVAP Leak Detection Pump Control Circuit High Conditions:**<br>Engine started, battery voltage must be at least 11.5v, all electrical components must be off, parking brake must be engaged (to keep daytime driving lights off), automatic transmission selector must be in park, the exhaust system must be properly sealed between the catalytic converter and the cylinder head, coolant temperature must be at least 80 degrees Celsius and oxygen sensor heaters for oxygen sensors before the catalytic converter must be functioning properly and the ground between the engine and the chassis must be well connected. The DME detected voltage irregularity in the leak detection pump control circuit.<br>**Possible Causes:**<br>• EVAP LDP power supply circuit is open<br>• EVAP LDP solenoid valve is damaged or it has failed<br>• EVAP LDP canister has a leak or a poor seal<br>• EVAP canister system has an improper seal<br>• Evaporative Emission (EVAP) canister purge regulator valve 1 has failed<br>• Leak Detection Pump (LDP) is faulty<br>• Aftermarket EVAP parts that do not conform to specifications<br>• EVAP component seals leaking (i.e., leaks in the Purge valve, fuel tank pressure sensor, canister vent solenoid, fuel vapor control valve tube assembly or fuel vapor vent valve). |
| **DTC: P1475**<br>**2T CCM, MIL: Yes**<br>**Years:** 2007, 2008<br>**Models:** Cooper<br>**Engines:** 1.6L, 1.6L Turbo<br>**Transmissions:** All | **EVAP Emission Control LDP Circuit Malfunction Conditions:**<br>Key on, KOEO Self-Test enabled, and the DME detected an unexpected voltage condition on the EVAP emission control leak detection pump circuit. The reed switch level stays high after activation of solenoids within the time threshold of more than 0.5 seconds. The coolant temperature is less than 60 degrees Celsius, and the ambient pressure is greater than 76.2994kPa. The air intake temperature at start is between 9.04 and 16.04 degrees Celsius. The change in barometric pressure since engine start is less than 0.9998kPa. The vehicle speed is less than 74.56mph, and the purge valve has opened enough on previous driving cycle.<br>**Possible Causes:**<br>• Leak Detection Pump has failed<br>• EVAP canister system has an improper or broken seal<br>• Evaporative Emission (EVAP) canister purge regulator valve 1 is faulty<br>• Hoses between the fuel pump and the EVAP canister are faulty<br>• Fuel filler cap is loose<br>• Fuel pump seal is defective, faulty or otherwise leaking<br>• Hoses between the EVAP canister and the fuel flap unit are faulty<br>• Hoses between the EVAP canister and the evaporative emission canister purge regulator valve are faulty |

| DTC | Trouble Code Title, Conditions & Possible Causes |
|---|---|
| **DTC: P1476**<br>**2T CCM, MIL: Yes**<br>**Years:** 2007, 2008<br>**Models:** Cooper<br>**Engines:** 1.6L, 1.6L Turbo<br>**Transmissions:** All | **EVAP Emission Control LDP Circuit Malfunction/Insufficient Vacuum Conditions:**<br>Key on, KOEO Self-Test enabled, and the DME detected an unexpected voltage condition on the EVAP emission control leak detection pump circuit. There is a clamped tube during the time period of any of the five first pump cycles. The coolant temperature is less than 60 degrees Celsius, and the ambient pressure is greater than 76.2994kPa. The air intake temperature at start is between 9.04 and 16.04 degrees Celsius. The change in barometric pressure since engine start is less than 0.9998kPa. The vehicle speed is less than 74.56mph, and the purge valve has opened enough on previous driving cycle.<br>**Possible Causes:**<br>• Leak Detection Pump has failed<br>• EVAP canister system has an improper or broken seal<br>• Evaporative Emission (EVAP) canister purge regulator valve 1 is faulty<br>• Hoses between the fuel pump and the EVAP canister are faulty<br>• Fuel filler cap is loose<br>• Fuel pump seal is defective, faulty or otherwise leaking<br>• Hoses between the EVAP canister and the fuel flap unit are faulty<br>• Hoses between the EVAP canister and the evaporative emission canister purge regulator valve are faulty |
| **DTC: P1477**<br>**2T CCM, MIL: Yes**<br>**Years:** 2007, 2008<br>**Models:** Cooper<br>**Engines:** 1.6L, 1.6L Turbo<br>**Transmissions:** All | **EVAP Emission Control LDP Circuit Malfunction Conditions:**<br>Key on, KOEO Self-Test enabled, and the DME detected an unexpected voltage condition on the EVAP emission control leak detection pump circuit. The reed switch level stays continuously low after activation of solenoids within the time threshold of more than 1 second. The coolant temperature is less than 60 degrees Celsius, and the ambient pressure is greater than 76.2994kPa. The air intake temperature at start is between 9.04 and 16.04 degrees Celsius. The change in barometric pressure since engine start is less than 0.9998kPa. The vehicle speed is less than 74.56mph, and the purge valve has opened enough on previous driving cycle.<br>**Possible Causes:**<br>• Leak Detection Pump has failed<br>• EVAP canister system has an improper or broken seal<br>• Evaporative Emission (EVAP) canister purge regulator valve 1 is faulty<br>• Hoses between the fuel pump and the EVAP canister are faulty<br>• Fuel filler cap is loose<br>• Fuel pump seal is defective, faulty or otherwise leaking<br>• Hoses between the EVAP canister and the fuel flap unit are faulty<br>• Hoses between the EVAP canister and the evaporative emission canister purge regulator valve are faulty |
| **DTC: P1481**<br>**MIL: No**<br>**Years:** 2007, 2008<br>**Models:** Cooper<br>**Engines:** 1.6L, 1.6L Turbo<br>**Transmissions:** All | **Cooling Fans Circuit Short to Ground or Open Circuit Conditions:**<br>The DME detected a high condition on the sensor.<br>**Possible Causes:**<br>• The cooling fan has failed.<br>• The DME has failed<br>• Check wiring |
| **DTC: P1482**<br>**MIL: No**<br>**Years:** 2007, 2008<br>**Models:** Cooper<br>**Engines:** 1.6L, 1.6L Turbo<br>**Transmissions:** All | **Cooling Fans Circuit Short to Battery Conditions:**<br>The DME detected a Low condition on the sensor.<br>**Possible Causes:**<br>• The cooling fan has failed.<br>• The DME has failed<br>• Check wiring |
| **DTC: P1484**<br>**MIL: No**<br>**Years:** 2007, 2008<br>**Models:** Cooper<br>**Engines:** 1.6L, 1.6L Turbo<br>**Transmissions:** All | **Cooling Fans Circuit Short to Ground or Open Circuit Conditions:**<br>The DME detected a high condition on the sensor.<br>**Possible Causes:**<br>• The cooling fan has failed.<br>• The DME has failed<br>• Check wiring |
| **DTC: P1485**<br>**MIL: No**<br>**Years:** 2007, 2008<br>**Models:** Cooper<br>**Engines:** 1.6L, 1.6L Turbo<br>**Transmissions:** All | **Cooling Fans Circuit Short to Battery Conditions:**<br>The DME detected a Low condition on the sensor.<br>**Possible Causes:**<br>• The cooling fan has failed.<br>• The DME has failed<br>• Check wiring |

| DTC | Trouble Code Title, Conditions & Possible Causes |
|---|---|
| **DTC: P1496**<br>**1T CCM, MIL: Yes**<br>**Years:** 2007, 2008<br>**Models:** Cooper<br>**Engines:** 1.6L, 1.6L Turbo<br>**Transmissions:** All | **Air Intake System Leak (Block 3) Conditions:**<br>Engine speed greater than 704rpm. The manifold pressure is greater than 15.002kPa. The throttle position is less than 89.98 percent. The DME detected that a comparison of the modeled mass airflow at the cylinder and the mass airflow at the throttle exceeds the threshold relative to the throttle opening by more than 1.3.<br>**Possible Causes:**<br>• Charge air system leaks<br>• Recirculation valve for turbocharger is faulty<br>• Turbocharging system is damaged<br>• Vacuum diaphragm for turbocharger needs adjusting<br>• Wastegate bypass regulator valve is faulty |
| **DTC: P1572**<br>**1T CCM, MIL: Yes**<br>**Years:** 2007, 2008<br>**Models:** Cooper<br>**Engines:** 1.6L, 1.6L Turbo<br>**Transmissions:** All | **ECM Sensor Supply "A" Noisy Signal:**<br>Engine speed greater than 704rpm. The manifold pressure is greater than 15.002kPa. The throttle position is less than 89.98 percent. The DME detected that a comparison of the modeled mass airflow at the cylinder and the mass airflow at the throttle exceeds the threshold relative to the throttle opening by more than 1.3.<br>**Possible Causes:**<br>• Charge air system leaks<br>• Recirculation valve for turbocharger is faulty<br>• Turbocharging system is damaged<br>• Vacuum diaphragm for turbocharger needs adjusting<br>• Wastegate bypass regulator valve is faulty |
| **DTC: P1575**<br>**1T CCM, MIL: Yes**<br>**Years:** 2007, 2008<br>**Models:** Cooper<br>**Engines:** 1.6L, 1.6L Turbo<br>**Transmissions:** All | **ECM Sensor Supply "B" Noisy Signal:**<br>Engine speed greater than 704rpm. The manifold pressure is greater than 15.002kPa. The throttle position is less than 89.98 percent. The DME detected that a comparison of the modeled mass airflow at the cylinder and the mass airflow at the throttle exceeds the threshold relative to the throttle opening by more than 1.3.<br>**Possible Causes:**<br>• Charge air system leaks<br>• Recirculation valve for turbocharger is faulty<br>• Turbocharging system is damaged<br>• Vacuum diaphragm for turbocharger needs adjusting<br>• Wastegate bypass regulator valve is faulty |
| **DTC: P1600**<br>**2T CCM, MIL: Yes**<br>**Years:** 2007, 2008<br>**Models:** Cooper<br>**Engines:** 1.6L, 1.6L Turbo<br>**Transmissions:** All | **Internal Control Module Random Access Memory (RAM) Error Conditions:**<br>Key on, and the DME detected an internal memory fault. This code will set if KAPWR to the DME is interrupted (at the initial key on). Watchdog on.<br>**Possible Causes:**<br>• Battery terminal corrosion, or loose battery connection<br>• Connection to the DME interrupted, or the circuit has been opened<br>• Reprogramming error has occurred<br>and needs replacement. Remember to check for Aftermarket Performance Products before replacing a DME. |
| **DTC: P1607**<br>**2T CCM, MIL: Yes**<br>**Years:** 2007, 2008<br>**Models:** Cooper<br>**Engines:** 1.6L, 1.6L Turbo<br>**Transmissions:** All | **CAN Bus Error Conditions:**<br>Engine started, VSS over 1 mph, and the DME detected a problem in the CAN Bus system during the self-test.<br>**Possible Causes:**<br>• Open/short circuit to ground in the communication wire from the transmission to the DME.<br>• The DME has failed |
| **DTC: P1611**<br>**2T CCM, MIL: Yes**<br>**Years:** 2007, 2008<br>**Models:** Cooper<br>**Engines:** 1.6L, 1.6L Turbo<br>**Transmissions:** All | **MIL Call-Up Circuit, Transmission Control Module Short to Ground Conditions:**<br>Engine started, VSS over 1 mph, and the DME detected a problem in the Transmission Control system during the self-test.<br>**Possible Causes:**<br>• Open/short circuit to ground in the communication wire from the transmission to the DME.<br>• The DME has failed |
| **DTC: P1612**<br>**2T CCM, MIL: Yes**<br>**Years:** 2007, 2008<br>**Models:** Cooper<br>**Engines:** 1.6L, 1.6L Turbo<br>**Transmissions:** All | **INSTR Module Error Conditions:**<br>Engine started, VSS over 1 mph, and the DME detected a problem in the INSTR Module system during the self-test.<br>**Possible Causes:**<br>• Open/short circuit to ground in the communication wire from the transmission to the DME.<br>• The DME has failed |

| DTC | Trouble Code Title, Conditions & Possible Causes |
|---|---|
| **DTC: P1613**<br>**2T CCM, MIL: Yes**<br>**Years:** 2007, 2008<br>**Models:** Cooper<br>**Engines:** 1.6L, 1.6L Turbo<br>**Transmissions:** All | **ASC Error Conditions:**<br>Engine started, VSS over 1 mph, and the DME detected a problem in the ASC system during the self-test.<br>**Possible Causes:**<br>• Open/short circuit to ground in the communication wire from the transmission to the DME.<br>• The DME has failed |
| **DTC: P1615**<br>**2T CCM, MIL: Yes**<br>**Years:** 2007, 2008<br>**Models:** Cooper<br>**Engines:** 1.6L, 1.6L Turbo<br>**Transmissions:** All | **SPI-Bus Error Conditions:**<br>Engine started, VSS over 1 mph, and the DME detected a problem in the SPI Bus system during the self-test.<br>**Possible Causes:**<br>• Open/short circuit to ground in the communication wire from the transmission to the DME.<br>• The DME has failed |
| **DTC: P1656**<br>**MIL: No**<br>**Years:** 2007, 2008<br>**Models:** Cooper<br>**Engines:** 1.6L, 1.6L Turbo<br>**Transmissions:** All | **Timeout EWS (Electronic Immobilizer) Telegram Conditions:**<br>Key on the DME detected an electrical malfunction regarding the EWS. The wrong message was received. Check the EWS fuse, the ECU or CAN for fault, turn the ignition off and then on to repeat start calibration.<br>**Possible Causes:**<br>• Circuit from the MIL to the DME<br>• ECU Failure, BUS system failure<br>• Circuit from the EPC to the DME<br>• Defective CAS |
| **DTC: P1661**<br>**MIL: No**<br>**Years:** 2007, 2008<br>**Models:** Cooper<br>**Engines:** 1.6L, 1.6L Turbo<br>**Transmissions:** All | **Timeout EWS (Electronic Immobilizer) Telegram Conditions:**<br>Key on the DME detected an electrical malfunction regarding the EWS. Check the EWS fuse, the ECU or CAN for fault, turn the ignition off and then on to repeat start calibration.<br>**Possible Causes:**<br>• Circuit from the MIL to the DME<br>• ECU Failure, BUS system failure<br>• Circuit from the EPC to the DME<br>• Defective CAS |
| **DTC: P1679**<br>**1T CCM, MIL: Yes**<br>**Years:** 2007, 2008<br>**Models:** Cooper<br>**Engines:** 1.6L, 1.6L Turbo<br>**Transmissions:** All | **Monitoring of Torque Losses Conditions:**<br>Key on, engine running, the DME has detected that there is an error in the torque loss calculation. The limit was exceeded in the threshold map during the first 360 ms of operation.<br>**Possible Causes:**<br>• Battery terminal corrosion, loose battery connection, or faulty<br>• Connection to the DME interrupted, or the circuit has been opened<br>• Reprogramming error has occurred<br>and needs replacement.<br>• Voltage supply for Engine Control Module (DME) is faulty |
| **DTC: P1680**<br>**1T CCM, MIL: Yes**<br>**Years:** 2007, 2008<br>**Models:** Cooper<br>**Engines:** 1.6L, 1.6L Turbo<br>**Transmissions:** All | **Monitoring of A to D Conversion Conditions:**<br>Key on, engine running to at least 1200rpm, the DME has detected that the PVS ratio differences exceeds the threshold greater than 0.273 volts.<br>**Possible Causes:**<br>• Battery terminal corrosion, loose battery connection, or faulty<br>• Connection to the DME interrupted, or the circuit has been opened<br>• Reprogramming error has occurred<br>and needs replacement.<br>• Voltage supply for Engine Control Module (DME) is faulty |
| **DTC: P1681**<br>**1T CCM, MIL: Yes**<br>**Years:** 2007, 2008<br>**Models:** Cooper<br>**Engines:** 1.6L, 1.6L Turbo<br>**Transmissions:** All | **Monitoring of Engine Speed Conditions:**<br>Key on, engine running to at least 1200rpm, the DME has detected that the engine speed difference exceeds the threshold of 576rpm<br>**Possible Causes:**<br>• Battery terminal corrosion, loose battery connection, or faulty<br>• Connection to the DME interrupted, or the circuit has been opened<br>• Reprogramming error has occurred<br>and needs replacement.<br>• Voltage supply for Engine Control Module (DME) is faulty |

| DTC | Trouble Code Title, Conditions & Possible Causes |
|---|---|
| **DTC: P1682**<br>**1T CCM, MIL: Yes**<br>**Years:** 2007, 2008<br>**Models:** Cooper<br>**Engines:** 1.6L, 1.6L Turbo<br>**Transmissions:** All | **Idle Speed Control, Monitoring of the Proportional Derivative Conditions:**<br>Key on, engine running to at least 1200rpm, the DME has detected that there is an error in the torque demand from the proportional derivative part. The maximum limit has been exceeded.<br>**Possible Causes:**<br>• Battery terminal corrosion, loose battery connection, or faulty<br>• Connection to the DME interrupted, or the circuit has been opened<br>• Reprogramming error has occurred<br>and needs replacement.<br>• Voltage supply for Engine Control Module (DME) is faulty |
| **DTC: P1683**<br>**1T CCM, MIL: Yes**<br>**Years:** 2007, 2008<br>**Models:** Cooper<br>**Engines:** 1.6L, 1.6L Turbo<br>**Transmissions:** All | **Idle Speed Control, Monitoring of the Integral Part Conditions:**<br>Key on, engine running to at least 1200rpm, the DME has detected that there is an error in the torque demand from the integral part is greater than 25NM.<br>**Possible Causes:**<br>• Battery terminal corrosion, loose battery connection, or faulty<br>• Connection to the DME interrupted, or the circuit has been opened<br>• Reprogramming error has occurred<br>and needs replacement.<br>• Voltage supply for Engine Control Module (DME) is faulty |
| **DTC: P1684**<br>**1T CCM, MIL: Yes**<br>**Years:** 2007, 2008<br>**Models:** Cooper<br>**Engines:** 1.6L, 1.6L Turbo<br>**Transmissions:** All | **Monitoring of Minimum Torque at Clutch Conditions:**<br>Key on, engine running, the DME has detected that there is an error in the minimum torque at the clutch calculation. The limit was exceeded in the threshold map.<br>**Possible Causes:**<br>• Battery terminal corrosion, loose battery connection, or faulty<br>• Connection to the DME interrupted, or the circuit has been opened<br>• Reprogramming error has occurred<br>and needs replacement.<br>• Voltage supply for Engine Control Module (DME) is faulty |
| **DTC: P1685**<br>**1T CCM, MIL: Yes**<br>**Years:** 2007, 2008<br>**Models:** Cooper<br>**Engines:** 1.6L, 1.6L Turbo<br>**Transmissions:** All | **Monitoring of Maximum Torque at Clutch Conditions:**<br>Key on, engine running, the DME has detected that there is an error in the maximum torque at the clutch calculation. The limit was exceeded in the threshold map.<br>**Possible Causes:**<br>• Battery terminal corrosion, loose battery connection, or faulty·<br>• Connection to the DME interrupted, or the circuit has been opened<br>• Reprogramming error has occurred<br>and needs replacement.<br>• Voltage supply for Engine Control Module (DME) is faulty |
| **DTC: P1686**<br>**1T CCM, MIL: Yes**<br>**Years:** 2007, 2008<br>**Models:** Cooper<br>**Engines:** 1.6L, 1.6L Turbo<br>**Transmissions:** All | **Monitoring of Pedal Values Conditions:**<br>Key on, engine running, the DME has detected that there is an error in pedal value checks. The difference exceeds the threshold map by 15.23 to 28.91 percent.<br>**Possible Causes:**<br>• Battery terminal corrosion, loose battery connection, or faulty<br>• Connection to the DME interrupted, or the circuit has been opened<br>• Reprogramming error has occurred<br>and needs replacement.<br>• Voltage supply for Engine Control Module (DME) is faulty |
| **DTC: P1687**<br>**1T CCM, MIL: Yes**<br>**Years:** 2007, 2008<br>**Models:** Cooper<br>**Engines:** 1.6L, 1.6L Turbo<br>**Transmissions:** All | **Monitoring of Throttle Position Conditions:**<br>Key on, engine running, the DME has detected that there is an error in the throttle position sensor ratio calculation by greater than 0.313 volts.<br>**Possible Causes:**<br>• Battery terminal corrosion, loose battery connection, or faulty<br>• Connection to the DME interrupted, or the circuit has been opened<br>• Reprogramming error has occurred<br>and needs replacement.<br>• Voltage supply for Engine Control Module (DME) is faulty |

| DTC | Trouble Code Title, Conditions & Possible Causes |
|---|---|
| **DTC: P1688**<br>**1T CCM, MIL: Yes**<br>**Years:** 2007, 2008<br>**Models:** Cooper<br>**Engines:** 1.6L, 1.6L Turbo<br>**Transmissions:** All | **Monitoring of Mass Airflow Conditions:**<br>Key on, engine running, the DME has detected that there is an error in the MAF calculation. The limit was exceeded in the threshold map by 0.044 to 0.218g/rev.<br>**Possible Causes:**<br>• Battery terminal corrosion, loose battery connection, or faulty<br>• Connection to the DME interrupted, or the circuit has been opened<br>• Reprogramming error has occurred<br>and needs replacement.<br>• Voltage supply for Engine Control Module (DME) is faulty |
| **DTC: P1689**<br>**1T CCM, MIL: Yes**<br>**Years:** 2007, 2008<br>**Models:** Cooper<br>**Engines:** 1.6L, 1.6L Turbo<br>**Transmissions:** All | **Monitoring of Actual Indicated Engine Torque Conditions:**<br>Key on, engine running, the DME has detected that there is an error in the maximum torque at the clutch calculation. The limit was exceeded in the threshold map by 30 to 38NM.<br>**Possible Causes:**<br>• Battery terminal corrosion, loose battery connection, or faulty<br>• Connection to the DME interrupted, or the circuit has been opened<br>• Reprogramming error has occurred<br>and needs replacement.<br>• Voltage supply for Engine Control Module (DME) is faulty |
| **DTC: P1691**<br>**1T CCM, MIL: Yes**<br>**Years:** 2007, 2008<br>**Models:** Cooper<br>**Engines:** 1.6L, 1.6L Turbo<br>**Transmissions:** All | **Monitoring of Engine Speed Limit in Limp Home Conditions:**<br>Key on, engine running, the DME has detected that monitoring of the engine speed limit in limp home condition exceeds the threshold map by greater than 2656rpm.<br>**Possible Causes:**<br>• Battery terminal corrosion, loose battery connection, or faulty<br>• Connection to the DME interrupted, or the circuit has been opened<br>• Reprogramming error has occurred<br>and needs replacement.<br>• Voltage supply for Engine Control Module (DME) is faulty |
| **DTC: P1692**<br>**1T CCM, MIL: Yes**<br>**Years:** 2007, 2008<br>**Models:** Cooper<br>**Engines:** 1.6L, 1.6L Turbo<br>**Transmissions:** All | **Monitoring of Processor Calculations Conditions:**<br>Key on, engine running, the DME has detected that there is an error in the for the final request for disabled power stages of MTC and IV.<br>**Possible Causes:**<br>• Battery terminal corrosion, loose battery connection, or faulty<br>• Connection to the DME interrupted, or the circuit has been opened<br>• Reprogramming error has occurred<br>and needs replacement.<br>• Voltage supply for Engine Control Module (DME) is faulty |
| **DTC: P1693**<br>**1T CCM, MIL: Yes**<br>**Years:** 2007, 2008<br>**Models:** Cooper<br>**Engines:** 1.6L, 1.6L Turbo<br>**Transmissions:** All | **Monitoring of Processor Calculations Conditions:**<br>Key on, engine running, the DME has detected that there is an error in the for the temporary request for disabled power stages of MTC and IV.<br>**Possible Causes:**<br>• Battery terminal corrosion, loose battery connection, or faulty<br>• Connection to the DME interrupted, or the circuit has been opened<br>• Reprogramming error has occurred<br>and needs replacement.<br>• Voltage supply for Engine Control Module (DME) is faulty |
| **DTC: P1698**<br>**1T CCM, MIL: Yes**<br>**Years:** 2007, 2008<br>**Models:** Cooper<br>**Engines:** 1.6L, 1.6L Turbo<br>**Transmissions:** A/T | **ECU Functionality Incorrect Conditions:**<br>The ECU Functionality is in error as there are internal errors. This test is performed by the GIB (Gearbox Interface Box), a system dedicated to low level control of the transmission control unit.<br>**Possible Causes:**<br>• Short to battery<br>• Short to ground<br>• Open circuit<br>• CAN data bus wires have short circuited to each other<br>• ECU has failed<br>• BUS system failure<br>• Defective bus controller (EGS) |

| DTC | Trouble Code Title, Conditions & Possible Causes |
|---|---|
| **DTC: P1699**<br>**1T CCM, MIL: Yes**<br>**Years:** 2007, 2008<br>**Models:** Cooper<br>**Engines:** All<br>**Transmissions:** A/T | **EPROM Checksum Incorrect Conditions:**<br>The EPROM Checksum is incorrect. This test is performed by the GIB (Gearbox Interface Box), a system dedicated to low level control of the transmission control unit.<br>**Possible Causes:**<br>• Short to battery<br>• Short to ground<br>• Open circuit<br>• CAN data bus wires have short circuited to each other<br>• ECU has failed<br>• BUS system failure<br>• Defective bus controller (EGS) |
| **DTC: P1705**<br>**2T CCM, MIL: Yes**<br>**Years:** 2007, 2008<br>**Models:** Cooper<br>**Engines:** 1.6L, 1.6L Turbo<br>**Transmissions:** A/T | **LED Drives Plausibility Conditions:**<br>Key on or engine running; and the DME detected an implausible signal (fault performed by the Gearbox Interface Box). The battery voltage is greater than 9 volts and the CAN Bus is operational.<br>**Possible Causes:**<br>• Short to battery<br>• Short to ground<br>• Open circuit<br>• CAN data bus wires have short circuited to each other<br>• ECU has failed<br>• BUS system failure<br>• Defective bus controller (EGS) |
| **DTC: P1706**<br>**2T CCM, MIL: Yes**<br>**Years:** 2007, 2008<br>**Models:** Cooper<br>**Engines:** 1.6L, 1.6L Turbo<br>**Transmissions:** A/T | **LED Drives Short Circuit Conditions:**<br>Key on or engine running; and the DME detected short circuit (fault performed by the Gearbox Interface Box). The battery voltage is greater than 9 volts and the CAN Bus is operational.<br>**Possible Causes:**<br>• Short to battery<br>• Short to ground<br>• Open circuit<br>• CAN data bus wires have short circuited to each other<br>• ECU has failed<br>• BUS system failure<br>• Defective bus controller (EGS) |
| **DTC: P1739**<br>**2T CCM, MIL: Yes**<br>**Years:** 2007, 2008<br>**Models:** Cooper<br>**Engines:** 1.6L, 1.6L Turbo<br>**Transmissions:** A/T | **Clutch Solenoid Circuit Communication Error Conditions:**<br>The clutch solenoid circuit signal is implausible or missing. This test is performed by the GIB (Gearbox Interface Box), a system dedicated to low level control of the transmission control unit.<br>**Possible Causes:**<br>• Short to battery<br>• Short to ground<br>• Open circuit |
| **DTC: P1741**<br>**2T CCM, MIL: Yes**<br>**Years:** 2007, 2008<br>**Models:** Cooper<br>**Engines:** 1.6L, 1.6L Turbo<br>**Transmissions:** A/T | **Clutch Solenoid Circuit Open Circuit Conditions:**<br>The clutch solenoid circuit continuity is in error. This test is performed by the GIB (Gearbox Interface Box), a system dedicated to low level control of the transmission control unit.<br>**Possible Causes:**<br>• Short to battery<br>• Short to ground<br>• Open circuit |
| **DTC: P1742**<br>**2T CCM, MIL: Yes**<br>**Years:** 2007, 2008<br>**Models:** Cooper<br>**Engines:** 1.6L, 1.6L Turbo<br>**Transmissions:** A/T | **Clutch Solenoid Circuit Short Circuit Conditions:**<br>The clutch solenoid circuit continuity is in error. This test is performed by the GIB (Gearbox Interface Box), a system dedicated to low level control of the transmission control unit.<br>**Possible Causes:**<br>• Short to battery<br>• Short to ground<br>• Open circuit |

| DTC | Trouble Code Title, Conditions & Possible Causes |
|---|---|
| **DTC: P1749**<br>**2T CCM, MIL: Yes**<br>**Years:** 2007, 2008<br>**Models:** Cooper<br>**Engines:** 1.6L, 1.6L Turbo<br>**Transmissions:** A/T | **Secondary Pressure Solenoid Circuit Communication Error Conditions:**<br>The Secondary Pressure circuit signal is implausible or missing. This test is performed by the GIB (Gearbox Interface Box), a system dedicated to low level control of the transmission control unit.<br>**Possible Causes:**<br>• Short to battery<br>• Short to ground<br>• Open circuit |
| **DTC: P1751**<br>**2T CCM, MIL: Yes**<br>**Years:** 2007, 2008<br>**Models:** Cooper<br>**Engines:** 1.6L, 1.6L Turbo<br>**Transmissions:** A/T | **Secondary Pressure Circuit Open Circuit Conditions:**<br>The Secondary Pressure circuit continuity is in error. This test is performed by the GIB (Gearbox Interface Box), a system dedicated to low level control of the transmission control unit.<br>**Possible Causes:**<br>• Short to battery<br>• Short to ground<br>• Open circuit |
| **DTC: P1752**<br>**2T CCM, MIL: Yes**<br>**Years:** 2007, 2008<br>**Models:** Cooper<br>**Engines:** 1.6L, 1.6L Turbo<br>**Transmissions:** A/T | **Secondary Pressure Solenoid Circuit Short Circuit Conditions:**<br>The Secondary Pressure circuit continuity is in error. This test is performed by the GIB (Gearbox Interface Box), a system dedicated to low level control of the transmission control unit.<br>**Possible Causes:**<br>• Short to battery<br>• Short to ground<br>• Open circuit |
| **DTC: P1785**<br>**2T CCM, MIL: Yes**<br>**Years:** 2007, 2008<br>**Models:** Cooper<br>**Engines:** 1.6L, 1.6L Turbo<br>**Transmissions:** A/T | **Secondary Pressure Solenoid Circuit Short Circuit Conditions:**<br>The Secondary Pressure circuit continuity is in error. This test is performed by the GIB (Gearbox Interface Box), a system dedicated to low level control of the transmission control unit.<br>**Possible Causes:**<br>• Short to battery<br>• Short to ground<br>• Open circuit |
| **DTC: P1786**<br>**2T CCM, MIL: Yes**<br>**Years:** 2007, 2008<br>**Models:** Cooper<br>**Engines:** 1.6L, 1.6L Turbo<br>**Transmissions:** A/T | **Secondary Pressure Solenoid Circuit Short Circuit Conditions:**<br>The Secondary Pressure circuit continuity is in error. This test is performed by the GIB (Gearbox Interface Box), a system dedicated to low level control of the transmission control unit.<br>**Possible Causes:**<br>• Short to battery<br>• Short to ground<br>• Open circuit |
| **DTC: P1787**<br>**2T CCM, MIL: Yes**<br>**Years:** 2007, 2008<br>**Models:** Cooper<br>**Engines:** 1.6L, 1.6L Turbo<br>**Transmissions:** A/T | **Secondary Pressure Solenoid Circuit Short Circuit Conditions:**<br>The Secondary Pressure circuit continuity is in error. This test is performed by the GIB (Gearbox Interface Box), a system dedicated to low level control of the transmission control unit.<br>**Possible Causes:**<br>• Short to battery<br>• Short to ground<br>• Open circuit |
| **DTC: P1788**<br>**2T CCM, MIL: Yes**<br>**Years:** 2007, 2008<br>**Models:** Cooper<br>**Engines:** 1.6L, 1.6L Turbo<br>**Transmissions:** A/T | **Secondary Pressure Solenoid Circuit Short Circuit Conditions:**<br>The Secondary Pressure circuit continuity is in error. This test is performed by the GIB (Gearbox Interface Box), a system dedicated to low level control of the transmission control unit.<br>**Possible Causes:**<br>• Short to battery<br>• Short to ground<br>• Open circuit |
| **DTC: P1789**<br>**2T CCM, MIL: Yes**<br>**Years:** 2007, 2008<br>**Models:** Cooper<br>**Engines:** 1.6L, 1.6L Turbo<br>**Transmissions:** A/T | **Secondary Pressure Solenoid Circuit Short Circuit Conditions:**<br>The Secondary Pressure circuit continuity is in error. This test is performed by the GIB (Gearbox Interface Box), a system dedicated to low level control of the transmission control unit.<br>**Possible Causes:**<br>• Short to battery<br>• Short to ground<br>• Open circuit |

| DTC | Trouble Code Title, Conditions & Possible Causes |
|---|---|
| **DTC: P1815**<br>**2T CCM, MIL: Yes**<br>**Years:** 2007, 2008<br>**Models:** Cooper<br>**Engines:** 1.6L, 1.6L Turbo<br>**Transmissions:** A/T | **Secondary Pressure Solenoid Circuit Short Circuit Conditions:**<br>The Secondary Pressure circuit continuity is in error. This test is performed by the GIB (Gearbox Interface Box), a system dedicated to low level control of the transmission control unit.<br>**Possible Causes:**<br>• Short to battery<br>• Short to ground<br>• Open circuit |
| **DTC: P1816**<br>**2T CCM, MIL: Yes**<br>**Years:** 2007, 2008<br>**Models:** Cooper<br>**Engines:** 1.6L, 1.6L Turbo<br>**Transmissions:** A/T | **Secondary Pressure Solenoid Circuit Short Circuit Conditions:**<br>The Secondary Pressure circuit continuity is in error. This test is performed by the GIB (Gearbox Interface Box), a system dedicated to low level control of the transmission control unit.<br>**Possible Causes:**<br>• Short to battery<br>• Short to ground<br>• Open circuit |

## Gas Engine OBD II Trouble Code List (P2xxx Codes)

| DTC | Trouble Code Title, Conditions & Possible Causes |
|---|---|
| **DTC: P2065**<br>**MIL: No**<br>**Years:** 2007, 2008<br>**Models:** Cooper<br>**Engines:** 1.6L, 1.6L Turbo<br>**Transmissions:** All | **Fuel Level Sensor "B" Circuit:**<br>The DME detected a high condition on the pump.<br>**Possible Causes:**<br>• The fuel pump has failed.<br>• Short to positive<br>• Check wiring |
| **DTC: P2067**<br>**MIL: No**<br>**Years:** 2007, 2008<br>**Models:** Cooper<br>**Engines:** 1.6L, 1.6L Turbo<br>**Transmissions:** All | **Fuel Level Sensor "B" Circuit Low:**<br>The DME detected a high condition on the pump.<br>**Possible Causes:**<br>• The fuel pump has failed.<br>• Short to ground<br>• Check wiring |
| **DTC: P2088**<br>**MIL: No**<br>**Years:** 2007, 2008<br>**Models:** Cooper<br>**Engines:** 1.6L, 1.6L Turbo<br>**Transmissions:** All | **"A" Camshaft Position Actuator Control Circuit Low (Bank 1):**<br>The DME detected a high condition on the pump.<br>**Possible Causes:**<br>• Short to positive VANOS intake-side solenoid valve.<br>• Check wiring<br>• DME |
| **DTC: P2089**<br>**MIL: No**<br>**Years:** 2007, 2008<br>**Models:** Cooper<br>**Engines:** 1.6L, 1.6L Turbo<br>**Transmissions:** All | **"A" Camshaft Position Actuator Control Circuit/Open (Bank 1):**<br>The DME detected a high condition on the pump.<br>**Possible Causes:**<br>• Short to positive VANOS Intake-side solenoid valve.<br>• Check wiring<br>• DME |
| **DTC: P2090**<br>**MIL: No**<br>**Years:** 2007, 2008<br>**Models:** Cooper<br>**Engines:** 1.6L, 1.6L Turbo<br>**Transmissions:** All | **"B" Camshaft Position Actuator Control Circuit High (Bank 1):**<br>The DME detected a high condition on the pump.<br>**Possible Causes:**<br>• Short to positive VANOS exhaust-side solenoid valve.<br>• Check wiring<br>• DME |
| **DTC: P2091**<br>**MIL: No**<br>**Years:** 2007, 2008<br>**Models:** Cooper<br>**Engines:** 1.6L, 1.6L Turbo<br>**Transmissions:** All | **"B" Camshaft Position Actuator Control Circuit Low (Bank 1):**<br>The DME detected a high condition on the pump.<br>**Possible Causes:**<br>• Short to positive VANOS exhaust-side solenoid valve.<br>• Check wiring<br>• DME |

| DTC | Trouble Code Title, Conditions & Possible Causes |
|-----|--------------------------------------------------|
| **DTC: P2096**<br>**2T CCM, MIL: Yes**<br>**Years:** 2007, 2008<br>**Models:** Cooper<br>**Engines:** 1.6L, 1.6L Turbo<br>**Transmissions:** All | **Post Catalyst Fuel Trim System Too Lean (Bank 1) Conditions:**<br>Engine started, battery voltage must be at least 11.5v, all electrical components must be off, the ground between the engine and the chassis must be well connected, the exhaust system must be properly sealed between the catalytic converter and the cylinder head, and the oxygen sensor heater for oxygen sensor before the catalytic converter must be properly functioning. The DME detected a problem with the fuel mixture. Trim control 1 segment (precision controller with oxygen sensor behind cat.) below delta lambda threshold of less than −1.56. Coolant temperature greater than 45 degrees Celsius. O2 heaters ready, fuel system in a closed loop, but the rear O2 sensor is in voltage outside the parameters.<br>**Note: For resistance testing of sensor heating, oxygen sensor should be cooled to ambient temperature. High temperatures at oxygen sensor may lead to inaccurate measurements.**<br>**Possible Causes:**<br>• Oxygen sensor (before catalytic converter) is faulty<br>• Oxygen sensor (behind catalytic converter) is faulty<br>• Oxygen sensor heater (before catalytic converter) is faulty<br>• Oxygen sensor heater (behind catalytic converter) is faulty<br>• Check circuits for shorts to each other, ground or power |
| **DTC: P2097**<br>**2T CCM, MIL: Yes**<br>**Years:** 2007, 2008<br>**Models:** Cooper<br>**Engines:** 1.6L, 1.6L Turbo<br>**Transmissions:** All | **Post Catalyst Fuel Trim System Too Rich (Bank 1) Conditions:**<br>Engine started, battery voltage must be at least 11.5v, all electrical components must be off, the ground between the engine and the chassis must be well connected, the exhaust system must be properly sealed between the catalytic converter and the cylinder head, and the oxygen sensor heater for oxygen sensor before the catalytic converter must be properly functioning. The DME detected a problem with the fuel mixture. Trim control 1 segment (precision controller with oxygen sensor behind cat.) below delta lambda threshold of less than −1.56. Coolant temperature greater than 45 degrees Celsius. O2 heaters ready, fuel system in a closed loop, but the rear O2 sensor is in voltage outside the parameters.<br>**Note: For resistance testing of sensor heating, oxygen sensor should be cooled to ambient temperature. High temperatures at oxygen sensor may lead to inaccurate measurements.**<br>**Possible Causes:**<br>• Oxygen sensor (before catalytic converter) is faulty<br>• Oxygen sensor (behind catalytic converter) is faulty<br>• Oxygen sensor heater (before catalytic converter) is faulty<br>• Oxygen sensor heater (behind catalytic converter) is faulty<br>• Check circuits for shorts to each other, ground or power |
| **DTC: P2122**<br>**2T CCM, MIL: Yes**<br>**Years:** 2007, 2008<br>**Models:** Cooper<br>**Engines:** 1.6L, 1.6L Turbo<br>**Transmissions:** All | **Accelerator Pedal Position Sensor 'D' Circuit Low Input Conditions:**<br>Engine started, battery voltage at least 11.5v, all electrical components off, ground connections between engine and chassis well connected, the DME detected that the accelerator pedal position sensor signal was outside the parameters to function normally.<br>**Note: Both the Throttle Position (TP) Sensor and Accelerator Pedal Position Sensor are located at the accelerator pedal module and communicate the driver's intentions to the DME completely independently of each other. Both sensors are stored in one housing.**<br>**Possible Causes:**<br>• Ground between engine and chassis may be broken<br>• Throttle position sensor may have failed<br>• Accelerator Pedal Position Sensor has failed<br>• Throttle position sensor wiring may have shorted<br>• Throttle position sensor has failed<br>• Faulty voltage supply |
| **DTC: P2123**<br>**2T CCM, MIL: Yes**<br>**Years:** 2007, 2008<br>**Models:** Cooper<br>**Engines:** 1.6L, 1.6L Turbo<br>**Transmissions:** All | **Accelerator Pedal Position Sensor 'D' Circuit High Input Conditions:**<br>Engine started, battery voltage at least 11.5v, all electrical components off, ground connections between engine and chassis well connected, the DME detected that the accelerator pedal position sensor signal was outside the parameters to function normally.<br>**Note: Both the Throttle Position (TP) Sensor and Accelerator Pedal Position Sensor are located at the accelerator pedal module and communicate the driver's intentions to the DME completely independently of each other. Both sensors are stored in one housing.**<br>**Possible Causes:**<br>• Ground between engine and chassis may be broken<br>• Throttle position sensor may have failed<br>• Accelerator Pedal Position Sensor has failed<br>• Throttle position sensor wiring may have shorted<br>• Throttle position sensor has failed<br>• Faulty voltage supply |

| DTC | Trouble Code Title, Conditions & Possible Causes |
|---|---|
| **DTC: P2127**<br>**2T CCM, MIL: Yes**<br>**Years:** 2007, 2008<br>**Models:** Cooper<br>**Engines:** 1.6L, 1.6L Turbo<br>**Transmissions:** All | **Accelerator Pedal Position Sensor 'E' Circuit Low Input Conditions:**<br>Engine started, battery voltage at least 11.5v, all electrical components off, ground connections between engine and chassis well connected, the DME detected that the accelerator pedal position sensor signal was outside the parameters to function normally.<br>**Note: Both the Throttle Position (TP) Sensor and Accelerator Pedal Position Sensor are located at the accelerator pedal module and communicate the driver's intentions to the DME completely independently of each other. Both sensors are stored in one housing.**<br>**Possible Causes:**<br>    • Ground between engine and chassis may be broken<br>    • Throttle position sensor may have failed<br>    • Accelerator Pedal Position Sensor has failed<br>    • Throttle position sensor wiring may have shorted<br>    • Throttle position sensor has failed<br>    • Faulty voltage supply |
| **DTC: P2128**<br>**2T CCM, MIL: Yes**<br>**Years:** 2007, 2008<br>**Models:** Cooper<br>**Engines:** 1.6L, 1.6L Turbo<br>**Transmissions:** All | **Accelerator Pedal Position Sensor 'E' Circuit High Input Conditions:**<br>Engine started, battery voltage at least 11.5v, all electrical components off, ground connections between engine and chassis well connected, the DME detected that the accelerator pedal position sensor signal was outside the parameters to function normally.<br>**Note: Both the Throttle Position (TP) Sensor and Accelerator Pedal Position Sensor are located at the accelerator pedal module and communicate the driver's intentions to the DME completely independently of each other. Both sensors are stored in one housing.**<br>**Possible Causes:**<br>    • Ground between engine and chassis may be broken<br>    • Throttle position sensor may have failed<br>    • Accelerator Pedal Position Sensor has failed<br>    • Throttle position sensor wiring may have shorted<br>    • Throttle position sensor has failed<br>    • Faulty voltage supply |
| **DTC: P2138**<br>**2T CCM, MIL: Yes**<br>**Years:** 2007, 2008<br>**Models:** Cooper<br>**Engines:** 1.6L, 1.6L Turbo<br>**Transmissions:** All | **Throttle Position Sensor D/E Voltage Correlation Conditions:**<br>Engine started, battery voltage must be at least 11.5v, all electrical components must be off, parking brake must be engaged (to keep daytime driving lights off), automatic transmission selector must be in park; and the DME detected the Throttle Position 'D' (TPD) and Throttle Position 'B' (TPE) sensors disagreed, or that the TPD sensor should not be in its detected position, or that the TPE sensor should not be in its detected position during testing.<br>**Note: Both the Throttle Position (TP) Sensor and Accelerator Pedal Position Sensor are located at the accelerator pedal module and communicate the driver's intentions to the DME completely independently of each other. Both sensors are stored in one housing.**<br>**Possible Causes:**<br>    • ETC TP sensor connector is damaged or shorted<br>    • ETC TP sensor circuits shorted together in the wire harness<br>    • ETC TP sensor signal circuit is shorted to VREF (5v)<br>    • ETC TP sensor is damaged or the DME has failed |
| **DTC: P2177**<br>**2T CCM, MIL: Yes**<br>**Years:** 2007, 2008<br>**Models:** Cooper<br>**Engines:** 1.6L, 1.6L Turbo<br>**Transmissions:** All | **System Too Lean Off Idle (Bank 1):**<br>Engine started, battery voltage must be at least 11.5v, all electrical components must be off, parking brake must be engaged (to keep daytime driving lights off), automatic transmission selector must be in park; and the DME detected the Throttle Position 'D' (TPD) and Throttle Position 'B' (TPE) sensors disagreed, or that the TPD sensor should not be in its detected position, or that the TPE sensor should not be in its detected position during testing.<br>**Note: Both the Throttle Position (TP) Sensor and Accelerator Pedal Position Sensor are located at the accelerator pedal module and communicate the driver's intentions to the DME completely independently of each other. Both sensors are stored in one housing.**<br>**Possible Causes:**<br>    • ETC TP sensor connector is damaged or shorted<br>    • ETC TP sensor circuits shorted together in the wire harness<br>    • ETC TP sensor signal circuit is shorted to VREF (5v)<br>    • ETC TP sensor is damaged or the DME has failed |

| DTC | Trouble Code Title, Conditions & Possible Causes |
|---|---|
| **DTC: P2178**<br>**2T CCM, MIL: Yes**<br>**Years:** 2007, 2008<br>**Models:** Cooper<br>**Engines:** 1.6L, 1.6L Turbo<br>**Transmissions:** All | **System Too Rich Off Idle (Bank 1):**<br>Engine started, battery voltage must be at least 11.5v, all electrical components must be off, parking brake must be engaged (to keep daytime driving lights off), automatic transmission selector must be in park; and the DME detected the Throttle Position 'D' (TPD) and Throttle Position 'B' (TPE) sensors disagreed, or that the TPD sensor should not be in its detected position, or that the TPE sensor should not be in its detected position during testing.<br>**Note: Both the Throttle Position (TP) Sensor and Accelerator Pedal Position Sensor are located at the accelerator pedal module and communicate the driver's intentions to the DME completely independently of each other. Both sensors are stored in one housing.**<br>**Possible Causes:**<br>• ETC TP sensor connector is damaged or shorted<br>• ETC TP sensor circuits shorted together in the wire harness<br>• ETC TP sensor signal circuit is shorted to VREF (5v)<br>• ETC TP sensor is damaged or the DME has failed |
| **DTC: P2187**<br>**2T CCM, MIL: Yes**<br>**Years:** 2007, 2008<br>**Models:** Cooper<br>**Engines:** 1.6L, 1.6L Turbo<br>**Transmissions:** All | **System Too Lean At Idle (Bank 1):**<br>Engine started, battery voltage must be at least 11.5v, all electrical components must be off, parking brake must be engaged (to keep daytime driving lights off), automatic transmission selector must be in park; and the DME detected the Throttle Position 'D' (TPD) and Throttle Position 'B' (TPE) sensors disagreed, or that the TPD sensor should not be in its detected position, or that the TPE sensor should not be in its detected position during testing.<br>**Note: Both the Throttle Position (TP) Sensor and Accelerator Pedal Position Sensor are located at the accelerator pedal module and communicate the driver's intentions to the DME completely independently of each other. Both sensors are stored in one housing.**<br>**Possible Causes:**<br>• ETC TP sensor connector is damaged or shorted<br>• ETC TP sensor circuits shorted together in the wire harness<br>• ETC TP sensor signal circuit is shorted to VREF (5v)<br>• ETC TP sensor is damaged or the DME has failed |
| **DTC: P2188**<br>**2T CCM, MIL: Yes**<br>**Years:** 2007, 2008<br>**Models:** Cooper<br>**Engines:** 1.6L, 1.6L Turbo<br>**Transmissions:** All | **System Too Rich At Idle (Bank 1):**<br>Engine started, battery voltage must be at least 11.5v, all electrical components must be off, parking brake must be engaged (to keep daytime driving lights off), automatic transmission selector must be in park; and the DME detected the Throttle Position 'D' (TPD) and Throttle Position 'B' (TPE) sensors disagreed, or that the TPD sensor should not be in its detected position, or that the TPE sensor should not be in its detected position during testing.<br>**Note: Both the Throttle Position (TP) Sensor and Accelerator Pedal Position Sensor are located at the accelerator pedal module and communicate the driver's intentions to the DME completely independently of each other. Both sensors are stored in one housing.**<br>**Possible Causes:**<br>• ETC TP sensor connector is damaged or shorted<br>• ETC TP sensor circuits shorted together in the wire harness<br>• ETC TP sensor signal circuit is shorted to VREF (5v)<br>• ETC TP sensor is damaged or the DME has failed |
| **DTC: P2195**<br>**2T CCM, MIL: Yes**<br>**Years:** 2007, 2008<br>**Models:** Cooper<br>**Engines:** 1.6L, 1.6L Turbo<br>**Transmissions:** All | **O2S Signal Biased/Stuck Lean (Bank 1 Sensor 1):**<br>Engine running, battery voltage 11.5, all electrical components off, ground between engine and chassis well connected and the exhaust system must be properly sealed between catalytic converter and the cylinder head. The DME detected the HO2S signal was in a negative voltage range referred to as "character shift downward". This code sets when the HO2S signal remains in a low state for a measured period of time. In effect, it does not switch properly in the closed loop operation. Engine speed is less than 8000rpm.<br>**Possible Causes:**<br>• Wiring fault<br>• O2S befor catalytic converter defective<br>• Check wiring harness plug connections CMI-216, CMI-215, CMI-228, CMI-227, CMI-230<br>• Replace O2S |
| **DTC: P2196**<br>**2T CCM, MIL: Yes**<br>**Years:** 2007, 2008<br>**Models:** Cooper<br>**Engines:** 1.6L, 1.6L Turbo<br>**Transmissions:** All | **O2S Signal Biased/Stuck Rich (Bank 1 Sensor 1):**<br>Engine running, battery voltage 11.5, all electrical components off, ground between engine and chassis well connected and the exhaust system must be properly sealed between catalytic converter and the cylinder head. The DME detected the HO2S signal was in a negative voltage range referred to as "character shift downward". This code sets when the HO2S signal remains in a low state for a measured period of time. In effect, it does not switch properly in the closed loop operation. Engine speed is less than 8000rpm.<br>**Possible Causes:**<br>• Wiring fault<br>• O2S befor catalytic converter defective<br>• Check wiring harness plug connections CMI-216, CMI-215, CMI-228, CMI-227, CMI-230<br>• Replace O2S |

| DTC | Trouble Code Title, Conditions & Possible Causes |
|---|---|
| **DTC: P2237**<br>**2T CCM, MIL: Yes**<br>**Years:** 2007, 2008<br>**Models:** Cooper<br>**Engines:** 1.6L, 1.6L Turbo<br>**Transmissions:** All | **HO2S (Bank 1 Sensor 1) Circuits:**<br>Engine running, battery voltage 11.5, all electrical components off, ground between engine and chassis well connected and the exhaust system must be properly sealed between catalytic converter and the cylinder head. The DME detected the HO2S signal was in a negative voltage range referred to as "character shift downward". This code sets when the HO2S signal remains in a low state for a measured period of time. In effect, it does not switch properly in the closed loop operation. Engine speed is less than 8000rpm.<br>**Possible Causes:**<br>• HO2S is contaminated (due to presence of silicone in fuel)<br>• HO2S signal and ground circuit wires crossed in wiring harness<br>• HO2S signal circuit is shorted to sensor or chassis ground<br>• HO2S element has failed (internal short condition)<br>• Leaks present in the exhaust manifold or exhaust pipes |
| **DTC: P2243**<br>**2T CCM, MIL: Yes**<br>**Years:** 2007, 2008<br>**Models:** Cooper<br>**Engines:** 1.6L, 1.6L Turbo<br>**Transmissions:** All | **HO2S (Bank 1 Sensor 1) Circuit:**<br>Engine running, battery voltage 11.5, all electrical components off, ground between engine and chassis well connected and the exhaust system must be properly sealed between catalytic converter and the cylinder head. The DME detected the HO2S signal was in a negative voltage range referred to as "character shift downward". This code sets when the HO2S signal remains in a low state for a measured period of time. In effect, it does not switch properly in the closed loop operation. Engine speed is less than 8000rpm.<br>**Possible Causes:**<br>• HO2S is contaminated (due to presence of silicone in fuel)<br>• HO2S signal and ground circuit wires crossed in wiring harness<br>• HO2S signal circuit is shorted to sensor or chassis ground<br>• HO2S element has failed (internal short condition)<br>• Leaks present in the exhaust manifold or exhaust pipes |
| **DTC: P2251**<br>**2T CCM, MIL: Yes**<br>**Years:** 2007, 2008<br>**Models:** Cooper<br>**Engines:** 1.6L, 1.6L Turbo<br>**Transmissions:** All | **HO2S (Bank 1 Sensor 1) Circuit:**<br>Engine running, battery voltage 11.5, all electrical components off, ground between engine and chassis well connected and the exhaust system must be properly sealed between catalytic converter and the cylinder head. The DME detected the HO2S signal was in a negative voltage range referred to as "character shift downward". This code sets when the HO2S signal remains in a low state for a measured period of time. In effect, it does not switch properly in the closed loop operation. Engine speed is less than 8000rpm.<br>**Possible Causes:**<br>• HO2S is contaminated (due to presence of silicone in fuel)<br>• HO2S signal and ground circuit wires crossed in wiring harness<br>• HO2S signal circuit is shorted to sensor or chassis ground<br>• HO2S element has failed (internal short condition)<br>• Leaks present in the exhaust manifold or exhaust pipes |
| **DTC: P2270**<br>**2T CCM, MIL: Yes**<br>**Years:** 2007, 2008<br>**Models:** Cooper<br>**Engines:** 1.6L, 1.6L Turbo<br>**Transmissions:** All | **Rear O2 Sensor Signal Stuck Lean Bank 1 Sensor 2 Conditions:**<br>Engine started, battery voltage must be at least 11.5v, all electrical components must be off, parking brake must be engaged (to keep daytime driving lights off), automatic transmission selector must be in park. The DME detected an unexpected voltage condition, or it detected an unexpected current draw in the heater circuit during the CCM test. Coolant temperature must been at least 80.25 degrees Celsius. The vehicle speed is greater than 27.96 and less than 80.76. The engine speed is between 1984 and 3647rpm. Ambient pressure is greater than 75.001kPa and the engine stability load is 6.94g/s.<br>**Note: Vehicle must be raised before connector for oxygen sensors is accessible.**<br>**Possible Causes:**<br>• Rear Oxygen sensor voltage exceeds a calibrated threshold<br>• Rear Oxygen sensor signal permanently lies below or above the set point<br>• Rear Oxygen sensor A/F mixture ia above the set point value |
| **DTC: P2271**<br>**2T CCM, MIL: Yes**<br>**Years:** 2007, 2008<br>**Models:** Cooper<br>**Engines:** 1.6L, 1.6L Turbo<br>**Transmissions:** All | **Rear O2 Sensor Signal Stuck Lean Bank 1 Sensor 2 Conditions:**<br>Engine started, battery voltage must be at least 11.5v, all electrical components must be off, parking brake must be engaged (to keep daytime driving lights off), automatic transmission selector must be in park. The DME detected an unexpected voltage condition, or it detected an unexpected current draw in the heater circuit during the CCM test. Coolant temperature must been at least 80.25 degrees Celsius. The vehicle speed is greater than 27.96 and less than 80.76. The engine speed is between 1984 and 3647rpm. Ambient pressure is greater than 75.001kPa and the engine stability load is 6.94g/s.<br>**Note: Vehicle must be raised before connector for oxygen sensors is accessible.**<br>**Possible Causes:**<br>• Rear Oxygen sensor voltage exceeds a calibrated threshold<br>• Rear Oxygen sensor signal permanently lies below or above the set point<br>• Rear Oxygen sensor A/F mixture ia above the set point value |

| DTC | Trouble Code Title, Conditions & Possible Causes |
|---|---|
| **DTC: P2300**<br>**2T CCM, MIL:** Yes<br>**Years:** 2007, 2008<br>**Models:** Cooper<br>**Engines:** 1.6L, 1.6L Turbo<br>**Transmissions:** All | **Ignition Coilpack A Primary/Secondary Circuit Malfunction Open Circuit/Short to Ground Conditions:**<br>Engine started, battery voltage must be at least 11.5v, all electrical components must be off, parking brake must be engaged (to keep daytime driving lights off), automatic transmission selector must be in park and the ground between the engine and the chassis must be well connected. The DME did not receive any valid pulses from the ignition module for the Ignition Coilpack A primary circuit.<br>**Note: Ignition coils and power output stages are one component and cannot be replaced individually.**<br>**Possible Causes:**<br>• Engine speed (RPM) sensor has failed<br>• Camshaft Position (CMP) sensor has failed<br>• Power Supply Relay is shorted to an open circuit<br>• There is a malfunction in voltage supply<br>• Ignition coilpack is damaged or it has failed<br>• Cylinder 1 to 4 Fuel Injector(s) have failed |
| **DTC: P2301**<br>**2T CCM, MIL:** Yes<br>**Years:** 2007, 2008<br>**Models:** Cooper<br>**Engines:** 1.6L, 1.6L Turbo<br>**Transmissions:** All | **Ignition Coilpack A Primary/Secondary Circuit Malfunction Short to Battery Conditions:**<br>Engine started, battery voltage must be at least 11.5v, all electrical components must be off, parking brake must be engaged (to keep daytime driving lights off), automatic transmission selector must be in park and the ground between the engine and the chassis must be well connected. The DME did not receive any valid pulses from the ignition module for the Ignition Coilpack A primary circuit.<br>**Note: Ignition coils and power output stages are one component and cannot be replaced individually.**<br>**Possible Causes:**<br>• Engine speed (RPM) sensor has failed<br>• Camshaft Position (CMP) sensor has failed<br>• Power Supply Relay is shorted to an open circuit<br>• There is a malfunction in voltage supply<br>• Ignition coilpack is damaged or it has failed<br>• Cylinder 1 to 4 Fuel Injector(s) have failed |
| **DTC: P2303**<br>**2T CCM, MIL:** Yes<br>**Years:** 2007, 2008<br>**Models:** Cooper<br>**Engines:** 1.6L, 1.6L Turbo<br>**Transmissions:** All | **Ignition Coilpack A Primary/Secondary Circuit Malfunction Open Circuit/Short to Ground Conditions:**<br>Engine started, battery voltage must be at least 11.5v, all electrical components must be off, parking brake must be engaged (to keep daytime driving lights off), automatic transmission selector must be in park and the ground between the engine and the chassis must be well connected. The DME did not receive any valid pulses from the ignition module for the Ignition Coilpack A primary circuit.<br>**Note: Ignition coils and power output stages are one component and cannot be replaced individually.**<br>**Possible Causes:**<br>• Engine speed (RPM) sensor has failed<br>• Camshaft Position (CMP) sensor has failed<br>• Power Supply Relay is shorted to an open circuit<br>• There is a malfunction in voltage supply<br>• Ignition coilpack is damaged or it has failed<br>• Cylinder 1 to 4 Fuel Injector(s) have failed |
| **DTC: P2304**<br>**2T CCM, MIL:** Yes<br>**Years:** 2007, 2008<br>**Models:** Cooper<br>**Engines:** 1.6L, 1.6L Turbo<br>**Transmissions:** All | **Ignition Coilpack A Primary/Secondary Circuit Malfunction Short to Battery Conditions:**<br>Engine started, battery voltage must be at least 11.5v, all electrical components must be off, parking brake must be engaged (to keep daytime driving lights off), automatic transmission selector must be in park and the ground between the engine and the chassis must be well connected. The DME did not receive any valid pulses from the ignition module for the Ignition Coilpack A primary circuit.<br>**Note: Ignition coils and power output stages are one component and cannot be replaced individually.**<br>**Possible Causes:**<br>• Engine speed (RPM) sensor has failed<br>• Camshaft Position (CMP) sensor has failed<br>• Power Supply Relay is shorted to an open circuit<br>• There is a malfunction in voltage supply<br>• Ignition coilpack is damaged or it has failed<br>• Cylinder 1 to 4 Fuel Injector(s) have failed |

| DTC | Trouble Code Title, Conditions & Possible Causes |
|---|---|
| **DTC: P240A**<br>**2T CCM, MIL: Yes**<br>**Years:** 2007, 2008<br>**Models:** Cooper<br>**Engines:** 1.6L, 1.6L Turbo<br>**Transmissions:** All | **EVAP Leak Detection Pump (LDP) Heater Control Circuit/Open:**<br>Engine started, battery voltage must be at least 11.5v, all electrical components must be off, parking brake must be engaged (to keep daytime driving lights off), automatic transmission selector must be in park, the exhaust system must be properly sealed between the catalytic converter and the cylinder head, coolant temperature must be at least 80 degrees Celsius and oxygen sensor heaters for oxygen sensors before the catalytic converter must be functioning properly and the ground between the engine and the chassis must be well connected. The DME detected voltage irregularity in the leak detection pump control circuit.<br>**Possible Causes:**<br>• EVAP LDP power supply circuit is open<br>• EVAP LDP solenoid valve is damaged or it has failed<br>• EVAP LDP canister has a leak or a poor seal<br>• EVAP canister system has an improper seal<br>• Evaporative Emission (EVAP) canister purge regulator valve 1 has failed<br>• Leak Detection Pump (LDP) is faulty<br>• Aftermarket EVAP parts that do not conform to specifications<br>• EVAP component seals leaking (i.e., leaks in the Purge valve, fuel tank pressure sensor, canister vent solenoid, fuel vapor control valve tube assembly or fuel vapor vent valve). |
| **DTC: P240B**<br>**2T CCM, MIL: Yes**<br>**Years:** 2007, 2008<br>**Models:** Cooper<br>**Engines:** 1.6L, 1.6L Turbo<br>**Transmissions:** All | **EVAP Leak Detection Pump (LDP) Heater Control Circuit Low:**<br>Engine started, battery voltage must be at least 11.5v, all electrical components must be off, parking brake must be engaged (to keep daytime driving lights off), automatic transmission selector must be in park, the exhaust system must be properly sealed between the catalytic converter and the cylinder head, coolant temperature must be at least 80 degrees Celsius and oxygen sensor heaters for oxygen sensors before the catalytic converter must be functioning properly and the ground between the engine and the chassis must be well connected. The DME detected voltage irregularity in the leak detection pump control circuit.<br>**Possible Causes:**<br>• EVAP LDP power supply circuit is open<br>• EVAP LDP solenoid valve is damaged or it has failed<br>• EVAP LDP canister has a leak or a poor seal<br>• EVAP canister system has an improper seal<br>• Evaporative Emission (EVAP) canister purge regulator valve 1 has failed<br>• Leak Detection Pump (LDP) is faulty<br>• Aftermarket EVAP parts that do not conform to specifications<br>• EVAP component seals leaking (i.e., leaks in the Purge valve, fuel tank pressure sensor, canister vent solenoid, fuel vapor control valve tube assembly or fuel vapor vent valve). |
| **DTC: P240C**<br>**2T CCM, MIL: Yes**<br>**Years:** 2007, 2008<br>**Models:** Cooper<br>**Engines:** 1.6L, 1.6L Turbo<br>**Transmissions:** All | **EVAP Leak Detection Pump (LDP) Heater Control Circuit High:**<br>Engine started, battery voltage must be at least 11.5v, all electrical components must be off, parking brake must be engaged (to keep daytime driving lights off), automatic transmission selector must be in park, the exhaust system must be properly sealed between the catalytic converter and the cylinder head, coolant temperature must be at least 80 degrees Celsius and oxygen sensor heaters for oxygen sensors before the catalytic converter must be functioning properly and the ground between the engine and the chassis must be well connected. The DME detected voltage irregularity in the leak detection pump control circuit.<br>**Possible Causes:**<br>• EVAP LDP power supply circuit is open<br>• EVAP LDP solenoid valve is damaged or it has failed<br>• EVAP LDP canister has a leak or a poor seal<br>• EVAP canister system has an improper seal<br>• Evaporative Emission (EVAP) canister purge regulator valve 1 has failed<br>• Leak Detection Pump (LDP) is faulty<br>• Aftermarket EVAP parts that do not conform to specifications<br>• EVAP component seals leaking (i.e., leaks in the Purge valve, fuel tank pressure sensor, canister vent solenoid, fuel vapor control valve tube assembly or fuel vapor vent valve). |

| DTC | Trouble Code Title, Conditions & Possible Causes |
| --- | --- |
| **DTC: P2400**<br>**2T CCM, MIL: Yes**<br>**Years:** 2007, 2008<br>**Models:** Cooper<br>**Engines:** 1.6L, 1.6L Turbo<br>**Transmissions:** All | **EVAP Leak Detection Pump (LDP) Control Circuit Open Conditions:**<br>Engine started, battery voltage must be at least 11.5v, all electrical components must be off, parking brake must be engaged (to keep daytime driving lights off), automatic transmission selector must be in park, the exhaust system must be properly sealed between the catalytic converter and the cylinder head, coolant temperature must be at least 80 degrees Celsius and oxygen sensor heaters for oxygen sensors before the catalytic converter must be functioning properly and the ground between the engine and the chassis must be well connected. The DME detected voltage irregularity in the leak detection pump control circuit.<br>**Possible Causes:**<br>• EVAP LDP power supply circuit is open<br>• EVAP LDP solenoid valve is damaged or it has failed<br>• EVAP LDP canister has a leak or a poor seal<br>• EVAP canister system has an improper seal<br>• Evaporative Emission (EVAP) canister purge regulator valve 1 has failed<br>• Leak Detection Pump (LDP) is faulty<br>• Aftermarket EVAP parts that do not conform to specifications<br>• EVAP component seals leaking (i.e., leaks in the Purge valve, fuel tank pressure sensor, canister vent solenoid, fuel vapor control valve tube assembly or fuel vapor vent valve). |
| **DTC: P2401**<br>**2T CCM, MIL: Yes**<br>**Years:** 2007, 2008<br>**Models:** Cooper<br>**Engines:** 1.6L, 1.6L Turbo<br>**Transmissions:** All | **EVAP Leak Detection Pump Control Circuit Low Conditions:**<br>Engine started, battery voltage must be at least 11.5v, all electrical components must be off, parking brake must be engaged (to keep daytime driving lights off), automatic transmission selector must be in park, the exhaust system must be properly sealed between the catalytic converter and the cylinder head, coolant temperature must be at least 80 degrees Celsius and oxygen sensor heaters for oxygen sensors before the catalytic converter must be functioning properly and the ground between the engine and the chassis must be well connected. The DME detected voltage irregularity in the leak detection pump control circuit.<br>**Possible Causes:**<br>• EVAP LDP power supply circuit is open<br>• EVAP LDP solenoid valve is damaged or it has failed<br>• EVAP LDP canister has a leak or a poor seal<br>• EVAP canister system has an improper seal<br>• Evaporative Emission (EVAP) canister purge regulator valve 1 has failed<br>• Leak Detection Pump (LDP) is faulty<br>• Aftermarket EVAP parts that do not conform to specifications<br>• EVAP component seals leaking (i.e., leaks in the Purge valve, fuel tank pressure sensor, canister vent solenoid, fuel vapor control valve tube assembly or fuel vapor vent valve). |
| **DTC: P2402**<br>**2T CCM, MIL: Yes**<br>**Years:** 2007, 2008<br>**Models:** Cooper<br>**Engines:** 1.6L, 1.6L Turbo<br>**Transmissions:** All | **EVAP Leak Detection Pump Control Circuit High Conditions:**<br>Engine started, battery voltage must be at least 11.5v, all electrical components must be off, parking brake must be engaged (to keep daytime driving lights off), automatic transmission selector must be in park, the exhaust system must be properly sealed between the catalytic converter and the cylinder head, coolant temperature must be at least 80 degrees Celsius and oxygen sensor heaters for oxygen sensors before the catalytic converter must be functioning properly and the ground between the engine and the chassis must be well connected. The DME detected voltage irregularity in the leak detection pump control circuit.<br>**Possible Causes:**<br>• EVAP LDP power supply circuit is open<br>• EVAP LDP solenoid valve is damaged or it has failed<br>• EVAP LDP canister has a leak or a poor seal<br>• EVAP canister system has an improper seal<br>• Evaporative Emission (EVAP) canister purge regulator valve 1 has failed<br>• Leak Detection Pump (LDP) is faulty<br>• Aftermarket EVAP parts that do not conform to specifications<br>• EVAP component seals leaking (i.e., leaks in the Purge valve, fuel tank pressure sensor, canister vent solenoid, fuel vapor control valve tube assembly or fuel vapor vent valve). |

| DTC | Trouble Code Title, Conditions & Possible Causes |
|---|---|
| **DTC: P2404**<br>**2T CCM, MIL: Yes**<br>**Years:** 2007, 2008<br>**Models:** Cooper<br>**Engines:** 1.6L, 1.6L Turbo<br>**Transmissions:** All | **EVAP Emission Control LDP Circuit Malfunction Pump Problem Conditions:**<br>Key on, KOEO Self-Test enabled, and the DME detected an unexpected voltage condition on the EVAP emission control leak detection pump circuit. The reed switch level stays low after activation of solenoids within the time threshold of more than 1 second. The coolant temperature is less than 60 degrees Celsius, and the ambient pressure is greater than 76.2994kPa. The air intake temperature at start is between 9.04 and 16.04 degrees Celsius. The change in barometric pressure since engine start is less than 0.9998kPa. The vehicle speed is less than 74.56mph, and the purge valve has opened enough on previous driving cycle.<br>**Possible Causes:**<br>• Leak Detection Pump has failed<br>• EVAP canister system has an improper or broken seal<br>• Evaporative Emission (EVAP) canister purge regulator valve 1 is faulty<br>• Hoses between the fuel pump and the EVAP canister are faulty<br>• Fuel filler cap is loose<br>• Fuel pump seal is defective, faulty or otherwise leaking<br>• Hoses between the EVAP canister and the fuel flap unit are faulty<br>• Hoses between the EVAP canister and the evaporative emission canister purge regulator valve are faulty |
| **DTC: P2414**<br>**2T CCM, MIL: Yes**<br>**Years:** 2007, 2008<br>**Models:** Cooper<br>**Engines:** 1.6L, 1.6L Turbo<br>**Transmissions:** All | **O2 SensorVoltage Check:**<br>Engine started, battery voltage must be at least 11.5v, all electrical components must be off, parking brake must be engaged (to keep daytime driving lights off), automatic transmission selector must be in park. The DME detected an unexpected voltage condition, or it detected an unexpected current draw in the heater circuit during the CCM test. Coolant temperature must been at least 80.25 degrees Celsius. The vehicle speed is greater than 27.96 and less than 80.76. The engine speed is between 1984 and 3647rpm. Ambient pressure is greater than 75.001kPa and the engine stability load is 6.94g/s.<br>**Note: Vehicle must be raised before connector for oxygen sensors is accessible.**<br>**Possible Causes:**<br>• Oxygen sensor not properly mounted in the exhaust system<br>• Oxygen sensor VA voltage lies below the maximum or above the calibrated value |
| **DTC: P2418**<br>**2T CCM, MIL: Yes**<br>**Years:** 2007, 2008<br>**Models:** Cooper<br>**Engines:** 1.6L, 1.6L Turbo<br>**Transmissions:** All | **EVAP Emission System Switching Valve Control Circuit/Open:**<br>Engine started, battery voltage must be at least 11.5v, all electrical components must be off, parking brake must be engaged (to keep daytime driving lights off), automatic transmission selector must be in park, the exhaust system must be properly sealed between the catalytic converter and the cylinder head, coolant temperature must be at least 80 degrees Celsius and oxygen sensor heaters for oxygen sensors before the catalytic converter must be functioning properly and the ground between the engine and the chassis must be well connected. The DME detected voltage irregularity in the leak detection pump control circuit.<br>**Possible Causes:**<br>• EVAP LDP power supply circuit is open<br>• EVAP LDP solenoid valve is damaged or it has failed<br>• EVAP LDP canister has a leak or a poor seal<br>• EVAP canister system has an improper seal<br>• Evaporative Emission (EVAP) canister purge regulator valve 1 has failed<br>• Leak Detection Pump (LDP) is faulty<br>• Aftermarket EVAP parts that do not conform to specifications<br>• EVAP component seals leaking (i.e., leaks in the Purge valve, fuel tank pressure sensor, canister vent solenoid, fuel vapor control valve tube assembly or fuel vapor vent valve). |
| **DTC: P2419**<br>**2T CCM, MIL: Yes**<br>**Years:** 2007, 2008<br>**Models:** Cooper<br>**Engines:** 1.6L, 1.6L Turbo<br>**Transmissions:** All | **EVAP Emission System Switching Valve Control Circuit Low:**<br>Engine started, battery voltage must be at least 11.5v, all electrical components must be off, parking brake must be engaged (to keep daytime driving lights off), automatic transmission selector must be in park, the exhaust system must be properly sealed between the catalytic converter and the cylinder head, coolant temperature must be at least 80 degrees Celsius and oxygen sensor heaters for oxygen sensors before the catalytic converter must be functioning properly and the ground between the engine and the chassis must be well connected. The DME detected voltage irregularity in the leak detection pump control circuit.<br>**Possible Causes:**<br>• EVAP LDP power supply circuit is open<br>• EVAP LDP solenoid valve is damaged or it has failed<br>• EVAP LDP canister has a leak or a poor seal<br>• EVAP canister system has an improper seal<br>• Evaporative Emission (EVAP) canister purge regulator valve 1 has failed<br>• Leak Detection Pump (LDP) is faulty<br>• Aftermarket EVAP parts that do not conform to specifications<br>• EVAP component seals leaking (i.e., leaks in the Purge valve, fuel tank pressure sensor, canister vent solenoid, fuel vapor control valve tube assembly or fuel vapor vent valve). |

| DTC | Trouble Code Title, Conditions & Possible Causes |
|---|---|
| **DTC: P2420**<br>**2T CCM, MIL: Yes**<br>**Years:** 2007, 2008<br>**Models:** Cooper<br>**Engines:** 1.6L, 1.6L Turbo<br>**Transmissions:** All | **EVAP Emission System Switching Valve Control Circuit High:**<br>Engine started, battery voltage must be at least 11.5v, all electrical components must be off, parking brake must be engaged (to keep daytime driving lights off), automatic transmission selector must be in park, the exhaust system must be properly sealed between the catalytic converter and the cylinder head, coolant temperature must be at least 80 degrees Celsius and oxygen sensor heaters for oxygen sensors before the catalytic converter must be functioning properly and the ground between the engine and the chassis must be well connected. The DME detected voltage irregularity in the leak detection pump control circuit.<br>**Possible Causes:**<br>• EVAP LDP power supply circuit is open<br>• EVAP LDP solenoid valve is damaged or it has failed<br>• EVAP LDP canister has a leak or a poor seal<br>• EVAP canister system has an improper seal<br>• Evaporative Emission (EVAP) canister purge regulator valve 1 has failed<br>• Leak Detection Pump (LDP) is faulty<br>• Aftermarket EVAP parts that do not conform to specifications<br>• EVAP component seals leaking (i.e., leaks in the Purge valve, fuel tank pressure sensor, canister vent solenoid, fuel vapor control valve tube assembly or fuel vapor vent valve). |
| **DTC: P2626**<br>**2T CCM, MIL: Yes**<br>**Years:** 2007, 2008<br>**Models:** Cooper<br>**Engines:** 1.6L, 1.6L Turbo<br>**Transmissions:** All | **HO2S (Bank 1 Sensor 1) Circuit:**<br>Engine running, battery voltage 11.5, all electrical components off, ground between engine and chassis well connected and the exhaust system must be properly sealed between catalytic converter and the cylinder head. The DME detected the HO2S signal was in a negative voltage range referred to as "character shift downward". This code sets when the HO2S signal remains in a low state for a measured period of time. In effect, it does not switch properly in the closed loop operation. Engine speed is less than 8000rpm.<br>**Possible Causes:**<br>• HO2S is contaminated (due to presence of silicone in fuel)<br>• HO2S signal and ground circuit wires crossed in wiring harness<br>• HO2S signal circuit is shorted to sensor or chassis ground<br>• HO2S element has failed (internal short condition)<br>• Leaks present in the exhaust manifold or exhaust pipes |

## Gas Engine OBD II Trouble Code List (P3xxx Codes)

| DTC | Trouble Code Title, Conditions & Possible Causes |
|---|---|
| **DTC: P3012**<br>**2T CCM, MIL: Yes**<br>**Years:** 2007, 2008<br>**Models:** Cooper<br>**Engines:** 1.6L, 1.6L Turbo<br>**Transmissions:** All | **HO2S (Bank 1 Sensor 1) Circuit:**<br>Engine running, battery voltage 11.5, all electrical components off, ground between engine and chassis well connected and the exhaust system must be properly sealed between catalytic converter and the cylinder head. The DME detected the HO2S signal was in a negative voltage range referred to as "character shift downward". This code sets when the HO2S signal remains in a low state for a measured period of time. In effect, it does not switch properly in the closed loop operation. Engine speed is less than 8000rpm.<br>**Possible Causes:**<br>• HO2S is contaminated (due to presence of silicone in fuel)<br>• HO2S signal and ground circuit wires crossed in wiring harness<br>• HO2S signal circuit is shorted to sensor or chassis ground<br>• HO2S element has failed (internal short condition)<br>• Leaks present in the exhaust manifold or exhaust pipes |
| **DTC: P3014**<br>**2T CCM, MIL: Yes**<br>**Years:** 2007, 2008<br>**Models:** Cooper<br>**Engines:** 1.6L, 1.6L Turbo<br>**Transmissions:** All | **HO2S (Bank 1 Sensor 1) Circuit:**<br>Engine running, battery voltage 11.5, all electrical components off, ground between engine and chassis well connected and the exhaust system must be properly sealed between catalytic converter and the cylinder head. The DME detected the HO2S signal was in a negative voltage range referred to as "character shift downward". This code sets when the HO2S signal remains in a low state for a measured period of time. In effect, it does not switch properly in the closed loop operation. Engine speed is less than 8000rpm.<br>**Possible Causes:**<br>• HO2S is contaminated (due to presence of silicone in fuel)<br>• HO2S signal and ground circuit wires crossed in wiring harness<br>• HO2S signal circuit is shorted to sensor or chassis ground<br>• HO2S element has failed (internal short condition)<br>• Leaks present in the exhaust manifold or exhaust pipes |

| DTC | Trouble Code Title, Conditions & Possible Causes |
|---|---|
| **DTC: P3016**<br>**2T CCM, MIL: Yes**<br>**Years:** 2007, 2008<br>**Models:** Cooper<br>**Engines:** All<br>**Transmissions:** All | **HO2S (Bank 1 Sensor 1) Heater Circuit Malfunction Conditions:**<br>Engine running, battery voltage is between 11 and 16 volts, all electrical components off, ground between engine and chassis well connected and the exhaust system must be properly sealed between catalytic converter and the cylinder head. The DME detected an unexpected voltage condition, or it detected excessive current draw in the heater circuit during the CCM test. The engine load is 25 to 160kg/h. The exhaust gas temperature is between 450 and 700 degrees Celsius.<br>**Possible Causes:**<br>• HO2S heater power circuit is open or heater ground circuit open<br>• HO2S signal tracking (due to oil or moisture in the connector)<br>• HO2S is damaged or has failed |
| **DTC: P3018**<br>**2T CCM, MIL: Yes**<br>**Years:** 2007, 2008<br>**Models:** Cooper<br>**Engines:** 1.6L, 1.6L Turbo<br>**Transmissions:** All | **HO2S (Bank 1 Sensor 1) Circuit:**<br>Engine running, battery voltage 11.5, all electrical components off, ground between engine and chassis well connected and the exhaust system must be properly sealed between catalytic converter and the cylinder head. The DME detected the HO2S signal was in a negative voltage range referred to as "character shift downward". This code sets when the HO2S signal remains in a low state for a measured period of time. In effect, it does not switch properly in the closed loop operation. Engine speed is less than 8000rpm.<br>**Possible Causes:**<br>• HO2S is contaminated (due to presence of silicone in fuel)<br>• HO2S signal and ground circuit wires crossed in wiring harness<br>• HO2S signal circuit is shorted to sensor or chassis ground<br>• HO2S element has failed (internal short condition)<br>• Leaks present in the exhaust manifold or exhaust pipes |
| **DTC: P3020**<br>**2T CCM, MIL: Yes**<br>**Years:** 2007, 2008<br>**Models:** Cooper<br>**Engines:** 1.6L, 1.6L Turbo<br>**Transmissions:** All | **HO2S (Bank 1 Sensor 1) Circuit:**<br>Engine running, battery voltage 11.5, all electrical components off, ground between engine and chassis well connected and the exhaust system must be properly sealed between catalytic converter and the cylinder head. The DME detected the HO2S signal was in a negative voltage range referred to as "character shift downward". This code sets when the HO2S signal remains in a low state for a measured period of time. In effect, it does not switch properly in the closed loop operation. Engine speed is less than 8000rpm.<br>**Possible Causes:**<br>• HO2S is contaminated (due to presence of silicone in fuel)<br>• HO2S signal and ground circuit wires crossed in wiring harness<br>• HO2S signal circuit is shorted to sensor or chassis ground<br>• HO2S element has failed (internal short condition)<br>• Leaks present in the exhaust manifold or exhaust pipes |
| **DTC: P3022**<br>**2T CCM, MIL: Yes**<br>**Years:** 2007, 2008<br>**Models:** Cooper<br>**Engines:** 1.6L, 1.6L Turbo<br>**Transmissions:** All | **HO2S (Bank 1 Sensor 1) Circuit:**<br>Engine running, battery voltage 11.5, all electrical components off, ground between engine and chassis well connected and the exhaust system must be properly sealed between catalytic converter and the cylinder head. The DME detected the HO2S signal was in a negative voltage range referred to as "character shift downward". This code sets when the HO2S signal remains in a low state for a measured period of time. In effect, it does not switch properly in the closed loop operation. Engine speed is less than 8000rpm.<br>**Possible Causes:**<br>• HO2S is contaminated (due to presence of silicone in fuel)<br>• HO2S signal and ground circuit wires crossed in wiring harness<br>• HO2S signal circuit is shorted to sensor or chassis ground<br>• HO2S element has failed (internal short condition)<br>• Leaks present in the exhaust manifold or exhaust pipes |
| **DTC: P3024**<br>**2T CCM, MIL: Yes**<br>**Years:** 2007, 2008<br>**Models:** Cooper<br>**Engines:** 1.6L, 1.6L Turbo<br>**Transmissions:** All | **HO2S (Bank 1 Sensor 1) Circuit:**<br>Engine running, battery voltage 11.5, all electrical components off, ground between engine and chassis well connected and the exhaust system must be properly sealed between catalytic converter and the cylinder head. The DME detected the HO2S signal was in a negative voltage range referred to as "character shift downward". This code sets when the HO2S signal remains in a low state for a measured period of time. In effect, it does not switch properly in the closed loop operation. Engine speed is less than 8000rpm.<br>**Possible Causes:**<br>• HO2S is contaminated (due to presence of silicone in fuel)<br>• HO2S signal and ground circuit wires crossed in wiring harness<br>• HO2S signal circuit is shorted to sensor or chassis ground<br>• HO2S element has failed (internal short condition)<br>• Leaks present in the exhaust manifold or exhaust pipes |

| DTC | Trouble Code Title, Conditions & Possible Causes |
|---|---|
| **DTC: P3026**<br>**2T CCM, MIL: Yes**<br>**Years:** 2007, 2008<br>**Models:** Cooper<br>**Engines:** All<br>**Transmissions:** All | **HO2S (Bank 1 Sensor 1) Heater Circuit Malfunction Conditions:**<br>Engine running, battery voltage is between 11 and 16 volts, all electrical components off, ground between engine and chassis well connected and the exhaust system must be properly sealed between catalytic converter and the cylinder head. The DME detected an unexpected voltage condition, or it detected excessive current draw in the heater circuit during the CCM test. The engine load is 25 to 160kg/h. The exhaust gas temperature is between 450 and 700 degrees Celsius.<br>**Possible Causes:**<br>• HO2S heater power circuit is open or heater ground circuit open<br>• HO2S signal tracking (due to oil or moisture in the connector)<br>• HO2S is damaged or has failed |
| **DTC: P2404**<br>**2T CCM, MIL: Yes**<br>**Years:** 2007, 2008<br>**Models:** Cooper<br>**Engines:** 1.6L, 1.6L Turbo<br>**Transmissions:** All | **EVAP Emission Control LDP Circuit Malfunction Pump Problem Conditions:**<br>Key on, KOEO Self-Test enabled, and the DME detected an unexpected voltage condition on the EVAP emission control leak detection pump circuit. The reed switch level stays low after activation of solenoids within the time threshold of more than 1 second. The coolant temperature is less than 60 degrees Celsius, and the ambient pressure is greater than 76.2994kPa. The air intake temperature at start is between 9.04 and 16.04 degrees Celsius. The change in barometric pressure since engine start is less than 0.9998kPa. The vehicle speed is less than 74.56mph, and the purge valve has opened enough on previous driving cycle.<br>**Possible Causes:**<br>• Leak Detection Pump has failed<br>• EVAP canister system has an improper or broken seal<br>• Evaporative Emission (EVAP) canister purge regulator valve 1 is faulty<br>• Hoses between the fuel pump and the EVAP canister are faulty<br>• Fuel filler cap is loose<br>• Fuel pump seal is defective, faulty or otherwise leaking<br>• Hoses between the EVAP canister and the fuel flap unit are faulty<br>• Hoses between the EVAP canister and the evaporative emission canister purge regulator valve are faulty |

# SAAB

9-2x • 9-3 • 9-5

# 11

## SPECIFICATIONS AND MAINTENANCE CHARTS

### ENGINE AND VEHICLE IDENTIFICATION CHART

| Code ① | Liters (cc) | Cu. In. | Cyl. | Fuel Sys. | Engine Type | Eng. Mfg. |
|---|---|---|---|---|---|---|
| 2 ③ | 2.0 (1994) | 122 | H4 | MFI-Turbo | DOHC | Subaru |
| S | 2.0 (1985) | 121 | I4 | MFI-Turbo | DOHC | Saab |
| Y | 2.0 (1985) | 121 | I4 | MFI-Turbo | DOHC | Saab |
| E | 2.3 (2290) | 140 | I4 | MFI-Turbo | DOHC | Saab |
| G | 2.3 (2290) | 140 | I4 | MFI-Turbo | DOHC | Saab |
| 6 ③ | 2.5 (2457) | 150 | H4 | MFI | SOHC | Subaru |
| 7 ③ | 2.5 (2457) | 150 | H4 | MFI-Turbo | DOHC | Subaru |
| U | 2.8 (2792) | 170 | V6 | MFI-Turbo | DOHC | Saab |

**Model Year**

| Code ② | Year |
|---|---|
| 5 | 2005 |
| 6 | 2006 |
| 7 | 2007 |
| 8 | 2008 |

CPC: Chevrolet/Pontiac/Canada

MFI: Multi-port Fuel Injection

SFI: Sequential Fuel Injection

① 8th position of VIN

② 10th position of VIN

③ 6th position of VIN

22205_SAAB_C0001

### GENERAL ENGINE SPECIFICATIONS

| Year | Model | Engine Displacement Liters | Engine ID/VIN | Net Horsepower @ rpm | Net Torque @ rpm (ft. lbs.) | Bore x Stroke (in.) | Compression Ratio | Oil Pressure @ rpm |
|---|---|---|---|---|---|---|---|---|
| 2005 | 9-2x | 2.0 | 2 | 227@6000 | 165@5600 | 3.62x2.95 | 8.0:1 | 43@5000 |
| | | 2.5 | 6 | 217@4000 | 166@4000 | 3.92x3.11 | 10.0:1 | 43@5001 |
| | 9-3 | 2.0 | S | 125@5500 | 125@5500 | 3.39x3.39 | 9.5:1 | 29@1000 |
| | | 2.0 | Y | 185@5500 | 185@5500 | 3.39x3.39 | 9.5:1 | 29@1000 |
| | 9-5 | 2.3 | E | 185@5300 | 207@1800 | 3.54x3.54 | 9.3:1 | 36@2000 |
| | | 2.3 | G | 250@5300 | 258@1900 | 3.54x3.54 | 9.3:1 | 36@2000 |
| 2006 | 9-2x | 2.5 | 6 | 173@6000 | 166@4400 | 3.92x3.11 | 10.0:1 | 43@5001 |
| | | 2.5 | 7 | 230@5600 | 235@3600 | 3.92x3.11 | 10.0:1 | 43@5001 |
| | 9-3 | 2.0 | S | 210@5500 | 221@2500 | 3.38x3.38 | 9.5:1 | 29@1000 |
| | | 2.0 | Y | 210@5500 | 221@2500 | 3.38x3.38 | 9.5:1 | 29@1000 |
| | | 2.8 | U | 250@5500 | 258@2000 | 3.50x2.94 | 9.5:1 | 138@2000 |
| | 9-5 | 2.3 | E | 260@5300 | 258@1900 | 3.54x3.54 | 9.3:1 | 36@2000 |
| | | 2.3 | G | 260@5300 | 258@1900 | 3.54x3.54 | 9.3:1 | 36@2000 |
| 2007 | 9-3 | 2.0 | S | 210@5500 | 221@2500 | 3.38x3.38 | 9.5:1 | 29@1000 |
| | | 2.0 | Y | 210@5500 | 221@2500 | 3.38x3.38 | 9.5:1 | 29@1000 |
| | | 2.8 | U | 250@5500 | 258@2000 | 3.50x2.94 | 9.5:1 | 138@2000 |
| | 9-5 | 2.3 | E | 260@5300 | 258@1900 | 3.54x3.54 | 9.3:1 | 36@2000 |
| | | 2.3 | G | 260@5300 | 258@1900 | 3.54x3.54 | 9.3:1 | 36@2000 |
| 2008 | 9-3 | 2.0 | S | 210@5500 | 221@2500 | 3.38x3.38 | 9.5:1 | 29@1000 |
| | | 2.0 | Y | 210@5500 | 221@2500 | 3.38x3.38 | 9.5:1 | 29@1000 |
| | | 2.8 | U | 255@5500 | 258@2000 | 3.50x2.94 | 9.5:1 | 138@2000 |
| | 9-5 | 2.3 | E | 260@5300 | 258@1900 | 3.54x3.54 | 9.3:1 | 36@2000 |
| | | 2.3 | G | 260@5300 | 258@1900 | 3.54x3.54 | 9.3:1 | 36@2000 |

22205_SAAB_C0002

## TUNE-UP SPECIFICATIONS

| Year | Engine Displacement Liters | Engine ID/VIN | Spark Plugs Gap (in.) | Ignition Timing (deg.) ① MT | AT | Fuel Pump (psi) | Idle Speed (rpm) MT | AT | Valve Clearance In. | Ex. |
|------|------|------|------|------|------|------|------|------|------|------|
| **2005** | 2.0 | 2 | 0.028-0.031 | ① | ① | 43.5 | 750 | 750 | 0.0079 | 0.0136 |
| | 2.0 | S | 0.035-0.039 | ① | ① | 43 ② | 850 | 720 | HYD | HYD |
| | 2.0 | Y | 0.035-0.039 | ① | ① | 43 ② | 720 | 720 | HYD | HYD |
| | 2.3 | E | 0.039-0.043 | ① | ① | 43 ② | 825 | 825 ③ | HYD | HYD |
| | 2.3 | G | 0.035-0.039 | ① | ① | 43 ② | 825 | 825 ③ | HYD | HYD |
| | 2.5 | 6 | 0.039-0.044 | ① | ① | 43.5 | 650 | 700 | 0.0079 | 0.0098 |
| **2006** | 2.0 | S | 0.035-0.039 | ① | ① | 43 ② | 850 | 720 | HYD | HYD |
| | 2.0 | Y | 0.035-0.039 | ① | ① | 43 ② | 720 | 720 | HYD | HYD |
| | 2.3 | E | 0.039-0.043 | ① | ① | 43 ② | 825 | 825 ③ | HYD | HYD |
| | 2.3 | G | 0.035-0.039 | ① | ① | 43 ② | 825 | 825 ③ | HYD | HYD |
| | 2.5 | 6 | 0.039-0.043 | ① | ① | 43.5 | 650 | 700 | 0.0079 | 0.0098 |
| | 2.5 | 7 | 0.028-0.031 | ① | ① | 43.5 | 650 | 700 ⑥ | 0.0079 | 0.0098 |
| | 2.8 | U | 0.035-0.039 | ① | ① | 43 ② | 700 | 700 | HYD | HYD |
| **2007** | 2.0 | S | 0.035-0.039 | ① | ① | 43 ② | 850 | 720 | HYD | HYD |
| | 2.0 | Y | 0.035-0.039 | ① | ① | 43 ② | 720 | 720 | HYD | HYD |
| | 2.3 | E | 0.039-0.043 | ① | ① | 43 ② | 825 | 825 ③ | HYD | HYD |
| | 2.3 | G | 0.039-0.043 | ① | ① | 43 ② | 825 | 825 ③ | HYD | HYD |
| | 2.8 | U | 0.035-0.039 | ① | ① | 43 ② | 700 | 700 | HYD | HYD |
| **2008** | 2.0 | S | 0.035-0.039 | ① | ① | 43 ② | 850 | 720 | HYD | HYD |
| | 2.0 | Y | 0.035-0.039 | ① | ① | 43 ② | 720 | 720 | HYD | HYD |
| | 2.3 | E | 0.039-0.043 | ① | ① | 43 ② | 825 | 825 ③ | HYD | HYD |
| | 2.3 | G | 0.039-0.043 | ① | ① | 43 ② | 825 | 825 ③ | HYD | HYD |
| | 2.8 | U | 0.035-0.039 | ① | ① | 43 ② | 700 | 700 | HYD | HYD |

NOTE: The Vehicle Emission Control Information label often reflects specification changes made during production.

The label figures must be used if they differ from figures in this chart.

HYD: Hydraulic

① Pre-programmed in ECU and cannot be adjusted

② With engine warm and operating at 2000 rpm.

③ Idle speed given is with transmission in Neutral; with engine in Drive: 860 rpm.

## CAPACITIES

| Year | Model | Engine Displacement Liters | Engine ID/VIN | Engine Oil with Filter (qts.) | Transmission (pts.) 5-Spd | 6-Spd | Auto. | Transfer Case (pts.) | Drive Axle Front (pts.) | Rear (pts.) | Fuel Tank (gal.) | Cooling System (qts.) |
|---|---|---|---|---|---|---|---|---|---|---|---|---|
| 2005 | 9-2x | 2.0 | 2 | 4.5 | 13.4 | — | 19.6 | NA | 2.6 | 1.6 | 16.4 | 8.0 |
|  |  | 2.5 | 6 | 4.4 | 13.4 | — | 19.6 | NA | 2.6 | 1.6 | 16.4 | 7.4 |
|  | 9-3 | 2.0 | S | 6.4 | 4.0 | 5.1 | ① | NA | NA | NA | 16.3 ② | 8.0 |
|  |  | 2.0 | Y | 6.4 | 4.0 | 5.1 | ① | NA | NA | NA | 16.3 ② | 8.0 |
|  | 9-5 | 2.3 | E | 4.2 | 4.2 | — | 6.8 | NA | NA | NA | 18.5 | 7.8 |
|  |  | 2.3 | G | 4.2 | 4.2 | — | 6.8 | NA | NA | NA | 18.5 | 7.8 |
| 2006 | 9-2x | 2.5 | 6 | 4.4 | 13.4 | — | 19.6 | NA | 2.6 | 1.6 | 16.4 | 7.4 |
|  |  | 2.5 | 7 | 4.4 | 13.4 | — | 19.6 | NA | 2.6 | 1.6 | 16.4 | 7.4 |
|  | 9-3 | 2.0 | S | 6.4 | 4.0 | 5.1 | ① | NA | NA | NA | 16.3 ② | 8.0 |
|  |  | 2.0 | Y | 6.4 | 4.0 | 5.1 | ① | NA | NA | NA | 16.3 ② | 8.0 |
|  |  | 2.8 | U | 6.4 | 3.8 | 5.1 | ③ | NA | NA | NA | 16.4 | 10.2 |
|  | 9-5 | 2.3 | E | 4.2 | 4.2 | — | 6.8 | NA | NA | NA | 18.0 | 7.8 |
|  |  | 2.3 | G | 4.2 | 4.2 | — | 6.8 | NA | NA | NA | 18.0 | 7.8 |
| 2007 | 9-3 | 2.0 | S | 6.4 | 4.0 | 5.1 | ① | NA | NA | NA | 16.4 | 8.0 |
|  |  | 2.0 | Y | 6.4 | 4.0 | 5.1 | ① | NA | NA | NA | 16.4 | 8.0 |
|  |  | 2.8 | U | 6.4 | 3.8 | 5.1 | ③ | NA | NA | NA | 16.4 | 10.2 |
|  | 9-5 | 2.3 | E | 4.2 | 4.2 | — | 6.8 | NA | NA | NA | 18.0 | 7.8 |
|  |  | 2.3 | G | 4.2 | 4.2 | — | 6.8 | NA | NA | NA | 18.0 | 7.8 |
| 2008 | 9-3 | 2.0 | S | 6.4 | 4.0 | 4.7 | ④ | NA | NA | NA | 16.4 | 8.0 |
|  |  | 2.0 | Y | 6.4 | 4.0 | 4.7 | ④ | NA | NA | NA | 16.4 | 8.0 |
|  |  | 2.8 | U | 6.4 | 4.0 | 4.7 | ⑤ | 1.4 | NA | ⑥ | 16.4 | 10.2 |
|  | 9-5 | 2.3 | E | 4.2 | 4.2 | — | 6.8 | NA | NA | NA | 18.0 | 7.8 |
|  |  | 2.3 | G | 4.2 | 4.2 | — | 6.8 | NA | NA | NA | 18.0 | 7.8 |
|  | 9-7x | 4.2 | S | 7.0 | — | — | 10.0 | 4.0 | 1.7 | 3.6 | 22.0 | 13.9 |

NOTE: All capacities are approximate. Add fluid gradually and check to be sure a proper fluid level is obtained.

NA: Not available.

① 5-Speed: 6.5 pts, 6-Speed: 6.0 pts.

② Sedan fuel tank capacity; For convertible: 17.0 gal.

③ 5-Speed: 6.5 pts, 6-Speed: 5.4 pts.

④ 5-Speed: 6.9 pts, 6-Speed: 6.0 pts.

⑤ 5-Speed: 6.9 pts, 6-Speed: 5.4 pts.

⑥ Rear Axle: 1.3 pts w/limited slip, 1.5 pts w/o limited slip.

Differential Cluch: 1.2 pts w/limited slip, 1.1 pts w/o limited slip.

22205_SAAB_C0004

## FLUID SPECIFICATIONS

| Year | Model | Engine Displacement Liters | Engine ID/VIN | Engine Oil | Auto. Trans. | Manual Trans. | Drive Axle | Transfer Case | Power Steering Fluid | Brake Master Cylinder |
|------|-------|------|------|------|------|------|------|------|------|------|
| 2005 | 9-2x | 2.0 | 2 | 5W-30 | Dexron III | 75W-90 | 75W-90 | Dexron III | Dexron III | DOT 4 |
|  |  | 2.5 | 6 | 5W-30 | Dexron III | 75W-90 | 75W-90 | Dexron III | Dexron III | DOT 4 |
|  | 9-3 | 2.0 | S | 5W-30 | ① | Saab MTF0063 | ② | ③ | Pentosin CHF 11 S | DOT 4 |
|  |  | 2.0 | Y | 5W-30 | ① | Saab MTF0063 | ② | ③ | Pentosin CHF 11 S | DOT 4 |
|  | 9-5 | 2.3 | E | 5W-30 | ① | Saab MTF0063 | — | — | Pentosin CHF 202 | DOT 4 |
|  |  | 2.3 | G | 5W-30 | ① | Saab MTF0063 | — | — | Pentosin CHF 202 | DOT 4 |
| 2006 | 9-2x | 2.5 | 6 | 5W-30 | Dexron III | 75W-90 | 75W-90 | Dexron III | Dexron III | DOT 4 |
|  |  | 2.5 | 7 | 5W-30 | Dexron III | 75W-90 | 75W-90 | Dexron III | Dexron III | DOT 4 |
|  | 9-3 | 2.0 | S | 5W-30 | ① | Saab MTF0063 | ② | ③ | Pentosin CHF 11 S | DOT 4 |
|  |  | 2.0 | Y | 5W-30 | ① | Saab MTF0063 | ② | ③ | Pentosin CHF 11 S | DOT 4 |
|  | 9-5 | 2.3 | E | 5W-30 | ① | Saab MTF0063 | — | — | Pentosin CHF 202 | DOT 4 |
|  |  | 2.3 | G | 5W-30 | ① | Saab MTF0063 | — | — | Pentosin CHF 202 | DOT 4 |
|  |  | 2.8 | U | 5W-30 | ① | Saab MTF0063 | — | — | Pentosin CHF 202 | DOT 4 |
| 2007 | 9-3 | 2.0 | S | 5W-30 | ① | Saab MTF0063 | ② | ③ | Pentosin CHF 11 S | DOT 4 |
|  |  | 2.0 | Y | 5W-30 | ① | Saab MTF0063 | ② | ③ | Pentosin CHF 11 S | DOT 4 |
|  | 9-5 | 2.3 | E | 5W-30 | ① | Saab MTF0063 | — | — | Pentosin CHF 202 | DOT 4 |
|  |  | 2.3 | G | 5W-30 | ① | Saab MTF0063 | — | — | Pentosin CHF 202 | DOT 4 |
|  |  | 2.8 | U | 5W-30 | ① | Saab MTF0063 | — | — | Pentosin CHF 202 | DOT 4 |
| 2008 | 9-3 | 2.0 | S | 5W-30 | ① | Saab MTF0063 | ② | ③ | Pentosin CHF 11 S | DOT 4 |
|  |  | 2.0 | Y | 5W-30 | ① | Saab MTF0063 | ② | ③ | Pentosin CHF 11 S | DOT 4 |
|  | 9-5 | 2.3 | E | 5W-30 | ① | Saab MTF0063 | — | — | Pentosin CHF 202 | DOT 4 |
|  |  | 2.3 | G | 5W-30 | ① | Saab MTF0063 | — | — | Pentosin CHF 202 | DOT 4 |
|  |  | 2.8 | U | 5W-30 | ① | Saab MTF0063 | — | — | Pentosin CHF 202 | DOT 4 |

DOT: Department Of Transpotation

① 5-Speed: Saab 3309, 6-Speed Saab AW-1

② Differential Clutch: Saab part # 93165387, Limited Slip Differential: Saab part # 93165388

③ Transfer Case: Saab part # 93165383

## VALVE SPECIFICATIONS

| Year | Engine Displacement Liters | Engine ID/VIN | Seat Angle (deg.) | Face Angle (deg.) | Spring Test Pressure (lbs. @ in.) | Spring Installed Height (in.) | Stem-to-Guide Clearance (in.) | | Stem Diameter (in.) | |
|---|---|---|---|---|---|---|---|---|---|---|
| | | | | | | | Intake | Exhaust | Intake | Exhaust |
| **2005** | 2.0 | 2 | 45 | NS | 46.3-53.1@ 1.42 | 1.42 | 0.0012- 0.0022 | 0.0016- 0.0026 | 0.2344- 0.2350 | 0.2341- 0.2346 |
| | 2.0 | S | 45 | NS | NS | 0.89- 1.28 | 0.0012- 0.0022 | 0.0016- 0.0026 | 0.2322- 0.2328 | 0.2319- 0.2325 |
| | 2.0 | Y | 45 | ① | NS | 0.89- 1.28 | 0.0012- 0.0022 | 0.0016- 0.0026 | 0.2322- 0.2328 | 0.2319- 0.2325 |
| | 2.3 | E | 45 | ① | 138-150@ 1.25 | 1.48 | 0.0067 | 0.0087 | 0.1988- 0.1994 | 0.1949- 0.1955 ② |
| | 2.3 | G | 45 | ① | 138-150@ 1.25 | 1.48 | 0.0067 | 0.0087 | 0.1988- 0.1994 | 0.1949- 0.1955 ② |
| | 2.5 | 6 | 45 | 45 | 48-55@ 1.77 | 1.77 | 0.0014- 0.0024 | 0.0016- 0.0026 | 0.2343- 0.2348 | 0.2341- 0.2346 |
| **2006** | 2.0 | S | 45 | NS | NS | 0.89- 1.28 | 0.0012- 0.0022 | 0.0016- 0.0026 | 0.2322- 0.2328 | 0.2319- 0.2325 |
| | 2.0 | Y | 45 | NS | NS | 0.89- 1.28 | 0.0012- 0.0022 | 0.0016- 0.0026 | 0.2322- 0.2328 | 0.2319- 0.2325 |
| | 2.3 | E | 45 | ① | 138-150@ 1.25 | 1.48 | 0.0067 | 0.0087 | 0.1988- 0.1994 | 0.1949- 0.1955 ② |
| | 2.3 | G | 45 | ① | 138-150@ 1.25 | 1.48 | 0.0067 | 0.0087 | 0.1988- 0.1994 | 0.1949- 0.1955 ② |
| | 2.5 | 6 | 45 | 45 | 46-53@ 1.417 | 1.77 | 0.0014- 0.0024 | 0.0016- 0.0026 | 0.2343- 0.2348 | 0.2341- 0.2346 |
| | 2.5 | 7 | 45 | 45 | 48-55@ 1.77 | 1.77 | 0.0012- 0.0022 | 0.0016- 0.0026 | 0.2344- 0.2350 | 0.2341- 0.2346 |
| | 2.8 | U | 45 | NS | NS | 1.37 | 0.0010- 0.0025 | 0.0014- 0.0029 | 0.2322- 0.2330 | 0.2318- 0.2326 |
| **2007** | 2.0 | S | 45 | NS | NS | 0.89- 1.28 | 0.0012- 0.0022 | 0.0016- 0.0026 | 0.2322- 0.2328 | 0.2319- 0.2325 |
| | 2.0 | Y | 45 | NS | NS | 0.89- 1.28 | 0.0012- 0.0022 | 0.0016- 0.0026 | 0.2322- 0.2328 | 0.2319- 0.2325 |
| | 2.3 | E | 45 | ① | 138-150@ 1.25 | 1.48 | 0.0067 | 0.0087 | 0.1988- 0.1994 | 0.1949- 0.1955 ② |
| | 2.3 | G | 45 | ① | 138-150@ 1.25 | 1.48 | 0.0067 | 0.0087 | 0.1988- 0.1994 | 0.1949- 0.1955 ② |
| | 2.8 | U | 45 | NS | NS | 1.37 | 0.0010- 0.0025 | 0.0014- 0.0029 | 0.2322- 0.2330 | 0.2318- 0.2326 |
| **2008** | 2.0 | S | 45 | NS | NS | 0.89- 1.28 | 0.0012- 0.0022 | 0.0016- 0.0026 | 0.2322- 0.2328 | 0.2319- 0.2325 |
| | 2.0 | Y | 45 | NS | NS | 0.89- 1.28 | 0.0012- 0.0022 | 0.0016- 0.0026 | 0.2322- 0.2328 | 0.2319- 0.2325 |
| | 2.3 | E | 45 | ① | 138-150@ 1.25 | 1.48 | 0.0067 | 0.0087 | 0.1988- 0.1994 | 0.1949- 0.1955 ② |
| | 2.3 | G | 45 | ① | 138-150@ 1.25 | 1.48 | 0.0067 | 0.0087 | 0.1988- 0.1994 | 0.1949- 0.1955 ② |
| | 2.8 | U | 45 | NS | NS | 1.37 | 0.0010- 0.0025 | 0.0014- 0.0029 | 0.2322- 0.2330 | 0.2318- 0.2326 |

NS - Not Specified by manufacturer.

① Exhaust: 44.5

   Intake: 45.3

② Nimonic engine: 0.1986-0.1992 inches.

# CAMSHAFT AND BEARING SPECIFICATIONS CHART

All measurements are given in inches.

| Year | Engine Displ. Liters | Engine ID/VIN | Journal Dia. | Brg. Oil Clearance | Shaft End-play | Runout | Journal Bore | Lobe Height Intake | Exhaust |
|---|---|---|---|---|---|---|---|---|---|
| **2005** | 2.0 | 2 | ① | 0.0015-0.0028 | 0.0015-0.0028 | 0.0079 | NS | 1.8210-1.8250 | 1.8210-1.8250 |
| | 2.0 | S | 1.0774-1.0784 | NS | ② | NS | NS | 0.2321 | 0.2348 |
| | 2.0 | Y | 1.0774-1.0784 | NS | ② | NS | NS | 0.2302 | 0.2348 |
| | 2.3 | E | 1.1568-1.1574 | NS | 0.0032-0.0140 | NS | NS | 0.3324 | 0.3324 |
| | 2.3 | G | 1.1568-1.1574 | NS | 0.0032-0.0140 | NS | NS | 0.3324 | 0.3324 |
| | 2.5 | 6 | 1.2570-1.2577 | 0.0022-0.0035 | 0.0012-0.0035 | 0.0010 | 1.2598-1.2650 | 1.5545-1.5585 | 1.5491-1.5530 |
| | 2.5 | 7 | ① | 0.0015-0.0028 | 0.0027-0.0046 | 0.0008 | NS | 1.833-1.8370 | 1.8410-1.8440 |
| | 2.8 | U | ③ | 0.0016-0.0034 | 0.0018-0.0086 | 0.0002 | NS | NS | NS |
| **2006** | 2.0 | S | 1.0774-1.0784 | NS | ② | NS | NS | 0.2321 | 0.2348 |
| | 2.0 | Y | 1.0774-1.0784 | NS | ② | NS | NS | 0.2302 | 0.2348 |
| | 2.3 | E | 1.1568-1.1574 | NS | 0.0032-0.0140 | NS | NS | 0.3324 | 0.3324 |
| | 2.3 | G | 1.1568-1.1574 | NS | 0.0032-0.0140 | NS | NS | 0.3324 | 0.3324 |
| | 2.5 | 6 | 1.2570-1.2577 | 0.0022-0.0035 | 0.0012-0.0035 | 0.0010 | 1.2598-1.2650 | 1.5545-1.5585 | 1.5491-1.5530 |
| | 2.5 | 7 | ① | 0.0015-0.0028 | 0.0027-0.0046 | 0.0008 | NS | 1.833-1.8370 | 1.8410-1.8440 |
| | 2.8 | U | ③ | 0.0016-0.0034 | 0.0018-0.0086 | 0.0002 | NS | NS | NS |
| **2007** | 2.0 | S | 1.0774-1.0784 | NS | ② | NS | NS | 0.2321 | 0.2348 |
| | 2.0 | Y | 1.0774-1.0784 | NS | ② | NS | NS | 0.2302 | 0.2348 |
| | 2.3 | E | 1.1568-1.1574 | NS | 0.0032-0.0140 | NS | NS | 0.3324 | 0.3324 |
| | 2.3 | G | 1.1568-1.1574 | NS | 0.0032-0.0140 | NS | NS | 0.3324 | 0.3324 |
| | 2.8 | U | ③ | 0.0016-0.0034 | 0.0018-0.0086 | 0.0002 | NS | NS | NS |

## CAMSHAFT AND BEARING SPECIFICATIONS CHART

All measurements are given in inches.

| Year | Engine Displ. Liters | Engine ID/VIN | Journal Dia. | Brg. Oil Clearance | Shaft End-play | Runout | Journal Bore | Lobe Height | |
|------|------|------|------|------|------|------|------|------|------|
| | | | | | | | | Intake | Exhaust |
| **2008** | 2.0 | S | 1.0774-1.0784 | NS | ② | NS | NS | 0.2321 | 0.2348 |
| | 2.0 | Y | 1.0774-1.0784 | NS | ② | NS | NS | 0.2302 | 0.2348 |
| | 2.3 | E | 1.1568-1.1574 | NS | 0.0032-0.0140 | NS | NS | 0.3324 | 0.3324 |
| | 2.3 | G | 1.1568-1.1574 | NS | 0.0032-0.0140 | NS | NS | 0.3324 | 0.3324 |
| | 2.8 | U | ③ | 0.0016-0.0034 | 0.0018-0.0086 | 0.0002 | NS | NS | NS |

NS: Not specified by manufacturer

① Front: 1.14939-1.4946 in.
Center/Rear: 1.1790-1.1796

② Without camshaft sprocket installed: 0.0044-0.0074 in.
With camshaft sprocket installed: 0.0042-0.0070 in.

③ Front: 1.4000-1.4008 in.
Center/Rear: 1.0774-1.0784

22205_SAAB_C0016

## CRANKSHAFT AND CONNECTING ROD SPECIFICATIONS

All measurements are given in inches.

| Year | Engine Displacement Liters | Engine ID/VIN | Crankshaft | | | | Connecting Rod | | |
|------|------|------|------|------|------|------|------|------|------|
| | | | Main Brg. Journal Dia. | Main Brg. Oil Clearance | Shaft End-play | Thrust on No. | Journal Diameter | Oil Clearance | Side Clearance |
| 2005 | 2.0 | 2 | 2.205 | 0.0004-0.0012 | 0.001-0.005 | NS | 2.046 | 0.0010-0.002 | 0.002-0.013 |
| | 2.0 | S | 1.960 | 0.0012-0.0029 | 0.002-0.003 | NS | 2.239 | 0.0013-0.0026 | NS |
| | 2.0 | Y | 1.960 | 0.0012-0.0029 | 0.002-0.003 | NS | 2.239 | 0.0013-0.0026 | NS |
| | 2.3 | E | 2.319 | 0.0005-0.0025 | 0.003-0.014 | NS | 2.079 | 0.0008-0.0027 | NS |
| | 2.3 | G | 2.319 | 0.0005-0.0025 | 0.003-0.014 | NS | 2.079 | 0.0008-0.0027 | NS |
| | 2.5 | 6 | 2.362 | 0.0001-0.0012 | 0.001-0.005 | NS | 2.046 | 0.0006-0.0017 | 0.003-0.013 |
| 2006 | 2.0 | S | 1.960 | 0.0012-0.0029 | 0.002-0.003 | NS | 2.239 | 0.0013-0.0026 | NS |
| | 2.0 | Y | 1.960 | 0.0012-0.0029 | 0.002-0.003 | NS | 2.239 | 0.0013-0.0026 | NS |
| | 2.3 | E | 2.319 | 0.0005-0.0025 | 0.003-0.014 | NS | 2.079 | 0.0008-0.0027 | NS |
| | 2.3 | G | 2.319 | 0.0005-0.0025 | 0.003-0.014 | NS | 2.079 | 0.0008-0.0027 | NS |
| | 2.5 | 6 | 2.362 | 0.0001-0.0012 | 0.001-0.005 | NS | 2.046 | 0.0006-0.0017 | 0.003-0.013 |
| | 2.5 | 7 | 2.362 | 0.0001-0.0012 | 0.001-0.005 | NS | 2.046 | 0.0006-0.0017 | 0.003-0.013 |
| | 2.8 | U | 2.239 | 0.0004-0.0028 | 0.004-0.013 | NS | 2.720 | 0.0004-0.0024 | NS |
| 2007 | 2.0 | S | 1.960 | 0.0012-0.0029 | 0.002-0.003 | NS | 2.239 | 0.0013-0.0026 | NS |
| | 2.0 | Y | 1.960 | 0.0012-0.0029 | 0.002-0.003 | NS | 2.239 | 0.0013-0.0026 | NS |
| | 2.3 | E | 2.319 | 0.0005-0.0025 | 0.003-0.014 | NS | 2.079 | 0.0008-0.0027 | NS |
| | 2.3 | G | 2.319 | 0.0005-0.0025 | 0.003-0.014 | NS | 2.079 | 0.0008-0.0027 | NS |
| | 2.8 | U | 2.239 | 0.0004-0.0028 | 0.004-0.013 | NS | 2.720 | 0.0004-0.0024 | NS |
| 2008 | 2.0 | S | 1.960 | 0.0012-0.0029 | 0.002-0.003 | NS | 2.239 | 0.0013-0.0026 | NS |
| | 2.0 | Y | 1.960 | 0.0012-0.0029 | 0.002-0.003 | NS | 2.239 | 0.0013-0.0026 | NS |
| | 2.3 | E | 2.319 | 0.0005-0.0025 | 0.003-0.014 | NS | 2.079 | 0.0008-0.0027 | NS |
| | 2.3 | G | 2.319 | 0.0005-0.0025 | 0.003-0.014 | NS | 2.079 | 0.0008-0.0027 | NS |
| | 2.8 | U | 2.239 | 0.0004-0.0028 | 0.004-0.013 | NS | 2.720 | 0.0004-0.0024 | NS |

NS: Not specified by manufacturer.

22205_SAAB_C0006

## PISTON AND RING SPECIFICATIONS
All measurements are given in inches.

| Year | Engine Size Liters | Engine ID/VIN | Piston Clearance | Ring Gap | | | Ring Side Clearance | | |
| | | | | Top Compression | Bottom Compression | Oil Control | Top Compression | Bottom Compression | Oil Control |
|---|---|---|---|---|---|---|---|---|---|
| **2005** | 2.0 | 2 | 0.0004-0.0012 | 0.008-0.0100 | 0.016-0.0200 | 0.008-0.0200 | 0.0016-0.0031 | 0.0012-0.0027 | NS |
| | 2.0 | S | 0.0005-0.0018 | 0.006-0.0140 | 0.016-0.0240 | 0.010-0.0300 | 0.0014-0.0032 | 0.0006-0.0024 | 0.002-0.006 |
| | 2.0 | Y | 0.0005-0.0018 | 0.006-0.0140 | 0.016-0.0240 | 0.010-0.0300 | 0.0014-0.0032 | 0.0006-0.0024 | 0.002-0.006 |
| | 2.3 | E | 0.001-0.0020 | 0.012-0.0190 | 0.012-0.0190 | 0.030-0.0390 | 0.0014-0.0031 | 0.0016-0.0030 | 0.0985 |
| | 2.3 | G | 0.001-0.0020 | 0.012-0.0190 | 0.012-0.0190 | 0.030-0.0390 | 0.0014-0.0031 | 0.0016-0.0030 | 0.0985 |
| | 2.5 | 6 | 0.0039-0.0039 | 0.007-0.0130 | 0.014-0.0200 | 0.008-0.0200 | 0.0016-0.0031 | 0.0012-0.0028 | NS |
| **2006** | 2.0 | S | 0.0005-0.0018 | 0.006-0.0140 | 0.016-0.0240 | 0.010-0.0300 | 0.0014-0.0032 | 0.0006-0.0024 | 0.002-0.006 |
| | 2.0 | Y | 0.0005-0.0018 | 0.006-0.0140 | 0.016-0.0240 | 0.010-0.0300 | 0.0014-0.0032 | 0.0006-0.0024 | 0.002-0.006 |
| | 2.3 | E | 0.001-0.0020 | 0.012-0.0190 | 0.012-0.0190 | 0.030-0.0390 | 0.0014-0.0031 | 0.0016-0.0030 | 0.0985 |
| | 2.3 | G | 0.001-0.0020 | 0.012-0.0190 | 0.012-0.0190 | 0.030-0.0390 | 0.0014-0.0031 | 0.0016-0.0030 | 0.0985 |
| | 2.5 | 6 | 0.0039-0.0039 | 0.007-0.0130 | 0.014-0.0200 | 0.008-0.0200 | 0.0016-0.0031 | 0.0012-0.0028 | NS |
| | 2.5 | 7 | 0.0004-0.0004 | 0.008-0.0120 | 0.015-0.0200 | 0.008-0.0200 | 0.0016-0.0031 | 0.0012-0.0028 | NS |
| | 2.8 | U | NS | NS | NS | NS | NS | NS | NS |
| **2007** | 2.0 | S | 0.0005-0.0018 | 0.006-0.0140 | 0.016-0.0240 | 0.010-0.0300 | 0.0014-0.0032 | 0.0006-0.0024 | 0.002-0.006 |
| | 2.0 | Y | 0.0005-0.0018 | 0.006-0.0140 | 0.016-0.0240 | 0.010-0.0300 | 0.0014-0.0032 | 0.0006-0.0024 | 0.002-0.006 |
| | 2.3 | E | 0.001-0.0020 | 0.012-0.0190 | 0.012-0.0190 | 0.030-0.0390 | 0.0014-0.0031 | 0.0016-0.0030 | 0.0985 |
| | 2.3 | G | 0.001-0.0020 | 0.012-0.0190 | 0.012-0.0190 | 0.030-0.0390 | 0.0014-0.0031 | 0.0016-0.0030 | 0.0985 |
| | 2.8 | U | NS | NS | NS | NS | NS | NS | NS |
| **2008** | 2.0 | S | 0.0005-0.0018 | 0.006-0.0140 | 0.016-0.0240 | 0.010-0.0300 | 0.0014-0.0032 | 0.0006-0.0024 | 0.002-0.006 |
| | 2.0 | Y | 0.0005-0.0018 | 0.006-0.0140 | 0.016-0.0240 | 0.010-0.0300 | 0.0014-0.0032 | 0.0006-0.0024 | 0.002-0.006 |
| | 2.3 | E | 0.001-0.0020 | 0.012-0.0190 | 0.012-0.0190 | 0.030-0.0390 | 0.0014-0.0031 | 0.0016-0.0030 | 0.0985 |
| | 2.3 | G | 0.001-0.0020 | 0.012-0.0190 | 0.012-0.0190 | 0.030-0.0390 | 0.0014-0.0031 | 0.0016-0.0030 | 0.0985 |
| | 2.8 | U | NS | NS | NS | NS | NS | NS | NS |

NS: Not specified by manufacturer

22205_SAAB_C0007

## TORQUE SPECIFICATIONS
All readings in ft. lbs.

| Year | Engine Displacement Liters | Engine ID/VIN | Cylinder Head Bolts | Main Bearing Bolts | Rod Bearing Bolts | Crankshaft Damper Bolts | Flywheel Bolts | Manifold Intake | Manifold Exhaust | Spark Plugs | Oil Pan Drain Plug |
|---|---|---|---|---|---|---|---|---|---|---|---|
| 2005 | 2.0 | 2 | ① | ② | 33 | 94 | 53 | 18 | 26 | 15 | 33 |
| | 2.0 | S | ③ | ④ | ⑤ | ⑥ | ⑦ | 7 | ⑧ | 21 | 18 |
| | 2.0 | Y | ③ | ④ | ⑤ | ⑥ | ⑦ | 7 | ⑧ | 21 | 18 |
| | 2.3 | E | ⑨ | ⑩ | ⑪ | 130 | 87 | 15 | 18 | 21 | 13 |
| | 2.3 | G | ⑨ | ⑩ | ⑪ | 130 | 59 | 15 | 13 | 21 | 19 |
| | 2.5 | 6 | ① | ② | 33 | 123-137 | 53 | 18 | 26 | 15 | 33 |
| 2006 | 2.0 | S | ③ | ④ | ⑤ | ⑥ | ⑦ | 7 | ⑧ | 21 | 18 |
| | 2.0 | Y | ③ | ④ | ⑤ | ⑥ | ⑦ | 7 | ⑧ | 21 | 18 |
| | 2.3 | E | ⑨ | ⑩ | ⑪ | 130 | 87 | 15 | 18 | 21 | 13 |
| | 2.3 | G | ⑨ | ⑩ | ⑪ | 130 | 59 | 15 | 13 | 21 | 19 |
| | 2.5 | 6 | ① | ② | 33 | 123-137 | 53 | 18 | 26 | 15 | 33 |
| | 2.5 | 7 | ① | ② | 33 | 123-137 | 53 | 18 | 26 | 15 | 33 |
| | 2.8 | U | ③ | ⑩ | ⑫ | ⑬ | ⑭ | 17 | 15 | 21 | 19 |
| 2007 | 2.0 | S | ③ | ④ | ⑤ | ⑥ | ⑦ | 7 | ⑦ | 21 | 18 |
| | 2.0 | Y | ⑮ | ④ | ⑤ | ⑥ | ⑦ | 7 | 12 | 21 | 18 |
| | 2.3 | E | ⑨ | ⑩ | ⑪ | 130 | 59 | 16 | 18 | 21 | 15 |
| | 2.3 | G | ⑨ | ⑩ | ⑪ | 130 | 59 | 16 | 18 | 21 | 19 |
| | 2.8 | U | ③ | ⑩ | ⑫ | ⑬ | ⑭ | 17 | 15 | 21 | 19 |
| 2008 | 2.0 | S | ③ | ④ | ⑤ | ⑥ | ⑦ | 7 | ⑦ | 21 | 18 |
| | 2.0 | Y | ⑮ | ④ | ⑤ | ⑥ | ⑦ | 7 | 12 | 21 | 18 |
| | 2.3 | E | ⑨ | ⑩ | ⑪ | 130 | 59 | 16 | 18 | 21 | 15 |
| | 2.3 | G | ⑨ | ⑩ | ⑪ | 130 | 59 | 16 | 18 | 21 | 19 |
| | 2.8 | U | ③ | ⑩ | ⑫ | ⑬ | ⑭ | 17 | 15 | 21 | 19 |

NS: Not specified by manufacturer

① Step 1: Tighten bolts in sequence to 22 ft. lbs.
Step 2: Tighten bolts in sequence to 51ft. lbs.
Step 3: Loosen all bolts by 180 degrees in reverse order
Step 4: Loosen all bolts an additional 180 degrees in reverse order
Step 5: Tighten bolts in sequence to 36ft. lbs.
Step 6: Tighten bolts in sequence 80-90 degrees.
Step 7: Tighten bolts in sequence 40-45 degrees.
Step 8: Tighten center two bolts 40-45 degrees.

② Note assembly procedure for torqe sequence and specification.

③ Step 1: 22 ft. lbs.
Step 2: Tighten each bolt an additional 150 degrees
Step 3: Tighten each bolt an additional 15 degrees

④ Step 1: 37 ft. lbs.
Step 2: Tighten each bolt an additional 45 degrees
Step 3: Tighten each bolt an additional 15 degrees

⑤ Step 1: 18 ft. lbs.
Step 2: Tighten each bolt an additional 30 degrees

⑥ M14 Bolt
Step 1: 74 ft. lbs.
Step 2: Tighten each bolt an additional 75 degrees

⑦ Step 1: 48 ft. lbs.
Step 2: Tighten each bolt an additional 40 degrees

⑧ Step 1: 18 ft. lbs.
Step 2: 24 ft. lbs.

⑨ Step 1: 30 ft. lbs.
Step 2: 44 ft. lbs.
Step 3: Tighten each bolt an additional 90 degrees

⑩ Step 1: 18 ft. lbs.
Step 2: Tighten each bolt an additional 100 degrees

⑪ Step 1: 18 ft. lbs.
Step 2: Tighten each bolt an additional 60 degrees

⑫ Step 1: 18 ft. lbs.
Step 2: Tighten each bolt an additional 60 degrees

⑬ Step 1: 74 ft. lbs.
Step 2: Tighten an additional 150 degrees

⑭ Step 1: 22 ft. lbs.
Step 2: Tighten each bolt an additional 45 degrees

⑮ Step 1: 18.5 ft. lbs.
Step 2: Tighten each bolt an additional 90 degrees
Step 3: Tighten each bolt an additional 90 degrees
Step 4: Tighten each bolt an additional 90 degrees
Step 5: Tighten each bolt an additional 45 degrees

## WHEEL ALIGNMENT

| Year | Model | | Caster Range (+/-Deg.) | Caster Preferred Setting (Deg.) | Camber Range (+/-Deg.) | Camber Preferred Setting (Deg.) | Toe-in (in.) |
|------|-------|---|---|---|---|---|---|
| 2005 | 9-2x ① | F | NS | +3.50 | 0.50 | -0.33 | 0.08+/-0.12 |
| | | R | — | — | 0.50 | -1.33 | 0.00+/-0.12 |
| | 9-2x ② | F | NS | +3.41 | 0.50 | -0.16 | 0.08+/-0.12 |
| | | R | — | — | 0.50 | -1.25 | 0.00+/-0.12 |
| | 9-3 ① | F | 0.50 | +2.90 | 0.50 | -0.80 | ③ |
| | | R | — | — | 0.30 | -0.70 | ④ |
| | 9-3 ② | F | 0.50 | +2.90 | 0.50 | -0.90 | ③ |
| | | R | — | — | 0.30 | -1.00 | ④ |
| | 9-5 ① | F | 0.50 | +2.90 | 0.50 | -0.80 | ③ |
| | | R | — | — | 0.30 | -0.70 | ④ |
| | 9-5 ② | F | 0.50 | +2.90 | 0.50 | -0.90 | ③ |
| | | R | — | — | 0.30 | -1.00 | ④ |
| 2006 | 9-2x ① | F | NS | +3.50 | 0.50 | -0.33 | 0.08+/-0.12 |
| | | R | — | — | 0.50 | -1.33 | 0.00+/-0.12 |
| | 9-2x ② | F | NS | +3.41 | 0.50 | -0.16 | 0.08+/-0.12 |
| | | R | — | — | 0.50 | -1.25 | 0.00+/-0.12 |
| | 9-3 ① | F | 0.50 | +2.90 | 0.50 | -0.80 | ③ |
| | | R | — | — | 0.30 | -0.70 | ④ |
| | 9-3 ② | F | 0.50 | +2.90 | 0.50 | -0.90 | ③ |
| | | R | — | — | 0.30 | -1.00 | ④ |
| | 9-5 ① | F | 0.50 | +2.90 | 0.50 | -0.80 | ③ |
| | | R | — | — | 0.30 | -0.70 | ④ |
| | 9-5 ② | F | 0.50 | +2.90 | 0.50 | -0.90 | ③ |
| | | R | — | — | 0.30 | -1.00 | ④ |
| 2007 | 9-2x ① | F | NS | +3.50 | 0.50 | -0.33 | 0.08+/-0.12 |
| | | R | — | — | 0.50 | -1.33 | 0.00+/-0.12 |
| | 9-2x ② | F | NS | +3.41 | 0.50 | -0.16 | 0.08+/-0.12 |
| | | R | — | — | 0.50 | -1.25 | 0.00+/-0.12 |
| | 9-3 ① | F | 0.50 | +2.90 | 0.50 | -0.80 | ③ |
| | | R | — | — | 0.30 | -0.70 | ④ |
| | 9-3 ② | F | 0.50 | +2.90 | 0.50 | -0.90 | ③ |
| | | R | — | — | 0.30 | -1.00 | ④ |
| | 9-5 ① | F | 0.50 | +2.90 | 0.50 | -0.80 | ③ |
| | | R | — | — | 0.30 | -0.70 | ④ |
| | 9-5 ② | F | 0.50 | +2.90 | 0.50 | -0.90 | ③ |
| | | R | — | — | 0.30 | -1.00 | ④ |

22205_SAAB_C0009

## WHEEL ALIGNMENT

| Year | Model | | Caster | | Camber | | Toe-in (in.) |
|------|-------|---|Range (+/-Deg.)|Preferred Setting (Deg.)|Range (+/-Deg.)|Preferred Setting (Deg.)| |
| 2008 | 9-2x ① | F | NS | +3.50 | 0.50 | -0.33 | 0.08+/-0.12 |
| | | R | — | — | 0.50 | -1.33 | 0.00+/-0.12 |
| | 9-2x ② | F | NS | +3.41 | 0.50 | -0.16 | 0.08+/-0.12 |
| | | R | — | — | 0.50 | -1.25 | 0.00+/-0.12 |
| | 9-3 ① | F | 0.50 | +2.90 | 0.50 | -0.80 | ③ |
| | | R | — | — | 0.30 | -0.70 | ④ |
| | 9-3 ② | F | 0.50 | +2.90 | 0.50 | -0.90 | ③ |
| | | R | — | — | 0.30 | -1.00 | ④ |
| | 9-3 ⑤ | F | 0.50 | +2.90 | 0.50 | -1.00 | ① |
| | | R | — | — | 0.30 | -1.00 | — |
| | 9-5 ① | F | 0.50 | +2.90 | 0.50 | -0.80 | ③ |
| | | R | — | — | 0.30 | -0.70 | ④ |
| | 9-5 ② | F | 0.50 | +2.90 | 0.50 | -0.90 | ③ |
| | | R | — | — | 0.30 | -1.00 | ④ |

NS: Not specified by manufacturer

① Standard Suspension

② Sport Suspension

③ With 16 inch wheels: 0.079 inch (+/-0.022 inch)

   With 17 inch wheels: 0.087 inch (+/- 0.025 inch)

   With 18 inch wheels : 0.094 inch (+/- 0.027 inch)

④ With 16 inch wheels: 0.067 inch (+/-0.022 inch)

   With 17 inch wheels: 0.074 inch (+/- 0.025 inch)

   With 18 inch wheels : 0.081 inch (+/- 0.027 inch)

⑤ AWD

22205_SAAB_C0017

## TIRE, WHEEL AND BALL JOINT SPECIFICATIONS

| Year | Model | OEM Tires | | Tire Pressures (psi) | | Wheel Size | Ball Joint Inspection | Lug Nut Torque (ft. lbs.) |
| | | Standard | Optional | Front | Rear | | | |
|---|---|---|---|---|---|---|---|---|
| 2005 | 9-2x Aero | 205/55-16 | — | 32 | 29 | 6.5 | NP | 81 |
| | | — | 215/45-17 | 33 | 32 | 7.0 | NP | 81 |
| | 9-2x Linear | 205/55-16 | None | 32 | 29 | 7.0 | NP | 81 |
| | 9-3 Aero | 225/45-17 | None | 35 | 35 | 7.0 | NP | 81 |
| | 9-3 Arc | 215/55-16 | — | 32 | 32 | 6.5 | NP | 81 |
| | | — | 225/45-17 | 35 | 35 | 7.0 | NP | 81 |
| | 9-3 Linear | 215/55-16 | None | 32 | 32 | 6.5 | NP | 81 |
| | 9-5 Aero | 225/45-17 | None | 36 | 35 | 7.0 | NP | 80 |
| | 9-5 Arc | 215/55-16 | — | 35 | 32 | 6.5 | NP | 80 |
| | | — | 225/45-17 | 36 | 35 | 7.0 | NP | 80 |
| | 9-5 Linear | 215/55-16 | — | 35 | 32 | 6.5 | NP | 80 |
| | | — | 225/45-17 | 36 | 35 | 7.0 | NP | 80 |
| 2006 | 9-2x 2.5i | 205/55-16 | — | 32 | 29 | 6.5 | NP | 81 |
| | | — | 215/45-17 | 33 | 32 | 7.0 | NP | 81 |
| | 9-2x Aero | 205/55-16 | None | 32 | 29 | 6.5 | NP | 81 |
| | 9-3 2.0T | 215/55-16 | None | 32 | 32 | 6.5 | NP | 81 |
| | 9-3 Aero | 235/45-17 | None | 35 | 35 | 7.0 | NP | 81 |
| | 9-5 2.3T ① | 215/55-16 | None | 35 | 32 | 6.5 | NP | 80 |
| | 9-5 2.3T ② | 225/45-17 | None | 36 | 35 | 7.0 | NP | 80 |
| | 9-5 Arc ② | 215/55-16 | — | 35 | 32 | 6.5 | NP | 80 |
| | | — | 225/45-17 | 36 | 35 | 7.0 | NP | 80 |
| | 9-5 Linear ② | 215/55-16 | — | 35 | 32 | 6.5 | NP | 80 |
| | | — | 225/45-17 | 36 | 35 | 7.0 | NP | 80 |
| 2007 | 9-3 2.0T | 215/55-16 | None | 32 | 32 | 6.5 | NP | 81 |
| | 9-3 Aero | 235/45-17 | None | 35 | 35 | 7.0 | NP | 81 |
| | 9-5 | 235/45-17 | None | 41 | 41 | 7.0 | NP | 80 |
| 2008 | 9-3 2.0T | 215/55-16 | None | 32 | 32 | 6.5 | NP | 81 |
| | 9-3 Aero | 235/45-17 | None | 35 | 35 | 7.0 | NP | 81 |
| | 9-3 TurboX | 235/45-17 | None | 35 | 35 | 7.0 | NP | 81 |
| | 9-5 | 235/45-17 | None | 41 | 41 | 7.0 | NP | 80 |

NP: No play visible upon inspection

① Sedan

② Wagon

# BRAKE SPECIFICATIONS
All measurements are in inches unless noted

| Year | Model | | Brake Disc | | | Lining | Brake Caliper | |
| | | | Original Thickness | Minimum Thickness | Maximum Runout | Minimum Thickness | Bracket Bolts (ft. lbs.) | Mounting Bolts (ft. lbs.) |
|------|-------|---|------|------|------|------|------|------|
| 2005 | 9-2x | F | 0.940 | 0.870 | 0.003 | 0.430 | 59 | 19 |
| | | R | 0.039 | 0.034 | 0.002 | 0.354 | 38 | 27 |
| | 9-3 | F | ① | ① | 0.003 | 0.078 | ② | 21 |
| | | R | ③ | ③ | 0.003 | 0.078 | ④ | 21 |
| | 9-5 | F | 0.980 | 0.870 | 0.003 | 0.200 | — | ⑤ |
| | | R | 0.390 | 0.320 | 0.003 | NS | — | 59 |
| 2006 | 9-2x | F | 0.940 | 0.870 | 0.003 | 0.430 | 59 | 19 |
| | | R | 0.039 | 0.034 | 0.002 | 0.354 | 38 | 27 |
| | 9-3 | F | ① | ① | 0.003 | 0.078 | ② | 21 |
| | | R | ③ | ③ | 0.003 | 0.078 | ④ | 21 |
| | 9-5 | F | 0.980 | 0.870 | 0.003 | 0.200 | — | ⑤ |
| | | R | 0.390 | 0.320 | 0.003 | NS | — | 59 |
| 2007 | 9-3 | F | ⑥ | ⑥ | 0.003 | 0.078 | ② | 21 |
| | | R | ⑦ | ⑦ | 0.003 | 0.078 | ④ | 21 |
| | 9-5 | F | 0.980 | 0.870 | 0.003 | 0.200 | — | ⑤ |
| | | R | 0.390 | 0.320 | 0.003 | NS | — | 59 |
| 2008 | 9-3 | F | ⑥ | ⑥ | 0.003 | 0.078 | ② | 21 |
| | | R | ⑦ | ⑦ | 0.003 | 0.078 | ④ | 21 |
| | 9-5 | F | 0.980 | 0.870 | 0.003 | 0.200 | — | ⑤ |
| | | R | 0.390 | 0.320 | 0.003 | NS | — | 59 |

NS: Not specified by manufacturer

① Level 1: Original thickness: 0.980 inch; minimum thickness: 0.870 inch

Levels 2, 3 and 4: Original thickness: 1.100 inch; minimum thickness: 0.980 inch

② Step 1: 155 ft. lbs.

Step 2: Tighten bolts an additional 30 degrees

③ Levels 1 and 4: Original thickness: 0.470 inch; minimum thickness: 0.390 inch

Levels 2 and 3: Original thickness: 0.790 inch; minimum thickness: 0.730 inch

④ Step 1: 96 ft. lbs.

Step 2: Tighten bolts an additional 45 degrees

⑤ Step 1: 103 ft. lbs.

Step 2: Tighten bolts an additional 45 degrees

⑥ 15" Rotor: Original thickness: 0.980 inch; minimum thickness: 0.870 inch

16" Rotor: Original thickness: 1.100 inch; minimum thickness: 0.980 inch

17" Rotor: Original thickness: 1.200 inch; minimum thickness: 1.100 inch

⑦ 15" Rotor: Original thickness: 0.470 inch; minimum thickness: 0.390 inch

16" Rotor: Original thickness: 0.790 inch; minimum thickness: 0.710 inch

22205_SAAB_C0011

## SCHEDULED MAINTENANCE INTERVALS
### SAAB 9-2x

| TO BE SERVICED | SERVICE | VEHICLE MILEAGE INTERVAL (x1000) | | | | | | | | | | | | |
|---|---|---|---|---|---|---|---|---|---|---|---|---|---|---|
| | | 7.5 | 15 | 22.5 | 30 | 37.5 | 45 | 52.5 | 60 | 67.5 | 75 | 82.5 | 90 | 97.5 |
| Cabin air filter | Replace | ✓ | ✓ | ✓ | ✓ | ✓ | ✓ | ✓ | ✓ | ✓ | ✓ | ✓ | ✓ | ✓ |
| Engine oil & filter | Replace | ✓ | ✓ | ✓ | ✓ | ✓ | ✓ | ✓ | ✓ | ✓ | ✓ | ✓ | ✓ | ✓ |
| Tires | Inspect/Rotate | ✓ | ✓ | ✓ | ✓ | ✓ | ✓ | ✓ | ✓ | ✓ | ✓ | ✓ | ✓ | ✓ |
| Brake lines | Inspect/Service | | ✓ | | ✓ | | ✓ | | ✓ | | ✓ | | ✓ | |
| Clutch system | Inspect/Service | | ✓ | | ✓ | | ✓ | | ✓ | | ✓ | | ✓ | |
| Disc brake pads & discs | Inspect/Service | | ✓ | | ✓ | | ✓ | | ✓ | | ✓ | | ✓ | |
| Axle boots & axle shaft joints | Inspect/Service | | ✓ | | ✓ | | ✓ | | ✓ | | ✓ | | ✓ | |
| Parking brake | Inspect/Service | | ✓ | | ✓ | | ✓ | | ✓ | | ✓ | | ✓ | |
| Power steering system | Inspect/Service | | ✓ | | ✓ | | ✓ | | ✓ | | ✓ | | ✓ | |
| Steering & suspension | Inspect/Service | | ✓ | | ✓ | | ✓ | | ✓ | | ✓ | | ✓ | |
| Air filter element | Replace | | | | ✓ | | | | ✓ | | | | ✓ | |
| Automatic transmission fluid & filter | Inspect/Service | | | | ✓ | | | | ✓ | | | | ✓ | |
| Brake fluid | Replace | | | | ✓ | | | | ✓ | | | | ✓ | |
| Camshaft drive belt | Inspect/Service | | | | ✓ | | | | ✓ | | | | ✓ | |
| Coolant system, level, hoses & clamps | Inspect/Service | | | | ✓ | | | | ✓ | | | | ✓ | |
| Dfferential gear fluid | Inspect/Service | | | | ✓ | | | | ✓ | | | | ✓ | |
| Drive belts | Inspect/Service | | | | ✓ | | | | ✓ | | | | ✓ | |
| Engine coolant | Replace | | | | ✓ | | | | ✓ | | | | ✓ | |
| Fuel system, hoses & connections | Inspect/Service | | | | ✓ | | | | ✓ | | | | ✓ | |
| Spark plugs (2.0L) | Replace | | | | ✓ | | | | ✓ | | | | | |
| Transmission fluid | Inspect/Service | | | | ✓ | | | | ✓ | | | | ✓ | |
| Spark plugs (2.5L) | Replace | | | | | | | | ✓ | | | | | |
| Front & rear wheel bearing | Inspect/Service | | | | | | | | ✓ | | | | | |
| Fuel filter | Replace | | | | | | | | ✓ | | | | ✓ | |

22205_SAAB_C0013

## PRECAUTIONS

Before servicing any vehicle, please be sure to read all of the following precautions, which deal with personal safety, prevention of component damage, and important points to take into consideration when servicing a motor vehicle:

• Never open, service or drain the radiator or cooling system when the engine is hot; serious burns can occur from the steam and hot coolant.

• Observe all applicable safety precautions when working around fuel. Whenever servicing the fuel system, always work in a well-ventilated area. Do not allow fuel spray or vapors to come in contact with a spark, open flame, or excessive heat (a hot drop light, for example). Keep a dry chemical fire extinguisher near the work area. Always keep fuel in a container specifically designed for fuel storage; also, always properly seal fuel containers to avoid the possibility of fire or explosion. Refer to the additional fuel system precautions later in this section.

• Fuel injection systems often remain pressurized, even after the engine has been turned **OFF**. The fuel system pressure must be relieved before disconnecting any fuel lines. Failure to do so may result in fire and/or personal injury.

• Brake fluid often contains polyglycol ethers and polyglycols. Avoid contact with the eyes and wash your hands thoroughly after handling brake fluid. If you do get brake fluid in your eyes, flush your eyes with clean, running water for 15 minutes. If eye irritation persists, or if you have taken brake fluid internally, IMMEDIATELY seek medical assistance.

• The EPA warns that prolonged contact with used engine oil may cause a number of skin disorders, including cancer. You should make every effort to minimize your exposure to used engine oil. Protective gloves should be worn when changing oil. Wash your hands and any other exposed skin areas as soon as possible after exposure to used engine oil. Soap and water, or waterless hand cleaner should be used.

• All new vehicles are now equipped with an air bag system, often referred to as a Supplemental Restraint System (SRS) or Supplemental Inflatable Restraint (SIR) system. The system must be disabled before performing service on or around system components, steering column, instrument panel components, wiring and sensors. Failure to follow safety and disabling procedures could result in accidental air bag deployment, possible personal injury and unnecessary system repairs.

• Always wear safety goggles when working with, or around, the air bag system. When carrying a non-deployed air bag, be sure the bag and trim cover are pointed away from your body. When placing a non-deployed air bag on a work surface, always face the bag and trim cover upward, away from the surface. This will reduce the motion of the module if it is accidentally deployed. Refer to the additional air bag system precautions later in this section.

• Clean, high quality brake fluid from a sealed container is essential to the safe and proper operation of the brake system. You should always buy the correct type of brake fluid for your vehicle. If the brake fluid becomes contaminated, completely flush the system with new fluid. Never reuse any brake fluid. Any brake fluid that is removed from the system should be discarded. Also, do not allow any brake fluid to come in contact with a painted surface; it will damage the paint.

• Never operate the engine without the proper amount and type of engine oil; doing so WILL result in severe engine damage.

• Timing belt maintenance is extremely important. Many models utilize an interference-type, non-freewheeling engine. If the timing belt breaks, the valves in the cylinder head may strike the pistons, causing potentially serious (also time-consuming and expensive) engine damage. Refer to the maintenance interval charts for the recommended replacement interval for the timing belt, and to the timing belt section for belt replacement and inspection.

• Disconnecting the negative battery cable on some vehicles may interfere with the functions of the on-board computer system(s) and may require the computer to undergo a relearning process once the negative battery cable is reconnected.

• When servicing drum brakes, only disassemble and assemble one side at a time, leaving the remaining side intact for reference.

• Only an MVAC-trained, EPA-certified automotive technician should service the air conditioning system or its components.

## BRAKES

## ANTI-LOCK BRAKE SYSTEM (ABS)

### GENERAL INFORMATION

*PRECAUTIONS*

• Certain components within the ABS system are not intended to be serviced or repaired individually.

• Do not use rubber hoses or other parts not specifically specified for and ABS system. When using repair kits, replace all parts included in the kit. Partial or incorrect repair may lead to functional problems and require the replacement of components.

• Lubricate rubber parts with clean, fresh brake fluid to ease assembly. Do not use shop air to clean parts; damage to rubber components may result.

• Use only DOT 3 brake fluid from an unopened container.

• If any hydraulic component or line is removed or replaced, it may be necessary to bleed the entire system.

• A clean repair area is essential. Always clean the reservoir and cap thoroughly before removing the cap. The slightest amount of dirt in the fluid may plug an orifice and impair the system function. Perform repairs after components have been thoroughly cleaned; use only denatured alcohol to clean components. Do not allow ABS components to come into contact with any substance containing mineral oil; this includes used shop rags.

• The Anti-Lock control unit is a microprocessor similar to other computer units in the vehicle. Ensure that the ignition switch is **OFF** before removing or installing controller harnesses. Avoid static electricity discharge at or near the controller.

• If any arc welding is to be done on the vehicle, the control unit should be unplugged before welding operations begin.

**BRAKES**                                            **BLEEDING THE BRAKE SYSTEM**

## BLEEDING PROCEDURE

### BLEEDING PROCEDURE

#### 9-2x

*See Figures 1 and 2.*

1. Raise the vehicle and safely support it.

2. Remove both front and rear wheels.

3. Draw out the brake fluid from master cylinder with syringe.

4. Refill reservoir tank with recommended brake fluid.

5. Install one end of a vinyl tube onto the air bleeder and insert the other end of the tube into a container to collect the brake fluid.

6. Instruct your co-worker to depress the brake pedal slowly two or three times and then hold it depressed.

7. Loosen bleeder screw approximately 1/4 turn until a small amount of brake fluid drains into container, and then quickly tighten screw.

8. Repeat again from the two former procedures above until there are no air bubbles in drained brake fluid and new fluid flows through vinyl tube.

➡**Add brake fluid as necessary while performing the air bleed operation, in order to prevent the tank from running short of brake fluid.**

9. After completing the bleeding operation, hold brake pedal depressed and tighten screw and install bleeder cap.

10. Bleed air from each wheel cylinder using the same procedures as described above.

11. Depress brake pedal and hold it there for approximately 1 minute. At this time check pedal to see if it shows any

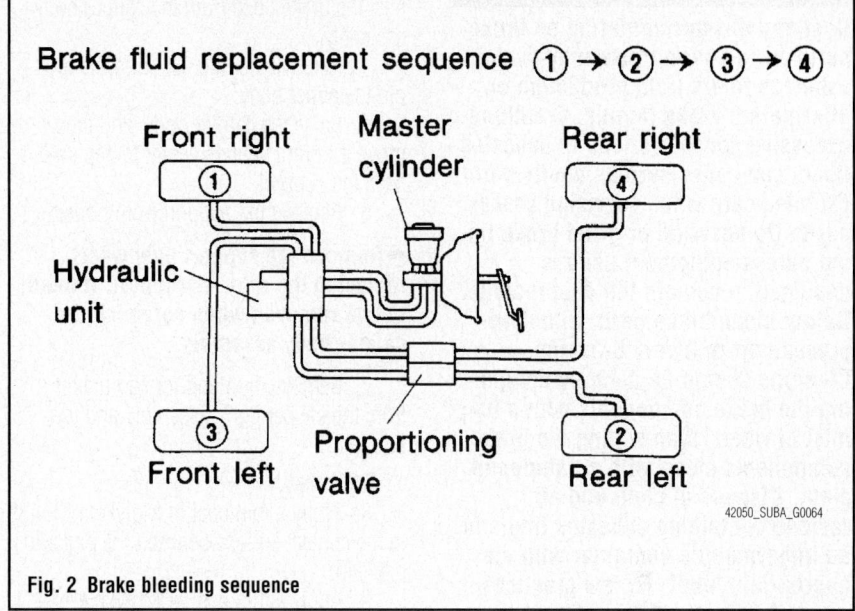

Fig. 2 Brake bleeding sequence

unusual movement. Visually inspect bleeder screws and brake pipe joints to make sure that there is no fluid leakage.

12. Install the wheels, and drive car for a short distance (between 1–2 miles to make sure that brakes are operating properly.

#### 9-3 and 9-5

*See Figure 3.*

1. Before servicing the vehicle, refer to the Precautions Section.

➡**Top up each time fluid is drained.**

2. Open the cap on the brake fluid reservoir and top up with brake fluid.

3. Raise the vehicle. Connect the bleeder hose to the air nipple on the front left brake caliper and open the nipple. Drain about 3.38 oz. (100 ml) brake fluid. Use 88 19 096 Bleeding equipment.

4. Top up with fluid in the brake fluid reservoir.

5. Tighten the nipple, install the rubber plug, move the brake bleeder to the air nipple on the rear right brake caliper and open the nipple. Drain about 1.7 oz. (50 ml) brake fluid.

6. Tighten the nipple, install the rubber plug, move the brake bleeder to the air nipple on the front right brake caliper and open the nipple. Drain about 3.38 oz. (100 ml) brake fluid.

7. Tighten the nipple, install the rubber plug, move the brake bleeder to the air nipple on the rear left brake caliper and open

the nipple. Drain about 1.7 oz. (50 ml) brake fluid.

8. Tighten the nipple, install the rubber plug and Lower the vehicle to the floor.

9. Check the operation of the brakes. Depress the brake pedal and check that it does not feel spongy or go all the way down to the floor.

10. Check the brake fluid level and adjust as necessary.

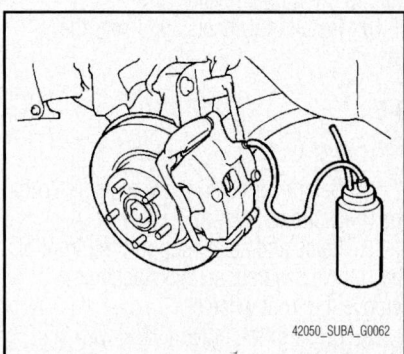

**Fig. 1 Connect one end of the tube onto the air bleeder.**

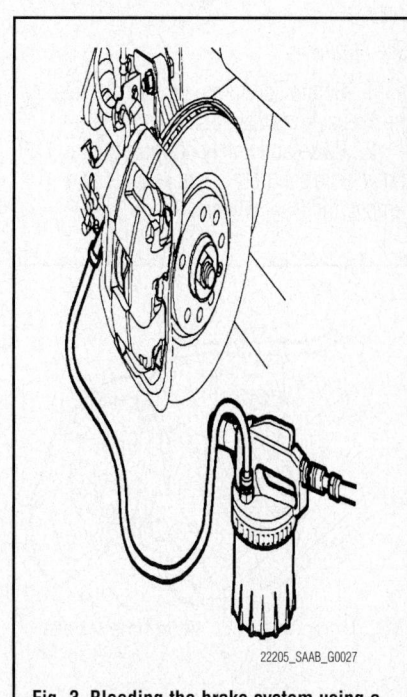

**Fig. 3 Bleeding the brake system using a vacuum bleeder—9-3 and 9-5**

## ✲✲ CAUTION

**Dust and dirt accumulating on brake parts during normal use may contain asbestos fibers from production or aftermarket brake linings. Breathing excessive concentrations of asbestos fibers can cause serious bodily harm. Exercise care when servicing brake parts. Do not sand or grind brake lining unless equipment used is designed to contain the dust residue. Do not clean brake parts with compressed air or by dry brushing. Cleaning should be done by dampening the brake components with a fine mist of water, then wiping the brake components clean with a dampened cloth. Dispose of cloth and all residue containing asbestos fibers in an impermeable container with the appropriate label. Follow practices prescribed by the Occupational Safety and Health Administration (OSHA) and the Environmental Protection Agency (EPA) for the handling, processing, and disposing of dust or debris that may contain asbestos fibers.**

## BRAKE CALIPER

### REMOVAL & INSTALLATION

#### 9-2x

*See Figure 4.*

1. Before servicing the vehicle, refer to the Precautions Section.
2. Raise and safely support the front of the vehicle securely on jackstands and remove the front wheels.

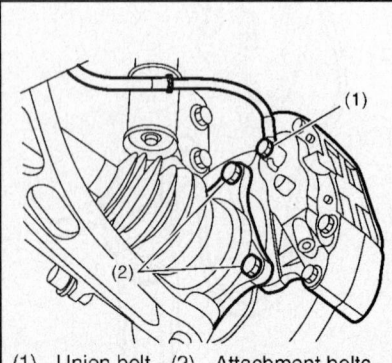

(1) Union bolt   (2) Attachment bolts

22140_SUBA_G0045

**Fig. 4 Rear brake caliper view—9-2x**

3. Remove the union bolt, and disconnect the brake hose from the caliper body assembly.
4. Remove the bolt securing the lock pin to caliper body.
5. Raise the caliper body, and then move it toward vehicle center to separate it from the support.
6. Remove the support from housing.

➡**Remove the support only when replacing the rotor or support. It need not be removed when servicing the caliper body assembly.**

7. Remove mud and foreign matter from the caliper body assembly and the support.

#### To install:

8. Apply a thin coat of Molykote M7439 to the contact surface between the pad and pad clip.
9. Apply a thin coat of Molykote AS-880N to both surfaces of the inner shim.
10. Install the pad to support.
11. Tightening specifications are as follows:
   - 16—inch type caliper body support housing to 59 ft. lbs. (80 Nm)
   - 17—inch type install the caliper body assembly to the housing and tighten to 114 ft. lbs. (155 Nm).
12. Connect the brake hose using a new brake hose gasket and tighten to 13 ft. lbs. (18 Nm).
13. Bleed the hydraulic system.

#### 9-3

*See Figure 5.*

1. Before servicing the vehicle, refer to the Precautions Section.
2. Raise and safely support the front of the vehicle securely on jackstands and remove the front wheels.
3. Press caliper piston back into caliper.
4. Use a brake pedal clamp to hold the pedal down.
5. Remove fuse No. 6 from electrical center.
6. Disconnect the brake hose from the brake caliper.
7. Remove the brake caliper mounting bolts and remove the brake caliper.

#### To install:

8. Install the brake caliper onto the strut. Apply thread-locking compound to the mounting bolts and tighten the caliper mounting bolts to 155 ft. lbs. (210 Nm) plus 30°.

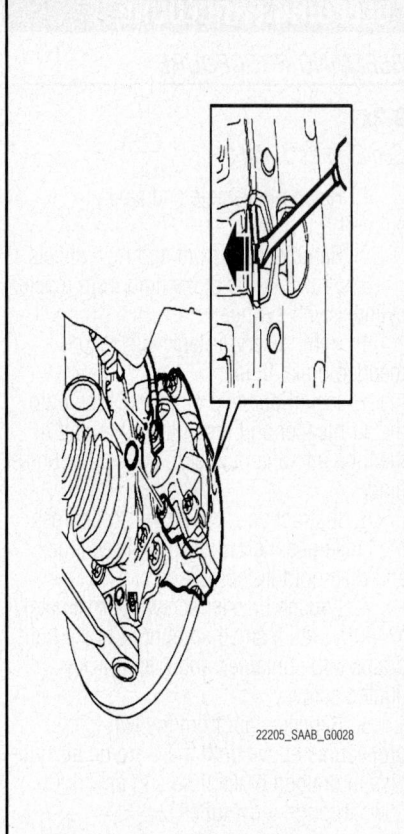

22205_SAAB_G0028

**Fig. 5 Press caliper piston back into caliper using a screwdriver—9-3**

9. Install the brake hose onto the caliper, with new seals, if equipped. Torque connection to 30 ft. lbs. (40 Nm).

## ✲✲ CAUTION

**Be careful to install the brake hose in its original position.**

10. Release the brake pedal clamp.
11. Install fuse No. 6 into electrical center.
12. Bleed the brake system and fill the fluid to the proper level.
13. Install the wheels and lower the vehicle.

#### 9-5

*See Figure 6.*

1. Before servicing the vehicle, refer to the Precautions Section.
2. Raise and safely support the front of the vehicle securely on jackstands and remove the front wheels.
3. Press in the brake piston with slip joint pliers.
4. Depress the brake pedal slightly with a brake pedal clamp.

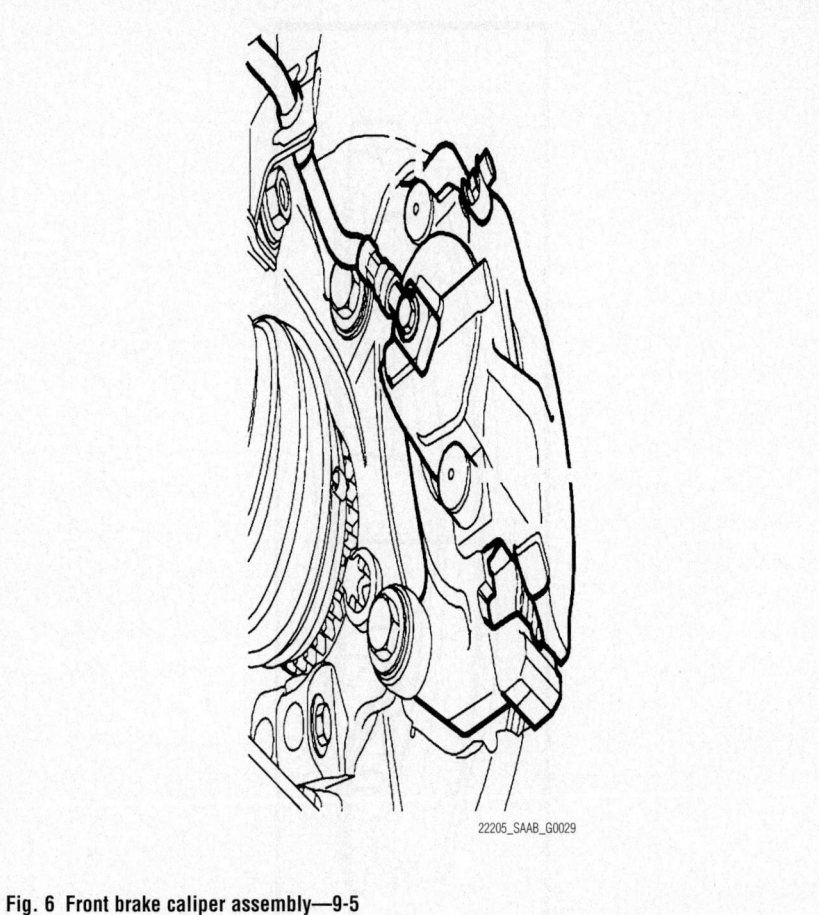

Fig. 6 Front brake caliper assembly—9-5

5. Remove fuse No. 1.

6. Remove the brake hose from the caliper.

7. Remove the caliper retaining bolts and remove the caliper.

**To install:**

8. Install the brake caliper and torque the retaining bolts to 86 ft. lbs. (117 Nm).

9. Install the brake hose to the caliper and torque it to 29 ft. lbs. (40 Nm).

10. Remove the brake pedal clamp and replace the fuse

11. Bleed the brake system and top off the fluid, if necessary.

12. Install the front wheels and lower the vehicle.

## DISC BRAKE PADS

*REMOVAL AND INSTALLATION*

### 9-2x

*See Figures 7 and 8.*

1. Before servicing the vehicle, refer to the Precautions Section.

2. Raise and safely support the vehicle. Remove the wheels

3. On 16 inch type, remove the "M" clip. Remove the pad pins and cross spring. Expand the pads and then push the piston back.

4. On 17 inch type, remove the clip. Remove the pad pins and cross spring. Expand the pads and then push the piston back.

5. Remove the disc brake pads.

**To install:**

6. Installation is the reverse of the removal procedure.

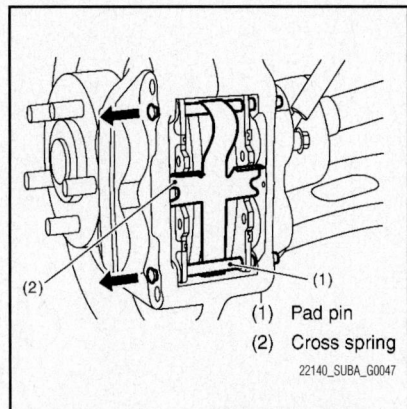

(1) Pad pin
(2) Cross spring

22140_SUBA_G0047

Fig. 7 16 inch type pads shown—9-2x

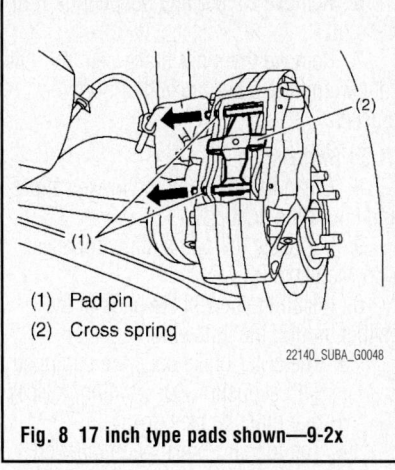

(1) Pad pin
(2) Cross spring

22140_SUBA_G0048

Fig. 8 17 inch type pads shown—9-2x

7. Apply a thin coat of Molykote AS-880N or equivalent to the frictional portion between the pad and pad inner shim.

8. Check the brake fluid level, correct as required.

9. Bleed the hydraulic system, as required.

### 9-3

*See Figure 9.*

1. Before servicing the vehicle, refer to the Precautions Section.

2. Raise and safely support the front of the vehicle securely on jackstands. Remove the wheels.

3. Remove the retaining clip from the brake caliper.

4. Push caliper piston back.

5. Remove the protective covers and caliper guide pins.

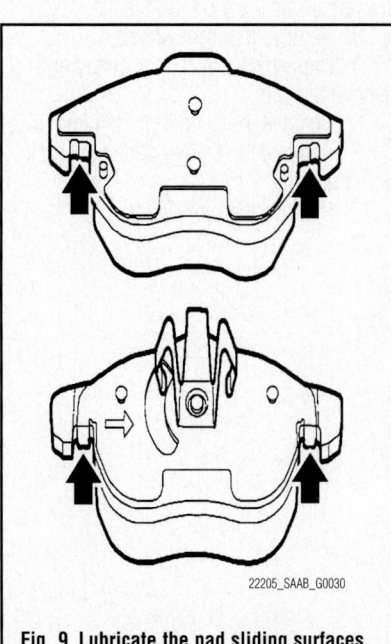

22205_SAAB_G0030

Fig. 9 Lubricate the pad sliding surfaces where indicated—9-3

6. Remove caliper and suspend it from the strut.

7. Remove the inner brake pad from the caliper and the outer brake pad from the bracket.

### To install:

8. Clean the inside of the brake caliper and check the dust covers.

9. Lubricate the pad sliding surfaces with Molykote P37.

10. Install the new brake pads in the caliper, noting the following:

- The outer brake pads are equipped with acoustic wear warning devices that must be face down.
- The inboard brake pads must be installed with the arrows pointing in the direction of rotation of the brake disc when the car is driven in a forward direction.

11. Install or connect the following:

- Brake caliper back into its original position on clean guide pins; torque to 21 ft. lbs. (28 Nm)
- Dust caps and retaining spring

12. Install the wheels and lower the vehicle.

13. Pump the brake pedal a few times to seat the brake pads.

14. Adjust brake fluid level, as necessary.

### 9-5

*See Figure 10.*

1. Before servicing the vehicle, refer to the Precautions Section.

2. Raise and safely support the front of the vehicle securely on jackstands.

3. Remove the front wheels.

4. Press back the caliper piston and remove the clip.

5. Remove the caliper guide pins.

6. Remove the caliper and suspend it from the strut.

7. Remove the brake pads.

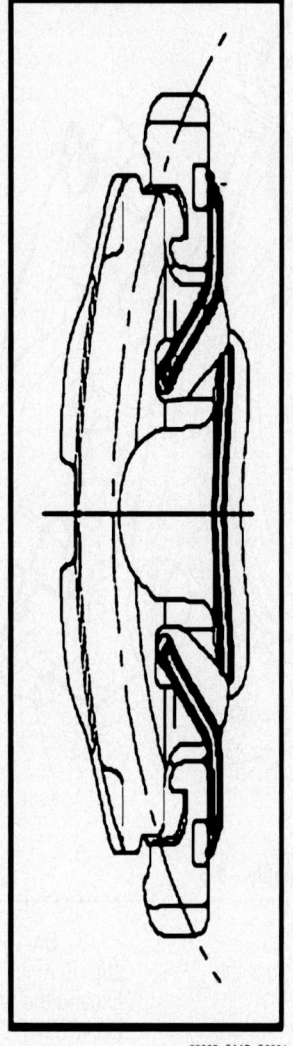

22205_SAAB_G0031

**Fig. 10 Make sure the clips are positioned as indicated—9-5**

### To install:

8. Install the brake pads in the caliper.

9. Install the brake caliper and torque the bolts to 21 ft. lbs. (28 Nm).

10. Install the brake clip depress the brake pedal to force out the pistons.

11. Install the front wheels.

12. Check and top off the brake fluid, if necessary.

**BRAKES**

### ✴✴ CAUTION

Dust and dirt accumulating on brake parts during normal use may contain asbestos fibers from production or aftermarket brake linings. Breathing excessive concentrations of asbestos fibers can cause serious bodily harm. Exercise care when servicing brake parts. Do not sand or grind brake lining unless equipment used is designed to contain the dust residue.

Do not clean brake parts with compressed air or by dry brushing. Cleaning should be done by dampening the brake components with a fine mist of water, then wiping the brake components clean with a dampened cloth. Dispose of cloth and all residue containing asbestos fibers in an impermeable container with the appropriate label. Follow practices prescribed by the Occupational Safety and Health Administration

(OSHA) and the Environmental Protection Agency (EPA) for the handling, processing, and disposing of dust or debris that may contain asbestos fibers.

### BRAKE CALIPER

*REMOVAL & INSTALLATION*

**9-2x**

*See Figure 11.*

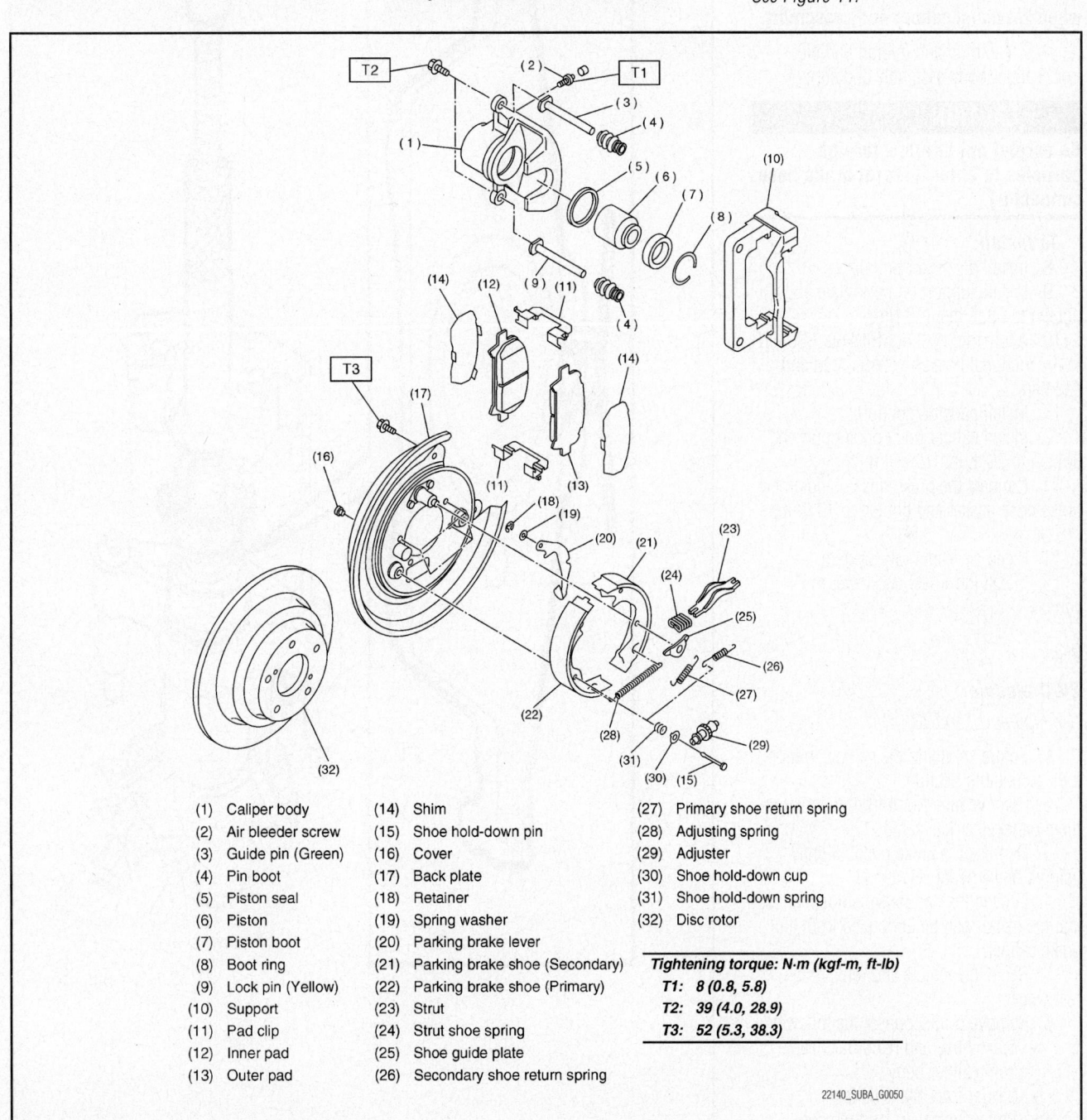

|  |  |  |  |  |  |
|---|---|---|---|---|---|
| (1) | Caliper body | (14) | Shim | (27) | Primary shoe return spring |
| (2) | Air bleeder screw | (15) | Shoe hold-down pin | (28) | Adjusting spring |
| (3) | Guide pin (Green) | (16) | Cover | (29) | Adjuster |
| (4) | Pin boot | (17) | Back plate | (30) | Shoe hold-down cup |
| (5) | Piston seal | (18) | Retainer | (31) | Shoe hold-down spring |
| (6) | Piston | (19) | Spring washer | (32) | Disc rotor |
| (7) | Piston boot | (20) | Parking brake lever |  |  |
| (8) | Boot ring | (21) | Parking brake shoe (Secondary) |  |  |
| (9) | Lock pin (Yellow) | (22) | Parking brake shoe (Primary) |  |  |
| (10) | Support | (23) | Strut |  |  |
| (11) | Pad clip | (24) | Strut shoe spring |  |  |
| (12) | Inner pad | (25) | Shoe guide plate |  |  |
| (13) | Outer pad | (26) | Secondary shoe return spring |  |  |

*Tightening torque: N·m (kgf-m, ft-lb)*

**T1:** *8 (0.8, 5.8)*
**T2:** *39 (4.0, 28.9)*
**T3:** *52 (5.3, 38.3)*

22140_SUBA_G0050

**Fig. 11 Exploded view of rear disc brake system—9-2x**

1. Before servicing the vehicle, refer to the Precautions Section.

2. Raise and safely support the rear of the vehicle securely on jackstands and remove wheels.

3. Disconnect brake hose from caliper body assembly.

4. Remove bolt securing lock pin to caliper body.

5. Raise caliper body and move it toward vehicle center to separate it from support.

6. Remove support from back plate.

➡**Remove support only when replacing it or the rotor. It need not be removed when servicing caliper body assembly.**

7. Clean mud and foreign particles from caliper body assembly and support.

### ※※ CAUTION

**Be careful not to allow foreign particles to enter inlet (at brake hose connector).**

#### To install:

8. Install disc rotor on hub.

9. Install support on back plate and tighten to 58 ft. lbs. (78 Nm).

10. Apply thin coat of Molykote AS880N to the frictional portion between pad and pad clip.

11. Install pads on support.

12. Install caliper body on support and tighten to 29 ft. lbs. (39 Nm).

13. Connect the brake hose using a new brake hose gasket and tighten to 13 ft. lbs. (18 Nm).

14. Bleed air from brake system.

15. Install the wheel and lower the vehicle.

### 9-3

#### 2WD Models

*See Figures 12 and 13.*

1. Before servicing the vehicle, refer to the Precautions Section.

2. Remove fuse No. 6 from the instrument panel electrical center.

3. Depress the brake pedal slightly using a brake pedal clamp.

4. Loosen the handbrake adjustment so that the cables can be unhooked from the brake caliper.

5. Raise the vehicle and remove the rear wheel.

6. Remove or disconnect the following:
   • Brake hose and handbrake cable from caliper body
   • Carrier from brake caliper
   • Protective covers and remove caliper body

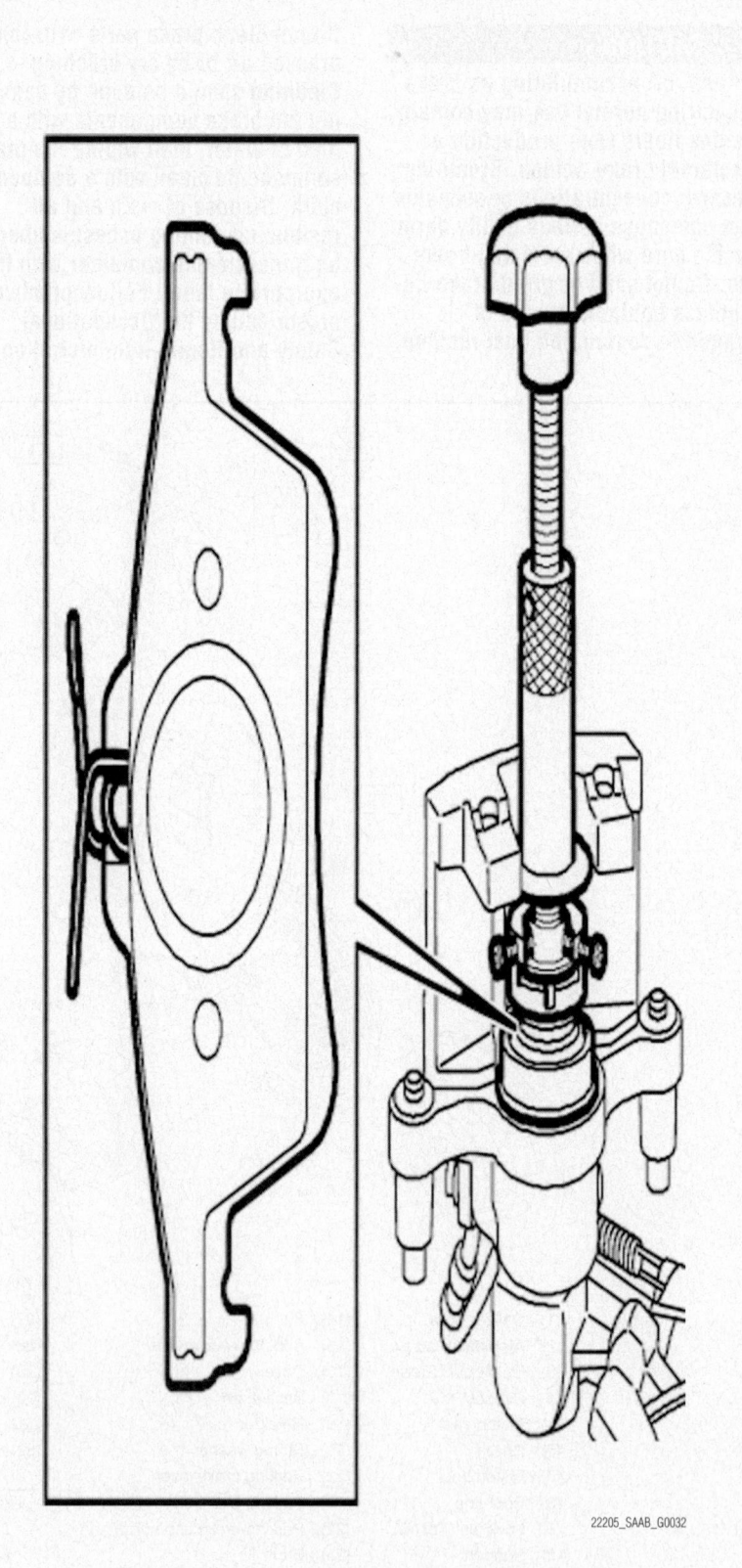

22205_SAAB_G0032

**Fig. 12 Screwing in the brake piston using the resetting tool—9-3**

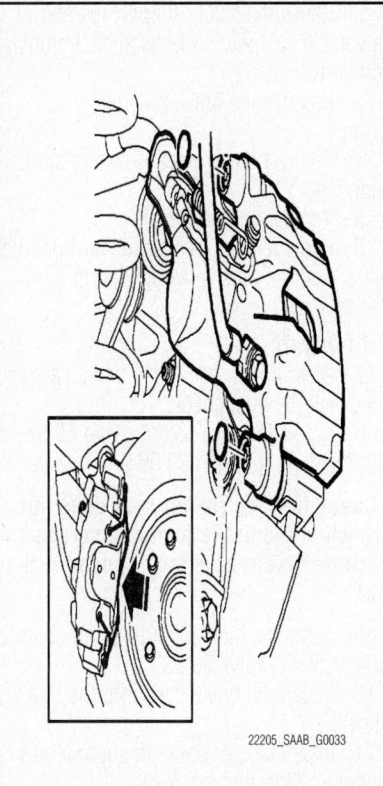

**Fig. 13 Make sure the clips are positioned as illustrated—9-3**

- Brake pads
- Carrier from steering knuckle.

***To install:***

7. Clean the contact surfaces between the steering knuckle and the brake caliper carrier.

8. Install or connect the following:
- Brake caliper carrier to steering knuckle, 74 96 268 thread lock adhesive to bolts and torque to 96 ft. lbs. (130 Nm) plus 45°.
- Screw in brake piston using 89 96 969 Resetting tool and 89 96 977 Adapter.
- Brake pads

➡**Make sure the springs on the inner pad enter the groove in the piston.**

- Caliper brake caliper; torque bolts to 21 ft. lbs. (28 Nm)
- Protective covers
- Retaining spring to the brake caliper
- Brake hose, with new seals, and torque connection to 30 ft. lbs. (40 Nm)

9. Remove the brake pedal clamp and Install fuse No. 6 in the instrument panel electrical center.

10. Bleed the brake system.

11. Adjust the handbrake.

12. Install the rear wheels and Lower the vehicle.

13. Depress the brake pedal repeatedly in order to seat brake piston and activate the handbrake self adjustment.

14. Check the operation of the foot brake and handbrake.

### 4WD Models

*See Figures 12 through 13.*

1. Before servicing the vehicle, refer to the Precautions Section.

2. Raise and safely support the rear of the vehicle securely on jackstands.

3. Remove the bolts for the bushings and fold down the anti-roll bar.

4. Remove the retaining spring from the brake caliper.

5. Remove the protective covers and the guide pins.

6. Hang up the brake caliper using a cable tie.

7. Remove the outer and inner brake pads.

***To install:***

8. Clean the inside of the brake caliper and check the dust covers.

9. Screw in brake piston using 89 96 969 Resetting tool and 89 96 977 Adapter.

➡**Check that the rubber seal does not turn when the brake piston is screwed in. Otherwise cracks may appear in the seal.**

10. Lubricate the pad sliding and contact surfaces with Molykote P37.

11. Install the brake pads to the holder and cut off the cable tie. Check that the inner pad locates in the piston's groove.

12. Install the brake caliper with the guide pins and the protective covers. Tighten the pins to 21 ft. lbs. (28 Nm).

13. Install the retaining spring to the brake caliper.

14. Position the stabilizer bar and Install the bolts for the bushings. Tighten bolts to 23 ft. lbs. (30 Nm).

15. Install the wheels and lower the vehicle.

16. Depress the brake pedal repeatedly in order to seat brake piston and activate the handbrake self adjustment.

17. Check the operation of the foot brake and handbrake.

### 9-5

*See Figure 14.*

1. Before servicing the vehicle, refer to the Precautions Section.

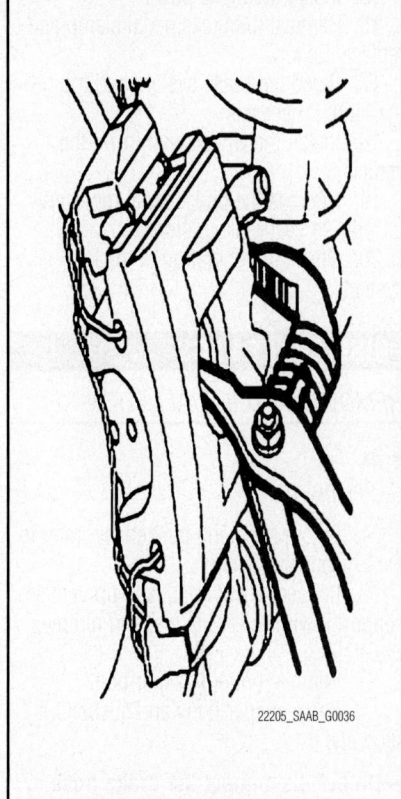

**Fig. 14 Pressing the brake caliper back using adjustable pliers—9-5**

2. Raise and safely support the rear of the vehicle securely on jackstands. Remove the rear wheels.

3. Remove the brake pads.

4. Depress the brake pedal slightly with a brake pedal clamp.

5. Remove fuse No. 1.

6. Remove the clip from the brake pedal pipe.

7. Remove the hose from the brake pedal pipe.

8. Remove the caliper guide pins and remove the brake caliper, swinging caliper only.

9. Remove the brake hose from the caliper.

10. Remove the caliper.

***To install:***

11. Install the brake lines to the brake caliper, swinging caliper only. Torque the line to 32 ft. lbs. (43 Nm).

12. Install the brake caliper and torque the bolts to 32 ft. lbs. (43 Nm).

13. Install the caliper and torque the bolts to 59 ft. lbs. (80 Nm).

14. Install the brake hose to the pipe and install the clip.

➡**Position the hose properly so it is not twisted.**

15. Install the brake pads.

16. Remove the brake pedal clamp and replace the fuse.

17. Bleed the brake system and top off the fluid, if necessary.

18. Install the wheels and lower the vehicle.

19. Depress the brake pedal repeatedly in order to seat brake pistons.

20. Check the operation of the system.

## DISC BRAKE PADS

### REMOVAL AND INSTALLATION

#### 9-2x

*See Figure 15.*

1. Before servicing the vehicle, refer to the Precautions Section.

2. Loosen wheel nuts, jack-up vehicle, support it with safety stands, and remove wheel.

3. Remove bottom caliper bolt.

4. Raise caliper body and suspend it securely.

➡**Do not disconnect the brake hose from caliper body.**

5. Remove pad from support.

6. If the brake pad is difficult to remove, proceed as follows:

  a. Remove the caliper body from support.

  b. Remove the support.

  c. Place a support in a vise between wooden blocks.

  d. Attach a rod of less than 0.47 in. (12 mm) dia. to the shaded area of brake pad, and strike the rod with a hammer to drive brake pad out of place.

➡**If it is difficult to push the piston during pad replacement, loosen air bleeder screw to assist in retracting the piston.**

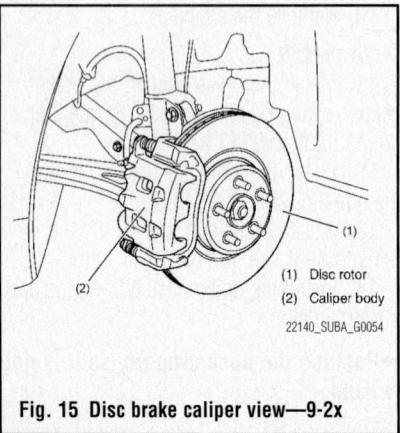

**Fig. 15 Disc brake caliper view—9-2x**

(1) Disc rotor
(2) Caliper body

22140_SUBA_G0054

#### To install:

7. Screw in brake piston using 89 96 969 Resetting tool and 89 96 977 Adapter.

➡**Check that the rubber seal does not turn when the brake piston is screwed in. Otherwise cracks may appear in the seal.**

8. Apply thin coat of Molykote M7439 to the frictional portion between pad and pad clip.

9. Install pad on support.

10. Install caliper body on support and tighten to 28 ft. lbs. (37 Nm).

11. Depress the brake pedal repeatedly in order to seat brake pads.

12. Check that brake fluid level is at max. line.

13. Install the wheels and lower the vehicle

14. Check the operation of the system.

#### 9-3

*See Figure 16.*

#### 2WD Models

1. Before servicing the vehicle, refer to the Precautions Section.

2. Raise and safely support the rear of the vehicle securely on jackstands. Remove the rear wheels.

3. Remove the brake pad retaining spring.

4. Remove the protective covers and guide pins.

5. Remove the caliper.

6. Remove the inner brake pad from the caliper and the outer brake pad from the bracket.

#### To install:

7. Clean the inside of the brake caliper and check the dust covers.

8. Screw in brake piston using 89 96 969 Resetting tool and 89 96 977 Adapter.

➡**Check that the rubber seal does not turn when the brake piston is screwed in. Otherwise cracks may appear in the seal.**

9. Lubricate the pad sliding and contact surfaces with Molykote P37.

10. Install the brake pads into the brake caliper.

11. Install caliper body on support and tighten to 28 ft. lbs. (37 Nm).

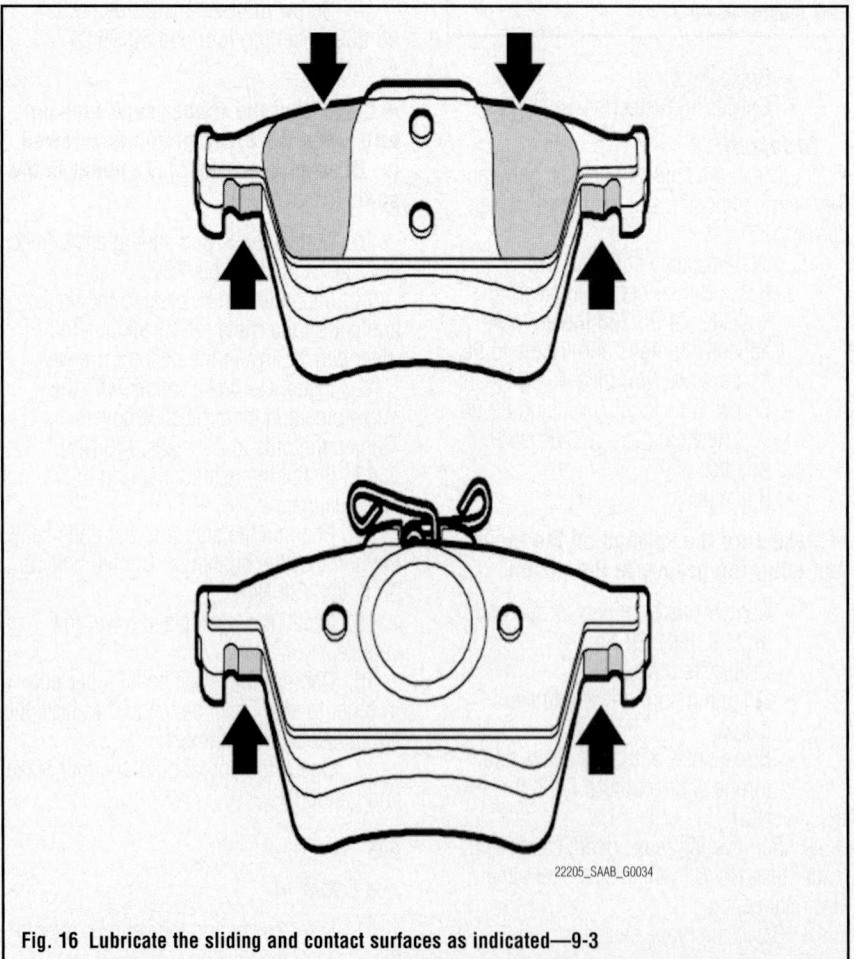

**Fig. 16 Lubricate the sliding and contact surfaces as indicated—9-3**

22205_SAAB_G0034

12. Install the protective covers and the retaining spring on the caliper.

13. Install the wheels and lower the vehicle

14. Press the brake pedal a few times to press out the brake pistons and adjust the handbrake.

15. Check that brake fluid level and adjust as necessary.

16. Check the operation of the system.

### 4WD Models

1. Before servicing the vehicle, refer to the Precautions Section.

2. Raise and safely support the rear of the vehicle securely on jackstands.

3. Remove the bolts for the bushings and fold down the anti-roll bar.

4. Remove the retaining spring from the brake caliper.

5. Remove the protective covers and the guide pins.

6. Hang up the brake caliper using a cable tie.

7. Remove the outer and inner brake pads.

**To install:**

8. Clean the inside of the brake caliper and check the dust covers.

9. Screw in brake piston using 89 96 969 Resetting tool and 89 96 977 Adapter.

> ❊❊ **WARNING**
>
> **Check that the rubber seal does not turn when the brake piston is screwed in. Otherwise cracks may appear in the seal.**

10. Lubricate the pad sliding and contact surfaces with Molykote P37.

11. Install the brake pads to the holder and cut off the cable tie. Check that the inner pad locates in the piston's groove.

12. Install the brake caliper with the guide pins and the protective covers. Tighten the pins to 21 ft. lbs. (28 Nm).

13. Install the retaining spring to the brake caliper.

14. Position the stabilizer bar and Install the bolts for the bushings. Tighten bolts to 23 ft. lbs. (30 Nm).

15. Install the wheels and lower the vehicle.

16. Depress the brake pedal repeatedly in order to seat brake piston and activate the handbrake self adjustment.

17. Check the operation of the foot brake and handbrake.

### 9-5

*See Figure 17.*

1. Before servicing the vehicle, refer to the Precautions Section.

2. Raise and safely support the rear of the vehicle securely on jackstands. Remove the rear wheels.

3. Press back the piston using slip joint pliers and remove the caliper clip.

4. Remove the dust caps and caliper guide pins.

5. Remove the brake caliper and suspend it from the suspension.

6. Remove the brake pads.

**To install:**

7. Clean the inside of the brake caliper with a soft wire brush and inspect the dust caps.

8. Install the brake pads. The inner and outer pads are different. The spring-

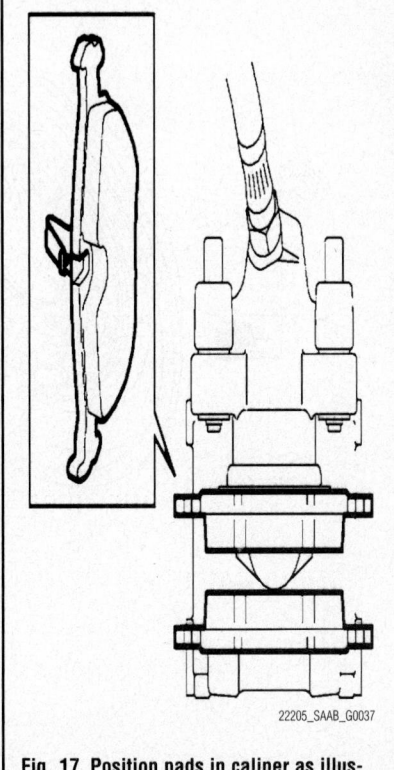

**Fig. 17 Position pads in caliper as illustrated—9-5**

installed pad must be installed against the piston.

9. Install the caliper and torque the pins to 20 ft. lbs. (28 Nm). Install the dust caps.

10. Install the rear wheels and lower the vehicle.

11. Depress the brake pedal to seat the brake pads.

12. Check and top off the brake fluid, as necessary.

## BRAKES

### PARKING BRAKE SHOES

*REMOVAL & INSTALLATION*

#### 9-2x

*See Figure 18.*

1. Before servicing the vehicle, refer to the Precautions Section.

2. Release the parking brake.

3. Remove the two mounting bolts and remove the brake caliper assembly.

4. Suspend the brake caliper assembly so that the hose is not stretched.

5. Remove the disc rotor.

6. If disc rotor is seized on the hub, drive the disc rotor out by pushing two 8 mm bolts in holes B on the rotor.

7. Remove the shoe return spring from the parking brake assembly.

8. Remove the front shoe hold down spring and pin.

9. Remove the strut and strut spring.

10. Remove the adjuster assembly from the parking brake assembly.

11. Remove the brake shoe.

12. Remove the rear shoe hold down spring and pin with pliers.

13. Remove the parking brake cable from lever.

14. Using a flat tip screwdriver, raise the retainer. Remove the parking lever and washer from brake shoe.

**To install:**

15. Apply brake grease to the following locations:

## PARKING BRAKE

- Six contact surfaces of the shoe rim and back plate gasket
- Contact surface of the shoe wave and the anchor pin
- Contact surface of the lever and strut
- Contact surface of the shoe wave and the adjuster assembly
- Contact surface of the shoe wave and the strut
- Contact surface of the lever and the shoe wave

16. Insert the primary side brake shoe into the anchor pin groove.

17. Secure the brake shoe with the shoe hold-down pin and cup.

18. Install the plate to the anchor pin, then install the primary return spring.

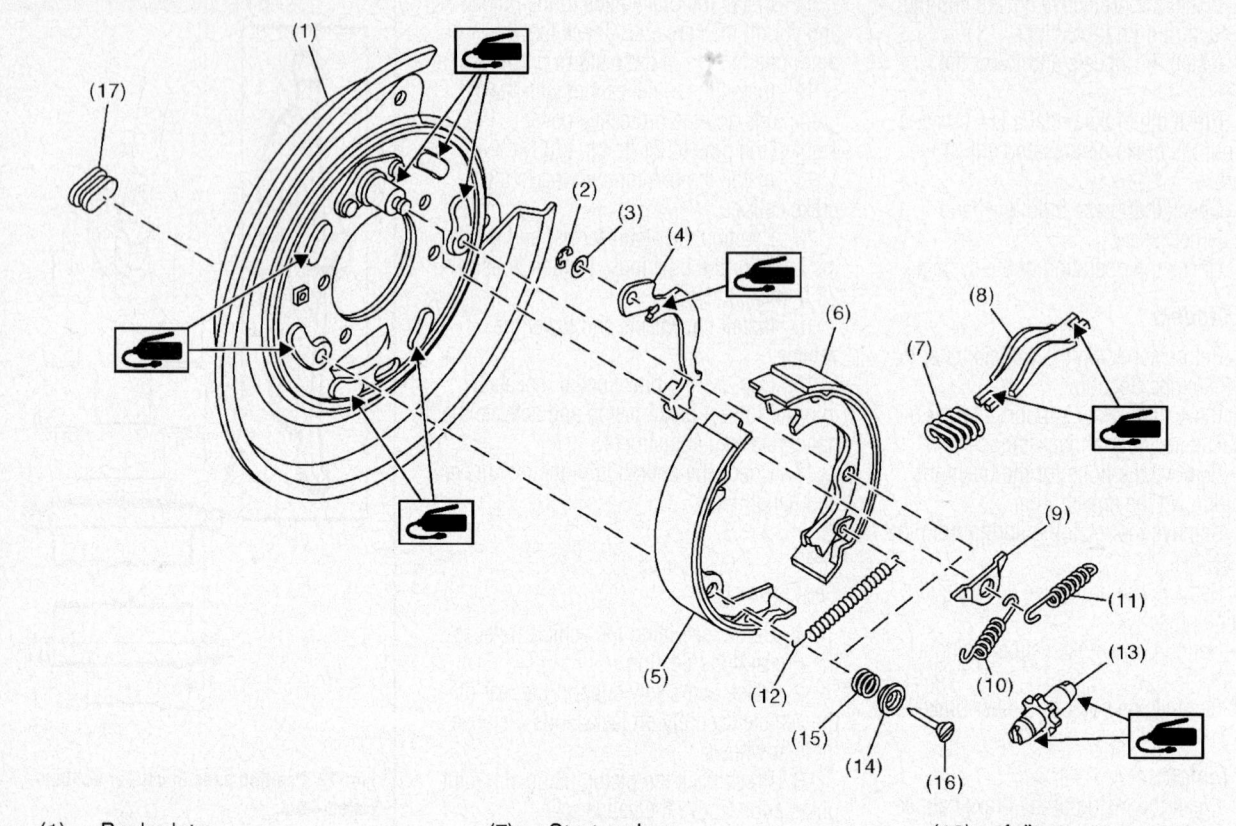

| | | | | |
|---|---|---|---|---|
| (1) | Back plate | (7) | Strut spring | (13) Adjuster |
| (2) | Retainer | (8) | Strut | (14) Shoe hold-down cup |
| (3) | Spring washer | (9) | Shoe guide plate | (15) Shoe hold-down spring |
| (4) | Lever | (10) | Primary return spring | (16) Shoe hold-down pin |
| (5) | Parking brake shoe (Primary) | (11) | Secondary return spring | (17) Adjusting hole cover |
| (6) | Parking brake shoe (Secondary) | (12) | Adjusting spring | |

22140_SUBA_G0059

**Fig. 18 Exploded view of parking brake components—9-2x**

19. Install the parking brake cable to the lever.

20. Install the strut and adjuster, then secure the secondary side brake shoe with the shoe hold-down pin and cup.

➡**Install the strut spring of both right and left wheel facing vehicle front. Install the adjuster assembly with screw section on the left side.**

21. Install the secondary return spring and the adjusting spring.

22. Adjust the parking brake.

23. Drive the vehicle to break-in the parking brake lining.

   a. Drive the vehicle at about 22 mph (35 km/h).

   b. With the parking brake release button pushed in, pull the parking brake lever gently.

   c. Drive the vehicle for about 0.12 miles (200 m) in this condition.

   d. Wait 5 to 10 minute s for the parking brake to cool down. Repeat again from step (A).

   e. After breaking-in, re—adjust the parking brakes.

### 9-5

*See Figure 19.*

1. Before servicing the vehicle, refer to the Precautions Section.

2. Remove the brake disc and swinging caliper.

3. Remove the return spring.

4. Unhook the cable from the actuating lever.

5. Remove the inner return springs, using tool 89 95 607, and the retaining pins using tool 89 96 647.

6. Remove the brake shoes and adjusting unit.

7. Inspect the actuating levers and adjusting unit for wear and replace as necessary.

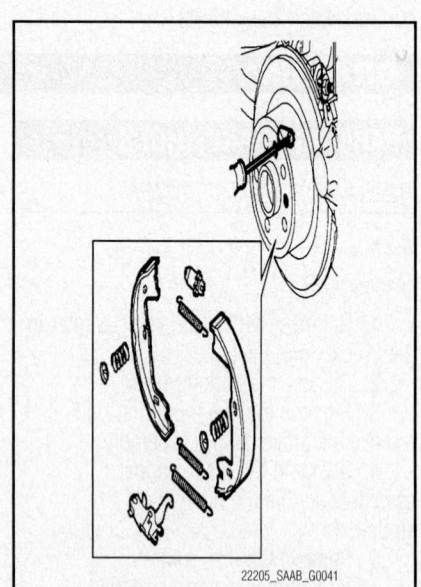

22205_SAAB_G0041

**Fig. 19 Exploded view of the parking brake shoes—9-5**

*To install:*

8. Position the brake shoes, actuating lever and adjusting unit on the backing plate.

9. Install the springs for the retaining pins, using tool 89 96 647, and the return springs using tool 89 95 607.

10. Install the brake disc and swinging caliper.

11. Hook the cable in place and install the return spring.

12. Adjust the handbrake.

13. Lower the vehicle.

14. Depress the brake pedal to seat the brake pistons.

15. Adjust the brake fluid level as necessary.

16. Check for proper brake operation.

## CHASSIS ELECTRICAL

### GENERAL INFORMATION

#### ✳✳ CAUTION

**These vehicles are equipped with an air bag system. The system must be disarmed before performing service on, or around, system components, the steering column, instrument panel components, wiring and sensors. Failure to follow the safety precautions and the disarming procedure could result in accidental air bag deployment, possible injury and unnecessary system repairs.**

#### *SERVICE PRECAUTIONS*

Disconnect and isolate the battery negative cable before beginning any airbag system component diagnosis, testing, removal, or installation procedures. Allow system capacitor to discharge for two minutes before beginning any component service. This will disable the airbag system. Failure to disable the airbag system may result in accidental airbag deployment, personal injury, or death.

Do not place an intact undeployed airbag face down on a solid surface. The airbag will propel into the air if accidentally deployed and may result in personal injury or death.

When carrying or handling an undeployed airbag, the trim side (face) of the airbag should be pointing towards the body to minimize possibility of injury if accidental deployment occurs. Failure to do this may result in personal injury or death.

Replace airbag system components with OEM replacement parts. Substitute parts may appear interchangeable, but internal differences may result in inferior occupant protection. Failure to do so may result in occupant personal injury or death.

Wear safety glasses, rubber gloves, and long sleeved clothing when cleaning powder residue from vehicle after an airbag deployment. Powder residue emitted from a deployed airbag can cause skin irritation. Flush affected area with cool water if irritation is experienced. If nasal or throat irritation is experienced, exit the vehicle for fresh air until the irritation ceases. If irritation continues, see a physician.

## AIR BAG (SUPPLEMENTAL RESTRAINT SYSTEM)

Do not use a replacement airbag that is not in the original packaging. This may result in improper deployment, personal injury, or death.

The factory installed fasteners, screws and bolts used to fasten airbag components have a special coating and are specifically designed for the airbag system. Do not use substitute fasteners. Use only original equipment fasteners listed in the parts catalog when fastener replacement is required.

During, and following, any child restraint anchor service, due to impact event or vehicle repair, carefully inspect all mounting hardware, tether straps, and anchors for proper installation, operation, or damage. If a child restraint anchor is found damaged in any way, the anchor must be replaced. Failure to do this may result in personal injury or death.

Deployed and non-deployed airbags may or may not have live pyrotechnic material within the airbag inflator.

Do not dispose of driver/passenger/curtain airbags or seat belt tensioners unless you are sure of complete deployment. Refer to the Hazardous Substance Control System for proper disposal.

Dispose of deployed airbags and tensioners consistent with state, provincial, local, and federal regulations.

After any airbag component testing or service, do not connect the battery negative cable. Personal injury or death may result if the system test is not performed first.

If the vehicle is equipped with the Occupant Classification System (OCS), do not connect the battery negative cable before performing the OCS Verification Test using the scan tool and the appropriate diagnostic information. Personal injury or death may result if the system test is not performed properly.

Never replace both the Occupant Restraint Controller (ORC) and the Occupant Classification Module (OCM) at the same time. If both require replacement, replace one, then perform the Airbag System test before replacing the other.

Both the ORC and the OCM store Occupant Classification System (OCS) calibration data, which they transfer to one another when one of them is replaced. If both are

replaced at the same time, an irreversible fault will be set in both modules and the OCS may malfunction and cause personal injury or death.

If equipped with OCS, the Seat Weight Sensor is a sensitive, calibrated unit and must be handled carefully. Do not drop or handle roughly. If dropped or damaged, replace with another sensor. Failure to do so may result in occupant injury or death.

If equipped with OCS, the front passenger seat must be handled carefully as well. When removing the seat, be careful when setting on floor not to drop. If dropped, the sensor may be inoperative, could result in occupant injury, or possibly death.

If equipped with OCS, when the passenger front seat is on the floor, no one should sit in the front passenger seat. This uneven force may damage the sensing ability of the seat weight sensors. If sat on and damaged, the sensor may be inoperative, could result in occupant injury, or possibly death.

#### *DISARMING THE SYSTEM*

#### ✳✳ CAUTION

**The air bag system must be disarmed before performing service around air bag system components or system wiring. Failure to do so may cause accidental deployment of the air bag, resulting in unnecessary air bag system repairs and/or personal injury.**

Always disconnect the battery cables (negative cable first) and wait 20 minute s prior to performing service around air bag system components or system wiring. Tape the battery cable for added protection.

#### ✳✳ CAUTION

**Do not use any diagnostic instruments that are battery powered, such as buzzers, ohmmeters or diode testers, to diagnose faults in the steering wheel or electronic control unit. Using such devices may trigger the air bag. Also, ensure that the battery cables cannot accidentally come into contact with the battery terminals.**

## ARMING THE SYSTEM

To rearm the air bag system, reconnect the battery cables.

## CLOCKSPRING CENTERING

### 9-2x

*See Figure 20.*

> ✳✳ **CAUTION**
>
> **When servicing a vehicle, be sure to turn the ignition switch to OFF, disconnect the ground cable from battery, and wait for more than 1 minute before starting work. The airbag system is installed with a backup power source. After disconnecting the battery ground cable, the airbag may deploy if you do not wait for more than 1 minute before starting the service of airbag system.**

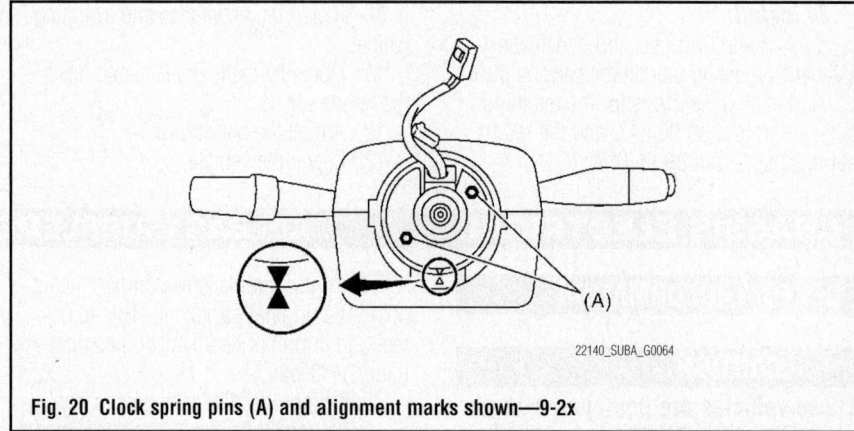

22140_SUBA_G0064

**Fig. 20 Clock spring pins (A) and alignment marks shown—9-2x**

1. Turn the ignition switch to **OFF**.
2. Disconnect the ground cable from battery and wait for at least 1 minute before starting work.
3. Check that front wheels are positioned in straight ahead direction.

4. Turn the clock spring pin (A) clockwise until it stops.
5. Turn the clock spring connector pins (A) approx. 3.25 turns until the marks are aligned.

# DRIVETRAIN

## AUTOMATIC TRANSAXLE ASSEMBLY

### REMOVAL & INSTALLATION

#### 9-2x

*See Figure 21.*

1. Before servicing the vehicle, refer to the Precautions Section.
2. Set the vehicle on a lift.
3. Open the front hood and support with the hood stay.
4. Disconnect the ground cable from the battery.
5. Remove the collector cover.
6. Remove the intercooler. (Turbo models)
7. Remove the air intake chamber. (Non-turbo models)

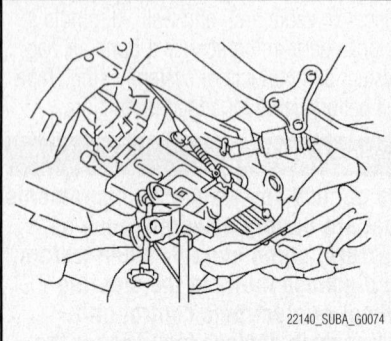

22140_SUBA_G0074

**Fig. 21 Transmission jack set into place—9-2x**

8. Remove the air cleaner case. (Non-turbo models)
9. Remove the air breather hose.
10. Remove the starter.
11. Front oxygen (A/F) sensor.
12. Transmission harness connectors.
13. Remove the intercooler stay and engine hanger rear. (Turbo models)
14. Disconnect the engine harness connectors, and then remove the engine hanger rear. (Non-turbo model)
15. Remove the water by-pass pipe. (Turbo models)
16. Separate the torque converter from flexplate as follows:
   - Remove the service hole plug.
   - Remove the bolts which hold torque converter to flexplate.
   - Remove the four bolts by rotating the clamp pulley a little at a time.
   - Make sure the torque converter moves freely by rotating with finger through the starter installation hole.
17. Attach the ST-498277200 converter holder to the converter case.
18. Remove the pitching stopper.
19. Remove the pitching stopper bracket.
20. Install the ST-41099AC000 engine support assembly.
21. Remove the air intake duct. (Turbo models)
22. Remove the air cleaner case. (Turbo models)
23. Remove the transmission mounting bolt (upper side).
24. Lift-up the vehicle. (Turbo models)

25. Remove the undercover. (Turbo models)
26. Remove the center and rear exhaust pipes and the muffler. (Turbo models)
27. Remove the front exhaust pipe, rear exhaust pipe and muffler. (Non-turbo model)
28. Remove the heat shield cover.
29. Remove the drain plug to drain transmission fluid.
30. Remove the oil charge pipe.
31. Disconnect the connector from turbine speed sensor 1.
32. Remove the turbine speed sensor 1 connector mounting bolt and rotate the sensor by 180°.

> ✳✳ **CAUTION**
>
> **Failure to follow this procedure may cause the interference between vehicle body and sensor while removing/installing transmission, and resulting in damage.**

33. Remove the driveshaft.
34. Remove the shift select cable.
35. Disconnect the hose from the ATF inlet and outlet pipes.
36. Remove the front crossmember support plate.
37. Remove the two clutch housing cover securing bolts.
38. Remove the front stabilizer bracket.
39. Remove the bolts which secure front ball joint to the housing.
40. Pull out the driveshaft from transmission.

41. Set the transmission jack under the transmission.

42. Remove the rear crossmember.

43. Remove the transmission mounting bolt (lower side).

44. Remove the transmission.

➡**Turn the engine support assembly from the vehicle under body to the left (to shorten the engine support length), and lower the rear of the engine for easy disassembly. Be careful not to allow breather pipe and etc. to touch the vehicle body when detaching the automatic transmission assembly by pulling it backward.**

*To install:*

45. Attach the ST-498277200 to the converter case.

46. Install the transmission onto the engine.

47. Lift-up the transmission gradually using transmission jack. Engage at the splined.

48. Install the engine mounting bolt (lower side) and tighten to 37 ft. lbs. (50 Nm).

49. Install the transmission rear crossmember and tighten the mounting bolts to 55 ft. lbs. (75 Nm).

50. Remove the transmission jack.

51. Lower the vehicle.

52. Install the engine mounting bolt (upper side) and tighten to 37 ft. lbs. (50 Nm).

53. Remove the ST from converter case.

54. Install the starter.

55. Install the torque converter to flexplate.

   a. Install the bolts which hold torque converter to flexplate.

   b. Install all four bolts by rotating the crank pulley a little at a time. Tighten to 18 ft. lbs. (25 Nm).

   c. Install the service hole.

56. Remove the ST-41099AC000 engine support assembly.

57. Install the pitching stopper.

58. Install the pitching stopper bracket and tighten to 30 ft. lbs. (41 Nm).

59. Install the pitching stopper and tighten to 37 ft. lbs. (50 Nm).

60. Lift-up the vehicle.

61. Replace the front differential side retainer oil seal.

62. Replace the circlip of the front driveshaft with a new part.

63. Apply grease to the oil seal lip.

64. Attach the ST-28399SA010 seal protector to side retainer.

65. Align and insert the splined of the front driveshaft to the splines of the differ-

ential bevel gear, and remove the ST-28399SA010.

66. Install the inlet and outlet hoses to the ATF inlet and outlet pipes.

67. Insert the ball joint into housing.

68. Install the front stabilizer bracket.

69. Install the clutch housing cover securing bolts.

70. Install the front cross support plate.

71. Install the driveshaft.

72. Install the shift select cable.

73. Install the turbine speed sensor 1 and harness, and then connect the connector. Tighten the mounting bolt to 5 ft. lbs. (7 Nm).

74. Install the oil charge pipe and tighten to 30 ft. lbs. (41 Nm).

75. Install the heat shield cover.

76. Install the center, rear exhaust pipes and the muffler. (Turbo models)

77. Install the front exhaust pipe, rear exhaust pipe and muffler. (Non-turbo models)

78. Install the undercover.

79. Lower the vehicle.

80. Install the air cleaner case.

81. Install the air intake duct.

82. Connect the following connectors:
- Transmission harness connectors
- Front oxygen (A/F) sensor

83. Install the intercooler stay RH and engine hanger rear. (Turbo models)

84. Install the engine hanger rear, and then connect the engine harness connector. (Non-turbo models)

85. Install the water by-pass pipe. (Turbo models)

86. Pour ATF from the oil charge pipe.

87. Install the air breather hose.

88. Install the intercooler. (Turbo models)

89. Install the air intake chamber. (Non-turbo models)

90. Install the air cleaner case. (Non-turbo models)

91. Install the collector cover.

92. Connect the ground cable to battery.

93. Perform Clear Memory 2 operation.

   a. Connect the Tech 2 to the data link connector.

   b. Turn the ignition switch to ON (engine OFF) and turn Tech 2 switch to ON.

   c. Ensure that the select lever is in the "P" range.

   d. On the "Main Menu" display screen, select the {Each System Check} and press the [YES] key.

   e. On the "System Selection Menu" display screen, select the {Transmission} and press the [YES] key.

   f. Press the [YES] key after the information of transmission type is displayed.

   g. On the "Transmission Diagnosis" display screen, select the {Clear Memory 2} and press the [YES] key.

94. Perform the inspection with driving the vehicle at the end of repair work, and make sure there is no faulty as below:
- Excessive shift shock
- Oil leakage from the transmission body, etc.
- Occurrence of noise caused by interference etc.

➡**If excessive shift shock is felt, execute advance operation of learning control.**

**9-3**

1. Before servicing the vehicle, refer to the Precautions Section.

2. Remove or disconnect the following:
- Upper engine cover
- Battery cover
- Battery with coolant pipe, hood switch and fuse box
- Control module connector
- Battery tray
- Connector under control module
- Ventilation hose from transmission
- Gear cable from selector lever arm (selector lever must be in N)
- Cable from the cable retainer (pull back locking sleeve)
- Ground cable from transmission casing
- Upper bolts on transmission casing
- Expansion tank from its holder
- Install engine lifting eyes. Attach a lifting beam and hoist to lifting eyes.
- Dipstick

3. With the hoist, take weight off the engine and transmission.

4. Remove the upper bumper shell mountings.

5. Attach 2 Straps above the radiator core so they are accessible from below (these will be used to restrain the radiator).

6. Remove or disconnect the following:
- Front wheels
- Lower front cover, then secure radiator core with Straps
- Radiator brackets from subframe
- Exhaust pipe (cut about 3.5 inches (87mm) from front of muffler)
- Engine torque rod from subframe
- Bolt and nut holding transverse link to steering knuckle on both sides
- Stabilizer bar from link arm
- Connector and cable clip of headlamp angle sensor (if equipped)

- Steering gear from subframe (retain nuts and washers); leave steering gear hanging
- Power steering pipe from subframe
- Front part of fender liner (bend it out of the way)

7. Position a trolley lift with jig underneath. Remove the subframe bolts and the rear brackets. Lower the subframe slightly.

8. Separate the steering arm ball joints from the steering knuckles.

9. Remove the stabilizer bar from the subframe (protect the steering gear boot). Install the bolts loosely in the link arms so the stabilizer bar is kept hanging in place.

10. Lower the subframe.

11. Remove the torque rod bracket. Undo the bolt for the torque rod for easier installing.

12. Remove or disconnect the following:
- Starter motor
- Torque converter from flexplate (6 bolts; turn crankshaft to access bolts)
- Undo plug and press torque converter towards transmission; install holding tool
- Transmission fluid
- Oil cooling hoses; plug holes and hoses
- Bolts in charge air pipe bracket and move pipe aside
- Lower transmission bolts; leave one bolt loosely in place
- Ground lead from transmission
- ABS cable from clip
- Left driveshaft; suspend driveshaft by means of a cable tie

13. Lower the vehicle, mark the location of the bolts on the left engine mount with a marker pen to ensure correct reinstalling. Remove the engine mount bracket.

14. Lower the powertrain with the lifting beam to facilitate removal of the transmission.

15. Undo the clip securing the power steering hose.

16. Raise the vehicle and move the power steering cooling coil aside and restrain with a Strap.

17. Install a holding device, 87 92 608, onto a column jack. Adjust and tighten the tool on the transmission. Use a bolt with 8.8 grade that is approx. 0.8 inch (20mm) longer than the removed bolt as well as a bolt (8.8 grade) with nut on the rear lifting eye.

18. Remove the remaining bolts securing the transmission to the engine.

19. Remove the last bolt, pull out the transmission and lower it.

20. Lift the transmission down from the column jack with an engine hoist.

## To install:

➡ **If installing a new transmission, transfer the torque converter holder to the new transmission from the old one.**

21. Turn the torque converter so that the bolt holes line up with the holes in the flexplate. Install torque converter holder during bolt installation.

22. Make sure the 2 guide sleeves are on the engine and apply anti-corrosion agent to the sleeves.

23. Install new driveshaft seals (lubricate them before installation).

24. Install the holder, 87 92 608, onto the column jack. Secure the tool to the transmission as for removal.

25. Install a protective collar, 83 95 162, in the right shaft seal in the transmission. This is done to protect the seal while the transmission is being installed. Lubricate the seal.

26. Place the transmission in position. Slide in the gearbox until about 0.8 inch (20mm) are remaining and remove the tool.

27. Push in the transmission until it is against the mating face. Tighten the bolts between the engine and the transmission that are accessible from below to 52 ft. lbs. (70 Nm).

28. Remove the lifting tool from the transmission and remove the jack. Install the bolt into the transmission and torque it to 16 ft. lbs. (22 Nm).

29. Undo the Strap and position the power steering cooling coil in place. Make sure it is installed on the correct side of the rubber seal.

30. Lower the vehicle and install the mounting for the engine bracket onto the transmission. Lift up the unit and install the bolts for the engine bracket as marked earlier. Tighten bolts to 69 ft. lbs. (93 Nm).

31. Remove the lifting beam and undo the lifting eyes.

32. Install the expansion tank in its holder.

33. Install the remaining upper bolts in the transmission and torque to 52 ft. lbs.(70 Nm).

34. Install a protective collar, 83 95 162, into the sealing ring. Lubricate the sealing ring.

35. Make sure the driveshaft is clean and align it with the tool. Slide in the gearbox until about 0.8 inch (20mm) are remaining and remove the tool. Push in the rest of the shaft until the circlip clicks in.

36. Attach the ABS cable to the clip.

37. Raise the vehicle and remove the torque converter holding tool. Press the torque converter against the flexplate. Install the plug.

38. Apply thread lock to the torque converter-to-flexplate bolts. Use the original bolts and washers. Using longer bolts will damage the torque converter. Install the 6 bolts without tightening.

39. Rotate the engine clockwise and tighten the bolts once they are all in place. Torque bolts to 22 ft. lbs. (30 Nm).

40. Install the starter motor with electrical connections. Torque bolts to 35 ft. lbs. (47 Nm).

41. Install the charge air pipe bracket. Torque the bolts to 15 ft. lbs. (20 Nm).

42. If original transmission is installed, change oil cooler hose seals. Install new seals lubricated with Vaseline on the oil cooler hoses. Lubricate the pipes and install the oil cooler hoses. The pipes will penetrate the seals when installed.

### ✳✳ CAUTION
**Ensure that the surfaces in the gearbox gasket opening are not damaged.**

43. Install or connect the following:
- Ground cable to transmission; torque to 7 ft. lbs. (10 Nm)
- Bracket for torque arm on transmission; torque to 69 ft. lbs. (93 Nm).
- Place subframe on trolley lift with jig; lift the frame
- Power steering pipe
- Transverse link ball joints in steering knuckle
- Stabilizer bar to subframe (opening of rubber bushings must face front); torque to 47 ft. lbs. (64 Nm)
- Subframe into position
- Guide pins in fixture to holes in body
- Subframe bolts and rear brackets
- Subframe against body so bushings do not rotate when bolts are tightened (remove lift); torque to 55 ft. lbs. (75 Nm), plus an additional 90° and to 60 ft. lbs. (90 Nm), plus an additional 45° for rear bracket bolts
- Steering gear bolts, washers and nuts to 37 ft. lbs. (50 Nm), plus an additional 60°
- Engine torque rod to subframe; torque to 52 ft. lbs. (70 Nm), plus an additional 90°
- Transverse links to steering knuckles; torque to 37 ft. lbs. (50 Nm)

### ✳✳ CAUTION
**Ensure that the steering knuckle stub is visible on the top of the steering knuckle housing before the bolt is installed.**

- Radiator brackets to subframe; torque to 35 ft. lbs. (47 Nm); remove the Straps

44. Set up the engine centering kit, 83 96 152, and check that engine is located correctly in relation to the subframe. Remove the centering tool.

45. Install the power steering pipe to the subframe. Insert all bolts first and then tighten.

46. Install a joint clamp on the front pipe and install the front exhaust pipe to the catalytic converter. Use a new gasket and new nuts. Torque nuts to 18 ft. lbs. (25 Nm). Adjust the joint clamp so that the pipe ends are in the middle. Tighten the joint clamp nuts to 30 ft. lbs. (40 Nm).

47. Connect or install the following:
- Fender liner section on both sides
- Front wheels
- Lower front cover
- Upper bumper shell mountings
- Cable to retainer (by pulling back the locking sleeve)
- Gear cable to selector lever arm

48. Carry out selector lever cable adjustment as follows:
a. Engage P with the gear selector on the transmission.
b. Rock the car until the parking lock engages.
c. Press down the locking brace by the selector lever housing securing the adjuster.
d. Check the shifting positions.
e. Install the plastic cover for the selector lever and the rubber mat in the storage compartment.

49. Install or connect the following:
- Transmission ventilation hose (lightly lubricated)
- Connectors under control module
- Battery tray
- Control module connector
- Hood switch
- Cooling pipe
- Main fuse box
- Battery, cables and cover
- Upper engine cover
- Dipstick
- Transmission fluid

50. Test drive the car with varying engine loads and speeds. Check for trouble codes. Also check the position of the steering wheel when driving straight ahead on a level road. Adjust if needed.

## 9-5

1. Before servicing the vehicle, refer to the Precautions Section.

2. Remove or disconnect the following:
- Intake manifold cover
- Battery and tray
- MAXI fuse board
- 16 pin and 10 pin wire connectors
- Transmission breather hose
- Gear selector arm from the transmission (place lever in L position)
- Shifting cable
- 3 upper bolts from the transmission
- Dipstick tube (plug hole in block)
- Oxygen sensor connectors
- Rear engine mount nut
- Top 2 bolts from the rear engine cushion, loosen only

3. Install a lifting beam to the engine and relieve the weight on the engine and transmission.

4. Remove or disconnect the following:
- Both front wheels
- Lower engine cover
- Front exhaust system
- Rear engine bracket and pad
- Steering gear bolts
- Rear clamps securing the power steering delivery pipe to the subframe
- A/C pipes from the subframe holder
- Air cleaner casing from the subframe
- Engine oil cooler from the charge air cooler

5. Attach a Strap around the radiator and crossmember to restrain the radiator.

6. Remove or disconnect the following:
- Outer ball joint to steering knuckle bolts
- Upper stabilizer bar ball joints
- Torque arm from the subframe
- Rear support plates

7. Position a lifting trolley with a holder under the vehicle. Align the trolley to the subframe.

8. Remove or disconnect the following:
- Remaining bolts from the subframe and slightly lower the subframe
- Power steering delivery pipe clamps. Lower the lifting trolley and move the subframe aside
- Splash plate
- Bolts securing the torque converter to the flexplate. Rotate the plate and pulley together

➡**The crankshaft must be rotated to gain access for all the bolts.**

9. Press the torque converter against the transmission to keep the converter in place during transmission removal.

10. Remove or disconnect the following:
- Ground leads
- Torque arm and bracket
- Transmission fluid
- Oil cooler inlet and outlet hoses (plug openings)
- Left driveshaft and suspend it
- Left side engine pad; lower transmission to clear structural member
- Transmission bracket
- Transmission from the engine assembly using a single-column jack
- Transmission from the vehicle

**To install:**

11. Rotate the torque converter so that the bolt holes align with the flexplate.

12. Make certain that the 2 guide sleeves are on the engine.

13. Lubricate and install new driveshaft seals.

14. Raise the transmission into position under the vehicle.

15. Install the transmission and torque the bottom bolts between the engine and transmission to 55 ft. lbs. (74 Nm) and the bolts between the transmission and oil pan to 34 ft. lbs. (47 Nm)

16. Remove the lifting beam from the transmission.

17. Install or connect the following:
- Transmission bolts and torque the bolts to 18 ft. lbs. (24 Nm)
- Engine pad mount to the transmission and torque the bolts to 62 ft. lbs. (84 Nm)
- Bolts securing the engine pad to the body and torque the bolts to 46 ft. lbs. (63 Nm)
- Protective sleeve in driveshaft seal
- Driveshaft and make certain that the circlip snaps into position (remove protective sleeve just before driveshaft is fully into transmission)
- Torque converter to flexplate and torque the bolts to 22 ft. lbs. (30 Nm)
- Splash plate
- Cooler hoses and torque the fasteners to 20 ft. lbs. (27 Nm)
- Torque arm bracket
- Ground cable to the bracket and raise the subframe into position
- Power steering delivery pipe clamps
- Outer ball joints and A/C pipes
- Subframe and rear support plates and torque the subframe bolts to 74 ft. lbs. (100 Nm) plus 45° and the support plate bolts to 44 ft. lbs. (60 Nm)

- Outer ball joints to the steering knuckles and torque the bolts to 63 ft. lbs. (85 Nm)
- Stabilizer bar link and torque the fastener to 68 ft. lbs. (92 Nm)
- Steering gear and torque the bolts to 66 ft. lbs. (90 Nm)
- Engine oil cooler and air filter housing
- Rear engine cushion
- Rear engine mount; torque bolts to 55 ft. lbs. (70 Nm)
- Torque arm-to-subframe; torque bolts to 18 ft. lbs. (25 Nm)
- Front exhaust system; torque bolts to 18 ft. lbs. (24 Nm) on 2.3L engine.
- Oxygen sensor wiring connectors and remove lifting beam
- Torque rear engine cushion bolts to 18 ft. lbs. (24 Nm) and nut to 33 ft. lbs. (45 Nm).
- Engine to transmission; torque bolts to 55 ft. lbs. (70 Nm)
- Dipstick tube
- Transmission breather hose
- Shifter cable to bracket
- Shifter cable to selector lever; adjust if necessary
- Torque rod and torque bolt to 34 ft. lbs. (47 Nm)
- MAXI fuse board
- Battery tray and battery
- Battery cables and cover
- Intake manifold cover
- Front wheels

18. Fill the transmission to the proper level.

19. Start the vehicle, check for leaks and repair if necessary.

## MANUAL TRANSAXLE ASSEMBLY

### REMOVAL & INSTALLATION

#### 9-2x

##### 5-Speed

*See Figures 22 through 24.*

1. Before servicing the vehicle, refer to the Precautions Section.
2. Open the hood to the full open position.
3. Disconnect the negative battery cable.
4. Drain the transmission fluid.
5. On non turbocharged engine, remove the air intake duct and cleaner case. Remove the air cleaner case stay.
6. On turbocharged engine, remove the intercooler assembly.

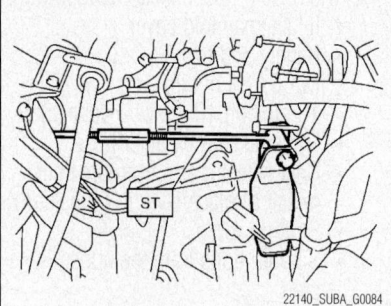

**Fig. 22 ST-41099AC000 engine support—9-2x 5-Speed**

7. Disconnect the neutral position switch connector and the backup light switch connector.
8. Disconnect the VSS sensor electrical connector.
9. Remove the starter.
10. Remove the clutch operating cylinder from the transmission and suspend on a wire.
11. Remove the pitching stopper. Position engine support tool ST-41099AC000, or equivalent to hold the engine assembly in place.
12. Remove the bolts which hold upper side of transmission to engine.
13. Lift the vehicle.
14. Remove the front and center exhaust pipes. (Non–turbo models)
15. Remove the center exhaust pipe. (Turbo models)
16. Remove the rear exhaust pipe and muffler.
17. Remove the heat shield cover. (If equipped)
18. Remove the driveshaft.
19. Remove the gear shift rod and the stay from the transmission.
   a. Disconnect the stay from the transmission.
   b. Remove the gear shift rod from the transmission.

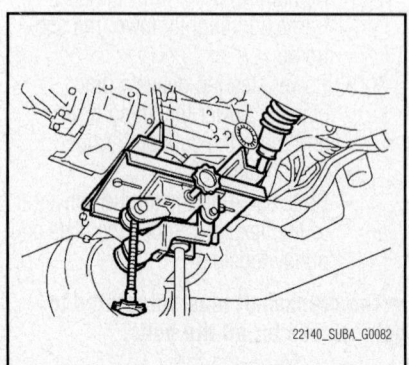

**Fig. 23 Transmission jack supporting the transmission—9-2x 5-Speed**

20. Remove the stabilizer link from the front arm.
21. Remove the bolt securing the ball joint of the front arm to the housing, then separate the front arms and the housing.
22. Using a crowbar, remove the left and right front driveshaft from the transmission.
23. Remove the bolts and nuts which hold lower side of transmission to engine.
24. Place the transmission jack under the transmission.

### ✳✳ CAUTION

**Always support the transmission case with a transmission jack.**

25. Remove the transmission rear crossmember from the vehicle.
26. Tighten the turnbuckle of the ST-41099AC000 while lowering the transmission jack to tilt the engine assembly towards the back.
27. Remove the transmission.

➡**Move the transmission jack towards the rear until the main shaft is withdrawn from the clutch disc.**

28. Separate the transmission assembly from the rear cushion rubber.

#### *To install:*

29. Replace the differential side retainer oil seal.
30. Install the rear cushion rubber to the transmission assembly and tighten to 26 ft. lbs. (35 Nm).
31. Install the transmission onto the engine.
   a. Lift-up the transmission gradually using a transmission jack.
   b. Engage at the splined section.
32. Loosen the turnbuckle of the ST-41099AC000 while raising the transmission jack to return the engine to its original position.

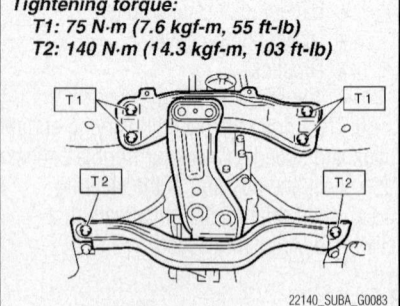

Tightening torque:
T1: 75 N·m (7.6 kgf-m, 55 ft-lb)
T2: 140 N·m (14.3 kgf-m, 103 ft-lb)

**Fig. 24 Transmission rear crossmember and mounting bolts—9-2x 5-Speed**

33. Install the transmission rear cross-member.

34. Take out the transmission jack.

35. Tighten the bolts and nuts which hold the lower side of transmission to the engine to 37 ft. lbs. (50 Nm).

36. Connect the transmission to the engine.

37. Install the starter

38. Tighten the bolts which hold the upper side of the transmission to the engine to 37 ft. lbs. (50 Nm).

39. Remove the ST-41099AC000 engine support tool, or equivalent.

40. Install the pitching stopper. Tighten to 37 ft. lbs. (50 Nm).

41. Lift the vehicle.

42. Install the front driveshaft into the transmission. Use seal protector ST-28399SA010 or equivalent.

43. Insert the ball joints of the front arm into the housing, then tighten the installing bolts to 37 ft. lbs. (50 Nm).

44. Attach the stabilizer link to the front arm and tighten to 33 ft. lbs. (45 Nm).

45. Attach the gear shift rod and stay.

   a. Attach the gear shift rod to the transmission.

   b. Attach the stay to the transmission. Tighten to 13 ft. lbs. (18 Nm).

46. Install the driveshaft.

47. Install the heat shield cover. (If equipped)

48. Install the rear exhaust pipe and muffler.

49. Install the front exhaust pipe and the center exhaust pipe. (Non-turbo models)

50. Install the center exhaust pipe. (Turbo models)

51. Install the operating cylinder and tighten to 27 ft. lbs. (37 Nm).

52. Connect the following connectors.

- Transmission ground cable, tighten to 9 ft. lbs. (13 Nm).
- Neutral position switch connector
- Back-up light switch connector

53. Fill transmission gear oil through the transmission level gauge hole.

54. Install the air intake chamber stay and tighten to 12 ft. lbs. (16 Nm). (Non-turbo models)

55. Install the air intake chamber and air cleaner case. (Non-turbo model)

56. Install the intercooler. (Turbo models)

57. Connect the battery ground cable to the battery.

58. Take off the vehicle from lift arms.

59. Start the engine and check for leaks, correct as required.

60. Road test the vehicle.

## 6-Speed

*See Figure 25.*

1. Before servicing the vehicle, refer to the Precautions Section.

2. Set the vehicle on a lift. Open the front hood, and support it with the stay.

3. Disconnect the ground cable from battery.

4. Remove the collector cover.

5. Remove the intercooler.

6. Remove the front wheels.

7. Disconnect the following harness connectors:

- Neutral position switch backup light switch connector
- Rear oxygen sensor connector

8. Remove the engine hanger rear.

9. Disconnect the ground cable on the upper side of the transmission case and body.

10. Remove the starter assembly.

11. Remove the operating cylinder from the transmission.

➡**Hang the removed operating cylinder with a piece of wire.**

12. Remove the pitching stopper and pitching stopper bracket.

13. Install the ST-41099AC000 engine support tool, or equivalent.

14. Remove the clutch release shaft.

   a. Remove the plug using a hexagon wrench.

   b. Attach a 6 mm (0.24 in) bolt to the release shaft, and pull out the release shaft.

   c. Lift the release fork, and remove from the claw of the release bearing. Pull the release fork to the engine side, and make it so that it moves freely.

15. Remove the bolts which hold upper side of transmission to engine.

16. Lift the vehicle.

17. Remove the center exhaust pipe.

18. Remove the rear exhaust pipe and muffler.

19. Remove the heat shield cover. (If equipped)

20. Remove the driveshaft.

21. Remove the front stabilizer link.

22. Remove the ball joint of front arm from the housing.

23. Remove the front drive shaft.

24. Set the transmission jack under the transmission, and remove the front cross-member and rear crossmember.

25. Move the transmission to the right side of the vehicle, and remove the joint COMPL, stay bolts and reverse check cable.

➡**If the transmission is not moved aside, the joint COMPL and stay bolts**

may contact the body and cause damage.

26. Tighten the turnbuckle of the ST-41099AC000 to tilt the engine assembly towards the back.

27. Remove the bolts holding the bottom of transmission to the engine, and remove the transmission from the vehicle.

### *To install:*

28. Set the release fork, release bearing and release shaft to the transmission.

29. Replace the front differential side retainer oil seal.

30. Remove the oil seal by using flat tip screwdriver etc.

31. Apply gear oil to the lip of new oil seals.

32. Install a new oil seal using ST-18675AA000.

➡**Be sure to replace the differential side oil seal after the procedure of removing front driveshaft from transmission.**

33. Loosen the turnbuckle of ST-41099AC000 to return the engine to its original position.

34. Install the transmission.

35. Tighten the bolts and nuts which hold the lower side of transmission to the engine to 37 ft. lbs. (50 Nm).

36. Move the transmission to the right side of the vehicle, and attach the joint COMPL, stay bolts and reverse check cable.

37. Install the front crossmember and rear crossmember.

38. Tighten the bolts which hold the upper side of the transmission to the engine to 37 ft. lbs. (50 Nm).

39. Make sure that the release bearing is completely inserted.

➡**Push the release fork towards the operating cylinder side until a clicking sound is heard. Pull the release fork towards the engine side. If the release**

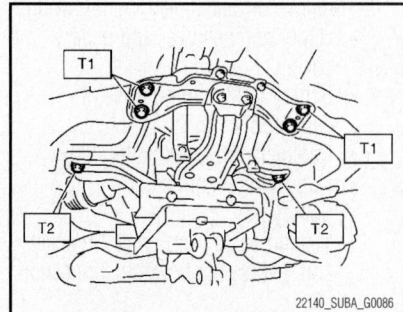

**Fig. 25 Transmission front and rear cross-member and mounting bolts—9-2x 6-Speed**

**fork is not in contact with the case, the setting is complete. Confirm that the boot cover is set securely.**

40. Install the pitching stopper bracket and tighten to 30 ft. lbs. (41 Nm).

41. Install the pitching stopper and tighten to 37 ft. lbs. (50 Nm).

42. Install the clutch slave cylinder. Tighten mounting bolts to 30 ft. lbs. (41 Nm).

43. Install the starter assembly.

44. Attach the ground cable to the transmission and body.

45. Connect the following harness connectors:

- Neutral position switch backup light switch connector
- Rear oxygen sensor connector

46. Attach the engine hanger rear.

47. Install the front drive shafts into the transmission. Use a seal protector ST-28399SA010 and replace the circlip of driveshaft with a new part.

48. Install the ball joint of the front arm. Tighten bolt to 37 ft. lbs. (50 Nm).

49. Install the front stabilizer links and tighten nuts to 33 ft. lbs. (45 Nm).

50. Install the heat shield cover. (If equipped)

51. Install the driveshaft.

52. Install the rear exhaust pipe and muffler.

53. Install the center exhaust pipe.

54. Fill the transmission gear oil.

55. Install the intercooler.

56. Install the collector cover.

57. Connect the battery ground cable to the battery.

58. Take off the vehicle from lift arms.

59. Start the engine and check for leaks, correct as required.

60. Road test the vehicle.

## 9-3

*See Figures 26 and 27.*

1. Before servicing the vehicle, refer to the Precautions Section.

2. Remove or disconnect the following:

- Upper engine cover and battery cover
- Battery (and cooling hose, if equipped)
- Cable clamp under battery tray
- Hood switch connector
- Battery tray
- Ground cable from engine bracket and reverse light switch connection (7)
- Gear cables from the transmission by pulling locking sleeves back, lifting up cables from levers, and carefully moving them aside (8)

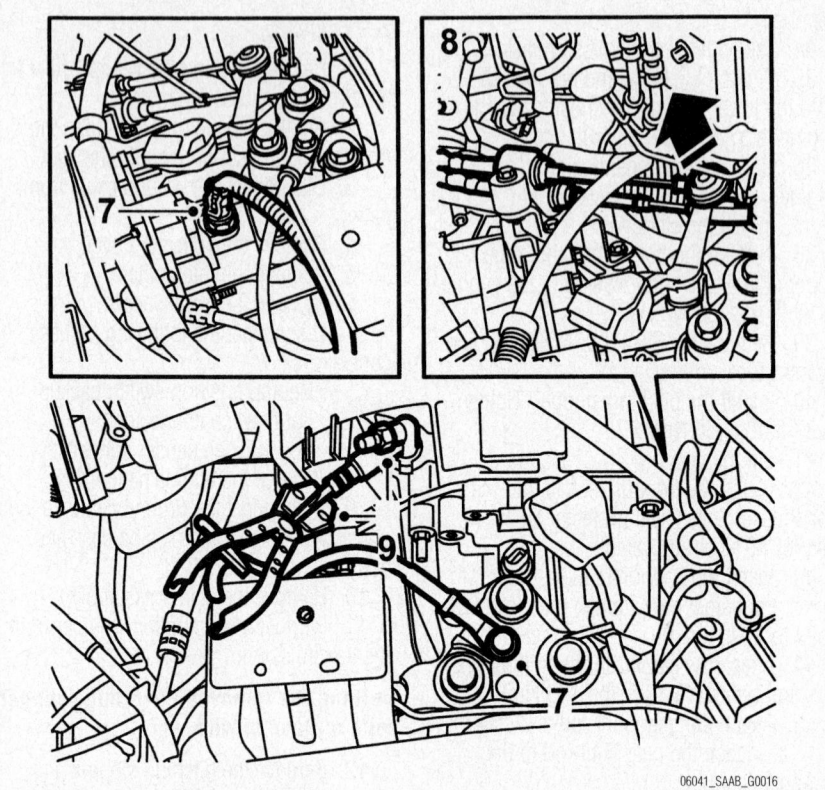

**Fig. 26 Disconnecting reverse light switch, removing gearbox cables, and pinching off clutch hoses—9-3 5–Speed**

06041_SAAB_G0016

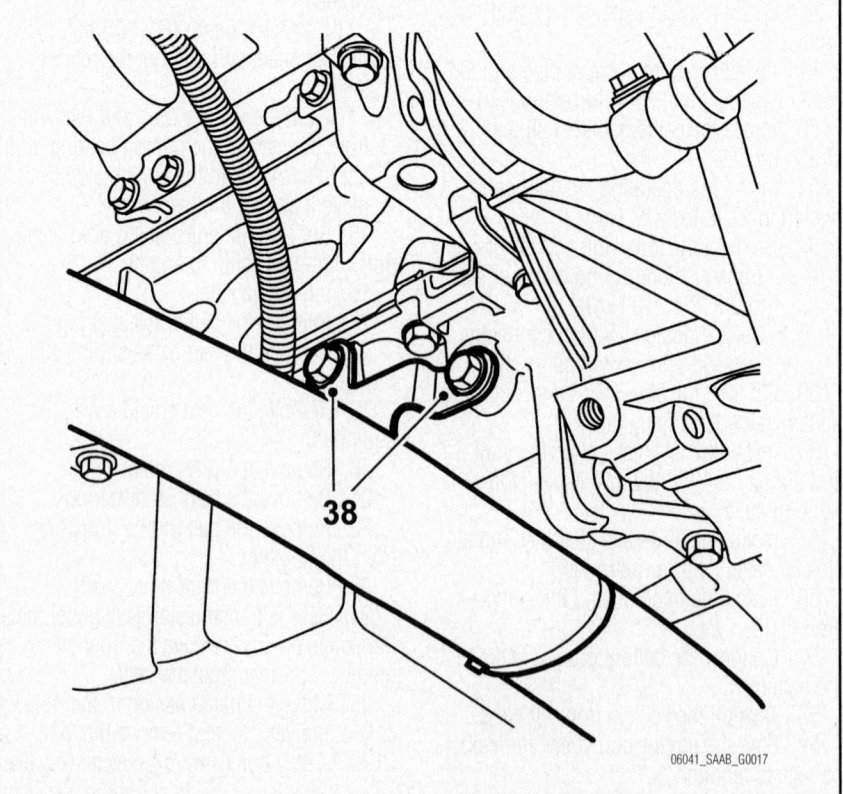

**Fig. 27 Showing the location of the lower transmission bolts—9-3 5–Speed**

06041_SAAB_G0017

- Pinch off clutch hose and undo quick-release coupling from clutch slave cylinder (9)
- Coolant reservoir (6-speed)
- Upper gearbox bolts

3. Install lift eyes, 83 96 178, to the engine and lifting beam 83 94 850 with holder 83 95 287 holder.
4. Remove the upper bumper shell mountings.
5. Suspend 2 Straps in place above the radiator core so they are accessible from below.

Remove or disconnect the following:

- Front wheels
- Front spoiler shield
- Hose from the headlamp washers and unplug and remove connector
- Secure radiator core with the Straps
- Radiator brackets from subframe
- Front part of exhaust pipe by cutting 3.43 inches (87mm) from front end of muffler
- Engine torque rod from subframe (6-speed)
- Transverse link from the steering knuckle on both sides
- Stabilizer bar from link arm.
- Headlamp angle sensor connector and cable clip (if equipped)
- Engine torque rod from subframe
- Steering gear from subframe (keep nuts and washers); leave steering gear hanging
- Clamps holding power steering pipe to subframe
- Front part of fender liner (bend out of way)

6. Place a trolley lift, 83 95 311, attaching fixture, 83 94 801, underneath. Position the guide pins and adjust the height with spacers to keep it level.
7. Remove the subframe bolts and the rear brackets.
8. Lower the subframe slightly.
9. Remove the stabilizer bar from the subframe, while protecting the steering gear boot.
10. Install bolts loosely in the link arms so the stabilizer bar is kept hanging in place.
11. Pull out the steering arm ball joints from the steering knuckles.
12. Lower the subframe.
13. Remove or disconnect the following:
    - Torque rod bracket (remove torque rod bolt for easier installing)
    - Gearbox oil (Install the oil plug)
    - Bolts in charge air pipe bracket (move pipe away)
    - Lower transmission bolts (leave one bolt in place loosely) (38)
    - Ground lead
    - ABS cable from clip
    - Left-hand driveshaft (suspend shaft by cables)
14. Lower the vehicle. Mark the bolt positions on the left engine bracket for correct reinstalling, and then remove the bolts from the mount.
15. Lower the powertrain about 2.75 inches (70mm), with the lifting beam, to facilitate removal of the gearbox.
16. Raise the vehicle. Move the power steering cooling coil aside and suspend it with a Strap.
17. Install single-column lift holder, 87 92 608, onto a column jack. Adjust and secure the tool in the transmission. Use bolt of 8.8 grade that are about 0.8 inch (20mm) longer than the bolts that were removed.
18. Remove the last bolt, pull out the transmission and lower it.
19. Lift the transmission down from the column jack with an engine lift and chain.
20. Remove the lifting tool from the transmission.

***To install:***

21. Bleed the slave cylinder, if necessary.
22. Lubricate the primary shaft splines.
23. Lubricate the guide sleeves on the engine with anti-corrosion agent.
24. Install a protective collar, 83 95 162, in the right shaft seal in the transmission. This is done to protect the seal while the transmission is being installed. Lubricate the seal.

**⁂ CAUTION**

**Always replace the shaft seals in the gearbox. Lubricate with gearbox oil.**

25. Install the gearbox, with the lifting tool, on the column jack.
26. Slide in the gearbox until approx. 0.8 inch (20mm) are remaining and remove the tool.
27. Push in the rest of the gearbox. Turn the crankshaft if necessary to get the gearbox in place.
28. Tighten all bolts, except the top one between the engine and the gearbox, to 30 ft. lbs. (40 Nm) for M10 bolts, and 52 ft. lbs. (70 Nm) for M12 bolts.
29. Install the charge air pipe bracket.
30. Remove the lifting tool from the gearbox and move the jack out of the way.
31. Install the gearbox bolts and torque to 18 ft. lbs. (24 Nm).
32. Undo the Straps and position the power steering cooling coil. Make sure it is installed on the correct side of the rubber seal.
33. Lower the vehicle. Lift in place the powertrain with the lifting beam until it meets the engine mounting.
34. Install the bracket for the engine mount according to the marks made earlier. Torque the bolts to 52 ft. lbs. (70 Nm), plus an additional 45°.
35. Remove the holder and the lifting beam.
36. Install or connect the following:
    - 3 upper gearbox bolts to 52 ft. lbs. (70 Nm).
    - Protective collar, 83 95 162, in left driveshaft seal (remove Strap, ensure driveshaft is clean, lubricate it and then align it with tool).
    - Driver in gearbox until approx.0.8 inch (20mm) are remaining and pull out tool before sealing surface of shaft reaches shaft seal; push in rest of driveshaft into gearbox until circlip clicks in.
    - ABS cable to the clip
    - Ground cable
    - Refill gearbox oil until level with plug hole.
    - Filler plug to 37 ft. lbs. (50 Nm) on 5-speed, or to 7 ft. lbs. (10 Nm) on 6-speed
    - Torque rod bracket, torque bolts to 52 ft. lbs. (70 Nm), plus an additional 90°
37. Position the subframe using the fixture. Also use a 3/8 inch square extension, 82 93 102, to guide the subframe. Lift until the stabilizer bar can be installed to the subframe.
38. Remove the link arm nuts and install the stabilizer bar to the subframe. Torque bolts to 13 ft. lbs. (18 Nm).
39. Install the transverse link ball joints to the steering knuckle and torque bolts to 37 ft. lbs. (50 Nm).

**⁂ CAUTION**

**Ensure that the steering knuckle stub is visible on the top of the steering knuckle housing before the bolt is installed.**

40. Check that the guide pins install into the reference holes, adjust the subframe until the guide pins go in easily and install the subframe bolts and brackets. Torque as follows:
    - Subframe bolts (except rear): 52 ft. lbs. (70 Nm), plus as an additional 90°
    - Rear subframe bolts: 66 ft. lbs. (90 Nm), plus an additional 45°
41. Install the stabilizer bar link arms and torque bolts to 47 ft. lbs. (64 Nm).

42. Move the lift and jig away.

43. Install the torque rods to the subframe and torque bolts to 52 ft. lbs. (70 Nm), plus an additional 90°.

44. Set up the Engine Centering Tool Kit, 83 96 152, and check that the engine is located correctly in relation to the subframe. Remove the centering tool.

45. Install or connect the following:

- Radiator brackets to subframe; remove Straps and torque bracket bolts to 35 ft. lbs. (47 Nm)
- Clamps holding power steering pipe to subframe
- Steering gear to subframe, torque bolts to 37 ft. lbs. (50 Nm), plus an additional 60°
- Joint clamp on front exhaust pipe and install it to catalytic converter; use a new gasket and new nuts, tightening to 18 ft. lbs. (25 Nm)
- Joint clamp where pipe was cut (place pipe ends in middle; torque joint clamp nuts to 30 ft. lbs. (40 Nm)
- Connector and cable clip to the headlamp angle sensor (if equipped)
- Coolant reservoir (6-speed)
- Left-hand side cover
- Fender liner section
- Connector and hose to headlamp washer
- Front spoiler shield
- Front wheels
- Bumper shell upper mountings
- Quick-release coupling to clutch slave cylinder (remove hose pinch-off pliers); ensure connection is in correct position
- Bleed the clutch
- Gear cables to gearbox
- Reverse light switch connection and ground cable to engine mount; torque bolt to 13 ft. lbs. (18 Nm)
- Battery tray and hood switch connector
- Battery (and cooling hose, if equipped)
- Battery cover
- Upper engine cover

46. Check the gear positions and adjust if necessary as described below:

a. Engage 4th gear on the gearbox.

b. Undo the gear lever cover with rubber mat and lift up the cover.

c. Undo the cable adjusters using a screwdriver.

d. Secure the gear lever in its adjusted position by lifting the adjusting sleeve while pressing in the two catches. Use a pair of pliers. Let the catch engage in the adjusting position. Check that 4th gear is engaged in the gearbox.

e. Press on the gear adjusters with a screwdriver.

f. Lift up the adjuster sleeve to its normal position.

g. Check the shifting positions.

47. Install the gear lever cover and rubber mat.

## 9-5

*See Figure 28.*

1. Before servicing the vehicle, refer to the Precautions Section.

2. Drain the transmission.

3. Remove or disconnect the following:
- Intake manifold cover
- Battery cover
- Battery and tray

4. Place the vehicle in 4th gear. Remove the plastic plug from the gearbox, and lock 4th gear in position with tool 87 92 335.

5. Remove the clip from the selector rod.

6. Raise gear selector boot. Engage 3rd gear so the selector rod disengages from the linkage. Install locking pin 87 92 335 in the lever housing.

7. Clamp brake hose or depress brake pedal about 2 inches (50mm) with brake clamp to prevent brake fluid from overflowing reservoir.

8. Remove or disconnect the following:
- Clip from slave cylinder and disconnect delivery line
- Reverse light connector
- 3 upper gearbox bolts and install locating studs in engine upper, outer guide holes
- Oxygen sensor
- Selector linkage from the gearbox
- Rear engine mount nut
- 3 bolts from rear engine pad (loosen but do not remove)
- Slightly lift engine and gearbox with a lifting beam on engine
- Both front wheels, wheel well covers, and under covers
- Headlamp leveling sensor brackets (if equipped); bend sensors aside

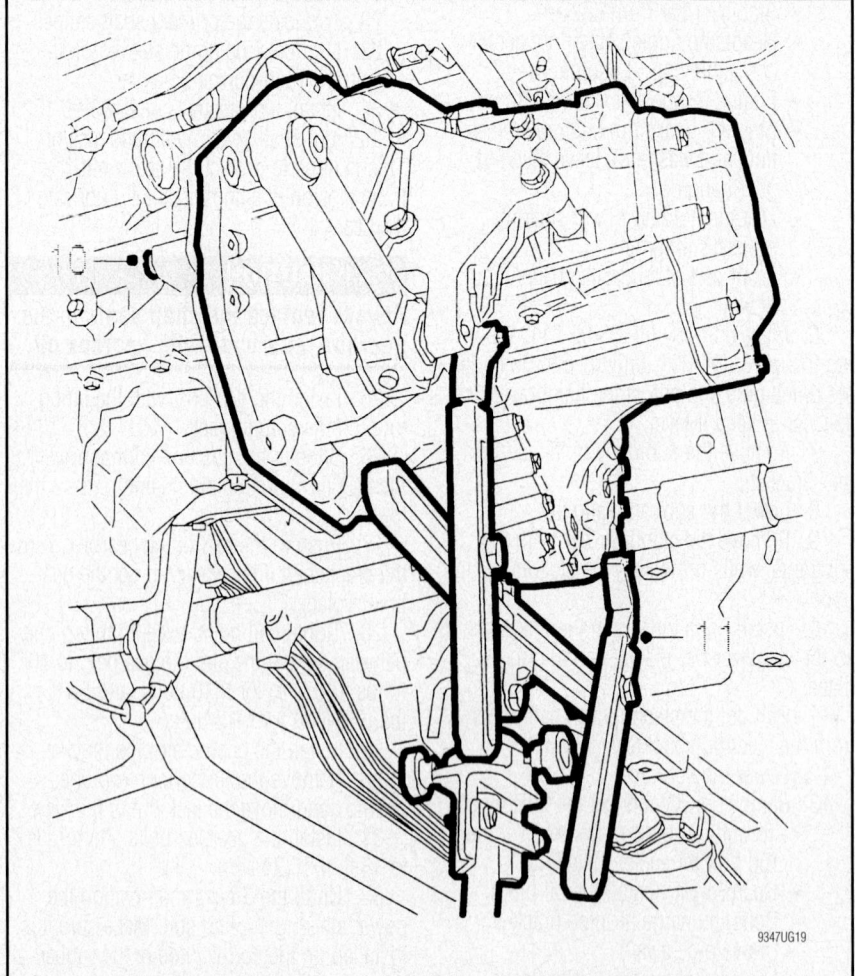

**Fig. 28 Install a lifting tool to the transmission—9-5**

9347UG19

- Front exhaust pipe
- Stay between the engine mount and engine
- Rear engine mount and pad
- Bolt securing torque arm to subframe (if equipped)
- Steering gear bolts
- Clips securing steering servo line to subframe
- A/C lines from the subframe
- Engine oil cooler from charge air cooler
- Place Strap around radiator and crossmember to retain radiator
- Bolts securing outer ball joints to steering swivel
- Loosen upper ball joint (both sides) from anti-roll bar
- Flywheel cover plate
- Bolts between gearbox and oil pan
- Bolts securing the rear support plate to subframe and place a lifting trolley under subframe
- Subframe and front mounts for the steering servo delivery line
- Gearbox oil
- Left driveshaft
- Ground cables from the gearbox
- Left engine pad
- Lower transmission so it clears structural member
- Transmission mounting

9. Connect a lifting tool to the transmission

10. Remove the transmission from the vehicle

### To install:

11. Lubricate the primary shaft splines.

12. Install a protective collar in the right shaft seal in the transmission.

13. Position the lifting trolley under the vehicle

14. Install or connect the following:

- Transmission and turn the engine shaft if needed to install the transmission
- Flywheel cover plate
- Bolts between the transmission and the oil pan and torque the bolts to 30 ft. lbs. (40 Nm)
- Screws between the engine and transmission and torque to 50 ft. lbs. (70 Nm). Remove the lifting trolley
- Transmission bolts and torque the bolts to 18 ft. lbs. (24 Nm)
- Ground cables to the transmission
- Shaft until the circlip snaps into position
- Transmission mountings and torque the bolts to 30 ft. lbs. (40 Nm)

- Clutch delivery line and clips
- Left engine pad. Torque the engine pad-to-transmission bolts to 62 ft. lbs. (85 Nm) and the engine pad to body bolts to 45 ft. lbs. (60 Nm)
- Reverse light electrical connector

15. Position the lifting trolley under the vehicle with the subframe properly aligned.

- Front ball joints and tighten the steering servo delivery lines
- Radiator journals to the subframe
- A/C lines into the brackets on the subframe
- Screws and rear support plate for the subframe. Move the lifting trolley aside. Torque the subframe bolts to 74 ft. lbs. (100 Nm) plus an additional 45°
- Torque the support plate bolts to 44 ft. lbs. (60 Nm)
- Torque rod-to-subframe bolts and torque the bolts to 22 ft. lbs. (30 Nm)
- Engine oil cooler
- Rear attaching clamps for the steering servo line to the subframe
- Steering gear and torque the bolts to 66 ft. lbs. (90 Nm)
- Outer ball joints to the steering knuckle and torque the bolts to 64 ft. lbs. 85 Nm)
- Stabilizer bar stay and torque the fastener to 64 ft. lbs. (85 Nm)
- Cover plate between the engine and transmission
- Rear engine pad and engine mount and torque the bolts to 50 ft. lbs. (70 Nm)
- Engine mount stay and torque the fastener to 16 ft. lbs. (22 Nm)
- Front exhaust pipe and torque the flange to the turbocharger to 18 ft. lbs. (25 Nm)
- Engine stay and torque the fastener to 16 ft. lbs. (22 Nm)
- Inner wheel well covers and remove the lifting devise from the top of the engine
- Upper engine bolts and torque the bolts to 50 ft. lbs. (70 Nm)
- Front torque arm to transmission and torque the arm to 34 ft. lbs. (47 Nm)
- Selector linkage to the transmission and torque the bolt to 18 ft. lbs. (24 Nm)
- Gear linkage to the selector rod. Place the vehicle in 4th gear and secure with a locking pin. Torque the clamp on the linkage to 16 ft. lbs. (22 Nm)
- Oxygen sensor connectors

- Battery and tray. Connect the battery cables
- Battery cover
- Front wheels
- Throttle body cover
- Grille

16. Fill and bleed the clutch system.

17. Fill the transmission fluid to the proper level.

18. Start the vehicle, check for leaks and proper operation. Repair if necessary.

## CLUTCH

### REMOVAL & INSTALLATION

#### 9-2x

1. Before servicing the vehicle, refer to the Precautions Section.

2. Disconnect the negative battery cable.

3. Remove the transmission.

4. Install the Clutch Disc Guide ST499747100, or equivalent on the flywheel.

➡**Take care not to allow oil on the clutch disc facing.**

5. Remove the clutch cover and clutch disc. Do not disassemble

➡**Be sure to put alignment marks on the flywheel and clutch cover before removing the clutch cover.**

#### To install:

6. Note the front and rear of the clutch disc when installing.

7. Insert the Clutch Disc Guide into clutch disc and install them on the flywheel by inserting the ST end into pilot bearing.

8. When installing a new clutch cover, position the clutch cover so that there is a gap of 120° or more between "0" marks on the flywheel and clutch cover. "0" marks indicate the directions of residual unbalance.

9. Align the alignment marks and install the clutch cover on flywheel. Tighten the bolts to 12 ft. lbs. (16 Nm). Tighten the clutch cover installing bolts gradually in a crisscross order.

10. Remove the Clutch Disc Guide.

11. Install the transmission.

#### 9-3 and 9-5

*See Figure 29.*

1. Before servicing the vehicle, refer to the Precautions Section.

2. Remove the transmission.

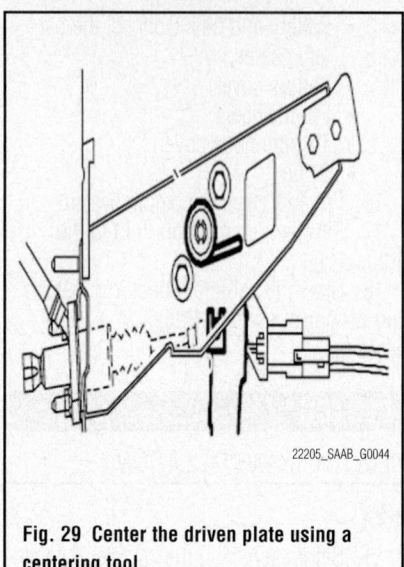

**Fig. 29 Center the driven plate using a centering tool**

3. Install flywheel holder tool 83 94 868.

4. Remove the pressure plate retaining nuts.

5. Remove the pressure plate and clutch disc.

### To install:

6. Check the flywheel surface in contact with the clutch plate. A blue tone and hairline cracks are no cause for concern. If the surface is deeply scored however, the flywheel should be changed.

7. Check the pressure plate for scratches and warping.

8. Check the release bearing for noise, wear and the like.

9. Check driven plate wear and change if necessary.

➡ **On vehicles equipped with self-adjusting pressure plate, the clutch plate and pressure plate are matched in pairs and must therefore be replaced as a pair.**

10. Position the driven plate and the pressure plate on the flywheel and loosely tighten the bolts.

➡ **On vehicles with dual mass flywheel, the convex side of the driven plate should face the gearbox.**

11. Center the driven plate using 87 92 327 Centering Tool.

12. Tighten the driven plate bolts to 22 ft. lbs. (30 Nm) on all transmission except the Z18XE where the Tighten to is 11 ft. lbs. (15 Nm)

13. Remove the Flywheel Locking Attachment.

14. Lubricate the primary shaft splines and install the transmission.

## BLEEDING

### 9-2x

### 5-Speed

See Figure 30.

1. On non turbocharged engine, remove the air intake chamber.

2. On turbocharged engine, remove the intercooler.

3. Connect a vinyl tube to the air bleeder on the master cylinder. Put the other end in a jar with clean clutch fluid.

4. Slowly depress the clutch pedal and keep it depressed. Open the air bleeder to discharge air and fluid.

5. Release the air bleeder for one or two seconds. With the bleeder closed, slowly release the clutch pedal.

6. Repeat the procedure until there are no more air bubbles in the vinyl tube.

7. Tighten the air bleeder.

8. Connect a vinyl tube to the air bleeder on the clutch operating (slave) cylinder. Put the other end in a jar with clean clutch fluid.

9. Slowly depress the clutch pedal and keep it depressed. Open the air bleeder to discharge air and fluid.

10. Release the air bleeder for one or two seconds. With the bleeder closed, slowly release the clutch pedal.

11. Repeat the procedure until there are no more air bubbles in the vinyl tube.

12. Tighten the air bleeder to 6 ft. lbs. (8 Nm).

13. After depressing the clutch pedal, make sure that there are no leaks in the entire system

14. Recheck to ensure that the clutch is operating correctly.

15. On non turbocharged engine, install the air intake chamber.

16. On turbocharged engine, install the intercooler.

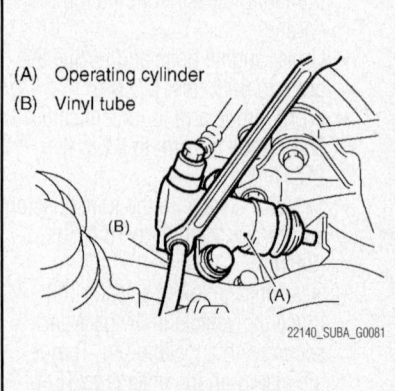

(A) Operating cylinder
(B) Vinyl tube

22140_SUBA_G0081

**Fig. 30 5-Speed bleeding—9-2x**

### 6-Speed

See Figure 31.

1. Remove the intercooler.

2. Remove the clutch operating cylinder. Do not remove the clutch hose.

3. Using a service clamp, fix the piston to avoid the piston from jumping out.

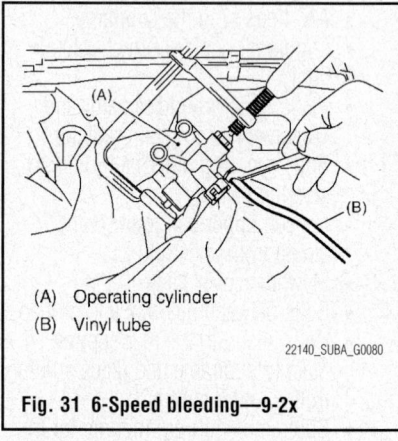

(A) Operating cylinder
(B) Vinyl tube

22140_SUBA_G0080

**Fig. 31 6-Speed bleeding—9-2x**

4. Connect a vinyl tube to the air bleeder on the clutch operating (slave) cylinder. Put the other end in a jar with clean clutch fluid.

5. Slowly depress the clutch pedal and keep it depressed. Open the air bleeder to discharge air and fluid.

6. Release the air bleeder for one or two seconds. With the bleeder closed, slowly release the clutch pedal.

➡ **Set the air breather screw position higher than the tip of the operating cylinder when performing this procedure.**

7. Repeat the procedure until there are no more air bubbles in the vinyl tube.

8. Tighten the air bleeder.

9. After depressing the clutch pedal, make sure that there are no leaks in the entire system

10. Recheck to ensure that the clutch is operating correctly.

### 9-3 and 9-5

### Slave Cylinder

See Figure 32.

Perform this procedure if the transmission is removed.

1. Connect a 17.5 in. (450 mm) long piece of clean, transparent 0.31 in. (8 mm) plastic hose to the slave cylinder delivery pipe connection.

2. Fully depress the release bearing once and then release it. The slave cylinder seal remains in its inner position.

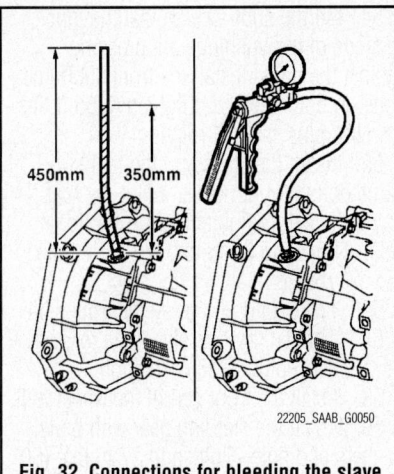

**Fig. 32 Connections for bleeding the slave cylinder in the vehicle**

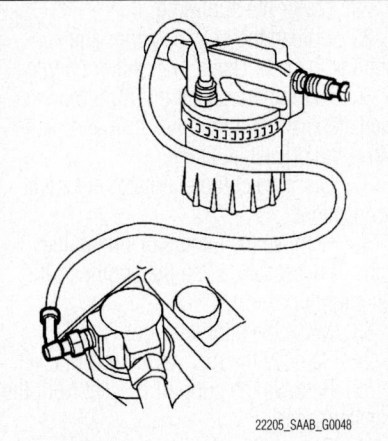

**Fig. 33 Connecting the hydraulic bleeder to the bleed nipple**

3. Fill the plastic hose with brake fluid to a level of about 13.65 in. (350 mm).

4. Connect a pressure/vacuum pump to the plastic hose. Use a 0.23 in. (6 mm) plastic hose (this is used to provide a tight seal).

5. Pump up pressure until the slave cylinder seal moves out and brake fluid runs down into the slave cylinder. The pressure increases when the seal has reached its farthest point.

6. Remove the pressure/vacuum pump and carefully press in the release bearing to its stop. Note the air bubbles in the plastic hose.

7. Repeat the preceding steps a few times until no air bubbles are visible in the hose.

8. Leave the slave cylinder seal in its innermost position. Drain and remove the plastic hose.

### Clutch Hydraulic System

*See Figures 33 and 34.*

Perform this procedure if the transmission is installed in the vehicle.

1. Before servicing the vehicle, refer to the precautions in the beginning of this section.

2. Fill the master cylinder with brake fluid.

3. Open the bleeder valve on the slave cylinder ½ turn.

4. Connect a 88 19 096 Vacuum Powered Brake Bleeder to the bleeder valve.

5. Bleed the clutch until clear fluid runs from the nipple.

6. Install a 30 05 451 Cooling System Tester to the brake fluid reservoir.

7. Hold the brake bleeder hose to the nipple and bleed the reservoir for about 1 minute or so long as bubbles rise to the surface.

8. Close the slave cylinder bleeder valve.

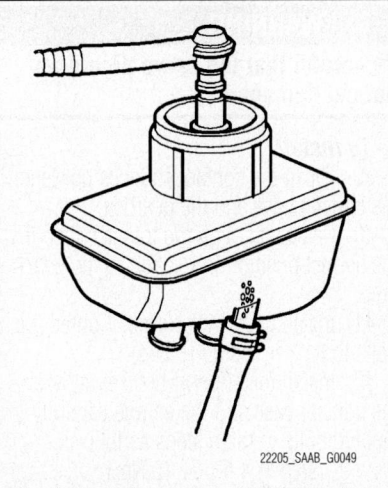

**Fig. 34 Install a cooling system tester to the brake fluid reservoir**

9. Check that all air was removed from the system and the clutch is functioning properly.

10. Adjust the fluid level, as required.

### TRANSFER CASE ASSEMBLY

#### REMOVAL & INSTALLATION

##### 9-2x

The transfer case is removed as an assembly with the transmission. Refer to "Transmission, Transmission Assembly, Removal & Installation".

##### 9-3

##### 2.0L Engine

1. Raise and safely support the vehicle securely on jackstands.

2. Remove the driveshaft. Refer to "Rear Drive Axle, Driveshaft, Removal & Installation".

3. On automatic transmission equipped vehicles, drain the transfer case fluid.

4. On manual transmission equipped vehicles, drain the gearbox of oil.

5. Unload the belt tensioner and unhook the belt. Use a 1/2" puller handle.

6. Tap the driveshaft out from the intermediate shaft using a brass drift and mallet. Move the driveshaft aside.

7. Disconnect the alternator electrical connections

8. Remove the alternator's retaining bolts. If necessary, push the engine to the left slightly.

9. Move the alternator out

10. Remove the rear torque rod bracket.

11. Remove the support bracket from the transfer case.

12. Support the transfer case with a column jack, and secure with cable ties.

13. Remove the transfer case.

14. Lower and ease out the transfer case.

### ❋❋ WARNING

**Be careful that the servo pipe does not get damaged.**

#### To install:

15. Clean the contact surfaces between the transfer case and the gearbox.

16. Raise and align the transfer case in the correct position in relation to the gearbox.

17. Install the transfer case. Tighten the bolts to 81.1 ft. lbs. (110 Nm).

18. Install the support bracket between the transfer case and the engine carefully Tighten bolts in three steps as follows:
 a. Step 1: 4 ft. lbs. (5 Nm)
 b. Step 2: 15ft. lbs. (20 Nm)
 c. Step 3: 44 ft. lbs. (60 Nm)

19. Install the rear torque rod bracket and tighten to 59 ft. lbs. (80 Nm)

20. Move the alternator in and install the retaining bolts. Tighten to 18 ft. lbs. (24 Nm).

21. Plug in the alternator electrical connector

22. Install the right-hand drive shaft.

23. Install the driveshaft.

24. Install the wheel well.

25. Check and adjust the transfer case oil level.

26. On automatic transmission equipped vehicles, fill the transfer case fluid.

27. On manual transmission equipped vehicles, fill the gearbox oil.

##### 2.8L Engine

1. On vehicles equipped with 2.8L engine, remove the subframe as follows:

2. Remove the upper bumper shell attachments.

3. Remove the air filter cover.

### ✳✳ WARNING

**Be careful of the electrical connection to the brake vacuum valve.**

4. Unplug the mass air flow sensor connector.

5. Detach the secondary air pipe and delivery hose from the turbocharger intake manifold and crankcase ventilation pipe.

6. Detach the turbocharger intake manifold from the air cleaner housing cover and the turbocharger.

7. Hang two 83 95 212 Straps in place over the radiator member and the radiator unit so that they are accessible from below.

8. Raise and safely support the vehicle securely on jackstands.

9. On automatic transmission equipped vehicles, drain the transfer case fluid.

10. On manual transmission equipped vehicles, drain the gearbox of oil.

11. Remove the front wheels

12. Remove the lower front cover.

13. Bind the radiator unit tightly with the Straps.

14. Remove the radiator brackets from the subframe.

15. Remove the driveshaft. Refer to "Rear Drive Axle, Driveshaft, Removal & Installation".

16. Remove the engine's front torque rod from the subframe.

17. Remove the bolt and nut which attach the transverse link to the steering knuckle on both sides.

18. On vehicles equipped with Xenon headlights, remove the level sensor on the left-hand transverse link.

19. Remove the steering gear from the subframe. Look after the nuts and washers. Allow the steering gear to remain suspended.

20. Open the clips that secure the power steering pipe to the subframe.

21. Position a 83 95 311 Trolley Lift with a 83 94 801 Carriage and 83 96 137 Subframe Centering Tool.

22. Remove the front part of the inner wheel well and fold away slightly.

23. Remove the subframe bolts on both sides. Remove the stays from the body.

24. Remove the bolts securing the rear torque rod to the subframe.

25. Lower the subframe slightly.

26. Pull out the transverse links ball-and-socket joints from the steering knuckles.

27. Remove the stabilizer bar from the subframe.

28. Lower the subframe.

29. Unload the belt tensioner and unhook the belt. Use a 1/2" puller handle.

30. Tap the driveshaft out from the intermediate shaft using a brass drift and mallet. Move the driveshaft aside.

31. Disconnect the alternator electrical connections

32. Remove the alternator's retaining bolts. If necessary, push the engine to the left slightly.

33. Move the alternator out

34. Remove the rear torque rod bracket.

35. Remove the support bracket from the transfer case.

36. Support the transfer case with a column jack, and secure with cable ties.

37. Remove the transfer case.

38. Lower and ease out the transfer case.

### ✳✳ WARNING

**Be careful that the servo pipe does not get damaged.**

#### *To install:*

39. Clean the contact surfaces between the transfer case and the gearbox.

40. Raise and align the transfer case in the correct position in relation to the gearbox.

41. Install the transfer case. Tighten the bolts to 81.1 ft. lbs. (110 Nm).

42. Install the support bracket between the transfer case and the engine carefully Tighten bolts in three steps as follows:
    a. Step 1: 4 ft. lbs. (5 Nm)
    b. Step 2: 15 ft. lbs. (20 Nm)
    c. Step 3: 44 ft. lbs. (60 Nm)

43. Install the rear torque rod bracket and tighten to 59 ft. lbs. (80 Nm)

44. Move the alternator in and install the retaining bolts. Tighten to 18 ft. lbs. (24 Nm).

45. Plug in the alternator electrical connector

46. Install the right-hand drive shaft.

47. Position the frame on a 83 95 311 Trolley Lift with a 83 94 801 Carriage and 83 96 137 Subframe Centering Tool.

48. Lift the frame.

49. Install the stabilizer bar to the subframe.

50. Install the bolts securing the rear torque rod to the subframe. Tighten M10 bolts to 48 ft. lbs. (65 Nm) and M12 bolts to 88 ft. lbs. (120 Nm).

51. Install the ball joints for the transverse links in the steering knuckle.

52. Lift the subframe into place and align 83 96 137 Subframe Centering Tool with the locating holes in the body.

53. Attach the power steering pipe to the plastic clips on the subframe.

54. Lift the subframe so that the inner sections of the bushings are pressing against the body. Install the front subframe bolts on both sides and tighten to 55 ft. lbs. (75 Nm) plus 135° of rotation.

55. Install the stays and the rear subframe bolts on both sides. Enter the rear stay bolts before tightening the subframe bolts. Tighten to 55 ft. lbs. (75 Nm) plus 135° of rotation.

56. Lower and pull away the trolley lift.

57. Tighten the rear stay bolts to 55 ft. lbs. (75 Nm) plus 135° of rotation

58. Install the front part of the wheel well.

59. Install the steering gear with bolts, washers and nuts. Tighten to 37 ft. lbs. (50 Nm) plus 60° of rotation.

60. On vehicles equipped with Xenon headlights, install the level sensor on the left-hand transverse link.

### ✳✳ WARNING

**The groove in the pin must be visible in the screw hole in the spindle housing. If the boot is pressed down, it will not seal properly to the swivel pin.**

61. Install the bolt and nut which attach the transverse link to the steering knuckle on both sides. Tighten to 37 ft. lbs. (50 Nm)

62. Install the engine's front torque rod to the subframe. Tighten to 44 ft. lbs. (60 Nm) plus 90° of rotation.

63. Set up the 83 96 137 Subframe Centering Tool and check that the engine is positioned correctly in relation to the subframe. Remove the centering tool.

64. Install the driveshaft.

65. Install the radiator brackets to the subframe and tighten to 35 ft. lbs. (47 Nm).

➡**Check that the upper radiator unit support is placed correctly.**

66. Remove the Straps.

67. Install the lower front cover.

68. Install the wheels and lower the vehicle.

69. Attach the turbocharger intake manifold to the air cleaner housing cover and the turbocharger.

70. Attach the secondary air pipe and delivery hose to the turbocharger intake manifold and crankcase ventilation pipe.

71. Plug in the mass air flow sensor connector.

72. Install the air filter cover.

73. Install the upper bumper shell attachments.

74. Check the straight-ahead position of the steering wheel when driving on a level road. Adjust if necessary.

## FRONT HALFSHAFT

### REMOVAL & INSTALLATION

#### 9-2x

*See Figure 35.*

1. Before servicing the vehicle, refer to the Precautions Section.
2. Disconnect ground cable from battery.
3. Jack-up vehicle, support it with safety stands, and remove front wheels.
4. Unlock axle nut.
5. Remove axle nut while depressing brake pedal to prevent front driveshaft from turning.

**✳✳ CAUTION**

**Be sure to loose and retighten axle nut after removing wheel from vehicle. Failure to follow this rule may damage wheel bearings.**

6. Remove stabilizer link.
7. Remove disc brake caliper from housing, and suspend it from strut using a wire.
8. Remove disc rotor from hub.
9. If disc rotor seizes up within hub, drive disc rotor out by installing an 8-mm bolt in screw hole on the rotor.
10. Remove cotter pin and castle nut which secure tie-rod end to housing knuckle arm.
11. Using a puller, remove tie-rod ball joint from knuckle arm.
12. Remove ABS sensor assembly and harness.
13. Remove bolt which secures sensor harness to strut.
14. Remove transverse link ball joint from housing.

15. Remove inner joint from transmission spindle.
16. Remove front driveshaft assembly from hub. If it is hard to remove using a ST-926470000 puller.

**✳✳ CAUTION**

**Be careful not to damage oil seal lip and tone wheel when removing front drive shaft. If front driveshaft is removed, replace inner oil seal with new one.**

**To install:**

17. Install front drive shaft.
18. Install transverse link ball joint to housing and tighten to 36 ft. lbs. (49 Nm).
19. Install ABS sensor harness on strut.
20. Install ABS sensor on housing. Tighten the mounting bolt to 24 ft. lbs. (32 Nm).
21. Install disc rotor on hub.
22. Install disc brake caliper on housing and tighten to 58 ft. lbs. (78 Nm).
23. Connect stabilizer link.
24. Install tie-rod end and tighten castle nut to 20 ft. lbs. (27 Nm).
25. After tightening castle nut to specified torque, retighten it further within 60° until a slot in castle nut is aligned with the ball joint hole. Insert cotter pin, and then bend the cotter pin around castle nut to secure it.
26. While depressing brake pedal to prevent front driveshaft from turning, tighten axle nut to 162 ft. lbs. (220 Nm).

**✳✳ CAUTION**

**When axle nut is removed, replace it with new one. Be sure to tighten axle nut to specified torque. Do not over tighten it as this may damage wheel bearing.**

27. After tightening axle nut, lock it securely.
28. Install wheel and tighten wheel nuts.

#### 9-3

*See Figure 36.*

1. Before servicing the vehicle, refer to the Precautions Section.
2. Raise and safely support the front of the vehicle securely on jackstands.
3. Remove the wheel hub center nut.
4. Remove the lower engine cover.
5. On convertible models, remove the bolts holding the front chassis reinforcement and remove the reinforcement.

**➡The front bolts are shorter than the others.**

6. Remove the ball joint bolt, lower the transverse link and place a wedge between the swing arm and the anti-roll bar.
7. Detach the brake hose from the clip and the ABS cable holder. Turn the wheel for better access.

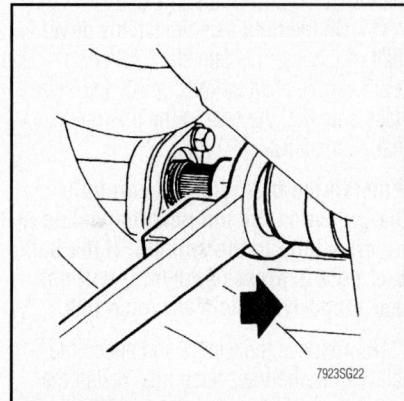

7923SG22

**Fig. 36 Removing the halfshaft joint from the intermediate shaft—9-3**

---

**2. FRONT DRIVE SHAFT ASSEMBLY**

| Model | Type of drive shaft | Axle diameterφ D mm (in) | Axle length L mm (in) |
|---|---|---|---|
| Turbo 5MT, 6MT | EBJ + PTJ | 26 (1.0) | 332.5 (13.09) |
| Other than above | EBJ + PTJ | 26 (1.0) | 349.6 (13.76) |

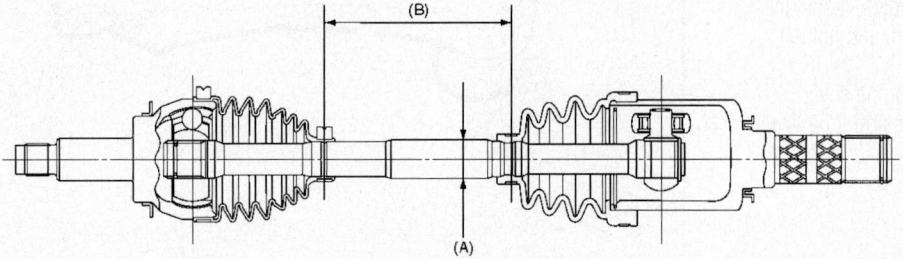

(A)  Axle diameter  (B)  Axle length

22140_SUBA_G0075

**Fig. 35 Front driveshaft assembly—9-2x**

8. Press out the driveshaft from the wheel hub about .39 in. (10 mm) using 89 96 951 Halfshaft Puller.

9. On the left side, pull out the driveshaft from the gearbox. Press with a 87 92 616 Halfshaft Removal Tool. Lift out the drive shaft. Collect the oil in a receptacle.

10. On the right side, tap out the driveshaft from the intermediate shaft with a brass drift and mallet. Lift out the drive shaft.

### To install:

11. On the left side, change the shaft seal in the gearbox using 87 92 350 Halfshaft Seal Puller with a standard gear puller.

12. Tap on the new seal using a drift to install. Lubricate the seal with gearbox oil.

13. Install the driveshaft into the wheel hub.

14. On the left side, install an 83 95 162 Protective Collar and install the driveshaft into the gearbox. Lubricate the shaft splines with gearbox oil. Pull out the collar when about 0.78 in. (20 mm_ is left. Press in the shaft until the clip snaps in place.

15. On the right side, install the driveshaft in the intermediate shaft splines. Lubricate the shaft splines using anti-seize. Make sure that the seal to the intermediate shaft is installed.

➡ **Install the ball joint pin carefully. The groove in the pin must be visible in the crew hole in the spindle. If the ball joint boot is pressed down, it will not seal properly to the transverse link.**

16. Remove the wedge and attach the ball joint to the transverse link. Install the bolt and nut. Tighten the nut to 37 ft. lbs. (50 Nm).

17. Install the brake hose in the clip and the ABS cable in the holder.

18. Install a new center hub nut loosely and install the wheels. Lower the vehicle and tighten the center hub nut to 170 ft. lbs. (230 Nm).

19. Press the protective washer securely on the wheel.

20. On convertible models, install the front chassis reinforcement and tighten bolts to 37 ft. lbs. (50 Nm)

21. Install the lower engine cover.

22. Check the transmission fluid level and top off if necessary.

### 9-5

*See Figure 37.*

1. Raise and safely support the front of the vehicle securely on jackstands.

2. Remove the wheels.

3. If equipped, remove any fasteners for the headlamp position sensor. Push the sensor to one side.

4. Remove the side cover.

5. Remove the protective cap from the hub center nut, if equipped. Then remove the center nut.

6. Remove the bolt securing the ball joint to the steering knuckle. Press loose and lower the transverse link. Fixate it in position with a wooden block placed between the transverse link and anti-roll bar.

7. Undo the clips on the brake hose and the ABS cable mountings.

➡ **Place a receptacle under the gearbox in order to avoid oil spillage on the floor.**

8. Tap the universal joint out of the wheel hub. Use a plastic mallet.

9. Tilt out the MacPherson strut. Pull the MacPherson strut backwards and fasten with a Strap.

10. On the left side, pull the driveshaft from the gearbox using 87 92 616 Halfshaft Removal Tool.

11. On the right side, tap out the driveshaft from the intermediate shaft with a brass drift and mallet. Lift out the drive shaft.

### To install:

12. On the left side, install an 83 95 162 Protective Collar and install the driveshaft into the gearbox. Lubricate the shaft splines with gearbox oil. Pull out the collar when about 0.78 in. (20 mm_ is left. Press in the shaft until the clip snaps in place.

13. On the right side, install the driveshaft in the intermediate shaft splines. Lubricate the shaft splines using anti-seize. Make sure that the seal to the intermediate shaft is installed.

14. Push in the rest of the shaft until the clip clicks in.

15. Tilt out the MacPherson strut and connect the driveshaft to the hub.

16. Make sure the driveshaft reaches its end position in the wheel hub. Install the ball joint to the transverse link and tighten to 37 ft. lbs. (50 Nm).

17. Install the side cover in the wheel housing.

18. Install the brake hose to the bracket on the steering knuckle and install the clips as well as the two mountings for the ABS lead.

19. If equipped, position the headlamp position sensor and install the fasteners.

20. Install a new center hub nut loosely and install the wheels. Lower the vehicle and tighten the center hub nut to 170 ft. lbs. (230 Nm).

21. Check and fill the transmission fluid to the proper level.

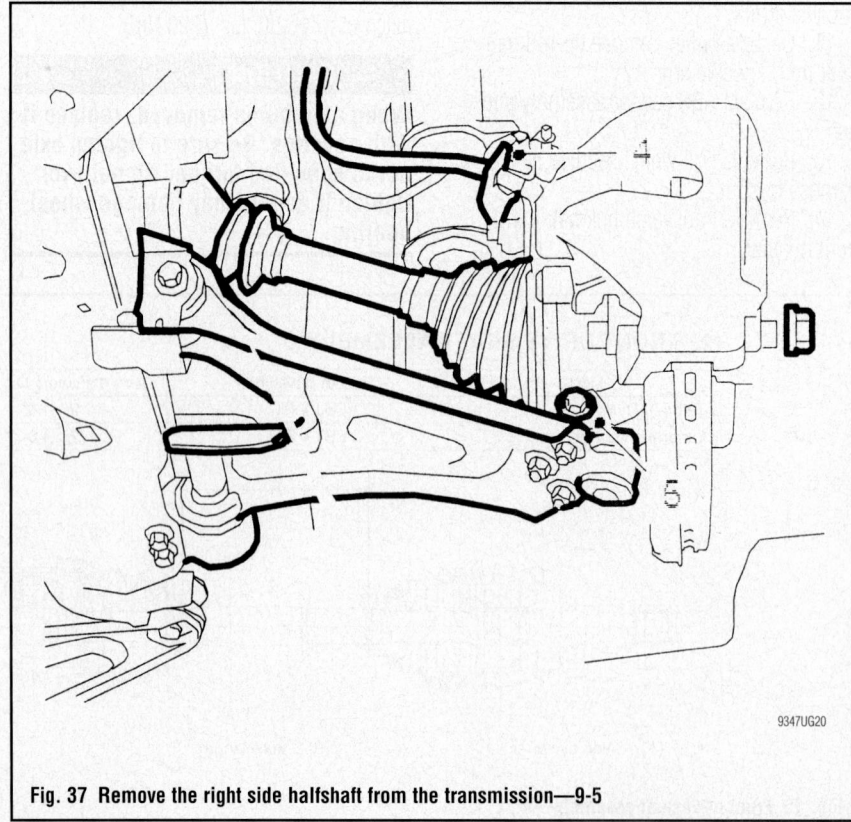

9347UG20

**Fig. 37 Remove the right side halfshaft from the transmission—9-5**

## REAR HALFSHAFT

### REMOVAL & INSTALLATION

#### 9-2x

*See Figure 38.*

1. Before servicing the vehicle, refer to the Precautions Section.

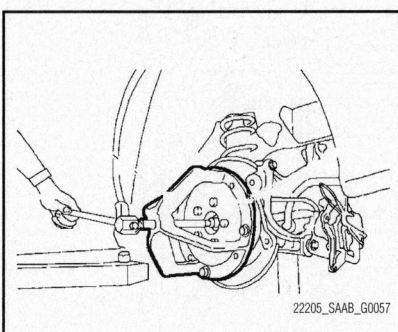

**Fig. 38 Remove halfshaft assembly from hub using puller ST-926470000—9-2x**

2. Disconnect ground cable from battery.
3. Jack-up vehicle, support it with safety stands, and remove front wheels.
4. Unlock axle nut.
5. Remove axle nut while depressing brake pedal to prevent front driveshaft from turning.

### ✻✻ CAUTION

**Be sure to loose and retighten axle nut after removing wheel from vehicle. Failure to follow this rule may damage wheel bearings.**

6. Return the parking brake lever and loosen the adjusting nut.
7. Remove disc brake caliper from housing, and suspend it from strut using a wire.
8. Remove disc rotor from hub.
9. If disc rotor seizes up within hub, drive disc rotor out by installing an 8-mm bolt in screw hole on the rotor.
10. Disconnect the parking brake cable end.
11. Disconnect the rear stabilizer from rear lateral link.
12. Remove the bolts which secure the trailing link assembly to rear housing.
13. Remove the bolts which secure the lateral assembly to rear housing.

14. Remove ABS sensor assembly and harness.
15. Remove bolt which secures sensor harness to strut.
16. Remove halfshaft assembly from hub. If it is hard to remove using a ST-926470000 puller.

### ✻✻ CAUTION

**Be careful not to damage oil seal lip and tone wheel when removing front drive shaft. If front driveshaft is removed, replace inner oil seal with new one.**

17. Remove the bolts which secure the rear housing to strut, and separate the two.

#### To install:

18. Temporarily tighten the rear axle to strut.
19. Insert the rear driveshaft into rear axle.

➡**Be careful not to damage the inner oil seal lip.**

20. Temporarily tighten the axle nut.
21. Using a new self-locking nut, temporarily tighten the rear housing assembly and lateral link assembly.
22. Using a new self-locking nut, temporarily tighten the rear housing assembly and trailing link assembly.
23. Tighten the rear housing assembly and strut assembly to 148 ft. lbs. (200 Nm) using a new self-locking nut.
24. Using a new self-locking nut, install the rear stabilizer and rear lateral link. Tighten to 33 ft. lbs. (44 Nm).
25. Connect the parking brake cable to parking brake.
26. Install the disc rotor on rear housing assembly.
27. Install the disc brake caliper on back plate. Tighten to 38 ft. lbs. (52 Nm).
28. Adjust the parking brake lever stroke by turning the adjuster.
29. While applying the parking brake, tighten a new axle nut using a socket wrench. Lock the axle nut after tightening to 140 ft. lbs. (190 Nm).

### ✻✻ WARNING

**Do not overtighten it as this may damage the wheel bearing.**

30. Install rear ABS wheel speed sensor.
31. Install the wheel and tighten the wheel nuts 66 ft. lbs. (90 Nm).
32. Whenever tightening bushing it is important that the tires be in contact with the ground fully and the vehicle be in curb weight. Lower the vehicle so the contact the ground fully.
33. Tighten the installation bolt of rear housing assembly and lateral link assembly to 103 ft. lbs. (140 Nm).
34. Tighten the installation bolt of rear housing assembly and trailing link assembly to 66 ft. lbs. (90Nm).

#### 9-3

*See Figure 39.*

1. Before servicing the vehicle, refer to the Precautions Section.
2. Raise and safely support the rear of the vehicle securely on jackstands.
3. Remove the wheels.
4. Remove the rear wheel hub. Refer to "Rear Suspension, Wheel Hub and Bearing, Removal and installation".
5. Tap out the driveshaft from the final drive gear using a brass drift and a mallet.
6. Remove the drive shaft.

#### To install:

7. Install the drive shaft. Lubricate the shaft's splines with gear oil. Press in the shaft until the engages.
8. Check/adjust the oil level in the final drive gear.
9. Install the wheel hub.

**Fig. 39 Tap out the driveshaft from the final drive gear using a brass drift and a mallet—9-3**

## ENGINE COOLING

### THERMOSTAT

*REMOVAL & INSTALLATION*

#### 2.0L I4 Engine

*See Figures 40 and 41.*

1. Before servicing the vehicle, refer to the Precautions Section.
2. Disconnect the negative battery cable.
3. Remove the expansion tank cap.
4. Raise the vehicle and remove front spoiler shield.
5. Drain and recycle the engine coolant. Close the drain cock.
6. Install the remove front spoiler shield and Lower the vehicle.
7. Disconnect the level sensor connector (8) and the expansion tank from the mounting on the body and set aside.
8. Disconnect the vacuum hose (9) from the vacuum pump.
9. Remove the cover on the thermostat housing and set aside.
10. Remove the thermostat.

#### To install:

11. Clean the sealing surfaces. Lubricate the new thermostat seal sparingly with Vaseline, non-acidic and install the thermostat to the thermostat housing.
12. Install thermostat and tighten bolts to 7 ft. lbs. (10 Nm).
13. To complete installation, reverse remaining removal procedure.
14. Fill cooling system and check for leaks.

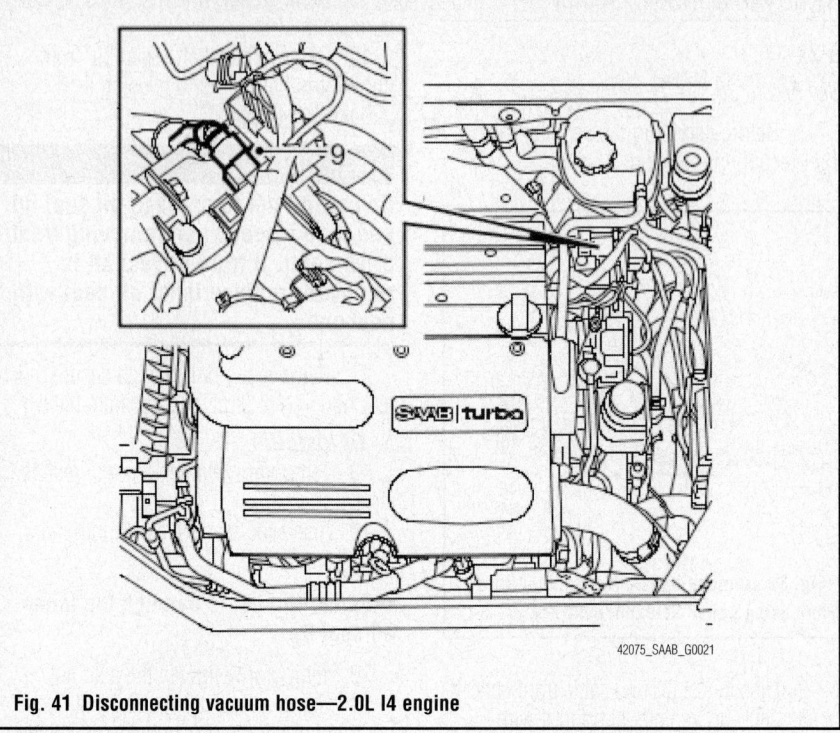

**Fig. 41 Disconnecting vacuum hose—2.0L I4 engine**

#### 2.0L H4 and 2.5L Engines

*See Figures 42 and 43.*

1. Before servicing the vehicle, refer to the Precautions Section.
2. Raise and safely support the vehicle.
3. Remove the engine undercover.
4. Drain the engine coolant.
5. Loosen the hose clamp and disconnect the lower radiator hose from the thermostat cover.

6. Remove the thermostat cover and gasket, and pull out the thermostat.

#### To install:

7. Install the thermostat.

➡**2.5L SOHC engine—The thermostat must be installed with the jiggle pin upward.**

➡**2.5L DOHC engines—The thermostat must be install with the jiggle pin facing front.**

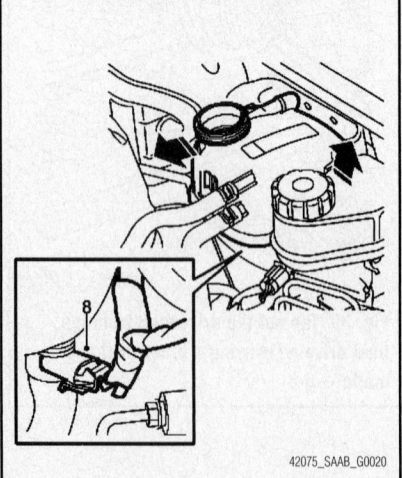

**Fig. 40 Disconnecting level sensor—2.0L I4 engine**

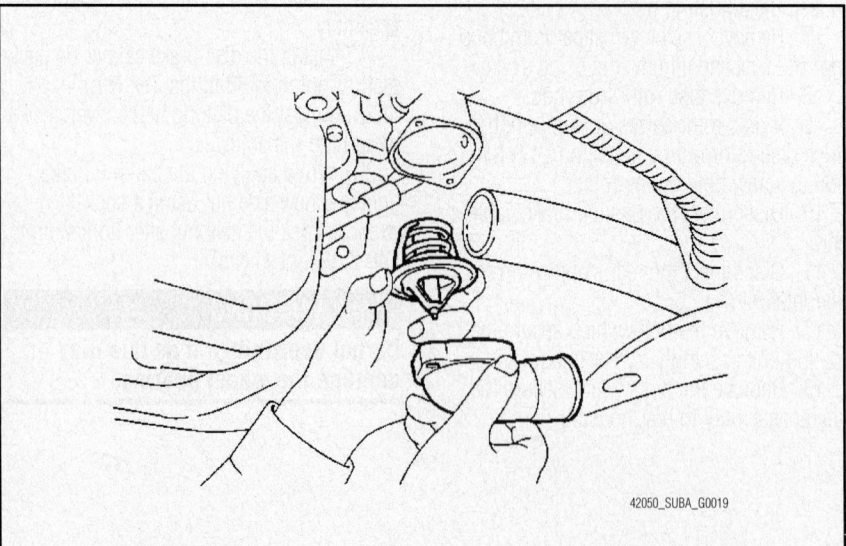

**Fig. 42 Remove the thermostat cover, and pull out the thermostat—2.5L SOHC engine shown**

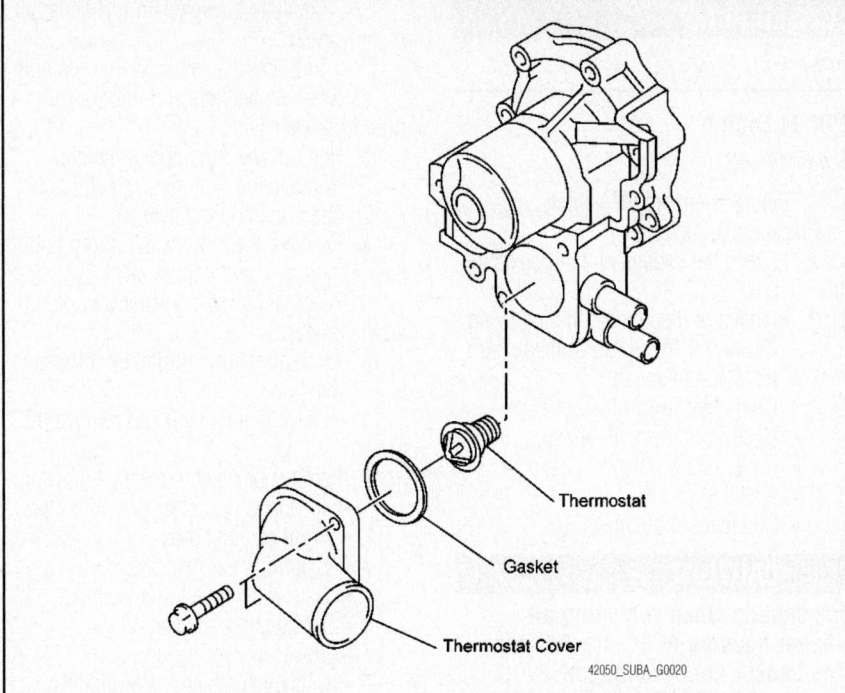

Fig. 43 Exploded view of the thermostat and related components—2.5L DOHC engine shown

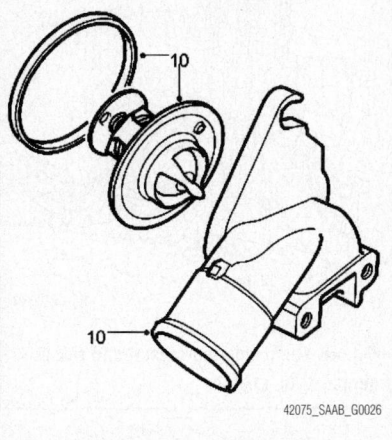

Fig. 46 Removing the thermostat housing (10) and thermostat (10)—2.3L engine

8. Install the thermostat cover together with a new gasket. Tighten to 9 ft. lbs. (12 Nm).

9. Install the lower radiator hose to the thermostat cover and tighten the hose clamp.

10. Install the engine undercover.

11. Lower the vehicle.

12. Refill the engine cooling system to the correct level.

13. Start the engine and check for leaks.

## 2.3L Engine

*See Figures 44 through 46.*

1. Before servicing the vehicle, refer to the Precautions Section.

2. Disconnect the negative battery cable.

3. Remove the expansion tank cap.

4. Raise the vehicle and remove front cover.

5. Drain and recycle the engine coolant.

6. Lower the vehicle to the floor. Loosen the hose clamp on the upper radiator hose mounting at the thermostat housing. Move the hose to one side.

7. Undo the ground connection and bracket (8) for the coolant pipe by the thermostat housing.

8. Slacken the upper bolt (9) for the thermostat housing and remove the lower bolt (9).

9. Remove the thermostat housing and thermostat.

### To install:

10. Check and clean the sealing surfaces. Grease the rubber seal lightly with non-acidic Vaseline.

11. When installing the thermostat, ensure that the vent hole is at the top (thermostat marking "TOP").

12. Install thermostat housing, tighten bolts to 16 ft. lbs. (22 Nm).

13. Install the ground cable and the coolant pipe bracket. Apply acid-free Vaseline to both sides of the ground connection.

14. To complete installation, reverse remaining removal procedure.

15. Fill with coolant. Do not forget to close the drain plug on the radiator first.

16. Check for coolant leakage. Start the engine, set the heat to max. and run the engine until warm.

17. Switch off the engine. Check the coolant level.

## 2.8L Engine

*See Figures 47 and 48.*

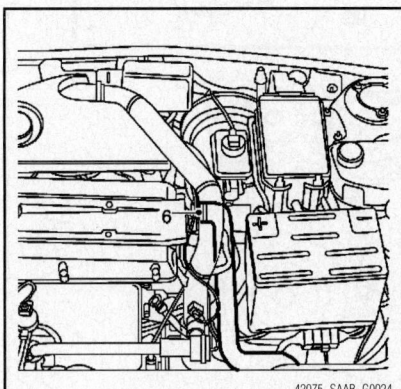

Fig. 44 Loosen the hose clamp on the upper radiator hose (6) mounting—2.3L engine

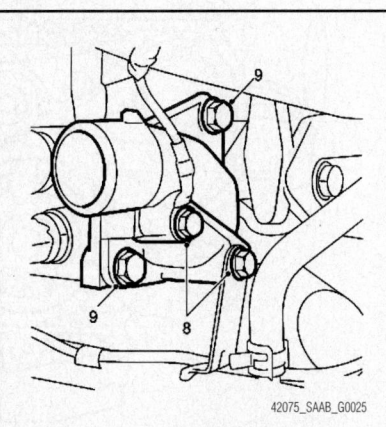

Fig. 45 Removing thermostat housing components—2.3L engine

Fig. 47 Disconnecting the cable duct—2.8L engine

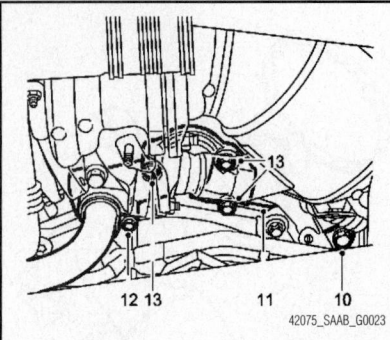

**Fig. 48 Removing thermostat and components—2.8L engine**

1. Before servicing the vehicle, refer to the Precautions Section.

2. Disconnect the negative battery cable.

3. Drain and recycle the engine coolant.

4. Raise the vehicle. Remove the bolt from the coolant pipe mounting on the front of the engine.

5. Slightly loosen the rear bolt holding the coolant pipe to the engine block.

6. Lower the vehicle. Remove the battery and battery cover.

7. Move the coolant reservoir aside. Remove the upper and lower heat shield of the catalytic converter.

8. Disconnect the cable duct (9) from the gearbox and move the duct aside.

9. Remove the bolt (10) holding the coolant pipe to the radiator assembly.

10. Disconnect the pipe (11) from the thermostat housing. Collect any spill.

11. Detach the coolant pipe (12) running from the thermostat housing to the radiator.

12. Remove the thermostat housing (13). For A/T vehicles: At the same time, press the coolant pipe to the radiator forward.

### To install:

13. Clean gasket residue from the sealing surface of the engine block.

14. Install the thermostat housing using a new seal. Fasten the seal with Vaseline on the engine block. Tighten bolts to 7 ft. lbs. (10 Nm).

15. Attach the coolant pipe using a new O-ring lightly coated with Vaseline. Do not tighten the bolt until the front mounting is installed.

16. Attach the pipe to the heater assembly using a new O-ring lightly coated with Vaseline.

17. To complete installation, reverse remaining removal procedure.

18. Fill cooling system and check for leaks.

## WATER PUMP

### REMOVAL & INSTALLATION

#### 2.0L I4 Engine

*See Figure 49.*

1. Before servicing the vehicle, refer to the Precautions Section.

2. Loosen the expansion tank pressure cap.

3. Remove or disconnect the following:
- Connector for oxygen sensors and temperature sensor
- Connector from MAF sensor
- Hose from the MAF sensor
- Cover on air cleaner housing
- Air filter
- Air cleaner housing

### ✳✳ CAUTION

**Use caution when removing air cleaner housing to ensure A/C pressure sensor is not damaged.**

- Turbocharger inlet hose (plug turbocharger opening)
- Upper bolt for water pump in timing cover and engine block
- Turbocharger heat shield
- Upper oxygen sensor

- Catalytic converter-to-turbocharger nuts
- Turbocharger oil delivery pipe bolt

4. Raise the vehicle and remove the right front wheel.

5. Remove the front spoiler shield.

6. Remove the right fender well liner.

7. Drain the cooling system.

8. Remove the lower water pump bolt in the timing cover and engine block.

9. Remove plug from water pump and drain coolant.

10. Mark direction of rotation of the auxiliary drive belt.

11. Relieve belt tension and remove the belt.

12. Remove or disconnect the following:
- Turbocharger lower pressure pipe
- Front exhaust pipe
- Steering gear heat shield
- Catalytic converter heat shield
- Temperature sensor
- Lower oxygen sensor
- Catalytic converter and bracket from engine block
- Turbocharger coolant pipe from thermostat housing and turbocharger
- Thermostat housing bolts

13. Remove the water pump cover from the timing cover (34).

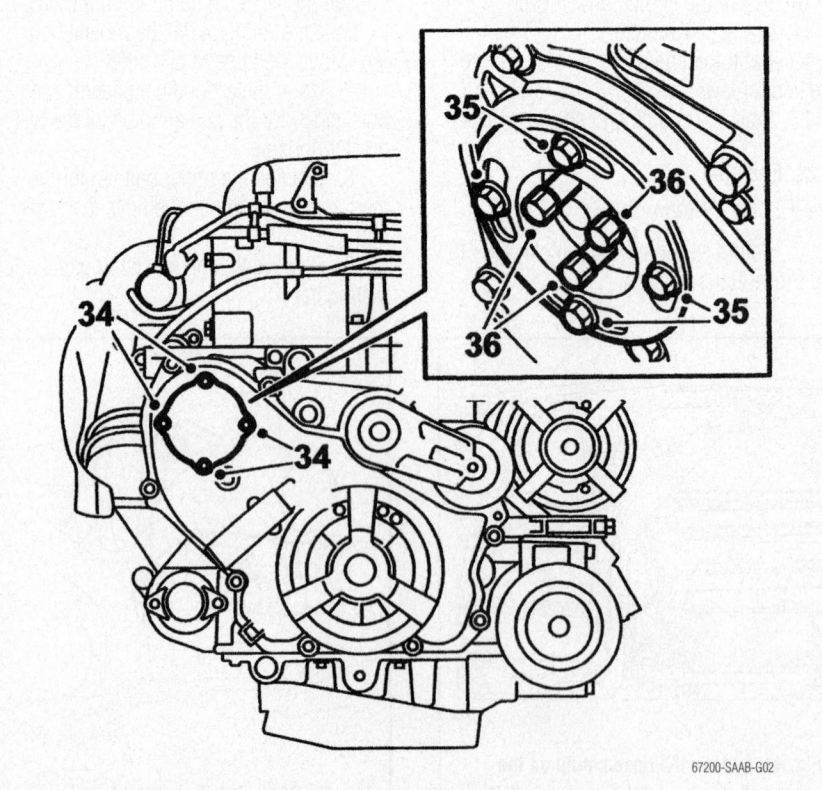

**Fig. 49 Showing water pump holding tool for securing balancer shaft pinion—2.0L engine**

14. Install water pump holder tool, 2.0L, to secure the balancer shaft pinion (35).

15. Remove the pinion bolts (36). Use a magnetic socket to remove the bolts from the holder tool.

16. Remove the bolts on the back of the water pump.

17. Move aside the thermostat housing, with the pipe, and remove the water pump.

18. Remove the water pipe from the thermostat housing.

***To install:***

19. Clean water pipe fittings. Install new O-rings, lightly coated with Vaseline.

20. Clean the thermostat housing connections, install a new seal and install the water pipe.

21. Install the stud to the water pump driver using a holding tool, 83 96 103.

22. Install the water pump, with a new seal coated lightly with Vaseline. Guide the stud through one of the holes in the balancer shaft pinion.

23. Install the 2 rear water pump bolts and the lower bolt through the timing cover.

24. Lower the vehicle.

25. Torque the water pump bolts to 15 ft. lbs. (20 Nm) for short bolts and 18 ft. lbs. (25 Nm) for the long bolts.

26. Install the coolant pipe to the water pump and install the thermostat.

27. Install the coolant pipe between the turbocharger and thermostat with a new seal. Tighten pipe to 15 ft. lbs. (20 Nm).

28. Lift catalytic converter and position its bracket to studs on the engine block. Position the catalytic converter to the turbocharger connection.

29. Insert the catalytic converter bracket bolts into the engine block and new nuts on the studs.

30. Lower the vehicle. Lubricate the turbocharger studs with screw thread paste 30 20 971.

31. Install the new nuts so they are firmly against the turbocharger, then torque to 7 ft. lbs. (10 Nm).

32. Raise the vehicle.

33. Tighten the catalytic converter nut and bolts one last time to 18 ft. lbs. (25 Nm).

34. Lubricate the lower oxygen sensor threads with screw thread paste and install the sensor and tighten to 30 ft. lbs. (40 Nm).

**✳✳ CAUTION**

**Be sure oxygen sensor cables are not twisted or damaged.**

35. Lubricate the temperature sensor threads with screw thread paste and install the sensor and tighten to 33 ft. lbs. (45 Nm).

36. Lift the oxygen sensor and temperature sensor cables to check for proper routing.

37. Install the heat shields on the catalytic converter and steering gear.

38. Clean the catalytic converter joint at pipe connection with front muffler. Install clamp and position front pipe, with new gasket, mounting nuts and rubber mountings. Apply thread paste to studs.

39. Position joint clamp so pipe ends are in middle of the clamp. Torque joint at catalytic converter to 18 ft. lbs. (25 Nm) and clamp nuts to 30 ft. lbs. (40 Nm).

40. Remove the stud and install bolts to the balancer shaft pinion, using a 10mm magnetic socket.

➡**It may be necessary to remove the Fixing Tool for access before finally tightening the bolts.**

41. Install the timing cover with a new gasket.

42. Install the auxiliary drive belt over the pulleys, making sure it is in proper direction of rotation.

43. Install the lower turbocharger delivery pipe.

44. Close the radiator drain and Lower the vehicle.

45. Apply thread paste and install upper oxygen sensor. Tighten it to 30 ft. lbs. (40 Nm).

46. Plug in and secure the oxygen sensor and temperature sensor connectors.

47. Install the turbocharger heat shield. Tighten bolts to 15 ft. lbs. (20 Nm).

48. Install the air cleaner assembly and hoses, including to the turbocharger.

49. Install the hose on the MAF sensor

50. Plug in the MAF sensor.

51. Fill cooling system and bleed system by starting and running the engine at varying speeds until the cooling fan comes on. Then, check and top off the coolant.

52. Install remaining components and check for any signs of leaks.

### 2.0L H4 and 2.5L Engines

*See Figures 50 through 52.*

1. Before servicing the vehicle, refer to the Precautions Section.

2. Remove or disconnect the following:
• Negative battery cable
• Engine undercover, if equipped

3. Drain the coolant into a suitable container.

4. Remove or disconnect the following:
• Radiator fan connectors
• Radiator outlet and heater hoses
• Heater bypass hose or overflow hose, if equipped
• Reservoir tank, on Legacy
• Radiator fan motor assembly
• Accessory drive belts
• Timing belt
• Belt tension adjuster
• Belt idler No. 2
• Camshaft Position (CMP) sensor
• Left side camshaft pulley
• Left side rear timing belt cover
• Tensioner bracket
• Radiator and heater hoses from water pump
• Water pump retainer bolts
• Water pump

5. Inspect the radiator hoses for deterioration and replace as necessary.

***To install:***

6. Clean the gasket mating surfaces thoroughly. Always use new gaskets during installation.

7. Install or connect the following:

8. Tighten the pump bolts in sequence to 8.9 ft. lbs. (12 Nm). After tightening the bolts once, retighten to the same specification again.
• Radiator heater hoses to water pump
• Tensioner bracket and tighten to 18 ft. lbs. (25 Nm)
• Left side rear timing belt cover
• Left side camshaft pulley(s). Tighten to 58 ft. lbs. (78 Nm) on non turbocharged engine and 72 ft. lbs. (98 Nm) on turbocharged engine.
• CMP sensor
• Belt idler No. 2 and tighten to 29 ft. lbs. (39 Nm)
• Belt tension adjuster
• Timing belt
• Accessory drive belts
• Radiator fan assembly
• Reservoir tank, if removed
• Heater bypass hose or overflow hose, if equipped
• Air intake duct
• Radiator outlet and heater hoses
• Radiator fan connectors
• Engine undercover, if removed

9. Fill the system with coolant and connect the negative battery cable.

10. Start the engine and allow it to reach operating temperature.

11. Check for leaks.

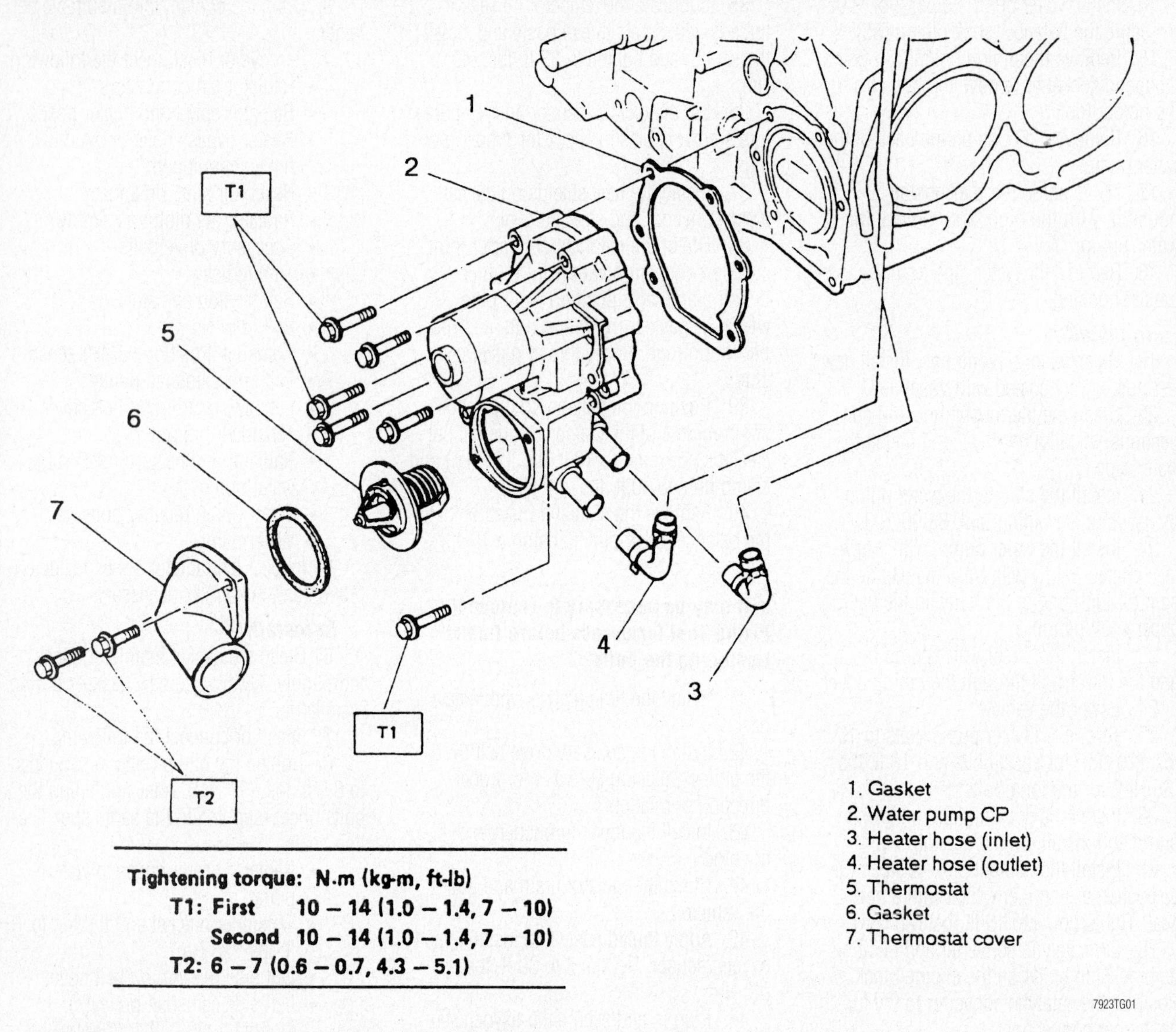

1
2
T1
5
6
7
T1
T2
4
3
Tightening torque: N.m (kg-m, ft-lb)
T1: First 10 – 14 (1.0 – 1.4, 7 – 10)
Second 10 – 14 (1.0 – 1.4, 7 – 10)
T2: 6 – 7 (0.6 – 0.7, 4.3 – 5.1)
1. Gasket
2. Water pump CP
3. Heater hose (inlet)
4. Heater hose (outlet)
5. Thermostat
6. Gasket
7. Thermostat cover
7923TG01

Fig. 50 Water pump and related components—9-2x

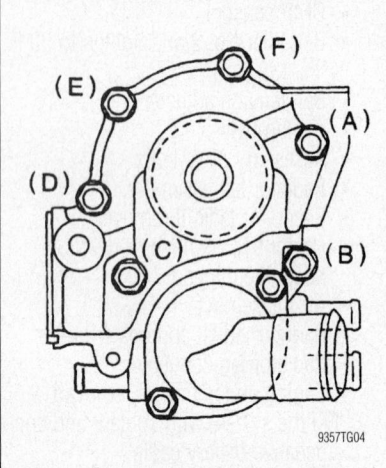

(F)
(E)
(A)
(D)
(C)
(B)
9357TG04

Fig. 51 Tighten the water pump bolts in two steps using the following sequence—2.0L engine

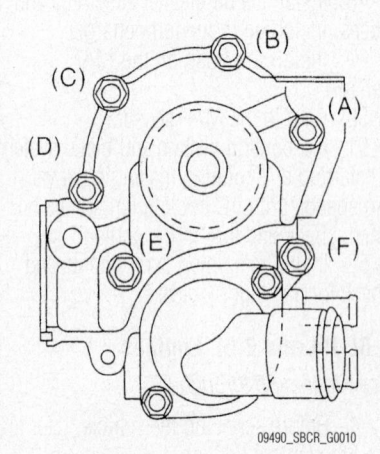

(B)
(C)
(A)
(D)
(E)
(F)
09490_SBCR_G0010

Fig. 52 Tighten the water pump bolts in two steps using the following sequence—2.5L engine

## 2.3L Engine

*See Figure 53.*

1. Before servicing the vehicle, refer to the Precautions Section.
2. Open cap of expansion tank and release any pressure.
3. Raise the vehicle and remove the lower front cover.
4. Drain the cooling system.
5. Lower the vehicle and remove or disconnect the following:
   - Negative battery cable
   - Mass Air Flow (MAF) sensor electrical connector
   - MAF sensor and air hose
   - V-belt tension with 83 95 254 Hydraulic Belt Tensioner Drawbar Handle
   - V-belt from the power steering pump and water pump

- Crankcase breather pipe
- Camshaft cover
- Boost pressure control valve connector
- Engine lifting eye
- Turbocharger wastegate valve hoses, loosen only
- Bypass valve and pipe
- Exhaust manifold heat shield
- Quick release coupling on the vent hose with tool 83 95 261
- Turbocharger intake pipe (plug turbocharger opening)
- Power steering pump and move it aside
- Water pump inlet hoses
- 2 longitudinal coolant pipes from the engine block and water pump
- Water pump and connecting piece

### To install:

6. Clean all mating surfaces and connecting piece seals. Replace the seals if necessary

7. Install or connect the following:
- Connecting piece to the engine block
- Water pump and torque the bolts to 16 ft. lbs. (22 Nm)
- Coolant pipe to the water pump and turbocharger. Torque the water pump bolt to 15 ft. lbs. (20 Nm) and the turbocharger bolt to 18 ft. lbs. (24 Nm)
- Longitudinal pipes to the water pump and torque the bolts to 7 ft. lbs. (10 Nm)
- Water pump inlet hoses
- Power steering pump and torque the bolts to 18 ft. lbs. (24 Nm)
- Turbocharger intake pipe

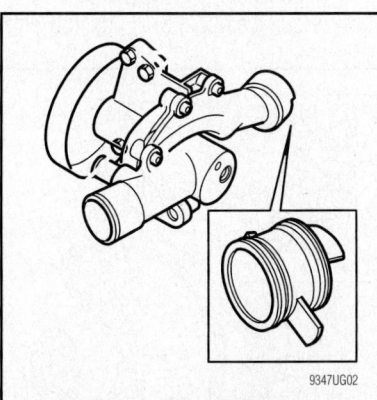

**Fig. 53 Remove the water pump and engine block connecting piece—2.3L engine**

- Turbocharger vent hose
- Exhaust manifold heat shield and torque the bolts to 15 ft. lbs. (20 Nm)
- Bypass pipe and valve and torque the bolts to 6 ft. lbs. (8 Nm)
- Turbocharger wastegate valve hoses
- Engine lifting eye
- Boost pressure control valve connector
- Crankcase breather pipe fastened to turbocharger inlet pipe and camshaft cover, torque to 18 ft. lbs. (24 Nm)
- V-belt
- MAF sensor and air hose
- Negative battery cable

8. Fill the cooling system.

9. Start the vehicle, check for leaks and repair if necessary.

### 2.8L Engine

*See Figure 54.*

1. Before servicing the vehicle, refer to the Precautions Section.

2. Disconnect the negative battery cable.

3. Take the cap off the expansion tank to release any pressure.

4. Raise and safely support the vehicle.

5. Undo the right-hand side of the spoiler shield. Place a suitable receptacle under the radiator. Connect a hose to the radiator and drain the coolant. Close the cock and Install the spoiler shield.

6. Remove the chassis reinforcement from the front subframe (convertible model).

7. Install 83 96 145 Centering tool, subframe - engine and 83 96 152 Centering

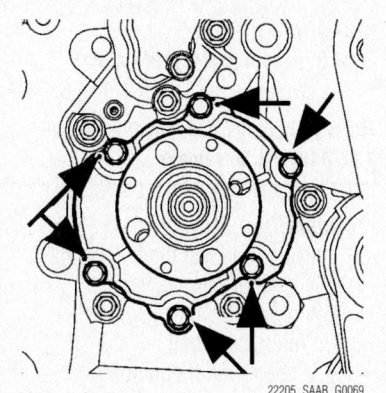

**Fig. 54 Water pump bolt locations—2.8L engine**

tool, power train with 32 025 059 Support. First remove the two long supports. Relieve the load on the engine.

8. Lower the vehicle.

9. Disconnect the turbocharger intake hose.

10. Remove the air cleaner casing cover.

11. Remove the air cleaner. Detach the intake hose and remove the air cleaner casing.

12. Remove the engine pad from the body and bracket.

13. Remove the engine bracket. The lower hole is slotted.

14. Loosen the bolts of the water pump pulley slightly.

15. Unload the drive belt tensioner using a ½ inch puller and remove the belt from the water pump pulley.

16. Remove the pulley.

17. Remove the water pump.

### To install:

18. Clean the sealing surfaces.

19. Install the water pump and tighten the bolts to 7 ft. lbs. (10 Nm).

20. Install the pulley.

21. Install the drive belt.

22. Tighten the bolts on the pulley to 9 ft. lbs. (12 Nm).

23. Install the engine bracket. Tighten the M10 size bolts to 31 ft. lbs. (42 Nm), and the M12 size bolts to 69 ft. lbs. (93 Nm).

24. Install the engine pad to the body and bracket using 32 025 063 Special box spanner. Tighten the bolts to the body to 37 ft. lbs. (50 Nm), plus an additional 180°. Tighten the bolts to the engine mounting to 33 ft. lbs. (45 Nm), plus an additional 90°.

25. Install the air cleaner casing and connect the intake hose. Install the air cleaner.

26. Install the air cleaner casing cover.

27. Install the turbocharger inlet hose.

28. Raise and safely support the vehicle.

29. Remove 83 96 145 Centering tool, subframe - engine and 83 96 152 Centering tool, power train with 32 025 059 Support.

30. Install the chassis reinforcement to the front subframe (convertible model).

31. Install the spoiler shield.

32. Lower the vehicle.

33. Connect the negative battery cable.

34. Fill cooling system and bleed system by starting and running the engine at varying speeds until the cooling fan comes on.

35. Stop the engine and top off the coolant.

36. Install remaining components and check for any signs of leaks.

## ENGINE ELECTRICAL

## CHARGING SYSTEM

### ALTERNATOR

*REMOVAL & INSTALLATION*

#### 2.0L I4 Engine

*See Figure 55.*

1. Before servicing the vehicle, refer to the Precautions Section.
2. Remove or disconnect the following:
   - Negative battery cable
   - Lower engine cover
   - Drain coolant
   - Plastic nuts from wheel well liner
   - Drive belt (through wheel well liner)
   - Upper engine cover
   - Connector(s) from ECM
   - ECM from mounting
   - Air hose from air filter housing
   - Secondary air injection hose
   - Air filter housing cover and air filter
   - Pressure sensor connector
   - Air filter housing
   - Radiator hose
   - Vent hose from air hose to throttle body
   - Loosen fastener in camshaft cover
   - Bend vent hose aside
   - Alternator electrical connections
   - Alternator retaining bolts
   - Alternator (lift upward)

*To install:*

3. Install or connect the following:
   - Alternator and retaining bolts
   - Tighten bolts to 18 ft. lbs. (24 Nm)
   - Vent hose fastener to camshaft cover
   - Vent hose to throttle body air hose
   - Radiator hose

- Intake air pipe to air filter housing
- Pressure sensor connector
- Air filter and housing cover
- Secondary air injection hose
- Air hose to air filter housing
- ECM and connections
- Upper engine cover
- Auxiliary drive belt (through wheel well liner)
- Plastic nuts on wheel well liner
- Lower vehicle to floor
- Negative battery cable
- Battery cover
- Refill cooling system

4. Reset the clock and radio, if necessary.

#### 2.0L H4 and 2.5L Engines

*See Figure 56.*

1. Before servicing the vehicle, refer to the Precautions Section.
2. Remove or disconnect the following:
   - Negative battery cable
   - Connector and terminal from the alternator
   - V-belt cover, if equipped
   - Front side V-belt
   - Alternator to bracket bolts
   - Alternator from the vehicle

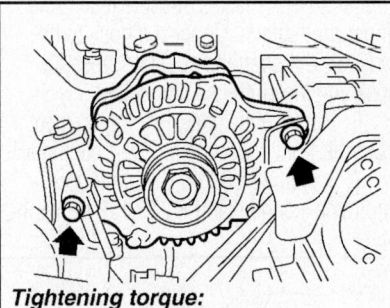

**Tightening torque:**
**13 N·m (1.3 kgf-m, 9.6 ft-lb)**

22140_SUBA_G0104

**Fig. 56 Alternator and mounting bolts—2.0L H4 and 2.5L engines**

*To install:*

3. Install or connect the following:
   - Alternator into the vehicle
   - Alternator to bracket bolts
   - Front side V-belt
   - V-belt cover, if equipped
   - Connector and terminal to the alternator
   - Negative battery cable

4. Check and adjust the belt tension.

#### 2.3L Engine

*See Figure 57.*

1. Before servicing the vehicle, refer to the Precautions Section.
2. Remove or disconnect the following:
   - Negative battery cable
   - Intake manifold shield
   - Crankcase ventilation solenoid and constant pressure valves
   - Belt tensioner with special tool 83 95 254
   - Alternator upper mounting bolt
   - Right side engine mount nut
   - Rear engine mount nut
   - Front exhaust pipe
   - Hose between the oil trap and the oil pan
   - Alternator electrical connectors
   - Alternator lower bolt and insert a bar between the gearbox and subframe
   - Alternator (tip and lower downward)

*To install:*

3. Install or connect the following:

➡**To make it easier to Install alternator, tap bushings until they are level with the mountings.**

- Insert bar between gearbox and subframe.
- Alternator and loosely install the lower retaining bolt
- Alternator electrical connectors
- Hose between the oil trap and oil pan
- Front exhaust pipe
- Upper retaining bolts and torque the bolts to 33 ft. lbs. (45 Nm)
- Auxiliary drive belt and tensioner
- Crankcase ventilation solenoid and constant pressure valves
- Right side engine mount and torque the nut to 37 ft. lbs. (50 Nm)

06041_SAAB_G0001

**Fig. 55 Showing location of ECM connectors—2.0L I4 engine**

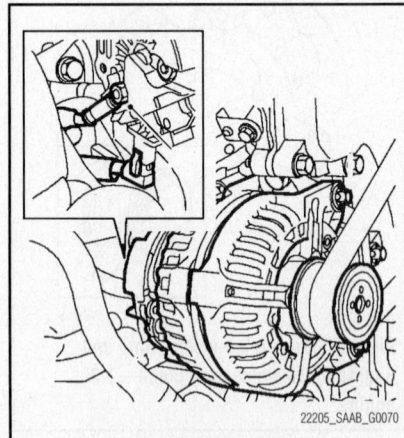

22205_SAAB_G0070

**Fig. 57 To make is easier to refit the generator, tap the bushings until they are level with the mountings—2.3L engine**

- Rear engine mount and torque the nut to 18 ft. lbs. (25 Nm)
- Intake manifold shield
- Negative battery cable

### 2.8L Engine

*See Figure 58.*

1. Before servicing the vehicle, refer to the Precautions Section.
2. Remove the battery cover.
3. Disconnect the negative battery cable.
4. Raise and safely support the front of the vehicle.
5. Remove the front right wheel.
6. Remove the right wheel well liner.
7. Relieve the belt tensioner and unhook the belt. Use a ½ inch drawbar handle.
8. Remove the bolt from the steering knuckle, lower the transverse link and place a 83 95 238 Wedge between the transverse link and the anti-roll bar.
9. Tap the driveshaft out of the intermediate shaft with a brass drift and large hammer. Move aside the drive shaft. Use protective goggles to protect against flying splinters of metal.

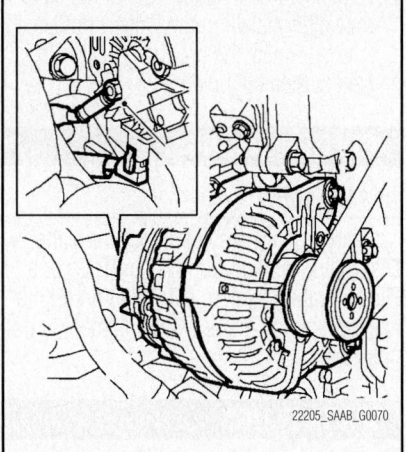

**Fig. 58 Alternator and attaching fastener location—2.8L engine**

10. Remove the alternator electrical connections.
11. Remove the alternator retaining bolts.
12. Remove the alternator.

**To install:**

13. Install the alternator. Torque the retaining bolts to 18 ft. lbs. (24 Nm).

14. Connect the alternator electrical connections.
15. Grease the driveshaft splines and make sure the rubber seal is mounted on the shaft.
16. Install the driveshaft to the intermediate shaft. Make sure the lock ring clicks into place.
17. Remove the wedge and connect the ball joint to the steering knuckle with a nut and bolt. The pin must be visible above the steering knuckle before the bolt is installed. Tighten the nut to 37 ft. lbs. (50 Nm).

➡**Press up the pin carefully. The groove in the pin must be visible in the screw hole in the spindle housing. If the rubber seal is pressed down, it will not seal properly to the swivel pin.**

18. Relieve the belt tensioner using a ½ inch drawbar handle and mount the belt.
19. Install the right wheel well liner.
20. Install the front wheel.
21. Lower the vehicle.
22. Connect the negative battery cable.
23. Install the battery cover.
- Reset clock and radio

## ENGINE ELECTRICAL

### IGNITION COIL

*REMOVAL & INSTALLATION*

#### 2.5L SOHC Engine

*See Figure 59.*

1. Before servicing the vehicle, refer to the Precautions Section.
2. Disconnect the negative battery cable.
3. Disconnect the spark plug cables from the ignition coil pack.
4. Disconnect the electrical connector from the ignition coil pack.

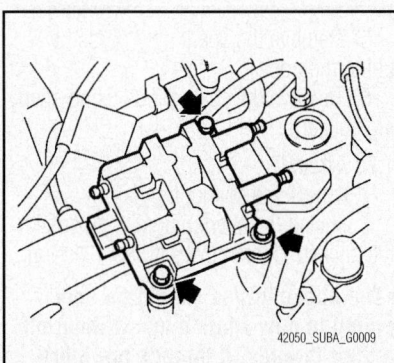

**Fig. 59 Location of the ignition coil pack mounting bolts—2.5L SOHC engine**

5. Remove the ignition coil pack assembly.

**To install:**

6. Installation is the reverse order of removal.
7. Tighten the ignition coil pack mounting bolts to 5 ft. lbs. (6 Nm).

#### 2.0L I4 Engine

*See Figure 60.*

1. Before servicing the vehicle, refer to the Precautions Section.

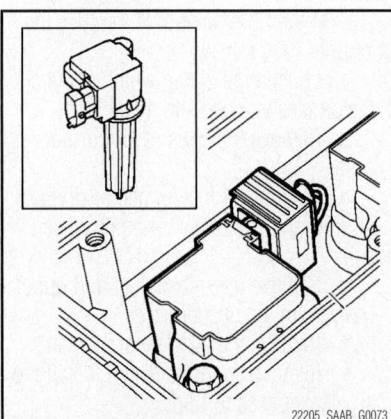

**Fig. 60 Ignition coil with integrated power module—2.0L engine**

## IGNITION SYSTEM

2. Remove the cover over the ignition coils.
3. Remove the ignition coil.
4. Unplug the ignition coil connector.

**To install:**

5. Install the ignition coil. Tighten to 15 ft. lbs. (20 Nm).
6. Make sure to locate the ignition coil cables so they are not pinched.
7. Install the cover over the ignition coils. Tighten to 6 ft. lbs. (8 Nm).

#### 2.0L H4 and 2.5L DOHC Engines

*Right Side*

1. Before servicing the vehicle, refer to the Precautions Section.
2. Disconnect the negative battery cable.
3. Remove the air cleaner lower case as follows:
   a. Disconnect the connector from mass air flow and intake air temperature sensor.
   b. Loosen the clamp which connects the air intake boot and intake duct.
   c. Remove the clip from air cleaner upper cover.
   d. Remove the air cleaner upper cover.
   e. Remove the air cleaner element.
   f. Remove the air cleaner lower case.

4. Disconnect the connector from ignition coil and igniter assembly.

5. Remove the ignition coil and igniter assembly.

### To install:

6. Install the ignition coil and igniter assembly.

7. Plug in the connector on ignition coil and igniter assembly.

8. Install the air cleaner lower case as follows:

    a. Install the air cleaner lower case.

    b. Install the air cleaner element.

    c. Install the air cleaner upper cover.

    d. Install the clip from air cleaner upper cover.

    e. Tighten the clamp which connects the air intake boot and intake duct.

    f. Plug in the connector on mass air flow and intake air temperature sensor.

9. Connect the negative battery cable.

### Left Side

1. Before servicing the vehicle, refer to the Precautions Section.

2. Remove the battery and battery carrier as follows:

    a. Disconnect the positive (+) cable after disconnecting the ground (-) cable of battery.

    b. Remove the flange nuts from battery rods, and then take off the battery holder.

    c. Remove the battery.

3. Remove the secondary air pump as follows:

    a. Disconnect the connector from secondary air pump

    b. Disconnect the hose from secondary air pump.

    c. Remove the bolt which secures the secondary air pump to the vehicle.

4. Disconnect the connector from ignition coil and igniter assembly.

5. Remove the ignition coil and igniter assembly.

### To install:

6. Install the ignition coil and igniter assembly.

7. Plug in the connector from ignition coil and igniter assembly.

8. Install the secondary air pump as follows:

    a. Install the bolt which secures the secondary air pump to the vehicle.

    b. connect the hose to secondary air pump.

    c. Plug in the connector on secondary air pump

9. Install the battery and battery carrier as follows:

    a. Install the battery.

    b. Install the flange nuts on battery rods after installing the battery holder.

    c. Connect the ground (-) cable of battery then the positive

## IGNITION TIMING

### ADJUSTMENT

Saab vehicles are equipped with either a Distributorless Ignition System (DIS) or a Direct Ignition System (DI). Ignition timing is controlled by the ECM. Adjustment is not possible.

## SPARK PLUGS

### REMOVAL & INSTALLATION

#### 2.0L I4 Engine

1. Before servicing the vehicle, refer to the Precautions Section.

2. Remove the cover over the ignition coils.

3. Remove the bolts, lift up each ignition coil and move aside. Start with the ignition coil for cylinder 1.

4. Remove the spark plugs using 83 94 785 Spark Plug Socket.

### To install:

5. Check the spark plug gap.

6. Install the spark plugs, keeping the socket straight so the spark plugs are not damaged. Tighten to 21 ft. lbs. (28 Nm).

7. Install the respective ignition coils starting with the one for cylinder #4. Tighten to 15 ft. lbs. (20 Nm).

8. Make sure to locate the ignition coil cables so they are not pinched.

9. Install the cover over the ignition coils. Tighten to 6 ft. lbs. (8 Nm).

#### 2.0L H4 and 2.5L DOHC engines

1. Disconnect the negative battery cable.

2. On the right hand side, remove the air cleaner case as follows:

    a. Loosen the clamp which connects the air cleaner case and intake duct.

    b. Remove the clips of air cleaner case.

    c. Disconnect the connector of mass air flow and intake air temperature sensor.

    d. Remove the intake duct and upper cover from air cleaner case.

    e. Remove the air cleaner element.

    f. Remove the bolts which install the air cleaner case to body.

    g. Remove the air cleaner case.

3. On the left hand side, remove the battery and carrier.

4. Disconnect the connector from ignition coil and igniter assembly.

5. Remove the ignition coil and igniter assembly.

6. Remove the spark plugs using spark plug socket.

### To install:

7. Check the spark plug gap.

8. Install the spark plugs using spark plug socket. Tighten to 15 ft. lbs. (21 Nm).

➡The above torque should be only applied to new spark plugs without oil on their threads. If threads are lubricated, the torque should be reduced to 10 ft. lbs. (14 Nm) in order to avoid over-stressing.

9. The remainder of installation is in the reverse order of removal procedure.

10. Connect the negative battery cable.

#### 2.5L SOHC Engine

1. Before servicing the vehicle, refer to the Precautions Section.

2. Disconnect the negative battery cable.

3. On the right hand side, remove the air cleaner case as follows:

    a. Loosen the clamp which connects the air cleaner case and intake duct.

    b. Remove the clips of air cleaner case.

    c. Disconnect the connector of mass air flow and intake air temperature sensor.

    d. Remove the intake duct and upper cover from air cleaner case.

    e. Remove the air cleaner element.

    f. Remove the bolts which install the air cleaner case to body.

    g. Remove the air cleaner case.

4. On the left hand side, remove the battery.

### ✳✳ WARNING

**Do not pull the spark plug cable itself.**

5. Remove the spark plug cords by pulling the boot.

6. Remove the spark plugs using spark plug socket.

### To install:

7. Check the spark plug gap.

8. Install the spark plugs using spark plug socket. Tighten to 15 ft. lbs. (21 Nm).

➡The above torque should be only applied to new spark plugs without oil on their threads. If threads are lubricated, the torque should be reduced to 10 ft. lbs. (14 Nm) in order to avoid over-stressing.

9. The remainder of installation is in the reverse order of removal procedure.

### 2.3L Engine

1. Before servicing the vehicle, refer to the Precautions Section.

2. Unplug the 10-pin ignition discharge module connector.

3. Remove the 4 ignition discharge module screws.

4. Remove the spark plugs.

#### To install:

5. Coat the spark plug threads with anti-seize.

6. Coat the rubber seals with Netzmittel K8 lubricant.

7. Check the spark gap and install the spark plugs. Tighten to 21 ft. lbs. (28 Nm).

8. Install the 4 ignition discharge module screws. Tighten to 8 ft. lbs. (11 Nm).

9. Plug in the 10-pin ignition discharge module connector.

### 2.8L Engine

*See Figure 61.*

1. Before servicing the vehicle, refer to the Precautions Section.

2. Remove the ignition coil and igniter assembly. Refer to "Distributorless Ignition System, Ignition Coil and Igniter Assembly, Removal & Installation".

3. Remove the spark plugs using 83 94 785 Spark Plug Socket.

#### To install:

4. Check the spark plug gap.

5. Install the spark plugs using 83 94 785 Spark Plug Socket. Tighten to 21 ft. lbs. (28 Nm).

6. Install the ignition coil and igniter assembly.

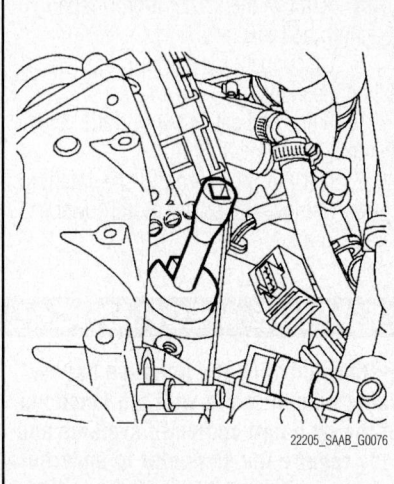

22205_SAAB_G0076

**Fig. 61 Removing the spark plugs using 83 94 785 Spark Plug Socket—2.8L engine**

---

## ENGINE ELECTRICAL                                    STARTING SYSTEM

### STARTER

#### *REMOVAL & INSTALLATION*

##### 2.0L H4 and 2.5L Engines

1. Before servicing the vehicle, refer to the Precautions Section.

2. Disconnect the negative battery cable.

3. Remove the air intake chamber, on non turbocharged engine.

4. Remove the intercooler, on turbocharged engine.

5. Remove the air intake chamber stay, on non turbocharged engine.

6. Disconnect the electrical connectors from the starter.

7. Remove the starter retaining bolts. Remove the starter from the vehicle.

#### To install:

8. Installation is the reverse of the removal procedure.

9. Torque the starter retaining bolts to 37 ft. lbs. (50 Nm).

##### 2.0L I4 and 2.8L Engine

*See Figures 62 and 63.*

1. Before servicing the vehicle, refer to the Precautions Section.

2. Remove the battery cover.

3. Disconnect the negative battery cable.

4. Raise and safely support the vehicle securely on jackstands.

5. On 2.8L engines, remove the secondary air pump.

6. Disconnect the starter motor electrical connections

7. Remove the starter motor retaining bolts.

8. Lower the starter motor and lift it out.

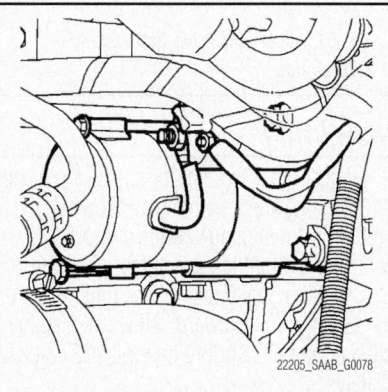

22205_SAAB_G0078

**Fig. 62 Starter location—2.0L I4 engine**

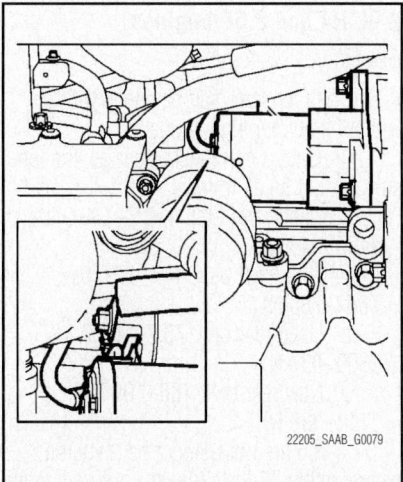

22205_SAAB_G0079

**Fig. 63 Starter location—2.8L engine**

#### To install:

9. Lift the starter motor into place.

10. Install the starter motor retaining bolts. Tighten to 35 ft. lbs. (47 Nm).

11. Clean, apply Vaseline and connect the starter motor connecting cables.

12. On 2.8L engines, install the secondary air pump.

13. Lower the vehicle.

14. Connect the negative battery cable.

15. Install the battery cover.

##### 2.3L Engine

*See Figure 64.*

1. Before servicing the vehicle, refer to the Precautions Section.

2. Remove the battery cover and the shield on the intake manifold.

3. Disconnect the negative battery cable.

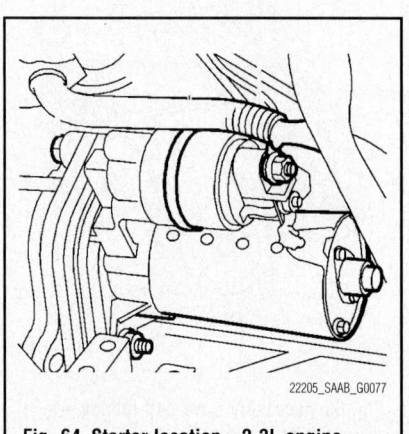

22205_SAAB_G0077

**Fig. 64 Starter location—2.3L engine**

4. Remove the starter motor upper retaining bolt (18 mm).

5. Disconnect the starter motor electrical connections

6. Raise and safely support the vehicle securely on jackstands.

7. Cut the Strap around the solenoid.

8. Remove the starter motor lower retaining nut.

9. Lower the starter motor and lift it out.

**To install:**

10. Clean the starter motor mounting surfaces and electrical contacts and apply Vaseline.

11. Screw on the starter motor's retaining nut.

12. Install a new Strap.

13. Lower the vehicle.

14. Clean the starter motor connecting cables, apply Vaseline and reconnect them.

15. Screw on the starter motor's upper retaining bolt.

16. Reconnect the negative cable to the battery.

17. Install the cover and shield.

# ENGINE MECHANICAL

➡ Disconnecting the negative battery cable may interfere with the functions of the on board computer systems and may require the computer to undergo a relearning process, once the negative battery cable is reconnected.

## ACCESSORY DRIVE BELTS

### ACCESSORY BELT ROUTING

*See Figures 65 through 67.*

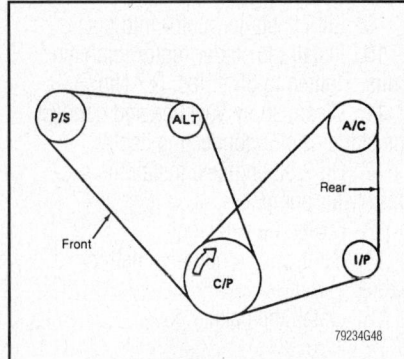

**Fig. 65 Accessory drive belt routing—2.0L H4 engine and 2.5L engines**

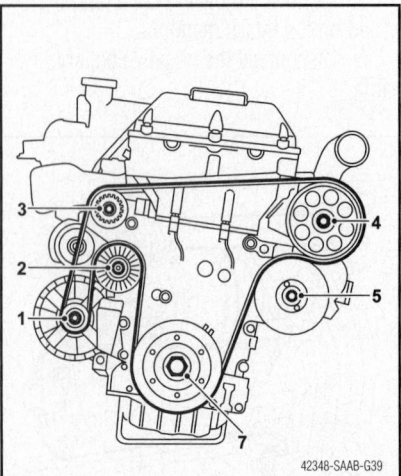

**Fig. 66 Accessory drive belt routing—9-3 and 9-5 2.0L and 2.3L engines**

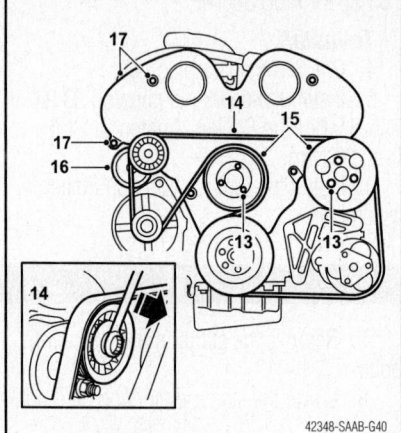

**Fig. 67 Accessory drive belt routing—2.8L engine**

### INSPECTION

Inspect the drive belt for signs of glazing or cracking. A glazed belt will be perfectly smooth from slippage, while a good belt will have a slight texture of fabric visible. Cracks will usually start at the inner edge of the belt and run outward. All worn or damaged drive belts should be replaced immediately.

### ADJUSTMENT

#### 2.0L H4 and 2.5L Engines

*See Figures 68 and 69.*

The belt tension may be check with or without a belt tension gauge.

1. If using a belt tension gauge, the tension should be as follows:
   a. Used front belt: 110.2–143.9 lbs. (490–640 N)
   b. New front belt: 143–175 lbs. (640–780 N)
   c. Used rear belt: 78.7–101.2 lbs. (350–450 N)
   d. New rear belt: 166–198 lbs. (740–880 N)

2. If you are not using a belt tension gauge, using 22 lbs. (98 Nm) of force, push on the belt as shown in the illustration. The

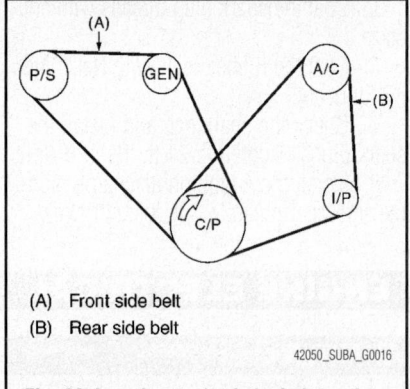

(A) Front side belt
(B) Rear side belt

**Fig. 68 Location to check the belt tension if you are using a belt tension gauge—9-2x**

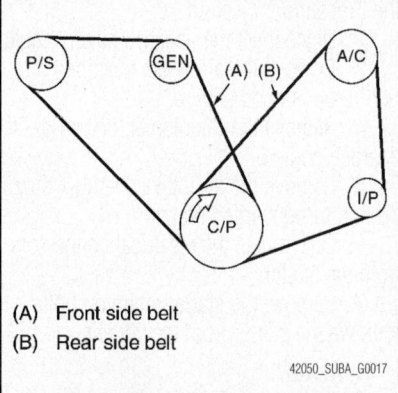

(A) Front side belt
(B) Rear side belt

**Fig. 69 Location to check to the belt tension if you are not using a belt tension gauge—9-2x**

total distance the belt travels up and down should be:
   a. Used front belt: 0.354–0.433 in. (9–11 mm)
   b. New front belt: 0.276–0.354 in. (7–9 mm)
   c. Used rear belt: 0.354–0.394 in. (9.0–10.0 mm)
   d. New rear belt: 0.295–0.335 in. (7.5–8.5 mm)

#### 2.0L I4, 2.3L and 2.8L Engines

These vehicles use a self tensioning pulley. Adjustment is not possible.

## REMOVAL & INSTALLATION

### 2.0L I4 Engine

*See Figure 66.*

1. Before servicing the vehicle, refer to the Precautions Section.

2. Raise and safely support the vehicle securely on jackstands.

3. Disconnect the front part of the right wheel well and pull it back to gain access to the belt tensioner.

4. Relieve the tension on the belt tensioner with 83 96 095 Drivebelt Relieving Tool.

5. Remove the drive belt from the pulleys.

*To install:*

6. To install, reverse removal procedure.

7. Check belt position on all pulleys.

### 2.0L H4 and 2.5L Engines

*See Figure 65.*

1. Remove the accessory drive belt cover, if equipped.

2. Loosen the lock bolt (bottom) of the front belt tensioner.

3. Loosen the slider bolt (top) of the front belt tensioner.

4. Remove the front side belt.

5. Loosen the lock bolt (bottom) of the rear belt tensioner.

6. Loosen the slider bolt (top) of the rear belt tensioner.

7. Remove the rear side belt.

*To install:*

➡**Wipe off any oil or water on the belt and pulley.**

8. Install the rear side belt, then tighten the slider bolt until the belt reaches the specified tension. Refer to "Engine Mechanical Components, Accessory Drive Belts, Adjustment".

9. Tighten the lock nut of the rear belt tensioner to 18 ft. lbs. (25 Nm).

10. Install the front side belt, then tighten the slider bolt until the belt reaches the specified tension. Refer to "Engine Mechanical Components, Accessory Drive Belts, Adjustment".

11. Tighten the lock nut of the front belt tensioner to 17 ft. lbs. (23 Nm).

### 2.3L Engine

*See Figure 66.*

1. Before servicing the vehicle, refer to the Precautions Section.

2. Cover the fenders to prevent damage and remove the cover on the intake manifold.

3. Safely raise and support the vehicle.

4. Turn the right front wheel outwards and disconnect the power steering pump pipe from the subframe.

5. Drive a wooden wedge between the oil sump and subframe.

6. Disconnect the mass air flow sensor's rubber boot and move it aside.

7. Remove the right-hand engine mount.

8. Take the strain off the belt tensioner with the aid of a ratchet handle extension and insert a 3 mm hexagon key in the hole.

### ❄❄ WARNING

**Exercise the utmost care to ensure that the belt tensioner does not break at its end position!**

9. Remove the belt. If the old belt is to be reinstalled, mark the direction of rotation.

*To install:*

10. Install the drive belt round the pulleys, check its position on all the pulleys and tension it.

11. Install the right-hand engine mount. Tighten to 36 ft. lbs. (50 Nm).

12. Connect the mass air flow sensor's rubber boot.

13. Remove the wedge and attach the pipe to the subframe.

14. Lower the vehicle.

15. Start the engine and run it at idling speed for a while. Switch off the engine and check that the belt is positioned correctly.

### 2.8L Engine

*See Figure 67.*

1. Before servicing the vehicle, refer to the Precautions Section.

2. On convertible models, remove the bolts holding the front chassis reinforcement and remove the reinforcement.

➡**The front bolts (of the rear mounting) are shorter than the others**

3. Raise and safely support the vehicle.

4. Remove the chassis reinforcement from the front subframe (convertible model).

5. Install 83 96 145 Centering tool, subframe - engine and 83 96 152 Centering tool, power train with 32 025 059 Support. First remove the two long supports. Relieve the load on the engine.

6. Remove the front spoiler shield.

7. Lower the vehicle.

8. Remove the upper then the lower section of the air cleaner.

9. Remove the engine pad.

10. Remove the power steering pipe clip from the engine bracket.

11. Remove the engine bracket. The lower hole is slotted.

12. Loosen the bolts of the water pump pulley slightly.

13. Relieve the tension in the belt with a ½" handle and remove the belt. Guide the belt behind the belt tensioners gear. Mark the direction of rotation if the belt is to be re-installed.

*To install:*

14. Position the belt, relieve the load on the belt tensioner and install the belt. Start by installing the belt on the belt tensioner.

15. Check the belt position on all pulleys.

16. Install the engine bracket. Tighten the M10 size bolts to 31 ft. lbs. (42 Nm), and the M12 size bolts to 69 ft. lbs. (93 Nm).

17. Install the engine pad to the body and bracket using 32 025 063 Special box spanner. Tighten the bolts to the body to 37 ft. lbs. (50 Nm), plus an additional 180°. Tighten the bolts to the engine mounting to 33 ft. lbs. (45 Nm), plus an additional 90°.

18. Install the air cleaner casing and connect the intake hose. Install the air cleaner.

19. Install the air cleaner casing cover.

20. Raise and safely support the vehicle.

21. Remove 83 96 145 Centering tool, subframe - engine and 83 96 152 Centering tool, power train with 32 025 059 Support.

22. Install the chassis reinforcement to the front subframe (convertible model).

23. Install the spoiler shield.

24. Lower the vehicle.

### CAMSHAFT AND VALVE LIFTERS

## REMOVAL & INSTALLATION

### 2.0L I4 Engine

*See Figures 70 and 71.*

1. Before servicing the vehicle, refer to the Precautions Section.

2. Remove the upper engine cover.

3. Unplug the MAF sensor connector. Detach the SAI hose. Undo the turbocharger inlet hose and remove the filter casing cover. Remove the filter element.

4. Detach the fuel pipes from their snap fasteners and remove the bracket bolts from the camshaft cover.

5. Remove or disconnect the following:
   • Cover over the ignition coils.
   • Ignition coil bolts, lift up each ignition coil and move aside, starting with the ignition coil for cylinder No. 1.

- Crankcase ventilation hose from the camshaft cover
- Cable duct from the cylinder head and camshaft cover and move it carefully aside
- Ground lead from the camshaft cover
- Heat shield over the turbocharger
- Camshaft cover
- Right-hand wheel
- Right fender liner

6. Lower the vehicle and zero the engine by turning the crankshaft in the direction of rotation of the engine until the marking on the crankshaft pulley is aligned with the marking on the timing cover. The cam lobes on cylinder No. 1 intake and exhaust camshafts should be pointing up and in.

7. Remove chain tensioner.

8. Remove the lock ring from the chain tensioner and remove the piston.

9. Remove the upper timing chain guide rail.

10. Remove the camshaft sprockets. Grip the camshaft flats with a spanner when loosening the bolts.

11. Remove the power steering pump bolts and shift the pump outward.

12. Remove the vacuum pump bolts and shift the pump outward.

13. Slacken the camshaft bearing caps evenly in the indicated order. Carefully detach the camshaft guide bearings with a rubber mallet.

14. Note the bearing cap designations.

➡**The cylinder head must be replaced in the event of a defective camshaft bearing. The rocker arms must be replaced when replacing the camshafts.**

15. Lift off the rocker arms.

### To install:

16. Install the rocker arms. Dip the rocker arms in oil before installing. Make sure that the rocker arms make contact with the valve stem and tappet.

17. Lubricate the contact surfaces with engine oil.

18. Apply a thin film of sealant on the sealing surfaces of the guide bearing.

19. Note the bearing cap designations.

20. Install the camshaft bearing caps in the specified order. Tighten to 7 ft. lbs. (10 Nm).

21. Install the guide bearings. Tighten to 16 ft. lbs. (22 Nm).

22. Install the vacuum pump bolts. Tighten to 16 ft. lbs. (22 Nm).

23. Install the power steering pump bolts 16 ft. lbs. (22 Nm).

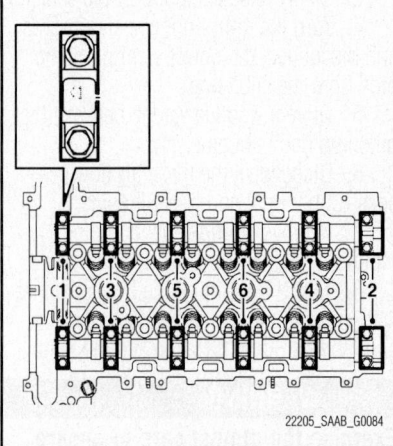

**Fig. 70 Loosen and tighten the camshaft bearing caps in the specified order—2.0L engine**

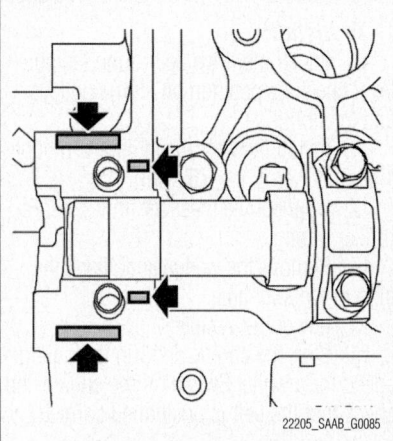

**Fig. 71 Apply a thin film of sealant on the sealing surfaces of the guide bearing—2.0L engine**

24. Clean the camshaft and sprocket contact surfaces from oil and grease.

25. Align the chain on the pinions, install the pinions on the camshaft and install the bolts. Turn the camshaft drive tensioner clockwise with a screwdriver until it engages in the tensioned position.

26. Place the plunger in the chain tensioner sleeve. Install the circlip and check that it is installed correctly in the groove.

➡**Check that the O-ring and gasket are intact. If damaged, replace the chain tensioner assembly.**

27. Install the chain tensioner. Tighten to 55 ft. lbs. (75 Nm).

28. Release/activate the chain tensioner by carefully pressing on the timing chain or chain guide (as shown in the illustration). Use a screwdriver. Check that the tensioner has released.

29. Remove the exhaust camshaft and intake camshaft bearing caps no. 2 and no. 7. Position EN-48368 Camshaft Adjustment Tool and EN-48366 Camshaft Adjustment Tool by manually pressing the tools down onto the camshafts without tightening the bolts. Turn the camshaft until the tool aligns with the spanner on the camshaft flats. Tighten the Adjustment Tools' bolts. Tighten to 7 ft. lbs. (10 Nm)

30. Check that the markings on the crankshaft pulley and the timing cover are aligned. Tighten the camshaft gears a first time. Use a wrench to grip the camshaft flats. Tighten to 22 ft. lbs. (30 Nm).

31. Remove the Adjustment Tools and install the bearing caps. Tighten to 6 ft. lbs. (8 Nm)

32. Continue to tighten the camshaft gears to the final torque. Use a wrench to grip the camshaft flats.

Tighten to 63 ft. lbs. (85 Nm) plus 30°.

33. Rotate the crankshaft 2 turns in the direction of engine rotation until the mark on the crankshaft pulley agrees with the mark on the timing cover. Remove the bearing caps and install EN-48368 Camshaft Adjustment Tool and EN-48366 Camshaft Adjustment Tool again to check that the camshaft setting is correct. Remove the Adjustment Tools and Install the bearing caps. Tighten to 6 ft. lbs. (8 Nm)

34. Install the upper timing chain guide rail.

35. Install the camshaft cover with new gaskets for the bolts. Check that the gaskets remain in place during installing. Tighten to 8 ft. lbs. (10 Nm)

36. Install or connect the following:
- Heat shield over the turbocharger
- Ground lead from the camshaft cover
- Cable duct from the cylinder head and camshaft cover
- Crankcase ventilation hose from the camshaft cover
- Ignition coil bolts, lift up each ignition coil and move aside, starting with the ignition coil for cylinder #4. Tighten to 6 ft. lbs. (8 Nm)
- Cover over the ignition coils.
- Camshaft cover

37. Connect the fuel pipes on their snap fasteners.

38. Plug in the MAF sensor connector.

39. Connect the inlet hose.

40. Remove the upper engine cover.

41. Install the right fender liner and wheel.

42. Lower the vehicle completely. Check the engine's oil level.

43. Top up as necessary.

## 2.0L H4 and 2.5L DOHC Engines

*See Figures 72 through 75.*

1. Before servicing the vehicle, refer to the Precautions Section.

2. Disconnect the negative battery cable.

3. Remove the collector cover.

4. To remove the front side V belt, remove the belt covers. Loosen the lock bolt. Loosen the slider bolt. Remove the front side belt.

5. To remove the rear side V belt, remove the belt covers. Loosen the lock bolt. Loosen the slider bolt. Remove the rear side belt. Remove the belt tensioner.

6. Remove the crankshaft pulley bolt.

7. Lock the crankshaft in place using tool ST499977100, or equivalent.

8. Remove the crankshaft pulley.

9. Remove the left side timing belt cover.

10. Remove the right side timing belt cover.

11. Remove the front timing belt cover.

12. Remove the timing belt.

13. Remove the camshaft position sensor.

14. Remove the camshaft sprockets.

➡**Be sure to lock the camshaft in place using tool ST499207400, or equivalent.**

15. Lock the crankshaft in place using tool ST499977100, or equivalent.

16. Remove the crankshaft pulley.

17. Remove the tensioner bracket. Remove the right and left timing belt No. 2 covers.

18. Remove the spark plug wires. Remove the oil level gauge, on the left side.

19. Remove the rocker cover and gasket. Remove the oil pipe.

20. Loosen the oil flow control solenoid valve assembly and the intake camshaft cap bolts equally and in the proper sequence.

21. Loosen the exhaust camshaft cap bolts equally and in the proper sequence.

22. Remove the oil flow control solenoid valve assembly, intake camshaft cap and camshaft.

23. Remove the exhaust camshaft caps and camshaft.

➡**Arrange the camshafts caps so that they can be installed in their original positions.**

**To install:**

➡**Lubricate the camshaft journals with clean engine oil prior to installation.**

24. Install the camshaft so that the valves are closed or in contact with the "base circle" of the cam lobe.

25. If the camshafts are positioned as shown in the, the camshafts need to be rotated at a minimum to align the timing belt during installation.

26. The right hand camshaft need not be rotated when set at the position illustrated, the left hand intake camshaft should be rotated 80° clockwise. The left hand exhaust camshaft should be rotated 45° counterclockwise.

27. To install the camshaft cap and oil flow control solenoid valve, apply a small amount of liquid gasket to the mating surface of the cap.

➡**Do not apply an excessive amount of sealant, as it will squish out and flow toward the seal resulting in an oil leak.**

28. Apply a thin coat of engine oil to the cap bearing surface and install the cap according to the cap identification mark. Gradually tighten the cap bolts in two stages, first to 7.2 ft. lbs. and then to 14.5 ft. lbs. in the proper sequence.

➡**After tightening the camshaft cap, ensure that the camshaft rotates slightly while holding it at base circle.**

29. Using tools ST49587600 and ST499597200, install a new seal on the camshaft. Be sure to coat the seal with clean engine oil before installation. Use tool ST499587700 and install the plug.

30. Install a new gasket on the rocker cover. Apply liquid gasket to the cylinder head (see illustration).

➡**Apply an extra amount of liquid gasket around the semicircular plugs, 5mm or more.**

31. Temporarily tighten the rocker cover retaining bolts, in the proper sequence, and then tighten to 4.7 ft. lbs. Install the oil pipe.

32. Continue the installation in the reverse order of the removal procedure.

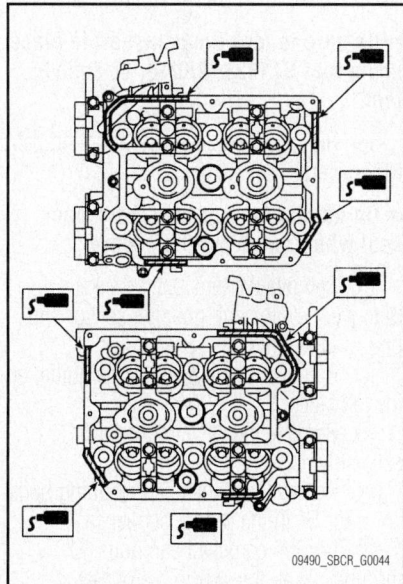

Fig. 74 Rocker cover liquid gasket application points—2.5L DOHC engine

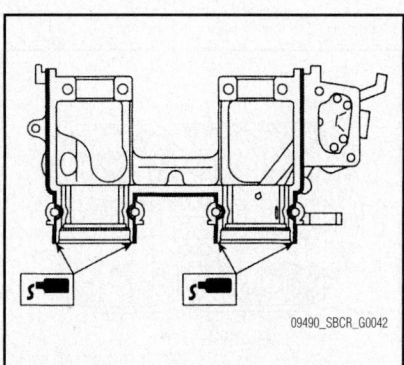

Fig. 72 Camshaft cap liquid gasket application points—2.5L DOHC engine

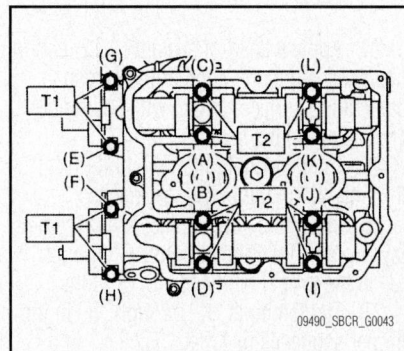

Fig. 73 Camshaft bolt tightening sequence—2.5L DOHC engine

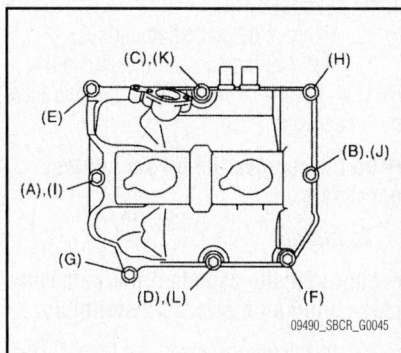

Fig. 75 Rocker cover tightening sequence—2.5L DOHC engine

### 2.5L SOHC Engine

*See Figures 76 through 83.*

1. Before servicing the vehicle, refer to the Precautions Section.

2. Disconnect the negative battery cable.

3. To remove the front side V belt, remove the belt covers. Loosen the lock bolt. Loosen the slider bolt. Remove the front side belt.

4. To remove the rear side V belt, remove the belt covers. Loosen the lock bolt. Loosen the slider bolt. Remove the rear side belt. Remove the belt tensioner.

5. Remove the crankshaft pulley bolt.

6. Lock the crankshaft in place using tool ST499977100, or equivalent.

7. Remove the crankshaft pulley.

8. Remove the left side timing belt cover.

9. Remove the front timing belt cover.

10. Remove the timing belt.

11. Remove the camshaft position sensor.

12. Remove the camshaft sprockets.

➡**Be sure to lock the camshaft in place using tool ST18231AA010, or equivalent.**

13. Remove the left and right timing belt NO2 covers.

➡**Do not damage or lose the rubber seal when removing the covers.**

14. Remove the tensioner bracket. Remove the camshaft position sensor support, on the left side.

15. Remove the oil level gauge guide, on the left side.

16. Remove the valve rocker arm assembly.

17. Remove camshaft cap retaining bolts "A" and "B" in the proper sequence.

18. Loosen camshaft cap bolts "C" through "J" all the way in the proper sequence.

19. Remove camshaft cap bolts "K" through "P" in the proper sequence using a Torx® head bit.

20. Remove the camshaft caps.

21. Remove the camshaft. Remove the oil seal. Remove the plug from the rear side of the camshaft.

➡**Do not remove the oil seal unless necessary.**

### To install:

➡**Lubricate the camshaft journals with clean engine oil prior to installation.**

22. Install the camshaft into the cylinder head.

23. Apply liquid gasket to the mating surfaces of the camshaft cap.

24. Apply a bead of sealant (0.12 inch in diameter) along the edge of the camshaft cap mating surface. Install with 20 minute s after applying the sealant.

25. Temporarily tighten the bolts "A" through "D" in the proper sequence.

26. Install the valve rocker arm assembly. Tighten Torx® head bolts "E" through "J" in the proper sequence to 13 ft. lbs.

27. Tighten bolts "K" through "R" in the proper sequence to 7.2 ft. lbs.

28. Tighten bolts "S" through "T" in the proper sequence to 7.2 ft. lbs.

➡**Be sure to use a new seal washer.**

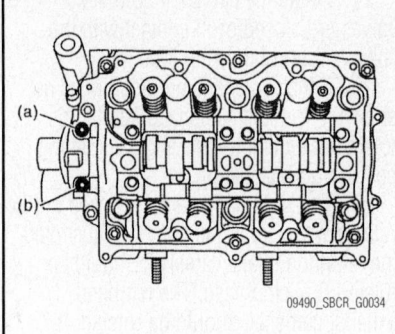

**Fig. 76 Camshaft bolt loosening sequence—2.5L SOHC engine**

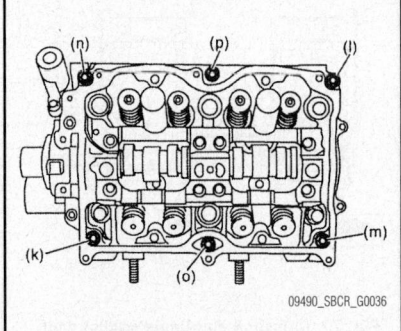

**Fig. 77 Camshaft bolt loosening sequence—2.5L SOHC engine**

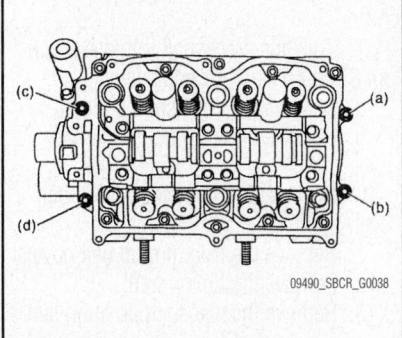

**Fig. 78 Camshaft bolt loosening sequence—2.5L SOHC engine**

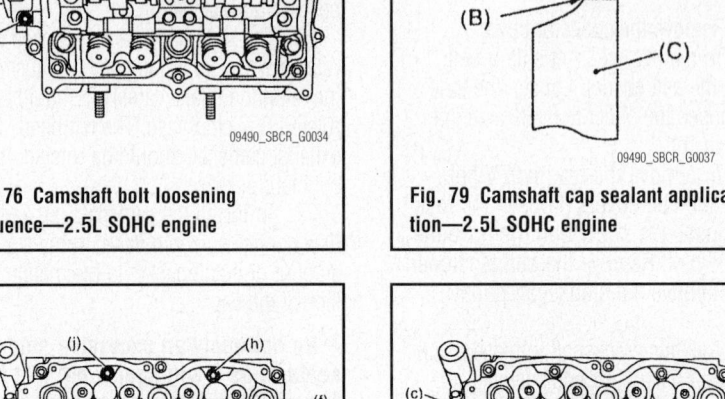

**Fig. 79 Camshaft cap sealant application—2.5L SOHC engine**

**Fig. 80 Camshaft bolt tightening sequence—2.5L SOHC engine**

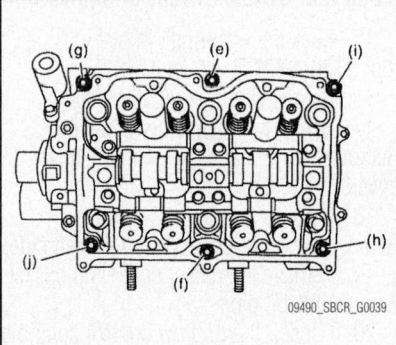

**Fig. 81 Camshaft bolt tightening sequence—2.5L SOHC engine**

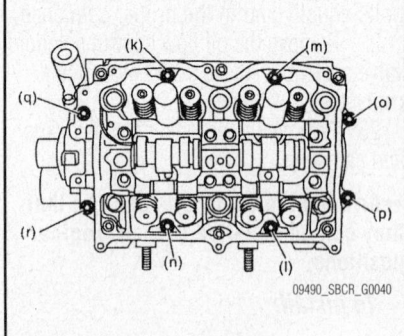

**Fig. 82 Camshaft bolt tightening sequence—2.5L SOHC engine**

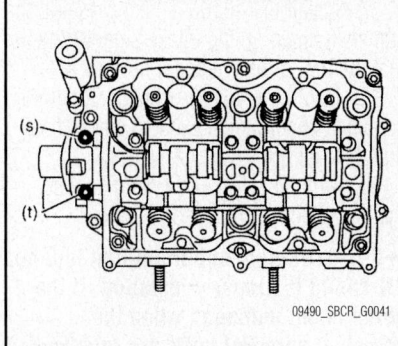

**Fig. 83 Camshaft bolt tightening sequence—2.5L SOHC engine**

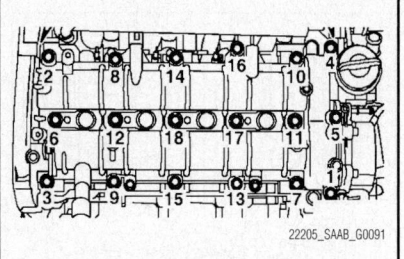

**Fig. 84 Remove and install the camshaft shaft housing and the guide sleeves in the order shown—2.3L engine**

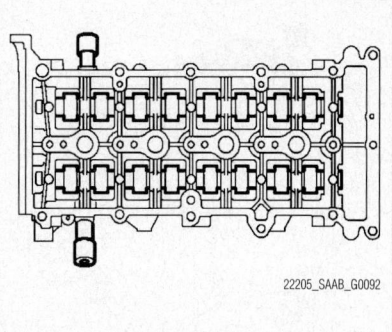

**Fig. 85 Lock both camshafts with 32 025 008 camshaft fixing tool—2.3L engine**

29. Using tools ST499597000 and ST499587500, install a new seal on the camshaft. Be sure to coat the seal with clean engine oil before installation. Use tool ST499587700 and install the plug.

30. Adjust the valve clearance.

31. Continue the installation in the reverse order of the removal procedure.

### 2.3L Engines

*See Figures 84 and 85.*

1. Before servicing the vehicle, refer to the Precautions Section.

2. Remove the upper engine cover.

3. Unplug the MAF sensor connector. Detach the SAI hose. Undo the turbocharger inlet hose and remove the filter casing cover. Remove the filter element.

4. Detach the fuel pipes from their snap fasteners and remove the bracket bolts from the camshaft cover.

5. Remove or disconnect the following:
   - Cover over the ignition coils.
   - Ignition coil bolts, lift up each ignition coil and move aside, starting with the ignition coil for cylinder #1.
   - Crankcase ventilation hose from the camshaft cover
   - Cable duct from the cylinder head and camshaft cover and move it carefully aside
   - Ground lead from the camshaft cover
   - Heat shield over the turbocharger
   - Camshaft cover
   - Right-hand wheel
   - Right fender liner

6. Relieve the pressure on the belt tensioner and remove the belt. Use 83 96 095 drivebelt relieving tool. Mark the belt's direction of rotation.

7. Lower the vehicle and zero the engine by turning the crankshaft in the direction of rotation of the engine until the marking on the crankshaft pulley is aligned

with the marking on the timing cover. The cam lobes on cylinder No. 1 intake and exhaust camshafts should be pointing up and in.

8. Remove the exhaust camshaft and intake camshaft bearing shells No. 2 and No. 7. Position the 83 96 046 or 83 96 079 Camshaft Adjustment Tool on camshaft by pressing down the tool by hand onto the camshafts without tightening the bolts.

9. If the Adjustment Tool does not install the cam lobes undo the camshaft gear bolts slightly, gripping the camshaft with a spanner on the flats. Tighten the Adjustment Tool bolts to 7 ft. lbs. (10 Nm).

10. Check that the markings on the crankshaft pulley and the timing cover are in agreement. Tighten the camshaft gear a first step, gripping the camshaft with a wrench on the flats. Tighten to is 22 ft. lbs. (30 Nm).

11. Remove the Adjustment Tools and install the bearing caps. Torque to 6 ft. lbs. (8 Nm). Tighten the camshaft gear to its final torque. Grip the camshaft with a wrench on its flats. Tighten to 63 ft. lbs. (85 Nm), plus an additional 30°.

12. Rotate the crankshaft 2 turns in the direction of engine rotation until the mark on the crankshaft pulley agrees with the mark on the timing cover. Remove the bearing caps and install 83 96 046 or 83 96 079 Adjustment Tool again to check the camshaft setting. Remove the Adjustment Tool and Install the bearing caps. Torque to 6 ft. lbs. (8 Nm).

**To install:**

13. Raise the vehicle, relieve the belt tensioner and install the belt in the marked direction of rotation. Install the right fender liner. Install the front wheels.

14. Lower the vehicle and install the camshaft cover with new seals. Take care not to disturb the position of the seals when

installing. Torque cover bolts to 8 ft. lbs. (10 Nm).

15. Install or connect the following:
   - Heat shield over turbocharger (make sure clip under heat shield snaps onto holder)
   - Ground cable to camshaft cover
   - Cable duct on camshaft cover
   - Crankcase ventilation hose to the camshaft cover
   - Ignition coils (start with coil for cylinder No. 4); torque to 6 ft. lbs. (8 Nm)
   - Cover over the ignition coils
   - Bolts for bracket to camshaft cover; install fuel pipes in snap fasteners
   - Filter element
   - Cover on air cleaner casing
   - MAF sensor connector
   - SAI hose
   - Turbocharger inlet hose to the air cleaner
   - Upper engine cover

### 2.8L Engine

#### 2006-07 Front (Old Version)

*See Figures 86 through 88.*

1. Before servicing the vehicle, refer to the Precautions Section.

2. Disconnect the negative battery cable.

3. Remove the heat shield between the engine and turbo.

4. Remove the bleeder pipe.

5. Remove the front secondary air pipe complete with valve and bracket.

6. Remove the power steering fluid reservoir's bolt.

7. Remove the left grounding point.

8. Unplug the connectors and remove ignition coils 2, 4 and 6.

9. Raise the heat shield and unplug the coolant temperature sensor connector.

10. Release the right ground cable from the wiring harness.

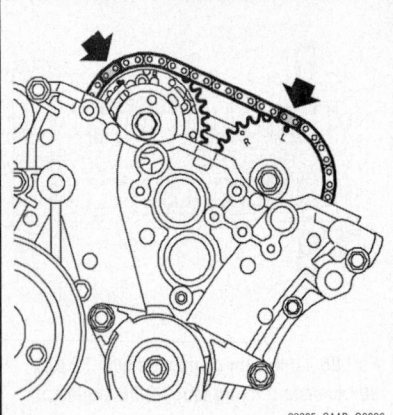

**Fig. 86 Rotate the crankshaft so that the markings on the camshaft sprockets are visible and the indentations align with the cylinder head—2006–07 2.8L front**

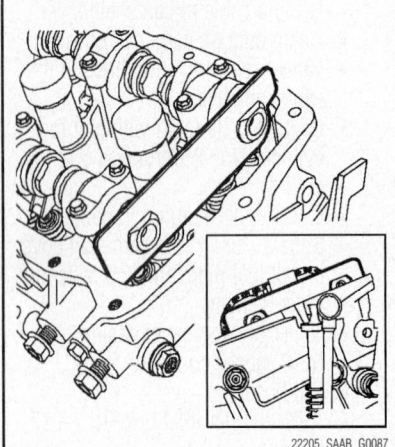

**Fig. 87 Use EN-46105-1 adjustment tool as a gauge to determine if the camshafts are now in their neutral position—2006–07 2.8L front**

11. Remove the wiring harness from the camshaft cover.

12. Unplug the connectors from the camshaft position sensor and the camshaft solenoid valve.

13. Raise the wiring harness slightly.

14. Remove the ground cable of the engine control module.

15. Unplug the engine control module connectors.

16. Detach the cable duct from the intake manifold.

17. Unplug the atmospheric pressure sensor connector.

18. Detach the corner bracket and move it aside.

19. Remove the bolt holding the injector bracket in place and unplug the connector.

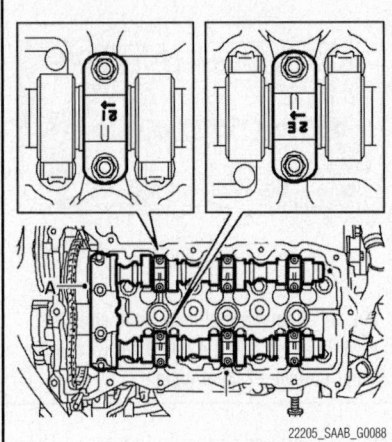

**Fig. 88 Remove the camshaft bearing caps by loosening the bolts in stages - 1/2 to 1 turn. It is important that removal ends at the bearing cap where the tappets are loaded. Note the markings on the bearing caps when refitting to ensure that they are in the correct spots—2006–07 2.8L front**

20. Detach the brake vacuum hose from the intake manifold.

21. Detach the crankcase ventilation from the intake manifold.

22. Unplug the fuel pressure sensor connector.

23. Detach the fuel line, wiping up any fuel spill with a cloth.

24. Remove the intake manifold.

25. Loosen the lifting eye. Remove one of the bolts and fold the lifting eye down.

26. Remove the camshaft cover bolts and seals.

27. Remove the camshaft cover.

28. Remove the camshaft cover gasket and seals around the spark plug holes.

29. Remove the spark plugs.

30. Remove the coolant port.

31. Remove the camshaft solenoid valve.

32. Remove the camshaft position sensor.

33. Rotate the crankshaft so that the markings on the camshaft sprocket are visible and the indentations align with the cylinder head. Use EN-46105-1 Adjustment Tool as a gauge (the camshafts are now in their neutral position).

34. Install EN-46106 Flywheel locking attachment.

35. Mark the chain and camshaft sprocket so they can be reinstallted in the same position.

36. Reduce the tightening torque of the bolts holding the camshaft sprockets. Counterhold using an open wrench on the camshaft flats (do not remove the bolts).

37. Install EN-46108-1 Fixirg Tool to the camshaft chain on the intake side. Check that the tool grips the chain firmly.

38. Install EN-46108-2 Fixing Tool to the camshaft chain on the exhaust side. Check that the tool grips the chain firmly.

39. Tighten EN-46108 Fixing Tool so that the chain is unloaded.

→**It is extremely important that tool no. EN-46108 is properly installed. If the chain is not unloaded when the camshaft sprocket bolts are removed, the automatic chain tensioner is triggered. If this occurs, the timing cover must be removed to adjust the chain tensioner.**

40. Remove the camshaft sprocket bolts.

41. Remove the sprockets from the camshafts.

42. Remove the camshaft bearing caps by loosening the bolts in stages—1/2 to 1 turn. It is important that removal ends at the bearing cap where the tappets are loaded. Note the markings on the bearing caps when reinstalling to ensure that they are in the correct spots.

43. Lift out the camshafts and note the rear marking on the camshafts.

44. Remove the roller rockers and the hydraulic adjuster elements. Store them in order so they do not get mixed up.

45. Cover all return oil ducts in the cylinder head so that the valve cone or contaminants cannot fall down into the engine.

46. Install EN-46110-eu Valve Spring Tool.

47. Install 83 94 173 Spark Plug Hole Air Nipple. Connect compressed air and put the piston and valves under pressure.

48. Press down the valve plate with EN-46110-eu Valve Spring Tool.

49. Remove the valve cone with a magnet. The compressed air must not be disconnected when the valve cone is removed. Do not press on the valve stem either as counter pressure in the cylinder could reduce and the valve could fall down into the cylinder.

50. Lift out the spring plate and valve spring.

51. Lift out the valve guide seal using 83 94 157 Valve Guide Seal Pliers.

***To install:***

52. Clean all parts and check contact surfaces and bearing surfaces for wear.

53. Install the valve guide seal using 83 94 157 Valve Guide Seal Pliers.

54. Install the valve spring and spring plate.

55. Press down the valve plate with EN-46110-eu Valve Spring Tool.

56. Install the valve cone using EN-46117 Installing tool.

57. Unload the valve and remove EN-46110-eu Valve Spring Tool.

58. Disconnect the compressed air and remove the air nipple.

59. Position the camshafts and use EN-46105 Adjustment Tool to check that the camshafts are mounted at the correct angle.

60. Install the bearing caps in order (markings on caps). Start where the cams point down and load the tappets. Tighten the bolts in stages—1/2 to 1 turn. Tighten to 7 ft. lbs. (10 Nm).

61. Install the sprockets with chain on the camshafts. Tighten the bolts by hand.

62. Check that the marks on the chain match the markings on the sprockets.

63. Remove EN-46108-2 and EN-46108-1 Fixing Tools.

64. Tighten the bolts of both camshaft sprockets, counter holding with an open wrench on the camshaft flats. Tighten to 48 ft. lbs. (65 Nm).

65. Remove EN-46106 Flywheel locking attachment.

66. Install the camshaft position sensor using a new gasket. Torque to 7 ft. lbs. (10 Nm).

67. Install the camshaft solenoid valve using a new seal. Torque to 7 ft. lbs. (10 Nm).

68. Install the coolant port using a new seal and gasket.

69. Install the spark plugs.

70. Install new seals on the camshaft cover.

71. Add a spot of sealant to each mating surface between the cylinder head and timing cover.

72. Position the camshaft cover.

73. Install the bolts using new seals and tighten to 7 ft. lbs. (10 Nm).

74. Fold the eye up and install the bolt. Tighten the bolt to 48 ft. lbs. (65 Nm).

75. Install the intake manifold.

76. Install the fuel line.

77. Install the fuel pressure sensor connector.

78. Install the crankcase ventilation to the intake manifold.

79. Install the brake vacuum hose to the intake manifold.

80. Plug in the injector connectors and install the bracket.

81. Reposition the corner bracket and install the bolts.

82. Plug in the atmospheric pressure sensor connector.

83. Install the cable duct to the intake manifold.

84. Plug in the engine control module connectors.

85. Install the ground cable of the engine control module.

86. Plug in the connectors of the camshaft position sensor and the camshaft solenoid valve.

87. Install the wiring harness to the camshaft cover using new cable ties.

88. Secure the ground cable to the wiring harness.

89. Plug in the coolant temperature sensor connector and press down the heat shield.

90. Install ignition coils 2, 4 and 6. Plug in the connectors.

91. Install the left grounding point.

92. Install the power steering fluid reservoir's bolt and tighten to 7 ft. lbs. (10 Nm).

93. Install the secondary air pipe complete with valve and bracket. Use a new gasket. Tighten the bolts to the exhaust manifold to 16 ft. lbs. (22 Nm), and the bracket bolts to 48 ft. lbs. (65 Nm).

94. Install the bleeder pipe and tighten fasteners to 7 ft. lbs. (10 Nm).

95. Install the heat shield between the engine and turbo.

96. Connect the negative battery cable.

### 2008 Front (New Version)

*See Figures 89 through 91.*

1. Before servicing the vehicle, refer to the Precautions Section.

2. Disconnect the negative battery cable.

3. Remove the battery cover and the coolant pipe.

4. Remove the upper engine cover.

> ❋❋ **WARNING**
>
> **The cooling system is under pressure. Hot coolant and steam can escape. Open the cap slowly to release the pressure. Open the expansion tank cap and relieve any excess pressure.**

➥**When removing the air cleaner casing cover, be careful not to damage the brake vacuum pump sensor.**

5. Remove the upper section of the air cleaner.

> ❋❋ **WARNING**
>
> **Take care when releasing the locking mechanism on the connector so as not to damage the connector. Pull the connector straight out when unplug-ging avoid bending the pins. For further information regarding connectors, refer to Connectors, handling and inspection.**

6. Remove the secondary air pipe , the hose clip, the wiring harness from the clip and remove the intake manifold between the air filter and the mass air flow sensor.

7. Unplug the mass air flow sensor connector, and remove the hose from the turbocharger's intake manifold and the crankcase ventilation line.

8. Undo the hose clip and remove the intake hose and the mass air flow sensor from the turbocharger.

9. Remove the charge air pipe by first removing the clips and then unplug the boost pressure sensor connector. Remove the power steering line from the clips.

10. Remove the secondary air injection pump.

11. Remove the electrical connections and the cable clip from the turbocharger.

12. Remove the crankcase ventilation hose from the turbocharger.

13. Cut off the cable tie and fold the wiring harness and the crankcase ventilation hose aside

14. Remove both of the secondary air injection pump's pipes and unplug the connector.

15. Remove the front lifting eye.

16. Remove the bolts and remove the breather pipe from the coolant port. Plug the pipe and fold the line aside.

17. Detach the vacuum hose and the fuel hose from the power steering fluid reservoir and remove the bolt.

18. Remove the ignition coils for cylinders 2, 4 and 6.

19. Remove the left-hand grounding point.

20. Lift up the heat shield and unplug the coolant temperature sensor connector.

21. Detach the wiring harness from the camshaft cover.

22. Unplug the connectors from the camshaft position sensor and from the camshaft solenoid valve.

23. Release the hook and lift up the wiring harness holder slightly.

24. Release the right-hand ground cable from the wiring harness. Carefully cut the tape.

25. Remove the engine control module's ground cables.

26. Unplug the engine control module connectors.

27. Remove the cable duct from the intake manifold.

28. Unplug the connectors for the atmospheric pressure sensors.

29. Detach the corner bracket and move it aside.

30. Remove the bolt holding the injectors' bracket and unplug the connector.

31. Remove the brake vacuum hose from the intake manifold.

32. Remove the crankcase ventilation from the intake manifold.

33. Unplug the fuel pressure sensor connector.

34. Remove the fuel line using 83 95 261 Fuel Line tool. Collect any spilled fuel with a rag.

35. Remove the upper intake manifold bolts.

36. Lift up the upper intake manifold and unplug the throttle body connector. Lift away the intake manifold.

37. Remove the gasket and cover the inlet ducts.

38. Remove the fuel rail with injectors and wiring harness. The injector for cylinder 1 may need to be removed from the fuel rail.

39. Remove the lower part of the intake manifold. Seal the inlet ducts to the cylinder head using lint-free rags.

40. Remove the secondary air system's valve for cylinders 2-4-6.

41. Detach the turbocharger diaphragm unit by removing the bolts so that the unit can be moved slightly to the side.

42. Remove the secondary air pipe.

43. Remove the upper bolts for the turbocharger delivery pipe.

44. Remove the camshaft cover bolts and rubber bushings.

45. Remove the camshaft cover for cylinders 2-4-6.

46. Remove the camshaft cover gasket and seals around the spark plug holes.

47. Undo the right-hand side of the spoiler shield.

48. Place a suitable receptacle under the radiator. Connect a hose to the radiator and drain the coolant.

49. Close the cock and Install the spoiler shield.

50. Remove the starter motor.

51. Remove the right engine mount.

52. Lower the vehicle.

53. Remove the spark plugs.

54. Remove the coolant port.

55. Remove the camshaft solenoid valve.

56. Remove the camshaft position sensor.

57. Rotate the crankshaft so that the markings on the camshaft sprockets are visible and the indentations align with the cylinder head. Use EN-48383-3 Adjustment Tool as a gauge (the camshafts are now in their neutral position).

58. Raise the vehicle.

59. Install EN-46106 Flywheel locking attachment.

60. Lower the vehicle.

➡ **Store all removed valve parts in the Valve Stand 83 93 787.**

61. Mark the chain and camshaft sprockets so they can be reinstalled in the same position.

62. Loosen the torque on the bolts holding the camshaft sprockets. Counterhold using an open spanner on the camshaft flats (do not remove the bolts).

63. Install EN-48313 Timing Chain Holding Tool to the timing chain. Check that the tool grips the chain securely and that the lower groove of the tool aligns with the top of the cylinder head (the timing cover is removed in the illustration to better depict tool installing).

64. Tighten EN-48313 Fixing Tool manually so that the chain is unloaded. Counterhold using an open spanner so that the tool does not turn.

### ✳✳ WARNING

**It is vital that EN-48313 is properly installed. If tension is not relieved from the chain when the camshaft sprocket is removed, the chain tensioner will trigger automatically. To then adjust the chain tensioner, the timing cover must be removed.**

65. Remove the camshaft sprocket bolts.

66. Remove the sprockets from the camshafts.

67. Remove the camshaft bearing caps by loosening the bolts in stages - 1/2 to 1 turn. It is important that the removal ends at the bearing cap where the tappets are loaded. Note the markings on the bearing caps when reinstalling to ensure that they are in the correct locations.

68. Lift out the camshafts and note the rear marking on the camshafts. The last letters in the marking: LI = intake camshaft, front cylinder bank. LE = exhaust camshaft, front cylinder bank.

### *To install:*

69. Position the camshafts and use EN-48383-3 Adjustment Tool to check that the camshafts are mounted at the correct angle. The last letters in the marking: LI = intake camshaft, front cylinder bank. LE = exhaust camshaft, front cylinder bank.

70. Install the bearing caps as illustrated (markings on caps). Start where the cams

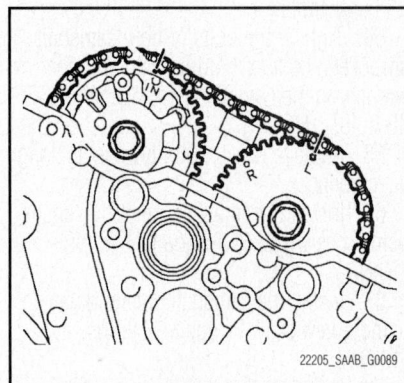

**Fig. 90 Mark the chain and camshaft sprockets so they can be refitted in the same position—2008 2.8L front**

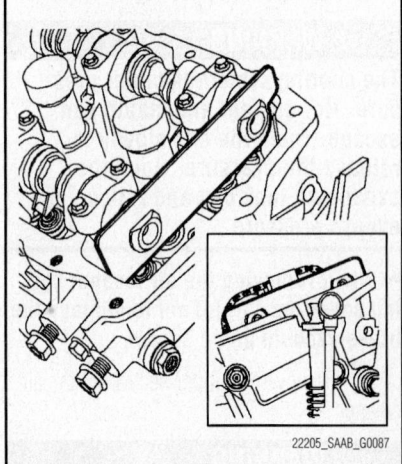

**Fig. 89 Use EN-46105-1 adjustment tool as a gauge to determine if the camshafts are now in their neutral position— 2006–07 2.8L front**

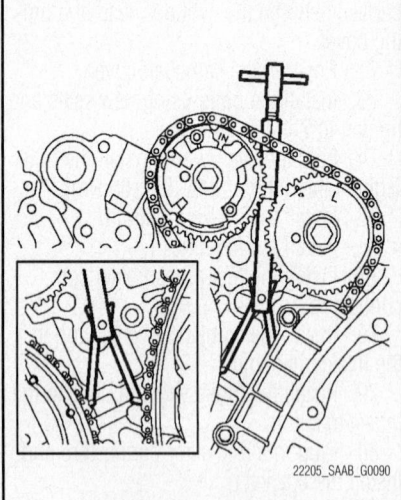

**Fig. 91 Timing chain holding tool—2008 2.8L**

point down and load the tappets. Tighten the bolts in stages - 1/2 to 1 turn at a time to 7 ft. lbs. (10 Nm).

71. Install the sprockets with chain on the camshafts. Tighten the bolts by hand.

72. Check that the marks on the chain match the markings on the sprockets.

73. Remove EN-48313 Fixing Tool.

74. Tighten the bolts of both camshaft sprockets (B), counterholding with an open spanner on the camshaft flats. Tighten to: 48 ft. lbs. (65 Nm)

75. Install the camshaft position sensor using a new gasket. Tighten to: 7 ft. lbs. (10 Nm).

76. Install the camshaft solenoid valve using a new seal. Tighten to 7 ft. lbs. (10 Nm).

77. Install the coolant port using a new seal and gasket. Tighten to 7 ft. lbs. (10 Nm).

78. Install the spark plugs.

79. Install right engine mount.

80. Raise the vehicle.

81. Remove EN-46106 Flywheel Locking Attachment.

82. Install the Starter motor.

83. Lower the vehicle.

84. Clean the sealing surfaces of the camshaft cover.

85. Install a new camshaft cover gasket and new seals for the camshaft cover. Use J-5590 Drift and J-24254-A Installing Tool.

86. Add a dot of 93 160 951 Sealant on each mating face between the cylinder head and the timing cover.

87. Position the camshaft cover.

88. Install the bolts with new rubber bushings. Tighten to 8 ft. lbs. (10 Nm).

89. Install the upper bolts to the turbocharger delivery pipe.

90. Install the secondary air pipe. Use a new seal. Tighten bolts to 8 ft. lbs. (10 Nm). Tighten exhaust manifold to 16 ft. lbs. (22 Nm)

91. Install the secondary air system's check valve. Use a new seal.

92. Install the turbocharger's diaphragm unit by installing the bolts.

93. Clean the sealing surfaces and install the lower section of the intake manifold with a new seal. Remove the rags from the inlet ducts. Tighten to 17 ft. lbs. (23 Nm).

94. Install the fuel rail with injectors and wiring harness. Tighten to 8 ft. lbs. (10 Nm).

95. Clean the sealing surfaces and install the intake manifold upper section in place with new seals and plug in the throttle body connector.

96. Install the intake manifold bolts. Tighten to 17 ft. lbs. (23 Nm)

97. Install the fuel line.

98. Plug in the fuel pressure sensor connector.

99. Install the crankcase ventilation to the intake manifold. Tighten to 8 ft. lbs. (10 Nm).

100. Install the brake vacuum hose to the intake manifold.

101. Plug in the injector connectors and install the bracket. Tighten to 8 ft. lbs. (10 Nm).

102. Reposition the corner bracket and install the bolts.

103. Plug in the atmospheric pressure sensor connector.

104. Install the cable duct to the intake manifold.

105. Plug in the engine control module connectors.

106. Install the engine control module's ground cables, one on each side.

107. Secure the ground cable to the wiring harness.

108. Plug in the connectors for the camshaft position sensor and for the camshaft solenoid valve and install the holder for the camshaft cover.

109. Attach the wiring harness to the camshaft cover using new cable ties.

110. Plug in the coolant temperature sensor connector and press down the heat shield.

111. Install the ignition coils for cylinders 2-4-6. Plug in the connectors. Tighten to 8 ft. lbs. (10 Nm).

112. Install the left-hand grounding point.

113. Install the power steering fluid reservoir's bolt. Tighten to 7 ft. lbs. (10 Nm).

114. Install the vacuum hose and the fuel hose to the power steering fluid reservoir.

115. Install the Secondary air injection pump.

116. Plug in the connector and install both of the secondary air injection pump's pipes.

117. Install the engine's lifting eye. Tighten to 48 ft. lbs. (65 Nm).

118. Attach the crankcase ventilation line to the turbocharger. Tighten to 8 ft. lbs. (10 Nm).

119. Install the turbocharger wiring harness. Plug in the connectors and install the cable clip. Install the cable tie.

120. Install the coolant return line. Install the clip and the bolts. Tighten to 8 ft. lbs. (10 Nm).

121. Install the charge air pipe. Install the clips. Tighten to 2.5 ft. lbs. (3.5 Nm).

122. Plug in the boost pressure sensor connector. Secure the power steering line in the clips.

123. Install the intake hose and the mass air flow sensor to the turbocharger. Tighten to 2.5 ft. lbs. (3.5 Nm).

124. Plug in the mass air flow sensor connector and install the hose for the turbocharger intake manifold and install the crankcase ventilation line to the dipstick tube.

125. Install the turbocharger intake manifold between the air filter and the mass air flow sensor, and the secondary air pipe and wiring harness to the clip. Tighten clip to 2.5 ft. lbs. (3.5 Nm). Tighten bolt to 1.1 ft. lbs. (1.5 Nm).

126. Install the top section of the air filter.

127. Replace the upper engine cover.

128. Connect the negative cable and install the coolant pipe and the battery cover. Tighten to 2.5 ft. lbs. (3.5 Nm).

129. Check the coolant level.

130. Connect the negative battery cable.

### 2006-07 Rear (Old Version)

*See Figures 92 through 94.*

1. Before servicing the vehicle, refer to the Precautions Section.

2. Disconnect the negative battery cable.

3. Remove the heat shield between the engine and turbo.

4. Remove the bleeder pipe.

5. Release the quick-coupling of the secondary air pipe and put the pipe aside.

6. Remove the power steering fluid reservoir's bolt.

7. Unplug the connectors from ignition coils 2, 4 and 6.

8. Remove the left grounding point.

9. Raise the heat shield and unplug the coolant temperature sensor connector.

10. Release the right ground cable from the wiring harness.

11. Remove the wiring harness from the camshaft cover.

12. Unplug the connectors from the camshaft position sensor and the camshaft solenoid valve.

13. Raise the wiring harness slightly.

14. Remove the ground cable of the engine control module.

15. Unplug the engine control module connectors.

16. Detach the cable duct from the intake manifold.

17. Unplug the atmospheric pressure sensor connectors.

18. Detach the corner bracket and move it aside.

19. Remove the bolt holding the injector bracket in place and unplug the connector.

20. Detach the brake vacuum hose from the intake manifold.

21. Detach the crankcase ventilation from the intake manifold.

22. Unplug the fuel pressure sensor connector.

23. Detach the fuel line, wiping up any fuel spill with a cloth.

24. Remove the intake manifold.

25. Unplug the connectors from the camshaft position sensor and the camshaft solenoid valve.

26. Raise the wiring harness slightly.

27. Remove the connector and pipe from the fuel bleeder valve. Lift away the valve.

28. Detach the vacuum pipe from the venturi valve.

29. Unplug the 7-pin contact connector.

30. Unplug the lower oxygen sensor connector and the connectors of ignition coils 1, 3 and 5.

31. Remove ignition coils 1, 3 and 5.

32. Remove the rear cable duct from the camshaft cover. Use an angled screwdriver to release the 3 clips from the cable duct. Press the cable duct back toward the bulkhead.

33. Remove the camshaft cover bolts.

34. Raise the wiring harness and servo reservoir. Remove the camshaft cover.

35. Remove the camshaft cover gasket and seals around the spark plug holes.

36. Remove the spark plugs.

37. Remove the coolant port.

38. Remove the camshaft solenoid valve.

39. Remove the camshaft position sensor.

40. Rotate the crankshaft so that the markings on the camshaft sprocket are visible and the indentations align with the cylinder head. Use EN-46105-1 Adjustment Tool as a gauge (the camshafts are now in their neutral position).

41. Install EN-46106 Flywheel locking attachment.

42. Mark the chain and camshaft sprocket so they can be reinstallted in the same position.

43. Reduce the tightening torque of the bolts holding the camshaft sprockets. Counterhold using an open wrench on the camshaft flats (do not remove the bolts).

44. Install EN-46108-1 Fixing Tool to the camshaft chain on the intake side. Check that the tool grips the chain firmly.

45. Install EN-46108-2 Fixing Tool to the camshaft chain on the exhaust side. Check that the tool grips the chain firmly.

46. Tighten EN-46108 Fixing Tool so that the chain is unloaded.

➡️It is extremely important that tool no. EN-46108 is properly installed. If the chain is not unloaded when the

**camshaft sprocket bolts are removed, the automatic chain tensioner is triggered. If this occurs, the timing cover must be removed to adjust the chain tensioner.**

47. Remove the camshaft sprocket bolts.

48. Remove the sprockets from the camshafts.

49. Remove the camshaft bearing caps by loosening the bolts in stages—1/2 to 1 turn. It is important that removal ends at the bearing cap where the tappets are loaded. Note the markings on the bearing caps when reinstalling to ensure that they are in the correct spots.

50. Lift out the camshafts and note the rear marking on the camshafts.

51. Remove the roller rockers and the hydraulic adjuster elements. Store them in order so they do not get mixed up.

52. Cover all return oil ducts in the cylinder head so that the valve cone or contaminants cannot fall down into the engine.

53. Install EN-46110-eu Valve Spring Tool.

54. Install 83 94 173 Spark Plug Hole Air Nipple. Connect compressed air and put the piston and valves under pressure.

55. Press down the valve plate with EN-46110-eu Valve Spring Tool.

56. Remove the valve cone with a magnet. The compressed air must not be disconnected when the valve cone is removed. Do not press on the valve stem either as counter pressure in the cylinder could reduce and the valve could fall down into the cylinder.

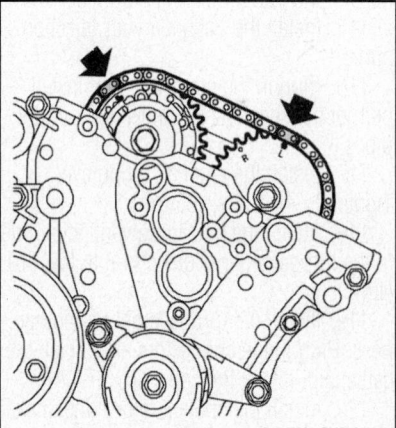

**Fig. 92 Rotate the crankshaft so that the markings on the camshaft sprockets are visible and the indentations align with the cylinder head—2006–07 2.8L front**

57. Lift out the spring plate and valve spring.

58. Lift out the valve guide seal using 83 94 157 Valve Guide Seal Pliers.

### To install:

59. Clean all parts and check contact surfaces and bearing surfaces for wear.

60. Install the valve guide seal using 83 94 157 Valve Guide Seal Pliers.

61. Install the valve spring and spring plate.

62. Press down the valve plate with EN-46110-eu Valve Spring Tool.

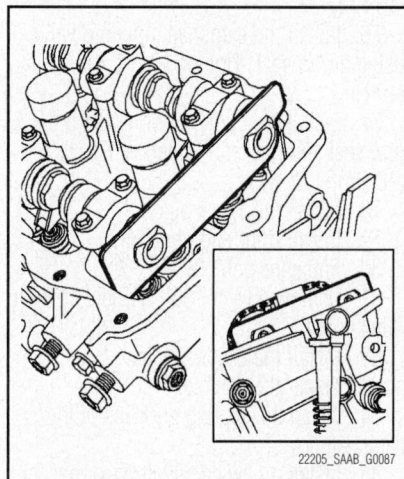

**Fig. 93 Use EN-46105-1 adjustment tool as a gauge to determine if the camshafts are now in their neutral position—2006–07 2.8L front**

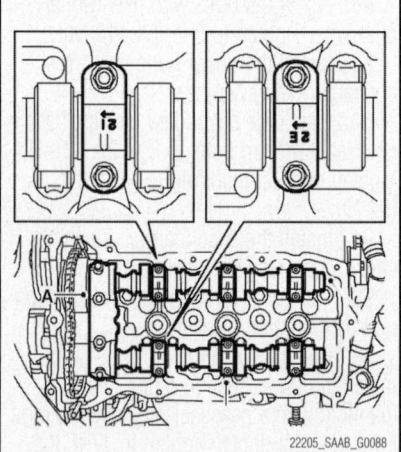

**Fig. 94 Remove the camshaft bearing caps by loosening the bolts in stages - 1/2 to 1 turn. It is important that removal ends at the bearing cap where the tappets are loaded. Note the markings on the bearing caps when refitting to ensure that they are in the correct spots.—2006–07 2.8L front**

63. Install the valve cone using EN-46117 Installing tool.

64. Unload the valve and remove EN-46110-eu Valve Spring Tool.

65. Disconnect the compressed air and remove the air nipple.

66. Position the camshafts and use EN-46105 Adjustment Tool to check that the camshafts are mounted at the correct angle.

67. Install the bearing caps in order (markings on caps). Start where the cams point down and load the tappets. Tighten the bolts in stages—1/2 to 1 turn. Tighten to 7 ft. lbs. (10 Nm).

68. Install the sprockets with chain on the camshafts. Tighten the bolts by hand.

69. Check that the marks on the chain match the markings on the sprockets.

70. Remove EN-46108-2 and EN-46108-1 Fixing Tools.

71. Tighten the bolts of both camshaft sprockets, counter holding with an open wrench on the camshaft flats. Tighten to 48 ft. lbs. (65 Nm).

72. Remove EN-46106 Flywheel locking attachment.

73. Install the camshaft position sensor using a new gasket. Torque to 7 ft. lbs. (10 Nm).

74. Install the camshaft solenoid valve using a new seal. Torque to 7 ft. lbs. (10 Nm).

75. Install the coolant port using a new seal and gasket.

76. Install the spark plugs.

77. Install new seals on the camshaft cover.

78. Add a spot of sealant to each mating surface between the cylinder head and timing cover.

79. Position the camshaft cover.

80. Install the bolts using new seals and tighten to 7 ft. lbs. (10 Nm).

81. Install the cable duct on the camshaft cover.

82. Install ignition coils 1, 3 and 5.

83. Install the connectors of the lower oxygen sensor and ignition coils 1, 3 and 5.

84. Plug in the 7-pin contact connector.

85. Install the vacuum pipe to the venturi valve.

86. Install the connector and pipe to the fuel bleeder valve.

87. Install the wiring harness to the camshaft cover.

88. Plug in the connectors of the camshaft position sensor and the camshaft solenoid valve.

89. Install the intake manifold.

90. Install the fuel line.

91. Install the fuel pressure sensor connector.

92. Install the crankcase ventilation to the intake manifold.

93. Install the brake vacuum hose to the intake manifold.

94. Plug in the injector connectors and install the bracket.

95. Reposition the corner bracket and install the bolts.

96. Plug in the atmospheric pressure sensor connector.

97. Install the cable duct to the intake manifold.

98. Plug in the engine control module connectors.

99. Install the ground cable of the engine control module.

100. Plug in the connectors of the camshaft position sensor and the camshaft solenoid valve.

101. Install the wiring harness to the camshaft cover using new cable ties.

102. Secure the ground cable to the wiring harness.

103. Plug in the coolant temperature sensor connector and press down the heat shield.

104. Install ignition coils 2, 4 and 6. Plug in the connectors.

105. Install the left grounding point.

106. Install the power steering fluid reservoir's bolt and tighten to 7 ft. lbs. (10 Nm).

107. Install the secondary air pipe and check that the quick-coupling engages.

108. Install the bleeder pipe and tighten fasteners to 7 ft. lbs. (10 Nm).

109. Install the heat shield between the engine and turbo.

110. Connect the negative battery cable.

### 2008 Rear (New Version)
*See Figures 95 through 97.*

1. Before servicing the vehicle, refer to the Precautions Section.

2. Disconnect the negative battery cable.

3. Remove the battery cover and the coolant pipe.

4. Remove the upper engine cover.

⁑ WARNING

**The cooling system is under pressure. Hot coolant and steam can escape. Open the cap slowly to release the pressure.**

5. Open the expansion tank cap and relieve any excess pressure.

➡ When removing the air cleaner casing cover, be careful not to damage the brake vacuum pump sensor.

6. Remove the upper section of the air cleaner.

⁑ WARNING

**Take care when releasing the locking mechanism on the connector so as not to damage the connector. Pull the connector straight out when unplugging avoid bending the pins. For further information regarding connectors, refer to Connectors, handling and inspection.**

7. Remove the secondary air pipe , the hose clip, the wiring harness from the clip and remove the intake manifold between the air filter and the mass air flow sensor.

8. Unplug the mass air flow sensor connector, and remove the hose from the turbocharger's intake manifold and the crankcase ventilation line.

9. Undo the hose clip and remove the intake hose and the mass air flow sensor from the turbocharger.

10. Remove the charge air pipe by first removing the clips and then unplug the boost pressure sensor connector. Remove the power steering line from the clips.

11. Remove the secondary air injection pump.

12. Remove the electrical connections and the cable clip from the turbocharger.

13. Remove the crankcase ventilation hose from the turbocharger.

14. Cut off the cable tie and fold the wiring harness and the crankcase ventilation hose aside

15. Remove both of the secondary air injection pump's pipes and unplug the connector.

16. Remove the front lifting eye.

17. Remove the bolts and remove the breather pipe from the coolant port. Plug the pipe and fold the line aside.

18. Detach the vacuum hose and the fuel hose from the power steering fluid reservoir and remove the bolt.

19. Remove the ignition coils for cylinders 2, 4 and 6.

20. Remove the left-hand grounding point.

21. Lift up the heat shield and unplug the coolant temperature sensor connector.

22. Detach the wiring harness from the camshaft cover.

23. Unplug the connectors from the camshaft position sensor and from the camshaft solenoid valve.

24. Release the hook and lift up the wiring harness holder slightly.

25. Release the right-hand ground cable from the wiring harness. Carefully cut the tape.

26. Remove the engine control module's ground cables.

27. Unplug the engine control module connectors.

28. Remove the cable duct from the intake manifold.

29. Unplug the connectors for the atmospheric pressure sensors.

30. Detach the corner bracket and move it aside.

31. Remove the bolt holding the injectors' bracket and unplug the connector.

32. Remove the brake vacuum hose from the intake manifold.

33. Remove the crankcase ventilation from the intake manifold.

34. Unplug the fuel pressure sensor connector.

35. Remove the fuel line using 83 95 261 Fuel Line tool. Collect any spilled fuel with a rag.

36. Remove the upper intake manifold bolts.

37. Lift up the upper intake manifold and unplug the throttle body connector. Lift away the intake manifold.

38. Remove the gasket and cover the inlet ducts.

39. Unplug the connectors from the camshaft position sensor and from the camshaft solenoid valve.

40. Release the hook and lift up the wiring harness holder slightly.

41. Release the right-hand ground cable from the wiring harness. Carefully cut the tape and cut off the cable tie.

42. Unplug the connector, detach the pipe from the fuel bleeder valve and lift away the valve. Detach the vacuum pipe from the venturi valve.

43. Unplug the lower oxygen sensor connector and the connectors for ignition coils 1, 3 and 5. Remove ignition coils 1, 3 and 5.

44. Remove the rear cable duct from the camshaft cover. Use an angle screw driver to release the 3 clips from the cable duct. Press the cable duct back toward the bulkhead.

45. Remove the camshaft cover bolts and rubber bushings. Raise the wiring harness and power steering fluid reservoir. Remove the camshaft cover.

46. Remove the camshaft cover gasket and seals around the spark plug holes.

47. Remove the spark plugs.

48. Remove the camshaft solenoid valve.

49. Remove the camshaft position sensor.

50. Rotate the crankshaft so that the markings on the camshaft sprockets are visible and the indentations align with the cylinder head. Use EN-48383-3 Adjustment Tool as a gauge (the camshafts are now in their neutral position).

51. Raise the vehicle.

52. Remove the starter motor.

53. Install EN-46106 Flywheel locking attachment.

54. Lower the vehicle.

➡ **Store all removed valve parts in the Valve Stand 83 93 787.**

55. Mark the chain and camshaft sprockets so they can be reinstalled in the same position.

56. Slightly loosen the torque on the bolts holding the camshaft sprockets. Counterhold using an open spanner on the camshaft flats (do not remove the bolts).

57. Install EN-48313 Timing Chain Holding Tool to the timing chain. Check that the tool grips the chain securely and that the lower groove of the tool aligns with the top of the cylinder head (the timing cover is removed in the illustration to better depict tool installing).

58. Tighten EN-48313 Fixing Tool manually so that the chain is unloaded. Counterhold using an open spanner so that the tool does not turn.

**※※ WARNING**

**It is vital that EN-48313 is properly installed. If tension is not relieved from the chain when the camshaft sprocket is removed, the chain tensioner will trigger automatically. To then adjust the chain tensioner, the timing cover must be removed.**

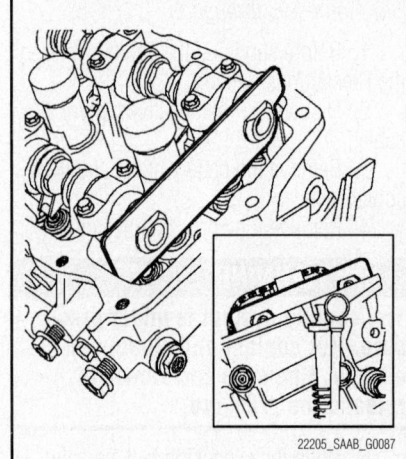

**Fig. 95 Use EN-46105-1 adjustment tool as a gauge to determine if the camshafts are now in their neutral position—2006–07 2.8L front**

59. Remove the camshaft sprocket bolts.

60. Remove the sprockets from the camshafts.

61. Remove the camshaft bearing caps by loosening the bolts in stages - 1/2 to 1 turn. It is important that the removal ends at the bearing cap where the tappets are loaded. Note the markings on the bearing caps when reinstalling to ensure that they are in the correct locations.

62. Lift out the camshafts and note the rear marking on the camshafts. The last letters in the marking: LI = intake camshaft, front cylinder bank. LE = exhaust camshaft, front cylinder bank.

***To install:***

63. Position the camshafts and use EN-48383-3 Adjustment Tool to check that the camshafts are mounted at the correct angle. The last letters in the marking: LI = intake camshaft, front cylinder bank. LE = exhaust camshaft, front cylinder bank.

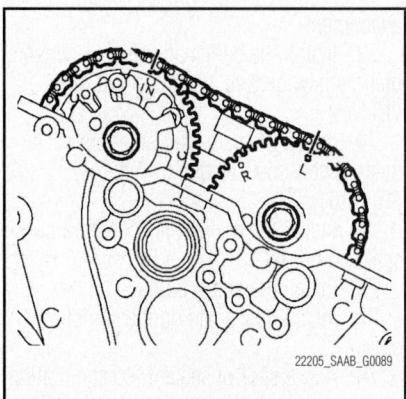

**Fig. 96 Mark the chain and camshaft sprockets so they can be refitted in the same position—2008 2.8L front**

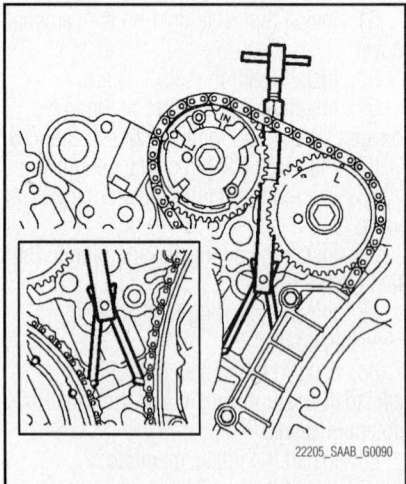

**Fig. 97 Timing chain holding tool—2008 2.8L**

64. Install the bearing caps as illustrated (markings on caps). Start where the cams point down and load the tappets. Tighten the bolts in stages - 1/2 to 1 turn at a time to 7 ft. lbs. (10 Nm).

65. Install the sprockets with chain on the camshafts. Tighten the bolts by hand.

66. Check that the marks on the chain match the markings on the sprockets.

67. Remove EN-48313 Fixing Tool.

68. Tighten the bolts of both camshaft sprockets (B), counterholding with an open spanner on the camshaft flats. Tighten to: 48 ft. lbs. (65 Nm)

69. Install the camshaft position sensor using a new gasket. Tighten to: 7 ft. lbs. (10 Nm).

70. Install the camshaft solenoid valve using a new seal. Tighten to 7 ft. lbs. (10 Nm).

71. Install the spark plugs.

72. Raise the vehicle.

73. Remove EN-46106 Flywheel Locking Attachment.

74. Install the Starter motor.

75. Lower the vehicle.

76. Clean the sealing surfaces of the camshaft cover.

77. Install a new camshaft cover gasket and new seals for the camshaft cover. Use J-5590 Drift and J-24254-A Installing Tool.

78. Add a dot of 93 160 951 Sealant on each mating face between the cylinder head and the timing cover.

79. Position the camshaft cover.

80. Install the bolts with new rubber bushings. Tighten to 8 ft. lbs. (10 Nm).

81. Remove the clips from the camshaft cover and install them in the grooves on the cable duct.

82. Install the cable duct on the camshaft cover.

83. Install ignition coils 1, 3 and 5.

84. Plug in the lower oxygen sensor connector and ignition coils 1, 3 and 5.

85. Install the vacuum pipe to the venturi valve.

86. Plug in the connector and attach the pipe to the fuel bleeder valve (A).

87. Secure the ground cable to the wiring harness and install the wiring harness to the camshaft cover. Plug in the connectors for the camshaft position sensor and the camshaft solenoid valve.

88. Plug in the connectors for the camshaft position sensor and for the camshaft solenoid valve and install the holder for the camshaft cover.

89. Clean the sealing surfaces and install the intake manifold upper section in place with new seals and plug in the throttle body connector.

90. Install the intake manifold bolts. Tighten to 17 ft. lbs. (23 Nm).

91. Install the fuel line.

92. Plug in the fuel pressure sensor connector.

93. Install the crankcase ventilation to the intake manifold. Tighten to 8 ft. lbs. (10 Nm).

94. Install the brake vacuum hose to the intake manifold.

95. Plug in the injector connectors and install the bracket. Tighten to 8 ft. lbs. (10 Nm).

96. Reposition the corner bracket and install the bolts.

97. Plug in the atmospheric pressure sensor connector.

98. Install the cable duct to the intake manifold.

99. Plug in the engine control module connectors.

100. Install the engine control module's ground cables, one on each side.

101. Secure the ground cable to the wiring harness.

102. Plug in the connectors for the camshaft position sensor and for the camshaft solenoid valve and install the holder for the camshaft cover.

103. Attach the wiring harness to the camshaft cover using new cable ties.

104. Plug in the coolant temperature sensor connector and press down the heat shield.

105. Install the ignition coils for cylinders 2-4-6. Plug in the connectors. Tighten to 8 ft. lbs. (10 Nm).

106. Install the left-hand grounding point.

107. Install the power steering fluid reservoir's bolt. Tighten to 8 ft. lbs. (10 Nm).

108. Install the Secondary air injection pump.

109. Plug in the connector and install both of the secondary air injection pump's pipes.

110. Attach the crankcase ventilation line to the turbocharger. Tighten to 8 ft. lbs. (10 Nm).

111. Install the turbocharger wiring harness. Plug in the connectors and install the cable clip. Install the cable tie.

112. Install the coolant return line. Install the clip and the bolts. Tighten to 8 ft. lbs. (10 Nm).

113. Install the charge air pipe. Install the clips. Tighten to 2.5 ft. lbs. (3.5 Nm).

114. Plug in the boost pressure sensor connector. Secure the power steering line in the clips.

115. Install the intake hose and the mass air flow sensor to the turbocharger. Tighten to 2.5 ft. lbs. (3.5 Nm).

116. Plug in the mass air flow sensor connector and install the hose for the turbocharger intake manifold and install the crankcase ventilation line to the dipstick tube.

117. Install the turbocharger intake manifold between the air filter and the mass air flow sensor, and the secondary air pipe and wiring harness to the clip. Tighten clip to 2.5 ft. lbs. (3.5 Nm). Tighten bolt to 1.1 ft. lbs. (1.5 Nm).

118. Install the top section of the air filter.

119. Replace the upper engine cover.

120. Connect the negative cable and install the coolant pipe and the battery cover. Tighten to 2.5 ft. lbs. (3.5 Nm).

121. Check the coolant level.

122. Connect the negative battery cable.

## CRANKSHAFT FRONT SEAL

### REMOVAL & INSTALLATION

#### 2.0L H4 Engine

Refer to "Engine Mechanical Components, Timing Belt and Sprockets, Removal & Installation"

#### 2.0L I4 Engine

*See Figures 98 and 99.*

1. Raise the vehicle, remove the right front wheel.

2. Remove the right wheel well.

3. Mark the belt direction of rotation. Relieve the belt tensioner using 83 96 095 Drivebelt Relieving Tool and remove the poly-V-belt.

4. Remove the crankshaft pulley using 83 95 360 and 83 96 210 Holding Tool. Fold up the plate and carefully hold the engine

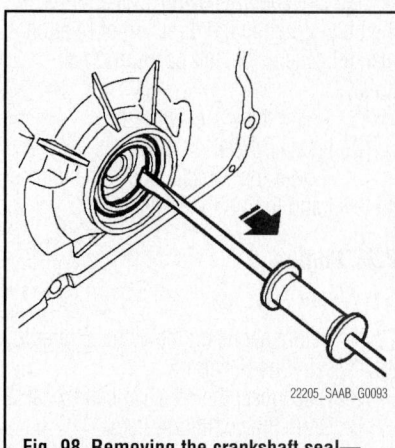

22205_SAAB_G0093

**Fig. 98 Removing the crankshaft seal— 2.0L I4 engine**

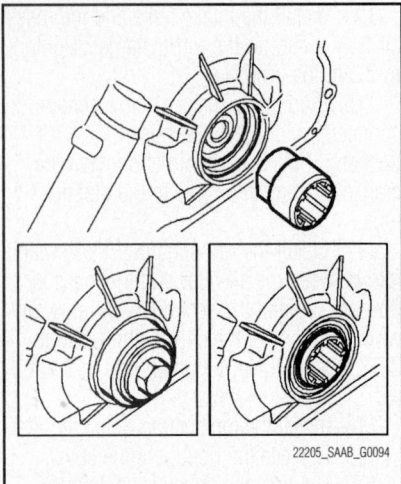

**Fig. 99 Installing the crankshaft seal— 2.0L I4 engine**

down slightly with a bracket to gain access to the bolt.

5. Remove the gasket using a 87 91 360 Extractor

### To install:

6. Clean the sealing surfaces.

7. Position the 83 96 202 Front Crankshaft Seal Installing Tool protective collar on the crankshaft. Lubricate the new sealing ring with non-acidic Vaseline and position it on the Installing Tool.

8. Position the Installing Tool on the crankshaft. Screw in the sealing ring using the crankshaft pulley bolt so that it is flush with the timing cover. Remove the tool.

9. Install the crankshaft pulley with a new bolt. Use 83 95 360 and 83 96 210Holding Tools. Hold the engine down slightly with a bracket to gain access to the bolt. Bend the plate back and spray Teroson Terotex HV 200 Extra T108 to prevent rust damage. Tighten to 74 ft. lbs. (100 Nm) plus 75° rotation.

10. Relieve the belt tensioner using 83 96 095 Drivebelt Relieving Tool and Install the poly-V-belt in the direction of rotation marked. Check the belt position on all pulleys.

11. Install the wheel well.

12. Install Wheel.

13. Lower the vehicle, check the engine oil level and top up if necessary.

### 2.5L Engine

*See Figure 100.*

1. Before servicing the vehicle, refer to the Precautions Section.

2. Disconnect the negative battery cable.

3. Drain the cooling system.

4. On the DOHC engine, remove the collector cover.

5. Raise and support the vehicle safely.

6. Remove the undercover.

7. On the DOHC engine, remove the bolts which retain the water pipe of the oil cooler to the oil pump. Remove the water pipe and hoses between the oil cooler and the water pump.

8. Lower the vehicle. Remove the radiator.

9. To remove the front side V belt, remove the belt covers. Loosen the lock bolt. Loosen the slider bolt. Remove the front side belt.

10. To remove the rear side V belt, remove the belt covers. Loosen the lock bolt. Loosen the slider bolt. Remove the rear side belt.

11. On the SOHC engine remove the belt tensioner.

12. On the DOHC engine remove the rear side V belt tensioner

13. Remove the crankshaft position sensor.

14. Remove the crankshaft pulley bolt.

15. Lock the crankshaft in place using tool ST499977100 for the SOHC engine and tool ST499207400 for the DOHC engine, or equivalent.

16. Remove the crankshaft pulley.

17. Remove the water pump.

18. If equipped, remove the timing belt guide. Remove the crankshaft sprocket.

19. Remove the oil pump retaining bolts.

➡**When disassembling and checking the oil pump, loosen the relief valve plug before removing the oil pump from its mounting.**

20. Using a flat tip tool remove the oil pump from the engine.

21. Carefully remove the front oil seal.

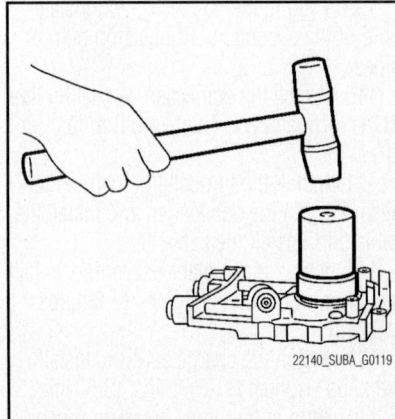

**Fig. 100 Front seal installation using ST499587100—2.5L engine**

### To install:

22. Replace the front oil seal with a new part using ST499587100.

23. Be sure all mating surfaces are clean and free of dirt.

24. Apply liquid gasket part number 004403007, or equivalent to the mating surfaces of the oil pump.

25. Be sure to replace the O-ring with a new one.

26. Apply a thin coat of clean engine oil to the inside of the oil seal.

27. Position the oil pump to its mounting, aligning the notched area with the crankshaft and push the pump straight.

➡**Be sure that the oil seal lip is not folded.**

28. Install the oil pump. Apply liquid gasket part number 004403042, or equivalent to the three retaining bolt threads. Install the bolts and tighten to 4.7 ft. lbs.

29. Continue the installation in the reverse order of the removal procedure.

30. Be sure to fill the cooling system with the proper grade and type coolant.

31. Start the engine and check for leaks. Correct, as required.

### 2.3L Engine

*See Figure 101.*

1. Before servicing the vehicle, refer to the Precautions Section.

2. Raise and safely support the vehicle securely on jackstands.

3. Remove the right-hand front wheel and detach the power steering pump from the subframe.

4. Remove the cover for the drivebelt. Install a wedge between the oil sump and the subframe so that it can be driven in to raise the engine for installing.

5. Remove the right-hand engine mounting with yoke.

6. Take the strain off the belt tensioner with the aid of a ratchet handle extension and insert a 3 mm hexagon key in the hole.

### ✳✳ WARNING

**Exercise the utmost care to ensure that the belt tensioner does not break at its end position!**

7. Remove the flywheel cover plate and install 83 94 868 Flywheel Locking Tool.

8. Remove the crankshaft pulley.

9. Use a screwdriver to pry out the old seal, taking care not to damage the sealing surface.

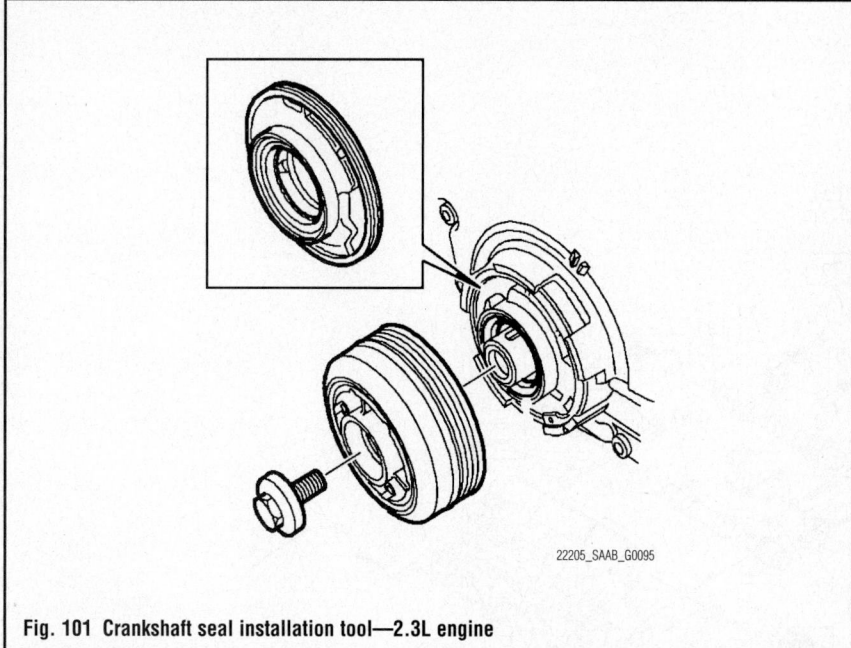

22205_SAAB_G0095

**Fig. 101 Crankshaft seal installation tool—2.3L engine**

### To install:

10. Install a new seal using 83 94 876 Crankshaft Seal Installing Tool.

11. Install the crankshaft pulley. Tighten to 130 ft. lbs. (175 Nm).

12. Remove the flywheel locking segment and install the flywheel cover plate.

13. Drive in the wedge to raise the engine.

14. Install the drive belt, check its position on all the pulleys and remove the strain-relieving hexagon key.

15. Install the RH engine mounting. Tighten to 37 ft. lbs. (50 Nm).

16. Remove the wedge and install the power steering pump pipe.

17. Install the drivebelt cover and the wheel. Tighten to 80 ft. lbs. (110 Nm).

18. Start the engine and run it at idling speed for a while.

19. Switch off the engine and check that the belt is correctly positioned.

### 2.8L Engine

1. Before servicing the vehicle, refer to the Precautions Section.

2. Disconnect the negative battery cable.

3. Raise and safely support the vehicle.

4. Remove the right front wheel.

5. Remove the right wheel well.

6. Mark the belt's direction of rotation. Unload the belt tensioner. Use a ½ inch puller and remove the poly-V-belt from the crankshaft pulley.

7. Remove the bolt holding the crankshaft pulley using 83 95 360 Holding tool, crankshaft pulley (handle only) and EN-47981 Holding tool, pulley.

➡**The pulley is pressed onto the crankshaft.**

8. Remove the pulley using EN-47982 Puller and EN-47981 Holding tool, pulley.

9. Remove the front crankshaft seal using 87 91 360 Extractor.

### To install:

10. Clean the sealing surfaces.

11. Lubricate the new seal with non-acidic Vaseline.

12. Install the seal using EN-47635 Installing tool and a plastic mallet.

13. Clean the sealing surface of the pulley.

14. Press the pulley onto the crankshaft using KM-J 41998 Installing tool.

15. Install the bolt using 83 95 360 Holding tool, crankshaft pulley (handle only) and EN-47981 Holding tool, pulley. Tighten the crankshaft pulley bolt to 74 ft. lbs. (100 Nm), plus an additional 150°.

16. Install the drive belt.

17. Install the right wheel well.

18. Install the right front wheel.

19. Lower the vehicle.

20. Connect the negative battery cable.

## CYLINDER HEAD

### REMOVAL & INSTALLATION

### ✸✸ CAUTION

**The fuel system pressure must be relieved before disconnecting any fuel lines. Failure to do so may result in personal injury.**

### 2.0L H4 Engine

*See Figures 102 through 105.*

1. Remove or disconnect the following:
   - Negative battery cable
   - Drive belt
   - Crankshaft pulley
   - Belt cover
   - Timing belt assembly
   - Camshaft sprocket
   - Intake manifold
   - Bolt that attaches the A/C compressor bracket to the head
   - Camshaft
   - Cylinder head bolts in the proper sequence. Leave bolts A and D installed loosely to prevent the cylinder head from falling.
   - Cylinder head from the block using a plastic-faced hammer, if needed, to separate the head from the cylinder block
   - Bolts A and D
   - Cylinder head and gasket

2. Clean all gasket material from both mating surfaces.

### To install:

3. Inspect the cylinder head for warpage. Warpage should not exceed 0.0020 in. (0.05mm).

4. Install a new head gasket and the cylinder head.

5. Secure the head in place with the mounting bolts. Coat each bolts with clean engine oil, and hand-tighten. Tighten the cylinder head bolts, in sequence, to the following specifications:
   a. Step 1: 22 ft. lbs. (29 Nm).
   b. Step 2: 51 ft. lbs. (69 Nm).
   c. Step 3: loosen all bolts by 180°, then loosen an additional 180°.
   d. Step 4: 31 ft. lbs. (42 Nm).
   e. Step 5: all bolts plus 80–90°.
   f. Step 6: center bolts plus 40–45°.

### ✸✸ WARNING

**Do not exceed 90° total tightening on the last two steps.**

6. Install or connect the following:
   - Camshaft
   - Bolt that attaches the A/C compressor bracket to the head
   - Camshaft sprocket
   - Intake manifold
   - Timing belt assembly
   - Belt cover
   - Crankshaft pulley
   - Drive belt
   - Negative battery cable

7. Start the engine and allow it to reach operating temperature. Check for leaks.

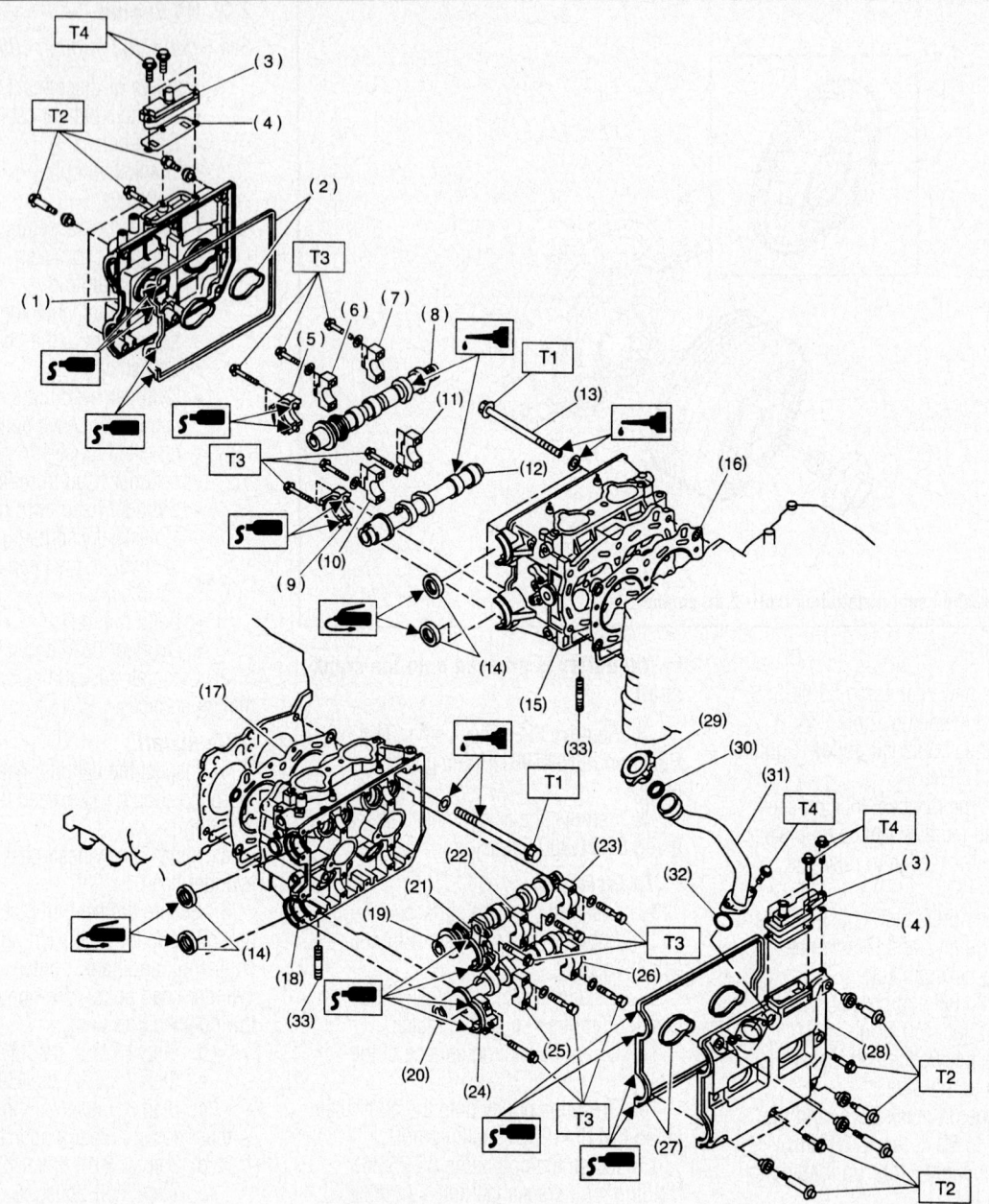

| | | | | | | |
|---|---|---|---|---|---|---|
| (1) | Rocker cover (RH) | (15) | Cylinder head (RH) | (29) | Oil filler cap | |
| (2) | Rocker cover gasket (RH) | (16) | Cylinder head gasket (RH) | (30) | Gasket | |
| (3) | Oil separator cover | (17) | Cylinder head gasket (LH) | (31) | Oil filler duct | |
| (4) | Gasket | (18) | Cylinder head (LH) | (32) | O-ring | |
| (5) | Intake camshaft cap (Front RH) | (19) | Intake camshaft (LH) | (33) | Stud bolt | |
| (6) | Intake camshaft cap (Center RH) | (20) | Exhaust camshaft (LH) | | | |
| (7) | Intake camshaft cap (Rear RH) | (21) | Intake camshaft cap (Front LH) | | | |
| (8) | Intake camshaft (RH) | (22) | Intake camshaft cap (Center LH) | | | |
| (9) | Exhaust camshaft cap (Front RH) | (23) | Intake camshaft cap (Rear LH) | | | |
| (10) | Exhaust camshaft cap (Center RH) | (24) | Exhaust camshaft (Front LH) | | | |
| (11) | Exhaust camshaft cap (Rear RH) | (25) | Exhaust camshaft cap (Center LH) | | | |
| (12) | Exhaust camshaft (RH) | (26) | Exhaust camshaft cap (Rear LH) | | | |
| (13) | Cylinder head bolt | (27) | Rocker cover gasket (LH) | | | |
| (14) | Oil seal | (28) | Rocker cover (LH) | | | |

**Tightening torque: N·m (kgf-m, ft-lb)**

T1: *<Ref. to ME(DOHC TURBO)-64, INSTALLATION, Cylinder Head Assembly.>*

T2: *5 (0.5, 3.6)*

T3: *10 (1.0, 7)*

T4: *6.4 (0.65, 4.7)*

9357TG05

**Fig. 102 Cylinder head and related components—2.0L H4 engine**

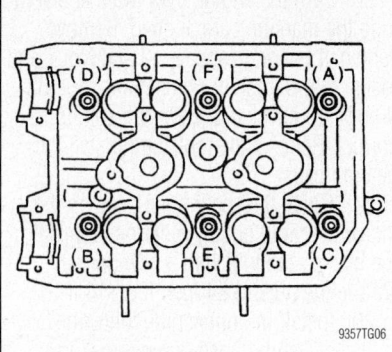

**Fig. 103 Cylinder head bolt loosening sequence (except bolts A and D which are left in place at this time)—2.0L H4 engine**

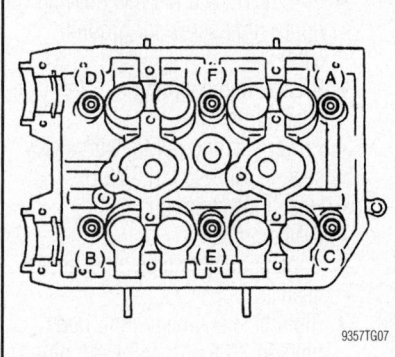

**Fig. 104 Tap on the block with a rubber mallet prior to removing cylinder head bolts A and D—2.0L H4 engine**

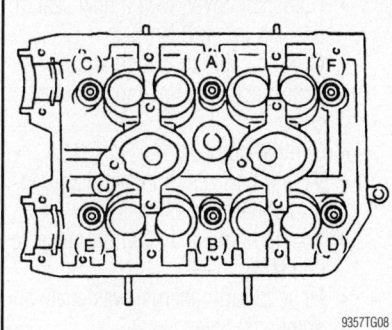

**Fig. 105 Cylinder head bolt tightening sequence—2.0L engine**

## 2.0L I4 Engine

*See Figures 106 through 109.*

1. Before servicing the vehicle, refer to the Precautions Section.

2. Remove or disconnect the following:
- Cap from expansion tank
- Oxygen sensor connector
- Temperature sensor connector
- Heat shield over turbocharger
- Turbocharger oil delivery pipe
- Upper oxygen sensor
- Catalytic converter connection from turbocharger
- Right front wheel and wheel well
- Lower spoiler shield
- Hose and connector from headlamp washers
- Drain the cooling system
- Lower turbocharger pressure pipe
- Front exhaust pipe from muffler (cut about 3.5 inches (87 mm) from the end of the muffler)
- Front pipe from catalytic converter
- Steering gear heat shield
- Catalytic converter heat shield
- Temperature sensor
- Lower oxygen sensor
- Bracket from catalytic converter and block
- Catalytic converter
- Drain remaining coolant
- Turbocharger oil return pipe
- Stay from intake manifold and cylinder block
- Starter positive cable
- Power steering pipe clamps on subframe
- Rear torsion bar bolts (loosen slightly)

3. Install an Engine Centering Tool Kit, 83 96 152, to subframe. Adjust the centering tool bolts so adjuster screw slots are completely visible.

4. Lower the vehicle and remove the upper engine cover.

5. Remove or disconnect the following:
- Turbocharger delivery hose from throttle body and pipe
- Turbocharger inlet pipe (plug openings)
- MAF sensor connector
- Air cleaner assembly

6. Release any fuel pressure by carefully pressing the service valve needle. Collect spilled fuel.

7. Detach both fuel lines for the fuel rail while gripping the lower nut. Plug the fuel line openings.

8. Remove or disconnect the following:
- Quick-release coupling for ventilation line
- Fuel lines and vent lines from camshaft cover (bend up hose and out of the way)
- Ignition coil cover
- Ignition coil connectors
- Turbocharger solenoid connector
- Connectors for A/C pressure sensor connector, coolant temperature sensor, engine control module, bypass solenoid valve, throttle body, MAP sensor, atmospheric pressure sensor, IDM module
- Ground connection on ECM
- Hose from turbocharger wastegate and solenoid valve
- Wiring harness duct and bracket from camshaft cover (bend wiring harness aside)
- Wiring harness from oil filter housing and inlet pipe
- Ignition coils
- Crankcase vent hose from camshaft cover
- Ground lead from camshaft cover
- Camshaft cover

9. Zero the engine by turning the crankshaft clockwise until the marking on the crankshaft pulley with aligned with the marking on the timing cover. The No. 1 cylinder cams on the intake and exhaust camshafts should be pointing up and inward.

10. Remove or disconnect the following:
- Coolant hose from cylinder head
- Right engine mount
- IDM module
- Couplings from vacuum pump
- Power steering pump
- EVAP canister purge valve and bracket
- Throttle body
- ECM
- Fuel injector connectors
- Dipstick tube bolt
- Vacuum hose and purge hose from intake manifold
- Vacuum hose from fuel pressure regulator
- Intake manifold
- Timing chain tensioner
- Camshaft pinions (hold camshaft flats with wrench)

11. Install a cable tie to the exhaust camshaft pinion and chain. Remove the camshaft pinions and lower the exhaust camshaft pinion with the chain.

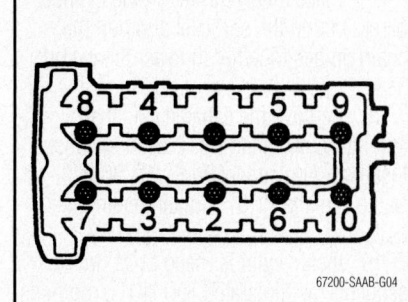

**Fig. 106 Remove the cylinder head bolts in sequence—2.0L engine**

26. Turn the engine over twice and zero it so the markings are aligned. Remove camshaft bearing caps No. 2 and No. 7 and install that Camshaft Adjustment Tool, 83 96 046, to check that camshaft setting is correct. Remove the tools and install the bearing caps.

27. Clean the sealing surfaces on the intake manifold and cylinder head. Install the intake manifold with a new gasket. Torque the nuts to 84 inch lbs. (10 Nm).

28. Install or connect the following:
- Vacuum hose to fuel pressure regulator
- Dipstick tube bolt
- Vacuum hose and purge hose to intake manifold
- Fuel injector connectors
- Throttle body and ECM (leaving upper right screw for ground cable)
- Bracket and coupling to EVAP canister purge valve
- Power steering pump, with a new seal, to 16 ft. lbs. (22 Nm)
- Hose to vacuum pump
- IDM module
- Coolant hose to pipe on cylinder head
- Right side engine mount: body bolts to 30 ft. lbs. (40 Nm), plus an additional 60° and bracket bolts to 52 ft. lbs. (70 Nm), plus an additional 60°
- Remaining coolant hose to cylinder head pipe
- Camshaft cover, with a new seal, to 84 inch lbs. (10 Nm)
- Ground cable to camshaft cover
- Crankcase vent hose to camshaft cover
- Ignition coils
- Wiring harness to intake manifold and oil filter housing
- Wiring harness duct and bracket to camshaft cover
- Hose to turbocharger wastegate and solenoid valve
- Connectors for A/C pressure sensor, coolant temperature sensor, ECM, wastegate solenoid, throttle body, MAP sensor, barometric pressure sensor, IDM module
- Ground cable to ECM
- Ignition module connectors (ensure cables are not pinched by cover)
- Turbocharger solenoid valve connectors with cable clips
- Cover over ignition coils
- Fuel lines and vent lines into clips on camshaft cover
- Coupling for vent line

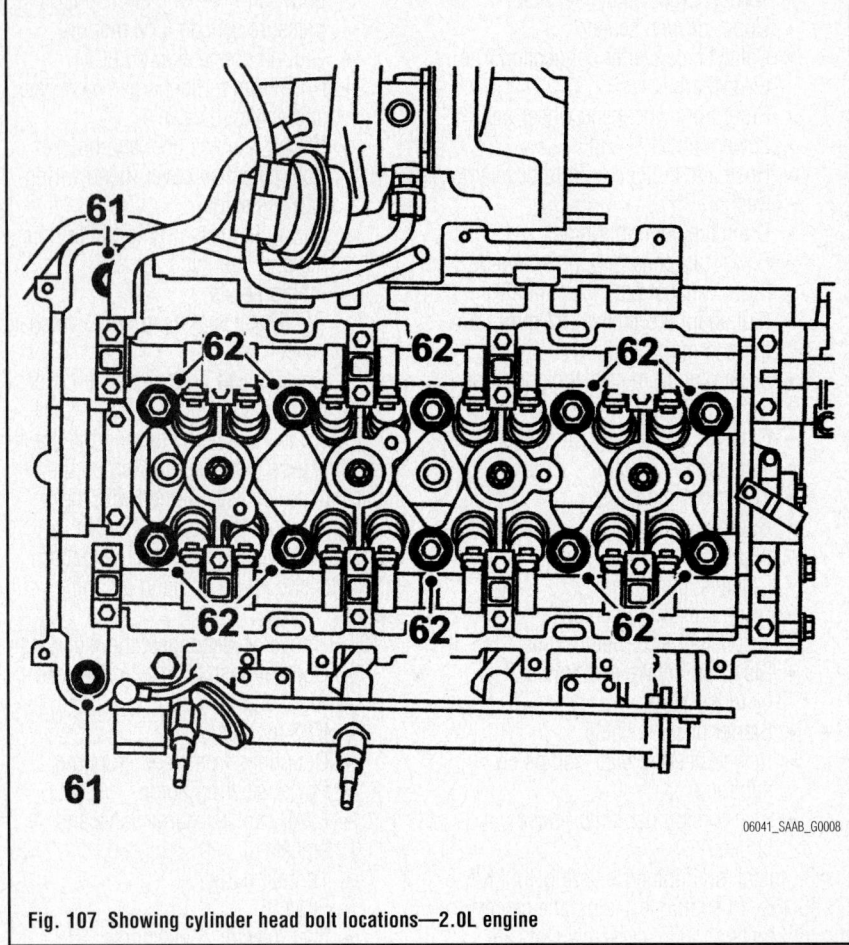

06041_SAAB_G0008

**Fig. 107 Showing cylinder head bolt locations—2.0L engine**

12. Remove the cylinder head plug.

13. Remove the bolts (61) for the chain guide on the intake side. Move the guide away and install the bolt again.

14. Remove the cylinder head. Remove the cylinder head bolts in an alternating pattern.

### To install:

15. Thoroughly clean all mating surfaces. Blow out bolt holes. Check for any warping or damage.

16. Check that marks on the pulley and timing cover are aligned.

17. Place a new gasket on the cylinder block. Lift up the sprocket and pull the chain guides together to lower the cylinder head.

18. Remove the exhaust camshaft and intake camshaft No. 2 bearing caps. Install an Adjustment Tools, 83 96 046, to the camshafts (5). Tighten retaining bolts.

19. Install cylinder head bolts, in sequence as illustrated. DO NOT drop bolts into place or threads will be damaged. Tighten cylinder head bolts in 3 steps as follows:

a. Step 1: 22 ft. lbs. (30 Nm)

b. Step 2: Each bolt, in sequence, an additional 150°

c. Step 3: Each bolt, in sequence, an additional 15°

d. Install and tighten timing section bolts to 26 ft. lbs. (35 Nm).

20. Remove the bolt for the chain guide. Put the guide in place an install and tighten the bolt.

21. Install the plug on the front of the cylinder head, with a new sealing washer. Tighten the plug to 18 ft. lbs. (25 Nm).

22. Install the pinion on the exhaust camshaft and intake camshaft without tightening. Verify that the engine is still "zeroed".

23. Install the chain tensioner reset with a new seal. The groove on the end of the chain tensioner should be vertical when installed. Torque tensioner reset to 55 ft. lbs. (75 Nm).

24. Torque the camshaft pinions while gripping the respective camshaft with a wrench on the flats. Torque to 63 ft. lbs. (85 Nm), plus an additional 30°. Remove the cable tie.

25. Remove the Camshaft Adjustment Tool and install the camshaft No. 2 bearing caps. Torque the bolts to 6 ft. lbs. (8 Nm).

- Both fuel lines, with new seals, to fuel rail while gripping lower nut to 84 inch lbs. (10 Nm)
- Air cleaner assembly

### ✳✳ WARNING

**To prevent hoses on delivery side of turbocharger from coming loose under pressure, thoroughly clean all mating areas on inside of hoses and pipes before attaching.**

- Air cleaner cover
- MAF sensor connector
- Turbocharger inlet pipe to air cleaner
- Hoses to inlet pipe
- Turbocharger hose to throttle body and turbo delivery pipe

29. Raise the vehicle and remove the engine centering tool from the subframe.

30. Torque the rear torsion arm bolts to 52 ft. lbs. (70 Nm), plus an additional 90°.

31. Install or connect the following:
- Bolts on power steering pipe clamps on subframe
- Positive cable to starter

- Stay between intake manifold and cylinder block to 16 ft. lbs. (22 Nm)
- Oil return line to turbocharger with new gasket and O-ring; torque to 11 ft. lbs. (15 Nm)
- Stay between exhaust manifold and cylinder block to 16 ft. lbs. (22 Nm)
- New seals on turbocharger coolant pipe; torque to 30 ft. lbs. (40 Nm)
- Catalytic converter and bracket to studs on engine block and to turbocharger connection
- Catalytic converter bracket bolts into engine block and new nuts on studs

32. Lower the vehicle and install the catalytic converter with new nuts so it is evenly positioned to the turbocharger. Use thread paste on the studs.

33. Raise the vehicle and tighten the catalytic converter bracket nuts and bolts to 18 ft. lbs. (25 Nm).

34. Sparingly apply thread paste to the lower oxygen sensor threads and install it.

Be sure wiring is not twisted or damaged. Torque oxygen sensor to 30 ft. lbs. (40 Nm).

35. Install or connect the following:
- Catalytic converter heat shield
- Steering gear heat shield
- Front exhaust pipe to catalytic converter with new gasket and nuts torque to 18 ft. lbs. (25 Nm)
- Joining clamp centered on exhaust pipe at cut; torque nuts to 30 ft. lbs. (40 Nm)
- Lower turbocharger delivery pipe
- Headlamp washer hose and connector
- Lower spoiler shield
- Chassis reinforcement front support frame, if equipped
- Right front wheel well and front wheel
- Upper oxygen sensor, with thread paste sparingly, torque to 30 ft. lbs. (40 Nm)
- Turbocharger oil delivery pipe, with new seals, to 16 ft. lbs. (22 Nm)
- Turbocharger heat shield

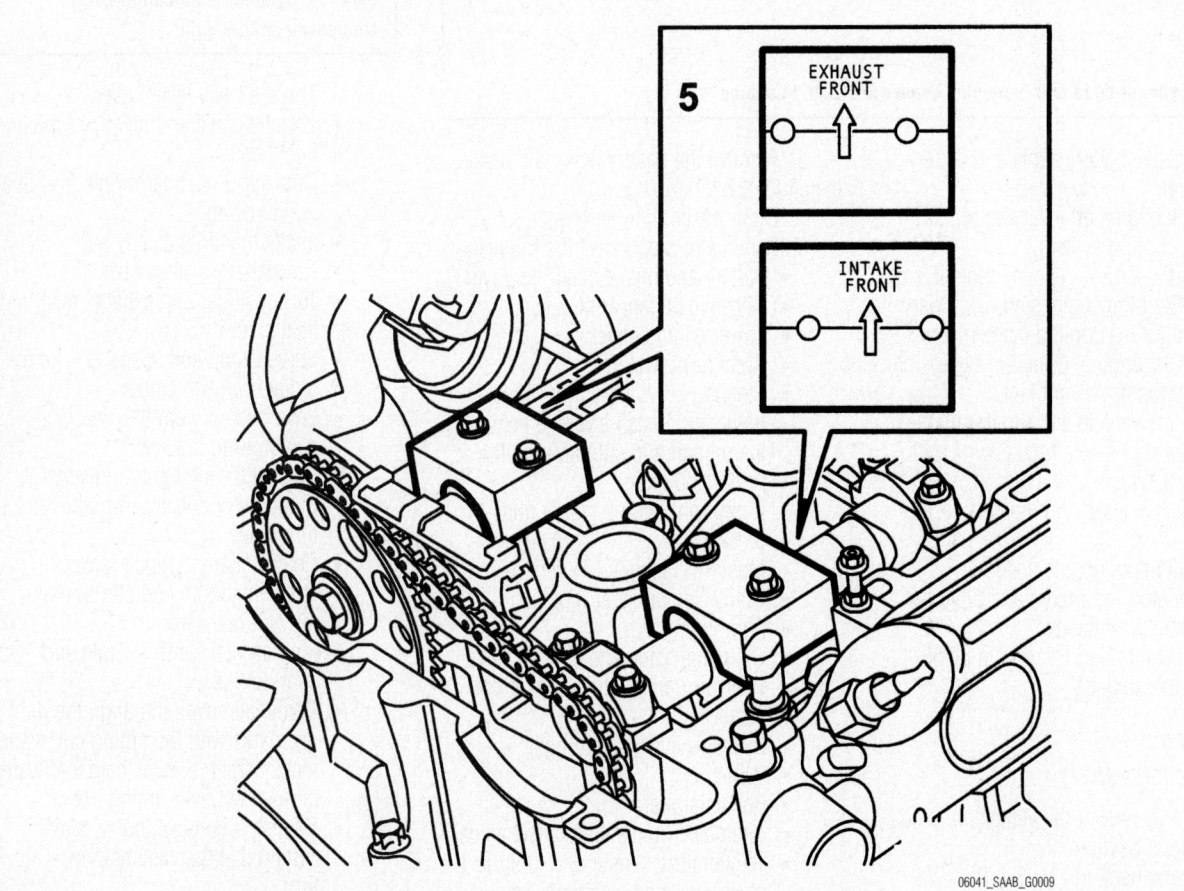

06041_SAAB_G0009

**Fig. 108 Using camshaft adjusting tools—2.0L I4 engine**

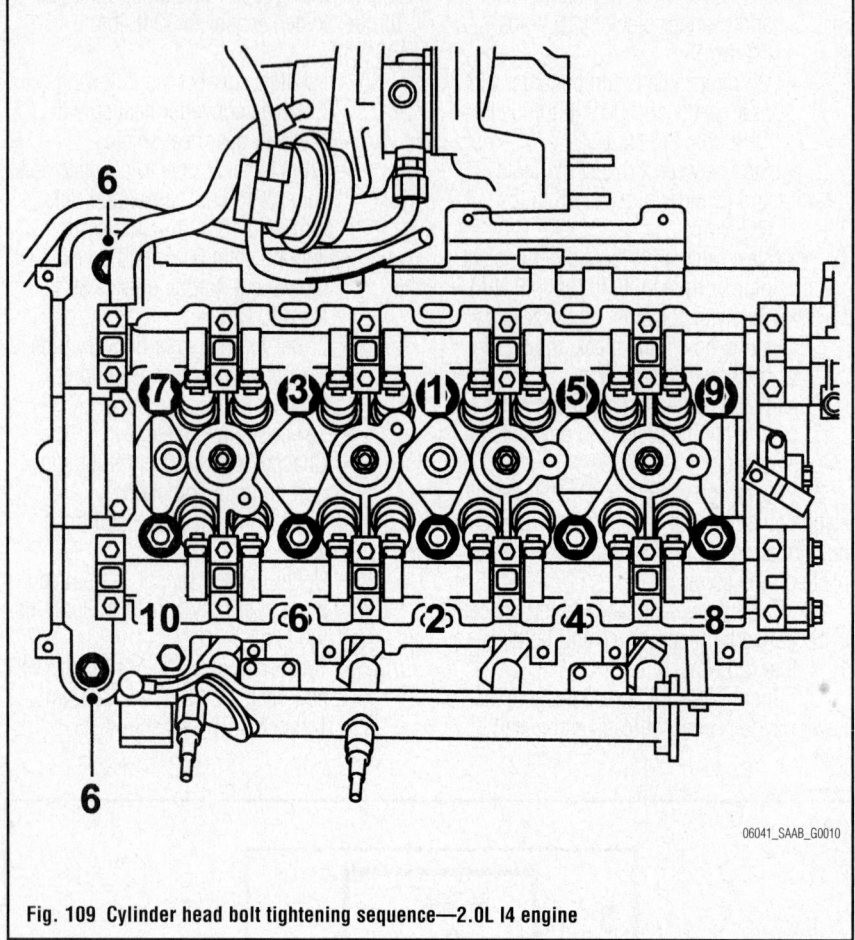

Fig. 109 Cylinder head bolt tightening sequence—2.0L I4 engine

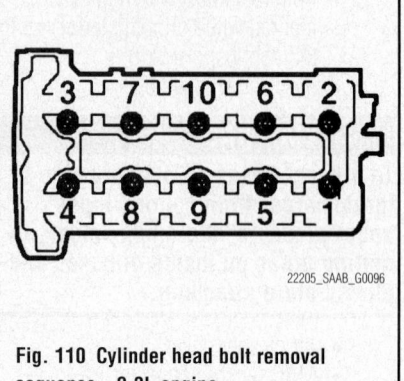

Fig. 110 Cylinder head bolt removal sequence—2.3L engine

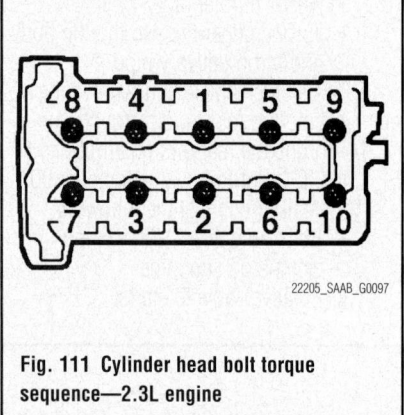

Fig. 111 Cylinder head bolt torque sequence—2.3L engine

• Negative battery cable and battery cover

36. Check engine oil level and top as needed. Fill cooling system.

37. With the A/C off, ensure coolant is at MAX level. Start the engine and run at varying speeds until the cooling fan comes on.

38. Stop the engine. Open the expansion tank cap and top up the coolant.

39. Start the vehicle and run it until the thermostat opens. Turn off the engine and check coolant level.

40. Check for leaks and repair if necessary.

41. Install upper engine cover.

42. Reset clock and radio and other electrical accessories as needed.

43. Use scan tool to reprogram ECM. Follow tool instructions.

## 2.3L Engine

*See Figures 110 and 111.*

1. Before servicing the vehicle, refer to the Precautions Section.

2. Run engine at idle speed. Remove fuse No 19 from fuse panel. When engine stops, turn off ignition. Re-install the fuse.

3. Remove the battery cover and intake manifold cover.

4. Drain the cooling system.

5. Remove or disconnect the following:
• Battery and intake manifold cover
• Negative battery cable
• Lower engine cover
• Right front wheel
• Steering servo pipe from the subframe and install a wedge between the oil pan and subframe on the right hand side
• Right front engine mount and bracket
• Mass Air Flow (MAF) sensor
• Crankcase breather pipe
• Front lifting eye
• Charge air pipe and bypass valve
• Pressure/temperature sensor connector
• Bypass valve vacuum hose
• V-belt
• Belt tensioner
• Alternator (move the bracket aside)
• Temperature sensor and ignition discharge module connectors
• Coolant hoses from the cylinder head

• Throttle body lever cover
• Throttle cable and dipstick tube
• Fuel hoses
• Crankcase breather nipple from the timing cover
• Intake manifold steady bar
• Turbocharger steady bar
• Turbocharger and exhaust manifold heat shield
• Servo pump and suspend it on the radiator crossmember
• Lower screw from the steering servo pump bracket
• Bracket for the ignition discharge module's connector from the rear lifting eye
• Heat exchanger pipe bracket
• Intake manifold and partition and move it rearward
• Ignition discharge module and spark plugs
• Camshaft cover and align the pulley marks with the timing cover and make certain that the camshafts are in line with their timing marks
• Camshaft sprocket bolts. Make certain that the camshaft does not turn
• Idler sprocket bolt and remove the chain tensioner

- Camshaft sprockets and place a rubber band between the chain guides
- Cylinder head by first removing the bolts from the timing cover

### To install:

6. Clean all mating surfaces.
7. Install or connect the following:
   - Inner gasket for the camshaft cover partition
   - Cylinder head gasket and turn the crankshaft 45° in the rotational direction of the engine to lower pistons
8. Bind chain guides with Strap or rubber band and install the new cylinder head gasket and position cylinder head. Make sure chain is not trapped.
9. Install the cylinder head bolts. Torque the bolts in sequence shown as follows:
   a. 30 ft. lbs. (40 Nm).
   b. 44 ft. lbs. (60 Nm).
   c. Plus an additional 90°.
10. Install or connect the following:
    - 2 bolts between the timing cover and cylinder head and torque the bolts to 16 ft. lbs. (22 Nm). Make certain that the camshafts are aligned with the timing marks and reset the crankshaft to the **0** mark.
    - Camshaft sprocket and chain starting with the exhaust camshaft. Do not tighten the bolts
    - Timing chain tensioner and torque the bolts to 47 ft. lbs. (63 Nm)
    - Timing chain tensioner plug, push rod and spring and torque the bolt to 16 ft. lbs. (22 Nm)
    - Torque the idler pulley and camshaft sprocket bolts to 47 ft. lbs. (63 Nm). Rotate the crankshaft 2 revolutions and make check the settings of the crankshaft pulley and camshafts
    - Camshaft cover after lightly coating the opening with clean oil and torque the bolts, in sequence from the front at the timing chain end and working around, to 11 ft. lbs. (15 Nm)
    - Spark plugs and torque the plugs to 21 ft. lbs. (28 Nm)
    - Ignition discharge module and torque the bolt to 8 ft. lbs. (11 Nm)
    - Intake manifold and intermediate partition. Remove the Straps and torque the bolts to 16 ft. lbs. (22 Nm).
    - Alternator bracket and make certain that the adjuster sleeve is tapped out slightly

- Belt tensioner and tighten the ignition discharge module connector bracket
- Coolant hoses to the cylinder head and install the bolt securing the pipe to the thermostat housing cover
- Ignition discharge module electrical connector
- Temperature sensor electrical connector
- Turbocharger retaining nuts between the exhaust manifold and turbocharger and torque the bolts to 18 ft. lbs. (25 Nm)
- Turbocharger heat shield
- Dipstick tube
- Turbocharger and intake manifold steady bars
- Fuel hoses and rubber protectors
- Crankcase ventilation nipple on the valve cover
- Throttle cable and adjust as necessary
- Ground connections to the cylinder head
- Throttle body cover
- Servo pump and install the V-belt. Torque the pulley bolts to 15 ft. lbs. (20 Nm)
- Lifting eye
- Solenoid valve connector
- Charge air pipe and bypass valve
- Bypass valve vacuum hose and pressure/temperature sensor connector
- MAF sensor
- Right side engine mount and bracket. Torque the bolts to 39 ft. lbs. (47 Nm) and the nuts to 78 ft. lbs. (105 Nm). Remove the wedge and secure the servo pump pipe to the subframe
- Right front wheel
- Crankcase breather pipe and torque the banjo bolt to 18 ft. lbs. (24 Nm)
- Lower engine cover
- Negative battery cable
- Intake manifold and battery cover

11. Fill the cooling system and check all fluid levels.
12. Start the vehicle, check for leaks and repair if necessary.

### 2.5L Engines

*See Figures 112 through 117.*

1. Before servicing the vehicle, refer to the Precautions Section.
2. Disconnect the negative battery cable.
3. If equipped with DOHC engine, remove the collector cover.

4. To remove the front side V belt, remove the belt covers. Loosen the lock bolt. Loosen the slider bolt. Remove the front side belt.
5. To remove the rear side V belt, remove the belt covers. Loosen the lock bolt. Loosen the slider bolt. Remove the rear side belt. Remove the belt tensioner.
6. Remove the crankshaft pulley bolt.
7. Lock the crankshaft in place using tool ST499977100, or equivalent.
8. Remove the crankshaft pulley.
9. Remove the left side timing belt cover.
10. If equipped with DOHC engine, remove the right side timing belt cover.
11. Remove the front timing belt cover.
12. Remove the timing belt.
13. Remove the camshaft position sensor.
14. Remove the camshaft sprockets.

➡**Be sure to lock the camshaft in place using tool ST18231AA010 (SOHC engine) and ST499207400 (DOHC engine), or equivalent.**

15. Remove the intake manifold.
16. If equipped with SOHC engine, remove the bolt that retains the air conditioning compressor bracket to the cylinder head.
17. Remove the rocker cover retaining bolts. Remove the rocker cover.
18. If equipped with SOHC engine, remove the rocker arm assembly.
19. If equipped with SOHC engine remove the camshaft. If equipped with DOHC engine, remove the camshafts.
20. Remove the cylinder head, in the proper sequence. On the DOHC engine leave bolts A and D installed loosely to prevent the cylinder head from falling. On the SOHC engine leave bolts A and C installed loosely to prevent the cylinder head from falling.
21. Loosen the cylinder head from the block using a plastic-faced hammer, if needed.
22. Remove bolts A and D on the DOHC engine and bolts A and C on the SOHC engine. Remove the cylinder head from the engine. Discard the gasket
23. Clean all gasket material from both mating surfaces.

### To install:

24. Installation is the reverse of the removal procedure.
25. Apply a thin coat of clean engine oil to the washers and cylinder head bolts.

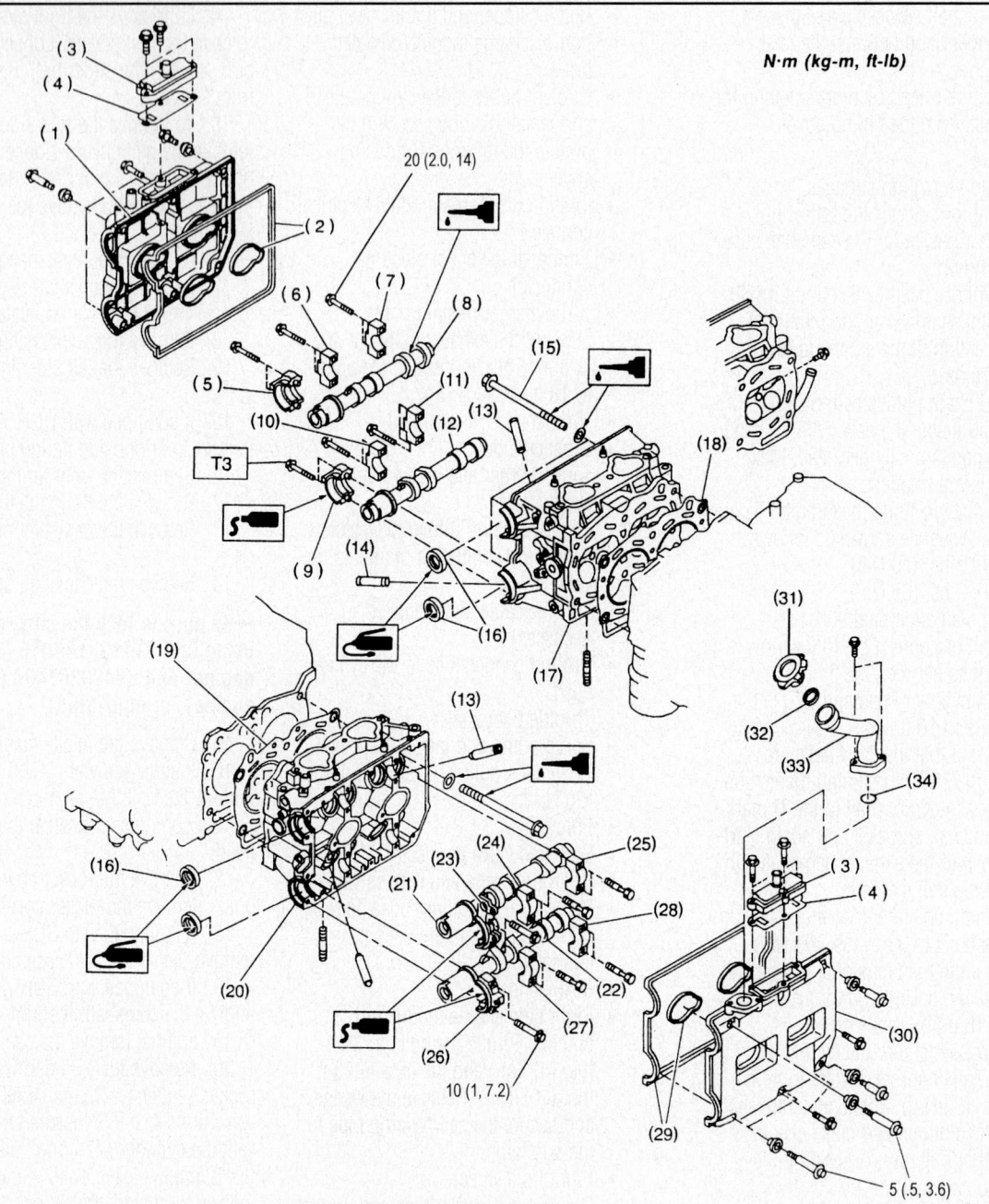

N·m (kg-m, ft-lb)

20 (2.0, 14)

T3

10 (1, 7.2)

5 (.5, 3.6)

(1)  Rocker cover (RH)
(2)  Rocker cover gasket (RH)
(3)  Oil separator cover
(4)  Gasket
(5)  Intake camshaft cap (Front RH)
(6)  Intake camshaft cap (Center RH)
(7)  Intake camshaft cap (Rear RH)
(8)  Intake camshaft (RH)
(9)  Exhaust camshaft cap (Front RH)
(10) Exhaust camshaft cap (Center RH)
(11) Exhaust camshaft cap (Rear RH)
(12) Exhaust camshaft (RH)
(13) Intake valve guide
(14) Exhaust valve guide

(15) Cylinder head bolt
(16) Oil seal
(17) Cylinder head (RH)
(18) Cylinder head gasket (RH)
(19) Cylinder head gasket (LH)
(20) Cylinder head (LH)
(21) Intake camshaft (LH)
(22) Exhaust camshaft (LH)
(23) Intake camshaft cap (Front LH)
(24) Intake camshaft cap (Center LH)
(25) Intake camshaft cap (Rear LH)
(26) Exhaust camshaft (Front LH)
(27) Exhaust camshaft cap (Center LH)
(28) Exhaust camshaft cap (Rear LH)

(29) Rocker cover gasket (LH)
(30) Rocker cover (LH)
(31) Oil filler cap
(32) Gasket
(33) Oil filler duct
(34) O-ring

7923TG07

**Fig. 112 Cylinder head and related components—2.5L DOHC engine**

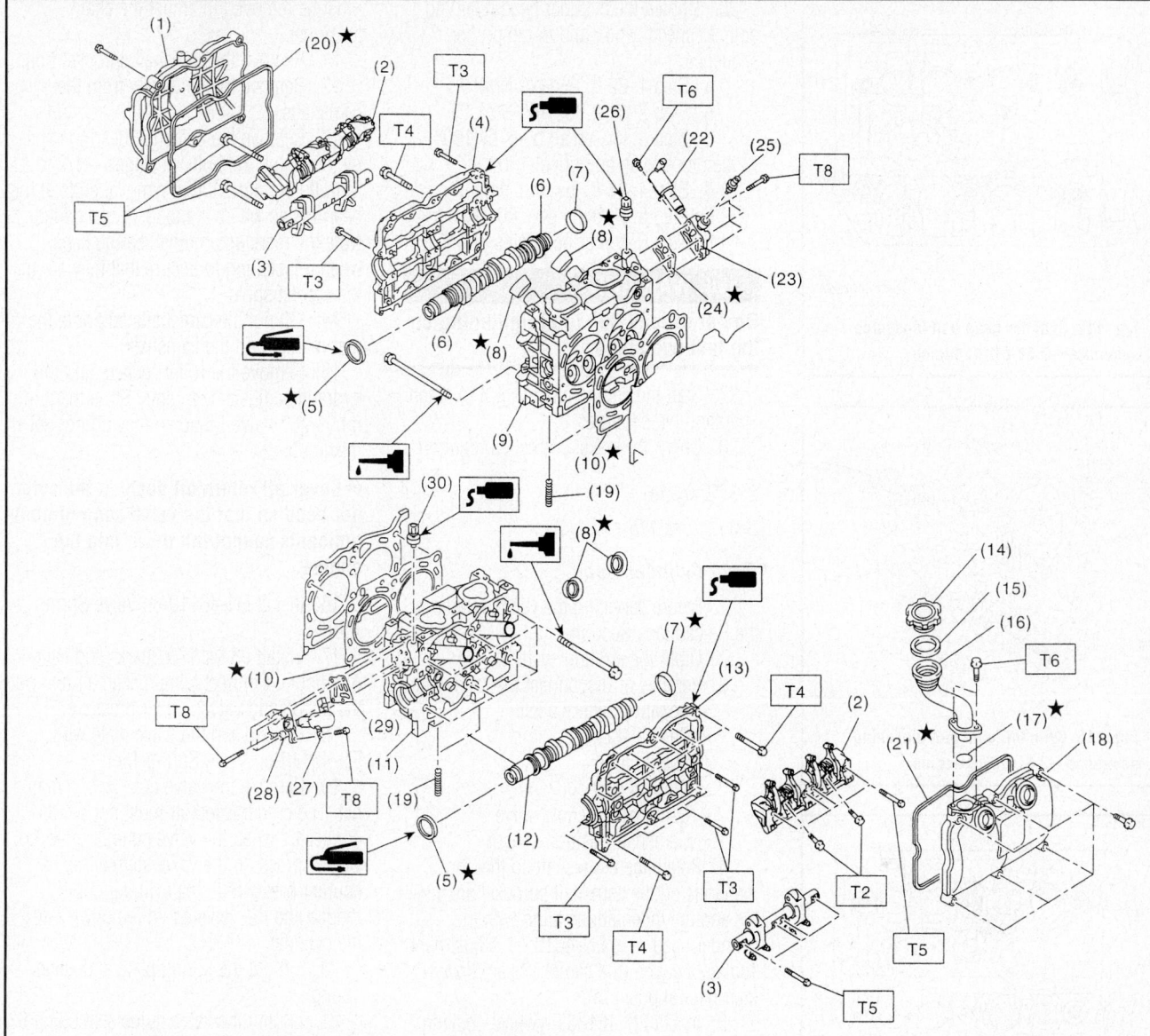

| (1) | Rocker cover (RH) | (17) | O-ring | (30) | Variable valve lift diagnosis oil pressure switch (LH) |
|---|---|---|---|---|---|
| (2) | Intake valve rocker assembly | (18) | Rocker cover (LH) | | |
| (3) | Exhaust valve rocker assembly | (19) | Stud bolt | | |
| (4) | Camshaft cap (RH) | (20) | Rocker cover gasket (RH) | | |
| (5) | Oil seal | (21) | Rocker cover gasket (LH) | | |
| (6) | Camshaft (RH) | (22) | Oil switching solenoid valve (RH) | | |
| (7) | Plug | (23) | Oil switching solenoid valve holder (RH) | | |
| (8) | Spark plug pipe gasket | | | | |
| (9) | Cylinder head (RH) | (24) | Gasket | | |
| (10) | Cylinder head gasket | (25) | Oil temperature sensor | | |
| (11) | Cylinder head (LH) | (26) | Variable valve lift diagnosis oil pressure switch (RH) | | |
| (12) | Camshaft (LH) | | | | |
| (13) | Camshaft cap (LH) | (27) | Oil switching solenoid valve (LH) | | |
| (14) | Oil filler cap | (28) | Oil switching solenoid valve holder (LH) | | |
| (15) | Gasket | | | | |
| (16) | Oil filler duct | (29) | Gasket | | |

**Tightening torque: N·m (kgf-m, ft-lb)**

| | |
|---|---|
| T3: | 9.75 (1.0, 7.2) |
| T4: | 18 (1.8, 13.0) |
| T5: | 25 (2.5, 18.1) |
| T6: | 6.4 (0.65, 4.7) |
| T7: | 8 (0.8, 5.9) |
| T8: | 10 (1.0, 7.4) |

09490_SBCR_G0016

**Fig. 113 Cylinder head and related components—2.5L SOHC engine**

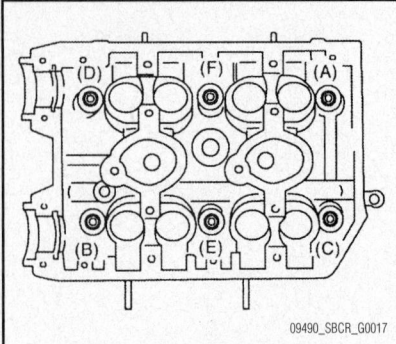

**Fig. 114 Cylinder head bolt loosening sequence—2.5L DOHC engine**

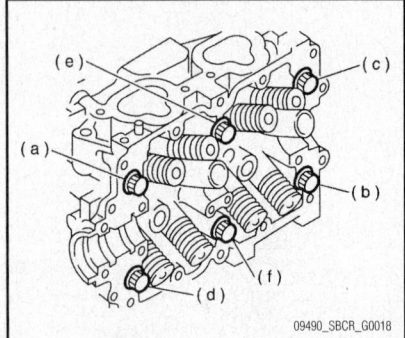

**Fig. 115 Cylinder head bolt loosening sequence—2.5L SOHC engine**

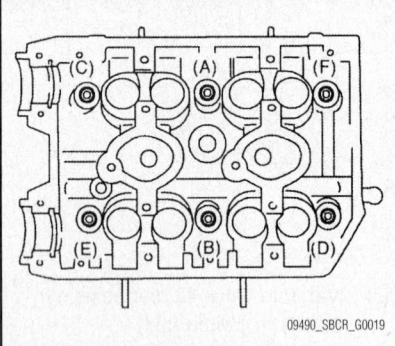

**Fig. 116 Cylinder head bolt tightening sequence—2.5L DOHC engine**

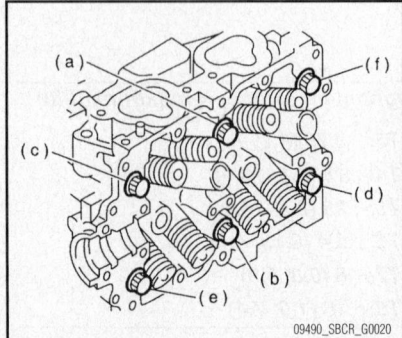

**Fig. 117 Cylinder head bolt tightening sequence—2.5L SOHC engine**

26. Tighten the cylinder head retaining bolts to specification and in the proper sequence.
   a. Step 1: 22 ft. lbs. (29 Nm).
   b. Step 2: 51 ft. lbs. (69 Nm).
   c. Step 3: loosen all bolts by 180°, then loosen an additional 180°.
   d. Step 4: 31 ft. lbs. (42 Nm).
   e. Step 5: all bolts plus 80–90°.
   f. Step 6: center bolts plus 40–45°.

### ☀ WARNING

**Do not exceed 90° total tightening on the last two steps.**

27. Start the engine and allow it to reach operating temperature.
28. Check for leaks, correct as required.

### 2.8L Engine

*See Figures 118 and 119.*

### Front Cylinder Head

1. Before servicing the vehicle, refer to the Precautions Section.
2. Drain the cooling system.
3. Remove or disconnect the following:
   - Negative battery cable
   - Camshaft cover
   - Spark plugs
   - Coolant port
   - Camshaft solenoid valve
   - Camshaft position sensor
4. Rotate the crankshaft so that the markings on the camshaft sprocket are visible and the indentations align with the cylinder head. Use EN-46105-1 Adjustment Tool as a gauge (the camshafts are now in their neutral position).
5. Install EN-46106 Flywheel Locking Attachment.
6. Mark the chain and camshaft sprocket so they can be reinstalled in the same position.
7. Reduce the tightening torque of the bolts holding the camshaft sprockets. Counterhold using an open wrench on the camshaft flats (do not remove the bolts).
8. Install EN-46108-1 Fixing Tool to the camshaft chain on the intake side. Check that the tool grips the chain firmly.
9. Install EN-46108-2 Fixing Tool to the camshaft chain on the exhaust side. Check that the tool grips the chain firmly.
10. Tighten EN-46108 Fixing Tool so that the chain is unloaded. It is extremely important that tool no. EN-46108 is properly installed. If the chain is not unloaded when the camshaft sprocket bolts are removed, the automatic chain tensioner is triggered. If this occurs, the timing cover

must be removed to adjust the chain tensioner.
11. Remove the camshaft sprocket bolts.
12. Remove the sprockets from the camshafts.
13. Remove the camshaft bearing caps by loosening the bolts in stages—1/2 to 1 turn. It is important that removal ends at the bearing cap where the tappets are loaded. Note the markings on the bearing caps when reinstalling to ensure that they are in the correct spots.
14. Lift out the camshafts and note the rear marking on the camshafts.
15. Remove the roller rockers and the hydraulic adjuster elements. Store them in 83 93 787 Valve Stand so they do not get mixed up.

➡ **Cover all return oil ducts in the cylinder head so that the valve cone or contaminants cannot fall down into the engine.**

16. Install EN-46110-eu Valve Spring Tool.
17. Install 83 94 173 Spark Plug Hole Air Nipple. Connect compressed air and put the piston and valves under pressure.
18. Press down the valve plate with EN-46110-eu Valve Spring Tool.
19. Remove the valve cone with a magnet. The compressed air must not be disconnected when the valve cone is removed. Do not press on the valve stem either as counter pressure in the cylinder could reduce and the valve could fall down into the cylinder.
20. Lift out the spring plate and valve spring.
21. Lift out the valve guide seal using 83 94 157 Valve Guide Seal Pliers.
22. Remove the cylinder head bolts in an alternating sequence by loosening them ¼ turn then ½ turn
23. Remove the cylinder head.

### *To install:*

24. Clean all mating surfaces of any residual gasket material.
25. Install the cylinder head. Torque the new bolts in an alternating sequence starting with the center bolts and radiating out towards the ends:
   a. Step 1: 18.5 ft. lbs. (22 Nm).
   b. Step 2: Plus an additional 90°.
   c. Step 3: Plus an additional 90°.
   d. Step 4: Plus an additional 90°.
   e. Step 5: Plus an additional 45°.
26. Clean all parts and check contact surfaces and bearing surfaces for wear.
27. Install the valve guide seal using 83 94 157 Valve Guide Seal Pliers.

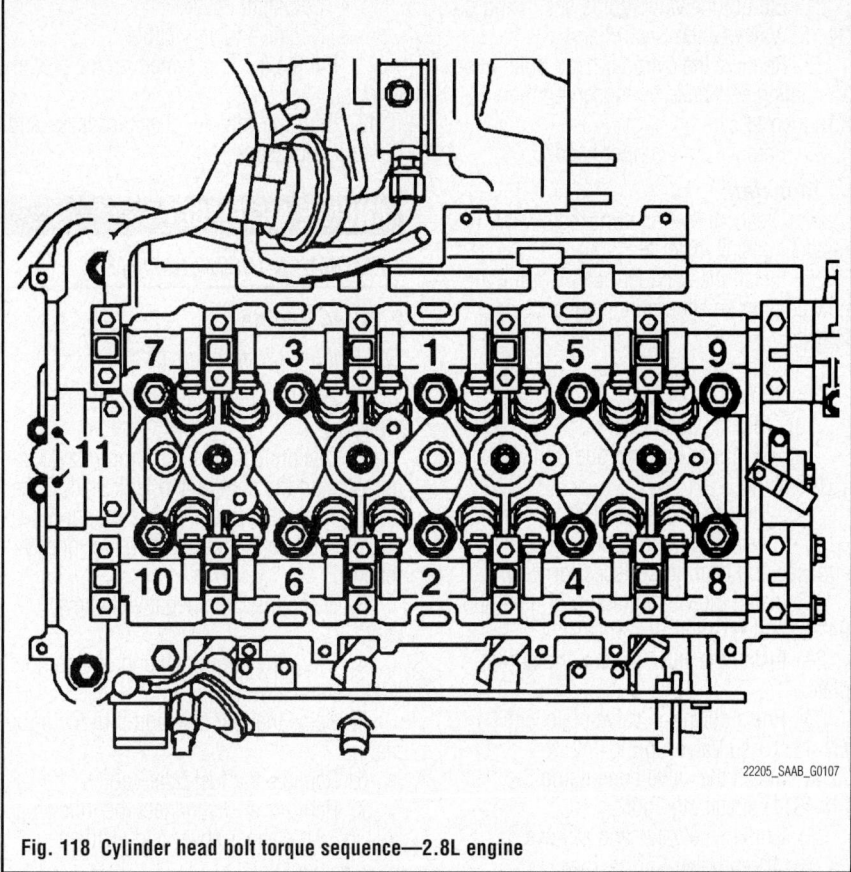

**Fig. 118 Cylinder head bolt torque sequence—2.8L engine**

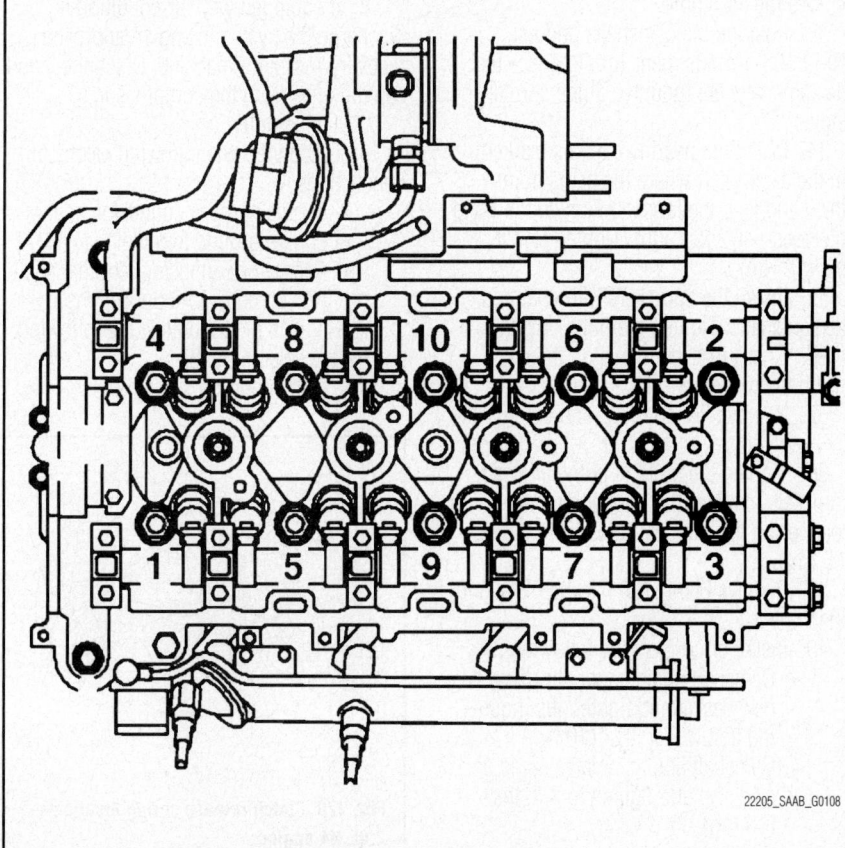

**Fig. 119 Cylinder head bolt removal sequence—2.8L engine**

28. Install the valve spring and spring plate.

29. Press down the valve plate with EN-46110-eu Valve Spring Tool.

30. Install the valve cone using EN-46117 Installing Tool.

31. Unload the valve and remove EN-46110-eu Valve Spring Tool.

32. Disconnect the compressed air and remove the air nipple.

33. Position the camshafts and use EN-46105 Adjustment Tool to check that the camshafts are mounted at the correct angle.

34. Install the bearing caps as marked on the caps. Start where the cams point down and load the tappets. Tighten the bolts in stages—1/2 to 1 turn. Tighten to 7 ft. lbs. (10 Nm).

35. Install the sprockets with chain on the camshafts. Tighten the bolts by hand.

36. Check that the marks on the chain match the markings on the sprockets.

37. Remove EN-46108-2 and EN-46108-1 Fixing Tools.

38. Tighten the bolts of both camshaft sprockets, counter holding with an open spanner on the camshaft flats to 48 ft. lbs. (65 Nm).

39. Remove EN-46106 Flywheel Locking Attachment.

40. Install or connect the following:
- Camshaft position sensor (555F) using a new gasket and tighten the mounting bolt to 7 ft. lbs. (10 Nm)
- Camshaft solenoid valve (695F) using a new seal and tighten to 7 ft. lbs. (10 Nm)
- Coolant port using a new seal and gasket. Tighten the mounting bolts to 7 ft. lbs. (10 Nm)
- Spark plugs
- Camshaft cover.
- Negative battery cable

41. Fill the cooling system to the proper level.

42. Start the vehicle, check for leaks and repair if necessary.

### Rear Cylinder Head

1. Before servicing the vehicle, refer to the Precautions Section.

2. Drain the cooling system.

3. Remove or disconnect the following:
- Camshaft cover
- Spark plugs
- Camshaft solenoid valve
- Camshaft position sensor

4. Rotate the crankshaft so that the markings on the camshaft sprocket are visible and the indentations align with the cylinder head. Use EN-46105 Adjustment

Tool as a gauge (the camshafts are now in their neutral position).

5. Install EN-46106 Flywheel Locking Attachment.

6. Mark the chain and camshaft sprocket so they can be reinstalled in the same position.

7. Slightly loosen the bolts holding the camshaft sprockets. Counterhold using an open wrench on the camshaft flats (do not remove the bolts).

8. Install EN-46108-2 Fixing Tool to the camshaft chain on the intake side. Check that the tool grips the chain firmly.

9. Install EN-46108-1 Fixing Tool to the camshaft chain on the exhaust side. Check that the tool grips the chain firmly.

10. Tighten EN-46108 Fixing Tool so that the chain is unloaded. It is extremely important that tool no. EN-46108 is properly installed. If the chain is not unloaded when the camshaft sprocket bolts are removed, the automatic chain tensioner is triggered. If this occurs, the timing cover must be removed to adjust the chain tensioner.

11. Remove the camshaft sprocket bolts.

12. Remove the sprockets from the camshafts.

13. Remove the camshaft bearing caps by loosening the bolts in stages—1/2 to 1 turn. It is important that removal ends at the bearing cap where the tappets are loaded. Note the markings on the bearing caps when reinstalling to ensure that they are in the correct spots.

14. Lift out the camshafts and note the rear marking on the camshafts.

15. Remove the roller rockers and the hydraulic adjuster elements. Store them in 83 93 787 Valve Stand so they do not get mixed up.

➡️**Cover all return oil ducts in the cylinder head so that the valve cone or contaminants cannot fall down into the engine.**

16. Install EN-46110-eu Valve Spring Tool.

17. Install 83 94 173 Spark Plug Hole Air Nipple. Connect compressed air and put the piston and valves under pressure.

18. Press down the valve plate with EN-46110-eu Valve Spring Tool.

19. Remove the valve cone with a magnet. The compressed air must not be disconnected when the valve cone is removed. Do not press on the valve stem either as counter pressure in the cylinder could reduce and the valve could fall down into the cylinder.

20. Lift out the spring plate and valve spring.

21. Lift out the valve guide seal using 83 94 157 Valve Guide Seal Pliers.

22. Remove the cylinder head bolts in an alternating sequence by loosening them ¼ turn then ½ turn

23. Remove the cylinder head.

**To install:**

24. Clean all mating surfaces of any residual gasket material.

25. Install the cylinder head. Torque the new bolts in an alternating sequence as follows:

   a. M8 size bolts torque 11 ft. lbs. (15 Nm).

   b. Plus an additional 60°.

   c. M11 size bolts torque 33 ft. lbs. (45 Nm).

   d. Plus an additional 120°.

26. Clean all parts and check contact surfaces and bearing surfaces for wear.

27. Install the valve guide seal using 83 94 157 Valve Guide Seal Pliers.

28. Install the valve spring and spring plate.

29. Press down the valve plate with EN-46110-eu Valve Spring Tool.

30. Install the valve cone using EN-46117 Installing Tool.

31. Unload the valve and remove EN-46110-eu Valve Spring Tool.

32. Disconnect the compressed air and remove the air nipple.

33. Position the camshafts and use EN-46105-1 Adjustment Tool to check that the camshafts are mounted at the correct angle.

34. Install the bearing caps as marked on the caps. Start where the cams point down and load the tappets. Tighten the bolts in stages—1/2 to 1 turn. Tighten to 7 ft. lbs. (10 Nm).

35. Install the sprockets with chain on the camshafts. Tighten the bolts by hand.

36. Check that the marks on the chain match the markings on the sprockets.

37. Remove EN-46108-2 and EN-46108-1 Fixing Tools.

38. Tighten the bolts of both camshaft sprockets, counter holding with an open spanner on the camshaft flats to 48 ft. lbs. (65 Nm).

39. Remove EN-46106 Flywheel Locking Attachment.

40. Install or connect the following:

- Camshaft position sensor using a new gasket and tighten the mounting bolt to 7 ft. lbs. (10 Nm)
- Camshaft solenoid valve using a new seal and tighten to 7 ft. lbs. (10 Nm)
- Spark plugs

- Camshaft cover.
- Negative battery cable.

41. Fill the cooling system to the proper level.

42. Start the vehicle, check for leaks and repair if necessary.

## ENGINE ASSEMBLY

### REMOVAL & INSTALLATION

**2.0L H4 Engine**

*See Figures 120 through 122.*

1. Properly relieve the fuel system pressure.

2. Disconnect the fuel pump relay connector, start the engine and let it stall. Once the engine stalls, crank it for a further 5 seconds to ensure the fuel system is properly relieved.

3. Disconnect the negative battery cable.

4. Drain the engine oil and coolant into suitable containers.

5. Raise the rear seat and turn the floor mat up.

6. Remove the fuel filler cap.

7. Remove or disconnect the following:
- Air cleaner cover and element
- Radiator
- Coolant filler tank

8. If equipped with air conditioning, discharge the system using an approved recovery/recycling machine. Disconnect and cap the lines from the compressor.
- Intercooler

9. Disconnect the following electrical connections:
- Engine harness connector
- Engine ground terminal
- Alternator connector, terminal and A/C compressor connections

10. Remove or disconnect the following:
- Accelerator cable
- Clutch release spring

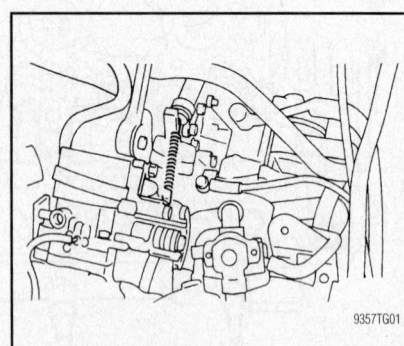

9357TG01

**Fig. 120 Clutch release spring location— 2.0L H4 engine**

- Brake booster hose
- Heater inlet and outlet hoses

11. Remove the power steering pump from the bracket by performing the following steps:

   a. Loosen the lock and slider bolts.

   b. Remove the V-belt.

   c. Disconnect the power steering switch connection.

   d. Remove the pipe with bracket from the intake manifold.

   e. Remove the power steering pump from the engine.

   f. Remove the power steering tank from the bracket by pulling it upwards.

   g. Place the power steering pump on the wheel apron on the right.

12. Remove or disconnect the following:

- Center exhaust pipe
- Nuts that attach the lower side of the engine to the transmission
- Nuts that attach the front cushion rubber onto the crossmember

13. Disconnect the clutch release fork from the release bearing as follows:

   a. Remove the clutch cylinder from the transmission.

   b. Using a 10mm wrench, remove the plug.

   c. Screw a 6mm diameter bolt into the release fork and remove it.

   d. Raise the release fork and unfasten the release tabs to free the release fork.

14. Disconnect the torque converter clutch from the flexplate if equipped with automatic transmission as follows:

   a. Remove the service hole plug.

   b. Remove the torque converter clutch-to-flexplate bolts.

   c. Remove the remaining bolts while rotating the engine using a crankshaft pulley wrench.

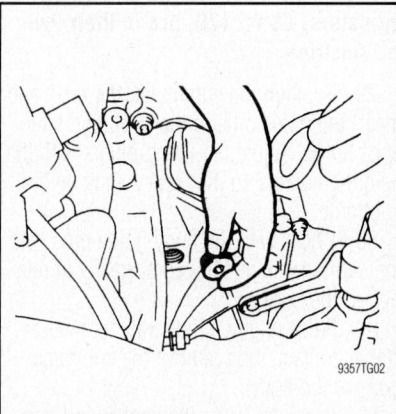

**Fig. 121 Using a 10mm wrench, remove the plug—2.0L H4 engine**

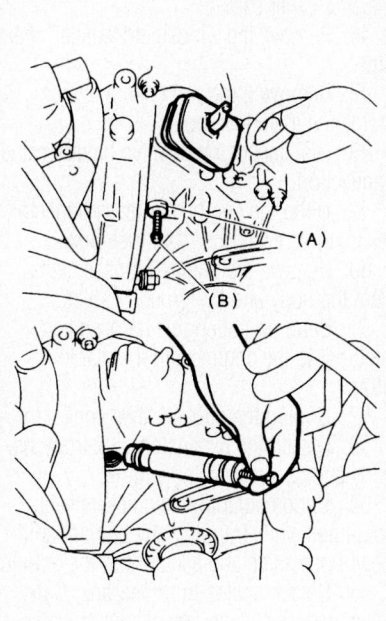

(A) Shaft
(B) Bolt

9357TG03

**Fig. 122 Screw a 6mm diameter bolt into the release fork and remove it—2.0L H4 engine**

15. Remove or disconnect the following:

- Pitching stopper
- Fuel delivery, return and evaporation hoses
- Fuel filter and bracket

16. Attach a lifting device to the engine.

17. Using a floor jack, support the transmission.

18. Remove the starter.

19. Separate the engine from the transmission.

20. Remove the upper right transmission-to-engine bolts.

21. Remove the engine as follows:

   a. Raise the engine slightly.

   b. Using the floor jack, raise the transmission.

   c. Move the engine horizontally until the mainshaft is withdrawn from the clutch cover.

   d. Remove the engine.

### To install:

22. Installation is the reverse of removal, please note the following torques:

- Clutch release fork plug: 32 ft. lbs. (44 Nm)
- Front cushion rubbers: 25 ft. lbs. (34 Nm)
- Bolts attaching the upper right side of the transmission to the engine: 37 ft. lbs. (50 Nm)
- Pitching stopper-to-fender bolt: 37 ft. lbs. (50 Nm)
- Pitching stopper-to-engine bolt: 43 ft. lbs. (58 Nm)
- Torque converter clutch-to-flexplate bolts, while rotating the engine: 18 ft. lbs. (25 Nm)
- Power steering pump bolts: 15 ft. lbs. (20 Nm)
- Bolts attaching the lower side of the transmission to the engine: 37 ft. lbs. (50 Nm)
- Front cushion rubber-to-crossmember bolts: 61 ft. lbs. (83 Nm)

23. Fill the engine with the recommended oil.

24. Fill and bleed the cooling system.

25. Charge the air conditioning system using an approved recovery/recycling machine.

26. Adjust the clutch cable.

27. If equipped, check the automatic transmission fluid level and add Dexron®II if necessary.

28. Start the engine and allow it to reach normal operating temperature. Check for leaks.

### 2.0L I4 Engine

*See Figures 123 through 125.*

1. Before servicing the vehicle, refer to the Precautions Section.

2. Open cap on expansion tank.

3. On lift, raise car slightly and remove front wheels.

4. Remove lower spoiler shield. Detach hose to headlamp washers, unplug connector and remove it.

5. Drain cooling system.

6. Remove wheel well liners.

7. Remove cover by the gearbox.

8. Detach the hose for the headlamp washers and drain washer fluid reservoir.

9. Lower the vehicle slightly and remove the front bumper.

10. Remove the cover on the left side and the hose between the charge air pipe and cooler.

11. Remove the lower charge air pipe bracket from the fan cowling.

12. Remove cover on right side and hose between charge air cooler and pipe.

13. Remove receiver-drier bolts from the radiator.

14. Remove the lower seal between the charge air cooler and the radiator.

15. Lower the vehicle and remove the upper engine cover.

16. Remove the battery cover and battery cooler pipe.

17. Remove the hood release cable coupling by first undoing the clip in the body and then separating the quick-release couple using a screwdriver.

18. Remove the upper radiator support.

19. Unplug the pressure-temperature sensor connector.

20. Disconnect the bypass valve hose.

21. Detach the charge air hose from the throttle body and release the upper charge air pipe bracket from the fan cowling. Remove the pipe and hose.

22. Remove the battery.

23. Detach the ventilation hose from the radiator.

24. Detach the vacuum hose quick-release coupling from the vacuum pump.

25. With automatic transmission, unplug the connector from the TCM. Remove the control module.

26. Unplug the connector under the control module.

27. Release the main fuse box in front of the battery tray and move it aside.

28. Unplug the hood switch connector, release the cable clip and remove the battery tray.

29. Detach the connector and release the clip holding the cables on the fan cowling.

30. Unplug the reverse light switch connector.

31. Remove the seal over the radiator core.

32. Remove the upper radiator brackets from the body.

33. Unplug both connectors from the left side structural member.

34. With automatic transmission, detach the pipes from the fluid cooler.

35. Remove the fan cowling bolts and loosen cowling slightly.

36. Secure the condenser and charge air cooler to the body with cable ties.

37. Remove the fan cowling. Carefully detach A/C pipes from retaining clip on right side structural member.

38. With receptacle under radiator, remove lower radiator hose.

39. Remove upper radiator hose.

40. Remove the radiator.

41. Unplug the A/C pressure sensor connector.

42. Detach the hoses from the turbocharger inlet pipe and remove the pipe.

43. Unplug the mass air flow (MAF) sensor connector. Cap the turbocharger inlet.

44. Remove the air cleaner cover and the air cleaner.

45. Detach the intake hose and remove the air cleaner casing.

46. Remove the windshield washer filler pipe.

47. Remove 2 mounting screws from inside the main fuse box.

48. Disconnect the positive battery cable connection.

49. Undo the engine harness connector retaining screw in the main fuse box.

50. Remove the engine harness clamp from the body and the ground cables.

51. Bend up the engine harness and secure it to the engine with a suitable Strap.

52. Unplug the coolant level connector.

53. Detach the expansion tank from the body and secure it to the engine.

54. Undo coolant hose quick-release couplings while trapping any coolant spill. Bend hose aside and secure it to the engine.

55. Detach cables from gearbox. Carefully bend then aside and secure to expansion tank bracket with cable ties.

56. With pliers, pinch off clutch hose and disconnect quick-release coupling from clutch slave cylinder.

57. Release any pressure in the fuel system by carefully pressing the service valve needle. Collect any fuel spillage.

58. Detach both fuel lines from fuel rail while gripping the lower nut. Plug the fuel line openings.

59. Disconnect quick-release coupling for ventilation line and detach fuel lines and vent lines from clips on camshaft cover. Bend up vent hose from intake manifold and place it on camshaft cover.

60. Be sure steering wheel and front wheels are in the straight-ahead position.

### ❋❋ WARNING

**To prevent contact roller for airbag from breaking due to twisting, steering wheel should be fixed in position, such as taping it to the instrument panel.**

61. Detach the steering shaft from the steering gear.

62. Raise the vehicle. Mark the rotation direction of the auxiliary drive belt, then remove it.

63. Unplug the A/C compressor connector and remove the compressor retaining bolts.

64. Remove the power steering pipe clamp bolts from the subframe.

65. Slightly loosen rear torsion arm bolts.

66. Install an Engine Centering Tool Kit, 83 96 152, to subframe. Adjust the centering tool bolts so adjuster screw slots are completely visible.

➡**Because of narrow tolerances on the driveshafts, centering tool must be used to carefully install the powertrain to the subframe and body during installation.**

67. Lower the vehicle and re-install the radiator support member.

68. Attach a Strap, 83 95 212, to radiator support, lower the Strap and wind it an extra turn around the A/C compressor to relieve any strain.

69. Remove left and right side engine mounts.

70. Raise the vehicle and disconnect the driveshaft from the hub by removing the center nuts. It may be necessary to use a Driveshaft Puller, 89 96 951, or a brass drift and mallet.

71. Remove outboard steering link nuts and separate links from steering knuckle.

72. Grip stabilizer bar link flats with a wrench, then disconnect the stabilizer bar links.

73. Undo the lower swivel joints from the steering knuckles and lower the transverse links.

74. Disconnect the driveshafts and move them away.

75. Measure a distance of about 8 inches (87 mm) from front end of muffler and cut the exhaust pipe between the muffler and flexpipe.

76. Detach front pipe from the catalytic converter.

77. Disconnect the ground cable from the gearbox and the connector for the angle sensor (if equipped).

78. Position a Subframe-To-Engine Centering Tool, 83 96 145, on engine lift.

➡**Ensure adjuster screws on height adjusters, 83 95 170, are in their lowest position.**

79. Position the engine lift, the raise and insert the guide pins in the subframe reference holes. Adjust the lifting pillars with the height adjusters so they rest evenly on the subframe.

80. Check with the body guide pins that the subframe is positioned correctly in relation to the body.

81. Raise engine lift slightly to ensure stable contact, then remove the subframe bolts in the body.

82. Carefully lower the engine and transmission assembly on the engine lift, checking that no components a caught or damaged during removal.

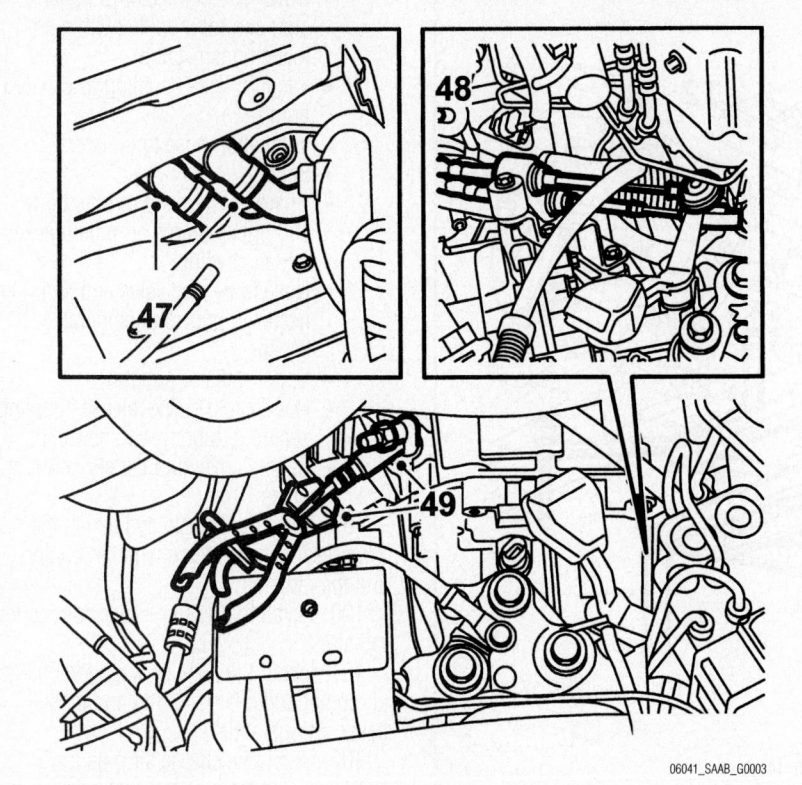

**Fig. 123 Showing location of coolant quick-release connections, gearbox cables and clutch hose—2.0L I4 engine**

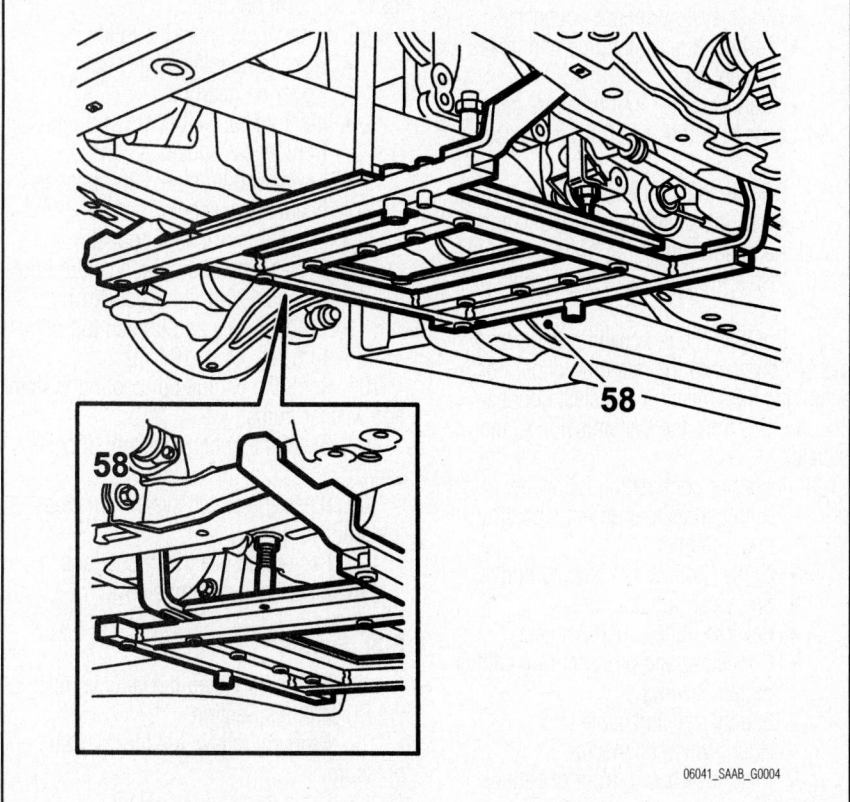

**Fig. 124 Showing engine centering tool installed—2.0L I4 engine**

*To install:*

83. Position the engine lift under the vehicle, moving the engine and transmission in position.

84. Lift the powertrain until the inboard driveshaft U-joints are level with the wheel hubs, then insert the driveshafts into the hubs.

85. Lift the powertrain a little higher and position the lower swivel joints to steering knuckles. If necessary, adjust the engine lift screws to provide even contact with the body. Ensure guide pins are correctly positioned.

86. Install the subframe-to-body bolts. Raise the engine lift fully and tighten the subframe bolts to 55 ft. lbs. (75 Nm), plus and additional 90°.

87. Install or connect the following:

- Lower swivel joint bolts to 37 ft. lbs. (50 Nm); ensure steering knuckle stub is visible on top of steering knuckle housing before bolt is installed.
- Stabilizer bar link bolts to 47 ft. lbs. (64 Nm)
- Outboard steering links and bolts to 26 ft. lbs. (35 Nm)
- Ground cable to gearbox and connector for angle sensor (if equipped).
- Position joint clamp on front exhaust pipe
- Catalytic converter to front exhaust pipe, tightening nuts to 18 ft. lbs. (25 Nm)
- Adjust joint clamp so pipe ends are in the middle, then tighten clamp nuts to 30 ft. lbs. (40 Nm)
- Right engine mount, after lowering car: right mount to 52 ft. lbs. (70 Nm), plus an additional 60°
- Left engine mount to 52 ft. lbs. (70 Nm), plus an additional 45°
- Right engine pad-to-body to 30 ft. lbs. (40 Nm), plus an additional 60°
- Release Strap from radiator member and A/C compressor
- Raise car and remove engine centering kit
- Rear torsion arm bolts to 52 ft. lbs. (70 Nm), plus an additional 90°
- Bolts to power steering pipe clamps on subframe
- A/C compressor and connector
- Auxiliary drive belt (in proper direction of rotation)
- Steering shaft to steering gear. Use new clamp bolt and tighten to 20 ft. lbs. (27 Nm)

88. Lower the vehicle and remove the restraint from steering wheel

**Fig. 125 Showing engine lift equipment installed—2.0L I4 engine**

89. Connect the quick-release coupling on the vent line and press fuel lines and vent lines into clips on camshaft cover.

90. Unplug fuel lines and connect to fuel rail while gripping lower nut. Use new seals.

91. Connect the quick-release coupling to the clutch slave cylinder and release the pliers.

92. Connection at clutch slave cylinder must be mounted at the correct angle, as illustrated.

93. Install or connect the following:
- Cables to gearbox
- Coolant hoses at quick-release couplings
- Expansion tank to body
- Coolant level sensor connector
- Ground cables and clamp for engine harness to body
- Retaining screw for engine harness connector in main fuse box
- Positive cable to battery
- 2 retaining screws for main fuse box and box cover
- Windshield washer filler pipe
- Air cleaner housing, intake hose and air cleaner cover

### ✳✳ CAUTION

**To reduce risk of hoses coming loose from delivery side of turbocharger, hoses and connecting pieces must be cleaned thoroughly with a cleaning agent, 30 15 815. Use new hose clamps.**

- Mass airflow sensor connector
- Turbocharger inlet pipe and hoses to inlet pipe
- Connector for A/C pressure sensor

94. Install the radiator. Cut off the cable ties.

95. Be sure the condenser and charge air cooler hooks are positioned correctly and install the radiator retaining bolts.

96. Install the upper and lower radiator hoses.

97. Position the fan cowling and press the A/C pipes into the retaining clips. For automatic transmission vehicles, connect the oil pipes from the transmission to the radiator.

98. Install or connect the following:
- Both connectors on left side structural member
- Upper radiator brackets to body
- Seal over radiator core
- Reverse light switch connector
- Connector and clip securing cables on fan cowling
- Battery tray and cable clip
- Hood switch connector
- Main fuse box in front of battery tray
- Connector to control module

- Cover and connector to transmission control module (with automatic transmission)
- Vacuum hose coupling to vacuum pump
- Ventilation hose to radiator
- Battery
- Charge air hose to throttle body and upper charge air pipe bracket on fan cowling
- Hose to bypass valve and connector to pressure/temperature sensor
- Upper radiator support
- Hood cable quick-release coupling; secure it to body with new clips
- Battery cover and battery cooler pipes

99. Raise the vehicle and install the lower seal between the charge air cooler and the radiator.

100. Install the receiver-drier bolts to the radiator.

101. Install the hose between the charge air cooler and charge air pip and install cover on right side.

102. Install the charge air pipe lower bracket on the fan cowling (left side).

103. Install the charge air hose between the charge air pip and charge air cooler and install cover (left side).

104. Lower the vehicle slightly and install the front bumper.

105. Install or connect the following:
- Hose to headlamp washers
- Cover on gearbox
- Inner wheel well liners and both front wheel housings
- Lower spoiler shield and hose to headlamp washers
- Headlamp washer connector
- Driveshafts and tighten center hub nuts to 170 ft. lbs. (230 Nm)
- Front wheels and tighten lug nuts to 81 ft. lbs. (110 Nm)

106. Refill the engine oil, cooling system and washer fluids.

107. Bleed the cooling system, if necessary.

108. Bleed the clutch system (manual transmission).

109. Restore all electrical functions.

110. Install upper engine cover.

### 2.3L Engine

1. Before servicing the vehicle, refer to the Precautions Section.

2. Properly relieve the fuel system pressure.

3. Drain the engine coolant.

4. Drain the engine oil.

5. Remove or disconnect the following:
- Battery and tray
- Ground cable
- Breather hose from under the fuse box and cables to the gearbox, automatic transmission only
- Reverse light switch and remove the lower engine cover, manual transmission only
- Steering column locking bolt and move the steering column upward
- Throttle cable
- Evaporative Emissions (EVAP) purge valve vacuum hose from the intake manifold
- Brake servo vacuum hose
- Fuel connections and plug the lines
- Positive battery cable at the distribution terminal
- Negative battery cable at the gearbox
- Clutch slave cylinder hose, manual transmission only
- Selector rod after placing the vehicle in 3rd gear, manual transmission only
- Selector lever cable, automatic transmission only
- Upper radiator hose
- Bypass valve vacuum hose
- Hose between the throttle body and charge air cooler
- Pressure/temperature sensor electrical connector
- Bypass valve and intake manifold
- Radiator fan electrical connector
- Upper radiator hose to the oil cooler, if equipped
- Fan cowling
- Grille
- Hose between the turbocharger and charge air cooler
- MAF sensor
- Lower radiator from the water pump
- Heat exchanger hoses
- Vacuum hose from the bypass, if equipped with ACC
- Radiator
- Engine wire harness and bracket
- Wiper arms and covers
- Control module electrical connectors
- Hub nuts and raise the vehicle
- Both front wheels. Drive a wedge between the gearbox and subframe, and between the oil pan and subframe
- A/C compressor retaining bolts
- Air filter housing retaining nuts and air hose. Lower the vehicle
- Condenser cooler, charge air cooler and A/C compressor and suspend them to the radiator member

- A/C compressor connector
- Quick release coupling to the wastegate
- Right side engine mount and yoke
- Belt tensioner and belt
- Left side engine mount
- Power steering reservoir and hoses
- Separate the quick release couplings to the gearbox oil cooler, if equipped
- Steering swivel joint
- Hardware securing the stabilizer bar supports to the strut
- Outer steering links
- Halfshafts from the hub
- Oil cooler
- Separate the exhaust system behind the catalytic converter
- Temperature sensor and place a lifting trolley under the subframe
- Subframe supporting plate bolts

6. Lower the trolley slightly and unhook the oil cooler hose from the bracket

7. Move the trolley to the rear to allow access for the A/C compressor and remove the powertrain assembly.

### To install:

8. Position the trolley under the vehicle with the engine in position.

9. Install or connect the following:
- Driveshafts to the steering swivel joint and torque the bolts to 22 ft. lbs. (30 Nm) plus 90°
- Subframe and supporting plate. Torque the subframe bolts to 84 ft. lbs. (115 Nm) and the supporting plate to 48 ft. lbs. (65 Nm). Move the trolley away from the vehicle.
- Exhaust pipe and temperature sensor and torque the bolts to 18 ft. lbs. (25 Nm)
- Air cleaner
- Outer steering knuckles to the stabilizer bar support
- Hub nuts but do not tighten them
- A/C compressor and air filter housing and torque the bolt to 35 ft. lbs. (47 Nm)
- Steering gear into position. Torque the locking nut to 18 ft. lbs. (25 Nm).
- Left side engine mount
- Right side engine mount and yoke and remove the wedges. Torque the mounting bolts to 36 ft. lbs. (50 Nm)
- Selector lever cable, if equipped
- Selector rod on manual transaxles and remove the locking pins
- Bleeder hose under the fuse box
- Hoses to the heat exchanger

- Coolant hose to the expansion tank
- Engine wire harness to the bulkhead partition
- Engine control module
- Wiper arms and cover plate
- Ground cable to the gearbox
- Positive cable to the positive terminal
- Connector with electrical leads to the gear box, automatic transmission only
- Reverse light connector, manual transmission only
- Quick release couplings to the automatic transmission fluid cooler
- Fuel connections and rubber protectors
- Vacuum hose for the brake servo and intake manifold
- EVAP canister purge valve vacuum hose
- Throttle cable
- Hose between the turbocharger and charge air cooler
- Engine oil cooler and cover
- Radiator fans
- Front grille
- Upper and lower radiator hoses
- Power steering reservoir
- Power steering pump pipes
- Quick release coupling for the turbocharger wastegate vacuum hose
- A/C compressor electrical connectors
- MAF sensor connector
- Bypass valve and intake manifold
- Pressure pipe and charge air cooler hose
- Pressure/temperature sensor connector
- Front wheels and torque the hub nuts to 215 ft. lbs. (290 Nm)
- Battery tray and battery
- Intake manifold cover and torque the bolts to 16 ft. lbs. (22 Nm)
- Shut off valve vacuum hose

10. Fill the engine with coolant
11. Fill the engine with clean oil.
12. Fill the transmission to the proper level.
13. Start the vehicle, check for leaks and repair if necessary.

### 2.5L Engines

*See Figures 126 through 129.*

1. Before servicing the vehicle, refer to the Precautions Section.
2. Open the front hood fully and support with the front food stay.
3. Collect the refrigerant from the A/C system.

4. Release the fuel pressure.

5. Disconnect the ground cable from the battery.

6. Open the fuel filler flap lid, and remove the fuel filler cap.

7. Remove the air intake duct, air cleaner case and air intake chamber.

8. Remove the undercover.

9. Remove the radiator from the vehicle.

10. Disconnect the A/C pressure hoses from A/C compressor.

11. Remove the air intake chamber stay.

12. Disconnect the following connectors and cables.

- Front oxygen (A/F) sensor connector
- Rear oxygen sensor connector
- Engine ground cable
- Engine harness connectors
- Alternator connector and terminal
- A/C compressor connector
- Power steering switch connector

13. Disconnect the following hoses.

- Brake booster vacuum hose
- Heater inlet and outlet hoses

14. Remove the power steering pump.

15. Remove the bolts which secure the power steering pump to the bracket. Place the power steering pump on the right side wheel apron.

16. Remove the front side V—belts.

17. Remove the front and center exhaust pipes.

18. Remove the bolts and nuts which hold lower side of transmission to engine.

19. Remove the nuts which install front cushion rubber onto front crossmember.

20. Separate the torque converter clutch from flexplate as follows: (AT models)

- Lower the vehicle.
- Remove the service hole plug.
- Remove the bolts which hold torque converter clutch to flexplate.
- Remove all bolts while turning the crankshaft with a socket wrench.

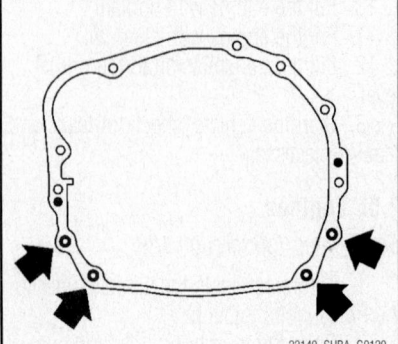

Fig. 126 Lower side of transmission bolts and nuts to engine shown—2.5L engines

Fig. 127 Engine supported with lifting device—2.5L engines

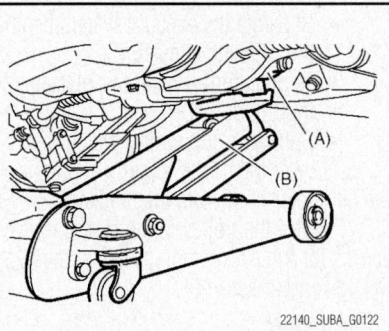

Fig. 128 Transmission supported with garage jack—2.5L engines

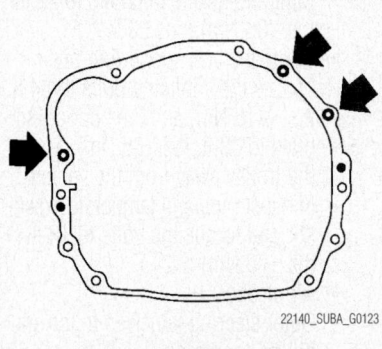

Fig. 129 Upper side of transmission bolts and nuts to engine—2.5L engines

21. Remove the pitching stopper.

22. Disconnect the fuel hoses from fuel pipe.

23. Remove the clip and disconnect the evaporation hose from the pipe.

24. Support the engine with a lifting device.

25. Support the transmission with a garage jack.

### ✳✳ CAUTION

**Before removing the engine away from transmission, check to be sure no work has been overlooked.**

26. Remove the starter.

27. Remove the bolts which hold upper side of transmission to engine.

28. Attach the ST498277200 to the converter case to keep converter from slipping out. (AT models)

29. Carefully remove the engine from vehicle as follows:

- Slightly raise the engine.
- Raise the transmission with garage jack.
- Move the engine horizontally until main shaft is withdrawn from clutch cover.
- Slowly move the engine away from engine compartment.

➡ Be careful not to damage adjacent parts or body panels with crank pulley, oil level gauge, etc.

30. Remove the front cushion rubber engine mounts.

### To install:

31. Install the front cushion rubber mounts to engine, tighten to 26 ft. lbs. (35 Nm).

32. Position the engine in engine compartment and align it with transmission.

33. Apply a small amount of grease to splines of main shaft. (MT models)

34. Tighten the bolts which hold upper side of transmission to engine, to 37 ft. lbs. (50 Nm).

35. Remove the lifting.

36. Remove the garage jack.

37. Install the pitching stopper.

38. Remove the ST498277200 from converter case. (AT models)

39. Install the starter.

40. Tighten the bolts which hold torque converter clutch to flexplate (AT models).

41. Tighten other bolts while rotating the crankshaft using a socket wrench.

42. Tighten the bolts to 18 ft. lbs. (25 Nm).

43. Attach the service hole plug to prevent entry of foreign objects.

44. Install the power steering pump.

45. Install the power steering pump to the bracket, and tighten the bolts to 16 ft. lbs. (22 Nm).

46. Connect the power steering switch connector.

47. Install and adjust the front side belt.

48. Lift up the vehicle.

49. Tighten the bolts and nuts which hold the lower side of the transmission to engine. Tighten to 37 ft. lbs. (50 Nm).

50. Tighten the nuts which install the front cushion rubber onto crossmember to 55 ft. lbs. (75 Nm).

➡**Make sure the front cushion rubber mounting bolts and locator are securely installed.**

51. Install the front and center exhaust pipe.

52. Lower the vehicle.

53. Connect the following hoses:
   - Fuel delivery hose and evaporation hose
   - Heater inlet and outlet hoses
   - Brake booster vacuum hose

54. Connect the following connectors:
   - Front oxygen (A/F) sensor connector
   - Rear oxygen sensor connector
   - Engine harness connectors
   - Alternator connector and terminal
   - A/C compressor connector

55. Install the air intake chamber stay and tighten to 12 ft. lbs. (16 Nm).

56. Tighten the engine ground cable to 10 ft. lbs. (14 Nm).

57. Install the A/C pressure hoses.

58. Install the radiator to vehicle.

59. Install the air intake duct, air cleaner case and air intake chamber.

60. Install the undercover.

61. Install the battery in the vehicle, and connect cables.

62. Fill engine coolant and bleed the cooling system.

63. Check the ATF level and replenish it if necessary.

64. Vacuum and charge the A/C system with refrigerant. Test for leaks at connections.

65. Remove the front hood stay, and close the front hood.

66. Lower the vehicle from the lift.

67. Road test and check for leaks.

## 2.8L Engine

1. Before servicing the vehicle, refer to the Precautions Section.

2. Disconnect the negative battery cable.

3. Drain the oil and coolant from the engine and the fluid from the gearbox.

4. Remove the alternator.

5. Remove the right-hand driveshaft with intermediate shaft.

6. Remove the bearing anchorage of the intermediate shaft.

7. Remove the engine mounting from the engine.

8. Remove the drive belt.

9. Remove the power steering pump. Let the servo hoses remain on the pump.

10. Unplug the control module connectors and disconnect the ground cable.

11. Unhook the power steering reservoir and servo hose clips.

12. Remove the A/C bracket and the turbocharger delivery pipe mounting.

13. Remove the power steering pump bracket.

14. Remove the secondary air injection pump complete with connectors, hoses and mountings.

15. Detach the turbocharger delivery hose from the turbocharger.

16. Detach the coolant hoses from the oil cooler. Collect any fluid spill.

17. Detach the coolant pipe with hoses from the thermostat housing.

18. Remove the starter motor.

19. If equipped with automatic transaxle, remove the bolts holding the torque converter in place.

20. Remove the wiring harness between the starter motor and alternator as well as the one from the rear of the intake manifold. Release the two clips from the back of the engine and the cable duct on the gearbox.

21. If equipped with manual transaxle, remove the cable mounting from the gearbox.

22. If equipped with automatic transaxle, remove the clamp and ground cable from the plate bracket and unplug the connector from the gear selector position sensor/control module.

23. Detach the 7-pin connector and the cable duct from the intake manifold.

24. Remove the upper and lower heat shields from the turbocharger.

25. Unplug the connectors of the heated oxygen sensors.

26. Detach the upper oxygen sensor from the catalytic converter.

27. Remove the catalytic converter with mounting.

28. Remove the upper mounting of the catalytic converter from the engine.

29. Remove the coolant pipe between the heat exchanger and thermostat housing.

30. Unplug the connectors and remove the engine control module with mounting.

31. Connect 83 92 409 Lifting Yoke to the engine and unload with an engine lift.

32. Remove the bolts holding the torque arms to the subframe.

33. Raise the engine from the subframe.

34. Remove the 4 lower bolts of the transaxle.

35. Place the engine on the floor.

36. Remove the transaxle bolts and lift away the transaxle.

37. If equipped with automatic transaxle, press the torque converter against the gearbox.

38. If equipped with automatic transaxle, install 87 92 574 Holder to hold the torque converter in place.

39. Transfer the crankcase ventilation hose.

### To install:

➡**Read the engine designation (L or E) and the serial number on the old engine and stamp in the equivalent on the new one.**

40. If equipped with automatic transaxle, turn the torque converter so that the bolt holes align with the holes on the driver plate and turn the engine so that the oval hole in the driver is centered at the starter motor opening. Install 87 92 574 Holder to hold the torque converter in place during installing. Lubricate the torque converter guide pin with grease.

41. Make sure the two guide sleeves are installed on the engine and apply grease to the guide sleeves.

42. Install the transaxle on the engine. If equipped with automatic transaxle, remove 87 92 574 Holder just before the gearbox is in place.

43. Tighten the upper bolts between the engine and transaxle to 14 ft. lbs. (19 Nm). If equipped with automatic transaxle, install the plate bracket for the ground cable.

44. Connect 83 92 409 Lifting Yoke, raise the engine with an engine lift and place it on the subframe.

45. Install the bolts that secure the torque arms to the subframe. Tighten the rear torque arm bolts to 59 ft. lbs. (80 Nm), and the front torque arm bolts to 44 ft. lbs. (60 Nm) plus an additional 90°.

46. Install the lower bolts between the engine and transaxle.

47. If equipped with automatic transaxle, perform the following:

   a. Press the torque converter against the driver plate. Install the plug.

   b. Apply a thread locking adhesive to the bolts holding the torque converter to the flexplate. Use the original bolts with corresponding washers. The torque converter will be ruined if the bolts that are used are too long.

   c. Turn the crankshaft clockwise with the pulley and tighten the bolts one at a time to 46 ft. lbs. (62 Nm), starting with the oval hole in the flexplate.

48. Install the coolant pipe between the heat exchanger and thermostat housing. Use a new O-ring coated with acid-free Vaseline.

49. Position the wiring harness.

50. If equipped with automatic transaxle, plug in the connector of the gear selector position sensor/control module. Install the clamp and ground cable to the plate bracket.

51. If equipped with manual transaxle, install the cable mounting to the transaxle.

52. Connect the two clips on the rear of the engine and install the cable duct on the transaxle.

53. Install the engine control module with mounting.

54. Install the cable duct of the intake manifold and plug in the 7-pin connector.

55. Install the starter.

56. Install the coolant pipe to the thermostat housing. Use a new O-ring coated with acid-free Vaseline.

57. Attach the coolant hoses to the oil cooler.

58. Attach the turbocharger delivery hose to the turbocharger.

59. Install the secondary air injection pump complete with connectors, hoses and mountings.

60. Install the upper mounting of the catalytic converter to the engine and tighten the M10 size bolt to 30 ft. lbs. (42 Nm).

61. Install the catalytic converter with bracket, using Anti-Seize on the nut threads. Install the upper heated oxygen sensor. Tighten the flange-to-turbocharger bolts to 18 ft. lbs. (25 Nm), mounting-to-oil sump bolts to 14 ft. lbs. (19 Nm), and the mounting-to-exhaust flange bolts to 30 ft. lbs. (40 Nm).

62. Plug in the connectors of the heated oxygen sensors.

63. Install the upper and lower heat shields to the turbocharger.

64. Install the power steering pump bracket.

65. Install the A/C bracket and the turbocharger delivery pipe mounting. Tighten the A/C bracket bolts to 28 ft. lbs. (38 Nm).

66. Hook on the power steering fluid reservoir; secure the hose with the clip.

67. Plug in the engine control module connectors and connect the ground cable.

68. Install the power steering pump and tighten the mounting bolts to 14 ft. lbs. (19 Nm).

69. Install the drive belt.

70. Install the engine mounting to the engine and tighten the mounting bolts to 69 ft. lbs. (93 Nm).

71. Install the bearing anchorage of the intermediate shaft and tighten the bolts to 18 ft. lbs. (24 Nm).

72. Install the alternator bracket. Tighten the M8 size bolt to 16 ft. lbs. (22 Nm), and the M10 size bolt to 28 ft. lbs. (38 Nm).

73. Install the alternator.

74. Install the right-hand driveshaft with intermediate shaft.

75. Fill the engine with coolant

76. Fill the engine with clean oil.

77. Fill the transmission to the proper level.

78. Connect the negative battery cable.

79. Start the vehicle, check for leaks and repair if necessary.

80. Bleed the cooling system, if necessary.

81. Bleed the clutch system (manual transmission).

82. Restore all electrical functions.

## EXHAUST MANIFOLD

### REMOVAL & INSTALLATION

#### 2.0L H4 Engine

*See Figure 130.*

1. Remove or disconnect the following:
   - Negative battery cable
   - Front Oxygen Sensor
   - Front undercover, if equipped
   - Lower exhaust manifold cover on the right hand side
   - Upper and lower exhaust manifold covers on the left hand side
   - Nuts and bolts that attach the front exhaust pipe to the turbocharger joint pipe
   - Nuts that attach the front exhaust pipe to the to the cylinder head while holding the front pipe
   - Front exhaust pipe assembly
   - Covers from the front exhaust pipe and manifold
   - Front exhaust pipe from the manifolds and discard the gaskets

*To install:*

2. Clean all gasket surfaces completely.
3. Install or connect the following:
   - New gaskets
   - Front exhaust pipe to the manifolds and tighten the retainers to 26 ft. lbs. (35 Nm)
   - Covers to the front exhaust pipe and tighten to 18 ft. lbs. (25 Nm)

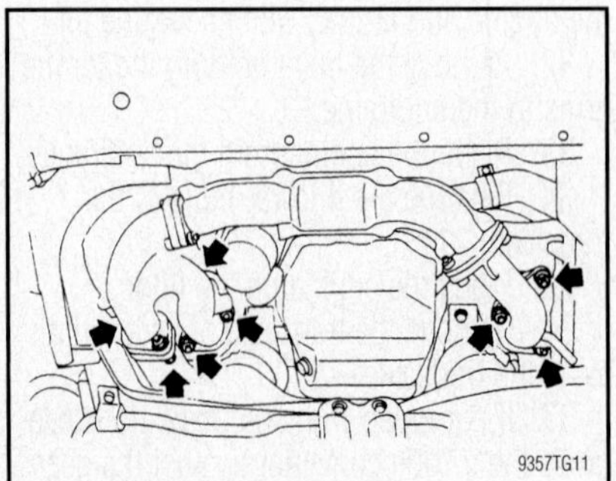

**Fig. 130 Location of the front exhaust pipe retainers—2.0L H4 engine**

9357TG11

- Upper exhaust manifold cover on the right hand side and tighten the retainers to 13 ft. lbs. (19 Nm)
- Front exhaust pipe assembly and tighten the retainers to 26 ft. lbs. (35 Nm)
- Right hand side manifold to the turbocharger joint pipe and tighten the retainers to 13 ft. lbs. (19 Nm)
- Upper and lower manifold covers on the left hand side to 13 ft. lbs. (19 Nm)
- Front Oxygen Sensor
- Front undercover, if equipped
- Negative battery cable

#### 2.0L I4 Engine

1. Before servicing the vehicle, refer to the Precautions Section.
2. Drain the cooling system.
3. Remove or disconnect the following:
   - Negative battery cable
   - Turbocharger
   - Heat Shield
   - Exhaust manifold and the stay.

*To install:*

4. Clean all mating surfaces of any residual gasket material.
5. Install or connect the following:
   - New gaskets.
   - Exhaust manifold and torque the bolts to 18 ft. lbs. (25 Nm), then to 23 ft. lbs. (32 Nm).
   - Exhaust manifold stay. Tighten to 35 ft. lbs. (48 Nm)
   - Heat shield. Tighten to 16 ft. lbs. (22 Nm)
   - Turbocharger
   - Negative battery cable
6. Start the vehicle, check for leaks and repair if necessary.

#### 2.3L Engine

1. Before servicing the vehicle, refer to the Precautions Section.
2. Drain the cooling system.
3. Remove or disconnect the following:
   - Negative battery cable
   - Lower front cover
   - Turbocharger-to-block mounting bolt
   - Turbocharger upper mounting bolt
   - Turbocharger coolant pipe
   - Turbocharger return hose
   - Oil pipe from the turbocharger to the oil filter adapter
   - Bypass pipe and valve
   - Exhaust manifold heat shield
   - Solenoid connector
   - MAF sensor-to-turbocharger connecting hose

- Crankcase breather pipe (bend pipe aside)
- Solenoid valve hoses (note positions)
- Engine lifting eye
- Turbocharger intake pipe from V-clamp (plug turbocharger inlet opening)
- 2 upper nuts between turbocharger and front exhaust pipe (move exhaust aside)
- Turbocharger coolant return pipe from cylinder head and pressure sensor mount
- Wastegate from turbocharger
- Nuts holding turbocharger to exhaust manifold (slightly lower turbocharger)
- Relieve auxiliary drive belt tensioner
- Auxiliary drive belt from power steering pump pulley
- Through-bolts, lower bolt and 2 upper bolts holding power steering pump (lift pump and bracket aside)
- Exhaust manifold

### To install:

4. Install or connect the following:
- Exhaust manifold with a new gasket and torque the bolts to 18 ft. lbs. (25 Nm)
- Power steering pump and bracket. Torque the upper left mounting bolt to 18 ft. lbs. (24 Nm) first then torque the remaining bolts to the same specification
- Auxiliary drive belt to power steering pump pulley and tighten it with the tensioner

➡**Fill the turbo inlet with oil. Rotate the compressor wheel by hand several times to assure the bearings are well lubricated.**

- Turbocharger to the exhaust manifold and torque the lock nuts to 18 ft. lbs. (24 Nm) and the studs to 16 ft. lbs. (22 Nm)
- Front exhaust pipe to the turbocharger and torque the bolts to 18 ft. lbs. (24 Nm)
- Wastegate
- Coolant return pipe to turbocharger with new copper washers and torque the bolts to 18 ft. lbs. (25 Nm)
- Coolant return pipe to cylinder head and torque bolts to 18 ft. lbs. (25 Nm)
- Turbocharger intake pipe with a new lubricated O-ring to the power steering pump bracket. Torque the

bolt and adjusting screw to 18 ft. lbs. (24 Nm).
- Coolant pipe to turbocharger, with new sealing washers, and torque to 18 ft. lbs. (25 Nm)
- Oil pipe from turbo to oil filter adapter, with new washers, and torque to 18 ft. lbs. (25 Nm)
- Turbocharger-to-block upper bolt then remaining bolt torque to 18 ft. lbs. (24 Nm)
- Lower nut on joint between turbocharger and exhaust pipe
- Intake pipe with lifting eye to turbocharger
- Solenoid valve hoses
- Solenoid valve connector
- Crankcase-to-camshaft cover and turbo intake pipe and torque the bolts to 18 ft. lbs. (24 Nm)
- MAF sensor-to-turbocharger intake pipe hose
- Bypass hose and valve with new O-ring
- Exhaust manifold heat shield
- Negative battery cable

5. Fill the cooling system and bleed by running engine (A/C off), at varying speeds, until cooling fan comes on. Stop the engine and top up the coolant.
6. Check and adjust level of engine oil
7. Start the vehicle, check for leaks and repair if necessary.

## 2.5L Engines

Due to the unique design of the 2.5L engine, an exhaust manifold is not used. The exhaust enters directly into the front Y-pipe.

## 2.8L Engine

### Front

1. Before servicing the vehicle, refer to the Precautions Section.
2. Disconnect the negative battery cable.
3. Remove or disconnect the following:
- Fan cowling
- Intake manifold of secondary air injection pump from pump
- Discharge pipe of secondary air injection pump from pump
- Intake air sensor connector
- Turbocharger delivery pipe
- Dipstick tube
- Front exhaust manifold heat shield
- Intermediate exhaust pipe from turbocharger and front exhaust manifold
- Secondary air pipe bolts to raise it slightly

4. Lubricate the studs and nuts with rust removal oil and carefully loosen.

5. Remove the exhaust manifold.

### To install:

6. Clean the contact surfaces from gasket residue and soot.
7. Install the exhaust manifold using new bolts and a new gasket. Tighten the bolts to 15 ft. lbs. (20 Nm).
8. Install the secondary air pipe using a new gasket and tighten the bolts to 17 ft. lbs. (23 Nm).
9. Connect the intermediate exhaust pipe to the front exhaust manifold and the turbocharger. Use a new gasket and coat the studs with Anti-Seize. Install all nuts.

➡**The conical side of the nut should face the center exhaust pipe.**

10. Tighten the mounting nuts first on the exhaust manifold and then on the turbocharger to 22 ft. lbs. (30 Nm). Tighten in an alternating fashion on the turbocharger. The surfaces should align with one another when the nuts are installed. Check using a feeler gauge. If the surfaces are not flat or the pipe is deformed, the center exhaust pipe and/or exhaust manifold must be replaced.
11. Install or connect the following:
- Front exhaust manifold heat shield. Tighten fasteners to 7 ft. lbs. (10 Nm).
- Dipstick tube with new O-rings
- Turbocharger delivery pipe
- Intake air sensor connector
- Discharge pipe of secondary air injection pump to the pump
- Intake manifold of secondary air injection pump to the pump
- Fan cowling

12. Connect the negative battery cable.
13. Start the engine, check for leaks and repair if necessary.

### Rear

1. Before servicing the vehicle, refer to the Precautions Section.
2. Disconnect the negative battery cable.
3. Remove or disconnect the following:
- Front catalytic converter
- Drive belt
- Alternator

4. Raise and safely support the vehicle.
5. Remove the exhaust manifold heat shield.
6. Lower the vehicle.
7. Remove the intermediate exhaust pipe.
8. Raise and safely support the vehicle.
9. Detach the rear secondary air pipe between the exhaust manifold and the check valve.
10. Remove the exhaust manifold.

## To install:

11. Clean the contact surfaces from gasket residue and soot.

12. Install new gaskets. Coat the studs with Anti-Seize.

13. Install the exhaust manifold using new bolts and a new gasket. Tighten the bolts to 15 ft. lbs. (20 Nm).

14. Attach the rear secondary air pipe between the exhaust manifold and the check valve.

15. Lower the vehicle.

16. Connect the intermediate exhaust pipe to the front exhaust manifold and the turbocharger. Use a new gasket and coat the studs with Anti-Seize. Install all nuts.

➡**The conical side of the nut should face the center exhaust pipe.**

17. Tighten the mounting nuts first on the exhaust manifold and then on the turbocharger to 22 ft. lbs. (30 Nm). Tighten in an alternating fashion on the turbocharger. The surfaces should align with one another when the nuts are installed. Check using a feeler gauge. If the surfaces are not flat or the pipe is deformed, the center exhaust pipe and/or exhaust manifold must be replaced.

18. Raise and safely support the vehicle.

19. Install the exhaust manifold heat shield. Tighten fasteners to 7 ft. lbs. (10 Nm).

20. Lower the vehicle.

21. Install the generator to the bracket.

22. Install the drive belt.

23. Install the front catalytic converter.

24. Connect the negative battery cable.

25. Start the engine, check for leaks and repair if necessary.

## INTAKE MANIFOLD

### REMOVAL & INSTALLATION

#### 2.0L H4 Engine

*See Figures 131 and 132.*

1. Release the fuel system pressure.

2. Remove or disconnect the following:
- Negative battery cable
- Engine cover, if necessary
- Air cleaner upper cover and boot
- Air cleaner filter
- Intercooler
- Accelerator cable
- Coolant filler tank

3. Remove the power steering pump from the bracket by performing the following steps:

a. Loosen the lock and slider bolts.

b. Remove the V-belt.

c. Disconnect the power steering switch connection.

d. Remove the bolts that attach the power steering pump pipe brackets to the intake manifold. Do not disconnect the hose.

e. Bolts that attach the power steering pump bracket.

f. Remove the power steering tank from the bracket by pulling it upwards.

g. Place the power steering pump on the wheel apron on the right.

4. Remove or disconnect the following:
- Emission hose from the Positive Crankcase Ventilation (PCV) valve
- Engine coolant temperature hoses from the throttle body
- Brake booster hose
- Pressure hose from the intake duct
- Engine harness connectors from the bulkhead connections

5. Disconnect the following electrical connections:
- Engine Coolant Temperature (ECT) sensor
- Oil pressure switch
- Crankshaft Position (CKP) sensor
- Knock Sensor (KS)
- Camshaft Position (CMP) sensor
- Ignition coil

6. Remove or disconnect the following:
- Engine harness fixed clip from the bracket
- Fuel delivery, return and evaporative hoses
- Intake manifold bolts
- Intake manifold and gasket

## To install:

7. Install or connect the following:
- Intake manifold and gasket
- Intake manifold bolts and tighten to 18 ft. lbs. (25 Nm)
- Fuel delivery, return and evaporative hoses
- Engine harness fixed clip from the bracket

8. Connect the following electrical connections:
- Oil pressure switch
- CKP sensor
- ECT sensor
- KS sensor
- CMP sensor
- Ignition coil
- Engine harness connectors to the bulkhead connections

9. Install or connect the following:
- Brake booster hose
- Engine coolant temperature hoses to the throttle body
- Emission hose to the PCV valve
- Pressure hose to the intake duct

10. Install the power steering pump on the bracket by performing the following steps:

a. Install the power steering tank on the bracket.

b. Attach the power steering switch connection.

c. Install the bolts that attach the power steering pump bracket and tighten to 16 ft. lbs. (22 Nm).

d. Attach the power steering pump pipe brackets to the intake manifold.

e. Install the V-belt.

11. Install or connect the following:
- Coolant filler tank
- Accelerator cable
- Intercooler
- Air cleaner filter
- Air cleaner upper cover and boot
- Engine undercover
- Negative battery cable

12. Fill the cooling system.

13. Start the engine and allow it to reach operating temperature. Check for leaks and test drive the vehicle.

#### 2.0L I4 Engine

1. Remove throttle body actuator unit.

2. Remove Trionic T8 control module.

3. Remove the solenoid valve's hose from the intake manifold.

4. Remove the atmospheric pressure sensor from the intake manifold. Allow the sensor to hang in the connector.

5. Undo the nut for the cable duct.

6. Undo the fuel pressure regulator's hose from the intake manifold.

7. Undo the EVAP canister purge valve's hose from the intake manifold and fold aside.

8. Undo the brake servo's plastic hose from the intake manifold.

9. Unplug the wiring harness' connector for the injectors.

10. Undo the wiring harness' connector from the clip under the intake manifold.

11. Unplug the injector connectors.

12. Lift up the wiring harness to the injectors.

13. Remove the upper bracket to the dipstick and undo it from the sump and fold aside.

14. Remove the intake manifold's stay.

15. Remove the intake manifold.

## To install:

16. Install a new gasket.

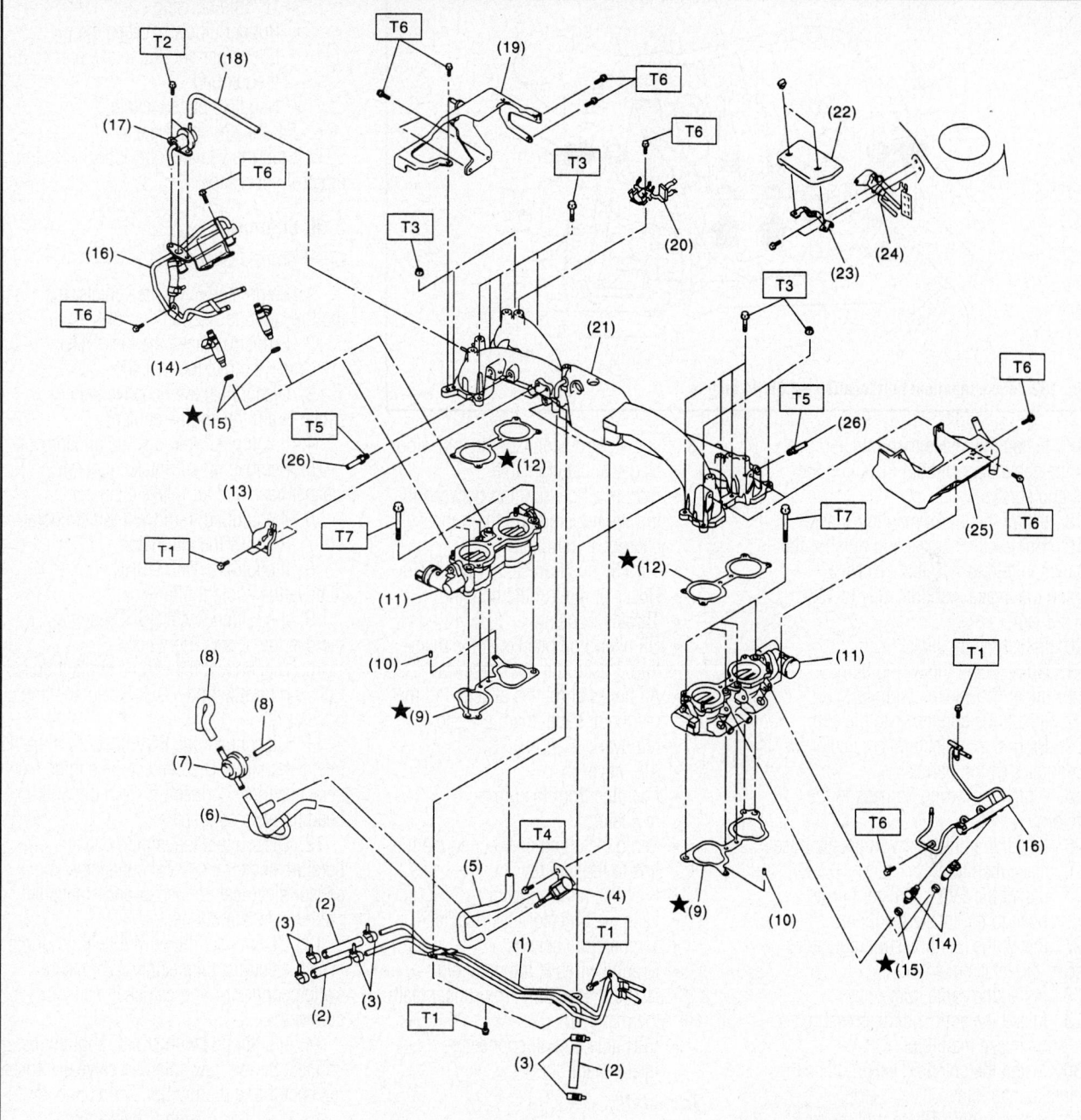

| (1) | Fuel pipe ASSY | (13) | Accelerator cable bracket | (25) | Fuel pipe protector LH |
| --- | --- | --- | --- | --- | --- |
| (2) | Fuel hose | (14) | Fuel injector | (26) | Nipple |
| (3) | Clip | (15) | Insulator | | |
| (4) | Purge control solenoid valve | (16) | Fuel injector pipe | | |
| (5) | Vacuum hose | (17) | Pressure regulator | | |
| (6) | Vacuum control hose | (18) | Pressure regulator hose | | |
| (7) | Purge valve | (19) | Fuel pipe protector RH | | |
| (8) | Purge hose | (20) | Blow-by hose stay | | |
| (9) | Intake manifold gasket | (21) | Intake manifold | | |
| (10) | Guide pin | (22) | Solenoid valve cover | | |
| (11) | Tumble generator valve ASSY | (23) | Solenoid valve cover stay | | |
| (12) | Tumble generator valve gasket | (24) | Wastegate control solenoid valve ASSY | | |

*Tightening torque: N·m (kgf-m, ft-lb)*

*T1:  4.9 (0.5, 3.6)*
*T2:  6.4 (0.65, 4.7)*
*T3:  8.25 (0.84, 6.1)*
*T4:  16 (1.6, 11.8)*
*T5:  17 (1.73, 12.5)*
*T6:  19 (1.94, 13.7)*
*T7:  25 (2.5, 18.1)*

9357TG09

**Fig. 131 Intake manifold and related components—2.0L H4 engine**

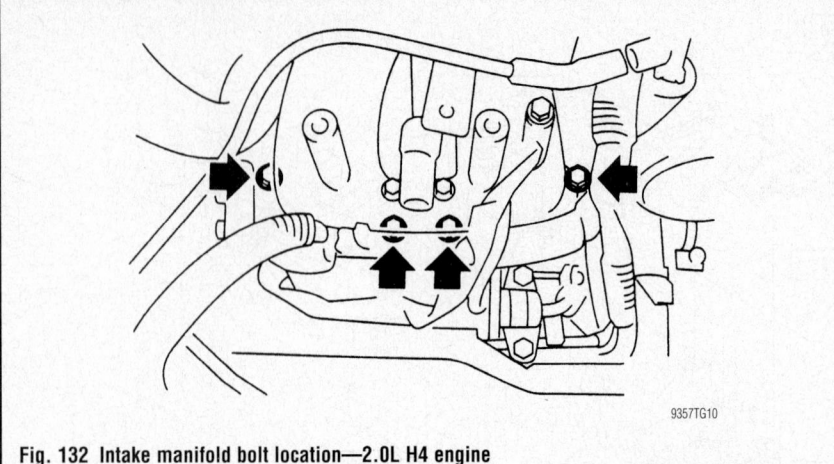

**Fig. 132 Intake manifold bolt location—2.0L H4 engine**

9357TG10

17. Install the intake manifold and position the cable duct without installing the nuts and screws.

18. Install the intake manifold's stay.

19. Tighten the screws and nuts for the intake manifold to 7 ft. lbs. (10 Nm). Tighten the intake manifold stay to 16 ft. lbs. (22 Nm).

20. Install the dipstick.

21. Guide down the wiring harness' connector under the intake manifold.

22. Install the connector to the clip.

23. Plug in the injectors' connector for the engine's wiring harness.

24. Attach the wiring harness to the injectors.

25. Install the brake servo's plastic hose to the intake manifold.

26. Install the EVAP canister purge valve's hose to the intake manifold.

27. Install the fuel pressure regulator's hose to the intake manifold.

28. Install the cable duct's nut.

29. Install the atmospheric pressure sensor to the intake manifold.

30. Install the solenoid valve's hose to the intake manifold.

31. Install Trionic T8 Control module.

32. Install Throttle body actuator unit.

## 2.3L Engine

**⁂ CAUTION**

**The fuel injection system remains under pressure, even after the engine has been turned OFF. The fuel system pressure must be relieved before disconnecting any fuel lines. Failure to do so may result in fire or personal injury.**

1. Before servicing the vehicle, refer to the Precautions Section.

2. Drain the cooling system.

3. Remove or disconnect the following:
- Negative battery cable
- Rubber elbow running between the throttle housing and the turbocharger, if equipped
- Throttle position sensor connector
- Hoses at the throttle housing
- Throttle housing
- Oil filler pipe bracket at the manifold. Position it out of the way
- All hoses and lines attached to the manifold. Label them prior to removal
- AIC valve
- Fuel line from the pressure regulator
- Banjo installing connecting the fuel line to the fuel rail
- Fuel line and regulator
- Each fuel injector electrical lead
- Temperature sensor
- Ground wires at the manifold
- Harness assembly from underneath the manifold
- EGR pipe and all connectors
- Intake manifold

**To install:**

4. Scrape off any excess gasket material

5. Install or connect the following;
- Intake manifold with a new gasket and torque the bolts in a crisscross pattern to 15 ft. lbs. (22 Nm)
- Wire harness
- EGR pipe
- Ground wires
- Temperature sensor
- Injector leads
- Fuel line to the pressure regulator
- Fuel line/regulator to the fuel rail. Secure it with a plastic tie
- Oil filler pipe bracket to the manifold
- AIC valve

- Throttle housing
- Rubber elbow between the turbocharger and the intake manifold, if equipped
- Negative battery cable

6. Fill the cooling system.

7. Start the vehicle, check for leaks and repair if necessary.

## 2.5L Engines

*See Figures 134 and 135.*

1. Before servicing the vehicle, refer to the Precautions Section.

2. Properly relieve the fuel system pressure. Remove the fuel cap.

3. Disconnect the negative battery cable. Drain the engine coolant.

4. If equipped, remove the undercover.

5. Remove the air intake duct, air cleaner case and air intake chamber.

6. If equipped, remove the intercooler.

7. Remove the alternator.

8. If equipped with DOHC engine, remove the coolant filler tank.

9. If equipped with SOHC engine, remove the spark plug wires.

10. Disconnect the engine coolant hoses from the throttle body. Disconnect the brake booster hose.

11. Disconnect the PCV hose from the intake manifold. Disconnect the engine harness electrical connectors from the bulkhead harness connectors.

12. Disconnect the engine coolant temperature sensor electrical connector, knock sensor electrical connector and crankshaft position sensor connector.

13. Disconnect the power steering pump switch electrical connector, oil pressure switch connector and camshaft sensor connector.

14. If equipped with DOHC engine, disconnect the oil flow solenoid valve electrical connector and the ignition coil connector.

15. If equipped with SOHC engine, remove the EGR pipe from the intake manifold and disconnect the fuel lines from the fuel pipe.

16. If equipped with DOHC engine, disconnect the fuel delivery hose, return hose and evaporation hose.

17. Remove the intake manifold retaining bolts. Remove the intake manifold from the engine.

**To install:**

18. Installation is the reverse of the removal procedure.

19. Be sure to use new intake manifold gaskets. Tighten the manifold retaining bolts to 18 ft. lbs. (25 Nm). and in alternating sequence.

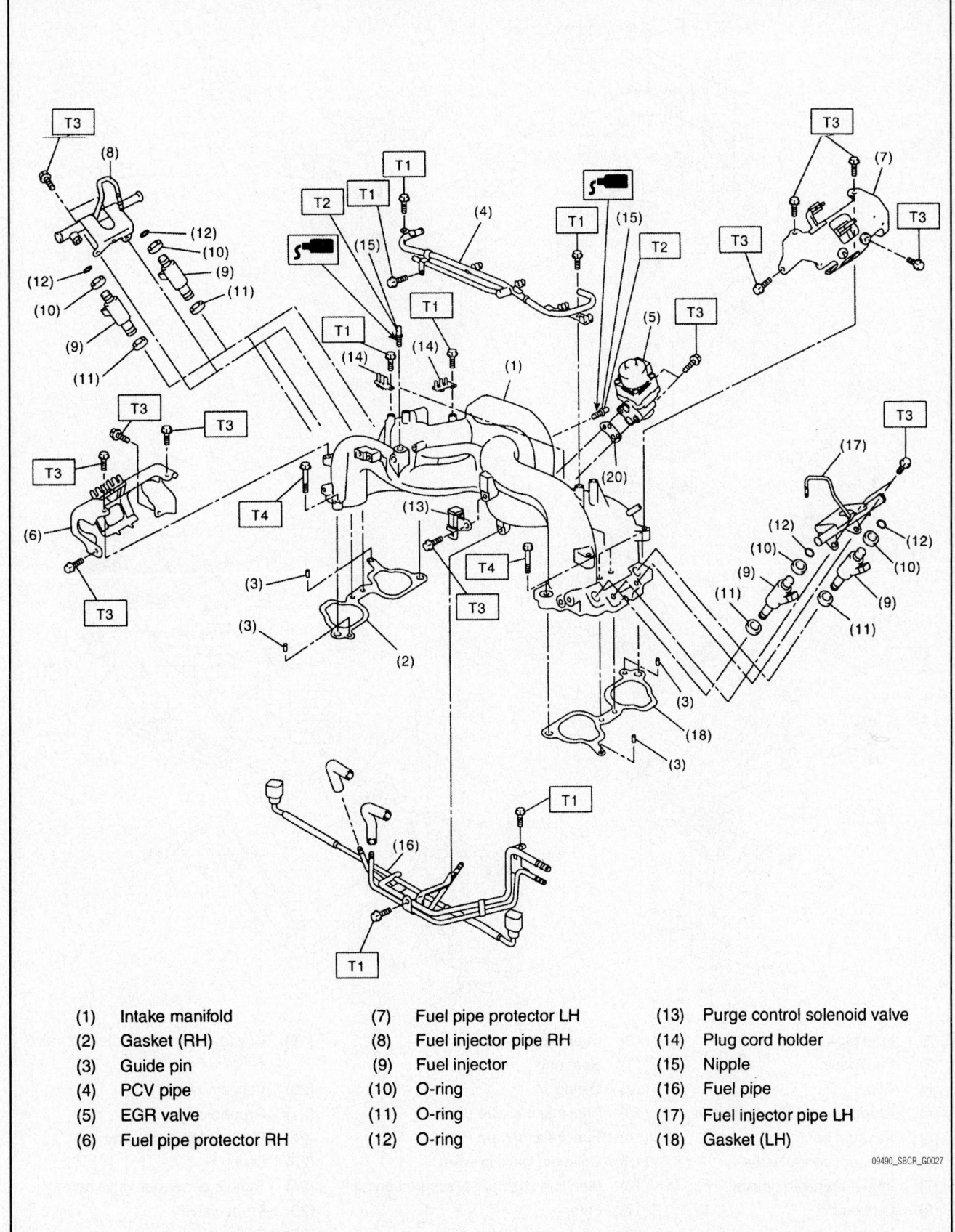

| | | | | |
|---|---|---|---|---|
| (1) | Intake manifold | (7) | Fuel pipe protector LH | (13) Purge control solenoid valve |
| (2) | Gasket (RH) | (8) | Fuel injector pipe RH | (14) Plug cord holder |
| (3) | Guide pin | (9) | Fuel injector | (15) Nipple |
| (4) | PCV pipe | (10) | O-ring | (16) Fuel pipe |
| (5) | EGR valve | (11) | O-ring | (17) Fuel injector pipe LH |
| (6) | Fuel pipe protector RH | (12) | O-ring | (18) Gasket (LH) |

09490_SBCR_G0027

Fig. 134 Intake manifold and related components—2.5L SOHC engine

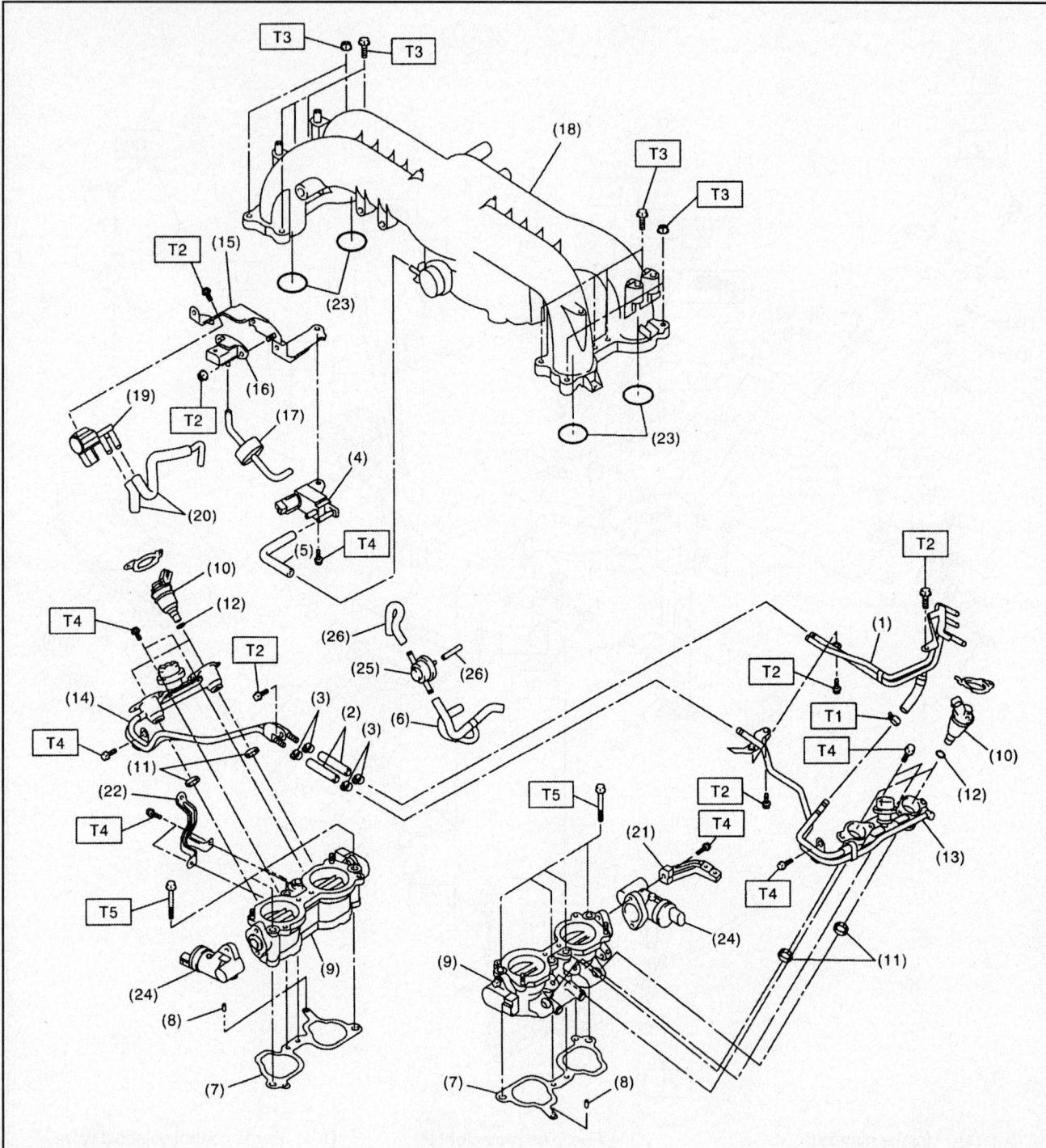

| | | | | | |
|---|---|---|---|---|---|
| (1) | Fuel pipe ASSY | (10) | Fuel injector | (19) | Wastegate control solenoid valve ASSY |
| (2) | Fuel hose | (11) | Seal ring | (20) | Vacuum hose |
| (3) | Clip | (12) | O-ring | (21) | Ground stay |
| (4) | Purge control solenoid valve | (13) | Fuel injector pipe LH | (22) | Coolant filler tank stay |
| (5) | Vacuum hose | (14) | Fuel injector pipe RH | (23) | O-ring |
| (6) | Vacuum control hose | (15) | Solenoid valve bracket | (24) | Tumble generator valve actuator |
| (7) | Intake manifold gasket | (16) | Manifold absolute pressure sensor | (25) | Purge valve |
| (8) | Guide pin | (17) | Filter | (26) | Purge hose |
| (9) | Intake manifold (lower) | (18) | Intake manifold | | |

09490_SBCR_G0028

**Fig. 135 Intake manifold and related components—2.5L DOHC engine**

20. Be sure to fill and bleed the cooling system with the proper grade and type engine coolant.

21. Start the engine and check for leaks, correct as required.

### 2.8L Engine

*See Figures 136 and 137.*

**✳✳ CAUTION**

**The fuel injection system remains under pressure, even after the engine has been turned OFF. The fuel system pressure must be relieved before disconnecting any fuel lines. Failure to do so may result in fire or personal injury.**

1. Before servicing the vehicle, refer to the Precautions Section.

2. Drain the cooling system.

3. Remove or disconnect the following:
- Negative battery cable
- Engine cover
- Power steering fluid reservoir bolt and place reservoir aside
- Connector of intake air sensor
- Upper charge air pipe
- Throttle body
- Ground cable of the control module
- Engine control module connectors
- Engine breather pipe
- Cable duct and move it aside
- Atmospheric pressure sensor bracket
- Injectors' wiring harness bracket
- Brake servo pipe from intake manifold
- Crankcase ventilation pipes from intake manifold
- Upper intake manifold bolts
- Upper intake manifold and gaskets
- Injectors' wiring harness from engine

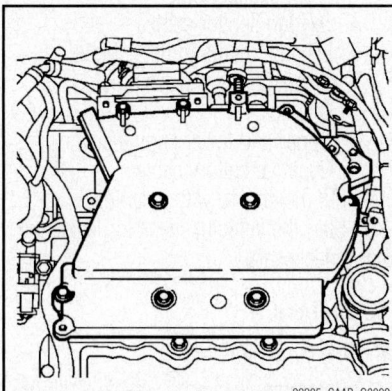

**Fig. 136 Upper intake manifold bolt locations—2.8L engine**

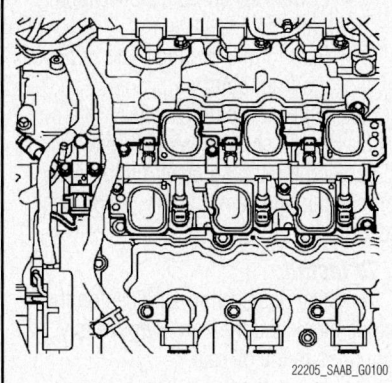

22205_SAAB_G0100

**Fig. 137 Lower intake manifold bolt locations—2.8L engine**

➡**Plug the intake ducts and blow the intake manifold clean.**

- Fuel pressure sensor connector.

4. Release any fuel pressure by pressing in the needle valve on the fuel rail. Catch any fuel spill with a cloth.
- Fuel line from fuel rail using 83 95 261 Fuel Line Tool
- Lower intake manifold bolts
- Lower intake manifold and gaskets

**To install:**

5. Plug the ducts and clean the sealing surfaces. Remove the plugs.

6. Install new gaskets and position the lower intake manifold. Install and tighten the bolts to 17 ft. lbs. (23 Nm).

7. Remove or disconnect the following:
- Fuel line to fuel rail
- Fuel pressure sensor connector
- Injectors' wiring harness to the engine

8. Clean the sealing surfaces and install new gaskets. Carefully position the upper intake manifold.

9. Install and tighten the upper intake manifold bolts to 17 ft. lbs. (23 Nm).
- Crankcase ventilation pipes to intake manifold and install bolt
- Brake servo pipe to intake manifold
- Injectors' wiring harness bracket
- Atmospheric pressure sensor bracket
- Cable duct.
- Engine breather pipe
- Engine control module connectors
- Ground cable of the control module
- Throttle body. Tighten throttle body bolts to 7 ft. lbs. (9 Nm).
- Charge air pipe
- Intake air sensor

- Power steering fluid reservoir and install bolt
- Engine cover
- Negative battery cable

10. Fill the cooling system.

11. Start the vehicle, check for leaks and repair if necessary.

### OIL PAN

*REMOVAL & INSTALLATION*

#### 2.0L H4 Engine

1. Before servicing the vehicle, refer to the Precautions Section.

2. Raise and safely support the vehicle securely on jackstands.

3. Disconnect the negative battery cable.

4. Disconnect the connector from mass air flow sensor.

5. Remove the air intake boot and air cleaner upper cover.

6. Remove the intercooler.

7. Remove the engine torque rod.

8. Remove the radiator upper brackets.

9. Support the engine with a lifting device and wire ropes.

10. Lift-up the engine.

11. Remove the undercover.

12. Drain the engine oil.

13. Remove the nuts which install the front cushion rubber onto front crossmember.

14. Remove the bolts which install the oil pan on cylinder block while raising up engine.

15. Insert the oil pan cutter blade between cylinder block-to-oil pan clearance.

**✳✳ WARNING**

**Do not use a screwdriver or similar tool in place of oil pan cutter.**

16. Remove the oil strainer.

17. Remove the baffle plate.

**To install:**

18. Before installing the oil pan, clean sealant from oil pan and engine block.

19. Install the baffle plate. Tighten to 5 ft. lbs. (6 Nm).

20. Replace the O-ring with a new one.

21. Install the oil strainer onto baffle plate. Tighten to 7 ft. lbs. (10 Nm).

22. Apply Three Bond 1207C Sealant to the mating surfaces, and then install the oil pan.

23. Tighten the bolts which install the oil pan onto engine block. Tighten to 4 ft. lbs. (5 Nm).

24. Lower the engine onto front crossmember.

25. Tighten the nuts which install the front cushion rubber onto front crossmember. Tighten to 61 ft. lbs. (83 Nm).

26. Install the undercover.

27. Lower the vehicle.

28. Remove the lifting device and steel cables.

29. Install the torque rod. Tighten engine side bolt to 37 ft. lbs. (50 Nm) and chassis side bolt to 43 ft. lbs. (58 Nm).

30. Install the radiator upper brackets.

31. Install the intercooler.

32. Install the air intake boot and air cleaner upper cover.

33. Plug in the connector to mass air flow sensor.

34. Install the front wheels.

35. Connect the battery ground cable to battery.

36. Fill engine oil.

### 2.0L I4 Engine

*See Figure 138.*

1. Before servicing the vehicle, refer to the Precautions Section.

2. Remove or disconnect the following:
- Negative battery cable.
- Front right wheel
- Right fender liner
- Drain the engine oil
- Install oil plug with a new seal and Lower the vehicle
- Upper engine cover
- SAI pump inlet hose from the pump and air cleaner
- Dipstick and dipstick pipe retaining screw
- Connector from mounting on dipstick (undo clip and pull up dipstick from oil pan)
- Oil level sensor connector
- Lower charge air pipe

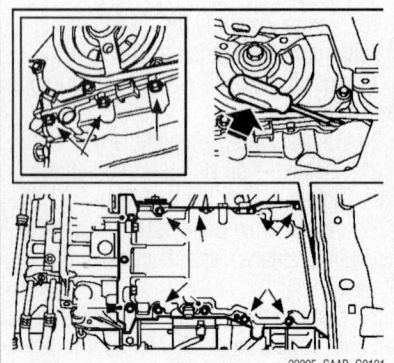

**Fig. 138 Oil pan bolt locations—2.0L I4 engine**

- Lower A/C compressor retaining bolt
- Rear torque arm bolt and remove the front bolts
- Oil pan retaining bolts

3. Place a screwdriver between the oil pan and the timing cover by the A/C compressor and carefully pry the oil pan loose.

### To install:

4. Clean the sealing surfaces on the engine block and the pan. Remove any impurities in the oil pan.

5. Apply a 2mm thick bead of 83 95 691 Sealant on the oil pan sealing surface and on the connection to the oil strainer suction pipe.

6. Install the oil pan carefully so that the sealing compound is not disturbed. Torque oil pan bolts to 16 ft. lbs. (22 Nm) and bolts in gearbox to 52 ft. lbs. (70 Nm).

7. Install the front torque arm bolts to the oil pan and tighten to 27 ft. lbs. (37 Nm). Torque the rear bolt to 52 ft. lbs. (70 Nm), plus an additional 90°.

8. Install the lower A/C compressor retaining bolt and torque to 18 ft. lbs. (24 Nm).

9. Install or connect the following:
- Charge air pipe
- Oil level sensor connector
- Oil plug with a new O-ring; torque to 18 ft. lbs. (25 Nm)
- New fastening clip on connector
- Right fender liner
- Right front wheel.
- Remove connector fastening clip and install dipstick pipe with new O-rings lubricated with 30 15 286 Vaseline
- Connector clip to the dipstick; plug in connector
- Ventilation line clip to dipstick
- Dipstick
- SAI hose to pump and filter element

10. Fill engine with the specified engine oil.

11. Start the engine and idle. Switch off the engine and wait 2-5 minutes. Check the oil level and adjust as necessary.

12. Install the upper engine cover.

### 2.3L Engine

*See Figure 139.*

1. Before servicing the vehicle, refer to the Precautions Section.

2. Drain the engine oil.

3. Remove or disconnect the following:
- Negative battery cable
- Upper engine cover

**Fig. 139 Exploded view of the oil pan and splash guard—2.3L engine**

- Dipstick
- Oxygen sensor cables
- Turbocharger bypass pipe
- Exhaust manifold heat shield
- Lower engine cover
- Front exhaust pipe
- Gearbox cover plate
- Crankcase breather hose from the oil pan
- Oil pan

### To install:

4. Transfer the splash guard and pie to a new oil pan, if replacing the pan.

5. Clean all mating surfaces of gasket material.

6. Apply an even bead of Sealant to the mating surface on the oil pan.

7. Install or connect the following:
- Oil pan and make certain that the pipe to the oil filter adapter is properly positioned in the oil pan and torque the oil pan bolts evenly to 16 ft. lbs. (22 Nm)
- Crankcase ventilation hose
- Gearbox cover plate
- Front exhaust pipe
- Oxygen sensor cables
- Exhaust manifold heat shield
- Turbocharger bypass pipe
- Dipstick
- Upper and lower engine covers
- Negative battery cable

8. Fill the engine with clean oil.

9. Start the vehicle, check for leaks and repair if necessary.

### 2.5L Engines

*See Figures 140 and 141.*

1. Before servicing the vehicle, refer to the Precautions Section.

2. Disconnect the negative battery cable.

3. Raise and support the vehicle safely.

4. Remove the front tires and wheels.

5. Lower the vehicle.

6. Remove the air intake duct and the air cleaner case. Remove the air intake chamber.

7. Remove the pitching stopper.

8. Remove the hood stay holder and the radiator upper brackets.

9. Properly support the engine with a lifting device and wire ropes.

10. Lift the vehicle and support it safely.

➡**When lifting the vehicle, raise the wire ropes at the same time.**

11. Remove the undercover.

12. Drain the engine oil.

13. Remove the front and center exhaust pipes.

14. Remove the nuts which retain the front cushion rubber onto the front crossmember.

15. Remove the bolts that retain the oil pan to the cylinder block, with the engine in the raised position.

16. Insert an oil pan gasket cutter tool into the gap between the cylinder block and the oil pan. Remove the oil pan from the engine.

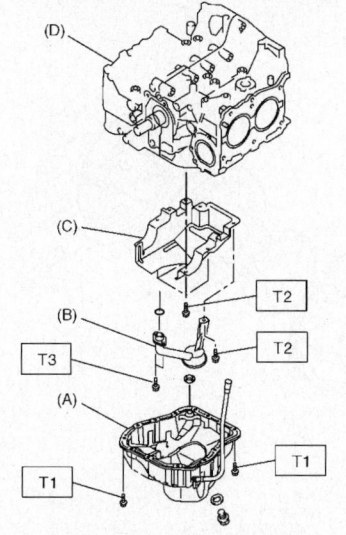

**Tightening torque:**
**T1: 5 N·m (0.5 kgf-m, 3.6 ft-lb)**
**T2: 6.4 N·m (0.65 kgf-m, 4.7 ft-lb)**
**T3: 10 N·m (1.0 kgf-m, 7.2 ft-lb)**

(A) Oil pan    (C) Baffle plate
(B) Oil strainer    (D) Cylinder block

22140_SUBA_G0125

**Fig. 140 Exploded view of oil pan, strainer, baffle plate and cylinder block—2.5L engine**

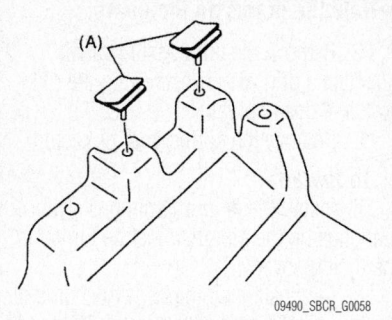

09490_SBCR_G0058

**Fig. 141 Oil pan baffle plate seal location and positioning—2.5L engine**

➡**Do not use a screwdriver or similar tool in place of the cutter tool.**

17. Remove the oil strainer, if required. Remove the baffle plate, if required.

*To install:*

18. Be sure to clean the old gasket material from the mating surfaces.

19. Apply a continuous bead of sealer to a new oil pan gasket.

20. Make sure that the seals (A) are installed securely on the baffle plate and in the direction shown in the illustration. Install the baffle plate; tighten the retaining bolts to 4.7 ft. lbs.

21. Replace the O-ring and install the oil strainer. Tighten the bolt to 7.2 ft. lbs.

22. Apply liquid gasket, part number 004403012 or equivalent, to the oil pan mating surface. Install the oil pan. Torque the retaining bolts to specification.

23. Continue the installation in the reverse order of the removal procedure.

24. Tighten the front cushion mounting bolts to 63 ft. lbs.

25. Be sure to fill the engine with the correct grade and type engine oil.

26. Start the engine and check for leaks. Correct as required.

## 2.8L Engine

*See Figures 142 and 143.*

1. Before servicing the vehicle, refer to the Precautions Section.

2. Disconnect the negative battery cable.

3. Drain the cooling system.

4. Drain the engine oil.

5. Remove or disconnect the following:
- Engine assembly
- Camshaft covers
- Timing cover
- Oil cooler assembly and gasket from cylinder block
- Oil pan. Carefully remove with a pry bar.

6. Clean sealing surfaces of gasket residue and the like.

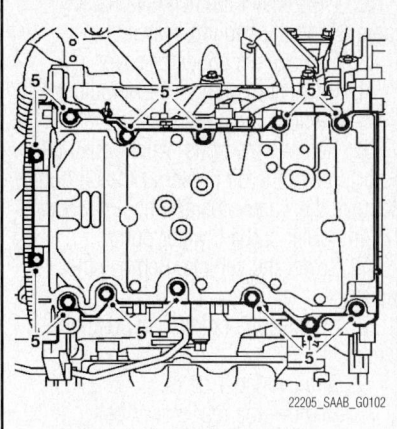

22205_SAAB_G0102

**Fig. 142 Oil pan bolt locations—2.8L engine**

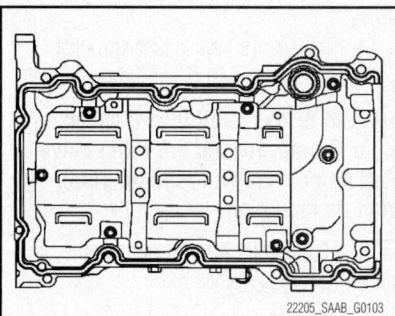

22205_SAAB_G0103

**Fig. 143 Apply a 2 mm thick bead of flange sealant on the oil sump sealing surface—2.8L engine**

*To install:*

7. Apply a 2mm thick bead of Sealant on the oil pan's sealing surface and install a new seal on the oil pan intake pipe.

8. Install the oil pan. Tighten the M6 size bolts to 7 ft. lbs. (10 Nm), and the M8 size bolts to 18 ft. lbs. (25 Nm).

9. Install a new gasket and install the oil cooler to the cylinder block. Tighten the mounting bolts to 18 ft. lbs. (25 Nm).

10. Install or connect the following:
- Timing cover
- Camshaft covers
- Engine assembly

11. Fill the engine cooling system.

12. Fill the engine with clean oil.

13. Connect the negative battery cable.

14. Start the vehicle, check for leaks and repair if necessary

## OIL PUMP

### REMOVAL & INSTALLATION

#### 2.0L H4 Engine

1. Before servicing the vehicle, refer to the Precautions Section.

2. Disconnect the negative battery cable. Drain the cooling system.

3. Remove the collector cover.

4. Raise and support the vehicle safely.

5. Remove the undercover.

6. Remove the bolts which retain the water pipe of the oil cooler to the oil pump. Remove the water pipe and hoses between the oil cooler and the water pump.

7. Lower the vehicle. Remove the radiator.

8. Remove the crankshaft position sensor.

9. Remove the V belts.

10. Remove the rear side V belt tensioner.

11. Remove the crankshaft pulley.

12. Remove the water pump.

13. If equipped, remove the timing belt guide.

14. Remove the crankshaft sprocket.

15. Remove the oil pump retaining bolts.

➡**When disassembling and checking the oil pump, loosen the relief valve plug before removing the oil pump from its mounting.**

16. Using a flat tip tool remove the oil pump from the engine.

### To install:

17. Be sure all mating surfaces are clean and free of dirt.

18. Apply three Bond 1215 Sealant to the mating surfaces of the oil pump.

19. Be sure to replace the O-ring with a new one.

20. Apply a thin coat of clean engine oil to the inside of the oil seal.

21. Position the oil pump to its mounting, aligning the notched area with the crankshaft and push the pump straight.

➡**Be sure that the oil seal lip is not folded.**

22. Install the oil pump. Tighten to 5 ft. lbs (6 Nm).

23. Continue the installation in the reverse order of the removal procedure.

24. Be sure to fill the cooling system with the proper grade and type coolant.

25. Start the engine and check for leaks. Correct, as required.

### 2.0L I4 Engine

*See Figure 144.*

1. Before servicing the vehicle, refer to the Precautions Section.

2. Disconnect the negative battery cable.

3. Remove the timing cover.

4. Remove the oil pump bolts. Lift off the cover and take out the pinions.

➡**Note the marks on the gears.**

5. Remove the oil pressure valve. Maintain a grip when unscrewing the plug so that it does not fall.

6. Remove the spring and the piston.

### To install:

7. Install the oil pump pinions. The markings on the pinions must be turned to the timing cover.

8. Lubricate the pinions with engine oil.

9. Install the oil pump cover. Tighten to 5 ft. lbs (6 Nm).

10. Install the oil pressure valve with the piston and the spring. Replace the gasket. Tighten to 29 ft. lbs (40 Nm).

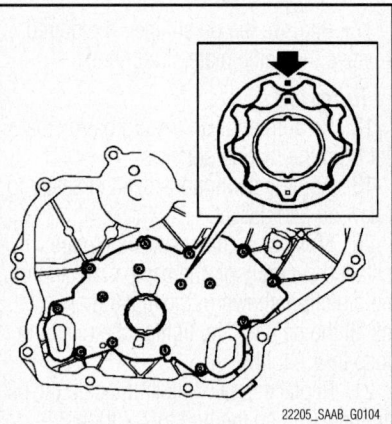

22205_SAAB_G0104

**Fig. 144 Oil pump bolt locations and gear alignment—2.0L I4 engine**

11. Install the timing cover.

12. Connect the negative battery cable.

### 2.3L Engine

*See Figure 145.*

1. Before servicing the vehicle, refer to the Precautions Section.

2. Disconnect the negative battery cable.

3. Remove the timing belt.

4. Use a lifting yoke to raise the engine approximately 1.5 in. (40 mm).

5. Raise the vehicle.

6. Remove oil pan.

7. Remove the oil suction pipe.

8. Remove the belt pulley using 32 025 006 Crankshaft Holder and 83 95 360 Crankshaft Pulley Holding Tool.

➡**Loosen the bolt by turning it CLOCK-WISE. The bolt is left-threaded.**

9. Remove the oil pump seal ring.

➡**Pry loose the seal ring from the oil pump using a suitable tool. Do not damage the sealing surfaces.**

10. Remove the oil pump.

### To install:

11. Clean the treads and sealing surfaces.

12. Install the oil pump using a new gasket. Use Thread locking adhesive, Loctite® 242 for M6x20 bolts. Install the stud of 32 025 009 Crankshaft Fixing Tool.

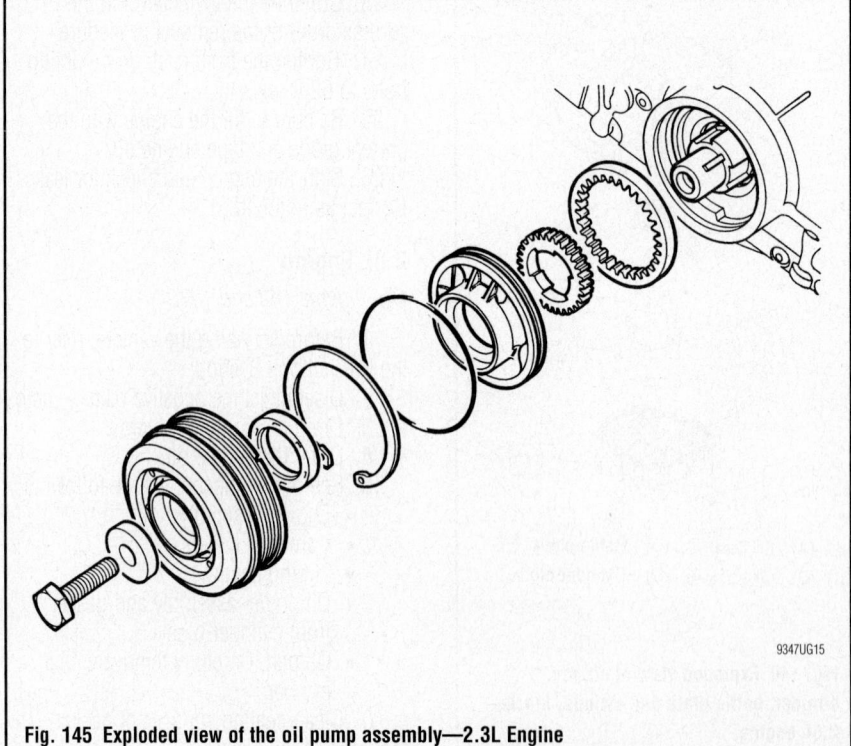

9347UG15

**Fig. 145 Exploded view of the oil pump assembly—2.3L Engine**

13. Install the oil pump seal using 32 025 014 Oil Pump Seal Installing Tool.

➡ **Moisten the gasket's seal lip with a little 90 167 353 Silicone Paste.**

14. Install the belt pulley using 32 025 006 Crankshaft Holder and 83 95 360 Crankshaft Pulley Holding Tool. Tighten to 245 ft. lbs (340 Nm).

➡ **Tighten the bolt by turning it COUNTERCLOCKWISE. The bolt is left-threaded.**

15. Remove Crankshaft holder and Crankshaft Pulley Holding Tool.
16. Install the oil intake pipe.
17. Install oil pan.
18. Lower the vehicle to the floor.
19. Install the timing belt.
20. Fill the engine with clean oil.
21. Connect the negative battery cable.
22. Start the vehicle, check for leaks and repair if necessary.

### 2.5L Engines

*See Figure 146.*

1. Before servicing the vehicle, refer to the Precautions Section.
2. Disconnect the negative battery cable. Drain the cooling system.
3. On the DOHC engine, remove the collector cover.
4. Raise and support the vehicle safely.
5. Remove the undercover.
6. On the DOHC engine, remove the bolts which retain the water pipe of the oil cooler to the oil pump. Remove the water pipe and hoses between the oil cooler and the water pump.
7. Lower the vehicle. Remove the radiator.
8. To remove the front side V belt, remove the belt covers. Loosen the lock bolt. Loosen the slider bolt. Remove the front side belt.
9. To remove the rear side V belt, remove the belt covers. Loosen the lock bolt. Loosen the slider bolt. Remove the rear side belt.
10. On the SOHC engine remove the belt tensioner.
11. On the DOHC engine remove the rear side V belt tensioner
12. Remove the crankshaft position sensor.
13. Remove the crankshaft pulley bolt.
14. Lock the crankshaft in place using tool ST499977100 for the SOHC engine and tool ST499207400 for the DOHC engine, or equivalent.
15. Remove the crankshaft pulley.

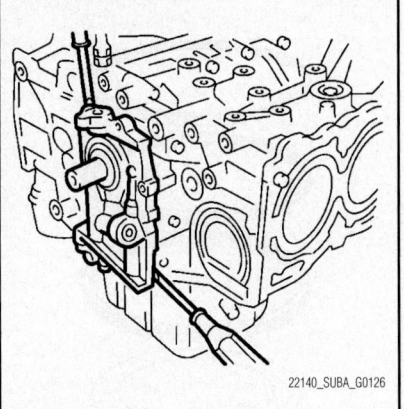

Fig. 146 Oil pump removal shown—2.5L engine

22140_SUBA_G0126

16. Remove the water pump.
17. If equipped, remove the timing belt guide. Remove the crankshaft sprocket.
18. Remove the oil pump retaining bolts.

➡ **When disassembling and checking the oil pump, loosen the relief valve plug before removing the oil pump from its mounting.**

19. Using a flat tip tool remove the oil pump from the engine.

### To install:
20. Be sure all mating surfaces are clean and free of dirt.
21. Apply liquid gasket part number 004403007, or equivalent to the mating surfaces of the oil pump.
22. Be sure to replace the O-ring with a new one.
23. Apply a thin coat of clean engine oil to the inside of the oil seal.
24. Position the oil pump to its mounting, aligning the notched area with the crankshaft and push the pump straight.

➡ **Be sure that the oil seal lip is not folded.**

25. Install the oil pump. Apply liquid gasket part number 004403042, or equivalent to the three retaining bolt threads. Install the bolts and tighten to 5 ft. lbs (6 Nm)
26. Continue the installation in the reverse order of the removal procedure.
27. Be sure to fill the cooling system with the proper grade and type coolant.
28. Start the engine and check for leaks. Correct, as required.

### 2.8L Engine
*See Figure 147.*

1. Before servicing the vehicle, refer to the Precautions Section.

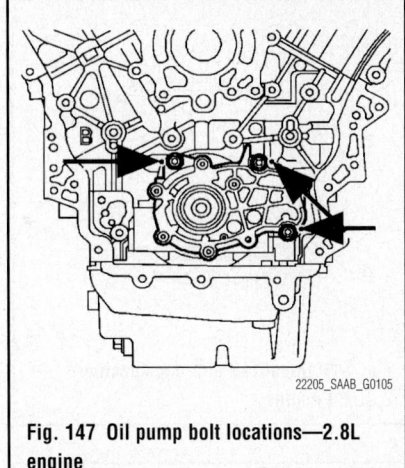

Fig. 147 Oil pump bolt locations—2.8L engine

22205_SAAB_G0105

2. Disconnect the negative battery cable.
3. Drain the cooling system.
4. Drain the engine oil.
5. Remove the timing chain, sprockets and front cover.
6. Remove the oil pump.

### To install:
7. Install the oil pump against the flat surface at the crankshaft.
8. Install the oil pump bolts and tighten to 17 ft. lbs. (23 Nm).
9. Install the timing chain, sprockets and front cover.
10. Fill the cooling system to the proper level.
11. Fill the engine with clean oil.
12. Connect the negative battery cable.
13. Start the vehicle, check for leaks and repair if necessary.

### PISTON AND RING

*POSITIONING*

*See Figures 148 through 155.*

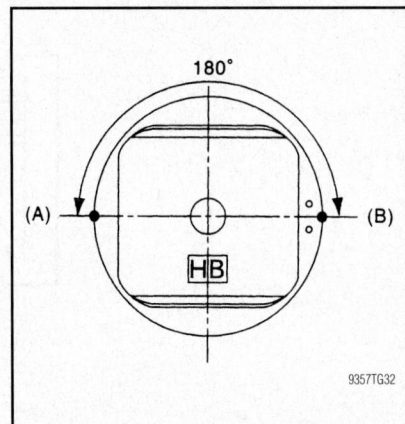

Fig. 148 Top ring end-gap spacing—2.0L H4 engine

9357TG32

(E)
(C)
35°
25°
HB
(D)
(F)
9357TG33

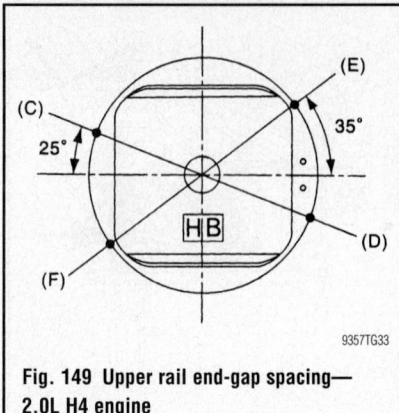

Fig. 149 Upper rail end-gap spacing—2.0L H4 engine

HB
(A)
(A) Front mark
9357TG34

Fig. 150 Piston front mark faces towards the front of the engine—2.0L H4 engine

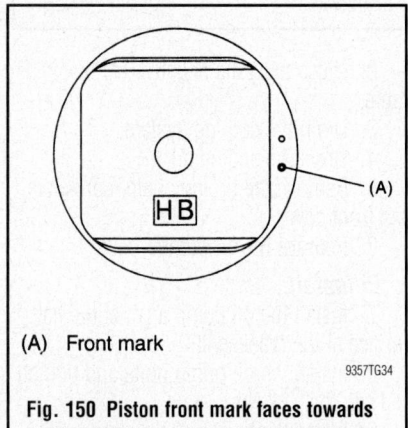

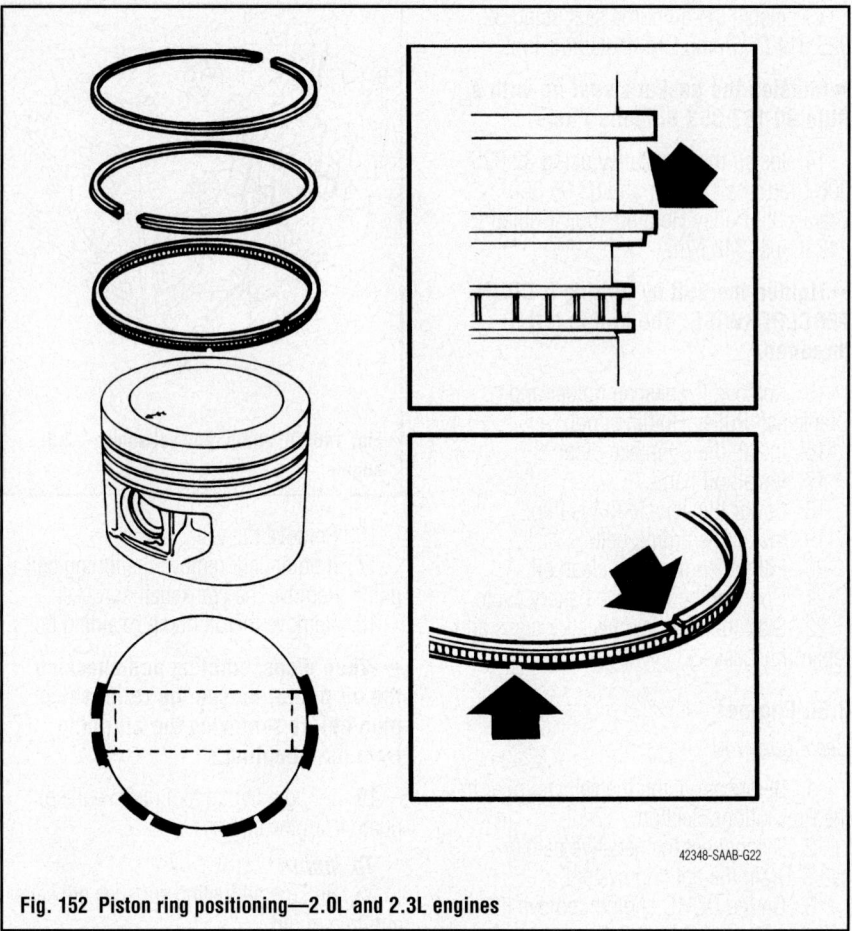

42348-SAAB-G22

Fig. 152 Piston ring positioning—2.0L and 2.3L engines

1
2
3
4
2
1. Piston, pin & connecting rod
2. Piston pin snap ring
3. Connecting rod upper bearing
4. Connecting rod bolt sleeves
42348-SAAB-G21

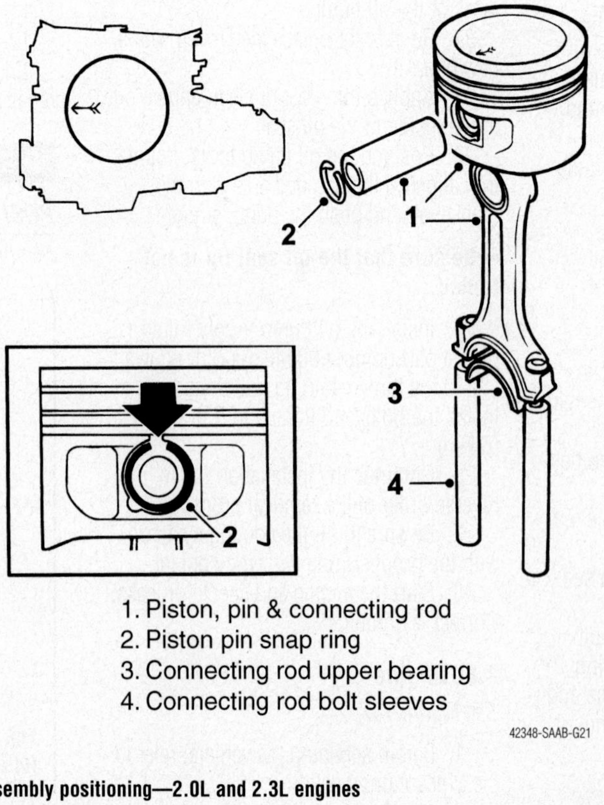

Fig. 151 Piston and connecting rod assembly positioning—2.0L and 2.3L engines

Position the top ring gap at (A) or (B) in the figure.

Position the second ring gap at 180° on the reverse side the top ring gap.

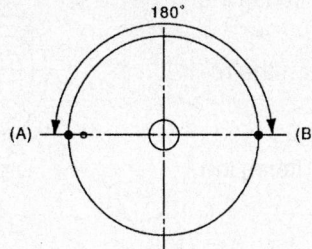

Position the upper rail gap at (C) in the figure.

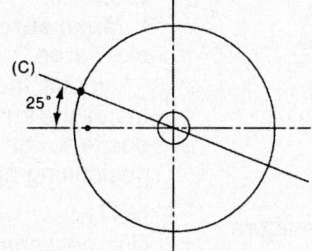

Align the upper rail spin stopper (D) to the side hole (E) on the piston.

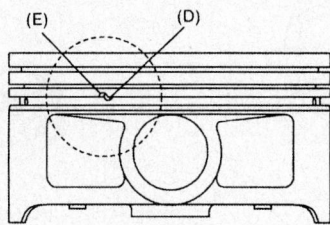

Position the expander gap at (F) in the figure on the 180° opposite direction of (C).

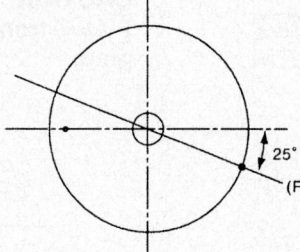

Position the lower rail gap at (G) in the figure.

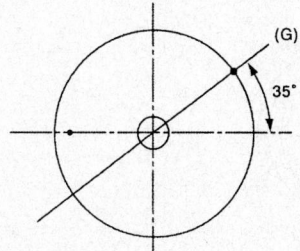

09490_SBCR_G0079

**Fig. 153 Piston ring Alignment and positioning—2.5L SOHC engine**

Position the top ring gap at (A) or (B) in the figure.

NOTE:
Assemble so that the piston ring mark "R" faces the upper side of the piston.

Position the second ring gap at 180° on the reverse side the top ring gap.

NOTE:
Assemble so that the piston ring mark "R" faces the upper side of the piston.

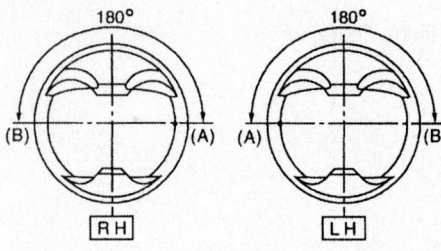

Position the upper rail gap at (C) in the figure.

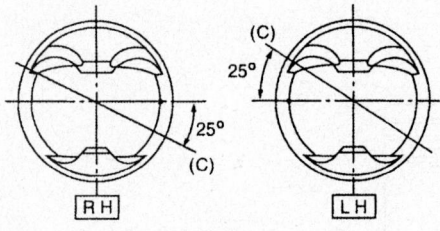

Align the upper rail spin stopper (E) to the side hole (D) on the piston.

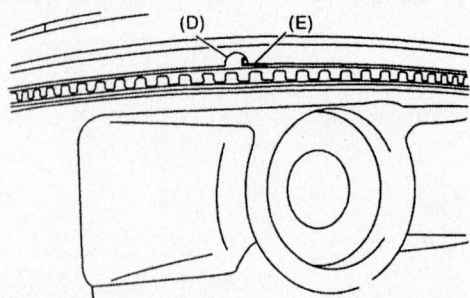

Position the expander gap at (F) in the figure.

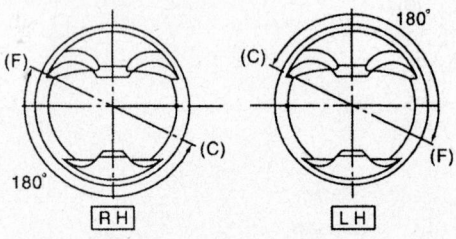

Position the lower rail gap at (G) in the figure.

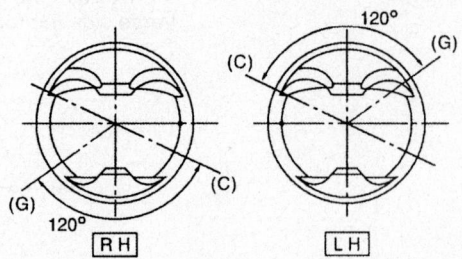

**CAUTION:**
- **Make sure ring gaps do not face the same direction.**
- **Make sure ring gaps are not within the piston skirt area.**

Install the snap ring.
Install the snap rings in the piston holes located opposite to the service holes in cylinder block when positioning all pistons in corresponding cylinders.

NOTE:
Use new snap rings.

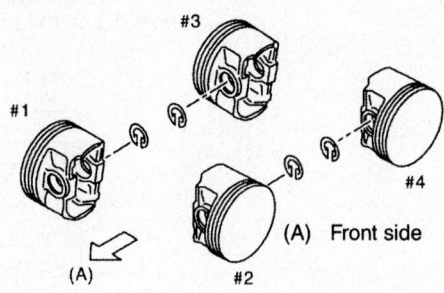

**CAUTION:**
**Piston front mark faces towards the front of engine.**

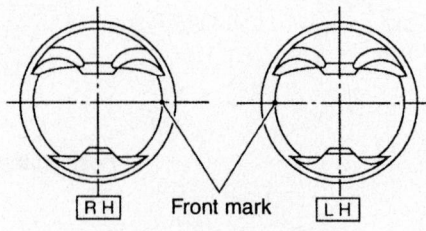

09490_SBCR_G0080

**Fig. 154 Piston ring Alignment and positioning—2.5L DOHC engine**

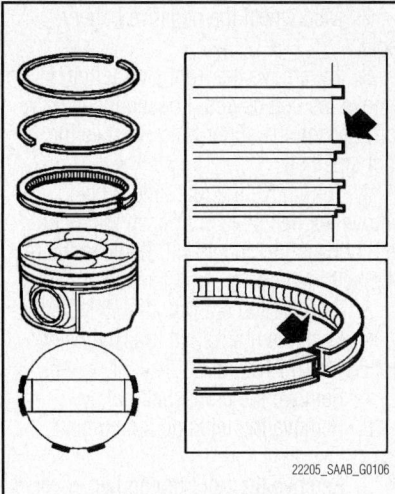

**Fig. 155 Piston ring Alignment and positioning—2.8L engine**

## REAR MAIN SEAL

*REMOVAL & INSTALLATION*

### 2.0L H4 and 2.5L Engines

*See Figure 156.*

➡**This procedure reflects Saab's recommendation that the Removal and installation procedure be performed with the engine removed from the vehicle. It may be possible to perform this procedure with the engine installed in the vehicle.**

1. Before servicing the vehicle, refer to the Precautions Section.
2. Remove or disconnect the following:
   - Engine from the vehicle
   - Clutch assembly/flywheel using the Clutch Disc Guide tool 499747000, if equipped with a manual transmission

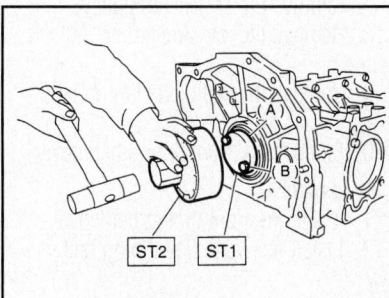

(A)   Rear oil seal
(B)   Drive plate attaching bolt

42356-SBCR-G28

**Fig. 156 Installing the rear main seal using oil seal guide ST1 499597100 and ST2 499598200**

- Torque converter flexplate from the crankshaft, if equipped with an automatic transmission
- Oil seal from the cylinder block using a small pry bar

*To install:*

3. Install or connect the following:
   - New oil seal by pressing it into the cylinder block using the appropriate driver and hammer
   - Flywheel housing using new gaskets and sealant where necessary.
   - Flywheel and tighten the bolts to specification.
   - Engine

### 2.0L I4 and 2.3L Engines

*See Figures 157 through 159.*

1. Before servicing the vehicle, refer to the Precautions Section.
2. Disconnect the negative battery cable.
3. Remove the transmission.
4. On vehicles equipped with a manual transmission, remove the pressure plate and clutch flexplate.
5. Remove the flexplate or the flywheel.
6. Carefully pry out the seal using a screwdriver.

*To install:*

7. Clean the sealing surfaces of gasket residue.
8. Position the Rear Crankshaft Seal Protective Collar 83 94 975 on the crankshaft. Lubricate the new sealing ring with non-acidic Vaseline and position it on the Installing Tool.
9. Position the Installing Tool on the crankshaft. Drive in the sealing ring with the

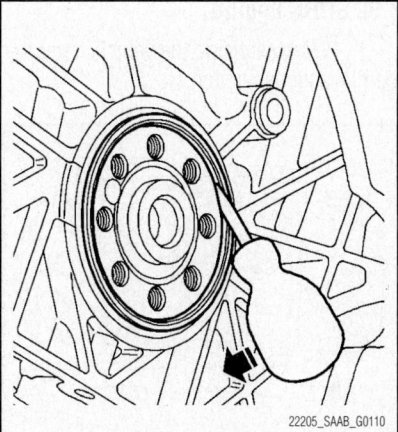

**Fig. 157 Carefully pry out the seal using a screwdriver—2.0L I4 and 2.3L engines**

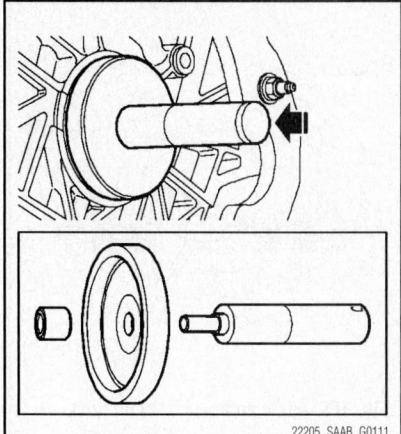

22205_SAAB_G0111

**Fig. 158 Position the rear crankshaft seal protective collar fitting tool, on the crankshaft. Lubricate the new sealing ring with non-acidic Vaseline and drive in the sealing ring with the shaft and the narrow guide sleeve until it is flush with the timing cover—2.0L I4 and 2.3L engines**

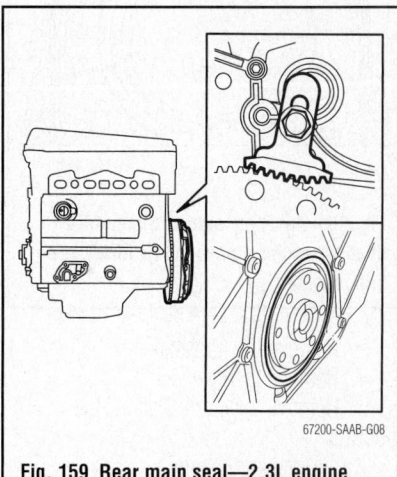

67200-SAAB-G08

**Fig. 159 Rear main seal—2.3L engine**

shaft and the narrow guide sleeve until it is flush with the timing cover. Remove the tool.

10. Install the flywheel or flexplate with new bolts.
11. On vehicles equipped with a manual transmission, install the pressure plate and the clutch flexplate.
12. Install the transmission.
13. Connect the negative battery cable.
14. Check the oil level in the engine. Top up as necessary.

### 2.8L Engine

*See Figures 160 and 161.*

1. Before servicing the vehicle, refer to the Precautions Section.
2. Disconnect the negative battery cable.

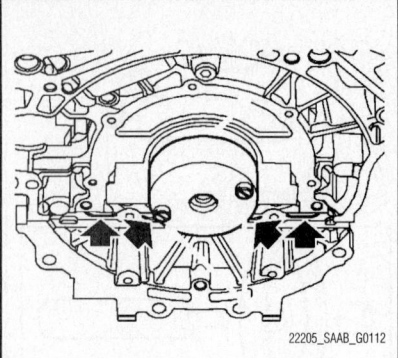

**Fig. 160 Apply an approx. 3 mm thick bead of sealing compound on the oil sump's grooves (arrows)—2.8L engine**

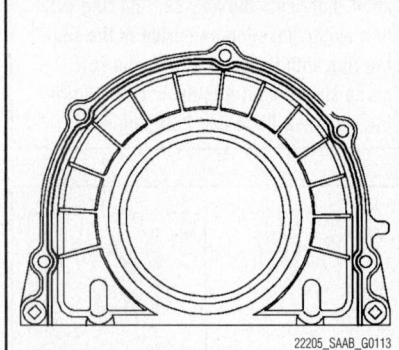

**Fig. 161 Apply an approx. 3 mm thick bead of sealing compound as illustrated— 2.8L engine**

3. Remove the transaxle.
4. Remove the flywheel.
5. Remove the crankshaft seal as follows:

a. Remove the bolts.

b. Pry off the crankshaft seal using a suitable tool from the cylinder block and the oil sump. The crankshaft sealing ring is only supplied complete with housing.

c. Clean the sealing surfaces and oil sump grooves.

*To install:*

6. Apply sealing compound and install the special tool as follows:

a. Apply an approx. 3mm thick bead of sealing compound on the oil sump's grooves.

b. Press gently so that the sealing string becomes flat.

c. Apply sealing compound on the edges to the left and right of the sealing surface.

d. Install EN-47839 Installing Tool onto the crankshaft.

e. Tighten the bolts.

➡ Check that the oil sump's grooves are filled with sealing compound and that the sealing string is applied consistently across the whole width of the oil sump's sealing surface. The component may only be touched on its metal housing - the sealing ring must not be touched!

7. Apply an approx. 3mm thick bead of sealing compound

8. Install the crankshaft seal as follows:

a. Press the crankshaft seal into the cylinder block over EN-47839 Installing Tool.

b. Install the bolts and tighten to 7 ft. lbs. (10 Nm).

9. Remove EN-47839 Installing Tool.
10. Install the flywheel with new bolts.
11. Install the transaxle.
12. Connect the negative battery cable.

## TIMING BELT FRONT COVER

### REMOVAL & INSTALLATION

#### 2.0L H4 engine

*See Figure 162.*

1. Disconnect the negative battery cable.
2. Remove the V-belt.
3. Remove the crankshaft pulley.
4. Remove the belt cover as follows:

a. Crankshaft pulley.
b. Left hand belt cover.
c. Right hand belt cover.
d. Front hand belt cover.

*To install:*

5. Install the belt covers and tighten to 3.5 ft. lbs. (5 Nm).

6. Install the crankshaft pulley and tighten the bolt to 94 ft. lbs. (127 Nm).

7. Install the V-belt.

#### 2.5L SOHC Engine

1. Before servicing the vehicle, refer to the Precautions Section.

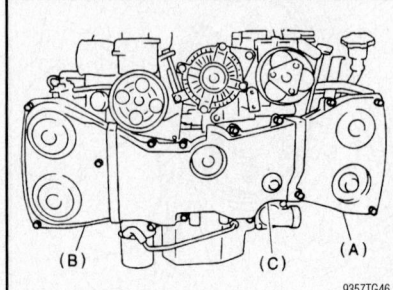

**Fig. 162 Remove the left hand (A), right hand (B) and front (C) belt covers—2.0L H4 engine**

2. Disconnect the negative battery cable.

3. To remove the front side V belt, remove the belt covers. Loosen the lock bolt. Loosen the slider bolt. Remove the front side belt.

4. To remove the rear side V belt, remove the belt covers. Loosen the lock bolt. Loosen the slider bolt. Remove the rear side belt. Remove the belt tensioner.

5. Remove the crankshaft pulley bolt.

6. Lock the crankshaft in place using tool ST499977100, or equivalent.

7. Remove the crankshaft pulley.

8. Remove the left side timing belt cover.

9. Remove the front timing belt cover.

*To install:*

10. Install the front timing belt cover.
11. Install the left side timing belt cover.
12. Install the crankshaft pulley.
13. Lock the crankshaft in place using tool ST499977100, or equivalent.
14. Install the crankshaft pulley bolt.
15. Install the rear side belt.
16. Install the belt tensioner.
17. Install the belt covers.
18. Connect the negative battery cable.

#### 2.5L DOHC Engine

1. Disconnect the negative battery cable.

2. Remove the collector cover.

3. To remove the front side V belt, remove the belt covers. Loosen the lock bolt. Loosen the slider bolt. Remove the front side belt.

4. To remove the rear side V belt, remove the belt covers. Loosen the lock bolt. Loosen the slider bolt. Remove the rear side belt. Remove the belt tensioner.

5. Remove the crankshaft pulley bolt.

6. Lock the crankshaft in place using tool ST499977100, or equivalent.

7. Remove the crankshaft pulley.

8. Remove the left side timing belt cover.

9. Remove the right side timing belt cover.

10. Remove the front timing belt cover.

*To install:*

11. Install the front timing belt cover.
12. Install the right side timing belt cover.
13. Install the left side timing belt cover.
14. Install the crankshaft pulley.
15. Lock the crankshaft in place using tool ST499977100, or equivalent.
16. Install the crankshaft pulley bolt.
17. Install the rear side belt.
18. Install the belt tensioner.

19. Install the belt covers.
20. Install the collector cover.
21. Connect the negative battery cable.

## TIMING BELT AND SPROCKETS

### REMOVAL & INSTALLATION

#### 2.0L H4 engine

*See Figures 163 through 179.*

1. Disconnect the negative battery cable.
2. Remove the V-belt.
3. Remove the crankshaft pulley.
4. Remove the belt cover as follows:
   a. Crankshaft pulley.
   b. Left hand belt cover.
   c. Right hand belt cover.
   d. Front hand belt cover.
5. Remove the timing belt guides on vehicles equipped with a manual transmission.
6. If the alignment marks that indicate rotation are faded, put new marks on the belt before removal as follows:
   a. Turn the crankshaft using crankshaft sprocket tool 499987500 and a breaker bar to align the crankshaft sprocket, left hand intake camshaft sprocket, left hand exhaust camshaft, right hand intake camshaft sprocket and right hand exhaust camshaft sprocket with the on the cover and cylinder block.
   b. Using white paint such as white out, place alignment marks on the belts in relation to the sprockets.
7. Remove the belt idler (A), illustrated in the accompanying illustration.
8. Remove the timing belt.
9. If necessary, remove belt idlers (B) and (C).
10. Remove the belt idler 2.
11. Remove the automatic belt tension adjuster assembly.

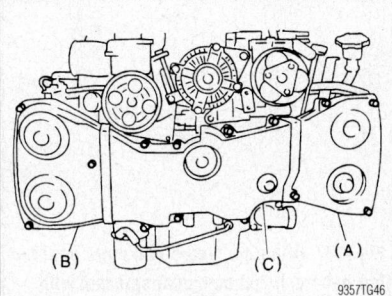

**Fig. 163 Remove the left hand (A), right hand (B) and front (C) belt covers–2.0L H4 engine**

**Fig. 164 Location of the upper belt guide–2.0L H4 engine**

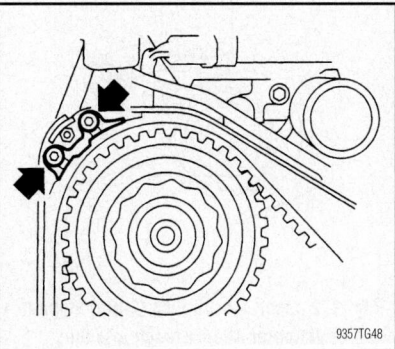

**Fig. 165 Location of the upper left belt guide–2.0L H4 engine**

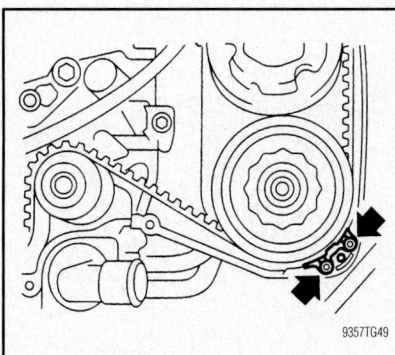

**Fig. 166 Location of the lower right belt guide–2.0L H4 engine**

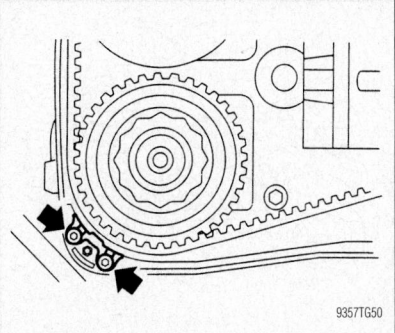

**Fig. 167 Location of the lower left belt guide–2.0L H4 engine**

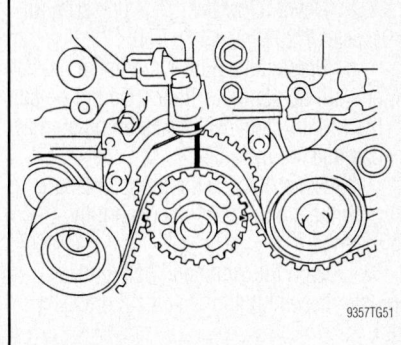

**Fig. 168 Mark the upper belt-to-sprocket alignment–2.0L H4 engine**

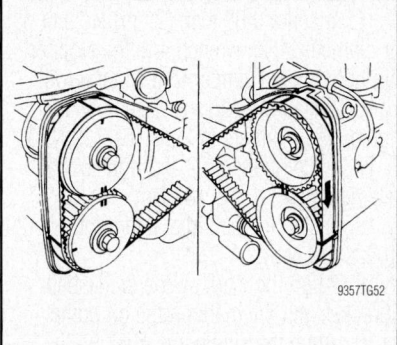

**Fig. 169 Mark the lower belt-to-sprocket alignment–2.0L H4 engine**

**Fig. 170 Remove the belt idler (A)–2.0L H4 engine**

### To install:

12. To prepare the automatic belt tensioner for assembly, perform the following steps:
   a. Always use a vertical type pressing tool to move the adjuster rod down.
   b. Do not use a lateral type vise.
   c. Always push the adjuster rod vertically.
   d. Make sure to slowly move the adjuster rod down applying a pressure of 66 lbs. (294 N).
   e. Press in the push adjuster rod gradually taking more than 3 minutes.

f. Never allow the press pressure to exceed 2,205 lbs. (9,807 N).

g. Press the adjuster rod as far as the end surface of the cylinder. Do not press the rod into the cylinder as doing so may damage the cylinder.

h. Never release the press pressure until the stopper pin has been fully inserted.

13. Attach the automatic belt tension adjuster assembly to the vertical pressing tool.

14. Move the adjuster rod down slowly using a pressure of 66 lbs. (294 N) until the rod is aligned with the stopper pin hole in the cylinder.

15. Insert a 0.08 inch (2mm) stopper pin or diameter Allen wrench into the stopper pin hole in the cylinder to retain the rod.

16. Install the adjuster assembly and tighten the retainers to 29 ft. lbs. (39 Nm).

17. Install belt idle 2 and tighten the retainers to 29 ft. lbs. (39 Nm).

18. Install the belt idler and tighten to 29 ft. lbs. (39 Nm).

19. Align the mark on the crankshaft sprocket with the mark on the oil pump.

20. Align the single line mark on the right hand exhaust camshaft sprocket with the notch on the belt cover.

21. Align the single line mark on the right hand intake camshaft with the notch on the belt cover. Make sure the double lines on the intake camshaft and exhaust sprockets are aligned as shown in the accompanying illustration.

22. Align the single line mark on the left hand exhaust camshaft sprocket with the notch on the belt cover by turning the sprocket counterclockwise (as viewed from the front of the engine).

23. Align the single line mark on the left hand intake camshaft sprocket with the

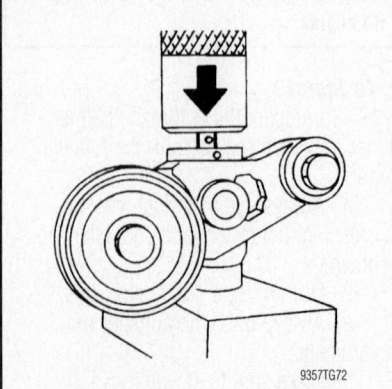

Fig. 171 Attach the automatic belt tension adjuster assembly to the vertical pressing tool–2.0L H4 engine

notch on the belt cover by turning the sprocket counterclockwise (as viewed from the front of the engine). Make sure the double lines on the intake camshaft and exhaust sprockets are aligned as shown in the accompanying illustration.

➡Make sure the camshaft and crankshaft sprockets are positioned correctly. The intake and exhaust camshafts on this engine can be

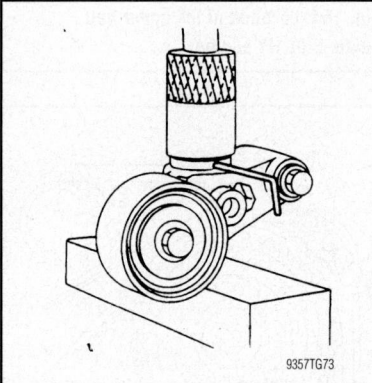

Fig. 172 Insert a 0.08 inch (2mm) stopper pin or diameter Allen wrench into the stopper pin hole in the cylinder to retain the rod–2.0L H4 engine

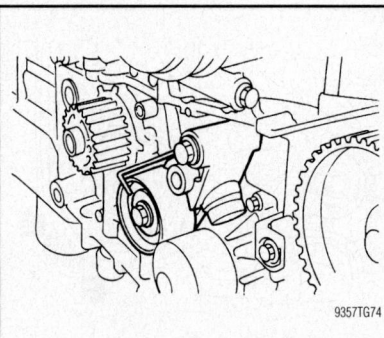

Fig. 173 Install the adjuster assembly–2.0L H4 engine

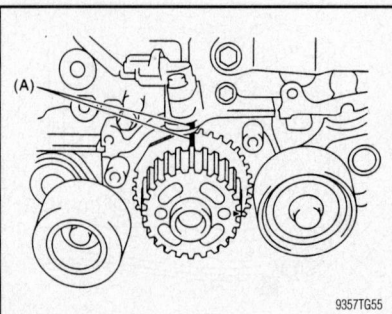

Fig. 174 Align the mark on the crankshaft sprocket with the mark on the oil pump–2.0L H4 engine

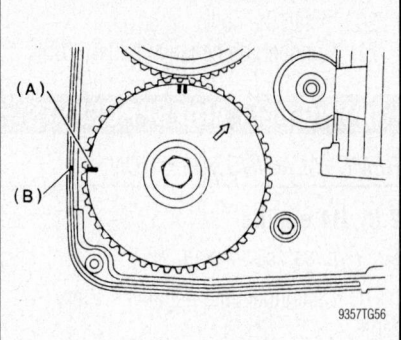

Fig. 175 Align the single line mark on the right hand exhaust camshaft sprocket with the notch on the belt cover–2.0L H4 engine

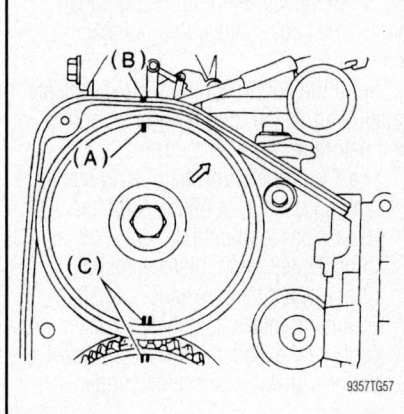

Fig. 176 Align the single line mark on the right hand intake camshaft with the notch on the belt cover. Make sure the double lines on the intake camshaft and exhaust sprockets are aligned–2.0L H4 engine

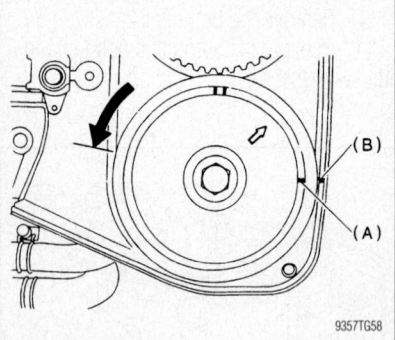

Fig. 177 Align the single line mark on the left hand exhaust camshaft sprocket with the notch on the belt cover by turning the sprocket counterclockwise (as viewed from the front of the engine)–2.0L H4 engine

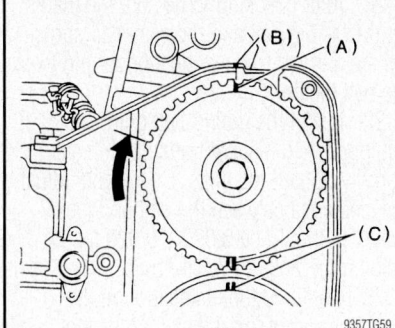

**Fig. 178 Align the single line mark on the left hand intake camshaft sprocket with the notch on the belt cover by turning the sprocket counterclockwise (as viewed from the front of the engine). Make sure the double lines on the intake camshaft and exhaust sprockets are aligned—2.0L H4 engine**

rotated independently with the timing belt removed. By looking at the illustration it will show you that if the intake and exhaust valve are lift together the heads will hit each other and bend.

➡When the timing belts are not installed, 4 camshafts are held at "zero lift" position, where all cams on the camshafts do not push the intake and exhaust valves down (under this condition all valves remain unlifted). When the camshafts are rotated to install the timing belts, No. 2 intake and No. 4 exhaust cam of the left hand camshafts are held to push their corresponding valves down. Under this condition these valves are held lifted. The right side camshafts are held in so that their cams do not push the valves down. The left hand camshafts must be rotated from the "zero lift" position to the position where the timing belt is to be installed at as small an angle as possible, in order to prevent mutual interference of intake and exhaust valve heads. Do not allow the camshafts to rotate in the direction illustrated as this causes both the intake and exhaust valves to lift off at the same time with will cause valve damage.

24. When installing the belt, make sure to align the marks made during removal or if using a new belt, align the in alphabetical order as shown in the illustration.

**⁑ WARNING**

**Disengagement of more than 3 timing belt teeth may result in contact between the valve and piston. Always make sure the belts rotation is correct.**

25. Install the belt idlers and tighten to 29 ft. lbs. (39 Nm).

**⁑ WARNING**

**Make sure the marks on the belt and sprockets are properly aligned.**

26. Once the marks on the belt and sprockets are aligned, remove the stopper pin from the tensioner adjuster.

27. Install the timing belt guide on vehicles with manual transmission. Measure the clearance between the belt and guide. the clearance should be 0.019–0.059 inch (0.5–1.5mm) and tighten the retainers to 7 ft. lbs. (10 Nm).

28. Install the belt covers and tighten to 3.5 ft. lbs. (5 Nm).

29. Install the crankshaft pulley and tighten the bolt to 94 ft. lbs. (127 Nm).

30. Install the V-belt.

### 2.5L SOHC Engine

*See Figures 180 through 184.*

1. Before servicing the vehicle, refer to the Precautions Section.

2. Disconnect the negative battery cable.

3. To remove the front side V belt, remove the belt covers. Loosen the lock bolt. Loosen the slider bolt. Remove the front side belt.

4. To remove the rear side V belt, remove the belt covers. Loosen the lock

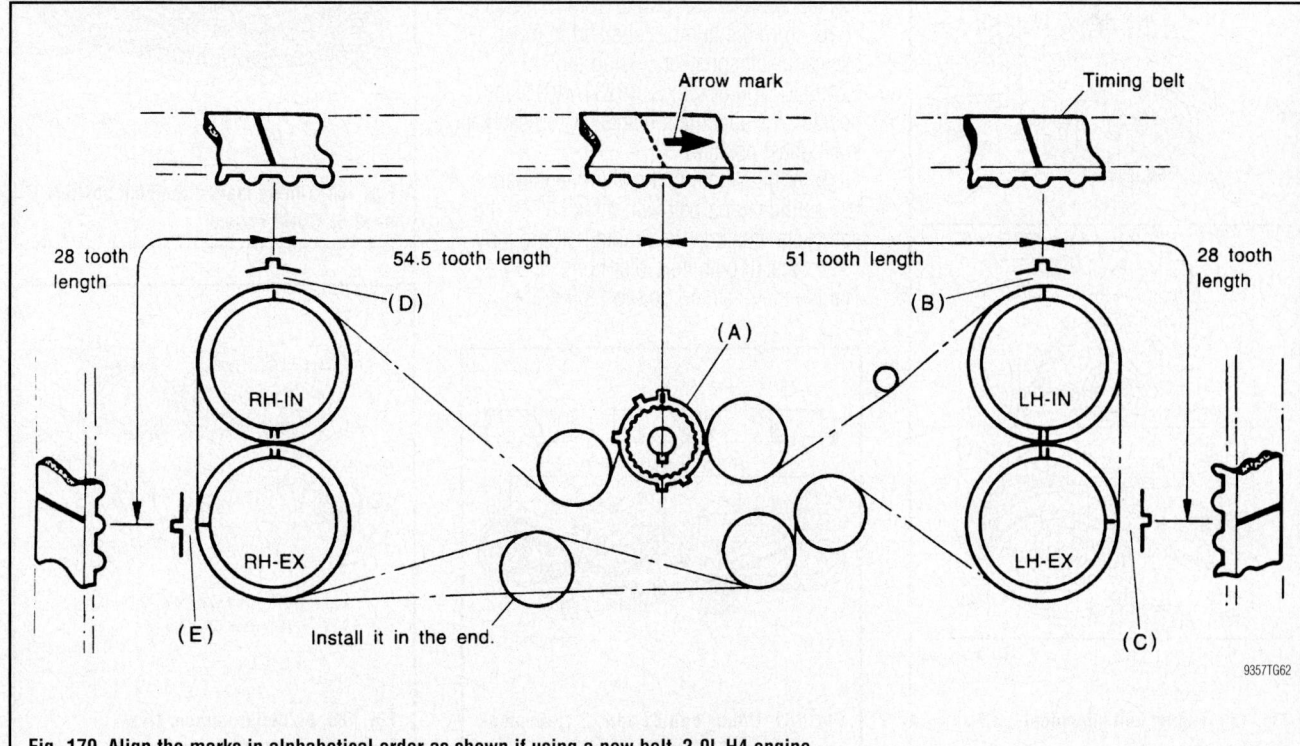

**Fig. 179 Align the marks in alphabetical order as shown if using a new belt—2.0L H4 engine**

bolt. Loosen the slider bolt. Remove the rear side belt. Remove the belt tensioner.

5. Remove the crankshaft pulley bolt.

6. Lock the crankshaft in place using tool ST499977100, or equivalent.

7. Remove the crankshaft pulley.

8. Remove the left side timing belt cover.

9. Remove the front timing belt cover.

10. If equipped with manual transmission, remove the timing belt guide.

➡**If the belt is going to be reused and the alignment mark on the belt is not readable, put a new mark on the belt to indicate the direction of rotation. Using tool ST499987500, turn the crankshaft to align the mark of the sprocket "A" to the cylinder mark notch "B". Ensure that the right side cam sprocket mark "C", cam cap and cylinder head matching surface "D" or left side cam sprocket mark "E", timing belt cover notch "F" are properly aligned. Paint an alignment mark on the belt in relation to the crankshaft sprocket and camshaft sprockets. Z1 measurement is 46.8 teeth. Z2 measurement is 43.7 teeth.**

11. Remove both the number two belt idlers. Remove the timing belt from the engine.

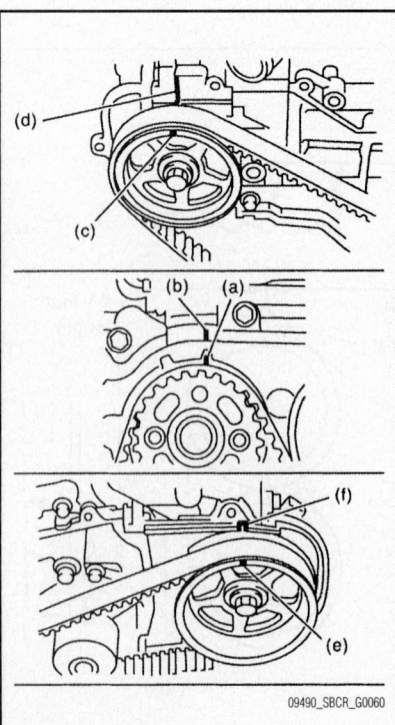

**Fig. 180 Timing belt alignment—2.5L SOHC engine**

12. Remove the number one belt idler. Remove the automatic belt tension adjuster assembly.

### To install:

13. Attach the automatic belt tension adjuster assembly to a vertical pressing tool.

➡**Always use a vertical type pressing tool to move the adjuster rod downward. Do not use a lateral type vise. Push the adjuster rod vertically. Press in the push adjuster rod gradually, which should take three minutes or more. Do not allow pressure to exceed 2,205 lb. force.**

14. Slowly move the adjuster rod down until the adjuster rod is aligned with the stopper pin hole in the cylinder.

➡**Press the adjuster rod as far as the end surface of the cylinder. Do not press the adjuster rod into the cylinder. Doing so may damage the cylinder.**

15. Using a 0.08 inch stopper pin, insert it into the stopper pin hole in the cylinder. Secure the adjuster rod.

➡**Do not release the press pressure until the stopper pin is completely inserted in the hole.**

16. Install the automatic belt tensioner assembly. Tighten the retaining bolt to 28.9 ft. lbs.

17. Install the belt idler number one. Tighten the retaining bolt to 28.9 ft. lbs.

18. Turn the number one and number two camshaft sprockets, using tool ST499207100 or tool ST18231AA010 and position the alignment marks "A" on each at the highest position.

19. While aligning the alignment mark "B" on the timing belt with mark "A" on the sprockets, position the timing belt properly.

20. Install both belt idler number two's. Tighten the retaining bolt to 28.9 ft. lbs.

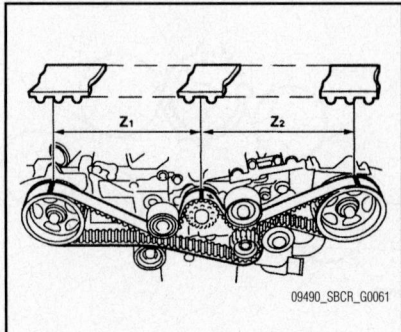

**Fig. 181 Timing belt Z1 and Z2 teeth measurement—2.5L SOHC engine**

21. After checking to be sure the marks on the timing belt and the camshaft sprockets are aligned remove the stopper pin from the belt tension adjuster.

22. Install the timing belt guide, if equipped with manual transmission. Temporarily tighten the bolts. Check and adjust the clearance between the belt and the guide. It should be 0.039 ± 0.020 inch. Tighten the bolts to 7.2 ft. lbs.

23. Continue the installation in the reverse order of the removal procedure.

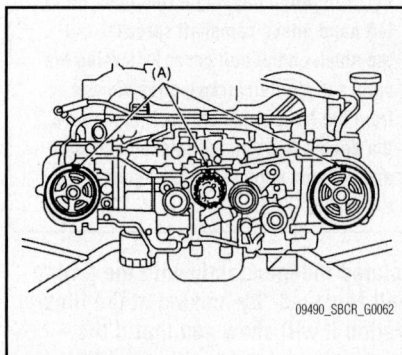

**Fig. 182 Timing mark alignment position A—2.5L SOHC engine**

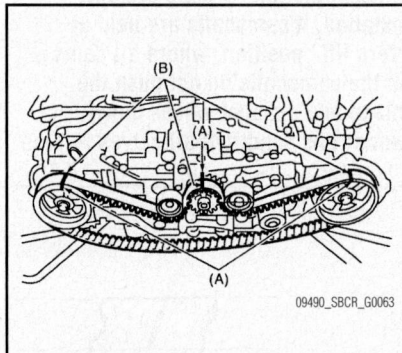

**Fig. 183 Timing mark alignment position B—2.5L SOHC engine**

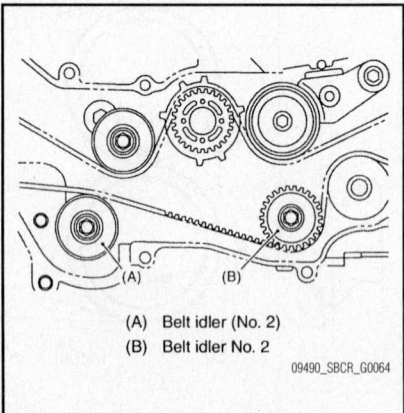

(A) Belt idler (No. 2)
(B) Belt idler No. 2

**Fig. 184 Belt idler number two locations—2.5L SOHC engine**

## 2.5L DOHC Engine

*See Figures 185 through 196.*

1. Disconnect the negative battery cable.

2. Remove the collector cover.

3. To remove the front side V belt, remove the belt covers. Loosen the lock bolt. Loosen the slider bolt. Remove the front side belt.

4. To remove the rear side V belt, remove the belt covers. Loosen the lock bolt. Loosen the slider bolt. Remove the rear side belt. Remove the belt tensioner.

5. Remove the crankshaft pulley bolt.

6. Lock the crankshaft in place using tool ST499977100, or equivalent.

7. Remove the crankshaft pulley.

8. Remove the left side timing belt cover.

9. Remove the right side timing belt cover.

10. Remove the front timing belt cover.

11. Remove the timing belt guide, if equipped.

➡ If the belt is going to be reused and the alignment mark on the belt is not readable, put a new mark on the belt to indicate the direction of rotation. Using tool ST499987500, turn the crankshaft to align the mark on the crankshaft sprocket, intake camshaft sprocket (left), exhaust camshaft sprocket (left), intake camshaft sprocket (right), exhaust camshaft sprocket (right) with the notches of the timing belt cover and cylinder block. Paint an alignment mark on the belts in relation to the camshaft sprockets. $Z1$ measurement

is 54.4 teeth. $Z2$ measurement is 51.0 teeth and $Z3$ measurement is 28.0 teeth.

12. Remove the belt idler belt "A". Remove the timing belt.

13. Remove the belt idlers belt "B" and "C".

14. Remove the belt idler number two. Remove the automatic belt tension adjuster assembly.

### To install:

15. Attach the automatic belt tension adjuster assembly to a vertical pressing tool.

➡ Always use a vertical type pressing tool to move the adjuster rod downward. Do not use a lateral type vise. Push the adjuster rod vertically. Press in the push adjuster rod gradually, which should take three minutes or more. Do not allow pressure to exceed 2,205 lb. force.

16. Slowly move the adjuster rod down until the adjuster rod is aligned with the stopper pin hole in the cylinder.

➡ Press the adjuster rod as far as the end surface of the cylinder. Do not press the adjuster rod into the cylinder. Doing so may damage the cylinder.

17. Using a 0.08 inch stopper pin, insert it into the stopper pin hole in the cylinder. Secure the adjuster rod.

➡ Do not release the press pressure until the stopper pin is completely inserted in the hole.

18. Install the automatic belt tensioner assembly. Tighten the retaining bolt to 28.9 ft. lbs.

19. Install the belt idler number two. Tighten the retaining bolt to 28.9 ft. lbs.

20. Install the belt idlers. Tighten the retaining bolts to 28.9 ft. lbs.

21. Align the mark "A" on the crankshaft sprocket with the mark on the oil pump at the cylinder block.

Align the single line mark "A" on the right exhaust camshaft sprocket with the notch "B" on the timing belt cover.

22. Align single line mark "A" on the right intake camshaft sprocket with the notch "B" on the timing cover. Ensure that the double lines "C" on the intake and exhaust camshaft sprockets are aligned.

23. Align single line mark "A" on the left exhaust camshaft sprocket with the notch "B" on the timing cover by turning the sprocket counterclockwise as viewed from the front of the engine.

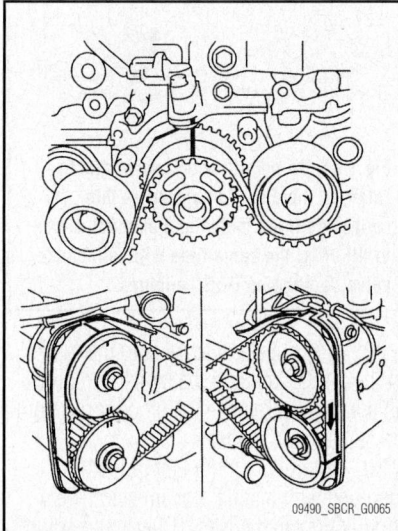

**Fig. 185 Timing belt alignment—2.5L DOHC engine**

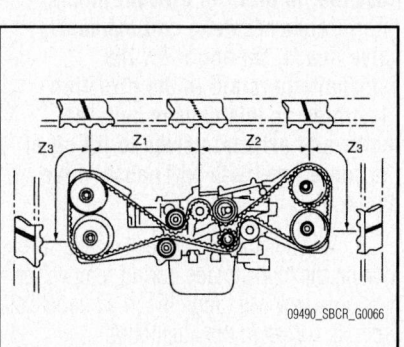

**Fig. 186 Timing belt Z1, Z2 and Z3 teeth measurement—2.5L DOHC engine**

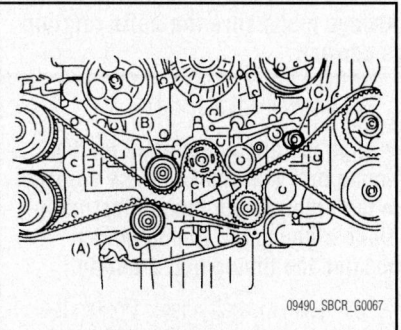

**Fig. 187 Belt idler identification and location—2.5L DOHC engine**

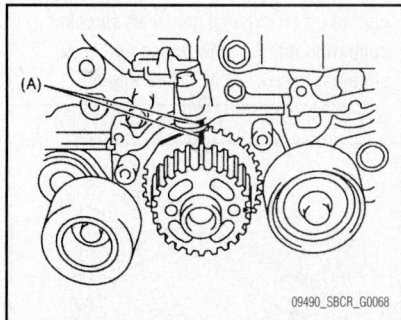

**Fig. 188 Crankshaft sprocket mark A to oil pump cover alignment—2.5L DOHC engine**

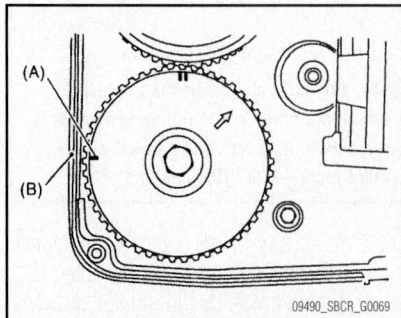

**Fig. 189 Right exhaust camshaft sprocket alignment mark A with timing belt cover B alignment mark—2.5L DOHC engine**

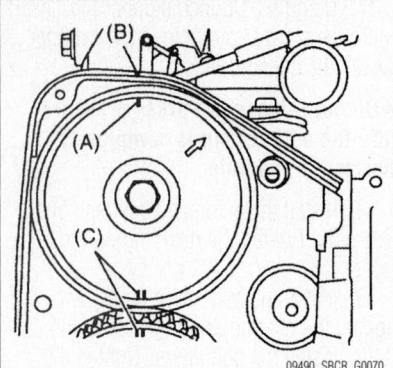

Fig. 190 Right intake camshaft sprocket alignment mark A with timing belt cover B alignment mark an double line C alignment mark—2.5L DOHC engine

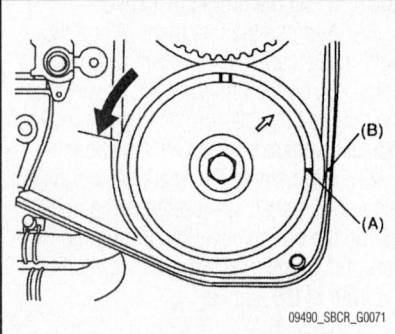

Fig. 191 Left exhaust camshaft sprocket alignment mark A with timing belt cover B alignment mark—2.5L DOHC engine

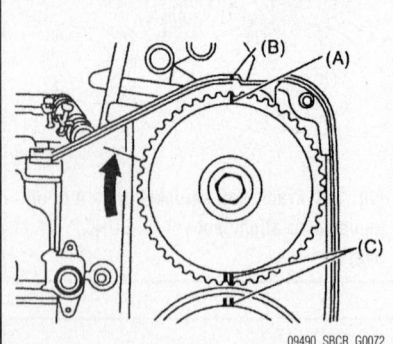

Fig. 192 Left intake camshaft sprocket alignment mark A with timing belt cover B alignment mark an double line C alignment mark—2.5L DOHC engine

24. Align single line mark "A" on the left intake camshaft sprocket with the notch "B" on the timing cover, by turning the sprocket clockwise as viewed from the front of the engine. Ensure that the double lines "C" on the intake and exhaust camshaft sprockets are aligned.

25. Make sure that the camshaft and crankshaft sprockets are positioned properly.

➡The intake and exhaust camshafts on this engine can be rotated independently with the timing belt removed. By looking at the illustration it will show you that if the intake and exhaust valve are lift together the heads will hit each other and bend.

➡When the timing belts are not installed, 4 camshafts are held at "zero lift" position, where all cams on the camshafts do not push the intake and exhaust valves down (under this condition all valves remain unlifted). When the camshafts are rotated to install the timing belts, No. 2 intake and No. 4 exhaust cam of the left hand camshafts are held to push their corresponding valves down. Under this condition these valves are held lifted. The right side camshafts are held in so that their cams do not push the valves down. The left hand camshafts must be rotated from the "zero lift" position to the position where the timing belt is to be installed at as small an angle as possible, in order to prevent mutual interference of intake and exhaust valve heads. Do not allow the camshafts to rotate in the direction illustrated as this causes both the intake and exhaust valves to lift off at the same time with will cause valve damage.

26. When installing the belt, make sure to align the marks made during removal or if using a new belt, align the in alphabetical order as shown in the illustration.

### ✴✴ WARNING
**Disengagement of more than 3 timing belt teeth may result in contact between the valve and piston. Always make sure the belts rotation is correct.**

27. Install the timing belt.

➡Align the alignment mark on the timing belt with marks on the sprocket in the order shown in the illustration. While aligning the timing marks, position the timing belt properly.

28. Install the belt idlers. Tighten the retaining bolts to 28.9 ft. lbs. (39 Nm).

➡Make sure that the marks on the timing belt and sprockets are aligned.

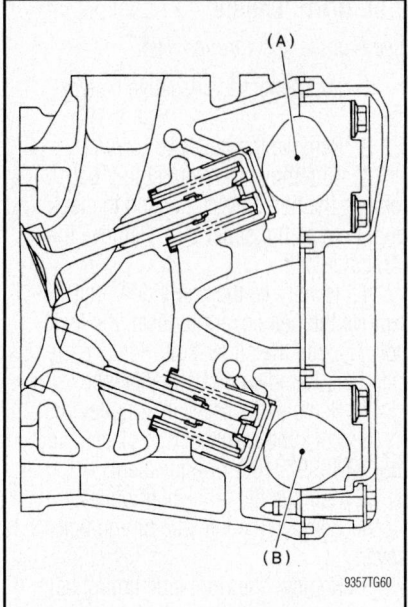

Fig. 193 If the intake and exhaust valve are lift together the heads will hit each other and bend —2.5L DOHC engine

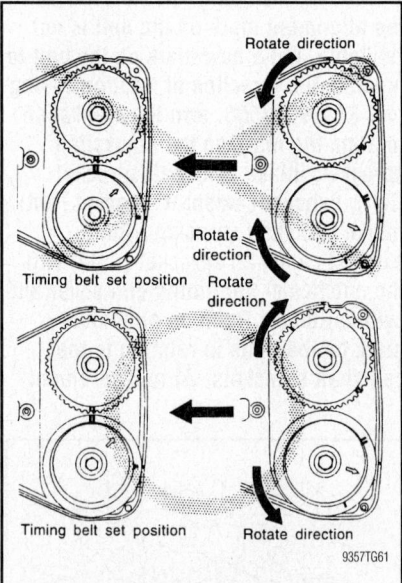

Fig. 194 Do not allow the camshafts to rotate in the direction shown as this causes both the intake and exhaust valves to lift off at the same time with will cause valve damage on DOHC engines

29. After checking to be sure the marks on the timing belt and the camshaft sprockets are aligned remove the stopper pin from the belt tension adjuster.

30. Install the timing belt guide, if equipped with manual transmission. Temporarily tighten the bolts. Check and adjust the clearance between the belt and the guide. It should be 0.039 ± 0.020 inch.

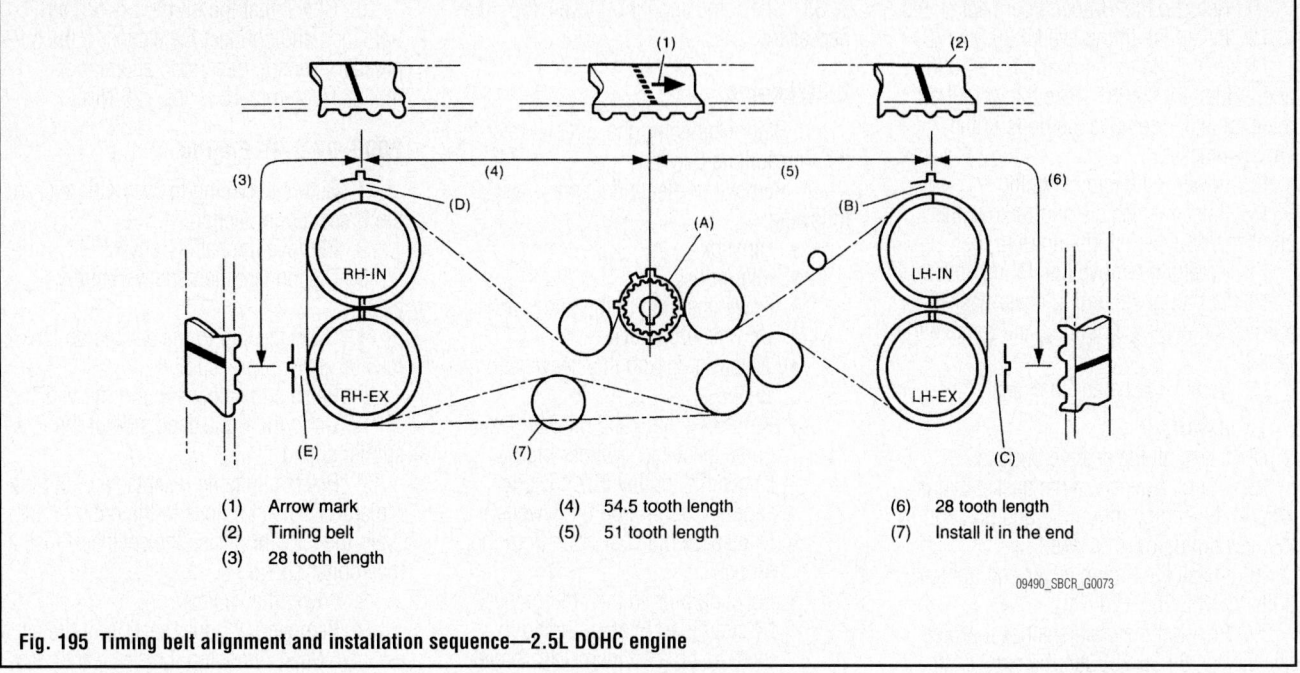

| | | | |
|---|---|---|---|
| (1) | Arrow mark | (4) | 54.5 tooth length |
| (2) | Timing belt | (5) | 51 tooth length |
| (3) | 28 tooth length | | |

| | |
|---|---|
| (6) | 28 tooth length |
| (7) | Install it in the end |

09490_SBCR_G0073

**Fig. 195 Timing belt alignment and installation sequence—2.5L DOHC engine**

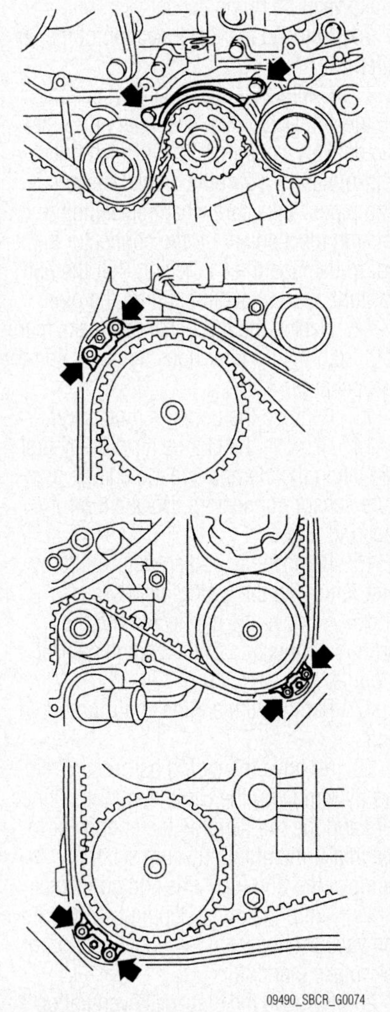

31. Install the timing belt cover.
32. Install the crank pulley.
33. Install the V-belts.

## TIMING CHAIN COVER AND SEAL

### REMOVAL & INSTALLATION

#### 2.0L I4 Engine

*See Figure 197.*

1. Before servicing the vehicle, refer to the Precautions Section.
2. Raise the vehicle and remove the front right wheel.

3. Remove the right wheel well.
4. Relieve the belt tensioner, using 83 96 095, and remove the belt. Mark the direction of rotation of the belt.
5. Remove the belt tensioner.
6. Lower the vehicle and remove the upper engine cover.
7. Unplug the mass air flow sensor connector and detach the inlet hose from the air cleaner casing cover. Remove the cover and the air filter. Remove the SAI hose.
8. Detach the inlet hose and remove the air cleaner casing. Unplug the connector for the A/C pressure sensor.

09490_SBCR_G0074

**Fig. 196 Timing belt guide bolt location—2.5L DOHC engine**

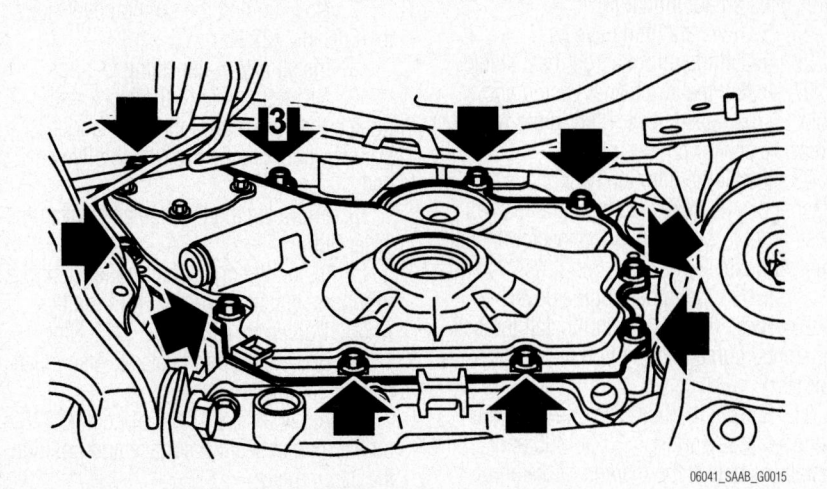

06041_SAAB_G0015

**Fig. 197 Showing the location of timing cover bolts (start loosening or tightening with the bolt shown with the number)—2.0L engine**

9. Remove the turbocharger heat shield.

10. Install the lifting eye kit 83 96 178.

11. Install the lifting beam, 83 94 850, and holder, 83 95 287. Take the weight off the engine and remove the bolts in the engine bracket.

12. Lower the engine slightly.

13. Raise the vehicle and remove the crankshaft pulley using a holding tool.

14. Carefully remove the timing cover.

15. Cut away the timing cover gasket around the engine mounting and remove the gasket.

16. Remove the crankshaft seal.

### To install:

17. Clean all the sealing surfaces.

18. Cut off the part of the gasket that is around the engine mounting and position the new timing cover gasket.

19. Install the timing cover and tighten bolts to 15 ft. lbs. (20 Nm).

20. Position the crankshaft seal protective sleeve, 83 96 202, on the crankshaft. Lubricate the new seal with non-acidic Vaseline and position it on the tool.

21. Position the tool on the crankshaft. Screw in the seal, using the crankshaft pulley bolt so that it is flush with the timing cover. Remove the tool.

22. Install the crankshaft pulley with a new bolt, while using a holding tool to prevent crankshaft rotation. Torque pulley bolt to 74 ft. lbs. (100 Nm), plus an additional 75°.

23. Lower the vehicle and raise the engine with the lifting beam until it rests against the engine mount.

24. Install the bolts to the engine bracket. Remove the lifting beam with holder. Tighten the bolts to 52 ft. lbs. (70 Nm), plus an additional 60°.

25. Remove the lifting eye kit.

26. Install the turbocharger heat shield.

27. Install the air cleaner casing and connect the intake hose. Plug in the A/C pressure sensor connector.

28. Install the filter element and the air cleaner casing cover. Connect the intake hose and the MAF sensor connector. Install the SAI hose.

29. Install the upper engine cover.

30. Raise the vehicle and install the belt tensioner, tightening the bolt to 37 ft. lbs. (50 Nm).

31. Relieve the belt tensioner with the drivebelt relieving tool, 83 96 095 and install the belt in the marked direction of rotation. Make sure the belt is located correctly on all the pulleys.

32. Install the right fender liner and the front wheel.

33. Check the engine oil level. Top up as necessary.

## 2.3L Engine

1. Before servicing the vehicle, refer to the Precautions Section.

2. Remove or disconnect the following:

- Dipstick
- Idler pulley
- Power steering pump and bracket with the lifting eye
- Water pump and the sleeve with O-rings
- Protective plate and oil pan (leave guide sleeve in cylinder block)
- Crankshaft pulley (use locking segment 83 94 868 on flywheel)
- Crankcase breather hose from the oil pan
- Locating pins in the timing cover by cutting an internal thread in them using a ⅜ inch UNC thread tap and withdraw them with sliding hammer 83 90 270.
- Timing cover retaining bolts

3. Pull the timing cover away, starting at the bottom, then lift the cover outwards/downwards to avoid damaging the gasket at the cylinder head.

### To install:

4. Thoroughly remove all remains of sealant on all surfaces.

5. Apply a bead of Loctite® 518 about 0.40 inch (1mm) thick along the middle of the sealing surfaces.

6. Position the timing cover, carefully turning it into position. Install the retaining bolts, but do not tighten them. Tap the locating pins in place.

7. Now, tighten the timing cover bolts to 16 ft. lbs. (22 Nm).

8. Install new water pump O-rings

9. Clean the hole in the block and tighten the sleeve until its large fender is pointing horizontally towards the flywheel end.

10. Lubricate the O-rings and install the water pump.

11. Install the crankshaft pulley, using a locking segment, 83 94 868, on the flywheel. Tighten to 130 ft. lbs. (175 Nm).

12. Be sure oil pan is clean. Also clean mating surfaces.

13. Apply an even bead of Loctite®518 on the oil pan sealing surface and position the pan in place.

Tighten bolts to 16 ft. lbs. (22 Nm).

14. Install the oil pan protective plate.

15. Plug in the oil level sensor connector and press the cable back into its clamps.

16. Check that the oil plug is properly tightened and connect the cables to the oil pressure sensor, generator, and starter motor. Tighten to 18 ft. lbs. (25 Nm).

## 2006–07 2.8L Engine

1. Before servicing the vehicle, refer to the Precautions Section.

2. Remove the battery cover.

3. Disconnect the negative battery cable.

4. Open the cap to the expansion tank, release any overpressure.

5. Raise and safely support the vehicle.

6. Undo the right-hand side of the spoiler shield.

7. Place a suitable receptacle under the radiator. Connect a hose to the radiator and drain the coolant. Close the cock and Install the spoiler shield.

8. Lower the vehicle.

9. Remove the upper section of the air cleaner. When removing the air cleaner casing cover, be careful not to damage the brake vacuum pump sensor.

10. Remove the lower section of the air filter.

11. Unplug the mass air flow sensor connector. Detach the secondary air pipe, the delivery hose from the turbocharger intake manifold and the crankcase ventilation pipes. Take care when releasing the locking mechanism on the connector so as not to damage the connector. Pull the halves straight apart to avoid bending the pins.

12. Detach the turbocharger intake manifold from the air filter housing cover and the turbocharger.

13. Remove the upper engine cover.

14. Remove the charge air pipe by first removing the clamps and the charge pressure sensor connectors. Detach the power steering line.

15. Remove the coolant return line by first removing the wiring harnesses.

16. Remove the connector and the ground connection to the engine control module.

17. Remove the engine control module bolts.

18. Remove the wiring harnesses holders by removing the cable duct, the connectors and the nut. Unplug the connector to the intake manifold pressure sensor. Remove the bolt, cable tie and connector.

19. Undo the quick coupling and remove the vacuum line from the upper section of the intake manifold.

20. Remove the crankcase ventilation lines by removing the bolt and removing the holder.

21. Remove the holder to the injector's connector.

22. Remove the throttle body connector and remove the upper section of the intake manifold.

23. Undo the quick couplings and remove the bleeder valve plus lines.

24. Place a collecting pan under the vehicle.

25. Detach the fuel line, wiping up any fuel spill with a cloth.

26. Remove the power steering reservoir by removing the bolt, return hose and clamp. Seal the hose and reservoir. Turn the reservoir to the side.

27. Unplug the connectors to the fuel pressure sensor.

28. Remove the wiring harnesses from the cylinder head for cylinders 1-3-5. Unplug the camshaft position sensor connector, the connector to the camshaft setting valve and the holder.

29. Remove the wiring harnesses from the cylinder head for cylinders 2-4-6. Unplug the camshaft position sensor connector, the connector to the camshaft setting valve and the holder.

30. Remove the wiring harness for the injectors by first unplugging the injectors' connectors.

31. Remove the fuel rail with injectors.

32. Remove the lower section of the intake manifold. Seal the inlet ducts to the cylinder head with lintless rags.

33. Remove the ignition coils for cylinders 2-4-6 by first unplugging the connectors.

34. Remove the wiring harness holder from the camshaft cover for cylinders 1-3-5.

35. Remove the ignition coils for cylinders 1-3-5 by first unplugging the connectors.

36. Remove the ground cable from the wiring harnesses by carefully cutting open the tape strip (arrows). Place the ground cable to one side.

37. Remove the wiring harnesses plus connectors. Remove the cable tie.

38. Remove the battery cover bottom section.

39. Loosen the expansion tank from the mounting.

40. Remove the turbocharger heat shield panel.

41. Remove the crankcase ventilation line from the turbocharger.

42. Remove the crankcase ventilation check valve from the camshaft cover for cylinders 1-3-5.

43. Undo the quick coupling and remove the secondary air system's valve for cylinders 2-4-6.

44. Undo the upper quick coupling by removing the bolt and removing the secondary air system's delivery line.

45. Remove the turbocharger diaphragm unit by undoing the bolt and pressing it out.

46. Remove the secondary air system's delivery line for cylinders 1-3-5.

47. Remove the secondary air system's delivery line for cylinders 2-4-6.

48. Remove the ventilation line to the oil dipstick's guide tube by undoing the quick coupling.

49. Remove the engine's wiring harness from cylinders 2-4-6 by unplugging the connectors, and by removing the head shield and the ground cable from the cylinder head. Fold the wiring harnesses to the side.

50. Remove the engine's transport lug by removing the bolts.

51. Remove the camshaft cover for cylinders 2-4-6 by removing the bolts.

52. Remove the camshaft cover for cylinders 1-3-5 by removing the bolts.

53. Raise the vehicle to half height.

54. Remove the front wheels.

55. Remove the right wheel well.

56. Remove the power steering hydraulic line from the holder on the front axle beam.

57. Remove the right engine mounting.

58. Relieve the load on the belt tensioner. Use a ¼ inch puller handle and remove the driver belt from the A/C compressor pulley.

59. Remove the bolt holding the crankshaft pulley using 83 95 360 Holding tool, crankshaft pulley (handle only) and EN-47981 Holding tool, pulley.

60. Remove the pulley using EN-47982 Puller and EN-47981 Holding tool, pulley.

61. Remove the transaxle housing at the bottom by removing the bolts.

62. Remove the wiring harness holder from the bottom of the transaxle housing by removing the bolt.

63. Lower the vehicle.

64. Remove the coolant line from the coolant outlet pipe by removing the bolt.

65. Remove the coolant outlet pipe by removing the coolant hose, remove the clamp and remove the bolts.

66. Remove the coolant pump pulley by removing the bolts. Hold firm with a screwdriver.

67. Remove the coolant pump by removing the bolts.

68. Remove the power steering pump at the top by removing the bolted joint.

69. Raise and safely support the vehicle.

70. Remove the power steering pump at the bottom by removing the bolt. Move the power steering pump towards the front of the car

71. Lower the vehicle.

72. Remove the drive belt tensioner by removing the bolt.

73. Remove the camshaft setting valve for cylinders 1-3-5 by removing the bolt.

74. Remove the camshaft setting valve for cylinders 2-4-6 by removing the bolt.

75. Remove the upper bolts of the transaxle housing.

76. Remove the timing cover. Carefully pry the cover loose with a pry bar and a screw that is carefully screwed in (to press the cover loose).

### To install:

77. Clean the sealing surfaces.

78. Install EN-46109-2 Centering tool in the engine block and cylinder head.

79. Apply 93 165 267 Sealant. Apply an approx. 3mm thick bead on the sealing surfaces of the transaxle housing.

80. Install the transaxle housing with help from an assistant. Check that the sealing bead remains undamaged during the Alignment of the transaxle housing.

81. Install the upper bolts for the transmission housing and tighten to 17 ft. lbs. (23 Nm). Remove the centering tools.

82. Secure the starter motor's wiring harness holder on the wiring harnesses.

83. Raise and safely support the vehicle.

84. Install the lower bolts of the transaxle housing and tighten to 17 ft. lbs. (23 Nm).

85. Install the lower bolt to the power steering pump and tighten to 16 ft. lbs. (22 Nm).

86. Install the pulley to the crankshaft using J-41998 Installing Tool to press on the pulley.

87. Install the pulley bolt using 83 95 360 Holding tool, crankshaft pulley (handle only) and EN-47981 Pulley holding tool. Tighten the pulley bolt to 74 ft. lbs. (100 Nm), plus an additional 150°.

88. Secure the starter motor's wiring harness holder at the bottom. Tighten the bolt.

89. Lower the vehicle.

90. Install the camshaft setting solenoid valves Tighten to 8 ft. lbs. (10 Nm).

91. Install the belt tensioner and the bolt. Tighten the bolt to 16 ft. lbs. (22 Nm).

92. Secure the starter motor's wiring harness holder at the top. Tighten the bolt.

93. Secure the power steering pump at the top. Tighten the fasteners to 16 ft. lbs. (22 Nm).

94. Install the mounting bolts to the engine pad's attaching plate and tighten to 32 ft. lbs. (44 Nm).

95. Install the coolant pump. Use a new gasket and tighten the bolts to 8 ft. lbs. (10 Nm).

96. Install the coolant pump's belt pulley. Counterhold with a screwdriver. Tighten the bolts to 9 ft. lbs. (12 Nm).

97. Install the coolant outlet pipe. Use a new gasket and tighten the bolts to 9 ft. lbs. (12 Nm).

98. Install the coolant hose and secure the clamp.

99. Unload the belt tensioner. Use a 1/4 inch puller and install the drive belt.

100. Install the right engine mounting.

101. Install the holder to the power steering line and tighten the bolt to 6 ft. lbs. (8 Nm).

102. Install the coolant line to the coolant outlet pipe. Use a new sealing ring. Tighten to 6 ft. lbs. (8 Nm).

103. Raise and safely support the vehicle.

104. Secure the power steering hydraulic line in the holder on the front axle beam.

105. Lower the vehicle to half height.

106. Install the right-hand wheel well.

107. Install the wheels.

108. Clean the camshaft cover sealing surfaces.

109. Replace the camshaft cover gasket for cylinders 2-4-6.

110. Apply an approx. 3mm thick bead of sealing compound on the separating points of the cylinder head and transmission housing.

111. Install the camshaft cover for cylinders 2-4-6 and tighten the bolts to 8 ft. lbs. (10 Nm).

112. Install the engine's transport lug and tighten the bolt to 48 ft. lbs. (65 Nm).

113. Clean the camshaft cover sealing surfaces.

114. Replace the camshaft cover gasket for cylinders 1-3-5. Use new gaskets.

115. Apply an approx. 3mm thick bead of sealing compound on the separating points of the cylinder head and transmission housing.

116. Install the camshaft cover for cylinders 1-3-5 and tighten the bolts to 8 ft. lbs. (10 Nm).

117. Install the engine wiring harnesses to the cylinder head for cylinders 2-4-6 by plugging in the connectors on the cylinder head for cylinders 2-4-6. Install the heat shield and install the ground cable on the cylinder head for cylinders 2-4-6. Install the bolts. Secure the wiring harnesses. Secure the engine control module's ground cable in the wiring harnesses. Install the cable ties.

118. Take care when plugging in the connector so as not to damage or press out the pins/sleeves in the connector.

119. Install the secondary air system line for cylinders 2-4-6. Install the exhaust pipe and tighten the bolts to 8 ft. lbs. (10 Nm).

120. Install the ventilation line to the oil dipstick's guide tube. Connect the quick coupling.

121. Install the turbocharger diaphragm unit by installing the bolts.

122. Install the secondary air system's delivery line. Connect the quick couplings.

123. Install the secondary air system's check valve for cylinders 2-4-6. Use a new gasket. Tighten the bolts and connect the quick coupling.

124. Install the secondary air system's delivery line and bolt.

125. Install the check valve to the crankcase ventilation on the camshaft cover for cylinders 1-3-5.

126. Install the crankcase ventilation line on the turbocharger and tighten the bolt to 8 ft. lbs. (10 Nm).

127. Install the turbocharger's upper heat shield panel. Install the nut and tighten to 8 ft. lbs. (10 Nm).

128. Install the expansion tank in the holder.

129. Install the battery cover bottom section.

130. Install the charge pressure regulator's wiring harness. Plug in the connector. Install the cable tie.

131. Install the ignition coils for cylinders 2-4-6. Plug in the connectors and install the bolts. Tighten the bolts to 8 ft. lbs. (10 Nm).

132. Install the wiring harness holder on the camshaft cover for cylinders 1-3-5. Secure in the holder.

133. Install the ignition coils for cylinders 1-3-5. Plug in the connectors and install the bolts. Tighten the bolts to 8 ft. lbs. (10 Nm).

134. Install the lower section of the intake manifold with a new seal. Remove the rags from the inlet ducts.

135. Install the fuel rail with injectors and tighten the bolts to 8 ft. lbs. (10 Nm).

136. Install the injectors' wiring harness. Plug in the connectors.

137. Plug in the connectors to the fuel pressure sensor.

138. Install the wiring harness to the camshaft guide for cylinders 2-4-6. Plug in the connectors and secure in the holder.

139. Connect the wiring harness to the camshaft guide for cylinders 1-3-5. Plug in the connectors and secure in the holder.

140. Install the power steering reservoir. Install the return hose. Secure the clamp and tighten to 7 ft. lbs. (9 Nm).

141. Connect the fuel delivery line.

142. Install the bleeder valve plus lines. Connect the quick couplings. Plug in the connectors to the bleeder valve.

143. Install the upper section of the intake manifold. Plug in the throttle body's connector. Use new gaskets and tighten the fasteners to 17 ft. lbs. (23 Nm).

144. Install the holder to the injector's connector.

145. Install the lines for the crankcase ventilation.

146. Install the vacuum line on the upper section of the intake manifold. Connect the quick coupling.

147. Install the wiring harnesses' holders. Install the bolt. Plug in the connectors for the intake manifold's pressure sensor. Install the cable cables. Secure the connectors. Install the cable duct. Plug in the connectors. Install the nut.

148. Install the engine control module. Install the bolts and tighten to 8 ft. lbs. (10 Nm).

149. Install the ground connection.

150. Plug in the engine control module connector.

151. Install the coolant return line. Install the clamps and tighten the bolts to 8 ft. lbs. (10 Nm).

152. Install the charge air pipe. Install the clamp.

153. Plug in the charge pressure sensor connector. Secure the power steering line with clips.

154. Install the air intake pipe.

155. Install the air filter housing with the mass air flow sensor as follows:
   a. Plug in the mass air flow sensor connector.
   b. Secure the cable ties.
   c. Install the secondary air system's intake manifold. Secure the line.
   d. Connect the quick coupling.

156. Install the upper engine cover. Remove the oil filler opening's screw cap. Install the protective cover. Install the oil filler opening's screw cap.

157. Fill and bleed the cooling system.

158. Connect the negative battery cable and install the battery cover.

## 2008 2.8L Engine

1. Before servicing the vehicle, refer to the Precautions Section.

2. Remove the camshaft covers. Refer to Engine Mechanical Components, Camshaft Covers, Removal & Installation.

3. Remove the coolant port. Remove the seal and O-ring.

4. Remove the front and rear solenoid valves of the variable camshaft. Remove the seals.

5. Remove the front and rear position sensor of the camshaft.

6. Remove the coolant pump pulley.

7. Remove the belt tensioner and screw.

8. Remove the bolt holding the crankshaft pulley using 83 95 360 Crankshaft Pulley Holding Tool.

➡**The pulley is pressed onto the crankshaft.**

9. Remove the pulley using EN-47982 Puller.

10. Remove the timing cover. Use a pry bar to carefully pry loose the cover plus a screw that is carefully screwed in as illustrated (to press the cover loose).

### To install:

11. Clean sealing surfaces of gasket residue and the like.

12. Apply a 2 mm thick bead of 93 160 951 Flange sealant on the sealing surface as the timing cover as illustrated. Fit a new seal at the coolant pump.

13. Install the timing cover. Tighten to 18 ft. lbs. (25 Nm).

14. Install the pulley to the crankshaft. Use KM-J 41998 Fitting tool to press the pulley into place.

15. Install the bolt using 83 95 360 Crankshaft Pulley Holding Tool. Tighten to 74 ft. lbs. (100 Nm) plus 150° rotation.

16. Install the belt tensioner and the screw

17. Install the coolant pump pulley using a screwdriver as a holding tool. Tighten to 9 ft. lbs. (12 Nm).

18. Install the front and rear camshaft position sensors using new seals. Tighten to 7 ft. lbs. (10 Nm).

19. Install the seals and the front and rear solenoid valves of the variable camshaft. Tighten to 7 ft. lbs. (10 Nm).

20. Install the coolant port using a new seal and O-ring. Tighten to 7 ft. lbs. (10 Nm).

21. Install camshaft covers.

### TIMING CHAIN AND SPROCKETS

#### REMOVAL & INSTALLATION

#### 2.0L I4 Engine

1. Before servicing the vehicle, refer to the Precautions Section.

2. Remove the upper engine cover.

3. Detach the mass air flow sensor connector and the intake hose from the air filter housing cover.

4. Remove the cover and the air filter.

5. Detach the intake hose and remove the air filter housing. Detach the connector for the A/C pressure sensor.

6. Undo the fuel pipes from the snap fasteners and remove the bolts for the bracket from the camshaft cover.

7. Remove the cover over the ignition coils.

8. Unplug the ignition coil connectors. Unplug the connector to the turbocharger solenoid valve and undo the cable clips.

9. Remove the bolts, pick up and move aside each ignition coil. Start with ignition coil for cylinder #1.

10. Detach the crankcase ventilation hose from camshaft cover.

11. Undo and carefully move away the cable duct from the cylinder head and camshaft cover.

12. Remove the turbocharger heat shield.

13. Undo ground cable from camshaft cover.

14. Remove the camshaft cover.

15. Raise and safely support the vehicle securely on jackstands.

16. Remove the front right wheel.

17. Remove the right wheel well.

18. On convertible models, remove the front subframe chassis reinforcement.

19. Relieve the belt tensioner with 83 96 095 Drivebelt Relieving tool and remove the belt. Mark the direction of rotation on the belt.

20. Remove the belt tensioner.

21. Lower the vehicle and zero the engine by turning the crankshaft in the direction of engine rotation until the mark on the crankshaft pulley agrees with the mark on the timing cover. The cams on cylinder 1 intake and exhaust camshaft must be pointing in/up.

22. Remove the chain tensioner. Use 83 96 129 Oil Filter Tool and camshaft drive chain tensioner.

23. Remove the lock ring from the chain tensioner and remove the piston.

24. Install 83 96 178 Lifting Eyes.

25. Install 83 94 850 Lifting Beam and 83 95 287 Holder. Relieve the engine and remove the engine bracket.

26. Lower the engine about 3 cm using the lifting beam.

27. Raise the car and remove the crankshaft pulley with 83 95 360 Crankshaft Pulley Holding tool.

28. Place a receptacle under the car to collect any oil and carefully remove the timing cover.

29. Cut off the part of the timing cover gasket that goes round the engine mounting and remove the gasket.

30. Remove the crankshaft seal. Lower the vehicle.

31. Remove the plug from the cylinder head.

32. Remove the upper bolt from the chain guide on the inlet side.

33. Remove the upper timing chain guide rail.

34. Remove the lower chain guide bolt. Pull down the chain guide. Detach the lower section of the chain.

35. Remove the camshaft sprockets. Use a spanner to grip the camshaft flats when the bolts are loosened. Lift up the chain.

36. Remove the tensioner guide by lifting it up. Inspect the condition of the guide. Replace if necessary.

### To install:

37. Clean all sealing surfaces.

38. Clean the camshaft and sprocket contact surfaces from oil and grease.

39. Install the tensioner guide.

40. Lay the chain on the intake sprocket and lower the chain. Fit the sprocket on the camshaft and insert the bolt.

41. Raise and safely support the vehicle securely on jackstands.

42. Install the chain on the crankshaft sprocket and install the chain guide. Fit the lower bolt. Tighten to 7 ft. lbs. (10 Nm).

43. Lower the vehicle.

44. Install the exhaust sprocket and enter the bolt.

45. Install the upper bolt. Fit the plug on the cylinder head.

46. Turn the camshaft drive tensioner clockwise with a screwdriver until it engages in the tensioned position.

47. Place the plunger in the chain tensioner sleeve. Fit the circlip and check that it is fitted correctly in the groove.

➡**Check that the O-ring and gasket are intact. If damaged, replace the chain tensioner assembly.**

48. Install the chain tensioner. Tighten to 55 ft. lbs. (75 Nm).

49. Release/activate the chain tensioner by carefully pressing on the transmission chain or chain control. Use a screwdriver. Check that the tensioner releases.

50. Remove the exhaust camshaft and intake camshaft bearing caps no. 2 and no. 7. Position EN-48368 or EN-48366 Camshaft Adjustment Tool by manually pressing the tools down onto the camshafts without tightening the bolts. Turn the camshaft until the tool aligns with the spanner on the camshaft flats. Tighten the adjustment tools' bolts. Tighten to 7 ft. lbs. (10 Nm).

51. Cut off the part of the gasket that is round the engine mounting and position the new timing cover gasket.

52. Install the timing cover. Tighten to 15 ft. lbs. (20 Nm).

53. Position the 83 96 202 Front Crankshaft Seal Fitting Tools protective collar on the crankshaft. Lubricate the new sealing ring with non-acidic Vaseline and position it on the fitting tool.

54. Position the fitting tool on the crankshaft. Tighten the sealing ring using the crankshaft pulley bolt until it is flush with the timing cover. Remove the tool.

55. Install the crankshaft pulley with a new bolt. Use 83 95 360 Crankshaft Pulley Holding Tool. Tighten to 74 ft. lbs. (100 Nm) plus 75° rotation.

56. Check that the markings on the crankshaft pulley and the timing cover are aligned. Tighten the camshaft gears a first time. Use a spanner to grip the camshaft flats. Tighten to 22 ft. lbs. (30 Nm).

57. Remove the adjustment tools and install the bearing caps. Tighten to 6 ft. lbs. (8 Nm).

58. Continue to tighten the camshaft gears to the final torque. Use a spanner to grip the camshaft flats.

Tighten to 63 ft. lbs. (85 Nm) plus 30° rotation.

59. Rotate the crankshaft 2 turns in the direction of engine rotation until the mark on the crankshaft pulley agrees with the mark on the timing cover. Remove the bearing caps and install EN-48368 Camshaft Adjustment Tool again to check that the camshaft setting is correct. Remove the adjustment tools and refit the bearing caps. Tighten to 6 ft. lbs. (8 Nm).

60. Install the upper timing chain guide rail.

61. Lower the vehicle and lift the engine with the lifting beam. Fit the engine bracket.

62. Install the bolts to the engine bracket. Tighten to 52 ft. lbs. (70 Nm) plus 60° rotation.

63. Remove 83 94 850 Lifting Beam with holder. Remove 83 96 178 Lifting Eye.

64. Install the camshaft cover with new seal. Tighten to 7 ft. lbs. (10 Nm).

65. Connect the ground cable to the camshaft cover.

66. Connect the crankcase ventilation hose to the camshaft cover.

67. Install the cable duct and bracket to the camshaft cover.

68. Install the ignition coils. Tighten to 6 ft. lbs. (8 Nm).

69. Plug in the ignition coil connectors. Make sure the cable is lying correctly and does not get pinched by the cover. Plug in the connector to the turbocharger solenoid valve and fasten the cable clips.

70. Install the cover over the ignition coils.

71. Press the fuel lines and breather lines into the clips on the camshaft cover. Connect the quick release coupling for the breather line.

72. Connect both fuel lines to the fuel rail while restraining the lower nut. Use new seals. Tighten to 7 ft. lbs. (10 Nm).

73. Install the turbocharger heat shield.

74. Install the air filter housing and connect the intake hose. Plug in the connector for the A/C pressure sensor.

75. Install the air filter and air filter housing cover. Connect the intake hose and the mass air flow sensor connector.

76. Install the upper engine cover.

77. Raise the car and install the belt tensioner. Tighten to 37 ft. lbs. (50 Nm).

78. Relieve the belt tensioner with 83 96 095 Drivebelt Relieving tool and install the belt in the marked direction of rotation. Check that the belt is positioned correctly over all the pulleys.

79. On convertible models, install the front subframe chassis reinforcement.

80. Lower the vehicle slightly and install the right wheel well.

81. Install the front right wheel. See Wheels.

82. Lower the vehicle completely.

83. Check engine oil level. Top up as necessary.

## 2.3L Engine

1. Before servicing the vehicle, refer to the Precautions Section.

2. Raise and safely support the vehicle securely on jackstands.

3. Remove the drivebelt cover.

4. Lower the vehicle and remove the upper engine cover.

5. Remove the ignition discharge module and spark plugs.

6. Remove the crankcase ventilation pipe and hose. Cut the cable tie on the check valve. Unplug the solenoid valve connector and remove the camshaft cover.

7. Align the crankshaft and camshafts with their respective setting marks by turning the crankshaft clockwise.

8. Relieve the tension from the multi-groove belt and undo the idler sprocket so that the chain tensioner sleeve is freed.

9. Check chain wear by removing the chain tensioner plug, spring and pushrod, and removing the chain tensioner without altering the position of the pistons.

10. The protruding part of the chain tensioner must not be longer than 15 mm. If it is, the chain must be changed.

11. Check the wear on chain sprockets and chain guide.

12. Chain guide wear must not be so great that the chain acts on the surface between the outer tracks.

13. If the surface between the outer tracks shows signs of wear, they must also be replaced and the chain changed in the usual manner.

### To install:

14. Change the chain as follows:

a. Cover the area round the chain with a cloth and also secure the chain with cable ties on both sides.

b. Split the chain by pressing out a link using 83 94 637 Removal tool and then removing the link with a pair of pliers. Attach a cable tie to the end of the old chain.

c. Couple together the new chain with the old one using 83 94 660 Chain locking link.

➥Take care not to let the chain fall down.

d. Make sure that the new chain lies over the camshaft sprockets on the intake side.

e. Install the chain support in 83 94 652 Fixture and remove the cloth and the cable tie on the camshaft sprocket.

f. With the old chain in your hand and the new one resting over your hand, carefully feed in the new chain while a helper turns the crankshaft.

**❋❋ WARNING**

**Be sure to keep the old chain under tension as it is fed forward. It might otherwise fold itself double down at the crankshaft.**

g. When the new chain has been fed forward to such an extent that only a few links still remain on the inlet side's camshaft sprocket, cover the area round the chain with a cloth once again. Secure the chain with a cable tie, and remove the chain link and the old chain.

h. Connect the ends of the new chain together with a chain link and remove the cable tie and cloth. Pull the chain round until the link is midway between the shanks of the chain support.

i. Cover the opening with a cloth, remove the chain link and install a new chain lock.

j. Check that the inserts marked 2 are mounted in the tool. Then place the chain link's plate in the tool and align it over the link on the chain.

k. Press the plate in place.

l. Reverse the insert in the tool and position the tool with the V notch directly over one of the shanks of the link.

m. Rivet the shanks. Repeat for the other shank.

n. Check the riveting by checking the compressed shank diameter with a Vernier caliper. The correct diameter is 3.4-3.6 mm.

15. Remove the cloth and shank support.

16. Reassemble and install the chain tensioner using a new washer. Fit the plug with a new O-ring sparingly greased with Vaseline. Tighten the chain tensioner to 46 ft. lbs. (63 Nm). Tighten the plug to 16 ft. lbs. (22 Nm).

17. Tighten the chain tensioner bolt and make sure the multi-groove belt is correctly positioned on all pulleys. Tighten to 47 ft. lbs. (63 Nm).

18. Turn the engine over two turns and check the 0 marks of the camshafts and crankshaft.

19. Wash the cylinder head gasket surface with benzene and oil the four recesses with engine oil.

20. Install the camshaft cover. Tighten to 11 ft. lbs. (15 Nm).

21. Install the spark plugs.

22. Install the ignition discharge module and connector.

23. Install the crankcase ventilation hose and pipe. Secure the check valve with a cable tie.

24. Raise the vehicle and install the drivebelt cover.

25. Lower the vehicle and install the upper engine cover.

### 2006–07 2.8L Engine

1. Before servicing the vehicle, refer to the Precautions Section.

2. Remove the battery cover.

3. Disconnect the negative battery cable.

4. Open the cap to the expansion tank, release any overpressure.

5. Raise and safely support the vehicle.

6. Undo the right-hand side of the spoiler shield.

7. Place a suitable receptacle under the radiator. Connect a hose to the radiator and drain the coolant. Close the cock and Install the spoiler shield.

8. Lower the vehicle.

9. Remove the upper section of the air cleaner. When removing the air cleaner casing cover, be careful not to damage the brake vacuum pump sensor.

10. Remove the lower section of the air filter.

11. Unplug the mass air flow sensor connector. Detach the secondary air pipe, the delivery hose from the turbocharger intake manifold and the crankcase ventilation pipes. Take care when releasing the locking mechanism on the connector so as not to damage the connector. Pull the halves straight apart to avoid bending the pins.

12. Detach the turbocharger intake manifold from the air filter housing cover and the turbocharger.

13. Remove the upper engine cover.

14. Remove the charge air pipe by first removing the clamps and the charge pressure sensor connectors. Detach the power steering line.

15. Remove the coolant return line by first removing the wiring harnesses.

16. Remove the connector and the ground connection to the engine control module.

17. Remove the engine control module bolts.

18. Remove the wiring harnesses holders by removing the cable duct, the connectors and the nut. Unplug the connector to the intake manifold pressure sensor. Remove the bolt, cable tie and connector.

19. Undo the quick coupling and remove the vacuum line from the upper section of the intake manifold.

20. Remove the crankcase ventilation lines by removing the bolt and removing the holder.

21. Remove the holder to the injector's connector.

22. Remove the throttle body connector and remove the upper section of the intake manifold.

23. Undo the quick couplings and remove the bleeder valve plus lines.

24. Place a collecting pan under the vehicle.

25. Detach the fuel line, wiping up any fuel spill with a cloth.

26. Remove the power steering reservoir by removing the bolt, return hose and clamp. Seal the hose and reservoir. Turn the reservoir to the side.

27. Unplug the connectors to the fuel pressure sensor.

28. Remove the wiring harnesses from the cylinder head for cylinders 1-3-5. Unplug the camshaft position sensor connector, the connector to the camshaft setting valve and the holder.

29. Remove the wiring harnesses from the cylinder head for cylinders 2-4-6. Unplug the camshaft position sensor connector, the connector to the camshaft setting valve and the holder.

30. Remove the wiring harness for the injectors by first unplugging the injectors' connectors.

31. Remove the fuel rail with injectors.

32. Remove the lower section of the intake manifold. Seal the inlet ducts to the cylinder head with lintless rags.

33. Remove the ignition coils for cylinders 2-4-6 by first unplugging the connectors.

34. Remove the wiring harness holder from the camshaft cover for cylinders 1-3-5.

35. Remove the ignition coils for cylinders 1-3-5 by first unplugging the connectors.

36. Remove the ground cable from the wiring harnesses by carefully cutting open the tape strip (arrows). Place the ground cable to one side.

37. Remove the wiring harnesses plus connectors. Remove the cable tie.

38. Remove the battery cover bottom section.

39. Loosen the expansion tank from the mounting.

40. Remove the turbocharger heat shield panel.

41. Remove the crankcase ventilation line from the turbocharger.

42. Remove the crankcase ventilation check valve from the camshaft cover for cylinders 1-3-5.

43. Undo the quick coupling and remove the secondary air system's valve for cylinders 2-4-6.

44. Undo the upper quick coupling by removing the bolt and removing the secondary air system's delivery line.

45. Remove the turbocharger diaphragm unit by undoing the bolt and pressing it out.

46. Remove the secondary air system's delivery line for cylinders 1-3-5.

47. Remove the secondary air system's delivery line for cylinders 2-4-6.

48. Remove the ventilation line to the oil dipstick's guide tube by undoing the quick coupling.

49. Remove the engine's wiring harness from cylinders 2-4-6 by unplugging the connectors, and by removing the head shield and the ground cable from the cylinder head. Fold the wiring harnesses to the side.

50. Remove the engine's transport lug by removing the bolts.

51. Remove the camshaft cover for cylinders 2-4-6 by removing the bolts.

52. Remove the camshaft cover for cylinders 1-3-5 by removing the bolts.

53. Raise the vehicle to half height.

54. Remove the front wheels.

55. Remove the right wheel well.

56. Remove the power steering hydraulic line from the holder on the front axle beam.

57. Remove the right engine mounting.

58. Relieve the load on the belt tensioner. Use a ¼ inch puller handle and remove the driver belt from the A/C compressor pulley.

59. Remove the bolt holding the crankshaft pulley using 83 95 360 Holding tool, crankshaft pulley (handle only) and EN-47981 Holding tool, pulley.

60. Remove the pulley using EN-47982 Puller and EN-47981 Holding tool, pulley.

61. Remove the transaxle housing at the bottom by removing the bolts.

62. Remove the wiring harness holder from the bottom of the transaxle housing by removing the bolt.

63. Lower the vehicle.

64. Remove the coolant line from the coolant outlet pipe by removing the bolt.

65. Remove the coolant outlet pipe by removing the coolant hose, remove the clamp and remove the bolts.

66. Remove the coolant pump pulley by removing the bolts. Hold firm with a screwdriver.

67. Remove the coolant pump by removing the bolts.

68. Remove the power steering pump at the top by removing the bolted joint.

69. Raise and safely support the vehicle.

70. Remove the power steering pump at the bottom by removing the bolt. Move the power steering pump towards the front of the car

71. Lower the vehicle.

72. Remove the drive belt tensioner by removing the bolt.

73. Remove the camshaft setting valve for cylinders 1-3-5 by removing the bolt.

74. Remove the camshaft setting valve for cylinders 2-4-6 by removing the bolt.

75. Remove the upper bolts of the transaxle housing.

76. Remove the timing cover. Carefully pry the cover loose with a pry bar and a screw that is carefully screwed in (to press the cover loose).

77. Remove the starter motor's wiring harness holder from the wiring harnesses.

➡**There is no marking to check valve timing on the camshafts and cylinder heads. The timing chains have silver links that are to align with the marks on the sprockets. The links of the primary chain are marked with yellow.**

78. Turn the crankshaft so that the cams of cylinder 1 point upward and inward at an angle, thus aligning the installation marks on the crankshaft sprocket and the oil

pump. When the engine is approx. 15° ATDC using EN-46111 Sleeve to turn the engine.

79. Remove the timing chain tensioner of the rear cylinder bank.

80. Remove the chain tensioner gasket.

➡**The exhaust camshaft turns somewhat when tension is relieved from the chain.**

81. Remove the guides and lift off the chain.

82. Remove the chain tensioner of the primary circuit and remove the gasket.

83. Remove the guides and lift off the chain.

84. Remove the rear primary sprocket.

85. Remove the timing chain tensioner of the front cylinder bank and remove the gasket.

86. Remove the guides and lift off the front primary sprocket. Lift off the chain.

87. Remove the crankshaft sprocket, counter holding with a spanner on the camshaft flats.

88. Remove the crankshaft sprocket, counter holding with a fixed wrench on the camshaft flats.

### ❋ CAUTION

**The camshafts must not be rotated as the valves may collide with the pistons or other valves and be damaged.**

*To install:*

89. Install the crankshaft sprocket, counter holding with a fixed wrench on the camshaft flats. Tighten the bolt to 48 ft. lbs. (65 Nm).

90. Install the sprocket on the crankshaft. The marking should be visible. The intake sprocket is variable.

91. Check that the installation marks on the crankshaft sprocket and the oil pump are aligned. Use EN-46111 Sleeve to adjust the position of the crankshaft. The crankshaft must not be rotated. Otherwise the pistons could damage the valves.

92. Install the camshaft drive of the front cylinder bank. Position the chain on the camshaft sprockets so that the silver links align with the marks on the sprockets. The front cylinder bank sprockets are marked with an L.

93. Install the primary sprocket so that the lower silver link is visible through the hole in the primary sprocket and tighten the bolt to 48 ft. lbs. (65 Nm).

94. Install the chain guides and tighten the fasteners to 18 ft. lbs. (25 Nm).

95. Restore the chain tensioner to its previous position by removing the piston

and turning the spring clockwise until it engages in the inner position. Use a screwdriver. Install the piston in the tensioner and check that the groove in the piston is vertical so that is installs with the tensioner guide.

96. Install the chain tensioner using a new gasket and tighten the fasteners to 18 ft. lbs. (25 Nm).

97. Trigger/activate the chain tensioner by carefully pressing on the timing chain or guide.

98. Install the rear primary sprocket. Tighten the bolt to 48 ft. lbs. (65 Nm).

99. Install the primary chain so that the yellow links align with the marks on the primary sprockets and crankshaft sprocket.

100. Install the front chain guide and tighten the fasteners to 7 ft. lbs. (10 Nm).

101. Install the upper guide and tighten the fasteners to 18 ft. lbs. (25 Nm).

102. Restore the chain tensioner to its previous position by removing the piston and turning the spring clockwise until it engages in the inner position. Use a screwdriver. Install the piston in the tensioner. Check that the groove in the piston is vertical so that is installs with the tensioner guide.

103. Install the chain tensioner using a new gasket and tighten the fasteners to 18 ft. lbs. (25 Nm).

104. Trigger/activate the chain tensioner by carefully pressing on the timing chain or guide.

105. Position the chain on the camshaft sprockets so that the silver links align with the marks on the sprockets. The rear cylinder bank sprockets are marked with an R.

106. Rotate the crankshaft approx. ¼ turn approx. 100° ATDC so that the lower silver link aligns with the marking on the primary sprocket.

107. Install the chain guides and tighten the fasteners to 18 ft. lbs. (25 Nm).

108. Restore the chain tensioner to its previous position by removing the piston and turning the spring clockwise until it engages in the inner position. Use a screwdriver. Install the piston in the tensioner. Check that the groove in the piston is vertical so that is installs with the tensioner guide.

109. Install the chain tensioner using a new gasket and tighten the fasteners to 18 ft. lbs. (25 Nm). Trigger/activate the chain tensioner by carefully pressing on the timing chain or guide.

110. Rotate the crankshaft one turn to check and ensure that all chain tensioners have been triggered and that the chain is correctly seated in the guide.

111. Clean the sealing surfaces.

112. Install EN-46109-2 Centering tool in the engine block and cylinder head.

113. Apply 93 165 267 Sealant. Apply an approx. 3mm thick bead on the sealing surfaces of the transaxle housing.

114. Install the transaxle housing with help from an assistant. Check that the sealing bead remains undamaged during the Alignment of the transaxle housing.

115. Install the upper bolts for the transmission housing and tighten to 17 ft. lbs. (23 Nm). Remove the centering tools.

116. Secure the starter motor's wiring harness holder on the wiring harnesses.

117. Raise and safely support the vehicle.

118. Install the lower bolts of the transaxle housing and tighten to 17 ft. lbs. (23 Nm).

119. Install the lower bolt to the power steering pump and tighten to 16 ft. lbs. (22 Nm).

120. Install the pulley to the crankshaft using J-41998 Installing Tool to press on the pulley.

121. Install the pulley bolt using 83 95 360 Holding tool, crankshaft pulley (handle only) and EN-47981 Pulley holding tool. Tighten the pulley bolt to 74 ft. lbs. (100 Nm), plus an additional 150°.

122. Secure the starter motor's wiring harness holder at the bottom. Tighten the bolt.

123. Lower the vehicle.

124. Install the camshaft setting solenoid valves Tighten to 8 ft. lbs. (10 Nm).

125. Install the belt tensioner and the bolt. Tighten the bolt to 16 ft. lbs. (22 Nm).

126. Secure the starter motor's wiring harness holder at the top. Tighten the bolt.

127. Secure the power steering pump at the top. Tighten the fasteners to 16 ft. lbs. (22 Nm).

128. Install the mounting bolts to the engine pad's attaching plate and tighten to 32 ft. lbs. (44 Nm).

129. Install the coolant pump. Use a new gasket and tighten the bolts to 8 ft. lbs. (10 Nm).

130. Install the coolant pump's belt pulley. Counterhold with a screwdriver. Tighten the bolts to 9 ft. lbs. (12 Nm).

131. Install the coolant outlet pipe. Use a new gasket and tighten the bolts to 9 ft. lbs. (12 Nm).

132. Install the coolant hose and secure the clamp.

133. Unload the belt tensioner. Use a 1/4 inch puller and install the drive belt.

134. Install the right engine mounting.

135. Install the holder to the power steering line and tighten the bolt to 6 ft. lbs. (8 Nm).

136. Install the coolant line to the coolant outlet pipe. Use a new sealing ring. Tighten to 6 ft. lbs. (8 Nm).

137. Raise and safely support the vehicle.

138. Secure the power steering hydraulic line in the holder on the front axle beam.

139. Lower the vehicle to half height.

140. Install the right-hand wheel well.

141. Install the wheels.

142. Clean the camshaft cover sealing surfaces.

143. Replace the camshaft cover gasket for cylinders 2-4-6.

144. Apply an approx. 3mm thick bead of sealing compound on the separating points of the cylinder head and transmission housing.

145. Install the camshaft cover for cylinders 2-4-6 and tighten the bolts to 8 ft. lbs. (10 Nm).

146. Install the engine's transport lug and tighten the bolt to 48 ft. lbs. (65 Nm).

147. Clean the camshaft cover sealing surfaces.

148. Replace the camshaft cover gasket for cylinders 1-3-5. Use new gaskets.

149. Apply an approx. 3mm thick bead of sealing compound on the separating points of the cylinder head and transmission housing.

150. Install the camshaft cover for cylinders 1-3-5 and tighten the bolts to 8 ft. lbs. (10 Nm).

151. Install the engine wiring harnesses to the cylinder head for cylinders 2-4-6 by plugging in the connectors on the cylinder head for cylinders 2-4-6. Install the heat shield and install the ground cable on the cylinder head for cylinders 2-4-6. Install the bolts. Secure the wiring harnesses. Secure the engine control module's ground cable in the wiring harnesses. Install the cable ties.

152. Take care when plugging in the connector so as not to damage or press out the pins/sleeves in the connector.

153. Install the secondary air system line for cylinders 2-4-6. Install the exhaust pipe and tighten the bolts to 8 ft. lbs. (10 Nm).

154. Install the ventilation line to the oil dipstick's guide tube. Connect the quick coupling.

155. Install the turbocharger diaphragm unit by installing the bolts.

156. Install the secondary air system's delivery line. Connect the quick couplings.

157. Install the secondary air system's check valve for cylinders 2-4-6. Use a new gasket. Tighten the bolts and connect the quick coupling.

158. Install the secondary air system's delivery line and bolt.

159. Install the check valve to the crankcase ventilation on the camshaft cover for cylinders 1-3-5.

160. Install the crankcase ventilation line on the turbocharger and tighten the bolt to 8 ft. lbs. (10 Nm).

161. Install the turbocharger's upper heat shield panel. Install the nut and tighten to 8 ft. lbs. (10 Nm).

162. Install the expansion tank in the holder.

163. Install the battery cover bottom section.

164. Install the charge pressure regulator's wiring harness. Plug in the connector. Install the cable tie.

165. Install the ignition coils for cylinders 2-4-6. Plug in the connectors and install the bolts. Tighten the bolts to 8 ft. lbs. (10 Nm).

166. Install the wiring harness holder on the camshaft cover for cylinders 1-3-5. Secure in the holder.

167. Install the ignition coils for cylinders 1-3-5. Plug in the connectors and install the bolts. Tighten the bolts to 8 ft. lbs. (10 Nm).

168. Install the lower section of the intake manifold with a new seal. Remove the rags from the inlet ducts.

169. Install the fuel rail with injectors and tighten the bolts to 8 ft. lbs. (10 Nm).

170. Install the injectors' wiring harness. Plug in the connectors.

171. Plug in the connectors to the fuel pressure sensor.

172. Install the wiring harness to the camshaft guide for cylinders 2-4-6. Plug in the connectors and secure in the holder.

173. Connect the wiring harness to the camshaft guide for cylinders 1-3-5. Plug in the connectors and secure in the holder.

174. Install the power steering reservoir. Install the return hose. Secure the clamp and tighten to 7 ft. lbs. (9 Nm).

175. Connect the fuel delivery line.

176. Install the bleeder valve plus lines. Connect the quick couplings. Plug in the connectors to the bleeder valve.

177. Install the upper section of the intake manifold. Plug in the throttle body's connector. Use new gaskets and tighten the fasteners to 17 ft. lbs. (23 Nm).

178. Install the holder to the injector's connector.

179. Install the lines for the crankcase ventilation.

180. Install the vacuum line on the upper section of the intake manifold. Connect the quick coupling.

181. Install the wiring harnesses' holders. Install the bolt. Plug in the connectors for the intake manifold's pressure sensor. Install the cable cables. Secure the connectors. Install the cable duct. Plug in the connectors. Install the nut.

182. Install the engine control module. Install the bolts and tighten to 8 ft. lbs. (10 Nm).

183. Install the ground connection.

184. Plug in the engine control module connector.

185. Install the coolant return line. Install the clamps and tighten the bolts to 8 ft. lbs. (10 Nm).

186. Install the charge air pipe. Install the clamp.

187. Plug in the charge pressure sensor connector. Secure the power steering line with clips.

188. Install the air intake pipe.

189. Install the air filter housing with the mass air flow sensor as follows:

   a. Plug in the mass air flow sensor connector.

   b. Secure the cable ties.

   c. Install the secondary air system's intake manifold. Secure the line.

   d. Connect the quick coupling.

190. Install the upper engine cover. Remove the oil filler opening's screw cap. Install the protective cover. Install the oil filler opening's screw cap.

191. Fill and bleed the cooling system.

192. Connect the negative battery cable and install the battery cover.

## 2008 2.8L Engine

➡**There is no marking for checking valve timing on the camshafts and cylinder heads. The timing chains and primary chain have white markings on the links which should correspond with the markings on the sprockets.**

1. Remove the Timing cover.

2. Rotate the crankshaft so that the cams of cylinder 1 point upward and inward at an angle, thus aligning the installation marks on the crankshaft sprocket and the oil pump. When the engine is approx. 15 degrees ATDC, use EN-46111 Sleeve to turn the engine.

3. Remove the timing chain tensioner for the rear cylinder bank.

4. Remove the chain tensioner gasket.

➡**The exhaust camshaft turns somewhat when tension is relieved from the chain.**

5. Remove the guides and lift off the chain.

6. Remove the chain tensioner for the primary circuit and remove the gasket.

7. Remove the guides and lift off the chain.

8. Remove the rear primary sprocket.

9. Remove the timing chain tensioner for the front cylinder bank and remove the gasket.

10. Remove the guides and lift off the front primary sprocket. Lift off the chain.

11. Remove the crankshaft sprocket.

12. Remove the camshafts' bolts (A) and sprockets (B) (x 4), counterholding with a fixed spanner on the camshaft flats.

### ※※ WARNING

**The camshafts must not be rotated as the valves may collide with the pistons or other valves and be damaged.**

*To install:*

13. Install the camshafts' sprockets (B) and bolts (A) (x 4), counterholding with a fixed spanner on the camshaft flats. Tighten to 48 ft. lbs. (65 Nm).

➡**The intake sprocket is variable.**

14. Install the sprocket on the crankshaft. The marking must be turned outwards. Check that the fitting markings on the crankshaft sprocket and oil pump are aligned. Use EN-46111 Sleeve to adjust the position of the crankshaft.

### ※※ WARNING

**The crankshaft must not be rotated. Otherwise the pistons could damage the valves.**

15. Install the front primary sprocket. Tighten to 48 ft. lbs. (65 Nm).

16. Install the camshaft drive for the front cylinder bank. Position the chain on the camshaft sprockets so that the links with white marking align with the markings on the sprockets. The front cylinder bank sprockets are marked with an L.

17. Install the chain guide. The link with white marking must be visible through the hole (arrow) in the primary sprocket. The link must be right in front of the hole.

18. Install the thread adhesive on the moving guide's bolt. Tighten to 17 ft. lbs. (23 Nm).

19. Restore the chain tensioner to its previous position by removing the piston and turning the spring clockwise until it engages in the inner position. Use a screwdriver. Refit the piston in the tensioner and

check that the groove in the piston is vertical so that is fits with the tensioner guide.

20. Install the chain tensioner with a new gasket. Note the position of the gasket. Tighten to 17 ft. lbs. (23 Nm).

21. Trigger/activate the chain tensioner by carefully pressing on the timing chain or guide.

22. Install the rear primary sprocket. Tighten to 48 ft. lbs. (65 Nm).

23. Install the primary chain so that the white links align with the markings on the primary sprockets and crankshaft sprocket.

24. Install the front chain guide. Tighten to 10 ft. lbs. (13 Nm).

25. Install the upper guide. Tighten to 18 ft. lbs. (23 Nm).

26. Restore the chain tensioner to its previous position by removing the piston and turning the spring clockwise until it engages in the inner position. Use a screwdriver. Refit the piston in the tensioner. Check that the groove in the piston is vertical so that is fits with the tensioner guide.

27. Install the chain tensioner with new gasket. Note the position of the gasket. Tighten to 17 ft. lbs. (23 Nm).

28. Trigger/activate the chain tensioner by carefully pressing on the timing chain or guide.

29. Rotate the camshafts of the rear cylinder bank forward slightly and install EN-48383-3 Adjustment tool on the camshafts.

30. Rotate the crankshaft forward approx. ¼ turn, approx. 100° ATDC, so that the marking on the crankshaft sprocket is aligned with the marking on the oil pump.

31. Fit the chain guides.

32. Install the thread adhesive on the moving guide's bolt. Tighten to 17 ft. lbs (23 Nm).

33. Restore the chain tensioner to its previous position by removing the piston and turning the spring clockwise until it engages in the inner position. Use a screwdriver. Refit the piston in the tensioner. Check that the groove in the piston is vertical so that is fits with the tensioner guide.

34. Install the chain tensioner (A) with a new gasket. Note the position of the gasket. Tighten to 17 ft. lbs (23 Nm).

35. Trigger/activate the chain tensioner by carefully pressing on the timing chain or guide.

36. Remove EN-48383-3 and rotate the crankshaft 1 3/4 turns to check and ensure that all chain tensioners have been triggered and that the chain is correctly seated in the guides.

37. Set the crankshaft sprocket's marking against the oil pump's marking.

38. Check that EN-48383-2 Adjustment tool fits on the rear bank camshafts and EN-48383-3 Adjustment Tool fits on the front bank camshafts. Remove the Adjustment Tool

39. Install the Timing cover.

## VALVE LASH

### *ADJUSTMENT*

#### 2.0L H4 Engine

*See Figures 198 through 210.*

➡ **Inspection and adjustment of the valve clearance should be performed with the engine cold.**

1. Before servicing the vehicle, refer to the Precautions Section.
2. Remove or disconnect the following:
   - Negative battery cable
   - Air intake duct
   - Bolt that attaches the right hand timing cover
   - Engine undercover
   - Remaining bolts attaching the right hand timing belt cover and the cover
3. When inspecting the No. 1 and No. 3 cylinders:
   a. Pull out the engine harness connector with the bracket from the air cleaner upper cover.
   b. Remove the air cleaner case.
   c. Disconnect the spark plug wires from the No. 1 and No. 3 cylinders.
   d. Disconnect the Positive Crankcase Ventilation (PCV) hose from the right hand rocker cover.
   e. Remove the right hand rocker cover.
4. When inspecting the No. 2 and No. 4 cylinders:
   a. Remove the battery and tray.
   b. Remove the bolt that attaches the engine harness onto the body.
   c. Disconnect the washer motor connectors.
   d. Remove the washer tank bolts and lift the tank upwards.
   e. Disconnect the spark plug wires from the No. 2 and No. 4 cylinders.
   f. Disconnect the PCV hose from the left hand rocker cover.
   g. Remove the left hand rocker cover.
   h. Turn the crankshaft pulley clockwise until the arrow mark on the camshaft is positioned as shown in the illustration to measure the No. 1 intake and No. 3 exhaust valves.
5. Using a suitable feeler gauge, measure the No. 1 and No. 3 cylinder exhaust valve clearance. Insert the gauge in as horizontal a direction with respect to the shim. Make sure to measure the exhaust valve clearances while lifting up the vehicle.
6. The intake valve clearance should be 0.0071–0.0087 inch (0.18–0.22mm). The exhaust valve clearance should be 0.0090–0.0106 inch (0.23–0.27mm).
7. If not within specification, adjust the valve as outlined in the adjustment steps.
8. Turn the crankshaft pulley clockwise to measure the valve clearance for the No. 2 exhaust and No. 3 intake valves as shown in the illustration.
9. Turn the crankshaft pulley clockwise to measure the valve clearance for the No. 2 intake and No. 4 exhaust valves as shown in the illustration.
10. Turn the crankshaft pulley clockwise to measure the valve clearance for the No. 1 exhaust and No. 4 intake valves as shown in the illustration.
11. Adjust the valve clearance as follows:
    a. Measure and record all valve clearances using the procedures outlined in the inspection steps in this section.
    b. Prepare shim replacer tool 498187200.
    c. Rotate the notch of the valve lifter outwards 45 degrees.
    d. Adjust the shim replacer tool notch to the lifter and set it.

➡ **Make sure when setting the tool that the edge does not touch the shim.**

    e. Tighten bolt "A" and attach it to the cylinder head. Refer to the illustration for bolt locations.
    f. Tighten bolt "B" and insert the lifter. Refer to the illustration for bolt locations.

    g. Use tweezers and remove the shim from the lifter. A magnet can also be used to remove the shim.
    h. Measure the shim thickness using a micrometer.
    i. Using the table supplied, select a suitable shim using measured valve clearance and shim thickness.
    j. Install the replacement shim to the lifter.

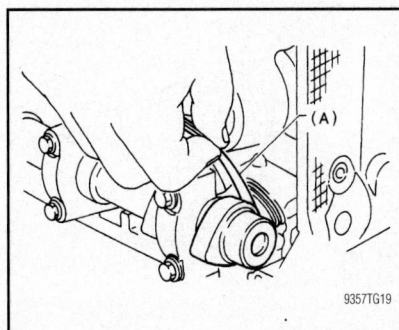

**Fig. 199 Use a feeler gauge to inspect the valve clearance—2.0L H4 engines**

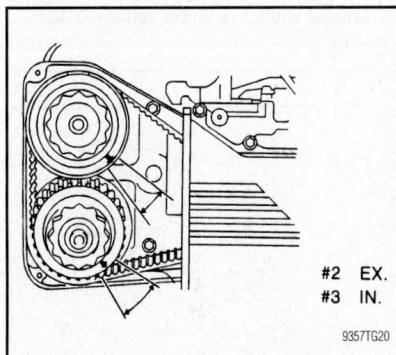

#2 EX.
#3 IN.

**Fig. 200 Turn the crankshaft pulley clockwise until the arrow mark on the camshaft is positioned as shown to measure the No. 2 exhaust and No. 3 intake valves—2.0L H4 engines**

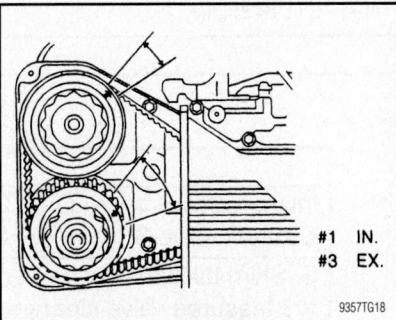

#1 IN.
#3 EX.

**Fig. 198 Turn the crankshaft pulley clockwise until the arrow mark on the camshaft is positioned as shown to measure the No. 1 intake and No. 3 exhaust valves—2.0L H4 engines**

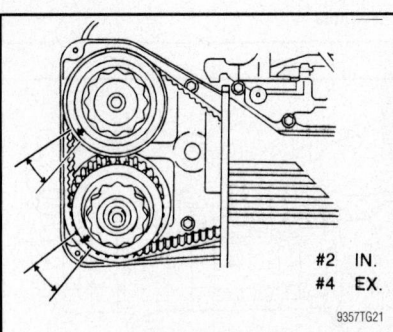

#2 IN.
#4 EX.

**Fig. 201 Turn the crankshaft pulley clockwise until the arrow mark on the camshaft is positioned as shown to measure the No. 2 intake and No. 4 exhaust valves—2.0L H4 engines**

k. After all shims have been adjusted, inspect the valve clearances again.

l. After completion, install all removed components.

## 2.0L I4 and 2.3L Engines

The hydraulic cam followers used in Saab engines do not require adjusting. The cam followers keep the valve clearance

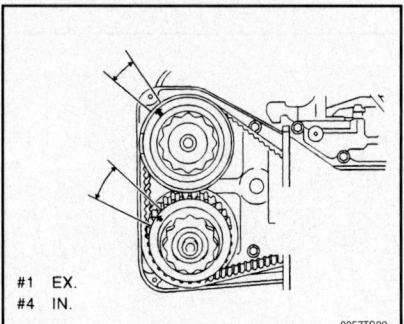

Fig. 202 Turn the crankshaft pulley clockwise until the arrow mark on the camshaft is positioned as shown to measure the No. 1 exhaust and No. 4 intake valves—2.0L H4 engines

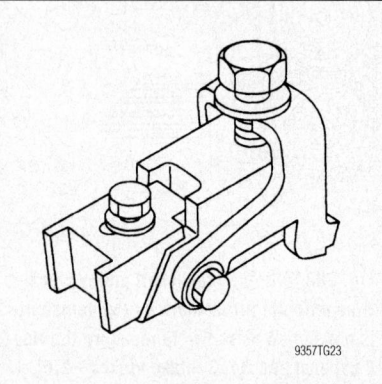

Fig. 203 Shim replacer tool 498187200 is required to adjust the valves—2.0L H4 engines

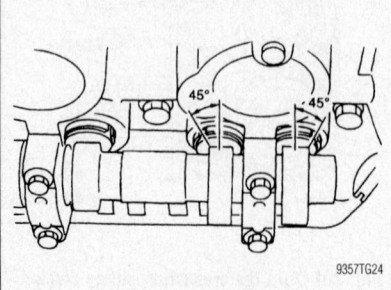

Fig. 204 Rotate the notch of the valve lifter outwards 45 degrees—2.0L H4 engines

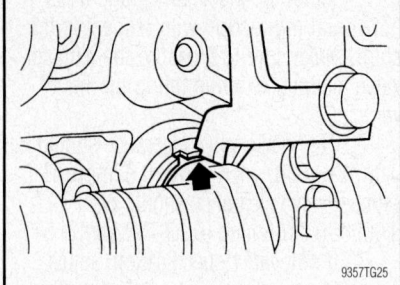

Fig. 205 Adjust the shim replacer tool notch to the lifter and set it—2.0L H4 engines

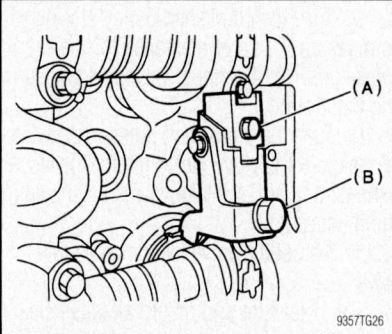

Fig. 206 Location of bolts "A" and "B" on the shim replacer tool—2.0L H4 engines

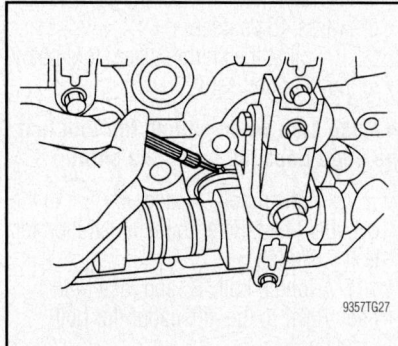

Fig. 207 Remove the shim from the lifter—2.0L H4 engine

within specification. However, if the cam followers are making excessive noise or are diagnosed to be defective, perform the following procedure:

1. Before servicing the vehicle, refer to the precautions in the beginning of this section.

2. Disconnect the negative battery cable.

3. If a cam follower is noisy, it can be found by removing the valve cover and, using a screwdriver, gently pushing down on each cam follower until the defective follower(s) is found by exhibiting a spongy feeling.

4. Replace the defective cam follower(s); first removing the camshaft(s).

5. Install the camshaft(s) and the valve cover.

## 2.5L SOHC Engine

*See Figure 211.*

➡**The valve adjustment should be performed while the engine is cold.**

1. Before servicing the vehicle, refer to the Precautions Section.

2. Raise and support the vehicle safely.

3. Remove the undercover.

4. Lower the vehicle.

5. Disconnect the negative battery cable.

6. To remove the front side V belt, remove the belt covers. Loosen the lock bolt. Loosen the slider bolt. Remove the front side belt.

7. To remove the rear side V belt, remove the belt covers. Loosen the lock bolt. Loosen the slider bolt. Remove the rear side belt. Remove the belt tensioner.

8. Remove the crankshaft pulley bolt.

9. Lock the crankshaft in place using tool ST499977100, or equivalent.

10. Remove the crankshaft pulley.

11. Remove the left side timing belt cover.

| | Unit: mm |
|---|---|
| Intake valve:$S = (V + T) - 0.20$ | |
| Exhaust valve:$S = (V + T) - 0.25$ | |
| S: Shim thickness to be used | |
| V: Measured valve clearance | |
| T: Shim thickness required | |

Fig. 208 Use this table to help you select a suitable shim—2.0L H4 engine

| Part No. | Thickness mm (in) |
|---|---|
| 13218 AK010 | 2.00 (0.0787) |
| 13218 AK020 | 2.02 (0.0795) |
| 13218 AK030 | 2.04 (0.0803) |
| 13218 AK040 | 2.06 (0.0811) |
| 13218 AK050 | 2.08 (0.0819) |
| 13218 AK060 | 2.10 (0.0827) |
| 13218 AK070 | 2.12 (0.0835) |
| 13218 AK080 | 2.14 (0.0843) |
| 13218 AK090 | 2.16 (0.0850) |
| 13218 AK100 | 2.18 (0.0858) |
| 13218 AK110 | 2.20 (0.0866) |
| 13218 AE710 | 2.22 (0.0874) |
| 13218 AE730 | 2.24 (0.0882) |
| 13218 AE750 | 2.26 (0.0890) |
| 13218 AE770 | 2.28 (0.0898) |
| 13218 AE790 | 2.30 (0.0906) |
| 13218 AE810 | 2.32 (0.0913) |
| 13218 AE830 | 2.34 (0.0921) |
| 13218 AE850 | 2.36 (0.0929) |
| 13218 AE870 | 2.38 (0.0937) |
| 13218 AE890 | 2.40 (0.0945) |
| 13218 AE910 | 2.42 (0.0953) |
| 13218 AE920 | 2.43 (0.0957) |
| 13218 AE930 | 2.44 (0.0961) |
| 13218 AE940 | 2.45 (0.0965) |
| 13218 AE950 | 2.46 (0.0969) |
| 13218 AE960 | 2.47 (0.0972) |
| 13218 AE970 | 2.48 (0.0976) |
| 13218 AE980 | 2.49 (0.0980) |
| 13218 AE990 | 2.50 (0.0984) |
| 13218 AF000 | 2.51 (0.0988) |
| 13218 AF010 | 2.52 (0.0992) |
| 13218 AF020 | 2.53 (0.0996) |
| 13218 AF030 | 2.54 (0.1000) |
| 13218 AF040 | 2.55 (0.1004) |
| 13218 AF050 | 2.56 (0.1008) |
| 13218 AF060 | 2.57 (0.1012) |
| 13218 AF070 | 2.58 (0.1016) |
| 13218 AF090 | 2.60 (0.1024) |
| 13218 AF110 | 2.62 (0.1031) |
| 13218 AF130 | 2.64 (0.1039) |
| 13218 AF150 | 2.66 (0.1047) |
| 13218 AF170 | 2.68 (0.1055) |
| 13218 AF190 | 2.70 (0.1063) |

9357TG30

Fig. 209 Valve adjusting shim chart—2.0L H4 engine

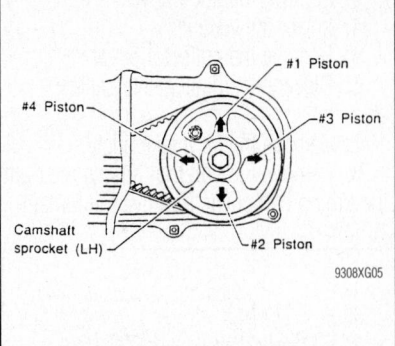

Fig. 210 Position the camshaft for adjustment to valves—2.0L H4 engine

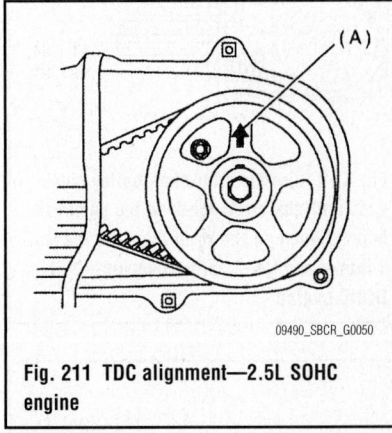

Fig. 211 TDC alignment—2.5L SOHC engine

12. Remove the fuel injector.

13. Remove the rocker cover.

14. Position the number one piston at TDC of the compression stroke.

➡**When the arrow (see illustration) on the camshaft sprocket (left side) comes exactly to the top, number one cylinder piston is at TDC of the compression stroke.**

15. Measure the valve clearance, using a feeler gauge.

16. If adjustment is needed, loosen the valve rocker nut and screw. Position the feeler gauge.

➡**Insert the feeler gauge in a horizontally as possible with respect to the valve stem end face. Adjust the exhaust valve clearance while lifting up the vehicle.**

17. While noting the valve clearance, tighten the rocker adjusting screw.

18. When the proper valve clearance is obtained, tighten the valve rocker nut to 7.2 ft. lbs.

19. Adjust the valve clearance on the remaining cylinders, following the above procedure.

➡**Be sure to position the pistons to their respective TDC positions on the compression stroke, before checking and adjusting the valves. By rotating the crankshaft pulley clockwise every 180 degrees from the state that number one piston is on TDC of the compression stroke, the remaining pistons come to TDC of the compression stroke in the following order, #3, #2, and #4.**

20. After adjustment, replace any removed components.

21. Be sure to use new gaskets and seals, as required.

### 2.5L DOHC Engine

*See Figures 212 through 218.*

➡**The valve adjustment should be performed while the engine is cold.**

1. Before servicing the vehicle, refer to the Precautions Section.

2. Raise and support the vehicle safely.

3. Remove the undercover.

4. Lower the vehicle.

5. Remove the collector cover.

6. Disconnect the negative battery cable.

7. Remove the air intake duct.

8. Remove the bolt that retains the right side timing belt cover. Remove the remain- ing bolts and remove the right side timing belt cover.

9. Disconnect the ignition coil electrical connector. Remove the ignition coil.

10. Position a suitable container under the vehicle.

11. Disconnect the PCV hose from the rocker cover. Remove the rocker cover retaining bolts. Remove the rocker cover from the vehicle.

12. Position the number one piston at TDC of the compression stroke.

13. Using a feeler gauge, measure and record the clearance of the number one cylinder intake and the number three cylin- der exhaust valves.

➡**Insert the feeler gauge in a horizon- tally as possible with respect to the valve lifter. Measure and record the exhaust valve clearance while lifting up the vehicle.**

14. Rotate the crankshaft pulley clockwise until the arrow mark on the camshaft is positioned as shown to measure and record the clearance on the number two exhaust and number three intake valves.

15. Rotate the crankshaft pulley clock- wise until the arrow mark on the camshaft is positioned as shown to measure and record the number two intake and number four exhaust valves.

16. Rotate the crankshaft pulley clock- wise until the arrow mark on the camshaft is positioned as shown to measure and record the number one exhaust and number four intake valves.

17. If adjustment is required, remove the camshafts.

18. Remove and measure the thickness of the valve lifter. Select a suitable shim, using the shim selection chart.

19. Install the replacement shim to the lifter.

20. After all shims have been adjusted, inspect the valve clearances again.

21. After completion, install all removed components.

### 2.8L Engine

1. Before servicing the vehicle, refer to the Precautions Section.

2. Remove the cylinder head.

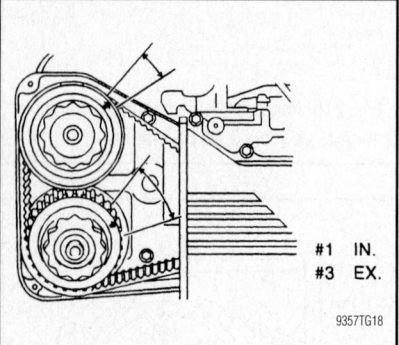

**Fig. 212 Turn the crankshaft pulley clock- wise until the arrow mark on the camshaft is positioned as shown to measure the No. 1 intake and No. 3 exhaust valves—2.5L DOHC engine**

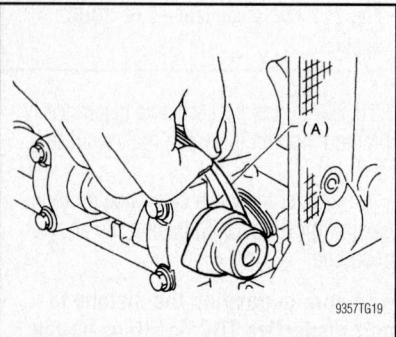

**Fig. 213 Use a feeler gauge to inspect the valve clearance—2.5L DOHC engine**

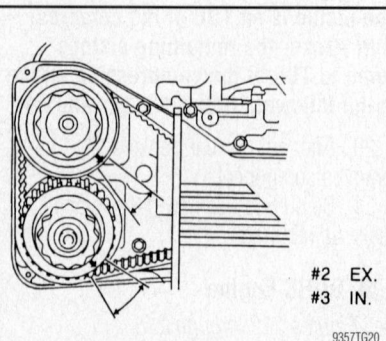

**Fig. 214 Turn the crankshaft pulley clock- wise until the arrow mark on the camshaft is positioned as shown to measure the No. 2 exhaust and No. 3 intake valves—2.5L DOHC engine**

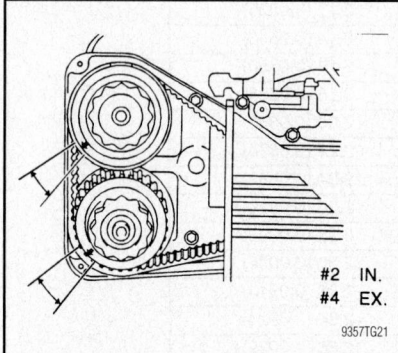

**Fig. 215 Turn the crankshaft pulley clock- wise until the arrow mark on the camshaft is positioned as shown to measure the No. 2 intake and No. 4 exhaust valves—2.5L DOHC engine**

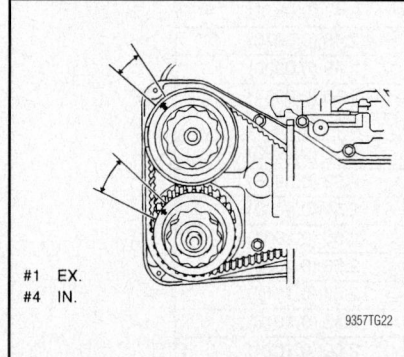

**Fig. 216 Turn the crankshaft pulley clock- wise until the arrow mark on the camshaft is positioned as shown to measure the No. 1 exhaust and No. 4 intake valves—2.5L DOHC engine**

| Unit: (mm) | | |
|---|---|---|
| Intake valve: $S = (V + T) - 0.20$ | | |
| Exhaust valve: $S = (V + T) - 0.35$ | | |
| S: Valve lifter thickness required | | |
| V: Measured valve clearance | | |
| T: Valve lifter thickness to be used | | |

09490_SBCR_G0051

**Fig. 217 Use this table to help you select a suitable shim—2.5L DOHC engine**

| Part No. | Thickness mm (in) |
|---|---|
| 13228 AB102 | 4.68 (0.1843) |
| 13228 AB112 | 4.69 (0.1846) |
| 13228 AB122 | 4.70 (0.1850) |
| 13228 AB132 | 4.71 (0.1854) |
| 13228 AB142 | 4.72 (0.1858) |
| 13228 AB152 | 4.73 (0.1862) |
| 13228 AB162 | 4.74 (0.1866) |
| 13228 AB172 | 4.75 (0.1870) |
| 13228 AB182 | 4.76 (0.1874) |
| 13228 AB192 | 4.77 (0.1878) |
| 13228 AB202 | 4.78 (0.1882) |
| 13228 AB212 | 4.79 (0.1886) |
| 13228 AB222 | 4.80 (0.1890) |
| 13228 AB232 | 4.81 (0.1894) |
| 13228 AB242 | 4.82 (0.1898) |
| 13228 AB252 | 4.83 (0.1902) |
| 13228 AB262 | 4.84 (0.1906) |
| 13228 AB272 | 4.85 (0.1909) |
| 13228 AB282 | 4.86 (0.1913) |
| 13228 AB292 | 4.87 (0.1917) |
| 13228 AB302 | 4.88 (0.1921) |
| 13228 AB312 | 4.89 (0.1925) |
| 13228 AB322 | 4.90 (0.1929) |
| 13228 AB332 | 4.91 (0.1933) |
| 13228 AB342 | 4.92 (0.1937) |
| 13228 AB352 | 4.93 (0.1941) |
| 13228 AB362 | 4.94 (0.1945) |
| 13228 AB372 | 4.95 (0.1949) |
| 13228 AB382 | 4.96 (0.1953) |
| 13228 AB392 | 4.97 (0.1957) |
| 13228 AB402 | 4.98 (0.1961) |
| 13228 AB412 | 4.99 (0.1965) |
| 13228 AB422 | 5.00 (0.1969) |
| 13228 AB432 | 5.01 (0.1972) |
| 13228 AB442 | 5.02 (0.1976) |
| 13228 AB452 | 5.03 (0.1980) |
| 13228 AB462 | 5.04 (0.1984) |
| 13228 AB472 | 5.05 (0.1988) |
| 13228 AB482 | 5.06 (0.1992) |
| 13228 AB492 | 5.07 (0.1996) |
| 13228 AB502 | 5.08 (0.2000) |
| 13228 AB512 | 5.09 (0.2004) |
| 13228 AB522 | 5.10 (0.2008) |
| 13228 AB532 | 5.11 (0.2012) |
| 13228 AB542 | 5.12 (0.2016) |
| 13228 AB552 | 5.13 (0.2020) |
| 13228 AB562 | 5.14 (0.2024) |
| 13228 AB572 | 5.15 (0.2028) |
| 13228 AB582 | 5.16 (0.2031) |
| 13228 AB592 | 5.17 (0.2035) |
| 13228 AB602 | 5.18 (0.2039) |
| 13228 AB612 | 5.19 (0.2043) |

| Part No. | Thickness mm (in) |
|---|---|
| 13228 AB622 | 5.20 (0.2047) |
| 13228 AB632 | 5.21 (0.2051) |
| 13228 AB642 | 5.22 (0.2055) |
| 13228 AB652 | 5.23 (0.2059) |
| 13228 AB662 | 5.24 (0.2063) |
| 13228 AB672 | 5.25 (0.2067) |
| 13228 AB682 | 5.26 (0.2071) |
| 13228 AB692 | 5.27 (0.2075) |
| 13228 AB702 | 4.38 (0.1724) |
| 13228 AB712 | 4.40 (0.1732) |
| 13228 AB722 | 4.42 (0.1740) |
| 13228 AB732 | 4.44 (0.1748) |
| 13228 AB742 | 4.46 (0.1756) |
| 13228 AB752 | 4.48 (0.1764) |
| 13228 AB762 | 4.50 (0.1771) |
| 13228 AB772 | 4.52 (0.1780) |
| 13228 AB782 | 4.54 (0.1787) |
| 13228 AB792 | 4.56 (0.1795) |
| 13228 AB802 | 4.58 (0.1803) |
| 13228 AB812 | 4.60 (0.1811) |
| 13228 AB822 | 4.62 (0.1819) |
| 13228 AB832 | 4.64 (0.1827) |
| 13228 AB842 | 4.66 (0.1835) |
| 13228 AB852 | 5.29 (0.2083) |
| 13228 AB862 | 5.31 (0.2091) |
| 13228 AB872 | 5.33 (0.2098) |
| 13228 AB882 | 5.35 (0.2106) |
| 13228 AB892 | 5.37 (0.2114) |
| 13228 AB902 | 5.39 (0.2122) |
| 13228 AB912 | 5.41 (0.2123) |
| 13228 AB922 | 5.43 (0.2138) |
| 13228 AB932 | 5.45 (0.2146) |
| 13228 AB942 | 5.47 (0.2154) |
| 13228 AB952 | 5.49 (0.2161) |
| 13228 AB962 | 5.51 (0.2169) |
| 13228 AB972 | 5.53 (0.2177) |
| 13228 AB982 | 5.55 (0.2185) |
| 13228 AB992 | 5.57 (0.2193) |
| 13228 AC002 | 5.59 (0.2201) |
| 13228 AC012 | 5.61 (0.2209) |
| 13228 AC022 | 5.63 (0.2217) |
| 13228 AC032 | 5.65 (0.2224) |

09490_SBCR_G0052

Fig. 218 Valve adjusting shim chart—2.5L DOHC engine

3. Check and adjust the valve clearance relative to the working range of the tappet.

4. The tolerances for valve clearance are 0.76–0.80 in. (19.5–20.5mm).

5. The valve clearance settings are 0.78–0.80 in. (20.0–20.4mm) with a nominal clearance of 0.79 in. (20.2 mm).

6. Valve clearance is equivalent to the distance between the end of the valve stem and the camshaft bearing seat.

7. Before valve clearances can be checked, the camshafts and valve tappets must first be removed.

8. The valve clearance is checked with Valve Clearance Gauge 83 93 753 as follows:

a. Place valve clearance gauge across two of the camshaft bearing seats with the depth gauge against the end of the valve stem.

b. Check that the maximum gauge depth of 20.5 mm actually reaches down to the end of the valve stem (noticed by the valve clearance gauge not bottoming against the bearing seat closest to the depth gauge).

c. Then check that the min depth 19.5 mm does not reach the end of the valve stem spindle. Correct valve position should be between the depth measurement min and max values. If the valve position deviates from the given measurements, adjustments are done on the valve spindle or valve seat. The valve spindle is shortened if measurement is below the min value and the valve seat is machined if the max value is exceeded. When adjusting the valve position, set this at nominal value 20.2 mm.

# ENGINE PERFORMANCE & EMISSION CONTROL

## CAMSHAFT POSITION (CMP) SENSOR

### LOCATION

#### 2.0L H4 and 2.5L Engines

The camshaft position sensor is located on the left (driver's) side of the engine at the camshaft sprocket.

#### 2.3L Engine

The camshaft position sensor is located under the crankcase ventilation pipe on the engine.

#### 2.8L Engine

The front cylinder bank camshaft position sensor is located on the front cylinder bank's drivebelt end near the rear bank. The rear cylinder band camshaft position sensor is located on the rear cylinder bank's drivebelt end near the front bank.

### REMOVAL & INSTALLATION

#### 2.0L H4 Engine

*See Figure 219.*

1. Before servicing the vehicle, refer to the Precautions Section.

Fig. 219 Camshaft position sensor location—2.0L H4 engine

2. Disconnect the negative battery cable.
3. Disconnect the connector from the camshaft position sensor.
4. Remove the camshaft position sensor from the LH camshaft support.

*To install:*

5. Install in the reverse order of removal.
6. Tighten bolt to 5 ft. lbs. (6 Nm).

#### 2.3L Engine

*See Figure 220.*

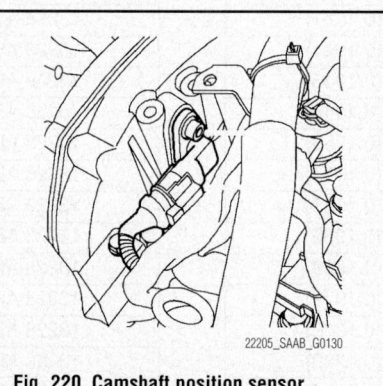

Fig. 220 Camshaft position sensor location—2.3L engine

1. Remove the upper engine cover and insulation.
2. Disconnect the crankcase ventilation pipe 2 screws and a hose clip and raise it slightly.

➡On vehicles equipped with preheated crankcase ventilation, take care not to damage the heating circuit in the hose.

3. Unplug the position sensor connector.

### ※※ WARNING

Take care when releasing the locking mechanism on the connector so as not to damage the connector. Pull the halves straight apart to avoid bending the pins. For further information

regarding connectors, refer to Connectors, handling and inspection.

4. Remove the position sensor.

*To install:*

5. Install the position sensor. Tighten to 7 ft. lbs. (9 Nm).
6. Install the crankcase ventilation pipe.
7. Plug in the position sensor.
8. Install the insulation and the upper engine cover.

#### 2.5L SOHC Engine

*See Figure 221.*

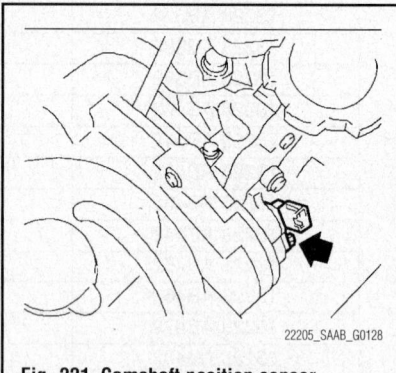

Fig. 221 Camshaft position sensor location—2.5L engines

1. Before servicing the vehicle, refer to the Precautions Section.
2. Disconnect the negative battery cable.
3. Disconnect the connector from the sensor.
4. Remove the bolt that retains the sensor to the sensor support.
5. Remove the bolt that retains the sensor support to the camshaft cap.
6. Remove the sensor and the sensor support as a unit.
7. Separate the sensor from the support.

*To install:*

8. Installation is the reverse of the removal procedure.

9. Tighten the sensor support to 4.7 ft. lbs. (6.4 Nm).

10. Tighten the sensor to 4.7 ft. lbs. (6.4 Nm).

### 2.5L DOHC Engine

*See Figure 222.*

1. Before servicing the vehicle, refer to the Precautions Section.

2. Disconnect the negative battery cable.

3. Remove the collector cover.

4. Disconnect the connector from camshaft position sensor RH.

5. Remove the camshaft position sensor RH from the rear side of the cylinder head.

6. Remove the cam shaft position sensor LH in the same way as RH.

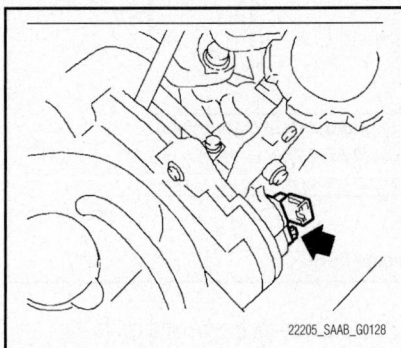

**Fig. 222 Camshaft position sensor location—2.5L engines**

*To install:*

7. Installation is the reverse of the removal procedure.

8. Tighten the sensor to 4.7 ft. lbs. (6.4 Nm).

### 2.8L Engine

#### Front

*See Figure 223.*

1. Before servicing the vehicle, refer to the Precautions Section.

2. Drain the coolant

3. Remove the upper engine cover.

4. Detach the turbocharger intake hose.

### ❋❋ WARNING

**When removing the air cleaner casing cover, be careful not to damage the brake vacuum pump sensor.**

5. Remove the air cleaner casing cover and the air cleaner.

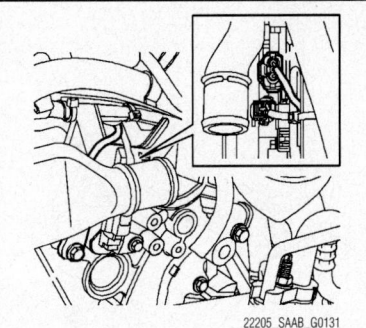

**Fig. 223 Camshaft position sensor location—2.8L engine front**

6. Detach the intake hose and remove the air cleaner casing.

7. Remove the coolant pipe from the port.

8. Position a jack under the oil sump and relieve the load on the engine mounting.

9. Remove the engine pad from the body and bracket.

10. Unplug the connector from the intake air sensor.

11. Detach the upper turbocharger delivery pipe from the throttle body and detach the lower delivery pipe.

12. Remove the P-clamp of the servo pipe.

13. Remove the engine bracket.

14. Detach the coolant hose from the port.

15. Remove the position sensor connector.

16. Remove the position sensor.

*To install:*

17. Install the position sensor using a new O-ring coated with a thin layer of acid-free Vaseline.

18. Tighten the sensor to 7.5 ft. lbs. (10 Nm).

19. Plug in the connector.

20. Install the coolant hose to the port.

21. Install the engine bracket. Tighten M12 bolts to 69 ft. lbs. (93 Nm) and M10 bolts to 31 ft. lbs. (42 Nm).

22. Install the P-clamp of the servo pipe.

23. Install the upper turbocharger delivery pipe to the throttle body and the lower delivery pipe.

24. Plug in the connector to the intake air sensor.

25. Install and adjust the engine pad to the body. Tighten the bolts between the engine pad and the bracket. The engine pad and bracket should be aligned with each other. Tighten engine bolts to 33 ft. lbs. (45 Nm) plus 90° rotation. Tighten body bolts to 37 ft. lbs. (50 Nm) plus 150° rotation.

26. Remove the jack.

27. Install the coolant pipe to the port using a new O-ring coated with a thin layer of acid-free Vaseline.

28. Install the air cleaner casing and intake hose.

29. Fit the air cleaner and the air cleaner casing cover.

30. Install the turbocharger inlet hose.

31. Install the upper engine cover.

32. Fill with coolant.

33. Pressure test the cooling system.

#### Rear

*See Figure 224.*

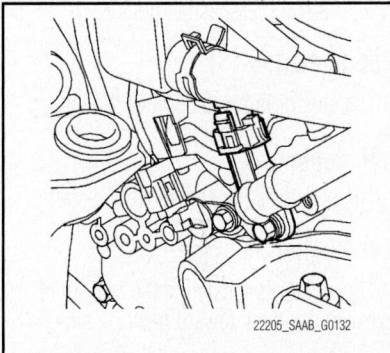

**Fig. 224 Camshaft position sensor location—2.8L engine rear**

1. Before servicing the vehicle, refer to the Precautions Section.

2. Drain the coolant

3. Remove the upper engine cover.

4. Detach the turbocharger intake hose.

### ❋❋ WARNING

**When removing the air cleaner casing cover, be careful not to damage the brake vacuum pump sensor.**

5. Remove the air cleaner casing cover and the air cleaner.

6. Detach the intake hose and remove the air cleaner casing.

7. Detach the coolant hose from the engine and move it aside.

8. Remove the position sensor connector.

9. Remove the sensor.

*To install:*

10. Install the position sensor using a new O-ring coated with a thin layer of acid-free Vaseline.

11. Tighten the sensor to 7.5 ft. lbs. (10 Nm).

12. Plug in the connector.

13. Reposition and install the coolant hose.

14. Install the air cleaner casing and intake hose.

15. Install the air cleaner and the air cleaner casing cover.

16. Install the turbocharger inlet hose.

17. Install the upper engine cover.

18. Fill with coolant.

19. Pressure test the cooling system.

## CRANKSHAFT POSITION (CKP) SENSOR

### LOCATION

#### 2.0L H4 and 2.5L Engines

The sensor is located at the front of the engine near the crankshaft pulley.

#### 2.0L I4 Engine

The sensor is located above the starter.

#### 2.3L Engine

The sensor is located above the starter.

#### 2.8L Engine

The sensor is located at the bottom of rear cylinder bank toward gearbox side.

### REMOVAL & INSTALLATION

#### 2.0L H4 and 2.5L Engines

*See Figures 225 and 226.*

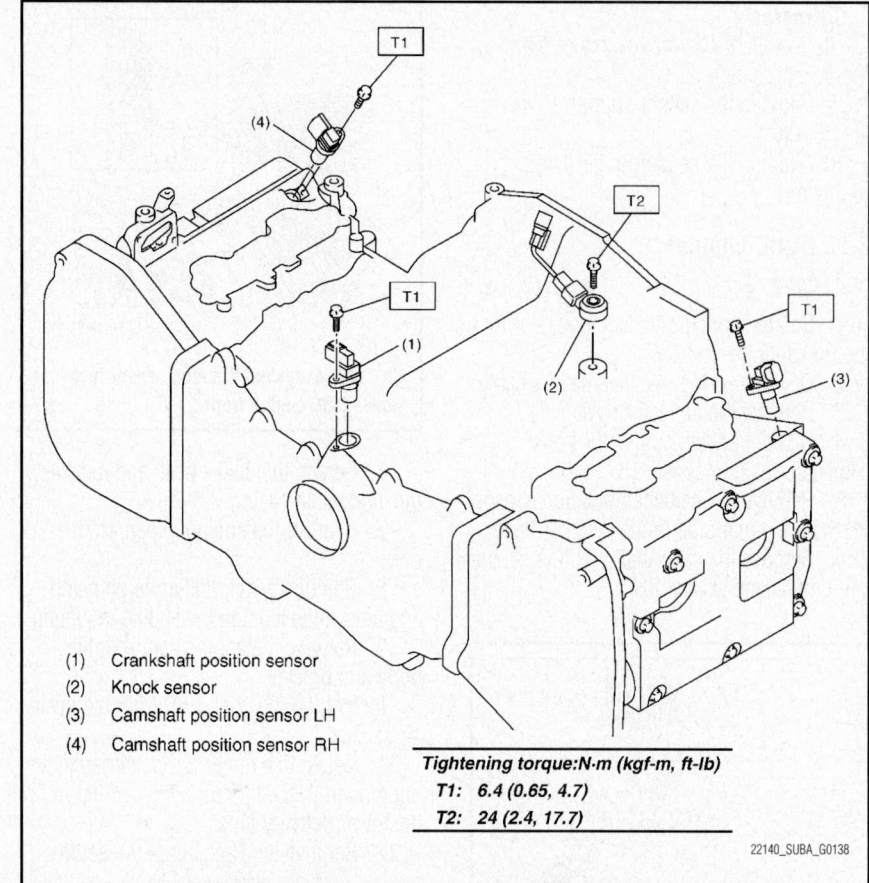

| (1) | Crankshaft position sensor |
| (2) | Knock sensor |
| (3) | Camshaft position sensor LH |
| (4) | Camshaft position sensor RH |

**Tightening torque:N·m (kgf-m, ft-lb)**
T1:  6.4 (0.65, 4.7)
T2:  24 (2.4, 17.7)

22140_SUBA_G0138

**Fig. 226 Crankshaft Position (CKP) Sensor—2.5L engine DOHC**

1. Disconnect the ground cable from battery.

2. Remove the collector cover.

3. Remove the alternator, as required.

4. Remove the bolt which installs crankshaft position sensor to cylinder block.

5. Remove the crankshaft position sensor, and then disconnect the connector from it.

**To install:**

6. Installation is the reverse of the removal procedure.

7. Tighten the sensor to 4.7 ft. lbs. (6.4 Nm).

#### 2.0L I4 Engine

*See Figure 227.*

1. Remove the upper engine cover.

2. Remove the starter motor, as necessary.

3. Unplug the crankshaft position sensor connector.

4. Remove the crankshaft position sensor.

**To install:**

5. Plug in the connector and install the crankshaft position sensor with a new O-ring sparingly lubricated with Vaseline.

6. Install the upper engine cover.

**Tightening torque:N·m (kgf-m, ft-lb)**
T1:  6.4 (0.65, 4.7)
T2:  24 (2.4, 17.7)

| (1) | Crankshaft position sensor |
| (2) | Knock sensor |
| (3) | Camshaft position sensor |
| (4) | Camshaft position sensor support |

22140_SUBA_G0136

**Fig. 225 Crankshaft Position (CKP) Sensor—2.5L engine SOHC**

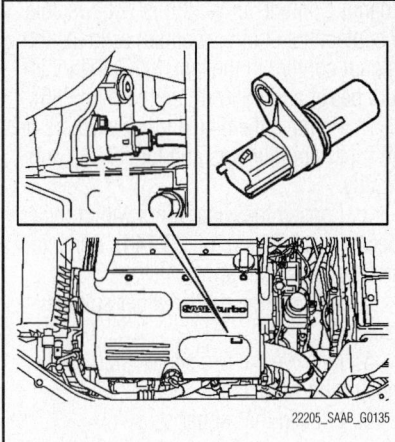

**Fig. 227 Crankshaft position sensor location—2.0L I4 engine**

### 2.3L Engine

*See Figure 228.*

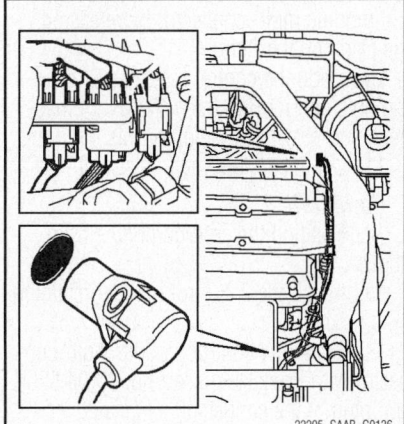

**Fig. 228 Crankshaft position sensor location—2.3L engine**

1. Raise and safely support the vehicle securely on jackstands.
2. Remove the lower engine cover.
3. Unplug the position sensor connector.
4. Remove the position sensor.

**To install:**

5. Grease the position sensor O-ring with acid-free Vaseline and install it. Take care not to damage the O-ring.
6. Install the position sensor. Tighten to 7 ft. lbs. (9 Nm).
7. Plug in the position sensor.
8. Install the lower engine cover.
9. Lower the vehicle.

### 2.8L Engine

*See Figure 229.*

1. Raise and safely support the vehicle securely on jackstands.

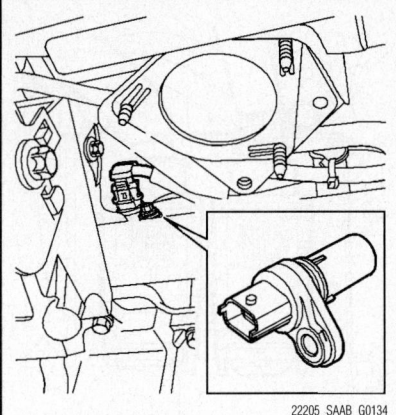

**Fig. 229 Crankshaft position sensor location—2.8L engine**

2. Remove the rear catalytic converter.
3. Remove the bracket between the front exhaust pipe and the engine using 32 025 061 Crow Foot Spanner Wrench.
4. Undo the cable fixing from the heat shield.
5. Remove the heat shield.
6. Unplug the crankshaft sensor connector. Remove the sensor.

**To install:**

7. Install the sensor using a new O-ring lightly coated with acid-free Vaseline.
8. Tighten to 7 ft. lbs. (10 Nm).
9. Plug in the connector.
10. Install the heat shield.
11. Secure the cable to the heat shield.

12. Install the bracket to the engine and front exhaust pipe.
13. Tighten the engine bolt to 17 ft. lbs. (23 Nm), the exhaust pipe bolt to 30 ft. lbs. (40 Nm) and the heat shield bolt to M6 6ft. lbs. (9 Nm) and M10 28 ft. lbs. (38 Nm).
14. Install rear catalytic converter using a new gasket.
15. Lower the vehicle.

## ELECTRONIC CONTROL MODULE (ECM)

### *LOCATION*

#### 2.0L H4 and 2.5L Engines

*See Figure 230.*

The ECM is located on the passenger's side of the vehicle, underneath the floor mat.

#### 2.0L I4 Engine

The ECM is located at the front of the engine.

#### 2.3L Engine

The ECM is located at the rear of the engine compartment adjacent to the MacPherson strut tower.

#### 2.8L Engine

The ECM is located at the top of the rear cylinder bank.

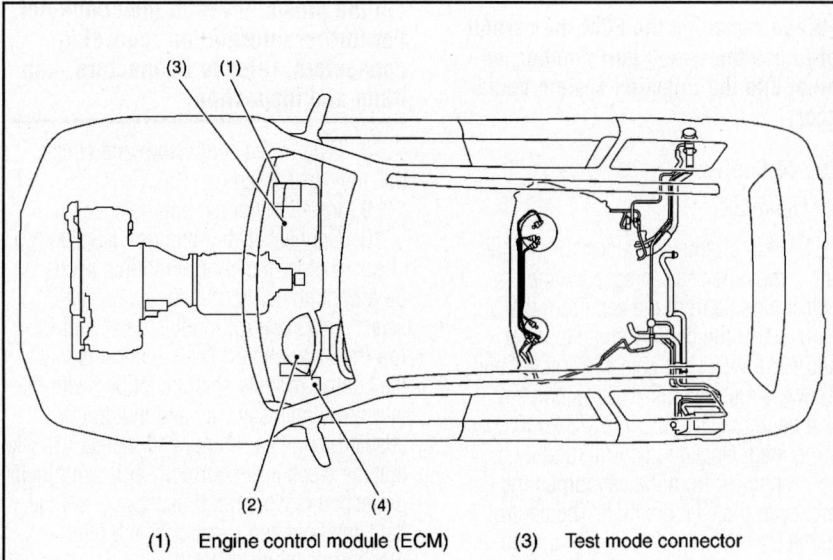

| (1) | Engine control module (ECM) | (3) | Test mode connector |
|-----|------------------------------|-----|---------------------|
| (2) | Malfunction indicator light  | (4) | Data link connector |

**Fig. 230 ECM and related components**

*REMOVAL & INSTALLATION*

## 2.0L H4 and 2.5L Engines

*See Figure 231.*

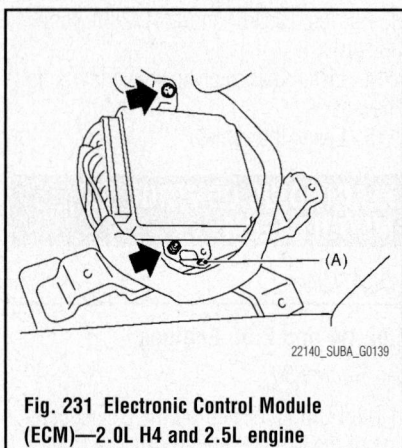

22140_SUBA_G0139

**Fig. 231 Electronic Control Module (ECM)—2.0L H4 and 2.5L engine**

1. Disconnect the negative battery cable.
2. Remove the lower inner trim on the passenger's side of the vehicle.
3. Detach the floor mat. Remove the protective cover.
4. Remove the Electronic Control Module (ECM) bracket retaining nuts. Remove the clip (A) from the bracket.
5. Disconnect the connectors.
6. Remove the ECM from the vehicle.

**To install:**

7. Installation is the reverse of the removal procedure.
8. Tighten the retaining screws to 5.5 ft. lbs. (7.5 Nm).

➡ When replacing the ECM, be careful not to use the wrong part number, as damage to the injection system could occur.

## 2.0L I4 Engine

*See Figure 232.*

1. Before removing a control module, Tech 2 must be used to separate the control module from the car. From the "All" menu, select the control module under "And/Remove". Then select "Remove" and follow the instructions. The ignition key must be in the ON position. TIS 2000 may be required. Once the control module has been separated from the car, turned the ignition key to the OFF position. The control module can then be removed. Carry out the "Remove" action with Tech2 even if the control module cannot be contacted or is completely missing because data in other control modules must be changed.
2. Turn the ignition key to the **LOCK** position.

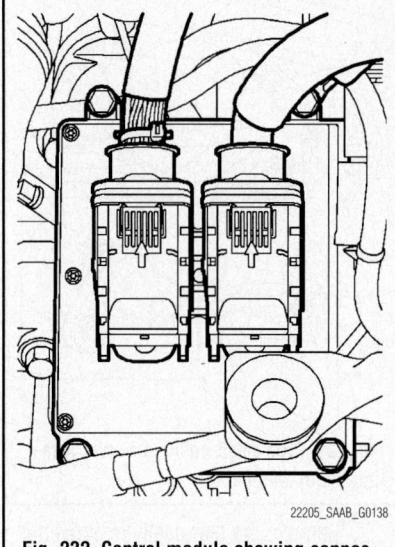

22205_SAAB_G0138

**Fig. 232 Control module showing connector removal—2.0L I4 engine**

3. Remove the upper engine cover.
4. Carefully press down the catches and carefully unplug the connectors from the control module one at a time so that the connection pins, etc. are not damaged.
5. Remove the control module.

**To install:**

6. Install the control module. Do not forget the ground cable.
7. Tighten to 7 ft. lbs. (10 Nm)

### ✳✳ WARNING

**Take care when plugging in the connector so as not to damage or press out the pins/sleeves in the connector. For further information regarding connectors, refer to Connectors, handling and inspection.**

8. Plug in the connectors and check that they are locked.
9. Install the upper engine cover.
10. Connect Tech 2 and use TIS 2000 to check whether the control module needs to be programmed. Access the menu "SPS", select "Read control module data" and follow the instructions. The intention is that the control module shall be loaded with the latest available software and in addition adapted to the car model and market. Check that the clock is set correctly and that pinch protection is working. If necessary, set the date and time and carry out calibration of pinch protection.

## 2.3L Engine

*See Figure 233.*

1. Connect Tech2 and contact Trionic. Select BioPower E85. Note "Adapted

Ethanol Content" in %. If it is not possible to contact the control module, estimate the ethanol content in the tank (0% ethanol is pure petrol and 85% ethanol is pure E85).

2. Remove the blanking-off washers on the wiper spindle nuts and loosen the nuts slightly.
3. Loosen the wiper arms with a puller, special tool (part no. 85 80 144). Remove the nuts and wiper arms.
4. Remove the rubber wiper spindle seals.
5. Remove the two screw clips on the ends of the windscreen cover.
6. Loosen the rubber seal.
7. Remove the windscreen cover by grasping the front edge and lifting upward/forward. Be careful with the bonnet release wire on the left-hand side.
8. Undo the protective cover on the control module.
9. Tilt up the cover and unplug the control module multi-connector by releasing the catch on the right.
10. Undo the control module's 2 retaining nuts and remove them. A 10 mm magnetic socket will be useful for this.
11. Lift the control module straight up.

**To install:**

12. Position the control module from above.
13. Install the 2 control module retaining nuts.
14. Spray the control module multi-connector with Kontakt 61 (part no. 30 04 520) and plug in the connector.
15. Install the protective cover on the control module.
16. Install the windscreen cover. Be careful with the bonnet release wire on the left-hand side.

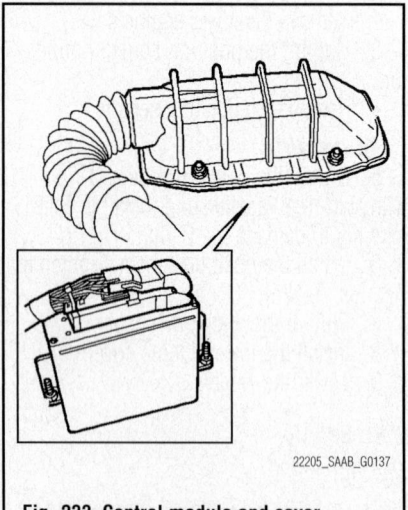

22205_SAAB_G0137

**Fig. 233 Control module and cover assembly—2.3L engine**

17. Install the rubber seal.

18. Fasten the screw clips on the ends of the windscreen cover.

19. Install the rubber wiper spindle seals.

20. Install the nuts on the wiper spindles with the blanking-off washers on the nuts.

21. Perform Measures after changing a control module.

22. Connect Tech 2 and use TIS 2000 to check whether the control module requires programming. Go to the "SPS" menu and select "Read ECU data". Follow the instructions. The control module should be programmed with the latest available software and be adapted to the vehicle variant and market.

23. Reset immobilization. Connect Tech2 and turn the ignition to **ON**. Select Body and TWICE. Select Immobilization. After 10 seconds the immobilizer is programmed.

### 2.8L Engine

*See Figure 234.*

1. Before removing a control module, Tech 2 must be used to separate the control module from the car. From the "All" menu, select the control module under "And/Remove". Then select "Remove" and follow the instructions. The ignition key must be in the ON position. TIS 2000 may be required. Once the control module has been separated from the car, turned the ignition key to the OFF position. The control module can then be removed. Carry out the "Remove" action with Tech2 even if the control module cannot be contacted or is completely missing because data in other control modules must be changed.

> **✳✳ WARNING**
>
> **It is important to follow the order in Tech2 as some data is loaded into the new control module.**

2. Move the ignition key to the **LOCK** position.

3. Remove the upper engine cover.

4. Detach the ground cables from the control module.

5. Carefully press down the catches and carefully unplug the connectors from the control module one at a time so that the connection pins are not damaged.

> **✳✳ WARNING**
>
> **Take care when releasing the locking mechanism on the connector so as not to damage the connector. Pull the halves straight apart to avoid bending the pins.**

6. Remove the control module.

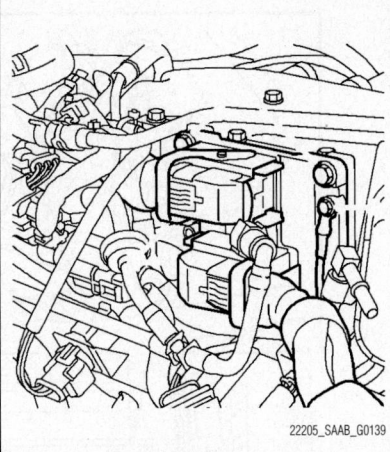

**Fig. 234 Control module and connector assembly—2.8L engine**

22205_SAAB_G0139

### To install:

7. Install the control module. Tighten to 7 ft. lbs. (10 Nm).

8. Plug in the connectors and check that they are locked.

9. Attach the ground cables.

10. Install the upper engine cover.

11. Connect Tech 2 and use TIS 2000 to check whether the control module needs to be programmed. Access the menu "SPS", select "Read control module data" and follow the instructions. The intention is that the control module shall be loaded with the latest available software and in addition adapted to the car model and market. Check that the clock is set correctly and that pinch protection is working. If necessary, set the date and time and carry out calibration of pinch protection.

## ENGINE COOLANT TEMPERATURE (ECT) SENSOR

*LOCATION*

### 2.0L H4 and 2.5L Engines

The sensor is located by the heater outlet or in a cooling passage on the engine, depending upon the particular vehicle. Refer below for engine location view.

### 2.0L I4 Engine

The coolant temperature sensor is located on the engine's top right (passenger side) front corner.

### 2.3L Engine

The sensor is located by the heater outlet.

### 2.8L Engine

The coolant temperature sensor is located below oil pressure switch.

*REMOVAL & INSTALLATION*

### 2.0L H4 and 2.5L Engines

*See Figure 235.*

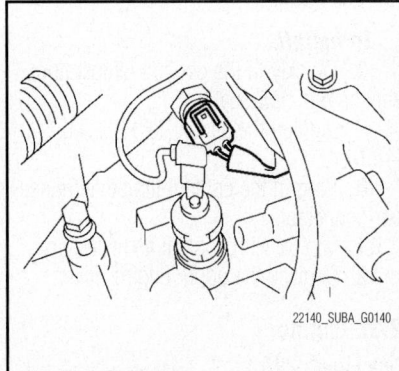

**Fig. 235 Coolant temperature sensor location—2.0L H4 and 2.5L engines**

22140_SUBA_G0140

1. Disconnect the negative battery cable.

2. Remove the alternator, as required.

3. Remove the air intake duct and air cleaner case, as required.

4. Disconnect the connector from the sensor.

5. Drain the cooling system, as required.

6. Remove the sensor from its mounting.

### To install:

7. Installation is the reverse of the removal procedure.

8. Tighten the sensor to 13.3 ft. lbs. (18 Nm).

### 2.0L I4 Engine

*See Figure 236.*

1. Before servicing the vehicle, refer to the Precautions Section.

2. Drain and recycle the engine coolant.

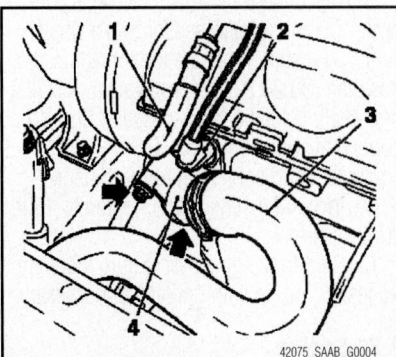

**Fig. 236 Removing coolant temperature sensor and components**

42075_SAAB_G0004

3. Remove the upper engine cover.

4. Undo the cap on the expansion tank to release pressure, retighten the cap.

5. Unplug the connector to the coolant temperature sensor.

6. Have some paper ready to collect any spilled coolant.

7. Remove the coolant temperature sensor. Use 83 96 087 Coolant Temperature Sensor Socket.

### To install:

8. Quickly fit the coolant temperature sensor with new seal.

9. Tighten the sensor to 11 ft. lbs. (15 Nm).

10. Plug in the coolant temperature sensor connector.

11. Top up with coolant as necessary.

12. Replace the upper engine cover.

## 2.3L Engine

*See Figure 237.*

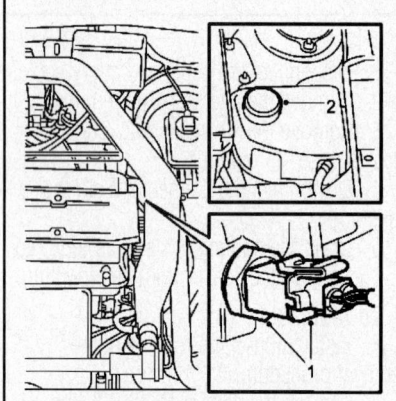

**Fig. 237 Identifying and locating coolant temperature sensor**

1. Before servicing the vehicle, refer to the Precautions Section.

2. Loosen coolant reservoir (1) cap and re-tighten.

3. Disconnect the sensor electrical connector (2) and remove sensor with a wrench.

4. To install, lubricate the threads with Vaseline (30 20 271) or equivalent. Install a new sealing washer as necessary. Spray the connector with Kontakt 61 (30 04 520) or equivalent and connect it to the sensor.

5. Install the sensor and tighten to 10 ft. lbs. (13 Nm).

6. Start the engine. Make sure there are no leaks. Top up with coolant as necessary.

## 2.8L Engine

*See Figure 238.*

1. Before servicing the vehicle, refer to the Precautions Section.

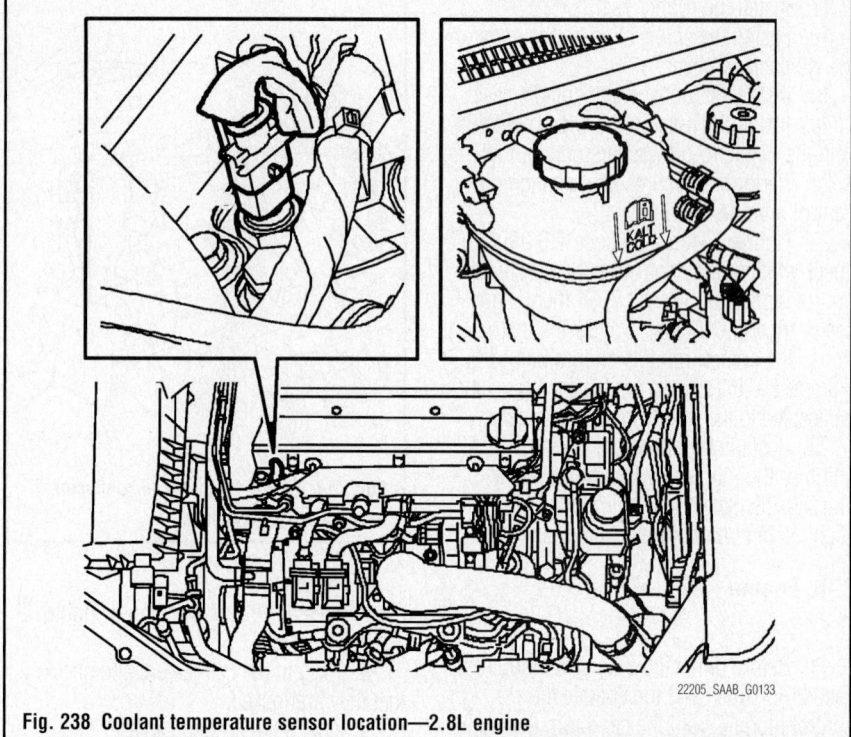

**Fig. 238 Coolant temperature sensor location—2.8L engine**

2. Remove the upper engine cover.

3. Undo the cap on the expansion tank to release pressure, retighten the cap.

### ✳✳ WARNING

**Take care when releasing the locking mechanism on the connector so as not to damage the connector. Pull the halves straight apart to avoid bending the pins.**

4. Detach the secondary air hose from the turbocharger intake manifold.

5. Remove the air cleaner casing cover and the bolts on the radiator member.

6. Detach the turbocharger intake manifold at the mass air flow sensor and remove it.

7. Lift up the heat shield and unplug the coolant temperature sensor connector.

8. Remove the sensor.

### To install:

9. Quickly refit the temperature sensor using a new seal. Tighten to 16 ft. lbs. (22 Nm).

10. Plug in the sensor connector and press down the heat shield.

11. Refit the turbocharger intake manifold.

12. Install the air cleaner casing cover and the radiator member bolts.

13. Attach the secondary air hose to the turbocharger intake manifold.

14. Replace the upper engine cover.

15. Top up coolant as necessary. Bleed the cooling system.

## HEATED OXYGEN (HO2S) SENSOR

### LOCATION

#### 2.0L H4 and 2.5L Engines

On the SOHC engine the front oxygen sensor is located in the front section of the front catalytic converter. The rear oxygen sensor is located in the rear section of the front catalytic converter. On the DOHC engines the front oxygen sensor is located in the front section of the exhaust system, just past the crossover pipe. The rear oxygen sensor is located in the front section of the catalytic converter.

#### 2.0L I4 and 2.8L Engines

The front heated oxygen sensor is located before the catalytic converter. The blue connector is located on the console between the power steering pump and the vacuum pump

The rear heated oxygen sensor is located after the catalytic converter. The connector is brown and located on the bracket between the power steering pump and the vacuum pump

#### 2.3L Engine

The front heated oxygen sensor is located in the exhaust system under the bypass pipe

The rear heated oxygen sensor is located after the catalytic converter.

*REMOVAL & INSTALLATION*

### 2.0L and 2.5L SOHC Engines

*See Figures 239 through 241.*

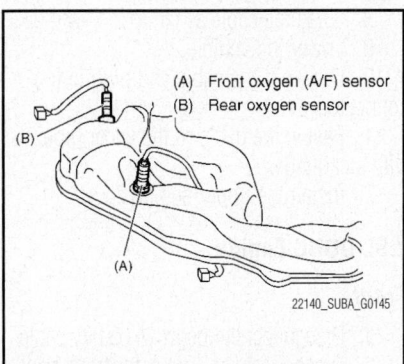

**Fig. 239 Heated Oxygen (HO2S) Sensor—2.5L SOHC engine**

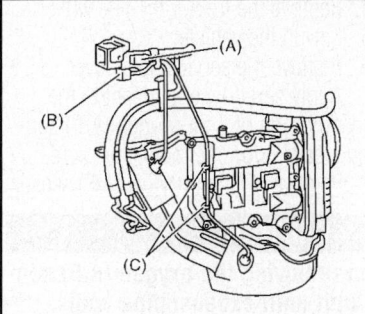

(A) Front oxygen (A/F) sensor connector
(B) Exhaust temperature sensor connector
(C) Clip

**Fig. 240 Front Heated Oxygen (HO2S) Sensor—2.5L DOHC engine**

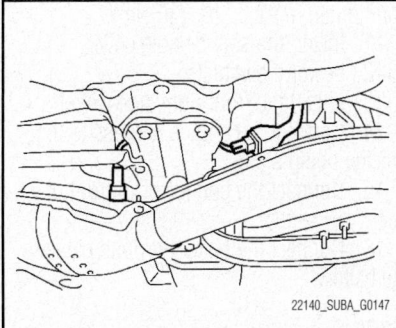

**Fig. 241 Rear Heated Oxygen (HO2S) Sensor—2.5L DOHC engine**

### Front

1. Disconnect the negative battery cable.
2. Remove the clip fastening the harness and disconnect the front oxygen (A/F) sensor connector.

3. Lift-up the vehicle.
4. Remove the undercover.
5. Apply spray-type lubricant to the threaded portion of front oxygen (A/F) sensor, and leave it for one minute or more.
6. Remove the front oxygen (A/F) sensor.

> ❊❊ **CAUTION**
>
> **When removing the front oxygen (A/F) sensor, wait until exhaust pipe cools, because it can damage the exhaust pipe.**

### To install:

7. Before installing front oxygen (A/F) sensor, apply anti-seize compound only to the threaded portion of front oxygen (A/F) sensor to make the next removal easier.
8. Install the front oxygen (A/F) sensor and tighten to 15 ft. lbs. (21 Nm).
9. Install the undercover.
10. Lower the vehicle.
11. Connect the connector of front oxygen (A/F) sensor connector and fasten the harness with clips.
12. Connect the battery ground cable to the battery.

### Rear

1. Disconnect the negative battery cable.
2. Remove the clip fastening the harness and disconnect the rear oxygen sensor connector.
3. Lift-up the vehicle.
4. Remove the undercover.
5. Apply spray-type lubricant to the threaded portion of rear oxygen sensor, and leave it for one minute or more.
6. Remove the rear oxygen sensor.

### To install:

7. Before installing rear oxygen sensor, apply the anti-seize compound only to the threaded portion of rear oxygen sensor to make the next removal easier.
8. Install the rear oxygen sensor and tighten to 15 ft. lbs. (21 Nm).
9. Install the undercover.
10. Lower the vehicle.
11. Connect the connector to rear oxygen sensor and mount the harness clips to the bracket for fastening.
12. Connect the battery ground cable to the battery.

### 2.0L I4 Engine

### Upper

*See Figure 242.*

1. Unplug the oxygen sensor and carefully lower the cable.

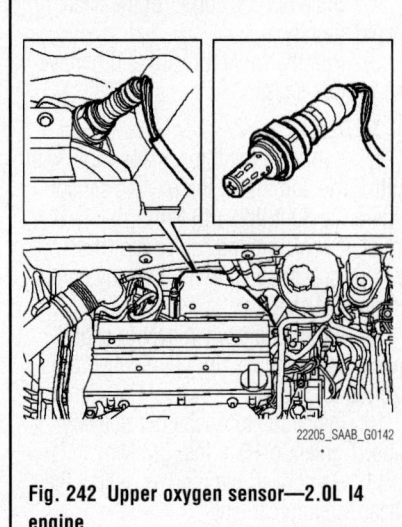

**Fig. 242 Upper oxygen sensor—2.0L I4 engine**

2. Remove the heat shield over the turbocharger (note the clip underneath the heat shield).
3. Remove the oxygen sensor.

### To install:

4. Lubricate the threads sparingly with Anti-Seize and install the oxygen sensor. Tighten to 30 ft. lbs. (40 Nm).
5. Plug in the oxygen sensor connector and secure it.
6. Fit the turbocharger heat shield.

### Lower

*See Figure 243.*

1. Unplug the connector for the oxygen sensor and carefully lower the cable.
2. Raise and safely support the vehicle securely on jackstands.
3. On convertible models, remove the chassis reinforcement.
4. Remove the lower turbocharger pressure pipe.

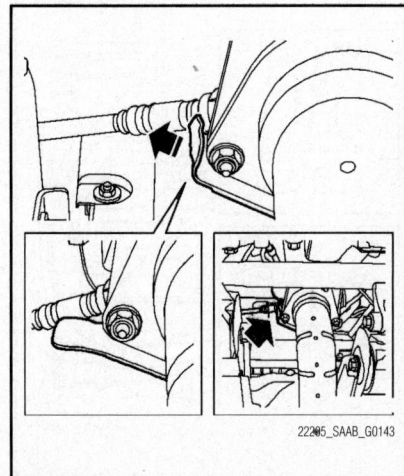

**Fig. 243 Lower oxygen sensor—2.0L I4 engine**

5. Bend out the corner of the heat shield slightly.

6. Run the cable forward and remove the oxygen sensor.

### To install:

7. Lubricate the threads sparingly with Anti-Seize and install the oxygen sensor. Tighten to 30 ft. lbs. (40 Nm).

8. Check that the cables are NOT twisted or damaged. Bend back the corner of the heat shield.

9. Lift up the oxygen sensor cable, making sure it does not get pinched or is exerted to chafing.

10. Install the turbocharger delivery pipe. Tighten to 16 ft. lbs. (22 Nm).

11. On convertible models, install the chassis reinforcement.

12. Lower the vehicle.

13. Plug in the oxygen sensor connector and secure it.

### 2.3L Engine

#### Front

*See Figure 244.*

> **⁂ WARNING**
>
> **The cables must not be twisted. Twisted cables exposed to vibration can break. The oxygen sensor is sensitive to impacts and jolting and must be handled with care.**

1. Remove the engine cover.

2. Remove the oxygen sensor connector from the holder by pressing the lugs together.

3. Remove the bypass pipe and heat shield.

4. Uncut the cable tie to the oxygen sensor and draw out the cables.

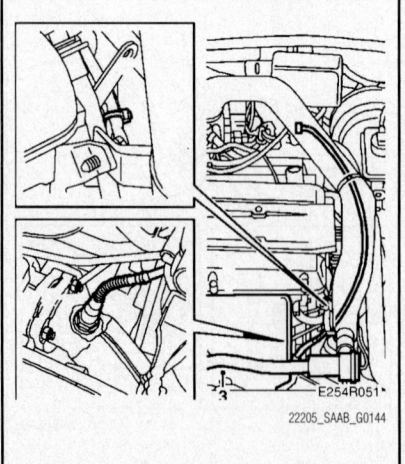

**Fig. 244 Front oxygen sensor—2.3L engine**

5. Remove the oxygen sensor from the turbo outlet pipe.

### To install:

6. Coat the oxygen sensor threads with Molykote 1000 (30 20 971) to prevent binding.

7. Install the oxygen sensor. Tighten to 41 ft. lbs. (55 Nm).

> **⁂ WARNING**
>
> **Chemicals such as contact spray and grease must not come into contact with the oxygen sensor connectors.**

8. Draw out the cable and plug in the connector.

9. Fasten the cable to the water pipe with a cable tie.

10. Install the heat shield and bypass pipe. Tighten to 15 ft. lbs. (20 Nm).

11. Install the upper engine cover.

#### Rear

*See Figure 245.*

> **⁂ WARNING**
>
> **The cables must not be twisted. Twisted cables exposed to vibration can break. The oxygen sensor is sensitive to impacts and jolting and must be handled with care.**

1. Remove the engine cover.

2. Remove the oxygen sensor connector from the holder by pressing the lugs together.

3. Cut the cable tie to the oxygen sensor and draw out the cables.

4. Raise and safely support the vehicle securely on jackstands.

5. Pull down the cable and remove the oxygen sensor.

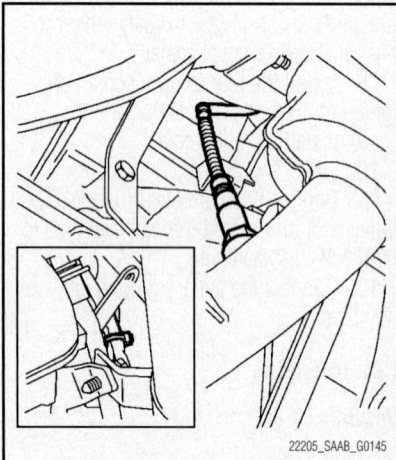

**Fig. 245 Rear oxygen sensor—2.3L engine**

6. Coat the oxygen sensor threads with Molykote 1000 (30 20 971) to prevent binding.

7. Install the oxygen sensor. Tighten to 41 ft. lbs. (55 Nm).

8. Lead the cable as far up as it will go.

9. Lower the vehicle.

10. Draw out the cable and plug in the connector.

11. Fasten the cable to the water pipe with a cable tie.

12. Install the upper engine cover.

### 2.5L DOHC Engine

#### Front

1. Disconnect the negative battery cable.

2. Disconnect the connector from front oxygen (A/F) sensor.

3. Disconnect the engine harness fixed by a clip, from the bracket.

4. Remove the front right side wheel.

5. Lift-up the vehicle.

6. Remove the service hole cover.

7. Apply spray-type lubricant to the threaded portion of front oxygen (A/F) sensor, and leave it for one minute or more.

8. Remove the front oxygen (A/F) sensor.

> **⁂ CAUTION**
>
> **When removing the oxygen (A/F) sensor, wait until exhaust pipe cools, otherwise it will damage the exhaust pipe.**

### To install:

9. Before installing front oxygen (A/F) sensor, apply the anti-seize compound only to the threaded portion of front oxygen (A/F) sensor. This facilitates the next removal.

10. Install the front oxygen (A/F) sensor and tighten to 22 ft. lbs. (30 Nm).

11. Install the service hole cover.

12. Lower the vehicle.

13. Install the front right side wheel.

14. Connect the engine harness to the bracket using a clip.

15. Connect the connector of front oxygen (A/F) sensor.

16. Connect the battery ground cable to the battery.

#### Rear

1. Disconnect the negative battery cable.

2. Lift-up the vehicle.

3. Disconnect the connector from the rear oxygen sensor.

4. Apply spray-type lubricant to the threaded portion of rear oxygen sensor, and leave it for one minute or more.

5. Remove the rear oxygen sensor.

## CAUTION

**When removing the rear oxygen sensor, wait until exhaust pipe cools, otherwise it may damage the exhaust pipe.**

### *To install:*

6. Before installing rear oxygen sensor, apply the anti-seize compound only to the threaded portion of rear oxygen sensor to make the next removal easier.

7. Install the rear oxygen sensor and tighten to 15 ft. lbs. (21 Nm).

8. Connect the connector to rear oxygen sensor.

9. Lower the vehicle.

10. Connect the battery ground cable to the battery.

### 2.8L Engine

#### *Front*
*See Figure 246.*

## WARNING

**The oxygen sensor cables must not be twisted. Twisted cables in combination with vibrations can lead to cable breaks. The oxygen sensor is sensitive to knocks and jolts and must be handled with care.**

1. Remove the upper engine cover.

2. Remove the battery and battery cover bottom.

3. Remove the heat shield from the turbocharger and catalytic converter.

4. Detach the upper oxygen sensor connector and remove the cable fixings.

5. Carefully raise the heat shield slightly.

6. Remove the oxygen sensor using 83 96 350 Oxygen Sensor Removal Socket.

7. Detach the ventilation pipe from the intake manifold and remove the oxygen sensor cable.

### *To install:*

8. Lightly coat the threads with Anti-Seize and install the oxygen sensor. Tighten 33 ft. lbs. (45 Nm).

9. Carefully press down the heat shield.

10. Position the cable and attach the ventilation pipe to the intake manifold.

11. Secure the cable and plug in the connector.

12. Install the heat shield to the turbocharger and catalytic converter.

13. Install the battery and battery cover bottom.

14. Replace the upper engine cover.

#### *Rear*
*See Figure 247.*

## WARNING

**The oxygen sensor cables must not be twisted. Twisted cables in combination with vibrations can lead to cable breaks. The oxygen sensor is sensitive to knocks and jolts and must be handled with care.**

1. Remove the upper engine cover.

2. Detach the lower oxygen sensor connector and remove the cable fixings.

3. Carefully raise the heat shield slightly.

4. Remove the oxygen sensor using 83 96 350 Oxygen Sensor Removal Socket.

### *To install:*

5. Lightly coat the threads with Anti-Seize and install the oxygen sensor. Tighten 33 ft. lbs. (45 Nm).

6. Carefully press down the heat shield.

7. Secure the cable and plug in the connector.

8. Replace the upper engine cover.

## INTAKE AIR TEMPERATURE (IAT) SENSOR

### *LOCATION*

#### 2.0L I4 and 2.8L Engine

The intake air temperature sensor is integrated in the intake air sensor on the charge air pipe.

#### 2.3L Engine

The intake air temperature sensor is located on the charge air pipe.

#### 2.5L SOHC engines

The sensor is mounted in the intake air hose of the air cleaner assembly. On the 2.0L H4 and 2.5L DOHC engine this sensor is combined with the mass air flow sensor.

### *REMOVAL & INSTALLATION*

#### 2.0L I4 and 2.8L Engine
*See Figure 248.*

1. Remove the upper engine cover and insulation.

## WARNING

**Take care when releasing the locking mechanism on the connector so as not to damage the connector. Pull the halves straight apart to avoid bending the pins.**

2. Unplug the sensor connector.

3. Remove the sensor.

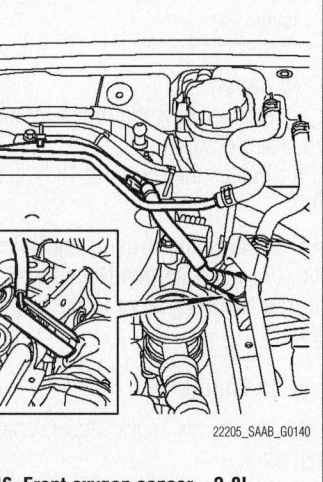
22205_SAAB_G0140

**Fig. 246 Front oxygen sensor—2.8L engine**

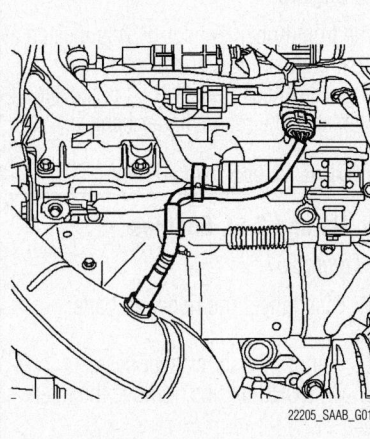
22205_SAAB_G0141

**Fig. 247 Rear oxygen sensor—2.8L engine**

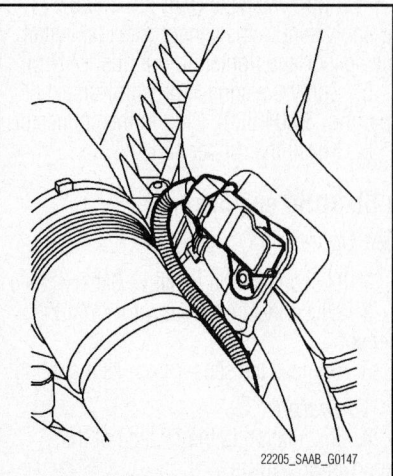
22205_SAAB_G0147

**Fig. 248 Intake air temperature sensor location—2.0L I4 and 2.8L engine**

*To install:*

4. Lubricate the sensor O-ring with acid-free Vaseline.

## ✳✳ WARNING

**Be careful not to damage the O-ring.**

5. Install the sensor. Tighten to 7 ft. lbs. (9 Nm).
6. Install the sensor connector.
7. Fit the insulation and the upper engine cover.

### 2.3L Engine

*See Figure 249.*

1. Remove the rubber cap and unplug the connector.
2. Remove the sensor.

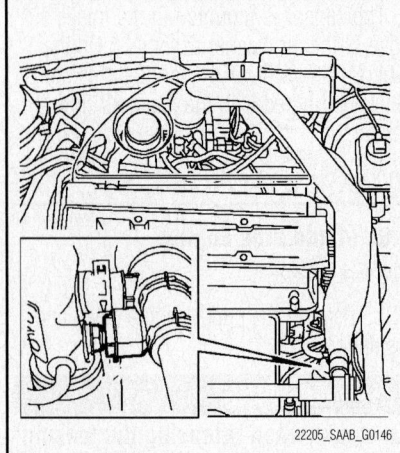

**Fig. 249 Intake air temperature sensor location—2.3L engine**

*To install:*

3. Check the sealing washer and install a new one if necessary.
4. Apply Vaseline (part no. 30 20 271) or equivalent to the sensor threads. Install the sensor and tighten to 5 ft. lbs. (7 Nm).
5. Spray the connector with Kontakt 61 (part no. 30 04 520). Plug in the connector.
6. Install the rubber cap.

### 2.5L SOHC engines

*See Figure 250.*

1. Disconnect the negative battery cable.
2. Disconnect the connector from the sensor.
3. Remove the sensor from its mounting.

*To install:*

4. Installation is the reverse of removal procedure.
5. On the DOHC engine, torque the retaining screw to 0.8 ft. lbs.

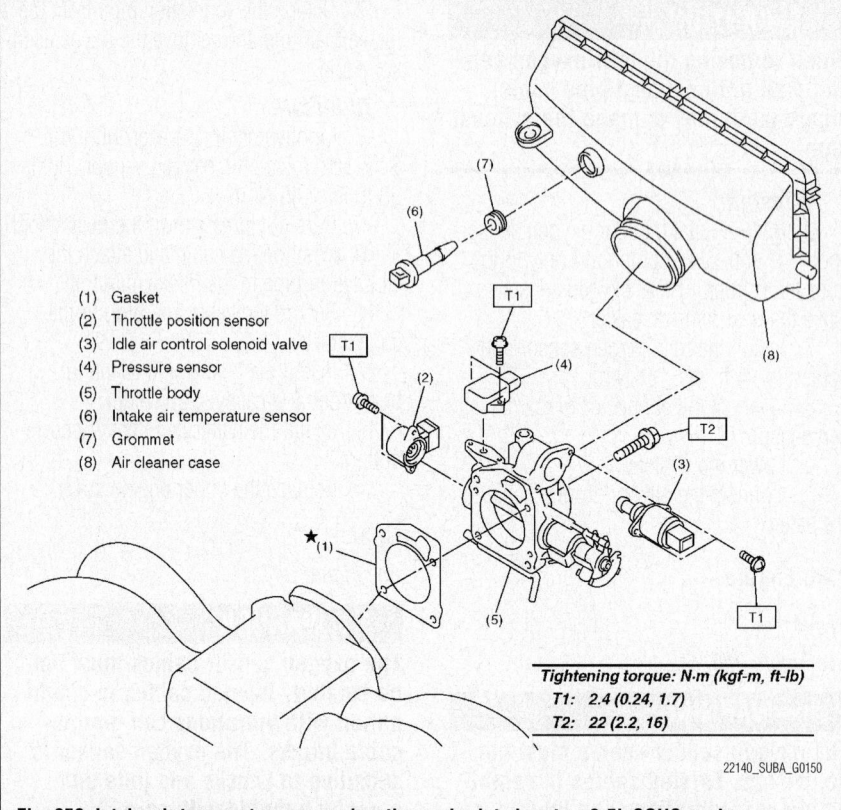

(1) Gasket
(2) Throttle position sensor
(3) Idle air control solenoid valve
(4) Pressure sensor
(5) Throttle body
(6) Intake air temperature sensor
(7) Grommet
(8) Air cleaner case

Tightening torque: N·m (kgf-m, ft-lb)
T1: 2.4 (0.24, 1.7)
T2: 22 (2.2, 16)

22140_SUBA_G0150

**Fig. 250 Intake air temperature sensor location and related parts—2.5L SOHC engine**

## KNOCK SENSOR (KS)

### LOCATION

#### 2.0L H4 and 2.5L Engines

The Knock Sensor (KS) is located at the top right (driver's) side of the engine and is positioned on the cylinder block.

#### 2.0L I4 Engine

The Knock Sensor (KS) is located in the center of the rear side of the engine.

#### 2.8L Engine

The front Knock Sensor (KS) is located at the rear edge of the starter motor.
The rear Knock Sensor (KS) is located at the center of the rear cylinder bank.

### REMOVAL & INSTALLATION

#### 2.0L H4 and 2.5L Engines

*See Figure 251.*

1. Disconnect the negative battery cable.
2. Remove the air cleaner case.
3. On DOHC engine, remove the intercooler.
4. Disconnect the sensor connector.
5. Remove the sensor from its mounting.

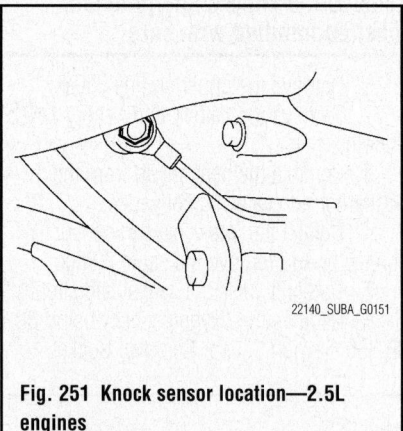

22140_SUBA_G0151

**Fig. 251 Knock sensor location—2.5L engines**

*To install:*

6. Installation is the reverse of the removal procedure.
7. Tighten the sensor to 18 ft. lbs. (24 Nm).

➡**The extraction area of the knock sensor wire must be positioned at a 60 degree angle relative to the engine rear.**

#### 2.0L I4 Engine

1. Unplug the knock sensor cable set connector.
2. Remove the knock sensor from the engine block.

*To install:*

**✳✳ WARNING**

**The knock sensor tightening torque must be maintained under all circumstances or operating disturbances may arise.**

3. Install the knock sensor on the engine block. Tighten to 15 ft. lbs. (20 Nm).

4. Connect the knock sensor cable set connector and properly route the cable.

### 2.8L Engine

#### *Front*

*See Figure 252.*

1. Raise and safely support the vehicle securely on jackstands.

2. Remove the secondary air injection pump.

3. Remove the knock sensor.

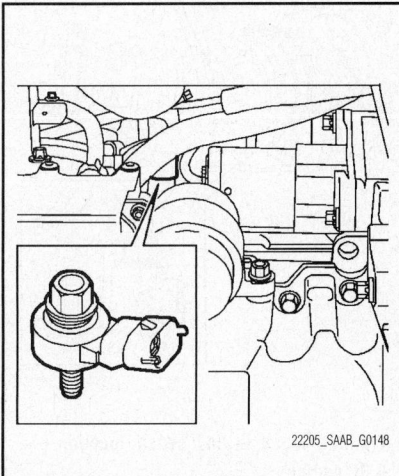

**Fig. 252 Front knock sensor location— 2.8L engine**

*To install:*

4. Plug in the connector.

5. Install the knock sensor and tighten to 17 ft. lbs. (23 Nm).

6. Install the secondary air injection pump.

7. Lower the vehicle.

#### *Rear*

*See Figure 253.*

1. Raise and safely support the vehicle securely on jackstands.

2. Remove the rear catalytic converter.

3. Remove the bracket between the front exhaust pipe and the engine using 32 025 061 Crow Foot Spanner Wrench.

4. Undo the cable fixing from the heat shield.

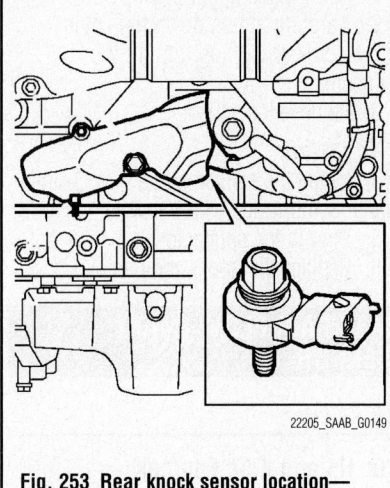

**Fig. 253 Rear knock sensor location— 2.8L engine**

5. Remove the heat shield.

6. Unplug the knock sensor connector.

7. Remove the knock sensor.

*To install:*

8. Install the knock sensor and tighten to 17 ft. lbs. (23 Nm).

9. Plug in the connector.

10. Install the heat shield.

11. Secure the cable to the heat shield.

12. Install the bracket to the engine and front exhaust pipe.

13. Tighten the engine bolt to 17 ft. lbs. (23 Nm), the exhaust pipe bolt to 30 ft. lbs. (40 Nm) and the heat shield bolt to M6 6ft. lbs. (9 Nm) and M10 28 ft. lbs. (38 Nm).

14. Install rear catalytic converter using a new gasket.

15. Lower the vehicle.

### MANIFOLD ABSOLUTE PRESSURE (MAP) SENSOR

#### LOCATION

##### 2.0L H4 and 2.5L SOHC Engines

*See Figure 254.*

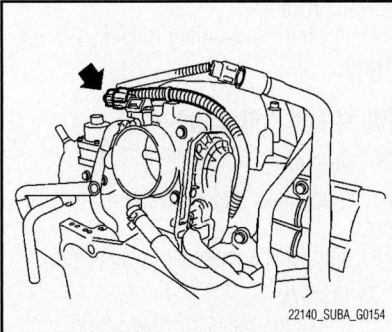

**Fig. 254 The Manifold Absolute Pressure (MAP) Sensor location SOHC 2.5L engine**

The Manifold Absolute Pressure (MAP) Sensor location for the SOHC 2.5L engine is located on the throttle body unit.

##### 2.3L Engine

The Manifold Absolute Pressure (MAP) Sensor is located at the top of the engine, under the engine cover.

##### 2.5L DOHC Engine

*See Figure 255.*

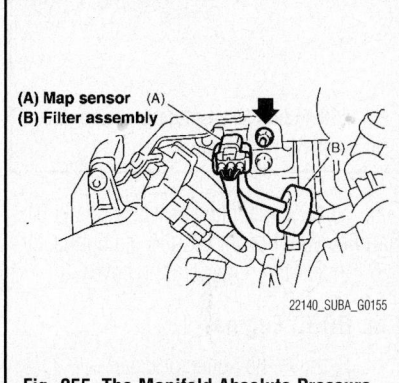

(A) Map sensor (A)
(B) Filter assembly

**Fig. 255 The Manifold Absolute Pressure (MAP) Sensor location DOHC 2.5L engine**

The Manifold Absolute Pressure (MAP) Sensor location for the DOHC 2.5L engine is located on the solenoid valve bracket.

##### 2.0L I4 and 2.8L Engines

The Manifold Absolute Pressure (MAP) Sensor location is located next to the throttle body.

#### REMOVAL & INSTALLATION

##### 2.0L H4 and 2.5L SOHC Engines

1. Disconnect the ground cable from the battery.

2. Disconnect the connector from manifold absolute pressure sensor.

3. Remove the manifold absolute pressure sensor from throttle body.

*To install:*

4. Install in the reverse order of removal.

5. Use new O-rings.

6. Tighten the manifold absolute pressure sensor to 1.5 ft. lbs. (2 Nm).

##### 2.3L Engine

*See Figure 256.*

1. Remove the upper engine cover.

2. Unplug the connector.

3. Remove the sensor.

*To install:*

4. Apply Vaseline the sensor threads. Install the sensor.

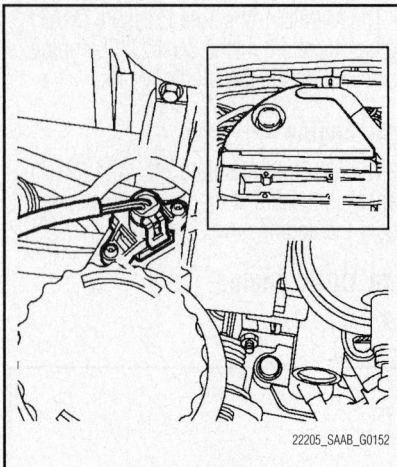

**Fig. 256 Manifold absolute pressure sensor location—2.3L engine**

5. Spray the connector with Kontakt 61 (part no. 30 04 520). Plug in the connector.
6. Install the upper engine cover.

### 2.5L DOHC Engine

1. Remove the collector cover.
2. Disconnect the ground cable from battery.
3. Disconnect the connector from manifold absolute pressure sensor, and remove the filter assembly from intake manifold.
4. Remove the manifold absolute pressure sensor from the solenoid valve bracket.

#### To install:
5. Install in the reverse order of removal.
6. Tighten the manifold absolute pressure sensor to 4.7 ft. lbs. (6.4 Nm).

### 2.0L I4 and 2.8L Engines

*See Figure 257.*

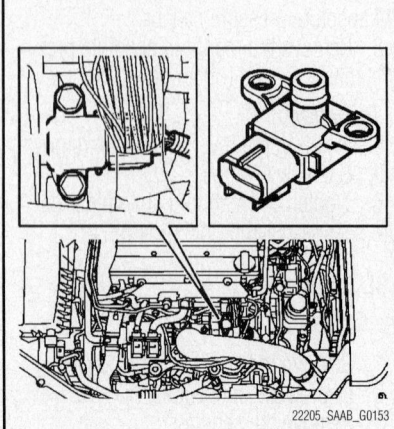

**Fig. 257 Manifold absolute pressure sensor location—2.0L I4 and 2.8L engines**

1. Remove the upper engine cover.
2. Carefully press down the catch with a screwdriver and unplug the connector.
3. Remove the sensor.

#### To install:
4. Install the sensor with a new O-ring sparingly lubricated with Vaseline.
5. Plug in the connector.
6. Replace the upper engine cover.

## MASS AIR FLOW (MAF) SENSOR

### LOCATION

### 2.0L H4 and 2.5L Engines

The MAF is mounted in the intake air hose of the air cleaner assembly. This sensor is combined with the intake air temperature sensor.

### 2.0L I4 and 2.8L Engine

The MAF is mounted on right-hand (passenger's side) MacPherson strut tower

### 2.3L Engine

The MAF is mounted in the intake air hose of the air cleaner assembly.

### REMOVAL & INSTALLATION

### 2.0L H4 and 2.5L Engines

1. Disconnect the negative battery cable.
2. Remove the collector cover, if equipped.
3. Disconnect the connector from the sensor.
4. Remove the filter assembly from the intake manifold, if equipped.
5. Remove the sensor from its mounting.

#### To install:
6. Installation is the reverse of the removal procedure.
7. Tighten the retaining bolts to 1 ft. lbs. (2 Nm).

### 2.0L I4 and 2.8L Engine

*See Figure 258.*

1. Disconnect the connector from the sensor.
2. Remove the mass air flow sensor.

#### To install:
3. Install the mass air flow sensor with a new O-ring, sparingly lubricated with Vaseline.
4. Plug in the connector.

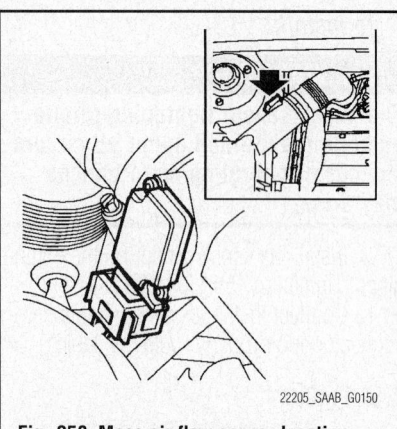

**Fig. 258 Mass air flow sensor location— 2.0L I4 and 2.8L engine**

### 2.3L Engine

*See Figure 259.*

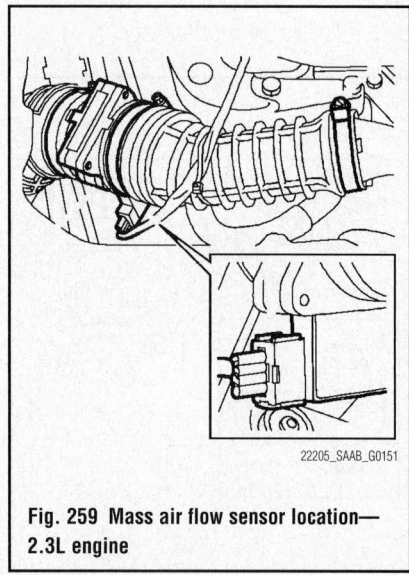

**Fig. 259 Mass air flow sensor location— 2.3L engine**

1. Unplug the connector.
2. Loosen the hose clips on the connecting hoses.
3. Remove the mass air flow sensor.

#### To install:
4. The mass air flow sensor must be fitted in the intake air pipe with the arrows in the direction of flow.
5. Put the mass air flow sensor in place.
6. Tighten the hose clips on the connecting hoses.
7. Spray the connector with Kontakt 61 (part no. 30 04 520) and plug it in.

## THROTTLE POSITION SENSOR (TPS)

### LOCATION

The throttle position sensor is located at the throttle body. On some engines the throttle position sensor is an integral part of the throttle body and cannot be serviced separately.

## REMOVAL & INSTALLATION

### 2.0L H4 Engine

1. Disconnect the negative battery cable.
2. Remove the intercooler, as necessary.
3. Disconnect the sensor connector.
4. Remove the sensor retaining screws.
5. Remove the sensor from its mounting.

**To install:**

6. Installation is the reverse of the removal procedure.
7. Tighten the sensor to 1 ft. lbs. (2 Nm).

### 2.0L I4 Engine

*See Figures 260 and 261.*

1. Remove the upper engine cover.
2. Detach the turbocharger delivery hose from the throttle body and bend it aside.
3. Carefully bend up the bracket catch and remove the EVAP canister purge valve from the bracket.
4. Remove the bracket from the throttle body and carefully bend it aside.
5. Unplug the connector from the throttle body.
6. Remove the throttle body.
7. Remove the old seal and clean the sealing surfaces.

**To install:**

8. Install the throttle body with a new seal.
9. Plug in the connector.
10. Install the bracket.
11. Install the EVAP canister purge valve and bend the catch back in place.

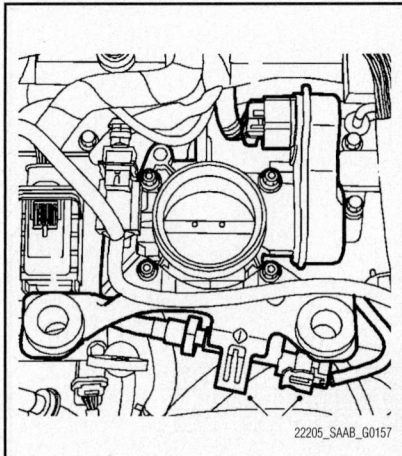

**Fig. 260 Throttle position sensor location—2.0L I4 engine**

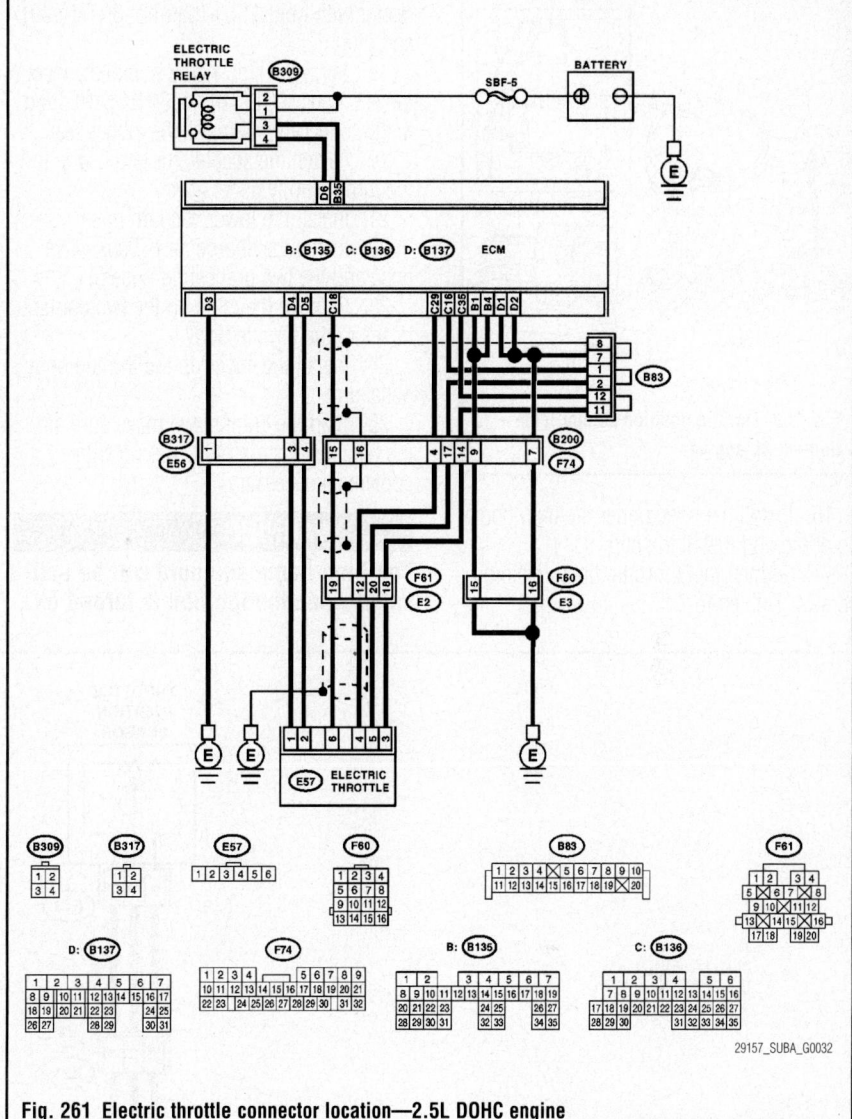

**Fig. 261 Electric throttle connector location—2.5L DOHC engine**

12. Install the turbocharger delivery hose to the throttle body.
13. Install the upper engine cover.
14. If a new throttle body has been installed, connect the Tech 2 diagnostic tool for programming.

### 2.3L Engine

*See Figure 262.*

1. Turn the ignition to the **OFF** position.
2. Remove the upper engine cover.
3. Undo the coolant expansion tank cap to release any pressure and then retighten it.
4. Remove the cover on the throttle body pedal arm.
5. Pinch together the 2 coolant hoses to the throttle body.
6. Undo the charge air bypass valve hose and the two preheating hoses.
7. Detach the lower vacuum hose.
8. Undo the turbocharger delivery pipe retaining screw in the cylinder head.
9. Undo the hose clip and carefully tilt up the turbocharger delivery pipe.
10. Detach the accelerator pedal wire from the throttle body pedal arm.
11. Lift off the rubber seal and unplug the limp-home solenoid connector.
12. Unplug the throttle body 10-pin connector.
13. Undo the 3 throttle body retaining screws.
14. Lift up the throttle body and detach the hose under the limp-home solenoid.

**To install:**

15. Position the throttle body and screw it on with the 3 retaining screws. Fit a new seal if necessary and lubricate it with a thin coat of Vaseline.

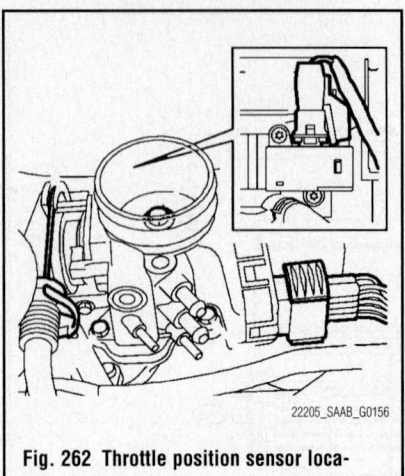

**Fig. 262 Throttle position sensor location—2.3L engine**

16. Install the hose under the limp home solenoid and install the clip.

17. Tighten the 3 throttle body retaining screws. Tighten to 7 ft. lbs. (10 Nm).

18. Spray the throttle body 10-pin connector with Kontakt 61 (part no. 30 04 520) and plug it in.

19. Spray the limp-home solenoid connector with Kontakt 61 (part no. 30 04 520). Plug in the connector and install the rubber seal.

20. Fasten the accelerator pedal wire to the throttle body pedal arm.

21. Install the lower vacuum hose.

22. Install the charge air bypass valve hose and the two preheating hoses.

23. Remove the clips on the two coolant hoses to the throttle body.

24. Install the cover on the throttle body pedal arm.

25. Start the engine and make sure all the connections are tight. Top up with coolant as necessary.

## ❄ WARNING

**The limp-home solenoid will be activated when the ignition is turned on.**

**The limp-home mechanism must therefore be reset after turning on the ignition.**

26. Attach female connector for throttle.

27. Turn the ignition ON.

28. Clear diagnostic trouble codes.

29. Turn the ignition OFF.

30. Carefully slide the end of the spring towards the throttle body using the point of a pen or similar.

31. At the same time, rotate the black tooth disc anticlockwise with a screwdriver. A click will be heard in its end position.

32. Turn the pedal arm with throttle cable clockwise. Make sure that the throttle arm does not follow the movement of the pedal arm. Refer to the illustration of a reset throttle.

33. Refit the engine cover.

### 2.5L Engines

*See Figure 263.*

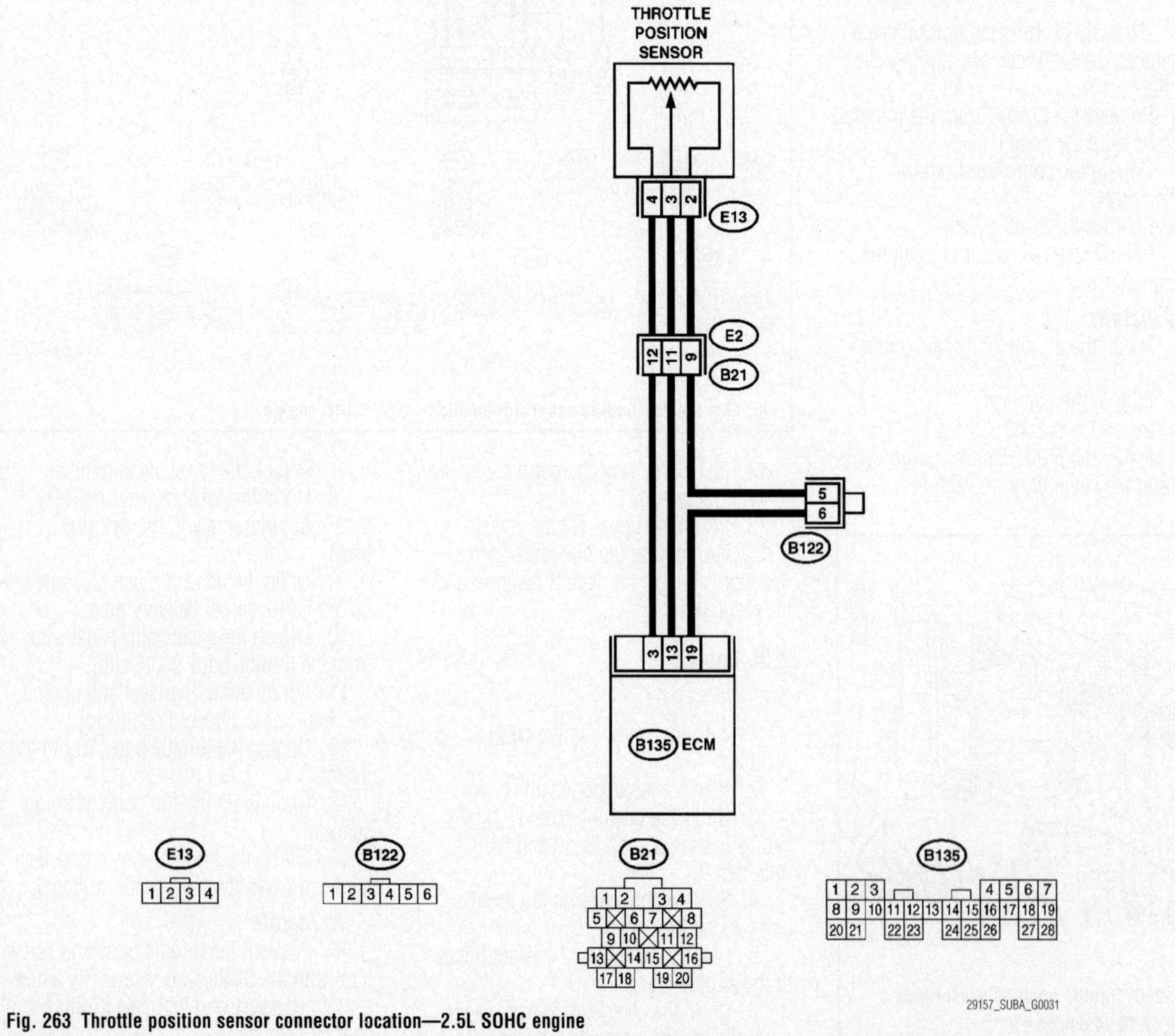

**Fig. 263 Throttle position sensor connector location—2.5L SOHC engine**

The throttle position sensor is an integral part of the throttle body. Do not attempt to remove the throttle position sensor from throttle body.

### 2.8L Engine

*See Figure 264.*

1. Before removing a control module, Tech 2 must be used to separate the control module from the car. From the "All" menu, select the control module under "And/Remove". Then select "Remove" and follow the instructions. The ignition key must be in the ON position. TIS 2000 may be required. Once the control module has been separated from the car, turned the ignition key to the OFF position. The control module can then be removed. Carry out the "Remove" action with Tech2 even if the control module cannot be contacted or is completely missing because data in other control modules must be changed.

2. Remove the engine cover.

3. Remove the power steering fluid reservoir bolt. Put the reservoir aside.

4. Unplug the connector of the intake air sensor.

5. Remove the charge air pipe.

6. Remove the throttle body bolts.

7. Unplug the connector and lift away the throttle body.

**To install:**

8. Clean the sealing surfaces. Position the throttle body using a new seal. Plug in the connector.

9. Install the throttle body bolts. Tighten to 7 ft. lbs. (9 Nm).

10. Replace the charge air pipe.

11. Plug in the connector of the intake air sensor.

12. Position the power steering fluid reservoir and install the bolt.

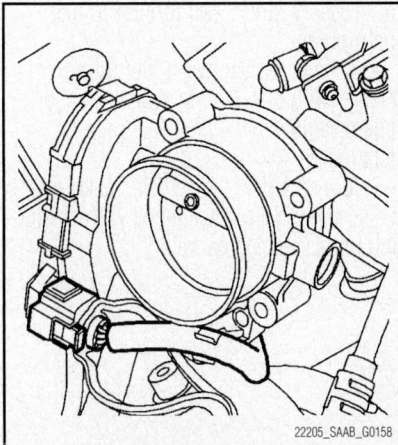

**Fig. 264 Throttle position sensor location—2.8L engine**

13. Install the engine cover.

14. If a new throttle body has been installed, connect the Tech 2 diagnostic tool for programming.

### VEHICLE SPEED SENSOR (VSS)

#### LOCATION

#### 2.0L H4 and 2.5L Engines

This vehicle uses a front and rear sensor. Both sensors are mounted on the transaxle.

#### 2.0L I4 and 2.8L Engines

This vehicle uses a front and rear sensor. Both sensors are mounted on the transaxle.

#### 2.3L Engine

This vehicle uses a front and rear sensor. Both sensors are mounted on the transaxle.

#### REMOVAL & INSTALLATION

#### 2.0L H4 and 2.5L Engines

*See Figures 265 and 266.*

1. Raise and support the vehicle safely.

2. Remove the exhaust, as required.

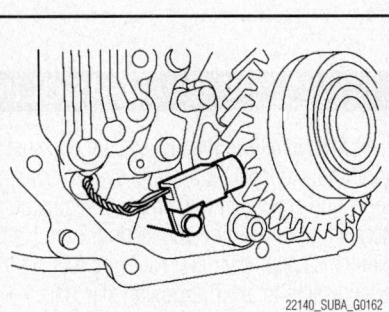

**Fig. 265 Front Vehicle Speed Sensor (VSS) location view–2.0L H4 and 2.5L engines**

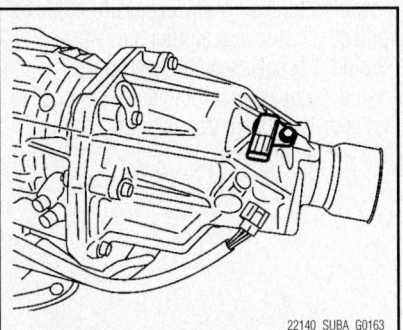

**Fig. 266 Rear Vehicle Speed Sensor (VSS) location view–2.0L H4 and 2.5L engines**

3. Place a drip pan below the speed sensor to catch any spilled fluid.

4. Disconnect the connector.

5. Remove the sensor from its mounting.

**To install:**

6. Installation is the reverse of the removal procedure.

7. Replace any lost fluid.

#### 2.0L I4 and 2.8L Engines

*See Figures 267 and 268.*

1. Remove the battery cover.

2. Remove the battery, coolant pipe, main fuse box and hood switch.

3. Unplug the control module connector. Unplug the connectors under the control module.

4. Remove the battery tray.

5. Unplug the speed sensor connector.

6. Remove the input shaft speed sensor.

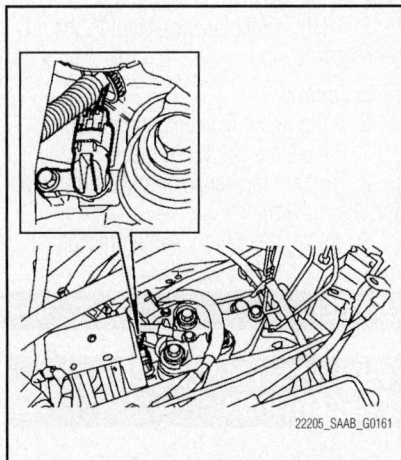

**Fig. 267 Input speed sensor—2.0L I4 and 2.8L engines**

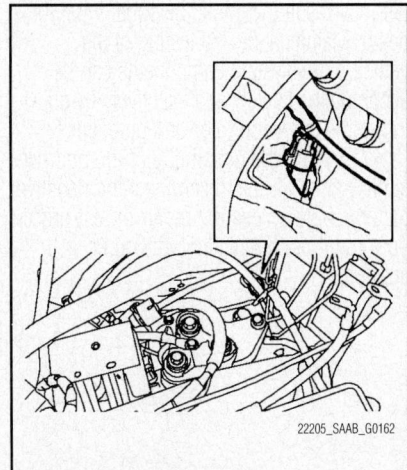

**Fig. 268 Output speed sensor—2.0L I4 and 2.8L engines**

*To install:*

7. Install the speed sensor and plug in the connector.

8. Plug in the connectors under the control module.

9. Plug in the control module connector and install the battery tray.

10. Install the main fuse box and coolant pipe.

11. Install the battery.

12. Install the battery cover.

### 2.3L Engine

#### Input

*See Figure 269.*

1. Remove the battery cover and the battery.

2. Remove the MAXI fuse holder from the battery tray and place it to one side. Remove the battery tray.

3. Undo the battery cable bracket.

4. Clean the area round the sensor. Remove the sensor's retaining bolt.

5. Remove the sensor. Unplug the connector.

*To install:*

6. Plug in the sensor's connector.

7. Install the sensor in place.

8. Tighten the sensor's retaining bolt to 5 ft. lbs. (6 Nm).

9. Install the battery cable bracket.

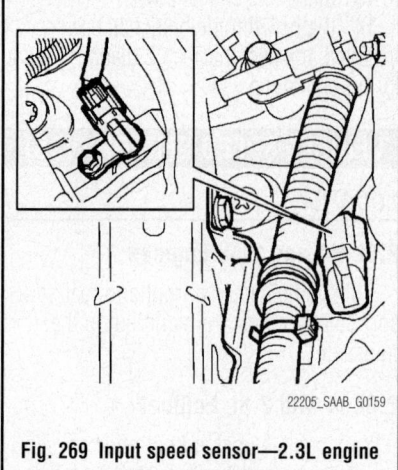

22205_SAAB_G0159

**Fig. 269 Input speed sensor—2.3L engine**

10. Install the battery tray, MAXI fuse board, battery and battery cover.

11. Set the right time on the clock.

#### Output

*See Figure 270.*

1. Remove the battery cover and the battery.

2. Remove the MAXI fuse holder from the battery tray and place it to one side. Remove the battery tray.

3. Remove the MAXI fuse board and place it to one side. Remove the battery tray.

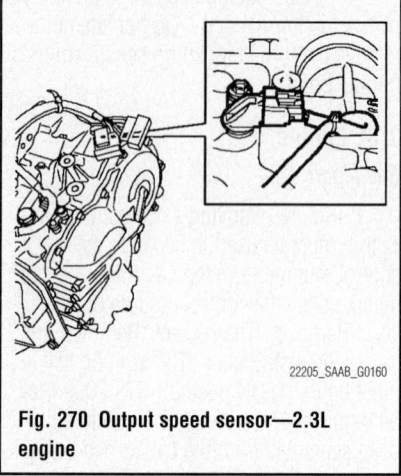

22205_SAAB_G0160

**Fig. 270 Output speed sensor—2.3L engine**

4. Disconnect the electrical distribution unit and move it aside.

5. Clean the area round the sensor. Remove the retaining bolt and withdraw the sensor.

6. Unplug the sensor's connector.

*To install:*

7. Plug in the sensor's connector.

8. Install the sensor in place and tighten the retaining bolt to 5 ft. lbs. (6 Nm).

9. Install the electrical distribution unit.

10. Install the battery tray and MAXI fuse board.

11. Install the battery and battery cover.

---

## FUEL                    GASOLINE FUEL INJECTION SYSTEM

### FUEL SYSTEM SERVICE PRECAUTIONS

Safety is the most important factor when performing not only fuel system maintenance but any type of maintenance. Failure to conduct maintenance and repairs in a safe manner may result in serious personal injury or death. Maintenance and testing of the vehicle's fuel system components can be accomplished safely and effectively by adhering to the following rules and guidelines.

• To avoid the possibility of fire and personal injury, always disconnect the negative battery cable unless the repair or test procedure requires that battery voltage be applied.

• Always relieve the fuel system pressure prior to disconnecting any fuel system component (injector, fuel rail, pressure regulator, etc.), fitting or fuel line connection. Exercise extreme caution whenever relieving fuel system pressure to avoid exposing skin, face and eyes to fuel spray. Please be advised that fuel under pressure may penetrate the skin or any part of the body that it contacts.

• Always place a shop towel or cloth around the fitting or connection prior to loosening to absorb any excess fuel due to spillage. Ensure that all fuel spillage (should it occur) is quickly removed from engine surfaces. Ensure that all fuel soaked cloths or towels are deposited into a suitable waste container.

• Always keep a dry chemical (Class B) fire extinguisher near the work area.

• Do not allow fuel spray or fuel vapors to come into contact with a spark or open flame.

• Always use a back-up wrench when loosening and tightening fuel line connection fittings. This will prevent unnecessary stress and torsion to fuel line piping.

• Always replace worn fuel fitting O-rings with new Do not substitute fuel hose or equivalent where fuel pipe is installed.

Before servicing the vehicle, make sure to also refer to the precautions in the beginning of this section as well.

**FUEL**

**GASOLINE FUEL INJECTION SYSTEM**

## FUEL SYSTEM SERVICE PRECAUTIONS

Safety is the most important factor when performing not only fuel system maintenance but any type of maintenance. Failure to conduct maintenance and repairs in a safe manner may result in serious personal injury or death. Maintenance and testing of the vehicle's fuel system components can be accomplished safely and effectively by adhering to the following rules and guidelines.

• To avoid the possibility of fire and personal injury, always disconnect the negative battery cable unless the repair or test procedure requires that battery voltage be applied.

• Always relieve the fuel system pressure prior to disconnecting any fuel system component (injector, fuel rail, pressure regulator, etc.), fitting or fuel line connection. Exercise extreme caution whenever relieving fuel system pressure to avoid exposing skin, face and eyes to fuel spray. Please be advised that fuel under pressure may penetrate the skin or any part of the body that it contacts.

• Always place a shop towel or cloth around the fitting or connection prior to loosening to absorb any excess fuel due to spillage. Ensure that all fuel spillage (should it occur) is quickly removed from engine surfaces. Ensure that all fuel soaked cloths or towels are deposited into a suitable waste container.

• Always keep a dry chemical (Class B) fire extinguisher near the work area.

• Do not allow fuel spray or fuel vapors to come into contact with a spark or open flame.

• Always use a back-up wrench when loosening and tightening fuel line connection fittings. This will prevent unnecessary stress and torsion to fuel line piping.

• Always replace worn fuel fitting O-rings with new Do not substitute fuel hose or equivalent where fuel pipe is installed.

Before servicing the vehicle, make sure to also refer to the precautions in the beginning of this section as well.

## RELIEVING FUEL SYSTEM PRESSURE

### ✳✳ CAUTION

The fuel injection system remains under pressure, even after the engine has been tuned OFF. The fuel system pressure must be relieved before

disconnecting any fuel lines. Failure to do so may result in fire and/or personal injury.

### 2.0L H4 and 2.5L Engines

➡This procedure must be performed prior to servicing any component of the fuel injection system.

1. Remove the fuel pump fuse from the main fuse box.
2. Start the engine and run until it stalls.
3. Crank the engine for 5 seconds or more to ensure the fuel pressure is properly relieved. If the engine starts during this time, allow it to run until it stalls.
4. Turn the ignition switch to the OFF position. Remove the key.
5. Disconnect the negative battery cable.
6. Remove the fuel cap.

### 2.0L I4, 2.3L and 2.8L Engines
See Figure 271.

### ✳✳ CAUTION

The fuel injection system remains under pressure, even after the engine has been turned OFF. The fuel system pressure must be relieved before disconnecting any fuel lines. Failure to do so may result in fire and/or personal injury.

1. Before servicing the vehicle, refer to the Precautions Section.
2. Remove fuel pump fuse from fuse panel while the engine is running.
3. Switch the ignition **OFF** once the engine has stopped.
4. Install fuel pump fuse.

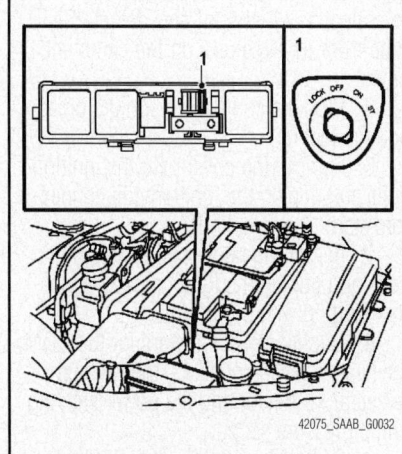

42075_SAAB_G0032

**Fig. 271 Locating the fuel pump fuse**

## FUEL FILTER

*REMOVAL & INSTALLATION*

### ✳✳ WARNING

Removal and installation involves partial dismantling of the car's fuel system. Work only in a well-ventilated area! If approved equipment for the extraction of fuel vapor is available then make sure it is used. Wear suitable gloves! Prolonged contact with fuel can cause skin irritation. Keep a class BE fire extinguisher close by! Be mindful of the danger of sparks caused by short circuits and when connecting and disconnecting leads in electrical circuits. Do not smoke in work area. Wear protective goggles.

➡Be particularly observant regarding cleanliness when working on the fuel system. Loss of function may occur due to very small particles. Prevent dirt and grime from entering the fuel system by cleaning the connections and plugging pipes and lines during disassembly.

### ✳✳ CAUTION

The fuel injection system remains under pressure, even after the engine has been tuned OFF. The fuel system pressure must be relieved before disconnecting any fuel lines. Failure to do so may result in fire and/or personal injury.

### 2.0L H4 and 2.5L Engines
See Figure 272.

The fuel filter is an integral component of the fuel pump/fuel sender assembly. However, it can be serviced separately.

1. Reduce the fuel pressure. Refer to "Gasoline Fuel Injection System, Relieving Fuel System Pressure, Procedure".
2. Remove the fuel pump assembly.
3. Remove the fuel pump.
4. Separate the fuel filter from fuel pump.
5. Turn the filter holder around to the arrow direction, and then remove the filter.

**To install:**
6. Install the fuel filter and turn the filter holder around to tighten.
7. Install the Separate the fuel filter from fuel pump.

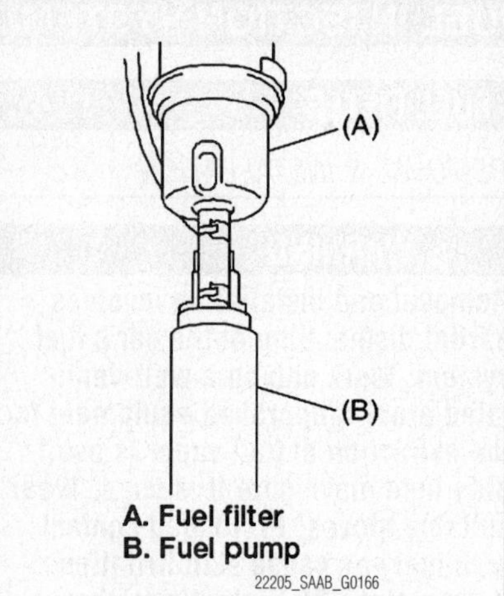
(A)
(B)
A. Fuel filter
B. Fuel pump
22205_SAAB_G0166

Fig. 272 Locating the fuel pump fuse

8. Install the fuel pump.
9. Install the fuel pump assembly.

### 2.0L I4, 2.3L and 2.8L Engines

The fuel filter is an integral component of the fuel pump/fuel sender assembly and cannot be serviced separately.

## FUEL INJECTORS

### *REMOVAL & INSTALLATION*

**⁂ WARNING**

**Removal and installation involves partial dismantling of the car's fuel system. Work only in a well-ventilated area! If approved equipment for the extraction of fuel vapor is available then make sure it is used. Wear suitable gloves! Prolonged contact with fuel can cause skin irritation. Keep a class BE fire extinguisher close by! Be mindful of the danger of sparks caused by short circuits and when connecting and disconnecting leads in electrical circuits. Do not smoke in work area. Wear protective goggles.**

**➡Be particularly observant regarding cleanliness when working on the fuel system. Loss of function may occur due to very small particles. Prevent dirt and grime from entering the fuel system by cleaning the connections and plugging pipes and lines during disassembly.**

**⁂ CAUTION**

**The fuel injection system remains under pressure, even after the engine has been tuned OFF. The fuel system pressure must be relieved before disconnecting any fuel lines. Failure to do so may result in fire and/or personal injury.**

### 2.0L H4 Engine

1. Before servicing the vehicle, refer to the Precautions Section.
2. Reduce the fuel pressure. Refer to "Gasoline Fuel Injection System, Relieving Fuel System Pressure, Procedure".
3. Open the fuel filler flap lid and remove the fuel filler cap.
4. Disconnect the ground cable from battery.
5. Remove the intake manifold.
6. Remove the fuel pipe protector.
7. Disconnect the connector from the fuel injector.
8. Remove the bolts which hold the injector pipe to intake manifold.
9. Remove the fuel injector pipe and fuel injectors.

***To install:***

10. Installation procedure is the reverse of removal.
11. Replace the O-ring and insulators with new ones.
12. Install the bolts which hold the injector pipe to intake manifold. Tighten to 14 ft. lbs. (19 Nm).

### 2.0L I4 Engine

*See Figure 273.*

1. Before servicing the vehicle, refer to the Precautions Section.
2. Reduce the fuel pressure. Refer to "Gasoline Fuel Injection System, Relieving Fuel System Pressure, Procedure".
3. Remove the upper engine cover.
4. Remove the inlet and return fuel pipes from the fuel rail. Use a support so that the lower connection on the fuel inlet pipe does not come loose. Use absorbent to collect any fuel spill and detach the pipe from its fasteners on the camshaft cover.
5. Plug the connections on the pipes and the fuel rail.
6. Remove the cover over the ignition coils and unplug the ignition coil connectors starting with cylinder 4.
7. Remove the cover over the ignition coils and unplug the ignition coil connectors starting with cylinder 4.
8. Unplug the connectors for the CDM, the heated oxygen sensors, the exhaust temperature sensor and the power steering pump pressure sensor.
9. Detach the cable duct and fold it forward.
10. Detach the vacuum hose from the fuel pressure regulator.

Remove the bolts, loosen the fuel rail and pull it up together with the injectors. Plug the holes in the cylinder head.

11. Unplug the connector from the injector(s).

Remove the injector fastening clips. Press the clip link to open it and detach with a screwdriver.

12. Remove the injector(s). Collect any fuel spill.

***To install:***

13. Install the injector(s) with new O-rings, sparingly lubricated with Vaseline, to the fuel rail. Adjust their positions so that the fastening clip is located in the correct groove on the injector. Check the fastening.
14. Plug in the injector connectors, making sure they are locked.
15. Remove the plugs and Install the fuel rail to the cylinder head. Tighten to 7 ft. lbs. (10 Nm).
16. Install the vacuum hose to the fuel pressure regulator.
17. Install the cable duct and plug in the connectors for the CDM , the heated oxygen sensors and the connector for the exhaust temperature sensor, and the pressure sensor for the power steering.
18. Install the cable duct and plug in the connectors to the throttle body and coolant temperature sensor. Make sure the coolant hose is not under the cable duct.
19. Plug in the connectors to the ignition coils. Make sure to locate the ignition coil cables so they are not trapped.
20. Install the cover over the ignition coils.
21. Install new O-rings , sparingly lubricated with Vaseline, to the fuel inlet and return pips.
22. Connect the fuel inlet and return pipes to the fuel rail. Secure the pipes to the mounting on the camshaft cover.

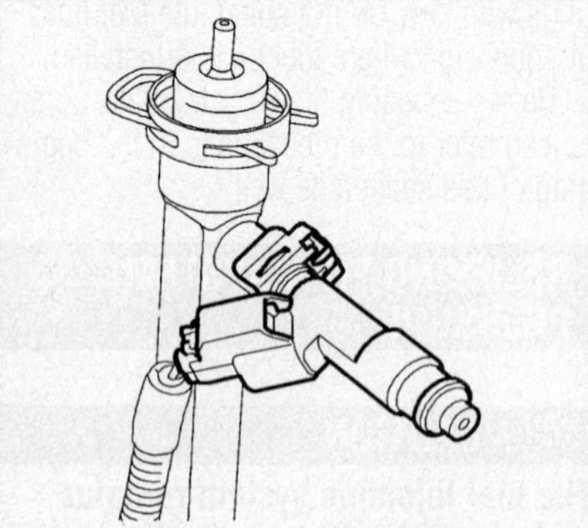
22205_SAAB_G0168

Fig. 273 Fuel rail and injector

23. Connect an exhaust extractor. Start the engine and check for fuel leaks.

24. Install the upper engine cover.

### 2.3L Engine

*See Figure 274.*

1. Before servicing the vehicle, refer to the Precautions Section.

2. Reduce the fuel pressure. Refer to "Gasoline Fuel Injection System, Relieving Fuel System Pressure, Procedure".

3. Remove or disconnect the following:
- Negative battery cable
- Upper engine cover
- Crankcase ventilation hoses
- Dipstick with the oil filler pipe
- Cover and detach the throttle cable from the spindle
- Turbocharger delivery pipe
- Fuel injector connectors
- Ignition discharge module
- Manifold Absolute Pressure (MAP) sensor
- Boost pressure control valve
- Turbocharger pressure sensor and slacken the upper bolt on the wiring holder
- Fuel rail retaining bolts and cable ties
- Fuel lines
- Pressure regulator vacuum hose
- Fuel rail with the injectors
- Fuel injector locking clips
- Fuel injectors

**To install:**

4. Lubricate the O-rings on the fuel injectors

5. Install or connect the following:
- Fuel injectors to the fuel rail and make certain that the injectors are installed to the proper cables
- Fuel rail and torque the bolts to 6 ft. lbs. (8 Nm)
- Vacuum hose to the pressure regulator

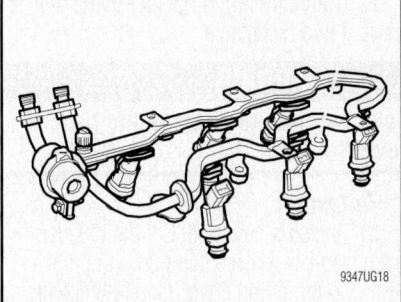

**Fig. 274 Remove the fuel rail and injectors as an assembly**

9347UG18

- Fuel injector hoses
- Screws for the wiring holder
- Throttle cable and cover and adjust if necessary
- Dipstick and filler pipe
- Crankcase ventilation hoses
- Upper engine cover
- Negative battery cable

6. Start the vehicle, check for leaks and repair if necessary.

### 2.5L Engines

1. Before servicing the vehicle, refer to the Precautions Section.

2. Reduce the fuel pressure. Refer to "Gasoline Fuel Injection System, Relieving Fuel System Pressure, Procedure".

3. Open the fuel filler flap lid, and remove the fuel filler cap.

4. Disconnect the ground cable from battery.

5. On the right hand (passenger's) side of the engine:

a. Remove the air intake chamber and air cleaner case.

b. Remove the power steering pump and tank from the brackets.

c. Remove the front side V-belt.

d. Remove the bolts which hold the power steering pipes onto the intake manifold protector.

e. Remove the bolts which install the power steering pump to the bracket.

f. Disconnect the power steering pump switch connector.

g. Remove the reservoir tank from the bracket by pulling it upwards.

h. Place the power steering pump and tank on the right side wheel apron.

6. On the left hand (driver's) side of the engine:

a. Remove the battery.

7. Remove the spark plug cords from spark plugs.

8. Remove the fuel pipe protector.

9. Disconnect the connector from fuel injector.

10. Remove the harness band which holds engine harness to injector pipe.

11. Remove the bolts which hold fuel injector pipe onto intake manifold.

12. Remove the fuel injector while lifting up the fuel injector pipe.

**To install:**

13. Installation procedure is the reverse of removal.

14. Replace the O-ring and insulators with new ones.

15. Install the bolts which hold the injector pipe to intake manifold. Tighten to 14 ft. lbs. (19 Nm).

### 2.8L Engine

*See Figure 275.*

1. Before servicing the vehicle, refer to the Precautions Section.

2. Reduce the fuel pressure. Refer to "Gasoline Fuel Injection System, Relieving Fuel System Pressure, Procedure".

3. Remove the engine cover.

4. Remove the upper intake manifold.

5. Remove the intake manifold gaskets.

6. Unplug the injector connectors using a small screwdriver.

7. Detach the injectors' wiring harness.

8. Unplug the fuel pressure sensor connector.

9. Release any fuel pressure by pressing in the needle valve on the fuel rail. Catch any fuel spill with a cloth.

10. Detach the fuel line from the fuel rail using 83 95 261 Fuel Line Tool. Catch any fuel spill.

11. Remove the bolts. Remove the fuel rail with injectors from the engine.

12. Remove the clips from the valves.

13. Remove the valves from the fuel rail.

**To install:**

14. Install the injectors to the fuel rail using new seals coated with acid-free Vaseline. Install the clips on the injectors and press them into the fuel rail.

15. Install the fuel rail with injectors to the engine. Install the bolts and tighten to 7 ft. lbs. (10 Nm).

16. Connect the fuel line to the fuel rail.

17. Plug in the fuel pressure sensor connector.

18. Connect the injectors' wiring harness.

19. Plug in the injector connectors.

20. Install the intake manifold gaskets.

21. Install the upper intake manifold.

22. Connect the negative battery cable.

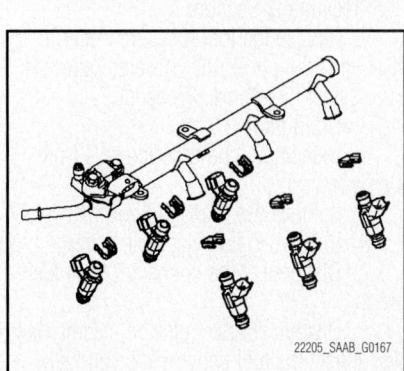

**Fig. 275 Fuel rail and injector exploded view**

22205_SAAB_G0167

23. Start the vehicle, check for leaks and repair if necessary.

## FUEL PUMP

### REMOVAL & INSTALLATION

#### ✳✳ WARNING

Removal and installation involves partial dismantling of the car's fuel system. Work only in a well-ventilated area! If approved equipment for the extraction of fuel vapor is available then make sure it is used. Wear suitable gloves! Prolonged contact with fuel can cause skin irritation. Keep a class BE fire extinguisher close by! Be mindful of the danger of sparks caused by short circuits and when connecting and disconnecting leads in electrical circuits. Do not smoke in work area. Wear protective goggles.

➡Be particularly observant regarding cleanliness when working on the fuel system. Loss of function may occur due to very small particles. Prevent dirt and grime from entering the fuel system by cleaning the connections and plugging pipes and lines during disassembly.

#### ✳✳ CAUTION

The fuel injection system remains under pressure, even after the engine has been tuned OFF. The fuel system pressure must be relieved before disconnecting any fuel lines. Failure to do so may result in fire and/or personal injury.

### 2.0L H4 and 2.5L Engines

*See Figure 276.*

1. Before servicing the vehicle, refer to the Precautions Section.
2. Reduce the fuel pressure. Refer to "Gasoline Fuel Injection System, Relieving Fuel System Pressure, Procedure".
3. Drain fuel.
4. Disconnect the ground cable from the battery.
5. Remove the luggage floor mat.
6. Remove the service hole cover.
7. Disconnect the connector from fuel pump.
8. Disconnect the quick connector, then disconnect the fuel delivery tube and jet pump tube.
9. Remove the nuts which install fuel pump assembly onto fuel tank.

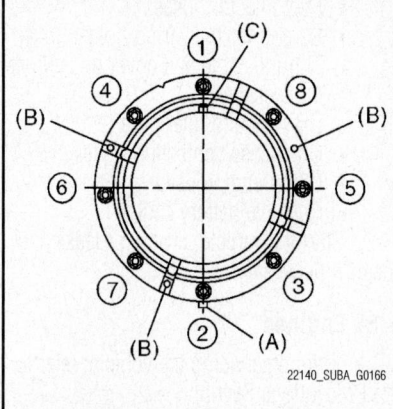

**Fig. 276 Fuel pump assembly tightening procedure**

22140_SUBA_G0166

10. Remove the fuel pump assembly from the fuel tank.

### To install:

11. Install in the reverse order of removal while being careful of the following:

- Make sure the sealing portion is free from fuel or foreign matter before installation.
- When assembling, point the protrusion of the gasket towards the front of the vehicle.
- Insert the protrusion of the gasket into the upper plate.
- Align the protrusion of the fuel pump assembly to the cut out in the upper plate.

12. Tighten the nuts to 3.2 ft. lbs. (4.4 Nm) in the order as shown in the figure below.

### 2.0L I4 Engine

*See Figure 277.*

1. Before servicing the vehicle, refer to the Precautions Section.
2. Reduce the fuel pressure. Refer to "Gasoline Fuel Injection System, Relieving Fuel System Pressure, Procedure".
3. Drain the fuel tank.
4. Remove the tank cap.
5. Raise and safely support the vehicle securely on jackstands.
6. On convertible models, remove the rear subframe chassis reinforcement on right-hand side.
7. Remove the rubber mountings and carefully lower the exhaust system, support it with a strap.
8. Disconnect the fuel and purge connections on the EVAP emission canister. Collect any fuel spill.
9. Seal the connections to the outlet line from the fuel pump.

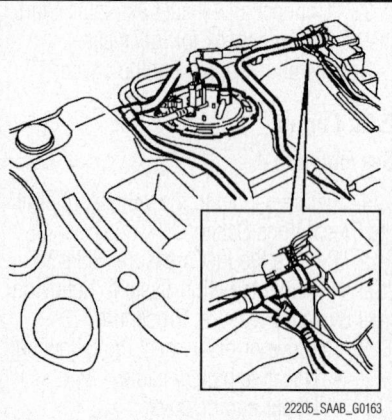

**Fig. 277 Fuel pump and sender in fuel tank—2.0L I4 and 2.8L engines**

22205_SAAB_G0163

10. Unplug and detach the connector on the EVAP emission canister.
11. Remove the fuel filler pipe bolt and detach the hose between filler pipe and tank, and disengage the quick coupling for the bleeder. Plug the tank's filler port.
12. Place a column jack under the tank, remove the tank straps and fuel tank.

#### ✳✳ WARNING

Be careful when removing the tank. Ensure that no components or connections are damaged. To prevent dirt entering the fuel system, clean the outside of the tank if necessary before removing the fuel pump.

13. Unplug the fuel pump connector.
14. Remove the fuel pump. Use tool 83 96 194 Fuel Armature Installing Tool to undo the fuel pump lock ring.
15. Undo the quick couplings on the EVAP canister and the hoses from the tank's fastenings.
16. Raise the fuel pump slightly, press in the catch and undo the quick coupling for the tank bleeder from the fuel pump.
17. Carefully lift up the fuel pump and place it in a receptacle.

#### ✳✳ WARNING

Take care not to damage the fuel level sensor.

### To install:

18. Install a new seal. Lift the bleeder hose, semi-insert the fuel pump and connect it to the pump cover. Carefully lower the fuel pump into the tank.
19. Make sure the recesses in the tank connection are in line with the corresponding heels on the pump cover.

20. Make sure the seal is located correctly. Tighten the fuel pump's fastening using 83 96 194 Fuel Armature Installing Tool.

21. Connect the quick-release couplings to the EVAP emission canister.

22. Attach the fuel lines to the fasteners on the fuel tank.

23. Secure the ventilation line to the outlet in the tank.

24. Plug in the fuel pump connector.

25. Install the fuel tank.

26. Replace the connecting hose between the tank and the filler pipe.

27. Use Vaseline to lubricate the filler pipe connection to the filler cap opening.

28. Install the filler pipe and tighten the bolts. Check the rubber seal in the filler cap opening, and adjust if necessary.

29. Raise the tank and Install the tank straps. Tighten to 18 ft. lbs. (25 Nm).

30. Connect the fuel filler pipe hose and tighten the clamp. Tighten to 3 ft. lbs. (4 Nm).

31. Tighten the bolt on the fuel filler tube. Tighten to 18 ft. lbs. (25 Nm).

32. Connect vent hose.

33. Attach the connector to the evaporative emission canister.

34. Connect the vent and fuel lines.

35. Install the rubber mountings and install the exhaust system.

36. On convertible models, install the rear subframe chassis reinforcement on right-hand side.

37. Lower the vehicle.

38. Fill with fuel and install the fuel filler flap.

39. Start the engine and test for proper function.

## 2.3L Engine

*See Figure 278.*

1. Before servicing the vehicle, refer to the Precautions Section.

2. Reduce the fuel pressure. Refer to "Gasoline Fuel Injection System, Relieving Fuel System Pressure, Procedure".

3. Raise the rear seat cushions and fold the carpeting out of the way.

4. Remove the fuel pump cover.

5. Unplug the upper connector.

### ❄❄ WARNING

**The connector on the pump must not be removed.**

6. Carefully work loose the check valves with fuel lines from the pump (move the yellow hooks to one side with a screwdriver). The check valves are connected to

the pump with quick-release couplings. The white ones are the delivery side and marked "Pressure" on the pump. The black ones are the return side and are marked "Return" on the pump. Move them out of the way and secure them under the plate seam.

7. Remove the screw ring using the 83 94 462 Fuel Pump Tool.

8. Lift up the pump until the top is approximately 2 in. (50 mm) above the fuel tank. Turn the pump clockwise approximately 80° and remove it carefully. Have a cloth or paper ready to collect any fuel that spills out.

9. Transfer the pump to a suitable receptacle and pour off the fuel.

### To install:

➡**Spray the connectors with Kontakt 61, part no. 30 04 520, before connecting them.**

10. Clean sealing surfaces.

11. Install a new O-ring in the tank groove.

12. Carefully lower the pump into its correct position, making sure that the alignment marks on the tank and fuel pump are opposite one other. Carefully press the pump downward into position.

### ❄❄ WARNING

**It is essential to the fuel level sensor function that the pump unit be positioned correctly.**

13. Lubricate the screw threads on the pump cover and the upper edge of the pump with acid-free Vaseline to prevent the pump from turning when the screw ring is tightened.

14. Attach the screw ring with the 83 94 462 Fuel pump tool. Tighten to 55 ft. lbs. (75 Nm).

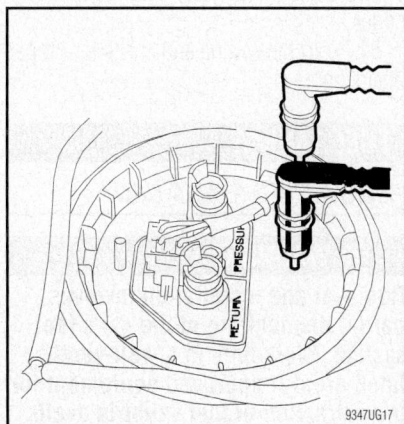

9347UG17

**Fig. 278 Remove the pressure and return valves from the top of the fuel pump—2.3L engine**

15. Inspect and lubricate the O-rings on the fuel line connections with acid-free Vaseline. Connect the fuel line connections to the pump.

16. Make sure that the yellow hooks have snapped in place over the connections.

17. Plug in the connector.

18. Check that the pump functions properly and that it does not leak.

19. Install the fuel pump cover, replace the rear seat cushions and fold back the carpeting.

## 2.8L Engine

*See Figure 279.*

1. Before servicing the vehicle, refer to the Precautions Section.

2. Reduce the fuel pressure. Refer to "Gasoline Fuel Injection System, Relieving Fuel System Pressure, Procedure".

3. Drain the fuel tank.

4. Remove the tank cap.

5. Raise and safely support the vehicle securely on jackstands.

6. Remove the right rear wheel.

7. Remove the bolts and nuts to the fender liner.

8. Release the fender liner from the stud bolts.

9. Release the fender liner from the liner edge starting at the back. Then release the entire liner edge.

10. Remove the fender liner.

11. Undo the rear spoiler shield.

12. On convertible models, remove the rear subframe chassis reinforcement on right-hand side.

13. Remove the rubber mountings and carefully lower the exhaust system, support it with a strap.

14. Disconnect the fuel and purge connections on the EVAP emission canister. Collect any fuel spill.

15. Seal the connections to the outlet line from the fuel pump.

16. Unplug and detach the connector on the EVAP emission canister.

17. Remove the fuel filler pipe bolt and detach the hose between filler pipe and tank, and disengage the quick coupling for the bleeder. Plug the tank's filler port.

18. Place a column jack under the tank, remove the tank straps and fuel tank.

### ❄❄ WARNING

**Be careful when removing the tank. Ensure that no components or connections are damaged. To prevent dirt entering the fuel system, clean the outside of the tank if necessary before removing the fuel pump.**

19. Unplug the fuel pump connector.

20. On vehicles with ORVR, unplug the connector for the pressure sensor.

21. Remove the fuel pump. Use tool 83 96 194 Fuel Armature Installing Tool to undo the fuel pump lock ring.

22. Undo the quick couplings on the EVAP canister and the hoses from the tank's fastenings.

23. Raise the fuel pump slightly, press in the catch and undo the quick coupling for the tank bleeder from the fuel pump.

24. Carefully lift up the fuel pump and place it in a receptacle.

### ❋❋ WARNING

**Take care not to damage the fuel level sensor.**

### To install:

25. Install a new seal. Lift the bleeder hose, semi-insert the fuel pump and connect it to the pump cover. Carefully lower the fuel pump into the tank.

26. Make sure the recesses in the tank connection are in line with the corresponding heels on the pump cover.

27. Make sure the seal is located correctly. Tighten the fuel pump's fastening using 83 96 194 Fuel Armature Installing Tool.

28. Connect the quick-release couplings to the EVAP emission canister.

29. Attach the fuel lines to the fasteners on the fuel tank.

30. Secure the ventilation line to the outlet in the tank.

31. Plug in the fuel pump connector.

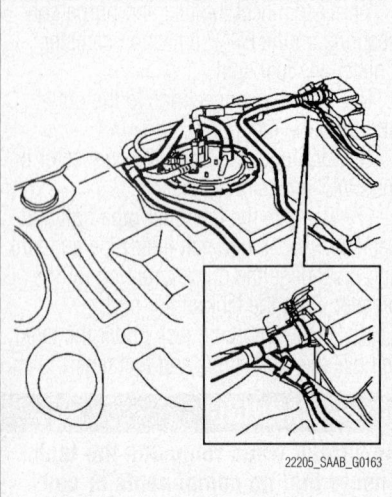

**Fig. 279 Fuel pump and sender in fuel tank—2.0L I4 and 2.8L engines**

22205_SAAB_G0163

32. On cars with ORVR, plug in the connector to the pressure sensor.

33. Install the fuel tank.

34. Replace the connecting hose between the tank and the filler pipe.

35. Use Vaseline to lubricate the filler pipe connection to the filler cap opening.

36. Install the filler pipe and tighten the bolts. Check the rubber seal in the filler cap opening, and adjust if necessary.

37. Raise the tank and Install the tank straps. Tighten to 18 ft. lbs. (25 Nm).

38. Connect the fuel filler pipe hose and tighten the clamp. Tighten to 3 ft. lbs. (4 Nm).

39. Tighten the bolt on the fuel filler tube. Tighten to 18 ft. lbs. (25 Nm).

40. Connect vent hose.

41. Attach the connector to the evaporative emission canister.

42. Connect the vent and fuel lines.

43. Fold in the fender liner. Press the lower liner edge against the wheel housing. Guide the liner onto the upper stud bolts. Then install onto the other stud bolts.

44. Install the fender liner to the liner edge and bumper cavity. The fender liner should lie behind the bumper cavity.

45. Check the install of the fender liner.

46. Attach the rear spoiler shield.

47. Install all bolts and nuts.

48. Install the wheel.

49. Install the rubber mountings and install the exhaust system.

50. On convertible models, install the rear subframe chassis reinforcement on right-hand side.

51. Lower the vehicle.

52. Fill with fuel and install the fuel filler flap.

53. Connect a TECH2 scan tool, or equivalent and erase any diagnostic codes.

54. Start the engine and check for proper operation.

## FUEL TANK

### REMOVAL & INSTALLATION

### ❋❋ WARNING

**Removal and installation involves partial dismantling of the car's fuel system. Work only in a well-ventilated area! If approved equipment for the extraction of fuel vapor is available then make sure it is used. Wear suitable gloves! Prolonged contact with fuel can cause skin irritation. Keep a class BE fire extinguisher close by! Be mindful of the danger of sparks caused by short circuits and when connecting and disconnecting leads in electrical circuits. Do not smoke in work area. Wear protective goggles.**

➡Be particularly observant regarding cleanliness when working on the fuel system. Loss of function may occur due to very small particles. Prevent dirt and grime from entering the fuel system by cleaning the connections and plugging pipes and lines during disassembly.

### ❋❋ CAUTION

**The fuel injection system remains under pressure, even after the engine has been tuned OFF. The fuel system pressure must be relieved before disconnecting any fuel lines. Failure to do so may result in fire and/or personal injury.**

### 2.0L H4 and 2.5L Engines

1. Before servicing the vehicle, refer to the Precautions Section.

2. Reduce the fuel pressure. Refer to "Gasoline Fuel Injection System, Relieving Fuel System Pressure, Procedure".

3. Drain the fuel tank.

4. Remove the tank cap.

5. Remove the rear seat.

6. Disconnect the connector of fuel tank cord to the rear harness.

7. Push the grommet which holds the fuel tank cord on floor panel into under the body.

8. Remove the rear crossmember.

9. Remove the canister.

10. Disconnect the connector from pressure control solenoid valve.

11. Loosen the clamp and disconnect the fuel filler hose (A) and evaporation hose (B) from the fuel filler pipe.

12. Move the clips, and disconnect the quick connector.

13. Disconnect the fuel hoses.

14. Support the fuel tank with a transmission jack, remove the bolts from bands and dismount the fuel tank from the vehicle.

➡An assistant is required to perform this work. Fuel may be left in the side, which has no drain plug, of the fuel tank. In this case, the tank is imbalanced between right and left sides. Be careful not to drop it when removing.

### To install:

15. Support the fuel tank with a transmission jack and push the fuel tank harness into the access hole with the grommet.

16. Set the fuel tank and temporarily tighten the bolts of fuel tank bands.

17. Insert the fuel filler hose approx. 1.38 to 1.57 in (35 to 40 mm) over the lower end of fuel filler pipe, and tighten the clamp.

18. Insert the evaporation hose to the lower end of evaporation pipe, and hold the clamp and clip.

### ✶✶ WARNING

**Do not allow clips to touch hose and rear suspension crossmember.**

19. Connect the fuel hoses, and hold them with clips and quick connector.

20. Connect the connector to the pressure control solenoid valve.

21. Install the canister. Tighten the band mounting bolts to 25 ft. lbs. (33 Nm).

22. Install the rear crossmember.

23. Connect the connectors to the fuel tank cord and plug the service hole with grommet.

24. Set the rear seat and floor mat.

25. Connect the connector to the fuel pump relay.

### 2.0L I4 Engine

1. Before servicing the vehicle, refer to the Precautions Section.

2. Reduce the fuel pressure. Refer to "Gasoline Fuel Injection System, Relieving Fuel System Pressure, Procedure".

3. Drain the fuel tank.

4. Remove the tank cap.

5. Raise and safely support the vehicle securely on jackstands.

6. On convertible models, remove the rear subframe chassis reinforcement on right-hand side.

7. Remove the rubber mountings and carefully lower the exhaust system, support it with a strap.

8. Disconnect the fuel and purge connections on the EVAP emission canister. Collect any fuel spill.

9. Seal the connections to the outlet line from the fuel pump.

10. Unplug and detach the connector on the EVAP emission canister.

11. Remove the fuel filler pipe bolt and detach the hose between filler pipe and tank, and disengage the quick coupling for the bleeder. Plug the tank's filler port.

12. Place a column jack under the tank, remove the tank straps and fuel tank.

### ✶✶ WARNING

**Be careful when removing the tank. Ensure that no components or connections are damaged. To prevent**

dirt entering the fuel system, clean the outside of the tank if necessary before removing the fuel pump.

13. Unplug the fuel pump connector.

*To install:*

14. Plug in the fuel pump connector.

15. Install the fuel tank.

16. Replace the connecting hose between the tank and the filler pipe.

17. Use Vaseline to lubricate the filler pipe connection to the filler cap opening.

18. Install the filler pipe and tighten the bolts. Check the rubber seal in the filler cap opening, and adjust if necessary.

19. Raise the tank and Install the tank straps. Tighten to 18 ft. lbs. (25 Nm).

20. Connect the fuel filler pipe hose and tighten the clamp. Tighten to 3 ft. lbs. (4 Nm).

21. Tighten the bolt on the fuel filler tube. Tighten to 18 ft. lbs. (25 Nm).

22. Connect vent hose.

23. Attach the connector to the evaporative emission canister.

24. Connect the vent and fuel lines.

25. Install the rubber mountings and install the exhaust system.

26. On convertible models, install the rear subframe chassis reinforcement on right-hand side.

27. Lower the vehicle.

28. Fill with fuel and install the fuel filler flap.

29. Start the engine and test for proper function.

### 2.3L Engine

1. Before servicing the vehicle, refer to the Precautions Section.

2. Reduce the fuel pressure. Refer to "Gasoline Fuel Injection System, Relieving Fuel System Pressure, Procedure".

3. Drain the fuel tank.

4. Remove the tank cap.

5. Raise and safely support the vehicle securely on jackstands.

6. Raise and safely support the vehicle securely on jackstands.

7. Remove the rear exhaust system.

8. Remove both handbrake cables from the handbrake lever and suspension straps.

9. Remove the filler pipe screw.

10. Disconnect the bleeder hose with tool 83 95 261 Fuel Line Tool.

11. Remove the filler pipe hose from the tank.

12. Carefully move aside the filler pipe.

13. Remove the feed and return lines and tank bleeder line. Use 83 95 261 Fuel Line Tool for one of the lines.

14. Remove the heat shield from the front of the tank.

15. Place a pillar jack under the tank.

16. Remove the front bolts of the strap holding the tank.

17. Lower the tank slightly and unplug the connector on the top of the tank.

18. Lower the tank.

*To install:*

### ✶✶ WARNING

**When installing the filler pipe it is important to replace the rubber hoses between the filler pipe and the fuel tank. There is a risk of cracks forming on the hoses after repeated tightening of the hose clips and this can lead to fuel leakage.**

➡Spray the connector with anti-corrosion agent before connecting.

19. Carefully jack up the tank.

20. Plug in the connector above the tank.

Raise the tank as far as possible, insert the handbrake cables into the grooves on the tank and tighten the straps that hold the tank. Make sure that the cables do not get trapped.

21. Install the heat shield.

22. Install the bleeder hose.

23. Install the filler pipe hose to the tank.

24. Install the filler pipe screw.

25. Connect the feed and return lines and the tank bleeder line.

26. Install both handbrake cables to the handbrake lever and suspension straps.

27. Install the rear exhaust system.

28. Lower the vehicle.

29. Fill with any fuel that was drained.

30. Start the engine and check the integrity of the system.

### 2.8L Engine

1. Before servicing the vehicle, refer to the Precautions Section.

2. Reduce the fuel pressure. Refer to "Gasoline Fuel Injection System, Relieving Fuel System Pressure, Procedure".

3. Drain the fuel tank.

4. Remove the tank cap.

5. Raise and safely support the vehicle securely on jackstands.

6. Remove the right rear wheel.

7. Remove the bolts and nuts to the fender liner.

8. Release the fender liner from the stud bolts.

9. Release the fender liner from the liner edge starting at the back. Then release the entire liner edge.

10. Remove the fender liner.

11. Undo the rear spoiler shield.

12. On convertible models, remove the rear subframe chassis reinforcement on right-hand side.

13. Remove the rubber mountings and carefully lower the exhaust system, support it with a strap.

14. Disconnect the fuel and purge connections on the EVAP emission canister. Collect any fuel spill.

15. Seal the connections to the outlet line from the fuel pump.

16. Unplug and detach the connector on the EVAP emission canister.

17. Remove the fuel filler pipe bolt and detach the hose between filler pipe and tank, and disengage the quick coupling for the bleeder. Plug the tank's filler port.

18. Place a column jack under the tank, remove the tank straps and fuel tank.

### ❋❋ WARNING

**Be careful when removing the tank. Ensure that no components or connections are damaged. To prevent dirt entering the fuel system, clean the outside of the tank if necessary before removing the fuel pump.**

19. Unplug the fuel pump connector.

*To install:*

20. Plug in the fuel pump connector.

21. On cars with ORVR, plug in the connector to the pressure sensor.

22. Install the fuel tank.

23. Replace the connecting hose between the tank and the filler pipe.

24. Use Vaseline to lubricate the filler pipe connection to the filler cap opening.

25. Install the filler pipe and tighten the bolts. Check the rubber seal in the filler cap opening, and adjust if necessary.

26. Raise the tank and Install the tank straps. Tighten to 18 ft. lbs. (25 Nm).

27. Connect the fuel filler pipe hose and tighten the clamp. Tighten to 3 ft. lbs. (4 Nm).

28. Tighten the bolt on the fuel filler tube. Tighten to 18 ft. lbs. (25 Nm).

29. Connect vent hose.

30. Attach the connector to the evaporative emission canister.

31. Connect the vent and fuel lines.

32. Fold in the fender liner. Press the lower liner edge against the wheel housing. Guide the liner onto the upper stud bolts. Then install onto the other stud bolts.

33. Install the fender liner to the liner edge and bumper cavity. The fender liner should lie behind the bumper cavity.

34. Check the install of the fender liner.

35. Attach the rear spoiler shield.

36. Install all bolts and nuts.

37. Install the wheel.

38. Install the rubber mountings and install the exhaust system.

39. On convertible models, install the rear subframe chassis reinforcement on right-hand side.

40. Lower the vehicle.

41. Fill with fuel and install the fuel filler flap.

42. Connect a TECH2 scan tool, or equivalent and erase any diagnostic codes.

43. Start the engine and check for proper operation.

## IDLE SPEED

### ADJUSTMENT

Idle speed is maintained by the Powertrain Control Module (PCM). Adjustment is not possible.

## THROTTLE BODY

### REMOVAL & INSTALLATION

#### 2.0L H4 Engine

*See Figure 280.*

1. Before servicing the vehicle, refer to the Precautions Section.

2. Disconnect the negative battery cable.

3. Remove the intercooler.

4. Disconnect the connectors from the throttle position sensor, idle air control solenoid and manifold absolute pressure sensor.

5. Remove the accelerator cable.

6. Disconnect the coolant hoses from the throttle body.

7. Remove the bolts which secure the throttle body to the intake manifold, and remove the throttle body.

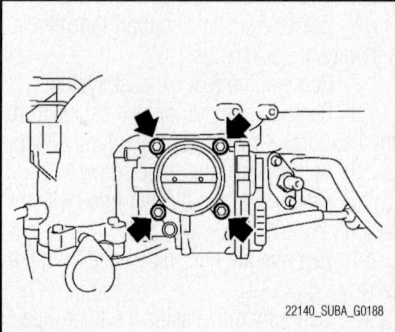

**Fig. 280 Remove the bolts which secure the throttle body**

*To install:*

8. The installation procedure is in the reverse order of removal.

9. Install the throttle body with a new gasket.

10. Tighten the mounting bolts to 15 ft. lbs. (22 Nm).

#### 2.0L I4 Engine

*See Figure 281.*

1. Before servicing the vehicle, refer to the Precautions Section.

2. Disconnect the negative battery cable.

3. Remove the upper engine cover.

4. Detach the turbocharger delivery hose from the throttle body and bend it aside.

5. Carefully bend up the bracket catch and remove the EVAP canister purge valve from the bracket.

6. Remove the bracket from the throttle body and carefully bend it aside.

7. Unplug the connector from the throttle body.

8. Remove the throttle body.

9. Remove the old seal and clean the sealing surfaces.

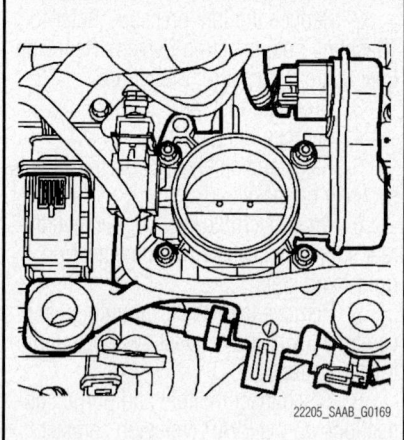

**Fig. 281 Throttle body assembly—2.0L I4 engine**

*To install:*

10. Install the throttle body with a new seal.

11. Plug in the connector.

12. Install the bracket.

13. Install the EVAP canister purge valve and bend the catch back in place.

14. Install the turbocharger delivery hose to the throttle body.

15. Install the upper engine cover.

#### 2.3L Engine

*See Figure 282.*

1. Before servicing the vehicle, refer to the Precautions Section.

2. Disconnect the negative battery cable.

3. Remove the engine protection cover.

4. Undo the coolant expansion tank cap to release any pressure and then retighten it.

5. Remove the cover on the throttle body pedal arm.

6. Pinch together the 2 coolant hoses to the throttle body.

7. Undo the charge air bypass valve hose and the two preheating hoses.

8. Detach the lower vacuum hose.

9. Undo the turbocharger delivery pipe retaining screw in the cylinder head.

10. Undo the hose clip and carefully tilt up the turbocharger delivery pipe.

11. Detach the accelerator pedal wire from the throttle body pedal arm.

12. Lift off the rubber seal and unplug the limp-home solenoid connector.

13. Unplug the throttle body 10-pin connector.

14. Undo the 3 throttle body retaining screws.

15. Lift up the throttle body and detach the hose under the limp-home solenoid.

### To install:

16. Position the throttle body and screw it on with the 3 retaining screws. Install a new seal if necessary and lubricate it with a thin coat of Vaseline (part no. 30 20 271).

17. Install the hose under the limp home solenoid and install the clip.

18. Tighten the 3 throttle body retaining screws to 6 ft. lbs. (8 Nm).

19. Spray the throttle body 10-pin connector with Kontakt 61 and plug it in.

20. Spray the limp-home solenoid connector with Kontakt 61. Plug in the connector and install the rubber seal.

**Fig. 282 Throttle body assembly—2.3L engine**

21. Fasten the accelerator pedal wire to the throttle body pedal arm.

22. Install the lower vacuum hose.

23. Install the charge air bypass valve hose and the two preheating hoses.

24. Remove the clips on the two coolant hoses to the throttle body.

25. Install the cover on the throttle body pedal arm.

➡ **The limp-home solenoid will be activated when the ignition is turned on. The limp-home mechanism must therefore be reset after turning on the ignition.**

26. Reset the limp home mode as follows:
   a. Remove the upper engine cover.
   b. Rectify the fault
   c. Attach female connector for throttle.
   d. Turn the ignition ON.
   e. Clear diagnostic trouble codes.
   f. Turn the ignition OFF.
   g. Carefully slide the end of the spring towards the throttle body using the point of a pen or similar.
   h. At the same time, rotate the black tooth disc anticlockwise with a screwdriver. A click will be heard in its end position.
   i. Turn the pedal arm with throttle cable clockwise. Make sure that the throttle arm does not follow the movement of the pedal arm. Refer to the illustration of a reset throttle.
27. Install the upper engine cover.

### 2.5L Engine

*See Figure 283.*

1. Before servicing the vehicle, refer to the Precautions Section.

2. Disconnect the negative battery cable.

3. On SOHC, remove the air intake chamber.

4. On DOHC, remove the intercooler.

5. Disconnect the connectors from the throttle position sensor, idle air control solenoid and manifold absolute pressure sensor.

6. Disconnect the coolant hoses from the throttle body.

7. Remove the bolts which secure the throttle body to the intake manifold, and remove the throttle body.

### To install:

8. The installation procedure is in the reverse order of removal.

9. Install the throttle body with a new gasket.

10. Tighten the mounting bolts to 6 ft. lbs. (8 Nm).

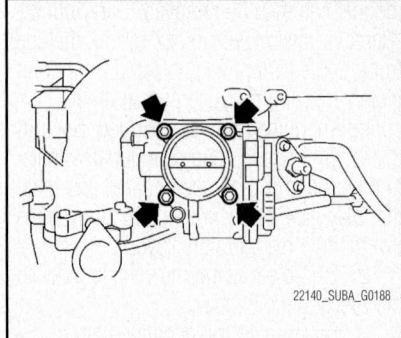

**Fig. 283 Remove the bolts which secure the throttle body**

### 2.8L Engine

*See Figure 284.*

⁕⁕ **WARNING**

**The control module is sensitive to electrostatic discharges. Never touch the pins on a control module with your hands or clothes. Ground yourself by touching the car body/engine when unplugging or plugging in the connector on the car's control module. When installing, place the old control module in the return packaging without touching its pins. Keep the new control module in its packaging as long as possible.**

1. Before removing a control module, Tech 2 must be used to separate the control module from the car. From the "All" menu, select the control module under "And/Remove". Then select "Remove" and follow the instructions. The ignition key must be in the ON position. TIS 2000 may be required. Once the control module has been separated from the car, turned the ignition key to the OFF position. The control

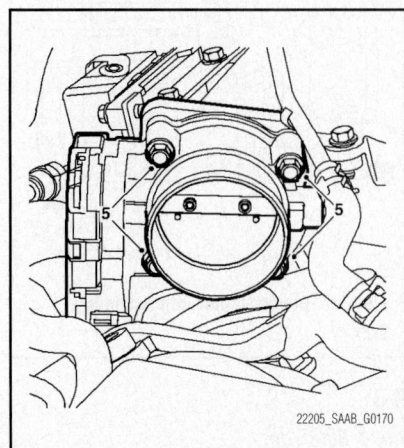

**Fig. 284 Throttle body assembly—2.8L engine**

module can then be removed. Carry out the "Remove" action with Tech2 even if the control module cannot be contacted or is completely missing because data in other control modules must be changed. Security codes will be reset, bus lists updated, data in other control modules updated and data that may have to be written into a new control module read and displayed.

2. Before servicing the vehicle, refer to the Precautions Section.

3. Disconnect the negative battery cable.

4. Remove the engine cover.

5. Remove the power steering fluid reservoir bolt. Put the reservoir aside.

6. Unplug the connector of the intake air sensor.

7. Remove the charge air pipe.

8. Remove the throttle body bolts.

9. Unplug the connector and lift away the throttle body.

**To install:**

10. Clean the sealing surfaces. Position the throttle body using a new seal. Plug in the connector.

11. Install the throttle body bolts. Tighten to 6 ft. lbs (9 Nm).

12. Replace the charge air pipe.

13. Plug in the connector of the intake air sensor (688).

14. Position the power steering fluid reservoir and install the bolt.

15. Install the engine cover.

16. If a new throttle body has been installed, connect the diagnostic tool for programming.

# HEATING & AIR CONDITIONING SYSTEM

## BLOWER MOTOR

### REMOVAL & INSTALLATION

#### 9-2x

See Figure 285.

1. Before servicing the vehicle, refer to the Precautions Section.

2. Disconnect the negative battery cable.

3. Remove the glove box.

4. Loosen the nut to remove the support beam stay.

5. Disconnect the blower motor wiring harness.

6. Disconnect the blower resistor connector.

7. Loosen the bolt and nut to remove the blower motor unit assembly.

8. Installation is the reverse order of removal.

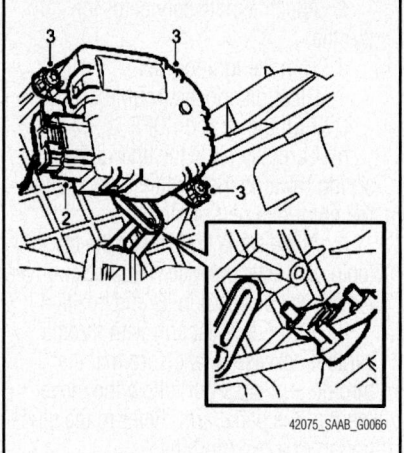
42075_SAAB_G0066

**Fig. 286 Disconnecting electrical connector (2) and removing blower motor (3)—9-3**

42075_SAAB_G0067

**Fig. 287 Locating and removing blower motor—9-5**

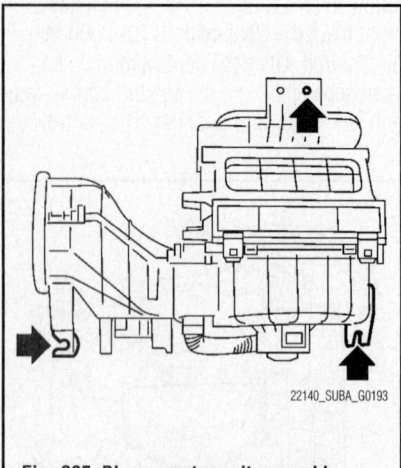
22140_SUBA_G0193

**Fig. 285 Blower motor unit assembly—9-2x**

#### 9-3

See Figure 286.

1. Before servicing the vehicle, refer to the Precautions Section.

2. Disconnect the negative battery cable.

3. Remove the glove box.

4. The blower motor is located on the fan housing's outer end.

5. Disconnect the blower motor connector by pressing in the lock and pulling out the connector.

6. Dismantle the blower motor.

7. To install, reverse removal procedure.

#### 9-5

See Figure 287.

1. Before servicing the vehicle, refer to the Precautions Section.

2. Disconnect the negative battery cable.

3. Remove the glove box.

4. Remove the blower motor and 2 bolts.

5. Unplug the connector.

6. To install, reverse removal procedure.

## HEATER CORE

### REMOVAL & INSTALLATION

#### 9-2x

See Figure 288.

1. Before servicing the vehicle, refer to the Precautions Section.

2. Disconnect the negative battery cable.

3. Discharge the air conditioning system.

4. Drain the cooling system.

5. Remove the bolts securing the expansion valve and pipe in the engine compartment. Release the heater hose clamps in the engine compartment to remove the heater hoses.

6. Remove the instrument panel.

7. Remove the support beam.

8. Remove the blower motor assembly.

9. Loosen the bolts and nuts to remove the heater and cooling unit.

**Fig. 288 Removing the heater core from the heater and cooling unit—9-2x**

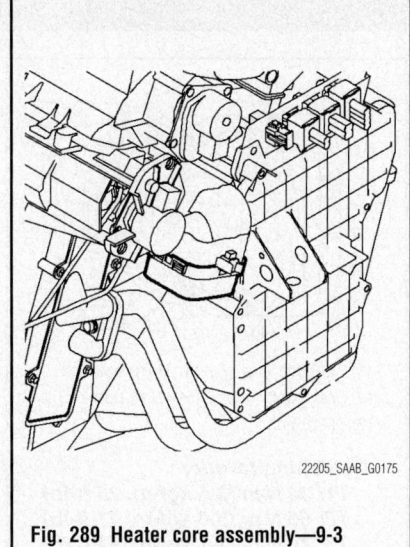

**Fig. 289 Heater core assembly—9-3**

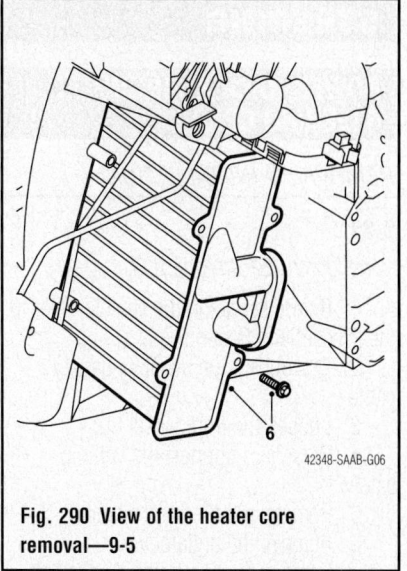

**Fig. 290 View of the heater core removal—9-5**

10. Loosen the screws to remove the heater core cover.

11. Remove the heater core.

12. Installation is the reverse order of removal.

### 9-3

*See Figure 289.*

1. Before servicing the vehicle, refer to the Precautions Section.

2. Disconnect the negative battery cable.

3. Drain the cooling system into a clean container for reuse.

4. Disconnect the hoses from the heater core fittings at the firewall by pulling up the clasp and pulling the hoses forward.

5. Blow any remaining coolant from the heater core with compressed air.

6. Remove or disconnect the following:
• Glove box
• Front side console panel on passenger side.
• Passenger sound shield.
• ACC or MCC module.
• Ashtray
• Instrument panel right side end cover
• Passenger side A-pillar trim
• Nuts holding dashboard on passenger side (through end opening)
• 4 nuts holding instrument panel (through center console opening)

7. Remove heater core cover.

8. Remove the pipe clamps from the heater core.

9. Pull the pipes out from the heater core.

10. Carefully pull of the instrument panel on the passenger side and pull out the heater core.

*To install:*

11. Place O-rings and insert the heater core into position.

12. Reposition instrument panel.

13. Install pipes into heater core and install clamps.

14. Install heater core cover.

15. Install or connect the following:
• Nuts in center of instrument panel
• Nuts on end of instrument panel
• Instrument panel end cover
• A-pillar trim
• Ashtray
• ACC or MCC module
• Passenger sound shield
• Floor console side panel
• Glove box
• Heater hoses at firewall

16. Refill the cooling system.

17. Start and run the engine at varying speeds until the cooling fan comes on. Top up the coolant, then continue to run the engine until the cooling fan comes on 3 more times.

18. Stop the engine and check coolant level.

19. Check for leaks.

### 9-5

*See Figure 290.*

1. Before servicing the vehicle, refer to the Precautions Section.

2. Disconnect the negative battery cable.

3. Drain the cooling system into a clean container for reuse.

4. Disconnect the heater hoses at the firewall in the engine compartment. Plug the ends of the fittings on the valve.

5. Blow any remaining coolant from the heater core with compressed air.

6. Remove the glove compartment retaining screws, bolt, quick-release pin and bracket catch. Pull the glove compartment out partway to disconnect the lamp; the remove the glove compartment.

7. Remove the trim from the side of the center console.

8. Remove the air vent from the air duct. Do not remove the seal.

9. Detach the pad connector for the piping to the heater core.

10. Remove the 4 heater core housing bolts.

11. Remove the heater core.

*To install:*

12. Install the heater core and 4 mounting bolts.

13. Replace and lubricate the O-rings with synthetic Vaseline. Place O-rings on heater core pipes.

14. Connect the pipes to the heater core and screw on the pad connector.

15. Install the air vent to the air duct.

16. Replace the side trim on the center console.

17. Install the glove compartment.

18. Connect the heater core hoses to the engine bulkhead connections.

19. Refill the cooling system.

20. Run engine at varying speeds until the cooling fan comes on.

21. Open the expansion tank cap and top up the coolant.

22. Let the engine run until the cooling fan has started 3 more times.

23. Stop the engine and top up the coolant.

24. Check for leaks.

## STEERING

### POWER RACK & PINION STEERING GEAR

#### REMOVAL & INSTALLATION

#### 9-2x

*See Figures 291 through 293.*

1. Before servicing the vehicle, refer to the Precautions Section.
2. Disconnect the negative battery cable.
3. Loosen the front wheel nut.
4. Raise and support the vehicle safely.
5. Remove the tires and wheels.
6. Remove the undercover.
7. Remove the sub frame. Leave bolt (1) connected by a few threads and remove the bolts in the sequence illustrated. Once the other bolts are removed, remove bolt (1) and the sub frame.
8. Remove the front exhaust pipe.
9. Remove the cotter pin and castle nut. Using a puller, remove the tie rod end from the knuckle arm.
10. Remove the jack up plate. Remove the front stabilizer.
11. Disconnect the power steering fluid pipe at the center of the gearbox and attach a vinyl hose. Discharge the fluid into a suitable container by turning the steering wheel fully clockwise and counterclockwise. Disconnect the other fluid line, and repeat the discharge procedure.
12. Remove the steering wheel.
13. Make a match mark on the universal joint. Remove the universal joint bolts and remove the joint from the vehicle.
14. Disconnect the fluid lines from the steering gear, pressure hose first.
15. Remove the steering gear retaining bolts and clamps securing the steering gear to the crossmember.

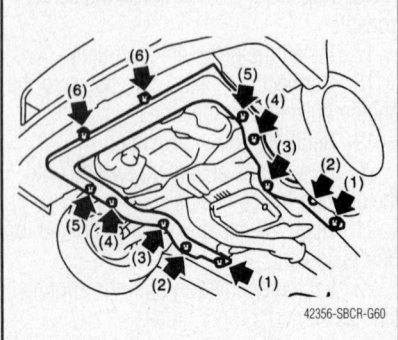

**Fig. 291 Sub frame bolt removal sequence—9-2x**

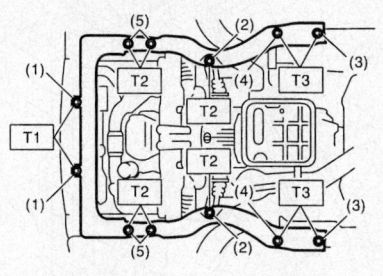

| (1) M8 bolt | (4) M10 bolt |
| (2) M12 bolt | (5) M12 bolt |
| (3) M10 bolt | |

**Tightening torque:**
**T1: 34 N·m (3.5 kgf-m, 25 ft-lb)**
**T2: 55 N·m (5.6 kgf-m, 41 ft-lb)**
**T3: 70 N·m (7.1 kgf-m, 52 ft-lb)**

09490_SBCR_G0095

**Fig. 292 Sub frame bolt tightening sequence—9-2x**

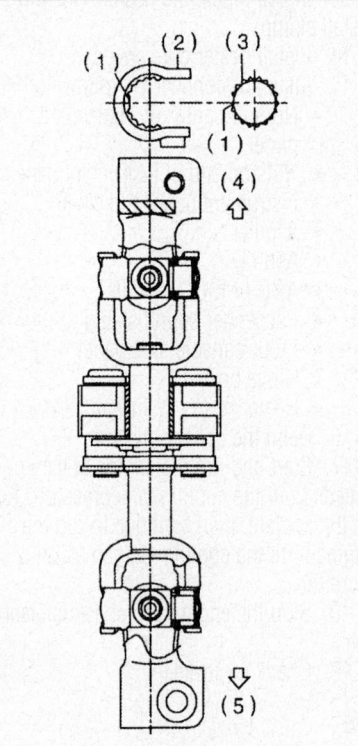

| (1) | Cutout portion |
| (2) | Yoke |
| (3) | Column shaft |
| (4) | Column shaft side |
| (5) | Gearbox side |

09490_SBCR_G0013

**Fig. 293 Steering column joint alignment—9-2x**

16. Remove the steering gear from the vehicle.

#### To install:

17. Insert the steering gear into the cross-member. Be careful not to damage the gearbox boot.
18. Tighten the steering gear to the cross-member bracket to 44.3 ft. lbs. (60 Nm).
19. Connect the fluid lines.
20. Align the cutout at the serrated section of the column shaft and yoke. Insert the universal joint into the column shaft.
21. Align the mating marks and insert the universal joint to serrated section of the steering gear assembly. Tighten the bolt to 17.4 ft. lbs. (24 Nm).
22. After adjusting toe-in and steering angle, tighten the lock nut on tie-rod end.
23. Continue the installation in the reverse order of the removal procedure.
24. When installing the sub frame be sure to torque the bolts to specification and in the proper sequence.
25. Fill the power steering system with the proper grade and type fluid.
26. Start the engine and check for leaks. Correct as required.

#### 9-3

*See Figure 294.*

1. Before servicing the vehicle, refer to the Precautions Section.
2. Clamp the return hose using suitable pinch-off pliers.
3. Position the steering wheel and the wheels so that they face straight ahead. Use woven tape to secure the steering wheel to the dashboard.
4. Raise the vehicle and remove the front wheels.
5. Remove the nuts securing the tie rod ends to the steering knuckles. Remove the track rod.
6. Remove the stabilizer bar links from the anti-roll bar.
7. Remove the steering shaft joint from the steering gear.
8. Detach the delivery line and the return line from the steering gear. Plug the lines.
9. Remove the heat shield from the steering gear.
10. Remove the steering gear bolts, nuts and washers.
11. Twist up the stabilizer bar as high as possible and lift out the steering gear through the wheel housing on the passenger side.

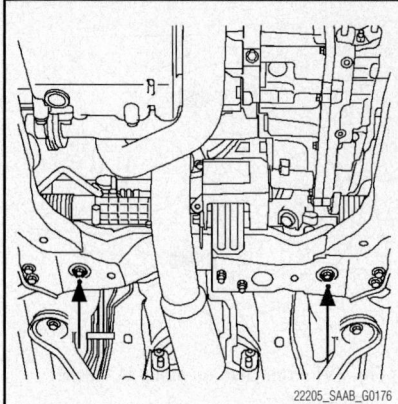

**Fig. 294 Steering gear mounting locations—9-3**

***To install:***

12. Lift the steering gear into place through the wheel housing on the passenger side.

13. Install the steering gear to the subframe and tighten mounting bolts to 37 ft. lbs. (50 Nm), plus an additional 60°.

14. Install the heat shield to the steering gear.

15. Attach the delivery line and return line to the steering gear. Install new O-rings. Tighten fittings to 21 ft. lbs. (28 Nm).

16. Twist the stabilizer bar down into position. Install the stabilizer bar links to the anti-roll bar. Hold with a thin 17mm open wrench so that the ball joint does not turn. Torque to 47 ft. lbs. (64 Nm).

17. Install the tie rod ends to the steering knuckles. Torque nuts to 26 ft. lbs. (35 Nm).

18. Install the steering shaft joint to the steering gear. Torque pinch bolt to 20 ft. lbs. (27 Nm).

19. Lower the vehicle and install the wheels.

20. Remove the hose pinch-off pliers and fill steering reservoir with power steering fluid.

21. Remove the tape used to prevent the steering wheel from moving.

22. Bleed the power steering system and check for leaks.

23. Carry out a wheel alignment .

## 9-5

*See Figure 295.*

1. Before servicing the vehicle, refer to the Precautions Section.

**✴✴ CAUTION**

**Carefully clean area around all fittings for steering gear and hoses. Immediately plug openings.**

2. Drain the power steering reservoir.

3. Remove or disconnect the following:
- Power steering reservoir and move it aside
- Return hose from the power steering reservoir (place end of hose in large container)

4. With wheels clear of floor, start the engine and turn steering wheel from lock-to-lock until flow of power steering fluid slows down into container. Turn engine off (do not allow pump to run dry).

5. For all engines, set wheels in straight-ahead position and secure steering wheel so it will not turn.

6. Remove or disconnect the following:
- Steering column shaft from the steering gear
- Intake manifold cover
- Rear engine cushion to the subframe
- Engine cushion to the engine mount
- Turbocharger bypass pipe, bend pipe aside and plug opening
- Loosen, but do not remove, exhaust pipe from flange
- Attach engine lift and hoist
- Both front wheels
- Tie rod lock nut
- Tie rod end from the steering swivel
- Tie rod end from the tie rod (count number of turns)
- Reinforcement from the rear attaching point on the subframe
- Exhaust pipe between the catalytic converter and the silencer (cut pipe)
- Exhaust pipe at rear exhaust manifold

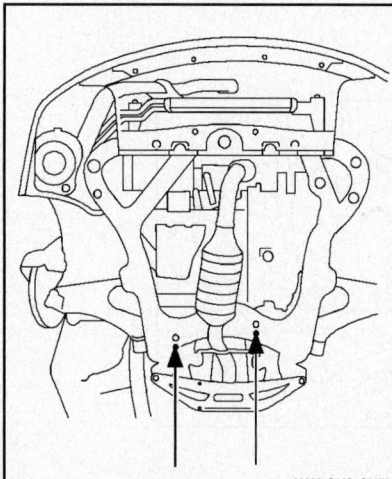

**Fig. 295 Steering gear mounting location—9-5**

- Subframe center attaching point, lower the subframe
- Engine cushion
- Delivery and return pipes from the valve body
- Steering gear retaining bolts
- Steering gear through the passenger side wheel well housing

***To install:***

7. Turn pinion shaft until rack is in center position.

8. Install or connect the following:
- Steering gear into position. Hand tighten the bolts at this time
- Valve body seal after lubricating
- Delivery and return pipes to the valve body, with new sealing rings; do not torque the bolts
- Delivery pipe to steering gear with clamp
- Steering gear bolts, with bracket on right side, but do not torque bolts
- Torque power steering pipes to steering gear to 25 ft. lbs. (30 Nm)
- Return hose and oxygen sensor cables to delivery pipe
- Torque steering gear retaining bolts to 70 ft. lbs. (95 Nm)
- Engine cushion; install bolt only snug
- Raise the subframe and torque the center attaching bolts to 75 ft. lbs. (100 Nm) plus 45°
- Exhaust system between the catalytic converter and the silencer; torque clamp bolts to 30 ft. lbs. (40 Nm)
- Subframe rear attaching points to the reinforcement and torque the bolts to 75 ft. lbs. (100 Nm) plus 45°
- Reinforcement at subframe rear attachment point; torque to 50 ft. lbs. (65 Nm)
- Tie rod ends to the track rods same number of turns as removed. Hand tighten the lock nuts
- Tie rod ends on the steering knuckles and torque the lock nuts to 45 ft. lbs. (60 Nm)
- Both front wheels
- Rear engine cushion to the subframe and torque the bolts to 20 ft. lbs. (25 Nm)
- Rear engine cushion to engine mount and torque the bolts to 35 ft. lbs. (50 Nm)
- Intake manifold cover
- Steering column shaft with the steering gear and torque the bolts to 25 ft. lbs. (30 Nm)
- Release steering wheel

- Return hose to the power steering reservoir (4-cylinder)
- Power steering fluid reservoir (4-cylinder)

9. Fill and bleed the power steering system by starting engine turning steering wheel lock-to-lock 2–3 times. Turn the engine off and top up the fluid.

10. Check the toe-in and adjust if necessary.

11. Tighten the tie rod lock nuts to 55 ft. lbs. (70 Nm).

## POWER STEERING PUMP

### REMOVAL & INSTALLATION

#### 9-2x

*See Figures 296 and 297.*

1. Before servicing the vehicle, refer to the Precautions Section.

2. Disconnect the ground cable from the battery.

3. Remove the air intake duct.

4. Remove the pulley belt cover.

5. Loosen the belt tension securing bolt and generator securing bolt, then remove the power steering pump V-belt.

6. Disconnect the connector from the power steering pressure switch.

7. Disconnect the pressure hose and suction hose from the oil pump.

#### ✳✳ CAUTION

**Prevent foreign matter from entering the hose and pipe, cover the open ends with clean cloth.**

8. Remove the installation bolt of the power steering pump bracket.

9. Place the oil pump bracket in a vise, and remove the two bolts from the front side of the oil pump.

10. Remove the bolt from the rear side of oil pump.

11. Disassemble the oil pump and bracket by inserting a flat tip screwdriver as shown in the figure.

12. Remove the oil pump.

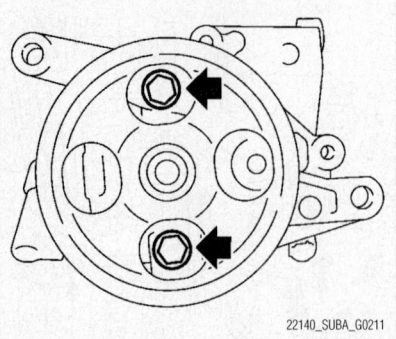

**Fig. 297 Front side oil pump to bracket bolts—9-2x**

#### To install:

13. Install the oil pump to bracket.

14. Place the oil pump bracket in a vise.

15. Tighten the bushing using a 12.7 mm (½) type, 14 and 21 mm box wrench until it is in contact with the oil pump mounting surface.

#### ✳✳ CAUTION

**When securing the oil pump bracket in a vice, hold the oil pump bracket with the least possible force between two pieces of wood.**

16. Tighten the two front bolts which hold the oil pump to the bracket to 12 ft. lbs. (16 Nm).

17. Tighten the rear pump to bracket bolt to 27.5 ft. lbs. (37 Nm).

18. Attach the installation bolts of the power steering pump bracket.

19. Connect the pressure hose and suction hose. Tighten the pressure hose eye bolt to 29.5 ft. lbs. (40 Nm).

20. Connect the power steering pressure switch connector.

21. Install the V-belts to the oil pump.

22. Check the tension of the V-belt.

23. Tighten the belt tension bolt to 18.4 ft. lbs. (25 Nm).

24. Install the pulley belt cover.

25. Install the air intake duct.

26. Connect the battery ground cable to the battery.

27. Fill with the specified power steering fluid. (ATF DEXRON III®).

#### ✳✳ CAUTION

**Never start the engine before feeding the fluid otherwise the vane pump might be seized.**

28. Bleed the power steering system.

29. Start the engine and check for leaks.

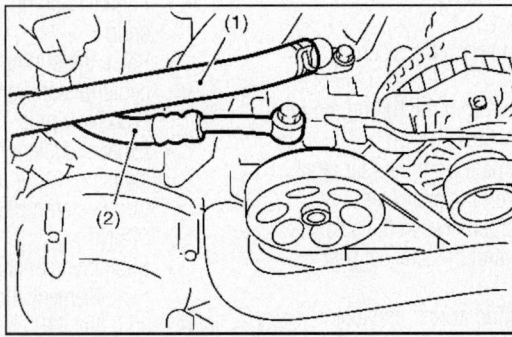

Non-turbo model

(1) Suction hose
(2) Pressure hose

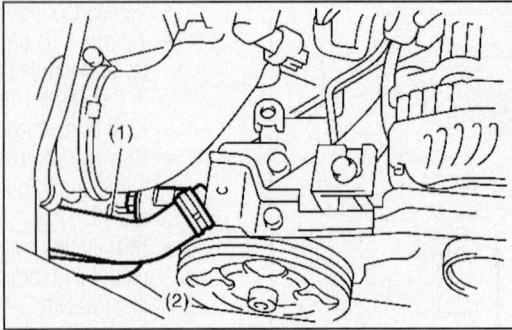

Turbo model

(1) Suction hose
(2) Pressure hose

22140_SUBA_G0210

**Fig. 296 Pressure hose and suction hose—9-2x**

### 9-3

#### 2.0L Engine

*See Figure 298.*

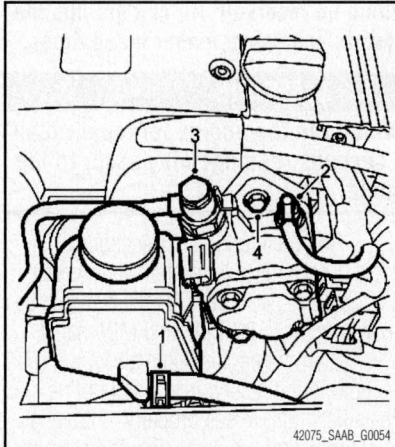

Fig. 298 Removing power steering pump and components—9-3 with 2.0L engine

1. Before servicing the vehicle, refer to the Precautions Section.
2. Disconnect the negative battery cable.
3. Remove drive belt.
4. Detach the return hose (1), wipe up any fluid spill with a cloth and plug the reservoir pipe using plugs (82 92 955). Suspend the hose so that fluid does not run out.
5. Remove the electrical connection (2) of the pressure sensor.
6. Remove the delivery pipe (3) from the pump. Wipe up any oil spill with a cloth.
7. Remove the pump (4). Some engine oil may run out.
8. Empty the fluid from the pump.

#### To install:

9. Install the pump using a new gasket. Tighten bolts to 16 ft. lbs. (22 Nm).
10. Connect the return hose.
11. To compete installation, reverse remaining removal procedure.
12. Fill up the system with fresh fluid as per specification.
13. Check the pump connections for leaks.
14. Bleed power steering system.

#### 2.8L Engine

*See Figure 299.*

1. Before servicing the vehicle, refer to the Precautions Section.
2. Disconnect the negative battery cable.
3. Remove drive belt.
4. Remove pipe, mass air flow sensor and hose from the turbo.

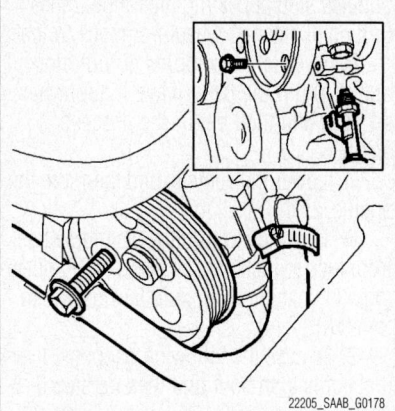

Fig. 299 The power steering pump bolt is removed through the hole in the pulley— 2.8L engine

5. Remove the upper engine cover.
6. Remove the turbo delivery pipe.
7. Raise the vehicle.
8. Remove the turbo delivery pipe and the bolt securing the turbo delivery pipe to the AC compressor.
9. Lower the vehicle to the floor.
10. Remove the turbo delivery pipe bolts on the front of the engine.
11. Remove the battery cover and radiator pipe.
12. Remove the upper radiator mounting from the radiator member.
13. Press forward the radiator and remove the turbo delivery pipe.
14. Clamp the supply hose with pinch-off pliers (30 07 730).
15. Remove the delivery hose from the power steering pump, using wrench (32 025 061)
16. Remove the supply hose from the power steering pump.
17. Remove the power steering pump, press forward the radiator and lift out the pump.

#### To install:

18. Install the power steering pump and tighten bolts to 16 ft. lbs. (22 Nm).
19. To complete installation, reverse removal procedure.
20. Tighten delivery hose to 22 ft. lbs. (30 Nm).
21. Fill up the system with fresh fluid as per specification.
22. Check the pump connections for leaks.
23. Bleed power steering system.

### 9-5

*See Figure 300.*

1. Before servicing the vehicle, refer to the Precautions Section.

2. Disconnect the negative battery cable.
3. Drain the power steering fluid reservoir, using oil suction equipment.
4. Detach the reservoir and lift it aside.
5. Carefully clean the area round the return hose's connection to the power steering fluid reservoir.
6. Detach the return hose from the power steering fluid reservoir.
7. Place the end of the hose in a receptacle that holds at least 1 liter.
8. Raise the front assembly so that the wheels are clear of the floor and start the engine so that the power steering fluid is pumped out of the steering gear. Turn the steering wheel to full right and full left lock. Switch off the engine when the flow of fluid diminishes, as the pump should not be allowed to run dry.
9. Connect the return hose to the power steering fluid reservoir and install the reservoir back in place.

➡The position of the reservoir is fixed.

10. Detach the intake hose with mass air flow sensor and turn it to one side.
11. Use a ratchet handle extension to relieve the strain on the belt tensioner and remove the belt.
12. Remove the suction hose from the pump.
13. Slacken the delivery pipe's hose clip and detach the pipe from the pump.
14. Remove the power steering pump.

#### To install:

15. Install pump, tighten bolts to 20 ft. lbs. (25 Nm).
16. Connect the delivery pipe to the pump and tighten the hose clip to 25 ft. lbs. (30 Nm).
17. Fill up the system with fresh fluid as per specification.

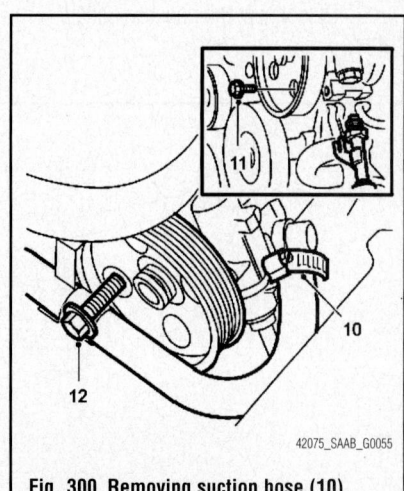

Fig. 300 Removing suction hose (10), delivery hose (11) and bolts (12)—9-5

18. Check the pump connections for leaks.

19. Bleed power steering system.

### BLEEDING

#### 9-2x

1. Fill the power steering fluid reservoir about half way with the specified fluid.

2. Continue to turn the steering wheel slowly from lock to lock until bubbles stop appearing on oil surface while keeping the fluid at that level.

3. If turning the steering wheel in low fluid level condition, air will be sucked in pipe. In this case, leave it about half an hour and then repeat the previous step.

4. Lift up the vehicle, start the engine and let it idle.

5. Continue to turn the steering wheel slowly from lock to lock again until bubbles stop appearing on oil surface while keeping the fluid at that level. It is normal that bubbles stop appearing after three times turning of steering wheel from lock to lock.

6. In case the bubbles do not stop appearing in the tank, leave it about half an hour and then begin the process again.

7. Lower the vehicle, and then idle the engine.

8. Continue to turn the steering wheel from lock to lock until bubbles stop appearing and change of the fluid level is within 3 mm (0.12 in).

9. In case the following happens, leave it about half an hour and then do step 5–8 again.

   a. The fluid level changes over 3 mm (0.12 in).

   b. Bubbles remain on the upper surface of the fluid.

   c. Grinding noise is generated from oil pump.

10. Check the fluid leakage after turning steering wheel from lock to lock with engine running.

#### 9-3 and 9-5

1. Fill up the system with fresh fluid as per specification.

➡️Do not turn the steering wheel when filling up reservoir. Air can get into the system and cause longer bleed times.

#### ❊❊ WARNING

**Do not run the engine for longer than 5 seconds, damage may occur to the pump.**

2. Run the engine for 5 seconds.

3. Check the fluid level in the reservoir and top up as necessary. At 68°, it should be midway between the MAX and MIN marks.

4. Start the engine again and run it until the fluid in the reservoir stops bubbling. Use a flashlight to see properly. If there is an abnormally loud noise, this indicates that there is still air in the system.

5. Stop the engine and check the fluid level. Top up as necessary.

## SUSPENSION

### COIL SPRING

#### REMOVAL & INSTALLATION

The coil spring Removal & Installation procedure is covered in the Strut Removal & Installation procedure.

### LOWER BALL JOINT

#### REMOVAL & INSTALLATION

#### 9-2x

See Figure 301.

1. Before servicing the vehicle, refer to the Precautions Section.

2. Disconnect the negative battery cable.

3. Raise and support the vehicle safely.

4. Remove the tire and wheel.

5. Remove the cotter pin from the ball stud.

6. Remove the castle nut.

7. Extract the ball stud from the transverse link.

8. Remove the bolt securing the ball joint to the housing.

9. Extract the ball joint from the housing.

#### To install:

10. Installation is the reverse of the removal procedure.

11. Install the ball joint to the transverse link arm and tighten to 37 ft. lbs. (50 Nm).

12. Install the castle nut and tighten to 29 ft. lbs. (39 Nm). Tighten the castle nut an

## FRONT SUSPENSION

additional 60 degrees until the slot in the castle nut is aligned with the cotter pin hole in the ball joint.

13. For Sedan turbo and STI models, tighten the castle nut to 22 ft. lbs. (30 Nm). Tighten the castle nut an additional 60 degrees until the slot in the castle nut is aligned with the cotter pin hole in the ball joint.

14. Check and adjust alignment , as required.

15. Install the front wheel.

#### 9-3

See Figure 302.

➡️The ball joint cannot be removed from the transverse link. To replace the ball joint, the transverse link must be replaced.

1. Before servicing the vehicle, refer to the Precautions Section.

2. Raise the vehicle.

3. Remove or disconnect the following:
   • Wheel
   • Sway bar link bolt
   • Ball joint out of the steering knuckle
   • Retaining nut at the support arm
   • Retaining bolt at the subframe

#### To install:

4. Install or connect the following:
   • Arm and bolt at the subframe and torque the retaining bolt at the subframe to 85 ft. lbs. (115 Nm)

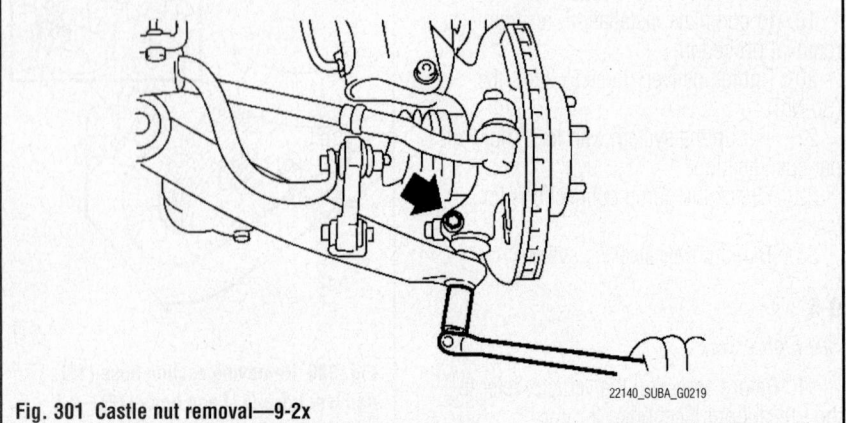

**Fig. 301 Castle nut removal—9-2x**

22140_SUBA_G0219

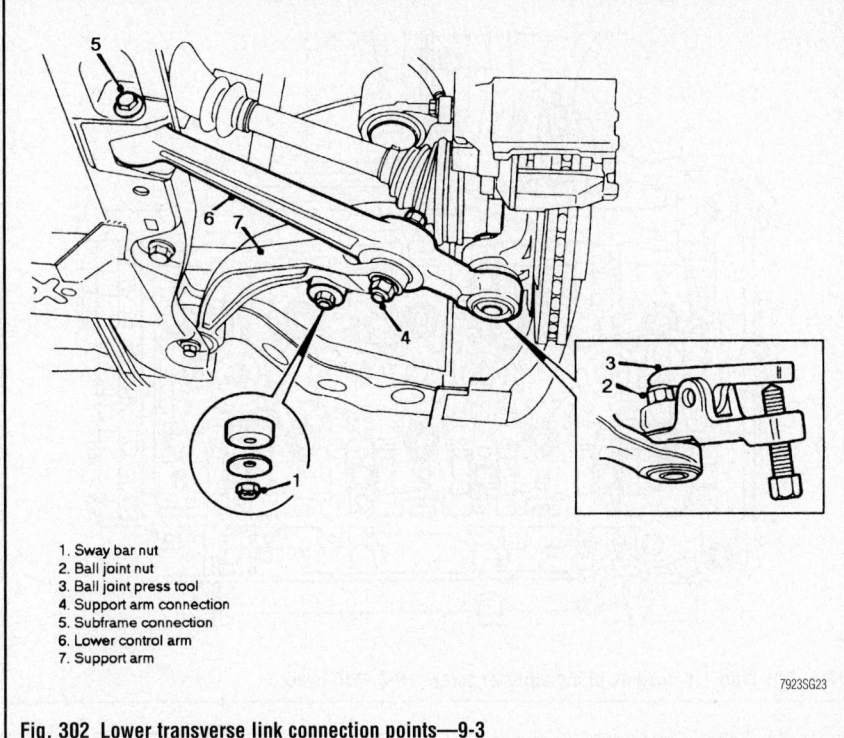

1. Sway bar nut
2. Ball joint nut
3. Ball joint press tool
4. Support arm connection
5. Subframe connection
6. Lower control arm
7. Support arm

7923SG23

**Fig. 302 Lower transverse link connection points—9-3**

- Bolt at the support arm and torque the bolt to 68 ft. lbs. (92 Nm)
- Sway bar link and torque the link to 89 inch lbs. (10 Nm)
- Ball joint and torque to 55 ft. lbs. (75 Nm)
- Wheel

5. Check front wheel Alignment .

**9-5**

See Figure 303.

1. Before servicing the vehicle, refer to the Precautions Section.

2. Remove or disconnect the following:

3. Remove the rear suspension arm mounting from the subframe.

4. Remove the front suspension arm mounting from the subframe.

5. Remove the ball joint from the knuckle.

6. Remove the ball joint from the suspension arm.

7. Press out the ball joint with tool 89 96 761

**To install:**

8. Press in the ball joint with tool 89 96 761

9. Install the rear bushing in line with the suspension arm. Tighten to 77 ft. lbs. (105 Nm).

10. Install the ball joint to the suspension arm. Tighten to 23 ft. lbs. (60 Nm) plus 30°.

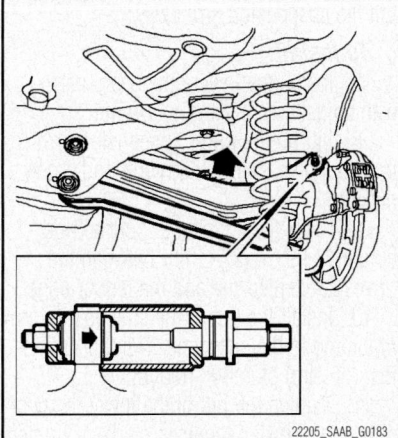

22205_SAAB_G0183

**Fig. 303 Pressing out the ball joint using the special tool—9-5**

11. Install the ball joint to the knuckle. Make sure that the journal on the ball joint sticks up above the knuckle mounting.

12. Position the suspension arm against the subframe and insert the bolts in the front and rear attachment points.

13. Raise the suspension arm with a jack to the same level as to when the car is on its wheels.

14. Tighten the bolts securing the front suspension mounting to the subframe. Tighten to 81 ft. lbs. (110 Nm) plus 90°.

15. Tighten the bolts securing the rear suspension mounting to the subframe to 88 ft. lbs. (120 Nm) plus 90°.

16. Tighten the bolt securing the ball joint to the knuckle to 36 ft. lbs. (49 Nm) plus 90°.

17. Install the wheel.

18. Lower the vehicle.

19. Check the wheel alignment and adjust if necessary.

## LOWER CONTROL ARM

*REMOVAL & INSTALLATION*

**9-3**

See Figure 304.

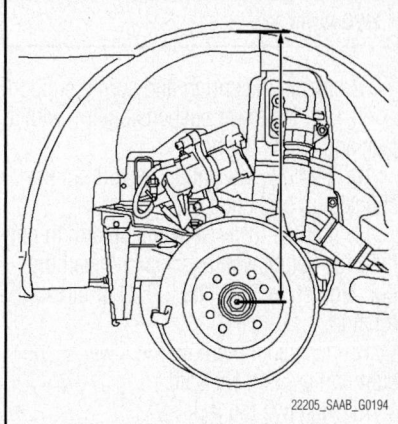

22205_SAAB_G0194

**Fig. 304 Measuring the normal ride height between the between the driveshaft center and the fender edge.**

### *2WD Models Lower*

See Figure 305.

1. Before servicing the vehicle, refer to the Precautions Section.

2. Raise and safely support the vehicle securely on jackstands. Remove the wheel.

3. On vehicles equipped with Xenon headlights, when working on the left (driver's) side suspension arm remove the level sensor.

4. Relieve the weight on the lower suspension arm with a jack and remove the bolt.

### ☀☀ **WARNING**

**The bolt must be screwed out carefully or there is risk that the hole will be reamed by the threads.**

5. Pull the suspension arm down and remove the spring and the spring support.

6. Mark the position of the adjustment bolt. Remove the adjustment bolt and nut. Lift the link arm away.

**To install:**

7. Install the link arm with the adjustment bolt and install a new nut.

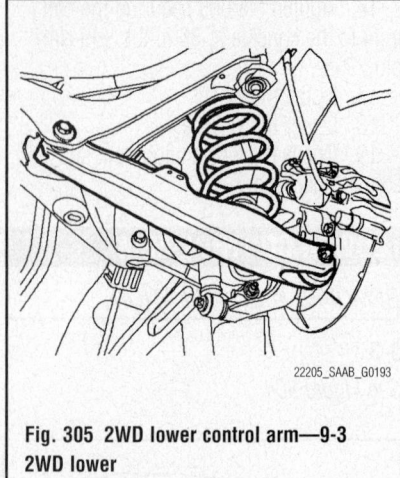

**Fig. 305 2WD lower control arm—9-3 2WD lower**

8. Install the spring and spring support.

9. Lift the lower suspension arm with a jack and install the bolt.

10. Remove the stabilizer bar from the knuckle on each side.

11. Lift the lower suspension arm to normal ride height with a jack and install the nut. Tighten to 55 ft. lbs. (75 Nm) plus 60° rotation.

12. Tighten the nut on the lower suspension arm adjustment bolt.

13. Remove the jack.

14. Install the stabilizer bar to the knuckle. Tighten to 39 ft. lbs. (53 Nm)

15. On vehicles equipped with Xenon headlights, when working on the left (driver's) side suspension arm install the level sensor.

16. Install the wheel and lower the vehicle.

17. Perform a four wheel alignment.

18. On vehicles equipped with Xenon headlights, perform an light alignment/calibration.

### 4WD Models Lower

*See Figure 306.*

1. Before servicing the vehicle, refer to the Precautions Section.

2. Raise and safely support the vehicle securely on jackstands. Remove the wheel.

3. Remove the spring.

4. On vehicles equipped with Xenon headlights, when working on the left (driver's) side suspension arm remove the level sensor.

5. Remove the lower suspension arm mounting from the knuckle.

### ✳✳ WARNING

**The bolt must be screwed out carefully or there is risk that the hole will be reamed by the threads.**

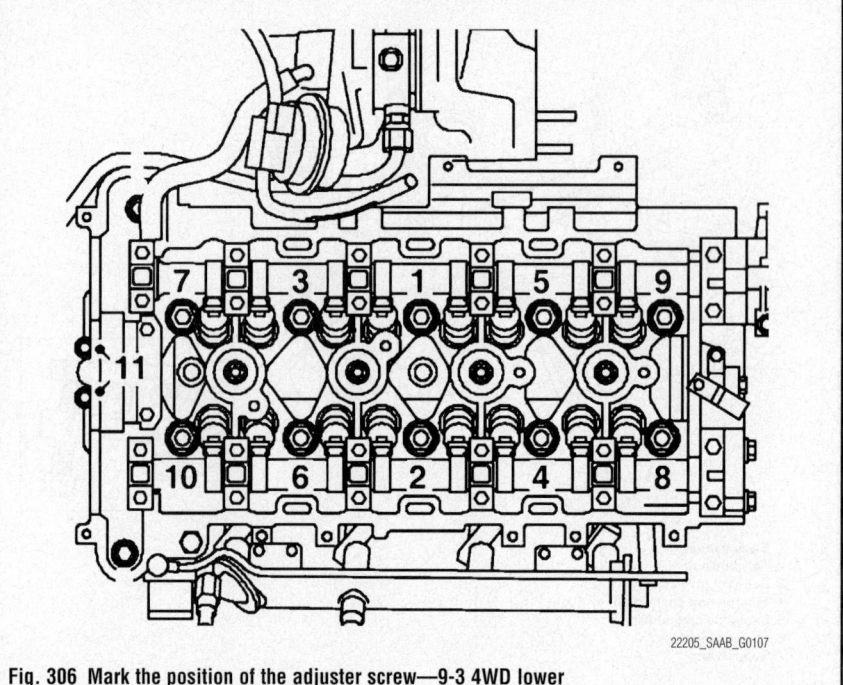

**Fig. 306 Mark the position of the adjuster screw—9-3 4WD lower**

6. Mark the position of the adjuster screw. Remove the adjuster screw and nut. Lift the suspension arm away.

### To install:

7. Install the lower suspension arm with adjuster screw and a new nut.

8. Lift up the lower suspension arm toward the knuckle. Fit the bolt and a new nut.

9. Raise the knuckle to normal ride height 14.59 in. (374 mm) between the driveshaft center line and the fender edge.

10. Install the nut of the suspension arm mounting to the subframe. Tighten to 55 ft. lbs. (75 Nm) plus 90° rotation.

11. Tighten the nut of the lower suspension arm adjuster screw as marked. Tighten to 55 ft. lbs. (75 Nm) plus 60° rotation.

12. Install the coil spring

13. On vehicles equipped with Xenon headlights, when working on the left (driver's) side suspension arm install the level sensor.

14. Install the wheel and lower the vehicle.

15. Perform a four wheel alignment.

### MACPHERSON STRUT

*REMOVAL & INSTALLATION*

#### 9-2x

*See Figure 307.*

1. Remove the strut cap on the quarter trim.

2. Loosen the rear wheel lug nuts.

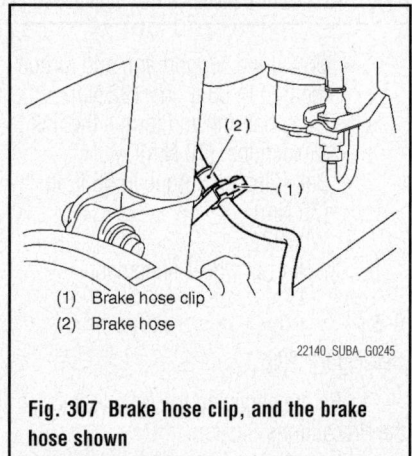

(1) Brake hose clip
(2) Brake hose

**Fig. 307 Brake hose clip, and the brake hose shown**

3. Raise and support the vehicle safely.

4. Remove the tire and wheel.

5. Remove the brake hose clip, and then remove the brake hose from the rear strut.

6. Remove the bolts that retain the strut to the housing.

7. Remove the nuts retaining the strut to the body.

8. Remove the strut from the vehicle.

### To install:

9. Installation is the reverse of the removal procedure.

10. Be sure to use new locknuts, as required.

11. Do not subject the ABS wheel speed sensor to excessive tension.

12. Check and adjust the wheel alignment, as necessary.

## OVERHAUL

1. Remove the strut from the vehicle.

2. Using a coil spring compressor tool, carefully compress the spring. Remove the self locking nut.

3. Remove the strut mount, upper spring and rubber seat from the strut.

4. Gradually decrease the compression force of the spring compressor tool. Remove the coil spring.

5. Remove the dust cover and helper spring.

6. Check for the presence of air in the damping force generating mechanism.

7. Using the spring compression tool, compress the coil spring.

➡**Be sure to properly install the coil spring.**

8. Position the coil spring so that its end face fits good into the spring seat.

9. Install the helper spring and dust cover to the piston rod.

10. Pull the piston rod fully upward, and install the rubber seat and spring seat.

11. Install the strut mount to the piston rod, and then tighten the self locking nut, temporarily. Be sure to use a new self locking nut.

12. Use a hexagon wrench to prevent the strut rod from turning. Tighten the self locking nut.

13. Carefully loosen the coil spring.

## STABILIZER BAR

### REMOVAL & INSTALLATION

#### 9-2x

*See Figure 308.*

1. Before servicing the vehicle, refer to the Precautions Section.

2. Loosen the rear wheel lug nuts.

3. Raise and support the vehicle safely.

4. Remove the tire and wheel.

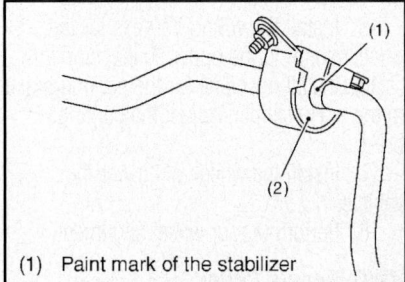

(1)  Paint mark of the stabilizer

(2)  Stabilizer bushing identification color

22140_SUBA_G0246

**Fig. 308 Stabilizer bar and bushing identification markings**

5. Remove the bolts that secure the stabilizer link to the rear arm.

6. Remove the bolts which secure the stabilizer bar to the sub frame.

7. Remove the stabilizer bar from the vehicle.

#### *To install:*

8. Installation is the reverse of the removal procedure.

9. Be sure that the stabilizer bar and the bushings have the same identification markings and/or colors.

10. Be sure to use new bolts and nuts, as required.

11. Always fully tighten the rubber bushings when the wheels are in full contact with the ground and the vehicle is at curb height.

12. Check and adjust the wheel alignment, as necessary.

### 9-3

#### *2WD Models*

*See Figure 309.*

1. Before servicing the vehicle, refer to the Precautions Section.

2. Raise and safely support the vehicle securely on jackstands. Remove the rear wheels.

3. On vehicle equipped with tire pressure monitoring: Remove the rear RH wheel well. Unplug the connector of the signal detector, release the wheel housing clips and fold down the wiring harness.

4. On convertible models, remove the chassis reinforcement from the rear subframe.

5. Cut the front pipe between the flexhose and the silencer, 87 mm above the silencer. Use 83 95 667 Pipe cutter/exhaust system.

6. Lift down the rear section of the exhaust system.

7. Remove the retaining spring from the brake caliper.

8. Remove the protective covers.

9. Remove the brake caliper and suspend it with a hook in the brake pipe holder.

➡**Make sure the brake pipe is not damaged.**

10. Remove the outer brake pad.

11. Remove the brake caliper.

12. Relieve the weight on the shocks on both sides.

13. Clean the bolts holding the shocks to the steering knuckle and lubricate the threads. Remove the bolts.

14. Open the protective case and unplug the connection for the electrical circuit.

15. Place a pillar jack under the centre of the subframe.

16. Remove the subframe bolts from the body

17. Lower the subframe and lift the springs away.

➡**Do not lower the lower edge of the subframe more than 7.8 in. (200 mm).**

18. Remove the stabilizer bar from the steering knuckles.

19. Remove the stabilizer bar mountings from the subframe.

20. Lift out the stabilizer bar towards the rear between the subframe and the body

#### *To install:*

21. Lift the stabilizer bar into place between the subframe and the body.

22. Install the bushings and caps on the subframe.

23. Install the stabilizer bar to the subframe. Tighten the 8·8 (flange diameter 15.2 mm) bolt to 13 ft. lbs. (18 Nm) using Loctite® 242. Tighten the 10·9 (flange diameter 16.7 mm) bolt to 23 ft. lbs. (31 Nm).

24. Install the stabilizer bar link to the steering knuckle on both sides. Tighten to 39 ft. lbs. (53 Nm).

25. Install the spring supports on the springs. Place the springs on the lower suspension arms.

26. Raise the subframe with a pillar jack, pushing it forward slightly.

➡**Make sure the dampers are positioned in front of the anti-roll bar.**

27. Install the subframe to the body. Tighten to 55 ft. lbs. (75 Nm) plus 135° rotation.

28. Remove the jack.

29. Connect the wiring harness, plug in the connector and close the protective case.

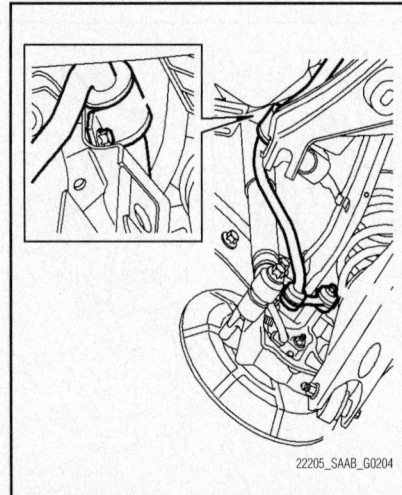

22205_SAAB_G0204

**Fig. 309 Stabilizer bar mounting points— 9-3**

30. Lift the steering knuckles and install the dampers on both sides. Tighten to 110 ft. lbs. (150 Nm)

31. Remove the inner brake pad. Screw in the brake piston with 89 96 969 Resetting Tool and 89 96 977 Adapter.

32. Install the brake pads.

33. Install the brake caliper.

34. Install the protective covers.

35. Install the retaining spring to the hydraulic body.

36. Clean the exhaust pipe joints and fittings. Fit the pipes with joint clamps. Tighten to 30 ft. lbs. (40 Nm).

37. On convertible models, install the chassis reinforcement to the rear subframe.

38. On vehicle equipped with tire pressure monitoring, install the wiring harness and plug in the connector for the signal detector. Fit the wheel well.

39. Install the rear wheels and lower the vehicle.

40. Depress the brake pedal several times to press out the self adjustment of the brake pistons and the parking brake.

### 4WD Models

*See Figure 310.*

1. Before servicing the vehicle, refer to the Precautions Section.

2. Raise and safely support the vehicle securely on jackstands. Remove the rear wheels.

3. Remove the bolts of the anti-roll bar.

4. Lift down the anti-roll bar.

### To install:

5. Installation is the reverse of the removal procedure.

6. Tighten stabilizer bar bolts to 23 ft. lbs. (30 Nm).

7. Lower the vehicle.

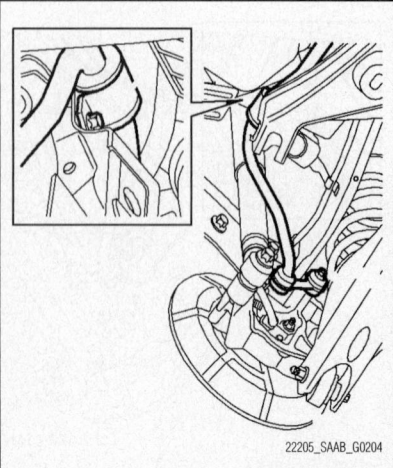

**Fig. 310 Stabilizer bar mounting points—9-3**

### 9-5
*See Figure 311.*

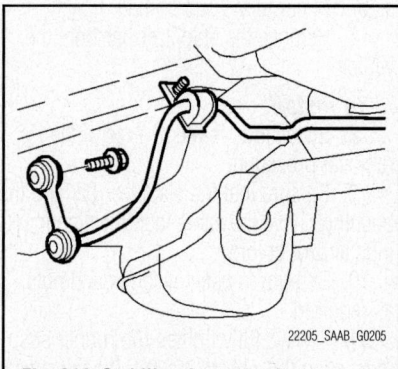

**Fig. 311 Stabilizer bar mounting points—9-5**

1. Before servicing the vehicle, refer to the Precautions Section.

2. Raise and safely support the vehicle securely on jackstands. Remove the rear wheels.

3. Remove the stabilizer bar retaining nuts and bolts.

4. Remove the anti-roll bar.

5. Tap out the clamp's screw.

6. Dismantle the clamp and remove the bushing.

### To install:

➡ **If necessary, lubricate the stabilizer bar bushings with Molykote 33 (bushings on new anti-roll bars are already lubricated), position the stabilizer bar and install all the bolts and nuts.**

7. Install the bushing in place and mount the clamp.

8. Tap in the clamp's bolt.

9. Tighten the nuts and bolts to 40 ft. lbs. (50 Nm).

10. Install the wheels and lower the vehicle.

## UPPER CONTROL ARM

### REMOVAL & INSTALLATION

### 9-3
*See Figure 312.*

### 2WD Models Upper

1. Before servicing the vehicle, refer to the Precautions Section.

2. Slacken the parking brake lever adjustment slightly.

3. Measure the normal ride height between the between the driveshaft center and the fender edge.

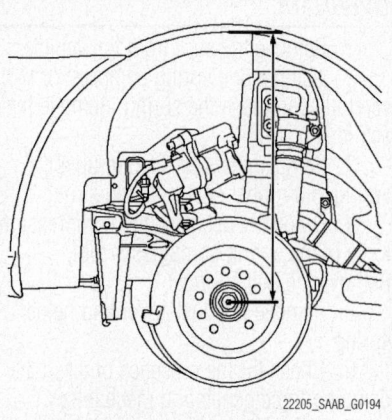

**Fig. 312 Measuring the normal ride height between the between the driveshaft center and the fender edge.**

4. Raise and safely support the vehicle securely on jackstands. Remove the wheel.

5. Remove the parking brake cable from the brake caliper.

6. Remove the wiring harness for the wheel sensor from the upper suspension arm.

7. Remove the upper suspension arm mounting from the subframe.

8. Lift the lower suspension arm with a jack in order to relieve the upper suspension arm mounting to the knuckle from bearing weight.

9. Remove the upper suspension arm mounting from the knuckle.

10. Remove the upper suspension arm.

### To install:

11. Install the upper suspension arm into place.

12. Install the bolts for the upper suspension arm mounting to the subframe.

13. Install the upper suspension arm to the swivel member. Tighten to 92 ft. lbs. (125 Nm) plus 135° rotation.

14. Lift the upper suspension arm to normal ride height and Install the nut to the suspension arm mounting to the subframe. Tighten to 55 ft. lbs. (75 Nm) plus 90° rotation.

15. Install the wiring harness for the wheel sensor to the upper suspension arm.

16. Install the cable for the parking brake to the brake caliper. Adjust the parking brake.

17. Install the wheel and lower the vehicle.

18. Perform a four wheel alignment.

### 4WD Models Upper

1. Before servicing the vehicle, refer to the Precautions Section.

2. Raise and safely support the vehicle securely on jackstands. Remove the wheel.

3. Remove the coil spring

4. Remove the wheel sensor wiring harness from the upper suspension arm.

5. Remove the upper suspension arm mounting from the subframe.

6. Remove the upper suspension arm mounting from the knuckle. Remove the ball joint from the knuckle using 87 91 287 Puller, 150 mm.

7. Remove the upper suspension arm.

**To install:**

8. Install the upper suspension arm into place.

9. Install the bolts for the upper suspension arm mounting to the subframe.

10. Install the upper suspension arm mounting (A) to the knuckle. Tighten to 44 ft. lbs. (60 Nm) plus 60° rotation.

11. Raise the knuckle to normal ride height 14.59 in. (374 mm) between the driveshaft center line and the fender edge.

12. Install the nut of the suspension arm mounting to the subframe. Tighten to 55 ft. lbs. (75 Nm) plus 90° rotation.

13. Install the wheel sensor wiring harness to the upper suspension arm.

14. Install the coil spring

15. Install the wheel and lower the vehicle.

16. Perform a four wheel alignment.

## SUSPENSION                                    REAR SUSPENSION

### COIL SPRING

*REMOVAL & INSTALLATION*

**9-3**

*See Figure 313.*

1. Before servicing the vehicle, refer to the Precautions Section.

2. Raise the vehicle and safely support it on jackstands. Do not place the stands under the rear axle assembly.

3. Remove the wheel.

4. Remove the spring from the brake caliper.

5. Remove the protective covers.

6. Remove and move aside the brake caliper.

7. On 4WD models, remove the bushing bolts and move aside the anti-roll bar.

8. Remove the outer brake pad.

9. Compress the spring and lift out the

spring. Remove the spring from the compression tool.

**To install:**

10. Compress the spring and place the spring support in the spring.

11. Lift the spring into position. Remove the spring compressor.

12. Remove the inner brake pad and screw in the brake piston using 89 96 969 Resetting Tool and 89 96 977 Adapter.

13. Install the brake pads.

14. Make sure the springs on the inner pad rest in the piston groove.

15. Install the brake caliper. Torque the bolts to 21 ft. lbs. (28 Nm).

16. Install the protective covers.

17. Install the spring to the brake caliper.

18. On 4WD models, install anti-roll bar. Tighten to 23 ft. lbs. (30 Nm).

19. Install the rear wheel and lower the vehicle

20. Depress the brake pedal several times to press out the self adjustment of the brake pistons and the parking brake.

21. On 4WD vehicles equipped with Xenon headlights, calibrate the headlight control unit and align the headlights.

**9-5**

*See Figure 314.*

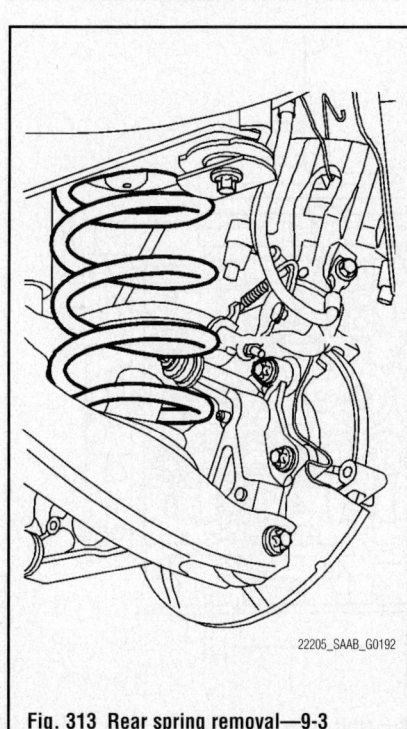

22205_SAAB_G0192

**Fig. 313 Rear spring removal—9-3**

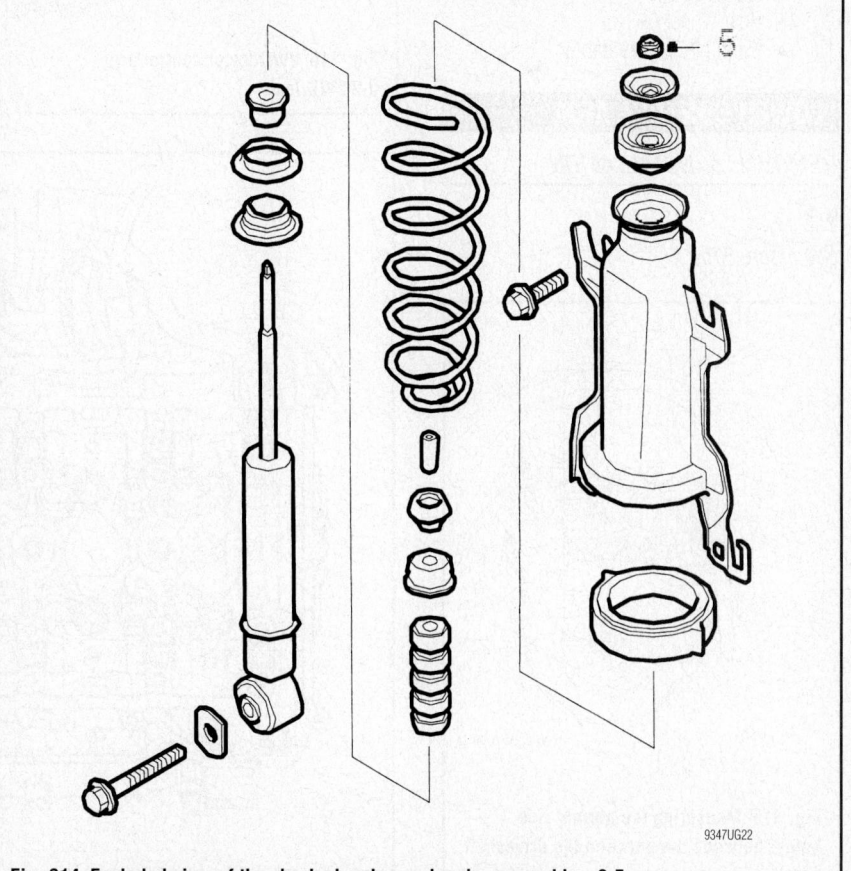

9347UG22

**Fig. 314 Exploded view of the shock absorber and spring assembly—9-5**

1. Before servicing the vehicle, refer to the Precautions Section.

2. Remove or disconnect the following:

- Negative battery cable
- Rear wheel
- Spring bracket lower bolts and loosen the upper bolts
- Lower shock absorber retaining bolt
- Shock absorber and spring assembly and loosen the lock nut

3. Press the spring bracket down to relieve the load. Remove the center nut, washer and bushing. If necessary using a spring compressor tool 88 18 791.

4. Remove the shock absorber.

### To install:

5. Install or connect the following:

- Spring, spacer ring and bracket on the shock absorber
- Press the bracket down to relieve the shock absorber load and install the bushing and washer
- New locknut and torque the nut to 15 ft. lbs. (20 Nm)
- Shock absorber and torque the bolts to 40 ft. lbs. (55 Nm)
- Lower mounting bolt on the rear axle and torque the bolt to 140 ft. lbs. (190 Nm)
- Rear wheel
- Negative battery cable

## LOWER CONTROL ARM

### REMOVAL & INSTALLATION

#### 9-3

See Figure 315.

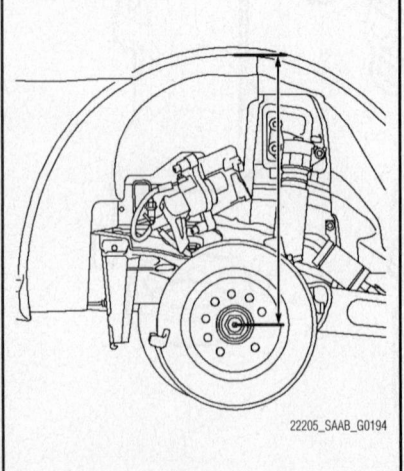

**Fig. 315 Measuring the normal ride height between the between the driveshaft center and the fender edge.**

### 2WD Models Lower

See Figure 316.

1. Before servicing the vehicle, refer to the Precautions Section.

2. Raise and safely support the vehicle securely on jackstands. Remove the wheel.

3. On vehicles equipped with Xenon headlights, when working on the left (driver's) side suspension arm remove the level sensor.

4. Relieve the weight on the lower suspension arm with a jack and remove the bolt.

### ✳✳ WARNING

**The bolt must be screwed out carefully or there is risk that the hole will be reamed by the threads.**

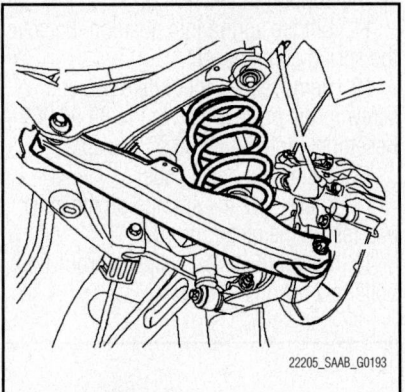

**Fig. 316 2WD lower control arm— 9-3 2WD lower**

5. Pull the suspension arm down and remove the spring and the spring support.

6. Mark the position of the adjustment bolt. Remove the adjustment bolt and nut. Lift the link arm away.

### To install:

7. Install the link arm with the adjustment bolt and install a new nut.

8. Install the spring and spring support.

9. Lift the lower suspension arm with a jack and install the bolt.

10. Remove the stabilizer bar from the knuckle on each side.

11. Lift the lower suspension arm to normal ride height with a jack and install the nut. Tighten to 55 ft. lbs. (75 Nm) plus 60° rotation.

12. Tighten the nut on the lower suspension arm adjustment bolt.

13. Remove the jack.

14. Install the stabilizer bar to the knuckle. Tighten to 39 ft. lbs. (53 Nm)

15. On vehicles equipped with Xenon headlights, when working on the left (driver's) side suspension arm install the level sensor.

16. Install the wheel and lower the vehicle.

17. Perform a four wheel alignment.

18. On vehicles equipped with Xenon headlights, perform an light alignment/calibration.

### 4WD Models Lower

See Figure 317.

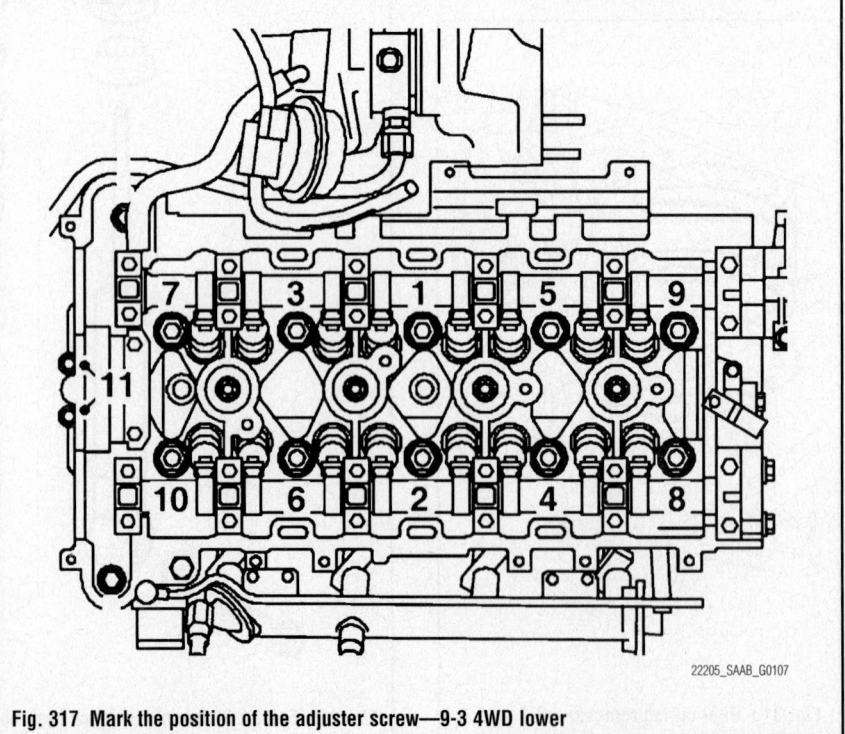

**Fig. 317 Mark the position of the adjuster screw—9-3 4WD lower**

1. Before servicing the vehicle, refer to the Precautions Section.

2. Raise and safely support the vehicle securely on jackstands. Remove the wheel.

3. Remove the spring.

4. On vehicles equipped with Xenon headlights, when working on the left (driver's) side suspension arm remove the level sensor.

5. Remove the lower suspension arm mounting from the knuckle.

### ✳✳ WARNING

**The bolt must be screwed out carefully or there is risk that the hole will be reamed by the threads.**

6. Mark the position of the adjuster screw. Remove the adjuster screw and nut. Lift the suspension arm away.

### To install:

7. Install the lower suspension arm with adjuster screw and a new nut.

8. Lift up the lower suspension arm toward the knuckle. Fit the bolt and a new nut.

9. Raise the knuckle to normal ride height 14.59 in. (374 mm) between the driveshaft center line and the fender edge.

10. Install the nut of the suspension arm mounting to the subframe. Tighten to 55 ft. lbs. (75 Nm) plus 90° rotation.

11. Tighten the nut of the lower suspension arm adjuster screw as marked. Tighten to 55 ft. lbs. (75 Nm) plus 60° rotation.

12. Install the coil spring

13. On vehicles equipped with Xenon headlights, when working on the left (driver's) side suspension arm install the level sensor.

14. Install the wheel and lower the vehicle.

### STABILIZER BAR

#### REMOVAL & INSTALLATION

#### 9-2x

*See Figure 318.*

1. Before servicing the vehicle, refer to the Precautions Section.

2. Loosen the rear wheel lug nuts.

3. Raise and support the vehicle safely.

4. Remove the tire and wheel.

5. Remove the bolts that secure the stabilizer link to the rear arm.

6. Remove the bolts which secure the stabilizer bar to the sub frame.

7. Remove the stabilizer bar from the vehicle.

### To install:

8. Installation is the reverse of the removal procedure.

9. Be sure that the stabilizer bar and the bushings have the same identification markings and/or colors.

10. Be sure to use new bolts and nuts, as required.

11. Always fully tighten the rubber bushings when the wheels are in full contact with the ground and the vehicle is at curb height.

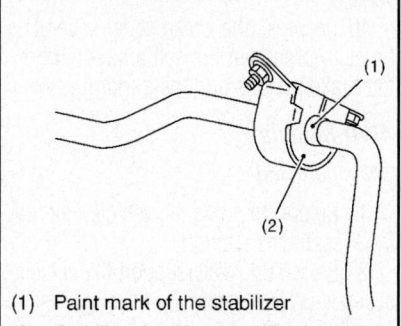

(1) Paint mark of the stabilizer
(2) Stabilizer bushing identification color

22140_SUBA_G0246

**Fig. 318 Stabilizer bar and bushing identification markings**

12. Check and adjust the wheel alignment, as necessary.

#### 9-3

#### *2WD Models*

*See Figure 319.*

1. Before servicing the vehicle, refer to the Precautions Section.

2. Raise and safely support the vehicle securely on jackstands. Remove the rear wheels.

3. On vehicle equipped with tire pressure monitoring: Remove the rear RH wheel well. Unplug the connector of the signal detector, release the wheel housing clips and fold down the wiring harness.

4. On convertible models, remove the chassis reinforcement from the rear subframe.

5. Cut the front pipe between the flexhose and the silencer, 87 mm above the silencer. Use 83 95 667 Pipe cutter/exhaust system.

6. Lift down the rear section of the exhaust system.

7. Remove the retaining spring from the brake caliper.

8. Remove the protective covers.

9. Remove the brake caliper and suspend it with a hook in the brake pipe holder.

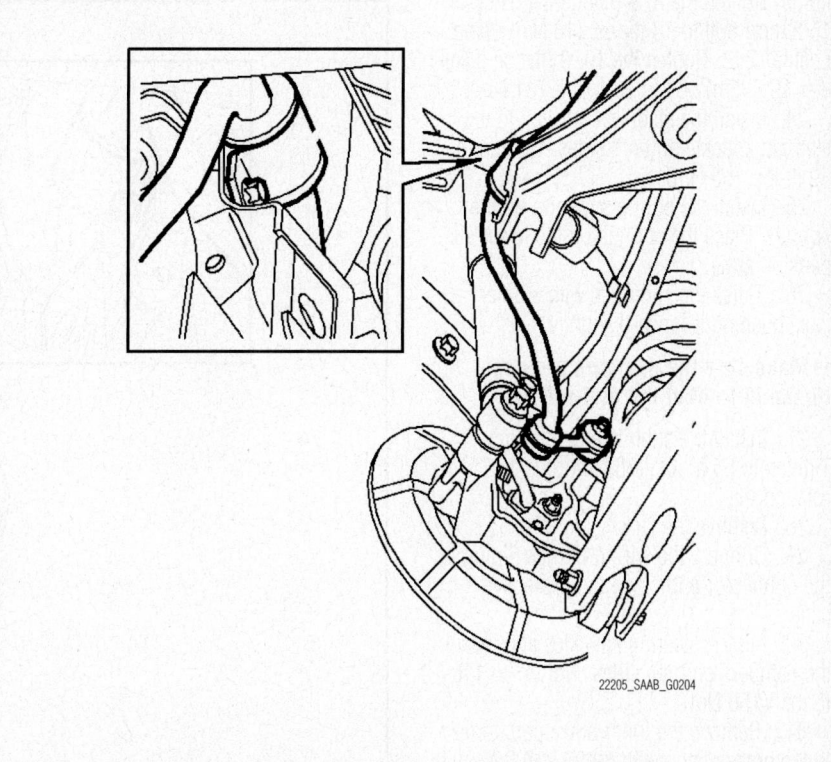

22205_SAAB_G0204

**Fig. 319 Stabilizer bar mounting points—9-3**

➡**Make sure the brake pipe is not damaged.**

10. Remove the outer brake pad.

11. Remove the brake caliper.

12. Relieve the weight on the shocks on both sides.

13. Clean the bolts holding the shocks to the steering knuckle and lubricate the threads. Remove the bolts.

14. Open the protective case and unplug the connection for the electrical circuit.

15. Place a pillar jack under the centre of the subframe.

16. Remove the subframe bolts from the body

17. Lower the subframe and lift the springs away.

➡**Do not lower the lower edge of the subframe more than 7.8 in. (200 mm).**

18. Remove the stabilizer bar from the steering knuckles.

19. Remove the stabilizer bar mountings from the subframe.

20. Lift out the stabilizer bar towards the rear between the subframe and the body

### To install:

21. Lift the stabilizer bar into place between the subframe and the body.

22. Install the bushings and caps on the subframe.

23. Install the stabilizer bar to the subframe. Tighten the 8·8 (flange diameter 15.2 mm) bolt to 13 ft. lbs. (18 Nm) using Loctite® 242. Tighten the 10·9 (flange diameter 16.7 mm) bolt to 23 ft. lbs. (31 Nm).

24. Install the stabilizer bar link to the steering knuckle on both sides. Tighten to 39 ft. lbs. (53 Nm).

25. Install the spring supports on the springs. Place the springs on the lower suspension arms.

26. Raise the subframe with a pillar jack, pushing it forward slightly.

➡**Make sure the dampers are positioned in front of the anti-roll bar.**

27. Install the subframe to the body. Tighten to 55 ft. lbs. (75 Nm) plus 135° rotation.

28. Remove the jack.

29. Connect the wiring harness, plug in the connector and close the protective case.

30. Lift the steering knuckles and install the dampers on both sides. Tighten to 110 ft. lbs. (150 Nm)

31. Remove the inner brake pad. Screw in the brake piston with 89 96 969 Resetting Tool and 89 96 977 Adapter.

32. Install the brake pads.

33. Install the brake caliper.

34. Install the protective covers.

35. Install the retaining spring to the hydraulic body.

36. Clean the exhaust pipe joints and installtings. Fit the pipes with joint clamps. Tighten to 30 ft. lbs. (40 Nm).

37. On convertible models, install the chassis reinforcement to the rear subframe.

38. On vehicle equipped with tire pressure monitoring, install the wiring harness and plug in the connector for the signal detector. Fit the wheel well.

39. Install the rear wheels and lower the vehicle.

40. Depress the brake pedal several times to press out the self adjustment of the brake pistons and the parking brake.

### 4WD Models

*See Figure 320.*

1. Before servicing the vehicle, refer to the Precautions Section.

2. Raise and safely support the vehicle securely on jackstands. Remove the rear wheels.

3. Remove the bolts of the anti-roll bar.

4. Lift down the anti-roll bar.

### To install:

5. Installation is the reverse of the removal procedure.

6. Tighten stabilizer bar bolts to 23 ft. lbs. (30 Nm).

7. Lower the vehicle.

### 9-5

*See Figure 321.*

1. Before servicing the vehicle, refer to the Precautions Section.

2. Raise and safely support the vehicle securely on jackstands. Remove the rear wheels.

3. Remove the stabilizer bar retaining nuts and bolts.

4. Remove the anti-roll bar.

5. Tap out the clamp's screw.

6. Dismantle the clamp and remove the bushing.

### To install:

➡**If necessary, lubricate the stabilizer bar bushings with Molykote 33 (bushings on new anti-roll bars are already lubricated), position the stabilizer bar and install all the bolts and nuts.**

7. Install the bushing in place and mount the clamp.

8. Tap in the clamp's bolt.

9. Tighten the nuts and bolts to 40 ft. lbs. (50 Nm).

10. Install the wheels and lower the vehicle.

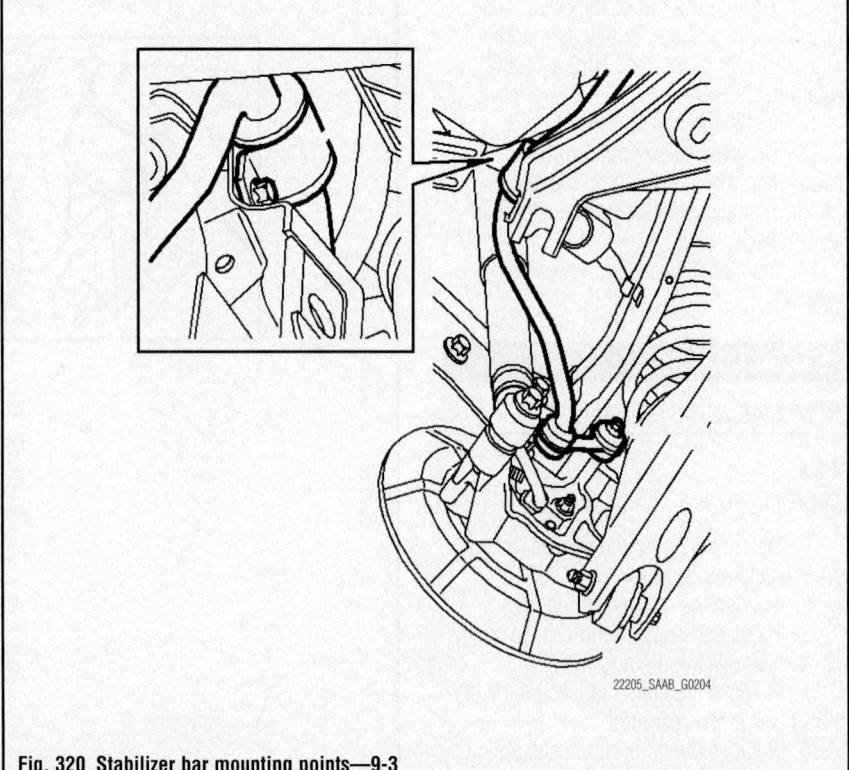

22205_SAAB_G0204

**Fig. 320 Stabilizer bar mounting points—9-3**

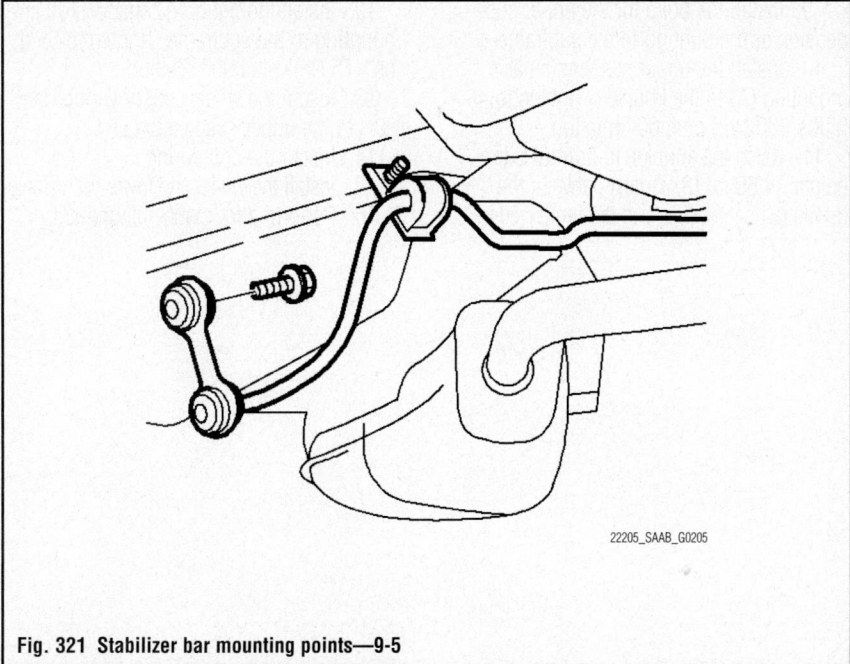

**Fig. 321 Stabilizer bar mounting points—9-5**

## STRUT & SPRING ASSEMBLY

### REMOVAL & INSTALLATION

#### 9-2x

*See Figure 322.*

1. Remove the strut cap on the quarter trim.
2. Loosen the rear wheel lug nuts.
3. Raise and support the vehicle safely.
4. Remove the tire and wheel.
5. Remove the brake hose clip, and then remove the brake hose from the rear strut.
6. Remove the bolts that retain the strut to the housing.
7. Remove the nuts retaining the strut to the body.
8. Remove the strut from the vehicle.

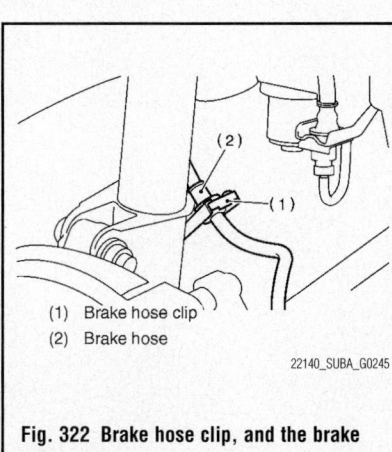

(1) Brake hose clip
(2) Brake hose

**Fig. 322 Brake hose clip, and the brake hose shown**

**To install:**

9. Installation is the reverse of the removal procedure.
10. Be sure to use new locknuts, as required.
11. Do not subject the ABS wheel speed sensor to excessive tension.
12. Check and adjust the wheel alignment, as necessary

## UPPER CONTROL ARM

### REMOVAL & INSTALLATION

#### 9-3

*See Figure 323.*

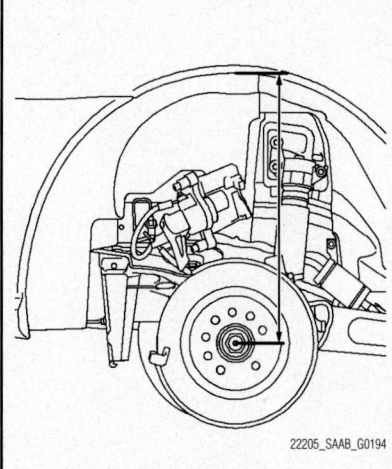

**Fig. 323 Measuring the normal ride height between the between the driveshaft center and the fender edge.**

#### 2WD Models Upper

1. Before servicing the vehicle, refer to the Precautions Section.
2. Slacken the parking brake lever adjustment slightly.
3. Measure the normal ride height between the between the driveshaft center and the fender edge.
4. Raise and safely support the vehicle securely on jackstands. Remove the wheel.
5. Remove the parking brake cable from the brake caliper.
6. Remove the wiring harness for the wheel sensor from the upper suspension arm.
7. Remove the upper suspension arm mounting from the subframe.
8. Lift the lower suspension arm with a jack in order to relieve the upper suspension arm mounting to the knuckle from bearing weight.
9. Remove the upper suspension arm mounting from the knuckle.
10. Remove the upper suspension arm.

#### To install:

11. Install the upper suspension arm into place.
12. Install the bolts for the upper suspension arm mounting to the subframe.
13. Install the upper suspension arm to the swivel member. Tighten to 92 ft. lbs. (125 Nm) plus 135° rotation.
14. Lift the upper suspension arm to normal ride height and Install the nut to the suspension arm mounting to the subframe. Tighten to 55 ft. lbs. (75 Nm) plus 90° rotation.
15. Install the wiring harness for the wheel sensor to the upper suspension arm.
16. Install the cable for the parking brake to the brake caliper. Adjust the parking brake.
17. Install the wheel and lower the vehicle.
18. Perform a four wheel alignment.

#### 4WD Models Upper

1. Before servicing the vehicle, refer to the Precautions Section.
2. Raise and safely support the vehicle securely on jackstands. Remove the wheel.
3. Remove the coil spring
4. Remove the wheel sensor wiring harness from the upper suspension arm.
5. Remove the upper suspension arm mounting from the subframe.

6. Remove the upper suspension arm mounting from the knuckle. Remove the ball joint from the knuckle using 87 91 287 Puller, 150 mm.

7. Remove the upper suspension arm.

**To install:**

8. Install the upper suspension arm into place.

9. Install the bolts for the upper suspension arm mounting to the subframe.

10. Install the upper suspension arm mounting (A) to the knuckle. Tighten to 44 ft. lbs. (60 Nm) plus 60° rotation.

11. Raise the knuckle to normal ride height 14.59 in. (374 mm) between the driveshaft center line and the fender edge.

12. Install the nut of the suspension arm mounting to the subframe. Tighten to 55 ft. lbs. (75 Nm) plus 90° rotation.

13. Install the wheel sensor wiring harness to the upper suspension arm.

14. Install the coil spring

15. Install the wheel and lower the vehicle.

16. Perform a four wheel alignment.

# SAAB

## 9-7x

# 12

## SPECIFICATIONS AND MAINTENANCE CHARTS

### ENGINE AND VEHICLE IDENTIFICATION CHART

| Engine | | | | | | | Model Year | |
|---|---|---|---|---|---|---|---|---|
| Code ① | Liters (cc) | Cu. In. | Cyl. | Fuel Sys. | Engine Type | Eng. Mfg. | Code ② | Year |
| S | 4.2 (4200) | 256 | V6 | MFI | DOHC | CPC | 5 | 2005 |
| P | 5.3 (5326) | 325 | V8 | SFI | OHV | CPC | 6 | 2006 |
| M | 5.3 (5326) | 325 | V8 | SFI | OHV | CPC | 7 | 2007 |
| H | 6.0 (5967) | 364 | V8 | SFI | OHV | CPC | 8 | 2008 |

CPC: Chevrolet/Pontiac/Canada

MFI: Multi-port Fuel Injection

SFI: Sequential Fuel Injection

① 6th position of VIN

② 10th position of VIN

22205_S97X_C0001

### GENERAL ENGINE SPECIFICATIONS

| Year | Model | Engine Displacement Liters | Engine ID/VIN | Net Horsepower @ rpm | Net Torque @ rpm (ft. lbs.) | Bore x Stroke (in.) | Compression Ratio | Oil Pressure @ rpm |
|---|---|---|---|---|---|---|---|---|
| 2005 | 9-7x | 4.2 | S | 275@6000 | 275@3600 | 3.66x4.02 | 10.0:1 | 12@1200 |
| | | 5.3 | P | 300@5300 | 325@4000 | 3.75x3.62 | 9.95:1 | 18@2000 |
| 2006 | 9-7x | 4.2 | S | 291@6000 | 277@4800 | 3.66x4.02 | 10.3:1 | 12@1200 |
| | | 5.3 | M | 300@5200 | 330@4000 | 3.75x3.62 | 9.95:1 | 18@2000 |
| 2007 | 9-7x | 4.2 | S | 291@6000 | 277@4800 | 3.66x4.02 | 10.3:1 | 12@1200 |
| | | 5.3 | M | 300@5200 | 330@4000 | 3.75x3.62 | 9.95:1 | 18@2000 |
| 2008 | 9-7x | 4.2 | S | 291@6000 | 277@4800 | 3.66x4.02 | 10.3:1 | 12@1200 |
| | | 5.3 | M | 300@5200 | 330@4000 | 3.75x3.62 | 9.95:1 | 18@2000 |
| | | 6.0 | H | 395@6000 | 400@4000 | 4.00x3.62 | 10.8:1 | 18@2000 |

22205_S97X_C0002

### TUNE-UP SPECIFICATIONS

| Year | Engine Displacement Liters | Engine ID/VIN | Spark Plugs Gap (in.) | Ignition Timing (deg.) MT | Ignition Timing (deg.) AT | Fuel Pump (psi) | Idle Speed (rpm) MT | Idle Speed (rpm) AT | Valve Clearance In. | Valve Clearance Ex. |
|---|---|---|---|---|---|---|---|---|---|---|
| 2005 | 4.2 | S | 0.042 | — | ① | 50-57 ② | — | ① | HYD | HYD |
| | 5.3 | P | 0.040 | — | ① | 55-62 ② | — | ① | HYD | HYD |
| 2006 | 4.2 | S | 0.042 | — | ① | 50-57 ② | — | ① | HYD | HYD |
| | 5.3 | M | 0.040 | — | ① | 50-60 ② | — | ① | HYD | HYD |
| 2007 | 4.2 | S | 0.042 | — | ① | 50-57 ② | — | ① | HYD | HYD |
| | 5.3 | M | 0.040 | — | ① | 50-60 ② | — | ① | HYD | HYD |
| 2008 | 4.2 | S | 0.042 | — | ① | 50-57 ② | — | ① | HYD | HYD |
| | 5.3 | M | 0.040 | — | ① | 50-60 ② | — | ① | HYD | HYD |
| | 6.0 | H | 0.040 | — | ① | 50-60 ② | — | ① | HYD | HYD |

NOTE: The Vehicle Emission Control Information label often reflects specification changes made during production.

The label figures must be used if they differ from figures in this chart.

HYD: Hydraulic

① Pre-programmed in ECU and cannot be adjusted

② With key ON and engine OFF

22205_S97X_C0003

## CAPACITIES

| Year | Model | Engine Displacement Liters | Engine ID/VIN | Engine Oil with Filter (qts.) | Transmission (pts.) 5-Spd | 6-Spd | Auto. | Transfer Case (pts.) | Drive Axle Front (pts.) | Rear (pts.) | Fuel Tank (gal.) | Cooling System (qts.) |
|---|---|---|---|---|---|---|---|---|---|---|---|---|
| 2005 | 9-7x | 4.2 | S | 7.0 | — | — | 10.0 | 4.0 | 1.7 | 3.6 | 22.0 | 13.9 |
|  |  | 5.3 | P | 6.0 | — | — | 10.0 | 4.0 | 1.7 | 4.3 | 22.0 | 15.2 |
| 2006 | 9-7x | 4.2 | S | 7.0 | — | — | 10.0 | 4.0 | 1.7 | 3.6 | 22.0 | 13.9 |
|  |  | 5.3 | M | 6.0 | — | — | 10.0 | 4.0 | 1.7 | 4.3 | 22.0 | 15.2 |
| 2007 | 9-7x | 4.2 | S | 7.0 | — | — | 10.0 | 4.0 | 1.7 | 3.6 | 22.0 | 13.9 |
|  |  | 5.3 | M | 6.0 | — | — | 10.0 | 4.0 | 1.7 | 4.3 | 22.0 | 15.2 |
| 2008 | 9-7x | 4.2 | S | 7.0 | — | — | 10.0 | 4.0 | 1.7 | 3.6 | 22.0 | 13.9 |
|  |  | 5.3 | M | 6.0 | — | — | 10.0 | 4.0 | 1.7 | 4.3 | 22.0 | 15.2 |
|  |  | 6.0 | H | 6.0 | — | — | 10.0 | 4.0 | 1.7 | 4.3 | 22.0 | 15.2 |

NOTE: All capacities are approximate. Add fluid gradually and check to be sure a proper fluid level is obtained.

22205_S97X_C0012

## FLUID SPECIFICATIONS

| Year | Model | Engine Displacement Liters | Engine ID/VIN | Engine Oil | Auto. Trans. | Drive Axle | Transfer Case | Power Steering Fluid | Brake Master Cylinder |
|---|---|---|---|---|---|---|---|---|---|
| 2005 | 9-7x | 4.2 | S | 5W-30 | Dexron-III | 75W-90 | Auto-Trak II | GM PS Fluid | DOT 3 |
|  |  | 5.3 | P | 5W-30 | Dexron-III | 75W-90 | Auto-Trak II | GM PS Fluid | DOT 3 |
| 2006 | 9-7x | 4.2 | S | 5W-30 | Dexron-VI | 75W-90 | Auto-Trak II | GM PS Fluid | DOT 3 |
|  |  | 5.3 | M | 5W-30 | Dexron-VI | 75W-90 | Auto-Trak II | GM PS Fluid | DOT 3 |
| 2007 | 9-7x | 4.2 | S | 5W-30 | Dexron-VI | 75W-90 | Auto-Trak II | GM PS Fluid | DOT 3 |
|  |  | 5.3 | M | 5W-30 | Dexron-VI | 75W-90 | Auto-Trak II | GM PS Fluid | DOT 3 |
| 2008 | 9-7x | 4.2 | S | 5W-30 | Dexron-VI | 75W-90 | Auto-Trak II | GM PS Fluid | DOT 3 |
|  |  | 5.3 | M | 5W-30 | Dexron-VI | 75W-90 | Auto-Trak II | GM PS Fluid | DOT 3 |
|  |  | 6.0 | H | 5W-30 ① | Dexron-VI | 75W-90 ② | Auto-Trak II | GM PS Fluid | DOT 3 |

DOT: Department Of Transpotation
① Mobil 1 synthetic oil
② Plus a limited slip axle additive

22205_S97X_C0012

## VALVE SPECIFICATIONS

| Year | Engine Displacement Liters | Engine ID/VIN | Seat Angle (deg.) | Face Angle (deg.) | Spring Test Pressure (lbs. @ in.) | Spring Installed Height (in.) | Stem-to-Guide Clearance (in.) | | Stem Diameter (in.) | |
|---|---|---|---|---|---|---|---|---|---|---|
| | | | | | | | Intake | Exhaust | Intake | Exhaust |
| 2005 | 4.2 | S | NS | NS | 130-142@1.26 | NS | 0.0011-0.0025 | 0.0015-0.0030 | NS | NS |
| | 5.3 | P | 46 | 45 | 220@1.32 | 1.80 | 0.0010-0.0026 | 0.0010-0.0026 | 0.313-0.314 | 0.313-0.314 |
| 2006 | 4.2 | S | NS | NS | 130-142@1.26 | NS | 0.0011-0.0025 | 0.0015-0.0030 | NS | NS |
| | 5.3 | M | 46 | 45 | 220@1.32 | 1.80 | 0.0010-0.0026 | 0.0010-0.0026 | 0.313-0.314 | 0.313-0.314 |
| 2007 | 4.2 | S | NS | NS | 130-142@1.26 | NS | 0.0011-0.0025 | 0.0015-0.0030 | NS | NS |
| | 5.3 | M | 46 | 45 | 220@1.32 | 1.80 | 0.0010-0.0026 | 0.0010-0.0026 | 0.313-0.314 | 0.313-0.314 |
| 2008 | 4.2 | S | NS | NS | 130-142@1.26 | NS | 0.0011-0.0025 | 0.0015-0.0030 | NS | NS |
| | 5.3 | M | 46 | 45 | 220@1.32 | 1.80 | 0.0010-0.0026 | 0.0010-0.0026 | 0.313-0.314 | 0.313-0.314 |
| | 6.0 | H | 46 | 45 | 220@1.32 | 1.80 | 0.0010-0.0026 | 0.0010-0.0026 | 0.313-0.314 | 0.313-0.314 |

NS - Not Specified by manufacturer.

22205_S97X_C0005

## CAMSHAFT AND BEARING SPECIFICATIONS CHART

All measurements are given in inches.

| Year | Engine Displ. Liters | Engine ID/VIN | Journal Dia. | Brg. Oil Clearance | Shaft End-play | Runout | Journal Bore | Lobe Height | |
|------|---|---|---|---|---|---|---|---|---|
| | | | | | | | | Intake | Exhaust |
| **2005** | 4.2 | S | 1.1794-1.1804 ① | 0.0015-0.0033 | 0.0017-0.0084 ② | NS | NS | NS | NS |
| | 5.3 | P | 2.1640-2.1660 | | 0.0010-0.0120 | NS | NS | 0.2890 ③ | 0.2890 ③ |
| **2006** | 4.2 | S | 1.1794-1.1804 ① | 0.0015-0.0033 | 0.0017-0.0084 ② | NS | NS | NS | NS |
| | 5.3 | M | 2.1640-2.1660 | NS | 0.0010-0.0120 | NS | ④ | 0.2890 ③ | 0.2890 ③ |
| **2007** | 4.2 | S | 1.1794-1.1804 ① | 0.0015-0.0033 | 0.0017-0.0084 ② | NS | NS | NS | NS |
| | 5.3 | M | 2.1640-2.1660 | NS | 0.0010-0.0120 | NS | ④ | 0.2890 ③ | 0.2890 ③ |
| **2008** | 4.2 | S | 1.1794-1.1804 ① | 0.0015-0.0033 | 0.0017-0.0084 ② | NS | NS | NS | NS |
| | 5.3 | M | 2.1640-2.1660 | NS | 0.0010-0.0120 | NS | ④ | 0.2890 ③ | 0.2890 ③ |
| | 6.0 | H | 2.1640-2.1660 | NS | 0.0010-0.0120 | NS | ④ | 0.2890 ③ | 0.2890 ③ |

NS: Not specified by manufacturer

① Exhaust journal #1, all others 1.0612-1.0622 in,

② Exhaust; Intake 0.0020-0.079 in.

③ Non Displacement -on-Demand Cylinders 0.0283 in.

④ Bore 1 & 5: 2.345-2.347

   Bore 2 & 4: 2.325-2.327

   Bore 3: 2.306-2.308

22205_S97X_C0014

## CRANKSHAFT AND CONNECTING ROD SPECIFICATIONS

All measurements are given in inches.

| Year | Engine Displacement Liters | Engine ID/VIN | Crankshaft | | | | Connecting Rod | | |
|------|------|------|------|------|------|------|------|------|------|
| | | | Main Brg. Journal Dia. | Main Brg. Oil Clearance | Shaft End-play | Thrust on No. | Journal Diameter | Oil Clearance | Side Clearance |
| 2005 | 4.2 | S | 2.757 | 0.0004-0.0025 | 0.0044-0.0153 | 4 | 2.237 | 0.0008-0.0025 | 0.0019-0.0137 |
| | 5.3 | P | 2.559 | 0.0008-0.0021 | 0.0015-0.0078 | 4 | 2.099 | 0.0009-0.0025 | 0.0043-0.0200 |
| 2006 | 4.2 | S | 2.757 | 0.0004-0.0025 | 0.0044-0.0153 | 4 | 2.237 | 0.0008-0.0025 | 0.0019-0.0137 |
| | 5.3 | M | 2.559 | 0.0008-0.0021 | 0.0015-0.0078 | 4 | 2.099 | 0.0009-0.0025 | 0.0043-0.0200 |
| 2007 | 4.2 | S | 2.757 | 0.0004-0.0025 | 0.0044-0.0153 | 4 | 2.237 | 0.0008-0.0025 | 0.0019-0.0137 |
| | 5.3 | M | 2.559 | 0.0008-0.0021 | 0.0015-0.0078 | 4 | 2.099 | 0.0009-0.0025 | 0.0043-0.0200 |
| 2007 | 4.2 | S | 2.757 | 0.0004-0.0025 | 0.0044-0.0153 | 4 | 2.237 | 0.0008-0.0025 | 0.0019-0.0137 |
| | 5.3 | M | 2.559 | 0.0008-0.0021 | 0.0015-0.0078 | 4 | 2.099 | 0.0009-0.0025 | 0.0043-0.0200 |
| | 6 | H | 2.559 | 0.0008-0.0021 | 0.0015-0.0078 | 4 | 2.099 | 0.0009-0.0025 | 0.0043-0.0200 |

NS: Not specified by manufacturer.

22205_S97X_C0006

## PISTON AND RING SPECIFICATIONS

All measurements are given in inches.

| Year | Engine Displacement Liters | Engine ID/VIN | Piston Clearance | Ring Gap | | | Ring Side Clearance | | |
|------|------|------|------|------|------|------|------|------|------|
| | | | | Top Compression | Bottom Compression | Oil Control | Top Compression | Bottom Compression | Oil Control |
| 2005 | 4.2 | S | 0.0006-0.0014 | 0.0059-0.0118 | 0.0142-0.0201 | 0.0098-0.0299 | 0.0017-0.0037 | 0.0017-0.0037 | 0.0023-0.0085 |
| | 5.3 | P | 0.0006-0.0014 | 0.0090-0.0173 | 0.0173-0.0275 | 0.0070-0.0295 | 0.0015-0.0033 | 0.0015-0.0031 | 0.0005-0.0078 |
| 2006 | 4.2 | S | 0.0006-0.0014 | 0.0059-0.0118 | 0.0142-0.0201 | 0.0098-0.0299 | 0.0017-0.0037 | 0.0017-0.0037 | 0.0023-0.0085 |
| | 5.3 | M | 0.0006-0.0014 | 0.0090-0.0173 | 0.0173-0.0275 | 0.0070-0.0295 | 0.0015-0.0033 | 0.0015-0.0031 | 0.0005-0.0078 |
| 2007 | 4.2 | S | 0.0006-0.0014 | 0.0059-0.0118 | 0.0142-0.0201 | 0.0098-0.0299 | 0.0017-0.0037 | 0.0017-0.0037 | 0.0023-0.0085 |
| | 5.3 | M | 0.0006-0.0014 | 0.0090-0.0173 | 0.0173-0.0275 | 0.0070-0.0295 | 0.0015-0.0033 | 0.0015-0.0031 | 0.0005-0.0078 |
| 2008 | 4.2 | S | 0.0006-0.0014 | 0.0059-0.0118 | 0.0142-0.0201 | 0.0098-0.0299 | 0.0017-0.0037 | 0.0017-0.0037 | 0.0023-0.0085 |
| | 5.3 | M | 0.0006-0.0014 | 0.0090-0.0173 | 0.0173-0.0275 | 0.0070-0.0295 | 0.0015-0.0033 | 0.0015-0.0031 | 0.0005-0.0078 |
| | 6.0 | H | 0.0009-0.0012 | 0.0080-0.016 | 0.0150-0.027 | 0.0090-0.031 | 0.0012-0.004 | 0.0014-0.0031 | 0.0005-0.0079 |

NS: Not specified by manufacturer

22205_S97X_C0007

## TORQUE SPECIFICATIONS
All readings in ft. lbs.

| Year | Engine Displacement Liters | Engine ID/VIN | Cylinder Head Bolts | Main Bearing Bolts | Rod Bearing Bolts | Crankshaft Damper Bolts | Flywheel Bolts | Manifold Intake* | Manifold Exhaust | Spark Plugs | Oil Pan Drain Plug |
|------|------|------|------|------|------|------|------|------|------|------|------|
| 2005 | 4.2 | S | ① | ② | ③ | ④ | ⑤ | ⑥ | ⑦ | 13 | 19 |
|      | 5.3 | P | ⑧ | ⑨ | ⑩ | ⑪ | ⑫ | ⑬ | ⑭ | 11 | 18 |
| 2006 | 4.2 | S | ① | ② | ③ | ④ | ⑤ | ⑥ | ⑦ | 13 | 19 |
|      | 5.3 | M | ⑧ | ⑨ | ⑩ | ⑪ | ⑫ | ⑬ | ⑭ | 11 | 18 |
| 2007 | 4.2 | S | ① | ② | ③ | ④ | ⑤ | ⑥ | ⑦ | 13 | 19 |
|      | 5.3 | M | ⑧ | ⑨ | ⑩ | ⑪ | ⑫ | ⑬ | ⑭ | 11 | 18 |
| 2008 | 4.2 | S | ① | ② | ③ | ④ | ⑤ | ⑥ | ⑦ | 13 | 19 |
|      | 5.3 | M | ⑧ | ⑨ | ⑩ | ⑪ | ⑫ | ⑬ | ⑭ | 11 | 18 |
|      | 6.0 | H | ⑧ | ⑨ | ⑩ | ⑪ | ⑫ | ⑬ | ⑭ | 11 | 18 |

* NOTE: Applies to Lower Manifold only.

① Cylinder head bolts (14)

1st pass: 22 ft. lbs.

2nd pass: Plus 155 degrees

2 short end bolts: 62 INCH lbs.

2nd pass: plus 60 degrees

1 long end bolt: 62 INCH lbs.

2nd pass: plus 120 degrees

② 18 ft. lbs., plus 180 depress

③ 18 ft. lbs., plus 110 degrees

④ 110 ft. lbs., plus 180 degrees

⑤ 18 ft. lbs., plus 50 degrees

⑥ 89 inch lbs.

⑦ 1st pass: 15 ft. lbs.

2nd pass: 15 ft. lbs.

3rd pass: 15 ft. lbs.

⑧ M11 bolts: 22 ft. lbs.

2nd pass: Plus 90 degrees

3rd pass: Plus 70 degrees

M8 bolts: 22 ft. lbs.

⑨ Inner bolts:

1st pass: 15 ft. lbs.

Final pass: Plus 80 degrees

Outer bolts:

1st pass: 15 ft. lbs.

Final pass: Plus 51 degrees

M8 bolts: 18 ft. lbs.

⑩ 15 ft. lbs. plus 85 degrees

⑪ Installation pass: 240 ft. lbs. (discard bolt)

First pass: 37 ft. lbs. (new bolt)

Final pass: Plus 140 degrees

⑫ 1st pass: 15 ft. lbs.

2nd pass: 37 ft. lbs.

3rd pass: 74 ft. lbs.

⑬ 1st pass: 44 inch lbs.

2nd pass: 89 inch lbs.

⑭ 1st pass: 11 ft. lbs.

2nd pass: 18 ft. lbs.

22205_S97X_C0008

## WHEEL ALIGNMENT

| Year | Model | | Caster Range (+/-Deg.) | Caster Preferred Setting (Deg.) | Camber Range (+/-Deg.) | Camber Preferred Setting (Deg.) | Toe-in (in.) |
|------|-------|---|------|------|------|------|------|
| 2005 | 9-7x | F | 0.60 | +4.00 | 0.60 | 0.00 | -0.10+/-0.20 |
| 2006 | 9-7x | F | 0.60 | +4.00 | 0.60 | 0.00 | -0.10+/-0.20 |
| 2007 | 9-7x | F | 0.60 | +4.00 | 0.60 | 0.00 | -0.10+/-0.20 |
| 2008 | 9-7x | F | 0.60 | +4.00 | 0.60 | 0.00 | -0.10+/-0.20 |

22205_S97X_C0009

## TIRE, WHEEL AND BALL JOINT SPECIFICATIONS

| Year | Model | OEM Tires Standard | OEM Tires Optional | Tire Pressures (psi) Front | Tire Pressures (psi) Rear | Wheel Size | Ball Joint Inspection | Lug Nut Torque (ft. lbs.) |
|------|-------|------|------|------|------|------|------|------|
| 2005 | 9-7x | 255/55-18 | None | 36 | 36 | 7-JJ | L ① | 103 |
| 2006 | 9-7x | 255/55-18 | None | 36 | 36 | 7-JJ | L ① | 103 |
| 2007 | 9-7x | 255/55-18 | None | 36 | 36 | 7-JJ | L ① | 103 |
| 2007 | 9-7x | 255/55-18 | None | 36 | 36 | 7-JJ | L ① | 103 |

OEM: Original Equipment Manufacturer

PSI: Pounds Per Square Inch

L: Lower (ball joint)

① Do not lift truck. Inspect the boss into which the grease fitting is threaded. Replace if the boss is flush or receded below the surface of the ball joint

22205_S97X_C0010

# BRAKE SPECIFICATIONS

All measurements are in inches unless noted

| Year | Model | | Brake Disc Original Thickness | Brake Disc Minimum Thickness | Brake Disc Maximum Runout | Lining Minimum Thickness | Brake Caliper Bracket Bolts (ft. lbs.) | Brake Caliper Mounting Bolts (ft. lbs.) |
|------|-------|---|------|------|------|------|------|------|
| 2005 | 9-7x | F | 1.14 | 1.08 | 0.002 | NS | 110 | 31 |
|      |      | R | 0.787 | 0.728 | 0.002 | NS | 148 | 23 |
| 2006 | 9-7x | F | 1.14 | 1.08 | 0.002 | NS | 118 | 31 |
|      |      | R | 0.787 | 0.728 | 0.002 | NS | 148 | 23 |
| 2007 | 9-7x | F | 1.14 | 1.08 | 0.002 | NS | 118 | 31 |
|      |      | R | 0.787 | 0.728 | 0.002 | NS | 148 | 23 |
| 2008 | 9-7x | F | 1.14 | 1.08 | 0.002 | NS | 118 | 31 |
|      |      | R | 0.787 | 0.728 | 0.002 | NS | 148 | 23 |

NS: Not Specified by Manufacturer

22205_S97X_C0011

# SCHEDULED MAINTENANCE INTERVALS

## SAAB 9-7X

| TO BE SERVICED | TYPE OF SERVICE | 10 | 20 | 30 | 40 | 50 | 60 | 70 | 80 |
|----------------|-----------------|----|----|----|----|----|----|----|----|
| Air filter | Inspect/Service | | ✓ | | ✓ | | ✓ | | ✓ |
| Air filter | Replace | | | | | ✓ | | | |
| Automatic transmission fluid | Inspect/Service | | ✓ | | ✓ | | ✓ | | ✓ |
| Automatic transmission fluid & filter | Replace | Every 100,000 miles | | | | | | | |
| Body | Inspect/Service | ✓ | ✓ | ✓ | ✓ | ✓ | ✓ | ✓ | ✓ |
| Brake system | Inspect/Service | ✓ | ✓ | ✓ | ✓ | ✓ | ✓ | ✓ | ✓ |
| Chassis | Inspect/Service | ✓ | ✓ | ✓ | ✓ | ✓ | ✓ | ✓ | ✓ |
| Coolant | Inspect/Service | ✓ | ✓ | ✓ | ✓ | ✓ | ✓ | ✓ | ✓ |
| Cooling system | Inspect/Service | | ✓ | | ✓ | | ✓ | | ✓ |
| Engine oil & filter | Inspect/Service | ✓ | ✓ | ✓ | ✓ | ✓ | ✓ | ✓ | ✓ |
| Exhaust system | Inspect/Service | | | | | ✓ | | | |
| Fuel filter | Replace | | | | | ✓ | | | |
| Fuel system | Inspect/Service | | | | | ✓ | | | |
| Parking brake cable guides | Inspect/Lubricate | ✓ | ✓ | ✓ | ✓ | ✓ | ✓ | ✓ | ✓ |
| Radiator core & A/C condenser | Clean | Every 150,000 miles | | | | | | | |
| Seat belts | Inspect/Service | | ✓ | | ✓ | | ✓ | | ✓ |
| Shift linkage | Inspect/Service | ✓ | ✓ | ✓ | ✓ | ✓ | ✓ | ✓ | ✓ |
| Spark plugs | Replace | Every 100,000 miles | | | | | | | |
| Steering & suspension | Inspect/Service | ✓ | ✓ | ✓ | ✓ | ✓ | ✓ | ✓ | ✓ |
| Throttle system | Inspect/Service | ✓ | ✓ | ✓ | ✓ | ✓ | ✓ | ✓ | ✓ |
| Transfer case fluid | Replace | | | | | ✓ | | | |
| Wheels & tires | Inspect/Rotate | ✓ | ✓ | ✓ | ✓ | ✓ | ✓ | ✓ | ✓ |
| Windshield washer | Inspect/Service | ✓ | ✓ | ✓ | ✓ | ✓ | ✓ | ✓ | ✓ |
| Windshield wiper | Inspect/Service | ✓ | ✓ | ✓ | ✓ | ✓ | ✓ | ✓ | ✓ |

22205_S97X_C0013

## PRECAUTIONS

Before servicing any vehicle, please be sure to read all of the following precautions, which deal with personal safety, prevention of component damage, and important points to take into consideration when servicing a motor vehicle:

• Never open, service or drain the radiator or cooling system when the engine is hot; serious burns can occur from the steam and hot coolant.

• Observe all applicable safety precautions when working around fuel. Whenever servicing the fuel system, always work in a well-ventilated area. Do not allow fuel spray or vapors to come in contact with a spark, open flame, or excessive heat (a hot drop light, for example). Keep a dry chemical fire extinguisher near the work area. Always keep fuel in a container specifically designed for fuel storage; also, always properly seal fuel containers to avoid the possibility of fire or explosion. Refer to the additional fuel system precautions later in this section.

• Fuel injection systems often remain pressurized, even after the engine has been turned **OFF**. The fuel system pressure must be relieved before disconnecting any fuel lines. Failure to do so may result in fire and/or personal injury.

• Brake fluid often contains polyglycol ethers and polyglycols. Avoid contact with the eyes and wash your hands thoroughly after handling brake fluid. If you do get brake fluid in your eyes, flush your eyes with clean, running water for 15 minutes. If eye irritation persists, or if you have taken brake fluid internally, IMMEDIATELY seek medical assistance.

• The EPA warns that prolonged contact with used engine oil may cause a number of skin disorders, including cancer. You should make every effort to minimize your exposure to used engine oil. Protective gloves should be worn when changing oil. Wash your hands and any other exposed skin areas as soon as possible after exposure to used engine oil. Soap and water, or waterless hand cleaner should be used.

• All new vehicles are now equipped with an air bag system, often referred to as a Supplemental Restraint System (SRS) or Supplemental Inflatable Restraint (SIR) system. The system must be disabled before performing service on or around system components, steering column, instrument panel components, wiring and sensors. Failure to follow safety and disabling procedures could result in accidental air bag deployment, possible personal injury and unnecessary system repairs.

• Always wear safety goggles when working with, or around, the air bag system. When carrying a non-deployed air bag, be sure the bag and trim cover are pointed away from your body. When placing a non-deployed air bag on a work surface, always face the bag and trim cover upward, away from the surface. This will reduce the motion of the module if it is accidentally deployed. Refer to the additional air bag system precautions later in this section.

• Clean, high quality brake fluid from a sealed container is essential to the safe and proper operation of the brake system. You should always buy the correct type of brake fluid for your vehicle. If the brake fluid becomes contaminated, completely flush the system with new fluid. Never reuse any brake fluid. Any brake fluid that is removed from the system should be discarded. Also, do not allow any brake fluid to come in contact with a painted surface; it will damage the paint.

• Never operate the engine without the proper amount and type of engine oil; doing so WILL result in severe engine damage.

• Timing belt maintenance is extremely important. Many models utilize an interference-type, non-freewheeling engine. If the timing belt breaks, the valves in the cylinder head may strike the pistons, causing potentially serious (also time-consuming and expensive) engine damage. Refer to the maintenance interval charts for the recommended replacement interval for the timing belt, and to the timing belt section for belt replacement and inspection.

• Disconnecting the negative battery cable on some vehicles may interfere with the functions of the on-board computer system(s) and may require the computer to undergo a relearning process once the negative battery cable is reconnected.

• When servicing drum brakes, only disassemble and assemble one side at a time, leaving the remaining side intact for reference.

• Only an MVAC-trained, EPA-certified automotive technician should service the air conditioning system or its components.

## BRAKES

### GENERAL INFORMATION

#### PRECAUTIONS

• Certain components within the ABS system are not intended to be serviced or repaired individually.

• Do not use rubber hoses or other parts not specifically specified for and ABS system. When using repair kits, replace all parts included in the kit. Partial or incorrect repair may lead to functional problems and require the replacement of components.

• Lubricate rubber parts with clean, fresh brake fluid to ease assembly. Do not use shop air to clean parts; damage to rubber components may result.

• Use only DOT 3 brake fluid from an unopened container.

• If any hydraulic component or line is removed or replaced, it may be necessary to bleed the entire system.

• A clean repair area is essential. Always clean the reservoir and cap thoroughly before removing the cap. The slightest amount of dirt in the fluid may plug an orifice and impair the system function. Perform repairs after components have been thoroughly cleaned; use only denatured alcohol

## ANTI-LOCK BRAKE SYSTEM (ABS)

to clean components. Do not allow ABS components to come into contact with any substance containing mineral oil; this includes used shop rags.

• The Anti-Lock control unit is a microprocessor similar to other computer units in the vehicle. Ensure that the ignition switch is **OFF** before removing or installing controller harnesses. Avoid static electricity discharge at or near the controller.

• If any arc welding is to be done on the vehicle, the control unit should be unplugged before welding operations begin.

## BRAKES

## BLEEDING THE BRAKE SYSTEM

### BLEEDING PROCEDURE

#### BLEEDING PROCEDURE

1. Before servicing the vehicle, refer to the Precautions Section.
2. Raise the vehicle in order to access the system bleed screws.
3. Bleed the system at the right rear wheel first.
4. Install a clear hose on the bleed screw.
5. Immerse the opposite end of the hose into a container partially filled with clean DOT 3 brake fluid.

6. Open the bleed screw ½ to 1 full turn.
7. Slowly depress the brake pedal. While the pedal is depressed to its full extent, tighten the bleed screw.
8. Release the brake pedal and wait 10–15 seconds for the master cylinder pistons to return to the home position.
9. Repeat the previous steps for the remaining wheels. The brake fluid which is present at each bleed screw should be clean and free of air.
10. This procedure may use more than a pint of fluid per wheel. Check the master

cylinder fluid level every four to six strokes of the brake pedal in order to avoid running the system dry.
11. Press the brake pedal firmly and run the Scan Tool Automated Bleed Procedure. Release the brake pedal between each test.
12. Bleed all four wheels again using Steps 3–9. This will remove the remaining air from the brake system.
13. Evaluate the feel of the brake pedal before attempting to drive the vehicle.
14. Bleed the system as many times as necessary in order to obtain the appropriate feel of the pedal.

## BRAKES

## FRONT DISC BRAKES

### ✳✳ CAUTION

**Dust and dirt accumulating on brake parts during normal use may contain asbestos fibers from production or aftermarket brake linings. Breathing excessive concentrations of asbestos fibers can cause serious bodily harm. Exercise care when servicing brake parts. Do not sand or grind brake lining unless equipment used is designed to contain the dust residue. Do not clean brake parts with compressed air or by dry brushing. Cleaning should be done by dampening the brake components with a fine mist of water, then wiping the brake components clean with a dampened cloth. Dispose of cloth and all residue containing asbestos fibers in an impermeable container with the appropriate label. Follow practices prescribed by the Occupational Safety and Health Administration (OSHA) and the Environmental Protection Agency (EPA) for the handling, processing, and disposing of dust or debris that may contain asbestos fibers.**

### BRAKE CALIPER

#### REMOVAL & INSTALLATION

See Figure 1.

1. Before servicing the vehicle, refer to the Precautions Section.
2. Raise and safely support the vehicle securely on jackstands.
3. Remove the tire and wheel assembly.
4. Remove the front brake hose fitting bolt.

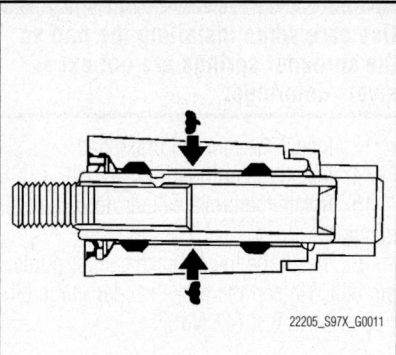

22205_S97X_G0011

**Fig. 1 Apply high temperature silicone brake lubricant to the brake caliper guide pin as illustrated**

5. Cap the brake hose fitting to prevent brake fluid loss and contamination.
6. Discard the copper gaskets. do not reuse them replace with new copper gaskets.
7. Remove the front brake caliper guide pin bolts.
8. Remove the front brake caliper from the bracket.

#### To install:

9. If the brake caliper guide pin is to be reused, clean the brake caliper guide pin using denatured alcohol, or equivalent.
10. Dry the brake caliper guide pin using non-lubricated, filtered air.
11. Apply high temperature silicone brake lubricant to the brake caliper guide pin. do not apply lubricant to the brake pad hardware. Tighten to 31 ft. lbs. (42 Nm).
12. Install the front brake caliper to the bracket.

### ✳✳ WARNING

**Tighten the leading bolt, closest to the bleed screw, first.**

13. Install the new copper gaskets to the front brake hose fitting.
14. Install the front brake hose fitting bolt and gaskets. Tighten to 30 ft. lbs. (40 Nm).
15. Refill the master cylinder to the correct level.
16. Bleed the brake system if the fluid lines were disconnected from the caliper.
17. Lower the vehicle.

### DISC BRAKE PADS

#### REMOVAL & INSTALLATION

See Figures 2 and 3.

1. Before servicing the vehicle, refer to the Precautions Section.
2. Raise and safely support the vehicle securely on jackstands.
3. Remove the wheels.
4. Remove ⅔ of the brake fluid from the master cylinder
5. Place a C-clamp around the outer pad and caliper; tighten the C-clamp until the piston is fully compressed in the caliper.
6. Remove the lower mounting bolt from the guide pin.
7. Rotate the brake caliper upward until it stops.
8. Remove the outboard brake pad.
9. Remove the inboard brake pad.
10. Remove the brake pad retaining clips from the brake caliper mounting bracket and discard.
11. Remove all foreign material from the brake caliper using denatured alcohol.

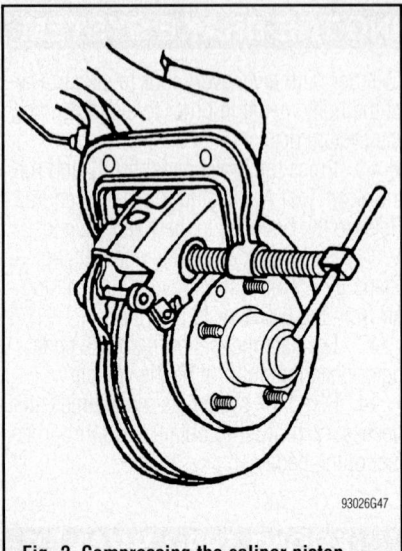

**Fig. 2 Compressing the caliper piston with a C-clamp**

*To install:*

### ✳✳ WARNING

**When installing new brake pads, do not reuse the old retaining clips for the brake pad. Use only new brake pad retaining clips.**

12. Install the brake pad retaining clips to the brake caliper mounting bracket.

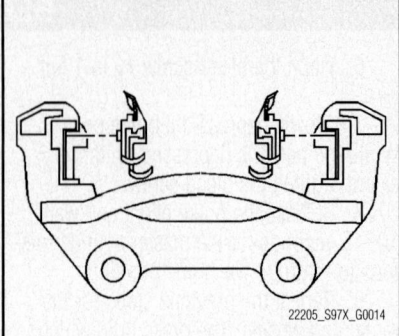

**Fig. 3 When installing new brake pads, do not reuse the old retaining clips for the brake pad. Use only new brake pad retaining clips**

### ✳✳ WARNING

**Use care when installing the pad so the spreader springs are not excessively deformed.**

13. Install the inboard brake pad.
14. Install the outboard brake pad.
15. Rotate the brake caliper down into position.
16. Install the lower brake caliper guide pin bolt. Tighten the brake caliper guide pin bolt to 31 ft. lbs. (42 Nm).

17. Install the tire and wheel assembly.
18. Lower the vehicle.
19. Fill the brake master cylinder reservoir.

### ✳✳ CAUTION

**Do not move the vehicle until a firm brake pedal is obtained. Failure to obtain a firm pedal before moving vehicle may result in personal injury.**

20. Pump the brake pedal slowly and firmly to seat the brake pads.
21. Burnish the brake pads and rotors as follows:
    a. Select a smooth road with little or no traffic.
    b. Accelerate the vehicle to 48 km/h (30 mph).

### ✳✳ WARNING

**Use care to avoid overheating the brakes while performing this step.**

    c. Using moderate to firm pressure, apply the brakes to bring the vehicle to a stop. Do not allow the brakes to lock.
    d. Repeat the previous two steps until approximately 20 stops have been completed. Allow sufficient cooling periods between stops in order to properly burnish the brake pads and rotors.

## BRAKES

### ✳✳ CAUTION

**Dust and dirt accumulating on brake parts during normal use may contain asbestos fibers from production or aftermarket brake linings. Breathing excessive concentrations of asbestos fibers can cause serious bodily harm. Exercise care when servicing brake parts. Do not sand or grind brake lining unless equipment used is designed to contain the dust residue. Do not clean brake parts with compressed air or by dry brushing. Cleaning should be done by dampening the brake components with a fine mist of water, then wiping the brake components clean with a dampened cloth. Dispose of cloth and all residue containing asbestos fibers in an impermeable container with the appropriate label. Follow practices prescribed by the Occupational Safety and Health Administration (OSHA) and the Environmental Protection Agency (EPA) for the handling, processing, and disposing of dust or debris that may contain asbestos fibers.**

### BRAKE CALIPER

*REMOVAL & INSTALLATION*
*See Figures 4 and 5.*

1. Before servicing the vehicle, refer to the Precautions Section.
2. Raise and safely support the vehicle.
3. Remove the tire and wheel assembly.
4. Remove the rear brake hose fitting bolt from the rear brake caliper.
5. Cap the brake hose fitting to prevent brake fluid loss and contamination.

### ✳✳ WARNING

**Discard the copper gaskets. do not reuse the copper, replace them.**

6. Remove the copper gaskets from the brake hose fitting bolt.
7. Remove the rear brake caliper guide pin bolts.

## REAR DISC BRAKES

8. Remove the rear brake caliper from the brake caliper bracket.

*To install:*
9. If the brake caliper guide pin is to be reused, clean the brake caliper guide pin using denatured alcohol, or equivalent.
10. Dry the brake caliper guide pin using non-lubricated, filtered air.
11. Apply high temperature silicone brake lubricant to the brake caliper guide pin. do not apply lubricant to the brake pad hardware.
12. Install the rear brake caliper assembly to the bracket.
13. Install the rear brake caliper guide pin bolts. Tighten to 23 ft. lbs. (31 Nm).
14. Install the new copper gaskets on the brake hose fitting.
15. Remove the plug from the brake hose end.
16. Install the brake hose fitting and fitting bolt to the rear brake caliper. Tighten to 40 ft. lbs. (54 Nm)
17. Bleed the hydraulic brake system.
18. Install the tire and wheel assembly.
19. Lower the vehicle.

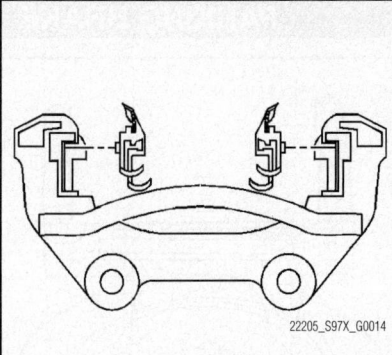

22205_S97X_G0014

**Fig. 4 When installing new brake pads, do not reuse the old retaining clips for the brake pad. Use only new brake pad retaining clips**

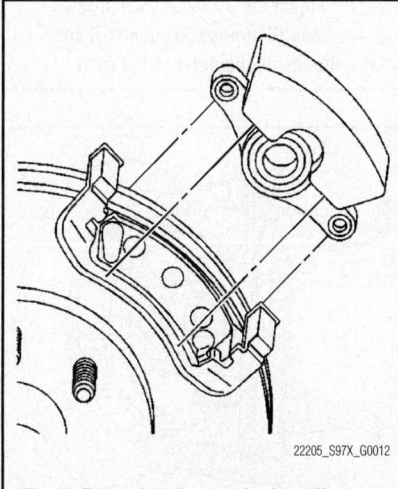

22205_S97X_G0012

**Fig. 5 Removing the rear brake caliper**

### DISC BRAKE PADS

*REMOVAL & INSTALLATION*

*See Figures 6 and 7.*

1. Before servicing the vehicle, refer to the Precautions Section.
2. Remove ⅔ of the brake fluid from the master cylinder
3. Raise and safely support the vehicle.
4. Remove the tire and wheel assembly.
5. Place a C-clamp around the outer pad and caliper; tighten the C-clamp until the piston is fully compressed in the caliper.
6. Remove the upper brake caliper mounting bolt.

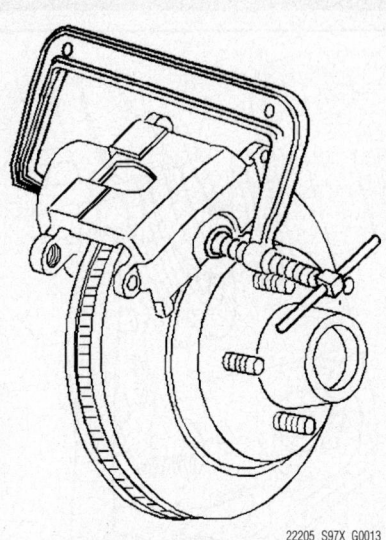

22205_S97X_G0013

**Fig. 6 Place a C-clamp around the outer pad and caliper and tighten the C-clamp until the piston is fully compressed in the caliper**

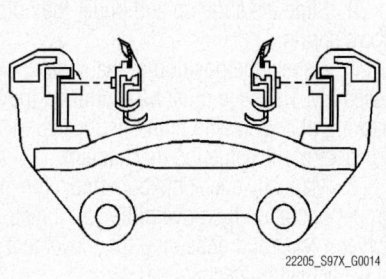

22205_S97X_G0014

**Fig. 7 When installing new brake pads, do not reuse the old retaining clips for the brake pad. Use only new brake pad retaining clips**

7. Rotate the rear brake caliper downward until it stops.
8. Remove the rear brake pads.
9. Remove the brake pad retaining clips from the brake caliper mounting bracket.
10. Remove the retaining clip from the brake caliper.
11. Clean the brake caliper mounting bracket with denatured alcohol.
12. Using non-lubricated, filtered air, dry the brake caliper mounting bracket.

*To install:*

### ✳✳ WARNING

**When installing new brake pads, do not reuse the old retaining clips for the brake pad. Use only new brake pad retaining clips.**

13. Install the brake pad retaining clips on the brake caliper mounting bracket.
14. Install the brake pads to the brake caliper mounting bracket.

### ✳✳ WARNING

**Ensure that the retaining clip for the brake pads is properly seated in the brake caliper.**

15. Install the retaining clip from the brake caliper.
16. Rotate the rear brake caliper upward until the brake caliper assembly is in the proper position.
17. Install the upper brake caliper mounting bolt. Tighten the brake caliper bolt to 23 ft. lbs. (31 Nm).
18. Install the tire and wheel assembly.
19. Lower the vehicle.
20. Fill the brake master cylinder reservoir, if needed.

### ✳✳ CAUTION

**Do not move the vehicle until a firm brake pedal is obtained. Failure to obtain a firm pedal before moving vehicle may result in personal injury.**

21. Pump the brake pedal slowly and firmly in order to seat the brake pads.
22. Burnish the new brake pads as follows:
    a. Select a smooth road with little or no traffic.
    b. Accelerate the vehicle to 48 km/h (30 mph).

### ✳✳ WARNING

**Use care to avoid overheating the brakes while performing this step.**

c. Using moderate to firm pressure, apply the brakes to bring the vehicle to a stop. Do not allow the brakes to lock.
d. Repeat the previous two steps until approximately 20 stops have been completed. Allow sufficient cooling periods between stops in order to properly burnish the brake pads and rotors.

## PARKING BRAKE SHOES

### REMOVAL & INSTALLATION

*See Figure 8.*

1. Before servicing the vehicle, refer to the Precautions Section.
2. Release the park brake, if applied.
3. Raise and the vehicle.
4. Remove the wheel and the tire.

➡**In the following service procedure, the brake caliper and mounting bracket does not have to be separated. Relocate the brake caliper and bracket to the side and secure.**

5. Remove the brake caliper and bracket.
6. Remove the brake rotor.
7. Remove the park brake cable from the park brake lever.
8. Slide the park brake shoe down until it is disengaged from the hold down spring.
9. Lift the shoe away from the backing plate and slide the shoe up, off of the actuation mechanism.
10. Remove the shoe over the axle flange and from the vehicle.
11. Clean the debris and the dust from the park brake components using a clean towel.
12. Turn the adjustment screw to the fully home position in the notched adjustment nut, then back it off 1/4 turn.
13. Align the slots in both the adjusting screw and tappet to be parallel with the backing plate face.
14. Using denatured alcohol, clean the rear backing plate of dirt and foreign materials.
15. Using non-lubricated, filtered air, dry the backing plate.

**To install:**

16. Install a new park brake shoe over the axle flange.
17. Position the shoe on the actuation mechanism.
18. Holding the lower end of the shoe away from the backing plate, slide the shoe down, over the top of the hold down spring.

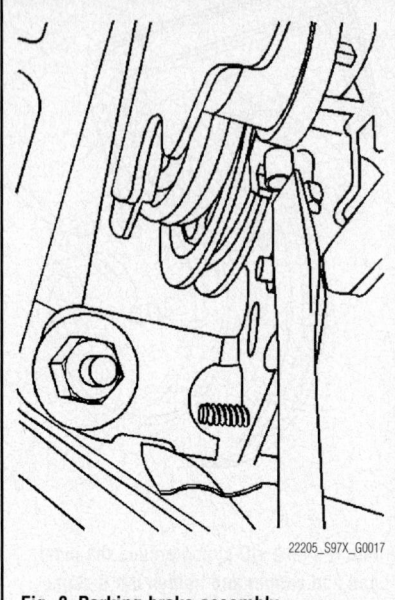

**Fig. 8 Parking brake assembly**

19. Place the lower end of the shoe against the backing plate.
20. Slide the shoe up and under the hold down spring.
21. Inspect the position of the shoe assembly. The shoe must be central on the backing plate and with both tips located in the slots of the actuation mechanism.
22. Manually check the park brake for proper operation by moving the park brake actuator lever and observing the movement of the actuation mechanism.
23. Install the park brake cable to the park brake lever.
24. Adjust the park brake shoe.
25. Install the brake rotor.
26. Install the brake caliper and mounting bracket.
27. Install the tire and wheel.
28. Lower the vehicle.

### ADJUSTMENT

*See Figures 9 and 10.*

1. Adjust the J 21177-A Drum-to-Brake Shoe Clearance Gage until it contacts the inside diameter of the rotor.

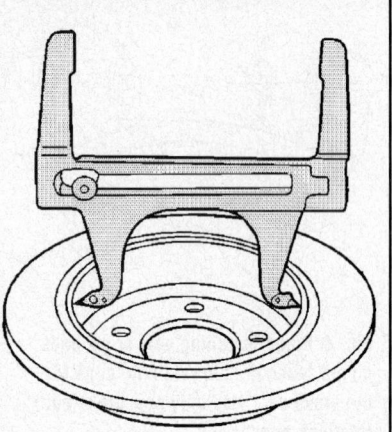

**Fig. 9 Adjust the J 21177-A Drum-to-Brake Shoe Clearance Gage until it contacts the inside diameter of the rotor**

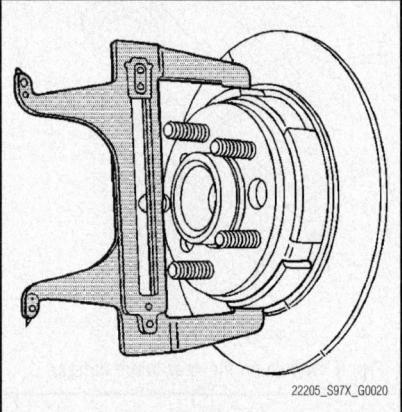

**Fig. 10 Position the gauge over the park shoe lining, left side shown, at the widest point**

2. Position the gauge over the park shoe lining, left side shown, at the widest point.
3. Turn the adjuster nut until the lining just contacts the gauge.
4. Repeat steps to adjust the right rear park brake.
5. The clearance between the rear park brake shoe lining and the rotor should be 0.010–0.020 in (0.020–0.25 mm).

## CHASSIS ELECTRICAL    AIR BAG (SUPPLEMENTAL RESTRAINT SYSTEM)

### GENERAL INFORMATION

#### ✷✷ CAUTION

**These vehicles are equipped with an air bag system. The system must be disarmed before performing service on, or around, system components, the steering column, instrument panel components, wiring and sensors. Failure to follow the safety precautions and the disarming procedure could result in accidental air bag deployment, possible injury and unnecessary system repairs.**

#### SERVICE PRECAUTIONS

Disconnect and isolate the battery negative cable before beginning any airbag system component diagnosis, testing, removal, or installation procedures. Allow system capacitor to discharge for two minutes before beginning any component service. This will disable the airbag system. Failure to disable the airbag system may result in accidental airbag deployment, personal injury, or death.

Do not place an intact undeployed airbag face down on a solid surface. The airbag will propel into the air if accidentally deployed and may result in personal injury or death.

When carrying or handling an undeployed airbag, the trim side (face) of the airbag should be pointing towards the body to minimize possibility of injury if accidental deployment occurs. Failure to do this may result in personal injury or death.

Replace airbag system components with OEM replacement parts. Substitute parts may appear interchangeable, but internal differences may result in inferior occupant protection. Failure to do so may result in occupant personal injury or death.

Wear safety glasses, rubber gloves, and long sleeved clothing when cleaning powder residue from vehicle after an airbag deployment. Powder residue emitted from a deployed airbag can cause skin irritation. Flush affected area with cool water if irritation is experienced. If nasal or throat irritation is experienced, exit the vehicle for fresh air until the irritation ceases. If irritation continues, see a physician.

Do not use a replacement airbag that is not in the original packaging. This may result in improper deployment, personal injury, or death.

The factory installed fasteners, screws and bolts used to fasten airbag components

have a special coating and are specifically designed for the airbag system. Do not use substitute fasteners. Use only original equipment fasteners listed in the parts catalog when fastener replacement is required.

During, and following, any child restraint anchor service, due to impact event or vehicle repair, carefully inspect all mounting hardware, tether straps, and anchors for proper installation, operation, or damage. If a child restraint anchor is found damaged in any way, the anchor must be replaced. Failure to do this may result in personal injury or death.

Deployed and non-deployed airbags may or may not have live pyrotechnic material within the airbag inflator.

Do not dispose of driver/passenger/curtain airbags or seat belt tensioners unless you are sure of complete deployment. Refer to the Hazardous Substance Control System for proper disposal.

Dispose of deployed airbags and tensioners consistent with state, provincial, local, and federal regulations.

After any airbag component testing or service, do not connect the battery negative cable. Personal injury or death may result if the system test is not performed first.

If the vehicle is equipped with the Occupant Classification System (OCS), do not connect the battery negative cable before performing the OCS Verification Test using the scan tool and the appropriate diagnostic information. Personal injury or death may result if the system test is not performed properly.

Never replace both the Occupant Restraint Controller (ORC) and the Occupant Classification Module (OCM) at the same time. If both require replacement, replace one, then perform the Airbag System test before replacing the other.

Both the ORC and the OCM store Occupant Classification System (OCS) calibration data, which they transfer to one another when one of them is replaced. If both are replaced at the same time, an irreversible fault will be set in both modules and the OCS may malfunction and cause personal injury or death.

If equipped with OCS, the Seat Weight Sensor is a sensitive, calibrated unit and must be handled carefully. Do not drop or handle roughly. If dropped or damaged, replace with another sensor. Failure to do so may result in occupant injury or death.

If equipped with OCS, the front passenger seat must be handled carefully as well. When removing the seat, be careful when

setting on floor not to drop. If dropped, the sensor may be inoperative, could result in occupant injury, or possibly death.

If equipped with OCS, when the passenger front seat is on the floor, no one should sit in the front passenger seat. This uneven force may damage the sensing ability of the seat weight sensors. If sat on and damaged, the sensor may be inoperative, could result in occupant injury, or possibly death.

#### DISARMING THE SYSTEM

There are two procedures for disarming the system.

Use the Negative Battery Cable procedure under the following circumstances:
- Vehicle involved in an accident with air bag deployment
- Removing or replacing components
- Vehicle is suspected of having shorted electrical wires

Use the Air Bag Fuse method only when performing electrical diagnosis on components other than Supplemental Inflatable Restraint (SIR/Airbag) components

#### Air Bag Fuse

#### ✷✷ WARNING

**Only use this procedure when performing electrical diagnosis on components other than Supplemental Inflatable Restraint (SIR) or Airbag components.**

1. Turn the steering wheel so that the vehicles wheels are pointing straight ahead.
2. Place the ignition in the OFF position.

➡**The Sensing and Diagnostic Module (SDM) may have more than one fused power input. To ensure there is no unwanted SIR deployment, personal injury, or unnecessary SIR system repairs, remove all fuses supplying power to the SDM. With all SDM fuses removed and the ignition switch in the ON position, the AIR BAG warning indicator illuminates. This is normal operation, and does not indicate a SIR system malfunction.**

3. Locate and remove the fuse(s) supplying power to the SDM.
4. Wait 1 minute before working on the system.

#### Negative Battery Cable

1. Turn the steering wheel so that the vehicles wheels are pointing straight ahead.
2. Place the ignition in the OFF position.

3. Disconnect the negative battery cable.
4. Wait 1 minute before working on system.

## ARMING THE SYSTEM

### Air Bag Fuse

**✳✳ WARNING**

**Only use this procedure when performing electrical diagnosis on components other than Supplemental Inflatable Restraint (SIR) or Airbag components.**

1. Place the ignition in the OFF position.
2. Install the fuse(s) supplying power to the SDM.
3. Turn the ignition switch to the ON position. The AIR BAG indicator will flash then turn OFF.

### Negative Battery Cable

1. Place the ignition in the OFF position.
2. Connect the negative battery cable.
3. Turn the ignition switch to the ON position. The AIR BAG indicator will flash then turn OFF.

## INFLATABLE RESTRAINT MODULE COIL CENTERING

*See Figures 11 through 15.*

**✳✳ WARNING**

**A new inflatable restraint steering wheel module coil is pre-centered. Do not remove the centering tab from the new inflatable restraint steering wheel module coil until installation is complete.**

➡ The new SIR coil assembly will be centered. Improper alignment of the

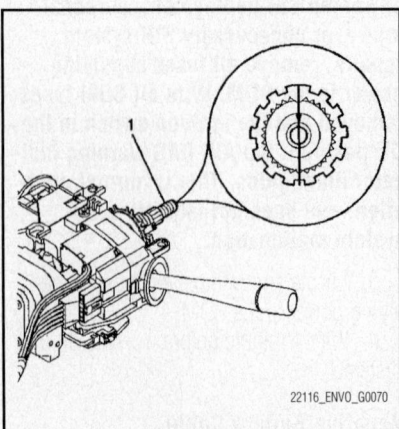

**Fig. 11 Verify that the block tooth of the steering shaft assembly is in the 12 o'clock position**

SIR coil assembly may damage the unit, causing an inflatable restraint malfunction.

1. Verify the following conditions before centering the SIR coil:
- The wheels on the vehicle are straight ahead
- The block tooth of the steering shaft assembly is in the 12 o'clock position
- The ignition switch assembly is in the LOCK position

2. If the front of the SIR coil has a centering window, and the back side has a spring service lock , perform the following steps:
   a. Hold the coil with the face up.
   b. While depressing the spring service lock, rotate the coil hub clockwise until the coil ribbon stops.
   c. Rotate the coil hub slowly, counter-clockwise, until the centering window appears yellow and both arrows line up.
   d. Release the spring service lock between the locking tab. The SIR coil is now centered.
   e. Align the centered SIR coil with the horn tower and slide onto the steering shaft assembly.

3. If the front of the SIR coil has a centering window and the back side has

NO spring service lock, perform the following steps:
   a. Hold the coil with the face up.
   b. Rotate the coil hub clockwise until the coil ribbon stops.
   c. Rotate the coil hub slowly, counter-clockwise until the centering window appears yellow and both arrows line up. This is the CENTER position.
   d. While holding the coil hub in the CENTER position, align the coil with the horn tower and slide the coil onto the steering shaft assembly.

4. If no centering window is present on the front side of the SIR coil, but a spring service lock is on the back side , perform the following steps:
   a. Hold the coil with the back side up.
   b. While depressing the spring service lock, rotate the coil hub in the direction of the arrow until the coil ribbon stops.
   c. Still pressing the spring service lock, rotate the coil hub in the opposite direction 21/2 revolutions.
   d. Release the spring service lock between the locking tabs. The SIR coil is now centered.
   e. Align the centered coil with the horn tower and slide the coil onto the steering shaft assembly.

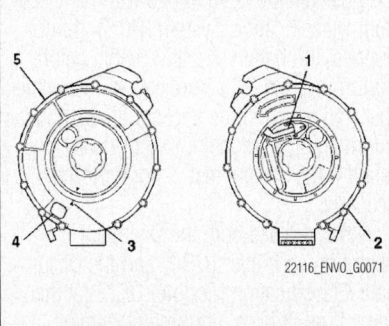

22116_ENVO_G0071

**Fig. 12 Spring service lock, back side, alignment arrows, centering window and front of the SIR coil**

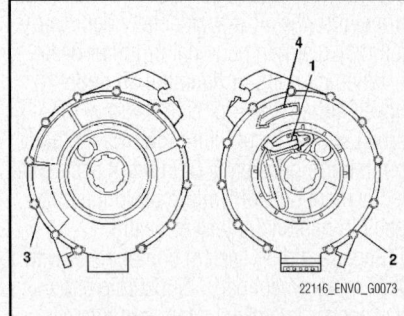

22116_ENVO_G0073

**Fig. 14 Spring service lock, back side, front side and directional arrow of the SIR coil**

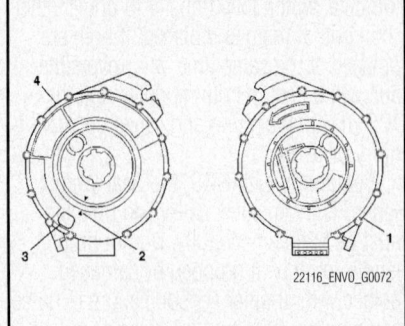

22116_ENVO_G0072

**Fig. 13 Back side, alignment arrows , centering window and front of the SIR coil**

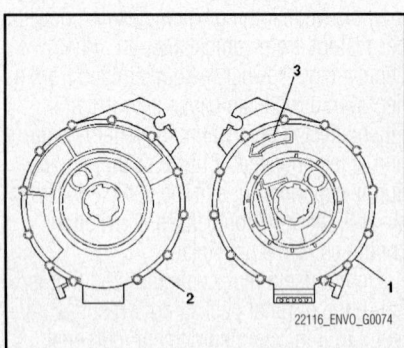

22116_ENVO_G0074

**Fig. 15 Back side, front side and directional arrow of the SIR coil**

5. If no centering window appears on the front side of the SIR coil and no spring service lock exists on the back side, perform the following steps:

a. Hold the coil with the face up.

b. Rotate the coil hub in the direction of the arrow until the coil ribbon stops.

c. Rotate the coil hub, slowly, counterclockwise, for 2 1/2 revolutions. This is the CENTER position.

d. While maintaining the coil hub in the CENTER position, align the centered coil with the horn tower and slide the coil onto the steering shaft assembly.

## DRIVETRAIN

### AUTOMATIC TRANSMISSION ASSEMBLY

*REMOVAL & INSTALLATION*

**4.2L Engine**

➡**This procedure requires the use of a Converter Holding Strap tool No. J 21366 Converter Holding Strap to secure the torque converter to the transmission during removal and installation.**

1. Disconnect the negative battery.
2. Drain the transmission fluid.
3. Remove the filler tube nut and stud located on the right side of the engine.
4. Raise the vehicle.
5. Remove the transfer case.
6. Support the transmission with a transmission jack.
7. Remove the fuel tank shield if equipped.
8. Remove the transmission support.
9. Remove the transmission mount bolts and mount.
10. Remove the front exhaust pipe assembly.
11. Lower the transmission for access to the top and sides of the transmission.
12. Remove the range selector cable end from the transmission range selector lever ball stud and bracket.
13. Remove the transmission heat shield, transmission vent hose park/neutral position switch connector, and main connector from the transmission.
14. Remove the bolt that secures the fuel line bracket to the left side of the transmission.
15. Remove the flywheel-to-torque converter bolts. Be careful not to drop the bolts into the bell housing.
16. Disconnect the transmission oil cooler lines from the transmission. Plug the transmission oil cooler lines connectors in the transmission case.
17. Install a safety chain around the transmission.
18. Remove the bolt that secures the fuel line bracket to the bell housing.
19. Remove the bolts that secure the coolant pipe to the bell housing.

20. Remove the remaining nuts, studs and/or bolts that secure the transmission to the engine.
21. Install Converter Holding Strap tool No. J 21366 Converter Holding Strap onto the transmission bell housing to hold the torque converter.
22. Pull the transmission straight back and remove it from the vehicle.

*To install:*

Installation is the reverse of removal, but please note the following important steps.
23. Make sure the torque converter is fully seated in the pump drive. If not, the transmission will not fit tightly to the rear of the engine block.
24. Raise the transmission into position and remove the torque converter holding strap. Carefully slide the transmission forward until the dowel pins are engaged while lining up the marks on the flywheel made during removal.
25. The torque converter should be flush with the flywheel and turn freely by hand.
26. Tighten the torque converter-to-flywheel bolts to 44 ft. lbs. (66 Nm).
27. Install the transmission-to-engine nuts, studs and or bolts. Tighten the studs and/or bolts to 37 ft. lbs. (50 Nm).
28. Tighten the bolts securing the heat shield to the transmission to 13 ft. lbs. (17 Nm).
29. Tighten the bolts and washers securing the transmission mount to 18 ft. lbs. (25 Nm).
30. Tighten the nut and washer securing the transmission mount to the transmission support to 35 ft. lbs. (46 Nm).
31. Refill the transmission with the proper amount and type of fluid.
32. Connect the negative battery cable. Start the vehicle and allow to warm while checking for leaks. Road test the vehicle to check for shift quality.

**5.3L and 6.0L Engines**

➡**This procedure requires the use of a Converter Holding Strap tool No. J 21366 Converter Holding Strap to secure the torque converter to the transmission during removal and installation.**

1. Disconnect the negative battery.
2. Drain the transmission fluid.
3. Raise and support the vehicle.
4. Remove the rear propeller shaft.
5. Support the transmission with a jack.
6. Remove the nuts securing the transmission mount to the transmission support.
7. Remove the transmission support from the vehicle.
8. Remove the transmission mount.
9. Remove the front exhaust pipe assembly.
10. Lower the transmission to gain access to the top and sides of the transmission.
11. Remove the transfer case, if equipped.
12. Remove the range selector cable end from the transmission range selector lever ball stud and the bracket.
13. Remove the transmission heat shield.
14. Disconnect the transmission vent hose, the park/neutral position switch connectors, and the main electrical connector from the transmission.
15. Remove the transmission harness from the retainers.
16. Remove the bolt that secures the fuel line bracket to the left side of the transmission.
17. Remove the torque converter access plug.
18. Mark the flywheel and torque converter orientation for reassembly.
19. Remove the flywheel to torque converter bolts. Use care not to drop the bolts into the bell housing.
20. Disconnect the transmission oil cooler lines from the transmission.
21. Plug the transmission oil cooler line connectors in the transmission case.
22. Install a safety chain around the transmission.
23. Remove the nut that secures the filler tube to the bell housing.
24. Remove the transmission filler tube.
25. Remove the remaining nuts, studs and/or bolts that secure the transmission to the engine.
26. Install the J-21366 onto the transmission bell housing to retain the torque converter.

27. Pull the transmission straight back.
28. Remove the transmission from the vehicle.

### To install:

29. Raise the transmission into place and remove the torque converter holding tool.

30. Slide the transmission straight onto the locating pins while lining up the marks on the flywheel and the torque converter made during removal. The torque converter must be flush onto the flywheel and rotate freely by hand.

31. Install nuts, studs and/or bolts securing the transmission to the engine and tighten to 37 ft. lbs. (50 Nm).

32. Install the fuel line retaining bracket to the transmission.

33. Install the flywheel-to-torque converter bolts and tighten to 44 ft. lbs. (66 Nm).

34. Install the torque converter access plug.

35. Remove the safety chain from the transmission.

36. Install the transmission filler tube.

37. Install the filler tube nut.

38. Install the transmission vent hose, fuel lines, and the wiring harness to the transmission.

39. Install the transmission harness to the retainers.

40. Install the heat shield to the transmission.

41. Install the bolts securing the heat shield to the transmission and tighten to 13 ft. lbs. (17 Nm).

42. Install the shift cable end to the transmission shift lever ball stud and bracket.

43. Install the transfer case, if equipped.

44. Install the front exhaust pipe assembly.

45. Install the transmission mount to the vehicle.

46. Install the bolts securing the transmission mount to the transmission and tighten to 18 ft. lbs. (25 Nm).

47. Install the transmission support to the vehicle.

48. Lower the transmission and remove the transmission jack.

49. Install the nuts securing the transmission mount to the transmission support and tighten to 35 ft. lbs. (46 Nm).

50. Install the rear propeller shaft.

51. Flush the transmission oil cooler and cooling lines at this time, if necessary.

52. Connect the transmission oil cooler lines to the transmission.

53. Lower the vehicle.

54. Connect the battery cable.

55. Fill the transmission to the proper level with DEXRON® III transmission fluid and check for leaks.

56. Road test the vehicle and check for proper operation.

## TRANSFER CASE ASSEMBLY

### REMOVAL & INSTALLATION

*See Figure 16.*

1. Before servicing the vehicle, refer to the Precautions Section.

2. Raise and safely support the vehicle securely on jackstands.

3. Remove the rear propeller shaft.

4. Remove the front propeller shaft.

5. Remove the electrical harness from the retainer.

6. Disconnect the electrical connector from the speed sensor.

7. Disconnect the vent hose.

8. Install a transmission jack on the transfer case.

9. Remove the transfer case adapter to transfer case nuts.

10. Lower the transfer case.

### ✳✳ WARNING

**The transfer case gasket may become damaged during the removal procedure.**

11. Disconnect the electrical harness from the top of the transfer case.

12. Remove the transfer case.

13. Remove the transfer case gasket.

### To install:

14. If the transfer case gasket is damaged, replace it.

### ✳✳ WARNING

**DO NOT use silicone sealant in place or with the transfer case gasket.**

15. Install or replace the transfer case gasket.

16. Raise the transfer case and install the electrical harness.

17. Install the transfer case in the vehicle.

18. Install the transfer case adapter to transfer case nuts. Tighten the nuts to 35 ft. lbs. (47 Nm).

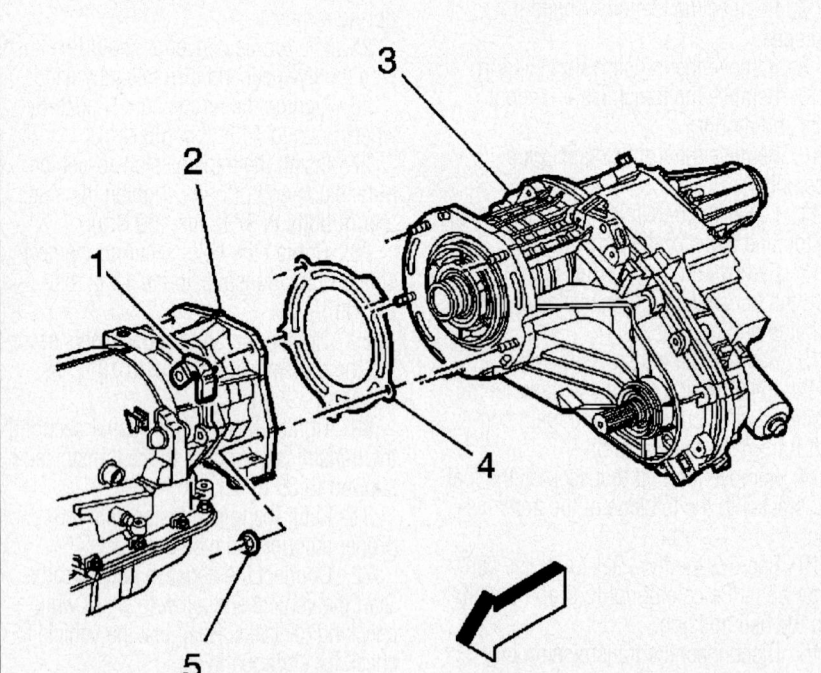

1. Fuel line retainer
2. Transfer case adapter
3. Transfer case
4. Transfer case gasket
5. Transfer case nuts

22205_S97X_G0026

**Fig. 16 Exploded view of the transmission/transfer case assembly**

19. Ensure the fuel line retainer is in place.

20. Remove the transmission jack stand.

21. Connect the electrical connector to the speed sensor.

22. Connect the vent hose.

23. Install the transfer case harness to the transfer case.

24. Install the rear propeller shaft.

25. Install the front propeller shaft.

26. Fill the transfer case with fluid.

27. Lower the vehicle.

## FRONT AXLE TUBE BEARING

### REMOVAL & INSTALLATION

The axle shaft bearing and seal are a sealed unit on this vehicle. Refer to Front Suspension, Wheel Bearing and Hub, Removal & Installation.

## FRONT HALFSHAFT

### REMOVAL & INSTALLATION

See Figures 17 through 19.

1. Raise and safely support the vehicle securely on jackstands.

2. Remove the tire and wheel assembly.

3. Remove the engine protection shield.

4. Remove the wheel speed sensor wiring harness from the retainers.

5. Disconnect the wheel speed sensor from the harness.

6. Remove the retaining bolt for the front brake hose.

7. Remove the front stabilizer bar link from the lower control arm.

8. Remove the upper shock module retaining from the shock tower.

9. Remove the tie rod end from the steering knuckle.

10. Remove the left and right upper ball joint pinch bolt and nut.

11. Remove the shock module from the shock tower.

12. Remove the steering knuckle from the upper control arm.

13. Remove the front wheel drive axle from the steering knuckle.

14. Using mechanics wire or hook, support the front shock module/steering knuckle to the frame.

15. Disconnect the left side wheel drive shaft from the differential carrier assembly by placing a brass drift against the tripod housing. Firmly strike the brass drift outward from the case with a hammer. Strike hard enough to overcome the snap ring pressure holding in the shaft.

16. Disconnect the right side wheel drive shaft from the clutch fork housing assembly by placing a brass drift against the tripod housing. Firmly strike the brass drift outward from the case with a hammer.

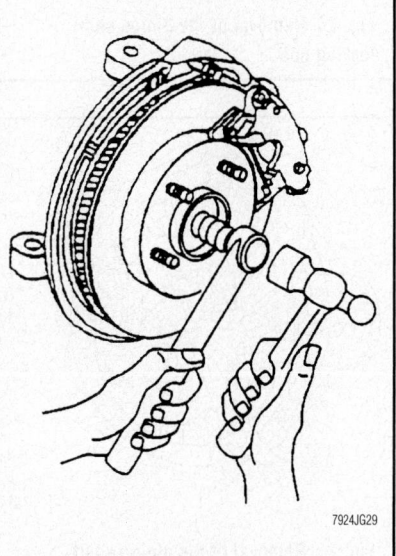

**Fig. 18 Tap the halfshaft out of the hub without damaging the threads**

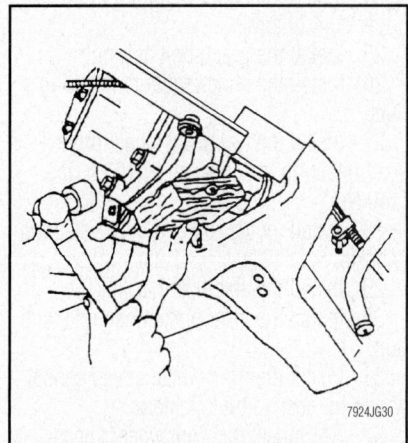

**Fig. 19 Disengage the halfshaft from the differential assembly**

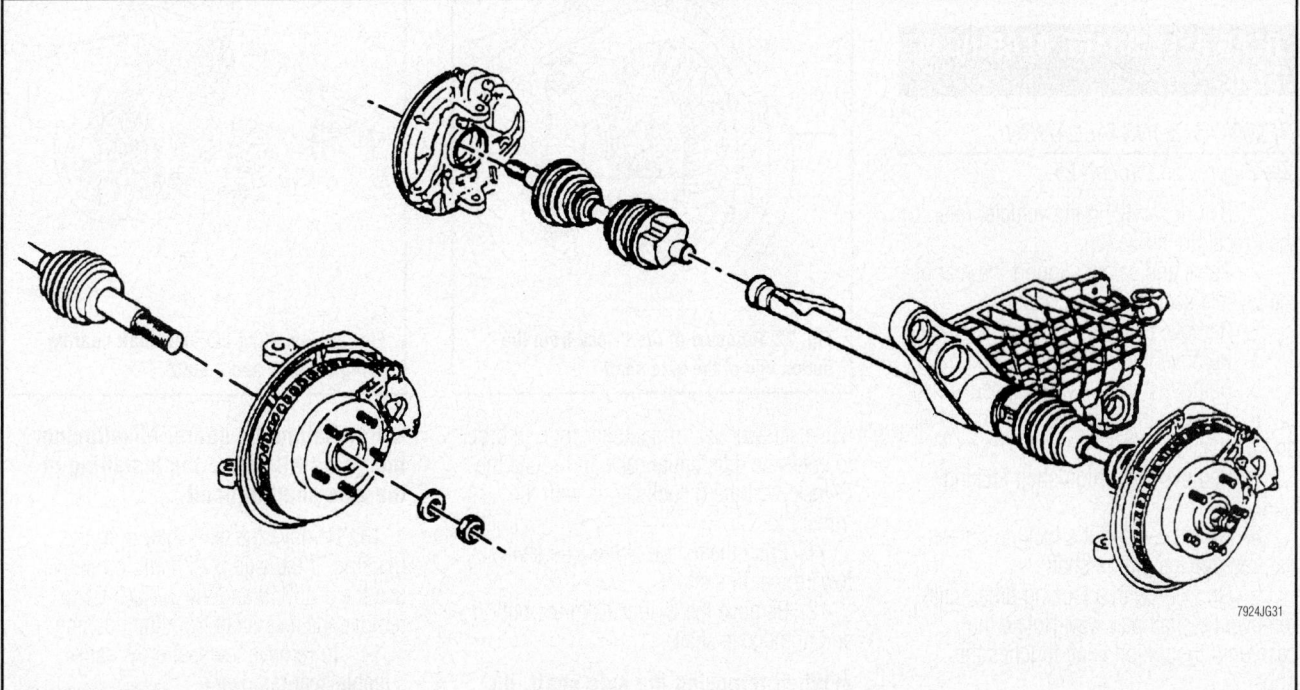

**Fig. 17 Halfshafts and related components**

Strike hard enough to overcome the snap ring pressure holding in the shaft.

17. Pull the wheel drive shaft straight out from the differential carrier assembly or the clutch fork housing assembly.

18. Remove the halfshaft.

### To install:

19. Install the front wheel drive axle front differential assembly.

20. Remove the mechanics wire or hook from the front shock module/steering

21. Install the front wheel drive axle in the steering knuckle.

22. Position the shock module in the shock tower.

23. Install the shock module into the shock tower

24. Install the upper ball joint in the upper control arm.

25. Install the pinch bolt and nut.

26. Install the shock module retaining nuts.

27. Install the new front wheel drive shaft retaining nut. Tighten to 103 ft. lbs. (140 Nm)

28. Install the tie rod end in the steering knuckle.

29. Install the stabilizer bar link.

30. Install the front brake hose retaining bolt.

31. Install the front wheel speed sensor wiring harness in the retainers.

32. Reconnect the front wheel speed sensors electrical connector.

33. Install the engine protection shield.

34. Install the tire and wheel assembly.

35. Lower the vehicle.

## REAR AXLE SHAFT, BEARING & SEAL

### REMOVAL & INSTALLATION

*See Figures 20 through 25.*

1. Before servicing the vehicle, refer to the Precautions Section.

2. Raise and safely support the rear of the vehicle securely on jackstands.

3. Remove the tire and wheel assembly.

4. Remove the brake caliper.

5. Remove the rear wheel speed sensor.

6. Remove the rear axle housing cover and the gasket.

7. Remove the pinion shaft locking bolt.

8. On axles without a locking differential, remove the pinion shaft.

9. On axles with a locking differential, remove the shaft part way. Rotate the case until the pinion shaft touches the housing.

10. On axles with a locking differential,

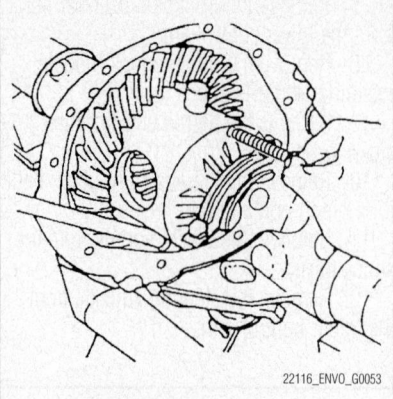

Fig. 20 Removal of the pinion shaft locking bolt

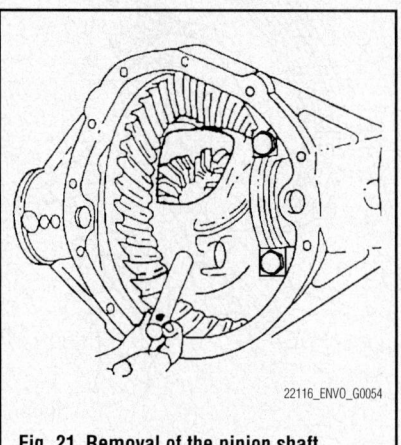

Fig. 21 Removal of the pinion shaft

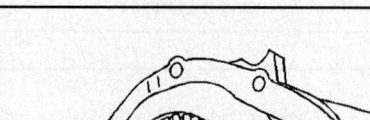
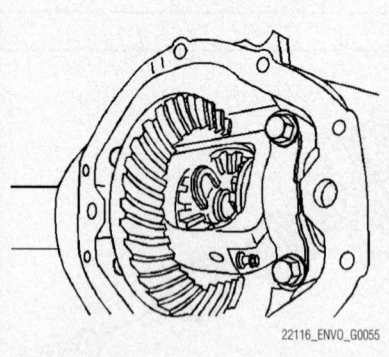

Fig. 22 Removal of the C-lock from the button end of the axle shaft

use a screwdriver, or a similar tool, in order to enter the differential case and rotate the C-lock until the C-lock aligns with the thrust block.

11. Push the flange of the axle shaft toward the differential.

12. Remove the C-lock from the button end of the axle shaft.

➡**When removing the axle shaft, do not rotate the shaft. Rotating the shaft**

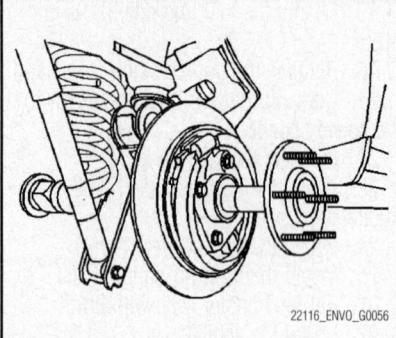

Fig. 23 Removal of the axle shaft from the housing

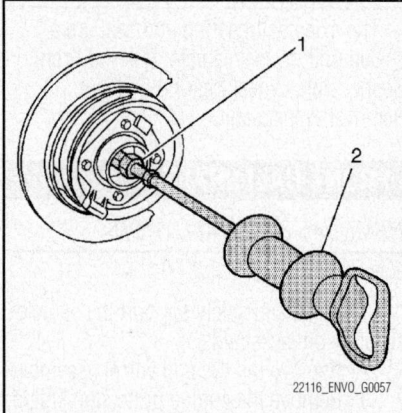

Fig. 24 Removal of the axle shaft seal and bearing together using special tools J-45857 and J-2619-01

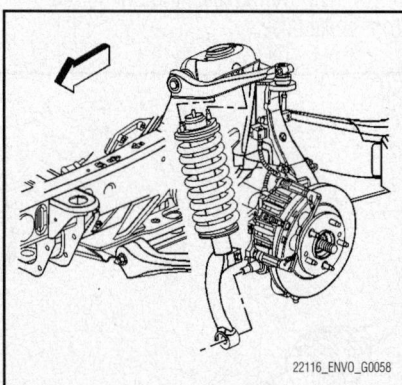

Fig. 25 Installing the axle shaft bearing using J-23690 and J-8092

**will misalign the gears. Misaligning the gears will make the installing of the axle shaft difficult.**

13. Remove the axle shaft from the housing. If the axle is difficult to remove, use the J-45859 and the J-2619-01 to remove the axle shaft from the housing.

14. To remove the seal only, use a suitable seal remover.

15. To remove the axle shaft seal and

the bearing together from the axle housing, use special tools J-45857 and J-2619-01.

### To install:

16. Using the J-23690 and the J-8092, install the axle shaft bearing.

17. Drive the axle shaft bearing into the axle housing until the tool bottoms against the tube.

18. Using the J-21128, install the axle shaft seal.

19. Drive the tool into the bore until the axle shaft seal bottoms flush with the tube.

### ✳✳ WARNING

**Carefully insert the axle shaft in order to not damage the seal.**

20. Install the axle shaft into the rear axle housing.

21. Slide the axle shaft into place allowing the splines to engage the differential side gear.

22. On axles without a locking differential, place the C-lock on the button end of the axle shaft.

23. On axles with a locking differential, keep the pinion shaft partially withdrawn.

24. On axles with a locking differential, place the C-lock on the axle shaft so that the ends are flush with the thrust block.

25. Pull the shaft flange outward in order to seat the C-lock in the differential gear.

26. Align the hole in the pinion shaft with the bolt hole in the differential case.

27. Install the new pinion shaft locking bolt:

- For the 8.0/8.6 inch axle, tighten the pinion shaft locking bolt to 27 ft. lbs. (36 Nm).
- For the 9.5 LD inch axle, tighten the pinion shaft locking bolt to 37 ft. lbs. (50 Nm).

➡**The axle housing gasket is reusable. Replace only if damaged.**

28. Install the axle housing cover gasket and axle housing cover.

29. Install the mounting bolts:

- For the 9.5 inch axle, tighten the rear axle housing cover bolts in a crosswise pattern to 30 ft. lbs. (40 Nm).
- For the 8.0 inch axle, tighten the rear axle housing cover bolts in a crosswise pattern to 20 ft. lbs. (30 Nm).
- For the 8.6 inch axle, tighten the rear axle housing cover bolts in a crosswise pattern to 18 ft. lbs. (25 Nm).

30. Install the drain plug and tighten to 24 inch lbs. (33 Nm).

31. Fill the rear axle with the proper axle lubricant as follows:

- For the 8.0 and 8.6 inch axles, the lubricant level should be between 0–0.4 inch (0–10mm) below the fill plug opening.
- For the 9.5 inch axle, the lubricant level should be between 0–0.5 inch (0–13mm) below the fill plug opening.

32. Install the brake caliper.

33. Install the rear wheel speed sensor.

34. Install the tire and wheel assembly.

35. Fill the rear axle with axle lubricant. Use the proper fluid.

36. Lower the vehicle.

### REAR HALFSHAFT

#### REMOVAL & INSTALLATION

*See Figures 26 through 28.*

1. Raise and safely support the vehicle securely on jackstands.

2. Remove the tire and wheel assembly.

3. Remove the engine protection shield.

4. Remove the wheel speed sensor wiring harness from the retainers.

5. Disconnect the wheel speed sensor from the harness.

6. Remove the retaining bolt for the front brake hose.

7. Remove the front stabilizer bar link from the lower control arm.

8. Remove the upper shock module retaining from the shock tower.

9. Remove the tie rod end from the steering knuckle.

10. Remove the left and right upper ball joint pinch bolt and nut.

11. Remove the shock module from the shock tower.

12. Remove the steering knuckle from the upper control arm.

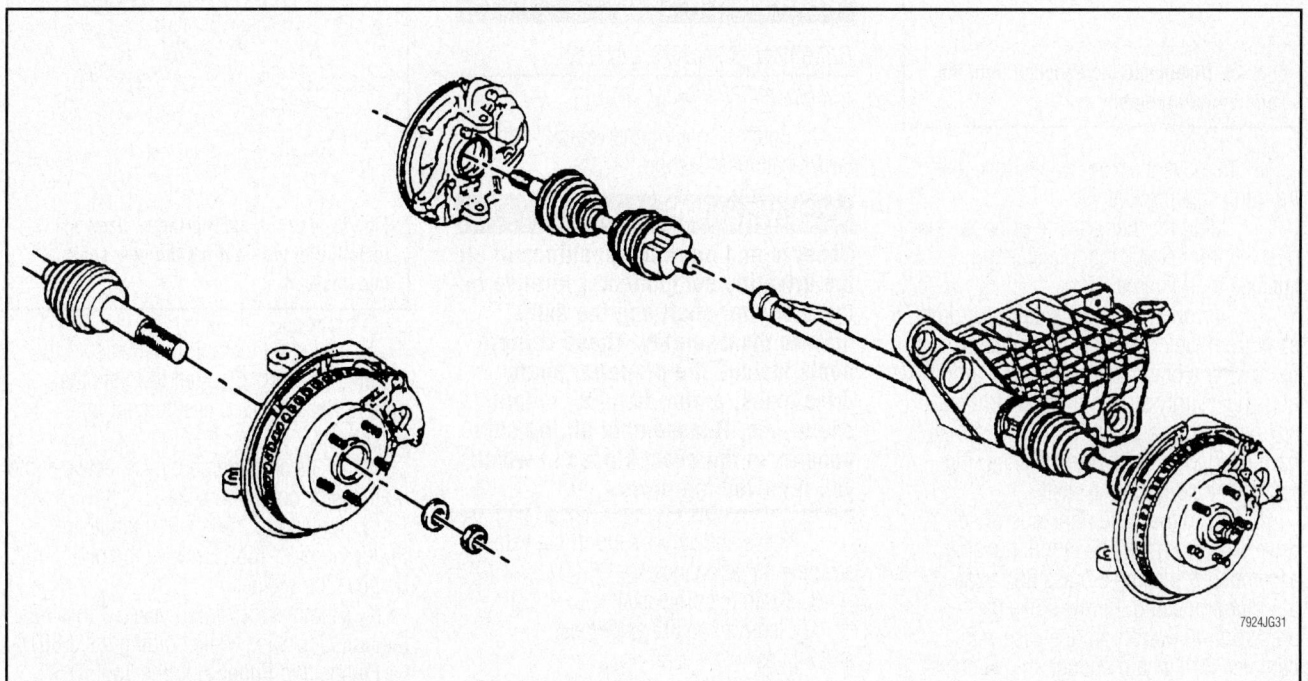

**Fig. 26 Halfshafts and related components**

7924JG31

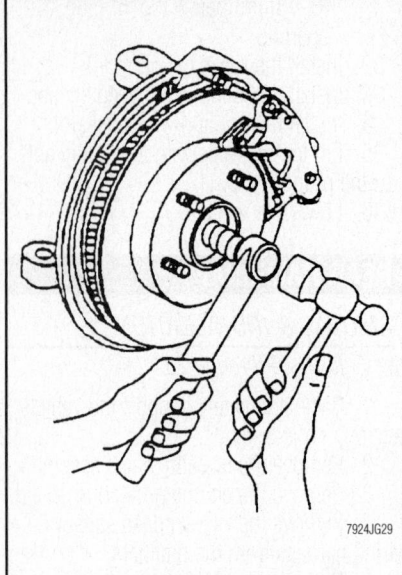

**Fig. 27 Tap the halfshaft out of the hub without damaging the threads**

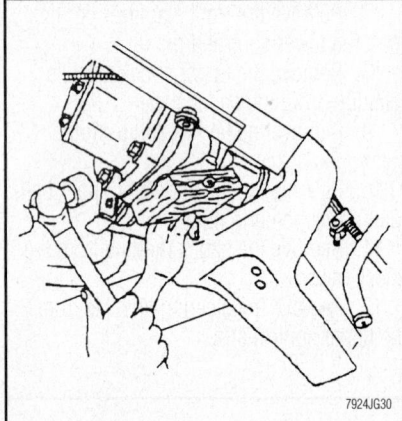

**Fig. 28 Disengage the halfshaft from the differential assembly**

13. Remove the front wheel drive axle from the steering knuckle.

14. Using mechanics wire or hook, support the front shock module/steering knuckle to the frame.

15. Disconnect the left side wheel drive shaft from the differential carrier assembly by placing a brass drift against the tripod housing. Firmly strike the brass drift outward from the case with a hammer. Strike hard enough to overcome the snap ring pressure holding in the shaft.

16. Disconnect the right side wheel drive shaft from the clutch fork housing assembly by placing a brass drift against the tripod housing. Firmly strike the brass drift outward from the case with a hammer. Strike hard enough to overcome the snap ring pressure holding in the shaft.

17. Pull the wheel drive shaft straight out from the differential carrier assembly or the clutch fork housing assembly.

18. Remove the halfshaft.

*To install:*

19. Install the front wheel drive axle front differential assembly.

20. Remove the mechanics wire or hook from the front shock module/steering

21. Install the front wheel drive axle in the steering knuckle.

22. Position the shock module in the shock tower.

23. Install the shock module into the shock tower

24. Install the upper ball joint in the upper control arm.

25. Install the pinch bolt and nut.

26. Install the shock module retaining nuts.

27. Install the new front wheel drive shaft retaining nut. Tighten to 103 ft. lbs. (140 Nm)

28. Install the tie rod end in the steering knuckle.

29. Install the stabilizer bar link.

30. Install the front brake hose retaining bolt.

31. Install the front wheel speed sensor wiring harness in the retainers.

32. Reconnect the front wheel speed sensors electrical connector.

33. Install the engine protection shield.

34. Install the tire and wheel assembly.

35. Lower the vehicle.

## REAR PINION SEAL

### REMOVAL & INSTALLATION

*See Figures 29 through 31.*

1. Before servicing the vehicle, refer to the Precautions Section.

### ※※ WARNING

**Observe and mark the positions of all the driveline components, relative to the propeller shaft and the axles, prior to disassembly. These components include the propeller shafts, drive axles, pinion flanges, output shafts, etc. Reassemble all the components in the exact places in which you removed the parts.**

2. Raise and safely support the vehicle securely on jackstands.

3. Drain the drive axle.

4. Remove the tire and wheel assemblies.

5. Remove the brake rotors.

6. Remove the propeller shaft.

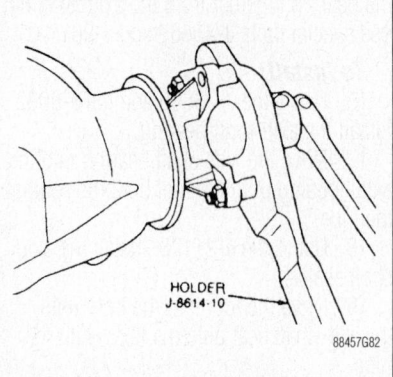

**Fig. 29 Removing the pinion nut using a J 8614-01 Flange and Pulley Holding Tool**

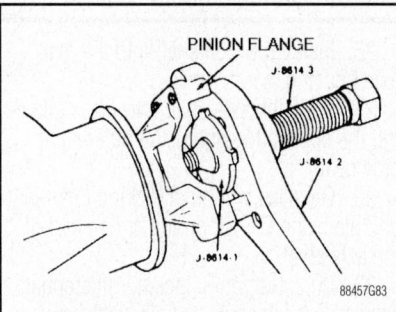

**Fig. 30 A puller and adapter should be used to withdraw the pinion from the housing**

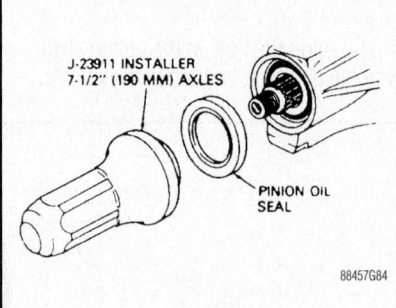

**Fig. 31 Use the appropriately sized installation tool to drive the new seal into position**

7. Using an inch pound, measure the amount of torque required to rotate the pinion. Record this measurement for reassembly.

8. Place an alignment mark between the pinion and the pinion yoke.

9. Install the J 8614-01 Flange and Pulley Holding Tool. Remove the pinion nut. Remove the washer.

10. Remove the pinion yoke by turning the bolt clockwise while holding the J 8614-01 Flange and Pulley Holding Tool. Use a container in order to retrieve the lubricant.

11. Using a suitable seal remover, remove the pinion oil seal. Do not damage the housing.

*To install:*

12. Examine the seal surface of pinion flange for tool marks, nicks or damage, such as a groove worn by the seal. If damaged, replace flange.

13. Using the J 33782 Seal Installer, install a new pinion oil seal.

14. Apply sealant GM P/N 12346004 or equivalent to the splines of the pinion yoke.

**✱✱ WARNING**

**Align the marks made during removal.**

15. Install the pinion yoke.

16. Seat the pinion yoke onto the pinion shaft by tapping it with a soft-faced hammer until a few pinion shaft threads show through the yoke.

17. Install the washer and a new pinion nut.

18. Install the J 8614-01 Flange and Pulley Holding Tool onto the pinion yoke.

**✱✱ WARNING**

**If the rotating torque is exceeded, the pinion will have to be removed and a new collapsible spacer installed.**

19. Using the J 8614-01 to hold the pinion nut, tighten the pinion nut until the pinion end play is just taken up. Rotate the pinion while tightening the nut to seat the bearings.

20. Measure the rotating torque of the pinion. Compare this measurement with the rotating torque recorded during removal.

21. Tighten the nut in small increments, as needed, until the rotating torque is 3–5 inch lbs. (0.40-0.57 Nm) greater than the rotating torque recorded during removal.

22. Once the specified torque is obtained, rotate the pinion several times to ensure the bearings have seated. Recheck the rotating torque and adjust if necessary.

23. Install the propeller shaft.

24. Install the brake rotors.

25. Install the tire and wheel assemblies.

26. Fill the drive axle.

27. Remove the support and lower the vehicle.

# ENGINE COOLING

## THERMOSTAT

*REMOVAL & INSTALLATION*

### 4.2L Engine

*See Figure 32.*

1. Before servicing the vehicle, refer to the Precautions Section.

2. Drain the cooling system to a level below the thermostat.

3. Remove the alternator, as outlined in the Engine Electrical Section.

4. Loosen the outlet hose clamp at the thermostat housing. Remove the outlet hose from the thermostat housing.

5. Remove the thermostat housing bolts.

6. Remove the thermostat housing from the engine block.

7. Clean all of the surfaces of the thermostat housing.

8. Clean the sealing surface of the engine block.

*To install:*

9. Install the thermostat housing to the engine block.

10. Install the thermostat housing bolts and tighten to 89 inch lbs. (10 Nm).

11. Lubricate the inner diameter of the radiator hose with engine coolant.

12. Install the outlet hose to the thermostat housing. Secure the hose with the clamp.

13. Install the alternator.

14. Fill the cooling system with specified coolant and concentration.

15. Inspect all sealing surfaces for leaks after starting the engine.

### 5.3L and 6.0L Engines

*See Figure 33.*

➥The thermostat is not serviceable separately. The water pump inlet and thermostat must be replaced as an assembly.

1. Before servicing the vehicle, refer to the Precautions Section.

2. Drain the cooling system to a level below the thermostat.

3. Remove the radiator outlet hose.

4. Remove the water pump inlet bolts.

5. Remove the water pump inlet and thermostat from the water pump.

*To install:*

6. Install the thermostat and thermostat housing to the water pump.

7. Install the thermostat housing bolts. Tighten the bolts to 11 ft. lbs. (15 Nm).

8. Install the radiator outlet hose.

9. Properly fill the engine cooling system and check for leaks.

## WATER PUMP

*REMOVAL & INSTALLATION*

*See Figures 34 and 35.*

1. Before servicing the vehicle, refer to the Precautions Section.

2. Drain the engine cooling system.

3. Remove the fan and shroud.

4. On 5.3L and 6.0L engines, loosen the air cleaner outlet duct clamps and remove the air cleaner outlet duct.

5. Relieve the belt tension and remove the accessory drive belts or the serpentine drive belt, as applicable.

6. Remove the water pump pulley using a suitable tool to hold the pulley while removing the bolts

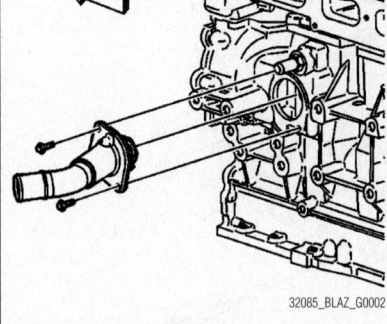

32085_BLAZ_G0002

**Fig. 32 Thermostat mounting—4.2L engine**

22205_S97X_G0029

**Fig. 33 Thermostat mounting—5.3L and 6.0L engines**

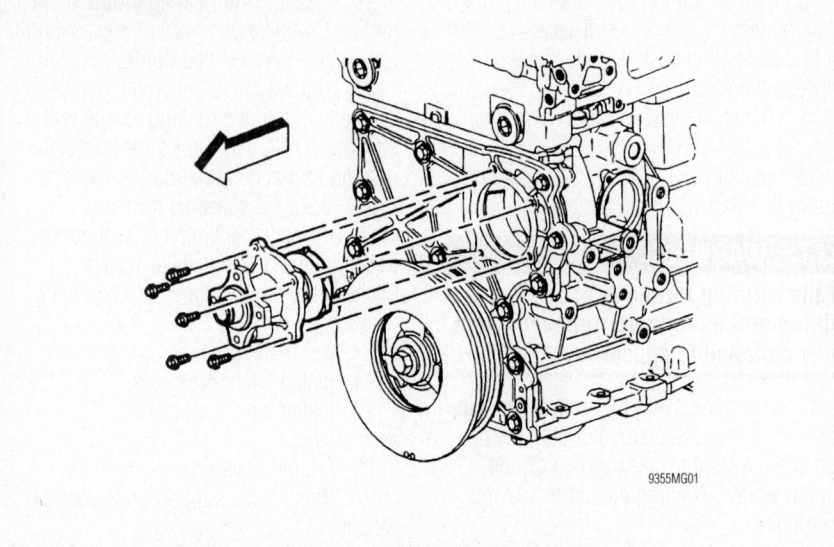

**Fig. 34 Exploded view of the water pump assembly mounting—4.2L engine**

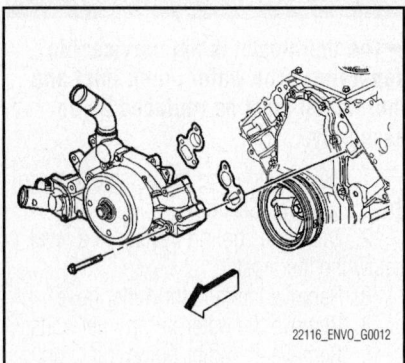

**Fig. 35 Exploded view of the water pump assembly mounting—5.3L and 6.0L engines**

7. Remove the coolant hose(s) from the water pump.

➡For the hoses on some engines, removal may be easier if the hose is left attached until the pump is free from the block. Once the pump is removed from the engine, the pump may be pulled (giving a better grip and greater leverage) from the tight hose connection.

8. Remove the water pump bolts, then the water pump.

### ✳✳ WARNING

Note the positions of all retainers as some engines will utilize different length fasteners in different locations and/or bolts and studs in different locations.

*To install:*

9. Clean the gasket mounting surfaces.

➡The water pumps on some of the engines covered may have been installed using sealer only, no gasket, at the factory. If a gasket is supplied with the replacement part, it should be used. Otherwise, a ⅛ in. (3mm) bead of RTV sealer should be used around the sealing surface of the pump.

10. Apply sealant to the water pump retainer threads.

11. Install water pump using a new gasket. Tighten the water pump retainers in two stages to 35 inch lbs. (4 Nm) and then 89 inch lbs. (10 Nm).

12. Install the water pump pulley. Tighten the pulley bolts to 18 ft. lbs. (25 Nm).

13. Install the coolant hose(s).

14. Install the serpentine drive belt (if equipped) by positioning the belt over the pulleys and carefully allow the tensioner back into contact with the belt. As required, install the V-belts (if equipped) and adjust the tension.

15. On 5.3L and 6.0L engines, install the air cleaner outlet duct and tighten the air cleaner outlet duct clamps.

16. Install the fan and the shroud.

17. Refill the engine cooling system.

18. Run the engine and check for leaks.

# ENGINE ELECTRICAL

## ALTERNATOR

### REMOVAL & INSTALLATION

#### 4.2L Engine

*See Figure 36.*

1. Before servicing the vehicle, refer to the Precautions Section.

2. Disconnect the negative battery cable.

3. Remove the accessory drive belt.

4. Remove the A/C line mounting bracket bolt at the engine lift hook.

5. Remove the right engine lift hook bolts. Remove the engine lift hook.

6. Remove the 3 alternator mounting bolts and remove the alternator.

7. Disconnect the battery positive cable nut on the alternator.

8. Remove the alternator.

# CHARGING SYSTEM

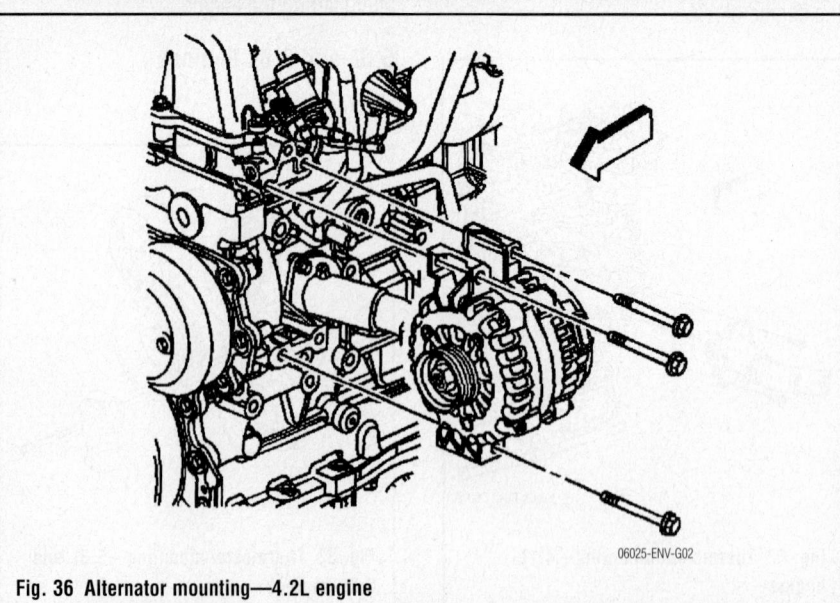

**Fig. 36 Alternator mounting—4.2L engine**

*To install:*

9. Connect the battery positive cable nut on the alternator. tighten the nut to 80 inch lbs. (9 Nm).

10. Install the alternator. Tighten the bolts to 37 ft. lbs. (50 Nm).

11. Install the right engine lift hook. Tighten to 37 ft. lbs. (50 Nm)

12. Install the A/C line mounting bracket bolt at the engine lift hook. Tighten the retaining bolt to 89 inch lbs. (10 Nm).

13. Install the accessory drive belt.

14. Connect the negative battery cable.

### 5.3L and 6.0L Engines

*See Figure 37.*

1. Before servicing the vehicle, refer to the Precautions Section.

2. Disconnect the negative battery cable.

3. Disconnect the alternator electrical connector.

4. Remove the alternator cable from the alternator, perform the following:

    a. Slide the boot down revealing the terminal stud.

    b. Remove the alternator cable nut from the terminal stud.

    c. Remove the alternator cable.

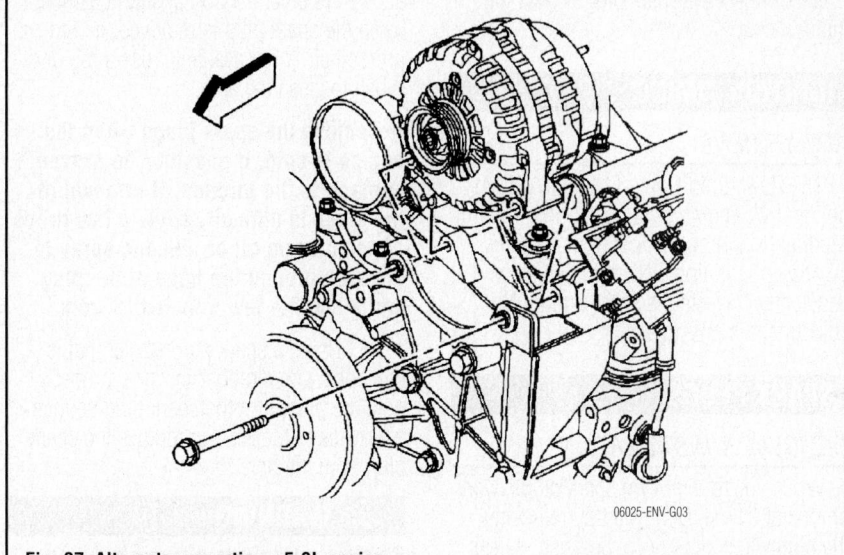

06025-ENV-G03

**Fig. 37 Alternator mounting—5.3L engine**

5. Remove the alternator bolts, then the alternator.

*To install:*

6. Install the alternator and tighten the bolts to 37 ft. lbs. (50 Nm).

7. Install the alternator cable, perform the following:

    a. Install the alternator cable.

    b. Install the alternator cable nut on the terminal stud. Tighten to 80 inch lbs. (9 Nm)

    c. Slide the boot down covering the terminal stud.

8. Connect the alternator electrical connector.

9. Connect the negative battery cable.

---

## ENGINE ELECTRICAL

### FIRING ORDER

The firing order for the 4.2L engine is 1-5-3-6-2-4.

The firing order for the 5.3L and 6.0L engines is 1-8-7-2-6-5-4-3.

### IGNITION COIL

*REMOVAL & INSTALLATION*

#### 4.2L Engine

*See Figure 38.*

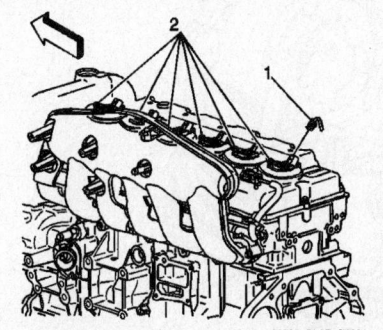

32085_BLAZ_G0001

**Fig. 38 Detach the connectors and remove the bolts from the ignition coils**

1. Before servicing the vehicle, refer to the Precautions Section.

2. Remove the air cleaner outlet resonator.

3. Disconnect the ignition coil connectors from the ignition coils.

4. Remove the retaining bolts from the ignition coils.

5. Remove the ignition coils from the engine.

*To install:*

#### ※※ WARNING

**Make sure that the ignition coil seals are properly seated to the valve cover.**

6. Install the ignition coil.

7. Install the ignition coil retaining bolts and tighten to 89 inch lbs. (10 Nm).

8. Replace the ignition coil connectors.

9. Install the air cleaner outlet resonator.

#### 5.3L and 6.0L Engines

*See Figure 39.*

1. Before servicing the vehicle, refer to the Precautions Section.

2. Remove the spark plug wire from the ignition coil.

## IGNITION SYSTEM

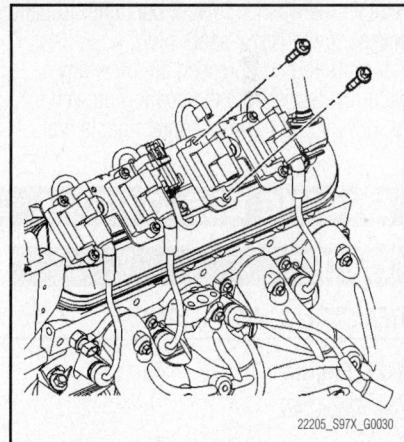

22205_S97X_G0030

**Fig. 39 Detach the connectors and remove the bolts from the ignition coils**

3. Disconnect the ignition coil electrical connector.

4. Remove the ignition coil bolts.

5. Remove the ignition coil.

*To install:*

6. Install the ignition coil.

7. Install the ignition coil bolts and tighten to 71 inch lbs. (8 Nm).

8. Connect the ignition coil electrical connector.

9. Connect the spark plug wire to the ignition coil.

## IGNITION TIMING

### ADJUSTMENT

The Powertrain Control Module (PCM) on the 4.2L engine or the Electronic Control Module (ECM) on the 5.3L/6.0L engines controls all ignition system functions, and constantly corrects the spark timing. No adjustment is possible.

## SPARK PLUGS

### REMOVAL & INSTALLATION

When you're removing spark plugs, work on one at a time. Don't start by removing the plug wires all at once, because, unless you number them, they may become mixed up. Take a minute before you begin and number the wires with tape.

1. Before servicing the vehicle, refer to the Precautions Section.

2. On the 4.2L engine, remove the ignition coils. On the 5.3L and 6.0L engines, carefully twist the spark plug wire boot to loosen it, then remove the boot from the plug. Be sure to pull on the boot and not on the wire, otherwise the connector located inside the boot may become separated.

3. On the 5.3L and 6.0L engines, remove the washer solvent container to gain access to the No. 2 spark plug.

4. Using compressed air, blow any water or debris from the spark plug well to assure that no harmful contaminants are allowed to enter the combustion chamber when the spark plug is removed. If compressed air is not available, use a rag or a brush to clean the area.

➡Remove the spark plugs when the engine is cold, if possible, to prevent damage to the threads. If removal of the plugs is difficult, apply a few drops of penetrating oil or silicone spray to the area around the base of the plug, and allow it a few minutes to work.

5. Using a spark plug socket that is equipped with a rubber insert to properly hold the plug, turn the spark plug counterclockwise to loosen and remove the spark plug from the bore.

### ❋❋ WARNING

**Be sure not to use a flexible extension on the socket. Use of a flexible extension may allow a shear force to be applied to the plug. A shear force could break the plug off in the cylinder head, leading to costly and frustrating repairs.**

### To install:

6. Inspect the spark plug boot for tears or damage. If a damaged boot is found, the spark plug wire must be replaced.

7. Using a wire feeler gauge, check and adjust the spark plug gap. When using a gauge, the proper size should pass between the electrodes with a slight drag. The next larger size should not be able to pass while the next smaller size should pass freely.

8. Carefully thread the plug into the bore by hand. If resistance is felt before the plug is almost completely threaded, back the plug out and begin threading again. In small, hard to reach areas, an old spark plug wire and boot could be used as a threading tool. The boot will hold the plug while you twist the end of the wire and the wire is supple enough to twist before it would allow the plug to crossthread.

### ❋❋ WARNING

**Do not use the spark plug socket to thread the plugs. Always carefully thread the plug by hand or using an old plug wire to prevent the possibility of crossthreading and damaging the cylinder head bore.**

9. Carefully tighten the spark plug. If the plug you are installing is equipped with a crush washer, seat the plug, then tighten about ¼ turn to crush the washer. If you are installing a tapered seat plug, tighten the plug to specifications provided by the vehicle or plug manufacturer.

10. On the 5.3L and 6.0L engines, install the washer solvent container to gain access to the No. 2 spark plug.

11. On the 4.2L engine, install the ignition coils. On the 5.3L and 6.0L engines, apply a small amount of silicone dielectric compound to the end of the spark plug lead or inside the spark plug boot to prevent sticking, then install the boot to the spark plug and push until it clicks into place. The click may be felt or heard, then gently pull back on the boot to assure proper contact.

## ENGINE ELECTRICAL

## STARTING SYSTEM

## STARTER

### REMOVAL & INSTALLATION

#### 4.2L Engine
*See Figure 40.*

1. Before servicing the vehicle, refer to the Precautions Section.

2. Disconnect the negative battery cable

3. Remove the vacuum brake booster hose.

4. Remove the battery positive lead from the solenoid.

5. Remove the starter solenoid S-terminal lead nut from the solenoid.

6. Remove the starter solenoid S-terminal lead from the solenoid.

7. Remove the starter mount bolt and nut.

8. Remove the starter motor.

06025-ENV-G06

**Fig. 40 Starter mounting—4.2L engine**

*To install:*

9. Install the starter motor.

10. Install the starter mount bolt and nut. Tighten to 37 ft. lbs. (50 Nm).

11. Install the starter solenoid S-terminal lead from the solenoid. Tighten the nut to 20 inch lbs. (2 Nm).

12. Install the starter solenoid S-terminal lead nut from the solenoid. Tighten the nut to 20 inch lbs. (2 Nm).

13. Install the battery positive lead from the solenoid. Tighten the nut to 80 inch lbs. (9 Nm).

14. Install the vacuum brake booster hose.

15. Connect the negative battery cable

16. Install the vacuum brake booster hose

17. Connect the negative battery cable.

### 5.3L and 6.0L Engines

*See Figure 41.*

1. Before servicing the vehicle, refer to the Precautions Section.

2. Disconnect the negative battery cable

3. Raise and safely support the vehicle securely on jackstands.

4. Remove the rear steering gear crossmember.

5. Remove the wire harness from the wire harness retaining clips on the transmission oil cooler line bracket.

6. Remove the transmission oil cooler line bracket bolt.

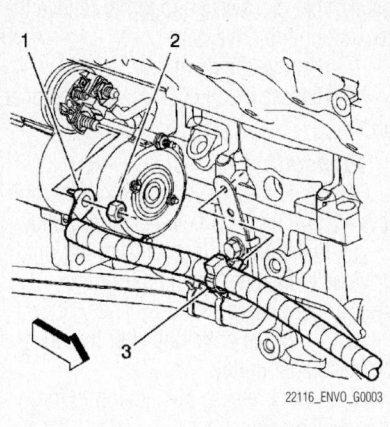

**Fig. 41 View of the starter, positive cable and starter lead nut—5.3L and 6.0L engines**

7. Remove the right transmission cover bolt.

8. Remove the starter bolts.

9. Move the starter toward the front of the vehicle, and remove the transmission cover.

10. Remove the starter solenoid heat shield.

11. Tilt and rotate the starter in order to pass the starter between the transmission oil cooler lines and the engine oil pan.

12. Remove the starter solenoid nut.

13. Remove the starter lead from the solenoid stud.

14. Remove the battery positive cable nut.

15. Remove the battery positive cable from the starter solenoid.

16. Finish removing the starter from the vehicle.

*To install:*

17. Begin installing the starter between the transmission oil cooler lines and the engine oil pan.

18. Install the battery positive cable to the starter stud. Tighten nut to 80 inch lbs. (9 Nm).

19. Install the starter solenoid lead to the solenoid stud. Tighten nut to 30 inch lbs. (3 Nm).

20. Install the starter solenoid heat shield.

21. Slide the starter toward the front of the vehicle.

22. Position the transmission cover to the transmission.

23. Position the starter to the engine. Install the starter bolts and tighten to 37 ft. lbs. (50 Nm).

24. Install right transmission cover bolt and tighten to 80 inch lbs. (9 Nm).

25. Install right transmission oil cooler line bracket bolt and tighten to 80 inch lbs. (9 Nm).

26. Attach the wire harness to the wire harness retaining clips on the transmission oil cooler line bracket.

27. Install the rear steering gear crossmember.

28. Lower the vehicle.

29. Connect the negative battery cable.

## ENGINE MECHANICAL

➡ **Disconnecting the negative battery cable may interfere with the functions of the on board computer systems and may require the computer to undergo a relearning process, once the negative battery cable is reconnected.**

### ACCESSORY DRIVE BELTS

*ACCESSORY BELT ROUTING*

*See Figures 42 and 43.*

*INSPECTION*

Inspect the drive belt for signs of glazing or cracking. A glazed belt will be perfectly smooth from slippage, while a good belt will have a slight texture of fabric visible. Cracks will usually start at the inner edge of the belt and run outward. All worn or damaged drive belts should be replaced immediately.

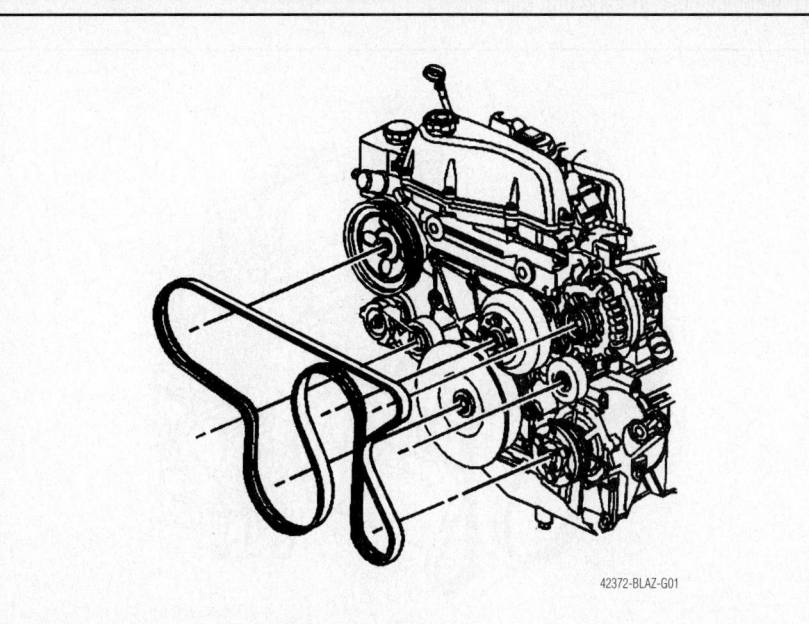

**Fig. 42 Accessory serpentine belt routing—4.2L engines**

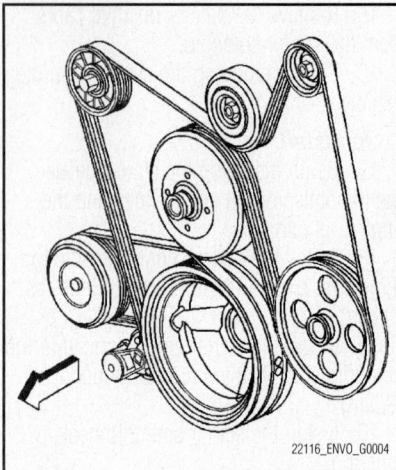

**Fig. 43 Accessory drive belt and A/C belt routing—5.3L and 6.0L engines**

## ADJUSTMENT

Serpentine belts are automatically tensioned by a system of idler and tensioner pulleys and require no adjustment. The serpentine belt tension can be checked by simply observing the belt acceptable belt wear range indicator located on the tensioner spindle. If the belt does not meet the specified range, it must be replaced.

➡**A belt is considered "used" after 15 minutes of operation.**

## REMOVAL & INSTALLATION

### 4.2L Engine

*See Figure 44.*

1. Install ⅜ inch breaker bar on the drive belt tensioner arm and turn the breaker bar clockwise enough to relieve the tension on the drive belt.
2. Remove the drive belt.
3. Release the tension on the tensioner arm.

*To install:*

4. Route the drive belt over all the pulleys except the drive belt tensioner pulley.
5. Install the ⅜ inch breaker bar on the drive belt tensioner arm and turn the breaker bar clockwise.
6. Install the drive belt over the drive belt tensioner pulley.
7. Slowly release the tension to the drive belt tensioner arm.
8. Inspect for proper installation of the drive belt on the pulleys.

### 5.3L and 6.0L Engines

#### Accessory Drive Belt

*See Figure 43.*

1. Remove the air cleaner outlet duct.
2. Install a breaker bar with hex-head socket to the drive belt tensioner bolt.
3. Rotate the drive belt tensioner clockwise in order to relieve tension on the belt.
4. Remove the belt from the alternator pulley.
5. Slowly release the tension on the drive belt tensioner.
6. Remove the breaker bar and socket and from the drive belt tensioner bolt.
7. Remove the belt from the remaining pulleys.
8. Clean and inspect the belt surfaces of all the pulleys.

*To install:*

9. Route the drive belt around all the pulleys except the alternator pulley.
10. Install the breaker bar with hex-head socket to the belt tensioner bolt.
11. Rotate the belt tensioner clockwise in order to relieve the tension on the belt.
12. Install the drive belt on the alternator pulley.
13. Slowly release the tension on the belt tensioner.
14. Remove the breaker bar and socket from the belt tensioner bolt.
15. Inspect the drive belt for proper installation and alignment.
16. Install the air cleaner outlet duct.

### A/C Compressor Belt

*See Figure 43.*

1. Remove the accessory drive belt.
2. Raise the vehicle.
3. Install a ratchet into the square opening of the air conditioning (A/C) belt tensioner.
4. Rotate the A/C belt tensioner clockwise in order to relieve tension on the belt.
5. Remove the A/C belt from the pulleys.
6. Slowly release the tension on the A/C belt tensioner.
7. Remove the ratchet from the A/C belt tensioner.
8. Clean and inspect the belt surfaces of all the pulleys.

*To install:*

9. Install the A/C belt around the crankshaft balancer.
10. Install a ratchet into the square opening of the A/C drive belt tensioner.
11. Rotate the A/C belt tensioner clockwise in order to relieve tension on the belt.
12. Install the A/C belt over the idler pulley.
13. Install the A/C belt around the A/C compressor pulley.
14. Slowly release the tension on the A/C belt tensioner.
15. Remove the ratchet from the A/C belt tensioner.
16. Inspect the A/C belt for proper installation and alignment.
17. Lower the vehicle.
18. Install the accessory drive belt.

## CAMSHAFT AND VALVE LIFTERS

### REMOVAL & INSTALLATION

#### 4.2L Engine

1. Before servicing the vehicle, refer to the Precautions Section.

**Fig. 44 Accessory serpentine belt routing—4.2L engines**

2. Disconnect the negative battery cable.

3. Discharge and recover the refrigerant from the air conditioning system, using the proper equipment.

4. Remove or disconnect the following:
- Intake manifold
- A/C line from the oil level indicator tube
- A/C line from the accumulator
- A/C bracket bolt from the engine lift hook
- Engine lift bracket
- Ignition control module electrical connectors
- Ignition control module bolts and module

### ❈❈ WARNING

**Be careful not to damage the clips that hold the harness housing in place.**

- Engine electrical harness housing from the camshaft cover
- Fuel injection harness electrical connector
- Camshaft cover bolts and cover
- Exhaust and intake sprocket bolts

5. Install a suitable sprocket holding tool onto the cylinder head and adjust the horizontal bolts into the camshaft sprockets to maintain timing chain tension and avoid disturbing the timing chain components.

6. Carefully move the sprockets with the timing chain off of the camshafts.

➡**Make sure to place the camshaft caps in a rack to keep them in order, so they may be installed in their original locations.**

7. Remove or disconnect the following:
- Camshaft cap bolts and caps
- Camshafts

### *To install:*

8. Coat the camshaft journals, camshaft journal thrust face, and camshaft lobes with clean engine oil.

9. Install the holding tool with the camshaft flats up and the number 1 cylinder at top dead center.

10. Install the camshafts, in their original position

11. Install the camshaft caps in their original locations. Tighten the bolts to 106 inch lbs. (12 Nm).

12. Carefully place the camshaft sprockets back onto the camshafts and remove the holding tool.

13. Install the intake camshaft sprocket washer and bolt and the exhaust camshaft actuator bolt. Tighten the intake camshaft

sprocket bolt to 22 ft. lbs. (30 Nm), plus an additional 135 degrees and the exhaust camshaft actuator bolt to 18 ft. lbs. (25 Nm), plus an additional 135 degrees.

14. Install or connect the following:
- New camshaft cover seal
- New rubber ignition control module seals
- Camshaft cover and bolts. Tighten the bolts to 89 inch lbs. (10 Nm).
- Ignition control module. Tighten the bolts to 89 inch lbs. (10 Nm).
- Ignition control module electrical connectors
- Fuel injector electrical connectors
- Engine electrical harness housing
- A/C line bracket to the oil level indicator tube stud and secure with the nut. Tighten the nut to 62 inch lbs. (7 Nm).
- Engine lift bracket and secure the lift hook with the bolts. Tighten the bolts to 37 ft. lbs. (50 Nm).
- A/C line bracket to the engine lift bracket. Tighten the bolt to 89 inch lbs. (10 Nm).
- Intake manifold

15. Using the proper equipment, recharge the A/C system.

### 5.3L and 6.0L Engines

*See Figures 45 through 47.*

1. Before servicing the vehicle, refer to the Precautions Section.

2. Disconnect the negative battery cable.

3. Discharge and recover the refrigerant from the air conditioning system, using the proper equipment.

4. Remove or disconnect the following:
- Condenser
- Cylinder head and gasket

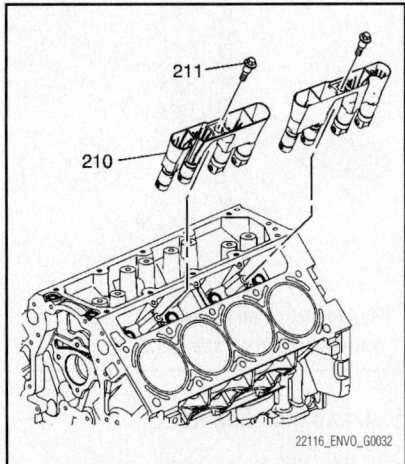

**Fig. 45 Valve lifters and guides—5.3L and 6.0L engines**

- Valve lifter guide bolts
- Valve lifters and guide

➡**If the lifters are stuck in the bores due to built up deposits, use Valve Lifter Remover tool No. J 3049-A or equivalent to remove the lifters**

- Valve lifters from the guide

➡**Make sure to keep the lifters in order as you are removing them. They must be installed in their original locations.**

5. Clean and inspect the lifters for damage.
- Camshaft sensor bolt and sensor

6. Rotate the crankshaft until the timing marks on the crankshaft and camshaft sprockets are aligned.
- Camshaft sprocket bolts

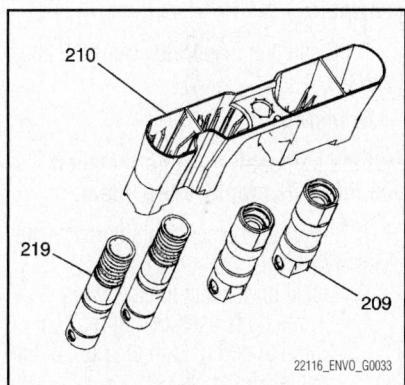

**Fig. 46 Remove the lifters from the guides, making sure to keep them in order—5.3L and 6.0L engines**

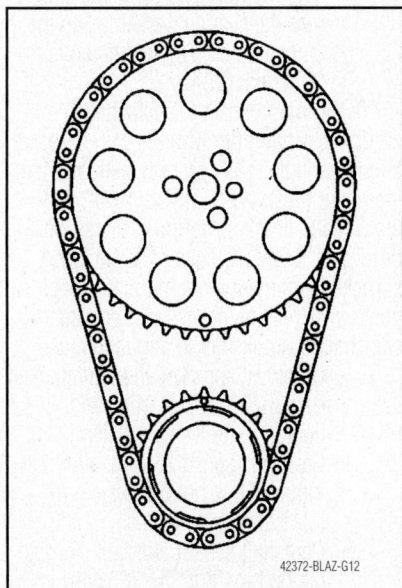

**Fig. 47 Make sure the crankshaft and camshaft timing marks are aligned**

## ✳✳ WARNING

**Do NOT turn the crankshaft after the timing chain has been removed to avoid damaging the pistons or valves!**

- Camshaft sprocket and reposition the timing chain
- Camshaft retaining bolts and retainer
- Camshaft by installing three M8-1.25 x 4.0 in. (M8-1.25 x 1.00mm) bolts in the front of the camshaft to act as a handle; then, remove the camshaft while turning slightly from side to side, as necessary. Remove the bolts from the camshaft.

➡**Take care not to damage the camshaft bearings when removing the camshaft.**

7. Clean and inspect the camshaft and bearings.

### To install:

➡**If the camshaft must be replaced, you must also replace the lifters.**

8. Lubricate the camshaft journals with clean engine oil.
9. Install or connect the following:
- Three bolts used during removal into the bolt hold in the front of the camshaft
- Camshaft carefully into the engine block, using the bolts as a handle. Remove the bolts.
- Camshaft retainer and bolts. Make sure the retaining plate is installed with the sealing gasket facing the engine block. Tighten the bolts to 18 ft. lbs. (25 Nm).
10. Properly locate the camshaft sprocket locating pin with the cam sprocket alignment hole. The sprocket teeth and timing chain must mesh. The camshaft and crankshaft sprocket alignment marks MUST be aligned properly. Locate the camshaft sprocket alignment mark in the 6 o'clock position. It may be necessary to rotate the camshaft or crankshaft to align the marks.
- Camshaft sprocket and timing chain
- Camshaft sprocket bolts and tighten to 26 ft. lbs. (35 Nm)
- Camshaft sensor O-ring, after making sure it is not damaged and lubricating it with clean engine oil
- Camshaft sensor and bolt. Torque the bolt to 18 ft. lbs. (25 Nm).
11. Lubricate the valve lifters and engine block lifter bores with clean engine oil.

12. Install or connect the following:
- Lifters into the lifter guides. Align the area on top of the lifter with the flat area in the lifter guide bore. Push the lifter completely into the guide bore.
- Valve lifters and guide to the engine block
- Valve lifter guide bolt and tighten to 106 inch lbs. (12 Nm)
- Cylinder head and gasket
- Condenser
13. Using the proper equipment, recharge the A/C system.

## CRANKSHAFT FRONT SEAL

### REMOVAL & INSTALLATION

#### 4.2L Engine

*See Figure 48.*

1. Before servicing the vehicle, refer to the Precautions Section.
2. Remove the crankshaft damper (balancer).
3. Pry out the crankshaft front oil seal using a suitable tool. Use the provided slots for prying out the seal.

➡**Do not damage the engine front cover or the crankshaft.**

### To install:

4. Apply the engine oil to the outside diameter of the crankshaft front oil seal.
5. Use the special tool J44218 to install the front oil seal. Remove the J44218.
6. Install the crankshaft damper.

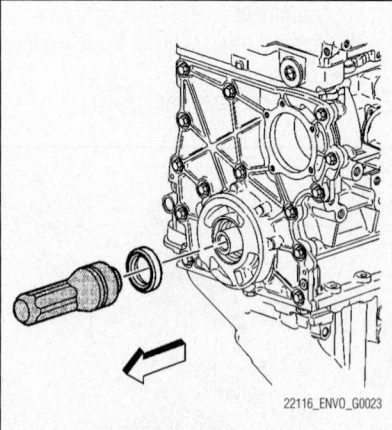

**Fig. 48 Using the proper tool to install the crankshaft front oil seal—4.2L engine**

#### 5.3L and 6.0L Engines

*See Figure 49.*

1. Before servicing the vehicle, refer to the Precautions Section.

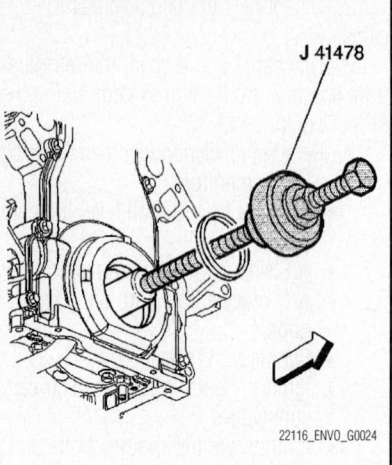

**Fig. 49 Using the proper tool to install the crankshaft front oil seal—5.3L and 6.0L engines**

2. Remove the radiator.
3. Remove the crankshaft damper (balancer).
4. Remove the crankshaft oil seal from the front cover.

### To install:

➡**Do not lubricate the oil seal sealing surface. Do not reuse the crankshaft oil seal.**

5. Lubricate the outer edge of the oil seal with clean engine oil.
6. Lubricate the front cover oil seal bore with clean engine oil.
7. Install the crankshaft front oil seal onto the J41478 guide.
8. Install J41478 threaded rod with nut, washer, guide, and oil seal into the end of the crankshaft.
9. Use J41478 in order to install the oil seal into the cover bore.
   a. Use a wrench and hold the hex on the installer bolt.
   b. Use a second wrench and rotate the installer nut clockwise until the seal bottoms in the cover bore.
   c. Remove J41478.
   d. Inspect the oil seal for proper installation.
10. The oil seal should be installed evenly and completely into the front cover bore.
11. Install the crankshaft damper.
12. Install the radiator.

## CYLINDER HEAD

### REMOVAL & INSTALLATION

#### 4.2L Engine

*See Figures 50 through 55.*

1. Before servicing the vehicle, refer to the Precautions Section.

2. Remove the air cleaner element.

3. Remove the air cleaner outlet resonator.

4. Remove the powertrain control module (ECM) and engine wire harness bracket and related hoses and connections.

5. Remove the alternator.

6. Remove the intake manifold.

7. Remove the exhaust manifold. Do not remove the exhaust pipe from the manifold. Only have the manifold pushed off to the side of the engine.

8. Position the A/C line out of the way towards the front of the vehicle.

9. Disconnect the following cross-vehicle engine wiring harness connectors:

- Engine coolant temperature sensor
- Manifold absolute pressure (MAP) sensor
- Ignition coils
- Harness clamps at power steering pump
- Wiring harness fastener at the right front inner fender
- Throttle body
- Camshaft sensors
- Camshaft actuators
- Fuel rail
- Heated oxygen sensor (HO2S)

10. Set aside the cross-vehicle engine wiring harness on the left side of the vehicle.

11. Remove the camshaft cover.

12. Partially drain the cooling system as follows:

   a. Raise and support the vehicle only high enough to access the thermostat housing through the wheelhouse. Refer to Lifting and Jacking the Vehicle.

   b. Place an approved container under the thermostat housing.

➡**Do not completely remove the thermostat housing. Complete removal of the thermostat housing will not provide steady drain path and will increase clean up time.**

   c. Loosen the thermostat housing bolts and slowly pull the thermostat housing back away from the engine. This will allow for a steady drain path for coolant.

   d. Once coolant is drained remove the thermostat housing bolts, thermostat housing, and thermostat. Inspect and replace as necessary.

   e. Clean and inspect the O-ring. Replace as necessary.

   f. Clean and inspect the sealing surface of the engine block.

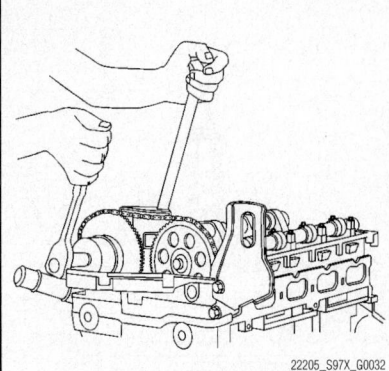

Fig. 50 Before performing one of the top dead center (TDC) procedures, break loose both the exhaust and intake camshaft sprocket bolts

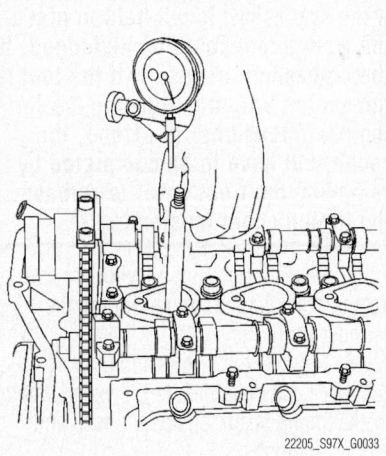

Fig. 51 Rotate the engine clockwise by hand to TDC on the compression stroke by using a piston TDC indicator tool and/or dial indicator in the number 1 cylinder

Fig. 52 When the piston is at TDC, the flats at the rear of the camshafts will be facing up and level when using a straight edge across the camshaft flats

13. Before performing one of the top dead center (TDC) procedures, break loose both the exhaust and intake camshaft sprocket bolts. Use a 25 mm (1 in) open end wrench on the camshaft hexes to hold the camshaft from turning. DO NOT remove the bolts.

14. Perform one of the following methods for the service timing procedure.

15. First Method:

- Rotate the engine clockwise by hand to TDC on the compression stroke by using a piston TDC indicator tool and/or dial indicator in the number 1 cylinder.
- The TDC indicator tool graduation marks on the shaft should note top of the piston stroke.
- When the piston is at TDC, the flats at the rear of the camshafts will be facing up and level when using a straight edge across the camshaft flats.

16. Second Method:

- Rotate the crankshaft in the engine rotational direction clockwise until the number 1 piston is at TDC on the compression stroke. The word Delphi on the exhaust camshaft position actuator will be parallel with the cylinder head to cam cover mating surface. When the piston is at TDC, the flats at the rear of the camshafts will be facing up and level when using a straight edge across the camshaft flats. A 0.005 inch feeler gage should not slide under the straight edge.

17. Once TDC is located for the number 1 cylinder using above methods, raise the vehicle and lock the flywheel with the J 44226 Crankshaft Balancer Remover.

18. Use a white paint pen or equivalent to place a reference mark on the harmonic balancer to the front cover for alignment purposes.

19. Lower the vehicle.

**✷✷ CAUTION**

**The camshaft holding tools must be installed on the camshafts to prevent camshaft rotation. When performing service to the valve train and/or timing components, valve spring pressure can cause the camshafts to rotate unexpectedly and can cause personal injury.**

➡**If the timing is correct—TDC compression stroke number 1 cylinder— the camshaft flats will be in the up position.**

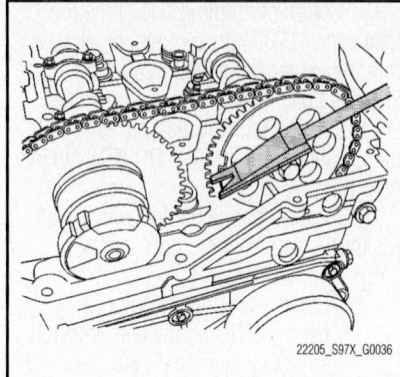

22205_S97X_G0036

**Fig. 53 DO NOT use excessive force to seat the wedge tool**

20. Install J 44221 Camshaft Holding Tool to the back of the camshafts.

21. Remove the upper timing chain guide to the cylinder head.

22. Clean the timing chain and gears with brake cleaner or suitable solvent. Use a white paint pen or equivalent to place a reference mark on both timing gear sprockets and the timing chain to mark location prior to disassembly. It is recommended that the paint marks be in the 12 o'clock position.

### ✳✳ WARNING

**DO NOT use excessive force to seat the wedge tool. If excessive force is used, you may damage the timing chain tensioner or break the front cover bolt requiring complete disassembly of the front engine.**

23. Install EN-48464 Lower Timing Gear Tensioner Holding Tool. It is important to install the tool with the proper orientation and to ensure that it is seated square against the timing chain and against the timing cover center bolt.

24. The narrow ramp of the wedge tool needs to be placed so that it faces the timing chain. Front cover removed for illustration purposes.

25. The wedge tool should be lightly seated using a couple of very light taps with a small plastic or brass hammer.

26. Once the tool is correctly installed, unscrew the handle and remove the handle.

### ✳✳ WARNING

**Use a 25 mm (1 in) open end wrench on the camshaft hexes to hold the camshaft from turning. It is critical that the crankshaft does not move and is held at TDC when the intake and exhaust camshaft sprocket bolts are removed.**

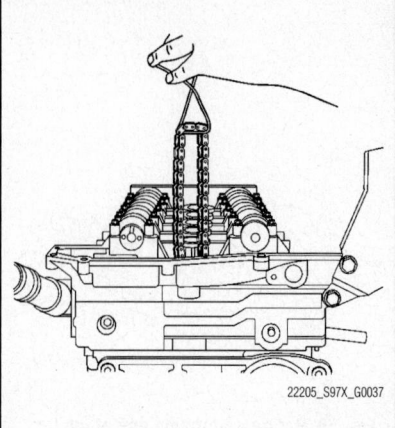

22205_S97X_G0037

**Fig. 54 Tie a piece of mechanics wire on the timing chain and let it drop**

### ✳✳ WARNING

**If the crankshaft is not held in place, the wedge tool could be dislodged. If the crankshaft moves, or if the tool is not seated properly allowing the timing chain tensioner to extend, the repair will have to be completed by removing the front cover to release the timing chain tensioner.**

27. Remove both upper cylinder head access hole plugs from the front of the cylinder head.

28. Remove the 1 long and 2 short cylinder head bolts next to the exhaust and intake timing chain tensioner shoes and discard the bolts.

29. Remove both upper timing chain tensioner shoe bolts.

30. Remove the exhaust and the intake camshaft sprocket bolts. Discard the bolts.

31. Carefully remove the exhaust and intake camshaft sprockets with the timing chain from the exhaust and intake camshafts. The illustration shows the exhaust camshaft sprocket already removed.

32. Remove the sprockets from the chain, tie a piece of mechanics wire on the timing chain and let it drop.

33. Before removing the cylinder head bolts, use a drift punch and hammer to shock the bolts. This will ensure that the cylinder head bolts will not strip out the threads in the engine block or break. If a bolt breaks during engine disassembly, EN-47702 Bolt Extractor Kit is available to assist in the removal of the remaining bolt segment.

34. Remove the cylinder head bolts. Discard the bolts.

35. Remove the cylinder head.

36. Place the cylinder head on a flat, clean surface with the combustion chambers face up, in order to prevent damage to the deck face.

37. Remove the cylinder head gasket. Discard the gasket.

38. Remove all remaining gasket material from the engine block.

39. Inspect the cylinder head gasket mating surface on the engine block.

40. Clean and inspect the cylinder head.

### To install:

41. Install the dowel pins, cylinder head locator, if necessary.

42. Position a NEW cylinder head gasket to the engine block.

➡ **Ensure all wires, components, etc. are out of the way when installing the cylinder head.**

43. Install the cylinder head.

### ✳✳ WARNING

**This engine uses torque-to-yield bolts. When servicing this component do not reuse the bolts, New torque-to-yield bolts must be installed. Reusing used torque-to-yield bolts will not provide proper bolt torque and clamp load. Failure to install NEW torque-to-yield bolts may lead to engine damage.**

44. Install NEW cylinder head bolts.

45. Tighten the NEW cylinder head bolts in the following sequence:

   a. Tighten the long bolts (1-14), in sequence, to 30 ft. lbs. (40 Nm).

   b. Tighten the long bolts, in sequence, an additional 90 degrees.

   c. Tighten the long bolts, in sequence, an additional 60 degrees.

   d. Tighten the long end bolts to 15 ft. lbs. (20 Nm).

   e. Tighten the 1 short end bolt to 13 ft. lbs. (18 Nm).

46. Install the camshafts with the flats up using J 44221 Camshaft Holding Tool.

➡ **Tension must be always kept on the intake side of the timing chain to properly keep the engine in time. If the chain is loose the timing will be off, which may cause internal engine damage or set DTC P0017.**

➡ **The exhaust camshaft actuator must be fully advanced during installation. Engine damage may occur if the camshaft actuator is not fully advanced.**

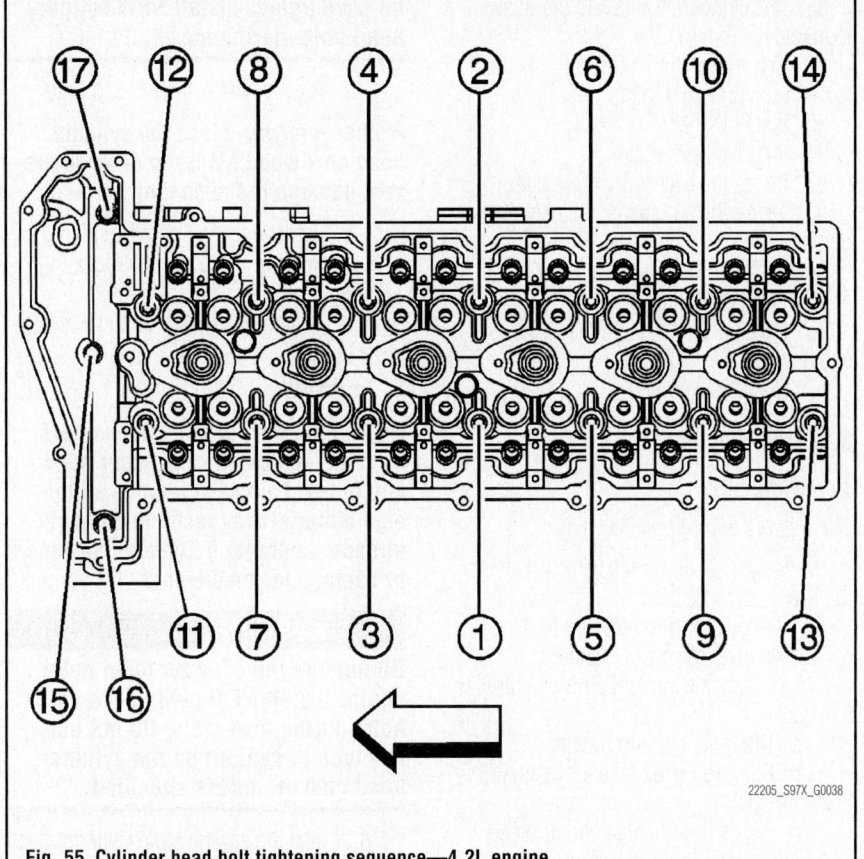

Fig. 55 Cylinder head bolt tightening sequence—4.2L engine

47. Ensure that the camshaft position actuator is in the fully advanced position.

> ❋❋ **WARNING**
>
> **To aid in aligning the actuator to the camshaft, use a 25 mm (1 in) open end wrench on the hex of the camshaft to rotate. This will ensure the alignment pin is properly engaged with the camshaft and hand tighten the new exhaust camshaft sprocket bolt.**

48. Install the exhaust camshaft actuator/sprocket and chain onto the exhaust camshaft. Use the paint marks as an alignment guide.

> ❋❋ **WARNING**
>
> **To aid in aligning the intake sprocket to the camshaft, use a 25 mm (1 in) open end wrench on the hex of the camshaft to rotate. This will ensure the alignment pin is properly engaged with the camshaft and hand tighten the new intake camshaft sprocket bolt.**

49. Install the intake camshaft sprocket and chain onto the intake camshaft. Use paint marks as alignment guide.

50. Position the timing chain tensioner shoe to the engine.

51. Install the timing chain tensioner shoe bolt. Tighten the timing chain tensioner shoe bolt to 18 ft. lbs. (25 Nm).

52. Position the lower timing chain guide to the engine.

53. Install the lower timing chain guide bolts. Tighten the lower timing chain guide bolts to 107 inch lbs. (12 Nm).

54. Install both upper timing chain tensioner shoe bolts. Tighten the tensioner shoe bolts to 18 ft. lbs. (25 Nm).

55. Install both upper cylinder head access hole plugs to the front of the cylinder head. Tighten the plugs to 44 inch lbs. (5 Nm).

56. Tighten the new intake camshaft sprocket bolt. Tighten the intake camshaft sprocket bolt to 15 ft. lbs. (20 Nm) plus 100° rotation.

57. Tighten the new exhaust camshaft actuator sprocket bolt. Tighten the exhaust camshaft actuator sprocket bolt to 18 ft. lbs. (25 Nm) plus 135° rotation.

58. Lift the vehicle and remove the J 44226 Crankshaft Balancer Remover.

59. Lower the vehicle.

60. Remove the J 44221 Camshaft Holding Tool from the back of the camshafts.

➡ Ensure that the wedge tool is removed from engine prior to rotation. If the wedge tool is not removed, engine damage will result.

61. Install the handle of EN-48464 Lower Timing Gear Tensioner Holding Tool and remove the wedge portion of the tool from the engine.

> ❋❋ **WARNING**
>
> **It is critical that the engine is at TDC and not a couple of degrees off. If in doubt, repeat this step.**

62. Rotate the engine clockwise by hand two complete revolutions to TDC number 1 on the compression stroke. Refer to First Method or Second Method for TDC. If you go past TDC, rotate the engine back approximately 45 degrees before TDC and then rotate clockwise up to TDC to ensure that the timing chain is tight (no slack) between the crank sprocket and the timing gears.

> ❋❋ **WARNING**
>
> **DO NOT use the J 44221 Camshaft Holding Tool installed to the back of the camshafts, as a method to verify timing.**

63. Both intake and exhaust camshaft flats should be facing up and flat and level with the cylinder head. If J 44221 Camshaft Holding Tool is used to verify cam timing, you could be off approximately one tooth and cause DTC P0017 to set. If a worn or new J 44221 Camshaft Holding Tool is used to verify timing, the timing will be off.

64. To verify timing, set a straight edge across the flats of the camshafts.

65. A 0.005 inch feeler gage should not be able to slip under the straight edge. If the feeler gage slips under one or both camshaft flats, then the timing is off. Repeat step 20 and recheck. If the camshaft flats are still not flat, the camshaft timing will have to be reset. This may require removal and reinstallation of one or both camshaft sprockets.

66. Install the 1 long and 2 short cylinder head bolts next to the exhaust and intake timing chain tensioner shoes and tighten the bolts.

67. Position the upper timing chain guide to the cylinder head. Apply thread locker GM P/N 89021297 (Canadian P/N 10953488) to the upper timing chain guide bolt threads.

68. Install the upper timing chain guide bolts. Tighten the bolts to 89 inch lbs. (10 Nm).

69. Install the radiator inlet hose and clamp to the cylinder head.

70. Clean and inspect the camshaft cover.

71. Install a NEW camshaft cover seal and NEW ignition control module seals to the cam cover. Position the camshaft cover to the cylinder head.

72. Install the camshaft cover bolts. Tighten the bolts to 89 inch lbs. (10 Nm).

73. Check the gap on all of the spark plugs. Install and tighten all of the spark plugs.

74. Install the ignition coils into the camshaft cover.

75. Install the ignition coil bolts. Tighten the bolts to 89 inch lbs. (10 Nm).

76. Reposition the exhaust manifold to cylinder head and install the exhaust manifold bolts to the cylinder head.

77. If equipped, install a NEW AIR injection gasket, then the cover and pipe studs to the cylinder head.

Tighten the pipe studs to 18 ft. lbs. (25 Nm).

78. Install the exhaust manifold heat shield to the exhaust manifold.

79. Apply anti-seize to the exhaust manifold heat shield nuts.

80. Install the exhaust manifold heat shield nuts. Tighten the nuts to 89 inch lbs. (10 Nm).

81. Install the intake manifold to the cylinder head.

82. Raise the vehicle and install the blind intake manifold bolts from the left front wheel well access.

83. Reposition the engine wiring harness bracket to the engine and harnesses. Install the engine wiring harness bracket bolts. Tighten the bracket bolts to 89 inch lbs. (10 Nm).

84. Install the left front wheel well panel and the left wheel and tire.

85. Drain the engine oil again.

86. If removed, install the radiator outlet hose.

87. Install the cross-vehicle wiring harness connectors to the following components:

- ECM
- Map sensor
- Ignition coils
- Harness clamps at power steering pump
- Wiring harness fastener at right front inner fender
- Throttle body
- Camshaft sensors
- Exhaust camshaft actuator
- Fuel injectors
- HO2S
- AIR valve and connectors

88. Install the PCV pipes to the intake manifold.

89. Reposition the Fuel/EVAP lines to the intake manifold retainer.

90. Install the thermostat.

91. Install the alternator.

92. Install the accessory drive belt.

93. Install the air cleaner element and resonator.

94. Install NEW engine oil.

95. Install NEW coolant.

96. Install a scan tool and start the engine. Check for DTCs.

97. Road test vehicle.

### 5.3L and 6.0L Engines

#### *Left Side*

*See Figures 56 through 58.*

1. Before servicing the vehicle, refer to the Precautions Section.
2. Remove the alternator bracket.
3. Remove the intake manifold.
4. Remove the coolant air bleed pipe as follows:
   a. Drain the cooling system.
   b. Remove the air cleaner resonator outlet duct.
   c. Reposition the coolant air bleed hose clamp at the coolant air bleed pipe.
   d. Remove the coolant air bleed hose from the coolant air bleed pipe.
   e. Remove the coolant air bleed pipe bolts.
   f. Remove the coolant air bleed pipe with seals.
   g. Remove the coolant air bleed pipe cover bolts, if necessary.
   h. Remove the coolant air bleed covers with seals, if necessary.
   i. Remove and discard the seals from the coolant air bleed pipe and covers.
5. Remove the left exhaust manifold.
6. Remove the pushrods as follows:
   a. Remove the valve cover.

**➡Place the valve rocker arms, pushrods, and pivot support, in a rack so that they can be installed in the same location from which they were removed.**

   b. Remove the valve rocker arm bolts.
   c. Remove the valve rocker arms.
   d. Remove the valve rocker arm pivot support.
   e. Remove the pushrods.
7. Remove the cylinder head as follows:

**WARNING**

**The cylinder head bolts are of a torque-to-yield design and are NOT to be used again. Install NEW cylinder head bolts during assembly.**

   a. Remove the cylinder head bolts.

**➡After removal, place the cylinder head on 2 wood blocks in order to prevent damage to the sealing surfaces.**

   b. Remove the cylinder head.
   c. Remove the gasket and locating pins.
   d. Discard the gasket and cylinder head bolts.

#### *To install:*

**➡Clean all dirt, debris, and coolant from the engine block cylinder head bolt holes. Failure to remove all foreign material may result in damaged threads, improperly tightened fasteners or damage to components.**

**WARNING**

**Do not use the cylinder head bolts again. Install NEW cylinder head bolts during assembly. Do not use any type of sealant on the cylinder head gasket, unless specified.**

8. Clean the engine block cylinder head bolt holes. Thread Repair Tool J 42385-107 may be used to clean the threads of old threadlocking material. Spray cleaner GM P/N 12346139 or equivalent, into the hole. Clean the cylinder head bolt holes with compressed air.
9. Install the cylinder head locating pins.
10. Inspect the displacement markings on the gasket for proper usage. Install the NEW cylinder head gasket onto the locating pins.

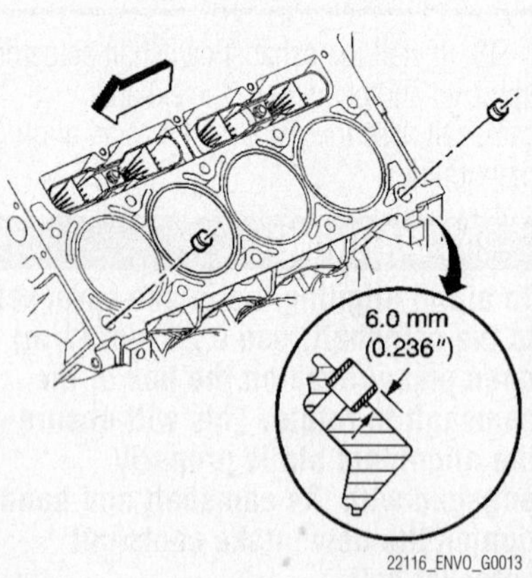
6.0 mm
(0.236")
22116_ENVO_G0013

**Fig. 56 Make sure the cylinder head locating pins are properly installed—5.3L and 6.0L engines**

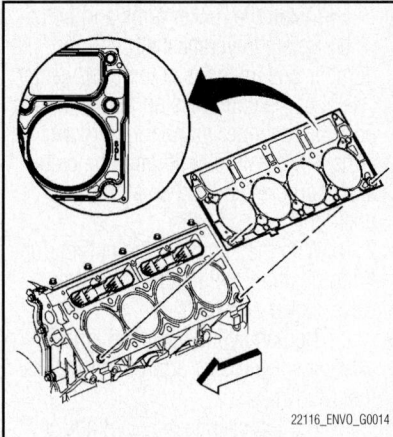

**Fig. 57 Proper cylinder head gasket installation—5.3L and 6.0L engines**

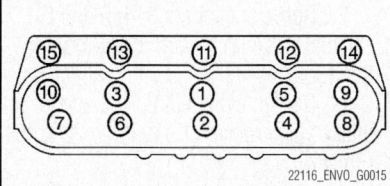

**Fig. 58 Cylinder head bolt torque sequence—5.3L and 6.0L engines**

11. Install the cylinder head as follows:

a. Install the cylinder head onto the locating pins and the gasket.

b. Install the NEW cylinder head bolts.

c. Tighten the cylinder head bolts in sequence as follows:

- Tighten the M11 bolts to 22 ft. lbs. (30 Nm).
- Tighten the M11 an additional 90 degrees.
- Tighten M11 bolts, an additional 70 degrees.
- Tighten the M8 bolts to 22 ft. lbs. (30 Nm). Tighten all the bolts beginning with the center bolt and working outward, alternating sides

12. Install the pushrods as follows:

### ✳✳ WARNING

**When reusing the valve train components, always install the components to the original location and position.**

➥**The valve lash is net build, no valve adjustment is required.**

a. Lubricate the valve rocker arms and pushrods with clean engine oil.

b. Lubricate the flange of the valve rocker arm bolts with clean engine oil.

c. Lubricate the flange or washer surface of the bolt that will contact the valve rocker arm.

d. Install the valve rocker arm pivot support.

➥**Make sure that the pushrods seat properly to the valve lifter sockets.**

e. Install the pushrods.

➥**Verify that the pushrods seat properly to the ends of the rocker arms. DO NOT tighten the rocker arm bolts at this time.**

f. Install the rocker arms and bolts.

g. Rotate the crankshaft until the number one piston is at top dead center (TDC) of the compression stroke. In this position, cylinder number one rocker arms will be off lobe lift, and the crankshaft sprocket key will be at the 1:30 position.

h. With the engine in the number one firing position, tighten the following valve rocker arm bolts:

i. Tighten cylinders 1, 2, 7 and 8 exhaust valve rocker arm bolts to 22 ft. lbs. (30 Nm).

j. Tighten cylinders 1, 3, 4 and 5 intake valve rocker arm bolts to 22 ft. lbs. (30 Nm).

k. Rotate the crankshaft 360 degrees. Tighten the following valve rocker arm bolts:

l. Tighten cylinders 3, 4, 5 and 6 exhaust valve rocker arm bolts to 22 ft. lbs. (30 Nm).

m. Tighten cylinders 2, 6, 7 and 8 intake valve rocker arm bolts to 22 ft. lbs. (30 Nm).

13. Install the valve cover.

14. Install the exhaust manifold.

15. Install the coolant air bleed pipe assembly as follows:

a. Position the gasket O-ring seal onto the nipple portion of the pipe.

b. Install the seals onto the coolant air bleed pipe and covers.

c. Install the coolant air bleed pipe with seals.

d. Install the coolant air bleed pipe bolts. Tighten the bolts to 106 inch lbs. (12 Nm).

e. Install the coolant air bleed covers with seals, if necessary.

f. Remove the coolant air bleed pipe cover bolts, if necessary. Tighten the bolts to 106 inch lbs. (12 Nm).

g. Install the coolant air bleed hose to the coolant air bleed pipe.

h. Position the coolant air bleed hose clamp at the coolant air bleed pipe.

i. Fill the cooling system, if necessary.

j. Install the air cleaner resonator outlet duct.

16. Install the alternator bracket.

17. Install the intake manifold.

### Right Side

See Figures 59 through 61.

1. Before servicing the vehicle, refer to the Precautions Section.

2. Remove the intake manifold.

3. Remove the oil level indicator.

4. Remove the coolant air bleed pipe a follows:

a. Drain the cooling system.

b. Remove the air cleaner resonator outlet duct.

c. Reposition the coolant air bleed hose clamp at the coolant air bleed pipe.

d. Remove the coolant air bleed hose from the coolant air bleed pipe.

e. Remove the coolant air bleed pipe bolts.

f. Remove the coolant air bleed pipe with seals.

g. Remove the coolant air bleed pipe cover bolts, if necessary.

h. Remove the coolant air bleed covers with seals, if necessary.

i. Remove and discard the seals from the coolant air bleed pipe and covers.

5. Remove the right exhaust manifold.

6. Remove the pushrods as follows:

a. Remove the valve cover.

➥**Place the valve rocker arms, pushrods, and pivot support, in a rack so that they can be installed in the same location from which they were removed.**

b. Remove the valve rocker arm bolts.

c. Remove the valve rocker arms.

d. Remove the valve rocker arm pivot support.

e. Remove the pushrods.

7. Remove the cylinder head as follows:

### ✳✳ WARNING

**The cylinder head bolts are of a torque-to-yield design and are NOT to be used again. Install NEW cylinder head bolts during assembly.**

a. Remove the cylinder head bolts.

➥**After removal, place the cylinder head on 2 wood blocks in order to prevent damage to the sealing surfaces.**

b. Remove the cylinder head.

c. Remove the gasket and locating pins.

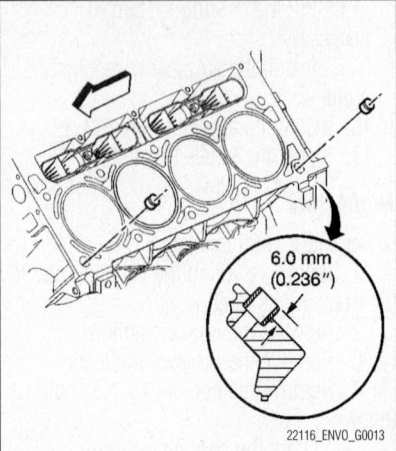

**Fig. 59 Make sure the cylinder head locating pins are properly installed—5.3L and 6.0L engines**

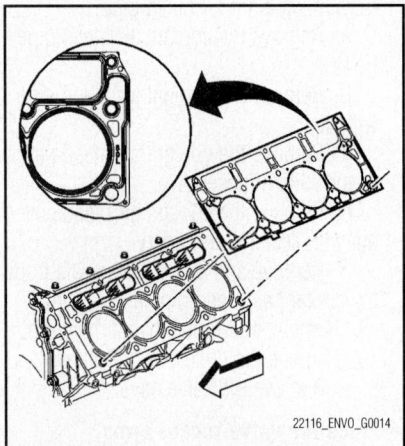

**Fig. 60 Proper cylinder head gasket installation—5.3L and 6.0L engines**

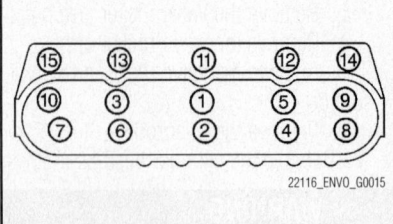

**Fig. 61 Cylinder head bolt torque sequence—5.3L and 6.0L engines**

d. Discard the gasket and cylinder head bolts.

**To install:**

➡Clean all dirt, debris, and coolant from the engine block cylinder head bolt holes. Failure to remove all foreign material may result in damaged threads, improperly tightened fasteners or damage to components.

## ※※ WARNING

**Do not use the cylinder head bolts again. Install NEW cylinder head bolts during assembly. Do not use any type of sealant on the cylinder head gasket, unless specified.**

8. Clean the engine block cylinder head bolt holes. Thread Repair Tool J 42385-107 may be used to clean the threads of old threadlocking material. Spray cleaner GM P/N 12346139 or equivalent, into the hole. Clean the cylinder head bolt holes with compressed air.

9. Install the cylinder head locating pins.

10. Inspect the displacement markings on the gasket for proper usage. Install the NEW cylinder head gasket onto the locating pins.

11. Install the cylinder head as follows:

a. Install the cylinder head onto the locating pins and the gasket.

b. Install the NEW cylinder head bolts.

c. Tighten the cylinder head bolts in sequence as follows:

- Tighten the M11 bolts to 22 ft. lbs. (30 Nm).
- Tighten the M11 an additional 90 degrees.
- Tighten M11 bolts, an additional 70 degrees.
- Tighten the M8 bolts to 22 ft. lbs. (30 Nm). Tighten all the bolts beginning with the center bolt and working outward, alternating sides

12. Install the pushrods as follows:

## ※※ WARNING

**When reusing the valve train components, always install the components to the original location and position.**

➡The valve lash is net build, no valve adjustment is required.

a. Lubricate the valve rocker arms and pushrods with clean engine oil.

b. Lubricate the flange of the valve rocker arm bolts with clean engine oil.

c. Lubricate the flange or washer surface of the bolt that will contact the valve rocker arm.

d. Install the valve rocker arm pivot support.

➡Make sure that the pushrods seat properly to the valve lifter sockets.

e. Install the pushrods.

➡Verify that the pushrods seat properly to the ends of the rocker arms. **DO NOT** tighten the rocker arm bolts at this time.

f. Install the rocker arms and bolts.

g. Rotate the crankshaft until the number one piston is at top dead center (TDC) of the compression stroke. In this position, cylinder number one rocker arms will be off lobe lift, and the crankshaft sprocket key will be at the 1:30 position.

h. With the engine in the number one firing position, tighten the following valve rocker arm bolts:

i. Tighten cylinders 1, 2, 7 and 8 exhaust valve rocker arm bolts to 22 ft. lbs. (30 Nm).

j. Tighten cylinders 1, 3, 4 and 5 intake valve rocker arm bolts to 22 ft. lbs. (30 Nm).

k. Rotate the crankshaft 360 degrees. Tighten the following valve rocker arm bolts:

l. Tighten cylinders 3, 4, 5 and 6 exhaust valve rocker arm bolts to 22 ft. lbs. (30 Nm).

m. Tighten cylinders 2, 6, 7 and 8 intake valve rocker arm bolts to 22 ft. lbs. (30 Nm).

13. Install the valve cover.

14. Install the intake manifold.

15. Install the exhaust manifold.

16. Install the coolant air bleed pipe assembly as follows:

a. Position the gasket O-ring seal onto the nipple portion of the pipe.

b. Install the seals onto the coolant air bleed pipe and covers.

c. Install the coolant air bleed pipe with seals.

d. Install the coolant air bleed pipe bolts. Tighten the bolts to 106 inch lbs. (12 Nm).

e. Install the coolant air bleed covers with seals, if necessary.

f. Remove the coolant air bleed pipe cover bolts, if necessary. Tighten the bolts to 106 inch lbs. (12 Nm).

g. Install the coolant air bleed hose to the coolant air bleed pipe.

h. Position the coolant air bleed hose clamp at the coolant air bleed pipe.

i. Fill the cooling system.

j. Install the air cleaner resonator outlet duct.

17. Remove the oil level indicator.

## ENGINE ASSEMBLY

### REMOVAL & INSTALLATION

#### 4.2L Engine

1. Remove the hood.

2. Disconnect the negative battery cable.

3. Drain engine coolant.

4. Recover the refrigerant.

➡ **Keep drain plug removed during engine removal and installation.**

5. Remove the oil pan drain plug and drain the oil.

6. Install a suitable plug into the oil pan after draining to prevent any oil leakage during the rest of the procedure.

7. Remove the air cleaner assembly.

8. Remove the throttle body.

9. Remove the manifold absolute pressure sensor.

10. Remove the washer solvent container.

11. Remove the grille.

12. Remove the headlamp housing.

13. Remove the radiator support brace.

14. Remove the hood latch.

15. Disconnect A/C lines at the condenser.

16. Disconnect the transmission cooler lines at the engine, not the radiator.

17. Remove the cooling fan and the shroud tilting the radiator forward, and the cooling fan and the shroud rearward for clearance.

18. Remove the radiator with condenser and transmission cooler lines.

19. Remove the drive belt.

20. Remove the power steering pump bolts and lay the pump aside.

21. Ensure the heater hoses (1, 2) at the heater core are disconnected.

22. Remove the secondary air injection reaction solenoid valve.

23. Install the Lift Hook J 44220 to the AIR port on the engine head.

24. Disconnect the oxygen sensor electrical connector.

25. Disconnect the A/C line at the accumulator.

26. Disconnect the front axle actuator electrical connector.

27. Disconnect the camshaft phase actuator valve electrical connector.

28. Unclip the transmission cooler lines from right side of the engine block.

29. Disconnect the ignition coil harness connectors.

30. Carefully disconnect harness retainer at clips and remove.

31. Remove power brake hose at booster.

32. Remove the engine control module.

33. Remove all harnesses from the engine harness bracket.

34. Disconnect the front differential vent hose from the engine harness bracket.

35. Remove the engine harness bracket bolt and remove bracket.

36. Disconnect starter electrical connections.

37. Disconnect the A/C pressure sensor and clutch electrical connector.

38. Disconnect alternator electrical connector and battery lead.

39. Disconnect the knock sensor electrical connector.

40. Disconnect the crankshaft sensor electrical connector.

41. Disconnect the camshaft sensor electrical connector.

42. Remove 4 grounds on the left side of the block.

43. Raise the vehicle.

44. Remove the wheel drive shafts, left and right.

45. Remove the propeller shaft from the front axle pinion yoke. Remove the engine protection shield.

46. Disconnect the exhaust pipe from the exhaust manifold and slide the exhaust pipe backward slightly.

47. Remove the fuel tank shield, if equipped.

48. Disconnect the AIR pipes from the AIR pump.

49. Remove the torque converter bolt access cover.

50. Remove the torque converter bolts.

51. Place a jack on the transmission oil pan for support.

52. Remove the transmission support.

53. Lower the transmission enough to reach the top bell housing bolts.

54. Remove the top 4 bell housing bolts. There may be 2 harness clips that will need to be removed in order to have access to 2 of the top bolts.

55. Raise the transmission.

56. Reinstall the transmission support using only 2 through bolts.

57. Remove the remaining bell housing bolts, for a total of 11 bolts.

58. Remove the left and right engine lower mount nuts.

59. Disconnect the oil pressure switch electrical connector.

60. Lower the vehicle.

61. Remove the left upper engine mount nut.

62. Remove the right upper engine mount nut.

63. Install the engine hoist.

64. Raise the engine out of the compartment slowly keeping the transmission supported.

65. Remove both engine mounts for clearance.

66. Remove the fasteners securing the AIR pipes to the back of the engine head.

67. Remove the AIR pipes from the vehicle.

68. Continue raising the engine out of the vehicle.

69. Install the engine to the engine stand.

***To install:***

70. Remove the engine from the engine stand.

71. Slowly install the engine into the engine compartment aligning the engine mounts with the brackets.

72. Install the AIR pipes to the engine.

73. Install the fasteners securing the AIR pipes to the back of the engine head.

74. When the engine mounts are aligned, install the engine mounts, putting the mount up through the engine mount brackets before inserting into the chassis mount brackets.

75. Lower the engine onto the mounts and install the upper engine mount nuts. Tighten the nuts to 51 ft. lbs. (70 Nm).

76. Remove the engine hoist.

77. Lay the radiator into the radiator support, but do not install the radiator completely.

78. Raise and safely support the vehicle securely on jackstands.

79. Install all of the lower bell housing bolts, excluding the top 4.

80. Remove the through bolts securing the transmission support.

81. Lower the transmission.

82. Install the top 4 bell housing bolts. Tighten all 11 bell housing bolts to 37 ft. lbs. (50 Nm).

83. Raise the transmission.

84. Install the transmission support.

85. Install the 3 torque converter bolts. Tighten the torque converter bolts to 44 ft. lbs. (60 Nm).

86. Install the torque converter bolt cover.

87. Install the fuel tank shield, if removed.

88. Connect the AIR pipes to the AIR pump.

89. Install the engine protection shield.

90. Install the propeller shaft to the front axle pinion yoke.

91. Connect the exhaust pipe to the manifold and secure the pipe with the bolts. Tighten the exhaust pipe bolts to 37 ft. lbs. (50 Nm).

92. Connect the oil pressure sensor electrical connector.

93. Install the oil pan drain plug. Tighten the plug to 19 ft. lbs. (26 Nm).

94. Install the lower radiator hose.

95. Install the left and right wheel drive shafts.

96. Lower the vehicle.

97. Install the 4 grounds on the left side of the block.

98. Install the camshaft sensor electrical connectors.

99. Install the crankshaft sensor electrical connector.

100. Install the knock sensor electrical connector.

101. Install the alternator electrical connector and battery lead. Tighten the alternator battery lead nut to 80 inch lbs. (9 Nm).

102. Install the A/C pressure sensor and clutch electrical connector.

103. Install starter electrical connectors and battery lead. Tighten the starter battery lead to 80 inch lbs. (9 Nm).

104. Install the engine harness bracket and bolt. Tighten the engine harness bracket bolt to 37 ft. lbs. (50 Nm).

105. Install the front differential vent hose, to the engine harness bracket.

106. Install all the harnesses to the engine harness bracket.

107. Install the ECM.

108. Install the power brake hose at booster.

109. Install the harness retainer to the original location.

110. Install the ignition coil harness electrical connectors.

111. Clip the transmission cooler lines to the right side of the engine block.

112. Connect the camshaft actuator valve electrical connector.

113. Connect the front axle actuator electrical connector.

114. Connect the A/C line at the accumulator.

115. Install the oxygen sensor electrical connector.

116. Remove the lift hook

117. Install the secondary AIR solenoid valve.

118. Install the heater hoses to the heater core.

119. Install the power steering pump.

120. Install the power steering pump bolts. Tighten the bolts to 18 ft. lbs. (25 Nm).

121. Install the drive belt.

122. Install cooling fan and shroud, tilting the radiator forward for clearance.

123. Finish installing the radiator.

124. Install transmission cooler lines together.

125. Install the MAP sensor.

126. Install the throttle body.

127. Install the hood latch.

128. Install the head lamp housing.

129. Install the grill.

130. Install the washer solvent container.

131. Install the air cleaner assembly.

132. Connect the negative battery cable.

133. Install the hood.

134. Service the engine oil.

135. Fill the cooling system.

136. Recharge the refrigerant.

137. After an overhaul, the engine should be tested. Use the following procedure after the engine is installed in the vehicle.

138. Test the vehicle using the following procedure:

   a. Disable the ignition system.

   b. Crank the engine several times. Listen for any unusual noises or evidence that parts are binding.

   c. Enable the ignition system.

   d. Start the engine and listen for unusual noises.

   e. Check the vehicle oil pressure gauge or light and confirm that the engine has acceptable oil pressure.

   f. Run the engine speed at about 1,000 RPM until the engine has reached normal operating temperature.

   g. Inspect for fuel, oil and/or coolant leaks while the engine is running.

### 5.3L and 6.0L Engines

1. Remove the hood.
2. Disconnect the negative battery cable.
3. Remove the intake manifold sight shield.
4. Recover the refrigerant.
5. Remove the radiator.
6. Remove the radiator support brace.
7. Remove the front differential drive axle.
8. Remove the wheel drive shafts.
9. Remove the intake manifold.
10. Disconnect the oil pressure sensor, oxygen sensor and camshaft position sensor wiring harnesses.
11. Remove the air conditioning compressor hose.
12. Disconnect the rear auxiliary A/C compressor pipe fitting.
13. Remove the rear auxiliary A/C compressor pipe nut and bolt.
14. Tie the pipe assembly out of the way.
15. Disconnect the engine coolant temperature sensor.
16. Remove the ground terminal bolt.
17. Remove the retaining clips from the brackets.
18. Disconnect the A/C pressure switch electrical connector.
19. Remove the retaining clip from the cylinder head.
20. Raise and suitably support the vehicle.
21. Remove the ground terminal bolts.
22. Remove the starter.
23. Remove the battery cable channel bolt.

24. Remove the battery cable channel from the oil pan.

25. Disconnect the A/C compressor electrical connector.

26. Lower the vehicle.

27. Gather all branches of the engine wiring harness and reposition the harness off to the side.

28. Remove the generator cable from the generator. Perform the following:

29. Slide the boot down revealing the terminal stud.

30. Remove the generator cable nut from the terminal stud.

31. Remove the generator cable.

32. Remove the generator bracket bolts.

33. Position the bracket with generator aside.

34. Using hose clamp pliers, remove the inlet hose from the water outlet.

35. Using hose clamp pliers, remove the outlet hose from the water outlet.

36. Disconnect the auxiliary heater inlet and outlet hose/pipe assembly from the heater water shutoff valve pipes.

37. Using hose clamp pliers, remove the auxiliary heater inlet and outlet hoses/pipes from the water pump.

38. Remove the ignition coils as required for the proper fit of J 41798 Engine Lift Bracket.

39. Install J 41798 Engine Lift Bracket. Tighten the M8 bolts to 18 ft. lbs. (25 Nm). Tighten the M10 bolts to 37 ft. lbs. (50 Nm).

40. Raise the vehicle.

41. Remove the catalytic converter. Refer to Catalytic Converter Replacement.

42. Remove the 3 bracket bolts from both the right and the left sides of the frame engine mount.

43. Remove the torque converter bolts.

44. Remove the transmission oil level indicator tube nut.

45. Remove the transmission oil level indicator tube.

46. Remove the transmission bolt and stud on the right side.

47. Remove the lower transmission bolt/studs.

48. Lower the vehicle.

49. Remove the 3 upper transmission bolts/studs.

50. Install an engine hoist to the J 41798 Engine Lift Bracket.

51. Install a floor jack under the transmission for support.

52. Separate the engine from the transmission.

53. Remove the engine.

54. Install the engine to an engine stand.

55. Install the J 21366 Converter Holding Strap.

**To install:**

56. Remove the J 21366 Converter Holding Strap.

57. Install an engine hoist to the J 41798 Engine Lift Bracket.

58. Remove the engine from the engine stand.

59. Install the engine.

60. Mate the engine to the transmission.

61. Remove the floor jack from under the transmission for support.

62. Install the 3 upper transmission bolts/studs. Tighten the bolts/studs to 37 ft. lbs. (50 Nm).

63. Raise the vehicle.

64. Install the lower transmission bolt/studs. Tighten the bolts/studs to 37 ft. lbs. (50 Nm).

65. Install the transmission bolt and stud on the right side. Tighten the bolts/studs to 37 ft. lbs. (50 Nm).

66. Install the transmission oil level indicator tube.

67. Install the transmission oil level indicator tube nut. Tighten the nut to 89 inch lbs. (10 Nm).

68. Install the torque converter bolts. Tighten the bolts to 44 ft. lbs. (60 Nm).

69. Install the 3 bracket bolts to both the right and the left sides of the frame engine mount. Tighten the bolts to 37 ft. lbs. (50 Nm).

70. Install the catalytic converter. Refer to Catalytic Converter Replacement.

71. Lower the vehicle.

72. Remove J 41798 Engine Lift Bracket.

73. Install the ignition coils, as required. Tighten the bolts to 71 inch lbs. (8 Nm).

74. Install the auxiliary heater inlet and outlet hoses/pipes to the water pump using hose clamp pliers.

75. Connect the auxiliary heater inlet and outlet hose/pipe assembly to the heater water shutoff valve pipes.

76. Install the outlet hose to the water outlet using hose clamp pliers.

77. Install the inlet hose to the water outlet using hose clamp pliers.

78. Position the bracket with generator to the engine.

79. Install the generator bracket bolts. Tighten the bolts to 37 ft. lbs. (50 Nm).

80. Install the generator cable to the generator, perform the following procedure:

81. Install the generator cable.

82. Install the generator cable nut to the terminal stud. Tighten the nut to 80 inch lbs. (9 Nm).

83. Slide the boot down covering the terminal stud.

84. Gather all branches of the engine wiring harness and position the harness over the engine.

85. Raise the vehicle.

86. Connect the A/C compressor electrical connector.

87. Install the battery cable channel to the oil pan.

88. Install the battery cable channel bolt. Tighten the bolt to 106 inch lbs. (12 Nm).

89. Install the starter.

90. Install the ground terminal bolts. Tighten the bolt to 18 ft. lbs. (25 Nm).

91. Lower the vehicle.

92. Install the retaining clip to the cylinder head.

93. Connect the A/C pressure switch electrical connector.

94. Install the retaining clips to the brackets.

95. Install the ground terminal bolt. Tighten the bolt to 18 ft. lbs. (25 Nm).

96. Connect the ECT sensor.

97. Install the rear auxiliary A/C compressor pipe nut and bolt. Tighten the nut/bolt to 15 ft. lbs. (20 Nm).

98. Connect the rear auxiliary A/C compressor pipe fitting. Tighten the nut to 12 ft. lbs. (16 Nm).

99. Install the A/C compressor hose.

100. Connect the following oil pressure sensor, oxygen sensor and camshaft position sensor electrical connectors.

101. Install the intake manifold..

102. Install the wheel drive shafts.

103. Install the front differential drive axle.

104. Install the radiator support brace.

105. Install the radiator.

106. Recharge the refrigerant.

107. Connect the negative battery cable.

108. Pre-lube the engine.

109. Perform the CKP system variation learn procedure.

110. Install the hood.

111. After an overhaul, the engine should be tested. Use the following procedure after the engine is installed in the vehicle.

112. Test the vehicle using the following procedure:

    a. Disable the ignition system.

    b. Crank the engine several times. Listen for any unusual noises or evidence that parts are binding.

    c. Enable the ignition system.

    d. Start the engine and listen for unusual noises.

    e. Check the vehicle oil pressure gauge or light and confirm that the engine has acceptable oil pressure.

    f. Run the engine speed at about 1,000 RPM until the engine has reached normal operating temperature.

    g. Inspect for fuel, oil and/or coolant leaks while the engine is running.

    h. Install the intake manifold sight shield.

## EXHAUST MANIFOLD

### REMOVAL & INSTALLATION

**4.2L Engine**

*See Figure 62.*

1. Before servicing the vehicle, refer to the Precautions Section.

2. Raise and safely support the vehicle securely on jackstands.

3. Loosen and remove the exhaust pipe bolts from the exhaust manifold.

4. Lower the vehicle.

5. Remove the manifold heat shield as follows:

    a. Remove the air cleaner resonator outlet duct.

    b. Remove the secondary air injection (AIR) solenoid valve.

    c. Remove the oil level indicator tube.

    d. Remove the oxygen sensor from the exhaust manifold.

    e. Remove the 4 manifold heat shield nuts and remove the heat shield.

6. Loosen and remove the exhaust manifold bolts.

7. Remove the exhaust manifold.

8. Remove the exhaust manifold gasket.

**To install:**

9. Using a putty knife, clean the gasket mounting surfaces. Inspect the exhaust manifold for distortion, cracks or damage; replace if necessary.

10. Install a new exhaust manifold gasket.

11. Install the exhaust manifold.

12. Using GM P/N 12345493 Threadlock on the manifold bolts, install the bolts onto the manifold. Tighten the exhaust manifold bolts in three passes to 15 ft. lbs. (20 Nm) each pass.

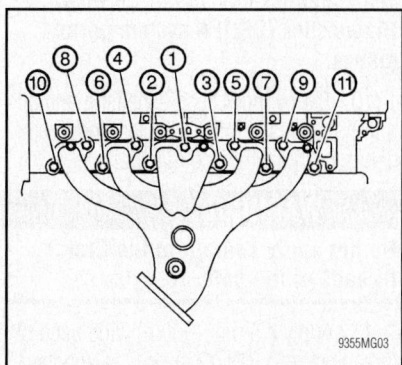

**Fig. 62 Exhaust manifold bolt tightening sequence—4.2L engine**

13. Install the exhaust manifold heat shield studs. Tighten the heat shield studs to 89 inch lbs. (10 Nm).

14. Install the manifold heat shield as follows:

a. Install the exhaust manifold heat shield with the 4 nuts. Tighten the exhaust manifold heat shield nuts to 89 inch lbs. (10 Nm).

15. Install the oxygen sensor.

16. Install the oil level indicator.

17. Install the AIR solenoid valve.

18. Install the air cleaner resonator outlet duct.

19. Raise the vehicle.

20. Install the exhaust pipe to the manifold with seal and secure the pipe with the nuts. Tighten the exhaust pipe nuts to 37 ft. lbs. (50 Nm).

21. Lower the vehicle.

### 5.3L and 6.0L Engines

1. Before servicing the vehicle, refer to the Precautions Section.

2. Raise and suitably support the vehicle. Refer to Lifting and Jacking the Vehicle.

3. Remove the nuts securing the catalytic converter to the left exhaust manifold.

4. Lower the vehicle.

5. Remove the spark plugs.

6. Remove the heat shield bolts, and shield from the exhaust manifold.

7. Remove the exhaust manifold bolts, manifold, and gasket.

8. Discard the gasket.

#### To install:

9. Clean and inspect the exhaust manifold. Refer to Exhaust Manifold Cleaning and Inspection.

➡**Tighten the exhaust manifold bolts as specified in the service procedure. Improperly installed and/or leaking exhaust manifold gaskets may affect vehicle emissions and/or On-Board Diagnostics (OBD) II system performance.**

10. The cylinder head exhaust manifold bolt hole threads must be clean and free of debris or threadlocking material.

#### ✳✳ WARNING
**Do not apply sealant to the first 3 threads of the bolt.**

11. Apply a 5 mm (0.2 in) wide band of threadlock GM P/N 12345493, (Canadian P/N 10953488), or equivalent to the threads of the exhaust manifold bolts.

12. Install the exhaust manifold NEW gasket and bolts.

13. Tighten the bolts a first pass to 11 ft. lbs. (15 Nm). Tighten the bolts beginning with the center 2 bolts. Alternate from side-to-side, and work toward the outside bolts.

14. Tighten the bolts a final pass to 18 ft. lbs. (25 Nm). Tighten the bolts beginning with the center 2 bolts. Alternate from side-to-side, and work toward the outside bolts.

15. Install the heat shield, and bolts to the exhaust manifold. Tighten the bolts to 80 inch lbs. (9 Nm).

16. Using a flat punch, bend over the exposed edge of the exhaust manifold gasket at the rear of the left cylinder head.

17. Install the spark plugs.

18. Raise the vehicle.

19. Install the nuts that secure the catalytic converter to the left exhaust manifold. Tighten the nuts to 37 ft. lbs. (50 Nm).

20. Lower the vehicle.

## INTAKE MANIFOLD

### REMOVAL & INSTALLATION

#### 4.2L Engine

1. Before servicing the vehicle, refer to the Precautions Section.

2. Disconnect the negative battery cable.

3. Relieve the fuel pressure.

4. Remove the throttle body.

5. Remove the powertrain control module retaining bolts and nuts.

6. Remove the PCM mounting studs and position the PCM out of the way.

7. Disconnect the engine coolant temperature sensor electrical connector.

8. Disconnect the fuel feed and fuel return pipes from the fuel rail.

9. Disconnect the integral clip from the wire harness bracket.

10. Remove the engine wire harness bracket bolt.

11. Position the engine electrical wire harness bracket with wires attached out of the way.

12. Disconnect the manifold absolute pressure sensor electrical connector.

13. Disconnect the crankcase ventilation hose from the intake manifold.

14. Disconnect the vacuum brake booster hose at the intake manifold.

15. Remove the generator.

#### ✳✳ WARNING
**The intake manifold bolts are captured within the intake manifold. Do not attempt to remove the bolts from the intake manifold.**

16. Loosen the intake manifold bolts.

17. Remove the intake manifold.

#### To install:

18. Install a new intake manifold gasket to the intake manifold.

19. Install the intake manifold onto the engine and secure the manifold with the bolts. Tighten the intake manifold bolts to 89 inch lbs. (10 Nm).

20. Install the generator.

21. Install the vacuum brake booster hose to the intake manifold.

22. Lubricate the inner diameter of the crankcase ventilation hose.

23. Install the crankcase ventilation hose.

24. Connect the MAP sensor electrical connector.

25. Properly position the engine electrical harness bracket to the intake manifold.

26. Install the engine electrical harness bracket bolt. Tighten the bolt to 89 inch lbs. (10 Nm).

27. Connect the integral clip to the wire harness bracket.

28. Connect the fuel feed and fuel return pipes to the fuel rail.

29. Connect the ECT sensor electrical connector.

30. Install the PCM mounting studs to the intake manifold. Tighten the studs to 53 inch lbs. (6 Nm).

31. Install the PCM onto the studs.

32. Install the PCM retaining bolts. Tighten the bolts to 71 inch lbs. (8 Nm).

33. Install the PCM retaining nuts. Tighten the nuts to 71 inch lbs. (8 Nm).

34. Install the throttle body.

35. Connect the negative battery cable.

36. Inspect for leaks using the following procedure:

37. Turn ON the ignition, with the engine OFF for 2 seconds.

38. Turn OFF the ignition, for 10 seconds.

39. Turn ON the ignition, with the engine OFF.

40. Inspect for fuel leaks.

#### 5.3L Engine

*See Figures 63 through 65.*

1. Before servicing the vehicle, refer to the Precautions Section.

2. Properly relieve the fuel system pressure.

3. Disconnect the negative battery cable.

➡**The intake manifold, throttle body, fuel rail and injectors can be removed as an assembly. If you are not servicing** these components individually, remove the intake manifold as a complete assembly.

4. Drain the engine cooling system.
5. Remove or disconnect the following:
   - Air cleaner outlet duct
   - A/C compressor pressure switch electrical connector
   - Harness clip from the cylinder head and fuel rail
   - Mass Airflow/Intake Air Temperature sensor connector
6. Disconnect the electrical connectors from the following:
   a. Main coil
   b. Electronic Throttle Control (ETC)
   c. Fuel injectors. Matchmark the connectors, pull the Connector Position Assurance (CPA) retainer up 1 click. Push the tab on the connector in, then detach the injector connector.
   - Alternator connector
   - Evaporative emission (EVAP) purge solenoid electrical connector
   - Knock Sensor (KS) electrical connector
   - Main coil
   - Fuel injector electrical connector
   - Electrical harness clips from the fuel rail
   - KS harness electrical connector from the intake manifold
   - Positive Crankcase Ventilation (PCV) valve hose and valve
   - Heater water shutoff valve actuator inlet hose from the intake manifold
   - EVAP purge solenoid vent tube
   - Vacuum brake booster hose from the rear of the intake manifold
   - Upper engine wire harness retainer nut. Position the wire harness aside.
   - Intake manifold bolts

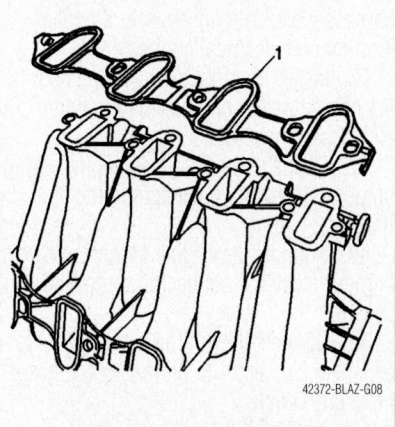

**Fig. 63 Make sure to use new intake manifold gaskets—5.3L engine**

- Intake manifold and gaskets. Discard the gaskets.

*To install:*

7. Clean the gasket mounting surfaces. Be sure to inspect the manifold for warpage and/or cracks. If necessary, replace it.
8. Properly position a new intake manifold gasket.
9. Apply a 0.20 in. (5mm) band of a suitable threadlocking material to the intake manifold bolt threads.
10. Install or connect the following:
   - Intake manifold and bolts. Torque the bolts, in sequence to 44 inch lbs. (5 Nm), then to 89 inch lbs. (10 Nm).
   - Route the electrical harness into position over the engine.
   - Engine harness bracket nut and tighten to 89 inch lbs. (10 Nm)
   - Vacuum brake booster hose to the rear of the intake manifold
   - EVAP purge solenoid valve
   - Heater water shutoff valve actuator inlet hose to the intake manifold
   - PCV valve and hose

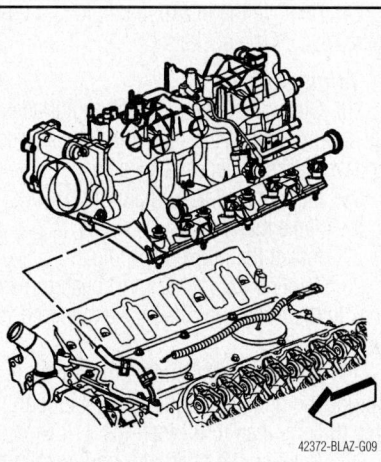

**Fig. 64 Exploded view of the intake manifold—5.3L engine**

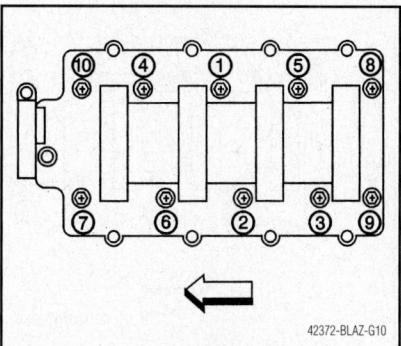

**Fig. 65 Intake manifold bolt tightening sequence—5.3L engine**

- EVAP purge solenoid, KS, MAP sensor, main coil & fuel injector electrical connectors
- Harness clips to the fuel rail
- Alternator electrical connector
- Main coil, ETC, fuel injector electrical connectors
- Electrical harness clips to the fuel rail
- A/C compressor pressure switch electrical connector
- Harness clip to the cylinder head
- Mass Airflow/Intake Air Temperature sensor connector
- Air cleaner outlet duct
- Fuel fill cap
- Negative battery cable
11. Refill the engine cooling system.

### 6.0L Engine

*See Figures 66 through 68.*

➡The intake manifold, throttle body, fuel rail and injectors can be removed as an assembly. If you are not servicing these components individually, remove the intake manifold as a complete assembly.

1. Before servicing the vehicle, refer to the Precautions Section.

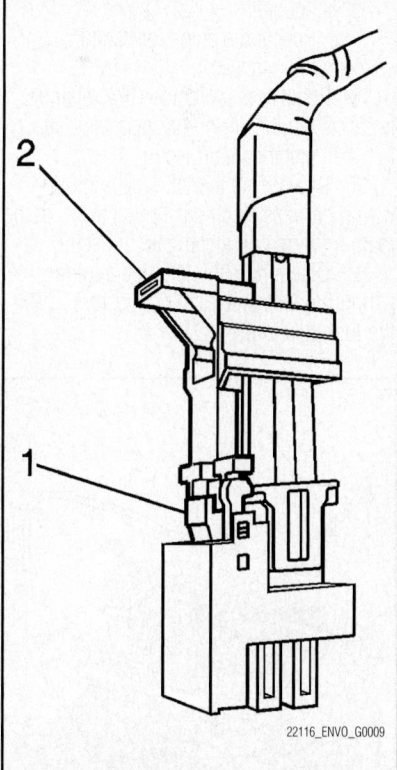

**Fig. 66 Pull the CPA retainer on the connector up one click, then push the tab on the connector in to disconnect the fuel injector electrical connector—6.0L engine**

2. Properly relieve the fuel system pressure.

3. Disconnect the negative battery cable.

4. Drain the engine cooling system.

5. Remove or disconnect the following:
- A/C compressor pressure switch electrical connector
- Mass Airflow/Intake Air Temperature sensor connector
- Electronic Throttle Control (ETC)

6. Remove the right side connector position assurance (CPA) retainer from the engine wiring harness main ignition coil electrical connector.

7. Disconnect the right side engine wiring harness electrical connector from the main ignition coil electrical connector.

8. Disconnect the right side engine wiring harness electrical connectors from the fuel injectors.

9. Perform the following steps (for the left and right sides) in order to disconnect the fuel injector electrical connectors.

   a. Mark the connectors to their corresponding injectors to ensure correct reassembly.

   b. Pull the connector position assurance (CPA) retainer on the connector up one click.

   c. Push the tab on the connector in.

   d. Disconnect the fuel injector electrical connector.

   e. Repeat the steps for each injector electrical connector.

10. Remove the left side CPA retainer from the engine wiring harness main ignition coil electrical connector.

11. Disconnect the left side engine wiring harness electrical connector from the main ignition coil electrical connector.

12. Disconnect the left side engine wiring harness electrical connectors from the fuel injectors.

13. Disconnect the engine wiring harness electrical connector from the alternator.

14. Disconnect the engine wiring harness electrical connector from the manifold absolute pressure (MAP) sensor.

15. Disconnect the engine wiring harness electrical connector from the evaporative emission (EVAP) canister purge solenoid valve.

16. Disconnect the positive crankcase ventilation (PCV) hose.

17. Disconnect the EVAP purge solenoid vent tubes.

18. Disconnect the fuel feed pipe from the fuel rail.

19. Reposition the vacuum brake booster hose clamp at the brake booster and disconnect the vacuum brake booster hose from the brake booster.

20. Remove the engine wire harness retainer nut and reposition the upper engine wire harness aside.

21. Remove the drive belt.

22. Remove the right alternator bolt, then loosen the left alternator bolt and reposition the alternator to the left.

23. Remove the intake manifold bolts.

24. Remove the intake manifold and gaskets. Discard the gaskets.

### To install:

25. Clean the gasket mounting surfaces. Be sure to inspect the manifold for warpage and/or cracks. If necessary, replace it.

26. Install new intake manifold gaskets to the intake manifold.

27. Install the intake manifold.

28. Apply a 0.2 inch (5mm) bead threadlock to the threads of the intake manifold bolts.

29. Install the intake manifold bolts. Torque the bolts, in sequence to 44 inch lbs. (5 Nm), then to 89 inch lbs. (10 Nm).

30. Position the alternator and install the right alternator bolt and tighten the left alternator bolt to 37 ft. lbs. (50 Nm).

31. Install the drive belt.

32. Route the electrical harness into position over the engine. Install the engine harness bracket nut and tighten to 89 inch lbs. (10 Nm).

33. Connect the vacuum brake booster hose to the brake booster and position the vacuum brake booster hose clamp at the brake booster.

34. Connect the fuel feed pipe to the fuel rail.

35. Install the EVAP purge solenoid vent tubes.

36. Install the PCV hose.

37. Connect the engine wiring harness electrical connector to the MAP sensor.

38. Connect the engine wiring harness electrical connector to the EVAP canister purge solenoid valve.

39. Connect the engine wiring harness electrical connector to the alternator.

40. Connect the left side engine wiring harness electrical connector to the main ignition coil electrical connector.

41. Install the left side CPA retainer to the engine wiring harness main ignition coil electrical connector.

42. Connect the left side engine wiring harness electrical connectors to the fuel injectors.

43. Perform the following steps (for the left and right sides) in order to connect the fuel injector electrical connectors:

   a. Install the connectors to their corresponding injectors to ensure correct reassembly.

   b. Connect the fuel injector electrical connector.

   c. Push the CPA retainer on the connector in one click.

   d. Repeat the steps for each injector electrical connector.

44. Connect the engine wiring harness electrical connector to the ETC.

45. Connect the right side engine wiring harness electrical connector to the main ignition coil electrical connector.

46. Install the right side CPA retainer to the engine wiring harness main ignition coil electrical connector.

47. Connect the right side engine wiring harness electrical connectors to the fuel injectors.

48. Connect the engine harness wiring harness electrical connector to the MAF/IAT sensor.

49. Connect the engine wiring harness electrical connector to the A/C compressor pressure switch.

50. Install the fuel fill cap.

51. Connect the negative battery cable.

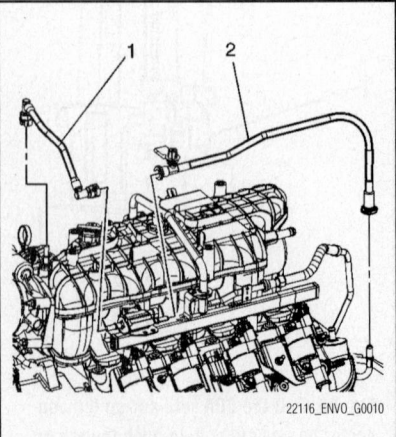

22116_ENVO_G0010

**Fig. 67 Disconnect the EVAP purge solenoid vent tubes (1 and 2)—6.0L engine**

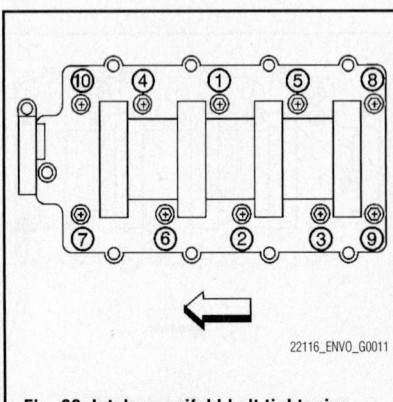

22116_ENVO_G0011

**Fig. 68 Intake manifold bolt tightening sequence—6.0L engine**

52. Use the following procedure in order to inspect for leaks:

    a. Turn the ignition ON, with the engine OFF, for 2 seconds.

    b. Turn the ignition OFF for 10 seconds.

    c. Turn the ignition ON, with the engine OFF.

    d. Inspect for fuel leaks.

## OIL PAN

*REMOVAL & INSTALLATION*

### 4.2L Engine

*See Figure 69.*

1. Before servicing the vehicle, refer to the Precautions Section.

2. Disconnect the negative battery cable.

3. Remove the A/C compressor bottom bolts and loosen the top bolts.

4. Remove the oil level indicator and tube.

5. Remove the stabilizer bar.

6. Remove the front differential and secure to the frame.

7. Remove the front drive axle intermediate shaft bearing assembly.

8. Drain the engine oil.

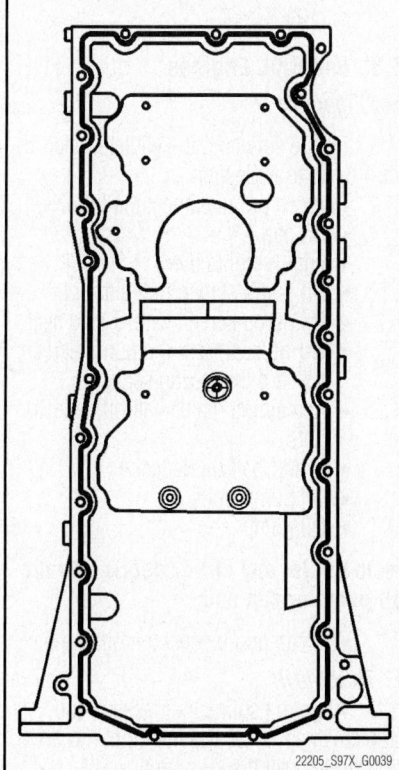

**Fig. 69 Apply a 0.12 in (3 mm) bead of sealer to the block, rather than the oil pan**

9. Unclip the transmission cooler lines from the engine block.

10. Remove 4 transmission bell housing bolts that are attached to the oil pan.

11. Remove the remaining oil pan bolts.

12. Place 2 oil pan bolts in the jack screws on the oil pan and tighten evenly to release the oil pan from the engine.

13. Clean and inspect the oil pan.

**To install:**

➡**The oil pan must be installed within 10 minutes from when sealer was applied.**

14. Apply a 3 mm (0.12 in) bead of sealer to the block, rather than the oil pan.

### ✹✹ WARNING

**When you install the oil pan, it could be shifted front or back a little which could cause a transmission alignment problem. The back of the oil pan needs to be flush with the block.**

15. Install the oil pan, maneuvering the oil pan to clear the oil pump and screen assembly.

16. Install the oil pan bolts.

17. Inspect the oil pan alignment. Use a straight edge on the back of the block and the oil pan transmission mounting surface.

18. Tighten the oil pan side bolts to 18 ft. lbs. (25 Nm). Tighten the oil pan end bolts to 89 inch lbs. (10 Nm).

19. Install the 4 transmission bell housing bolts that attach to the oil pan. Tighten bolts to 35 ft. lbs. (47 Nm).

20. Clip transmission cooler lines to the engine block.

21. Install the front drive axle intermediate shaft bearing assembly.

22. Install the oil drain plug and filter.

23. Install the front differential to the engine.

24. Install the stabilizer bar.

25. Install the A/C compressor 2 bottom bolts. Tighten bolts to 37 ft. lbs. (50 Nm).

26. Install the oil level indicator and tube.

27. Connect the negative battery cable.

28. Fill the engine with oil.

29. Start the engine and inspect for oil leaks in order to ensure all sealing surfaces are sealed.

### 5.3L and 6.0L Engines

*See Figures 70 through 72.*

1. Before servicing the vehicle, refer to the Precautions Section.

2. Disconnect the negative battery cable.

3. Drain the engine crankcase oil and differential oil.

4. Remove or disconnect the following:

5. Remove the oil level indicator tube.

6. Remove the front differential and secure to the frame.

7. Drain the engine oil.

8. Remove the transmission oil cooler lines from the retainer.

9. Remove the transmission oil cooler line retaining bracket bolt and bracket.

10. Remove the starter.

11. Remove the flywheel inspection cover from the left side of the transmission.

12. Remove the battery cable channel bolt from the front of the oil pan.

13. Remove the battery cable channel from the oil pan.

14. Loosen the upper air conditioning (A/C) compressor bracket bolts.

15. Remove the lower A/C compressor bracket bolts.

16. Remove the lower bellhousing bolts.

17. Remove the oil pan bolts.

18. Remove the oil pan by tilting the rear of the oil pan down to clear the transmission, pull the oil pan rearward past the front wire harness, then lower the oil pan clear of the vehicle.

### ✹✹ WARNING

**DO NOT allow foreign material to enter the oil passages of the oil pan, cap or cover the openings as required.**

19. Drill out the oil pan gasket retaining rivets, if required.

➡**The oil pan gasket is reusable. It is NOT necessary to remove the oil pan gasket unless damaged.**

20. Remove the gasket from the pan.

21. Discard the gasket and rivets.

22. Clean and inspect the oil pan.

**To install:**

➡**The alignment of the structural oil pan is critical. The rear bolt hole locations of the oil pan provide mounting points for the transmission bellhousing. To ensure the rigidity of the powertrain and correct transmission alignment, it is important that the rear of the block and the rear of the oil pan must NEVER protrude beyond the engine block and transmission bellhousing plane.**

➡**If replacing the oil pan gasket it is not necessary to rivet the NEW gasket to the oil pan.**

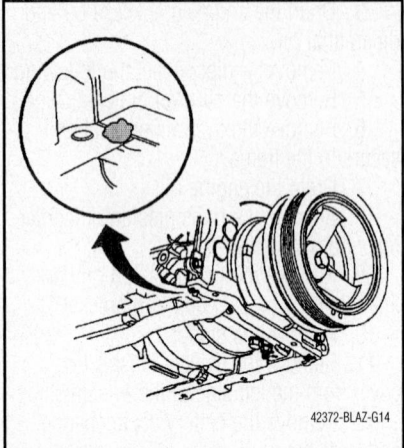

**Fig. 70 Proper sealant application to the front cover gasket**

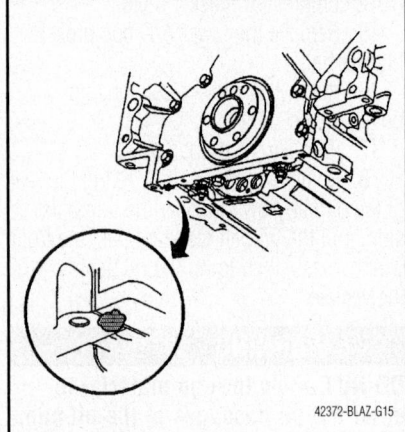

**Fig. 71 Proper sealant application to the rear cover gasket**

23. Apply a 0.20 in. (5mm) bead of sealant 0.80 in. (20mm) long to the engine block. Apply the sealant directly onto the tabs of the front cover gasket that protrudes into the oil pan surface.

24. Apply a 0.20 in. (5mm) bead of sealant 0.80 in. (20mm) long to the engine block. Apply the sealant directly onto the tabs of the rear cover gasket that protrudes into the oil pan surface.

25. Pre-assemble the oil pan gasket and bolts to the pan. Install the gasket onto the pan. Install the oil pan bolts to the pan and through the gasket.

26. Install the oil pan, oil pan gasket and bolts to the engine block as an assembly.

27. Hand-start the bolts into the engine block snug-tight. Do not fully tighten yet.

28. Install the lower bellhousing bolts and tighten to 37 ft. lbs. (50 Nm).

29. Tighten the rear oil pan-to-rear cover bolts to 106 inch lbs. (12 Nm) and the remaining oil pan bolts to 18 ft. lbs. (25 Nm).

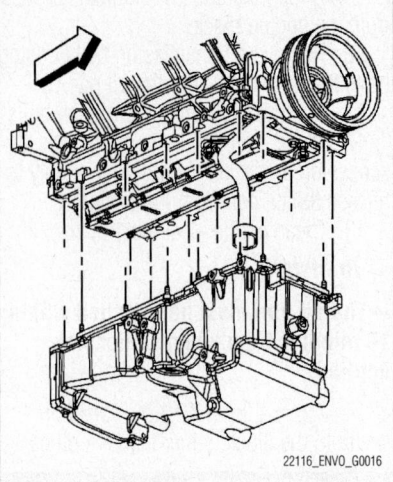

**Fig. 72 Oil pan mounting—5.3L and 6.0L engines**

30. Install the lower A/C compressor bracket bolts. Tighten the bolts to 37 ft. lbs. (50 Nm).

31. Tighten the upper A/C compressor bracket bolts. Tighten the bolts to 37 ft. lbs. (50 Nm).

32. Install the battery cable channel to the oil pan.

33. Install the battery cable channel bolt to the oil pan. Tighten the bolt to 106 inch lbs. (12 Nm).

34. Install the flywheel inspection cover to the left side of the transmission.

35. Install the starter.

36. Install the inner axle shaft.

37. Install the transmission oil cooler line retaining bracket and bolt. Tighten the bolt to 80 inch lbs. (9 Nm).

38. Install the transmission oil cooler line retaining bracket and bolt. Torque the bolt to 80 inch lbs. (9 Nm).

39. Install the oil level indicator tube.

40. Fill the engine with oil.

41. Install the front differential.

42. Connect the negative battery cable.

## OIL PUMP

### REMOVAL & INSTALLATION

#### 4.2L Engine

*See Figure 73.*

1. Before servicing the vehicle, refer to the Precautions Section.

2. Remove or disconnect the following:
- Engine front cover
- Oil pump cover bolts
- Oil pump cover. Mark the inner and outer gears in relation to the pump housing.

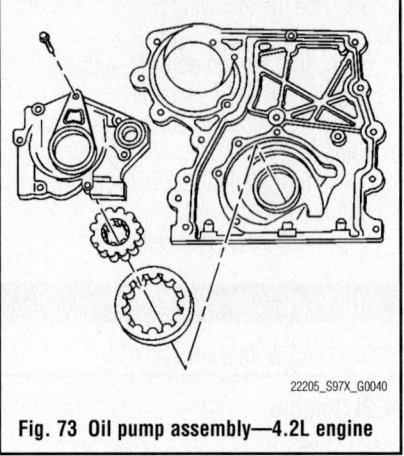

**Fig. 73 Oil pump assembly—4.2L engine**

- Inner and outer pump gears
- Oil pump pressure relief valve plug
- Oil pump pressure relief valve and spring

*To install:*

3. Install or connect the following:
- Oil pump pressure relief valve and spring
- Oil pump pressure relief valve plug. Tighten to 10 ft. lbs. (14 Nm).
- Oil pump outer and inner gears, as marked during removal
- Oil pump cover and bolts. Tighten the bolts to 89 inch lbs. (10 Nm).
- Front cover

#### 5.3L and 6.0L Engines

*See Figure 74.*

1. Before servicing the vehicle, refer to the Precautions Section.

2. Remove or disconnect the following:
- Oil pan
- Engine front cover
- Oil pump screen bolt and nuts
- Oil pump screen with O-ring seal
- O-ring seal from the pump screen. Discard the O-ring seal.
- Remaining crankshaft oil deflector nuts
- Crankshaft oil deflector
- Oil pump bolts
- Oil pump

➡ **Do not let any dirt or debris into the oil pump or cap end.**

- Clean and inspect the oil pump.

*To install:*

3. Align the splined surfaces of the crankshaft sprocket and the oil pump drive gear and install the oil pump.

4. Install or connect the following:
- Oil pump onto the crankshaft sprocket until the pump housing

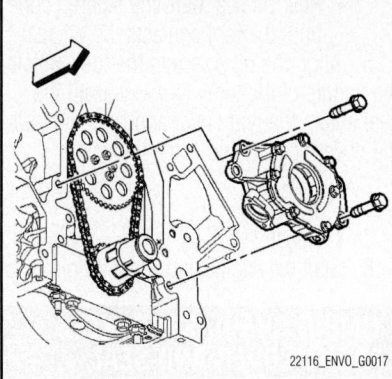

**Fig. 74 Exploded view of the oil pump mounting—5.3L and 6.0L engines**

contacts the face of the engine block
- Oil pump bolts and tighten to 18 ft. lbs. (25 Nm)
- Crankshaft oil deflector and nuts until snug
- New oil pump screen O-ring seal into the oil pump screen, after lubricating with clean engine oil

➡ **Push the oil pump screen tube completely into the oil pump prior to tightening the bolt. Do not let the bolt pull the tube into the pump.**

5. Align the oil pump screen mounting brackets with the correct crankshaft bearing cap studs.
- Oil pump screen
- Oil pump screen bolts and nuts. Tighten the bolts to 106 inch lbs. (12 Nm) and the nuts to 18 ft. lbs. (25 Nm).
- Engine front cover
- Oil pan

## PISTON AND RING

### POSITIONING

*See Figures 75 and 76.*

## REAR MAIN SEAL

### REMOVAL & INSTALLATION

#### 4.2L Engine

##### 2006 Models

*See Figure 77.*

Please note that the transmission assembly must be removed to perform this procedure.

1. Before servicing the vehicle, refer to the precautions in the beginning of this section.

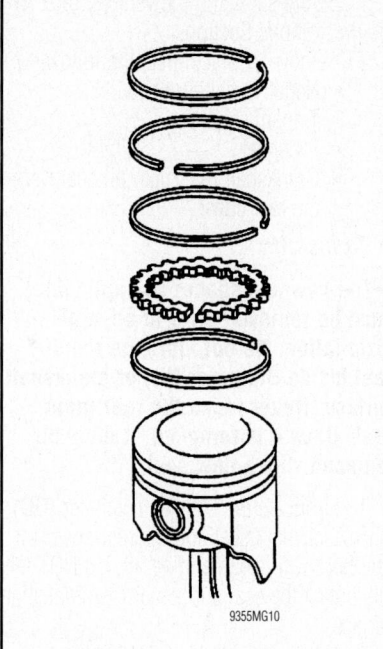

**Fig. 75 Piston ring positioning—4.2L engine**

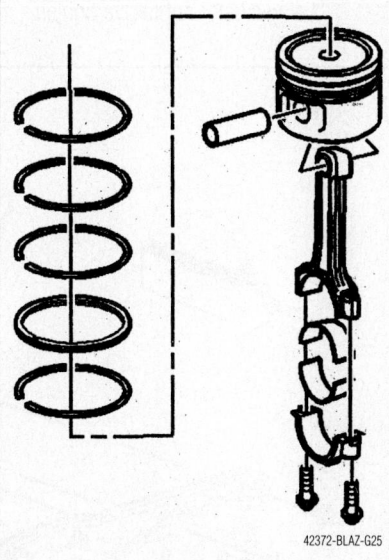

**Fig. 76 Piston ring positioning—5.3L engine**

2. Remove or disconnect the following:
- Negative battery cable
- Transmission
- Flywheel
- Crankshaft rear main seal housing bolts. Install 2 bolts into the jackscrew holes to release the cover from the block
- Crankshaft and rear main seal housing
- Rear main seal from the crankshaft snout

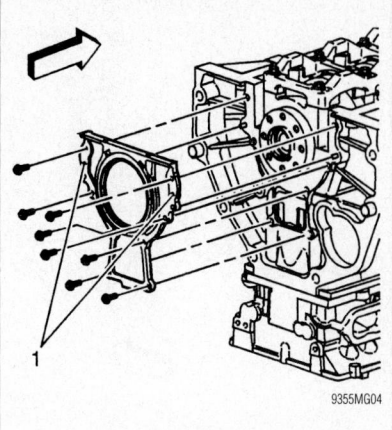

**Fig. 77 Install 2 bolts into the jackscrew holes (1) to push the cover off of the block**

#### To install:
3. Install or connect the following:
- Rear main seal, using a suitable seal installation tool, then remove the tool
- Apply a 0.12 in. (3mm) bead of 12378521, or equivalent sealant to the rear mail seal housing
- Suitable cover alignment pins into the block

➡ **When you install a new seal, make sure to use the plastic installation sleeve supplies with the new seal. The sleeve should come off and be discarded after the seal is installed.**

4. Slide the crankshaft rear main seal housing over the alignment pins and crankshaft.

5. Install the crankshaft rear main seal housing bolts, except the 2 in place of the guide pins.

6. Remove the guide pins.

7. Install or connect the following:
- Remaining 2 crankshaft rear main seal housing bolts and tighten to 89 inch lbs. (10 Nm). Wipe off any excess sealant.
- Flywheel
- Transmission

##### 2007–08 Models

*See Figure 78.*

1. Before servicing the vehicle, refer to the Precautions Section.

2. Remove or disconnect the following:
- Negative battery cable
- Transmission
- Flywheel

**✳✳ WARNING**

**Do not damage the crankshaft or seal bore.**

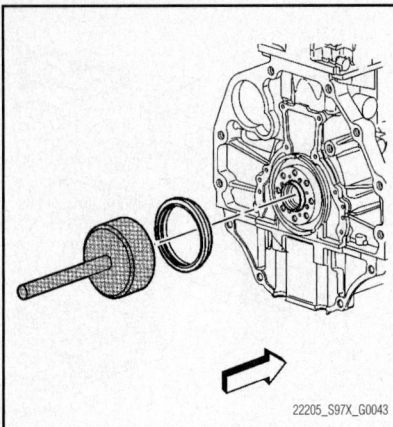

**Fig. 78 Install 2 bolts into the jackscrew holes to push the cover off of the block**

- Pry the crankshaft rear oil seal out of the rear oil seal housing using a suitable tool.

#### To install:

3. Install or connect the following:
- Rear main seal, using a suitable seal installation tool, then remove the tool

➡ **When you install a new seal, make sure to use the plastic installation sleeve supplies with the new seal. The sleeve should come off and be discarded after the seal is installed.**

4. Install or connect the following:
- Flywheel
- Transmission
- Negative battery cable

### 5.3L and 6.0L Engines

*See Figure 79.*

Please note that the transmission assembly must be removed to perform this procedure.

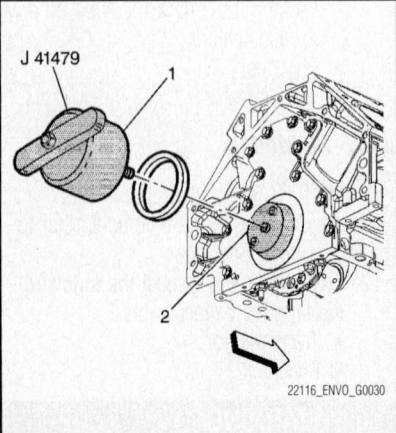

**Fig. 79 View of the rear main seal installation—5.3L and 6.0L engines**

1. Before servicing the vehicle, refer to the Precautions Section.
2. Remove or disconnect the following:
- Negative battery cable
- Transmission
- Flywheel
- Crankshaft rear main oil seal from the rear cover

#### To install:

➡ **The flywheel spacer (if applicable) must be removed prior to oil seal installation. Do not lubricate the oil seal Inside Diameter (ID) or crankshaft surface. Never reuse the rear main seal. Once it is removed, it must be replaced with a new seal.**

3. Lubricate the Outside Diameter (OD) of the rear main seal and the rear cover oil seal bore with clean engine oil. Do NOT let oil contact the seal surface or the crankshaft surface.
4. Install or connect the following:
- Crankshaft Rear Oil Seal Installer Tool No. J 41479 tapered cone and bolts onto the rear of the crankshaft. Tighten the bolts until just snug, being careful not to overtighten.

- Rear oil seal onto the tapered cone until the tool contacts the oil seal
5. Align the oil seal into the tool, Rotate the handle of the tool clockwise until the seal enters the rear cover and bottoms into the cover bore. Remove the tool.
- Flywheel
- Transmission
- Negative battery cable
6. Start the engine and verify no oil leaks.

## TIMING CHAIN, SPROCKETS, FRONT COVER AND SEAL

### REMOVAL & INSTALLATION

#### 4.2L Engine

*See Figures 80 through 83.*

1. Remove the camshaft cover.
2. Before servicing the vehicle, refer to the Precautions Section.
3. Drain the engine coolant.
4. Remove the cooling fan and the shroud.
5. Remove the drive belt.
6. Remove the water pump.
7. Remove the crankshaft balancer.
8. Remove the power steering pump.

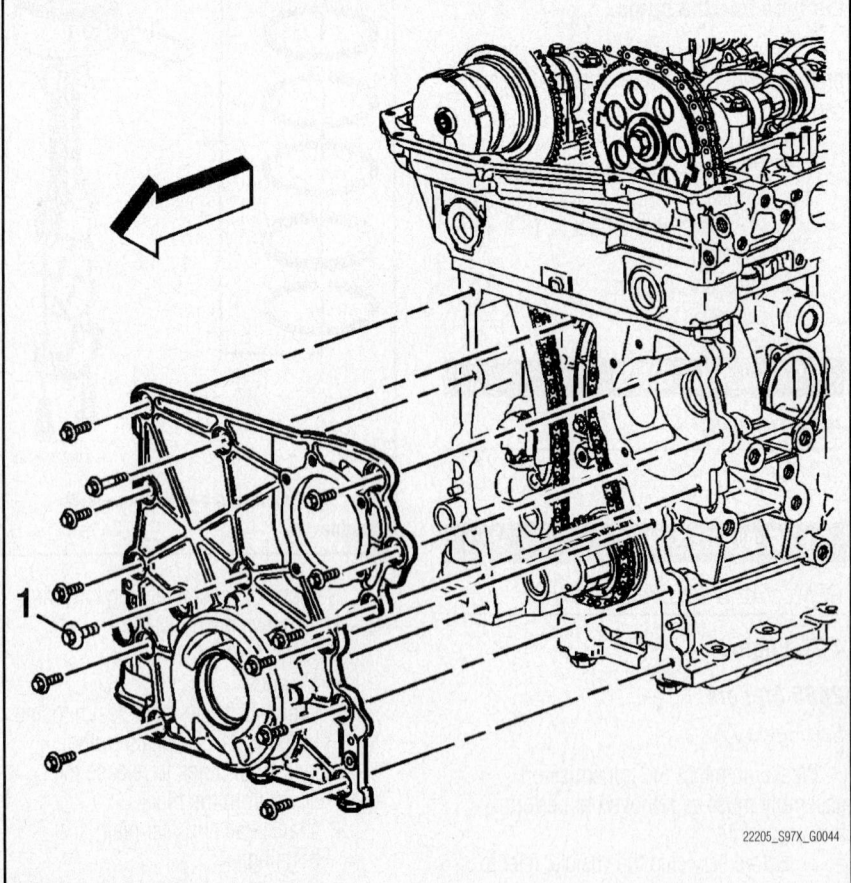

**Fig. 80 Engine front cover bolt locations**

9. Raise and safely support the vehicle securely on jackstands.

10. Remove the oil pan.

11. Lower the vehicle.

12. Remove the 7 mm center bolt.

13. Loosen and remove the remaining engine front cover bolts.

14. Place 2 of the front cover bolts in the jack screw holes on the front cover and tighten the bolts evenly to release the front cover from the engine.

15. Remove the 2 bolts from the front cover.

16. Remove the oil pump.

17. Remove the tension on the timing chain by moving the tensioner shoe in. Place a tee into the tension to hold the shoe in place.

18. Remove the exhaust camshaft position actuator bolt and actuator.

19. Remove the intake camshaft sprocket bolt and sprocket.

20. Remove the timing chain.

21. Remove the crankshaft sprocket.

22. Remove the cylinder head access hole plugs.

23. Remove the timing chain tensioner shoe bolt and shoe.

24. Remove the timing chain tensioner guide bolts and guide.

25. Remove the timing chain tensioner bolts and tensioner.

**To install:**

➡**Every seventh link of the timing chain is darkened to help in aligning the timing marks.**

26. Install timing chain tensioner and bolts. Tighten to 18 ft. lbs. (25 Nm).

27. Install timing chain guide and bolts. Tighten to 89 inch lbs. (10 Nm).

28. Install timing chain tensioner shoe and bolt. Tighten to 19 ft. lbs. (26 Nm).

29. Install cylinder head access hole plugs and tighten to 44 inch lbs. (5 Nm).

30. Install Crankshaft Holding tool No. J-44221, or equivalent with the camshaft flats up and the No. 1 cylinder at Top Dead Center (TDC).

31. Install crankshaft sprocket.

32. Install intake camshaft sprocket into the timing chain.

33. Align the dark link of the timing chain with the timing mark on the intake camshaft sprocket.

34. Feed the timing chain down through the opening in the head.

35. Install the timing chain onto the crankshaft sprocket. Align the dark link of the timing chain with the timing mark on the crankshaft sprocket.

➡**It may be necessary to remove the crankshaft holding tool to rotate and hold the camshaft hex to align the pin to the camshaft sprocket**

36. Install the intake camshaft sprocket onto the intake camshaft.

37. Install the intake camshaft washer and bolt.

38. Install the exhaust camshaft actuator into the timing chain. Align the dark link of the timing chain with the timing mark on the exhaust camshaft actuator.

➡**It may be necessary to remove the crankshaft holding tool to rotate and hold the camshaft hex to align the pin to the camshaft sprocket**

39. Install the exhaust camshaft actuator onto the exhaust camshaft.

➡**Rotate the camshaft actuator clockwise relative to the camshaft prior to tightening the bolt.**

40. Rotate the camshaft actuator clockwise (as seen from the front of the vehicle).

※※ **WARNING**

**The camshaft actuator must be fully advanced during installation. Engine damage may occur if the camshaft actuator is not fully advanced.**

41. Install the exhaust camshaft actuator bolt and tighten to 18 ft. lbs. (25 Nm), plus an additional 135° rotation, using a torque angle meter.

42. Tighten the intake camshaft sprocket bolt to 22 ft. lbs. (30 Nm), plus an additional 100° rotation, using a torque angle meter.

43. Remove the tee from the timing chain tensioner to regain tension on the timing chain.

44. Remove the crankshaft holding tool. The dark lines on the timing chain should be aligned with the marks on the sprockets.

45. Clean the gasket mating surfaces of the engine and cover of all remaining gasket or sealer material. Be careful not to score or damage the surfaces.

46. Install J 44219 Cover Alignment Pins, onto the engine.

➡**The front cover MUST be installed within 10 minutes of applying the sealant.**

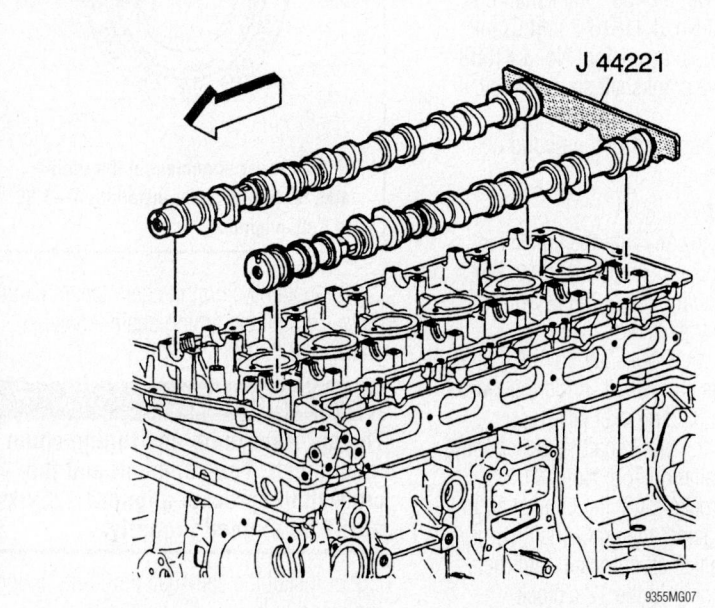

Fig. 81 Proper installation of the crankshaft holding tool with the No. 1 cylinder at TDC

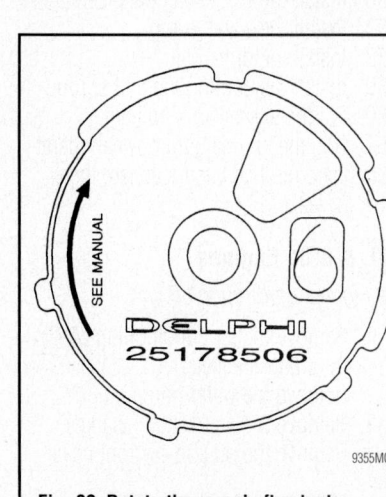

Fig. 82 Rotate the camshaft actuator clockwise

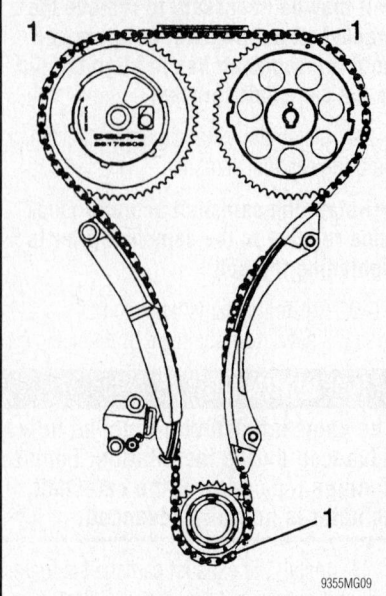

**Fig. 83 The dark lines on the timing chain should be aligned with the marks on the sprockets**

47. Apply a 0.12 in. (3mm) beat of 12378521 or equivalent sealant to the trace grooves on the back side of the engine front cover. Apply sealant on the inside 3 bolt hole bosses on the cover also.

48. Align the oil pump to the crankshaft sprocket splines.

49. Install the front cover and bolts, tighten the center bolt last. Tighten to 89 inch lbs. (10 Nm).

50. Remove the alignment pins.

51. Raise and safely support the vehicle securely on jackstands.

52. Install the oil pan.

53. Lower the vehicle.

54. Install the camshaft cover.

55. Install the power steering pump.

56. Install the crankshaft balancer.

57. Install the water pump.

58. Install the drive belt.

59. Install the cooling fan and shroud.

60. Fill the engine with coolant.

61. Run the engine until normal operating temperature has been reached, then check for leaks.

## 5.3L & 6.0L Engines

*See Figures 84 through 88.*

1. Remove the air conditioning (A/C) compressor and bracket.

2. Remove the water pump.

3. Remove the crankshaft balancer.

4. Remove the oil pan-to-front cover bolts.

5. Remove the front cover bolts.

6. Remove the front cover and gasket.

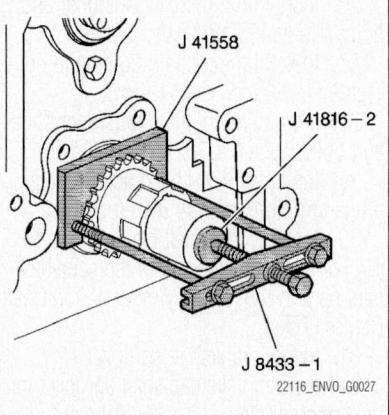

**Fig. 84 Use the proper tools to remove the crankshaft sprocket—5.3L and 6.0L engines**

7. Discard the front cover gasket.

8. Remove the oil pump.

9. Rotate the crankshaft until the timing marks on the crankshaft and the camshaft sprockets are aligned.

> ❊❊ **WARNING**
> **Do NOT turn the crankshaft after the timing chain has been removed to prevent damage to the pistons and valves.**

10. Remove and discard the camshaft sprocket bolts

11. Remove camshaft sprocket and timing chain

12. Remove the bolts and timing chain tensioner.

13. Remove crankshaft sprocket using Pulley Puller No. J 8433, Crankshaft End Protector Tool No. J 41816-2 and Crankshaft Sprocket Removal Tool No. J 41558

14. Remove crankshaft sprocket key, if necessary

15. Clean and inspect the timing chain and sprockets.

### To install:

16. Install the key into the crankshaft keyway, if removed. Tap the key into the keyway until both ends of the key bottom into the crankshaft.

17. Install the crankshaft sprocket onto the front of the crankshaft. Align the crankshaft key with the sprocket keyway.

18. Install the crankshaft sprocket using Sprocket Installation Tool No. J 41665. Install the sprocket onto the crankshaft until fully seated against the crankshaft flange. Rotate the crankshaft sprocket until the alignment mark is in the 12 o'clock position.

19. Compress the timing chain tensioner guide and install the EN 46330 Holding Pin.

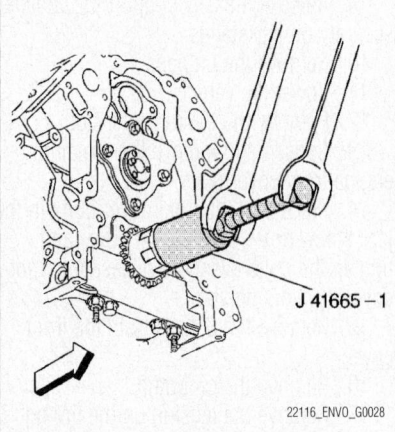

**Fig. 85 Crankshaft sprocket installation—5.3L and 6.0L engines**

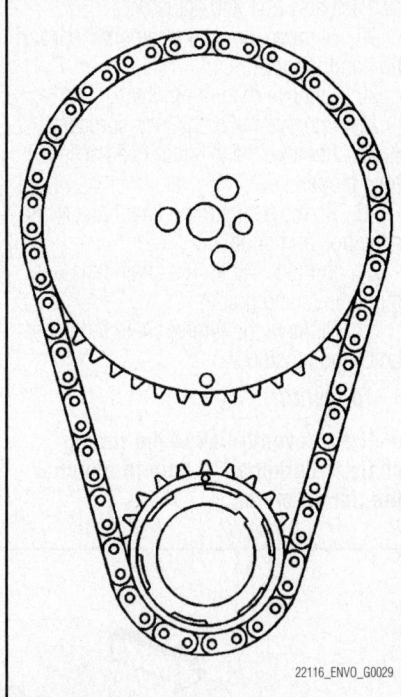

**Fig. 86 Proper alignment of the timing marks for timing chain installation—5.3L and 6.0L engines**

20. Install the timing chain tensioner and bolts. Tighten the timing chain tensioner bolts to 18 ft. lbs. (25 Nm).

> ❊❊ **WARNING**
> **The sprocket teeth and timing chain must mesh. The camshaft and the crankshaft sprocket alignment marks MUST be aligned properly.**

21. Install the camshaft sprocket, timing chain and bolt.

22. Inspect the sprockets for proper alignment. The mark on the camshaft

sprocket should be located in the 6 o'clock position and the mark on the crankshaft sprocket should be located in the 12 o'clock position.

23. Remove the EN 46330 Holding Pin.

24. Install the camshaft sprocket bolts and tighten to 26 ft. lbs. (35 Nm).

25. Install the oil pump.

26. Clean and inspect the engine front cover.

### ✳✳ WARNING

**Do not reuse the crankshaft oil seal or front cover gasket. Do not apply any type of sealant to the front cover gasket, unless specified. A special tool is used to properly center the front crankshaft front oil seal. All gasket surfaces should be free of oil or other foreign material during assembly. The crankshaft front oil seal MUST be centered in relation to the crankshaft. An improperly aligned front cover may cause premature front oil seal wear and/or engine oil leaks.**

27. Apply a 0.20 in. (5mm) bead of sealant 0.80 in. (20mm) long to the oil pan-to-engine block junction.

28. Install the front cover gasket and cover.

29. Install the front cover bolts until snug. Do not over tighten.

30. Install the oil pan-to-front cover bolts until snug. Do not over tighten.

31. Install the J 41476 Front and Rear Cover Alignment Tool. Align the tapered legs of the tool with the machined alignment surfaces on the front cover.

32. Install the crankshaft balancer bolt, finger-tight.

33. Install the oil pan-to-front cover bolts to 18 ft. lbs. (25 Nm).

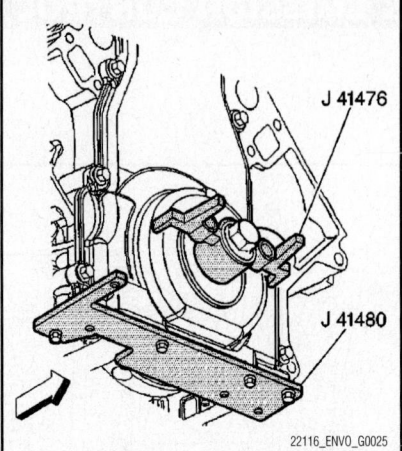

**Fig. 87 Align the tapered legs of the tool with the machined alignment surfaces on the front cover—5.3L and 6.0L engines**

34. Install the front cover bolts to 18 ft. lbs. (25 Nm).

35. Remove the tool.

36. Install a new crankshaft front oil seal as follows:

a. Remove the radiator for access.

b. Remove the crankshaft balancer.

c. Remove the crankshaft oil seal.

d. Lubricate the outer edge ONLY of the new crankshaft oil seal with clean engine oil.

e. Install the crankshaft front oil seal into the Crankshaft Front Seal Installation Tool No. J 41478 guide.

f. Install the J 41478 threaded rod (with nut, washer, guide and oil seal) into the end of the crankshaft.

g. Use J 41478 to install the oil seal into the cover bore. Use a wrench and hold the hex on the installer bolt. Use a second wrench to rotate the installer nut clockwise until the seal bottoms in the cover bore. Remove the tool.

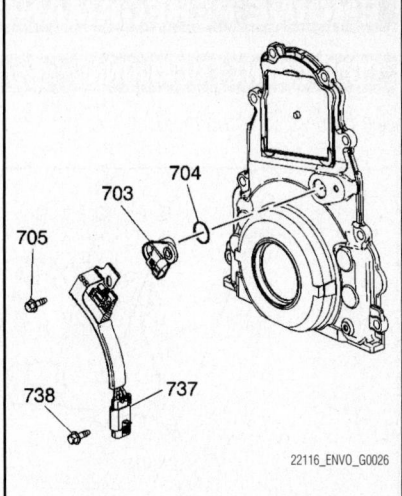

**Fig. 88 Front cover seal installation using the proper tool—5.3L and 6.0L engines**

h. Check the seal for proper installation. It should be installed evenly and completely into the front cover bore.

37. Install the crankshaft balancer. Tighten the bolt to 37 ft. lbs. (50 Nm), plus an additional 140 degrees using a torque angle meter.

38. Install the radiator.

39. Install the water pump.

40. Install the A/C compressor and bracket

41. Fill the cooling system with coolant

42. Run the engine until normal operating temperature has been reached, then check for leaks.

### VALVE LASH

#### ADJUSTMENT

The 4.2L, 5.3L and 6.0L engines do not require periodic valve lash adjustment.

# ENGINE PERFORMANCE & EMISSION CONTROL

## COMPONENT LOCATIONS

*See Figures 89 through 91.*

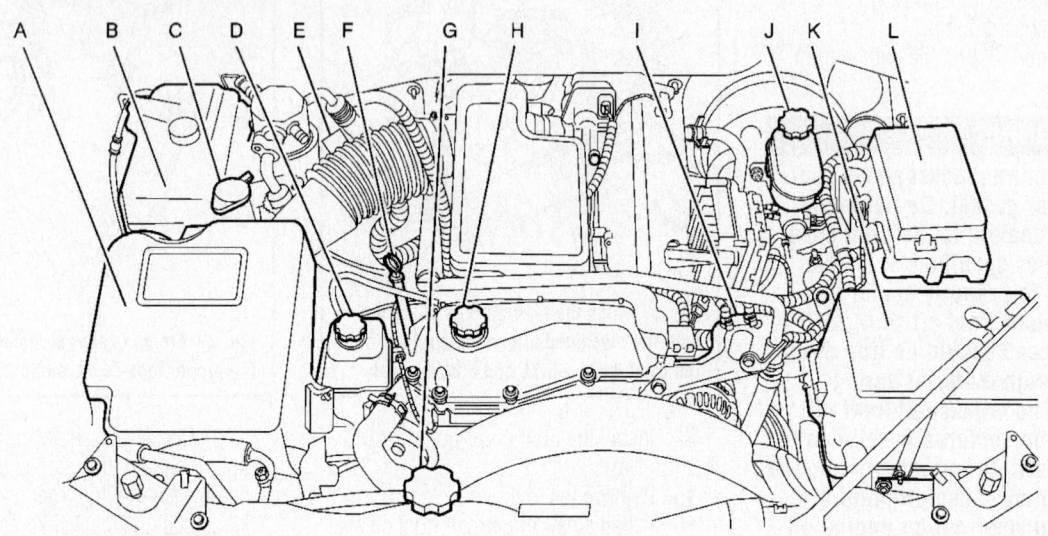

A. Engine air filter
B. Coolant recovery tank
C. Washer fluid reservoir
D. Power steering fluid reservoir
E. Transmission fluid dipstick (out of view)
F. Engine oil dipstick
G. Radiator cap
H. Engine oil fill cap
I. Remote negative (-) battery terminal (marked GND)
J. Brake master cylinder reservoir
K. Battery
L. Engine compartment fuse block

22116_ENVO_G0141

**Fig. 89 Engine compartment component locations—4.2L engine**

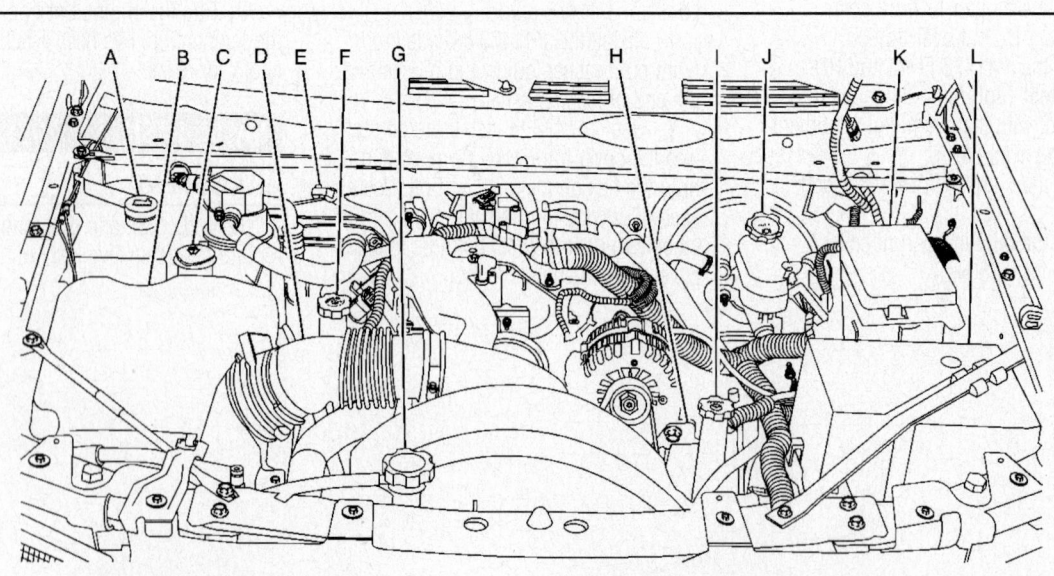

A. Engine coolant recovery tank
B. Engine air filter
C. Washer fluid reservoir
D. Engine oil dipstick
E. Transmission fluid dipstick
F. Engine oil fill cap
G. Radiator cap
H. Remote negative (-) terminal (marked GND)
I. Power steering fluid reservoir
J. Brake master cylinder reservoir
K. Engine compartment fuse block
L. Battery

22116_ENVO_G0142

**Fig. 90 Engine compartment component locations—5.3L engine**

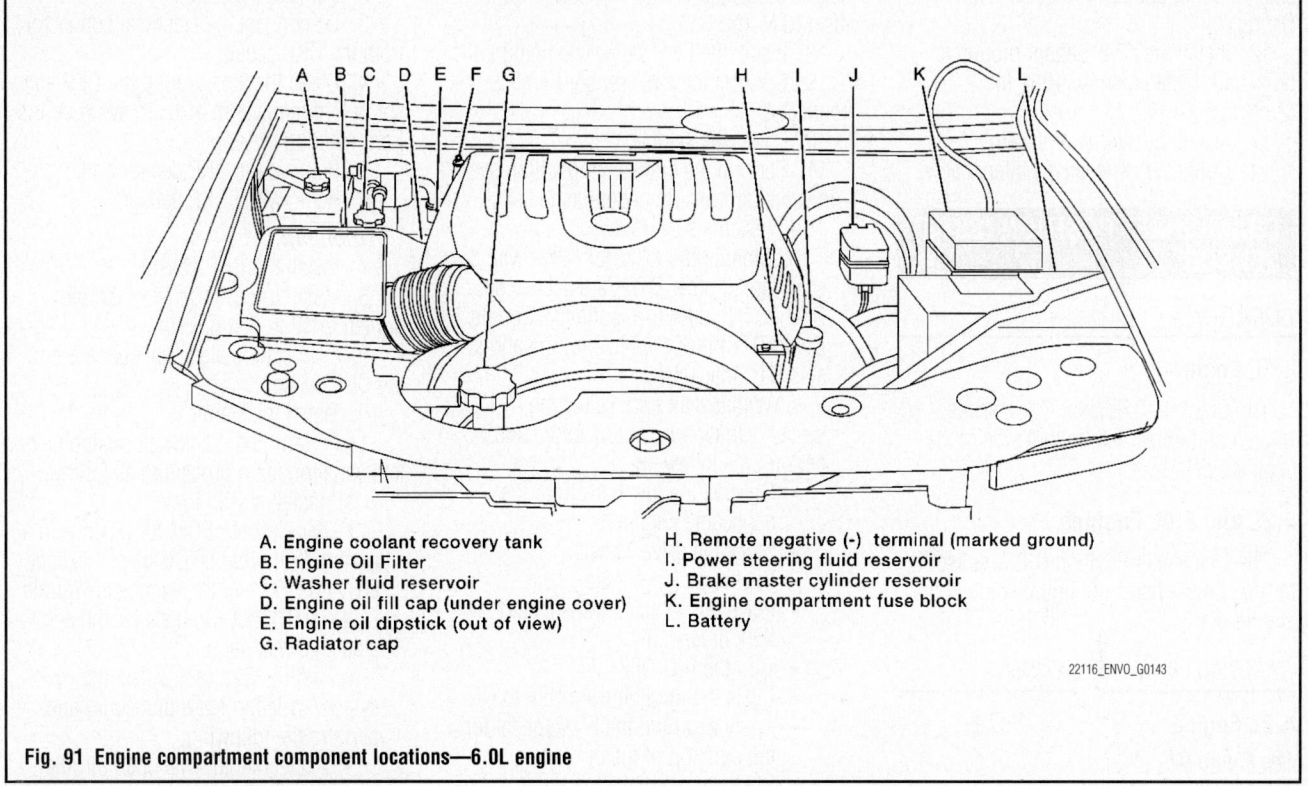

A. Engine coolant recovery tank
B. Engine Oil Filter
C. Washer fluid reservoir
D. Engine oil fill cap (under engine cover)
E. Engine oil dipstick (out of view)
G. Radiator cap

H. Remote negative (-) terminal (marked ground)
I. Power steering fluid reservoir
J. Brake master cylinder reservoir
K. Engine compartment fuse block
L. Battery

22116_ENVO_G0143

**Fig. 91 Engine compartment component locations—6.0L engine**

## CAMSHAFT POSITION (CMP) SENSOR

### LOCATION

#### 4.2L Engine

The Camshaft Position (CMP) sensor is located on the front right corner of the engine cylinder head.

#### 5.3L and 6.0L Engines

The Camshaft Position (CMP) sensor is located on the front of the engine block, just above the crankshaft.

### REMOVAL & INSTALLATION

#### 4.2L Engine

*See Figure 92.*

1. Disconnect the negative battery cable.
2. Remove the Camshaft Position (CMP) sensor electrical connector.
3. Remove the CMP sensor retaining bolt.

***To install:***

4. Install the CMP sensor and tighten the CMP sensor bolt to 89 inch lbs. (10 Nm).
5. Install the CMP sensor electrical connector.
6. Connect the negative battery cable.

22116_ENVO_G0112

**Fig. 92 Location of the Camshaft Position (CMP) sensor—4.2L engine**

#### 5.3L and 6.0L Engines

*See Figure 93.*

1. Disconnect the negative battery cable.
2. Remove the alternator bracket assembly.
3. Remove the Camshaft Position (CMP) sensor mounting bolts.
4. Remove the CMP sensor assembly (4, 5, 6) from the front cover.
5. Disconnect the CMP sensor jumper harness and the engine harness electrical connectors.

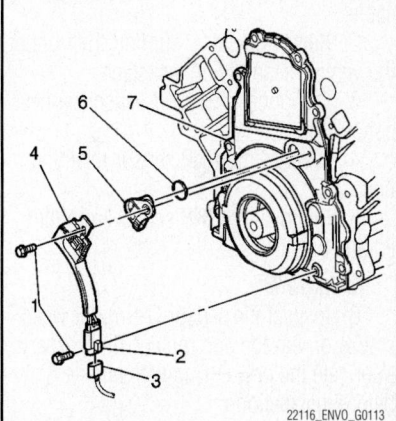

22116_ENVO_G0113

**Fig. 93 Location of the Camshaft Position (CMP) sensor and related components—5.3L and 6.0L engines**

6. Remove the CMP sensor assembly (4, 5, 6).
7. Disconnect CMP sensor from the jumper harness.

***To install:***

8. Reconnect the CMP sensor and the jumper harness.
9. Install the O-ring on the CMP sensor assembly (4, 5).
10. Reconnect the CMP sensor assembly (4, 5, 6) and the engine harness connector.
11. Install the CMP sensor assembly (4, 5, 6) in the front cover. Apply a

small amount of clean motor oil to the O-ring.

12. Install the CMP sensor mounting bolts and tighten them to 18 ft. lbs. (25 Nm).

13. Install the alternator assembly.

14. Connect the negative battery cable.

## CRANKSHAFT POSITION (CKP) SENSOR

### LOCATION

#### 4.2L Engine

The Crankshaft Position (CKP) sensor is located on the rear left bottom side of the engine block.

#### 5.3L and 6.0L Engines

The Crankshaft Position (CKP) sensor is located on the rear right bottom side of the engine block.

### REMOVAL & INSTALLATION

#### 4.2L Engine

See Figure 94.

1. Disconnect the negative battery cable.

2. Raise and safely support the front of the vehicle securely on jackstands.

3. Disconnect the CKP sensor harness connector.

4. Remove the CKP sensor retaining bolt.

5. Remove the CKP sensor from the engine block.

**To install:**

6. Inspect the sensor O-ring for wear cracks or leakage and replace if necessary. Lubricate the new O-ring with engine oil before installation.

7. Install the CKP sensor into the

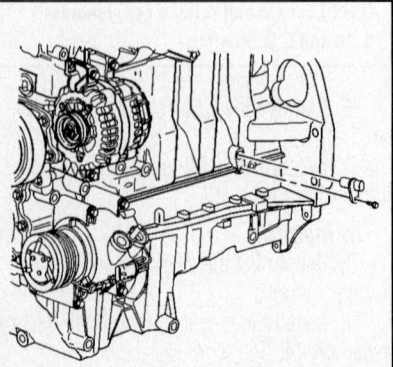

**Fig. 94 Location of the Crankshaft Position (CKP) sensor—4.2L engine**

22116_ENVO_G0114

engine block and tighten the bolt to 89 inch lbs. (10 Nm).

8. Install the CKP sensor retaining bolt.

9. Connect the CKP sensor harness connector.

10. Lower the vehicle.

11. Perform the crankshaft position system variation learn procedure as follows:

    a. Install a scan tool.

    b. Monitor the ECM for DTCs with a scan tool. If other DTCs are set, except DTC P0315, refer to Diagnostic Trouble Code (DTC) List - Vehicle for the applicable DTC that set.

    c. With a scan tool, select the CKP system variation learn procedure and perform the following:

- Observe the fuel cut-off for the applicable engine
- Block the drive wheels
- Set the parking brake
- Place the vehicle's transmission in Park or Neutral
- Turn the A/C OFF
- Cycle the ignition from OFF to ON
- Apply and hold the brake pedal for the duration of the procedure
- Start and idle the engine
- Accelerate to wide open throttle (WOT). The engine should not accelerate beyond the calibrated fuel cut-off RPM value noted earlier. Release the throttle immediately if the value is exceeded.
- While the learn procedure is in progress, release the throttle immediately when the engine starts to decelerate. The engine control is returned to the operator and the engine responds to throttle position after the learn procedure is complete.
- Release the throttle when fuel cut-off occurs

    d. The scan tool displays Learn Status: Learned this Ignition. If the scan tool indicates that DTC P0315 ran and passed, the CKP variation learn procedure is complete. If the scan tool indicates DTC P0315 failed or did not run, refer to DTC P0315. If any other DTCs set, refer to Diagnostic Trouble Code (DTC) List - Vehicle for the applicable DTC that set.

    e. Turn OFF the ignition for 30 seconds after the learn procedure is completed successfully.

12. Connect the negative battery cable.

#### 5.3L and 6.0L Engines

See Figure 95.

1. Disconnect the negative battery cable.

2. Remove the starter.

3. Disconnect the electrical connector from the CKP sensor.

4. Clean the area around the CKP sensor before removal in order to avoid debris from entering the engine.

5. Remove the CKP sensor bolt.

6. Remove the CKP sensor.

**To install:**

7. Install the CKP sensor.

8. Install the CKP sensor bolt and tighten to 18 ft. lbs. (25 Nm).

9. Connect the electrical connector to the CKP sensor.

10. Install the starter.

11. Perform the crankshaft position system variation learn procedure as follows:

    a. Install a scan tool.

    b. Monitor the ECM for DTCs with a scan tool. If other DTCs are set, except DTC P0315, refer to Diagnostic Trouble Code (DTC) List - Vehicle for the applicable DTC that set.

    c. With a scan tool, select the CKP system variation learn procedure and perform the following:

- Accelerate to wide open throttle (WOT)
- Release throttle when fuel cut-off occurs
- Observe fuel cut-off for applicable engine
- Engine should not accelerate beyond calibrated RPM value
- Release throttle immediately if value is exceeded
- Block drive wheels
- Set parking brake
- do not apply brake pedal
- Cycle ignition from OFF to ON
- Apply and hold brake pedal
- Start and idle engine
- Turn A/C OFF

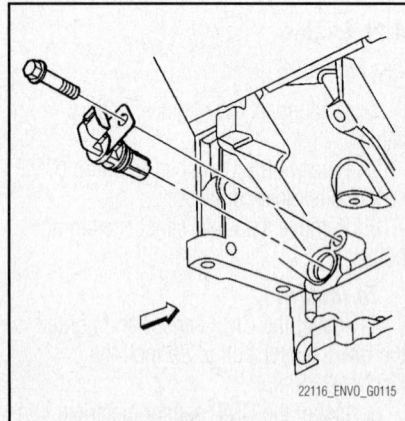

**Fig. 95 Location CKP sensor—5.3L and 6.0L engines**

22116_ENVO_G0115

- Vehicle must remain in Park or Neutral

d. The scan tool displays Learn Status: Learned this Ignition. If the scan tool indicates that DTC P0315 ran and passed, the CKP variation learn procedure is complete. If the scan tool indicates DTC P0315 failed or did not run, refer to DTC P0315. If any other DTCs set, refer to Diagnostic Trouble Code (DTC) List - Vehicle for the applicable DTC that set.

e. Turn OFF the ignition for 30 seconds after the learn procedure is completed successfully.

12. Connect the negative battery cable.

## ELECTRONIC CONTROL MODULE (ECM)

### LOCATION

The Engine Control Module (ECM) is located in the front portion of the engine compartment

### REMOVAL & INSTALLATION

#### 4.2L Engine

Refer to Powertrain Control Module (PCM).

#### 5.3L and 6.0L Engines

*See Figure 96.*

➡ **It is necessary to record the remaining engine oil life. If the replacement module is not programmed with the remaining engine oil life, the engine oil life will default to 100%. If the replacement module is not programmed with the remaining engine oil life, the engine oil will need to be changed at 5000 km (3,000 mi) from the last engine oil change.**

1. Using a scan tool, retrieve the percentage of remaining engine oil. Record the remaining engine oil life.
2. Disconnect the negative battery cable.
3. Disconnect the cooling fan electrical connector for additional clearance while removing the ECM.
4. Depress the ECM/Transmission Control Module (TCM) cover retainers.
5. Remove the ECM/TCM cover from the ECM/TCM bracket.

### ✳ WARNING

**Do not touch the connector pins or soldered components on the circuit board in order to prevent possible ElectroStatic Discharge (ESD) damage to the ECM.**

➡ **It is not necessary to disconnect the ECM electrical connectors in order to remove the ECM from the ECM/TCM bracket. Only disconnect the electrical connectors if servicing of component requires disconnecting of the electrical connectors.**

### ✳ WARNING

**Remove any debris from around the ECM connector surfaces before servicing the ECM. Inspect the ECM module connector gaskets when diagnosing or replacing the ECM. Ensure that the gaskets are installed correctly. The gaskets prevent contaminant intrusion into the ECM.**

6. Disconnect the ECM electrical connectors from the ECM.
7. Release the bracket ECM retainers.
8. Tilt the ECM away from the ECM/TCM bracket.
9. Remove the ECM from the ECM bracket.
10. Only when replacement of the ECM/TCM bracket is necessary, remove the TCM.
11. Remove the ECM/TCM bracket retaining bolts.
12. Remove the ECM/TCM bracket from the vehicle frame.

#### To install:

13. If the ECM/TCM bracket was previously removed, install the ECM/TCM bracket to the vehicle frame.
14. Install the ECM/TCM bracket retaining bolts.
15. Tighten the ECM/TCM bracket bolts to 89 inch lbs. (10 Nm).
16. If the TCM was previously removed from the ECM/TCM bracket, install the TCM.

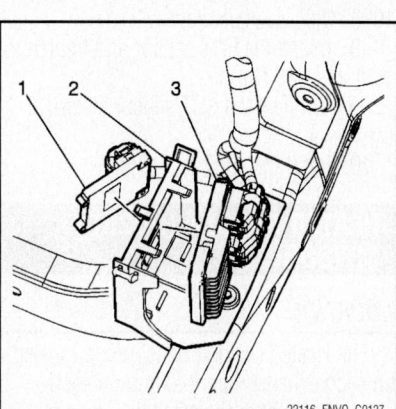

22116_ENVO_G0127

**Fig. 96 Transmission Control Module (TCM) , ECM/TCM bracket and Engine Control Module—5.3L and 6.0L engines**

17. Insert the ECM into the retaining slots of the ECM/TCM bracket.
18. Secure the ECM to the ECM/TCM mounting bracket ensuring the ECM retaining tabs are fully engaged.
19. Connect the ECM electrical connectors to the ECM if previously removed.
20. Install the ECM/TCM cover to the ECM/TCM bracket.
21. Ensure the ECM/TCM cover retainers are fully engaged with the ECM/TCM bracket.
22. Connect the cooling fan electrical connector.
23. Connect the negative battery cable.
24. If the ECM was replaced the replacement ECM must be programmed.

## ENGINE COOLANT TEMPERATURE (ECT) SENSOR

### LOCATION

#### 4.2L Engine

The Engine Coolant Temperature (ECT) sensor is mounted on the top front of the engine cylinder head.

#### 5.3L and 6.0L Engines

The Engine Coolant Temperature (ECT) sensor is mounted on the top front of the engine's left cylinder head.

### OPERATION

The Engine Coolant Temperature (ECT) sensor is used to monitor the temperature of the engine coolant. Its resistance changes in proportion to coolant temperature. Input from the coolant sensor tells the computer when the engine is warm so the PCM (4.2L) or ECM (5.3L and 6.0L) can go into closed loop feedback fuel control and handle other emission functions (EGR, canister purge, etc.) that may be temperature dependent.

### REMOVAL & INSTALLATION

#### 4.2L Engine

*See Figure 97.*

### ✳ WARNING

**Use care when handling the coolant sensor. Damage to the coolant sensor will affect the operation of the fuel control system.**

1. Turn the engine OFF.
2. Disconnect the negative battery terminal.
3. Drain coolant below the level of the Engine Coolant Temperature (ECT) sensor.

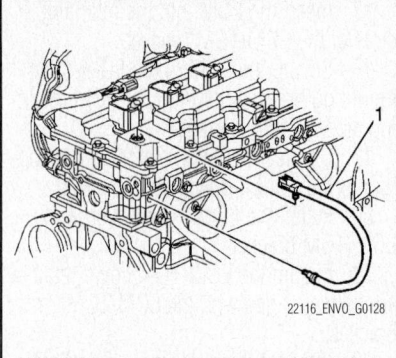

**Fig. 97 Engine Coolant Temperature (ECT) sensor—4.2L engine**

4. Disconnect the ECT sensor electrical connector.
5. Carefully remove the ECT sensor.

**To install:**

6. If installing the original sensor or a new sensor without sealant, apply thread sealer P/N 12346004 or equivalent.
7. Install the ECT sensor and tighten to 12 ft. lbs. (16 Nm).
8. Connect the ECT electrical connector.
9. Connect the negative battery terminal.
10. Refill the engine coolant.

### 5.3L and 6.0L Engines

*See Figures 98 through 100.*

**✳✳ WARNING**

**Use care when handling the coolant sensor. Damage to the coolant sensor will affect the operation of the fuel control system.**

1. Turn OFF the ignition.
2. Raise and suitably support the vehicle.

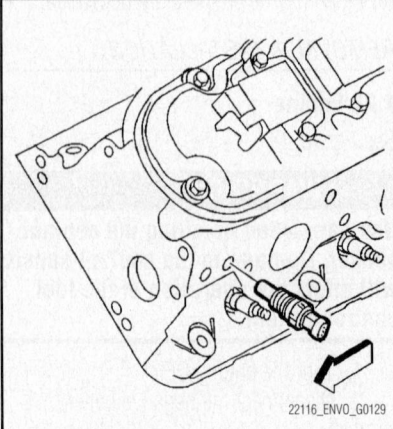

**Fig. 98 Engine Coolant Temperature (ECT) sensor—5.3L and 6.0L engines**

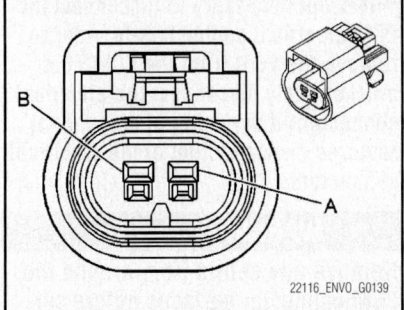

**Fig. 99 Engine Coolant Temperature (ECT) sensor connector: ECT sensor signal, low reference—4.2L engine**

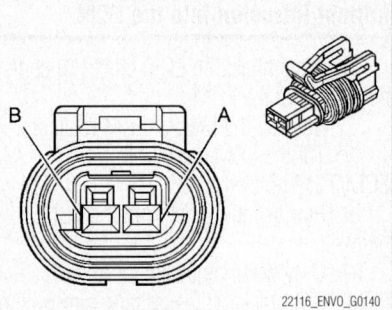

**Fig. 100 Engine Coolant Temperature (ECT) sensor connector: low reference, ECT sensor signal—5.3L and 6.0L engines**

3. Drain the cooling system below the level of the Engine Coolant Temperature (ECT) sensor.
4. Lower the vehicle.
5. Disconnect the ECT sensor electrical connector.
6. Remove the ECT sensor.

**To install:**

7. Coat the ECT sensor threads with sealer GM P/N 12346004 (Canadian P/N 10953480), or equivalent.
8. Install the ECT sensor and tighten to 15 ft. lbs. (20 Nm).
9. Connect the ECT sensor electrical connector.
10. Refill the engine coolant.

### HEATED OXYGEN (HO2S) SENSOR

#### LOCATION

The Heated Oxygen Sensors are located on each exhaust manifold and on each exhaust pipe after the catalytic converter.

#### OPERATION

Heated Oxygen Sensors (HO2S) are used for fuel control and post catalyst monitoring.

Each HO2S compares the oxygen content of the surrounding air with the oxygen content in the exhaust stream. The HO2S must reach operating temperature to provide an accurate voltage signal. A heating element inside the HO2S minimizes the time required for the sensor to reach operating temperature. Voltage is provided to the heater by the ignition 1 voltage circuit through a fuse. With the engine running, ground is provided to the heater by the HO2S heater low control circuit, through a low side driver within the Engine Control Module (ECM).

The ECM commands the heater ON or OFF to maintain a specific HO2S operating temperature range. The ECM monitors the voltage on the HO2S heater low control circuit for heater fault diagnosis. If the ECM detects that the HO2S heater low control circuit voltage is not within a specified range.

#### REMOVAL & INSTALLATION

*See Figure 101.*

1. Disconnect the negative battery cable.
2. Raise and safely support the vehicle, if necessary.
3. Disconnect the Heated Oxygen Sensor (HO2S) electrical connector.
4. Remove the HO2S using a J-39194-B.

**To install:**

5. Coat the threads of the heated oxygen sensor with the anti-seize compound P/N 5613695, or the equivalent if necessary.
6. Install the heated oxygen sensor using a J-39194-B and tighten to 30 ft. lbs. (41 Nm).
7. Connect the HO2S electrical connector.
8. Lower the vehicle, if raised.
9. Connect the negative battery cable.

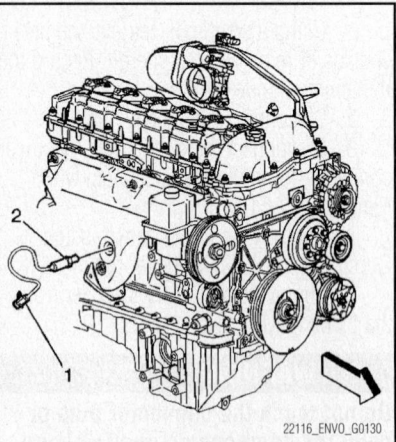

**Fig. 101 Heated Oxygen Sensor (HO2S) 1 and electrical connector—4.2L engine shown**

## INTAKE AIR TEMPERATURE (IAT) SENSOR

Refer to the Mass Air Flow/Intake Air Temperature (MAF/IAT) sensor procedure.

## KNOCK SENSOR (KS)

### LOCATION

#### 4.2L Engine

The knock sensors (KS) are located on the left side of the engine block.

#### 5.3L and 6.0L Engines

There are two knock sensors (KS), one each located on the left and right middle sides of the engine block.

### REMOVAL & INSTALLATION

#### 4.2L Engine

*See Figure 102.*

1. Disconnect the negative battery cable.
2. Raise and safely support the front of the vehicle securely on jackstands.
3. Remove the knock sensor harness connector.
4. Remove the knock sensor retaining bolt.
5. Remove the appropriate knock sensor (1 or 2).

#### To install:

6. Install the knock sensor (1 or 2) and the bolt, then tighten the sensor to 18 ft. lbs. (25 Nm).
7. Connect the knock sensor harness connector.
8. Lower the vehicle.
9. Connect the negative battery cable.

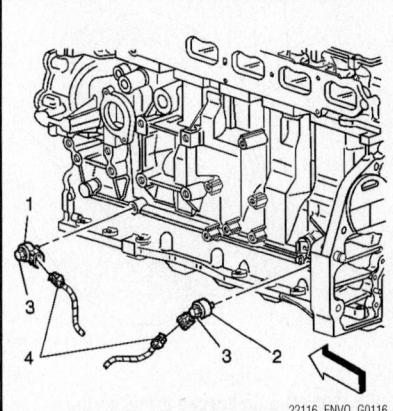

**Fig. 102 Location of the knock sensors (1 and 2)—4.2L engine**

#### 5.3L and 6.0L Engines

### Left Side

*See Figure 103.*

1. Disconnect the negative battery cable.
2. Remove the mounting bolt for the knock sensor 1.
3. Disconnect the electrical connector of the knock sensor from the engine harness.
4. Remove the knock sensor from the engine block.

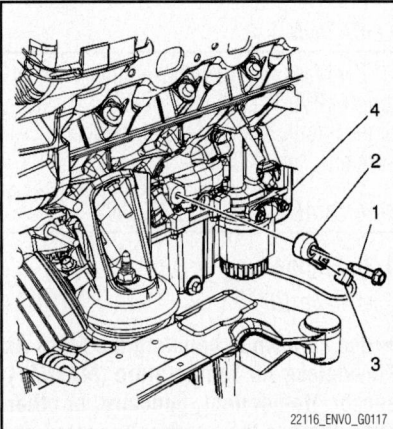

**Fig. 103 Location of the left side knock sensor—5.3L and 6.0L engines**

#### To install:

5. Reconnect the engine harness and the knock sensor electrical connectors.
6. Position the knock sensor 2 on the engine block.
7. Install the mounting bolt for the knock sensor 2.
8. Tighten the knock sensor mounting bolt to 15 ft. lbs. (20 Nm).
9. Connect the negative battery cable.

### Right Side

*See Figure 104.*

1. Disconnect the negative battery cable.
2. Disconnect the electrical connector from the knock sensor.
3. Remove the knock sensor bolt.
4. Remove the knock sensor from the engine block.

#### To install:

5. Position the knock sensor on the engine block.
6. Install the knock sensor bolt and tighten to 15 ft. lbs. (20 Nm).
7. Connect the electrical connector to the knock sensor.
8. Connect the negative battery cable.

**Fig. 104 Location of the right side knock sensor—5.3L and 6.0L engines**

### TESTING

1. Connect the Digital Multi-Meter (DMM) from the Knock Sensor (KS) signal circuit to the KS low reference circuit on the sensor side of the KS harness connector.
2. Set the DMM to the 400 mV AC hertz scale and wait for the DMM to stabilize at 0 Hz.

➡**do not tap on plastic engine components.**

3. Tap on the engine block with a non-metallic object near the KS while observing the signal indicated on the DMM.
4. The DMM should display a fluctuating frequency while tapping on the engine block.

## MANIFOLD ABSOLUTE PRESSURE (MAP) SENSOR

### LOCATION

#### 4.2L Engine

The Manifold Absolute Pressure (MAP) sensor is located on top of the engine, near the firewall.

#### 5.3L and 6.0L Engines

The Manifold Absolute Pressure (MAP) sensor is located on top of the engine, on top of the intake manifold plenum.

### REMOVAL & INSTALLATION

#### 4.2L Engine

*See Figure 105.*

1. Disconnect the negative battery cable.
2. Turn OFF the ignition.
3. Disconnect the Manifold Absolute Pressure (MAP) sensor electrical connector.

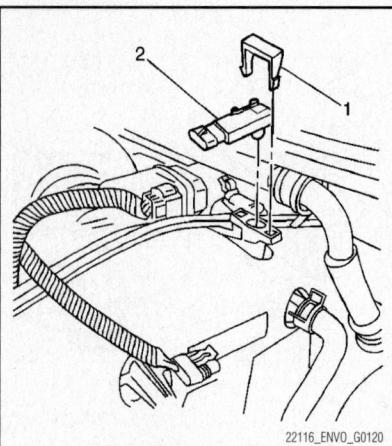

**Fig. 105 Location of the retainer and the Manifold Absolute Pressure (MAP) sensor—4.2L engine**

4. Press the retainer locking tabs inward, then pull the retainer up to remove it.

5. Remove the MAP sensor.

6. Inspect the MAP sensor seal for damage, and replace as necessary.

**To install:**

7. Install the MAP sensor.

8. Install the MAP sensor retainer.

9. Connect the electrical connector.

10. Connect the negative battery cable.

### 5.3L and 6.0L Engines

*See Figure 106.*

1. Disconnect the negative battery cable.

2. Disconnect the manifold absolute pressure (MAP) sensor electrical connector.

3. Remove the MAP sensor retaining clip from the intake manifold.

4. Remove the MAP sensor from the intake manifold.

**To install:**

5. Lightly coat the MAP sensor seal with clean engine oil before installing the sensor.

6. Install the MAP sensor. Push the MAP sensor into the intake manifold.

7. Install the MAP sensor retainer to the intake manifold.

8. Connect the MAP sensor electrical connector.

9. Connect the negative battery cable.

### MASS AIR FLOW (MAF) SENSOR

#### LOCATION

The Mass Air Flow/Intake Air Temperature (MAF/IAT) sensor is mounted ahead of the throttle body on the air cleaner assembly.

#### REMOVAL & INSTALLATION

#### 4.2L Engine

*See Figure 107.*

➡ Use care when handling the Mass Air Flow/Intake Air Temperature (MAF/IAT) sensor. Do not dent, puncture, or otherwise damage the honeycell located at the air inlet end of the MAF/IAT. Do not touch the sensing elements or allow anything including cleaning solvents and lubricants to come in contact with them. Use a small amount of a non-silicone based lubricant, on the air duct only, to aid in installation.

1. Disconnect the negative battery cable.

2. Disconnect the engine harness electrical connector from the MAF/IAT sensor.

3. Remove the MAF/IAT sensor screws.

4. Remove the MAF/IAT sensor.

**To install:**

5. Install the MAF/IAT sensor.

6. Install the MAF/IAT sensor screws and tighten to 5 inch lbs. (0.6 Nm).

7. Connect the engine harness electrical connector to the MAF/IAT sensor.

8. Connect the negative battery cable.

### 5.3L and 6.0L Engines

*See Figure 108.*

➡ Use care when handling the Mass Air Flow/Intake Air Temperature (MAF/IAT) sensor. Do not dent, puncture, or otherwise damage the honeycell located at the air inlet end of the MAF/IAT. Do not touch the sensing elements or allow anything including cleaning solvents and lubricants to come in contact with them. Use a small amount of a non-silicone based lubricant, on the air duct only, to aid in installation.

1. Disconnect the negative battery cable.

2. Disconnect the MAF/IAT sensor electrical connector.

3. Loosen the clamps at the MAF/IAT sensor and the throttle body.

4. Remove the air cleaner outlet duct bolt.

5. Remove the air cleaner outlet duct.

6. Loosen the clamp attaching the MAF/IAT sensor to the air cleaner housing.

7. Remove the MAF/IAT sensor from the air cleaner housing.

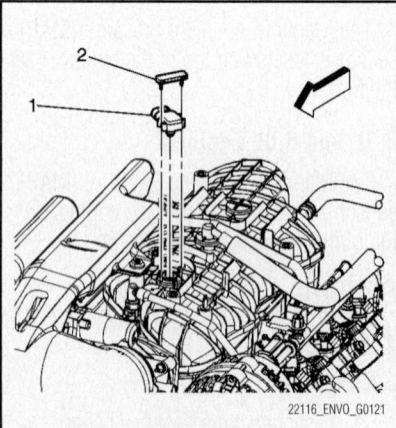

**Fig. 106 Location of the Manifold Absolute Pressure (MAP) sensor and retaining clip—5.3L and 6.0L engines**

**Fig. 107 Location of the Mass Air Flow/Intake Air Temperature (MAF/IAT) sensor—4.2L engine**

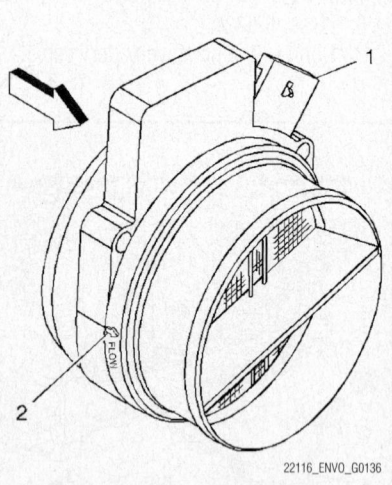

**Fig. 108 The embossed arrow on the MAF/IAT sensor indicates the proper air flow direction. The arrow must point toward the engine—5.3L and 6.0L engines**

### To install:

→ **The embossed arrow on the MAF/IAT sensor indicates the proper air flow direction. The arrow must point toward the engine.**

8. Locate the air flow direction arrow on the MAF/IAT sensor.

9. Install the MAF/IAT sensor on to the air cleaner housing.

10. Tighten the clamp securing the MAF/IAT sensor to the air cleaner housing and tighten the clamp to 62 inch lbs. (7 Nm).

11. Install the air cleaner outlet duct.

12. Install the air cleaner outlet duct bolt and tighten to 89 inch lbs. (10 Nm).

13. Tighten the clamps at the MAF/IAT sensor and the throttle body, then tighten the clamps to 62 inch lbs. (7 Nm).

14. Connect the MAF/IAT electrical connector.

15. Connect the negative battery cable.

## THROTTLE POSITION SENSOR (TPS)

### LOCATION

The Throttle Position Sensor (TPS) is an integral component of the throttle body assembly and cannot be serviced separately.

### REMOVAL & INSTALLATION

Refer to the Throttle Body R&I procedure.

## VEHICLE SPEED SENSOR (VSS)

### LOCATION

The vehicle Speed Sensor (VSS) is located on the right rear side of the transmission case.

### REMOVAL & INSTALLATION

See Figure 109.

1. Disconnect the negative battery cable.

2. Remove the harness connector.

3. Remove the bolt.

4. Remove the vehicle speed sensor.

5. Remove the O-ring seal.

### To install:

6. Install the O-ring seal on the vehicle speed sensor.

7. Coat the O-ring seal with a thin film of transmission fluid.

8. Install the vehicle speed sensor into the transmission case.

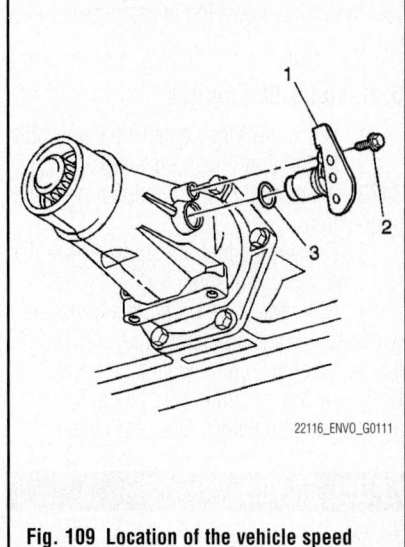

22116_ENVO_G0111

**Fig. 109 Location of the vehicle speed sensor, bolt and O-ring seal**

9. Install the bolt and tighten to 97 inch lbs. (11 Nm).

10. Connect the wiring harness electrical connector to the vehicle speed sensor.

11. Refill the fluid as required.

12. Connect the negative battery cable.

---

## FUEL | GASOLINE FUEL INJECTION SYSTEM

## FUEL SYSTEM SERVICE PRECAUTIONS

Safety is the most important factor when performing not only fuel system maintenance but any type of maintenance. Failure to conduct maintenance and repairs in a safe manner may result in serious personal injury or death. Maintenance and testing of the vehicle's fuel system components can be accomplished safely and effectively by adhering to the following rules and guidelines.

• To avoid the possibility of fire and personal injury, always disconnect the negative battery cable unless the repair or test procedure requires that battery voltage be applied.

• Always relieve the fuel system pressure prior to disconnecting any fuel system component (injector, fuel rail, pressure regulator, etc.), fitting or fuel line connection. Exercise extreme caution whenever relieving fuel system pressure to avoid exposing skin, face and eyes to fuel spray. Please be advised that fuel under pressure may penetrate the skin or any part of the body that it contacts.

• Always place a shop towel or cloth around the fitting or connection prior to loosening to absorb any excess fuel due to spillage. Ensure that all fuel spillage (should it occur) is quickly removed from engine surfaces. Ensure that all fuel soaked cloths or towels are deposited into a suitable waste container.

• Always keep a dry chemical (Class B) fire extinguisher near the work area.

• Do not allow fuel spray or fuel vapors to come into contact with a spark or open flame.

• Always use a back-up wrench when loosening and tightening fuel line connection fittings. This will prevent unnecessary stress and torsion to fuel line piping.

• Always replace worn fuel fitting O-rings with new Do not substitute fuel hose or equivalent where fuel pipe is installed.

Before servicing the vehicle, make sure to also refer to the precautions in the beginning of this section as well.

## RELIEVING FUEL SYSTEM PRESSURE

The fuel systems operate under high fuel pressures. It is very important that the pressure be properly relieved prior to servicing the system or any of its components.

**4.2L Engine**

1. Before servicing the vehicle, refer to the Precautions Section.

### ❋❋ WARNING

**Do not perform this procedure for more than 2 minutes to avoid damaging the catalytic converter.**

2. Loosen the fuel filler cap to release the fuel tank pressure.

3. Remove the fuel pump relay from the junction block.

4. Crank the engine, allowing it to start and stall.

5. Crank the engine for an additional 3 seconds to relieve any remaining fuel pressure.

6. Disconnect the negative battery cable to avoid repressurizing the fuel system.

7. Install the fuel pump relay in the junction block.

8. Tighten the fuel filler cap.

9. After you are finished working on the fuel system, connect the negative battery cable.

### 5.3L and 6.0L Engines

1. Disconnect the negative battery cable.
2. Install Fuel Pressure Gauge J 34730-1A or equivalent to the fuel pressure connection.
3. Loosen the fuel fill cap to relieve the fuel tank vapor pressure.
4. Open the valve on the fuel pressure gauge to bleed the system pressure. The fuel connections are now safe for servicing. Drain any fuel remaining in the gauge into an approved container. Once the system

## FUEL FILTER

### REMOVAL & INSTALLATION

The fuel filter is contained in the fuel sender assembly inside the fuel tank. The paper filter element traps particles in the fuel that may damage the fuel injection system. The filter housing is made to withstand maximum fuel system pressure, exposure to fuel additives, and changes in temperature.

1. Before servicing the vehicle, refer to the Precautions Section.
2. Properly relieve the fuel system pressure.
3. Disconnect the negative battery cable.
4. Remove the fuel filler cap, if not already done.
5. Remove the fuel pump assembly.
6. Remove the fuel filter from the fuel pump assembly. Replace the seals or O-rings.

**To install:**

7. Install the new fuel filter to the fuel pump assembly along with new seals or O-rings.
8. Install the fuel pump assembly.
9. Install the fuel filler cap.
10. Connect the negative battery cable.
11. Start the engine and check for leaks.

### DRAINING WATER FROM THE SYSTEM

1. Remove the fuel sender assembly.
2. Inspect the fuel pump inlet for dirt and debris. Replace the fuel pump if you find dirt or debris in the fuel pump inlet.

➡**When flushing the fuel tank, handle the fuel and water mixture as a hazardous material. Handle the fuel and water mixture in accordance with all applicable local, state, and federal laws and regulations.**

3. Flush the fuel tank with hot water.
4. Pour the water out of the fuel sender assembly opening. Rock the tank to be sure that removal of the water from the tank is complete.
5. Install the fuel sender assembly.

## FUEL INJECTORS

### REMOVAL & INSTALLATION

#### 4.2L Engine

*See Figure 110.*

1. Before servicing the vehicle, refer to the Precautions Section.
2. Relieve the fuel system pressure.
3. Remove or disconnect the following:
   - Fuel feed pipe from the fuel rail
   - Intake manifold

➡**Clean the fuel rail assembly with a suitable spray cleaner before proceeding. Never soak the fuel rail in a cleaning solvent.**

   - Fuel pressure regulator vacuum line
   - Fuel feed and return pipes
   - Fuel injector in-line electrical connector
   - Fuel rail attaching bolts and fuel rail
   - Fuel injector harness connector from the fuel injectors
   - Injector retaining clip
   - Injector from the fuel rail
   - Retainer clip and O-ring seals from each end of the injector and discard

**To install:**

➡**Each injector is calibrated. When replacing the fuel injectors, be sure to replace it with the correct injector.**

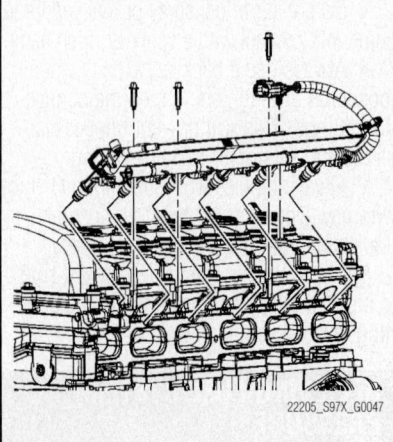

**Fig. 110 View of the fuel rail mounting— 4.2L engine**

22205_S97X_G0047

4. Lubricate the new injector O-ring seats with engine oil.
5. Install or connect the following:
   - O-rings on the injector
   - New retainer clip on the injector
6. Push the fuel injector into the fuel rail socket, making sure the connector faces outward. The retainer clip locks to a flange on the fuel rail injector socket.
   - Fuel rail assembly. Tighten the bolts to 89 inch lbs. (10 Nm).
   - Fuel feed and return lines to the rail
   - Fuel injector electrical connectors
   - Fuel pressure regulator vacuum line
   - Intake manifold
7. Turn the ignition **ON** for 2 seconds and then turn it **OFF** for 10 seconds. Again turn the ignition **ON** and check for leaks.

#### 5.3L and 6.0L Engines

*See Figures 111 and 112.*

1. Before servicing the vehicle, refer to the Precautions Section.
2. Relieve the fuel system pressure. Refer to the fuel system relief procedure in this section.
3. Remove or disconnect the following:
   - Negative battery cable, if not done already
   - Positive crankcase ventilation (PCV) foul air hose.
   - A/C compressor pressure switch electrical connector
   - Wire harness from the clip on the cylinder head
   - Mass Airflow/Intake Air Temperature (MAF/IAT) sensor connector
   - Alternator electrical connector
   - Electrical connectors from the coil main electrical harness, Electronic Throttle Control (ETC) and fuel injectors.
4. To detach the injector connector: Matchmark the connectors, pull the Connector Position Assurance (CPA) retainer up 1 click. Push the tab on the connector in, then detach the injector connector.
   - Electrical harness from the clips on the ignition coil bracket
   - Evaporative emission (EVAP) purge solenoid electrical connector
   - Knock Sensor (KS) electrical connector
   - Manifold Absolute Pressure (MAP) electrical connector
   - Main coil
   - Fuel injector electrical connector (right side)
   - Electrical harness from the clips on the ignition coil bracket

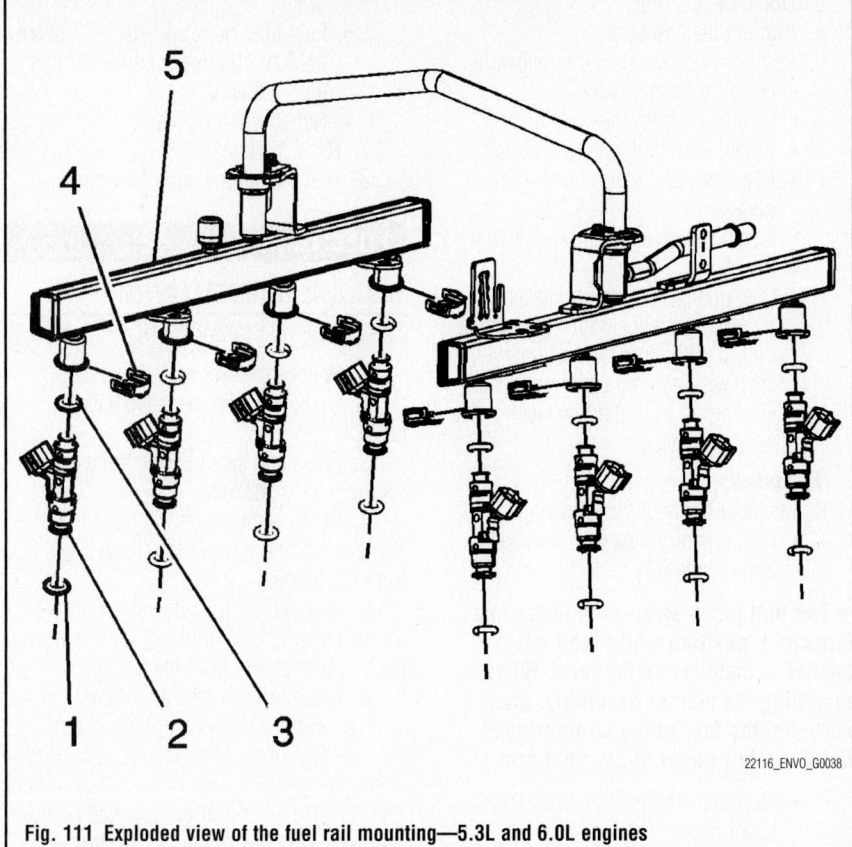

**Fig. 111 Exploded view of the fuel rail mounting—5.3L and 6.0L engines**

- Upper engine wire harness retainer nut. Position the wire harness aside.
- Fuel feed and return pipes from the rail
- Fuel pressure regulator vacuum line
- Fuel rail bolts
- Fuel rail, after cleaning with a spray-type cleaner

### ❄ WARNING

**Be very careful when removing the fuel rail and injectors not to damage the connector terminals or injector spray tips**

- Fuel injector from the fuel rail
- Fuel injector retainer clip and discard
- Fuel injector lower O-ring seals and discard

**To install:**

5. Install or connect the following:
- New O-ring seals on the injectors, after lubricating with clean engine oil
- New retainer clip on the injector
- Fuel injector by pushing it into the fuel rail socket
- Fuel rail

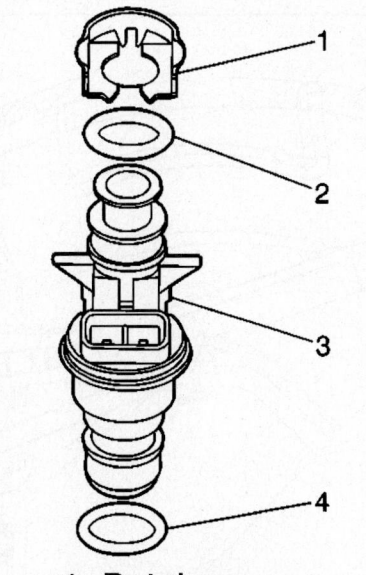

1. **Retainer**
2. **O-ring seal**
3. **Fuel injector**
4. **O-ring seal**

22116_ENVO_G0039

**Fig. 112 Exploded view of the fuel injector, retainer and O-ring seals (2, 4)—5.3L and 6.0L engines**

- Apply 0.20 (5mm) band of thread-lock to the threads of the fuel rail bolts
- Fuel rail bolts and tighten to 89 inch lbs. (10 Nm)
- Fuel pressure regulator vacuum line
- Fuel feel and return pipes
- Route the upper electrical harness into position over the engine.
- Engine harness bracket nut and tighten to 89 inch lbs. (10 Nm)
- PCV valve and hose
- EVAP purge solenoid, KS, MAP sensor, main coil & fuel injector electrical connectors
- Harness to the clips on the ignition coil bracket
- Main coil, ETC, fuel injector electrical connectors
- Harness to the clips on the ignition coil bracket
- Alternator electrical connector
- MAF/IAT sensor connector
- Wire harness to the clip on the cylinder head
- A/C compressor switch electrical connector
- Positive crankcase ventilation (PCV) foul air hose.
- Air cleaner outlet duct
- Fuel fill cap
- Negative battery cable
6. Refill the engine cooling system.

### FUEL PRESSURE REGULATOR

#### REMOVAL & INSTALLATION

*See Figure 113.*

The fuel system is a returnless, on-demand design. The fuel pressure regulator is contained in the fuel sender assembly inside the fuel tank.

1. Before servicing the vehicle, refer to the Precautions Section.
2. Properly relieve the fuel system pressure.
3. Disconnect the negative battery cable.
4. Remove the fuel filler cap, if not already done.
5. Remove the fuel pump assembly.
6. Remove the fuel pressure regulator from the fuel pump assembly. Replace the seal(s) or O-ring(s).

**To install:**

7. Install the new fuel pressure regulator to the fuel pump assembly along with new seal(s) or O-ring(s).
8. Install the fuel pump assembly.
9. Install the fuel filler cap.
10. Connect the negative battery cable.

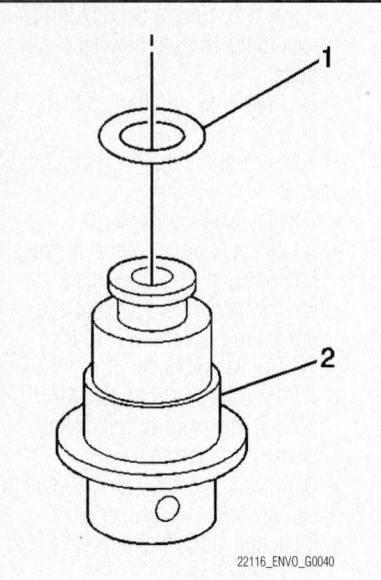

**Fig. 113 Fuel pressure regulator and O-ring assembly—4.2L engine shown, others similar**

11. Start the engine and check for leaks.

## FUEL PUMP

### REMOVAL & INSTALLATION

*See Figure 114.*

1. Before servicing the vehicle, refer to the Precautions Section.
2. Properly relieve the fuel system pressure.

3. Drain the fuel tank.
4. Support the fuel tank.
5. Remove or disconnect the following:
   - Negative battery cable
   - Filler neck from the tank
   - Shield from tank and tank straps
   - Fuel lines and vapor hose from pump
   - Electrical connection from fuel pump
   - Fuel tank
   - Fuel pump/sending unit assembly by turning the locking ring (located on top of the fuel tank) counterclockwise using a spanner wrench
   - Fuel pump from the fuel lever sending device

### To install:

6. Install or connect the following:
   - Fuel pump in tank with new seal around opening

➡ **The fuel pump strainer must be in a horizontal position when the fuel sender is installed in the tank. When installing the sender assembly, make sure that the fuel pump strainer does not block full travel of the float arm.**

   - Tank and connect fuel lines and vapor hose
   - Tank to the frame. Torque the fasteners to 24 ft. lbs. (32 Nm).

   - Shield
   - Fuel filler neck and clamp. Tighten the fuel fill hose clamp to 22 inch lbs. (2.5 Nm).
   - Negative battery cable
7. Refill the tank.
8. Run the engine and check for leaks.

## FUEL TANK

### REMOVAL & INSTALLATION

*See Figures 115 through 117.*

1. Relieve the fuel system pressure.
2. Disconnect the negative battery cable.
3. Raise and safely support the vehicle securely on jackstands.
4. Remove the mounting bolts from the frame brace and remove the frame brace from the frame.
5. Remove the fuel tank shield to the frame retaining bolts and nut, then remove the fuel tank shield from the frame.
6. Drain the fuel tank as follows:
   a. Loosen the fuel fill hose clamp.
   b. Disconnect the fuel fill hose from the fuel tank.
   c. Use a hand or air operated pump device in order to drain as much fuel from the fuel tank as possible.

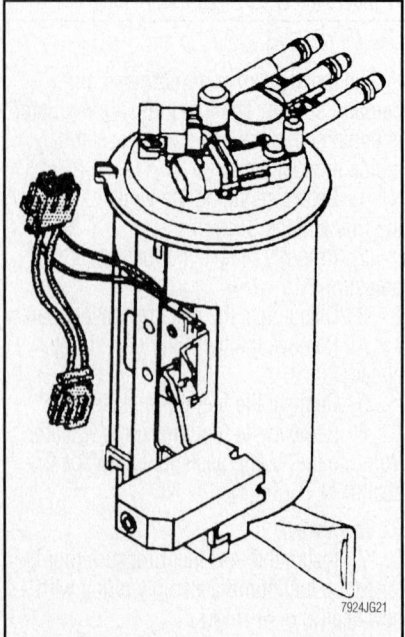

**Fig. 114 View of the in-tank fuel pump assembly**

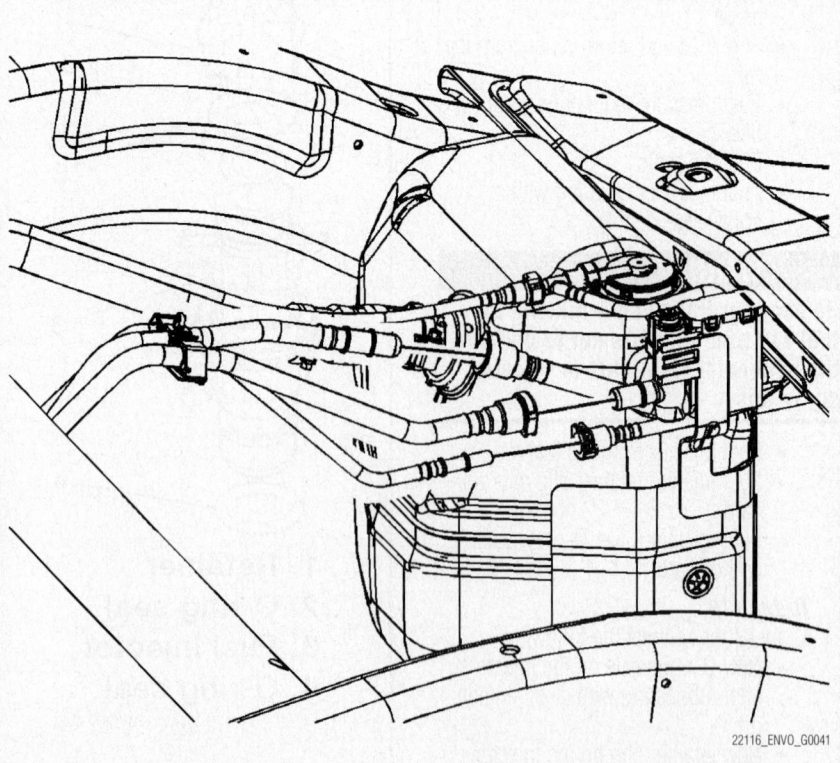

**Fig. 115 Disconnect the EVAP canister fresh air pipe, EVAP canister solenoid pipe and EVAP purge pipe from the fuel tank**

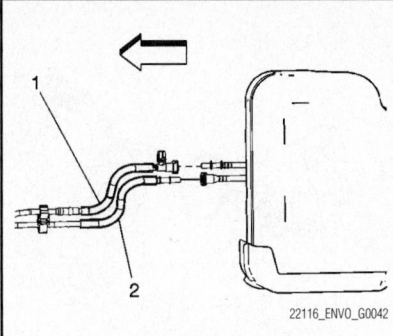

**Fig. 116 Disconnect the EVAP pipe and the fuel feed pipe from the fuel tank**

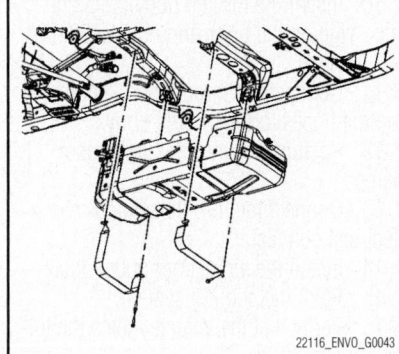

**Fig. 117 Remove the fuel tank straps and carefully lower the fuel tank**

7. Disconnect the evaporative emission (EVAP) canister fresh air pipe.

8. Disconnect the EVAP canister solenoid pipe.

9. Disconnect the EVAP purge pipe.

10. Disconnect the fuel filler pipe recirculation hose from the fuel tank.

11. Loosen the clamp securing the fuel fill pipe to the fuel tank.

12. Disconnect the fuel fill pipe from the fuel tank.

13. Disconnect the fuel feed pipe and EVAP pipe from the fuel tank.

14. Cap the fuel and EVAP pipes in order to prevent possible fuel system contamination.

15. Support the fuel tank.

16. Remove the fuel tank strap attaching bolts.

17. Remove the fuel tank straps.

18. Carefully lower the fuel tank.

19. Disconnect the EVAP vent valve electrical connector.

20. Disconnect the fuel tank pressure sensor electrical connector.

21. Disconnect the fuel sender electrical connector.

22. Remove the fuel tank.

23. Place the fuel tank in a suitable work area.

**To install:**

24. Support the fuel tank.

25. Connect the fuel sender electrical connector.

26. Connect the EVAP vent valve electrical connector.

27. Connect the fuel pressure sensor electrical connector.

28. Install the fuel tank straps.

29. Install the fuel tank strap attaching bolts and tighten to 24 ft. lbs. (32 Nm).

30. Remove the caps from the fuel and EVAP pipes.

31. Connect the fuel feed pipe and the EVAP pipe as follows:

a. Apply a few drops of clean engine oil to the male connection end.

b. Push both sides of the quick-connect fitting together in order to cause the retaining feature to snap into place.

c. Once installed, pull on both sides of the quick-connect fitting in order to make sure the connection is secure.

32. Connect the fuel fill pipe to the fuel tank and tighten the fuel fill hose clamp to 22 inch lbs. (2.5 Nm).

33. Connect the fuel filler pipe recirculation hose to the fuel tank as follows:

a. Apply a few drops of clean engine oil to the male connection end.

b. Push both sides of the quick-connect fitting together in order to cause the retaining feature to snap into place.

c. Once installed, pull on both sides of the quick-connect fitting in order to make sure the connection is secure.

34. Connect the EVAP purge pipe as follows:

a. Apply a few drops of clean engine oil to the male connection end.

b. Push both sides of the quick-connect fitting together in order to cause the retaining feature to snap into place.

c. Once installed, pull on both sides of the quick-connect fitting in order to make sure the connection is secure.

35. Connect the EVAP canister solenoid pipe as follows:

a. Apply a few drops of clean engine oil to the male connection end.

b. Push both sides of the quick-connect fitting together in order to cause the retaining feature to snap into place.

c. Once installed, pull on both sides of the quick-connect fitting in order to make sure the connection is secure.

36. Connect the EVAP canister fresh air pipe as follows:

a. Apply a few drops of clean engine oil to the male connection end.

b. Push both sides of the quick-connect fitting together in order to cause the retaining feature to snap into place.

c. Once installed, pull on both sides of the quick-connect fitting in order to make sure the connection is secure.

37. Lower the vehicle.

38. Refill the fuel tank.

39. Install the fuel filler cap.

40. Connect the negative battery cable.

41. Raise the vehicle.

42. Inspect for leaks as follows:

a. Turn ON the ignition, with the engine OFF for 10 seconds.

b. Turn OFF the ignition for 10 seconds.

c. Turn ON the ignition, with the engine OFF.

d. Inspect for fuel leaks.

43. Install the fuel tank shield, if equipped, to the frame and tighten the retaining bolts and nut to 24 ft. lbs. (32 Nm).

44. Install the frame brace to the frame using the mounting bolts and tighten to 37 ft. lbs. (50 Nm).

45. Lower the vehicle.

## IDLE SPEED

### ADJUSTMENT

Idle speed is maintained by the Powertrain Control Module (PCM). No adjustment is possible.

## THROTTLE BODY

### REMOVAL & INSTALLATION

#### 4.2L Engine

*See Figure 118.*

1. Remove the resonator assembly.

2. Remove the evaporative emission (EVAP) canister purge line from the throttle body.

3. Disconnect the throttle body electrical connector.

4. Remove the throttle body assembly retaining bolts.

5. Remove the throttle body assembly and the gasket from the intake manifold.

6. Clean the gasket surface.

**To install:**

7. Install the throttle body assembly to the intake manifold with the gasket.

8. Add sealer GM P/N 12346004 (Canadian P/N 10953480) to the throttle control module bolt threads.

9. Install the throttle body assembly retaining bolts. Tighten the bolts to 89 inch lbs. (10 Nm).

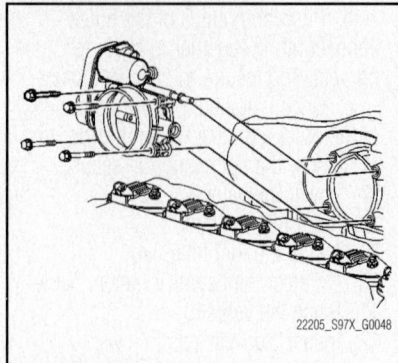

**Fig. 118 Throttle body assembly—4.2L engine**

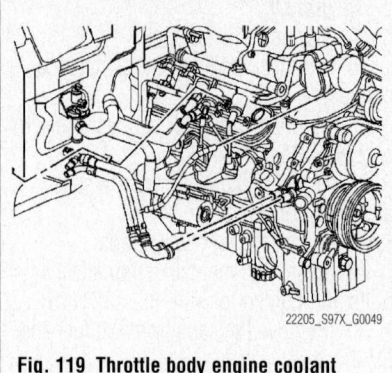

**Fig. 119 Throttle body engine coolant hoses—5.3L and 6.0L engines**

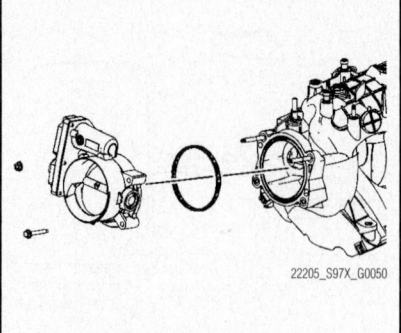

**Fig. 120 Throttle body assembly—5.3L and 6.0L engines**

10. Connect the throttle body electrical connector.

11. Install the EVAP canister purge line to the throttle body.

12. Install the resonator assembly.

### 5.3L and 6.0L Engines

*See Figures 119 and 120.*

> ✳✳ **WARNING**
>
> **Handle the electronic throttle control components carefully. Use cleanliness in order to prevent damage. Do not drop the electronic throttle control components. Do not roughly handle the electronic throttle control components. Do not immerse the electronic throttle control components in cleaning solvents of any type.**

> ✳✳ **WARNING**
>
> **Do not for any reason, insert a screwdriver or other small hand tools into the throttle body to hold open the throttle plate, as the throttle body could be damaged.**

➡ **An eight digit part identification number is stamped on the throttle body casting. Refer to this number if servicing, or part replacement is required.**

1. Partially drain the cooling system in order to allow the hose at the throttle body to be removed.

2. Remove the air cleaner outlet duct.

3. Disconnect the throttle actuator motor electrical connector.

4. Reposition the throttle body hose clamp.

5. Remove both of the throttle body engine coolant hoses from the throttle body.

6. Remove the throttle body bolts and nuts.

➡ **Do not reuse the throttle body gasket. Install a new gasket during assembly.**

7. Remove the throttle body and gasket. Discard the gasket.

#### To install:

8. Install a new throttle body gasket.

9. Install the throttle body.

10. Install the throttle body bolts and nuts. Tighten the bolts and nuts to 53 inch lbs. (6 Nm).

11. Connect the throttle body engine coolant hoses to the throttle body.

12. Position the throttle body hose clamps.

13. Connect the throttle actuator motor electrical connector.

14. Install the air cleaner outlet duct.

15. Refill the cooling system.

16. Verify that the vehicle meets the following conditions:

   a. The vehicle is not in a reduced engine power mode.

   b. The ignition is **ON**.

   c. The engine is OFF.

17. Connect a scan tool in order to test for a proper throttle-opening and throttle-closing range.

18. Operate the accelerator pedal and monitor the throttle angles. The accelerator pedal should operate freely, without binding, between a closed throttle, and a wide open throttle (WOT).

19. Start the engine.

20. Inspect for coolant leaks.

# HEATING & AIR CONDITIONING SYSTEM

## BLOWER MOTOR

### REMOVAL & INSTALLATION
*See Figure 121.*

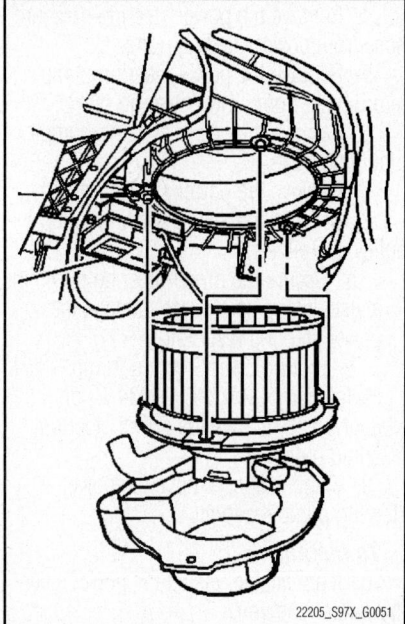

**Fig. 121 Exploded view of blower motor assembly**

1. Before servicing the vehicle, refer to the Precautions Section.
2. Remove the right closeout/insulator panel.
3. Remove the instrument panel storage compartment door.
4. Disconnect the blower motor electrical connector.
5. Remove the blower motor mounting screws.
6. Remove the blower motor cooling tube.
7. Remove the blower motor.

#### To install:
8. Install the blower motor.
9. Install the blower motor cooling tube.
10. Install the blower motor mounting screws and tighten to 18 inch lbs. (2 Nm).
11. Connect the blower motor electrical connector.
12. Install the right closeout/insulator panel.

## HEATER CORE

### REMOVAL & INSTALLATION
*See Figure 122.*

1. Before servicing the vehicle, refer to the Precautions Section.

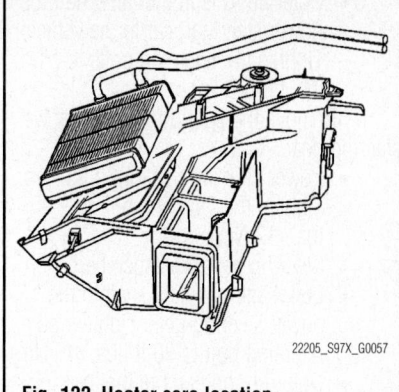

**Fig. 122 Heater core location**

2. Remove the HVAC module assembly.
3. Remove the heater core access cover screws.
4. Remove the heater core access cover.
5. Remove the heater core from the HVAC module assembly.

#### To install:
6. Install the heater core to the HVAC module assembly.
7. Install the heater core access cover.
8. Install the heater core access cover screws. Tighten the screws to 17 inch lbs. (1.9 Nm).
9. Install the HVAC module assembly.

# STEERING

## POWER STEERING GEAR

### REMOVAL & INSTALLATION
*See Figures 123 and 124.*

1. Before servicing the vehicle, refer to the Precautions Section.
2. Raise and support the vehicle.
3. Position a fluid catch pan under the power steering gear.
4. Remove or disconnect the following:
   - Front tire and wheel assemblies
   - Outer tie rod retaining nuts

### ✳ WARNING

**Do not try to separate a steering linkage joint by driving a wedge between the joint and the attached part. Doing this can cause seal damage and premature failure of the part.**

   - Outer tie rods from the steering knuckles using a suitable steering linkage and tie rod puller
   - Lower intermediate shaft retaining bolt and shaft from the power steering gear

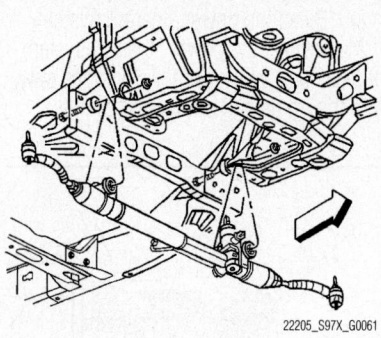

**Fig. 123 Steering gear attaching bolt locations**

   - Steering gear crossmember
   - Feed and return fluid hoses from the steering gear. Immediately cap or plug all openings to prevent system contamination or excessive fluid loss.
5. Support the power steering gear.
   - Power steering gear mounting bolts, then remove the gear from the vehicle

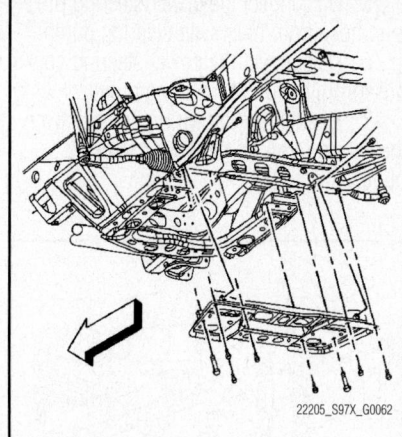

**Fig. 124 Steering gear crossmember mounting bolt locations**

6. Loosen the outer tie rod jam nuts, then remove the outer tie rods from the inner tie rods and discard the jam nut.

#### To install:
7. Lubricate the inner tie rod threads with a suitable lubricant before installing the outer tie rod.

8. Install or connect the following:
- New jam nuts to the outer tie rods
- Outer tie rods to the inner tie rods
- Power steering gear to the vehicle. Tighten the retaining bolts to 81 ft. lbs. (110 Nm).

9. Remove the support from the power steering gear.
- Power steering hose(s) to the gear. Tighten the retaining bolt to 9 ft. lbs. (12 Nm).
- Steering gear crossmember
- Lower intermediate shaft to the power steering gear. Tighten the retaining bolt to 30 ft. lbs. (40 Nm).
- Outer tie rod ends to the steering knuckles. Tighten the retaining nuts to 33 ft. lbs. (45 Nm).
- Front tire and wheel assemblies

10. Remove the drain pan, then lower the vehicle.

11. Bleed the power steering system and adjust the front toe as necessary.

## POWER STEERING PUMP

### REMOVAL & INSTALLATION

#### 4.2L Engine

*See Figures 125 and 126.*

1. Before servicing the vehicle, refer to the Precautions Section.
2. Remove the air cleaner assembly.
3. Remove the drive belt.
4. Install a drain pan under the vehicle.
5. Disconnect the power steering pressure hose from the power steering pump.
6. Disconnect the power steering cooler hose from the power steering pump.
7. Disconnect the wiring harness from the wiring loom on the power steering pump.

8. Remove the power steering pump mounting bolts.
9. Remove the power steering pump.
10. Remove the power steering pump pulley.
11. Remove the power steering pump pulley using Power Steering Pump Pulley Remover tool no. J 25034-C, or equivalent.

#### To install:

12. Install the power steering pump pulley, as follows:
   a. Install the power steering pump pulley to the end of the power steering pump shaft.
   b. Install the power steering pump pulley to the power steering pump using Power Steering Pump Pulley Installer tool no. J 25033-C, or equivalent.
   c. Install the power steering pump pulley flush against the end of the power steering pump shaft , with an allowable variance of 0.010 in. (0.25mm).

13. Install the power steering pump.
14. Install the power steering pump mounting bolts. Tighten the power steering pump mounting bolts to 18 ft. lbs. (25 Nm).
15. Install the power steering cooler hose to the power steering pump.
16. Install the power steering pressure hose to the power steering pump. Tighten the power steering pressure hose to 18 ft. lbs. (25 Nm).
17. Remove the drain pan from under the vehicle.
18. Install the drive belt.
19. Install the air cleaner assembly.
20. Bleed the power steering system.
21. Inspect the power steering system for leaks and the hoses for clearance away from the frame and other components.

#### 5.3L and 6.0L Engines

*See Figures 126 and 127.*

1. Before servicing the vehicle, refer to the Precautions Section.
2. Remove the drive belt.
3. Remove the PCM from PCM mounting bracket and move to the side.
4. Remove the power steering pressure hose from power steering pump.
5. Remove the power steering pump return hose from power steering pump.
6. Remove the power steering pump mounting bolts.
7. Remove the power steering pump.
8. Remove the power steering pump pulley, as follows:
   a. Secure the power steering pump in a vise, taking care not to damage the power steering reservoir.
   b. Using Power Steering Pump Pulley Removal Tool J 25034 C, or equivalent, remove the power steering pump pulley.

9. If applicable, remove the power steering pump reservoir.

#### To install:

10. If applicable, install the power steering pump reservoir.
11. Install the power steering pump pulley, as follows:
   a. Install the power steering pump pulley to the end of the power steering pump shaft.
   b. Install the power steering pump pulley to the power steering pump using Power Steering Pump Pulley Installer tool no. J 25033-C, or equivalent.
   c. Install the power steering pump pulley flush against the end of the power

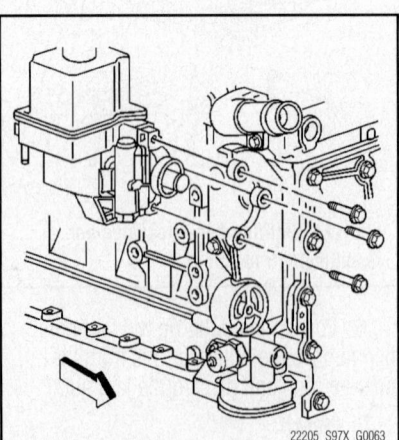

Fig. 125 Power steering pump mounting bolt locations—4.2L engine

22205_S97X_G0063

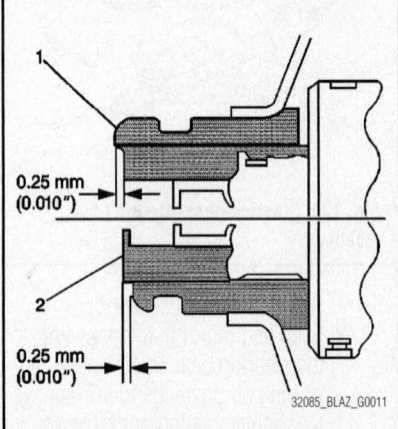

Fig. 126 Install the power steering pump pulley flush against the end of the power steering pump shaft, with an allowable variance of 0.010 in. (0.25mm)

32085_BLAZ_G0011

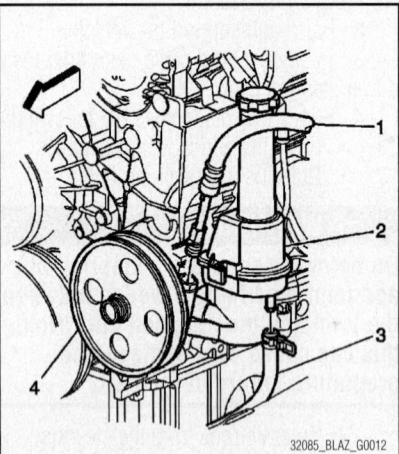

Fig. 127 Remove the power steering pressure hose and return hose from power steering pump

32085_BLAZ_G0012

steering pump shaft , with an allowable variance of 0.010 in. (0.25mm).

12. Align the power steering pump with mounting bolt holes on engine block.

13. Install the power steering pump mounting bolts. Tighten the bolts to 18 ft. lbs. (25 Nm).

14. Attach the power steering pump return hose to power steering pump.

15. Attach the power steering pump pressure hose to power steering pump.

16. Tighten the fittings to 18 ft. lbs. (25 Nm).

17. Install the PCM to PCM mounting bracket.

18. Install the drive belt.

19. Bleed the power steering system.

### BLEEDING

➡**Make sure to use clean, new power steering fluid type only. Hoses touching the frame, body or engine may cause system noise. Verify that the hoses do** not touch any other part of the vehicle. Loose connections may not leak, but could allow air into the steering system. Verify that all hose connections are tight.

### ✳✳ WARNING

**Power steering fluid level must be maintained throughout bleed procedure.**

1. Before servicing the vehicle, refer to the Precautions Section.

2. Fill pump reservoir with fluid to minimum system level, FULL COLD level, or middle of hash mark on cap stick fluid level indicator.

➡**With hydro-boost only, the oil level will appear falsely high if the hydro-boost accumulator is not fully charged. Do not apply the brake pedal with the engine OFF. This will discharge the hydro-boost accumulator.**

3. If equipped with hydro-boost, fully charge the hydro-boost accumulator using the following procedure:
   a. Start the engine.
   b. Firmly apply the brake pedal 10-15 times.
   c. Turn the engine OFF.

4. Raise the vehicle until the front wheels are off the ground.

5. Key on engine OFF, turn the steering wheel from stop to stop 12 times. Vehicles equipped with hydro-boost systems or longer length power steering hoses may require turns up to 15 to 20 stop to stops.

6. Verify power steering fluid level per operating specification.

7. Start the engine. Rotate steering wheel from left to right. Check for sign of cavitation or fluid aeration (pump noise/whining).

8. Verify the fluid level. Repeat the bleed procedure, if necessary.

## SUSPENSION

### COIL SPRING

#### REMOVAL & INSTALLATION

*See Figure 128.*

➡**This procedure requires the use of a suitable spring compressor.**

1. Before servicing the vehicle, refer to the Precautions Section.

2. Remove or disconnect the following:
   • Wheel
   • Shock module
   • Shock module yoke-to-shock absorber pinch bolt and nut

3. Spread the shock module yoke at the pinch bolt using a suitable flat-bladed tool.

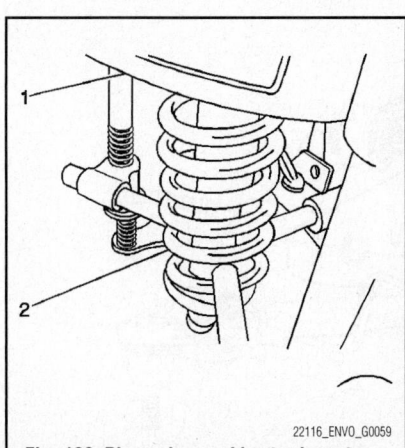

22116_ENVO_G0059

**Fig. 128 Place pieces of heater hose to the spring where the compressor contacts the lower part of the spring**

   • Shock module yoke from the shock absorber

4. Install pieces of heater hose or equivalent material to the shock module spring where the spring compressor contacts the lower part of the spring.

5. Install the shock module into the spring compressor.

➡**The spring is compressed when the shock absorber moves freely.**

6. Turn the spring compressor forcing screw until the coil spring is compressed.

7. Remove or disconnect the following:
   • Shock absorber upper retaining nut
   • Shock absorber from the shock module

8. Loosen the compressor forcing screw until the upper mounting plate and coil spring can be removed.
   • Upper mounting plate and coil spring from the spring compressor

#### To install:

9. Install or connect the following:
   • Coil spring and upper mounting plate to the spring compressor

10. Turn the compressor forcing screw until the coil spring is compressed.
   • Shock absorber to the shock module. Tighten the retaining nut to 33 ft. lbs. (45 Nm)

11. Remove the shock module from the spring compressor. Remove the pieces of heater hose from the spring.

## FRONT SUSPENSION

   • Shock module yoke to the shock absorber
   • Shock module yoke-to-shock pinch bolt and nut and tighten to 52 ft. lbs. (70 Nm)
   • Shock module to the vehicle
   • Tire and wheel

12. Lower the vehicle

### LOWER BALL JOINT

#### REMOVAL & INSTALLATION

*See Figures 129 through 131.*

1. Before servicing the vehicle, refer to the Precautions Section.

2. Remove the wheel center cap and drive axle nut.

3. Raise and support the vehicle.

4. Remove or disconnect the following:
   • Tire and wheel
   • Wheel hub and bearing, if necessary
   • Outer tie rod retaining nut
   • Out tie rod from the steering knuckle using a suitable puller
   • Brake hose bracket retaining bolts and bracket
   • Upper control arm-to-steering knuckle pinch bolt and nut
   • Upper control arm from the steering knuckle
   • Lower ball joint retaining nut
   • Steering knuckle from the lower control arm using a suitable ball joint removal tool

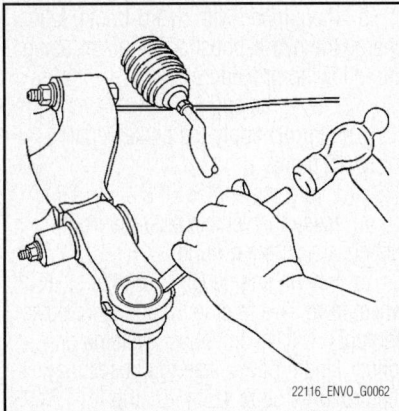

**Fig. 129 Remove the lower ball joint flange with a chisel**

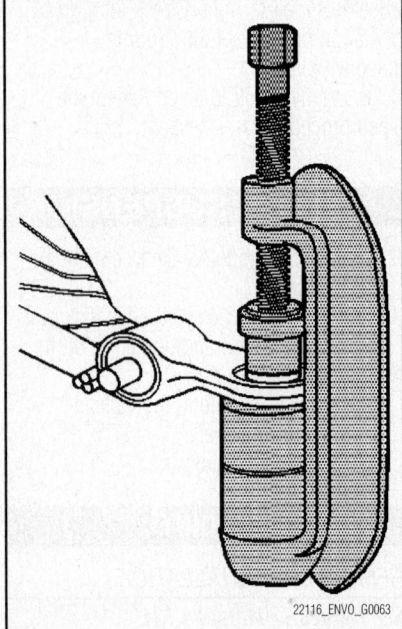

**Fig. 130 Driving the lower joint from the control arm**

- Steering knuckle from the vehicle
- Lower ball joint flange with a chisel

5. Install tools J 9519-E Lower Ball Joint Remover and Installer and J 34874 Booster Seal Remover/Installer to the lower ball joint, then use those tools to remove the lower ball joint from the lower control arm.

**To install:**

6. Install or connect the following:
- Lower ball joint to the lower control arm, using tools J 9519-E Lower Ball Joint Remover and Installer, J 41435 Ball Joint Installer and J 45105-2 Receiver

7. Remove the tools from the lower control arm.

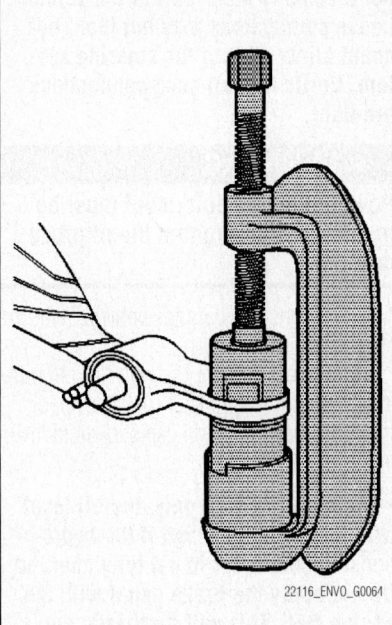

**Fig. 131 Installing a new ball joint**

- Tools J 9519-E Lower Ball Joint Remover and Installer and J 45105-1 Ball Joint Flaring Adapter to the lower ball joint

8. Flare the lower ball joint flange, then remove the tools from the lower ball joint.
- Steering knuckle to the lower control arm
- Lower ball joint retaining nut and tighten to 81 ft. lbs. (110 Nm)
- Upper control arm to the steering knuckle
- Upper control arm pinch bolt and nut and tighten to 30 ft. lbs. (41 Nm)

- Brake hose bracket to the steering knuckle
- Brake hose bracket retaining nuts and tighten to 7 ft. lbs. (10 Nm)
- Outer tie rod to the steering knuckle
- Outer tie rod retaining nut and tighten to 33 ft. lbs. (45 Nm)
- Wheel hub and bearing, if removed
- Tire and wheel

9. Lower the vehicle
- Drive axle nut and tighten to 103 ft. lbs. (140 Nm)
- Wheel center cap, if removed

10. Check the front wheel alignment.

## LOWER CONTROL ARM

### REMOVAL & INSTALLATION

*See Figure 132.*

1. Before servicing the vehicle, refer to the Precautions Section.
2. Raise and support the vehicle.
3. Remove the wheel and tire.
4. Remove the outer tie rod retaining nut.
5. Disconnect the outer tie rod from the steering knuckle using a tie rod puller.
6. Remove the stabilizer bar link lower retaining nut.
7. Disconnect the stabilizer bar link and washer from the lower control arm.
8. Remove the shock module yoke lower mounting nut.
9. Disconnect the shock module yoke from the lower control arm using a tie rod puller.
10. Remove the lower ball joint retaining nut.

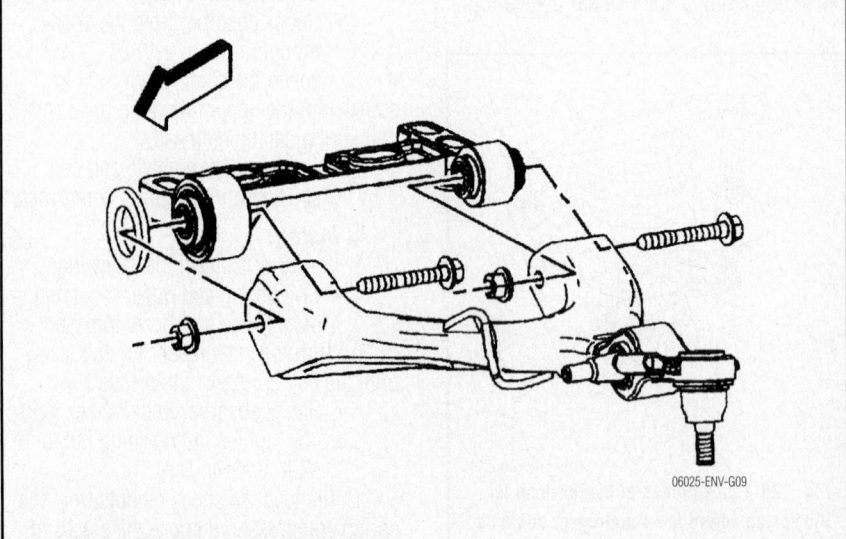

**Fig. 132 Front lower control arm mounting**

11. Disconnect the lower ball joint from steering knuckle using ball joint remover.

12. Remove the lower control arm-to-lower control arm bracket mounting nuts.

13. Note the direction the bolts are removed for installation.

14. Remove the lower control arm to lower control arm bracket mounting bolts.

15. Take care not to disengage the axle shaft from the transmission.

16. Pivot the lower control arm outward and downward in order to disconnect the lower control arm from the lower control arm bracket.

17. Ensure that the spacer stays in position on the front control arm bracket front bushing.

18. Remove the lower control arm from the vehicle.

### To install:

19. Position the lower control arm ball joint stud to the steering knuckle.

20. Ensure that the spacer stays in position on the front control arm bracket front bushing.

21. Pivot the lower control arm outward and upward in order to connect the lower control arm to the lower control arm bracket.

22. Install the lower control arm to lower control arm bracket mounting bolts.

➡**Ensure that the lower control arm is parallel to the lower control arm bracket during the installation and tightening of the lower control arm mounting bolts and nuts. This will ensure correct alignment of the lower control arm bushings.**

23. Install the lower control arm to lower control arm bracket mounting nuts and tighten to 96 ft. lbs. (130 Nm).

24. Connect the shock module yoke to the lower control arm.

25. Install the shock module yoke lower mounting nut and tighten to 81 ft. lbs. (110 Nm).

26. Install the lower ball joint retaining nut and tighten to 81 ft. lbs. (110 Nm).

27. Install the stabilizer bar link and washer to the lower control arm.

28. Install the stabilizer bar link retaining nut and tighten to 114 ft. lbs. (155 Nm).

29. Install the outer tie rod to the steering knuckle.

30. Install the outer tie rod retaining nut and tighten to 33 ft. lbs. (45 Nm).

31. Install the tire and wheel.

32. Lower the vehicle.

33. Check the front wheel alignment.

## SHOCK ABSORBER MODULE

### REMOVAL & INSTALLATION

*See Figures 133 and 134.*

A "shock module", similar to a McPherson strut is used on these vehicles.

1. Before servicing the vehicle, refer to the Precautions Section.

2. Remove the shock module upper retaining nuts.

➡**Use care when handling the coil springs in order to avoid chipping or scratching the coating. Damage to the coating will result in premature failure of the coil springs.**

3. Raise and support the vehicle.

4. Remove the tire and wheel.

5. Remove the shock module yoke from the lower control arm.

6. Remove the shock module from the shock tower and lower control arm.

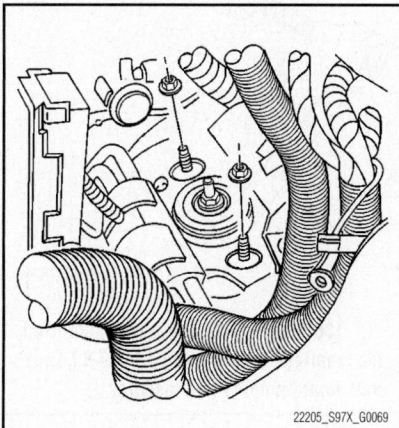

22205_S97X_G0069

**Fig. 133 Removing the shock module upper retaining nuts**

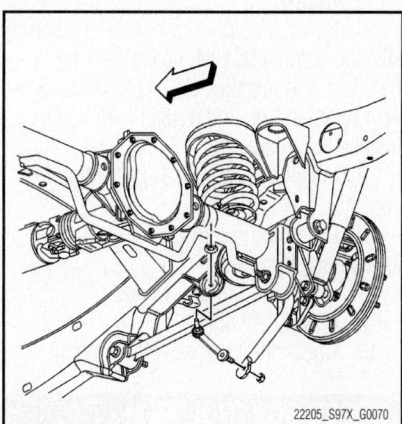

22205_S97X_G0070

**Fig. 134 Remove the shock module from the shock tower and lower control arm**

### To install:

7. Install the shock module to the shock tower and lower control arm.

8. Install the shock module yoke to the lower control arm.

9. Lower the vehicle.

10. Install the shock module upper retaining nuts. Tighten the shock module upper retaining nuts to 33 ft. lbs. (45 Nm).

11. Raise the vehicle.

12. Install the tire and wheel.

13. Remove the support and lower the vehicle.

## STABILIZER BAR

### REMOVAL & INSTALLATION

1. Before servicing the vehicle, refer to the Precautions Section.

2. Raise and support the vehicle.

3. Remove the stabilizer bar links to the stabilizer bar retaining nuts.

4. Remove the stabilizer bar insulator clamp mounting bolts.

5. Remove the stabilizer bar insulator clamp from the stabilizer bar insulator.

6. Remove the stabilizer bar insulators from the stabilizer bar.

7. Remove the stabilizer bar from the vehicle.

### To install:

8. Install the stabilizer bar to the vehicle.

9. Install the stabilizer bar insulators to the stabilizer bar.

10. Install the stabilizer bar insulator clamp to the stabilizer bar insulator.

11. Install the stabilizer bar insulator clamp mounting bolts and tighten to 41 ft. lbs. (55 Nm).

12. Install the stabilizer bar links to the stabilizer bar and tighten to 74 ft. lbs. (100 Nm).

13. Lower the vehicle.

## STEERING KNUCKLE

### REMOVAL & INSTALLATION

1. Before servicing the vehicle, refer to the Precautions Section.

2. Raise and support the vehicle.

3. Remove the tire and wheel.

4. Remove wheel center cap and the drive axle nut and washer

5. Remove the wheel hub and bearing.

6. Remove the outer tie rod retaining nut.

7. Disconnect the outer tie rod from the steering knuckle using a tie rod puller.

8. Remove the brake hose bracket retaining bolts.

9. Remove the brake hose bracket from the steering knuckle.

10. Disconnect the ABS wheel speed sensor wiring harness bracket from the steering knuckle.

11. Remove the upper control arm to the steering knuckle pinch bolt and nut.

12. Disconnect the upper control arm from the steering knuckle.

13. Remove the lower ball joint retaining nut.

14. Remove the steering knuckle from the lower control arm.

15. Remove the steering knuckle from the vehicle.

### To install:

16. Install the steering knuckle to the lower control arm.

17. Install the lower ball joint retaining nut and tighten to 81 ft. lbs. (110 Nm).

18. Connect the upper control arm to the steering knuckle.

19. Install upper control arm pinch bolt and nut and tighten to 30 ft. lbs. (40 Nm).

20. Connect the ABS wheel speed sensor wiring harness bracket to the steering knuckle.

21. Install the brake hose bracket to the steering knuckle.

22. Install the brake hose bracket retaining bolts and tighten to 89 inch lbs. (10 Nm).

23. Install the outer tie rod to the steering knuckle.

24. Install the new outer tie rod retaining nut and tighten to 44 ft. lbs. (60 Nm).

25. Install the wheel hub and bearing.

26. Install the drive axle nut and tighten to 103 ft. lbs. (140 Nm), then install the center cap.

27. Install the tire and wheel.

28. Lower the vehicle.

29. Adjust the front toe.

## UPPER BALL JOINT

### REMOVAL & INSTALLATION

*See Figures 135 and 136.*

1. Before servicing the vehicle, refer to the Precautions Section.

2. Raise and safely support the front of the vehicle securely on jackstands.

3. Remove the tire and wheel.

4. Remove the steering knuckle with wheel hub attached.

5. Remove the upper ball joint retaining clip.

6. Remove the upper ball joint boot.

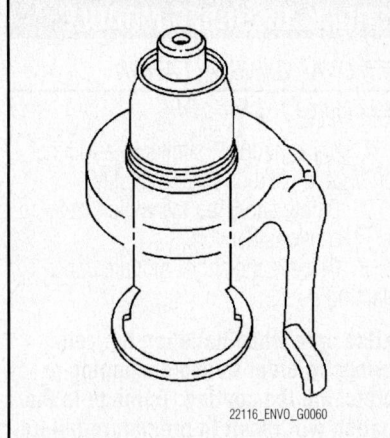

**Fig. 135 Remove the upper ball joint boot**

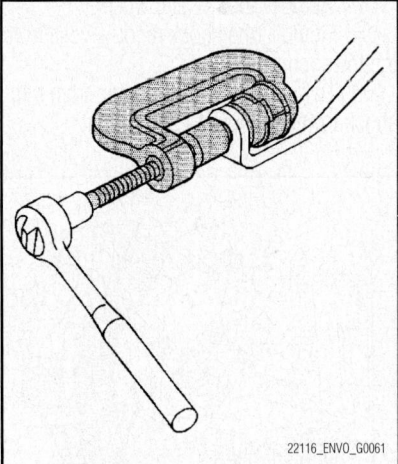

**Fig. 136 Remove the upper ball joint from the steering knuckle using J 9519-E Lower Ball Joint Remover and Installer**

7. Remove the upper ball joint from the steering knuckle using J 9519-E Lower Ball Joint Remover and Installer.

### To install:

8. Install the upper ball joint to steering knuckle using J 9519-E Lower Ball Joint Remover and Installer, J 21474-01 Control Arm Bushing Set, and J 45117 Ball Joint Installation Spacer.

9. Install the upper ball joint retaining clip.

10. Install the steering knuckle with wheel hub attached.

11. Install the tire and wheel.

12. Lower the vehicle.

13. Check the front wheel alignment.

## UPPER CONTROL ARM

### REMOVAL & INSTALLATION

1. Before servicing the vehicle, refer to the Precautions Section.

2. Remove or disconnect the following:
   - Tire and wheel assembly
   - Upper ball joint-to-upper control arm pinch bolt and nut
   - Upper control arm from the knuckle
   - Anti-lock Brake System (ABS) wheel speed sensor wiring harness
   - If servicing the left upper control arm, remove the battery tray

➡️**Gently pry out on inner fender body panel to access forward facing bolt.**

   - Upper control arm mounting bolts
   - Upper control arm

### To install:

3. Install or connect the following:
   - Upper control arm and tighten the bolts to 108 ft. lbs. (146 Nm)
   - ABS wheel speed sensor wiring harness
   - Battery tray
   - Upper control arm to the steering knuckle
   - Upper ball joint-to-upper control arm pinch bolt and nut and tighten to 30 ft. lbs. (40 Nm)
   - Tire and wheel

4. Check the front wheel alignment.

## WHEEL BEARINGS

### REMOVAL & INSTALLATION

*See Figure 137.*

1. Remove the wheel center cap.

2. Remove the drive axle nut.

3. Raise and safely support the vehicle securely on jackstands.

4. Remove the tire and wheel.

5. Remove the brake rotor.

6. Disengage the wheel drive shaft from the wheel hub and bearing. Place a brass drift against the outer end of the wheel drive

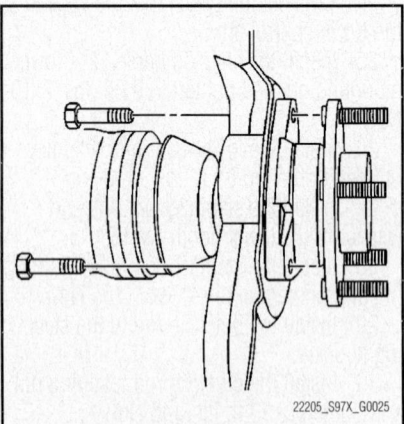

**Fig. 137 Removing the wheel hub and bearing to the steering knuckle mounting bolts**

shaft in order to protect the wheel drive shaft threads. Sharply strike the brass drift with the hammer. Do not attempt to remove the wheel drive shaft from the wheel hub and bearing at this time.

7. Remove the ABS sensor mounting bolt from the wheel hub and bearing.

8. Remove the ABS sensor from the wheel hub and bearing.

9. Remove the wheel hub and bearing to the steering knuckle mounting bolts.

10. Remove the wheel hub and bearing from the steering knuckle.

11. Remove the splash shield from the steering knuckle.

**To install:**

12. Install the splash shield to the steering knuckle. Align the splash shield to the steering knuckle threaded holes.

13. Install the wheel hub and bearing to the steering knuckle. Align the threaded holes.

14. Install the wheel hub and bearing to the steering knuckle mounting bolts. Tighten the wheel hub and bearing mounting bolts to 77 ft. lbs. (105 Nm).

15. Install the ABS sensor to the wheel hub and bearing.

16. Install the ABS sensor mounting bolt to the wheel hub and bearing. Tighten the

ABS sensor to the wheel hub and bearing mounting bolt to 13 ft. lbs. (18 Nm).

17. Install the brake rotor.

18. Install the tire and wheel.

19. Lower the vehicle.

20. Install the drive axle nut and draw the hub and bearing onto the axle. Tighten the drive axle nut to 103 ft. lbs. (140 Nm).

21. Install the tire and wheel center cap.

## ADJUSTMENT

The wheel bearings on these vehicles are not adjustable. If the bearings become loose or make noise, they must be replaced.

# SUSPENSION                                        REAR SUSPENSION

## COIL SPRING

### REMOVAL & INSTALLATION

1. Raise and support the vehicle. Refer to Lifting and Jacking the Vehicle.

2. Install the rear axle support.

3. Remove the shock absorber lower mounting bolts.

➡**Do not lower the rear axle so that the upper control arms contact the frame. Damage to the upper control arms will result.**

4. Lower the rear axle.

➡**Use care when handling the coil springs in order to avoid chipping or scratching the coating. Damage to the coating will result in premature failure of the coil springs.**

5. Remove the rear coil springs.

**To install:**

6. Install the rear coil springs.

7. Raise the rear axle.

8. Install the shock absorber lower mounting bolts. Tighten the shock absorber lower mounting bolts to 63 ft. lbs. (85 Nm).

9. Remove the rear axle support.

10. Lower the vehicle.

## LOWER CONTROL ARM

### REMOVAL & INSTALLATION

*See Figure 138.*

1. Raise and support the vehicle.

2. Remove the wheel and tire.

3. Raise and support the rear axle at the trim height of 6.12 in. (155.4mm).

4. If equipped with air suspension, depressurize the air suspension system.

5. Remove the rear axle lower control arm to the axle mounting nut and bolt.

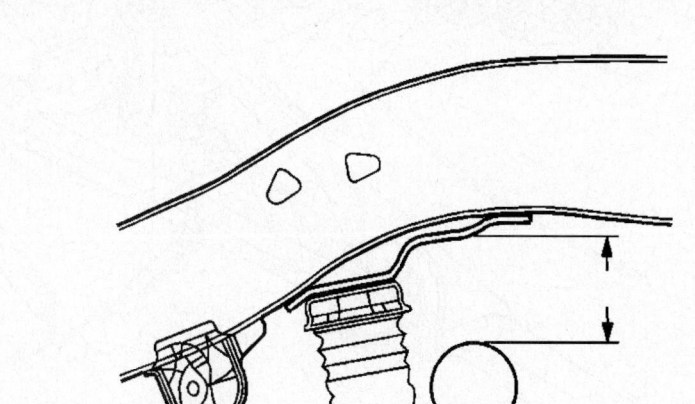

22205_S97X_G0067

Fig. 138 Raise or lower the vehicle so that the trim height (distance between the arrows) is at specification

6. Remove the rear axle lower control arm to the frame mounting nut and bolt.

7. Remove the lower control arm.

**To install:**

8. Install the lower control arm.

9. Install the rear axle lower control arm to the frame mounting nut and bolt.

10. Install the rear axle lower control arm to the axle mounting bolt and nut and tighten to 74 ft. lbs. (100 Nm).

11. Remove the rear axle support.

12. Lower the vehicle.

## STABILIZER BAR

### REMOVAL & INSTALLATION

*See Figure 139.*

1. Before servicing the vehicle, refer to the Precautions Section.

2. Raise and support the vehicle.

3. Remove the tire and wheel.

4. Remove the stabilizer shaft links to the stabilizer shaft retaining nuts.

### ❋❋ WARNING

**Do not pry on the stabilizer shaft link. Use care when removing or installing the stabilizer shaft link in order to avoid tearing or puncturing the stabilizer shaft link boot. Damage to the stabilizer shaft link boot will lead to damage to the stabilizer shaft link.**

5. Disconnect the stabilizer shaft links from the stabilizer shaft.

6. Remove the stabilizer shaft insulator clamp mounting nuts.

7. Remove the stabilizer shaft.

8. Remove the stabilizer shaft insulators.

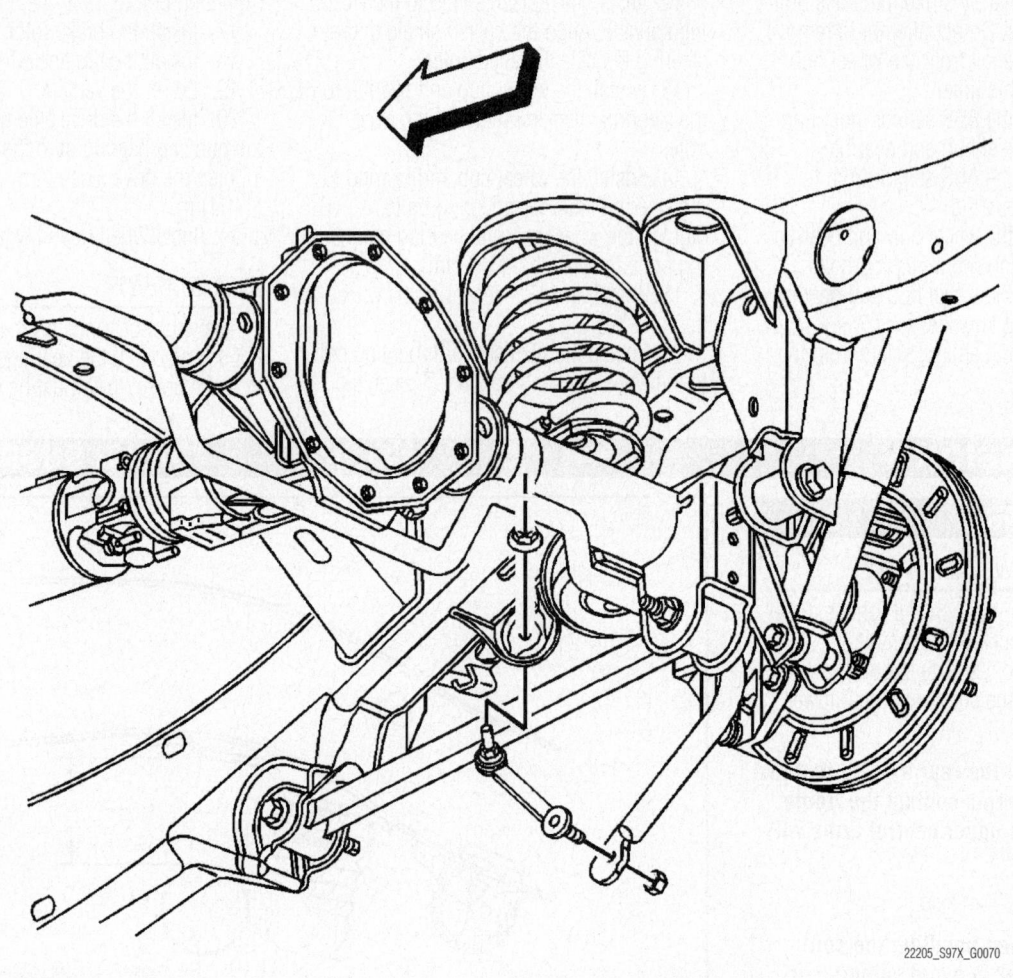

**Fig. 139 Rear stabilizer bar assembly**

22205_S97X_G0070

*To install:*

9. Install the stabilizer shaft insulators.

10. Install the stabilizer shaft.

11. Install the stabilizer shaft insulator clamp mounting nuts. Tighten the stabilizer shaft insulator clamp mounting nuts to 55 ft. lbs. (75 Nm).

12. Connect the stabilizer shaft links to the stabilizer shaft. Hand tighten only.

13. Install the stabilizer shaft links to stabilizer shaft retaining nuts. Tighten the stabilizer shaft links to stabilizer shaft retaining nuts to 66 ft. lbs. (90 Nm).

14. Install the tire and wheel.

15. Lower the vehicle.

## SHOCK ABSORBER

### REMOVAL & INSTALLATION

1. Before servicing the vehicle, refer to the Precautions Section.

2. Raise and support the vehicle.

### ✳✳ WARNING

**The rear axle must not be allowed to hang freely while servicing rear suspension components. Not supporting the rear axle will result in damage to the upper control arm and/or to the air suspension components.**

3. Support the rear axle.

4. Remove the shock absorber upper mounting bolt.

5. Remove the shock absorber lower mounting bolt.

6. Remove the shock absorber from the vehicle.

*To install:*

7. Install the shock absorber to the vehicle.

8. Install the shock absorber lower mounting bolt. Tighten the shock absorber lower mounting bolt to 63 ft. lbs. (85 Nm).

Install the shock absorber upper mounting bolt. Tighten the shock absorber upper mounting nut to 63 ft. lbs. (85 Nm).

9. Remove the support from the rear axle.

## UPPER CONTROL ARM

### REMOVAL & INSTALLATION

*See Figures 140 and 141.*

1. Before servicing the vehicle, refer to the Precautions Section.

2. Raise and safely support the rear of the vehicle securely on jackstands.

3. Remove the tire and wheel.

4. Remove the wheel well.

5. Raise and support the rear axle at the trim height of 5.17–5.49 inches (131.4–139.4mm).

6. Depressurize the air suspension system.

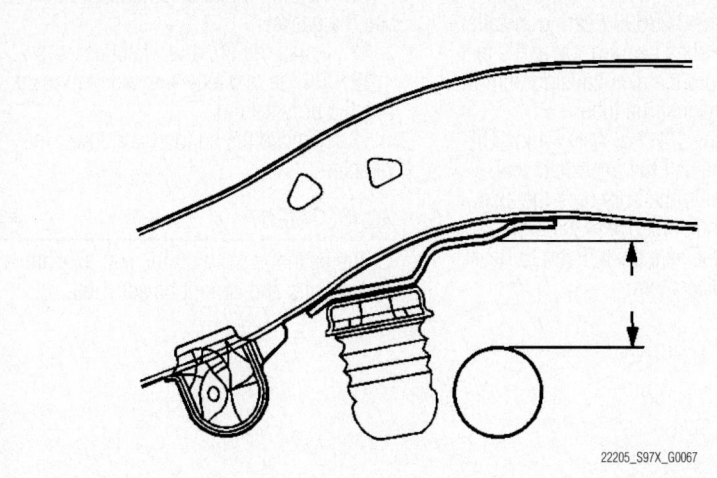

**Fig. 140 Measure the trim height by measuring vertical distance between the top surface of the axle and to the side of the hole center on the jounce bumper reinforcement bracket**

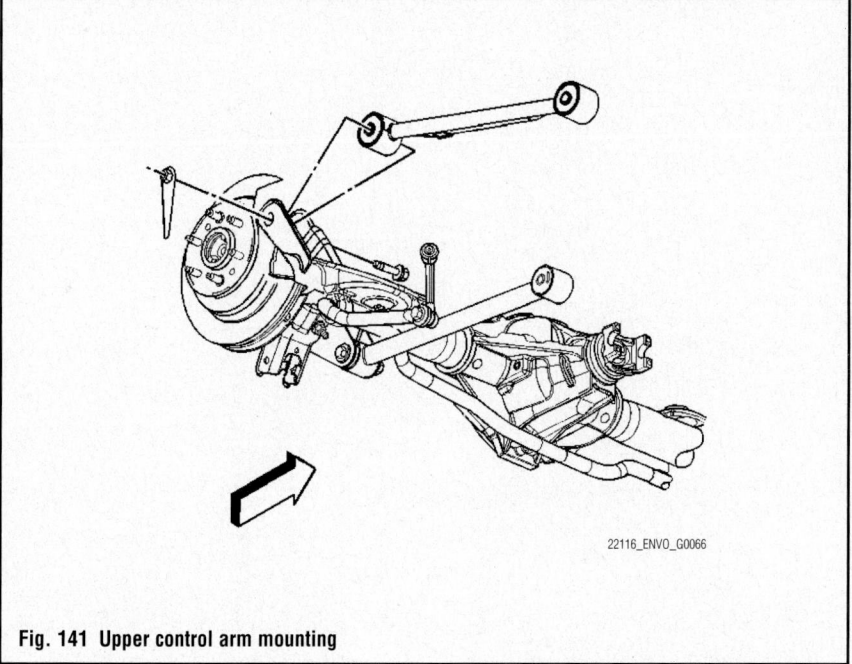

**Fig. 141 Upper control arm mounting**

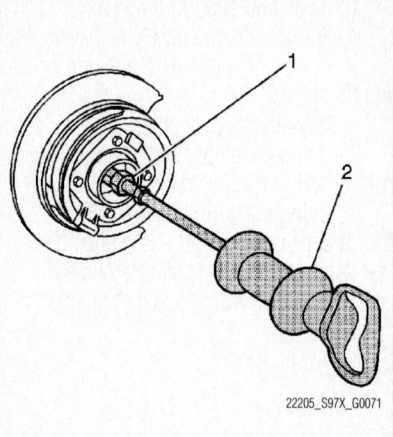

**Fig. 142 Using the J 45857 tone wheel and/or Bearing remover and a slide hammer, remove the axle shaft seal and the bearing together from the axle housing**

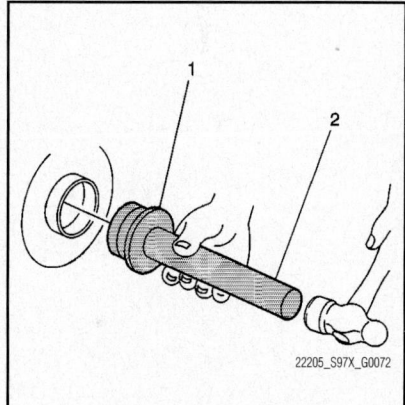

**Fig. 143 Using the J 23690 bearing installer, install the axle shaft bearing**

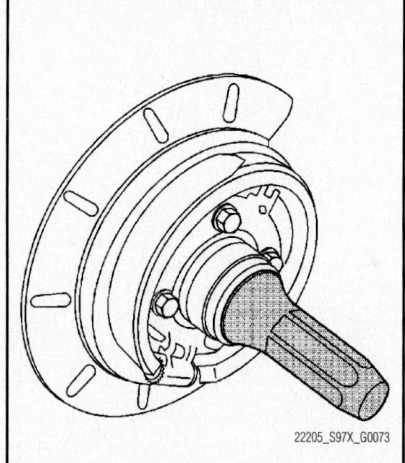

**Fig. 144 Using the J 21128 axle pinion oil seal installer, install the axle shaft seal**

7. Disconnect the air suspension leveling sensor link from the rear axle upper control arm.

8. Remove the rear axle upper control arm to axle mounting bolt and nut.

9. Remove the rear axle upper control arm to frame mounting bolt.

10. Remove the rear axle upper control arm.

**To install:**

11. Install the rear axle upper control arm.

12. Install the rear axle upper control arm to frame mounting bolt.

13. Install the rear axle upper control arm to axle mounting nut and bolt and tighten the bolts to 97 ft. lbs. (131 Nm).

14. Connect the air suspension leveling sensor link to the rear axle upper control arm.

15. Remove the rear axle support.

16. Install the wheel well.

17. Install the tire and wheel.

18. Lower the vehicle.

## WHEEL BEARINGS

### REMOVAL & INSTALLATION

*See Figures 142 through 144.*

1. Raise and support the vehicle.
2. Remove the tire and wheel assembly.
3. Remove the rear axle housing cover and the gasket.
4. Remove the axle shaft.
5. Using a suitable seal remover, remove the rear axle shaft seal.
6. Using the J 45857 Tone Wheel and/or Bearing Remover and a slide hammer, remove the axle shaft seal and the bearing together from the axle housing.

**To install:**

7. Using the J 23690 Bearing Installer, install the axle shaft bearing. Drive the axle shaft bearing into the axle housing until the tool bottoms against the tube.
8. Using the J 21128 Axle Pinion Oil Seal Installer, install the axle shaft seal. Drive the tool into the bore until the axle shaft seal bottoms flush with the tube.
9. Install the axle shaft. Refer to Rear Axle Shaft Replacement.

10. Install the rear axle housing cover and the gasket.
11. Install the tire and wheel assembly.
12. Fill the rear axle with axle lubricant. Use the proper fluid.
13. Remove the support and lower the vehicle.

## ADJUSTMENT

The bearings used on the rear axle are sealed units and cannot be adjusted.

# SAAB

## Diagnostic Trouble Codes

## DIAGNOSTIC TROUBLE CODES

### OBD II VEHICLE APPLICATIONS

*SAAB*

**9-2x**
2005–06
- 2.0L H4 MFI Turbo (SOHC) . . . . VIN 2
2005–08
- 2.5L H4 MFI (SOHC) . . . . . . . . . VIN 6
2006–08
- 2.5L H4 MFI Turbo (DOHC) . . . . VIN 7

**9-3**
2005-08
- 2.0L I4 MFI Turbo (DOHC) . . . . . VIN S
2005–08
- 2.0L I4 MFI Turbo (DOHC) . . . . . VIN Y
2006–08
- 2.8L V6 MFI Turbo (DOHC) . . . . VIN U

**9-5**
2005-08
- 2.3L I4 MFI Turbo (DOHC) . . . . . VIN E

2005–08
- 2.3L I4 MFI Turbo (DOHC) . . . . . VIN G

**9-7X**
2005–08
- 4.2L V6 MFI (DOHC) . . . . . . . . . VIN S
2005
- 5.3L V8 MFI (OHV) . . . . . . . . . . VIN P
2006–08
- 5.3L V8 MFI (OHV) . . . . . . . . . . VIN M
2008
- 6.0L V8 MFI (OHV) . . . . . . . . . . VIN H

## OBD II Trouble Code List (P0xxx Codes)

| DTC | Trouble Code Title, Conditions & Possible Causes |
|---|---|
| **DTC: P0008**<br>**2T CCM, MIL: Yes**<br>**Years:** 2005, 2006, 2007, 2008<br>**Engines:** 2.8L VIN U<br>**Models:** All<br>**Transmissions:** All | **Camshaft Bank 1, Misaligned to Crankshaft**<br>Control module did not Control module detected any Camshaft Position (Camshaft Position ) sensor signals, or the Camshaft Position sensor signal was interrupted after the engine was running during the CCM test.<br>**Note: The engine may not start without a proper Camshaft Position sensor signal.**<br>**Possible Causes:**<br>&bull; Check any other DTC on display<br>&bull; Check power supply<br>&bull; Check harness between sensor connector and Control module<br>&bull; Check sensor installation and condition<br>&bull; Check the sensor<br>&bull; Check for poor contact<br>&bull; Faulty Camshaft Position sensor |
| **DTC: P0009**<br>**2T CCM, MIL: Yes**<br>**Years:** 2005, 2006, 2007, 2008<br>**Engines:** 2.8L VIN U<br>**Models:** All<br>**Transmissions:** All | **Camshaft Bank 2, Misaligned to Crankshaft**<br>Control module did not Control module detected any Camshaft Position (Camshaft Position ) sensor signals, or the Camshaft Position sensor signal was interrupted after the engine was running during the CCM test.<br>**Note: The engine may not start without a proper Camshaft Position sensor signal.**<br>**Possible Causes:**<br>&bull; Check any other DTC on display<br>&bull; Check power supply<br>&bull; Check harness between sensor connector and Control module<br>&bull; Check sensor installation and condition<br>&bull; Check the sensor<br>&bull; Check for poor contact<br>&bull; Faulty Camshaft Position sensor |
| **DTC: P0010**<br>**2T CCM, MIL: Yes**<br>**Years:** 2005, 2006, 2007, 2008<br>**Engines:** 2.8L VIN U<br>**Models:** All<br>**Transmissions:** All | **Intake Cam Angle Solenoid Circuit Bank 1, Open**<br>Control module did not Control module detected any Camshaft Position (Camshaft Position ) sensor signals, or the Camshaft Position sensor signal was interrupted after the engine was running during the CCM test.<br>**Note: The engine may not start without a proper Camshaft Position sensor signal.**<br>**Possible Causes:**<br>&bull; Check any other DTC on display<br>&bull; Check power supply<br>&bull; Check harness between sensor connector and Control module<br>&bull; Check sensor installation and condition<br>&bull; Check the sensor<br>&bull; Check for poor contact<br>&bull; Faulty Camshaft Position sensor |
| **DTC: P0011**<br>**2T CCM, MIL: Yes**<br>**Years:** 2005, 2006, 2007, 2008<br>**Engines:** 2.8L VIN U<br>**Models:** All<br>**Transmissions:** All | **Intake Camshaft Position Timing (over advanced) Bank 1**<br>Control module detects the camshaft timing system malfunction. Engine stalls. Erroneous idling.<br>**Possible Causes:**<br>&bull; Check any other DTC on display<br>&bull; Using a scan tool, check current data.<br>&bull; Oil pipe clog<br>&bull; Clogged oil flow control solenoid valve<br>&bull; Intake camshaft damaged<br>&bull; Timing chain problem |
| **DTC: P0013**<br>**2T CCM, MIL: Yes**<br>**Years:** 2005, 2006, 2007, 2008<br>**Engines:** 4.2L VIN S<br>**Models:** All<br>**Transmissions:** All | **Camshaft Position Actuator Circuit Malfunction**<br>Control module detected an unexpected voltage condition on the Camshaft Position actuator high control or low reference circuit for 250 milliseconds.<br>**Note: A 50% duty cycle is used to maintain a steady retard angle.**<br>**Possible Causes**<br>&bull; Camshaft Position actuator high control circuit is open or shorted to ground<br>&bull; Camshaft Position actuator low reference circuit is open<br>&bull; Camshaft Position actuator is damaged or has failed<br>&bull; Control module has failed |

| DTC | Trouble Code Title, Conditions & Possible Causes |
|---|---|
| **DTC: P00014**<br>**2T CCM, MIL: Yes**<br>**Years:** 2005, 2006, 2007, 2008<br>**Engines:** 4.2L VIN S<br>**Models:** All<br>**Transmissions:** All | **Camshaft Phasing System Malfunction**<br>Control module detected the difference between the Actual and Desired Cam Phase angle was over 1.5 degrees with the Cam Phaser steady for 20 seconds (50% duty cycle signal achieves a steady retard angle).<br>**Possible Causes**<br>&bull; CKP or Camshaft Position sensor loose/damaged (causes signal variation)<br>&bull; Contamination or debris interfering with the actuator operation<br>&bull; Timing chain and gear assembly has excessive free-play<br>&bull; TSB 01-06-04-052 contains a repair procedure for this code |
| **DTC: P0016**<br>**2T CCM, MIL: Yes**<br>**Years:** 2005, 2006, 2007, 2008<br>**Engines:** 2.8L VIN U, 5.3L VIN M/P, 6.0L VIN H<br>**Models:** All<br>**Transmissions:** All | **Crankshaft Position (camshaft position correlation) Bank 1**<br>Control module detects the camshaft timing system malfunction. Engine stalls. Erroneous idling.<br>Engine stalls. Erroneous idling.<br>**Possible Causes:**<br>&bull; Check any other DTC on display<br>&bull; Using a scan tool, check current data.<br>&bull; Oil pipe clog<br>&bull; Clogged oil flow control solenoid valve<br>&bull; Intake camshaft damaged<br>&bull; Timing chain problem |
| **DTC: P0017**<br>**2T CCM, MIL: Yes**<br>**Years:** 2005, 2006, 2007, 2008<br>**Engines:** 4.2L VIN S<br>**Models:** All<br>**Transmissions:** All | **Crankshaft Position Camshaft Position Correlation (Bank 1 Sensor B)**<br>Control module did not Control module detected any Camshaft Position (Camshaft Position ) sensor signals, or the Camshaft Position sensor signal was interrupted after the engine was running during the CCM test.<br>**Note: The engine may not start without a proper Camshaft Position sensor signal.**<br>**Possible Causes:**<br>&bull; Check any other DTC on display<br>&bull; Check power supply<br>&bull; Check harness between sensor connector and Control module<br>&bull; Check sensor installation and condition<br>&bull; Check the sensor<br>&bull; Check for poor contact<br>&bull; Faulty Camshaft Position sensor |
| **DTC: P0018**<br>**2T CCM, MIL: Yes**<br>**Years:** 2005, 2006, 2007, 2008<br>**Engines:** 2.8L VIN U<br>**Models:** All<br>**Transmissions:** All | **Crankshaft Position (camshaft position correlation) Bank 2**<br>Control module detects the camshaft timing system malfunction. Engine stalls. Erroneous idling.<br>**Possible Causes:**<br>&bull; Check any other DTC on display<br>&bull; Using a scan tool, check current data.<br>&bull; Oil pipe clog<br>&bull; Clogged oil flow control solenoid valve<br>&bull; Intake camshaft damaged<br>&bull; Timing chain problem |
| **DTC: P0020**<br>**2T CCM, MIL: Yes**<br>**Years:** 2005, 2006, 2007, 2008<br>**Engines:** 2.8L VIN U<br>**Models:** All<br>**Transmissions:** All | **Camshaft Position Actuator Circuit (Bank 2)**<br>Control module did not Control module detected any Camshaft Position (Camshaft Position ) sensor signals, or the Camshaft Position sensor signal was interrupted after the engine was running during the CCM test.<br>**Note: The engine may not start without a proper Camshaft Position sensor signal.**<br>**Possible Causes:**<br>&bull; Check any other DTC on display<br>&bull; Check power supply<br>&bull; Check harness between sensor connector and Control module<br>&bull; Check sensor installation and condition<br>&bull; Check the sensor<br>&bull; Check for poor contact<br>&bull; Faulty Camshaft Position sensor |
| **DTC: P0021**<br>**2T CCM, MIL: Yes**<br>**Years:** 2005, 2006, 2007, 2008<br>**Engines:** 2.8L VIN U<br>**Models:** All<br>**Transmissions:** All | **Intake Camshaft Position Timing (over advanced) Bank 2**<br>Control module detects the camshaft timing system malfunction. Engine stalls. Erroneous idling.<br>**Possible Causes:**<br>&bull; Check any other DTC on display<br>&bull; Using a scan tool, check current data.<br>&bull; Oil pipe clog<br>&bull; Clogged oil flow control solenoid valve<br>&bull; Dirty engine oil<br>&bull; Intake camshaft damaged<br>&bull; Timing chain problem |

| DTC | Trouble Code Title, Conditions & Possible Causes |
|---|---|
| **DTC: P0026**<br>**2T CCM, MIL: Yes**<br>**Years:** 2005, 2006, 2007, 2008<br>**Engines:** 2.5L SOHC VIN 6<br>**Models:** All<br>**Transmissions:** All | **Intake Valve Control Solenoid Circuit Range/Performance Bank 1**<br>Variable lift control is operational. Erroneous idling.<br>**Possible Causes:**<br>• Check any other DTC on display<br>• Check harness between Control module and variable valve lift diagnosis oil pressure switch connector<br>• Check harness between Control module and variable valve lift diagnosis oil pressure switch connector<br>• Check oil switching solenoid valve |
| **DTC: P0028**<br>**2T CCM, MIL: Yes**<br>**Years:** 2005, 2006, 2007, 2008<br>**Engines:** 2.5L SOHC VIN 6<br>**Models:** All<br>**Transmissions:** All | **Intake Valve Control Solenoid Circuit Range/Performance Bank 2**<br>Variable lift control is operational. Erroneous idling.<br>**Possible Causes:**<br>• Check any other DTC on display<br>• Check harness between Control module and variable valve lift diagnosis oil pressure switch connector<br>• Check harness between Control module and variable valve lift diagnosis oil pressure switch connector<br>• Check oil switching solenoid valve |
| **DTC: P0030**<br>**2T CCM, MIL: Yes**<br>**Years:** 2005, 2006, 2007, 2008<br>**Engines:** 2.0L VIN 2, 2.0L VIN S/Y, 2.5L SOHC, DOHC VIN 6, 2.8L VIN U, 4.2L VIN S, 5.3L VIN M/P, 6.0L VIN H<br>**Models:** All<br>**Transmissions:** All | **O2S Heater Control Circuit**<br>Control module detected functional errors of the oxygen sensor heater. Poor drive ability.<br>**Possible Causes:**<br>• Check harness between Control module and front oxygen sensor connector<br>• Check harness between main relay and front oxygen sensor connector<br>• Check front oxygen sensor<br>• Check poor contact |
| **DTC: P0031**<br>**2T CCM, MIL: Yes**<br>**Years:** 2005, 2006, 2007, 2008<br>**Engines:** 2.0L VIN 2, 2.0L VIN S/Y, 2.5L SOHC, DOHC VIN 6, 2.8L VIN U<br>**Models:** All<br>**Transmissions:** All | **O2S Heater Control Circuit Low**<br>Control module detected the open or short circuit of heater. The heater performs duty control, and the output terminal voltage at ON is 0 V and the output terminal voltage at OFF is the battery voltage. Control module detects when the terminal voltage remains Low.<br>**Possible Causes:**<br>• Check power supply to front oxygen sensor<br>• Check ground circuit for Control module<br>• Using a scan tool check current data<br>• Check output signal of Control module<br>• Check front oxygen sensor |
| **DTC: P0032**<br>**2T CCM, MIL: Yes**<br>**Years:** 2005, 2006, 2007, 2008<br>**Engines:** 2.0L VIN 2, 2.0L VIN S/Y, 2.5L SOHC, DOHC VIN 6, 2.8L VIN U<br>**Models:** All<br>**Transmissions:** All | **O2S Heater Control Circuit High**<br>Control module detected the open or short circuit of heater. The heater performs duty control, and the output terminal voltage at ON is 0 V and the output terminal voltage at OFF is the battery voltage. Control module detects when the terminal voltage remains High.<br>**Possible Causes:**<br>• Check output signal of Control module<br>• Check front oxygen sensor heater current<br>• Faulty Control module |
| **DTC: P0036**<br>**2T CCM, MIL: Yes**<br>**Years:** 2005, 2006, 2007, 2008<br>**Engines:** 2.0L VIN S/Y, 2.8L VIN U, 4.2L VIN S, 5.3L VIN M/P, 6.0L VIN H<br>**Models:** All<br>**Transmissions:** All | **O2S Heater Low Side Open Circuit**<br>Control module detected that the O2S heater low control circuit signal indicating an open O2S heater circuit condition.<br>**Possible Causes:**<br>• O2S heater low side control circuit open or shorted to ground<br>• O2S heater power circuit is open (check the PRE O2 fuse)<br>• O2S heater is damaged or it has failed<br>• Control module has failed |
| **DTC: P0037**<br>**2T CCM, MIL: Yes**<br>**Years:** 2005, 2006, 2007, 2008<br>**Engines:** 2.0L VIN 2, 2.0L VIN S/Y, 2.5L SOHC, DOHC VIN 6, 2.8L VIN U<br>**Models:** All<br>**Transmissions:** All | **O2S Heater Control Circuit Low**<br>Control module detected the open or short circuit of heater. The heater performs duty control, and the output terminal voltage at ON is 0 V and the output terminal voltage at OFF is the battery voltage. Control module detects when the terminal voltage remains Low.<br>**Possible Causes:**<br>• Check power supply to rear oxygen sensor<br>• Check ground circuit for Control module<br>• Using a scan tool check current data<br>• Check output signal of Control module<br>• Check rear oxygen sensor |

| DTC | Trouble Code Title, Conditions & Possible Causes |
|---|---|
| **DTC: P0038**<br>**Years:** 2005, 2006, 2007, 2008<br>**Engines:** 2.0L VIN 2, 2.0L VIN S/Y, 2.5L SOHC, DOHC VIN 6, 2.8L VIN U<br>**Models:** All<br>**Transmissions:** All | **O2S Heater Control Circuit High**<br>Control module detected the open or short circuit of heater. The heater performs duty control, and the output terminal voltage at ON is 0 V and the output terminal voltage at OFF is the battery voltage. Control module detects when the terminal voltage remains High.<br>**Possible Causes:**<br>• Check input signal of Control module<br>• Using a scan tool check current data<br>• Check poor contact |
| **DTC: P0050**<br>**2T CCM, MIL: Yes**<br>**Years:** 2005, 2006, 2007, 2008<br>**Engines:** 2.5L SOHC, DOHC VIN 6, 5.3L VIN M/P, 6.0L VIN H<br>**Models:** All<br>**Transmissions:** All | **O2S Heater Control Circuit**<br>Control module detected functional errors of the oxygen sensor heater. Poor drive ability.<br>**Possible Causes:**<br>• Check harness between Control module and front oxygen sensor connector<br>• Check harness between main relay and front oxygen sensor connector<br>• Check front oxygen sensor<br>• Check poor contact |
| **DTC: P0051**<br>**2T CCM, MIL: Yes**<br>**Years:** 2005, 2006, 2007, 2008<br>**Engines:** 2.5L SOHC, DOHC VIN 6<br>**Models:** All<br>**Transmissions:** All | **O2S Heater Control Circuit Low**<br>Control module detected the open or short circuit of heater. The heater performs duty control, and the output terminal voltage at ON is 0 V and the output terminal voltage at OFF is the battery voltage. Control module detects when the terminal voltage remains Low.<br>**Possible Causes:**<br>• Check power supply to front oxygen sensor<br>• Check ground circuit for Control module<br>• Using a scan tool check current data<br>• Check output signal of Control module<br>• Check front oxygen sensor |
| **DTC: P0052**<br>**2T CCM, MIL: Yes**<br>**Years:** 2005, 2006, 2007, 2008<br>**Engines:** 2.5L SOHC, DOHC VIN 6<br>**Models:** All<br>**Transmissions:** All | **O2S Heater Control Circuit High**<br>Control module detected the open or short circuit of heater. The heater performs duty control, and the output terminal voltage at ON is 0 V and the output terminal voltage at OFF is the battery voltage. Control module detects when the terminal voltage remains High.<br>**Possible Causes:**<br>• Check output signal of Control module<br>• Check front oxygen sensor heater current<br>• Check output signal of Control module<br>• Faulty Control module |
| **DTC: P0053**<br>**2T CCM, MIL: Yes**<br>**Years:** 2005, 2006, 2007, 2008<br>**Engines:** 2.8L VIN U, 4.2L VIN S, 5.3L VIN M/P, 6.0L VIN H<br>**Models:** All<br>**Transmissions:** All | **O2S Heater Resistance**<br>Control module detected the open or short circuit of heater. The heater performs duty control, and the output terminal voltage at ON is 0 V and the output terminal voltage at OFF is the battery voltage. Control module detects when the terminal voltage remains High.<br>**Possible Causes:**<br>• Check output signal of Control module<br>• Check front oxygen sensor heater current<br>• Check output signal of Control module<br>• Faulty Control module |
| **DTC: P0054**<br>**2T CCM, MIL: Yes**<br>**Years:** 2005, 2006, 2007, 2008<br>**Engines:** 4.2L VIN S, 5.3L VIN M/P, 6.0L VIN H<br>**Models:** All<br>**Transmissions:** All | **O2S Heater Resistance**<br>Control module detected the open or short circuit of heater. The heater performs duty control, and the output terminal voltage at ON is 0 V and the output terminal voltage at OFF is the battery voltage. Control module detects when the terminal voltage remains High.<br>**Possible Causes:**<br>• Check output signal of Control module<br>• Check front oxygen sensor heater current<br>• Check output signal of Control module<br>• Faulty Control module |
| **DTC: P0056**<br>**2T CCM, MIL: Yes**<br>**Years:** 2005, 2006, 2007, 2008<br>**Engines:** 5.3L VIN M/P, 6.0L VIN H<br>**Models:** All<br>**Transmissions:** All | **O2S Heater Low Side Open Circuit**<br>Engine started and Control module detected that the O2S heater low control circuit had an open O2S heater circuit condition.<br>**Possible Causes**<br>• O2S heater low side control circuit open or shorted to ground<br>• O2S heater power circuit is open (check the POST O2 fuse)<br>• O2S heater is damaged or it has failed<br>• Control module has failed |

| DTC | Trouble Code Title, Conditions & Possible Causes |
|---|---|
| **DTC: P0057**<br>**2T CCM, MIL: Yes**<br>**Years:** 2005, 2006, 2007, 2008<br>**Engines:** 2.5L SOHC, DOHC VIN 6<br>**Models:** All<br>**Transmissions:** All | **O2S Heater Control Circuit Low**<br>Control module detected the open or short circuit of heater. The heater performs duty control, and the output terminal voltage at ON is 0 V and the output terminal voltage at OFF is the battery voltage. Control module detects when the terminal voltage remains Low. Poor drive ability.<br>**Possible Causes:**<br>• Check power supply to rear oxygen sensor<br>• Check ground circuit for Control module<br>• Using a scan tool check current data<br>• Check output signal of Control module<br>• Check rear oxygen sensor |
| **DTC: P0058**<br>**2T CCM, MIL: Yes**<br>**Years:** 2005, 2006, 2007, 2008<br>**Engines:** 2.5L SOHC, DOHC VIN 6<br>**Models:** All<br>**Transmissions:** All | **O2S Heater Control Circuit High**<br>Control module detected the open or short circuit of heater. The heater performs duty control, and the output terminal voltage at ON is 0 V and the output terminal voltage at OFF is the battery voltage. Control module detects when the terminal voltage remains High. Poor drive ability.<br>**Possible Causes:**<br>• Check input signal of Control module<br>• Using a scan tool check current data<br>• Check poor contact |
| **DTC: P0059**<br>**2T CCM, MIL: Yes**<br>**Years:** 2005, 2006, 2007, 2008<br>**Engines:** 5.3L VIN M/P, 6.0L VIN H<br>**Models:** All<br>**Transmissions:** All | **O2S Heater Resistance**<br>Control module detected the open or short circuit of heater. The heater performs duty control, and the output terminal voltage at ON is 0 V and the output terminal voltage at OFF is the battery voltage. Control module detects when the terminal voltage remains High.<br>**Possible Causes:**<br>• Check output signal of Control module<br>• Check front oxygen sensor heater current<br>• Check output signal of Control module<br>• Faulty Control module |
| **DTC: P0060**<br>**2T CCM, MIL: Yes**<br>**Years:** 2005, 2006, 2007, 2008<br>**Engines:** 5.3L VIN M/P, 6.0L VIN H<br>**Models:** All<br>**Transmissions:** All | **O2S Heater Resistance**<br>Control module detected the open or short circuit of heater. The heater performs duty control, and the output terminal voltage at ON is 0 V and the output terminal voltage at OFF is the battery voltage. Control module detects when the terminal voltage remains High.<br>**Possible Causes:**<br>• Check output signal of Control module<br>• Check front oxygen sensor heater current<br>• Check output signal of Control module<br>• Faulty Control module |
| **DTC: P0068**<br>**2T CCM, MIL: Yes**<br>**Years:** 2005, 2006, 2007, 2008<br>**Engines:** 2.0L VIN 2, 2.5L SOHC, DOHC VIN 6, 4.2L VIN S, 5.3L VIN M/P, 6.0L VIN H<br>**Models:** All<br>**Transmissions:** All | **MAP Sensor Range/Performance**<br>Control module detected problems in the intake manifold pressure sensor output properties. Determining from the engine condition, if the intake manifold pressure AD value is small even when operating conditions suggest it should be large, or if the intake manifolds pressure AD value is large even though the operating condition suggests it should be small, this is judged as being NG.<br>**Possible Causes:**<br>• Check any other DTC on display<br>• Check condition of the sensor<br>• Check condition of the throttle body |
| **DTC: P0076**<br>**1T CCM, MIL: Yes**<br>**Years:** 2005, 2006, 2007, 2008<br>**Engines:** 2.5L SOHC, DOHC VIN 6<br>**Models:** All<br>**Transmissions:** All | **Intake Valve Control Circuit Low Bank 1**<br>Control module detected the open circuit of the oil switching solenoid valve. Control module detects when the current is small even though the output duty is large. Control module detects when the current is small even though the output duty is large. Poor idling condition<br>**Possible Causes:**<br>• Check harness between Control module and oil switching solenoid valve<br>• Check oil switching solenoid valve |
| **DTC: P0077**<br>**1T CCM, MIL: Yes**<br>**Years:** 2005, 2006, 2007, 2008<br>**Engines:** 2.5L SOHC, DOHC VIN 6<br>**Models:** All<br>**Transmissions:** All | **Intake valve Control Circuit High Bank 1**<br>Control module detected short circuits of the oil switching solenoid valve. Judge as a short NG when the current is large even though the output duty is small. Poor idling condition.<br>**Possible Causes:**<br>• Check harness between Control module and oil switching solenoid valve<br>• Check oil switching solenoid valve |

| DTC | Trouble Code Title, Conditions & Possible Causes |
|---|---|
| **DTC: P0082**<br>**1T CCM, MIL: Yes**<br>**Years:** 2005, 2006, 2007, 2008<br>**Engines:** 2.5L SOHC, DOHC VIN 6<br>**Models:** All<br>**Transmissions:** All | **Intake Valve Control Circuit Low Bank 2**<br>Control module detected the open circuit of the oil switching solenoid valve. Control module detects when the current is small even though the output duty is large. Control module detects when the current is small even though the output duty is large. Poor idling condition.<br>**Possible Causes:**<br>• Check harness between Control module and oil switching solenoid valve<br>• Check oil switching solenoid valve |
| **DTC: P0083**<br>**1T CCM, MIL: Yes**<br>**Years:** 2005, 2006, 2007, 2008<br>**Engines:** 2.5L SOHC, DOHC VIN 6<br>**Models:** All<br>**Transmissions:** All | **Intake Valve Control Circuit High Bank 2**<br>Control module detected short circuits of the oil switching solenoid valve. Judge as a short NG when the current is large even though the output duty is small. Poor idling condition.<br>**Possible Causes:**<br>• Check harness between Control module and oil switching solenoid valve<br>• Check oil switching solenoid valve |
| **DTC: P0087**<br>**1T CCM, MIL: Yes**<br>**Years:** 2005, 2006, 2007, 2008<br>**Engines:** 2.8L VIN U<br>**Models:** All<br>**Transmissions:** All | **Fuel Rail/System Pressure - Too Low**<br>Control module monitors the fuel rail pressure. If the sensor indicates a pressure different than the commanded rail pressure plus a possible transitional overshoot, Control module will set a DTC for fuel rail pressure.<br>**Possible Causes:**<br>• Check the engine oil for fuel contamination<br>• Fuel filter is clogged or restricted<br>• Fuel supply lines between fuel tank and injector pump clogged<br>• Fuel rail pressure sensor signal circuit is shorted to ground<br>• Fuel rail pressure sensor is out-of-calibration or it has failed<br>• Control module has failed |
| **DTC: P0088**<br>**1T CCM, MIL: Yes**<br>**Years:** 2005, 2006, 2007, 2008<br>**Engines:** 2.8L VIN U<br>**Models:** All<br>**Transmissions:** All | **Fuel Rail/System Pressure - Too High**<br>Control module monitors the fuel rail pressure. If the sensor indicates a pressure different than the commanded rail pressure plus a possible transitional overshoot, Control module will set a DTC for fuel rail pressure.<br>**Possible Causes:**<br>• Pressure sensor signal circuit is open<br>• Sensor signal circuit has a high resistance condition<br>• Sensor is damaged or it has failed<br>• Regulator is damaged or it has failed<br>• Control module has failed |
| **DTC: P0089**<br>**1T CCM, MIL: Yes**<br>**Years:** 2005, 2006, 2007, 2008<br>**Engines:** 2.8L VIN U<br>**Models:** All<br>**Transmissions:** All | **Fuel Pressure Regulator 1 Performance**<br>Control module monitors the fuel rail pressure. If the sensor indicates a pressure different than the commanded rail pressure plus a possible transitional overshoot, Control module will set a DTC for fuel rail pressure.<br>**Possible Causes:**<br>• Pressure sensor signal circuit is shorted to ground<br>• Sensor is damaged or it has failed<br>• Regulator is damaged or it has failed<br>• Control module has failed |
| **DTC: P0090**<br>**1T CCM, MIL: Yes**<br>**Years:** 2005, 2006, 2007, 2008<br>**Engines:** 2.8L VIN U<br>**Models:** All<br>**Transmissions:** All | **Fuel Pressure Regulator 1 Control Circuit**<br>Control module monitors the fuel rail pressure. If the sensor indicates a pressure different than the commanded rail pressure plus a possible transitional overshoot, Control module will set a DTC for fuel rail pressure.<br>**Possible Causes:**<br>• Fuel rail pressure regulator control circuit is open or grounded<br>• Fuel rail pressure regulator supply circuit is open or grounded<br>• Fuel rail pressure regulator is damaged or it has failed<br>• Control module has failed |
| **DTC: P0091**<br>**1T CCM, MIL: Yes**<br>**Years:** 2005, 2006, 2007, 2008<br>**Engines:** 2.8L VIN U<br>**Models:** All<br>**Transmissions:** All | **Fuel Pressure Regulator 1 Control Circuit Low**<br>Control module monitors the fuel rail pressure. If the sensor indicates a pressure different than the commanded rail pressure plus a possible transitional overshoot, Control module will set a DTC for fuel rail pressure.<br>**Possible Causes:**<br>• Fuel rail pressure regulator control circuit is open or grounded<br>• Fuel rail pressure regulator supply circuit is open or grounded<br>• Fuel rail pressure regulator is damaged or it has failed<br>• Control module has failed |

| DTC | Trouble Code Title, Conditions & Possible Causes |
|---|---|
| **DTC: P0092**<br>**1T CCM, MIL: Yes**<br>**Years:** 2005, 2006, 2007, 2008<br>**Engines:** 2.8L VIN U<br>**Models:** All<br>**Transmissions:** All | **Fuel Pressure Regulator 1 Control Circuit High**<br>Control module monitors the fuel rail pressure. If the sensor indicates a pressure different than the commanded rail pressure plus a possible transitional overshoot, Control module will set a DTC for fuel rail pressure.<br>**Possible Causes:**<br>• Fuel rail pressure regulator control circuit is open or grounded<br>• Fuel rail pressure regulator supply circuit is open or grounded<br>• Fuel rail pressure regulator is damaged or it has failed<br>• Control module has failed |
| **DTC P0100**<br>**2T CCM, MIL: Yes**<br>**Years:** 2005, 2006, 2007, 2008<br>**Engines:** 2.3L VIN E/G<br>**Models:** All<br>**Transmissions:** All | **Mass or Volume Air Flow Circuit Malfunction**<br>Engine speed over 500 rpm, system voltage over 11v, and Control module detected the MAF sensor was less than a preset minimum voltage amount for 1 second. The Control module uses Speed Density fuel calculation when this code sets.<br>**Possible Causes**<br>• MAF sensor signal circuit is open or shorted to ground<br>• MAF sensor power circuit is open<br>• MAF sensor is damaged (it may have dirt on it) or has failed<br>• Control module has failed |
| **DTC P0101**<br>**2T CCM, MIL: Yes**<br>**Years:** 2005, 2006, 2007, 2008<br>**Engines:** All<br>**Models:** All<br>**Transmissions:** All | **Mass or Volume Air Flow Circuit Range/Performance**<br>Engine coolant temperature is 167°F. (75°C). Control module detected for abnormalities in the air flow sensor output properties. Judge as a low side NG when the air flow voltage indicates a small value regardless of running in a state where the air flow voltage increases. Judge as a high side NG when the air flow voltage indicates a large value regardless of running in a state where the air flow voltage decreases. Judge air flow sensor property NG when the Low side or High side becomes NG. Poor idling condition. Engine stalls. Poor driving performance.<br>**Possible Causes:**<br>• Check any other DTC on display<br>• Faulty Mass Air Flow (MAF) sensor |
| **DTC: P0102**<br>**2T CCM, MIL: Yes**<br>**Years:** 2005, 2006, 2007, 2008<br>**Engines:** All<br>**Models:** All<br>**Transmissions:** All | **Mass or Volume Air Flow Circuit Low Input**<br>Control module detects open or short circuits of the air flow sensor. Engine stalls. Poor idling condition. Poor driving performance.<br>**Possible Causes:**<br>• Using a scan tool, check current data<br>• Check input signal of Control module<br>• Check power supply to sensor<br>• Check harness between Control module and sensor connector<br>• Check for poor contacts<br>• Faulty Mass Air Flow (MAF) sensor |
| **DTC: P0103**<br>**1T CCM, MIL: Yes**<br>**Years:** 2005, 2006, 2007, 2008<br>**Engines:** All<br>**Models:** All<br>**Transmissions:** All | **Mass or Volume Air Flow Circuit High Input**<br>Control module detects open or short circuits of the air flow sensor. Engine stalls. Poor idling condition. Poor driving performance.<br>**Possible Causes:**<br>• Using a scan tool, check current data<br>• Check harness between Control module and sensor connector<br>• Check for poor contact<br>• Faulty Mass Air Flow (MAF) sensor |
| **DTC: P0106**<br>**2T CCM, MIL: Yes**<br>**Years:** 2005, 2006, 2007, 2008<br>**Models:** All<br>**Engines:** 2.0L VIN S/Y, 2.3L VIN E/G, 4.2L VIN S, 5.3L VIN M/P, 6.0L VIN H<br>**Transmissions:** All | **Manifold Absolute Pressure (MAP) Sensor Performance**<br>The Engine Control Module detects that the MAP sensor pressure is not within range of the calculated pressure that is derived from the system of models for more than 0.5 second.<br>**Possible Causes:**<br>• MAP sensor VREF circuit shorted to voltage or high resistance<br>• MAP sensor signal circuit shorted to voltage or high resistance<br>• MAP sensor low reference open or high resistance<br>• MAP sensor is damaged or has failed<br>• Control module has failed |
| **DTC: P0107**<br>**1T CCM, MIL: Yes**<br>**Years:** 2005, 2006, 2007, 2008<br>**Engines:** 2.0L VIN 2, 2.0L VIN S/Y, 2.3L VIN E/G, 2.5L SOHC, DOHC VIN 6, 4.2L VIN S, 5.3L VIN M/P, 6.0L VIN H<br>**Models:** All<br>**Transmissions:** All | **Manifold Absolute pressure/Barometric Pressure Circuit Low Input**<br>Control module detects the open or short circuit of intake manifold pressure sensor.<br>**Possible Causes:**<br>• Using a scan tool, check current data<br>• Check for poor contacts<br>• Check output signal of Control module<br>• Check input signal of Control module<br>• Check harness between Control module and sensor connector<br>• Faulty MAP sensor |

| DTC | Trouble Code Title, Conditions & Possible Causes |
|---|---|
| **DTC: P0108**<br>**1T CCM, MIL: Yes**<br>**Years:** 2005, 2006, 2007, 2008<br>**Engines:** 2.0L VIN 2, 2.0L VIN S/Y, 2.3L VIN E/G, 2.5L SOHC, DOHC VIN 6, 4.2L VIN S, 5.3L VIN M/P, 6.0L VIN H<br>**Models:** All<br>**Transmissions:** All | **Manifold Absolute pressure/Barometric Pressure Circuit High Input**<br>Control module detects the open or short circuit of intake manifold pressure sensor.<br>**Possible Causes:**<br>• Check output signal of Control module<br>• Check input signal of Control module<br>• Check harness between Control module and sensor connector<br>• Check harness between the sensor connector<br>• Check poor contact<br>• Faulty MAP sensor |
| **DTC: P0111**<br>**2T CCM, MIL: Yes**<br>**Years:** 2005, 2006, 2007, 2008<br>**Engines:** 2.0L VIN 2, 2.0L VIN S/Y, 2.3L VIN E/G, 2.5L SOHC, DOHC VIN 6, 2.8L VIN U<br>**Models:** All<br>**Transmissions:** All | **Intake Air Temperature Circuit Range/Performance**<br>Control module detects the malfunction of intake air temperature sensor output property. Control module detects when the intake air temperature is not varied whereas it seemed to be varied from the viewpoint of engine condition. Poor idling condition. Poor driving performance.<br>**Possible Causes:**<br>• Check any other DTC on display<br>• Check engine coolant temperature |
| **DTC: P0112**<br>**2T CCM, MIL: Yes**<br>**Years:** 2005, 2006, 2007, 2008<br>**Engines:** All<br>**Models:** All<br>**Transmissions:** All | **Intake Air Temperature Circuit Low Input**<br>Control module detected open or short circuit of the intake air temperature sensor. Poor idling condition. Poor driving performance.<br>**Possible Causes:**<br>• Using a scan tool, check current data<br>• Check the harness between the sensor and Control module connector<br>• Check for poor contacts<br>• Faulty IAT sensor |
| **DTC: P0113**<br>**1T CCM, MIL: Yes**<br>**Years:** 2005, 2006, 2007, 2008<br>**Engines:** 2.0L VIN S/Y, 2.3L VIN E/G<br>**Models:** All<br>**Transmissions:** All | **Intake Air Temperature Circuit High Input**<br>Control module detected open or short circuit of the intake air temperature sensor. Poor idling condition. Poor driving performance.<br>**Possible Causes:**<br>• Using a scan tool, check current data<br>• Check the harness between the sensor and Control module connector<br>• Check for poor contacts<br>• Faulty IAT sensor |
| **DTC: P0115**<br>**1T CCM, MIL: Yes**<br>**Years:** 2005, 2006, 2007, 2008<br>**Engines:** 2.0L VIN S/Y, 2.5L SOHC, DOHC VIN 6, 2.3L VIN E/G<br>**Models:** All<br>**Transmissions:** All | **Engine Coolant Temperature Circuit Problem**<br>Control module detects the open or short circuit of the engine coolant temperature sensor. Hard starting condition. Poor idling condition. Poor driving performance.<br>**Possible Causes:**<br>• Using a scan tool, check current data<br>• Check harness between sensor and Control module connector<br>• Check for poor contacts<br>• Faulty ECT sensor |
| **DTC: P0116**<br>**2T CCM, MIL: Yes**<br>**Years: 2005, 2006, 2007**<br>**Models:** All<br>**Engines:** 2.3L VIN E/G, 2.8L VIN U, 4.2L VIN S, 5.3L VIN M/P, 6.0L VIN H<br>**Transmissions:** All | **Engine Coolant Temperature (ECT) Sensor Performance**<br>Control module detects a temperature difference at power-up that indicates that the ECT sensor is more than the IAT sensor.<br>OR. Control module detects a temperature difference at power-up that indicates that the ECT sensor is more than the IAT sensor<br>**Possible Causes:**<br>• ECT sensor circuit has an intermittent high resistance condition<br>• ECT sensor circuit has an intermittent grounded condition<br>• ECT sensor is out of calibration or "skewed" high<br>• IAT sensor is out of calibration or it is "skewed" high or low<br>• Control module has failed |
| **DTC: P0117**<br>**1T CCM, MIL: Yes**<br>**Years:** 2005, 2006, 2007, 2008<br>**Engines:** All<br>**Models:** All<br>**Transmissions:** All | **Engine Coolant Temperature Sensor Circuit Low Input**<br>Control module detects the open or short circuit of the engine coolant temperature sensor. Hard starting condition. Poor idling condition. Poor driving performance.<br>**Possible Causes:**<br>• Using a scan tool, check current data<br>• Check harness between sensor and Control module connector<br>• Check for poor contacts<br>• Faulty ECT sensor |

| DTC | Trouble Code Title, Conditions & Possible Causes |
|---|---|
| **DTC: P0118**<br>**1T CCM, MIL: Yes**<br>**Years:** 2005, 2006, 2007, 2008<br>**Engines:** All<br>**Models:** All<br>**Transmissions:** All | **Engine Coolant Temperature Sensor Circuit High Input**<br>Control module detects the open or short circuit of the engine coolant temperature sensor. Hard starting condition. Poor idling condition. Poor driving performance.<br>**Possible Causes:**<br>• Using a scan tool, check current data<br>• Check harness between sensor and Control module connector<br>• Check for poor contacts<br>• Faulty ECT sensor |
| **DTC: P0119**<br>**1T CCM, MIL: Yes**<br>**Years:** 2005, 2006, 2007, 2008<br>**Engines:** 2.0L VIN S/Y, 2.3L VIN E/G<br>**Models:** All<br>**Transmissions:** All | **Engine Coolant Temperature Sensor Circuit Intermittent**<br>Control module detects the open or short circuit of the engine coolant temperature sensor. Hard starting condition. Poor idling condition. Poor driving performance.<br>**Possible Causes:**<br>• Using a scan tool, check current data<br>• Check harness between sensor and Control module connector<br>• Check for poor contacts<br>• Faulty ECT sensor |
| **DTC: P0120**<br>**1T CCM, MIL: Yes**<br>**Years:** 2005, 2006, 2007<br>**Models:** All<br>**Engines:** 4.2L VIN S, 5.3L VIN M/P, 6.0L VIN H<br>**Transmissions:** All | **Throttle Position (TP) Sensor Performance**<br>The predicted air flow and the predicted MAP combined are outside a calibrated range for more than 3 seconds.<br>**Possible Causes:**<br>• Vacuum hoses for splits, kinks, and proper connections<br>• Throttle body for dirt, debris, and coking<br>• Air leaks at throttle body mounting area and intake manifold sealing surfaces<br>• Faulty MAP sensor or circuit<br>• Faulty MAF sensor or circuit<br>• Faulty throttle body assembly |
| **DTC: P0121**<br>**1T CCM, MIL: Yes**<br>**Years:** 2005, 2006, 2007<br>**Models:** All<br>**Engines:** 2.0L VIN 2, 2.0L VIN S/Y, 2.8L VIN U, 4.2L VIN S, 5.3L VIN M/P, 6.0L VIN H<br>**Transmissions:** All | **Throttle Position (TP) Sensor Performance**<br>The predicted air flow and the predicted MAP combined are outside a calibrated range for more than 3 seconds.<br>**Possible Causes:**<br>• Vacuum hoses for splits, kinks, and proper connections<br>• Throttle body for dirt, debris, and coking<br>• Air leaks at throttle body mounting area and intake manifold sealing surfaces<br>• Faulty MAP sensor or circuit<br>• Faulty MAF sensor or circuit<br>• Faulty throttle body assembly |
| **DTC: P0122**<br>**1T CCM, MIL: Yes**<br>**Years:** 2005, 2006, 2007, 2008<br>**Engines:** 2.0L VIN 2, 2.0L VIN S/Y, 2.5L SOHC, DOHC VIN 6, 2.8L VIN U, 4.2L VIN S, 5.3L VIN M/P, 6.0L VIN H<br>**Models:** All<br>**Transmissions:** All | **Throttle/Pedal Position Sensor Circuit Low Input**<br>Control module detects the open or short circuit of throttle position Sensor. Poor idling condition. Engine stalls Poor driving performance<br>**Possible Causes:**<br>• Check sensor output<br>• Check for poor contact<br>• Check harness between Control module and electronic throttle control<br>• Faulty electronic throttle control<br>• Faulty Control module |
| **DTC: P0123**<br>**1T CCM, MIL: Yes**<br>**Years:** 2005, 2006, 2007, 2008<br>**Engines:** 2.0L VIN 2, 2.0L VIN S/Y, 2.5L SOHC, DOHC VIN 6, 2.8L VIN U, 4.2L VIN S, 5.3L VIN M/P, 6.0L VIN H<br>**Models:** All<br>**Transmissions:** All | **Throttle/Pedal Position Sensor Circuit Low Input**<br>Control module detects the open or short circuit of throttle position Sensor. Poor idling condition. Engine stalls Poor driving performance.<br>**Possible Causes:**<br>• Check sensor output<br>• Check for poor contact<br>• Check harness between Control module and electronic throttle control<br>• Faulty electronic throttle control<br>• Faulty Control module |
| **DTC: P0125**<br>**2T CCM, MIL: Yes**<br>**Years:** 2005, 2006, 2007, 2008<br>**Engines:** 2.0L VIN 2, 2.0L VIN S/Y, 2.3L VIN E/G, 2.5L SOHC, DOHC VIN 6<br>**Models:** All<br>**Transmissions:** All | **Insufficient Coolant Temperature For Closed loop Fuel Control**<br>Control module detects the malfunction of engine coolant temperature output property. Engine does not return to normal idle.<br>**Possible Causes:**<br>• Check any other DTC on display<br>• Check tire size<br>• Check engine coolant<br>• Check thermostat<br>• Faulty ECT sensor |

| DTC | Trouble Code Title, Conditions & Possible Causes |
|---|---|
| **DTC: P0126**<br>**2T CCM, MIL: Yes**<br>**Years:** 2005, 2006, 2007, 2008<br>**Engines:** 2.0L VIN 2, 2.0L VIN S/Y, 2.3L VIN E/G, 2.5L SOHC, DOHC VIN 6<br>**Models:** All<br>**Transmissions:** All | **Insufficient Coolant Temperature For Stable Operation**<br>Control module detected the malfunction of the engine coolant temperature sensor characteristics.<br>**Possible Causes:**<br>• Check any other DTC on display<br>• Check Engine Coolant Temperature (ECT) sensor and Control module connectors<br>• Faulty ECT sensor |
| **DTC: P0128**<br>**2T CCM, MIL: Yes**<br>**Years:** 2005, 2006, 2007, 2008<br>**Engines:** 2.0L VIN 2, 2.0L VIN S/Y, 2.5L SOHC, DOHC VIN 6, 2.8L VIN U<br>**Models:** All<br>**Transmissions:** All | **Coolant Temperature Thermostat (coolant temperature below thermostat regulating temperature)**<br>Control module detects malfunctions of the thermostat function.<br>**Possible Causes:**<br>• Open or missing thermostat<br>• Check any other DTC on display<br>• Check engine coolant for proper mixture<br>• Check radiator fan (Stuck On) |
| **DTC: P0130**<br>**1T CCM, MIL: Yes**<br>**Years:** 2005, 2006, 2007, 2008<br>**Engines:** 2.0L VIN 2, 2.5L SOHC, DOHC VIN 6, 2.8L VIN U<br>**Models:** All<br>**Transmissions:** All | **O2 Sensor Circuit**<br>Control module detected an unexpected voltage condition on the A/F sensor signal circuit for over 6 seconds during the CCM test.<br>**Possible Causes:**<br>• Check harness between Control module and sensor<br>• Shorted ground circuit of harness between Control module and front oxygen A/F sensor connector<br>• Check for poor contacts<br>• Faulty front oxygen A/F sensor |
| **DTC: P0131**<br>**1T CCM, MIL: Yes**<br>**Years:** 2005, 2006, 2007, 2008<br>**Engines:** All<br>**Models:** All<br>**Transmissions:** All | **O2 Sensor Circuit Low Voltage**<br>Control module detected an unexpected "low" voltage condition on the A/F sensor signal circuit for over 6 seconds during the CCM test.<br>**Possible Causes:**<br>• Check harness between Control module and sensor<br>• Shorted ground circuit of harness between Control module and front oxygen A/F sensor connector<br>• Check for poor contacts<br>• Faulty front oxygen A/F sensor |
| **DTC: P0132**<br>**2T CCM, MIL: Yes**<br>**Years:** 2005, 2006, 2007, 2008<br>**Engines:** 2.0L VIN S/Y, 2.3L VIN E/G, 2.5L SOHC, DOHC VIN 6, 2.8L VIN U<br>**Models:** All<br>**Transmissions:** All | **O2 Sensor Circuit High Voltage**<br>Control module detected an unexpected "high" voltage condition on the A/F sensor signal circuit for over 6 seconds during the CCM test.<br>**Possible Causes:**<br>• Check harness between Control module and sensor<br>• Shorted power circuit of harness between Control module and front oxygen A/F sensor connector<br>• Check for poor contacts<br>• Faulty front oxygen A/F sensor |
| **DTC: P0133**<br>**2T CCM, MIL: Yes**<br>**Years:** 2005, 2006, 2007, 2008<br>**Engines:** All<br>**Models:** All<br>**Transmissions:** All | **O2 Sensor Circuit Slow Response**<br>Control module detected the number of A/F rich-to-lean or lean-to-rich switches was less than a calibrated amount.<br>**Possible Causes:**<br>• Check any other DTC on display<br>• Check exhaust system<br>• Faulty front oxygen A/F sensor |
| **DTC: P0134**<br>**1T CCM, MIL: Yes**<br>**Years:** 2005, 2006, 2007, 2008<br>**Engines:** All<br>**Models:** All<br>**Transmissions:** All | **O2 Sensor Circuit No Activity Detected**<br>Control module detects open circuits of the sensor. Control module detects when the impedance of the element is large. Poor drive ability.<br>**Possible Causes:**<br>• Check harness between Control module and sensor connector<br>• Check poor contact<br>• Faulty front oxygen A/F sensor |
| **DTC: P0135**<br>**1T CCM, MIL: Yes**<br>**Years:** 2005, 2006, 2007, 2008<br>**Engines:** All<br>**Models:** All<br>**Transmissions:** All | **O2 Sensor Circuit No Activity Detected**<br>Control module detects open circuits of the sensor. Control module detects when the impedance of the element is large. Poor drive ability.<br>**Possible Causes:**<br>• Check harness between Control module and sensor connector<br>• Check poor contact<br>• Faulty front oxygen A/F sensor |

| DTC | Trouble Code Title, Conditions & Possible Causes |
|---|---|
| **DTC: P0137**<br>**2T CCM, MIL: Yes**<br>**Years:** 2005, 2006, 2007, 2008<br>**Engines:** All<br>**Models:** All<br>**Transmissions:** All | **O2 Sensor Circuit Low Voltage**<br>Control module detected an unexpected low voltage condition on the O2S circuit during the CCM test period.<br>**Possible Causes:**<br>• Check any other DTC on display<br>• Using the a scan tool, check current data<br>• Check Harness between Control module and rear oxygen sensor connector<br>• Check exhaust system<br>• Faulty rear oxygen sensor |
| **DTC: P0138**<br>**2T CCM, MIL: Yes**<br>**Years:** 2005, 2006, 2007, 2008<br>**Engines:** All<br>**Models:** All<br>**Transmissions:** All. | **O2 Sensor Circuit High Voltage**<br>Control module detected an unexpected high voltage condition on the O2S circuit during the CCM test period.<br>**Possible Causes:**<br>• Check any other DTC on display<br>• Using a scan tool, check current data<br>• Check Harness between Control module and rear oxygen sensor connector<br>• Check exhaust system<br>• Faulty rear oxygen sensor |
| **DTC: P0139**<br>**2T CCM, MIL: Yes**<br>**Years:** 2005, 2006, 2007, 2008<br>**Engines:** 2.0L VIN 2, 2.5L SOHC, DOHC VIN 6, 2.8L VIN U<br>**Models:** All<br>**Transmissions:** All | **O2 Sensor Circuit Slow Response**<br>Control module detected the number of O2S rich-to-lean or lean-to-rich switches was less than a calibrated amount.<br>**Possible Causes:**<br>• Check any other DTC on display<br>• Check Harness between Control module and rear oxygen sensor connector<br>• Check sensor<br>• Faulty rear oxygen sensor |
| **DTC: P0140**<br>**2T CCM, MIL: Yes**<br>**Years:** 2005, 2006, 2007, 2008<br>**Engines:** All<br>**Models:** All<br>**Transmissions:** All | **O2 Sensor Circuit No Activity Detected**<br>Control module detects open circuits of the sensor. Control module detects when the impedance of the element is large. Poor drive ability.<br>**Possible Causes:**<br>• Check any other DTC on display<br>• Using a scan tool, check current data<br>• Check Harness between Control module and rear oxygen sensor connector<br>• Check exhaust system<br>• Faulty rear oxygen sensor |
| **DTC: P0141**<br>**2T CCM, MIL: Yes**<br>**Years:** 2005, 2006, 2007, 2008<br>**Engines:** 2.0L VIN 2, 2.0L VIN S/Y, 2.5L SOHC, DOHC VIN 6, 2.8L VIN U, 4.2L VIN S, 5.3L VIN M/P, 6.0L VIN H<br>**Models:** All<br>**Transmissions:** All | **O2 Sensor Heater Circuit**<br>Engine started and Control module detected that the O2S heater circuit signal was between 2.6v to 4.6v with the heater commanded off, indicating an open O2S heater circuit condition.<br>**Possible Causes:**<br>• O2S heater low side control circuit open or shorted to ground<br>• O2S heater power circuit is open (check the PRE O2 fuse)<br>• O2S heater is damaged or it has failed<br>• Control module has failed |
| **DTC: P0151**<br>**1T CCM, MIL: Yes**<br>**Years:** 2005, 2006, 2007, 2008<br>**Engines:** 5.3L VIN M/P, 6.0L VIN H<br>**Models:** All<br>**Transmissions:** All | **O2 Sensor Circuit Low Voltage**<br>Control module detected an unexpected "low" voltage condition on the front A/F sensor signal circuit for over 6 seconds during the CCM test.<br>**Possible Causes:**<br>• Check any other DTC on display<br>• Using a scan tool, check current data<br>• Check the harness between Control module and the front sensor connector<br>• Faulty front oxygen A/F sensor |
| **DTC: P0152**<br>**1T CCM, MIL: Yes**<br>**Years:** 2005, 2006, 2007, 2008<br>**Engines:** 5.3L VIN M/P, 6.0L VIN H<br>**Models:** All<br>**Transmissions:** All | **O2 Sensor Circuit High Voltage**<br>Control module detected an unexpected "high" voltage condition on the front A/F sensor signal circuit for over 6 seconds in the CCM test.<br>**Possible Causes:**<br>• Check any other DTC on display<br>• Using a scan tool, check current data<br>• Check the harness between Control module and the front sensor connector<br>• Faulty front oxygen A/F sensor |

| DTC | Trouble Code Title, Conditions & Possible Causes |
|---|---|
| **DTC: P0153**<br>**2T CCM, MIL: Yes**<br>**Years:** 2005, 2006, 2007, 2008<br>**Engines:** 5.3L VIN M/P, 6.0L VIN H<br>**Models:** All<br>**Transmissions:** All | **O2 Sensor Circuit Slow Response**<br>Control module detected the number of A/F S-21 rich-to-lean or lean-to-rich switches was less than a calibrated amount.<br>**Possible Causes:**<br>• Check any other DTC on display<br>• Check exhaust system<br>• Faulty front oxygen A/F sensor |
| **DTC: P0154**<br>**1T CCM, MIL: Yes**<br>**Years:** 2005, 2006, 2007, 2008<br>**Engines:** 5.3L VIN M/P, 6.0L VIN H<br>**Models:** All<br>**Transmissions:** All | **O2 Sensor Circuit No Activity Detected**<br>Control module detects open circuits of the sensor. Control module detects when the impedance of the element is large. Poor drive ability.<br>**Possible Causes:**<br>• Check the harness between Control module and the front sensor<br>• Check poor contact<br>• Faulty front oxygen A/F sensor |
| **DTC: P0155**<br>**2T CCM, MIL: Yes**<br>**Years:** 2005, 2006, 2007, 2008<br>**Engines:** 5.3L VIN M/P, 6.0L VIN H<br>**Models:** All<br>**Transmissions:** All | **O2 Sensor Heater Circuit**<br>Engine started and Control module detected that the O2S heater circuit signal was between 2.6v to 4.6v with the heater commanded off, indicating an open O2S heater circuit condition.<br>**Possible Causes:**<br>• O2S heater low side control circuit open or shorted to ground<br>• O2S heater power circuit is open (check the PRE O2 fuse)<br>• O2S heater is damaged or it has failed<br>• Control module has failed |
| **DTC: P0157**<br>**2T CCM, MIL: Yes**<br>**Years:** 2005, 2006, 2007, 2008<br>**Engines:** 5.3L VIN M/P, 6.0L VIN H<br>**Models:** All<br>**Transmissions:** All | **O2 Sensor Circuit Low Voltage**<br>Control module detected an unexpected low voltage condition on the O2S circuit during the CCM test period Poor drive ability.<br>**Possible Causes:**<br>• Check any other DTC on display<br>• Using a scan tool, check data<br>• Check harness between Control module and rear sensor connector<br>• Check harness between rear sensor and Control module connector<br>• Check exhaust system<br>• Faulty rear oxygen sensor |
| **DTC: P0158**<br>**2T CCM, MIL: Yes**<br>**Years:** 2005, 2006, 2007, 2008<br>**Engines:** 5.3L VIN M/P, 6.0L VIN H<br>**Models:** All<br>**Transmissions:** All | **O2 Sensor Circuit High Voltage**<br>Control module detected an unexpected high voltage condition on the O2S circuit during the CCM test period.<br>**Possible Causes:**<br>• Check any other DTC on display<br>• Using a scan tool, check data<br>• Check harness between Control module and rear sensor connector<br>• Check harness between rear sensor and Control module connector<br>• Check exhaust system<br>• Faulty rear oxygen sensor |
| **DTC: P0159**<br>**2T CCM, MIL: Yes**<br>**Years:** 2005, 2006, 2007, 2008<br>**Engines:** 5.3L VIN M/P, 6.0L VIN H<br>**Models:** All<br>**Transmissions:** All | **O2 Sensor Circuit Slow Response**<br>Control module detected the number of O2S rich-to-lean or lean-to-rich switches was less than a calibrated amount. Poor drive ability.<br>**Possible Causes:**<br>• Check any other DTC on display<br>• Check harness between Control module and rear sensor connector<br>• Faulty rear oxygen sensor |
| **DTC: P0160**<br>**2T CCM, MIL: Yes**<br>**Years:** 2005, 2006, 2007, 2008<br>**Engines:** 5.3L VIN M/P, 6.0L VIN H<br>**Models:** All<br>**Transmissions:** All | **O2 Sensor Circuit No Activity Detected**<br>Control module detects open circuits of the sensor. Control module detects when the impedance of the element is large. Poor drive ability.<br>**Possible Causes:**<br>• Check any other DTC on display<br>• Using a scan tool, check data<br>• Check harness between Control module and rear sensor connector<br>• Check harness between rear sensor and Control module connector<br>• Check exhaust system<br>• Faulty rear oxygen sensor |

| DTC | Trouble Code Title, Conditions & Possible Causes |
|---|---|
| **DTC: P0171**<br>**2T CCM, MIL: Yes**<br>**Years:** 2005, 2006, 2007, 2008<br>**Engines:** All<br>**Models:** All<br>**Transmissions:** All | **System Too Lean Bank 1**<br>The Fuel system is diagnosed by comparing the target air fuel ratio calculated by Control module with the actual air fuel ratio measured by sensor. Improper idling. Engine stalls. Poor driving performance.<br>**Possible Causes:**<br>• Check any other DTC on display<br>• Check exhaust system<br>• Check air intake system<br>• Check fuel pressure<br>• Check fuel injector<br>• Check coolant temperature<br>• Check intake manifold pressure sensor<br>• Check intake air temperature sensor |
| **DTC: P0172**<br>**2T CCM, MIL: Yes**<br>**Years:** 2005, 2006, 2007, 2008<br>**Engines:** All<br>**Models:** All<br>**Transmissions:** All | **System Too Rich Bank 1**<br>The Fuel system is diagnosed by comparing the target air fuel ratio calculated by Control module with the actual air fuel ratio measured by sensor. Improper idling. Engine stalls. Poor driving performance.<br>**Possible Causes:**<br>• Check any other DTC on display<br>• Check exhaust system<br>• Check air intake system<br>• Check fuel pressure<br>• Check fuel injector<br>• Check coolant temperature<br>• Check intake manifold pressure sensor<br>• Check intake air temperature sensor |
| **DTC: P0174**<br>**2T CCM, MIL: Yes**<br>**Years:** 2005, 2006, 2007, 2008<br>**Engines:** 4.2L VIN S, 5.3L VIN M/P, 6.0L VIN H<br>**Models:** All<br>**Transmissions:** All | **System Too Lean Bank 2**<br>The Fuel system is diagnosed by comparing the target air fuel ratio calculated by Control module with the actual air fuel ratio measured by sensor. Improper idling. Engine stalls. Poor driving performance.<br>**Possible Causes:**<br>• Check any other DTC on display<br>• Check exhaust system<br>• Check air intake system<br>• Check any other DTC on display<br>• Check exhaust system<br>• Check air intake system<br>• Check fuel pressure<br>• Check fuel injector<br>• Check coolant temperature<br>• Check intake manifold pressure sensor<br>• Check intake air temperature sensor |
| **DTC: P0175**<br>**2T CCM, MIL: Yes**<br>**Years:** 2005, 2006, 2007, 2008<br>**Engines:** 4.2L VIN S, 5.3L VIN M/P, 6.0L VIN H<br>**Models:** All<br>**Transmissions:** All | **System Too Rich Bank 2**<br>The Fuel system is diagnosed by comparing the target air fuel ratio calculated by Control module with the actual air fuel ratio measured by sensor. Improper idling. Engine stalls. Poor driving performance.<br>**Possible Causes:**<br>• Check any other DTC on display<br>• Check exhaust system<br>• Check air intake system<br>• Check fuel pressure<br>• Check fuel injector<br>• Check coolant temperature<br>• Check intake manifold pressure sensor<br>• Check intake air temperature sensor |
| **DTC: P0181**<br>**2T CCM, MIL: Yes**<br>**Years:** 2005, 2006, 2007, 2008<br>**Engines:** 2.0L VIN 2, 2.5L SOHC, DOHC VIN 6<br>**Models:** All<br>**Transmissions:** All | **Fuel Temperature Sensor Circuit Range/Performance**<br>Control module detected the Fuel Temperature Sensor 'A' signal was out-of-range, or the signal was not plausible.<br>**Possible Causes:**<br>• Check any other DTC on display<br>• Faulty fuel temperature sensor |

| DTC | Trouble Code Title, Conditions & Possible Causes |
|---|---|
| **DTC: P0182**<br>**1T CCM, MIL: Yes**<br>**Years:** 2005, 2006, 2007, 2008<br>**Engines:** 2.0L VIN 2, 2.5L SOHC, DOHC VIN 6<br>**Models:** All<br>**Transmissions:** All | **Fuel Temperature Sensor Circuit Low Input**<br>Control module detected the Fuel Temperature Sensor signal was less than 0.10 volts.<br>**Possible Causes:**<br>&bull; Using a scan tool, check data<br>&bull; Shorted ground circuit of harness between Control module and fuel pump<br>&bull; Faulty fuel temperature sensor |
| **DTC: P0183**<br>**1T CCM, MIL: Yes**<br>**Years:** 2005, 2006, 2007, 2008<br>**Engines:** 2.0L VIN 2, 2.5L SOHC, DOHC VIN 6<br>**Models:** All<br>**Transmissions:** All | **Fuel Temperature Sensor Circuit High Input**<br>Control module detected the Fuel Temperature Sensor signal was more than 4.80 volts.<br>**Possible Causes:**<br>&bull; Using a scan tool, check data<br>&bull; Shorted ground circuit of harness between Control module and fuel pump<br>&bull; Faulty fuel temperature sensor |
| **DTC: P0192**<br>**1T CCM, MIL: Yes**<br>**Years:** 2005, 2006, 2007, 2008<br>**Engines:** 2.8L VIN U<br>**Models:** All<br>**Transmissions:** All | **Fuel Rail Pressure Sensor Circuit Input Low**<br>If the fuel rail pressure increases, the FRP signal voltage increases. The FICM monitors the FRP sensor and communicates the data to the Control module by a dedicated pulse width modulated (pulse width modulation) circuit.<br>**Possible Causes**<br>&bull; FRP sensor 5-volt power circuit is open or shorted to ground<br>&bull; FRP Sensor signal circuit is shorted to ground<br>&bull; FRP Sensor is damaged or has failed<br>&bull; Control module has failed |
| **DTC: P0193**<br>**1T CCM, MIL: Yes**<br>**Years:** 2005, 2006, 2007, 2008<br>**Engines:** 2.8L VIN U<br>**Models:** All<br>**Transmissions:** All | **Fuel Rail Pressure Sensor Circuit Input High**<br>Control module detected the Fuel Rail Pressure (FRP) sensor was more than 4.95v for 5 seconds.<br>**Possible Causes**<br>&bull; FRP sensor signal circuit is open between sensor and the Control module<br>&bull; FRP Sensor ground circuit is open between sensor and Control module<br>&bull; FRP sensor signal circuit is shorted to VREF or system power<br>&bull; FRP Sensor is damaged or has failed<br>&bull; Control module has failed |
| **DTC: P0196**<br>**2T CCM, MIL: Yes**<br>**Years:** 2005, 2006, 2007, 2008<br>**Engines:** 2.5L SOHC, VIN 6<br>**Models:**<br>**Transmissions:** All | **Engine Oil Temperature Sensor Circuit Range/Performance**<br>Detected when two consecutive driving cycles with fault occur. Hard starting condition. Engine stalls. Erroneous idling. Poor driving performance.<br>**Possible Causes:**<br>&bull; Check any other DTC on display<br>&bull; Faulty oil temperature sensor |
| **DTC: P0197**<br>**1T CCM, MIL: Yes**<br>**Years:** 2005, 2006, 2007, 2008<br>**Engines:** 2.5L SOHC, VIN 6,<br>**Models:**<br>**Transmissions:** All | **Engine Oil Temperature Sensor Circuit Low**<br>Control module detected the open or short circuit of the oil temperature sensor Hard starting condition. Engine stalls. Erroneous idling. Poor driving performance.<br>**Possible Causes:**<br>&bull; Check harness between sensor and Control module<br>&bull; Check poor contact<br>&bull; Faulty oil temperature sensor |
| **DTC: P0198**<br>**1T CCM, MIL: Yes**<br>**Engines:** 2.5L SOHC, VIN 6,<br>**Models:**<br>**Transmissions:** All | **Engine Oil Temperature Sensor Circuit High**<br>Control module detects the open or short circuit of the oil temperature sensor Hard starting condition. Engine stalls. Improper idling. Poor driving performance.<br>**Possible Causes:**<br>&bull; Check harness between sensor and Control module<br>&bull; Check poor contact<br>&bull; Faulty oil temperature sensor |
| **DTC: P0201**<br>**2T CCM, MIL: Yes**<br>**Years:** 2005, 2006, 2007, 2008<br>**Models:** All<br>**Engines:** 2.0L VIN S/Y, 2.8L VIN U, 4.2L VIN S, 5.3L VIN M/P, 6.0L VIN H<br>**Transmissions:** All | **Fuel Injector 1 Control Circuit**<br>Control module detects an incorrect voltage on the fuel injector control circuit. The above condition is met for 1 second.<br>**Possible Causes:**<br>&bull; Injector 1 control circuit is open between injector and Control module<br>&bull; Injector 1 control circuit is grounded between injector and Control module<br>&bull; Injector 1 power circuit is open (test INJ fuse in fuse block)<br>&bull; Injector 1 is damaged or it has failed<br>&bull; Control module is damaged |

| DTC | Trouble Code Title, Conditions & Possible Causes |
|---|---|
| **DTC: P0202**<br>**2T CCM, MIL: Yes**<br>**Years:** 2005, 2006, 2007, 2008<br>**Models:** All<br>**Engines:** 2.0L VIN S/Y, 2.8L VIN U, 4.2L VIN S, 5.3L VIN M/P, 6.0L VIN H<br>**Transmissions:** All | **Fuel Injector 2 Control Circuit**<br>Control module detects an incorrect voltage on the fuel injector control circuit. The above condition is met for 1 second<br>**Possible Causes:**<br>• Injector 2 control circuit is open between injector and Control module<br>• Injector 2 control circuit is grounded between injector and Control module<br>• Injector 2 power circuit is open (test INJ fuse in fuse block)<br>• Injector 2 is damaged or it has failed<br>• Control module is damaged |
| **DTC: P0203**<br>**2T CCM, MIL: Yes**<br>**Years:** 2005, 2006, 2007, 2008<br>**Models:** All<br>**Engines:** 2.0L VIN S/Y, 2.8L VIN U, 4.2L VIN S, 5.3L VIN M/P, 6.0L VIN H<br>**Transmissions:** All | **Fuel Injector 3 Control Circuit**<br>Control module detects an incorrect voltage on the fuel injector control circuit. The above condition is met for 1 second.<br>**Possible Causes:**<br>• Injector 3 control circuit is open between injector and Control module<br>• Injector 3 control circuit is grounded between injector and Control module<br>• Injector 3 power circuit is open (test INJ fuse in fuse block)<br>• Injector 3 is damaged or it has failed<br>• Control module is damaged |
| **DTC: P0204**<br>**2T CCM, MIL: Yes**<br>**Years:** 2005, 2006, 2007, 2008<br>**Models:** All<br>**Engines:** 2.0L VIN S/Y, 2.8L VIN U, 4.2L VIN S, 5.3L VIN M/P, 6.0L VIN H<br>**Transmissions:** All | **Fuel Injector 4 Control Circuit**<br>Control module detects an incorrect voltage on the fuel injector control circuit. The above condition is met for 1 second.<br>**Possible Causes:**<br>• Injector 4 control circuit is open between injector and Control module<br>• Injector 4 control circuit is grounded between injector and Control module<br>• Injector 4 power circuit is open (test INJ fuse in fuse block)<br>• Injector 4 is damaged or it has failed<br>• Control module is damaged |
| **DTC: P0205**<br>**2T CCM, MIL: Yes**<br>**Years:** 2005, 2006, 2007, 2008<br>**Models:** All<br>**Engines:** 2.8L VIN U, 4.2L VIN S, 5.3L VIN M/P, 6.0L VIN H<br>**Transmissions:** All | **Fuel Injector 5 Control Circuit**<br>Control module detects an incorrect voltage on the fuel injector control circuit. The above condition is met for 1 second.<br>**Possible Causes:**<br>• Injector 5 control circuit is open between injector and Control module<br>• Injector 5 control circuit is grounded between injector and Control module<br>• Injector 5 power circuit is open (test INJ fuse in fuse block)<br>• Injector 5 is damaged or it has failed<br>• Control module is damaged |
| **DTC: P0206**<br>**2T CCM, MIL: Yes**<br>**Years:** 2005, 2006, 2007, 2008<br>**Models:** All<br>**Engines:** 2.8L VIN U, 4.2L VIN S, 5.3L VIN M/P, 6.0L VIN H<br>**Transmissions:** All | **Fuel Injector 6 Control Circuit**<br>Control module detects an incorrect voltage on the fuel injector control circuit. The above condition is met for 1 second.<br>**Possible Causes:**<br>• Injector 6 control circuit is open between injector and Control module<br>• Injector 6 control circuit is grounded between injector and Control module<br>• Injector 6 power circuit is open (test INJ fuse in fuse block)<br>• Injector 6 is damaged or it has failed<br>• Control module is damaged |
| **DTC: P0207**<br>**2T CCM, MIL: Yes**<br>**Years:** 2005, 2006, 2007, 2008<br>**Models:** All<br>**Engines:** 4.2L VIN S, 5.3L VIN M/P, 6.0L VIN H<br>**Transmissions:** All | **Fuel Injector 7 Control Circuit**<br>Control module detects an incorrect voltage on the fuel injector control circuit. The above condition is met for 1 second.<br>**Possible Causes:**<br>• Injector 7 control circuit is open between injector and Control module<br>• Injector 7 control circuit is grounded between injector and Control module<br>• Injector 7 power circuit is open (test INJ fuse in fuse block)<br>• Injector 7 is damaged or it has failed<br>• Control module is damaged |
| **DTC: P0208**<br>**2T CCM, MIL: Yes**<br>**Years:** 2005, 2006, 2007, 2008<br>**Models:** All<br>**Engines:** 4.2L VIN S, 5.3L VIN M/P, 6.0L VIN H<br>**Transmissions:** All | **Fuel Injector 8 Control Circuit**<br>Control module detects an incorrect voltage on the fuel injector control circuit. The above condition is met for 1 second.<br>**Possible Causes:**<br>• Injector 8 control circuit is open between injector and Control module<br>• Injector 8 control circuit is grounded between injector and Control module<br>• Injector 8 power circuit is open (test INJ fuse in fuse block)<br>• Injector 8 is damaged or it has failed<br>• Control module is damaged |

| DTC | Trouble Code Title, Conditions & Possible Causes |
|---|---|
| **DTC: P0219**<br>**2T CCM, MIL: Yes**<br>**Years:** 2005, 2006, 2007, 2008<br>**Models:** All<br>**Engines:** 2.8L VIN U<br>**Transmissions:** All | **Engine Overspeed Condition**<br>Control module detected five (5) Engine Shutoff (ESO) cycles occurred with a resulting engine speed change from 800-1200 rpm occurred continuously.<br>**Possible Causes**<br>• Fuel injection pump may need to be replaced<br>• Fuel injection pump timing adjustment needs to be performed<br>• Control module has failed |
| **DTC: P0220**<br>**1T CCM, MIL: No**<br>**Years:** 2005, 2006, 2007<br>**Models:** All<br>**Engines:** 4.2L VIN S, 5.3L VIN M/P, 6.0L VIN H<br>**Transmissions:** All | **Throttle Position (TP) Sensor Circuit**<br>Control module detects the TP Sensor voltage is less than 0.25volt or more than 4.59 volts for more than 1 second.<br>**Possible Causes:**<br>• TP Sensor signal circuit for an open or high resistance<br>• Short between the TP Sensor signal circuit and all other TAC module circuits<br>• APP Sensor 5-volt reference circuit shorted to voltage<br>• TP Sensor 5-volt reference circuit open or high resistance<br>• TP Sensor 5-volt reference circuit shorted to ground<br>• APP Sensor 5-volt reference circuit for a short to ground<br>• Faulty throttle body assembly<br>• Faulty APP sensor.<br>• Control module is defective |
| **DTC: P0221**<br>**1T CCM, MIL: No**<br>**Years:** 2005, 2006, 2007<br>**Models:** All<br>**Engines:** 2.0L VIN S/Y, 2.8L VIN U<br>**Transmissions:** All | **Throttle/Pedal Position Sensor Circuit Range/Performance**<br>Control module detects the open or short circuit of throttle position Sensor. Improper idling. Poor driving performance. Engine stalls.<br>**Possible Causes:**<br>• Check sensor output<br>• Check poor contact<br>• Check harness between Control module and electronic throttle control<br>• Check sensor power supply<br>• Check for short circuit in Control module |
| **DTC: P0222**<br>**1T CCM, MIL: Yes**<br>**Years:** 2005, 2006, 2007, 2008<br>**Engines:** 2.0L VIN S/Y, 2.5L SOHC, DOHC VIN 6, 2.8L VIN U<br>**Models:** All<br>**Transmissions:** All | **Throttle/Pedal Position Sensor Circuit Low Input**<br>Control module detects the open or short circuit of throttle position Sensor. Improper idling. Poor driving performance. Engine stalls.<br>**Possible Causes:**<br>• Check sensor output<br>• Check poor contact<br>• Check harness between Control module and electronic throttle control<br>• Check sensor power supply<br>• Check for short circuit in Control module |
| **DTC: P0223**<br>**1T CCM, MIL: Yes**<br>**Years:** 2005, 2006, 2007, 2008<br>**Engines:** All<br>**Models:** All<br>**Transmissions:** All | **Throttle/Pedal Position Sensor Circuit High Input**<br>Control module detects the open or short circuit of throttle position Sensor. Improper idling. Poor driving performance. Engine stalls.<br>**Possible Causes:**<br>• Check sensor output<br>• Check poor contact<br>• Check harness between Control module and electronic throttle control<br>• Check for short circuit in Control module |
| **DTC: P0230**<br>**2T CCM, MIL: Yes**<br>**Years:** 2005, 2006, 2007, 2008<br>**Engines:** 2.0L VIN 2, 2.5L SOHC, DOHC VIN 6, 4.2L VIN S, 5.3L VIN M/P, 6.0L VIN H<br>**Models:** All<br>**Transmissions:** All | **Fuel Pump Primary Circuit**<br>Fuel pump control unit detects the open or short circuit malfunction for each line, and then sends NG signals if one of them is found NG. Poor drive ability.<br>**Possible Causes:**<br>• Check power supply to fuel pump control unit<br>• Check ground circuit of fuel pump control unit<br>• Check harness between the fuel pump control unit and fuel pump connector<br>• Check poor contact<br>• Check if vehicle has previously ran out of fuel<br>• Faulty fuel pump control unit |
| **DTC: P0234**<br>**2T CCM, MIL: Yes**<br>**Years:** 2005, 2006, 2007, 2008<br>**Engines:** 2.8L VIN U<br>**Models:** All<br>**Transmissions:** All | **Turbocharger Boost Circuit Range/Performance Conditions**<br>Boost Pressure sensor was above the expected range by 35 kPa or more for 12 seconds.<br>**Possible Causes**<br>• Boost sensor signal, VREF or ground circuit connection faults<br>• Boost sensor is damaged or has failed<br>• Pressure hose from the charged air tube to the Wastegate actuator is disconnected or ruptured<br>• Wastegate or Turbocharger is damaged or has failed |

| DTC | Trouble Code Title, Conditions & Possible Causes |
|---|---|
| **DTC: P0236**<br>**2T CCM, MIL: Yes**<br>**Years:** 2005, 2006, 2007, 2008<br>**Engines:** 2.8L VIN U<br>**Models:** All<br>**Transmissions:** All | **Turbocharger Boost System Performance Conditions**<br>Boost Pressure sensor was above the expected range by 35 kPa or more for 12 seconds.<br>**Possible Causes**<br>• Boost sensor signal, VREF or ground circuit connection faults<br>• Boost sensor is damaged or has failed<br>• Pressure hose from the charged air tube to the Wastegate actuator is disconnected or ruptured<br>• Wastegate or Turbocharger is damaged or has failed |
| **DTC: P0237**<br>**2T CCM, MIL: Yes**<br>**Years:** 2005, 2006, 2007, 2008<br>**Engines:** 2.0L VIN S/Y, 2.8L VIN U<br>**Models:** All<br>**Transmissions:** All | **Charge Air Absolute Pressure Sensor Circuit Input Low**<br>Control module detected the Boost Pressure sensor was less than 0.10v (Scan Tool reads less than 40 kPa) for 2 seconds during the CCM continuous test.<br>**Possible Causes**<br>• MAP sensor signal circuit shorted to sensor or chassis ground<br>• MAP sensor VREF circuit open between the sensor and Control module<br>• MAP sensor is damaged or has failed<br>• Control module has failed |
| **DTC: P0238**<br>**2T CCM, MIL: Yes**<br>**Years:** 2005, 2006, 2007, 2008<br>**Engines:** 2.0L VIN S/Y, 2.8L VIN U<br>**Models:** All<br>**Transmissions:** All | **Charge Air Absolute Pressure Sensor Circuit Input High**<br>Control module detected the Turbocharger Boost Pressure sensor was more than, or was equal to 4.80v (202S kPa) for 2 seconds.<br>**Possible Causes**<br>• MAP sensor signal circuit is open or it is shorted to VREF<br>• MAP sensor ground circuit open between sensor and the Control module<br>• MAP sensor is damaged or has failed<br>• Control module has failed |
| **DTC: P0244**<br>**2T CCM, MIL: Yes**<br>**Years:** 2005, 2006, 2007, 2008<br>**Engines:** 2.0L VIN S/Y, 2.8L VIN U<br>**Models:** All<br>**Transmissions:** All | **Charge Air Control System, Performance Problem**<br>Control module detected an unexpected voltage on the Boost Control Solenoid control circuit for 30 seconds.<br>**Possible Causes**<br>• Boost solenoid circuit is open between the solenoid and Control module<br>• Boost solenoid circuit is shorted to ground<br>• Boost solenoid power circuit is open to system power<br>• Boost solenoid is damaged or has failed<br>• Control module has failed |
| **DTC: P0245**<br>**2T CCM, MIL: Yes**<br>**Years:** 2005, 2006, 2007, 2008<br>**Engines:** 2.0L VIN S/Y, 2.8L VIN U<br>**Models:** All<br>**Transmissions:** All | **Charge Air Control Valve Circuit Low Input**<br>Control module detected an unexpected voltage on the Boost Control Solenoid control circuit for 30 seconds.<br>**Possible Causes**<br>• Boost solenoid circuit is open between the solenoid and Control module<br>• Boost solenoid circuit is shorted to ground<br>• Boost solenoid power circuit is open to system power<br>• Boost solenoid is damaged or has failed<br>• Control module has failed |
| **DTC: P0246**<br>**2T CCM, MIL: Yes**<br>**Years:** 2005, 2006, 2007, 2008<br>**Engines:** 2.0L VIN S/Y, 2.8L VIN U<br>**Models:** All<br>**Transmissions:** All | **Charge Air Control Valve Circuit High Input**<br>Control module detected an unexpected voltage on the Boost Control Solenoid control circuit for 30 seconds.<br>**Possible Causes**<br>• Boost solenoid circuit is open between the solenoid and Control module<br>• Boost solenoid circuit is shorted to ground<br>• Boost solenoid power circuit is open to system power<br>• Boost solenoid is damaged or has failed<br>• Control module has failed |
| **DTC: P0261**<br>**2T CCM, MIL: Yes**<br>**Years:** 2005, 2006, 2007, 2008<br>**Engines:** 2.8L VIN U<br>**Models:** All<br>**Transmissions:** All | **Fuel Injector 1 Control Circuit Input Low**<br>Control module detected a low voltage condition on the Fuel Injector 1 circuit for 30 seconds.<br>**Possible Causes**<br>• Fuel injector 1 connector is damaged or shorted<br>• Fuel injector 1 power circuit is open (check the INJ/Coil fuse)<br>• Fuel injector 1 control circuit is shorted to ground<br>• Fuel injector 1 is damaged or it has failed<br>• Control module is damaged |

| DTC | Trouble Code Title, Conditions & Possible Causes |
|---|---|
| **DTC: P0262**<br>**2T CCM, MIL: Yes**<br>**Years:** 2005, 2006, 2007, 2008<br>**Engines:** 2.8L VIN U<br>**Models:** All<br>**Transmissions:** All | **Fuel Injector 1 Control Circuit Input High**<br>Control module detected a high voltage condition on the Fuel Injector 1 circuit for 30 seconds.<br>**Possible Causes**<br>• Fuel injector 1 connector is damaged or shorted<br>• Fuel injector 1 control circuit is shorted to system power (B+)<br>• Fuel injector 1 is damaged or it has failed<br>• Control module is damaged |
| **DTC: P0264**<br>**2T CCM, MIL: Yes**<br>**Years:** 2005, 2006, 2007, 2008<br>**Engines:** 2.8L VIN U<br>**Models:** All<br>**Transmissions:** All | **Fuel Injector 2 Control Circuit Input Low**<br>Control module detected a low voltage condition on the Fuel Injector 2 circuit for 30 seconds.<br>**Possible Causes**<br>• Fuel injector 2 connector is damaged or shorted<br>• Fuel injector 2 power circuit is open (check the INJ/Coil fuse)<br>• Fuel injector 2 control circuit is shorted to ground<br>• Fuel injector 2 is damaged or it has failed<br>• Control module is damaged |
| **DTC: P0265**<br>**2T CCM, MIL: Yes**<br>**Years:** 2005, 2006, 2007, 2008<br>**Engines:** 2.8L VIN U<br>**Models:** All<br>**Transmissions:** All | **Fuel Injector 2 Control Circuit Input High**<br>Control module detected a high voltage condition on the Fuel Injector 2 circuit for 30 seconds.<br>**Possible Causes**<br>• Fuel injector 2 connector is damaged or shorted<br>• Fuel injector 2 control circuit is shorted to system power (B+)<br>• Fuel injector 2 is damaged or it has failed<br>• Control module is damaged |
| **DTC: P0267**<br>**2T CCM, MIL: Yes**<br>**Years:** 2005, 2006, 2007, 2008<br>**Engines:** 2.8L VIN U<br>**Models:** All<br>**Transmissions:** All | **Fuel Injector 3 Control Circuit Input Low**<br>Control module detected a low voltage condition on the Fuel Injector 3 circuit for 30 seconds.<br>**Possible Causes**<br>• Fuel injector 3 connector is damaged or shorted<br>• Fuel injector 3 power circuit is open (check the INJ/Coil fuse)<br>• Fuel injector 3 control circuit is shorted to ground<br>• Fuel injector 3 is damaged or it has failed<br>• Control module is damaged |
| **DTC: P0268**<br>**2T CCM, MIL: Yes**<br>**Years:** 2005, 2006, 2007, 2008<br>**Engines:** 2.8L VIN U<br>**Models:** All<br>**Transmissions:** All | **Fuel Injector 3 Control Circuit Input High**<br>Control module detected a high voltage condition on the Fuel Injector 3 circuit for 30 seconds.<br>**Possible Causes**<br>• Fuel injector 3 connector is damaged or shorted<br>• Fuel injector 3 control circuit is shorted to system power (B+)<br>• Fuel injector 3 is damaged or it has failed<br>• Control module is damaged |
| **DTC: P0270**<br>**2T CCM, MIL: Yes**<br>**Years:** 2005, 2006, 2007, 2008<br>**Engines:** 2.8L VIN U<br>**Models:** All<br>**Transmissions:** All | **Fuel Injector 4 Control Circuit Input Low**<br>Control module detected a low voltage condition on the Fuel Injector 4 circuit for 30 seconds.<br>**Possible Causes**<br>• Fuel injector 4 connector is damaged or shorted<br>• Fuel injector 4 power circuit is open (check the INJ/Coil fuse)<br>• Fuel injector 4 control circuit is shorted to ground<br>• Fuel injector 4 is damaged or it has failed<br>• Control module is damaged |
| **DTC: P0271**<br>**2T CCM, MIL: Yes**<br>**Years:** 2005, 2006, 2007, 2008<br>**Engines:** 2.8L VIN U<br>**Models:** All<br>**Transmissions:** All | **Fuel Injector 4 Control Circuit Input High**<br>Control module detected a high voltage condition on the Fuel Injector 4 circuit for 30 seconds.<br>**Possible Causes**<br>• Fuel injector 4 connector is damaged or shorted<br>• Fuel injector 4 control circuit is shorted to system power (B+)<br>• Fuel injector 4 is damaged or it has failed<br>• Control module is damaged |

| DTC | Trouble Code Title, Conditions & Possible Causes |
|---|---|
| **DTC: P0273**<br>**2T CCM, MIL: Yes**<br>**Years:** 2005, 2006, 2007, 2008<br>**Engines:** 2.8L VIN U<br>**Models:** All<br>**Transmissions:** All | **Fuel Injector 5 Control Circuit Input Low**<br>Control module detected a low voltage condition on the Fuel Injector 5 circuit for 30 seconds.<br>**Possible Causes**<br>• Fuel injector 5 connector is damaged or shorted<br>• Fuel injector 5 power circuit is open (check the INJ/Coil fuse)<br>• Fuel injector 5 control circuit is shorted to ground<br>• Fuel injector 5 is damaged or it has failed<br>• Control module is damaged |
| **DTC: P0274**<br>**2T CCM, MIL: Yes**<br>**Years:** 2005, 2006, 2007, 2008<br>**Engines:** 2.8L VIN U<br>**Models:** All<br>**Transmissions:** All | **Fuel Injector 5 Control Circuit Input High**<br>Control module detected a high voltage condition on the Fuel Injector 5 circuit for 30 seconds.<br>**Possible Causes**<br>• Fuel injector 5 connector is damaged or shorted<br>• Fuel injector 5 control circuit is shorted to system power (B+)<br>• Fuel injector 5 is damaged or it has failed<br>• Control module is damaged |
| **DTC: P0276**<br>**2T CCM, MIL: Yes**<br>**Years:** 2005, 2006, 2007, 2008<br>**Engines:** 2.8L VIN U<br>**Models:** All<br>**Transmissions:** All | **Fuel Injector 6 Control Circuit Input Low**<br>Control module detected a low voltage condition on the Fuel Injector 6 circuit for 30 seconds.<br>**Possible Causes**<br>• Fuel injector 6 connector is damaged or shorted<br>• Fuel injector 6 power circuit is open (check the INJ/Coil fuse)<br>• Fuel injector 6 control circuit is shorted to ground<br>• Fuel injector 6 is damaged or it has failed<br>• Control module is damaged |
| **DTC: P0277**<br>**2T CCM, MIL: Yes**<br>**Years:** 2005, 2006, 2007, 2008<br>**Engines:** 2.8L VIN U<br>**Models:** All<br>**Transmissions:** All | **Fuel Injector 6 Control Circuit Input High**<br>Control module detected a high voltage condition on the Fuel Injector 6 circuit for 30 seconds.<br>**Possible Causes**<br>• Fuel injector 6 connector is damaged or shorted<br>• Fuel injector 6 control circuit is shorted to system power (B+)<br>• Fuel injector 6 is damaged or it has failed<br>• Control module is damaged |
| **DTC: P0299**<br>**2T CCM, MIL: Yes**<br>**Years:** 2005, 2006, 2007, 2008<br>**Engines:** 2.8L VIN U<br>**Models:** All<br>**Transmissions:** All | **Turbo Charger Under Boost Detected**<br>Control module detected the Actual Boost Pressure was higher than, lower than or equal to 20 kPa of the Desired value.<br>**Possible Causes:**<br>• Engine vacuum line is loose, open or pinched, or misrouted<br>• Engine source vacuum restricted where vacuum line connects<br>• Turbocharger solenoid vent filter or vent orifice is restricted (the vacuum to the solenoid should fluctuate in normal operation) |
| **DTC: P0300**<br>**2T CCM, MIL: Yes**<br>**Years:** 2005, 2006, 2007, 2008<br>**Engines:** All<br>**Models:** All<br>**Transmissions:** All | **Random/Multiple Cylinder Misfire Detected**<br>Control module detected a misfire in Random/Multiple Cylinders in the 200 (Catalyst) or 1000 rpm (High Emissions) revolution range. Engine stalls. Improper idling. Rough driving condition.<br>**Note: If the misfire is severe, the MIL will flash on/off on the 1st trip!**<br>**Possible Causes:**<br>• Check any other DTC on display<br>• Check output signal of Control module<br>• Check harness between injector and Control module connector<br>• Check fuel injector<br>• Check power supply line<br>• Check power supply to mass air flow sensor<br>• Check harness between Control module and sensor connector<br>• Check installation of camshaft and crankshaft sensors<br>• Check crank plate<br>• Check condition of timing chain<br>• Check fuel level<br>• Check status of MIL light<br>• Check air intake system<br>• Check cause of misfire (engine running)<br>• Check engine condition<br>• Check all cylinders individually, in pairs, in groups and at random |

| DTC | Trouble Code Title, Conditions & Possible Causes |
|---|---|
| **DTC: P0301**<br>**2T CCM, MIL: Yes**<br>**Years:** 2005, 2006, 2007, 2008<br>**Engines:** All<br>**Models:** All<br>**Transmissions:** All | **Cylinder No. 1 Misfire Detected**<br>Control module detected a misfire in Cylinder 1 in the 200 (Catalyst) or 1000 rpm (High Emissions) revolution range. Engine stalls. Improper idling. Rough driving condition.<br>**Note: If the misfire is severe, the MIL will flash on/off on the 1st trip!**<br>**Possible Causes:**<br>• Check any other DTC on display<br>• Check output signal of Control module<br>• Check harness between injector and Control module connector<br>• Check fuel injector<br>• Check power supply line<br>• Check power supply to mass air flow sensor<br>• Check harness between Control module and sensor connector<br>• Check installation of camshaft and crankshaft sensors<br>• Check crank plate<br>• Check condition of timing chain<br>• Check fuel level<br>• Check status of MIL light<br>• Check air intake system<br>• Check cause of misfire (engine running)<br>• Check engine condition<br>• Check all cylinders individually, in pairs, in groups and at random |
| **DTC P0302**<br>**2T CCM, MIL: Yes**<br>**Years:** 2005, 2006, 2007, 2008<br>**Engines:** All<br>**Models:** All<br>**Transmissions:** All | **Cylinder No. 2 Misfire Detected**<br>Control module detected a misfire in Cylinder 2 in the 200 (Catalyst) or 1000 rpm (High Emissions) revolution range. Engine stalls. Improper idling. Rough driving condition.<br>**Note: If the misfire is severe, the MIL will flash on/off on the 1st trip!**<br>**Possible Causes:**<br>• Check any other DTC on display<br>• Check output signal of Control module<br>• Check harness between injector and Control module connector<br>• Check fuel injector<br>• Check power supply line<br>• Check power supply to mass air flow sensor<br>• Check harness between Control module and sensor connector<br>• Check installation of camshaft and crankshaft sensors<br>• Check crank plate<br>• Check condition of timing chain<br>• Check fuel level<br>• Check status of MIL light<br>• Check air intake system<br>• Check cause of misfire (engine running)<br>• Check engine condition<br>• Check all cylinders individually, in pairs, in groups and at random |
| **DTC: P0303**<br>**2T CCM, MIL: Yes**<br>**Years:** 2005, 2006, 2007, 2008<br>**Engines:** All<br>**Models:** All<br>**Transmissions:** All | **Cylinder No. 3 Misfire Detected**<br>Control module detected a misfire in Cylinder 3 in the 200 (Catalyst) or 1000 rpm (High Emissions) revolution range. Engine stalls. Improper idling. Rough driving condition.<br>**Note: If the misfire is severe, the MIL will flash on/off on the 1st trip!**<br>**Possible Causes:**<br>• Check any other DTC on display<br>• Check output signal of Control module<br>• Check harness between injector and Control module connector<br>• Check fuel injector<br>• Check power supply line<br>• Check power supply to mass air flow sensor<br>• Check harness between Control module and sensor connector<br>• Check installation of camshaft and crankshaft sensors<br>• Check crank plate<br>• Check condition of timing chain<br>• Check fuel level<br>• Check status of MIL light<br>• Check air intake system<br>• Check cause of misfire (engine running)<br>• Check engine condition<br>• Check all cylinders individually, in pairs, in groups and at random |

| DTC | Trouble Code Title, Conditions & Possible Causes |
|---|---|
| **DTC: P0304**<br>**2T CCM, MIL: Yes**<br>**Years:** 2005, 2006, 2007, 2008<br>**Engines:** All<br>**Models:** All<br>**Transmissions:** All | **Cylinder No. 4 Misfire Detected**<br>Control module detected a misfire in Cylinder 4 in the 200 (Catalyst) or 1000 rpm (High Emissions) revolution range. Engine stalls. Improper idling. Rough driving condition.<br>**Note: If the misfire is severe, the MIL will flash on/off on the 1st trip!**<br>**Possible Causes:**<br>• Check any other DTC on display<br>• Check output signal of Control module<br>• Check harness between injector and Control module connector<br>• Check fuel injector<br>• Check power supply line<br>• Check power supply to mass air flow sensor<br>• Check harness between Control module and sensor connector<br>• Check installation of camshaft and crankshaft sensors<br>• Check crank plate<br>• Check condition of timing chain<br>• Check fuel level<br>• Check status of MIL light<br>• Check air intake system<br>• Check cause of misfire (engine running)<br>• Check engine condition<br>• Check all cylinders individually, in pairs, in groups and at random |
| **DTC: P0305**<br>**2T CCM, MIL: Yes**<br>**Years:** 2005, 2006, 2007, 2008<br>**Engines:** 2.8L VIN U<br>**Models:** All<br>**Transmissions:** All | **Cylinder No. 5 Misfire Detected**<br>Control module detected a misfire in Cylinder 5 in the 200 (Catalyst) or 1000 rpm (High Emissions) revolution range. Engine stalls. Improper idling. Rough driving condition.<br>**Note: If the misfire is severe, the MIL will flash on/off on the 1st trip!**<br>**Possible Causes:**<br>• Check any other DTC on display<br>• Check output signal of Control module<br>• Check harness between injector and Control module connector<br>• Check fuel injector<br>• Check power supply line<br>• Check power supply to mass air flow sensor<br>• Check harness between Control module and sensor connector<br>• Check installation of camshaft and crankshaft sensors<br>• Check crank plate<br>• Check condition of timing chain<br>• Check fuel level<br>• Check status of MIL light<br>• Check air intake system<br>• Check cause of misfire (engine running)<br>• Check engine condition<br>• Check all cylinders individually, in pairs, in groups and at random |
| **DTC: P0306**<br>**2T CCM, MIL: Yes**<br>**Years:** 2005, 2006, 2007, 2008<br>**Engines:** 2.8L VIN U<br>**Models:** All<br>**Transmissions:** All | **Cylinder No. 6 Misfire Detected**<br>Control module detected a misfire in Cylinder 6 in the 200 (Catalyst) or 1000 rpm (High Emissions) revolution range. Engine stalls. Improper idling. Rough driving condition.<br>**Note: If the misfire is severe, the MIL will flash on/off on the 1st trip!**<br>**Possible Causes:**<br>• Check any other DTC on display<br>• Check output signal of Control module<br>• Check harness between injector and Control module connector<br>• Check fuel injector<br>• Check power supply line<br>• Check power supply to mass air flow sensor<br>• Check harness between Control module and sensor connector<br>• Check installation of camshaft and crankshaft sensors<br>• Check crank plate<br>• Check condition of timing chain<br>• Check fuel level<br>• Check status of MIL light<br>• Check air intake system<br>• Check cause of misfire (engine running)<br>• Check engine condition<br>• Check all cylinders individually, in pairs, in groups and at random |

| DTC | Trouble Code Title, Conditions & Possible Causes |
|---|---|
| **DTC: P0307**<br>**2T CCM, MIL: Yes**<br>**Years:** 2005, 2006, 2007, 2008<br>**Engines:** 4.2L VIN S, 5.3L VIN M/P, 6.0L VIN H<br>**Models:** All<br>**Transmissions:** All | **Cylinder No. 7 Misfire Detected**<br>Control module detected a misfire in Cylinder 7 in the 200 (Catalyst) or 1000 rpm (High Emissions) revolution range. Engine stalls. Improper idling. Rough driving condition.<br>**Note: If the misfire is severe, the MIL will flash on/off on the 1st trip!**<br>**Possible Causes:**<br>    • Check any other DTC on display<br>    • Check output signal of Control module<br>    • Check harness between injector and Control module connector<br>    • Check fuel injector<br>    • Check power supply line<br>    • Check power supply to mass air flow sensor<br>    • Check harness between Control module and sensor connector<br>    • Check installation of camshaft and crankshaft sensors<br>    • Check crank plate<br>    • Check condition of timing chain<br>    • Check fuel level<br>    • Check status of MIL light<br>    • Check air intake system<br>    • Check cause of misfire (engine running)<br>    • Check engine condition<br>    • Check all cylinders individually, in pairs, in groups and at random |
| **DTC: P0308**<br>**2T CCM, MIL: Yes**<br>**Years:** 2005, 2006, 2007, 2008<br>**Engines:** 4.2L VIN S, 5.3L VIN M/P, 6.0L VIN H<br>**Models:** All<br>**Transmissions:** All | **Cylinder No. 8 Misfire Detected**<br>Control module detected a misfire in Cylinder 8 in the 200 (Catalyst) or 1000 rpm (High Emissions) revolution range. Engine stalls. Improper idling. Rough driving condition.<br>**Note: If the misfire is severe, the MIL will flash on/off on the 1st trip!**<br>**Possible Causes:**<br>    • Check any other DTC on display<br>    • Check output signal of Control module<br>    • Check harness between injector and Control module connector<br>    • Check fuel injector<br>    • Check power supply line<br>    • Check power supply to mass air flow sensor<br>    • Check harness between Control module and sensor connector<br>    • Check installation of camshaft and crankshaft sensors<br>    • Check crank plate<br>    • Check condition of timing chain<br>    • Check fuel level<br>    • Check status of MIL light<br>    • Check air intake system<br>    • Check cause of misfire (engine running)<br>    • Check engine condition<br>    • Check all cylinders individually, in pairs, in groups and at random |
| **DTC: P0313**<br>**2T CCM, MIL: Yes**<br>**Years:** 2005, 2006, 2007, 2008<br>**Engines:** 2.0L VIN S/Y, 2.8L VIN U<br>**Models:** All<br>**Transmissions:** All | **Misfire Detected, Lean Air Fuel Condition Present**<br>Control module detected a lean condition or a misfire condition for 10 seconds.<br>**Possible Causes**<br>    • Verify that there is an adequate fuel supply in the fuel tank<br>    • If this code is set and a misfire or lean condition is not currently present, the vehicle has run low on fuel and caused this code. |
| **DTC: P0315**<br>**2T CCM, MIL: Yes**<br>**Years:** 2005, 2006, 2007, 2008<br>**Engines:** 4.2L VIN S, 5.3L VIN M/P, 6.0L VIN H<br>**Models:** All<br>**Transmissions:** All | **Crankshaft Position (CKP) System Variation Not Learned**<br>Control module detects that the CKP system variation values are not stored in memory<br>**Possible Causes:**<br>    • Worn crankshaft main bearings<br>    • A damaged reluctor wheel<br>    • Excessive crankshaft runout<br>    • A damaged crankshaft<br>    • Interference in the signal circuit of the CKP sensor<br>    • Foreign material passing between the CKP sensor and the reluctor wheel<br>    • Coolant temperature that is not within the Conditions For Running the DTC<br>    • Control module has failed. |

| DTC | Trouble Code Title, Conditions & Possible Causes |
|---|---|
| **DTC: P0318**<br>**2T CCM, MIL: Yes**<br>**Years:** 2005, 2006, 2007, 2008<br>**Engines:** 2.8L VIN U<br>**Models:** All<br>**Transmissions:** All | **Misfire Detected, Lean Air Fuel Condition Present**<br>Control module detected it lost communication with the Econtrol module for 10 seconds.<br>**Possible Causes**<br>• Perform the Diagnostic Circuit Check for the Control module<br>• Perform the Diagnostic Circuit Check for the Econtrol module<br>• Check the serial data connections to the Econtrol module and the Control module |
| **DTC: P0324**<br>**2T CCM, MIL: Yes**<br>**Years:** 2005, 2006, 2007, 2008<br>**Engines:** 2.8L VIN U , 4.2L VIN S,<br>5.3L VIN M/P, 6.0L VIN H<br>**Models:** All<br>**Transmissions:** All | **Knock Sensor Circuit Malfunction**<br>Control module detected a fault the KS diagnostic circuit that did not allow proper diagnosis of the KS system.<br>**Possible Causes**<br>• Clear the codes and then recheck for the same code to reset.<br>• Control module has failed |
| **DTC: P0325**<br>**2T CCM, MIL: Yes**<br>**Years:** 2005, 2006, 2007, 2008<br>**Engines:** 2.0L VIN S/Y, 2.8L<br>VIN U , 4.2L VIN S, 5.3L VIN M/P,<br>6.0L VIN H<br>**Models:** All<br>**Transmissions:** All | **Knock Sensor Circuit Malfunction**<br>Control module detected the Knock Sensor signal was below 0.8v or above 2.0v for 20 seconds during the CCM test.<br>**Possible Causes**<br>• Knock sensor signal circuit is open, shorted to ground or power<br>• Knock sensor ground circuit is open (i.e., not mounted properly)<br>• Knock sensor is damaged or has failed<br>• Control module has failed |
| **DTC: P0326**<br>**2T CCM, MIL: Yes**<br>**Years:** 2005, 2006, 2007, 2008<br>**Engines:** 2.8L VIN U , 4.2L VIN S,<br>5.3L VIN M/P, 6.0L VIN H<br>**Models:** All<br>**Transmissions:** All | **Knock Sensor Excessive Spark Retard**<br>Control module detected an unexpected voltage condition on KS diagnostic circuit.<br>**Possible Causes**<br>• Clear the codes and retest for this same code. If it resets, the Control module has failed and needs to be replaced. |
| **DTC: P0327**<br>**1T CCM, MIL: Yes**<br>**Years:** 2005, 2006, 2007, 2008<br>**Engines:** All<br>**Models:** All<br>**Transmissions:** All | **Knock Sensor Circuit Low Input Bank 1 or Single Sensor**<br>Control module detected an unexpected "low" voltage condition on the Knock Sensor (KS) 1 circuit during the CCM test. Poor driving condition. Knocking occurs.<br>**Possible Causes:**<br>• Check harness between sensor and Control module connector<br>• Check sensor installation<br>• Faulty KS |
| **DTC: P0328**<br>**1T CCM, MIL: Yes**<br>**Years:** 2005, 2006, 2007, 2008<br>**Engines:** All<br>**Models:** All<br>**Transmissions:** All | **Knock Sensor Circuit High Input Bank 1 or Single Sensor**<br>Control module detected an unexpected "high" voltage condition on the Knock Sensor (KS) 1 circuit during the CCM test. Poor driving condition. Knocking occurs.<br>**Possible Causes:**<br>• Check harness between sensor and Control module connector<br>• Check sensor<br>• Check input signal of Control module<br>• Faulty KS |
| **DTC: P0330**<br>**1T CCM, MIL: Yes**<br>**Years:** 2005, 2006, 2007, 2008<br>**Engines:** 4.2L VIN S, 5.3L VIN<br>M/P, 6.0L VIN H<br>**Models:** All<br>**Transmissions:** All | **Knock Sensor Circuit Malfunction (Bank 2)**<br>Control module did not Control module detected any change between the KS noise level at idle and noise level at 2500-3000 rpm for 5 seconds.<br>**Possible Causes**<br>• Knock sensor signal circuit is open, shorted to ground or power<br>• Knock sensor ground circuit is open (i.e., not mounted properly)<br>• Knock sensor is damaged or has failed<br>• Control module has failed |
| **DTC: P0332**<br>**1T CCM, MIL: Yes**<br>**Years:** 2005, 2006, 2007, 2008<br>**Engines:** 2.5L SOHC, DOHC VIN<br>6, 2.8L VIN U, 4.2L VIN S, 5.3L VIN<br>M/P, 6.0L VIN H<br>**Models:** All<br>**Transmissions:** All | **Knock Sensor Circuit Low Input Bank 2**<br>Control module detected an unexpected "low" voltage condition on the Knock Sensor (KS) 2 circuit during the CCM test. Poor driving condition. Knocking occurs.<br>**Possible Causes:**<br>• Check harness between sensor and Control module connector<br>• Check sensor installation<br>• Faulty KS |

| DTC | Trouble Code Title, Conditions & Possible Causes |
|---|---|
| **DTC: P0333**<br>**1T CCM, MIL: Yes**<br>**Years:** 2005, 2006, 2007, 2008<br>**Engines:** 2.5L SOHC, DOHC VIN 6, 2.8L VIN U, 4.2L VIN S, 5.3L VIN M/P, 6.0L VIN H<br>**Models:** All<br>**Transmissions:** All | **Knock Sensor Circuit High Input Bank 2**<br>Control module detected an unexpected "high" voltage condition on the Knock Sensor (KS) 2 circuit during the CCM test.<br>**Possible Causes:**<br>• Check harness between sensor and Control module connector<br>• Check sensor installation<br>• Faulty KS<br>• Check input signal of Control module |
| **DTC: P0335**<br>**1T CCM, MIL: Yes**<br>**Years:** 2005, 2006, 2007, 2008<br>**Engines:** All<br>**Models:** All<br>**Transmissions:** All | **Crankshaft Position Sensor Circuit**<br>Control module did not Control module detected any Crankshaft Position (CKP) sensor signals, or the CKP sensor signal was interrupted after then engine was running during the CCM test.<br>**Note: The engine will not start without a proper CKP sensor signal.**<br>**Possible Causes:**<br>• Check harness between the sensor and Control module connector<br>• Check condition of the sensor<br>• Faulty CKP sensor |
| **DTC: P0336**<br>**2T CCM, MIL: Yes**<br>**Years:** 2005, 2006, 2007, 2008<br>**Engines:** All<br>**Models:** All<br>**Transmissions:** All | **Crankshaft Position Sensor Circuit range/Performance**<br>Control module did not Control module detected any Crankshaft Position Sensor (CKP) signals, or the CKP sensor signal was interrupted after the engine was running during the CCM test.<br>**Note: The engine may not start, or it may stall if it loses the proper CKP sensor signal after is has been running.**<br>**Possible Causes:**<br>• Check any other DTC on display<br>• Check condition of the sensor<br>• Check the crankshaft plate<br>• Check timing chain<br>• Faulty CKP sensor |
| **DTC: P0337**<br>**2T CCM, MIL: Yes**<br>**Years:** 2005, 2006, 2007, 2008<br>**Engines:** 2.0L VIN S/Y, 2.3L VIN E/G<br>**Models:** All<br>**Transmissions:** All | **Crankshaft Position Sensor Circuit Low**<br>Control module detected less than 58 CKP sensor reference signals.<br>**Possible Causes**<br>• CKP sensor wires routed close to other wiring or components<br>• CKP sensor positive (+) signal circuit is open<br>• CKP sensor is damaged or it has failed<br>• Excessive air gap between CKP sensor and reluctor ring, or material lodged in the ring<br>• Control module has failed |
| **DTC: P0338**<br>**2T CCM, MIL: Yes**<br>**Years:** 2005, 2006, 2007, 2008<br>**Engines:** 2.8L VIN U<br>**Models:** All<br>**Transmissions:** All | **Crankshaft Position Sensor Circuit High**<br>Control module detected more than 58 CKP sensor reference signals.<br>**Possible Causes**<br>• CKP sensor wires routed close to other wiring or components<br>• CKP sensor resistance out of specification (700 00 ohms)<br>• CKP sensor is damaged or it has failed<br>• Excessive air gap between the CKP sensor and reluctor ring<br>• Foreign material lodged between CKP sensor and reluctor ring.<br>• Control module has failed |
| **DTC: P0339**<br>**2T CCM, MIL: Yes**<br>**Years:** 2005, 2006, 2007, 2008<br>**Engines:** 2.0L VIN S/Y<br>**Models:** All<br>**Transmissions:** All | **Crankshaft Position Sensor Circuit Intermittent**<br>Control module did not Control module detected any Camshaft Position sensor (Hall Effect) pulses at least once during 2 complete crankshaft revolutions during the CCM test.<br>**Possible Causes**<br>• Camshaft Position sensor signal circuit is open or shorted to ground<br>• Camshaft Position sensor VREF circuit is open between sensor and Control module<br>• Camshaft Position sensor ground (Low Reference) circuit is open<br>• Camshaft Position sensor is damaged or it failed (check the reluctor wheel)<br>• Control module has failed |

| DTC | Trouble Code Title, Conditions & Possible Causes |
|---|---|
| **DTC: P0340**<br>**1T CCM, MIL: Yes**<br>**Years:** 2005, 2006, 2007, 2008<br>**Engines:** All<br>**Models:** All<br>**Transmissions:** All | **Camshaft Position Sensor Circuit Bank 1 or Single Sensor**<br>Control module did not Control module detected any Camshaft Position (Camshaft Position ) sensor signals, or the Camshaft Position sensor signal was interrupted after the engine was running during the CCM test.<br>**Note: The engine may not start without a proper Camshaft Position sensor signal.**<br>**Possible Causes:**<br><ul><li>Check any other DTC on display</li><li>Check power supply</li><li>Check harness between sensor connector and Control module</li><li>Check sensor installation and condition</li><li>Check the sensor</li><li>Check for poor contact</li><li>Faulty Camshaft Position sensor</li></ul> |
| **DTC: P0341**<br>**1T CCM, MIL: Yes**<br>**Years:** 2005, 2006, 2007, 2008<br>**Engines:** All<br>**Models:** All<br>**Transmissions:** All | **Camshaft Position Sensor Signal Range/Performance**<br>Control module detected more than 15 Camshaft Position sensor resynchronizations during a 4 minute 16 second period.<br>**Possible Causes**<br><ul><li>Camshaft Position sensor signal circuit is open, shorted to ground or VREF</li><li>Camshaft Position sensor signal wire is routed to close to the Generator, spark plug wires or any other possible cause of EMI/RFI under the hood (check for high power receivers causing interference)</li><li>Camshaft Position sensor "shield" ground circuit is open (intermittent fault)</li><li>Camshaft Position sensor is cracked or damaged (check the reluctor wheel)</li><li>Control module has failed</li></ul> |
| **DTC: P0342**<br>**1T CCM, MIL: Yes**<br>**Years:** 2005, 2006, 2007, 2008<br>**Engines:** 2.8L VIN U<br>**Models:** All<br>**Transmissions:** All | **Camshaft Position Sensor Circuit Input Low**<br>Control module detected the Camshaft Position sensor signal was in a low state (when the signal should have been in a high state) for 1.5 seconds in the test.<br>**Possible Causes**<br><ul><li>Camshaft reluctor wheel is damaged or foreign material present</li><li>Camshaft Position sensor signal circuit is open, shorted to ground or VREF</li><li>Camshaft Position sensor is contacting the reluctor wheel or is damaged</li><li>Control module has failed</li></ul> |
| **DTC: P0343**<br>**1T CCM, MIL: Yes**<br>**Years:** 2005, 2006, 2007, 2008<br>**Engines:** 2.8L VIN U<br>**Models:** All<br>**Transmissions:** All | **Camshaft Position Sensor Circuit Input High**<br>Control module detected the Camshaft Position sensor signal was stuck high (when the signal should have been in a low state) for 1.5 seconds in the CCM test.<br>**Possible Causes**<br><ul><li>Camshaft Position sensor connector is damaged, loose or shorted</li><li>Camshaft Position sensor low reference circuit is open or shorted to VREF</li><li>Camshaft reluctor wheel is damaged or foreign material present</li><li>Camshaft Position sensor is contacting the reluctor wheel or is damaged</li><li>Control module has failed</li></ul> |
| **DTC: P0345**<br>**1T CCM, MIL: Yes**<br>**Years:** 2005, 2006, 2007, 2008<br>**Engines:** 2.5L DOHC VIN 6, 2.8L VIN U<br>**Models:** All<br>**Transmissions:** All | **Camshaft Position Sensor Circuit**<br>Control module detected the open or short circuit of the camshaft position sensor. Control module detects when the number of camshaft signals remains abnormal. Engine stalls. Failure of engine to start.<br>**Possible Causes:**<br><ul><li>Check any other DTC on display</li><li>Check power supply</li><li>Check harness between sensor connector and Control module</li><li>Check the sensor</li><li>Check poor contact</li><li>Check the installation and condition of the sensor</li></ul> |
| **DTC: P0346**<br>**1T CCM, MIL: Yes**<br>**Years:** 2005, 2006, 2007, 2008<br>**Engines:** 2.8L VIN U<br>**Models:** All<br>**Transmissions:** All | **Camshaft Position Sensor Circuit Range/Performance**<br>Control module detected the open or short circuit of the camshaft position sensor. Control module detects when the number of camshaft signals remains abnormal. Engine stalls. Failure of engine to start.<br>**Possible Causes:**<br><ul><li>Check any other DTC on display</li><li>Check power supply</li><li>Check harness between sensor connector and Control module</li><li>Check the sensor</li><li>Check poor contact</li><li>Check the installation and condition of the sensor</li></ul> |

| DTC | Trouble Code Title, Conditions & Possible Causes |
|---|---|
| **DTC: P0347**<br>**1T CCM, MIL: Yes**<br>**Years:** 2005, 2006, 2007, 2008<br>**Engines:** 2.8L VIN U<br>**Models:** All<br>**Transmissions:** All | **Camshaft Position Sensor Circuit Low**<br>Control module detected the open or short circuit of the camshaft position sensor. Control module detects when the number of camshaft signals remains abnormal. Engine stalls. Failure of engine to start.<br>**Possible Causes:**<br>• Check any other DTC on display<br>• Check power supply<br>• Check harness between sensor connector and Control module<br>• Check the sensor<br>• Check poor contact<br>• Check the installation and condition of the sensor |
| **DTC: P0348**<br>**1T CCM, MIL: Yes**<br>**Years:** 2005, 2006, 2007, 2008<br>**Engines:** 2.8L VIN U<br>**Models:** All<br>**Transmissions:** All | **Camshaft Position Sensor Circuit High**<br>Control module detected the open or short circuit of the camshaft position sensor. Control module detects when the number of camshaft signals remains abnormal. Engine stalls. Failure of engine to start.<br>**Possible Causes:**<br>• Check any other DTC on display<br>• Check power supply<br>• Check harness between sensor connector and Control module<br>• Check the sensor<br>• Check poor contact<br>• Check the installation and condition of the sensor |
| **DTC: P0351**<br>**2T CCM, MIL: Yes**<br>**Years:** 2005, 2006, 2007, 2008<br>**Models:** All<br>**Engines:** 2.8L VIN U, 4.2L VIN S, 5.3L VIN M/P, 6.0L VIN H<br>**Transmissions:** All | **Ignition Coil 1 Control Circuit**<br>Control module detects the IC circuit is grounded, open, or shorted to voltage for less than 1 second.<br>**Possible Causes:**<br>• IC circuit for a short to ground<br>• IC circuit for a short to voltage<br>• IC circuit for open condition<br>• Intermittent or poor connection at the ignition coil<br>• Intermittent or poor connection at the Control module<br>• Open in the ignition voltage circuit<br>• Faulty ignition coil<br>• Control module has failed |
| **DTC: P0352**<br>**2T CCM, MIL: Yes**<br>**Years:** 2005, 2006, 2007, 2008<br>**Models:** All<br>**Engines:** 2.8L VIN U, 4.2L VIN S, 5.3L VIN M/P, 6.0L VIN H<br>**Transmissions:** All | **Ignition Coil 2 Control Circuit**<br>Control module detects the IC circuit is grounded, open, or shorted to voltage for less than 1 second.<br>**Possible Causes:**<br>• IC circuit for a short to ground<br>• IC circuit for a short to voltage<br>• IC circuit for open condition<br>• Intermittent or poor connection at the ignition coil<br>• Intermittent or poor connection at the Control module<br>• Open in the ignition voltage circuit<br>• Faulty ignition coil<br>• Control module has failed |
| **DTC: P0353**<br>**2T CCM, MIL: Yes**<br>**Years:** 2005, 2006, 2007, 2008<br>**Models:** All<br>**Engines:** 2.8L VIN U, 4.2L VIN S, 5.3L VIN M/P, 6.0L VIN H<br>**Transmissions:** All | **Ignition Coil 3 Control Circuit**<br>Control module detects the IC circuit is grounded, open, or shorted to voltage for less than 1 second.<br>**Possible Causes:**<br>• IC circuit for a short to ground<br>• IC circuit for a short to voltage<br>• IC circuit for open condition<br>• Intermittent or poor connection at the ignition coil<br>• Intermittent or poor connection at the Control module<br>• Open in the ignition voltage circuit<br>• Faulty ignition coil<br>• Control module has failed |

| DTC | Trouble Code Title, Conditions & Possible Causes |
|---|---|
| **DTC: P0354**<br>**2T CCM, MIL: Yes**<br>**Years:** 2005, 2006, 2007, 2008<br>**Models:** All<br>**Engines:** 2.8L VIN U, 4.2L VIN S, 5.3L VIN M/P, 6.0L VIN H<br>**Transmissions:** All | **Ignition Coil 4 Control Circuit**<br>Control module detects the IC circuit is grounded, open, or shorted to voltage for less than 1 second.<br>**Possible Causes:**<br>• IC circuit for a short to ground<br>• IC circuit for a short to voltage<br>• IC circuit for open condition<br>• Intermittent or poor connection at the ignition coil<br>• Intermittent or poor connection at the Control module<br>• Open in the ignition voltage circuit<br>• Faulty ignition coil<br>• Control module has failed |
| **DTC: P0355**<br>**2T CCM, MIL: Yes**<br>**Years:** 2005, 2006, 2007, 2008<br>**Models:** All<br>**Engines:** 2.8L VIN U, 4.2L VIN S, 5.3L VIN M/P, 6.0L VIN H<br>**Transmissions:** All | **Ignition Coil 5 Control Circuit**<br>Control module detects the IC circuit is grounded, open, or shorted to voltage for less than 1 second.<br>**Possible Causes:**<br>• IC circuit for a short to ground<br>• IC circuit for a short to voltage<br>• IC circuit for open condition<br>• Intermittent or poor connection at the ignition coil<br>• Intermittent or poor connection at the Control module<br>• Open in the ignition voltage circuit<br>• Faulty ignition coil<br>• Control module has failed |
| **DTC: P0356**<br>**2T CCM, MIL: Yes**<br>**Years:** 2005, 2006, 2007, 2008<br>**Models:** All<br>**Engines:** 2.8L VIN U, 4.2L VIN S, 5.3L VIN M/P, 6.0L VIN H<br>**Transmissions:** All | **Ignition Coil 6 Control Circuit**<br>Control module detects the IC circuit is grounded, open, or shorted to voltage for less than 1 second.<br>**Possible Causes:**<br>• IC circuit for a short to ground<br>• IC circuit for a short to voltage<br>• IC circuit for open condition<br>• Intermittent or poor connection at the ignition coil<br>• Intermittent or poor connection at the Control module<br>• Open in the ignition voltage circuit<br>• Faulty ignition coil<br>• Control module has failed |
| **DTC: P0357**<br>**2T CCM, MIL: Yes**<br>**Years:** 2005, 2006, 2007, 2008<br>**Models:** All<br>**Engines:** 4.2L VIN S, 5.3L VIN M/P, 6.0L VIN H<br>**Transmissions:** All | **Ignition Coil 7 Control Circuit**<br>Control module detects the IC circuit is grounded, open, or shorted to voltage for less than 1 second.<br>**Possible Causes:**<br>• IC circuit for a short to ground<br>• IC circuit for a short to voltage<br>• IC circuit for open condition<br>• Intermittent or poor connection at the ignition coil<br>• Intermittent or poor connection at the Control module<br>• Open in the ignition voltage circuit<br>• Faulty ignition coil<br>• Control module has failed |
| **DTC: P0358**<br>**2T CCM, MIL: Yes**<br>**Years:** 2005, 2006, 2007, 2008<br>**Models:** All<br>**Engines:** 4.2L VIN S, 5.3L VIN M/P, 6.0L VIN H<br>**Transmissions:** All | **Ignition Coil 8 Control Circuit**<br>Control module detects the IC circuit is grounded, open, or shorted to voltage for less than 1 second.<br>**Possible Causes:**<br>• IC circuit for a short to ground<br>• IC circuit for a short to voltage<br>• IC circuit for open condition<br>• Intermittent or poor connection at the ignition coil<br>• Intermittent or poor connection at the Control module<br>• Open in the ignition voltage circuit<br>• Faulty ignition coil<br>• Control module has failed |

| DTC | Trouble Code Title, Conditions & Possible Causes |
|---|---|
| **DTC: P0365**<br>**1T CCM, MIL: Yes**<br>**Years:** 2008<br>**Engines:** 4.2L VIN S<br>**Models:** All<br>**Transmissions:** All | **Camshaft Position Sensor Circuit**<br>Control module detected the open or short circuit of the camshaft position sensor. Control module detects when the number of camshaft signals remains abnormal. Engine stalls. Failure of engine to start.<br>**Possible Causes:**<br>• Check any other DTC on display<br>• Check power supply<br>• Check harness between sensor connector and Control module<br>• Check the sensor<br>• Check poor contact<br>• Check the installation and condition of the sensor |
| **DTC: P0366**<br>**1T CCM, MIL: Yes**<br>**Years:** 2008<br>**Engines:** 4.2L VIN S<br>**Models:** All<br>**Transmissions:** All | **Camshaft Position Sensor Circuit Range/Performance**<br>Control module detected the open or short circuit of the camshaft position sensor. Control module detects when the number of camshaft signals remains abnormal. Engine stalls. Failure of engine to start.<br>**Possible Causes:**<br>• Check any other DTC on display<br>• Check power supply<br>• Check harness between sensor connector and Control module<br>• Check the sensor<br>• Check poor contact<br>• Check the installation and condition of the sensor |
| **DTC: P0390**<br>**1T CCM, MIL: Yes**<br>**Years:** 2008<br>**Engines:** 2.5L SOHC, DOHC VIN 6<br>**Models:** All<br>**Transmissions:** All | **Camshaft Position Sensor Circuit Bank 2**<br>Control module detected the open or short circuit of the camshaft position sensor. Control module detects when the number of camshaft signals remains abnormal. Engine stalls. Failure of engine to start.<br>**Possible Causes:**<br>• Check any other DTC on display<br>• Check power supply<br>• Check harness between sensor connector and Control module<br>• Check the sensor<br>• Check poor contact<br>• Check the installation and condition of the sensor |
| **DTC: P0400**<br>**2T CCM, MIL: Yes**<br>**Years:** 2005, 2006, 2007, 2008<br>**Engines:** 2.5L SOHC, VIN 6,<br>**Models:** All<br>**Transmissions:** All | **Exhaust Gas Recirculation Flow**<br>Movement performance problem when engine is low speed. Erroneous idling. Movement performance problem.<br>**Possible Causes:**<br>• Plugged piping or foreign objects caught in the EGR system<br>• Manifold absolute pressure sensor and throttle body improperly installed<br>• Faulty EGR valve |
| **DTC: P0410**<br>**2T CCM, MIL: Yes**<br>**Years:** 2005, 2006, 2007, 2008<br>**Engines:** 2.5L DOHC<br>**Models:** All<br>**Transmissions:** All | **Secondary Air Injection System**<br>Faulty secondary air delivery pipe pressure, pulse of secondary air delivery pipe pressure and secondary air pipe airflow amount.<br>**Possible Causes:**<br>• Blown fuse<br>• Check harness between fuse, pump harness and ground<br>• Damage, clogged or disconnected duct<br>• Faulty secondary air combination valve<br>• Open circuit between air pump relay and pump<br>• Faulty air pump relay<br>• Faulty air pump relay power supply circuit<br>• Faulty air pump |
| **DTC: P0411**<br>**2T CCM, MIL: Yes**<br>**Years:** 2005, 2006, 2007, 2008<br>**Engines:** 2.5L DOHC, 2.8L VIN U,<br>4.2L VIN S<br>**Models:** All<br>**Transmissions:** All | **Secondary Air Injection System Incorrect Flow Detected**<br>Faulty secondary air delivery pipe pressure, pulse of secondary air delivery pipe pressure and secondary air pipe airflow amount.<br>**Possible Causes:**<br>• Blown fuse<br>• Check harness between fuse, pump harness and ground<br>• Damage, clogged or disconnected duct<br>• Faulty secondary air combination valve<br>• Open circuit between air pump relay and pump<br>• Faulty air pump relay<br>• Faulty air pump relay power supply circuit<br>• Faulty air pump |

| DTC | Trouble Code Title, Conditions & Possible Causes |
|---|---|
| **DTC: P0411**<br>**2T CCM, MIL: Yes**<br>**Years:** 2005, 2006, 2007, 2008<br>**Engines:** 4.2L VIN S<br>**Models:** All<br>**Transmissions:** All | **Secondary Air System Control Circuit Malfunction**<br>Control module detected the Actual and Commanded state of the AIR solenoid driver did not match for 5 seconds during the test.<br>**Possible Causes**<br>• AIR solenoid control circuit is open, shorted to ground or B+<br>• AIR solenoid power circuit is open (test power from IGN fuse)<br>• AIR solenoid is damaged or has failed, or Control module has failed |
| **DTC: P0413**<br>**1T CCM, MIL: Yes**<br>**Years:** 2005, 2006, 2007, 2008<br>**Engines:** 2.5L DOHC<br>**Models:** All<br>**Transmissions:** All | **Secondary Air Injection System Switching Valve Circuit Open**<br>Control module output level differs from the actual terminal level.<br>**Possible Causes:**<br>• Open circuit of harness between Control module and secondary air combination valve relay<br>• Ground short circuit of harness between Control module and secondary air combination valve relay |
| **DTC: P0414**<br>**1T CCM, MIL: Yes**<br>**Years:** 2005, 2006, 2007, 2008<br>**Engines:** 2.5L DOHC<br>**Models:** All<br>**Transmissions:** All | **Secondary Air Injection System Switching Valve Circuit Shorted**<br>Control module output level differs from the actual terminal level.<br>**Possible Causes:**<br>• Short circuit to power in the harness between Control module and secondary air combination valve relay 1<br>• Poor contact of Control module connector |
| **DTC: P0416**<br>**1T CCM, MIL: Yes**<br>**Years:** 2005, 2006, 2007, 2008<br>**Engines:** 2.5L DOHC<br>**Models:** All<br>**Transmissions:** All | **Secondary Air Injection System Switching Valve Circuit Open**<br>Control module output level differs from the actual terminal level.<br>**Possible Causes:**<br>• Open circuit of harness between Control module and secondary air combination valve relay<br>• Ground short circuit of harness between Control module and secondary air combination valve relay. |
| **DTC: P0417**<br>**1T CCM, MIL: Yes**<br>**Years:** 2005, 2006, 2007, 2008<br>**Engines:** 2.5L DOHC<br>**Models:** All<br>**Transmissions:** All | **Secondary Air Injection System Switching Valve Circuit Shorted**<br>Control module output level differs from the actual terminal level.<br>**Possible Causes:**<br>• Short circuit to power in the harness between Control module and secondary air combination valve relay 1<br>• Poor contact of Control module connector |
| **DTC: P0418**<br>**1T CCM, MIL: Yes**<br>**Years:** 2005, 2006, 2007, 2008<br>**Engines:** 2.5L DOHC, 2.8L VIN U, 4.2L VIN S<br>**Models:** All<br>**Transmissions:** All | **Secondary Air Injection System Control Circuit**<br>Control module output level differs from the actual terminal level.<br>**Possible Causes:**<br>• Short circuit of harness between Control module and secondary air pump relay<br>• Open circuit of harness between Control module and secondary air pump relay |
| **DTC: P0420**<br>**2T CCM, MIL: Yes**<br>**Years:** 2005, 2006, 2007, 2008<br>**Engines:** All<br>**Models:** All<br>**Transmissions:** All | **Catalyst System Efficiency Below Threshold Bank 1**<br>Control module detected the amplitudes of the O2S and O2S signals were too similar during the test. Engine stalls. Idle mixture is out of specifications.<br>**Possible Causes:**<br>• Check any other DTC on display<br>• Check harness between fuse and rear oxygen sensor connector<br>• Check exhaust system<br>• Faulty catalytic converter |
| **DTC: P0430**<br>**2T CAT, MIL: Yes**<br>**Years:** 2005, 2006, 2007, 2008<br>**Models:** All<br>**Engines:** 4.2L VIN S, 5.3L VIN M/P, 6.0L VIN H<br>**Transmissions:** All | **Catalyst System Efficiency Below Threshold Bank 2**<br>Control module has determined the catalyst efficiency has degraded below a calibrated threshold. This diagnostic may conclude in one test attempt. However, this diagnostic may require as many as 18 test attempts, which would require at least 3 drive cycles. Each test attempt may conclude within approximately 1 minute.<br>**Possible Causes:**<br>• Air leaks at the exhaust manifold or in the exhaust pipes<br>• Base engine problems (i.e., high engine oil or coolant usage)<br>• Catalytic converter is damaged, contaminated or has failed<br>• Continuous engine misfire conditions, or weak or low coil output<br>• Front O2S or rear O2S is contaminated with fuel or moisture<br>• Rear O2S is loose in the mounting hole (check it for a leak)<br>• Front O2S older (aged) than the rear O2S (O2Sis lazy) |

| DTC | Trouble Code Title, Conditions & Possible Causes |
|---|---|
| **DTC: P0441**<br>**2T CAT, MIL: Yes**<br>**Years:** 2005, 2006, 2007, 2008<br>**Models:** All<br>**Engines:** 2.0L VIN S/Y, 2.3L VIN E/G, 4.2L VIN S, 5.3L VIN M/P, 6.0L VIN H<br>**Transmissions:** All | **EVAP System No Flow During Purge**<br>Control module detected the EVAP Vacuum switch indicated low with Purge enabled for 5 seconds.<br>**Possible Causes**<br>• Charcoal canister is loaded with fuel or moisture<br>• EVAP switch is damaged, disconnected or it failed<br>• Fuel filler cap is loose, cross-threaded, damaged or wrong part<br>• Fuel tank vapor line(s) is clogged, damaged or disconnected<br>• Purge valve vapor line is clogged, damaged, or disconnected<br>• Purge solenoid is damaged or sticking (it may be stuck closed)<br>• Purge solenoid power circuit is open (check the fuse)<br>• Control module has failed |
| **DTC: P0442**<br>**2T CCM, MIL: Yes**<br>**Years:** 2005, 2006, 2007, 2008<br>**Engines:** All<br>**Models:** All<br>**Transmissions:** All | **Evaporative Emission Control System Leak Detected (small leak)**<br>Control module detected a vacuum decaying condition existed during the test. There is a hole of more than 0.04 inch (1.0 mm) diameter in evaporation system or fuel tank. Possible fuel odor.<br>**Possible Causes:**<br>• Check any other DTC on display<br>• Check fuel filler cap<br>• Check fuel filler pipe packing<br>• Check the drain valve<br>• Check the purge control solenoid valve<br>• Check the pressure control solenoid valve<br>• Check evaporative emission control system line<br>• Check canister<br>• Check fuel tank<br>• Check any other mechanical components in the evaporative control system |
| **DTC: P0443**<br>**2T EVAP, MIL: Yes**<br>**Years:** 2005, 2006, 2007, 2008<br>**Models:** All<br>**Engines:** 2.8L VIN U, 4.2L VIN S, 5.3L VIN M/P, 6.0L VIN H<br>**Transmissions:** All | **Evaporative Emission (EVAP) Purge Solenoid Control Circuit**<br>Control module detects that the commanded state of the driver and the actual state of the control circuit do not match for a minimum of 5 seconds.<br>**Possible Causes:**<br>• EVAP control circuit of the canister purge solenoid valve for an open or short to voltage<br>• EVAP control circuit of the canister purge solenoid valve for a short to ground<br>• Inspect for poor connections<br>• Open or short to ground in the ignition 1 voltage circuit<br>• Faulty EVAP canister purge solenoid valve<br>• Control module has failed |
| **DTC: P0444**<br>**2T EVAP, MIL: Yes**<br>**Years:** 2005, 2006, 2007, 2008<br>**Models:** All<br>**Engines:** 2.0L VIN S/Y, 2.3L VIN E/G<br>**Transmissions:** All | **EVAP Purge Solenoid Control Circuit Input Low**<br>Control module detected 0v on the EVAP purge valve control circuit with the control driver commanded "off" for over 50 seconds.<br>**Possible Causes**<br>• Purge solenoid connector is damaged or shorted<br>• Purge solenoid control circuit is shorted to ground<br>• Control module has failed |
| **DTC: P0445**<br>**2T EVAP, MIL: Yes**<br>**Years:** 2005, 2006, 2007, 2008<br>**Models:** All<br>**Engines:** 2.0L VIN S/Y, 2.3L VIN E/G<br>**Transmissions:** All | **EVAP Purge Solenoid Control Circuit Input High**<br>Control module detected 12v on the EVAP purge valve control circuit with the control driver commanded "on" for over 50 seconds.<br>**Possible Causes**<br>• Purge solenoid connector is damaged or shorted<br>• Purge solenoid control circuit is shorted to system power (B+)<br>• Purge solenoid valve is damaged or it has failed<br>• Control module has failed |
| **DTC: P0446**<br>**2T EVAP, MIL: Yes**<br>**Years:** 2005, 2006, 2007, 2008<br>**Models:** All<br>**Engines:** All<br>**Transmissions:** All | **Evaporative Emission (EVAP) Vent System Performance**<br>Control module detected EVAP purge valve control circuit performance problem.<br>**Possible Causes:**<br>• EVAP vent fresh air hose is clogged, kinked or restricted<br>• EVAP Vent solenoid is contaminated, damaged or has failed<br>• EVAP Canister plugged or severely restricted<br>• Fuel Cap or EVAP Service Port leaking<br>• Fuel vapor lines or purge lines damaged or leaking<br>• FTP sensor is out-of-calibration, damaged or "skewed"<br>• Control module has failed |

| DTC | Trouble Code Title, Conditions & Possible Causes |
|---|---|
| **DTC: P0447**<br>**2T CCM, MIL: Yes**<br>**Years:** 2005, 2006, 2007, 2008<br>**Engines:** 2.0L VIN 2, 2.5L SOHC, DOHC VIN 6<br>**Models:** All<br>**Transmissions:** All | **Evaporative Emission Control System Vent Control Circuit Open**<br>Control module detected an open or short circuit of the drain valve.<br>**Possible Causes:**<br>• Check output signal from Control module<br>• Check poor contact<br>• Check harness between drain valve and Control module connector<br>• Check drain valve<br>• Check power supply to drain valve |
| **DTC: P0448**<br>**2T CCM, MIL: Yes**<br>**Years:** 2005, 2006, 2007, 2008<br>**Engines:** 2.0L VIN 2, 2.5L SOHC, DOHC VIN 6<br>**Models:** All<br>**Transmissions:** All | **Evaporative Emission Control System Vent Control Circuit Shorted**<br>Control module detected an open or short circuit of the drain valve.<br>**Possible Causes:**<br>• Check input signal from Control module<br>• Check poor contact<br>• Check harness between drain valve and Control module connector<br>• Check drain valve |
| **DTC: P0449**<br>**2T EVAP, MIL: Yes**<br>**Years:** 2005, 2006, 2007, 2008<br>**Models:** All<br>**Engines:** 2.8L VIN U, 4.2L VIN S, 5.3L VIN M/P, 6.0L VIN H<br>**Transmissions:** All | **Evaporative Emission (EVAP) Vent Solenoid Control Circuit**<br>Control module detects that the commanded state of the driver and the actual state of the control circuit do not match for a minimum of 5 seconds.<br>**Possible Causes:**<br>• EVAP control circuit of the canister vent solenoid shorted to voltage or open<br>• EVAP control circuit of the canister vent solenoid shorted to ground<br>• Inspect for poor connections<br>• Open or short to ground in the battery positive voltage circuit.<br>• Faulty EVAP canister vent solenoid valve<br>• Control module has failed |
| **DTC: P0451**<br>**2T CCM, MIL: Yes**<br>**Years:** 2005, 2006, 2007, 2008<br>**Engines:** All<br>**Models:** All<br>**Transmissions:** All | **Evaporative Emission Control System Pressure Sensor Malfunction**<br>Control module detected the tank pressure sensor output property abnormality. Control module detects when there is no pressure variation, which should exist in the tank, considering the engine status.<br>**Possible Causes:**<br>• Check any other DTC on display<br>• Check fuel filler cap<br>• Check pressure vacuum line<br>• Faulty fuel tank pressure sensor |
| **DTC: P0452**<br>**2T CCM, MIL: Yes**<br>**Years:** 2005, 2006, 2007, 2008<br>**Engines:** All<br>**Models:** All<br>**Transmissions:** All | **Evaporative Emission Control System Pressure Sensor Low Input**<br>Control module detected Control module detected an open or short circuit of the fuel tank pressure sensor.<br>**Possible Causes:**<br>• Using a scan tool, check current data<br>• Check power supply to fuel tank pressure sensor<br>• Check input signal to Control module<br>• Check harness between Control module and coupling connector in rear wiring harness<br>• Check fuel tank cord<br>• Check the purge control solenoid valve<br>• Check poor contact<br>• Faulty fuel tank pressure sensor |
| **DTC P0453**<br>**1T CCM, MIL: Yes**<br>**Years:** 2005, 2006, 2007, 2008<br>**Engines:** All<br>**Models:** All<br>**Transmissions:** All | **Evaporative Emission Control System Pressure Sensor High Input**<br>Control module detected Control module detected an open or short circuit of the fuel tank pressure sensor.<br>**Possible Causes:**<br>• Using a scan tool, check current data<br>• Check power supply to fuel tank pressure sensor<br>• Check input signal to Control module<br>• Check harness between Control module and coupling connector in rear wiring harness<br>• Check fuel tank cord<br>• Check poor contact<br>• Check harness between Control module and fuel tank pressure sensor connector<br>• Faulty fuel tank pressure sensor |

| DTC | Trouble Code Title, Conditions & Possible Causes |
|---|---|
| **DTC: P0454**<br>**2T CCM, MIL: Yes**<br>**Years:** 2005, 2006, 2007, 2008<br>**Models:** All<br>**Engines:** 4.2L VIN S, 5.3L VIN M/P, 6.0L VIN H<br>**Transmissions:** All | **Fuel Tank Pressure (FTP) Sensor Circuit Intermittent**<br>Control module detects an abrupt FTP signal change, other than a refueling event, this DTC will set.<br>**Possible Causes:**<br>• Intermittent or poor connection at the FTP sensor<br>• Faulty FTP sensor |
| **DTC: P0455**<br>**2T CCM, MIL: Yes**<br>**Years:** 2005, 2006, 2007, 2008<br>**Models:** All<br>**Engines:** All<br>**Transmissions:** All | **Evaporative Emission (EVAP) System Large Leak Detected**<br>The EVAP system is not able to achieve or maintain vacuum during the diagnostic test.<br>**Possible Causes:**<br>• Loose, missing, incorrect, or damaged fuel fill cap<br>• Vacuum lines<br>• EVAP hoses<br>• EVAP vent solenoid valve<br>• FTP sensor is damaged or it has failed |
| **DTC: P0456**<br>**2T CCM, MIL: Yes**<br>**Years:** 2005, 2006, 2007, 2008<br>**Engines:** All<br>**Models:** All<br>**Transmissions:** All | **Evaporative Emission Control System Pressure Sensor Very Small Leak Detected**<br>Control module detected a vacuum decaying condition existed during the test. There is a hole of more than 0.04 inch (1.0 mm) diameter in evaporation system or fuel tank. Possible fuel odor.<br>**Possible Causes:**<br>• Check any other DTC on display<br>• Check fuel filler cap<br>• Check fuel filler pipe packing<br>• Check drain valve<br>• Check purge control solenoid valve<br>• Check pressure control solenoid valve<br>• Check harness between Control module and fuel tank pressure sensor connector<br>• Check evaporative emission control system line<br>• Check canister<br>• Check any other mechanical component in the system |
| **DTC: P0457**<br>**2T CCM, MIL: Yes**<br>**Years:** 2005, 2006, 2007, 2008<br>**Engines:** 2.0L VIN 2, 2.5L SOHC, DOHC VIN 6,<br>**Models:** All<br>**Transmissions:** All | **Evaporative Emission Control System Leak Detected (Fuel Cap Loose/Off)**<br>Control module detected an unexpected low voltage condition on the EVAP Purge Controls solenoid control circuit.<br>**Possible Causes:**<br>• Check any other DTC on display<br>• Check fuel filler cap<br>• Check fuel filler pipe packing<br>• Check drain valve<br>• Check purge control solenoid valve<br>• Check pressure control solenoid valve<br>• Check canister<br>• Check fuel tank<br>• Check any other mechanical component in the system |
| **DTC: P0458**<br>**2T CCM, MIL: Yes**<br>**Years:** 2005, 2006, 2007, 2008<br>**Engines:** 2.0L VIN 2, 2.5L SOHC, DOHC VIN 6, 2.8L VIN U<br>**Models:** All<br>**Transmissions:** All | **Evaporative Emission System Purge Control Valve Circuit Low**<br>Control module detected open or short circuit of the purge control solenoid valve. Elapsed time after start up 1 second. Improper idling.<br>**Possible Causes:**<br>• Check output signal of Control module<br>• Check harness between purge control solenoid valve and Control module connector<br>• Check purge control solenoid valve<br>• Check power supply to purge control solenoid valve<br>• Check poor contact |
| **DTC: P0459**<br>**2T CCM, MIL: Yes**<br>**Years:** 2005, 2006, 2007, 2008<br>**Engines:** 2.0L VIN 2, 2.5L SOHC, DOHC VIN 6, 2.8L VIN U<br>**Models:** All<br>**Transmissions:** All | **Evaporative Emission System Purge Control Valve Circuit High**<br>Control module detected open or short circuit of the purge control solenoid valve. Elapsed time after start up 1 second. Improper idling.<br>**Possible Causes:**<br>• Check output signal of Control module<br>• Check harness between purge control solenoid valve and Control module connector<br>• Check purge control solenoid valve<br>• Check poor contact<br>• Faulty purge control solenoid valve |

| DTC | Trouble Code Title, Conditions & Possible Causes |
|---|---|
| **DTC: P0460**<br>**2T CCM, MIL: Yes**<br>**Years:** 2005, 2006, 2007, 2008<br>**Engines:** 2.0L VIN 2, 2.0L VIN S/Y, 2.5L SOHC, DOHC VIN 6, 2.8L VIN U<br>**Models:** All<br>**Transmissions:** All | **Fuel Level Sensor Circuit**<br>Control module detected the Fuel Level sensor signal was too low or tool high during the CCM test.<br>**Possible Causes:**<br>• Check any other DTC on display<br>• Faulty fuel level sensor and fuel sub level sensor |
| **DTC: P0461**<br>**2T CCM, MIL: Yes**<br>**Years:** 2005, 2006, 2007, 2008<br>**Engines:** 2.0L VIN 2, 2.0L VIN S/Y, 2.5L SOHC, DOHC VIN 6, 2.8L VIN U<br>**Models:** All<br>**Transmissions:** All | **Fuel Level Sensor Circuit Range/Performance**<br>Control module detected the Fuel Level sensor signal was too low or tool high during the CCM test.<br>**Possible Causes:**<br>• Check any other DTC on display<br>• Faulty fuel level sensor and fuel sub level sensor |
| **DTC: P0462**<br>**2T CCM, MIL: Yes**<br>**Years:** 2005, 2006, 2007, 2008<br>**Engines:** 2.0L VIN 2, 2.0L VIN S/Y, 2.5L SOHC, DOHC VIN 6, 2.8L VIN U<br>**Models:** All<br>**Transmissions:** All | **Fuel Level Sensor Circuit Low**<br>Control module detected the Fuel Level sensor signal indicated less than 0.10 volt at any time during the CCM continuous test.<br>**Possible Causes:**<br>• Check speedometer and tachometer operation<br>• Check input signal of Control module<br>• Check input voltage from Control module<br>• Check harness between Control module and combination meter connector<br>• Check fuel tank cord<br>• Check fuel level sensor<br>• Check fuel sub level sensor |
| **DTC: P0463**<br>**2T CCM, MIL: Yes**<br>**Years:** 2005, 2006, 2007, 2008<br>**Engines:** 2.0L VIN 2, 2.0L VIN S/Y, 2.5L SOHC, DOHC VIN 6, 2.8L VIN U<br>**Models:** All<br>**Transmissions:** All | **Fuel Level Sensor Circuit High**<br>Control module detected the Fuel Level sensor signal indicated more than 4.60 volts at any time during the CCM continuous test.<br>**Possible Causes:**<br>• Check any other DTC on display<br>• Check harness between Control module and combination meter connector |
| **DTC: P0464**<br>**2T CCM, MIL: Yes**<br>**Years:** 2005, 2006, 2007, 2008<br>**Engines:** 2.0L VIN 2, 2.5L SOHC, DOHC VIN 6<br>**Models:** All<br>**Transmissions:** All | **Fuel level Sensor Circuit Intermittent**<br>Control module detected the unstable output faults from the fuel level sensor caused by noise. Control module detects when the Max value and cumulative value of output voltage variation of the fuel level sensor is larger than the threshold value.<br>**Possible Causes:**<br>• Check any other DTC on display<br>• Check harness between Control module and combination meter connector |
| **DTC: P0480**<br>**2T CCM, MIL: Yes**<br>**Years:** 2005, 2006, 2007, 2008<br>**Models:** All<br>**Engines:** 2.8L VIN U<br>**Transmissions:** All | **Cooling Fan Relay 1 Control Circuit**<br>The commanded state of the Output Driver Module (ODM) and the actual state of the control circuit do not match. The condition is present for more than 5 seconds.<br>**Possible Causes:**<br>• Fan control relay control circuit is open or shorted to ground<br>• Fan control relay control circuit is shorted to system power<br>• Fan control relay power circuit is open (check Cool Fan 1 fuse)<br>• Fan control relay is damaged or has failed<br>• Control module has failed |
| **DTC: P0481**<br>**2T CCM, MIL: Yes**<br>**Years:** 2005, 2006, 2007, 2008<br>**Models:** All<br>**Engines:** 2.8L VIN U<br>**Transmissions:** All | **Cooling Fan Relay 2 Control Circuit**<br>The commanded state of the Output Driver Module (ODM) and the actual state of the control circuit do not match. The condition is present for more than 5 seconds.<br>**Possible Causes:**<br>• Fan control relay control circuit is open or shorted to ground<br>• Fan control relay control circuit is shorted to system power<br>• Fan control relay power circuit is open (check Cool Fan 1 fuse)<br>• Fan control relay is damaged or has failed<br>• Control module has failed |

| DTC | Trouble Code Title, Conditions & Possible Causes |
|---|---|
| **DTC: P0482**<br>**2T CCM, MIL: Yes**<br>**Years:** 2005, 2006, 2007, 2008<br>**Models:** All<br>**Engines:** 2.8L VIN U<br>**Transmissions:** All | **Cooling Fan Relay 3 Control Circuit**<br>The commanded state of the Output Driver Module (ODM) and the actual state of the control circuit do not match. The condition is present for more than 5 seconds.<br>**Possible Causes:**<br>• Fan control relay control circuit is open or shorted to ground<br>• Fan control relay control circuit is shorted to system power<br>• Fan control relay power circuit is open (check Cool Fan 1 fuse)<br>• Fan control relay is damaged or has failed<br>• Control module has failed |
| **DTC: P0483**<br>**2T CCM, MIL: Yes**<br>**Years:** 2005, 2006, 2007, 2008<br>**Engines:** 2.0L VIN 2, 2.5L SOHC, DOHC VIN 6<br>**Models:** All<br>**Transmissions:** All | **Fan Rationality Check**<br>Control module detected the function abnormality of the fan operation. Control module detects when the engine coolant temperature slowly decreases even when the radiator fan is rotating.<br>**Possible Causes:**<br>• Check any other DTC on display<br>• Check radiator fan and fan motor |
| **DTC: P0496**<br>**2T CCM, MIL: Yes**<br>**Years:** 2005, 2006, 2007, 2008<br>**Engines:** 2.8L VIN U, 4.2L VIN S, 5.3L VIN M/P, 6.0L VIN H<br>**Models:** All<br>**Transmissions:** All | **Evaporative Emission (EVAP) System Flow During Non-Purge**<br>A continuous open purge flow condition is detected during the diagnostic test. The fuel tank pressure decreases to less than a calibrated value.<br>**Possible Causes:**<br>• Faulty EVAP purge solenoid valve<br>• Faulty FTP sensor |
| **DTC: P0498**<br>**2T CCM, MIL: Yes**<br>**Years:** 2005, 2006, 2007, 2008<br>**Engines:** 2.0L VIN S/Y, 2.8L VIN U<br>**Models:** All<br>**Transmissions:** All | **Canister Close Valve Circuit Open / Short to Ground**<br>The control module detects that the commanded state of the driver and the actual state of the control circuit do not match.<br>**Possible Causes:**<br>• EVAP control circuit of the canister purge solenoid valve for a short to ground<br>• Inspect for poor connections<br>• Open or short to ground in the ignition 1 voltage circuit<br>• Faulty EVAP canister purge solenoid valve<br>• PCM has failed |
| **DTC: P0499**<br>**2T CCM, MIL: Yes**<br>**Years:** 2005, 2006, 2007, 2008<br>**Engines:** 2.0L VIN S/Y, 2.8L VIN U<br>**Models:** All<br>**Transmissions:** All | **Canister Close Valve Circuit Short to B+**<br>The control module detects that the commanded state of the driver and the actual state of the control circuit do not match.<br>**Possible Causes:**<br>• EVAP control circuit of the canister purge solenoid valve for an open or short to voltage<br>• Inspect for poor connections<br>• Open or short to ground in the ignition 1 voltage circuit<br>• Faulty EVAP canister purge solenoid valve<br>• PCM has failed |
| **DTC: P0500**<br>**1T CCM, MIL: Yes**<br>**Years:** 2005, 2006, 2007, 2008<br>**Engines:** 2.5L SOHC, DOHC VIN 6, 2.8L VIN U<br>**Models:** All<br>**Transmissions:** All | **Vehicle Speed Sensor**<br>Control module did not Control module detected any VSS signals during the CCM Rationality test.<br>**Possible Causes:**<br>• Check DTC of ABS<br>• Check poor contact of Control module connector. |
| **DTC: P0501**<br>**2T CCM, MIL: Yes**<br>**Years:** 2005, 2006, 2007, 2008<br>**Models:** All<br>**Engines:** 2.0L VIN S/Y, 2.3L VIN E/G<br>**Transmissions:** All | **Vehicle Speed Sensor Range/Performance**<br>Control module detects range/performance issue with VSS.<br>**Possible Causes:**<br>• High circuit of the VSS for a short to voltage<br>• High circuit of the VSS for a short to ground or open<br>• Low circuit of the VSS for an open.<br>• High circuit and low circuit of the VSS shorted together<br>• Output shaft speed sensor rotor for damage or misalignment<br>• Faulty VSS/OSS<br>• Control module has failed |

| DTC | Trouble Code Title, Conditions & Possible Causes |
|---|---|
| **DTC: P0502**<br>**2T CCM, MIL: Yes**<br>**Years:** 2005, 2006, 2007, 2008<br>**Models:** All<br>**Engines:** 2.0L VIN S/Y, 2.3L VIN E/G<br>**Transmissions:** All | **Vehicle Speed Sensor Circuit Low Input**<br>Control module detects vehicle speed sensor circuit input low.<br>**Possible Causes:**<br>• High circuit of the VSS for a short to voltage<br>• High circuit of the VSS for a short to ground or open<br>• Low circuit of the VSS for an open.<br>• High circuit and low circuit of the VSS shorted together<br>• Output shaft speed sensor rotor for damage or misalignment<br>• Faulty VSS/OSS<br>• Control module has failed |
| **DTC: P0503**<br>**2T CCM, MIL: Yes**<br>**Years:** 2005, 2006, 2007, 2008<br>**Models:** All<br>**Engines:** 2.0L VIN S/Y<br>**Transmissions:** All | **Vehicle Speed Sensor Intermittent/Erratic/High**<br>Control module detects vehicle speed sensor circuit input intermittent/erratic/high.<br>**Possible Causes:**<br>• High circuit of the VSS for a short to voltage<br>• High circuit of the VSS for a short to ground or open<br>• Low circuit of the VSS for an open.<br>• High circuit and low circuit of the VSS shorted together<br>• Output shaft speed sensor rotor for damage or misalignment<br>• Faulty VSS/OSS<br>• Control module has failed |
| **DTC: P0504**<br>**2T CCM, MIL: Yes**<br>**Years:** 2005, 2006, 2007, 2008<br>**Models:** All<br>**Engines:** 2.8L VIN U<br>**Transmissions:** All | **Brake Switch / Correlation**<br>Control module detects a incorrect voltage signal on the brake signal circuit.<br>**Possible Causes:**<br>• Stop lamp switch voltage circuit shorted to ground, open or high resistance<br>• TCC brake signal circuit for an open or high resistance<br>• Faulty stop lamp switch |
| **DTC: P0506**<br>**2T CCM, MIL: Yes**<br>**Engines:** All<br>**Models:** All<br>**Transmissions:** All | **Idle Air Control System RPM Lower Than Expected**<br>Engine started. Control module Control module detects a malfunction in actual engine speed is not close to target engine speed during idling. The engine may be hard to start, stall or not start when this code is set. Improper idling could also be caused.<br>**Possible Causes:**<br>• Check any other DTC on display<br>• Clogged air cleaner element<br>• Check electronic throttle control |
| **DTC: P0507**<br>**2T CCM, MIL: Yes**<br>**Engines:** All<br>**Models:** All<br>**Transmissions:** All | **Idle Air Control System RPM Higher Than Expected**<br>Engine startedControl module detected the Actual idle speed was 100-200 rpm more than the Target idle speed for over 10 seconds.<br>**Note: Improper idling could also be caused.**<br>**Possible Causes:**<br>• Check any other DTC on display<br>• Check air intake system<br>• Check electronic throttle control |
| **DTC: P0508**<br>**2T CCM, MIL: Yes**<br>**Engines:** 2.0L VIN 2<br>**Models:** All<br>**Transmissions:** All | **Idle Control System Circuit Low**<br>Control module detected open or short circuit in idle air control solenoid valve. Period of idle air control solenoid valve is 4 milliseconds and it is too short. Judge OK/NG in accordance with the number of change of OFF to ON signal.<br>**Possible Causes:**<br>• Check any other DTC on display<br>• Check idle air control valve<br>• Check control module |
| **DTC: P0509**<br>**2T CCM, MIL: Yes**<br>**Engines:** 2.0L VIN 2<br>**Models:** All<br>**Transmissions:** All | **Idle Control System Circuit High**<br>Control module detected open or short circuit in idle air control solenoid valve. Period of idle air control solenoid valve is 4 milliseconds and it is too short. Judge OK/NG in accordance with the number of change of OFF to ON signal.<br>**Possible Causes:**<br>• Check any other DTC on display<br>• Check idle air control valve<br>• Check control module |

| DTC | Trouble Code Title, Conditions & Possible Causes |
|---|---|
| **DTC: P0512**<br>**2T CCM, MIL: No**<br>**Engines:** 2.0L VIN 2, 2.5L SOHC, DOHC VIN 6<br>**Models:** All<br>**Transmissions:** All | **Starter Request Circuit**<br>Control module detected the open or short circuit of starter SW. Judge as ON NG when the starter SW signal remains ON. Engine Fails to start<br>**Possible Causes:**<br>• Check any other DTC on display<br>• Short circuit to power in the harness between Control module and ignition switch |
| **DTC: P0513**<br>**2T CCM, MIL: No**<br>**Engines:** 2.8L VIN U<br>**Models:** All<br>**Transmissions:** All | **Control module or CIM Not Added**<br>The PCM/ECM detects an internal failure or incomplete programming for more than 5 seconds.<br>**Possible Causes:**<br>• PCM/ECM is not programmed<br>• PCM/ECM has failed |
| **DTC: P0513**<br>**2T CCM, MIL: No**<br>**Engines:** 2.5L SOHC, DOHC VIN 6<br>**Models:** All<br>**Transmissions:** All | **Incorrect Immobilizer Key**<br>Use of unregistered key in the body integrated unit<br>**Possible Causes:**<br>• Perform teaching operation of ignition key<br>• Check the ignition keys (including transponder) which cannot be registered<br>• Replace the body integrated unit and replace all the ignition keys (including transponder).<br>• Execute the registration procedure next |
| **DTC: P0519**<br>**1T CCM, MIL: Yes**<br>**Years:** 2005, 2006, 2007, 2008<br>**Engines:** 2.0L VIN 2, 2.5L SOHC, DOHC VIN 6<br>**Models:** All<br>**Transmissions:** All | **Idle Air Control System Performance**<br>Control module detected malfunctions in which the engine RPM continues to rise during idling. Engine keeps running at higher speed than specified idle speed.<br>**Possible Causes:**<br>• Check any other DTC on display<br>• Check air intake system<br>• Vacuum leaks<br>• Check electronic throttle control |
| **DTC: P0532**<br>**1T CCM, MIL: No**<br>**Years:** 2005, 2006, 2007, 2008<br>**Models:** All<br>**Engines:** 2.0L VIN S/Y, 2.8L VIN U<br>**Transmissions:** All | **Air Conditioning (A/C) Refrigerant Pressure Sensor Circuit Low Voltage**<br>A/C pressure of less than 0.1 volt or more than 4.92 volts. The condition must be present for more than 5 seconds.<br>**Possible Causes:**<br>• A/C refrigerant pressure<br>• Pressure sensor VREF circuit open shorted to ground or high resistance<br>• Pressure sensor signal circuit open shorted to ground or high resistance<br>• A/C refrigerant pressure sensor.<br>• Control module has failed |
| **DTC: P0533**<br>**1T CCM, MIL: No**<br>**Years:** 2005, 2006, 2007, 2008<br>**Models:** All<br>**Engines:** 2.0L VIN S/Y, 2.8L VIN U<br>**Transmissions:** All | **Air Conditioning (A/C) Refrigerant Pressure Sensor Circuit High Voltage**<br>The engine is running. A/C pressure of less than 0.1 volt or more than 4.92 volts. The condition must be present for more than 5 seconds.<br>**Possible Causes:**<br>• A/C refrigerant pressure<br>• Pressure sensor VREF circuit open shorted to ground or high resistance<br>• Pressure sensor signal circuit open shorted to ground or high resistance<br>• A/C refrigerant pressure sensor.<br>• Control module has failed |
| **DTC: P0545**<br>**1T CCM, MIL: No**<br>**Years:** 2005, 2006, 2007, 2008<br>**Models:** All<br>**Engines:** 2.0L VIN 2<br>**Transmissions:** All | **Exhaust Gas Temperature Sensor Circuit Low**<br>Control module detected the open or short circuit of exhaust temperature sensor. Control module detects when out of the standard value.<br>**Possible Causes:**<br>• Check sensor harness<br>• Check sensor output voltage |
| **DTC: P0546**<br>**1T CCM, MIL: No**<br>**Years:** 2005, 2006, 2007, 2008<br>**Models:** All<br>**Engines:** 2.0L VIN 2<br>**Transmissions:** All | **Exhaust Gas Temperature Sensor Circuit High**<br>Control module detected the open or short circuit of exhaust temperature sensor. Control module detects when out of the standard value.<br>**Possible Causes:**<br>• Check sensor harness<br>• Check sensor output voltage |

| DTC | Trouble Code Title, Conditions & Possible Causes |
|---|---|
| **DTC: P0551**<br>**1T CCM, MIL: No**<br>**Years:** 2005, 2006, 2007, 2008<br>**Models:** All<br>**Engines:** 2.0L VIN S/Y<br>**Transmissions:** All | **Power Steering Pressure Sensor/Switch Circuit Range/Performance**<br>Control module detected the open or short circuit of exhaust temperature sensor. Control module detects when out of the standard value.<br>**Possible Causes:**<br>• Check sensor harness<br>• Check sensor output voltage |
| **DTC: P0556**<br>**1T CCM, MIL: No**<br>**Years:** 2005, 2006, 2007, 2008<br>**Models:** All<br>**Engines:** 5.3L VIN M/P, 6.0L VIN H<br>**Transmissions:** All | **Brake Booster Pressure Sensor Circuit Range/Performance**<br>The control module detects brake booster vacuum as being less than the intake manifold vacuum for more than 100 milliseconds.<br>**Possible Causes:**<br>• Check sensor harness<br>• Check sensor resistance |
| **DTC: P0557**<br>**1T CCM, MIL: No**<br>**Years:** 2005, 2006, 2007, 2008<br>**Models:** All<br>**Engines:** 5.3L VIN M/P, 6.0L VIN H<br>**Transmissions:** All | **Brake Booster Pressure Sensor Circuit Low Input**<br>The control module detects less than 0.04 volts on the brake booster sensor signal circuit for more than 12.5 milliseconds.<br>**Possible Causes:**<br>• Check sensor harness<br>• Check sensor resistance |
| **DTC: P0558**<br>**1T CCM, MIL: No**<br>**Years:** 2005, 2006, 2007, 2008<br>**Models:** All<br>**Engines:** 5.3L VIN M/P, 6.0L VIN H<br>**Transmissions:** All | **Brake Booster Pressure Sensor Circuit High Input**<br>Control module detects greater than 4.89 volts on the brake booster sensor signal circuit for more than 12.5 milliseconds.<br>**Possible Causes:**<br>• Check sensor harness<br>• Check sensor resistance |
| **DTC: P0560**<br>**1T CCM, MIL: No**<br>**Years:** 2005, 2006, 2007, 2008<br>**Models:** All<br>**Engines:** 2.8L VIN U<br>**Transmissions:** All | **System Voltage Malfunction**<br>Control module detected the system voltage was less than 9v, or that it was more than 18v for 25 seconds during the CCM test.<br>**Possible Causes**<br>• Check for high resistance at the battery connections or at the Underhood Fuse Block power circuit connection to the Control module<br>• Check the drive belt for excessive wear and the proper tension<br>• Check the condition of the battery and the Generator output<br>• Vehicle may have been used to jump-start another vehicle |
| **DTC: P0562**<br>**1T CCM, MIL: No**<br>**Years:** 2005, 2006, 2007, 2008<br>**Models:** All<br>**Engines:** 2.8L VIN U<br>**Transmissions:** All | **System Voltage Low**<br>Control module detects an improper voltage below 11 volts for 5 seconds.<br>**Possible Causes:**<br>• Check for high resistance at battery connections<br>• Check the drive belt for excessive wear and the proper tension<br>• Test the operation of the alternator (it may be undercharging)<br>• Control module has failed |
| **DTC: P0563**<br>**1T CCM, MIL: No**<br>**Years:** 2005, 2006, 2007, 2008<br>**Models:** All<br>**Engines:** 2.8L VIN U<br>**Transmissions:** All | **System Voltage High**<br>Control module detects a system voltage above 16 volts for less than 1 second.<br>**Possible Causes:**<br>• Check the condition of the battery (it may be worn out)<br>• Test the operation of the alternator (it may be overcharging)<br>• Control module has failed |
| **DTC: P0565**<br>**1T CCM, MIL: No**<br>**Years:** 2005, 2006, 2007, 2008<br>**Models:** All<br>**Engines:** 2.0L VIN 2<br>**Transmissions:** All | **Cruise Control On Signal**<br>Control module detects cruise control set signal is continued to be ON for a certain period of time at the vehicle speed less than 20 km/h (12 MPH).<br>**Possible Causes:**<br>• Check harness between transmission module and control module<br>• Check harness connectors<br>• Check input signal for transmission module<br>• Check for poor contact |

| DTC | Trouble Code Title, Conditions & Possible Causes |
|---|---|
| **DTC: P0572**<br>**1T CCM, MIL: No**<br>**Years:** 2005, 2006, 2007, 2008<br>**Models:** All<br>**Engines:** 2.0L VIN S/Y<br>**Transmissions:** All | **Brake Switch Circuit Low Voltage**<br>Control module detects a low voltage signal on the TCC brake signal circuit when the serial data message from the control module indicates the brakes are applied.<br>**Possible Causes:**<br>• Stop lamp switch voltage circuit shorted to ground open or high resistance<br>• TCC brake signal circuit for an open or high resistance<br>• Faulty stop lamp switch |
| **DTC: P0573**<br>**1T CCM, MIL: No**<br>**Years:** 2005, 2006, 2007, 2008<br>**Models:** All<br>**Engines:** 2.0L VIN S/Y<br>**Transmissions:** All | **Brake Switch Circuit High Voltage**<br>Control module detects a high voltage on the TCC brake signal circuit when the serial data message from the control module indicates the brakes are not applied.<br>**Possible Causes:**<br>• TCC brake signal circuit for a short to voltage<br>• Faulty stop lamp switch |
| **DTC: P0600**<br>**1T CCM, MIL: Yes**<br>**Years:** 2005, 2006, 2007, 2008<br>**Engines:** 2.0L VIN S/Y, 2.5L SOHC, DOHC VIN 6, 4.2L VIN S, 5.3L VIN M/P, 6.0L VIN H<br>**Models:** All<br>**Transmissions:** All | **Serial communication link**<br>Tech 2 is required for reading DTC, performing diagnosis and reading current data.<br>**Note: Check harness for broken wires or short circuits, shake trouble spot or connector.**<br>**Possible Causes:**<br>• Check the harness between Control module and transmission module which might affect body control<br>• Check resistance between Control module connector and chassis<br>• Faulty Control module |
| **DTC: P0601**<br>**1T CCM, MIL: No**<br>**Years:** 2005, 2006, 2007, 2008<br>**Models:** All<br>**Engines:** 2.0L VIN S/Y, 4.2L VIN S, 5.3L VIN M/P, 6.0L VIN H<br>**Transmissions:** All | **Control Module Read Only Memory (ROM)**<br>Control module detects that the check sum value is incorrect for 0.05 second. This diagnostic runs continuous.<br>**Possible Causes:**<br>• Ground circuits open high resistance or short<br>• Voltage supply circuits open high resistance or short<br>• Control module has failed |
| **DTC: P0602**<br>**1T CCM, MIL: Yes**<br>**Years:** 2006<br>**Engines:** 2.0L VIN S/Y, 2.5L SOHC, 2.8L VIN U, 4.2L VIN S, 5.3L VIN M/P, 6.0L VIN H<br>**Models:** All<br>**Transmissions:** All | **Control Module Programming Error**<br>Engine keeps running at higher speed than specified idle speed. Engine keeps running at lower speed than specified idle speed. Engine stalls.<br>**Possible Causes:**<br>• Check any other DTC on display<br>• Check engine oil level<br>• Check exhaust system<br>• Check all mechanical parts<br>• Check all electrical out put sensors for open or short to ground<br>• Check for poor connections at Control module and sensors<br>• Faulty Control module |
| **DTC: P0603**<br>**1T CCM, MIL: No**<br>**Years:** 2005, 2006, 2007, 2008<br>**Models:** All<br>**Engines:** 4.2L VIN S, 5.3L VIN M/P, 6.0L VIN H<br>**Transmissions:** All | **Control Module Long Term Memory Reset**<br>Control module detects an internal failure or incomplete programming for more than 10 seconds.<br>**Possible Causes:**<br>• Open high resistance or shorted voltage and ground inputs to Control module<br>• Attempt to program Control module before replacing Control module If DTC P0602S resets, replace Control module. |
| **DTC: P0604**<br>**1T CCM, MIL: No**<br>**Years:** 2005, 2006, 2007, 2008<br>**Models:** All<br>**Engines:** All<br>**Transmissions:** All | **Transmission Control Module Random Access Memory (RAM)**<br>The transmission module has detected an internal malfunction.<br>**Possible Causes:**<br>• transmission module is not programmed<br>• transmission module has failed |
| **DTC: P0605**<br>**1T CCM, MIL: Yes**<br>**Years:** 2005, 2006, 2007, 2008<br>**Engines:** 2.3L VIN E/G, 2.5L SOHC, DOHC VIN 6, 2.8L VIN U<br>**Models:** All<br>**Transmissions:** All | **Internal Control Module Read Only Memory (ROM) Error**<br>SUM value of ROM is outside the standard value. Ignition switch is on.<br>**Possible Causes:**<br>• Check any other DTC on display<br>• There may be a temporary connector contact failure<br>• Faulty control module |

| DTC | Trouble Code Title, Conditions & Possible Causes |
|---|---|
| **DTC: P0606**<br>**1T CCM, MIL: No**<br>**Years:** 2005, 2006, 2007, 2008<br>**Models:** All<br>**Engines:** All<br>**Transmissions:** All | **Control Module Internal Performance**<br>Control module detects an internal failure or incomplete programming for more than 5 seconds.<br>**Possible Causes:**<br>• Control module is not programmed<br>• Control module has failed |
| **DTC: P0607**<br>**2T CCM, MIL: Yes**<br>**Years:** 2005, 2006, 2007, 2008<br>**Engines:** All<br>**Models:** All<br>**Transmissions:** All | **Control Module Performance**<br>Control module detects a performance issue for more than 5 seconds.<br>**Possible Causes:**<br>• Open, grounded shorted circuits of power supply circuit<br>• Open, grounded shorted circuits of ground supply circuit<br>• Faulty control module connections<br>• Faulty control module |
| **DTC: P0610**<br>**2T CCM, MIL: Yes**<br>**Years:** 2005, 2006, 2007, 2008<br>**Engines:** 2.0L VIN S/Y, 2.8L VIN U<br>**Models:** All<br>**Transmissions:** All | **Control Module Performance**<br>Control module detected it was not programmed for the correct application.<br>**Possible Causes:**<br>• Control module must be reprogrammed to correct this problem. Once this step is done, recheck the code to verify the repair is done. |
| **DTC: P0614**<br>**1T CCM, MIL: No**<br>**Years:** 2005, 2006, 2007<br>**Models:** All<br>**Engines:** 2.8L VIN U<br>**Transmissions:** All | **Control Module / Transmission Module Incompatible**<br>Control module detected it was not programmed for the correct application.<br>**Possible Causes:**<br>• Control module must be reprogrammed to correct this problem. Once this step is done, recheck the code to verify the repair is done. |
| **DTC: P0615**<br>**1T CCM, MIL: No**<br>**Years:** 2005, 2006, 2007<br>**Models:** All<br>**Engines:** 2.8L VIN U<br>**Transmissions:** All | **Starter Relay Control Circuit**<br>Control module detects an improper voltage level on the output circuit that controls the starter relay. The condition exists for at least 2 seconds.<br>**Possible Causes:**<br>• Control circuit of the start 1 relay for a short to voltage or an open<br>• Poor connections at the start 1 relay or Control module<br>• Faulty starter relay<br>• Control module has failed |
| **DTC: P0616**<br>**1T CCM, MIL: No**<br>**Years:** 2005, 2006, 2007<br>**Models:** All<br>**Engines:** 2.0L VIN S/Y, 2.8L VIN U<br>**Transmissions:** All | **Starter Relay Control Circuit Input Low**<br>Control module detected the voltage on the Starter Relay control circuit did not match the commanded state for at least two seconds.<br>**Possible Causes**<br>• Starter relay connector is damaged or shorted to ground<br>• Starter relay control circuit is shorted to ground<br>• Control module has failed |
| **DTC: P0617**<br>**1T CCM, MIL: No**<br>**Years:** 2005, 2006, 2007<br>**Models:** All<br>**Engines:** 2.0L VIN S/Y, 2.8L VIN U<br>**Transmissions:** All | **Starter Relay Control Circuit Input Low**<br>Control module detected the voltage on the Starter Relay control circuit did not match the commanded state for at least two seconds.<br>**Possible Causes**<br>• Starter relay connector is damaged or shorted to power<br>• Starter relay control circuit is shorted to system power<br>• Control module has failed |
| **DTC: P0620**<br>**1T CCM, MIL: No**<br>**Years:** 2005, 2006, 2007<br>**Models:** All<br>**Engines:** 2.8L VIN U<br>**Transmissions:** All | **Generator Signal Range/Performance**<br>Control module detected the 'L' terminal voltage was low with the Generator commanded "on", or the 'F' terminal pulse width modulation was less than 5% with the engine speed below 2500 rpm for 30 seconds.<br>**Note: Refer to the Freeze Frame Records for additional information.**<br>**Possible Causes**<br>• Generator 'L' terminal circuit is open, shorted to ground or B+<br>• Generator 'F' terminal circuit is open, shorted to ground or B+<br>• Control module has failed |

| DTC | Trouble Code Title, Conditions & Possible Causes |
|---|---|
| **DTC: P0621**<br>**1T CCM, MIL: No**<br>**Years:** 2005, 2006, 2007<br>**Models:** All<br>**Engines:** 2.0L VIN S/Y<br>**Transmissions:** All | **Generator 'L' Terminal Circuit Malfunction**<br>Control module detected the Generator 'L' Terminal voltage was high for 5 seconds; or with the engine running or the Control module detected the Generator 'L' Terminal voltage was low for over 15 seconds.<br>**Possible Causes**<br>• Generator 'L' terminal circuit is shorted to system power (B+)<br>• Generator is damaged or has failed<br>• Control module has failed |
| **DTC: P0625**<br>**1T CCM, MIL: No**<br>**Years:** 2005, 2006, 2007<br>**Models:** All<br>**Engines:** 2.0L VIN S/Y, 2.8L VIN U<br>**Transmissions:** All | **Generator Control Circuit Input Low**<br>Control module detected the Generator pulse width modulation signal was less than 5% for 15 seconds.<br>**Possible Causes**<br>• Battery positive cable is open or has a high resistance condition<br>• Generator connector is damaged or shorted to power<br>• Generator control circuit is shorted to ground<br>• Control module has failed |
| **DTC: P0626**<br>**1T CCM, MIL: No**<br>**Years:** 2005, 2006, 2007<br>**Models:** All<br>**Engines:** 2.0L VIN S/Y, 2.8L VIN U<br>**Transmissions:** All | **Generator Control Circuit Input High**<br>Control module detected the Generator pulse width modulation signal was more than 5% for 15 seconds.<br>**Possible Causes**<br>• Generator connector is damaged or shorted to system power<br>• Generator control circuit is shorted to system power (B+)<br>• Control module has failed |
| **DTC: P0628**<br>**1T CCM, MIL: No**<br>**Years:** 2005, 2006, 2007<br>**Models:** All<br>**Engines:** 2.0L VIN S/Y<br>**Transmissions:** All | **Fuel Pump Control Circuit Input Low**<br>Control module detected an unexpected low voltage on the Fuel Pump control circuit for 1 second.<br>**Possible Causes**<br>• Fuel pump relay connector is damaged or shorted<br>• Fuel pump relay control circuit is shorted to ground<br>• Fuel pump relay is damaged or it has failed<br>• Control module is damaged |
| **DTC: P0629**<br>**1T CCM, MIL: No**<br>**Years:** 2005, 2006, 2007<br>**Models:** All<br>**Engines:** 2.0L VIN S/Y<br>**Transmissions:** All | **Fuel Pump Control Circuit Input High**<br>Control module detected an unexpected high voltage on the Fuel Pump control circuit for 1 second.<br>**Possible Causes**<br>• Fuel pump relay connector is damaged or open<br>• Fuel pump relay control circuit is open or shorted to power (B+)<br>• Fuel pump relay is damaged or it has failed<br>• Control module is damaged |
| **DTC: P0630**<br>**1T CCM, MIL: No**<br>**Years:** 2005, 2006, 2007<br>**Models:** All<br>**Engines:** 2.0L VIN S/Y, 2.8L VIN U<br>**Transmissions:** All | **VIN Not Programmed or Incompatible - Control module**<br>Control module detected it was not programmed for the correct application.<br>**Possible Causes**<br>• Control module must be reprogrammed to correct this problem. Once this step is done, recheck the code to verify the repair is done. |
| **DTC: P0632**<br>**1T CCM, MIL: No**<br>**Years:** 2005, 2006, 2007<br>**Models:** All<br>**Engines:** 2.0L VIN S/Y, 2.8L VIN U<br>**Transmissions:** All | **Odometer Not Programmed - Control module**<br>Control module detected it was not programmed for the correct application.<br>**Possible Causes**<br>• Control module must be reprogrammed to correct this problem. Once this step is done, recheck the code to verify the repair is done. |
| **DTC: P0633**<br>**1T CCM, MIL: No**<br>**Years:** 2005, 2006, 2007<br>**Models:** All<br>**Engines:** 2.0L VIN S/Y, 2.8L VIN U<br>**Transmissions:** All | **Theft Deterrent Key Not Programmed**<br>Control module detected it was not programmed for the correct application.<br>**Possible Causes**<br>• Control module must be reprogrammed to correct this problem. Once this step is done, recheck the code to verify the repair is done. |

| DTC | Trouble Code Title, Conditions & Possible Causes |
|---|---|
| **DTC: P0638**<br>**1T CCM, MIL: Yes**<br>**Years:** 2005, 2006, 2007, 2008<br>**Engines:** 2.0L VIN S/Y, 2.5L SOHC, DOHC VIN 6, 2.8L VIN U<br>**Models:** All<br>**Transmissions:** All | **Throttle Actuator Control Range/Performance Bank 1**<br>Control module detects when the target opening angle and actual opening angle is mismatched or the current to motor is the specified duty or more for specified time continuously. Improper idling. Engine stalls. Poor driving performance.<br>**Possible Causes:**<br>• Check electronic throttle control relay<br>• Check power supply of electronic control relay<br>• Check harness between Control module and electronic throttle control relay<br>• Check sensor output<br>• Check poor contact<br>• Check harness between Control module and electronic throttle control<br>• Check sensor power supply<br>• Check for short in Control module<br>• Check sensor output<br>• Check harness between Control module and electronic throttle control motor<br>• Check electronic throttle control motor harness<br>• Check ground circuit<br>• Check electronic throttle control |
| **DTC: P0641**<br>**1T CCM, MIL: No**<br>**Years:** 2005, 2006, 2007<br>**Models:** All<br>**Engines:** 2.0L VIN S/Y, 4.2L VIN S, 5.3L VIN M/P, 6.0L VIN H<br>**Transmissions:** All | **5-Volt Reference 1 Circuit**<br>Control module detects a voltage out of tolerance condition on the 5-volt reference circuit for more than 2 seconds.<br>**Possible Causes:**<br>• 5-volt reference circuits for a short to voltage<br>• MAP sensor signal circuit for a short to voltage<br>• Faulty MAP sensor<br>• Faulty EOP sensor<br>• Control module has failed |
| **DTC: P0645**<br>**1T CCM, MIL: No**<br>**Years:** 2005, 2006, 2007<br>**Models:** All<br>**Engines:** 2.8L VIN U<br>**Transmissions:** All | **Air Conditioning (A/C) Clutch Relay Control Circuit**<br>Control module detects an open on the control circuit of the A/C compressor clutch relay.<br>**Possible Causes:**<br>• Control circuit of the A/C compressor clutch relay for a short to ground, short to<br>• Voltage, or an open<br>• Poor connections at the A/C compressor clutch relay or Control module<br>• Faulty A/C relay<br>• Control module has failed |
| **DTC: P0646**<br>**1T CCM, MIL: No**<br>**Years:** 2005, 2006, 2007<br>**Models:** All<br>**Engines:** 2.0L VIN S/Y, 2.8L VIN U<br>**Transmissions:** All | **A/C Clutch Relay Control Circuit Low**<br>Control module detects an open on the control circuit of the A/C compressor clutch relay.<br>**Possible Causes:**<br>• Control circuit of the A/C compressor clutch relay for a short to ground, short to<br>• Voltage, or an open<br>• Poor connections at the A/C compressor clutch relay or Control module<br>• Faulty A/C relay<br>• Control module has failed |
| **DTC: P0647**<br>**1T CCM, MIL: No**<br>**Years:** 2005, 2006, 2007<br>**Models:** All<br>**Engines:** 2.0L VIN S/Y, 2.8L VIN U<br>**Transmissions:** All | **A/C Clutch Relay Control Circuit High**<br>Control module detects an short to voltage on the control circuit of the A/C compressor clutch relay.<br>**Possible Causes:**<br>• Control circuit of the A/C compressor clutch relay for a short to ground, short to<br>• Voltage, or an open<br>• Poor connections at the A/C compressor clutch relay or Control module<br>• Faulty A/C relay<br>• Control module has failed |
| **DTC: P0650**<br>**1T CCM, MIL: No**<br>**Years:** 2005, 2006, 2007<br>**Models:** All<br>**Engines:** 2.8L VIN U, 4.2L VIN S, 5.3L VIN M/P, 6.0L VIN H<br>**Transmissions:** All | **Malfunction Indicator Lamp (MIL) Control Circuit**<br>Control module detects that the commanded state of the MIL driver and the actual state of the control circuit do not match for more than 5 seconds.<br>**Possible Causes:**<br>• Open fuse that supplies voltage to the MIL<br>• MIL control circuit for an open or high resistance.<br>• Intermittent or poor connection at the IPC<br>• Intermittent or poor connection at the Control module.<br>• Short to ground in the MIL control circuit<br>• Short to voltage in the MIL control circuit<br>• Faulty IPC<br>• Control module has failed |

| DTC | Trouble Code Title, Conditions & Possible Causes |
|---|---|
| **DTC: P0651**<br>**1T CCM, MIL: No**<br>**Years:** 2005, 2006, 2007<br>**Models:** All<br>**Engines:** 2.0L VIN S/Y, 4.2L VIN S, 5.3L VIN M/P, 6.0L VIN H<br>**Transmissions:** All | **5-Volt Reference 2 Circuit**<br>The Control module detects a voltage out of tolerance condition on the 5-volt reference circuit for more than 2 seconds. A short to voltage on the signal circuit of the FTP sensor will back-feed through the sensor into the 5-volt reference circuit and set this DTC.<br>**Possible Causes:**<br>• 5-volt reference circuit for a short to ground or any sensor low reference circuit<br>• 5-volt reference circuits for a short to voltage<br>• FTP sensor signal circuit for a short to voltage<br>• Faulty FTP sensor<br>• Faulty A/C pressure sensor<br>• Control module has failed |
| **DTC: P0685**<br>**1T CCM, MIL: No**<br>**Years:** 2005, 2006, 2007<br>**Models:** All<br>**Engines:** 2.0L VIN S/Y, 4.2L VIN S, 5.3L VIN M/P, 6.0L VIN H<br>**Transmissions:** All | **Control Module Power Relay Control Circuit /Open**<br>Control module detects the commanded state of the ODM and the actual state of the control circuit do not match. The condition is present for more than 5 seconds.<br>**Possible Causes:**<br>• Relay coil control open or high resistance<br>• Relay coil control shorted to ground or voltage<br>• Relay coil B+ control shorted to ground or high resistance<br>• Control module has failed |
| **DTC: P0686**<br>**1T CCM, MIL: No**<br>**Years:** 2005, 2006, 2007<br>**Models:** All<br>**Engines:** 2.0L VIN S/Y, 4.2L VIN S, 5.3L VIN M/P, 6.0L VIN H<br>**Transmissions:** All | **Control Module Power Relay Control Circuit /Low**<br>Control module detects the commanded state of the ODM and the actual state of the control circuit do not match. The condition is present for more than 5 seconds.<br>**Possible Causes:**<br>• Relay coil control open or high resistance<br>• Relay coil control shorted to ground or voltage<br>• Relay coil B+ control shorted to ground or high resistance<br>• Control module has failed |
| **DTC: P0687**<br>**1T CCM, MIL: No**<br>**Years:** 2005, 2006, 2007<br>**Models:** All<br>**Engines:** 2.0L VIN S/Y, 4.2L VIN S, 5.3L VIN M/P, 6.0L VIN H<br>**Transmissions:** All | **Control Module Power Relay Control Circuit /High**<br>Control module detects the commanded state of the ODM and the actual state of the control circuit do not match. The condition is present for more than 5 seconds.<br>**Possible Causes:**<br>• Relay coil control open or high resistance<br>• Relay coil control shorted to ground or voltage<br>• Relay coil B+ control shorted to ground or high resistance<br>• Control module has failed |
| **DTC: P0689**<br>**1T CCM, MIL: No**<br>**Years:** 2005, 2006, 2007<br>**Models:** All<br>**Engines:** 4.2L VIN S, 5.3L VIN M/P, 6.0L VIN H<br>**Transmissions:** All | **Engine Controls Relay Feedback Circuit Low Voltage**<br>Control module detects less than 5 volts on the ignition 1 voltage circuit to Control module. The condition is present for more than 5 seconds.<br>**Possible Causes:**<br>• Relay switch B+ shorted to ground open or high resistance<br>• Relay ignition output ignition 1 voltage shorted to ground open or high resistance<br>• Control module has failed |
| **DTC: P0690**<br>**1T CCM, MIL: No**<br>**Years:** 2005, 2006, 2007<br>**Models:** All<br>**Engines:** 4.2L VIN S, 5.3L VIN M/P, 6.0L VIN H<br>**Transmissions:** All | **Engine Controls Relay Feedback Circuit High Voltage**<br>Control module detects more than 16 volts on the ignition 1 voltage circuit to Control module when the relay is commanded ON. Control module detects more than 2 volts on the ignition 1 voltage circuit to Control module when the relay is commanded OFF. The condition is present for more than 2 seconds.<br>**Possible Causes:**<br>• Relay ignition output ignition 1 voltage shorted to voltage<br>• Control module has failed |
| **DTC: P0691**<br>**2T CCM, MIL: No**<br>**Years:** 2005, 2006, 2007, 2008<br>**Engines:** 2.0L VIN 2, 2.0L VIN S/Y, 2.5L SOHC, DOHC VIN 6, 2.8L VIN U<br>**Models:** All<br>**Transmissions:** All | **Cooling Fan 1 Control Circuit Low**<br>Control module detects the open or short circuit of radiator fan circuit. Radiator fan does not operate properly. Overheating condition.<br>**Possible Causes:**<br>• Check any other DTC on display<br>• Check radiator fan system<br>• Check for temporary poor contact |

| DTC | Trouble Code Title, Conditions & Possible Causes |
|---|---|
| **DTC: P0692**<br>**2T CCM, MIL: No**<br>**Years:** 2005, 2006, 2007, 2008<br>**Engines:** 2.0L VIN 2, 2.0L VIN S/Y, 2.5L SOHC, DOHC VIN 6, 2.8L VIN U<br>**Models:** All<br>**Transmissions:** All | **Cooling Fan 1 Control Circuit High**<br>Control module detects the open or short circuit of radiator fan circuit. Radiator fan does not operate properly. Overheating condition.<br>**Possible Causes:**<br>• Check any other DTC on display<br>• Check radiator fan system<br>• Check for temporary poor contact |
| **DTC: P0693**<br>**2T CCM, MIL: No**<br>**Years:** 2005, 2006, 2007, 2008<br>**Engines:** 2.0L VIN 2, 2.0L VIN S/Y, 2.5L SOHC, DOHC VIN 6, 2.8L VIN U<br>**Models:** All<br>**Transmissions:** All | **Cooling Fan 2 Control Circuit Low**<br>Control module detects the open or short circuit of radiator fan circuit. Radiator fan does not operate properly. Overheating condition.<br>**Possible Causes:**<br>• Check any other DTC on display<br>• Check radiator fan system<br>• Check for temporary poor contact |
| **DTC: P0694**<br>**2T CCM, MIL: No**<br>**Years:** 2005, 2006, 2007, 2008<br>**Engines:** 2.0L VIN S/Y, 2.5L SOHC, DOHC VIN 6, 2.8L VIN U<br>**Models:** All<br>**Transmissions:** All | **Cooling Fan 1 Control Circuit High**<br>Control module detects the open or short circuit of radiator fan circuit. Radiator fan does not operate properly. Overheating condition.<br>**Possible Causes:**<br>• Check any other DTC on display<br>• Check radiator fan system<br>• Check for temporary poor contact |
| **DTC: P0695**<br>**2T CCM, MIL: No**<br>**Years:** 2005, 2006, 2007, 2008<br>**Engines:** 2.0L VIN S/Y, 2.5L SOHC, DOHC VIN 6, 2.8L VIN U<br>**Models:** All<br>**Transmissions:** All | **Cooling Fan 1 Control Circuit Low**<br>Control module detects the open or short circuit of radiator fan circuit. Radiator fan does not operate properly. Overheating condition.<br>**Possible Causes:**<br>• Check any other DTC on display<br>• Check radiator fan system<br>• Check for temporary poor contact |
| **DTC: P0696**<br>**2T CCM, MIL: No**<br>**Years:** 2005, 2006, 2007, 2008<br>**Engines:** 2.0L VIN S/Y, 2.5L SOHC, DOHC VIN 6, 2.8L VIN U<br>**Models:** All<br>**Transmissions:** All | **Cooling Fan 1 Control Circuit High**<br>Control module detects the open or short circuit of radiator fan circuit. Radiator fan does not operate properly. Overheating condition.<br>**Possible Causes:**<br>• Check any other DTC on display<br>• Check radiator fan system<br>• Check for temporary poor contact |
| **DTC: P0700**<br>**1T CCM, MIL: Yes**<br>**Years:** 2005, 2006, 2007, 2008<br>**Engines:** 2.5L SOHC, DOHC VIN 6, 2.8L VIN U, 4.2L VIN S, 5.3L VIN M/P, 6.0L VIN H<br>**Models:** All<br>**Transmissions:** All | **Transaxle Control System (MIL Request)**<br>There is CAN communication with the transaxle module and there is a MIL lighting request. Vehicle performance issue.<br>**Possible Causes:**<br>• Road Test<br>• Inhibitor Switch<br>• Time Lag Test<br>• Transfer Clutch Pressure Test<br>• Line pressure test<br>• Stall speed test<br>• Oil leakage<br>• Check engine start failure (does engine run)<br>• Check illumination of MIL light<br>• Check indication of DTC on display<br>• Using the a scan tool, perform diagnosis |
| **DTC: P0703**<br>**1T CCM, MIL: Yes**<br>**Years:** 2005, 2006, 2007<br>**Models:** All<br>**Engines:** 2.0L VIN S/Y, 2.8L VIN U<br>**Transmissions:** All | **Brake Switch Circuit**<br>Control module receives an invalid brake pedal status serial data message from the control module. This diagnostic runs continuously.<br>**Possible Causes:**<br>• Brake Pedal Position (BPP) sensor<br>• Control module is damaged<br>• control module has failed |

| DTC | Trouble Code Title, Conditions & Possible Causes |
|---|---|
| **DTC: P0704**<br>**1T CCM, MIL: Yes**<br>**Years:** 2005, 2006, 2007<br>**Models:** All<br>**Engines:** 2.0L VIN S/Y, 2.8L VIN U<br>**Transmissions:** All | **Clutch Switch Circuit Malfunction**<br>Control module detected any change in the Clutch switch status.<br>**Possible Causes**<br>&bull; Clutch switch circuit is open, shorted to ground or to power<br>&bull; Clutch switch power circuit is open (test the ENG IGN fuse)<br>&bull; Clutch switch is out of adjustment, damaged or has failed<br>&bull; Control module has failed |
| **DTC: P0705**<br>**2T CCM, MIL: Yes**<br>**Years:** 2005, 2006, 2007, 2008<br>**Engines:** 2.0L VIN 2, 2.5L SOHC, DOHC VIN 6<br>**Models:** All<br>**Transmissions:** All | **Transmission Range Sensor Circuit (PRNDL Input)**<br>The inhibitor switch is open or short. Shift characteristics are erroneous. Shift indicator light does not match with select lever. Shift indicator light does not illuminate. N-D, N-R shock occurs.<br>**Possible Causes:**<br>&bull; Short circuit of harness between transmission module connector and chassis ground<br>&bull; Open circuit of harness between transmission module connector and transmission connector<br>&bull; Short circuit of harness between control valve body connector and transmission connector<br>&bull; Faulty transmission module<br>&bull; Control valve body |
| **DTC: P0710**<br>**2T CCM, MIL: Yes**<br>**Years:** 2005, 2006, 2007, 2008<br>**Engines:** 2.0L VIN 2, 2.5L SOHC, DOHC VIN 6<br>**Models:** All<br>**Transmissions:** All | **Transmission Fluid Temperature Sensor Circuit**<br>Input signal circuit to ATF temperature Sensor is incorrect.<br>**Possible Causes:**<br>&bull; Open circuit of harness between transmission module and transmission connector<br>&bull; Open circuit of harness between transmission connector and control valve body connector<br>&bull; Poor contact in ATF temperature Sensor circuit<br>&bull; Faulty Sensor<br>&bull; Faulty transmission module |
| **DTC: P0712**<br>**2T CCM, MIL: Yes**<br>**Years:** 2005, 2006, 2007, 2008<br>**Engines:** 2.5L SOHC, DOHC VIN 6<br>**Models:** All<br>**Transmissions:** All | **Transmission Fluid Temperature Sensor Circuit Low Input**<br>Input signal circuit to ATF temperature Sensor is open. Excessive shift shock<br>**Possible Causes:**<br>&bull; Open circuit of harness between transmission module and transmission connector<br>&bull; Open circuit of harness between transmission connector and control valve body connector<br>&bull; Poor contact in ATF temperature Sensor circuit<br>&bull; Faulty Sensor<br>&bull; Faulty transmission module |
| **DTC: P0713**<br>**2T CCM, MIL: Yes**<br>**Years:** 2005, 2006, 2007, 2008<br>**Engines:** 2.5L SOHC, DOHC VIN 6<br>**Models:** All<br>**Transmissions:** All | **Transmission Fluid Temperature Sensor Circuit High Input**<br>Input signal circuit to ATF temperature Sensor is shorted. Excessive shift shock.<br>**Possible Causes:**<br>&bull; Short circuit of harness between transmission module and transmission connector<br>&bull; Short circuit of harness between transmission connector and control valve body connector<br>&bull; Poor contact in ATF temperature Sensor circuit<br>&bull; Faulty Sensor<br>&bull; Faulty transmission module |
| **DTC: P0715**<br>**2T CCM, MIL: Yes**<br>**Years:** 2005, 2006, 2007, 2008<br>**Engines:** 2.5L SOHC, DOHC VIN 6<br>**Models:** All<br>**Transmissions:** All | **Input/Turbine Speed Sensor Circuit**<br>Input signal circuit of transmission module is open or shorted. Excessive shift shock. Does not shift into 5th.<br>**Possible Causes:**<br>&bull; Open circuit of harness between transmission module and transmission connector<br>&bull; Short circuit of harness between transmission module and chassis ground<br>&bull; Open or short circuit for power supply and ground<br>&bull; Open circuit of harness between transmission module and transmission connector, or poor contact of connector.<br>&bull; Faulty turbine speed Sensor<br>&bull; Check for mechanical problems<br>&bull; Faulty transmission module |
| **DTC: P0716**<br>**2T CCM, MIL: Yes**<br>**Years:** 2005, 2006, 2007, 2008<br>**Engines:** 2.0L VIN 2<br>**Models:** All<br>**Transmissions:** All | **Input/Turbine Speed Sensor Circuit Range/Performance**<br>Input signal circuit of transmission module is out of specification.<br>**Possible Causes:**<br>&bull; Open circuit of harness between transmission module and transmission connector<br>&bull; Short circuit of harness between transmission module and chassis ground<br>&bull; Open or short circuit for power supply and ground<br>&bull; Open circuit of harness between transmission module and transmission connector, or poor contact of connector.<br>&bull; Faulty turbine speed Sensor<br>&bull; Check for mechanical problems<br>&bull; Faulty transmission module |

| DTC | Trouble Code Title, Conditions & Possible Causes |
|---|---|
| **DTC: P0719**<br>**2T CCM, MIL: Yes**<br>**Years:** 2005, 2006, 2007, 2008<br>**Engines:** 2.0L VIN 2, 2.0L VIN S/Y, 2.5L SOHC, DOHC VIN 6<br>**Models:** All<br>**Transmissions:** All | **Torque Converter/Brake Switch Circuit Low**<br>Brake switch malfunction, open input signal circuit. Brake down control is not operated at SPORT mode.<br>**Possible Causes:**<br>• Poor contact of connector or harness may be the cause<br>• Faulty transmission module<br>• Open circuit of harness between body integrated unit and stop light switch<br>• Short circuit of harness between body integrated unit and stop light switch<br>• Faulty Switch |
| **DTC: P0720**<br>**2T CCM, MIL: Yes**<br>**Years:** 2005, 2006, 2007, 2008<br>**Engines:** 2.0L VIN 2<br>**Models:** All<br>**Transmissions:** All | **Output Speed Sensor Circuit**<br>Vehicle speed signal is abnormal. The harness connector between transmission module and vehicle speed sensor is shorted or open. Deterioration of shifting quality. Driving performance is poor.<br>**Possible Causes:**<br>• Poor contact of connector or harness may be the cause<br>• Open or short circuit for power supply and ground<br>• Open circuit of harness between control valve body connector and transmission connector<br>• Short circuit of harness between transmission connector and transmission ground<br>• Control valve body<br>• Faulty transmission module<br>• Vehicle speed sensor |
| **DTC: P0724**<br>**2T CCM, MIL: Yes**<br>**Years:** 2005, 2006, 2007, 2008<br>**Engines:** 2.0L VIN 2, 2.0L VIN S/Y, 2.5L SOHC, DOHC VIN 6<br>**Models:** All<br>**Transmissions:** All | **Torque Converter/Brake Switch Circuit High**<br>Brake switch malfunction, open input signal circuit. Gear is not shifted down when climbing a hill.<br>**Possible Causes:**<br>• Temporary poor contact of connector or harness may be the cause<br>• Faulty stop light switch<br>• Short circuit of harness between transmission module and stop light switch<br>• Faulty transmission module |
| **DTC: P0725**<br>**2T CCM, MIL: Yes**<br>**Years:** 2005, 2006, 2007, 2008<br>**Engines:** 2.5L SOHC, DOHC VIN 6<br>**Models:** All<br>**Transmissions:** All | **Engine Speed Input Circuit**<br>Information of engine speed is not correctly received from Control module. No lock-up occurs. (After engine is warmed-up)<br>**Possible Causes:**<br>• Temporary poor contact of connector or harness may be the cause<br>• Check indication of DTC on display<br>• Faulty transmission module |
| **DTC: P0726**<br>**2T CCM, MIL: Yes**<br>**Years:** 2005, 2006, 2007, 2008<br>**Engines:** 2.0L VIN 2<br>**Models:** All<br>**Transmissions:** All | **Engine Speed Input Circuit Range/Performance**<br>Information of engine speed is not correctly received from Control module. No lock-up occurs. (After engine is warmed-up)<br>**Possible Causes:**<br>• Temporary poor contact of connector or harness may be the cause<br>• Check indication of DTC on display<br>• Faulty transmission module |
| **DTC: P0731**<br>**2T CCM, MIL: Yes**<br>**Years:** 2005, 2006, 2007, 2008<br>**Engines:** 2.0L VIN 2, 2.5L SOHC, DOHC VIN 6<br>**Models:** All<br>**Transmissions:** All | **Gear 1 Incorrect Ratio**<br>Target gear ratio and actual gear ratio do not match. Shift point is too high or too low. Excessive shift shock. Gear is not changed. The vehicle does not move in D or R range with the engine running at high speed.<br>**Possible Causes:**<br>• Using the a scan tool, perform diagnosis<br>• Temporary poor contact of connector or harness may be the cause<br>• Faulty transmission assembly |
| **DTC: P0732**<br>**2T CCM, MIL: Yes**<br>**Years:** 2005, 2006, 2007, 2008<br>**Engines:** 2.0L VIN 2, 2.5L SOHC, DOHC VIN 6<br>**Models:** All<br>**Transmissions:** All | **Gear 2 Incorrect Ratio**<br>Target gear ratio and actual gear ratio do not match. Shift point is too high or too low. Excessive shift shock. Gear is not changed. The vehicle does not move in D or R range with the engine running at high speed.<br>**Possible Causes:**<br>• Using the a scan tool, perform diagnosis<br>• Temporary poor contact of connector or harness may be the cause<br>• Faulty transmission assembly |
| **DTC: P0733**<br>**2T CCM, MIL: Yes**<br>**Years:** 2005, 2006, 2007, 2008<br>**Engines:** 2.0L VIN 2, 2.5L SOHC, DOHC VIN 6<br>**Models:** All<br>**Transmissions:** All | **Gear 3 Incorrect Ratio**<br>Target gear ratio and actual gear ratio do not match. Shift point is too high or too low. Excessive shift shock. Gear is not changed. The vehicle does not move in D or R range with the engine running at high speed.<br>**Possible Causes:**<br>• Using the a scan tool, perform diagnosis<br>• Temporary poor contact of connector or harness may be the cause<br>• Faulty transmission assembly |

| DTC | Trouble Code Title, Conditions & Possible Causes |
|---|---|
| **DTC: P0734**<br>**2T CCM, MIL: Yes**<br>**Years:** 2005, 2006, 2007, 2008<br>**Engines:** 2.0L VIN 2, 2.5L SOHC, DOHC VIN 6<br>**Models:** All<br>**Transmissions:** All | **Gear 4 Incorrect Ratio**<br>Target gear ratio and actual gear ratio do not match. Shift point is too high or too low. Excessive shift shock. Gear is not changed. The vehicle does not move in D or R range with the engine running at high speed.<br>**Possible Causes:**<br>• Using the a scan tool, perform diagnosis<br>• Temporary poor contact of connector or harness may be the cause<br>• Faulty transmission assembly |
| **DTC: P0735**<br>**2T CCM, MIL: Yes**<br>**Years:** 2005, 2006, 2007, 2008<br>**Engines:** 2.5L SOHC, DOHC VIN 6<br>**Models:** All<br>**Transmissions:** All | **Gear 5 Incorrect Ratio**<br>Target gear ratio and actual gear ratio do not match. Shift point is too high or too low. Excessive shift shock. Gear is not changed. The vehicle does not move in D or R range with the engine running at high speed.<br>**Possible Causes:**<br>• Using the a scan tool, perform diagnosis<br>• Temporary poor contact of connector or harness may be the cause<br>• Faulty transmission assembly |
| **DTC: P0736**<br>**2T CCM, MIL: Yes**<br>**Years:** 2005, 2006, 2007, 2008<br>**Engines:** 2.5L SOHC, DOHC VIN 6<br>**Models:** All<br>**Transmissions:** All | **Reverse Incorrect Ratio**<br>Target gear ratio and actual gear ratio do not match. Shift point is too high or too low. Excessive shift shock. Gear is not changed. The vehicle does not move in D or R range with the engine running at high speed.<br>**Possible Causes:**<br>• Using the a scan tool, perform diagnosis<br>• Temporary poor contact of connector or harness may be the cause<br>• Faulty transmission assembly |
| **DTC: P0741**<br>**2T CCM, MIL: Yes**<br>**Years:** 2005, 2006, 2007, 2008<br>**Engines:** 2.0L VIN 2, 2.5L SOHC, DOHC VIN 6<br>**Models:** All<br>**Transmissions:** All | **Torque Converter Clutch Circuit Performance Or Stuck Off**<br>Defective lock-up clutch or torque converter assembly. Defective control valve. Defective turbine speed Sensor or 2. No lock-up occurs. (After engine is warmed-up)<br>**Possible Causes:**<br>• Using the a scan tool, perform diagnosis<br>• Temporary poor contact of connector or harness may be the cause<br>• Faulty transmission assembly<br>• Suspect the transmission assembly when DTC P0741 is displayed |
| **DTC: P0743**<br>**2T CCM, MIL: Yes**<br>**Years:** 2005, 2006, 2007, 2008<br>**Engines:** 2.0L VIN 2, 2.5L SOHC, DOHC VIN 6<br>**Models:** All<br>**Transmissions:** All | **Torque Converter Clutch Circuit Electrical**<br>The output signal circuit of lock up solenoid is open or shorted. No lock-up occurs. (After engine is warmed-up)<br>**Possible Causes:**<br>• Open circuit of harness between transmission module connector and transmission connector<br>• Short circuit of harness between transmission module connector and transmission connector<br>• Open circuit of harness between transmission connector and control valve body connector<br>• Short circuit of harness between control valve body connector and transmission ground<br>• Faulty control valve body<br>• Temporary poor contact of connector or harness may be the cause<br>• Faulty transmission module |
| **DTC: P0748**<br>**2T CCM, MIL: Yes**<br>**Years:** 2005, 2006, 2007, 2008<br>**Engines:** 2.0L VIN 2, 2.5L SOHC, DOHC VIN 6<br>**Models:** All<br>**Transmissions:** All | **Pressure Control Solenoid Electrical**<br>Output signal circuit of line pressure solenoid is open or shorted. Excessive shift shock.<br>**Possible Causes:**<br>• Temporary poor contact of connector or harness may be the cause<br>• Open circuit of harness between transmission module connector and transmission connector<br>• Short circuit of harness between transmission module connector and transmission connector<br>• Open circuit of harness between transmission connector and control valve body connector<br>• Short circuit of harness between control valve body connector and transmission ground<br>• Faulty control valve body<br>• Faulty transmission module |
| **DTC: P0751**<br>**2T CCM, MIL: Yes**<br>**Years:** 2005, 2006, 2007, 2008<br>**Engines:** 2.5L SOHC, DOHC VIN 6<br>**Models:** All<br>**Transmissions:** All | **Shift Solenoid Performance Or Stuck Off**<br>Output signal of front brake solenoid does not match with oil pressure. Locked to 4th or 5th gear.<br>**Possible Causes:**<br>• Open circuit of harness between transmission module and transmission connector<br>• Short circuit of harness between transmission module and transmission connector<br>• Faulty control valve body<br>• Transmission harness assembly |

| DTC | Trouble Code Title, Conditions & Possible Causes |
|---|---|
| **DTC: P0753**<br>**2T CCM, MIL:** Yes<br>**Years:** 2005, 2006, 2007, 2008<br>**Engines:** 2.0L VIN 2, 2.5L SOHC, DOHC VIN 6<br>**Models:** All<br>**Transmissions:** All | **Shift Solenoid Electrical**<br>Output signal circuit of front brake solenoid is open or shorted. Locked to 4th or 5th gear.<br>**Possible Causes:**<br>• Open circuit of harness between transmission module connector and transmission connector<br>• Short circuit of harness between transmission module connector and transmission connector<br>• Open circuit of harness between transmission connector and control valve body connector<br>• Short circuit of harness between control valve body and transmission connector<br>• Temporary poor contact of connector or harness may be the cause<br>• Faulty control valve body<br>• Faulty transmission module |
| **DTC: P0756**<br>**2T CCM, MIL:** Yes<br>**Years:** 2005, 2006, 2007, 2008<br>**Engines:** 2.5L SOHC, DOHC VIN 6<br>**Models:** All<br>**Transmissions:** All | **Shift Solenoid Performance Or Stuck Off**<br>Output signal value of input clutch solenoid and oil pressure does not match. Locked to 4th gear.<br>**Possible Causes:**<br>• Open circuit of harness between transmission module and transmission connector<br>• Short circuit of harness between transmission module and transmission connector<br>• Temporary poor contact of connector or harness may be the cause<br>• Faulty control valve body<br>• Transmission harness assembly |
| **DTC: P0758**<br>**2T CCM, MIL:** Yes<br>**Years:** 2005, 2006, 2007, 2008<br>**Engines:** 2.0L VIN 2, 2.5L SOHC, DOHC VIN 6<br>**Models:** All<br>**Transmissions:** All | **Shift Solenoid Electrical**<br>Output signal circuit of input clutch solenoid is open or shorted. Locked to 4th gear.<br>**Possible Causes:**<br>• Open circuit of harness between transmission module and transmission connector<br>• Short circuit of harness between transmission module and transmission connector<br>• Temporary poor contact of connector or harness may be the cause<br>• Transmission harness assembly<br>• Faulty transmission module |
| **DTC: P0761**<br>**2T CCM, MIL:** Yes<br>**Years:** 2005, 2006, 2007, 2008<br>**Engines:** 2.5L SOHC, DOHC VIN 6<br>**Models:** All<br>**Transmissions:** All | **Shift Solenoid Performance Or Stuck Off**<br>Output signal value of high & low reverse clutch solenoid and oil pressure does not match. Locked to 4th gear.<br>**Possible Causes:**<br>• Open circuit of harness between transmission module and transmission connector<br>• Short circuit of harness between transmission module and transmission connector<br>• Faulty control valve body<br>• Transmission harness assembly |
| **DTC: P0763**<br>**2T CCM, MIL:** Yes<br>**Years:** 2005, 2006, 2007, 2008<br>**Engines:** 2.5L SOHC, DOHC VIN 6<br>**Models:** All<br>**Transmissions:** All | **Shift Solenoid Electrical**<br>Output signal circuit of input clutch solenoid is open or shorted. Locked to 4th gear.<br>**Possible Causes:**<br>• Open circuit of harness between transmission module and transmission connector<br>• Short circuit of harness between transmission module and transmission connector<br>• Temporary poor contact of connector or harness may be the cause<br>• Transmission harness assembly<br>• Faulty transmission module |
| **DTC: P0766**<br>**2T CCM, MIL:** Yes<br>**Years:** 2005, 2006, 2007, 2008<br>**Engines:** 2.5L SOHC, DOHC VIN 6<br>**Models:** All<br>**Transmissions:** All | **Shift Solenoid Performance Or Stuck Off**<br>Output signal of front brake solenoid does not match with oil pressure. Locked to 4th or 5th gear.<br>**Possible Causes:**<br>• Open circuit of harness between transmission module and transmission connector<br>• Short circuit of harness between transmission module and transmission connector<br>• Faulty control valve body<br>• Transmission harness assembly |
| **DTC: P0768**<br>**2T CCM, MIL:** Yes<br>**Years:** 2005, 2006, 2007, 2008<br>**Engines:** 2.5L SOHC, DOHC VIN 6<br>**Models:** All<br>**Transmissions:** All | **Shift Solenoid Electrical**<br>The output signal circuit of direct clutch solenoid is open or shorted. Locked to 4th gear.<br>**Possible Causes:**<br>• Open circuit of harness between transmission module and transmission connector<br>• Short circuit of harness between transmission module and transmission connector<br>• Temporary poor contact of connector or harness may be the cause<br>• Transmission harness assembly<br>• Faulty transmission module |

| DTC | Trouble Code Title, Conditions & Possible Causes |
|---|---|
| **DTC: P0771**<br>**2T CCM, MIL: Yes**<br>**Years:** 2005, 2006, 2007, 2008<br>**Engines:** 2.0L VIN 2, 2.5L SOHC, DOHC VIN 6<br>**Models:** All<br>**Transmissions:** All | **Shift Solenoid Performance Or Stuck Off**<br>Output signal value of low coast brake solenoid and oil pressure does not match. Locked to 2nd gear. Engine brake does not function at 1st or 2nd of manual mode.<br>**Possible Causes:**<br>• Open circuit of harness between transmission module and transmission connector<br>• Short circuit of harness between transmission module and transmission connector<br>• Temporary poor contact of connector or harness may be the cause<br>• Transmission harness assembly<br>• Faulty control valve body |
| **DTC: P0773**<br>**2T CCM, MIL: Yes**<br>**Years:** 2005, 2006, 2007, 2008<br>**Engines:** 2.5L SOHC, DOHC VIN 6<br>**Models:** All<br>**Transmissions:** All | **Shift Solenoid Electrical**<br>Output signal circuit of low coast brake solenoid is open or shorted. Locked to 2nd 3rd or 4th gear. Engine brake does not function at 1st or 2nd of manual mode.<br>**Possible Causes:**<br>• Open circuit of harness between transmission module connector and transmission connector<br>• Short circuit of harness between transmission module connector and transmission connector<br>• Open circuit of harness between transmission connector and control valve body connector<br>• Short circuit of harness between control valve body connector and transmission ground<br>• Faulty control valve body<br>• Faulty transmission module |
| **DTC: P0775**<br>**2T CCM, MIL: Yes**<br>**Years:** 2005, 2006, 2007, 2008<br>**Engines:** 2.5L SOHC, DOHC VIN 6<br>**Models:** All<br>**Transmissions:** All | **Shift Solenoid Electrical**<br>Output signal circuit of low coast brake solenoid is open or shorted. Locked to 2nd 3rd or 4th gear. Engine brake does not function at 1st or 2nd of manual mode.<br>**Possible Causes:**<br>• Open circuit of harness between transmission module connector and transmission connector<br>• Short circuit of harness between transmission module connector and transmission connector<br>• Open circuit of harness between transmission connector and control valve body connector<br>• Short circuit of harness between control valve body connector and transmission ground<br>• Faulty control valve body<br>• Faulty transmission module |
| **DTC: P0778**<br>**2T CCM, MIL: Yes**<br>**Years:** 2005, 2006, 2007, 2008<br>**Engines:** 2.0L VIN 2<br>**Models:** All<br>**Transmissions:** All | **Pressure Control Solenoid Electrical**<br>Output signal circuit of low coast brake solenoid is open or shorted.<br>**Possible Causes:**<br>• Open circuit of harness between transmission module connector and transmission connector<br>• Short circuit of harness between transmission module connector and transmission connector<br>• Open circuit of harness between transmission connector and control valve body connector<br>• Short circuit of harness between control valve body connector and transmission ground<br>• Faulty control valve body<br>• Faulty transmission module |
| **DTC: P0801**<br>**2T CCM, MIL: Yes**<br>**Years:** 2005, 2006, 2007, 2008<br>**Engines:** 2.5L SOHC, DOHC VIN 6<br>**Models:** All<br>**Transmissions:** All | **Reverse Inhibitor Control Circuit**<br>Shift lock solenoid malfunction, open or short reverse inhibitor control circuit<br>Gear is shifted from "N" range to "R" range during driving at 12MPH or more. Gear can not be shifted from "N" range to "R" range though the vehicle is parked.<br>**Possible Causes:**<br>• Blown fuse number 32<br>• Short circuit of harness between fuse number 32 and transmission module<br>• Open circuit of harness between transmission module and shift lock solenoid connector<br>• Short circuit of harness between transmission module and shift lock solenoid connector<br>• Open circuit of harness between chassis ground and shift lock solenoid connector<br>• Faulty shift lock solenoid<br>• Poor contact in the reverse inhibitor control circuit<br>• Faulty transmission module |

| DTC | Trouble Code Title, Conditions & Possible Causes |
|---|---|
| **DTC: P0817**<br>**2T CCM, MIL:** Yes<br>**Years:** 2005, 2006, 2007, 2008<br>**Engines:** 2.5L SOHC, DOHC VIN 6<br>**Models:** All<br>**Transmissions:** All | **Starter Disable Circuit**<br>Open or short in P/N signal output circuit. Engine can be started on other than "P" or "N" range. Engine can not be started on "P" or "N" range.<br>**Possible Causes:**<br>• Blown fuse number 32<br>• Short circuit of harness between fuse number 32 and transmission module<br>• Open circuit of harness between transmission module and transmission connector, or poor contact of connector<br>• Short circuit of harness between transmission connector and chassis ground<br>• Temporary poor contact of connector or harness may be the cause<br>• Neutral switch of Control module<br>• Faulty transmission module |
| **DTC: P0851**<br>**2T CCM, MIL:** No<br>**Years:** 2005, 2006, 2007, 2008<br>**Engines:** 2.0L VIN 2, 2.5L SOHC, DOHC VIN 6<br>**Models:** All<br>**Transmissions:** A/T | **Neutral Switch Input Circuit Low (Automatic Transaxle)**<br>Erroneous idling.<br>**Possible Causes:**<br>• Check any other DTC on display<br>• Check input signal of Control module<br>• Check harness between Control module and transmission harness connector<br>• Check transmission harness connector<br>• Check inhibitor switch<br>• Check selector cable connection |
| **DTC: P0851**<br>**2T CCM, MIL:** No<br>**Years:** 2005, 2006, 2007, 2008<br>**Engines:** 2.0L VIN 2, 2.5L SOHC, DOHC VIN 6<br>**Models:** All<br>**Transmissions:** M/T | **Neutral Switch Input Circuit Low (Manual Transaxle)**<br>Control module detected an unexpected low voltage condition on the Neutral Position switch circuit during the CCM test.<br>**Possible Causes:**<br>• Check input signal of Control module<br>• Check poor contact<br>• Check neutral safety switch<br>• Check harness between Control module and neutral safety switch connector<br>• Check neutral safety switch ground |
| **DTC: P0852**<br>**2T CCM, MIL:** No<br>**Years:** 2005, 2006, 2007, 2008<br>**Engines:** 2.0L VIN 2, 2.5L SOHC, DOHC VIN 6<br>**Models:** All<br>**Transmissions:** A/T | **Neutral Switch Input Circuit High (Automatic Transaxle)**<br>Control module detected an unexpected low voltage condition on the Neutral Position switch circuit during the CCM test.<br>**Possible Causes:**<br>• Check any other DTC on display<br>• Check input signal of Control module<br>• Check poor contact<br>• Check harness between Control module and inhibitor switch connector<br>• Check inhibitor switch ground<br>• Check inhibitor switch<br>• Check selector cable connection |
| **DTC: P0852**<br>**2T CCM, MIL:** No<br>**Years:** 2005, 2006, 2007, 2008<br>**Engines:** 2.0L VIN 2, 2.5L SOHC, DOHC VIN 6<br>**Models:** All<br>**Transmissions:** M/T | **Neutral Switch Input Circuit High (Manual Transaxle)**<br>Control module detected an unexpected low voltage condition on the Neutral Position switch circuit during the CCM test.<br>**Possible Causes:**<br>• Check input signal of Control module<br>• Check poor contact<br>• Check harness between Control module and transmission harness connector<br>• Check neutral safety switch ground<br>• Check neutral safety switch |
| **DTC: P0864**<br>**2T CCM, MIL:** No<br>**Years:** 2005, 2006, 2007, 2008<br>**Engines:** 2.0L VIN 2, 2.5L SOHC, DOHC VIN 6<br>**Models:** All<br>**Transmissions:** A/T | **Transmission Module Communication Circuit Range/Performance**<br>Control module detected an unexpected voltage condition on the A/T Diagnosis circuit during the CCM test.<br>**Possible Causes:**<br>• Check driving condition (is AT shift control functioning)<br>• Check accessory |
| **DTC: P0865**<br>**2T CCM, MIL:** No<br>**Years:** 2005, 2006, 2007, 2008<br>**Engines:** 2.0L VIN 2, 2.5L SOHC, DOHC VIN 6<br>**Models:** All<br>**Transmissions:** All | **Transmission Module Communication Circuit Low**<br>Control module detected an unexpected low voltage condition on the A/T Diagnosis circuit during the CCM test.<br>**Possible Causes:**<br>• Check harness between Control module and transmission module Connector<br>• Check output signal for Control module<br>• Check trouble code for automatic transaxle |

| DTC | Trouble Code Title, Conditions & Possible Causes |
|---|---|
| **DTC: P0866**<br>**2T CCM, MIL: No**<br>**Years:** 2005, 2006, 2007, 2008<br>**Engines:** 2.0L VIN 2, 2.5L SOHC, DOHC VIN 6<br>**Models:** All<br>**Transmissions:** All | **Transmission Module Communication Circuit High**<br>Control module detected an unexpected high voltage condition on the A/T Diagnosis circuit during the CCM test.<br>**Possible Causes:**<br>• Check harness between Control module and transmission module Connector<br>• Check poor contact |
| **DTC: P0882**<br>**2T CCM, MIL: Yes**<br>**Years:** 2005, 2006, 2007, 2008<br>**Engines:** 2.5L SOHC, DOHC VIN 6<br>**Models:** All<br>**Transmissions:** All | **Transmission Module Power Input Signal Low**<br>Malfunction of PVIGN power supply relay or open, short circuit of PVIGN power supply circuit. Gear is not changed.<br>**Possible Causes:**<br>• MAIN SBF, SBF 8 or fuse number 12 blown out<br>• Open circuit of harness between fuse (No. 12) and PVIGN relay<br>• Temporary poor contact of connector or harness may be the cause<br>• Faulty PVIGN relay<br>• Faulty transmission module |
| **DTC: P0957**<br>**2T CCM, MIL: Yes**<br>**Years:** 2005, 2006, 2007, 2008<br>**Engines:** 2.5L SOHC, DOHC VIN 6<br>**Models:** All<br>**Transmissions:** All | **Backup Light Relay Circuit Low**<br>Shorted circuits of back-up light relay output circuit. Back-up light does not illuminate in "R" range<br>**Possible Causes:**<br>• Open circuit of harness between transmission module and transmission connector, or poor contact of connector<br>• Short circuit of harness between transmission module and transmission connector<br>• Faulty back-up light relay<br>• Open or short circuit of harness between fuse number 18 and back-up light relay<br>• Faulty transmission module |
| **DTC: P0958**<br>**2T CCM, MIL: Yes**<br>**Years:** 2005, 2006, 2007, 2008<br>**Engines:** 2.5L SOHC, DOHC VIN 6<br>**Models:** All<br>**Transmissions:** All | **Backup Light Relay Circuit High**<br>Backup light relay malfunction, or open/short circuit in back-up light relay output circuit. Back-up light does not illuminate in "R" range.<br>• Back-up light always illuminate except in "R" range.<br>**Possible Causes:**<br>• Open circuit of harness between transmission module and transmission connector, or poor contact of connector<br>• Short circuit of harness between transmission module and transmission connector<br>• Faulty back-up light relay<br>• Open or short circuit of harness between fuse number 18 and back-up light relay<br>• Faulty transmission module |

## OBD II Trouble Code List (P1xxx Codes)

| DTC | Trouble Code Title, Conditions & Possible Causes |
|---|---|
| **DTC: P1086**<br>**1T CCM, MIL: Yes**<br>**Years:** 2005, 2006, 2007, 2008<br>**Engines:** 2.5L DOHC<br>**Models:** All<br>**Transmissions:** All | **Tumble Generated Valve Position Sensor Circuit Low**<br>Engine stalls. Erroneous idling. Poor driving performance.<br>**Possible Causes:**<br>• Using a scan tool, check current data<br>• Check input signal for Control module<br>• Check harness between Control module and tumble generator valve position sensor connector<br>• Check poor contact |
| **DTC: P1087**<br>**1T CCM, MIL: Yes**<br>**Years:** 2005, 2006, 2007, 2008<br>**Engines:** 2.5L DOHC<br>**Models:** All<br>**Transmissions:** All | **Tumble Generated Valve Position Sensor Circuit High**<br>Engine stalls. Erroneous idling. Poor driving performance.<br>**Possible Causes:**<br>• Using a scan tool, check current data<br>• Check harness between tumble generator valve position sensor and Control module connector<br>• Check harness between throttle position sensor and Control module connector |
| **DTC: P1088**<br>**1T CCM, MIL: Yes**<br>**Years:** 2005, 2006, 2007, 2008<br>**Engines:** 2.5L DOHC<br>**Models:** All<br>**Transmissions:** All | **Tumble Generated Valve Position Sensor Circuit Low**<br>Engine stalls. Erroneous idling. Poor driving performance.<br>**Possible Causes:**<br>• Using a scan tool, check current data<br>• Check input signal for Control module<br>• Check harness between Control module and tumble generator valve position sensor connector<br>• Check poor contact |

| DTC | Trouble Code Title, Conditions & Possible Causes |
|---|---|
| **DTC: P1089**<br>**1T CCM, MIL: Yes**<br>**2002-04**<br>**Engines:** 2.5L DOHC<br>**Models:** All<br>**Transmissions:** All | **Tumble Generated Valve Position Sensor Circuit High**<br>Engine stalls. Erroneous idling. Poor driving performance.<br>**Possible Causes:**<br>• Using a scan tool, check current data<br>• Check harness between tumble generator valve position sensor and Control module connector |
| **DTC: P1090**<br>**1T CCM, MIL: Yes**<br>**Years:** 2005, 2006, 2007, 2008<br>**Engines:** 2.5L DOHC<br>**Models:** All<br>**Transmissions:** All | **Tumble Generated Valve System 1 (Valve Open)**<br>Engine stalls. Erroneous idling. Poor driving performance.<br>**Possible Causes:**<br>• Check any other DTC on display<br>• Check tumble generator valve (RH) |
| **DTC: P1091**<br>**1T CCM, MIL: Yes**<br>**2002-04**<br>**Engines:** 2.5L DOHC<br>**Models:** All<br>**Transmissions:** All | **Tumble Generated Valve System 1 (Valve Closed)**<br>Engine stalls. Erroneous idling. Poor driving performance.<br>**Possible Causes:**<br>• Check any other DTC on display<br>• Check tumble generator valve (RH) |
| **DTC: P1092**<br>**1T CCM, MIL: Yes**<br>**Years:** 2005, 2006, 2007, 2008<br>**Engines:** 2.5L DOHC<br>**Models:** All<br>**Transmissions:** All | **Tumble Generated Valve System 2 (Valve Open)**<br>Engine stalls. Erroneous idling. Poor driving performance.<br>**Possible Causes:**<br>• Check any other DTC on display<br>• Check tumble generator valve (RH) |
| **DTC: P1093**<br>**1T CCM, MIL: Yes**<br>**2002-04**<br>**Engines:** 2.5L DOHC<br>**Models:** All<br>**Transmissions:** All | **Tumble Generated Valve System 2 (Valve Closed)**<br>Engine stalls. Erroneous idling. Poor driving performance.<br>**Possible Causes:**<br>• Check any other DTC on display<br>• Check tumble generator valve (RH) |
| **DTC: P1094**<br>**1T CCM, MIL: Yes**<br>**Years:** 2005, 2006, 2007, 2008<br>**Engines:** 2.5L DOHC<br>**Models:** All<br>**Transmissions:** All | **Tumble Generated Valve Signal 1 Circuit Malfunction**<br>Engine stalls. Erroneous idling. Poor driving performance.<br>**Possible Causes:**<br>• Check harness between Control module and tumble generator valve actuator connector<br>• Check poor contact |
| **DTC: P1095**<br>**1T CCM, MIL: Yes**<br>**Years:** 2005, 2006, 2007, 2008<br>**Engines:** 2.5L DOHC<br>**Models:** All<br>**Transmissions:** All | **Tumble Generated Valve Signal 1 Circuit Malfunction (Short)**<br>Engine stalls. Erroneous idling. Poor driving performance.<br>**Possible Causes:**<br>• Check harness between Control module and tumble generator valve actuator connector<br>• Check poor contact |
| **DTC: P1096**<br>**1T CCM, MIL: Yes**<br>**Years:** 2005, 2006, 2007, 2008<br>**Engines:** 2.5L DOHC<br>**Models:** All<br>**Transmissions:** All | **Tumble Generated Valve Signal 2 Circuit Malfunction (Open)**<br>Engine stalls. Erroneous idling. Poor driving performance.<br>**Possible Causes:**<br>• Check harness between Control module and tumble generator valve actuator connector<br>• Check poor contact |
| **DTC: P1097**<br>**1T CCM, MIL: Yes**<br>**Years:** 2005, 2006, 2007, 2008<br>**Engines:** 2.5L DOHC<br>**Models:** All<br>**Transmissions:** All | **Tumble Generated Valve Signal 2 Circuit Malfunction (Short)**<br>Engine stalls. Erroneous idling. Poor driving performance.<br>**Possible Causes:**<br>• Check harness between Control module and tumble generator valve actuator connector<br>• Check poor contact |

| DTC | Trouble Code Title, Conditions & Possible Causes |
|---|---|
| **DTC: P1105**<br>**1T CCM, MIL: Yes**<br>**Years:** 2005, 2006, 2007, 2008<br>**Engines:** 2.3L VIN E/G<br>**Models:** All<br>**Transmissions:** All | **Charge Air Absolute Pressure Sensor Circuit Performance Problem**<br>Intake manifold pressure exceeding a value that is 5 kPa lower than atmospheric pressure. Charge air pressure below a value 5 kPa higher than atmospheric pressure.<br>**Possible Causes**<br>• Defective turbo unit, e.g. open wastegate valve or sticking turbine shaft<br>• Major air leak between turbocharger and throttle body<br>• Catalytic converter or exhaust system blockage |
| **DTC: P1106**<br>**1T CCM, MIL: Yes**<br>**Years:** 2005, 2006, 2007, 2008<br>**Engines:** x<br>**Models:** All<br>**Transmissions:** All | **Charge Air Absolute Pressure Sensor Circuit Performance Problem**<br>Intake manifold pressure exceeding a value that is 5 kPa lower than atmospheric pressure. Charge air pressure below a value 5 kPa higher than atmospheric pressure.<br>**Possible Causes**<br>• Defective turbo unit, e.g. open wastegate valve or sticking turbine shaft<br>• Major air leak between turbocharger and throttle body<br>• Catalytic converter or exhaust system blockage |
| **DTC: P1107**<br>**1T CCM, MIL: Yes**<br>**Years:** 2005, 2006, 2007, 2008<br>**Engines:** 2.3L VIN E/G<br>**Models:** All<br>**Transmissions:** All | **Charge Air Absolute Pressure Sensor Circuit Short to Ground / Open**<br>Engine speed drops slower than normal when accelerator pedal is released. Density compensation of the throttle angle is blocked. The charge air adaptation is blocked and zeroed, as a substitute value the barometric pressure sensor is used. Voltage below 0.185 V for more than 10 second(s).<br>**Possible Causes**<br>• Check connector<br>• Check wiring<br>• Check charge air sensor |
| **DTC: P1108**<br>**1T CCM, MIL: Yes**<br>**Years:** 2005, 2006, 2007, 2008<br>**Engines:** 2.3L VIN E/G<br>**Models:** All<br>**Transmissions:** All | **Charge Air Absolute Pressure Sensor Circuit Short to B+**<br>Engine speed drops slower than normal when accelerator pedal is released. Density compensation of the throttle angle is blocked. The charge air adaptation is blocked and zeroed, as a substitute value the barometric pressure sensor is used. Voltage higher than 4.815 V for more than 10 second(s).<br>**Possible Causes**<br>• Check connector<br>• Check wiring<br>• Check charge air sensor |
| **DTC: P1110**<br>**1T CCM, MIL: Yes**<br>**Years:** 2005, 2006, 2007, 2008<br>**Engines:** 2.3L VIN E/G<br>**Models:** All<br>**Transmissions:** All | **Charge Air Bypass Valve Performance Problems**<br>The inlet pressure drops rapidly by more than 3 kPa whereupon the accelerator pedal is fully released. The charge air pressure exceeds 120 kPa. The solenoid valve for the by-pass valve has been deactivated.<br>**Possible Causes**<br>• Defective turbo unit, e.g. open wastegate valve or sticking turbine shaft<br>• Major air leak between turbocharger and throttle body<br>• Catalytic converter or exhaust system blockage |
| **DTC: P1131**<br>**2T CCM, MIL: Yes**<br>**Years:** 2005, 2006, 2007, 2008<br>**Engines:** 2.0L VIN S/Y, 2.3L VIN E/G<br>**Models:** All<br>**Transmissions:** All | **O2S Sensor Circuit Lean**<br>Detected when two consecutive driving cycles with fault occur. Poor drive ability.<br>**Possible Causes:**<br>• Open circuit in harness between Control module and front oxygen A/F sensor connector<br>• Poor contact in front oxygen A/F sensor connector<br>• Poor contact in Control module connector<br>• Poor contact of coupling connector<br>• Faulty front oxygen A/F sensor |
| **DTC: P1132**<br>**2T CCM, MIL: Yes**<br>**Years:** 2005, 2006, 2007, 2008<br>**Engines:** 2.0L VIN S/Y, 2.3L VIN E/G<br>**Models:** All<br>**Transmissions:** All | **O2S Sensor Circuit Rich**<br>Detected when two consecutive driving cycles with fault occur. Poor drive ability.<br>**Possible Causes:**<br>• Open circuit in harness between Control module and front oxygen A/F sensor connector<br>• Poor contact in front oxygen A/F sensor connector<br>• Poor contact in Control module connector<br>• Poor contact of coupling connector<br>• Faulty front oxygen A/F sensor |
| **DTC: P1134**<br>**2T CCM, MIL: Yes**<br>**Years:** 2005, 2006, 2007, 2008<br>**Engines:** 2.0L VIN 2<br>**Models:** All<br>**Transmissions:** All | **A/F Sensor Micro-Computer Problem**<br>Detect the malfunction of IC communication. Control module detects when the IC communication malfunction occurs.<br>**Possible Causes:**<br>• Check for other DTC |

| DTC | Trouble Code Title, Conditions & Possible Causes |
|---|---|
| **DTC: P1135**<br>**2T CCM, MIL: Yes**<br>**Years:** 2005, 2006, 2007, 2008<br>**Engines:** 2.3L VIN E/G<br>**Models:** All<br>**Transmissions:** All | **O2S Heater Circuit Current Low**<br>Detected when two consecutive driving cycles with fault occur. Poor drive ability.<br>**Possible Causes:**<br>• Open circuit in harness between Control module and front oxygen A/F sensor connector<br>• Poor contact in front oxygen A/F sensor connector<br>• Poor contact in Control module connector<br>• Poor contact of coupling connector<br>• Faulty front oxygen A/F sensor |
| **DTC: P1136**<br>**2T CCM, MIL: Yes**<br>**Years:** 2005, 2006, 2007, 2008<br>**Engines:** 2.3L VIN E/G<br>**Models:** All<br>**Transmissions:** All | **O2S Heater Circuit Current High**<br>Detected when two consecutive driving cycles with fault occur. Poor drive ability.<br>**Possible Causes:**<br>• Poor contact in front oxygen A/F sensor connector<br>• Shorted ground circuit of harness between Control module and front oxygen A/F sensor connector<br>• Short circuit to power in the harness between Control module and front oxygen A/F sensor connector<br>• Faulty front oxygen A/F sensor |
| **DTC: P1137**<br>**2T CCM, MIL: Yes**<br>**Years:** 2005, 2006, 2007, 2008<br>**Engines:** 2.3L VIN E/G, 2.5L SOHC<br>**Models:** All<br>**Transmissions:** All | **O2S Low Voltage During Power Enrichment**<br>Control module detected the O2S signal was above 700 mv while the rear O2S was less than 399 mv for 9.5 seconds.<br>**Possible Causes**<br>• Air leaks present in exhaust manifold, exhaust pipes, or in the O2S mounting location<br>• O2S may be contaminated (due to improper fuel or silicone)<br>• Control module has failed |
| **DTC: P1141**<br>**2T CCM, MIL: Yes**<br>**Years:** 2005, 2006, 2007, 2008<br>**Engines:** 2.3L VIN E/G<br>**Models:** All<br>**Transmissions:** All | **O2S Heater Circuit Malfunction**<br>Control module detected an unexpected voltage condition on the O2S heater circuit during the CCM test.<br>**Possible Causes**<br>• Air leaks present in the exhaust manifold or the exhaust pipes<br>• O2S heater control circuit is open or shorted to ground<br>• O2S heater power circuit is open (test power circuit to O2S)<br>• O2S is damaged or has failed |
| **DTC: P1142**<br>**2T CCM, MIL: Yes**<br>**Years:** 2005, 2006, 2007, 2008<br>**Engines:** 2.3L VIN E/G<br>**Models:** All<br>**Transmissions:** All | **O2S Heater Circuit Malfunction**<br>Control module detected an unexpected voltage condition on the O2S heater circuit during the CCM test.<br>**Possible Causes**<br>• Air leaks present in the exhaust manifold or the exhaust pipes<br>• O2S heater control circuit is open or shorted to ground<br>• O2S heater power circuit is open (test power circuit to O2S)<br>• O2S is damaged or has failed |
| **DTC: P1152**<br>**2T CCM, MIL: Yes**<br>**Years:** 2005, 2006, 2007, 2008<br>**Engines:** 2.0L VIN 2, 2.5L SOHC, DOHC VIN 6<br>**Models:** All<br>**Transmissions:** All | **O2S Sensor Circuit Range/Performance Low**<br>Detected when two consecutive driving cycles with fault occur. Poor drive ability.<br>**Possible Causes:**<br>• Open circuit in harness between Control module and front oxygen A/F sensor connector<br>• Poor contact in front oxygen A/F sensor connector<br>• Poor contact in Control module connector<br>• Poor contact of coupling connector<br>• Faulty front oxygen A/F sensor |
| **DTC: P1153**<br>**2T CCM, MIL: Yes**<br>**Years:** 2005, 2006, 2007, 2008<br>**Engines:** 2.0L VIN 2, 2.5L SOHC, DOHC VIN 6<br>**Models:** All<br>**Transmissions:** All | **O2S Sensor Circuit Range/Performance High**<br>Detected when two consecutive driving cycles with fault occur. Poor drive ability.<br>**Possible Causes:**<br>• Poor contact in front oxygen A/F sensor connector<br>• Shorted ground circuit of harness between Control module and front oxygen A/F sensor connector<br>• Short circuit to power in the harness between Control module and front oxygen A/F sensor connector<br>• Faulty front oxygen A/F sensor |
| **DTC: P1154**<br>**2T CCM, MIL: Yes**<br>**Years:** 2005, 2006, 2007, 2008<br>**Engines:** 2.5L SOHC, DOHC VIN 6<br>**Models:** All<br>**Transmissions:** All | **O2S Sensor circuit Range/Performance Low**<br>Detected when two consecutive driving cycles with fault occur. Poor drive ability.<br>**Possible Causes:**<br>• Open circuit in harness between Control module and front oxygen A/F sensor connector<br>• Poor contact in front oxygen A/F sensor connector<br>• Poor contact in Control module connector<br>• Poor contact of coupling connector<br>• Faulty front oxygen A/F sensor |

| DTC | Trouble Code Title, Conditions & Possible Causes |
|---|---|
| **DTC: P1155**<br>**2T CCM, MIL: Yes**<br>**Years:** 2005, 2006, 2007, 2008<br>**Engines:** 2.5L SOHC, DOHC VIN 6<br>**Models:** All<br>**Transmissions:** All | **O2S Sensor circuit Range/Performance High**<br>Detected when two consecutive driving cycles with fault occur. Poor drive ability.<br>**Possible Causes:**<br>• Poor contact in front oxygen A/F sensor connector<br>• Shorted ground circuit of harness between Control module and front oxygen A/F sensor connector<br>• Short circuit to power in the harness between Control module and front oxygen A/F sensor connector<br>• Faulty front oxygen A/F sensor |
| **DTC: P1160**<br>**1T CCM, MIL: Yes**<br>**Years:** 2005, 2006, 2007, 2008<br>**Engines:** 2.5L DOHC<br>**Models:** All<br>**Transmissions:** All | **Return Spring Failure**<br>Control module detects when the valve is opened more than the default opening angle, but does not move to the close direction with the motor power stopped. Improper idling. Poor driving performance. Engine stalls.<br>**Possible Causes:**<br>• Check electronic throttle control relay<br>• Check power supply of electronic control relay<br>• Check harness between Control module and electronic throttle control relay<br>• Check sensor output<br>• Check poor contact<br>• Check harness between Control module and electronic throttle control<br>• Check sensor power supply<br>• Check for short in Control module<br>• Check sensor output<br>• Check harness between Control module and electronic throttle control motor<br>• Check electronic throttle control motor harness<br>• Check ground circuit |
| **DTC: P1171**<br>**2T CCM, MIL: Yes**<br>**Years:** 2005, 2006, 2007, 2008<br>**Engines:** 2.3L VIN E/G<br>**Models:** All<br>**Transmissions:** All | **Short Term Fuel Trim Max Value Air/Fuel Lean**<br>Control module can identify if the fuel delivery system can supply enough fuel during heavy acceleration. When this mode is requested during closed loop operation, the Control module provides extra fuel to the engine, and the Control module should Control module detected a rich condition. If it does not Control module detected a rich condition at this time, it will set this code.<br>**Possible Causes**<br>• Fuel filter clogged or restricted<br>• Fuel injectors dirty or restricted<br>• Fuel level too low during heavy acceleration events<br>• Fuel pressure regulator has failed (not supplying enough fuel)<br>• Water or alcohol in fuel (low O2S signal during acceleration) |
| **DTC: P1172**<br>**2T CCM, MIL: Yes**<br>**Years:** 2005, 2006, 2007, 2008<br>**Engines:** 2.3L VIN E/G<br>**Models:** All<br>**Transmissions:** All | **Short Term Fuel Trim Max Value Air/Fuel Rich**<br>Engine running and closed loop active. One multiplicative adaptation has been made. No change in status of brake light switch. Provided the fault criteria are not fulfilled during the driving cycle the OK report occurs at ignition before the engine stalls. Two types of OK reports can be given: Type 1: If a DTC is already stored for the diagnosis and similar operating conditions have occurred (coolant temperature above or below 71°C, engine speed ±375 rpm and load ±10%). Type 2: If there is no DTC stored for the diagnosis or similar operating conditions have not occurred<br>**Possible Causes**<br>• Manifold absolute pressure<br>• Charge air absolute pressure<br>• Atmosphere absoslute pressure<br>• EVAP canister purge valve |
| **DTC: P1174**<br>**1T CCM, MIL: Yes**<br>**Years:** 2005, 2006, 2007, 2008<br>**Engines:** 4.2L VIN S<br>**Models:** All<br>**Transmissions:** All | **Fuel Trim Cylinder Balance**<br>The Fuel Trim Cylinder Balance diagnostic detects a rich or a lean cylinder-to-cylinder air/fuelratio imbalance. The Fuel Trim Cylinder Balance diagnostic is very sensitive to O2S design. A non-OE sensor or an incorrect part number may set a false DTC. Monitoring the misfire current counters, or misfire graph, may help to isolate the cylinder that is causing the condition.<br>**Possible Causes:**<br>• Check MAP sensor<br>• Check air induction system<br>• Check restricted exhaust<br>• Fuel injectors<br>• Fuel contamination<br>• Ignition system |

| DTC | Trouble Code Title, Conditions & Possible Causes |
|---|---|
| **DTC: P1181**<br>**2T CCM, MIL: Yes**<br>**Years:** 2005, 2006, 2007, 2008<br>**Engines:** 2.3L VIN E/G<br>**Models:** All<br>**Transmissions:** All | **Long Term Fuel Trim Max Value Air/Fuel Lean**<br>The Control module can identify if the fuel delivery system can supply enough fuel during heavy acceleration. When this mode is requested during closed loop operation, the Control module provides extra fuel to the engine, and the Control module should Control module detected a rich condition. If it does not Control module detected a rich condition at this time, it will set this code.<br>**Possible Causes**<br>• Fuel filter clogged or restricted<br>• Fuel injectors dirty or restricted<br>• Fuel level too low during heavy acceleration events<br>• Fuel pressure regulator has failed (not supplying enough fuel)<br>• Water or alcohol in fuel (low O2S signal during acceleration) |
| **DTC: P1182**<br>**2T CCM, MIL: Yes**<br>**Years:** 2005, 2006, 2007, 2008<br>**Engines:** 2.3L VIN E/G<br>**Models:** All<br>**Transmissions:** All | **Long Term Fuel Trim Max Value Air/Fuel Lean**<br>The Control module can identify if the fuel delivery system can supply enough fuel during heavy acceleration. When this mode is requested during closed loop operation, the Control module provides extra fuel to the engine, and the Control module should Control module detected a rich condition. If it does not Control module detected a rich condition at this time, it will set this code.<br>**Possible Causes**<br>• Fuel filter clogged or restricted<br>• Fuel injectors dirty or restricted<br>• Fuel level too low during heavy acceleration events<br>• Fuel pressure regulator has failed (not supplying enough fuel)<br>• Water or alcohol in fuel (low O2S signal during acceleration) |
| **DTC: P1230**<br>**2T CCM, MIL: Yes**<br>**Years:** 2005, 2006, 2007, 2008<br>**Engines:** 2.3L VIN E/G<br>**Models:** All<br>**Transmissions:** All | **Throttle Position Sensor 1 and 2 Circuit Sum Out of Range**<br>Ignition in ON position. Continuous. But interrupted when fault criteria fulfilled and will not restart until next driving cycle. The sum of voltages from throttle position sensors 1 and 2 exceeds 5.70 V for more than 0.8 s or is less than 4.59 V for more than 0.3 s.<br>**Possible Causes:**<br>• Check sensor output<br>• Check poor contact<br>• Check harness between Control module and electronic throttle control<br>• Check sensor power supply<br>• Check for short circuit in Control module<br>• Check sensor output<br>• Check electronic throttle control harness |
| **DTC: P1231**<br>**2T CCM, MIL: Yes**<br>**Years:** 2005, 2006, 2007, 2008<br>**Engines:** 2.3L VIN E/G<br>**Models:** All<br>**Transmissions:** All | **Throttle Position Sensor 1 and 2 Circuit Sum Out of Range (No Limp Home)**<br>Ignition in ON position. Continuous. But interrupted when fault criteria fulfilled and will not restart until next driving cycle. The sum of voltages from throttle position sensors 1 and 2 exceeds 5.70 V for more than 0.8 s or is less than 4.59 V for more than 0.3 s.<br>**Possible Causes:**<br>• Check sensor output<br>• Check poor contact<br>• Check harness between Control module and electronic throttle control<br>• Check sensor power supply<br>• Check for short circuit in Control module<br>• Check sensor output<br>• Check electronic throttle control harness |
| **DTC: P1240**<br>**1T CCM, MIL: Yes**<br>**Years:** 2005, 2006, 2007, 2008<br>**Engines:** 2.3L VIN E/G<br>**Models:** All<br>**Transmissions:** All | **Throttle Motor Circuit Shorted**<br>Motor current exceeding 8A during the positive or negative section of the pulse width modulation period. Criteria fulfilled for 3 min.<br>**Possible Causes:**<br>• Check sensor output<br>• Check poor contact<br>• Check harness between Control module and electronic throttle control<br>• Check sensor power supply<br>• Check for short circuit in Control module<br>• Check sensor output |
| **DTC: P1251**<br>**1T CCM, MIL: Yes**<br>**Years:** 2005, 2006, 2007, 2008<br>**Engines:** 2.3L VIN E/G<br>**Models:** All<br>**Transmissions:** All | **Throttle Motor Full Pulse Width Modulation in Closing Direction**<br>The system cannot close the throttle disc. Throttle control goes into limp-home mode. The limp-home solenoid will be activated 5 times in succession each time the ignition is turned on until the fault has been rectified and the trouble codes cleared.<br>**Possible Causes:**<br>• Open and close the flap valve by hand.<br>• The flap valve should not bind in any position.<br>• The spring should return the flap valve to the stop screw.<br>• No external objects, hoses, cables, etc. must obstruct the movement of the throttle arm. |

| DTC | Trouble Code Title, Conditions & Possible Causes |
|---|---|
| **DTC: P1252**<br>**1T CCM, MIL: Yes**<br>**Years:** 2005, 2006, 2007, 2008<br>**Engines:** 2.3L VIN E/G<br>**Models:** All<br>**Transmissions:** All | **Throttle Motor Full Pulse Width Modulation in Open Direction and No Motor Current**<br>Throttle motor receiving 100% pulse width modulation in open direction but there is no current reaching the motor. Throttle control goes into limp-home mode. The limp-home solenoid will be activated 5 times in succession each time the ignition is turned on until the fault has been rectified and the trouble codes cleared.<br>**Possible Causes:**<br>• Check sensor output<br>• Check poor contact<br>• Check harness between Control module and electronic throttle control<br>• Check sensor power supply<br>• Check for short circuit in Control module<br>• Check sensor output |
| **DTC: P1253**<br>**1T CCM, MIL: Yes**<br>**Years:** 2005, 2006, 2007, 2008<br>**Engines:** 2.3L VIN E/G<br>**Models:** All<br>**Transmissions:** All | **Throttle Motor Full Pulse Width Modulation in Open Direction During Cranking**<br>Throttle motor receiving 100% pulse width modulation in opening direction. Throttle control goes into limp-home mode. The limp-home solenoid will be activated 5 times in succession each time the ignition is turned on until the fault has been rectified and the trouble codes cleared.<br>**Possible Causes:**<br>• Open and close the flap valve by hand.<br>• The flap valve should not bind in any position.<br>• The spring should return the flap valve to the stop screw.<br>• No external objects, hoses, cables, etc. must obstruct the movement of the throttle |
| **DTC: P1260**<br>**1T CCM, MIL: Yes**<br>**Years:** 2005, 2006, 2007, 2008<br>**Engines:** 2.3L VIN E/G<br>**Models:** All<br>**Transmissions:** All | **Throttle Return Spring Too Low Force**<br>Throttle motor receiving close to 0% pulse width modulation in opening direction for 4 s. Throttle return spring is too weak.<br>**Possible Causes:**<br>• Open and close the flap valve by hand.<br>• The flap valve should not bind in any position.<br>• The spring should return the flap valve to the stop screw.<br>• No external objects, hoses, cables, etc. must obstruct the movement of the throttle |
| **DTC: P1261**<br>**1T CCM, MIL: Yes**<br>**Years:** 2005, 2006, 2007, 2008<br>**Engines:** 2.3L VIN E/G<br>**Models:** All<br>**Transmissions:** All | **Throttle Binding**<br>Engine running and throttle control is already in limp-home mode due to another diagnostic trouble code. Air mass per combustion is greater than requested.<br>**Possible Causes:**<br>• Open and close the flap valve by hand.<br>• The flap valve should not bind in any position.<br>• The spring should return the flap valve to the stop screw.<br>• No external objects, hoses, cables, etc. must obstruct the movement of the throttle |
| **DTC: P1263**<br>**1T CCM, MIL: Yes**<br>**Years:** 2005, 2006, 2007, 2008<br>**Engines:** 2.3L VIN E/G<br>**Models:** All<br>**Transmissions:** All | **Throttle Moved Manually When Engine Was Running**<br>Engine running for at least 5 s. No vehicle speed. Throttle motor receiving 100% pulse width modulation in closing direction.<br>**Possible Causes:**<br>• Open and close the flap valve by hand.<br>• The flap valve should not bind in any position.<br>• The spring should return the flap valve to the stop screw.<br>• No external objects, hoses, cables, etc. must obstruct the movement of the throttle |
| **DTC: P1264**<br>**1T CCM, MIL: Yes**<br>**Years:** 2005, 2006, 2007, 2008<br>**Engines:** 2.3L VIN E/G<br>**Models:** All<br>**Transmissions:** All | **Throttle Open When Pedal Is Released**<br>Engine running. Throttle position exceeds the maximum permissible with pedal released.<br>**Possible Causes:**<br>• Open and close the flap valve by hand.<br>• The flap valve should not bind in any position.<br>• The spring should return the flap valve to the stop screw.<br>• No external objects, hoses, cables, etc. must obstruct the movement of the throttle |

| DTC | Trouble Code Title, Conditions & Possible Causes |
|---|---|
| **DTC: P1300**<br>**1T CCM, MIL: Yes**<br>**Years:** 2005, 2006, 2007, 2008<br>**Engines:** 2.3L VIN E/G<br>**Models:** All<br>**Transmissions:** All | **Misfire Detected (High Temperature Exhaust Gas)**<br>Control module detected whether the misfire occurred or not. (Exhaust temperature method) Control module detects when the exhaust temperature is high. Poor driving performance. Engine stalls. Poor Idling condition.<br>**Possible Causes:**<br>• Check any other DTC on display<br>• Check output signal of Control module<br>• Check harness between injector and Control module connector<br>• Check fuel injector<br>• Check power supply line<br>• Check power supply to mass air flow sensor<br>• Check harness between Control module and sensor connector<br>• Check installation of camshaft and crankshaft sensors<br>• Check crank plate<br>• Check condition of timing chain<br>• Check fuel level<br>• Check status of MIL light<br>• Check air intake system<br>• Check cause of misfire (engine running)<br>• Check engine condition<br>• Check all cylinders individually, in pairs, in groups and at random |
| **DTC: P1301**<br>**1T CCM, MIL: Yes**<br>**Years:** 2005, 2006, 2007, 2008<br>**Engines:** 2.0L VIN 2, 2.5L DOHC<br>**Models:** All<br>**Transmissions:** All | **Cylinder 1 Misfire Detected**<br>Control module detected whether the misfire occurred or not. (Exhaust temperature method) Control module detects when the exhaust temperature is high. Poor driving performance. Engine stalls. Poor Idling condition.<br>**Possible Causes:**<br>• Check any other DTC on display<br>• Check output signal of Control module<br>• Check harness between injector and Control module connector<br>• Check fuel injector<br>• Check power supply line<br>• Check power supply to mass air flow sensor<br>• Check harness between Control module and sensor connector<br>• Check installation of camshaft and crankshaft sensors<br>• Check crank plate<br>• Check condition of timing chain<br>• Check fuel level<br>• Check status of MIL light<br>• Check air intake system<br>• Check cause of misfire (engine running)<br>• Check engine condition<br>• Check all cylinders individually, in pairs, in groups and at random |
| **DTC: P1302**<br>**1T CCM, MIL: Yes**<br>**Years:** 2005, 2006, 2007, 2008<br>**Engines:** 2.3L VIN E/G, 2.5L DOHC<br>**Models:** All<br>**Transmissions:** All | **Cylinder 2 Misfire Detected**<br>Control module detected whether the misfire occurred or not. (Exhaust temperature method) Control module detects when the exhaust temperature is high. Poor driving performance. Engine stalls. Poor Idling condition.<br>**Possible Causes:**<br>• Check any other DTC on display<br>• Check output signal of Control module<br>• Check harness between injector and Control module connector<br>• Check fuel injector<br>• Check power supply line<br>• Check power supply to mass air flow sensor<br>• Check harness between Control module and sensor connector<br>• Check installation of camshaft and crankshaft sensors<br>• Check crank plate<br>• Check condition of timing chain<br>• Check fuel level<br>• Check status of MIL light<br>• Check air intake system<br>• Check cause of misfire (engine running)<br>• Check engine condition<br>• Check all cylinders individually, in pairs, in groups and at random |

| DTC | Trouble Code Title, Conditions & Possible Causes |
|---|---|
| **DTC: P1303**<br>**1T CCM, MIL: Yes**<br>**Years:** 2005, 2006, 2007, 2008<br>**Engines:** 2.3L VIN E/G, 2.5L DOHC<br>**Models:** All<br>**Transmissions:** All | **Cylinder 3 Misfire Detected**<br>Control module detected whether the misfire occurred or not. (Exhaust temperature method) Control module detects when the exhaust temperature is high. Poor driving performance. Engine stalls. Poor Idling condition.<br>**Possible Causes:**<br>• Check any other DTC on display<br>• Check output signal of Control module<br>• Check harness between injector and Control module connector<br>• Check fuel injector<br>• Check power supply line<br>• Check power supply to mass air flow sensor<br>• Check harness between Control module and sensor connector<br>• Check installation of camshaft and crankshaft sensors<br>• Check crank plate<br>• Check condition of timing chain<br>• Check fuel level<br>• Check status of MIL light<br>• Check air intake system<br>• Check cause of misfire (engine running)<br>• Check engine condition<br>• Check all cylinders individually, in pairs, in groups and at random |
| **DTC: P1304**<br>**1T CCM, MIL: Yes**<br>**Years:** 2005, 2006, 2007, 2008<br>**Engines:** 2.3L VIN E/G, 2.5L DOHC<br>**Models:** All<br>**Transmissions:** All | **Cylinder 4 Misfire Detected**<br>Control module detected whether the misfire occurred or not. (Exhaust temperature method) Control module detects when the exhaust temperature is high. Poor driving performance. Engine stalls. Poor Idling condition.<br>**Possible Causes:**<br>• Check any other DTC on display<br>• Check output signal of Control module<br>• Check harness between injector and Control module connector<br>• Check fuel injector<br>• Check power supply line<br>• Check power supply to mass air flow sensor<br>• Check harness between Control module and sensor connector<br>• Check installation of camshaft and crankshaft sensors<br>• Check crank plate<br>• Check condition of timing chain<br>• Check fuel level<br>• Check status of MIL light<br>• Check air intake system<br>• Check cause of misfire (engine running)<br>• Check engine condition<br>• Check all cylinders individually, in pairs, in groups and at random |
| **DTC: P1310**<br>**1T CCM, MIL: Yes**<br>**Years:** 2005, 2006, 2007, 2008<br>**Engines:** 2.3L VIN E/G, 2.5L DOHC<br>**Models:** All<br>**Transmissions:** All | **Ignition Discharge Module Not Powered, Bank 1**<br>Crankshaft rotating. Main relay voltage above 6V. Voltage on all ignition trigger outputs below 2V for 1 second(s).<br>**Possible Causes:**<br>• Check module output<br>• Check poor contact<br>• Check harness between Control module and module<br>• Check power supply<br>• Check for short circuit in Control module |
| **DTC: P1312**<br>**1T CCM, MIL: Yes**<br>**Years:** 2005, 2006, 2007, 2008<br>**Engines:** 2.0L VIN 2<br>**Models:** All<br>**Transmissions:** All | **Combustion Detection Cylinder 1/2 Open Circuit / Short to B+**<br>Atmospheric pressure must exceed 72 kPa and the engine coolant and intake air temperatures must exceed −7°C. Voltage exceeding 5V continuously during 200 combustions.<br>**Possible Causes:**<br>• Check output signal from Control module<br>• Check for poor contact in Control module connector<br>• Check harness between module and Control module |

| DTC | Trouble Code Title, Conditions & Possible Causes |
|---|---|
| **DTC: P1334**<br>**1T CCM, MIL: Yes**<br>**Years:** 2005, 2006, 2007, 2008<br>**Engines:** 2.3L VIN E/G, 2.5L DOHC<br>**Models:** All<br>**Transmissions:** All | **Combustion Detection Cylinder 3/4 Open Circuit / Short to B+**<br>Atmospheric pressure must exceed 72 kPa and the engine coolant and intake air temperatures must exceed −7°C. Voltage exceeding 5V continuously during 200 combustions.<br>**Possible Causes:**<br>• Check output signal from Control module<br>• Check for poor contact in Control module connector<br>• Check harness between module and Control module |
| **DTC: P1380**<br>**1T CCM, MIL: Yes**<br>**Years:** 2005, 2006, 2007, 2008<br>**Engines:** 4.2L VIN S, 5.3L VIN M/P, 6.0L VIN H<br>**Models:** All<br>**Transmissions:** All | **Misfire Detected - No Communication With Brake Control Module**<br>The engine control module detects engine misfire by detecting variations in crankshaft deceleration between firing strokes. An ABS malfunction exists for more than 10 seconds, preventing Control module from receiving rough road detection data. Engine misfire is detected and DTC P0300 set.<br>**Possible Causes:**<br>• Check for other DTC<br>• Check Control module for communication<br>• Check faulty connectors |
| **DTC: P1381**<br>**1T CCM, MIL: Yes**<br>**Years:** 2005, 2006, 2007, 2008<br>**Engines:** 4.2L VIN S, 5.3L VIN M/P, 6.0L VIN H<br>**Models:** All<br>**Transmissions:** All | **Misfire Detected - No Communication With Brake Control Module**<br>The engine control module detects engine misfire by detecting variations in crankshaft deceleration between firing strokes. An ABS malfunction exists for more than 10 seconds, preventing Control module from receiving rough road detection data. Engine misfire is detected and DTC P0300 set.<br>**Possible Causes:**<br>• Check for other DTC<br>• Check Control module for communication<br>• Check faulty connectors |
| **DTC: P1390**<br>**1T CCM, MIL: Yes**<br>**Years:** 2005, 2006, 2007, 2008<br>**Engines:** 2.3L VIN E/G<br>**Models:** All<br>**Transmissions:** All | **Random Misfire Detected Catalyst Damaging Fuel Level Low**<br>Running unevenly and misfiring. Engine started. Fuel level below 4 litres. Main relay voltage exceeding 10V. Ignition discharge module combustion signals registered. The number of missing combustion signals is accumulated for each cylinder during 200 engine revolutions. The total number of accepted misfires during 200 engine revolutions is specified in a load and engine speed-dependent matrix.<br>**Possible Causes:**<br>• Check for other DTC<br>• Check Control module for communication<br>• Check faulty connectors<br>• Check the ignition coil<br>• Check the injector<br>• Check the fuel supply |
| **DTC: P1391**<br>**1T CCM, MIL: Yes**<br>**Years:** 2005, 2006, 2007, 2008<br>**Engines:** 2.3L VIN E/G<br>**Models:** All<br>**Transmissions:** All | **Cylinder 1 Misfire Detected Catalyst Damaging. Fuel Level Low**<br>Running unevenly and misfiring. Engine started. Fuel level below 4 litres. Main relay voltage exceeding 10V. Ignition discharge module combustion signals registered. The number of missing combustion signals is accumulated for each cylinder during 200 engine revolutions. The total number of accepted misfires during 200 engine revolutions is specified in a load and engine speed-dependent matrix.<br>**Possible Causes:**<br>• Check for other DTC<br>• Check Control module for communication<br>• Check faulty connectors<br>• Check the ignition coil<br>• Check the injector<br>• Check the fuel supply |
| **DTC: P1392**<br>**1T CCM, MIL: Yes**<br>**Years:** 2005, 2006, 2007, 2008<br>**Engines:** 2.3L VIN E/G<br>**Models:** All<br>**Transmissions:** All | **Cylinder 2 Misfire Detected Catalyst Damaging. Fuel Level Low**<br>Running unevenly and misfiring. Engine started. Fuel level below 4 litres. Main relay voltage exceeding 10V. Ignition discharge module combustion signals registered. The number of missing combustion signals is accumulated for each cylinder during 200 engine revolutions. The total number of accepted misfires during 200 engine revolutions is specified in a load and engine speed-dependent matrix.<br>**Possible Causes:**<br>• Check for other DTC<br>• Check Control module for communication<br>• Check faulty connectors<br>• Check the ignition coil<br>• Check the injector<br>• Check the fuel supply |

| DTC | Trouble Code Title, Conditions & Possible Causes |
|---|---|
| **DTC: P1393**<br>**1T CCM, MIL: Yes**<br>**Years:** 2005, 2006, 2007, 2008<br>**Engines:** 2.3L VIN E/G<br>**Models:** All<br>**Transmissions:** All | **Cylinder 3 Misfire Detected Catalyst Damaging. Fuel Level Low**<br>Running unevenly and misfiring. Engine started. Fuel level below 4 litres. Main relay voltage exceeding 10V. Ignition discharge module combustion signals registered. The number of missing combustion signals is accumulated for each cylinder during 200 engine revolutions. The total number of accepted misfires during 200 engine revolutions is specified in a load and engine speed-dependent matrix.<br>**Possible Causes:**<br>• Check for other DTC<br>• Check Control module for communication<br>• Check faulty connectors<br>• Check the ignition coil<br>• Check the injector<br>• Check the fuel supply |
| **DTC: P1394**<br>**1T CCM, MIL: Yes**<br>**Years:** 2005, 2006, 2007, 2008<br>**Engines:** 2.3L VIN E/G<br>**Models:** All<br>**Transmissions:** All | **Cylinder 4 Misfire Detected Catalyst Damaging. Fuel Level Low**<br>Running unevenly and misfiring. Engine started. Fuel level below 4 litres. Main relay voltage exceeding 10V. Ignition discharge module combustion signals registered. The number of missing combustion signals is accumulated for each cylinder during 200 engine revolutions. The total number of accepted misfires during 200 engine revolutions is specified in a load and engine speed-dependent matrix.<br>**Possible Causes:**<br>• Check for other DTC<br>• Check Control module for communication<br>• Check faulty connectors<br>• Check the ignition coil<br>• Check the injector<br>• Check the fuel supply |
| **DTC: P1400**<br>**1T CCM, MIL: Yes**<br>**Years:** 2005, 2006, 2007, 2008<br>**Engines:** 2.0L VIN 2, 2.0L VIN S/Y, 4.2L VIN S, 5.3L VIN M/P, 6.0L VIN H<br>**Models:** All<br>**Transmissions:** All | **Cold Start Emission Reduction Control System**<br>The catalytic converter must be warmed to efficiently reduce the emissions. The cold start strategy is to reduce the amount of time it takes to warm the catalytic converter. During a cold start, the engine idle speed is elevated and spark timing is retarded to allow the catalyst to warm quickly. Any loading of the engine that lowers engine RPM, such as partial application of the clutch, A/C cycling, etc. during the first 120 seconds of the engine run time may set this DTC.<br>**Possible Causes:**<br>• Check air induction system<br>• Check exhaust system<br>• Check engine mechanical<br>• Check Control module for communication<br>• Check faulty connectors |
| **DTC: P1400**<br>**2T CCM, MIL: Yes**<br>**Years:** 2005, 2006, 2007, 2008<br>**Engines:** 2.5L DOHC<br>**Models:** All, All<br>**Transmissions:** All | **Fuel Tank Pressure Control Solenoid Valve Circuit**<br>Control module detected open or short circuit of pressure control solenoid valve. Control module detects when Control module output level is different from actual terminal level.<br>**Possible Causes:**<br>• Check output signal from Control module<br>• Check for poor contact in Control module connector<br>• Check harness between pressure control solenoid valve and Control module<br>• Faulty pressure control solenoid valve |
| **DTC: P1410**<br>**1T CCM, MIL: Yes**<br>**Years:** 2005, 2006, 2007, 2008<br>**Engines:** 2.5L DOHC<br>**Models:** All, All<br>**Transmissions:** All | **Secondary Air Injection System Switching Valve Stuck Open**<br>Continually detects for a combination solenoid valve and lead valve stuck open condition. Calculate the integrated value of secondary air supply piping pressure sensor output voltage maximum/minimum values and output voltage deviation for a constant time period after engine start, and if the difference between the maximum/minimum is large and the integrated value is also large, judge as NG.<br>**Possible Causes:**<br>• Check secondary combination valve<br>• Check for poor contact of connector<br>• Faulty secondary air combination valve |
| **DTC: P1418**<br>**1T CCM, MIL: Yes**<br>**Years:** 2005, 2006, 2007, 2008<br>**Engines:** 2.5L DOHC<br>**Models:** All<br>**Transmissions:** All | **Secondary Air Injection System Control Circuit Shorted**<br>Control module output level is different from the actual terminal level.<br>**Possible Causes:**<br>• Check harness between Control module and secondary air pump relay<br>• Short circuit in harness between Control module and secondary air pump relay terminal. |

| DTC | Trouble Code Title, Conditions & Possible Causes |
|---|---|
| **DTC: P1420**<br>**2T CCM, MIL: Yes**<br>**Years:** 2005, 2006, 2007, 2008<br>**Engines:** 2.0L VIN 2, 2.5L SOHC, DOHC<br>**Models:** All, All<br>**Transmissions:** All | **Fuel Tank Pressure Control Solenoid Valve Circuit High**<br>Control module detected open or short circuit of pressure control solenoid valve. Control module detects when Control module output level is different from actual terminal level.<br>**Possible Causes:**<br>• Check input signal for Control module<br>• Check for poor contact in Control module connector<br>• Check harness between pressure control solenoid valve and Control module<br>• Faulty Control module |
| **DTC: P1441**<br>**1T CCM, MIL: Yes**<br>**Years:** 2005, 2006, 2007, 2008<br>**Engines:** 2.3L VIN E/G<br>**Models:** All,<br>**Transmissions:** All | **EVAP Purge Valve Malfunction and Malfunctional Fuel Level Sensor**<br>Control module detected the abnormal function (stuck closed) of the drain valve. Control module detects when fuel tank pressure is low.<br>**Possible Causes:**<br>• Check any other DTC on display<br>• Check vent line hoses<br>• Check drain valve operation |
| **DTC: P1442**<br>**1T CCM, MIL: Yes**<br>**Years:** 2005, 2006, 2007, 2008<br>**Engines:** 2.0L VIN S/Y<br>**Models:** All,<br>**Transmissions:** All | **EVAP System, Fuel Level Sensor Failure and Small Leak Detected**<br>Control module detected the EVAP system was not able to achieve or maintain a vacuum during the diagnostic test period.<br>**Possible Causes:**<br>• Fuel tank vapor line(s) is clogged, damaged or disconnected<br>• Fuel tank pressure sensor low reference circuit is open<br>• Fuel tank pressure sensor is damaged or it has failed<br>• EVAP charcoal canister is clogged or loaded with fuel or water<br>• EVAP purge or EVAP vent valve is damaged or it has failed<br>• PCM has failed |
| **DTC: P1443**<br>**1T CCM, MIL: Yes**<br>**Years:** 2005, 2006, 2007, 2008<br>**Engines:** 2.0L VIN 2, 2.5L SOHC, DOHC VIN 6<br>**Models:** All,<br>**Transmissions:** All | **Vent Control Solenoid Valve Function Problem**<br>Control module detected the abnormal function (stuck closed) of the drain valve. Control module detects when fuel tank pressure is low.<br>**Possible Causes:**<br>• Check any other DTC on display<br>• Check vent line hoses<br>• Check drain valve operation |
| **DTC: P1444**<br>**1T CCM, MIL: Yes**<br>**Years:** 2005, 2006, 2007, 2008<br>**Engines:** 2.3L VIN E/G<br>**Models:** All,<br>**Transmissions:** All | **Canister Close Valve Circuit Open / Short to Ground**<br>Engine running. Main relay voltage 10 V – 16 V. Output not active. Voltage below 2V for 1 s.<br>**Possible Causes:**<br>• Check input signal for Control module<br>• Check for poor contact in Control module connector<br>• Check harness between solenoid valve and Control module<br>• Faulty Control module |
| **DTC: P1445**<br>**1T CCM, MIL: Yes**<br>**Years:** 2005, 2006, 2007, 2008<br>**Engines:** 2.3L VIN E/G<br>**Models:** All,<br>**Transmissions:** All | **Canister Close Valve Circuit Open / Short to B+**<br>Engine running. Main relay voltage 10 V – 16 V. Output not active. Voltage exceeds 2V for 1 s.<br>**Possible Causes:**<br>• Check input signal for Control module<br>• Check for poor contact in Control module connector<br>• Check harness between solenoid valve and Control module<br>• Faulty Control module |
| **DTC: P1446**<br>**2T CCM, MIL: Yes**<br>**Years:** 2006<br>**Engines:** 2.0L VIN 2, 2.5L SOHC, DOHC VIN 6<br>**Models:** All,<br>**Transmissions:** All | **Fuel Tank Sensor Control Valve Circuit Low**<br>Control module detected the open or short circuit of tank pressure switching solenoid. Control module detects when Control module output level is different from actual terminal level.<br>**Possible Causes:**<br>• Check output signal for Control module<br>• Check for poor contact in Control module connector<br>• Check harness between fuel tank sensor control valve and Control module connector<br>• Check fuel tank sensor control valve<br>• Check power supply to sensor control valve<br>• Check for poor contact at sensor control valve connector |

| DTC | Trouble Code Title, Conditions & Possible Causes |
|---|---|
| **DTC: P1447**<br>**2T CCM, MIL: Yes**<br>**Years:** 2006<br>**Engines:** 2.0L VIN 2, 2.5L SOHC, DOHC VIN 6<br>**Models:** All,<br>**Transmissions:** All | **Fuel Tank Sensor Control Valve Circuit High**<br>Control module detected the open/short circuit of fuel tank sensor control valve. Control module detects when Control module output level is different from actual terminal level.<br>**Possible Causes:**<br>• Check any other DTC on display<br>• Check fuel filler cap<br>• Check EVAP hoses and pipes<br>• Faulty fuel tank pressure sensor |
| **DTC: P1448**<br>**2T CCM, MIL: Yes**<br>**Years:** 2006<br>**Engines:** 2.0L VIN 2, 2.5L SOHC, DOHC VIN 6<br>**Models:** All,<br>**Transmissions:** All | **Fuel Tank Sensor Control Valve Range/Performance**<br>Control module detects the tank pressure switching solenoid function abnormality. The tank pressure sensor is a relative pressure sensor, which normally compares the pressure with the atmospheric pressure. The tank pressure switching solenoid is a solenoid, which shifts the compare space from opening to closed during the EVAP diagnosis. Control module detected the malfunction that the compare space remains closed. (Not Control module detects after enable condition completed but assume NG before enable condition completed.)<br>**Possible Causes:**<br>• Check any other DTC on display<br>• Check fuel filler cap<br>• Check EVAP hoses and pipes<br>• Faulty fuel tank pressure sensor |
| **DTC: P1451**<br>**2T CCM, MIL: Yes**<br>**Years:** 2006<br>**Engines:** 2.0L VIN S/Y, 2.3L VIN E/G<br>**Models:** All,<br>**Transmissions:** All | **EVAP System Fuel Level Sensor Failure and Pressure Sensor Performance Problem**<br>Control module detected the open or short circuit of tank pressure sensor. Control module detects when Control module output level is different from actual terminal level.<br>**Possible Causes:**<br>• Check output signal for Control module<br>• Check for poor contact in Control module connector<br>• Check harness between sensor and Control module connector<br>• Check sensor<br>• Check power supply to sensor<br>• Check for poor contact at sensor |
| **DTC: P1452**<br>**2T CCM, MIL: Yes**<br>**Years:** 2006<br>**Engines:** 2.0L VIN S/Y, 2.3L VIN E/G<br>**Models:** All,<br>**Transmissions:** All | **EVAP System Tank Pressure Sensor Value Low**<br>Control module detected the open or short circuit of tank pressure sensor. Control module detects when Control module output level is different from actual terminal level.<br>**Possible Causes:**<br>• Check output signal for Control module<br>• Check for poor contact in Control module connector<br>• Check harness between sensor and Control module connector<br>• Check sensor<br>• Check power supply to sensor<br>• Check for poor contact at sensor |
| **DTC: P1453**<br>**2T CCM, MIL: Yes**<br>**Years:** 2006<br>**Engines:** 2.0L VIN S/Y, 2.3L VIN E/G<br>**Models:** All,<br>**Transmissions:** All | **EVAP System Tank Pressure Sensor Value High**<br>Control module detected the open or short circuit of tank pressure sensor. Control module detects when Control module output level is different from actual terminal level.<br>**Possible Causes:**<br>• Check output signal for Control module<br>• Check for poor contact in Control module connector<br>• Check harness between sensor and Control module connector<br>• Check sensor<br>• Check power supply to sensor<br>• Check for poor contact at sensor |
| **DTC: P1491**<br>**1T CCM, MIL: Yes**<br>**Years:** 2005, 2006, 2007, 2008<br>**Engines:** 2.0L VIN 2, 2.3L VIN E/G, 2.5L SOHC, DOHC VIN 6<br>**Models:** All<br>**Transmissions:** All | **Positive Crankcase Ventilation (Blow-By) Function Problem**<br>Control module detected the blow-by hose release abnormality. Control module detects when the diagnosis terminal voltage is high. Poor idling condition.<br>**Possible Causes:**<br>• Check the blow-by hose for disconnection or cracks<br>• Check harness between PCV diagnosis connector and Control module connector<br>• Check between PCV diagnosis connector and engine ground<br>• Check for poor contact in Control module and PCV diagnosis connector<br>• Faulty PCV diagnosis connector |

| DTC | Trouble Code Title, Conditions & Possible Causes |
|---|---|
| **DTC: P1492**<br>**1T CCM, MIL: Yes**<br>**Years:** 2005, 2006, 2007, 2008<br>**Engines:** 2.0L VIN S/Y, 2.3L VIN E/G, 2.5L SOHC, DOHC VIN 6<br>**Models:** All<br>**Transmissions:** All | **EGR Solenoid Valve Signal 1 Circuit Low Input**<br>Control module detects open or short circuit of EGR. Control module detects when Control module output level differs from the actual terminal level. EGR target position 0 step. 1 second after startup. Improper idling. Poor driving performance. Engine breathing<br>**Possible Causes:**<br>• Open circuit in harness between EGR valve and main relay connector<br>• Poor contact of coupling connector<br>• Open circuit in harness between Control module and EGR valve connector<br>• Ground short in harness between Control module and EGR valve connector<br>• Poor contact in Control module or EGR valve connector<br>• Faulty EGR valve |
| **DTC: P1493**<br>**1T CCM, MIL: Yes**<br>**Years:** 2005, 2006, 2007, 2008<br>**Engines:** 2.0L VIN S/Y, 2.3L VIN E/G, 2.5L SOHC, DOHC VIN 6<br>**Models:** All<br>**Transmissions:** All | **EGR Solenoid Valve Signal 1 Circuit High Input**<br>Control module detects open or short circuit of EGR. Control module detects when Control module output level differs from the actual terminal level. EGR target position 0 step. 1 second after startup. Improper idling. Poor driving performance. Engine breathing<br>**Possible Causes:**<br>• Short circuit to power in the harness between Control module and EGR valve connectors<br>• Poor contact of Control module connector |
| **DTC: P1494**<br>**1T CCM, MIL: Yes**<br>**Years:** 2005, 2006, 2007, 2008<br>**Engines:** 2.5L SOHC, DOHC VIN 6<br>**Models:** All<br>**Transmissions:** All | **EGR Solenoid Valve Signal 2 Circuit Low Input**<br>Control module detects open or short circuit of EGR. Control module detects when control module output level differs from the actual terminal level. EGR target position 0 step. 1 second after startup. Improper idling. Poor driving performance. Engine breathing<br>**Possible Causes:**<br>• Open circuit in harness between EGR valve and main relay connector<br>• Poor contact of coupling connector<br>• Open circuit in harness between Control module and EGR valve connector<br>• Ground short in harness between Control module and EGR valve connector<br>• Poor contact in Control module or EGR valve connector<br>• Faulty EGR valve |
| **DTC: P1495**<br>**1T CCM, MIL: Yes**<br>**Years:** 2005, 2006, 2007, 2008<br>**Engines:** 2.5L SOHC, DOHC VIN 6<br>**Models:** All<br>**Transmissions:** All | **EGR Solenoid Valve Signal 2 Circuit High Input**<br>Control module detects open or short circuit of EGR. Control module detects when control module output level differs from the actual terminal level. EGR target position 0 step. 1 second after startup. Improper idling. Poor driving performance. Engine breathing<br>**Possible Causes:**<br>• Short circuit to power in the harness between Control module and EGR valve connectors<br>• Poor contact of Control module connector |
| **DTC: P1496**<br>**1T CCM, MIL: Yes**<br>**Years:** 2005, 2006, 2007, 2008<br>**Engines:** 2.5L SOHC, DOHC VIN 6<br>**Models:** All<br>**Transmissions:** All | **EGR Solenoid Valve Signal 3 Circuit Low Input**<br>Control module detects open or short circuit of EGR. Control module detects when control module output level differs from the actual terminal level. EGR target position 0 step. 1 second after startup. Improper idling. Poor driving performance. Engine breathing<br>**Possible Causes:**<br>• Open circuit in harness between EGR valve and main relay connector<br>• Poor contact of coupling connector<br>• Open circuit in harness between Control module and EGR valve connector<br>• Ground short in harness between Control module and EGR valve connector<br>• Poor contact in Control module or EGR valve connector<br>• Faulty EGR valve |
| **DTC: P1497**<br>**1T CCM, MIL: Yes**<br>**Years:** 2005, 2006, 2007, 2008<br>**Engines:** 2.5L SOHC, DOHC VIN 6<br>**Models:** All<br>**Transmissions:** All | **EGR Solenoid Valve Signal 3 Circuit High Input**<br>Control module detects open or short circuit of EGR. Control module detects when Control module output level differs from the actual terminal level. EGR target position 0 step. 1 second after startup. Improper idling. Poor driving performance. Engine breathing<br>**Possible Causes:**<br>• Short circuit to power in the harness between Control module and EGR valve connectors<br>• Poor contact of Control module connector |
| **DTC: P1498**<br>**1T CCM, MIL: Yes**<br>**Years:** 2005, 2006, 2007, 2008<br>**Engines:** 2.5L SOHC, DOHC VIN 6<br>**Models:** All<br>**Transmissions:** All | **EGR Solenoid Valve Signal 4 Circuit Low Input**<br>Control module detects open or short circuit of EGR. Control module detects when control module output level differs from the actual terminal level. EGR target position 0 step. 1 second after startup. Improper idling. Poor driving performance. Engine breathing<br>**Possible Causes:**<br>• Open circuit in harness between EGR valve and main relay connector<br>• Poor contact of coupling connector<br>• Open circuit in harness between Control module and EGR valve connector<br>• Ground short in harness between Control module and EGR valve connector<br>• Poor contact in Control module or EGR valve connector<br>• Faulty EGR valve |

| DTC | Trouble Code Title, Conditions & Possible Causes |
|---|---|
| **DTC: P1499**<br>**1T CCM, MIL: Yes**<br>**Years:** 2005, 2006, 2007, 2008<br>**Engines:** 2.5L SOHC, DOHC VIN 6<br>**Models:** All<br>**Transmissions:** All | **EGR Solenoid Valve Signal 4 Circuit High Input**<br>Control module detects open or short circuit of EGR. Control module detects when control module output level differs from the actual terminal level. EGR target position 0 step. 1 second after startup. Improper idling. Poor driving performance. Engine breathing<br>**Possible Causes:**<br>• Short circuit to power in the harness between Control module and EGR valve connectors<br>• Poor contact of Control module connector |
| **DTC: P1510**<br>**1T CCM, MIL: Yes**<br>**Years:** 2006<br>**Engines:** 2.5L SOHC Vin 6<br>**Models:** All<br>**Transmissions:** All | **ISC Solenoid Valve Signal No. 1 Circuit Malfunction (low input)**<br>Control module detected an unexpected "low" condition on the IAC solenoid Signal 1 circuit during the test.<br>**Possible Causes:**<br>• Check power supply to idle air control solenoid valve<br>• Check harness between Control module and idle control solenoid valve connector<br>• Check poor contact |
| **DTC: P1511**<br>**1T CCM, MIL: Yes**<br>**Years:** 2006<br>**Engines:** 2.5L SOHC Vin 6<br>**Models:** All<br>**Transmissions:** All | **ISC Solenoid Valve Signal No. 1 Circuit Malfunction (high input)**<br>Control module detected an unexpected "low" condition on the IAC solenoid Signal 1 circuit during the test.<br>**Possible Causes:**<br>• Check any other DTC on display<br>• Check ground circuit for Control module<br>• Check harness between Control module and idle control solenoid valve connector |
| **DTC: P1512**<br>**1T CCM, MIL: Yes**<br>**Years:** 2006<br>**Engines:** 2.5L SOHC Vin 6<br>**Models:** All<br>**Transmissions:** All | **ISC Solenoid Valve Signal No. 2 Circuit Malfunction (low input)**<br>Control module detected an unexpected "low" condition on the IAC solenoid Signal 2 circuit during the test.<br>**Possible Causes:**<br>• Check power supply to idle air control solenoid valve<br>• Check harness between Control module and idle control solenoid valve connector<br>• Check poor contact |
| **DTC: P1513**<br>**2T CCM, MIL: Yes**<br>**1998-06**<br>**Engines:** 2.5L SOHC Vin 6<br>**Models:** All<br>**Transmissions:** All | **ISC Solenoid Valve Signal No. 2 Circuit Malfunction (high input)**<br>Control module detected an unexpected "low" condition on the IAC solenoid Signal 2 circuit during the test.<br>**Possible Causes:**<br>• Check any other DTC on display<br>• Check ground circuit for Control module<br>• Check harness between Control module and idle control solenoid valve connector |
| **DTC: P1514**<br>**2T CCM, MIL: Yes**<br>**Years:** 2006<br>**Engines:** 2.5L SOHC Vin 6<br>**Models:** All<br>**Transmissions:** All | **ISC Solenoid Valve Signal No. 3 Circuit Malfunction (low input)**<br>Control module detected an unexpected "low" condition on the IAC solenoid Signal 3 circuit during the test.<br>**Possible Causes:**<br>• Check power supply to idle air control solenoid valve<br>• Check harness between Control module and idle control solenoid valve connector<br>• Check poor contact |
| **DTC: P1515**<br>**2T CCM, MIL: Yes**<br>**Years:** 2006<br>**Engines:** 2.5L SOHC Vin 6<br>**Models:** All<br>**Transmissions:** All | **ISC Solenoid Valve Signal No. 3 Circuit Malfunction (high input)**<br>Control module detected an unexpected "low" condition on the IAC solenoid Signal 3 circuit during the test.<br>**Possible Causes:**<br>• Check any other DTC on display<br>• Check ground circuit for Control module<br>• Check harness between Control module and idle control solenoid valve connector |
| **DTC: P1516**<br>**2T CCM, MIL: Yes**<br>**1998-06**<br>**Engines:** 2.5L SOHC Vin 6<br>**Models:** All<br>**Transmissions:** All | **ISC Solenoid Valve Signal No. 4 Circuit Malfunction (low input)**<br>Control module detected an unexpected "low" condition on the IAC solenoid Signal 4 circuit during the test.<br>**Possible Causes:**<br>• Check power supply to idle air control solenoid valve<br>• Check harness between Control module and idle control solenoid valve connector<br>• Check poor contact |
| **DTC: P1517**<br>**2T CCM, MIL: Yes**<br>**Years:** 2006<br>**Engines:** 2.5L SOHC Vin 6<br>**Models:** All<br>**Transmissions:** All | **ISC Solenoid Valve Signal No. 4 Circuit Malfunction (high input)**<br>Control module detected an unexpected "low" condition on the IAC solenoid Signal 4 circuit during the test.<br>**Possible Causes:**<br>• Check any other DTC on display<br>• Check ground circuit for Control module<br>• Check harness between Control module and idle control solenoid valve connector |

| DTC | Trouble Code Title, Conditions & Possible Causes |
|---|---|
| **DTC: P1518**<br>**1T CCM, MIL: No**<br>**Years:** 2005, 2006, 2007, 2008<br>**Engines:** 2.0L VIN 2, 2.5L SOHC, DOHC VIN 6<br>**Models:** All<br>**Transmissions:** All | **Starter Switch Circuit Low Input**<br>Control module detected an unexpected "low" voltage condition on the Starter Switch circuit during the test.<br>**Note: The engine will not start with this condition present.**<br>**Possible Causes:**<br>• Check operation of starter motor circuit<br>• Open or ground short circuit of harness between Control module and starter motor connector |
| **DTC: P1523**<br>**1T CCM, MIL: No**<br>**Years:** 2005, 2006, 2007, 2008<br>**Engines:** 2.0L VIN S/Y<br>**Models:** All<br>**Transmissions:** All | **A/C Pressure Sensor Circuit Open/Short to Ground**<br>A/C pressure of less than 0.1 volt or more than 4.92 volts. The condition must be present for more than 5 seconds.<br>**Possible Causes:**<br>• A/C refrigerant pressure<br>• Pressure sensor reference circuit open shorted to ground or high resistance<br>• Pressure sensor signal circuit open shorted to ground or high resistance<br>• A/C refrigerant pressure sensor.<br>• Control module has failed |
| **DTC: P1530**<br>**1T CCM, MIL: No**<br>**Years:** 2005, 2006, 2007, 2008<br>**Engines:** 2.3L VIN E/G<br>**Models:** All<br>**Transmissions:** All | **Pedal Position Sensor 1 and 2 Circuit Sum Out of Range**<br>Control module detected the out of range circuit of accelerator pedal position. Control module detects if out of specification. Improper idling. Poor driving performance.<br>**Possible Causes:**<br>• Check accelerator pedal position sensor output<br>• Check poor contact<br>• Check harness between Control module and accelerator pedal position sensor<br>• Check power supply of accelerator pedal position sensor<br>• Check sensor |
| **DTC: P1531**<br>**1T CCM, MIL: No**<br>**Years:** 2005, 2006, 2007, 2008<br>**Engines:** 2.3L VIN E/G<br>**Models:** All<br>**Transmissions:** All | **Pedal Position Sensor 1 and 2 Circuit Adapted Sum Out of Range**<br>Control module detected the out of range circuit of accelerator pedal position. Control module detects if out of specification. Improper idling. Poor driving performance.<br>**Possible Causes:**<br>• Check accelerator pedal position sensor output<br>• Check poor contact<br>• Check harness between Control module and accelerator pedal position sensor<br>• Check power supply of accelerator pedal position sensor<br>• Check sensor |
| **DTC: P1532**<br>**1T CCM, MIL: No**<br>**Years:** 2005, 2006, 2007, 2008<br>**Engines:** 2.3L VIN E/G<br>**Models:** All<br>**Transmissions:** All | **Pedal Position Sensor 1 and 2 Circuit Control Module Inputs Shorted to Each Other**<br>Control module detected the out of range circuit of accelerator pedal position. Control module detects if out of specification. Improper idling. Poor driving performance.<br>**Possible Causes:**<br>• Check accelerator pedal position sensor output<br>• Check poor contact<br>• Check harness between Control module and accelerator pedal position sensor<br>• Check power supply of accelerator pedal position sensor<br>• Check sensor |
| **DTC: P1544**<br>**1T CCM, MIL: No**<br>**Years:** 2006<br>**Engines:** 2.0L VIN 2, 2.5L SOHC, DOHC VIN 6<br>**Models:** All<br>**Transmissions:** All | **Exhaust Gas Temperature To High**<br>Control module detected the malfunction of high exhaust gas temperature. Control module detects when the exhaust gas becomes too high. Erroneous idling. Poor driving performance.<br>**Possible Causes:**<br>• Check any other DTC on display<br>• Check exhaust system |
| **DTC: P1549**<br>**1T CCM, MIL: No**<br>**Years:** 2005, 2006, 2007, 2008<br>**Engines:** 2.3L VIN E/G<br>**Models:** All<br>**Transmissions:** All | **Charge Air Malfunction**<br>Control module detected an unexpected voltage on the Boost Control Solenoid control circuit for 30 seconds.<br>**Possible Causes**<br>• Boost solenoid circuit is open between the solenoid and Control module<br>• Boost solenoid circuit is shorted to ground<br>• Boost solenoid power circuit is open to system power<br>• Boost solenoid is damaged or has failed<br>• Control module has failed |

| DTC | Trouble Code Title, Conditions & Possible Causes |
|---|---|
| **DTC: P1551**<br>**1T CCM, MIL: Yes**<br>**Years:** 2005, 2006, 2007, 2008<br>**Engines:** 2.8L VIN U<br>**Models:** All<br>**Transmissions:** All | **Throttle Mechanical Rest Position Not Reached During Learn**<br>Control module detects when the motor current becomes large or drive circuit is heated. Improper idling. Engine stalls. Poor driving performance.<br>**Possible Causes:**<br>• Check electronic throttle control relay<br>• Check power supply of electronic control relay<br>• Check harness between Control module and electronic throttle control relay<br>• Check sensor output<br>• Check poor contact<br>• Check harness between Control module and electronic throttle control<br>• Check sensor power supply<br>• Check for short in Control module<br>• Check sensor output<br>• Check harness between Control module and electronic throttle control motor<br>• Check electronic throttle control motor harness<br>• Check ground circuit<br>• Check electronic throttle control |
| **DTC: P1560**<br>**1T CCM, MIL: Yes**<br>**Years:** 2005, 2006, 2007, 2008<br>**Engines:** 2.0L VIN 2, 2.5L SOHC, DOHC VIN 6<br>**Models:** All<br>**Transmissions:** All | **Back-Up Voltage Circuit Malfunction**<br>Control module detected an unexpected "low" voltage condition on the Battery Backup circuit during the CCM test.<br>**Note: The engine will not start with this condition present.**<br>**Possible Causes:**<br>• Check input signal of Control module<br>• Check harness between Control module and main fuse box connector<br>• Check fuse number 13 |
| **DTC: P1570**<br>**1T CCM, MIL: Yes**<br>**Years:** 2005, 2006, 2007, 2008<br>**Engines:** 2.5L SOHC, DOHC VIN 6<br>**Models:** All<br>**Transmissions:** All | **Antenna Incorrect Immobilizer Key**<br>When the engine is started. Improper antenna<br>**Possible Causes:**<br>• Perform teaching operation of ignition key<br>• Check the ignition keys (including transponder) which cannot be registered<br>• Replace the body integrated unit and replace all the ignition keys (including transponder).<br>• Execute the registration procedure next |
| **DTC: P1572**<br>**1T CCM, MIL: Yes**<br>**Years:** 2005, 2006, 2007, 2008<br>**Engines:** 2.5L SOHC, DOHC VIN 6<br>**Models:** All<br>**Transmissions:** All | **IMM Circuit Failure (Except Antenna Circuit)**<br>When starting the engine. Communication failure between body integrated unit and Control module. Incorrect immobilizer key (Use of unregistered key in body integrated unit.<br>**Possible Causes:**<br>• Poor connections<br>• Wrong key<br>• Faulty body integrated unit<br>• Faulty Control module |
| **DTC: P1574**<br>**1T CCM, MIL: Yes**<br>**Years:** 2005, 2006, 2007, 2008<br>**Engines:** 2.5L SOHC, DOHC VIN 6<br>**Models:** All<br>**Transmissions:** All | **Key Communication Failure**<br>When starting the engine. Incorrect immobilizer key (Use of unregistered key in body integrated unit.<br>**Possible Causes:**<br>• Poor connections<br>• Wrong key |
| **DTC: P1576**<br>**1T CCM, MIL: No**<br>**Years:** 2005, 2006, 2007, 2008<br>**Engines:** 2.3L VIN E/G<br>**Models:** All<br>**Transmissions:** All | **Brake Light Switch Circuit Open / Short to B+**<br>Control module detected brake Switch signals did not cycle open and closed during trip.<br>**Possible Causes**<br>• Brake switch B+ circuit is open<br>• Brake switch is damaged (closed) or has failed |
| **DTC: P1577**<br>**1T CCM, MIL: No**<br>**Years:** 2005, 2006, 2007, 2008<br>**Engines:** 2.3L VIN E/G<br>**Models:** All<br>**Transmissions:** All | **Brake Light Switch Circuit Short to Ground**<br>Control module detected brake Switch signals did not cycle open and closed during trip.<br>**Possible Causes**<br>• Brake switch signal circuit is shorted to ground<br>• Brake switch is damaged or has failed |

| DTC | Trouble Code Title, Conditions & Possible Causes |
|---|---|
| **DTC: P1601**<br>**2T CCM, MIL: Yes**<br>**Years:** 2005, 2006, 2007, 2008<br>**Engines:** 2.3L VIN E/G, 2.5L SOHC, DOHC VIN 6<br>**Models:** All<br>**Transmissions:** All | **Internal Control Module Malfunction**<br>Communication does not complete between control valve memory box. Shifting quality malfunction.<br>**Possible Causes:**<br>• Check loose connection on transmission module connector<br>• Check transmission module output signal<br>• Check harness between transmission module connector and transmission connector<br>• Check circuit of harness between transmission connector and control valve body connector<br>• Faulty transmission assembly |
| **DTC: P1602**<br>**2T CCM, MIL: Yes**<br>**Years:** 2005, 2006, 2007, 2008<br>**Engines:** 2.3L VIN E/G, 2.5L SOHC, DOHC VIN 6<br>**Models:** All<br>**Transmissions:** All | **Internal Control Module Malfunction**<br>Communication does not complete between control valve memory box. Shifting quality malfunction.<br>**Possible Causes:**<br>• Check loose connection on transmission module connector<br>• Check transmission module output signal<br>• Check harness between transmission module connector and transmission connector<br>• Check circuit of harness between transmission connector and control valve body connector<br>• Faulty transmission assembly |
| **DTC: P1603**<br>**2T CCM, MIL: Yes**<br>**Years:** 2005, 2006, 2007, 2008<br>**Engines:** 2.3L VIN E/G<br>**Models:** All<br>**Transmissions:** All | **Internal Control Module Malfunction**<br>Control module detects an internal failure or incomplete programming for more than 5 seconds.<br>**Possible Causes:**<br>• Control module is not programmed<br>• Control module has failed |
| **DTC: P1604**<br>**2T CCM, MIL: Yes**<br>**Years:** 2005, 2006, 2007, 2008<br>**Engines:** 2.3L VIN E/G<br>**Models:** All<br>**Transmissions:** All | **Internal Control Module Malfunction**<br>Control module detects an internal failure or incomplete programming for more than 5 seconds.<br>**Possible Causes:**<br>• Control module is not programmed<br>• Control module has failed |
| **DTC: P1605**<br>**2T CCM, MIL: Yes**<br>**Years:** 2005, 2006, 2007, 2008<br>**Engines:** 2.3L VIN E/G<br>**Models:** All<br>**Transmissions:** All | **Internal Control Module Malfunction**<br>Control module detects an internal failure or incomplete programming for more than 5 seconds.<br>**Possible Causes:**<br>• Control module is not programmed<br>• Control module has failed |
| **DTC: P1606**<br>**2T CCM, MIL: Yes**<br>**Years:** 2005, 2006, 2007, 2008<br>**Engines:** 2.0L VIN S/Y, 2.3L VIN E/G<br>**Models:** All<br>**Transmissions:** All | **Control Module Internal Fault**<br>Control module detects an internal failure or incomplete programming for more than 5 seconds.<br>**Possible Causes:**<br>• Control module is not programmed<br>• Control module has failed |
| **DTC: P1608**<br>**2T CCM, MIL: Yes**<br>**Years:** 2005, 2006, 2007, 2008<br>**Engines:** 2.3L VIN E/G<br>**Models:** All<br>**Transmissions:** All | **Internal Control Module Malfunction**<br>Control module detects an internal failure or incomplete programming for more than 5 seconds.<br>**Possible Causes:**<br>• Control module is not programmed<br>• Control module has failed |
| **DTC: P1609**<br>**2T CCM, MIL: Yes**<br>**Years:** 2005, 2006, 2007, 2008<br>**Engines:** 2.0L VIN S/Y, 2.3L VIN E/G<br>**Models:** All<br>**Transmissions:** All | **Internal Control Module Malfunction**<br>Control module detects an internal failure or incomplete programming for more than 5 seconds.<br>**Possible Causes:**<br>• Control module is not programmed<br>• Control module has failed |

| DTC | Trouble Code Title, Conditions & Possible Causes |
|---|---|
| **DTC: P1610**<br>**2T CCM, MIL: Yes**<br>**Years:** 2005, 2006, 2007, 2008<br>**Engines:** 2.0L VIN S/Y, 2.3L VIN E/G<br>**Models:** All<br>**Transmissions:** All | **ECU Internal Fault Security Function Disable**<br>Control module detects an internal failure or incomplete programming for more than 5 seconds.<br>**Possible Causes:**<br>   • Control module is not programmed<br>   • Control module has failed |
| **DTC: P1611**<br>**2T CCM, MIL: Yes**<br>**Years:** 2005, 2006, 2007, 2008<br>**Engines:** 2.0L VIN S/Y, 2.3L VIN E/G<br>**Models:** All<br>**Transmissions:** All | **Control Module Not Added**<br>Control module detects an internal failure or incomplete programming for more than 5 seconds.<br>**Possible Causes:**<br>   • Control module is not programmed<br>   • Control module has failed |
| **DTC: P1613**<br>**2T CCM, MIL: Yes**<br>**Years:** 2005, 2006, 2007, 2008<br>**Engines:** 2.0L VIN S/Y, 2.3L VIN E/G<br>**Models:** All<br>**Transmissions:** All | **Control Module Not Added or Missing on Bus**<br>Control module detected it was not programmed for the correct application.<br>**Possible Causes**<br>   • Control module must be reprogrammed to correct this problem. Once this step is done, recheck the code to verify the repair is done. |
| **DTC: P1614**<br>**2T CCM, MIL: Yes**<br>**Years:** 2005, 2006, 2007, 2008<br>**Engines:** 2.0L VIN S/Y, 2.3L VIN E/G<br>**Models:** All<br>**Transmissions:** All | **Control Module Not Added**<br>Control module detected it was not programmed for the correct application.<br>**Possible Causes**<br>   • Control module must be reprogrammed to correct this problem. Once this step is done, recheck the code to verify the repair is done. |
| **DTC: P1615**<br>**2T CCM, MIL: Yes**<br>**Years:** 2005, 2006, 2007, 2008<br>**Engines:** 2.0L VIN S/Y<br>**Models:** All<br>**Transmissions:** All | **Control Module Not Added**<br>Control module detected it was not programmed for the correct application.<br>**Possible Causes**<br>   • Control module must be reprogrammed to correct this problem. Once this step is done, recheck the code to verify the repair is done. |
| **DTC: P1616**<br>**2T CCM, MIL: Yes**<br>**Years:** 2005, 2006, 2007, 2008<br>**Engines:** 2.0L VIN S/Y<br>**Models:** All<br>**Transmissions:** All | **Control Module or I Bus ECU Not Added**<br>Control module detected it was not programmed for the correct application.<br>**Possible Causes**<br>   • Control module must be reprogrammed to correct this problem. Once this step is done, recheck the code to verify the repair is done. |
| **DTC: P1619**<br>**2T CCM, MIL: Yes**<br>**Years:** 2005, 2006, 2007, 2008<br>**Engines:** 2.0L VIN S/Y<br>**Models:** All<br>**Transmissions:** All | **ECU Attributes Not Programmed/Internal Fault**<br>Control module detects an internal failure or incomplete programming for more than 5 seconds.<br>**Possible Causes:**<br>   • Control module is not programmed<br>   • Control module has failed |
| **DTC: P1621**<br>**2T CCM, MIL: Yes**<br>**Years:** 2005, 2006, 2007, 2008<br>**Engines:** 2.3L VIN E/G<br>**Models:** All<br>**Transmissions:** All | **Internal Control Module Malfunction**<br>Control module detects an internal failure or incomplete programming for more than 5 seconds.<br>**Possible Causes:**<br>   • Control module is not programmed<br>   • Control module has failed |
| **DTC: P1622**<br>**2T CCM, MIL: Yes**<br>**Years:** 2005, 2006, 2007, 2008<br>**Engines:** 2.3L VIN E/G<br>**Models:** All<br>**Transmissions:** All | **Internal Control Module Malfunction**<br>Control module detects an internal failure or incomplete programming for more than 5 seconds.<br>**Possible Causes:**<br>   • Control module is not programmed<br>   • Control module has failed |

| DTC | Trouble Code Title, Conditions & Possible Causes |
|---|---|
| **DTC: P1623**<br>**2T CCM, MIL: Yes**<br>**Years:** 2005, 2006, 2007, 2008<br>**Engines:** 2.0L VIN S/Y, 2.3L VIN E/G<br>**Models:** All<br>**Transmissions:** All | **Transmission Control Module Missing on Bus**<br>Control module detected it was not programmed for the correct application.<br>**Possible Causes**<br>• Control module must be reprogrammed to correct this problem. Once this step is done, recheck the code to verify the repair is done. |
| **DTC: P1624**<br>**2T CCM, MIL: Yes**<br>**Years:** 2005, 2006, 2007, 2008<br>**Engines:** 2.0L VIN S/Y, 2.3L VIN E/G<br>**Models:** All<br>**Transmissions:** All | **Control Module Missing on Bus**<br>Control module detected it was not programmed for the correct application.<br>**Possible Causes**<br>• Control module must be reprogrammed to correct this problem. Once this step is done, recheck the code to verify the repair is done. |
| **DTC: P1625**<br>**2T CCM, MIL: Yes**<br>**Years:** 2005, 2006, 2007, 2008<br>**Engines:** 2.0L VIN S/Y, 2.3L VIN E/G<br>**Models:** All<br>**Transmissions:** All | **Traction Control Module Missing on Bus**<br>Control module detected it was not programmed for the correct application.<br>**Possible Causes**<br>• Control module must be reprogrammed to correct this problem. Once this step is done, recheck the code to verify the repair is done. |
| **DTC: P1629**<br>**2T CCM, MIL: Yes**<br>**Years:** 2005, 2006, 2007, 2008<br>**Engines:** 2.8L VIN U<br>**Models:** All<br>**Transmissions:** All | **Control Module Not Added or Missing on Bus**<br>Control module detected it was not programmed for the correct application.<br>**Possible Causes**<br>• Control module must be reprogrammed to correct this problem. Once this step is done, recheck the code to verify the repair is done. |
| **DTC: P1631**<br>**2T CCM, MIL: Yes**<br>**Years:** 2005, 2006, 2007, 2008<br>**Engines:** 2.3L VIN E/G<br>**Models:** All<br>**Transmissions:** All | **Internal Control Module Malfunction**<br>Control module detects an internal failure or incomplete programming for more than 5 seconds.<br>**Possible Causes:**<br>• Control module is not programmed<br>• Control module has failed |
| **DTC: P1632**<br>**2T CCM, MIL: Yes**<br>**Years:** 2005, 2006, 2007, 2008<br>**Engines:** 2.3L VIN E/G , 2.8L VIN U<br>**Models:** All<br>**Transmissions:** All | **Internal Control Module Malfunction**<br>Control module detects an internal failure or incomplete programming for more than 5 seconds.<br>**Possible Causes:**<br>• Control module is not programmed<br>• Control module has failed |
| **DTC: P1632**<br>**2T CCM, MIL: Yes**<br>**Years:** 2005, 2006, 2007, 2008<br>**Engines:** 2.3L VIN E/G<br>**Models:** All<br>**Transmissions:** All | **Internal Control Module Malfunction**<br>Control module detects an internal failure or incomplete programming for more than 5 seconds.<br>**Possible Causes:**<br>• Control module is not programmed<br>• Control module has failed |
| **DTC: P1640**<br>**2T CCM, MIL: Yes**<br>**Years:** 2005, 2006, 2007, 2008<br>**Engines:** 2.3L VIN E/G<br>**Models:** All<br>**Transmissions:** All | **Main Relay Circuit No Voltage to Control Module**<br>Control module detects no voltage.<br>**Possible Causes:**<br>• Check input signal for Control module<br>• Check for poor contact in Control module connector<br>• Check harness between relay and Control module<br>• Faulty Control module |
| **DTC: P1641**<br>**2T CCM, MIL: Yes**<br>**Years:** 2005, 2006, 2007, 2008<br>**Engines:** 2.3L VIN E/G<br>**Models:** All<br>**Transmissions:** All | **Fuel Pump Relay No Voltage to Fuel Pump**<br>Control module detects no voltage.<br>**Possible Causes:**<br>• Check input signal for Control module<br>• Check for poor contact in Control module connector<br>• Check harness between relay and Control module<br>• Faulty Control module |

| DTC | Trouble Code Title, Conditions & Possible Causes |
|---|---|
| **DTC: P1648**<br>**2T CCM, MIL: Yes**<br>**Years:** 2005, 2006, 2007, 2008<br>**Engines:** 2.8L VIN U<br>**Models:** All<br>**Transmissions:** All | **Control Module Missing on Bus**<br>Control module detected it was not programmed for the correct application.<br>**Possible Causes**<br>&bull; Control module must be reprogrammed to correct this problem. Once this step is done, recheck the code to verify the repair is done. |
| **DTC: P1652**<br>**2T CCM, MIL: Yes**<br>**Years:** 2005, 2006, 2007, 2008<br>**Engines:** 2.3L VIN E/G<br>**Models:** All<br>**Transmissions:** All | **Main Relay Coil Circuit Open / Short to Ground**<br>Control module detects voltage below specification.<br>**Possible Causes:**<br>&bull; Check input signal for Control module<br>&bull; Check for poor contact in Control module connector<br>&bull; Check harness between solenoid valve and Control module<br>&bull; Faulty Control module |
| **DTC: P1653**<br>**2T CCM, MIL: Yes**<br>**Years:** 2005, 2006, 2007, 2008<br>**Engines:** 2.3L VIN E/G<br>**Models:** All<br>**Transmissions:** All | **Main Relay Coil Circuit Short to B+**<br>Control module detects voltage above specification.<br>**Possible Causes:**<br>&bull; Check input signal for Control module<br>&bull; Check for poor contact in Control module connector<br>&bull; Check harness between solenoid valve and Control module<br>&bull; Faulty Control module |
| **DTC: P1654**<br>**2T CCM, MIL: Yes**<br>**Years:** 2005, 2006, 2007, 2008<br>**Engines:** 2.3L VIN E/G<br>**Models:** All<br>**Transmissions:** All | **Fuel Pump Relay Coil Circuit Open / Short to Ground**<br>Control module detects voltage below specification.<br>**Possible Causes:**<br>&bull; Check input signal for Control module<br>&bull; Check for poor contact in Control module connector<br>&bull; Check harness between solenoid valve and Control module<br>&bull; Faulty Control module |
| **DTC: P1655**<br>**2T CCM, MIL: Yes**<br>**Years:** 2005, 2006, 2007, 2008<br>**Engines:** 2.3L VIN E/G<br>**Models:** All<br>**Transmissions:** All | **Fuel Pump Relay Coil Circuit Short to B+**<br>Control module detects voltage above specification.<br>**Possible Causes:**<br>&bull; Check input signal for Control module<br>&bull; Check for poor contact in Control module connector<br>&bull; Check harness between solenoid valve and Control module<br>&bull; Faulty Control module |
| **DTC: P1656**<br>**2T CCM, MIL: Yes**<br>**Years:** 2005, 2006, 2007, 2008<br>**Engines:** 2.3L VIN E/G<br>**Models:** All<br>**Transmissions:** All | **A/C Relay Coil Circuit Open / Short to Ground**<br>Control module detects voltage below specification.<br>**Possible Causes:**<br>&bull; Check input signal for Control module<br>&bull; Check for poor contact in Control module connector<br>&bull; Check harness between solenoid valve and Control module<br>&bull; Faulty Control module |
| **DTC: P1657**<br>**2T CCM, MIL: Yes**<br>**Years:** 2005, 2006, 2007, 2008<br>**Engines:** 2.3L VIN E/G<br>**Models:** All<br>**Transmissions:** All | **A/C Relay Coil Circuit Short to B+**<br>Control module detects voltage above specification.<br>**Possible Causes:**<br>&bull; Check input signal for Control module<br>&bull; Check for poor contact in Control module connector<br>&bull; Check harness between solenoid valve and Control module<br>&bull; Faulty Control module |
| **DTC: P1658**<br>**2T CCM, MIL: Yes**<br>**Years:** 2005, 2006, 2007, 2008<br>**Engines:** 2.3L VIN E/G<br>**Models:** All<br>**Transmissions:** All | **Charge Air Bypass Valve Circuit Open / Short to Ground**<br>Control module detects voltage below specification.<br>**Possible Causes:**<br>&bull; Check input signal for Control module<br>&bull; Check for poor contact in Control module connector<br>&bull; Check harness between solenoid valve and Control module<br>&bull; Faulty Control module |

| DTC | Trouble Code Title, Conditions & Possible Causes |
|---|---|
| **DTC: P1659**<br>**2T CCM, MIL:** Yes<br>**Years:** 2005, 2006, 2007, 2008<br>**Engines:** 2.3L VIN E/G<br>**Models:** All<br>**Transmissions:** All | **Charge Air Bypass Valve Circuit Short to B+**<br>Control module detects voltage above specification.<br>**Possible Causes:**<br>• Check input signal for Control module<br>• Check for poor contact in Control module connector<br>• Check harness between solenoid valve and Control module<br>• Faulty Control module |
| **DTC: P1662**<br>**2T CCM, MIL:** Yes<br>**Years:** 2005, 2006, 2007, 2008<br>**Engines:** 2.3L VIN E/G<br>**Models:** All<br>**Transmissions:** All | **Charge Air Control Valve Circuit Open / Short to Ground**<br>Control module detects voltage below specification.<br>**Possible Causes:**<br>• Check input signal for Control module<br>• Check for poor contact in Control module connector<br>• Check harness between solenoid valve and Control module<br>• Faulty Control module |
| **DTC: P1663**<br>**2T CCM, MIL:** Yes<br>**Years:** 2005, 2006, 2007, 2008<br>**Engines:** 2.3L VIN E/G<br>**Models:** All<br>**Transmissions:** All | **Charge Air Control Valve Circuit Short to B+**<br>Control module detects voltage above specification.<br>**Possible Causes:**<br>• Check input signal for Control module<br>• Check for poor contact in Control module connector<br>• Check harness between solenoid valve and Control module<br>• Faulty Control module |
| **DTC: P1670**<br>**2T CCM, MIL:** Yes<br>**Years:** 2005, 2006, 2007, 2008<br>**Engines:** 2.3L VIN E/G<br>**Models:** All<br>**Transmissions:** All | **Limp-Home Relay Open Circuit or Short to Ground**<br>Control module detects voltage below specification.<br>**Possible Causes:**<br>• Check input signal for Control module<br>• Check for poor contact in Control module connector<br>• Check harness between solenoid valve and Control module<br>• Faulty Control module |
| **DTC: P1671**<br>**2T CCM, MIL:** Yes<br>**Years:** 2005, 2006, 2007, 2008<br>**Engines:** 2.3L VIN E/G<br>**Models:** All<br>**Transmissions:** All | **Limp-Home Relay Short to B+**<br>Control module detects voltage above specification.<br>**Possible Causes:**<br>• Check input signal for Control module<br>• Check for poor contact in Control module connector<br>• Check harness between solenoid valve and Control module<br>• Faulty Control module |
| **DTC: P1676**<br>**2T CCM, MIL:** Yes<br>**Years:** 2005, 2006, 2007, 2008<br>**Engines:** 2.3L VIN E/G<br>**Models:** All<br>**Transmissions:** All | **Injector Circuit Open / Short to Ground or B+**<br>Control module detects voltage above specification.<br>**Possible Causes:**<br>• Check input signal for Control module<br>• Check for poor contact in Control module connector<br>• Check harness between solenoid valve and Control module<br>• Faulty Control module |
| **DTC: P1677**<br>**2T CCM, MIL:** Yes<br>**Years:** 2005, 2006, 2007, 2008<br>**Engines:** 2.8L VIN U<br>**Models:** All<br>**Transmissions:** All | **Control Module Internal Fault**<br>Control module detects an internal failure or incomplete programming for more than 5 seconds.<br>**Possible Causes:**<br>• Control module is not programmed<br>• Control module has failed |
| **DTC: P1678**<br>**2T CCM, MIL:** Yes<br>**Years:** 2005, 2006, 2007, 2008<br>**Engines:** 2.8L VIN U<br>**Models:** All<br>**Transmissions:** All | **Control Module Not Added**<br>Control module detected it was not programmed for the correct transmission application.<br>**Possible Causes**<br>• Control module must be reprogrammed to correct this problem. Once this step is done, recheck the code to verify the repair is done. |

| DTC | Trouble Code Title, Conditions & Possible Causes |
|---|---|
| **DTC: P1679**<br>**2T CCM, MIL: Yes**<br>**Years:** 2005, 2006, 2007, 2008<br>**Engines:** 2.8L VIN U<br>**Models:** All<br>**Transmissions:** All | **Control Module Not Added**<br>Control module detected it was not programmed for the correct transmission application.<br>**Possible Causes**<br>• Control module must be reprogrammed to correct this problem. Once this step is done, recheck the code to verify the repair is done. |
| **DTC: P1680**<br>**2T CCM, MIL: Yes**<br>**Years:** 2005, 2006, 2007, 2008<br>**Engines:** 2.0L VIN S/Y<br>**Models:** All<br>**Transmissions:** All | **Control Module Internal Fault**<br>Control module detects an internal failure or incomplete programming for more than 5 seconds.<br>**Possible Causes:**<br>• Control module is not programmed<br>• Control module has failed |
| **DTC: P1681**<br>**2T CCM, MIL: Yes**<br>**Years:** 2005, 2006, 2007, 2008<br>**Engines:** 2.0L VIN S/Y<br>**Models:** All<br>**Transmissions:** All | **Control Module Internal Fault**<br>Control module detects an internal failure or incomplete programming for more than 5 seconds.<br>**Possible Causes:**<br>• Control module is not programmed<br>• Control module has failed |
| **DTC: P1682**<br>**1T CCM, MIL: Yes**<br>**Years:** 2005, 2006, 2007, 2008<br>**Engines:** 4.2L VIN S, 5.3L VIN M/P, 6.0L VIN H<br>**Models:** All<br>**Transmissions:** All | **Ignition 1 Switch Circuit**<br>There are two ignition 1 voltage circuits supplied to the engine control module (Control module). The first ignition 1 voltage circuit is provided by the powertrain relay through a fuse. This ignition 1 voltage circuit supplies power to all the internal Control module circuits associated with the throttle actuator control (TAC) operation. The ignition main relay provides the second ignition 1 voltage circuit to Control module through a fuse. This ignition 1 voltage provides power to other internal Control module circuits, except those associated with TAC operation. Control module continuously monitors the voltage level difference between the 2 circuits.<br>**Possible Causes:**<br>• Check for other DTC<br>• Check power train relay<br>• Check coil voltage<br>• Check poor contact |
| **DTC: P1698**<br>**2T CCM, MIL: Yes**<br>**Years:** 2005, 2006, 2007, 2008<br>**Engines:** 2.5L SOHC, DOHC VIN 6<br>**Models:** All<br>**Transmissions:** All | **Engine Torque Control Cut Signal Circuit Malfunction (Low Input)**<br>Control module detected an unexpected "low" voltage condition on the Engine Torque Control Cut Signal circuit (5-volt) during the CCM test.<br>**Possible Causes:**<br>• Check output signal for Control module<br>• Check harness between Control module and transmission module connector |
| **DTC: P1699**<br>**2T CCM, MIL: Yes**<br>**Years:** 2005, 2006, 2007, 2008<br>**Engines:** 2.5L SOHC, DOHC VIN 6<br>**Models:** All<br>**Transmissions:** All | **Engine Torque Control Cut Signal Circuit Malfunction (High Input)**<br>Control module detected an unexpected "high" voltage condition on the Engine Torque Control Cut Signal circuit (5-volt) during the CCM test<br>**Possible Causes:**<br>• Check output signal for Control module<br>• Check harness between Control module and transmission module connector |
| **DTC: P1700**<br>**2T CCM, MIL: No**<br>**Years:** 2006<br>**Engines:** 2.0L VIN 2, 2.5L SOHC, DOHC VIN 6<br>**Models:** All<br>**Transmissions:** All | **Throttle Position Sensor Circuit Malfunction**<br>Control module detected an unexpected "low" or "high" voltage condition on the TP Sensor circuit.<br>**Note: The TP sensor signal is shared with Control module on this circuit.**<br>**Possible Causes:**<br>• Check any other DTC on display<br>• Check throttle position sensor circuit |
| **DTC: P1706**<br>**1T CCM, MIL: Yes**<br>**Years:** 2005, 2006, 2007, 2008<br>**Engines:** 2.5L SOHC, DOHC VIN 6<br>**Models:** All<br>**Transmissions:** All | **Vehicle Speed Sensor Circuit (Rear Wheels)**<br>Input signal circuit of transmission module is open or shorted. Shifting quality malfunction. Tight corner braking phenomenon is occurred.<br>**Possible Causes:**<br>• Open or short circuit for power supply and ground<br>• Open circuit of harness between transmission module and transmission connector<br>• Short circuit of harness between transmission module and transmission connector<br>• Poor contact of harness in ATF temperature sensor and transmission connector<br>• Faulty transmission harness<br>• Faulty transmission assembly |

| DTC | Trouble Code Title, Conditions & Possible Causes |
|---|---|
| **DTC: P1707**<br>**1T CCM, MIL: Yes**<br>**Years:** 2005, 2006, 2007, 2008<br>**Engines:** 2.5L SOHC, DOHC VIN 6<br>**Models:** All<br>**Transmissions:** All | **AWD Solenoid Valve Circuit**<br>Output signal circuit of transfer solenoid is open or shorted. Tight corner braking phenomenon is occurred. Drivability getting worse.<br>**Possible Causes:**<br> • Open circuit of harness between transmission module connector and transmission connector<br> • Short circuit in accelerator pedal position sensor<br> • Open circuit of harness between transmission connector and control valve body connector<br> • Short circuit of harness between control valve body connector and transmission ground<br> • Faulty control valve body<br> • Faulty transmission module |
| **DTC: P1708**<br>**1T CCM, MIL: Yes**<br>**Years:** 2006<br>**Engines:** 2.5L SOHC DOHC Vin 6<br>**Models:** All<br>**Transmissions:** All | **Throttle Position Circuit (Low Input)**<br>The input signal circuit of accelerator pedal position sensor is open or shorted. Shift point too high or too low. Excessive shift shock. Excessive tight corner "braking".<br>**Possible Causes:**<br> • Check engine ground connections<br> • Check Control module ground<br> • Check accelerator pedal position sensor connector receptacle's terminals<br> • Check harness between transmission module and accelerator pedal position sensor<br> • Check input signal of transmission module using a scan tool<br> • Faulty transmission module |
| **DTC: P1709**<br>**1T CCM, MIL: Yes**<br>**Years:** 2006<br>**Engines:** 2.5L SOHC DOHC Vin 6<br>**Models:** All<br>**Transmissions:** All | **Throttle Position Circuit (High Input)**<br>The input signal circuit of accelerator pedal position sensor is shorted. Shift point too high or too low. Excessive shift shock. Excessive tight corner "braking".<br>**Possible Causes:**<br> • Check engine ground connections<br> • Check Control module ground<br> • Check accelerator pedal position sensor connector receptacle's terminals<br> • Check harness between transmission module and accelerator pedal position sensor<br> • Check input signal of transmission module using a scan tool<br> • Faulty transmission module |
| **DTC: P1710**<br>**1T CCM, MIL: Yes**<br>**Years:** 2005, 2006, 2007, 2008<br>**Engines:** 2.5L SOHC, DOHC VIN 6,<br>**Models:** All<br>**Transmissions:** All | **Torque Converter Turbine 2 Speed Signal Circuit**<br>Input signal circuit of transmission module is open or shorted. Excessive shift shock. Does not shift to 5th<br>**Possible Causes:**<br> • Open or short circuit for power supply and ground<br> • Open circuit of harness between transmission module and transmission connector<br> • Short circuit of harness between transmission module and transmission connector<br> • Faulty transmission module |
| **DTC: P1711**<br>**1T CCM, MIL: Yes**<br>**Years:** 2005, 2006, 2007, 2008<br>**Engines:** 2.0L VIN 2<br>**Models:** All<br>**Transmissions:** All | **Engine Torque Control Signal #1 Circuit Malfunction**<br>Excessive shift shock. Control module detects the difference 250 times in a row by comparing the module output with signal line output every 10 milliseconds.<br>**Possible Causes:**<br> • Check input signal for control module<br> • Check for poor contact<br> • Check harness |
| **DTC: P1712**<br>**1T CCM, MIL: Yes**<br>**Years:** 2005, 2006, 2007, 2008<br>**Engines:** 2.0L VIN 2<br>**Models:** All<br>**Transmissions:** All | **Engine Torque Control Signal #2 Circuit Malfunction**<br>Excessive shift shock. Control module detects the difference 250 times in a row by comparing the module output with signal line output every 10 milliseconds.<br>**Possible Causes:**<br> • Check input signal for control module<br> • Check for poor contact<br> • Check harness |
| **DTC: P1762**<br>**1T CCM, MIL: Yes**<br>**Years:** 2005, 2006, 2007, 2008<br>**Engines:** 4.2L VIN S, 5.3L VIN M/P, 6.0L VIN H<br>**Models:** All<br>**Transmissions:** All | **Transmission Mode Switch Signal Circuit**<br>The Transmission Mode Switch Signal enables the operator to achieve enhanced shift performance when towing or hauling a load. When tow/haul mode is selected, the tow/haul switch input signal to the body control module is momentarily toggled to zero volts. This code sets when the rolling count of the transmission control module does not match the same value sent from the control module.<br>**Possible Causes:**<br> • Check for communication<br> • Check communication references |

| DTC | Trouble Code Title, Conditions & Possible Causes |
|---|---|
| **DTC: P1810**<br>**1T CCM, MIL: Yes**<br>**Years:** 2005, 2006, 2007, 2008<br>**Engines:** 4.2L VIN S, 5.3L VIN M/P, 6.0L VIN H<br>**Models:** All<br>**Transmissions:** All | **Transmission Fluid Pressure (TFP) Position Switch Circuit**<br>The automatic transmission fluid pressure (TFP) manual valve position switch consists of 5 pressure switches and a transmission fluid temperature (TFT) sensor combined into one unit. The combined unit mounts on the valve body. The transmission control module supplies ignition voltage for each range signal. By grounding one or more of these circuits through various combinations of the pressure switches, the transmission module detects which manual valve position you select. The transmission module compares the actual voltage combination of the switches to a TFP manual valve position switch combination chart stored in memory.<br>**Possible Causes:**<br>• Check transmission fluid level<br>• Check transmission fluid pressure manual valve<br>• Check communication references |
| **DTC: P1815**<br>**1T CCM, MIL: Yes**<br>**Years:** 2005, 2006, 2007, 2008<br>**Engines:** 4.2L VIN S, 5.3L VIN M/P, 6.0L VIN H<br>**Models:** All<br>**Transmissions:** All | **Transmission Fluid Pressure (TFP) Valve Position Switch - Start in Wrong Range**<br>The automatic transmission fluid pressure (TFP) manual valve position switch consists of 5 pressure switches and a transmission fluid temperature (TFT) sensor combined into one unit. The combined unit mounts on the valve body. The transmission control module supplies ignition voltage for each range signal. By grounding one or more of these circuits through various combinations of the pressure switches, the transmission module detects which manual valve position you select. The transmission module compares the actual voltage combination of the switches to a TFP manual valve position switch combination chart stored in memory.<br>**Possible Causes:**<br>• Check transmission fluid level<br>• Check transmission fluid pressure manual valve<br>• Check communication references |
| **DTC: P1816**<br>**1T CCM, MIL: Yes**<br>**Years:** 2005, 2006, 2007, 2008<br>**Engines:** 4.2L VIN S, 5.3L VIN M/P, 6.0L VIN H<br>**Models:** All<br>**Transmissions:** All | **Transmission Fluid Pressure (TFP) Valve Position Switch Indicates Park/Neutral (P/N) with Drive Ratio**<br>The automatic transmission fluid pressure (TFP) manual valve position switch consists of 5 pressure switches and a transmission fluid temperature (TFT) sensor combined into one unit. The combined unit mounts on the valve body. The transmission control module supplies ignition voltage for each range signal. By grounding one or more of these circuits through various combinations of the pressure switches, the transmission module detects which manual valve position you select. The transmission module compares the actual voltage combination of the switches to a TFP manual valve position switch combination chart stored in memory.<br>**Possible Causes:**<br>• Check transmission fluid level<br>• Check transmission fluid pressure manual valve<br>• Check communication references |
| **DTC: P1817**<br>**1T CCM, MIL: Yes**<br>**Years:** 2005, 2006, 2007, 2008<br>**Engines:** 2.5L SOHC, DOHC VIN 6<br>**Models:** All<br>**Transmissions:** All | **Sport Mode Manual Switch Circuit (Manual Switch)**<br>Input signal circuit of manual mode switch is open or shorted. Manual mode can not be set. When shifting to "N", the SPORT shift indicator light illuminates.<br>**Possible Causes:**<br>• Faulty select lever assembly<br>• Short circuit of harness between body integrated unit and manual mode<br>• Temporary poor contact of connector or harness may be the cause<br>• Faulty transmission module |
| **DTC: P1818**<br>**1T CCM, MIL: Yes**<br>**Years:** 2005, 2006, 2007, 2008<br>**Engines:** 4.2L VIN S, 5.3L VIN M/P, 6.0L VIN H<br>**Models:** All<br>**Transmissions:** All | **Transmission Fluid Pressure (TFP) Valve Position Switch Indicates Drive without Drive Ratio**<br>The automatic transmission fluid pressure (TFP) manual valve position switch consists of 5 pressure switches and a transmission fluid temperature (TFT) sensor combined into one unit. The combined unit mounts on the valve body. The transmission control module supplies ignition voltage for each range signal. By grounding one or more of these circuits through various combinations of the pressure switches, the transmission module detects which manual valve position you select. The transmission module compares the actual voltage combination of the switches to a TFP manual valve position switch combination chart stored in memory.<br>**Possible Causes:**<br>• Check transmission fluid level<br>• Check transmission fluid pressure manual valve<br>• Check communication references |
| **DTC: P1840**<br>**1T CCM, MIL: Yes**<br>**Years:** 2005, 2006, 2007, 2008<br>**Engines:** 2.5L SOHC, DOHC VIN 6<br>**Models:** All<br>**Transmissions:** All | **Transmission Fluid Pressure Sensor/Switch A Circuit**<br>Output signal of front brake solenoid does not match with oil pressure. Locked to 1st gear.<br>**Possible Causes:**<br>• Open circuit of harness between transmission module and transmission connector<br>• Short circuit of harness between transmission module and transmission connector<br>• Transmission harness assembly<br>• Faulty control valve body |

| DTC | Trouble Code Title, Conditions & Possible Causes |
|---|---|
| **DTC: P1841**<br>**1T CCM, MIL: Yes**<br>**Years:** 2005, 2006, 2007, 2008<br>**Engines:** 2.5L SOHC, DOHC VIN 6<br>**Models:** All<br>**Transmissions:** All | **Transmission Fluid Pressure Sensor/Switch B Circuit**<br>Output signal of forward brake solenoid does not match the oil pressure. Locked to 2nd, 3rd, 4th gear.<br>**Possible Causes:**<br>• Open circuit of harness between transmission module and transmission connector<br>• Short circuit of harness between transmission module and transmission connector<br>• Transmission harness assembly<br>• Faulty control valve body |
| **DTC: P1842**<br>**1T CCM, MIL: Yes**<br>**Years:** 2005, 2006, 2007, 2008<br>**Engines:** 2.5L SOHC, DOHC VIN 6<br>**Models:** All<br>**Transmissions:** All | **Transmission Fluid Pressure Sensor/Switch C Circuit**<br>Output signal value of input clutch solenoid and oil pressure does not match. Locked to 1st or 4th gear.<br>**Possible Causes:**<br>• Open circuit of harness between transmission module and transmission connector<br>• Short circuit of harness between transmission module and transmission connector<br>• Transmission harness assembly<br>• Faulty control valve body |
| **DTC: P1843**<br>**1T CCM, MIL: Yes**<br>**Years:** 2005, 2006, 2007, 2008<br>**Engines:** 2.5L SOHC, DOHC VIN 6<br>**Models:** All<br>**Transmissions:** All | **Transmission Fluid Pressure Sensor/Switch D Circuit**<br>Output signal value of direct clutch solenoid and oil pressure does not match. Locked to 1st or 4th gear.<br>**Possible Causes:**<br>• Open circuit of harness between transmission module and transmission connector<br>• Short circuit of harness between transmission module and transmission connector<br>• Transmission harness assembly<br>• Faulty control valve body |
| **DTC: P1844**<br>**1T CCM, MIL: Yes**<br>**Years:** 2005, 2006, 2007, 2008<br>**Engines:** 2.5L SOHC, DOHC VIN 6<br>**Models:** All<br>**Transmissions:** All | **Transmission Fluid Pressure Sensor/Switch E Circuit**<br>Output signal value of high & low reverse clutch solenoid and oil pressure does not match. Locked to 1st gear.<br>**Possible Causes:**<br>• Open circuit of harness between transmission module and transmission connector<br>• Short circuit of harness between transmission module and transmission connector<br>• Transmission harness assembly<br>• Faulty control valve body |

## OBD II Trouble Code List (P2xxx Codes)

| DTC | Trouble Code Title, Conditions & Possible Causes |
|---|---|
| **DTC: P2004**<br>**1T CCM, MIL: Yes**<br>**Years:** 2005, 2006, 2007, 2008<br>**Engines:** 2.0L VIN 2, 2.5L DOHC VIN 6<br>**Models:** All<br>**Transmissions:** All | **Intake Manifold Runner Control Stuck Open (Bank 1)**<br>Vehicle performance issue.<br>**Possible Causes:**<br>• Check any other DTC on display<br>• Check tumble generator valve (RH) |
| **DTC: P2005**<br>**1T CCM, MIL: Yes**<br>**Years:** 2005, 2006, 2007, 2008<br>**Engines:** 2.0L VIN 2, 2.5L DOHC VIN 6<br>**Models:** All<br>**Transmissions:** All | **Intake Manifold Runner Control Stuck Open (Bank 2)**<br>Vehicle performance issue.<br>**Possible Causes:**<br>• Check any other DTC on display<br>• Check tumble generator valve (RH) |
| **DTC: P2006**<br>**1T CCM, MIL: Yes**<br>**Years:** 2005, 2006, 2007, 2008<br>**Engines:** 2.0L VIN 2, 2.5L DOHC VIN 6<br>**Models:** All<br>**Transmissions:** All | **Intake Manifold Runner Control Stuck Closed (Bank 1)**<br>Vehicle performance issue.<br>**Possible Causes:**<br>• Check any other DTC on display<br>• Check tumble generator valve (RH) |
| **DTC: P2007**<br>**1T CCM, MIL: Yes**<br>**Years:** 2005, 2006, 2007, 2008<br>**Engines:** 2.0L VIN 2, 2.5L DOHC VIN 6<br>**Models:** All<br>**Transmissions:** All | **Intake Manifold Runner Control Stuck Closed (Bank 2)**<br>Vehicle performance issue.<br>**Possible Causes:**<br>• Check any other DTC on display<br>• Check tumble generator valve (RH) |

| DTC | Trouble Code Title, Conditions & Possible Causes |
|---|---|
| **DTC: P2008**<br>**1T CCM, MIL: Yes**<br>**Years:** 2005, 2006, 2007, 2008<br>**Engines:** 2.0L VIN 2, 2.5L DOHC VIN 6<br>**Models:** All<br>**Transmissions:** All | **Intake Manifold Runner Control Circuit Open (Bank 1)**<br>Vehicle performance issue.<br>**Possible Causes:**<br>• Check harness between Control module and tumble generator valve actuator connector<br>• Check poor contact |
| **DTC: P2009**<br>**1T CCM, MIL: Yes**<br>**Years:** 2005, 2006, 2007, 2008<br>**Engines:** 2.0L VIN 2, 2.5L DOHC VIN 6<br>**Models:** All<br>**Transmissions:** All | **Intake Manifold Runner Control Circuit Low (Bank 1)**<br>Vehicle performance issue.<br>**Possible Causes:**<br>• Check harness between Control module and tumble generator valve actuator connector |
| **DTC: P2011**<br>**1T CCM, MIL: Yes**<br>**Years:** 2005, 2006, 2007, 2008<br>**Engines:** 2.0L VIN 2, 2.5L DOHC VIN 6<br>**Models:** All<br>**Transmissions:** All | **Intake Manifold Runner Control Circuit Open (Bank 2)**<br>Vehicle performance issue.<br>**Possible Causes:**<br>• Check harness between Control module and tumble generator valve actuator connector |
| **DTC: P2012**<br>**1T CCM, MIL: Yes**<br>**Years:** 2005, 2006, 2007, 2008<br>**Engines:** 2.0L VIN 2, 2.5L DOHC VIN 6<br>**Models:** All<br>**Transmissions:** All | **Intake Manifold Runner Control Circuit Low (Bank 2)**<br>Vehicle performance issue.<br>**Possible Causes:**<br>• Check harness between Control module and tumble generator valve actuator connector |
| **DTC: P2016**<br>**1T CCM, MIL: Yes**<br>**Years:** 2005, 2006, 2007, 2008<br>**Engines:** 2.0L VIN 2, 2.5L DOHC VIN 6<br>**Models:** All<br>**Transmissions:** All | **Intake Manifold Runner Position Sensor Circuit Low (Bank 1)**<br>Engine stalls. Erroneous idling. Poor driving performance.<br>**Possible Causes:**<br>• Using the a scan tool, check current data<br>• Check input signal for Control module<br>• Check harness between Control module and tumble generator valve position sensor connector<br>• Check harness between Control module and throttle position sensor connector<br>• Check poor contact |
| **DTC: P2017**<br>**1T CCM, MIL: Yes**<br>**Years:** 2005, 2006, 2007, 2008<br>**Engines:** 2.0L VIN 2, 2.5L DOHC VIN 6<br>**Models:** All<br>**Transmissions:** All | **Intake Manifold Runner Position Sensor Circuit High (Bank 1)**<br>Engine stalls. Erroneous idling. Poor driving performance.<br>**Possible Causes:**<br>• Using a scan tool, check current data<br>• Check harness between tumble generator valve position sensor and Control module connector |
| **DTC: P2021**<br>**1T CCM, MIL: Yes**<br>**Years:** 2005, 2006, 2007, 2008<br>**Engines:** 2.0L VIN 2, 2.5L DOHC VIN 6<br>**Models:** All<br>**Transmissions:** All | **Intake Manifold Runner Position Sensor Circuit Low (Bank 2)**<br>Engine stalls. Erroneous idling. Poor driving performance.<br>**Possible Causes:**<br>• Using a scan tool, check current data<br>• Check input signal for Control module<br>• Check harness between Control module and tumble generator valve position sensor connector<br>• Check poor contact |
| **DTC: P2022**<br>**1T CCM, MIL: Yes**<br>**Years:** 2005, 2006, 2007, 2008<br>**Engines:** 2.0L VIN 2, 2.5L DOHC VIN 6<br>**Models:** All<br>**Transmissions:** All | **Intake Manifold Runner Position Sensor Circuit High (Bank 2)**<br>Engine stalls. Erroneous idling. Poor driving performance.<br>**Possible Causes:**<br>• Using a scan tool, check current data<br>• Check harness between tumble generator valve position sensor and Control module connector<br>• Check poor contact |

| DTC | Trouble Code Title, Conditions & Possible Causes |
|---|---|
| **DTC: P2088**<br>**1T CCM, MIL:** No<br>**Years:** 2005, 2006, 2007, 2008<br>**Engines:** 2.8L VIN U<br>**Models:** All<br>**Transmissions:** All | **Intake Camshaft Position Actuator Control Circuit Low (Bank 1 )**<br>Control module detected open or short circuit of the oil flow control solenoid valve. Control module detects when the current is small even though the duty signal is large Improper idling<br>**Possible Causes:**<br>• Open circuit of the harness between Control module and oil flow control solenoid valve connector<br>• Ground short circuit of harness between Control module and oil flow control solenoid valve connector<br>• Poor contact of Control module or oil flow control solenoid valve connector<br>• Faulty oil flow control solenoid valve |
| **DTC: P2089**<br>**1T CCM, MIL:** No<br>**Years:** 2005, 2006, 2007, 2008<br>**Engines:** 2.8L VIN U<br>**Models:** All<br>**Transmissions:** All | **Intake Camshaft Position Actuator Control Circuit High (Bank 1 )**<br>Control module detected open or short circuit of the oil flow control solenoid valve. Control module detects when the current is small even though the duty signal is large Improper idling<br>**Possible Causes:**<br>• Short circuit to power in the harness between Control module and oil flow control solenoid valve connector<br>• Open circuit of the harness between Control module and oil flow control solenoid valve connector<br>• Poor contact of Control module or oil flow control solenoid valve connector<br>• Faulty oil flow control solenoid valve |
| **DTC: P2090**<br>**1T CCM, MIL:** No<br>**Years:** 2005, 2006, 2007, 2008<br>**Engines:**<br>**Models:** All<br>**Transmissions:** All | **Exhaust Camshaft Position Actuator Control Circuit Low (Bank 1 )**<br>Control module detected open circuit of the oil flow control solenoid valve. Control module detected open circuit of the oil flow control solenoid valve. Improper idling<br>**Possible Causes:**<br>• Open circuit of the harness between Control module and oil flow control solenoid valve connector<br>• Poor contact of coupling connector<br>• Short circuit to ground in harness between Control module and oil flow control solenoid valve connector<br>• Poor contact of Control module or oil flow control solenoid valve connector<br>• Faulty oil flow control solenoid valve |
| **DTC: P2091**<br>**1T CCM, MIL:** No<br>**Years:** 2005, 2006, 2007, 2008<br>**Engines:**<br>**Models:** All<br>**Transmissions:** All | **Exhaust Camshaft Position Actuator Control Circuit High (Bank 1 )**<br>Control module detected short circuit of oil flow control solenoid valve. Judge as short NG when the current is large even though the duty signal is small. Improper idling.<br>**Possible Causes:**<br>• Short circuit to power in the harness between Control module and oil flow control solenoid valve connector<br>• Open circuit of the harness between Control module and oil flow control solenoid valve connector<br>• Poor contact of Control module or oil flow control solenoid valve connector<br>• Faulty oil flow control solenoid valve |
| **DTC: P2092**<br>**2T CCM, MIL:** No<br>**Years:** 2005, 2006, 2007, 2008<br>**Engines:** 2.5L DOHC, Vin 6, 2.8L VIN U<br>**Models:** All<br>**Transmissions:** All | **OCV Solenoid Valve Signal Circuit Open (Bank 2)**<br>Erroneous idling.<br>**Possible Causes:**<br>• Check harness between Control module and oil flow control solenoid valve<br>• Check oil flow control solenoid valve |
| **DTC: P2093**<br>**2T CCM, MIL:** No<br>**Years:** 2005, 2006, 2007, 2008<br>**Engines:** 2.5L SOHC, DOHC VIN 6, 2.8L VIN U<br>**Models:** All<br>**Transmissions:** All | **OCV Solenoid Valve Signal Circuit Short (Bank 2)**<br>Erroneous idling.<br>**Possible Causes:**<br>• Check harness between Control module and oil flow control solenoid valve<br>• Check oil flow control solenoid valve |
| **DTC: P2094**<br>**1T CCM, MIL:** No<br>**Years:** 2005, 2006, 2007, 2008<br>**Engines:**<br>**Models:** All<br>**Transmissions:** All | **Exhaust Camshaft Position Actuator Control Circuit Low (Bank 1 )**<br>Control module detected open circuit of the oil flow control solenoid valve. Control module detected open circuit of the oil flow control solenoid valve. Improper idling<br>**Possible Causes:**<br>• Open circuit of the harness between Control module and oil flow control solenoid valve connector<br>• Poor contact of coupling connector<br>• Short circuit to ground in harness between Control module and oil flow control solenoid valve connector<br>• Poor contact of Control module or oil flow control solenoid valve connector<br>• Faulty oil flow control solenoid valve |

| DTC | Trouble Code Title, Conditions & Possible Causes |
| --- | --- |
| **DTC: P2095**<br>**1T CCM, MIL: No**<br>**Years:** 2005, 2006, 2007, 2008<br>**Engines:**<br>**Models:** All<br>**Transmissions:** All | **Exhaust Camshaft Position Actuator Control Circuit High (Bank 1 )**<br>Control module detected short circuit of oil flow control solenoid valve. Judge as short NG when the current is large even though the duty signal is small. Improper idling.<br>**Possible Causes:**<br>• Short circuit to power in the harness between Control module and oil flow control solenoid valve connector<br>• Open circuit of the harness between Control module and oil flow control solenoid valve connector<br>• Poor contact of Control module or oil flow control solenoid valve connector<br>• Faulty oil flow control solenoid valve |
| **DTC: P2096**<br>**2T CCM, MIL: Yes**<br>**Years:** 2005, 2006, 2007, 2008<br>**Engines:** 2.0L VIN 2, 2.5L SOHC, DOHC VIN 6, 2.8L VIN U<br>**Models:** All<br>**Transmissions:** All | **Post Catalyst Fuel Trim System Too Lean (Bank 1)**<br>Poor Drive ability.<br>**Possible Causes:**<br>• Check any other DTC on display<br>• Using the a scan tool, check front oxygen sensor data<br>• Using the Subaru scan tool, check rear oxygen sensor data<br>• Check exhaust system<br>• Check air intake system<br>• Check fuel pressure<br>• Check engine coolant temperature sensor<br>• Check Mass air flow and intake air temperature<br>• Check harness between Control module and front oxygen sensor connector<br>• Using harness between Control module and rear oxygen sensor connector |
| **DTC: P2097**<br>**2T CCM, MIL: Yes**<br>**Years:** 2005, 2006, 2007, 2008<br>**Engines:** 2.0L VIN 2, 2.5L SOHC, DOHC VIN 6, 2.8L VIN U<br>**Models:** All<br>**Transmissions:** All | **Post Catalyst Fuel Trim System Too Rich (Bank 1)**<br>Poor Drive ability.<br>**Possible Causes:**<br>• Check any other DTC on display<br>• Using a scan tool, check front oxygen sensor data<br>• Using a scan tool, check rear oxygen sensor data<br>• Check exhaust system<br>• Check air intake system<br>• Check fuel pressure<br>• Check engine coolant temperature sensor<br>• Check Mass air flow and intake air temperature<br>• Check harness between Control module and front oxygen sensor connector<br>• Using harness between Control module and rear oxygen sensor connector |
| **DTC: P2098**<br>**2T CCM, MIL: Yes**<br>**Years:** 2005, 2006, 2007, 2008<br>**Engines:** 2.5L SOHC, DOHC Vin 6<br>**Models:** All<br>**Transmissions:** All | **Post Catalyst Fuel Trim System Too Lean (Bank 2)**<br>Poor Drive ability.<br>**Possible Causes:**<br>• Check any other DTC on display<br>• Using a scan tool, check front oxygen sensor data<br>• Using a scan tool, check rear oxygen sensor data<br>• Check exhaust system<br>• Check air intake system<br>• Check fuel pressure<br>• Check engine coolant temperature sensor<br>• Check Mass air flow and intake air temperature<br>• Check harness between Control module and front oxygen sensor connector<br>• Using harness between Control module and rear oxygen sensor connector |
| **DTC: P2099**<br>**2T CCM, MIL: Yes**<br>**Years:** 2005, 2006, 2007, 2008<br>**Engines:**<br>**Models:** All<br>**Transmissions:** All | **Post Catalyst Fuel Trim System Too Rich (Bank 2)**<br>Poor Drive ability.<br>**Possible Causes:**<br>• Check any other DTC on display<br>• Using a scan tool, check front oxygen sensor data<br>• Using a scan tool, check rear oxygen sensor data<br>• Check exhaust system<br>• Check air intake system<br>• Check fuel pressure<br>• Check engine coolant temperature sensor<br>• Check Mass air flow and intake air temperature<br>• Check harness between Control module and front oxygen sensor connector<br>• Using harness between Control module and rear oxygen sensor connector |

| DTC | Trouble Code Title, Conditions & Possible Causes |
|---|---|
| **DTC: P2100**<br>**1T CCM, MIL: Yes**<br>**Years:** 2005, 2006, 2007, 2008<br>**Engines:** 2.8L VIN U<br>**Models:** All<br>**Transmissions:** All | **Throttle Actuator Control Motor Circuit**<br>Control module detects when the motor current becomes large or drive circuit is heated. Improper idling. Engine stalls. Poor driving performance.<br>**Possible Causes:**<br>• Check electronic throttle control relay<br>• Check power supply of electronic control relay<br>• Check harness between Control module and electronic throttle control relay<br>• Check sensor output<br>• Check poor contact<br>• Check harness between Control module and electronic throttle control<br>• Check sensor power supply<br>• Check for short in Control module<br>• Check sensor output<br>• Check harness between Control module and electronic throttle control motor<br>• Check electronic throttle control motor harness<br>• Check ground circuit<br>• Check electronic throttle control |
| **DTC: P2101**<br>**1T CCM, MIL: Yes**<br>**Years:** 2005, 2006, 2007, 2008<br>**Engines:** 2.5L SOHC, DOHC VIN 6, 4.2L VIN S, 5.3L VIN M/P, 6.0L VIN H<br>**Models:** All<br>**Transmissions:** All | **Throttle Actuator Control Motor Circuit Range/Performance**<br>Control module detects when the motor current becomes large or drive circuit is heated. Improper idling. Engine stalls. Poor driving performance.<br>**Possible Causes:**<br>• Check electronic throttle control relay<br>• Check power supply of electronic control relay<br>• Check harness between Control module and electronic throttle control relay<br>• Check sensor output<br>• Check poor contact<br>• Check harness between Control module and electronic throttle control<br>• Check sensor power supply<br>• Check for short in Control module<br>• Check sensor output<br>• Check harness between Control module and electronic throttle control motor<br>• Check electronic throttle control motor harness<br>• Check ground circuit<br>• Check electronic throttle control |
| **DTC: P2102**<br>**1T CCM, MIL: Yes**<br>**Years:** 2005, 2006, 2007, 2008<br>**Engines:** 2.5L SOHC, DOHC VIN 6,<br>**Models:** All<br>**Transmissions:** All | **Throttle Actuator Control Motor Circuit Low**<br>Control module detects when the electronic throttle control power is not supplied even when Control module sets the electronic throttle control relay to ON Improper idling. Engine stalls. Poor driving performance.<br>**Possible Causes:**<br>• Check electronic throttle control relay<br>• Check power supply of electronic control relay<br>• Check harness between Control module and electronic throttle control relay |
| **DTC: P2103**<br>**1T CCM, MIL: Yes**<br>**Years:** 2005, 2006, 2007, 2008<br>**Engines:** 2.5L SOHC, DOHC VIN 6,<br>**Models:** All<br>**Transmissions:** All | **Throttle Actuator Control Motor Circuit High**<br>Control module detects when the electronic throttle control power is supplied even when Control module sets the electronic throttle control relay to OFF. Poor drive ability.<br>**Possible Causes:**<br>• Check electronic throttle control relay<br>• Check for short circuit of the relay and power supply<br>• Check harness between Control module and electronic throttle control relay |

| DTC | Trouble Code Title, Conditions & Possible Causes |
|---|---|
| **DTC: P2107**<br>**1T CCM, MIL: Yes**<br>**Years:** 2005, 2006, 2007, 2008<br>**Engines:** 2.8L VIN U<br>**Models:** All<br>**Transmissions:** All | **Throttle Actuator Control Motor Circuit Range/Performance**<br>Control module detects when the motor current becomes large or drive circuit is heated. Improper idling. Engine stalls. Poor driving performance.<br>**Possible Causes:**<br>• Check electronic throttle control relay<br>• Check power supply of electronic control relay<br>• Check harness between Control module and electronic throttle control relay<br>• Check sensor output<br>• Check poor contact<br>• Check harness between Control module and electronic throttle control<br>• Check sensor power supply<br>• Check for short in Control module<br>• Check sensor output<br>• Check harness between Control module and electronic throttle control motor<br>• Check electronic throttle control motor harness<br>• Check ground circuit<br>• Check electronic throttle control |
| **DTC: P2109**<br>**1T CCM, MIL: Yes**<br>**Years:** 2005, 2006, 2007, 2008<br>**Engines:** 2.5L SOHC, DOHC VIN 6,<br>**Models:** All<br>**Transmissions:** All | **Throttle/Pedal Position Sensor Minimum Stop Performance**<br>Control module detects when full close point learning cannot be conducted or abnormal value is detected Improper idling. Engine stalls. Poor driving performance.<br>**Possible Causes:**<br>• Check electronic throttle control relay<br>• Check power supply of electronic control relay<br>• Check harness between Control module and electronic throttle control relay<br>• Check sensor output<br>• Check poor contact<br>• Check harness between Control module and electronic throttle control<br>• Check sensor power supply<br>• Check for short in Control module<br>• Check sensor output<br>• Check harness between Control module and electronic throttle control motor<br>• Check electronic throttle control motor harness<br>• Check ground circuit<br>• Check electronic throttle control |
| **DTC: P2119**<br>**1T CCM, MIL: Yes**<br>**Years:** 2005, 2006, 2007, 2008<br>**Engines:** 2.5L SOHC, DOHC VIN 6, 2.8L VIN U, 4.2L VIN S, 5.3L VIN M/P, 6.0L VIN H<br>**Models:** All<br>**Transmissions:** All | **Throttle Closed Position Performance**<br>Control module detects when the motor current becomes large or drive circuit is heated. Improper idling. Engine stalls. Poor driving performance.<br>**Possible Causes:**<br>• Check electronic throttle control relay<br>• Check power supply of electronic control relay<br>• Check harness between Control module and electronic throttle control relay<br>• Check sensor output<br>• Check poor contact<br>• Check harness between Control module and electronic throttle control<br>• Check sensor power supply<br>• Check for short in Control module<br>• Check sensor output<br>• Check harness between Control module and electronic throttle control motor<br>• Check electronic throttle control motor harness<br>• Check ground circuit<br>• Check electronic throttle control |
| **DTC: P2120**<br>**1T CCM, MIL: Yes**<br>**Years:** 2005, 2006, 2007, 2008<br>**Engines:** 4.2L VIN S, 5.3L VIN M/P, 6.0L VIN H<br>**Models:** All<br>**Transmissions:** All | **Throttle/Pedal Position Sensor Circuit**<br>Control module detected the open or short circuit of accelerator pedal position. Control module detects if out of specification. Improper idling. Poor driving performance.<br>**Possible Causes:**<br>• Check accelerator pedal position sensor output<br>• Check poor contact<br>• Check harness between Control module and accelerator pedal position sensor<br>• Check power supply of accelerator pedal position sensor<br>• Check sensor |

| DTC | Trouble Code Title, Conditions & Possible Causes |
|---|---|
| **DTC: P2121**<br>**1T CCM, MIL: Yes**<br>**Years:** 2005, 2006, 2007, 2008<br>**Engines:** 2.0L VIN S/Y<br>**Models:** All<br>**Transmissions:** All | **Pedal Position Sensor Circuit Out of Range**<br>Control module detects when the signal level of throttle position Sensor is different from the throttle position Sensor. Improper idling. Poor driving performance.<br>**Possible Causes:**<br>• Check sensor output<br>• Check poor contact<br>• Check harness between Control module and electronic throttle control<br>• Check sensor power supply<br>• Check for short circuit in Control module<br>• Check sensor output<br>• Check electronic throttle control harness |
| **DTC: P2122**<br>**1T CCM, MIL: Yes**<br>**Years:** 2005, 2006, 2007, 2008<br>**Engines:** 2.0L VIN S/Y, 2.8L VIN U<br>**Models:** All<br>**Transmissions:** All | **Pedal Position Sensor Circuit 1 Open Circuit/Short to Ground**<br>Control module detects when the signal level of throttle position Sensor is different from the throttle position Sensor. Improper idling. Poor driving performance.<br>**Possible Causes:**<br>• Check sensor output<br>• Check poor contact<br>• Check harness between Control module and electronic throttle control<br>• Check sensor power supply<br>• Check for short circuit in Control module<br>• Check sensor output<br>• Check electronic throttle control harness |
| **DTC: P2123**<br>**1T CCM, MIL: Yes**<br>**Years:** 2005, 2006, 2007, 2008<br>**Engines:** 2.0L VIN S/Y, 2.8L VIN U<br>**Models:** All<br>**Transmissions:** All | **Pedal Position Sensor Circuit 1 Short to B+**<br>Control module detects when the signal level of throttle position Sensor is different from the throttle position Sensor. Improper idling. Poor driving performance.<br>**Possible Causes:**<br>• Check sensor output<br>• Check poor contact<br>• Check harness between Control module and electronic throttle control<br>• Check sensor power supply<br>• Check for short circuit in Control module<br>• Check sensor output<br>• Check electronic throttle control harness |
| **DTC: P2126**<br>**1T CCM, MIL: Yes**<br>**Years:** 2005, 2006, 2007, 2008<br>**Engines:** 2.0L VIN S/Y<br>**Models:** All<br>**Transmissions:** All | **Pedal Position Sensor Circuit Out of Range**<br>Control module detects when the signal level of throttle position Sensor is different from the throttle position Sensor. Improper idling. Poor driving performance.<br>**Possible Causes:**<br>• Check sensor output<br>• Check poor contact<br>• Check harness between Control module and electronic throttle control<br>• Check sensor power supply<br>• Check for short circuit in Control module<br>• Check sensor output<br>• Check electronic throttle control harness |
| **DTC: P2127**<br>**1T CCM, MIL: Yes**<br>**Years:** 2005, 2006, 2007, 2008<br>**Engines:** 2.0L VIN S/Y, 2.8L VIN U<br>**Models:** All<br>**Transmissions:** All | **Pedal Position Sensor Circuit 2 Short to Ground/Open Circuit**<br>Control module detects when the signal level of throttle position Sensor is different from the throttle position Sensor. Improper idling. Poor driving performance.<br>**Possible Causes:**<br>• Check sensor output<br>• Check poor contact<br>• Check harness between Control module and electronic throttle control<br>• Check sensor power supply<br>• Check for short circuit in Control module<br>• Check sensor output<br>• Check electronic throttle control harness |

| DTC | Trouble Code Title, Conditions & Possible Causes |
|---|---|
| **DTC: P2128**<br>**1T CCM, MIL: Yes**<br>**Years:** 2005, 2006, 2007, 2008<br>**Engines:** 2.0L VIN S/Y, 2.8L VIN U<br>**Models:** All<br>**Transmissions:** All | **Pedal Position Sensor Circuit 2 Short to B+**<br>Control module detects when the signal level of throttle position Sensor is different from the throttle position Sensor. Improper idling. Poor driving performance.<br>**Possible Causes:**<br>&bull; Check sensor output<br>&bull; Check poor contact<br>&bull; Check harness between Control module and electronic throttle control<br>&bull; Check sensor power supply<br>&bull; Check for short circuit in Control module<br>&bull; Check sensor output<br>&bull; Check electronic throttle control harness |
| **DTC: P2122**<br>**1T CCM, MIL: Yes**<br>**Years:** 2005, 2006, 2007, 2008<br>**Engines:** 2.5L SOHC, DOHC VIN 6, 4.2L VIN S, 5.3L VIN M/P, 6.0L VIN H<br>**Models:** All<br>**Transmissions:** All | **Throttle/Pedal Position Sensor Circuit Low Input**<br>Control module detected the open or short circuit of accelerator pedal position Sensor. Control module detects if out of specification. Improper idling. Poor driving performance.<br>**Possible Causes:**<br>&bull; Check accelerator pedal position sensor output<br>&bull; Check poor contact<br>&bull; Check harness between Control module and accelerator pedal position sensor<br>&bull; Check power supply of accelerator pedal position sensor<br>&bull; Check sensor |
| **DTC: P2123**<br>**1T CCM, MIL: Yes**<br>**Years:** 2005, 2006, 2007, 2008<br>**Engines:** 2.5L SOHC, DOHC VIN 6, 4.2L VIN S, 5.3L VIN M/P, 6.0L VIN H<br>**Models:** All<br>**Transmissions:** All | **Throttle/Pedal Position Sensor Circuit High Input**<br>Control module detected the open or short circuit of accelerator pedal position Sensor. Control module detects if out of specification. Improper idling. Poor driving performance.<br>**Possible Causes:**<br>&bull; Check accelerator pedal position sensor output<br>&bull; Check poor contact<br>&bull; Check harness between Control module and accelerator pedal position sensor |
| **DTC: P2125**<br>**1T CCM, MIL: Yes**<br>**Years:** 2005, 2006, 2007, 2008<br>**Engines:** 2.5L SOHC, DOHC VIN 6, 4.2L VIN S, 5.3L VIN M/P, 6.0L VIN H<br>**Models:** All<br>**Transmissions:** All | **Throttle/Pedal Position Sensor/Switch Circuit**<br>Control module detected the open or short circuit of accelerator pedal position Sensor Control module detects if out of specification. Improper idling. Poor driving performance.<br>**Possible Causes:**<br>&bull; Check accelerator pedal position sensor output<br>&bull; Check poor contact<br>&bull; Check harness between Control module and accelerator pedal position sensor<br>&bull; Check sensor |
| **DTC: P2127**<br>**1T CCM, MIL: Yes**<br>**Years:** 2005, 2006, 2007, 2008<br>**Engines:** 2.5L SOHC, DOHC VIN 6, 4.2L VIN S, 5.3L VIN M/P, 6.0L VIN H<br>**Models:** All<br>**Transmissions:** All | **Throttle/Pedal Position Sensor Circuit Low Input**<br>Control module detected the open or short circuit of accelerator pedal position Sensor Control module detects if out of specification. Improper idling. Poor driving performance.<br>**Possible Causes:**<br>&bull; Check accelerator pedal position sensor output<br>&bull; Check poor contact<br>&bull; Check harness between Control module and accelerator pedal position sensor<br>&bull; Check sensor |
| **DTC: P2128**<br>**1T CCM, MIL: Yes**<br>**Years:** 2005, 2006, 2007, 2008<br>**Engines:** 2.5L SOHC, DOHC VIN 6, 4.2L VIN S, 5.3L VIN M/P, 6.0L VIN H<br>**Models:** All<br>**Transmissions:** All | **Throttle/Pedal Position Sensor Circuit High Input**<br>Control module detected the open or short circuit of accelerator pedal position Sensor. Control module detects if out of specification. Improper idling. Poor driving performance.<br>**Possible Causes:**<br>&bull; Check accelerator pedal position sensor output<br>&bull; Check poor contact<br>&bull; Check harness between Control module and accelerator pedal position sensor<br>&bull; Check sensor |

| DTC | Trouble Code Title, Conditions & Possible Causes |
|---|---|
| **DTC: P2135**<br>**Years:** 2005, 2006, 2007, 2008<br>**Engines:** 2.5L SOHC, DOHC VIN 6,<br>**Models:** All<br>**Transmissions:** All | **Throttle/Pedal Position Sensor/Switch/ Voltage Correlation**<br>Control module detects when the signal level of throttle position Sensor is different from the throttle position Sensor. Improper idling. Poor driving performance.<br>**Possible Causes:**<br>• Check sensor output<br>• Check poor contact<br>• Check harness between Control module and electronic throttle control<br>• Check sensor power supply<br>• Check for short circuit in Control module<br>• Check sensor output<br>• Check electronic throttle control harness |
| **DTC: P2138**<br>**1T CCM, MIL: Yes**<br>**Years:** 2005, 2006, 2007, 2008<br>**Engines:** 2.0L VIN S/Y, 2.5L SOHC, DOHC VIN 6, 2.8L VIN U, 4.2L VIN S, 5.3L VIN M/P, 6.0L VIN H<br>**Models:** All<br>**Transmissions:** All | **Throttle/Pedal Position Sensor/Switch / Voltage Correlation**<br>Control module detects when the signal level of throttle position Sensor is different from the throttle position Sensor. Improper idling. Poor driving performance.<br>**Possible Causes:**<br>• Check accelerator pedal position sensor output<br>• Check poor contact<br>• Check harness between Control module and sensor<br>• Check sensor<br>• Check sensor output |
| **DTC: P2176**<br>**1T CCM, MIL: Yes**<br>**Years:** 2005, 2006, 2007, 2008<br>**Engines:** 2.0L VIN S/Y, 2.5L SOHC, DOHC VIN 6, 2.8L VIN U, 4.2L VIN S, 5.3L VIN M/P, 6.0L VIN H<br>**Models:** All<br>**Transmissions:** All | **Minimum Throttle Position Not Learned**<br>Control module detects when the motor current becomes large or drive circuit is heated. Improper idling. Engine stalls. Poor driving performance.<br>**Possible Causes:**<br>• Check electronic throttle control relay<br>• Check power supply of electronic control relay<br>• Check harness between Control module and electronic throttle control relay<br>• Check sensor output<br>• Check poor contact<br>• Check harness between Control module and electronic throttle control<br>• Check sensor power supply<br>• Check for short in Control module<br>• Check sensor output<br>• Check harness between Control module and electronic throttle control motor<br>• Check electronic throttle control motor harness<br>• Check ground circuit<br>• Check electronic throttle control |
| **DTC: P2187**<br>**1T CCM, MIL: Yes**<br>**Years:** 2005, 2006, 2007, 2008<br>**Engines:** 2.8L VIN U<br>**Models:** All<br>**Transmissions:** All | **Long Term Fuel Trim Additive Max Value Air/Fuel Lean**<br>The Control module can identify if the fuel delivery system can supply enough fuel during heavy acceleration. When this mode is requested during closed loop operation, the Control module provides extra fuel to the engine, and the Control module should detect a rich condition. If it does not Control module detected a rich condition at this time, it will set this code.<br>**Possible Causes**<br>• Fuel filter clogged or restricted<br>• Fuel injectors dirty or restricted<br>• Fuel level too low during heavy acceleration events<br>• Fuel pressure regulator has failed (not supplying enough fuel)<br>• Water or alcohol in fuel (low O2S signal during acceleration) |
| **DTC: P2188**<br>**1T CCM, MIL: Yes**<br>**Years:** 2005, 2006, 2007, 2008<br>**Engines:** 2.8L VIN U<br>**Models:** All<br>**Transmissions:** All | **Long Term Fuel Trim Additive Max Value Air/Fuel Rich**<br>The long term fuel trim makes coarse adjustments in order to maintain an optimum air/fuel ratio. The fuel trim diagnostic will conduct a test to determine if a rich failure actually exists.<br>**Possible Causes:**<br>• Check any other DTC on display<br>• Check exhaust system<br>• Check air intake system<br>• Check fuel pressure<br>• Check fuel injector<br>• Check coolant temperature<br>• Check intake manifold pressure sensor<br>• Check intake air temperature sensor |

| DTC | Trouble Code Title, Conditions & Possible Causes |
|---|---|
| **DTC: P2191**<br>**1T CCM, MIL: Yes**<br>**Years:** 2005, 2006, 2007, 2008<br>**Engines:** 2.8L VIN U<br>**Models:** All<br>**Transmissions:** All | **Long Term Fuel Trim Additive Max Value Air/Fuel Lean**<br>The Control module can identify if the fuel delivery system can supply enough fuel during heavy acceleration. When this mode is requested during closed loop operation, the Control module provides extra fuel to the engine, and the Control module should detect a rich condition. If it does not Control module detected a rich condition at this time, it will set this code.<br>**Possible Causes**<br>  • Fuel filter clogged or restricted<br>  • Fuel injectors dirty or restricted<br>  • Fuel level too low during heavy acceleration events<br>  • Fuel pressure regulator has failed (not supplying enough fuel)<br>  • Water or alcohol in fuel (low O2S signal during acceleration) |
| **DTC: P2192**<br>**1T CCM, MIL: Yes**<br>**Years:** 2005, 2006, 2007, 2008<br>**Engines:** 2.8L VIN U<br>**Models:** All<br>**Transmissions:** All | **Long Term Fuel Trim Multiplicative Max Value Air/Fuel Rich**<br>The long term fuel trim makes coarse adjustments in order to maintain an optimum air/fuel ratio. The fuel trim diagnostic will conduct a test to determine if a rich failure actually exists.<br>**Possible Causes:**<br>  • Check any other DTC on display<br>  • Check exhaust system<br>  • Check air intake system<br>  • Check fuel pressure<br>  • Check fuel injector<br>  • Check coolant temperature<br>  • Check intake manifold pressure sensor<br>  • Check intake air temperature sensor |
| **DTC: P2195**<br>**1T CCM, MIL: Yes**<br>**Years:** 2005, 2006, 2007, 2008<br>**Engines:** 2.0L VIN S/Y, 2.8L VIN U<br>**Models:** All<br>**Transmissions:** All | **Short Term Fuel Trim Additive Max Value Air/Fuel Lean**<br>The Control module can identify if the fuel delivery system can supply enough fuel during heavy acceleration. When this mode is requested during closed loop operation, the Control module provides extra fuel to the engine, and the Control module should detect a rich condition. If it does not Control module detected a rich condition at this time, it will set this code.<br>**Possible Causes**<br>  • Fuel filter clogged or restricted<br>  • Fuel injectors dirty or restricted<br>  • Fuel level too low during heavy acceleration events<br>  • Fuel pressure regulator has failed (not supplying enough fuel)<br>  • Water or alcohol in fuel (low O2S signal during acceleration) |
| **DTC: P2196**<br>**1T CCM, MIL: Yes**<br>**Years:** 2005, 2006, 2007, 2008<br>**Engines:** 2.0L VIN S/Y, 2.8L VIN U<br>**Models:** All<br>**Transmissions:** All | **Short Term Fuel Trim Min Value Air/Fuel too Rich**<br>The short term fuel trim makes coarse adjustments in order to maintain an optimum air/fuel ratio. The fuel trim diagnostic will conduct a test to determine if a rich failure actually exists.<br>**Possible Causes:**<br>  • Check any other DTC on display<br>  • Check exhaust system<br>  • Check air intake system<br>  • Check fuel pressure<br>  • Check fuel injector<br>  • Check coolant temperature<br>  • Check intake manifold pressure sensor<br>  • Check intake air temperature sensor |
| **DTC: P2227**<br>**2T CCM, MIL: Yes**<br>**Years:** 2005, 2006, 2007, 2008<br>**Engines:** 2.0L VIN 2, 2.5L SOHC, DOHC VIN 6, 2.8L VIN U<br>**Models:** All<br>**Transmissions:** All | **Barometric Pressure Circuit Range/Performance**<br>Control module detected the malfunction of barometric pressure sensor output property Control module detects when the barometric pressure sensor output is largely different from the intake manifold pressure at engine start Poor drive ability.<br>**(Note: The barometric pressure sensor is built into Control module).**<br>**Possible Causes:**<br>  • Check any other DTC on display<br>  • Faulty Control module |
| **DTC: P2228**<br>**1T CCM, MIL: Yes**<br>**Years:** 2005, 2006, 2007, 2008<br>**Engines:** 2.0L VIN 2, 2.5L SOHC, DOHC VIN 6, 2.8L VIN U<br>**Models:** All<br>**Transmissions:** All | **Barometric Pressure Circuit Low Input**<br>Control module detected the open/short circuit of the barometric pressure sensor. Control module detects if out of specification. Poor drive ability.<br>**(Note: The barometric pressure sensor is built into Control module).**<br>**Possible Causes:**<br>  • Check any other DTC on display<br>  • Faulty Control module |

| DTC | Trouble Code Title, Conditions & Possible Causes |
|---|---|
| **DTC: P2229**<br>**1T CCM, MIL: Yes**<br>**Years:** 2005, 2006, 2007, 2008<br>**Engines:** 2.0L VIN 2, 2.5L SOHC, DOHC VIN 6, 2.8L VIN U<br>**Models:** All<br>**Transmissions:** All | **Barometric Pressure Circuit High Input**<br>Control module detected the open/short circuit of the barometric pressure sensor. Control module detects if out of specification. Poor drive ability.<br>**(Note: The barometric pressure sensor is built into Control module).**<br>**Possible Causes:**<br>• Check any other DTC on display<br>• Faulty Control module |
| **DTC: P2300**<br>**1T CCM, MIL: Yes**<br>**Years:** 2005, 2006, 2007, 2008<br>**Engines:** 2.0L VIN S/Y, 2.8L VIN U<br>**Models:** All<br>**Transmissions:** All | **Ignition Coil Trigger Cylinder 1 Circuit Short to Ground**<br>Control module detects voltage below specification.<br>**Possible Causes:**<br>• Check input signal for Control module<br>• Check for poor contact in Control module connector<br>• Check harness between solenoid valve and Control module<br>• Faulty Control module |
| **DTC: P2301**<br>**1T CCM, MIL: Yes**<br>**Years:** 2005, 2006, 2007, 2008<br>**Engines:** 2.0L VIN S/Y, 2.8L VIN U<br>**Models:** All<br>**Transmissions:** All | **Ignition Coil Trigger Cylinder 1 Circuit Open/Short to B+**<br>Control module detects voltage above specification.<br>**Possible Causes:**<br>• Check input signal for Control module<br>• Check for poor contact in Control module connector<br>• Check harness between solenoid valve and Control module<br>• Faulty Control module |
| **DTC: P2303**<br>**1T CCM, MIL: Yes**<br>**Years:** 2005, 2006, 2007, 2008<br>**Engines:** 2.0L VIN S/Y, 2.8L VIN U<br>**Models:** All<br>**Transmissions:** All | **Ignition Coil Trigger Cylinder 2 Circuit Short to Ground**<br>Control module detects voltage below specification.<br>**Possible Causes:**<br>• Check input signal for Control module<br>• Check for poor contact in Control module connector<br>• Check harness between solenoid valve and Control module<br>• Faulty Control module |
| **DTC: P2304**<br>**1T CCM, MIL: Yes**<br>**Years:** 2005, 2006, 2007, 2008<br>**Engines:** 2.0L VIN S/Y, 2.8L VIN U<br>**Models:** All<br>**Transmissions:** All | **Ignition Coil Trigger Cylinder 2 Circuit Open/Short to B+**<br>Control module detects voltage above specification.<br>**Possible Causes:**<br>• Check input signal for Control module<br>• Check for poor contact in Control module connector<br>• Check harness between solenoid valve and Control module<br>• Faulty Control module |
| **DTC: P2306**<br>**1T CCM, MIL: Yes**<br>**Years:** 2005, 2006, 2007, 2008<br>**Engines:** 2.0L VIN S/Y, 2.8L VIN U<br>**Models:** All<br>**Transmissions:** All | **Ignition Coil Trigger Cylinder 3 Circuit Short to Ground**<br>Control module detects voltage below specification.<br>**Possible Causes:**<br>• Check input signal for Control module<br>• Check for poor contact in Control module connector<br>• Check harness between solenoid valve and Control module<br>• Faulty Control module |
| **DTC: P2307**<br>**1T CCM, MIL: Yes**<br>**Years:** 2005, 2006, 2007, 2008<br>**Engines:** 2.0L VIN S/Y, 2.8L VIN U<br>**Models:** All<br>**Transmissions:** All | **Ignition Coil Trigger Cylinder 3 Circuit Open/Short to B+**<br>Control module detects voltage above specification.<br>**Possible Causes:**<br>• Check input signal for Control module<br>• Check for poor contact in Control module connector<br>• Check harness between solenoid valve and Control module<br>• Faulty Control module |
| **DTC: P2309**<br>**1T CCM, MIL: Yes**<br>**Years:** 2005, 2006, 2007, 2008<br>**Engines:** 2.0L VIN S/Y, 2.8L VIN U<br>**Models:** All<br>**Transmissions:** All | **Ignition Coil Trigger Cylinder 4 Circuit Short to Ground**<br>Control module detects voltage below specification.<br>**Possible Causes:**<br>• Check input signal for Control module<br>• Check for poor contact in Control module connector<br>• Check harness between solenoid valve and Control module<br>• Faulty Control module |

| DTC | Trouble Code Title, Conditions & Possible Causes |
|---|---|
| **DTC: P2310**<br>**1T CCM, MIL: Yes**<br>**Years:** 2005, 2006, 2007, 2008<br>**Engines:** 2.0L VIN S/Y, 2.8L VIN U<br>**Models:** All<br>**Transmissions:** All | **Ignition Coil Trigger Cylinder 4 Circuit Open/Short to B+**<br>Control module detects voltage above specification.<br>**Possible Causes:**<br>• Check input signal for Control module<br>• Check for poor contact in Control module connector<br>• Check harness between solenoid valve and Control module<br>• Faulty Control module |
| **DTC: P2312**<br>**1T CCM, MIL: Yes**<br>**Years:** 2005, 2006, 2007, 2008<br>**Engines:** 2.8L VIN U<br>**Models:** All<br>**Transmissions:** All | **Ignition Coil Trigger Cylinder 5 Circuit Short to Ground**<br>Control module detects voltage below specification.<br>**Possible Causes:**<br>• Check input signal for Control module<br>• Check for poor contact in Control module connector<br>• Check harness between solenoid valve and Control module<br>• Faulty Control module |
| **DTC: P2313**<br>**1T CCM, MIL: Yes**<br>**Years:** 2005, 2006, 2007, 2008<br>**Engines:** 2.8L VIN U<br>**Models:** All<br>**Transmissions:** All | **Ignition Coil Trigger Cylinder 5 Circuit Open/Short to B+**<br>Control module detects voltage above specification.<br>**Possible Causes:**<br>• Check input signal for Control module<br>• Check for poor contact in Control module connector<br>• Check harness between solenoid valve and Control module<br>• Faulty Control module |
| **DTC: P2315**<br>**1T CCM, MIL: Yes**<br>**Years:** 2005, 2006, 2007, 2008<br>**Engines:** 2.8L VIN U<br>**Models:** All<br>**Transmissions:** All | **Ignition Coil Trigger Cylinder 6 Circuit Short to Ground**<br>Control module detects voltage below specification.<br>**Possible Causes:**<br>• Check input signal for Control module<br>• Check for poor contact in Control module connector<br>• Check harness between solenoid valve and Control module<br>• Faulty Control module |
| **DTC: P2316**<br>**1T CCM, MIL: Yes**<br>**Years:** 2005, 2006, 2007, 2008<br>**Engines:** 2.8L VIN U<br>**Models:** All<br>**Transmissions:** All | **Ignition Coil Trigger Cylinder 6 Circuit Open/Short to B+**<br>Control module detects voltage above specification.<br>**Possible Causes:**<br>• Check input signal for Control module<br>• Check for poor contact in Control module connector<br>• Check harness between solenoid valve and Control module<br>• Faulty Control module |
| **DTC: P2430**<br>**2T CCM, MIL: Yes**<br>**2006**<br>**Engines:** 2.5L DOHC, 4.2L VIN S<br>**Models:** All<br>**Transmissions:** All | **Secondary Air Injection System Air Flow/Pressure Sensor Circuit**<br>Poor Drive ability.<br>**Possible Causes:**<br>• Check any other DTC on display<br>• Using a scan tool, check front oxygen sensor data |
| **DTC: P2431**<br>**2T CCM, MIL: Yes**<br>**2006**<br>**Engines:** 2.5L DOHC, 4.2L VIN S<br>**Models:** All<br>**Transmissions:** All | **Secondary Air Injection System Air Flow/Pressure Sensor Circuit Range/Performance**<br>Poor Drive ability.<br>**Possible Causes:**<br>• Check any other DTC on display<br>• Using a scan tool, check front oxygen sensor data |
| **DTC: P2432**<br>**1T CCM, MIL: Yes**<br>**Years:** 2005, 2006, 2007, 2008<br>**Engines:** 2.5L SOHC, DOHC VIN 6, 4.2L VIN S<br>**Models:** All<br>**Transmissions:** All | **Secondary Air Injection System Air Flow/Pressure Sensor Circuit Low**<br>Poor Drive ability.<br>**Possible Causes:**<br>• Check harness between Control module and valve LH connector |

| DTC | Trouble Code Title, Conditions & Possible Causes |
|---|---|
| **DTC: P2433**<br>**1T CCM, MIL: Yes**<br>**Years:** 2005, 2006, 2007, 2008<br>**Engines:** 2.5L SOHC, DOHC VIN 6, 4.2L VIN S<br>**Models:** All<br>**Transmissions:** All | **Secondary Air Injection System Air Flow/Pressure Sensor Circuit High**<br>Poor Drive ability.<br>**Possible Causes:**<br>• Check harness between Control module and valve LH connector |
| **DTC: P2440**<br>**2T CCM, MIL: Yes**<br>**Years:** 2005, 2006, 2007, 2008<br>**Engines:** 2.5L SOHC, DOHC VIN 6, 4.2L VIN S<br>**Models:** All<br>**Transmissions:** All | **Secondary Air Injection System Switching Valve Stock Open (bank 1)**<br>Poor Drive ability.<br>**Possible Causes:**<br>• Check secondary air combination valve operation<br>• Check duct between secondary air pump and secondary air combination valve<br>• Check pipe between secondary air combination valve and cylinder head<br>• Check power supply to valve<br>• Check harness between secondary air combination valve relay and secondary air combination valve connector terminal<br>• Check valve relay<br>• Check harness between Control module and valve relay connector |
| **DTC: P2441**<br>**2T CCM, MIL: Yes**<br>**Years:** 2005, 2006, 2007, 2008<br>**Engines:** 2.5L SOHC, DOHC VIN 6,<br>**Models:** All<br>**Transmissions:** All | **Secondary Air Injection System Switching Valve Stock Closed (bank 1)**<br>Poor Drive ability.<br>**Possible Causes:**<br>• Check secondary air combination valve operation<br>• Check duct between secondary air pump and secondary air combination valve<br>• Check pipe between secondary air combination valve and cylinder head<br>• Check power supply to valve<br>• Check harness between secondary air combination valve relay and secondary air combination valve connector terminal<br>• Check valve relay<br>• Check harness between Control module and valve relay connector |
| **DTC: P2442**<br>**2T CCM, MIL: Yes**<br>**Years:** 2005, 2006, 2007, 2008<br>**Engines:** 2.5L SOHC, DOHC VIN 6,<br>**Models:** All<br>**Transmissions:** All | **Secondary Air Injection System Switching Valve Stock Open (bank 2)**<br>Poor Drive ability.<br>**Possible Causes:**<br>• Check secondary air combination valve operation<br>• Check duct between secondary air pump and secondary air combination valve<br>• Check pipe between secondary air combination valve and cylinder head<br>• Check power supply to valve<br>• Check harness between secondary air combination valve relay and secondary air combination valve connector terminal<br>• Check valve relay<br>• Check harness between Control module and valve relay connector |
| **DTC: P2443**<br>**2T CCM, MIL: Yes**<br>**Years:** 2005, 2006, 2007, 2008<br>**Engines:** 2.5L SOHC, DOHC VIN 6,<br>**Models:** All<br>**Transmissions:** All | **Secondary Air Injection System Switching Valve Stock Closed (bank 2)**<br>Poor Drive ability.<br>**Possible Causes:**<br>• Check secondary air combination valve operation<br>• Check duct between secondary air pump and secondary air combination valve<br>• Check pipe between secondary air combination valve and cylinder head<br>• Check power supply to valve<br>• Check harness between secondary air combination valve relay and secondary air combination valve connector terminal<br>• Check valve relay<br>• Check harness between Control module and valve relay connector |
| **DTC: P2444**<br>**1T CCM, MIL: Yes**<br>**Years:** 2005, 2006, 2007, 2008<br>**Engines:** 2.5L SOHC, DOHC VIN 6, 2.8L VIN U, 4.2L VIN S<br>**Models:** All<br>**Transmissions:** All | **Secondary Air Injection Relay Circuit, Short to B+**<br>Poor Drive ability.<br>**Possible Causes:**<br>• Check secondary air piping pressure<br>• Check power supply to valve<br>• Check valve relay |
| **DTC: P2445**<br>**1T CCM, MIL: Yes**<br>**Years:** 2005, 2006, 2007, 2008<br>**Engines:** 2.8L VIN U<br>**Models:** All<br>**Transmissions:** All | **Secondary Air Injection Relay Circuit, Short to Ground**<br>Poor Drive ability.<br>**Possible Causes:**<br>• Check secondary air piping pressure<br>• Check power supply to valve<br>• Check valve relay |

| DTC | Trouble Code Title, Conditions & Possible Causes |
|---|---|
| **DTC: P2503**<br>**1T CCM, MIL: No**<br>**Years:** 2005, 2006, 2007, 2008<br>**Engines:** All<br>**Models:** All<br>**Transmissions:** All | **Charging System Voltage Low**<br>Vehicle performance issue.<br>**Possible Causes:**<br>&bull; Check harness between alternator and Control module connector |
| **DTC: P2504**<br>**1T CCM, MIL: No**<br>**Years:** 2005, 2006, 2007, 2008<br>**Engines:** 2.5L SOHC, DOHC VIN 6,<br>**Models:** All<br>**Transmissions:** All | **Charging System Voltage High**<br>Vehicle performance issue.<br>**Possible Causes:**<br>&bull; Check harness between alternator and Control module connector |
| **DTC: P2544**<br>**1T CCM, MIL: No**<br>**Years:** 2005, 2006, 2007, 2008<br>**Engines:** 2.5L SOHC, DOHC VIN 6,<br>**Models:** All<br>**Transmissions:** All | **Transmission Torque Request Circuit**<br>To improve shift feel, the transmission control module may request that the engine control module reduce engine torque during shift events. When such a request is received, Control module responds by retarding the base ignition timing and notifying the transmission module that the request has succeeded. If Control module is unable to comply with the request, Control module sends the transmission module a message that the request has failed.<br>**Possible Causes:**<br>&bull; Check other DTC<br>&bull; Check low system voltage<br>&bull; Check other control modules references |
| **DTC: P3400**<br>**1T CCM, MIL: No**<br>**Years:** 2005, 2006, 2007, 2008<br>**Engines:** 2.5L SOHC, DOHC VIN 6, 5.3L VIN M/P, 6.0L VIN H<br>**Models:** All<br>**Transmissions:** All | **Cylinder Deactivation System Performance**<br>The Cylinder Deactivation System Performance diagnostic is a test of the internal mechanical condition of the engine, during operation in V4 mode. The engine control module determines the internal mechanical condition of the engine through the actual measured values of the mass airflow (MAF) sensor, the intake manifold absolute pressure (MAP) sensor, and the throttle position (TP) sensor. The actual measured values of the 3 sensors are then compared to the calculated values for V4 mode, stored in Control module. If the measured values are incorrect for V4 mode, Control module will set DTC P3400, and will return to V8 mode operation.<br>**Possible Causes:**<br>&bull; Check other DTC<br>&bull; Check system voltage<br>&bull; Check valve lifter oil manifold (VLOM) solenoids |
| **DTC: P3401**<br>**1T CCM, MIL: No**<br>**Years:** 2005, 2006, 2007, 2008<br>**Engines:** 2.5L SOHC, DOHC VIN 6, 5.3L VIN M/P, 6.0L VIN H<br>**Models:** All<br>**Transmissions:** All | **Cylinder Deactivation Circuit Solenoid 1**<br>The Cylinder Deactivation System Performance diagnostic is a test of the internal mechanical condition of the engine, during operation in V4 mode. The engine control module determines the internal mechanical condition of the engine through the actual measured values of the mass airflow (MAF) sensor, the intake manifold absolute pressure (MAP) sensor, and the throttle position (TP) sensor. The actual measured values of the 3 sensors are then compared to the calculated values for V4 mode, stored in Control module. If the measured values are incorrect for V4 mode, Control module will set DTC P3400, and will return to V8 mode operation.<br>**Possible Causes:**<br>&bull; Check other DTC<br>&bull; Check system voltage<br>&bull; Check valve lifter oil manifold (VLOM) solenoids |
| **DTC: P3425**<br>**1T CCM, MIL: No**<br>**Years:** 2005, 2006, 2007, 2008<br>**Engines:** 2.5L SOHC, DOHC VIN 6, 5.3L VIN M/P, 6.0L VIN H<br>**Models:** All<br>**Transmissions:** All | **Cylinder Deactivation Circuit Solenoid 2**<br>The Cylinder Deactivation System Performance diagnostic is a test of the internal mechanical condition of the engine, during operation in V4 mode. The engine control module determines the internal mechanical condition of the engine through the actual measured values of the mass airflow (MAF) sensor, the intake manifold absolute pressure (MAP) sensor, and the throttle position (TP) sensor. The actual measured values of the 3 sensors are then compared to the calculated values for V4 mode, stored in Control module. If the measured values are incorrect for V4 mode, Control module will set DTC P3400, and will return to V8 mode operation.<br>**Possible Causes:**<br>&bull; Check other DTC<br>&bull; Check system voltage<br>&bull; Check valve lifter oil manifold (VLOM) solenoids |

| DTC | Trouble Code Title, Conditions & Possible Causes |
|---|---|
| **DTC: P3441**<br>**1T CCM, MIL: No**<br>**Years:** 2005, 2006, 2007, 2008<br>**Engines:** 2.5L SOHC, DOHC VIN 6, 5.3L VIN M/P, 6.0L VIN H<br>**Models:** All<br>**Transmissions:** All | **Cylinder Deactivation Circuit Solenoid 3**<br>The Cylinder Deactivation System Performance diagnostic is a test of the internal mechanical condition of the engine, during operation in V4 mode. The engine control module determines the internal mechanical condition of the engine through the actual measured values of the mass airflow (MAF) sensor, the intake manifold absolute pressure (MAP) sensor, and the throttle position (TP) sensor. The actual measured values of the 3 sensors are then compared to the calculated values for V4 mode, stored in Control module. If the measured values are incorrect for V4 mode, Control module will set DTC P3400, and will return to V8 mode operation.<br>**Possible Causes:**<br>• Check other DTC<br>• Check system voltage<br>• Check valve lifter oil manifold (VLOM) solenoids |
| **DTC: P3449**<br>**1T CCM, MIL: No**<br>**Years:** 2005, 2006, 2007, 2008<br>**Engines:** 2.5L SOHC, DOHC VIN 6, 5.3L VIN M/P, 6.0L VIN H<br>**Models:** All<br>**Transmissions:** All | **Cylinder Deactivation Circuit Solenoid 4**<br>The Cylinder Deactivation System Performance diagnostic is a test of the internal mechanical condition of the engine, during operation in V4 mode. The engine control module determines the internal mechanical condition of the engine through the actual measured values of the mass airflow (MAF) sensor, the intake manifold absolute pressure (MAP) sensor, and the throttle position (TP) sensor. The actual measured values of the 3 sensors are then compared to the calculated values for V4 mode, stored in Control module. If the measured values are incorrect for V4 mode, Control module will set DTC P3400, and will return to V8 mode operation.<br>**Possible Causes:**<br>• Check other DTC<br>• Check system voltage<br>• Check valve lifter oil manifold (VLOM) solenoids |

# VOLKSWAGEN

Golf • GTI • Jetta • New Beetle • Passat • Rabbit

**14**

## SPECIFICATIONS AND MAINTENANCE CHARTS

### ENGINE AND VEHICLE IDENTIFICATION

| | | Engine | | | | | | Model Year | |
|---|---|---|---|---|---|---|---|---|---|
| ENG Co | Liters (cc) | Cu. In. | Cyl. | Fuel Sys. ① | Engine Type | Eng. Mfg. | | Code ② | Year |
| BGP | 2.5 (2480) | 151 | 5 | ME 7.1.1 | DOHC | Volkswagen | | 6 | 2006 |
| BGQ | 2.5 (2480) | 151 | 5 | ME 7.1.1 | DOHC | Volkswagen | | 7 | 2007 |
| BLV | 3.6 (3580) | 218 | 6 | MED 9.1 | DOHC | Volkswagen | | 8 | 2008 |
| BPR | 2.5 (2480) | 151 | 5 | ME 7.1.1 | DOHC | Volkswagen | | | |
| BPS | 2.5 (2480) | 151 | 5 | ME 7.1.1 | DOHC | Volkswagen | | | |
| BPY | 2.0 (1984) | 121 | 4 | MED 9.1 | DOHC | Volkswagen | | | |
| CBFA | 2.0 (1984) | 121 | 4 | MED 12.5 | DOHC | Volkswagen | | | |
| CBUA | 2.5 (2480) | 151 | 5 | ME 7.1.1 | DOHC | Volkswagen | | | |
| CBTA | 2.5 (2480) | 151 | 5 | ME 7.1.1 | DOHC | Volkswagen | | | |
| CCTA | 2.0 (1984) | 121 | 4 | MED 12.5 | DOHC | Volkswagen | | | |

DOHC: Double Overhead Camshafts

SOHC: Single Overhead Camshaft

① Bosch Motronic Version

② 10th digit of VIN

22205_VWCA_C0001

### GENERAL ENGINE SPECIFICATIONS

| Year | Model | Engine Displacement Liters | Engine ID/VIN | Net Horsepower @ rpm | Net Torque@rpm (ft. lbs.) | Bore x Stroke (in.) | Compression Ratio | Oil Pressure @ rpm |
|---|---|---|---|---|---|---|---|---|
| **2006** | Rabbit | 2.5 | BGP/BGQ | 150@5000 | 170@3750 | 3.25x3.65 | 9.5:1 | 39@2000 |
| | GTI | 2.0 | BPY | 200@5700 | 207@2000 | 3.25x3.65 | 10.5:1 | 39@2000 |
| | Jetta | 2.0 | BPY | 200@5700 | 207@2000 | 3.25x3.65 | 10.5:1 | 39@2000 |
| | Jetta | 2.5 | BGP/BGQ | 150@5000 | 170@3750 | 3.25x3.65 | 9.5:1 | 39@2000 |
| | New Beetle | 2.5 | BPR/BPS | 150@5000 | 170@3750 | 3.25x3.65 | 10:01 | 39@2000 |
| | Passat | 2.0 | BPY | 200@5700 | 207@2000 | 3.25x3.65 | 10.5:1 | 39@2000 |
| | Passat | 3.6 | BLV | 280@6200 | 265@2800 | 3.25x3.40 | 12.0:1 | 43@2000 |
| **2007** | Rabbit | 2.5 | BGP/BGQ | 150@5000 | 170@3750 | 3.25x3.65 | 9.5:1 | 39@2000 |
| | GTI | 2.0 | BPY | 200@5700 | 207@2000 | 3.25x3.65 | 10.5:1 | 39@2000 |
| | Jetta | 2.0 | BPY | 200@5700 | 207@2000 | 3.25x3.65 | 10.5:1 | 39@2000 |
| | Jetta | 2.5 | BGP/BGQ | 150@5000 | 170@3750 | 3.25x3.65 | 9.5:1 | 39@2000 |
| | New Beetle | 2.5 | BPR/BPS | 150@5000 | 170@3750 | 3.25x3.65 | 10:01 | 39@2000 |
| | Passat | 2.0 | BPY | 200@5700 | 207@2000 | 3.25x3.65 | 10.5:1 | 39@2000 |
| | Passat | 3.6 | BLV | 280@6200 | 265@2800 | 3.25x3.40 | 12.0:1 | 43@2000 |
| **2008** | Rabbit | 2.5 | CBTA/CBUA | 170@5700 | 177@4250 | 3.25x3.65 | 9.5:1 | 39@2000 |
| | GTI | 2.0 | BPY | 200@5700 | 207@2000 | 3.25x3.65 | 10.5:1 | 39@2000 |
| | Jetta | 2.0 | BPY | 200@5700 | 207@2000 | 3.25x3.65 | 10.5:1 | 39@2000 |
| | Jetta | 2.5 | CBTA/CBUA | 170@5700 | 177@4250 | 3.25x3.65 | 9.5:1 | 39@2000 |
| | New Beetle | 2.5 | BPR/BPS | 150@5000 | 170@3750 | 3.25x3.65 | 10:01 | 39@2000 |
| | Passat | 2.0 | BPY | 200@5700 | 207@2000 | 3.25x3.65 | 10.5:1 | 39@2000 |
| | Passat | 3.6 | BLV | 280@6200 | 265@2800 | 3.25x3.40 | 12.0:1 | 43@2000 |

22205_VWCA_C0002

## GASOLINE ENGINE TUNE-UP SPECIFICATIONS

| Year | Engine Displacement Liters | Engine ID/VIN | Spark Plug Gap (in.) | Ignition Timing (deg.) A MT | AT | Fuel Pump (psi) | Idle Speed (rpm) MT | AT | Valve Clearance Intake | Exhaust |
|------|------|------|------|------|------|------|------|------|------|------|
| 2006 | 2.5 (2480) | BGP | 0.039-0.043 | ① | ① | 58 | 640-800 | 640-800 | HYD | HYD |
| | 2.5 (2480) | BGQ | 0.039-0.043 | ① | ① | 58 | 680-780 | 680-780 | HYD | HYD |
| | 3.6 (3580) | BLV | 0.039-0.043 | ① | ① | 81-87 ② | 600-800 | 600-800 | HYD | HYD |
| | 2.5 (2480) | BPR | 0.039-0.044 | ① | ① | 51-73 ③ | 680 | 680 | HYD | HYD |
| | 2.5 (2480) | BPS | 0.039-0.045 | ① | ① | 51-73 ③ | 680 | 680 | HYD | HYD |
| | 2.0 (1984) | BPY | 0.028-0.031 | ① | ① | 51-73 ③ | 620-800 | 620-800 | HYD | HYD |
| 2007 | 2.5 (2480) | BGP | 0.039-0.043 | ① | ① | 58 | 640-800 | 640-800 | HYD | HYD |
| | 2.5 (2480) | BGQ | 0.039-0.043 | ① | ① | 58 | 680-780 | 680-780 | HYD | HYD |
| | 3.6 (3580) | BLV | 0.039-0.043 | ① | ① | 81-87 ② | 600-800 | 600-800 | HYD | HYD |
| | 2.5 (2480) | BPR | 0.039-0.044 | ① | ① | 51-73 ③ | 680 | 680 | HYD | HYD |
| | 2.5 (2480) | BPS | 0.039-0.045 | ① | ① | 51-73 ③ | 680 | 680 | HYD | HYD |
| | 2.0 (1984) | BPY | 0.028-0.031 | ① | ① | 51-73 ③ | 620-800 | 620-800 | HYD | HYD |
| | 2.5 (2480) | CBTA | 0.039-0.043 | ① | ① | 58 | 680-780 | 680-780 | HYD | HYD |
| | 2.5 (2480) | CBUA | 0.039-0.043 | ① | ① | 58 | 680-780 | 680-780 | HYD | HYD |
| 2008 | 2.5 (2480) | BPR | 0.039-0.043 | ① | ① | 51-73 ③ | 680 | 680 | HYD | HYD |
| | 2.5 (2480) | BPS | 0.039-0.043 | ① | ① | 51-73 ③ | 680 | 680 | HYD | HYD |
| | 2.0 (1984) | BPY | 0.028-0.031 | ① | ① | 51-73 ③ | 620-800 | 620-800 | HYD | HYD |
| | 3.6 (3580) | BLV | 0.039-0.043 | ① | ① | 81-87 ② | 600-800 | 600-800 | HYD | HYD |
| | 2.5 (2480) | CBTA | 0.039-0.043 | ① | ① | 58 | 680-780 | 680-780 | HYD | HYD |
| | 2.5 (2480) | CBUA | 0.039-0.043 | ① | ① | 58 | 680-780 | 680-780 | HYD | HYD |

Note: The Vehicle Emission Control Information label reflects specification changes made during production.

The label figures must be used if they differ from those in this chart.

HYD: Hydraulic

① Ignition timing controlled electronically. Specification no longer provided.

② Engine OFF. Ignition ON. Fuel Pump connector pin 4 jumpered to harness pin 4. FP pin 1 jumpered to harness pin 3 with remote starter switch.

③ Keep Engine OFF. Cycle Ignition switch until fuel pressure no longer rises.

22205_VWCA_C0003

## CAPACITIES

| Year | Model | Engine Displacement Liters | Engine ID/VIN | Engine Oil with Filter | Transmission Manual (pts.) | Auto. (qts.) | Final Drive (pts.) | Fuel Tank (gal.) | Cooling System (qts.) |
|------|-------|-----|-----|-----|-----|-----|-----|-----|-----|
| **2006** | Rabbit | 2.5 | BGP/BGQ | 6.3 | 2.0 | 7.4 ① | — | 14.5 | 10.0 |
| | GTI | 2.0 | BPY | 4.9 | 2.4 | 5.5 | — | 14.5 | 8.0 |
| | Jetta | 2.5 | BGP/BGQ | 6.3 | 2.0 | 7.4 ① | — | 14.5 | 10.0 |
| | Jetta | 2.0 | BPY | 4.9 | 2.4 | 5.5 | — | 14.5 | 8.5 |
| | New Beetle | 2.5 | BPR/BPS | 6.3 | 2.0 | 7.4 ① | — | 14.5 | 10.0 |
| | Passat | 3.6 | BLV | 5.8 | — | 7.4 ① | — | 18.5 | 8.5 |
| | Passat | 2.0 | BPY | 4.9 | 2.4 | 5.5 | — | 18.5 | 7.6 |
| **2007** | Rabbit | 2.5 | BGP/BGQ | 6.3 | 2.0 | 7.4 ① | — | 14.5 | 10.0 |
| | GTI | 2.0 | BPY | 4.9 | 2.4 | 5.5 | — | 14.5 | 8.0 |
| | Jetta | 2.5 | BGP/BGQ | 6.3 | 2.0 | 7.4 ① | — | 14.5 | 10.0 |
| | Jetta | 2.0 | BPY | 4.9 | 2.4 | 5.5 | — | 14.5 | 8.5 |
| | New Beetle | 2.5 | BPR/BPS | 6.3 | 2.0 | 7.4 ① | — | 14.5 | 10.0 |
| | Passat | 3.6 | BLV | 5.8 | — | 7.4 ① | — | 18.5 | 8.5 |
| | Passat | 2.0 | BPY | 4.9 | 2.4 | 5.5 | — | 18.5 | 7.6 |
| **2008** | Rabbit | 2.5 | CBTA/CBUA | 5.8 | 2.0 | 7.4 ① | — | 14.5 | 10.0 |
| | GTI | 2.0 | BPY | 4.9 | 2.4 | 5.5 | — | 14.5 | 8.5 |
| | Jetta | 2.0 | BPY | 4.9 | 2.4 | 5.5 | — | 14.5 | 8.5 |
| | Jetta | 2.5 | CBTA/CBUA | 5.8 | 2.0 | 7.4 ① | — | 14.5 | 10.0 |
| | New Beetle | 2.5 | BPR/BPS | 6.3 | 2.0 | 7.4 ① | — | 14.5 | 10.0 |
| | Passat | 3.6 | BLV | 5.8 | — | 7.4 ① | — | 18.5 | 8.5 |
| | Passat | 2.0 | BPY | 4.4 | 2.4 | 7.4 ① | — | 18.5 | 7.6 |

Note: All capacities are approximate. Add fluid gradually and check often to avoid overfilling or underfilling.

① Fill amount for new or rebuit trans. Be careful not to overfill.

22205_VWCA_C0004

## VALVE SPECIFICATIONS

| Year | Engine Displacement Liters | Engine ID/VIN | Seat Angle (deg.) | Face Angle (deg.) | Spring Test Pressure (lbs. @ in.) | Spring Installed Height (in.) | Stem-to-Guide Clearance (in.) | | Stem Diameter (in.) | |
|---|---|---|---|---|---|---|---|---|---|---|
| | | | | | | | Intake | Exhaust | Intake | Exhaust |
| 2006 | 2.5 | BGP | 45 | 45 | NA | NA | 0.031 | 0.031 | 0.2346 ① | 0.2340 ② |
| | 2.5 | BGQ | 45 | 45 | NA | NA | 0.031 | 0.031 | 0.2346 ① | 0.2340 ② |
| | 3.6 | BLV | 45 | 45 | NA | NA | 0.031 | 0.031 | 0.2354 | 0.2350 |
| | 2.5 | BPR | 45 | 45 | NA | NA | 0.031 | 0.031 | 0.2346 ① | 0.2340 ② |
| | 2.5 | BPS | 45 | 45 | NA | NA | 0.031 | 0.031 | 0.2346 ① | 0.2340 ② |
| | 2.0 | BPY | 45 | 45 | NA | NA | 0.031 | 0.031 | 0.2354 | 0.2342 |
| 2007 | 2.5 | BGP | 45 | 45 | NA | NA | 0.031 | 0.031 | 0.2346 ① | 0.2340 ② |
| | 2.5 | BGQ | 45 | 45 | NA | NA | 0.031 | 0.031 | 0.2346 ① | 0.2340 ② |
| | 3.6 | BLV | 45 | 45 | NA | NA | 0.031 | 0.031 | 0.2354 | 0.2350 |
| | 2.5 | BPR | 45 | 45 | NA | NA | 0.031 | 0.031 | 0.2346 ① | 0.2340 ② |
| | 2.5 | BPS | 45 | 45 | NA | NA | 0.031 | 0.031 | 0.2346 ① | 0.2340 ② |
| | 2.0 | BPY | 45 | 45 | NA | NA | 0.031 | 0.031 | 0.2354 | 0.2342 |
| | 2.5 | CBTA | 45 | 45 | NA | NA | 0.031 | 0.031 | 0.2346 ① | 0.2340 ② |
| | 2.5 | CBUA | 45 | 45 | NA | NA | 0.031 | 0.031 | 0.2346 ① | 0.2340 ② |
| 2008 | 2.5 | BPR | 45 | 45 | NA | NA | 0.031 | 0.031 | 0.2346 ① | 0.2340 ③ |
| | 2.5 | BPS | 45 | 45 | NA | NA | 0.031 | 0.031 | 0.2346 ① | 0.2340 ② |
| | 2.0 | BPY | 45 | 45 | NA | NA | 0.031 | 0.031 | 0.2354 | 0.2342 |
| | 3.6 | BLV | 45 | 45 | NA | NA | 0.031 | 0.031 | 0.2354 | 0.2350 |
| | 2.5 | CBTA | 45 | 45 | NA | NA | 0.031 | 0.031 | 0.2346 ① | 0.2340 ② |
| | 2.5 | CBUA | 45 | 45 | NA | NA | 0.031 | 0.031 | 0.2346 ① | 0.2340 ② |

NA: Not Available

① +/- 0.0004 in.

② +/- 0.0002 in.

## CRANKSHAFT AND CONNECTING ROD SPECIFICATIONS
All measurements are given in inches.

| Year | Engine Displacement Liters | Engine ID/VIN | Crankshaft | | | | Connecting Rod | | |
| | | | Main Brg. Journal Dia. | Main Brg. Oil Clearance | Shaft End-play | Thrust on No. | Journal Diameter | Oil Clearance | Side Clearance |
|---|---|---|---|---|---|---|---|---|---|
| **2006** | 2.5 | BGP | 2.2834 | 0.0009-0.0017 ① | 0.0009-0.0017 ② | 3 | 1.8819 | 0.0008-0.0024 ③ | 0.0039-0.0138 ④ |
| | 2.5 | BGQ | 2.2834 | 0.0009-0.0017 ① | 0.0009-0.0017 ② | 3 | 1.8819 | 0.0008-0.0024 ③ | 0.0039-0.0138 ④ |
| | 3.6 | BLV | 2.3605-2.3613 | 0.0008-0.0024 ⑤ | 0.0028-0.0091 ⑥ | 5 | 2.1243-2.1251 | 0.0008-0.0028 ⑤ | 0.0020-0.0122 ④ |
| | 2.5 | BPR | 2.2834 | 0.0009-0.0017 ① | 0.0009-0.0017 ② | 3 | 1.8819 | 0.0008-0.0024 ③ | 0.0039-0.0138 ④ |
| | 2.5 | BPS | 2.2834 | 0.0009-0.0017 ① | 0.0009-0.0017 ② | 3 | 1.8819 | 0.0008-0.0024 ③ | 0.0039-0.0138 ④ |
| | 2.0 | BPY | 2.1260 | 0.0007-0.0015 ① | 0.0028-0.0090 ② | 3 | 1.8819 | 0.0008-0.0024 ③ | 0.0039-0.0138 ④ |
| **2007** | 2.5 | BGP | 2.2834 | 0.0009-0.0017 ① | 0.0009-0.0017 ② | 3 | 1.8819 | 0.0008-0.0024 ③ | 0.0039-0.0138 ④ |
| | 2.5 | BGQ | 2.2834 | 0.0009-0.0017 ① | 0.0009-0.0017 ② | 3 | 1.8819 | 0.0008-0.0024 ③ | 0.0039-0.0138 ④ |
| | 3.6 | BLV | 2.3605-2.3613 | 0.0008-0.0024 ⑤ | 0.0028-0.0091 ⑥ | 5 | 2.1243-2.1251 | 0.0008-0.0028 ⑤ | 0.0020-0.0122 ④ |
| | 2.5 | BPR | 2.2834 | 0.0009-0.0017 ① | 0.0009-0.0017 ② | 3 | 1.8819 | 0.0008-0.0024 ③ | 0.0039-0.0138 ④ |
| | 2.5 | BPS | 2.2834 | 0.0009-0.0017 ① | 0.0009-0.0017 ② | 3 | 1.8819 | 0.0008-0.0024 ③ | 0.0039-0.0138 ④ |
| | 2.0 | BPY | 2.1260 | 0.0007-0.0015 ① | 0.0028-0.0090 ② | 3 | 1.8819 | 0.0008-0.0024 ③ | 0.0039-0.0138 ④ |
| | 2.5 | CBTA | 2.2834 | 0.0009-0.0017 ① | 0.0009-0.0017 ② | 3 | 1.8819 | 0.0008-0.0024 ③ | 0.0039-0.0138 ④ |
| | 2.5 | CBUA | 2.2834 | 0.0009-0.0017 ① | 0.0009-0.0017 ② | 3 | 1.8819 | 0.0008-0.0024 ③ | 0.0039-0.0138 ④ |
| **2008** | 2.5 | BPR | 2.2834 | 0.0009-0.0017 ① | 0.0009-0.0017 ② | 3 | 1.8819 | 0.0008-0.0024 ③ | 0.0039-0.0138 ④ |
| | 2.5 | BPS | 2.2834 | 0.0009-0.0017 ① | 0.0009-0.0017 ② | 3 | 1.8819 | 0.0008-0.0024 ③ | 0.0039-0.0138 ④ |
| | 2.0 | BPY | 2.1260 | 0.0007-0.0015 ① | 0.0028-0.0090 ② | 3 | 1.8819 | 0.0008-0.0024 ③ | 0.0039-0.0138 ④ |
| | 3.6 | BLV | 2.3605-2.3613 | 0.0008-0.0024 ⑤ | 0.0028-0.0091 ⑥ | 5 | 2.1243-2.1251 | 0.0008-0.0028 ⑤ | 0.0020-0.0122 ④ |
| | 2.5 | CBTA | 2.2834 | 0.0009-0.0017 ① | 0.0009-0.0017 ② | 3 | 1.8819 | 0.0008-0.0024 ③ | 0.0039-0.0138 ④ |
| | 2.5 | CBUA | 2.2834 | 0.0009-0.0017 ① | 0.0009-0.0017 ② | 3 | 1.8819 | 0.0008-0.0024 ③ | 0.0039-0.0138 ④ |

NA: Not Available

① Measurement with new parts. Wear limit is 0.0028 in.
② Measurement with new parts. Wear limit is 0.0118 in.
③ Measurement with new parts. Wear limit is 0.0035 in.
④ Measurement with new parts. Wear limit is 0.0157 in.
⑤ Measurement with new parts. Wear limit is 0.0039 in.
⑥ Measurement with new parts. Wear limit is 0.0003 in.

## PISTON AND RING SPECIFICATIONS

All measurements are given in inches.

| Year | Engine Size Liters | Engine ID/VIN | Piston Clearance | Ring Gap ② Top Compression | Ring Gap ② Bottom Compression | Ring Gap ② Oil Control | Ring Side Clearance ③ Top Compression | Ring Side Clearance ③ Bottom Compression | Ring Side Clearance ③ Oil Control |
|------|------|------|------|------|------|------|------|------|------|
| 2006 | 2.5 | BGP | 0.0177 ① | 0.008-0.016 ② | 0.008-0.016 ② | 0.010-0.020 ② | 0.0024-0.0035 ③ | 0.0024-0.0035 ③ | 0.0012-0.0024 ④ |
| | 2.5 | BGQ | 0.0177 ① | 0.008-0.016 ② | 0.008-0.016 ② | 0.010-0.020 ② | 0.0024-0.0035 ③ | 0.0024-0.0035 ③ | 0.0012-0.0024 ④ |
| | 3.6 | BLV | 0.0026 | 0.008-0.016 ⑤ | 0.008-0.016 ⑤ | 0.010-0.020 ⑤ | 0.0016-0.0035 ④ | 0.0012-0.0024 ④ | 0.0008-0.0024 ④ |
| | 2.5 | BPR | 0.0177 ① | 0.008-0.016 ② | 0.008-0.016 ② | 0.010-0.020 ② | 0.0024-0.0035 ③ | 0.0024-0.0035 ③ | 0.0012-0.0024 ④ |
| | 2.5 | BPS | 0.0177 ① | 0.008-0.016 ② | 0.008-0.016 ② | 0.010-0.020 ② | 0.0024-0.0035 ③ | 0.0024-0.0035 ③ | 0.0012-0.0024 ④ |
| | 2.0 | BPY | 0.0177 ① | 0.008-0.016 ② | 0.008-0.016 ② | 0.010-0.020 ② | 0.0024-0.0035 ③ | 0.0024-0.0035 ③ | 0.0012-0.0024 ④ |
| 2007 | 2.5 | BGP | 0.0177 ① | 0.008-0.016 ② | 0.008-0.016 ② | 0.010-0.020 ② | 0.0024-0.0035 ③ | 0.0024-0.0035 ③ | 0.0012-0.0024 ④ |
| | 2.5 | BGQ | 0.0177 ① | 0.008-0.016 ② | 0.008-0.016 ② | 0.010-0.020 ② | 0.0024-0.0035 ③ | 0.0024-0.0035 ③ | 0.0012-0.0024 ④ |
| | 3.6 | BLV | 0.0026 | 0.008-0.016 ⑤ | 0.008-0.016 ⑤ | 0.010-0.020 ⑤ | 0.0016-0.0035 ④ | 0.0012-0.0024 ④ | 0.0008-0.0024 ④ |
| | 2.5 | BPR | 0.0177 ① | 0.008-0.016 ② | 0.008-0.016 ② | 0.010-0.020 ② | 0.0024-0.0035 ③ | 0.0024-0.0035 ③ | 0.0012-0.0024 ④ |
| | 2.5 | BPS | 0.0177 ① | 0.008-0.016 ② | 0.008-0.016 ② | 0.010-0.020 ② | 0.0024-0.0035 ③ | 0.0024-0.0035 ③ | 0.0012-0.0024 ④ |
| | 2.0 | BPY | 0.0177 ① | 0.008-0.016 ② | 0.008-0.016 ② | 0.010-0.020 ② | 0.0024-0.0035 ③ | 0.0024-0.0035 ③ | 0.0012-0.0024 ④ |
| | 2.5 | CBTA | 0.0177 ① | 0.008-0.016 ② | 0.008-0.016 ② | 0.010-0.020 ② | 0.0024-0.0035 ③ | 0.0024-0.0035 ③ | 0.0012-0.0024 ④ |
| | 2.5 | CBUA | 0.0177 ① | 0.008-0.016 ② | 0.008-0.016 ② | 0.010-0.020 ② | 0.0024-0.0035 ③ | 0.0024-0.0035 ③ | 0.0012-0.0024 ④ |
| 2008 | 2.5 | BPR | 0.0177 ① | 0.008-0.016 ② | 0.008-0.016 ② | 0.010-0.020 ② | 0.0024-0.0035 ③ | 0.0024-0.0035 ③ | 0.0012-0.0024 ④ |
| | 2.5 | BPS | 0.0177 ① | 0.008-0.016 ② | 0.008-0.016 ② | 0.010-0.020 ② | 0.0024-0.0035 ③ | 0.0024-0.0035 ③ | 0.0012-0.0024 ④ |
| | 2.0 | BPY | 0.0177 ① | 0.008-0.016 ② | 0.008-0.016 ② | 0.010-0.020 ② | 0.0024-0.0035 ③ | 0.0024-0.0035 ③ | 0.0012-0.0024 ④ |
| | 3.6 | BLV | 0.0026 | 0.008-0.016 ⑤ | 0.008-0.016 ⑤ | 0.010-0.020 ⑤ | 0.0016-0.0035 ④ | 0.0012-0.0024 ④ | 0.0008-0.0024 ④ |
| | 2.5 | CBTA | 0.0177 ① | 0.008-0.016 ② | 0.008-0.016 ② | 0.010-0.020 ② | 0.0024-0.0035 ③ | 0.0024-0.0035 ③ | 0.0012-0.0024 ④ |
| | 2.5 | CBUA | 0.0177 ① | 0.008-0.016 ② | 0.008-0.016 ② | 0.010-0.020 ② | 0.0024-0.0035 ③ | 0.0024-0.0035 ③ | 0.0012-0.0024 ④ |

NA: Not Available

① Measurement does not include graphite coating of new piston (0.0007 in.) Graphite coating wears off.

② Measurement with new parts. Wear limit is 0.031 in.

③ Measurement with new parts. Wear limit is 0.008 in.

④ Measurement with new parts. Wear limit is 0.0059 in.

⑤ Measurement with new parts. Wear limit is 0.0394 in.

## TORQUE SPECIFICATIONS
All readings in ft. lbs.

| Year | Engine Displacement Liters | Engine ID/VIN | Cylinder Head Bolts | Main Bearing Bolts | Rod Bearing Bolts | Crankshaft Damper Bolt | Flywheel Bolts | Manifold Intake | Manifold Exhaust | Spark Plugs | Oil Pan Drain Plug |
|---|---|---|---|---|---|---|---|---|---|---|---|
| 2006 | 2.5 | BGP | ① | ② | ③ | ④ | ⑤ | 7 | 18 ⑥ | 18 | 22 |
| | 2.5 | BGQ | ① | ② | ③ | ④ | ⑤ | 7 | 18 ⑥ | 18 | 22 |
| | 3.6 | BLV | ⑦ | ⑧ | ③ | ⑨ | ⑤ | 6 | 18 ⑥ | 13 | 22 |
| | 2.5 | BPR | ① | ② | ③ | ④ | ⑤ | 7 | 18 ⑥ | 18 | 22 |
| | 2.5 | BPS | ① | ② | ③ | ⑧ | ⑤ | 7 | 18 ⑥ | 18 | 22 |
| | 2.0 | BPY | ① | ⑩ | ③ | ⑪ | ⑤ | 7 | 15 ⑥ | 18 | 22 |
| 2007 | 2.5 | BGP | ① | ② | ③ | ④ | ⑤ | 7 | 18 ⑥ | 18 | 22 |
| | 2.5 | BGQ | ① | ② | ③ | ④ | ⑤ | 7 | 18 ⑥ | 18 | 22 |
| | 3.6 | BLV | ⑦ | ⑧ | ③ | ⑨ | ⑤ | 6 | 18 ⑥ | 13 | 22 |
| | 2.5 | BPR | ① | ② | ③ | ④ | ⑤ | 7 | 18 ⑥ | 18 | 22 |
| | 2.5 | BPS | ① | ② | ③ | ④ | ⑤ | 7 | 18 ⑥ | 18 | 22 |
| | 2.0 | BPY | ① | ⑩ | ③ | ⑪ | ⑤ | 7 | 15 ⑥ | 18 | 22 |
| | 2.5 | CBTA | ① | ② | ③ | ④ | ⑤ | 7 | 18 ⑥ | 18 | 22 |
| | 2.5 | CBUA | ① | ② | ③ | ④ | ⑤ | 7 | 18 ⑥ | 18 | 22 |
| 2008 | 2.5 | BPR | ① | ② | ③ | ④ | ⑤ | 7 | 18 ⑥ | 18 | 22 |
| | 2.5 | BPS | ① | ② | ③ | ④ | ⑤ | 7 | 18 ⑥ | 18 | 22 |
| | 2.0 | BPY | ① | ⑩ | ③ | ⑫ | ⑤ | 7 | 15 ⑥ | 18 | 22 |
| | 3.6 | BLV | ⑦ | ⑧ | ③ | ⑨ | ⑤ | 6 | 18 ⑥ | 13 | 22 |
| | 2.5 | CBTA | ① | ② | ③ | ④ | ⑤ | 7 | 18 ⑥ | 18 | 22 |
| | 2.5 | CBUA | ① | ② | ③ | ④ | ⑤ | 7 | 18 ⑥ | 18 | 22 |

① Torque in three steps. Use new bolts.
  Step 1:  30 ft. lbs.
  Step 2:  plus 90 degrees
  Step 3:  plus 90 degrees

② Two Steps. 30 ft. lbs. plus 90 degrees. Use new bolts.

③ Two Steps. 22 ft. lbs. plus 90 degrees. Use new bolts.

④ Two Steps. 37 ft. lbs. plus 90 degrees. Use new bolts.

⑤ Two Steps. 44 ft. lbs. plus 90 degrees. Use new bolts.

⑥ Use new bolts.

⑦ Torque in three steps. Use new bolts.
  Step 1:  11 ft. lbs.
  Step 2:  22 ft. lbs.
  Step 3:  plus 180 degrees

⑧ Two Steps. 22 ft. lbs. plus 180 degrees. Use new bolts.

⑨ Two Steps. 74 ft. lbs. plus 180 degrees. Use new bolt.

⑩ Two Steps. 66 ft. lbs. Plus 90 degrees. Use new bolt.

⑪ Two Steps. 66 ft. lbs. Plus 90 degrees. Use new bolt.

22205_VWCA_C0010

## WHEEL ALIGNMENT

| Year | Model | | Caster | | Camber | | Toe-in |
| | | | Range (+/-Deg.) | Preferred Setting (Deg.) | Range (+/-Deg.) | Preferred Setting (Deg.) | (Deg.) |
|---|---|---|---|---|---|---|---|
| 2006 | Rabbit | F | 0.50 | +7.57 | 0.50 | -0.50 | 1.63+/-0.33 ① |
| | Standard | R | — | — | 0.50 | -1.33 | 0.17+/-0.21 ② |
| | Sport w/o 18" | F | 0.50 | +7.83 | 0.50 | -0.68 | 1.67+/-0.33 ① |
| | wheels | R | — | — | 0.50 | -1.33 | 0.17+/-0.21 ② |
| | Sport w/ 18" | F | 0.50 | +7.83 | 0.50 | -0.68 | 1.67+/-0.33 ① |
| | wheels | R | — | — | 0.50 | -1.75 | 0.17+/-0.21 ② |
| | GTI | F | 0.50 | +7.78 | 0.50 | -0.73 | 1.63+/-0.33 ① |
| | Standard | R | — | — | 0.17 | -1.45 | 0.17+/-0.21 ② |
| | Jetta | F | 0.50 | +7.57 | 0.50 | -0.50 | 1.63+/-0.33 ① |
| | Standard | R | — | — | 0.50 | -1.33 | 0.17+/-0.21 ② |
| | Jetta | F | 0.50 | +7.83 | 0.50 | -0.68 | 1.67+/-0.33 ① |
| | Sport | R | — | — | 0.50 | -1.33 | 0.17+/-0.21 ② |
| | GTI | F | 0.50 | +7.83 | 0.50 | -0.58 | 1.67+/-0.33 ① |
| | | R | — | — | 0.50 | -1.33 | 0.17+/-0.21 ② |
| | New Beetle | F | 0.50 | +7.67 | 0.50 | -0.50 | 1.50+/-0.33 ① |
| | Standard | R | — | — | 0.16 | -1.45 | 0.33+/-0.17 ② |
| | New Beetle | F | 0.50 | +7.83 | 0.50 | -0.55 | 1.52+/-0.33 ① |
| | Sport | R | — | — | 0.16 | -1.45 | 0.42+/-0.17 ② |
| | Passat | F | 0.50 | 7.53 | 0.50 | -0.50 | 1.31+/-0.33 ① |
| | Standard | R | — | — | 0.50 | -1.33 | 0.17+/-0.17 ② |
| | Sport w/0 18" | F | 0.50 | 7.73 | 0.50 | -0.68 | 1.35+/-0.33 ① |
| | wheels | R | — | — | 0.50 | -1.33 | 0.17+/-0.17 ② |
| | Sport w 18" | F | 0.50 | 7.73 | 0.50 | -0.68 | 1.35+/-0.33 ① |
| | wheels | R | — | — | 0.50 | -1.45 | 0.17+/-0.17 ② |
| 2007 | Rabbit | F | 0.50 | +7.57 | 0.50 | -0.50 | 1.63+/-0.33 ① |
| | Standard | R | — | — | 0.50 | -1.33 | 0.17+/-0.21 ② |
| | Sport w/o 18" | F | 0.50 | +7.83 | 0.50 | -0.68 | 1.67+/-0.33 ① |
| | wheels | R | — | — | 0.50 | -1.33 | 0.17+/-0.21 ② |
| | Sport w 18" | F | 0.50 | +7.83 | 0.50 | -0.68 | 1.67+/-0.33 ① |
| | wheels | R | — | — | 0.50 | -1.75 | 0.17+/-0.21 ② |
| | GTI | F | 0.50 | +7.78 | 0.50 | -0.73 | 1.63+/-0.33 ① |
| | | R | — | — | 0.17 | -1.45 | 0.17+/-0.21 ② |
| | Jetta | F | 0.50 | +7.57 | 0.50 | -0.50 | 1.63+/-0.33 ① |
| | Standard | R | — | — | 0.50 | -1.33 | 0.17+/-0.21 ② |
| | Jetta | F | 0.50 | +7.83 | 0.50 | -0.68 | 1.67+/-0.33 ① |
| | Sport | R | — | — | 0.50 | -1.33 | 0.17+/-0.21 ② |
| | Jetta | F | 0.50 | +7.83 | 0.50 | -0.58 | 1.67+/-0.33 ① |
| | GLI | R | — | — | 0.50 | -1.33 | 0.17+/-0.21 ② |
| | New Beetle | F | 0.50 | +7.67 | 0.50 | -0.50 | 1.50+/-0.33 ① |
| | Standard | R | — | — | 0.16 | -1.45 | 0.33+/-0.17 ② |
| | New Beetle | F | 0.50 | +7.83 | 0.50 | -0.55 | 1.52+/-0.33 ① |
| | Sport | R | — | — | 0.16 | -1.45 | 0.42+/-0.17 ② |
| | Passat | F | 0.50 | 7.53 | 0.50 | -0.50 | 1.31+/-0.33 ① |
| | Standard | R | — | — | 0.50 | -1.33 | 0.17+/-0.17 ② |
| | Sport w/o 18" | F | 0.50 | 7.73 | 0.50 | -0.68 | 1.35+/-0.33 ① |
| | wheels | R | — | — | 0.50 | -1.33 | 0.17+/-0.17 ② |
| | Sport w 18" | F | 0.50 | 7.73 | 0.50 | -0.68 | 1.35+/-0.33 ① |
| | wheels | R | — | — | 0.50 | -1.45 | 0.17+/-0.17 ② |

## WHEEL ALIGNMENT

| Year | Model | | Caster | | Camber | | Toe-in (Deg.) |
|---|---|---|---|---|---|---|---|
| | | | Range (+/-Deg.) | Preferred Setting (Deg.) | Range (+/-Deg.) | Preferred Setting (Deg.) | |
| 2008 | Rabbit | F | 0.50 | +7.57 | 0.50 | -0.50 | 1.63+/-0.33 ① |
| | Standard | R | — | — | 0.50 | -1.33 | 0.17+/-0.21 ② |
| | Sport w/o 18" | F | 0.50 | +7.83 | 0.50 | -0.68 | 1.67+/-0.33 ① |
| | wheels | R | — | — | 0.50 | -1.33 | 0.17+/-0.21 ② |
| | Sport w 18" | F | 0.50 | +7.83 | 0.50 | -0.68 | 1.67+/-0.33 ① |
| | wheels | R | — | — | 0.50 | -1.75 | 0.17+/-0.21 ② |
| | GTI | F | 0.50 | +7.78 | 0.50 | -0.73 | 1.63+/-0.33 ① |
| | Standard | R | — | — | 0.17 | -1.45 | 0.17+/-0.21 ② |
| | Jetta | F | 0.50 | +7.57 | 0.50 | -0.50 | 1.63+/-0.33 ① |
| | Standard | R | — | — | 0.50 | -1.33 | 0.17+/-0.21 ② |
| | Sport | F | 0.50 | +7.83 | 0.50 | -0.68 | 1.67+/-0.33 ① |
| | | R | — | — | 0.50 | -1.33 | 0.17+/-0.21 ② |
| | Sport GLI | F | 0.50 | +7.83 | 0.50 | -0.58 | 1.67+/-0.33 ① |
| | | R | — | — | 0.50 | -1.33 | 0.17+/-0.21 ② |
| | New Beetle | F | 0.50 | +7.67 | 0.50 | -0.50 | 1.50+/-0.33 ① |
| | Standard | R | — | — | 0.16 | -1.45 | 0.33+/-0.17 ② |
| | Sport | F | 0.50 | +7.83 | 0.50 | -0.55 | 1.52+/-0.33 ① |
| | | R | — | — | 0.16 | -1.45 | 0.42+/-0.17 ② |
| | Passat | F | 0.50 | 7.53 | 0.50 | -0.50 | 1.31+/-0.33 ① |
| | Standard | R | — | — | 0.50 | -1.33 | 0.17+/-0.17 ② |
| | Sport w/o 18" | F | 0.50 | 7.73 | 0.50 | -0.68 | 1.35+/-0.33 ① |
| | wheels | R | — | — | 0.50 | -1.33 | 0.17+/-0.17 ② |
| | Sport w 18" | F | 0.50 | 7.73 | 0.50 | -0.68 | 1.35+/-0.33 ① |
| | wheels | R | — | — | 0.50 | -1.45 | 0.17+/-0.17 ② |

① Toe differential angle with 20 deg. steering lock to left and right

② Total toe at specified camber.

22205_VWCA_C0012

## TIRE, WHEEL AND BALL JOINT SPECIFICATIONS

| Year | Model | OEM Tires | | Tire Pressures (psi) | | Wheel Size | Lug Nut (ft. lbs.) |
| | | Standard | Optional | Front | Rear | | |
|------|-------|-----------|----------|-------|------|------------|--------------------|
| 2006 | Golf | P205/55HR16 | None | ① | ① | 6.5-J | 89 |
| | GTI | P225/4HR17 | 225/40HR18 | ① | ① | 7-J ② | 89 |
| | Jetta | P205/55HR16 | P225/4HR17 | ① | ① | 7-J | 89 |
| | New Beetle | 185/60HR14 | 195/60HR14 | ① | ① | 6-J | 89 |
| | Rabbit | P205/55HR16 | P225/4HR17 | ① | ① | 7-J | 89 |
| | Passat | P205/55HR16 | P235/45HR17 | ① | ① | 6.5-J ② | 89 |
| 2007 | GTI | P225/4HR17 | 225/40HR18 | ① | ① | 7-J ② | 89 |
| | Jetta | P205/55HR16 | P225/4HR17 | ① | ① | 7-J | 89 |
| | New Beetle | P205/55HR16 | P225/4HR17 | ① | ① | 7-J | 89 |
| | Rabbit | P205/55HR16 | P225/4HR17 | ① | ① | 7-J | 89 |
| | Passat | 195/65R15 | None | ① | ① | 6.5-J | 89 |
| 2008 | GTI | P225/4HR17 | 225/40HR18 | ① | ① | 7-J ② | 89 |
| | Jetta | P205/55HR16 | P225/4H R17 | ① | ① | 7-J | 89 |
| | New Beetle | P205/55HR16 | P225/4H R17 | ① | ① | 7-J | 89 |
| | Rabbit | P205/55HR16 | P225/4H R17 | ① | ① | 7-J | 89 |
| | Passat | 195/65R15 | None | ① | ① | 6.5-J | 89 |

① See Vehicle Tire Information Sticker for correct inflation values.

② Optional rim size is 7.5-J.

22205_VWCA_C0013

# BRAKE SPECIFICATIONS

All measurements in inches unless noted

| Year | Model | | Brake Disc Original Thickness | Brake Disc Minimum Thickness | Brake Disc Maximum Run-out | Drum Diameter Original Inside Diameter | Drum Diameter Maximum Machine Diameter | Minimum Lining Thickness | Brake Caliper Guide Pins (ft. lbs.) | Brake Caliper Mounting Bolts (ft. lbs.) |
|---|---|---|---|---|---|---|---|---|---|---|
| **2006** | GTI | F | 0.980 | 0.870 | 0.002 | — | — | 0.070 ① | 22 | — |
| | | R | 0.472 | 0.390 | 0.002 | — | — | 0.070 ① | — | 35 ② |
| | Jetta | F | 0.866 | 0.748 | 0.002 | — | — | 0.276 ① | 22 | — |
| | Standard | R | 0.354 | 0.276 | 0.002 | — | — | 0.295 ③ | — | 48 |
| | Jetta | F | 0.980 | 0.870 | 0.002 | — | — | 0.070 ① | 22 | — |
| | GLI | R | 0.472 | 0.390 | 0.002 | — | — | 0.070 ① | — | 35 ② |
| | New Beetle | F | 0.866 | 0.748 | 0.002 | — | — | 0.276 ③ | 22 | — |
| | Standard | R | 0.390 | 0.315 | 0.002 | — | — | 0.275 ③ | 26 | 48 |
| | Rabbit | F | 0.866 | 0.748 | 0.002 | — | — | 0.276 ③ | 22 | — |
| | | R | 0.354 | 0.276 | 0.002 | — | — | 0.295 ③ | — | 48 |
| | Passat | F | 0.980 | 0.870 | 0.002 | — | — | 0.070 ① | 22 | — |
| | | R | 0.472 | 0.390 | 0.002 | — | — | 0.070 ① | — | 35 ② |
| **2007** | GTI | F | 0.980 | 0.870 | 0.002 | — | — | 0.070 ① | 22 | — |
| | Standard | R | 0.472 | 0.390 | 0.002 | — | — | 0.070 ① | — | 35 ② |
| | Jetta | F | 0.866 | 0.748 | 0.002 | — | — | 0.276 ③ | 22 | — |
| | Standard | R | 0.354 | 0.276 | 0.002 | — | — | 0.295 ③ | — | 48 |
| | Jetta | F | 0.980 | 0.870 | 0.002 | — | — | 0.070 ① | 22 | — |
| | GLI | R | 0.472 | 0.390 | 0.002 | — | — | 0.070 ① | — | 35 ② |
| | New Beetle | F | 0.866 | 0.748 | 0.002 | — | — | 0.276 ③ | 22 | — |
| | Standard | R | 0.390 | 0.315 | 0.002 | — | — | 0.275 ③ | 26 | 48 |
| | Rabbit | F | 0.866 | 0.748 | 0.002 | — | — | 0.276 ③ | 22 | — |
| | | R | 0.354 | 0.276 | 0.002 | — | — | 0.295 ③ | — | 48 |
| | Passat | F | 0.980 | 0.870 | 0.002 | — | — | 0.070 ① | 22 | — |
| | | R | 0.472 | 0.390 | 0.002 | — | — | 0.070 ① | — | 35 ② |
| **2008** | GTI | F | 0.980 | 0.870 | 0.002 | — | — | 0.070 ① | 22 | — |
| | Standard | R | 0.472 | 0.390 | 0.002 | — | — | 0.070 ① | — | 35 ② |
| | Jetta | F | 0.866 | 0.748 | 0.002 | — | — | 0.276 ③ | 22 | — |
| | Standard | R | 0.354 | 0.276 | 0.002 | — | — | 0.295 ③ | — | 48 |
| | Jetta | F | 0.980 | 0.870 | 0.002 | — | — | 0.070 ① | 22 | — |
| | GLI | R | 0.472 | 0.390 | 0.002 | — | — | 0.070 ① | — | 35 ② |
| | New Beetle | F | 0.866 | 0.748 | 0.002 | — | — | 0.276 ③ | 22 | — |
| | Standard | R | 0.390 | 0.315 | 0.002 | — | — | 0.275 ③ | 26 | 48 |
| | Rabbit | F | 0.866 | 0.748 | 0.002 | — | — | 0.276 ③ | 22 | — |
| | | R | 0.354 | 0.276 | 0.002 | — | — | 0.295 ③ | — | 48 |
| | Passat | F | 0.980 | 0.870 | 0.002 | — | — | 0.070 ① | 22 | — |
| | | R | 0.472 | 0.390 | 0.002 | — | — | 0.070 ① | — | 35 ② |

① Measurement does not include backing plate.

② Use new bolts.

③ Measurement does not include backing plate.

## SCHEDULED MAINTENANCE INTERVALS
### VOLKSWAGEN—GOLF, GTI, JETTA, NEW BEETLE, RABBIT & PASSAT

| TO BE SERVICED | TYPE OF SERVICE | VEHICLE MILEAGE INTERVAL (x1000) | | | | | | | | | | | | |
|---|---|---|---|---|---|---|---|---|---|---|---|---|---|---|
| | | 5 | 10 | 20 | 30 | 40 | 50 | 60 | 70 | 80 | 90 | 100 | 110 | 120 |
| Engine Oil & Filter - replace | R | ✓ | ✓ | ✓ | ✓ | ✓ | ✓ | ✓ | ✓ | ✓ | ✓ | ✓ | ✓ | ✓ |
| Brake Pad thickness - check | S/I | ✓ | ✓ | ✓ | ✓ | ✓ | ✓ | ✓ | ✓ | ✓ | ✓ | ✓ | ✓ | ✓ |
| Service Interval Display - reset | S/I | ✓ | ✓ | ✓ | ✓ | ✓ | ✓ | ✓ | ✓ | ✓ | ✓ | ✓ | ✓ | ✓ |
| Washer Fluid - level check | S/I | | ✓ | ✓ | ✓ | ✓ | ✓ | ✓ | ✓ | ✓ | ✓ | ✓ | ✓ | ✓ |
| Brake system - inspection | S/I | | ✓ | ✓ | ✓ | ✓ | ✓ | ✓ | ✓ | ✓ | ✓ | ✓ | ✓ | ✓ |
| Rotate wheels | S/I | | ✓ | ✓ | ✓ | ✓ | ✓ | ✓ | ✓ | ✓ | ✓ | ✓ | ✓ | ✓ |
| Auto Shift Lock - inspection | S/I | | ✓ | ✓ | ✓ | ✓ | ✓ | ✓ | ✓ | ✓ | ✓ | ✓ | ✓ | ✓ |
| Engine - leak check | S/I | | | ✓ | | ✓ | | ✓ | | ✓ | | ✓ | | ✓ |
| Engine OBD - check & purge | S/I | | | ✓ | | ✓ | | ✓ | | ✓ | | ✓ | | ✓ |
| Auto Trans - leak check | S/I | | | ✓ | | ✓ | | ✓ | | ✓ | | ✓ | | ✓ |
| Battery - check | S/I | | | ✓ | | ✓ | | ✓ | | ✓ | | ✓ | | ✓ |
| Door Arresters - lubricate | S/I | | | ✓ | | ✓ | | ✓ | | ✓ | | ✓ | | ✓ |
| Sunroof Frame - lubricate | S/I | | | ✓ | | ✓ | | ✓ | | ✓ | | ✓ | | ✓ |
| Cooling System - level check | S/I | | | ✓ | | ✓ | | | | ✓ | | ✓ | | ✓ |
| Exhaust System - inspection | S/I | | | ✓ | | ✓ | | | | ✓ | | ✓ | | ✓ |
| Tires/Spare - inspection | S/I | | | ✓ | | ✓ | | | | ✓ | | ✓ | | ✓ |
| Front Suspension - inspection | S/I | | | ✓ | | ✓ | | | | ✓ | | ✓ | | ✓ |
| Headlights - Check and adjust | S/I | | | ✓ | | ✓ | | | | ✓ | | ✓ | | ✓ |
| Drive Shafts - check boots | S/I | | | ✓ | | ✓ | | | | ✓ | | ✓ | | ✓ |
| Manual Trans - level check | S/I | | | ✓ | | ✓ | | | | ✓ | | ✓ | | ✓ |
| Road Test - general insp. | S/I | | | | ✓ | ✓ | | | | ✓ | | ✓ | | ✓ |
| Cabin Air Filter - replace | R | | | | | ✓ | | | | ✓ | | | | ✓ |
| Air Cleaner (Beetle) - replace | R | | | | | ✓ | | | | ✓ | | | | ✓ |
| Wiper Blades - check | S/I | | | | | ✓ | | | | ✓ | | | | ✓ |
| Interior lighting - check | S/I | | | | | ✓ | | | | ✓ | | | | ✓ |
| Underbody - inspection | S/I | | | | | ✓ | | | | ✓ | | | | ✓ |
| Serpentine Belt - inspection | S/I | | | | | ✓ | | | | ✓ | | | | ✓ |
| Auto Trans - level check | S/I | | | | | ✓ | | | | ✓ | | | | ✓ |
| Final Drive - level check | S/I | | | | | ✓ | | | | ✓ | | | | ✓ |
| Spark Plugs - replace | R | | | | | ✓ | | | | ✓ | | | | ✓ |
| DSG Trans - replace fluid & filter | R | | | | ✓ | | | | | ✓ | | | | ✓ |
| Power Steering - level check | S/I | | | | | ✓ | | | | ✓ | | | | ✓ |
| Air Cleaner - replace | R | | | | | | | ✓ | | | | | | ✓ |
| Check Airbag System | | Every 12 months | | | | | | | | | | | | |
| Change Brake Fluid | | Every 24 months | | | | | | | | | | | | |
| Check Convertible Top Latches | | Every 24 months | | | | | | | | | | | | |
| Replace Tire Filler Bottle | | Every 48 months | | | | | | | | | | | | |
| Replace Tire Pressure Sensors | | Every 72 months | | | | | | | | | | | | |

R: Replace     S/I: Service or Inspect

22205_VWCA_C0015

## PRECAUTIONS

Before servicing any vehicle, please be sure to read all of the following precautions, which deal with personal safety, prevention of component damage, and important points to take into consideration when servicing a motor vehicle:

• Never open, service or drain the radiator or cooling system when the engine is hot; serious burns can occur from the steam and hot coolant.

• Observe all applicable safety precautions when working around fuel. Whenever servicing the fuel system, always work in a well-ventilated area. Do not allow fuel spray or vapors to come in contact with a spark, open flame, or excessive heat (a hot drop light, for example). Keep a dry chemical fire extinguisher near the work area. Always keep fuel in a container specifically designed for fuel storage; also, always properly seal fuel containers to avoid the possibility of fire or explosion. Refer to the additional fuel system precautions later in this section.

• Fuel injection systems often remain pressurized, even after the engine has been turned **OFF**. The fuel system pressure must be relieved before disconnecting any fuel lines. Failure to do so may result in fire and/or personal injury.

• Brake fluid often contains polyglycol ethers and polyglycols. Avoid contact with the eyes and wash your hands thoroughly after handling brake fluid. If you do get brake fluid in your eyes, flush your eyes with clean, running water for 15 minutes. If eye irritation persists, or if you have taken brake fluid internally, IMMEDIATELY seek medical assistance.

• The EPA warns that prolonged contact with used engine oil may cause a number of skin disorders, including cancer. You should make every effort to minimize your exposure to used engine oil. Protective gloves should be worn when changing oil. Wash your hands and any other exposed skin areas as soon as possible after exposure to used engine oil. Soap and water, or waterless hand cleaner should be used.

• All new vehicles are now equipped with an air bag system, often referred to as a Supplemental Restraint System (SRS) or Supplemental Inflatable Restraint (SIR) system. The system must be disabled before performing service on or around system components, steering column, instrument panel components, wiring and sensors. Failure to follow safety and disabling procedures could result in accidental air bag deployment, possible personal injury and unnecessary system repairs.

• Always wear safety goggles when working with, or around, the air bag system. When carrying a non-deployed air bag, be sure the bag and trim cover are pointed away from your body. When placing a non-deployed air bag on a work surface, always face the bag and trim cover upward, away from the surface. This will reduce the motion of the module if it is accidentally deployed. Refer to the additional air bag system precautions later in this section.

• Clean, high quality brake fluid from a sealed container is essential to the safe and proper operation of the brake system. You should always buy the correct type of brake fluid for your vehicle. If the brake fluid becomes contaminated, completely flush the system with new fluid. Never reuse any brake fluid. Any brake fluid that is removed from the system should be discarded. Also, do not allow any brake fluid to come in contact with a painted surface; it will damage the paint.

• Never operate the engine without the proper amount and type of engine oil; doing so WILL result in severe engine damage.

• Timing belt maintenance is extremely important. Many models utilize an interference-type, non-freewheeling engine. If the timing belt breaks, the valves in the cylinder head may strike the pistons, causing potentially serious (also time-consuming and expensive) engine damage. Refer to the maintenance interval charts for the recommended replacement interval for the timing belt, and to the timing belt section for belt replacement and inspection.

• Disconnecting the negative battery cable on some vehicles may interfere with the functions of the on-board computer system(s) and may require the computer to undergo a relearning process once the negative battery cable is reconnected.

• When servicing drum brakes, only disassemble and assemble one side at a time, leaving the remaining side intact for reference.

• Only an MVAC-trained, EPA-certified automotive technician should service the air conditioning system or its components.

## BRAKES

## ANTI-LOCK BRAKE SYSTEM (ABS)

### GENERAL INFORMATION

*PRECAUTIONS*

• Certain components within the ABS system are not intended to be serviced or repaired individually.

• Do not use rubber hoses or other parts not specifically specified for and ABS system. When using repair kits, replace all parts included in the kit. Partial or incorrect repair may lead to functional problems and require the replacement of components.

• Lubricate rubber parts with clean, fresh brake fluid to ease assembly. Do not use shop air to clean parts; damage to rubber components may result.

• Use only DOT 3 brake fluid from an unopened container.

• If any hydraulic component or line is removed or replaced, it may be necessary to bleed the entire system.

• A clean repair area is essential. Always clean the reservoir and cap thoroughly before removing the cap. The slightest amount of dirt in the fluid may plug an orifice and impair the system function. Perform repairs after components have been thoroughly cleaned; use only denatured alcohol to clean components. Do not allow ABS components to come into contact with any substance containing mineral oil; this includes used shop rags.

• The Anti-Lock control unit is a microprocessor similar to other computer units in the vehicle. Ensure that the ignition switch is **OFF** before removing or installing controller harnesses. Avoid static electricity discharge at or near the controller.

• If any arc welding is to be done on the vehicle, the control unit should be unplugged before welding operations begin.

## BRAKES                                    BLEEDING THE BRAKE SYSTEM

### BLEEDING PROCEDURE

*BLEEDING PROCEDURE*

**Without Pressure Bleeder**

1. Before servicing the vehicle, refer to the Precautions Section.
2. Build-up brake pressure in brake system by pumping brake pedal.
3. Connect the hose from bleeder bottle to brake bleeder valve at each wheel in the following sequence:
   a. 1–Right rear caliper
   b. 2–Left rear caliper
   c. 3–Right front caliper
   d. 4–Left front caliper
4. Open bleeder valve.
5. Hold brake pedal down, close bleeder screw.

➡The brake fluid level in the reservoir must not fall below the MIN mark during bleeding.

6. Release brake pedal.
7. Repeat operation until brake fluid flows without air bubbles.

**With Brake Filler and Bleeder Unit VAS 5234**

*Vehicles with ABS Mark 20*

See Figure 1.

➡This procedure requires the use of brake filler and bleeder unit VAS 5234, or equivalent brake bleeding equipment.

➡Bleeding the brake system on vehicles with ABS is the same as for vehicles with a conventional brake system, with the exception of using scan tool function 04 to bleed systems that have been completely empty.

1. Before servicing the vehicle, refer to the Precautions Section.
2. On vehicles with a brake pressure regulator valve on the rear axle, move the regulator lever when bleeding rear brakes.
3. Connect brake filling and bleeding unit.
4. Open bleeder valves in specified sequence and bleed brake calipers in the following sequence:
   a. 1–Right rear caliper
   b. 2–Left rear caliper
   c. 3–Right front caliper
   d. 4–Left front caliper
5. On vehicles with ABS or ABS/EDL, test drive the vehicle and perform at least one brake application sufficient to engage ABS regulation.

*Vehicles with ABS Mark 60*

See Figure 1.

➡This procedure requires the use of brake filler and bleeder unit VAS 5234, or equivalent brake bleeding equipment.

➡Please note the following important points about vehicles with ABS/EDL, ABS/EDL/ASR and ABS/EDL/ASR/ESP:

• When at least one chamber in the brake fluid reservoir is completely empty (e.g. leaks in braking system) the hydraulic unit must be bled with scan tool VAG 1551 in function "basic setting".
• Brakes must be bled in the proper sequence
• Then the hydraulic unit must be bled with scan tool VAG 1551 in function "basic setting".
• After bleeding the brake system, perform a zero compensation of the brake pressure sender.

• Once the bleeding sequence is completed, carry out a test drive and perform at least one ABS application when doing this.

### ✳✳ WARNING

**Make sure that a filling pressure of 1 bar (14.5 psi) is not exceeded, when filling brake fluid using brake filler and bleeder unit VAS 5234. The brake fluid pressure must be reduced to 1 bar (14.5 psi) on brake filler and bleeder unit VAS 5234. Refer to the equipment operating instructions**

### ✳✳ WARNING

**Brake fluid is poisonous. Brake fluid must not come into contact with paintwork as it is very corrosive. Brake fluid is hygroscopic, which means it absorbs moisture from the ambient air and should always be stored in air tight containers. Rinse off any spilled brake fluid using plenty of water.**

1. Connect brake filler and bleeder unit VAS 5234.
2. Open bleeder screws in the following sequence and bleed brake caliper/wheel cylinder:
   a. 1–Left front brake caliper
   b. 2–Right front brake caliper
   c. 3–Left rear wheel cylinder/brake caliper
   d. 4–Right rear wheel cylinder/brake caliper
3. With hose still connected, leave the respective bleeder screw open until brake fluid is free from air bubbles out of the bleeder screw.
4. A test drive must be carried out after bleeding brakes. When doing this an ABS application must be performed at least once!
5. Pre-bleed the brake system when at least one chamber in the brake fluid reservoir is completely empty (e.g. leaks in braking system). Pre-bleeding sequence
   a. 1–Bleed left front and right front brake calipers simultaneously
   b. 2–Bleed left rear and right rear brake calipers simultaneously
   c. With hose still connected, leave the respective bleeder screw open until brake fluid is free from air bubbles flowing out of the bleeder screw.

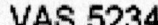
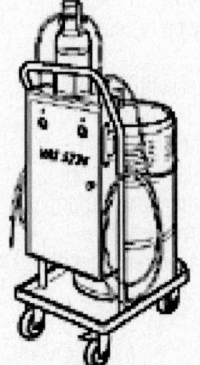

VAS 5234

42075_VWCA_G0124

**Fig. 1 View of the brake filler and bleeder unit needed to bleed the brakes**

**BRAKES**                                    **FRONT DISC BRAKES**

### ✷✷ CAUTION

Dust and dirt accumulating on brake parts during normal use may contain asbestos fibers from production or aftermarket brake linings. Breathing excessive concentrations of asbestos fibers can cause serious bodily harm. Exercise care when servicing brake parts. Do not sand or grind brake lining unless equipment used is designed to contain the dust residue. Do not clean brake parts with compressed air or by dry brushing. Cleaning should be done by dampening the brake components with a fine mist of water, then wiping the brake components clean with a dampened cloth. Dispose of cloth and all residue containing asbestos fibers in an impermeable container with the appropriate label. Follow practices prescribed by the Occupational Safety and Health Administration (OSHA) and the Environmental Protection Agency (EPA) for the handling, processing, and disposing of dust or debris that may contain asbestos fibers.

### BRAKE CALIPER

#### REMOVAL & INSTALLATION

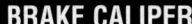

*See Figures 2 and 3.*

1. Before servicing the vehicle, refer to the Precautions Section.
2. Remove the wheels.
3. Loosen the hydraulic line at the caliper, then remove the caliper from the carrier. With guide pin calipers, be sure to hold the pin with a back-up wrench when removing the caliper bolts.
4. Remove the caliper from the hydraulic line.
5. The carrier can be removed by removing the 2 bolts.

#### To install:

6. If removed, install the carrier. On standard brakes, torque the carrier bolts to 52 ft. lbs. (70 Nm). On ABS brakes, torque the carrier bolts to 92 ft. lbs. (125 Nm).
7. Thread the caliper onto the hydraulic line and hand-tighten it. Fit the caliper into place on the carrier.
8. On calipers with guide pins, torque the bolts to 25 ft. lbs. (35 Nm). On calipers with sleeves and bushings, torque the bolts to 18 ft. lbs. (25 Nm).

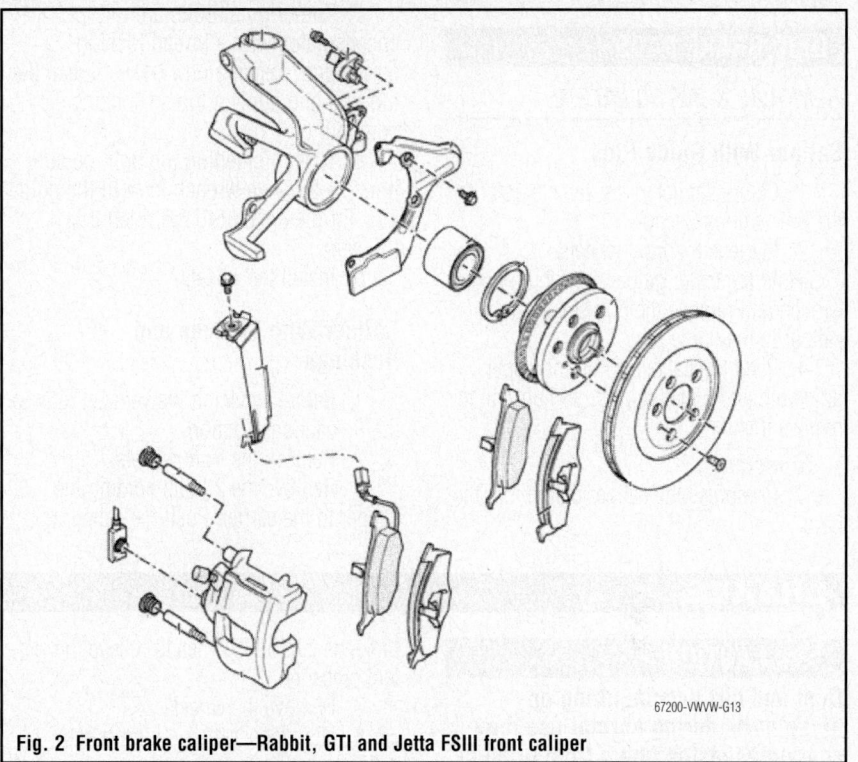

67200-VWVW-G13

**Fig. 2 Front brake caliper—Rabbit, GTI and Jetta FSIII front caliper**

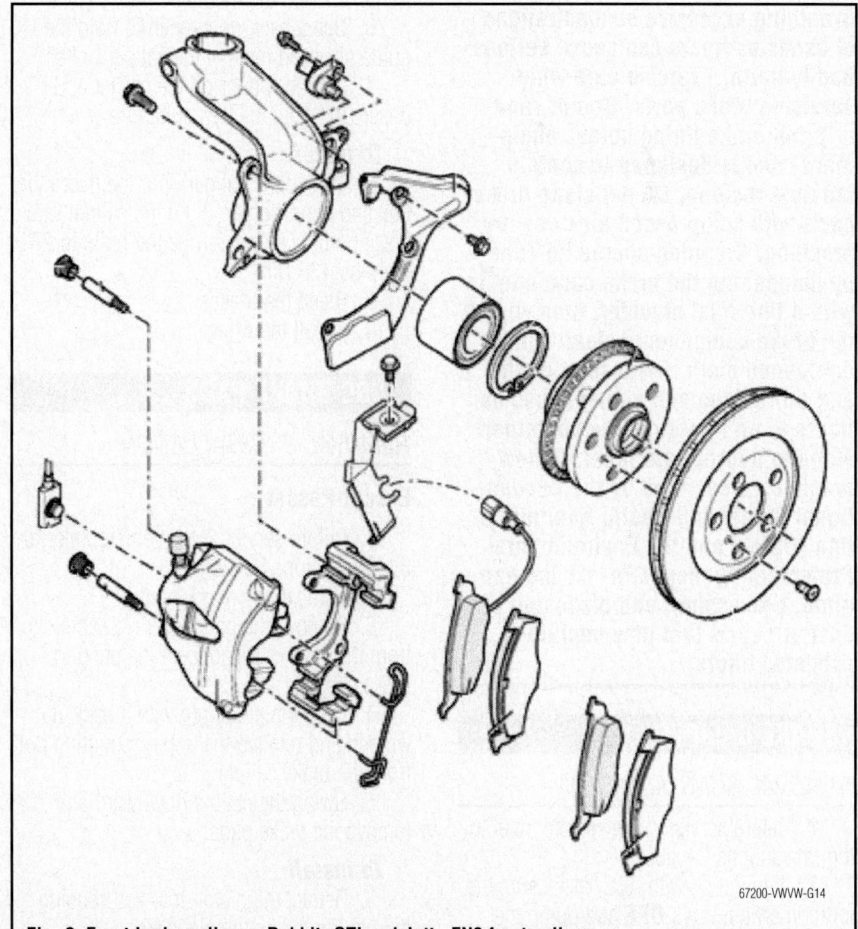

67200-VWVW-G14

**Fig. 3 Front brake caliper—Rabbit, GTI and Jetta FN3 front caliper**

9. Tighten the hydraulic line and bleed the brakes.

## DISC BRAKE PADS

### REMOVAL & INSTALLATION

#### Caliper With Guide Pins

1. Before servicing the vehicle, refer to the Precautions Section.
2. Remove the front wheels.
3. Hold the lower guide pin with an open wrench and remove the bolt securing the caliper to the guide pin.
4. Pivot the caliper up on the upper guide pin and slide the pads straight out to remove them.

**To install:**

5. Compress the caliper piston into the bore.

6. Fit the new pads into the carrier and pivot the caliper into place.
7. The original bolts are micro-encapsulated with a thread locking compound. Install a new bolt or clean the old bolt and apply a thread-locking compound.
8. When tightening the bolt, be sure to use a back-up wrench to hold the guide pin. Torque the bolt to 271 inch lbs. (35 Nm).
9. Install the wheels.

#### Caliper With Sleeves and Bushings

1. Before servicing the vehicle, refer to the Precautions Section.
2. Remove the front wheels.
3. Remove the 2 bolts holding the caliper to the carrier. Push the caliper up

and pivot the bottom of the caliper out of the carrier.
4. Remove the anti-rattle springs and the pads from the carrier and note their location.

**To install:**

5. Fit the anti-rattle springs into place and slide the new pads onto the carrier.
6. Fit the caliper into place at the top and push up so it can be pivoted into place at the bottom. The tabs on the anti-rattle springs should be pushing against the inside of the caliper.
7. Make sure the caliper mounting bolts are clean. Install the bolts and torque them to 18 ft. lbs. (25 Nm).
8. Install the wheels.

## BRAKES

### ❊ CAUTION

**Dust and dirt accumulating on brake parts during normal use may contain asbestos fibers from production or aftermarket brake linings. Breathing excessive concentrations of asbestos fibers can cause serious bodily harm. Exercise care when servicing brake parts. Do not sand or grind brake lining unless equipment used is designed to contain the dust residue. Do not clean brake parts with compressed air or by dry brushing. Cleaning should be done by dampening the brake components with a fine mist of water, then wiping the brake components clean with a dampened cloth. Dispose of cloth and all residue containing asbestos fibers in an impermeable container with the appropriate label. Follow practices prescribed by the Occupational Safety and Health Administration (OSHA) and the Environmental Protection Agency (EPA) for the handling, processing, and disposing of dust or debris that may contain asbestos fibers.**

## BRAKE CALIPER

### REMOVAL & INSTALLATION

1. Before servicing the vehicle, refer to the Precautions Section.
2. If equipped with ABS, make sure the ignition switch stays **OFF** and pump the

brake pedal 25–35 times to relieve the system pressure.
3. Remove the wheels.
4. Disconnect the parking brake cable.
5. Loosen the hydraulic line.
6. Use a back-up wrench to hold the guide pins and remove the caliper bolts.
7. Lift the caliper off the carrier and unscrew it from the hydraulic line.

**To install:**

8. Thread the caliper onto the hydraulic line and hand-tighten it. Fit the caliper into place on the carrier. Torque the bolts to 271 inch lbs. (35 Nm).
9. Bleed the brakes.
10. Install the wheels.

## DISC BRAKE PADS

### REMOVAL & INSTALLATION

#### Except Passat

1. Before servicing the vehicle, refer to the Precautions Section.
2. Remove the rear wheels.
3. Remove the parking brake cable clip from the caliper. Disconnect the parking brake cable.
4. Hold the guide pin with a back-up wrench and remove the upper mounting bolt from the brake caliper.
5. Swing the caliper downward and remove the brake pads.

**To install:**

6. Retract the piston into the housing by rotating the piston clockwise.

## REAR DISC BRAKES

7. Install the new brake pads onto the pad carrier.
8. Install the caliper to the pad carrier using a new self locking bolt or a thread locking compound and torque to 271 inch lbs. (35 Nm).
9. Attach the hand brake cable to the caliper.
10. Check the parking brake operation and adjust the cable if necessary.
11. Install the wheels.

#### Passat

➡ **Rear brake service requires the factory scan tool VAS 5051 or equivalent. Do not disconnect the connectors from the parking brake motors or you will set trouble codes.**

1. With the parking brake in the **Released** position, raise and safely support the vehicle and remove the rear wheels.
2. The pistons of the parking brake must be driven back using VAS 5051. Before doing this, draw off brake fluid from reservoir using a bleeder bottle to avoid overflowing the reservoir.
3. Connect the scan tool, select electromechanical parking brake and function "Moving piston of parking brake motor out and in."
4. Drive pistons back using VAS 5051.
5. Remove bolts from brake caliper, while counter-holding guide pins.
6. Remove brake caliper downward from brake carrier.

➡Resetting the piston with VAS 5051 is often not sufficient but necessary! The pressure nut in the piston is mounted floating, therefore the piston can only be pressed and not pulled back. Only the spindle with the pressure nut is moved back.

**✳✳ WARNING**

The piston must first be reset with VAS 5051.

7. Now press piston all the way back with Piston Resetting Tool T10145. Do not use a clamp on the motor.

8. Secure brake caliper with wire so weight of caliper does not load or damage brake line.

9. Remove brake pads and anti-rattle springs.

*To install:*

10. Thoroughly clean contact surfaces for brake pads at brake carrier, remove corrosion.

11. Insert brake pad retention plates and brake pads in brake carrier.

12. Make sure that the brake pads are located correctly in anti-rattle springs.

13. Secure brake caliper using new self-locking bolts. Torque the bolts to 26 ft. lbs. (35 Nm).

14. Drive the pistons out with VAS 5051, then perform the basic setting function.

15. Install wheels and test drive the vehicle.

## BRAKES

**✳✳ CAUTION**

Dust and dirt accumulating on brake parts during normal use may contain asbestos fibers from production or aftermarket brake linings. Breathing excessive concentrations of asbestos fibers can cause serious bodily harm. Exercise care when servicing brake parts. Do not sand or grind brake lining unless equipment used is designed to contain the dust residue. Do not clean brake parts with compressed air or by dry brushing. Cleaning should be done by dampening the brake components with a fine mist of water, then wiping the brake components clean with a dampened cloth. Dispose of cloth and all residue containing asbestos fibers in an impermeable container with the appropriate label. Follow practices prescribed by the Occupational Safety and Health Administration (OSHA) and the Environmental Protection Agency (EPA) for the handling, processing, and disposing of dust or debris that may contain asbestos fibers.

### BRAKE DRUM

*REMOVAL & INSTALLATION*

1. Before servicing the vehicle, refer to the Precautions Section.

2. Remove the rear wheels.

3. Insert a small pry tool through a wheel bolt hole and push up on the adjusting wedge to slacken the rear brake adjustment.

4. Remove the grease cap, cotter pin, locking ring, axle nut, thrust washer, and outer bearing.

5. Remove the drum.

*To install:*

6. Install the drum. Install the outer bearing, washer, and nut.

7. Install the lock ring and cotter pin. Install the grease cap.

8. Install the wheels.

### BRAKE SHOES

*REMOVAL & INSTALLATION*

1. Before servicing the vehicle, refer to the Precautions Section.

2. Remove the rear wheels.

3. Remove the rear brake drum.

4. Remove the brake shoe hold-down spring retainers.

5. Remove the shoes from the back plate by pulling first 1 shoe, then the other against the upper spring and from its wheel cylinder slot. Detach the parking brake cable from the brake lever. Remove the brake shoe assembly from the vehicle.

6. Clamp the pushrod that holds the shoes apart at the top in a vise and begin removing the springs. Start with the lower return spring, adjusting wedge spring,

## REAR DRUM BRAKES

upper return spring and then the tensioning spring and adjusting wedge.

7. On most vehicles, the parking brake lever must be removed from the old shoes and reused.

*To install:*

8. Clean the back plate and lubricate the shoe contact points with a suitable brake lubricant.

9. With the push rod clamped in a vise, attach the front brake shoe and tensioning spring.

10. Insert the adjusting wedge between the front shoe and pushrod so its lug is pointing toward the backing plate.

11. Remove the parking brake lever from the old shoe and attach it onto the new rear brake shoe.

12. Put the rear brake shoe and parking brake lever assembly onto the pushrod and hook up the spring.

13. Connect the parking brake cable to the lever and place the whole assembly onto the backing plate.

14. Install the hold-down springs.

15. Install the upper and lower return springs.

16. Install the adjusting wedge spring.

17. Center the brake shoes on the backing plate, making sure the adjusting wedge is fully released (all the way up) before installing the drum.

18. Install the drum and wheel assembly.

19. Apply the brake pedal a few times to bring the brake shoe into adjustment.

## CHASSIS ELECTRICAL | AIR BAG (SUPPLEMENTAL RESTRAINT SYSTEM)

### GENERAL INFORMATION

#### ※※ CAUTION

**These vehicles are equipped with an air bag system. The system must be disarmed before performing service on, or around, system components, the steering column, instrument panel components, wiring and sensors. Failure to follow the safety precautions and the disarming procedure could result in accidental air bag deployment, possible injury and unnecessary system repairs.**

#### SERVICE PRECAUTIONS

Disconnect and isolate the battery negative cable before beginning any airbag system component diagnosis, testing, removal, or installation procedures. Allow system capacitor to discharge for two minutes before beginning any component service. This will disable the airbag system. Failure to disable the airbag system may result in accidental airbag deployment, personal injury, or death.

Do not place an intact undeployed airbag face down on a solid surface. The airbag will propel into the air if accidentally deployed and may result in personal injury or death.

When carrying or handling an undeployed airbag, the trim side (face) of the airbag should be pointing towards the body to minimize possibility of injury if accidental deployment occurs. Failure to do this may result in personal injury or death.

Replace airbag system components with OEM replacement parts. Substitute parts may appear interchangeable, but internal differences may result in inferior occupant protection. Failure to do so may result in occupant personal injury or death.

Wear safety glasses, rubber gloves, and long sleeved clothing when cleaning powder residue from vehicle after an airbag deployment. Powder residue emitted from a deployed airbag can cause skin irritation. Flush affected area with cool water if irritation is experienced. If nasal or throat irritation is experienced, exit the vehicle for fresh air until the irritation ceases. If irritation continues, see a physician.

Do not use a replacement airbag that is not in the original packaging. This may result in improper deployment, personal injury, or death.

The factory installed fasteners, screws and bolts used to fasten airbag components have a special coating and are specifically designed for the airbag system. Do not use substitute fasteners. Use only original equipment fasteners listed in the parts catalog when fastener replacement is required.

During, and following, any child restraint anchor service, due to impact event or vehicle repair, carefully inspect all mounting hardware, tether straps, and anchors for proper installation, operation, or damage. If a child restraint anchor is found damaged in any way, the anchor must be replaced. Failure to do this may result in personal injury or death.

Deployed and non-deployed airbags may or may not have live pyrotechnic material within the airbag inflator.

Do not dispose of driver/passenger/curtain airbags or seat belt tensioners unless you are sure of complete deployment. Refer to the Hazardous Substance Control System for proper disposal.

Dispose of deployed airbags and tensioners consistent with state, provincial, local, and federal regulations.

After any airbag component testing or service, do not connect the battery negative cable. Personal injury or death may result if the system test is not performed first.

If the vehicle is equipped with the Occupant Classification System (OCS), do not connect the battery negative cable before performing the OCS Verification Test using the scan tool and the appropriate diagnostic information. Personal injury or death may result if the system test is not performed properly.

Never replace both the Occupant Restraint Controller (ORC) and the Occupant Classification Module (OCM) at the same time. If both require replacement, replace one, then perform the Airbag System test before replacing the other.

Both the ORC and the OCM store Occupant Classification System (OCS) calibration data, which they transfer to one another when one of them is replaced. If both are replaced at the same time, an irreversible fault will be set in both modules and the OCS may malfunction and cause personal injury or death.

If equipped with OCS, the Seat Weight Sensor is a sensitive, calibrated unit and must be handled carefully. Do not drop or handle roughly. If dropped or damaged, replace with another sensor. Failure to do so may result in occupant injury or death.

If equipped with OCS, the front passenger seat must be handled carefully as well. When removing the seat, be careful when setting on floor not to drop. If dropped, the sensor may be inoperative, could result in occupant injury, or possibly death.

If equipped with OCS, when the passenger front seat is on the floor, no one should sit in the front passenger seat. This uneven force may damage the sensing ability of the seat weight sensors. If sat on and damaged, the sensor may be inoperative, could result in occupant injury, or possibly death.

#### DISARMING THE SYSTEM

Disconnect the negative battery cable. No waiting time is necessary before beginning work.

#### ARMING THE SYSTEM

After reconnecting the airbag wiring, turn the ignition switch **ON** and make sure no one is inside the vehicle, then connect the battery cable.

#### CLOCKSPRING CENTERING

*See Figures 4 through 6.*

➡**There are two different manufacturers of the spiral spring. On each one, the color band appears in the window when the spring is centered.**

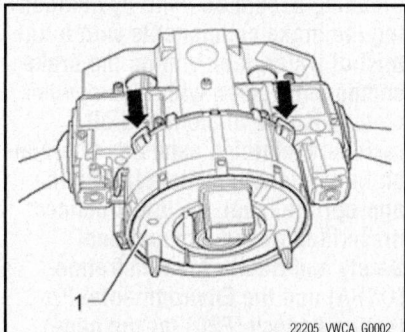

22205_VWCA_G0002

**Fig. 4 These clips hold the clockspring (spiral spring) in place—2006–08 Rabbit, GTI**

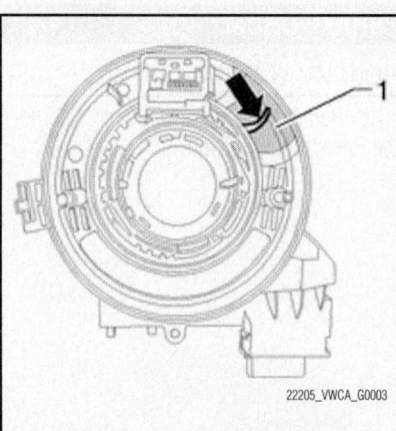

22205_VWCA_G0003

**Fig. 5 The black color band appears in the window when the spiral spring is centered**

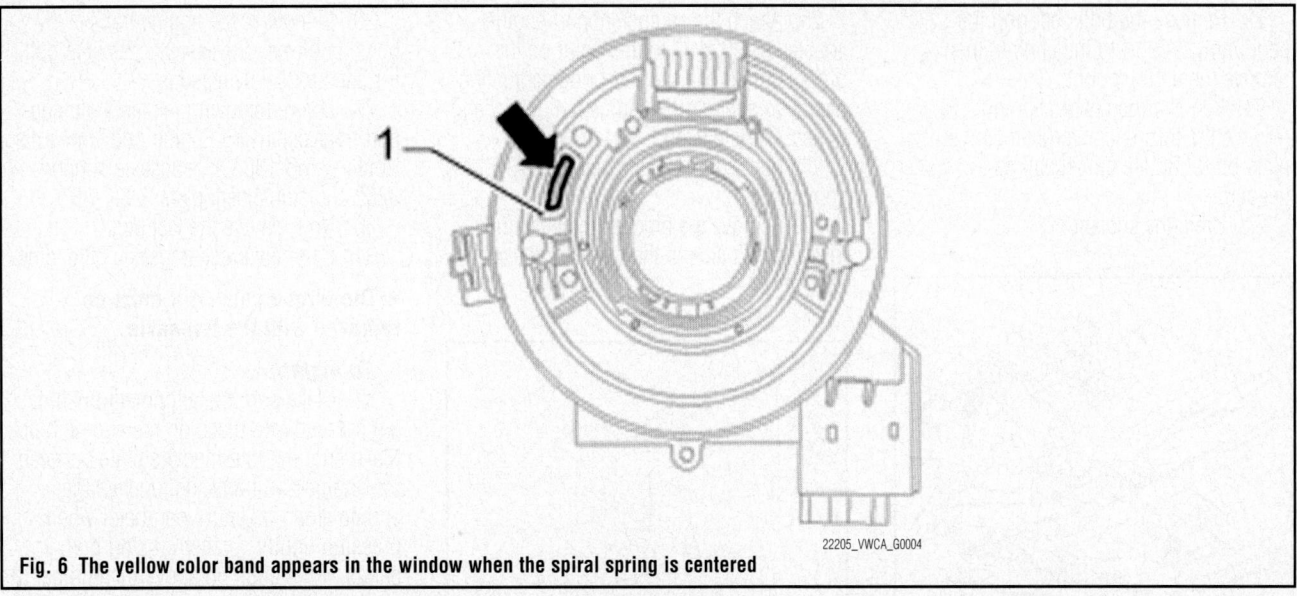

Fig. 6 The yellow color band appears in the window when the spiral spring is centered

## DRIVETRAIN

### AUTOMATIC TRANSAXLE ASSEMBLY

*REMOVAL & INSTALLATION*

**Rabbit, Jetta and GTI**

*See Figures 7 through 14.*

➡ **The transaxle can be lowered from the vehicle without removing the engine. Several new nuts and bolts are required.**

1. Before servicing the vehicle, refer to the Precautions Section.
2. Loosen the front axle bolts ¼ with the vehicle on the ground. Do not loosen any more or roll the vehicle or the front wheel bearings will be damaged.
3. Raise and safely support the vehicle and remove the front wheels.

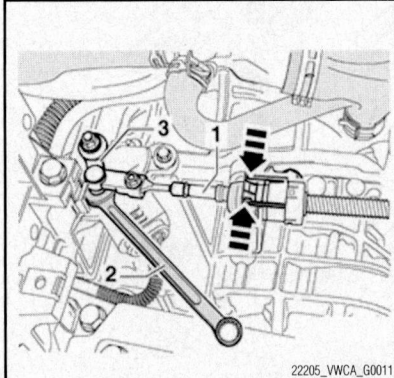

Fig. 7 Squeeze the clips to pull the cable from the bracket, then pry the cable end from the ball

4. Remove engine cover/air filter assembly.
5. Remove intake hose from throttle valve unit.
6. Remove battery and battery housing.
7. Disengage selector lever cable from bracket. Handle with care and make sure not to kink cable. Do not use pliers or the retainer straps may break off.
8. Carefully pry the cable end from the ball.
9. Disconnect wiring from transaxle and starter.
10. Remove wiring harness retainer from starter bolt and remove upper starter bolt.
11. Remove upper engine/transaxle bolts.
12. To install the engine support, disconnect lines (1) from transport strap (3). Remove transport tab from engine and pull out of eye (2). Insert a shackle (10–222 A /12) in this eye.
13. Install engine support bridge 10–222 A. Extend right hook 10–222 A /10 with

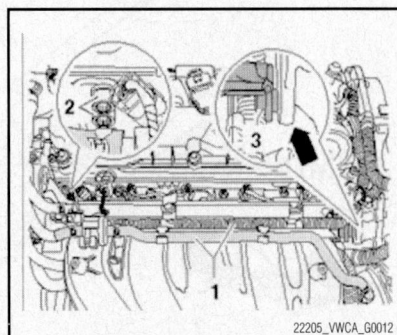

Fig. 8 Attachment points (1 and 3) for engine support

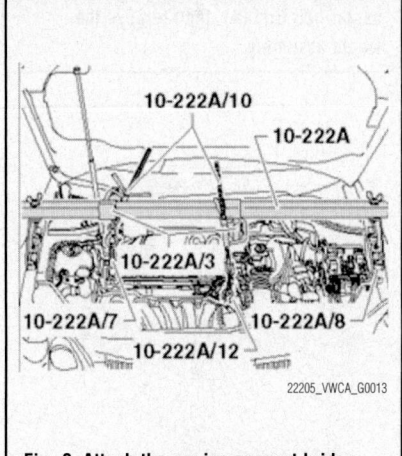

Fig. 9 Attach the engine support bridge

adapter 10–222 A /7. Hook points downward and will be engaged on engine block later.

14. Remove the lower engine cover and left lower part of wheel housing liner.
15. Remove heat shield over right half-shaft.
16. Remove the nuts that secure the ball joints to the lower control arms.
17. Carefully pry both halfshafts out of transaxle. Remove left shaft.
18. Move right shaft upward as far as possible and secure in this position. Do not damage protective coating on shaft.
19. Unplug the connectors from the transaxle.
20. Pull out connector under starter and disconnect.
21. Remove bracket from lower starter bolt. Remove lower starter bolt and remove starter.

22. Remove the bolt securing the pendulum support to the engine, then remove the entire mount.

23. Remove the cover cap and remove the torque converter-to-drive plate bolts. Rotate crankshaft as needed.

24. Drain the coolant.

**Fig. 10 Remove the pendulum support center bolt first (A), then remove the whole assembly**

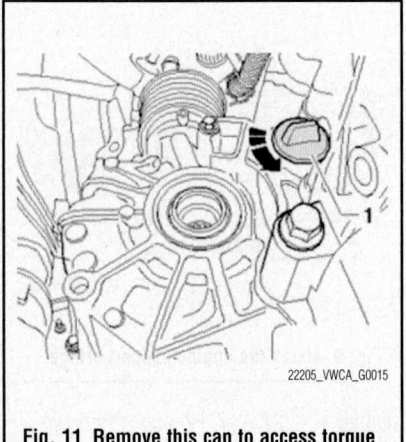

**Fig. 11 Remove this cap to access torque converter bolts**

**Fig. 12 Support engine with Adapter 10-222 A/7 at this point on the block**

25. Attach the engine support engine adapter 10-222 A/7 to this point on the block. Tighten the support bridge adjusting handle to support the engine but do not add unnecessary tension.

26. Remove coolant hoses from ATF cooler.

27. Remove the bracket, then carefully lower the left side of the engine/transaxle assembly.

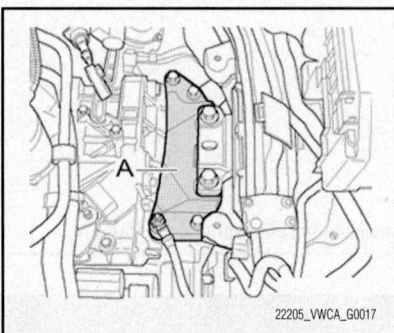

**Fig. 13 Remove this bracket, then carefully lower the left side of the engine/transaxle assembly**

28. Remove lower engine/transaxle bolts, but leave one easily accessible bolt installed for safety reasons.

29. Before removing the last bolt, support transaxle using Engine And Transaxle Holder V.A.G 1383 A , transaxle support 3282 and adjustment plate 3282 /36 .

30. Now remove the last bolt.

31. Carefully press transaxle off engine.

➡ **The torque converter must be removed with the transaxle.**

*To install:*

32. Make sure the alignment bushings are pressed into place on the engine block. Make sure the intermediate plate between the engine and transaxle touches the engine along its entire perimeter when pressing lightly against it using both hands. If it stands off above the engine, it is not positioned correctly behind the drive plate.

33. Carefully fit transaxle to engine and install the bolts hand tight. Torque the 12 mm bolts to 60 ft. lbs. (80 Nm) and the 10 mm bolts to 30 ft. lbs. (40 Nm).

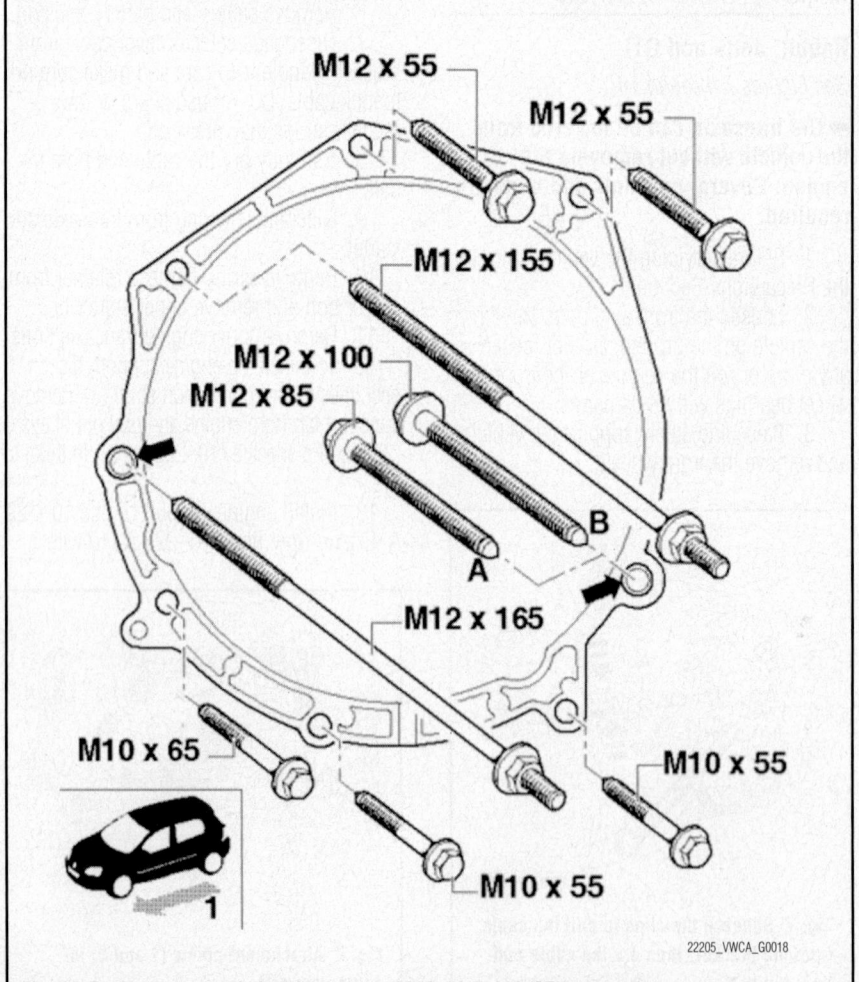

**Fig. 14 Engine-to-transaxle bolt locations—2006–08 Rabbit/GTI automatic**

34. Position the engine/transaxle assembly and begin installing the mounts and brackets. Install each bolt only hand tight until all are installed.
- Torque the two pendulum support bracket bolts to 30 ft. lbs. (40 Nm) plus 90°. Install a new pendulum support center bolt and torque it to 75 ft. lbs. (100 Nm) plus 90°.
- Install four new mount-to-transaxle bolts and torque to 30 ft. lbs. (40 Nm) plus 90°.
- Install two new mount-to-body bolts and torque to 45 ft. lbs. (60 Nm) plus 90°.

35. Remove the engine support bridge and adapter brackets.

36. Install the torque converter-to-drive plate bolts and torque to 45 ft. lbs. (60 Nm). Install the cap.

37. Connect ATF cooler hoses.

38. Install the starter and brackets.

39. Reconnect starter and transaxle wiring.

40. Install the halfshafts. Do not tighten the hub bolts yet.

41. Install new nuts that secure the ball joints to the lower control arms and torque to 45 ft. lbs. (60 Nm).

42. Install the halfshaft heat shield.

43. Install remaining components and connect all wiring, hoses and control cables. The shift cable will be adjusted later.

### ✳✳ WARNING

**Do not roll vehicle before tightening the front hub bolt or the wheel bearings will be damaged.**

44. Install the wheels and lower the vehicle to the ground.

45. Install new axle hub bolts and have a helper hold the brake pedal to tighten them.
- With 6-point bolts, torque to 150 ft. lbs. (200 Nm) plus 180°
- With 12-point bolts, torque to 50 ft. lbs. (70 Nm) plus 90°

46. To adjust the shift cable
- Place gear selector lever in Park.
- Loosen the adjuster bolt at the transaxle end of the cable.
- Lightly tap the shift knob forward and backward. Make sure to keep knob in Park.
- Make sure the transaxle really is in Park. The vehicle should not roll.
- Tighten the bolt.

47. Install the battery and refill the transaxle with fluid.

### Passat

*See Figures 15 and 16.*

➡**The transmission is removed downward without the engine. Two people are required for this procedure.**

1. Raise and safely support the vehicle.

2. Disconnect negative battery cable with ignition switched **OFF**.

3. Remove the air filter.

4. Disconnect the shift cable from the transaxle.

➡**Handle the cable with care, making sure not to kink the cable. Do not use pliers, otherwise retainer straps may break off.**

5. Remove the cable out of the auxiliary bracket on the transmission.

6. Disconnect the starter wiring.

7. Disconnect the wiring form the transaxle.

8. Remove the coolant lines from the engine and disengage the wiring harness.

9. Install the engine support bridge 10–222 A. Support the engine/transaxle. Do not lift yet.

10. Remove the transaxle mount bracket bolts.

11. Install the hose clamps 3094 onto the hoses and disconnect the hoses from the ATF cooler.

12. Remove both upper engine/transmission bolts.

13. Press the pedal brake in order to remove the left halfshaft hub bolt (second technician required).

➡**After this, the vehicle must not contact the ground again or the wheel bearing will be damaged.**

14. Remove engine undercover.

15. Remove the right halfshaft heat shield from the engine.

16. Remove the left lower part of the wheel housing liner.

17. Remove the pendulum support (transaxle mount underneath).

**Fig. 15 Pendulum support**

18. Remove the nuts to disengage the ball joints from both lower control arms.

19. Separate both halfshafts from the transaxle. Remove the left one.

20. Secure the right halfshaft up as high as possible on the strut.

21. Remove the lower engine/transaxle bolts. Start with two the lower bolts and keep an easily accessible bolt installed for safety reasons.

22. Remove the six torque converter nuts using socket wrench insert V/175 . While doing so, counter hold at starter ring gear using screwdriver.

23. Clamp off the ATF lines.

24. Remove the bolt and pull the ATF lines off of the transaxle. Plug the lines using clean plugs.

25. Align the transmission support 3282 with adjustment plate 3282/56.

26. Support the transaxle with the engine and transmission jack V.A.G 1383 A, transmission support 3282, do not raise.

27. Remove the last bolt.

28. Remove the transaxle from the engine.

➡**The torque converter must be removed with transaxle.**

29. Remember that the following components must be reassembled when installing new a transmission:
- Starter with bracket for wiring harness
- Transaxle mount bracket
- Brackets for shift lever cable

*To install:*

30. Make sure the alignment bushings are pressed into place on the engine block.

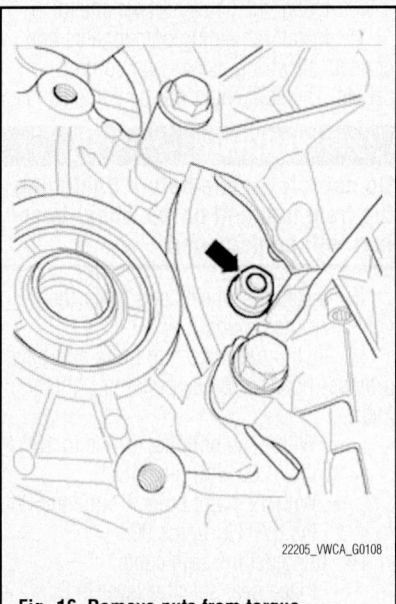

22205_VWCA_G0108

**Fig. 16 Remove nuts from torque converter—Passat**

Make sure the intermediate plate between the engine and transaxle touches the engine along its entire perimeter when pressing lightly against it using both hands. If it stands off above the engine, it is not positioned correctly behind the drive plate.

31. Carefully fit transaxle to engine and install the bolts hand tight. Torque the 12 mm bolts to 60 ft. lbs. (80 Nm) and the 10 mm bolts to 30 ft. lbs. (40 Nm).

32. Position the engine/transaxle assembly and begin installing the mounts and brackets. Install each bolt only hand tight until all are installed.

- Torque the two pendulum support bracket bolts to 30 ft. lbs. (40 Nm) plus 90°. Install a new pendulum support center bolt and torque it to 75 ft. lbs. (100 Nm) plus 90°.
- Install four new mount-to-transaxle bolts and torque to 30 ft. lbs. (40 Nm) plus 90°.
- Install two new mount-to-body bolts and torque to 45 ft. lbs. (60 Nm) plus 90°.

33. Remove the engine support bridge and adapter brackets.

34. Install the torque converter-to-drive plate bolts and torque to 45 ft. lbs. (60 Nm). Install the cap.

35. Connect ATF cooler hoses.

36. Install the starter and brackets.

37. Reconnect starter and transaxle wiring.

38. Install the halfshafts. Do not tighten the hub bolts yet.

39. Install new nuts that secure the ball joints to the lower control arms and torque to 45 ft. lbs. (60 Nm).

40. Install the halfshaft heat shield.

41. Install remaining components and connect all wiring, hoses and control cables. The shift cable will be adjusted later.

## ✳✳ WARNING

**Do not roll vehicle before tightening the front hub bolt or the wheel bearings will be damaged.**

42. Install the wheels and lower the vehicle to the ground.

43. Install new axle hub bolts and have a helper hold the brake pedal to tighten them.

- With 6-point bolts, torque to 150 ft. lbs. (200 Nm) plus 180°
- With 12-point bolts, torque to 50 ft. lbs. (70 Nm) plus 90°

44. To adjust the shift cable

- Place gear selector lever in Park.
- Loosen the adjuster bolt at the transaxle end of the cable.

- Lightly tap the shift knob forward and backward. Make sure to keep knob in Park.
- Make sure the transaxle really is in Park. The vehicle should not roll.
- Tighten the bolt.

45. Install the battery and refill the transaxle with fluid.

## MANUAL TRANSAXLE ASSEMBLY

### REMOVAL & INSTALLATION

### Rabbit, Jetta and GTI
*See Figures 17 through 19.*

➡**The transaxle can be lowered from the vehicle without removing the engine.**

1. Before servicing the vehicle, refer to the Precautions Section.

2. Do not raise the vehicle yet. If equipped with a coded radio, obtain the anti-theft code.

3. Disconnect the negative battery cable with ignition switched **OFF**.

4. Remove the engine cover/air filter housing.

5. Remove the battery.

6. To disconnect the shift cable, remove the lock washer and remove the cable from the pin.

7. To disconnect the relay lever cable:

- If equipped with a metal relay lever, remove the selector cable lock washer and removed the relay cable from the pin.
- If equipped with a plastic relay lever, pull securing mechanism forward to its stop (arrow 1), unlock by twisting the cable (arrow 2), then press the relay lever forward (arrow 3). Press the tab or remove the clip to slide the lever off.

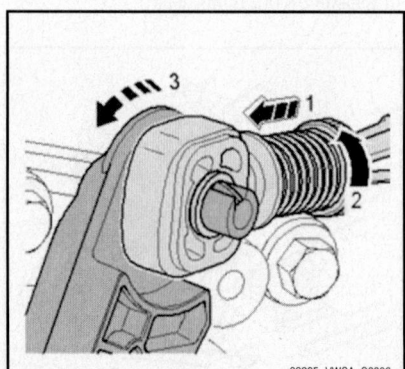

22205_VWCA_G0006

**Fig. 17 Disconnecting the relay lever cable with a plastic relay lever**

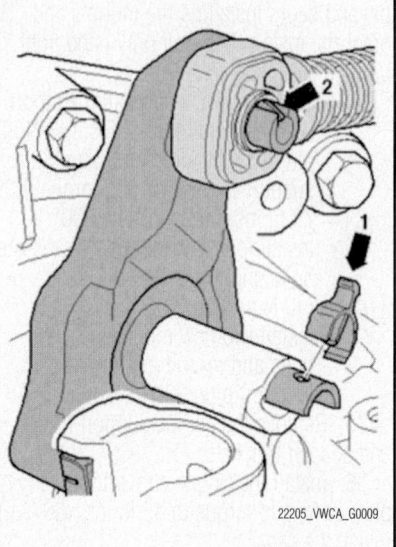

22205_VWCA_G0009

**Fig. 18 Press the tab or remove the clip to slide the cable or relay lever off the pin**

8. Remove cable mounting bracket from transaxle and tie it up out of the way.

9. Remove the transaxle support near the clutch slave cylinder.

10. If the clutch slave cylinder is accessible, remove the bolts to remove the slave cylinder without disconnecting the fluid line. Secure it out of the way.

11. If the slave cylinder is not accessible, clamp the fluid supply line and disconnect it.

12. Remove the ground cable at the starter.

13. Disconnect the wiring from the transaxle and starter.

14. Remove the upper engine/transaxle bolts and the upper bolt on starter.

15. Remove filler pieces from upper edge of left and right fenders.

16. Disconnect any hoses and cables in the area of the lifting eyes on the engine.

17. Position engine support bridge 10-222 A with the engine support feet in front of hood gas struts. Slightly pre-tension engine/transaxle assembly on spindle.

18. Raise and safely support the vehicle and remove the front wheels.

19. Remove the engine undercover and the left inner fender.

20. Separate the exhaust system and remove the exhaust pipe bracket from the front subframe.

21. If there is a CV-joint shield on the engine, remove it now.

22. Disconnect the halfshaft flanges from the transaxle and secure the halfshafts out of the way with wire. Do not let them hang from the transaxle.

23. Remove the pendulum support (rear engine/transaxle mount).

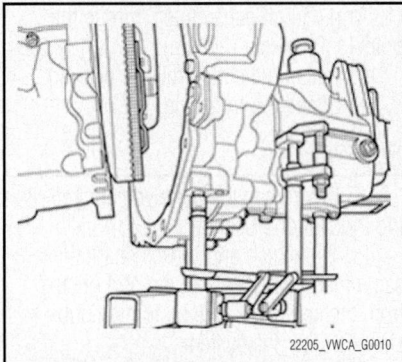

**Fig. 19 Transaxle properly secured to the support bracket and transaxle jack**

24. Remove the upper transaxle mount.

25. Make sure all necessary wires, cables, brackets and hoses are disconnected and out of the way and carefully lower the support bridge to tilt the transaxle out below the vehicle.

26. Remove the small flywheel cover, if equipped.

27. Remove the exhaust system bracket.

28. Remove the starter.

29. Secure the transaxle to a transaxle jack (support bracket 3282 and transaxle holder V.A.G 1383A ). Make sure it's secured firmly on the jack.

30. Remove the remaining bolts and carefully separate the transaxle from the engine.

31. Lower the transaxle out of the vehicle.

**To install:**

32. Carefully fit the transaxle into place on the engine. Make sure the transaxle input shaft fits properly into the clutch.

### ✳✳ WARNING

**Do not use the engine/transaxle bolts to draw the two together.**

33. Install the lower engine/transaxle bolts and torque as follows:
   • M6 bolts–7 ft. lbs. (10 Nm)
   • M7 bolts–11 ft. lbs. (15 Nm)
   • M8 bolts–18 ft. lbs. (25 Nm)
   • M10 bolts–30 ft. lbs. (40 Nm)
   • M12 bolts–45 ft. lbs. (60 Nm)

34. Raise the engine support bridge enough to install the engine/transaxle mounts.

35. Remove the transaxle jack and install the starter, exhaust system brace and the small flywheel cover, if equipped.

36. Attach the halfshafts and the CV-joint shield.

37. Connect the exhaust system and install the exhaust pipe bracket.

38. Install the starter.

39. Reconnect the shift linkage.

40. Reconnect wires, hoses and cables. Install the clutch slave cylinder.

41. Install all remaining components and install the battery.

### Passat

*See Figures 20 through 23.*

➡ **The transaxle can be lowered from the vehicle without removing the engine.**

1. Before servicing the vehicle, refer to the Precautions Section.

2. Do not raise the vehicle yet. If equipped with a coded radio, obtain the anti-theft code.

3. Disconnect the negative battery cable with ignition switched **OFF**.

4. Remove the engine cover/air filter housing.

5. Remove the battery.

6. To disconnect the shift cable, remove the lock washer and remove the cable from the pin.

7. To disconnect the relay lever cable:
   • If equipped with a metal relay lever, remove the selector cable lock washer and removed the relay cable from the pin.
   • If equipped with a plastic relay lever, pull securing mechanism forward to its stop (arrow 1), unlock by twisting the cable (arrow 2), then press the relay lever forward (arrow 3). Press the tab or remove the clip to slide the lever off.

8. Remove cable mounting bracket from transaxle and tie it up out of the way.

9. Remove the transaxle support near the clutch slave cylinder.

10. If the clutch slave cylinder is accessible, remove the bolts to remove the slave cylinder without disconnecting the fluid line. Secure it out of the way.

**Fig. 20 Disconnecting the relay lever cable with a plastic relay lever**

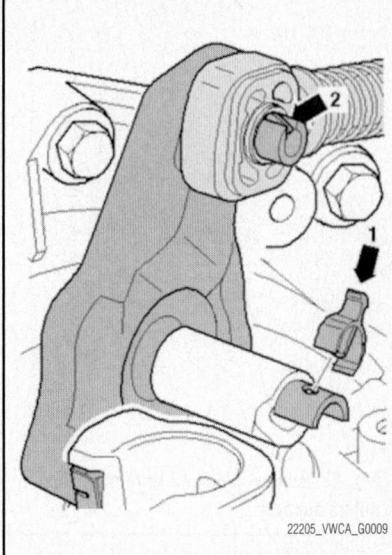

**Fig. 21 Press the tab or remove the clip to slide the cable or relay lever off the pin**

11. If the slave cylinder is not accessible, clamp the fluid supply line and disconnect it.

12. Remove the ground cable at the starter.

13. Disconnect the wiring from the transaxle and starter.

14. Remove the upper engine/transaxle bolts and the upper bolt on starter.

15. Disconnect any hoses and cables in the area of the lifting eyes on the engine.

16. Position engine support bridge 10-222 A with the engine support feet in front of hood gas struts. Slightly pre-tension engine/transaxle assembly on spindle.

17. Raise and safely support the vehicle and remove the front wheels.

18. Remove the engine undercover and the left inner fender.

19. Remove charge air tube.

20. Separate the exhaust system and remove the exhaust pipe bracket from the front subframe.

21. If there is a CV-joint shield on the engine, remove it now.

22. Disconnect the halfshaft flanges from the transaxle and secure the halfshafts out of the way with wire. Do not let them hang from the transaxle.

23. Remove the subframe:
   • Remove heat shield from subframe.
   • Remove coupling rod from stabilizer bar.
   • Remove pendulum support from transaxle.
   • Remove the three bolts securing the ball joints to the control arm and
   • Remove bolts 3 and 6.

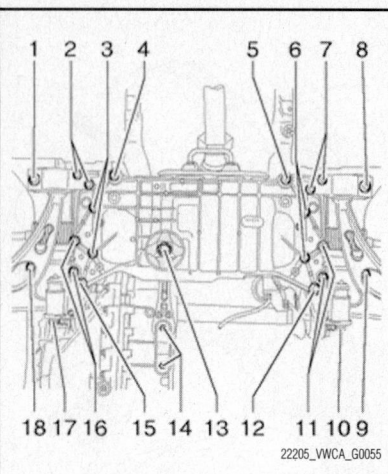

**Fig. 22 Remove bolts 13 to remove pendulum support**

- Remove bolts 11 and 14.
- remove bolts 4 and 5.
- Lower subframe using engine/transmission jack V.A.G 1383 A .
- Secure steering gear to body.

24. Remove the upper transaxle mount.

25. Make sure all necessary wires, cables, brackets and hoses are disconnected and out of the way and carefully lower the support bridge to tilt the transaxle out below the vehicle.

26. Remove the small flywheel cover, if equipped.

27. Remove the exhaust system bracket.

28. Remove the starter.

29. Secure the transaxle to a transaxle jack (support bracket 3282 and transaxle holder V.A.G 1383A). Make sure it's secured firmly on the jack.

30. Remove the remaining bolts and carefully separate the transaxle from the engine.

31. Lower the transaxle out of the vehicle.

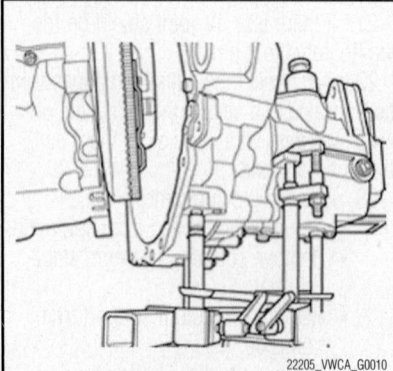

**Fig. 23 Transaxle properly secured to the support bracket and transaxle jack**

### To install:

32. Carefully fit the transaxle into place on the engine. Make sure the transaxle input shaft fits properly into the clutch.

### ✳✳ WARNING

**Do not use the engine/transaxle bolts to draw the two together.**

33. Install the lower engine/transaxle bolts and torque as follows:
- M6 bolts–7 ft. lbs. (10 Nm)
- M7 bolts–11 ft. lbs. (15 Nm)
- M8 bolts–18 ft. lbs. (25 Nm)
- M10 bolts–30 ft. lbs. (40 Nm)
- M12 bolts–45 ft. lbs. (60 Nm)

34. Raise the engine support bridge enough to install the engine/transaxle mounts.

35. Remove the transaxle jack and install the starter, exhaust system brace and the small flywheel cover, if equipped.

36. Attach the halfshafts and the CV-joint shield.

37. Connect the exhaust system and install the exhaust pipe bracket.

38. Install the starter.

39. Reconnect the shift linkage.

40. Reconnect wires, hoses and cables. Install the clutch slave cylinder.

41. Install all remaining components and install the battery.

### CLUTCH

#### REMOVAL & INSTALLATION

1. Before servicing the vehicle, refer to the Precautions Section.

2. Remove the engine and separate it from the transaxle.

3. Matchmark the flywheel and pressure plate if the pressure plate is going to be reused.

4. Gradually loosen the pressure plate bolts 1–2 turns at a time in a crisscross pattern to prevent distortion.

5. Remove the pressure plate and disc.

6. Check the clutch disc for uneven or excessive lining wear. Examine the pressure plate for cracking, scorching or scoring. Replace any questionable components.

### To install:

7. Install the clutch disc with the word "Getriebeseite" (transaxle side) faces towards the transaxle.

8. Install the pressure plate and use an alignment tool to keep the clutch disc centered.

9. Gradually tighten the pressure plate-to-flywheel bolts in a crisscross pattern. On a LuK clutch, torque the bolts to 10 ft. lbs.

(13 Nm). On a Sachs clutch, torque the bolts to 15 ft. lbs. (20 Nm).

10. Install the clutch release bearing.

11. Install the transaxle.

#### BLEEDING

1. Before servicing the vehicle, refer to the Precautions Section.

2. The clutch and brakes share the same reservoir. Clean all dirt and grease from the cap to be sure no foreign substances enter the system.

3. Remove the cap and diaphragm and fill the reservoir to the top with the approved DOT 3 or 4 brake fluid. Fully loosen the bleed screw which is in the slave cylinder body next to the inlet connection.

4. At this point bubbles of air will appear at the bleed screw outlet. When the slave cylinder is full and a steady stream of fluid comes out of the slave cylinder bleeder, tighten the bleed screw.

5. Refill the reservoir and cap it. Exert a light load of about 20 lbs. (9 kg) to the slave cylinder piston by pushing the release lever towards the cylinder and loosen the bleed screw. Maintain a constant light load; fluid and any air that is left will be expelled through the bleed port. Tighten the bleed screw when a steady flow of fluid with no air is being expelled.

6. Fill the reservoir fluid level back to normal capacity, if necessary repeat Step 4.

7. Exert a light load to the release lever but do not open the bleeder screw as the piston in the slave cylinder will move slowly down the bore. Repeat this operation 2–3 times; the fluid movement will force any air left in the system into the reservoir. The hydraulic system should now be fully bled.

8. Check the operation of the clutch hydraulic system and repeat this procedure, if necessary.

### FRONT HALFSHAFT

#### REMOVAL & INSTALLATION

### ✳✳ CAUTION

**When loosening or tightening axle bolts, be sure the vehicle is on the ground. Axle nut torque is high enough that attempting to loosen it may cause the vehicle to fall.**

1. Before servicing the vehicle, refer to the Precautions Section.

2. With the vehicle on the ground, loosen the front axle bolt or nut.

3. Raise and safely support the vehicle and remove the front wheels.

4. Remove the socket head bolts retaining the halfshaft to the transaxle flange.

5. Remove the transaxle side of the halfshaft from the drive flange and secure it out of the way. Do not let it hang unsupported.

6. Remove the three nuts to detach the ball joint from the control arm.

7. Push the halfshaft out of the hub. A wheel puller may be required.

***To install:***

8. Fit the halfshaft to the drive flange and install the bolt. Do not tighten it yet.

9. Fit ball joint into control arm. Install new self-locking nuts and torque nut to 45 ft. lbs. (60 Nm)

10. Install the wheel and hold it to keep the axle from turning. Torque the inner axle bolts to 33 ft. lbs. (45 Nm).

11. With the vehicle on the ground, tighten the axle nut as follows:

- Rabbit, GTI and Jetta with 6-point bolts, torque to 150 ft. lbs. (200 Nm) plus 180°
- Rabbit, GTI and Jetta with 12-point bolts, torque to 50 ft. lbs. (70 Nm) plus 90°

12. Check and adjust the front wheel alignment.

## REAR AXLE SHAFT, BEARING & SEAL

### *REMOVAL & INSTALLATION*

**Passat with All Wheel Drive**

> ❋❋ **WARNING**
>
> **With axle hub bolt loosened, wheel bearing can be damaged by vehicle's own weight. If a vehicle needs to be moved when the halfshaft is removed, an outer CV-joint must be installed and tightened to 90 ft. lbs. (120 Nm).**

1. With the vehicle sitting on the ground, loosen the axle hub bolt no more than 90°.

2. Raise and safely support the vehicle.

3. Have a helper hold brake pedal to hold the wheel and remove the axle hub bolt.

4. Remove wheel.

5. Remove coil spring.

6. Remove bolts for tie rod and lower transverse link from bottom of wheel bearing housing.

7. Remove lower shock absorber mount bolt.

8. Loosen the halfshaft at rear final drive flange.

9. Swing wheel bearing housing upward and remove halfshaft from inner splines.

***To install:***

10. Fit the halfshaft into the wheel bearing and attach it to the final drive flange. Install new bolts and torque them to 30 ft. lbs. in a diagonal pattern in two steps.

11. Install a new hub bolt and just snug for now.

12. Install new tie rod and transverse link bolts hand tight for now.

13. Install the shock absorber and tighten the nut to 90 ft. lbs. (120 Nm).

14. Install the wheel and have an assistant hold the brake while tightening the axle hub bolt:

- 6-point bolt: 150 ft. lbs. (200 Nm) plus an additional 180°
- 12-point bolt: 52 ft. lbs. (70 Nm) plus an additional 90°

15. With the vehicle sitting on its wheels, torque the transverse link bolt to 70 ft. lbs. (90 Nm) plus an additional 90° and the tie rod bolt to 96 ft. lbs. (130 Nm) plus an additional 90°.

16. Check rear wheel alignment.

# ENGINE COOLING

## THERMOSTAT

### *REMOVAL & INSTALLATION*

**2.0L Engine**

1. Before servicing the vehicle, refer to the Precautions Section.

2. Remove the alternator.

3. Disconnect the hose from the coolant distribution housing.

4. Remove the bolts and remove the coolant distribution housing with thermostat from the engine.

***To install:***

5. Clean sealing surface for O-ring.

6. Replace O-ring and coat it with coolant.

7. Fasten the coolant thermostat housing to the engine block and tighten the bolts evenly to 11 ft. lbs. (15 Nm).

8. Install the alternator.

9. Refill the coolant reservoir with G 12 coolant and install the cap. Run the engine at 2000 rpm for three minutes or until the fan runs and check coolant level.

**2.5L Engine**

*See Figure 24.*

1. Before servicing the vehicle, refer to the Precautions Section.

2. Drain the cooling system

3. Remove the engine cover/air filter housing.

4. Remove the intake manifold.

5. Reinstall the oil dipstick tube so no escaping coolant can run into the engine.

6. Place a suitable container under coolant regulator housing to catch coolant flowing out.

7. Remove the bolts to remove the coolant regulator and thermostat.

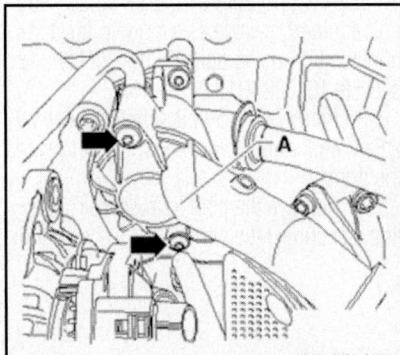

22205_VWCA_G0021

**Fig. 24 Remove these bolts to remove the thermostat**

***To install:***

8. Fit the thermostat into place with the jiggle valve at the top.

9. Install the housing and bolts. Torque the bolts to 44 in. lbs. (5 Nm)

10. Install the intake manifold and dipstick tube.

11. Refill the cooling system with G 12 coolant and install the cap. Run the engine at 2000 rpm for three minutes or until the fan runs and check coolant level.

## WATER PUMP

### *REMOVAL & INSTALLATION*

**2.0L Engine**

*See Figure 25.*

1. Before servicing the vehicle, refer to the Precautions Section.

2. Remove or disconnect:

- Negative battery cable.
- Coolant.
- Accessory drive belt and tensioner.
- Upper and center timing belt covers.

3. Position the engine so that the No. 1 cylinder is at TDC.

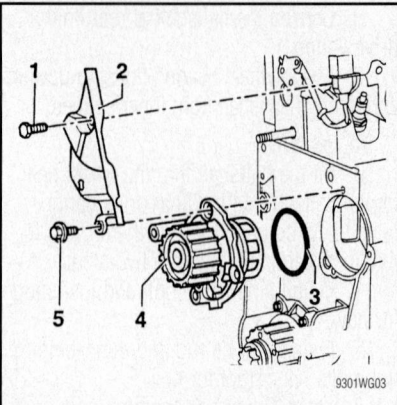

**Fig. 25 Exploded view of the water pump mounting—2.0L (AEG/AVH/AZG/BDF) engine**

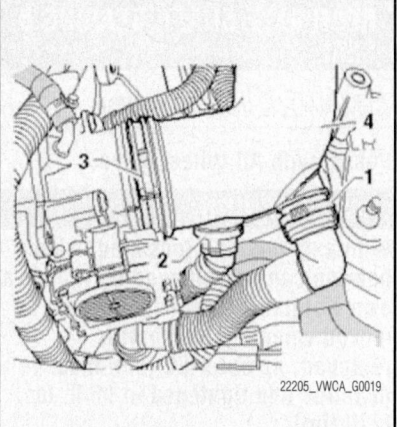

**Fig. 26 Release the spring clamps to remove the air hoses**

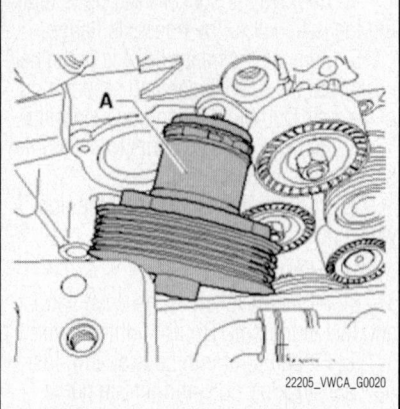

**Fig. 27 Swivel the pump to remove it from the engine compartment—2.5L engine**

### ✳✳ WARNING

**Cover the timing belt to protect it from being contaminated with coolant.**

4. Loosen the timing belt tension and slide the belt off the water pump sprocket.
5. Remove or disconnect:
- Timing belt guard (2) mounting bolt (1), if equipped.
- Water pump mounting bolts (5), then the water pump (4).
- O-ring (3) and clean the seating area.

*To install:*
6. Install or connect:
- O-ring, moistened with coolant.
- Water pump so that the plug in the housing faces downward. Mounting bolts: 11 ft. lbs. (15 Nm).
- Timing belt guard, if equipped. Mounting bolt: 15 ft. lbs. (20 Nm).
- Timing belt.
- Upper and center timing belt covers.
- Accessory drive belt tensioner. Mounting bolt: 18 ft. lbs. (25 Nm).
- Accessory drive belt.
- Coolant.
- Negative battery cable.
7. Check and clear any DTCs, then match the ECM to the TCM.

### 2.5L Engine

*See Figures 26 and 27.*

➡This procedure requires removal of the front engine mount. New bolts will be needed.

1. Before servicing the vehicle, refer to the Precautions Section.
2. Remove the engine cover/air filter assembly.
3. Remove the battery and battery box.
4. Drain the cooling system by disconnecting the bottom radiator hose.
5. Remove air inlet tube guide above the fans.
6. Remove the air intake hose (4) by disconnecting the smaller air hoses (1 and 2) and removing the spring clamp (3).
7. Remove right inner fender.
8. Remove the coolant pump drive belt.
9. Disconnect the exhaust pipe from the manifold and remove the bracket bolts. Tie the pipe out of the way. The flex joint in the front exhaust pipe must not be bent more than 10 degrees, otherwise it may be damaged.
10. Remove the pendulum support (under car engine mount).
11. Disconnect the shift control cables at the transaxle.
12. Install engine support bridge 10-222A with adapters 10-222A/8 and support engine/transaxle assembly in installation position using adapter 10-222A/3 and shackle 10-222 A/12.
13. Remove the bolt that holds the windshield washer reservoir and move it towards the front.
14. Remove the coolant reservoir bolts and disconnect the wiring. Set the reservoir

out of the way without disconnecting the hoses.
15. Remove the bolts to remove the front engine mount. The rear bolt in cylinder block is accessible above wheel housing liner.
16. Remove the two transaxle mount bolts.
17. Slide the engine as far as possible toward the front and left.
18. Remove the three mounting bolts and swivel the coolant pump out.

*To install:*
19. Install the pump and torque the bolts to 7 ft. lbs. (10 Nm).
20. Install the engine mounts using new bolts.
- Torque the two pendulum support bracket bolts to 30 ft. lbs. (40 Nm) plus 90°. Install a new pendulum support center bolt and torque it to 75 ft. lbs. (100 Nm) plus 90°.
- Install two new transaxle mount bolts and torque to 45 ft. lbs. (60 Nm) plus 90°.
21. Remove the engine support bridge and adapter brackets.
22. Connect the shift cables and adjust as necessary (See Transaxle Removal and Installation).
23. Connect the exhaust pipe and install the bracket bolts.
24. Install the remaining components and connect all wiring and hoses.
25. Refill the coolant reservoir with G 12 coolant and install the cap. Run the engine at 2000 rpm for three minutes or until the fan runs and check coolant level.

**ENGINE ELECTRICAL** | **CHARGING SYSTEM**

## ALTERNATOR

### REMOVAL & INSTALLATION

#### 2.0L Engine

*See Figure 28.*

Before purchasing a replacement alternator, read the specification plate on the housing. The number 14V will appear to indicate maximum voltage rating. On the same line will be two more digits followed by the letter **A**. This is the maximum amperage output. Be sure to purchase an alternator with the same rating. The regulator can be replaced without removing the alternator.

1. Before servicing the vehicle, refer to the Precautions Section.

2. Disconnect the negative battery cable.

3. Put a wrench on the belt tensioner lug and pivot the tensioner.

4. Install a pin to keep the tensioner in place.

5. Remove the bolts to loosen the coolant pipe. The pipe can remain connected.

6. Remove the alternator bolts and carefully lift the alternator from the bracket.

7. Disconnect the wiring from the alternator. If necessary, mark the wires with tape or other means to ensure they are connected properly upon installation.

#### To install:

8. On the rear alternator mounting holes, drive the threaded sleeves towards the rear of the housing approximately 0.160 in. (4 mm).

9. Attach the wire harness bracket and fit the alternator into place.

10. Connect the alternator wiring.

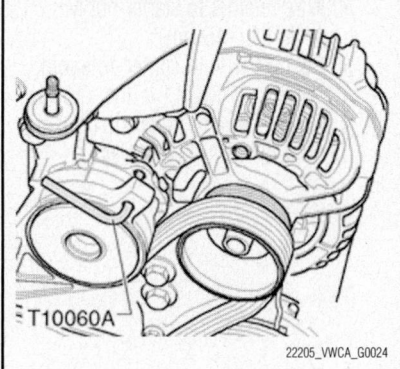

**Fig. 28 Pin the tensioner in place to remove the belt—2.0L engine**

11. Fit the alternator into place and install the mounting bolts. Torque to 18 ft. lbs. (25 Nm).

12. Install the drive belt so it runs in the original direction and remove the pin from the tensioner.

13. Install remaining components and connect the battery cable.

#### 2.5L Engine

*See Figures 29 and 30.*

1. Before servicing the vehicle, refer to the Precautions Section.

2. Disconnect the negative battery cable.

➡**Air intake and engine cover component layout may differ between vehicles.**

3. Remove screws, clips or clamps for air intake to air filter/engine cover from lock carrier.

4. Disconnect the Mass Airflow (MAF) sensor (if necessary).

5. Remove engine cover/air filter housing assembly.

6. Remove the accessory drive belt.
- Raise and safely support the front of the vehicle.
- Remove front part of right front inner fender.
- Put a wrench on the A/C belt tensioner and push towards the rear of the vehicle.
- Remove the compressor belt and mark its direction of rotation for installation.

7. Unplug the A/C compressor connector (1) and remove the rear compressor mounting bolt (2) .

8. Remove both front compressor mounting bolts. Without disconnecting the refrigerant hoses, carefully lower the compressor and hang it from the body with wire or rope. Do not let it hang by the refrigerant hoses.

9. Remove five bolts to remove the idler pulley. Two are below the alternator.

10. Carefully disconnect the alternator wiring.

11. Remove the mounting bolts and remove the alternator. Don't forget the wiring harness bracket at the rear.

#### To install:

12. On the rear alternator mounting holes, drive the threaded sleeves towards

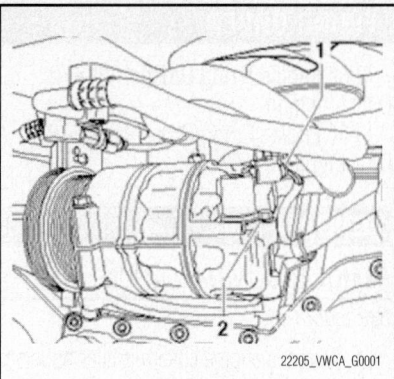

**Fig. 29 Unplug the connector, remove the bolts and hang the A/C compressor from the car with wire. Do not let it hang by the hoses—2.5L engine**

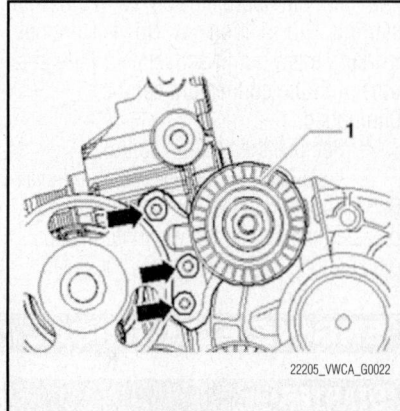

**Fig. 30 Remove these three bolts and two more underneath to remove the idler pulley—2.5L engine**

the rear of the housing approximately 0.160 in. (4 mm).

13. Attach the wire harness bracket and fit the alternator into place.

14. Install the two mounting bolts and torque to 18 ft. lbs. (25 Nm).

15. Connect the alternator wiring.

16. Install the idler pulley and torque the bolts to 18 ft. lbs. (25 Nm).

17. Before installing the A/C compressor, make sure both centering sleeves are inserted in the sub-assembly bracket threaded holes. Torque the mounting bolts to 18 ft. lbs. (25 Nm).

18. Install the drive belt so it runs in the original direction.

19. Install remaining components and connect the battery cable.

## ENGINE ELECTRICAL — IGNITION SYSTEM

### FIRING ORDER

2.0L Engine Firing Order: 1–3–4–2
Direct Ignition
2.5L Engine Firing Order: 1–2–4–5–3
Direct Ignition

### IGNITION COIL

*REMOVAL & INSTALLATION*

*See Figure 31.*

1. Remove engine cover/air filter housing.
2. If present, disconnect harness connector from Secondary Air Injection (AIR) solenoid valve.
3. In order to prevent damage to the cable guide, remove all the ignition coil assemblies together. Using puller T40039, pull them out approximately ⅜ in. (10 mm) starting with ignition coil No. 1. Continue working them out in sequence a little at a time until the connectors can be unplugged.
4. Remove the ignition coil(s).
5. Installation is the reverse of removal. Remember to connect the wiring before pressing the coil all the way onto the spark plug.

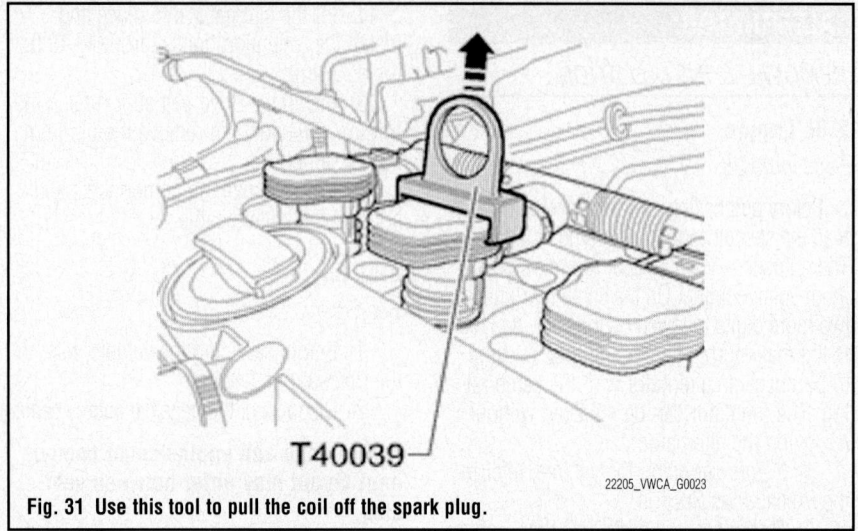

**Fig. 31 Use this tool to pull the coil off the spark plug.**

22205_VWCA_G0023

### IGNITION TIMING

*ADJUSTMENT*

➡The ignition timing is controlled by the engine control module and is not adjustable. However the timing can be monitored on a scan tool connected to the DLC in the vehicle. No specification has been given by the manufacturer.

### SPARK PLUGS

*REMOVAL & INSTALLATION*

1. Remove ignition coils.
2. Remove spark plugs
3. Installation is the reverse of removal. Torque spark plugs to 22 ft. lbs. (30 Nm).

## ENGINE ELECTRICAL — STARTING SYSTEM

### STARTER

*REMOVAL & INSTALLATION*

➡On vehicles with electric power steering, after connecting the battery and turning the ignition switch ON, the ASR/ESP Control Lamp lights up continuously. The lamp goes out automatically when the vehicle is driven a distance straight ahead at 15 to 20 km/h. This re-activates Steering Angle Sensor.

1. Disconnect the battery ground cable.
2. Remove the engine cover/air filter housing.
3. Slide protective cap downward and remove the positive wire from starter terminal 50.
4. Remove the ground strap from starter mounting bolt.
5. Remove mounting bolt.
6. Raise and safely support the vehicle and remove the engine under cover.
7. Remove the nuts and bolts to remove the starter.

8. Installation is the reverse of removal. Torque the fasteners as follows:
   - Starter to transaxle bolt M12: 59 ft. lbs. (80 Nm)
   - Starter to transaxle bolt M10: 30 ft. lbs. (40 Nm)
   - Ground wire to starter nut M8: 11 ft. lbs. (15 Nm)
   - Wire retainer to starter nut M8: 11 ft. lbs. (15 Nm)
   - Positive wire to starter solenoid switch nut M8: 11 ft. lbs. (15 Nm)
   - Ground wire to transaxle housing nut M8: 11 ft. lbs. (15 Nm)

## ENGINE MECHANICAL

➡Disconnecting the negative battery cable may interfere with the functions of the on board computer systems and may require the computer to undergo a relearning process, once the negative battery cable is reconnected.

### ACCESSORY DRIVE BELTS

#### ACCESSORY BELT ROUTING

*See Figures 32 through 34.*

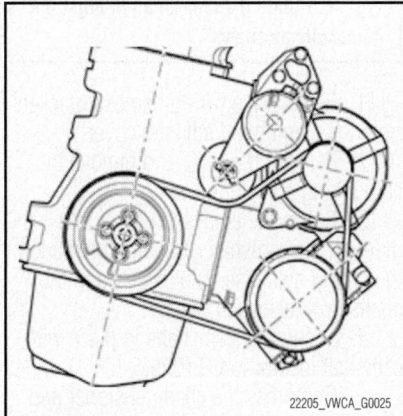

**Fig. 32 Accessory drive belt routing: 2.0L Engine**

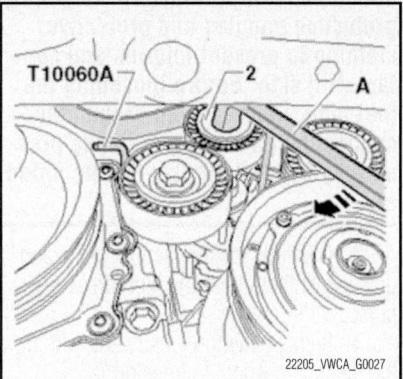

**Fig. 34 Releasing belt tensioner—2.5L engine**

#### INSPECTION

1. Turn the engine at vibration damper/ crankshaft pulley with a suitable socket wrench.
2. Raise the vehicle if necessary. Check the drive belt for:
   a. Sub-surface (deep) cracks
   b. Layer separation (top layer, cord strands)
   c. Traces of oil and grease
3. Replace the belt if any damage is found or if contaminated with oil or grease.

#### ADJUSTMENT

All models use an automatic (spring powered) tensioner. No adjustment is required.

#### REMOVAL & INSTALLATION

**2.0L Engine**

*See Figure 35.*

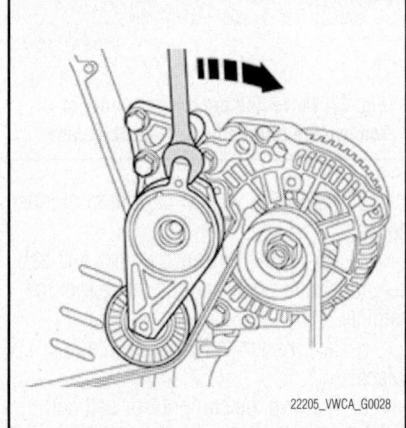

**Fig. 35 Releasing belt tensioner—2.0L engine**

1. Put a wrench on the belt tensioner lug and pivot the tensioner.
2. Install a pin to keep the tensioner in place.
3. Do not leave the pin in place if moving on to other work.

**2.5L Engine**

*See Figures 36 and 37.*

1. Before servicing the vehicle, refer to the Precautions Section.
2. Raise and safely support the front of the vehicle.

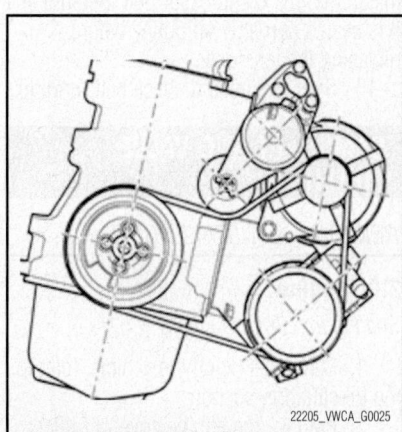

**Fig. 36 Move and pin this tensioner to remove the A/C compressor belt—2.5L engine**

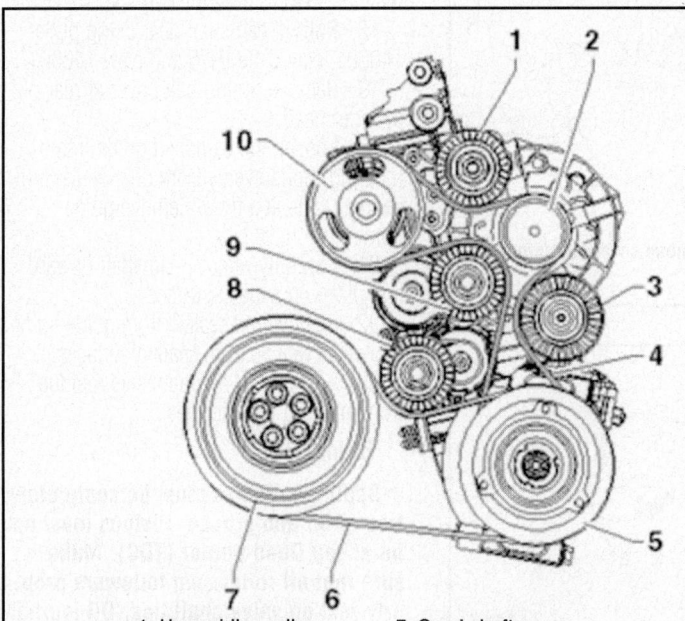

1. Upper idler pulley
2. Alternator
3. Lower idler pulley
4. Ribbed belt for generator and coolant pump
5. A/C compressor
6. Belt for A/C compressor
7. Crankshaft
8. Lower tensioner for ribbed belt for A/C compressor
9. Upper tensioner for ribbed belt for generator and coolant pump
10. Coolant pump

**Fig. 33 Accessory drive belt routing: 2.5L Engine**

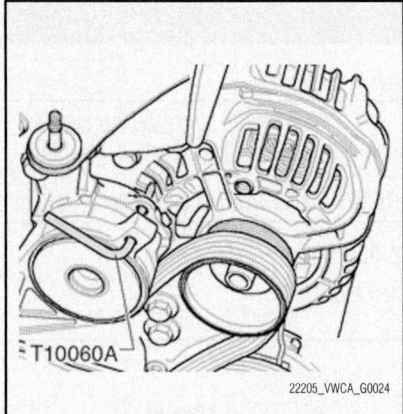

**Fig. 37 Move and pin this tensioner to remove the alternator belt—2.5L engine**

3. Remove front part of right front inner fender.

4. Put a 15mm wrench on the A/C belt tensioner and push towards the rear of the vehicle.

5. Insert a pin to hold the wrench in place.

6. Remove the compressor belt and mark its running direction for installation. Do not leave the wrench and pin in place if moving on to other work.

7. Mark the running direction of the alternator belt.

8. Move the tensioner and insert a pin to hold it in place.

9. Remove the belt. Do not leave the wrench and pin in place if moving on to other work.

### To install:

10. Place the alternator belt on the alternator and coolant pump pulleys, then lastly on the idler pulley.

11. Before releasing the tensioner, check ribbed for correct seating in the pulleys.

12. Secure the other tensioner and install the A/C compressor belt. Be sure that it is seated correctly on pulley before releasing the tensioner.

13. Start engine and check belt running.

## CAMSHAFT AND VALVE LIFTERS

### REMOVAL & INSTALLATION

#### 2.0L Engine

*See Figures 38 through 43.*

1. Before servicing the vehicle, refer to the Precautions Section.

2. Remove the engine cover/air filter housing.

3. Disconnect the wiring from the high pressure fuel pump.

### ✳✳ CAUTION

**Fuel pipes are pressurized! Wear protective goggles and protective clothing to prevent injuries and contact with skin. Before loosening the fuel pipe, place a cloth around the connection point. Then release pressure by carefully loosening the union nut.**

4. Carefully loosen the fuel pipes and disconnect them from the high pressure pump.

5. Remove the three bolts and withdraw the fuel pump from the housing.

6. Remove the ignition coils.

7. Remove the upper timing belt cover.

8. Disconnect the vent hoses from the cylinder head cover. Remove the 4 screws to remove the valve assembly from the front of the cylinder head cover.

9. Loosen the cylinder head cover bolts in reverse of tightening order and remove the cover.

10. Disconnect the coolant hoses and remove the bracket bolt to remove the coolant pipe from the heat shield.

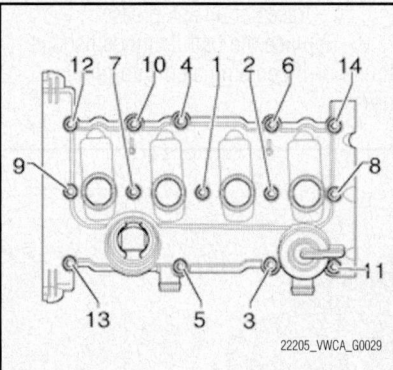

**Fig. 38 Cylinder head cover bolt torque sequence**

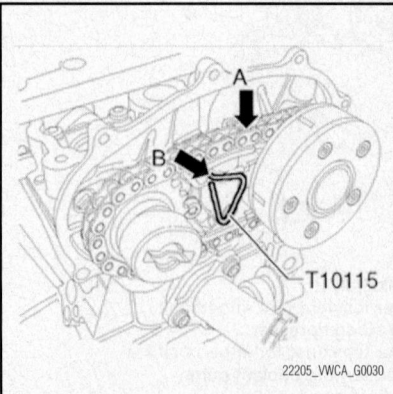

**Fig. 39 Compress the chain tensioner and pin it in place**

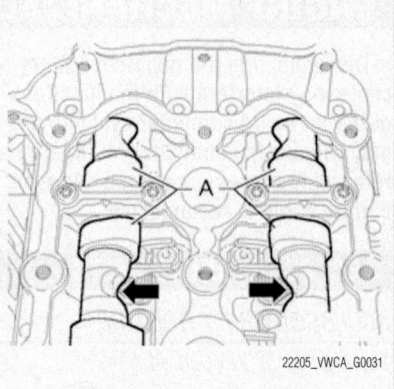

**Fig. 40 Rotate the crankshaft to align the camshafts as shown**

11. Remove the wiring harness brackets from the cam timing adjuster cover.

12. Remove the bolts and remove the cam timing adjuster cover.

13. Rotate the crankshaft to align the mark on the camshaft gear with the mark on the timing belt case. The recesses in the shafts now face each other.

14. Secure the camshafts in place with camshaft locator tool T10252.

15. Compress the chain tensioner and remove the center bolt to remove the camshaft adjuster and chain.

16. Remove the timing belt.

17. Loosen camshaft sprocket using retainer 3036 to hold the camshaft.

18. Pull off camshaft gear using puller T40001, claw T40001/6 and claw T40001/7.

19. Remove timing belt cover at rear of cylinder head.

20. Loosen the camshaft guide frame (bearing) bolts evenly from outside toward inside a little at a time. Remove guide frame.

21. Carefully remove camshaft upward and place on a clean surface.

22. Remove old sealant from guide frame groove as well as from sealing surfaces.

23. Prevent dirt and adhesive residue from entering cylinder head.

### To install:

➡️**Sealing surfaces must be completely free of oil and grease. Pistons must not be at Top Dead Center (TDC). Make sure that all roller cam followers properly rest on valve shaft tips. Oil journal surfaces of camshafts.**

24. Carefully fit camshafts into the cylinder head. Cams of cylinder 4 must face each other. Recesses must point to each other.

25. Apply an even, slightly raised bead of sealant D 188 800 A1 into clean grooves

of guide frame. Apply sealant uniformly on sealing surface. Sealant must not be applied too thick, wipe off any excess.

26. Place guide frame onto cylinder head.

27. Tighten the bolts in sequence one turn at a time. The final torque it 71 inch lbs. (8 Nm) plus 90°.

28. Install a new sealing cap and camshaft seal.

29. Install timing belt case.

30. Insert the sprocket key and the sprocket with a new bolt. Hold the sprocket and torque the bolt to 37 ft. lbs. (50 Nm) plus 180°.

### ☀☀ WARNING

**This is an interference engine. When turning camshaft, pistons may not be at TDC for any cylinder. Valves and/or pistons may be damaged.**

31. Install timing belt .

32. Install camshaft adjuster .

- Make sure the camshafts are aligned and secured using camshaft locator T10252.
- Place chain on to camshaft adjuster.
- Hold the camshaft adjuster in front of the exhaust camshaft so that the notch aligns with the pin.

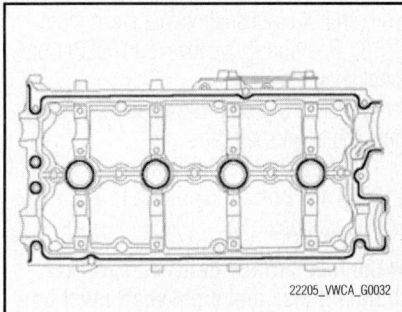

**Fig. 41 Apply a thin bead of sealant as shown to the guide frame**

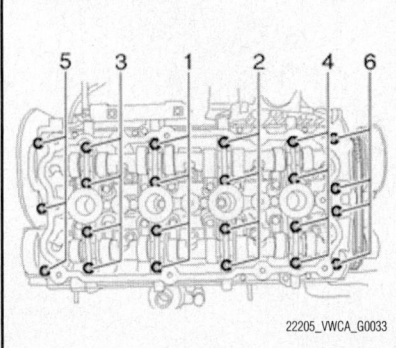

**Fig. 42 Guide frame bolt torque sequence**

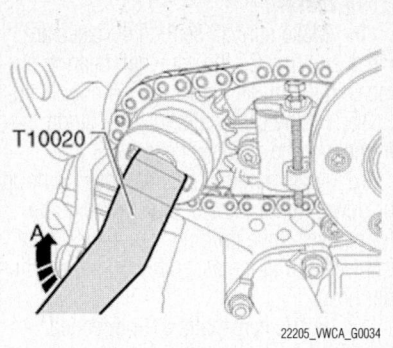

**Fig. 43 Hold the adjuster on the camshaft while turning the other camshaft clockwise. The pin on the adjuster should click into place in the notch on the camshaft**

- Fit chain into place on the chain sprocket of intake camshaft with all the slack at the bottom.
- Slowly turn the intake camshaft clockwise with tool T10020 until the camshaft adjuster fits onto the camshaft.

➡ **If the adjuster pin does not fit into the notch on the camshaft, remove the chain and start over.**

33. Install camshaft adjuster bolt and torque to 15 ft. lbs. (20 Nm) plus 45°.

34. Remove the pin from the chain tensioner.

35. Rotate the crankshaft two full turns to make sure all timing marks align properly.

36. Install the cylinder head cover and torque the bolts in sequence to 7 ft. lbs. (10 Nm).

37. Install remaining components and connect all wiring and hoses.

38. To install the high pressure fuel pump:

- Fit a new O-ring and coat it lightly with clean engine oil.
- Check plunger for damage and insert it into cylinder head.
- Fit the pump carefully into place and install the bolts hand-tight.
- Install the fuel pipes hand-tight.
- Torque the mounting bolts to 7 ft. lbs. (10 Nm) and the union bolts to 18 ft. lbs. (25 Nm).

### 2.5L Engine

*See Figures 44 through 49.*

1. Before servicing the vehicle, refer to the Precautions Section.

2. Remove the battery and battery holder.

3. Remove engine cover/air filter assembly.

4. Disconnect PCV hose from cylinder head cover. If equipped, remove the secondary air tube.

5. Remove the ignition coils.

6. Remove bolts for cylinder head cover in reverse of the tightening sequence.

7. Remove the timing chain cover. This requires removing the intake manifold.

8. Remove the brake booster vacuum pump.

### ☀☀ WARNING

**The 4 cover bolts must not be loosened under any circumstances.**

9. If valve timing is correct:

- Secure the crankshaft as if for removing the timing chain. See Timing Chain Removal and Installation.
- Install the camshaft locator T40070 on the camshafts and tighten bolts to 15 ft. lbs. (20 Nm). If the bolts could not be screwed in easily, rotate exhaust camshaft slightly using an open-end wrench.
- Relieve tension on timing chain. To do so, pry between piston of chain tensioner and tensioning track. Secure completely pressed in piston using Locking Pins T03006.

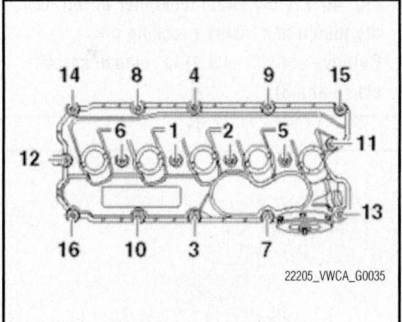

**Fig. 44 Cylinder head cover bolt tightening sequence—2.5L engine**

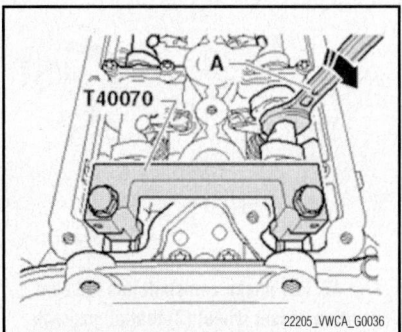

**Fig. 45 Lock the camshafts in place with this tool. Rock camshaft slightly if needed.**

Locking pin must be inserted until it stops.

10. If valve timing is not correct:
  - Remove timing chain case cover
  - Rotate crankshaft to TDC cylinder 5 but do not lock it into place.
  - Rotate crankshaft so that camshaft locator T40070 can be screwed easily on to camshafts.

11. Remove the bolts to remove the sprockets. It may be necessary to gently pry them off.

12. Remove the camshaft locking tool and loosen the guide frame bolts a little at a time from the outside towards the center.

13. Remove the bolts and guide frame and lift the camshafts out of the cylinder head.

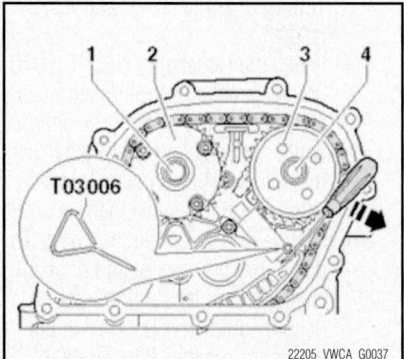

**Fig. 46 Pry the chain tensioner to retract the piston and insert a locking pin. Remove bolts (1 and 4) to remove sprockets (2 and 3)**

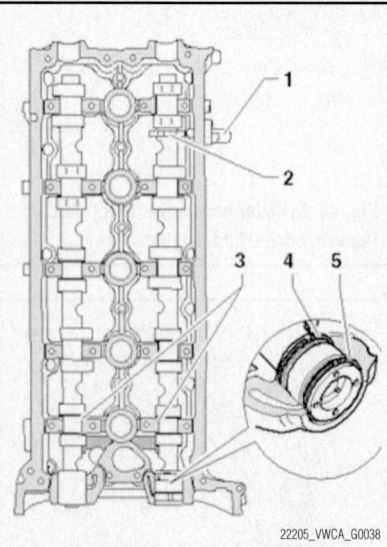

**Fig. 47 The intake camshaft has the cam position sensor wheel (2) that aligns with the sensor (1)—make sure the sealing ring ends (4 and 5) point up or down, not towards the sides**

### To install:

14. Make sure all sealing surfaces are clean and dry. Oil the camshaft bearing surfaces in the guide frame and cylinder head.

15. Fit the camshafts into the guide frame, not the cylinder head.

16. Make sure the sealing ends point up or down, not to the sides. Turn the guide frame over while holding camshafts firmly in place and fit the assembly onto the cylinder head.

17. Rotate camshafts until threaded holes point upward.

18. Check whether camshafts still lie correctly in the bearings, then install the holding tool T40070.

19. Carefully apply sealant as indicated. Do not use too much sealant or it will get into the cam bearings and prevent proper lubrication.

20. Install the bolts and tighten them in several stages. Torque all bolts in sequence to 71 in. lbs. (8 Nm), then repeat the sequence turning each bolt an additional 90°.

21. Sealant must bulge outward slightly in the timing chain compartment. Wipe off any excess sealant.

22. Carefully press in new sealing plugs until they reach end of chamfer.

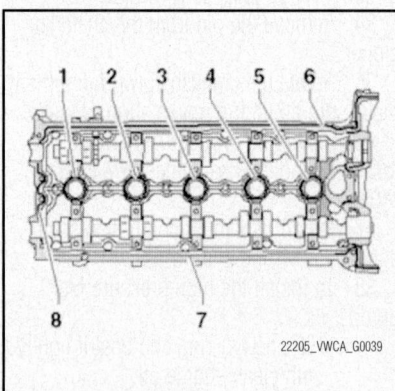

**Fig. 48 Apply sealant as indicated**

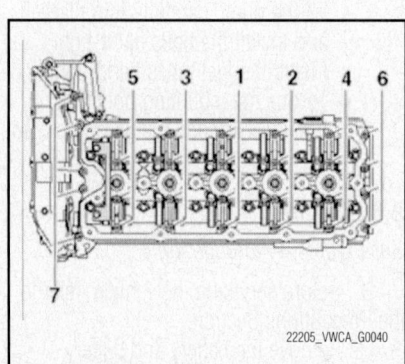

**Fig. 49 Camshaft guide frame bolt torque sequence**

➡ **If sealing plug was pressed in too far, it must be pressed through and pressed in again up to the marking.**

23. Install the timing chain and remove the crankshaft locking pin from the cylinder block.

24. Install the remaining components.

## CRANKSHAFT FRONT SEAL

### REMOVAL & INSTALLATION

#### 2.0L Engine

*See Figure 50.*

1. Remove the timing belt.

2. Hold the crankshaft sprocket with tool 3145 and remove sprocket bolt and sprocket.

3. To guide seal puller, thread central bolt into crankshaft by hand until stop.

4. Remove inner portion of seal puller 3203 nine turns from outer portion and secure with knurled-head bolt.

5. Grease threaded head of seal puller and install into oil seal as far as possible with forced pressure.

6. Loosen knurled bolt and turn inner portion against crankshaft until the oil seal is pulled out.

### To install:

7. Before installing, remove oil remains from end of crankshaft with a clean cloth.

8. Position guide sleeve T10053/1 on crankshaft pin.

9. Slide dry oil seal over guide sleeve onto the crankshaft pin.

10. Press in oil seal using assembly tool T10053 and bolt T10053/2 (M16 x 1.5 x 60) up to stop.

➡ **Contact surface between sprocket, diamond disc and crankshaft must be free of oil.**

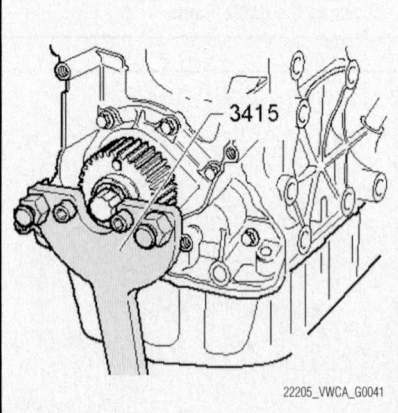

**Fig. 50 Hold sprocket with tool 3145 to loosen and tighten bolt**

11. Install the sprocket and washer and install a new bolt hand tight.

12. Secure the sprocket with counterholder tool 3415.

13. Torque the bolt to 66 ft. lbs. (90 Nm) plus an additional 90°

14. Install the timing belt.

### 2.5L Engine

*See Figures 51 through 53.*

1. Raise and safely support the vehicle and remove the engine undercover.

2. Remove the front part of the right inner fender.

3. Remove the A/C compressor drive belt. Mark the belt's direction of rotation for reinstallation.

4. Remove the belt tensioner.

5. Set the engine to TDC of cylinder No. 5 and insert the crankshaft locking tool at the rear of the engine block.

6. Remove the bolts that hold the front seal flange to the engine block.

7. Starting at the alignment pins, carefully pry off the sealing flange. Take care to prevent damaging the cylinder block. The sealing flange will be damaged while removing.

> **❄❄ WARNING**
>
> **To prevent injuries from shavings, wear protective goggles and protective clothing.**

8. Remove sealant from cylinder block. Make sure that no sealant residue enters the engine.

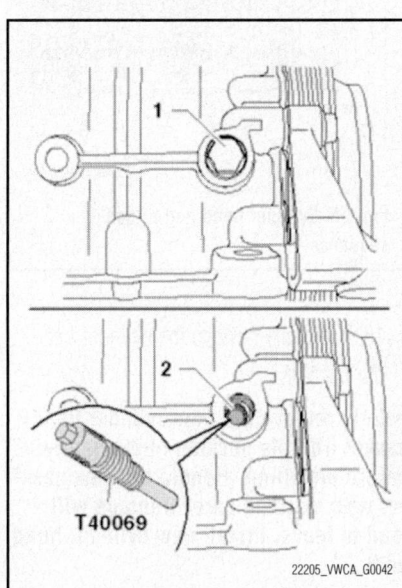

**Fig. 51 With the engine at TDC No. 5, remove the plug and use this tool to lock the crankshaft**

9. Clean sealing surface of cylinder block and the crankshaft; they must be clean and dry.

### *To install:*

➡**The following steps must be followed so that the sealing lip of sealing flange does not roll itself up when installing. Do not use any lubricants!**

10. Widen sealing lip of new sealing flange using the tapered end of assembly sleeve T03004.

11. After a short time, remove the assembly sleeve and insert the wide end into sealing ring. Assembly sleeve must protrude approximately 0.020 in. (3 mm) on the engine side of the seal.

12. Apply sealant bead into groove of sealing flange. Insert sealing flange sleeve T03004 over the crankshaft and press the flange plate uniformly into place.

13. Start all the bolts, then tighten uniformly in diagonal sequence. Torque to 7 ft. lbs. (10 Nm).

14. Install the crankshaft pulley and torque the bolt to 37 ft. lbs. (50 Nm) plus 90°.

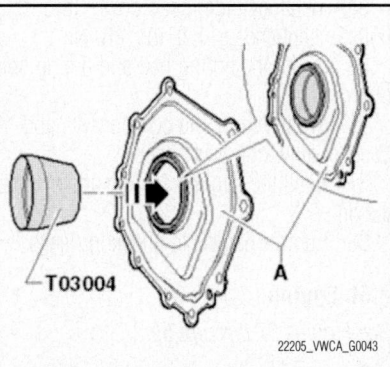

**Fig. 52 Use the assembly sleeve to widen the seal enough to fit over the crankshaft—do not use any lubricants**

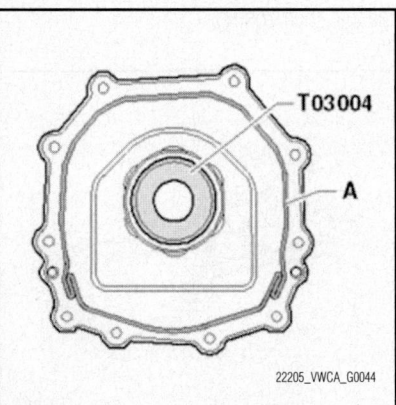

**Fig. 53 Apply a bead of sealant and immediately install the sealing flange**

15. Remove the crankshaft locking tool and install the plug.

16. Install the belt tensioner and A/C drive belt.

17. Install remaining components and run the engine to check for leaks.

## CYLINDER HEAD

### *REMOVAL & INSTALLATION*

#### 2.0L Engine

*See Figures 54 through 56.*

1. Before servicing the vehicle, refer to the Precautions Section.

2. If a coded radio is installed, obtain the anti-theft coding.

3. With ignition switched **OFF**, disconnect battery cables.

4. On GTI, remove plenum chamber facing wall (below the windshield).

5. Remove intake manifold.

6. Drain the cooling system.

7. Remove the engine cover/air filter housing.

8. Remove front part of right wheel housing liner.

9. Remove catalytic converter with front exhaust pipe.

10. Remove charge air hose from charge air cooler.

11. Remove bolts and disconnect charge air pipe.

12. Pull off/disconnect all electrical lines from turbocharger and set aside.

13. Remove oil supply line and coolant supply line at turbocharger.

14. Disconnect oil return line from turbocharger.

15. Remove the bolts and remove support for turbocharger.

16. Remove the battery and battery holder.

17. Disconnect the coolant hoses and remove the coolant pipe at the end of the cylinder head.

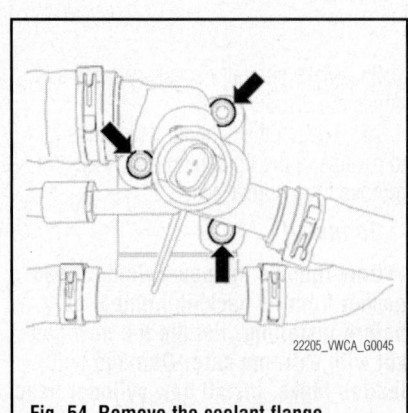

**Fig. 54 Remove the coolant flange**

18. Disconnect the wiring from the high pressure fuel pump.

19. Remove the coolant flange.

20. Remove the accessory drive belt and belt tensioner.

21. Remove the timing belt and the timing belt housing:

22. Remove the ignition coils.

23. Disconnect the vent hoses from the cylinder head cover. Remove the 4 screws to remove the valve assembly from the front of the cylinder head cover.

24. Loosen the cylinder head cover bolts in reverse of tightening order and remove the cover.

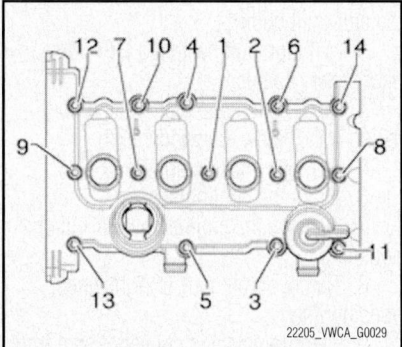

**Fig. 55 Cylinder head cover bolt torque sequence**

25. Make sure all necessary wires and hoses are disconnected.

26. Loosen the cylinder head bolts in reverse of the tightening sequence and remove the cylinder head.

**To install:**

➡Only remove the new cylinder head gasket from its packing immediately before installing. Handle the new gasket with extreme care. Damage will lead to leaks. Install new cylinder head bolts.

27. Clean all cylinder block sealing surfaces. There must be no oil or coolant in the blind holes for the cylinder head bolts in the cylinder block. Stuff clean rags into cylinders so no dirt or abrasive material can get between cylinder wall and piston. Do not allow dirt or abrasives to get into coolant either.

28. Carefully clean cylinder head and avoid introducing scratches or scoring (do not use sandpaper with grit below 100).

29. Carefully remove metal particles, emery remains and cloths.

30. In the event the crankshaft has been rotated, set piston of cylinder 1 to TDC and turn crankshaft back again slightly.

31. Install new cylinder head gasket. Replacement part number must be up.

32. Install cylinder head. Insert cylinder head bolts and tighten them hand-tight.

33. Tighten cylinder head bolts in sequence as follows:
- Step 1: torque to 30 ft. lbs. (40 Nm).
- Step 2: tighten in sequence an additional 90°
- Step 3: tighten in sequence an additional 90°

34. Install cylinder head cover. Torque bolts in sequence to 7 ft. lbs. (10 Nm).

35. Install the timing belt and timing belt cover.

36. Install remaining components and connect all wiring and hoses.

37. Refill the cooling system and change the oil.

38. Run the engine to check for leaks.

### 2.5L Engine

*See Figures 57 through 59.*

➡Two people are needed to remove the cylinder head.

1. Before servicing the vehicle, refer to the Precautions Section.

2. If a coded radio is installed, obtain the anti-theft coding.

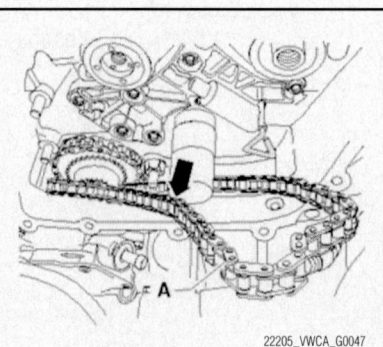

**Fig. 57 Hold timing chain at "A" and lay it under the pipe connection**

3. Remove the battery and battery holder.

4. Drain the cooling system.

5. Remove the engine cover/air filter housing.

6. Remove the intake manifold.

7. Install transport strap on cylinder head to better hold the head during removal.

8. Remove timing chain case cover.

9. Remove cylinder head cover.

10. Secure camshafts and remove chain sprockets for camshafts. See Camshaft Removal and Installation.

11. Disconnect the exhaust pipe and remove the bracket bolts. Move the pipe aside.

➡Flex joint must not be bent more than 10° or it will be damaged.

12. Disconnect the oxygen sensor wiring at the bulkhead.

13. Remove cable bracket at air injection valve.

14. Remove cylinder head bolts in reverse of tightening sequence.

➡If bolt 12 could not be pulled out using a magnet, loosen bolts of camshaft locator T40070 one rotation, slide camshaft locator T40070 toward right front of vehicle and tighten bolts again.

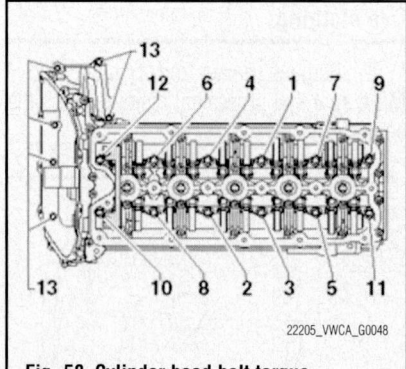

**Fig. 58 Cylinder head bolt torque sequence**

15. Carefully remove the cylinder head.

**To install:**

➡Only remove the new cylinder head gasket from its packing immediately before installing. Handle the new gasket with extreme care. Damage will lead to leaks. Install new cylinder head bolts.

16. Clean all cylinder block sealing surfaces. There must be no oil or coolant in the blind holes for the cylinder head bolts in the cylinder block. Stuff clean rags into cylinders

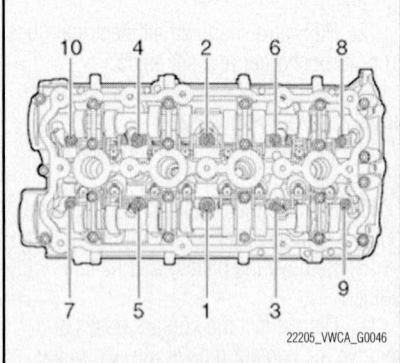

**Fig. 56 Cylinder head bolt torque sequence**

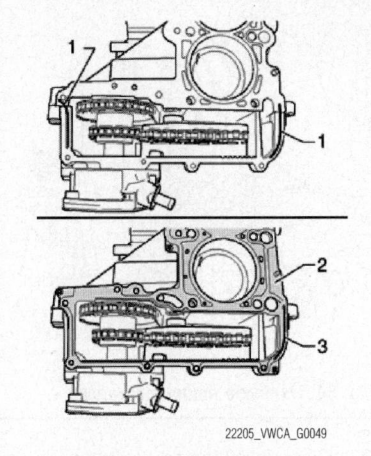

**Fig. 59 Apply sealant to the block at location 1, then install the gasket (2) and apply sealant to location 3**

so no dirt or abrasive material can get between cylinder wall and piston. Do not allow dirt or abrasives to get into coolant either.

17. Carefully clean cylinder head and avoid introducing scratches or scoring (do not use sandpaper with grit below 100).

18. Carefully remove metal particles, emery remains and cloths.

19. Apply a bead of sealant on the block as illustrated.

20. Install new cylinder head gasket and apply sealant to the same places on the gasket.

21. Install cylinder head.

22. Guide timing chain over pipe connection.

23. Insert cylinder head bolts and tighten them hand-tight.

24. Torque cylinder head bolts in sequence shown as follows:
- Step 1: torque bolts 1–12 in sequence to 30 ft. lbs.
- Step 2: tighten bolts 1–12 in sequence an additional 90°
- Step 3: tighten bolts 1–12 in sequence an additional 90°
- Step 4: torque bolts 13 to 7 ft. lbs. (10 Nm).

25. Wipe off any excess sealant.

26. Install camshaft sprockets and rotate crankshaft to check timing mark alignment.

27. Install cylinder head cover.

28. Install remaining components and connect all wiring and hoses.

29. Refill with new coolant and engine oil.

30. Run the engine to check for leaks.

## ENGINE ASSEMBLY

### REMOVAL & INSTALLATION

**2.0L Engine**

*See Figures 60 through 64.*

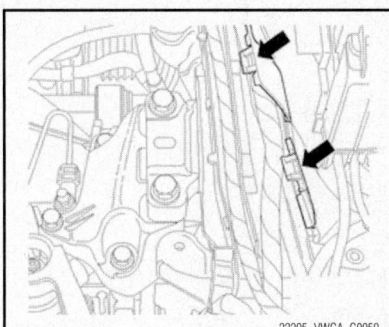

**Fig. 60 Release the wiring harness from the cable guide and set the harness on top of the engine**

➡ **The engine and transaxle are lowered from the vehicle as an assembly. A scan tool and several new bolts are required for installation.**

1. Before servicing the vehicle, refer to the Precautions Section.

2. If a coded radio is installed, obtain the anti-theft coding.

3. Remove the battery and battery holder.

4. Remove the wiper arms, the plenum chamber cover and the bulkhead plenum chamber.

5. Disconnect engine wiring harness connector from Engine Control Module (ECM).

6. Release pass-through for engine wiring harness and pull off upward. The wiring harness will remain with the engine.

7. Disconnect alternator wire from fuse box.

8. Remove the main ground strap.

9. Open locking mechanisms of cable guide. Disconnect all harness connectors of engine wiring harness/body and set down engine wiring harness on engine.

### ✷✷ WARNING

**Fuel supply line is under pressure! Wear protective goggles and protective clothing to prevent injuries and contact with skin. Before removing from hose connection wrap a cloth around the connection. Then release pressure by carefully pulling hose off connection.**

10. Disconnect fuel supply hoses.

11. Remove the EVAP canister.

12. Raise and safely support the vehicle and remove the engine under cover and the front of both inner fender liners.

13. Disconnect charge air hoses.

14. Remove heat shield for right half-shaft.

15. Remove both halfshafts.

16. Remove front exhaust pipe with catalytic converter.

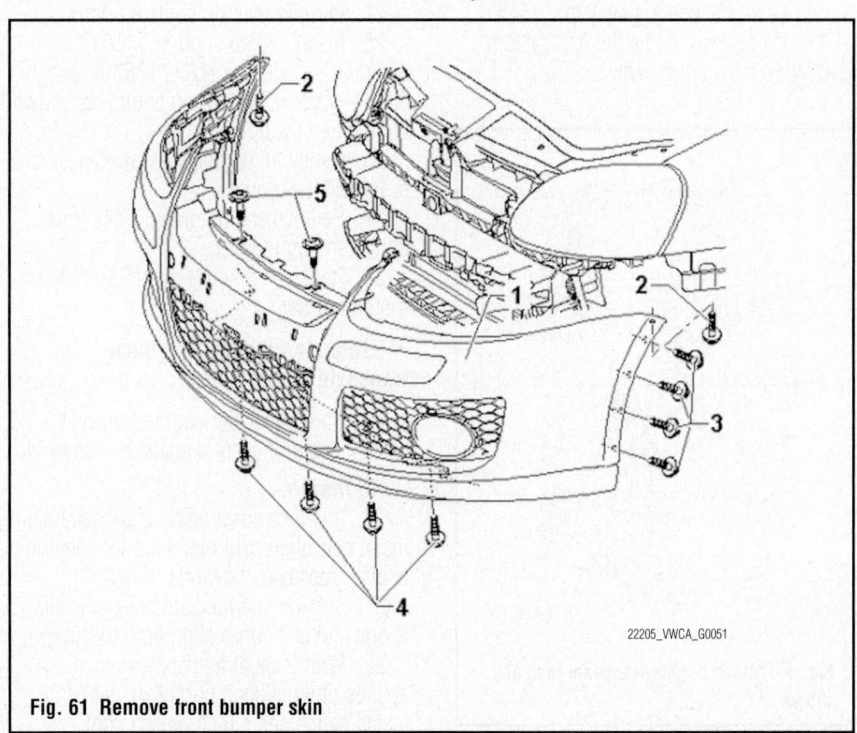

**Fig. 61 Remove front bumper skin**

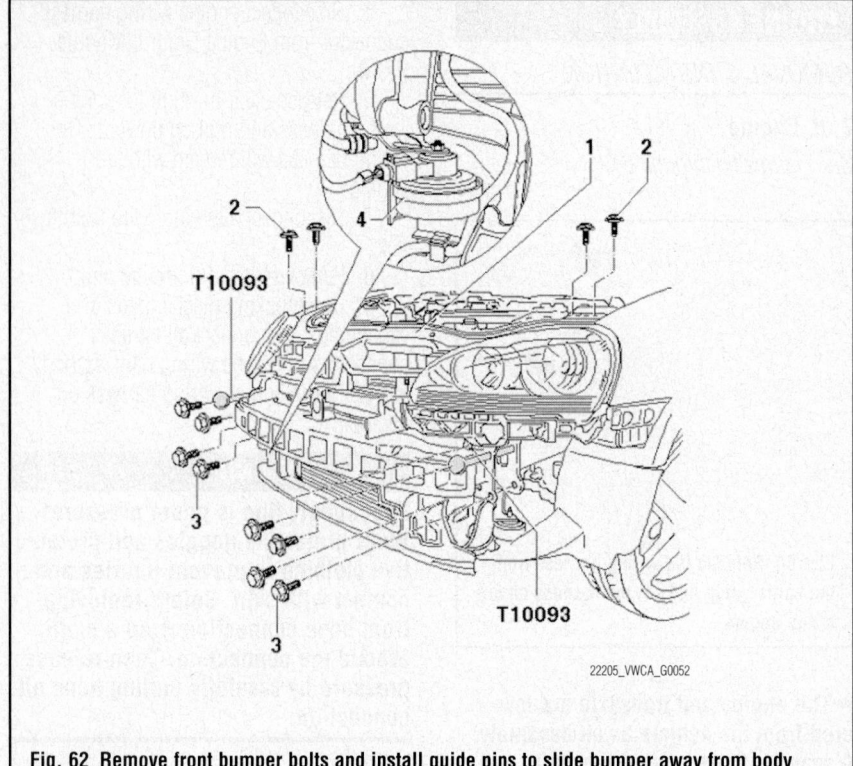

**Fig. 62 Remove front bumper bolts and install guide pins to slide bumper away from body**

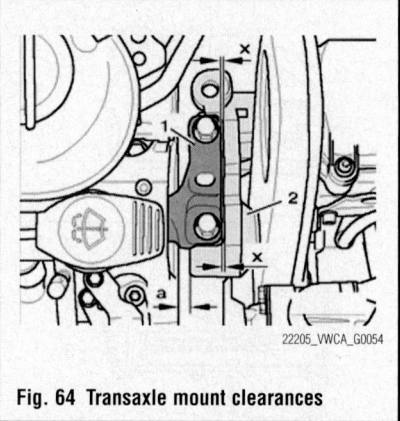

**Fig. 64 Transaxle mount clearances**

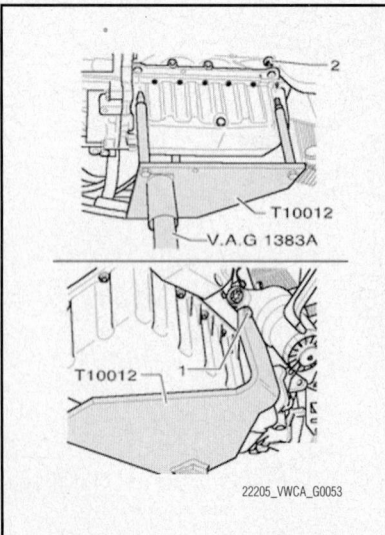

**Fig. 63 Attach engine/transaxle jack as shown**

17. Bring the lock carrier into service position:
  • Remove the front bumper skin and grille
  • Remove the front bumper bolts and install guide pin tools T10093. The bumper and lock carrier can now be slid out to provide more room to work

18. Drain the cooling system.

19. Properly evacuate the A/C system and recover the refrigerant

20. Disconnect the high pressure line from A/C compressor and from the condenser.

21. Remove the pendulum support (engine/trans mount under the vehicle).

22. Remove all remaining wires and hoses from engine/transaxle and set aside.

23. Disconnect transaxle shift mechanism. See Transaxle Removal and Installation.

24. Remove electric coolant pump.

25. Insert engine support T10012 in engine/transaxle jack V.A.G 1383 A. Attach engine support T10012 on engine as shown and tighten the bolts.

26. Gently lift engine/transaxle assembly to unload the mounts.

27. Remove engine mount bolts from engine mount bracket .

28. Remove transaxle mount bolts from mount bracket.

➡**Use a step ladder to remove transaxle mount bolts.**

29. Carefully lower engine/transaxle assembly and prevent damage to bodywork.

*To install:*

30. If engine and transaxle are separated, insert new alignment bushings for centering engine/transaxle in cylinder block.

31. Engage intermediate plate at sealing flange and slide onto alignment bushings.

32. With manual transaxle, grease splines of input shaft lightly. Inspect and install clutch and clutch mechanism.

33. When installing engine/transaxle assembly, check for clearance to subframe as well as to radiator. Align engine mount as follows:
  • There must be a distance of at least ⅜ in. (10 mm) between engine mount bracket and right longitudinal member.
  • Casting edge on engine mount bracket must stand parallel to engine mount.

34. Install pendulum support with a new center mount bolt. Torque the bracket bolts to 30 ft. lbs. (40 Nm) plus 90° the new center mount bolt to 75 ft. lbs. (100 Nm) plus 90°.

35. Install four new mount-to-transaxle bolts and torque to 30 ft. lbs. (40 Nm) plus 90°.

36. Install two new mount-to-body bolts and torque to 45 ft. lbs. (60 Nm) plus 90°.

37. Reassemble the A/C system components.

38. Install front exhaust pipe.

39. Install halfshafts.

40. Attach/connect shift mechanism and adjust if necessary.

41. Begin connecting or installing wires and hoses. Pay careful attention to routing.

42. Reassemble body work.

43. Install the battery. Make sure no one is inside the vehicle when connecting the battery cables.

44. Fill with fresh coolant and oil.

45. Connect scan tool and check fault code memory, erase all fault codes which may have occurred during the assembly.

46. Perform test drive, then recharge the A/C system

### 2.5L Engine

*See Figures 65 and 66.*

➡**The engine and transaxle are lowered from the vehicle as an assembly after removing the front suspension subframe. A scan tool and several new mounting bolts are required.**

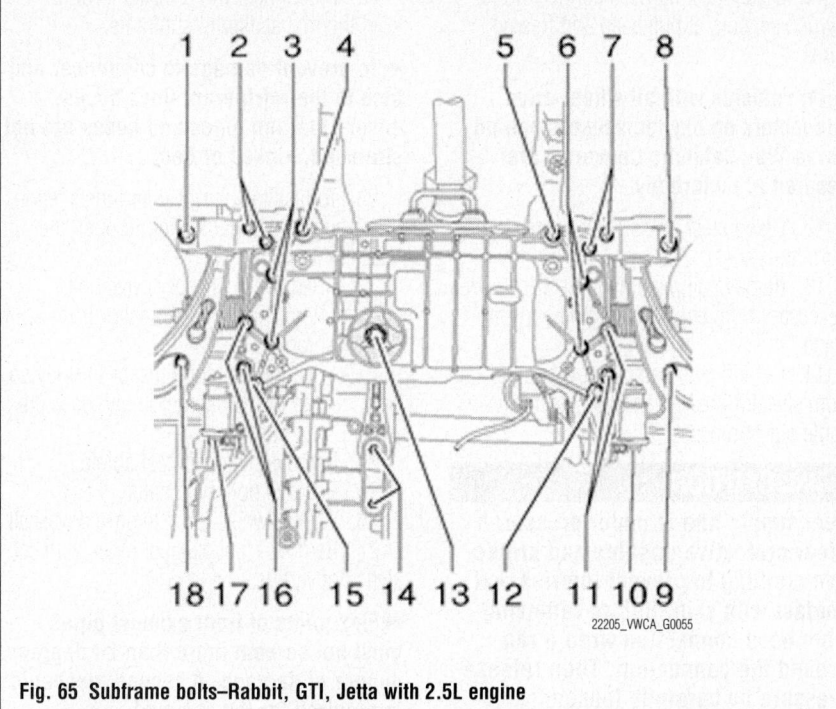

Fig. 65 Subframe bolts–Rabbit, GTI, Jetta with 2.5L engine

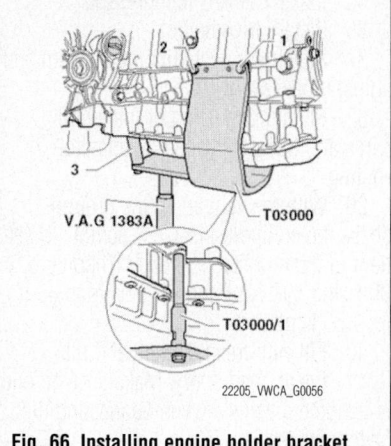

Fig. 66 Installing engine holder bracket T03000

1. Before servicing the vehicle, refer to the Precautions Section.

2. If a coded radio is installed, obtain the anti-theft coding.

3. Remove the battery and battery holder.

4. Disconnect engine wiring harness connector from Engine Control Module (ECM).

5. Release the wiring harness from all its holding fixtures and lay it up on the engine.

6. Disconnect the starter wiring and remove the main ground strap.

7. Disconnect the transaxle wiring and shift cables.

8. On manual transaxle, remove the clutch slave cylinder from the transaxle without disconnecting the hydraulic line. Secure the cylinder out of the way.

9. Disconnect the vacuum hose from the brake booster.

10. Disconnect the oxygen sensor at the bulkhead.

11. Remove the EVAP canister.

12. Raise and safely support the vehicle and remove the engine under cover and the front of both inner fender liners.

13. Drain the cooling system.

14. Remove the accessory drive belt.

15. Remove the A/C compressor without disconnecting any hoses and secure it out of the way. Do not let it hang by the hoses.

16. Disconnect the alternator wiring.

17. Disconnect the exhaust pipe from the exhaust manifold. Do not bend the flex joint more than 10°.

18. Remove both halfshafts.

19. Remove the front suspension subframe:
- Remove pendulum support bolts 13 and 14.
- Remove bolts 3 and 6.
- Remove stabilizer bar bolts 11 and 16.
- Remove subframe bolts 4, 5, 12 and 15.

### ✳✳ WARNING

**Fuel supply line is under pressure! Wear protective goggles and protective clothing to prevent injuries and contact with skin. Before removing from hose connection wrap a cloth around the connection. Then release pressure by carefully pulling hose off connection.**

20. Disconnect fuel supply and vent hoses. Seal the lines to prevent contamination.

21. Remove the coolant reservoir and windshield washer reservoir without disconnecting the hoses. Secure them out of the way.

22. Disconnect the radiator hoses.

23. Place engine/transaxle jack V.A.G 1383 A on Engine Holder Bracket T03000 and lift engine/transaxle assembly slightly.

24. Remove engine mount from above. The rear bolt in cylinder block is accessible above wheel housing liner.

25. Remove transaxle mount bolts.

➡Use a step ladder to remove mounting bolts.

26. Carefully lower engine/transaxle assembly and prevent damage to bodywork.

### To install:

27. If engine and transaxle are separated, insert new alignment bushings for centering engine/transaxle in cylinder block.

28. Engage intermediate plate at sealing flange and slide onto alignment bushings.

29. With manual transaxle, grease splines of input shaft lightly. Inspect and install clutch and clutch mechanism.

30. Carefully lift engine/transaxle into place and install new mounting bolts finger tight. Final tightening will be done with the vehicle on the ground.

31. Install subframe:
- Install new subframe-to-body bolts and torque to 52 ft. lbs. (70 Nm) plus an additional 90°
- Install new subframe-to-bracket bolts and torque to 52 ft. lbs. (70 Nm) plus an additional 90°
- Install new stabilizer bar-to-subframe bolts and torque to 15 ft. lbs. (20 Nm) plus an additional 90°
- Install a new stabilizer bar-to-connecting link new nut. Counterhold at joint pin inner fitting and torque to 48 ft. lbs. (65 Nm)
- Install new steering gear-to-subframe bolts and torque to 37 ft. lbs. (50 Nm) plus an additional 90°

32. Install pendulum support with a new center mount bolt. Torque the bracket bolts to 30 ft. lbs. (40 Nm) plus 90° the new center mount bolt to 75 ft. lbs. (100 Nm) plus 90°.

33. Install the exhaust system components.

34. Install the halfshafts.

35. Install the A/C compressor.

36. Install drive belt(s).

37. Attach/connect shift mechanism and adjust if necessary

38. Begin connecting or installing wires and hoses. Pay careful attention to routing.

39. With the vehicle on the ground, shake the engine/transaxle assembly to settle it in the mounts. Torque M10 bolts to 30 ft. lbs. (40 Nm) and M12 bolts to 45 ft. lbs. (60 Nm).

40. Fill with fresh coolant and oil.

41. Install the battery. Make sure no one is inside the vehicle when connecting the battery cables.

42. Connect scan tool and perform vehicle system test using "Guided Fault Finding"

43. Perform test drive, then repair any system malfunctions.

44. Recharge the A/C system

### 3.6L Engine

*See Figures 67 and 68.*

The engine and transaxle are lowered out of the vehicle together. Check DTC memories of all control modules, before removing engine Leave vehicle key in the ignition lock to prevent steering wheel lock from locking. When the engine is installed, some components cannot be removed or can only be removed with great difficulty. Therefore determine which components are faulty before removing engine.

1. With ignition switched turned **OFF**, remove the battery.

2. Completely remove air filter with connecting hose to throttle valve control module.

3. Remove bracket for air filter.

4. Remove windshield wiper arms and plenum chamber cover.

5. With Engine Control Module (ECM) Equipped with Anti-Theft Immobilizer, remove locking plate or locking bracket from Engine Control Module (ECM).

6. Disconnect engine wiring harness connector from Engine Control Module (ECM).

7. Release pass-through for engine wiring harness and pull off upward.

8. Unscrew wire for generator at fuse holder.

9. Unscrew ground (GND) cable on longitudinal member.

10. Open locking mechanisms of cable guide on longitudinal member. Disconnect all harness connectors of engine wiring harness/body and set down engine wiring harness on engine.

11. Disconnect harness connectors for oxygen sensors at bulkhead and free up wires.

➡ **On vehicles with all wheel drive, connectors on oxygen sensors behind Three Way Catalytic Converter are secured at underbody.**

12. Disconnect all harness connectors from transaxle.

13. Remove all remaining electrical wires necessary from engine/transaxle and set aside.

14. Pull off gear selector lever cable from selector shaft, then pull selector lever cable out of bracket.

### ❋❋ CAUTION

**Fuel supply line is under pressure! Wear protective goggles and protective clothing to prevent injuries and contact with skin. Before removing from hose connection wrap a rag around the connection. Then release pressure by carefully loosening union nut.**

15. Disconnect the fuel tank breather line, vacuum line and fuel supply line. To do so, press release buttons.

16. Disconnect vacuum line at brake booster.

17. Disconnect all other connections, vacuum and intake hoses from engine. Seal the lines so that the fuel system is not contaminated by dirt etc.

18. Open and close the cap on the coolant expansion tank to relieve the cooling system pressure.

19. Remove engine undercover.

20. Remove lower sections of front wheel housing liners.

21. Disconnect harness connector from oil level thermal sensor.

22. Disconnect wire bracket from oil level thermal sensor at subframe.

➡ **To prevent damage to condenser and also to the refrigerant lines/hoses, ensure that the pipes and hoses are not stretched, kinked or bent.**

23. To facilitate removal and installation of the engine without having to open the refrigerant circuit:
- Remove accessory drive belt.
- Remove A/C compressor from bracket.
- Secure A/C compressor to body so that the refrigerant lines/hoses are not burdened.

24. Remove halfshaft heat shield.

25. Remove both halfshafts.

26. With all wheel drive, remove driveshaft.

27. Remove front exhaust pipes with catalytic converters.

➡ **Flex joints of front exhaust pipes must not be bent more than 20 degrees danger of damage. A second mechanic is required for the removal.**

28. Remove pendulum support (transaxle mount underneath).

29. Drain the cooling system.

30. Disconnect coolant hoses for radiator at engine using spring clamp pliers.

31. Disconnect transaxle oil lines from transaxle.

32. Insert engine support T40074 in engine/transmission jack V.A.G 1383 A.

33. Install the engine support T40074 on the engine as shown and tighten the mounting nut–1–to approximately 20 Nm on the cylinder block.

➡ **On vehicles with all wheel drive, it is advisable to first attached adapter T40074/1 to cylinder block and then secure it to engine support T40074.**

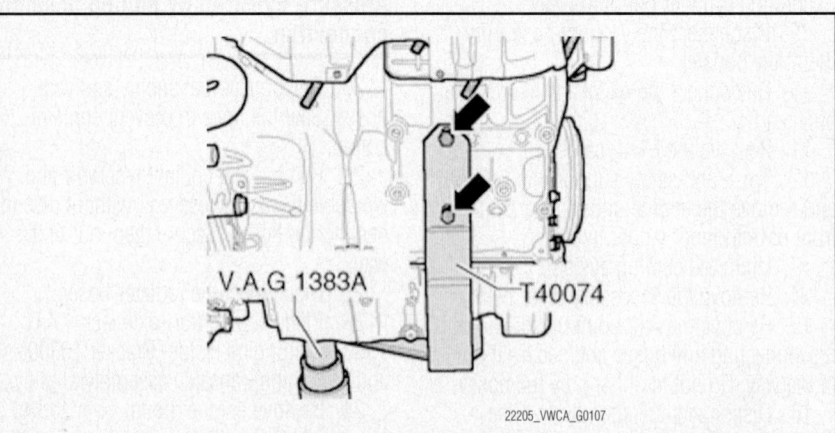

22205_VWCA_G0107

**Fig. 67 With all wheel drive, attach adapter T40074 to secure assembly engine/transmission jack**

34. Tighten the engine support T40074 bolts on the exhaust side to the cylinder block.

35. Gently lift engine and transmission.

36. Unbolt engine side of engine mount from engine mount bracket.

37. Unbolt transaxle mount from mount bracket on transaxle side from above. Use a step ladder to remove bolts.

38. Carefully lower engine/transaxle assembly from the vehicle.

### To install:

39. When installing engine/transaxle assembly, check for clearance to subframe as well as to radiator.

40. Align engine mount as follows:
- There must be a distance (a) of at least ⅜ in. (10 mm) between engine mount and right longitudinal member.
- Casting edge on engine mount bracket (2) must stand parallel to engine mount (1).

**Fig. 68 Align engine mount as shown—3.6L engine**

41. Install new engine mount bolts and torque to 45 ft. lbs. (60 Nm) plus an additional 90°.

42. Install pendulum support using new bolts. Torque the two bracket bolts to 30 ft. lbs. (40 Nm) plus 90° and the center bolt to 75 ft. lbs. (100 Nm) plus 90°.

43. Install front exhaust pipes with catalytic converters.

44. Install halfshafts and heat shield. With all four wheel drive models, install driveshaft.

45. Install A/C compressor and accessory drive belt.

46. Reconnect all wiring, hoses and fuel lines.

47. Refill the cooling system and check for leaks.

48. Refill other fluids as needed.

49. Before installing the battery and providing power to the ECM, check carefully for any disconnected wiring.

50. Connect vehicle diagnosis, testing and information system VAS 5051B. Check all DTC memories and erase all DTC entries which may have occurred during the assembly.

51. Perform a drive test and generate the readiness code "Guided Function". Then perform vehicle system test and repair any occurring malfunctions.

## EXHAUST MANIFOLD

### REMOVAL & INSTALLATION

#### 2.0L Engine

The exhaust manifold and turbocharger are one assembly. See Turbocharger Removal and Installation.

## INTAKE MANIFOLD

### REMOVAL & INSTALLATION

#### 2.0L Engine

*See Figures 69 and 70.*

**Fig. 69 Press the buttons to release the wiring connectors**

➡️**If a fuel injector is pulled out of cylinder head when removing the intake manifold, the Teflon sealing ring must be replaced.**

1. Before servicing the vehicle, refer to the Precautions Section.

2. Remove the engine cover/air filter housing assembly.

3. Remove throttle valve assembly:
- Raise and safely support the vehicle and remove the engine undercover.
- Remove the air tube between the charge air cooler and throttle body.

4. Unplug the wiring connectors near the coolant reservoir.

5. Disconnect all electrical harness connectors necessary.

6. Disconnect vacuum hose between intake manifold and vacuum pump from intake manifold.

### ❋❋ CAUTION

**Fuel supply line is under pressure! Wear protective goggles and protective clothing to prevent injuries and contact with skin. Before removing from hose connection wrap a rag around the connection. Then release pressure by carefully pulling hose off connection.**

7. Disconnect both fuel pipes from high pressure fuel pump.

8. Disconnect coolant pipe and oil dipstick tube from intake manifold. Pull oil dipstick tube upward from engine.

9. Remove intake manifold support and disconnect pressure sensor.

10. Remove all bolts from intake manifold.

11. Carefully pull off intake manifold with fuel rail from cylinder head.

### To install:

12. If any fuel injectors were pulled out when removing the intake manifold, replace Teflon sealing rings. Replace O-rings between fuel injectors and fuel rail and coat them lightly with clean motor oil.

13. Fit a new gasket into place.

14. Position intake manifold with fuel rail on cylinder head and press evenly onto fuel injectors.

15. Torque mounting bolts for intake manifold to 7 ft. lbs. (10 Nm).

16. Install support. Torque bolts to 18 ft. lbs. (25 Nm) and nut to 7 ft. lbs. (10 Nm).

17. Connect fuel supply line to high pressure pump and torque fittings to 18 ft. lbs. (25 Nm).

18. Connect the fuel return line using a new banjo fitting.

19. Install the throttle valve assembly. Replace the seal if damaged.

20. Connect all wiring, hoses, tubes and vacuum lines.

21. If intake flap motor or intake manifold was replaced, intake manifold runner position sensor must be adapted to ECM using "Guided Functions" on scan tool.

#### 2.5L Engine

*See Figures 71 through 73.*

1. Before servicing the vehicle, refer to the Precautions Section.

2. First, check whether a coded radio is installed. If necessary, obtain the anti-theft coding.

3. Disconnect the negative battery cable.

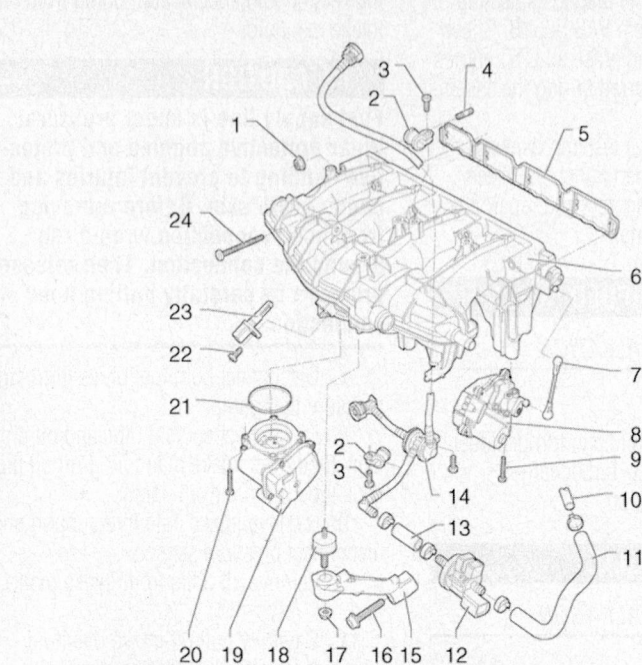

1. Manifold nut
2. Retaining clip
3. Retaining clip screw
4. Stud bolt
5. Gasket
6. Intake manifold
7. Coupling rod
8. Intake flap motor with intake manifold runner position sensor
9. Intake flap motor screw
10. To EVAP canister
11. Hose
12. Evaporative emission (EVAP) canister purge regulator valve
13. Hose
14. Double check-valve
15. Intake manifold support
16. Support bolt
17. Bushing nut
18. Bonded rubber bushing
19. Throttle valve control module
20. Throttle valve control module screw
21. Seal: replace if damaged
22. IAT sensor screw
23. Intake Air Temperature (IAT) sensor
24. Manifold bolt

22205_VWCA_G0061

**Fig. 70 Intake manifold assembly—2.5L Engine**

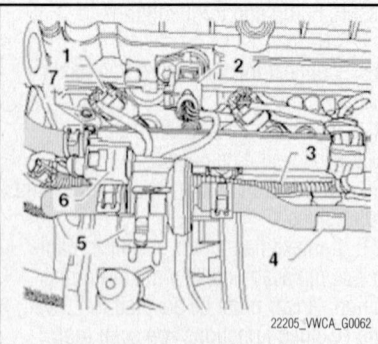

22205_VWCA_G0062

**Fig. 71 Disconnect wiring and clamps as indicated**

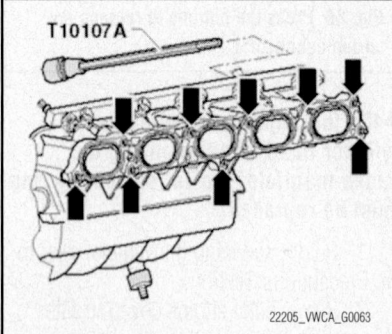

22205_VWCA_G0063

**Fig. 72 This long multi point socket is needed to remove the manifold bolts**

4. Remove engine cover/air filter assembly.

### ❊❊ CAUTION

**Fuel supply line is under pressure! Wear protective goggles and protective clothing to prevent injuries and contact with skin. Before loosening the fuel lines, place a rag around the connection point. Then release pressure by carefully pulling hose off connection.**

5. Disconnect the three under-hood fuel lines. To release wires, press the circlip in.

On the fuel supply line, the retainer must be pressed upward in the housing.

6. Disconnect connectors (1), (2) and (6). Remove wiring harness (3) from transport strap. Pull clamps (4) and retaining ring (5) out of locking mechanism. Remove bolts (7) and remove transport strap.

7. Remove throttle valve control module (throttle body). The coolant hoses remain attached.

8. Disconnect manifold wiring and hose for crankcase ventilation.

9. Remove wiring harness by carefully pressing off clips.

10. Pull oil dipstick out and press dipstick tube retaining ring downward.

11. Raise and safely support the vehicle and remove the engine undercover.

12. Loosen bolts or nuts on the bottom side of the intake manifold.

13. Loosen bolt for intake manifold support and dipstick tube. Lay tube aside.

14. Open clip on leak detection pump (LDP) vacuum hose.

15. Loosen intake manifold bolts using Tool T10107A. Bolts remain in intake manifold.

16. Remove intake manifold upward at an angle. Make sure that no bolts fall out.

17. Seal intake ports in cylinder head using a clean rag.

18. If manifold must be replaced:
   • Remove fuel rail with injectors.
   • Disconnect vacuum hose for Leak Detection Pump (LDP).
   • Remove Manifold Absolute Pressure (MAP) sensor.

### To install:

19. Make sure all sealing surfaces are clean and dry. Fit new sealing rings to the intake manifold runners.

20. If the injectors were removed, fit new O-rings, lubricate them with engine oil and install the injectors.

21. Replace oil dipstick tube seal.

22. Fit the manifold into place and torque the screws to 6 ft. lbs. (9 Nm) working from the middle and working towards the ends in a diagonal sequence.

23. Install mounting supports and torque bolts to 18 ft. lbs. (25 Nm).

24. Install the dipstick tube and insert the dipstick.

25. Install the throttle body.

26. Connect all wires, hoses and tubes using new seals or gaskets as needed.

27. Connect the battery.

28. Bleed fuel supply system.

29. Run the engine to check for leaks.

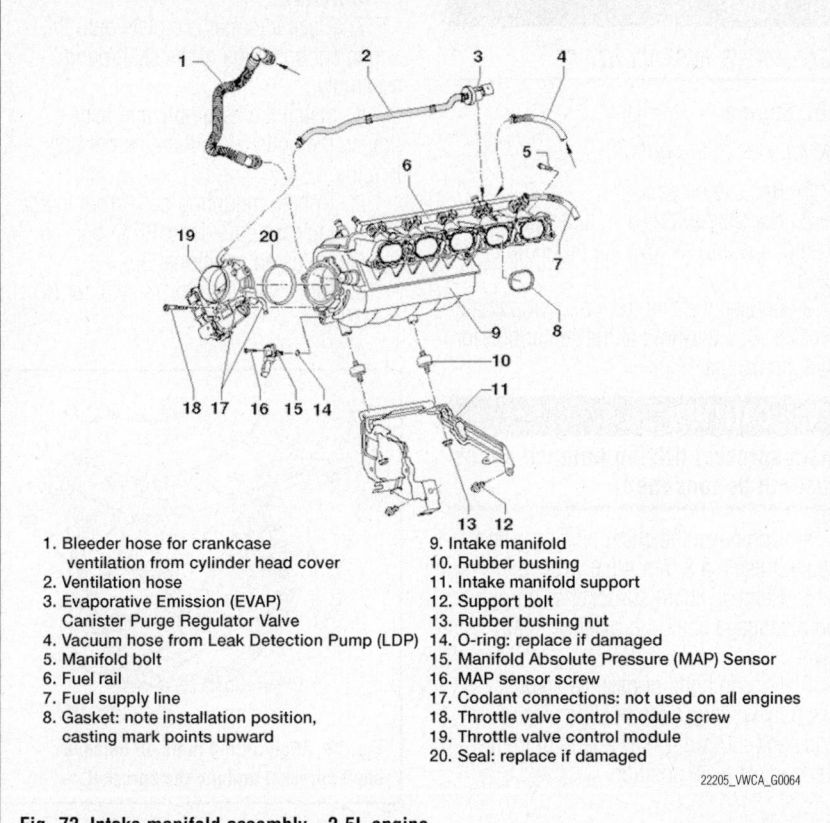

1. Bleeder hose for crankcase ventilation from cylinder head cover
2. Ventilation hose
3. Evaporative Emission (EVAP) Canister Purge Regulator Valve
4. Vacuum hose from Leak Detection Pump (LDP)
5. Manifold bolt
6. Fuel rail
7. Fuel supply line
8. Gasket: note installation position, casting mark points upward
9. Intake manifold
10. Rubber bushing
11. Intake manifold support
12. Support bolt
13. Rubber bushing nut
14. O-ring: replace if damaged
15. Manifold Absolute Pressure (MAP) Sensor
16. MAP sensor screw
17. Coolant connections: not used on all engines
18. Throttle valve control module screw
19. Throttle valve control module
20. Seal: replace if damaged

22205_VWCA_G0064

**Fig. 73 Intake manifold assembly—2.5L engine**

## OIL PAN

### REMOVAL & INSTALLATION

#### 2.0L Engine

1. Before servicing the vehicle, refer to the Precautions Section.
2. Raise and safely support the vehicle and remove the engine undercover.
3. Drain engine oil.
4. Remove the bolts holding the air tube below the crankshaft pulley.
5. On the side of the oil pan, remove oil return line bolt and unplug the sensor connector.
6. Remove three bolts that hold the transaxle to the oil pan.
7. Remove remaining bolts and remove oil pan. Loosen oil pan with light blows of a rubber hammer if necessary.

**To install:**
8. Make sure all sealing surfaces are clean and dry.
9. Apply silicone sealant to oil pan. The bead must run inside the bolt holes.

➡**Sealant bead must not be more than 0.120 in. (3 mm) thick to prevent excess sealant from getting into the oil pan and clogging the oil pump intake screen.**

10. Fit the pan into place and start all the bolts.
11. If the transaxle is mounted on the engine, tighten the three pan-to-transaxle bolts first to 30 ft. lbs. (40 Nm).
12. Torque the remaining bolts in three steps in a diagonal sequence to 11 ft. lbs. (15 Nm).
13. After installing oil pan, allow sealant to dry for approximately 30 minutes before refilling the engine with oil.

#### 2.5L Engine

The lower oil pan can be removed easily. The upper oil pan must be removed to remove the oil pump.

##### Lower Oil Pan

1. Raise and safely support the vehicle and remove the engine undercover.
2. Drain the oil.
3. Remove the bolts to remove the lower oil pan. There are several recesses around the edge to pry without damaging the sealing surfaces.

**To install:**
4. Make sure all sealing surfaces are clean and dry.
5. Apply a bead of silicone sealant to the lower oil pan. The bead should run inside the bolt holes.

➡**Sealant bead must not be more than 0.080 in. (2 mm) thick to prevent excess sealant from getting into the oil pan and clogging the oil pump intake screen.**

6. Fit the pan into place and start all the bolts.
7. Torque the bolts in a diagonal sequence to 7 ft. lbs. (10 Nm).
8. After installing oil pan, allow sealant to dry for approximately 30 minutes before refilling the engine with oil.

##### Upper Oil Pan

*See Figures 74 and 75.*

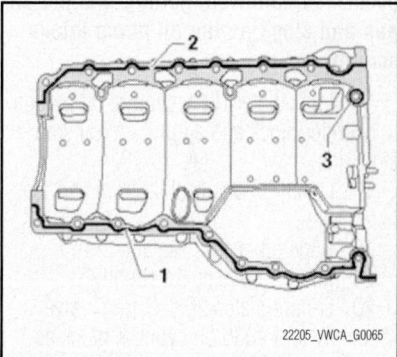

22205_VWCA_G0065

**Fig. 74 Apply sealant to upper oil pan as shown—2.5L engine**

Removing the upper oil pan requires removing the flywheel.

1. Before servicing the vehicle, refer to the Precautions Section.
2. Raise and safely support the vehicle and drain engine oil.
3. Remove the transaxle or engine/transaxle assembly.
4. Remove the flywheel.
5. Remove the timing chain cover and cylinder head cover.
6. Hold the camshafts and loosen the camshaft sprockets.
7. Remove the control housing cover from the end of the cylinder block.
8. Compress the timing chain tensioner and secure it with a pin.
9. Remove chain guide.
10. Remove crankshaft pulley sealing flange.
11. Remove lower section of oil pan.
12. Remove bolts to remove oil intake.
13. Remove bolts and carefully pry upper section of oil pan from cylinder block at designated points using a suitable screwdriver.

*To install:*

## ❊❊ CAUTION

**To prevent injuries from shavings, wear protective goggles and protective clothing. Make sure that no sealant residue enters the engine.**

14. Remove remainder of sealant from upper section of oil pan and cylinder block e.g. using a rotating plastic brush. Clean sealing surfaces so they are completely free of any oil or grease.

15. Apply sealant bead to upper section of oil pan.

➡**Sealant bead must not be more than 0.080 in. (2 mm) thick to prevent excess sealant from getting into the oil pan and clogging the oil pump intake screen.**

16. Immediately fit section of oil pan on to cylinder block and align it on transmission side.

17. Install 2 bolts each at front and rear hand-tight.

18. Wipe off excess sealant.

19. Loosen bolts again slightly.

20. Secure plates 2036/1 from valve assembly tool 2036 on cylinder block as depicted.

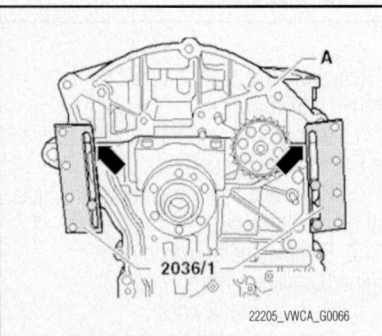

**Fig. 75 Oil pan alignment plates—2.5L engine**

21. Press upper section of oil pan firmly on to plates 2036/1 and tighten bolts again hand-tight.

22. Install remaining bolts and tighten by hand.

23. Make sure that upper section of oil pan makes contact on plates 2036/1 .

24. Tighten all bolts in diagonal sequence from inside working toward outside. Torque bolts to 18 ft. lbs. (25 Nm).

25. Insert new seal into oil pump.

26. Install oil pump intake and bracket.

27. Install chain guide and release the chain tensioner.

28. Adjust valve timing.

## OIL PUMP

*REMOVAL & INSTALLATION*

### 2.0L Engine

*See Figures 76 through 78.*

1. Remove oil pan

2. Use a screwdriver to disengage three locking lugs and remove the oil pump chain guard.

3. Loosen the bolt from oil pump chain sprocket. Counter-hold at the center bolt for vibration damper.

## ❊❊ CAUTION

**Chain sprocket driving balance shafts must not be loosened!**

4. Compress the chain tensioner and secure it using a 3 mm Allen wrench.

5. Remove chain sprocket of oil pump and disengage chain on balance shaft drive.

6. Loosen bolts of balance shaft assembly working from outside toward inside and remove it. Pay attention to the intermediate plate position.

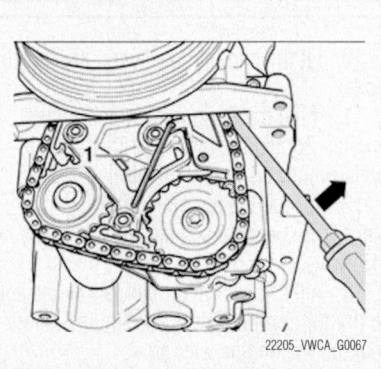

**Fig. 76 Compress the chain tensioner and lock it in place.**

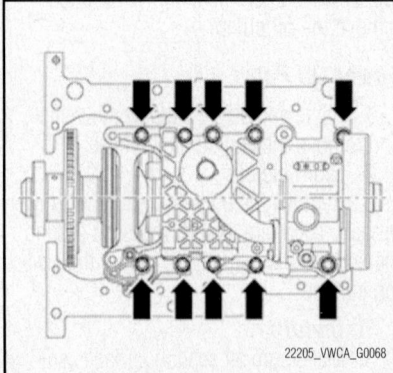

**Fig. 77 Remove these bolts to remove the oil pump/balance shaft assembly**

*To install:*

7. Place intermediate plate onto the alignment bushings of the shaft/pump assembly.

8. Install the assembly and hand-tighten the bolts. Note the different bolt lengths.

9. Tighten mounting bolts from inside working toward outside to 11 ft. lbs. (15 Nm) plus an additional 90°

10. Turn crankshaft to set cylinder No. 1 at TDC.

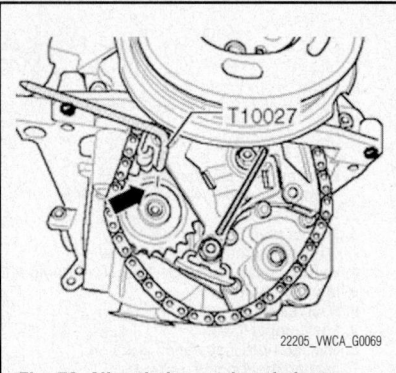

**Fig. 78 Align timing mark on balance shaft sprocket and pin the sprocket**

11. Set timing mark on balance shaft sprocket opposite alignment bore. Secure sprocket in this position using connecting pin T10027. Fit the chain onto the sprocket.

12. Install oil pump chain sprocket and tighten bolt by hand.

➡**Oil pump chain sprocket fits only in one position. To install, only the oil pump must be turned.**

13. Remove connecting pin T10027 and Allen wrench.

14. Hold the crankshaft from turning and torque the oil pump sprocket bolt to 15 ft. lbs. (20 Nm) plus an additional 90°

15. Install oil pan.

### 2.5L Engine

The flywheel must be removed to remove the upper oil pan.

1. Before servicing the vehicle, refer to the Precautions Section.

2. Remove upper oil pan.

3. Tension chain tensioner, secure it with locking pin T10115 and remove it.

4. Remove bolts and hold chain sprocket tightly with counterhold tool T10172.

5. Remove chain sprocket from oil pump and remove oil pump.

*To install:*

6. Secure crankshaft.

7. Replace O-ring and hand tighten bolts for oil pump to cylinder block.

8. Set chain sprocket onto oil pump shaft with text facing outward and fasten it using a new bolt (drive chain still not installed).

9. Tighten bolts to 14 ft. lbs. (20 Nm), plus an additional 90°.

10. Loosen the oil pump bolt closest to the crankshaft and the one closest to the chain sprocket; oil pump must be easy to move.

11. Make sure that no shavings are found on magnets of oil pump locating tool T03005. Contact surfaces of crankshaft and chain sprocket must be clean.

12. Place oil pump locating tool T03005 onto end of crankshaft and fasten it with two of the vibration damper/belt pulley bolts.

13. Tighten to 21 ft. lbs. (30 Nm). Oil pump is pulled into place by magnets.

14. Remove locking pin T40069.

15. Push crankshaft, as for checking axial bearing play, toward pulley drive.

➡**This work step is important to guarantee the correct positioning of the chain sprockets to one another.**

16. In this position, first tighten the bolt closest to the crankshaft, the one closest to the chain sprocket, and then the remaining bolt to 17 ft. lbs. (25 Nm).

17. Install locking pin T40069 again. Crankshaft may only be moved minimally around TDC point. Otherwise, there is a risk that valves will contact piston.

18. Remove oil pump locating tool T03005.

19. If a new oil pump is installed, fill oil pump with some clean engine oil via suction pipe and turn pump shaft a several times.

20. Place drive chain onto chain sprocket of oil pump.

21. Install oil pan.

22. Install guide rail, release tension on chain tensioner and pull out locking pin T10115.

➡**Make sure that the drive chain is correctly positioned in the guide rail and in the tensioning rail.**

23. Adjust valve timing.

24. Further assembly is performed in reverse order of removal. Note the following:

a. Remove locking pin T40069 from rear of cylinder block and screw in locking bolt. Tighten to 21 ft. lbs. (30 Nm).

b. Refill the engine cooling system.

## PISTON AND RING

### POSITIONING

*See Figure 79.*

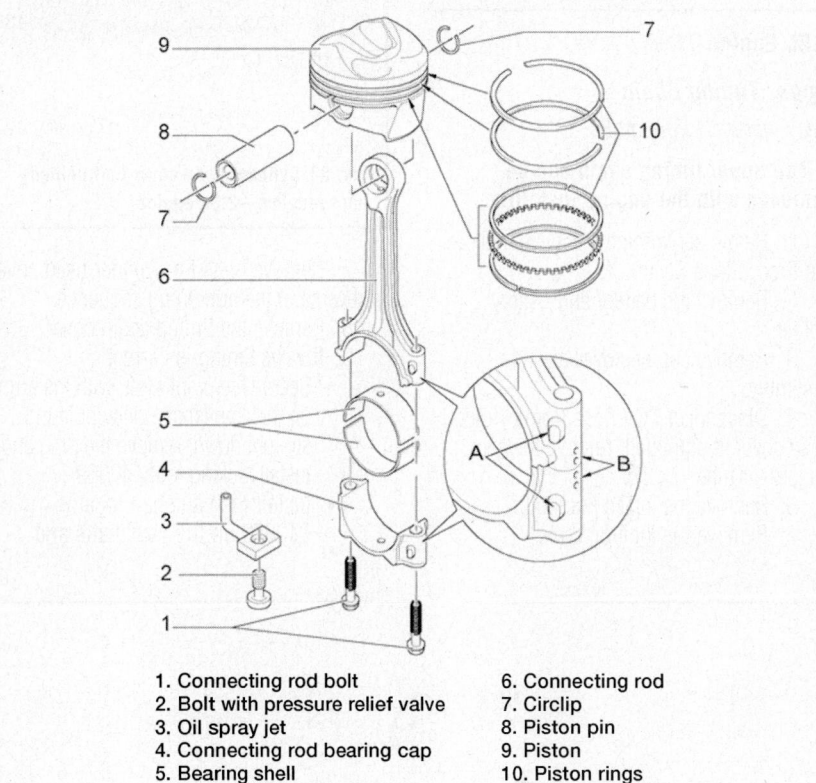

1. Connecting rod bolt
2. Bolt with pressure relief valve
3. Oil spray jet
4. Connecting rod bearing cap
5. Bearing shell
6. Connecting rod
7. Circlip
8. Piston pin
9. Piston
10. Piston rings

22205_VWCA_G0070

**Fig. 79 2.0L and 2.5L engines—piston ring end-gap spacing**

## REAR MAIN SEAL

### REMOVAL & INSTALLATION

#### 2.0L Engine

1. Remove transaxle and flywheel.
2. Remove oil pan.
3. Remove sealing flange bolts.

➡**Do not lubricate oil seal!**

4. Clean sealing surfaces. They must be free of oil and grease.

5. To install, use the support sleeve provided with the new part. Supporting sleeve may only be removed after the sealing flange has been slid onto the crankshaft pin. Sealing flange sits on alignment pins.

6. Torque bolts to 11 ft. lbs. (15 Nm).

7. Install oil pan

#### 2.5L Engine

*See Figure 80.*

1. Remove the transaxle and flywheel.
2. Pull out old seal using pulling hook T20143/2.

➡**Do not lubricate sealing lip of oil seal!**

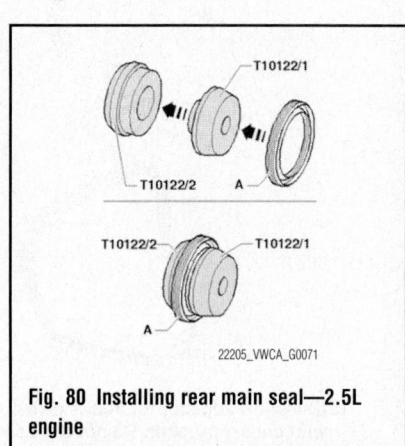

22205_VWCA_G0071

**Fig. 80 Installing rear main seal—2.5L engine**

3. Clean sealing surfaces. They must be free of oil and grease.

4. Insert assembly device T10122/1 onto pull sleeve T10122/2 and slide seal onto pull sleeve. Remove assembly device T10122/1.

5. Install pull sleeve T10122/2 with seal onto crankshaft.

6. Press in seal all around evenly and flush using pressure sleeve T10122/3.

## TIMING CHAIN, SPROCKETS, FRONT COVER AND SEAL

### REMOVAL & INSTALLATION

#### 2.5L Engine

#### Upper Timing Chain

*See Figures 81 through 85.*

➡ **The upper timing chain can be removed with the engine installed.**

1. Before servicing the vehicle, refer to the Precautions Section.
2. Remove the battery and battery holder.
3. Remove engine cover/air filter assembly.
4. Disconnect PCV hose from cylinder head cover. If equipped, remove the secondary air tube.
5. Remove the intake manifold.
6. Remove the ignition coils.

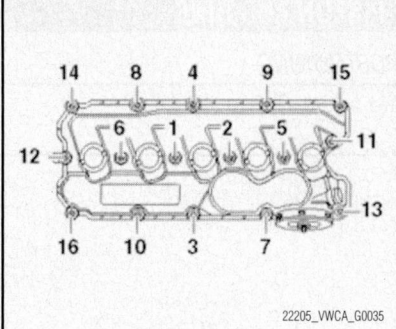

Fig. 82 Cylinder head cover bolt tightening sequence—2.5L engine

7. Remove bolts for cylinder head cover in reverse of the tightening sequence.
8. Remove the timing chain cover.
9. If valve timing is correct:
   - Secure the crankshaft: with the arrow on the crankshaft pulley pointing straight down, remove the plug and install locking tool T40069.
   - Install the camshaft locator T40070 on the camshafts and

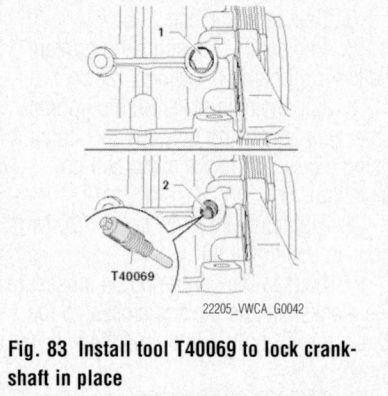

Fig. 83 Install tool T40069 to lock crankshaft in place

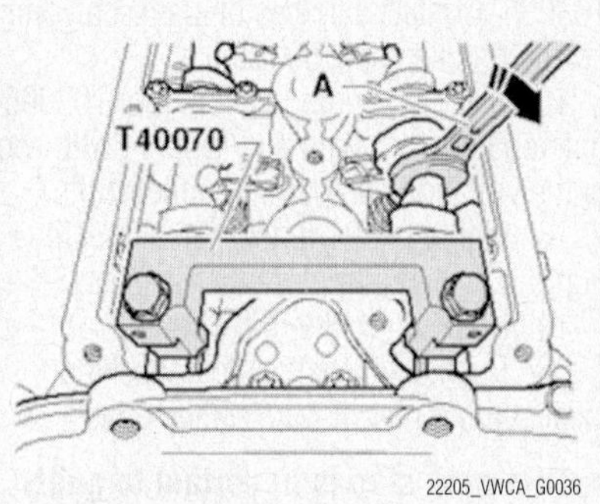

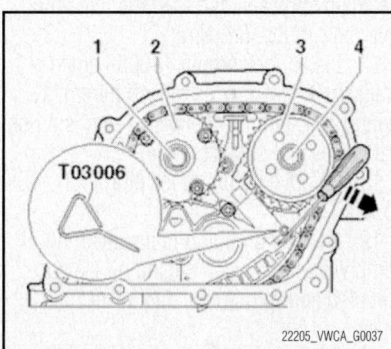

Fig. 84 Lock the camshafts in place with this tool. Rock camshaft slightly if needed.

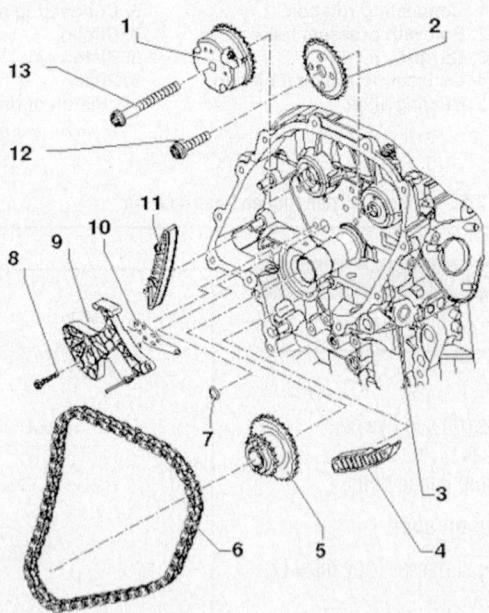

1. Camshaft adjuster for intake camshaft with chain sprocket: Do not disassemble
2. Chain sprocket for exhaust camshaft
3. Cylinder Head
4. Tensioning rail
5. Double chain sprocket (drive wheel)
6. Timing Chain
7. Strainer
8. Chain tensioner screw
9. Chain tensioner
10. Gasket
11. Guide rail
12. Sprocket bolt: always replace
13. Sprocket bolt: always replace

Fig. 81 Upper timing chain components—2.5L engine

Fig. 85 Pry the chain tensioner to retract the piston and insert a locking pin. Remove bolts (1 and 4) to remove sprockets (2 and 3)

tighten bolts to 15 ft. lbs. (20 Nm). If the bolts could not be screwed in easily, rotate exhaust camshaft slightly using an open-end wrench.
   - Relieve tension on timing chain. To do so, pry between piston of chain tensioner and tensioning track. Secure completely pressed in piston using Locking Pins T03006. Locking pin must be inserted until it stops.

10. If valve timing is not correct:
- Remove timing chain case cover
- Rotate crankshaft to TDC cylinder 5 but do not lock it into place.
- Rotate crankshaft so that camshaft locator T40070 can be screwed easily on to camshafts.

11. Mark the direction of travel and remove the timing chain.

12. If necessary, remove the bolts to remove the sprockets. It may be necessary to gently pry them off. New bolts are required for installation.

### To install:

13. Make sure all sealing surfaces are clean and dry.

14. Install the timing chain in the original direction of travel.

15. If removed, install the tensioner. Remove the tensioner lock pin.

16. Remove the camshaft locator tool.

17. Remove the crankshaft locking pin from the cylinder block. Rotate the crankshaft two full turns to make sure camshaft timing marks still align properly.

18. Install the remaining components.

### Lower Timing Chain

*See Figure 86.*

1. Remove engine.
2. Remove control housing cover.
3. Remove timing chain. Mark direction of travel.
4. Remove chain tensioner.

➡️When installing, place on in original direction of travel.

## TIMING BELT, FRONT COVER AND SEAL

### *REMOVAL & INSTALLATION*

#### 2.0L Engine

*See Figures 87 through 91.*

➡️To remove the engine mount, the engine must be lifted quite far using engine support bridge 10-222A. To prevent damage to halfshafts from significant bend angles, it is necessary to disengage them from the transaxle. New engine mount bolts are required for installation.

1. Before servicing the vehicle, refer to the Precautions Section.
2. Remove engine cover/air filter housing.

#### ✳️✳️ CAUTION

**Fuel supply line is under pressure! Wear protective goggles and protective clothing to prevent injuries and contact with skin. Before removing from hose connection wrap a rag around the connection. Then release pressure by carefully pulling hose off connection.**

3. For safety reasons, remove the fuel pump control module fuse from the fuse panel before opening the fuel system. This prevents the fuel pump from being activated when opening the driver side door.

4. Disconnect fuel lines 1, 2 and 3. To do so, press release buttons.

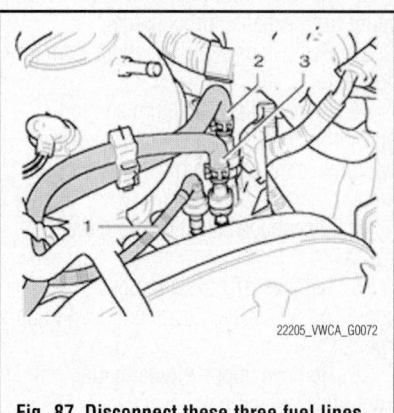
22205_VWCA_G0072

**Fig. 87 Disconnect these three fuel lines**

5. Disconnect ventilation line and remove the EVAP canister
6. Open coolant reservoir to release any pressure, then seal it again.
7. Loosen clamp on the coolant tube in front of the front cover and disconnect the hose. Collect escaping coolant with a cloth.
8. Without disconnecting the hoses, unplug the wiring and remove the coolant reservoir. Lay it aside.
9. Remove the accessory drive belt and the belt tensioner.
10. Attach bracket T10339 at both upper belt tensioner holes.
11. Position engine support bridge 10-222A with base 10-222A/1 and adapter 10–222 A /18 on fender edges.
12. Insert fender insert pads T40045 under right and left fender edge. Support engine in installation location.
13. Raise and safely support the vehicle and remove the engine undercover.
14. Remove right front part of wheel housing liners.
15. With two-piece timing belt cover, remove the upper cover. With one-piece cover, remove timing belt cover inspection cover.

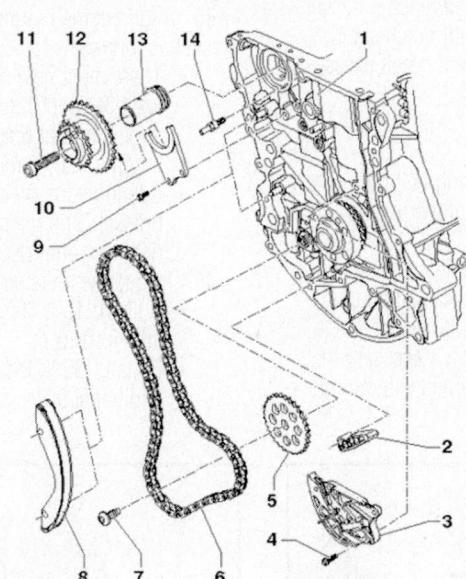

1. Cylinder block
2. Guide rail
3. Chain tensioner
4. Chain tensioner screw
5. Chain sprocket of oil pump
6. Power Take-Off Drive Chain
7. Sprocket bolt: always replace
8. Guide rail
9. Thrust bearing screw
10. Axial (thrust) bearing disc
11. Sprocket bolt: always replace
12. Double chain sprocket (drive wheel)
13. Journal for double chain sprocket (drive wheel)
14. Pin for tensioning track

22205_VWCA_G0078

**Fig. 86 Lower timing chain (power take-off chain)—2.5L engine**

16. Bring camshaft gear to marking for TDC cylinder No. 1 by turning crankshaft. Marking on camshaft gear must align with arrow on timing belt guard.

17. Remove vibration damper/belt pulley.

18. Remove bolts from lower section of belt guard.

19. Remove bottom engine bracket bolt (near crankshaft).

20. Loosen bolts of right wheel housing liner and push it aside.

21. Remove bolt through hole in wheel housing.

➡️**The next repair steps are necessary in order to be able to lift the engine far enough.**

22. Remove front pipe with catalytic converter from exhaust turbocharger

23. Remove catalytic converter from turbocharger.

24. Remove the pendulum support bracket bolts (bottom engine mount). Do not remove the bolt from the center of the mount.

25. Disengage halfshafts from the transaxle.

26. Remove A/C system pipes from body. The A/C system does not need to be opened.

27. Remove upper engine mount from in front of the timing belt cover.

### ✳✳ WARNING

**When lifting the engine with the engine support bridge 10-222A, make sure that no components/hoses are damaged, overstressed or torn off.**

28. Lift engine using engine support bridge 10-222A far enough until upper bolt of engine mount bracket can be loosened and removed.

29. Continue to raise engine until engine mount bracket can be removed upward.

30. Remove remaining bolts from timing belt guard and remove guard from engine.

31. Mark rotational direction of timing belt.

32. Loosen tensioning roller and remove timing belt.

33. Then, turn the crankshaft back slightly to lower pistons from TDC.

### *To install:*

➡️**This is an interference engine. When turning camshaft, crankshaft must not be at TDC. Valves and/or pistons may be damaged.**

34. Place the timing belt onto the crankshaft sprocket (observe direction of rotation).

35. Secure the lower timing belt guard with the two lower bolts.

36. Install vibration damper/belt pulley with new bolts. Torque to 7 ft. lbs. (10 Nm) plus an additional 90°

37. Set the crankshaft and camshaft to TDC at cylinder 1 using timing marks.

38. Route timing belt in sequence: Tensioning roller, camshaft gear, coolant pump and last over the idler pulley.

➡️**Be sure tensioning roller is properly seated in cylinder head.**

39. Tension timing belt:
- Turn socket head wrench on eccentric clockwise until notch stands above the tab (timing belt excessively tensioned).
- Release tension on timing belt again.
- Now tension timing belt until notch and tab align. Tighten nuts to 18 ft. lbs. (25 Nm).
- Turn the crankshaft 2 revolutions in direction of engine rotation until the engine is positioned at TDC again. For this it is necessary that the last 45° is turned without interruption.

**Fig. 90 Engine mount/bracket alignment**

- Check tension of timing belt again. Tab and notch must align.
- Check the valve timing again.

40. Install the timing belt cover.

41. Install engine bracket on cylinder block from above and tighten bolts hand-tight. The bottom bolt is shorter than the other two. One bolt can be installed through hole in wheel housing and tightened.

42. Lower the engine far enough until lower bracket bolt can be tightened. Torque all bolts to 33 ft. lbs. (45 Nm).

43. Install the engine mount with new bolts. Make bolts only hand tight at this point.

44. Lower the engine and install the engine mount-to-bracket bolts hand-tight.

45. Align engine mount/engine mount bracket as follows:
- There must be a distance (a) of at least ⅜ in. (10 mm) between engine mount bracket and right longitudinal member.
- Casting edge on engine mount bracket (2) must stand parallel to engine mount (1).

46. Torque the bolts as follows:
- Bolts A: 15 ft. lbs. (20 Nm) plus an additional 90°
- Bolts B: 30 ft. lbs. (40 Nm) plus an additional 90°

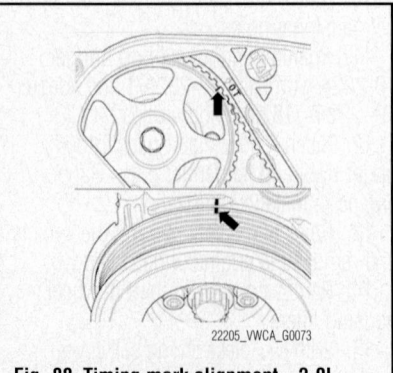

**Fig. 88 Timing mark alignment—2.0L Engine**

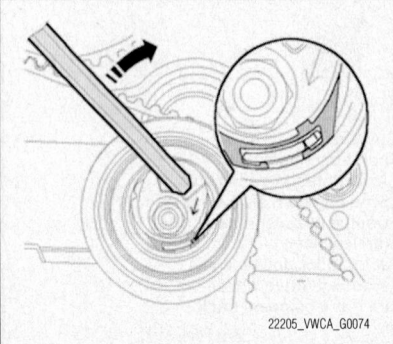

**Fig. 89 Timing belt tensioner adjustment showing the notch standing above the tab**

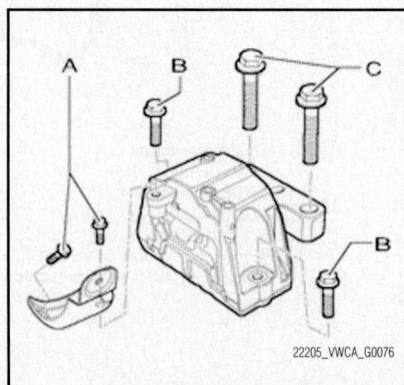

**Fig. 91 Engine mount bolt installation**

- Bolts C: 45 ft. lbs. (60 Nm) plus an additional 90°

47. Remove engine support bridge.

48. Remove bracket T10339 and install accessory belt tensioner and the belt.

49. Reconnect fuel and ventilation lines. Make sure connector couplings are securely fastened.

50. Install coolant expansion tank and all other components.

51. Insert the fuel pump control module fuse and run the pump to check for leaks. The pump should run for two seconds when the driver's door is opened.

## TURBOCHARGER

### REMOVAL & INSTALLATION

#### 2.0L Engine

*See Figures 92 and 93.*

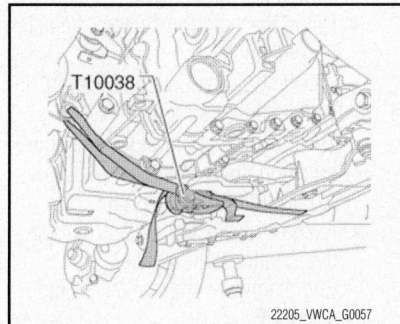

22205_VWCA_G0057

**Fig. 92 With the pendulum support (lower mount) bracket bolts removed, pull the engine towards the rear to gain clearance**

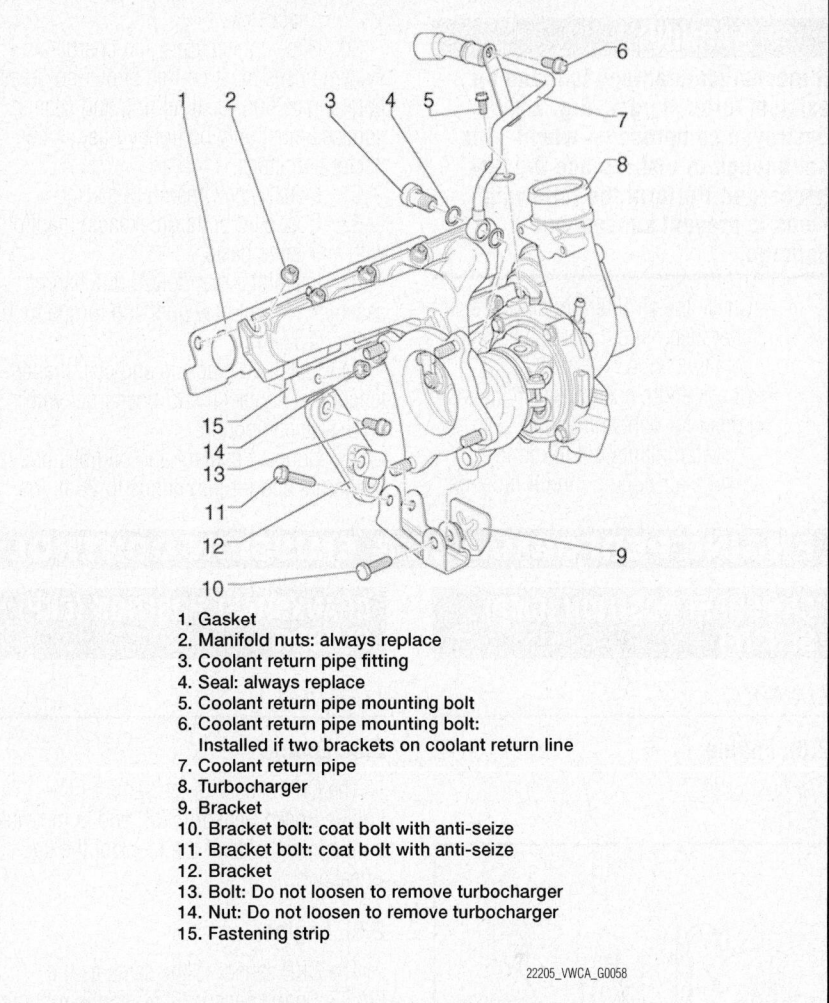

1. Gasket
2. Manifold nuts: always replace
3. Coolant return pipe fitting
4. Seal: always replace
5. Coolant return pipe mounting bolt
6. Coolant return pipe mounting bolt: Installed if two brackets on coolant return line
7. Coolant return pipe
8. Turbocharger
9. Bracket
10. Bracket bolt: coat bolt with anti-seize
11. Bracket bolt: coat bolt with anti-seize
12. Bracket
13. Bolt: Do not loosen to remove turbocharger
14. Nut: Do not loosen to remove turbocharger
15. Fastening strip

22205_VWCA_G0058

**Fig. 93 Turbocharger assembly–2.0L Engine**

1. Remove engine cover/air filter housing and remove intake hose from turbocharger.

2. Drain the cooling system.

3. Remove right front part of wheel housing liner.

4. Remove charge air hose from charge air cooler.

5. Remove bolts below the crankshaft pulley and disconnect charge air pipe.

6. Unplug the connectors and unclip wiring at turbocharger inlet.

7. Disconnect oxygen sensor connector above the brake booster.

8. Disconnect ignition coils and set wiring harness aside.

9. Disconnect the coolant line to coolant expansion tank.

10. Disconnect coolant hoses from the coolant pipe around the end of the engine.

11. Disconnect the heater hoses at the firewall.

12. Remove the heat shield above the exhaust manifold with coolant pipe.

13. Remove line for crankcase ventilation with heat shield from turbocharger.

14. Disconnect line for crankcase ventilation from cylinder head cover.

15. Disconnect EVAP canister line to turbocharger from cylinder head cover.

16. Loosen oil supply line at turbocharger.

17. Remove both upper nuts of front exhaust pipe/turbocharger.

18. Remove heat shield for right half-shaft.

19. Remove both lower nuts of front exhaust pipe/turbocharger.

20. Remove bracket for exhaust pipe.

21. Disconnect the exhaust system at the clamping sleeve.

22. Disconnect front exhaust pipe from turbocharger and slide it toward rear slightly.

➡**Catalytic converter is not removed and oxygen sensor connector is not disconnected.**

23. Remove oil supply line and coolant supply line at turbocharger.

24. Disconnect oil return line from turbocharger.

25. Remove bolts and remove support for turbocharger.

➡**The following work step is necessary to achieve some more room between cylinder head and bulkhead.**

26. Remove the pendulum support bracket bolts. Do not remove the bolt from the center of the mount.

27. Pull the engine approximately ¾ in. (20 mm) toward rear using tensioning strap.

28. Disconnect coolant pipe in front of the timing belt cover.

➡**Retaining strip nuts do not have to be loosened.**

29. Remove exhaust manifold nuts and remove turbocharger/exhaust manifold upward.

*To install:*

> ❄ **WARNING**
>
> **If mechanical damage is found on exhaust turbocharger, e.g. a destroyed compression wheel, it is not enough to just replace the turbocharger. Perform the following steps to prevent subsequent damage.**

- Check the air filter housing, the air filter insert and the intake hoses for contamination.
- Check entire charge air circuit and cooler for contamination.
- If contaminants are found in charge air circuit, circuit must be cleaned and cooler replaced if necessary.

30. Hose connections and charge air system hoses must be free of oil and grease before installing. Sealing ring and sealing surfaces must only be lightly oiled at connector couplings.

31. Install a new manifold gasket.

32. Coat stud bolts on exhaust manifold with anti-seize paste.

33. Install the manifold/turbocharger assembly. Install new nuts and torque to 15 ft. lbs. (20 Nm).

34. Install all brackets and coat the fasteners that contact the turbocharger with anti-seize compound.

35. Connect coolant and oil lines using new seals and torque fittings to 26 ft. lbs.

(35 Nm). Before connecting the oil supply line, pour fresh oil into the turbocharger.

36. Install new pendulum support bracket bolts and torque to 30 ft. lbs. (40 Nm) plus an additional 90°.

37. Install remaining components and reconnect all wiring and hoses.

38. Fill the cooling system.

39. After starting, let engine idle for approximately 1 minute to ensure adequate oil supply to the turbocharger.

## VALVE LASH

### ADJUSTMENT

All engines are equipped with hydraulic valve lash adjusters. No periodic valve lash adjustment is necessary.

# ENGINE PERFORMANCE & EMISSION CONTROL

## CAMSHAFT POSITION (CMP) SENSOR

### LOCATION

#### 2.0L Engine

*See Figure 94.*

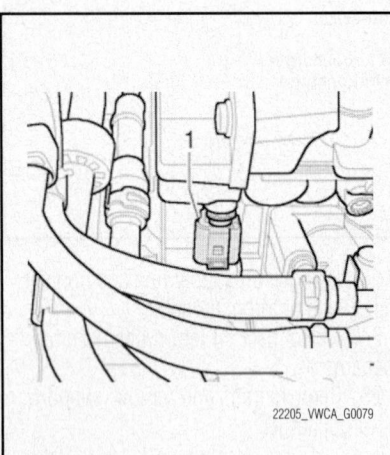

22205_VWCA_G0079

**Fig. 94 Camshaft position sensor—2.0L engine**

#### 2.5L Engine

The CMP sensor is on the intake side of the cylinder head above the cylinder No. 1 intake runner.

### REMOVAL & INSTALLATION

1. Unplug the connector.
2. Remove the screw and the sensor and cover the hole to prevent dirt from getting into the engine.
3. When installing, use a new seal.

## CRANKSHAFT POSITION (CKP) SENSOR

### LOCATION

#### 2.0L Engine

The CKP sensor is the same as the Engine Speed Sensor (G28) and is mounted on the intake side of the block at the flywheel end.

#### 2.5L Engine

The CKP sensor is the same as the Engine Speed Sensor (G28) and is mounted on the timing chain cover at the flywheel end.

### REMOVAL & INSTALLATION

1. Unplug the connector.
2. Remove the screw and the sensor and cover the hole to prevent dirt from getting into the engine.
3. When installing, use a new seal.

## ELECTRONIC CONTROL MODULE (ECM)

### REMOVAL & INSTALLATION

#### Rabbit, GTI and Jetta

*See Figures 95 through 97.*

➡ **When the Motronic Engine Control Module (ECM) J623 electrical harness connectors are disconnected, the adaptation values are erased and the DTC memory content remains intact.**

1. Check the identification of the previous Motronic Engine Control Module (ECM) J623 as follows:

- Connect the scan tool.
- Switch the ignition **ON**.
- Using the scan tool, select "Vehicle information".
- Select "Calibration Identification" in vehicle information.
- The electronic control module identification number will be displayed, e.g.: 06A906032NA 4983
- Record the electronic control module identification number.
- End diagnosis and switch the ignition **OFF**.

2. Switch the ignition **ON**.

3. Actuate the touch-wipe function to allow the wipers to move to the service position.

4. Remove the wiper arms caps.

5. Loosen the wiper arm nuts several turns.

6. Loosen the wiper arms from the wiper axle by rocking slightly.

7. Remove the wiper arms nuts and the wiper arms from the wiper axles.

8. Press the spray nozzles with the water lines attached through the plenum chamber opening.

9. Remove the rubber seal and the plenum chamber cover.

10. With 2.0L engine
- Saw off two rounded sides of the shear bolt heads so that two flat parallel surfaces result.
- Using locking pliers, remove the shear bolts and discard them. The threads of the shear bolts have been coated with a locking compound.
- Using a screwdriver, pry the protective housing upward and sideways to remove from the retaining plate.

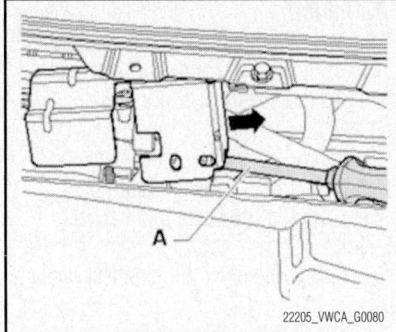

**Fig. 95 After removing the shear bolts, carefully pry off the retaining plate to access the ECM—2.0L engine**

- Disconnect the forward electrical harness connector.
- Lift up the locking mechanism slightly.
- Remove the Motronic Engine Control Module (ECM) J623 from the retainer.
- Disconnect the rear electrical harness connector.
11. With 2.5L engine
- Remove the screws and cover from the E-Box in the plenum chamber.
- Using a screwdriver, remove the retainer bar and Engine Control Module (ECM) J623 from the E-box.

**❋❋ WARNING**

**To prevent damage (burning) to the wiring and harness connections, insulation and control module, perform the following work procedures exactly! Observe operating instructions for heat gun.**

**❋❋ CAUTION**

**When heating the shear bolts, parts of the protective housing will become extremely hot. Do not burn yourself on this! Make sure that only the shear bolts are heated as much as possible, and not any of the surrounding parts.**

➡The threads of shear bolts (that are not screwed into ECM) are coated with a locking compound. For this reason, the threads must be heated with the heat gun to remove both bolts. The threads of both shear bolts (that are screwed into ECM) are not coated with a locking compound. The threads in the ECM housing must not be heated and do not require to be heated (unintentional heating of ECM).

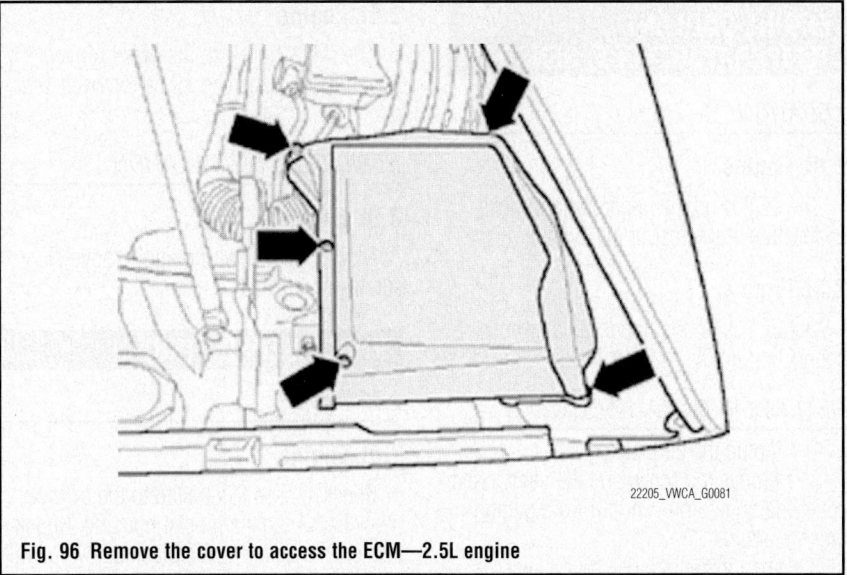

**Fig. 96 Remove the cover to access the ECM—2.5L engine**

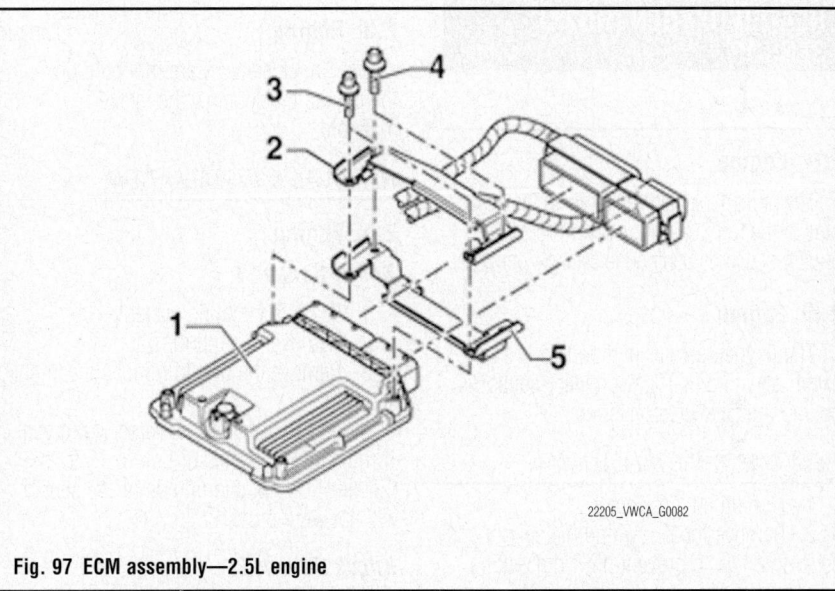

**Fig. 97 ECM assembly—2.5L engine**

- Using heat gun, direct the heat gun nozzle at shear bolt of retaining tab for approx. 20 to 25 seconds.
- Using locking pliers, remove shear bolt.
- The procedure for second shear bolt is exactly the same.
- Remove the protective housing from the ECM.
- Disengage the electrical connector retainers and disconnect electrical connectors.
- Remove the ECM.

*To install:*

➡The Engine Control Module must be installed with the protective housing.

12. Fit the ECM into place with the protective housing. Connect the rear wiring harness connector.

13. Install the ECM and attach the other connector.

14. With 2.0L engine, secure the locking mechanism.

15. Install the protective housing using new shear bolts. Tighten the bolts until the heads shear off.

16. The new ECM and immobilizer must be activated. Refer to the Ebahn website for ECM and immobilizer activation instructions.

17. After completing the repair work, the following work steps must be performed in the following sequence:
- Check the DTC memory.
- If necessary, erase the DTC memory.
- If the DTC memory was erased, generate readiness code.

## ENGINE COOLANT TEMPERATURE (ECT) SENSOR

### LOCATION

#### 2.0L Engine

The ECT is on the upper radiator hose connection at the rear of the cylinder head.

#### 2.5L Engine

The ECT is on the upper radiator hose connection on the cylinder head.

### REMOVAL & INSTALLATION

1. Unplug the connector.
2. Remove the screw and the sensor and cover the hole to prevent dirt from getting into the engine.
3. When installing, use a new seal.

## HEATED OXYGEN (HO2S) SENSOR

### LOCATION

#### 2.0L Engine

The primary oxygen sensor is on the exhaust pipe at the turbocharger outlet. The secondary sensor is behind the catalytic converter.

#### 2.5L Engine

The oxygen sensor is mounted in the exhaust manifold. The secondary sensor is behind the catalytic converter.

### REMOVAL & INSTALLATION

1. Unplug the connector.
2. Remove the screw and the sensor and cover the hole to prevent dirt from getting into the engine.
3. When installing, apply anti-seize compound to the threads. Be careful not to get any on the sensor element.

## INTAKE AIR TEMPERATURE (IAT) SENSOR

### LOCATION

#### 2.0L Engine

1. The IAT is mounted low in the center of the intake manifold plenum.

#### 2.5L Engine

The IAT is built into the Mass Airflow Sensor (MAF) and cannot be removed separately.

### REMOVAL & INSTALLATION

#### 2.0L Engine

Disconnect the wiring and remove the screw to remove the sensor.

## KNOCK SENSOR (KS)

### LOCATION

#### 2.0L Engine

Knock sensor 1 is bolted to the cylinder block below the thermostat housing. Knock sensor 2 is bolted to the block below the oil filter bracket.

#### 2.5L Engine

Two knock sensors are mounted on the engine block below the intake manifold.

### REMOVAL & INSTALLATION

#### 2.0L Engine
#### Knock Sensor 1

1. Drain the cooling system.
2. Remover the thermostat.
3. Remove the bolt to remove the knock sensor.
4. Installation is the reverse of removal. Torque the knock sensor bolt to 15 ft. lbs. (20 Nm). Do not overtighten or the sensor will not work properly.

#### Knock Sensor 2

1. Drain the cooling system.
2. Remove the intake manifold.
3. Disconnect the coolant hoses from the oil cooler.
4. Remove the oil cooler.
5. Disconnect the oil pressure switch.
6. Remove the four bolts to remove the oil filter bracket.
7. Remove the bolt to remove the knock sensor.

#### To install:

8. Install the sensor and torque the bolt to 15 ft. lbs. (20 Nm). Do not overtighten or the sensor will not work properly.
9. Install the baffle plate, then install the oil filter bracket using new gaskets. Torque the bolts to 11 ft. lbs. (15 Nm).
10. Install oil cooler and connect coolant hoses.
11. Install the intake manifold.
12. Refill the cooling system and run the engine to check for leaks.

#### 2.5L Engine

1. Remove the intake manifold.
2. Remove the bolt to remove the knock sensor.
3. Installation is the reverse of removal. Torque the knock sensor bolt to 15 ft. lbs. (20 Nm). Do not overtighten or the sensor will not work properly.

## MASS AIR FLOW (MAF) SENSOR

### LOCATION

#### 2.0 Engine

The MAF is mounted in the engine cover/air filter housing assembly behind the opening for the oil filler cap.

#### 2.5L Engine

The MAF is in the intake air tube next to the battery.

### REMOVAL & INSTALLATION

Unplug the connector and remove the screw to remove the sensor. Do not blow pressurized air at the sensor element.

## THROTTLE POSITION SENSOR (TPS)

### LOCATION

The throttle position sensor is built into the electronic throttle valve control module (throttle body).

**FUEL**                                    GASOLINE FUEL INJECTION SYSTEM

### FUEL SYSTEM SERVICE PRECAUTIONS

Safety is the most important factor when performing not only fuel system maintenance but any type of maintenance. Failure to conduct maintenance and repairs in a safe manner may result in serious personal injury or death. Maintenance and testing of the vehicle's fuel system components can be accomplished safely and effectively by adhering to the following rules and guidelines.

• To avoid the possibility of fire and personal injury, always disconnect the negative battery cable unless the repair or test procedure requires that battery voltage be applied.

• Always relieve the fuel system pressure prior to disconnecting any fuel system component (injector, fuel rail, pressure regulator, etc.), fitting or fuel line connection. Exercise extreme caution whenever relieving fuel system pressure to avoid exposing skin, face and eyes to fuel spray. Please be advised that fuel under pressure may penetrate the skin or any part of the body that it contacts.

• Always place a shop towel or cloth around the fitting or connection prior to loosening to absorb any excess fuel due to spillage. Ensure that all fuel spillage (should it occur) is quickly removed from engine surfaces. Ensure that all fuel soaked cloths or towels are deposited into a suitable waste container.

• Always keep a dry chemical (Class B) fire extinguisher near the work area.

• Do not allow fuel spray or fuel vapors to come into contact with a spark or open flame.

• Always use a back-up wrench when loosening and tightening fuel line connection fittings. This will prevent unnecessary stress and torsion to fuel line piping.

• Always replace worn fuel fitting O-rings with new Do not substitute fuel hose or equivalent where fuel pipe is installed.

Before servicing the vehicle, make sure to also refer to the precautions in the beginning of this section as well.

### RELIEVING FUEL SYSTEM PRESSURE

The fuel injection system operates under high pressure. This makes it necessary to first relieve the system of pressure before servicing. The pressurized fuel, when released, may ignite or cause personal injury.

➡On most models, the fuel pump runs for two seconds when the driver door is opened from the outside.

1. Before servicing the vehicle, refer to the Precautions Section.
2. Remove or disconnect:
   • Power to the fuel pump by removing the relay or the fuel pump fuse. Check the list on the fuse box lid to be sure. The fuse can be removed to stop the fuel pump from running. With the engine operating at idle, wait until the engine stalls from fuel starvation.
3. Carefully loosen the fuel line at the high pressure fuel pump. Wrap a clean rag around the connection while loosening to catch any fuel.
4. After service is complete, discard the fuel soaked rag in the proper manner and reconnect the negative battery cable, relay or fuses.

### FUEL FILTER

#### REMOVAL & INSTALLATION

Most vehicles use a fuel filter mounted under the vehicle near the fuel tank. An arrow should be on the filter indicating fuel flow direction. Install with arrow pointing to engine. Use care not to mix up fuel supply or return lines. Fuel pressure applied to the return side of the system will cause damage.

1. Before servicing the vehicle, refer to the Precautions Section.
2. For safety reasons, remove the fuel pump control module fuse from the fuse panel before opening the fuel system. This prevents the fuel pump from being activated when opening the driver side door.
3. Relieve the fuel system pressure.
4. Disconnect the fuel lines leading into and out of the fuel filter.
5. Unscrew the filter retaining bracket and remove the filter.
6. Install a new filter in the bracket and reattach the bracket. Be sure the arrows are pointing in the direction of the fuel flow.
7. Reconnect the fuel lines, start the engine and check for leaks.

### FUEL INJECTORS

#### REMOVAL & INSTALLATION

**2.0L Engine**

*See Figures 98 through 102.*

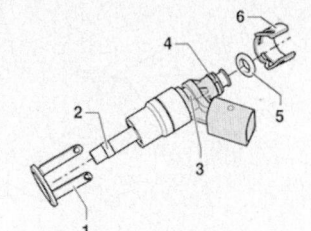

1. Radial compensation: always replace
2. Combustion chamber sealing ring
3. Groove on fuel injector
4. Support washer: always replace
5. O-ring: always replace
6. Support ring

22205_VWCA_G0083

**Fig. 98 Fuel injector assembly**

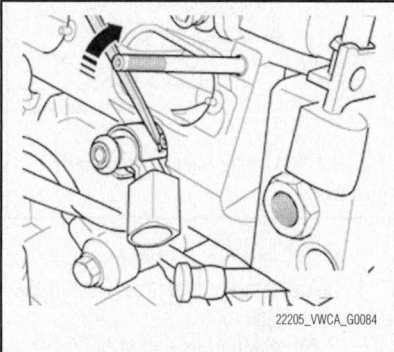

22205_VWCA_G0084

**Fig. 99 Pry the radial compensation retaining tab to remove it**

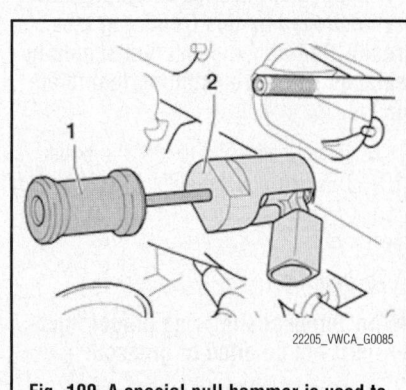

22205_VWCA_G0085

**Fig. 100 A special pull hammer is used to remove the injector**

1. Before servicing the vehicle, refer to the Precautions Section.
2. For safety reasons, remove the fuel pump control module fuse from the fuse panel before opening the fuel system. This prevents the fuel pump from being activated when opening the driver side door.

**Fig. 101 A special press tool is used to install the fuel injector**

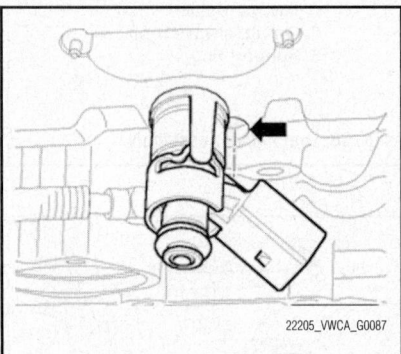

**Fig. 102 Make sure injector is properly positioned**

3. Relieve the fuel system pressure.

4. Remove intake manifold. Seal intake ports with a clean cloth.

5. Bend retaining tabs of radial compensation to the side using a screwdriver and pull off support ring from fuel injector.

➡Radial compensation retaining tabs are destroyed by this (retaining tabs break). Radial compensation should be replaced before reinstalling fuel injector.

6. Install hammer T10133/3 to puller T10133/2 . Then guide puller T10133/2 into groove on fuel injector and carefully tap fuel injector out.

### To install:

➡The Teflon sealing ring of fuel injector must not be oiled or greased.

7. Thoroughly clean bores for high pressure fuel injectors in cylinder head using nylon brush T10133/4. Possibly an opened intake valve hinders the cleaning. In this case, engine must be turned further by hand using a box wrench on the crankshaft.

8. Replace O-ring, radial compensation, support washer and Teflon seal on fuel injector

9. Using the assembly drift T10133/9 , press the fuel injector into the hole in the cylinder head until it stops. Make sure fuel injectors are positioned correctly in cylinder head.

10. Install intake manifold.

11. Start the engine, check for leaks and repair if necessary.

### 2.5L Engine

*See Figures 103 and 104.*

1. Before servicing the vehicle, refer to the Precautions Section.

2. First, check whether a coded radio is installed. If necessary, obtain the anti-theft coding.

3. Disconnect the negative battery cable.

4. Remove engine cover/air filter housing.

**✳✳ CAUTION**

**Fuel supply line is under pressure! Wear protective goggles and protective clothing to prevent injuries and contact with skin. Before loosening the fuel lines, place a rag around the connection point. Then release pressure by carefully pulling hose off connection.**

5. Disconnect the fuel tank vent line and the fuel supply line. To release the connectors, press the circlip in. On the fuel

supply line, the retainer must be pressed upward in the housing.

6. Disconnect connectors 1, 2 and 6.

7. Remove wiring harness (3) from transport strap.

8. Pull clamps (4) and retaining ring (5) out of locking mechanism.

9. Remove bolts (7) and remove transport strap.

10. Remove bolts and pull fuel rail with fuel injectors evenly out of intake manifold.

11. Seal or cover openings in intake manifold.

12. Pull of retaining clips and then remove fuel injectors from fuel rail.

### To install:

13. Install new O-rings for fuel injectors and coat them lightly with clean engine oil.

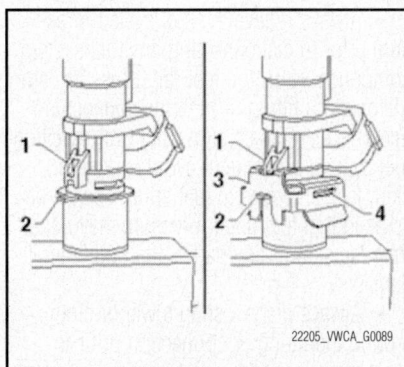

**Fig. 104 The clip (3) must engage both tabs (1 and 2) and the collar (4)**

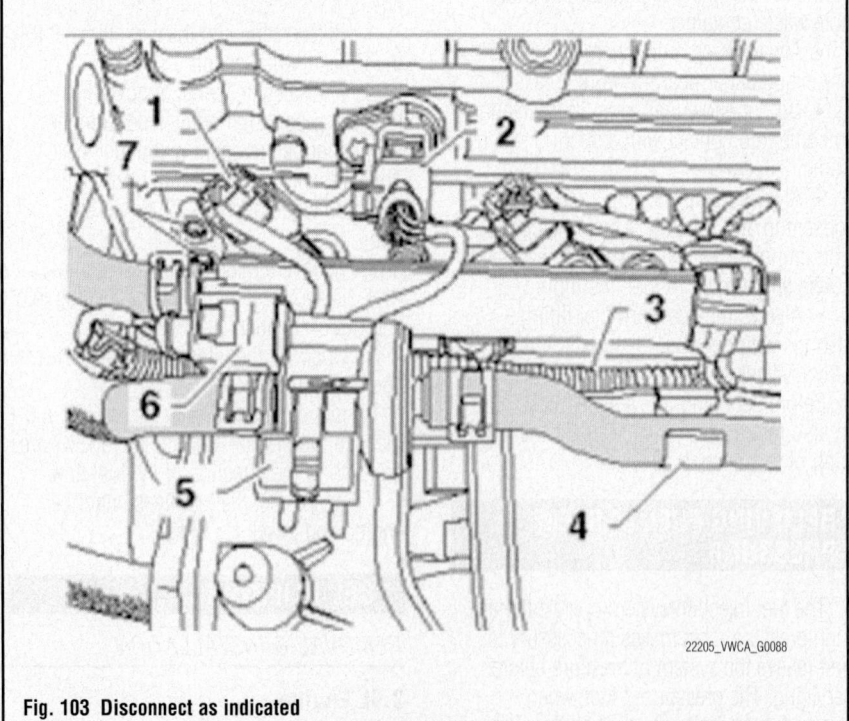

**Fig. 103 Disconnect as indicated**

14. Press fuel injectors into fuel rail so that tabs (1 and 2) align.

15. Slide retaining clip (3) into groove of fuel injector. Collar (4) must be located correctly in cutout of retaining clip on both sides.

16. After assembling, check all fuel injectors for correct fitting.

17. Fit fuel rail/injector assembly on intake manifold and press it in uniformly.

18. Bolt fuel rail to intake manifold.

19. Connect all wiring and the fuel lines.

20. Connect the battery and bleed the fuel system.

### BLEEDING

**2.5L Engine**
*See Figure 105.*

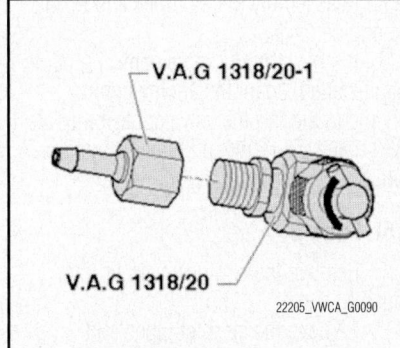

Fig. 105 Valve and hose adapter used to bleed the fuel system—2.5L engine

1. Remove cover cap from bleeder valve on the fuel rail.

2. Install adapter V.A.G 1318/20-1 on to adapter V.A.G 1318/20.

3. Screw adapter V.A.G 1318/20 hand-tight on to bleeder valve.

4. Turn valve at T-piece counterclockwise until it is completely open.

5. Connect a hose to the adapter fitting.

6. Connect vehicle diagnosis, testing and information system VAS 5051B to the Data Link Connector (DLC) in drivers footwell.

7. Press the following buttons on the display one after another:
- Vehicle Self-Diagnosis
- On Board Diagnosis (OBD) micro
- Engine Electronics micro
- Output Diagnostic Test Mode micro
- Press micro until Fuel Pump (FP) Relay J17 is activated. This activates the fuel pump.

8. Let the diagnostic run until fuel without bubbles runs out of the bleeder valve. Then end output diagnostic test mode.

➡ **If the output diagnostic test mode is interrupted, the engine must be started for a short time before the mode can be accessed again. Output diagnostic test mode is automatically cancelled after 60 seconds.**

9. Turn valve at T-piece counterclockwise until it is completely open again.

10. Clamp the vent hose and disconnect it from adapter V.A.G 1318/20-1.

11. Remove adapter from bleeder valve.

12. Screw on cover cap on bleeder valve.

### FUEL PUMP

### REMOVAL & INSTALLATION

**Rabbit, GTI and Jetta**
*See Figure 106.*

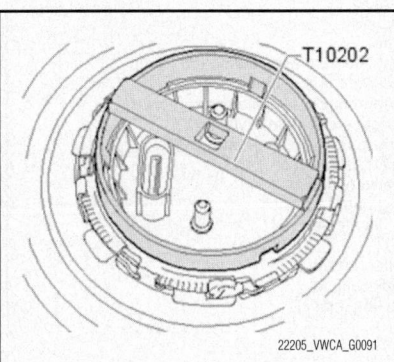

Fig. 106 Tool used to remove/install fuel pump—Rabbit, GTI, Jetta

1. Before servicing the vehicle, refer to the Precautions Section.

2. Relieve the fuel system pressure.

3. Disconnect the negative battery cable or remove the fuel pump fuse.

4. Remove the rear seat cushion.

5. Remove the access cover.

6. Disconnect the wiring and fuel lines.

7. Using spanner tool T10202, remove the fuel pump flange nut.

8. Remove the fuel pump unit and seal.

### *To install:*

9. Installation is the reverse of removal. Be sure to use new sealing rings and/or gaskets.

10. Torque the flange nut to 80 ft. lbs. (110 Nm).

11. The fuel supply line is black (or has black markings) and the return line is blue (or has blue markings).

12. If fuel delivery unit was replaced, adapt Engine Control Module to Fuel Pump (FP) using the "Guided Functions" procedure on the factory scan tool.

### FUEL TANK

### REMOVAL & INSTALLATION

**Rabbit, GTI, Jetta**
*See Figures 107 and 108.*

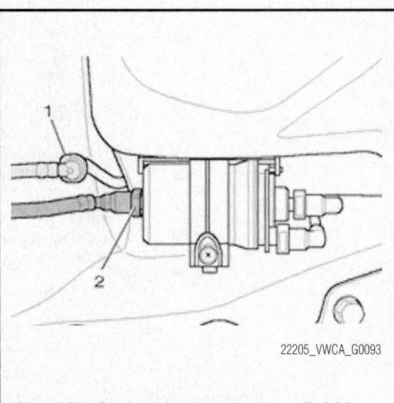

Fig. 107 Fuel tank components—Rabbit, GTI and Jetta

➡ **New fuel tank strap bolts are required.**

1. Empty the fuel tank if possible using a fuel extracting device such as VAS 519

2. Remove rear seat bench.

3. Remove the pump access cover and disconnect the wiring and fuel lines.

4. Raise and safely support the vehicle and remove the right rear wheel.

5. Remove right rear fender liner.

6. Unscrew mounting bolt for fuel flap unit and remove fuel flap unit.

7. Disconnect the vent hoses.

8. Remove filler tube bolts.

9. Unclip electrical wire on filler connection.

10. Remove center and rear mufflers.

11. Remove heat shield from center and rear mufflers.

### ❋❋ CAUTION

**Fuel supply line is under pressure! Wear protective goggles and protective clothing to prevent injuries and contact with skin.**

12. Before removing from hose connection wrap a rag around the connection. Then release pressure by carefully pulling hose off connection.

13. Disconnect vent line (1, white) and fuel line (2, black).

14. Remove tension strap and bolts. Support fuel tank using engine/transmission jack V.A.G 1383 A while doing so.

15. Carefully lower fuel tank.

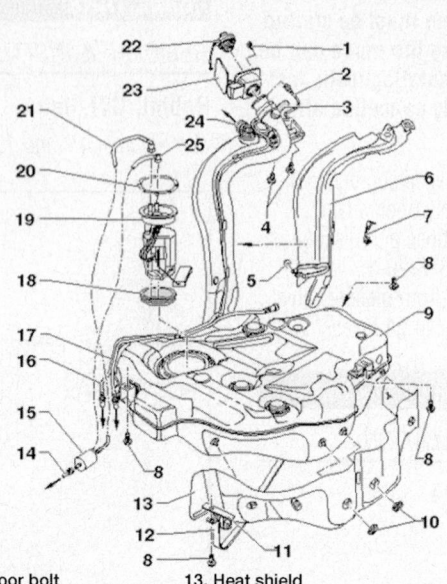

1. Filler door bolt
2. Ground connection
3. Vacuum line to Leak
   Detection Pump (LDP) V144
4. Protective plate bolts
5. Rivet
6 Protective plate: riveted to
   lower clamp at factory
7. Cable bracket
8. Bolt: always replace
9. Fuel tank
10. Lock washer
11. Bracket for exhaust system
12. Securing strap:
    note installation position

13. Heat shield
14. Supply line
15. Fuel filter
16. Vacuum line from Leak Detection Pump
    (LDP) V144 to intake manifold
17. Vent line from EVAP canister to
    evaporative emission (EVAP)
    canister purge regulator valve N80
18. Seal: always replace
19. Fuel delivery unit
20. Locking ring, 110 Nm
21. Return line (blue)
22. Cap
23. Fuel filler door unit
24. To EVAP canister
25. Supply line (black)

22205_VWCA_G0092

**Fig. 108 Disconnect fuel lines 1 and 2**

**To install:**

16. Fit the fuel tank into place using the jack and start the tension strap bolts. Use new M8 (larger) bolts.

17. Make sure the tank and straps are properly positioned. Torque the M6 bolts to 7 ft. lbs (10 Nm) and the M8 bolts to 18 ft. lbs. (25 Nm).

18. Install the remaining fuel tank components and reconnect all wiring and fuel lines. Make sure the lines are not kinked and pay close attention to the chassis ground connection at the filler tube.

19. Install the heat shields and exhaust system components.

20. If fuel delivery unit was replaced, adapt engine control module to Fuel Pump (FP) using "Guided Functions" procedure on the factory scan tool.

## IDLE SPEED

### ADJUSTMENT

Idle speed is maintained by the Powertrain Control Module (PCM). No adjustment is necessary or possible.

## THROTTLE BODY

### REMOVAL & INSTALLATION

#### 2.0L Engine

1. Remove the engine cover/air filter housing.

2. Remove the air inlet pipes and hoses

3. Raise and safely support the vehicle and remove the engine undercover.

4. Disconnect electrical connector and remove the bolt to pull the guide pipe off the charge air cooler and remove it downward.

5. Remove four bolts on throttle valve control module and remove it.

**To install:**

6. Install the throttle valve with a new seal.

7. Reassembly the air tubes and install the engine cover.

8. If a new throttle valve control module was installed, adapt the engine control module to the throttle valve control module J338 using the "Guided Functions" procedure on the factory scan tool.

#### 2.5L Engine

1. Remove the engine cover/air filter housing.

2. Remove the air inlet pipes and hoses

3. Disconnect the wiring and coolant hoses (if equipped).

4. Remove the screws to remove the throttle body assembly.

**To install:**

5. Install the throttle valve with a new seal.

6. Reassembly the air tubes and install the engine cover.

7. If a new throttle valve control module was installed, adapt the engine control module to the throttle valve control module J338 using the "Guided Functions" procedure on the factory scan tool.

## HEATING & AIR CONDITIONING SYSTEM

### HEATER CORE

*REMOVAL & INSTALLATION*

**Rabbit, GTI and Jetta**

*See Figures 109 and 110.*

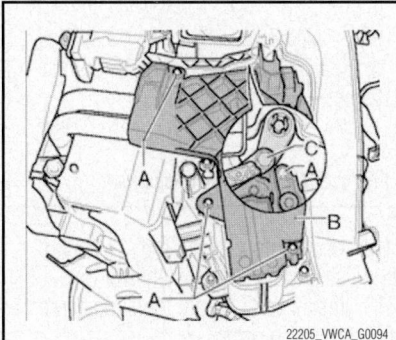

**Fig. 109 Remove covers to access heater controls—Rabbit, GTI, Jetta**

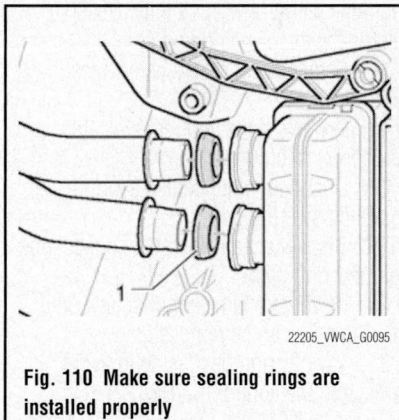

**Fig. 110 Make sure sealing rings are installed properly**

1. Place a drip tray under the engine and disconnect the heater hoses at the firewall.

2. Connect a section of hose onto upper connection of heater core and carefully blow coolant out of heater core with compressed air.

3. Depending on the engine, additional parts may need to be removed such as the charge air pipe.

4. Slightly loosen but do not remove the screw on connecting flange between heater core connections.

➡**By loosening the screw, the coolant pipes are loosened and it is easier to remove the heater core.**

5. Remove driver side footwell trim.

6. Remove left footwell vent.

7. Remove screws (A) and remove cover (B).

➡**There are different versions of cover (B). The illustration shows the version with Auxiliary Air Heater Heating Element.**

8. If the temperature door lever (C) is a position that hinders access to the upper screw (A), change the position of the temperature door using the controls for hot air and fresh on, on vehicles with Climatronic with operating and display unit (e.g. "Hi" setting).

9. Cover carpet in area under heat exchanger with waterproof foil and water absorbing paper.

10. Open pipe clamps and disconnect coolant pipes from heater core.

11. Remove heater core from heating unit.

**To install:**

12. Check the seals on the heater core. Only install a heater core with undamaged seals.

➡**An incorrectly glued seal can roll up into heating unit when sliding in the heater core. Cold air may flow past heat exchanger if seal is damaged or not properly fitted.**

13. With the heater core removed, check heating unit for dirt or debris in the heater core opening. If necessary, remove dirt or coolant from heating unit, for example after removing leaky heater core.

14. Push heater core into heating unit.

15. Coat new sealing rings with coolant and insert sealing rings into connection on heater core. Ensure sealing rings are installed on the proper side, as shown in the illustration. If pipe clamps are deformed, replace them. Do not reuse old sealing rings.

16. Connect coolant pipes to heater core. Clamps must be able to be twisted slightly when installing onto the coolant pipes.

17. Tighten the pipe clamps. Check seating of both clamps after tightening screws. They must completely enclose the flange on the heat exchanger and coolant pipe and must not come in contact with other components.

18. Tighten the connecting flange screw between heater core connections under the hood.

19. Make sure the grommet in firewall is properly seated.

20. Seal the flanges on the coolant pipes leading to the heater core and for the expansion valve to the evaporator (with A/C) at the grommet with silicon adhesive sealant to prevent water from penetrating.

21. After replacing heater core, refill the cooling system with new coolant

22. Check coolant circuit for leaks, pay particular attention between coolant hoses and heater core.

## STEERING

### POWER STEERING GEAR

*REMOVAL & INSTALLATION*

**Rabbit, GTI and Jetta**

*See Figures 111 through 113.*

➡**The subframe must be lowered to remove the steering gear. New mounting bolts are required for installation.**

1. Before servicing the vehicle, refer to the Precautions Section.
2. Disconnect negative battery cable.
3. Remove driver side foot well trim.
4. Remove bolt from steering shaft universal joint.
5. Raise and safely support the vehicle and remove front wheels.
6. Loosen tie rod end nut, but do not remove yet.

### ❄ WARNING

**To protect thread, screw nut on pin a few turns.**

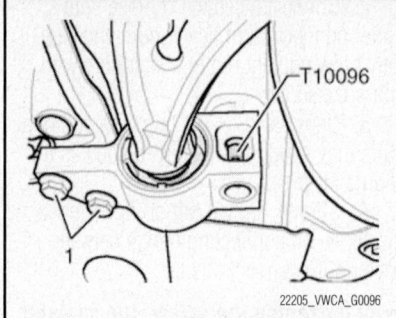

**Fig. 112 Subframe locating pins hold subframe in place while bolts are removed—Rabbit, GTI and Jetta**

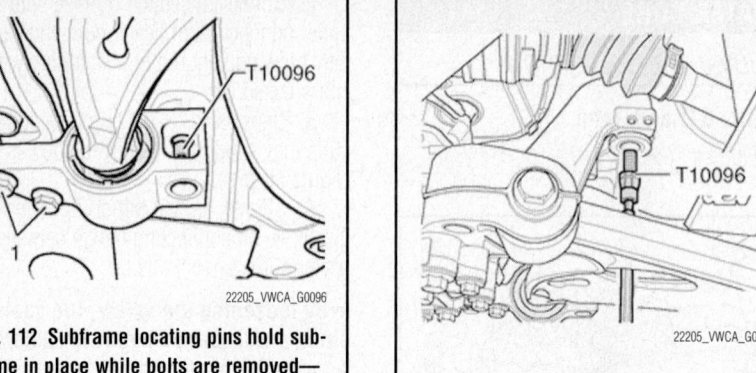

**Fig. 113 Subframe locating pins**

7. Press tie rod end off of steering knuckle using Ball Joint Puller 3287A.
8. Remove engine undercover.
9. Remove pendulum support bracket bolts (bottom transaxle mount).
10. Remove bracket for exhaust system from subframe.
11. Remove bolts from heat shield and remove heat shield from subframe.

12. Now remove bolts 3, 6, 11 and 16 for the steering gear and stabilizer bar.
13. Install locating pin tools to secure the subframe in place.
14. Place Engine/Transmission Jack VAG1383A under subframe. Place a block of wood between the jack and the subframe.
15. Loosen bolts (4) and (5) and lower subframe with brackets slightly. Observe electrical wires, when doing so.
16. Remove heat shield above steering gear.
17. Remove bolts and remove cable guide from subframe.
18. Unclip all other cable securing points on steering gear.
19. Disconnect all electrical connections from steering gear.
20. Lower subframe with the jack until steering gear can be removed.
21. Lay steering gear aside to avoid damage to the control module.

*To install:*

➡**Coat seal on steering gear with lubricant, e.g. soft soap, before installing steering gear.**

22. Make sure sealing surfaces are clean.
23. Before fastening the bolts for subframe, position steering gear on subframe and fasten bolts for steering gear and stabilizer.
24. Fit the steering gear into place taking care not to damage the control module. The steering gear threaded sleeves must be seated in the bracket holes.
25. After attaching steering gear to steering shaft, make sure that seal on steering gear is positioned to mounting plate without kinks and opening to foot well is sealed correctly. Ingress of water and/or noises may be the result.

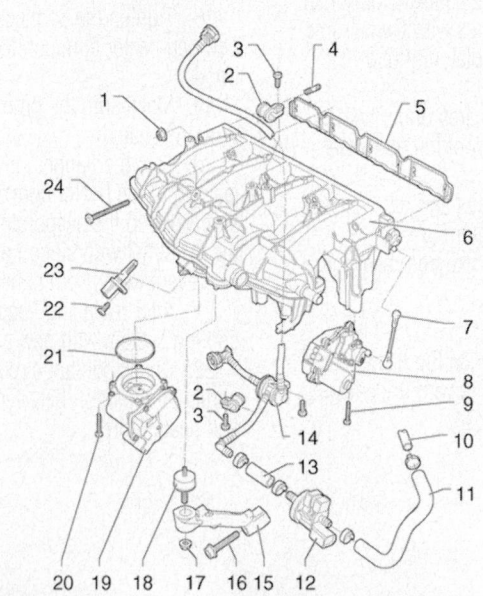

1. Manifold nut
2. Retaining clip
3. Retaining clip screw
4. Stud bolt
5. Gasket
6. Intake manifold
7. Coupling rod
8. Intake flap motor with intake manifold runner position sensor
9. Intake flap motor screw
10. To EVAP canister
11. Hose
12. Evaporative emission (EVAP) canister purge regulator valve
13. Hose
14. Double check-valve
15. Intake manifold support
16. Support bolt
17. Bushing nut
18. Bonded rubber bushing
19. Throttle valve control module
20. Throttle valve control module screw
21. Seal: replace if damaged
22. IAT sensor screw
23. Intake Air Temperature (IAT) sensor
24. Manifold bolt

**Fig. 111 Subframe bolt numbers—Rabbit, GTI and Jetta**

26. Connect steering gear wiring.

27. Install new steering gear-to-subframe bolts and torque to 37 ft. lbs. plus an additional 90°

28. Install the new subframe bolts hand tight.

29. Install new stabilizer bolts hand tight.

30. Torque the subframe-to-body bolts to 52 ft. lbs (70 Nm) plus an additional 90°.

31. Torque the stabilizer bar-to-subframe bolts to 15 ft. lbs. (20 Nm) plus an additional 90°.

32. Install new stabilizer bar-to-connecting link nuts. Counterhold at joint pin inner multi-point fitting and torque to 48 ft. lbs. (65 Nm).

33. Install new pendulum support bracket bolts and torque to 30 ft. lbs. (40 Nm) plus an additional 90°.

34. Install a new nut on the tie rod end and torque to 15 ft. lbs. (20 Nm) plus an additional 90°.

35. Install remaining components and connect all wiring.

36. Perform Steering Angle Sensor G85 basic setting with the Vehicle Diagnosis, Testing and Information System VAS 5051.

37. If new steering gear was installed, Power Steering Control Module J500 must be adapted using Vehicle Diagnosis, Testing and Information System VAS 5051. If Parking Assist 2 is installed in the vehicle, the Power Steering Control Module J500 must be recoded after installing new steering gear.

38. After installation, position of steering wheel must be checked with a road test.

39. If the steering wheel is crooked or new steering gear was installed, check front wheel alignment.

## SUSPENSION

## COIL SPRING

### REMOVAL & INSTALLATION

*See Figure 114.*

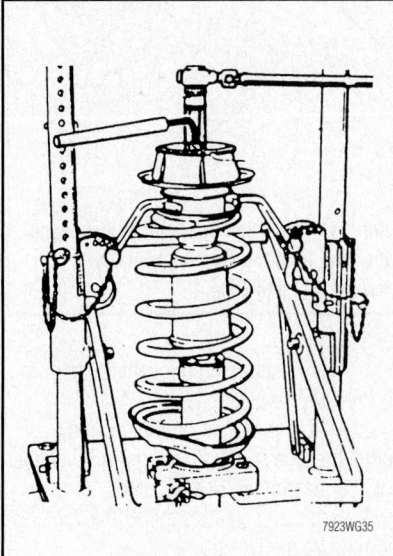

7923WG35

**Fig. 114 Compress the coil spring before removing the upper strut rod nut**

1. Before servicing the vehicle, refer to the Precautions Section.

2. Remove the strut from the vehicle.

3. Clamp the Spring Compressor VAG 1752/2 or equivalent in a vise.

4. Install the strut into the spring compressor.

5. Pry off the mounting bolt cap.

6. Compress the coil spring and remove the self-locking nut from the piston rod.

7. Matchmark the position of the spring retainer and spring mount.

8. Remove:
• Spring seat and related components noting the order of removal.
• Strut from the spring compressor.
• Spring out of the compressor.

**To install:**

9. Install the new spring into the compressor.

10. Compress the spring and insert the strut through the spring.

11. Install the spring seat and related components in the reverse order as they were removed and aligning the matchmarks.

12. Install a new self-locking nut.

13. Reinstall the mounting bolt cap.

14. Release the spring compressor and install the strut into the vehicle.

## LOWER BALL JOINT

### REMOVAL & INSTALLATION

#### Rabbit, GTI and Jetta

*See Figure 115.*

1. With the vehicle sitting on the ground, loosen drive axle hub bolt.

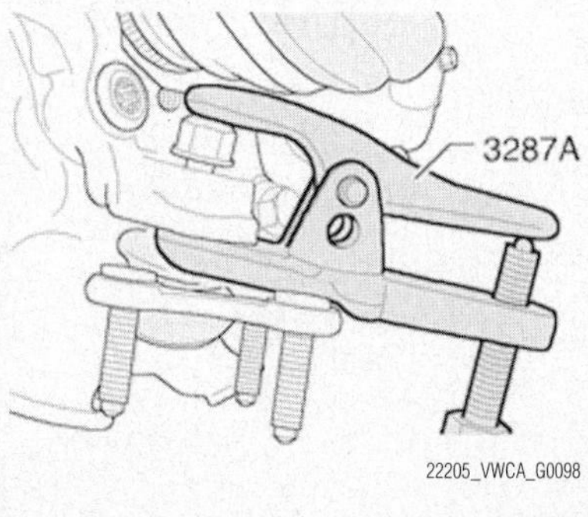

3287A

22205_VWCA_G0098

**Fig. 115 Pressing ball joint out of steering knuckle—Rabbit, GTI, Jetta**

## FRONT SUSPENSION

### ✳ WARNING

Vehicle must not be sitting on the wheels when tightening or loosening. With bolt loosened, wheel bearing can be damaged by vehicles own weight. If a vehicle needs to be moved when the drive axle is removed, an outer joint must be installed and tightened to 37 ft. lbs. (50 Nm).

2. Raise and safely support the vehicle and remove the wheel.

3. Remove the nuts securing the ball joint to the control arm.

4. Pull drive axle slightly out of wheel hub.

5. Pull ball joint out of control arm.

6. Move control arm downward as far as possible.

7. Loosen the ball joint nut. To protect ball joint thread, leave nut on a few turns.

8. Attach ball joint puller 3287 A and press out ball joint.

### ✳ CAUTION

Place engine/transmission jack V.A.G 1383 A or similar underneath (risk of accident if components fall off when pressing out ball joint).

**To install:**

9. Insert ball joint into steering knuckle. Install a new nut, ensure that boot is not damaged or twisted and torque nut to 45 ft. lbs. (60 Nm).

10. Install drive axle in wheel hub. Do not tighten bolt yet.

11. Fit ball joint into control arm. Install new self-locking nuts and torque nut to 45 ft. lbs. (60 Nm)

12. Install any remaining components and install the wheel.

13. With the vehicle sitting on the ground, torque the front axle bolt:
- With 6-point bolt, torque to 150 ft. lbs. (plus an additional 180°.
- With 12-point bolt, torque to 52 ft. lbs. (70 Nm) plus an additional 90°.

## LOWER CONTROL ARM

### REMOVAL & INSTALLATION

**Rabbit, GTI and Jetta**

*See Figures 116 and 117.*

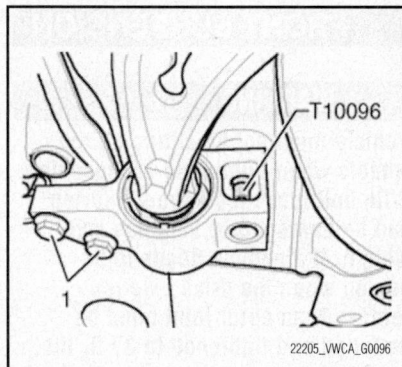

**Fig. 116 Locating pins keep suspension bracket in place with bolts removed**

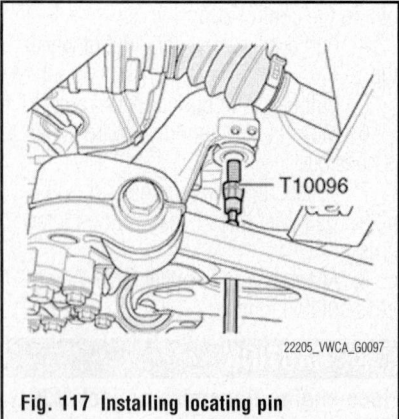

**Fig. 117 Installing locating pin**

➡ **This procedure requires lowering the subframe slightly. New mounting bolts are required for installation.**

1. Raise and safely support the vehicle and remove the wheels and engine undercover.

2. Remove bracket for exhaust system from subframe.

3. If equipped, remove heat shield from subframe.

4. Remove pendulum support bracket (bottom transaxle mount). Do not remove the large bolt from the center of the mount

5. To remove the left side control arm, secure the control arm mounting bracket using locating pins T10096.

6. Remove the nuts to disconnect the ball joint from the control arm.

7. Remove the bolts for steering gear, stabilizer bar and subframe on right side.

8. Place engine/transmission jack V.A.G 1383 A below subframe. Place a piece of wood between engine/transmission jack and subframe.

9. Remove the four corner bolts and lower the subframe with brackets as far as necessary. When doing this, pry alignment bushings of steering gear out of left bracket.

10. Remove bolt to remove control arm from bracket.

### To install:

11. Fit the control arm into place and start the new bolts.

12. Raise the subframe and start all the new bolts.

13. Torque the new subframe-to-body bolts to 52 ft. lbs. (70 Nm) plus an additional 90°.

14. Torque the new control arm brake bolts to 52 ft. lbs. (70 Nm) plus an additional 90°.

15. Torque the new pendulum support bracket bolts to 30 ft. lbs. (40 Nm) plus an additional 90°.

16. Torque new ball joint-to-control arm nuts to 45 ft. lbs. (60 Nm).

17. Torque new stabilizer bar bolts to 15 ft. lbs. (20 Nm) plus an additional 90°.

18. Install remaining components.

### BUSHING REPLACEMENT

**Rabbit, GTI and Jetta**

*See Figure 118.*

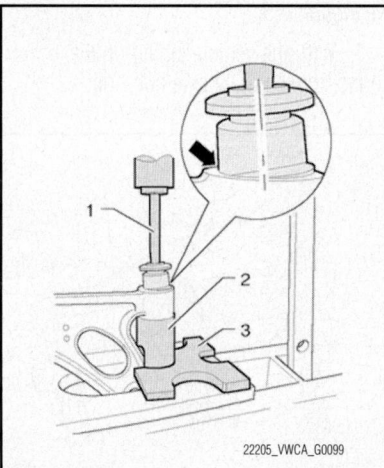

**Fig. 118 Fit bushing into control arm at an angle**

1. Remove the control arm.

2. Position the control arm on a press. Carefully push the bushing out of the control arm using the press.

### To install:

3. Lightly lubricate, then position the bushing on the control arm.

4. Carefully press the bushing into the control arm. When completely pressed in, the bushing must protrude evenly at both sides.

5. Install the control arm.

## MACPHERSON STRUT

### REMOVAL & INSTALLATION

**Rabbit, GTI and Jetta**

*See Figures 119 and 120.*

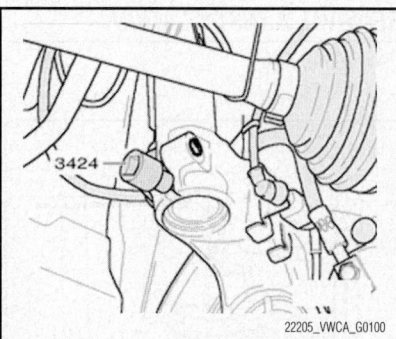

**Fig. 119 Use a spreader tool to disengage the strut from the steering knuckle— Rabbit, GTI and Jetta**

1. Before servicing the vehicle, refer to the Precautions Section.

2. With the wheels on the ground, loosen the axle hub bolt. Do not roll vehicle or wheel bearing will be damaged.

3. Raise and safely support the vehicle and remove the wheels.

4. Remove coupling rod bolt from suspension strut.

5. Disengage wire for wheel speed sensor from suspension strut.

6. Remove the three lower ball joint nuts and pull the knuckle with ball joint out of control arm.

7. Pull drive axle outer joint out of wheel hub. Secure drive axle to body using wire.

### ✷✷ WARNING

**Drive axle must not hang down, otherwise inner joint may be damaged when bent too far.**

8. Secure the steering knuckle to engine/transmission jack V.A.G 1383 A with Wheel Hub Support T10149 to wheel hub

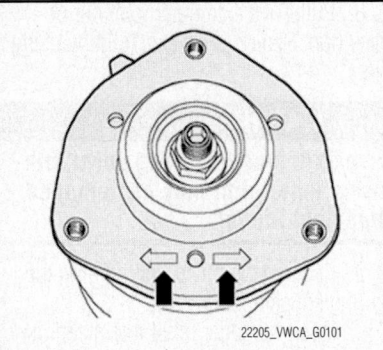

**Fig. 120 One of these arrows must point forward when installing strut**

using one wheel bolt. The jack must prevent the knuckle from falling when the strut is removed.

9. Remove the pinch bolt holding the steering knuckle to the strut.

10. Insert spreader 3424 into wheel bearing housing slot and turn it to open the slot. Press brake disc in direction of suspension strut by hand. Otherwise strut tube may be canted in knuckle.

11. Pull knuckle downward off strut tube and lower using engine/transmission jack V.A.G 1383 A until strut tube hangs free.

12. Tie wheel bearing housing onto bracket/subframe using wire. Do not let it hang by the brake hose.

13. Remove wiper arms.

14. Remove plenum chamber cover.

15. Remove bolts for upper strut mounting and remove suspension strut.

***To install:***

16. Insert suspension strut, one of the two markings–arrows–must point in direction of travel when doing this.

17. Install new bolts and torque to 11 ft. lbs. (15 Nm) plus an additional 90°

18. Install plenum chamber cover.

19. Install wiper arms.

20. Secure engine/transmission jack V.A.G 1383 A with wheel hub support T10149 to wheel hub using a wheel bolt.

21. Position suspension strut on steering knuckle.

22. Carefully lift the knuckle with the jack and engage the suspension strut. Install a new bolt hand tight.

23. Remove spreader tool and torque the bolt to 52 ft. lbs. (70 Nm) plus an additional 90°.

24. Remove the jack.

25. Install drive axle in wheel hub.

26. Fit the ball joint into the control arm and install new bolts hand tight. Ensure that boot is not damaged or twisted.

27. Torque the ball joint bolts to 45 ft. lbs. (60 Nm).

28. Install or connect remaining components and install the wheels.

29. With the vehicle on the ground, torque the axle hub bolt.
- Six-point bolt: 150 ft. lbs. (200 Nm) plus an additional 180°
- Twelve-point bolt: 52 ft. lbs. (70 Nm) plus an additional 90°

## STABILIZER BAR

*REMOVAL & INSTALLATION*

**Rabbit, GTI and Jetta**
*See Figures 121 through 123.*

➡The subframe must be lowered to remove the steering gear. New mounting bolts are required for installation.

1. Before servicing the vehicle, refer to the Precautions Section.

2. Disconnect negative battery cable.

3. Remove driver side foot well trim.

4. Remove bolt from steering shaft universal joint.

5. Raise and safely support the vehicle and remove front wheels.

6. Loosen tie rod end nut, but do not remove yet.

**✳✳ WARNING**

**To protect thread, screw nut on pin a few turns.**

7. Press tie rod end off of steering knuckle using Ball Joint Puller 3287A.

8. Remove engine undercover.

9. Remove nuts 11 and 16 to release stabilizer bar from subframe.

10. Remove pendulum support bracket bolts (bottom transaxle mount).

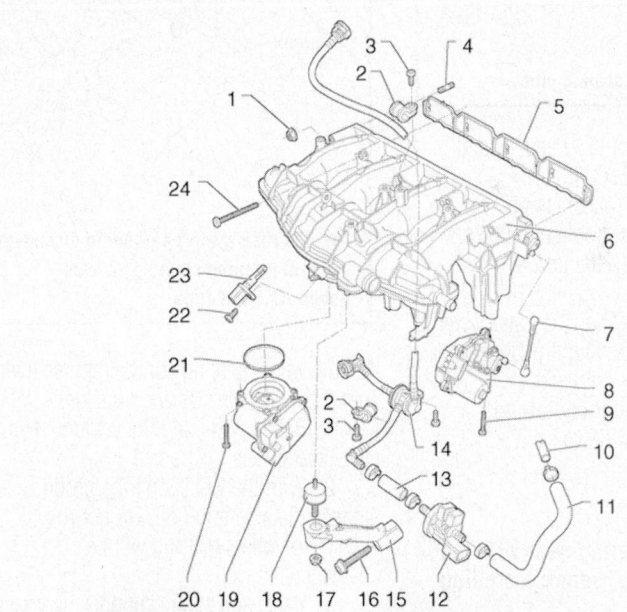

1. Manifold nut
2. Retaining clip
3. Retaining clip screw
4. Stud bolt
5. Gasket
6. Intake manifold
7. Coupling rod
8. Intake flap motor with intake manifold runner position sensor
9. Intake flap motor screw
10. To EVAP canister
11. Hose
12. Evaporative emission (EVAP) canister purge regulator valve
13. Hose
14. Double check-valve
15. Intake manifold support
16. Support bolt
17. Bushing nut
18. Bonded rubber bushing
19. Throttle valve control module
20. Throttle valve control module screw
21. Seal: replace if damaged
22. IAT sensor screw
23. Intake Air Temperature (IAT) sensor
24. Manifold bolt

**Fig. 121 Subframe bolt numbers—Rabbit, GTI and Jetta**

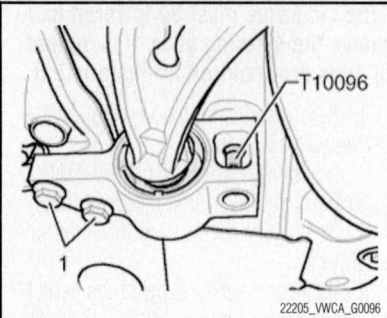

**Fig. 122 Subframe locating pins hold subframe in place while bolts are removed—Rabbit, GTI and Jetta**

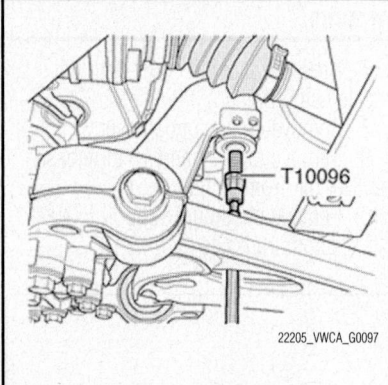

**Fig. 123 Subframe locating pins**

11. Remove bolts 1, 8, 9 and 18 and install locating pins T10096.

12. Place Engine/Transmission Jack VAG1383A under subframe. Place a block of wood between the jack and the subframe.

13. Loosen bolts (4) and (5) and lower subframe with brackets slightly. Observe electrical wires, when doing so.

14. Now lift stabilizer bar toward front over bracket and down from subframe.

### To install:

➡ **Coat seal on steering gear with lubricant, e.g. soft soap, before installing steering gear.**

15. Position stabilizer bar on subframe and install new nuts hand tight.

16. Attach steering gear to steering shaft and make sure that seal on steering gear is positioned to mounting plate without kinks and opening to footwell is sealed correctly. Ingress of water and/or noises may be the result.

17. Install the new subframe bolts hand tight.

18. Torque the subframe-to-body bolts to 52 ft. lbs (70 Nm) plus an additional 90°

19. Torque the stabilizer bar-to-subframe nuts to 15 ft. lbs. (20 Nm) plus an additional 90°

20. Install new stabilizer bar-to-connecting link nuts. Counterhold at joint pin inner multi-point fitting and torque to 48 ft. lbs. (65 Nm).

21. Install new pendulum support bracket bolts and torque to 30 ft. lbs. (40 Nm) plus an additional 90°.

22. Install a new nut on the tie rod end and torque to 15 ft. lbs. (20 Nm) plus an additional 90°.

23. Install remaining components and connect all wiring.

## STEERING KNUCKLE

### REMOVAL & INSTALLATION

#### Rabbit, GTI and Jetta
*See Figure 124.*

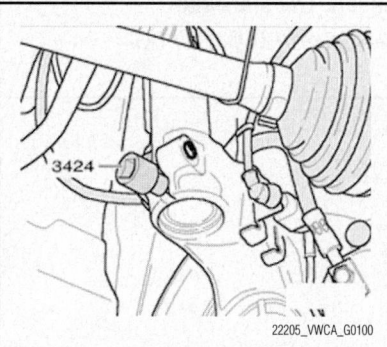

**Fig. 124 Use a spreader tool to disengage the strut from the steering knuckle—Rabbit, GTI and Jetta**

The wheel bearing can be removed without removing the steering knuckle.

1. Before servicing the vehicle, refer to the Precautions Section.

2. With the wheels on the ground, loosen the axle hub bolt. Do not roll vehicle or wheel bearing will be damaged.

3. Raise and safely support the vehicle and remove the wheels.

4. Remove the brake caliper and brake disc.

5. Remove the brake backing plate.

6. Loosen tie rod end nut, but do not remove yet.

### ✴ WARNING

**To protect thread, screw nut on pin a few turns.**

7. Press tie rod end off of steering knuckle using Ball Joint Puller 3287A.

8. Pull drive axle outer joint out of wheel hub. Secure drive axle to body using wire.

### ✴ WARNING

**Drive axle must not hang down, otherwise inner joint may be damaged when bent too far.**

9. Remove the pinch bolt holding the steering knuckle to the strut.

10. Insert spreader 3424 into steering knuckle slot and turn it to open the slot. Press brake disc in direction of suspension strut by hand. Otherwise strut tube may be canted in knuckle.

11. Remove the three lower ball joint nuts and pull the knuckle with ball joint out of control arm.

12. Place engine/transmission jack V.A.G 1383 A beneath wheel bearing housing.

13. First, pull ball joint from control arm, then pull off steering knuckle from suspension strut.

### To install:

14. Fit the steering knuckle onto the strut. Support it with a jack if necessary.

15. Remove the spreader tool and install a new pinch bolt finger tight.

16. Install drive axle in wheel hub and install a new bolt finger tight.

17. Fit the ball joint into the control arm and install new bolts hand tight. Ensure that boot is not damaged or twisted. Install new nuts finger tight.

18. Remove the jack.

19. Torque the ball joint bolts to 45 ft. lbs. (60 Nm).

20. Torque the steering knuckle pinch bolt to 52 ft. lbs. (70 Nm) plus an additional 90°.

21. Engage the tie rod in the steering knuckle, install a new nut and torque to 15 ft. lbs. (20 Nm) plus an additional 90°.

22. Install the brake plate, rotor and brake caliper.

23. Install or connect remaining components and install the wheels.

24. With the vehicle on the ground, torque the axle hub bolt.

- Six-point bolt: 150 ft. lbs. (200 Nm) plus an additional 180°
- Twelve-point bolt: 52 ft. lbs. (70 Nm) plus an additional 90°

## WHEEL BEARINGS

### REMOVAL & INSTALLATION

#### Rabbit, GTI and Jetta
*See Figure 125.*

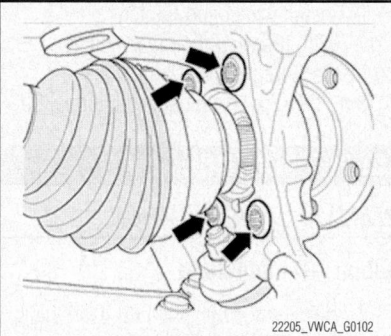

**Fig. 125 Remove these bolts to remove the wheel bearing—Rabbit, GTI and Jetta**

The wheel bearing can be removed without removing the steering knuckle.

1. Before servicing the vehicle, refer to the Precautions Section.
2. With the wheels on the ground, loosen the axle hub bolt. Do not roll vehicle or wheel bearing will be damaged.
3. Raise and safely support the vehicle and remove the wheels.
4. Remove the wheel speed sensor, brake caliper and brake disc.
5. Remove the brake backing plate.

*To install:*

6. Fit the new bearing into place and install new bolts. Torque the bolts to 52 ft. lbs. (70 Nm) plus an additional 90°.
7. Fit the axle into the hub and install a new bolt finger tight.
8. Install the brake backing plate and brake assembly.
9. With the vehicle on the ground, torque the axle hub bolt.
   - Six-point bolt: 150 ft. lbs. (200 Nm) plus an additional 180°
   - Twelve-point bolt: 52 ft. lbs. (70 Nm) plus an additional 90°

## SUSPENSION

### COIL SPRING

*REMOVAL & INSTALLATION*

**Rabbit, GTI and Jetta**

*See Figures 126 and 127.*

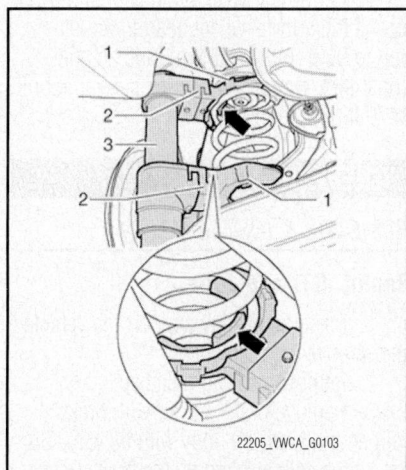

**Fig. 126 Removing rear coil spring with a spring compressor—Rabbit, GTI and Jetta**

1. Raise and safely support the vehicle and remove the rear wheel.
2. Insert spring compressor.

### ✳✳ CAUTION

**Be sure coil spring is properly seated in spring holder. Accidental release may cause serious injury.**

3. Use wrench or ratchet for tightening spring compressor. Compress coil spring until it can be removed.
4. Remove the lower spring seat.

*To install:*

5. Fit the spring and spring seat into the spring compressor. Make sure the spring aligns with the pin on the lower spring seat.
6. Fit the compressed spring assembly into place. Make sure the pin on the lower spring seat fits into the hole in the lower transverse link.
7. Make sure the spring fits properly in the upper seat and carefully release the spring compressor.
8. Install the wheel.

## REAR SUSPENSION

### LOWER CONTROL ARM

*REMOVAL & INSTALLATION*

**Rabbit, GTI and Jetta**

*See Figure 128.*

1. Raise and safely support the vehicle and remove the wheel.
2. Remove the coil spring
3. Remove any exhaust system components that may be in the way.
4. Mark the position of eccentric bolt to the subframe.
5. Remove the bolt to remove the lower transverse link.

*To install:*

6. Install lower transverse link in vehicle, install new nuts and bolts and hand-tighten them. Note position marking of eccentric bolt adjustment.

➡ **The transverse link bolts must be tightened with the suspension at normal loaded height.**

7. Install the exhaust system.
8. Install coil spring
9. Install wheel.
10. Perform wheel alignment. Torque the lower transverse link-to-wheel bearing housing bolt and nut to 66 ft. lbs. (90 Nm) plus an additional 90°.

*BUSHING REPLACEMENT*

1. Remove the control arm.
2. Position the control arm on a press. Carefully push the bushing out of the control arm using the press.

*To install:*

3. Lightly lubricate, then position the bushing on the control arm.
4. Carefully press the bushing into the control arm.
5. Install the control arm.

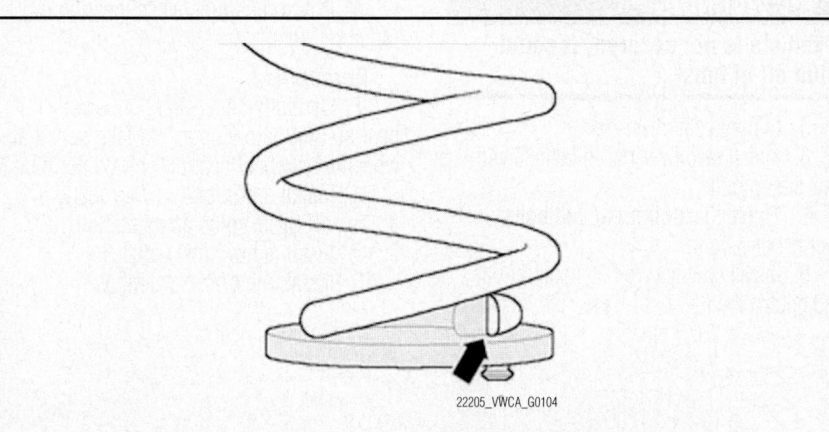

**Fig. 127 Make sure the spring aligns with the pin on the lower spring seat—Rabbit, GTI and Jetta**

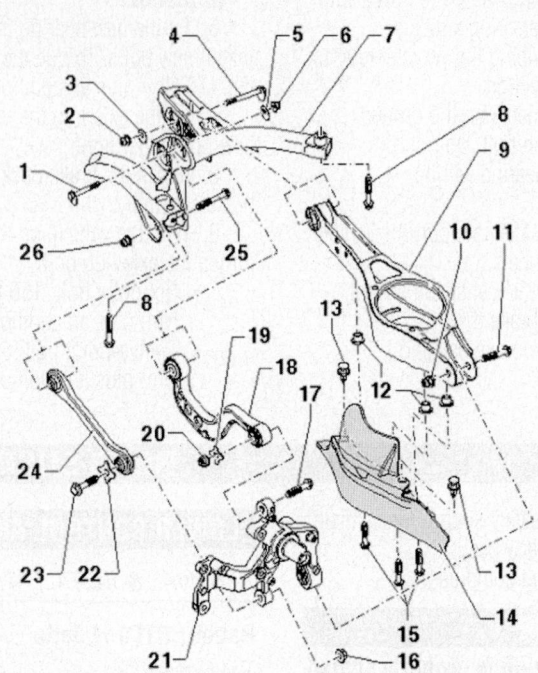

1. Eccentric bolt
2. Self-locking nut: always replace
3. Eccentric washer
4. Eccentric bolt
5. Eccentric washer
6. Self-locking nut: always replace
7. Subframe
8. Bolt: always replace
9. Lower transverse link
10. Self-locking nut: always replace
11. Bolt: always replace
12. Threaded rivet
13. Expanding rivet
14. Stone protection plate
15. Protection plate bolts
16. Self-locking nut: always replace
17. Bolt: always replace
18. Upper control arm
19. Washer
20. Self-locking nut: always replace
21. Wheel Bearing Housing
22. Washer
23. Bolt: always replace
24. Tie Rod
25. Bolt: always replace
26. Self-locking nut: always replace

22205_VWCA_G0105

**Fig. 128 Exploded view of rear suspension—Rabbit, GTI and Jetta**

## STABILIZER BAR

### REMOVAL & INSTALLATION

#### Rabbit, GTI and Jetta

➡The following work steps are described for left side of vehicle. These work steps apply also to right side of vehicle.

1. Raise and safely support the vehicle and remove rear wheels.
2. Remove the nut and remove coupling rod from stabilizer bar.
3. Remove stabilizer bar clip bolts.
4. If upper bolts on stabilizer bar clamp on right side of the vehicle cannot be removed, use a jack to raise the suspension enough to access the bolts and/or remove the bar. Make sure the vehicle is secured so it can't fall off the lift or jackstands.

### ❋❋ WARNING

**If vehicle is not secured, it could slide off of hoist.**

5. Remove stabilizer bar.
6. Install stabilizer bar in vehicle with new clamp bolts.
7. Tighten stabilizer bar clip bolts evenly hand tight.
8. Install coupling rod in stabilizer bar and tighten the new nut to 33 ft. lbs. (45 Nm).

9. Install wheel and tighten.
10. With the weight on the vehicle, torque the stabilizer bar clamp bolts to 18 ft. lbs. (25 Nm) plus an additional 90°.

## UPPER CONTROL ARM

### REMOVAL & INSTALLATION

#### Rabbit, GTI and Jetta

1. Raise and safely support the vehicle and remove the rear wheel.
2. Remove the coil spring.
3. Disengage wire for speed sensor from upper control arm.
4. Mark the position of the eccentric adjusters and remove the bolts from the upper control arm.
5. Remove the upper control arm.

#### *To install:*

6. Fit the control arm into place with new bolts. Make sure the washer is installed between bolt and wheel bearing housing. Make the bolts only finger tight.
7. Perform a wheel alignment and torque the arm-to-wheel bearing housing bolt to 96 ft. lbs. (130 Nm) plus an additional 90°. Torque the arm-to-subframe bolt to 70 ft. lbs. (95 Nm).

## WHEEL BEARINGS

### REMOVAL & INSTALLATION

#### Rabbit, GTI and Jetta

1. Raise and safely support the vehicle and remove the rear wheel.
2. Remove bearing dust cap.
3. Remove brake carrier with brake caliper and hang on body with tie wire. Do not let the caliper hang by the brake hose.
4. Remove the brake disc.
5. Remove socket head bolt using bit XZN 18 T10162.
6. Pull wheel hubs/wheel bearing unit off stub axle.

#### *To install:*

7. Carefully slide wheel hubs/wheel bearing unit onto stub axle. Make sure that wheel hubs/wheel bearing unit do not tilt.
8. Install a new bolt and torque to 133 ft. lbs., (180 Nm) plus an additional 180°.
9. Install a new dust cap.
10. Install the brake assembly.

# VOLKSWAGEN

**Touareg**

## SPECIFICATIONS AND MAINTENANCE CHARTS

### ENGINE AND VEHICLE IDENTIFICATION

| VIN ① | Cu. In. | Cyl. | Fuel Sys. | Engine Type | Eng. Mfg. | Code ② | Year |
|---|---|---|---|---|---|---|---|
| | | | Engine | | | Model Year | |
| E | 220 | 6 | Motronic | DOHC | Volkswagen | 7 | 2007 |
| B | 256 | 8 | Motronic | DOHC | Volkswagen | 8 | 2008 |
| T | 305 | 10 | TDI | DOHC | Volkswagen | | |

DOHC: Double Overhead Camshafts

① 5th digit of VIN

② 10th digit of VIN

22205_TOUA_C0001

### GENERAL ENGINE SPECIFICATIONS

| Year | Model | Engine Displacement Liters | Engine VIN | Net Horsepower @ rpm | Net Torque@rpm (ft. lbs.) | Bore x Stroke (in.) | Compression Ratio | Oil Pressure @ rpm |
|---|---|---|---|---|---|---|---|---|
| **2007** | Touareg | 3.6 | E | 276@6200 | 266@2800 | 3.50x3.80 | 10.85:1 | 44-80@2000 |
| | Touareg | 4.2 | B | 306@6200 | 302@3200 | 3.33x3.66 | 11.0:1 | 44-80@2000 |
| | Touareg | 5.0 | T | 308@3750 | 553@2000 | 3.19x3.76 | 18.5:1 | 29@2000 |
| **2008** | Touareg | 3.6 | E | 276@6200 | 266@2800 | 3.50x3.80 | 10.85:1 | 44-80@2000 |
| | Touareg | 4.2 | B | 306@6200 | 302@3200 | 3.33x3.66 | 11.0:1 | 44-80@2000 |
| | Touareg | 5.0 | T | 308@3750 | 553@2000 | 3.19x3.76 | 18.5:1 | 29@2000 |

22205_TOUA_C0002

### ENGINE TUNE-UP SPECIFICATIONS

| Year | Engine Displacement Liters | Engine VIN | Spark Plug Gap (in.) | Ignition Timing (deg.) MT | AT | Fuel Pump (psi) | Idle Speed (rpm) MT | AT | Valve Clearance Intake | Exhaust |
|---|---|---|---|---|---|---|---|---|---|---|
| **2007** | 3.6 | E | 0.043 | — | ① | 58 ② | — | 600-800 | HYD | HYD |
| | 4.2 ③ | B | 0.043 | — | ① | 58 ② | — | 670-730 | HYD | HYD |
| | 4.2 ④ | B | 0.043 | — | ① | 58 ② | — | 670-730 | HYD | HYD |
| | 5.0 | T | — | — | — | — | — | — | ⑤ | ⑤ |
| **2008** | 3.6 | E | 0.043 | — | ① | 58 ② | — | 600-800 | HYD | HYD |
| | 4.2 ③ | B | 0.043 | — | ① | 58 ② | — | 670-730 | HYD | HYD |
| | 4.2 ④ | B | 0.043 | — | ① | 58 ② | — | 670-730 | HYD | HYD |
| | 5.0 | T | — | — | — | — | — | — | ⑤ | ⑤ |

Note: The Vehicle Emission Control Information label reflects specification changes made during production.

The label figures must be used if they differ from those in this chart.

HYD: Hydraulic

NA: Information not available

① The ignition timing is controlled by the ECM and is not adjustable.

② System pressure at idle.

③ Engine codes AXQ, BHX

④ Engine code BAR

⑤ Uses Camshaft Adjuster Valves

22205_TOUA_C0003

## CAPACITIES

| Year | Model | Engine Displacement Liters | Engine VIN | Engine Oil with Filter | Transmission (qts.) Manual | Transmission (qts.) Automatic | Final Drive (pts.) | Fuel Tank (gal.) | Cooling System (qts.) |
|------|-------|------|------|------|------|------|------|------|------|
| 2007 | Touareg | 3.6 | E | 7.3 | NA | 9.5 | ① | ② | 9.5 |
| | Touareg | 4.2 ③ | B | 7.9 | NA | 9.5 | ① | ② | 9.5 |
| | Touareg | 4.2 ④ | B | 6.7 | NA | 9.5 | ① | ② | 12.7 |
| | Touareg | 5.0 | T | 12.1 | NA | 9.5 | ① | ② | 9.5 |
| 2008 | Touareg | 3.6 | E | 7.3 | NA | 9.5 | ① | ② | 9.5 |
| | Touareg | 4.2 ③ | B | 7.9 | NA | 9.5 | ① | ② | 9.5 |
| | Touareg | 4.2 ④ | B | 6.7 | NA | 9.5 | ① | ② | 12.7 |
| | Touareg | 5.0 | T | 12.1 | NA | 9.5 | ① | ② | 9.5 |

Note: All capacities are approximate. Add fluid gradually and check often to avoid overfilling.

NA: Information not available

① Transfer case: 0.9 qts.; Front final drive: 1.0 qts.; Rear final drive: 1.7 qts.

② Refer to vehicle owner's manual for specification.

③ With engine code AXQ or BHX

④ With engine code BAR

22205_TOUA_C0004

## FLUID SPECIFICATIONS

| Year | Model | Engine Displacement Liters | Engine VIN | Engine Oil | Auto. Trans. | Manual Trans. | Power Steering Fluid | Brake Master Cylinder |
|------|-------|------|------|------|------|------|------|------|
| 2007 | Touareg | 3.6 | E | 5W-40 | VW ATF | — | ① | DOT 4 |
| | Touareg | 4.2 | B | 5W-40 | VW ATF | — | ① | DOT 4 |
| | Touareg | 5.0 | T | 5W-40 ② | VW ATF | — | ① | DOT 4 |
| 2008 | Touareg | 3.6 | E | 5W-40 | VW ATF | — | ① | DOT 4 |
| | Touareg | 4.2 | B | 5W-40 | VW ATF | — | ① | DOT 4 |
| | Touareg | 5.0 | T | 5W-40 ② | VW ATF | — | ① | DOT 4 |

DOT: Department Of Transportation

① VW G002 000 (do not use ATF)

② Use only low-ash oils meeting VW specifications.

22205_TOUA_C0005

## VALVE SPECIFICATIONS

| Year | Engine Displacement Liters | Engine VIN | Seat Angle (deg.) | Face Angle (deg.) | Spring Test Pressure (lbs. @ in.) | Spring Installed Height (in.) | Stem-to-Guide Clearance (in.) | | Stem Diameter (in.) | |
|---|---|---|---|---|---|---|---|---|---|---|
| | | | | | | | Intake | Exhaust | Intake | Exhaust |
| 2007 | 3.6 | E | 45 | 44°40' | N/A | N/A | N/A | N/A | 0.235 | 0.235 |
| | 4.2 ③ | B | 45 | 45 | N/A | N/A | N/A | N/A | 0.235 | 0.235 |
| | 4.2 ④ | B | 45 | 45 | N/A | N/A | N/A | N/A | 0.235 | 0.235 |
| | 5.0 | T | 45 | 45 | N/A | N/A | N/A | N/A | 0.275 | 0.274 |
| 2008 | 3.6 | E | 45 | 44°40' | N/A | N/A | N/A | N/A | 0.235 | 0.235 |
| | 4.2 ③ | B | 45 | 45 | N/A | N/A | N/A | N/A | 0.235 | 0.235 |
| | 4.2 ④ | B | 45 | 45 | N/A | N/A | N/A | N/A | 0.235 | 0.235 |
| | 5.0 | T | 45 | 45 | N/A | N/A | N/A | N/A | 0.275 | 0.274 |

NA: Information not available

22205_TOUA_C0006

## CAMSHAFT AND BEARING SPECIFICATIONS CHART

All measurements are given in inches.

| Year | Engine Displ. Liters | Engine VIN | Journal Dia. | Brg. Oil Clearance | Shaft End-play | Runout | Journal Bore | Lobe Height | |
|---|---|---|---|---|---|---|---|---|---|
| | | | | | | | | Intake | Exhaust |
| 2007 | 3.6 | E | NA | NA | 0.0157 | NA | NA | NA | NA |
| | 4.2 | B | NA | NA | 0.0079 | NA | NA | NA | NA |
| | 5.0 | T | NA | NA | 0.0039-0.0075 | NA | NA | NA | NA |
| 2008 | 3.6 | E | NA | NA | 0.0157 | NA | NA | NA | NA |
| | 4.2 | B | NA | NA | 0.0079 | NA | NA | NA | NA |
| | 5.0 | T | NA | NA | 0.0039-0.0075 | NA | NA | NA | NA |

NA: Not Available

22205_TOUA_C0007

## CRANKSHAFT AND CONNECTING ROD SPECIFICATIONS

All measurements are given in inches.

| Year | Engine Displacement Liters | Engine VIN | Crankshaft | | | | Connecting Rod | | |
|---|---|---|---|---|---|---|---|---|---|
| | | | Main Brg. Journal Dia. | Main Brg. Oil Clearance | Shaft End-play | Thrust on No. | Journal Diameter | Oil Clearance | Side Clearance |
| 2007 | 3.6 | E | 2.3605 | 0.0007-0.0023 | 0.0027-0.0090 | 5 | 2.1243 | 0.0007-0.0027 | 0.0019-0.0122 |
| | 4.2 | B | 2.558-2.5600 | 0.0007-0.0017 | 0.0035-0.0100 | 4 | 2.1251-2.1276 | 0.0008-0.0027 | NA |
| | 5.0 | T | 2.5572-2.5603 | 0.0012-0.0031 | 0.0028-0.0067 | 3, 4 | 2.2843-2.2818 | 0.0008-0.0027 | 0.0015 |
| 2008 | 3.6 | E | 2.3605 | 0.0007-0.0023 | 0.0027-0.0090 | 5 | 2.1243 | 0.0007-0.0027 | 0.0019-0.0122 |
| | 4.2 | B | 2.558-2.5600 | 0.0007-0.0017 | 0.0035-0.0100 | 4 | 2.1251-2.1276 | 0.0008-0.0027 | NA |
| | 5.0 | T | 2.5572-2.5603 | 0.0012-0.0031 | 0.0028-0.0067 | 3, 4 | 2.2843-2.2818 | 0.0008-0.0027 | 0.0015 |

NA: Information not available

22205_TOUA_C0008

## PISTON AND RING SPECIFICATIONS

All measurements are given in inches.

| Year | Engine Displacement Liters | Engine VIN | Piston Clearance | Ring Gap | | | Ring Side Clearance | | |
|------|------|------|------|------|------|------|------|------|------|
| | | | | Top Compression | Bottom Compression | Oil Control | Top Compression | Bottom Compression | Oil Control |
| 2007 | 3.6 | E | 0.0025 | 0.008-0.015 | 0.008-0.015 | 0.009-0.019 | 0.0015-0.0035 | 0.0012-0.0024 | 0.0008-0.0024 |
| | 4.2 | B | 0.0027 | 0.008-0.014 | 0.008-0.016 | 0.008-0.016 | 0.0014-0.0034 | 0.0002-0.0018 | 0.0004-0.0020 |
| | 5.0 | T | 0.0017 | 0.008-0.0160 | 0.012-0.0200 | 0.010-0.0200 | 0.0031-0.0039 | 0.0012-0.0019 | 0.0008-0.0019 |
| 2008 | 3.6 | E | 0.0025 | 0.008-0.015 | 0.008-0.015 | 0.009-0.019 | 0.0015-0.0035 | 0.0012-0.0024 | 0.0008-0.0024 |
| | 4.2 | B | 0.0027 | 0.008-0.014 | 0.008-0.016 | 0.008-0.016 | 0.0014-0.0034 | 0.0002-0.0018 | 0.0004-0.0020 |
| | 5.0 | T | 0.0017 | 0.008-0.0160 | 0.012-0.0200 | 0.010-0.0200 | 0.0031-0.0039 | 0.0012-0.0019 | 0.0008-0.0019 |

NA: Information not available

22205_TOUA_C0009

## TORQUE SPECIFICATIONS

All readings in ft. lbs.

| Year | Engine Displacement Liters | Engine VIN | Cylinder Head Bolts | Main Bearing Bolts | Rod Bearing Bolts | Crankshaft Damper Bolt | Flywheel Bolts | Manifold | | Spark Plugs | Oil Pan Drain Plug |
|------|------|------|------|------|------|------|------|------|------|------|------|
| | | | | | | | | Intake | Exhaust | | |
| 2007 | 3.6 | E | ① | ② | ③ | ④ | ⑤ | ⑥ | NA | 15 | 22 |
| | 4.2 | B | ⑦ | ② | ⑧ | ⑨ | ⑨ | ⑥ | NA | 22 | 37 |
| | 5.0 | T | ⑩ | ② | ③ | ④ | ⑤ | N/A | ⑪ | 11 | 22 |
| 2008 | 3.6 | E | ① | ② | ③ | ④ | ⑤ | ⑥ | NA | 15 | 22 |
| | 4.2 | B | ⑦ | ② | ⑧ | ⑨ | ⑨ | ⑥ | NA | 22 | 37 |
| | 5.0 | T | ⑩ | ② | ③ | ④ | ⑤ | N/A | ⑪ | 11 | 22 |

NA: Information not available

① Torque in four steps:  (use new bolts)

  Step 1:  22 ft. lbs.

  Step 2:  37 ft. lbs.

  Step 3:  plus 90 degrees

  Step 4:  plus 90 degrees

② Use new bolts: 22 ft. lbs. plus 180 degrees

③ Use new bolts: 22 ft. lbs. plus 90 degrees

④ Use new bolts: 74 ft. lbs. plus 90 degrees

⑤ Use new bolts: 44 ft. lbs. plus 90 degrees

⑥ Upper intake manifold bolts: 7 ft. lbs.

  Lower intake manifold bolts: 6 ft. lbs.

⑦ Torque in three steps:  (use new bolts)

  Step 1:  11 ft. lbs.

  Step 2:  22 ft. lbs.

  Step 3:  plus 180 degrees

⑧ M8 nuts: 18 ft. lbs

  M10 nuts:  30 ft. lbs.

⑨ Use new bolts: 37 ft. lbs. plus 90 degrees

⑩ Torque in three steps: (use new bolts)

  Step 1:  26 ft. lbs.

  Step 2:  44 ft. lbs.

  Step 3:  plus 90 degrees

⑪ Use new bolts: 44 ft. lbs. plus 90 degrees

22205_TOUA_C0010

## WHEEL ALIGNMENT

| Year | Model | | Caster Range (+/-Deg.) | Caster Preferred Setting (Deg.) | Camber Range (+/-Deg.) | Camber Preferred Setting (Deg.) | Toe-in (Deg.) |
|------|-------|---|---------|---------|--------|---------|---------|
| 2007 | Standard | F | 0.16 | 8.50 | 0.33 | 0.00 | 0.04 +/- 0.04 |
| | Suspension | R | — | — | 0.33 | 1.00 | 0.16 +/- 0.08 |
| | Sport | F | 0.16 | 8.58 | 0.33 | -0.16 | 0.08 +/- 0.04 |
| | Suspension | R | — | — | 0.33 | 1.00 | 0.16 +/- 0.08 |
| | Air Springs | F | 0.50 | 8.45 | 0.1 | -0.20 | 0.10 +/- 0.30 |
| | Suspension | R | — | — | 0.2 | -1.20 | 0.10 +/- 0.05 |
| | Standard | F | 0.16 | 8.16 | 0.33 | 0.00 | 0.04 +/- 0.04 |
| | Suspension | R | — | — | 0.33 | 1.00 | 0.16 +/- 0.08 |
| 2008 | Sport | F | 0.16 | 8.58 | 0.33 | -0.16 | 0.08 +/- 0.04 |
| | Suspension | R | — | — | 0.33 | 1.00 | 0.16 +/- 0.08 |
| | Air Springs | F | 0.50 | 8.45 | 0.1 | -0.20 | 0.10 +/- 0.30 |
| | Suspension | R | — | — | 0.2 | -1.20 | 0.10 +/- 0.05 |

22205_TOUA_C0011

## TIRE, WHEEL AND BALL JOINT SPECIFICATIONS

| Year | Model | OEM Tires ① Standard | OEM Tires ① Optional | Tire Pressures (psi) Front | Tire Pressures (psi) Rear | Wheel Size | Lug Nut (ft. lbs.) |
|------|-------|----------|----------|-------|------|------|------|
| 2007 | Touareg 3.6L | P235/65R17 | P235/60R18 | ① | ① | NA | 133 |
| | Touareg 4.2L | P235/65R17 | P235/60R18 | ① | ① | NA | 133 |
| | Touareg 5.0L | P235/65R17 | P235/60R18 | ① | ① | NA | 133 |
| 2008 | Touareg 3.6L | P235/65R17 | P235/60R18 | ① | ① | NA | 133 |
| | Touareg 4.2L | P235/65R17 | P235/60R18 | ① | ① | NA | 133 |
| | Touareg 5.0L | P235/65R17 | P235/60R18 | ① | ① | NA | 133 |

NA: Information not available

OEM: Original Equipment Manufacturer

PSI: Pounds Per Square Inch

① All models equipped with TPMS; consult vehicle tag or owner's handbook for specific pressures.

22205_TOUA_C0012

## BRAKE SPECIFICATIONS

All measurements in inches unless noted

| Year | Model | | Brake Disc Original Thickness | Brake Disc Minimum Thickness | Brake Disc Maximum Run-out | Minimum Lining Thickness | Brake Caliper Guide Pins (ft. lbs.) | Brake Caliper Mounting Bolts (ft. lbs.) |
|------|-------|---|-----------|-----------|----------|---------|-------|-------|
| 2007 | Touareg | F | ① | NA | NA | 0.07 | NA | 199 |
| | | R | ② | NA | NA | 0.07 | NA | 133 |
| 2008 | Touareg | F | ① | NA | NA | 0.07 | NA | 199 |
| | | R | ② | NA | NA | 0.07 | NA | 133 |

NA: Information not available

① 17" 1LC, 1LE: 1.26"

   18" 1LF: 1.34"

② 17" 1KF, 1KQ: 1.10"

22205_TOUA_C0013

## SCHEDULED MAINTENANCE INTERVALS
### VOLKSWAGEN TOUAREG - 3.6L & 4.2L

| TO BE SERVICED | TYPE OF SERVICE | VEHICLE MILEAGE INTERVAL (x1000) | | | | | | | | | | | | |
|---|---|---|---|---|---|---|---|---|---|---|---|---|---|---|
| | | 5 | 10 | 20 | 30 | 40 | 45 | 50 | 60 | 70 | 75 | 80 | 90 | 100 |
| Engine oil & filter | R | ✓ | ✓ | ✓ | ✓ | ✓ | | ✓ | ✓ | ✓ | | ✓ | ✓ | ✓ |
| Brake pad thickness | S/I | | ✓ | ✓ | ✓ | ✓ | | ✓ | ✓ | ✓ | | ✓ | ✓ | ✓ |
| Dust seals on ball joints, tie rod ends & tie rods | S/I | | | ✓ | | ✓ | | | ✓ | | | ✓ | | ✓ |
| Battery | S/I | | ✓ | ✓ | ✓ | ✓ | | ✓ | ✓ | ✓ | | ✓ | ✓ | ✓ |
| Brake system | S/I | | ✓ | ✓ | ✓ | ✓ | | ✓ | ✓ | ✓ | | ✓ | ✓ | ✓ |
| Cooling system | S/I | | | ✓ | | ✓ | | | ✓ | | | ✓ | | ✓ |
| Driveshaft boots | S/I | | | ✓ | | ✓ | | | ✓ | | | ✓ | | ✓ |
| Engine (check for leaks) | S/I | | | ✓ | | ✓ | | | ✓ | | | ✓ | | ✓ |
| Engine exhaust | S/I | | | ✓ | | ✓ | | | ✓ | | | ✓ | | ✓ |
| Power steering fluid level | S/I | | | ✓ | | ✓ | | | ✓ | | | ✓ | | ✓ |
| Underbody | S/I | | | | | | ✓ | | | | | ✓ | | |
| Timing belt (V8) | R | | | | | | | | | | | ✓ | | |
| Transmission fluid level | S/I | | | ✓ | | ✓ | | | ✓ | | | ✓ | | ✓ |
| Air filter element | R | | | | | ✓ | | | | | | ✓ | | |
| Spark plugs (V6) | R | | | | | ✓ | | | | | | ✓ | | |
| Spark plugs (V8) | R | | | | | | | | ✓ | | | | | |
| Passenger compartment air filter | R | | ✓ | | | ✓ | | | ✓ | | | ✓ | | ✓ |
| Drive belts | S/I | | | | | ✓ | | | | | | ✓ | | |
| Brake fluid ① | R | | | | | | | | | | | | | |

R: Replace   S/I: Service or Inspect

① Replace every two years regardless of mileage.

## FREQUENT OPERATION MAINTENANCE (SEVERE SERVICE)

If a vehicle is operated under any of the following conditions it is considered severe service:

- Extremely dusty areas.

- 50% or more of the vehicle operation is in 32°C (90°F) or higher temperatures, or constant operation in temperatures below 0°C (32°F).

- Prolonged idling (vehicle operation in stop and go traffic).

- Frequent short running periods (engine does not warm to normal operating temperatures).

- Police, taxi, delivery usage or trailer towing usage.

Oil & oil filter change: change every 3750 miles.

Air filter element: service or inspect every 15,000 miles.

Automatic transaxle fluid & filter: replace every 30,000 miles.

22205_TOUA_C0014

## SCHEDULED MAINTENANCE INTERVALS
### VOLKSWAGEN TOUAREG - 5.0L TDI

| TO BE SERVICED | TYPE | VEHICLE MILEAGE INTERVAL (x1000) | | | | | | | | | | | | |
|---|---|---|---|---|---|---|---|---|---|---|---|---|---|---|
| | | 5 | 10 | 20 | 30 | 40 | 50 | 60 | 70 | 80 | 90 | 100 | 110 | 120 |
| Engine oil & filter | R | ✓ | ✓ | ✓ | ✓ | ✓ | ✓ | ✓ | ✓ | ✓ | ✓ | ✓ | ✓ | ✓ |
| Brake pad thickness | S/I | ✓ | ✓ | ✓ | ✓ | ✓ | ✓ | ✓ | ✓ | ✓ | ✓ | ✓ | ✓ | ✓ |
| Dust seals on ball joints, tie rod ends & tie rods | S/I | | | ✓ | | ✓ | | ✓ | | ✓ | | ✓ | | ✓ |
| Battery | S/I | | | | | ✓ | | | | ✓ | | | | ✓ |
| Brake system | S/I | | ✓ | ✓ | ✓ | ✓ | ✓ | ✓ | ✓ | ✓ | ✓ | ✓ | ✓ | ✓ |
| Cooling system | S/I | | | ✓ | | ✓ | | ✓ | | ✓ | | ✓ | | ✓ |
| Driveshaft boots | S/I | | | ✓ | | ✓ | | ✓ | | ✓ | | ✓ | ✓ | ✓ |
| Engine (check for leaks) | S/I | | | ✓ | | ✓ | | ✓ | | ✓ | | ✓ | | ✓ |
| Engine exhaust | S/I | | | ✓ | | ✓ | | ✓ | | ✓ | | ✓ | ✓ | ✓ |
| Power steering fluid level | S/I | | | | | ✓ | | | | ✓ | | | | ✓ |
| Underbody | S/I | | | | | ✓ | | | | ✓ | | | | ✓ |
| Transmission fluid level | S/I | | | | | ✓ | | | | ✓ | | | | ✓ |
| Final driver | S/I | | | | | ✓ | | | | ✓ | | | | ✓ |
| Fuel filter | R | | ✓ | | | ✓ | | ✓ | | ✓ | | ✓ | | ✓ |
| Air cleaner fitler | S/I | | | | | | | | | ✓ | | | | |
| Air suspension (check for leaks) | S/I | | | | | ✓ | | | | ✓ | | | | ✓ |
| Passenger compartment air filter | R | | | ✓ | | ✓ | | ✓ | | ✓ | | ✓ | ✓ | ✓ |
| Drive belts | S/I | | | | | ✓ | | | | ✓ | | | | ✓ |
| Water separator - drain | S/I | ✓ | ✓ | ✓ | ✓ | ✓ | ✓ | ✓ | ✓ | ✓ | ✓ | ✓ | ✓ | ✓ |
| Diesel particle filter | R | | | | | | | | | | | | | ✓ |
| Brake fluid ① | R | | | | | | | | | | | | | |

R: Replace          S/I: Service or Inspect

① Replace every two years regardless of mileage.

## FREQUENT OPERATION MAINTENANCE (SEVERE SERVICE)
If a vehicle is operated under any of the following conditions it is considered severe service:

- Extremely dusty areas.

- 50% or more of the vehicle operation is in 32°C (90°F) or higher temperatures, or constant operation in temperatures below 0°C (32°F).

- Prolonged idling (vehicle operation in stop and go traffic).

- Frequent short running periods (engine does not warm to normal operating temperatures).

- Police, taxi, delivery usage or trailer towing usage.

Oil & oil filter change: change every 5000 miles.

Air filter element: service or inspect every 80,000 miles.

Automatic transmission fluid & filter: replace every 40,000 miles.

22205_TOUA_C0015

## PRECAUTIONS

Before servicing any vehicle, please be sure to read all of the following precautions, which deal with personal safety, prevention of component damage, and important points to take into consideration when servicing a motor vehicle:

• Never open, service or drain the radiator or cooling system when the engine is hot; serious burns can occur from the steam and hot coolant.

• Observe all applicable safety precautions when working around fuel. Whenever servicing the fuel system, always work in a well-ventilated area. Do not allow fuel spray or vapors to come in contact with a spark, open flame, or excessive heat (a hot drop light, for example). Keep a dry chemical fire extinguisher near the work area. Always keep fuel in a container specifically designed for fuel storage; also, always properly seal fuel containers to avoid the possibility of fire or explosion. Refer to the additional fuel system precautions later in this section.

• Fuel injection systems often remain pressurized, even after the engine has been turned **OFF**. The fuel system pressure must be relieved before disconnecting any fuel lines. Failure to do so may result in fire and/or personal injury.

• Brake fluid often contains polyglycol ethers and polyglycols. Avoid contact with the eyes and wash your hands thoroughly after handling brake fluid. If you do get brake fluid in your eyes, flush your eyes with clean, running water for 15 minutes. If eye irritation persists, or if you have taken brake fluid internally, IMMEDIATELY seek medical assistance.

• The EPA warns that prolonged contact with used engine oil may cause a number of skin disorders, including cancer. You should make every effort to minimize your exposure to used engine oil. Protective gloves should be worn when changing oil. Wash your hands and any other exposed skin areas as soon as possible after exposure to used engine oil. Soap and water, or waterless hand cleaner should be used.

• All new vehicles are now equipped with an air bag system, often referred to as a Supplemental Restraint System (SRS) or Supplemental Inflatable Restraint (SIR) system. The system must be disabled before performing service on or around system components, steering column, instrument panel components, wiring and sensors. Failure to follow safety and disabling procedures could result in accidental air bag deployment, possible personal injury and unnecessary system repairs.

• Always wear safety goggles when working with, or around, the air bag system. When carrying a non-deployed air bag, be sure the bag and trim cover are pointed away from your body. When placing a non-deployed air bag on a work surface, always face the bag and trim cover upward, away from the surface. This will reduce the motion of the module if it is accidentally deployed. Refer to the additional air bag system precautions later in this section.

• Clean, high quality brake fluid from a sealed container is essential to the safe and proper operation of the brake system. You should always buy the correct type of brake fluid for your vehicle. If the brake fluid becomes contaminated, completely flush the system with new fluid. Never reuse any brake fluid. Any brake fluid that is removed from the system should be discarded. Also, do not allow any brake fluid to come in contact with a painted surface; it will damage the paint.

• Never operate the engine without the proper amount and type of engine oil; doing so WILL result in severe engine damage.

• Timing belt maintenance is extremely important. Many models utilize an interference-type, non-freewheeling engine. If the timing belt breaks, the valves in the cylinder head may strike the pistons, causing potentially serious (also time-consuming and expensive) engine damage. Refer to the maintenance interval charts for the recommended replacement interval for the timing belt, and to the timing belt section for belt replacement and inspection.

• Disconnecting the negative battery cable on some vehicles may interfere with the functions of the on-board computer system(s) and may require the computer to undergo a relearning process once the negative battery cable is reconnected.

• When servicing drum brakes, only disassemble and assemble one side at a time, leaving the remaining side intact for reference.

• Only an MVAC-trained, EPA-certified automotive technician should service the air conditioning system or its components.

## BRAKES

### GENERAL INFORMATION

*PRECAUTIONS*

• Certain components within the ABS system are not intended to be serviced or repaired individually.

• Do not use rubber hoses or other parts not specifically specified for and ABS system. When using repair kits, replace all parts included in the kit. Partial or incorrect repair may lead to functional problems and require the replacement of components.

• Lubricate rubber parts with clean, fresh brake fluid to ease assembly. Do not use shop air to clean parts; damage to rubber components may result.

• Use only DOT 3 brake fluid from an unopened container.

• If any hydraulic component or line is removed or replaced, it may be necessary to bleed the entire system.

• A clean repair area is essential. Always clean the reservoir and cap thoroughly before removing the cap. The slightest amount of dirt in the fluid may plug an orifice and impair the system function. Perform repairs after components have been thoroughly cleaned; use only denatured alcohol

## ANTI-LOCK BRAKE SYSTEM (ABS)

to clean components. Do not allow ABS components to come into contact with any substance containing mineral oil; this includes used shop rags.

• The Anti-Lock control unit is a microprocessor similar to other computer units in the vehicle. Ensure that the ignition switch is **OFF** before removing or installing controller harnesses. Avoid static electricity discharge at or near the controller.

• If any arc welding is to be done on the vehicle, the control unit should be unplugged before welding operations begin.

## BRAKES

## BLEEDING THE BRAKE SYSTEM

### BLEEDING PROCEDURE

*BLEEDING PROCEDURE*

➡If one chamber of the brake fluid reservoir has run completely empty (e.g. leak in the brake system), the brake system must be pre-bled first.

**Pre-Bleeding**

➡There are two bleeder valves installed on Brembo brake calipers Always bleed at both bleeder valves.

1. Bleed at outer bleeder valve first.
2. Connect brake filling and bleeding tool (VAS 5234 or V.A.G 1869).
3. Note the bleeding sequence:
  - Bleed left front and right front brake caliper together simultaneously
  - Bleed left rear and right rear brake caliper together simultaneously
4. With bleeder bottle hoses attached, leave bleeder valves open long enough that brake fluid exits without bubbles.
5. Then hydraulic unit must be bled once more via function "Basic setting" using an approved tool (VAS 5051) or scan tool.
6. Initiate basic setting (to bleed brake system), then, brake system must be bled normally
7. After bleeding brake system, basic

setting for brake pressure sensor 1 (G201) must be performed.

**Bleeding**

➡Adhere strictly to work sequence when bleeding brake system.

➡There are two bleeder valves installed on Brembo brake calipers Always bleed at both bleeder valves.

1. Bleed at outer bleeder valve first.
2. Connect brake filling and bleeding tool (VAS 5234 or V.A.G 1869).
3. Open bleeder valves in the pre-scribed sequence and bleed brake caliper:
  - Right rear brake caliper
  - Left rear brake caliper
  - Front right brake caliper
  - Front left brake caliper
4. Use suitable bleeder hose. It must fit tightly on bleeder valve so that no air gets into brake system.
5. With bleeder bottle hose attached, leave bleeder valves open long enough that brake fluid exits without bubbles.

**Post-Bleeding**

➡A second mechanic is required to assist.

1. Depress brake pedal forcefully and hold.

2. Open bleeder valve at brake caliper.
3. Press brake pedal down onto stop.
4. Close bleeder valve with pedal depressed.
5. Release brake pedal slowly.

➡This bleeding procedure must be performed 5 times per brake caliper.

6. Note the proper bleeding sequence:
  - Right rear brake caliper
  - Left rear brake caliper
  - Front right brake caliper
  - Front left brake caliper
7. A road test must be performed after bleeding. During this, at least one ABS regulation must be performed.

*MASTER CYLINDER BLEEDING*

➡For these models, a complete brake bleeding procedure is required. See "Brake Bleeding" above.

*BRAKE LINE BLEEDING*

➡For these models, a complete brake bleeding procedure is required. See "Brake Bleeding" above.

*BLEEDING THE ABS SYSTEM*

➡For these models, a complete brake bleeding procedure is required. See "Brake Bleeding" above.

## BRAKES

## FRONT DISC BRAKES

### ✸✸ CAUTION

Dust and dirt accumulating on brake parts during normal use may contain asbestos fibers from production or aftermarket brake linings. Breathing excessive concentrations of asbestos fibers can cause serious bodily harm. Exercise care when servicing brake parts. Do not sand or grind brake lining unless equipment used is designed to contain the dust residue. Do not clean brake parts with compressed air or by dry brushing. Cleaning should be done by dampening the brake components with a fine mist of water, then wiping the brake components clean with a dampened cloth. Dispose of cloth and all residue containing asbestos fibers in an impermeable container with the appropriate label. Follow practices prescribed by the Occupational Safety and Health Administration (OSHA) and the Environmental Protection Agency (EPA) for

the handling, processing, and disposing of dust or debris that may contain asbestos fibers.

### BRAKE CALIPER

*REMOVAL & INSTALLATION*

*See Figure 1.*

1. Remove wheels.
2. Connect bleeder hose of bleeder bottle to bleeder valve of brake caliper and then open bleeder valve.
3. Insert A brake pedal depressor (V.A.G 1869/2).
4. Close bleeder valve and remove bleeder bottle.
5. Unscrew brake line from brake caliper.
6. Unscrew bracket from wheel bearing housing.
7. Remove the brake pads.
8. Remove brake caliper from wheel bearing housing.

**To install:**

9. Bolt brake caliper onto wheel bearing

housing. Tighten the bolt to 199 ft. lbs. (270 Nm).
10. Install brake pads.
11. Install brake line on brake caliper. Tighten to 10 ft. lbs. (14 Nm).
12. Install bracket to wheel bearing housing. Tighten bolt to 7 ft. lbs. (9 Nm).
13. Remove brake pedal loading device.
14. Bleed brake system. See "Bleeding Procedure."
15. Install the wheels. Tighten the wheel nuts to 133 ft. lbs. (180 Nm).
16. Before moving vehicle, depress brake pedal several times firmly to properly seat brake pads in their normal operating position.
17. Check brake fluid level.

### DISC BRAKE PADS

*REMOVAL & INSTALLATION*

*See Figure 1.*

1. Before servicing the vehicle, refer to the precautions in the beginning of this Section.

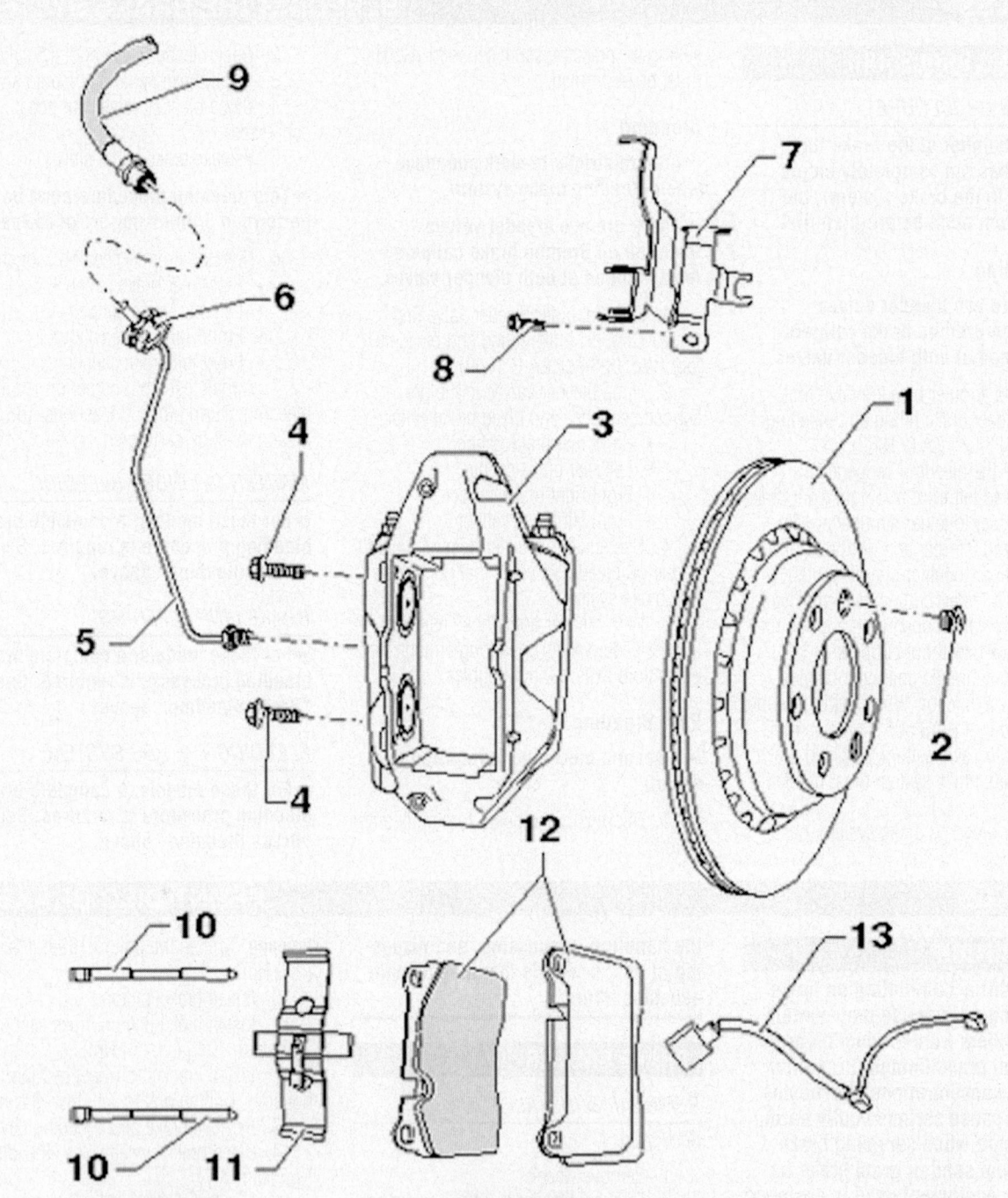

1. Disc
2. Bolt
3. Caliper
4. Bolt
5. Brake Line
6. Spring Clamp
7. Bracket
8. Bolt
9. Brake Hose
10. Retaining Pins
11. Retaining Spring
12. Brake Pads
13. Pad Wear Indicator Wire

22205_TOUA_G0002

**Fig. 1 Exploded view of the front brake system**

2. Raise the vehicle and remove the wheels.

3. Unplug the connector for the brake pad wear indicator.

4. Remove the cotter pin on the inner side of the caliper from the pad retaining pin.

5. Press retaining spring down and remove pad retaining pin at the same time.

6. Pull the wire for the brake pad wear indicator out of the brake caliper housing and out of the retaining spring.

7. Remove retaining spring.

8. Before pressing brake pads back, draw off brake fluid from reservoir using a bleeder bottle. Otherwise, especially if reservoir has been topped off, fluid will overflow and cause damage.

9. Press brake pads off brake disc and remove from brake caliper.

10. Carefully remove the contact sensor with the brake pad wear indicator wire from the brake pads and check for damage. (Reuse undamaged contact sensors and wires.)

### To install:

11. Clean the brake caliper with mineral spirits of any adhesive residue. The surface must be dry and clean.

12. Before pressing the piston back in its bore to insert the new brake pads, draw off brake fluid from the reservoir with a bleeder bottle.

13. Push the piston back in its bore. Carefully install the brake pad wear indicator wire contact sensor in the new brake pads.

14. Insert the new brake pads into the caliper.

15. Insert retaining spring and install the wire for the brake pad wear indicator below the tab of the retaining spring and in the brake caliper housing.

16. Press the retaining spring down and press the pad retaining pin in until the stop.

17. Secure the pad retaining pin using securing cotter pin.

18. Connect connectors of the brake pad wear indicator in the bracket of brake caliper housing.

19. Install the wheels.

20. After replacing the brake pads, depress brake pedal firmly several times with the vehicle stationary so that the brake pads are properly seated in their normal operating position.

21. Check brake fluid level after replacing brake pad

## BRAKES

### ✳✳ CAUTION

**Dust and dirt accumulating on brake parts during normal use may contain asbestos fibers from production or aftermarket brake linings. Breathing excessive concentrations of asbestos fibers can cause serious bodily harm. Exercise care when servicing brake parts. Do not sand or grind brake lining unless equipment used is designed to contain the dust residue. Do not clean brake parts with compressed air or by dry brushing. Cleaning should be done by dampening the brake components with a fine mist of water, then wiping the brake components clean with a dampened cloth. Dispose of cloth and all residue containing asbestos fibers in an impermeable container with the appropriate label. Follow practices prescribed by the Occupational Safety and Health Administration (OSHA) and the Environmental Protection Agency (EPA) for the handling, processing, and disposing of dust or debris that may contain asbestos fibers.**

### BRAKE CALIPER

#### REMOVAL & INSTALLATION

*See Figure 2.*

1. Remove wheels.

2. Connect bleeder hose of bleeder bottle to bleeder valve of brake caliper and then open bleeder valve.

3. Insert A brake pedal depressor (V.A.G 1869/2).

4. Close bleeder valve and remove bleeder bottle.

5. Unscrew brake line from brake caliper.

6. Unscrew bracket from wheel bearing housing.

7. Remove the brake pads.

8. Remove brake caliper from wheel bearing housing.

#### To install:

9. Bolt brake caliper onto wheel bearing housing. Tighten the bolt to 199 ft. lbs. (270 Nm).

10. Install brake pads.

11. Install brake line on brake caliper. Tighten to 10 ft. lbs. (14 Nm).

12. Install bracket to wheel bearing housing. Tighten bolt to 7 ft. lbs. (9 Nm).

13. Remove brake pedal loading device.

14. Bleed brake system. See "Bleeding Procedure."

15. Install the wheels. Tighten the wheel nuts to 133 ft. lbs. (180 Nm).

16. Before moving vehicle, depress brake pedal several times firmly to properly seat brake pads in their normal operating position.

17. Check brake fluid level.

### DISC BRAKE PADS

#### REMOVAL & INSTALLATION

*See Figure 2.*

1. Remove the wheels.

2. Separate the connector for brake pad wear indicator.

3. Drive out the pad retaining pins.

## REAR DISC BRAKES

4. Pull the wire for the brake pad wear indicator out of the brake caliper housing and out of retaining spring.

5. Remove the retaining spring.

### ✳✳ CAUTION

**Before pressing brake pads back, draw off brake fluid from the reservoir, using a bleeder bottle. Otherwise, especially if reservoir has been topped off, fluid will overflow and cause damage.**

6. Press brake pads off the brake disc and remove the pads from brake caliper.

7. Carefully remove the contact sensor with brake pad wear indicator wire from the brake pads and check for damage. Reuse undamaged contact sensors and wires.

#### To install:

8. Clean the brake caliper.

### ✳✳ CAUTION

**Before pressing piston into cylinder using piston resetting tool, brake fluid must be extracted from brake fluid reservoir. Otherwise, especially if reservoir has been topped off, fluid will overflow and cause damage.**

9. Press the pistons back.

10. Carefully install the brake pad wear indicator wire contact sensor in the new brake pads.

11. Insert the brake pads into the brake caliper housing.

12. Insert the retaining spring.

13. Install the wire for the brake pad wear indicator below the tab of the retaining spring and in the brake caliper housing.

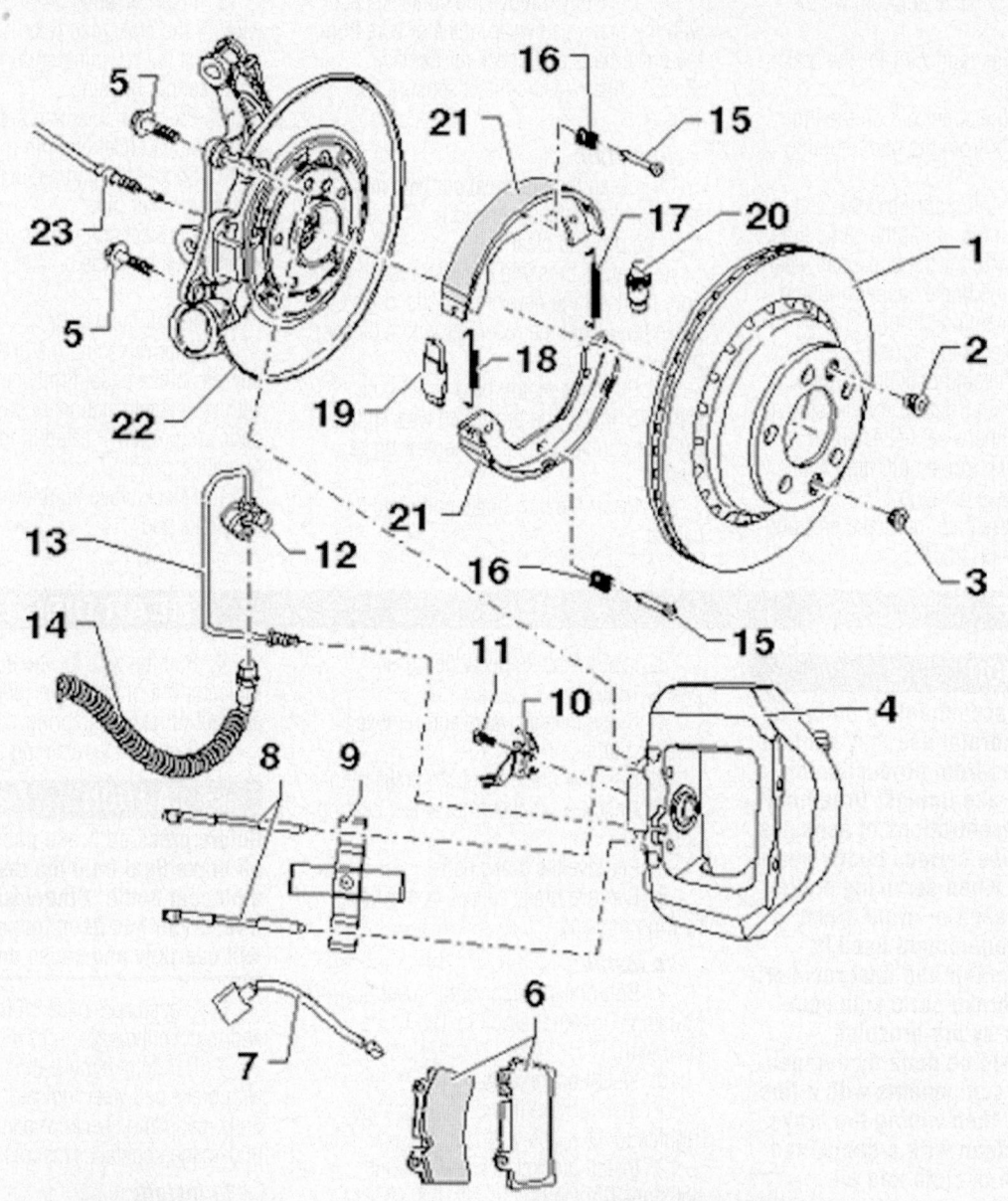

1. Disc/Drum
2. Bolt
3. Bolt
4. Caliper
5. Pads
6. Pad Wear Indicator Wire
7. Pad Retaining Spring
8. Retaining Pin
9. Bracket
10. Bolt
11. Spring Clamp
12. Brake Line
13. Brake Hose
14. Spring Dowel Sleeve
15. Compression Spring
16. Pull Spring
17. Pull Spring
18. Spreader
19. Adjustment Nut
20. Brake Shoe
21. Brake Carrier
22. Rear Brake Cable

22205_TOUA_G0003

**Fig. 2 Exploded view of the rear brake system**

14. Press the retaining spring down and place the pad retaining pin in the brake caliper.

15. Drive the pad retaining pin into the brake caliper as far as the stop.

16. Connect the connectors of the brake pad wear indicator in the bracket of brake caliper housing.

17. Install the wheels.

18. After replacing brake pads, depress brake pedal firmly several times with vehicle stationary so that the brake pads are properly seated in their normal operating position.

19. Check and adjust the brake fluid level after replacing brake pads.

## BRAKES

### PARKING BRAKE SHOES

#### REMOVAL & INSTALLATION

*See Figure 3.*

1. Remove wheels.
2. Remove brake pads
3. Pull wheel speed sensor wire out of bracket at brake caliper.
4. Remove mounting bolts of brake caliper.
5. Remove brake caliper housing and secure with wire so that the weight of the brake caliper does not burden or damage the brake hose.
6. Remove locking bolt (arrow) for adjustment nut of parking brake.
7. With a proper adjusting tool inserted

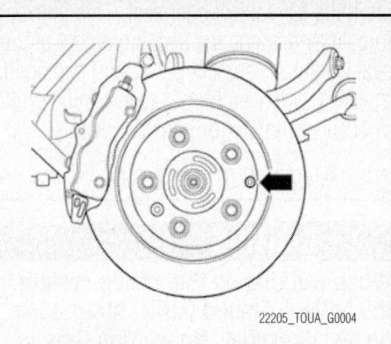

22205_TOUA_G0004

**Fig. 3 Showing location of the locking bolt**

through the disk, turn the adjustment nut against resistance from the pull-spring.

8. Remove the brake disc.
9. Unhook pull-springs.

## PARKING BRAKE

10. Remove spring dowel sleeves with compression springs and remove brake shoes.

#### To install:

11. Install the brake shoes.
12. Install the spring dowel sleeves with the compression springs.
13. Hook the pull-spring to its original location.
14. Install the brake disc.
15. Install the brake caliper to the disc and tighten the bolts.
16. Reposition the wheel speed sensor wire to the bracket.
17. Install the rear disc pads.
18. Install the wheels.
19. Adjust the parking brake and lever. See "Adjustment" in this section.

## CHASSIS ELECTRICAL

### GENERAL INFORMATION

#### ✳✳ CAUTION

**These vehicles are equipped with an air bag system. The system must be disarmed before performing service on, or around, system components, the steering column, instrument panel components, wiring and sensors. Failure to follow the safety precautions and the disarming procedure could result in accidental air bag deployment, possible injury and unnecessary system repairs.**

#### SERVICE PRECAUTIONS

Disconnect and isolate the battery negative cable before beginning any airbag system component diagnosis, testing, removal, or installation procedures. Allow system capacitor to discharge for two minutes before beginning any component service. This will disable the airbag system. Failure to disable the airbag system may result in accidental airbag deployment, personal injury, or death.

Do not place an intact undeployed airbag face down on a solid surface. The airbag will propel into the air if accidentally

## AIR BAG (SUPPLEMENTAL RESTRAINT SYSTEM)

deployed and may result in personal injury or death.

When carrying or handling an undeployed airbag, the trim side (face) of the airbag should be pointing towards the body to minimize possibility of injury if accidental deployment occurs. Failure to do this may result in personal injury or death.

Replace airbag system components with OEM replacement parts. Substitute parts may appear interchangeable, but internal differences may result in inferior occupant protection. Failure to do so may result in occupant personal injury or death.

Wear safety glasses, rubber gloves, and long sleeved clothing when cleaning powder residue from vehicle after an airbag deployment. Powder residue emitted from a deployed airbag can cause skin irritation. Flush affected area with cool water if irritation is experienced. If nasal or throat irritation is experienced, exit the vehicle for fresh air until the irritation ceases. If irritation continues, see a physician.

Do not use a replacement airbag that is not in the original packaging. This may result in improper deployment, personal injury, or death.

The factory installed fasteners, screws and bolts used to fasten airbag components have a special coating and are

specifically designed for the airbag system. Do not use substitute fasteners. Use only original equipment fasteners listed in the parts catalog when fastener replacement is required.

During, and following, any child restraint anchor service, due to impact event or vehicle repair, carefully inspect all mounting hardware, tether straps, and anchors for proper installation, operation, or damage. If a child restraint anchor is found damaged in any way, the anchor must be replaced. Failure to do this may result in personal injury or death.

Deployed and non-deployed airbags may or may not have live pyrotechnic material within the airbag inflator.

Do not dispose of driver/passenger/curtain airbags or seat belt tensioners unless you are sure of complete deployment. Refer to the Hazardous Substance Control System for proper disposal.

Dispose of deployed airbags and tensioners consistent with state, provincial, local, and federal regulations.

After any airbag component testing or service, do not connect the battery negative cable. Personal injury or death may result if the system test is not performed first.

If the vehicle is equipped with the Occupant Classification System (OCS), do not connect the battery negative cable before performing the OCS Verification Test using the scan tool and the appropriate diagnostic information. Personal injury or death may result if the system test is not performed properly.

Never replace both the Occupant Restraint Controller (ORC) and the Occupant Classification Module (OCM) at the same time. If both require replacement, replace one, then perform the Airbag System test before replacing the other.

Both the ORC and the OCM store Occupant Classification System (OCS) calibration data, which they transfer to one another when one of them is replaced. If both are replaced at the same time, an irreversible fault will be set in both modules and the OCS may malfunction and cause personal injury or death.

If equipped with OCS, the Seat Weight Sensor is a sensitive, calibrated unit and must be handled carefully. Do not drop or handle roughly. If dropped or damaged, replace with another sensor. Failure to do so may result in occupant injury or death.

If equipped with OCS, the front passenger seat must be handled carefully as well. When removing the seat, be careful when setting on floor not to drop. If dropped, the sensor may be inoperative, could result in occupant injury, or possibly death.

If equipped with OCS, when the passenger front seat is on the floor, no one should sit in the front passenger seat. This uneven force may damage the sensing ability of the seat weight sensors. If sat on and damaged, the sensor may be inoperative, could result in occupant injury, or possibly death.

### DISARMING THE SYSTEM

> ### ✳✳ CAUTION
>
> **When working on the airbag system, the battery ground (GND) strap must be disconnected. No waiting time is necessary after disconnecting battery. When connecting the airbag system to a voltage source, there must be no person present inside the vehicle.**

To avoid personal injury when working on vehicles equipped with an air bag, the negative battery cable must be disconnected before working on the system. Failure to do so may result in deployment of the air bag.

1. Before servicing the vehicle, refer to the precautions in the beginning of this section.

2. Turn the ignition switch to the **LOCK** position.

3. Disconnect the negative battery cable. Shield the cable by wrapping electrical tape around it. Work can begin immediately after disconnecting the battery, no waiting time is needed.

### ARMING THE SYSTEM

After repairs are completed, the negative battery cable is properly reconnected to the battery. Ensure there are no additional personnel in the vehicle when reconnecting the negative battery cable.

# DRIVETRAIN

## AUTOMATIC TRANSMISSION ASSEMBLY

### REMOVAL & INSTALLATION

*See Figures 4 through 6.*

The automatic transmission is removed together with the engine, front final drive and transfer case. The procedure for this is described in this section under Engine Assembly removal and installation.

➡ **It is advisable to remove the lower connecting bolts of the engine and transmission, even before the removal of the transfer case. When possible, the torque converter fasteners should also be mounted early on. This applies when removing as well as installing the transmission.**

1. Before servicing the vehicle, refer to the Precautions Section.

2. Remove the engine. See "Engine" section.

3. On V6 and V8 engines, remove the starter motor.

4. On V10 TDI engines, remove the starter cable bolt.

➡ **On V10 TDI engines, the torque converter fasteners and lower engine/transmission bolts should be installed with gear set mounted.**

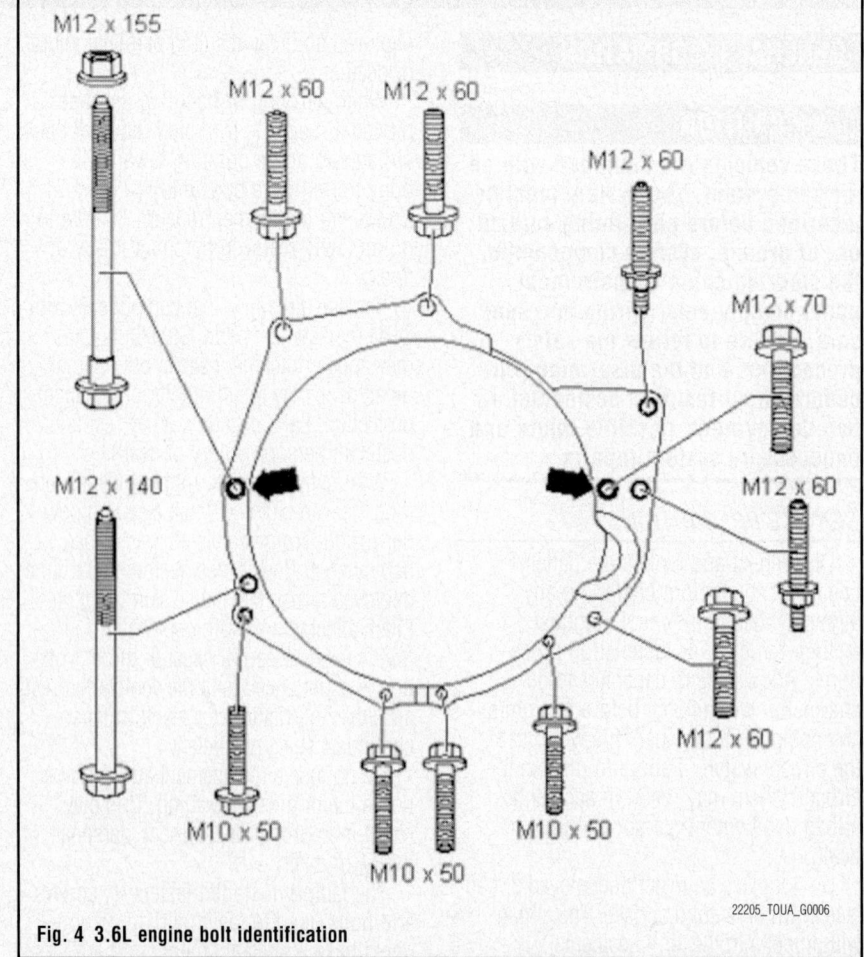

Fig. 4 3.6L engine bolt identification

22205_TOUA_G0006

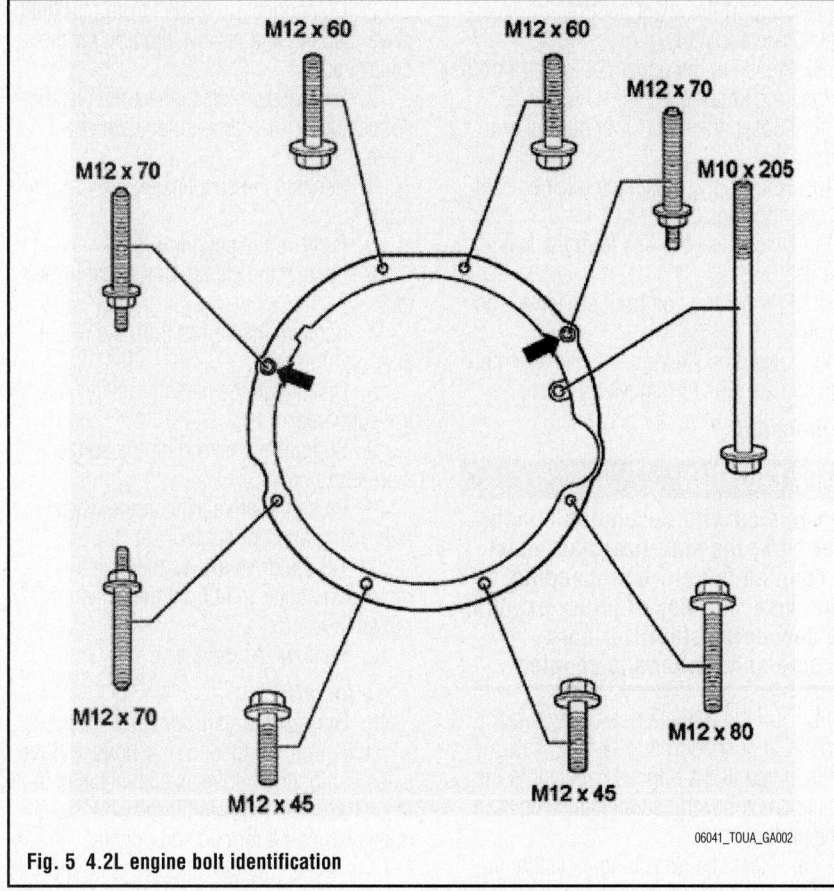

**Fig. 5 4.2L engine bolt identification**

06041_TOUA_GA002

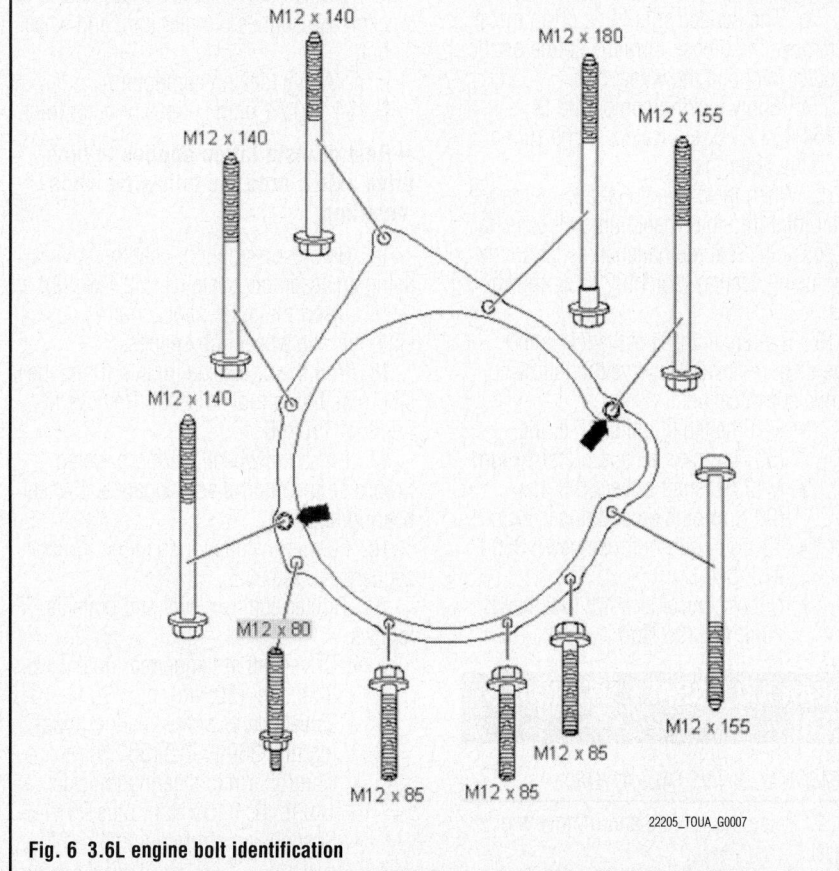

**Fig. 6 3.6L engine bolt identification**

22205_TOUA_G0007

5. Make sure that at least two plates on the lifting table are placed under the oil pan as support for the automatic transmission.

6. Remove the driveshaft between the transfer case and the front final drive.

7. Support the transfer case with a hoist.

8. Support the transmission with a hoist.

9. Loosen and detach the ATF lines.

10. Remove all the connecting bolts securing the transmission to the engine.

11. Carefully pull the transmission from the engine.

➡**Make sure that the converter is also removed from the drive plate with the transmission.**

*To install:*

12. Installation is the reverse of removal. Observe the following torque specifications (refer to the illustrations for bolt identification):

a. 3.6L engines:
- M12 × 60 (except starter): 59 ft. lbs. (80 Nm)
- M12 × 60 (starter): 48 ft. lbs. (65 Nm)
- M12 × 70: 59 ft. lbs. (80 Nm)
- M10 × 50: 33 ft. lbs. (45 Nm)
- M12 × 140: 59 ft. lbs. (80 Nm)
- M12 × 155: 59 ft. lbs. (80 Nm)

b. 4.2L engines:
- M12 × 60: 59 ft. lbs. (80 Nm)
- M12 × 70: 59 ft. lbs. (80 Nm)
- M10 × 205: 33 ft. lbs. (45 Nm)
- M12 × 80: 48 ft. lbs. (65 Nm)
- M12 × 45: 59 ft. lbs. (80 Nm)
- M12 × 70: 59 ft. lbs. (80 Nm)

c. V10 TDI engines:
- M12 × 140: 59 ft. lbs. (80 Nm)
- M12 × 180: 59 ft. lbs. (80 Nm)
- M12 × 155: 48 ft. lbs. (65 Nm)
- M12 × 85: 59 ft. lbs. (80 Nm)
- M12 × 80 (with bracket): 59 ft. lbs. (80 Nm)
- M12 × 140: 59 ft. lbs. (80 Nm)
- Drive plate to converter: 63 ft. lbs. (85 Nm)

## TRANSFER CASE ASSEMBLY

### REMOVAL & INSTALLATION

The transfer case is removed together with the engine, front final drive and transmission. The procedure for this is described in this section under Engine Assembly removal and installation.

1. Before servicing the vehicle, refer to the Precautions Section.

2. Remove the engine.

3. Remove the driveshaft between the transfer case and the front final drive.

Counterhold the driveshaft using tool T10172 or equivalent to ease removal of the bolts.

4. Support the transfer case with a hoist.

5. Remove the bolts, then separate the transfer case and set it aside.

*To install:*

6. Check if the alignment sleeves for centering the transmission/transfer case are installed in the transmission; install if necessary.

7. Always replace the O-ring seal between the transfer case and transmission; lubricate it lightly.

8. Lubricate the transfer case input shaft and transmission output shaft with grease G 000 100 or equivalent.

9. Push the transfer case completely onto the transmission, ensuring that transfer case input shaft splines are centered as they are guided onto the transmission output shaft.

➡**If the splines are correctly positioned, the transfer case will slide to a stop against the transmission. Do not use the bolts to pull the transfer case onto the transmission!**

10. The remaining installation is the reverse of removal. Observe the following torque specifications:

- Transfer case mounting bolts (use new): 33 ft. lbs. (45 Nm)
- Transfer case damper: 24 ft. lbs. (32 Nm)

## FRONT HALFSHAFT

### REMOVAL & INSTALLATION

1. Before servicing the vehicle, refer to the Precautions Section.

2. Loosen the 12-point hub nut on the wheel hub.

➡**Loosen and tighten the 12-point nut when the vehicle is on its wheels. Do not move the vehicle when the 12-point nut is loosened, otherwise the wheel bearing could be damaged. If the vehicle has to be moved with the halfshaft removed, an outer joint must be installed and the 12-point nut tightened to 111 ft. lbs. (150 Nm).**

3. Remove the wheel and raise the vehicle.

4. Remove the halfshaft bolts from the front final drive. Use socket T10099/1 or equivalent to loosen the bolts.

5. Press the tie rod from the steering knuckle.

6. Press the drive axle out from the wheel bearing hub.

7. Support the lower control arm/steering knuckle assembly.

8. Press out the upper control arm from the steering knuckle.

9. Loosen the strut lower control arm bolt.

10. Remove the brake line retainer from the steering knuckle.

11. Disconnect the link from the stabilizer bar.

12. Pull out the bolt from the lower control arm.

13. Lower the steering knuckle assembly until you are able to remove drive axle.

*To install:*

### ❄❄ CAUTION

**On vehicles with decouplable stabilizer bars, the stabilizer bars must be coupled in before proceeding. Otherwise, the risk of injury exists if the decoupled stabilizer bars become unintentionally coupled.**

14. During production, a change was made from bonded to non-bonded halfshafts. If you find a bonded halfshaft in the vehicle, the following additional steps have to be performed:

a. Clean the remaining locking compound from the splines on the outer joint and on the wheel hub.

b. The surface should be clean metal; remove the grease from the spline on the outer joint and the wheel hub.

c. Apply locking compound D 154 000 A1 or equivalent to the spline on the outer joint.

15. When installing the shafts, guide the outer joint into the wheel hub splines as far as possible. Seat the halfshaft as far as the stop using seating tool T10206 or equivalent.

16. The remaining installation is the reverse of removal. Observe the following torque specifications:

- M10 halfshaft bolts: 37 ft. lbs. (50 Nm) plus an additional ¼ turn
- M12 halfshaft bolts: 66 ft. lbs. (90 Nm) plus an additional ¼ turn
- 12-point hub nut (use new): 369 ft. lbs. (500 Nm)
- Upper control arm nut (use new): 70 ft. lbs. (95 Nm)

## REAR AXLE SHAFT, BEARING & SEAL

### REMOVAL & INSTALLATION

1. Raise the vehicle and remove the wheel.

2. Disconnect driveshaft from rear final drive. Loosen and remove the bolt for the control arm.

3. If necessary, disconnect the electrical connector for the damping adjustment valve.

4. Remove the suspension strut mounting bolt.

5. Remove the coupling rod.

6. Loosen the clamp on the stabilizer bar.

7. Remove the tie rod from the wheel bearing housing.

8. Disengage the parking brake cable at the subframe.

9. Support the tie rod on the suspension strut bolt.

10. Press the drive axle off the wheel hub, using appropriate tool set.

11. Tilt the drive axle to the side to remove the outer joint from the wheel bearing splines.

12. Remove the drive axle.

*To install:*

13. During start of production, there was a change from bonded to non-bonded drive axles. If a bonded drive axle is installed in the vehicle, the following three work steps must also be performed additionally:

a. Clean splines of outer joint and wheel hub of remaining locking fluid.

b. Clean the metallic surface and degrease splines of outer joint and wheel hub.

c. Apply locking compound D 154 000 A1 onto splines of outer joint.

➡**Rest of installation applies to both drive axles, note the following when installing:**

14. Before assembling, coat the splines, using an assembly paste (G 052 109 A2).

15. Insert the outer joint as far as possible into the wheel hub splines.

16. Pull the drive axle in until it hits the stop (use fitting tool T10206). Remove fitting tool T10206 .

17. Press the wheel bearing housing inward so the control arm engages at wheel bearing housing.

18. Further installation is in the reverse sequence to removal.

19. Tighten the new nuts and bolts as follows:

- Driveshaft at flange/rear final drive: 37 ft. lbs. (50 Nm), plus 90°
- Drive axle to wheel bearing housing nut: 369 ft. lbs. (500 Nm)
- Control arm to steering knuckle: 110 ft. lbs. (150 Nm), plus 90 °
- Suspension strut/coupling rod to

wheel bearing housing: 66 ft. lbs. (90 Nm), plus 90°
- Clamp for stabilizer bar: 30 ft. lbs. (40 Nm)
- Decoupling stabilizer bar clamp: 37 ft. lbs. (50 Nm)
- Stabilizer bar to connecting link: 74 ft. lbs. (100 Nm)
- Tie rod to steering knuckle: 110 ft. lbs. (150 Nm), plus 90 °

## PROPELLER SHAFT

### REMOVAL & INSTALLATION
*See Figure 7.*

### ✳✳ CAUTION

**Before removing, mark the positions of all parts in relation to each other. Reinstall in the same position otherwise imbalance will be excessive, the bearings could be damaged causing rumbling noises.**

➡**If droning noises occur when driving, unbolt front driveshaft from flange at front final drive, rotate one hole relative to flange and bolt on again. If droning noises are still audible, the front driveshaft can be rotated one hole relative to flange a total of 5 times.**

1. Select position "N" with selector lever.
2. Insert a brake pedal depressor (V.A.G 1869/2) .
3. Raise the vehicle.
4. Remove any sound insulation on bottom of transmission.
5. If an underbody impact guard is located beneath the driveshaft, remove it.
6. Check whether there is a colored marking on front driveshaft and at driveshaft flange at front final drive. If this marking is not present, then mark in color the position of driveshaft flange to front final drive.

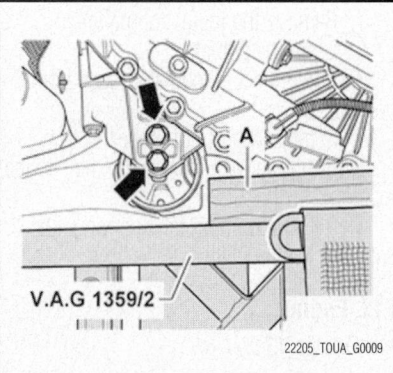

**Fig. 7 Do not loosen bolts (arrows) yet.**

7. Remove the lower three bolts from front driveshaft at front final drive.
8. Remove the lower three bolts from front driveshaft at output flange of transfer case.
9. Lower the vehicle.
10. Remove the brake pedal depressor.
11. Raise vehicle.
12. Rotate both front wheels in same direction, at the same time, so the front driveshaft rotates one-half turn (180 degrees).
13. Lower the vehicle.
14. Reinstall the brake pedal depressor.
15. Raise the vehicle.
16. Remove the remaining three bolts from the front driveshaft at the front final drive and at the transfer case.
17. Remove the transmission carrier as follows:
    a. Place an engine/transmission jack, with a universal adapter and a block of wood, under the transfer case.
    b. Do not loosen bolts (arrows) yet.
    c. Unbolt transmission carrier (crossmember) from the body.
    d. Remove the bolt for transmission carrier (crossmember) from the transfer case bracket.
18. Remove front driveshaft.

➡**If the front driveshaft cannot be removed, lower the engine/transmission assembly slightly. When doing this, the engine must not come to rest against the bulkhead**

#### To install:
19. Installation is in reverse order of removal. Note the following:
    a. Always replace bolts for front driveshaft.
    b. Markings between the driveshaft and front final drive must line up as much as possible.
20. Install transmission carrier as follows:
    a. Use new bolts.
    b. First tighten the transmission carrier to the transfer case bracket to specifications.
    c. Then, tighten the transmission carrier to the body to specifications.
21. Insert the bolts for the front driveshaft and tighten to specifications.
22. If an underbody impact guard is located beneath the driveshaft, install it.
23. Tighten the new nuts and bolts as follows:
- Bracket to transmission carrier: 37 ft. lbs. (50 Nm), plus 90°
- Transmission carrier to structure: 37 ft. lbs. (50 Nm), plus 90°

- Front driveshaft to front final drive flange: 22 ft. lbs. (30 Nm), plus 90°
- Front driveshaft to transfer case flange: 22 ft. lbs. (30 Nm), plus 90°

## REAR HALFSHAFT

### REMOVAL & INSTALLATION
*See Figures 8 and 9.*

1. Before servicing the vehicle, refer to the Precautions Section.
2. Loosen the 12-point hub nut on the wheel hub.

➡**Loosen and tighten the 12-point nut when the vehicle is on its wheels. Do not move the vehicle when the 12-point nut is loosened, otherwise the wheel bearing could be damaged. If the vehicle has to be moved with the halfshaft removed, an outer joint must be installed and the 12-point nut tightened to 111 ft. lbs. (150 Nm).**

3. Remove the wheel and raise the vehicle.
4. Remove the halfshaft bolts from the rear final drive. Use socket T10099 or equivalent to loosen the bolts.

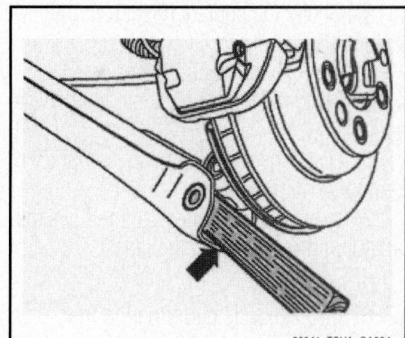

**Fig. 8 Insert a block of into the control arm**

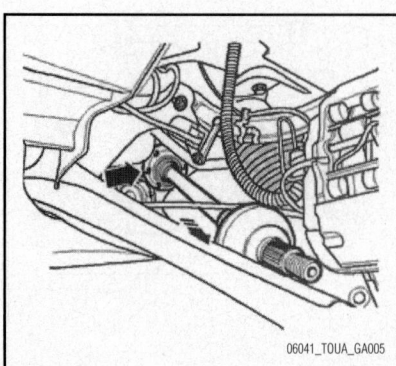

**Fig. 9 Pull the halfshaft out in the direction shown**

5. Remove the lower bolt from the control arm. Insert a block of wood between the control arm and the wheel bearing housing.

### ⁜ CAUTION

**If this block of wood is not inserted, there is a risk of injury!**

6. Press out the drive axle.

7. Push the drive axle as far as possible beyond the final drive flange so that the outer joint can be pulled out of the wheel bearing spline.

***To install:***

### ⁜⁜ CAUTION

**On vehicles with decouplable stabilizer bars, the stabilizer bars must be coupled in before proceeding. Otherwise, the risk of injury exists if the decoupled stabilizer bars become unintentionally coupled.**

8. During production, a change was made from bonded to non-bonded half-shafts. If you find a bonded halfshaft in the vehicle, the following additional steps have to be performed:

  a. Clean the remaining locking compound from the splines on the outer joint and on the wheel hub.

  b. The surface should be clean metal; remove the grease from the spline on the outer joint and the wheel hub.

  c. Apply locking compound (D 154 000 A1, or equivalent) to the spline on the outer joint.

9. Coat the spline with assembly paste (G 052 109 A2, or equivalent).

10. When installing the shafts, guide the outer joint into the wheel hub splines as far as possible. Seat the halfshaft as far as the stop using a proper seating tool (T10206, or equivalent). Make absolutely sure that you are pressing only on the wheel bearing housing.

11. The remaining installation is the reverse of removal. Observe the following torque specifications:

- Halfshaft bolts: 37 ft. lbs. (50 Nm), plus an additional ¼ turn
- 12-point hub nut (use new): 369 ft. lbs. (500 Nm)
- Control arm bolt and nut (use new): 111 ft. lbs. (150 Nm) plus an additional ¼ turn

## ENGINE COOLING

### ENGINE FAN

#### *REMOVAL & INSTALLATION*

*See Figure 10.*

#### 3.6L & 5.0L Engines

1. Remove the insulation panel.

2. Remove the front coolant pipe from the fan mount.

3. Unclip the coolant hose from the lock carrier.

4. Remove the front bumper.

5. Bring the lock carrier into the service position.

6. Disconnect the fan connector and free up the wiring.

7. Loosen the fasteners and remove the fan support upward with the fans.

***To install:***

8. Installation is performed in the reverse of the removal procedure.

#### 4.2L Engine

*See Figures 11 and 12.*

1. Before servicing the vehicle, refer to the Precautions Section.

2. Open the cap on the coolant expansion tank to relieve any residual pressure.

3. Push down on the retaining clips of the upper coolant hose. Loosen the spring clamps on the pipe, then rotate it upward to separate it from the cooling fan mount.

4. Detach the cooling fan connectors.

5. Pull off the securing ring from the brake system vacuum pump and lay it aside.

6. Release the retaining clips from the cooling fan mount, then lift the entire cooling fan assembly up and out.

7. If necessary, remove the bolts securing the fan and remove the fan from the mount.

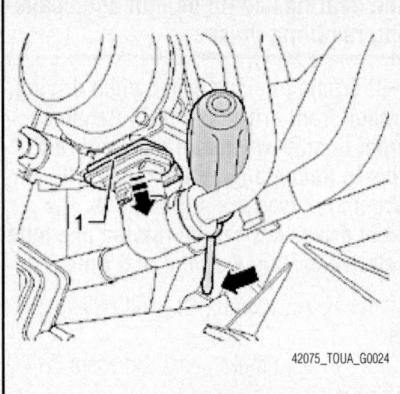

**Fig. 12 Pull off the securing ring (1) from the brake system vacuum pump**

8. Installation is the reverse of removal. Tighten the fan mounting bolts to 7 ft. lbs. (10 Nm).

## RADIATOR

#### *REMOVAL & INSTALLATION*

#### 3.6L Engine

1. Remove the insulation panel.

2. Drain the coolant.

3. Disconnect all coolant hoses from the radiator.

4. Remove the front bumper.

5. Bring the lock carrier into the service position.

6. Remove radiator fan support with fans. See "Engine Fan."

#### 4.2L Engine

*See Figures 13 and 14.*

1. Before servicing the vehicle, refer to the Precautions Section.

**Fig. 10 Showing the engine cooling fan and support assembly**

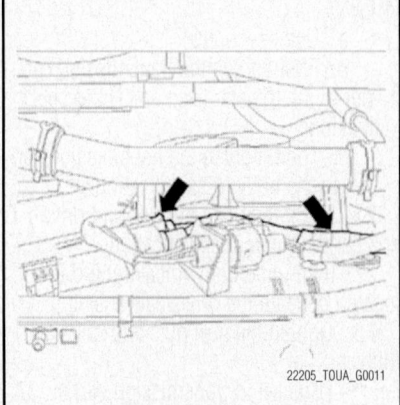

**Fig. 11 Detach the coolant pipe assembly from the cooling fan mount**

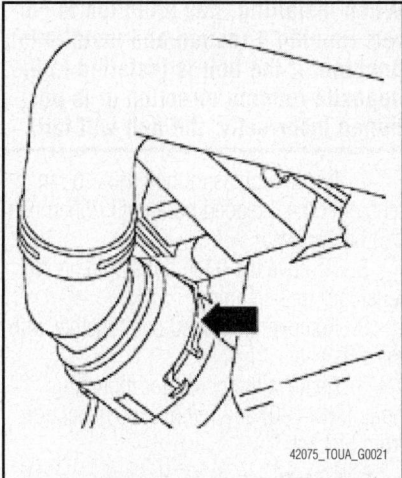

**Fig. 13 Pull the retaining clips securing the radiator hoses**

2. Drain the cooling system.

3. Pull the retaining clips out of the couplings for the upper and lower radiator hoses. Pull the hoses from the radiator.

4. Carefully unclip the VW emblem from the grille. Remove the grille securing screw behind it.

5. Carefully disengage the clips securing the grille, then remove it from the vehicle.

6. Remove the bolts securing the bumper to the vehicle. Once all the bolts are removed, release the retaining tabs under the headlights. With the aid of an assistant, pull the bumper away from the vehicle.

7. Bring the hood lock carrier up to the service position.

8. Remove the cooling fan assembly.

9. Detach the securing clips for the condenser and transmission oil cooler.

10. Remove the upper radiator securing

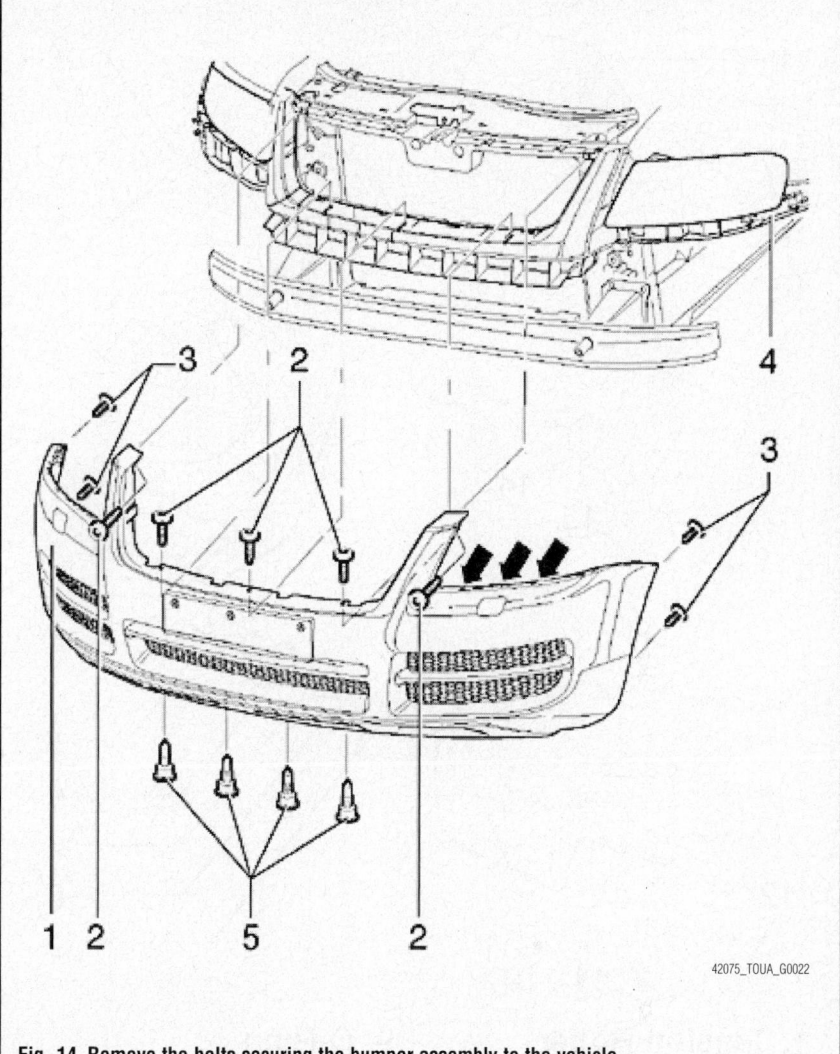

**Fig. 14 Remove the bolts securing the bumper assembly to the vehicle**

bolts and remove the radiator by pulling it straight up and out.

11. Installation is the reverse of removal.

Tighten radiator mounting bolts to 7 ft. lbs. (10 Nm). The bumper fasteners should be tightened until just snug.

## ENGINE ELECTRICAL

### ALTERNATOR

*REMOVAL & INSTALLATION*

**3.6L Engine**

*See Figures 15 and 16.*

**✳ WARNING**

**When disconnecting battery, procedure must always be followed as described. If the sequence is not followed, the pyrotechnic battery isolation system switch may trigger, which may damage electrical components in the vehicle.**

## CHARGING SYSTEM

1. Use the following procedures to properly disconnect the negative battery cable:

a. To disconnect battery, anti-theft warning system must be deactivated.

b. Switch off ignition, switch off all electrical consumers and remove ignition key.

c. Remove the left front seat rail cover.

d. Remove the left front seat trim.

e. Remove the screws and tilt the seat frame and seat to the rear.

f. Remove the screw and the air guide from the battery box.

g. Open the battery box clips and remove the cover.

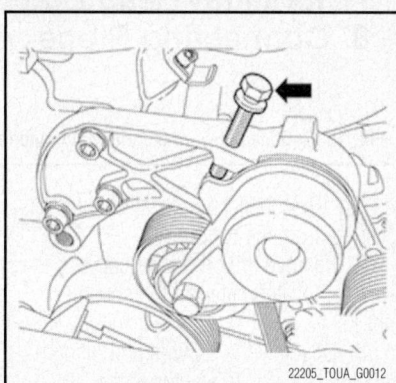

**Fig. 15 Install a pressure bolt M8×50 in tensioner the threaded bore until ribbed belt can be removed**

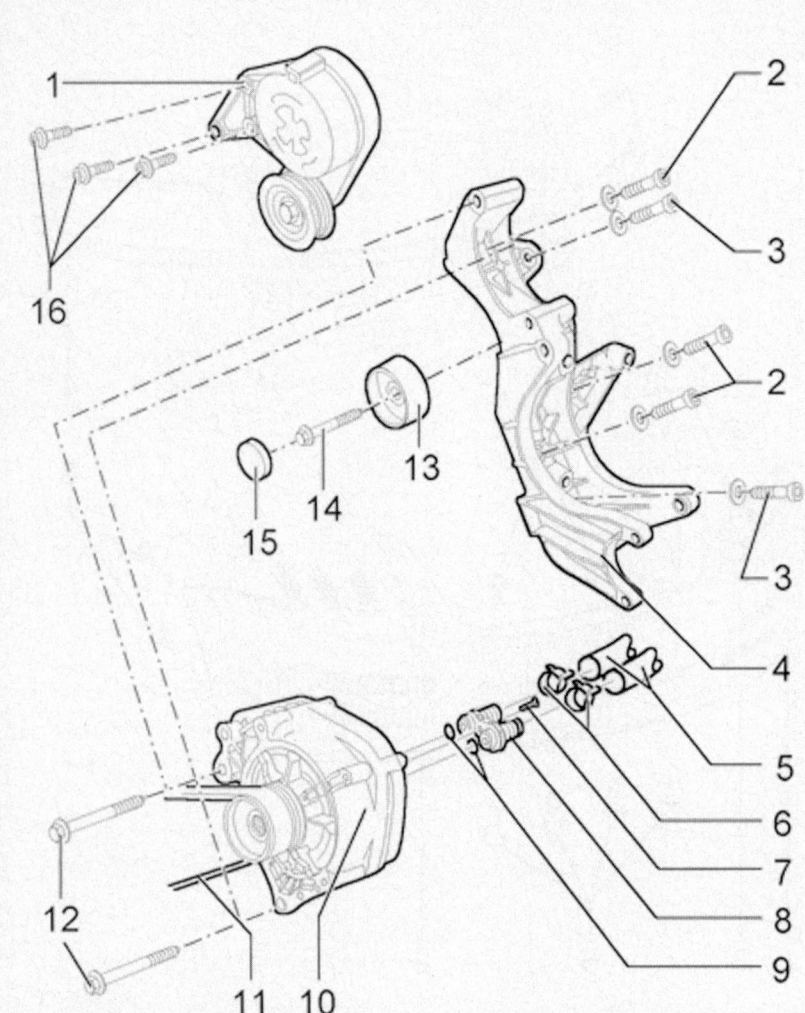

1. Tension Roller
2. Socket Head Bolts
3. Fitting Bolts
4. Bracket
5. Coolant Hose
6. Clamp
7. Torx Flat Head Bolt
8. Connecting Piece
9. O-Ring
10. Alternator
11. Ribbed Belt
12. Collar Bolts
13. Idler Pulley
14. Bolt
15. Cap
16. Socket Head Bolts

22205_TOUA_G0013

**Fig. 16 Exploded view of the alternator and related components—3.6L engines**

h. Disconnect the negative cable terminal from the battery.

2. Remove the protective cap of B+ wire, located between the coolant pipes.

3. Unscrew the B+ wire and disconnect the harness connector of the DF wire.

4. Clamp off the coolant hoses, using hose clamping pliers.

5. Remove the B+ wire bolt.

6. Pull the coolant pipes off of the generator.

➡The coolant hoses may remain connected to the coolant pipes.

**✳✳ CAUTION**

Before removing ribbed belt, mark the top side and direction of travel.

When installing, pay attention to correct running direction and installation position. If the belt is installed in the opposite running direction or is positioned incorrectly, the belt will fail!

7. Install a pressure bolt M8×50 in tensioner the threaded bore until ribbed belt can be removed.

8. Remove the three bolts holding the tensioner to the engine.

9. Remove the ribbed belt together with the tensioner.

10. Remove the generator mounting bolts (M8×90) and remove the generator from bracket.

***To install:***

11. Install in reverse order of removal, noting the following:

a. Tighten the threaded connections to the specification.

b. Remove pressure bolt M80×50 from tensioner.

c. Tighten the fasteners to the following:

- Socket head bolt (M8×30 mm): 15 ft. lbs. (20 Nm)
- Fitting bolts (M8×1.10 in.): 15 ft. lbs. (20 Nm)
- Torx flat head bolt (M6×16 mm): 7 ft. lbs. (9 Nm)
- Collar bolts (M8×3.54 in.): 15 ft. lbs. (20 Nm)
- Bolt (M10×1.77 in.): 15 ft. lbs. (20 Nm)

d. Fill with coolant .

### 4.2L Engine

*See Figure 17.*

1. Before servicing the vehicle, refer to the Precautions Section.

2. Disconnect the battery.

3. Remove the engine assembly.

4. Mark the direction of rotation on the belt.

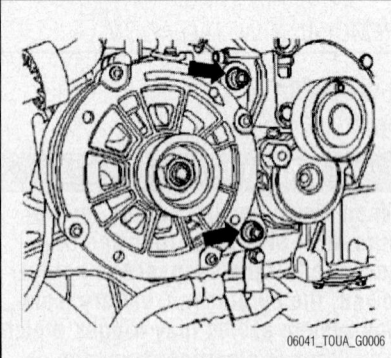

06041_TOUA_G0006

**Fig. 17 Alternator mounting bolts—4.2L engines**

5. Rotate the tensioner using a 19mm wrench and secure it in the released position using tool T10060A or a similar sized pin punch. Remove the belt.

6. Unplug the electrical connector from the back of the alternator.

7. Remove the cap from the positive terminal on the alternator and remove the nut securing the cable.

8. Remove the bolt securing the coolant lines on the alternator, then pull the lines away from it.

9. Remove the 2 bolts securing the alternator to the engine.

### To install:

10. Installation is the reverse of removal. Observe the following tightening torques:

- Alternator bracket and mounting bolts: 15 ft. lbs. (20 Nm)
- Tensioner mounting bolts: 15 ft. lbs. (20 Nm)
- Coolant pipe: 7 ft. lbs. (9 Nm)
- Positive terminal nut: 11 ft. lbs. (15 Nm)

➡**Be sure to reinstall the belt in the proper direction of rotation and to remove the belt tension release tool. Use new O-rings on the coolant pipes and check the coolant level.**

### 5.0L TDI Engine

*See Figures 18 through 20.*

➡**On generators with hard rubber disc clutch, this should generally be replaced with torsion-elastic clutch.**

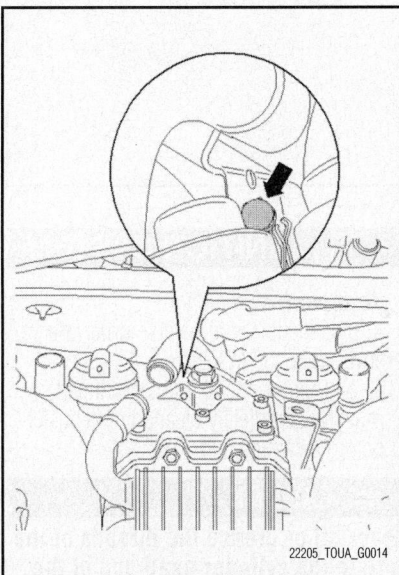

**Fig. 18 To remove EGR valve, remove coolant pipe bolt**

22205_TOUA_G0014

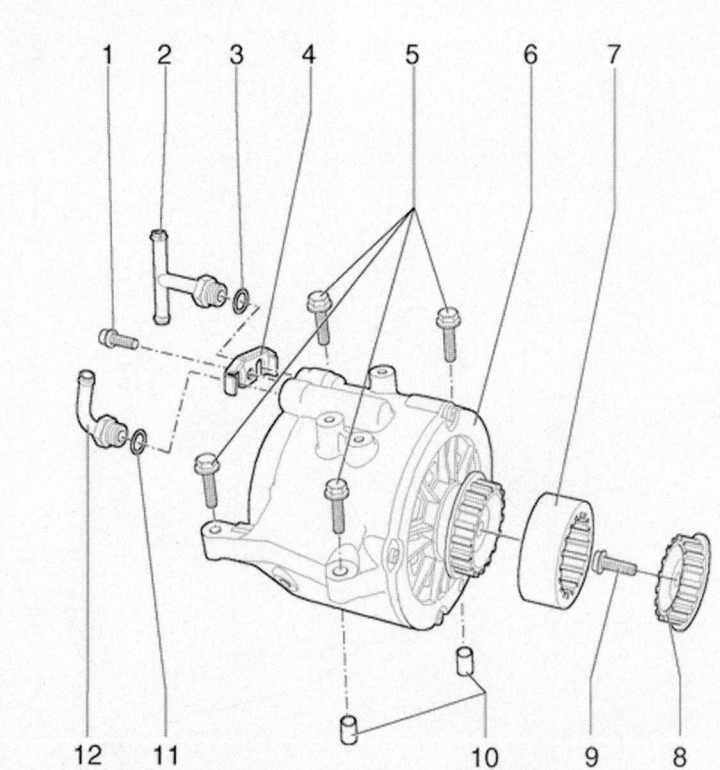

1. Socket Head Bolt with Washer
2. Coolant Piece/T-Piece
3. O-Ring
4. Angle Bracket
5. Bolts
6. Alternator
7. Torsion Elastic Clutch
8. Drive Gear
9. Bolt
10. Alignment Bushings
11. O-Ring
12. Coolant Pipe

22205_TOUA_G0015

**Fig. 19 Exploded view of alternator with torsion-elastic clutch**

1. Disconnect the battery.

2. Before removing generator, perform the following work procedures:
   a. Drain the coolant.
   b. Remove the intake manifolds.

3. To remove EGR valve, remove coolant pipe bolt.

4. Remove the bolt and remove the bracket for the coolant lines from the generator.

5. Remove the coolant lines from the generator.

➡**The coolant hoses may remain connected to the coolant pipes.**

6. Separate the connector for the DF lead and remove the protective cap.

7. Remove the B+ lead from the generator.

8. Remove the bolt and the bracket for the B+ lead from the generator.

9. Remove the mounting bolts for the generator.

10. Remove the generator. When doing this, the torsion-elastic clutch remains on the drive gear.

### To install:

11. Install in reverse order of removal, noting the following:

12. Check if alignment bushings are equipped and securely seated. If alignment bushings are loose, coat them with commercially available grease before installing.

13. Tighten to specifications as follows:

- Socket head bolt with washer (M6 × 0.79 in.): 6 ft. lbs. (8 Nm)
- Bolts (M8 × 1.38 in.): 15 ft. lbs. (20 Nm)
- Bolts (M10 × 1 × 30 mm): 37 ft. lbs. (50 Nm), plus 90°

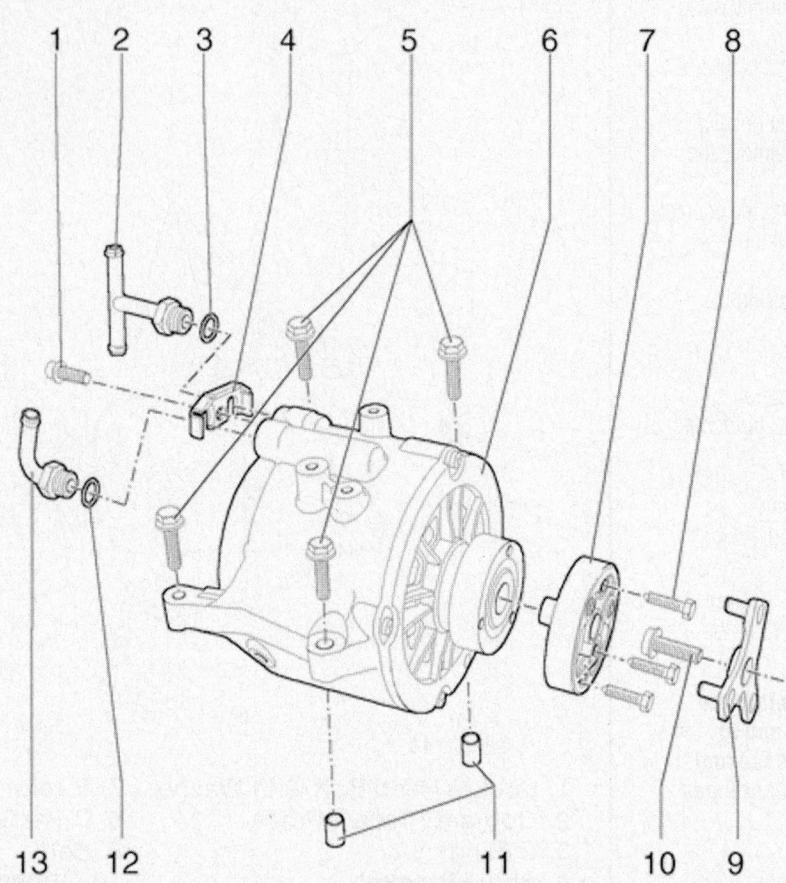

1. Socket Head Bolt with Washer
2. Coolant Piece/T-Piece
3. O-Ring
4. Angle Bracket
5. Bolts
6. Alternator
7. Hard Rubber Clutch
8. Hex Bolts with Socket Head and Points
9. Drive Hub
10. Bolt
11. Alignment Bushings
12. O-Ring
13. Coolant Pipe

22205_TOUA_G0016

**Fig. 20 Exploded view of alternator with hard rubber clutch**

## ENGINE ELECTRICAL

### FIRING ORDER

**3.6L Engine**

1–5–3–6–2–4.

**4.2L Engine**

1–5–4–8–6–3–7–2.

**5.0L TDI Engine**

1–6–5–10–2–7–3–8–4–9.

### GLOW PLUGS

**REMOVAL & INSTALLATION**

1. Remove the intake manifolds. See "Intake Manifold" under "ENGINE MECHANICAL."

2. Disconnect the connectors from the ceramic sheathed element glow plugs, using proper pliers (6275).

3. Remove the ceramic sheathed element glow plugs, using flex wrench (3220).

## IGNITION SYSTEM

**To install:**

4. Installation is in reverse order of removal, note the following:

   a. Cylinder head bore and thread must be completely cleaned of deposits before installing.

### ✳✳ CAUTION

**Never oil or grease the threads of the bore in the cylinder head and of the ceramic sheathed element glow plugs.**

b.  Install ceramic sheathed element glow plugs into cylinder head by hand, using flex wrench (3220).

c.  Tighten the ceramic sheathed element glow plugs to 12 ft. lbs. (15 Nm).

## IGNITION COIL

### REMOVAL & INSTALLATION

#### 3.6L Engine

*See Figure 21.*

1.  Remove the ignition coil wiring harness cover strip.

2.  Place assembly tool (T10118) on the locking button and carefully pull down on the harness connector.

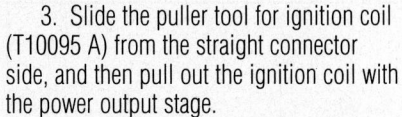

22205_TOUA_G0017

**Fig. 21  Place assembly tool (T10118_ on the locking button (arrow) and carefully pull down on the harness connector**

3.  Slide the puller tool for ignition coil (T10095 A) from the straight connector side, and then pull out the ignition coil with the power output stage.

4.  To install, insert ignition coil with power output stage into corresponding spark plug shaft so that straight connector sides fit with each other.

5.  Slide the puller for the ignition coil (T10095 A) from straight connector side and press the ignition coil, with power output stage, onto the spark plugs.

#### 4.2L Engine

*See Figure 22.*

1.  Press in the retaining tab of the corresponding connector to be removed, and then pull it off.

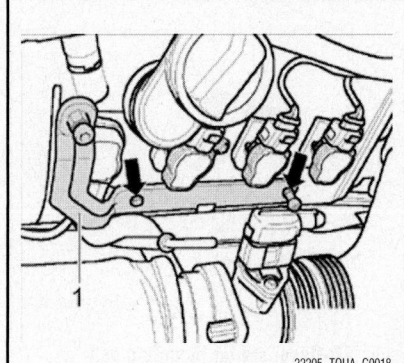

22205_TOUA_G0018

**Fig. 22  For connectors of cylinders 5 through 8, first remove the bracket (1) and bolts.**

2.  For connectors of cylinders 5 through 8, first remove bracket and bolts.

3.  Push the Ignition Coil Puller (T40039) onto the ignition coil with power output stage, and pull it off in the proper direction.

4.  To install, correctly position the ignition coil with power output stage in the spark plug shaft and press the ignition coil with power output stage onto the spark plug until it stops.

## IGNITION TIMING

### ADJUSTMENT

➡**Ignition timing is electronically controlled and cannot be adjusted manually.**

## SPARK PLUGS

### REMOVAL & INSTALLATION

#### 3.6L Engine

1.  Remove the ignition coils. See "Ignition Coils."

2.  Remove the spark plugs.

3.  Install the spark plugs and tighten to 15 ft. lbs. (20 Nm).

#### 4.2L Engine

1.  Remove the ignition coils. See "Ignition Coils."

2.  Remove the spark plugs.

3.  Install the spark plugs and tighten to 20 ft. lbs. (30 Nm).

# ENGINE ELECTRICAL

## STARTER

### REMOVAL & INSTALLATION

#### 3.6L Engine

1.  Disconnect the battery.

2.  Remove the noise insulation.

3.  Open snaps and remove the starter solenoid heat shield mat.

4.  Pry off the protective caps from the mounting nuts.

5.  Remove the nut at solenoid positive (B+) terminal.

6.  Remove the nut at solenoid Terminal 50.

### ✳✳ CAUTION

**Screw connections on magnetic switch can twist. The solenoid switch can be damaged. When removing and installing nuts at positive terminal and terminal 50, counterhold the**

**threaded connection at the solenoid switch using an open-end wrench.**

7.  Remove the lower starter bolt (M12 × 60) on engine side.

8.  Remove the bolt for the cable retainer and set aside. Remove the nuts to free the cable.

9.  Remove the bolt for the exhaust system bracket from the engine side and remove the bracket.

10.  Remove the bolt for the coolant pipe at engine sump.

11.  Remove the upper bolt for the starter on the transmission side.

12.  Remove the starter.

#### To install:

13.  Install the starter.

14.  Install and tighten the upper bolt for the starter on the transmission side.

15.  Install and tighten the bolt for the coolant pipe at the engine sump.

# STARTING SYSTEM

16.  Install the bolt for the exhaust system bracket from the engine side and remove the bracket.

17.  Install the lower starter bolt (M12 × 60) on engine side. Tighten it to 55 ft. lbs. (75 Nm).

18.  Install the remaining items in reverse of the removal procedure.

#### 4.2L Engines (5-Valve, Code AXQ, BHX)

*See Figures 23 through 27.*

1.  Before servicing the vehicle, refer to the Precautions Section.

2.  Remove the engine from the vehicle.

3.  Remove the exhaust manifold.

4.  Remove bolt (1). Unclip cable harness (2) and remove the cable retainer (3) from mounting.

5.  Remove the nuts and bolts from the engine mount (arrows).

6.  Lift the engine with a hoist to take

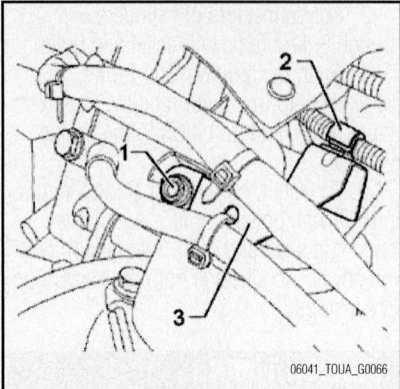

Fig. 23 Remove bolt (1), unclip cable harness (2) and remove the cable retainer (3)—4.2L engines

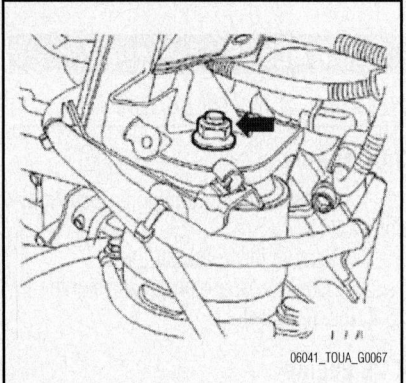

Fig. 24 Remove the nut from the mount—4.2L engines

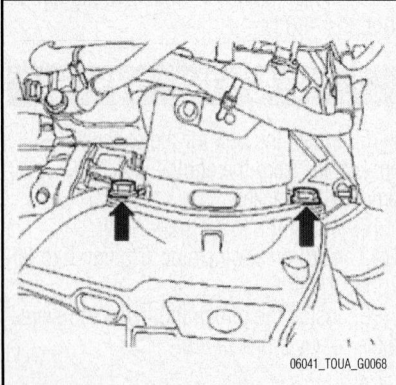

Fig. 25 Remove the bolts securing the mount to the carrier—4.2L engines

the weight off the mount. Remove the engine mount.

7. Remove the three bolts for the engine mount bracket at engine block, then remove the bracket.

8. Pry off the caps from the nuts at the positive (B+) terminal and at terminal 50 on the starter, if present. Remove the nuts.

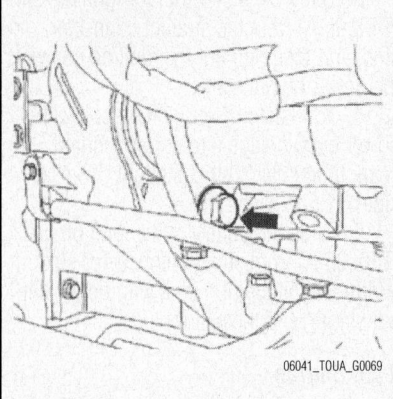

Fig. 26 Lower starter mounting bolt location—4.2L engines

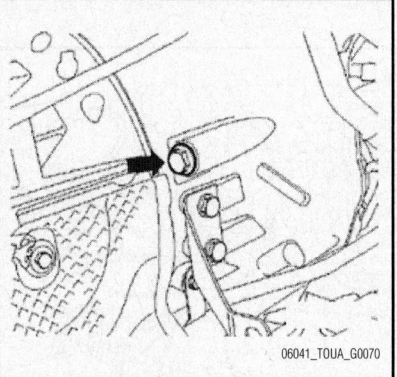

Fig. 27 Upper starter mounting bolt location—4.2L engines

➡When removing the nuts from the starter solenoid terminals, always counterhold the terminal stud in order to avoid damage to starter solenoid.

9. Remove the bolts securing the starter, then remove starter.

*To install:*

10. Installation is the reverse of removal. Note the following torque specifications:
- Starter mounting bolts: 55 ft. lbs. (75 Nm)
- B+ terminal nut: 11 ft. lbs. (15 Nm)
- Terminal 50 nut: 71 inch lbs. (8 Nm)
- Engine bracket to mount: 55 ft. lbs. (75 Nm)
- Engine mount to carrier: 44 ft. lbs. (60 Nm)

## 4.2L Engines (4-Valve, Code BAR)
*See Figure 28.*

1. Remove the engine. See "ENGINE MECHANICAL" section.

2. Remove the nut from the engine mount.

3. Remove the screws as shown in the graphic.

4. Engage a lifting chain in the lifting eyes on the cylinder heads, and hook the lifting tackle into the shop crane.

5. Lift the engine.

6. Remove the engine mount.

7. Remove the three bolts for the engine mount bracket at engine block.

8. Move the engine bracket slightly to side.

➡**Engine bracket must not be removed completely. Ground lines remain connected.**

9. Remove the upper starter bolt on the transmission side.

10. Remove the heat shield from the wiring retainer.

11. Remove the lower bolt for the starter on the engine side.

12. Remove the starter from the transmission housing, being careful of wires.

13. Disconnect the electrical connection from the starter.

14. Remove the protective cap and nut beneath.

15. Remove the starter.

*To install:*

16. Installation is the reverse of the removal procedure, noting the following torque specifications:
- Starter Bolt (M10 × 205): 33 ft. lbs. (45 Nm)
- Starter Bolt (M12 × 80): 48 ft. lbs. (65 Nm)
- Wire Terminal 50 to Starter Solenoid Switch: (8 Nm)
- Positive Wire to Solenoid Switch: 12 ft. lbs. (15 Nm)
- Engine Bracket to Cylinder Block: 44 ft. lbs. (60 Nm)
- Engine Bracket to Engine Mount: 55 ft. lbs. (75 Nm)

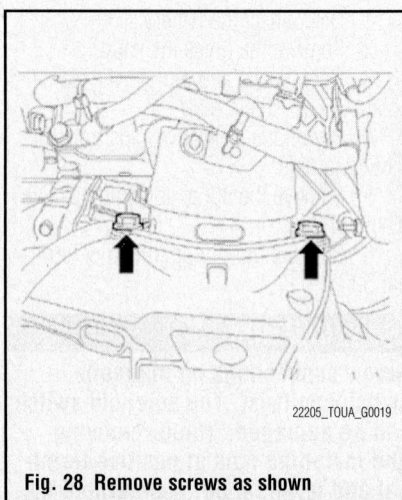

Fig. 28 Remove screws as shown

### 5.0L TDI Engine

*See Figure 29.*

1. Remove the engine. See "ENGINE MECHANICAL" section.

2. Remove the right turbocharger. See "Turbocharger" in 'DIESEL FUEL INJECTION" section.

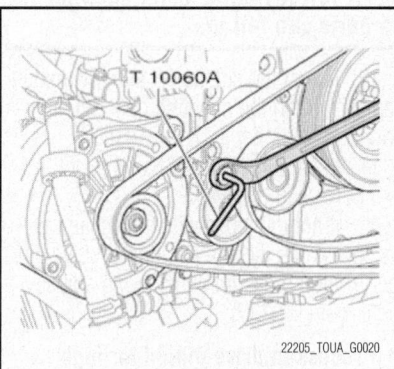

22205_TOUA_G0020

**Fig. 29 Remove nut (arrow) from engine mount**

3. Remove the starter solenoid heat shield.

4. Remove the nut from the engine mount as shown.

5. Remove 4 bolts M10 × 95 from right engine support and remove engine support.

6. Remove bolt for coolant pipe at engine block.

7. Remove bolt for coolant pipes at transmission.

8. Pry off cap from nut at positive (B+) terminal.

9. Remove nut at solenoid positive (B+) terminal.

10. Remove nut at solenoid Terminal 50.

> ✷✷ **CAUTION**
>
> **Screw connections on magnetic switch can twist. The solenoid switch can be damaged. When removing and installing nuts at positive terminal and terminal 50, counterhold the threaded connection at the solenoid switch using an open-end wrench.**

11. Remove bolt for cable retainer and set cables aside.

12. Remove lower starter bolt M12 × 165 on engine side.

13. Remove upper starter bolt M12 × 165 on transmission side.

14. Remove starter.

### To install:

15. Install in reverse order of removal, noting the following:

16. Tighten the fasteners to the specifications:

- Starter Bolts (M12 × 165): 48 ft. lbs. (65 Nm)
- Wire Terminal 50 to Starter Solenoid Switch: 6 ft. lbs. (8 Nm)
- Positive Wire to Solenoid Switch: 11 ft. lbs. (15 Nm)
- Engine Bracket to Cylinder Block (M10 × 95): 44 ft. lbs. (60 Nm)
- Engine Bracket to Engine Mount: 55 ft. lbs. (75 Nm)

# ENGINE MECHANICAL

➡**Disconnecting the negative battery cable may interfere with the functions of the on board computer systems and may require the computer to undergo a relearning process, once the negative battery cable is reconnected.**

## ACCESSORY DRIVE BELTS

### ACCESSORY BELT ROUTING

*See Figures 30 through 32.*

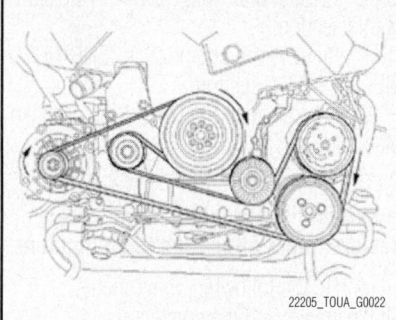

22205_TOUA_G0022

**Fig. 31 Showing accessory belt routing— 4.2L 5-Valve Engine (Code AXQ, BHX)**

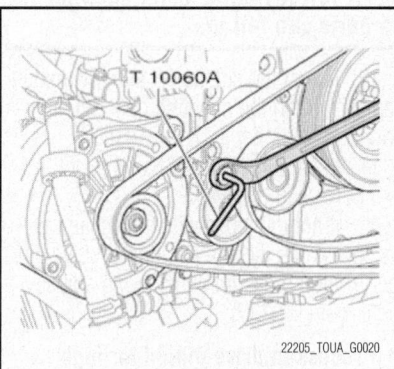

22205_TOUA_G0024

**Fig. 30 Showing accessory belt routing— 3.6L engine**

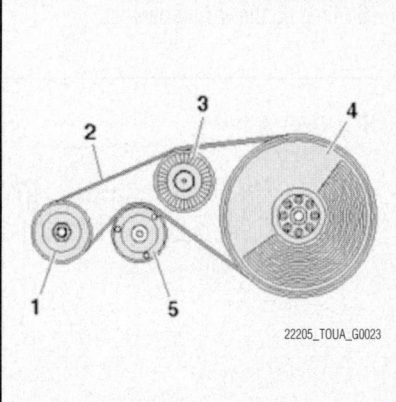

22205_TOUA_G0023

**Fig. 32 Showing accessory belt routing— 4.2L 4-Valve Engine (Code BAR)**

### REMOVAL & INSTALLATION

### 3.6L Engine

*See Figure 33.*

1. Install pressure bolt M8◊50 in the tensioner threaded bore until ribbed belt can be removed.

> ✷✷ **CAUTION**
>
> **Tensioner housing may be damaged. If hex bolt is threaded too far into tensioner housing, the housing may be damaged. Only screw the bolt in so far until the ribbed belt can be removed.**

2. Remove bolts for the tensioner.

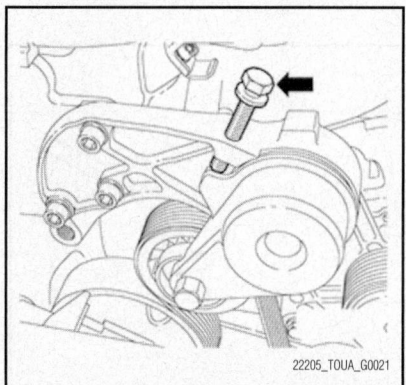

22205_TOUA_G0021

**Fig. 33 Removing the ribbed belt—3.6L engine**

3. Remove the ribbed belt together with the tensioner.

→ **If reusing the same belt, ensure the direction of rotation is maintained.**

4. Install and adjust tension.

### 4.2L 5-Valve Engine (Code AXQ, BHX)

*See Figure 34.*

1. Remove noise insulation pan.
2. If reusing the belt, mark the direction of rotation of ribbed belt.
3. Turn wrench clockwise until tensioning element can be locked in place using Locking Pin T10060 A .
4. Remove ribbed belt.

### To install:

→ **Ensure, before installing ribbed belt, that all ancillaries (generator, air conditioner compressor, power steering pump) are secured tightly.**

5. Check that idler roller turns easily.
6. Note previously marked direction of belt rotation and be sure that it is seated correctly on pulley.
7. Route ribbed belt in the proper direction and locations. See "Accessory Belt Routing" above.
8. After completing repairs, start the engine and check belt running.

### 4.2L 4-Valve Engine (Code BAR)

1. Pull front engine cover off.

→ **Before removing the ribbed belt, mark the turning direction on it with chalk or a felt tip pen. A reversed turning direction can cause damage to the ribbed belt under operating conditions.**

2. To release ribbed belt tension, swing the tensioner downward to relieve belt tension. Use a standard socket and ratchet.

3. Remove ribbed belt and release tensioning device.

### To install:

4. Place ribbed belt over belt pulley in specified sequence:
- Generator
- Ribbed belt
- Idler roller
- Vibration damper
- Tensioning roller

→ **When installing the ribbed belt, make sure it is seated correctly on the pulleys.**

5. The rest of installation is in reverse order of removal.
6. Start the engine and check running belt.

## CAMSHAFT AND VALVE LIFTERS

### REMOVAL & INSTALLATION

#### 3.6L Engine

*See Figures 35 through 44.*

1. Before starting this procedure, refer to the "Precautions" information at the start of this section.
2. Remove the engine. See "ENGINE MECHANICAL" section.
3. Remove intake manifold. See "Intake Manifold."
4. Remove the cylinder head cover.

→ **Before disconnecting harness connectors, mark allocation to component.**

5. On the following components, pull off the covers and the connectors:
- Camshaft adjustment valve (intake)
- Camshaft adjustment valve (exhaust)
- Camshaft Position (CMP) sensors 1 and 2
6. Free up the wiring harness.

7. Remove the vacuum pump, if equipped. See "Vacuum Pump."
8. Remove top and bottom coolant pipes on cover piece.
9. Remove the timing chain cover bolts from the cylinder head.

### ❊❊ CAUTION

**Cover lower timing chain opening so no parts can fall in.**

10. Adjust the crankshaft at the vibration damper bolt in the direction of engine rotation to the cylinder 1 TDC marking.
11. Insert the camshaft bar (T10068 A) into both shaft grooves.
12. If not equipped with a mechanical vacuum pump, secure position of drive pinion for high pressure pump using adjustment tool (T10332).

→ **If recess in drive pinion for high pressure pump does not point upward: Remove camshaft bar T10068 A. Rotate crankshaft further in direction of rotation until this position has been**

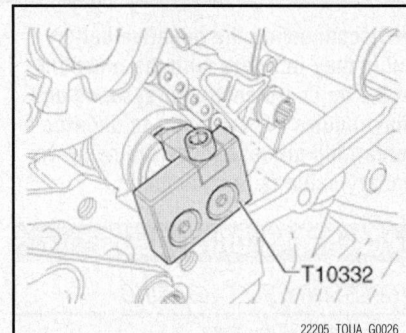

Fig. 36 If not equipped with a mechanical vacuum pump, secure position of drive pinion for high pressure pump using adjustment tool (T10332)

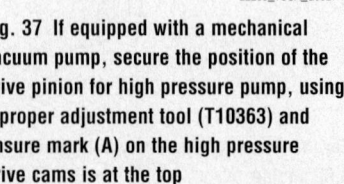

Fig. 37 If equipped with a mechanical vacuum pump, secure the position of the drive pinion for high pressure pump, using a proper adjustment tool (T10363) and ensure mark (A) on the high pressure drive cams is at the top

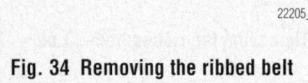

Fig. 34 Removing the ribbed belt

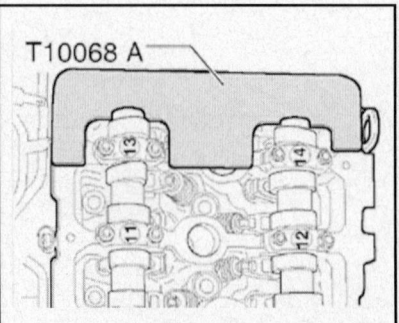

Fig. 35 Insert the camshaft bar (T10068 A) into both shaft grooves

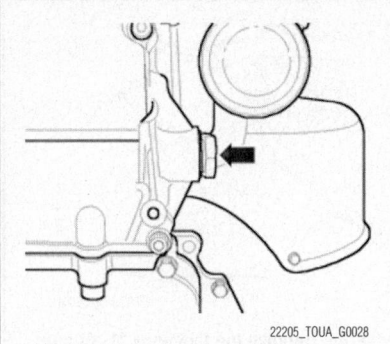

**Fig. 38 On all engines, remove the chain tensioner bolt for camshaft timing chain**

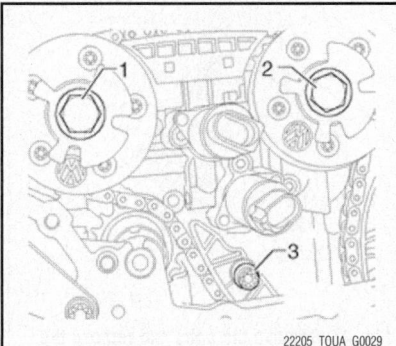

**Fig. 39 Loosen the bolts of the camshaft adjusters 1 and 2**

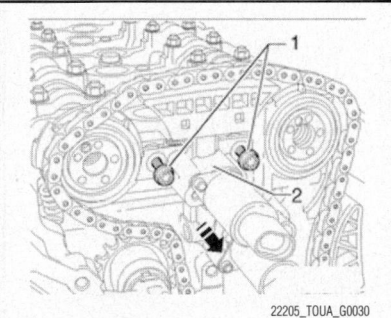

**Fig. 40 Remove the control housing bolts from the cylinder head and pull it from the camshafts**

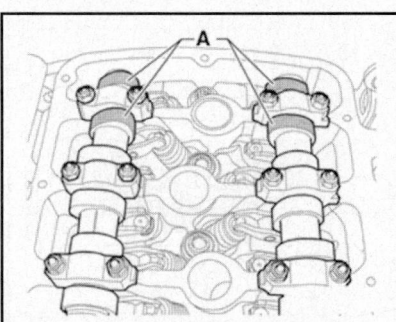

**Fig. 41 When installing the camshafts, the cam lobes for cylinder 1 must point upward**

**attained and camshaft bar T10068 A and adjustment tool T10332 can be inserted.**

13. If equipped with a mechanical vacuum pump, secure the position of the drive pinion for high pressure pump, using a proper adjustment tool (T10363). Mark "A" on the high pressure drive cams must be at the top.

➡**If the vacuum pump drive pins are not vertical, remove the camshaft bar (T10068 A), turn the crankshaft in the rotation direction until the pins are vertical and the camshaft bar (T10068 A) can be inserted.**

14. On all engines, remove the chain tensioner bolt for camshaft timing chain.

➡**Only counter-hold at camshaft using a spanner wrench (27 mm). Camshaft bar (T10068 A) must not be inserted when tightening or loosening the camshaft adjuster.**

15. Loosen the bolts of the camshaft adjusters 1 and 2.

16. Remove both camshaft adjusters.

17. Remove the control housing bolts from the cylinder head and pull it from the camshafts.

18. Place the camshaft timing chain to the side.

19. On the intake camshaft, perform the following:

  a. First remove bearing caps 1 and 13.

  b. Remove bearing caps 3 and 11.

  c. Remove bearing cap 7.

  d. Loosen and remove bearing caps 5 and 9 in alternation and in diagonal sequence.

20. On the exhaust camshaft, perform the following:

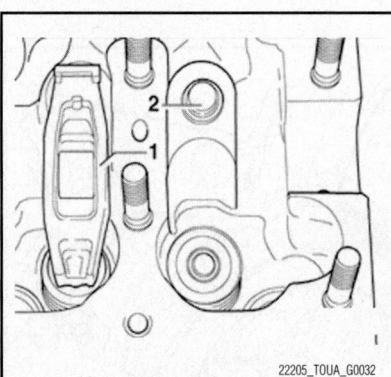

**Fig. 42 Make sure that all roller cam followers (1) properly contact the valve shaft tips and are clipped into the respective support elements (2)**

  a. First remove bearing caps 2 and 14.

  b. Remove bearing caps 4 and 12.

  c. Remove bearing cap 8.

  d. Loosen bearing caps 6 and 10 in alternation and in diagonal sequence.

21. On both camshafts, carefully remove the camshafts and place on a clean surface.

22. Remove roller rocker lever together with support elements and place on a clean surface.

23. Ensure that the roller rocker levers and the support elements are not interchanged.

*To install:*

24. When installing the camshafts, the cam lobes for cylinder 1 must point upward.

25. Coat the contact surface of bearing caps 7 and 8 lightly with grease before installing.

26. Insert the support element in cylinder head and install the roller rocker lever onto the respective valve stem end and support element.

27. Make sure that all roller cam followers properly contact the valve shaft tips and are clipped into the respective support elements.

28. Oil the running surfaces of both camshafts.

29. Place the respective camshaft carefully in the camshaft bearings of the cylinder head. While doing so, observe proper identification of camshafts.

30. Observe the installed position of the bearing caps: the points of the intake and exhaust camshaft bearing caps must face outwards.

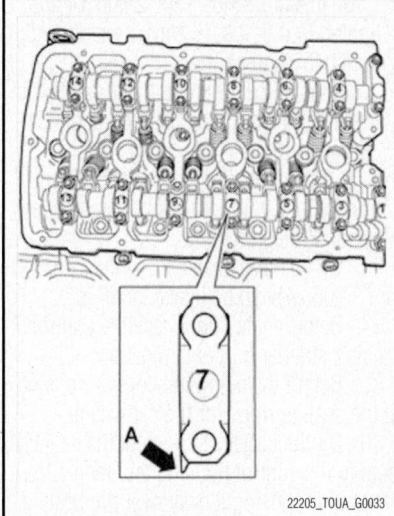

**Fig. 43 Observe the installed position of the bearing caps: the points of the intake and exhaust camshaft bearing caps must face outwards**

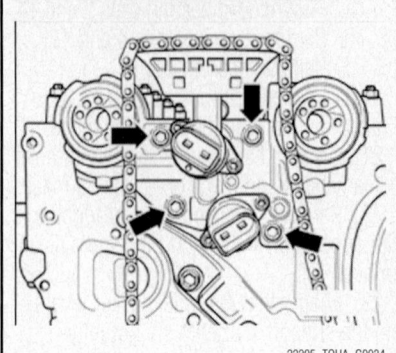

**Fig. 44 Install the control housing and the mounting bolts (arrows) using locking adhesive**

31. On the intake camshaft, install and tighten the bearing caps as follows:

a. Tighten bearing caps 5 and 9 in alternation and in diagonal sequence. Tighten them to 4 ft. lbs. (5 Nm), plus an additional 45 degree turn.

b. Install bearing caps 1 and 13 and tighten to 4 ft. lbs. (5 Nm), plus an additional 45 degree turn.

c. Install bearing cap 7 and tighten to 4 ft. lbs. (5 Nm), plus an additional 45 degree turn.

d. Install bearing caps 3 and 11 and tighten to 4 ft. lbs. (5 Nm), plus an additional 45 degree turn.

32. On the exhaust camshaft, install and tighten the bearing caps as follows:

a. Tighten bearing caps 6 and 10 in alternation and in diagonal sequence. Tighten them to 4 ft. lbs. (5 Nm), plus an additional 45 degree turn.

b. Install bearing caps 2 and 14 and tighten to 4 ft. lbs. (5 Nm), plus an additional 45 degree turn.

c. Install bearing cap 8 and tighten to 4 ft. lbs. (5 Nm), plus an additional 45 degree turn.

d. Install bearing caps 4 and 12 and tighten to 4 ft. lbs. (5 Nm), plus an additional 45 degree turn.

33. On both camshafts, insert the camshaft bar (T10068 A) into both shaft grooves.

34. Before installing, check the control housing strainer for contamination.

35. Before installing the control housing, oil the sealing rings for the camshafts.

36. Oil the contact surface of the sealing rings in the control housing and then slide the control housing slowly over the sealing rings for the camshafts.

37. Install the control housing and the mounting bolts, using locking adhesive on the threads. Tighten the bolts to 6 ft. lbs. (8 Nm).

38. Install the camshaft adjustor with the timing chain for the camshaft drive. See "Timing Belt, Sprockets, Front Cover and Seal."

39. Clean the sealing surface on the cover piece as well as on the cylinder head.

➡ If the sealing rings in the cover piece are to be replaced, see "Timing Belt, Sprockets, Front Cover and Seal" in this section.

40. Coat the sealing surface of the cover piece with sealant (D 176 501 A1) and install immediately.

### ✳✳ CAUTION

**Sealant D 176 501 A1 hardens quickly.**

41. First, install all timing chain cover bolts and tighten by hand. Now, tighten mounting bolts to 6 ft. lbs. (8 Nm).

42. Install the chain tensioner for the camshaft timing chain and tighten to 30 ft. lbs. (40 Nm).

43. Install the vacuum pump, if equipped. See "Mechanical Vacuum Pump."

44. Install the cylinder head cover.

45. Install the intake manifold. See "Intake Manifold."

46. Reset the service position.

### 4.2L 5-Valve Engine (Code AXQ, BHX)

*See Figures 45 through 52.*

➡ This procedure includes the timing chain adjuster.

1. Set the engine to TDC.
2. Remove the toothed belt and camshaft gear. See "Timing Belt, Sprockets, Front Cover and Seal."
3. On the left cylinder head, perform the following:

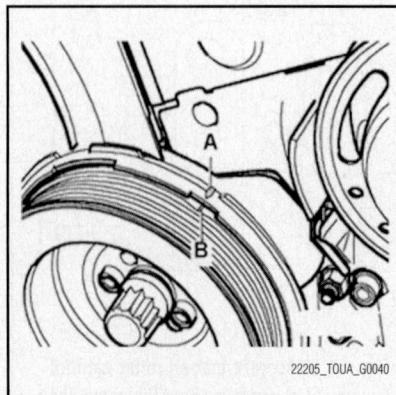

**Fig. 45 Setting the engine at TDC—4.2L 5-valve engine (AXQ, BHX)**

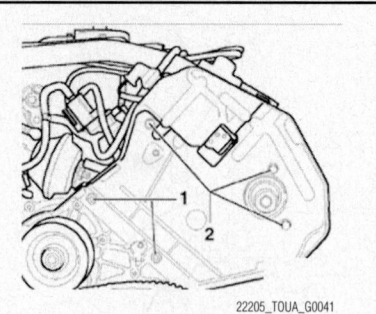

**Fig. 46 Remove the five bolts (1, 2) and remove the rear toothed belt cover—4.2L 5-valve engine (AXQ, BHX)**

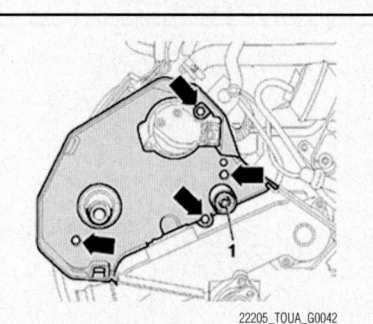

**Fig. 47 Remove the bolts and remove the rear toothed belt cover for the right cylinder head—4.2L 5-valve engine (AXQ, BHX)**

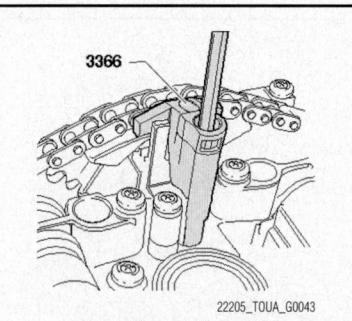

**Fig. 48 Secure camshaft adjuster using Bracket for Chain Adjustment (3366)—4.2L 5-valve engine (AXQ, BHX)**

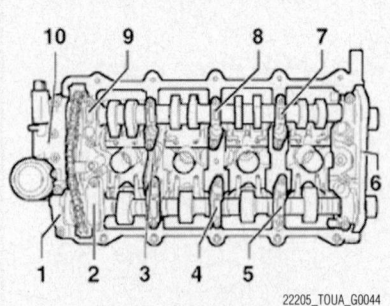

**Fig. 49 Camshaft bearing cap removal and installation sequence—4.2L 5-valve engine (AXQ, BHX)**

a. Remove the left cylinder head cover.

b. Remove the five bolts and remove the rear toothed belt cover.

4. On the right cylinder head, perform the following:

a. Remove the right cylinder head cover.

b. Remove the idler roller.

c. Remove the bolts and remove rear toothed belt cover.

5. Disconnect the connector from Camshaft Position (CMP) sensor and remove the sensor housing with the cover and cone.

6. Verify the TDC position of camshafts. The markings on the camshafts must be aligned with both arrows on the bearing cap.

7. When re-using camshaft roller chain, mark the roller chain before removing (e.g. with paint, make an arrow pointing in direction of rotation).

8. Secure the camshaft adjuster using a special Bracket For Chain Adjustment (3366).

**✳✳ CAUTION**

**If the bracket for chain adjustment is fastened too tightly, the camshaft adjuster can be damaged.**

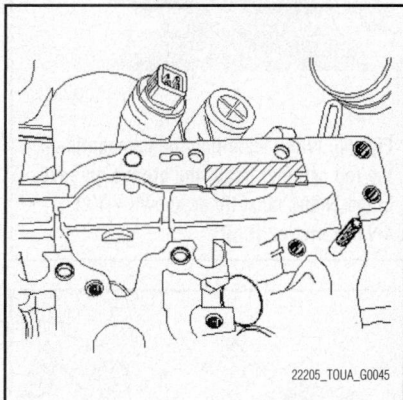

22205_TOUA_G0045

**Fig. 50 Lightly coat the shaded area with sealant**

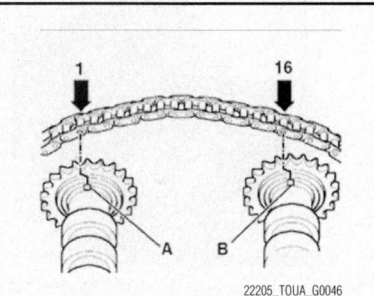

22205_TOUA_G0046

**Fig. 51 Showing where "1" and "16" drive chain rollers must be installed on the chain sprockets—4.2L 5-valve engine (AXQ, BHX)**

9. Independent of the existing markings, mark the installation position and sequence of all bearing caps as shown (e.g. using a water-proof felt pen).

a. Remove mounting bolts of camshaft adjuster.

b. Remove bearing cap 1.

c. Remove bearing caps 2, 4, 6, 8 and 9 and lay them on a clean surface in the correct order.

d. Loosen bearing caps 3, 5 and 7 alternately, in a diagonal sequence, and remove.

e. Remove both camshafts with camshaft adjusters and place on a clean surface.

*To install:*

➡**During installation, replace the half-round sealing plugs and the gasket for the camshaft adjuster.**

10. Lightly coat the shaded area with sealant (AMV 188 001 02).

11. Place the camshaft roller chain onto the camshaft drive sprockets as follows.

12. If installing the same chain, move the color markings to align with the marks made prior to removal.

13. If using a new chain, the distance between notches "A" and "B" must consist of 16 rollers on the drive chain. The illustration shows where "1" and "16" drive chain rollers must be installed on the chain sprockets. The chain rollers "1" and "16" are across from notches "A" and "B", and are respectively a one-half tooth-width offset to the left.

14. For all chain installations, push the camshaft adjuster in between the camshaft roller chain.

15. Install the camshafts with the camshaft roller chain and camshaft adjuster into the cylinder head.

16. Oil the journal surfaces of the camshafts.

➡**Alignment bushings for the bearing caps and camshaft adjusters must be installed in the cylinder head.**

17. Install bearing caps "3". "5" and "7" according to their markings and fasten new bolts alternately in a diagonal sequence to 4 ft. lbs. (5 Nm), plus an additional 90 degrees.

18. Tighten the camshaft adjuster "10" to 4 ft. lbs. (5 Nm), plus an additional 90 degrees.

19. Install the chain tensioner

20. Verify the TDC position of the camshafts. The markings on the camshafts must be aligned with both arrows on bearing caps.

➡**So that both markings line up, turn the camshaft slightly back and forth, if necessary.**

21. Lightly coat the double bearing cap "6" and bearing cap "1", using a proper sealant (AMV 188 001 02). Lightly coat the separation surfaces of the front and rear bearing caps before installing. Torque the bearing cap new bolts to 4 ft. lbs. (5 Nm), plus an additional 90 degrees.

22. Install the remaining bearing caps in the sequence shown. Torque the new bolts to 4 ft. lbs. (5 Nm), plus an additional 90 degrees.

23. Replace the camshaft seals.

24. Replace the seals for the Camshaft Position (CMP) Sensor.

25. Carefully drive in the sealing cover, using a proper driver (3202).

26. The rest of assembly is in reverse order of disassembling.

27. After installing the camshafts, the engine should not be started for at least 30 minutes. The hydraulic equalization elements must seat themselves (otherwise, the valves will crash into the pistons).

28. After working on the valve train and lifters, carefully rotate the crankshaft by hand at least 2 full revolutions before starting, to be sure that valves do not strike the pistons.

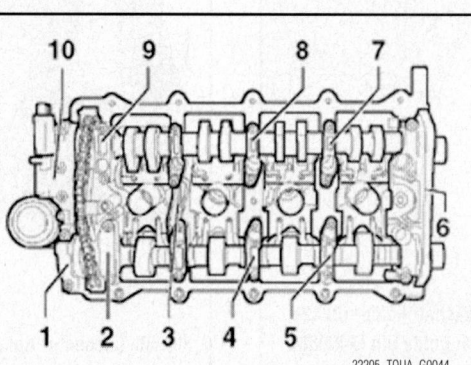

22205_TOUA_G0044

**Fig. 52 Camshaft bearing cap removal and installation sequence—4.2L 5-valve engine (AXQ, BHX)**

### 4.2L 4-Valve Engine (Code BAR)

*See Figures 53 through 59.*

1. Remove the engine. See "Engine Assembly" in this section.

2. Disconnect the electrical connector at the intake camshaft position (CMP) sensor.

3. Remove the high-pressure pump. See "GASOLINE FUEL INJECTION SYSTEM."

4. Remove the camshaft timing chains from camshafts. See "Timing Chain, Sprockets, Front Cover and Seal."

5. Remove the crankshaft holder (3242) from the upper part of the oil pan.

➡ **This holder tool (3242) is installed during the timing chain removal procedure.**

6. Into the crankshaft holder (3242), insert a proper socket guide pin (T40058) so that the large diameter points to the

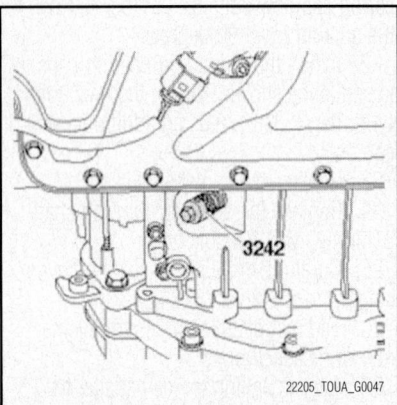

**Fig. 53 Remove the crankshaft holder (3242) from the upper part of the oil pan—4.2L 4-valve engine (BAR)**

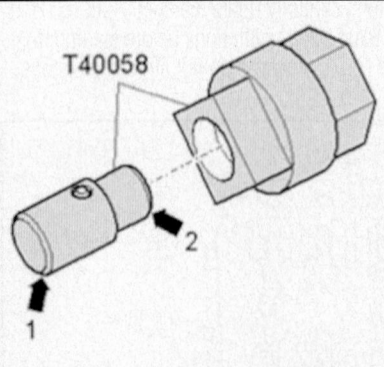

**Fig. 54 Into the crankshaft holder (3242), insert a proper socket guide pin (T40058) so that the large diameter (1) points to the socket and the small diameter (2) points to engine—4.2L 4-valve engine (BAR)**

socket and the small diameter points to engine.

> ※※ **CAUTION**
>
> **To prevent valves from contacting pistons during the following steps, crankshaft must be turned so that no pistons are in "TDC" position.**

7. Rotate the crankshaft, opposite the engine normal rotation direction, so the pointer mark is 40 degrees out of "TDC" position.

8. Remove the camshaft clamp (T40070) from the cylinder head.

➡ **This clamp (T40070) was installed during timing chain removal at the beginning of this procedure.**

9. Loosen the 24 bearing bracket bolts in the sequence shown for each cylinder head.

10. Mark the position and location of each camshaft and remove from the cylinder head.

➡ **On engines with alignment pins for bearing bracket, these must be driven out with a cotter pin driver before cleaning the sealing surface and then reinserted securely before installing the camshaft assembly bracket.**

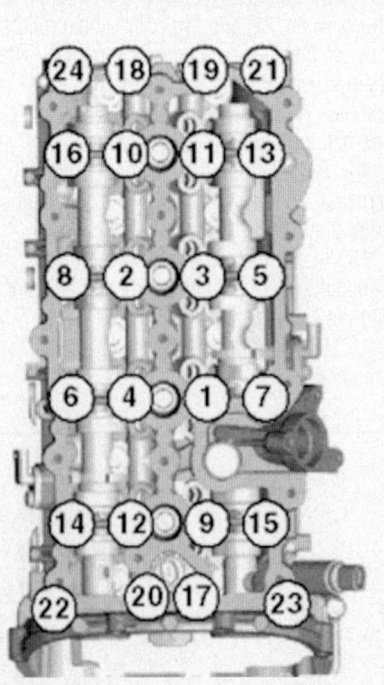

**Fig. 55 Loosen or tighten the 24 bearing bracket bolts in the sequence shown for each cylinder head—4.2L 4-valve engine (BAR)**

*To install:*

➡ **Always replace gaskets and seals. Wear safety glasses.**

11. Remove all sealant residue on the cylinder head and other mating surfaces with a plastic brush.

> ※※ **CAUTION**
>
> **Make sure that no sealant residue enters the cylinder head and bearings.**

12. Clean all sealing surfaces so they are free of oil and grease.

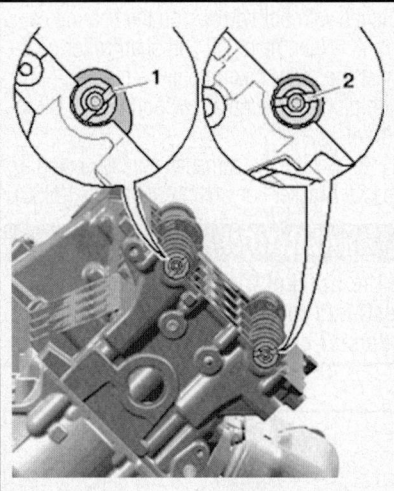

**Fig. 56 When installing the camshafts on the left cylinder head, the groove on ends of the shafts must lie as shown—4.2L 4-valve engine (BAR)**

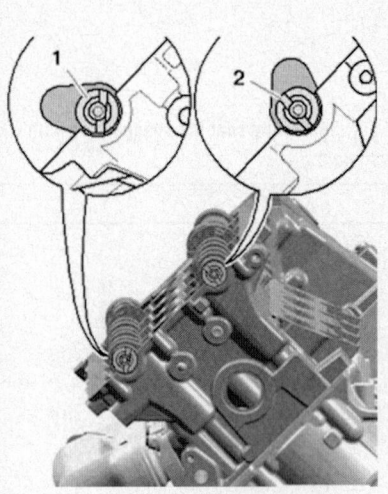

**Fig. 57 When installing the camshafts on the right cylinder head, the groove on ends of the shafts must lie as shown—4.2L 4-valve engine (BAR)**

13. Oil the journal surfaces of the camshafts.

14. Place the camshafts in the cylinder head, noting the position of camshafts so that the bearing bracket can be installed without tension.

15. When installing the camshafts on the left cylinder head, the groove on ends of the shafts must lie as shown.

16. When installing the camshafts on the right cylinder head, the groove on ends of the shafts must lie as shown.

17. Lay a new seal in the bearing bracket groove around the edge and down the center of the cylinder head. Thickness of sealant beads is 2.5 mm.

➡ **Sealant beads must not be thicker than specified, otherwise extra sealant can enter the camshaft bearing.**

18. Immediately place the bearing bracket on cylinder head. Ensure camshafts can be inserted in bearing bracket axial bearing without force.

19. Insert the bearing bracket locating pins (T40116) in the bearing bracket and cylinder head.

➡ **Placing the bearing bracket in place and tightening it should follow sealant application immediately, since the sealant begins to harden immediately.**

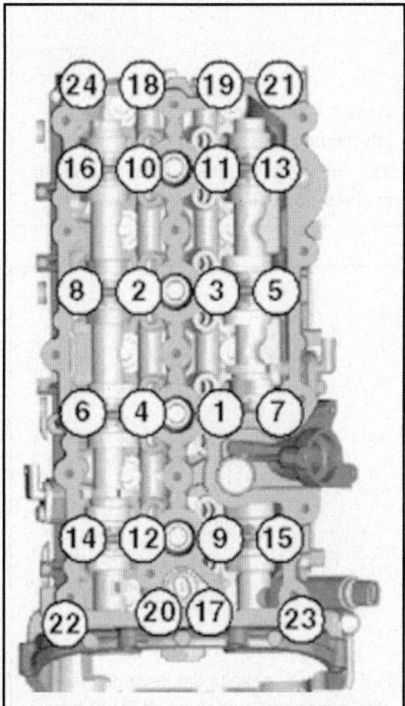

**Fig. 58 Loosen or tighten the 24 bearing bracket bolts in the sequence shown for each cylinder head—4.2L 4-valve engine (BAR)**

22205_TOUA_G0049

20. After installing the bearing bracket, the sealant must cure for approximately 30 minutes.

21. Hand-tighten the 24 bearing bracket bolts equally, in sequence shown. Then, fasten the bolts in sequence to 6 ft. lbs. (8 Nm).

22. Drive the new sealing plugs in flush in the cylinder head bores at the end of the camshafts.

23. Remove the bearing bracket locating pins (T40116). Pins that are too secure can be removed with an impact puller (T10133/3).

24. Rotate the intake camshaft to "TDC" and tighten the camshaft adjuster screw with a socket (SW 24) inserted between.

➡ **When tightening the bolt, counterhold the socket with pliers, if necessary.**

25. Position a lever or ratchet with a socket (T10035) on the bolt and rotate the camshaft until the threaded holes for the camshaft clamp (T40070) face up.

26. Next, loosely fasten the camshaft clamp (T40070) to the intake camshaft.

27. The camshaft locating tool (T40070) is correctly positioned when the holes for the cylinder head bolts remain free.

28. Position the camshaft adjuster screw and socket to the exhaust camshaft.

29. Rotate the exhaust camshaft until the threaded hole for the camshaft clamp (T40070) faces up.

30. At the same time, position a counterhole (SW 24) on the camshaft clamp

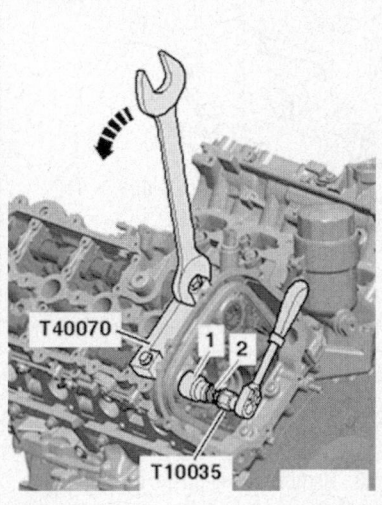

**Fig. 59 Rotate the camshaft to "TDC" and complete camshaft installation using the indicated tools—4.2L 4-valve engine (BAR)**

22205_TOUA_G0052

(T40070) and swing the camshaft clamp (T40070) against the exhaust camshaft to install.

31. Tighten the camshaft clamp (T40070) by hand onto the exhaust camshaft to avoid damaging the threads (a second technician is needed).

32. Tighten the camshaft clamp (T40070) bolts to 18 ft. lbs. (25 Nm).

33. Using a proper socket (T40058), rotate the crankshaft in the direction of engine rotation to "TDC".

34. Install the locking pin (3242) into the crankcase hole. If necessary, rotate the crankshaft very slightly back and forth to completely center the bolt. Tighten the crankshaft holder (3242) to 15 ft. lbs. (20 Nm).

⁑ **CAUTION**

**Do not turn the crankshaft while touching the "TDC" hole with your finger.**

35. The rest of the installation is in the reverse order of removal, note the following:

⁑ **CAUTION**

**After installing the camshafts, do not crank the engine for at least 30 minutes. The hydraulic equalization elements must seat themselves (otherwise the valves will crash into the pistons).**

a. After working on the valve train, carefully rotate the engine by hand at least two full revolutions to ensure that the valves do not strike the pistons when starting.

b. Position the camshaft timing chains on camshafts. See "Timing Chain, Sprockets, Front Cover and Seal."

c. Install the left and right timing chain covers.

d. Install the cylinder head covers.

e. Install the engine. See "Engine Assembly."

### 5.0L TDI Engine

1. Remove the engine and transmission assembly. See "Engine Assembly."

2. Remove the drive gear for the camshaft. See "Camshaft Compensation Wheel and Input Gear."

3. Remove the rocker lever shafts.

4. Loosen the camshaft bearing caps, starting from the outside and working inward so the bolts of the last two bearing caps are loosened in an alternating diagonal sequence.

*To install:*

5. During installation, note the following:

   a. Make sure that no valve is pressed onto the piston when installing the camshaft. Bring the pistons evenly to just below Top Dead Center (TDC) position.

   b. Do not interchange run-in connecting rod bearing shells (mark).

   c. When installing the camshaft, make sure the bearing shell retaining tabs are seated properly in the bearing caps and cylinder head.

   d. Before installing the bearing caps, make sure that washers for cylinder head bolts are inserted in the cylinder head.

6. Be sure that all oil bearing cap contact surfaces are clean.

7. Tighten the inner two bearing cap new bolts, alternating in a diagonal sequence, and tighten to 6 ft. lbs. (8 Nm), plus an additional 90 degrees.

8. Install the remaining bearing caps and also tighten the new bolts to 6 ft. lbs. (8 Nm), plus an additional 90 degrees.

9. Seal the joint surfaces of the outer bearing caps with a proper sealant (AMV 176 501).

10. Install the rocker arm axles and tighten first the inner and then the outer mounting new bolts evenly, in a diagonal sequence, to 15 ft. lbs. (20 Nm), plus an additional 90 degrees.

11. Re-install the drive gear for the camshaft See "Camshaft Compensation Wheel and Input Gear."

12. Adjust the valve timing. See "Valve Timing" in this section.

13. Install the acoustic cover, as well as the cylinder head cover of the respective cylinder.

14. Install the engine and transmission. See "Engine Assembly."

## ✽✽ CAUTION

**After installing the new valve lifters, the engine must not be started for 30 minutes. The hydraulic equalization elements must seat themselves (otherwise the valves will crash into the pistons).**

## CAMSHAFT COMPENSATION WHEEL AND INPUT GEAR

### REMOVAL & INSTALLATION

#### 5.0L TDI Engine

*See Figures 60 through 65.*

1. Remove the engine and transmission. See "Engine Assembly."

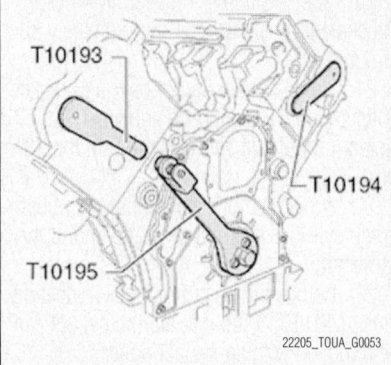

**Fig. 60 Installing the Camshaft Tool to the crankshaft stub—5.0L TDI engine**

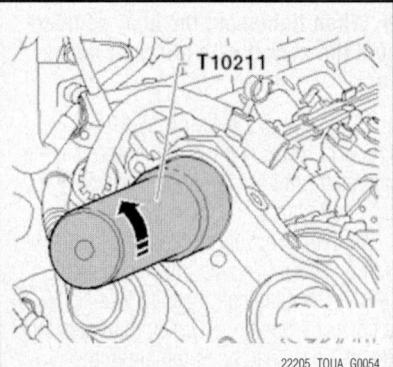

**Fig. 61 Position the special wrench (T10211), as shown, and loosen the adapter ring on the pump injection unit central connector—5.0L TDI engine**

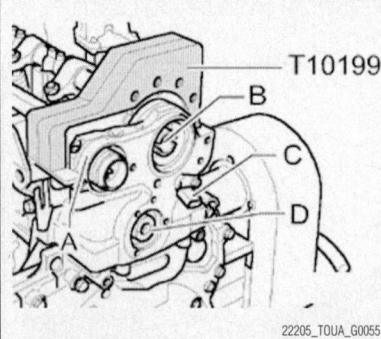

**Fig. 62 Removing the outer bear cap and tandem pump drive—5.0L TDI engine**

2. Remove the intake manifold flap and the crankcase ventilation line from the respective cylinder.

3. Remove the acoustic cover and the cylinder head cover of the applicable cylinder head.

4. Check the rocker arm shaft running surfaces for wear.

5. Remove the coolant pipe from below the tandem fuel pump.

6. Remove the vibration damper. Use a proper counter-hold tool (V10 T10172) with threading pins (T10172/1).

7. Install the Camshaft Tool (T10195) onto the crankshaft stub. Observe the alignment pin while doing so.

8. Now carefully rotate the crankshaft until the pin of the Camshaft Tool (T10195) is inserted into the alignment bore.

9. Remove the corresponding tandem pump or fuel pump. See "DIESEL FUEL INJECTION SYSTEM" section.

10. Remove the oil supply line to the exhaust turbocharger.

11. Position the special wrench (T10211), as shown, and loosen the adapter ring on the pump injection unit central connector.

12. Remove the Camshaft Position (CMP) Sensor "C".

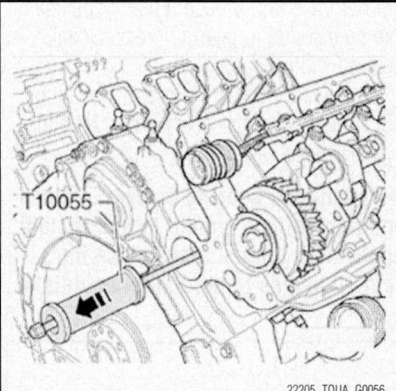

**Fig. 63 Install a puller (T10055) into the full-floating axle and pull the full-floating axle out of its seat, using careful knocking motions—5.0L TDI engine**

**Fig. 64 First, engage the inner, engine-side compensation linkage at the bottom (if necessary, guide it through the compensating piston hole with your finger) and then slide the top onto the sleeve—5.0L TDI engine**

13. Remove the outer bearing cap "A", place the clamping device (T10199) on the camshaft gear and tighten the bolt to 30 ft. lbs. (40 Nm).

14. Loosen the mounting bolt "B" for the camshaft gear using a socket insert (T10198) and remove the bolt together with tandem pump drive.

15. Loosen the bolts on the Clamping Device (T10199) and remove it.

16. Remove the compensating piston for the full-floating axle of the compensation linkage.

17. Remove the pull-off head for the pump-injector unit from the puller (T10055).

18. Install a puller (T10055) into the full-floating axle and pull the full-floating axle out of its seat, using careful knocking motions.

19. Now, pull out the outer compensation linkage on the transmission side, along with its sleeve. If necessary, loosen it by hitting lightly with a plastic hammer.

20. Remove the camshaft gear.

21. Now, remove the compensation wheel and inner compensation linkage on the engine side.

22. Remove the diamond disc from the camshaft stub.

**To install:**

23. Clean the bearing cap contact surfaces of any sealant residue.

24. Oil the compensation linkage sleeve guide surfaces.

➡️**If there is marked crimping or deposits on the compensation linkage sleeve guide surfaces, these must be replaced.**

25. First, engage the inner, engine-side compensation linkage at the bottom (if necessary, guide it through the compensating

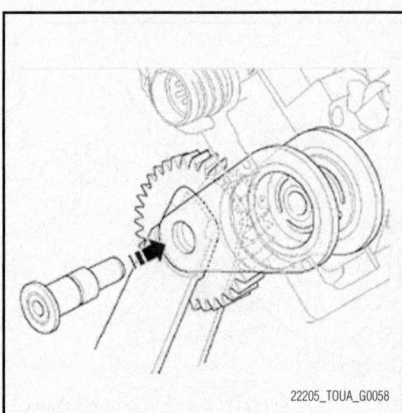

22205_TOUA_G0058

**Fig. 65 Showing the proper compensation linkage position—5.0L TDI engine**

piston hole with your finger) and then slide the top onto the sleeve.

26. Place a new diamond disc on the camshaft stub.

27. Bring the lower, transmission-side compensation linkage into position and insert the compensation wheel with phase in direction of transmission, as follows:

a. Lubricate the free-floating axle with oil, then carefully slide it in by hand, with as even a force as possible. Correct seat of the full-floating axle faces toward a clear gap to the cylinder head bore.

b. Remove the tandem or fuel pump drive.

c. Position the bearing cap flush with the cylinder head edge, using a proper sealant (AMV 176 501) and tighten to 6 ft. lbs. (8 Nm), plus an addition 90 degrees.

d. Install new compensating pistons, tightening to 81 ft. lbs. (110 Nm).

e. Positioning the Clamping Device (T10199) and tighten camshaft gear bolt to 44 ft. lbs. (60 Nm).

28. When inserting the compensation linkage, turn the camshaft gear so the marking on the sensor wheel aligns with the upper edge of the sealing surface.

29. Secure the installation by carefully sliding the full-floating axle on by hand, with force as even as possible.

⚜️ **CAUTION**

**There is the risk that a linkage may pivot to the side out of the wheel cassette and will not engage when full-floating axle is inserted, or it may be distorted when inserting the full-floating axle.**

30. Insert the guide sleeve in the transmission-side compensation linkage.

31. Carefully pull the full-floating axle back out of its seat. When doing so, ensure the compensation linkages, which have already been positioned, remain in correct location.

32. Install the transmission-side compensation linkage with its writing toward the transmission. When doing so, support it if necessary with a light tapping on the guide sleeve with a plastic hammer.

33. Control the correct seating of the compensation linkage with a finger placed through the compensating piston hole.

34. The compensation wheel must be aligned so that the holes in four linkages and in the sprocket are positioned flush behind one another.

### CRANKSHAFT FRONT SEAL

*REMOVAL & INSTALLATION*

### 3.6L Engine

1. The engine must be removed. See "Engine Assembly."

2. Remove ribbed accessory drive belt.

3. Remove the vibration damper. To do so, lock the vibration damper in place, using a counter-holder tool (T10069).

4. Remove the oil pan.

5. Unscrew the crankshaft front sealing flange.

6. Remove the sealant residue on sealing surfaces.

**To install:**

➡️**Before installing, remove any oil remains from the end of crankshaft with a clean cloth.**

7. Apply a sealant bead of approximately 2 to 3 mm on the clean sealing surface of sealing flange.

8. Cover the sealing ring with a clean cloth before applying the sealing bead.

➡️**Sealant bead must not be thicker than 2 to 3 mm, otherwise excess sealant could get into oil pan and clog oil intake pipe strainer. The sealing flange must be installed within 5 minutes after application of silicon sealant.**

9. Insert a guide sleeve (T10215/1) at the front on to the crankshaft pin.

10. Now, slide the sealing flange (with sealing ring) carefully over the guide sleeve.

11. Install the sealing flange bolts to cylinder crankshaft housing.

12. Install the oil pan. Tighten the oil pan bolts to 9 ft. lbs. (12 Nm).

13. Install the vibration damper. Tighten the damper bolt to 74 ft. lbs. (100 Nm), plus an additional 180 degrees.

14. Install accessory drive ribbed belt.

### 4.2L 5-Valve Engine (Code AXQ, BHX)

1. Remove noise insulation pan.

2. Remove the toothed timing belt. See "Timing Belt, Sprockets, Front Cover and Seal" in this section.

3. Remove the centered bolt of the toothed belt crankshaft sprocket, and remove the sprocket from the end of the crankshaft.

4. Then, pull out the crankshaft seal, using a proper pulling hook (T20143/1).

5. Remove the thrust collar from the end of the crankshaft and mark the thrust

collar on its face surface, using a waterproof felt pen, for proper reinstallation orientation.

### To install:

6. Before installing, remove any oil remains from the end of the crankshaft and thrust collar with a clean cloth.

7. Press in a new seal, up to the stop, using a pressure piece (T40007).

8. Turn the thrust collar and press it onto the end of the crankshaft so that the marked surface faces the engine.

9. Install the toothed belt crankshaft sprocket, using a new centered bolt. Torque the bolt to 148 ft. lbs. (200 Nm), plus an additional 180 degrees.

10. Install the toothed timing belt. See "Timing Belt, Sprockets, Front Cover and Seal" in this section.

11. Install the accessory drive ribbed belt.

12. Install the noise insulation pan.

### 4.2L 4-Valve Engine (Code BAR)

*See Figure 66.*

1. Remove the vibration damper.

2. Place the inner part of an oil seal extractor (T40019) flush with the outer part and secure it using a knurled-head screw.

3. Lubricate the threaded head of the seal remover, place it against the seal, and with strong force, screw it into the seal as far as possible.

4. Loosen the knurled thumb screw and turn the inner portion against the crankshaft until seal is pulled out.

5. Clamp the seal extractor at it mounting points in a vise. Remove the seal using pliers.

### To install:

6. Clean the operating and sealing surfaces.

7. Position an assembly device (T40048/1) on a pull sleeve (T40048/2) and slide the new seal onto the pull sleeve.

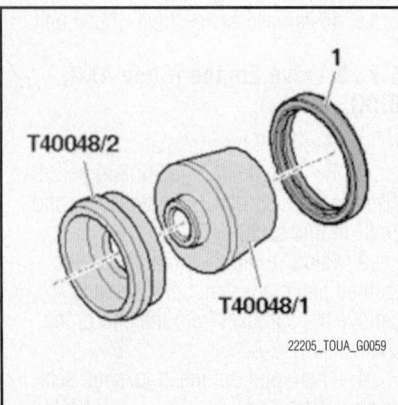

**Fig. 66 Showing the assembly device and pull sleeve with a new seal for installation**

8. Remove the assembly device.

9. Place the pull sleeve (T40048/2) on the crankshaft and slide the new seal into the sealing surface on the engine. The pull sleeve remains on crankshaft for pressing in.

10. Position the pressure sleeve (T40048/3) with two M8 × 55 mm bolts on the crankshaft. Then, install bolts by hand. Tighten the bolts one-half turn each by alternating sides to press the seal in until it reaches its stop.

11. The rest of installation is in reverse order of removal, note the following:

a. Install vibration damper.

### 5.0L TDI Engine

*See Figure 67.*

1. Bring the lock carrier into the service position.

2. Remove the vibration damper.

#### ✳✳ CAUTION

**Be careful not to damage the sealing surface when removing the sealing ring.**

3. Carefully remove the old sealing ring from its seat, using a hook (T20143/2).

4. Before installing the new seal, remove any oil remains from the end of crankshaft with a clean cloth.

➡**Do not remove the support ring out of the seal until immediately before installation.**

5. Join assembly sleeves (T10196/1 and T10196/2). Slide the sealing ring to its stop on the assembly sleeve.

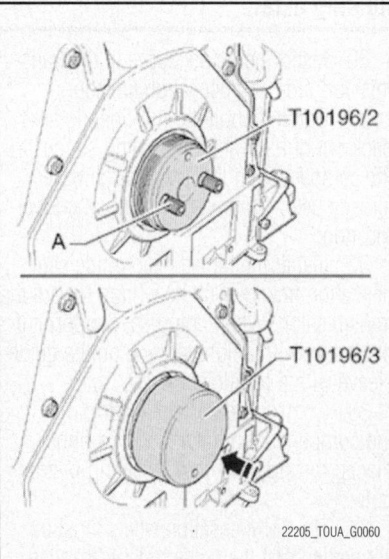

**Fig. 67 Installing the seal with the assembly sleeve tool set—5.0L TDI engine**

6. Separate both assembly sleeves.

7. Place assembly sleeve (T10196/2), together with the sealing ring, on the crankshaft flange and tighten the knurled bolts hand-tight.

8. Press the seal up to its stop, using pressure sleeve (T1096/3).

9. The rest of assembly is basically a reverse of disassembling sequence.

### CYLINDER HEAD

*REMOVAL & INSTALLATION*

#### 3.6L Engine

*See Figures 68 through 73.*

1. Remove the engine. See "Engine Assembly" in this section.

➡**All cable ties which are opened or cut open when removing, must be replaced in the same position when installing.**

➡**When the engine is installed in the engine compartment, some components cannot be removed or can only be removed with great difficulty.**

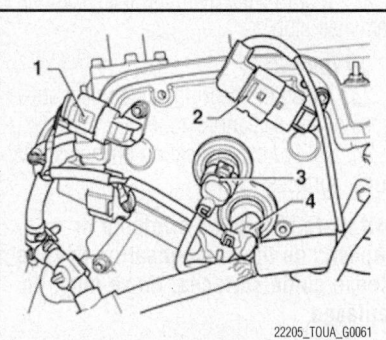

**Fig. 68 Showing the sensors and other connectors on the cylinder head front—3.6L engine**

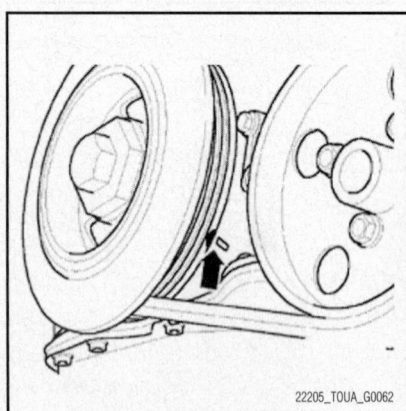

**Fig. 69 Showing cylinder No. 1 TDC marking alignment—3.6L engine**

**Therefore determine which components are faulty before removing engine.**

2. If equipped with a one-piece intake manifold, remove the intake manifold. See "Intake Manifold."

3. If equipped with a two-piece intake manifold, remove the intake manifold upper section. See "Intake Manifold."

4. For all configurations, remove the pump plunger.

5. Unscrew the heat shield, with the intake manifold support, from the cylinder head.

6. Disconnect the connectors from the Camshaft Position (CMP) sensor 1 and Camshaft Position (CMP) sensor 2.

7. Disconnect the harness connector from the camshaft adjustment valve 1 and the camshaft adjustment valve 1 (exhaust).

8. Mark all wiring prior to removal, then pull off or disconnect all remaining electrical wires required from cylinder head and set aside.

9. Remove the bracket from cover.

10. Remove the vacuum pump, if equipped.

11. Remove the three bolts on the end of the coolant connection.

12. Remove the cylinder head cover.

13. Rotate the crankshaft at the vibration damper bolt, in the direction of engine rotation, to align the cylinder 1 TDC marking.

14. The cams of cylinder 1 must face each other (pointing slightly inward).

15. Remove the chain tensioner (bolt) for camshaft timing chain from the side of the cylinder head.

16. Remove both bolts from the sealing flange and then all bolts from the timing chain cover.

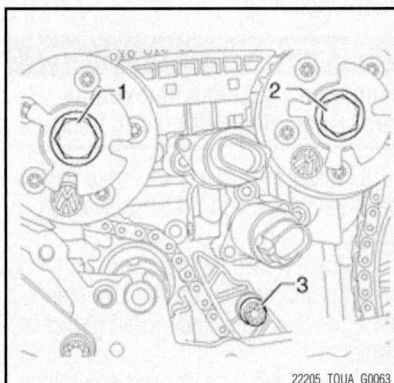

**Fig. 70 Remove bolts "1" and "2" and then remove the bolts of the guide rail "3". Remove the camshaft adjuster—3.6L engine**

➡ **Only counter-hold at the camshaft with a spanner wrench (27 mm). The camshaft bar (T10068 A) must not be inserted when tightening or loosening the camshaft adjuster.**

17. Remove bolts "1" and "2" and then remove the bolts of the guide rail "3". Remove the camshaft adjuster.

18. Using a screwdriver at cylinder head, pry the bearing shaft forward slightly and pull it out.

19. Now, the drive pinion for the high pressure pump can be removed.

20. Remove the cylinder head bolts, in the shown sequence, starting from the outside and working toward the inside.

21. Remove the four bolts from the control housing and pull down the control housing from camshafts.

22. Cover the camshaft ends with clean paper strips and wrap ends with adhesive tape.

23. Assemble lifting tackle and carefully lift off the cylinder head.

24. When the head is removed, note the following:

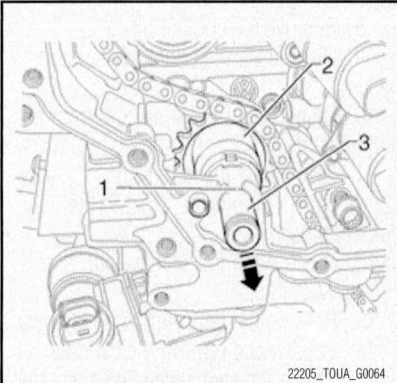

**Fig. 71 Using a screwdriver at cylinder head "1", pry the bearing shaft "3" forward slightly and pull it out, then the drive pinion for the high pressure pump "2" can be removed—3.6L engine**

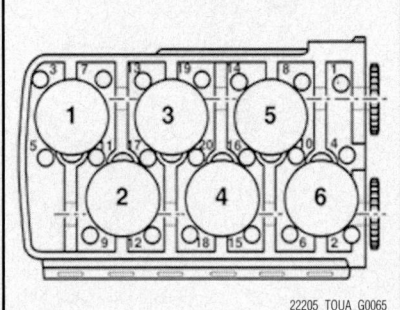

**Fig. 72 Cylinder head bolt removal/ installation sequence—3.6L engine**

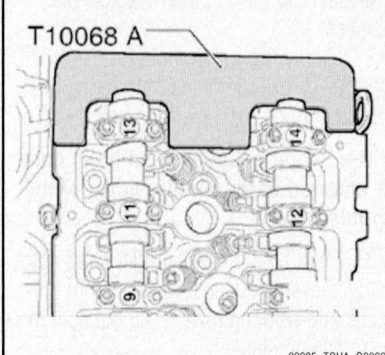

**Fig. 73 It must be possible to insert the camshaft bar (T10068 A) into both shaft grooves**

a. Stuff clean cloths into the cylinders so that no dirt or abrasive powder can get between cylinder wall and piston.

b. Do not allow dirt or abrasive powder to get into coolant either.

c. Carefully clean cylinder head and cylinder block sealing surfaces. Avoid introducing scratches or scoring (do not use sandpaper with grit below "100").

d. Clean all threaded bores for cylinder head bolts.

e. Check the cylinder head for warping

### To install:

25. If needed, adjust the crankshaft at the vibration damper bolt in the direction of engine rotation to the cylinder 1 TDC marking. Have a second technician guide the camshaft timing chain by hand when doing this.

26. Make sure that the alignment bushings are inserted in bores 12 and 20 in the cylinder block.

27. Place a 2 mm thick sealant bead of sealant (D 176 501 A1) on to the partition of the cylinder block and sealing flange.

➡ **Only remove the new cylinder head gasket from its packing immediately before installing. Handle the new gasket with extreme care. Damaging will lead to leaks.**

28. Immediately place the new cylinder head gasket in position. The text side (replacement part number) must be visible.

29. Also fill some more sealant (D 176 501 A1) into both 3 mm bores, which lie on the sealant bead.

30. Position the camshafts in the cylinder head so TDC is set for No. 1 cylinder. 1. The cams of cylinder 1 must face each other (point slightly inward).

31. It must be possible to insert the camshaft bar (T10068 A) into both shaft

grooves. Now the cylinder head can be installed.

32. Apply locking compound (D 197 300 A2) to each cylinder head bold in the head contact surface and insert the bolts.

➡ **The longer cylinder head bolts must be inserted in the middle holes of the cylinder head.**

33. Tighten the new cylinder head bolts, in the shown sequence, starting from the inside and working toward the outside, in the following steps:

a. Pre-tighten all the bolts to 11 ft. lbs. (15 Nm).

b. Tighten all the bolts to 22 ft. lbs. (30 Nm).

c. Tighten all bolts an additional 180 degrees.

34. The rest of the assembly is basically a reverse of the disassembling sequence.

➡ **There is no requirement to retighten the cylinder head bolts after repairs.**

## 4.2L 5-Valve Engine (Code AXQ, BHX)

*See Figures 74 through 78.*

1. Remove the engine. See "Engine Assembly" in this section.

➡ **All cable ties which are opened or cut open when removing, must be replaced in the same position when installing.**

2. Remove the front exhaust pipe from the exhaust manifold.

3. Remove the oil dipstick guide tube.

4. Disconnect the connectors "1", "2" and "3".

5. Remove the toothed belt and camshaft gear. See "Timing Belt, Sprockets, Front Cover and Seal" in this section.

6. Remove the rear toothed belt cover.

7. Disconnect the vacuum hose from the combination valve. Remove the bolts,

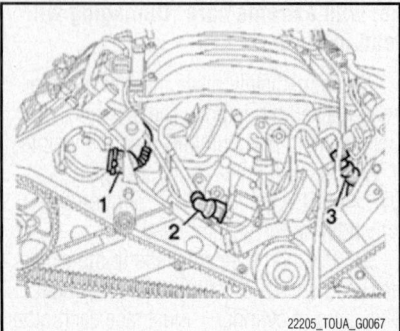

**Fig. 74 Disconnect the connectors "1", "2" and "3"—4.2L 5-valve engine**

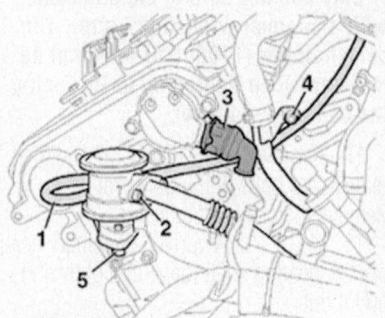

1. Vacuum Hose
2. Bolts
3. Connector
4. Bolt/Oil Pipe
5. Bolts/Combination Valve

22205_TOUA_G0068

**Fig. 75 Disconnect the vacuum hose from the combination valve. Remove the bolts, disconnect the connector, remove the oil pipe, and then remove the combination valve—4.2L 5-valve engine**

disconnect the connector, remove the oil pipe, and then remove the combination valve.

8. Remove the coolant pipe.

9. Disconnect the knock sensor connectors.

10. Disconnect the connectors from all fuel injectors.

11. Disconnect the connector "1", hose "2", bolts "8", connector "9", and the coolant pipe.

12. Remove the intake manifold. See "Intake Manifold."

13. Remove the left cylinder head cover.

14. Loosen and remove the cylinder head bolts, in the sequence shown, starting from the outside working toward the inside.

15. Carefully lift the cylinder head off.

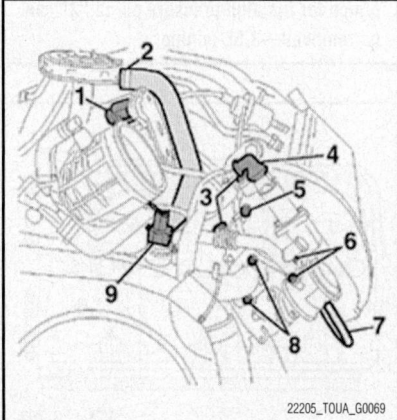

22205_TOUA_G0069

**Fig. 76 Disconnect the connector "1", hose "2", bolts "8", connector "9", and the coolant pipe**

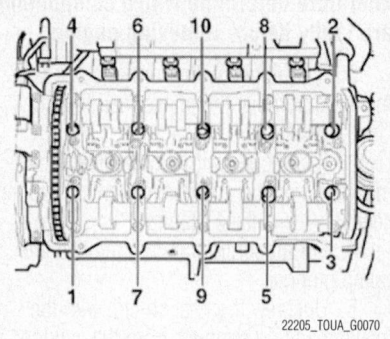

22205_TOUA_G0070

**Fig. 77 Cylinder head bolt loosening sequence—4.2L 5-valve engine**

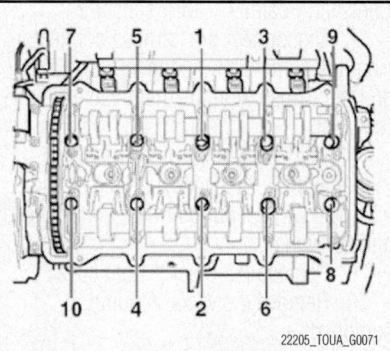

22205_TOUA_G0071

**Fig. 78 Cylinder head bolt tightening sequence—4.2L 5-valve engine**

16. Stuff clean cloths into the cylinders so that no dirt or abrasive powder can get between cylinder wall and piston.

17. Carefully clean the cylinder head and cylinder block sealing surfaces. This ensures that no scoring or scratches are formed (when using abrasive paper, the grade must not be less than 100).

*To install:*

➡ **Only remove the new cylinder head gasket from its packing immediately before installing.**

### ✳✳ CAUTION
**Handle new gasket with extreme care. Damaging will lead to leaks.**

18. Install the new cylinder head gasket. The inscription (Part No.) must be facing upward.

19. Ensure that the alignment bushings are inserted into the cylinder block and that the cylinder head gasket is seated.

20. Install the cylinder head. Insert the new cylinder head bolts and hand tighten.

21. Tighten the cylinder head bolts for each step, in the tightening sequence shown, working from inside to outside. Tighten the bolts in four steps:

- Step 1: Pre-tighten all bolts to 26 ft. lbs. (35 Nm).
- Step 2: Tighten all bolts to 44 ft. lbs. (60 Nm).
- Step 3: Tighten all bolts and additional 90 degrees, using a rigid wrench.
- Step 4: Tighten all bolts again an additional 90 degrees.

22. Adjusting the valve timing. See "Valve Timing."

23. Install the cylinder head cover.

24. Install the intake manifold. See "Intake Manifold."

25. The rest of assembly is in reverse order of disassembling.

➡ There is no requirement to re-tighten the cylinder head bolts after repairs.

## 4.2L 4-Valve Engine (Code BAR)

*See Figures 79 through 85.*

➡ All cable ties opened or cut during engine removal must be reinstalled at the same locations during installation.

1. Drain the coolant.
2. Remove the engine. See "Engine Assembly."
3. Separate the engine and transmission.

4. If removing the left cylinder head, remove the left primary catalytic converter.

5. Remove the electrical connectors for the knock sensors 3 and 4 from the bracket.

6. Remove the air guide hose from the AIR combination valves.

7. Remove the intake manifold. See "Intake Manifold."

8. Remove left and right coolant pipes.

9. Remove the rear coolant pipe.

10. Remove the camshaft timing chains from the camshafts. See "Timing Chain, Sprocket, Front Cover and Seal" in this section.

11. Remove the right chain tensioner.

12. Disconnect the electrical connectors at fuel injectors.

13. Remove the high pressure line from the connector on the fuel rail. To do this, counter-hold at the hex head with an open-end wrench and loosen the union nut.

➡ Do not change bent shape of high pressure lines.

14. Remove the locking bolt as indicated.

15. Loosen the cylinder head bolts with a polydrive bit and drive socket (T10070) in the sequence shown.

16. Remove the cylinder head bolts and carefully remove cylinder head. Lay cylinder head on a soft surface (foam).

### To install:

Cylinder head installation notes:
- Replace cylinder head bolts.
- During assembly, replace self-locking nuts and bolts.
- Always replace bolts that are tightened to torque as well as O-rings and gaskets.
- Carefully remove sealant residue from

Fig. 83 Install the locking bolt as indicated

Fig. 84 Check whether the camshaft timing chain tensioner guide rail is secured, with a locking pin (T40071)—4.2L 4-valve engine

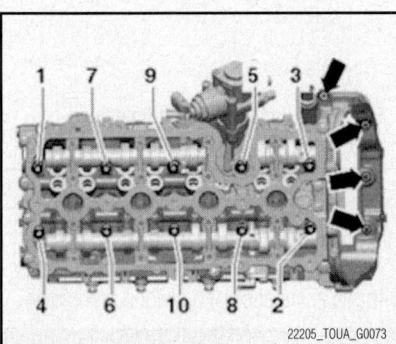

Fig. 79 Remove the locking bolt as indicated

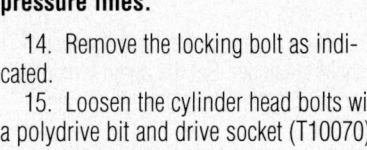

Fig. 81 Crankshaft holder (3242) must be installed

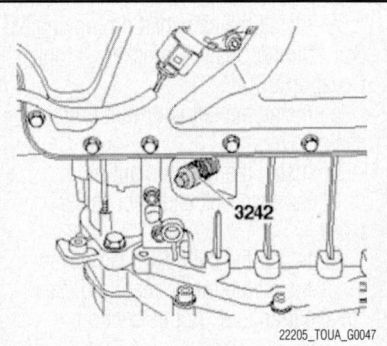

Fig. 80 Cylinder head bolt removal sequence—4.2L 4-valve engine

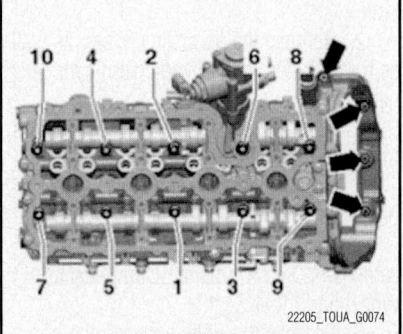

Fig. 82 Cylinder head bolt tightening sequence—4.2L 4-valve engine

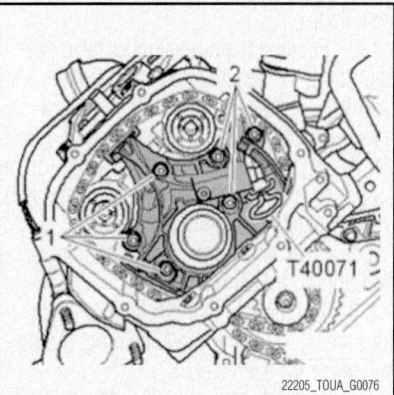

Fig. 85 Tighten bolts "1" and "2"—4.2L 4-valve engine

cylinder head and cylinder block. Make sure that no long scrapes or scratches result.

• Carefully remove all grinding and sanding residue.

• There must be no oil or coolant in the blind holes for the cylinder head bolts in the cylinder block.

• Only unpack new cylinder head gasket immediately prior to installation.

• Handle gasket carefully. Damages to the silicone layer and in areas of recesses may result in leaks.

• Cylinder heads with cracks between the valve seats, or between the valve seat and the spark plug threads, can continue to be used without reducing the service life, as long as the cracks have a width of max. 0.3 mm, or only the first 4 threads of the spark plug threads are cracked.

• After installing a replacement cylinder head with camshafts installed, oil contact surfaces between roller rocker levers and cam lubricating surfaces after installing cylinder head.

• Do not remove plastic bases protecting exposed valves until immediately before installing cylinder head.

• Secure all hose connections with hose clamps appropriate for this model.

• During installation, all cable ties must be reinstalled at the same location.

• After working on the valve train and lifters, carefully rotate the crankshaft by hand at least 2 full revolutions before starting to be sure that valves do not strike the pistons.

• When replacing the cylinder head or cylinder head gasket, coolant must be completely replaced.

17. Ensure that the camshafts of both cylinder heads are in "TDC" position.

18. Camshaft clamp (T40070) must be installed on both cylinder heads and tightened to 18 ft. lbs. (25 Nm).

19. Crankshaft holder (3242) must be installed.

20. Position the new cylinder head gasket. Pay close attention to the alignment bushings in the cylinder block. Also, pay attention to the installation position of cylinder head gasket, marking "oben" (top) (or the part number must face toward cylinder head).

21. Install cylinder head.

22. Insert all new cylinder head bolts and tighten by hand.

23. Tighten the cylinder head bolts in four steps, following the shown tightening sequence for each step:

• Step 1: Using a torque wrench, tighten to 22 ft. lbs. (30 Nm).

• Step 2: Using a torque wrench, tighten to 44 ft. lbs. (60 Nm).

• With a Torx key, tighten an additional 90 degrees.

• With a Torx key, tighten an additional 90 degrees.

➡ **There is no requirement to retighten the cylinder head bolts after repairs.**

24. Insert bolts around the bellhousing after applying locking compound. Tighten to 6 ft. lbs. (8 Nm).

25. Again, tighten the cylinder head bolts an additional 90 degrees, using a ratchet.

26. Tighten the locking bolt.

27. Check whether the camshaft timing chain tensioner guide rail is secured, with a locking pin (T40071).

a. If the tensioning element is to be removed from the chain tensioner, observe the installed position:

b. The hole in the housing floor faces toward chain tensioner and the piston faces toward the tensioning rail.

c. Clean chain tensioner oil screen, if necessary.

28. Place a new gasket onto the rear of the chain tensioner. Set the chain tensioner in place and install the camshaft timing chain.

29. Tighten bolts "1" and "2".

30. The rest of the installation is in reverse order of removal, note the following:

a. Install the camshaft timing chains. See "Timing Chain, Sprockets, Front Cover and Seal".

b. Install the left and right timing chain covers.

c. Install the rear coolant pipe.

d. Install the left and right coolant pipes.

e. Install the cylinder head cover.

f. Install the intake manifold. See "Intake Manifold" in this section.

### 5.0L TDI Engine

*See Figures 86 through 88.*

1. Remove the engine and transmission.

2. Remove the acoustic covers, as well as the cylinder head covers, from cylinder heads.

3. Remove the tandem and fuel pumps from the cylinder heads.

4. Set the crankshaft to TDC for Cylinder 1.

5. Remove the drive gears for the camshafts. See "Camshaft Compensation Wheel and Input Gear."

6. Remove the bolts between the control housing and the cylinder head of cylinder bank 1 and 2.

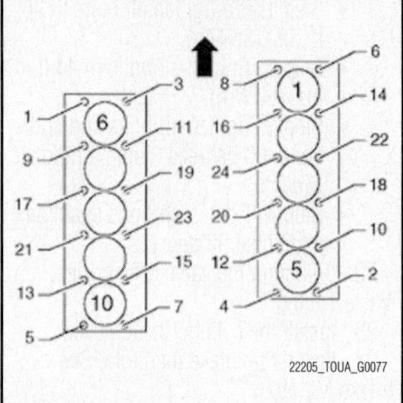

**Fig. 86 Cylinder head bolt loosening sequence—5.0L TDI Engine**

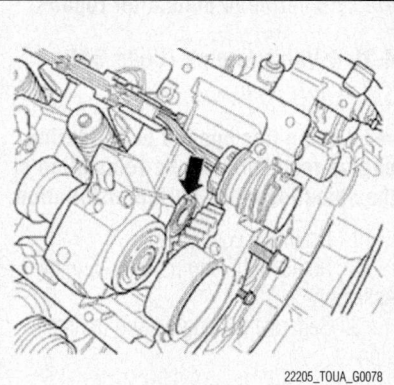

**Fig. 87 Install the cylinder head according to the position of the compensation linkage—5.0L TDI Engine**

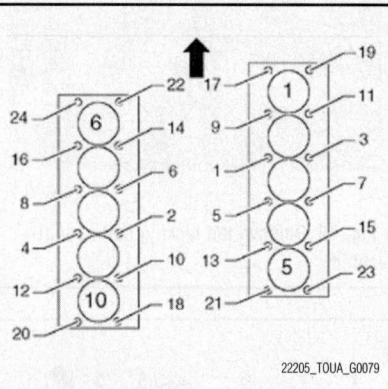

**Fig. 88 Cylinder head bolt tightening sequence—5.0L TDI Engine**

7. Loosen and remove the cylinder head bolts in the given sequence.

8. Carefully remove the cylinder heads with the assistance of a second technician.

❖❖ **CAUTION**

**The cylinder heads must be carefully guided to prevent damages.**

9. Place clean rags into cylinders so that no dirt or emery cloth particles can get in between cylinder wall and piston. Do not allow dirt or abrasive powder to get into coolant either.

10. Carefully clean cylinder head and cylinder block sealing surfaces. Avoid introducing scratches or scoring (do not use sandpaper with grit below 100).

**To install:**

11. Note the following during installation:

    a. Always replace the cylinder head bolts.

    b. Only unpack the new cylinder head gasket immediately prior to installation.

    c. Install the cylinder head gasket with the same "thickness" index as the one removed.

    d. Handle the gasket carefully. Damages to the silicone layer and in areas of recesses may result in leaks.

12. Before installing the cylinder heads, position the crankshaft to TDC for cylinder 1.

13. Turn back the crankshaft in the opposite direction of engine rotation direction until all pistons stand almost evenly below TDC.

    a. Set cylinder head gasket in place.

14. Install the cylinder head according to the position of the compensation linkage.

15. Insert the cylinder head bolts with their associated washers and tighten hand-tight.

➡**All cylinder head bolts, except number "1", "2", "17" and "19" in the tightening sequence must be installed with special notched washers.**

16. Tighten the cylinder head bolts, in the sequence indicated, in four steps as follows:

- Step 1: (30 Nm), using a torque wrench
- Step 2: (60 Nm), using a torque wrench
- Step 3: Plus 180 degrees additional turn, using a solid wrench
- Step 4: Plus 180 degrees additional turn, using a solid wrench

17. Install the drive gears for camshafts. See "Camshaft Compensation Wheel and Input Gear."

18. Adjust the valve timing. See "Valve Timing" in this section.

19. Install the tandem and fuel pumps from the cylinder heads.

20. Install the acoustic covers, as well as the cylinder head covers, to the cylinder heads.

21. Install the engine and transmission. See "Engine Assembly."

## ENGINE ASSEMBLY

### REMOVAL & INSTALLATION

**3.6L Engine**

*See Figures 89 through 95.*

1. Note the following prior to and during removal:

    a. To allow free rotation of the driveshaft, move the selector lever in the "N" position.

    b. Leave the vehicle key in the ignition lock to prevent steering wheel lock from locking.

    c. It is advisable to remove the front wheels before removing the engine/transmission assembly. The vehicle can be lowered on the hoist until the cover plates of the brake discs are just above the floor. This enables the most ergonomic work position possible regarding accessibility of components in the engine compartment.

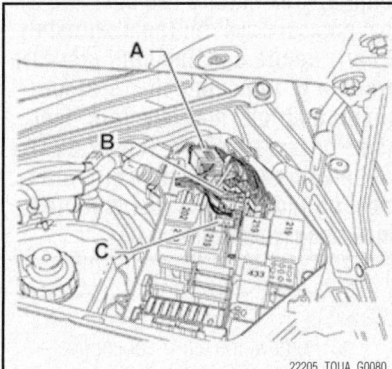

Fig. 89 Open the fuse box cover on left side of the plenum chamber and separate connectors "A", "B", and "C"—3.6L Engine

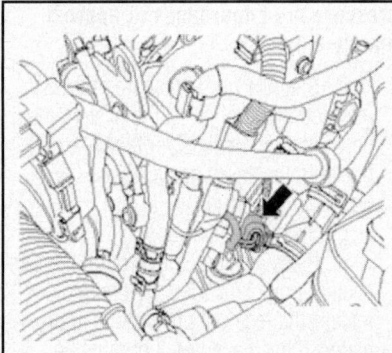

Fig. 90 Remove the electrical connector from fuel pressure regulator valve—3.6L Engine

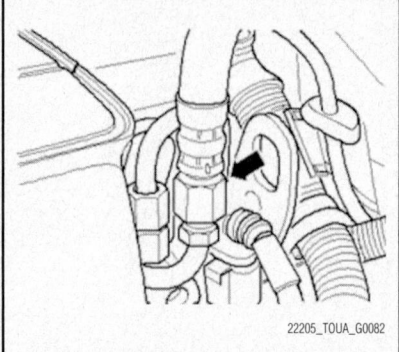

Fig. 91 Lay a cloth around the threaded connection and loosen the union nut from the fuel supply hose—3.6L Engine

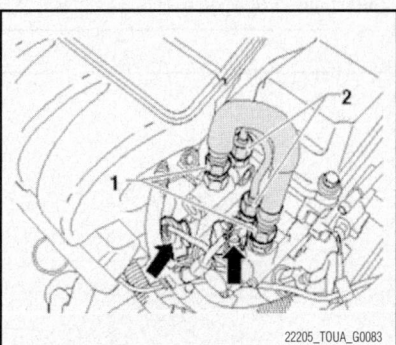

Fig. 92 Remove the connector and remove the low pressure fuel line and the high pressure line—3.6L Engine

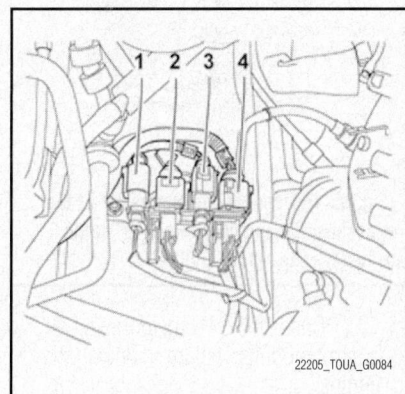

Fig. 93 Disconnect the oxygen sensor connectors "1" through "4" and lay them on the engine—3.6L Engine

    d. To prevent damage to the removed components, use a proper parts container (V.A.G 1698) for storage.

    e. When the engine is installed in the engine compartment some components cannot be removed or can only be removed with great difficulty. Therefore determine which components are faulty before removing engine.

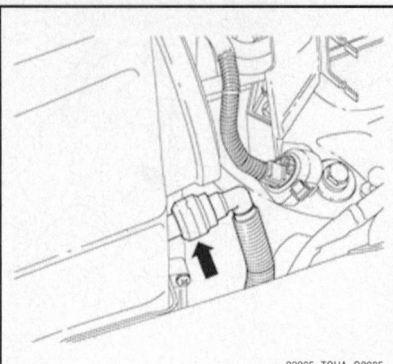

**Fig. 94 Carefully pry off the green circlip "1" using a screwdriver, and then, press the clamping ring "2" in the arrow direction—3.6L Engine**

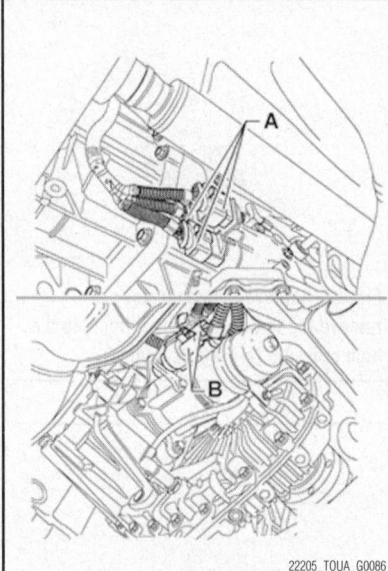

**Fig. 95 Disconnect connectors "A" at the transmission and connectors "B" at the transfer case—3.6L Engine**

f. Check the DTC memories of all control modules, before removing the engine.

2. Turn off the ignition and all electrical consumers. Disconnect the battery under the driver's seat.

3. Remove the front wheels.

4. Remove the windshield wiper arms.

5. Remove the plenum chamber cover.

6. Pull the engine compartment cover seal off the bulkhead.

➡**All cable ties which are opened or cut open when removing engine, must be replaced in the same position when installing engine.**

7. Remove the covers at each end above the bulkhead.

8. Disconnect the small connector of the engine control module and disconnect the ground connection of the wiring harness.

9. Open the fuse box cover on left side of the plenum chamber and separate connectors "A", "B", and "C".

10. Remove the wiring harness from the plenum chamber and place it on the engine.

11. Disconnect the starter and generator ground wires and place them on the engine.

**⁑ WARNING**

**Fuel supply lines are under pressure! Wear protective goggles and protective gloves to avoid damage and contact with skin. Before removing from any hose connection, wrap a cloth around the connection, and then release the pressure by carefully opening the fuel line.**

**⁑ CAUTION**

**The fuel injection system is divided into a high pressure section (maximum approximately 120 bar) and a low pressure system (approximately 6 bar). Before opening high pressure area - e.g. removing high pressure pump, fuel rail, fuel injectors, fuel pipes or fuel pressure Sensor G247 - fuel pressure in high pressure area must be reduced to a residual pressure of approximately 6 bar.**

12. Release the fuel system pressure as follows:

a. Remove the electrical connector from fuel pressure regulator valve.

b. Allow the engine to idle approximately 10 seconds.

➡**When the fuel pressure regulator valve electrical connector is disconnected during idle, pressure in the high pressure area decreases to approximately 6 bar.**

c. After the high pressure has been released, the high pressure system must be opened immediately; otherwise, the pressure increases again due to the warming of the fuel.

d. Switch off ignition.

e. Place a clean cloth around the connection point and carefully open to release the residual pressure of approximately 6 bar. Escaping fuel must be absorbed.

f. To conclude this procedure, check the DTC memory of the Engine Control Module (ECM); erase all DTC entries

which may have occurred from removing the connector.

➡**If the DTC memory was erased, generate readiness code "Guided Function", using a scan tool.**

13. Lay a cloth around the threaded connection and loosen the union nut from the fuel supply hose. To do so, counter-hold on the fuel supply line using a wrench.

14. Remove the connector and remove the low pressure fuel line and the high pressure line. When doing this, counter-hold on the connection of the high pressure pump using a wrench.

15. Remove the vacuum hoses from the intake manifold.

16. Remove the intake hose from the Mass Air Flow (MAF) sensor and throttle valve control module.

17. Remove the throttle valve control module.

18. Disconnect the oxygen sensor connectors "1" through "4" and lay them on the engine.

➡**Cylinders 1 through 3 have black connectors; cylinders 4 through 6 have brown connectors.**

19. Separate the connecting line to the air suspension compressor at the air filter as follows:

a. Carefully pry off the green circlip "1" using a screwdriver. Then, press the clamping ring "2" in the arrow direction.

b. Pull only the slackened line from the connection at the air filter.

20. Now remove the upper part of air filter housing with the Mass Air Flow (MAF) sensor.

21. Unclip the transmission breather line from the air filter housing.

22. Evacuate refrigerant from air conditioning system, using an approved recovery/recycling A/C service center.

23. Remove the noise insulation tray.

24. Remove the heat shield the on steering mechanism and loosen the universal joint at the lower end of the steering column.

25. Unclip the selector lever cable for the transmission.

26. Drain the cooling system.

27. Pull upper coolant hose off the radiator, and lower the coolant hose off the coolant pipe.

28. Disconnect the lines from the transmission oil cooler at the lower right. Catch any fluid which runs out.

29. Disconnect the line from the power steering fluid cooler below on the left. Catch any fluid which runs out.

30. Now, remove the mount from the catalytic converter on the vehicle floor.

31. Loosen the double clamp between the catalytic converter and the center muffler and slide back to the center muffler.

**➡The center muffler can remain installed.**

32. Disconnect connectors "A" at the transmission and connectors "B" at the transfer case.

33. Remove the driveshaft from the transmission.

34. Remove the front wheel housing liners.

35. In the wheel housings, separate all connectors between the body and front axle.

36. Loosen the brake line bracket and remove the brake carrier from the wheel bearing housing.

37. Suspend the brake caliper with wire in wheel housing without bending brake lines.

38. Remove the stabilizer bar bolts.

39. Remove the transmission transverse support (crossmember).

40. Remove the ground wire on the front right longitudinal member.

41. Remove one of the rear longitudinal member mounting bolts.

42. Use jack stands and wooden blocks to fully support the subframe at both sides of the vehicle body, and then remove the other rear longitudinal member mounting bolt.

43. Using a suitable lift table, mount the front supports on the lift table. Support both axle supports to the lift table. Rotate the turntables of the axle supports downward. Place the supports for the subframe and transmission console into the corresponding positions on the lift table.

44. Remove the subframe bolts and the transmission transverse support (crossmember).

45. Slowly lower the engine/transmission assembly, constantly observing clearance.

46. Now, install a bolt and nut from the subframe fastener through the empty bore on both sides. Tighten the engine carrier to the subframe. This prevents the engine carrier from slipping.

***To install:***

## ❄❄ CAUTION

**When positioning the engine/transmission assembly into the body, it is always necessary to bring the engine carrier into contact on the body using the scissor lift table. Pulling the**

**engine carrier up with bolt torque damages thread inserts!**

47. Installation is performed in the reverse order. When doing this note the following:

a. Make sure the alignment sleeves for centering the engine/transmission are installed in the cylinder block. Install if necessary.

b. If the subframe with suspension assembly has been removed from the assembly platform, position the axle assembly with subframe onto the prepared scissor lift table.

c. Set engine on assembly carrier.

d. If transmission is not yet bolted on, support the engine properly and install the transmission to the engine.

e. When installing the engine, use scissor lift table to lift so far that the selector cable can be attached.

f. Note the following torque specifications during installation:

- Bolts and Nuts M6: 8 ft. lbs. (10 Nm)
- Bolts and Nuts M7: 11 ft. lbs. (15 Nm)
- Bolts and Nuts M8: 18 ft. lbs. (25 Nm)
- Bolts and Nuts M10: 30 ft. lbs. (40 Nm)
- Bolts and Nuts M12: 44 ft. lbs. (60 Nm)
- Engine bracket to engine mount (nut), M10: 55 ft. lbs. (75 Nm)
- Engine mount to engine subframe (bolt), M10: 44 ft. lbs. (60 Nm)
- Subframe to chassis (bolt), M12: 74 ft. lbs. (100 Nm), plus an additional 180 degrees
- Driveshaft to transmission rear final drive: 29 ft. lbs. (39 Nm)
- Transmission carrier to body: 27 ft. lbs. (37 Nm)

g. Top off the ATF level.

h. Reconnect all lines, hoses and connections that were disconnected for the removal sequence.

i. Connect the A/C lines and properly recharge the A/C system.

j. Fill up the power steering fluid.

k. Fill the engine oil.

l. Install the selector lever cable connection.

m. Ensure the proper electrical connections and routing.

n. Fill the coolant system.

o. After installing the assembly, perform a vehicle alignment.

p. Perform a road test and check all DTC memories.

## 4.2L 5-Valve Engine (Code AXQ, BHX)

*See Figures 96 through 101.*

1. Check DTC memories of all control modules, before removing engine.

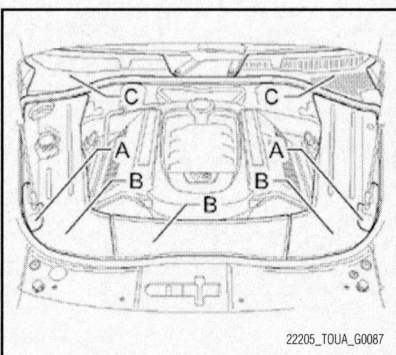

22205_TOUA_G0087

**Fig. 96 Remove covers "A", "B" and "C" in the engine compartment—4.2L 5-valve engine**

22205_TOUA_G0088

**Fig. 97 Disconnect fuel supply line, then, from the other line, disconnect breather line to solenoid valve in engine compartment—4.2L 5-valve engine**

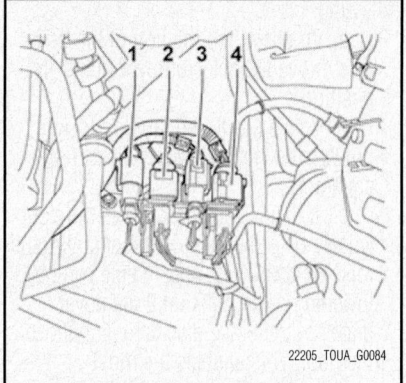

22205_TOUA_G0084

**Fig. 98 Disconnect the connectors "1" through "4" for the oxygen sensors and lay the connectors aside on the engine—4.2L 5-valve engine**

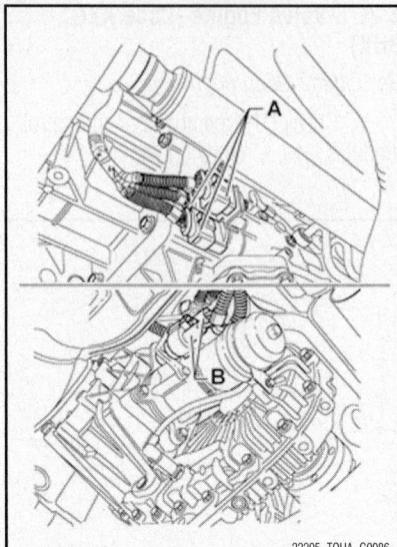

**Fig. 99 Separate connectors "A" on the transmission and "B" on the transfer case. Loosen the selector lever cable—4.2L 5-valve engine**

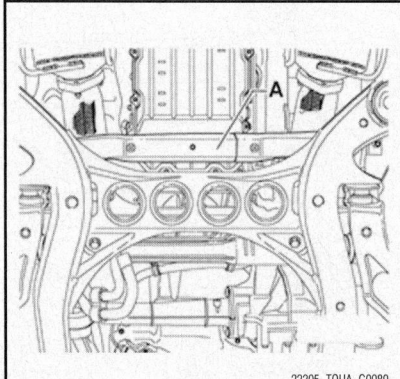

**Fig. 100 Remove transmission transverse support "A"—4.2L 5-valve engine**

2. Before and during removal, note the following:

   a. To allow free rotation of the drive axle, move the selector lever in the "N" position.

   b. Leave key in the ignition lock to prevent the steering wheel lock from engaging.

   c. It is advisable to remove the front wheels before removing the engine/transmission assembly. The vehicle can be lowered on the hoist until the cover plates of the brake discs are just above the floor. This enables the most ergonomic work position possible regarding accessibility of components in the engine compartment.

   d. When the engine is installed in the engine compartment some components cannot be removed or can only be removed with great difficulty. Therefore determine which components are faulty before removing engine.

   e. To prevent damage to the removed components, use a proper container for removed parts as storage.

   f. All cable ties which are opened or cut open when removing engine, must be replaced in the same position when installing engine.

3. Switch off all electrical accessories. Turn the ignition switch to OFF. Disconnect the battery under the driver's seat.

4. Remove the left and right wiper arms.

5. Pull the engine compartment cover seal off bulkhead.

6. Remove covers "A", "B" and "C" in the engine compartment.

### ☀☀ WARNING

**Fuel supply lines are under pressure! Wear protective eye wear and gloves to prevent injury and skin contact. Before loosening the hose connections, place a rag around the connection point. Then release pressure by carefully pulling hose off connection.**

7. Place a rag around connection to catch escaping fuel.

8. Disconnect fuel supply line, then, from the other line, disconnect breather line to solenoid valve in engine compartment.

9. Disconnect the small connector from the engine control module (ECM) and disconnect the ground connection of the wiring harness.

10. Open the fuse box cover on the left in the plenum chamber and disconnect the three main wiring connectors. Remove the

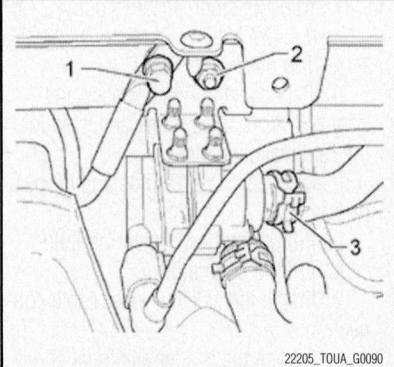

**Fig. 101 Remove the ground wire "1" from the long member at right front. Pull off the coolant hose "3" and unfasten the coolant pump with bracket "2"—4.2L 5-valve engine**

wiring harness from the plenum chamber and set it on the engine.

11. Disconnect the starter ground wire and lay it aside on the engine.

12. Disconnect the connector from the left mass air flow (MAF) sensor, and remove the intake hose and left air filter housing.

13. Disconnect the connectors "1" through "4" for the oxygen sensors and lay the connectors aside on the engine.

➡**Connectors for cylinder bank 1 are Black, and connectors for cylinder bank 2 are Brown.**

14. Remove the right intake hose to the throttle valve control module.

15. Pull the intake hose to Secondary Air Injection (AIR) Pump Motor off from the upper part of the air filter and remove the upper part of air filter.

16. Remove the air filter element.

17. Pull off the connecting wire to air suspension compressor on the right of the air filter housing.

18. Unclip the transmission breather line from the air filter housing.

19. Remove the bracket for the right engine cover.

20. Pull the lower part of the air filter off to the right from the air duct and remove it.

21. Discharge and recover the A/C system refrigerant.

22. Remove the sound insulation tray.

23. Loosen the double pipe clamp between the catalytic converter-center muffler and push the clamp forward.

24. Support the exhaust system between the rear muffler and center muffler with a transmission jack. Remove the mountings and lower the exhaust system.

25. Remove the driveshaft from the transmission and rear final drive.

26. Remove the front wheels.

27. Remove the front wheel housing liners.

28. Separate the hose connection to the vacuum reservoir, in front of the left wheel housing.

29. Remove the heat shield on the steering gear and loosen the universal joint from the lower end of the steering column.

30. Separate connectors "A" on the transmission and "B" on the transfer case. Loosen the selector lever cable.

31. Disconnect the brake lines in the wheel housing at the brake hose and catch the brake fluid which runs out.

32. In wheel housings, separate all connectors between the body and front axle.

33. Remove the connection lines for the air suspension at the strut.

34. Remove transmission transverse support "A".

35. Open and close the cap on the expansion tank, to relieve cooling system pressure.

36. Drain the coolant.

37. From the area of the left wheel housing, separate the hose connections for the heater core and catch escaping coolant.

38. Pull the upper coolant hose off radiator, and lower coolant hose off the coolant pipe.

39. Disconnect the lines from the transmission oil cooler at lower right. Catch fluid which runs out.

40. Disconnect the line from the power steering fluid cooler below on left. Catch fluid which runs out.

41. Remove the ground wire "1" from the long member at right front. Pull off the coolant hose "3" and unfasten the coolant pump with bracket "2".

42. Remove one of the rear mounting bolts from the longitudinal member.

43. Support the subframe and suspension with jacks and wooden blocks, then remove the other rear mounting bolt. From the area of the plenum chamber, remove two least accessible bolts from suspension strut on each side of vehicle.

44. If available, prepare a lift table under the engine/transmission assembly. When lift table or other support device is properly in place, remove all of the subframe bolts and transmission crossmember bolts.

45. Carefully lower the engine/transmission assembly from the vehicle.

### To install:

46. Installation is in reverse order of removal. When doing this, note the following:

a. If the engine was separated from transmission. check whether the alignment bushings for centralizing the engine/transmission are equipped in cylinder block, install if necessary.

b. Position the engine/transmission assembly into the vehicle, making sure to avoid damaging any lines, hoses or wiring. At first, raise the assembly only far enough to clip in the selector lever cable, then position it all the way into the vehicle.

c. Reconnect all lines, hoses and electrical connections that were disconnected for the removal procedure.

d. Note the torque specifications as follows:

- Bolts and nuts, M6: 8 ft. lbs. (10 Nm)
- Bolts and nuts, M7: 11 ft. lbs. (15 Nm)

- Bolts and nuts, M8: 18 ft. lbs. (25 Nm)
- Bolts and nuts, M10: 30 ft. lbs. (40 Nm)
- Bolts and nuts, M12: 44 ft. lbs. (60 Nm)
- Engine bracket to engine mount (nut), M10: 55 ft. lbs. (75 Nm)
- Engine mount to engine carrier (bolt), M10: 44 ft. lbs. (60 Nm)
- Engine carrier to chassis (bolt), M12: 74 ft. lbs. (100 Nm), plus an additional 180 degrees
- Drive axles to transmission: 30 ft. lbs. (40 Nm)
- Transmission carrier to body (rear final drive): 25 ft. lbs. (34 Nm)

e. Top off the ATF level.

f. Connect the air conditioning lines and recharge the air conditioning system with refrigerant.

g. Fill up oil for power steering.

h. Fill the coolant system.

i. Fill up with engine oil.

j. Connect the battery.

k. Perform a wheel alignment check.

l. Perform a vehicle system test (on-board diagnostic) at the "Guided Fault Finding" screen.

m. Erase any DTC entries.

n. Generate a readiness code in conjunction with a road test.

o. Perform a road test.

p. After the road test, perform a vehicle on-board system test again and repair any occurring malfunctions.

### 4.2L 4-Valve Engine (Code BAR)

*See Figures 102 through 107.*

1. Check DTC memories of all control modules, before removing engine.

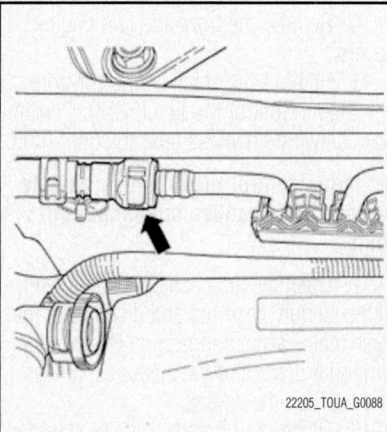

**Fig. 102 Disconnect the breather line to solenoid valve in engine compartment—4.2L 4-valve engine**

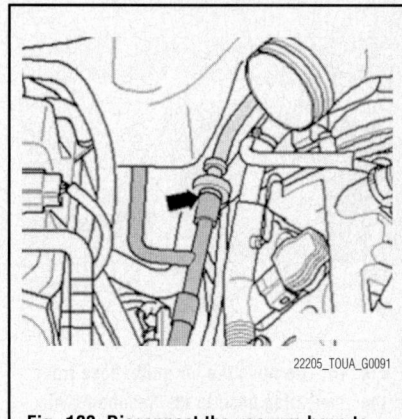

**Fig. 103 Disconnect the vacuum hose to the suction jet pump—4.2L 4-valve engine**

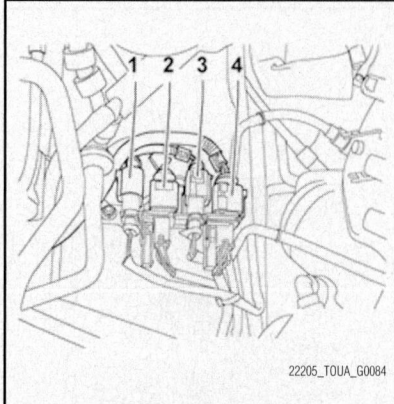

**Fig. 104 Disconnect the connectors "1" through "4" for the oxygen sensors and lay the connectors aside on the engine—4.2L 4-valve engine**

2. Before and during removal, note the following:

a. To allow free rotation of the drive axle, move the selector lever in the "N" position.

b. Leave key in the ignition lock to prevent the steering wheel lock from engaging.

c. It is advisable to remove the front wheels before removing the engine/transmission assembly. The vehicle can be lowered on the hoist until the cover plates of the brake discs are just above the floor. This enables the most ergonomic work position possible regarding accessibility of components in the engine compartment.

d. When the engine is installed in the engine compartment some components cannot be removed or can only be removed with great difficulty. Therefore determine which components are faulty before removing engine.

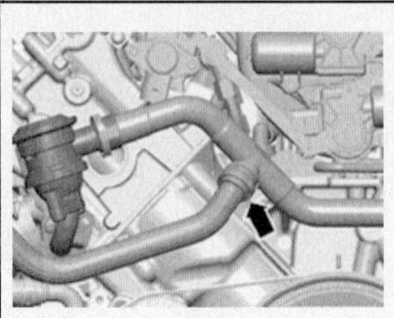

**Fig. 105 Remove the air guide hose from the connecting hose to the Secondary Air Injection (AIR) combination valves—4.2L 4-valve engine**

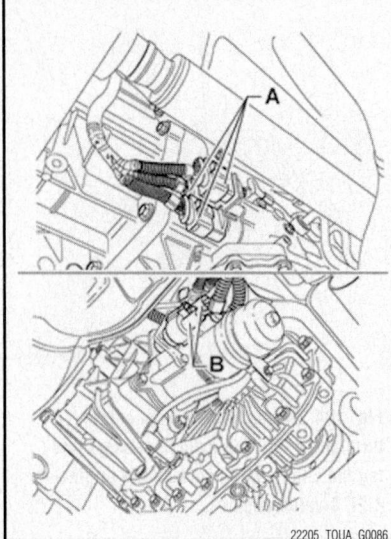

**Fig. 106 Separate connectors "A" on the transmission and "B" on the transfer case—4.2L 4-valve engine**

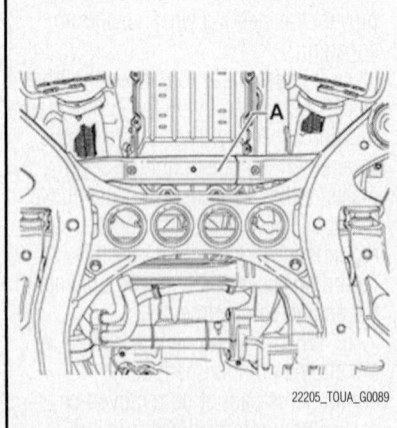

**Fig. 107 Remove transmission transverse support "A"—4.2L 4-valve engine**

e. To prevent damage to the removed components, use a proper container for removed parts as storage.

f. All cable ties which are opened or cut open when removing engine, must be replaced in the same position when installing engine.

3. Relieve the fuel system pressure as follows:

a. Disconnect connector from fuel metering valve on right high pressure pump.

b. Disconnect connector from fuel metering valve 2 on left high pressure pump.

c. Allow engine to idle approximately 10 seconds.

➡ **If the fuel metering valve electrical connectors are disconnected at idle, the pressure in the high pressure area drops to approximately 6 bar. After high pressure has been released, high pressure system must be opened immediately. Otherwise, the pressure increases again due to the warming of the fuel.**

d. Switch off ignition.

e. Place a clean cloth around connection point and carefully open to release the residual pressure of approximately 6 bar. Escaping fuel must be absorbed.

4. Switch off all electrical accessories. Turn the ignition switch to OFF. Disconnect the battery under the driver's seat.

5. Place a rag around connection to catch escaping fuel. Disconnect fuel supply line.

6. Evacuate the A/C system.

7. Pull the engine compartment cover seal off bulkhead.

8. Remove the wiper arms.

9. Remove the plenum chamber cover.

10. Remove the front and rear engine covers.

11. Remove the small connector from the Engine Control Module (ECM). Disconnect the wiring harness ground connection.

➡ **Engine control module may have to be removed because of space restrictions**

12. Open the fuse box cover on the left in the plenum chamber and disconnect the three main wiring connectors. Remove the wiring harness from the plenum chamber and set it on the engine.

13. Disconnect breather line to solenoid valve in engine compartment.

14. Disconnect the starter ground wire and lay it aside on the engine.

15. Extract the hydraulic oil from the

power steering reservoir, using a suitable extracting device.

16. Remove the hose from the power steering fluid reservoir.

17. Disconnect connector from left and right Mass Air Flow (MAF) sensors and remove intake hoses and left and right air filter housing.

18. Disconnect the connecting wire to the air suspension compressor on the right of the air filter housing.

19. Disconnect the vacuum hose to the suction jet pump.

20. Disconnect the connectors "1" through "4" for the oxygen sensors and lay the connectors aside on the engine.

➡ **Connectors for cylinder bank 1 are Black, and connectors for cylinder bank 2 are Brown.**

21. Remove the air guide hose from the connecting hose to the Secondary Air Injection (AIR) combination valves.

22. Unclip the transmission ventilation hose on the brake system vacuum pump.

➡ **In order to reach the coolant lines better, remove left front wheel housing liner if necessary.**

23. Remove the left lower part of the air filter.

24. Remove the refrigerant lines from A/C compressor. Seal open lines and connections at A/C compressor with suitable sealing caps (to prevent dirt and moisture from entering).

25. Remove the front and rear noise insulation.

26. Separate connectors "A" on the transmission and "B" on the transfer case.

27. Loosen the double pipe clamps between the catalytic converter/front muffler assembly and slide them forward.

28. Support the exhaust system between front muffler and rear muffler, using engine/transmission jack, and then remove mountings and lower exhaust system.

29. Remove the rear driveshaft.

30. Remove the ground (GND) cable from the longitudinal frame member at the right front.

31. Remove the selector lever cable heat shield. Press the selector lever cable ball head from the selector shaft lever, using a pry bar.

32. Remove the securing clamps and remove the selector lever cable from the transmission.

33. Disconnect the electrical connector at the steering gear.

34. Remove the heat shield from the steering gear.

35. Separate universal joint from steering gear only when the front wheels are in a straight-ahead position.

➡**Do not change steering wheel position and steering gear position any more, secure steering wheel with adhesive tape.**

36. Remove the universal joint bolt at the lower end of the steering column. Press the universal joint off of the steering gear.

37. Place an collecting and extracting device under the left cooler.

38. Remove the hydraulic line hose from the power steering cooler at left of lock carrier.

39. In the wheel housings, separate all connectors between the body and front axle.

40. Loosen the brake line bracket and remove brake caliper from the wheel bearing housing.

41. Suspend the brake caliper with wire in the wheel housing, without bending the brake lines.

42. Remove the right and left lower bolt on the stabilizer bar. The other bolt remains fastened.

43. Remove transmission transverse support "A".

44. Open and close the cap on the expansion tank, to relieve cooling system pressure.

45. Drain the coolant. Remove both hoses from the coolant reservoir.

46. From the area of the left wheel housing, separate the hose connections for the heater core and catch escaping coolant.

47. Pull the upper coolant hose off radiator, and lower coolant hose off the coolant pipe.

48. Disconnect the lines from the transmission oil cooler at lower right. Catch fluid which runs out.

49. Remove one of the rear mounting bolts from the longitudinal member.

50. Using locating pins (T10300), secure the subframe to the body.

51. Remove the other rear mounting bolt.

52. If available, prepare a lift table under the engine/transmission assembly. When lift table or other support device is properly in place, remove all of the subframe bolts and transmission crossmember bolts.

53. Now, remove the remaining lower bolt on the right and left suspension strut.

54. Remove the upper right and left control arms from wheel bearing housings.

55. Loosen the subframe bolts, as well as the transmission transverse support bolts.

56. Carefully lower the engine/transmission assembly from the vehicle.

57. Install a bolt with nut on the right and left side to prevent the engine carrier from slipping. Secure the engine carrier with the subframe.

*To install:*

**✳✳ CAUTION**

**When installing powertrain in body, it is absolutely necessary to rest engine carrier against body using the lift table (VAS 6131) or other suitable lifting/jacking equipment. By pulling up engine carrier via the mounting bolts, the threaded inserts will be damaged!**

58. Installation is in reverse order of removal. When doing this, note the following:

a. If the engine was separated from transmission. check whether the alignment bushings for centralizing the engine/transmission are equipped in cylinder block, install if necessary.

b. Position the engine/transmission assembly into the vehicle, making sure to avoid damaging any lines, hoses or wiring. At first, raise the assembly only far enough to clip in the selector lever cable, then position it all the way into the vehicle.

c. Reconnect all lines, hoses and electrical connections that were disconnected for the removal procedure.

59. Install the plenum chamber cover.

60. Install the left and right wiper arms.

a. Note the torque specifications as follows:

- Bolts and nuts, M6: 8 ft. lbs. (10 Nm)
- Bolts and nuts, M7: 11 ft. lbs. (15 Nm)
- Bolts and nuts, M8: 18 ft. lbs. (25 Nm)
- Bolts and nuts, M10: 30 ft. lbs. (40 Nm)
- Bolts and nuts, M12: 44 ft. lbs. (60 Nm)
- Engine bracket to cylinder block, M10: 44 ft. lbs. (60 Nm)
- Engine bracket to engine mount (nut), M10: 55 ft. lbs. (75 Nm)
- Engine mount to engine carrier (bolt), M10: 44 ft. lbs. (60 Nm)
- Engine carrier to chassis (bolt), M12: 74 ft. lbs. (100 Nm), plus an additional 180 degrees
- Driveshafts to transmission: 30 ft. lbs. (40 Nm)
- Transmission carrier to body (rear final drive): 25 ft. lbs. (34 Nm)

b. Top off the ATF level.

c. Connect the air conditioning lines and recharge the air conditioning system with refrigerant.

d. Fill up oil for power steering.

e. Fill the coolant system.

f. Fill up with engine oil.

g. Connect the battery.

h. Perform a wheel alignment check.

i. Perform a vehicle system test (on-board diagnostic) at the "Guided Fault Finding" screen.

j. Erase any DTC entries.

k. Generate a readiness code in conjunction with a road test.

l. Perform a road test.

m. After the road test, perform a vehicle on-board system test again and repair any occurring malfunctions.

### 5.0L TDI Engine

*See Figures 108 through 113.*

1. Note the following specifics during engine removal:

a. Before removing engine, check DTC memories of all control modules or scan tool functions

b. To allow free rotation of the drive axle, move the selector lever in the "N" position.

c. Leave key in the ignition lock to prevent the steering wheel lock from engaging.

d. It is advisable to remove the front wheels before the disassembly. The vehicle can be lowered on the hoist until the cover plates of the brake discs are just above the floor. This enables the most ergonomic work position possible regarding accessibility of components in the engine compartment.

e. When the engine is installed in the

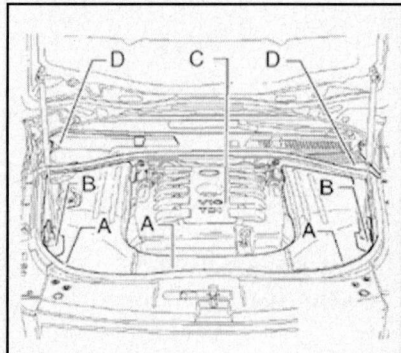

22205_TOUA_G0093

**Fig. 108 Remove covers "A", "B" (if equipped) and "C" in engine compartment and "C" for plenum chamber—5.0L TDI engine**

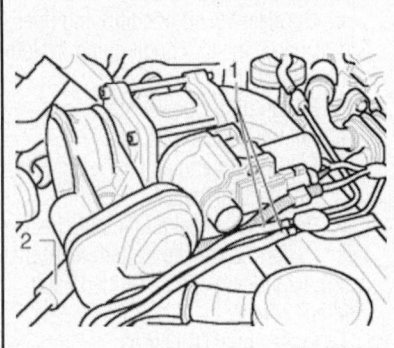

**Fig. 109 Disconnect vacuum lines "1" for vacuum actuators of the bypass flap (mark lines before separating) and brake booster "2"—5.0L TDI engine**

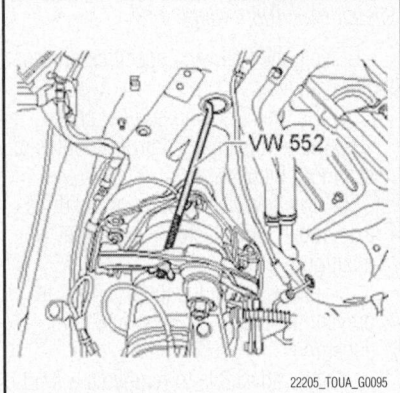

**Fig. 110 Secure suspension struts using Tensioner Hooks (VW 552) or heavy wiring—5.0L TDI engine**

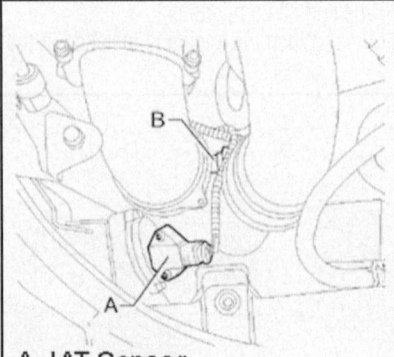

A. IAT Sensor
B. MAF Sensor

**Fig. 111 Disconnect connectors from Intake Air Temperature (IAT) sensor and from Mass Air Flow (MAF) sensors for cylinder both banks—bank 1 shown; bank 2 similar—5.0L TDI engine**

engine compartment some components cannot be removed or can only be removed with great difficulty. Therefore determine which components are faulty before removing engine.

f. To prevent damage to the removed components, use a proper part container (V.A.G 1698) for storage.

g. All cable ties which are opened or cut open when removing engine, must be replaced in the same position when installing engine.

2. Switch off ignition and all electrical accessories. Disconnect batter under driver' seat. Then, disconnect the auxiliary battery, in spare wheel well of luggage compartment.

3. Remove left and right windshield wiper arms.

4. Pull engine compartment cover seal off bulkhead.

5. Remove covers "A", "B" (if equipped) and "C" in engine compartment and "C" for plenum chamber.

6. Disconnect smaller connector from Engine Control Modules (ECM) and remove Ground (GND) connection for wiring harness.

7. Open fuse box cover on left in plenum chamber and disconnect the three wiring harness connectors.

8. Remove wiring harness from plenum chamber and place it on engine.

9. Remove intake line on both sides between air filter and intake connection.

10. Disconnect vacuum lines "1" for vacuum actuators of the bypass flap (mark lines before separating) and brake booster "2".

11. Disconnect the following connections and free up wiring harness to the following components:
- 1 - 6-pin harness connector, black: For Heated Oxygen Sensor (HO2S)
- 2 - 6-pin harness connector, brown: For Heated Oxygen Sensor (HO2S) 2
- 3 - 2-pin harness connector, black: for Exhaust Gas Temperature (EGT) Sensor 1
- 4 - 2-pin harness connector, brown: for Bank 2 Exhaust Gas Temperature (EGT) Sensor 1

12. Disconnect harness connectors for glow plugs in cylinder bank 1 and cylinder bank 2.

13. Secure suspension struts using Tensioner Hooks (VW 552) or heavy wiring.

14. Remove upper suspension strut upper bolts.

15. Remove noise insulation.

16. Discharge and recover refrigerant from air conditioning system.

17. Drain coolant

18. Disconnect line from transmission oil cooler below on right. Catch fluid which runs out.

19. Disconnect line from cooler for servo oil at lower left. Catch fluid which runs out.

20. Remove charge air connecting pipes for both banks.

21. Loosen clamps from connecting pipes to air filter for cylinder bank 1 and cylinder bank 2 and remove.

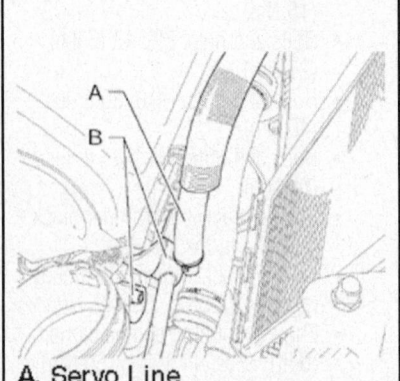

A. Servo Line
B. A/C Compressor Lines

**Fig. 112 Open clamp and pull off servo-line "A", then remove A/C lines from A/C compressor "B"—5.0L TDI engine**

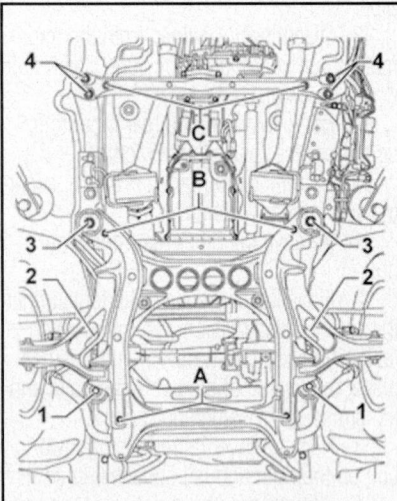

**Fig. 113 With proper engine/transmission support assembly in place, loosen and remove all subframe and member bolts—5.0L TDI engine**

22. Disconnect coolant hoses and lay then so that they cannot be pinched when removing and installing engine.

23. Disconnect starter and generator wire connections at body and place them on engine.

24. Unclip transmission ventilation line.

25. Disconnect connectors from Intake Air Temperature (IAT) sensor and from Mass Air Flow (MAF) sensors for cylinder both banks.

26. Remove wire connection of compressor for air suspension on air filter housing.

27. Remove both top sections of air filter upward.

28. Open clamp and pull off servo-line. Catch fluid which runs out. Remove A/C lines from A/C compressor.

29. Disconnect fuel lines from fuel filter.

30. Separate connection of particle filter/exhaust pipe and remove exhaust system.

31. Remove heat shield on steering gear and loosen universal joint.

32. Separate connectors on transmission, as well as transfer case, and loosen selector lever cable.

33. Remove rear drive shaft (propeller shaft).

34. Remove transmission transverse support (crossmember).

35. Disconnect brake hoses where they join brake lines. Catch escaping brake fluid.

36. In wheel housings, separate all connectors between body and front axle.

➡**Manufacturer recommends the use of a scissor-type lift table (VAS 6131) for removal of the engine/transmission assembly.**

37. With proper engine/transmission support assembly in place, loosen and remove all subframe and member bolts. Slowly lower assembly, watching for possible interference or potential damage to components.

***To install:***

38. Installation is in reverse order of removal. When doing this note the following:

a. Make sure centering sleeves for engine to transmission are installed in cylinder block. Install if necessary.

b. During installation, note the following torque specifications:

- Engine to transmission, M10: 33 ft. lbs. (45 Nm)
- Except starter, M12: 59 ft. lbs. (80 Nm)
- Engine support to cylinder block: 37 ft. lbs. (50 Nm)

- Engine support to engine mount: 55 ft. lbs. (75 Nm)
- Engine mount to engine carrier: 44 ft. lbs. (60 Nm)
- Engine carrier to body: 74 ft. lbs. (100 Nm), plus an additional 180 degrees
- Transmission carrier to body: 37 ft. lbs. (50 Nm), plus an additional 90 degrees

c. Reconnect all lines, hoses and connections that were disconnected for the removal sequence.

d. Connect air conditioning lines and properly recharge the A/C system.

e. Top off oil level for power steering system.

f. Fill with coolant.

g. Check headlight adjustment, correct if necessary.

h. Install new oil and filter.

i. Perform test drive and check for DTCs.

## EXHAUST MANIFOLD

### *REMOVAL & INSTALLATION*

**3.6L Engine**

*See Figure 114.*

1. Disconnect batteries.

2. If equipped with one-piece intake manifold:

a. Remove the four vacuum hoses from intake manifold.

b. Remove air filter housing with intake hose to throttle valve control module.

c. Remove throttle valve control module from intake manifold.

3. If equipped with a two-piece intake manifold, remove the upper manifold. See "Intake Manifold" in this section.

4. Disconnect the four oxygen sensor connectors.

5. Using a ring spanner (3337), remove the heated oxygen sensors (HO2S) in front of catalytic converters.

6. Remove heat shield with intake manifold support.

7. Remove the oxygen sensor (O2S) behind both three-way catalytic (TWC) converters.

8. Raise the vehicle.

9. Identify both exhaust pipe flanges, this makes assembly later easier.

10. Remove nuts on flanges.

11. Remove support to transmission .

12. Unbolt flange to exhaust manifold.

➡**To loosen or tighten nuts more easily, shorten a commercially available**

**16 mm open end wrench to approximately 11 cm handle length.**

13. First, remove exhaust pipe with catalytic converter from cylinders 4 to 6, then from cylinders 1 to 3.

14. Remove exhaust manifold.

***To install:***

15. When installing exhaust pipes, ensure that the flange connection after the catalytic converter seals tightly. Leaks in this area produce pulsations in the exhaust. This allows ambient air to reach the lambda probe after catalytic converter and the lambda regulation will be disturbed.

16. Tighten the new exhaust manifold nuts to:

- M8: 18 ft. lbs. (25 Nm)
- M10: 30 ft. lbs. (40 Nm)

17. Adjust the exhaust system so that there is sufficient clearance to the transmission and subframe.

18. The rest of the assembly is basically a reverse of the disassembling sequence.

**4.2L 5-Valve Engine (Code AXQ, BHX)**

*See Figures 115 and 116.*

1. On cylinder bank 2, unfasten bracket for oil dipstick guide tube.

2. Disconnect refrigerant line and close off connections with plugs or a clean rag.

3. Disconnect the exhaust pipe flange from the manifold(s).

4. Remove exhaust manifold nuts and remove the manifold(s).

5. Installation is the reverse of the removal procedure.

➡**Always use new gaskets and new retaining nuts and all exhaust locations.**

6. Tighten new exhaust manifold nuts, in an alternating sequence, to 18 ft. lbs. (25 Nm).

7. Install new exhaust pipe flange nuts and tighten to:

- M8: 18 ft. lbs. (25 Nm)
- M10: 30 ft. lbs. (40 Nm)

**4.2L 4-Valve Engine (Code BAR)**

1. Remove the applicable main catalytic converter.

2. Remove the applicable primary catalytic converter.

3. Drain coolant.

4. Remove coolant pipe.

5. Remove six nuts and remove exhaust manifold.

➡**Both securing strips can remain installed when removing exhaust manifold.**

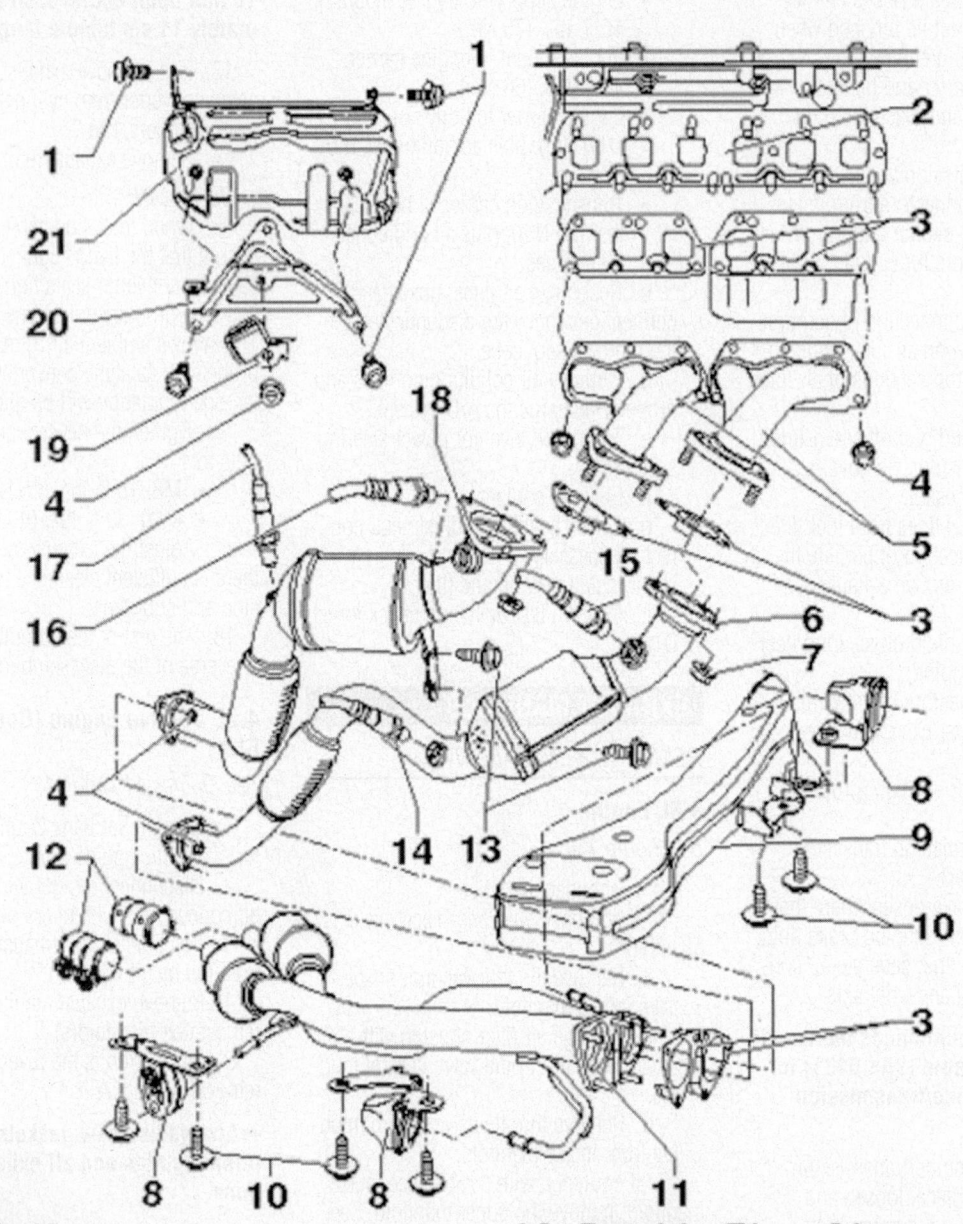

1. Bolts
2. Cylinder Head
3. Gaskets
4. Securing Nuts
5. Exhaust Manifold
6. Primary Catalytic Converter
7. Securing Nuts
8. Suspended Mount
9. Transmission Bracket
10. Bolts
11. Catalytic Converters
12. Double Pipe Clamp
13. Bolts
14. Oxygen Sensor
15. Heated Oxygen Sensor
16. Heated Oxygen Sensor
17. Oxygen Sensor
18. Primary Catalytic Converter
19. Bracket
20. Intake Manifold Support
21. Heat Shield

22205_TOUA_G0099

**Fig. 114 Exploded view of the exhaust manifold and related components—3.6L engine**

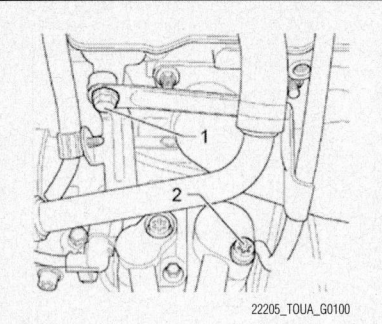

**Fig. 115 Showing bracket "1" and refrigerant line "2" to disconnect—4.2L 5-valve engine**

**To install:**

6. Installation is in reverse order of removal, note the following:

   a. Replace gasket and self-locking nuts.

   b. Grease studs and bolts with hot-bolt paste.

   c. Place new gasket onto stud bolts.

   d. Insert exhaust manifold in securing strips and hand tighten nuts.

   e. Tighten exhaust manifold nuts in two steps as follows:

- Step 1: Tighten nuts, from back to front, to 15 ft. lbs. (20 Nm).
- Step 2: Tighten nuts in same sequence 22 ft. lbs. (30 Nm).

   f. Install left coolant pipe.

   g. Install left primary catalytic converter.

   h. Align exhaust system free of tension.

   i. Fill with coolant.

### 5.0L TDI Engine

*See Figure 117.*

➡**Manufacturer does not provide a step-by-step procedure. Use the illustration as a guide for removal and installation.**

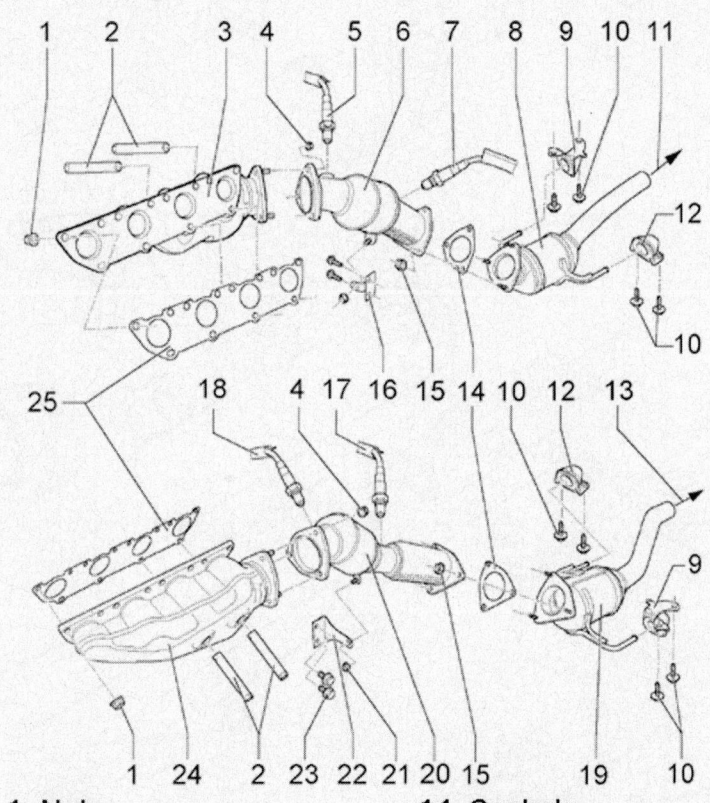

1. Nuts
2. Nuts
3. Exhaust Manifold
4. Nuts
5. Heated Oxygen Sensor
6. Primary Catalytic Converter
7. Oxygen Sensor
8. Catalytic Converter
9. Suspended Mount
10. Bolts
11. To Middle Muffler
12. Suspended Mount
13. To Middle Muffler
14. Gaskets
15. Nuts
16. Bracket
17. Oxygen Sensor
18. Heated Oxygen Sensor
19. Catalytic Converter
20. Primary Catalytic Converter
21. Nut
22. Bracket
23. Bolts
24. Exhaust Manifolds
25. Gaskets

**Fig. 116 Exploded view of the exhaust manifold and related components—4.2L 5-valve engine**

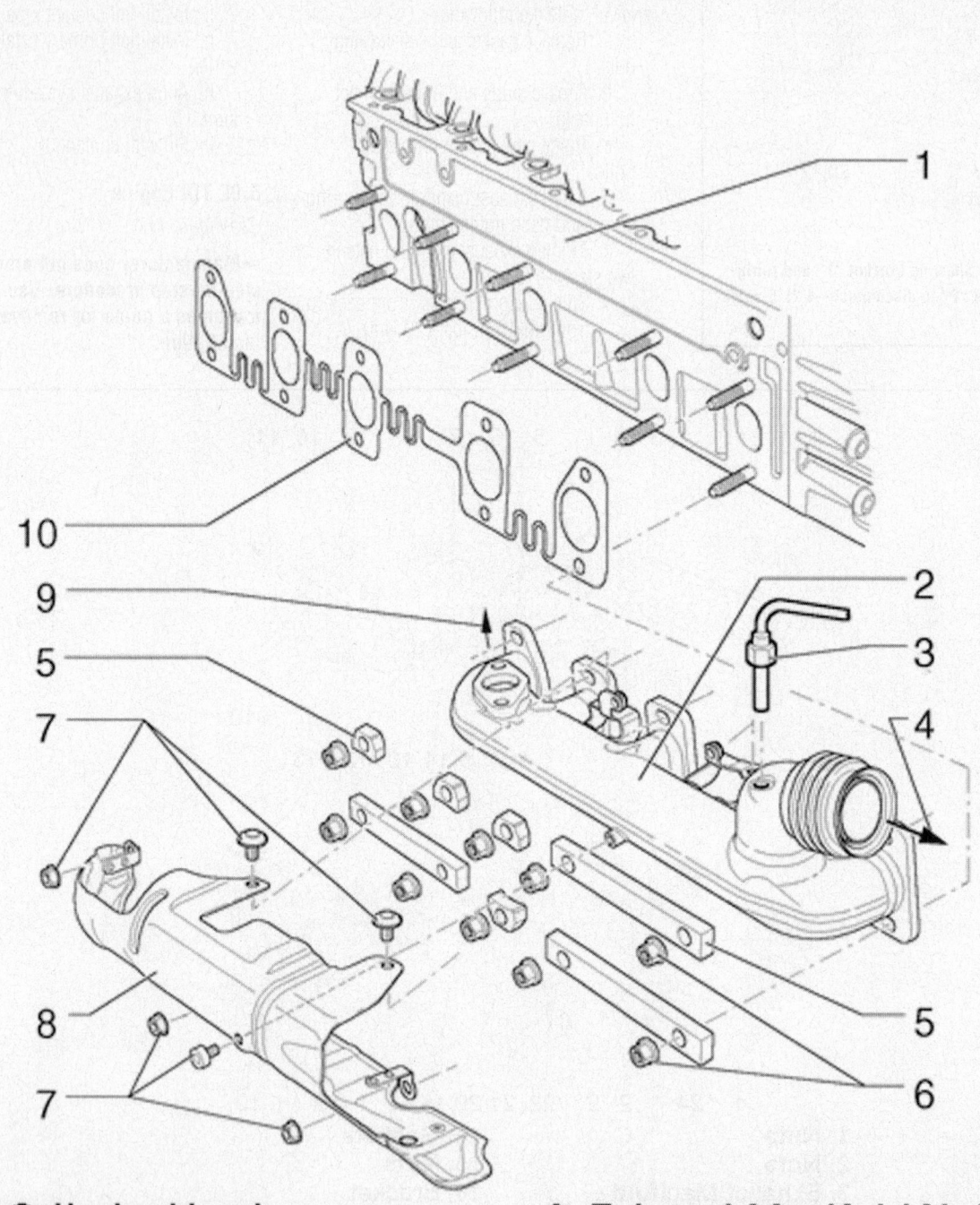

1. Cylinder Head
2. Exhaust Manifold
3. Turbocharger Intake Sensor
4. To Turbocharger
5. Spacer Piece
6. Exhaust Manifold Nuts
7. Heat Shield Fasteners
8. To Connecting Pipe
9. Gasket

22205_TOUA_G0102

**Fig. 117 Exploded view of the exhaust manifold and related components—5.0L TDI engine**

➡During installation, replace the gasket and tighten new exhaust manifold nuts to 17 ft. lbs. (23 Nm), in an alternating sequence.

## INTAKE MANIFOLD

### REMOVAL & INSTALLATION

#### 3.6L Engine

*See Figures 118 through 123.*

➡If a fuel injector is removed, Teflon sealing ring of combustion chamber as well as O-ring must always be replaced. If fuel rail is removed, spring elements for fuel injectors must be replaced.

1. Disconnect battery ground (GND) strap with ignition switched off.
2. Drain the coolant.
3. Remove coolant hoses from the straps on the manifold.

4. Remove bolt from oil dipstick guide tube. Remove guide tube with oil dipstick.
5. Remove accessory drive ribbed belt.
6. Remove ribbed belt tensioner.
7. Remove generator.
8. Remove coolant hoses from coolant pipe on the intake manifold/cylinder head.

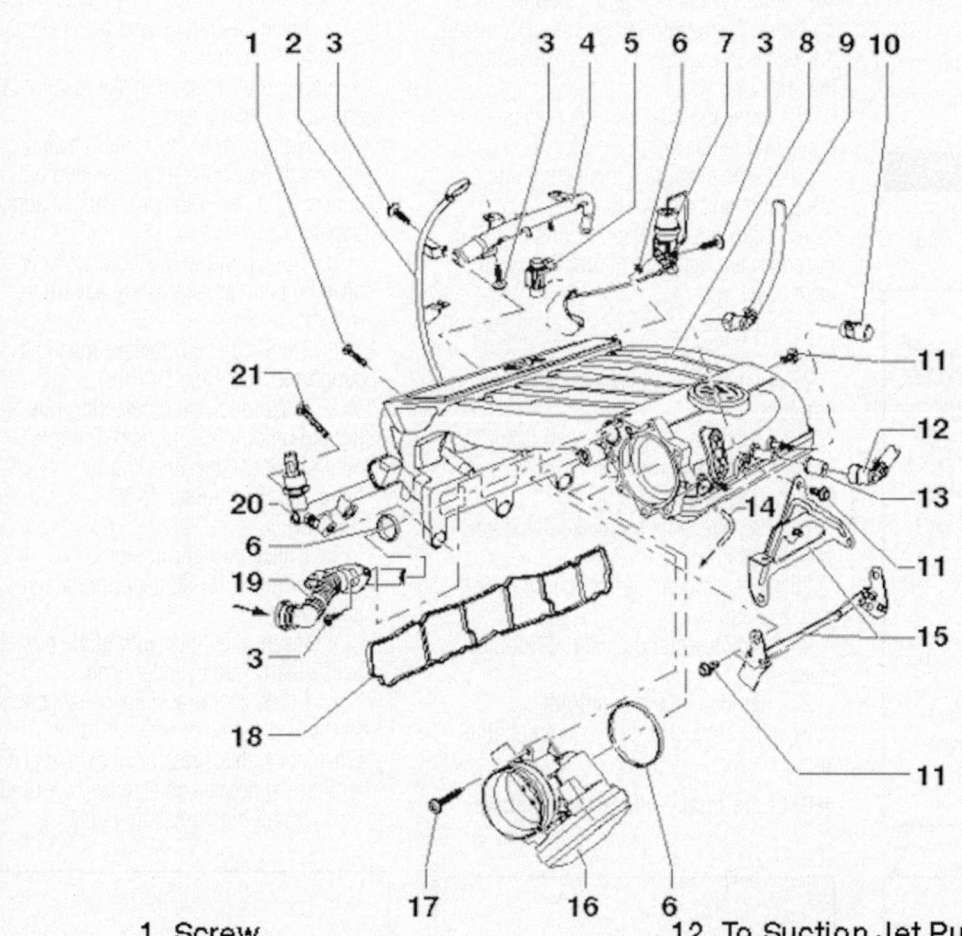

1. Screw
2. Guide Tube
3. Screw
4. Coolant Line
5. Intake Manifold Runner Control (IMRC) Valve
6. Seal
7. Vacuum Actuator
8. Intake Manifold
9. From EVAP Purge Regulator Valve
10. Sealing Plugs
11. Bolt
12. To Suction Jet Pump
13. Grommet
14. To IMRC Valve
15. Intake Manifold Support
16. Throttle Valve Control Module
17. Bolt
18. Gasket
19. Connection
20. Fuel Rail
21. Bolt

22205_TOUA_G0103

**Fig. 118 Exploded view of the intake manifold and related components—3.6L engine**

9. Remove vacuum hoses from Intake Manifold Runner Control (IMRC) valve.

10. Loosen bolts "2" from coolant pipe approximately 1 turn, then remove bolt "3" and remove coolant pipe "1" from intake manifold.

11. Remove connecting hose to throttle valve control module.

12. Disconnect harness connector from throttle valve control module.

13. Disconnect harness connectors from ignition coil with power output stage.

14. Remove ignition coils with power output stage.

15. Remove wiring harness of ignition coils with power output stage from intake manifold.

### ✳✳ WARNING

**Fuel supply lines are under pressure! Wear protective goggles and**

**Fig. 119 Loosen bolts "2" from coolant pipe approximately 1 turn, then remove bolt "3" and remove coolant pipe "1" from intake manifold—3.6L engine**

**protective clothing to prevent injuries and contact with skin. Before loosening the fuel lines, place a cloth around the connection point. Then release pressure by carefully loosening the union nuts.**

16. Lay a cloth around threaded connection and loosen union nut from fuel supply hose. To do so, counter hold on fuel supply line using a wrench.

17. Disconnect high pressure line "1" and low pressure line "2" from high pressure pump. When doing this, counter hold on connection of high pressure pump using a wrench. Disconnect harness connectors in this area (arrows).

18. Remove bolts , of high pressure pump and remove high pressure pump.

19. Loosen fuel rail connecting line union nuts and remove the line.

20. Disconnect the connecting hose for the crankcase ventilation from the cylinder head cover.

21. Remove rear intake manifold support bolt.

22. Remove front intake manifold support bolt.

23. Disconnect lines (arrows) from intake manifold. Remove right intake manifold support bolt "1".

24. Remove bolts for intake manifold/ cylinder head.

25. Remove bolts of fuel rail for cylinders 2, 4 and 6.

26. Carefully pull off fuel rail from fuel injectors.

27. Remove intake manifold downward and set it down on a suitable surface.

➡Seal the intake channels in intake

manifold and in cylinder head using clean cloths.

### To install:

28. Installation is performed in the reverse order of removal, noting the following:

a. Replace O-rings between fuel injectors and fuel rail and coat them lightly with clean motor oil.

b. Replace spring elements at all fuel injectors.

c. Position the intake manifold on the cylinder head.

d. Place fuel rail on and press evenly on to fuel injectors.

e. Insert all bolts of intake manifold and of fuel rail by hand.

f. Evenly tighten fuel rail to cylinder head with new bolts from inside to outside to 22 ft. lbs. (30 Nm), plus an additional 90 degrees.

g. Tighten intake manifold/cylinder head bolts, in an alternating pattern to 6 ft. lbs. (8 Nm).

h. Tighten bolts on intake manifold support to 15 ft. lbs. (20 Nm).

i. Install fuel rail connecting line, tighten union nuts. Tighten union nuts of connecting line first by hand and then tighten to 21 ft. lbs. (28 Nm).

j. Check plunger in side of cylinder head, for damage; replace if necessary.

k. Insert oiled lifter with guide perpendicularly into cylinder head.

l. Rotate engine at vibration damper slowly in direction of engine rotation. When doing this, press plunger in until it reaches the deepest point in cylinder head.

m. Install high pressure pump.

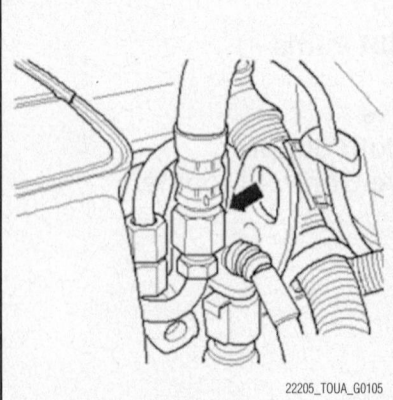

**Fig. 120 Lay a cloth around threaded connection and loosen union nut from fuel supply hose—3.6L engine**

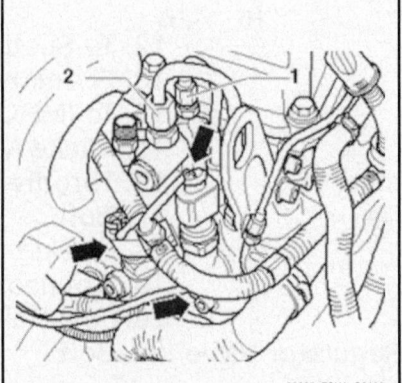

**Fig. 121 Disconnect high pressure line "1" and low pressure line "2" from high pressure pump—3.6L engine**

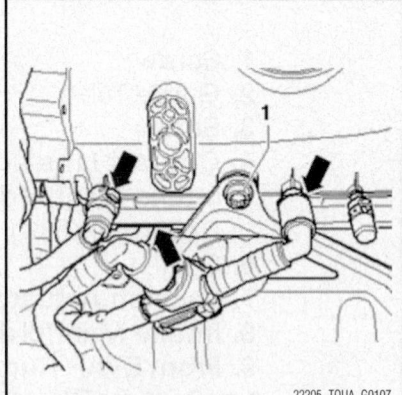

**Fig. 122 Disconnect lines (arrows) from intake manifold. Remove right intake manifold support bolt "1"—3.6L engine**

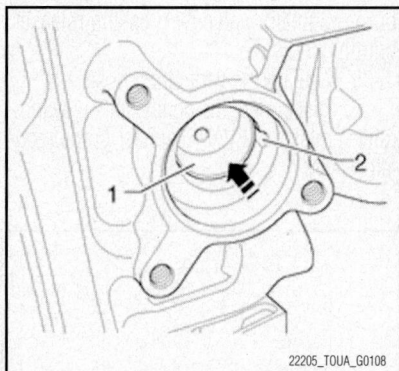

**Fig. 123 Insert oiled lifter "1" with guide "2" perpendicularly into cylinder head—3.6L engine**

➥ O-ring of high pressure pump must always be replaced.

n. Before installing fuel lines, first tighten connection for fuel lines on high pressure pump. Tighten as follows:
- Connection for high pressure line: 30 ft. lbs. (40 Nm)
- Connection for low pressure line: 21 ft. lbs. (28 Nm)

o. Tighten union nuts of fuel lines first by hand and then tighten to 21 ft. lbs. (28 Nm).

p. Install ribbed belt tensioner.

q. Install ribbed belt

r. Fill with coolant.

### 4.2L 5-Valve Engine (Code AXQ, BHX)

*See Figure 124.*

➥ Manufacturer does not provide a step-by-step procedure. Use the illustration as a guide for removal and installation.

1. Tighten retaining bolts as follows:
- Throttle valve control module bolts: 8 ft. lbs. (10 Nm)
- Crankcase vent valve connecting pipe bolts: 8 ft. lbs. (10 Nm)
- Intake manifold bolts: 8 ft. lbs. (10 Nm)

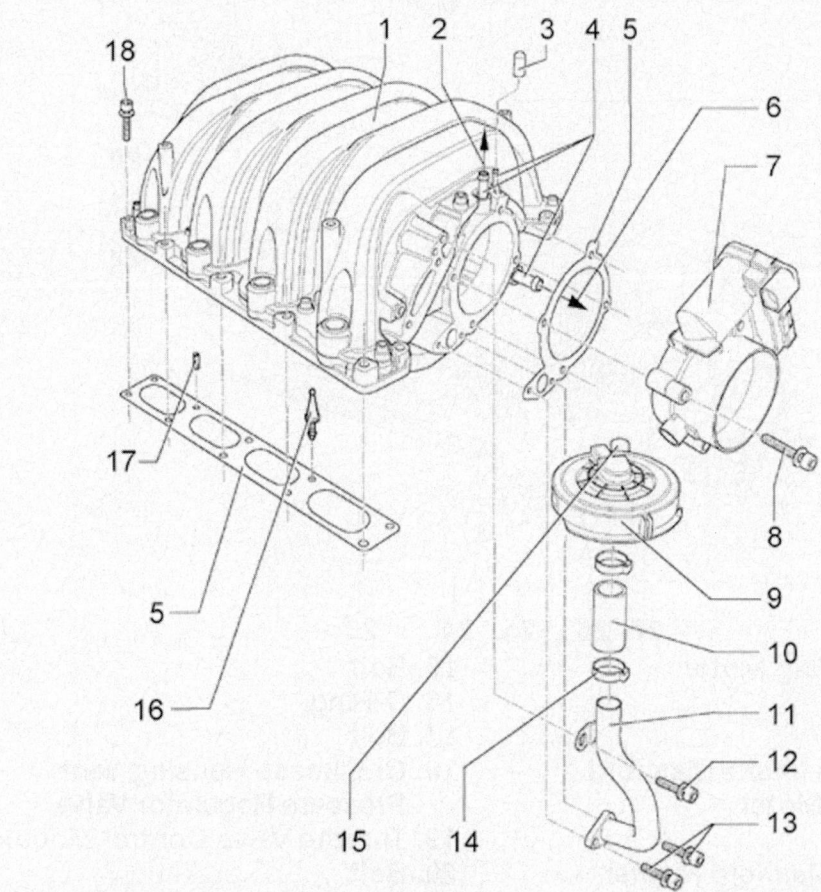

1. Intake Manifold Changeover
2. To EVAP Canister Purge Regulator Valve
3. Sealing Cap
4. Vacuum Connection
5. Gasket
6. To Brake Booster
7. Throttle Valve Control Module
8. Bolt
9. Crankcase Vent Valve
10. Connecting Hose
11. Connecting Pip
12. Bolt
13. Bolt
14. Connection
15. Alignment Cone
16. Alignment Pin
17. Bolts

**Fig. 124 Exploded view of the intake manifold and related components—4.2L 5-valve engine**

### 4.2L 4-Valve Engine (Code BAR)

*See Figures 125 and 126.*

➡**All cable ties opened or cut during engine removal must be reinstalled at the same locations during installation.**

1. With ignition switched off, disconnect battery ground (GND) cable.
2. Remove front engine cover.
3. Remove rear engine cover.
4. Disconnect vent hose to evaporative emission (EVAP) canister purge regulator valve by pressing release button.
5. Free up the ventilation hose.
6. Disconnect the connector on evaporative emission (EVAP) canister purge regulator valve.

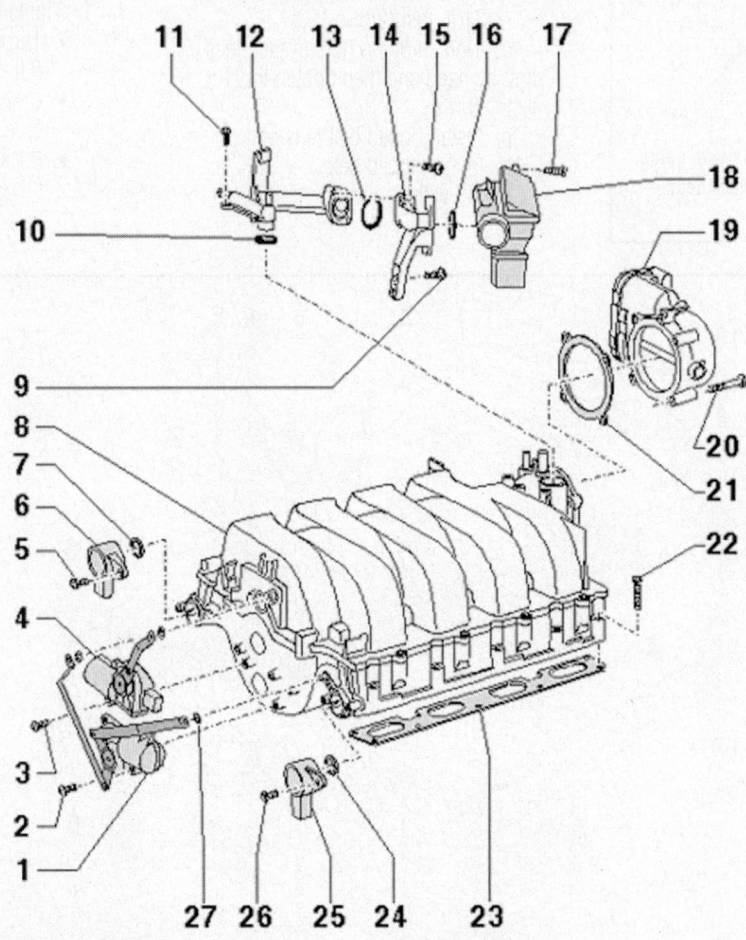

1. Intake Flap Motor
2. Bolt
3. Bolt
4. Variable Intake Manifold Runner Motor
5. Bolt
6. Intake Manifold Runner Position Sensor
7. Gasket
8. Intake Manifold
9. Bolt
10. O-Ring
11. Bolt
12. Coolant Connection Supports
13. O-Ring
14. Connection
15. Bolt
16. O-Ring
17. Bolt
18. Crankcase Housing Vent Pressure Regulator Valve
19. Throttle Valve Control Module
20. Bolt
21. Gasket
22. Bolt
23. Gasket
24. Gasket
25. Intake Manifold Runner Position Sensor 2
26. Bolt
27. Circlip

22205_TOUA_G0110

**Fig. 125 Exploded view of the intake manifold and related components—4.2L 4-valve engine**

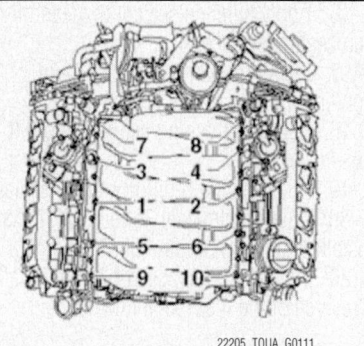

22205_TOUA_G0111

**Fig. 126 Loosening/tightening intake manifold bolt sequence—4.2L 4-valve engine**

7. Remove the ventilation hose from intake manifold.

8. Remove evaporative emission (EVAP) canister purge regulator valve down from bracket.

### ☼☼ WARNING

**Fuel system is under pressure! Always follow safety measures before opening the system.**

9. Place a rag around threaded connection and relieve fuel pressure by briefly opening the line.

10. Now remove fuel supply line on distribution piece.

11. Remove air filter pipe.

12. Remove bolts and place crankcase ventilation pressure regulation valve slightly to side. The hose can remain connected.

13. Clamp off coolant hose above intake manifold.

14. Clamp off coolant hose on crankcase housing ventilation valve.

15. Now, both coolant hoses can be removed from throttle valve control module.

16. Remove crankcase ventilation hoses together with coolant line.

17. Remove bolt and union nut and remove fuel supply line.

18. Disconnect electrical connector and remove the bracket.

19. Carefully, open union nuts at high pressure lines. Place a rag around each connection as it is opened.

20. Remove fuel rail with high pressure lines.

21. Remove vacuum hose from intake manifold.

22. Disconnect connector on throttle valve control module by pulling the rubber cover over release button.

23. Remove throttle valve control module, ensure seal does not fall down.

24. Remove cover for oil filter housing.

25. Remove oil filter insert. Cover filter housing.

26. Disconnect left and right electrical connectors on fuel metering valve and on fuel metering valve 2 at high pressure pump.

27. Disconnect left and right electrical connectors on Camshaft Position (CMP) sensor and on Camshaft Position (CMP) sensor 2 on cylinder head.

28. Remove ground (GND) cable - 3 - at right and left from cylinder head.

29. Disconnect electrical connectors on:
- Intake flap motor (brown connector)
- Intake manifold runner position sensor
- Variable intake manifold runner motor (black connector)
- Intake manifold runner position sensor

30. Mark installation location of pipe clamps for later assembly and remove them.

31. Remove all four engine lifting eyes, left and right front and left and right rear.

32. Remove intake manifold bolts in sequence.

➡**The intake manifold bolts have a large bolt head.**

➡**To avoid scratching intake manifold, apply adhesive tape in area of high pressure lines.**

33. Raise front of intake manifold upward.

34. Swing intake manifold to right side of vehicle, under high pressure fuel lines and out.

35. Remove intake manifold from engine compartment.

➡**Plug intake ports of the cylinder head with clean rags.**

#### To install:

36. Installation is in reverse order of removal, note the following:

a. Replace gaskets and O-rings.

b. Secure all hose connections with hose clamps appropriate for the model.

c. During installation, all cable ties must be reinstalled at the same location.

d. Replace intake manifold gaskets while checking thoroughly for proper contact and locating points.

e. Be aware of alignment pins when setting intake manifold in place.

f. Tighten intake manifold bolts in sequence to 6 ft. lbs. (9 Nm).

g. Ensure that the high pressure line connections do not show any signs of damage.

h. Do not change bend shape of high pressure lines.

i. Hand-tighten union nuts for high-pressure lines.

j. Make sure high-pressure lines are seated free of stress.

k. To tighten union nuts on high pressure lines, use a proper torque wrench with socket insert. Tighten to 18 ft. lbs. (25 Nm).

l. Check that all fuel rail and line fittings are tightened to 18 ft. lbs. (25 Nm).

➡**Only install retaining tabs after high pressure lines have been tightened.**

m. Replace engine lifting eyes and tighten bolts to 16 ft. lbs. (22 Nm)

n. Reconnect the battery.

o. Top off coolant if necessary.

### 5.0L TDI Engine

*See Figure 127.*

➡**Manufacturer does not provide a step-by-step procedure for this component. Refer to the illustration as a guide for removal and installation.**

1. During installation, note the following tightening specifications:
- Intake manifold bolts: 15 ft. lbs. (20 Nm)
- Fuel rail bolts: 6 ft. lbs. (8 Nm), plus an additional 90 degrees
- Glow plugs: 11 ft. lbs. (15 Nm)

➡**Replace all gaskets and seals.**

### OIL PAN

#### REMOVAL & INSTALLATION

### 3.6L Engine

1. Drain engine oil.

2. Disconnect wire connection to air suspension compressor at air filter.

3. Completely remove air filter with Mass Air Flow (MAF) sensor.

4. Remove nuts on right and left engine bracket.

5. Position an engine support/lifting device to the engine.

6. With the lifting device in place, pretension engine slightly.

7. Remove the subframe.

8. Remove engine carrier. Engine mounts can remain attached to engine carrier.

9. Disconnect 3-pin connector from oil level thermal sensor.

10. Remove the oil pan.

11. Loosen oil pan with light blows of a rubber headed hammer if necessary.

12. Remove sealant residue from cylinder block with a flat scraper.

13. Remove sealant residue at oil pan using a rotating brush, e.g. a drill with plastic brush attachment (wear protective glasses).

14. Clean the sealing surfaces, they must be free of oil and grease.

**To install:**

➡ The oil pan must be installed within 5 minutes after application of silicone sealant.

15. Apply silicone sealant to clean sealing surfaces of oil pan. The sealing compound bead must be about 2 to 3 mm thick, and running on the inside of bolt holes.

➡ Sealant bead must not be thicker than specified. Otherwise, excess sealant could get into oil pan and clog strainer in intake line of oil pump.

16. Apply silicone sealant to clean oil pan sealing surfaces.

17. Install oil pan immediately and tighten all oil pan bolts lightly.

18. Repeat tightening sequence to 9 ft. lbs. (12 Nm).

19. After installing oil pan, allow sealant to dry for approximately 30 minutes, before installing any new engine oil.

20. The rest of the assembly is basically a reverse of the disassembling sequence.

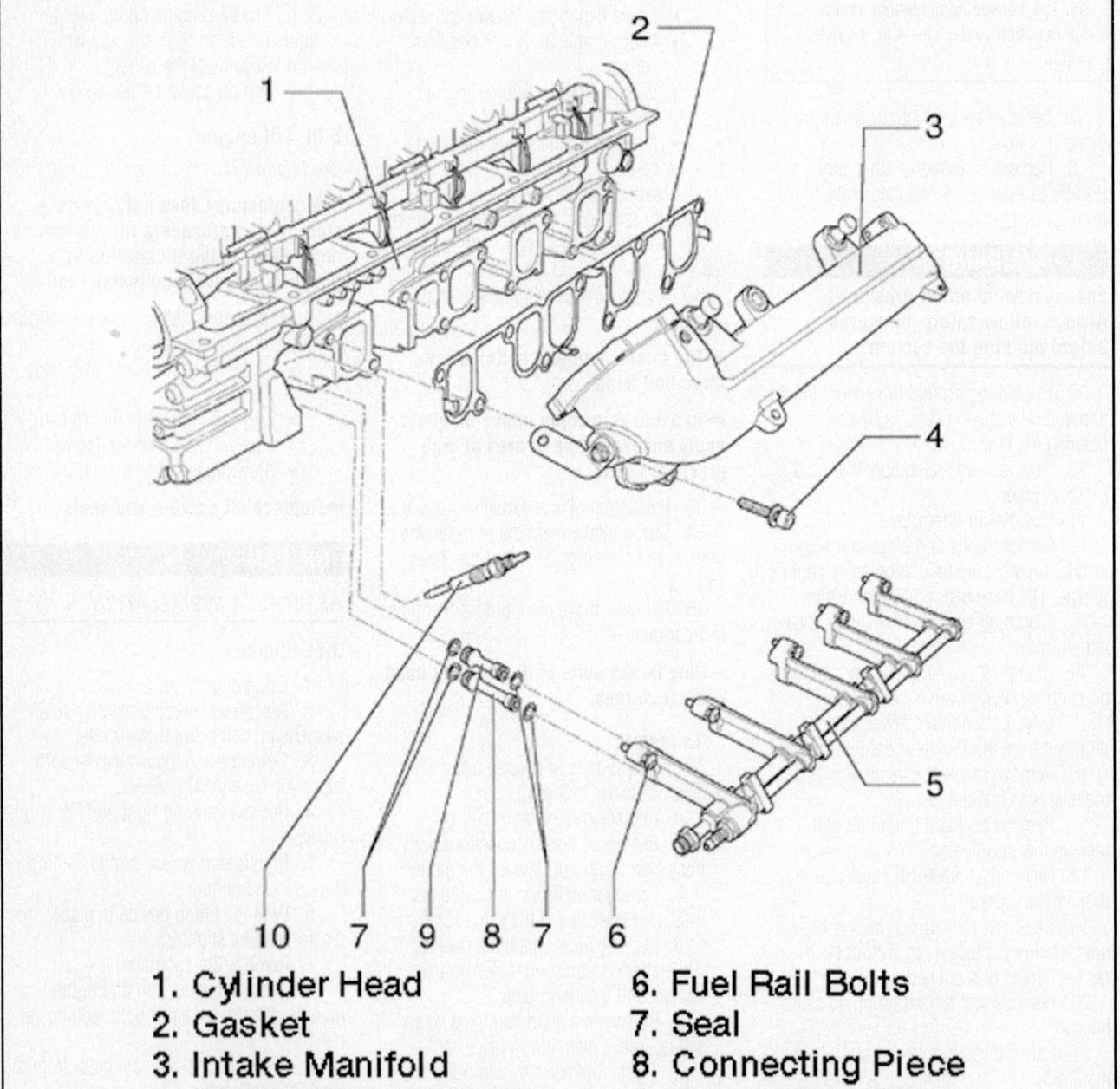

1. Cylinder Head
2. Gasket
3. Intake Manifold
4. Intake Manifold Bolts
5. Fuel Rail
6. Fuel Rail Bolts
7. Seal
8. Connecting Piece
9. Connecting Piece
10. Glow Plugs

22205_TOUA_G0112

**Fig. 127 Exploded view of intake manifold and related components—5.0L TDI engine**

**4.2L 5-Valve Engine (Code AXQ, BHX)**

*See Figure 128.*

➡**Manufacturer does not provide a specific step-by-step procedure for this component. Use the illustration as a guide for removal and installation.**

1. Note the following torque specification during installation:
- Pipe connection bolt: 7 ft. lbs. (10 Nm)
- Chain tensioner bolt: 7 ft. lbs. (10 Nm)
- Chain sprocket bolt: 25 ft. lbs. (34 Nm)

- Oil pump bolt: 22 ft. lbs. (30 Nm)
- Lower oil pan bolts: 7 ft. lbs. (10 Nm)
- Oil level thermal sensor bolt: 7 ft. lbs. (10 Nm)
- Drain plug (upper or lower): 37 ft. lbs. (50 Nm)
- Upper oil pan bolts: 16 ft. lbs. (22 Nm)

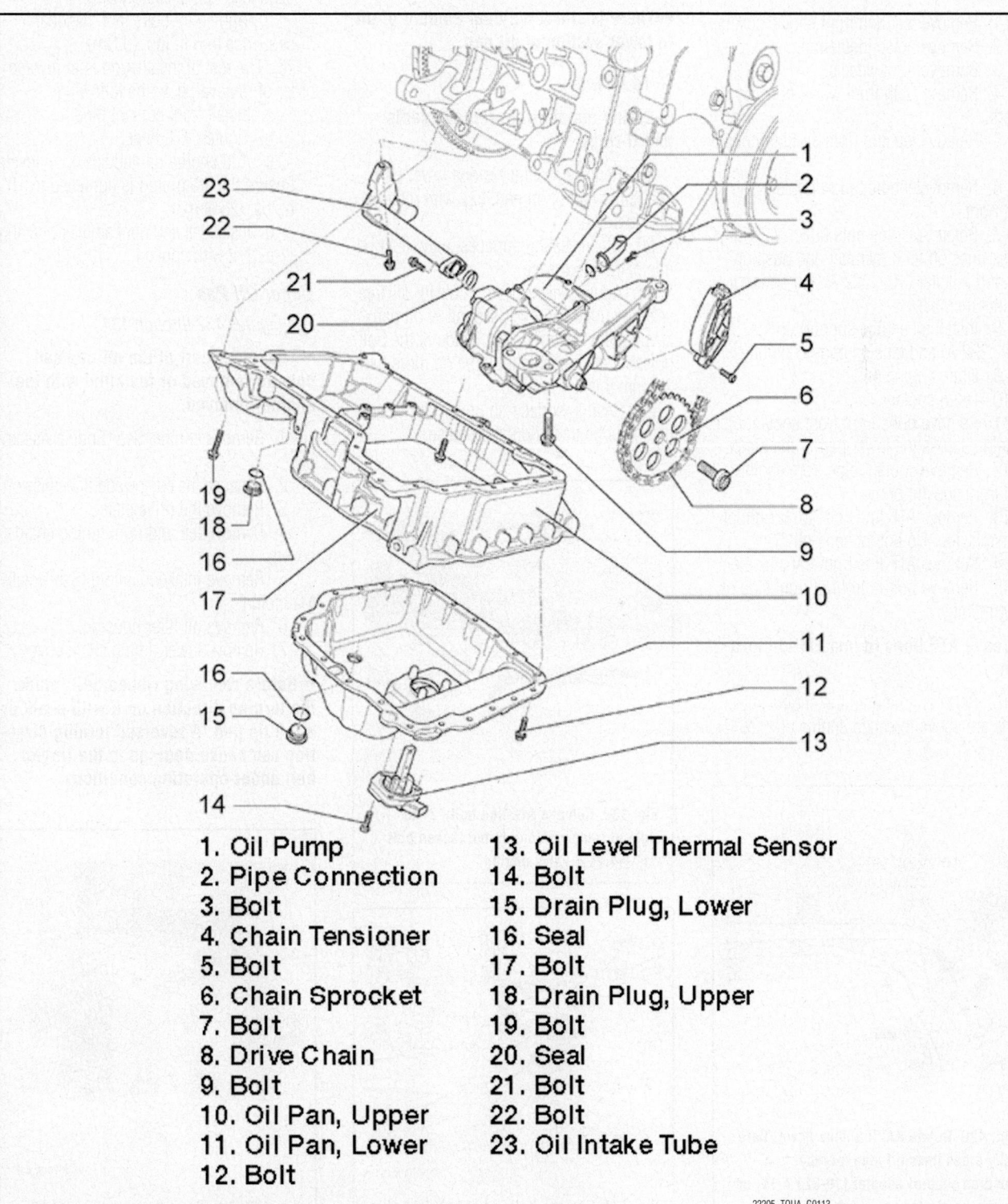

1. Oil Pump
2. Pipe Connection
3. Bolt
4. Chain Tensioner
5. Bolt
6. Chain Sprocket
7. Bolt
8. Drive Chain
9. Bolt
10. Oil Pan, Upper
11. Oil Pan, Lower
12. Bolt
13. Oil Level Thermal Sensor
14. Bolt
15. Drain Plug, Lower
16. Seal
17. Bolt
18. Drain Plug, Upper
19. Bolt
20. Seal
21. Bolt
22. Bolt
23. Oil Intake Tube

22205_TOUA_G0113

**Fig. 128 Exploded view of the oil pan and related components—4.2L 5-valve engine**

- Oil intake tube to oil pump bolt: 7 ft. lbs. (10 Nm)
- Oil intake tube to block bolt: 18 ft. lbs. (25 Nm)

### 4.2L 4-Valve Engine (Code BAR)

#### Lower Oil Pan

*See Figures 129 through 131.*

1. Remove rear and front engine covers.
2. Remove noise insulation.
3. Remove front wheels.
4. Remove both front wheel housing liners.
5. Remove left and right air filter housing.
6. Remove engine bracket nuts at left and right.
7. Rotate A/C line nuts down. Carefully press lines off long member and position support adapter (10 - 222 A /19) on longitudinal member.
8. Install an engine support bridge (10 - 222 A) and tension engine slightly.
9. Drain engine oil.
10. Drain coolant.
11. Remove bolts from front coolant pipe running across the lower part of the block.
12. Remove coolant pipe from engine and from coolant pump.
13. Remove ATF line bolt "2" at right of transmission. Do not loosen bolt "1".
14. Remove ATF line from transmission.
15. Remove bolt in front of right side of long member.

➡**Leave ATF lines in installation location.**

16. Raise engine until the engine supports are above threaded engine bracket

**Fig. 129 Rotate A/C line nuts down. Carefully press lines off long member and position support adapter (10-222 A/19) on longitudinal member—4.2L 4-valve engine**

pins and tension all spindles evenly to achieve even weight distribution.

17. Disconnect electrical connector at oil level thermal sensor on pan.
18. Remove oil pan (lower part) and pry out carefully.

➡**To remove lower part of oil pan, pull ATF lines down slightly.**

➡**There is still a residual amount of oil in lower section of oil pan.**

#### To install:

➡**During installation, replace seals and O-rings.**

19. Remove sealant residue lower part and upper part of oil pan, e.g. with rotating plastic brush.
20. Clean sealing surfaces, they must be free of oil and grease.
21. Apply sealant bead on clean, oil-free sealing surface of lower section of oil pan, keeping the sealant on the inside of the bolt holes. Sealant bead should be no more than 2.5 mm thick.
22. Sealing surface on upper part of oil pan must be free of oil and grease.

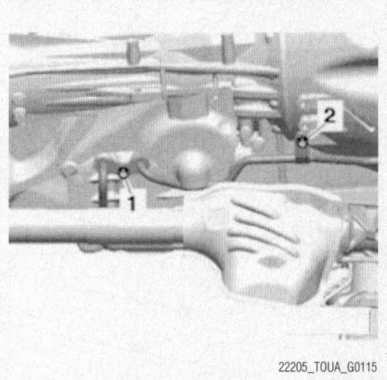

**Fig. 130 Remove ATF line bolt "2" at right of transmission; do not loosen bolt "1"—4.2L 4-valve engine**

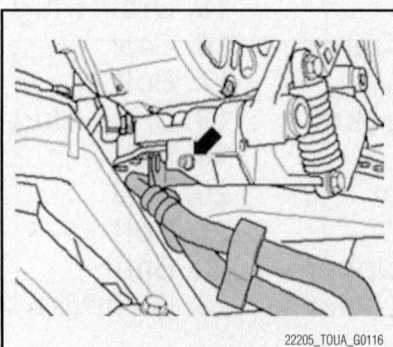

**Fig. 131 Remove bolt in front of right side of long member—4.2L 4-valve engine**

23. The oil pan (lower part) must be installed within 5 minutes after application of sealant.
24. Position lower part of oil pan and hand tighten all bolts.
25. Tighten bolts for lower part of oil pan in 2 stages as follows.
   a. Pre-tighten all bolts in a diagonal sequence to 3 ft. lbs. (5 Nm).
   b. Tighten all bolts in a diagonal sequence to 6 ft. lbs. (9 Nm).
26. The rest of installation is in reverse order of removal, note the following:
   a. Install front coolant pipe.
   b. Tighten ATF lines.
   c. Add engine oil and check oil level. Ensure the drain plug is tightened to 18 ft. lbs. (25 Nm).
   d. Install left and right air filter housing.
   e. Fill with coolant.

#### Upper Oil Pan

*See Figures 132 through 134.*

➡**The upper part of the oil pan can only be removed or installed with the engine removed.**

1. Remove engine. See "Engine Assembly."
2. Separate the engine and transmission.
3. Remove the drive plate.
4. Remove left and right timing chain covers.
5. Remove intake manifold. See "Intake Manifold."
6. Remove oil filter housing.
7. Remove lower timing chain cover.

➡**Before removing ribbed belt, mark the turning direction on it with chalk or a felt tip pen. A reversed turning direction can cause damage to the ribbed belt under operating conditions.**

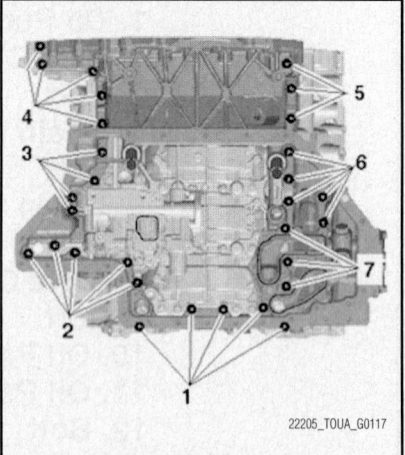

**Fig. 132 Showing locations of upper oil pan bolts—4.2L 4-valve engine**

8. Pivot the belt tensioning device downward to relieve tension on ribbed belt. Remove ribbed belt and release tensioning device.

9. Remove bolts holding tensioning device and remove ribbed belt tensioner from upper part of oil pan.

10. Remove electrical connections from rear of generator.

11. Remove the generator.

12. Remove bolts and remove air generator bracket.

Remove bolts and remove front coolant pipe from engine and from coolant pump.

13. Remove lower section of oil pan. See procedure above.

14. Remove oil pump.

15. Remove bolts (1 through 7) for upper section of oil pan.

16. Press upper part of oil pan from alignment pins of cylinder block.

### To install:

➡**During installation, replace seals and O-rings.**

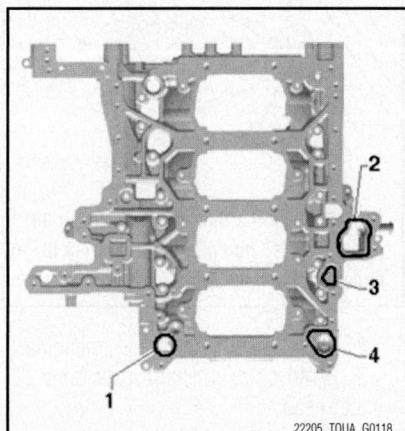

**Fig. 133 Insert new seals ("1" though "4") into grooves on cylinder block—4.2L 4-valve engine**

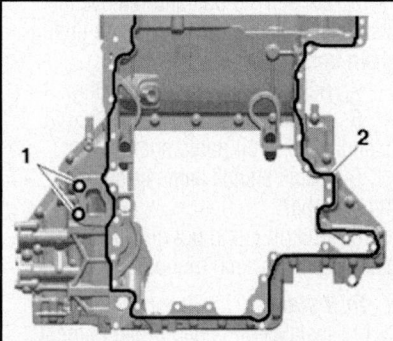

**Fig. 134 Apply sealant beads on clean sealing surfaces of upper part of oil pan, as shown—4.2L 4-valve engine**

17. Using rotating plastic brush, remove any remaining sealant from oil pan (upper part) and at cylinder block.

18. Clean sealing surfaces, they must be free of oil and grease.

19. Insert new seals ("1" though "4") into grooves on cylinder block.

20. Apply sealant beads on clean sealing surfaces of upper part of oil pan, as shown in illustration. Thickness of sealant beads: 2.5 mm.

➡**The oil pan (upper part) must be installed within 5 minutes after application of sealant.**

21. Position upper part of oil pan and tighten bolts in two steps, using a diagonal tightening sequence:

a. Pre-tighten bolts to diagonally to 3 ft. lbs. (5 Nm).

b. Tighten bolts diagonally to 10 ft. lbs. (14 Nm).

22. The rest of installation is in reverse order of removal, note the following:

a. Add engine oil and check oil level.

**5.0L TDI Engine**

*Lower Oil Pan*

See Figure 135.

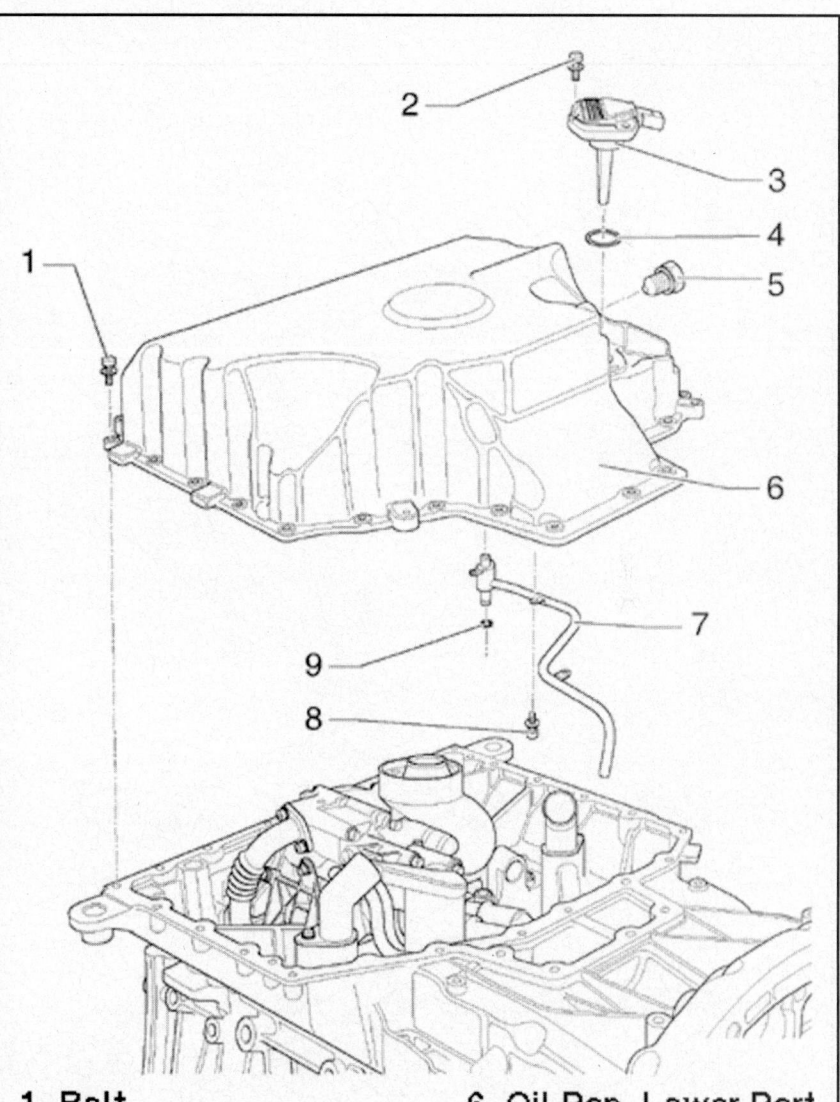

1. Bolt
2. Bolt
3. Oil Level Thermal Sensor
4. Seal
5. Drain Plug
6. Oil Pan, Lower Part
7. Oil Pipe
8. Bolt
9. O-Ring

**Fig. 135 Exploded view of the lower oil pan—5.0L TDI engine**

➡**Manufacturer does not provide a step-by-step procedure. Use the illustration as a guide for removal and installation.**

1. During installation, torque bolts as follows:
   - Except drain plug: 6 ft. lbs. (8 Nm)
   - Drain plug: 22 ft. lbs. (30 Nm)

## Upper Oil Pan

*See Figure 136.*

➡**Manufacturer does not provide a step-by-step procedure. Use the illustration as a guide for removal and installation.**

1. During installation, torque bolts as follows:
   - Upper oil pan bolts: 6 ft. lbs. (8 Nm)
   - Oil pipe bolts: 6 ft. lbs. (8 Nm)
   - Splash shield bolts: 7 ft. lbs. (10 Nm)

## OIL PUMP

### REMOVAL & INSTALLATION

#### 3.6L Engine

*See Figures 137 and 138.*

1. Remove engine. See "Engine Assembly" in this section.

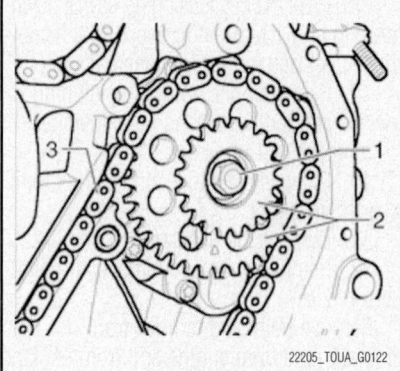

**Fig. 137 Loosen the bolt "1" and remove the chain sprockets "2", together with the timing chain "3" from the oil pump—3.6L engine**

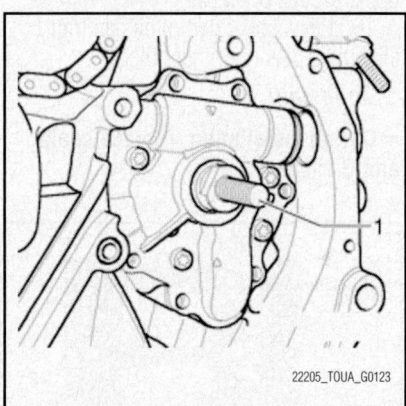

**Fig. 138 Install a threaded bolt M10 × 30 (standard) "1" into oil pump shaft—3.6L engine**

2. Separate transmission from engine.
3. Remove timing chain cover from cylinder head.
4. Remove drive plate and sealing flange from engine.
5. Remove the chain tensioner and then remove the camshaft control chain from the front chain sprocket for the oil pump.
6. Loosen the bolt and remove the chain sprockets, together with the timing chain from the oil pump.
7. Remove the bolts of oil pump.
8. Install a threaded bolt M10 × 30 (standard) into oil pump shaft.
9. Install a slide hammer onto the threaded bolt.
10. Pull oil pump out of cylinder block using light knocking motions.

#### To install:

11. Installation is performed in the reverse of removal, noting the following:
    a. O-ring as well as mounting bolts for oil pump and chain sprockets must always be replaced.

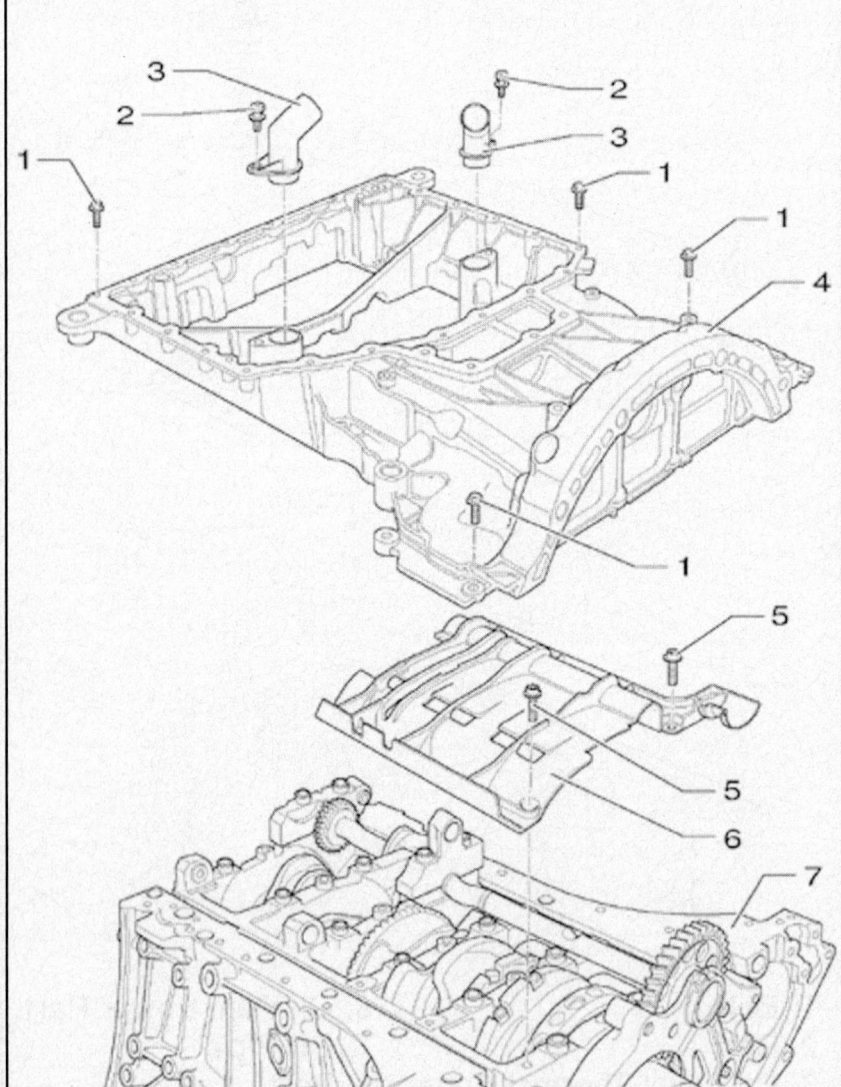

**Fig. 136 Exploded view of the upper oil pan—5.0L TDI engine**

b. Install oil pump with new bolts. Tighten the bolts to 6 ft. lbs. (8 Nm).

c. Install chain sprockets with a new bolt. Tighten the bolt to 47 ft. lbs. (60 Nm), plus an additional 90 degrees.

d. Install camshaft adjustor with timing chain for camshaft drive.

### 4.2L 5-Valve Engine (Code AXQ, BHX)

*See Figure 139.*

1. Remove the engine. See "Engine Assembly" in this section.

2. Drain engine oil from lower and upper part of oil pan.

3. Remove right engine mount.

4. Remove oil filter housing.

5. At left cylinder head, unfasten bracket from oil dipstick guide tube and pull guide tube out.

6. Unfasten bracket of coolant drain pipe at oil pan, upper part. If necessary, loosen upper banjo bolt slightly.

7. Remove lower and upper part of oil pan. See "Oil Pan."

8. If necessary, loosen oil pan using light blows of a rubber headed hammer or by lightly prying using a pry bar.

9. Compress chain tensioner in direction of arrow and secure it using a locking pin (T40011). Remove chain sprocket from oil pump.

10. Remove three bolts from oil pump and oil intake tube .

11. Remove oil pump.

12. Remove residual sealant on cylinder block.

13. Remove sealant residue from lower and upper part of oil pan using a rotating brush, e.g. a hand drill with a plastic brush attachment (wear eye protection).

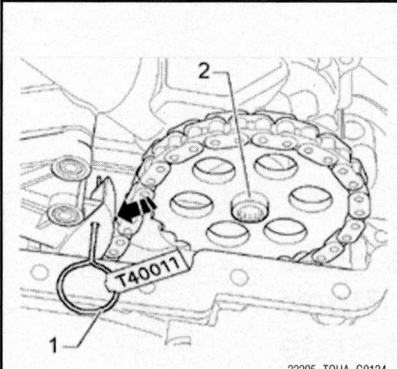

**Fig. 139 Compress chain tensioner in direction of arrow and secure it using a locking pin "1", then remove chain sprocket "2" from oil pump**

22205_TOUA_G0125

**Fig. 140 Remove three bolts indicated and remove intake tube—4.2L 4-valve engine**

14. Clean sealing surfaces. They must be free of oil and grease.

### *To install:*

15. Position the oil pump.

16. Bolt chain sprocket to oil pump. Tighten the bolt to 25 ft. lbs. (34 Nm).

17. Pull locking pin (T40011) out from chain tensioner.

18. Install upper part of oil pan as follows:

**➡Oil pan must be installed within 5 minutes of applying silicone sealing compound.**

19. Apply silicone sealing compound to the clean sealing surface of upper part of oil pan. Sealing compound bead must be: 1.5 mm thick and run on inside of bolt holes.

20. Install upper part of oil pan immediately and tighten all oil pan bolts lightly in a "diagonal sequence".

21. Then, fasten oil pan bolts "1" through "15" to 11 ft. lbs. (15 Nm) and bolts "2" through "22" to 16 ft. lbs. (22 Nm) in a "diagonal sequence" .

22. Apply a 1.5 mm thick bead of sealant onto clean sealing surface of lower part of oil pan.

23. Install lower part of oil pan immediately and tighten all oil pan bolts lightly in a "diagonal sequence". Then, tighten the oil pan bolts in a "diagonal sequence" to 7 ft. lbs. (10 Nm).

24. Set oil dipstick guide tube in place.

25. Fasten coolant drain pipe to upper part of oil pan to 7 ft. lbs. (10 Nm) and then to the cylinder block to 22 ft. lbs. (30 Nm).

**➡After installing the oil pan the sealant must be allowed to dry for approx. 30 minutes. Only after then may the engine oil be replenished.**

26. The rest of assembly is basically a reverse of disassembling sequence.

### 4.2L 4-Valve Engine (Code BAR)

*See Figure 140.*

1. Remove lower section of oil pan. See "Oil Pan."

2. Remove coolant (water) pump.

3. Remove drive shaft for coolant pump from bore in oil pump.

4. Remove bolts and remove oil pipe from the side of the engine.

5. Remove three bolts indicated and remove intake tube.

6. Remove oil pipe together with oil check valve housing.

7. Press back drive shaft for oil pump against spring force and clamp tightly, using long-nose pliers.

8. Remove bolts and remove oil pump.

### *To install:*

**➡During installation, replace all seals and O-rings.**

9. Check whether two alignment bushings are present in cylinder block, install if necessary.

10. Install the oil pump and tighten the bolts to 6 ft. lbs. (8 Nm), plus an additional 90 degrees.

11. Unlock long-nose gripping pliers and let drive shaft glide into oil pump.

12. Check whether drive shaft is friction locked to oil pump. To do so, reach into intake opening of oil pump and try to rotate oil pump gears. Toothed gears must not be able to be rotated.

13. The rest of installation is in reverse order of removal, note the following:

a. Install coolant pump. Tighten all bolts to 7 ft. lbs. (9 Nm).

b. Install lower section of oil pan. See "Oil Pan."

c. Add engine oil and check oil level.

d. Fill with coolant.

### 5.0L TDI Engine

*See Figure 141.*

➡**Manufacturer does not provide a step-by-step procedure. Use the illustration as a guide for removal and installation.**

1. During installation, torque bolts as follows:

- Oil return line bolt: 6 ft. lbs. (8 Nm), plus an additional 90 degrees
- Baffle plate bolts: 7 ft. lbs. (10 Nm)
- Oil pump bolts: 15 ft. lbs. (20 Nm), plus an additional 90 degrees

### PISTON AND RING

*POSITIONING*

### 3.6L Engine

*See Figure 142.*

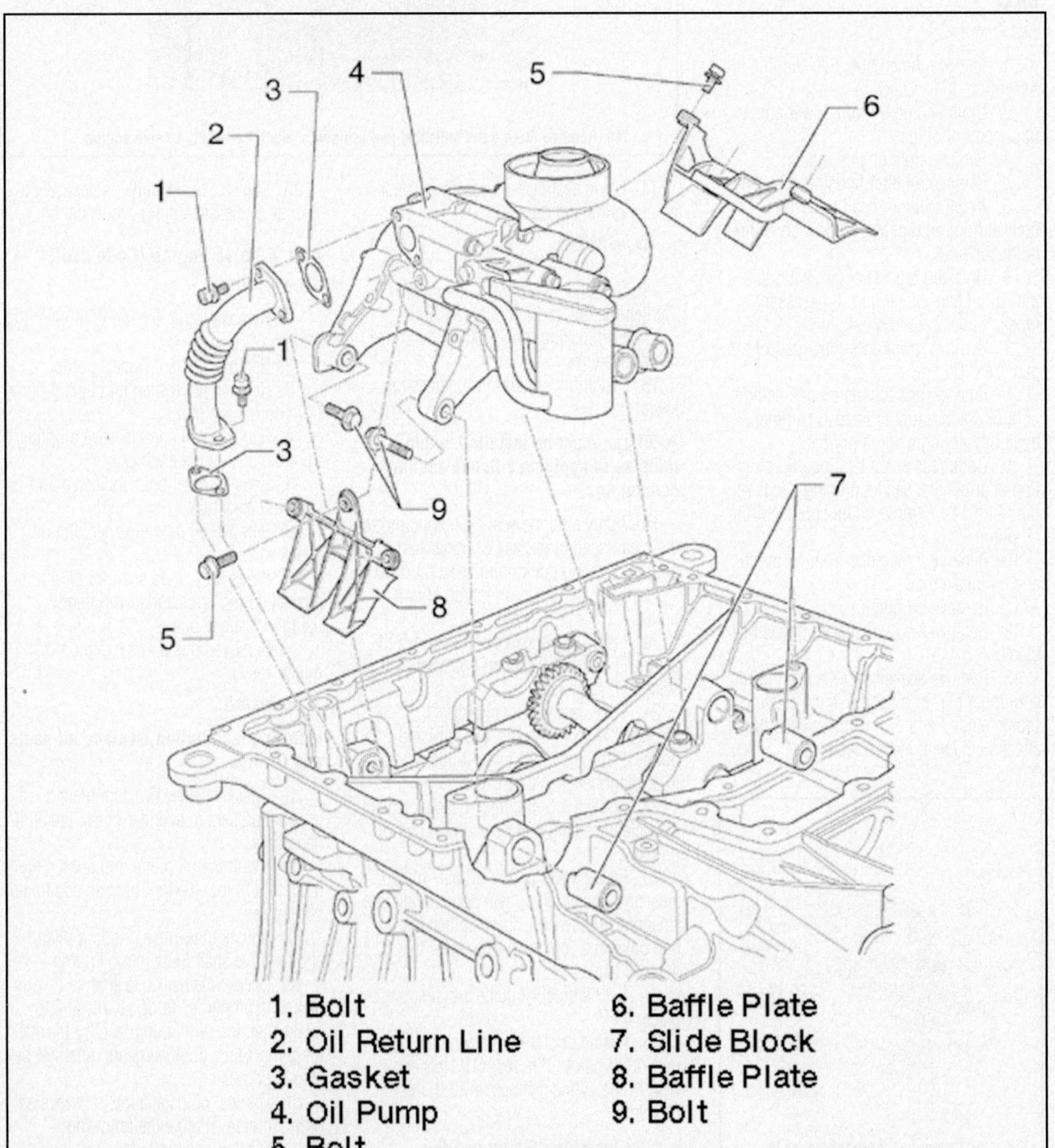

1. Bolt
2. Oil Return Line
3. Gasket
4. Oil Pump
5. Bolt
6. Baffle Plate
7. Slide Block
8. Baffle Plate
9. Bolt

22205_TOUA_G0126

**Fig. 141 Exploded view of the oil pump and related components—5.0L TDI engine**

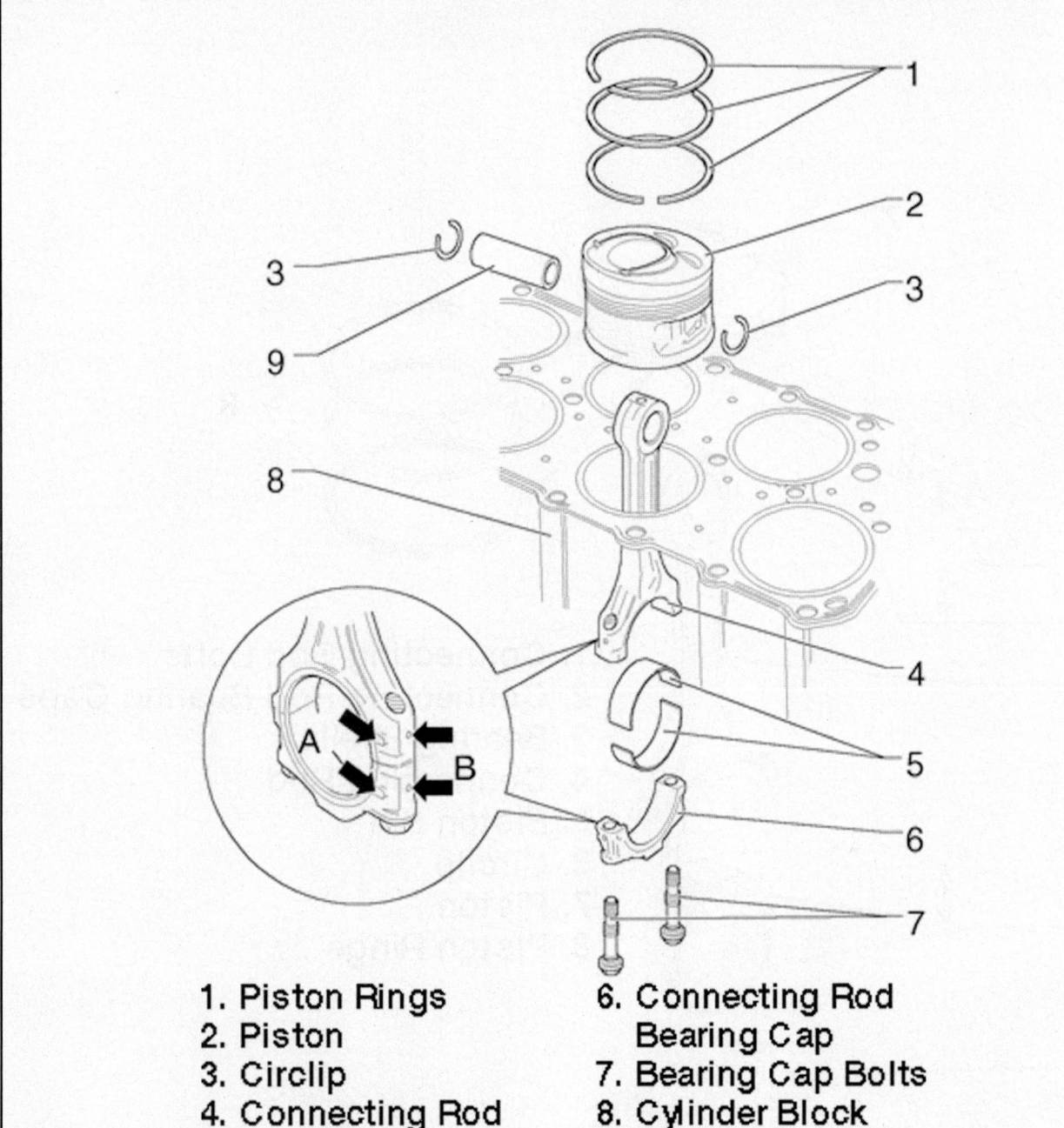

1. Piston Rings
2. Piston
3. Circlip
4. Connecting Rod
5. Bearing Shell
6. Connecting Rod
   Bearing Cap
7. Bearing Cap Bolts
8. Cylinder Block
9. Piston Pin

22205_TOUA_G0127

Fig. 142 Exploded view of the piston and rod assembly—note positions of ring end gaps on piston—3.6L engine

**4.2L 5-Valve Engine (Code AXQ, BHX)**

➡Manufacturer does not provide an illustration of this component.

**4.2L 4-Valve Engine (Code BAR)**

See Figure 143.

**5.0L TDI Engine**

See Figure 144.

## REAR MAIN SEAL

*REMOVAL & INSTALLATION*

### 3.6L Engine and 5.0L TDI Engine

1. Remove the engine and separate the transmission. See "Engine Assembly" in this section.
2. Place pulling hook (T20143/2) behind sealing lip of rear main sealing ring.

3. Support the pulling hook on sealing flange and pry out sealing ring.

*To install:*

4. Pull sealing ring with its outside over the sleeve and onto a proper assembly/installer tool .
5. Place assembly tool with dry sealing ring onto crankshaft pin.
6. Knock it into sealing flange until it stops.

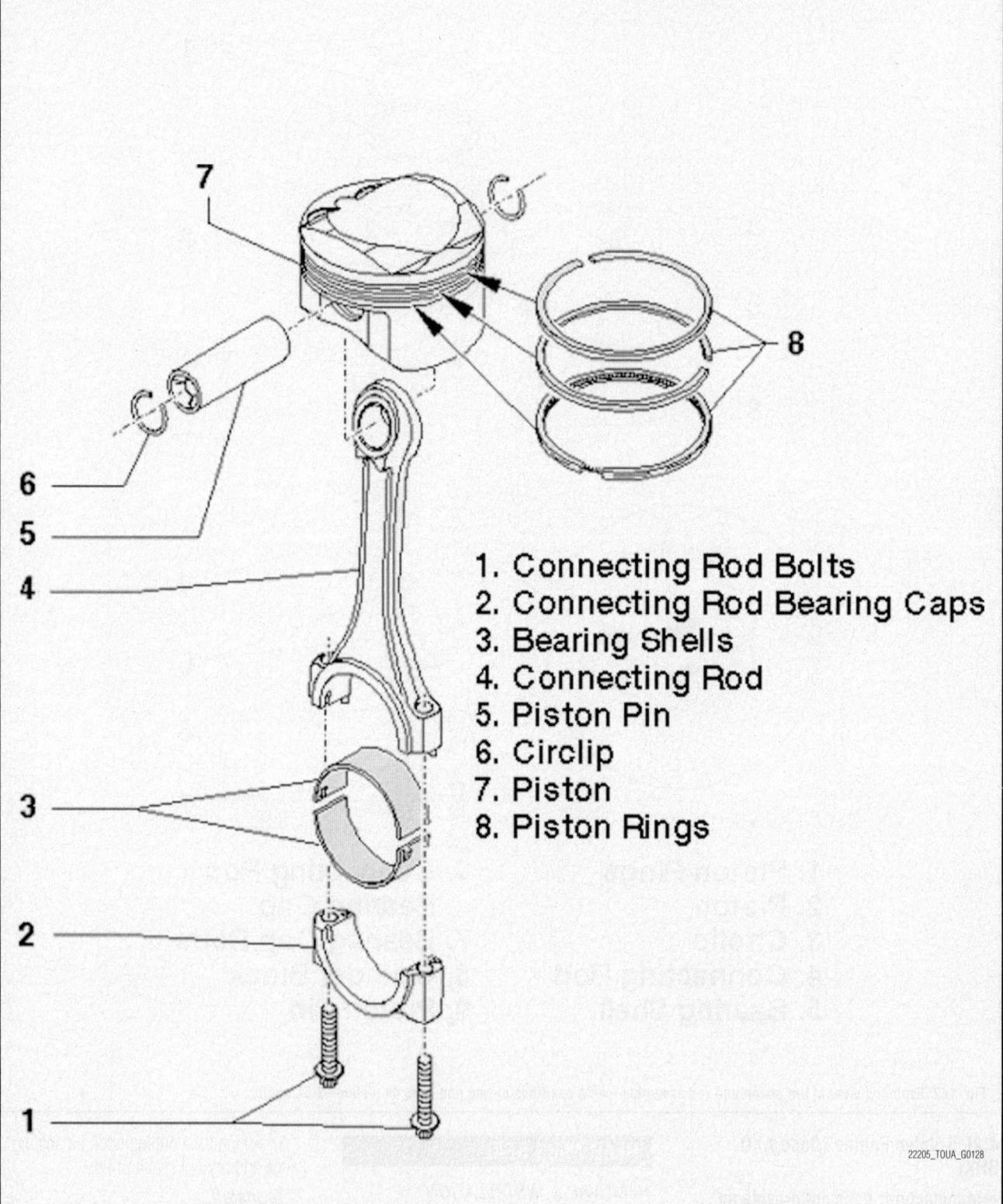

1. Connecting Rod Bolts
2. Connecting Rod Bearing Caps
3. Bearing Shells
4. Connecting Rod
5. Piston Pin
6. Circlip
7. Piston
8. Piston Rings

22205_TOUA_G0128

Fig. 143 Exploded view of the piston and rod assembly—note positions of ring end gaps on piston—4.2L 4-valve engine

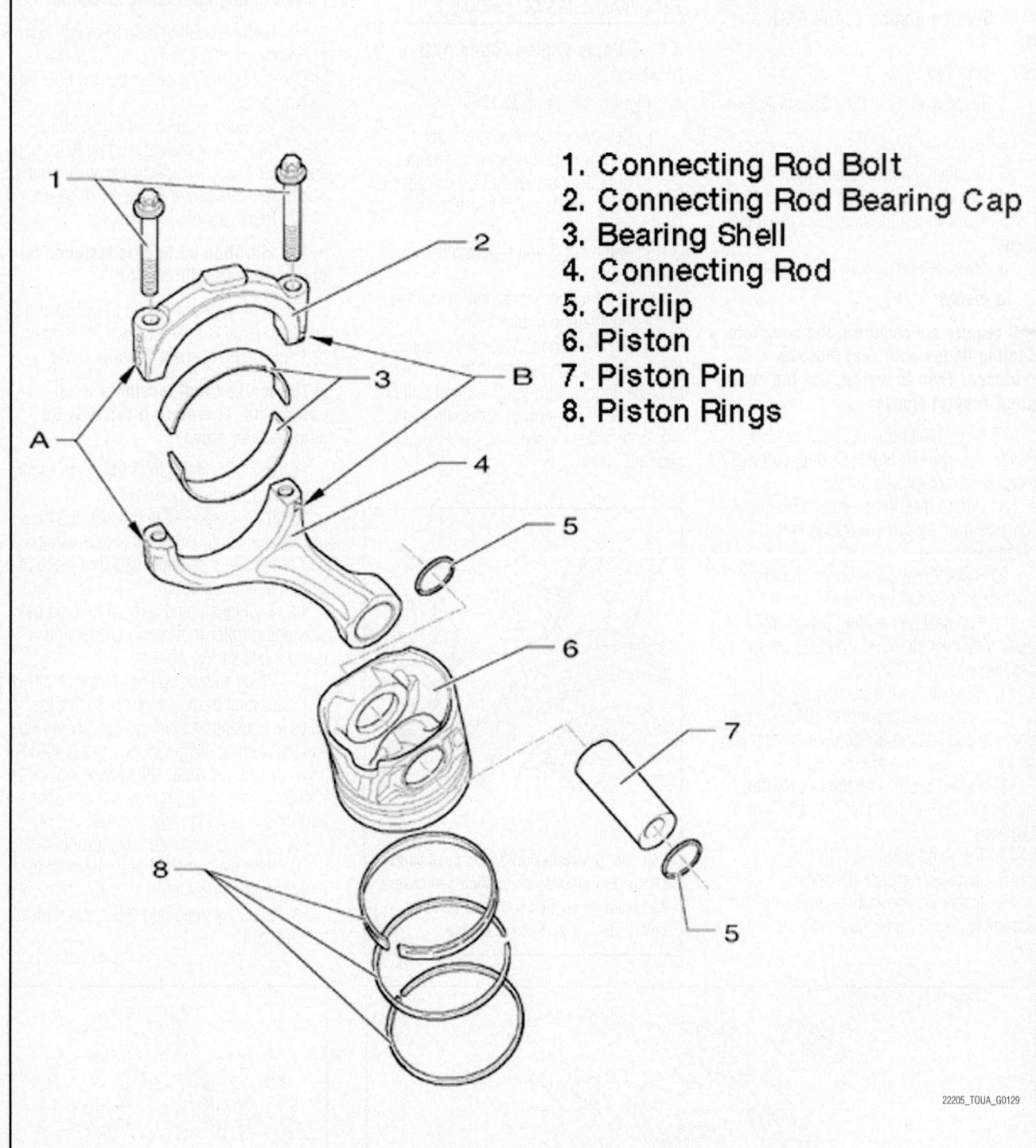

1. Connecting Rod Bolt
2. Connecting Rod Bearing Cap
3. Bearing Shell
4. Connecting Rod
5. Circlip
6. Piston
7. Piston Pin
8. Piston Rings

22205_TOUA_G0129

**Fig. 144 Exploded view of the piston and rod assembly—note positions of ring end gaps on piston—5.0L TDI engine**

7. Attach transmission to engine and install the assembly. See "Engine Assembly" in this section.

### 4.2L 5-Valve Engine (Code AXQ, BHX)

*See Figure 145.*

1. Remove engine. See "Engine Assembly."
2. Separate engine from transmission.
3. Remove the drive plate (flywheel).
4. Remove oil pan. See "Oil Pan."
5. Remove sealing flange from cylinder block.
6. Remove and discard old gasket.

### To install:

➡ **If repairs are required, the complete sealing flange with seal must be replaced. Then to install, use the supplied support sleeve.**

7. Clean sealing surfaces of sealing flange and cylinder block. Sealing surfaces must be free of oil and grease.
8. Before installing, remove oil remains from end of crankshaft with a clean cloth.
9. Place a new gasket onto alignment bushings of the cylinder head.
10. Pull seal and sealing flange with outer side over sleeve (T10122/1) onto assembly tool (T10122/2).
11. Separate both assembly sleeves.
12. Place assembly tool (T10122/2) with dry seal and sealing flange onto crankshaft flange.
13. Fasten sealing flange to cylinder block and tighten bolts to 7 ft. lbs. (10 Nm).
14. Install oil pan. See "Oil Pan."
15. Install drive plate (flywheel).
16. Install engine and transmission assembly. See "Engine Assembly" in this section.

## TIMING BELT, SPROCKETS, FRONT COVER AND SEAL

### *REMOVAL & INSTALLATION*

### 4.2L 5-Valve Engine (Code AXQ, BHX)

*See Figures 146 through 156.*

1. Remove noise insulation tray.
2. Remove coolant fan with mount.
3. Bring lock carrier into service position.
4. Mark accessory drive ribbed belt and remove it.
5. Remove left and right toothed (timing) belt cover.
6. Mark running direction of toothed belt using a felt tip marker.
7. Set crankshaft to top dead center (TDC). The marking on toothed belt cover must be aligned with notch on belt pulley.
8. Check position of camshaft gears. The large holes on securing plates must align on inside.

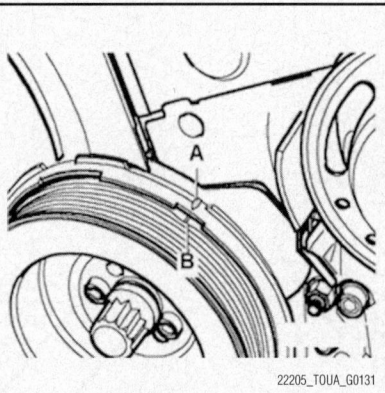

**Fig. 146 Set crankshaft to top dead center (TDC). The marking on toothed belt cover "A" must be aligned with notch on belt pulley "B"—4.2L 5-valve engine**

➡ **If the large holes are positioned on the outer side of the toothed belt gears, crankshaft must be turned one rotation further in engine running direction.**

9. Remove sealing plug from left-side of cylinder block. The TDC hole for crankshaft must be visible or felt behind hole for sealing plug.
10. Carefully install a crankshaft holder (3242) into hole of sealing plug up to its stop, and thereby secure crankshaft against turning.
11. Remove center toothed belt guard.
12. Remove vibration damper.

➡ **The vibration damper is fastened to the crankshaft with 8 bolts.**

13. Remove bolt from oil filter housing and then remove bolts of cover for toothed belt tensioning element. Remove cover.

➡ **The toothed belt tensioner is oil-dampened. Therefore it can only be compressed slowly.**

14. Use a locking pin (T40011) to secure tensioning element.
15. If necessary, align holes in the housing and tensioning element piston using a pair of needle nose pliers or a thin piece of wire, before tensioning.
16. Turn tensioning lever of toothed belt tensioning roller in downward direction using a hex socket wrench.
17. When tensioning lever pushes toothed belt tensioning element together so that the holes in housing and piston align, secure tensioning element using locking pin a (T40011).
18. Place camshaft adjustment tool (T40005) into securing plates of camshafts and loosen bolts approx. 5 turns.
19. Remove camshaft locator tool.
20. Remove camshaft gears from cone using proper puller tool.
21. Loosen tensioning roller and remove toothed belt.

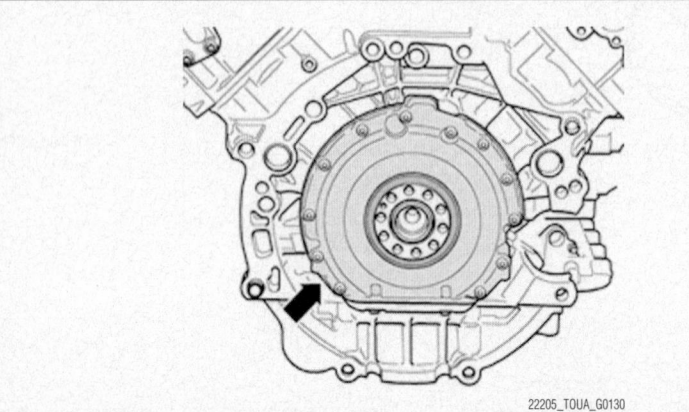

**Fig. 145 Remove sealing flange from cylinder block—4.2L 5-valve engine**

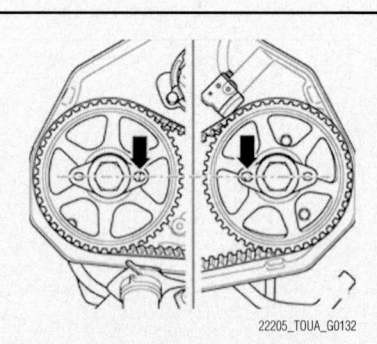

**Fig. 147 Check position of camshaft gears. The large holes on securing plates (arrows) must align on inside—4.2L 5-valve engine**

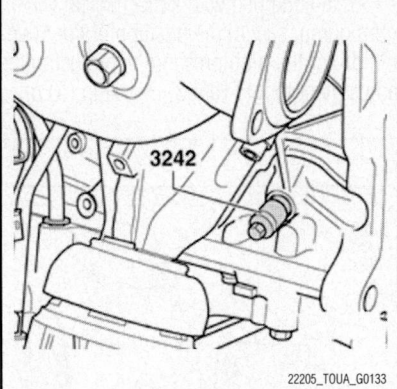

22205_TOUA_G0133

Fig. 148 Carefully install a crankshaft holder (3242) into hole up to stop, and thereby secure crankshaft against turning—4.2L 5-valve engine

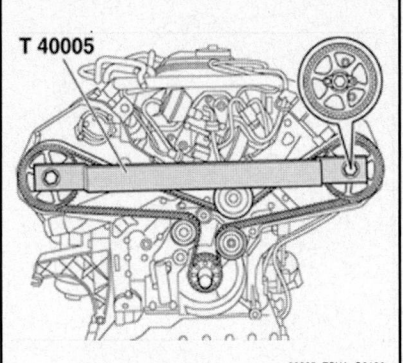

22205_TOUA_G0136

Fig. 151 Place camshaft adjustment tool (T40005) into securing plates of camshafts and loosen bolts approx. 5 turns—4.2L 5-valve engine

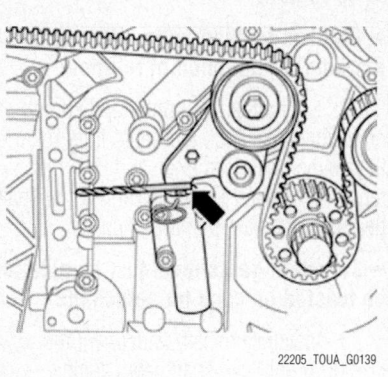

22205_TOUA_G0139

Fig. 154 Set a 5 mm drill between tensioning lever and piston of tensioning element—4.2L 5-valve engine

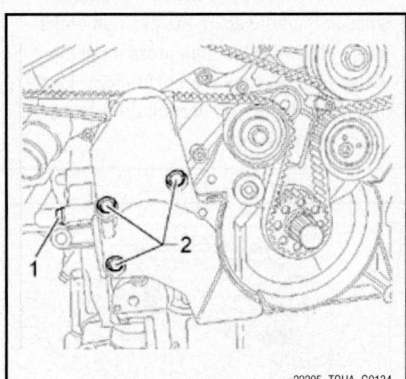

22205_TOUA_G0134

Fig. 149 Remove bolt from oil filter housing and then remove bolts of cover for toothed belt tensioning element—4.2L 5-valve engine

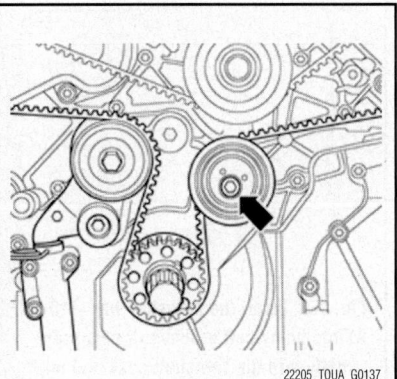

22205_TOUA_G0137

Fig. 152 Loosen tensioning roller (arrow) and remove toothed belt—4.2L 5-valve engine

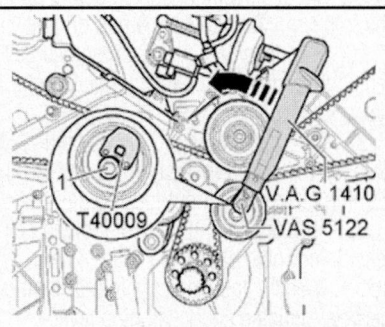

22205_TOUA_G0140

Fig. 155 Pretension still loosened tensioning roller using a torque wrench and a connected reversible ratchet with tensioning roller wrench counter-clockwise to 3 ft. lbs. (4 Nm)—4.2L 5-valve engine

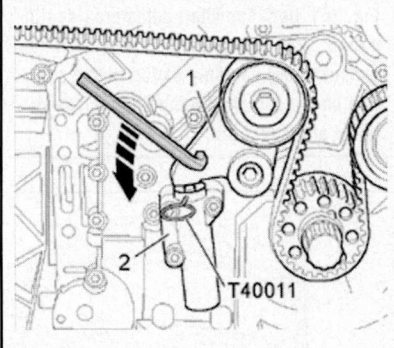

22205_TOUA_G0135

Fig. 150 Turn tensioning lever of toothed belt tensioning roller "1" in direction of arrow using a hex socket wrench, then when tensioning lever "2" pushes toothed belt tensioning element together so that the holes in housing and piston align, secure tensioning element using a locking pin (T40011)—4.2L 5-valve engine

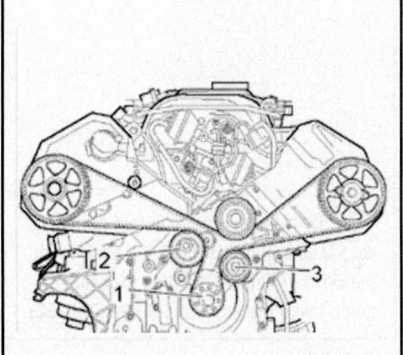

22205_TOUA_G0138

Fig. 153 Set toothed belt onto toothed belt crankshaft gear "1" first, then onto idler roller of the toothed belt tensioner "2", and then onto tensioning roller "3", and then, set it onto camshaft gears and coolant pump belt pulley—4.2L 5-valve engine

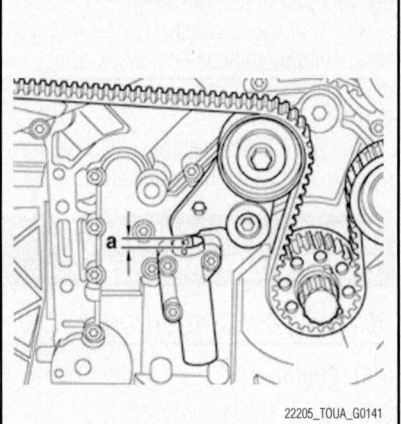

22205_TOUA_G0141

Fig. 156 Turn over crankshaft twice in direction of engine rotation and check adjustment dimension; it should be 0.197 in. (5 mm)—4.2L 5-valve engine

*To install:*

➡ **On used toothed belts, observe marking for direction of rotation.**

22. Set toothed belt onto toothed belt crankshaft gear first, then onto idler roller of the toothed belt tensioner, and then onto tensioning roller. Then, set it onto camshaft gears and coolant pump belt pulley.

➡ **Camshaft gears must just be able to be twisted on cone for camshafts.**

23. Install camshaft adjustment tool (T40005) onto camshaft gears again.

24. Set a 5 mm drill between tensioning lever and piston of tensioning element.

25. Pretension still loosened tensioning roller using a torque wrench and a connected reversible ratchet with tensioning roller wrench counter-clockwise to 3 ft. lbs. (4 Nm). In this position, tighten bolt to 33 ft. lbs. (45 Nm). Then remove 5 mm drill bit.

26. Turn tensioning lever of toothed belt tensioning roller counterclockwise using a hex socket wrench.

27. When tensioning lever has compressed the piston in the toothed belt tensioning element, pull out locking pin.

28. Turn tensioning lever of toothed belt tensioning roller clockwise using a hex socket wrench, and place a 7 mm drill bit between housing and tensioning lever. Tighten camshaft gears to 41 ft. lbs. (55 Nm).

29. Remove camshaft locator (T40005).

30. Remove 7 mm drill bit, that was inserted between housing and tensioning element.

31. Remove crankshaft holder (3242) from hole and install sealing plug and tighten to 22 ft. lbs. (30 Nm).

32. Turn over crankshaft twice in direction of engine rotation and check adjustment dimension. It should be 0.197 in. (5 mm).

33. The rest of assembly is basically a reverse of disassembling sequence.

## TIMING CHAIN, SPROCKETS, FRONT COVER AND SEAL

*REMOVAL & INSTALLATION*

### 3.6L Engine

*See Figures 157 through 166.*

1. Adjust the crankshaft at the vibration damper bolt in the direction of engine rotation to the cylinder 1 TDC marking.

2. Position the camshafts in the cylinder head to TDC No. 1 cylinder. 1. The cam lobes of cylinder 1 must face each other.

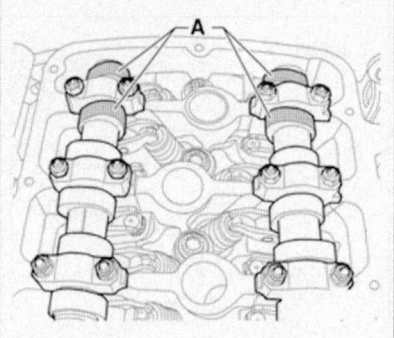

**Fig. 157 Position the camshafts in the cylinder head to TDC No. 1 cylinder. 1. The cam lobes of cylinder 1 must face each other**

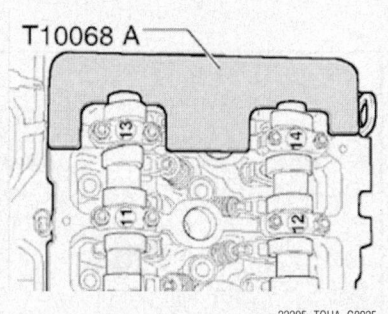

**Fig. 158 Insert the camshaft bar (T10068 A) into both shaft grooves. If necessary, slightly turn the camshaft back and forth using an open-end wrench**

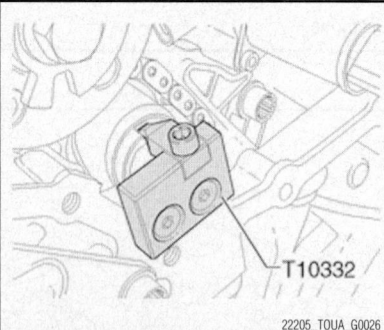

**Fig. 159 If not equipped with a mechanical vacuum pump, secure the position of the drive pinion for the high pressure pump using the adjustment tool (T10332)**

3. Insert the camshaft bar (T10068 A) into both shaft grooves. If necessary, slightly turn the camshaft back and forth using an open-end wrench.

4. If not equipped with a mechanical vacuum pump, secure the position of the drive pinion for the high pressure pump using the adjustment tool (T10332).

5. If equipped with a mechanical vacuum pump, secure the position of the drive pinion for the high pressure pump using the adjustment tool (T10363). Marking (A) on

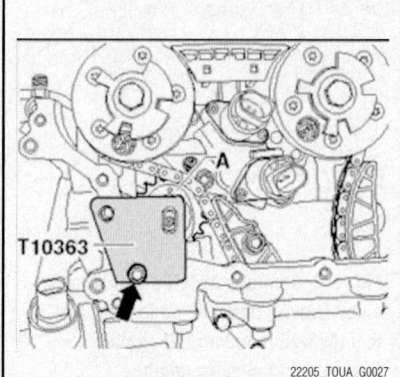

**Fig. 160 If equipped with a mechanical vacuum pump, secure the position of the drive pinion for the high pressure pump using the adjustment tool (T10363). Marking (A) on the high pressure drive cams must be at the top**

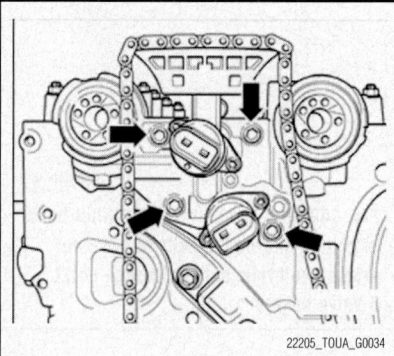

**Fig. 161 Both camshaft adjusters (identification: "24E" on intake side and "32A" on exhaust side) can only bolted in one position on the camshaft mountings (arrows) by an alignment pin**

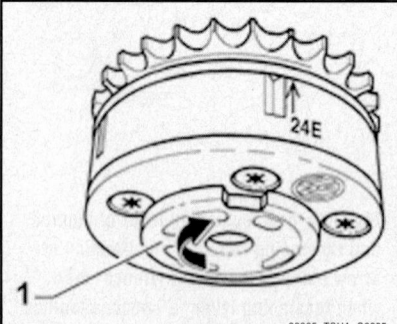

**Fig. 162 Turn the sensor wheel on the intake camshaft adjuster clockwise until it stops and hold the adjuster in this position**

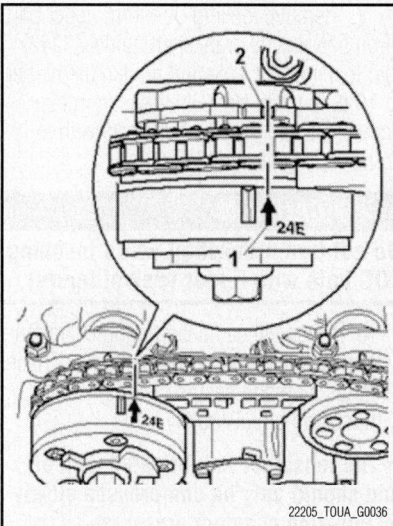

Fig. 163 Arrow on camshaft adjuster "24E" must align with the right notch of control housing. Markings on control housing

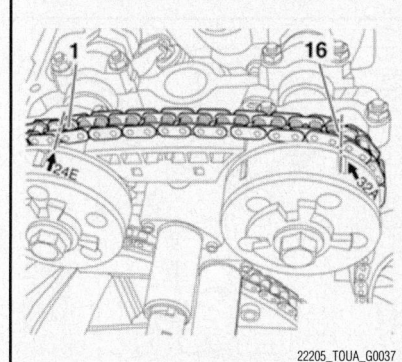

Fig. 164 Now count exactly 16 rollers on the timing chain (to the right from tooth with the arrow aligning with the notch)

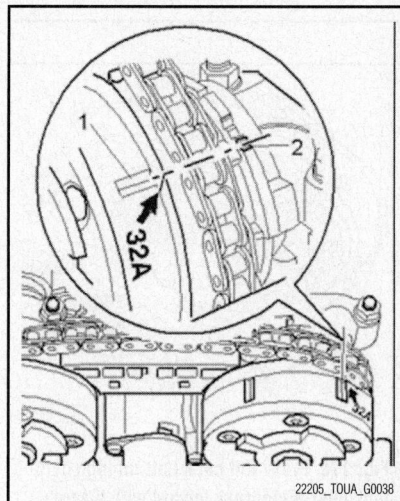

Fig. 165 Place the camshaft adjuster "32A" with camshaft timing chain installed on exhaust camshaft

the high pressure drive cams must be at the top.

➡ **Both camshaft adjusters (identification: "24E" on intake side and "32A" on exhaust side) can only bolted in one position on the camshaft mountings by an alignment pin.**

6. On the intake camshaft, turn the sensor wheel on the intake camshaft adjuster clockwise until it stops and hold the adjuster in this position.

➡ **If the intake camshaft adjuster is attached to the camshaft, the adjuster must be rotated left accordingly with the chain sprocket and then the camshaft timing chain must be routed.**

7. Place the intake camshaft adjuster with the timing chain in place on the camshaft. Note the following:

- The timing chain for the high pressure pump drive train sprocket must not hang through.
- It must be easy to mount the camshaft adjuster with the timing chain taut and can then be tighten by hand.
- Arrow on camshaft adjuster "24E" must align with the right notch of control housing. Markings on control housing.

a. Now count exactly 16 rollers on the timing chain (to the right from tooth with the arrow aligning with the notch). Mark this roller with a color marker.

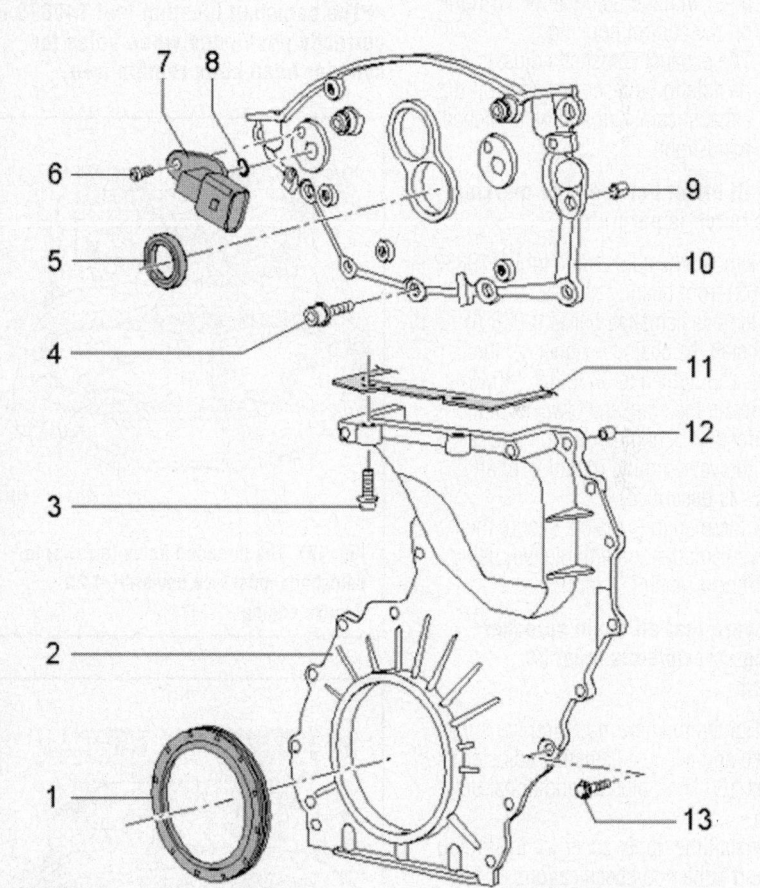

1. Seal
2. Sealing Flange
3. Bolt
4. Bolt
5. Seal
6. Bolt
7. Camshaft Position (CMP) Sensor
8. O-Ring
9. Alignment Pins
10. Cover
11. Cylinder Head Gasket
12. Alignment Pins
13. Bolt

Fig. 166 Exploded view of the timing chain cover—3.6L engines

➡The exhaust camshaft adjuster is locked in the rest state. Therefore, the sensor wheel cannot be rotated when adjusting the valve timing. If the locking mechanism in the rest state is not engaged (locked), turn the adjuster in both directions by hand until it locks. If that is not possible, replaced the camshaft adjuster.

8. Now insert the exhaust camshaft adjuster "32A" positioned with the tooth at arrow marking into the camshaft timing chain so that the exact, previously counted 16 rollers lie between markings "24E" and "32A".

9. Place the camshaft adjuster "32A" with camshaft timing chain installed on exhaust camshaft. Note the following:

- The mark on the camshaft adjuster (which the "arrow" points to), must align with the notch at the far right on the control housing.
- The exhaust camshaft adjuster must be able to be inserted easily on the exhaust camshaft and be tightened hand-tight.

➡A small offset between the marking and the notch is permitted.

10. Remove the adjustment tool (T10332 or T10363) from bearing shaft.

11. Remove camshaft bar (T10068 A).

12. Install the chain tensioner for the timing chain and tighten to 30 ft. lbs. (40 Nm).

13. Rotate the crankshaft two revolutions in the direction of engine rotation and recheck the valve timing (position of all markings as described).

14. If the markings match, secure the respective camshaft to be tightened, using an open-end wrench (27 mm).

➡Be aware that all chain sprocket securing screws/bolts must be replaced.

15. Tighten the new mounting bolts for the intake and exhaust camshaft adjuster to 44 ft. lbs. (60 Nm), plus an additional 90 degrees.

16. Install the upper cover for the timing chain, and tighten to specifications as follows (reference bolt numbering in the illustration):

- Bolt 3: 17 ft. lbs. (23 Nm)
- Bolt 4: 6 ft. lbs. (8 Nm)
- Bolt 6: 7 ft. lbs. (10 Nm)
- Bolt 13: 7 ft. lbs. (10 Nm)

## 4.2L 4-Valve Engine (Code BAR)

*See Figures 167 through 172.*

➡This description provides for the timing chains for the camshafts to remain on the engine. Even if work is performed only on one of the cylinder heads, the procedure described must be followed, since then the valve timing at both cylinder heads must be adjusted.

1. Remove engine.
2. Separate engine and transmission.
3. Remove left and right timing chain covers.
4. Remove cylinder head covers.
5. Using a proper socket and adapter, rotate crankshaft in direction of engine rotation to TDC.

➡The threaded holes in camshafts must face upward.

6. Mount camshaft locating tool (T40070) to both cylinder heads and tighten bolts to 18 ft. lbs. (25 Nm).

➡The camshaft locating tool T40070 is correctly positioned when holes for cylinder head bolts remain free.

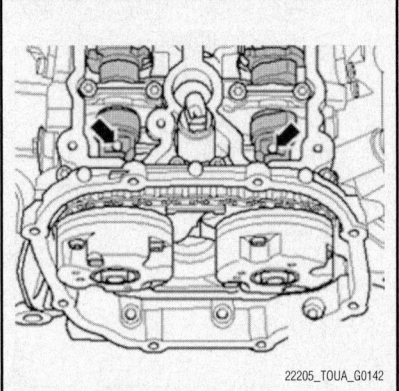

**Fig. 167 The threaded holes (arrows) in camshafts must face upward—4.2L 4-valve engine**

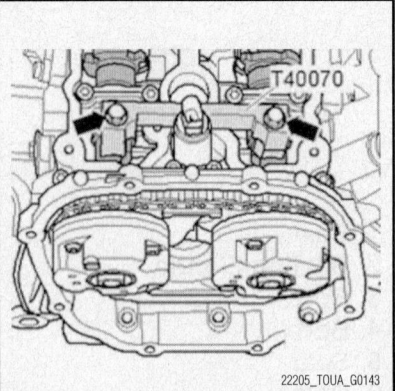

**Fig. 168 Mount camshaft locating tool (T40070) to both cylinder heads and tighten bolts to 18 ft. lbs. (25 Nm)—4.2L 4-valve engine**

7. Remove locking bolt from upper part of oil pan. Install crankshaft holder (3242) into locking bolt hole and tighten the holder to 15 ft. lbs. (20 Nm). If necessary rotate crankshaft very slightly back and forth to completely center holder.

### ✳✳ CAUTION

**Do not turn crankshaft while touching TDC hole with finger (risk of injury).**

8. Press left camshaft timing chain tensioner glide track inward with a screwdriver as far as stop and secure chain tensioner with locking pin (T40071).

➡The tensioner is lubricated with oil and should only be compressed slowly by applying constant pressure.

9. Press right camshaft timing chain tensioner glide track inward with a screwdriver as far as stop and secure chain tensioner with locking pin T40071 .

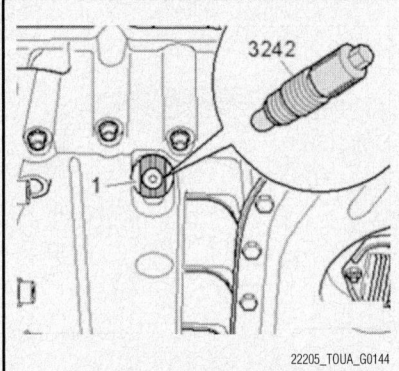

**Fig. 169 Remove locking bolt from upper part of oil pan and install crankshaft holder (3242) into locking bolt hole and tighten the holder—4.2L 4-valve engine**

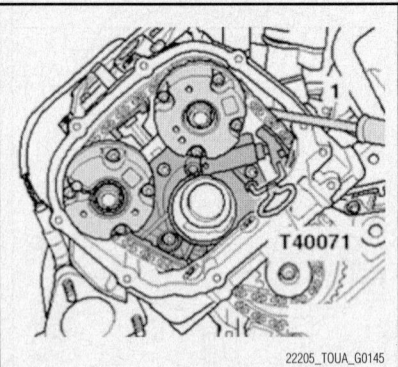

**Fig. 170 Press left camshaft timing chain tensioner glide track inward with a screwdriver "1" as far as stop and secure chain tensioner with locking pin—4.2L 4-valve engine**

10. Mark position of camshaft adjuster to camshaft for reinstallation.

11. Remove bolts remove both camshaft adjusters from both cylinder banks.

### To install:

➡️Replace bolts which have been tightened to torque.

### ✲✲ CAUTION

**When turning camshaft, crankshaft must not be at "TDC" for any cylinder. Valves and/or pistons may be damaged.**

12. If removed, install the drive chain for timing mechanism. See "Timing Mechanism Drive Chain" in this section.

13. Secure crankshaft in TDC position using crankshaft holder (3242).

14. Ensure camshaft locating tool (T40070) is mounted on both cylinder heads and fastened to 18 ft. lbs. (25 Nm).

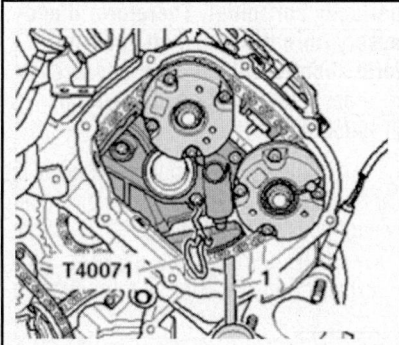

22205_TOUA_G0146

**Fig. 171 Press right camshaft timing chain tensioner glide track inward with a screwdriver "1" as far as stop and secure chain tensioner with locking pin—4.2L 4-valve engine**

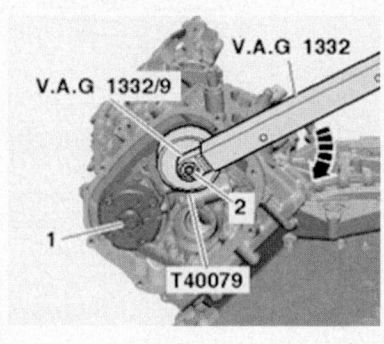

22205_TOUA_G0147

**Fig. 172 Positioning adapter to adjuster and setting pretension—left camshaft shown; right similar—4.2L 4-valve engine**

15. Reinstall left camshaft adjuster according to mark applied during removal.

➡️Replace camshaft bolts.

16. Place camshaft timing chain onto drive sprocket and onto camshaft adjusters and loosely thread in camshaft adjuster bolts.

➡️Both camshaft adjusters must be able to still be rotated on camshaft and must not tip.

17. Remove the locking pin (T40071).

18. Repeat for the right camshaft.

➡️A 2nd technician is needed for further work.

19. Position adapter (T40079) on intake camshaft adjuster at cylinder head.

20. Position torque wrench (VAG 1332) with open ring spanner insert (VAG 1332/9 on adapter T40079).

21. Pretension camshaft adjuster to 30 ft. lbs. (40 Nm) in clockwise direction and hold tension.

22. Tighten bolt on exhaust camshaft simultaneously to initial specification of 44 ft. lbs. (60 Nm).

23. Continue holding pretension on exhaust camshaft and pre-tighten bolt on exhaust camshaft to 44 ft. lbs. (60 Nm).

24. Repeat on the other cylinder bank camshaft.

25. Remove camshaft locating tool (T40070) from both cylinder heads.

26. First tighten outer camshaft bolt and then inner camshaft bolt to final torque of 59 ft. lbs. (80 Nm), plus an additional 90 degrees.

27. Remove the crankshaft holder (3242).

28. Using socket (T40058), turn crankshaft two complete rotations in direction of engine rotation until crankshaft stands at TDC again.

➡️If rotated unintentionally beyond TDC, turn back crankshaft again approximately 30 degree and set to TDC again.

29. Ensure the threaded holes in camshafts face upward.

30. Mount camshaft locating tools (T40070) to both cylinder heads and tighten bolts to 18 ft. lbs. (25 Nm).

➡️The camshaft locating tool (T40070) is correctly positioned when holes for cylinder head bolts remain free.

31. Install crankshaft holder (3242) in bore and tighten to 15 ft. lbs. (20 Nm).

➡️**The crankshaft holder (3242) must engage in locating hole of crankshaft, otherwise repeat adjustment.**

32. Remove camshaft locating tools from both cylinder heads.

33. Remove crankshaft holder (3242).

34. Install TDC mark sealing plug with new seal in upper part of oil pan. Tighten plug to 26 ft. lbs. (35 Nm).

35. The rest of installation is in reverse order of removal, note the following:

a. Install left and right timing chain covers.

b. Install cylinder head covers.

### TIMING MECHANISM DRIVE CHAIN

#### REMOVAL & INSTALLATION

#### 4.2L 4-Valve Engine (Code BAR)

*See Figure 173.*

1. Remove engine. See "Engine Assembly" in this section.

2. Separate engine and transmission.

3. Remove or disconnect the following:

- Drive plate
- Cylinder head covers
- Left and right timing chain covers
- Intake manifold. See "Intake Manifold."
- Oil filter housing
- Lower timing chain cover
- Camshaft timing chains
- Power take off drive chain. See "Power Take Off Drive Chain" in this section.

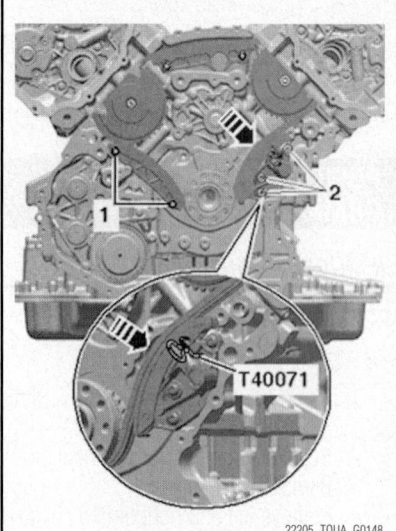

22205_TOUA_G0148

**Fig. 173 Showing the timing mechanism drive chain guide rail and tensioner—4.2L 4-valve engine**

4. Push drive chain tensioner guide rail outward from center and secure chain tensioner using locking pin (T40071).

5. Mark drive chain running direction with paint.

6. Remove bolts and remove guide rail.

7. Remove bolts and remove chain tensioner.

8. Remove timing mechanism drive chain.

### To install:

9. Installation is in reverse order of removal, note the following:

➡**Replace bolts which have been tightened to torque.**

a. Route timing mechanism drive chain according to marks applied to drive chain sprockets during removal.

b. Install guide rail and tighten bolts.

c. Install chain tensioner and tighten bolts.

d. Press drive chain tensioner guide rail outward from center and remove locking pin (T40071) from chain tensioner.

e. Install the following:
- Power take off drive chain. See "Power Take Off Drive Chain."
- Camshaft timing chains. See "Timing Chain, Sprockets, Front Cover and Seal."
- Lower timing chain cover
- Crankshaft seal, timing chain side. See "Rear Main Seal."
- Oil filter housing
- Intake manifold. See "Intake Manifold."
- Left and right timing chain covers
- Cylinder head covers
- Drive plate

f. Install engine. See "Engine Assembly."

g. Add engine oil and check oil level.

## POWER TAKE OFF DRIVE CHAIN

### REMOVAL & INSTALLATION

#### 4.2L 4-Valve Engine (Code BAR)

*See Figure 174.*

1. Remove engine. See "Engine Assembly."

2. Separate engine and transmission.

3. Remove drive plate.

4. Remove left and right timing chain covers.

5. Remove intake manifold. See "Intake Manifold."

6. Remove oil filter housing.

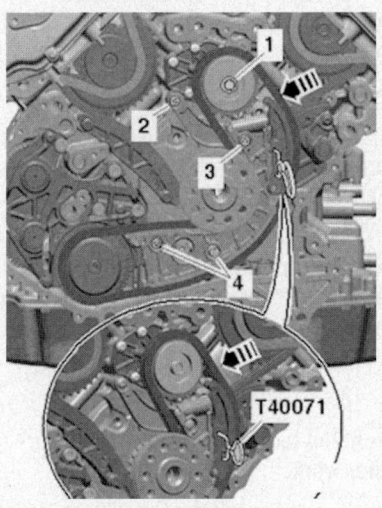

22205_TOUA_G0149

**Fig. 174 Press tensioning rail in direction of arrow and secure chain tensioner with locking pin (T40071). Remove bolt "1" and remove idler sprocket—4.2L 4-valve engine**

7. Remove lower timing chain cover.

8. Mark running direction of power take off chain with paint.

9. Press tensioning rail in direction of arrow and secure chain tensioner with locking pin (T40071). Remove bolt "1" and remove idler sprocket.

➡**When removing, be careful of spring in drive spur gear shaft.**

10. Remove bolts "2" through "4" (see above illustration) and remove chain tensioner.

11. Remove power take off drive chain.

### To install:

12. Installation is in reverse order of removal, note the following:

a. Replace gaskets.

b. Secure all hose connections with hose clamps appropriate for the model.

c. Note the following torque specifications during installation:
- Chain tensioner on cylinder block: 7 ft. lbs. (9 Nm)
- Idler pulley mounting pins to mounting bracket: 31 ft. lbs. (42 Nm)

d. Install the following:
- Lower timing chain cover
- Crankshaft seal, timing chain side. See "Rear Main Seal."
- Oil filter housing
- Intake manifold. See "Intake Manifold" in this section.
- Left and right timing chain covers
- Drive plate

13. Install engine. See "Engine Assembly."

14. Add engine oil and check oil level.

## VALVE TIMING

### ADJUSTMENT

#### 3.6L Engine

*See Figures 175 and 176.*

1. Remove insulation panel.

2. Remove intake manifold. See "Intake Manifold."

3. Remove cylinder head covers.

4. Adjust the crankshaft at the vibration damper bolt in the direction of engine rotation to the cylinder 1 TDC marking.

➡**Cams lobes of cylinder 1 must face inward toward each other.**

5. Insert the camshaft bar (T10068 A) onto both shaft grooves.

➡**Due to the camshaft adjuster function, the camshaft grooves may not be perfectly horizontal. Therefore, if necessary, turn the camshaft back and forth slightly with an open end wrench in order to insert the camshaft bar (T10068 A).**

6. Verify the installation markings of the camshaft adjuster with the markings on the control housing as follows:

a. The "arrow" on the intake or exhaust camshaft adjuster must align

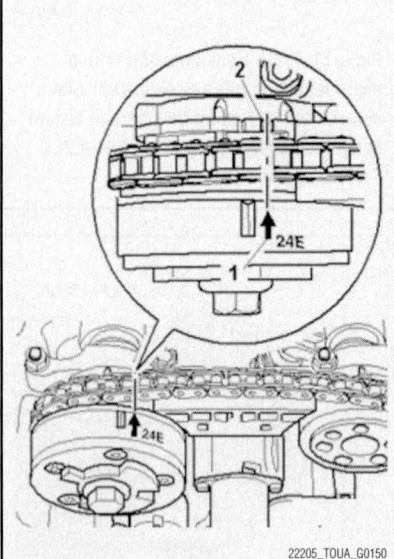

22205_TOUA_G0150

**Fig. 175 The "arrow" on the intake or exhaust camshaft adjuster must align with the notch on the far right of the control housing—3.6L engine**

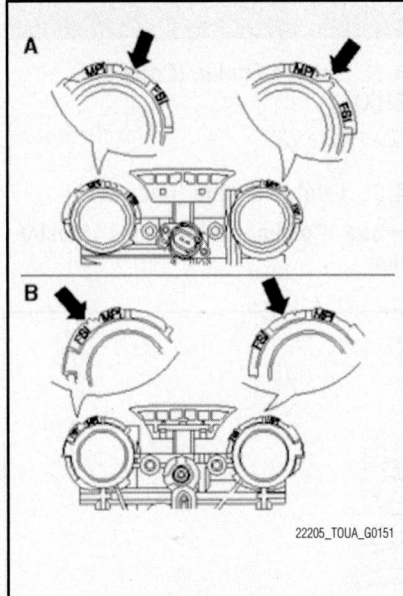

Fig. 176 Showing the markings on the
control housing from front and rear views:
view "A" flywheel side; view "B" vibration
damper side—3.6L engine

with the notch on the far right of the control housing.

➡A small offset between the marking
"1" and notch "2" is permitted.

7. Ensure the distance between the
markings on the camshaft adjuster are
exactly 16 rollers of camshaft timing chain.
8. If the markings do not match, reinstall camshaft adjuster and/or camshaft
drive timing chain.
9. If the markings match, install:
• Cylinder head cover
• Intake manifold

### 5.0L TDI Engine

*See Figures 177 through 179.*

1. Remove engine and transmission.
See "Engine Assembly."
2. Remove tandem and fuel pumps
from cylinder heads. See "DIESEL FUEL
INJECTION SYSTEM" section.
3. Remove noise insulation and covers
from cylinder heads.
4. Remove acoustic covers and cylinder
head cover.
5. Turn engine using counter-hold tool
(V10 T10172) until at least one camshaft
locator bar can be inserted.
6. Place clamping device (T10199)

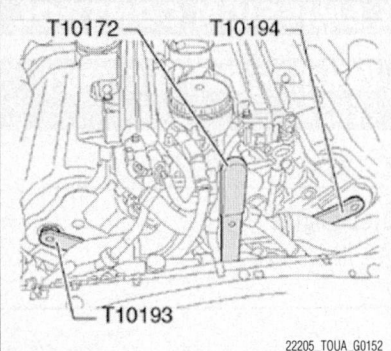

Fig. 177 Turn engine using counter-hold
tool (V10 T10172) until at least one
camshaft locator bar can be inserted—
5.0L TDI engine

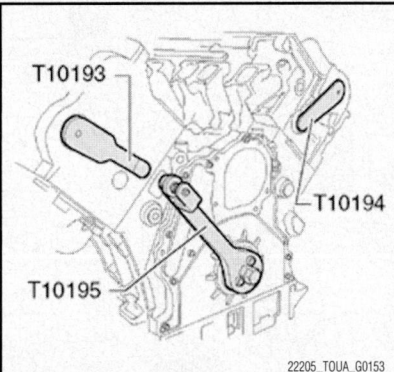

Fig. 178 Install Camshaft Tool (T10195)
onto crankshaft stub. Observe alignment
pin while doing so, and then carefully
rotate crankshaft until pin Camshaft Tool
is inserted into alignment bore—5.0L TDI
engine

onto wheel of secured camshaft and loosen
camshaft wheel bolt. Wheel can now be
turned freely on camshaft.
7. If only one camshaft locator bar
could be inserted:
a. Turn engine using counterhold tool
(V10 T10172) until camshaft locator bar
for other cylinder bank can be inserted.
b. Place clamping device (T10199) on
camshaft gear and loosen its bolt. Wheel
can also now be turned freely on
camshaft.
c. Remove vibration damper. For this,
use counterhold tool (V10 T10172) with
threading pins (T10172/1).
8. Install Camshaft Tool (T10195)
onto crankshaft stub. Observe alignment pin
while doing so.

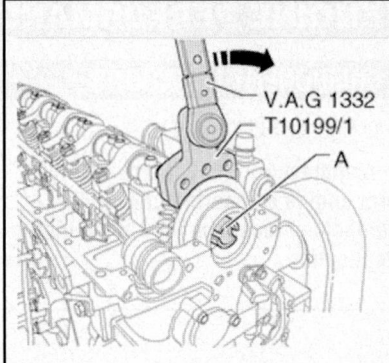

Fig. 179 Place clamping device
(T10199/1) as shown and tighten its
clamping bolts to 52 ft. lbs. (70 Nm), then
place a torque wrench into square of
clamping device and push with 59 ft. lbs.
(80 Nm) in opposite direction of engine
rotation direction to eliminate play from
gear drive—5.0L TDI engine

9. Now, carefully rotate crankshaft until
pin Camshaft Tool is inserted into alignment
bore.
10. Place clamping device (T10199/1) as
shown and tighten its clamping bolts to 52
ft. lbs. (70 Nm).
11. Place a torque wrench into square of
clamping device and push with 59 ft. lbs.
(80 Nm) in opposite direction of engine
rotation direction to eliminate play from
gear drive.
12. Maintain specified force and tighten
mounting bolt of camshaft gear to a tightening torque of 50 Nm using socket (T10198).
Remove clamping device (T10199/1).
13. Place clamping device (T10199) on
camshaft gear.

➡Make sure that clamping device
makes contact flush on cylinder head.

14. Tighten clamping device bolts to
30 ft. lbs. (40 Nm).
15. Tighten camshaft gear, using a
socket (T10198) to 111 ft. lbs. (150 Nm),
plus an additional 90 degrees.
16. Remove the clamping device.
17. Remove anchors from camshafts and
crankshaft.
18. Continue turning crankshaft two
rotations in direction of engine rotation until
crankshaft is again at TDC for cylinder 1.
19. Check whether anchors for
camshafts as well as for crankshaft can be
inserted simultaneously.

# ENGINE PERFORMANCE & EMISSION CONTROL

## COMPONENT LOCATIONS

*See Figures 180 through 183.*

➡Manufacturer did not provide a general engine compartment overview of component locations for 3.6L engine application.

## CAMSHAFT POSITION (CMP) SENSOR

### LOCATION

**3.6L Engine**
*See Figure 184.*

### 4.2L 5-Valve Engine (Code AXQ, BHX)

*See Figure 185.*

### 4.2L 4-Valve Engine (Code BAR)

➡See "Component Locations" illustration.

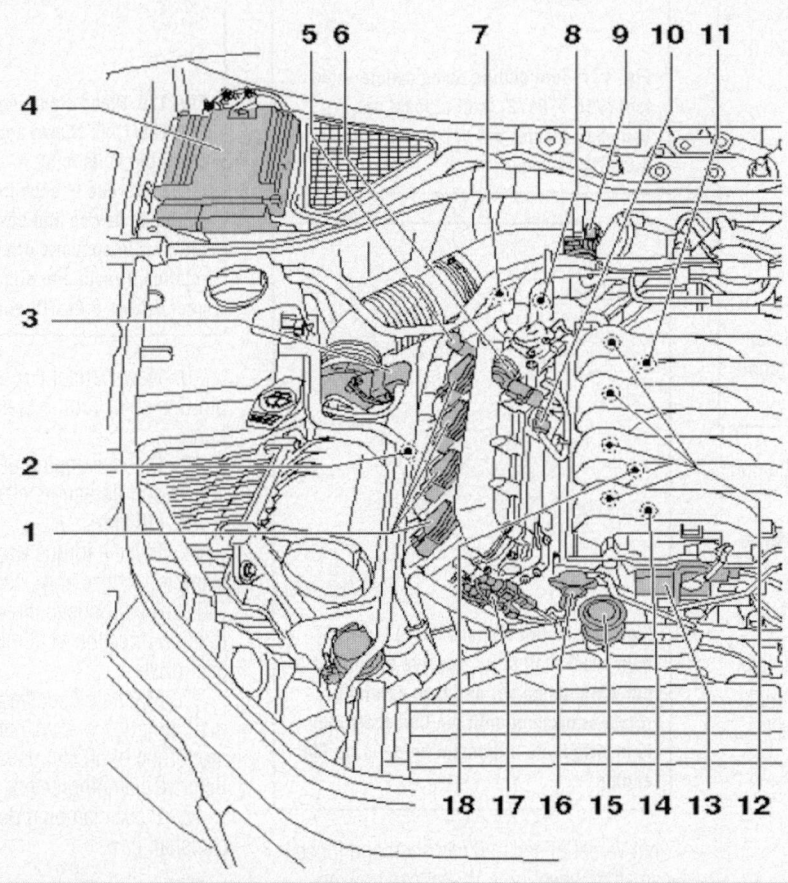

| 1 - Ignition coils, Cylinder bank 1: **N70, N127, N291 N292** | 10 - Camshaft Position (CMP) sensor -**G40**- |
| --- | --- |
| 2 - Camshaft position (CMP) sensor 3 -**G300**- | 11 - Knock sensor 2 -**G66**- |
| 3 - Mass Air Flow (MAF) Sensor -**G70**-/Intake Air Temperature (IAT) Sensor -**G42**- | 12 - Fuel injectors, Cylinder bank 1: **N30 N31 N32 N33** |
| 4 - Engine control module -**J623**- | 13 - Variable Intake Manifold Runner Motor -**V183**- |
| 5 - Camshaft Adjustment Valve 1 (exhaust) -**N318**- | 14 - Fuel Pressure Sensor -**G247**- |
| 6 - Fuel Metering Valve -**N290**- | 15 - Secondary Air Injection (AIR) combi-valve |
| 7 - Engine Coolant Temperature (ECT) Sensor -**G62**- | 16 - Intake Manifold Runner Position Sensor -**G336**- |
| 8 - Camshaft Adjustment Valve 1 -**N205**- | 17 - Electrical connectors: Black for Knock Sensor 1 -**G61**, Brown for **G66** |
| 9 - Evaporative Emission (EVAP) Canister Purge Solenoid Valve -**N80**- | 18 - Knock sensor 1 -**G61**- |

22205_TOUA_G0155

**Fig. 180 Engine compartment view of sensor component locations—4.2L 5-valve engine—1 of 2**

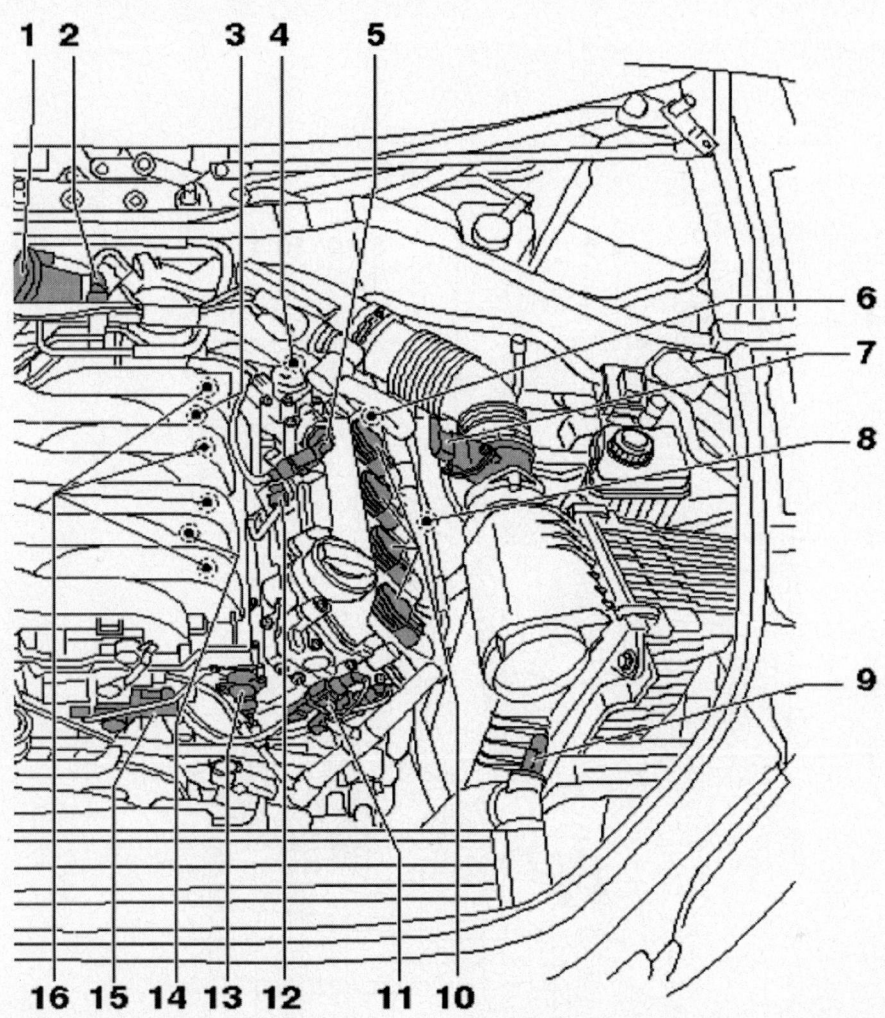

| 1 - | Throttle valve control module -J338- | 9 - | Engine Coolant Temperature (ECT) Sensor (on Radiator) -G83- |
|---|---|---|---|
| 2 - | Low Fuel Pressure Sensor -G410- | 10 - | Ignition coils, Cylinder bank 2: **N323 N324 N325 N326** |
| 3 - | Knock sensor 4 -G199- | 11 - | Electrical connectors<br>- Black for Knock Sensor 3 -G198-<br>- Brown for Knock sensor (KS) 4 -G199- |
| 4 - | Camshaft Adjustment Valve 2 -N208- | 12 - | Camshaft position (CMP) sensor 2 - G163- |
| 5 - | Fuel Metering Valve 2 -N402- | 13 - | Intake Manifold Runner Position Sensor 2 -G512- |
| 6 - | Camshaft Adjustment Valve 2 (exhaust) - N319- | 14 - | Knock sensor 3 -G198- |
| 7 - | Mass Air Flow (MAF) Sensor 2 -G246- | 15 - | Intake Flap Motor -V157- |
| 8 - | Camshaft position (CMP) sensor 4 - G301- | 16 - | Fuel injectors, Cylinder bank 2 |

22205_TOUA_G0156

**Fig. 181 Engine compartment view of sensor component locations—4.2L 5-valve engine—2 of 2**

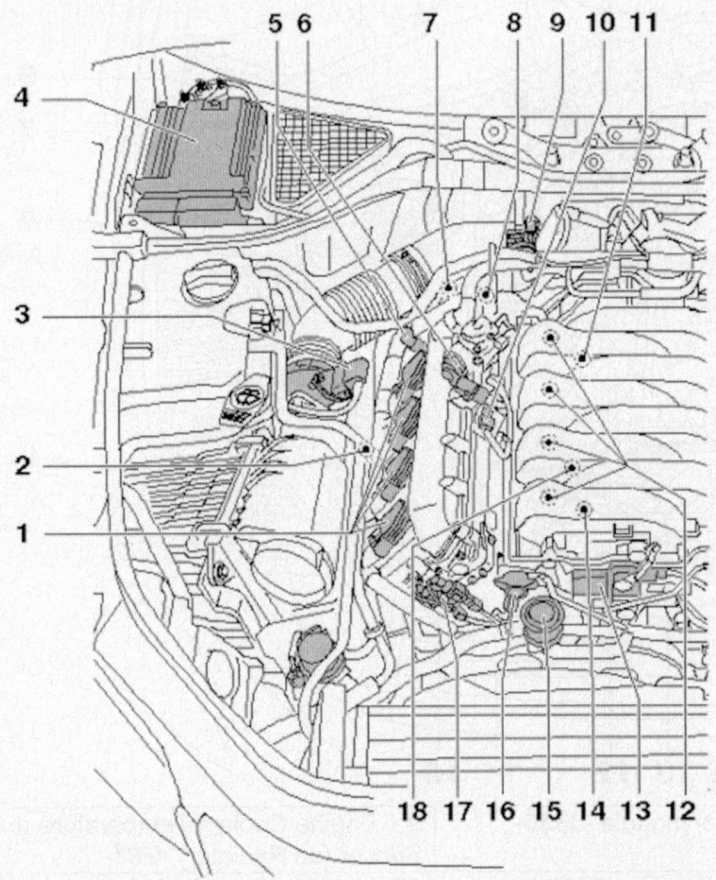

1. Ignition Coils 1, 2, 3, 4
2. Camshaft Position (CMP) Sensor 3
3. Mass Airflow (MAF) Sensor/Intake
   Air Temperature (IAT) Sensor
4. Engine Control Module
5. Camshaft Adjustment Valve 1 (Exhaust)
6. Fuel Metering Valve 1
7. Engine Coolant Temperature (ECT) Sensor
8. Camshaft Adjust Valve 1 (Intake)
9. Fuel Metering Valve

10. Camshaft Position (CMP) Sensor 1
11. Knock Sensor 2
12. Fuel Injectors
13. Variable Intake Manifold Runner Motor
14. Fuel Pressure Sensor
15. Secondary Air Injection (AIR)
    Combination Valve
16. Intake Manifold Runner Position Sensor 1
17. Electrical Connectors
18. Knock Sensor 1

22205_TOUA_G0159

**Fig. 182 Engine compartment overview of sensor component locations on left bank—4.2L 4-valve engine**

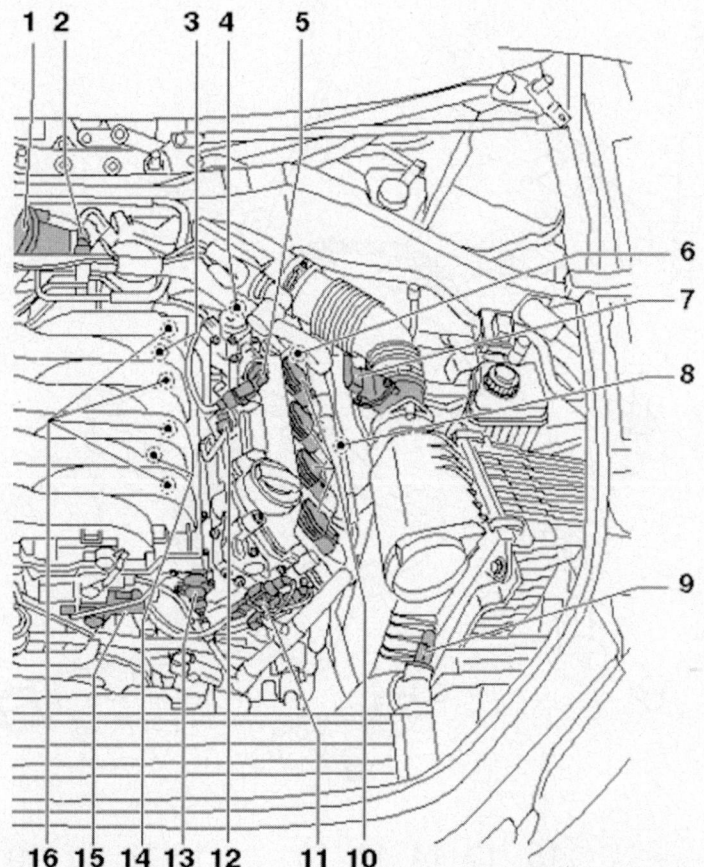

1. Throttle Valve Control Module
2. Low Fuel Pressure Sensor
3. Knock Sensor 4
4. Camshaft Adjustment Valve 2 (Intake)
5. Fuel Metering Valve 2
6. Camshaft Adjustment Valve 2 (Exhaust)
7. Mass Airflow (MAF) Sensor 2
8. Camshaft Position (CMP) Sensor 4

9. Engine Coolant Temperature (ECT) Sensor
10. Ignition Coils 5, 6, 7, 8
11. Electrical Connectors
12. Camshaft Position (CMP) Sensor 2
13. Intake Manifold Runner Position Sensor 2
14. Knock Sensor 3
15. Intake Flap Motor
16. Fuel Injectors

22205_TOUA_G0160

**Fig. 183 Engine compartment overview of sensor component locations on right bank—4.2L 4-valve engine**

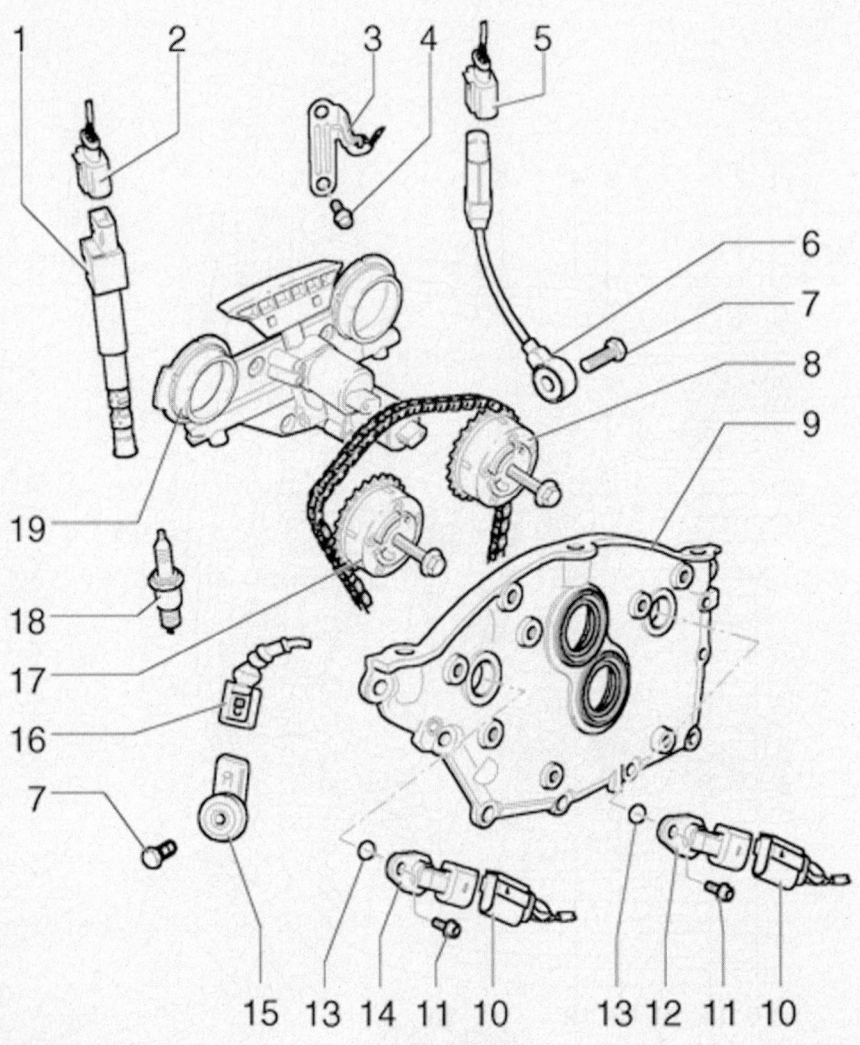

1. Ignition Coil
2. Connector
3. Bracket
4. Bolt
5. Connector
6. Knock Sensor 1
7. Bolt
8. Exhaust Camshaft Adjuster
9. Cover
10. Connector
11. Bolt
12. Camshaft Position (CMP) Sensor 2
13. Seal
14. Camshaft Position (CMP) Sensor 1
15. Knock Sensor 2
16. Connector
17. Intake Camshaft Adjuster
18. Spark Plug
19. Control Housing

22205_TOUA_G0157

**Fig. 184 Showing ignition system components with location of CMP sensor 1 and CMP sensor 2—3.6L engine**

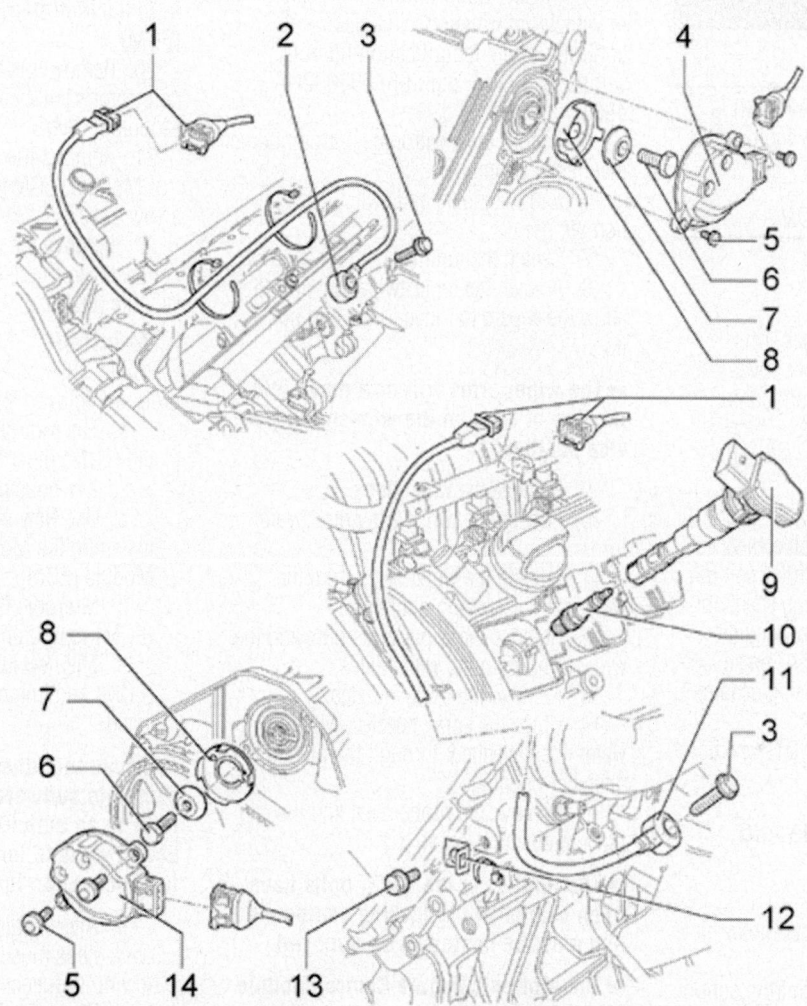

1. Connection for Knock Sensor
2. Knock Sensor 1
3. Bolt
4. Camshaft Position (CMP) Sensor 2
5. Bolt
6. Bolt
7. Washer
8. Hood
9. Ignition Coil
10. Spark Plug
11. Knock Sensor 2
12. Bracket
13. Bolt
14. Camshaft Position (CMP) Sensor 1

22205_TOUA_G0158

**Fig. 185 Showing ignition system components with location of CMP sensor 1 and CMP sensor 2—4.2L 5-valve engine**

## REMOVAL & INSTALLATION

➡️**Manufacturer does not provide step-by-step procedure for this component. Use illustrations as a guide for removal and installation.**

## ELECTRONIC CONTROL MODULE (ECM)

### LOCATION

The Engine Control Module (ECM) is located in the right side of the engine compartment, under the plenum cover.

### REMOVAL & INSTALLATION

**3.6L Engine**

1. Switch off ignition.
2. Remove windshield wiper arms.
3. Remove plenum chamber cover.
4. Disengage connector from control module and then disconnect it.
5. Remove the old control unit and insert the new one.
6. Recode the control unit and adapt to electronic immobilizer and throttle valve control unit. If necessary, enable cruise control system "Guided Function", using scan tool.
7. Read fault memory of new engine control module and, if necessary, erase fault memory "Guided Function", using scan tool.
8. Perform test drive.
9. Check control modules DTC memory again.

**4.2L 5-Valve Engine (Code AXQ, BHX)**

1. Switch off ignition.
2. Remove windshield wiper motor on the right side.
3. Release both connectors from control module and disconnect them.
4. Control module can now be removed.

*To install:*

5. Set control module into mounting frame.
6. Connect harness connectors and engage them.
7. Install windshield wiper motor on right side.
8. Install plenum chamber cover.
9. With scan tool and proper software, program the new Engine Control Module (ECM), using "Guided Functions".

**4.2L 4-Valve Engine (Code BAR)**

*See Figure 186.*

➡️**When the Motronic Engine Control Module (ECM) electrical harness**

connectors are disconnected, the adaptation values are erased and the DTC memory content remains intact.

1. Connect the scan tool.
2. Switch the ignition on.
3. Using the scan tool, select "Vehicle information".
4. Select "Calibration Identification" in vehicle information. The electronic control module identification number will be displayed, e.g.: 06A906032NA 4983.
5. Record the electronic control module identification number.
6. End diagnosis and switch the ignition off.
7. Switch the ignition on.
8. Actuate the touch-wipe function to allow the wipers to move to the end position.

➡️**The wiper arms will now move into the line of view on the windshield (service position).**

9. Remove the wiper arms caps.
10. Loosen the wiper arm nuts several turns.
11. Loosen the wiper arms from the wiper axle by rocking slightly.
12. Remove the wiper arms nuts and the wiper arms from the wiper axles.
13. Remove the spray nozzles.
14. Press the spray nozzles, with the water lines attached, through the plenum chamber opening.
15. Remove the rubber seal and the plenum chamber cover.

➡️**The threads of the shear bolts have been coated with a locking compound and must be heated to be removed.**

➡️**The Motronic Engine Control Module (ECM) is secured via a protective housing and shear bolts.**

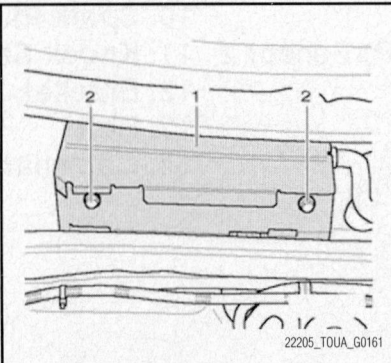

22205_TOUA_G0161

Fig. 186 The Motronic Engine Control Module (ECM) is secured via a protective housing "1" and shear bolts "2"

16. Using a Heat gun set at its lowest setting, heat the shear bolt for approx. 20 to 25 seconds.
17. Remove the shear bolt with locking pliers. Discard the used shear bolts.
18. Repeat the previous steps for the second shear bolt.
19. Remove the protective housing from the Motronic Engine Control Module (ECM).
20. Remove the front electrical harness connector from the Motronic Engine Control Module (ECM).
21. Remove the Motronic Engine Control Module (ECM) from the retainer - arrow - .
22. Remove the rear electrical harness connector from the Motronic Engine Control Module (ECM).

*To install:*

23. Installation is performed in reverse order of removal. Note the following:

a. The Motronic Engine Control Module (ECM) must be installed with the protective housing.

b. Use New shear bolts when installing the Motronic Engine Control Module (ECM).

c. Motronic Engine Control Module (ECM) reprogramming is required.

d. The new Engine Control Module (ECM) and immobilizer must be activated.

➡️**If scan tool does not have proper and complete software for this purpose, refer to an authorized dealer or to the Ebahn website for ECM and immobilizer activation instructions.**

e. After repair work, the following work steps must be performed in the following sequence:
- Check the DTC memory.
- If necessary, erase the DTC memory.
- If the DTC memory was erased, generate readiness code.

**5.0L TDI Engine**

➡️**This vehicle is equipped with two separate Engine Control Modules (ECM) which cannot be visually distinguished from one another. If engine control modules are installed interchanged, the DTC memories are not detected. Therefore it is recommended to mark the engine control modules, as well as their corresponding harness connectors, before removing in order to prevent interchanging the engine control modules.**

➡️Before removing Engine Control Modules (ECM), the control module identification and with it also the coding of the previous control module must be checked using an appropriate scan tool and software.

1. Switch off ignition.
2. Remove right plenum chamber cover.
3. Release and disconnect connector of engine control module and/or.
4. Remove old control module and install new control module.
5. Verify old coding and code new control module using an appropriate scan tool and software.
6. Check DTC memory of new Engine Control Module (ECM) and erase if necessary using an appropriate scan tool and software.
7. Perform test drive.
8. Again, check engine control module DTC memory

## ENGINE COOLANT TEMPERATURE (ECT) SENSOR

### LOCATION

#### 3.6L Engine and 4.2L 5-Valve Engine

The Engine Coolant Temperature (ECT) sensor is located on the lower radiator outlet.

#### 4.2L 4-Valve Engine

The Engine Coolant Temperature (ECT) sensor is located on the cooling line from the thermostat and is near the back side of the engine.

#### 5.0L TDI Engine

The Engine Coolant Temperature (ECT) sensor is located on the thermostat housing.

### REMOVAL & INSTALLATION

#### 3.6L Engine

1. Drain the cooling system.
2. Disconnect electrical connector from the ECT sensor.
3. Unscrew and remove the ECT sensor from the radiator lower outlet connection.
4. Installation is the reverse of the removal procedure.

#### 4.2L 4-Valve Engine

See Figures 187 through 189.

1. Remove rear engine cover.
2. Remove connector on Mass Air Flow (MAF) sensor on intake air assembly.
3. Release tension from spring clips and remove right air intake hose.

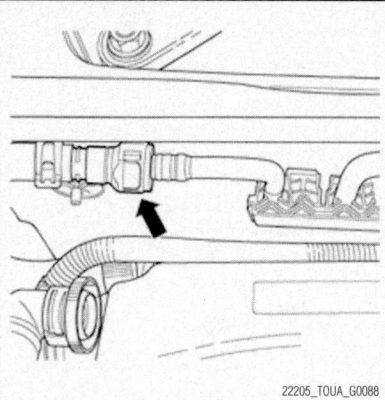

**Fig. 187 Disconnect breather line to evaporative emission (EVAP) canister purge regulator valve in engine compartment—4.2L 4-valve engine**

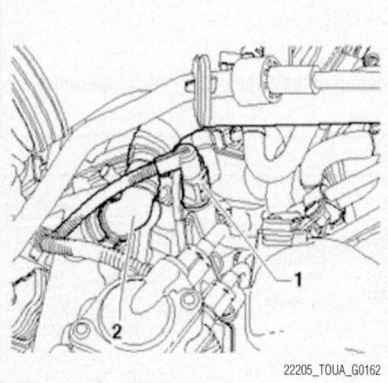

**Fig. 188 Remove connector "1" from camshaft adjustment valve 1, then remove right crankcase housing ventilation "2", unclip hose from clamp and pull it forward in direction of arrow, with coolant line attached—4.2L 4-valve engine**

4. Disconnect breather line to evaporative emission (EVAP) canister purge regulator valve in engine compartment.

5. Remove connector from camshaft adjustment valve 1, then remove right crankcase housing ventilation, unclip hose from clamp and pull it forward in direction shown, with coolant line attached.
6. Remove connector from evaporative emission (EVAP) canister purge regulator valve, pull valve down from bracket, and lay it aside.

### ❊❊ WARNING

**Hot steam may escape when opening expansion tank. Wear protective goggles and protective clothing to prevent damage to eyes and scalding. Cover cap with a rag and open carefully.**

7. Open and seal reservoir to remove pressure from cooling system.
8. Remove bracket to obtain better access. Now, remove Engine Coolant Temperature (ECT) sensor.

*To install:*
9. Installation is in reverse order of removal, note the following:
   a. Replace O-ring.
   b. Secure all hose connections with hose clamps appropriate for the model.
   c. Insert Engine Coolant Temperature (ECT) sensor as quickly as possible so that coolant does not escape unnecessarily.
   d. Ensure sensor retaining clip is seated securely.
   e. Check coolant level.

#### 5.0L TDI Engine

*See Figure 190.*

➡️Manufacturer did not provide a step-by-step procedure for this component. Use the illustration as a guide for removal and installation.

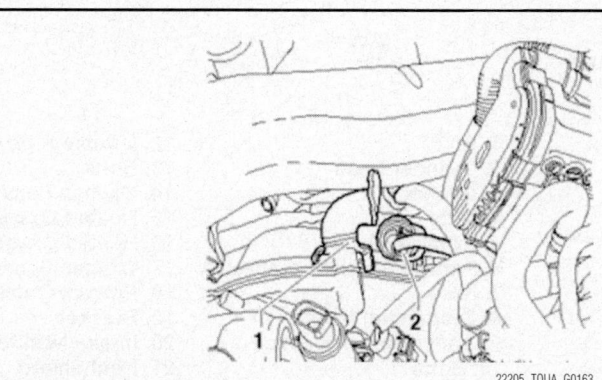

**Fig. 189 Remove bracket "1" to obtain better access, then, remove Engine Coolant Temperature (ECT) sensor "2"—4.2L 4-valve engine**

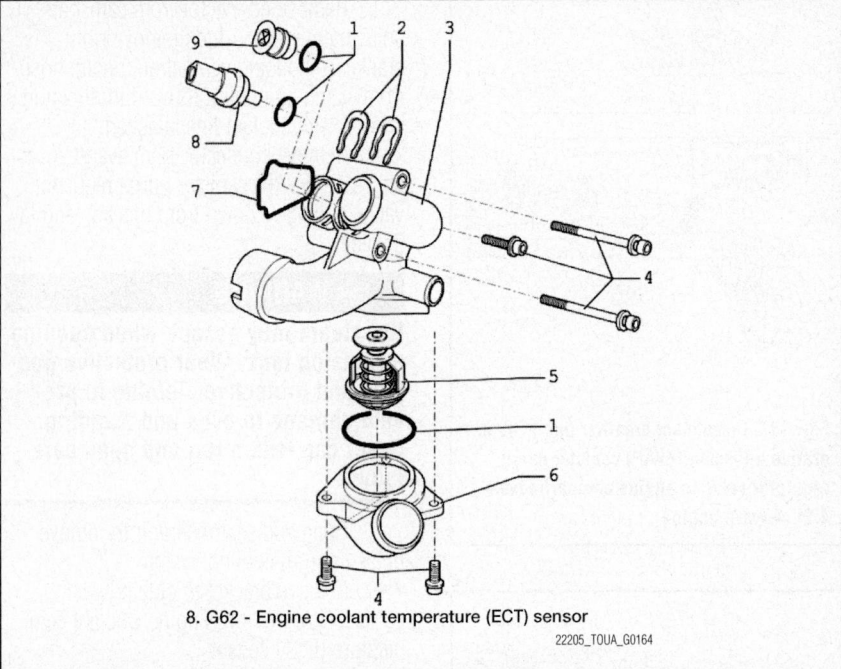

8. G62 - Engine coolant temperature (ECT) sensor

22205_TOUA_G0164

**Fig. 190 Showing the thermostat housing and the location of the Engine Coolant Temperature (ECT) sensor—5.0L TDI engine**

## HEATED OXYGEN (HO2S) SENSOR

### LOCATION

#### 3.6L Engine

*See Figures 191 and 192.*

#### 4.2L 5-Valve Engine (Code AXQ, BHX)

*See Figure 193.*

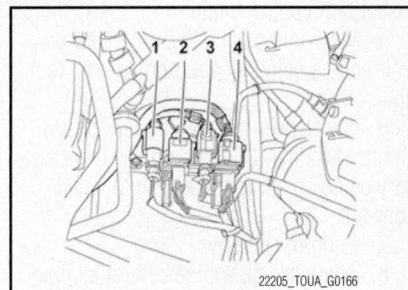

22205_TOUA_G0166

**Fig. 192 Showing the oxygen sensor electrical connectors "1" through "4"—3.6L engine**

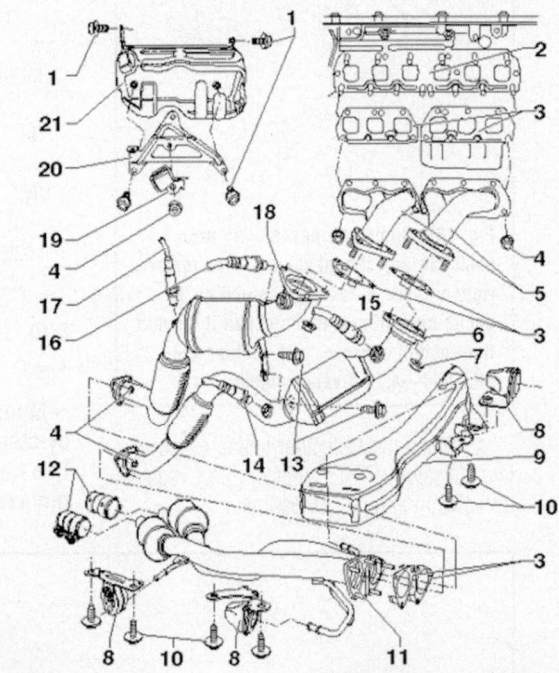

1. Bolt
2. Cylinder Head
3. Gasket
4. Nuts
5. Exhaust Manifold
6. Primary Catalytic Converter
7. Nuts
8. Suspended Mount
9. Transmission Bracket
10. Bolts
11. Catalytic Converters
12. Double Pipe Clamp
13. Bolts
14. Oxygen Sensor 1
15. Heated Oxygen Sensor 1
16. Heated Oxygen Sensor 2
17. Oxygen Sensor 2
18. Primary Catalytic Converter
19. Bracket
20. Intake Manifold Support
21. Heat Shield

22205_TOUA_G0165

**Fig. 191 Exploded view of the exhause system components, showing the location of the Heated Oxygen Sensors and the Oxygen Sensors—3.6L Engine**

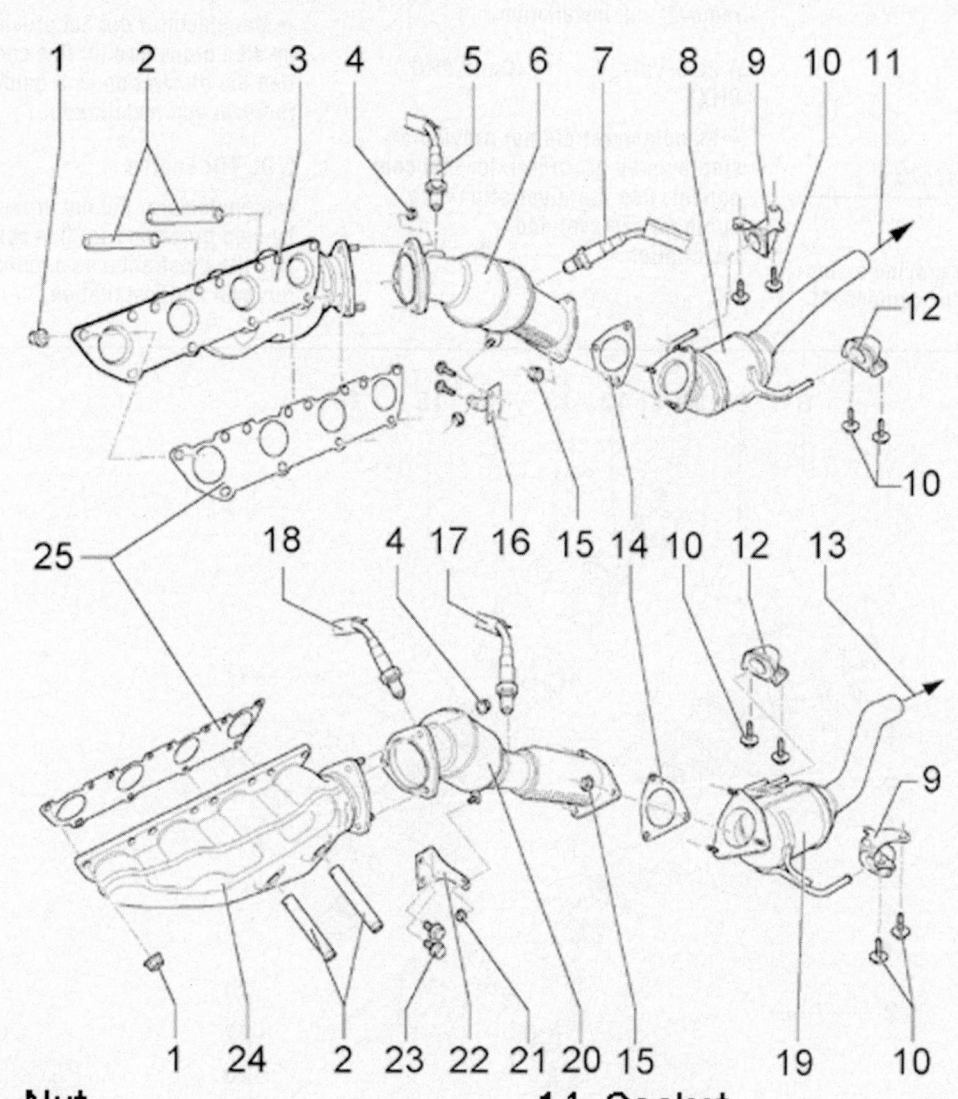

1. Nut
2. Nut
3. Exhaust Manifold
4. Bolt
5. Heated Oxygen Sensor 1
6. Primary Catalytic Converter
7. Oxygen Sensor 1
8. Catalytic Converter
9. Suspended Mount
10. Bolt
11. To Middle Muffler
12. Suspended Mount
13. To Middle Muffler
14. Gasket
15. Bolt
16. Bracket
17. Oxygen Sensor 2
18. Heated Oxygen Sensor 2
19. Catalytic Converter
20. Primary Catalytic Converter
21. Bolt
22. Bracket
23. Bolt
24. Exhaust Manifold
25. Gasket

22205_TOUA_G0167

**Fig. 193 Exploded view of the exhaust system components, showing the location of the Heated Oxygen Sensors and the Oxygen Sensors—4.2L 5-valve engine**

### 4.2L 4-Valve Engine (Code BAR)
*See Figure 194.*

### 5.0L TDI Engine
*See Figures 195 and 196.*

### REMOVAL & INSTALLATION

### 3.6L Engine
➡Manufacturer did not provide a step-by-step procedure for this component.

Use the illustration as a guide for removal and installation.

### 4.2L 5-Valve Engine (Code AXQ, BHX)
➡Manufacturer did not provide a step-by-step procedure for this component. Use the illustration as a guide for removal and installation.

### 4.2L 4-Valve Engine (Code BAR)
➡Manufacturer did not provide a step-by-step procedure for this component. Use the illustration as a guide for removal and installation.

### 5.0L TDI Engine
➡Manufacturer did not provide a step-by-step procedure for this component. Use the illustration as a guide for removal and installation.

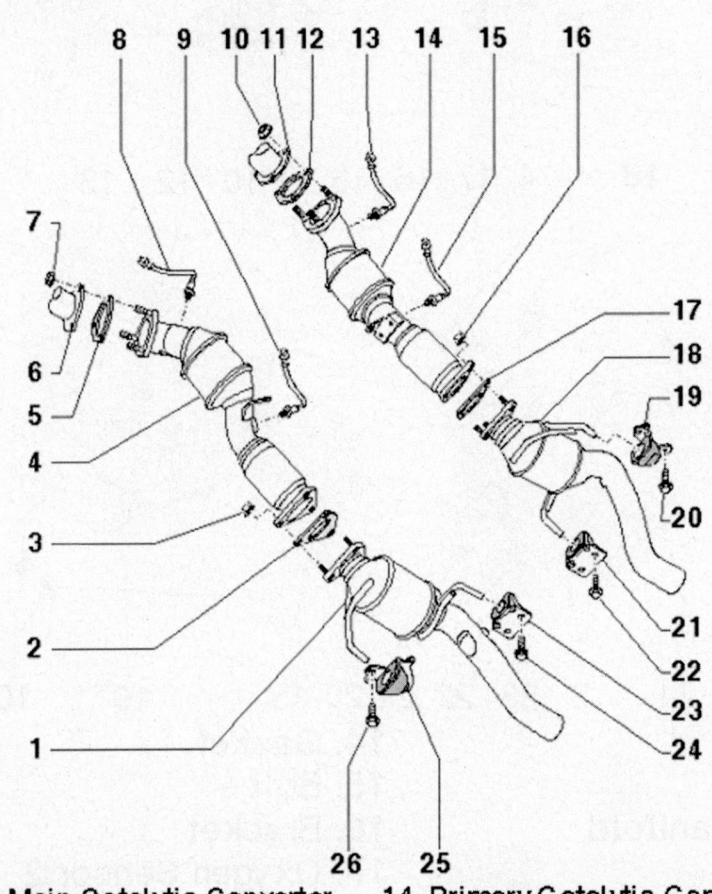

1. Main Catalytic Converter
2. Gasket
3. Bolt
4. Primary Catalytic Converter
5. Gasket
6. Exhaust Manifold
7. Bolt
8. Heated Oxygen Sensor 1
9. Oxygen Sensor 1
10. Bolt
11. Exhaust Manifold
12. Gasket
13. Heated Oxygen Sensor 2
14. Primary Catalytic Converter
15. Oxygen Sensor 2
16. Bolt
17. Gasket
18. Main Catalytic Converter
19. Suspended Mount
20. Bolt
21. Suspended Mount
22. Bolt
23. Suspended Mount
24. Bolt
25. Suspended Mount
26. Bolt

22205_TOUA_G0168

**Fig. 194 Exploded view of the catalytic converter components, showing the location of the Heated Oxygen Sensors and the Oxygen Sensors—4.2L 4-valve engine**

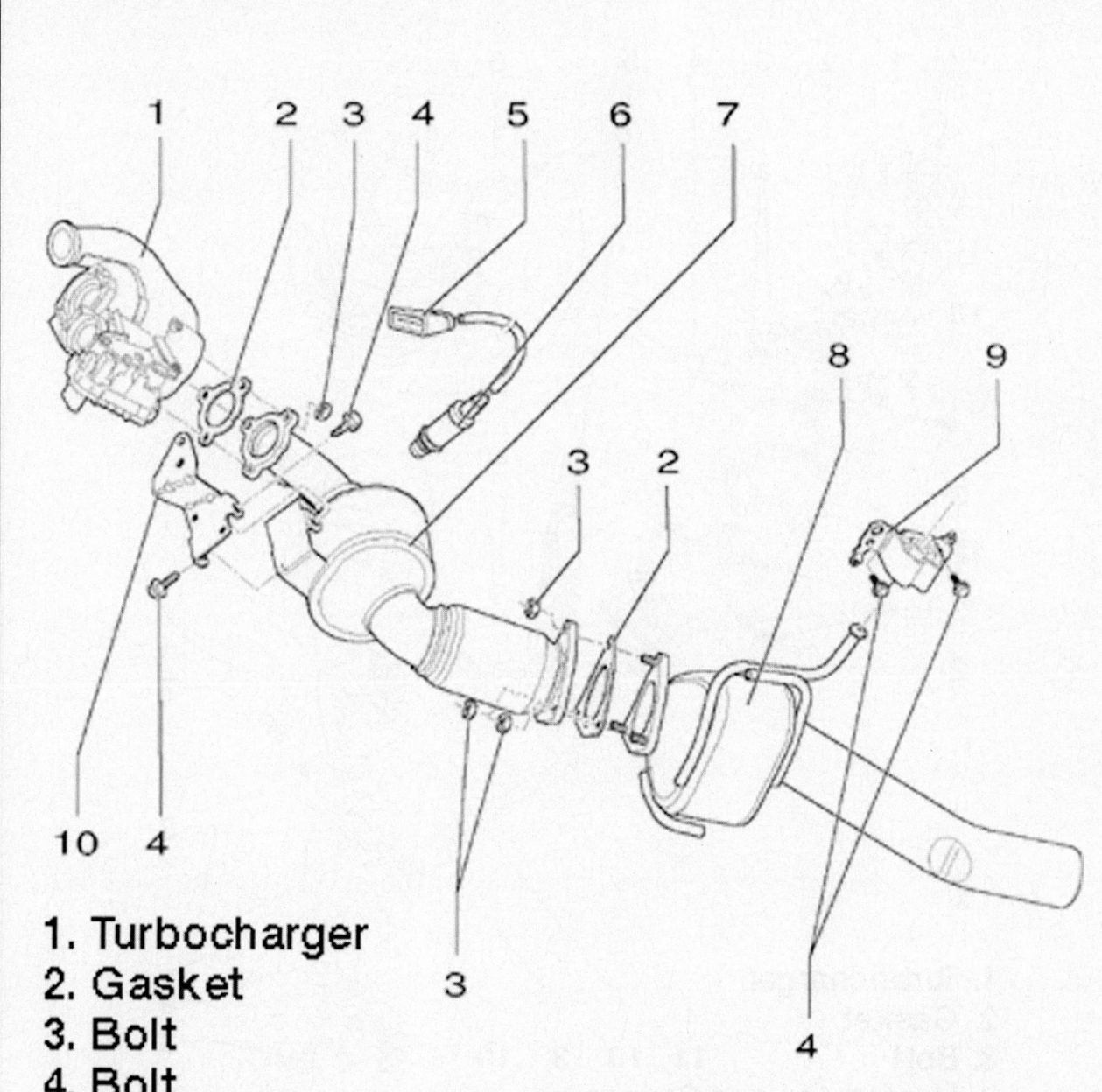

1. Turbocharger
2. Gasket
3. Bolt
4. Bolt
5. Connector
6. Heated Oxygen Sensor
7. Front Exhaust Pipe with Catalytic Converter
8. Catalytic Converter
9. Suspended Mount
10. Bracket

22205_TOUA_G0169

**Fig. 195 Exploded view of the exhaust system components, showing the location of the Heated Oxygen Sensors—5.0L TDI engine**

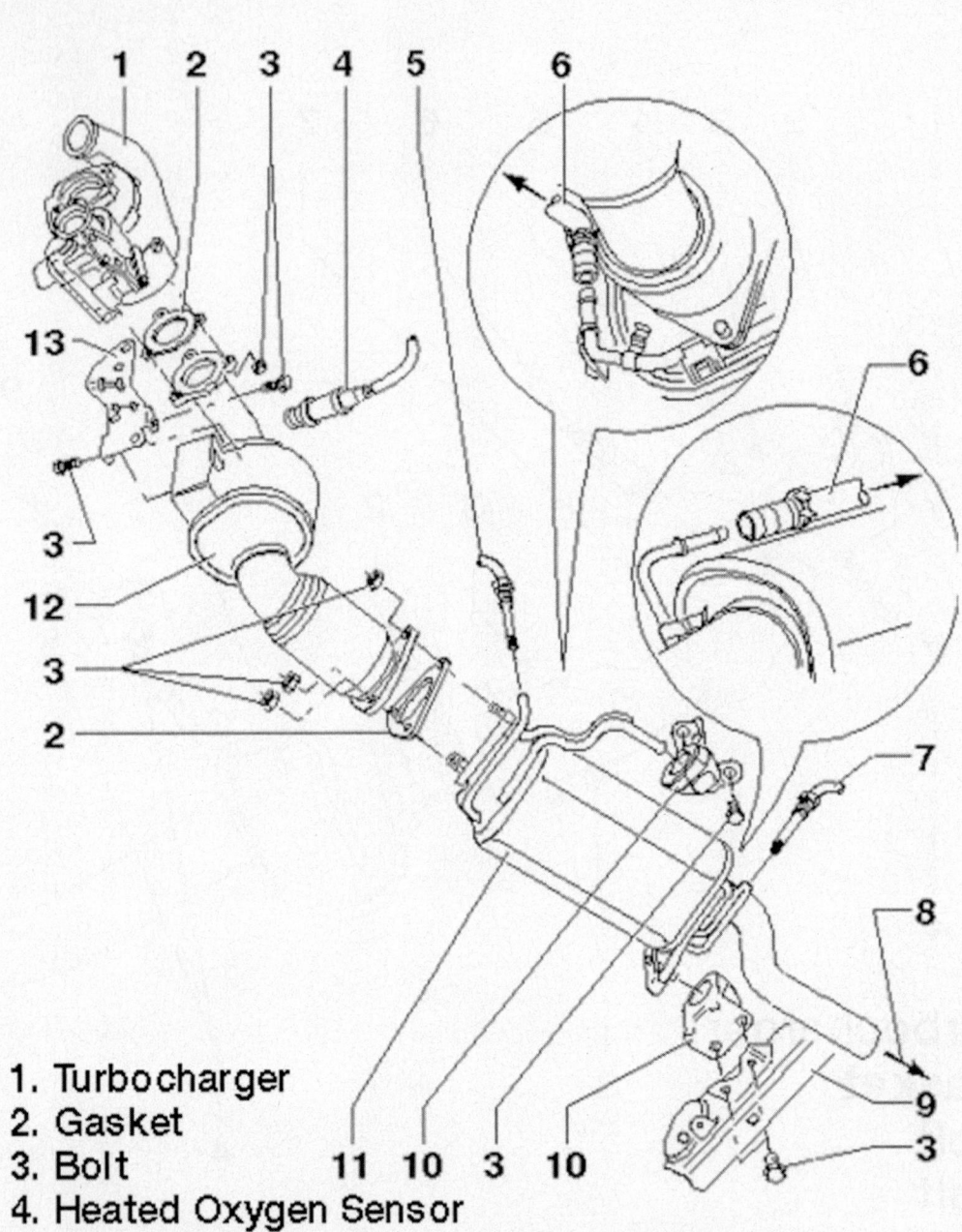

1. Turbocharger
2. Gasket
3. Bolt
4. Heated Oxygen Sensor
5. Pre-Particle Filter Temperature Sensor
6. Pressure Line
7. Post-Particle Filter Temperature Sensor
8. To Rear Muffler
9. Transmission Crossmember
10. Suspended Mount
11. Particle Filter
12. Front Exhaust Pipe with Catalytic Converter
13. Bracket

22205_TOUA_G0170

**Fig. 196 Exploded view of the catalytic converter components, showing the location of the Heated Oxygen Sensors with Particulate Filters—5.0L TDI engine**

## INTAKE AIR TEMPERATURE (IAT) SENSOR

### LOCATION

**3.6L Engine**

➡Intake air temperature sensor is not applicable on this engine.

**4.2L 5-Valve Engine (Code AXQ, BHX)**

*See Figure 197.*

**4.2L 4-Valve Engine (Code BAR)**

➡Intake air temperature sensor is not applicable on this engine.

**5.0L TDI Engine**

➡Intake air temperature sensor is not applicable on this engine.

### REMOVAL & INSTALLATION

➡Manufacturer did not provide a step-by-step procedure for this component. Use the illustration as a guide for removal and installation.

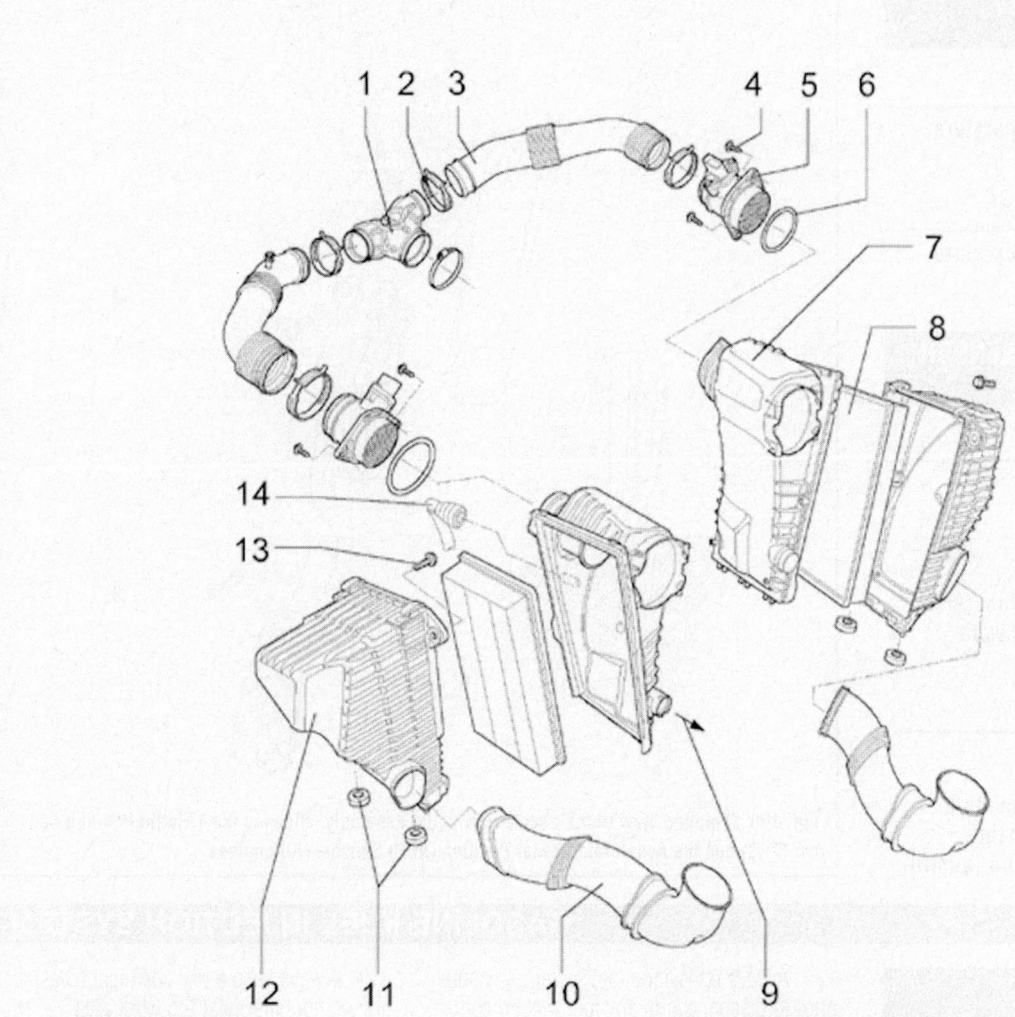

1. Intake Tube
2. Spring Clip
3. Intake Tube
4. Bolt
5. Mass Air Flow (MAF) Sensor with Intake Air Temperature (IAT) Sensor
6. Seal
7. Air Filter Upper Part
8. Filter Element
9. To AIR System Pump Motor
10. Air Duct
11. Rubber Bushing
12. Air Filter Lower Part
13. Bolt
14. To Compressor for Air Suspension

22205_TOUA_G0171

**Fig. 197 Exploded view of the air filter and air intake system, showing the location of the MAF/IAT Sensor—4.2L 5-valve engine**

## KNOCK SENSOR (KS)

### LOCATION

➡ Refer to "Camshaft Position (CMP) Sensor."

### REMOVAL & INSTALLATION

➡ Refer to "Camshaft Position (CMP) Sensor."

## MASS AIR FLOW (MAF) SENSOR

### LOCATION

➡ Refer to "Intake Air Temperature (IAT) Sensor."

### REMOVAL & INSTALLATION

➡ Refer to "Intake Air Temperature (IAT) Sensor."

## THROTTLE POSITION SENSOR (TPS)

### LOCATION

**All Engines**

*See Figure 198.*

The Throttle Position Sensor (TPS)/Accelerator Pedal Position (APP) Sensor is located on the accelerator pedal.

### REMOVAL & INSTALLATION

**All Engines**

➡ Manufacturer did not provide a step-by-step procedure for this component. Use the location illustration as a guide for removal and installation.

## VEHICLE SPEED SENSOR (VSS)

### LOCATION

#### 4.2L Engine

The Engine Speed Sensor (ESS) is located on the left side of the transmission, just rear of the engine flange.

➡ This component is commonly called the Vehicle Speed Sensor (VSS).

#### 5.0L TDI Engine

The Engine Speed Sensor (ESS) is located on the engine block below bank 1.

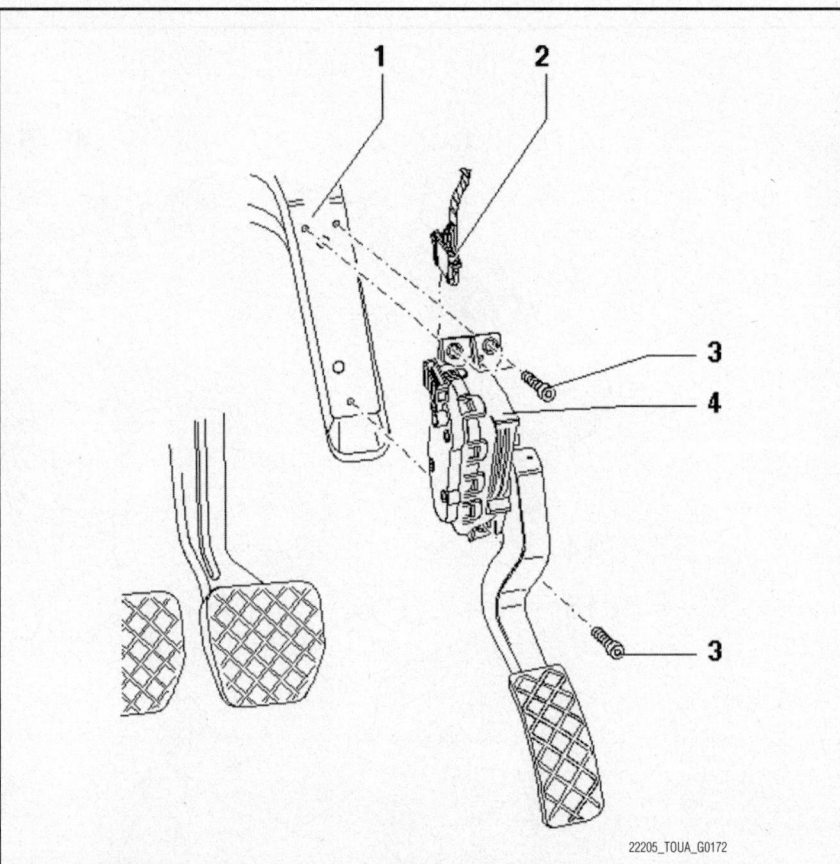

22205_TOUA_G0172

**Fig. 198 Exploded view of the accelerator pedal assembly, showing the Throttle Position Sensor (TPS) and the Accelerator Pedal Position (APP) Sensor—All engines**

---

# FUEL                                          GASOLINE FUEL INJECTION SYSTEM

## FUEL SYSTEM SERVICE PRECAUTIONS

Safety is the most important factor when performing not only fuel system maintenance but any type of maintenance. Failure to conduct maintenance and repairs in a safe manner may result in serious personal injury or death. Maintenance and testing of the vehicle's fuel system components can be accomplished safely and effectively by adhering to the following rules and guidelines.

• To avoid the possibility of fire and personal injury, always disconnect the negative battery cable unless the repair or test procedure requires that battery voltage be applied.

• Always relieve the fuel system pressure prior to disconnecting any fuel system component (injector, fuel rail, pressure regulator, etc.), fitting or fuel line connection. Exercise extreme caution whenever relieving fuel system pressure to avoid exposing skin, face and eyes to fuel spray. Please be advised that fuel under pressure may penetrate the skin or any part of the body that it contacts.

• Always place a shop towel or cloth around the fitting or connection prior to loosening to absorb any excess fuel due to spillage. Ensure that all fuel spillage (should it occur) is quickly removed from engine surfaces. Ensure that all fuel soaked cloths or towels are deposited into a suitable waste container.

• Always keep a dry chemical (Class B) fire extinguisher near the work area.

• Do not allow fuel spray or fuel vapors to come into contact with a spark or open flame.

• Always use a back-up wrench when loosening and tightening fuel line connection fittings. This will prevent unnecessary stress and torsion to fuel line piping.

• Always replace worn fuel fitting O-rings with new Do not substitute fuel hose or equivalent where fuel pipe is installed.

Before servicing the vehicle, make sure to also refer to the precautions in the beginning of this section as well.

## RELIEVING FUEL SYSTEM PRESSURE

### 3.6L Engine

*See Figure 199.*

1. Remove electrical connector from fuel pressure regulator valve.

2. Allow engine to idle approximately 10 seconds.

➡**When the fuel pressure regulator valve N276 electrical connector is disconnected during idle, pressure in high pressure area decreases to approximately 87 psi (6 bar).**

3. After high pressure has been released, high pressure system must be opened immediately. Otherwise, the pressure increases again due to the warming of the fuel.

4. Switch off ignition.

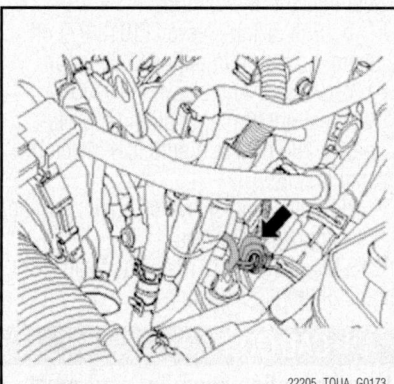

22205_TOUA_G0173

**Fig. 199 Remove electrical connector from fuel pressure regulator valve (arrow)— 3.6L engine**

❊❊ **WARNING**

**Fuel lines are pressurized! Wear protective goggles and protective clothing to prevent injuries and contact with skin. Before opening the high pressure system, place a cloth around the connection.**

5. Place a clean cloth around the connection point and carefully open to release the residual pressure of approximately 87 psi (6 bar). Escaping fuel must be absorbed.

6. To conclude work, check DTC memory of Engine Control Module (ECM), erase all DTC entries which may have occurred from removing the connector.

### 4.2L Engine

1. Disconnect connector from fuel metering valve on right high pressure

pump. Then, disconnect connector from fuel metering valve 2 on left high pressure pump.

2. Allow engine to idle approximately 10 seconds.

➡**If the fuel metering valve electrical connectors are disconnected at idle, the pressure in the high pressure area drops to approximately 87 psi (6 bar).**

3. After high pressure has been released, high pressure system must be opened immediately. Otherwise, the pressure increases again due to the warming of the fuel.

4. Switch off ignition.

❊❊ **WARNING**

**Fuel lines are pressurized! Wear protective goggles and protective clothing to prevent injuries and contact with skin. Before opening the high pressure system, place a cloth around the connection.**

5. Place a clean cloth around connection point and carefully open to release the residual pressure of approximately 87 psi (6 bar). Escaping fuel must be absorbed.

6. To conclude work, check DTC memory of Engine Control Module (ECM), erase all DTC entries which may have occurred from removing the connector.

## FUEL FILTER

### REMOVAL & INSTALLATION

#### 3.6L Engine and 4.2L Engine

Fuel filter is located in left fuel tank opening, when viewed in direction of travel.

1. Before beginning work, read the "Fuel System Service Precautions" in this section.

2. Locate and remove fuel system fuses from fuse panel.

3. Drain the fuel tank.

4. Remove rear seat bench.

5. Cut a circular opening in the carpet, at the pre-cut area, on the left side near the seat attaching bracket (feel through carpet to determine location of access cover).

6. Remove bolts for seat mounting bracket and press it up slightly. Loosen bolts for the fuel access cover and remove cover.

7. Remove vent line and connector from flange.

❊❊ **WARNING**

**Fuel supply lines are under pressure! Wear protective goggles and protec-**

tive gloves to avoid damage and contact with skin. Before removing from hose connection wrap a cloth around the connection. Then release pressure by carefully pulling hose off connection.

➡**When breather line is pulled off, hose coupling button often cannot be pressed in. In that case, use special assembly tool (T10118).**

8. Remove locking ring from left flange using locking ring wrench (T10202).

9. Remove flange with fuel filter housing from fuel tank.

10. Remove connector from fuel level sensor and fuel lines from fuel filter housing.

➡**Release line connections by pressing button on hose coupling.**

11. Empty fuel filter housing.

12. Mark installation position of filter cover, using colored felt tip marker pen.

13. Disconnect ground wire and unbolt filter cover from housing.

➡**If filter insert is to be replaced, ensure that the ground (GND) connection contacts are not bent and have sufficient pretension.**

## FUEL INJECTORS

### REMOVAL & INSTALLATION

#### 3.6L Engine

*See Figure 200.*

1. Before beginning work, read the "Fuel System Service Precautions" in this section.

2. Remove the intake manifold. See "Intake Manifold" in the "ENGINE MECHANICAL" section.

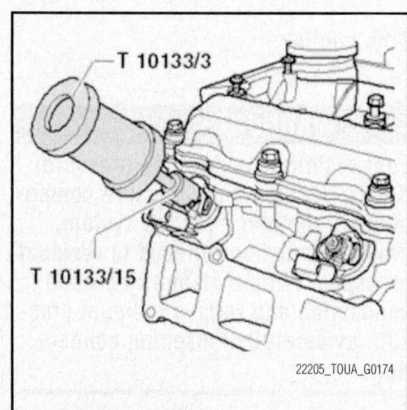

22205_TOUA_G0174

**Fig. 200 Using the hammer/puller to extract fuel injectors—3.6L engine**

3. If fuel injectors for cylinders 1, 3 and 5 are to be removed, remove fuel rail.

4. Push the O-ring upward by hand and remove it from fuel injector.

5. Assemble a slide hammer/puller assembly (T10133/3 and T10133/15).

6. Guide the puller onto the groove on the fuel injector.

→Spring element must not be removed prior to removing the injector valves.

7. Carefully remove fuel injector.

*To install:*

8. Thoroughly clean bores for fuel injectors in cylinder head using nylon brush T10133/4 .

9. Check plastic support washer for damage, replace if necessary.

→Replace spring element and Teflon sealing ring each time fuel injector is removed.

10. Replace O-rings between fuel injectors and fuel rail and coat them lightly with clean motor oil.

11. The Teflon sealing ring of fuel injector may not be oiled or greased.

12. Press fuel injector by hand into cylinder head bore until it stops.

13. Check fuel injectors for correct seating and installation position in cylinder head.

14. If fuel injectors for cylinders 1, 3 and 5 were removed, place fuel rail on and press evenly on to fuel injectors.

15. Tighten fuel rail with new bolts, uniformly, to 22 ft. lbs. (30 Nm), plus an additional 90 degrees.

16. Install union nut for fuel supply line tightly to supply line of fuel rail. To do so, counter hold on fuel supply line using a wrench. Tighten to 16 ft. lbs. (22 Nm).

17. Install intake manifold. See "Intake Manifold" in "ENGINE MECHANICAL" section.

## 4.2L Engine

*See Figure 201.*

### ✽✽ WARNING

**Fuel system is under high pressure! Before opening high pressure components of the fuel injection system, pressure must be relieved to residual pressure. Wrap a clean rag around connection and relieve residual pressure by carefully loosening connection.**

1. Remove intake manifold. See "Intake Manifold" in "ENGINE MECHANICAL" section.

2. Seal intake channels in cylinder heads with clean cloths.

3. Disconnect electrical harness connectors at fuel injectors.

4. Remove high pressure lines from connector on fuel rail. To do this, counter hold at hex head with and open-end wrench and loosen union nut.

5. Remove bolts "1", "2", "3", "4", "6" and "8" as shown.

6. Do not change bent shape of high pressure lines.

7. Remove fuel rail with fuel injectors.

8. If fuel injectors cannot be pulled out of cylinder head by hand, proceed as follows:

   a. Using a screwdriver, bend retaining tabs of radial adjustment aside and pull support ring from fuel injector.

   b. Remove O-ring from fuel injector.

   c. Attach slide wrench and adapter (T10133/3 and T10133/10) to injector.

   d. Guide puller adapter into groove on fuel injector and carefully drive fuel injector out.

→When setting the puller in place, the radial adjustment can be destroyed, because the retaining tabs break.

9. Carefully remove old combustion chamber seal by cutting seal open with a knife or spreading seal open with a small screwdriver and pulling it forward and off.

10. Make sure that the groove of fuel injector does not become damaged. If groove is damage, fuel injector must be replaced.

*To install:*

→During installation, replace combustion chamber seal and O-ring.

11. Replace spacer ring if damaged.

12. Lightly moisten fuel injector O-rings with clean engine oil.

13. Re-insert injector lines at same cylinder.

**Fig. 201 Remove bolts "1", "2", "3", "4", "6" and "8"—4.2L engine**

14. Clean bore in cylinder head with nylon cylinder brush T10133/4 .

15. Clip radial adjustment to support ring on injector.

16. When re-installing a fuel injector, use a clean cloth to clean combustion residue from groove for combustion chamber seal and shaft of fuel injector.

17. Place assembly tool (cone T10133/5), with new combustion chamber seal, on fuel injector.

18. Slide combustion chamber seal as far possible onto installer assembly tool (cone T10133/5).

19. Slide combustion chamber seal into groove.

20. When pushing combustion chamber seal onto the fuel injector, the seal spreads open. Therefore after pushing it on, it must be tightened again in 2 steps, as follows.

   a. Press a sleeve with a slight turning motion (approximately 180 degrees) onto fuel injector until it stops.

   b. Pull sizing sleeve (T10133/7) off again, turning it in opposite direction.

   c. Press sizing sleeve (T10133/8) with a slight turning motion (approximately 180 degrees) onto fuel injector until it stops.

   d. Pull sizing sleeve off again, turning it in opposite direction.

21. Moisten new O-ring with clean engine oil before installing.

### ✽✽ CAUTION

**The combustion chamber seal must not be oiled.**

22. Slide fuel injector into bore cylinder head as far as stop, using a proper assembly drift (T10133/9).

→The fuel injector must not be difficult to install. If necessary, wait as the combustion chamber seal continues to pull itself together.

23. Make sure fuel injectors are correctly positioned in cylinder head.

24. The electrical connection of fuel injector must engage in intended recess of cylinder head.

25. Press fuel rail onto fuel injectors with uniform pressure.

26. Tighten the removed bolts diagonally in stages to 6 ft. lbs. (9 Nm).

→High pressure line connections must not show any signs of damage.

### ✽✽ CAUTION

**Do not change bent shape of high pressure lines.**

27. Hand-tighten union nuts for high-pressure lines.

28. Make sure high-pressure lines are seated free of stress.

29. To tighten the union nut on the high pressure line, use a torque wrench (VAG 1331) with a socket insert (AF 14, open ring VAG 1331/8 or 1331/6).

30. Tighten high pressure line nuts to 18 ft. lbs. (25 Nm).

31. Only install retaining tabs after high pressure lines have been tightened.

32. Install intake manifold. See "Intake Manifold" in "ENGINE MECHANICAL" section.

33. Fill with coolant, if necessary.

## FUEL PUMP

*REMOVAL & INSTALLATION*

### 3.6L Engine

#### HIGH PRESSURE PUMP

*See Figure 202.*

### ☀ WARNING

**Follow safety measures for releasing fuel pressure in high pressure area. Fuel pipes are under pressure! Wear protective goggles and protective clothing to prevent injuries and contact with skin. Before loosening the fuel pipe, place a cloth around the connection point. Then release pressure by carefully loosening.**

1. Before beginning work, read the "Fuel System Service Precautions" in this section.

2. Lay a cloth around threaded connection and loosen union nut from fuel supply hose. To do so, counter hold on fuel supply line using a wrench.

3. Remove connector and remove low pressure line and high pressure line. When

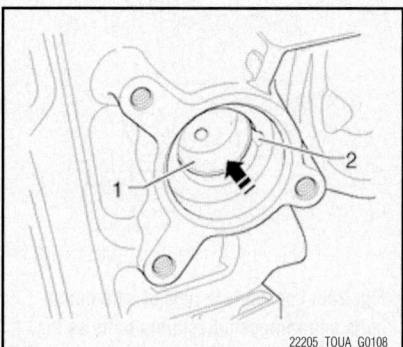

**Fig. 202 Insert oiled lifter with guide perpendicularly into cylinder head**

22205_TOUA_G0108

doing this, counter hold on connection of high pressure pump using a wrench.

4. Disconnect connector, remove bolts, and remove high pressure pump from engine.

### *To install:*

5. Check plunger , for damage; replace if necessary.

6. Insert oiled lifter with guide perpendicularly into cylinder head.

7. Rotate engine at vibration damper slowly in direction of engine rotation. When doing this, press plunger in until it reaches the deepest point in cylinder head.

8. Replace high pressure pump O-ring.

9. Install high pressure pump and evenly tighten bolts to 7 ft. lbs. (10 Nm).

➡**O-ring of high pressure pump must always be replaced.**

10. Before installing fuel lines, first tighten connection for fuel lines on high pressure pump.

- Connection for high pressure line: 30 ft. lbs. (40 Nm)
- Connection for low pressure line: 21 ft. lbs. (28 Nm)

### 4.2L Engine

1. Before beginning work, read the "Fuel System Service Precautions" in this section.

2. Before opening high pressure components of the fuel injection system, pressure must be relieved to residual pressure. See "Relieving Fuel System Pressure."

3. With ignition switched off, disconnect battery ground (GND) cable.

4. Remove rear engine cover.

5. Disconnect electrical connector, loosen union nuts, and remove mounting bolts.

➡**Do not change bent shape of high pressure lines.**

6. Carefully remove high pressure pump.

7. Remove roller tappet from cylinder head.

### *To install:*

8. Installation is in reverse order of removal, note the following:

   a. Replace O-ring.

   b. Insert roller tappet in cylinder head.

   c. To set high pressure pump in place, lift high pressure lines only slightly.

   d. Insert high pressure pump in cylinder head and tighten.

➡**High pressure line connections must not show any signs of damage. Do not**

change bent shape of high pressure lines.

   e. Hand-tighten union nuts for high-pressure lines.

   f. Make sure high-pressure lines are seated free of stress.

   g. Tighten union nuts on high pressure lines to 18 ft. lbs. (25 Nm).

   h. Reconnect the battery.

## FUEL TANK

*REMOVAL & INSTALLATION*

### 3.6L Engine and 4.2L Engine

1. Before beginning work, read the "Fuel System Service Precautions" in this section.

### ☀ CAUTION

**During all repair procedures on the fuel tank, be aware of the following:**

- Route all the various lines (e.g. for fuel, EVAP system, or vacuum) and electrical wiring so that the original routing positions are restored.
- Make sure that the ground (-) strap between the fuel filler tube and body is securely fastened, to prevent electrostatic charging.
- Ensure sufficient clearance to all moving or hot components.

2. Drain the fuel tank.

3. Remove rear seat bench.

4. Cut open carpet on right side in pre cut area (feel near seat bracket for fuel tank access panel before cutting).

5. Remove the nuts on the right side of the fuel delivery unit cover. If necessary, remove backrest support or mounting bracket.

### ☀ WARNING

**Fuel supply lines are under pressure! Wear protective goggles and protective gloves to avoid damage and contact with skin. Before removing from hose connection wrap a cloth around the connection. Then release pressure by carefully pulling hose off connection.**

6. Remove fuel supply line, auxiliary heater fuel line, fuel pump connector, vent line, and fuel pump (FP) control module connector from fuel sending unit.

7. Press in securing ring to disengage the fuel line.

8. Remove mufflers and mountings.

9. Remove driveshaft.

10. Remove rear axle.

11. Open fuel flap and remove fuel tank cap.

12. Pull rubber gasket off filler neck.

13. Remove bolts on filler neck and pull off ground wire.

14. Remove right rear wheel housing liner.

15. Unbolt fuel line cover plate.

16. Unclip fuel tank breather lines at securing clip attached on longitudinal member.

17. Remove bolts for filler pipe and EVAP canister in wheel housing.

18. Bend filler neck slightly downward and pull off breather line connections to EVAP canister.

➡ **Release connection by pressing button on hose coupling.**

19. Disconnect ground wire clipped to EVAP canister and remove canister.

20. Separate fuel pump connectors, on left next to fuel tank.

21. Remove securing straps with covers on left and right below fuel tank.

22. Support fuel tank using engine/transmission jack, and remove securing strap at center of fuel tank.

23. Carefully lower fuel tank about 12 inches (30 cm).

24. Grab between fuel tank and vehicle floor and disconnect vent line from left sensor flange.

➡ **This step eliminates having to cut open the carpeting in the vehicle interior in the vicinity of sender flange cover.**

25. Lower the fuel tank.

*To install:*

26. Installation is performed in the reverse order of removal, noting the following:

    a. Connections for breather and fuel lines must engage audibly when joined.

    b. Make sure ventilation and fuel lines are not kinked when installed.

    c. The flange seal should be replaced each time it is opened.

    d. Secure fuel hoses with spring-type clamps.

    e. Ensure fuel hoses are seated securely.

    f. The ground strap at the fuel filler tube must be securely connected to the body.

    g. Before fastening the fuel tank, check that the supply and ventilation lines are still clipped onto the fuel tank.

## IDLE SPEED

*ADJUSTMENT*

➡ **Idle speed is electronically monitored and controlled through the engine control module. Therefore, idle speed is not manually adjustable.**

## THROTTLE BODY

*REMOVAL & INSTALLATION*

➡ **The throttle body is removed as operational steps within the removal procedure for the intake manifold. See "Intake Manifold" in "ENGINE MECHANICAL" section.**

# FUEL                                           DIESEL FUEL INJECTION SYSTEM

## FUEL SYSTEM SERVICE PRECAUTIONS

Safety is the most important factor when performing not only fuel system maintenance but any type of maintenance. Failure to conduct maintenance and repairs in a safe manner may result in serious personal injury or death. Maintenance and testing of the vehicle's fuel system components can be accomplished safely and effectively by adhering to the following rules and guidelines.

    • To avoid the possibility of fire and personal injury, always disconnect the negative battery cable unless the repair or test procedure requires that battery voltage be applied.

    • Always relieve the fuel system pressure prior to disconnecting any fuel system component (injector, fuel rail, pressure regulator, etc.), fitting or fuel line connection. Exercise extreme caution whenever relieving fuel system pressure to avoid exposing skin, face and eyes to fuel spray. Please be advised that fuel under pressure may penetrate the skin or any part of the body that it contacts.

    • Always place a shop towel or cloth around the fitting or connection prior to loosening to absorb any excess fuel due to spillage. Ensure that all fuel spillage

(should it occur) is quickly removed from engine surfaces. Ensure that all fuel soaked cloths or towels are deposited into a suitable waste container.

    • Always keep a dry chemical (Class B) fire extinguisher near the work area.

    • Do not allow fuel spray or fuel vapors to come into contact with a spark or open flame.

    • Always use a back-up wrench when loosening and tightening fuel line connection fittings. This will prevent unnecessary stress and torsion to fuel line piping.

    • Always replace worn fuel fitting O-rings with new. Do not substitute fuel hose or equivalent where fuel pipe is installed.

Before servicing the vehicle, make sure to also refer to the precautions in the beginning of this section as well.

## INJECTION PUMP

*REMOVAL & INSTALLATION*

### 5.0L TDI Engine

*See Figures 203 through 205.*

1. Remove acoustic cover and cylinder head cover of applicable cylinder head.

2. Rotate crankshaft until cam pair of respective Pump-Injector Unit to be installed and removed face upward simultaneously.

3. Loosen lock nuts of adjustment bolts "1" and remove adjustment bolts as far until respective rocker arm makes contact at plunger spring of Pump-Injector Unit.

4. Loosen bolts "2" for rocker lever shaft from outside toward inside, using a proper socket insert (3410) and remove rocker lever shaft.

5. Loosen mounting bolt "3" for tension block, using a proper socket insert (T10054) and remove tension block.

6. Pry out connector from Pump-Injector Unit using a screwdriver. In order to

22205_TOUA_G0176

**Fig. 203 Loosen lock nuts of adjustment bolts and remove adjustment bolts as far until respective rocker arm makes contact at plunger spring of Pump-Injector Unit— 5.0L TDI engine**

avoid canting, support opposite side of connector with light finger pressure.

7. Set a puller (T10055) in place of tension block into side slit of pump-injector unit.

8. Using careful knocking motions, pull Pump-Injector Unit upward out of its cylinder head seat.

### To install:

9. Install a new pump-injector unit, also replace corresponding adjustment bolt in rocker arm.

→For every procedure that requires adjustment of the pump-injector unit, the adjustment screw in the rocker lever and ball pin of pump-injector unit must be replaced. New Pump-Injector Units are shipped with O-rings and heat protection seal.

10. Install pump-injector unit again, replace O-rings and heat protection seal.

11. Before installing Pump-Injector Unit, check that three sealing rings, heat protection seal and securing ring are seated properly.

→Sealing rings must not be twisted on themselves.

12. Oil the sealing rings and insert Pump-Injector Unit with great care into cylinder head seat.

13. Using uniform pressure, push pump-injector unit into cylinder head seat onto stop.

14. Insert tension block into lateral slot of Pump-Injector Unit.

→If Pump-Injector Unit does not stand at right angle to tension block, securing bolt can loosen and thereby cause damage to Pump-Injector Unit / cylinder head.

15. Align pump-injector unit as follows:
   a. Install new bolt into tension block so far that the pump-injector unit can still be lightly turned.
   b. Insert a gauge (T10210) between thrust bearing and Pump-Injector Unit as shown.
   c. Turn pump-injector unit against gauge T10210 by hand.

16. Re-align Pump-Injector Unit if necessary and tighten securing bolt to 10 ft. lbs. (12 Nm), plus an additional 270 degrees. The additional turn can be done in several stages, if needed.

17. Install rocker arm axle and tighten new mounting bolts as follows:
   a. Tighten first inner and then both outer bolts by hand.
   b. Tighten, in the same sequence, to 15 ft. lbs. (20 Nm), plus an additional 90 degrees.

18. Place a dial gauge onto adjustment screw of pump-injector unit.

19. Rotate crankshaft in direction of engine rotation until rocker arm roller stands at camshaft point.
   • Roller side "A" is positioned at highest point
   • Dial gauge "B" is positioned at lowest point

20. Remove dial gauge .

21. Now thread adjusting screw into rocker arm against spring force of pump-injector unit significant resistance is detected (pump-injector unit is positioned at stop).

22. Turn back adjustment bolt from stop approx. 180 degree

23. Hold adjustment bolt in this position and tighten lock nut to 22 ft. lbs. (30 Nm).

24. Connect connector of pump-injector unit and install cylinder head cover and acoustic cover.

*INJECTION TIMING*

→Injection timing is performed as a part of injection pump installation. See procedure above.

### TANDEM PUMP AND FUEL PUMP

*REMOVAL & INSTALLATION*
See Figure 206.

**✳✳ CAUTION**

The tandem pump must not be disassembled under any circumstances because it can cause the vacuum part to malfunction. The result would be brake booster failure.

→Work sequence for removing and installing tandem pump is described. Work sequence for Fuel Pump (FP) is identical up to disconnecting the vacuum line.

1. Remove bypass flap with cooler for Exhaust Gas Recirculation (EGR).

2. Remove intake manifold. See "Intake Manifold" in "ENGINE MECHANICAL" section.

3. Disconnect vacuum line from brake booster of tandem pump.

4. Disconnect central connector for Pump-Injector Units "3".

5. Disconnect fuel hose "2" at tandem pump.

6. Remove mounting bolts.

7. Remove tandem pump from cylinder head.

8. Pull tandem pump slightly upward, remove fuel hose and remove tandem pump.

### To install:

9. Installation is in reverse order of removal, note the following:

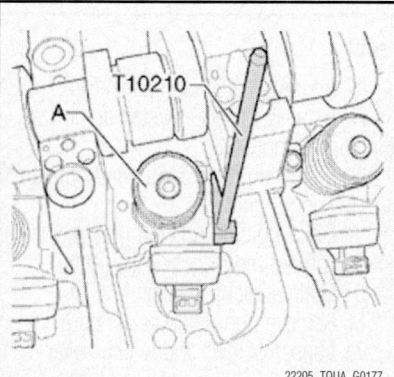

22205_TOUA_G0177

**Fig. 204 Insert a gauge (T10210) between thrust bearing and Pump-Injector Unit—5.0L TDI engine**

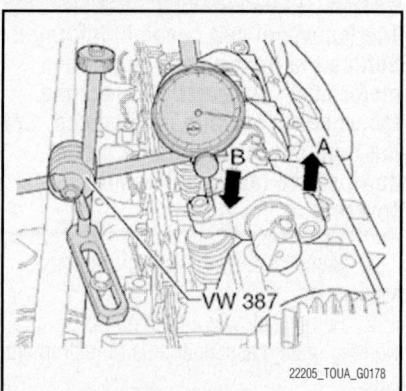

22205_TOUA_G0178

**Fig. 205 Place a dial gauge onto adjustment screw of pump-injector unit—5.0L TDI engine**

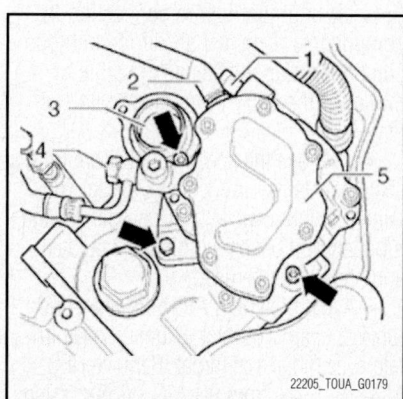

22205_TOUA_G0179

**Fig. 206 Disconnect vacuum line "1" from brake booster of tandem pump "5"—5.0L TDI engine**

a. Make sure that tandem pump coupling has proper seating in camshaft.

b. Tandem pump seals must always be replaced.

c. Connect fuel line with new sealing rings to tandem pump and tighten banjo bolt to 18 ft. lbs. (25 Nm).

10. Install tandem pump and tighten mounting bolts:
- M6: 6 ft. lbs. (8 Nm)
- M8: 15 ft. lbs. (20 Nm)

a. Connect fuel hose and vacuum line to tandem pump.

b. Connect central connector for Pump-Injector Units.

c. Install bypass flap with cooler for Exhaust Gas Recirculation (EGR).

d. Install intake manifold.

11. Perform test drive and check DTC memory.

➡ **By disconnecting the central connection for the pump-injector unit, DTCs will be stored. Query the fault memory and erase, if necessary.**

## GLOW PLUGS

*REMOVAL & INSTALLATION*

### ✳✳ CAUTION

**When removing and installing, do not cant ceramic sheathed element glow plugs.**

➡ **Remove components which hinder installation.**

1. Remove intake manifolds. See "Intake Manifold" in "ENGINE MECHANICAL" section.

2. Disconnect connectors from ceramic sheathed element glow plugs using proper pliers (6275).

3. Remove ceramic sheathed element glow plugs using a flex wrench (3220).

### *To install:*

4. Installation is in reverse order of removal, note the following:

a. Cylinder head bore and thread must be completely cleaned of deposits before installing.

b. Never oil or grease thread of bore in cylinder head and ceramic sheathed element glow plugs.

c. Install ceramic sheathed element glow plugs into cylinder head by hand using flex wrench (3220).

d. Tighten ceramic sheathed element glow plugs. Torque specification: 11 ft. lbs. (15 Nm)

e. After installing and before first engine start with a cold engine, always perform a resistance test at all ceramic sheathed element glow plugs. Specified value: 1.0 ohm maximum.

f. If specified value is exceeded, replace malfunctioning ceramic sheathed element glow plug.

g. If the malfunctioning ceramic sheathed element glow plug is broken, remove all broken pieces from engine, otherwise they can cause damage to engine.

# HEATING & AIR CONDITIONING SYSTEM

## PRECAUTIONS

Before servicing the air conditioning system on any vehicle, please be sure to read all of the following precautions, which deal with personal safety, prevention of component damage, and important points to take into consideration when servicing the air conditioning system:

- When removing refrigerant components from a vehicle, immediately cap (seal) the component to minimize the entry of moisture from the atmosphere.

- When installing refrigerant components to a vehicle, do not remove the caps (unseal) until just before connecting the components. Connect all refrigerant loop components as quickly as possible to minimize the entry of moisture into system.

- Only use the specified lubricant from a sealed container. Immediately reseal containers of lubricant. Without proper sealing, lubricant will become moisture saturated and should not be used.

- Avoid breathing A/C refrigerant and lubricant vapor or mist. Exposure may irritate eyes, nose and throat. Remove HFC-134a (R-134a) from the A/C system, using certified service equipment meeting requirements of SAE J2210 HFC-134a (R-134a) recycling equipment, or J2209 HFC-134a (R-134a) recovery equipment. If accidental system discharge occurs, ventilate work area before resuming service. Additional health and safety information may be obtained from refrigerant and lubricant manufacturers.

- Do not allow lubricant to come in contact with Styrofoam parts. Damage may result.

## BLOWER MOTOR

*REMOVAL & INSTALLATION*

### Front Blower Motor

### ✳✳ CAUTION

**The fan wheel can become deformed. Setting the front blower regulation motor down incorrectly causes the fan wheel to become unbalanced. Set the front blower regulation motor down so the fan wheel is not covered.**

1. Remove footwell trim on front passenger side.

2. Remove blower mounting bolts, working under passenger side of instrument panel.

3. Disconnect blower electrical connector.

4. Install in reverse order of removal.

### Rear Blower Motor

1. Remove the rear HVAC assembly.

2. Disconnect connector at rear blower motor.

3. Lift tab (on center plate of blower motor) slightly and rotate rear blower motor to the right.

4. Remove rear blower motor from heating and air conditioning unit

5. Install in reverse order.

➡ **Condensation water hose to valve must not be kinked or pinched**

## HEATER CORE

*REMOVAL & INSTALLATION*

### Front

1. Remove lower dash panel trim.

2. Remove the access/start authorization control module.

3. Removing relay carrier and onboard supply control unit.

4. Discharge the A/C refrigerant system using a proper recovery service center.

5. Drain coolant system.

6. Remove right-hand heater core cover.

7. Loosen screws and unclip cover from retainers.

8. Remove interior carpet from center tunnel in left-hand area.

9. Remove left-hand heat exchanger cover.

10. Remove screws and unclip cover.

➡**One of the screws is covered by the heater core cover.**

11. Remove footwell vent.

12. Cover carpet in area under heater core with waterproof foil and water absorbing paper.

13. Remove retaining clips from refrigerant lines.

14. Slowly remove refrigerant lines from heater core and allow escaping refrigerant to run into an accumulator reservoir.

➡**There is a retaining pin on the left side below the heater core. This connects the lower heater core cover with the air conditioner unit.**

15. Remove the retaining pin from the locking mechanism and then pull retaining pin out.

16. Fold cover down and remove heater core to left.

### To install:

17. Installation is carried out in the reverse order, when doing this note the following:

   a. Always replace O-rings.

   b. Secure all the connections with the same type of clamps used in series production.

c. After installing retaining clips, make sure they are securely seated on the heater core.

### With Rear A/C

1. Switch off ignition.

2. Release the pressure in the coolant circuit by opening the cap on the coolant reservoir.

3. Remove the rear heating A/C unit from into mounting position.

4. Disconnect the electrical connectors from the A/C unit.

5. Remove left rear air door motor.

6. Remove right rear air door motor.

7. Remove the retaining bolts for the upper cover.

8. Mark both coolant hose positions to rear heater core.

9. Clamp off both coolant hoses.

10. Remove the heater core from the HVAC unit.

11. Place a container under both connections to the heat exchanger and disconnect both coolant hoses.

### To install:

12. Clean the rear A/C unit.

13. Check the foam seals on the heater core. The foam seals must be attached without gaps and must not be damaged.

14. Insert the heater core into heating A/C unit.

15. Connect supply coolant hose to coolant pipe to heater core.

➡**Return coolant hose is not connected for the time being.**

16. Add coolant to the reservoir up to the upper marking.

17. Cover the area under the A/C unit with absorbent paper to prevent leaking coolant from running onto the A/C unit or carpet.

18. Place a container under connection to heater core.

19. Open the hose clamps on coolant supply hose enough so coolant can flow slowly into the heater core (to pre-bleed the heat exchanger).

➡**If the coolant does not flow into the heater core on its own, use a hand pump to build up pressure in the reservoir and increase the flow.**

20. Wait until coolant exits from the coolant return hose connection on the coolant pipe to the heater core.

21. Install the removed components again in reverse order.

22. Bleed coolant circuit.

---

## STEERING

### POWER STEERING GEAR

#### REMOVAL & INSTALLATION

1. Remove front wheels.

2. Pinch off steering intake line and return line with clamping pliers or hose clamps.

3. Press off tie rods from wheel bearing housing.

4. Remove the left tie rod from the steering gear.

5. Remove bracket for hydraulic lines.

6. Remove pressure line and return line from steering gear.

7. Rotate steering gear toward left until stop.

8. If present, remove shielding plate from steering gear (2 bolts).

9. Remove bolt for universal joint at steering gear and remove universal joint from steering gear.

10. Remove steering gear heat shield.

11. Remove the steering gear mounting bolts.

12. Slide steering gear to right side of vehicle.

13. Swing left tie rod downward.

14. Remove steering gear downward toward left side of vehicle.

### To install:

15. Installation is the reverse of removal. Note the following:

   a. Bleed steering system. See "Bleeding Procedure" below.

   b. Check steering system for leaks.

   c. Check hydraulic fluid level and top off, if necessary.

   d. After installation, a vehicle alignment must be performed.

   e. Note the following tightening specifications:

   • Pressure/return line to steering gear: 22 ft. lbs. (30 Nm)

   • Universal joint to steering gear (new bolt): 30 ft. lbs. (40 Nm), plus an additional 90 degrees

   • Shielding plate to steering gear (if present): 8 ft. lbs. (10 Nm)

   • Tie rod to steering knuckle (new nut): 66 ft. lbs. (90 Nm)

   • Tie rod to steering gear: 74 ft. lbs. (100 Nm)

   • Steering gear to subframe (new nuts and bolts): 66 ft. lbs. (90 Nm), plus an additional 90 degrees

### Bleeding Procedure

   a. Inspect hydraulic oil level and top off as needed.

   b. Raise vehicle until front wheels are off the ground.

   c. Start engine and let it run at idle for approximately 5 seconds.

   d. Switch off engine and check hydraulic fluid level.

   e. Repeat procedure once more.

   f. Start engine again and turn steering wheel 3 times from stop to stop at idle speed.

   g. Switch off engine and check hydraulic fluid level, top off if necessary.

   h. Repeat procedure 2 times.

   i. To dissipate gas of hydraulic fluid, let engine stand 2 to 3 minutes.

   j. Lower vehicle.

   k. Now, once more, turn steering wheel 5 times from stop to stop at idle speed.

   l. Steering system has been bled when air bubbles no longer rise to the surface in hydraulic fluid reservoir.

*REMOVAL & INSTALLATION*

### 3.6L Engine

1. Remove engine noise insulation.
2. Mark ribbed belt running direction
3. Thread bolt M8 × 50 into threaded hole of tensioning roller until ribbed belt is free of tension.
4. Remove ribbed belt pulley of power steering pump.
5. Remove the ribbed belt.

➡**When removing ribbed belt, observe adjustment shim located underneath. The shim is important for measurement and adjustment of power steering pump. Do not remove adjustment shim from power steering pump.**

6. Clamp off power steering intake hose.
7. Remove connection banjo bolt.
8. Seal pressurized line using a plastic bag or something similar.
9. Open spring clip and pull intake hose off power steering pump.
10. Remove the power steering pump.

*To install:*

11. Installation is the reverse of removal, with special attention to the following:

   a. Fill hydraulic oil into power steering pump. Fill oil at pump intake connection of vane pump.
   b. Turn the hub by hand until oil runs out of the pressure side.
   c. Install power steering pump.
   d. Bleed steering system. See "Bleeding Procedure" below.
   e. Check hydraulic fluid level.
   f. Check steering system for leaks.
   g. Note the following tightening specifications:
   • Power steering pump to bracket: 18 ft. lbs. (25 Nm)
   • Belt pulley to power steering pump: 18 ft. lbs. (25 Nm)
   • Banjo bolt to power steering pump: 30 ft. lbs. (40 Nm)

*BLEEDING PROCEDURE*

   a. Inspect hydraulic oil level and top off as needed.
   b. Raise vehicle until front wheels are off the ground.
   c. Start engine and let it run at idle for approximately 5 seconds.
   d. Switch off engine and check hydraulic fluid level.
   e. Repeat procedure once more.
   f. Start engine again and turn steer-

ing wheel 3 times from stop to stop at idle speed.
   g. Switch off engine and check hydraulic fluid level, top off if necessary.
   h. Repeat procedure 2 times.
   i. To dissipate gas of hydraulic fluid, let engine stand 2 to 3 minutes.
   j. Lower vehicle.
   k. Now, once more, turn steering wheel 5 times from stop to stop at idle speed.
   l. Steering system has been bled when air bubbles no longer rise to the surface in hydraulic fluid reservoir.

### 4.2L Engine

*See Figures 207 and 208.*

1. Remove fan mount.
2. Remove engine noise insulation.
3. Loosen attachment bolts for belt pulley.
4. Mark ribbed belt running direction.
5. Turn tensioner roller so tension is relieved from belt.

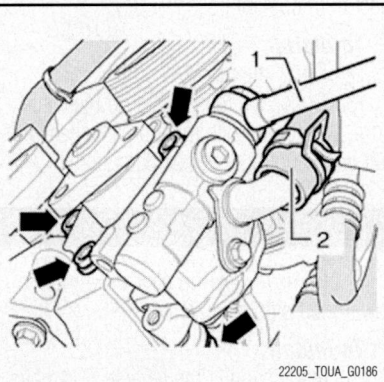

22205_TOUA_G0186

**Fig. 207 Remove banjo fitting and seal pressurized line "1"—4.2L engine**

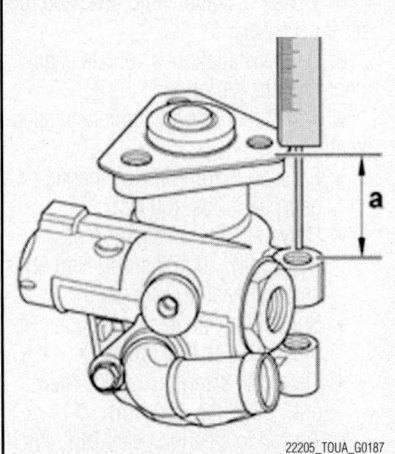

22205_TOUA_G0187

**Fig. 208 Measure dimension "a" (shim is included in measurement), and note measured value—4.2L engine**

6. Remove the ribbed belt.
7. Clamp off intake hose, using hose clamp (3094).
8. Remove belt drive pulley from the power steering pump.
9. Remove banjo fitting and seal pressurized line "1" using a plastic bag or something similar.
10. Open spring clip and pull intake hose "2" off power steering pump.
11. Remove the power steering pump (arrows).
12. Measure dimension "a". The shim is included in measurement. Note measured value.

*To install:*

13. Installation is the reverse of removal, with special attention to the following:

   a. Use the following torque specifications during installation:
   • Power steering pump to bracket: 18 ft. lbs. (25 Nm)
   • Belt pulley to power steering pump: 18 ft. lbs. (25 Nm)
   • Banjo bolt to power steering pump: 30 ft. lbs. (40 Nm)
   b. Using measured dimension, determine thickness of adjustment shim needed.
   • For example, if measured value of old power steering pump with adjustment shim was 33.65 mm and the measured value of new power steering pump without adjustment shim is 32.85 mm, then shim thickness should be 0.80 mm.
   • Shims are available in thickness of: 0.6 mm, 0.8 mm, 1.0 mm, 1.2 mm, and 1.4 mm.
   c. Select and install proper shim thickness.
   d. Fill hydraulic oil into power steering pump. Fill oil at pump intake connection of power steering pump.
   e. Turn the hub by hand until oil runs out of the pressure side.
   f. Install power steering pump.
   g. When tightening banjo fitting, strap of pressurized line must make contact with pump housing.
   h. Install ribbed belt on power steering pump.
   i. Using a proper gauge, determine if pulleys are aligned and ribs of the belt ride properly in the pulleys. If they do not align, a different suitable shim must be used.
   j. Install ribbed belt.
14. Bleed steering system.
15. Check hydraulic fluid level.
16. Check steering system for leaks.

**5.0L TDI Engine**

1. Clamp off intake hose, using hose clamp 3094.

2. Remove engine and transmission. See "Engine Assembly" in "ENGINE MECHANICAL" section.

3. Remove A/C compressor.

4. Remove charge line from cylinder block and from turbocharger and lay aside.

5. Remove pressurized line from vane pump.

6. Seal pressurized line using a plastic bag or something similar.

7. Remove intake line from power steering pump.

8. Remove power steering pump from cylinder block.

**To install:**

9. Installation is the reverse of removal, with special attention to the following:

a. Fill hydraulic oil into power steering pump. Fill oil at pump intake connection of power steering pump.

b. Turn the hub by hand until oil runs out of the pressure side.

c. Install power steering pump.

10. After installing engine, bleed the steering system.

11. Check hydraulic fluid level.

12. Check steering system for leaks.

## SUSPENSION

## FRONT SUSPENSION

### COIL SPRING

*REMOVAL & INSTALLATION*

➡See "MacPherson Strut."

### LOWER BALL JOINT

*REMOVAL & INSTALLATION*

The ball joints are not replaceable. If necessary, the entire control arm must be replaced.

### LOWER CONTROL ARM

*REMOVAL & INSTALLATION*

*See Figure 209.*

1. Before servicing the vehicle, refer to the precautions in the beginning of this Section.

2. Measure the distance from the center of the wheel hub to the lip of the front fender.

3. Raise the vehicle and remove the wheel.

4. Remove the wheel housing liner.

5. Using hooks VW 552 or equivalent, secure the upper control arm to the upper opening of the wheel housing.

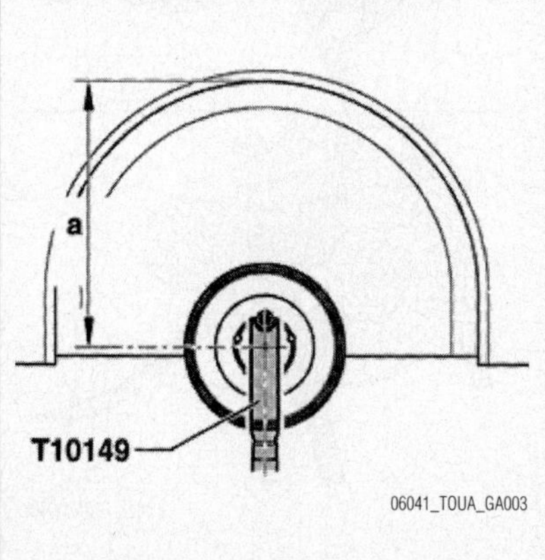

6. Raise the lower control arm slightly to preload it. This prevents damage to the stud of the ball joint.

7. Remove the suspension strut from the lower control arm.

8. Remove the ball joint nut from the lower control arm and press off the ball stud using a ball joint puller.

9. Remove the nuts, then remove the lower control arm from the subframe and steering knuckle.

**To install:**

10. Installation is the reverse of removal. When installing the strut, raise the control arm to the measurement taken in the beginning of the procedure. The nut must be tightened when in this position. Perform a wheel alignment and observe the following torque specifications (use new hardware):

- Lower control arm-to-subframe: 133 ft. lbs. (180 Nm)
- Suspension strut-to-lower control arm: 111 ft. lbs. (150 Nm) plus an additional ¼ turn
- Lower control arm-to-steering knuckle: 77 ft. lbs. (105 Nm)

### MACPHERSON STRUT

*REMOVAL & INSTALLATION*

*See Figure 210.*

1. Before servicing the vehicle, refer to the precautions in the beginning of this Section.

2. Remove the wheel and raise the vehicle.

3. Remove the nuts securing the suspension strut to the body.

4. Remove the halfshaft.

5. Make sure the lower control arm/steering knuckle assembly is properly supported.

6. Pull out bolt securing the strut to the lower control arm and remove suspension strut.

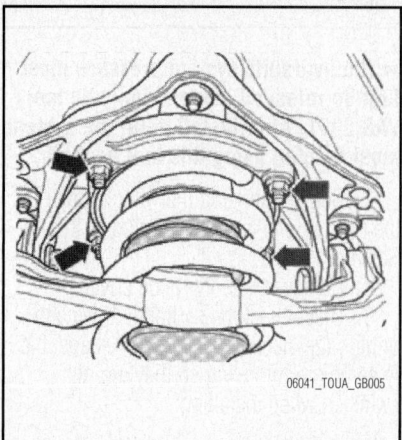

06041_TOUA_GB005

**Fig. 210 Remove the nuts securing the strut to the body**

**To install:**

7. Installation is the reverse of removal. Perform a wheel alignment and observe the following torque specifications (use new hardware):

- Suspension strut to body nuts: 22 ft. lbs. (30 Nm)
- Suspension strut to lower control arm: 111 ft. lbs. (150 Nm) plus an additional ¼ turn
- Upper control arm to steering knuckle: 70 ft. lbs. (95 Nm)
- Tie rod to steering knuckle: 66 ft. lbs. (90 Nm)

### STABILIZER BAR

*REMOVAL & INSTALLATION*
*See Figure 211.*

### ✳✳ CAUTION

**On vehicles with decouplable stabilizer bars, the stabilizer bars must be coupled in before proceeding. Otherwise, the risk of injury exists if the decoupled stabilizer bars become unintentionally coupled.**

06041_TOUA_GA003

**Fig. 209 Measure distance (A). The lower strut nut must be tightened in this position**

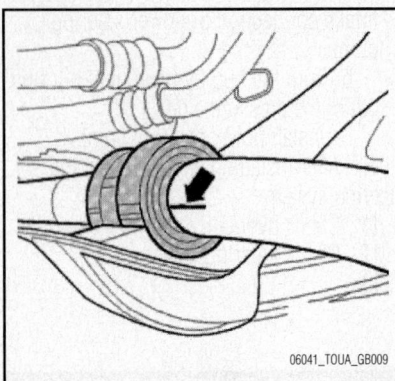

**Fig. 211 Mark the installed position of the bar**

➡**The hydraulic system pressure must first be released using diagnostic tool VAS 5051. After installation, the system must be bled using this tool as well.**

1. Before servicing the vehicle, refer to the precautions in the beginning of this Section.

2. Remove the underbody cover.

3. Disconnect the hydraulic lines and unplug the electrical connection. Label the lines before removing so they are not switched when installing.

➡**When removing or installing the hydraulic lines, counterhold them with an open-end wrench.**

4. Remove the stabilizer bar clamps and mark the installed location of the stabilizer bar mount on the bar.

5. Remove the stabilizer bar from the connecting links.

*To install:*

➡**The larger outside diameter of the stabilizer bar mount halves face toward the outside of vehicle.**

6. Installation is the reverse of removal. Align the markings made during removal. Observe the following torque specifications (use new hardware):
- Coupling rod to stabilizer bar: 81 ft. lbs. (110 Nm)
- Stabilizer bar to subframe: 44 ft. lbs. (60 Nm)
- Hydraulic lines: 11 ft. lbs. (15 Nm)

## STEERING KNUCKLE

### REMOVAL & INSTALLATION

1. Before servicing the vehicle, refer to the Precautions Section.

2. Raise and properly support the vehicle.

3. Remove the halfshaft from the steering knuckle assembly (refer to the Halfshaft removal and installation procedure).

4. Remove the ball joint nut from the lower control arm and press off the ball stud using a ball joint puller.

5. Remove the steering knuckle from the vehicle.

6. Installation is the reverse of removal. When installing the lower control arm ball joint nut (use a new one), tighten it to 77 ft. lbs. (105 Nm).

## STRUT & SPRING ASSEMBLY

### REMOVAL & INSTALLATION

➡**See "MacPherson Strut."**

## WHEEL BEARINGS

### REMOVAL & INSTALLATION

*See Figures 212 through 215.*

1. Before servicing the vehicle, refer to the precautions in the beginning of this Section.

2. Remove the halfshaft.

3. Attach the upper control arm to the steering knuckle.

4. Secure the brake rotor with a wheel bolt.

5. Remove the brake caliper and secure it with wire out of the way.

6. Remove the wheel bolt and the brake rotor. Remove the disc shield.

7. Unclip the wheel speed sensor wire from the retainer.

8. Remove the bolt and the ABS wheel speed sensor.

**❊❊ CAUTION**

**On vehicles with decouplable stabilizer bars, the stabilizer bars must be coupled in before proceeding. Otherwise, the risk of injury exists if the decoupled stabilizer bars become unintentionally coupled.**

9. The following tools or their commercially available equivalents can be used to remove and install the hub and bearing:
- Hydraulic press VAS 6178
- Threaded rod M20T10205/8-1
- Threaded nut M20T10205/8-2
- Cup T10205/2
- Puller T10205/1
- Press plates/supports T10205/3, 4, 6, 7, 9 and 10
- Mounting tube T10205/5

10. Attach the puller to the wheel hub using wheel bolts. Pull the wheel hub out.

11. Remove the circlip.

12. Assemble the tools as shown in illustration. Pull the wheel bearing out.

➡**The shoulder of press plate T10205/6 must point toward the final drive.**

13. To pull the inner race off the wheel hub, install the separating tool behind the inner race.

➡**The chamfers on the jaws face toward the inner race.**

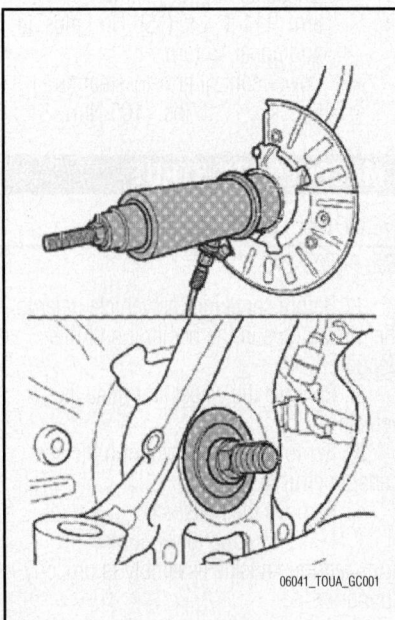

**Fig. 212 Pulling the bearing out of the housing**

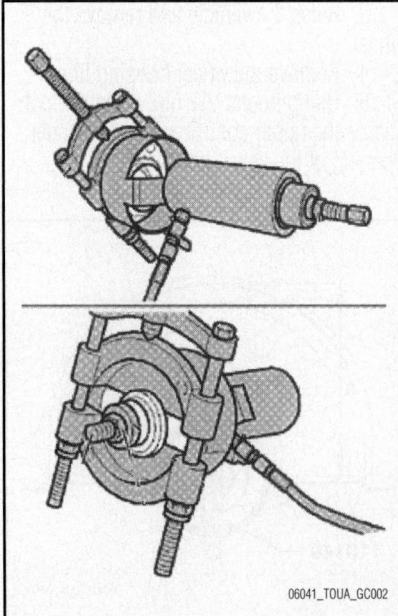

**Fig. 213 Removing the inner race from the hub**

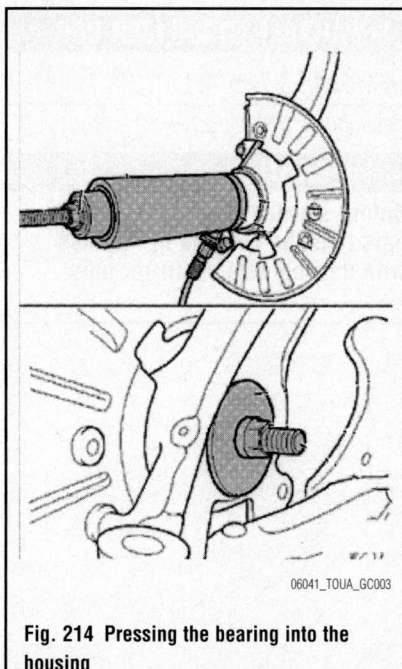

**Fig. 214 Pressing the bearing into the housing**

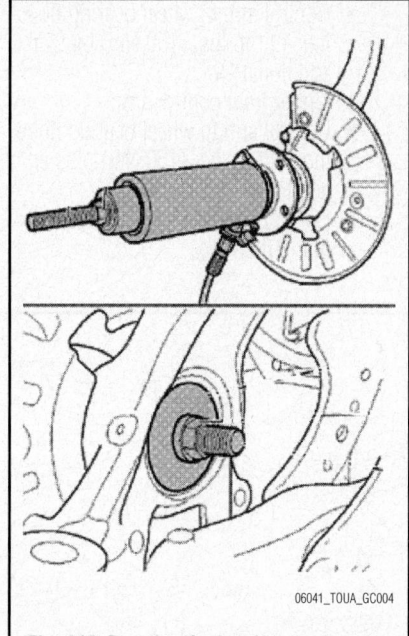

**Fig. 215 Pressing the hub into the housing**

15. Install the tools as shown in illustration to press the bearing in.

➡**The shoulder of the press plate T10205/9 must face toward wheel bearing housing.**

16. Press in the wheel bearing up to the stop.

➡**If using hydraulic press VAS 6178, the pressure reading must be between 100 and 190 bar shortly before the end of pressing in. The maximum pressure should not exceed 310 bar.**

17. Install the circlip.
18. Assemble the tools as shown in illustration to press the wheel hub in.

➡**The shoulder of the press plate T10205/6 must point toward the final drive.**

19. Press in the wheel bearing as far as the stop.

➡**If using hydraulic press VAS 6178, the pressure reading must be between 30 and 100 bar shortly before the end of pressing in. The maximum pressure should not exceed 140 bar.**

20. The remaining installation is the reverse of removal.

## ✳✳ CAUTION

**To prevent injury, hold the tool when pulling the bearing inner race off so that the race faces downward.**

**To install:**

14. Make sure that the rubberized ABS sensor ring faces towards the final drive. If no rubberized ring is visible, check using a paper clip which of the two sides is magnetic. When installed, this side must face toward final drive.

## SUSPENSION

### COIL SPRING

*REMOVAL & INSTALLATION*

*See Figures 216 and 217.*

1. Before servicing the vehicle, refer to the precautions in the beginning of this Section.
2. Remove the upper strut bolts.
3. Insert a length of wood between the underbody and the upper rear control arm. This will be used to press the control arm down to remove the strut.
4. Remove the lower strut nut.
5. Using the length of wood, press control arm down sufficiently to remove the bolt.
6. Remove the strut.

**To install:**

7. Installation is the reverse of removal. Observe the following torque specifications (use new hardware):

## REAR SUSPENSION

- Strut to body: 44 ft. lbs. (60 Nm)
- Strut to wheel bearing housing: 66 ft. lbs. (90 Nm) plus an additional ¼ turn

### LOWER CONTROL ARM

*REMOVAL & INSTALLATION*

➡**This procedure is for both the lower and upper control arms.**

The bushings of the control arms are not replaceable. If necessary, the entire control arm must be replaced

1. Before servicing the vehicle, refer to the precautions in the beginning of this Section.
2. Raise the vehicle and remove the wheels.
3. Remove the bolts securing the control arm and remove it from the vehicle.
4. Installation is the reverse of removal. Observe the following torque specifications (use new hardware):
   a. Lower control arm:
   - Control arm to subframe (eccentric): 133 ft. lbs. (180 Nm)

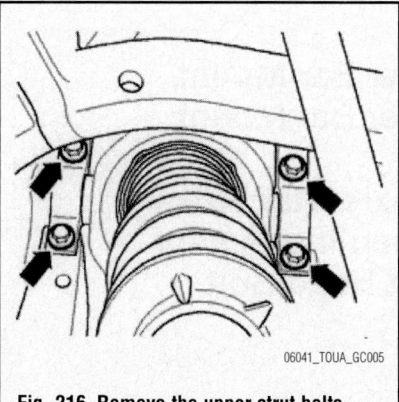

**Fig. 216 Remove the upper strut bolts**

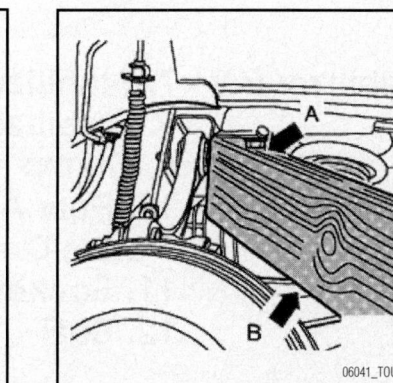

**Fig. 217 Insert the length of wood as shown**

- Control arm to subframe: 111 ft. lbs. (150 Nm) plus an additional ¼ turn
- Control arm to wheel bearing housing: 111 ft. lbs. (150 Nm) plus an additional ¼ turn
- b. Upper rear control arm:
- Control arm to subframe: 66 ft. lbs. (90 Nm) plus an additional ¼ turn

- Control arm to wheel bearing housing: 111 ft. lbs. (150 Nm) plus an additional ¼
- c. Upper front control arm:
- Control arm to wheel bearing housing: 111 ft. lbs. (150 Nm) plus an additional ¼
- Control arm to subframe: 133 ft. lbs. (180 Nm)

## SEPARABLE STABILIZER BAR

*REMOVAL & INSTALLATION*
See Figures 218 and 219.

### ✴✴ CAUTION

**Before starting work, stabilizer bars must be coupled in vehicles with decouplable stabilizer bars.**

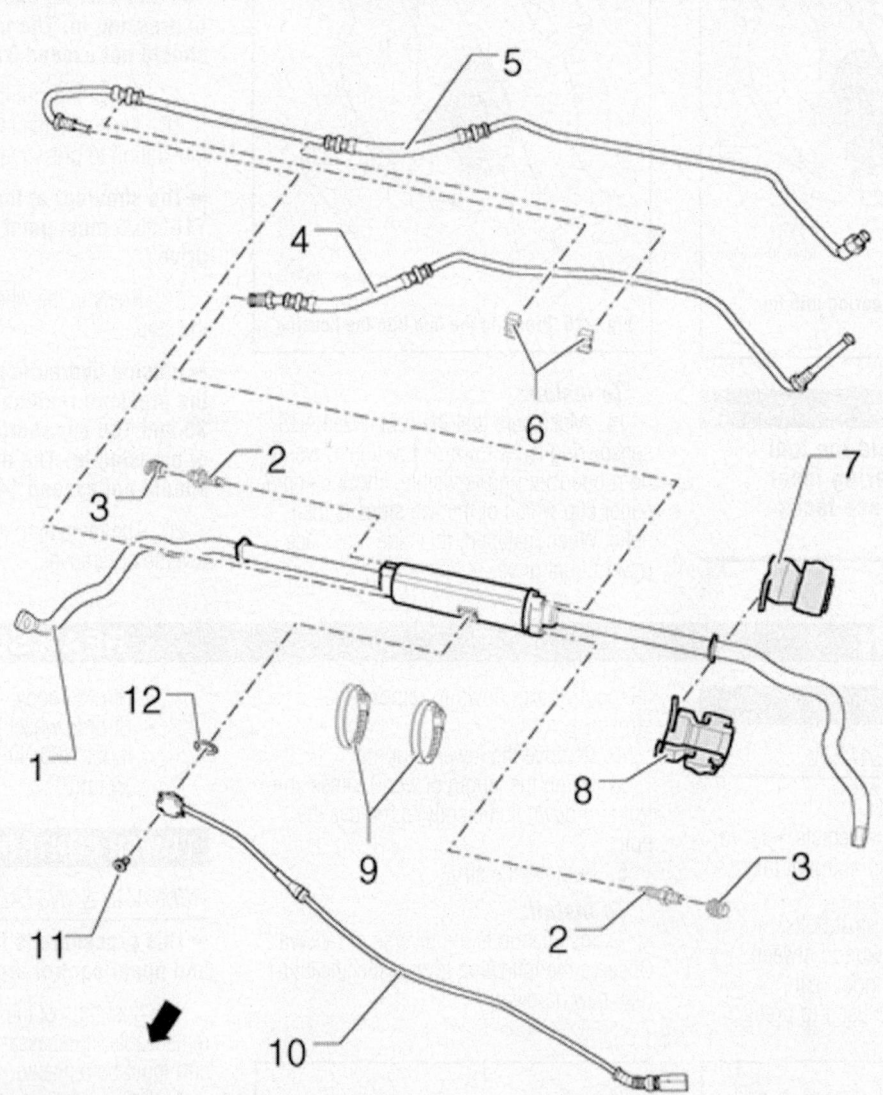

1. Rear Separable Stabilizer Bar
2. Bleeder Nipple
3. Dust Cap
4. Hydraulic Line
5. Hydraulic Line
6. Bracket
7. Stabilizer Bar Mount
8. Stabilizer Bar Mount
9. Clamp
10. Rear Axle Stabilizer De-Coupling Sensor
11. Socket Head Bolt
12. Seal

22205_TOUA_G0182

**Fig. 218 Exploded view of separable stabilizer bar and related components**

Otherwise, there is the danger of injury caused by unintentional coupling.

1. Depressurize hydraulic system.
2. Remove left rear wheel housing liner.
3. Disconnect both hydraulic lines and electrical connection for rear suspension.

### ✳✳ CAUTION

**Lines and wires must not be mixed up when connecting!**

➡**Check whether color marking is present on front in driving direction. If no longer present, apply a new one.**

4. When loosening hydraulic lines, counterhold using open-end wrench.
5. Remove coupling rods.
6. Remove tie rods on both sides of wheel bearing housing.
7. Back out bolt - 1 - to lock tie rod - 2 - .
8. Mark installation position of stabilizer bar bearings in relation to the stabilizer bar.
9. Remove clamps and stabilizer bar.

***To install:***
10. Installation is performed in the reverse order of removal.
11. Install stabilizer bar bearing halves on stabilizer bar. Note the following:
    a. Markings applied on stabilizer bar when removed must be aligned.
    b. Larger outer diameter of stabilizer bar bearing halves must face toward vehicle center
12. After installation, hydraulic system must be bled using "Separable Stabilizer Bar Bleeding" procedure below.
13. Note the following tightening specifications:
    • Coupling rod to stabilizer bar: 74 ft. lbs. (100 Nm)

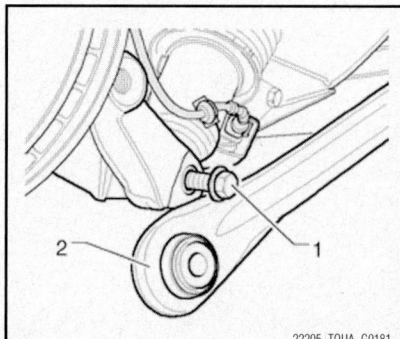

22205_TOUA_G0181

**Fig. 219 Back out bolt "1" to lock tie rod "2"**

• Stabilizer bar to subframe: 37 ft. lbs. (50 Nm)
• Connecting link to wheel bearing housing: 74 ft. lbs. (100 Nm)
• Tie rod to steering knuckle (new nuts and bolts): 111 ft. lbs. (150 Nm), plus an additional 90 degrees
• Hydraulic line connection: 11 ft. lbs. (15 Nm)
14. Install left rear wheel housing liner

### *SEPARABLE STABILIZER BAR BLEEDING*

1. Depressurize hydraulic system.
2. Check oil level in engine pump assembly.
3. Connect hose to bleeder nipple.
4. Secure hose on bleeder nipples using cable ties or hose clamp.

### ✳✳ CAUTION

**Due to high pressure, hose would slide from bleeder nipples during bleeding without this securing measure.**

5. Loosen bleeder nipple by approximately 1 turn.
6. Tighten bleeder nipple and remove hose.
7. Check oil level in engine pump assembly again

### STRUT & SPRING ASSEMBLY

### *REMOVAL & INSTALLATION*

1. Remove suspension strut upper mounting bolts.
2. Insert wood block between underbody and rear upper control arm.

➡**This is needed to press down wheel bearing housing.**

**Remove nut for lower strut retaining bolt.**

3. Using wood block, press down wheel bearing housing far enough to remove strut lower retaining bolt. Remove rear suspension strut.

***To install:***
4. Installation is in reverse order of removal.
5. Note the following tightening specifications:
   • Suspension strut to body: 44 ft. lbs. (60 Nm)
   • Suspension strut to wheel bearing housing (new bolt and nut): 66 ft. lbs. (90 Nm), plus an additional 90 degrees

### WHEEL BEARINGS

### *REMOVAL & INSTALLATION*

*See Figures 220 through 222.*

### ✳✳ CAUTION

**Before starting work, stabilizer bars must be coupled in vehicles with decouplable stabilizer bars. Otherwise, there is the danger of injury caused by unintentional coupling.**

1. Remove wheel.
2. Remove lower control arm from wheel bearing housing.
3. Install wood block between control arm and wheel bearing housing.

### ✳✳ WARNING

**Without this wood block inserted, there is the danger of injury when lower control arm moves back to initial position.**

4. Install tools onto wheel bearing housing, as shown in illustration.

### ✳✳ CAUTION

**Before pressing out, place engine/transmission jack V.A.G 1383 A with universal mount underneath cylinder, cylinder may fall down at end of pressing out procedure.**

5. Press out bonded rubber bushing.

***To install:***
6. Install tools and bonded rubber bushings onto wheel bearing housing, as shown.
7. Press in bonded rubber bushing until stop.

➡**By doing this, core of bonded rubber bushing is pressed in by thrust piece T10230/6 more than is required. However, this is necessary so that bonded**

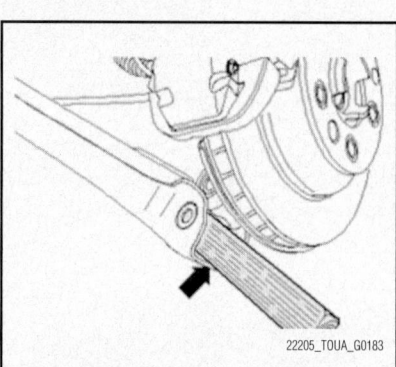

22205_TOUA_G0183

**Fig. 220 Install wood block between control arm and wheel bearing housing**

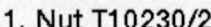

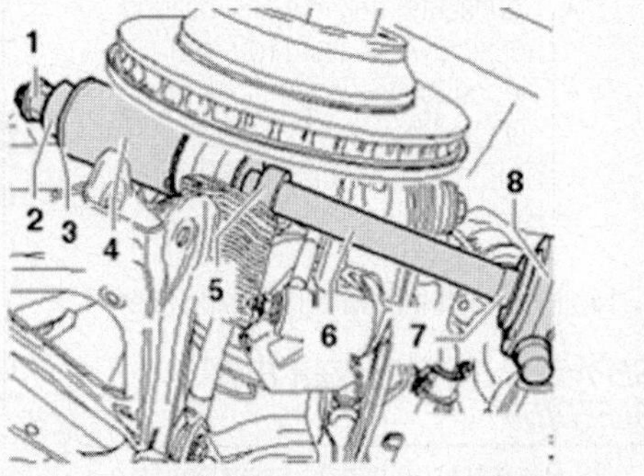

1. Nut T10230/2
2. Press piece T10230/13
3. Thrust pad T10230/6
4. Tube T10230/5
5. Press piece T10230/4
6. Tube T10230/3
7. Thrust plate T10230/7
8. Hydraulic Press VAS 6178 with Pressure Head T10205/13

22205_TOUA_G0184

**Fig. 221 Install tools onto wheel bearing housing, as shown**

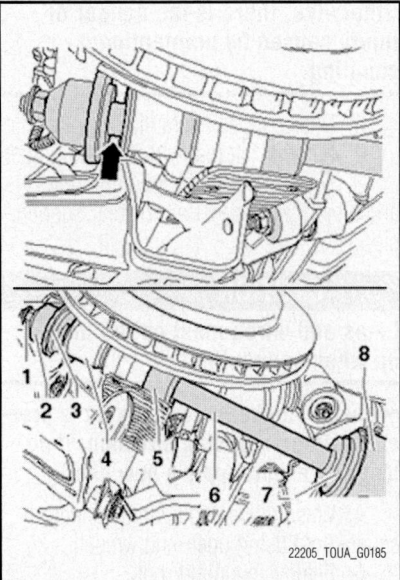

22205_TOUA_G0185

**Fig. 222 Install tools and bonded rubber bushings onto wheel bearing housing, as shown**

rubber bushing lip makes contact with control arm on outside. After loosening hydraulic cylinder VAS 6178, bonded rubber bushing core settles into correct position.

8. Remove wood block from between wheel bearing housing and lower control arm.

9. Press wheel bearing housing into position.

### ❋ CAUTION

**Make sure to press only at wheel bearing housing. There is the danger of injury at the connection point of control arm and wheel bearing housing!**

10. Ensure control arm engages at wheel bearing housing.

11. Further installation is in the reverse sequence to removal.

12. Note the following tightening specifications (tighten at curb weight position):
- Control arm to steering knuckle (new bolt and nut): 111 ft. lbs. (150 Nm), plus an additional 90 degrees
- Wheel to wheel hub: 133 ft. lbs. (180 Nm)

# VOLKSWAGEN

## Diagnostic Trouble Codes

# 16

# DIAGNOSTIC TROUBLE CODES

## OBD II VEHICLE APPLICATIONS

*VOLKSWAGEN*

**Golf**
2007–2008
- 2.0L MPI . . . . . . . Engine Codes: BPY

**GTI**
2007–2008
- 2.0L MPI . . . . . . . Engine Codes: BPY

**Jetta**
2007–2008
- 1.9L DFI . . . . . . . Engine Codes: BRM
- 2.0L MPI . . . . . . . Engine Codes: BPY
- 2.5L MPI . . Engine Codes: BGP, BGQ

**New Beetle**
2007–2008
- 1.9L DFI . . . . . . . Engine Codes: BEW
- 2.5L MPI . . .Engine Codes: BPR, BPS

**New Beetle Convertible**
2007–2008
- 2.5L MPI . . .Engine Codes: BPR, BPS

**Passat**
2007–2008
- 2.0L MPI . . . . . . . Engine Codes: BPY
- 2.5L, 3.6L MPI . . . Engine Codes: BLV

**Passat Wagon**
2007–2008
- 2.0L MPI . . . . . . . Engine Codes: BPY
- 2.5L, 3.6L MPI . . . Engine Codes: BLV

## Gas Engine OBD II Trouble Code List (P0xxx Codes)

| DTC | Trouble Code Title, Conditions & Possible Causes |
|---|---|
| **DTC: P0010**<br>**2T, MIL: Yes**<br>**Years:** 2007, 2008<br>**Models:** Passat, Jetta, Golf, GTI, New Beetle<br>**Engines:** 2.0L, 2.5L, 3.6L<br>**Transmissions:** All | **"A" Camshaft Position Actuator Circuit (Bank 1) Conditions:**<br>Key on or engine running; and the ECM detected an unexpected high voltage or low voltage condition on the camshaft position sensor. The relative position between the camshaft and crankshaft needs to be optimal so the engine has better torque, fuel economy and emissions.<br>**Note: The camshaft adjustment is load- and RPM dependant. The electrical camshaft adjustment valve 1 switches oil pressure onto camshaft adjuster (mechanical adjustment mechanism), which adjusts the camshaft.**<br>**Possible Causes:**<br>• Fuel pump has failed<br>• Actuator circuit is open<br>• ECM has failed<br>• Battery voltage below 11.5 volts<br>• Position actuator circuit may short to B+ or Ground |
| **DTC: P0011**<br>**2T, MIL: Yes**<br>**Years:** 2007, 2008<br>**Models:** Passat, Jetta, Golf, GTI, New Beetle<br>**Engines:** 2.0L, 2.5L, 3.6L<br>**Transmissions:** All | **"A" Camshaft Position Timing Over-Advanced (Bank 1) Conditions:**<br>Engine started and driven at an engine speed of more than 400 RPM; and the ECM detected the camshaft timing exceeded the maximum calibrated advance value, or the camshaft remained in an advanced position during the CCM test. The valve timing did not change from the current valve timing or it remained fixed during the testing.<br>**Note: The camshaft adjustment is load- and RPM dependant. The electrical camshaft adjustment valve 1 switches oil pressure onto camshaft adjuster (mechanical adjustment mechanism), which adjusts the camshaft.**<br>**Possible Causes:**<br>• Fuel pump has failed<br>• CPS circuit is open, shorted to ground or shorted to power<br>• ECM has failed<br>• Battery voltage below 11.5 volts<br>• Position actuator circuit may short to B+ or Ground<br>• Camshaft timing improperly set, or continuous oil flow to the VCT piston chamber<br>• Camshaft advance mechanism (the VCT unit) is sticking or binding mechanically<br>• VCT solenoid valve is stuck in open position |
| **DTC: P0012**<br>**2T, MIL: Yes**<br>**Years:** 2007, 2008<br>**Models:** Passat, Jetta, Golf, GTI, New Beetle<br>**Engines:** 2.0L, 2.5L, 3.6L<br>**Transmissions:** All | **"A" Camshaft Position Over-Retarded (Bank 1) Conditions:**<br>Engine started and driven at an engine speed of more than 400RPM; and the ECM detected the camshaft timing exceeded the minimu calibrated retarded value, or the camshaft remained in an retarted position during the CCM test. The valve timing did not change from the current valve timing or it remained fixed during the testing.<br>**Note: The camshaft adjustment is load- and RPM dependant. The electrical camshaft adjustment valve 1 switches oil pressure onto camshaft adjuster (mechanical adjustment mechanism), which adjusts the camshaft.**<br>**Possible Causes:**<br>• Fuel pump has failed<br>• CPS circuit is open, shorted to ground or shorted to power<br>• ECM has failed<br>• Battery voltage below 11.5 volts<br>• Position actuator circuit may short to B+ or Ground<br>• Camshaft timing improperly set, or continuous oil flow to the VCT piston chamber<br>• Camshaft advance mechanism (the VCT unit) is sticking or binding mechanically<br>• VCT solenoid valve is stuck in open position |
| **DTC: P0013**<br>**2T, MIL: Yes**<br>**Years:** 2007, 2008<br>**Models:** Passat, Jetta, Golf, New Beetle<br>**Engines:** 2.5L, 3.6L<br>**Transmissions:** All | **"B" Camshaft Position Actuator Circuit (Bank 1) Conditions:**<br>Key on or engine running; and the ECM detected an unexpected high voltage or low voltage condition on the camshaft position sensor. The relative position between the camshaft and crankshaft needs to be optimal so the engine has better torque, fuel economy and emissions.<br>**Note: The camshaft adjustment is load- and RPM dependant. The electrical camshaft adjustment valve 1 switches oil pressure onto camshaft adjuster (mechanical adjustment mechanism), which adjusts the camshaft.**<br>**Possible Causes:**<br>• Fuel pump has failed<br>• ECM has failed<br>• Battery voltage below 11.5 volts<br>• Position actuator circuit may short to B+ or Ground |

| DTC | Trouble Code Title, Conditions & Possible Causes |
|---|---|
| **DTC: P0014**<br>**2T, MIL: Yes**<br>**Years: 2007, 2008**<br>**Models:** Passat, Jetta, Golf, New Beetle<br>**Engines:** 2.5L, 3.6L<br>**Transmissions:** All | **"B" Camshaft Position Timing Over-Advanced (Bank 1) Conditions:**<br>Engine started and driven at an engine speed of more than 400RPM; and the ECM detected the camshaft timing exceeded the maximum calibrated advance value, or the camshaft remained in an advanced position during the CCM test. The valve timing did not change from the current valve timing or it remained fixed during the testing.<br>**Note: The camshaft adjustment is load- and RPM dependant. The electrical camshaft adjustment valve 1 switches oil pressure onto camshaft adjuster (mechanical adjustment mechanism), which adjusts the camshaft.**<br>**Possible Causes:**<br>• Fuel pump has failed<br>• CPS circuit is open, shorted to ground or shorted to power<br>• ECM has failed<br>• Battery voltage below 11.5 volts<br>• Position actuator circuit may short to B+ or Ground<br>• Camshaft timing improperly set, or continuous oil flow to the VCT piston chamber<br>• Camshaft advance mechanism (the VCT unit) is sticking or binding mechanically<br>• VCT solenoid valve is stuck in open position |
| **DTC: P0015**<br>**2T, MIL: Yes**<br>**Years: 2007, 2008**<br>**Models:** Passat, Jetta, Golf, New Beetle<br>**Engines:** 2.5L, 3.6L<br>**Transmissions:** All | **"B" Camshaft Position Over-Retarded (Bank 1) Conditions:**<br>Engine started and driven at an engine speed of more than 400 RPM; and the ECM detected the camshaft timing exceeded the minimu calibrated retarded value, or the camshaft remained in an retarded position during the CCM test. The valve timing did not change from the current valve timing or it remained fixed during the testing.<br>**Note: The camshaft adjustment is load- and RPM dependant. The electrical camshaft adjustment valve 1 switches oil pressure onto camshaft adjuster (mechanical adjustment mechanism), which adjusts the camshaft.**<br>**Possible Causes:**<br>• Fuel pump has failed<br>• CPS circuit is open, shorted to ground or shorted to power<br>• ECM has failed<br>• Battery voltage below 11.5 volts<br>• Position actuator circuit may short to B+ or Ground<br>• Camshaft timing improperly set, or continuous oil flow to the VCT piston chamber<br>• Camshaft advance mechanism (the VCT unit) is sticking or binding mechanically<br>• VCT solenoid valve is stuck in open position |
| **DTC: P0016**<br>**2T, MIL: Yes**<br>**Years: 2007, 2008**<br>**Models:** Passat, Jetta, Golf, New Beetle<br>**Engines:** 2.0L, 2.5L, 3.6L<br>**Transmissions:** All | **Crankshaft Position - Camshaft Position Correlation Bank 1 Sensor A Conditions:**<br>Engine started, engine running, and the ECM detected a diviation between the crankshaft position sensor signal and the camshaft position sensor. A rationality error has been detected for camshaft position out of phase with crankshaft.<br>**Possible Causes:**<br>• Camshaft Position (CMP) sensor is faulty<br>• CMP circuit short to ground, power or open<br>• Engine Speed (RPM) sensor is faulty<br>• ECM has failed |
| **DTC: P0030**<br>**2T, MIL: Yes**<br>**Years: 2007, 2008**<br>**Models:** Passat, Jetta, Golf, GTI, New Beetle, New Beetle Convertible,<br>**Engines:** 2.0L, 2.5L, 3.6L<br>**Transmissions:** All | **HO2S Heater (Bank 1 Sensor 1) Control Circuit Malfunction Conditions:**<br>Engine started, battery voltage must be at least 11.5v, all electrical components must be off, the ground between the engine and the chassis must be well connected, the exhaust system must be properly sealed between the catalytic converter and the cylinder head, the coolant temperature must be 80 degrees Celsius, and the oxygen sensor heater for oxygen sensor before the catalytic converter must be properly functioning. The ECM detected the HO2S signal was in a negative voltage range referred to as "character shift downward". This code sets when the HO2S signal remains in a low state (usually less than 156 mv). In effect, it does not switch properly between 0.1v and 1.1v in closed loop operation.<br>**Possible Causes:**<br>• HO2S is contaminated (due to presence of silicone in fuel)<br>• HO2S signal and ground circuit wires crossed in wiring harness<br>• HO2S signal circuit is shorted to sensor or chassis ground<br>• HO2S element has failed (internal short condition)<br>• ECM has failed |

| DTC | Trouble Code Title, Conditions & Possible Causes |
|---|---|
| **DTC: P0031**<br>**2T, MIL: Yes**<br>**Years: 2007, 2008**<br>**Models:** Passat, Jetta, Golf, GTI, New Beetle, New Beetle Convertible<br>**Engines:** 2.0L, 2.5L, 3.6L<br>**Transmissions:** All | **HO2S Heater (Bank 1 Sensor 1) Circuit Low Input Conditions:**<br>Engine started, battery voltage must be at least 11.5v, all electrical components must be off, the ground between the engine and the chassis must be well connected, the exhaust system must be properly sealed between the catalytic converter and the cylinder head, the coolant temperature must be 80 degrees Celsius, and the oxygen sensor heater for oxygen sensor before the catalytic converter must be properly functioning. The ECM detected the HO2S signal was in a negative voltage range referred to as "character shift downward". This code sets when the HO2S signal remains in a low state. In effect, it does not switch properly in the closed loop operation. The HO2S (before the three-way catalytic converter) has a short circuit to ground that has lasted longer than 200 seconds<br>**Possible Causes:**<br>• HO2S is contaminated (due to presence of silicone in fuel)<br>• HO2S signal and ground circuit wires crossed in wiring harness<br>• HO2S signal circuit is shorted to sensor or chassis ground<br>• HO2S element has failed (internal short condition)<br>• ECM has failed |
| **DTC: P0032**<br>**2T, MIL: Yes**<br>**Years: 2007, 2008**<br>**Models:** Passat, Jetta, Golf, GTI, New Beetle, New Beetle Convertible<br>**Engines:** 2.0L, 2.5L, 3.6L<br>**Transmissions:** All | **HO2S Heater (Bank 1 Sensor 1) Circuit High Input Conditions:**<br>Engine started, battery voltage must be at least 11.5v, all electrical components must be off, the ground between the engine and the chassis must be well connected, the exhaust system must be properly sealed between the catalytic converter and the cylinder head, the coolant temperature must be 80 degrees Celsius, and the oxygen sensor heater for oxygen sensor before the catalytic converter must be properly functioning. The ECM detected the HO2S signal remained in a high state.<br>**Note: The HO2S signal circuit may be shorted to the heater power circuit due to tracking inside of the HO2S connector. Remove the connector and visually inspect the connector for signs of oil or water.**<br>**Possible Causes:**<br>• HO2S signal shorted to heater power circuit inside connector<br>• HO2S signal circuit shorted to ground or to system voltage<br>• ECM has failed |
| **DTC: P0033**<br>**Years: 2007, 2008**<br>**Models:** Passat<br>**Engines:** 2.0L<br>**Transmissions:** All | **Turbocharger Bypass Valve Control Circuit Conditions:**<br>The ECM detected the turbocharger bypass valve control circuit signal was providing an invalid input.<br>**Possible Causes:**<br>• Leaks in the air charger system<br>• Turbocharger recirculating valve faulty<br>• Turbocharging system damaged<br>• Vacuum diaphragm out of adjustment<br>• Wastegate bypass valve regulator valve faulty<br>• ECM has failed |
| **DTC: P0034**<br>**Years: 2007, 2008**<br>**Models:** Passat<br>**Engines:** 2.0L<br>**Transmissions:** All | **Turbocharger Bypass Valve Control Circuit Low Conditions:**<br>The ECM detected the turbocharger bypass valve control circuit signal was exceeding the minimum threshold.<br>**Possible Causes:**<br>• Leaks in the air charger system<br>• Turbocharger recirculating valve faulty<br>• Turbocharging system damaged<br>• Vacuum diaphragm out of adjustment<br>• Wastegate bypass valve regulator valve faulty<br>• ECM has failed |
| **DTC: P0035**<br>**Years: 2007, 2008**<br>**Models:** Passat<br>**Engines:** 2.0L<br>**Transmissions:** All | **Turbocharger Bypass Valve Control Circuit High Conditions:**<br>The ECM detected the turbocharger bypass valve control circuit signal was exceeding the maximum threshold.<br>**Possible Causes:**<br>• Leaks in the air charger system<br>• Turbocharger recirculating valve faulty<br>• Turbocharging system damaged<br>• Vacuum diaphragm out of adjustment<br>• Wastegate bypass valve regulator valve faulty<br>• ECM has failed |

| DTC | Trouble Code Title, Conditions & Possible Causes |
|---|---|
| **DTC: P0036**<br>**2T, MIL: Yes**<br>**Years: 2007, 2008**<br>**Models:** Passat, Jetta, Golf, GTI, New Beetle, New Beetle Convertible<br>**Engines:** 2.0L, 2.5L, 3.6L<br>**Transmissions:** All | **HO2S Heater (Bank 1 Sensor 2) Control Circuit Malfunction Conditions:**<br>Engine started, battery voltage must be at least 11.5v, all electrical components must be off, the ground between the engine and the chassis must be well connected, the exhaust system must be properly sealed between the catalytic converter and the cylinder head, the coolant temperature must be 80 degrees Celsius, and the oxygen sensor heater for oxygen sensor before the catalytic converter must be properly functioning. The ECM detected the HO2S signal was in a negative voltage range referred to as "character shift downward". This code sets when the HO2S signal remains in a low state.<br>**Possible Causes:**<br>• HO2S is contaminated (due to presence of silicone in fuel)<br>• HO2S signal and ground circuit wires crossed in wiring harness<br>• HO2S signal circuit is shorted to sensor or chassis ground<br>• HO2S element has failed (internal short condition)<br>• ECM has failed |
| **DTC: P0037**<br>**2T, MIL: Yes**<br>**Years: 2007, 2008**<br>**Models:** Passat, Jetta, Golf, GTI, New Beetle, New Beetle Convertible<br>**Engines:** 2.0L, 2.5L, 3.6L<br>**Transmissions:** All | **HO2S Heater (Bank 1 Sensor 2) Circuit Low Input Conditions:**<br>Engine started, battery voltage must be at least 11.5v, all electrical components must be off, the ground between the engine and the chassis must be well connected, the exhaust system must be properly sealed between the catalytic converter and the cylinder head, the coolant temperature must be 80 degrees Celsius, and the oxygen sensor heater for oxygen sensor before the catalytic converter must be properly functioning. The ECM detected the HO2S signal was in a negative voltage range referred to as "character shift downward". This code sets when the HO2S signal remains in a low state. In effect, it does not switch properly in the closed loop operation. The HO2S (before the three-way catalytic converter) has a short circuit to ground that has lasted longer than 200 seconds<br>**Possible Causes:**<br>• HO2S is contaminated (due to presence of silicone in fuel)<br>• HO2S signal and ground circuit wires crossed in wiring harness<br>• HO2S signal circuit is shorted to sensor or chassis ground<br>• HO2S element has failed (internal short condition)<br>• ECM has failed |
| **DTC: P0038**<br>**2T, MIL: Yes**<br>**Years: 2007, 2008**<br>**Models:** Passat, Jetta, Golf, GTI, New Beetle, New Beetle Convertible<br>**Engines:** 2.0L, 2.5L, 3.6L<br>**Transmissions:** All | **HO2S Heater (Bank 1 Sensor 2) Circuit High Input Conditions:**<br>Engine started, battery voltage must be at least 11.5v, all electrical components must be off, the ground between the engine and the chassis must be well connected, the exhaust system must be properly sealed between the catalytic converter and the cylinder head, the coolant temperature must be 80 degrees Celsius, and the oxygen sensor heater for oxygen sensor before the catalytic converter must be properly functioning. The ECM detected the HO2S signal remained in a high state.<br>**Note: The HO2S signal circuit may be shorted to the heater power circuit due to tracking inside of the HO2S connector. Remove the connector and visually inspect the connector for signs of oil or water.**<br>**Possible Causes:**<br>• HO2S signal shorted to heater power circuit inside connector<br>• HO2S signal circuit shorted to ground or to system voltage<br>• ECM has failed |
| **DTC: P0042**<br>**Years: 2007, 2008**<br>**Models:** Jetta, New Beetle<br>**Engines:** 2.0L, 2.5L<br>**Transmissions:** All | **HO2S Heater Control Circuit Bank 1 Sensor 3 Conditions:**<br>Engine started, battery voltage must be at least 11.5v, all electrical components must be off, the ground between the engine and the chassis must be well connected, the exhaust system must be properly sealed between the catalytic converter and the cylinder head, and the coolant temperature must be 80 degrees Celsius. The ECM detected the HO2S signal was in a negative voltage range referred to as "character shift downward". This code sets when the HO2S signal remains in a low state. In effect, it does not switch properly in the closed loop operation. The HO2S (before the three-way catalytic converter) has a short circuit to ground that has lasted longer than 200 seconds.<br>**Possible Causes:**<br>• HO2S is contaminated (due to presence of silicone in fuel)<br>• HO2S signal and ground circuit wires crossed in wiring harness<br>• HO2S signal circuit is shorted to sensor or chassis ground<br>• HO2S element has failed (internal short condition)<br>• ECM has failed |

| DTC | Trouble Code Title, Conditions & Possible Causes |
|---|---|
| **DTC: P0043**<br>**Years: 2007, 2008**<br>**Models**: Jetta, New Beetle<br>**Engines**: 2.0L, 2.5L<br>**Transmissions**: All | **HO2S Heater Control Circuit Low Bank 1 Sensor 3 Conditions:**<br>Engine started, battery voltage must be at least 11.5v, all electrical components must be off, the ground between the engine and the chassis must be well connected, the exhaust system must be properly sealed between the catalytic converter and the cylinder head, and the coolant temperature must be 80 degrees Celsius. The ECM detected the HO2S signal was in a negative voltage range referred to as "character shift downward". This code sets when the HO2S signal remains in a low state. In effect, it does not switch properly in the closed loop operation. The HO2S (before the three-way catalytic converter) has a short circuit to ground that has lasted longer than 200 seconds.<br>**Possible Causes:**<br>&bull; HO2S is contaminated (due to presence of silicone in fuel)<br>&bull; HO2S signal and ground circuit wires crossed in wiring harness<br>&bull; HO2S signal circuit is shorted to sensor or chassis ground<br>&bull; HO2S element has failed (internal short condition)<br>&bull; ECM has failed |
| **DTC: P0044**<br>**Years: 2007, 2008**<br>**Models**: Jetta, New Beetle<br>**Engines**: 2.0L, 2.5L<br>**Transmissions**: All | **HO2S Heater Control Circuit High Bank 1 Sensor 3 Conditions:**<br>Engine started, battery voltage must be at least 11.5v, all electrical components must be off, the ground between the engine and the chassis must be well connected, the exhaust system must be properly sealed between the catalytic converter and the cylinder head, and the coolant temperature must be 80 degrees Celsius. The ECM detected the HO2S signal remained in a high state.<br>**Note: The HO2S signal circuit may be shorted to the heater power circuit due to tracking inside of the HO2S connector. Remove the connector and visually inspect the connector for signs of oil or water.**<br>**Possible Causes:**<br>&bull; HO2S signal shorted to heater power circuit inside connector<br>&bull; HO2S signal circuit shorted to ground or to system voltage<br>&bull; ECM has failed |
| **DTC: P0050**<br>**Years: 2007, 2008**<br>**Models**: Passat, Jetta, GTI<br>**Engines**: 2.5L, 3.6L<br>**Transmissions**: All | **HO2S Heater (Bank 2 Sensor 1) Control Circuit Malfunction Conditions:**<br>Engine started, battery voltage must be at least 11.5v, all electrical components must be off, the ground between the engine and the chassis must be well connected, the exhaust system must be properly sealed between the catalytic converter and the cylinder head, and the coolant temperature must be 80 degrees Celsius. The ECM detected the HO2S signal was in a negative voltage range referred to as "character shift downward".<br>**Possible Causes:**<br>&bull; HO2S is contaminated (due to presence of silicone in fuel)<br>&bull; HO2S signal and ground circuit wires crossed in wiring harness<br>&bull; HO2S signal circuit is shorted to sensor or chassis ground<br>&bull; HO2S element has failed (internal short condition)<br>&bull; ECM has failed |
| **DTC: P0051**<br>**Years: 2007, 2008**<br>**Models**: Passat, Jetta, GTI<br>**Engines**: 2.5L, 3.6L<br>**Transmissions**: All | **HO2S Heater (Bank 2 Sensor 1) Circuit Low Input Conditions:**<br>Engine started, battery voltage must be at least 11.5v, all electrical components must be off, the ground between the engine and the chassis must be well connected, the exhaust system must be properly sealed between the catalytic converter and the cylinder head, and the coolant temperature must be 80 degrees Celsius. The ECM detected the HO2S signal was in a negative voltage range referred to as "character shift downward". This code sets when the HO2S signal remains in a low state. In effect, it does not switch properly in the closed loop operation. The HO2S (before the three-way catalytic converter) has a short circuit to ground that has lasted longer than a specified time.<br>**Possible Causes:**<br>&bull; HO2S is contaminated (due to presence of silicone in fuel)<br>&bull; HO2S signal and ground circuit wires crossed in wiring harness<br>&bull; HO2S signal circuit is shorted to sensor or chassis ground<br>&bull; HO2S element has failed (internal short condition)<br>&bull; ECM has failed |
| **DTC: P0056**<br>**Years: 2007, 2008**<br>**Models**: Passat, Jetta, GTI<br>**Engines**: 2.5L, 3.6L<br>**Transmissions**: All | **HO2S Heater (Bank 2 Sensor 2) Circuit High Input Conditions:**<br>Engine started, battery voltage must be at least 11.5v, all electrical components must be off, the ground between the engine and the chassis must be well connected, the exhaust system must be properly sealed between the catalytic converter and the cylinder head, and the coolant temperature must be 80 degrees Celsius. The ECM detected the HO2S signal remained in a high state.<br>**Note: The HO2S signal circuit may be shorted to the heater power circuit due to tracking inside of the HO2S connector. Remove the connector and visually inspect the connector for signs of oil or water.**<br>**Possible Causes:**<br>&bull; HO2S signal shorted to heater power circuit inside connector<br>&bull; HO2S signal circuit shorted to ground or to system voltage<br>&bull; ECM has failed |

| DTC | Trouble Code Title, Conditions & Possible Causes |
|---|---|
| **DTC: P0057**<br>**Years: 2007, 2008**<br>**Models:** Passat, Jetta, GTI<br>**Engines:** 2.5L, 3.6L<br>**Transmissions:** All | **HO2S Heater (Bank 2 Sensor 2) Control Circuit Malfunction Conditions:**<br>Engine started, battery voltage must be at least 11.5v, all electrical components must be off, the ground between the engine and the chassis must be well connected, the exhaust system must be properly sealed between the catalytic converter and the cylinder head, and the coolant temperature must be 80 degrees Celsius. The ECM detected the HO2S signal was in a negative voltage range referred to as "character shift downward".<br>**Possible Causes:**<br>• HO2S is contaminated (due to presence of silicone in fuel)<br>• HO2S signal and ground circuit wires crossed in wiring harness<br>• HO2S signal circuit is shorted to sensor or chassis ground<br>• HO2S element has failed (internal short condition)<br>• ECM has failed |
| **DTC: P0058**<br>**Years: 2007, 2008**<br>**Models:** Passat, Jetta, GTI<br>**Engines:** 2.5L, 3.6L<br>**Transmissions:** All | **HO2S Heater (Bank 2 Sensor 2) Circuit Low Input Conditions:**<br>Engine started, battery voltage must be at least 11.5v, all electrical components must be off, the ground between the engine and the chassis must be well connected, the exhaust system must be properly sealed between the catalytic converter and the cylinder head, and the coolant temperature must be 80 degrees Celsius. The ECM detected the HO2S signal was in a negative voltage range referred to as "character shift downward". This code sets when the HO2S signal remains in a low state. In effect, it does not switch properly in the closed loop operation. The HO2S (before the three-way catalytic converter) has a short circuit to ground that has lasted longer than a specified time.<br>**Possible Causes:**<br>• HO2S is contaminated (due to presence of silicone in fuel)<br>• HO2S signal and ground circuit wires crossed in wiring harness<br>• HO2S signal circuit is shorted to sensor or chassis ground<br>• HO2S element has failed (internal short condition)<br>• ECM has failed |
| **DTC: P0068**<br>**Years: 2007, 2008**<br>**Models:** Passat, Jetta, New Beetle<br>**Engines:** 2.0L, 2.5L<br>**Transmissions:** All | **Mass Air Pressure and Mass Air Flow Miscommunication Conditions**<br>Engine running and the temperature must be at least 185-degrees (F) and all electrical equipment (A/C, lights, etc) must be off. The ECM detected a miscommunication between the mass air pressure and the mass air flow sensors<br>**Possible Causes:**<br>• Intake motor flap faulty<br>• Intake system is leaking<br>• Intake manifold runner position sensor is faulty<br>• Mass Air Flow sensor has failed<br>• Throttle valve control module has failed |
| **DTC: P0087**<br>**Years: 2007, 2008**<br>**Models:** Passat<br>**Engines:** 2.5L, 3.6L<br>**Transmissions:** All | **Fuel Rail/System Pressure Too Low Conditions**<br>Engine started, battery voltage must be at least 11.5v, all electrical components must be off, the ground between the engine and the chassis must be well connected, the exhaust system must be properly sealed between the catalytic converter and the cylinder head, and the coolant temperature must be 80 degrees Celsius. The ECM detected that the system's fuel pressure has fallen below the accepted normal calibrated value.<br>**Possible Causes:**<br>• Fuel Pressure Regulator Valve faulty<br>• Fuel Pressure Sensor faulty<br>• Fuel Pump (FP) Control Module faulty<br>• Fuel pump faulty<br>• Low fuel |
| **DTC: P0088**<br>**Years: 2007, 2008**<br>**Models:** Passat<br>**Engines:** 2.5L, 3.6L<br>**Transmissions:** All | **Fuel Rail/System Pressure Too High Conditions**<br>Engine started, battery voltage must be at least 11.5v, all electrical components must be off, the ground between the engine and the chassis must be well connected, the exhaust system must be properly sealed between the catalytic converter and the cylinder head, and the coolant temperature must be 80 degrees Celsius. The ECM detected that the system's fuel pressure has risen above the accepted normal calibrated value.<br>**Possible Causes:**<br>• Fuel Pressure Regulator Valve faulty<br>• Fuel Pressure Sensor faulty<br>• Fuel Pump (FP) Control Module faulty<br>• Fuel pump faulty |

| DTC | Trouble Code Title, Conditions & Possible Causes |
|---|---|
| **DTC: P0089**<br>**Years: 2007, 2008**<br>**Models:** Passat<br>**Engines:** 2.5L, 3.6L<br>**Transmissions**: All | **Fuel Pressure Regulator Range/Performance Conditions**<br>Engine started, battery voltage must be at least 11.5v, all electrical components must be off, the ground between the engine and the chassis must be well connected, the exhaust system must be properly sealed between the catalytic converter and the cylinder head, and the coolant temperature must be 80 degrees Celsius. The ECM detected that the system's fuel pressure sensor is providing a signal that is either outside the accepted normal values or is not receiving a signal at all.<br>**Possible Causes:**<br>• Fuel Pressure Regulator Valve faulty<br>• Fuel Pressure Sensor faulty<br>• Fuel Pump (FP) Control Module faulty<br>• Fuel pump faulty |
| **DTC: P0100**<br>**Years: 2007, 2008**<br>**Models:** Passat<br>**Engines:** 2.5L, 3.6L<br>**Transmissions**: All | **Mass Air Flow Circuit Range/Performance Conditions**<br>Engine running, with the system voltage more than 11.0v, and the temperature must be at least 185-degrees (F) and all electrical equipment (A/C, lights, etc) must be off. The ECM has detected that the MAF signal was less than the required minimum or out of a calculated range with the engine (or undetectable) for a certain period of time.<br>**Possible Causes:**<br>• Mass air flow (MAF) sensor has failed or is damaged<br>• MAF sensor signal circuit is open, shorted to ground or to B+<br>• ECM has failed |
| **DTC: P0101**<br>**2T, MIL: Yes**<br>**Years: 2007, 2008**<br>**Models:** Passat, Jetta, Golf, GTI, New Beetle, New Beetle Convertible<br>**Engines:** 2.0L, 2.5L, 3.6L<br>**Transmissions**: All | **Mass or Volume Air Flow Circuit Range/Performance Conditions**<br>Engine running, with the system voltage more than 11.0v, and the temperature must be at least 185-degrees (F) and all electrical equipment (A/C, lights, etc) must be off. The ECM has detected that the MAF signal was out of a calculated range with the engine (or undetectable) for a certain period of time.<br>**Possible Causes:**<br>• Mass air flow (MAF) sensor has failed or is damaged<br>• ECM has failed<br>• Signal and ground wires of Mass Air Flow (MAF) sensor has short circuited |
| **DTC: P0102**<br>**2T, MIL: Yes**<br>**Years: 2007, 2008**<br>**Models:** Passat, Jetta, Golf, GTI, New Beetle, New Beetle Convertible<br>**Engines:** 2.0L, 2.5L, 3.6L<br>**Transmissions**: All | **MAF Sensor Circuit Low Input Conditions:**<br>Key on, engine started, and the ECM detected the MAF sensor signal was less than the minimum calibrated value. The engine temperature must beat least 185-degrees (F) and all electrical equipment (A/C, lights, etc) must be off. The ECM has detected that the MAF signal was less than the required minimum.<br>**Possible Causes:**<br>• Check for leaks between MAF sensor and throttle valve control module<br>• Voltage supply faulty.<br>• Sensor power circuit open from fuel pump relay to MAF sensor<br>• Sensor signal circuit open (may be disconnected) from ECM and MAF<br>• Faulty ground cable resistance between connector terminal 1 and Ground<br>• MAF Sensor malfunction |
| **DTC: P0103**<br>**2T, MIL: Yes**<br>**Years: 2007, 2008**<br>**Models:** Passat, Jetta, Golf, GTI, New Beetle, New Beetle Convertible<br>**Engines:** 2.0L, 2.5L, 3.6L<br>**Transmissions**: All | **MAF Sensor Circuit High Input Conditions:**<br>Key on, engine started, and the ECM detected the MAF sensor signal was more than the minimum calibrated value. The engine temperature must beat least 185-degrees (F) and all electrical equipment (A/C, lights, etc) must be off. The ECM has detected that the MAF signal was more than the required minimum.<br>**Possible Causes:**<br>• Check for leaks between MAF sensor and throttle valve control module<br>• Voltage supply faulty.<br>• Sensor power circuit open from fuel pump relay to MAF sensor<br>• Sensor signal circuit open (may be disconnected) from ECM and MAF<br>• Faulty ground cable resistance between connector terminal 1 and Ground<br>• MAF Sensor malfunction |
| **DTC: P0106**<br>**2T, MIL: Yes**<br>**Years: 2007, 2008**<br>**Models:** Passat, Jetta, Golf, GTI, New Beetle, New Beetle Convertible<br>**Engines:** 2.0L, 2.5L, 3.6L<br>**Transmissions**: All | **Manifold Absolute Pressure/Barometric Pressure Sensor Circuit Performance Conditions:**<br>Engine started, the temperature must beat least 185-degrees (F) and all electrical equipment (A/C, lights, etc) must be off. The ECM detected the BARO sensor was out of range during the CCM test. The BARO sensor signal should be in 4.5v.<br>**Possible Causes:**<br>• Sensor has deteriorated (response time too slow) or has failed<br>• MAP sensor signal circuit is shorted to ground<br>• MAP sensor circuit (5v) is open<br>• MAP sensor is damaged or it has failed<br>• BARO sensor signal circuit is shorted to ground<br>• BARO sensor circuit (5v) is open<br>• BARO sensor is damaged or it has failed<br>• ECM is not connected properly<br>• ECM has failed |

| DTC | Trouble Code Title, Conditions & Possible Causes |
|---|---|
| **DTC: P0107**<br>**Years: 2007, 2008**<br>**Models:** Passat, Jetta, Golf, GTI,<br>**Engines:** 2.0L, 2.5L, 3.6L<br>**Transmissions**: All | **Manifold Absolute Pressure/Barometric Pressure Sensor Circuit Low Input Conditions:**<br>Engine started, the temperature must beat least 185-degrees (F) and all electrical equipment (A/C, lights, etc) must be off. The ECM detected the BARO sensor was out of range during the CCM test. The BARO sensor signal should be in 4.5v. The BARO sensor is a variable capacitance unit used to detect altitude.<br>**Possible Causes:**<br>• Sensor has deteriorated (response time too slow) or has failed<br>• MAP sensor signal circuit is shorted to ground<br>• MAP sensor circuit (5v) is open<br>• MAP sensor is damaged or it has failed<br>• BARO sensor signal circuit is shorted to ground<br>• BARO sensor circuit (5v) is open<br>• BARO sensor is damaged or it has failed<br>• ECM is not connected properly<br>• ECM has failed |
| **DTC: P0108**<br>**Years: 2007, 2008**<br>**Models:** Passat, Jetta, Golf, GTI,<br>**Engines:** 2.0L, 2.5L, 3.6L<br>**Transmissions**: All | **Manifold Absolute Pressure/Barometric Sensor Circuit High Input Conditions:**<br>Engine started, the temperature must beat least 185-degrees (F) and all electrical equipment (A/C, lights, etc) must be off. The ECM detected the BARO sensor was out of range during the CCM test. The BARO sensor signal should be in 4.5v. The BARO sensor is a variable capacitance unit used to detect altitude.<br>**Possible Causes:**<br>• Sensor has deteriorated (response time too slow) or has failed<br>• MAP sensor signal circuit is shorted to ground<br>• MAP sensor circuit (5v) is open<br>• MAP sensor is damaged or it has failed<br>• BARO sensor signal circuit is shorted to ground<br>• BARO sensor circuit (5v) is open<br>• BARO sensor is damaged or it has failed<br>• ECM is not connected properly<br>• ECM has failed |
| **DTC: P0111**<br>**Years: 2007, 2008**<br>**Models:** Passat, Jetta<br>**Engines:** 2.0L, 2.5L, 3.6L<br>**Transmissions**: All | **Intake Air Temperature Sensor Circuit Low Input Conditions:**<br>Key on or engine running, the temperature must beat least 185-degrees (F) and all electrical equipment (A/C, lights, etc) must be off; and the ECM detected the IAT sensor signal was less than the self-test minimum. This is a thermistor-type sensor with a variable resistance that changes when exposed to different temperatures. This means: the higher the temperature, the lower the resistance value.<br>**Possible Causes:**<br>• IAT sensor signal circuit is grounded (check wiring & connector)<br>• Resistance value between sockets 33 and 36 out of range<br>• IAT sensor has an open circuit<br>• IAT sensor is damaged or it has failed<br>• ECM has failed |
| **DTC: P0112**<br>**2T, MIL: Yes**<br>**Years: 2007, 2008**<br>**Models:** Passat, Jetta, Golf, GTI,<br>New Beetle, New Beetle Convertible<br>**Engines:** 2.0L, 2.5L, 3.6L<br>**Transmissions**: All | **Intake Air Temperature Sensor Circuit Low Input Conditions:**<br>Key on or Engine running, the temperature must beat least 185-degrees (F) and all electrical equipment (A/C, lights, etc) must be off; and the ECM detected the IAT sensor signal was less than the self-test minimum. This is a thermistor-type sensor with a variable resistance that changes when exposed to different temperatures. This means: the higher the temperature, the lower the resistance value.<br>**Possible Causes:**<br>• IAT sensor signal circuit is grounded (check wiring & connector)<br>• Resistance value between sockets 33 and 36 out of range<br>• IAT sensor has an open circuit<br>• IAT sensor is damaged or it has failed<br>• ECM has failed |
| **DTC: P0113**<br>**2T, MIL: Yes**<br>**Years: 2007, 2008**<br>**Models:** Passat, Jetta, Golf, GTI,<br>New Beetle, New Beetle Convertible<br>**Engines:** 2.0L, 2.5L, 3.6L<br>**Transmissions**: All | **Intake Air Temperature Sensor Circuit High Input Conditions:**<br>Key on or engine running, the temperature must beat least 185-degrees (F) and all electrical equipment (A/C, lights, etc) must be off; and the ECM detected the IAT sensor signal was more than the self-test maximum. This is a thermistor-type sensor with a variable resistance that changes when exposed to different temperatures. This means: the higher the temperature, the lower the resistance value.<br>**Possible Causes:**<br>• IAT sensor signal circuit is open (inspect wiring & connector)<br>• IAT sensor signal circuit is shorted<br>• Resistance value between sockets 33 and 36 out of range<br>• IAT sensor is damaged or it has failed<br>• ECM has failed |

| DTC | Trouble Code Title, Conditions & Possible Causes |
|---|---|
| **DTC: P0116**<br>**2T, MIL: Yes**<br>**Years: 2007, 2008**<br>**Models:** Passat, Jetta, Golf, GTI, New Beetle, New Beetle Convertible<br>**Engines:** 2.0L, 2.5L, 3.6L<br>**Transmissions:** All | **ECT Sensor / CHT Sensor Signal Range/Performance Conditions:**<br>Engine started (cold), battery voltage must be 11.5, and all equipment must be off. The ECM detected the ECT sensor exceeded the required calibrated value, or the engine is at idle and doesn't reach operating temperature quickly enough; the Catalyst, Fuel System, HO2S and Misfire Monitor did not complete, or the timer expired. Testing completion of procedure, the engine's temperature must rise uniformily during idle.<br>**Possible Causes:**<br>• Check for low coolant level or incorrect coolant mixture<br>• ECM detects a short circuit wiring in the ECT<br>• CHT sensor is out-of-calibration or it has failed<br>• ECT sensor is out-of-calibration or it has failed |
| **DTC: P0117**<br>**2T, MIL: Yes**<br>**Years: 2007, 2008**<br>**Models:** Passat, Jetta, Golf, GTI, New Beetle, New Beetle Convertible<br>**Engines:** 2.0L, 2.5L, 3.6L<br>**Transmissions:** All | **ECT Sensor Circuit Low Input Conditions:**<br>Engine started (cold), battery voltage must be 11.5, and all equipment must be off. The ECM detected the ECT sensor signal was less than the self-test minimum. This is a thermistor-type sensor with a variable resistance that changes when exposed to different temperatures<br>**Possible Causes:**<br>• ECT sensor signal circuit is grounded in the wiring harness<br>• ECT sensor doesn't react to changes in temperature<br>• ECT sensor is damaged or the ECM has failed |
| **DTC: P0118**<br>**2T, MIL: Yes**<br>**Years: 2007, 2008**<br>**Models:** Passat, Jetta, Golf, GTI, New Beetle, New Beetle Convertible<br>**Engines:** 2.0L, 2.5L, 3.6L<br>**Transmissions:** All | **ECT Sensor Circuit High Input Conditions:**<br>Engine started (cold), battery voltage must be 11.5, and all equipment must be off. The ECM detected the ECT sensor signal was more than the self-test maximum. This is a thermistor-type sensor with a variable resistance that changes when exposed to different temperatures<br>**Possible Causes:**<br>• ECT sensor signal circuit is open (inspect wiring & connector)<br>• ECT sensor signal circuit is shorted to ground<br>• ECT sensor is damaged or it has failed<br>• ECM has failed |
| **DTC: P0120**<br>**Years: 2007, 2008**<br>**Models:** Passat, Jetta, Golf, GTI, New Beetle, New Beetle Convertible<br>**Engines:** 2.0L, 2.5L, 3.6L<br>**Transmissions:** All | **Throttle/Pedal Position Sensor (A) Circuit Malfunction Conditions:**<br>Engine started, at idle, the temperature must be 80 degrees Celsius. The throttle position sensor supplies implausible signal to the ECM. The throttle valve activation occurs via an electric motor (throttle drive) in the throttle valve control module. It is activated by the Engine Control Module (ECM) according to specifications of the two sensors, Throttle Position (TP) Sensor and Accelerator Pedal Position Sensor 2.<br>**Possible Causes:**<br>• TP sensor signal circuit is open (inspect wiring & connector)<br>• TP sensor signal circuit is shorted to ground<br>• TP sensor or module is damaged or it has failed<br>• Throttle valve is damaged or dirty<br>• Throttle valve control module is faulty<br>• ECM has failed |
| **DTC: P0121**<br>**2T, MIL: Yes**<br>**Years: 2007, 2008**<br>**Models:** Passat, Jetta, Golf, GTI, New Beetle, New Beetle Convertible<br>**Engines:** 2.0L, 2.5L, 3.6L<br>**Transmissions:** All | **Throttle/Pedal Position Sensor Signal Range/Performance Conditions:**<br>Engine started; then immediately following a condition where the engine was running under at off-idle, the ECM detected the TP sensor signal indicated the throttle did not return to its previous closed position during the Rationality test.<br>**Possible Causes:**<br>• Throttle plate is binding, dirty or sticking<br>• Throttle valve is damaged or dirty<br>• Throttle valve control module is faulty<br>• TP sensor signal circuit open (inspect wiring & connector)<br>• TP sensor ground circuit open (inspect wiring & connector)<br>• TP sensor and/or control module is damaged or has failed<br>• MAF sensor signal is damaged, has failed or a short is present |
| **DTC: P0122**<br>**2T, MIL: Yes**<br>**Years: 2007, 2008**<br>**Models:** Passat, Jetta, Golf, GTI, New Beetle, New Beetle Convertible<br>**Engines:** 2.0L, 2.5L, 3.6L<br>**Transmissions:** All | **Throttle/Pedal Position Sensor Circuit Low Input Conditions:**<br>Engine started, at idle, the temperature must be at least 80 degrees Celsius. The throttle position sensor supplies implausible signal to the ECM.<br>**Possible Causes:**<br>• TP sensor signal circuit open (inspect wiring & connector)<br>• TP sensor signal shorted to ground (inspect wiring & connector)<br>• TP sensor is damaged or has failed<br>• Throttle control module's voltage supply is shorted or open<br>• ECM has failed |

| DTC | Trouble Code Title, Conditions & Possible Causes |
|---|---|
| **DTC: P0123**<br>**2T, MIL: Yes**<br>**Years: 2007, 2008**<br>**Models:** Passat, Jetta, Golf, GTI, New Beetle, New Beetle Convertible<br>**Engines:** 2.0L, 2.5L, 3.6L<br>**Transmissions:** All | **TP Sensor Circuit High Input Conditions:**<br>Engine started, at idle, the temperature must be at least 80 degrees Celsius. The ECM detected the TP sensor signal was more than the self-test maximum during testing.<br>**Possible Causes:**<br>• TP sensor not seated correctly in housing (may be damaged)<br>• TP sensor signal is circuit shorted to ground or system voltage<br>• TP sensor ground circuit is open (check the wiring harness)<br>• TP sensor and/or ECM has failed |
| **DTC: P0125**<br>**Years: 2007, 2008**<br>**Models:** Passat, Jetta, Golf, GTI, New Beetle,<br>**Engines:** 2.0L, 2.5L, 3.6L<br>**Transmissions:** All | **Insufficient Coolant Temperature For Closed Loop Conditions:**<br>Let engine run at idle until coolant temperature for oxygen sensor control has been reached. The ECM detected that the ECT sensor signal did not indicate the required engine temperature value to enter closed loop within a specified amount of time. The amount of time is calculated from the point at which the engine is started, and depends upon the ECT sensor signal value at startup.<br>**Possible Causes:**<br>• Check the coolant mixture for an incorrect mixture<br>• Check the operation of the thermostat (it may be stuck open or has failed)<br>• ECT sensor has failed or is disconnected<br>• Inspect for low coolant level |
| **DTC: P0130**<br>**2T, MIL: Yes**<br>**Years: 2007, 2008**<br>**Models:** Passat, Jetta, Golf, GTI, New Beetle, New Beetle Convertible<br>**Engines:** 2.0L, 2.5L, 3.6L<br>**Transmissions:** All | **O2 Sensor Circuit Bank 1 Sensor 1 Conditions:**<br>Engine running, battery voltage 11.5, all electrical components off, ground between engine and chassis well connected and the exhaust system must be properly sealed between catalytic converter and the cylinder head. The ECM detected the HO2S signal was implausible or not detected.<br>**Possible Causes:**<br>• Oxygen sensor heater for oxygen sensor (HO2S) before catalytic converter is faulty<br>• HO2S is contaminated (due to presence of silicone in fuel)<br>• HO2S signal and ground circuit wires crossed in wiring harness<br>• HO2S signal circuit is shorted to sensor or chassis ground<br>• HO2S element before the catalytic converter has failed (internal short condition)<br>• Leaks present in the exhaust manifold or exhaust pipes<br>• ECM has failed |
| **DTC: P0131**<br>**Years: 2007, 2008**<br>**Models:** Passat, Jetta, Golf, GTI, New Beetle, New Beetle Convertible<br>**Engines:** 2.0L, 2.5L, 3.6L<br>**Transmissions:** All | **HO2S (Bank 1 Sensor 1) Circuit Low Input Conditions:**<br>Engine running, battery voltage 11.5, all electrical components off, ground between engine and chassis well connected and the exhaust system must be properly sealed between catalytic converter and the cylinder head. The ECM detected the HO2S signal was in a negative voltage range referred to as "character shift downward". This code sets when the HO2S signal remains in a low state for a measured period of time. In effect, it does not switch properly in the closed loop operation.<br>**Possible Causes:**<br>• HO2S is contaminated (due to presence of silicone in fuel)<br>• HO2S signal and ground circuit wires crossed in wiring harness<br>• HO2S signal circuit is shorted to sensor or chassis ground<br>• HO2S element has failed (internal short condition)<br>• Leaks present in the exhaust manifold or exhaust pipes<br>• ECM has failed |
| **DTC: P0132**<br>**2T, MIL: Yes**<br>**Years: 2007, 2008**<br>**Models:** Passat, Jetta, Golf, GTI, New Beetle, New Beetle Convertible<br>**Engines:** 2.0L, 2.5L, 3.6L<br>**Transmissions:** All | **HO2S (Bank 1 Sensor 1) Circuit High Input Conditions:**<br>Engine running, battery voltage 11.5, all electrical components off, ground between engine and chassis well connected and the exhaust system must be properly sealed between catalytic converter and the cylinder head. The ECM detected the HO2S signal was in a high state. This code sets when the HO2S signal remains in a high state for a measured period of time. In effect, it does not switch properly in the closed loop operation.<br>**Note: The HO2S signal circuit may be shorted to the heater power circuit due to tracking inside of the HO2S connector. Remove the connector and visually inspect the connector for signs of oil or water.**<br>**Possible Causes:**<br>• HO2S is contaminated (due to presence of silicone in fuel)<br>• HO2S signal and ground circuit wires crossed in wiring harness<br>• HO2S signal circuit is shorted to sensor or chassis ground<br>• HO2S element has failed (internal short condition)<br>• Leaks present in the exhaust manifold or exhaust pipes<br>• ECM has failed |

| DTC | Trouble Code Title, Conditions & Possible Causes |
|---|---|
| **DTC: P0133**<br>**2T, MIL: Yes**<br>**Years: 2007, 2008**<br>**Models:** Passat, Jetta, Golf, GTI, New Beetle, New Beetle Convertible<br>**Engines:** 2.0L, 2.5L, 3.6L<br>**Transmissions:** All | **HO2S (Bank 1 Sensor 1) Circuit Slow Response Conditions:**<br>Engine running, battery voltage 11.5, all electrical components off, ground between engine and chassis well connected and the exhaust system must be properly sealed between catalytic converter and the cylinder head. The ECM detected the HO2S amplitude and frequency were out of the normal range (e.g., the HO2S rich to lean switch) during the HO2S Monitor test.<br>**Possible Causes:**<br>• HO2S before the three-way catalytic converter is contaminated (due to presence of silicone in fuel); Run the engine for three minutes at 3500 RPM as a self-cleaning effect<br>• HO2S signal circuit open<br>• Leaks present in the exhaust manifold or exhaust pipes<br>• HO2S is damaged or has failed<br>• ECM has failed |
| **DTC: P0134**<br>**2T, MIL: Yes**<br>**Years: 2007, 2008**<br>**Models:** Passat, Jetta, Golf, GTI, New Beetle, New Beetle Convertible<br>**Engines:** 2.0L, 2.5L, 3.6L<br>**Transmissions:** All | **HO2S (Bank 1 Sensor 1) Circuit No Activity Conditions:**<br>Engine running, battery voltage 11.5, all electrical components off, ground between engine and chassis well connected and the exhaust system must be properly sealed between catalytic converter and the cylinder head. The ECM detected the HO2S signal failed to meet the maximum or minimum voltage levels (i.e., it failed the voltage range check).<br>**Possible Causes:**<br>• Leaks present in the exhaust manifold or exhaust pipes<br>• HO2S signal wire and ground wire crossed in connector (voltage jumps)<br>• HO2S element is fuel contaminated or has failed<br>• ECM has failed |
| **DTC: P0135**<br>**2T, MIL: Yes**<br>**Years: 2007, 2008**<br>**Models:** Passat, Jetta, Golf, GTI, New Beetle, New Beetle Convertible<br>**Engines:** 2.0L, 2.5L, 3.6L<br>**Transmissions:** All | **HO2S (Bank 1 Sensor 1) Heater Circuit Malfunction Conditions:**<br>Engine running, battery voltage 11.5, all electrical components off, ground between engine and chassis well connected and the exhaust system must be properly sealed between catalytic converter and the cylinder head. The ECM detected an unexpected voltage condition, or it detected excessive current draw in the heater circuit during the CCM test.<br>**Possible Causes:**<br>• HO2S heater power circuit is open or heater ground circuit open<br>• HO2S signal tracking (due to oil or moisture in the connector)<br>• HO2S is damaged or has failed<br>• ECM has failed |
| **DTC: P0136**<br>**2T, MIL: Yes**<br>**Years: 2007, 2008**<br>**Models:** Passat, Jetta, Golf, GTI, New Beetle, New Beetle Convertible<br>**Engines:** 2.0L, 2.5L, 3.6L<br>**Transmissions:** All | **HO2S (Bank 1 Sensor 2) Circuit Malfunction Conditions:**<br>Engine running, battery voltage 11.5, all electrical components off, ground between engine and chassis well connected and the exhaust system must be properly sealed between catalytic converter and the cylinder head. The ECM detected the HO2S signal failed to meet the maximum or minimum voltage levels (i.e., it failed the voltage range check).<br>**Possible Causes:**<br>• Leaks present in the exhaust manifold or exhaust pipes<br>• HO2S signal wire and ground wire crossed in connector<br>• HO2S element is fuel contaminated or has failed<br>• ECM has failed |
| **DTC: P0137**<br>**2T, MIL: Yes**<br>**Years: 2007, 2008**<br>**Models:** Passat, Jetta, Golf, GTI, New Beetle, New Beetle Convertible<br>**Engines:** 2.0L, 2.5L, 3.6L<br>**Transmissions:** All | **HO2S (Bank 1 Sensor 2) Circuit Low Input Conditions:**<br>Engine running, battery voltage 11.5, all electrical components off, ground between engine and chassis well connected and the exhaust system must be properly sealed between catalytic converter and the cylinder head. The ECM detected the HO2S signal remained in a high state.<br>**Note: The HO2S signal circuit may be shorted to the heater power circuit due to "tracking inside of the HO2S connector. Remove the connector and visually inspect the connector for signs of oil or water.**<br>**Possible Causes:**<br>• HO2S signal shorted to heater power circuit in the connector<br>• HO2S signal circuit shorted to ground (for more than 200 seconds) or to system voltage<br>• ECM has failed |
| **DTC: P0138**<br>**2T, MIL: Yes**<br>**Years: 2007, 2008**<br>**Models:** Passat, Jetta, Golf, GTI, New Beetle, New Beetle Convertible<br>**Engines:** 2.0L, 2.5L, 3.6L<br>**Transmissions:** All | **HO2S (Bank 1 Sensor 2) Circuit High Input Conditions:**<br>Engine running, battery voltage 11.5, all electrical components off, ground between engine and chassis well connected and the exhaust system must be properly sealed between catalytic converter and the cylinder head. The ECM detected the HO2S signal remained in a high state.<br>**Note: The HO2S signal circuit may be shorted to the heater power circuit due to "tracking inside of the HO2S connector. Remove the connector and visually inspect the connector for signs of oil or water.**<br>**Possible Causes:**<br>• HO2S signal shorted to heater power circuit in the positive connector<br>• HO2S signal circuit shorted to ground or to system voltage<br>• HO2S has failed<br>• ECM has failed |

| DTC | Trouble Code Title, Conditions & Possible Causes |
|---|---|
| **DTC: P0139**<br>**2T, MIL: Yes**<br>**Years: 2007, 2008**<br>**Models:** Passat, Jetta, Golf, GTI, New Beetle, New Beetle Convertible<br>**Engines:** 2.0L, 2.5L, 3.6L<br>**Transmissions:** All | **HO2S (Bank 1 Sensor 2) Slow Response Conditions:**<br>Engine running, battery voltage 11.5, all electrical components off, ground between engine and chassis well connected and the exhaust system must be properly sealed between catalytic converter and the cylinder head. The ECM detected the HO2S amplitude and frequency were out of the normal range during the HO2S Monitor test.<br>**Possible Causes:**<br>• HO2S signal shorted to heater power circuit in the connector<br>• HO2S signal circuit shorted to VREF or to system voltage<br>• ECM has failed |
| **DTC: P0140**<br>**2T, MIL: Yes**<br>**Years: 2007, 2008**<br>**Models:** Passat, Jetta, Golf, GTI, New Beetle, New Beetle Convertible<br>**Engines:** 2.0L, 2.5L, 3.6L<br>**Transmissions:** All | **HO2S (Bank 1 Sensor 2) No Activity Conditions:**<br>Engine running, battery voltage 11.5, all electrical components off, ground between engine and chassis well connected and the exhaust system must be properly sealed between catalytic converter and the cylinder head. The ECM detected the HO2S signal failed to meet the maximum or minimum voltage levels (i.e., it failed the voltage range check).<br>**Possible Causes:**<br>• HO2S before the three-way catalytic converter is contaminated (due to presence of silicone in fuel); Run the engine for three minutes at 3500RPM as a self-cleaning effect<br>• Leaks present in the exhaust manifold or exhaust pipes<br>• HO2S signal wire and ground wire crossed in connector (voltage jumps)<br>• HO2S element is contaminated or has failed<br>• ECM has failed |
| **DTC: P0141**<br>**2T, MIL: Yes**<br>**Years: 2007, 2008**<br>**Models:** Passat, Jetta, Golf, GTI, New Beetle, New Beetle Convertible<br>**Engines:** 2.0L, 2.5L, 3.6L<br>**Transmissions:** All | **HO2S (Bank 1 Sensor 2) Malfunction Conditions:**<br>Engine running, battery voltage 11.5, all electrical components off, ground between engine and chassis well connected and the exhaust system must be properly sealed between catalytic converter and the cylinder head. The ECM detected the HO2S signal failed to meet the maximum or minimum voltage levels (i.e., it failed the voltage range check).<br>**Possible Causes:**<br>• Leaks present in the exhaust manifold or exhaust pipes<br>• HO2S signal wire and ground wire crossed in connector<br>• HO2S element is fuel contaminated or has failed<br>• ECM has failed |
| **DTC: P0142**<br>**Years: 2007, 2008**<br>**Models:** Jetta<br>**Engines:** 2.0L, 2.5L<br>**Transmissions:** All | **O2 Sensor Circuit Bank 1 Sensor 3 Conditions:**<br>Engine running, battery voltage 11.5, all electrical components off, ground between engine and chassis well connected and the exhaust system must be properly sealed between catalytic converter and the cylinder head. The ECM detected the HO2S amplitude and frequency were out of the normal range during the HO2S Monitor test.<br>**Possible Causes:**<br>• HO2S signal shorted to heater power circuit in the connector<br>• HO2S signal circuit shorted to VREF or to system voltage<br>• ECM has failed |
| **DTC: P0143**<br>**Years: 2007, 2008**<br>**Models:** Jetta<br>**Engines:** 2.0L, 2.5L<br>**Transmissions:** All | **O2 Sensor Circuit Low Voltage Bank 1 Sensor 3 Conditions:**<br>Engine running, battery voltage 11.5, all electrical components off, ground between engine and chassis well connected and the exhaust system must be properly sealed between catalytic converter and the cylinder head. The ECM detected the HO2S amplitude and frequency were out of the normal range during the HO2S Monitor test.<br>**Possible Causes:**<br>• HO2S signal shorted to heater power circuit in the connector<br>• HO2S signal circuit shorted to VREF or to system voltage<br>• ECM has failed |
| **DTC: P0144**<br>**Years: 2007, 2008**<br>**Models:** Jetta<br>**Engines:** 2.0L, 2.5L<br>**Transmissions:** All | **O2 Sensor Circuit High Voltage Bank 1 Sensor 3 Conditions:**<br>Engine running, battery voltage 11.5, all electrical components off, ground between engine and chassis well connected and the exhaust system must be properly sealed between catalytic converter and the cylinder head. The ECM detected the HO2S amplitude and frequency were out of the normal range during the HO2S Monitor test.<br>**Possible Causes:**<br>• HO2S signal shorted to heater power circuit in the connector<br>• HO2S signal circuit shorted to VREF or to system voltage<br>• ECM has failed |
| **DTC: P0145**<br>**Years: 2007, 2008**<br>**Models:** Jetta<br>**Engines:** 2.0L, 2.5L<br>**Transmissions:** All | **O2 Sensor Circuit Slow Response Bank 1 Sensor 3 Conditions:**<br>Engine running, battery voltage 11.5, all electrical components off, ground between engine and chassis well connected and the exhaust system must be properly sealed between catalytic converter and the cylinder head. The ECM detected the HO2S amplitude and frequency were out of the normal range during the HO2S Monitor test.<br>**Possible Causes:**<br>• HO2S signal shorted to heater power circuit in the connector<br>• HO2S signal circuit shorted to VREF or to system voltage<br>• ECM has failed |

| DTC | Trouble Code Title, Conditions & Possible Causes |
|---|---|
| **DTC: P0146**<br>**Years:** 2007, 2008<br>**Models:** Jetta<br>**Engines:** 2.0L, 2.5L<br>**Transmissions:** All | **O2 Sensor Circuit No Activity Detected Bank 1 Sensor 3 Conditions:**<br>Engine running, battery voltage 11.5, all electrical components off, ground between engine and chassis well connected and the exhaust system must be properly sealed between catalytic converter and the cylinder head. The ECM detected the HO2S amplitude and frequency were not detected during the HO2S Monitor test.<br>**Possible Causes:**<br>• HO2S signal shorted to heater power circuit in the connector<br>• HO2S signal circuit shorted to VREF or to system voltage<br>• ECM has failed |
| **DTC: P0147**<br>**Years:** 2007, 2008<br>**Models:** Jetta<br>**Engines:** 2.0L, 2.5L<br>**Transmissions:** All | **O2 Sensor Heater Circuit Bank 1 Sensor 3 Conditions:**<br>Engine running, battery voltage 11.5, all electrical components off, ground between engine and chassis well connected and the exhaust system must be properly sealed between catalytic converter and the cylinder head. The ECM detected the HO2S amplitude and frequency were out of the normal range during the HO2S Monitor test.<br>**Possible Causes:**<br>• HO2S signal shorted to heater power circuit in the connector<br>• HO2S signal circuit shorted to VREF or to system voltage<br>• ECM has failed |
| **DTC: P0150**<br>**Years:** 2007, 2008<br>**Models:** Passat, Jetta, GTI<br>**Engines:** 2.5L, 3.6L<br>**Transmissions:** All | **HO2S (Bank 2 Sensor 1) Circuit Malfunction Conditions:**<br>Engine running, battery voltage 11.5, all electrical components off, ground between engine and chassis well connected and the exhaust system must be properly sealed between catalytic converter and the cylinder head. The ECM detected the HO2S signal failed to meet the maximum or minimum voltage levels (i.e., it failed the voltage range check).<br>**Possible Causes:**<br>• Leaks present in the exhaust manifold or exhaust pipes<br>• HO2S signal wire and ground wire crossed in connector<br>• HO2S element is fuel contaminated or has failed<br>• ECM has failed |
| **DTC: P0151**<br>**Years:** 2007, 2008<br>**Models:** Passat, Jetta, GTI<br>**Engines:** 2.5L, 3.6L<br>**Transmissions:** All | **HO2S (Bank 2 Sensor 1) Low Input Conditions:**<br>Engine running, battery voltage 11.5, all electrical components off, ground between engine and chassis well connected and the exhaust system must be properly sealed between catalytic converter and the cylinder head. The ECM detected the HO2S signal remained in a high state.<br>**Note: The HO2S signal circuit may be shorted to the heater power circuit due to "tracking inside of the HO2S connector. Remove the connector and visually inspect the connector for signs of oil or water.**<br>**Possible Causes:**<br>• HO2S is contaminated (due to presence of silicone in fuel)<br>• HO2S signal tracking (due to oil or moisture in the connector)<br>• HO2S signal circuit is open or shorted to VREF<br>• ECM has failed |
| **DTC: P0152**<br>**Years:** 2007, 2008<br>**Models:** Passat, Jetta, GTI<br>**Engines:** 2.5L, 3.6L<br>**Transmissions:** All | **HO2S (Bank 2 Sensor 1) Circuit High Input Conditions:**<br>Engine running, battery voltage 11.5, all electrical components off, ground between engine and chassis well connected and the exhaust system must be properly sealed between catalytic converter and the cylinder head. The ECM detected the HO2S signal remained in a high state (more than 1.5v).<br>**Note: The HO2S signal circuit may be shorted to the heater power circuit due to "tracking inside of the HO2S connector. Remove the connector and visually inspect the connector for signs of oil or water.**<br>**Possible Causes:**<br>• HO2S is contaminated (due to presence of silicone in fuel)<br>• HO2S signal tracking (due to oil or moisture in the connector)<br>• HO2S signal circuit is open or shorted to VREF<br>• ECM has failed |
| **DTC: P0153**<br>**2T, MIL: Yes**<br>**Years:** 2007, 2008<br>**Models:** Passat, Jetta, GTI<br>**Engines:** 2.5L, 3.6L<br>**Transmissions:** All | **HO2S (Bank 2 Sensor 1) Circuit Slow Response Conditions:**<br>Engine running, battery voltage 11.5, all electrical components off, ground between engine and chassis well connected and the exhaust system must be properly sealed between catalytic converter and the cylinder head. The the ECM detected the HO2S amplitude and frequency were out of the normal range during the HO2S Monitor test.<br>**Possible Causes:**<br>• HO2S is contaminated (due to presence of silicone in fuel)<br>• Leaks present in the exhaust manifold or exhaust pipes<br>• HO2S is damaged or has failed<br>• ECM has failed |

| DTC | Trouble Code Title, Conditions & Possible Causes |
|---|---|
| **DTC: P0154**<br>**2T, MIL: Yes**<br>**Years: 2007, 2008**<br>**Models:** Passat, Jetta, GTI<br>**Engines:** 2.5L, 3.6L<br>**Transmissions:** All | **HO2S (Bank 2 Sensor 1) Circuit No Activity Conditions:**<br>Engine running, battery voltage 11.5, all electrical components off, ground between engine and chassis well connected and the exhaust system must be properly sealed between catalytic converter and the cylinder head. The ECM detected the HO2S signal failed to meet the maximum or minimum voltage (i.e., it failed the voltage check).<br>**Possible Causes:**<br>• Leaks present in the exhaust manifold or exhaust pipes<br>• HO2S signal wire and ground wire crossed in connector<br>• HO2S element is fuel contaminated or has failed<br>• ECM has failed |
| **DTC: P0155**<br>**2T, MIL: Yes**<br>**Years: 2007, 2008**<br>**Models:** Passat, Jetta, GTI<br>**Engines:** 2.5L, 3.6L<br>**Transmissions:** All | **HO2S (Bank 2 Sensor 1) Heater Circuit Malfunction Conditions:**<br>Engine running, battery voltage 11.5, all electrical components off, ground between engine and chassis well connected and the exhaust system must be properly sealed between catalytic converter and the cylinder head. The ECM detected an open or shorted condition, or excessive current draw in the heater circuit.<br>**Possible Causes:**<br>• HO2S heater power circuit is open<br>• HO2S heater ground circuit is open<br>• HO2S signal tracking (due to oil or moisture in the connector)<br>• HO2S is damaged or has failed<br>• ECM has failed |
| **DTC: P0156**<br>**2T, MIL: Yes**<br>**Years: 2007, 2008**<br>**Models:** Passat, Jetta, GTI<br>**Engines:** 2.5L, 3.6L<br>**Transmissions:** All | **HO2S (Bank 2 Sensor 2) Circuit No Activity Conditions:**<br>Engine running, battery voltage 11.5, all electrical components off, ground between engine and chassis well connected and the exhaust system must be properly sealed between catalytic converter and the cylinder head. The ECM detected the HO2S signal failed to meet the maximum or minimum voltage (i.e., it failed the voltage check).<br>**Possible Causes:**<br>• Leaks present in the exhaust manifold or exhaust pipes<br>• HO2S signal wire and ground wire crossed in connector<br>• HO2S element is fuel contaminated or has failed<br>• ECM has failed |
| **DTC: P0157**<br>**2T, MIL: Yes**<br>**Years: 2007, 2008**<br>**Models:** Passat, Jetta, GTI<br>**Engines:** 2.5L, 3.6L<br>**Transmissions:** All | **HO2S (Bank 2 Sensor 2) Circuit Low Voltage Conditions:**<br>Engine running, battery voltage 11.5, all electrical components off, ground between engine and chassis well connected and the exhaust system must be properly sealed between catalytic converter and the cylinder head. The ECM detected the HO2S signal remained in a high state.<br>**Note: The HO2S signal circuit may be shorted to the heater power circuit due to "tracking inside of the HO2S connector. Remove the connector and visually inspect the connector for signs of oil or water**<br>**Possible Causes:**<br>• HO2S is contaminated (due to presence of silicone in fuel)<br>• HO2S signal tracking (due to oil or moisture in the connector)<br>• HO2S signal circuit is open or shorted to VREF<br>• ECM has failed |
| **DTC: P0158**<br>**2T, MIL: Yes**<br>**Years: 2007, 2008**<br>**Models:** Passat, Jetta, GTI<br>**Engines:** 2.5L, 3.6L<br>**Transmissions:** All | **HO2S (Bank 2 Sensor 2) Circuit High Input Conditions:**<br>Engine running, battery voltage 11.5, all electrical components off, ground between engine and chassis well connected and the exhaust system must be properly sealed between catalytic converter and the cylinder head. The ECM detected the HO2S signal remained in a high state (i.e., more than 1.5v).<br>**Note: The HO2S signal circuit may be shorted to the heater power circuit due to "tracking inside of the HO2S connector. Remove the connector and visually inspect the connector for signs of oil or water.**<br>**Possible Causes:**<br>• HO2S signal shorted to the heater power circuit (due to oil or moisture in the connector)<br>• HO2S signal circuit shorted to VREF or to system voltage<br>• ECM has failed |
| **DTC: P0159**<br>**2T, MIL: Yes**<br>**Years: 2007, 2008**<br>**Models:** Passat, Jetta, GTI<br>**Engines:** 2.5L, 3.6L<br>**Transmissions:** All | **HO2S (Bank 2 Sensor 2) Circuit Slow Response Conditions:**<br>Engine running, battery voltage 11.5, all electrical components off, ground between engine and chassis well connected and the exhaust system must be properly sealed between catalytic converter and the cylinder head. The ECM detected the HO2S amplitude and frequency were out of the normal range during the HO2S Monitor test.<br>**Possible Causes:**<br>• HO2S is contaminated (due to presence of silicone in fuel)<br>• Leaks present in the exhaust manifold or exhaust pipes<br>• HO2S is damaged or has failed<br>• ECM has failed |

| DTC | Trouble Code Title, Conditions & Possible Causes |
|---|---|
| **DTC: P0160**<br>**2T, MIL: Yes**<br>**Years:** 2007, 2008<br>**Models:** Passat, Jetta, GTI<br>**Engines:** 2.5L, 3.6L<br>**Transmissions:** All | **HO2S (Bank 2 Sensor 2) Circuit No Activity Detected Conditions:**<br>Engine running, battery voltage 11.5, all electrical components off, ground between engine and chassis well connected and the exhaust system must be properly sealed between catalytic converter and the cylinder head. The ECM detected the HO2S signal failed to meet the maximum or minimum voltage (i.e., it failed the voltage check).<br>**Possible Causes:**<br>• Leaks present in the exhaust manifold or exhaust pipes<br>• HO2S signal wire and ground wire crossed in connector<br>• HO2S element is fuel contaminated or has failed<br>• ECM has failed |
| **DTC: P0161**<br>**2T, MIL: Yes**<br>**Years:** 2007, 2008<br>**Models:** Passat, Jetta, GTI<br>**Engines:** 2.5L, 3.6L<br>**Transmissions:** All | **HO2S (Bank 2 Sensor 2) Heater Circuit Malfunction Conditions:**<br>Engine running, battery voltage 11.5, all electrical components off, ground between engine and chassis well connected and the exhaust system must be properly sealed between catalytic converter and the cylinder head. The the ECM detected an open or shorted condition, or excessive current draw in the heater circuit.<br>**Possible Causes:**<br>• HO2S heater power circuit or the heater ground circuit is open<br>• HO2S signal tracking (due to oil or moisture in the connector)<br>• HO2S has failed, or the ECM has failed |
| **DTC: P0169**<br>**Years:** 2007, 2008<br>**Models:** Passat<br>**Engines:** 2.0L<br>**Transmissions:** All | **Incorrect Fuel Composition Conditions:**<br>The engine is running in a closed loop at a stable engine speed, and the ECM detected the lean or rich fuel trim correction valve was more than or less than a calibrated limit.<br>**Possible Causes:**<br>• One or more fuel injectors have failed |
| **DTC: P0170**<br>**Years:** 2007, 2008<br>**Models:** Passat, Jetta, Golf, GTI, New Beetle, New Beetle Convertible<br>**Engines:** 2.0L, 2.5L, 3.6L<br>**Transmissions:** All | **Fuel System Malfunction (Cylinder Bank 1) Conditions:**<br>The engine is running in a closed loop at a stable engine speed, and the ECM detected the lean or rich fuel trim correction valve was more than or less than a calibrated limit.<br>**Possible Causes:**<br>• Air leaks after the MAF sensor, or leaks in the PCV system<br>• Exhaust leaks before or near where the HO2S is mounted<br>• Fuel injector(s) restricted or not supplying enough fuel<br>• Fuel system not supplying enough fuel during high fuel demand conditions (e.g., the fuel pump may not supply enough fuel)<br>• Leaking EGR gasket, or leaking EGR valve diaphragm<br>• MAF sensor dirty (causes ECM to underestimate airflow)<br>• Vehicle running out of fuel or engine oil dip stick not seated |
| **DTC: P0171**<br>**2T, MIL: Yes**<br>**Years:** 2007, 2008<br>**Models:** Passat, Jetta, Golf, GTI, New Beetle, New Beetle Convertible<br>**Engines:** 2.0L, 2.5L, 3.6L<br>**Transmissions:** All | **Fuel System Too Lean (Cylinder Bank 1) Conditions:**<br>Key on or engine running, all electrical components off and coolant temperature at least 80 degrees Celsius; and the ECM detected the Bank 1 Adaptive Fuel Control System reached its rich correction limit (a lean A/F condition).<br>**Possible Causes:**<br>• Air leaks after the MAF sensor, or leaks in the PCV system<br>• Exhaust leaks before or near where the HO2S is mounted<br>• Fuel injector(s) restricted or not supplying enough fuel<br>• Fuel pump not supplying enough fuel during high fuel demand conditions<br>• Leaking EGR gasket, or leaking EGR valve diaphragm<br>• MAF sensor dirty (causes ECM to underestimate airflow)<br>• Vehicle running out of fuel or engine oil dip stick not seated |
| **DTC: P0172**<br>**2T, MIL: Yes**<br>**Years:** 2007, 2008<br>**Models:** Passat, Jetta, Golf, GTI, New Beetle, New Beetle Convertible<br>**Engines:** 2.0L, 2.5L, 3.6L<br>**Transmissions:** All | **Fuel System Too Rich (Cylinder Bank 1) Conditions:**<br>Key on or engine running, all electrical components off and coolant temperature at least 80 degrees Celsius; and the ECM detected the Bank 1 Adaptive Fuel Control System reached its rich correction limit (a rich A/F condition).<br>**Possible Causes:**<br>• Camshaft timing is incorrect, or the engine has an oil overfill condition<br>• EVAP vapor recovery system failure (may be pulling vacuum)<br>• Fuel pressure regulator is damaged or leaking<br>• HO2S element is contaminated with alcohol or water<br>• MAF or MAP sensor values are incorrect or out-of-range<br>• One of more fuel injectors is leaking |

| DTC | Trouble Code Title, Conditions & Possible Causes |
|---|---|
| **DTC: P0190**<br>**Years: 2007, 2008**<br>**Models:** Passat<br>**Engines:** 2.0L<br>**Transmissions**: All | **Fuel Rail Pressure Sensor Circuit Conditions**<br>Key on or engine running, all electrical components off and coolant temperature at least 80 degrees Celsius; and the ECM detected the fuel rail pressure sensor signal was outside the required voltage parameters in the self-test.<br>**Possible Causes:**<br>• Fuel Pressure Regulator Valve faulty<br>• Fuel Pressure Sensor faulty<br>• Fuel Pump (FP) Control Module faulty<br>• Fuel pump faulty |
| **DTC: P0191**<br>**Years: 2007, 2008**<br>**Models:** Passat<br>**Engines:** 2.0L<br>**Transmissions**: All | **Fuel Rail Pressure Sensor Circuit Range/Performance Conditions**<br>Key on or engine running; and the ECM detected the fuel rail pressure sensor signal was outside the required voltage parameters in the self-test.<br>**Possible Causes:**<br>• Fuel Pressure Regulator Valve faulty<br>• Fuel Pressure Sensor faulty<br>• Fuel Pump (FP) Control Module faulty<br>• Fuel pump faulty |
| **DTC: P0192**<br>**Years: 2007, 2008**<br>**Models:** Passat, Jetta<br>**Engines:** 2.0L<br>**Transmissions**: All | **Fuel Rail Pressure Sensor Circuit Low Conditions**<br>Key on or engine running, all electrical components off and coolant temperature at least 80 degrees Celsius; and the ECM detected the fuel rail pressure sensor signal was below the required voltage in the self-test.<br>**Possible Causes:**<br>• Fuel Pressure Regulator Valve faulty<br>• Fuel Pressure Sensor faulty<br>• Fuel Pump (FP) Control Module faulty<br>• Fuel pump faulty |
| **DTC: P0193**<br>**Years: 2007, 2008**<br>**Models:** Passat<br>**Engines:** 2.0L<br>**Transmissions**: All | **Fuel Rail Pressure Sensor Circuit High Conditions**<br>Key on or engine running, all electrical components off and coolant temperature at least 80 degrees Celsius; and the ECM detected the fuel rail pressure sensor signal was above the required voltage in the self-test.<br>**Possible Causes:**<br>• Fuel Pressure Regulator Valve faulty<br>• Fuel Pressure Sensor faulty<br>• Fuel Pump (FP) Control Module faulty<br>• Fuel pump faulty |
| **DTC: P0194**<br>**Years: 2007, 2008**<br>**Models:** Passat<br>**Engines:** 2.0L<br>**Transmissions**: All | **Fuel Rail Pressure Sensor Circuit Intermittent Conditions**<br>Key on or engine running, all electrical components off and coolant temperature at least 80 degrees Celsius; and the ECM detected the fuel rail pressure sensor signal was implausible.<br>**Possible Causes:**<br>• Fuel Pressure Regulator Valve faulty<br>• Fuel Pressure Sensor faulty<br>• Fuel Pump (FP) Control Module faulty<br>• Fuel pump faulty |
| **DTC: P0201**<br>**2T, MIL: Yes**<br>**Years: 2007, 2008**<br>**Models:** Passat, Jetta, Golf, GTI, New Beetle, New Beetle Convertible<br>**Engines:** 2.0L, 2.5L, 3.6L<br>**Transmissions**: All | **Cylinder 1 Injector Circuit Malfunction Conditions:**<br>Engine started, and the ECM detected the fuel injector "1" control circuit was in a high state when it should have been low, or in a low state when it should have been high (wiring harness & injector okay).<br>**Possible Causes:**<br>• Injector 1 connector is damaged, open or shorted<br>• Injector 1 control circuit is open, shorted to ground or to power<br>• ECM has failed (the injector driver circuit may be damaged) |
| **DTC: P0202**<br>**2T, MIL: Yes**<br>**Years: 2007, 2008**<br>**Models:** Passat, Jetta, Golf, GTI, New Beetle, New Beetle Convertible<br>**Engines:** 2.0L, 2.5L, 3.6L<br>**Transmissions**: All | **Cylinder 2 Injector Circuit Malfunction Conditions:**<br>Engine started, and the ECM detected the fuel injector "2" control circuit was in a high state when it should have been low, or in a low state when it should have been high (wiring harness & injector okay).<br>**Possible Causes:**<br>• Injector 2 connector is damaged, open or shorted<br>• Injector 2 control circuit is open, shorted to ground or to power<br>• ECM has failed (the injector driver circuit may be damaged) |

| DTC | Trouble Code Title, Conditions & Possible Causes |
|---|---|
| **DTC: P0203**<br>**2T, MIL**: Yes<br>**Years**: 2007, 2008<br>**Models**: Passat, Jetta, Golf, GTI, New Beetle, New Beetle Convertible<br>**Engines**: 2.0L, 2.5L, 3.6L<br>**Transmissions**: All | **Cylinder 3 Injector Circuit Malfunction Conditions:**<br>Engine started, and the ECM detected the fuel injector "3" control circuit was in a high state when it should have been low, or in a low state when it should have been high (wiring harness & injector okay).<br>**Possible Causes:**<br>• Injector 3 connector is damaged, open or shorted<br>• Injector 3 control circuit is open, shorted to ground or to power<br>• ECM has failed (the injector driver circuit may be damaged) |
| **DTC: P0204**<br>**2T, MIL**: Yes<br>**Years**: 2007, 2008<br>**Models**: Passat, Jetta, Golf, GTI, New Beetle, New Beetle Convertible<br>**Engines**: 2.0L, 2.5L, 3.6L<br>**Transmissions**: All | **Cylinder 4 Injector Circuit Malfunction Conditions:**<br>Engine started, and the ECM detected the fuel injector "4" control circuit was in a high state when it should have been low, or in a low state when it should have been high (wiring harness & injector okay).<br>**Possible Causes:**<br>• Injector 4 connector is damaged, open or shorted<br>• Injector 4 control circuit is open, shorted to ground or to power<br>• ECM has failed (the injector driver circuit may be damaged) |
| **DTC: P0207**<br>**2T, MIL**: Yes<br>**Years**: 2007, 2008<br>**Models**: Passat<br>**Engines**: 2.5L, 3.6L<br>**Transmissions**: All | **Cylinder 7 Injector Circuit Malfunction Conditions:**<br>Engine started, and the ECM detected the fuel injector "7" control circuit was in a high state when it should have been low, or in a low state when it should have been high (wiring harness & injector okay).<br>**Note: Monitor the INJIF PID Fault "flags" with the Scan Tool. The appropriate INJF PID "flag" will read Yes when this code is set.**<br>**Possible Causes:**<br>• Injector 7 connector is damaged, open or shorted<br>• Injector 7 control circuit is open, shorted to ground or to power<br>• ECM has failed (the injector driver circuit may be damaged) |
| **DTC: P0208**<br>**2T, MIL**: Yes<br>**Years**: 2007, 2008<br>**Models**: Passat<br>**Engines**: 2.5L, 3.6L<br>**Transmissions**: All | **Cylinder 8 Injector Circuit Malfunction Conditions:**<br>Engine started, and the ECM detected the fuel injector "8" control circuit was in a high state when it should have been low, or in a low state when it should have been high (wiring harness & injector okay).<br>**Note: Monitor the INJIF PID Fault "flags" with the Scan Tool. The appropriate INJF PID "flag" will read Yes when this code is set.**<br>**Possible Causes:**<br>• Injector 8 connector is damaged, open or shorted<br>• Injector 8 control circuit is open, shorted to ground or to power<br>• ECM has failed (the injector driver circuit may be damaged) |
| **DTC: P0219**<br>**Years**: 2007, 2008<br>**Models**: Passat, Jetta, Golf, GTI, New Beetle, New Beetle Convertible<br>**Engines**: 2.0L, 2.5L, 3.6L<br>**Transmissions**: All | **Engine Over-Speed Condition Conditions:**<br>Engine started, and the ECM determined the vehicle had been driven in a manner that caused the engine to over-speed, and to exceed the engine speed calibration limit stored in memory.<br>**Possible Causes:**<br>• Engine operated in the wrong transmission gear position<br>• Excessive engine speed with gear selector in Neutral position<br>• Wheel slippage due to wet, muddy or snowing conditions |
| **DTC: P0221**<br>**2T, MIL**: Yes<br>**Years**: 2007, 2008<br>**Models**: Passat, Jetta, Golf, GTI, New Beetle, New Beetle Convertible<br>**Engines**: 2.0L, 2.5L, 3.6L<br>**Transmissions**: All | **Throttle Position Sensor 'B' Signal Performance Conditions:**<br>Engine started, battery voltage at least 11.5v, all electrical components off, ground connections between engine and chassis well connected, coolant temperature at least 80-degrees Celicius and the throttle valve must not be damaged or dirty; and the ECM detected the TP Sensor 'B' circuit was out of its normal operating range during a condition with the throttle wide open, or with it completely closed. The throttle valve activation occurs via an electric motor (throttle drive) in the throttle valve control module. It is activated by the ECM according to specifications of the two sensors, Throttle Position Sensor and Accelerator Pedal Position Sensor 2. Slowly depress accelerator pedal up to Wide Open Throttle (WOT) stop while observing the percentage display on the PID data function of the scan tool. The percentage display must increase uniformly.<br>**Possible Causes:**<br>• Throttle body is damaged<br>• Throttle linkage is binding or sticking<br>• ETC TP Sensor 'B' signal circuit to the ECM is open<br>• ETC TP Sensor 'B' ground circuit is open<br>• ETC TP Sensor 'B' is damaged or it has failed |

| DTC | Trouble Code Title, Conditions & Possible Causes |
|---|---|
| **DTC: P0222**<br>**2T, MIL: Yes**<br>**Years: 2007, 2008**<br>**Models:** Passat, Jetta, Golf, GTI, New Beetle, New Beetle Convertible<br>**Engines:** 2.0L, 2.5L, 3.6L<br>**Transmissions:** All | **Throttle Position Sensor 'B' Circuit Low Input Conditions:**<br>Engine started, battery voltage at least 11.5v, all electrical components off, ground connections between engine and chassis well connected, coolant temperature at least 80-degrees Celicius and the throttle valve must not be damaged or dirty; and the ECM detected the TP Sensor 'B' circuit was out of its normal operating range during a condition with the throttle wide open, or with it completely closed. The throttle valve activation occurs via an electric motor (throttle drive) in the throttle valve control module. It is activated by the ECM according to specifications of the two sensors, Throttle Position Sensor and Accelerator Pedal Position Sensor 2. Slowly depress accelerator pedal up to Wide Open Throttle (WOT) stop while observing the percentage display on the PID data function of the scan tool. The percentage display must increase uniformly.<br>**Possible Causes:**<br>• ETC TP Sensor 'B' connector is damaged or shorted<br>• ETC TP Sensor 'B' signal circuit is shorted to ground<br>• ETC TP Sensor 'B' is damaged or it has failed<br>• ECM has failed |
| **DTC: P0223**<br>**2T, MIL: Yes**<br>**Years: 2007, 2008**<br>**Models:** Passat, Jetta, Golf, GTI, New Beetle, New Beetle Convertible<br>**Engines:** 2.0L, 2.5L, 3.6L<br>**Transmissions:** All | **Throttle Position Sensor 'B' Circuit High Input Conditions:**<br>Engine started, battery voltage at least 11.5v, all electrical components off, ground connections between engine and chassis well connected, coolant temperature at least 80-degrees Celicius and the throttle valve must not be damaged or dirty; and the ECM detected the TP Sensor 'B' circuit was out of its normal operating range during a condition with the throttle wide open, or with it completely closed. The throttle valve activation occurs via an electric motor (throttle drive) in the throttle valve control module. It is activated by the ECM according to specifications of the two sensors, Throttle Position Sensor and Accelerator Pedal Position Sensor 2. Slowly depress accelerator pedal up to Wide Open Throttle (WOT) stop while observing the percentage display on the PID data function of the scan tool. The percentage display must increase uniformly.<br>**Possible Causes:**<br>• ETC TP Sensor 'B' connector is damaged or open<br>• ETC TP Sensor 'B' signal circuit is open<br>• ETC TP Sensor 'B' signal circuit is shorted to VREF (5v)<br>• ETC TP Sensor 'B' is damaged or it has failed |
| **DTC: P0225**<br>**Years: 2007, 2008**<br>**Models:** New Beetle, New Beetle Convertible<br>**Engines:** 2.0L<br>**Transmissions:** All | **Throttle Position Sensor 'C' Signal Voltage Supply Conditions:**<br>Engine started, battery voltage at least 11.5v, all electrical components off, ground connections between engine and chassis well connected, coolant temperature at least 80-degrees Celicius and the throttle valve must not be damaged or dirty; and the ECM detected the TP Sensor 'B' circuit was out of its normal operating range during a condition with the throttle wide open, or with it completely closed. The throttle valve activation occurs via an electric motor (throttle drive) in the throttle valve control module. It is activated by the ECM according to specifications of the two sensors, Throttle Position Sensor and Accelerator Pedal Position Sensor 2. Slowly depress accelerator pedal up to Wide Open Throttle (WOT) stop while observing the percentage display on the PID data function of the scan tool. The percentage display must increase uniformly.<br>**Possible Causes:**<br>• Throttle body is damaged<br>• Throttle linkage is binding or sticking<br>• ETC TP Sensor 'C' signal circuit to the ECM is open<br>• ETC TP Sensor 'C' ground circuit is open<br>• ETC TP Sensor 'C' is damaged or it has failed |
| **DTC: P0226**<br>**Years: 2007, 2008**<br>**Models:** Passat, Jetta, Golf, GTI, New Beetle, New Beetle Convertible<br>**Engines:** 2.0L, 2.5L, 3.6L<br>**Transmissions:** All | **Throttle Position Sensor 'C' Signal Performance Conditions:**<br>Engine started, battery voltage at least 11.5v, all electrical components off, ground connections between engine and chassis well connected, coolant temperature at least 80-degrees Celicius and the throttle valve must not be damaged or dirty; and the ECM detected the TP Sensor 'B' circuit was out of its normal operating range during a condition with the throttle wide open, or with it completely closed. The throttle valve activation occurs via an electric motor (throttle drive) in the throttle valve control module. It is activated by the ECM according to specifications of the two sensors, Throttle Position Sensor and Accelerator Pedal Position Sensor 2. Slowly depress accelerator pedal up to Wide Open Throttle (WOT) stop while observing the percentage display on the PID data function of the scan tool. The percentage display must increase uniformly..<br>**Possible Causes:**<br>• Throttle body is damaged<br>• Throttle linkage is binding or sticking<br>• ETC TP Sensor 'C' signal circuit to the ECM is open<br>• ETC TP Sensor 'C' ground circuit is open<br>• ETC TP Sensor 'C' is damaged or it has failed |

| DTC | Trouble Code Title, Conditions & Possible Causes |
|---|---|
| **DTC: P0227**<br>**Years: 2007, 2008**<br>**Models:** Passat, Jetta, Golf, GTI, New Beetle, New Beetle Convertible<br>**Engines:** 2.0L, 2.5L, 3.6L<br>**Transmissions**: All | **Throttle Position Sensor 'C' Circuit Low Input Conditions:**<br>Engine started, battery voltage at least 11.5v, all electrical components off, ground connections between engine and chassis well connected, coolant temperature at least 80-degrees Celicius and the throttle valve must not be damaged or dirty; and the ECM detected the TP Sensor 'B' circuit was out of its normal operating range during a condition with the throttle wide open, or with it completely closed. The throttle valve activation occurs via an electric motor (throttle drive) in the throttle valve control module. It is activated by the ECM according to specifications of the two sensors, Throttle Position Sensor and Accelerator Pedal Position Sensor 2. Slowly depress accelerator pedal up to Wide Open Throttle (WOT) stop while observing the percentage display on the PID data function of the scan tool. The percentage display must increase uniformly.<br>**Possible Causes:**<br>• ETC TP Sensor 'C' connector is damaged or shorted<br>• ETC TP Sensor 'C' signal circuit is shorted to ground<br>• ETC TP Sensor 'C' is damaged or it has failed<br>• ECM has failed |
| **DTC: P0228**<br>**Years: 2007, 2008**<br>**Models:** Passat, Jetta, Golf, GTI, New Beetle, New Beetle Convertible<br>**Engines:** 2.0L, 2.5L, 3.6L<br>**Transmissions**: All | **Throttle Position Sensor 'C' Circuit High Input Conditions:**<br>Engine started, battery voltage at least 11.5v, all electrical components off, ground connections between engine and chassis well connected, coolant temperature at least 80-degrees Celicius and the throttle valve must not be damaged or dirty; and the ECM detected the TP Sensor 'B' circuit was out of its normal operating range during a condition with the throttle wide open, or with it completely closed. The throttle valve activation occurs via an electric motor (throttle drive) in the throttle valve control module. It is activated by the ECM according to specifications of the two sensors, Throttle Position Sensor and Accelerator Pedal Position Sensor 2. Slowly depress accelerator pedal up to Wide Open Throttle (WOT) stop while observing the percentage display on the PID data function of the scan tool. The percentage display must increase uniformly.<br>**Possible Causes:**<br>• ETC TP Sensor 'C' connector is damaged or open<br>• ETC TP Sensor 'C' signal circuit is open<br>• ETC TP Sensor 'C' signal circuit is shorted to VREF (5v)<br>• ETC TP Sensor 'C' is damaged or it has failed |
| **DTC: P0230**<br>**2T, MIL: Yes**<br>**Years: 2007, 2008**<br>**Models:** Passat, Jetta, Golf, GTI, New Beetle, New Beetle Convertible<br>**Engines:** 2.0L, 2.5L, 3.6L<br>**Transmissions**: All | **Fuel Pump Primary Circuit Malfunction Conditions:**<br>Engine started, battery voltage at least 11.5v, all electrical components off, ground connections between engine and chassis well connected, coolant temperature at least 80-degrees Celicius. The ECM detected high current in fuel pump or fuel shutoff valve (FSV) circuit, or it detected voltage with the valve off, or it did not detect voltage on the circuit. The circuit is used to energize the fuel pump relay at key on or while running. Fuel pressure value should be 3000 to 5000 kPa at idle.<br>**Possible Causes:**<br>• FP or FSV circuit is open or shorted<br>• Fuel pump relay VPWR circuit open<br>• Fuel pump relay is damaged or has failed<br>• Fuel pressure sensor has failed<br>• Fuel pump control module is faulty<br>• ECM has failed |
| **DTC: P0234**<br>**Years: 2007, 2008**<br>**Models:** Passat, Jetta, Golf, GTI, New Beetle<br>**Engines:** 2.0L<br>**Transmissions**: All | **Turbo/Supercharger Overboost Condition Conditions:**<br>Engine started, battery voltage at least 11.5v, all electrical components off, ground connections between engine and chassis well connected, coolant temperature at least 80-degrees Celicius. The ECM detected an operating condition that could harm the engine or automatic transmission.<br>**Possible Causes:**<br>• Ignition misfire condition exceeds the calibrated threshold<br>• Knock sensor circuit has failed, or excessive knock detected<br>• Low speed fuel pump relay not switching properly<br>• Transmission oil temperature beyond the calibrated threshold<br>• Shaft bearing of charge pressure regulator valve in turbocharger is blocked |

| DTC | Trouble Code Title, Conditions & Possible Causes |
|---|---|
| **DTC: P0235**<br>**Years: 2007, 2008**<br>**Models:** Passat, Jetta, Golf, GTI, New Beetle<br>**Engines:** 2.0L<br>**Transmissions:** All | **Turbocharger Boost Sensor (A) Circ Control Limit Not Reached Conditions:**<br>Engine started, battery voltage at least 11.5v, all electrical components off, ground connections between engine and chassis well connected, coolant temperature at least 80-degrees Celicius. The ECM detected an operating condition that could harm the engine or automatic transmission.<br>**Possible Causes:**<br>• Charge air pressure sensor is faulty<br>• Voltage supply to the charge air pressure sensor is open or shorted<br>• Check the charge air system for leaks<br>• Recirculating valve for turbocharger is faulty<br>• Turbocharging system is damaged or not functioning properly<br>• Turbocharger recirculating valve is faulty<br>• Vacuum diaphragm for turbocharger is out of adjustment<br>• Wastegate bypass regulator valve is faulty<br>• Boost sensor has failed<br>• ECM has failed |
| **DTC: P0236**<br>**Years: 2007, 2008**<br>**Models:** Passat, Jetta, Golf, GTI, New Beetle<br>**Engines:** 2.0L<br>**Transmissions:** All | **Turbocharger Boost Sensor (A) Circ Control Range/Performance Conditions:**<br>Engine started, battery voltage at least 11.5v, all electrical components off, ground connections between engine and chassis well connected, coolant temperature at least 80-degrees Celicius. The ECM detected an operating condition that could harm the engine or automatic transmission.<br>**Possible Causes:**<br>• Charge air pressure sensor is faulty<br>• Voltage supply to the charge air pressure sensor is open or shorted<br>• Check the charge air system for leaks<br>• Recirculating valve for turbocharger is faulty<br>• Turbocharging system is damaged or not functioning properly<br>• Turbocharger recirculating valve is faulty<br>• Vacuum diaphragm for turbocharger is out of adjustment<br>• Wastegate bypass regulator valve is faulty<br>• Boost sensor has failed<br>• ECM has failed |
| **DTC: P0237**<br>**2T, MIL: Yes**<br>**Years: 2007, 2008**<br>**Models:** Passat, Jetta, Golf, GTI, New Beetle<br>**Engines:** 2.0L<br>**Transmissions:** All | **Turbocharger Boost Sensor (A) Circ Low Input Conditions:**<br>Engine started, battery voltage at least 11.5v, all electrical components off, ground connections between engine and chassis well connected, coolant temperature at least 80-degrees Celicius. The ECM detected an operating condition that could harm the engine or automatic transmission.<br>**Possible Causes:**<br>• Charge air pressure sensor is faulty<br>• Voltage supply to the charge air pressure sensor is open or shorted<br>• Check the charge air system for leaks<br>• Recirculating valve for turbocharger is faulty<br>• Turbocharging system is damaged or not functioning properly<br>• Turbocharger recirculating valve is faulty<br>• Vacuum diaphragm for turbocharger is out of adjustment<br>• Wastegate bypass regulator valve is faulty<br>• Boost sensor has failed<br>• ECM has failed |
| **DTC: P0238**<br>**2T, MIL: Yes**<br>**Years: 2007, 2008**<br>**Models:** Passat, Jetta, Golf, GTI, New Beetle<br>**Engines:** 2.0L<br>**Transmissions:** All | **Turbocharger Boost Sensor (A) Circ High Input Conditions:**<br>Engine started, battery voltage at least 11.5v, all electrical components off, ground connections between engine and chassis well connected, coolant temperature at least 80-degrees Celicius. The ECM detected an operating condition that could harm the engine or automatic transmission.<br>**Possible Causes:**<br>• Charge air pressure sensor is faulty<br>• Voltage supply to the charge air pressure sensor is open or shorted<br>• Check the charge air system for leaks<br>• Recirculating valve for turbocharger is faulty<br>• Turbocharging system is damaged or not functioning properly<br>• Turbocharger recirculating valve is faulty<br>• Vacuum diaphragm for turbocharger is out of adjustment<br>• Wastegate bypass regulator valve is faulty<br>• Boost sensor has failed<br>• ECM has failed |

| DTC | Trouble Code Title, Conditions & Possible Causes |
|---|---|
| **DTC: P0243**<br>**Years: 2007, 2008**<br>**Models:** Passat, Jetta, Golf, GTI, New Beetle<br>**Engines:** 2.0L<br>**Transmissions:** All | **Turbocharger Boost Bypass Solenoid (A) Circuit Open/Short Circuit Conditions:**<br>Engine started, battery voltage at least 11.5v, all electrical components off, ground connections between engine and chassis well connected, coolant temperature at least 80-degrees Celicius. The ECM detected an unexpected voltage condition on the Bypass Solenoid control circuit<br>**Possible Causes:**<br>• Bypass solenoid power supply circuit is open<br>• Bypass solenoid control circuit is open, shorted to ground or system power<br>• Bypass solenoid assembly is damaged or has failed<br>• Charge air pressure sensor is faulty<br>• Voltage supply to the charge air pressure sensor is open or shorted<br>• Check the charge air system for leaks<br>• Recirculating valve for turbocharger is faulty<br>• Turbocharging system is damaged or not functioning properly<br>• Turbocharger recirculating valve is faulty<br>• Vacuum diaphragm for turbocharger is out of adjustment<br>• Wastegate bypass regulator valve is faulty<br>• Boost sensor has failed<br>• ECM has failed |
| **DTC: P0245**<br>**Years: 2007, 2008**<br>**Models:** Passat, Jetta, Golf, GTI, New Beetle<br>**Engines:** 2.0L<br>**Transmissions:** All | **Turbocharger Boost Bypass Solenoid (A) Circuit Low Input/Short to Ground Conditions:**<br>Engine started, battery voltage at least 11.5v, all electrical components off, ground connections between engine and chassis well connected, coolant temperature at least 80-degrees Celicius. The ECM detected an unexpected voltage condition on the Bypass Solenoid control circuit<br>**Possible Causes:**<br>• Bypass solenoid power supply circuit is open<br>• Bypass solenoid control circuit is open, shorted to ground or system power<br>• Bypass solenoid assembly is damaged or has failed<br>• Charge air pressure sensor is faulty<br>• Voltage supply to the charge air pressure sensor is open or shorted<br>• Check the charge air system for leaks<br>• Recirculating valve for turbocharger is faulty<br>• Turbocharging system is damaged or not functioning properly<br>• Turbocharger recirculating valve is faulty<br>• Vacuum diaphragm for turbocharger is out of adjustment<br>• Wastegate bypass regulator valve is faulty<br>• Boost sensor has failed<br>• ECM has failed |
| **DTC: P0246**<br>**Years: 2007, 2008**<br>**Models:** Passat, Jetta, Golf, GTI, New Beetle<br>**Engines:** 2.0L<br>**Transmissions:** All | **Turbocharger Boost Bypass Solenoid (A) Circuit High Input/Short to B+ Conditions:**<br>Engine started, battery voltage at least 11.5v, all electrical components off, ground connections between engine and chassis well connected, coolant temperature at least 80-degrees Celicius. The ECM detected an unexpected voltage condition on the Bypass Solenoid control circuit<br>**Possible Causes:**<br>• Bypass solenoid power supply circuit is open<br>• Bypass solenoid control circuit is open, shorted to ground or system power<br>• Bypass solenoid assembly is damaged or has failed<br>• Charge air pressure sensor is faulty<br>• Voltage supply to the charge air pressure sensor is open or shorted<br>• Check the charge air system for leaks<br>• Recirculating valve for turbocharger is faulty<br>• Turbocharging system is damaged or not functioning properly<br>• Turbocharger recirculating valve is faulty<br>• Vacuum diaphragm for turbocharger is out of adjustment<br>• Wastegate bypass regulator valve is faulty<br>• Boost sensor has failed<br>• ECM has failed |

| DTC | Trouble Code Title, Conditions & Possible Causes |
|---|---|
| **DTC: P0261**<br>**Years: 2007, 2008**<br>**Models:** Passat, Jetta, Golf, GTI, New Beetle, New Beetle Convertible<br>**Engines:** 2.0L, 2.5L, 3.6L<br>**Transmissions:** All | **Cylinder 1 Injector Circuit Low Input/Short to Ground Conditions:**<br>Key on or engine running, fuses in the instrument panel and the E-box in the engine compartment must be functioning, and the ground connections between the engine ad the chassis must be well connected; and the ECM detected an unexpected voltage condition on the injector circuit<br>**Possible Causes:**<br>• Injector 1 control circuit is open<br>• Injector 1 power circuit (B+) is open<br>• Injector 1 control circuit is shorted to chassis ground<br>• Injector 1 is damaged or has failed<br>• ECM is not connected or has failed |
| **DTC: P0262**<br>**2T, MIL: Yes**<br>**Years: 2007, 2008**<br>**Models:** Passat, Jetta, Golf, GTI, New Beetle, New Beetle Convertible<br>**Engines:** 2.0L, 2.5L, 3.6L<br>**Transmissions:** All | **Cylinder 1 Injector Circuit Low Input/Short to B+ Conditions:**<br>Key on or engine running, fuses in the instrument panel and the E-box in the engine compartment must be functioning, and the ground connections between the engine ad the chassis must be well connected; and the ECM detected an unexpected voltage condition on the injector circuit<br>**Possible Causes:**<br>• Injector control circuit is open<br>• Injector power circuit (B+) is open<br>• Injector control circuit is shorted to chassis ground<br>• Injector is damaged or has failed<br>• ECM is not connected or has failed<br>• Fuel pump relay has failed<br>• Fuel injectors may have malfunctioned<br>• Faulty engine speed sensor |
| **DTC: P0264**<br>**2T, MIL: Yes**<br>**Years: 2007, 2008**<br>**Models:** Passat, Jetta, Golf, GTI, New Beetle, New Beetle Convertible<br>**Engines:** 2.0L, 2.5L, 3.6L<br>**Transmissions:** All | **Cylinder 2 Injector Circuit Low Input/Short to Ground Conditions:**<br>Key on or engine running, fuses in the instrument panel and the E-box in the engine compartment must be functioning, and the ground connections between the engine ad the chassis must be well connected; and the ECM detected an unexpected voltage condition on the injector circuit<br>**Possible Causes:**<br>• Injector control circuit is open<br>• Injector power circuit (B+) is open<br>• Injector control circuit is shorted to chassis ground<br>• Injector is damaged or has failed<br>• ECM is not connected or has failed<br>• Fuel pump relay has failed<br>• Fuel injectors may have malfunctioned<br>• Faulty engine speed sensor |
| **DTC: P0265**<br>**2T, MIL: Yes**<br>**Years: 2007, 2008**<br>**Models:** Passat, Jetta, Golf, GTI, New Beetle, New Beetle Convertible<br>**Engines:** 2.0L, 2.5L, 3.6L<br>**Transmissions:** All | **Cylinder 2 Injector Circuit Low Input/Short to B+ Conditions:**<br>Key on or engine running, fuses in the instrument panel and the E-box in the engine compartment must be functioning, and the ground connections between the engine ad the chassis must be well connected; and the ECM detected an unexpected voltage condition on the injector circuit<br>**Possible Causes:**<br>• Injector control circuit is open<br>• Injector power circuit (B+) is open<br>• Injector control circuit is shorted to chassis ground<br>• Injector is damaged or has failed<br>• ECM is not connected or has failed<br>• Fuel pump relay has failed<br>• Fuel injectors may have malfunctioned<br>• Faulty engine speed sensor |

| DTC | Trouble Code Title, Conditions & Possible Causes |
|---|---|
| **DTC: P0267**<br>**2T, MIL: Yes**<br>**Years: 2007, 2008**<br>**Models**: Passat, Jetta, Golf, GTI, New Beetle, New Beetle Convertible<br>**Engines**: 2.0L, 2.5L, 3.6L<br>**Transmissions**: All | **Cylinder 3 Injector Circuit Low Input/Short to Ground Conditions:**<br>Key on or engine running, fuses in the instrument panel and the E-box in the engine compartment must be functioning, and the ground connections between the engine ad the chassis must be well connected; and the ECM detected an unexpected voltage condition on the injector circuit<br>**Possible Causes:**<br>• Injector control circuit is open<br>• Injector power circuit (B+) is open<br>• Injector control circuit is shorted to chassis ground<br>• Injector is damaged or has failed<br>• ECM is not connected or has failed<br>• Fuel pump relay has failed<br>• Fuel injectors may have malfunctioned<br>• Faulty engine speed sensor |
| **DTC: P0268**<br>**2T, MIL: Yes**<br>**Years: 2007, 2008**<br>**Models**: Passat, Jetta, Golf, GTI, New Beetle, New Beetle Convertible<br>**Engines**: 2.0L, 2.5L, 3.6L<br>**Transmissions**: All | **Cylinder 3 Injector Circuit Low Input/Short to B+ Conditions:**<br>Key on or engine running, fuses in the instrument panel and the E-box in the engine compartment must be functioning, and the ground connections between the engine ad the chassis must be well connected; and the ECM detected an unexpected voltage condition on the injector circuit<br>**Possible Causes:**<br>• Injector control circuit is open<br>• Injector power circuit (B+) is open<br>• Injector control circuit is shorted to chassis ground<br>• Injector is damaged or has failed<br>• ECM is not connected or has failed<br>• Fuel pump relay has failed<br>• Fuel injectors may have malfunctioned<br>• Faulty engine speed sensor |
| **DTC: P0270**<br>**2T, MIL: Yes**<br>**Years: 2007, 2008**<br>**Models**: Passat, Jetta, Golf, GTI, New Beetle, New Beetle Convertible<br>**Engines**: 2.0L, 2.5L, 3.6L<br>**Transmissions**: All | **Cylinder 4 Injector Circuit Low Input/Short to Ground Conditions:**<br>Key on or engine running, fuses in the instrument panel and the E-box in the engine compartment must be functioning, and the ground connections between the engine ad the chassis must be well connected; and the ECM detected an unexpected voltage condition on the injector circuit<br>**Possible Causes:**<br>• Injector control circuit is open<br>• Injector power circuit (B+) is open<br>• Injector control circuit is shorted to chassis ground<br>• Injector is damaged or has failed<br>• ECM is not connected or has failed<br>• Fuel pump relay has failed<br>• Fuel injectors may have malfunctioned<br>• Faulty engine speed sensor |
| **DTC: P0271**<br>**2T, MIL: Yes**<br>**Years: 2007, 2008**<br>**Models**: Passat, Jetta, Golf, GTI, New Beetle, New Beetle Convertible<br>**Engines**: 2.0L, 2.5L, 3.6L<br>**Transmissions**: All | **Cylinder 4 Injector Circuit Low Input/Short to B+ Conditions:**<br>Key on or engine running, fuses in the instrument panel and the E-box in the engine compartment must be functioning, and the ground connections between the engine ad the chassis must be well connected; and the ECM detected an unexpected voltage condition on the injector circuit<br>**Possible Causes:**<br>• Injector control circuit is open<br>• Injector power circuit (B+) is open<br>• Injector control circuit is shorted to chassis ground<br>• Injector is damaged or has failed<br>• ECM is not connected or has failed<br>• Fuel pump relay has failed<br>• Fuel injectors may have malfunctioned<br>• Faulty engine speed sensor |

| DTC | Trouble Code Title, Conditions & Possible Causes |
|---|---|
| **DTC: P0273**<br>**Years: 2007, 2008**<br>**Models:** Passat, Jetta, GTI<br>**Engines:** 2.5L, 3.6L<br>**Transmissions:** All | **Cylinder 5 Injector Circuit Low Input/Short to Ground Conditions:**<br>Key on or engine running, fuses in the instrument panel and the E-box in the engine compartment must be functioning, and the ground connections between the engine ad the chassis must be well connected; and the ECM detected an unexpected voltage condition on the injector circuit<br>**Possible Causes:**<br>• Injector control circuit is open<br>• Injector power circuit (B+) is open<br>• Injector control circuit is shorted to chassis ground<br>• Injector is damaged or has failed<br>• ECM is not connected or has failed<br>• Fuel pump relay has failed<br>• Fuel injectors may have malfunctioned<br>• Faulty engine speed sensor |
| **DTC: P0274**<br>**Years: 2007, 2008**<br>**Models:** Passat, Jetta, GTI<br>**Engines:** 2.5L, 3.6L<br>**Transmissions:** All | **Cylinder 5 Injector Circuit Low Input/Short to B+ Conditions:**<br>Key on or engine running, fuses in the instrument panel and the E-box in the engine compartment must be functioning, and the ground connections between the engine ad the chassis must be well connected; and the ECM detected an unexpected voltage condition on the injector circuit<br>**Possible Causes:**<br>• Injector control circuit is open<br>• Injector power circuit (B+) is open<br>• Injector control circuit is shorted to chassis ground<br>• Injector is damaged or has failed<br>• ECM is not connected or has failed<br>• Fuel pump relay has failed<br>• Fuel injectors may have malfunctioned<br>• Faulty engine speed sensor |
| **DTC: P0276**<br>**Years: 2007, 2008**<br>**Models:** Passat, Jetta, GTI<br>**Engines:** 2.5L, 3.6L<br>**Transmissions:** All | **Cylinder 6 Injector Circuit Low Input/Short to Ground Conditions:**<br>Key on or engine running, fuses in the instrument panel and the E-box in the engine compartment must be functioning, and the ground connections between the engine ad the chassis must be well connected; and the ECM detected an unexpected voltage condition on the injector circuit<br>**Possible Causes:**<br>• Injector control circuit is open<br>• Injector power circuit (B+) is open<br>• Injector control circuit is shorted to chassis ground<br>• Injector is damaged or has failed<br>• ECM is not connected or has failed<br>• Fuel pump relay has failed<br>• Fuel injectors may have malfunctioned<br>• Faulty engine speed sensor |
| **DTC: P0277**<br>**Years: 2007, 2008**<br>**Models:** Passat, Jetta, GTI<br>**Engines:** 2.5L, 3.6L<br>**Transmissions:** All | **Cylinder 6 Injector Circuit Low Input/Short to B+ Conditions:**<br>Key on or engine running, fuses in the instrument panel and the E-box in the engine compartment must be functioning, and the ground connections between the engine ad the chassis must be well connected; and the ECM detected an unexpected voltage condition on the injector circuit<br>**Possible Causes:**<br>• Injector control circuit is open<br>• Injector power circuit (B+) is open<br>• Injector control circuit is shorted to chassis ground<br>• Injector is damaged or has failed<br>• ECM is not connected or has failed<br>• Fuel pump relay has failed<br>• Fuel injectors may have malfunctioned<br>• Faulty engine speed sensor |

| DTC | Trouble Code Title, Conditions & Possible Causes |
|---|---|
| **DTC: P0299**<br>**Years:** 2007, 2008<br>**Models:** Passat, Jetta, Golf, GTI, New Beetle<br>**Engines:** 2.0L<br>**Transmissions:** All | **Turbocharger Underboost Conditions:**<br>Engine started, battery voltage at least 11.5v, all electrical components off, ground connections between engine and chassis well connected, coolant temperature at least 80-degrees Celicius. The ECM detected an operating condition that could harm the engine or automatic transmission.<br>**Possible Causes:**<br>• Charge air pressure sensor has failed<br>• Charge air system has leaks<br>• Recirculating valve for turbocharger is faulty<br>• Turbocharging system is faulty<br>• Vacuum diaphragm for turbocharger needs adjusting<br>• Wastegate bypass regulator valve is faulty |
| **DTC: P0300**<br>**2T, MIL: Yes**<br>**Years:** 2007, 2008<br>**Models:** Passat, Jetta, Golf, GTI, New Beetle, New Beetle Convertible<br>**Engines:** 2.0L, 2.5L, 3.6L<br>**Transmissions:** All | **Random/Multiple Misfire Detected Conditions:**<br>Engine running under positive torque conditions, and the ECM detected a misfire or uneven engine running in two or more cylinders.<br>**Note: If the misfire is severe, the MIL will flash on/off on the first trip!**<br>**Possible Causes:**<br>• Base engine mechanical fault that affects two or more cylinders<br>• Fuel metering fault that affects two or more cylinders<br>• Fuel pressure too low or too high, fuel supply contaminated<br>• EVAP system problem or the EVAP canister is fuel saturated<br>• EGR valve is stuck open or the PCV system has a vacuum leak<br>• Ignition system fault (coil, plugs) affecting two or more cylinders<br>• MAF sensor contamination (it can cause a very lean condition)<br>• Vehicle driven while very low on fuel (less than 1/8 of a tank) |
| **DTC: P0301**<br>**2T, MIL: Yes**<br>**Years:** 2007, 2008<br>**Models:** Passat, Jetta, Golf, GTI, New Beetle, New Beetle Convertible<br>**Engines:** 2.0L, 2.5L, 3.6L<br>**Transmissions:** All | **Cylinder Number 1 Misfire Detected Conditions:**<br>Engine running under positive torque conditions, and the ECM detected a misfire or uneven engine function.<br>**Note: If the misfire is severe, the MIL will flash on/off on the 1st trip!**<br>**Possible Causes:**<br>• Air leak in the intake manifold, or in the EGR or ECM system<br>• Base engine mechanical problem<br>• Fuel delivery component problem (i.e., a contaminated, dirty or sticking fuel injector)<br>• Fuel pump relay defective<br>• Ignition coil fuses have failed<br>• Ignition system problem (dirty damaged coil or plug)<br>• Engine speed (RPM) sensor has failed<br>• Camshaft position sensors have failed<br>• Ignition coil is faulty<br>• Spark plugs are not working properly or are not gapped properly |
| **DTC: P0302**<br>**2T, MIL: Yes**<br>**Years:** 2007, 2008<br>**Models:** Passat, Jetta, Golf, GTI, New Beetle, New Beetle Convertible<br>**Engines:** 2.0L, 2.5L, 3.6L<br>**Transmissions:** All | **Cylinder Number 2 Misfire Detected Conditions:**<br>Engine running under positive torque conditions, and the ECM detected a misfire or uneven engine function.<br>**Note: If the misfire is severe, the MIL will flash on/off on the 1st trip!**<br>**Possible Causes:**<br>• Air leak in the intake manifold, or in the EGR or ECM system<br>• Base engine mechanical problem<br>• Fuel delivery component problem (i.e., a contaminated, dirty or sticking fuel injector)<br>• Fuel pump relay defective<br>• Ignition coil fuses have failed<br>• Ignition system problem (dirty damaged coil or plug)<br>• Engine speed (RPM) sensor has failed<br>• Camshaft position sensors have failed<br>• Ignition coil is faulty<br>• Spark plugs are not working properly or are not gapped properly |

| DTC | Trouble Code Title, Conditions & Possible Causes |
|---|---|
| **DTC: P0303**<br>**2T, MIL: Yes**<br>**Years: 2007, 2008**<br>**Models:** Passat, Jetta, Golf, GTI, New Beetle, New Beetle Convertible<br>**Engines:** 2.0L, 2.5L, 3.6L<br>**Transmissions:** All | **Cylinder Number 3 Misfire Detected Conditions:**<br>Engine running under positive torque conditions, and the ECM detected a misfire or uneven engine function.<br>**Note: If the misfire is severe, the MIL will flash on/off on the 1st trip!**<br>**Possible Causes:**<br>• Air leak in the intake manifold, or in the EGR or ECM system<br>• Base engine mechanical problem<br>• Fuel delivery component problem (i.e., a contaminated, dirty or sticking fuel injector)<br>• Fuel pump relay defective<br>• Ignition coil fuses have failed<br>• Ignition system problem (dirty damaged coil or plug)<br>• Engine speed (RPM) sensor has failed<br>• Camshaft position sensors have failed<br>• Ignition coil is faulty<br>• Spark plugs are not working properly or are not gapped properly |
| **DTC: P0304**<br>**2T, MIL: Yes**<br>**Years: 2007, 2008**<br>**Models:** Passat, Jetta, Golf, GTI, New Beetle, New Beetle Convertible<br>**Engines:** 2.0L, 2.5L, 3.6L<br>**Transmissions:** All | **Cylinder Number 4 Misfire Detected Conditions:**<br>Engine running under positive torque conditions, and the ECM detected a misfire or uneven engine function.<br>**Note: If the misfire is severe, the MIL will flash on/off on the 1st trip!**<br>**Possible Causes:**<br>• Air leak in the intake manifold, or in the EGR or ECM system<br>• Base engine mechanical problem<br>• Fuel delivery component problem (i.e., a contaminated, dirty or sticking fuel injector)<br>• Fuel pump relay defective<br>• Ignition coil fuses have failed<br>• Ignition system problem (dirty damaged coil or plug)<br>• Engine speed (RPM) sensor has failed<br>• Camshaft position sensors have failed<br>• Ignition coil is faulty<br>• Spark plugs are not working properly or are not gapped properly |
| **DTC: P0305**<br>**Years: 2007, 2008**<br>**Models:** Passat, Jetta, GTI<br>**Engines:** 2.5L, 3.6L<br>**Transmissions:** All | **Cylinder Number 5 Misfire Detected Conditions:**<br>Engine running under positive torque conditions, and the ECM detected a misfire or uneven engine function.<br>**Note: If the misfire is severe, the MIL will flash on/off on the 1st trip!**<br>**Possible Causes:**<br>• Air leak in the intake manifold, or in the EGR or ECM system<br>• Base engine mechanical problem<br>• Fuel delivery component problem (i.e., a contaminated, dirty or sticking fuel injector)<br>• Fuel pump relay defective<br>• Ignition coil fuses have failed<br>• Ignition system problem (dirty damaged coil or plug)<br>• Engine speed (RPM) sensor has failed<br>• Camshaft position sensors have failed<br>• Ignition coil is faulty<br>• Spark plugs are not working properly or are not gapped properly |
| **DTC: P0306**<br>**Years: 2007, 2008**<br>**Models:** Passat, Jetta, GTI<br>**Engines:** 2.5L, 3.6L<br>**Transmissions:** All | **Cylinder Number 6 Misfire Detected Conditions:**<br>Engine running under positive torque conditions, and the ECM detected a misfire or uneven engine function.<br>**Note: If the misfire is severe, the MIL will flash on/off on the 1st trip!**<br>**Possible Causes:**<br>• Air leak in the intake manifold, or in the EGR or ECM system<br>• Base engine mechanical problem<br>• Fuel delivery component problem (i.e., a contaminated, dirty or sticking fuel injector)<br>• Fuel pump relay defective<br>• Ignition coil fuses have failed<br>• Ignition system problem (dirty damaged coil or plug)<br>• Engine speed (RPM) sensor has failed<br>• Camshaft position sensors have failed<br>• Ignition coil is faulty<br>• Spark plugs are not working properly or are not gapped properly |
| **DTC: P0318**<br>**Years: 2007, 2008**<br>**Models:** Passat, Jetta, Golf, GTI, New Beetle<br>**Engines:** 2.0L<br>**Transmissions:** All | **Rough Road Sensor Conditions:**<br>Engine running, and the ECM detected an implausible signal from the rough road sensor.<br>**Possible Causes:**<br>• Wire connection between Engine Control Module (ECM) and ABS Control Module |

| DTC | Trouble Code Title, Conditions & Possible Causes |
|---|---|
| **DTC: P0321**<br>**2T, MIL:** Yes<br>**Years:** 2007, 2008<br>**Models:** Passat, Jetta, Golf, GTI, New Beetle, New Beetle Convertible<br>**Engines:** 2.0L, 2.5L, 3.6L<br>**Transmissions:** All | **Ignition/Distributor Engine Speed Input Circuit Range/Performance Conditions:**<br>Engine started, vehicle driven, and the ECM detected the engine speed signal was more than the calibrated value.<br>**Note: The engine will not start if there is no speed signal. If the speed signal fails when the engine is running, it will cause the engine to stall immediately.**<br>**Possible Causes:**<br>• Engine speed sensor has failed or is damaged<br>• ECM has failed<br>• Sensor wheel is damaged or doesn't fit properly<br>• Sensor wheel spacer isn't seated properly |
| **DTC: P0322**<br>**2T, MIL:** Yes<br>**Years:** 2007, 2008<br>**Models:** Passat, Jetta, Golf, GTI, New Beetle, New Beetle Convertible<br>**Engines:** 2.0L, 2.5L, 3.6L<br>**Transmissions:** All | **Ignition/Distributor Engine Input Circuit No Signal Conditions:**<br>Key on, and the ECM could not detect the engine speed signal or the signal was erratic.<br>**Note: The engine will not start if there is no speed signal. If the speed signal fails when the engine is running, it will cause the engine to stall immediately.**<br>**Possible Causes:**<br>• Engine speed sensor has failed or is damaged<br>• ECM has failed<br>• Sensor wheel is damaged or doesn't fit properly<br>• Sensor wheel spacer isn't seated properly |
| **DTC: P0324**<br>**Years:** 2007, 2008<br>**Models:** Passat, Jetta, Golf, GTI, New Beetle, New Beetle Convertible<br>**Engines:** 2.0L, 2.5L, 3.6L<br>**Transmissions:** All | **Knock Control System Error Conditions:**<br>Engine started, vehicle driven, and the ECM detected the Knock Sensor 1 (KS1) signal was too low or not recognized by the ECM<br>**Possible Causes:**<br>• Knock sensor circuit is open<br>• Knock sensor is loose (tighten to 20 NM)<br>• Contact between the knock sensor and cylinder block is dirty, corroded or greasy<br>• Knock sensor circuit is shorted to ground, or shorted to power<br>• Knock sensor is damaged or it has failed<br>• Wrong kind of fuel used<br>• A component in the engine compartment is loose or not properly secured<br>• ECM has failed |
| **DTC: P0327**<br>**2T, MIL:** Yes<br>**Years:** 2007, 2008<br>**Models:** Passat, Jetta, Golf, GTI, New Beetle, New Beetle Convertible<br>**Engines:** 2.0L, 2.5L, 3.6L<br>**Transmissions:** All | **Knock Sensor 1 Signal Low Input Conditions:**<br>Engine started, vehicle driven, and the ECM detected the Knock Sensor 1 (KS1) signal was too low or not recognized by the ECM<br>**Possible Causes:**<br>• Knock sensor circuit is open<br>• Knock sensor is loose (tighten to 20 NM)<br>• Contact between the knock sensor and cylinder block is dirty, corroded or greasy<br>• Knock sensor circuit is shorted to ground, or shorted to power<br>• Knock sensor is damaged or it has failed<br>• Wrong kind of fuel used<br>• A component in the engine compartment is loose or not properly secured<br>• ECM has failed |
| **DTC: P0328**<br>**2T, MIL:** Yes<br>**Years:** 2007, 2008<br>**Models:** Passat, Jetta, Golf, GTI, New Beetle, New Beetle Convertible<br>**Engines:** 2.0L, 2.5L, 3.6L<br>**Transmissions:** All | **Knock Sensor 1 Signal High Input Conditions:**<br>Engine started, vehicle driven, and the ECM detected the Knock Sensor 1 (KS1) signal was too high<br>**Possible Causes:**<br>• Knock sensor circuit is open<br>• Knock sensor is loose (tighten to 20 NM)<br>• Contact between the knock sensor and cylinder block is dirty, corroded or greasy<br>• Knock sensor circuit is shorted to ground, or shorted to power<br>• Knock sensor is damaged or it has failed<br>• Wrong kind of fuel used<br>• A component in the engine compartment is loose or not properly secured<br>• ECM has failed |

| DTC | Trouble Code Title, Conditions & Possible Causes |
|---|---|
| **DTC: P0332**<br>**2T, MIL: Yes**<br>**Years: 2007, 2008**<br>**Models:** Passat, Jetta, Golf, GTI, New Beetle, New Beetle Convertible<br>**Engines:** 2.0L, 2.5L, 3.6L<br>**Transmissions:** All | **Knock Sensor 2 Signal Low Input Conditions:**<br>Engine started, vehicle driven, and the ECM detected the Knock Sensor 1 (KS1) signal was too low or not recognized by the ECM<br>**Possible Causes:**<br>• Knock sensor circuit is open<br>• Knock sensor is loose (tighten to 20 NM)<br>• Contact between the knock sensor and cylinder block is dirty, corroded or greasy<br>• Knock sensor circuit is shorted to ground, or shorted to power<br>• Knock sensor is damaged or it has failed<br>• Wrong kind of fuel used<br>• A component in the engine compartment is loose or not properly secured<br>• ECM has failed |
| **DTC: P0333**<br>**2T, MIL: Yes**<br>**Years: 2007, 2008**<br>**Models:** Passat, Jetta, Golf, GTI, New Beetle, New Beetle Convertible<br>**Engines:** 2.0L, 2.5L, 3.6L<br>**Transmissions:** All | **Knock Sensor 2 Signal High Input Conditions:**<br>Engine started, vehicle driven, and the ECM detected the Knock Sensor 1 (KS1) signal was too high<br>**Possible Causes:**<br>• Knock sensor circuit is open<br>• Knock sensor is loose (tighten to 20 NM)<br>• Contact between the knock sensor and cylinder block is dirty, corroded or greasy<br>• Knock sensor circuit is shorted to ground, or shorted to power<br>• Knock sensor is damaged or it has failed<br>• Wrong kind of fuel used<br>• A component in the engine compartment is loose or not properly secured<br>• ECM has failed |
| **DTC: P0340**<br>**2T, MIL: Yes**<br>**Years: 2007, 2008**<br>**Models:** Passat, Jetta, Golf, GTI, New Beetle, New Beetle Convertible<br>**Engines:** 2.0L, 2.5L, 3.6L<br>**Transmissions:** All | **Camshaft Position Sensor Circuit Malfunction Conditions:**<br>Engine started, battery voltage must be at least 11.5v, all electrical components must be off, parking brake must be engaged (to keep daytime driving lights off), automatic transmission selector must be in park and the ground between the engine and the chassis must be well connected. The ECM detected the CMP sensor signal was missing or it was erratic.<br>**Possible Causes:**<br>• CMP sensor circuit is open or shorted to ground<br>• CMP sensor circuit is shorted to power<br>• CMP sensor ground (return) circuit is open<br>• CMP sensor installation incorrect (Hall-effect type)<br>• CMP sensor is damaged or CMP sensor shielding damaged<br>• CMP sensor has failed<br>• ECM has failed |
| **DTC: P0341**<br>**2T, MIL: Yes**<br>**Years: 2007, 2008**<br>**Models:** Passat, Jetta, Golf, GTI, New Beetle, New Beetle Convertible<br>**Engines:** 2.0L, 2.5L, 3.6L<br>**Transmissions:** All | **Camshaft Position Sensor Circ Range/Performance Conditions:**<br>Engine started, battery voltage must be at least 11.5v, all electrical components must be off, parking brake must be engaged (to keep daytime driving lights off), automatic transmission selector must be in park and the ground between the engine and the chassis must be well connected. The ECM detected the CMP sensor signal was implausible.<br>**Possible Causes:**<br>• CMP sensor circuit is open or shorted to ground<br>• CMP sensor circuit is shorted to power<br>• CMP sensor ground (return) circuit is open<br>• CMP sensor installation incorrect (Hall-effect type)<br>• CMP sensor is damaged or CMP sensor shielding damaged<br>• ECM has failed |
| **DTC: P0342**<br>**2T, MIL: Yes**<br>**Years: 2007, 2008**<br>**Models:** Passat, Jetta, Golf, GTI, New Beetle, New Beetle Convertible<br>**Engines:** 2.0L, 2.5L, 3.6L<br>**Transmissions:** All | **Camshaft Position Sensor "A" Circuit (Bank 1 or Single Sensor) Low Input Conditions:**<br>Engine started, battery voltage must be at least 11.5v, all electrical components must be off, parking brake must be engaged (to keep daytime driving lights off), automatic transmission selector must be in park and the ground between the engine and the chassis must be well connected. The ECM detected the CMP sensor signal exceeded the bounds of the specified maximum limit.<br>**Possible Causes:**<br>• CMP sensor circuit is open or shorted to ground<br>• CMP sensor circuit is shorted to power<br>• CMP sensor ground (return) circuit is open<br>• CMP sensor installation incorrect (Hall-effect type)<br>• CMP sensor is damaged or CMP sensor shielding damaged<br>• ECM has failed |

| DTC | Trouble Code Title, Conditions & Possible Causes |
|---|---|
| **DTC: P0343**<br>**2T, MIL: Yes**<br>**Years:** 2007, 2008<br>**Models:** Passat, Jetta, Golf, GTI, New Beetle, New Beetle Convertible<br>**Engines:** 2.0L, 2.5L, 3.6L<br>**Transmissions:** All | **Camshaft Position Sensor "A" Circuit (Bank 1 or Single Sensor) High Input Conditions:**<br>Engine started, battery voltage must be at least 11.5v, all electrical components must be off, parking brake must be engaged (to keep daytime driving lights off), automatic transmission selector must be in park and the ground between the engine and the chassis must be well connected. The ECM detected the CMP sensor signal did not reach the specified minimum limit.<br>**Possible Causes:**<br>• CMP sensor circuit is open or shorted to ground<br>• CMP sensor circuit is shorted to power<br>• CMP sensor ground (return) circuit is open<br>• CMP sensor installation incorrect (Hall-effect type)<br>• CMP sensor is damaged or CMP sensor shielding damaged<br>• ECM has failed |
| **DTC: P0345**<br>**Years:** 2007, 2008<br>**Models:** Passat, Jetta, GTI<br>**Engines:** 2.5L, 3.6L<br>**Transmissions:** All | **Camshaft Position Sensor "A" Circuit (Bank 2) Malfunction Conditions:**<br>Engine started, battery voltage must be at least 11.5v, all electrical components must be off, parking brake must be engaged (to keep daytime driving lights off), automatic transmission selector must be in park and the ground between the engine and the chassis must be well connected. The ECM detected the CMP sensor signal was missing or it was erratic.<br>**Possible Causes:**<br>• CMP sensor circuit is open or shorted to ground<br>• CMP sensor circuit is shorted to power<br>• CMP sensor ground (return) circuit is open<br>• CMP sensor installation incorrect (Hall-effect type)<br>• CMP sensor is damaged or CMP sensor shielding damaged<br>• ECM has failed |
| **DTC: P0346**<br>**Years:** 2007, 2008<br>**Models:** Passat, Jetta, GTI<br>**Engines:** 2.5L, 3.6L<br>**Transmissions:** All | **Camshaft Position Sensor "A" Circuit (Bank 2) Range/Performance Conditions:**<br>Engine started, battery voltage must be at least 11.5v, all electrical components must be off, parking brake must be engaged (to keep daytime driving lights off), automatic transmission selector must be in park and the ground between the engine and the chassis must be well connected. The ECM detected the CMP sensor signal was implausible.<br>**Possible Causes:**<br>• CMP sensor circuit is open or shorted to ground<br>• CMP sensor circuit is shorted to power<br>• CMP sensor ground (return) circuit is open<br>• CMP sensor installation incorrect (Hall-effect type)<br>• CMP sensor is damaged or CMP sensor shielding damaged<br>• ECM has failed |
| **DTC: P0347**<br>**Years:** 2007, 2008<br>**Models:** Passat, Jetta, GTI<br>**Engines:** 2.5L, 3.6L<br>**Transmissions:** All | **Camshaft Position Sensor "A" Circuit (Bank 2) Low Input Conditions:**<br>Engine started, battery voltage must be at least 11.5v, all electrical components must be off, parking brake must be engaged (to keep daytime driving lights off), automatic transmission selector must be in park and the ground between the engine and the chassis must be well connected. The ECM detected the CMP sensor signal exceeded the bounds of the specified maximum limit.<br>**Possible Causes:**<br>• CMP sensor circuit is open or shorted to ground<br>• CMP sensor circuit is shorted to power<br>• CMP sensor ground (return) circuit is open<br>• CMP sensor installation incorrect (Hall-effect type)<br>• CMP sensor is damaged or CMP sensor shielding damaged<br>• ECM has failed |
| **DTC: P0348**<br>**Years:** 2007, 2008<br>**Models:** Passat, Jetta, GTI<br>**Engines:** 2.5L, 3.6L<br>**Transmissions:** All | **Camshaft Position Sensor "A" Circuit "A" Circuit (Bank 2) High Input Conditions:**<br>Engine started, battery voltage must be at least 11.5v, all electrical components must be off, parking brake must be engaged (to keep daytime driving lights off), automatic transmission selector must be in park and the ground between the engine and the chassis must be well connected. The ECM detected the CMP sensor signal did not reach the specified minimum limit.<br>**Possible Causes:**<br>• CMP sensor circuit is open or shorted to ground<br>• CMP sensor circuit is shorted to power<br>• CMP sensor ground (return) circuit is open<br>• CMP sensor installation incorrect (Hall-effect type)<br>• CMP sensor is damaged or CMP sensor shielding damaged<br>• ECM has failed |

| DTC | Trouble Code Title, Conditions & Possible Causes |
|---|---|
| **DTC: P0351**<br>**2T, MIL: Yes**<br>**Years: 2007, 2008**<br>**Models:** Passat, Jetta, Golf, GTI, New Beetle, New Beetle Convertible<br>**Engines:** 2.0L, 2.5L, 3.6L<br>**Transmissions:** All | **Ignition Coilpack A Primary/Secondary Circuit Malfunction Conditions:**<br>Engine started, battery voltage must be at least 11.5v, all electrical components must be off, parking brake must be engaged (to keep daytime driving lights off), automatic transmission selector must be in park and the ground between the engine and the chassis must be well connected. The ECM did not receive any valid pulses from the ignition module for the Ignition Coilpack A primary circuit.<br>**Note: Ignition coils and power output stages are one component and cannot be replaced individually.**<br>**Possible Causes:**<br>• Engine speed (RPM) sensor has failed<br>• Camshaft Position (CMP) sensor has failed<br>• Power Supply Relay is shorted to an open circuit<br>• There is a malfunction in voltage supply<br>• Ignition coilpack is damaged or it has failed<br>• Cylinder 1 to 4 Fuel Injector(s) have failed<br>• ECM has failed |
| **DTC: P0352**<br>**2T, MIL: Yes**<br>**Years: 2007, 2008**<br>**Models:** Passat, Jetta, Golf, GTI, New Beetle, New Beetle Convertible<br>**Engines:** 2.0L, 2.5L, 3.6L<br>**Transmissions:** All | **Ignition Coilpack B Primary/Secondary Circuit Malfunction Conditions:**<br>Engine started, battery voltage must be at least 11.5v, all electrical components must be off, parking brake must be engaged (to keep daytime driving lights off), automatic transmission selector must be in park and the ground between the engine and the chassis must be well connected. The ECM did not receive any valid pulses from the ignition module for the Ignition Coilpack B primary circuit.<br>**Note: Ignition coils and power output stages are one component and cannot be replaced individually.**<br>**Possible Causes:**<br>• Engine speed (RPM) sensor has failed<br>• Camshaft Position (CMP) sensor has failed<br>• Power Supply Relay is shorted to an open circuit<br>• There is a malfunction in voltage supply<br>• Ignition coilpack is damaged or it has failed<br>• Cylinder 1 to 4 Fuel Injector(s) have failed<br>• ECM has failed |
| **DTC: P0353**<br>**2T, MIL: Yes**<br>**Years: 2007, 2008**<br>**Models:** Passat, Jetta, Golf, GTI, New Beetle, New Beetle Convertible<br>**Engines:** 2.0L, 2.5L, 3.6L<br>**Transmissions:** All | **Ignition Coilpack C Primary/Secondary Circuit Malfunction Conditions:**<br>Engine started, battery voltage must be at least 11.5v, all electrical components must be off, parking brake must be engaged (to keep daytime driving lights off), automatic transmission selector must be in park and the ground between the engine and the chassis must be well connected. The ECM did not receive any valid pulses from the ignition module for the Ignition Coilpack C primary circuit.<br>**Note: Ignition coils and power output stages are one component and cannot be replaced individually.**<br>**Possible Causes:**<br>• Engine speed (RPM) sensor has failed<br>• Camshaft Position (CMP) sensor has failed<br>• Power Supply Relay is shorted to an open circuit<br>• There is a malfunction in voltage supply<br>• Ignition coilpack is damaged or it has failed<br>• Cylinder 1 to 4 Fuel Injector(s) have failed<br>• ECM has failed |
| **DTC: P0354**<br>**2T, MIL: Yes**<br>**Years: 2007, 2008**<br>**Models:** Passat, Jetta, Golf, GTI, New Beetle, New Beetle Convertible<br>**Engines:** 2.0L, 2.5L, 3.6L<br>**Transmissions:** All | **Ignition Coilpack D Primary/Secondary Circuit Malfunction Conditions:**<br>Engine started, battery voltage must be at least 11.5v, all electrical components must be off, parking brake must be engaged (to keep daytime driving lights off), automatic transmission selector must be in park and the ground between the engine and the chassis must be well connected. The ECM did not receive any valid pulses from the ignition module for the Ignition Coilpack D primary circuit.<br>**Note: Ignition coils and power output stages are one component and cannot be replaced individually.**<br>**Possible Causes:**<br>• Engine speed (RPM) sensor has failed<br>• Camshaft Position (CMP) sensor has failed<br>• Power Supply Relay is shorted to an open circuit<br>• There is a malfunction in voltage supply<br>• Ignition coilpack is damaged or it has failed<br>• Cylinder 1 to 4 Fuel Injector(s) have failed<br>• ECM has failed |

| DTC | Trouble Code Title, Conditions & Possible Causes |
|---|---|
| **DTC: P0355**<br>**Years: 2007, 2008**<br>**Models**: Passat, Jetta, GTI<br>**Engines**: 2.5L, 3.6L<br>**Transmissions**: All | **Ignition Coilpack E Primary/Secondary Circuit Malfunction Conditions:**<br>Engine started, battery voltage must be at least 11.5v, all electrical components must be off, parking brake must be engaged (to keep daytime driving lights off), automatic transmission selector must be in park and the ground between the engine and the chassis must be well connected. The ECM did not receive any valid pulses from the ignition module for the Ignition Coilpack E primary circuit.<br>**Note: Ignition coils and power output stages are one component and cannot be replaced individually.**<br>**Possible Causes:**<br>• Engine speed (RPM) sensor has failed<br>• Camshaft Position (CMP) sensor has failed<br>• Power Supply Relay is shorted to an open circuit<br>• There is a malfunction in voltage supply<br>• Ignition coilpack is damaged or it has failed<br>• Cylinder 1 to 4 Fuel Injector(s) have failed<br>• ECM has failed |
| **DTC: P0356**<br>**Years: 2007, 2008**<br>**Models**: Passat, Jetta, GTI<br>**Engines**: 2.5L, 3.6L<br>**Transmissions**: All | **Ignition Coilpack F Primary/Secondary Circuit Malfunction Conditions:**<br>Engine started, battery voltage must be at least 11.5v, all electrical components must be off, parking brake must be engaged (to keep daytime driving lights off), automatic transmission selector must be in park and the ground between the engine and the chassis must be well connected. The ECM did not receive any valid pulses from the ignition module for the Ignition Coilpack F primary circuit.<br>**Note: Ignition coils and power output stages are one component and cannot be replaced individually.**<br>**Possible Causes:**<br>• Engine speed (RPM) sensor has failed<br>• Camshaft Position (CMP) sensor has failed<br>• Power Supply Relay is shorted to an open circuit<br>• There is a malfunction in voltage supply<br>• Ignition coilpack is damaged or it has failed<br>• Cylinder 1 to 4 Fuel Injector(s) have failed<br>• ECM has failed |
| **DTC: P0411**<br>**2T, MIL: Yes**<br>**Years: 2007, 2008**<br>**Models**: Passat, Jetta, Golf, GTI, New Beetle, New Beetle Convertible<br>**Engines**: 2.0L, 2.5L, 3.6L<br>**Transmissions**: All | **Secondary Air Injection System Upstream Flow Detected Conditions:**<br>Engine started, battery voltage must be at least 11.5v, all electrical components must be off, parking brake must be engaged (to keep daytime driving lights off), automatic transmission selector must be in park and the ground between the engine and the chassis must be well connected. The ECM detected the Secondary AIR pump airflow was not diverted correctly when requested during the self-test. The pump is functioning but the quantity of air is recognized as insufficient by HO2S.<br>**Note: The solenoid valve is closed when no voltage is present.**<br>**Possible Causes:**<br>• Air pump output is blocked or restricted<br>• AIR bypass solenoid is leaking or it is restricted<br>• AIR bypass solenoid is stuck open or stuck closed<br>• Check valve (one or more) is damaged or leaking<br>• Electric air injection pump hose(s) leaking<br>• Electric air injection pump is damaged or faulty<br>• ECM has failed |
| **DTC: P0412**<br>**Years: 2007, 2008**<br>**Models**: Passat, Jetta, Golf, GTI, New Beetle, New Beetle Convertible<br>**Engines**: 2.0L, 2.5L, 3.6L<br>**Transmissions**: All | **Secondary Air Injection Solenoid Circuit Malfunction Conditions:**<br>Engine started, battery voltage must be at least 11.5v, all electrical components must be off, parking brake must be engaged (to keep daytime driving lights off), automatic transmission selector must be in park and the ground between the engine and the chassis must be well connected. The ECM detected an unexpected low or high voltage condition on the AIR solenoid control circuit during testing.<br>**Possible Causes:**<br>• AIR solenoid power circuit (B+) is open (check dedicated fuse)<br>• AIR bypass solenoid control circuit is open or shorted to ground<br>• AIR diverter solenoid control circuit open or shorted to ground<br>• AIR pump control circuit is open or shorted to ground<br>• Check valve (one or more) is damaged or leaking<br>• Solid State relay is damaged or it has failed<br>• ECM has failed |

| DTC | Trouble Code Title, Conditions & Possible Causes |
|---|---|
| **DTC: P0413**<br>**Years: 2007, 2008**<br>**Models:** Passat, Jetta, Golf, GTI, New Beetle, New Beetle Convertible<br>**Engines:** 2.0L, 2.5L, 3.6L<br>**Transmissions:** All | **Secondary Air Injection Solenoid Circuit Open Conditions:**<br>Engine started, battery voltage must be at least 11.5v, all electrical components must be off, parking brake must be engaged (to keep daytime driving lights off), automatic transmission selector must be in park and the ground between the engine and the chassis must be well connected. The ECM detected an unexpected low or high voltage condition on the AIR solenoid control circuit during testing.<br>**Possible Causes:**<br>• AIR solenoid power circuit (B+) is open (check dedicated fuse)<br>• AIR bypass solenoid control circuit is open or shorted to ground<br>• AIR diverter solenoid control circuit open or shorted to ground<br>• AIR pump control circuit is open or shorted to ground<br>• Check valve (one or more) is damaged or leaking<br>• Solid State relay is damaged or it has failed<br>• ECM has failed |
| **DTC: P0414**<br>**Years: 2007, 2008**<br>**Models:** Passat, Jetta, Golf, GTI, New Beetle, New Beetle Convertible<br>**Engines:** 2.0L, 2.5L, 3.6L<br>**Transmissions:** All | **Secondary Air Injection Solenoid Circuit Short Conditions:**<br>Engine started, battery voltage must be at least 11.5v, all electrical components must be off, parking brake must be engaged (to keep daytime driving lights off), automatic transmission selector must be in park and the ground between the engine and the chassis must be well connected. The ECM detected an unexpected low or high voltage condition on the AIR solenoid control circuit during testing.<br>**Possible Causes:**<br>• AIR solenoid power circuit (B+) is open (check dedicated fuse)<br>• AIR bypass solenoid control circuit is open or shorted to ground<br>• AIR diverter solenoid control circuit open or shorted to ground<br>• AIR pump control circuit is open or shorted to ground<br>• Check valve (one or more) is damaged or leaking<br>• Solid State relay is damaged or it has failed<br>• ECM has failed |
| **DTC: P0418**<br>**2T, MIL: Yes**<br>**Years: 2007, 2008**<br>**Models:** Passat, Jetta, Golf, GTI, New Beetle, New Beetle Convertible<br>**Engines:** 2.0L, 2.5L, 3.6L<br>**Transmissions:** All | **Secondary Air Injection Relay (A) Circuit Malfunction Conditions:**<br>Engine started, battery voltage must be at least 11.5v, all electrical components must be off, parking brake must be engaged (to keep daytime driving lights off), automatic transmission selector must be in park and the ground between the engine and the chassis must be well connected. The ECM detected an unexpected low or high voltage condition on the AIR solenoid control circuit during testing.<br>**Possible Causes:**<br>• AIR solenoid power circuit (B+) is open (check dedicated fuse)<br>• AIR bypass solenoid control circuit is open or shorted to ground<br>• AIR diverter solenoid control circuit open or shorted to ground<br>• AIR pump control circuit is open or shorted to ground<br>• Check valve (one or more) is damaged or leaking<br>• Solid State relay is damaged or it has failed<br>• ECM has failed |
| **DTC: P0420**<br>**2T, MIL: Yes**<br>**Years: 2007, 2008**<br>**Models:** Passat, Jetta, Golf, GTI, New Beetle, New Beetle Convertible<br>**Engines:** 2.0L, 2.5L, 3.6L<br>**Transmissions:** All | **Catalyst System Efficiency (Bank 1) Below Threshold Conditions:**<br>Engine started, battery voltage must be at least 11.5v, all electrical components must be off, parking brake must be engaged (to keep daytime driving lights off), automatic transmission selector must be in park, the exhaust system must be properly sealed between the catalytic converter and the cylinder head, coolant temperature must be at least 80 degrees Celsius and oxygen sensor heaters for oxygen sensors before the catalytic converter must be functioning properly and the ground between the engine and the chassis must be well connected. The ECM detected the switch rate of the rear HO2S-12 was close to the switch rate of front HO2S (it should be much slower).<br>**Possible Causes:**<br>• Air leaks at the exhaust manifold or in the exhaust pipes<br>• Catalytic converter is damaged, contaminated or it has failed<br>• ECT/CHT sensor has lost its calibration (the signal is incorrect)<br>• Engine cylinders misfiring, or the ignition timing is over retarded<br>• Engine oil is contaminated<br>• Front HO2S or rear HO2S is contaminated with fuel or moisture<br>• Front HO2S and/or the rear HO2S is loose in the mounting hole<br>• Front HO2S much older than the rear HO2S (HO2S-11 is lazy)<br>• Fuel system pressure is too high (check the pressure regulator)<br>• Rear HO2S wires improperly connected or the HO2S has failed |

| DTC | Trouble Code Title, Conditions & Possible Causes |
|---|---|
| **DTC: P0421**<br>**Years: 2007, 2008**<br>**Models:** Passat<br>**Engines:** 2.5L, 3.6L<br>**Transmissions:** All | **Warm Up Catalyst System Efficiency (Bank 1) Below Threshold Conditions:**<br>Engine started, battery voltage must be at least 11.5v, all electrical components must be off, parking brake must be engaged (to keep daytime driving lights off), automatic transmission selector must be in park, the exhaust system must be properly sealed between the catalytic converter and the cylinder head, coolant temperature must be at least 80 degrees Celsius and oxygen sensor heaters for oxygen sensors before the catalytic converter must be functioning properly and the ground between the engine and the chassis must be well connected. The ECM detected the switch rate of the rear HO2S-12 was close to the switch rate of front HO2S (it should be much slower).<br>**Possible Causes:**<br>• Air leaks at the exhaust manifold or in the exhaust pipes<br>• Catalytic converter is damaged, contaminated or it has failed<br>• ECT/CHT sensor has lost its calibration (the signal is incorrect)<br>• Engine cylinders misfiring, or the ignition timing is over retarded<br>• Engine oil is contaminated<br>• Front HO2S or rear HO2S is contaminated with fuel or moisture<br>• Front HO2S and/or the rear HO2S is loose in the mounting hole<br>• Front HO2S much older than the rear HO2S (HO2S-11 is lazy)<br>• Fuel system pressure is too high (check the pressure regulator)<br>• Rear HO2S wires improperly connected or the HO2S has failed |
| **DTC: P0422**<br>**2T, MIL: Yes**<br>**Years: 2007, 2008**<br>**Models:** Passat, Jetta, Golf, GTI, New Beetle, New Beetle Convertible<br>**Engines:** 2.0L, 2.5L, 3.6L<br>**Transmissions:** All | **Main Catalyst (Bank 1) Efficiency Below Threshold Conditions:**<br>Engine started, battery voltage must be at least 11.5v, all electrical components must be off, parking brake must be engaged (to keep daytime driving lights off), automatic transmission selector must be in park, the exhaust system must be properly sealed between the catalytic converter and the cylinder head, coolant temperature must be at least 80 degrees Celsius and oxygen sensor heaters for oxygen sensors before the catalytic converter must be functioning properly and the ground between the engine and the chassis must be well connected. The ECM detected the switch rate of the rear HO2S-12 was close to the switch rate of front HO2S (it should be much slower).<br>**Possible Causes:**<br>• Air leaks at the exhaust manifold or in the exhaust pipes<br>• Catalytic converter is damaged, contaminated or it has failed<br>• ECT/CHT sensor has lost its calibration (the signal is incorrect)<br>• Engine cylinders misfiring, or the ignition timing is over retarded<br>• Engine oil is contaminated<br>• Front HO2S or rear HO2S is contaminated with fuel or moisture<br>• Front HO2S and/or the rear HO2S is loose in the mounting hole<br>• Front HO2S much older than the rear HO2S<br>• Fuel system pressure is too high (check the pressure regulator)<br>• Rear HO2S wires improperly connected or the HO2S has failed |
| **DTC: P0430**<br>**Years: 2007, 2008**<br>**Models:** Passat, GTI<br>**Engines:** 2.5L, 3.6L<br>**Transmissions:** All | **Catalyst System Efficiency (Bank 2) Below Threshold Conditions:**<br>Engine started, battery voltage must be at least 11.5v, all electrical components must be off, parking brake must be engaged (to keep daytime driving lights off), automatic transmission selector must be in park, the exhaust system must be properly sealed between the catalytic converter and the cylinder head, coolant temperature must be at least 80 degrees Celsius and oxygen sensor heaters for oxygen sensors before the catalytic converter must be functioning properly and the ground between the engine and the chassis must be well connected. The ECM detected the switch rate of the rear HO2S-12 was close to the switch rate of front HO2S (it should be much slower).<br>**Possible Causes:**<br>• Air leaks at the exhaust manifold or in the exhaust pipes<br>• Catalytic converter is damaged, contaminated or it has failed<br>• ECT/CHT sensor has lost its calibration (the signal is incorrect)<br>• Engine cylinders misfiring, or the ignition timing is over retarded<br>• Engine oil is contaminated<br>• Front HO2S or rear HO2S is contaminated with fuel or moisture<br>• Front HO2S and/or the rear HO2S is loose in the mounting hole<br>• Front HO2S much older than the rear HO2S (HO2S-11 is lazy)<br>• Fuel system pressure is too high (check the pressure regulator)<br>• Rear HO2S wires improperly connected or the HO2S has failed |

| DTC | Trouble Code Title, Conditions & Possible Causes |
|---|---|
| **DTC: P0440**<br>**Years:** 2007, 2008<br>**Models:** Passat, Jetta, Golf, GTI, New Beetle, New Beetle Convertible<br>**Engines:** 2.0L, 2.5L, 3.6L<br>**Transmissions:** All | **EVAP System Malfunction Conditions:**<br>ECT sensor is cold during startup, engine started, battery voltage must be at least 11.5v, all electrical components must be off, parking brake must be engaged (to keep daytime driving lights off), automatic transmission selector must be in park, the exhaust system must be properly sealed between the catalytic converter and the cylinder head, coolant temperature must be at least 80 degrees Celsius and oxygen sensor heaters for oxygen sensors before the catalytic converter must be functioning properly and the ground between the engine and the chassis must be well connected. The ECM detected the switch rate of the rear HO2S-12 was close to the switch rate of front HO2S (it should be much slower).<br>ECM detected a problem in the EVAP system during the EVAP System Monitor test.<br>**Possible Causes:**<br>• EVAP canister purge valve is damaged<br>• EVAP canister has an improper seal<br>• Vapor line between purge solenoid and intake manifold vacuum reservoir is damaged, or vapor line between EVAP canister purge solenoid and charcoal canister is damaged<br>• Vapor line between charcoal canister and check valve, or vapor line between check valve and fuel vapor valves is damaged<br>• ECM has failed |
| **DTC: P0441**<br>**2T, MIL: Yes**<br>**Years:** 2007, 2008<br>**Models:** Passat, Jetta, Golf, GTI, New Beetle, New Beetle Convertible<br>**Engines:** 2.0L, 2.5L, 3.6L<br>**Transmissions:** All | **EVAP Control System Incorrect Purge Flow Conditions:**<br>ECT sensor is cold during startup, engine started, battery voltage must be at least 11.5v, all electrical components must be off, parking brake must be engaged (to keep daytime driving lights off), automatic transmission selector must be in park, the exhaust system must be properly sealed between the catalytic converter and the cylinder head, coolant temperature must be at least 80 degrees Celsius and oxygen sensor heaters for oxygen sensors before the catalytic converter must be functioning properly and the ground between the engine and the chassis must be well connected. The ECM detected the switch rate of the rear HO2S-12 was close to the switch rate of front HO2S (it should be much slower).<br>ECM detected a problem in the EVAP system during the EVAP System Monitor test.<br>**Possible Causes:**<br>• EVAP canister purge valve is damaged<br>• EVAP canister has an improper seal<br>• Vapor line between purge solenoid and intake manifold vacuum reservoir is damaged, or vapor line between EVAP canister purge solenoid and charcoal canister is damaged<br>• Vapor line between charcoal canister and check valve, or vapor line between check valve and fuel vapor valves is damaged<br>• ECM has failed |
| **DTC: P0442**<br>**2T, MIL: Yes**<br>**Years:** 2007, 2008<br>**Models:** Passat, Jetta, Golf, GTI, New Beetle, New Beetle Convertible<br>**Engines:** 2.0L, 2.5L, 3.6L<br>**Transmissions:** All | **EVAP Control System Small Leak Detected Conditions:**<br>Engine started, battery voltage must be at least 11.5v, all electrical components must be off, parking brake must be engaged (to keep daytime driving lights off), automatic transmission selector must be in park, the exhaust system must be properly sealed between the catalytic converter and the cylinder head, coolant temperature must be at least 80 degrees Celsius and oxygen sensor heaters for oxygen sensors before the catalytic converter must be functioning properly and the ground between the engine and the chassis must be well connected. The ECM detected a leak in the EVAP system as small as 0.040" during the EVAP Monitor Test.<br>**Possible Causes:**<br>• Aftermarket EVAP parts that do not conform to specifications<br>• CV solenoid remains partially open when commanded to close<br>• EVAP component seals leaking (i.e., leaks in the Purge valve, fuel tank pressure sensor, canister vent solenoid, fuel vapor control valve tube assembly or fuel vapor vent valve).<br>• Fuel filler cap damaged, cross-threaded or loosely installed<br>• Loose fuel vapor hose/tube connections to EVAP components<br>• Small holes or cuts in fuel vapor hoses or EVAP canister tubes |
| **DTC: P0443**<br>**Years:** 2007, 2008<br>**Models:** Passat, Jetta, Golf, GTI, New Beetle,<br>**Engines:** 2.0L<br>**Transmissions:** All | **EVAP Vapor Management Valve Circuit Malfunction Conditions:**<br>Engine started, battery voltage must be at least 11.5v, all electrical components must be off, parking brake must be engaged (to keep daytime driving lights off), automatic transmission selector must be in park, the exhaust system must be properly sealed between the catalytic converter and the cylinder head, coolant temperature must be at least 80 degrees Celsius and oxygen sensor heaters for oxygen sensors before the catalytic converter must be functioning properly and the ground between the engine and the chassis must be well connected. The ECM detected an unexpected high or low voltage condition on the Vapor Management Valve (VMV) circuit when the device was cycled On/Off during testing.<br>**Possible Causes:**<br>• EVAP power supply circuit is open<br>• EVAP solenoid control circuit is open or shorted to ground<br>• EVAP solenoid control circuit is shorted to power (B+)<br>• EVAP solenoid valve is damaged or it has failed<br>• ECM has failed |

| DTC | Trouble Code Title, Conditions & Possible Causes |
|---|---|
| **DTC: P0444**<br>**2T, MIL: Yes**<br>**Years: 2007, 2008**<br>**Models:** Passat, Jetta, Golf, GTI, New Beetle, New Beetle Convertible<br>**Engines:** 2.0L, 2.5L, 3.6L<br>**Transmissions:** All | **Evaporative Emission System Purge Control Valve Circuit Open Conditions:**<br>Engine started, battery voltage must be at least 11.5v, all electrical components must be off, parking brake must be engaged (to keep daytime driving lights off), automatic transmission selector must be in park, the exhaust system must be properly sealed between the catalytic converter and the cylinder head, coolant temperature must be at least 80 degrees Celsius and oxygen sensor heaters for oxygen sensors before the catalytic converter must be functioning properly and the ground between the engine and the chassis must be well connected. The ECM detected an unexpected voltage condition on the EVAP circuit when the device was cycled On/Off during testing.<br>**Possible Causes:**<br>• EVAP power supply circuit is open<br>• EVAP solenoid control circuit is open or shorted to ground<br>• EVAP solenoid control circuit is shorted to power (B+)<br>• EVAP solenoid valve is damaged or it has failed<br>• EVAP canister has a leak or a poor seal<br>• ECM has failed |
| **DTC: P0445**<br>**2T, MIL: Yes**<br>**Years: 2007, 2008**<br>**Models:** Passat, Jetta, Golf, GTI, New Beetle, New Beetle Convertible<br>**Engines:** 2.0L, 2.5L, 3.6L<br>**Transmissions:** All | **Evaporative Emission System Purge Control Valve Circuit Shorted Conditions:**<br>Engine started, battery voltage must be at least 11.5v, all electrical components must be off, parking brake must be engaged (to keep daytime driving lights off), automatic transmission selector must be in park, the exhaust system must be properly sealed between the catalytic converter and the cylinder head, coolant temperature must be at least 80 degrees Celsius and oxygen sensor heaters for oxygen sensors before the catalytic converter must be functioning properly and the ground between the engine and the chassis must be well connected. The ECM detected an unexpected voltage condition on the EVAP circuit when the device was cycled On/Off during testing.<br>**Possible Causes:**<br>• EVAP power supply circuit is open<br>• EVAP solenoid control circuit is open or shorted to ground<br>• EVAP solenoid control circuit is shorted to power (B+)<br>• EVAP solenoid valve is damaged or it has failed<br>• EVAP canister has a leak or a poor seal<br>• ECM has failed |
| **DTC: P0455**<br>**2T, MIL: Yes**<br>**Years: 2007, 2008**<br>**Models:** Passat, Jetta, Golf, GTI, New Beetle, New Beetle Convertible<br>**Engines:** 2.0L, 2.5L, 3.6L<br>**Transmissions:** All | **EVAP Control System Large Leak Detected Conditions:**<br>Engine started, battery voltage must be at least 11.5v, all electrical components must be off, parking brake must be engaged (to keep daytime driving lights off), automatic transmission selector must be in park, the exhaust system must be properly sealed between the catalytic converter and the cylinder head, coolant temperature must be at least 80 degrees Celsius and oxygen sensor heaters for oxygen sensors before the catalytic converter must be functioning properly and the ground between the engine and the chassis must be well connected. The ECM detected multiple small fuel vapor leaks; or it detected a large leak in the system during the leak test.<br>**Possible Causes:**<br>• Aftermarket EVAP hardware non-conforming to specifications<br>• EVAP canister tube, EVAP canister purge outlet tube or EVAP return tube disconnected or cracked, or canister is damaged<br>• EVAP canister purge valve stuck closed, or canister damaged<br>• Fuel filler cap missing, loose (not tightened) or the wrong part<br>• Loose fuel vapor hose/tube connections to EVAP components<br>• Canister vent (CV) solenoid stuck open<br>• Fuel tank pressure (FTP) sensor has failed mechanically |
| **DTC: P0456**<br>**2T, MIL: Yes**<br>**Years: 2007, 2008**<br>**Models:** Passat, Jetta, Golf, GTI, New Beetle, New Beetle Convertible<br>**Engines:** 2.0L, 2.5L, 3.6L<br>**Transmissions:** All | **EVAP Control System Small Leak Detected Conditions:**<br>Engine started, battery voltage must be at least 11.5v, all electrical components must be off, parking brake must be engaged (to keep daytime driving lights off), automatic transmission selector must be in park, the exhaust system must be properly sealed between the catalytic converter and the cylinder head, coolant temperature must be at least 80 degrees Celsius and oxygen sensor heaters for oxygen sensors before the catalytic converter must be functioning properly and the ground between the engine and the chassis must be well connected. The ECM detected multiple small fuel vapor leaks; or it detected a large leak in the system during the leak test.<br>**Possible Causes:**<br>• Aftermarket EVAP hardware non-conforming to specifications<br>• EVAP canister tube, EVAP canister purge outlet tube or EVAP return tube disconnected or cracked, or canister is damaged<br>• EVAP canister purge valve stuck closed, or canister damaged<br>• Fuel filler cap missing, loose (not tightened) or the wrong part<br>• Loose fuel vapor hose/tube connections to EVAP components<br>• Canister vent (CV) solenoid stuck open<br>• Fuel tank pressure (FTP) sensor has failed mechanically |

| DTC | Trouble Code Title, Conditions & Possible Causes |
|---|---|
| **DTC: P0458**<br>**Years:** 2007, 2008<br>**Models:** Passat, Jetta, Golf, GTI, New Beetle,<br>**Engines:** 2.0L , 2.5L, 3.6L<br>**Transmissions:** All | **Evaporative Emission System Purge Control Valve Circuit Low Conditions:**<br>Engine started, battery voltage must be at least 11.5v, all electrical components must be off, parking brake must be engaged (to keep daytime driving lights off), automatic transmission selector must be in park, the exhaust system must be properly sealed between the catalytic converter and the cylinder head, coolant temperature must be at least 80 degrees Celsius and oxygen sensor heaters for oxygen sensors before the catalytic converter must be functioning properly and the ground between the engine and the chassis must be well connected. The ECM detected an unexpected voltage condition on the EVAP circuit when the device was cycled On/Off during testing.<br>**Possible Causes:**<br>• EVAP power supply circuit is open<br>• EVAP solenoid control circuit is open or shorted to ground<br>• EVAP solenoid control circuit is shorted to power (B+)<br>• EVAP solenoid valve is damaged or it has failed<br>• EVAP canister has a leak or a poor seal<br>• ECM has failed |
| **DTC: P0459**<br>**Years:** 2007, 2008<br>**Models:** Passat, Jetta, Golf, GTI, New Beetle,<br>**Engines:** 2.0L , 2.5L, 3.6L<br>**Transmissions:** All | **Evaporative Emission System Purge Control Valve Circuit High Conditions:**<br>Engine started, battery voltage must be at least 11.5v, all electrical components must be off, parking brake must be engaged (to keep daytime driving lights off), automatic transmission selector must be in park, the exhaust system must be properly sealed between the catalytic converter and the cylinder head, coolant temperature must be at least 80 degrees Celsius and oxygen sensor heaters for oxygen sensors before the catalytic converter must be functioning properly and the ground between the engine and the chassis must be well connected. The ECM detected an unexpected voltage condition on the EVAP circuit when the device was cycled On/Off during testing.<br>**Possible Causes:**<br>• EVAP power supply circuit is open<br>• EVAP solenoid control circuit is open or shorted to ground<br>• EVAP solenoid control circuit is shorted to power (B+)<br>• EVAP solenoid valve is damaged or it has failed<br>• EVAP canister has a leak or a poor seal<br>• ECM has failed |
| **DTC: P0480**<br>**Years:** 2007, 2008<br>**Models:** Jetta<br>**Engines:** 2.0L, 2.5L<br>**Transmissions:** All | **Fan 1 Control Circuit Conditions:**<br>Engine running, battery voltage at least 11.5v, all electrical components off, and the ECM detected an problem with the fan control circuit<br>**Possible Causes:**<br>• Check connection of coolant fan control module according to wiring diagram<br>• Short to B+ or ground in the circuit<br>• Circuit open |
| **DTC: P0491**<br>**Years:** 2007, 2008<br>**Models:** Passat, Jetta, GTI<br>**Engines:** 2.5L, 3.6L<br>**Transmissions:** All | **Secondary Air Injection System Insufficient Flow (Bank 1) Conditions:**<br>Engine started, battery voltage must be at least 11.5v, all electrical components must be off, parking brake must be engaged (to keep daytime driving lights off), automatic transmission selector must be in park and the ground between the engine and the chassis must be well connected. The ECM detected the Secondary AIR pump airflow was not diverted correctly when requested during the self-test. The pump is functioning but the quantity of air is recognized as insufficient by HO2S<br>**Possible Causes:**<br>• Air pump output is blocked or restricted<br>• AIR bypass solenoid is leaking or it is restricted<br>• AIR bypass solenoid is stuck open or stuck closed<br>• Check valve (one or more) is damaged or leaking<br>• Electric air injection pump hose(s) leaking<br>• Electric air injection pump is damaged or faulty<br>• ECM has failed |
| **DTC: P0492**<br>**Years:** 2007, 2008<br>**Models:** Passat, Jetta, GTI<br>**Engines:** 2.5L, 3.6L<br>**Transmissions:** All | **Secondary Air Injection System Insufficient Flow (Bank 2) Conditions:**<br>Engine started, battery voltage must be at least 11.5v, all electrical components must be off, parking brake must be engaged (to keep daytime driving lights off), automatic transmission selector must be in park and the ground between the engine and the chassis must be well connected. The ECM detected the Secondary AIR pump airflow was not diverted correctly when requested during the self-test. The pump is functioning but the quantity of air is recognized as insufficient by HO2S<br>**Possible Causes:**<br>• Air pump output is blocked or restricted<br>• AIR bypass solenoid is leaking or it is restricted<br>• AIR bypass solenoid is stuck open or stuck closed<br>• Check valve (one or more) is damaged or leaking<br>• Electric air injection pump hose(s) leaking<br>• Electric air injection pump is damaged or faulty<br>• ECM has failed |

| DTC | Trouble Code Title, Conditions & Possible Causes |
|---|---|
| **DTC: P0501**<br>**2T, MIL: Yes**<br>**Years:** 2007, 2008<br>**Models:** Passat, Jetta, Golf, GTI, New Beetle, New Beetle Convertible<br>**Engines:** 2.0L, 2.5L, 3.6L<br>**Transmissions:** All | **Vehicle Speed Sensor or PSOM Range/Performance Conditions:**<br>Engine started; engine speed above the TCC stall speed, and the ECM detected a loss of the VSS signal over a period of time or the signal is not usable.<br>**Note: The ECM receives vehicle speed data from the VSS, TCSS, ABS module, CTM or GEM controller, depending up the application.**<br>**Possible Causes:**<br>• VSS signal circuit is open or shorted to ground<br>• VSS harness circuit is shorted to ground<br>• VSS harness circuit is shorted to power<br>• VSS circuit open between the ECM and related control module<br>• VSS or wheel speed sensors circuits are damaged<br>• Modules connected to VSC/VSS harness circuits are damaged<br>• Mechanical drive mechanism for the VSS is damaged |
| **DTC: P0506**<br>**2T, MIL: Yes**<br>**Years:** 2007, 2008<br>**Models:** Passat, Jetta, Golf, GTI, New Beetle, New Beetle Convertible<br>**Engines:** 2.0L, 2.5L, 3.6L<br>**Transmissions:** All | **Idle Air Control System RPM Lower Than Expected Conditions:**<br>Engine started, battery voltage must be at least 11.5v, all electrical components must be off, parking brake must be engaged (to keep daytime driving lights off), automatic transmission selector must be in park, the exhaust system must be properly sealed between the catalytic converter and the cylinder head, coolant temperature must be at least 80 degrees Celsius and oxygen sensor heaters for oxygen sensors before the catalytic converter must be functioning properly and the ground between the engine and the chassis must be well connected. The ECM detected it could not control the idle speed correctly, as it is constantly more than 100 RPM less than specification.<br>**Possible Causes:**<br>• Air inlet is plugged or the air filter element is severely clogged<br>• IAC circuit is open or shorted<br>• IAC circuit VPWR circuit is open<br>• IAC solenoid is damaged or has failed<br>• ECM has failed<br>• The VSS has failed |
| **DTC: P0507**<br>**2T, MIL: Yes**<br>**Years:** 2007, 2008<br>**Models:** Passat, Jetta, Golf, GTI, New Beetle, New Beetle Convertible<br>**Engines:** 2.0L, 2.5L, 3.6L<br>**Transmissions:** All | **Idle Air Control System RPM Higher Than Expected Conditions:**<br>Engine started, battery voltage must be at least 11.5v, all electrical components must be off, parking brake must be engaged (to keep daytime driving lights off), automatic transmission selector must be in park, the exhaust system must be properly sealed between the catalytic converter and the cylinder head, coolant temperature must be at least 80 degrees Celsius and oxygen sensor heaters for oxygen sensors before the catalytic converter must be functioning properly and the ground between the engine and the chassis must be well connected. The ECM detected it could not control the idle speed correctly, as it is constantly more than 200 RPM more than specification.<br>**Possible Causes:**<br>• Air intake leak located somewhere after the throttle body<br>• IAC control circuit is shorted to chassis ground<br>• IAC solenoid is damaged or has failed<br>• Throttle Valve Control module has failed or is clogged with carbon<br>• ECM has failed<br>• The VSS has failed |
| **DTC: P0510**<br>**Years:** 2007, 2008<br>**Models:** Passat, Jetta, Golf, GTI, New Beetle,<br>**Engines:** 2.0L, 2.5L, 3.6L<br>**Transmissions:** All | **Closed Throttle Position Switch Malfunction Conditions:**<br>The Engine Control Module is not receiving a usable signal from the Closed Throttle Position switch.<br>**Possible Causes:**<br>• Throttle Valve Control module has failed<br>• Throttle Valve Control module is shorted to ground<br>• Throttle Valve Control module is open<br>• ECM has failed |
| **DTC: P0560**<br>**Years:** 2007, 2008<br>**Models:** Passat, Jetta, Golf, GTI, New Beetle<br>**Engines:** 2.0L, 2.5L, 3.6L<br>**Transmissions:** All | **System Voltage Malfunction Conditions:**<br>Engine started, battery voltage must be at least 11.5v, all electrical components must be off, parking brake must be engaged (to keep daytime driving lights off), automatic transmission selector must be in park, and the ground between the engine and the chassis must be well connected. The ECM has detected a voltage value that is implausible or erratic.<br>**Possible Causes:**<br>• Alternator damaged or faulty<br>• Battery voltage low or insufficient<br>• Fuses blown or circuits open<br>• Battery connection to terminal not clean<br>• Voltage regulator has failed |

| DTC | Trouble Code Title, Conditions & Possible Causes |
|---|---|
| **DTC: P0562**<br>**Years: 2007, 2008**<br>**Models:** Passat, Jetta, Golf, GTI, New Beetle,<br>**Engines:** 2.0L, 2.5L, 3.6L<br>**Transmissions:** All | **System Voltage Low Conditions:**<br>Engine started, battery voltage must be at least 11.5v, all electrical components must be off, parking brake must be engaged (to keep daytime driving lights off), automatic transmission selector must be in park, and the ground between the engine and the chassis must be well connected. The ECM has detected a voltage value that is below the specified minimum limit for the system to function properly.<br>**Possible Causes:**<br>• Alternator damaged or faulty<br>• Battery voltage low or insufficient<br>• Fuses blown or circuits open<br>• Battery connection to terminal not clean<br>• Voltage regulator has failed |
| **DTC: P0563**<br>**Years: 2007, 2008**<br>**Models:** Passat, Jetta, Golf, GTI, New Beetle,<br>**Engines:** 2.0L, 2.5L, 3.6L<br>**Transmissions:** All | **System Voltage High Conditions:**<br>Engine started, battery voltage must be at least 11.5v, all electrical components must be off, parking brake must be engaged (to keep daytime driving lights off), automatic transmission selector must be in park, and the ground between the engine and the chassis must be well connected. The ECM has detected a voltage value that has exceeded the specified maximum limit for the system to function properly.<br>**Possible Causes:**<br>• Alternator damaged or faulty<br>• Battery voltage low or insufficient<br>• Fuses blown or circuits open<br>• Battery connection to terminal not clean<br>• Voltage regulator has failed |
| **DTC: P0568**<br>**Years: 2007, 2008**<br>**Models:** Passat, Jetta, Golf, GTI, New Beetle<br>**Engines:** 2.0L, 2.5L, 3.6L<br>**Transmissions:** All | **Cruise Control Set Signal Incorrect Signal Conditions:**<br>Engine started, battery voltage must be at least 11.5v, all electrical components must be off, parking brake must be engaged (to keep daytime driving lights off), automatic transmission selector must be in park, and the ground between the engine and the chassis must be well connected. The ECM has detected a voltage value that has exceeded the specified maximum limit for the system to function properly.<br>**Possible Causes:**<br>• Cruise control system is damaged<br>• Control circuit is shorted to chassis ground |
| **DTC: P0571**<br>**Years: 2007, 2008**<br>**Models:** Passat, Jetta, Golf, GTI, New Beetle<br>**Engines:** 2.0L, 2.5L, 3.6L<br>**Transmissions:** All | **Cruise/Brake Switch (A) Circuit Malfunction Conditions:**<br>Engine started, battery voltage must be at least 11.5v, all electrical components must be off, parking brake must be engaged (to keep daytime driving lights off), automatic transmission selector must be in park, and the ground between the engine and the chassis must be well connected. The ECM has detected a voltage value that is implausible or erratic.<br>**Possible Causes:**<br>• Brake light switch is faulty<br>• Control circuit is shorted to chassis ground |
| **DTC: P0600**<br>**2T, MIL: Yes**<br>**Years: 2007, 2008**<br>**Models:** Passat, Jetta, Golf, GTI, New Beetle<br>**Engines:** 2.0L, 2.5L, 3.6L<br>**Transmissions:** All | **Serial Communication Link (Data BUS) Message Missing Conditions:**<br>The Engine Control Module (ECM) communicates with all databus-capable control modules via a CAN databus. These databus-capable control modules are connected via two data bus wires which are twisted together (CAN_High and CAN_Low), and exchange information (messages). Missing information on the databus is recognized as a malfunction and stored. Trouble-free operation of the CAN-Bus requires that it have a terminal resistance. This central terminal resistor is located in the Engine Control Module (ECM).<br>**Possible Causes:**<br>• ECM has failed<br>• CAN data bus wires have short circuited to each other |
| **DTC: P0601**<br>**2T, MIL: Yes**<br>**Years: 2007, 2008**<br>**Models:** Passat, Jetta, Golf, GTI, New Beetle, New Beetle Convertible<br>**Engines:** 2.0L, 2.5L, 3.6L<br>**Transmissions:** All | **Internal Control Module Memory Check Sum Error Conditions:**<br>Key on, the ECM has detected a programming error<br>**Possible Causes:**<br>• Battery terminal corrosion, or loose battery connection<br>• Connection to the ECM interrupted, or the circuit has been opened<br>• Reprogramming error has occurred<br>• ECM has failed and needs replacement. Remember to check for Aftermarket Performance Products before replacing a ECM. |

| DTC | Trouble Code Title, Conditions & Possible Causes |
|---|---|
| **DTC: P0602**<br>**Years: 2007, 2008**<br>**Models:** Passat, Jetta, Golf, GTI, New Beetle, New Beetle Convertible<br>**Engines:** 2.0L, 2.5L, 3.6L<br>**Transmissions:** All | **Control Module Programming Error Conditions:**<br>Key on, and the ECM detected a programming error in the VID block. This fault requires that the VID Block be reprogrammed, or that the EEPROM be re-flashed.<br>**Possible Causes:**<br>• During the VID reprogramming function, the Vehicle ID (VID) data block failed during reprogramming wit the Scan Tool.<br>• Battery terminal corrosion, or loose battery connection<br>• Connection to the ECM interrupted, or the circuit has been opened<br>• Reprogramming error has occurred<br>• ECM has failed and needs replacement. Remember to check for Aftermarket Performance Products before replacing a ECM. |
| **DTC: P0604**<br>**2T, MIL: Yes**<br>**Years: 2007, 2008**<br>**Models:** Passat, Jetta, Golf, GTI, New Beetle, New Beetle Convertible<br>**Engines:** 2.0L, 2.5L, 3.6L<br>**Transmissions:** All | **Internal Control Module Random Access Memory (RAM) Error Conditions:**<br>Key on, and the ECM detected an internal memory fault. This code will set if KAPWR to the ECM is interrupted (at the initial key on).<br>**Possible Causes:**<br>• Battery terminal corrosion, or loose battery connection<br>• Connection to the ECM interrupted, or the circuit has been opened<br>• Reprogramming error has occurred<br>• ECM has failed and needs replacement. Remember to check for Aftermarket Performance Products before replacing a ECM. |
| **DTC: P0605**<br>**2T, MIL: Yes**<br>**Years: 2007, 2008**<br>**Models:** Passat, Jetta, Golf, GTI, New Beetle, New Beetle Convertible<br>**Engines:** 2.0L, 2.5L, 3.6L<br>**Transmissions:** All | **ECM Read Only Memory (ROM) Test Error Conditions:**<br>Key on, and the ECM detected a ROM test error (ROM inside ECM is corrupted). The ECM is normally replaced if this code has set.<br>**Possible Causes:**<br>• An attempt was made to change the module calibration, or a module programming error may have occurred<br>• Clear the trouble codes and then check for this trouble code. If it resets, the ECM has failed and needs replacement.<br>• Aftermarket performance products may have been installed.<br>• The Transmission Control Module (TCM) has failed. |
| **DTC: P0606**<br>**2T, MIL: Yes**<br>**Years: 2007, 2008**<br>**Models:** Passat, Jetta, Golf, GTI, New Beetle, New Beetle Convertible<br>**Engines:** 2.0L, 2.5L, 3.6L<br>**Transmissions:** All | **ECM Internal Communication Error Conditions:**<br>Key on, and the ECM detected an internal communications register read back error during the initial key on check period.<br>**Possible Causes:**<br>• Clear the trouble codes and then check for this trouble code. If it resets, the ECM has failed and needs replacement.<br>• Remember to check for signs of Aftermarket Performance Products installation before replacing the ECM. |
| **DTC: P0613**<br>**Years: 2007, 2008**<br>**Models:** New Beetle, New Beetle Convertible<br>**Engines:** 2.0L<br>**Transmissions:** All | **TCM Processor Conditions:**<br>Key on, and the ECM detected an internal communication error with the Transmission control module<br>**Possible Causes:**<br>• TCM failed<br>• ECM failed<br>• Replacement control module ID doesn't match old control module ID |
| **DTC: P0614**<br>**Years: 2007, 2008**<br>**Models:** Passat, Jetta, Golf, GTI, New Beetle, New Beetle Convertible<br>**Engines:** 2.0L, 2.5L, 3.6L<br>**Transmissions:** All | **ECM / TCM Incompatible Conditions:**<br>Key on, and the ECM detected a communication error between the Transmission control module and the ECM<br>**Possible Causes:**<br>• TCM failed<br>• ECM failed<br>• Circuit shorting between ECM and TCM<br>• Replacement control module ID doesn't match old control module ID |
| **DTC: P0627**<br>**Years: 2007, 2008**<br>**Models:** Passat, Jetta, New Beetle<br>**Engines:** 2.5L<br>**Transmissions:** All | **Fuel Pump "A" Control Circuit Open Conditions:**<br>Engine started, battery voltage must be at least 11.5v, all electrical components must be off, parking brake must be engaged (to keep daytime driving lights off), automatic transmission selector must be in park, and the ground between the engine and the chassis must be well connected. The ECM has detected a voltage value across the fuel pump control circuit that is out of the specified limits for the system to function properly.<br>**Possible Causes:**<br>• Fuel Pressure Regulator Valve is faulty<br>• Fuel Pressure Sensor is faulty<br>• Fuel Pump (FP) Control Module is faulty<br>• Fuel pump is faulty |

| DTC | Trouble Code Title, Conditions & Possible Causes |
|---|---|
| **DTC: P0628**<br>**Years:** 2007, 2008<br>**Models:** Passat, New Beetle<br>**Engines:** 2.0L<br>**Transmissions:** All | **Fuel Pump "A" Control Circuit Low Conditions:**<br>Engine started, battery voltage must be at least 11.5v, all electrical components must be off, parking brake must be engaged (to keep daytime driving lights off), automatic transmission selector must be in park, and the ground between the engine and the chassis must be well connected. The ECM has detected a voltage value across the fuel pump control circuit that is below the specified limit for the system to function properly.<br>**Possible Causes:**<br>• Fuel Pressure Regulator Valve is faulty<br>• Fuel Pressure Sensor is faulty<br>• Fuel Pump (FP) Control Module is faulty<br>• Fuel pump is faulty |
| **DTC: P0629**<br>**Years:** 2007, 2008<br>**Models:** Passat, Jetta, New Beetle<br>**Engines:** 2.5L<br>**Transmissions:** All | **Fuel Pump "A" Control Circuit High Conditions:**<br>Engine started, battery voltage must be at least 11.5v, all electrical components must be off, parking brake must be engaged (to keep daytime driving lights off), automatic transmission selector must be in park, and the ground between the engine and the chassis must be well connected. The ECM has detected a voltage value across the fuel pump control circuit that is above the specified limit for the system to function properly.<br>**Possible Causes:**<br>• Fuel Pressure Regulator Valve is faulty<br>• Fuel Pressure Sensor is faulty<br>• Fuel Pump (FP) Control Module is faulty<br>• Fuel pump is faulty |
| **DTC: P0638**<br>**2T, MIL: Yes**<br>**Years:** 2007, 2008<br>**Models:** Passat, Jetta, Golf, GTI, New Beetle, New Beetle Convertible<br>**Engines:** 2.0L, 2.5L, 3.6L<br>**Transmissions:** All | **Throttle Actuator Control Range/Performance Bank 1 Conditions:**<br>Engine started, battery voltage must be at least 11.5v, all electrical components must be off, parking brake must be engaged (to keep daytime driving lights off), automatic transmission selector must be in park, and the ground between the engine and the chassis must be well connected. The ECM has detected a voltage value across the throttle actuator control circuit that is out of the specified limit for the system to function properly. Both Throttle Position (TP) Sensor / Accelerator Pedal Position Sensor 2 are located at the accelerator pedal and communicate the driver's intentions to the Motronic engine control module (ECM) completely independently of each other. Both sensors are integrated into one housing.<br>**Possible Causes:**<br>• Throttle Position (TP) sensor is faulty<br>• Throttle valve control module is faulty<br>• ECM is faulty<br>• Circuit wires have short circuited to each other, to vehicle Ground (GND) or to B+.<br>• Accelerator pedal module is faulty |
| **DTC: P0639**<br>**Years:** 2007, 2008<br>**Models:** Jetta<br>**Engines:** 2.5L<br>**Transmissions:** All | **Throttle Actuator Control Range/Performance Bank 2 Conditions:**<br>Engine started, battery voltage must be at least 11.5v, all electrical components must be off, parking brake must be engaged (to keep daytime driving lights off), automatic transmission selector must be in park, and the ground between the engine and the chassis must be well connected. The ECM has detected a voltage value across the throttle actuator control circuit that is out of the specified limit for the system to function properly. Both Throttle Position (TP) Sensor / Accelerator Pedal Position Sensor 2 are located at the accelerator pedal and communicate the driver's intentions to the Motronic engine control module (ECM) completely independently of each other. Both sensors are integrated into one housing.<br>**Possible Causes:**<br>• Throttle Position (TP) sensor is faulty<br>• Throttle valve control module is faulty<br>• ECM is faulty<br>• Circuit wires have short circuited to each other, to vehicle Ground (GND) or to B+.<br>• Accelerator pedal module is faulty |
| **DTC: P0641**<br>**Years:** 2007, 2008<br>**Models:** Passat, Jetta<br>**Engines:** 2.0L, 2.5L<br>**Transmissions:** All | **Sensor Reference Voltage "A" Circuit Open Conditions:**<br>Engine started, battery voltage must be at least 11.5v, all electrical components must be off, parking brake must be engaged (to keep daytime driving lights off), automatic transmission selector must be in park, and the ground between the engine and the chassis must be well connected.<br>**Possible Causes:**<br>• Circuit harness connector contacts are corroded or ingressed of water<br>• Circuit wires have shorted to each other, to battery or ground<br>• Automatic Transmission Hydraulic Pressure Sensor 1 has failed<br>• Solenoid valves in valve body are faulty<br>• Transmission Control Module (TCM) needs replacing<br>• Transmission Input Speed (RPM) Sensor has failed<br>• Transmission Output Speed (RPM) Sensor has failed |

| DTC | Trouble Code Title, Conditions & Possible Causes |
|---|---|
| **DTC: P0642**<br>**Years:** 2007, 2008<br>**Models:** Passat, Jetta, Golf, GTI, New Beetle,<br>**Engines:** 2.0L, 2.5L, 3.6L<br>**Transmissions:** All | **Sensor Reference Voltage "A" Circuit Low Conditions:**<br>Engine started, battery voltage must be at least 11.5v, all electrical components must be off, parking brake must be engaged (to keep daytime driving lights off), automatic transmission selector must be in park, and the ground between the engine and the chassis must be well connected.<br>**Possible Causes:**<br>• Circuit harness connector contacts are corroded or ingressed of water<br>• Circuit wires have shorted to each other, to battery or ground<br>• Automatic Transmission Hydraulic Pressure Sensor 1 has failed<br>• Solenoid valves in valve body are faulty<br>• Transmission Control Module (TCM) needs replacing<br>• Transmission Input Speed (RPM) Sensor has failed<br>• Transmission Output Speed (RPM) Sensor has failed |
| **DTC: P0643**<br>**Years:** 2007, 2008<br>**Models:** Passat, Jetta<br>**Engines:** 2.0L<br>**Transmissions:** All | **Sensor Reference Voltage "A" Circuit High Conditions:**<br>Engine started, battery voltage must be at least 11.5v, all electrical components must be off, parking brake must be engaged (to keep daytime driving lights off), automatic transmission selector must be in park, and the ground between the engine and the chassis must be well connected.<br>**Possible Causes:**<br>• Circuit harness connector contacts are corroded or ingressed of water<br>• Circuit wires have shorted to each other, to battery or ground<br>• Automatic Transmission Hydraulic Pressure Sensor 1 has failed<br>• Solenoid valves in valve body are faulty<br>• Transmission Control Module (TCM) needs replacing<br>• Transmission Input Speed (RPM) Sensor has failed<br>• Transmission Output Speed (RPM) Sensor has failed |
| **DTC: P0651**<br>**Years:** 2007, 2008<br>**Models:** Passat, Jetta<br>**Engines:** 2.0L<br>**Transmissions:** All | **Sensor Reference Voltage "B" Circuit Open Conditions:**<br>Engine started, battery voltage must be at least 11.5v, all electrical components must be off, parking brake must be engaged (to keep daytime driving lights off), automatic transmission selector must be in park, and the ground between the engine and the chassis must be well connected.<br>**Possible Causes:**<br>• Circuit harness connector contacts are corroded or ingressed of water<br>• Circuit wires have shorted to each other, to battery or ground<br>• Automatic Transmission Hydraulic Pressure Sensor 1 has failed<br>• Solenoid valves in valve body are faulty<br>• Transmission Control Module (TCM) needs replacing<br>• Transmission Input Speed (RPM) Sensor has failed<br>• Transmission Output Speed (RPM) Sensor has failed |
| **DTC: P0652**<br>**Years:** 2007, 2008<br>**Models:** Passat, Jetta<br>**Engines:** 2.0L<br>**Transmissions:** All | **Sensor Reference Voltage "B" Circuit Low Conditions:**<br>Engine started, battery voltage must be at least 11.5v, all electrical components must be off, parking brake must be engaged (to keep daytime driving lights off), automatic transmission selector must be in park, and the ground between the engine and the chassis must be well connected.<br>**Possible Causes:**<br>• Circuit harness connector contacts are corroded or ingressed of water<br>• Circuit wires have shorted to each other, to battery or ground<br>• Automatic Transmission Hydraulic Pressure Sensor 1 has failed<br>• Solenoid valves in valve body are faulty<br>• Transmission Control Module (TCM) needs replacing<br>• Transmission Input Speed (RPM) Sensor has failed<br>• Transmission Output Speed (RPM) Sensor has failed |
| **DTC: P0653**<br>**Years:** 2007, 2008<br>**Models:** Passat, Jetta<br>**Engines:** 2.0L<br>**Transmissions:** All | **Sensor Reference Voltage "B" Circuit High Conditions:**<br>Engine started, battery voltage must be at least 11.5v, all electrical components must be off, parking brake must be engaged (to keep daytime driving lights off), automatic transmission selector must be in park, and the ground between the engine and the chassis must be well connected.<br>**Possible Causes:**<br>• Circuit harness connector contacts are corroded or ingressed of water<br>• Circuit wires have shorted to each other, to battery or ground<br>• Automatic Transmission Hydraulic Pressure Sensor 1 has failed<br>• Solenoid valves in valve body are faulty<br>• Transmission Control Module (TCM) needs replacing<br>• Transmission Input Speed (RPM) Sensor has failed<br>• Transmission Output Speed (RPM) Sensor has failed |

| DTC | Trouble Code Title, Conditions & Possible Causes |
|---|---|
| **DTC: P0654**<br>**Years:** 2007, 2008<br>**Models:** Jetta<br>**Engines:** 2.0L<br>**Transmissions:** All | **Engine RPM Output Circuit Conditions:**<br>Engine started, battery voltage must be at least 11.5v, all electrical components must be off, parking brake must be engaged (to keep daytime driving lights off), automatic transmission selector must be in park, and the ground between the engine and the chassis must be well connected.<br>**Possible Causes:**<br>• Circuit harness connector contacts are corroded or ingressed of water<br>• Circuit wires have shorted to each other, to battery or ground<br>• Automatic Transmission Hydraulic Pressure Sensor 1 has failed<br>• Solenoid valves in valve body are faulty<br>• Transmission Control Module (TCM) needs replacing<br>• Transmission Input Speed (RPM) Sensor has failed<br>• Transmission Output Speed (RPM) Sensor has failed |
| **DTC: P0657**<br>**Years:** 2007, 2008<br>**Models:** Passat<br>**Engines:** 2.0L<br>**Transmissions:** All | **Actuator Supply Voltage "A" Circuit Open Conditions:**<br>Engine started, battery voltage must be at least 11.5v, all electrical components must be off, parking brake must be engaged (to keep daytime driving lights off), automatic transmission selector must be in park, and the ground between the engine and the chassis must be well connected.<br>**Possible Causes:**<br>• Circuit harness connector contacts are corroded or ingressed of water<br>• Circuit wires have shorted to each other, to battery or ground<br>• Automatic Transmission Hydraulic Pressure Sensor 1 has failed<br>• Solenoid valves in valve body are faulty<br>• Transmission Control Module (TCM) needs replacing<br>• Transmission Input Speed (RPM) Sensor has failed<br>• Transmission Output Speed (RPM) Sensor has failed |
| **DTC: P0658**<br>**Years:** 2007, 2008<br>**Models:** Passat<br>**Engines:** 2.0L<br>**Transmissions:** All | **Actuator Supply Voltage "A" Circuit Low Conditions:**<br>Engine started, battery voltage must be at least 11.5v, all electrical components must be off, parking brake must be engaged (to keep daytime driving lights off), automatic transmission selector must be in park, and the ground between the engine and the chassis must be well connected.<br>**Possible Causes:**<br>• Circuit harness connector contacts are corroded or ingressed of water<br>• Circuit wires have shorted to each other, to battery or ground<br>• Automatic Transmission Hydraulic Pressure Sensor 1 has failed<br>• Solenoid valves in valve body are faulty<br>• Transmission Control Module (TCM) needs replacing<br>• Transmission Input Speed (RPM) Sensor has failed<br>• Transmission Output Speed (RPM) Sensor has failed |
| **DTC: P0659**<br>**Years:** 2007, 2008<br>**Models:** Passat<br>**Engines:** 2.0L<br>**Transmissions:** All | **Actuator Supply Voltage "A" Circuit High Conditions:**<br>Engine started, battery voltage must be at least 11.5v, all electrical components must be off, parking brake must be engaged (to keep daytime driving lights off), automatic transmission selector must be in park, and the ground between the engine and the chassis must be well connected.<br>**Possible Causes:**<br>• Circuit harness connector contacts are corroded or ingressed of water<br>• Circuit wires have shorted to each other, to battery or ground<br>• Automatic Transmission Hydraulic Pressure Sensor 1 has failed<br>• Solenoid valves in valve body are faulty<br>• Transmission Control Module (TCM) needs replacing<br>• Transmission Input Speed (RPM) Sensor has failed<br>• Transmission Output Speed (RPM) Sensor has failed |
| **DTC: P0685**<br>**Years:** 2007, 2008<br>**Models:** Passat, Jetta, Golf, GTI, New Beetle, New Beetle Convertible<br>**Engines:** 2.0L, 2.5L, 3.6L<br>**Transmissions:** All | **ECM Power Relay Control Circuit Open Conditions:**<br>Engine started, battery voltage must be at least 11.5v, all electrical components must be off, parking brake must be engaged (to keep daytime driving lights off), automatic transmission selector must be in park and the ground between the engine and the chassis must be well connected. The ECM detected the ECM power relay control circuit has a voltage outside requirement for proper function.<br>**Possible Causes:**<br>• Generator has failed or is damaged<br>• Fuel pump relay is faulty<br>• Circuit is grounded to power or chassis<br>• ECM has failed |

| DTC | Trouble Code Title, Conditions & Possible Causes |
|---|---|
| **DTC: P0686**<br>**Years: 2007, 2008**<br>**Models:** Passat, Jetta, Golf, GTI, New Beetle, New Beetle Convertible<br>**Engines:** 2.0L, 2.5L, 3.6L<br>**Transmissions:** All | **ECM/PCM Power Relay Control Circuit Low Conditions:**<br>Engine started, battery voltage must be at least 11.5v, all electrical components must be off, parking brake must be engaged (to keep daytime driving lights off), automatic transmission selector must be in park and the ground between the engine and the chassis must be well connected. The ECM detected the ECM power relay control circuit has a voltage outside requirement for proper function.<br>**Possible Causes:**<br>• Generator has failed or is damaged<br>• Fuel pump relay is faulty<br>• Circuit is grounded to power or chassis<br>• ECM has failed |
| **DTC: P0687**<br>**Years: 2007, 2008**<br>**Models:** Passat, Jetta, Golf, GTI, New Beetle, New Beetle Convertible<br>**Engines:** 2.0L, 2.5L, 3.6L<br>**Transmissions:** All | **ECM/PCM Power Relay Control Circuit High Conditions:**<br>Engine started, battery voltage must be at least 11.5v, all electrical components must be off, parking brake must be engaged (to keep daytime driving lights off), automatic transmission selector must be in park and the ground between the engine and the chassis must be well connected. The ECM detected the ECM power relay control circuit has a voltage outside requirement for proper function.<br>**Possible Causes:**<br>• Generator has failed or is damaged<br>• Fuel pump relay is faulty<br>• Circuit is grounded to power or chassis<br>• ECM has failed |
| **DTC: P0688**<br>**Years: 2007, 2008**<br>**Models:** Passat, Jetta, Golf, GTI, New Beetle<br>**Engines:** 2.0L, 2.5L, 3.6L<br>**Transmissions:** All | **ECM/PCM Power Relay Control Sense Circuit Open Conditions:**<br>Engine started, battery voltage must be at least 11.5v, all electrical components must be off, parking brake must be engaged (to keep daytime driving lights off), automatic transmission selector must be in park and the ground between the engine and the chassis must be well connected. The ECM detected the ECM power relay control circuit has a voltage outside requirement for proper function.<br>**Possible Causes:**<br>• Generator has failed or is damaged<br>• Fuel pump relay is faulty<br>• Circuit is grounded to power or chassis<br>• ECM has failed |
| **DTC: P0691**<br>**Years: 2007, 2008**<br>**Models:** Passat, Jetta<br>**Engines:** 2.0L, 2.5L<br>**Transmissions:** All | **Fan 1 Control Circuit Low Conditions:**<br>Engine running, battery voltage at least 11.5v, all electrical components off, and the ECM detected an problem with the fan control circuit<br>**Possible Causes:**<br>• Check connection of coolant fan control module according to wiring diagram<br>• Short to B+ or ground in the circuit<br>• Circuit open |
| **DTC: P0692**<br>**Years: 2007, 2008**<br>**Models:** Passat, Jetta<br>**Engines:** 2.0L, 2.5L<br>**Transmissions:** All | **Fan 1 Control Circuit High Conditions:**<br>Engine running, battery voltage at least 11.5v, all electrical components off, and the ECM detected an problem with the fan control circuit<br>**Possible Causes:**<br>• Check connection of coolant fan control module according to wiring diagram<br>• Short to B+ or ground in the circuit<br>• Circuit open |
| **DTC: P0697**<br>**Years: 2007, 2008**<br>**Models:** Passat, Jetta<br>**Engines:** 2.0L<br>**Transmissions:** All | **Sensor Reference Voltage "C" Circuit Open Conditions:**<br>Engine started, battery voltage must be at least 11.5v, all electrical components must be off, parking brake must be engaged (to keep daytime driving lights off), automatic transmission selector must be in park, and the ground between the engine and the chassis must be well connected.<br>**Possible Causes:**<br>• Circuit harness connector contacts are corroded or ingressed of water<br>• Circuit wires have shorted to each other, to battery or ground<br>• Automatic Transmission Hydraulic Pressure Sensor 1 has failed<br>• Solenoid valves in valve body are faulty<br>• Transmission Control Module (TCM) needs replacing<br>• Transmission Input Speed (RPM) Sensor has failed<br>• Transmission Output Speed (RPM) Sensor has failed |

| DTC | Trouble Code Title, Conditions & Possible Causes |
|---|---|
| **DTC: P0698**<br>**Years: 2007, 2008**<br>**Models:** Passat, Jetta<br>**Engines:** 2.0L<br>**Transmissions:** All | **Sensor Reference Voltage "C" Circuit Low Conditions:**<br>Engine started, battery voltage must be at least 11.5v, all electrical components must be off, parking brake must be engaged (to keep daytime driving lights off), automatic transmission selector must be in park, and the ground between the engine and the chassis must be well connected.<br>**Possible Causes:**<br>• Circuit harness connector contacts are corroded or ingressed of water<br>• Circuit wires have shorted to each other, to battery or ground<br>• Automatic Transmission Hydraulic Pressure Sensor 1 has failed<br>• Solenoid valves in valve body are faulty<br>• Transmission Control Module (TCM) needs replacing<br>• Transmission Input Speed (RPM) Sensor has failed<br>• Transmission Output Speed (RPM) Sensor has failed |
| **DTC: P0699**<br>**Years: 2007, 2008**<br>**Models:** Passat, Jetta<br>**Engines:** 2.0L<br>**Transmissions:** All | **Sensor Reference Voltage "C" Circuit High Conditions:**<br>Engine started, battery voltage must be at least 11.5v, all electrical components must be off, parking brake must be engaged (to keep daytime driving lights off), automatic transmission selector must be in park, and the ground between the engine and the chassis must be well connected.<br>**Possible Causes:**<br>• Circuit harness connector contacts are corroded or ingressed of water<br>• Circuit wires have shorted to each other, to battery or ground<br>• Automatic Transmission Hydraulic Pressure Sensor 1 has failed<br>• Solenoid valves in valve body are faulty<br>• Transmission Control Module (TCM) needs replacing<br>• Transmission Input Speed (RPM) Sensor has failed<br>• Transmission Output Speed (RPM) Sensor has failed |
| **DTC: P0700**<br>**Years: 2007, 2008**<br>**Models:** Passat, Jetta, Golf, GTI, New Beetle, New Beetle Convertible<br>**Engines:** 2.0L, 2.5L, 3.6L<br>**Transmissions:** All | **Transmission Control System Malfunction Conditions:**<br>Engine started, battery voltage must be at least 11.5v, all electrical components must be off, parking brake must be engaged (to keep daytime driving lights off), automatic transmission selector must be in park, and the ground between the engine and the chassis must be well connected. The ECM detected a malfunction int the transmission control system.<br>**Possible Causes:**<br>• Circuit harness connector contacts are corroded or ingressed of water<br>• Circuit wires have shorted to each other, to battery or ground<br>• Automatic Transmission Hydraulic Pressure Sensor 1 has failed<br>• Solenoid valves in valve body are faulty<br>• Transmission Input Speed (RPM) Sensor has failed<br>• Transmission Output Speed (RPM) Sensor has failed<br>• Engine Control Module (ECM) is faulty<br>• Voltage supply for Engine Control Module (ECM) is faulty<br>• Transmission Control Module (TCM) is faulty |
| **DTC: P0701**<br>**Years: 2007, 2008**<br>**Models:** Passat<br>**Engines:** 2.0L<br>**Transmissions:** All | **Transmission Control System Range/Performance Conditions:**<br>Engine started, battery voltage must be at least 11.5v, all electrical components must be off, parking brake must be engaged (to keep daytime driving lights off), automatic transmission selector must be in park, and the ground between the engine and the chassis must be well connected. The ECM detected a voltage outside the normal performance range to allow the system to properly function.<br>**Possible Causes:**<br>• Circuit harness connector contacts are corroded or ingressed of water<br>• Circuit wires have shorted to each other, to battery or ground<br>• Automatic Transmission Hydraulic Pressure Sensor 1 has failed<br>• Solenoid valves in valve body are faulty<br>• Transmission Input Speed (RPM) Sensor has failed<br>• Transmission Output Speed (RPM) Sensor has failed<br>• Engine Control Module (ECM) is faulty<br>• Voltage supply for Engine Control Module (ECM) is faulty<br>• Transmission Control Module (TCM) is faulty |

| DTC | Trouble Code Title, Conditions & Possible Causes |
|---|---|
| **DTC: P0702**<br>**Years: 2007, 2008**<br>**Models:** Passat, Jetta, Golf, GTI, New Beetle, New Beetle Convertible<br>**Engines:** 2.0L, 2.5L, 3.6L<br>**Transmissions:** All | **Transmission Control System Electrical Conditions:**<br>Engine started, battery voltage must be at least 11.5v, all electrical components must be off, parking brake must be engaged (to keep daytime driving lights off), automatic transmission selector must be in park, and the ground between the engine and the chassis must be well connected. The ECM detected a voltage outside the normal performance range to allow the system to properly function.<br>**Possible Causes:**<br>• Circuit harness connector contacts are corroded or ingressed of water<br>• Circuit wires have shorted to each other, to battery or ground<br>• Automatic Transmission Hydraulic Pressure Sensor 1 has failed<br>• Solenoid valves in valve body are faulty<br>• Transmission Input Speed (RPM) Sensor has failed<br>• Transmission Output Speed (RPM) Sensor has failed<br>• Engine Control Module (ECM) is faulty<br>• Voltage supply for Engine Control Module (ECM) is faulty<br>• Transmission Control Module (TCM) is faulty |
| **DTC: P0704**<br>**Years: 2007, 2008**<br>**Models:** Passat, Jetta, Golf, GTI, New Beetle, New Beetle Convertible<br>**Engines:** 2.0L, 2.5L, 3.6L<br>**Transmissions:** All | **Clutch Switch Input Circuit Malfunction Conditions:**<br>Engine started, battery voltage must be at least 11.5v, all electrical components must be off, parking brake must be engaged (to keep daytime driving lights off), automatic transmission selector must be in park, and the ground between the engine and the chassis must be well connected. The ECM detected a voltage outside the normal performance range to allow the system to properly function.<br>**Possible Causes:**<br>• Circuit harness connector contacts are corroded or ingressed of water<br>• Circuit wires have shorted to each other, to battery or ground<br>• Automatic Transmission Hydraulic Pressure Sensor 1 has failed<br>• Solenoid valves in valve body are faulty<br>• Transmission Input Speed (RPM) Sensor has failed<br>• Transmission Output Speed (RPM) Sensor has failed<br>• Engine Control Module (ECM) is faulty<br>• Voltage supply for Engine Control Module (ECM) is faulty<br>• Transmission Control Module (TCM) is faulty |
| **DTC: P0705**<br>**Years: 2007, 2008**<br>**Models:** Passat, Jetta, Golf, GTI, New Beetle, New Beetle Convertible<br>**Engines:** 2.0L, 2.5L, 3.6L<br>**Transmissions:** A/T | **TR Sensor Circuit Malfunction Conditions:**<br>Engine started, battery voltage must be at least 11.5v, all electrical components must be off, parking brake must be engaged (to keep daytime driving lights off), automatic transmission selector must be in park, and the ground between the engine and the chassis must be well connected. The ECM detected a voltage outside the normal performance range to allow the system to properly function.<br>**Possible Causes:**<br>• Circuit harness connector contacts are corroded or ingressed of water<br>• Circuit wires have shorted to each other, to battery or ground<br>• Automatic Transmission Hydraulic Pressure Sensor 1 has failed<br>• Solenoid valves in valve body are faulty<br>• Transmission Input Speed (RPM) Sensor has failed<br>• Transmission Output Speed (RPM) Sensor has failed<br>• Engine Control Module (ECM) is faulty<br>• Voltage supply for Engine Control Module (ECM) is faulty<br>• Transmission Control Module (TCM) is faulty |
| **DTC: P0706**<br>**Years: 2007, 2008**<br>**Models:** Passat, Jetta, Golf, GTI, New Beetle, New Beetle Convertible<br>**Engines:** 2.0L<br>**Transmissions:** A/T | **TR Sensor Circuit Range/Performance Conditions:**<br>Engine started, battery voltage must be at least 11.5v, all electrical components must be off, parking brake must be engaged (to keep daytime driving lights off), automatic transmission selector must be in park, and the ground between the engine and the chassis must be well connected. The ECM detected a voltage outside the normal performance range to allow the system to properly function.<br>**Possible Causes:**<br>• Circuit harness connector contacts are corroded or ingressed of water<br>• Circuit wires have shorted to each other, to battery or ground<br>• Automatic Transmission Hydraulic Pressure Sensor 1 has failed<br>• Solenoid valves in valve body are faulty<br>• Transmission Input Speed (RPM) Sensor has failed<br>• Transmission Output Speed (RPM) Sensor has failed<br>• Engine Control Module (ECM) is faulty<br>• Voltage supply for Engine Control Module (ECM) is faulty<br>• Transmission Control Module (TCM) is faulty |

| DTC | Trouble Code Title, Conditions & Possible Causes |
|---|---|
| **DTC: P0707**<br>**Years: 2007, 2008**<br>**Models:** Passat<br>**Engines:** 2.5L, 3.6L<br>**Transmissions:** A/T | **Transmission Range Sensor Circuit Low Input Conditions:**<br>Engine started, battery voltage must be at least 11.5v, all electrical components must be off, parking brake must be engaged (to keep daytime driving lights off), automatic transmission selector must be in park, and the ground between the engine and the chassis must be well connected. The ECM detected the Transmission Range sensor (TR) signal was less than the self-test minimum value in the test.<br>**Possible Causes:**<br>• Circuit harness connector contacts are corroded or ingressed of water<br>• Circuit wires have shorted to each other, to battery or ground<br>• Automatic Transmission Hydraulic Pressure Sensor 1 has failed<br>• Solenoid valves in valve body are faulty<br>• Transmission Input Speed (RPM) Sensor has failed<br>• Transmission Output Speed (RPM) Sensor has failed<br>• Engine Control Module (ECM) is faulty<br>• Voltage supply for Engine Control Module (ECM) is faulty<br>• Transmission Control Module (TCM) is faulty |
| **DTC: P0708**<br>**Years: 2007, 2008**<br>**Models:** Passat<br>**Engines:** 2.5L, 3.6L<br>**Transmissions:** A/T | **Transmission Range Sensor Circuit High Input Conditions:**<br>Engine started, battery voltage must be at least 11.5v, all electrical components must be off, parking brake must be engaged (to keep daytime driving lights off), automatic transmission selector must be in park, and the ground between the engine and the chassis must be well connected. The ECM detected the Transmission Range sensor (TR) input was more than the self-test maximum range in the test.<br>**Possible Causes:**<br>• Circuit harness connector contacts are corroded or ingressed of water<br>• Circuit wires have shorted to each other, to battery or ground<br>• Automatic Transmission Hydraulic Pressure Sensor 1 has failed<br>• Solenoid valves in valve body are faulty<br>• Transmission Input Speed (RPM) Sensor has failed<br>• Transmission Output Speed (RPM) Sensor has failed<br>• Engine Control Module (ECM) is faulty<br>• Voltage supply for Engine Control Module (ECM) is faulty<br>• Transmission Control Module (TCM) is faulty |
| **DTC: P0710**<br>**Years: 2007, 2008**<br>**Models:** Passat, Jetta, Golf, GTI, New Beetle, New Beetle Convertible<br>**Engines:** 2.0L, 2.5L, 3.6L<br>**Transmissions:** A/T | **Transmission Fluid Temperature Sensor Circuit Malfunction Conditions:**<br>Engine started, battery voltage must be at least 11.5v, all electrical components must be off, parking brake must be engaged (to keep daytime driving lights off), automatic transmission selector must be in park, and the ground between the engine and the chassis must be well connected. The ECM detected the Transmission fluid temperature sensor circuit was outside the normal range in the test to allow proper function.<br>**Possible Causes:**<br>• ATF is low, contaminated, dirty or burnt<br>• Circuit harness connector contacts are corroded or ingressed of water<br>• Circuit wires have shorted to each other, to battery or ground<br>• Automatic Transmission Hydraulic Pressure Sensor 1 has failed<br>• Solenoid valves in valve body are faulty<br>• Transmission Input Speed (RPM) Sensor has failed<br>• Transmission Output Speed (RPM) Sensor has failed<br>• Engine Control Module (ECM) is faulty<br>• Voltage supply for Engine Control Module (ECM) is faulty<br>• Transmission Control Module (TCM) is faulty |
| **DTC: P0711**<br>**Years: 2007, 2008**<br>**Models:** Passat, Jetta, Golf, GTI, New Beetle, New Beetle Convertible<br>**Engines:** 2.0L, 2.5L<br>**Transmissions:** A/T | **Transmission Fluid Temperature Sensor Signal Range/Performance Conditions:**<br>Engine started, battery voltage must be at least 11.5v, all electrical components must be off, parking brake must be engaged (to keep daytime driving lights off), automatic transmission selector must be in park, and the ground between the engine and the chassis must be well connected. The ECM detected the Transmission Fluid Temperature (TFT) sensor value was not close its normal operating temperature.<br>**Possible Causes:**<br>• ATF is low, contaminated, dirty or burnt<br>• TFT sensor signal circuit has a high resistance condition<br>• TFT sensor is out-of-calibration ("skewed") or it has failed<br>• ECM has failed |

| DTC | Trouble Code Title, Conditions & Possible Causes |
|---|---|
| **DTC: P0712**<br>**Years: 2007, 2008**<br>**Models:** Passat, Jetta, Golf, GTI, New Beetle, New Beetle Convertible<br>**Engines:** 2.0L, 2.5L<br>**Transmissions:** A/T | **Transmission Fluid Temperature Sensor Circuit Low Input Conditions:**<br>Engine started, battery voltage must be at least 11.5v, all electrical components must be off, parking brake must be engaged (to keep daytime driving lights off), automatic transmission selector must be in park, and the ground between the engine and the chassis must be well connected. The ECM detected the Transmission Fluid Temperature (TFT) sensor was less than its minimum self-test range in the test.<br>**Possible Causes:**<br>• TFT sensor signal circuit is shorted to chassis ground<br>• TFT sensor signal circuit is shorted to sensor ground<br>• TFT sensor is damaged, or out-of-calibration, or has failed<br>• ECM has failed |
| **DTC: P0713**<br>**Years: 2007, 2008**<br>**Models:** Passat, Jetta, Golf, GTI, New Beetle, New Beetle Convertible<br>**Engines:** 2.0L, 2.5L<br>**Transmissions:** A/T | **Transmission Fluid Temperature Sensor Circuit High Input Conditions:**<br>Engine started, battery voltage must be at least 11.5v, all electrical components must be off, parking brake must be engaged (to keep daytime driving lights off), automatic transmission selector must be in park, and the ground between the engine and the chassis must be well connected. The ECM detected the Transmission Fluid Temperature (TFT) sensor was more than its maximum self-test range in the test.<br>**Possible Causes:**<br>• TFT sensor signal circuit is open between the sensor and ECM<br>• TFT sensor ground circuit is open between sensor and ECM<br>• TFT sensor is damaged or has failed<br>• ECM has failed |
| **DTC: P0714**<br>**Years: 2007, 2008**<br>**Models:** Passat, Jetta, Golf, GTI, New Beetle, New Beetle Convertible<br>**Engines:** 2.0L, 2.5L<br>**Transmissions:** A/T | **Transmission Fluid Temperature Sensor Circuit Intermittent Conditions:**<br>Engine started, battery voltage must be at least 11.5v, all electrical components must be off, parking brake must be engaged (to keep daytime driving lights off), automatic transmission selector must be in park, and the ground between the engine and the chassis must be well connected. The ECM detected the Transmission Fluid Temperature (TFT) sensor was giving a false reading or was not reading at all.<br>**Possible Causes:**<br>• TFT sensor signal circuit is open between the sensor and ECM<br>• TFT sensor ground circuit is open between sensor and ECM<br>• TFT sensor is damaged or has failed<br>• ECM has failed |
| **DTC: P0715**<br>**Years: 2007, 2008**<br>**Models:** Passat, Jetta, Golf, GTI, New Beetle, New Beetle Convertible<br>**Engines:** 2.0L, 2.5L, 3.6L<br>**Transmissions:** A/T | **Input/Turbine Speed Sensor Circuit Malfunction Conditions:**<br>Engine started, vehicle driven with the vehicle speed sensor indicating more than 1 mph, and the ECM detected the Transmission Vehicle Speed Sensor signals were erratic, or that they were missing for a period of time.<br>**Possible Causes:**<br>• TVSS signal circuit is open<br>• TVSS signal is shorted to chassis ground<br>• TVSS signal is shorted to sensor ground<br>• TVSS assembly is damaged or it has failed<br>• ECM has failed |
| **DTC: P0716**<br>**Years: 2007, 2008**<br>**Models:** Passat, Jetta, Golf, GTI, New Beetle, New Beetle Convertible<br>**Engines:** 2.0L, 2.5L<br>**Transmissions:** A/T | **Input Turbine/Speed Sensor Circuit Range/Performance Conditions:**<br>Engine started, vehicle driven with the vehicle speed sensor indicating more than 1 mph, and the ECM detected the Transmission Vehicle Speed Sensor signals were erratic, or that they were missing for a period of time.<br>**Possible Causes:**<br>• TVSS signal circuit is open<br>• TVSS signal is shorted to chassis ground<br>• TVSS signal is shorted to sensor ground<br>• TVSS assembly is damaged or it has failed<br>• ECM has failed |
| **DTC: P0717**<br>**Years: 2007, 2008**<br>**Models:** Passat, Jetta, Golf, GTI, New Beetle, New Beetle Convertible<br>**Engines:** 2.0L, 2.5L<br>**Transmissions:** A/T | **Transmission Speed Shaft Sensor Signal Intermittent Conditions:**<br>Engine started, vehicle speed sensor indicating over 1 mph, and the ECM detected an intermittent loss of TSS signals (i.e., the TSS signals were erratic, irregular or missing).<br>**Possible Causes:**<br>• TSS connector is damaged, loose or shorted<br>• TSS signal circuit has an intermittent open condition<br>• TSS signal circuit has an intermittent short to ground condition<br>• TSS assembly is damaged or is has failed<br>• ECM has failed |

| DTC | Trouble Code Title, Conditions & Possible Causes |
|---|---|
| **DTC: P0721**<br>**Years:** 2007, 2008<br>**Models:** Passat, Jetta, Golf, GTI, New Beetle, New Beetle Convertible<br>**Engines:** 2.0L, 2.5L<br>**Transmissions:** A/T | **A/T Output Shaft Speed Sensor Noise Interference Conditions:**<br>Engine started, VSS signal more than 1 mph, and the ECM detected "noise" interference on the Output Shaft Speed (OSS) sensor circuit.<br>**Possible Causes:**<br>• After market add-on devices interfering with the OSS signal<br>• OSS connector is damaged, loose or shorted, or the wiring is misrouted or it is damaged<br>• OSS assembly is damaged or it has failed<br>• ECM has failed |
| **DTC: P0722**<br>**Years:** 2007, 2008<br>**Models:** Passat, Jetta, Golf, GTI, New Beetle, New Beetle Convertible<br>**Engines:** 2.0L, 2.5L, 3.6L<br>**Transmissions:** A/T | **A/T Output Speed Sensor No Signal Conditions:**<br>Engine started, and the ECM did not detect any Vehicle Speed Sensor (VSS) sensor signals upon initial vehicle movement.<br>**Possible Causes:**<br>• After market add-on devices interfering with the VSS signal<br>• VSS sensor wiring is misrouted, damaged or shorting<br>• ECM and/or TCM has failed |
| **DTC: P0725**<br>**Years:** 2007, 2008<br>**Models:** Passat, Jetta, Golf, GTI, New Beetle, New Beetle Convertible<br>**Engines:** 2.0L, 2.5L, 3.6L<br>**Transmissions:** A/T | **Engine Speed Input Circuit Malfunction Conditions:**<br>The Transmission Control Module (TCM) does not receive a signal from the Engine Control Module (ECM).<br>**Possible Causes:**<br>• The TCM circuit is shorting to ground, B+ or is open<br>• TCM has failed<br>• ECM has failed |
| **DTC: P0726**<br>**Years:** 2007, 2008<br>**Models:** Passat, Jetta, Golf, GTI, New Beetle,<br>**Engines:** 2.0L, 2.5L, 3.6L<br>**Transmissions:** A/T | **Engine Speed Input Circuit Range/Performance Conditions:**<br>The Engine Speed (RPM) Sensor detects engine speed and reference marks. Without an engine speed signal, the engine will not start. If the engine speed signal fails while the engine is running, the engine will stop immediately.<br>**Note: There is a larger-sized gap on the sensor wheel. This gap is the reference mark and does not mean that the sensor wheel is damaged.**<br>**Possible Causes:**<br>• Engine speed sensor has failed<br>• Circuit is shorting to ground, B+ or is open<br>• Sensor wheel is damaged, run out or not properly secured<br>• ECM has failed |
| **DTC: P0727**<br>**Years:** 2007, 2008<br>**Models:** Passat, Jetta, Golf, GTI, New Beetle,<br>**Engines:** 2.0L, 2.5L, 3.6L<br>**Transmissions:** A/T | **Engine Speed Input Circuit No Signal Conditions:**<br>The Engine Speed (RPM) Sensor detects engine speed and reference marks. Without an engine speed signal, the engine will not start. If the engine speed signal fails while the engine is running, the engine will stop immediately.<br>**Note: There is a larger-sized gap on the sensor wheel. This gap is the reference mark and does not mean that the sensor wheel is damaged.**<br>**Possible Causes:**<br>• Engine speed sensor has failed<br>• Circuit is shorting to ground, B+ or is open<br>• Sensor wheel is damaged, run out or not properly secured<br>• ECM has failed |
| **DTC: P0729**<br>**Years:** 2007, 2008<br>**Models:** Passat, Jetta, Golf, GTI, New Beetle, New Beetle Convertible<br>**Engines:** 2.0L, 2.5L, 3.6L<br>**Transmissions:** A/T | **Gear 6 Incorrect Ratio Conditions:**<br>Engine started, battery voltage must be at least 11.5v, all electrical components must be off, and the ground between the engine and the chassis must be well connected. The ECM detected an incorrect ratio within the sixth gear.<br>**Possible Causes:**<br>• ATF level is low<br>• Circuit harness connector contacts are corroded or ingressed of water<br>• Circuit wires have shorted to each other, to battery or ground<br>• Automatic Transmission Hydraulic Pressure Sensor 1 has failed<br>• Solenoid valves in valve body are faulty<br>• Transmission Control Module (TCM) needs replacing<br>• Transmission Input Speed (RPM) Sensor has failed<br>• Transmission Output Speed (RPM) Sensor has failed |

| DTC | Trouble Code Title, Conditions & Possible Causes |
|---|---|
| **DTC: P0730**<br>**Years:** 2007, 2008<br>**Models:** Passat, Jetta, Golf, GTI, New Beetle, New Beetle Convertible<br>**Engines:** 2.0L, 2.5L, 3.6L<br>**Transmissions:** A/T | **Gear Incorrect Ratio Conditions:**<br>Engine started, battery voltage must be at least 11.5v, all electrical components must be off, and the ground between the engine and the chassis must be well connected. The ECM detected an incorrect gear ratio.<br>**Possible Causes:**<br>• ATF level is low<br>• Circuit harness connector contacts are corroded or ingressed of water<br>• Circuit wires have shorted to each other, to battery or ground<br>• Automatic Transmission Hydraulic Pressure Sensor 1 has failed<br>• Solenoid valves in valve body are faulty<br>• Transmission Control Module (TCM) needs replacing<br>• Transmission Input Speed (RPM) Sensor has failed<br>• Transmission Output Speed (RPM) Sensor has failed |
| **DTC: P0731**<br>**Years:** 2007, 2008<br>**Models:** Passat, Jetta, Golf, GTI, New Beetle, New Beetle Convertible<br>**Engines:** 2.0L, 2.5L, 3.6L<br>**Transmissions:** A/T | **Incorrect First Gear Ratio Conditions:**<br>Engine started, vehicle operating with 1st gear commanded "on", and the ECM detected an incorrect 1st gear ratio during the test.<br>**Possible Causes:**<br>• 1st Gear solenoid harness connector not properly seated<br>• 1st Gear solenoid signal shorted to ground, or open<br>• 1st Gear solenoid wiring harness connector is damaged<br>• 1st Gear solenoid is damaged or not properly installed<br>• ATF level is low<br>• Circuit harness connector contacts are corroded or ingressed of water<br>• Circuit wires have shorted to each other, to battery or ground<br>• Automatic Transmission Hydraulic Pressure Sensor 1 has failed<br>• Transmission Control Module (TCM) needs replacing<br>• Transmission Input Speed (RPM) Sensor has failed<br>• Transmission Output Speed (RPM) Sensor has failed |
| **DTC: P0732**<br>**Years:** 2007, 2008<br>**Models:** Passat, Jetta, Golf, GTI, New Beetle, New Beetle Convertible<br>**Engines:** 2.0L, 2.5L, 3.6L<br>**Transmissions:** A/T | **Incorrect Second Gear Ratio Conditions:**<br>Engine started, vehicle operating with 2nd Gear commanded "on", and the ECM detected an incorrect 2nd gear ratio during the test.<br>**Possible Causes:**<br>• 2nd Gear solenoid harness connector not properly seated<br>• 2nd Gear solenoid signal shorted to ground, or open<br>• 2nd Gear solenoid wring harness connector is damaged<br>• 2nd Gear solenoid is damaged or not properly installed<br>• ATF level is low<br>• Circuit harness connector contacts are corroded or ingressed of water<br>• Circuit wires have shorted to each other, to battery or ground<br>• Automatic Transmission Hydraulic Pressure Sensor 1 has failed<br>• Transmission Control Module (TCM) needs replacing<br>• Transmission Input Speed (RPM) Sensor has failed<br>• Transmission Output Speed (RPM) Sensor has failed |
| **DTC: P0733**<br>**Years:** 2007, 2008<br>**Models:** Passat, Jetta, Golf, GTI, New Beetle, New Beetle Convertible<br>**Engines:** 2.0L, 2.5L, 3.6L<br>**Transmissions:** A/T | **Incorrect Third Gear Ratio Conditions:**<br>Engine started, vehicle operating with 3rd Gear commanded "on", and the ECM detected an incorrect 3rd gear ratio during the test.<br>**Possible Causes:**<br>• 3rd Gear solenoid harness connector not properly seated<br>• 3rd Gear solenoid signal shorted to ground, or open<br>• 3rd Gear solenoid wiring harness connector is damaged<br>• 3rd Gear solenoid is damaged or not properly installed<br>• ATF level is low<br>• Circuit harness connector contacts are corroded or ingressed of water<br>• Circuit wires have shorted to each other, to battery or ground<br>• Automatic Transmission Hydraulic Pressure Sensor 1 has failed<br>• Transmission Control Module (TCM) needs replacing<br>• Transmission Input Speed (RPM) Sensor has failed<br>• Transmission Output Speed (RPM) Sensor has failed |

| DTC | Trouble Code Title, Conditions & Possible Causes |
|---|---|
| **DTC: P0734**<br>**Years: 2007, 2008**<br>**Models:** Passat, Jetta, Golf, GTI, New Beetle, New Beetle Convertible<br>**Engines:** 2.0L, 2.5L, 3.6L<br>**Transmissions:** A/T | **Incorrect Fourth Gear Ratio Conditions:**<br>Engine started, vehicle operating with 4th Gear commanded "on", and the ECM detected an incorrect 4th gear ratio during the test.<br>**Possible Causes:**<br>• 4th Gear solenoid harness connector not properly seated<br>• 4th Gear solenoid signal shorted to ground, or open<br>• 4th Gear solenoid wiring harness connector is damaged<br>• 4th Gear solenoid is damaged or not properly installed<br>• ATF level is low<br>• Circuit harness connector contacts are corroded or ingressed of water<br>• Circuit wires have shorted to each other, to battery or ground<br>• Automatic Transmission Hydraulic Pressure Sensor 1 has failed<br>• Transmission Control Module (TCM) needs replacing<br>• Transmission Input Speed (RPM) Sensor has failed<br>• Transmission Output Speed (RPM) Sensor has failed |
| **DTC: P0740**<br>**Years: 2007, 2008**<br>**Models:** Passat, Jetta, Golf, GTI, New Beetle, New Beetle Convertible<br>**Engines:** 2.0L, 2.5L, 3.6L<br>**Transmissions:** A/T | **TCC Solenoid Circuit Malfunction Conditions:**<br>Engine started, KOER Self-Test enabled, vehicle driven at cruise speed, and the ECM did not detect any voltage drop across the TCC solenoid circuit during the test period.<br>**Possible Causes:**<br>• TCC solenoid control circuit is open or shorted to ground<br>• TCC solenoid wiring harness connector is damaged<br>• TCC solenoid is damaged or has failed<br>• ECM has failed |
| **DTC: P0743**<br>**Years: 2007, 2008**<br>**Models:** Passat, Jetta, Golf, GTI, New Beetle, New Beetle Convertible<br>**Engines:** 2.0L, 2.5L, 3.6L<br>**Transmissions:** A/T | **Torque Converter Clutch Circuit Electrical Malfunction Conditions:**<br>Engine started, KOER Self-Test enabled, vehicle driven at cruise speed, and the ECM did not detect any voltage drop across the TCC solenoid circuit during the test period.<br>**Possible Causes:**<br>• TCC solenoid control circuit is open or shorted to ground<br>• TCC solenoid wiring harness connector is damaged<br>• TCC solenoid is damaged or has failed<br>• ECM has failed |
| **DTC: P0746**<br>**Years: 2007, 2008**<br>**Models:** Passat, Jetta, Golf, GTI, New Beetle, New Beetle Convertible<br>**Engines:** 2.0L, 2.5L, 3.6L<br>**Transmissions:** A/T | **Pressure Control Solenoid "A" Performance or Stuck Off Conditions:**<br>Engine started, battery voltage must be at least 11.5v, all electrical components must be off, and the ground between the engine and the chassis must be well connected. The ECM detected the pressure control solenoid was in the "stuck off" position.<br>**Possible Causes:**<br>• ATF level is low<br>• Circuit harness connector contacts are corroded or ingressed of water<br>• Circuit wires have shorted to each other, to battery or ground<br>• Automatic Transmission Hydraulic Pressure Sensor 1 has failed<br>• Solenoid valves in valve body are faulty<br>• Transmission Control Module (TCM) needs replacing<br>• Transmission Input Speed (RPM) Sensor has failed<br>• Transmission Output Speed (RPM) Sensor has failed |
| **DTC: P0747**<br>**Years: 2007, 2008**<br>**Models:** Passat, Jetta, Golf, GTI, New Beetle, New Beetle Convertible<br>**Engines:** 2.0L, 2.5L, 3.6L<br>**Transmissions:** A/T | **Pressure Control Solenoid "A" Performance or Stuck On Conditions:**<br>Engine started, battery voltage must be at least 11.5v, all electrical components must be off, and the ground between the engine and the chassis must be well connected. The ECM detected the pressure control solenoid was in the "stuck on" position.<br>**Possible Causes:**<br>• ATF level is low<br>• Circuit harness connector contacts are corroded or ingressed of water<br>• Circuit wires have shorted to each other, to battery or ground<br>• Automatic Transmission Hydraulic Pressure Sensor 1 has failed<br>• Solenoid valves in valve body are faulty<br>• Transmission Control Module (TCM) needs replacing<br>• Transmission Input Speed (RPM) Sensor has failed<br>• Transmission Output Speed (RPM) Sensor has failed |

| DTC | Trouble Code Title, Conditions & Possible Causes |
|---|---|
| **DTC: P0748**<br>**Years: 2007, 2008**<br>**Models:** Passat, Jetta, Golf, GTI, New Beetle, New Beetle Convertible<br>**Engines:** 2.0L, 2.5L, 3.6L<br>**Transmissions:** A/T | **Pressure Control Solenoid Electrical Conditions:**<br>The valve body solenoid valve is not receiving a signal.<br>**Possible Causes:**<br>• Pressure control solenoid circuit is shorting to ground<br>• Pressure control solenoid circuit is open<br>• Valve has failed<br>• TCM has failed<br>• ECM has failed |
| **DTC: P0749**<br>**Years: 2007, 2008**<br>**Models:** Passat, Jetta, Golf, GTI, New Beetle, New Beetle Convertible<br>**Engines:** 2.0L, 2.5L, 3.6L<br>**Transmissions:** A/T | **Pressure Control Solenoid "A" Intermittent Conditions:**<br>The valve body solenoid valve is receiving an improper signal.<br>**Possible Causes:**<br>• Pressure control solenoid circuit is shorting to ground<br>• Pressure control solenoid circuit is open<br>• Valve has failed<br>• TCM has failed<br>• ECM has failed |
| **DTC: P0751**<br>**Years: 2007, 2008**<br>**Models:** Passat, Jetta, Golf, GTI, New Beetle, New Beetle Convertible<br>**Engines:** 2.0L, 2.5L, 3.6L<br>**Transmissions:** A/T | **Shift Solenoid "A" Performance or Stuck Off Conditions:**<br>Engine started, vehicle driven with the solenoid applied, and the ECM detected an unexpected voltage condition on the SS1/A solenoid circuit was incorrect during the test.<br>**Possible Causes:**<br>• Solenoid valves in valve body are faulty<br>• Solenoid circuit is shorting to ground<br>• Solenoid circuit is open<br>• TCM has failed or wiring is shorting<br>• ECM has failed |
| **DTC: P0752**<br>**Years: 2007, 2008**<br>**Models:** Passat, Jetta, Golf, GTI, New Beetle, New Beetle Convertible<br>**Engines:** 2.0L, 2.5L, 3.6L<br>**Transmissions:** A/T | **A/T Shift Solenoid 1/A Function Range/Performance Conditions:**<br>Engine started, vehicle driven with the solenoid applied, and the ECM detected a mechanical failure while operating the Shift Solenoid 1/A during the CCM test period.<br>**Possible Causes:**<br>• SS1/A solenoid is stuck in the "on" position<br>• SS1/A solenoid has a mechanical failure<br>• SS1/A solenoid has a hydraulic failure<br>• ECM has failed |
| **DTC: P0753**<br>**Years: 2007, 2008**<br>**Models:** Passat, Jetta, Golf, GTI, New Beetle, New Beetle Convertible<br>**Engines:** 2.0L, 2.5L, 3.6L<br>**Transmissions:** A/T | **A/T Shift Solenoid 1/A Circuit Malfunction Conditions:**<br>Engine started, vehicle driven with the solenoid applied, and the ECM detected an unexpected voltage condition on the SS1/A solenoid circuit was incorrect during the test.<br>**Possible Causes:**<br>• SS1/A solenoid control circuit is open<br>• SS1/A solenoid control circuit is shorted to ground<br>• SS1/A solenoid wiring harness connector is damaged<br>• SS1/A solenoid is damaged or has failed<br>• ECM has failed |
| **DTC: P0756**<br>**Years: 2007, 2008**<br>**Models:** Passat, Jetta, Golf, GTI, New Beetle, New Beetle Convertible<br>**Engines:** 2.0L, 2.5L, 3.6L<br>**Transmissions:** A/T | **A/T Shift Solenoid 2/B Function Range/Performance Conditions:**<br>Engine started, vehicle driven with the solenoid applied, and the ECM detected a mechanical failure while operating the Shift Solenoid 2/B during the CCM test period.<br>**Possible Causes:**<br>• SS2/B solenoid is stuck in the "on" position<br>• SS2/B solenoid has a mechanical failure<br>• SS2/B solenoid has a hydraulic failure<br>• ECM has failed |
| **DTC: P0757**<br>Years: 2007, 2008<br>**Models:** Passat, Jetta, Golf, GTI, New Beetle, New Beetle Convertible<br>**Engines:** 2.0L, 2.5L, 3.6L<br>**Transmissions:** A/T | **A/T Shift Solenoid 2/B Function Range/Performance Conditions:**<br>Engine started, vehicle driven with the solenoid applied, and the ECM detected a mechanical failure while operating the Shift Solenoid 2/B during the CCM test period.<br>**Possible Causes:**<br>• SS2/B solenoid is stuck in the "on" position<br>• SS2/B solenoid has a mechanical failure<br>• SS2/B solenoid has a hydraulic failure<br>• ECM has failed |

| DTC | Trouble Code Title, Conditions & Possible Causes |
|---|---|
| **DTC: P0758**<br>**Years: 2007, 2008**<br>**Models:** Passat, Jetta, Golf, GTI, New Beetle, New Beetle Convertible<br>**Engines:** 2.0L, 2.5L, 3.6L<br>**Transmissions**: A/T | **A/T Shift Solenoid 2/B Circuit Malfunction Conditions:**<br>Engine started, vehicle driven with the solenoid applied, and the ECM detected an unexpected voltage condition on the SS1/A solenoid circuit was incorrect during the test..<br>**Possible Causes:**<br>• Shift Solenoid 2/B connector is damaged, open or shorted<br>• Shift Solenoid 2/B control circuit is open<br>• Shift Solenoid 2/B control circuit is shorted to ground<br>• Shift Solenoid 2/B is damaged or it has failed<br>• ECM has failed |
| **DTC: P0761**<br>**Years: 2007, 2008**<br>**Models:** Passat, Jetta, Golf, GTI, New Beetle, New Beetle Convertible<br>**Engines:** 2.0L, 2.5L, 3.6L<br>**Transmissions**: A/T | **A/T Shift Solenoid 3/C Function Range/Performance Conditions:**<br>Engine started, vehicle driven with Shift Solenoid 3/C applied, and the ECM detected a mechanical failure occurred (stuck "off") while operating Shift Solenoid 3/C during the test.<br>**Possible Causes:**<br>• SS3/C solenoid may be stuck "off"<br>• SS3/C solenoid has a mechanical failure<br>• SS3/C solenoid has a hydraulic failure<br>• ECM has failed |
| **DTC: P0762**<br>**Years: 2007, 2008**<br>**Models:** Passat, Jetta, Golf, GTI, New Beetle, New Beetle Convertible<br>**Engines:** 2.0L, 2.5L, 3.6L<br>**Transmissions**: A/T | **A/T Shift Solenoid 3/C Function Range/Performance Conditions:**<br>Engine started, vehicle driven with Shift Solenoid 3/C applied, and the ECM detected a mechanical failure occurred (stuck "on") while operating Shift Solenoid 3/C during the test.<br>**Possible Causes:**<br>• SS3/C solenoid may be stuck "on"<br>• SS3/C solenoid has a mechanical failure<br>• SS3/C solenoid has a hydraulic failure<br>• ECM has failed |
| **DTC: P0763**<br>**Years: 2007, 2008**<br>**Models:** Passat, Jetta, Golf, GTI, New Beetle, New Beetle Convertible<br>**Engines:** 2.0L, 2.5L, 3.6L<br>**Transmissions**: A/T | **A/T Shift Solenoid 3/C Electrical Conditions:**<br>Engine started, vehicle driven with the solenoid applied, and the ECM detected an unexpected voltage condition on the SS3/C solenoid circuit was incorrect during the test..<br>**Possible Causes:**<br>• Shift Solenoid 3/C connector is damaged, open or shorted<br>• Shift Solenoid 3/C control circuit is open<br>• Shift Solenoid 3/C control circuit is shorted to ground<br>• Shift Solenoid 3/C is damaged or it has failed<br>• ECM has failed |
| **DTC: P0766**<br>**Years: 2007, 2008**<br>**Models:** Jetta, Golf, GTI, New Beetle, New Beetle Convertible<br>**Engines:** 2.0L, 2.5L, 3.6L<br>**Transmissions**: A/T | **A/T Shift Solenoid D Performance Conditions:**<br>Engine started, vehicle driven with the solenoid applied, and the ECM detected an unexpected voltage condition on the SS3/C solenoid circuit was incorrect during the test..<br>**Possible Causes:**<br>• Shift Solenoid D connector is damaged, open or shorted<br>• Shift Solenoid D control circuit is open<br>• Shift Solenoid D control circuit is shorted to ground<br>• Shift Solenoid D is damaged or it has failed<br>• ECM has failed |
| **DTC: P0768**<br>**Years: 2007, 2008**<br>**Models:** Passat, Jetta, Golf, GTI, New Beetle, New Beetle Convertible<br>**Engines:** 2.0L, 2.5L, 3.6L<br>**Transmissions**: A/T | **A/T Shift Solenoid 3/D Electrical Conditions:**<br>Engine started, vehicle driven with the solenoid applied, and the ECM detected an unexpected voltage condition on the SS3/D solenoid circuit was incorrect during the test..<br>**Possible Causes:**<br>• Shift Solenoid 3/D connector is damaged, open or shorted<br>• Shift Solenoid 3/D control circuit is open<br>• Shift Solenoid 3/D control circuit is shorted to ground<br>• Shift Solenoid 3/D is damaged or it has failed<br>• ECM has failed |

| DTC | Trouble Code Title, Conditions & Possible Causes |
|---|---|
| **DTC: P0771**<br>**Years: 2007, 2008**<br>**Models:** Jetta, Golf, GTI, New Beetle, New Beetle Convertible<br>**Engines:** 2.0L, 2.5L, 3.6L<br>**Transmissions:** A/ | **A/T Shift Solenoid E Performance Conditions:**<br>Engine started, vehicle driven with the solenoid applied, and the ECM detected an unexpected voltage condition on the SS3/C solenoid circuit was incorrect during the test..<br>**Possible Causes:**<br>• Shift Solenoid D connector is damaged, open or shorted<br>• Shift Solenoid D control circuit is open<br>• Shift Solenoid D control circuit is shorted to ground<br>• Shift Solenoid D is damaged or it has failed<br>• ECM has failed |
| **DTC: P0773**<br>**Years: 2007, 2008**<br>**Models:** Passat, Jetta, Golf, GTI, New Beetle, New Beetle Convertible<br>**Engines:** 2.0L, 2.5L, 3.6L<br>**Transmissions:** A/T | **A/T Shift Solenoid E Electrical Conditions:**<br>Engine started, vehicle driven with the solenoid applied, and the ECM detected an unexpected voltage condition on the SS3/D solenoid circuit was incorrect during the test..<br>**Possible Causes:**<br>• Shift Solenoid connector is damaged, open or shorted<br>• Shift Solenoid control circuit is open<br>• Shift Solenoid control circuit is shorted to ground<br>• Shift Solenoid is damaged or it has failed<br>• ECM has failed |
| **DTC: P0776**<br>**Years: 2007, 2008**<br>**Models:** Passat, Jetta, Golf, GTI, New Beetle, New Beetle Convertible<br>**Engines:** 2.0L, 2.5L, 3.6L<br>**Transmissions:** A/T | **Pressure Control Solenoid "B" Performance or Stuck Off Conditions:**<br>Engine started, vehicle driven with Shift Solenoid 3/C applied, and the ECM detected a mechanical failure occurred (stuck "off") while operating Shift Solenoid 3/C during the test.<br>**Possible Causes:**<br>• SS3/C solenoid may be stuck "off"<br>• SS3/C solenoid has a mechanical failure<br>• SS3/C solenoid has a hydraulic failure<br>• ECM has failed |
| **DTC: P0777**<br>**Years: 2007, 2008**<br>**Models:** Passat, Jetta, Golf, GTI, New Beetle, New Beetle Convertible<br>**Engines:** 2.0L, 2.5L, 3.6L<br>**Transmissions:** A/T | **Pressure Control Solenoid "B" Stuck On Conditions:**<br>Engine started, vehicle driven with Shift Solenoid 3/C applied, and the ECM detected a mechanical failure occurred (stuck "on") while operating Shift Solenoid 3/C during the test.<br>**Possible Causes:**<br>• SS3/C solenoid may be stuck "on"<br>• SS3/C solenoid has a mechanical failure<br>• SS3/C solenoid has a hydraulic failure<br>• ECM has failed |
| **DTC: P0778**<br>**Years: 2007, 2008**<br>**Models:** Passat, Jetta, Golf, GTI, New Beetle, New Beetle Convertible<br>**Engines:** 2.0L, 2.5L, 3.6L<br>**Transmissions:** A/T | **Pressure Control Solenoid "B" Electrical Conditions:**<br>Engine started, vehicle driven with the solenoid applied, and the ECM detected an unexpected voltage condition on the SS3/C solenoid circuit was incorrect during the test..<br>**Possible Causes:**<br>• Shift Solenoid connector is damaged, open or shorted<br>• Shift Solenoid control circuit is open<br>• Shift Solenoid control circuit is shorted to ground<br>• Shift Solenoid is damaged or it has failed<br>• ECM has failed |
| **DTC: P0785**<br>**Years: 2007, 2008**<br>**Models:** Passat, Jetta, Golf, GTI, New Beetle, New Beetle Convertible<br>**Engines:** 2.0L, 2.5L, 3.6L<br>**Transmissions:** A/T | **Shift/Timing Solenoid Conditions:**<br>Engine running and vehicle driven, the ECM detected a mechanical malfunction within the transmission<br>**Possible Causes:**<br>• Solenoid valves in valve body are faulty<br>• Solenoid circuit is shorting to ground<br>• Solenoid circuit is open<br>• TCM has failed or wiring is shorting<br>• ECM has failed<br>• Mechanical malfunction in transmission |

| DTC | Trouble Code Title, Conditions & Possible Causes |
|---|---|
| **DTC: P0796**<br>**Years: 2007, 2008**<br>**Models:** Passat, Jetta, Golf, GTI, New Beetle, New Beetle Convertible<br>**Engines:** 2.0L, 2.5L, 3.6L<br>**Transmissions:** A/T | **Pressure Solenoid "C" Performance or Stuck Off Conditions:**<br>Engine started, vehicle driven with the solenoid applied, and the ECM detected an unexpected voltage condition on the SS1/C solenoid circuit was incorrect during the test.<br>**Possible Causes:**<br>• Solenoid valves in valve body are faulty<br>• Solenoid circuit is shorting to ground<br>• Solenoid circuit is open<br>• TCM has failed or wiring is shorting<br>• ECM has failed |
| **DTC: P0797**<br>**Years: 2007, 2008**<br>**Models:** Passat, Jetta, Golf, GTI, New Beetle, New Beetle Convertible<br>**Engines:** 2.0L, 2.5L, 3.6L<br>**Transmissions:** A/T | **Pressure Solenoid "C" Performance or Stuck On Conditions:**<br>Engine started, vehicle driven with the solenoid applied, and the ECM detected an unexpected voltage condition on the SS1/C solenoid circuit was incorrect during the test.<br>**Possible Causes:**<br>• Solenoid valves in valve body are faulty<br>• Solenoid circuit is shorting to ground<br>• Solenoid circuit is open<br>• TCM has failed or wiring is shorting<br>• ECM has failed |
| **DTC: P0798**<br>**Years: 2007, 2008**<br>**Models:** Passat, Jetta, Golf, GTI, New Beetle, New Beetle Convertible<br>**Engines:** 2.0L, 2.5L, 3.6L<br>**Transmissions:** A/T | **Pressure Solenoid "C" Electrical Conditions:**<br>Engine started, vehicle driven with the solenoid applied, and the ECM detected an unexpected voltage condition on the SS1/C solenoid circuit was incorrect during the test.<br>**Possible Causes:**<br>• Solenoid valves in valve body are faulty<br>• Solenoid circuit is shorting to ground<br>• Solenoid circuit is open<br>• TCM has failed or wiring is shorting<br>• ECM has failed |
| **DTC: P0811**<br>**Years: 2007, 2008**<br>**Models:** Passat, Jetta, Golf, GTI, New Beetle, New Beetle Convertible<br>**Engines:** 2.0L, 2.5L, 3.6L<br>**Transmissions:** A/T | **Excessive Clutch Slippage Conditions:**<br>Engine started, vehicle driven and the ECM and/or TCM has detected that the clutch has slipped multiple times in a given time frame.<br>**Possible Causes:**<br>• Solenoid valves in valve body are faulty<br>• Solenoid circuit is shorting to ground<br>• Solenoid circuit is open<br>• TCM has failed or wiring is shorting<br>• ECM has failed |
| **DTC: P0840**<br>**Years: 2007, 2008**<br>**Models:** Jetta, New Beetle, New Beetle Convertible<br>**Engines:** 2.0L, 2.5L<br>**Transmissions:** A/T | **Transmission Fluid Pressure Sensor/Switch "A" Circuit Conditions:**<br>Engine started, vehicle driven with the solenoid applied, and the ECM detected an unexpected voltage condition on the pressure sensor/switch was incorrect during the test.<br>**Possible Causes:**<br>• Solenoid valves in valve body are faulty<br>• Solenoid circuit is shorting to ground<br>• Solenoid circuit is open<br>• TCM has failed or wiring is shorting<br>• Transmission Input Speed (RPM) Sensor has failed<br>• Transmission Output Speed (RPM) Sensor has failed<br>• ECM has failed |
| **DTC: P0841**<br>**Years: 2007, 2008**<br>**Models:** Jetta, New Beetle, New Beetle Convertible<br>**Engines:** 2.0L, 2.5L<br>**Transmissions:** A/T | **Transmission Fluid Pressure Sensor/Switch "A" Circuit Range/Performance Conditions:**<br>Engine started, vehicle driven with the solenoid applied, and the ECM detected an unexpected voltage condition on the pressure sensor/switch was incorrect during the test.<br>**Possible Causes:**<br>• Solenoid valves in valve body are faulty<br>• Solenoid circuit is shorting to ground<br>• Solenoid circuit is open<br>• TCM has failed or wiring is shorting<br>• Transmission Input Speed (RPM) Sensor has failed<br>• Transmission Output Speed (RPM) Sensor has failed<br>• ECM has failed |

| DTC | Trouble Code Title, Conditions & Possible Causes |
|---|---|
| **DTC: P0845**<br>**Years:** 2007, 2008<br>**Models:** Jetta, New Beetle, New Beetle Convertible<br>**Engines:** 2.0L, 2.5L<br>**Transmissions:** A/T | **Transmission Fluid Pressure Sensor/Switch "B" Circuit Conditions:**<br>Engine started, vehicle driven with the solenoid applied, and the ECM detected an unexpected voltage condition on the pressure sensor/switch was incorrect during the test.<br>**Possible Causes:**<br>• Solenoid valves in valve body are faulty<br>• Solenoid circuit is shorting to ground<br>• Solenoid circuit is open<br>• TCM has failed or wiring is shorting<br>• Transmission Input Speed (RPM) Sensor has failed<br>• Transmission Output Speed (RPM) Sensor has failed<br>• ECM has failed |
| **DTC: P0846**<br>**Years:** 2007, 2008<br>**Models:** Jetta, New Beetle, New Beetle Convertible<br>**Engines:** 2.0L, 2.5L<br>**Transmissions:** A/T | **Transmission Fluid Pressure Sensor/Switch "B" Circuit Range/Performance Conditions:**<br>Engine started, vehicle driven with the solenoid applied, and the ECM detected an unexpected voltage condition on the pressure sensor/switch was incorrect during the test.<br>**Possible Causes:**<br>• Solenoid valves in valve body are faulty<br>• Solenoid circuit is shorting to ground<br>• Solenoid circuit is open<br>• TCM has failed or wiring is shorting<br>• Transmission Input Speed (RPM) Sensor has failed<br>• Transmission Output Speed (RPM) Sensor has failed<br>• ECM has failed |
| **DTC: P0863**<br>**Years:** 2007, 2008<br>**Models:** Passat, Jetta, Golf, GTI, New Beetle, New Beetle Convertible<br>**Engines:** 2.0L, 2.5L, 3.6L<br>**Transmissions:** A/T | **TCM Communication Circuit Conditions:**<br>The Transmission Control Module (ECM) communicates with all databus-capable control modules via a CAN databus. These databus-capable control modules are connected via two data bus wires which are twisted together (CAN_High and CAN_Low), and exchange information (messages). Missing information on the databus is recognized as a malfunction and stored. Trouble-free operation of the CAN-Bus requires that it have a terminal resistance.<br>**Possible Causes:**<br>• ECM has failed<br>• Terminal resistance for CAN-bus are faulty<br>• Can data bus wires have short circuited to each other<br>• TCM has failed |
| **DTC: P0864**<br>**Years:** 2007, 2008<br>**Models:** Passat, Jetta, Golf, GTI, New Beetle, New Beetle Convertible<br>**Engines:** 2.0L, 2.5L, 3.6L<br>**Transmissions:** A/T | **TCM Communication Circuit Range/Performance Conditions:**<br>The Transmission Control Module (ECM) communicates with all databus-capable control modules via a CAN databus. These databus-capable control modules are connected via two data bus wires which are twisted together (CAN_High and CAN_Low), and exchange information (messages). Missing information on the databus is recognized as a malfunction and stored. Trouble-free operation of the CAN-Bus requires that it have a terminal resistance.<br>**Possible Causes:**<br>• ECM has failed<br>• Terminal resistance for CAN-bus are faulty<br>• Can data bus wires have short circuited to each other<br>• TCM has failed |
| **DTC: P0865**<br>**Years:** 2007, 2008<br>**Models:** Passat, Jetta, Golf, GTI, New Beetle, New Beetle Convertible<br>**Engines:** 2.0L, 2.5L, 3.6L<br>**Transmissions:** A/T | **TCM Communication Circuit Low Conditions:**<br>The Transmission Control Module (ECM) communicates with all databus-capable control modules via a CAN databus. These databus-capable control modules are connected via two data bus wires which are twisted together (CAN_High and CAN_Low), and exchange information (messages). Missing information on the databus is recognized as a malfunction and stored. Trouble-free operation of the CAN-Bus requires that it have a terminal resistance.<br>**Possible Causes:**<br>• ECM has failed<br>• Terminal resistance for CAN-bus are faulty<br>• Can data bus wires have short circuited to each other<br>• TCM has failed |

| DTC | Trouble Code Title, Conditions & Possible Causes |
|---|---|
| **DTC: P0884**<br>**Years: 2007, 2008**<br>**Models:** Passat, Jetta, Golf, GTI, New Beetle,<br>**Engines:** 2.0L<br>**Transmissions:** A/T | **TCM Power Input Signal Intermittent Conditions:**<br>The Transmission Control Module (ECM) communicates with all databus-capable control modules via a CAN databus. These databus-capable control modules are connected via two data bus wires which are twisted together (CAN_High and CAN_Low), and exchange information (messages). Missing information on the databus is recognized as a malfunction and stored. Trouble-free operation of the CAN-Bus requires that it have a terminal resistance.<br>**Possible Causes:**<br>• Solenoid valves in valve body are faulty<br>• Solenoid circuit is shorting to ground<br>• Solenoid circuit is open<br>• TCM has failed or wiring is shorting<br>• ECM has failed |
| **DTC: P0886**<br>**Years: 2007, 2008**<br>**Models:** Passat, Jetta, Golf, GTI, New Beetle,<br>**Engines:** 2.0L<br>**Transmissions:** A/T | **TCM Power Relay Control Circuit Low Conditions:**<br>The Transmission Control Module (ECM) communicates with all databus-capable control modules via a CAN databus. These databus-capable control modules are connected via two data bus wires which are twisted together (CAN_High and CAN_Low), and exchange information (messages). Missing information on the databus is recognized as a malfunction and stored. Trouble-free operation of the CAN-Bus requires that it have a terminal resistance.<br>**Possible Causes:**<br>• Solenoid valves in valve body are faulty<br>• Solenoid circuit is shorting to ground<br>• Solenoid circuit is open<br>• TCM has failed or wiring is shorting<br>• ECM has failed |
| **DTC: P0887**<br>**Years: 2007, 2008**<br>**Models:** Passat, Jetta, Golf, GTI, New Beetle,<br>**Engines:** 2.0L<br>**Transmissions:** A/T | **TCM Power Relay Control Circuit High Conditions:**<br>The Transmission Control Module (ECM) communicates with all databus-capable control modules via a CAN databus. These databus-capable control modules are connected via two data bus wires which are twisted together (CAN_High and CAN_Low), and exchange information (messages). Missing information on the databus is recognized as a malfunction and stored. Trouble-free operation of the CAN-Bus requires that it have a terminal resistance.<br>**Possible Causes:**<br>• Solenoid valves in valve body are faulty |
| **DTC: P0889**<br>**Years: 2007, 2008**<br>**Models:** Passat, Jetta, Golf, GTI, New Beetle, New Beetle Convertile<br>**Engines:** 2.0L, 2.5L, 3.6L<br>**Transmissions:** A/T | **TCM Power Relay Sense Circuit Range/Performance Conditions:**<br>The Transmission Control Module (ECM) communicates with all databus-capable control modules via a CAN databus. These databus-capable control modules are connected via two data bus wires which are twisted together (CAN_High and CAN_Low), and exchange information (messages). Missing information on the databus is recognized as a malfunction and stored. Trouble-free operation of the CAN-Bus requires that it have a terminal resistance.<br>**Possible Causes:**<br>• Solenoid valves in valve body are faulty<br>• Solenoid circuit is shorting to ground<br>• Solenoid circuit is open<br>• TCM has failed or wiring is shorting<br>• ECM has failed |
| **DTC: P0890**<br>**Years: 2007, 2008**<br>**Models:** Passat, Jetta, Golf, GTI, New Beetle, New Beetle Convertile<br>**Engines:** 2.0L, 2.5L, 3.6L<br>**Transmissions:** A/T | **TCM Power Relay Sense Circuit Low Conditions:**<br>The Transmission Control Module (ECM) communicates with all databus-capable control modules via a CAN databus. These databus-capable control modules are connected via two data bus wires which are twisted together (CAN_High and CAN_Low), and exchange information (messages). Missing information on the databus is recognized as a malfunction and stored. Trouble-free operation of the CAN-Bus requires that it have a terminal resistance.<br>**Possible Causes:**<br>• Solenoid valves in valve body are faulty<br>• Solenoid circuit is shorting to ground<br>• Solenoid circuit is open<br>• TCM has failed or wiring is shorting<br>• ECM has failed |

| DTC | Trouble Code Title, Conditions & Possible Causes |
|---|---|
| **DTC: P0891**<br>**Years: 2007, 2008**<br>**Models:** Passat, Jetta, Golf, GTI, New Beetle, New Beetle Convertile<br>**Engines:** 2.0L, 2.5L, 3.6L<br>**Transmissions:** A/T | **TCM Power Relay Sense Circuit High Conditions:**<br>The Transmission Control Module (ECM) communicates with all databus-capable control modules via a CAN databus. These databus-capable control modules are connected via two data bus wires which are twisted together (CAN_High and CAN_Low), and exchange information (messages). Missing information on the databus is recognized as a malfunction and stored. Trouble-free operation of the CAN-Bus requires that it have a terminal resistance.<br>**Possible Causes:**<br>• Solenoid valves in valve body are faulty<br>• Solenoid circuit is shorting to ground<br>• Solenoid circuit is open<br>• TCM has failed or wiring is shorting<br>• ECM has failed |
| **DTC: P0892**<br>**Years: 2007, 2008**<br>**Models:** Passat, Jetta, Golf, GTI, New Beetle, New Beetle Convertile<br>**Engines:** 2.0L, 2.5L, 3.6L<br>**Transmissions:** A/T | **TCM Power Relay Sense Circuit High Conditions:**<br>The Transmission Control Module (ECM) communicates with all databus-capable control modules via a CAN databus. These databus-capable control modules are connected via two data bus wires which are twisted together (CAN_High and CAN_Low), and exchange information (messages). Missing information on the databus is recognized as a malfunction and stored. Trouble-free operation of the CAN-Bus requires that it have a terminal resistance.<br>**Possible Causes:**<br>• Solenoid valves in valve body are faulty<br>• Solenoid circuit is shorting to ground<br>• Solenoid circuit is open<br>• TCM has failed or wiring is shorting<br>• ECM has failed |

## Gas Engine OBD II Trouble Code List (P1xxx Codes)

| DTC | Trouble Code Title, Conditions & Possible Causes |
|---|---|
| **DTC: P1025**<br>**Years: 2007, 2008**<br>**Models:** Jetta<br>**Engines:** 2.0L<br>**Transmissions:** All | **Fuel Pressure Regulator Valve Mechanical Malfunction Conditions:**<br>Engine started, battery voltage must be at least 11.5v, all electrical components must be off, the ground between the engine and the chassis must be well connected, the exhaust system must be properly sealed between the catalytic converter and the cylinder head, the coolant temperature must be 80 degrees Celsius, and the oxygen sensor heater for oxygen sensor before the catalytic converter must be properly functioning. The ECM detected a mechanical malfunction of the fuel pressure regulator valve.<br>**Possible Causes:**<br>• Fuel Pressure Regulator Valve is faulty<br>• Fuel Pressure Sensor is faulty<br>• Low Fuel Pressure Sensor is faulty |
| **DTC: P1093**<br>**Years: 2007, 2008**<br>**Models:** Jetta<br>**Engines:** 2.0L<br>**Transmissions:** All | **Fuel Trim 2, Bank 1 Malfunction Conditions:**<br>Engine started, battery voltage must be at least 11.5v, all electrical components must be off, the ground between the engine and the chassis must be well connected, the exhaust system must be properly sealed between the catalytic converter and the cylinder head, the coolant temperature must be 80 degrees Celsius, and the oxygen sensor heater for oxygen sensor before the catalytic converter must be properly functioning. The ECM detected a mechanical malfunction of the fuel pressure regulator valve.<br>**Possible Causes:**<br>• Fuel Pressure Regulator Valve is faulty<br>• Fuel Pressure Sensor is faulty<br>• Low Fuel Pressure Sensor is faulty |

| DTC | Trouble Code Title, Conditions & Possible Causes |
|---|---|
| **DTC: P1100**<br>**Years: 2007, 2008**<br>**Models:** Passat<br>**Engines:** 2.0L<br>**Transmissions:** All | **O2 Sensor Circuit (Bank 1-Sensor 2) Heating Circuit Voltage Too Low Conditions:**<br>Engine started, battery voltage must be at least 11.5v, all electrical components must be off, the ground between the engine and the chassis must be well connected, the exhaust system must be properly sealed between the catalytic converter and the cylinder head, and the oxygen sensor heater for oxygen sensor before the catalytic converter must be properly functioning. The ECM detected a voltage on the O2 sensor circuit that was below the parameters to function properly.<br>**Note: For resistance testing of sensor heating, oxygen sensor should be cooled to ambient temperature. High temperatures at oxygen sensor may lead to inaccurate measurements.**<br>**Possible Causes:**<br>• Oxygen sensor (before catalytic converter) is faulty<br>• Oxygen sensor (behind catalytic converter) is faulty<br>• Oxygen sensor heater (before catalytic converter) is faulty<br>• Oxygen sensor heater (behind catalytic converter) is faulty<br>• Circuit wiring has a short to power or ground<br>• Engine Component Power Supply Relay is faulty<br>• E-box fuses for oxygen sensor are faulty<br>• Leaks present in the exhaust manifold or exhaust pipes<br>• HO2S signal wire and ground wire crossed in connector<br>• HO2S element is fuel contaminated or has failed<br>• ECM has failed |
| **DTC: P1102**<br>**2T, MIL: Yes**<br>**Years: 2007, 2008**<br>**Models:** Passat, Jetta, Golf, GTI, New Beetle, New Beetle Convertible<br>**Engines:** 2.0L, 2.5L, 3.6L<br>**Transmissions:** A/T | **O2 Sensor Circuit (Bank 1-Sensor 1) Short to B+ Conditions:**<br>Engine started, battery voltage must be at least 11.5v, all electrical components must be off, the ground between the engine and the chassis must be well connected, the exhaust system must be properly sealed between the catalytic converter and the cylinder head, and the oxygen sensor heater for oxygen sensor before the catalytic converter must be properly functioning. The ECM detected a voltage on the O2 sensor circuit that was outside the parameters to function properly.<br>**Note: For resistance testing of sensor heating, oxygen sensor should be cooled to ambient temperature. High temperatures at oxygen sensor may lead to inaccurate measurements.**<br>**Possible Causes:**<br>• Oxygen sensor (before catalytic converter) is faulty<br>• Oxygen sensor (behind catalytic converter) is faulty<br>• Oxygen sensor heater (before catalytic converter) is faulty<br>• Oxygen sensor heater (behind catalytic converter) is faulty<br>• Circuit wiring has a short to power or ground<br>• Engine Component Power Supply Relay is faulty<br>• E-box fuses for oxygen sensor are faulty<br>• Leaks present in the exhaust manifold or exhaust pipes<br>• HO2S signal wire and ground wire crossed in connector<br>• HO2S element is fuel contaminated or has failed<br>• ECM has failed |
| **DTC: P1103**<br>**Years: 2007, 2008**<br>**Models:** Passat, Jetta, Golf, GTI, New Beetle, New Beetle Convertible<br>**Engines:** 2.0L, 2.5L, 3.6L<br>**Transmissions:** A/T | **O2 Sensor Circuit (Bank 1-Sensor 1) Output Too Low Conditions:**<br>Engine started, battery voltage must be at least 11.5v, all electrical components must be off, the ground between the engine and the chassis must be well connected, the exhaust system must be properly sealed between the catalytic converter and the cylinder head, and the oxygen sensor heater for oxygen sensor before the catalytic converter must be properly functioning. The ECM detected a voltage on the O2 sensor circuit that was outside the parameters to function properly.<br>**Note: For resistance testing of sensor heating, oxygen sensor should be cooled to ambient temperature. High temperatures at oxygen sensor may lead to inaccurate measurements.**<br>**Possible Causes:**<br>• Oxygen sensor (before catalytic converter) is faulty<br>• Oxygen sensor (behind catalytic converter) is faulty<br>• Oxygen sensor heater (before catalytic converter) is faulty<br>• Oxygen sensor heater (behind catalytic converter) is faulty<br>• Circuit wiring has a short to power or ground<br>• Engine Component Power Supply Relay is faulty<br>• E-box fuses for oxygen sensor are faulty<br>• Leaks present in the exhaust manifold or exhaust pipes<br>• HO2S signal wire and ground wire crossed in connector<br>• HO2S element is fuel contaminated or has failed<br>• ECM has failed |

| DTC | Trouble Code Title, Conditions & Possible Causes |
|---|---|
| **DTC: P1105**<br>**2T, MIL: Yes**<br>**Years: 2007, 2008**<br>**Models:** Passat, Jetta, Golf, GTI, New Beetle, New Beetle Convertible<br>**Engines:** 2.0L, 2.5L, 3.6L<br>**Transmissions:** A/T | **O2 Sensor Circuit (Bank 1-Sensor 2) Short to B+ Conditions:**<br>Engine started, battery voltage must be at least 11.5v, all electrical components must be off, the ground between the engine and the chassis must be well connected, the exhaust system must be properly sealed between the catalytic converter and the cylinder head, and the oxygen sensor heater for oxygen sensor before the catalytic converter must be properly functioning. The ECM detected a voltage on the O2 sensor circuit that was outside the parameters to function properly.<br>**Note: For resistance testing of sensor heating, oxygen sensor should be cooled to ambient temperature. High temperatures at oxygen sensor may lead to inaccurate measurements.**<br>**Possible Causes:**<br>• Oxygen sensor (before catalytic converter) is faulty<br>• Oxygen sensor (behind catalytic converter) is faulty<br>• Oxygen sensor heater (before catalytic converter) is faulty<br>• Oxygen sensor heater (behind catalytic converter) is faulty<br>• Circuit wiring has a short to power or ground<br>• Engine Component Power Supply Relay is faulty<br>• E-box fuses for oxygen sensor are faulty<br>• Leaks present in the exhaust manifold or exhaust pipes<br>• HO2S signal wire and ground wire crossed in connector<br>• HO2S element is fuel contaminated or has failed<br>• ECM has failed |
| **DTC: P1107**<br>**Years: 2007, 2008**<br>**Models:** Passat<br>**Engines:** 2.5L, 3.6L<br>**Transmissions:** A/T | **O2 Sensor Circuit (Bank 2-Sensor 1) Voltage Too Low Conditions:**<br>Engine started, battery voltage must be at least 11.5v, all electrical components must be off, the ground between the engine and the chassis must be well connected, the exhaust system must be properly sealed between the catalytic converter and the cylinder head, and the oxygen sensor heater for oxygen sensor before the catalytic converter must be properly functioning. The ECM detected a voltage on the O2 sensor circuit that was outside the parameters to function properly.<br>**Note: For resistance testing of sensor heating, oxygen sensor should be cooled to ambient temperature. High temperatures at oxygen sensor may lead to inaccurate measurements.**<br>**Possible Causes:**<br>• Oxygen sensor (before catalytic converter) is faulty<br>• Oxygen sensor (behind catalytic converter) is faulty<br>• Oxygen sensor heater (before catalytic converter) is faulty<br>• Oxygen sensor heater (behind catalytic converter) is faulty<br>• Circuit wiring has a short to power or ground<br>• Engine Component Power Supply Relay is faulty<br>• E-box fuses for oxygen sensor are faulty<br>• Leaks present in the exhaust manifold or exhaust pipes<br>• HO2S signal wire and ground wire crossed in connector<br>• HO2S element is fuel contaminated or has failed<br>• ECM has failed |
| **DTC: P1110**<br>**Years: 2007, 2008**<br>**Models:** Passat<br>**Engines:** 2.5L, 3.6L<br>**Transmissions:** A/T | **O2 Sensor Circuit (Bank 2-Sensor 2) Short to B+ Conditions:**<br>Engine started, battery voltage must be at least 11.5v, all electrical components must be off, the ground between the engine and the chassis must be well connected, the exhaust system must be properly sealed between the catalytic converter and the cylinder head, and the oxygen sensor heater for oxygen sensor before the catalytic converter must be properly functioning. The ECM detected a voltage on the O2 sensor circuit that was outside the parameters to function properly.<br>**Note: For resistance testing of sensor heating, oxygen sensor should be cooled to ambient temperature. High temperatures at oxygen sensor may lead to inaccurate measurements.**<br>**Possible Causes:**<br>• Oxygen sensor (before catalytic converter) is faulty<br>• Oxygen sensor (behind catalytic converter) is faulty<br>• Oxygen sensor heater (before catalytic converter) is faulty<br>• Oxygen sensor heater (behind catalytic converter) is faulty<br>• Circuit wiring has a short to power or ground<br>• Engine Component Power Supply Relay is faulty<br>• E-box fuses for oxygen sensor are faulty<br>• Leaks present in the exhaust manifold or exhaust pipes<br>• HO2S signal wire and ground wire crossed in connector<br>• HO2S element is fuel contaminated or has failed<br>• ECM has failed |

| DTC | Trouble Code Title, Conditions & Possible Causes |
|---|---|
| **DTC: P1111**<br>**Years: 2007, 2008**<br>**Models:** Passat, Jetta, Golf, GTI, New Beetle,<br>**Engines:** 2.0L<br>**Transmissions:** A/T | **O2 Control (Bank 1) System Too Lean Conditions:**<br>Engine started, battery voltage must be at least 11.5v, all electrical components must be off, the ground between the engine and the chassis must be well connected, the exhaust system must be properly sealed between the catalytic converter and the cylinder head, and the oxygen sensor heater for oxygen sensor before the catalytic converter must be properly functioning. The ECM detected a measurement on the O2 sensor circuit that was outside the parameters to function properly.<br>**Note: For resistance testing of sensor heating, oxygen sensor should be cooled to ambient temperature. High temperatures at oxygen sensor may lead to inaccurate measurements.**<br>**Note: When an O2S malfunction (P0131 to P0414) is also stored with this malfunction, the O2S malfunction(s) should be repaired first.**<br>**Possible Causes:**<br>• Oxygen sensor (before catalytic converter) is faulty<br>• Oxygen sensor (behind catalytic converter) is faulty<br>• Oxygen sensor heater (before catalytic converter) is faulty<br>• Oxygen sensor heater (behind catalytic converter) is faulty<br>• Circuit wiring has a short to power or ground<br>• Engine Component Power Supply Relay is faulty<br>• E-box fuses for oxygen sensor are faulty<br>• Leaks present in the exhaust manifold or exhaust pipes<br>• HO2S signal wire and ground wire crossed in connector<br>• HO2S element is fuel contaminated or has failed<br>• ECM has failed |
| **DTC: P1112**<br>**Years: 2007, 2008**<br>**Models:** Passat, Jetta, Golf, GTI, New Beetle,<br>**Engines:** 2.0L<br>**Transmissions:** A/T | **O2 Control (Bank 1) System Too Rich Conditions:**<br>Engine started, battery voltage must be at least 11.5v, all electrical components must be off, the ground between the engine and the chassis must be well connected, the exhaust system must be properly sealed between the catalytic converter and the cylinder head, and the oxygen sensor heater for oxygen sensor before the catalytic converter must be properly functioning. The ECM detected a measurement on the O2 sensor circuit that was outside the parameters to function properly.<br>**Note: For resistance testing of sensor heating, oxygen sensor should be cooled to ambient temperature. High temperatures at oxygen sensor may lead to inaccurate measurements.**<br>**Note: When an O2S malfunction (P0131 to P0414) is also stored with this malfunction, the O2S malfunction(s) should be repaired first.**<br>**Possible Causes:**<br>• Oxygen sensor (before catalytic converter) is faulty<br>• Oxygen sensor (behind catalytic converter) is faulty<br>• Oxygen sensor heater (before catalytic converter) is faulty<br>• Oxygen sensor heater (behind catalytic converter) is faulty<br>• Circuit wiring has a short to power or ground<br>• Engine Component Power Supply Relay is faulty<br>• E-box fuses for oxygen sensor are faulty<br>• Leaks present in the exhaust manifold or exhaust pipes<br>• HO2S signal wire and ground wire crossed in connector<br>• HO2S element is fuel contaminated or has failed<br>• ECM has failed |
| **DTC: P1113**<br>**2T, MIL: Yes**<br>**Years: 2007, 2008**<br>**Models:** Passat, Jetta, Golf, GTI, New Beetle,<br>**Engines:** 2.0L<br>**Transmissions:** A/T | **O2 Control (Bank 1 Sensor 1) Internal Resistance Too High Conditions:**<br>Engine started, battery voltage must be at least 11.5v, all electrical components must be off, the ground between the engine and the chassis must be well connected, the exhaust system must be properly sealed between the catalytic converter and the cylinder head, and the oxygen sensor heater for oxygen sensor before the catalytic converter must be properly functioning. The ECM detected a measurement on the O2 sensor circuit that was outside the parameters to function properly.<br>**Note: For resistance testing of sensor heating, oxygen sensor should be cooled to ambient temperature. High temperatures at oxygen sensor may lead to inaccurate measurements.**<br>**Possible Causes:**<br>• Oxygen sensor (before catalytic converter) is faulty<br>• Oxygen sensor (behind catalytic converter) is faulty<br>• Oxygen sensor heater (before catalytic converter) is faulty<br>• Oxygen sensor heater (behind catalytic converter) is faulty<br>• Circuit wiring has a short to power or ground<br>• Engine Component Power Supply Relay is faulty<br>• E-box fuses for oxygen sensor are faulty<br>• Leaks present in the exhaust manifold or exhaust pipes<br>• HO2S signal wire and ground wire crossed in connector<br>• HO2S element is fuel contaminated or has failed<br>• ECM has failed |

| DTC | Trouble Code Title, Conditions & Possible Causes |
|---|---|
| **DTC: P1114**<br>**2T, MIL:** Yes<br>**Years:** 2007, 2008<br>**Models:** Passat, Jetta, Golf, GTI, New Beetle,<br>**Engines:** 2.0L, 2.5L, 3.6L<br>**Transmissions:** All | **O2 Control (Bank 1 Sensor 2) Internal Resistance Too High Conditions:**<br>Engine started, battery voltage must be at least 11.5v, all electrical components must be off, the ground between the engine and the chassis must be well connected, the exhaust system must be properly sealed between the catalytic converter and the cylinder head, and the oxygen sensor heater for oxygen sensor before the catalytic converter must be properly functioning. The ECM detected a measurement on the O2 sensor circuit that was outside the parameters to function properly.<br>**Note: For resistance testing of sensor heating, oxygen sensor should be cooled to ambient temperature. High temperatures at oxygen sensor may lead to inaccurate measurements.**<br>**Possible Causes:**<br>• Oxygen sensor (before catalytic converter) is faulty<br>• Oxygen sensor (behind catalytic converter) is faulty<br>• Oxygen sensor heater (before catalytic converter) is faulty<br>• Oxygen sensor heater (behind catalytic converter) is faulty<br>• Circuit wiring has a short to power or ground<br>• Engine Component Power Supply Relay is faulty<br>• E-box fuses for oxygen sensor are faulty<br>• Leaks present in the exhaust manifold or exhaust pipes<br>• HO2S signal wire and ground wire crossed in connector<br>• HO2S element is fuel contaminated or has failed<br>• ECM has failed |
| **DTC: P1115**<br>**2T, MIL:** Yes<br>**Years:** 2007, 2008<br>**Models:** Passat, Jetta, Golf, GTI, New Beetle, New Beetle Convertible<br>**Engines:** 2.0L, 2.5L, 3.6L<br>**Transmissions:** All | **O2 Control (Bank 1 Sensor 1) Short to Ground Conditions:**<br>Engine started, battery voltage must be at least 11.5v, all electrical components must be off, the ground between the engine and the chassis must be well connected, the exhaust system must be properly sealed between the catalytic converter and the cylinder head, and the oxygen sensor heater for oxygen sensor before the catalytic converter must be properly functioning. The ECM detected a measurement on the O2 sensor circuit that was outside the parameters to function properly.<br>**Note: For resistance testing of sensor heating, oxygen sensor should be cooled to ambient temperature. High temperatures at oxygen sensor may lead to inaccurate measurements.**<br>**Possible Causes:**<br>• Oxygen sensor (before catalytic converter) is faulty<br>• Oxygen sensor (behind catalytic converter) is faulty<br>• Oxygen sensor heater (before catalytic converter) is faulty<br>• Oxygen sensor heater (behind catalytic converter) is faulty<br>• Circuit wiring has a short to power or ground<br>• Engine Component Power Supply Relay is faulty<br>• E-box fuses for oxygen sensor are faulty<br>• Leaks present in the exhaust manifold or exhaust pipes<br>• HO2S signal wire and ground wire crossed in connector<br>• HO2S element is fuel contaminated or has failed<br>• ECM has failed |
| **DTC: P1116**<br>**2T, MIL:** Yes<br>**Years:** 2007, 2008<br>**Models:** Passat, Jetta, Golf, GTI, New Beetle, New Beetle Convertible<br>**Engines:** 2.0L, 2.5L, 3.6L<br>**Transmissions:** All | **O2 Control (Bank 1 Sensor 1) Open Conditions:**<br>Engine started, battery voltage must be at least 11.5v, all electrical components must be off, the ground between the engine and the chassis must be well connected, the exhaust system must be properly sealed between the catalytic converter and the cylinder head, and the oxygen sensor heater for oxygen sensor before the catalytic converter must be properly functioning. The ECM detected a measurement on the O2 sensor circuit that was outside the parameters to function properly.<br>**Note: For resistance testing of sensor heating, oxygen sensor should be cooled to ambient temperature. High temperatures at oxygen sensor may lead to inaccurate measurements.**<br>**Possible Causes:**<br>• Oxygen sensor (before catalytic converter) is faulty<br>• Oxygen sensor (behind catalytic converter) is faulty<br>• Oxygen sensor heater (before catalytic converter) is faulty<br>• Oxygen sensor heater (behind catalytic converter) is faulty<br>• Circuit wiring has a short to power or ground<br>• Engine Component Power Supply Relay is faulty<br>• E-box fuses for oxygen sensor are faulty<br>• Leaks present in the exhaust manifold or exhaust pipes<br>• HO2S signal wire and ground wire crossed in connector<br>• HO2S element is fuel contaminated or has failed<br>• ECM has failed |

| DTC | Trouble Code Title, Conditions & Possible Causes |
|---|---|
| **DTC: P1117**<br>**2T, MIL: Yes**<br>**Years: 2007, 2008**<br>**Models:** Passat, Jetta, Golf, GTI, New Beetle, New Beetle Convertible<br>**Engines:** 2.0L, 2.5L, 3.6L<br>**Transmissions:** All | **O2 Control (Bank 1 Sensor 2) Open Conditions:**<br>Engine started, battery voltage must be at least 11.5v, all electrical components must be off, the ground between the engine and the chassis must be well connected, the exhaust system must be properly sealed between the catalytic converter and the cylinder head, and the oxygen sensor heater for oxygen sensor before the catalytic converter must be properly functioning. The ECM detected a measurement on the O2 sensor circuit that was outside the parameters to function properly.<br>**Note: For resistance testing of sensor heating, oxygen sensor should be cooled to ambient temperature. High temperatures at oxygen sensor may lead to inaccurate measurements.**<br>**Possible Causes:**<br>• Oxygen sensor (before catalytic converter) is faulty<br>• Oxygen sensor (behind catalytic converter) is faulty<br>• Oxygen sensor heater (before catalytic converter) is faulty<br>• Oxygen sensor heater (behind catalytic converter) is faulty<br>• Circuit wiring has a short to power or ground<br>• Engine Component Power Supply Relay is faulty<br>• E-box fuses for oxygen sensor are faulty<br>• Leaks present in the exhaust manifold or exhaust pipes<br>• HO2S signal wire and ground wire crossed in connector<br>• HO2S element is fuel contaminated or has failed<br>• ECM has failed |
| **DTC: P1118**<br>**2T, MIL: Yes**<br>**Years: 2007, 2008**<br>**Models:** Passat, Jetta, Golf, GTI, New Beetle, New Beetle Convertible<br>**Engines:** 2.0L, 2.5L, 3.6L<br>**Transmissions:** All | **O2 Sensor Heater Circ. (Bank 1-Sensor2) Open Conditions:**<br>Engine started, battery voltage must be at least 11.5v, all electrical components must be off, the ground between the engine and the chassis must be well connected, the exhaust system must be properly sealed between the catalytic converter and the cylinder head, and the oxygen sensor heater for oxygen sensor before the catalytic converter must be properly functioning. The ECM detected a measurement on the O2 sensor circuit that was outside the parameters to function properly.<br>**Note: For resistance testing of sensor heating, oxygen sensor should be cooled to ambient temperature. High temperatures at oxygen sensor may lead to inaccurate measurements.**<br>**Possible Causes:**<br>• Oxygen sensor (before catalytic converter) is faulty<br>• Oxygen sensor (behind catalytic converter) is faulty<br>• Oxygen sensor heater (before catalytic converter) is faulty<br>• Oxygen sensor heater (behind catalytic converter) is faulty<br>• Circuit wiring has a short to power or ground<br>• Engine Component Power Supply Relay is faulty<br>• E-box fuses for oxygen sensor are faulty<br>• Leaks present in the exhaust manifold or exhaust pipes<br>• HO2S signal wire and ground wire crossed in connector<br>• HO2S element is fuel contaminated or has failed<br>• ECM has failed |
| **DTC: P1127**<br>**2T, MIL: Yes**<br>**Years: 2007, 2008**<br>**Models:** Passat, Jetta, Golf, GTI, New Beetle, New Beetle Convertible<br>**Engines:** 2.0L, 2.5L, 3.6L<br>**Transmissions:** All | **Long Term Fuel Trim Add. Air. Bank 1 System Too Rich Conditions:**<br>Engine started, battery voltage must be at least 11.5v, all electrical components must be off, the ground between the engine and the chassis must be well connected, the exhaust system must be properly sealed between the catalytic converter and the cylinder head, and the oxygen sensor heater for oxygen sensor before the catalytic converter must be properly functioning. The fuel mixture is so rich that the O2S control is on lean limit.<br>**Note: After exhaust system repairs, make sure exhaust system is not under stress and that it has sufficient clearance from the bodywork. If necessary, loosen double clamps and align exhaust pipe so that sufficient clearance is maintained to the bodywork and support rings carry uniform loads. Do not use any silicone sealant. Traces of silicone components which are sucked into the engine are not burned there, and they damage the oxygen sensor.**<br>**Possible Causes:**<br>• MAF sensor circuit open<br>• MAF sensor circuit shorted to ground<br>• Air leak in the manifold<br>• Secondary air injection system combi-valve stuck open<br>• Secondary air injection system electrical short<br>• Fuel pressure too high, leaks in the vacuum hose to fuel pressure regulator<br>• Fuel pressure regulator has failed<br>• Fuel injectors are dirty, faulty or do not close properly<br>• ECM has failed |

| DTC | Trouble Code Title, Conditions & Possible Causes |
|---|---|
| **DTC: P1128**<br>**2T, MIL:** Yes<br>**Years:** 2007, 2008<br>**Models:** Passat, Jetta, Golf, GTI, New Beetle, New Beetle Convertible<br>**Engines:** 2.0L, 2.5L, 3.6L<br>**Transmissions:** All | **Long Term Fuel Trim Add. Air. Bank 1 System Too Lean Conditions:**<br>Engine started, battery voltage must be at least 11.5v, all electrical components must be off, the ground between the engine and the chassis must be well connected, the exhaust system must be properly sealed between the catalytic converter and the cylinder head, and the oxygen sensor heater for oxygen sensor before the catalytic converter must be properly functioning. The fuel mixture is so rich that the O2S control is on lean limit.<br>**Note: When an O2S malfunction (P0131 to P0414) is also stored with this malfunction, the O2S malfunction(s) should be repaired first.**<br>**Note: After exhaust system repairs, make sure exhaust system is not under stress and that it has sufficient clearance from the bodywork. If necessary, loosen double clamps and align exhaust pipe so that sufficient clearance is maintained to the bodywork and support rings carry uniform loads. Do not use any silicone sealant. Traces of silicone components which are sucked into the engine are not burned there, and they damage the oxygen sensor.**<br>**Possible Causes:**<br>• Fuel pressure is too low or fuel quantity supplied is too low<br>• Fuel filter faulty<br>• Transfer fuel pump has failed<br>• Fuel injector is faulty (sticking or not opening)<br>• Engine speed (RPM) sensor is faulty<br>• MAF sensor circuit open<br>• MAF sensor circuit shorted to ground<br>• Air leak in the manifold<br>• Secondary air injection system combi-valve stuck open<br>• Secondary air injection system electrical short<br>• ECM has failed |
| **DTC: P1129**<br>**Years:** 2007, 2008<br>**Models:** Passat<br>**Engines:** 2.5L, 3.6L<br>**Transmissions:** A/T | **Long Term Fuel Trim at Rich Limit Conditions:**<br>Engine started, battery voltage must be at least 11.5v, all electrical components must be off, the ground between the engine and the chassis must be well connected, the exhaust system must be properly sealed between the catalytic converter and the cylinder head, and the oxygen sensor heater for oxygen sensor before the catalytic converter must be properly functioning. The ECM detected the HO2S circuit was too rich, or that it could no longer change Fuel Trim because it was at its lean limit.<br>**Possible Causes:**<br>• Air intake system leaking, vacuum hoses leaking or damaged<br>• Air leaks located after the MAF sensor mounting location<br>• EGR valve sticking, EGR diaphragm leaking, or gasket leaking<br>• EVAP vapor recovery system has failed<br>• Excessive fuel pressure, leaking or contaminated fuel injectors<br>• Exhaust leaks before or near the HO2S(s) mounting location<br>• Fuel pressure regulator is leaking or damaged<br>• HO2S circuits wet or oily, corroded, or poor terminal contact<br>• HO2S is damaged or it has failed<br>• HO2S signal circuit open, shorted to ground, shorted to power<br>• Low fuel pressure or vehicle driven until it was out of fuel<br>• Oil dipstick not seated or engine oil level too high (overfilled) |
| **DTC: P1130**<br>**Years:** 2007, 2008<br>**Models:** Passat<br>**Engines:** 2.5L, 3.6L<br>**Transmissions:** A/T | **Long Term Fuel Trim at Lean Limit Conditions:**<br>Engine started, battery voltage must be at least 11.5v, all electrical components must be off, the ground between the engine and the chassis must be well connected, the exhaust system must be properly sealed between the catalytic converter and the cylinder head, and the oxygen sensor heater for oxygen sensor before the catalytic converter must be properly functioning. The ECM detected the HO2S circuit was too lean, or that it could no longer change Fuel Trim because it was at its lean limit.<br>**Possible Causes:**<br>• Air intake system leaking, vacuum hoses leaking or damaged<br>• Air leaks located after the MAF sensor mounting location<br>• EGR valve sticking, EGR diaphragm leaking, or gasket leaking<br>• EVAP vapor recovery system has failed<br>• Excessive fuel pressure, leaking or contaminated fuel injectors<br>• Exhaust leaks before or near the HO2S(s) mounting location<br>• Fuel pressure regulator is leaking or damaged<br>• HO2S circuits wet or oily, corroded, or poor terminal contact<br>• HO2S is damaged or it has failed<br>• HO2S signal circuit open, shorted to ground, shorted to power<br>• Low fuel pressure or vehicle driven until it was out of fuel<br>• Oil dipstick not seated or engine oil level too high (overfilled) |

| DTC | Trouble Code Title, Conditions & Possible Causes |
|---|---|
| **DTC: P1136**<br>**2T, MIL: Yes**<br>**Years: 2007, 2008**<br>**Models:** Passat, Jetta, Golf, GTI, New Beetle,<br>**Engines:** 2.0L, 2.5L, 3.6L<br>**Transmissions:** All | **Long Term Fuel Trim Add. Fuel, Bank 1 System Too Lean Conditions:**<br>Engine started, battery voltage must be at least 11.5v, all electrical components must be off, the ground between the engine and the chassis must be well connected, the exhaust system must be properly sealed between the catalytic converter and the cylinder head, and the oxygen sensor heater for oxygen sensor before the catalytic converter must be properly functioning. The ECM detected the HO2S circuit was too lean, or that it could no longer change Fuel Trim because it was at its lean limit.<br>**Possible Causes:**<br>• Air intake system leaking, vacuum hoses leaking or damaged<br>• Air leaks located after the MAF sensor mounting location<br>• EGR valve sticking, EGR diaphragm leaking, or gasket leaking<br>• EVAP vapor recovery system has failed<br>• Excessive fuel pressure, leaking or contaminated fuel injectors<br>• Exhaust leaks before or near the HO2S(s) mounting location<br>• Fuel pressure regulator is leaking or damaged<br>• HO2S circuits wet or oily, corroded, or poor terminal contact<br>• HO2S is damaged or it has failed<br>• HO2S signal circuit open, shorted to ground, shorted to power<br>• Low fuel pressure or vehicle driven until it was out of fuel<br>• Oil dipstick not seated or engine oil level too high (overfilled) |
| **DTC: P1137**<br>**2T, MIL: Yes**<br>**Years: 2007, 2008**<br>**Models:** Passat, Jetta, Golf, GTI, New Beetle,<br>**Engines:** 2.0L, 2.5L, 3.6L<br>**Transmissions:** All | **Long Term Fuel Trim Add. Fuel, Bank 1 System Too Rich Conditions:**<br>Engine started, battery voltage must be at least 11.5v, all electrical components must be off, the ground between the engine and the chassis must be well connected, the exhaust system must be properly sealed between the catalytic converter and the cylinder head, and the oxygen sensor heater for oxygen sensor before the catalytic converter must be properly functioning. The ECM detected the HO2S circuit was too rich, or that it could no longer change Fuel Trim because it was at its lean limit.<br>**Possible Causes:**<br>• Air intake system leaking, vacuum hoses leaking or damaged<br>• Air leaks located after the MAF sensor mounting location<br>• EGR valve sticking, EGR diaphragm leaking, or gasket leaking<br>• EVAP vapor recovery system has failed<br>• Excessive fuel pressure, leaking or contaminated fuel injectors<br>• Exhaust leaks before or near the HO2S(s) mounting location<br>• Fuel pressure regulator is leaking or damaged<br>• HO2S circuits wet or oily, corroded, or poor terminal contact<br>• HO2S is damaged or it has failed<br>• HO2S signal circuit open, shorted to ground, shorted to power<br>• Low fuel pressure or vehicle driven until it was out of fuel<br>• Oil dipstick not seated or engine oil level too high (overfilled) |
| **DTC: P1138**<br>**Years: 2007, 2008**<br>**Models:** Passat<br>**Engines:** 2.5L, 3.6L<br>**Transmissions:** A/T | **Long Term Fuel Trim Add. Fuel, Bank 2 System Too Lean Conditions:**<br>Engine started, battery voltage must be at least 11.5v, all electrical components must be off, the ground between the engine and the chassis must be well connected, the exhaust system must be properly sealed between the catalytic converter and the cylinder head, and the oxygen sensor heater for oxygen sensor before the catalytic converter must be properly functioning. The ECM detected the HO2S circuit was too lean, or that it could no longer change Fuel Trim because it was at its lean limit.<br>**Possible Causes:**<br>• Air intake system leaking, vacuum hoses leaking or damaged<br>• Air leaks located after the MAF sensor mounting location<br>• EGR valve sticking, EGR diaphragm leaking, or gasket leaking<br>• EVAP vapor recovery system has failed<br>• Excessive fuel pressure, leaking or contaminated fuel injectors<br>• Exhaust leaks before or near the HO2S(s) mounting location<br>• Fuel pressure regulator is leaking or damaged<br>• HO2S circuits wet or oily, corroded, or poor terminal contact<br>• HO2S is damaged or it has failed<br>• HO2S signal circuit open, shorted to ground, shorted to power<br>• Low fuel pressure or vehicle driven until it was out of fuel<br>• Oil dipstick not seated or engine oil level too high (overfilled) |

| DTC | Trouble Code Title, Conditions & Possible Causes |
|---|---|
| **DTC: P1139**<br>**Years: 2007, 2008**<br>**Models:** Passat<br>**Engines:** 2.5L, 3.6L<br>**Transmissions:** A/T | **Long Term Fuel Trim Add. Fuel, Bank 2 System Too Rich Conditions:**<br>Engine started, battery voltage must be at least 11.5v, all electrical components must be off, the ground between the engine and the chassis must be well connected, the exhaust system must be properly sealed between the catalytic converter and the cylinder head, and the oxygen sensor heater for oxygen sensor before the catalytic converter must be properly functioning. The ECM detected the HO2S circuit was too rich, or that it could no longer change Fuel Trim because it was at its lean limit.<br>**Possible Causes:**<br>• Air intake system leaking, vacuum hoses leaking or damaged<br>• Air leaks located after the MAF sensor mounting location<br>• EGR valve sticking, EGR diaphragm leaking, or gasket leaking<br>• EVAP vapor recovery system has failed<br>• Excessive fuel pressure, leaking or contaminated fuel injectors<br>• Exhaust leaks before or near the HO2S(s) mounting location<br>• Fuel pressure regulator is leaking or damaged<br>• HO2S circuits wet or oily, corroded, or poor terminal contact<br>• HO2S is damaged or it has failed<br>• HO2S signal circuit open, shorted to ground, shorted to power<br>• Low fuel pressure or vehicle driven until it was out of fuel<br>• Oil dipstick not seated or engine oil level too high (overfilled) |
| **DTC: P1141**<br>**Years: 2007, 2008**<br>**Models:** Passat, Jetta, Golf, GTI, New Beetle, New Beetle Convertible<br>**Engines:** 2.0L, 2.5L, 3.6L<br>**Transmissions:** All | **Load Calculation Cross Check Range/Performance Conditions:**<br>Engine started, battery voltage must be at least 11.5v, all electrical components must be off, the ground between the engine and the chassis must be well connected, the exhaust system must be properly sealed between the catalytic converter and the cylinder head, and the oxygen sensor heater for oxygen sensor before the catalytic converter must be properly functioning.<br>**Note: Vacuum in the intake system sucks in the leak detection spray with false air. Leak detection spray decreases ignition quality of the fuel mixture. This causes a drop in engine speed and changes the value produced by the Heated Oxygen Sensor.**<br>**Note: Both Throttle Position (TP) sensor and Sender 2 for accelerator pedal position are located at the accelerator pedal and communicate the driver's intentions to the ECM completely independently of each other. Both sensors are stored in one housing.**<br>**Possible Causes:**<br>• Intake system is leaking<br>• Signal is grounding<br>• ECM has failed<br>• Intake Manifold Runner Position Sensor is faulty<br>• Intake system for leaks (false air) is faulty<br>• Motor for intake flap is faulty<br>• Mass Air Flow (MAF) sensor is faulty<br>• Throttle Position (TP) sensor is faulty<br>• Throttle valve control module is faulty |
| **DTC: P1142**<br>**Years: 2007, 2008**<br>**Models:** Passat, Jetta, Golf, GTI, New Beetle, New Beetle Convertible<br>**Engines:** 2.0L, 2.5L, 3.6L<br>**Transmissions:** All | **Load Calculation Cross Check Lower Limit Conditions:**<br>Engine started, battery voltage must be at least 11.5v, all electrical components must be off, the ground between the engine and the chassis must be well connected, the exhaust system must be properly sealed between the catalytic converter and the cylinder head, and the oxygen sensor heater for oxygen sensor before the catalytic converter must be properly functioning.<br>**Note: Vacuum in the intake system sucks in the leak detection spray with false air. Leak detection spray decreases ignition quality of the fuel mixture. This causes a drop in engine speed and changes the value produced by the Heated Oxygen Sensor.**<br>**Note: Both Throttle Position (TP) sensor and Sender 2 for accelerator pedal position are located at the accelerator pedal and communicate the driver's intentions to the ECM completely independently of each other. Both sensors are stored in one housing.**<br>**Possible Causes:**<br>• Intake Manifold Runner Position Sensor is faulty<br>• Intake system for leaks (false air) is faulty<br>• Motor for intake flap is faulty<br>• Mass Air Flow (MAF) sensor is faulty<br>• Throttle Position (TP) sensor is faulty<br>• Throttle valve control module is faulty<br>• Intake system is leaking<br>• Signal is grounding<br>• ECM has failed |

| DTC | Trouble Code Title, Conditions & Possible Causes |
|---|---|
| **DTC: P1143**<br>**Years: 2007, 2008**<br>**Models:** Passat, Jetta, Golf, GTI, New Beetle, New Beetle Convertible<br>**Engines:** 2.0L, 2.5L, 3.6L<br>**Transmissions:** All | **Load Calculation Cross Check Upper Limit Conditions:**<br>Engine started, battery voltage must be at least 11.5v, all electrical components must be off, the ground between the engine and the chassis must be well connected, the exhaust system must be properly sealed between the catalytic converter and the cylinder head, and the oxygen sensor heater for oxygen sensor before the catalytic converter must be properly functioning.<br>**Note: Vacuum in the intake system sucks in the leak detection spray with false air. Leak detection spray decreases ignition quality of the fuel mixture. This causes a drop in engine speed and changes the value produced by the Heated Oxygen Sensor.**<br>**Note: Both Throttle Position (TP) sensor and Sender 2 for accelerator pedal position are located at the accelerator pedal and communicate the driver's intentions to the ECM completely independently of each other. Both sensors are stored in one housing.**<br>**Possible Causes:**<br>• Intake Manifold Runner Position Sensor is faulty<br>• Intake system for leaks (false air) is faulty<br>• Motor for intake flap is faulty<br>• Mass Air Flow (MAF) sensor is faulty<br>• Throttle Position (TP) sensor is faulty<br>• Throttle valve control module is faulty<br>• Intake system is leaking<br>• Signal is grounding<br>• ECM has failed |
| **DTC: P1149**<br>**Years: 2007, 2008**<br>**Models:** Passat, Jetta, Golf, GTI, New Beetle, New Beetle Convertible<br>**Engines:** 2.0L, 2.5L, 3.6L<br>**Transmissions:** All | **O2 Control (Bank 1) Out of Range Conditions:**<br>Engine started, battery voltage must be at least 11.5v, all electrical components must be off, the ground between the engine and the chassis must be well connected, the exhaust system must be properly sealed between the catalytic converter and the cylinder head, and the oxygen sensor heater for oxygen sensor before the catalytic converter must be properly functioning. The ECM detected a voltage on the O2 sensor circuit that was outside the parameters to function properly.<br>**Note: For resistance testing of sensor heating, oxygen sensor should be cooled to ambient temperature. High temperatures at oxygen sensor may lead to inaccurate measurements.**<br>**Possible Causes:**<br>• Oxygen sensor (before catalytic converter) is faulty<br>• Oxygen sensor (behind catalytic converter) is faulty<br>• Oxygen sensor heater (before catalytic converter) is faulty<br>• Oxygen sensor heater (behind catalytic converter) is faulty<br>• Circuit wiring has a short to power or ground<br>• Engine Component Power Supply Relay is faulty<br>• E-box fuses for oxygen sensor are faulty<br>• Leaks present in the exhaust manifold or exhaust pipes<br>• HO2S signal wire and ground wire crossed in connector<br>• HO2S element is fuel contaminated or has failed<br>• ECM has failed |
| **DTC: P1150**<br>**Years: 2007, 2008**<br>**Models:** Passat, Jetta, Golf, GTI, New Beetle, New Beetle Convertible<br>**Engines:** 2.0L, 2.5L, 3.6L<br>**Transmissions:** All | **Lack of HO2S-21 Switching, Fuel Trim At Rich/Lean Limit Conditions:**<br>Engine running in closed loop, and the ECM detected the HO2S circuit was too lean or too rich, or that it could no longer correct Fuel Trim (i.e., the Fuel Trim was at its calibrated rich limit or its calibrated lean limit).<br>**Possible Causes:**<br>• Air intake system leaking, vacuum hoses leaking or damaged<br>• Air leaks located after the MAF sensor mounting location<br>• EGR valve sticking, EGR diaphragm leaking, or gasket leaking<br>• EVAP vapor recovery system has failed<br>• Excessive fuel pressure, leaking or contaminated fuel injectors<br>• Exhaust leaks before or near the HO2S(s) mounting location<br>• Fuel pressure regulator is leaking or damaged<br>• HO2S circuits wet or oily, corroded, or poor terminal contact<br>• HO2S signal circuit open, shorted to ground, shorted to power, or the sensor has failed<br>• Low fuel pressure or vehicle driven until it was out of fuel<br>• Oil dipstick not seated or engine oil level too high (overfilled) |

| DTC | Trouble Code Title, Conditions & Possible Causes |
|---|---|
| **DTC: P1151**<br>**Years: 2007, 2008**<br>**Models:** Passat, Jetta, Golf, GTI, New Beetle, New Beetle Convertible<br>**Engines:** 2.0L, 2.5L, 3.6L<br>**Transmissions:** All | **Long Term Fuel Trim (Bank1, Range 1) Leanness Lower Limit Exceeded Conditions:**<br>Engine started, battery voltage must be at least 11.5v, all electrical components must be off, the ground between the engine and the chassis must be well connected, the exhaust system must be properly sealed between the catalytic converter and the cylinder head, and the oxygen sensor heater for oxygen sensor before the catalytic converter must be properly functioning. The fuel mixture is so lean that the O2S control is on rich limit.<br>**Note: After exhaust system repairs, make sure exhaust system is not under stress and that it has sufficient clearance from the bodywork. If necessary, loosen double clamps and align exhaust pipe so that sufficient clearance is maintained to the bodywork and support rings carry uniform loads. Do not use any silicone sealant. Traces of silicone components which are sucked into the engine are not burned there, and they damage the oxygen sensor.**<br>**Possible Causes:**<br>• Exhaust system is damaged<br>• Intake Manifold Runner Position Sensor is faulty<br>• Intake system for leaks (false air)<br>• Motor for intake flap has failed<br>• MAF sensor circuit open<br>• MAF sensor circuit shorted to ground<br>• Air leak in the manifold<br>• Secondary air injection system combi-valve stuck open<br>• Secondary air injection system electrical short<br>• Fuel pressure too high, leaks in the vacuum hose to fuel pressure regulator<br>• Fuel pressure regulator has failed<br>• Fuel injectors are dirty, faulty or do not close properly |
| **DTC: P1152**<br>**Years: 2007, 2008**<br>**Models:** Passat, Jetta, Golf, GTI, New Beetle, New Beetle Convertible<br>**Engines:** 2.0L, 2.5L, 3.6L<br>**Transmissions:** All | **Long Term Fuel Trim (Bank1, Range 2) Leanness Lower Limit Exceeded Conditions:**<br>Engine started, battery voltage must be at least 11.5v, all electrical components must be off, the ground between the engine and the chassis must be well connected, the exhaust system must be properly sealed between the catalytic converter and the cylinder head, and the oxygen sensor heater for oxygen sensor before the catalytic converter must be properly functioning. The fuel mixture is so lean that the O2S control is on rich limit.<br>**Note: After exhaust system repairs, make sure exhaust system is not under stress and that it has sufficient clearance from the bodywork. If necessary, loosen double clamps and align exhaust pipe so that sufficient clearance is maintained to the bodywork and support rings carry uniform loads. Do not use any silicone sealant. Traces of silicone components which are sucked into the engine are not burned there, and they damage the oxygen sensor.**<br>**Possible Causes:**<br>• Exhaust system is damaged<br>• Intake Manifold Runner Position Sensor is faulty<br>• Intake system for leaks (false air)<br>• Motor for intake flap has failed<br>• MAF sensor circuit open<br>• MAF sensor circuit shorted to ground<br>• Air leak in the manifold<br>• Secondary air injection system combi-valve stuck open<br>• Secondary air injection system electrical short<br>• Fuel pressure too high, leaks in the vacuum hose to fuel pressure regulator<br>• Fuel pressure regulator has failed<br>• Fuel injectors are dirty, faulty or do not close properly |

| DTC | Trouble Code Title, Conditions & Possible Causes |
|---|---|
| **DTC: P1165**<br>**Years: 2007, 2008**<br>**Models:** Passat, Jetta, Golf, GTI, New Beetle, New Beetle Convertible<br>**Engines:** 2.0L, 2.5L, 3.6L<br>**Transmissions:** All | **Bank 1, Long Term Fuel Trim, Range 1 Rich Limit Exceeded Conditions:**<br>Engine started, battery voltage must be at least 11.5v, all electrical components must be off, the ground between the engine and the chassis must be well connected, the exhaust system must be properly sealed between the catalytic converter and the cylinder head, and the oxygen sensor heater for oxygen sensor before the catalytic converter must be properly functioning. The fuel mixture is so rich that the O2S control is on lean limit.<br>**Note: After exhaust system repairs, make sure exhaust system is not under stress and that it has sufficient clearance from the bodywork. If necessary, loosen double clamps and align exhaust pipe so that sufficient clearance is maintained to the bodywork and support rings carry uniform loads. Do not use any silicone sealant. Traces of silicone components which are sucked into the engine are not burned there, and they damage the oxygen sensor.**<br>**Possible Causes:**<br>• Exhaust system is damaged<br>• Intake Manifold Runner Position Sensor is faulty<br>• Intake system for leaks (false air)<br>• Motor for intake flap has failed<br>• MAF sensor circuit open<br>• MAF sensor circuit shorted to ground<br>• Air leak in the manifold<br>• Secondary air injection system combi-valve stuck open<br>• Secondary air injection system electrical short<br>• Fuel pressure too high, leaks in the vacuum hose to fuel pressure regulator<br>• Fuel pressure regulator has failed<br>• Fuel injectors are dirty, faulty or do not close properly<br>• EVAP canister system lacks a proper seal<br>• Evaporative Emission (EVAP) canister purge regulator valve 1 is faulty<br>• Leak Detection Pump (LDP) is faulty<br>• Fuel injectors have failed<br>• Oxygen sensor (before catalytic converter) is faulty<br>• Oxygen sensor (behind catalytic converter) is faulty<br>• Oxygen sensor heater (before catalytic converter) is faulty<br>• Oxygen sensor heater (behind catalytic converter) is faulty |
| **DTC: P1166**<br>**Years: 2007, 2008**<br>**Models:** Passat, Jetta, Golf, GTI, New Beetle, New Beetle Convertible<br>**Engines:** 2.0L, 2.5L, 3.6L<br>**Transmissions:** All | **Bank 1, Long Term Fuel Trim, Range 2 Rich Limit Exceeded Conditions:**<br>Engine started, battery voltage must be at least 11.5v, all electrical components must be off, the ground between the engine and the chassis must be well connected, the exhaust system must be properly sealed between the catalytic converter and the cylinder head, and the oxygen sensor heater for oxygen sensor before the catalytic converter must be properly functioning. The fuel mixture is so rich that the O2S control is on lean limit.<br>**Note: After exhaust system repairs, make sure exhaust system is not under stress and that it has sufficient clearance from the bodywork. If necessary, loosen double clamps and align exhaust pipe so that sufficient clearance is maintained to the bodywork and support rings carry uniform loads. Do not use any silicone sealant. Traces of silicone components which are sucked into the engine are not burned there, and they damage the oxygen sensor.**<br>**Possible Causes:**<br>• Exhaust system is damaged<br>• Intake Manifold Runner Position Sensor is faulty<br>• Intake system for leaks (false air)<br>• Motor for intake flap has failed<br>• MAF sensor circuit open<br>• MAF sensor circuit shorted to ground<br>• Air leak in the manifold<br>• Secondary air injection system combi-valve stuck open<br>• Secondary air injection system electrical short<br>• Fuel pressure too high, leaks in the vacuum hose to fuel pressure regulator<br>• Fuel pressure regulator has failed<br>• Fuel injectors are dirty, faulty or do not close properly<br>• EVAP canister system lacks a proper seal<br>• Evaporative Emission (EVAP) canister purge regulator valve 1 is faulty<br>• Leak Detection Pump (LDP) is faulty<br>• Fuel injectors have failed<br>• Oxygen sensor (before catalytic converter) is faulty<br>• Oxygen sensor (behind catalytic converter) is faulty<br>• Oxygen sensor heater (before catalytic converter) is faulty<br>• Oxygen sensor heater (behind catalytic converter) is faulty |

| DTC | Trouble Code Title, Conditions & Possible Causes |
|---|---|
| **DTC: P1171**<br>**2T, MIL: Yes**<br>**Years: 2007, 2008**<br>**Models:** Passat, Jetta, Golf, GTI, New Beetle, New Beetle Convertible<br>**Engines:** 2.0L, 2.5L, 3.6L<br>**Transmissions:** All | **Throttle Actuation Potentiometer Sign.2 Range/Performance Conditions:**<br>Engine started, battery voltage must be at least 11.5v, all electrical components must be off, the ground between the engine and the chassis must be well connected, coolant temperature must be at least 80 degrees Celsius and the accelerator pedal must be properly adjusted. The ECM detected an incorrect singal from the throttle potentiometer.<br>**Note: If the complete throttle valve control module is current-less (e.g. connector disconnected) the throttle valve moves into a particular, specified mechanical position, which signals an increased idle speed with an engine at operating temperature. If only the Throttle Position (TP) actuator is current-less, the throttle valve also moves into the specified mechanical position (emergency running gap), however, since Closed Throttle Position (CTP) switch can still be recognized, an "almost normal idle RPM" is reached via the respective ignition angle retardation.**<br>**Note: Terminal assignment at throttle control module is different in vehicles with and without cruise control. Characteristic: Steering column switch with operating module for cruise control.**<br>**Possible Causes:**<br>• Throttle valve control module has failed<br>• Throttle valve is dirty or damaged<br>• Throttle valve is not in a closed position<br>• Voltage supply of throttle valve control module is shorted or open<br>• ECM has failed |
| **DTC: P1172**<br>**2T, MIL: Yes**<br>**Years: 2007, 2008**<br>**Models:** Passat, Jetta, Golf, GTI, New Beetle, New Beetle Convertible<br>**Engines:** 2.0L, 2.5L, 3.6L<br>**Transmissions:** All | **Throttle Actuation Potentiometer Sign.2 Signal Too Low Conditions:**<br>Engine started, battery voltage must be at least 11.5v, all electrical components must be off, the ground between the engine and the chassis must be well connected, coolant temperature must be at least 80 degrees Celsius and the accelerator pedal must be properly adjusted. The ECM detected an incorrect singal from the throttle potentiometer.<br>**Note: If the complete throttle valve control module is current-less (e.g. connector disconnected) the throttle valve moves into a particular, specified mechanical position, which signals an increased idle speed with an engine at operating temperature. If only the Throttle Position (TP) actuator is current-less, the throttle valve also moves into the specified mechanical position (emergency running gap), however, since Closed Throttle Position (CTP) switch can still be recognized, an "almost normal idle RPM" is reached via the respective ignition angle retardation.**<br>**Note: Terminal assignment at throttle control module is different in vehicles with and without cruise control. Characteristic: Steering column switch with operating module for cruise control.**<br>**Possible Causes:**<br>• Throttle valve control module has failed<br>• Throttle valve is dirty or damaged<br>• Throttle valve is not in a closed position<br>• Voltage supply of throttle valve control module is shorted or open<br>• ECM has failed |
| **DTC: P1173**<br>**2T, MIL: Yes**<br>**Years: 2007, 2008**<br>**Models:** Passat, Jetta, Golf, GTI, New Beetle, New Beetle Convertible<br>**Engines:** 2.0L, 2.5L, 3.6L<br>**Transmissions:** All | **Throttle Actuation Potentiometer Sign.2 Signal Too High Conditions:**<br>Engine started, battery voltage must be at least 11.5v, all electrical components must be off, the ground between the engine and the chassis must be well connected, coolant temperature must be at least 80 degrees Celsius and the accelerator pedal must be properly adjusted. The ECM detected an incorrect singal from the throttle potentiometer.<br>**Note: If the complete throttle valve control module is current-less (e.g. connector disconnected) the throttle valve moves into a particular, specified mechanical position, which signals an increased idle speed with an engine at operating temperature. If only the Throttle Position (TP) actuator is current-less, the throttle valve also moves into the specified mechanical position (emergency running gap), however, since Closed Throttle Position (CTP) switch can still be recognized, an "almost normal idle RPM" is reached via the respective ignition angle retardation.**<br>**Note: Terminal assignment at throttle control module is different in vehicles with and without cruise control. Characteristic: Steering column switch with operating module for cruise control.**<br>**Possible Causes:**<br>• Throttle valve control module has failed<br>• Throttle valve is dirty or damaged<br>• Throttle valve is not in a closed position<br>• Voltage supply of throttle valve control module is shorted or open<br>• ECM has failed |

| DTC | Trouble Code Title, Conditions & Possible Causes |
|---|---|
| **DTC: P1176**<br>**2T, MIL: Yes**<br>**Years: 2007, 2008**<br>**Models:** Passat, Jetta, Golf, GTI, New Beetle, New Beetle Convertible<br>**Engines:** 2.0L, 2.5L, 3.6L<br>**Transmissions**: All | **O2 Correction Behind Catalyst B1 Limit Attained Conditions:**<br>Engine started, battery voltage must be at least 11.5v, all electrical components must be off, the ground between the engine and the chassis must be well connected, the exhaust system must be properly sealed between the catalytic converter and the cylinder head, the coolant temperature must be at least 80 degrees Celsius, and the oxygen sensor heater for oxygen sensor before the catalytic converter must be properly functioning. The ECM has detected a malfunction of the oxygen sensor.<br>**Note: Vacuum in the intake system sucks in the leak detection spray with false air. Leak detection spray decreases ignition quality of the fuel mixture. This causes a drop in engine speed and changes the value produced by the Heated Oxygen Sensor (HO2S).**<br>**Note: Vehicle must be raised before connector for oxygen sensor is accessible.**<br>**Note: The oxygen sensor before catalytic converter has a static regulation and can be differentiated from the oxygen sensor behind catalytic converter via a 6-pin connector.**<br>**Possible Causes:**<br>• O2 sensor circuit has shorted to ground or B+<br>• O2 sensor circuit is open<br>• ECM has failed<br>• O2 sensor has failed<br>• Intake Manifold Runner Position Sensor is faulty<br>• Intake system for leaks (false air) is faulty<br>• Motor for intake flap is faulty |
| **DTC: P1177**<br>**Years: 2007, 2008**<br>**Models:** Passat, Jetta, GTI<br>**Engines:** 2.5L, 3.6L<br>**Transmissions**: All | **O2 Correction Behind Catalyst B2 Limit Attained Conditions:**<br>Engine started, battery voltage must be at least 11.5v, all electrical components must be off, the ground between the engine and the chassis must be well connected, the exhaust system must be properly sealed between the catalytic converter and the cylinder head, the coolant temperature must be at least 80 degrees Celsius, and the oxygen sensor heater for oxygen sensor before the catalytic converter must be properly functioning. The ECM has detected a malfunction of the oxygen sensor.<br>**Note: Vacuum in the intake system sucks in the leak detection spray with false air. Leak detection spray decreases ignition quality of the fuel mixture. This causes a drop in engine speed and changes the value produced by the Heated Oxygen Sensor (HO2S).**<br>**Note: Vehicle must be raised before connector for oxygen sensor is accessible.**<br>**Note: The oxygen sensor before catalytic converter has a static regulation and can be differentiated from the oxygen sensor behind catalytic converter via a 6-pin connector.**<br>**Possible Causes:**<br>• O2 sensor circuit has shorted to ground or B+<br>• O2 sensor circuit is open<br>• ECM has failed<br>• O2 sensor has failed<br>• Intake Manifold Runner Position Sensor is faulty<br>• Intake system for leaks (false air) is faulty<br>• Motor for intake flap is faulty |
| **DTC: P1178**<br>**Years: 2007, 2008**<br>**Models:** Jetta, Golf, GTI, New Beetle, New Beetle Convertible<br>**Engines:** 2.0L<br>**Transmissions**: All | **Linear O2 Sensor/Pump Current Open Circuit Conditions:**<br>Engine started, battery voltage must be at least 11.5v, all electrical components must be off, the ground between the engine and the chassis must be well connected, the exhaust system must be properly sealed between the catalytic converter and the cylinder head, the coolant temperature must be at least 80 degrees Celsius, and the oxygen sensor heater for oxygen sensor before the catalytic converter must be properly functioning. The ECM has detected a malfunction of the oxygen sensor.<br>**note: Vacuum in the intake system sucks in the leak detection spray with false air. Leak detection spray decreases ignition quality of the fuel mixture. This causes a drop in engine speed and changes the value produced by the Heated Oxygen Sensor (HO2S).**<br>**Note: Vehicle must be raised before connector for oxygen sensor is accessible.**<br>**Possible Causes:**<br>• O2 sensor circuit has shorted to ground or B+<br>• O2 sensor circuit is open<br>• ECM has failed<br>• O2 sensor has failed<br>• Intake Manifold Runner Position Sensor is faulty<br>• Intake system for leaks (false air) is faulty<br>• Motor for intake flap is faulty |

| DTC | Trouble Code Title, Conditions & Possible Causes |
|---|---|
| **DTC: P1179**<br>**Years: 2007, 2008**<br>**Models:** Jetta, Golf, GTI, New Beetle, New Beetle Convertible<br>**Engines:** 2.0L<br>**Transmissions:** All | **Linear O2 Sensor/Pump Current Short to Ground Conditions:**<br>Engine started, battery voltage must be at least 11.5v, all electrical components must be off, the ground between the engine and the chassis must be well connected, the exhaust system must be properly sealed between the catalytic converter and the cylinder head, the coolant temperature must be at least 80 degrees Celsius, and the oxygen sensor heater for oxygen sensor before the catalytic converter must be properly functioning. The ECM has detected a malfunction of the oxygen sensor.<br>**Note: Vacuum in the intake system sucks in the leak detection spray with false air. Leak detection spray decreases ignition quality of the fuel mixture. This causes a drop in engine speed and changes the value produced by the Heated Oxygen Sensor (HO2S).**<br>**Note: Vehicle must be raised before connector for oxygen sensor is accessible.**<br>**Possible Causes:**<br>&bull; O2 sensor circuit has shorted to ground or B+<br>&bull; O2 sensor circuit is open<br>&bull; ECM has failed<br>&bull; O2 sensor has failed<br>&bull; Intake Manifold Runner Position Sensor is faulty<br>&bull; Intake system for leaks (false air) is faulty<br>&bull; Motor for intake flap is faulty |
| **DTC: P1180**<br>**Years: 2007, 2008**<br>**Models:** Jetta, Golf, GTI, New Beetle, New Beetle Convertible<br>**Engines:** 2.0L<br>**Transmissions:** All | **Linear O2 Sensor/Pump Current Short to B+ Conditions:**<br>Engine started, battery voltage must be at least 11.5v, all electrical components must be off, the ground between the engine and the chassis must be well connected, the exhaust system must be properly sealed between the catalytic converter and the cylinder head, the coolant temperature must be at least 80 degrees Celsius, and the oxygen sensor heater for oxygen sensor before the catalytic converter must be properly functioning. The ECM has detected a malfunction of the oxygen sensor.<br>**Note: Vacuum in the intake system sucks in the leak detection spray with false air. Leak detection spray decreases ignition quality of the fuel mixture. This causes a drop in engine speed and changes the value produced by the Heated Oxygen Sensor (HO2S).**<br>**Note: Vehicle must be raised before connector for oxygen sensor is accessible.**<br>**Possible Causes:**<br>&bull; O2 sensor circuit has shorted to ground or B+<br>&bull; O2 sensor circuit is open<br>&bull; ECM has failed<br>&bull; O2 sensor has failed<br>&bull; Intake Manifold Runner Position Sensor is faulty<br>&bull; Intake system for leaks (false air) is faulty<br>&bull; Motor for intake flap is faulty |
| **DTC: P1181**<br>**Years: 2007, 2008**<br>**Models:** Jetta, Golf, GTI, New Beetle, New Beetle Convertible<br>**Engines:** 2.0L<br>**Transmissions:** All | **Linear O2 Sensor/Reference Voltage Open Circuit Conditions:**<br>Engine started, battery voltage must be at least 11.5v, all electrical components must be off, the ground between the engine and the chassis must be well connected, the exhaust system must be properly sealed between the catalytic converter and the cylinder head, the coolant temperature must be at least 80 degrees Celsius, and the oxygen sensor heater for oxygen sensor before the catalytic converter must be properly functioning. The ECM has detected a malfunction of the oxygen sensor.<br>**Note: Vacuum in the intake system sucks in the leak detection spray with false air. Leak detection spray decreases ignition quality of the fuel mixture. This causes a drop in engine speed and changes the value produced by the Heated Oxygen Sensor (HO2S).**<br>**Note: Vehicle must be raised before connector for oxygen sensor is accessible.**<br>**Possible Causes:**<br>&bull; O2 sensor circuit has shorted to ground or B+<br>&bull; O2 sensor circuit is open<br>&bull; ECM has failed<br>&bull; O2 sensor has failed<br>&bull; Intake Manifold Runner Position Sensor is faulty<br>&bull; Intake system for leaks (false air) is faulty<br>&bull; Motor for intake flap is faulty |

| DTC | Trouble Code Title, Conditions & Possible Causes |
|---|---|
| **DTC: P1182**<br>**Years: 2007, 2008**<br>**Models:** Jetta, Golf, GTI, New Beetle, New Beetle Convertible<br>**Engines:** 2.0L<br>**Transmissions:** All | **Linear O2 Sensor/Reference Voltage Short to Ground Conditions:**<br>Engine started, battery voltage must be at least 11.5v, all electrical components must be off, the ground between the engine and the chassis must be well connected, the exhaust system must be properly sealed between the catalytic converter and the cylinder head, the coolant temperature must be at least 80 degrees Celsius, and the oxygen sensor heater for oxygen sensor before the catalytic converter must be properly functioning. The ECM has detected a malfunction of the oxygen sensor.<br>**Note: Vacuum in the intake system sucks in the leak detection spray with false air. Leak detection spray decreases ignition quality of the fuel mixture. This causes a drop in engine speed and changes the value produced by the Heated Oxygen Sensor (HO2S).**<br>**Note: Vehicle must be raised before connector for oxygen sensor is accessible.**<br>**Possible Causes:**<br>&bull; O2 sensor circuit has shorted to ground or B+<br>&bull; O2 sensor circuit is open<br>&bull; ECM has failed<br>&bull; O2 sensor has failed<br>&bull; Intake Manifold Runner Position Sensor is faulty<br>&bull; Intake system for leaks (false air) is faulty<br>&bull; Motor for intake flap is faulty |
| **DTC: P1183**<br>**Years: 2007, 2008**<br>**Models:** Jetta, Golf, GTI, New Beetle, New Beetle Convertible<br>**Engines:** 2.0L<br>**Transmissions:** All | **Linear O2 Sensor / Reference Voltage Short to B+ Conditions:**<br>Engine started, battery voltage must be at least 11.5v, all electrical components must be off, the ground between the engine and the chassis must be well connected, the exhaust system must be properly sealed between the catalytic converter and the cylinder head, the coolant temperature must be at least 80 degrees Celsius, and the oxygen sensor heater for oxygen sensor before the catalytic converter must be properly functioning. The ECM has detected a malfunction of the oxygen sensor.<br>**Note: Vacuum in the intake system sucks in the leak detection spray with false air. Leak detection spray decreases ignition quality of the fuel mixture. This causes a drop in engine speed and changes the value produced by the Heated Oxygen Sensor (HO2S).**<br>**Note: Vehicle must be raised before connector for oxygen sensor is accessible.**<br>**Possible Causes:**<br>&bull; O2 sensor circuit has shorted to ground or B+<br>&bull; O2 sensor circuit is open<br>&bull; ECM has failed<br>&bull; O2 sensor has failed<br>&bull; Intake Manifold Runner Position Sensor is faulty<br>&bull; Intake system for leaks (false air) is faulty<br>&bull; Motor for intake flap is faulty |
| **DTC: P1184**<br>**Years: 2007, 2008**<br>**Models:** Jetta, Golf, GTI, New Beetle, New Beetle Convertible<br>**Engines:** 2.0L<br>**Transmissions:** All | **Linear O2 Sensor / Common Ground Wire Open Circuit Conditions:**<br>Engine started, battery voltage must be at least 11.5v, all electrical components must be off, the ground between the engine and the chassis must be well connected, the exhaust system must be properly sealed between the catalytic converter and the cylinder head, the coolant temperature must be at least 80 degrees Celsius, and the oxygen sensor heater for oxygen sensor before the catalytic converter must be properly functioning. The ECM has detected a malfunction of the oxygen sensor.<br>**Note: Vacuum in the intake system sucks in the leak detection spray with false air. Leak detection spray decreases ignition quality of the fuel mixture. This causes a drop in engine speed and changes the value produced by the Heated Oxygen Sensor (HO2S).**<br>**Note: Vehicle must be raised before connector for oxygen sensor is accessible.**<br>**Possible Causes:**<br>&bull; O2 sensor circuit has shorted to ground or B+<br>&bull; O2 sensor circuit is open<br>&bull; ECM has failed<br>&bull; O2 sensor has failed<br>&bull; Intake Manifold Runner Position Sensor is faulty<br>&bull; Intake system for leaks (false air) is faulty<br>&bull; Motor for intake flap is faulty |

| DTC | Trouble Code Title, Conditions & Possible Causes |
|---|---|
| **DTC: P1185**<br>**Years: 2007, 2008**<br>**Models:** Jetta, Golf, GTI, New Beetle, New Beetle Convertible<br>**Engines:** 2.0L<br>**Transmissions:** All | **Linear O2 Sensor/Common Ground Wire Short to Ground Conditions:**<br>Engine started, battery voltage must be at least 11.5v, all electrical components must be off, the ground between the engine and the chassis must be well connected, the exhaust system must be properly sealed between the catalytic converter and the cylinder head, the coolant temperature must be at least 80 degrees Celsius, and the oxygen sensor heater for oxygen sensor before the catalytic converter must be properly functioning. The ECM has detected a malfunction of the oxygen sensor.<br>**Note: Vacuum in the intake system sucks in the leak detection spray with false air. Leak detection spray decreases ignition quality of the fuel mixture. This causes a drop in engine speed and changes the value produced by the Heated Oxygen Sensor (HO2S).**<br>**Note: Vehicle must be raised before connector for oxygen sensor is accessible.**<br>**Possible Causes:**<br>• O2 sensor circuit has shorted to ground or B+<br>• O2 sensor circuit is open<br>• ECM has failed<br>• O2 sensor has failed<br>• Intake Manifold Runner Position Sensor is faulty<br>• Intake system for leaks (false air) is faulty<br>• Motor for intake flap is faulty |
| **DTC: P1186**<br>**Years: 2007, 2008**<br>**Models:** Jetta, Golf, GTI, New Beetle, New Beetle Convertible<br>**Engines:** 2.0L<br>**Transmissions:** All | **Linear O2 Sensor/Common Ground Wire Short to B+ Conditions:**<br>Engine started, battery voltage must be at least 11.5v, all electrical components must be off, the ground between the engine and the chassis must be well connected, the exhaust system must be properly sealed between the catalytic converter and the cylinder head, the coolant temperature must be at least 80 degrees Celsius, and the oxygen sensor heater for oxygen sensor before the catalytic converter must be properly functioning. The ECM has detected a malfunction of the oxygen sensor.<br>**Note: Vacuum in the intake system sucks in the leak detection spray with false air. Leak detection spray decreases ignition quality of the fuel mixture. This causes a drop in engine speed and changes the value produced by the Heated Oxygen Sensor (HO2S).**<br>**Note: Vehicle must be raised before connector for oxygen sensor is accessible.**<br>**Possible Causes:**<br>• O2 sensor circuit has shorted to ground or B+<br>• O2 sensor circuit is open<br>• ECM has failed<br>• O2 sensor has failed<br>• Intake Manifold Runner Position Sensor is faulty<br>• Intake system for leaks (false air) is faulty<br>• Motor for intake flap is faulty |
| **DTC: P1196**<br>**Years: 2007, 2008**<br>**Models:** Passat, Jetta, Golf, GTI, New Beetle, New Beetle Convertible<br>**Engines:** 2.0L, 2.5L, 3.6L<br>**Transmissions:** All | **O2 Sensor Heater Circuit (Bank 1-Sensor 1) Electrical Malfunction Conditions:**<br>Engine started, battery voltage must be at least 11.5v, all electrical components must be off, the ground between the engine and the chassis must be well connected, the exhaust system must be properly sealed between the catalytic converter and the cylinder head, and the oxygen sensor heater for oxygen sensor before the catalytic converter must be properly functioning.<br>**Note: For resistance testing of sensor heating, oxygen sensor should be cooled to ambient temperature. High temperatures at oxygen sensor may lead to inaccurate measurements. The ECM detected an open or shorted condition, or excessive current draw in the heater circuit.**<br>**Possible Causes:**<br>• HO2S heater power circuit is open<br>• HO2S heater ground circuit is open<br>• HO2S signal tracking (due to oil or moisture in the connector)<br>• HO2S is damaged or has failed<br>• ECM has failed<br>• Oxygen sensor (before catalytic converter) is faulty<br>• Oxygen sensor (behind catalytic converter) is faulty<br>• Oxygen sensor heater (before catalytic converter) is faulty<br>• Oxygen sensor heater (behind catalytic converter) is faulty |

| DTC | Trouble Code Title, Conditions & Possible Causes |
|-----|--------------------------------------------------|
| **DTC: P1197**<br>**Years: 2007, 2008**<br>**Models:** Passat<br>**Engines:** 2.5L, 3.6L<br>**Transmissions:** All | **O2 Sensor Heater Circuit (Bank 2-Sensor 1) Electrical Malfunction Conditions:**<br>Engine started, battery voltage must be at least 11.5v, all electrical components must be off, the ground between the engine and the chassis must be well connected, the exhaust system must be properly sealed between the catalytic converter and the cylinder head, and the oxygen sensor heater for oxygen sensor before the catalytic converter must be properly functioning.<br>**Note: For resistance testing of sensor heating, oxygen sensor should be cooled to ambient temperature. High temperatures at oxygen sensor may lead to inaccurate measurements. The ECM detected an open or shorted condition, or excessive current draw in the heater circuit.**<br>**Possible Causes:**<br>• HO2S heater power circuit is open<br>• HO2S heater ground circuit is open<br>• HO2S signal tracking (due to oil or moisture in the connector)<br>• HO2S is damaged or has failed<br>• ECM has failed<br>• Oxygen sensor (before catalytic converter) is faulty<br>• Oxygen sensor (behind catalytic converter) is faulty<br>• Oxygen sensor heater (before catalytic converter) is faulty<br>• Oxygen sensor heater (behind catalytic converter) is faulty |
| **DTC: P1198**<br>**Years: 2007, 2008**<br>**Models:** Passat, Jetta, Golf, GTI, New Beetle, New Beetle Convertible<br>**Engines:** 2.0L, 2.5L, 3.6L<br>**Transmissions:** All | **O2 Sensor Heater Circuit (Bank 1-Sensor 2) Electrical Malfunction Conditions:**<br>Engine started, battery voltage must be at least 11.5v, all electrical components must be off, the ground between the engine and the chassis must be well connected, the exhaust system must be properly sealed between the catalytic converter and the cylinder head, and the oxygen sensor heater for oxygen sensor before the catalytic converter must be properly functioning.<br>**Note: For resistance testing of sensor heating, oxygen sensor should be cooled to ambient temperature. High temperatures at oxygen sensor may lead to inaccurate measurements. The ECM detected an open or shorted condition, or excessive current draw in the heater circuit.**<br>**Possible Causes:**<br>• HO2S heater power circuit is open<br>• HO2S heater ground circuit is open<br>• HO2S signal tracking (due to oil or moisture in the connector)<br>• HO2S is damaged or has failed<br>• ECM has failed<br>• Oxygen sensor (before catalytic converter) is faulty<br>• Oxygen sensor (behind catalytic converter) is faulty<br>• Oxygen sensor heater (before catalytic converter) is faulty<br>• Oxygen sensor heater (behind catalytic converter) is faulty |
| **DTC: P1199**<br>**Years: 2007, 2008**<br>**Models:** Passat<br>**Engines:** 2.5L, 3.6L<br>**Transmissions:** All | **O2 Sensor Heater Circuit (Bank 2-Sensor 2) Electrical Malfunction Conditions:**<br>Engine started, battery voltage must be at least 11.5v, all electrical components must be off, the ground between the engine and the chassis must be well connected, the exhaust system must be properly sealed between the catalytic converter and the cylinder head, and the oxygen sensor heater for oxygen sensor before the catalytic converter must be properly functioning.<br>**Note: For resistance testing of sensor heating, oxygen sensor should be cooled to ambient temperature. High temperatures at oxygen sensor may lead to inaccurate measurements. The ECM detected an open or shorted condition, or excessive current draw in the heater circuit.**<br>**Possible Causes:**<br>• HO2S heater power circuit is open<br>• HO2S heater ground circuit is open<br>• HO2S signal tracking (due to oil or moisture in the connector)<br>• HO2S is damaged or has failed<br>• ECM has failed<br>• Oxygen sensor (before catalytic converter) is faulty<br>• Oxygen sensor (behind catalytic converter) is faulty<br>• Oxygen sensor heater (before catalytic converter) is faulty<br>• Oxygen sensor heater (behind catalytic converter) is faulty |

| DTC | Trouble Code Title, Conditions & Possible Causes |
|---|---|
| **DTC: P1200**<br>**Years:** 2007, 2008<br>**Models:** Passat<br>**Engines:**<br>**Transmissions:** All | **Turbocharger Bypass Valve Mechanical Malfunction Conditions:**<br>Engine started, battery voltage at least 11.5v, all electrical components off, ground connections between engine and chassis well connected, coolant temperature at least 80-degrees Celicius. The ECM detected an operating condition that could harm the engine or automatic transmission.<br>**Possible Causes:**<br>• Charge air pressure sensor is faulty<br>• Voltage supply to the charge air pressure sensor is open or shorted<br>• Check the charge air system for leaks<br>• Recirculating valve for turbocharger is faulty<br>• Turbocharging system is damaged or not functioning properly<br>• Turbocharger recirculating valve is faulty<br>• Vacuum diaphragm for turbocharger is out of adjustment<br>• Wastegate bypass regulator valve is faulty<br>• Boost sensor has failed<br>• ECM has failed |
| **DTC: P1201**<br>**Years:** 2007, 2008<br>**Models:** Passat, Jetta, Golf, GTI, New Beetle, New Beetle Convertible<br>**Engines:** 2.0L, 2.5L, 3.6L<br>**Transmissions:** All | **Cylinder 1 Fuel Injection Circuit Electrical Malfunction Conditions:**<br>Key on or engine running, fuses in the instrument panel and the E-box in the engine compartment must be functioning, and the ground connections between the engine ad the chassis must be well connected; and the ECM detected an unexpected voltage condition on the injector circuit<br>**Possible Causes:**<br>• Injector control circuit is open<br>• Injector power circuit (B+) is open<br>• Injector control circuit is shorted to chassis ground<br>• Injector is damaged or has failed<br>• ECM is not connected or has failed<br>• Fuel pump relay has failed<br>• Fuel injectors may have malfunctioned<br>• Faulty engine speed sensor |
| **DTC: P1202**<br>**Years:** 2007, 2008<br>**Models:** Passat, Jetta, Golf, GTI, New Beetle, New Beetle Convertible<br>**Engines:** 2.0L, 2.5L, 3.6L<br>**Transmissions:** All | **Cylinder 2 Fuel Injection Circuit Electrical Malfunction Conditions:**<br>Key on or engine running, fuses in the instrument panel and the E-box in the engine compartment must be functioning, and the ground connections between the engine ad the chassis must be well connected; and the ECM detected an unexpected voltage condition on the injector circuit<br>**Possible Causes:**<br>• Injector control circuit is open<br>• Injector power circuit (B+) is open<br>• Injector control circuit is shorted to chassis ground<br>• Injector is damaged or has failed<br>• ECM is not connected or has failed<br>• Fuel pump relay has failed<br>• Fuel injectors may have malfunctioned<br>• Faulty engine speed sensor |
| **DTC: P1203**<br>**Years:** 2007, 2008<br>**Models:** Passat, Jetta, Golf, GTI, New Beetle, New Beetle Convertible<br>**Engines:** 2.0L, 2.5L, 3.6L<br>**Transmissions:** All | **Cylinder 3 Fuel Injection Circuit Electrical Malfunction Conditions:**<br>Key on or engine running, fuses in the instrument panel and the E-box in the engine compartment must be functioning, and the ground connections between the engine ad the chassis must be well connected; and the ECM detected an unexpected voltage condition on the injector circuit<br>**Possible Causes:**<br>• Injector control circuit is open<br>• Injector power circuit (B+) is open<br>• Injector control circuit is shorted to chassis ground<br>• Injector is damaged or has failed<br>• ECM is not connected or has failed<br>• Fuel pump relay has failed<br>• Fuel injectors may have malfunctioned<br>• Faulty engine speed sensor |

| DTC | Trouble Code Title, Conditions & Possible Causes |
|---|---|
| **DTC: P1204**<br>**Years: 2007, 2008**<br>**Models:** Passat, Jetta, Golf, GTI, New Beetle, New Beetle Convertible<br>**Engines:** 2.0L, 2.5L, 3.6L<br>**Transmissions:** All | **Cylinder 4 Fuel Injection Circuit Electrical Malfunction Conditions:**<br>Key on or engine running, fuses in the instrument panel and the E-box in the engine compartment must be functioning, and the ground connections between the engine ad the chassis must be well connected; and the ECM detected an unexpected voltage condition on the injector circuit<br>**Possible Causes:**<br>• Injector control circuit is open<br>• Injector power circuit (B+) is open<br>• Injector control circuit is shorted to chassis ground<br>• Injector is damaged or has failed<br>• ECM is not connected or has failed<br>• Fuel pump relay has failed<br>• Fuel injectors may have malfunctioned<br>• Faulty engine speed sensor |
| **DTC: P1213**<br>**2T, MIL: Yes**<br>**Years: 2007, 2008**<br>**Models:** Passat, Jetta, Golf, GTI, New Beetle, New Beetle Convertible<br>**Engines:** 2.0L, 2.5L, 3.6L<br>**Transmissions:** All | **Cylinder 1 Fuel Injection Circuit Short to B+ Conditions:**<br>Key on or engine running, fuses in the instrument panel and the E-box in the engine compartment must be functioning, and the ground connections between the engine ad the chassis must be well connected; and the ECM detected an unexpected voltage condition on the injector circuit. Wiring or fuel injector has a short circuit to positive supply.<br>**Possible Causes:**<br>• Injector control circuit is open<br>• Injector power circuit (B+) is open<br>• Injector control circuit is shorted to chassis ground<br>• Injector is damaged or has failed<br>• ECM is not connected or has failed<br>• Fuel pump relay has failed<br>• Engine speed sensor has failed |
| **DTC: P1214**<br>**2T, MIL: Yes**<br>**Years: 2007, 2008**<br>**Models:** Passat, Jetta, Golf, GTI, New Beetle, New Beetle Convertible<br>**Engines:** 2.0L, 2.5L, 3.6L<br>**Transmissions:** All | **Cylinder 2 Fuel Injection Circuit Short to B+ Conditions:**<br>Key on or engine running, fuses in the instrument panel and the E-box in the engine compartment must be functioning, and the ground connections between the engine ad the chassis must be well connected; and the ECM detected an unexpected voltage condition on the injector circuit. Wiring or fuel injector has a short circuit to positive supply.<br>**Possible Causes:**<br>• Injector control circuit is open<br>• Injector power circuit (B+) is open<br>• Injector control circuit is shorted to chassis ground<br>• Injector is damaged or has failed<br>• ECM is not connected or has failed<br>• Fuel pump relay has failed<br>• Engine speed sensor has failed |
| **DTC: P1215**<br>**2T, MIL: Yes**<br>**Years: 2007, 2008**<br>**Models:** Passat, Jetta, Golf, GTI, New Beetle, New Beetle Convertible<br>**Engines:** 2.0L, 2.5L, 3.6L<br>**Transmissions:** All | **Cylinder 3 Fuel Injection Circuit Short to B+ Conditions:**<br>Key on or engine running, fuses in the instrument panel and the E-box in the engine compartment must be functioning, and the ground connections between the engine ad the chassis must be well connected; and the ECM detected an unexpected voltage condition on the injector circuit. Wiring or fuel injector has a short circuit to positive supply.<br>**Possible Causes:**<br>• Injector control circuit is open<br>• Injector power circuit (B+) is open<br>• Injector control circuit is shorted to chassis ground<br>• Injector is damaged or has failed<br>• ECM is not connected or has failed<br>• Fuel pump relay has failed<br>• Engine speed sensor has failed |
| **DTC: P1216**<br>**2T, MIL: Yes**<br>**Years: 2007, 2008**<br>**Models:** Passat, Jetta, Golf, GTI, New Beetle, New Beetle Convertible<br>**Engines:** 2.0L, 2.5L, 3.6L<br>**Transmissions:** All | **Cylinder 4 Fuel Injection Circuit Short to B+ Conditions:**<br>Key on or engine running, fuses in the instrument panel and the E-box in the engine compartment must be functioning, and the ground connections between the engine ad the chassis must be well connected; and the ECM detected an unexpected voltage condition on the injector circuit. Wiring or fuel injector has a short circuit to positive supply.<br>**Possible Causes:**<br>• Injector control circuit is open<br>• Injector power circuit (B+) is open<br>• Injector control circuit is shorted to chassis ground<br>• Injector is damaged or has failed<br>• ECM is not connected or has failed<br>• Fuel pump relay has failed<br>• Engine speed sensor has failed |

| DTC | Trouble Code Title, Conditions & Possible Causes |
|---|---|
| **DTC: P1217**<br>**Years: 2007, 2008**<br>**Models:** Passat, Jetta<br>**Engines:** 2.5L, 3.6L<br>**Transmissions:** All | **Cylinder 5 Fuel Injection Circuit Short to B+ Conditions:**<br>Key on or engine running, fuses in the instrument panel and the E-box in the engine compartment must be functioning, and the ground connections between the engine ad the chassis must be well connected; and the ECM detected an unexpected voltage condition on the injector circuit. Wiring or fuel injector has a short circuit to positive supply.<br>**Possible Causes:**<br>• Injector control circuit is open<br>• Injector power circuit (B+) is open<br>• Injector control circuit is shorted to chassis ground<br>• Injector is damaged or has failed<br>• ECM is not connected or has failed<br>• Fuel pump relay has failed<br>• Engine speed sensor has failed |
| **DTC: P1218**<br>**Years: 2007, 2008**<br>**Models:** Passat, Jetta<br>**Engines:** 2.5L, 3.6L<br>**Transmissions:** All | **Cylinder 6 Fuel Injection Circuit Short to B+ Conditions:**<br>Key on or engine running, fuses in the instrument panel and the E-box in the engine compartment must be functioning, and the ground connections between the engine ad the chassis must be well connected; and the ECM detected an unexpected voltage condition on the injector circuit. Wiring or fuel injector has a short circuit to positive supply.<br>**Possible Causes:**<br>• Injector control circuit is open<br>• Injector power circuit (B+) is open<br>• Injector control circuit is shorted to chassis ground<br>• Injector is damaged or has failed<br>• ECM is not connected or has failed<br>• Fuel pump relay has failed<br>• Engine speed sensor has failed |
| **DTC: P1225**<br>**2T, MIL: Yes**<br>**Years: 2007, 2008**<br>**Models:** Passat, Jetta, Golf, GTI, New Beetle, New Beetle Convertible<br>**Engines:** 2.0L, 2.5L, 3.6L<br>**Transmissions:** All | **Cylinder 1 Fuel Injection Circuit Short to Ground Conditions:**<br>Key on or engine running, fuses in the instrument panel and the E-box in the engine compartment must be functioning, and the ground connections between the engine ad the chassis must be well connected; and the ECM detected an unexpected voltage condition on the injector circuit. Wiring or fuel injector has a short circuit to ground.<br>**Possible Causes:**<br>• Injector control circuit is open<br>• Injector power circuit (B+) is open<br>• Injector control circuit is shorted to chassis ground<br>• Injector is damaged or has failed<br>• ECM is not connected or has failed<br>• Fuel pump relay has failed<br>• Engine speed sensor has failed |
| **DTC: P1226**<br>**2T, MIL: Yes**<br>**Years: 2007, 2008**<br>**Models:** Passat, Jetta, Golf, GTI, New Beetle, New Beetle Convertible<br>**Engines:** 2.0L, 2.5L, 3.6L<br>**Transmissions:** All | **Cylinder 2 Fuel Injection Circuit Short to Ground Conditions:**<br>Key on or engine running, fuses in the instrument panel and the E-box in the engine compartment must be functioning, and the ground connections between the engine ad the chassis must be well connected; and the ECM detected an unexpected voltage condition on the injector circuit. Wiring or fuel injector has a short circuit to ground.<br>**Possible Causes:**<br>• Injector control circuit is open<br>• Injector power circuit (B+) is open<br>• Injector control circuit is shorted to chassis ground<br>• Injector is damaged or has failed<br>• ECM is not connected or has failed<br>• Fuel pump relay has failed<br>• Engine speed sensor has failed |
| **DTC: P1227**<br>**2T, MIL: Yes**<br>**Years: 2007, 2008**<br>**Models:** Passat, Jetta, Golf, GTI, New Beetle, New Beetle Convertible<br>**Engines:** 2.0L, 2.5L, 3.6L<br>**Transmissions:** All | **Cylinder 3 Fuel Injection Circuit Short to Ground Conditions:**<br>Key on or engine running, fuses in the instrument panel and the E-box in the engine compartment must be functioning, and the ground connections between the engine ad the chassis must be well connected; and the ECM detected an unexpected voltage condition on the injector circuit. Wiring or fuel injector has a short circuit to ground.<br>**Possible Causes:**<br>• Injector control circuit is open<br>• Injector power circuit (B+) is open<br>• Injector control circuit is shorted to chassis ground<br>• Injector is damaged or has failed<br>• ECM is not connected or has failed<br>• Fuel pump relay has failed<br>• Engine speed sensor has failed |

| DTC | Trouble Code Title, Conditions & Possible Causes |
|---|---|
| **DTC: P1228**<br>**2T, MIL: Yes**<br>**Years:** 2007, 2008<br>**Models:** Passat, Jetta, Golf, GTI, New Beetle, New Beetle Convertible<br>**Engines:** 2.0L, 2.5L, 3.6L<br>**Transmissions:** All | **Cylinder 4 Fuel Injection Circuit Short to Ground Conditions:**<br>Key on or engine running, fuses in the instrument panel and the E-box in the engine compartment must be functioning, and the ground connections between the engine ad the chassis must be well connected; and the ECM detected an unexpected voltage condition on the injector circuit. Wiring or fuel injector has a short circuit to ground.<br>**Possible Causes:**<br>• Injector control circuit is open<br>• Injector power circuit (B+) is open<br>• Injector control circuit is shorted to chassis ground<br>• Injector is damaged or has failed<br>• ECM is not connected or has failed<br>• Fuel pump relay has failed<br>• Engine speed sensor has failed |
| **DTC: P1229**<br>**Years:** 2007, 2008<br>**Models:** Passat, Jetta<br>**Engines:** 2.5L, 3.6L<br>**Transmissions:** All | **Cylinder 5 Fuel Injection Circuit Short to Ground Conditions:**<br>Key on or engine running, fuses in the instrument panel and the E-box in the engine compartment must be functioning, and the ground connections between the engine ad the chassis must be well connected; and the ECM detected an unexpected voltage condition on the injector circuit. Wiring or fuel injector has a short circuit to ground.<br>**Possible Causes:**<br>• Injector control circuit is open<br>• Injector power circuit (B+) is open<br>• Injector control circuit is shorted to chassis ground<br>• Injector is damaged or has failed<br>• ECM is not connected or has failed<br>• Fuel pump relay has failed<br>• Engine speed sensor has failed |
| **DTC: P1230**<br>**Years:** 2007, 2008<br>**Models:** Passat, Jetta<br>**Engines:** 2.5L, 3.6L<br>**Transmissions:** All | **Cylinder 6 Fuel Injection Circuit Short to Ground Conditions:**<br>Key on or engine running, fuses in the instrument panel and the E-box in the engine compartment must be functioning, and the ground connections between the engine ad the chassis must be well connected; and the ECM detected an unexpected voltage condition on the injector circuit. Wiring or fuel injector has a short circuit to ground.<br>**Possible Causes:**<br>• Injector control circuit is open<br>• Injector power circuit (B+) is open<br>• Injector control circuit is shorted to chassis ground<br>• Injector is damaged or has failed<br>• ECM is not connected or has failed<br>• Fuel pump relay has failed<br>• Engine speed sensor has failed |
| **DTC: P1237**<br>**2T, MIL: Yes**<br>**Years:** 2007, 2008<br>**Models:** Passat, Jetta, Golf, GTI, New Beetle, New Beetle Convertible<br>**Engines:** 2.0L, 2.5L, 3.6L<br>**Transmissions:** All | **Cylinder 1 Fuel Injection Circuit Open Circuit Conditions:**<br>Key on or engine running, fuses in the instrument panel and the E-box in the engine compartment must be functioning, and the ground connections between the engine ad the chassis must be well connected; and the ECM detected an unexpected voltage condition on the injector circuit. Wiring or fuel injector has a short circuit that is open.<br>**Possible Causes:**<br>• Injector control circuit is open<br>• Injector power circuit (B+) is open<br>• Injector control circuit is shorted to chassis ground<br>• Injector is damaged or has failed<br>• ECM is not connected or has failed<br>• Fuel pump relay has failed<br>• Engine speed sensor has failed |
| **DTC: P1238**<br>**2T, MIL: Yes**<br>**Years:** 2007, 2008<br>**Models:** Passat, Jetta, Golf, GTI, New Beetle, New Beetle Convertible<br>**Engines:** 2.0L, 2.5L, 3.6L<br>**Transmissions:** All | **Cylinder 2 Fuel Injection Circuit Open Circuit Conditions:**<br>Key on or engine running, fuses in the instrument panel and the E-box in the engine compartment must be functioning, and the ground connections between the engine ad the chassis must be well connected; and the ECM detected an unexpected voltage condition on the injector circuit. Wiring or fuel injector has a short circuit that is open.<br>**Possible Causes:**<br>• Injector control circuit is open<br>• Injector power circuit (B+) is open<br>• Injector control circuit is shorted to chassis ground<br>• Injector is damaged or has failed<br>• ECM is not connected or has failed<br>• Fuel pump relay has failed<br>• Engine speed sensor has failed |

| DTC | Trouble Code Title, Conditions & Possible Causes |
|---|---|
| **DTC: P1239**<br>**2T, MIL**: Yes<br>**Years**: 2007, 2008<br>**Models**: Passat, Jetta, Golf, GTI, New Beetle, New Beetle Convertible<br>**Engines**: 2.0L, 2.5L, 3.6L<br>**Transmissions**: All | **Cylinder 3 Fuel Injection Circuit Open Circuit Conditions:**<br>Key on or engine running, fuses in the instrument panel and the E-box in the engine compartment must be functioning, and the ground connections between the engine ad the chassis must be well connected; and the ECM detected an unexpected voltage condition on the injector circuit. Wiring or fuel injector has a short circuit that is open.<br>**Possible Causes:**<br>• Injector control circuit is open<br>• Injector power circuit (B+) is open<br>• Injector control circuit is shorted to chassis ground<br>• Injector is damaged or has failed<br>• ECM is not connected or has failed<br>• Fuel pump relay has failed<br>• Engine speed sensor has failed |
| **DTC: P1240**<br>**2T, MIL**: Yes<br>**Years**: 2007, 2008<br>**Models**: Passat, Jetta, Golf, GTI, New Beetle, New Beetle Convertible<br>**Engines**: 2.0L, 2.5L, 3.6L<br>**Transmissions**: All | **Cylinder 4 Fuel Injection Circuit Open Circuit Conditions:**<br>Key on or engine running, fuses in the instrument panel and the E-box in the engine compartment must be functioning, and the ground connections between the engine ad the chassis must be well connected; and the ECM detected an unexpected voltage condition on the injector circuit. Wiring or fuel injector has a short circuit that is open.<br>**Possible Causes:**<br>• Injector control circuit is open<br>• Injector power circuit (B+) is open<br>• Injector control circuit is shorted to chassis ground<br>• Injector is damaged or has failed<br>• ECM is not connected or has failed<br>• Fuel pump relay has failed<br>• Engine speed sensor has failed |
| **DTC: P1241**<br>**Years**: 2007, 2008<br>**Models**: Passat, Jetta<br>**Engines**: 2.5L, 3.6L<br>**Transmissions**: All | **Cylinder 5 Fuel Injection Circuit Open Circuit Conditions:**<br>Key on or engine running, fuses in the instrument panel and the E-box in the engine compartment must be functioning, and the ground connections between the engine ad the chassis must be well connected; and the ECM detected an unexpected voltage condition on the injector circuit. Wiring or fuel injector has a short circuit that is open.<br>**Possible Causes:**<br>• Injector control circuit is open<br>• Injector power circuit (B+) is open<br>• Injector control circuit is shorted to chassis ground<br>• Injector is damaged or has failed<br>• ECM is not connected or has failed<br>• Fuel pump relay has failed<br>• Engine speed sensor has failed |
| **DTC: P1242**<br>**Years**: 2007, 2008<br>**Models**: Passat, Jetta<br>**Engines**: 2.5L, 3.6L<br>**Transmissions**: All | **Cylinder 6 Fuel Injection Circuit Open Circuit Conditions:**<br>Key on or engine running, fuses in the instrument panel and the E-box in the engine compartment must be functioning, and the ground connections between the engine ad the chassis must be well connected; and the ECM detected an unexpected voltage condition on the injector circuit. Wiring or fuel injector has a short circuit that is open.<br>**Possible Causes:**<br>• Injector control circuit is open<br>• Injector power circuit (B+) is open<br>• Injector control circuit is shorted to chassis ground<br>• Injector is damaged or has failed<br>• ECM is not connected or has failed<br>• Fuel pump relay has failed<br>• Engine speed sensor has failed |
| **DTC: P1250**<br>**Years**: 2007, 2008<br>**Models**: Passat, Jetta, Golf, GTI<br>**Engines**: 2.0L, 2.5L, 3.6L<br>**Transmissions**: All | **Fuel Pressure Regulator Control Circuit Malfunction (Fuel Level too Low) Conditions:**<br>KOEO or KOER Self-Test enabled, and the ECM detected a lack of power (VPWR) to the Fuel Pressure Regulator Control (FPRC) solenoid circuit.<br>**Possible Causes:**<br>• FPRC solenoid valve harness circuits are open or shorted<br>• FPRC input port or output port vacuum lines are damaged<br>• FRPC solenoid is damaged<br>• Fuel level is too low<br>• ECM has failed |

| DTC | Trouble Code Title, Conditions & Possible Causes |
|---|---|
| **DTC: P1255**<br>**Years: 2007, 2008**<br>**Models:** Passat, Jetta, Golf, GTI, New Beetle, New Beetle Convertible<br>**Engines:** 2.0L, 2.5L, 3.6L<br>**Transmissions:** All | **Engine Coolant Temperature Circuit Short to Ground Conditions:**<br>Key on, engine started, the ECM detected an unexpected voltage condition on the ECT circuit.<br>**Possible Causes:**<br>• Coolant Temperature Sensor has failed<br>• Circuit short to ground, open or other component<br>• ECM has failed |
| **DTC: P1256**<br>**Years: 2007, 2008**<br>**Models:** Passat, Jetta, Golf, GTI, New Beetle, New Beetle Convertible<br>**Engines:** 2.0L, 2.5L, 3.6L<br>**Transmissions:** All | **Engine Coolant Temperature Circuit Open/Short to B+ Conditions:**<br>Key on, engine started, the ECM detected an unexpected voltage condition on the ECT circuit.<br>**Possible Causes:**<br>• Coolant Temperature Sensor has failed<br>• Circuit short to ground, open or other component<br>• ECM has failed |
| **DTC: P1287**<br>**Years: 2007, 2008**<br>**Models:** Passat, Jetta, Golf, GTI, New Beetle<br>**Engines:** 2.0L, 2.5L, 3.6L<br>**Transmissions:** All | **Turbocharger Bypass Valve Open Conditions:**<br>Engine started, battery voltage at least 11.5v, all electrical components off, ground connections between engine and chassis well connected, coolant temperature at least 80-degrees Celicius. The ECM detected an unexpected voltage condition on the bypass valve control circuit<br>**Possible Causes:**<br>• Charge air system check for leaks<br>• Recirculating valve for turbocharger is faulty<br>• Turbocharging system may be damaged<br>• Vacuum diaphragm for turbocharger needs adjusting<br>• Wastegate bypass regulator valve is faulty<br>• Bypass solenoid power supply circuit is open<br>• Bypass solenoid control circuit is open, shorted to ground or system power<br>• Bypass solenoid assembly is damaged or has failed<br>• Charge air pressure sensor is faulty<br>• Voltage supply to the charge air pressure sensor is open or shorted<br>• Check the charge air system for leaks<br>• Recirculating valve for turbocharger is faulty<br>• Turbocharging system is damaged or not functioning properly<br>• Turbocharger recirculating valve is faulty<br>• Vacuum diaphragm for turbocharger is out of adjustment<br>• Wastegate bypass regulator valve is faulty<br>• ECM has failed |
| **DTC: P1288**<br>**Years: 2007, 2008**<br>**Models:** Passat, Jetta, Golf, GTI, New Beetle<br>**Engines:** 2.0L, 2.5L, 3.6L<br>**Transmissions:** All | **Turbocharger Bypass Valve Short to B+ Conditions:**<br>Engine started, battery voltage at least 11.5v, all electrical components off, ground connections between engine and chassis well connected, coolant temperature at least 80-degrees Celicius. The ECM detected an unexpected voltage condition on the bypass valve control circuit<br>**Possible Causes:**<br>• Charge air system check for leaks<br>• Recirculating valve for turbocharger is faulty<br>• Turbocharging system may be damaged<br>• Vacuum diaphragm for turbocharger needs adjusting<br>• Wastegate bypass regulator valve is faulty<br>• Bypass solenoid power supply circuit is open<br>• Bypass solenoid control circuit is open, shorted to ground or system power<br>• Bypass solenoid assembly is damaged or has failed<br>• Charge air pressure sensor is faulty<br>• Voltage supply to the charge air pressure sensor is open or shorted<br>• Check the charge air system for leaks<br>• Recirculating valve for turbocharger is faulty<br>• Turbocharging system is damaged or not functioning properly<br>• Turbocharger recirculating valve is faulty<br>• Vacuum diaphragm for turbocharger is out of adjustment<br>• Wastegate bypass regulator valve is faulty<br>• ECM has failedvalve is faulty |

| DTC | Trouble Code Title, Conditions & Possible Causes |
|---|---|
| **DTC: P1289**<br>**Years: 2007, 2008**<br>**Models:** Passat, Jetta, Golf, GTI, New Beetle<br>**Engines:** 2.0L, 2.5L, 3.6L<br>**Transmissions:** All | **Turbocharger Bypass Valve Short to Ground Conditions:**<br>Engine started, battery voltage at least 11.5v, all electrical components off, ground connections between engine and chassis well connected, coolant temperature at least 80-degrees Celicius. The ECM detected an unexpected voltage condition on the bypass valve control circuit<br>**Possible Causes:**<br>• Charge air system check for leaks<br>• Recirculating valve for turbocharger is faulty<br>• Turbocharging system may be damaged<br>• Vacuum diaphragm for turbocharger needs adjusting<br>• Wastegate bypass regulator valve is faulty<br>• Bypass solenoid power supply circuit is open<br>• Bypass solenoid control circuit is open, shorted to ground or system power<br>• Bypass solenoid assembly is damaged or has failed<br>• Charge air pressure sensor is faulty<br>• Voltage supply to the charge air pressure sensor is open or shorted<br>• Check the charge air system for leaks<br>• Recirculating valve for turbocharger is faulty<br>• Turbocharging system is damaged or not functioning properly<br>• Turbocharger recirculating valve is faulty<br>• Vacuum diaphragm for turbocharger is out of adjustment<br>• Wastegate bypass regulator valve is faulty<br>• ECM has failed |
| **DTC: P1295**<br>**Years: 2007, 2008**<br>**Models:** Passat, Jetta, Golf, GTI, New Beetle<br>**Engines:** 2.0L, 2.5L, 3.6L<br>**Transmissions:** All | **Turbocharger Bypass Valve Throughput Faulty Conditions:**<br>Engine started, battery voltage at least 11.5v, all electrical components off, ground connections between engine and chassis well connected, coolant temperature at least 80-degrees Celicius. The ECM detected an unexpected voltage condition on the bypass valve control circuit<br>**Possible Causes:**<br>• Charge air system check for leaks<br>• Recirculating valve for turbocharger is faulty<br>• Turbocharging system may be damaged<br>• Vacuum diaphragm for turbocharger needs adjusting<br>• Wastegate bypass regulator valve is faulty<br>• Bypass solenoid power supply circuit is open<br>• Bypass solenoid control circuit is open, shorted to ground or system power<br>• Bypass solenoid assembly is damaged or has failed<br>• Charge air pressure sensor is faulty<br>• Voltage supply to the charge air pressure sensor is open or shorted<br>• Check the charge air system for leaks<br>• Recirculating valve for turbocharger is faulty<br>• Turbocharging system is damaged or not functioning properly<br>• Turbocharger recirculating valve is faulty<br>• Vacuum diaphragm for turbocharger is out of adjustment<br>• Wastegate bypass regulator valve is faulty<br>• ECM has failed |
| **DTC: P1296**<br>**Years: 2007, 2008**<br>**Models:** Passat, Jetta, Golf, GTI, New Beetle, New Beetle Convertible<br>**Engines:** 2.0L, 2.5L, 3.6L<br>**Transmissions:** All | **Cooling System Malfunction Conditions:**<br>Key on, engine not running, the Engine Control Module (ECM) will use the intake air temperature as a replacement value for an engine start (start temperature replacement value) as soon as there is a Diagnostic Trouble Code (DTC) stored in DTC memory for the Engine Coolant Temperature (ECT) sensor. The temperature then rises according to a program stored in the ECM. When the engine has reached normal operating temperature a fixed replacement value will be displayed. This fixed value is also dependent upon the intake air temperature.<br>**Possible Causes:**<br>• Engine coolant temperature sensor has failed<br>• An open circuit or a short to B+ is present<br>• Sensor circuit is short to ground<br>• ECM has failed |

| DTC | Trouble Code Title, Conditions & Possible Causes |
|---|---|
| **DTC: P1297**<br>**Years:** 2007, 2008<br>**Models:** Passat, Jetta, Golf, GTI, New Beetle<br>**Engines:** 2.0L<br>**Transmissions**: All | **Connection Turbocharger/Throttle Valve Pressure Hose Conditions:**<br>Engine started, battery voltage at least 11.5v, all electrical components off, ground connections between engine and chassis well connected, coolant temperature at least 80-degrees Celicius. The ECM detected an unexpected voltage condition on the turbo valve pressure hose.<br>**Possible Causes:**<br>• Charge air system check for leaks<br>• Recirculating valve for turbocharger is faulty<br>• Turbocharging system may be damaged<br>• Vacuum diaphragm for turbocharger needs adjusting<br>• Wastegate bypass regulator valve is faulty<br>• Bypass solenoid power supply circuit is open<br>• Bypass solenoid control circuit is open, shorted to ground or system power<br>• Bypass solenoid assembly is damaged or has failed<br>• Charge air pressure sensor is faulty<br>• Voltage supply to the charge air pressure sensor is open or shorted<br>• Check the charge air system for leaks<br>• Recirculating valve for turbocharger is faulty<br>• Turbocharging system is damaged or not functioning properly<br>• Turbocharger recirculating valve is faulty<br>• Vacuum diaphragm for turbocharger is out of adjustment<br>• Wastegate bypass regulator valve is faulty<br>• ECM has failed |
| **DTC: P1300**<br>**Years:** 2007, 2008<br>**Models:** Passat, Jetta, Golf, GTI, New Beetle, New Beetle Convertible<br>**Engines:** 2.0L, 2.5L, 3.6L<br>**Transmissions:** All | **Misfire Detected Reason: Fuel Level Too Low Conditions:**<br>Engine running, the ECM detected a misfire because of lack of fuel<br>**Possible Causes:**<br>• Fuel level too low<br>• Fuel leak<br>• Fuel injector faulty |
| **DTC: P1325**<br>**Years:** 2007, 2008<br>**Models:** Passat, Jetta, Golf, GTI, New Beetle,<br>**Engines:** 2.0L, 2.5L, 3.6L<br>**Transmissions**: All | **Cylinder 1-Knock Control Limit Attained Conditions:**<br>Engine started, battery voltage at least 11.5v, all electrical components off, ground connections between engine and chassis well connected, and the ECM detected the Knock Sensor signal was more than the calibrated value.<br>**Possible Causes:**<br>• Knock sensor circuit is open<br>• Knock sensor circuit is shorted to ground, or shorted to power<br>• Knock sensor is damaged or it has failed<br>• Poor fuel quality<br>• Loosen knock sensors and tighten again to 20 Nm<br>• ECM has failed |
| **DTC: P1326**<br>**Years:** 2007, 2008<br>**Models:** Passat, Jetta, Golf, GTI, New Beetle,<br>**Engines:** 2.0L, 2.5L, 3.6L<br>**Transmissions**: All | **Cylinder 2-Knock Control Limit Attained Conditions:**<br>Engine started, battery voltage at least 11.5v, all electrical components off, ground connections between engine and chassis well connected, and the ECM detected the Knock Sensor signal was more than the calibrated value.<br>**Possible Causes:**<br>• Knock sensor circuit is open<br>• Knock sensor circuit is shorted to ground, or shorted to power<br>• Knock sensor is damaged or it has failed<br>• Poor fuel quality<br>• Loosen knock sensors and tighten again to 20 Nm<br>• ECM has failed |
| **DTC: P1327**<br>**Years:** 2007, 2008<br>**Models:** Passat, Jetta, Golf, GTI, New Beetle,<br>**Engines:** 2.0L, 2.5L, 3.6L<br>**Transmissions**: All | **Cylinder 3-Knock Control Limit Attained Conditions:**<br>Engine started, battery voltage at least 11.5v, all electrical components off, ground connections between engine and chassis well connected, and the ECM detected the Knock Sensor signal was more than the calibrated value.<br>**Possible Causes:**<br>• Knock sensor circuit is open<br>• Knock sensor circuit is shorted to ground, or shorted to power<br>• Knock sensor is damaged or it has failed<br>• Poor fuel quality<br>• Loosen knock sensors and tighten again to 20 Nm<br>• ECM has failed |

| DTC | Trouble Code Title, Conditions & Possible Causes |
|-----|---------------------------------------------------|
| **DTC: P1328**<br>**Years:** 2007, 2008<br>**Models:** Passat, Jetta, Golf, GTI, New Beetle,<br>**Engines:** 2.0L, 2.5L, 3.6L<br>**Transmissions:** All | **Cylinder 4-Knock Control Limit Attained Conditions:**<br>Engine started, battery voltage at least 11.5v, all electrical components off, ground connections between engine and chassis well connected, and the ECM detected the Knock Sensor signal was more than the calibrated value.<br>**Possible Causes:**<br>• Knock sensor circuit is open<br>• Knock sensor circuit is shorted to ground, or shorted to power<br>• Knock sensor is damaged or it has failed<br>• ECM has failed |
| **DTC: P1329**<br>**Years:** 2007, 2008<br>**Models:** Passat<br>**Engines:** 2.5L, 3.6L<br>**Transmissions:** All | **Cylinder 5-Knock Control Limit Attained Conditions:**<br>Engine started, battery voltage at least 11.5v, all electrical components off, ground connections between engine and chassis well connected, and the ECM detected the Knock Sensor signal was more than the calibrated value.<br>**Possible Causes:**<br>• Knock sensor circuit is open<br>• Knock sensor circuit is shorted to ground, or shorted to power<br>• Knock sensor is damaged or it has failed<br>• Poor fuel quality<br>• Loosen knock sensors and tighten again to 20 Nm<br>• ECM has failed |
| **DTC: P1330**<br>**Years:** 2007, 2008<br>**Models:** Passat<br>**Engines:** 2.5L, 3.6L<br>**Transmissions:** All | **Cylinder 6-Knock Control Limit Attained Conditions:**<br>Engine started, battery voltage at least 11.5v, all electrical components off, ground connections between engine and chassis well connected, and the ECM detected the Knock Sensor signal was more than the calibrated value.<br>**Possible Causes:**<br>• Knock sensor circuit is open<br>• Knock sensor circuit is shorted to ground, or shorted to power<br>• Knock sensor is damaged or it has failed<br>• Poor fuel quality<br>• Loosen knock sensors and tighten again to 20 Nm<br>• ECM has failed |
| **DTC: P1335**<br>**Years:** 2007, 2008<br>**Models:** Passat, Jetta, Golf, GTI, New Beetle, New Beetle Convertible<br>**Engines:** 2.0L, 2.5L, 3.6L<br>**Transmissions:** All | **Engine Torque Monitoring 2 Control Limit Exceeded Conditions:**<br>Engine cold, battery voltage at least 11.5v, all electrical components off, ground connections between engine and chassis well connected, the ECM detected a signal beyond the required limit.<br>**Possible Causes:**<br>• Engine Control Module (ECM) has failed<br>• Voltage supply for Engine Control Module (ECM) is shorted<br>• Engine Coolant Temperature (ECT) sensor is faulty<br>• Intake Air Temperature (IAT) sensor is faulty<br>• Intake Manifold Runner Position Sensor is faulty<br>• Intake system for leaks (false air) is faulty<br>• Motor for intake flap is faulty<br>• Mass Air Flow (MAF) sensor is faulty |
| **DTC: P1336**<br>**Years:** 2007, 2008<br>**Models:** Passat, Jetta, Golf, GTI, New Beetle, New Beetle Convertible<br>**Engines:** 2.0L, 2.5L, 3.6L<br>**Transmissions:** All | **Engine Torque Monitoring Control Limit Exceeded Conditions:**<br>Engine cold, battery voltage at least 11.5v, all electrical components off, ground connections between engine and chassis well connected, the ECM detected a signal beyond the required limit.<br>**Possible Causes:**<br>• Engine Control Module (ECM) has failed<br>• Voltage supply for Engine Control Module (ECM) is shorted<br>• Engine Coolant Temperature (ECT) sensor is faulty<br>• Intake Air Temperature (IAT) sensor is faulty<br>• Intake Manifold Runner Position Sensor is faulty<br>• Intake system for leaks (false air) is faulty<br>• Motor for intake flap is faulty<br>• Mass Air Flow (MAF) sensor is faulty |
| **DTC: P1337**<br>**2T, MIL: Yes**<br>**Years:** 2007, 2008<br>**Models:** Passat, Jetta, Golf, GTI, New Beetle<br>**Engines:** 2.0L, 2.5L, 3.6L<br>**Transmissions:** All | **Camshaft Position Sensor (Bank 1) Short to Ground Conditions:**<br>Engine started, battery voltage at least 11.5v, all electrical components off, ground connections between engine and chassis well connected, and the ECM detected an unexpected low or high voltage condition on the camshaft position sensor circuit<br>**Possible Causes:**<br>• Faulty CPM sensor<br>• ECM has failed |

| DTC | Trouble Code Title, Conditions & Possible Causes |
|---|---|
| **DTC: P1338**<br>**2T, MIL: Yes**<br>**Years: 2007, 2008**<br>**Models:** Passat, Jetta, Golf, GTI, New Beetle<br>**Engines:** 2.0L, 2.5L, 3.6L<br>**Transmissions:** All | **Camshaft Position Sensor (Bank 1) Open/Short to B+ Conditions:**<br>Engine started, battery voltage at least 11.5v, all electrical components off, ground connections between engine and chassis well connected, and the ECM detected an unexpected low or high voltage condition on the camshaft position sensor circuit<br>**Possible Causes:**<br>• Faulty CPM sensor<br>• ECM has failed |
| **DTC: P1340**<br>**2T, MIL: Yes**<br>**Years: 2007, 2008**<br>**Models:** Passat, Jetta, Golf, GTI, New Beetle, New Beetle Convertible<br>**Engines:** 2.0L, 2.5L, 3.6L<br>**Transmissions:** All | **Crankshaft Position/Camshaft Sensor Signal Out of Sequence Conditions:**<br>Engine started, battery voltage at least 11.5v, all electrical components off, ground connections between engine and chassis well connected, and the ECM detected the crankshaft position sensor and the camshaft sensor were out of sequence with each other.<br>**Note: The Engine Speed (RPM) Sensor detects engine speed and reference marks. Without an engine speed signal, the engine will not start. If the engine speed signal fails while the engine is running, the engine will stop immediately.**<br>**Possible Causes:**<br>• Engine speed sensor has failed or is contaminated (metal filings)<br>• Engine speed sensor's wheel is damaged<br>• Engine speed sensor circuit is shorted to the cable shield<br>• Engine speed sensor circuit is open<br>• ECM is faulty<br>• Canshaft position sensor is faulty |
| **DTC: P1355**<br>**Years: 2007, 2008**<br>**Models:** Passat, Jetta, Golf, GTI, New Beetle, New Beetle Convertible<br>**Engines:** 2.0L, 2.5L, 3.6L<br>**Transmissions:** All | **Cylinder 1 Ignition Circuit Open Circuit Conditions:**<br>Key on or Engine started, battery voltage at least 11.5v, all electrical components off, ground connections between engine and chassis well connected, and the ECM detected the voltage of the ignition was outside the designed parameters.<br>**Possible Causes:**<br>• Fuel pump relay faulty<br>• Canshaft position sensor has failed<br>• Engine speed sensor has failed<br>• Circuit wires have shorted to ground or are open<br>• Ignition coils with power output stages are faulty<br>• ECM has failed |
| **DTC: P1356**<br>**Years: 2007, 2008**<br>**Models:** Passat, Jetta, Golf, GTI, New Beetle,<br>**Engines:** 2.0L, 2.5L, 3.6L<br>**Transmissions:** All | **Cylinder 1 Ignition Circuit Short to B+ Conditions:**<br>Key on or Engine started, battery voltage at least 11.5v, all electrical components off, ground connections between engine and chassis well connected, and the ECM detected the voltage of the ignition was outside the designed parameters.<br>**Possible Causes:**<br>• Fuel pump relay faulty<br>• Canshaft position sensor has failed<br>• Engine speed sensor has failed<br>• Circuit wires have shorted to ground or are open<br>• Ignition coils with power output stages are faulty<br>• ECM has failed |
| **DTC: P1357**<br>**Years: 2007, 2008**<br>**Models:** Passat, Jetta, Golf, GTI, New Beetle,<br>**Engines:** 2.0L, 2.5L, 3.6L<br>**Transmissions:** All | **Cylinder 1 Ignition Circuit Short to Ground Conditions:**<br>Key on or Engine started, battery voltage at least 11.5v, all electrical components off, ground connections between engine and chassis well connected, and the ECM detected the voltage of the ignition was outside the designed parameters.<br>**Possible Causes:**<br>• Fuel pump relay faulty<br>• Canshaft position sensor has failed<br>• Engine speed sensor has failed<br>• Circuit wires have shorted to ground or are open<br>• Ignition coils with power output stages are faulty<br>• ECM has failed |
| **DTC: P1358**<br>**Years: 2007, 2008**<br>**Models:** Passat, Jetta, Golf, GTI, New Beetle, New Beetle Convertible<br>**Engines:** 2.0L, 2.5L, 3.6L<br>**Transmissions:** All | **Cylinder 2 Ignition Circuit Open Circuit Conditions:**<br>Key on or Engine started, battery voltage at least 11.5v, all electrical components off, ground connections between engine and chassis well connected, and the ECM detected the voltage of the ignition was outside the designed parameters.<br>**Possible Causes:**<br>• Fuel pump relay faulty<br>• Canshaft position sensor has failed<br>• Engine speed sensor has failed<br>• Circuit wires have shorted to ground or are open<br>• Ignition coils with power output stages are faulty<br>• ECM has failed |

| DTC | Trouble Code Title, Conditions & Possible Causes |
|---|---|
| **DTC: P1359**<br>**2T, MIL: Yes**<br>**Years: 2007, 2008**<br>**Models:** Passat, Jetta, Golf, GTI, New Beetle,<br>**Engines:** 2.0L, 2.5L, 3.6L<br>**Transmissions:** All | **Cylinder 2 Ignition Circuit Short to B+ Conditions:**<br>Key on or Engine started, battery voltage at least 11.5v, all electrical components off, ground connections between engine and chassis well connected, and the ECM detected the voltage of the ignition was outside the designed parameters.<br>**Possible Causes:**<br>• Fuel pump relay faulty<br>• Canshaft position sensor has failed<br>• Engine speed sensor has failed<br>• Circuit wires have shorted to ground or are open<br>• Ignition coils with power output stages are faulty<br>• ECM has failed |
| **DTC: P1360**<br>**Years: 2007, 2008**<br>**Models:** Passat, Jetta, Golf, GTI, New Beetle,<br>**Engines:** 2.0L, 2.5L, 3.6L<br>**Transmissions:** All | **Cylinder 2 Ignition Circuit Short to Ground Conditions:**<br>Key on or Engine started, battery voltage at least 11.5v, all electrical components off, ground connections between engine and chassis well connected, and the ECM detected the voltage of the ignition was outside the designed parameters.<br>**Possible Causes:**<br>• Fuel pump relay faulty<br>• Canshaft position sensor has failed<br>• Engine speed sensor has failed<br>• Circuit wires have shorted to ground or are open<br>• Ignition coils with power output stages are faulty<br>• ECM has failed |
| **DTC: P1361**<br>**Years: 2007, 2008**<br>**Models:** Passat, Jetta, Golf, GTI, New Beetle, New Beetle Convertible<br>**Engines:** 2.0L, 2.5L, 3.6L<br>**Transmissions:** All | **Cylinder 3 Ignition Circuit Open Circuit Conditions:**<br>Key on or Engine started, battery voltage at least 11.5v, all electrical components off, ground connections between engine and chassis well connected, and the ECM detected the voltage of the ignition was outside the designed parameters.<br>**Possible Causes:**<br>• Fuel pump relay faulty<br>• Canshaft position sensor has failed<br>• Engine speed sensor has failed<br>• Circuit wires have shorted to ground or are open<br>• Ignition coils with power output stages are faulty<br>• ECM has failed |
| **DTC: P1362**<br>**Years: 2007, 2008**<br>**Models:** Passat, Jetta, Golf, GTI, New Beetle,<br>**Engines:** 2.0L, 2.5L, 3.6L<br>**Transmissions:** All | **Cylinder 3 Ignition Circuit Short to B+ Conditions:**<br>Key on or Engine started, battery voltage at least 11.5v, all electrical components off, ground connections between engine and chassis well connected, and the ECM detected the voltage of the ignition was outside the designed parameters.<br>**Possible Causes:**<br>• Fuel pump relay faulty<br>• Canshaft position sensor has failed<br>• Engine speed sensor has failed<br>• Circuit wires have shorted to ground or are open<br>• Ignition coils with power output stages are faulty<br>• ECM has failed |
| **DTC: P1363**<br>**Years: 2007, 2008**<br>**Models:** Passat, Jetta, Golf, GTI, New Beetle,<br>**Engines:** 2.0L, 2.5L, 3.6L<br>**Transmissions:** All | **Cylinder 3 Ignition Circuit Short to Ground Conditions:**<br>Key on or Engine started, battery voltage at least 11.5v, all electrical components off, ground connections between engine and chassis well connected, and the ECM detected the voltage of the ignition was outside the designed parameters.<br>**Possible Causes:**<br>• Fuel pump relay faulty<br>• Canshaft position sensor has failed<br>• Engine speed sensor has failed<br>• Circuit wires have shorted to ground or are open<br>• Ignition coils with power output stages are faulty<br>• ECM has failed |
| **DTC: P1364**<br>**Years: 2007, 2008**<br>**Models:** Passat, Jetta, Golf, GTI, New Beetle, New Beetle Convertible<br>**Engines:** 2.0L, 2.5L, 3.6L<br>**Transmissions:** All | **Cylinder 4 Ignition Circuit Open Circuit Conditions:**<br>Key on or Engine started, battery voltage at least 11.5v, all electrical components off, ground connections between engine and chassis well connected, and the ECM detected the voltage of the ignition was outside the designed parameters.<br>**Possible Causes:**<br>• Fuel pump relay faulty<br>• Canshaft position sensor has failed<br>• Engine speed sensor has failed<br>• Circuit wires have shorted to ground or are open<br>• Ignition coils with power output stages are faulty<br>• ECM has failed |

| DTC | Trouble Code Title, Conditions & Possible Causes |
|---|---|
| **DTC: P1365**<br>**Years: 2007, 2008**<br>**Models:** Passat, Jetta, Golf, GTI, New Beetle,<br>**Engines:** 2.0L, 2.5L, 3.6L<br>**Transmissions:** All | **Cylinder 4 Ignition Circuit Short to B+ Conditions:**<br>Key on or Engine started, battery voltage at least 11.5v, all electrical components off, ground connections between engine and chassis well connected, and the ECM detected the voltage of the ignition was outside the designed parameters.<br>**Possible Causes:**<br>• Fuel pump relay faulty<br>• Canshaft position sensor has failed<br>• Engine speed sensor has failed<br>• Circuit wires have shorted to ground or are open<br>• Ignition coils with power output stages are faulty<br>• ECM has failed |
| **DTC: P1366**<br>**Years: 2007, 2008**<br>**Models:** Passat, Jetta, Golf, GTI, New Beetle,<br>**Engines:** 2.0L, 2.5L, 3.6L<br>**Transmissions:** All | **Cylinder 4 Ignition Circuit Short to Ground Conditions:**<br>Key on or Engine started, battery voltage at least 11.5v, all electrical components off, ground connections between engine and chassis well connected, and the ECM detected the voltage of the ignition was outside the designed parameters.<br>**Possible Causes:**<br>• Fuel pump relay faulty<br>• Canshaft position sensor has failed<br>• Engine speed sensor has failed<br>• Circuit wires have shorted to ground or are open<br>• Ignition coils with power output stages are faulty<br>• ECM has failed |
| **DTC: P1386**<br>**Years: 2007, 2008**<br>**Models:** Passat, Jetta, Golf, GTI, New Beetle<br>**Engines:** 2.0L, 2.5L, 3.6L<br>**Transmissions:** All | **Internal Control Module, Knock Control Circuit Error Conditions:**<br>Engine started, and the ECM detected a too high or too low voltage condition on the knock control circuits, or a miscommunication between the knock control and the ECM.<br>**Possible Causes:**<br>• ECM has failed |
| **DTC: P1387**<br>**2T, MIL: Yes**<br>**Years: 2007, 2008**<br>**Models:** Passat, Jetta, Golf, GTI, New Beetle,<br>**Engines:** 2.0L, 2.5L, 3.6L<br>**Transmissions:** All | **Internal Control Module Altitude Sensor Error Conditions:**<br>Ignition on, the ECM detected and altitude sensor error. To achieve optimal anti-theft protection for the vehicle, an anti-theft immobilizer is installed. The anti-theft immobilizer is a system for enabling and locking the Engine Control Module (ECM). So that this system cannot be circumvented, it is necessary to perform adaptation of the anti-theft immobilizer using the Vehicle Diagnostic and Information System VAS 5052 in the On Board Diagnostic (OBD) function. The great availability of equipment options makes it necessary to adapt the Engine Control Module (ECM) to the vehicle (e.g. throttle valve control module or cruise control system). This "writing" function is not possible with the generic scan tool.<br>**Possible Causes:**<br>• (If ECM was replaced) ECM ID not the same as the replaced unit<br>• ECM has failed<br>• Voltage supply for Engine Control Module (ECM) has shorted |
| **DTC: P1388**<br>**Years: 2007, 2008**<br>**Models:** Passat, Jetta, Golf, GTI, New Beetle,<br>**Engines:** 2.0L, 2.5L, 3.6L<br>**Transmissions:** All | **Internal Control Module Drive By Wire Error Conditions:**<br>Ignition on, the ECM detected and drive by wire error. To achieve optimal anti-theft protection for the vehicle, an anti-theft immobilizer is installed. The anti-theft immobilizer is a system for enabling and locking the Engine Control Module (ECM). So that this system cannot be circumvented, it is necessary to perform adaptation of the anti-theft immobilizer using the Vehicle Diagnostic and Information System VAS 5052 in the On Board Diagnostic (OBD) function. The great availability of equipment options makes it necessary to adapt the Engine Control Module (ECM) to the vehicle (e.g. throttle valve control module or cruise control system). This "writing" function is not possible with the generic scan tool.<br>**Possible Causes:**<br>• Engine Control Module (ECM) has failed<br>• Voltage supply for Engine Control Module (ECM) has shorted |
| **DTC: P1391**<br>**Years: 2007, 2008**<br>**Models:** Passat<br>**Engines:** 2.5L, 3.6L<br>**Transmissions:** All | **Camshaft Position Sensor (Bank 2) Short to Ground Conditions:**<br>Key on or Engine started, battery voltage must be at least 11.5v, all electrical components must be off, parking brake must be engaged (to keep daytime driving lights off), automatic transmission selector must be in park and the ground between the engine and the chassis must be well connected. The ECM detected an unexpected low or high voltage condition on the camshaft position sensor circuit.<br>**Possible Causes:**<br>• CMP sensor circuit is open or shorted to ground<br>• CMP sensor circuit is shorted to power<br>• CMP sensor ground (return) circuit is open<br>• CMP sensor installation incorrect (Hall-effect type)<br>• CMP sensor is damaged or CMP sensor shielding damaged<br>• ECM has failed |

| DTC | Trouble Code Title, Conditions & Possible Causes |
|---|---|
| **DTC: P1392**<br>**Years:** 2007, 2008<br>**Models:** Passat<br>**Engines:** 2.5L, 3.6L<br>**Transmissions:** All | **Camshaft Position Sensor (Bank 2) Open/Short to B+ Conditions:**<br>Key on or Engine started, battery voltage must be at least 11.5v, all electrical components must be off, parking brake must be engaged (to keep daytime driving lights off), automatic transmission selector must be in park and the ground between the engine and the chassis must be well connected. The ECM detected an unexpected low or high voltage condition on the camshaft position sensor circuit.<br>**Possible Causes:**<br>• CMP sensor circuit is open or shorted to ground<br>• CMP sensor circuit is shorted to power<br>• CMP sensor ground (return) circuit is open<br>• CMP sensor installation incorrect (Hall-effect type)<br>• CMP sensor is damaged or CMP sensor shielding damaged<br>• ECM has failed |
| **DTC: P1409**<br>**Years:** 2007, 2008<br>**Models:** Passat, Jetta, Golf, GTI, New Beetle<br>**Engines:** 2.0L, 2.5L, 3.6L<br>**Transmissions:** All | **Tank Ventilation Valve Circuit Malfunction Conditions**<br>Key on or engine running; and the ECM detected a too high or too low voltage level in the tank ventilation valve circuit.<br>**Possible Causes:**<br>• EVAP canister purge regulator valve has failed<br>• Activation wire is shorting to positive<br>• EVAP canister system has an improper or broken seal<br>• Evaporative Emission (EVAP) canister purge regulator valve 1 is faulty<br>• Leak Detection Pump (LDP) is faulty<br>• Fuel filler cap is not properly closed<br>• Lock ring on fuel pump not tightened<br>• Hoses between EVAP canister and purge regulator valve have failed<br>• ECM has failed |
| **DTC: P1410**<br>**2T, MIL: Yes**<br>**Years:** 2007, 2008<br>**Models:** Passat, Jetta, Golf, GTI, New Beetle, New Beetle Convertible<br>**Engines:** 2.0L, 2.5L, 3.6L<br>**Transmissions:** All | **Tank Ventilation Valve Circuit Short to B+:**<br>Key on or engine running; and the ECM detected a too high or too low voltage level in the tank ventilation valve circuit.<br>**Possible Causes:**<br>• EVAP canister purge regulator valve has failed<br>• Activation wire is shorting to positive<br>• EVAP canister system has an improper or broken seal<br>• Evaporative Emission (EVAP) canister purge regulator valve 1 is faulty<br>• Leak Detection Pump (LDP) is faulty<br>• Fuel filler cap is not properly closed<br>• Lock ring on fuel pump not tightened<br>• Hoses between EVAP canister and purge regulator valve have failed<br>• ECM has failed |
| **DTC: P1420**<br>**Years:** 2007, 2008<br>**Models:** Passat, Jetta, Golf, GTI, New Beetle, New Beetle Convertible<br>**Engines:** 2.0L, 2.5L, 3.6L<br>**Transmissions:** All | **Secondary Air Injector Valve Circuit Electrical Malfunction Conditions:**<br>The Engine Control Module activates the secondary air injection solenoid valve, but the Heated Oxygen Sensor (HO2S) does not detect secondary air injection.<br>**Note: Solenoid valve is closed when no voltage is present.**<br>**Possible Causes:**<br>• Connector to the secondary air injection valve is loose or disconnected<br>• Secondary air injector valve circuit short<br>• Secondary air injector valve circuit is open<br>• Faulty secondary air injector valve<br>• ECM has failed |
| **DTC: P1421**<br>**2T, MIL: Yes**<br>**Years:** 2007, 2008<br>**Models:** Passat, Jetta, Golf, GTI, New Beetle, New Beetle Convertible<br>**Engines:** 2.0L, 2.5L, 3.6L<br>**Transmissions:** All | **Secondary Air Injector Valve Circuit Short to Ground Conditions:**<br>The Engine Control Module detects a short circuit to ground when activating the secondary air injection solenoid valve.<br>**Note: Solenoid valve is closed when no voltage is present.**<br>**Possible Causes:**<br>• Connector to the secondary air injection valve is loose or disconnected<br>• Secondary air injector valve circuit short<br>• Secondary air injector valve circuit is open<br>• Faulty secondary air injector valve<br>• ECM has failed |

| DTC | Trouble Code Title, Conditions & Possible Causes |
|---|---|
| **DTC: P1422**<br>**2T, MIL: Yes**<br>**Years: 2007, 2008**<br>**Models:** Passat, Jetta, Golf, GTI, New Beetle, New Beetle Convertible<br>**Engines:** 2.0L, 2.5L, 3.6L<br>**Transmissions:** All | **Secondary Air Injector Valve Circuit Short to B+ Conditions:**<br>The Engine Control Module detects a short circuit to B+ when activating the secondary air injection solenoid valve.<br>**Note: Solenoid valve is closed when no voltage is present.**<br>**Possible Causes:**<br>• Connector to the secondary air injection valve is loose or disconnected<br>• Secondary air injector valve circuit short<br>• Secondary air injector valve circuit is open<br>• Faulty secondary air injector valve<br>• ECM has failed |
| **DTC: P1424**<br>**2T, MIL: Yes**<br>**Years: 2007, 2008**<br>**Models:** Passat, Jetta, Golf, GTI, New Beetle, New Beetle Convertible<br>**Engines:** 2.0L, 2.5L, 3.6L<br>**Transmissions:** All | **Secondary Air Injector System (Bank 1) Leak Detected Conditions:**<br>Ignition on or vehicle running, and the ECM detected a leak in the secondary air injector system.<br>**Possible Causes:**<br>• Poor hose/pipe connections between the secondary air injector pump motor and valve<br>• Faulty hoses or pipes<br>• Mechanical faults in the secondary air injector system |
| **DTC: P1425**<br>**2T, MIL: Yes**<br>**Years: 2007, 2008**<br>**Models:** Passat, Jetta, Golf, GTI, New Beetle, New Beetle Convertible<br>**Engines:** 2.0L, 2.5L, 3.6L<br>**Transmissions:** All | **Tank Ventilation Valve Short to Ground Conditions:**<br>Ignition off. The Evaporative Emission (EVAP) canister purge regulator valve in the tank venting system or activation wire has a short circuit to ground. Engine started, engine running at a steady cruise speed, canister vent solenoid enabled, and the ECM detected an unexpected voltage condition on the Canister Vent solenoid circuit.<br>**Note: Solenoid valve is closed when no voltage is present.**<br>**Possible Causes:**<br>• Activation wire has a short to ground<br>• ECM has failed<br>• EVAP canister has failed<br>• EVAP canister system has an improper or broken seal<br>• Evaporative Emission (EVAP) canister purge regulator valve is faulty<br>• Leak Detection Pump (LDP) is faulty |
| **DTC: P1426**<br>**2T, MIL: Yes**<br>**Years: 2007, 2008**<br>**Models:** Passat, Jetta, Golf, GTI, New Beetle, New Beetle Convertible<br>**Engines:** 2.0L, 2.5L, 3.6L<br>**Transmissions:** All | **Tank Ventilation Valve Open Conditions:**<br>Ignition off. The Evaporative Emission (EVAP) canister purge regulator valve in the tank venting system or activation wire has a short circuit to ground. Engine started, engine running at a steady cruise speed, canister vent solenoid enabled, and the ECM detected an unexpected voltage condition on the Canister Vent solenoid circuit.<br>**Possible Causes:**<br>• Activation wire has a short to ground<br>• ECM has failed<br>• EVAP canister has failed<br>• EVAP canister system has an improper or broken seal<br>• Evaporative Emission (EVAP) canister purge regulator valve 1 is faulty<br>• Leak Detection Pump (LDP) is faulty |
| **DTC: P1432**<br>**2T, MIL: Yes**<br>**Years: 2007, 2008**<br>**Models:** Passat, Jetta, Golf, GTI, New Beetle, New Beetle Convertible<br>**Engines:** 2.0L, 2.5L, 3.6L<br>**Transmissions:** All | **Secondary Air Injection Valve Open Conditions:**<br>The output Diagnostic Test Mode (DTM) can be activated only with the ignition switched on and the engine not running. The output DTM is interrupted if the engine is started, or if a rotary pulse from the ignition system is recognized..<br>**Possible Causes:**<br>• Fuel pump relays have failed<br>• Fuel injector has failed<br>• Hoses on the EVAP canister may be clogged<br>• EVAP canister purge regulator valve may be faulty<br>• ECM may have failed<br>• Manifold Tuning Valve (IMT) may have failed |
| **DTC: P1433**<br>**2T, MIL: Yes**<br>**Years: 2007, 2008**<br>**Models:** Passat, Jetta, Golf, GTI, New Beetle, New Beetle Convertible<br>**Engines:** 2.0L, 2.5L, 3.6L<br>**Transmissions:** All | **Secondary Air Injection System Pump Relay Circuit Open Conditions:**<br>The output Diagnostic Test Mode (DTM) can be activated only with the ignition switched on and the engine not running. The output DTM is interrupted if the engine is started, or if a rotary pulse from the ignition system is recognized..<br>**Possible Causes:**<br>• Fuel pump relays have failed<br>• Fuel injector has failed<br>• Hoses on the EVAP canister may be clogged<br>• EVAP canister purge regulator valve may be faulty<br>• ECM may have failed<br>• Manifold Tuning Valve (IMT) may have failed |

| DTC | Trouble Code Title, Conditions & Possible Causes |
|---|---|
| **DTC: P1434**<br>**2T, MIL: Yes**<br>**Years:** 2007, 2008<br>**Models:** Passat, Jetta, Golf, GTI, New Beetle, New Beetle Convertible<br>**Engines:** 2.0L, 2.5L, 3.6L<br>**Transmissions:** All | **Secondary Air Injection System Pump Relay Circuit Short to B+ Conditions:**<br>The output Diagnostic Test Mode (DTM) can be activated only with the ignition switched on and the engine not running. The output DTM is interrupted if the engine is started, or if a rotary pulse from the ignition system is recognized..<br>**Possible Causes:**<br>• Fuel pump relays have failed<br>• Fuel injector has failed<br>• Hoses on the EVAP canister may be clogged<br>• EVAP canister purge regulator valve may be faulty<br>• ECM may have failed<br>• Manifold Tuning Valve (IMT) may have failed |
| **DTC: P1435**<br>**2T, MIL: Yes**<br>**Years:** 2007, 2008<br>**Models:** Passat, Jetta, Golf, GTI, New Beetle, New Beetle Convertible<br>**Engines:** 2.0L, 2.5L, 3.6L<br>**Transmissions:** All | **Secondary Air Injection System Pump Relay Circuit Short to Ground Conditions:**<br>The output Diagnostic Test Mode (DTM) can be activated only with the ignition switched on and the engine not running. The output DTM is interrupted if the engine is started, or if a rotary pulse from the ignition system is recognized..<br>**Possible Causes:**<br>• Fuel pump relays have failed<br>• Fuel injector has failed<br>• Hoses on the EVAP canister may be clogged<br>• EVAP canister purge regulator valve may be faulty<br>• ECM may have failed<br>• Manifold Tuning Valve (IMT) may have failed |
| **DTC: P1436**<br>**Years:** 2007, 2008<br>**Models:** Passat, Jetta, Golf, GTI, New Beetle,<br>**Engines:** 2.0L, 2.5L, 3.6L<br>**Transmissions:** All | **A/C Evaporator Temperature (ACET) Circuit Low Input Conditions:**<br>Key on or engine running; and the ECM detected the ACET signal was less than the self-test minimum amount of in the self-test.<br>**Possible Causes:**<br>• ACET signal circuit shorted to sensor ground (return)<br>• ACET signal circuit shorted to chassis ground<br>• ACET sensor is damaged or has failed<br>• Check activation of Secondary Air Injection (AIR) Pump Relay<br>• ECM has failed |
| **DTC: P1450**<br>**Years:** 2007, 2008<br>**Models:** Passat, Jetta, Golf, GTI, New Beetle, New Beetle Convertible<br>**Engines:** 2.0L, 2.5L, 3.6L<br>**Transmissions:** All | **Secondary Air Injector Valve Circuit Short to B+ Conditions:**<br>The Engine Control Module detects a short circuit to positive (B+) when activating the secondary air injection solenoid valve.<br>**Possible Causes:**<br>• Connector to the secondary air injection valve is loose or disconnected<br>• Secondary air injector valve circuit short<br>• Secondary air injector valve circuit is open<br>• Faulty secondary air injector valve<br>• ECM has failed |
| **DTC: P1451**<br>**Years:** 2007, 2008<br>**Models:** Passat, Jetta, Golf, GTI, New Beetle, New Beetle Convertible<br>**Engines:** 2.0L, 2.5L, 3.6L<br>**Transmissions:** All | **Secondary Air Injector Valve Circuit Short to Ground Conditions:**<br>Engine started, engine running at a steady cruise speed, the Engine Control Module detects a short circuit open when activating the secondary air injection solenoid valve.<br>**Possible Causes:**<br>• Connector to the secondary air injection valve is loose or disconnected<br>• Secondary air injector valve circuit short<br>• Secondary air injector valve circuit is open<br>• Faulty secondary air injector valve<br>• ECM has failed |
| **DTC: P1452**<br>**Years:** 2007, 2008<br>**Models:** Passat, Jetta, Golf, GTI, New Beetle, New Beetle Convertible<br>**Engines:** 2.0L, 2.5L, 3.6L<br>**Transmissions:** All | **Secondary Air Injector Valve Circuit Open Conditions:**<br>Engine started, engine running at a steady cruise speed, the Engine Control Module detects a short circuit open when activating the secondary air injection solenoid valve.<br>**Possible Causes:**<br>• Connector to the secondary air injection valve is loose or disconnected<br>• Secondary air injector valve circuit short<br>• Secondary air injector valve circuit is open<br>• Faulty secondary air injector valve<br>• ECM has failed |

| DTC | Trouble Code Title, Conditions & Possible Causes |
|---|---|
| **DTC: P1470**<br>**Years: 2007, 2008**<br>**Models:** Passat, Jetta, Golf, GTI, New Beetle<br>**Engines:**<br>**Transmissions**: All | **EVAP Emission Control LDP Circuit Electrical Malfunction Conditions:**<br>Key on, KOEO Self-Test enabled, and the ECM detected an unexpected voltage condition on the EVAP emission control leak detection pump circuit.<br>**Possible Causes:**<br>• EVAP canister system has an improper or broken seal<br>• Evaporative Emission (EVAP) canister purge regulator valve 1 is faulty<br>• Leak Detection Pump (LDP) is faulty<br>• ECM has failed |
| **DTC: P1471**<br>**2T, MIL: Yes**<br>**Years: 2007, 2008**<br>**Models:** Passat, Jetta, Golf, GTI, New Beetle, New Beetle Convertible<br>**Engines:** 2.0L, 2.5L, 3.6L<br>**Transmissions**: All | **EVAP Emission Control Leak Detection Pump Circuit Short to B+ Conditions:**<br>Key on, KOEO Self-Test enabled, and the ECM detected an unexpected voltage condition on the EVAP emission control leak detection pump circuit.<br>**Possible Causes:**<br>• Leak Detection Pump has failed<br>• EVAP canister system has an improper or broken seal<br>• Evaporative Emission (EVAP) canister purge regulator valve 1 is faulty<br>• Hoses between the fuel pump and the EVAP canister are faulty<br>• Fuel filler cap is loose<br>• Fuel pump seal is defective, faulty or otherwise leaking<br>• Hoses between the EVAP canister and the fuel flap unit are faulty<br>• Hoses between the EVAP canister and the evaporative emission canister purge regulator valve are faulty<br>• ECM has failed |
| **DTC: P1472**<br>**2T, MIL: Yes**<br>**Years: 2007, 2008**<br>**Models:** Passat, Jetta, Golf, GTI, New Beetle, New Beetle Convertible<br>**Engines:** 2.0L, 2.5L, 3.6L<br>**Transmissions**: All | **EVAP Emission Control Leak Detection Pump Circuit Short to Ground Conditions:**<br>Key on, KOEO Self-Test enabled, and the ECM detected an unexpected voltage condition on the EVAP emission control leak detection pump circuit.<br>**Possible Causes:**<br>• Leak Detection Pump has failed<br>• EVAP canister system has an improper or broken seal<br>• Evaporative Emission (EVAP) canister purge regulator valve 1 is faulty<br>• Hoses between the fuel pump and the EVAP canister are faulty<br>• Fuel filler cap is loose<br>• Fuel pump seal is defective, faulty or otherwise leaking<br>• Hoses between the EVAP canister and the fuel flap unit are faulty<br>• Hoses between the EVAP canister and the evaporative emission canister purge regulator valve are faulty<br>• ECM has failed |
| **DTC: P1473**<br>**2T, MIL: Yes**<br>**Years: 2007, 2008**<br>**Models:** Passat, Jetta, Golf, GTI, New Beetle, New Beetle Convertible<br>**Engines:** 2.0L, 2.5L, 3.6L<br>**Transmissions**: All | **EVAP Emission Control Leak Detection Pump Circuit Open Conditions:**<br>Key on, KOEO Self-Test enabled, and the ECM detected an unexpected voltage condition on the EVAP emission control leak detection pump circuit.<br>**Possible Causes:**<br>• Leak Detection Pump has failed<br>• EVAP canister system has an improper or broken seal<br>• Evaporative Emission (EVAP) canister purge regulator valve 1 is faulty<br>• Hoses between the fuel pump and the EVAP canister are faulty<br>• Fuel filler cap is loose<br>• Fuel pump seal is defective, faulty or otherwise leaking<br>• Hoses between the EVAP canister and the fuel flap unit are faulty<br>• Hoses between the EVAP canister and the evaporative emission canister purge regulator valve are faulty<br>• ECM has failed |
| **DTC: P1475**<br>**2T, MIL: Yes**<br>**Years: 2007, 2008**<br>**Models:** Passat, Jetta, Golf, GTI, New Beetle, New Beetle Convertible<br>**Engines:** 2.0L, 2.5L, 3.6L<br>**Transmissions**: All | **EVAP Emission Control LDP Circuit Malfunction/Signal Circuit Open Conditions:**<br>Key on, KOEO Self-Test enabled, and the ECM detected an unexpected voltage condition on the EVAP emission control leak detection pump circuit.<br>**Possible Causes:**<br>• Leak Detection Pump has failed<br>• EVAP canister system has an improper or broken seal<br>• Evaporative Emission (EVAP) canister purge regulator valve 1 is faulty<br>• Hoses between the fuel pump and the EVAP canister are faulty<br>• Fuel filler cap is loose<br>• Fuel pump seal is defective, faulty or otherwise leaking<br>• Hoses between the EVAP canister and the fuel flap unit are faulty<br>• Hoses between the EVAP canister and the evaporative emission canister purge regulator valve are faulty<br>• ECM has failed |

| DTC | Trouble Code Title, Conditions & Possible Causes |
|---|---|
| **DTC: P1476**<br>**2T, MIL: Yes**<br>**Years: 2007, 2008**<br>**Models:** Passat, Jetta, Golf, GTI, New Beetle, New Beetle Convertible<br>**Engines:** 2.0L, 2.5L, 3.6L<br>**Transmissions:** All | **EVAP Emission Control LDP Circuit Malfunction/Insufficient Vacuum Conditions:**<br>Key on, KOEO Self-Test enabled, and the ECM detected an unexpected voltage condition on the EVAP emission control leak detection pump circuit.<br>**Possible Causes:**<br>• Leak Detection Pump has failed<br>• EVAP canister system has an improper or broken seal<br>• Evaporative Emission (EVAP) canister purge regulator valve 1 is faulty<br>• Hoses between the fuel pump and the EVAP canister are faulty<br>• Fuel filler cap is loose<br>• Fuel pump seal is defective, faulty or otherwise leaking<br>• Hoses between the EVAP canister and the fuel flap unit are faulty<br>• Hoses between the EVAP canister and the evaporative emission canister purge regulator valve are faulty<br>• ECM has failed |
| **DTC: P1477**<br>**Years: 2007, 2008**<br>**Models:** Passat, Jetta, Golf, GTI, New Beetle,<br>**Engines:** 2.0L, 2.5L, 3.6L<br>**Transmissions:** All | **EVAP Emission Control LDP Circuit Malfunction Conditions:**<br>Key on, KOEO Self-Test enabled, and the ECM detected an unexpected voltage condition on the EVAP emission control leak detection pump circuit.<br>**Possible Causes:**<br>• Leak Detection Pump has failed<br>• EVAP canister system has an improper or broken seal<br>• Evaporative Emission (EVAP) canister purge regulator valve 1 is faulty<br>• Hoses between the fuel pump and the EVAP canister are faulty<br>• Fuel filler cap is loose<br>• Fuel pump seal is defective, faulty or otherwise leaking<br>• Hoses between the EVAP canister and the fuel flap unit are faulty<br>• Hoses between the EVAP canister and the evaporative emission canister purge regulator valve are faulty<br>• ECM has failed |
| **DTC: P1478**<br>**Years: 2007, 2008**<br>**Models:** Passat, Jetta, Golf, GTI, New Beetle,<br>**Engines:** 2.0L, 2.5L, 3.6L<br>**Transmissions:** All | **EVAP Emission Control LDP Circuit Clamped Tube Detected Conditions:**<br>Key on, KOEO Self-Test enabled, and the ECM detected an unexpected voltage condition on the EVAP emission control leak detection pump circuit.<br>**Possible Causes:**<br>• Leak Detection Pump has failed<br>• EVAP canister system has an improper or broken seal<br>• Evaporative Emission (EVAP) canister purge regulator valve 1 is faulty<br>• Hoses between the fuel pump and the EVAP canister are faulty<br>• Fuel filler cap is loose<br>• Fuel pump seal is defective, faulty or otherwise leaking<br>• Hoses between the EVAP canister and the fuel flap unit are faulty<br>• Hoses between the EVAP canister and the evaporative emission canister purge regulator valve are faulty<br>• ECM has failed |
| **DTC: P1500**<br>**2T, MIL: Yes**<br>**Years: 2007, 2008**<br>**Models:** Passat, Jetta, Golf, GTI, New Beetle, New Beetle Convertible<br>**Engines:** 2.0L, 2.5L, 3.6L<br>**Transmissions:** All | **Fuel Pump Relay Circuit Electrical Malfunction Conditions:**<br>Engine running the ECM detected that the fuel pump relay signal was intermittent<br>**Possible Causes:**<br>• Fuel delivery unit connector is loose or not attached<br>• Fuse 18 cause a short to the transfer fuel pump or the O2S<br>• Fuel pump has failed<br>• Fuel pump relay circuit is shorted to ground, B+ or is open<br>• Fuel Pump (FP) Relay not activated<br>• ECM has failed |
| **DTC: P1501**<br>**Years: 2007, 2008**<br>**Models:** Passat, Jetta, Golf, GTI, New Beetle, New Beetle Convertible<br>**Engines:** 2.0L, 2.5L, 3.6L<br>**Transmissions:** All | **Fuel Pump Relay Circuit Electrical Short to Ground Conditions:**<br>Engine running the ECM detected that the fuel pump relay signal was intermittent<br>**Possible Causes:**<br>• Fuel delivery unit connector is loose or not attached<br>• Fuse 18 cause a short to the transfer fuel pump or the O2S<br>• Fuel pump has failed<br>• Fuel pump relay circuit is shorted to ground, B+ or is open<br>• Fuel Pump (FP) Relay not activated<br>• ECM has failed |

| DTC | Trouble Code Title, Conditions & Possible Causes |
|---|---|
| **DTC: P1502**<br>**2T, MIL: Yes**<br>**Years: 2007, 2008**<br>**Models:** Passat, Jetta, Golf, GTI, New Beetle, New Beetle Convertible<br>**Engines:** 2.0L, 2.5L, 3.6L<br>**Transmissions:** All | **Fuel Pump Relay Circuit Short to B+ Conditions:**<br>Engine running the ECM detected that the fuel pump relay signal was intermittent<br>**Possible Causes:**<br>• Fuel delivery unit connector is loose or not attached<br>• Fuse 18 cause a short to the transfer fuel pump or the O2S<br>• Fuel pump has failed<br>• Fuel pump relay circuit is shorted to ground, B+ or is open<br>• Fuel Pump (FP) Relay not activated<br>• ECM has failed |
| **DTC: P1512**<br>**Years: 2007, 2008**<br>**Models:** Passat<br>**Engines:** 2.5L, 3.6L<br>**Transmissions:** All | **Intake Manifold Changeover Valve Circuit Short to B+ Conditions:**<br>Engine started, and the ECM detected the changeover valve circuit was shorting to positive during the continuous self test.<br>**Possible Causes:**<br>• Leaky vacuum reservoir, vacuum lines loose or damaged<br>• Vacuum solenoid or vacuum actuator is damaged<br>• IMRC actuator cable/gears are seized, or the cables are improperly routed or seized<br>• IMRC housing return springs are damaged or disconnected<br>• Lever/shaft return stop may be obstructed or bent, or the lever/shaft wide open stop may be obstructed or bent, or the IMRC lever/shaft may be sticking, binding or disconnected<br>• IMRC control circuit open, shorted or the VPWR circuit is open<br>• ECM has failed |
| **DTC: P1515**<br>**Years: 2007, 2008**<br>**Models:** Passat<br>**Engines:** 2.5L, 3.6L<br>**Transmissions:** All | **Intake Manifold Changeover Valve Circuit Short to Ground Conditions:**<br>Engine started, and the ECM detected the changeover valve circuit was shorting to ground during the continuous self test.<br>**Possible Causes:**<br>• Leaky vacuum reservoir, vacuum lines loose or damaged<br>• Vacuum solenoid or vacuum actuator is damaged<br>• IMRC actuator cable/gears are seized, or the cables are improperly routed or seized<br>• IMRC housing return springs are damaged or disconnected<br>• Lever/shaft return stop may be obstructed or bent, or the lever/shaft wide open stop may be obstructed or bent, or the IMRC lever/shaft may be sticking, binding or disconnected<br>• IMRC control circuit open, shorted or the VPWR circuit is open<br>• ECM has failed |
| **DTC: P1516**<br>**Years: 2007, 2008**<br>**Models:** Passat<br>**Engines:** 2.5L, 3.6L<br>**Transmissions:** All | **Intake Manifold Runner Control Input Error (Bank 1) Conditions:**<br>Key on or engine running; and the ECM detected the IMRC Monitor signal for Bank 1 was outside of its expected calibrated range during the Continuous self test.<br>**Possible Causes:**<br>• IMRC mechanical fault-the linkage may be bound or seized<br>• Inspect for binding or improper routing. The cable core wire at the IMRC/IMSC housing attachment must have slack and lever must contact close plate stop screw |
| **DTC: P1517**<br>**Years: 2007, 2008**<br>**Models:** Passat, Jetta, Golf, GTI, New Beetle<br>**Engines:** 2.0L<br>**Transmissions:** All | **Main Relay Circuit Electrical Malfunction Conditions:**<br>The ECM detected an electrical malfunction on the main relay circuit<br>**Possible Causes:**<br>• Engine Control Module (ECM) has failed<br>• Voltage supply for Engine Control Module (ECM) is faulty<br>• Check activation of Motronic Engine Control Module (ECM) Power Supply Relay |
| **DTC: P1519**<br>**Years: 2007, 2008**<br>**Models:** Passat<br>**Engines:** 2.5L, 3.6L<br>**Transmissions:** All | **Intake Manifold Runner Control Stuck Closed Conditions:**<br>Key on, and the ECM detected the IMRC Monitor was more than the expected calibrated range at closed throttle.<br>**Possible Causes:**<br>• IMRC monitor signal circuit shorted to power ground<br>• IMRC Monitor signal circuit shorted to signal ground (return)<br>• IMRC actuator is damaged or has failed (e.g., there may be a small leak in the vacuum diaphragm of the actuator)<br>• ECM has failed |
| **DTC: P1522**<br>**Years: 2007, 2008**<br>**Models:** Passat<br>**Engines:** 2.5L, 3.6L<br>**Transmissions:** All | **Intake Camshaft Control (Bank 2) Malfunction Conditions:**<br>Key on or engine running; and the ECM detected the intake manifold control signal for was outside of its expected calibrated range.<br>**Possible Causes:**<br>• Camshaft control circuit is open or shorted to ground<br>• Camshaft sensor is damaged or the ECM has failed<br>• Camshaft out of adjustment |

| DTC | Trouble Code Title, Conditions & Possible Causes |
|---|---|
| **DTC: P1530**<br>**Years:** 2007, 2008<br>**Models:** Passat, Jetta<br>**Engines:** 2.0L<br>**Transmissions:** All | **Camshaft Control Circuit Short to Ground Conditions:**<br>Engine started and driven at an engine speed of more than 400RPM; and the ECM detected the camshaft timing exceeded the calibrated levels. The valve timing did not change from the current valve timing or it remained fixed during the testing.<br>**Note: The camshaft adjustment is load- and RPM dependant. The electrical camshaft adjustment valve 1 switches oil pressure onto camshaft adjuster (mechanical adjustment mechanism), which adjusts the camshaft.**<br>**Possible Causes:**<br>• Fuel pump has failed<br>• CPS circuit is open, shorted to ground or shorted to power<br>• ECM has failed<br>• Battery voltage below 11.5 volts<br>• Position actuator circuit may short to B+ or Ground<br>• Camshaft timing improperly set, or continuous oil flow to the VCT piston chamber<br>• Camshaft advance mechanism (the VCT unit) is sticking or binding mechanically<br>• VCT solenoid valve is stuck in open position |
| **DTC: P1531**<br>**Years:** 2007, 2008<br>**Models:** Passat, Jetta<br>**Engines:** 2.0L<br>**Transmissions:** All | **Camshaft Control Circuit Open Conditions:**<br>Engine started and driven at an engine speed of more than 400RPM; and the ECM detected the camshaft timing exceeded the calibrated levels. The valve timing did not change from the current valve timing or it remained fixed during the testing.<br>**Note: The camshaft adjustment is load- and RPM dependant. The electrical camshaft adjustment valve 1 switches oil pressure onto camshaft adjuster (mechanical adjustment mechanism), which adjusts the camshaft.**<br>**Possible Causes:**<br>• Fuel pump has failed<br>• CPS circuit is open, shorted to ground or shorted to power<br>• ECM has failed<br>• Battery voltage below 11.5 volts<br>• Position actuator circuit may short to B+ or Ground<br>• Camshaft timing improperly set, or continuous oil flow to the VCT piston chamber<br>• Camshaft advance mechanism (the VCT unit) is sticking or binding mechanically<br>• VCT solenoid valve is stuck in open position |
| **DTC: P1539**<br>**Years:** 2007, 2008<br>**Models:** Jetta, Golf, GTI, New Beetle<br>**Engines:**<br>**Transmissions:** All | **Clutch Vacuum Vent Valve Switch Incorrect Signal Conditions:**<br>Engine started, battery voltage must be at least 11.5v, all electrical components must be off, parking brake must be engaged (to keep daytime driving lights off), automatic transmission selector must be in park, and the ground between the engine and the chassis must be well connected. The ECM detected an incorrect signal from the clutch vacuum vent valve switch.<br>**Possible Causes:**<br>• Signal from clutch vacuum vent valve switch is faulty<br>• Clutch vacuum vent valve is faulty<br>• Circuit wires are short circuiting to each other, to vehicle ground or to B+<br>• ECM has failed |
| **DTC: P1541**<br>**Years:** 2007, 2008<br>**Models:** Passat, Jetta, Golf, GTI, New Beetle, New Beetle Convertible<br>**Engines:** 2.0L, 2.5L, 3.6L<br>**Transmissions:** All | **Fuel Pump Relay Circuit Open Conditions:**<br>The ECM detected an electrical malfunction on the fuel pump relay circuit<br>**Possible Causes:**<br>• Fuel pump relay not activiated |

| DTC | Trouble Code Title, Conditions & Possible Causes |
|---|---|
| **DTC: P1542**<br>**2T, MIL: Yes**<br>**Years: 2007, 2008**<br>**Models:** Passat, Jetta, Golf, GTI, New Beetle, New Beetle Convertible<br>**Engines:** 2.0L, 2.5L, 3.6L<br>**Transmissions:** All | **Throttle Actuation Potentiometer Range/Performance Conditions:**<br>Engine started, battery voltage must be at least 11.5v, all electrical components must be off, parking brake must be engaged (to keep daytime driving lights off), automatic transmission selector must be in park, the exhaust system must be properly sealed between the catalytic converter and the cylinder head, coolant temperature must be at least 80 degrees Celsius, and the ground between the engine and the chassis must be well connected. The signal from the Throttle Position Valve Module to the ECM detected was erratic, non existent or unreliable.<br>**Note: If the complete throttle valve control module is current-less (e.g. connector disconnected) the throttle valve moves into a particular, specified mechanical position, which signals an increased idle speed with an engine at operating temperature. If only the Throttle Position (TP) actuator –V60- is current-less, the throttle valve also moves into the specified mechanical position (emergency running gap), however, since Closed Throttle Position (CTP) switch –F60- can still be recognized, an "almost normal idle RPM" is reached via the respective ignition angle retardation. If the Engine Control Module (ECM) detects a malfunction at Throttle Position (TP) sensor –G69-, Throttle Position (TP) actuator –V60- is switched current-less by the Engine Control Module (ECM) and the throttle valve moves into the specified mechanical position (emergency running gap) again.**<br>**Note: Terminal assignment at throttle control module is different in vehicles with and without cruise control. Characteristic: Steering column switch with operating module for cruise control.**<br>**Possible Causes:**<br>• Throttle valve control module is faulty<br>• Throttle valve is damaged or dirty<br>• Throttle valve must be in closed throttle position<br>• Accelerator pedal is out of adjustment (AEG engines only)<br>• Throttle position actuator is shorting to ground or power |
| **DTC: P1543**<br>**2T, MIL: Yes**<br>**Years: 2007, 2008**<br>**Models:** Passat, Jetta, Golf, GTI, New Beetle, New Beetle Convertible<br>**Engines:** 2.0L, 2.5L, 3.6L<br>**Transmissions:** All | **Throttle Actuation Potentiometer Signal Too Low Conditions:**<br>Engine started, battery voltage must be at least 11.5v, all electrical components must be off, parking brake must be engaged (to keep daytime driving lights off), automatic transmission selector must be in park, the exhaust system must be properly sealed between the catalytic converter and the cylinder head, coolant temperature must be at least 80 degrees Celsius, and the ground between the engine and the chassis must be well connected. The signal from the Throttle Position Valve Module to the ECM detected was erratic, non existent or unreliable.<br>**Note: If the complete throttle valve control module is current-less (e.g. connector disconnected) the throttle valve moves into a particular, specified mechanical position, which signals an increased idle speed with an engine at operating temperature. If only the Throttle Position (TP) actuator –V60- is current-less, the throttle valve also moves into the specified mechanical position (emergency running gap), however, since Closed Throttle Position (CTP) switch –F60- can still be recognized, an "almost normal idle RPM" is reached via the respective ignition angle retardation. If the Engine Control Module (ECM) detects a malfunction at Throttle Position (TP) sensor –G69-, Throttle Position (TP) actuator –V60- is switched current-less by the Engine Control Module (ECM) and the throttle valve moves into the specified mechanical position (emergency running gap) again.**<br>**Note: Terminal assignment at throttle control module is different in vehicles with and without cruise control. Characteristic: Steering column switch with operating module for cruise control.**<br>**Possible Causes:**<br>• Throttle valve control module is faulty<br>• Throttle valve is damaged or dirty<br>• Throttle valve must be in closed throttle position<br>• Accelerator pedal is out of adjustment (AEG engines only)<br>• Throttle position actuator is shorting to ground or power |

| DTC | Trouble Code Title, Conditions & Possible Causes |
|---|---|
| **DTC: P1544**<br>**2T, MIL: Yes**<br>**Years: 2007, 2008**<br>**Models:** Passat, Jetta, Golf, GTI, New Beetle, New Beetle Convertible<br>**Engines:** 2.0L, 2.5L, 3.6L<br>**Transmissions:** All | **Throttle Actuation Potentiometer Signal Too High Conditions:**<br>Engine started, battery voltage must be at least 11.5v, all electrical components must be off, parking brake must be engaged (to keep daytime driving lights off), automatic transmission selector must be in park, the exhaust system must be properly sealed between the catalytic converter and the cylinder head, coolant temperature must be at least 80 degrees Celsius, and the ground between the engine and the chassis must be well connected. The signal from the Throttle Position Valve Module to the ECM detected was erratic, non existent or unreliable.<br>**Note: If the complete throttle valve control module is current-less (e.g. connector disconnected) the throttle valve moves into a particular, specified mechanical position, which signals an increased idle speed with an engine at operating temperature. If only the Throttle Position (TP) actuator –V60- is current-less, the throttle valve also moves into the specified mechanical position (emergency running gap), however, since Closed Throttle Position (CTP) switch –F60- can still be recognized, an "almost normal idle RPM" is reached via the respective ignition angle retardation. If the Engine Control Module (ECM) detects a malfunction at Throttle Position (TP) sensor –G69-, Throttle Position (TP) actuator –V60- is switched current-less by the Engine Control Module (ECM) and the throttle valve moves into the specified mechanical position (emergency running gap) again.**<br>**Note: Terminal assignment at throttle control module is different in vehicles with and without cruise control. Characteristic: Steering column switch with operating module for cruise control.**<br>**Possible Causes:**<br>• Throttle valve control module is faulty<br>• Throttle valve is damaged or dirty<br>• Throttle valve must be in closed throttle position<br>• Accelerator pedal is out of adjustment (AEG engines only)<br>• Throttle position actuator is shorting to ground or power |
| **DTC: P1545**<br>**2T, MIL: Yes**<br>**Years: 2007, 2008**<br>**Models:** Passat, Jetta, Golf, GTI, New Beetle, New Beetle Convertible<br>**Engines:** 2.0L, 2.5L, 3.6L<br>**Transmissions:** All | **Throttle Position Control Malfunction Conditions:**<br>Engine started, battery voltage must be at least 11.5v, all electrical components must be off, parking brake must be engaged (to keep daytime driving lights off), automatic transmission selector must be in park, the exhaust system must be properly sealed between the catalytic converter and the cylinder head, coolant temperature must be at least 80 degrees Celsius, and the ground between the engine and the chassis must be well connected. The signal from the Throttle Position Valve Module to the ECM detected was erratic, non existent or unreliable.<br>**Note: If the complete throttle valve control module is current-less (e.g. connector disconnected) the throttle valve moves into a particular, specified mechanical position, which signals an increased idle speed with an engine at operating temperature. If only the Throttle Position (TP) actuator is current-less, the throttle valve also moves into the specified mechanical position (emergency running gap), however, since Closed Throttle Position (CTP) switch – can still be recognized, an "almost normal idle RPM" is reached via the respective ignition angle retardation. If the Engine Control Module (ECM) detects a malfunction at Throttle Position (TP) sensor – Throttle Position (TP) actuator is switched current-less by the Engine Control Module (ECM) and the throttle valve moves into the specified mechanical position (emergency running gap) again.**<br>**Note: Terminal assignment at throttle control module is different in vehicles with and without cruise control. Characteristic: Steering column switch with operating module for cruise control.**<br>**Possible Causes:**<br>• Throttle valve control module is faulty<br>• Throttle valve is damaged or dirty<br>• Throttle valve must be in closed throttle position<br>• Accelerator pedal is out of adjustment (AEG engines only)<br>• Throttle position actuator is shorting to ground or power |
| **DTC: P1546**<br>**Years: 2007, 2008**<br>**Models:** Passat, Jetta, Golf, GTI, New Beetle<br>**Engines:** 2.0L<br>**Transmissions:** All | **Boost Pressure Control Valve Short to B+ Conditions:**<br>Engine started, battery voltage at least 11.5v, all electrical components off, ground connections between engine and chassis well connected, coolant temperature at least 80-degrees Celicius. The ECM detected an short in the boost pressure control valve.<br>**Possible Causes:**<br>• Charge air pressure sensor is faulty<br>• Voltage supply to the charge air pressure sensor is open or shorted<br>• Check the charge air system for leaks<br>• Recirculating valve for turbocharger is faulty<br>• Turbocharging system is damaged or not functioning properly<br>• Turbocharger recirculating valve is faulty<br>• Vacuum diaphragm for turbocharger is out of adjustment<br>• Wastegate bypass regulator valve is faulty<br>• Boost sensor has failed<br>• ECM has failed |

| DTC | Trouble Code Title, Conditions & Possible Causes |
|---|---|
| **DTC: P1547**<br>**Years:** 2007, 2008<br>**Models:** Passat, Jetta, Golf, GTI, New Beetle<br>**Engines:** 2.0L<br>**Transmissions:** All | **Boost Pressure Control Valve Short to Ground Conditions:**<br>Engine started, battery voltage at least 11.5v, all electrical components off, ground connections between engine and chassis well connected, coolant temperature at least 80-degrees Celicius. The ECM detected an short in the boost pressure control valve.<br>**Possible Causes:**<br>• Charge air pressure sensor is faulty<br>• Voltage supply to the charge air pressure sensor is open or shorted<br>• Check the charge air system for leaks<br>• Recirculating valve for turbocharger is faulty<br>• Turbocharging system is damaged or not functioning properly<br>• Turbocharger recirculating valve is faulty<br>• Vacuum diaphragm for turbocharger is out of adjustment<br>• Wastegate bypass regulator valve is faulty<br>• Boost sensor has failed<br>• ECM has failed |
| **DTC: P1548**<br>**Years:** 2007, 2008<br>**Models:** Passat, Jetta, Golf, GTI, New Beetle<br>**Engines:** 2.0L<br>**Transmissions:** All | **Boost Pressure Control Valve Open Conditions:**<br>Engine started, battery voltage at least 11.5v, all electrical components off, ground connections between engine and chassis well connected, coolant temperature at least 80-degrees Celicius. The ECM detected an short in the boost pressure control valve.<br>**Possible Causes:**<br>• Charge air pressure sensor is faulty<br>• Voltage supply to the charge air pressure sensor is open or shorted<br>• Check the charge air system for leaks<br>• Recirculating valve for turbocharger is faulty<br>• Turbocharging system is damaged or not functioning properly<br>• Turbocharger recirculating valve is faulty<br>• Vacuum diaphragm for turbocharger is out of adjustment<br>• Wastegate bypass regulator valve is faulty<br>• Boost sensor has failed<br>• ECM has failed |
| **DTC: P1555**<br>**Years:** 2007, 2008<br>**Models:** Passat, Jetta, Golf, GTI, New Beetle<br>**Engines:** 2.0L<br>**Transmissions:** All | **Charge Pressure Upper Limit Exceeded Conditions:**<br>Engine started, battery voltage at least 11.5v, all electrical components off, ground connections between engine and chassis well connected, coolant temperature at least 80-degrees Celicius. The ECM detected deviation from the normal operating parameters of the charge pressure sensor.<br>**Possible Causes:**<br>• Charge air system leaks<br>• Recirculating valve for turbocharger is faulty<br>• Turbocharging system is damaged<br>• Vacuum diaphragm for turbocharger needs adjusting<br>• Wastegate bypass regulator valve is faulty<br>• ECM has failed |
| **DTC: P1556**<br>**Years:** 2007, 2008<br>**Models:** Passat, Jetta, Golf, GTI, New Beetle<br>**Engines:** 2.0L<br>**Transmissions:** All | **Charge Pressure Control Negative Deviation Conditions:**<br>Engine started, battery voltage at least 11.5v, all electrical components off, ground connections between engine and chassis well connected, coolant temperature at least 80-degrees Celicius. The ECM detected deviation from the normal operating parameters of the charge pressure sensor.<br>**Possible Causes:**<br>• Charge air system leaks<br>• Recirculating valve for turbocharger is faulty<br>• Turbocharging system is damaged<br>• Vacuum diaphragm for turbocharger needs adjusting<br>• Wastegate bypass regulator valve is faulty<br>• ECM has failed |
| **DTC: P1557**<br>**Years:** 2007, 2008<br>**Models:** Passat, Jetta, Golf, GTI, New Beetle<br>**Engines:** 2.0L<br>**Transmissions:** All | **Charge Pressure Control Positive Deviation Conditions:**<br>Engine started, battery voltage at least 11.5v, all electrical components off, ground connections between engine and chassis well connected, coolant temperature at least 80-degrees Celicius. The ECM detected deviation from the normal operating parameters of the charge pressure sensor.<br>**Possible Causes:**<br>• Charge air system leaks<br>• Recirculating valve for turbocharger is faulty<br>• Turbocharging system is damaged<br>• Vacuum diaphragm for turbocharger needs adjusting<br>• Wastegate bypass regulator valve is faulty<br>• ECM has failed |

| DTC | Trouble Code Title, Conditions & Possible Causes |
|---|---|
| **DTC: P1558**<br>**2T, MIL: Yes**<br>**Years: 2007, 2008**<br>**Models:** Passat, Jetta, Golf, GTI, New Beetle, New Beetle Convertible<br>**Engines:** 2.0L, 2.5L, 3.6L<br>**Transmissions**: All | **Throttle Actuator Electrical Malfunction Conditions:**<br>Engine started, battery voltage at least 11.5v, all electrical components off, ground connections between engine and chassis well connected, coolant temperature at least 80-degrees Celicius and the throttle valve must not be damaged or dirty; and the ECM detected the signal from the Throttle Position Valve Module to the ECM detected was erratic, non existent or unreliable (too high or too low).<br>**Possible Causes:**<br>• Throttle valve control module has failed<br>• Throttle valve control module's circuit has shorted or is open<br>• The ECM has failed |
| **DTC: P1559**<br>**2T, MIL: Yes**<br>**Years: 2007, 2008**<br>**Models:** Passat, Jetta, Golf, GTI, New Beetle, New Beetle Convertible<br>**Engines:** 2.0L, 2.5L, 3.6L<br>**Transmissions**: All | **Idle Speed Control Throttle Position Adaptation Malfunction Conditions:**<br>Engine started, battery voltage at least 11.5v, all electrical components off, ground connections between engine and chassis well connected, coolant temperature at least 80-degrees Celicius and the throttle valve must not be damaged or dirty; and the ECM detected the signal from the Throttle Position Valve Module to the ECM detected was erratic, non existent or unreliable (too high or too low).<br>**Possible Causes:**<br>• Throttle valve control module has failed<br>• Throttle valve control module's circuit has shorted or is open<br>• The ECM has failed |
| **DTC: P1560**<br>**Years: 2007, 2008**<br>**Models:** Passat, Jetta, Golf, GTI, New Beetle<br>**Engines:** 2.0L<br>**Transmissions**: All | **Maximum Engine Speed Exceeded Conditions:**<br>Engine running, the ECM has detected that the maximum engine speed had been attained.<br>**Possible Causes:**<br>• Throttle valve control module has failed<br>• Throttle valve control module's circuit has shorted or is open<br>• The ECM has failed<br>• General engine damage |
| **DTC: P1564**<br>**Years: 2007, 2008**<br>**Models:** Passat, Jetta, Golf, GTI, New Beetle<br>**Engines:** 2.0L, 2.5L, 3.6L<br>**Transmissions**: All | **Idle Speed Control Throttle Position Low Voltage During Adaptation Conditions:**<br>Engine started, battery voltage at least 11.5v, all electrical components off, ground connections between engine and chassis well connected, coolant temperature at least 80-degrees Celicius and the throttle valve must not be damaged or dirty; and the ECM detected the signal from the Throttle Position Valve Module to the ECM detected was erratic, non existent or unreliable (too high or too low).<br>**Possible Causes:**<br>• Alternator failed<br>• ECM failed<br>• Fuses blown or open circuits<br>• Clean Throttle Valve Control Module<br>• Faulty battery<br>• Idle speed control throttle failed<br>• Wire connections to relay carrier and ground connection of ECM may have shorted |
| **DTC: P1565**<br>**2T, MIL: Yes**<br>**Years: 2007, 2008**<br>**Models:** Passat, Jetta, Golf, GTI, New Beetle<br>**Engines:** 2.0L, 2.0L<br>**Transmissions**: All | **Idle Speed Control Throttle Position Lower Limit Not Attainted Conditions:**<br>Engine started, battery voltage at least 11.5v, all electrical components off, ground connections between engine and chassis well connected, coolant temperature at least 80-degrees Celicius and the throttle valve must not be damaged or dirty; and the ECM detected the signal from the Throttle Position Valve Module to the ECM detected was erratic, non existent or unreliable (too high or too low).<br>**Possible Causes:**<br>• Alternator failed<br>• ECM failed<br>• Fuses blown or open circuits<br>• Clean Throttle Valve Control Module<br>• Accelerator cable not adjusted properly<br>• Idle speed control throttle failed<br>• Wire connections to relay carrier and ground connection of ECM may have shorted |

| DTC | Trouble Code Title, Conditions & Possible Causes |
|---|---|
| **DTC: P1568**<br>**2T, MIL: Yes**<br>**Years:** 2007, 2008<br>**Models:** Passat, Jetta, Golf, GTI, New Beetle<br>**Engines:** 2.0L<br>**Transmissions:** All | **Idle Speed Control Throttle Position Mechanical Malfunction Conditions:**<br>Engine started, battery voltage at least 11.5v, all electrical components off, ground connections between engine and chassis well connected, coolant temperature at least 80-degrees Celicius and the throttle valve must not be damaged or dirty; and the ECM detected the signal from the Throttle Position Valve Module to the ECM detected was erratic, non existent or unreliable (too high or too low) suggesting a mechanicl malfunction.<br>**Possible Causes:**<br>• Alternator failed<br>• ECM failed<br>• Fuses blown or open circuits<br>• Clean Throttle Valve Control Module<br>• Accelerator cable not adjusted properly<br>• Idle speed control throttle failed<br>• Wire connections to relay carrier and ground connection of ECM may have shorted |
| **DTC: P1569**<br>**Years:** 2007, 2008<br>**Models:** Passat, Jetta, Golf, GTI, New Beetle<br>**Engines:** 2.0L<br>**Transmissions:** All | **Cruise Control Switch Incorrect Signal Conditions:**<br>Key on or engine started and the ECM detected an incorrect signal from the cruise control switch<br>**Possible Causes:**<br>• Check Cruise Control System (CCS) wiring shorts to ground or B+ |
| **DTC: P1579**<br>**Years:** 2007, 2008<br>**Models:** Jetta, Golf, GTI<br>**Engines:** 2.0L<br>**Transmissions:** All | **Idle Speed Control Throttle Position Adaptation Not Started Conditions:**<br>Key on or engine started and the ECM detected an incorrect signal between the idle speed control and the throttle valve control module<br>**Possible Causes:**<br>• Adapt Engine Control Module (ECM) to throttle valve control module. |
| **DTC: P1580**<br>**Years:** 2007, 2008<br>**Models:** Passat, Jetta, Golf, GTI, New Beetle, New Beetle Convertible<br>**Engines:** 2.0L, 2.5L, 3.6L<br>**Transmissions:** All | **Throttle Actuator B1 Malfunction Conditions:**<br>Engine started, battery voltage at least 11.5v, all electrical components off, ground connections between engine and chassis well connected, coolant temperature at least 80-degrees Celicius. The ECM detected the throttle actuator B1 input failed the rationality test (i.e., the input did not change as expected by the ECM). The throttle valve activation occurs via an electric motor (throttle drive) in the throttle valve control module. It is activated by the Engine Control Module (ECM) according to specifications of the two sensors, Throttle Position (TP) Sensor –G79- and Accelerator Pedal Position Sensor 2<br>**Possible Causes:**<br>• The throttle position actuator B1 circuit is open<br>• The ECM has failed<br>• The throttle valve control module circuit is open<br>• The throttle position actuator has failed |
| **DTC: P1582**<br>**Years:** 2007, 2008<br>**Models:** Passat, Jetta, Golf, GTI, New Beetle, New Beetle Convertible<br>**Engines:** 2.0L, 2.5L, 3.6L<br>**Transmissions:** All | **Idle Adaptation at Limit**<br>Key on or engine running, the ECM detected that the idle adaptation reached its limit<br>**Possible Causes:**<br>• Crankcase oil is diluted (change the oil)<br>• Fuel injectors are worn<br>• Compression ratio is low<br>• Throttle valve control module is faulty<br>• Leak Detection Pump has failed<br>• EVAP canister system has an improper or broken seal<br>• Evaporative Emission (EVAP) canister purge regulator valve 1 is faulty<br>• Hoses between the fuel pump and the EVAP canister are faulty<br>• Fuel filler cap is loose<br>• Fuel pump seal is defective, faulty or otherwise leaking<br>• Hoses between the EVAP canister and the fuel flap unit are faulty<br>• Hoses between the EVAP canister and the evaporative emission canister purge regulator valve are faulty<br>• ECM has failed |

| DTC | Trouble Code Title, Conditions & Possible Causes |
|---|---|
| **DTC: P1602**<br>**Years:** 2007, 2008<br>**Models:** Passat, Jetta, Golf, GTI, New Beetle, New Beetle Convertible<br>**Engines:** 2.0L, 2.5L, 3.6L<br>**Transmissions:** All | **Power Supply (B+) Terminal 15 Low Voltage Conditions:**<br>Ignition on, the ECM detected a low voltage condition on the power supply terminal (15). To achieve optimal anti-theft protection for the vehicle, an anti-theft immobilizer is installed. The anti-theft immobilizer is a system for enabling and locking the Engine Control Module (ECM). So that this system cannot be circumvented, it is necessary to perform adaptation of the anti-theft immobilizer using the Vehicle Diagnostic and Information System VAS 5052 in the On Board Diagnostic (OBD) function. The great availability of equipment options makes it necessary to adapt the Engine Control Module (ECM) to the vehicle (e.g. throttle valve control module or cruise control system). This "writing" function is not possible with the generic scan tool.<br>**Possible Causes:**<br>    • (If ECM was replaced) ECM ID not the same as the replaced unit<br>    • ECM has failed<br>    • Voltage supply for Engine Control Module (ECM) has shorted |
| **DTC: P1603**<br>**Years:** 2007, 2008<br>**Models:** Passat, Jetta, Golf, GTI, New Beetle, New Beetle Convertible<br>**Engines:** 2.0L, 2.5L, 3.6L<br>**Transmissions:** All | **Internal Control Module Malfunction Conditions:**<br>Ignition on, the ECM detected a control module malfunction. To achieve optimal anti-theft protection for the vehicle, an anti-theft immobilizer is installed. The anti-theft immobilizer is a system for enabling and locking the Engine Control Module (ECM). So that this system cannot be circumvented, it is necessary to perform adaptation of the anti-theft immobilizer using the Vehicle Diagnostic and Information System VAS 5052 in the On Board Diagnostic (OBD) function. The great availability of equipment options makes it necessary to adapt the Engine Control Module (ECM) to the vehicle (e.g. throttle valve control module or cruise control system). This "writing" function is not possible with the generic scan tool.<br>**Possible Causes:**<br>    • (If ECM was replaced) ECM ID not the same as the replaced unit<br>    • ECM has failed<br>    • Voltage supply for Engine Control Module (ECM) has shorted |
| **DTC: P1604**<br>**2T, MIL: Yes**<br>**Years:** 2007, 2008<br>**Models:** Passat, Jetta, Golf, GTI, New Beetle, New Beetle Convertible<br>**Engines:** 2.0L, 2.5L, 3.6L<br>**Transmissions:** All | **Internal Control Module Driver Error Conditions:**<br>Ignition on, the ECM detected a control module malfunction. To achieve optimal anti-theft protection for the vehicle, an anti-theft immobilizer is installed. The anti-theft immobilizer is a system for enabling and locking the Engine Control Module (ECM). So that this system cannot be circumvented, it is necessary to perform adaptation of the anti-theft immobilizer using the Vehicle Diagnostic and Information System VAS 5052 in the On Board Diagnostic (OBD) function. The great availability of equipment options makes it necessary to adapt the Engine Control Module (ECM) to the vehicle (e.g. throttle valve control module or cruise control system). This "writing" function is not possible with the generic scan tool.<br>**Possible Causes:**<br>    • (If ECM was replaced) ECM ID not the same as the replaced unit<br>    • ECM has failed<br>    • Voltage supply for Engine Control Module (ECM) has shorted |
| **DTC: P1606**<br>**Years:** 2007, 2008<br>**Models:** Passat, Jetta, Golf, GTI, New Beetle<br>**Engines:** 2.0L, 2.5L, 3.6L<br>**Transmissions:** All | **Rough Road Spec Engine Torque ABS-ECU Electrical Malfunction Conditions:**<br>Ignition on, the ECM detected an electrical malfunction.<br>**Possible Causes:**<br>    • Check wire connection between Engine Control Module (ECM) and ABS Control Module |
| **DTC: P1609**<br>**Years:** 2007, 2008<br>**Models:** Passat, Jetta, Golf, GTI, New Beetle<br>**Engines:** 2.0L, 2.5L, 3.6L<br>**Transmissions:** All | **Crash Shut-Down Activated Conditions:**<br>The ECM detected that the car has been in an accident.<br>**Possible Causes:**<br>    • Check the vehicle for damage<br>    • Reset the ECU |
| **DTC: P1610**<br>**Years:** 2007, 2008<br>**Models:** Passat, Jetta, Golf, GTI, New Beetle,<br>**Engines:** 2.0L<br>**Transmissions:** All | **ECU Defective Conditions:**<br>To achieve optimal anti-theft protection for the vehicle, an anti-theft immobilizer is installed. The anti-theft immobilizer is a system for enabling and locking the Engine Control Module (ECM). So that this system cannot be circumvented, it is necessary to perform adaptation of the anti-theft immobilizer using the Vehicle Diagnostic and Information System VAS 5052 in the On Board Diagnostic (OBD) function. The great availability of equipment options makes it necessary to adapt the Engine Control Module (ECM) to the vehicle (e.g. throttle valve control module or cruise control system). This "writing" function is not possible with the generic scan tool.<br>**Possible Causes:**<br>    • (If ECM was replaced) ECM ID not the same as the replaced unit<br>    • ECM has failed<br>    • Voltage supply for Engine Control Module (ECM) has shorted |

| DTC | Trouble Code Title, Conditions & Possible Causes |
|---|---|
| **DTC: P1611**<br>**MIL: Yes**<br>**Years:** 2007, 2008<br>**Models:** Passat, Jetta, Golf, GTI,<br>**Engines:** 2.0L, 2.5L, 3.6L<br>**Transmissions:** All | **MIL Call-Up Circuit, Transmission Control Module Short to Ground Conditions:**<br>Engine started, VSS over 1 mph, and the ECM detected a problem in the Transmission Control system during the self-test.<br>**Possible Causes:**<br>&bull; Open/short circuit to ground in the communication wire from the transmission to the ECM.<br>&bull; The ECM has failed |
| **DTC: P1612**<br>**2T, MIL: Yes**<br>**Years:** 2007, 2008<br>**Models:** Passat, Jetta, Golf, GTI,<br>New Beetle<br>**Engines:** 2.0L, 2.5L, 3.6L<br>**Transmissions:** All | **Electronic Control Module Incorrect Coding Conditions:**<br>Ignition on, the ECM detected a control module malfunction. To achieve optimal anti-theft protection for the vehicle, an anti-theft immobilizer is installed. The anti-theft immobilizer is a system for enabling and locking the Engine Control Module (ECM). So that this system cannot be circumvented, it is necessary to perform adaptation of the anti-theft immobilizer using the Vehicle Diagnostic and Information System VAS 5052 in the On Board Diagnostic (OBD) function. The great availability of equipment options makes it necessary to adapt the Engine Control Module (ECM) to the vehicle (e.g. throttle valve control module or cruise control system). This "writing" function is not possible with the generic scan tool.<br>**Possible Causes:**<br>&bull; (If ECM was replaced) ECM ID not the same as the replaced unit<br>&bull; ECM has failed<br>&bull; Voltage supply for Engine Control Module (ECM) has shorted |
| **DTC: P1613**<br>**MIL: Yes**<br>**Years:** 2007, 2008<br>**Models:** Passat, Jetta, Golf, GTI,<br>**Engines:** 2.0L, 2.5L, 3.6L<br>**Transmissions:** All | **MIL Call-up Circuit Open/Short to B+ Conditions:**<br>Engine started, VSS over 1 mph, and the ECM detected a problem in the Transmission Control system during the self-test.<br>**Possible Causes:**<br>&bull; Open/short circuit to ground from the MIL to the ECM.<br>&bull; The ECM has failed<br>&bull; The MIL light has failed (check bulb) |
| **DTC: P1624**<br>**Years:** 2007, 2008<br>**Models:** Passat, Jetta, Golf, GTI,<br>New Beetle, New Beetle Convertible<br>**Engines:** 2.0L, 2.5L, 3.6L<br>**Transmissions:** All | **MIL Requested Signature Active Conditions:**<br>Ignition on, the ECM detected a control module malfunction. To achieve optimal anti-theft protection for the vehicle, an anti-theft immobilizer is installed. The anti-theft immobilizer is a system for enabling and locking the Engine Control Module (ECM). So that this system cannot be circumvented, it is necessary to perform adaptation of the anti-theft immobilizer using the Vehicle Diagnostic and Information System VAS 5052 in the On Board Diagnostic (OBD) function. The great availability of equipment options makes it necessary to adapt the Engine Control Module (ECM) to the vehicle (e.g. throttle valve control module or cruise control system). This "writing" function is not possible with the generic scan tool.<br>**Possible Causes:**<br>&bull; (If ECM was replaced) ECM ID not the same as the replaced unit<br>&bull; ECM has failed<br>&bull; Voltage supply for Engine Control Module (ECM) has shorted |
| **DTC: P1626**<br>**Years:** 2007, 2008<br>**Models:** Passat, Jetta, Golf, GTI,<br>New Beetle, New Beetle Convertible<br>**Engines:** 2.0L, 2.5L, 3.6L<br>**Transmissions:** All | **Data BUS Powertrain Missing Message From Transmission Control Conditions:**<br>Ignition on, the ECM detected a control module malfunction (Transmission). To achieve optimal anti-theft protection for the vehicle, an anti-theft immobilizer is installed. The anti-theft immobilizer is a system for enabling and locking the Engine Control Module (ECM). So that this system cannot be circumvented, it is necessary to perform adaptation of the anti-theft immobilizer using the Vehicle Diagnostic and Information System VAS 5052 in the On Board Diagnostic (OBD) function. The great availability of equipment options makes it necessary to adapt the Engine Control Module (ECM) to the vehicle (e.g. throttle valve control module or cruise control system). This "writing" function is not possible with the generic scan tool.<br>**Possible Causes:**<br>&bull; (If ECM was replaced) ECM ID not the same as the replaced unit<br>&bull; ECM has failed<br>&bull; Voltage supply for Engine Control Module (ECM) has shorted |
| **DTC: P1630**<br>**2T, MIL: Yes**<br>**Years:** 2007, 2008<br>**Models:** Passat, Jetta, Golf, GTI,<br>New Beetle, New Beetle Convertible<br>**Engines:** 2.0L, 2.5L, 3.6L<br>**Transmissions:** All | **Acceleration Pedal Position Sensor 1 Signal Too Low Conditions:**<br>Engine started, battery voltage at least 11.5v, all electrical components off, ground connections between engine and chassis well connected, the ECM detected that the accelerator pedal position sensor signal was too low.<br>**Note: Both the Throttle Position (TP) Sensor and Accelerator Pedal Position Sensor 2 are located at the accelerator pedal module and communicate the driver's intentions to the ECM completely independently of each other. Both sensors are stored in one housing.**<br>**Possible Causes:**<br>&bull; Ground between engine and chassis may be broken<br>&bull; Throttle position sensor may have failed<br>&bull; Accelerator Pedal Position Sensor 2 has failed<br>&bull; Throttle position sensor wiring may have shorted<br>&bull; Faulty voltage supply<br>&bull; ECM has failed |

| DTC | Trouble Code Title, Conditions & Possible Causes |
|---|---|
| **DTC: P1631**<br>**2T, MIL: Yes**<br>**Years: 2007, 2008**<br>**Models**: Passat, Jetta, Golf, GTI, New Beetle, New Beetle Convertible<br>**Engines**: 2.0L, 2.5L, 3.6L<br>**Transmissions**: All | **Acceleration Pedal Position Sensor 1 Signal Too High Conditions:**<br>Engine started, battery voltage at least 11.5v, all electrical components off, ground connections between engine and chassis well connected, the ECM detected that the accelerator pedal position sensor signal was too high.<br>**Note: Both the Throttle Position (TP) Sensor and Accelerator Pedal Position Sensor 2 are located at the accelerator pedal module and communicate the driver's intentions to the ECM completely independently of each other. Both sensors are stored in one housing.**<br>**Possible Causes:**<br>    • Ground between engine and chassis may be broken<br>    • Throttle position sensor may have failed<br>    • Accelerator Pedal Position Sensor 2 has failed<br>    • Throttle position sensor wiring may have shorted<br>    • Faulty voltage supply<br>    • ECM has failed |
| **DTC: P1633**<br>**2T, MIL: Yes**<br>**Years: 2007, 2008**<br>**Models**: Passat, Jetta, Golf, GTI, New Beetle, New Beetle Convertible<br>**Engines**: 2.0L, 2.5L, 3.6L<br>**Transmissions**: All | **Acceleration Pedal Position Sensor 2 Signal Too Low Conditions:**<br>Engine started, battery voltage at least 11.5v, all electrical components off, ground connections between engine and chassis well connected, the ECM detected that the accelerator pedal position sensor signal was too low.<br>**Note: Both the Throttle Position (TP) Sensor and Accelerator Pedal Position Sensor 2 are located at the accelerator pedal module and communicate the driver's intentions to the ECM completely independently of each other. Both sensors are stored in one housing.**<br>**Possible Causes:**<br>    • Ground between engine and chassis may be broken<br>    • Throttle position sensor may have failed<br>    • Accelerator Pedal Position Sensor 2 has failed<br>    • Throttle position sensor wiring may have shorted<br>    • Faulty voltage supply<br>    • ECM has failed |
| **DTC: P1634**<br>**2T, MIL: Yes**<br>**Years: 2007, 2008**<br>**Models**: Passat, Jetta, Golf, GTI, New Beetle, New Beetle Convertible<br>**Engines**: 2.0L, 2.5L, 3.6L<br>**Transmissions**: All | **Acceleration Pedal Position Sensor 2 Signal Too High Conditions:**<br>Engine started, battery voltage at least 11.5v, all electrical components off, ground connections between engine and chassis well connected, the ECM detected that the accelerator pedal position sensor signal was too high.<br>**Note: Both the Throttle Position (TP) Sensor and Accelerator Pedal Position Sensor 2 are located at the accelerator pedal module and communicate the driver's intentions to the ECM completely independently of each other. Both sensors are stored in one housing.**<br>**Possible Causes:**<br>    • Ground between engine and chassis may be broken<br>    • Throttle position sensor may have failed<br>    • Accelerator Pedal Position Sensor 2 has failed<br>    • Throttle position sensor wiring may have shorted<br>    • Faulty voltage supply<br>    • ECM has failed |
| **DTC: P1635**<br>**Years: 2007, 2008**<br>**Models**: Passat, Jetta, Golf, GTI, New Beetle, New Beetle Convertible<br>**Engines**: 2.0L, 2.5L, 3.6L<br>**Transmissions**: All | **Data BUS Powertrain Missing Message From Central A/C Control Conditions:**<br>Ignition off, the ECU is missing general Data BUS information from the A/C control. The Engine Control Module (ECM) communicates with all databus-capable control modules via a CAN databus. These databus-capable control modules are connected via two data bus wires which are twisted together (CAN_High and CAN_Low), and exchange information (messages). Missing information on the databus is recognized as a malfunction and stored. Trouble-free operation of the CAN-bus requires that it have a terminal resistance. This central terminal resistor is located in the Engine Control Module (ECM).<br>**Possible Causes:**<br>    • Check the Terminal resistance for CAN-bus<br>    • Data-Bus wires have short<br>    • Data-Bus components are malfunctioning<br>    • ECM has failed |
| **DTC: P1637**<br>**Years: 2007, 2008**<br>**Models**: Passat, Jetta<br>**Engines**: 2.0L<br>**Transmissions**: All | **Data BUS Powertrain Missing Message From Central Electrical Control Conditions:**<br>Ignition off, the ECU is missing general Data BUS information from the central electrical control. The Engine Control Module (ECM) communicates with all databus-capable control modules via a CAN databus. These databus-capable control modules are connected via two data bus wires which are twisted together (CAN_High and CAN_Low), and exchange information (messages). Missing information on the databus is recognized as a malfunction and stored. Trouble-free operation of the CAN-bus requires that it have a terminal resistance. This central terminal resistor is located in the Engine Control Module (ECM).<br>**Possible Causes:**<br>    • Check the Terminal resistance for CAN-bus<br>    • Data-Bus wires have short<br>    • Data-Bus components are malfunctioning<br>    • ECM has failed |

| DTC | Trouble Code Title, Conditions & Possible Causes |
|---|---|
| **DTC: P1639**<br>**2T, MIL: Yes**<br>**Years:** 2007, 2008<br>**Models:** Passat, Jetta, Golf, GTI, New Beetle, New Beetle Convertible<br>**Engines:** 2.0L, 2.5L, 3.6L<br>**Transmissions:** All | **Accelerator Pedal Position Sensor 1+2 Range/Performance Conditions:**<br>Engine started, battery voltage at least 11.5v, all electrical components off, ground connections between engine and chassis well connected, the ECM detected that the accelerator pedal position sensor signal was too high.<br>**Note: Both the Throttle Position (TP) Sensor and Accelerator Pedal Position Sensor 2 are located at the accelerator pedal module and communicate the driver's intentions to the ECM completely independently of each other. Both sensors are stored in one housing.**<br>**Possible Causes:**<br>• Ground between engine and chassis may be broken<br>• Throttle position sensor may have failed<br>• Accelerator Pedal Position Sensor 2 has failed<br>• Throttle position sensor wiring may have shorted<br>• Faulty voltage supply<br>• ECM has failed |
| **DTC: P1640**<br>**Years:** 2007, 2008<br>**Models:** Passat, Jetta, Golf, GTI, New Beetle, New Beetle Convertible<br>**Engines:** 2.0L, 2.5L, 3.6L<br>**Transmissions:** All | **Internal Control Module (EEPROM) Error Conditions:**<br>Ignition on, the ECM detected a control module malfunction (software). To achieve optimal anti-theft protection for the vehicle, an anti-theft immobilizer is installed. The anti-theft immobilizer is a system for enabling and locking the Engine Control Module (ECM). So that this system cannot be circumvented, it is necessary to perform adaptation of the anti-theft immobilizer using the Vehicle Diagnostic and Information System VAS 5052 in the On Board Diagnostic (OBD) function. The great availability of equipment options makes it necessary to adapt the Engine Control Module (ECM) to the vehicle (e.g. throttle valve control module or cruise control system). This "writing" function is not possible with the generic scan tool.<br>**Possible Causes:**<br>• Engine Control Module (ECM) has failed<br>• Voltage supply for Engine Control Module (ECM) has shorted |
| **DTC: P1647**<br>Passat: 2.0L (BPY); Jetta: 2.5L (BGP, BGQ), 2.8L (AFP, BDF); GTI: 2.8L (AFP, BDF)<br>Transmissions: All | **Please Check Coding of the ECUs in the Data Bus Powertrain Conditions:**<br>Ignition on, the ECM detected a control module malfunction (software). To achieve optimal anti-theft protection for the vehicle, an anti-theft immobilizer is installed. The anti-theft immobilizer is a system for enabling and locking the Engine Control Module (ECM). So that this system cannot be circumvented, it is necessary to perform adaptation of the anti-theft immobilizer using the Vehicle Diagnostic and Information System VAS 5052 in the On Board Diagnostic (OBD) function. The great availability of equipment options makes it necessary to adapt the Engine Control Module (ECM) to the vehicle (e.g. throttle valve control module or cruise control system). This "writing" function is not possible with the generic scan tool.<br>**Possible Causes:**<br>• Engine Control Module (ECM) has failed<br>• Voltage supply for Engine Control Module (ECM) has shorted |
| **DTC: P1648**<br>**Years:** 2007, 2008<br>**Models:** Passat, Jetta, Golf, GTI, New Beetle, New Beetle Convertible<br>**Engines:** 2.0L, 2.5L, 3.6L<br>**Transmissions:** All | **Data Bus Powertrain Malfunction Conditions:**<br>Ignition on, the ECM detected a data bus malfunction (software). To achieve optimal anti-theft protection for the vehicle, an anti-theft immobilizer is installed. The anti-theft immobilizer is a system for enabling and locking the Engine Control Module (ECM). So that this system cannot be circumvented, it is necessary to perform adaptation of the anti-theft immobilizer using the Vehicle Diagnostic and Information System VAS 5052 in the On Board Diagnostic (OBD) function. The great availability of equipment options makes it necessary to adapt the Engine Control Module (ECM) to the vehicle (e.g. throttle valve control module or cruise control system). This "writing" function is not possible with the generic scan tool.<br>**Possible Causes:**<br>• Ground between engine and chassis may be broken<br>• Throttle position sensor may have failed<br>• Accelerator Pedal Position Sensor 2 has failed<br>• Throttle position sensor wiring may have shorted<br>• Faulty voltage supply<br>• ECM has failed |

| DTC | Trouble Code Title, Conditions & Possible Causes |
|---|---|
| **DTC: P1649**<br>**2T, MIL: Yes**<br>**Years: 2007, 2008**<br>**Models:** Passat, Jetta, Golf, GTI, New Beetle, New Beetle Convertible<br>**Engines:** 2.0L, 2.5L, 3.6L<br>**Transmissions**: All | **Data Bus Powertrain Missing Message from ABS Control Module Conditions:**<br>Ignition off, the ECU is missing general Data BUS information from the central electrical control. The Engine Control Module (ECM) communicates with all databus-capable control modules via a CAN databus. These databus-capable control modules are connected via two data bus wires which are twisted together (CAN_High and CAN_Low), and exchange information (messages). Missing information on the databus is recognized as a malfunction and stored. Trouble-free operation of the CAN-bus requires that it have a terminal resistance. This central terminal resistor is located in the Engine Control Module (ECM).<br>**Possible Causes:**<br>• Ground between engine and chassis may be broken<br>• Throttle position sensor may have failed<br>• Accelerator Pedal Position Sensor 2 has failed<br>• Throttle position sensor wiring may have shorted<br>• Faulty voltage supply<br>• Check the Terminal resistance for CAN-bus<br>• Data-Bus wires have short<br>• Data-Bus components are malfunctioning<br>• ECM has failed |
| **DTC: P1650**<br>**Years: 2007, 2008**<br>**Models:** Passat, Jetta, Golf, GTI, New Beetle, New Beetle Convertible<br>**Engines:** 2.0L, 2.5L, 3.6L<br>**Transmissions**: All | **Data Bus Powertrain Missing Message from Instrument Panel ECU Conditions:**<br>Ignition off, the ECU is missing general Data BUS information from the instrument panel. The Engine Control Module (ECM) communicates with all databus-capable control modules via a CAN databus. These databus-capable control modules are connected via two data bus wires which are twisted together (CAN_High and CAN_Low), and exchange information (messages). Missing information on the databus is recognized as a malfunction and stored. Trouble-free operation of the CAN-bus requires that it have a terminal resistance. This central terminal resistor is located in the Engine Control Module (ECM).<br>**Possible Causes:**<br>• Ground between engine and chassis may be broken<br>• Faulty voltage supply<br>• ECM has failed |
| **DTC: P1653**<br>**Years: 2007, 2008**<br>**Models:** Passat<br>**Engines:** 2.0L<br>**Transmissions**: All | **Please Check DTC Memory of the ABS Control Module Conditions:**<br>Ignition off, the ECU detected a memory fault of the ABS control module. The Engine Control Module (ECM) communicates with all databus-capable control modules via a CAN databus. These databus-capable control modules are connected via two data bus wires which are twisted together (CAN_High and CAN_Low), and exchange information (messages). Missing information on the databus is recognized as a malfunction and stored. Trouble-free operation of the CAN-bus requires that it have a terminal resistance. This central terminal resistor is located in the Engine Control Module (ECM).<br>**Possible Causes:**<br>• Ground between engine and chassis may be broken<br>• Throttle position sensor may have failed<br>• Accelerator Pedal Position Sensor 2 has failed<br>• Throttle position sensor wiring may have shorted<br>• Faulty voltage supply<br>• ECM has failed |
| **DTC: P1654**<br>**Years: 2007, 2008**<br>**Models:** Passat<br>**Engines:**<br>**Transmissions**: All | **Please Check DTC Memory of the Control Panel ECU Conditions:**<br>The Engine Control Module (ECM) communicates with all databus-capable control modules via a CAN databus. These databus-capable control modules are connected via two data bus wires which are twisted together (CAN_High and CAN_Low), and exchange information (messages). Missing information on the databus is recognized as a malfunction and stored. Trouble-free operation of the CAN-bus requires that it have a terminal resistance. This central terminal resistor is located in the Engine Control Module (ECM).<br>**Possible Causes:**<br>• Ground between engine and chassis may be broken<br>• Faulty voltage supply<br>• ECM has failed |
| **DTC: P1655**<br>**Years: 2007, 2008**<br>**Models:** Passat<br>**Engines:** 2.0L<br>**Transmissions**: All | **Please Check DTC Memory of the ADR Control Module Conditions:**<br>The Engine Control Module (ECM) communicates with all databus-capable control modules via a CAN databus. These databus-capable control modules are connected via two data bus wires which are twisted together (CAN_High and CAN_Low), and exchange information (messages). Missing information on the databus is recognized as a malfunction and stored. Trouble-free operation of the CAN-bus requires that it have a terminal resistance. This central terminal resistor is located in the Engine Control Module (ECM).<br>**Possible Causes:**<br>• Ground between engine and chassis may be broken<br>• Throttle position sensor may have failed<br>• Accelerator Pedal Position Sensor 2 has failed<br>• Throttle position sensor wiring may have shorted<br>• Faulty voltage supply<br>• ECM has failed |

| DTC | Trouble Code Title, Conditions & Possible Causes |
|---|---|
| **DTC: P1676**<br>**Years:** 2007, 2008<br>**Models:** Passat, Jetta, Golf, GTI, New Beetle, New Beetle Convertible<br>**Engines:**<br>**Transmissions:** All | **Drive by Wire-MIL Circuit Electrical Malfunction Conditions:**<br>Key on or engine running, the ECM detected an electrical malfunction regarding the drive-by-wire circuit.<br>**Note: EPC" is an abbreviation and stands for Electronic Power Control and means "electronic engine load control". If malfunctions are recognized in the EPC system during operation of the engine, the Engine Control Module (ECM) switches on the EPC warning lamp. An entry is made in DTC memory at the same time. After a few seconds of the engine at idle, the EPC should extinguish itself.**<br>**Possible Causes:**<br>• Circuit from the MIL to the ECM<br>• ECM has failed<br>• Circuit from the EPC to the ECM |
| **DTC: P1677**<br>**Years:** 2007, 2008<br>**Models:** Passat, Jetta, Golf, GTI, New Beetle, New Beetle Convertible<br>**Engines:**<br>**Transmissions:** All | **Drive by Wire-MIL Circuit Short to B+ Conditions:**<br>Key on or engine running, the ECM detected an electrical malfunction regarding the drive-by-wire circuit.<br>**Note: EPC" is an abbreviation and stands for Electronic Power Control and means "electronic engine load control". If malfunctions are recognized in the EPC system during operation of the engine, the Engine Control Module (ECM) switches on the EPC warning lamp. An entry is made in DTC memory at the same time. After a few seconds of the engine at idle, the EPC should extinguish itself.**<br>**Possible Causes:**<br>• Circuit from the MIL to the ECM<br>• ECM has failed<br>• Circuit from the EPC to the ECM |
| **DTC: P1681**<br>**2T, MIL: Yes**<br>**Years:** 2007, 2008<br>**Models:** Passat, Jetta, Golf, GTI, New Beetle Convertible<br>**Engines:**<br>**Transmissions:** All | **Control Unit Programming, Programming Not Finished Conditions:**<br>The Engine Control Module (ECM) communicates with all databus-capable control modules via a CAN databus. These databus-capable control modules are connected via two data bus wires which are twisted together (CAN_High and CAN_Low), and exchange information (messages). Missing information on the databus is recognized as a malfunction and stored. Trouble-free operation of the CAN-bus requires that it have a terminal resistance. This central terminal resistor is located in the Engine Control Module (ECM).<br>**Possible Causes:**<br>• ECM has failed |
| **DTC: P1682**<br>**Years:** 2007, 2008<br>**Models:** Jetta<br>**Engines:** 2.0L<br>**Transmissions:** All | **Powertrain Data Bus Implausible Message from ABS Control Module Conditions:**<br>Ignition off, the ECU detected a I fault of the ABS control module. The Engine Control Module (ECM) communicates with all databus-capable control modules via a CAN databus. These databus-capable control modules are connected via two data bus wires which are twisted together (CAN_High and CAN_Low), and exchange information (messages). Missing information on the databus is recognized as a malfunction and stored. Trouble-free operation of the CAN-bus requires that it have a terminal resistance. This central terminal resistor is located in the Engine Control Module (ECM).<br>**Possible Causes:**<br>• Ground between engine and chassis may be broken<br>• Faulty voltage supply<br>• ECM has failed |
| **DTC: P1683**<br>**Years:** 2007, 2008<br>**Models:** Passat, Jetta<br>**Engines:** 2.0L<br>**Transmissions:** All | **Data Bus Powertrain Implausible Message from Airbag Control Conditions:**<br>Ignition off, the ECU detected a circuit fault of the airbag control module. The Engine Control Module (ECM) communicates with all databus-capable control modules via a CAN databus. These databus-capable control modules are connected via two data bus wires which are twisted together (CAN_High and CAN_Low), and exchange information (messages). Missing information on the databus is recognized as a malfunction and stored. Trouble-free operation of the CAN-bus requires that it have a terminal resistance. This central terminal resistor is located in the Engine Control Module (ECM).<br>**Possible Causes:**<br>• Ground between engine and chassis may be broken<br>• Throttle position sensor may have failed<br>• Accelerator Pedal Position Sensor 2 has failed<br>• Throttle position sensor wiring may have shorted<br>• Faulty voltage supply<br>• ECM has failed |
| **DTC: P1690**<br>**Years:** 2007, 2008<br>**Models:** Passat, Jetta, Golf, GTI, New Beetle<br>**Engines:** 2.5L, 3.6L<br>**Transmissions:** All | **Malfunction Indication Light Malfunction Conditions:**<br>The exhaust Malfunction Indicator Lamp (MIL) lights up when exhaust relevant malfunctions are recognized by the Engine Control Module (ECM). The Malfunction Indicator Lamp (MIL) can blink or remain lit continuously. Blinking: There is a malfunction that causes damage to the catalytic converter in this driving condition. In this case, vehicle must only be driven at reduced power! Continuously lit: There is a malfunction that causes increased emissions. Check DTC memory for Motronic control module. DTC memory must still be checked if there are driveability problems or customer complaints and the MIL is not lit, since malfunctions can be stored without causing the MIL to light immediately.<br>**Possible Causes:**<br>• Wire from ECM to MIL is shorted or grounded<br>• ECM has failed<br>• MIL has failed |

| DTC | Trouble Code Title, Conditions & Possible Causes |
|---|---|
| **DTC: P1691**<br>**Years: 2007, 2008**<br>**Models:** Passat, Jetta, GTI<br>**Engines:** 2.5L, 3.6L<br>**Transmissions:** All | **Malfunction Indication Light Open Conditions:**<br>The exhaust Malfunction Indicator Lamp (MIL) lights up when exhaust relevant malfunctions are recognized by the Engine Control Module (ECM). The Malfunction Indicator Lamp (MIL) can blink or remain lit continuously. Blinking: There is a malfunction that causes damage to the catalytic converter in this driving condition. In this case, vehicle must only be driven at reduced power! Continuously lit: There is a malfunction that causes increased emissions. Check DTC memory for Motronic control module. DTC memory must still be checked if there are driveability problems or customer complaints and the MIL is not lit, since malfunctions can be stored without causing the MIL to light immediately.<br>**Possible Causes:**<br>&bull; Wire from ECM to MIL is shorted or grounded<br>&bull; ECM has failed<br>&bull; MIL has failed |
| **DTC: P1692**<br>**Years: 2007, 2008**<br>**Models:** Passat, Jetta, GTI<br>**Engines:** 2.5L, 3.6L<br>**Transmissions:** All | **Malfunction Indication Light Short to Ground Conditions:**<br>The exhaust Malfunction Indicator Lamp (MIL) lights up when exhaust relevant malfunctions are recognized by the Engine Control Module (ECM). The Malfunction Indicator Lamp (MIL) can blink or remain lit continuously. Blinking: There is a malfunction that causes damage to the catalytic converter in this driving condition. In this case, vehicle must only be driven at reduced power! Continuously lit: There is a malfunction that causes increased emissions. Check DTC memory for Motronic control module. DTC memory must still be checked if there are driveability problems or customer complaints and the MIL is not lit, since malfunctions can be stored without causing the MIL to light immediately.<br>**Possible Causes:**<br>&bull; Wire from ECM to MIL is shorted or grounded<br>&bull; ECM has failed<br>&bull; MIL has failed |
| **DTC: P1693**<br>**Years: 2007, 2008**<br>**Models:** Passat, Jetta, Golf, GTI, New Beetle<br>**Engines:** 2.5L, 3.6L<br>**Transmissions:** All | **Malfunction Indication Light Short to B+ Conditions:**<br>The exhaust Malfunction Indicator Lamp (MIL) lights up when exhaust relevant malfunctions are recognized by the Engine Control Module (ECM). The Malfunction Indicator Lamp (MIL) can blink or remain lit continuously. Blinking: There is a malfunction that causes damage to the catalytic converter in this driving condition. In this case, vehicle must only be driven at reduced power! Continuously lit: There is a malfunction that causes increased emissions. Check DTC memory for Motronic control module. DTC memory must still be checked if there are driveability problems or customer complaints and the MIL is not lit, since malfunctions can be stored without causing the MIL to light immediately.<br>**Possible Causes:**<br>&bull; Wire from ECM to MIL is shorted or grounded<br>&bull; ECM has failed<br>&bull; MIL has failed |
| **DTC: P1698**<br>**Years: 2007, 2008**<br>**Models:** Jetta<br>**Engines:** 2.0L<br>**Transmissions:** All | **Check DTC Memory of Steering Column ECU Conditions:**<br>The Engine Control Module (ECM) communicates with all databus-capable control modules via a CAN databus. These databus-capable control modules are connected via two data bus wires which are twisted together (CAN_High and CAN_Low), and exchange information (messages). Missing information on the databus is recognized as a malfunction and stored. Trouble-free operation of the CAN-bus requires that it have a terminal resistance. This central terminal resistor is located in the Engine Control Module (ECM).<br>**Possible Causes:**<br>&bull; Ground between engine and chassis may be broken<br>&bull; Faulty voltage supply<br>&bull; ECM has failed |
| **DTC: P1778**<br>**Years: 2007, 2008**<br>**Models:** Passat, Jetta, Golf, GTI, New Beetle, New Beetle Convertible<br>**Engines:** 2.0L, 2.5L, 3.6L<br>**Transmissions:** All | **Solenoid EV7 Electrical Malfunction Conditions:**<br>Engine started, battery voltage must be at least 11.5v, all electrical components must be off, the ground between the engine and the chassis must be well connected, the exhaust system must be properly sealed between the catalytic converter and the cylinder head, and the oxygen sensor heater for oxygen sensor before the catalytic converter must be properly functioning. The ECM detected a loss of communication between the TCM and the valve body solenoid valve.<br>**Possible Causes:**<br>&bull; Valve body solenoid valve has a short<br>&bull; The TCM has failed<br>&bull; ECM has failed |
| **DTC: P1780**<br>**Years: 2007, 2008**<br>**Models:** Passat, Jetta, Golf, GTI, New Beetle<br>**Engines:** 2.5L, 3.6L<br>**Transmissions:** A/T | **Engine Intervention Readable Conditions:**<br>Key on or engine started, Self-Test enabled, and the ECM detected the Transmission Control Switch (TCS) was out of range during the test and a signal was not received.<br>**Note: The seal on the ATF level plug must always be replaced if the ATF level is checked.**<br>**Possible Causes:**<br>&bull; TCS circuit open or shorted in the wiring harness<br>&bull; TCS not cycled during the self-test<br>&bull; TCS is damaged or failed, or the ECM has failed<br>&bull; Check ATF level<br>&bull; Mechanical malfunction in transmission |

| DTC | Trouble Code Title, Conditions & Possible Causes |
|---|---|
| **DTC: P1823**<br>**Years: 2007, 2008**<br>**Models:** Passat, Jetta, Golf, GTI, New Beetle<br>**Engines:** 2.0L, 2.5L, 3.6L<br>**Transmissions**: All | **Pressure Control Solenoid 3 Electrical Conditions:**<br>Engine started, vehicle driven with the solenoid applied, and the ECM detected an unexpected voltage condition on the SS1/C solenoid circuit was incorrect during the test.<br>**Possible Causes:**<br>• Solenoid valves in valve body are faulty<br>• Solenoid circuit is shorting to ground<br>• Solenoid circuit is open<br>• TCM has failed or wiring is shorting<br>• Check harness connector for contact corrosion or water damage<br>• Check resistance of solenoid valves may not be up to specification<br>• Wires to Transmission Control Module (TCM) may have ground out or open<br>• Transmission Control Module (TCM) has failed<br>• ECM has failed |
| **DTC: P1828**<br>**Years: 2007, 2008**<br>**Models:** Passat, Jetta, Golf, GTI, New Beetle<br>**Engines:** 2.0L, 2.5L, 3.6L<br>**Transmissions**: All | **Pressure Control Solenoid 4 Electrical Conditions:**<br>Engine started, vehicle driven with the solenoid applied, and the ECM detected an unexpected voltage condition on the SS1/C solenoid circuit was incorrect during the test.<br>**Possible Causes:**<br>• Solenoid valves in valve body are faulty<br>• Solenoid circuit is shorting to ground<br>• Solenoid circuit is open<br>• TCM has failed or wiring is shorting<br>• Check harness connector for contact corrosion or water damage<br>• Check resistance of solenoid valves may not be up to specification<br>• Wires to Transmission Control Module (TCM) may have ground out or open<br>• Transmission Control Module (TCM) has failed<br>• ECM has failed |
| **DTC: P1847**<br>**Years: 2007, 2008**<br>**Models:** Passat, New Beetle<br>**Engines:** 2.0L<br>**Transmissions**: All | **Please Check DTC Memory of Brake System ECU Conditions:**<br>The ECU detected a memory fault of the brake system. The Engine Control Module (ECM) communicates with all databus-capable control modules via a CAN databus. These databus-capable control modules are connected via two data bus wires which are twisted together (CAN_High and CAN_Low), and exchange information (messages). Missing information on the databus is recognized as a malfunction and stored. Trouble-free operation of the CAN-bus requires that it have a terminal resistance. This central terminal resistor is located in the Engine Control Module (ECM).<br>**Possible Causes:**<br>• Fuses on E-box in engine compartment, left side, may be faulty<br>• Check harness connector for contact corrosion or water damage<br>• Check resistance of solenoid valves may not be up to specification<br>• Solenoid valve may be faulty<br>• Wires to Transmission Control Module (TCM) may have ground out or open<br>• Transmission Control Module (TCM) has failed |
| **DTC: P1850**<br>**Years: 2007, 2008**<br>**Models:** Passat, Jetta, Golf, GTI, New Beetle, New Beetle Convertible<br>**Engines:** 2.0L, 2.5L, 3.6L<br>**Transmissions**: All | **Data BUS Powertrain Missing Message from Engine Control Conditions:**<br>The Engine Control Module (ECM) communicates with all databus-capable control modules via a CAN databus. These databus-capable control modules are connected via two data bus wires which are twisted together (CAN_High and CAN_Low), and exchange information (messages). Missing information on the databus is recognized as a malfunction and stored. Trouble-free operation of the CAN-bus requires that it have a terminal resistance. This central terminal resistor is located in the Engine Control Module (ECM).<br>**Possible Causes:**<br>• Check the Terminal resistance for CAN-bus<br>• Data-Bus wires have short<br>• Data-Bus components are malfunctioning<br>• ECM has failed |
| **DTC: P1853**<br>**Years: 2007, 2008**<br>**Models:** Passat, Jetta, Golf, GTI, New Beetle<br>**Engines:** 2.0L<br>**Transmissions**: All | **Data BUS Powertrain Implausible Message from Brake Control Conditions:**<br>The Engine Control Module (ECM) communicates with all databus-capable control modules via a CAN databus. These databus-capable control modules are connected via two data bus wires which are twisted together (CAN_High and CAN_Low), and exchange information (messages). Missing information on the databus is recognized as a malfunction and stored. Trouble-free operation of the CAN-bus requires that it have a terminal resistance. This central terminal resistor is located in the Engine Control Module (ECM).<br>**Possible Causes:**<br>• Check the Terminal resistance for CAN-bus<br>• Data-Bus wires have short<br>• Data-Bus components are malfunctioning<br>• ECM has failed |

| DTC | Trouble Code Title, Conditions & Possible Causes |
|---|---|
| **DTC: P1854**<br>**Years:** 2007, 2008<br>**Models:** Passat, Jetta, Golf, GTI, New Beetle, New Beetle Convertible<br>**Engines:** 2.0L, 2.5L, 3.6L<br>**Transmissions:** All | **Data BUS Powertrain Hardware Defective Conditions:**<br>The Engine Control Module (ECM) communicates with all databus-capable control modules via a CAN databus. These databus-capable control modules are connected via two data bus wires which are twisted together (CAN_High and CAN_Low), and exchange information (messages). Missing information on the databus is recognized as a malfunction and stored. Trouble-free operation of the CAN-bus requires that it have a terminal resistance. This central terminal resistor is located in the Engine Control Module (ECM).<br>**Possible Causes:**<br>    • Check the Terminal resistance for CAN-bus<br>    • Data-Bus wires have short<br>    • Data-Bus components are malfunctioning<br>    • ECM has failed |
| **DTC: P1855**<br>**Years:** 2007, 2008<br>**Models:** Passat, Jetta, Golf, GTI, New Beetle, New Beetle Convertible<br>**Engines:** 2.0L, 2.5L, 3.6L<br>**Transmissions:** All | **Data BUS Powertrain Software Version Control Conditions:**<br>The Engine Control Module (ECM) communicates with all databus-capable control modules via a CAN databus. These databus-capable control modules are connected via two data bus wires which are twisted together (CAN_High and CAN_Low), and exchange information (messages). Missing information on the databus is recognized as a malfunction and stored. Trouble-free operation of the CAN-bus requires that it have a terminal resistance. This central terminal resistor is located in the Engine Control Module (ECM).<br>**Possible Causes:**<br>    • Check the Terminal resistance for CAN-bus<br>    • Data-Bus wires have short<br>    • Data-Bus components are malfunctioning<br>    • ECM has failed |
| **DTC: P1857**<br>**Years:** 2007, 2008<br>**Models:** Passat, Jetta, Golf, GTI, New Beetle, New Beetle Convertible<br>**Engines:** 2.0L, 2.5L, 3.6L<br>**Transmissions:** All | **Load Signal Error Message From Engine Control Conditions:**<br>The Engine Control Module (ECM) communicates with all databus-capable control modules via a CAN databus. These databus-capable control modules are connected via two data bus wires which are twisted together (CAN_High and CAN_Low), and exchange information (messages). Missing information on the databus is recognized as a malfunction and stored. Trouble-free operation of the CAN-bus requires that it have a terminal resistance. This central terminal resistor is located in the Engine Control Module (ECM).<br>**Possible Causes:**<br>    • Intake Manifold Runner Position Sensor is faulty<br>    • Intake system has leaks (false air)<br>    • Motor for intake flap is faulty<br>    • Mass Air Flow (MAF) sensor has failed<br>    • ECM has failed |
| **DTC: P1861**<br>**Years:** 2007, 2008<br>**Models:** Passat, Jetta, Golf, GTI, New Beetle, New Beetle Convertible<br>**Engines:** 2.0L 2.5L, 3.6L<br>**Transmissions:** All | **Throttle Position Sensor Message from ECM Conditions:**<br>The Engine Control Module (ECM) communicates with all databus-capable control modules via a CAN databus. These databus-capable control modules are connected via two data bus wires which are twisted together (CAN_High and CAN_Low), and exchange information (messages). Missing information on the databus is recognized as a malfunction and stored. Trouble-free operation of the CAN-bus requires that it have a terminal resistance. This central terminal resistor is located in the Engine Control Module (ECM).<br>**Note: Both the Throttle Position (TP) Sensor and Accelerator Pedal Position Sensor 2 are located at the accelerator pedal module and communicate the driver's intentions to the ECM completely independently of each other. Both sensors are stored in one housing.**<br>**Possible Causes:**<br>    • Throttle Position (TP) Sensor has failed<br>    • Accelerator Pedal Position Sensor 2 has failed<br>    • ECM has failed<br>    • Ground (GND) connections between engine and chassis must be OK<br>    • Engine Control Module (ECM) may not connected |
| **DTC: P1866**<br>**Years:** 2007, 2008<br>**Models:** Passat, Jetta, Golf, GTI, New Beetle, New Beetle Convertible<br>**Engines:** 2.0L, 2.5L, 3.6L<br>**Transmissions:** All | **Data Bus Powertrain Missing Messages Conditions:**<br>The Engine Control Module (ECM) communicates with all databus-capable control modules via a CAN databus. These databus-capable control modules are connected via two data bus wires which are twisted together (CAN_High and CAN_Low), and exchange information (messages). Missing information on the databus is recognized as a malfunction and stored. Trouble-free operation of the CAN-bus requires that it have a terminal resistance. This central terminal resistor is located in the Engine Control Module (ECM).<br>**Possible Causes:**<br>    • Check the Terminal resistance for CAN-bus<br>    • Data-Bus wires have short<br>    • Data-Bus components are malfunctioning<br>    • ECM has failed |
| **DTC: P1912**<br>**Years:** 2007, 2008<br>**Models:** Passat, Jetta, GTI<br>**Engines:** 2.5L, 3.6L<br>**Transmissions:** All | **Brake Booster Pressure Sensor Short Circuit to B+ Conditions:**<br>Key on or engine running, the ECM detected an error with the brake booster pressure sensor signal.<br>**Possible Causes:**<br>    • Circuit short to ground or open<br>    • Brake booster pressure sensor has failed or is dirty<br>    • ECM has failed |

| DTC | Trouble Code Title, Conditions & Possible Causes |
|---|---|
| **DTC: P1913**<br>**Years: 2007, 2008**<br>**Models:** Passat, Jetta, GTI<br>**Engines:** 2.5L, 3.6L<br>**Transmissions:** All | **Brake Booster Pressure Sensor Short Circuit to Ground Conditions:**<br>Key on or engine running, the ECM detected an error with the brake booster pressure sensor signal.<br>**Possible Causes:**<br>• Circuit short to ground or open<br>• Brake booster pressure sensor has failed or is dirty<br>• ECM has failed |
| **DTC: P2004**<br>**Years: 2007, 2008**<br>**Models:** Passat<br>**Engines:** 2.0L<br>**Transmissions:** All | **Intake Manifold Runner Control Stuck Open Bank 1 Conditions:**<br>Engine started, battery voltage must be at least 11.5v, all electrical components must be off, the ground between the engine and the chassis must be well connected. The ECM detected an unexpected voltage condition on the Intake Manifold Runner Control circuit during the CCM test period (i.e., the valve may be stuck open).<br>**Note: Intake Flap Motor and Intake Manifold Runner Position Sensor are one component and cannot be replaced individually.**<br>**Possible Causes:**<br>• Test for a sticking Accelerator or speed control cable condition: Turn the key off and disconnect accelerator and speed control cable from the throttle body. Rotate the throttle body linkage to determine if it rotates freely (the throttle body may have failed).<br>• Check the air cleaner and air inlet assembly for restrictions<br>• Check the IAC motor response (it may be damaged or sticking)<br>• Check the PCV system (valve and hoses) for leaks or plugging<br>• Check for signs of vacuum leaks in the engine or components<br>• Test TP sensor signal (due a sweep test at key on, engine off) |

## Gas Engine OBD II Trouble Code List (P2xxx Codes)

| DTC | Trouble Code Title, Conditions & Possible Causes |
|---|---|
| **DTC: P2008**<br>**Years: 2007, 2008**<br>**Models:** Passat, Jetta<br>**Engines:** 2.0L, 2.5L<br>**Transmissions:** All | **Intake Manifold Runner Control Circuit/Open Bank 1 Conditions:**<br>Engine started, battery voltage must be at least 11.5v, all electrical components must be off, the ground between the engine and the chassis must be well connected. The ECM detected an unexpected voltage condition on the Intake Manifold Runner Control circuit during the CCM test period (i.e., the valve may be stuck open).<br>**Note: Intake Flap Motor and Intake Manifold Runner Position Sensor are one component and cannot be replaced individually.**<br>**Possible Causes:**<br>• Accelerator or speed control cable sticking or binding. To test for this condition, turn the key off. Then disconnect the accelerator and speed control cable from the throttle body. Then rotate the throttle body linkage to determine if it rotates freely. If it is sticking, the throttle body may need replacement.<br>• Check the air cleaner and air inlet assembly for restrictions<br>• Check the IAC motor response (it may be damaged or sticking)<br>• Check the PCV system (valve and hoses) for leaks or plugging<br>• Check for signs of vacuum leaks in the engine or components<br>• Test TP sensor signal |
| **DTC: P2009**<br>**Years: 2007, 2008**<br>**Models:** Passat<br>**Engines:** 2.0L<br>**Transmissions:** All | **Intake Manifold Runner Control Circuit Low Bank 1 Conditions:**<br>Engine started, battery voltage must be at least 11.5v, all electrical components must be off, the ground between the engine and the chassis must be well connected. The ECM detected an unexpected voltage condition on the Intake Manifold Runner Control circuit during the CCM test period (i.e., the valve may be stuck open).<br>**Note: Intake Flap Motor and Intake Manifold Runner Position Sensor are one component and cannot be replaced individually.**<br>**Possible Causes:**<br>• Accelerator or speed control cable sticking or binding. To test for this condition, turn the key off. Then disconnect the accelerator and speed control cable from the throttle body. Then rotate the throttle body linkage to determine if it rotates freely. If it is sticking, the throttle body may need replacement.<br>• Check the air cleaner and air inlet assembly for restrictions<br>• Check the IAC motor response (it may be damaged or sticking)<br>• Check the PCV system (valve and hoses) for leaks or plugging<br>• Check for signs of vacuum leaks in the engine or components<br>• Test TP sensor signal |

| DTC | Trouble Code Title, Conditions & Possible Causes |
|-----|--------------------------------------------------|
| **DTC: P2014**<br>**Years:** 2007, 2008<br>**Models:** Passat<br>**Engines:** 2.0L<br>**Transmissions:** All | **Intake Manifold Runner Position Sensor/Switch Circuit Bank 1 Conditions:**<br>Engine started, battery voltage must be at least 11.5v, all electrical components must be off, the ground between the engine and the chassis must be well connected. The ECM detected an unexpected voltage condition on the Intake Manifold Runner Control circuit during the CCM test period (i.e., the valve may be stuck open).<br>**Note: Intake Flap Motor and Intake Manifold Runner Position Sensor are one component and cannot be replaced individually.**<br>**Possible Causes:**<br>• Accelerator or speed control cable sticking or binding. To test for this condition, turn the key off. Then disconnect the accelerator and speed control cable from the throttle body. Then rotate the throttle body linkage to determine if it rotates freely. If it is sticking, the throttle body may need replacement.<br>• Check the air cleaner and air inlet assembly for restrictions<br>• Check the IAC motor response (it may be damaged or sticking)<br>• Check the PCV system (valve and hoses) for leaks or plugging<br>• Check for signs of vacuum leaks in the engine or components<br>• Test TP sensor signal |
| **DTC: P2015**<br>**Years:** 2007, 2008<br>**Models:** Passat<br>**Engines:** 2.0L<br>**Transmissions:** All | **Intake Manifold Runner Position Sensor/Switch Circuit Range/Performance Bank 1 Conditions:**<br>Engine started, battery voltage must be at least 11.5v, all electrical components must be off, the ground between the engine and the chassis must be well connected. The ECM detected an unexpected voltage condition on the Intake Manifold Runner Control circuit during the CCM test period (i.e., the valve may be stuck open).<br>**Note: Intake Flap Motor and Intake Manifold Runner Position Sensor are one component and cannot be replaced individually.**<br>**Possible Causes:**<br>• Accelerator or speed control cable sticking or binding. To test for this condition, turn the key off. Then disconnect the accelerator and speed control cable from the throttle body. Then rotate the throttle body linkage to determine if it rotates freely. If it is sticking, the throttle body may need replacement.<br>• Check the air cleaner and air inlet assembly for restrictions<br>• Check the IAC motor response (it may be damaged or sticking)<br>• Check the PCV system (valve and hoses) for leaks or plugging<br>• Check for signs of vacuum leaks in the engine or components<br>• Test TP sensor signal |
| **DTC: P2016**<br>**Years:** 2007, 2008<br>**Models:** Passat<br>**Engines:** 2.0L<br>**Transmissions:** All | **Intake Manifold Runner Position Sensor/Switch Circuit Low Bank 1 Conditions:**<br>Engine started, battery voltage must be at least 11.5v, all electrical components must be off, the ground between the engine and the chassis must be well connected. The ECM detected an unexpected voltage condition on the Intake Manifold Runner Control circuit during the CCM test period (i.e., the valve may be stuck open).<br>**Note: Intake Flap Motor and Intake Manifold Runner Position Sensor are one component and cannot be replaced individually.**<br>Accelerator or speed control cable sticking or binding. To test for this condition, turn the key off. Then disconnect the accelerator and speed control cable from the throttle body. Then rotate the throttle body linkage to determine if it rotates freely. If it is sticking, the throttle body may need replacement.<br>**Possible Causes:**<br>• Check the air cleaner and air inlet assembly for restrictions<br>• Check the IAC motor response (it may be damaged or sticking)<br>• Check the PCV system (valve and hoses) for leaks or plugging<br>• Check for signs of vacuum leaks in the engine or components<br>• Test TP sensor signal |
| **DTC: P2017**<br>**Years:** 2007, 2008<br>**Models:** Passat<br>**Engines:** 2.0L<br>**Transmissions:** All | **Intake Manifold Runner Position Sensor/Switch Circuit High Bank 1 Conditions:**<br>Engine started, battery voltage must be at least 11.5v, all electrical components must be off, the ground between the engine and the chassis must be well connected. The ECM detected an unexpected voltage condition on the Intake Manifold Runner Control circuit during the CCM test period (i.e., the valve may be stuck open).<br>**Note: Intake Flap Motor and Intake Manifold Runner Position Sensor are one component and cannot be replaced individually.**<br>**Possible Causes:**<br>• Accelerator or speed control cable sticking or binding. To test for this condition, turn the key off. Then disconnect the accelerator and speed control cable from the throttle body. Then rotate the throttle body linkage to determine if it rotates freely. If it is sticking, the throttle body may need replacement.<br>• Check the air cleaner and air inlet assembly for restrictions<br>• Check the IAC motor response (it may be damaged or sticking)<br>• Check the PCV system (valve and hoses) for leaks or plugging<br>• Check for signs of vacuum leaks in the engine or components<br>• Test TP sensor signal |

| DTC | Trouble Code Title, Conditions & Possible Causes |
|---|---|
| **DTC: P2088**<br>**2T ECM, MIL: No**<br>**Years: 2007, 2008**<br>**Models:** Passat, New Beetle<br>**Engines:** 2.0L<br>**Transmissions:** All | **"A" Camshaft Position Control Circuit Low Bank 1 Conditions:**<br>Key on or engine running; and the ECM detected an unexpected voltage condition on the Camshaft Position Control circuit during the CCM test period. The relative position between the camshaft and crankshaft needs to be optimal so the engine has better torque, fuel economy and emissions.<br>**Note: camshaft adjustment is load- and RPM dependant. The electrical camshaft adjustment valve 1 switches oil pressure onto camshaft adjuster (mechanical adjustment mechanism), which adjusts the camshaft.**<br>**Possible Causes:**<br>• Camshaft position control wiring harness connector is damaged or open<br>• Camshaft adjustment valve has failed<br>• Circuit is open or grounded<br>• Assembly is damaged or it has failed (an open circuit)<br>• ECM power supply relay has failed<br>• ECM has failed |
| **DTC: P2089**<br>**2T ECM, MIL: No**<br>**Years: 2007, 2008**<br>**Models:** Passat, Jetta, New Beetle<br>**Engines:** 2.5L<br>**Transmissions:** All | **"A" Camshaft Position Control Circuit High Bank 1 Conditions:**<br>Key on or engine running; and the ECM detected an unexpected voltage condition on the Camshaft Position Control circuit during the CCM test period. The relative position between the camshaft and crankshaft needs to be optimal so the engine has better torque, fuel economy and emissions.<br>**Note: camshaft adjustment is load- and RPM dependant. The electrical camshaft adjustment valve 1 switches oil pressure onto camshaft adjuster (mechanical adjustment mechanism), which adjusts the camshaft.**<br>**Possible Causes:**<br>• Camshaft position control wiring harness connector is damaged or open<br>• Camshaft adjustment valve has failed<br>• Circuit is open or grounded<br>• Assembly is damaged or it has failed (an open circuit)<br>• ECM power supply relay has failed<br>• ECM has failed |
| **DTC: P2090**<br>**2T ECM, MIL: No**<br>**Years: 2007, 2008**<br>**Models:** Passat, Jetta, GTi<br>**Engines:** 2.0L, 2.5L, 3.6L<br>**Transmissions:** All | **"B" Camshaft Position Control Circuit Low Bank 1 Conditions:**<br>Key on or engine running; and the ECM detected an unexpected voltage condition on the Camshaft Position Control circuit during the CCM test period. The relative position between the camshaft and crankshaft needs to be optimal so the engine has better torque, fuel economy and emissions.<br>**Note: camshaft adjustment is load- and RPM dependant. The electrical camshaft adjustment valve 1 switches oil pressure onto camshaft adjuster (mechanical adjustment mechanism), which adjusts the camshaft.**<br>**Possible Causes:**<br>• Camshaft position control wiring harness connector is damaged or open<br>• Camshaft adjustment valve has failed<br>• Circuit is open or grounded<br>• Assembly is damaged or it has failed (an open circuit)<br>• ECM power supply relay has failed<br>• ECM has failed |
| **DTC: P2091**<br>**2T ECM, MIL: No**<br>**Years: 2007, 2008**<br>**Models:** Passat, Jetta, GTi<br>**Engines:** 2.0L, 2.5L, 3.6L<br>**Transmissions:** All | **"B" Camshaft Position Control Circuit High Bank 1 Conditions:**<br>Key on or engine running; and the ECM detected an unexpected voltage condition on the Camshaft Position Control circuit during the CCM test period. The relative position between the camshaft and crankshaft needs to be optimal so the engine has better torque, fuel economy and emissions.<br>**Note: camshaft adjustment is load- and RPM dependant. The electrical camshaft adjustment valve 1 switches oil pressure onto camshaft adjuster (mechanical adjustment mechanism), which adjusts the camshaft.**<br>**Possible Causes:**<br>• Camshaft position control wiring harness connector is damaged or open<br>• Camshaft adjustment valve has failed<br>• Circuit is open or grounded<br>• Assembly is damaged or it has failed (an open circuit)<br>• ECM power supply relay has failed<br>• ECM has failed |

| DTC | Trouble Code Title, Conditions & Possible Causes |
|---|---|
| **DTC: P2094**<br>**2T ECM, MIL:** No<br>**Years:** 2007, 2008<br>**Models:** Passat<br>**Engines:** 2.0L<br>**Transmissions:** All | **"B" Camshaft Position Control Circuit Low Bank 2 Conditions:**<br>Key on or engine running; and the ECM detected an unexpected voltage condition on the Camshaft Position Control circuit during the CCM test period. The relative position between the camshaft and crankshaft needs to be optimal so the engine has better torque, fuel economy and emissions.<br>**Note: camshaft adjustment is load- and RPM dependant. The electrical camshaft adjustment valve 1 switches oil pressure onto camshaft adjuster (mechanical adjustment mechanism), which adjusts the camshaft.**<br>**Possible Causes:**<br>• Camshaft position control wiring harness connector is damaged or open<br>• Camshaft adjustment valve has failed<br>• Circuit is open or grounded<br>• Assembly is damaged or it has failed (an open circuit)<br>• ECM power supply relay has failed<br>• ECM has failed |
| **DTC: P2095**<br>**2T ECM, MIL:** No<br>**Years:** 2007, 2008<br>**Models:** Passat<br>**Engines:** 2.0L<br>**Transmissions:** All | **"B" Camshaft Position Control Circuit High Bank 2 Conditions:**<br>Key on or engine running; and the ECM detected an unexpected voltage condition on the Camshaft Position Control circuit during the CCM test period. The relative position between the camshaft and crankshaft needs to be optimal so the engine has better torque, fuel economy and emissions.<br>**Note: camshaft adjustment is load- and RPM dependant. The electrical camshaft adjustment valve 1 switches oil pressure onto camshaft adjuster (mechanical adjustment mechanism), which adjusts the camshaft.**<br>**Possible Causes:**<br>• Camshaft position control wiring harness connector is damaged or open<br>• Camshaft adjustment valve has failed<br>• Circuit is open or grounded<br>• Assembly is damaged or it has failed (an open circuit)<br>• ECM power supply relay has failed<br>• ECM has failed |
| **DTC: P2096**<br>**Years:** 2007, 2008<br>**Models:** Passat, Jetta, Golf, GTI, New Beetle, New Beetle Convertible<br>**Engines:** 2.0L, 2.5L, 3.6L<br>**Transmissions:** All | **Post Catalyst Fuel Trim System Too Lean (Bank 1) Conditions:**<br>Engine started, battery voltage must be at least 11.5v, all electrical components must be off, the ground between the engine and the chassis must be well connected, the exhaust system must be properly sealed between the catalytic converter and the cylinder head, and the oxygen sensor heater for oxygen sensor before the catalytic converter must be properly functioning. The ECM detected a problem with the fuel mixture.<br>**Note: For resistance testing of sensor heating, oxygen sensor should be cooled to ambient temperature. High temperatures at oxygen sensor may lead to inaccurate measurements.**<br>**Possible Causes:**<br>• Oxygen sensor (before catalytic converter) is faulty<br>• Oxygen sensor (behind catalytic converter) is faulty<br>• Oxygen sensor heater (before catalytic converter) is faulty<br>• Oxygen sensor heater (behind catalytic converter) is faulty<br>• Check circuits for shorts to each other, ground or power<br>• ECM has failed |
| **DTC: P2097**<br>**Years:** 2007, 2008<br>**Models:** Passat, Jetta, Golf, GTI, New Beetle, New Beetle Convertible<br>**Engines:** 2.0L, 2.5L, 3.6L<br>**Transmissions:** All | **Post Catalyst Fuel Trim System Too Rich (Bank 1) Conditions:**<br>Engine started, battery voltage must be at least 11.5v, all electrical components must be off, the ground between the engine and the chassis must be well connected, the exhaust system must be properly sealed between the catalytic converter and the cylinder head, and the oxygen sensor heater for oxygen sensor before the catalytic converter must be properly functioning. The ECM detected a problem with the fuel mixture.<br>**Note: For resistance testing of sensor heating, oxygen sensor should be cooled to ambient temperature. High temperatures at oxygen sensor may lead to inaccurate measurements.**<br>**Possible Causes:**<br>• Oxygen sensor (before catalytic converter) is faulty<br>• Oxygen sensor (behind catalytic converter) is faulty<br>• Oxygen sensor heater (before catalytic converter) is faulty<br>• Oxygen sensor heater (behind catalytic converter) is faulty<br>• Check circuits for shorts to each other, ground or power<br>• ECM has failed |

| DTC | Trouble Code Title, Conditions & Possible Causes |
|---|---|
| **DTC: P2098**<br>**Years: 2007, 2008**<br>**Models:** Passat<br>**Engines:** 2.5L, 3.6L<br>**Transmissions:** All | **Post Catalyst Fuel Trim System Too Lean (Bank 2) Conditions:**<br>Engine started, battery voltage must be at least 11.5v, all electrical components must be off, the ground between the engine and the chassis must be well connected, the exhaust system must be properly sealed between the catalytic converter and the cylinder head, and the oxygen sensor heater for oxygen sensor before the catalytic converter must be properly functioning. The ECM detected a problem with the fuel mixture.<br>**Note: For resistance testing of sensor heating, oxygen sensor should be cooled to ambient temperature. High temperatures at oxygen sensor may lead to inaccurate measurements.**<br>**Possible Causes:**<br>• Oxygen sensor (before catalytic converter) is faulty<br>• Oxygen sensor (behind catalytic converter) is faulty<br>• Oxygen sensor heater (before catalytic converter) is faulty<br>• Oxygen sensor heater (behind catalytic converter) is faulty<br>• Check circuits for shorts to each other, ground or power<br>• ECM has failed |
| **DTC: P2099**<br>**Years: 2007, 2008**<br>**Models:** Passat<br>**Engines:** 2.5L, 3.6L<br>**Transmissions:** All | **Post Catalyst Fuel Trim System Too Rich (Bank 2) Conditions:**<br>Engine started, battery voltage must be at least 11.5v, all electrical components must be off, the ground between the engine and the chassis must be well connected, the exhaust system must be properly sealed between the catalytic converter and the cylinder head, and the oxygen sensor heater for oxygen sensor before the catalytic converter must be properly functioning. The ECM detected a problem with the fuel mixture.<br>**Note: For resistance testing of sensor heating, oxygen sensor should be cooled to ambient temperature. High temperatures at oxygen sensor may lead to inaccurate measurements.**<br>**Possible Causes:**<br>• Oxygen sensor (before catalytic converter) is faulty<br>• Oxygen sensor (behind catalytic converter) is faulty<br>• Oxygen sensor heater (before catalytic converter) is faulty<br>• Oxygen sensor heater (behind catalytic converter) is faulty<br>• Check circuits for shorts to each other, ground or power<br>• ECM has failed |
| **DTC: P2101**<br>**2T, MIL: Yes**<br>**Years: 2007, 2008**<br>**Models:** Passat, Jetta, Golf, GTI, New Beetle, New Beetle Convertible<br>**Engines:** 2.0L, 2.5L, 3.6L<br>**Transmissions:** All | **Throttle Actuator Control Motor Range/Performance Conditions:**<br>Engine started, battery voltage must be at least 11.5v, all electrical components must be off, parking brake must be engaged (to keep daytime driving lights off), automatic transmission selector must be in park, the exhaust system must be properly sealed between the catalytic converter and the cylinder head, coolant temperature must be at least 80 degrees Celsius. The ECM detected an unexpected low or high voltage condition on the Throttle Actuator Control Motor (TACM) circuit during the CCM test.<br>**Note: The throttle valve activation occurs via an electric motor (throttle drive) in the throttle valve control module. It is activated by the Engine Control Module (ECM) according to specifications of the two sensors, Throttle Position (TP) Sensor and Sender 2 for accelerator pedal position.**<br>**Possible Causes:**<br>• TACM wiring harness connector is damaged or open<br>• TACM wiring may be crossed in the wire harness assembly<br>• TACM (motor) circuit is open, or TACM assembly is damaged (possible open circuit)<br>• TACM or the Throttle Valve is dirty<br>• Throttle Position sensor has failed<br>• ECM has failed |
| **DTC: P2106**<br>**Years: 2007, 2008**<br>**Models:** Passat, Jetta, Golf, GTI, New Beetle, New Beetle Convertible<br>**Engines:** 2.0L, 2.5L, 3.6L<br>**Transmissions:** All | **Throttle Actuator Control System – Forced Limited Power Conditions**<br>Engine started, battery voltage must be at least 11.5v, all electrical components must be off, parking brake must be engaged (to keep daytime driving lights off), automatic transmission selector must be in park, the exhaust system must be properly sealed between the catalytic converter and the cylinder head, coolant temperature must be at least 80 degrees Celsius. The ECM detected an unexpected low or high voltage condition on the Throttle Actuator Control Motor (TACM) circuit during the CCM test.<br>**Note: The throttle valve activation occurs via an electric motor (throttle drive) in the throttle valve control module. It is activated by the Engine Control Module (ECM) according to specifications of the two sensors, Throttle Position (TP) Sensor and Sender 2 for accelerator pedal position.**<br>**Possible Causes:**<br>• TACM wiring harness connector is damaged or open<br>• TACM wiring may be crossed in the wire harness assembly<br>• TACM (motor) circuit is open, or TACM assembly is damaged (possible open circuit)<br>• TACM or the Throttle Valve is dirty<br>• Throttle Position sensor has failed<br>• ECM has failed |

| DTC | Trouble Code Title, Conditions & Possible Causes |
|---|---|
| **DTC: P2108**<br>**2T ECM, MIL: No**<br>**Years: 2007, 2008**<br>**Models:** Passat<br>**Engines:** 2.0L<br>**Transmissions**: All | **Throttle Actuator Control Motor Performance Conditions:**<br>Engine started, battery voltage must be at least 11.5v, all electrical components must be off, parking brake must be engaged (to keep daytime driving lights off), automatic transmission selector must be in park, the exhaust system must be properly sealed between the catalytic converter and the cylinder head, coolant temperature must be at least 80 degrees Celsius. The ECM detected an unexpected low or high voltage condition on the Throttle Actuator Control Motor (TACM) circuit during the CCM test.<br>**Note: The throttle valve activation occurs via an electric motor (throttle drive) in the throttle valve control module. It is activated by the Engine Control Module (ECM) according to specifications of the two sensors, Throttle Position (TP) Sensor and Sender 2 for accelerator pedal position.**<br>**Possible Causes:**<br>• TACM wiring harness connector is damaged or open<br>• TACM wiring may be crossed in the wire harness assembly<br>• TACM (motor) circuit is open, or TACM assembly is damaged (possible open circuit)<br>• TACM or the Throttle Valve is dirty<br>• Throttle Position sensor has failed<br>• ECM has failed |
| **DTC: P2110**<br>**2T ECM, MIL: No**<br>**Years: 2007, 2008**<br>**Models:** Passat<br>**Engines:** 2.0L<br>**Transmissions**: All | **Throttle Actuator Control System–Forced Limited RPM Conditions:**<br>Engine started, battery voltage must be at least 11.5v, all electrical components must be off, parking brake must be engaged (to keep daytime driving lights off), automatic transmission selector must be in park, the exhaust system must be properly sealed between the catalytic converter and the cylinder head, coolant temperature must be at least 80 degrees Celsius. The ECM detected an unexpected low or high voltage condition on the Throttle Actuator Control Motor (TACM) circuit during the CCM test.<br>**Note: The throttle valve activation occurs via an electric motor (throttle drive) in the throttle valve control module. It is activated by the Engine Control Module (ECM) according to specifications of the two sensors, Throttle Position (TP) Sensor and Sender 2 for accelerator pedal position.**<br>**Possible Causes:**<br>• TACM wiring harness connector is damaged or open<br>• TACM wiring may be crossed in the wire harness assembly<br>• TACM (motor) circuit is open, or TACM assembly is damaged (possible open circuit)<br>• TACM or the Throttle Valve is dirty<br>• Throttle Position sensor has failed<br>• ECM has failed |
| **DTC: P2122**<br>**2T, MIL: Yes**<br>**Years: 2007, 2008**<br>**Models:** Passat, Jetta, Golf, GTI, New Beetle, New Beetle Convertible<br>**Engines:** 2.0L, 2.5L, 3.6L<br>**Transmissions**: All | **Accelerator Pedal Position Sensor 'D' Circuit Low Input Conditions:**<br>Engine started, battery voltage at least 11.5v, all electrical components off, ground connections between engine and chassis well connected, the ECM detected that the accelerator pedal position sensor signal was outside the parameters to function normally.<br>**Note: Both the Throttle Position (TP) Sensor and Accelerator Pedal Position Sensor are located at the accelerator pedal module and communicate the driver's intentions to the ECM completely independently of each other. Both sensors are stored in one housing.**<br>**Possible Causes:**<br>• Ground between engine and chassis may be broken<br>• Throttle position sensor may have failed<br>• Accelerator Pedal Position Sensor has failed<br>• Throttle position sensor wiring may have shorted<br>• Throttle position sensor has failed<br>• Faulty voltage supply<br>• ECM has failed |
| **DTC: P2123**<br>**2T, MIL: Yes**<br>**Years: 2007, 2008**<br>**Models:** Passat, Jetta, Golf, GTI, New Beetle, New Beetle Convertible<br>**Engines:** 2.0L, 2.5L, 3.6L<br>**Transmissions**: All | **Accelerator Pedal Position Sensor 'D' Circuit High Input Conditions:**<br>Engine started, battery voltage at least 11.5v, all electrical components off, ground connections between engine and chassis well connected, the ECM detected that the accelerator pedal position sensor signal was outside the parameters to function normally.<br>**Note: Both the Throttle Position (TP) Sensor and Accelerator Pedal Position Sensor are located at the accelerator pedal module and communicate the driver's intentions to the ECM completely independently of each other. Both sensors are stored in one housing.**<br>**Possible Causes:**<br>• Ground between engine and chassis may be broken<br>• Throttle position sensor may have failed<br>• Accelerator Pedal Position Sensor has failed<br>• Throttle position sensor wiring may have shorted<br>• Throttle position sensor has failed<br>• Faulty voltage supply<br>• ECM has failed |

| DTC | Trouble Code Title, Conditions & Possible Causes |
|---|---|
| **DTC: P2127**<br>**2T, MIL: Yes**<br>**Years:** 2007, 2008<br>**Models:** Passat, Jetta, Golf, GTI, New Beetle, New Beetle Convertible<br>**Engines:** 2.0L, 2.5L, 3.6L<br>**Transmissions:** All | **Accelerator Pedal Position Sensor 'E' Circuit Low Input Conditions:**<br>Engine started, battery voltage at least 11.5v, all electrical components off, ground connections between engine and chassis well connected, the ECM detected that the accelerator pedal position sensor signal was outside the parameters to function normally.<br>**Note: Both the Throttle Position (TP) Sensor and Accelerator Pedal Position Sensor are located at the accelerator pedal module and communicate the driver's intentions to the ECM completely independently of each other. Both sensors are stored in one housing.**<br>**Possible Causes:**<br>    • Ground between engine and chassis may be broken<br>    • Throttle position sensor may have failed<br>    • Accelerator Pedal Position Sensor has failed<br>    • Throttle position sensor wiring may have shorted<br>    • Throttle position sensor has failed<br>    • Faulty voltage supply<br>    • ECM has failed |
| **DTC: P2128**<br>**2T, MIL: Yes**<br>**Years:** 2007, 2008<br>**Models:** Passat, Jetta, Golf, GTI, New Beetle, New Beetle Convertible<br>**Engines:** 2.0L, 2.5L, 3.6L<br>**Transmissions:** All | **Accelerator Pedal Position Sensor 'E' Circuit High Input Conditions:**<br>Engine started, battery voltage at least 11.5v, all electrical components off, ground connections between engine and chassis well connected, the ECM detected that the accelerator pedal position sensor signal was outside the parameters to function normally.<br>**Note: Both the Throttle Position (TP) Sensor and Accelerator Pedal Position Sensor are located at the accelerator pedal module and communicate the driver's intentions to the ECM completely independently of each other. Both sensors are stored in one housing.**<br>**Possible Causes:**<br>    • Ground between engine and chassis may be broken<br>    • Throttle position sensor may have failed<br>    • Accelerator Pedal Position Sensor has failed<br>    • Throttle position sensor wiring may have shorted<br>    • Throttle position sensor has failed<br>    • Faulty voltage supply<br>    • ECM has failed |
| **DTC: P2133**<br>**2T, MIL: Yes**<br>**Years:** 2007, 2008<br>**Models:** Jetta, Golf, GTI, New Beetle, New Beetle Convertible<br>**Engines:** 2.0L<br>**Transmissions:** All | **Accelerator Pedal Position Sensor 'F' Circuit High Input Conditions:**<br>Engine started, battery voltage at least 11.5v, all electrical components off, ground connections between engine and chassis well connected, the ECM detected that the accelerator pedal position sensor signal was outside the parameters to function normally.<br>**Note: Both the Throttle Position (TP) Sensor and Accelerator Pedal Position Sensor are located at the accelerator pedal module and communicate the driver's intentions to the ECM completely independently of each other. Both sensors are stored in one housing.**<br>**Possible Causes:**<br>    • Ground between engine and chassis may be broken<br>    • Throttle position sensor may have failed<br>    • Accelerator Pedal Position Sensor has failed<br>    • Throttle position sensor wiring may have shorted<br>    • Throttle position sensor has failed<br>    • Faulty voltage supply<br>    • ECM has failed |
| **DTC: P2138**<br>**2T, MIL: Yes**<br>**Years:** 2007, 2008<br>**Models:** Passat, Jetta, Golf, GTI, New Beetle, New Beetle Convertible<br>**Engines:** 2.0L, 2.5L, 3.6L<br>**Transmissions:** All | **Throttle Position Sensor D/E Voltage Correlation Conditions:**<br>Engine started, battery voltage must be at least 11.5v, all electrical components must be off, parking brake must be engaged (to keep daytime driving lights off), automatic transmission selector must be in park; and the ECM detected the Throttle Position 'D' (TPD) and Throttle Position 'B' (TPE) sensors disagreed, or that the TPD sensor should not be in its detected position, or that the TPE sensor should not be in its detected position during testing.<br>**Note: Both the Throttle Position (TP) Sensor and Accelerator Pedal Position Sensor are located at the accelerator pedal module and communicate the driver's intentions to the ECM completely independently of each other. Both sensors are stored in one housing.**<br>**Possible Causes:**<br>    • ETC TP sensor connector is damaged or shorted<br>    • ETC TP sensor circuits shorted together in the wire harness<br>    • ETC TP sensor signal circuit is shorted to VREF (5v)<br>    • ETC TP sensor is damaged or the ECM has failed |

| DTC | Trouble Code Title, Conditions & Possible Causes |
|---|---|
| **DTC: P2146**<br>**Years:** 2007, 2008<br>**Models:** Passat, Jetta<br>**Engines:** 2.0L<br>**Transmissions:** All | **Fuel Injector Group "A" Supply Voltage Circuit/Open Conditions:**<br>Engine started, battery voltage must be at least 11.5v, all electrical components must be off, the ground between the engine and the chassis must be well connected, the exhaust system must be properly sealed between the catalytic converter and the cylinder head, and the oxygen sensor heater for oxygen sensor before the catalytic converter must be properly functioning. The ECM detected the fuel injector supply voltage circuit was outside the normal range during the test period.<br>**Note:** For resistance testing of sensor heating, oxygen sensor should be cooled to ambient temperature. High temperatures at oxygen sensor may lead to inaccurate measurements.<br>**Possible Causes:**<br>• Oxygen sensor (before catalytic converter) is faulty<br>• Oxygen sensor (behind catalytic converter) is faulty<br>• Oxygen sensor heater (before catalytic converter) is faulty<br>• Oxygen sensor heater (behind catalytic converter) is faulty<br>• Check circuits for shorts to each other, ground or power<br>• Fuel Injector(s) may have failed<br>• ECM has failed |
| **DTC: P2149**<br>**Years:** 2007, 2008<br>**Models:** Passat, Jetta<br>**Engines:** 2.0L<br>**Transmissions:** All | **Fuel Injector Group "B" Supply Voltage Circuit/Open Conditions:**<br>Engine started, battery voltage must be at least 11.5v, all electrical components must be off, the ground between the engine and the chassis must be well connected, the exhaust system must be properly sealed between the catalytic converter and the cylinder head, and the oxygen sensor heater for oxygen sensor before the catalytic converter must be properly functioning. The ECM detected the fuel injector supply voltage circuit was outside the normal range during the test period.<br>**Note:** For resistance testing of sensor heating, oxygen sensor should be cooled to ambient temperature. High temperatures at oxygen sensor may lead to inaccurate measurements.<br>**Possible Causes:**<br>• Oxygen sensor (before catalytic converter) is faulty<br>• Oxygen sensor (behind catalytic converter) is faulty<br>• Oxygen sensor heater (before catalytic converter) is faulty<br>• Oxygen sensor heater (behind catalytic converter) is faulty<br>• Check circuits for shorts to each other, ground or power<br>• Fuel Injector(s) may have failed<br>• ECM has failed |
| **DTC: P2177**<br>**2T, MIL: Yes**<br>**Years:** 2007, 2008<br>**Models:** Jetta, Golf, GTI, New Beetle, New Beetle Convertible<br>**Engines:** 2.0L, 2.5L, 3.6L<br>**Transmissions:** All | **System Too Lean Off Idle Bank 1 Conditions:**<br>Engine started, battery voltage must be at least 11.5v, all electrical components must be off, the ground between the engine and the chassis must be well connected, the exhaust system must be properly sealed between the catalytic converter and the cylinder head, and the oxygen sensor heater for oxygen sensor before the catalytic converter must be properly functioning. The ECM detected the system indicated a lean signal, or it could no longer control bank 1 because it was at its lean limit.<br>**Possible Causes:**<br>• Intake Manifold Runner Position Sensor has failed<br>• Intake system has leaks (false air)<br>• Motor for intake flap is faulty<br>• Oxygen sensor (before catalytic converter) is faulty<br>• Oxygen sensor (behind catalytic converter) is faulty<br>• Oxygen sensor heater (before catalytic converter) is faulty<br>• Oxygen sensor heater (behind catalytic converter) is faulty<br>• Check circuits for shorts to each other, ground or power<br>• Fuel Injector(s) may have failed<br>• ECM has failed |
| **DTC: P2178**<br>**2T, MIL: Yes**<br>**Years:** 2007, 2008<br>**Models:** Jetta, Golf, GTI, New Beetle, New Beetle Convertible<br>**Engines:** 2.0L, 2.5L, 3.6L<br>**Transmissions:** All | **System Too Rich Off Idle Bank 1 Conditions:**<br>Engine started, battery voltage must be at least 11.5v, all electrical components must be off, the ground between the engine and the chassis must be well connected, the exhaust system must be properly sealed between the catalytic converter and the cylinder head, and the oxygen sensor heater for oxygen sensor before the catalytic converter must be properly functioning. The ECM detected the system indicated a rich signal, or it could no longer control bank 1 because it was at its rich limit.<br>**Possible Causes:**<br>• Intake Manifold Runner Position Sensor has failed<br>• Intake system has leaks (false air)<br>• Motor for intake flap is faulty<br>• Oxygen sensor (before catalytic converter) is faulty<br>• Oxygen sensor (behind catalytic converter) is faulty<br>• Oxygen sensor heater (before catalytic converter) is faulty<br>• Oxygen sensor heater (behind catalytic converter) is faulty<br>• Check circuits for shorts to each other, ground or power<br>• Fuel Injector(s) may have failed<br>• ECM has failed |

| DTC | Trouble Code Title, Conditions & Possible Causes |
|---|---|
| **DTC: P2181**<br>**2T, MIL: Yes**<br>**Years: 2007, 2008**<br>**Models:** Passat, Jetta, Golf, GTI, New Beetle, New Beetle Convertible<br>**Engines:** 2.0L, 2.5L, 3.6L<br>**Transmissions**: All | **Cooling System Performance Malfunction Conditions:**<br>Key on, engine cold; and the Engine Coolant Temperature (ECM) detected the ECT sensor signal was more or less than the self-test limits or has failed to gain a signal. This is a thermistor-type sensor with a variable resistance that changes when exposed to different temperatures<br>**Possible Causes:**<br>• ECT sensor has failed<br>• ECT Sensor (on Radiator) has failed<br>• ECT sensor signal circuit is open (inspect wiring & connector)<br>• ECT sensor signal circuit is shorted<br>• Cooling system malfunction, or the thermostat is stuck<br>• Engine not operating at normal operating temperature<br>• EOT sensor is damaged or it has failed |
| **DTC: P2184**<br>**Years: 2007, 2008**<br>**Models:** Passat, Jetta,<br>**Engines:** 2.0L, 2.5L, 3.6L<br>**Transmissions**: All | **Engine Coolant Temperature Sensor 2 Circuit Low Conditions:**<br>Key on or engine running; and the Engine Coolant Temperature (ECM) detected the ECT sensor signal was less than the self-test minimum. This is a thermistor-type sensor with a variable resistance that changes when exposed to different temperatures<br>**Possible Causes:**<br>• ECT sensor has failed<br>• ECT Sensor (on Radiator) has failed<br>• ECT sensor signal circuit is open (inspect wiring & connector)<br>• ECT sensor signal circuit is shorted<br>• Cooling system malfunction, or the thermostat is stuck<br>• Engine not operating at normal operating temperature<br>• EOT sensor is damaged or it has failed |
| **DTC: P2185**<br>**Years: 2007, 2008**<br>**Models:** Passat, Jetta,<br>**Engines:** 2.0L, 2.5L, 3.6L<br>**Transmissions**: All | **Engine Coolant Temperature Sensor 2 Circuit High Conditions:**<br>Key on or engine running; and the Engine Coolant Temperature (ECM) detected the ECT sensor signal was more than the self-test maximum. This is a thermistor-type sensor with a variable resistance that changes when exposed to different temperatures<br>**Possible Causes:**<br>• ECT sensor has failed<br>• ECT Sensor (on Radiator) has failed<br>• ECT sensor signal circuit is open (inspect wiring & connector)<br>• ECT sensor signal circuit is shorted<br>• Cooling system malfunction, or the thermostat is stuck<br>• Engine not operating at normal operating temperature<br>• EOT sensor is damaged or it has failed |
| **DTC: P2187**<br>**2T, MIL: Yes**<br>**Years: 2007, 2008**<br>**Models:** Passat, Jetta, Golf, GTI, New Beetle, New Beetle Convertible<br>**Engines:** 2.0L, 2.5L, 3.6L<br>**Transmissions**: All | **System Too Lean On Idle Bank 1 Conditions:**<br>Engine started, battery voltage must be at least 11.5v, all electrical components must be off, the ground between the engine and the chassis must be well connected, the exhaust system must be properly sealed between the catalytic converter and the cylinder head, and the oxygen sensor heater for oxygen sensor before the catalytic converter must be properly functioning. ECM detected the system indicated a lean signal, or it could no longer control bank 1 because it was at its lean limit.<br>**Possible Causes:**<br>• Evaporative Emission (EVAP) canister purge regulator valve is faulty<br>• Exhaust system components are damaged<br>• Fuel injectors are faulty<br>• Fuel pressure regulator and residual pressure have failed<br>• Fuel Pump (FP) in fuel tank is faulty<br>• Intake system has leaks (false air)<br>• Secondary Air Injection (AIR) system has an improper seal<br>• Intake Manifold Runner Position Sensor has failed<br>• Motor for intake flap is faulty<br>• Oxygen sensor (before catalytic converter) is faulty<br>• Oxygen sensor (behind catalytic converter) is faulty<br>• Oxygen sensor heater (before catalytic converter) is faulty<br>• Oxygen sensor heater (behind catalytic converter) is faulty<br>• Check circuits for shorts to each other, ground or power<br>• ECM has failed |

| DTC | Trouble Code Title, Conditions & Possible Causes |
|---|---|
| **DTC: P2188**<br>**2T, MIL: Yes**<br>**Years:** 2007, 2008<br>**Models:** Passat, Jetta, Golf, GTI, New Beetle, New Beetle Convertible<br>**Engines:** 2.0L, 2.5L, 3.6L<br>**Transmissions:** All | **System Too Rich On Idle Bank 1 Conditions:**<br>Engine started, battery voltage must be at least 11.5v, all electrical components must be off, the ground between the engine and the chassis must be well connected, the exhaust system must be properly sealed between the catalytic converter and the cylinder head, and the oxygen sensor heater for oxygen sensor before the catalytic converter must be properly functioning. ECM detected the system indicated a rich signal, or it could no longer control bank 1 because it was at its rich limit.<br><br>**Possible Causes:**<br>• Evaporative Emission (EVAP) canister purge regulator valve is faulty<br>• Exhaust system components are damaged<br>• Fuel injectors are faulty<br>• Fuel pressure regulator and residual pressure have failed<br>• Fuel Pump (FP) in fuel tank is faulty<br>• Intake system has leaks (false air)<br>• Secondary Air Injection (AIR) system has an improper seal<br>• Intake Manifold Runner Position Sensor has failed<br>• Motor for intake flap is faulty<br>• Oxygen sensor (before catalytic converter) is faulty<br>• Oxygen sensor (behind catalytic converter) is faulty<br>• Oxygen sensor heater (before catalytic converter) is faulty<br>• Oxygen sensor heater (behind catalytic converter) is faulty<br>• Check circuits for shorts to each other, ground or power<br>• ECM has failed |
| **DTC: P2191**<br>**Years:** 2007, 2008<br>**Models:** Passat, Jetta, Golf, GTI, New Beetle, New Beetle Convertible<br>**Engines:** 2.0L, 2.5L, 3.6L<br>**Transmissions:** All | **System Too Lean at Higher Load Bank 1 Conditions:**<br>Engine started, battery voltage must be at least 11.5v, all electrical components must be off, the ground between the engine and the chassis must be well connected, the exhaust system must be properly sealed between the catalytic converter and the cylinder head, and the oxygen sensor heater for oxygen sensor before the catalytic converter must be properly functioning. ECM detected the system indicated a lean signal, or it could no longer control bank 1 because it was at its lean limit.<br><br>**Possible Causes:**<br>• Evaporative Emission (EVAP) canister purge regulator valve is faulty<br>• Exhaust system components are damaged<br>• Fuel injectors are faulty<br>• Fuel pressure regulator and residual pressure have failed<br>• Fuel Pump (FP) in fuel tank is faulty<br>• Intake system has leaks (false air)<br>• Secondary Air Injection (AIR) system has an improper seal<br>• Intake Manifold Runner Position Sensor has failed<br>• Motor for intake flap is faulty<br>• Oxygen sensor (before catalytic converter) is faulty<br>• Oxygen sensor (behind catalytic converter) is faulty<br>• Oxygen sensor heater (before catalytic converter) is faulty<br>• Oxygen sensor heater (behind catalytic converter) is faulty<br>• Check circuits for shorts to each other, ground or power<br>• ECM has failed |
| **DTC: P2192**<br>**Years:** 2007, 2008<br>**Models:** Passat, Jetta, Golf, GTI, New Beetle, New Beetle Convertible<br>**Engines:** 2.0L, 2.5L, 3.6L<br>**Transmissions:** All | **System Too Rich at Higher Load Bank 1 Conditions:**<br>Engine started, battery voltage must be at least 11.5v, all electrical components must be off, the ground between the engine and the chassis must be well connected, the exhaust system must be properly sealed between the catalytic converter and the cylinder head, and the oxygen sensor heater for oxygen sensor before the catalytic converter must be properly functioning. ECM detected the system indicated a rich signal, or it could no longer control bank 1 because it was at its rich limit.<br><br>**Possible Causes:**<br>• Evaporative Emission (EVAP) canister purge regulator valve is faulty<br>• Exhaust system components are damaged<br>• Fuel injectors are faulty<br>• Fuel pressure regulator and residual pressure have failed<br>• Fuel Pump (FP) in fuel tank is faulty<br>• Intake system has leaks (false air)<br>• Secondary Air Injection (AIR) system has an improper seal<br>• Intake Manifold Runner Position Sensor has failed<br>• Motor for intake flap is faulty<br>• Oxygen sensor (before catalytic converter) is faulty<br>• Oxygen sensor (behind catalytic converter) is faulty<br>• Oxygen sensor heater (before catalytic converter) is faulty<br>• Oxygen sensor heater (behind catalytic converter) is faulty<br>• Check circuits for shorts to each other, ground or power<br>• ECM has failed |

| DTC | Trouble Code Title, Conditions & Possible Causes |
|---|---|
| **DTC: P2195**<br>**Years:** 2007, 2008<br>**Models:** Passat, Jetta, Golf, GTI, New Beetle, New Beetle Convertible<br>**Engines:** 2.0L, 2.5L, 3.6L<br>**Transmissions:** All | **O2 Sensor Signal Stuck Lean Bank 1 Sensor 1 Conditions:**<br>Engine running in closed loop, and the ECM detected the O2S indicated a lean signal, or it could no longer control Fuel Trim because it was at lean limit.<br>**Possible Causes:**<br>• Engine oil level high<br>• Camshaft timing error<br>• Cylinder compression low<br>• Exhaust leaks in front of O2S<br>• EGR valve is stuck open<br>• EGR gasket is leaking<br>• EVR diaphragm is leaking<br>• Damaged fuel pressure regulator or extremely low fuel pressure<br>• O2S circuit is open or shorted in the wiring harness<br>• Oxygen sensor (before catalytic converter) is faulty<br>• Oxygen sensor (behind catalytic converter) is faulty<br>• Oxygen sensor heater (before catalytic converter) is faulty<br>• Oxygen sensor heater (behind catalytic converter) is faulty<br>• Air leaks after the MAF sensor<br>• PCV system leaks<br>• Dip stick not seated properly |
| **DTC: P2196**<br>**Years:** 2007, 2008<br>**Models:** Passat, Jetta, Golf, GTI, New Beetle, New Beetle Convertible<br>**Engines:** 2.0L, 2.5L, 3.6L<br>**Transmissions:** All | **O2 Sensor Signal Stuck Rich Bank 1 Sensor 1 Conditions:**<br>Engine running in closed loop, and the ECM detected the O2S indicated a rich signal, or it could no longer control Fuel Trim because it was at its rich limit.<br>**Possible Causes:**<br>• Engine oil level high<br>• Camshaft timing error<br>• Cylinder compression low<br>• Exhaust leaks in front of O2S<br>• EGR valve is stuck open<br>• EGR gasket is leaking<br>• EVR diaphragm is leaking<br>• Damaged fuel pressure regulator or extremely low fuel pressure<br>• O2S circuit is open or shorted in the wiring harness<br>• Oxygen sensor (before catalytic converter) is faulty<br>• Oxygen sensor (behind catalytic converter) is faulty<br>• Oxygen sensor heater (before catalytic converter) is faulty<br>• Oxygen sensor heater (behind catalytic converter) is faulty<br>• Air leaks after the MAF sensor<br>• PCV system leaks<br>• Dip stick not seated properly |
| **DTC: P2231**<br>**Years:** 2007, 2008<br>**Models:** Passat, Jetta, Golf, GTI, New Beetle, New Beetle Convertible<br>**Engines:** 2.0L, 2.5L, 3.6L<br>**Transmissions:** All | **O2 Sensor Signal Circuit Shorted to Heater Circuit Bank 1 Sensor 1 Conditions:**<br>Engine started, battery voltage must be at least 11.5v, all electrical components must be off, parking brake must be engaged (to keep daytime driving lights off), automatic transmission selector must be in park. The ECM detected an unexpected voltage condition, or it detected an unexpected current draw in the sensor circuit during the CCM test.<br>**Note: Vehicle must be raised before connector for oxygen sensors is accessible.**<br>**Possible Causes:**<br>• Oxygen sensor (before catalytic converter) is faulty<br>• Oxygen sensor heater (before catalytic converter) is faulty<br>• Oxygen sensor heater (before catalytic converter) is faulty<br>• Oxygen sensor heater (behind catalytic converter) is faulty<br>• O2S circuit is open or shorted in the wiring harness<br>• ECM has failed |

| DTC | Trouble Code Title, Conditions & Possible Causes |
|---|---|
| **DTC: P2237**<br>**Years: 2007, 2008**<br>**Models:** Passat, Jetta, Golf, GTI, New Beetle, New Beetle Convertible<br>**Engines:** 2.0L, 2.5L, 3.6L<br>**Transmissions:** All | **O2 Sensor Positive Current Control Circuit/Open Bank 1 Sensor 1 Conditions:**<br>Engine started, battery voltage must be at least 11.5v, all electrical components must be off, parking brake must be engaged (to keep daytime driving lights off), automatic transmission selector must be in park. The ECM detected an unexpected voltage condition, or it detected an unexpected current draw in the sensor circuit during the CCM test.<br>**Note: Vehicle must be raised before connector for oxygen sensors is accessible.**<br>**Possible Causes:**<br>• Oxygen sensor (before catalytic converter) is faulty<br>• Oxygen sensor heater (before catalytic converter) is faulty<br>• Oxygen sensor heater (before catalytic converter) is faulty<br>• Oxygen sensor heater (behind catalytic converter) is faulty<br>• O2S circuit is open or shorted in the wiring harness<br>• ECM has failed |
| **DTC: P2240**<br>**Years: 2007, 2008**<br>**Models:** Passat, Jetta, GTI<br>**Engines:** 2.0L, 2.5L, 3.6L<br>**Transmissions:** All | **O2 Sensor Positive Current Control Circuit/Open Bank 2 Sensor 1 Conditions:**<br>Engine started, battery voltage must be at least 11.5v, all electrical components must be off, parking brake must be engaged (to keep daytime driving lights off), automatic transmission selector must be in park. The ECM detected an unexpected voltage condition, or it detected an unexpected current draw in the sensor circuit during the CCM test.<br>**Note: Vehicle must be raised before connector for oxygen sensors is accessible.**<br>**Possible Causes:**<br>• Oxygen sensor (before catalytic converter) is faulty<br>• Oxygen sensor heater (before catalytic converter) is faulty<br>• Oxygen sensor heater (before catalytic converter) is faulty<br>• Oxygen sensor heater (behind catalytic converter) is faulty<br>• O2S circuit is open or shorted in the wiring harness<br>• ECM has failed |
| **DTC: P2243**<br>**Years: 2007, 2008**<br>**Models:** Passat, Jetta, Golf, GTI, New Beetle, New Beetle Convertible<br>**Engines:** 2.0L, 2.5L, 3.6L<br>**Transmissions:** All | **O2 Sensor Reference Voltage Circuit/Open Bank 1 Sensor 1 Conditions:**<br>Engine started, battery voltage must be at least 11.5v, all electrical components must be off, parking brake must be engaged (to keep daytime driving lights off), automatic transmission selector must be in park. The ECM detected an unexpected voltage condition, or it detected an unexpected current draw in the sensor circuit during the CCM test.<br>**note: Vehicle must be raised before connector for oxygen sensors is accessible.**<br>**Possible Causes:**<br>• Oxygen sensor (before catalytic converter) is faulty<br>• Oxygen sensor heater (before catalytic converter) is faulty<br>• Oxygen sensor heater (before catalytic converter) is faulty<br>• Oxygen sensor heater (behind catalytic converter) is faulty<br>• O2S circuit is open or shorted in the wiring harness<br>• ECM has failed |
| **DTC: P2251**<br>**Years: 2007, 2008**<br>**Models:** Passat, Jetta, Golf, GTI, New Beetle, New Beetle Convertible<br>**Engines:** 2.0L, 2.5L, 3.6L<br>**Transmissions:** All | **O2 Sensor Negative Voltage Circuit/Open Bank 1 Sensor 1 Conditions:**<br>Engine started, battery voltage must be at least 11.5v, all electrical components must be off, parking brake must be engaged (to keep daytime driving lights off), automatic transmission selector must be in park. The ECM detected an unexpected voltage condition, or it detected an unexpected current draw in the sensor circuit during the CCM test.<br>**Note: Vehicle must be raised before connector for oxygen sensors is accessible.**<br>**Possible Causes:**<br>• Oxygen sensor (before catalytic converter) is faulty<br>• Oxygen sensor heater (before catalytic converter) is faulty<br>• Oxygen sensor heater (before catalytic converter) is faulty<br>• Oxygen sensor heater (behind catalytic converter) is faulty<br>• O2S circuit is open or shorted in the wiring harness<br>• ECM has failed |
| **DTC: P2257**<br>**Years: 2007, 2008**<br>**Models:** Passat, Jetta, Golf, GTI, New Beetle, New Beetle Convertible<br>**Engines:** 2.0L, 2.5L, 3.6L<br>**Transmissions:** All | **Secondary Air Injection System Control "A" Circuit Low Conditions:**<br>Engine started, battery voltage must be at least 11.5v, all electrical components must be off, parking brake must be engaged (to keep daytime driving lights off), automatic transmission selector must be in park and the ground between the engine and the chassis must be well connected. The ECM detected an unexpected voltage condition on the AIR system control circuit during testing.<br>**Possible Causes:**<br>• AIR solenoid power circuit (B+) is open (check dedicated fuse)<br>• AIR bypass solenoid control circuit is open or shorted to ground<br>• AIR diverter solenoid control circuit open or shorted to ground<br>• AIR pump control circuit is open or shorted to ground<br>• Check valve (one or more) is damaged or leaking<br>• Solid State relay is damaged or it has failed<br>• Check activation of Secondary Air Injection (AIR) Pump Relay<br>• ECM has failed |

| DTC | Trouble Code Title, Conditions & Possible Causes |
|---|---|
| **DTC: P2258**<br>**Years:** 2007, 2008<br>**Models:** Passat, Jetta, Golf, GTI, New Beetle, New Beetle Convertible<br>**Engines:** 2.0L, 2.5L, 3.6L<br>**Transmissions:** All | **Secondary Air Injection System Control "A" Circuit High Conditions:**<br>Engine started, battery voltage must be at least 11.5v, all electrical components must be off, parking brake must be engaged (to keep daytime driving lights off), automatic transmission selector must be in park and the ground between the engine and the chassis must be well connected. The ECM detected an unexpected voltage condition on the AIR system control circuit during testing.<br>**Possible Causes:**<br>• AIR solenoid power circuit (B+) is open (check dedicated fuse)<br>• AIR bypass solenoid control circuit is open or shorted to ground<br>• AIR diverter solenoid control circuit open or shorted to ground<br>• AIR pump control circuit is open or shorted to ground<br>• Check valve (one or more) is damaged or leaking<br>• Solid State relay is damaged or it has failed<br>• Check activation of Secondary Air Injection (AIR) Pump Relay<br>• ECM has failed |
| **DTC: P2270**<br>**Years:** 2007, 2008<br>**Models:** Passat, Jetta, Golf, GTI, New Beetle, New Beetle Convertible<br>**Engines:** 2.0L, 2.5L, 3.6L<br>**Transmissions:** All | **O2 Sensor Signal Stuck Lean Bank 1 Sensor 2 Conditions:**<br>Engine started, battery voltage must be at least 11.5v, all electrical components must be off, parking brake must be engaged (to keep daytime driving lights off), automatic transmission selector must be in park. The ECM detected an unexpected voltage condition, or it detected an unexpected current draw in the heater circuit during the CCM test.<br>**note: Vehicle must be raised before connector for oxygen sensors is accessible.**<br>**Possible Causes:**<br>• Oxygen sensor (before catalytic converter) is faulty<br>• Oxygen sensor heater (before catalytic converter) is faulty<br>• Oxygen sensor heater (before catalytic converter) is faulty<br>• Oxygen sensor heater (behind catalytic converter) is faulty<br>• O2S circuit is open or shorted in the wiring harness<br>• ECM has failed |
| **DTC: P2271**<br>**Years:** 2007, 2008<br>**Models:** Passat, Jetta, Golf, GTI, New Beetle, New Beetle Convertible<br>**Engines:** 2.0L, 2.5L, 3.6L<br>**Transmissions:** All | **O2 Sensor Signal Stuck Rich Bank 1 Sensor 2 Conditions:**<br>Engine started, battery voltage must be at least 11.5v, all electrical components must be off, parking brake must be engaged (to keep daytime driving lights off), automatic transmission selector must be in park. The ECM detected an unexpected voltage condition, or it detected an unexpected current draw in the heater circuit during the CCM test.<br>**Note: Vehicle must be raised before connector for oxygen sensors is accessible.**<br>**Possible Causes:**<br>• Oxygen sensor (before catalytic converter) is faulty<br>• Oxygen sensor heater (before catalytic converter) is faulty<br>• Oxygen sensor heater (before catalytic converter) is faulty<br>• Oxygen sensor heater (behind catalytic converter) is faulty<br>• O2S circuit is open or shorted in the wiring harness<br>• ECM has failed |
| **DTC: P2273**<br>**Years:** 2007, 2008<br>**Models:** Passat, Jetta, GTI<br>**Engines:** 2.5L, 3.6L<br>**Transmissions:** All | **O2 Sensor Signal Stuck Rich Bank 2 Sensor 2 Conditions:**<br>Engine started, battery voltage must be at least 11.5v, all electrical components must be off, parking brake must be engaged (to keep daytime driving lights off), automatic transmission selector must be in park. The ECM detected an unexpected voltage condition, or it detected an unexpected current draw in the heater circuit during the CCM test.<br>**Note: Vehicle must be raised before connector for oxygen sensors is accessible.**<br>**Possible Causes:**<br>• Oxygen sensor (before catalytic converter) is faulty<br>• Oxygen sensor heater (before catalytic converter) is faulty<br>• Oxygen sensor heater (before catalytic converter) is faulty<br>• Oxygen sensor heater (behind catalytic converter) is faulty<br>• O2S circuit is open or shorted in the wiring harness<br>• ECM has failed |
| **DTC: P2274**<br>**Years:** 2007, 2008<br>**Models:** Jetta, New Beetle<br>**Engines:** 2.0L, 2.5L<br>**Transmissions:** All | **O2 Sensor Signal Stuck Lean Bank 2 Sensor 3 Conditions:**<br>Engine started, battery voltage must be at least 11.5v, all electrical components must be off, parking brake must be engaged (to keep daytime driving lights off), automatic transmission selector must be in park. The ECM detected an unexpected voltage condition, or it detected an unexpected current draw in the heater circuit during the CCM test.<br>**Note: Vehicle must be raised before connector for oxygen sensors is accessible.**<br>**Possible Causes:**<br>• Oxygen sensor (before catalytic converter) is faulty<br>• Oxygen sensor heater (before catalytic converter) is faulty<br>• Oxygen sensor heater (before catalytic converter) is faulty<br>• Oxygen sensor heater (behind catalytic converter) is faulty<br>• O2S circuit is open or shorted in the wiring harness<br>• ECM has failed |

| DTC | Trouble Code Title, Conditions & Possible Causes |
|---|---|
| **DTC: P2275**<br>**Years: 2007, 2008**<br>**Models:** Jetta, New Beetle<br>**Engines:** 2.0L, 2.5L<br>**Transmissions:** All | **O2 Sensor Signal Stuck Rich Bank 2 Sensor 3 Conditions:**<br>Engine started, battery voltage must be at least 11.5v, all electrical components must be off, parking brake must be engaged (to keep daytime driving lights off), automatic transmission selector must be in park. The ECM detected an unexpected voltage condition, or it detected an unexpected current draw in the heater circuit during the CCM test.<br>**Note: Vehicle must be raised before connector for oxygen sensors is accessible.**<br>**Possible Causes:**<br>• Oxygen sensor (before catalytic converter) is faulty<br>• Oxygen sensor heater (before catalytic converter) is faulty<br>• Oxygen sensor heater (before catalytic converter) is faulty<br>• Oxygen sensor heater (behind catalytic converter) is faulty<br>• O2S circuit is open or shorted in the wiring harness<br>• ECM has failed |
| **DTC: P2279**<br>**Years: 2007, 2008**<br>**Models:** Passat, Jetta, Golf, GTI, New Beetle, New Beetle Convertible<br>**Engines:** 2.0L, 2.5L, 3.6L<br>**Transmissions:** All | **Intake Air System Leak Conditions:**<br>Engine running and the vehicle speed more than 25mph, the ECM detected the intake air system has a potential leak. The IAT sensor is a variable resistor that includes an IAT signal circuit and a low reference circuit to measure the temperature of the air entering the engine. The ECM supplies the sensor with a low voltage singal circuit and a low reference ground circuit. When the IAT sensor is cold, its resistence is high. When the air temperature increases, its resistence decreases. With high sensor resistance, the IAT sensor signal voltage is high. With lower sensor resistance, the IAT sensor signal voltage should be lower.<br>**Possible Causes:**<br>• Intake Manifold Runner Position Sensor is damaged or has failed<br>• Intake system has leaks (false air)<br>• Motor for intake flap is faulty<br>• ECM has failed<br>• IAT sensor signal circuit is shorted to sensor or chassis ground<br>• IAT sensor is damaged or has failed<br>• ECM has failed. |
| **DTC: P2301**<br>**Years: 2007, 2008**<br>**Models:** Passat, Jetta<br>**Engines:** 2.0L, 3.6L<br>**Transmissions:** All | **Ignition Coil "A" Primary Control Circuit High Conditions:**<br>Engine started, battery voltage must be at least 11.5v, all electrical components must be off, parking brake must be engaged (to keep daytime driving lights off), automatic transmission selector must be in park and the ground between the engine and the chassis must be well connected. The ECM detected voltage values from the ignition module for the Ignition coilpack primary circuit that were outside the normal values required for proper function.<br>**Note: Ignition coils and power output stages are one component and cannot be replaced individually.**<br>**Possible Causes:**<br>• Engine speed (RPM) sensor has failed<br>• Camshaft Position (CMP) sensor has failed<br>• Power Supply Relay is shorted to an open circuit<br>• There is a malfunction in voltage supply<br>• Ignition coilpack is damaged or it has failed<br>• Cylinder 1 to 4 Fuel Injector(s) have failed<br>• ECM has failed |
| **DTC: P2303**<br>**Years: 2007, 2008**<br>**Models:** Passat, Jetta<br>**Engines:** 2.0L, 3.6L<br>**Transmissions:** All | **Ignition Coil "B" Primary Control Circuit Low Conditions:**<br>Engine started, battery voltage must be at least 11.5v, all electrical components must be off, parking brake must be engaged (to keep daytime driving lights off), automatic transmission selector must be in park and the ground between the engine and the chassis must be well connected. The ECM detected voltage values from the ignition module for the Ignition coilpack primary circuit that were outside the normal values required for proper function.<br>**Note: Ignition coils and power output stages are one component and cannot be replaced individually.**<br>**Possible Causes:**<br>• Engine speed (RPM) sensor has failed<br>• Camshaft Position (CMP) sensor has failed<br>• Power Supply Relay is shorted to an open circuit<br>• There is a malfunction in voltage supply<br>• Ignition coilpack is damaged or it has failed<br>• Cylinder 1 to 4 Fuel Injector(s) have failed<br>• ECM has failed |

| DTC | Trouble Code Title, Conditions & Possible Causes |
|---|---|
| **DTC: P2304**<br>**Years: 2007, 2008**<br>**Models:** Passat, Jetta<br>**Engines:** 2.0L, 3.6L<br>**Transmissions:** All | **Ignition Coil "B" Primary Control Circuit High Conditions:**<br>Engine started, battery voltage must be at least 11.5v, all electrical components must be off, parking brake must be engaged (to keep daytime driving lights off), automatic transmission selector must be in park and the ground between the engine and the chassis must be well connected. The ECM detected voltage values from the ignition module for the Ignition coilpack primary circuit that were outside the normal values required for proper function.<br>**Note: Ignition coils and power output stages are one component and cannot be replaced individually.**<br>**Possible Causes:**<br>• Engine speed (RPM) sensor has failed<br>• Camshaft Position (CMP) sensor has failed<br>• Power Supply Relay is shorted to an open circuit<br>• There is a malfunction in voltage supply<br>• Ignition coilpack is damaged or it has failed<br>• Cylinder 1 to 4 Fuel Injector(s) have failed<br>• ECM has failed |
| **DTC: P2306**<br>**Years: 2007, 2008**<br>**Models:** Passat, Jetta<br>**Engines:** 2.0L, 3.6L<br>**Transmissions:** All | **Ignition Coil "C" Primary Control Circuit Low Conditions:**<br>Engine started, battery voltage must be at least 11.5v, all electrical components must be off, parking brake must be engaged (to keep daytime driving lights off), automatic transmission selector must be in park and the ground between the engine and the chassis must be well connected. The ECM detected voltage values from the ignition module for the Ignition coilpack primary circuit that were outside the normal values required for proper function.<br>**Note: Ignition coils and power output stages are one component and cannot be replaced individually.**<br>**Possible Causes:**<br>• Engine speed (RPM) sensor has failed<br>• Camshaft Position (CMP) sensor has failed<br>• Power Supply Relay is shorted to an open circuit<br>• There is a malfunction in voltage supply<br>• Ignition coilpack is damaged or it has failed<br>• Cylinder 1 to 4 Fuel Injector(s) have failed<br>• ECM has failed |
| **DTC: P2307**<br>**Years: 2007, 2008**<br>**Models:** Passat, Jetta<br>**Engines:** 2.0L, 3.6L<br>**Transmissions:** All | **Ignition Coil "C" Primary Control Circuit High Conditions:**<br>Engine started, battery voltage must be at least 11.5v, all electrical components must be off, parking brake must be engaged (to keep daytime driving lights off), automatic transmission selector must be in park and the ground between the engine and the chassis must be well connected. The ECM detected voltage values from the ignition module for the Ignition coilpack primary circuit that were outside the normal values required for proper function.<br>**Note: Ignition coils and power output stages are one component and cannot be replaced individually.**<br>**Possible Causes:**<br>• Engine speed (RPM) sensor has failed<br>• Camshaft Position (CMP) sensor has failed<br>• Power Supply Relay is shorted to open circuit<br>• There is a malfunction in voltage supply<br>• Cylinder 1 to 4 Fuel Injector(s) have failed<br>• ECM has failed |
| **DTC: P2309**<br>**Years: 2007, 2008**<br>**Models:** Passat, Jetta<br>**Engines:** 2.0L, 3.6L<br>**Transmissions:** All | **Ignition Coil "D" Primary Control Circuit Low Conditions:**<br>Engine started, battery voltage must be at least 11.5v, all electrical components must be off, parking brake must be engaged (to keep daytime driving lights off), automatic transmission selector must be in park and the ground between the engine and the chassis must be well connected. The ECM detected voltage values from the ignition module for the Ignition coilpack primary circuit that were outside the normal values required for proper function.<br>**Note: Ignition coils and power output stages are one component and cannot be replaced individually.**<br>**Possible Causes:**<br>• Engine speed (RPM) sensor has failed<br>• Camshaft Position (CMP) sensor has failed<br>• Power Supply Relay is shorted to an open circuit<br>• There is a malfunction in voltage supply<br>• Ignition coilpack is damaged or it has failed<br>• Cylinder 1 to 4 Fuel Injector(s) have failed<br>• ECM has failed |

| DTC | Trouble Code Title, Conditions & Possible Causes |
|---|---|
| **DTC: P2310**<br>**Years:** 2007, 2008<br>**Models:** Passat, Jetta<br>**Engines:** 2.0L, 3.6L<br>**Transmissions:** All | **Ignition Coil "D" Primary Control Circuit High Conditions:**<br>Engine started, battery voltage must be at least 11.5v, all electrical components must be off, parking brake must be engaged (to keep daytime driving lights off), automatic transmission selector must be in park and the ground between the engine and the chassis must be well connected. The ECM detected voltage values from the ignition module for the Ignition coilpack primary circuit that were outside the normal values required for proper function.<br>**Note: Ignition coils and power output stages are one component and cannot be replaced individually.**<br>**Possible Causes:**<br>• Engine speed (RPM) sensor has failed<br>• Camshaft Position (CMP) sensor has failed<br>• Power Supply Relay is shorted to an open circuit<br>• There is a malfunction in voltage supply<br>• Ignition coilpack is damaged or it has failed<br>• Cylinder 1 to 4 Fuel Injector(s) have failed<br>• ECM has failed |
| **DTC: P2400**<br>**2T, MIL: Yes**<br>**Years:** 2007, 2008<br>**Models:** Passat, Jetta, Golf, GTI, New Beetle, New Beetle Convertible<br>**Engines:** 2.0L, 2.5L, 3.6L<br>**Transmissions:** All | **EVAP Leak Detection Pump (LDP) Control Circuit Open Conditions:**<br>Engine started, battery voltage must be at least 11.5v, all electrical components must be off, parking brake must be engaged (to keep daytime driving lights off), automatic transmission selector must be in park, the exhaust system must be properly sealed between the catalytic converter and the cylinder head, coolant temperature must be at least 80 degrees Celsius and oxygen sensor heaters for oxygen sensors before the catalytic converter must be functioning properly and the ground between the engine and the chassis must be well connected. The ECM detected voltage irregularity in the leak detection pump control circuit.<br>**Possible Causes:**<br>• EVAP LDP power supply circuit is open<br>• EVAP LDP solenoid valve is damaged or it has failed<br>• EVAP LDP canister has a leak or a poor seal<br>• ECM has failed<br>• EVAP canister system has an improper seal<br>• Evaporative Emission (EVAP) canister purge regulator valve 1 has failed<br>• Leak Detection Pump (LDP) is faulty<br>• Aftermarket EVAP parts that do not conform to specifications<br>• EVAP component seals leaking (i.e., leaks in the Purge valve, fuel tank pressure sensor, canister vent solenoid, fuel vapor control valve tube assembly or fuel vapor vent valve). |
| **DTC: P2401**<br>**2T, MIL: Yes**<br>**Years:** 2007, 2008<br>**Models:** Passat, Jetta, Golf, GTI, New Beetle, New Beetle Convertible<br>**Engines:** 2.0L, 2.5L, 3.6L<br>**Transmissions:** All | **EVAP Leak Detection Pump Control Circuit Low Conditions:**<br>Engine started, battery voltage must be at least 11.5v, all electrical components must be off, parking brake must be engaged (to keep daytime driving lights off), automatic transmission selector must be in park, the exhaust system must be properly sealed between the catalytic converter and the cylinder head, coolant temperature must be at least 80 degrees Celsius and oxygen sensor heaters for oxygen sensors before the catalytic converter must be functioning properly and the ground between the engine and the chassis must be well connected. The ECM detected voltage irregularity in the leak detection pump control circuit.<br>**Possible Causes:**<br>• EVAP LDP power supply circuit is open<br>• EVAP LDP solenoid valve is damaged or it has failed<br>• EVAP LDP canister has a leak or a poor seal<br>• ECM has failed<br>• EVAP canister system has an improper seal<br>• Evaporative Emission (EVAP) canister purge regulator valve 1 has failed<br>• Leak Detection Pump (LDP) is faulty<br>• Aftermarket EVAP parts that do not conform to specifications<br>• EVAP component seals leaking (i.e., leaks in the Purge valve, fuel tank pressure sensor, canister vent solenoid, fuel vapor control valve tube assembly or fuel vapor vent valve). |

| DTC | Trouble Code Title, Conditions & Possible Causes |
|---|---|
| **DTC: P2402**<br>**2T, MIL: Yes**<br>**Years: 2007, 2008**<br>**Models:** Passat, Jetta, Golf, GTI, New Beetle, New Beetle Convertible<br>**Engines:** 2.0L, 2.5L, 3.6L<br>**Transmissions:** All | **EVAP Leak Detection Pump Control Circuit High Conditions:**<br>Engine started, battery voltage must be at least 11.5v, all electrical components must be off, parking brake must be engaged (to keep daytime driving lights off), automatic transmission selector must be in park, the exhaust system must be properly sealed between the catalytic converter and the cylinder head, coolant temperature must be at least 80 degrees Celsius and oxygen sensor heaters for oxygen sensors before the catalytic converter must be functioning properly and the ground between the engine and the chassis must be well connected. The ECM detected voltage irregularity in the leak detection pump control circuit.<br>**Possible Causes:**<br>• EVAP LDP power supply circuit is open<br>• EVAP LDP solenoid valve is damaged or it has failed<br>• EVAP LDP canister has a leak or a poor seal<br>• ECM has failed<br>• EVAP canister system has an improper seal<br>• Evaporative Emission (EVAP) canister purge regulator valve 1 has failed<br>• Leak Detection Pump (LDP) is faulty<br>• Aftermarket EVAP parts that do not conform to specifications<br>• EVAP component seals leaking (i.e., leaks in the Purge valve, fuel tank pressure sensor, canister vent solenoid, fuel vapor control valve tube assembly or fuel vapor vent valve). |
| **DTC: P2403**<br>**2T, MIL: Yes**<br>**Years: 2007, 2008**<br>**Models:** Passat, Jetta, Golf, GTI, New Beetle, New Beetle Convertible<br>**Engines:** 2.0L, 2.5L, 3.6L<br>**Transmissions:** All | **EVAP Leak Detection Pump Sense Circuit Open Conditions:**<br>Engine started, battery voltage must be at least 11.5v, all electrical components must be off, parking brake must be engaged (to keep daytime driving lights off), automatic transmission selector must be in park, the exhaust system must be properly sealed between the catalytic converter and the cylinder head, coolant temperature must be at least 80 degrees Celsius and oxygen sensor heaters for oxygen sensors before the catalytic converter must be functioning properly and the ground between the engine and the chassis must be well connected. The ECM detected voltage irregularity in the leak detection pump control circuit.<br>**Possible Causes:**<br>• EVAP LDP power supply circuit is open<br>• EVAP LDP solenoid valve is damaged or it has failed<br>• EVAP LDP canister has a leak or a poor seal<br>• ECM has failed<br>• EVAP canister system has an improper seal<br>• Evaporative Emission (EVAP) canister purge regulator valve 1 has failed<br>• Leak Detection Pump (LDP) is faulty<br>• Aftermarket EVAP parts that do not conform to specifications<br>• EVAP component seals leaking (i.e., leaks in the Purge valve, fuel tank pressure sensor, canister vent solenoid, fuel vapor control valve tube assembly or fuel vapor vent valve). |
| **DTC: P2404**<br>**2T, MIL: Yes**<br>**Years: 2007, 2008**<br>**Models:** Passat, Jetta, Golf, GTI, New Beetle, New Beetle Convertible<br>**Engines:** 2.0L, 2.5L, 3.6L<br>**Transmissions:** All | **EVAP Leak Detection Pump Sense Circuit Range/Performance Conditions:**<br>Engine started, battery voltage must be at least 11.5v, all electrical components must be off, parking brake must be engaged (to keep daytime driving lights off), automatic transmission selector must be in park, the exhaust system must be properly sealed between the catalytic converter and the cylinder head, coolant temperature must be at least 80 degrees Celsius and oxygen sensor heaters for oxygen sensors before the catalytic converter must be functioning properly and the ground between the engine and the chassis must be well connected. The ECM detected voltage irregularity in the leak detection pump control circuit.<br>**Possible Causes:**<br>• EVAP LDP power supply circuit is open<br>• EVAP LDP solenoid valve is damaged or it has failed<br>• EVAP LDP canister has a leak or a poor seal<br>• ECM has failed<br>• EVAP canister system has an improper seal<br>• Evaporative Emission (EVAP) canister purge regulator valve 1 has failed<br>• Leak Detection Pump (LDP) is faulty<br>• Aftermarket EVAP parts that do not conform to specifications<br>• EVAP component seals leaking (i.e., leaks in the Purge valve, fuel tank pressure sensor, canister vent solenoid, fuel vapor control valve tube assembly or fuel vapor vent valve). |

| DTC | Trouble Code Title, Conditions & Possible Causes |
|---|---|
| **DTC: P2414**<br>**Years:** 2007, 2008<br>**Models:** Passat, Jetta, Golf, GTI, New Beetle, New Beetle Convertible<br>**Engines:** 2.0L, 2.5L<br>**Transmissions:** All | **O2 Sensor Exhaust Sample Error Bank 1 Sensor 1 Conditions:**<br>Engine running (ground connections between the engine and the chassis must be well connected), and the ECM detected an error on the OS Sensor.<br>**Note: Intake Flap Motor and Intake Manifold Runner Position Sensor are one component and cannot be replaced individually.**<br>**Note: Vacuum in the intake system sucks in the leak detection spray with false air. Leak detection spray decreases ignition quality of the fuel mixture. This causes a drop in engine speed and changes the value produced by the Heated Oxygen Sensor (HO2S).**<br>**Possible Causes:**<br>• Intake Manifold Runner Position Sensor is damaged or has failed<br>• Intake system has leaks (false air)<br>• Motor for intake flap is faulty<br>• ECM has failed<br>• Oxygen sensor (before catalytic converter) is faulty<br>• Oxygen sensor heater (before catalytic converter) is faulty<br>• Oxygen sensor heater (before catalytic converter) is faulty<br>• Oxygen sensor heater (behind catalytic converter) is faulty<br>• O2S circuit is open or shorted in the wiring harness |
| **DTC: P2539**<br>**Years:** 2007, 2008<br>**Models:** Passat, Jetta<br>**Engines:** 2.0L, 3.6L<br>**Transmissions:** All | **Low Pressure Fuel System Sensor Circuit Conditions:**<br>Engine started, battery voltage must be at least 11.5v, all electrical components must be off, parking brake must be engaged (to keep daytime driving lights off), automatic transmission selector must be in park, the exhaust system must be properly sealed between the catalytic converter and the cylinder head, coolant temperature must be at least 80 degrees Celsius. The ECM detected an error on the fuel system sensor circuit.<br>**Note: The specified fuel pressure should be between 3000 to 5000 kPA**<br>**Possible Causes:**<br>• Fuel Pressure Regulator Valve has failed<br>• Fuel Pressure Sensor has failed<br>• Fuel Pump (FP) Control Module has failed<br>• Fuel pump has failed<br>• ECM has failed |
| **DTC: P2540**<br>**Years:** 2007, 2008<br>**Models:** Passat, Jetta<br>**Engines:** 2.0L, 3.6L<br>**Transmissions:** All | **Low Pressure Fuel System Sensor Circuit Range/Performance Conditions:**<br>Engine started, battery voltage must be at least 11.5v, all electrical components must be off, parking brake must be engaged (to keep daytime driving lights off), automatic transmission selector must be in park, the exhaust system must be properly sealed between the catalytic converter and the cylinder head, coolant temperature must be at least 80 degrees Celsius. The ECM detected an error on the fuel system sensor circuit.<br>**Note: The specified fuel pressure should be between 3000 to 5000 kPA**<br>**Possible Causes:**<br>• Fuel Pressure Regulator Valve has failed<br>• Fuel Pressure Sensor has failed<br>• Fuel Pump (FP) Control Module has failed<br>• Fuel pump has failed<br>• ECM has failed |
| **DTC: P2541**<br>**Years:** 2007, 2008<br>**Models:** Passat, Jetta<br>**Engines:** 2.0L, 3.6L<br>**Transmissions:** All | **Low Pressure Fuel System Sensor Circuit Low Conditions:**<br>Engine started, battery voltage must be at least 11.5v, all electrical components must be off, parking brake must be engaged (to keep daytime driving lights off), automatic transmission selector must be in park, the exhaust system must be properly sealed between the catalytic converter and the cylinder head, coolant temperature must be at least 80 degrees Celsius. The ECM detected an error on the fuel system sensor circuit.<br>**Note: The specified fuel pressure should be between 3000 to 5000 kPA**<br>**Possible Causes:**<br>• Fuel Pressure Regulator Valve has failed<br>• Fuel Pressure Sensor has failed<br>• Fuel Pump (FP) Control Module has failed<br>• Fuel pump has failed<br>• ECM has failed |
| **DTC: P2600**<br>**Years:** 2007, 2008<br>**Models:** Passat, Jetta<br>**Engines:** 2.0L, 3.6L<br>**Transmissions:** All | **Coolant Pump Control Circuit/Open Conditions:**<br>Key on, engine started, the ECM detected an unexpected voltage condition on the ECT circuit.<br>**Possible Causes:**<br>• Check activation of Recirculation Pump Relay<br>• Recirculation pump has failed<br>• Coolant Temperature Sensor has failed<br>• Circuit short to ground, open or other component<br>• ECM has failed |

| DTC | Trouble Code Title, Conditions & Possible Causes |
|---|---|
| **DTC: P2602**<br>**Years: 2007, 2008**<br>**Models:** Passat, Jetta<br>**Engines:** 2.0L, 3.6L<br>**Transmissions:** All | **Coolant Pump Control Circuit Low Conditions:**<br>Key on, engine started, the ECM detected an unexpected voltage condition on the ECT circuit.<br>**Possible Causes:**<br>• Check activation of Recirculation Pump Relay<br>• Recirculation pump has failed<br>• Coolant Temperature Sensor has failed<br>• Circuit short to ground, open or other component<br>• ECM has failed |
| **DTC: P2603**<br>**Years: 2007, 2008**<br>**Models:** Passat, Jetta<br>**Engines:** 2.0L, 3.6L<br>**Transmissions:** All | **Coolant Pump Control Circuit High Conditions:**<br>Key on, engine started, the ECM detected an unexpected voltage condition on the ECT circuit.<br>**Possible Causes:**<br>• Check activation of Recirculation Pump Relay<br>• Recirculation pump has failed<br>• Coolant Temperature Sensor has failed<br>• Circuit short to ground, open or other component<br>• ECM has failed |
| **DTC: P2626**<br>**Years: 2007, 2008**<br>**Models:** Passat, Jetta, Golf, GTI, New Beetle, New Beetle Convertible<br>**Engines:** 2.0L, 2.5L, 3.6L<br>**Transmissions:** All | **O2 Sensor Pumping Current Trim Circuit/Open Bank 1 Sensor 1 Conditions:**<br>Engine started, battery voltage must be at least 11.5v, all electrical components must be off, parking brake must be engaged (to keep daytime driving lights off), automatic transmission selector must be in park, the exhaust system must be properly sealed between the catalytic converter and the cylinder head, coolant temperature must be at least 80 degrees Celsius and oxygen sensor heaters for oxygen sensors before the catalytic converter must be functioning properly and the ground between the engine and the chassis must be well connected. The ECM detected a voltage value that doesn't fall within the desired parameters for a properly functioning O2 system.<br>**Possible Causes:**<br>• Check activation of Recirculation Pump Relay<br>• Oxygen sensor (before catalytic converter) is faulty<br>• Oxygen sensor (behind catalytic converter) is faulty<br>• Oxygen sensor heater (before catalytic converter) is faulty<br>• Oxygen sensor heater (behind catalytic converter) is faulty |
| **DTC: P2629**<br>**Years: 2007, 2008**<br>**Models:** Passat, Jetta<br>**Engines:** 2.0L, 3.6L<br>**Transmissions:** All | **O2 Sensor Pumping Current Trim Circuit/Open Bank 2 Sensor 1 Conditions:**<br>Engine started, battery voltage must be at least 11.5v, all electrical components must be off, parking brake must be engaged (to keep daytime driving lights off), automatic transmission selector must be in park, the exhaust system must be properly sealed between the catalytic converter and the cylinder head, coolant temperature must be at least 80 degrees Celsius and oxygen sensor heaters for oxygen sensors before the catalytic converter must be functioning properly and the ground between the engine and the chassis must be well connected. The ECM detected a voltage value that doesn't fall within the desired parameters for a properly functioning O2 system.<br>**Possible Causes:**<br>• Check activation of Recirculation Pump Relay<br>• Oxygen sensor (before catalytic converter) is faulty<br>• Oxygen sensor (behind catalytic converter) is faulty<br>• Oxygen sensor heater (before catalytic converter) is faulty<br>• Oxygen sensor heater (behind catalytic converter) is faulty |
| **DTC: P2637**<br>**Years: 2007, 2008**<br>**Models:** Passat, Jetta, Golf, GTI, New Beetle, New Beetle Convertible<br>**Engines:** 2.0L, 2.5L, 3.6L<br>**Transmissions:** All | **Torque Management Feedback Signal "A" Conditions:**<br>Engine started, battery voltage must be at least 11.5v, all electrical components must be off, parking brake must be engaged (to keep daytime driving lights off), automatic transmission selector must be in park, the exhaust system must be properly sealed between the catalytic converter and the cylinder head, coolant temperature must be at least 80 degrees Celsius and oxygen sensor heaters for oxygen sensors before the catalytic converter must be functioning properly and the ground between the engine and the chassis must be well connected. The ECM detected a voltage value on the torque management circuits that doesn't fall within the desired parameters<br>**Possible Causes:**<br>• Engine Control Module (ECM) has failed<br>• Voltage supply for Engine Control Module (ECM) is damaged<br>• Engine Coolant Temperature (ECT) sensor has failed<br>• Intake Air Temperature (IAT) sensor has failed<br>• Intake Manifold Runner Position Sensor has failed<br>• Intake system has leaks (false air)<br>• Motor for intake flap has failed<br>• Mass Air Flow (MAF) sensor has failed |

| DTC | Trouble Code Title, Conditions & Possible Causes |
|---|---|
| **DTC: P2714**<br>**Years: 2007, 2008**<br>**Models:** Passat, Jetta<br>**Engines:** 2.0L, 3.6L<br>**Transmissions:** All | **Pressure Control Solenoid "D" Performance or Stuck Off Conditions:**<br>Engine started, battery voltage must be at least 11.5v, all electrical components must be off, and the ground between the engine and the chassis must be well connected. The ECM detected the pressure control solenoid was in the "stuck off" position.<br>**Possible Causes:**<br>• ATF level is low<br>• Circuit harness connector contacts are corroded or ingressed of water<br>• Circuit wires have shorted to each other, to battery or ground<br>• Automatic Transmission Hydraulic Pressure Sensor 1 has failed<br>• Solenoid valves in valve body are faulty<br>• Transmission Control Module (TCM) needs replacing<br>• Transmission Input Speed (RPM) Sensor has failed<br>• Transmission Output Speed (RPM) Sensor has failed |
| **DTC: P2715**<br>**Years: 2007, 2008**<br>**Models:** Passat, Jetta<br>**Engines:** 2.0L, 3.6L<br>**Transmissions:** All | **Pressure Control Solenoid "D" Performance or Stuck On Conditions:**<br>Engine started, battery voltage must be at least 11.5v, all electrical components must be off, and the ground between the engine and the chassis must be well connected. The ECM detected the pressure control solenoid was in the "stuck on" position.<br>**Possible Causes:**<br>• ATF level is low<br>• Circuit harness connector contacts are corroded or ingressed of water<br>• Circuit wires have shorted to each other, to battery or ground<br>• Automatic Transmission Hydraulic Pressure Sensor 1 has failed<br>• Solenoid valves in valve body are faulty<br>• Transmission Control Module (TCM) needs replacing<br>• Transmission Input Speed (RPM) Sensor has failed<br>• Transmission Output Speed (RPM) Sensor has failed |
| **DTC: P2716**<br>**Years: 2007, 2008**<br>**Models:** Passat, Jetta, GTI, New Beetle, New Beetle Convertible<br>**Engines:** 2.0L, 2.5L<br>**Transmissions:** All | **Pressure Control Solenoid "D" Electrical Malfunction Conditions:**<br>Engine started, battery voltage must be at least 11.5v, all electrical components must be off, and the ground between the engine and the chassis must be well connected. The ECM detected the pressure control solenoid was experiencing electrical malfunctions.<br>**Possible Causes:**<br>• ATF level is low<br>• Circuit harness connector contacts are corroded or ingressed of water<br>• Circuit wires have shorted to each other, to battery or ground<br>• Automatic Transmission Hydraulic Pressure Sensor 1 has failed<br>• Solenoid valves in valve body are faulty<br>• Transmission Control Module (TCM) needs replacing<br>• Transmission Input Speed (RPM) Sensor has failed<br>• Transmission Output Speed (RPM) Sensor has failed |
| **DTC: P2717**<br>**Years: 2007, 2008**<br>**Models:** Passat, New Beetle<br>**Engines:** 2.0L<br>**Transmissions:** All | **Pressure Control Solenoid "D" Intermittent Conditions:**<br>Engine started, battery voltage must be at least 11.5v, all electrical components must be off, and the ground between the engine and the chassis must be well connected. The ECM detected the pressure control solenoid sending intermittent signals.<br>**Possible Causes:**<br>• ATF level is low<br>• Circuit harness connector contacts are corroded or ingressed of water<br>• Circuit wires have shorted to each other, to battery or ground<br>• Automatic Transmission Hydraulic Pressure Sensor 1 has failed<br>• Solenoid valves in valve body are faulty<br>• Transmission Control Module (TCM) needs replacing<br>• Transmission Input Speed (RPM) Sensor has failed<br>• Transmission Output Speed (RPM) Sensor has failed |
| **DTC: P2723**<br>**Years: 2007, 2008**<br>**Models:** Passat, Jetta, New Beetle, New Beetle Convertible<br>**Engines:** 2.0L<br>**Transmissions:** All | **Pressure Control Solenoid "E" Performance or Stuck Off Conditions:**<br>Engine started, battery voltage must be at least 11.5v, all electrical components must be off, and the ground between the engine and the chassis must be well connected. The ECM detected the pressure control solenoid was in the "stuck off" position.<br>**Possible Causes:**<br>• ATF level is low<br>• Circuit harness connector contacts are corroded or ingressed of water<br>• Circuit wires have shorted to each other, to battery or ground<br>• Automatic Transmission Hydraulic Pressure Sensor 1 has failed<br>• Solenoid valves in valve body are faulty<br>• Transmission Control Module (TCM) needs replacing<br>• Transmission Input Speed (RPM) Sensor has failed<br>• Transmission Output Speed (RPM) Sensor has failed |

| DTC | Trouble Code Title, Conditions & Possible Causes |
|---|---|
| **DTC: P2724**<br>**Years: 2007, 2008**<br>**Models:** Passat, Jetta, New Beetle, New Beetle Convertible<br>**Engines:** 2.0L, 2.5L<br>**Transmissions:** All | **Pressure Control Solenoid "E" Performance or Stuck On Conditions:**<br>Engine started, battery voltage must be at least 11.5v, all electrical components must be off, and the ground between the engine and the chassis must be well connected. The ECM detected the pressure control solenoid was in the "stuck on" position.<br>**Possible Causes:**<br>• ATF level is low<br>• Circuit harness connector contacts are corroded or ingressed of water<br>• Circuit wires have shorted to each other, to battery or ground<br>• Automatic Transmission Hydraulic Pressure Sensor 1 has failed<br>• Solenoid valves in valve body are faulty<br>• Transmission Control Module (TCM) needs replacing<br>• Transmission Input Speed (RPM) Sensor has failed<br>• Transmission Output Speed (RPM) Sensor has failed |
| **DTC: P2725**<br>**Years: 2007, 2008**<br>**Models:** Passat, Jetta, New Beetle, New Beetle Convertible<br>**Engines:** 2.0L, 2.5L<br>**Transmissions:** All | **Pressure Control Solenoid "E" Electrical Malfunction Conditions:**<br>Engine started, battery voltage must be at least 11.5v, all electrical components must be off, and the ground between the engine and the chassis must be well connected. The ECM detected the pressure control solenoid was experiencing electrical malfunctions.<br>**Possible Causes:**<br>• ATF level is low<br>• Circuit harness connector contacts are corroded or ingressed of water<br>• Circuit wires have shorted to each other, to battery or ground<br>• Automatic Transmission Hydraulic Pressure Sensor 1 has failed<br>• Solenoid valves in valve body are faulty<br>• Transmission Control Module (TCM) needs replacing<br>• Transmission Input Speed (RPM) Sensor has failed<br>• Transmission Output Speed (RPM) Sensor has failed |
| **DTC: P2726**<br>**Years: 2007, 2008**<br>**Models:** Passat, Jetta, New Beetle, New Beetle Convertible<br>**Engines:** 2.0L, 2.5L<br>**Transmissions:** All | **Pressure Control Solenoid "E" Intermittent Conditions:**<br>Engine started, battery voltage must be at least 11.5v, all electrical components must be off, and the ground between the engine and the chassis must be well connected. The ECM detected the pressure control solenoid sending intermittent signals.<br>**Possible Causes:**<br>• ATF level is low<br>• Circuit harness connector contacts are corroded or ingressed of water<br>• Circuit wires have shorted to each other, to battery or ground<br>• Automatic Transmission Hydraulic Pressure Sensor 1 has failed<br>• Solenoid valves in valve body are faulty<br>• Transmission Control Module (TCM) needs replacing<br>• Transmission Input Speed (RPM) Sensor has failed<br>• Transmission Output Speed (RPM) Sensor has failed |
| **DTC: P2732**<br>**Years: 2007, 2008**<br>**Models:** Passat, Jetta, New Beetle, New Beetle Convertible<br>**Engines:** 2.0L, 2.5L<br>**Transmissions:** All | **Pressure Control Solenoid "F" Performance or Stuck Off Conditions:**<br>Engine started, battery voltage must be at least 11.5v, all electrical components must be off, and the ground between the engine and the chassis must be well connected. The ECM detected the pressure control solenoid was in the "stuck off" position.<br>**Possible Causes:**<br>• ATF level is low<br>• Circuit harness connector contacts are corroded or ingressed of water<br>• Circuit wires have shorted to each other, to battery or ground<br>• Automatic Transmission Hydraulic Pressure Sensor 1 has failed<br>• Solenoid valves in valve body are faulty<br>• Transmission Control Module (TCM) needs replacing<br>• Transmission Input Speed (RPM) Sensor has failed<br>• Transmission Output Speed (RPM) Sensor has failed |
| **DTC: P2733**<br>**Years: 2007, 2008**<br>**Models:** Passat, Jetta, New Beetle, New Beetle Convertible<br>**Engines:** 2.0L, 2.5L<br>**Transmissions:** All | **Pressure Control Solenoid "F" Performance or Stuck On Conditions:**<br>Engine started, battery voltage must be at least 11.5v, all electrical components must be off, and the ground between the engine and the chassis must be well connected. The ECM detected the pressure control solenoid was in the "stuck on" position.<br>**Possible Causes:**<br>• ATF level is low<br>• Circuit harness connector contacts are corroded or ingressed of water<br>• Circuit wires have shorted to each other, to battery or ground<br>• Automatic Transmission Hydraulic Pressure Sensor 1 has failed<br>• Solenoid valves in valve body are faulty<br>• Transmission Control Module (TCM) needs replacing<br>• Transmission Input Speed (RPM) Sensor has failed<br>• Transmission Output Speed (RPM) Sensor has failed |

| DTC | Trouble Code Title, Conditions & Possible Causes |
|---|---|
| **DTC: P2734**<br>**Years:** 2007, 2008<br>**Models:** Passat, Jetta, New Beetle, New Beetle Convertible<br>**Engines:** 2.0L, 2.5L<br>**Transmissions:** All | **Pressure Control Solenoid "F" Electrical Malfunction Conditions:**<br>Engine started, battery voltage must be at least 11.5v, all electrical components must be off, and the ground between the engine and the chassis must be well connected. The ECM detected the pressure control solenoid was experiencing electrical malfunctions.<br>**Possible Causes:**<br>• ATF level is low<br>• Circuit harness connector contacts are corroded or ingressed of water<br>• Circuit wires have shorted to each other, to battery or ground<br>• Automatic Transmission Hydraulic Pressure Sensor 1 has failed<br>• Solenoid valves in valve body are faulty<br>• Transmission Control Module (TCM) needs replacing<br>• Transmission Input Speed (RPM) Sensor has failed<br>• Transmission Output Speed (RPM) Sensor has failed |
| **DTC: P2735**<br>**Years:** 2007, 2008<br>**Models:** Passat, Jetta, New Beetle, New Beetle Convertible<br>**Engines:** 2.0L, 2.5L<br>**Transmissions:** All | **Pressure Control Solenoid "F" Intermittent Conditions:**<br>Engine started, battery voltage must be at least 11.5v, all electrical components must be off, and the ground between the engine and the chassis must be well connected. The ECM detected the pressure control solenoid sending intermittent signals.<br>**Possible Causes:**<br>• ATF level is low<br>• Circuit harness connector contacts are corroded or ingressed of water<br>• Circuit wires have shorted to each other, to battery or ground<br>• Automatic Transmission Hydraulic Pressure Sensor 1 has failed<br>• Solenoid valves in valve body are faulty<br>• Transmission Control Module (TCM) needs replacing<br>• Transmission Input Speed (RPM) Sensor has failed<br>• Transmission Output Speed (RPM) Sensor has failed |

## Gas Engine OBD II Trouble Code List (P3xxx Codes)

| DTC | Trouble Code Title, Conditions & Possible Causes |
|---|---|
| **DTC: P3028**<br>**Years:** 2007, 2008<br>**Models:** Jetta<br>**Engines:** 2.5L<br>**Transmissions:** All | **Throttle Actuation 2 Potentiometer Sign.2 Range/Performance Conditions:**<br>Engine started, battery voltage must be at least 11.5v, all electrical components must be off, the ground between the engine and the chassis must be well connected, coolant temperature must be at least 80 degrees Celsius and the accelerator pedal must be properly adjusted. The ECM detected an incorrect singal from the throttle potentiometer.<br>**Note: If the complete throttle valve control module is current-less (e.g. connector disconnected) the throttle valve moves into a particular, specified mechanical position, which signals an increased idle speed with an engine at operating temperature. If only the Throttle Position (TP) actuator is current-less, the throttle valve also moves into the specified mechanical position (emergency running gap), however, since Closed Throttle Position (CTP) switch can still be recognized, an "almost normal idle RPM" is reached via the respective ignition angle retardation.**<br>**Note: Terminal assignment at throttle control module is different in vehicles with and without cruise control. Characteristic: Steering column switch with operating module for cruise control.**<br>**Possible Causes:**<br>• Throttle Position (TP) sensor is faulty<br>• Throttle valve control module is faulty<br>• ECM is fault<br>• Circuit wires have short circuited to each other, to vehicle Ground (GND) or to B+<br>• Accelerator pedal module is faulty |

| DTC | Trouble Code Title, Conditions & Possible Causes |
|---|---|
| **DTC: P3031**<br>**Years: 2007, 2008**<br>**Models:** Jetta<br>**Engines:** 2.5L<br>**Transmissions:** All | **Throttle Actuator 2 Electrical Malfunction Conditions:**<br>Engine started, battery voltage must be at least 11.5v, all electrical components must be off, the ground between the engine and the chassis must be well connected, coolant temperature must be at least 80 degrees Celsius and the accelerator pedal must be properly adjusted. The ECM detected an incorrect singal from the throttle potentiometer.<br>**Note: If the complete throttle valve control module is current-less (e.g. connector disconnected) the throttle valve moves into a particular, specified mechanical position, which signals an increased idle speed with an engine at operating temperature. If only the Throttle Position (TP) actuator is current-less, the throttle valve also moves into the specified mechanical position (emergency running gap), however, since Closed Throttle Position (CTP) switch can still be recognized, an "almost normal idle RPM" is reached via the respective ignition angle retardation.**<br>**Note: Terminal assignment at throttle control module is different in vehicles with and without cruise control. Characteristic: Steering column switch with operating module for cruise control.**<br>**Possible Causes:**<br>• Throttle Position (TP) sensor is faulty<br>• Throttle valve control module is faulty<br>• ECM is fault<br>• Circuit wires have short circuited to each other, to vehicle Ground (GND) or to B+<br>• Accelerator pedal module is faulty |
| **DTC: P3032**<br>**Years: 2007, 2008**<br>**Models:** Jetta<br>**Engines:** 2.5L<br>**Transmissions:** All | **Idle Speed Control Throttle Position 2 Adaptation Malfunction Conditions:**<br>Engine started, battery voltage at least 11.5v, all electrical components off, ground connections between engine and chassis well connected, coolant temperature at least 80-degrees Celicius and the throttle valve must not be damaged or dirty; and the ECM detected the signal from the Throttle Position Valve Module to the ECM detected was erratic, non existent or unreliable (too high or too low).<br>**Possible Causes:**<br>• Throttle valve control module has failed<br>• Throttle valve control module's circuit has shorted or is open<br>• The ECM has failed |
| **DTC: P3035**<br>**Years: 2007, 2008**<br>**Models:** Jetta<br>**Engines:** 2.5L<br>**Transmissions:** All | **Idle Speed Control Throttle Position 2 Mechanical Malfunction Conditions:**<br>Engine started, battery voltage at least 11.5v, all electrical components off, ground connections between engine and chassis well connected, coolant temperature at least 80-degrees Celicius and the throttle valve must not be damaged or dirty; and the ECM detected the signal from the Throttle Position Valve Module to the ECM detected was erratic, non existent or unreliable (too high or too low).<br>**Possible Causes:**<br>• Throttle valve control module has failed<br>• Throttle valve control module's circuit has shorted or is open<br>• The ECM has failed |
| **DTC: P3047**<br>**Years: 2007, 2008**<br>**Models:** Jetta, Golf, GTI, New Beetle<br>**Engines:** 2.0L<br>**Transmissions:** All | **Activation Starter Relay 2 Short Circuit to B+ Conditions:**<br>The ECM detected a short circuit on the starter relay circuit.<br>**Possible Causes:**<br>• Check activation of Starting interlock relay |
| **DTC: P3048**<br>**Years: 2007, 2008**<br>**Models:** Jetta, Golf, GTI, New Beetle<br>**Engines:** 2.0L<br>**Transmissions:** All | **Activation Starter Relay 2 Short Circuit to Ground Conditions:**<br>The ECM detected a short circuit on the starter relay circuit.<br>**Possible Causes:**<br>• Check activation of Starting interlock relay |
| **DTC: P3049**<br>**Years: 2007, 2008**<br>**Models:** Jetta, Golf, GTI, New Beetle<br>**Engines:** 2.0L<br>**Transmissions:** All | **Activation Starter Relay 2 Open Conditions:**<br>The ECM detected a short circuit on the starter relay circuit.<br>**Possible Causes:**<br>• Check activation of Starting interlock relay |
| **DTC: P3050**<br>**Years: 2007, 2008**<br>**Models:** Jetta, Golf, GTI, New Beetle<br>**Engines:** 2.0L<br>**Transmissions:** All | **Starter Relay 2 Electrical Malfunction in Circuit (Relay Stuck) Conditions:**<br>The ECM detected a malfunction of the starter relay circuit.<br>**Possible Causes:**<br>• Check activation of Starting interlock relay |

| DTC | Trouble Code Title, Conditions & Possible Causes |
|---|---|
| **DTC: P3081**<br>**2T, MIL: Yes**<br>**Years: 2007, 2008**<br>**Models:** Passat, Jetta, Golf, GTI, New Beetle, New Beetle Convertible<br>**Engines:** 2.0L, 2.5L, 3.6L<br>**Transmissions:** All | **Engine Temperature Too Low Conditions:**<br>Engine running and the ECM has detected that the engine temperature is too low.<br>**Possible Causes:**<br>&bull; Engine hasn't completely warmed up<br>&bull; Radiator malfunction<br>&bull; Thermostat malfunction<br>&bull; ECM failure |
| **DTC: P3096**<br>**Years: 2007, 2008**<br>**Models:** Passat, Jetta, Golf, GTI<br>**Engines:** 2.0L<br>**Transmissions:** All | **Internal Control Module Memory, Check Sum Error Conditions:**<br>Key on, the ECM has detected a programming error<br>**Possible Causes:**<br>&bull; Battery terminal corrosion, or loose battery connection<br>&bull; Connection to the ECM interrupted, or the circuit has been opened<br>&bull; Reprogramming error has occurred<br>&bull; ECM has failed and needs replacement.<br>&bull; Voltage supply for Engine Control Module (ECM) is faulty |
| **DTC: P3097**<br>**Years: 2007, 2008**<br>**Models:** Passat, Jetta, Golf, GTI<br>**Engines:** 2.0L<br>**Transmissions:** All | **Internal Control Module Memory, Check Sum Error Conditions:**<br>Key on, the ECM has detected a programming error<br>**Possible Causes:**<br>&bull; Battery terminal corrosion, or loose battery connection<br>&bull; Connection to the ECM interrupted, or the circuit has been opened<br>&bull; Reprogramming error has occurred<br>&bull; ECM has failed and needs replacement.<br>&bull; Voltage supply for Engine Control Module (ECM) is faulty |
| **DTC: P310A**<br>**Years: 2007, 2008**<br>**Models:** Passat, Jetta<br>**Engines:** 2.0L<br>**Transmissions:** All | **Low Fuel Pressure Regulation Coolant Fuel Pressure Outside Specification Conditions:**<br>Engine started, battery voltage must be at least 11.5v, all electrical components must be off, the ground between the engine and the chassis must be well connected, the exhaust system must be properly sealed between the catalytic converter and the cylinder head, and the coolant temperature must be 80 degrees Celsius. The ECM detected that the system's fuel pressure has fallen below the accepted normal calibrated value.<br>**Possible Causes:**<br>&bull; Fuel Pressure Regulator Valve faulty<br>&bull; Fuel Pressure Sensor faulty<br>&bull; Fuel Pump (FP) Control Module faulty<br>&bull; Fuel pump faulty<br>&bull; Low fuel |
| **DTC: P310B**<br>**Years: 2007, 2008**<br>**Models:** Passat, Jetta<br>**Engines:** 2.0L<br>**Transmissions:** All | **Low fuel Pressure Regulation Fuel Pressure Fluctuates Conditions:**<br>Engine started, battery voltage must be at least 11.5v, all electrical components must be off, the ground between the engine and the chassis must be well connected, the exhaust system must be properly sealed between the catalytic converter and the cylinder head, and the coolant temperature must be 80 degrees Celsius. The ECM detected that the system's fuel pressure is irratic.<br>**Possible Causes:**<br>&bull; Fuel Pressure Regulator Valve faulty<br>&bull; Fuel Pressure Sensor faulty<br>&bull; Fuel Pump (FP) Control Module faulty<br>&bull; Fuel pump faulty<br>&bull; Low fuel |
| **DTC: P310C**<br>**Years: 2007, 2008**<br>**Models:** Passat, Jetta<br>**Engines:** 2.0L<br>**Transmissions:** All | **Low fuel Pressure Regulation Fuel Pressure Breaks Down Conditions:**<br>Engine started, battery voltage must be at least 11.5v, all electrical components must be off, the ground between the engine and the chassis must be well connected, the exhaust system must be properly sealed between the catalytic converter and the cylinder head, and the coolant temperature must be 80 degrees Celsius. The ECM detected that the system's fuel pressure is irratic.<br>**Possible Causes:**<br>&bull; Fuel Pressure Regulator Valve faulty<br>&bull; Fuel Pressure Sensor faulty<br>&bull; Fuel Pump (FP) Control Module faulty<br>&bull; Fuel pump faulty<br>&bull; Low fuel |

| DTC | Trouble Code Title, Conditions & Possible Causes |
|---|---|
| **DTC: P3137**<br>**Years: 2007, 2008**<br>**Models:** Passat, Jetta<br>**Engines:** 2.0L<br>**Transmissions:** All | **Intake Manifold Runner Control (IMRC) Basic Setting Not Carried Out Conditions:**<br>Engine started, battery voltage must be at least 11.5v, all electrical components must be off, the ground between the engine and the chassis must be well connected, the exhaust system must be properly sealed between the catalytic converter and the cylinder head, and the coolant temperature must be 80 degrees Celsius.<br>**Note: The throttle valve activation occurs via an electric motor (throttle drive) in the throttle valve control module. It is activated by the Engine Control Module (ECM) according to specifications of the two sensors, Throttle Position (TP) Sensor and Accelerator Pedal Position Sensor.**<br>**Possible Causes:**<br>• |
| **DTC: P3211**<br>**2T, MIL: Yes**<br>**Years: 2007, 2008**<br>**Models:** Passat, Jetta, Golf, GTI, New Beetle, New Beetle Convertible<br>**Engines:** 2.0L, 2.5L, 3.6L<br>**Transmissions:** All | **Exhaust (Bank 1 Sensor 1) Heater Return Connection Conditions:**<br>Engine started, battery voltage must be at least 11.5v, all electrical components must be off, parking brake must be engaged (to keep daytime driving lights off), automatic transmission selector must be in park, the exhaust system must be properly sealed between the catalytic converter and the cylinder head, coolant temperature must be at least 80 degrees Celsius and oxygen sensor heaters for oxygen sensors before the catalytic converter must be functioning properly and the ground between the engine and the chassis must be well connected. The ECM detected a voltage value that doesn't fall within the desired parameters for a properly functioning exhaust system.<br>**Possible Causes:**<br>• Check activation of Recirculation Pump Relay<br>• Oxygen sensor (before catalytic converter) is faulty<br>• Oxygen sensor (behind catalytic converter) is faulty<br>• Oxygen sensor heater (before catalytic converter) is faulty<br>• Oxygen sensor heater (behind catalytic converter) is faulty |
| **DTC: P3255**<br>**Years: 2007, 2008**<br>**Models:** Passat, Jetta, Golf, GTI, New Beetle, New Beetle Convertible<br>**Engines:** 2.0L<br>**Transmissions:** All | **O2 Sensor Before Catalytic Converter (Bank 1), Heating Circuit Regulation at Upper Impact Conditions:**<br>Engine started, battery voltage must be at least 11.5v, all electrical components must be off, parking brake must be engaged (to keep daytime driving lights off), automatic transmission selector must be in park, the exhaust system must be properly sealed between the catalytic converter and the cylinder head, coolant temperature must be at least 80 degrees Celsius and oxygen sensor heaters for oxygen sensors before the catalytic converter must be functioning properly and the ground between the engine and the chassis must be well connected. The ECM detected a voltage value that doesn't fall within the desired parameters for a properly functioning O2 system.<br>**Possible Causes:**<br>• Check activation of Recirculation Pump Relay<br>• Oxygen sensor (before catalytic converter) is faulty<br>• Oxygen sensor (behind catalytic converter) is faulty<br>• Oxygen sensor heater (before catalytic converter) is faulty<br>• Oxygen sensor heater (behind catalytic converter) is faulty |
| **DTC: P3056**<br>**Years: 2007, 2008**<br>**Models:** Passat, Jetta, Golf, GTI, New Beetle, New Beetle Convertible<br>**Engines:** 2.0L<br>**Transmissions:** All | **O2 Sensor Before Catalytic Converter (Bank 1), Heating Circuit Regulation at Lower Impact Conditions:**<br>Engine started, battery voltage must be at least 11.5v, all electrical components must be off, parking brake must be engaged (to keep daytime driving lights off), automatic transmission selector must be in park, the exhaust system must be properly sealed between the catalytic converter and the cylinder head, coolant temperature must be at least 80 degrees Celsius and oxygen sensor heaters for oxygen sensors before the catalytic converter must be functioning properly and the ground between the engine and the chassis must be well connected. The ECM detected a voltage value that doesn't fall within the desired parameters for a properly functioning O2 system.<br>**Possible Causes:**<br>• Check activation of Recirculation Pump Relay<br>• Oxygen sensor (before catalytic converter) is faulty<br>• Oxygen sensor (behind catalytic converter) is faulty<br>• Oxygen sensor heater (before catalytic converter) is faulty<br>• Oxygen sensor heater (behind catalytic converter) is faulty |

| DTC | Trouble Code Title, Conditions & Possible Causes |
|---|---|
| **DTC: P3266**<br>**Years:** 2007, 2008<br>**Models:** Passat, Jetta, Golf, GTI, New Beetle, New Beetle Convertible<br>**Engines:** 2.0L<br>**Transmissions:** All | **Internal Resistance (Bank 1, Sensor 1) Too Large Conditions:**<br>Engine started, battery voltage must be at least 11.5v, all electrical components must be off, the ground between the engine and the chassis must be well connected, the exhaust system must be properly sealed between the catalytic converter and the cylinder head, and the oxygen sensor heater for oxygen sensor before the catalytic converter must be properly functioning. The ECM detected a measurement on the O2 sensor circuit that was outside the parameters to function properly.<br>**Note: For resistance testing of sensor heating, oxygen sensor should be cooled to ambient temperature. High temperatures at oxygen sensor may lead to inaccurate measurements.**<br>**Possible Causes:**<br>• Oxygen sensor (before catalytic converter) is faulty<br>• Oxygen sensor (behind catalytic converter) is faulty<br>• Oxygen sensor heater (before catalytic converter) is faulty<br>• Oxygen sensor heater (behind catalytic converter) is faulty<br>• Circuit wiring has a short to power or ground<br>• Engine Component Power Supply Relay is faulty<br>• E-box fuses for oxygen sensor are faulty<br>• Leaks present in the exhaust manifold or exhaust pipes<br>• HO2S signal wire and ground wire crossed in connector<br>• HO2S element is fuel contaminated or has failed<br>• ECM has failed |

## Gas Engine OBD II Trouble Code List (U1xxx Codes)

| DTC | Trouble Code Title, Conditions & Possible Causes |
|---|---|
| **DTC: U0001**<br>**Years:** 2007, 2008<br>**Models:** Passat, Jetta, Golf, GTI, New Beetle, New Beetle Convertible<br>**Engines:** 2.0L, 2.5L, 3.6L<br>**Transmissions:** All | **High Speed CAN Communication Bus Conditions:**<br>The Engine Control Module (ECM) communicates with all databus-capable control modules via a CAN databus. These databus-capable control modules are connected via two data bus wires which are twisted together (CAN_High and CAN_Low), and exchange information (messages). Missing information on the databus is recognized as a malfunction and stored. Trouble-free operation of the CAN-Bus requires that it have a terminal resistance. This central terminal resistor is located in the Engine Control Module (ECM).<br>**Possible Causes:**<br>• ECM has failed<br>• CAN data bus wires have short circuited to each other |
| **DTC: U0101**<br>**Years:** 2007, 2008<br>**Models:** Passat, Jetta, Golf, GTI, New Beetle, New Beetle Convertible<br>**Engines:** 2.0L, 2.5L, 3.6L<br>**Transmissions:** All | **Lost Communication With TCM Conditions:**<br>Key on, and the ECM detected that it has lost communication with the Transmission Control Module (TCM) during its initial startup. The Engine Control Module (ECM) communicates with all databus-capable control modules via a CAN databus. These databus-capable control modules are connected via two data bus wires which are twisted together (CAN_High and CAN_Low), and exchange information (messages). Missing information on the databus is recognized as a malfunction and stored. Trouble-free operation of the CAN-Bus requires that it have a terminal resistance.<br>**Possible Causes:**<br>• ECM has failed<br>• Terminal resistance for CAN-bus are faulty<br>• Can data bus wires have short circuited to each other<br>• TCM has failed |
| **DTC: U0104**<br>**Years:** 2007, 2008<br>**Models:** Passat, Jetta<br>**Engines:** 2.0L, 3.6L<br>**Transmissions:** All | **Lost Communication With Cruise Control Module Conditions:**<br>Key on, and the ECM detected that it has lost communication with the Cruise Control Module during its initial startup. The Engine Control Module (ECM) communicates with all databus-capable control modules via a CAN databus. These databus-capable control modules are connected via two data bus wires which are twisted together (CAN_High and CAN_Low), and exchange information (messages). Missing information on the databus is recognized as a malfunction and stored. Trouble-free operation of the CAN-Bus requires that it have a terminal resistance.<br>**Possible Causes:**<br>• ECM has failed<br>• Terminal resistance for CAN-bus are faulty<br>• Can data bus wires have short circuited to each other<br>• The cruise control module has failed |

| DTC | Trouble Code Title, Conditions & Possible Causes |
|---|---|
| **DTC: U0115**<br>**Years: 2007, 2008**<br>**Models:** Passat, Jetta<br>**Engines:** 2.0L, 3.6L<br>**Transmissions**: All | **Lost Communication With ECM "B" Conditions:**<br>Key on, and the ECM detected that it has lost communication during its initial startup. The Engine Control Module (ECM) communicates with all databus-capable control modules via a CAN databus. These databus-capable control modules are connected via two data bus wires which are twisted together (CAN_High and CAN_Low), and exchange information (messages). Missing information on the databus is recognized as a malfunction and stored. Trouble-free operation of the CAN-Bus requires that it have a terminal resistance.<br>**Possible Causes:**<br>• ECM has failed<br>• Terminal resistance for CAN-bus are faulty<br>• Can data bus wires have short circuited to each other |
| **DTC: U0121**<br>**Years: 2007, 2008**<br>**Models:** Passat, Jetta, Golf, GTI, New Beetle, New Beetle Convertible<br>**Engines:** 2.0L, 2.5L, 3.6L<br>**Transmissions**: All | **Lost Communication With Anti-Lock Brake System (ABS) Control Module Conditions:**<br>Key on, and the ECM detected that it has lost communication with the ABS Control Module during its initial startup. The Engine Control Module (ECM) communicates with all databus-capable control modules via a CAN databus. These databus-capable control modules are connected via two data bus wires which are twisted together (CAN_High and CAN_Low), and exchange information (messages). Missing information on the databus is recognized as a malfunction and stored. Trouble-free operation of the CAN-Bus requires that it have a terminal resistance.<br>**Possible Causes:**<br>• ECM has failed<br>• Terminal resistance for CAN-bus are faulty<br>• Can data bus wires have short circuited to each other<br>• There is a fault with the ABS control module |
| **DTC: U0126**<br>**Years: 2007, 2008**<br>**Models:** Passat, Jetta<br>**Engines:** 2.0L, 3.6L<br>**Transmissions**: All | **Lost Communication With Steering Angle Sensor Module Conditions:**<br>Key on, and the ECM detected that it has lost communication with the Steering Angle Sensor during its initial startup. The Engine Control Module (ECM) communicates with all databus-capable control modules via a CAN databus. These databus-capable control modules are connected via two data bus wires which are twisted together (CAN_High and CAN_Low), and exchange information (messages). Missing information on the databus is recognized as a malfunction and stored. Trouble-free operation of the CAN-Bus requires that it have a terminal resistance.<br>**Possible Causes:**<br>• ECM has failed<br>• Terminal resistance for CAN-bus are faulty<br>• Can data bus wires have short circuited to each other<br>• The steering angle sensor module is faulty |
| **DTC: U0128**<br>**Years: 2007, 2008**<br>**Models:** Passat, Jetta<br>**Engines:** 2.0L, 3.6L<br>**Transmissions**: All | **Lost Communication With Park Brake Control Module Conditions:**<br>Key on, and the ECM detected that it has lost communication with the Parking Brake Control Module during its initial startup. The Engine Control Module (ECM) communicates with all databus-capable control modules via a CAN databus. These databus-capable control modules are connected via two data bus wires which are twisted together (CAN_High and CAN_Low), and exchange information (messages). Missing information on the databus is recognized as a malfunction and stored. Trouble-free operation of the CAN-Bus requires that it have a terminal resistance.<br>**Possible Causes:**<br>• ECM has failed<br>• Terminal resistance for CAN-bus are faulty<br>• Can data bus wires have short circuited to each other<br>• The parking brake control module has failed |
| **DTC: U0146**<br>**Years: 2007, 2008**<br>**Models:** Passat, Jetta<br>**Engines:** 2.0L, 3.6L<br>**Transmissions**: All | **Lost Communication With Gateway "A" Conditions:**<br>Key on, and the ECM detected that it has lost communication with the Gateway "A" during its initial startup. The Engine Control Module (ECM) communicates with all databus-capable control modules via a CAN databus. These databus-capable control modules are connected via two data bus wires which are twisted together (CAN_High and CAN_Low), and exchange information (messages). Missing information on the databus is recognized as a malfunction and stored. Trouble-free operation of the CAN-Bus requires that it have a terminal resistance.<br>**Possible Causes:**<br>• ECM has failed or isn't properly coded<br>• Terminal resistance for CAN-bus are faulty<br>• Can data bus wires have short circuited to each other |

| DTC | Trouble Code Title, Conditions & Possible Causes |
|---|---|
| **DTC: U0151**<br>**Years:** 2007, 2008<br>**Models:** Passat, Jetta<br>**Engines:** 2.0L, 3.6L<br>**Transmissions:** All | **Lost Communication With Restraints Control Module Conditions:**<br>Key on, and the ECM detected that it has lost communication with the Restraints Control Module during its initial startup. The Engine Control Module (ECM) communicates with all databus-capable control modules via a CAN databus. These databus-capable control modules are connected via two data bus wires which are twisted together (CAN_High and CAN_Low), and exchange information (messages). Missing information on the databus is recognized as a malfunction and stored. Trouble-free operation of the CAN-Bus requires that it have a terminal resistance.<br>**Possible Causes:**<br>• ECM has failed or isn't properly coded<br>• Terminal resistance for CAN-bus are faulty<br>• Can data bus wires have short circuited to each other |
| **DTC: U0155**<br>**Years:** 2007, 2008<br>**Models:** Passat, Jetta, Golf, GTI, New Beetle, New Beetle Convertible<br>**Engines:** 2.0L, 2.5L, 3.6L<br>**Transmissions:** All | **Lost Communication With Instrument Cluster Conditions:**<br>Key on, and the ECM detected that it has lost communication with the Instrument Cluster Panel (I/P) during its initial startup. The Engine Control Module (ECM) communicates with all databus-capable control modules via a CAN databus. These databus-capable control modules are connected via two data bus wires which are twisted together (CAN_High and CAN_Low), and exchange information (messages). Missing information on the databus is recognized as a malfunction and stored. Trouble-free operation of the CAN-Bus requires that it have a terminal resistance.<br>**Possible Causes:**<br>• ECM has failed<br>• Terminal resistance for CAN-bus are faulty<br>• Can data bus wires have short circuited to each other |
| **DTC: U0164**<br>**Years:** 2007, 2008<br>**Models:** Passat, Jetta<br>**Engines:** 2.0L, 3.6L<br>**Transmissions:** All | **Lost Communication With HVAC Control Module Conditions:**<br>Key on, and the ECM detected that it has lost communication with the HVAC Control Module during its initial startup. The Engine Control Module (ECM) communicates with all databus-capable control modules via a CAN databus. These databus-capable control modules are connected via two data bus wires which are twisted together (CAN_High and CAN_Low), and exchange information (messages). Missing information on the databus is recognized as a malfunction and stored. Trouble-free operation of the CAN-Bus requires that it have a terminal resistance.<br>**Possible Causes:**<br>• ECM has failed<br>• Terminal resistance for CAN-bus are faulty<br>• Can data bus wires have short circuited to each other<br>• The HVAC control module has failed |
| **DTC: U0302**<br>**Years:** 2007, 2008<br>**Models:** Passat, Jetta, Golf, GTI, New Beetle, New Beetle Convertible<br>**Engines:** 2.0L, 2.5L, 3.6L<br>**Transmissions:** All | **Software Incompatibility with Transmission Control Module Conditions:**<br>Key on, and the ECM detected a software incompatibility condition with the Transmission Control Module during its initial startup. The Engine Control Module (ECM) communicates with all databus-capable control modules via a CAN databus. These databus-capable control modules are connected via two data bus wires which are twisted together (CAN_High and CAN_Low), and exchange information (messages). Missing information on the databus is recognized as a malfunction and stored. Trouble-free operation of the CAN-Bus requires that it have a terminal resistance.<br>**Possible Causes:**<br>• ECM or TCM has failed or is not properly coded<br>• Terminal resistance for CAN-bus are faulty<br>• Can data bus wires have short circuited to each other |
| **DTC: U0305**<br>**Years:** 2007, 2008<br>**Models:** Passat, Jetta<br>**Engines:** 2.0L, 3.6L<br>**Transmissions:** All | **Software Incompatibility with Cruise Control Module Conditions:**<br>Key on, and the ECM detected that it has lost communication with the Cruise Control Module during its initial startup. The Engine Control Module (ECM) communicates with all databus-capable control modules via a CAN databus. These databus-capable control modules are connected via two data bus wires which are twisted together (CAN_High and CAN_Low), and exchange information (messages). Missing information on the databus is recognized as a malfunction and stored. Trouble-free operation of the CAN-Bus requires that it have a terminal resistance.<br>**Possible Causes:**<br>• ECM has failed<br>• Terminal resistance for CAN-bus are faulty<br>• Can data bus wires have short circuited to each other<br>• The cruise control module has failed |

| DTC | Trouble Code Title, Conditions & Possible Causes |
|---|---|
| **DTC: U0315**<br>**Years: 2007, 2008**<br>**Models:** Passat, Jetta, Golf, GTI, New Beetle, New Beetle Convertible<br>**Engines:** 2.0L, 2.5L, 3.6L<br>**Transmissions:** All | **Software Incompatibility with Anti-Lock Brake System Control Module Conditions:**<br>Key on, and the ECM detected a software incompatibility condition with the Anti-Lock Brake System Control Module during its initial startup. The Engine Control Module (ECM) communicates with all databus-capable control modules via a CAN databus. These databus-capable control modules are connected via two data bus wires which are twisted together (CAN_High and CAN_Low), and exchange information (messages). Missing information on the databus is recognized as a malfunction and stored. Trouble-free operation of the CAN-Bus requires that it have a terminal resistance.<br>**Possible Causes:**<br>• ECM has failed<br>• Terminal resistance for CAN-bus are faulty<br>• Can data bus wires have short circuited to each other<br>• The AB S control module has failed |
| **DTC: U0402**<br>**Years: 2007, 2008**<br>**Models:** Passat, Jetta, Golf, GTI, New Beetle, New Beetle Convertible<br>**Engines:** 2.0L, 2.5L, 3.6L<br>**Transmissions:** All | **Invalid Data Received From Transmission Control Module Conditions:**<br>Key on, and the ECM detected a software invalid data from the Cruise Control Module during its initial startup. The Engine Control Module (ECM) communicates with all databus-capable control modules via a CAN databus. These databus-capable control modules are connected via two data bus wires which are twisted together (CAN_High and CAN_Low), and exchange information (messages). Missing information on the databus is recognized as a malfunction and stored. Trouble-free operation of the CAN-Bus requires that it have a terminal resistance.<br>**Possible Causes:**<br>• ECM or TCM has failed<br>• Terminal resistance for CAN-bus are faulty<br>• Can data bus wires have short circuited to each other |
| **DTC: U0404**<br>**Years: 2007, 2008**<br>**Models:** Jetta, GTI, New Beetle, New Beetle Convertible<br>**Engines:** 2.0L, 2.5L, 3.6L<br>**Transmissions:** All | **Invalid Data Received From Gear Shift Control Module Conditions:**<br>Key on, and the PCM detected a software invalid data from the Gear Shift Control Module during its initial startup. The Engine Control Module (ECM) communicates with all databus-capable control modules via a CAN databus. These databus-capable control modules are connected via two data bus wires which are twisted together (CAN_High and CAN_Low), and exchange information (messages). Missing information on the databus is recognized as a malfunction and stored. Trouble-free operation of the CAN-Bus requires that it have a terminal resistance.<br>**Possible Causes:**<br>• ECM has failed<br>• Terminal resistance for CAN-bus are faulty<br>• Can data bus wires have short circuited to each other<br>• Gear shift control module has failed |
| **DTC: U0405**<br>**Years: 2007, 2008**<br>**Models:** Passat, Jetta<br>**Engines:** 2.0L, 3.6L<br>**Transmissions:** All | **Invalid Data Received From Cruise Control Module Conditions:**<br>Key on, and the ECM detected that it is getting invalid data from the Cruise Control Module during its initial startup. The Engine Control Module (ECM) communicates with all databus-capable control modules via a CAN databus. These databus-capable control modules are connected via two data bus wires which are twisted together (CAN_High and CAN_Low), and exchange information (messages). Missing information on the databus is recognized as a malfunction and stored. Trouble-free operation of the CAN-Bus requires that it have a terminal resistance.<br>**Possible Causes:**<br>• ECM has failed<br>• Terminal resistance for CAN-bus are faulty<br>• Can data bus wires have short circuited to each other<br>• The cruise control module has failed |
| **DTC: U0415**<br>**Years: 2007, 2008**<br>**Models:** Passat, Jetta, GTI, New Beetle<br>**Engines:** 2.0L, 2.5L, 3.6L<br>**Transmissions:** All | **Software Incompatibility with Anti-Lock Brake System Control Module Conditions:**<br>Key on, and the ECM detected a software incompatibility condition with the Anti-Lock Brake System Control Module (FICM) during its initial startup. The Engine Control Module (ECM) communicates with all databus-capable control modules via a CAN databus. These databus-capable control modules are connected via two data bus wires which are twisted together (CAN_High and CAN_Low), and exchange information (messages). Missing information on the databus is recognized as a malfunction and stored. Trouble-free operation of the CAN-Bus requires that it have a terminal resistance.<br>**Possible Causes:**<br>• ECM has failed<br>• Terminal resistance for CAN-bus are faulty<br>• Can data bus wires have short circuited to each other<br>• The ABS control module has failed |

| DTC | Trouble Code Title, Conditions & Possible Causes |
|---|---|
| **DTC: U0417**<br>**Years:** 2007, 2008<br>**Models:** Passat, Jetta<br>**Engines:** 2.0L, 3.6L<br>**Transmissions:** All | **Invalid Data Received From Park Brake Control Module Conditions:**<br>Key on, and the ECM detected invalid data from the Park Brake Control Module during its initial startup. The Engine Control Module (ECM) communicates with all databus-capable control modules via a CAN databus. These databus-capable control modules are connected via two data bus wires which are twisted together (CAN_High and CAN_Low), and exchange information (messages). Missing information on the databus is recognized as a malfunction and stored. Trouble-free operation of the CAN-Bus requires that it have a terminal resistance.<br>**Possible Causes:**<br>• ECM has failed<br>• Terminal resistance for CAN-bus are faulty<br>• Can data bus wires have short circuited to each other<br>• Park brake control module has failed |
| **DTC: U0433**<br>**Years:** 2007, 2008<br>**Models:** Passat, Jetta<br>**Engines:** 2.0L, 3.6L<br>**Transmissions:** All | **Invalid Data Received from Cruise Control Front Distance Range Sensor Conditions:**<br>Key on, and the ECM detected invalid data from the Cruise Control Module during its initial startup. The Engine Control Module (ECM) communicates with all databus-capable control modules via a CAN databus. These databus-capable control modules are connected via two data bus wires which are twisted together (CAN_High and CAN_Low), and exchange information (messages). Missing information on the databus is recognized as a malfunction and stored. Trouble-free operation of the CAN-Bus requires that it have a terminal resistance.<br>**Possible Causes:**<br>• ECM has failed<br>• Terminal resistance for CAN-bus are faulty<br>• Can data bus wires have short circuited to each other<br>• Cruise control distance range sensor has failed |

*DIESEL ENGINE TROUBLE CODE LIST*

## Diesel Engine OBD II Trouble Code List (P0xxx Codes)

| DTC | Trouble Code Title, Conditions & Possible Causes |
|---|---|
| **DTC: P0030**<br>**Years:** 2007<br>**Models:** Jetta, New Beetle<br>**Engines:** 1.9L<br>**Transmissions:** All | **HO2S Heater (Bank 1 Sensor 1) Control Circuit Malfunction Conditions:**<br>Engine started, battery voltage must be at least 11.5v, all electrical components must be off, the ground between the engine and the chassis must be well connected, the exhaust system must be properly sealed between the catalytic converter and the cylinder head, the coolant temperature must be 80 degrees Celsius, and the oxygen sensor heater for oxygen sensor before the catalytic converter must be properly functioning. The ECM detected the HO2S signal was in a negative voltage range referred to as "character shift downward". This code sets when the HO2S signal remains in a low state (usually less than 156 mv). In effect, it does not switch properly between 0.1v and 1.1v in closed loop operation.<br>**Possible Causes:**<br>• HO2S is contaminated (due to presence of silicone in fuel)<br>• HO2S signal and ground circuit wires crossed in wiring harness<br>• HO2S signal circuit is shorted to sensor or chassis ground<br>• HO2S element has failed (internal short condition)<br>• ECM has failed |
| **DTC: P0031**<br>**Years:** 2007<br>**Models:** Jetta, New Beetle<br>**Engines:** 1.9L<br>**Transmissions:** All | **HO2S Heater (Bank 1 Sensor 1) Circuit Low Input Conditions:**<br>Engine started, battery voltage must be at least 11.5v, all electrical components must be off, the ground between the engine and the chassis must be well connected, the exhaust system must be properly sealed between the catalytic converter and the cylinder head, the coolant temperature must be 80 degrees Celsius, and the oxygen sensor heater for oxygen sensor before the catalytic converter must be properly functioning. The ECM detected the HO2S signal was in a negative voltage range referred to as "character shift downward". This code sets when the HO2S signal remains in a low state. In effect, it does not switch properly in the closed loop operation. The HO2S (before the three-way catalytic converter) has a short circuit to ground that has lasted longer than 200 seconds<br>**Possible Causes:**<br>• HO2S is contaminated (due to presence of silicone in fuel)<br>• HO2S signal and ground circuit wires crossed in wiring harness<br>• HO2S signal circuit is shorted to sensor or chassis ground<br>• HO2S element has failed (internal short condition)<br>• ECM has failed |

| DTC | Trouble Code Title, Conditions & Possible Causes |
|---|---|
| **DTC: P0032**<br>**Years:** 2007<br>**Models:** Jetta, New Beetle<br>**Engines:** 1.9L<br>**Transmissions:** All | **HO2S Heater (Bank 1 Sensor 1) Circuit High Input Conditions:**<br>Engine started, battery voltage must be at least 11.5v, all electrical components must be off, the ground between the engine and the chassis must be well connected, the exhaust system must be properly sealed between the catalytic converter and the cylinder head, the coolant temperature must be 80 degrees Celsius, and the oxygen sensor heater for oxygen sensor before the catalytic converter must be properly functioning. The ECM detected the HO2S signal remained in a high state.<br>**Note: The HO2S signal circuit may be shorted to the heater power circuit due to tracking inside of the HO2S connector. Remove the connector and visually inspect the connector for signs of oil or water.**<br>**Possible Causes:**<br>    • HO2S signal shorted to heater power circuit inside connector<br>    • HO2S signal circuit shorted to ground or to system voltage<br>    • ECM has failed |
| **DTC: P0100**<br>**Years:** 2007<br>**Models:** Jetta, New Beetle<br>**Engines:** 1.9L<br>**Transmissions:** All | **Mass Air Flow Circuit Range/Performance Conditions**<br>Engine running, with the system voltage more than 11.0v, and the temperature must be at least 185-degrees (F) and all electrical equipment (A/C, lights, etc) must be off. The ECM has detected that the MAF signal was less than the required minimum or out of a calculated range with the engine (or undetectable) for a certain period of time.<br>**Possible Causes:**<br>    • Mass air flow (MAF) sensor has failed or is damaged<br>    • MAF sensor signal circuit is open, shorted to ground or to B+<br>    • ECM has failed |
| **DTC: P0101**<br>**2T, MIL: Yes**<br>**Years:** 2007<br>**Models:** Jetta, New Beetle<br>**Engines:** 1.9L<br>**Transmissions:** All | **Mass or Volume Air Flow Circ. Range/Performance Conditions**<br>Engine running, with the system voltage more than 11.0v, and the temperature must be at least 185-degrees (F) and all electrical equipment (A/C, lights, etc) must be off. The ECM has detected that the MAF signal was out of a calculated range with the engine (or undetectable) for a certain period of time.<br>**Possible Causes:**<br>    • Mass air flow (MAF) sensor has failed or is damaged<br>    • ECM has failed<br>    • Signal and ground wires of Mass Air Flow (MAF) sensor has short circuited |
| **DTC: P0102**<br>**2T, MIL: Yes**<br>**Years:** 2007<br>**Models:** Jetta, New Beetle<br>**Engines:** 1.9L<br>**Transmissions:** All | **MAF Sensor Circuit Low Input Conditions:**<br>Key on, engine started, and the ECM detected the MAF sensor signal was less than the minimum calibrated value. The engine temperature must beat least 185-degrees (F) and all electrical equipment (A/C, lights, etc) must be off. The ECM has detected that the MAF signal was less than the required minimum.<br>**Possible Causes:**<br>    • Check for leaks between MAF sensor and throttle valve control module<br>    • Voltage supply faulty.<br>    • Sensor power circuit open from fuel pump relay to MAF sensor<br>    • Sensor signal circuit open (may be disconnected) from ECM and MAF<br>    • Faulty ground cable resistance between connector terminal 1 and Ground<br>    • MAF Sensor malfunction |
| **DTC: P0103**<br>**2T, MIL: Yes**<br>**Years:** 2007<br>**Models:** Jetta, New Beetle<br>**Engines:** 1.9L<br>**Transmissions:** All | **MAF Sensor Circuit High Input Conditions:**<br>Key on, engine started, and the ECM detected the MAF sensor signal was more than the minimum calibrated value. The engine temperature must beat least 185-degrees (F) and all electrical equipment (A/C, lights, etc) must be off. The ECM has detected that the MAF signal was more than the required minimum.<br>**Possible Causes:**<br>    • Check for leaks between MAF sensor and throttle valve control module<br>    • Voltage supply faulty.<br>    • Sensor power circuit open from fuel pump relay to MAF sensor<br>    • Sensor signal circuit open (may be disconnected) from ECM and MAF<br>    • Faulty ground cable resistance between connector terminal 1 and Ground<br>    • MAF Sensor malfunction |

| DTC | Trouble Code Title, Conditions & Possible Causes |
|---|---|
| **DTC: P0105**<br>**2T, MIL: Yes**<br>**Years:** 2007<br>**Models:** Jetta, New Beetle<br>**Engines:** 1.9L<br>**Transmissions:** All | **Barometric Pressure Sensor Circuit Conditions:**<br>Engine started, the temperature must beat least 185-degrees (F) and all electrical equipment (A/C, lights, etc) must be off. The ECM detected the BARO sensor was out of range during the CCM test. The BARO sensor signal should be in 4.5v.<br>**Possible Causes:**<br>• Sensor has deteriorated (response time too slow) or has failed<br>• MAP sensor signal circuit is shorted to ground<br>• MAP sensor circuit (5v) is open<br>• MAP sensor is damaged or it has failed<br>• BARO sensor signal circuit is shorted to ground<br>• BARO sensor circuit (5v) is open<br>• BARO sensor is damaged or it has failed<br>• ECM is not connected properly<br>• ECM has failed |
| **DTC: P0106**<br>**2T, MIL: Yes**<br>**Years:** 2007<br>**Models:** Jetta, New Beetle<br>**Engines:** 1.9L<br>**Transmissions:** A/T | **Manifold Absolute Pressure/Barometric Pressure Sensor Circuit Performance Conditions:**<br>Engine started, the temperature must beat least 185-degrees (F) and all electrical equipment (A/C, lights, etc) must be off. The ECM detected the BARO sensor was out of range during the CCM test. The BARO sensor signal should be in 4.5v.<br>**Possible Causes:**<br>• Sensor has deteriorated (response time too slow) or has failed<br>• MAP sensor signal circuit is shorted to ground<br>• MAP sensor circuit (5v) is open<br>• MAP sensor is damaged or it has failed<br>• BARO sensor signal circuit is shorted to ground<br>• BARO sensor circuit (5v) is open<br>• BARO sensor is damaged or it has failed<br>• ECM is not connected properly<br>• ECM has failed |
| **DTC: P0107**<br>**2T, MIL: Yes**<br>**Years:** 2007<br>**Models:** Jetta, New Beetle<br>**Engines:** 1.9L<br>**Transmissions:** All | **Manifold Absolute Pressure/Barometric Pressure Sensor Circuit Low Input Conditions:**<br>Engine started, the temperature must beat least 185-degrees (F) and all electrical equipment (A/C, lights, etc) must be off. The ECM detected the BARO sensor was out of range during the CCM test. The BARO sensor signal should be in 4.5v. The BARO sensor is a variable capacitance unit used to detect altitude.<br>**Possible Causes:**<br>• Sensor has deteriorated (response time too slow) or has failed<br>• MAP sensor signal circuit is shorted to ground<br>• MAP sensor circuit (5v) is open<br>• MAP sensor is damaged or it has failed<br>• BARO sensor signal circuit is shorted to ground<br>• BARO sensor circuit (5v) is open<br>• BARO sensor is damaged or it has failed<br>• ECM is not connected properly<br>• ECM has failed |
| **DTC: P0108**<br>**2T, MIL: Yes**<br>**Years:** 2007<br>**Models:** Jetta, New Beetle<br>**Engines:** 1.9L<br>**Transmissions:** All | **Manifold Absolute Pressure/Barometric Sensor Circuit High Input Conditions:**<br>Engine started, the temperature must beat least 185-degrees (F) and all electrical equipment (A/C, lights, etc) must be off. The ECM detected the BARO sensor was out of range during the CCM test. The BARO sensor signal should be in 4.5v. The BARO sensor is a variable capacitance unit used to detect altitude.<br>**Possible Causes:**<br>• Sensor has deteriorated (response time too slow) or has failed<br>• MAP sensor signal circuit is shorted to ground<br>• MAP sensor circuit (5v) is open<br>• MAP sensor is damaged or it has failed<br>• BARO sensor signal circuit is shorted to ground<br>• BARO sensor circuit (5v) is open<br>• BARO sensor is damaged or it has failed<br>• ECM is not connected properly<br>• ECM has failed |

| DTC | Trouble Code Title, Conditions & Possible Causes |
|---|---|
| **DTC: P0112**<br>**Years:** 2007<br>**Models:** Jetta, New Beetle<br>**Engines:** 1.9L<br>**Transmissions**: All | **Intake Air Temperature Sensor Circuit Low Input Conditions:**<br>Key on or engine running, the temperature must beat least 185-degrees (F) and all electrical equipment (A/C, lights, etc) must be off; and the ECM detected the IAT sensor signal was less than the self-test minimum. This is a thermistor-type sensor with a variable resistance that changes when exposed to different temperatures. This means: the higher the temperature, the lower the resistance value.<br>**Possible Causes:**<br>• IAT sensor signal circuit is grounded (check wiring & connector)<br>• Resistance value between sockets 33 and 36 out of range<br>• IAT sensor has an open circuit<br>• IAT sensor is damaged or it has failed<br>• ECM has failed |
| **DTC: P0113**<br>**Years:** 2007<br>**Models:** Jetta, New Beetle<br>**Engines:** 1.9L<br>**Transmissions**: All | **Intake Air Temperature Sensor Circuit High Input Conditions:**<br>Key on or engine running, the temperature must beat least 185-degrees (F) and all electrical equipment (A/C, lights, etc) must be off; and the ECM detected the IAT sensor signal was more than the self-test maximum. This is a thermistor-type sensor with a variable resistance that changes when exposed to different temperatures. This means: the higher the temperature, the lower the resistance value.<br>**Possible Causes:**<br>• IAT sensor signal circuit is open (inspect wiring & connector)<br>• IAT sensor signal circuit is shorted<br>• Resistance value between sockets 33 and 36 out of range<br>• IAT sensor is damaged or it has failed<br>• ECM has failed |
| **DTC: P0116**<br>**2T, MIL: Yes**<br>**Years:** 2007<br>**Models:** Jetta, New Beetle<br>**Engines:** 1.9L<br>**Transmissions**: All | **ECT Sensor / CHT Sensor Signal Range/Performance Conditions:**<br>Engine started (cold), battery voltage must be 11.5, and all equipment must be off. The ECM detected the ECT sensor exceeded the required calibrated value, or the engine is at idle and doesn't reach operating temperature quickly enough; the Catalyst, Fuel System, HO2S and Misfire Monitor did not complete, or the timer expired. Testing completion of procedure, the engine's temperature must rise uniformily during idle.<br>**Possible Causes:**<br>• Check for low coolant level or incorrect coolant mixture<br>• ECM detects a short circuit wiring in the ECT<br>• CHT sensor is out-of-calibration or it has failed<br>• ECT sensor is out-of-calibration or it has failed |
| **DTC: P0117**<br>**2T, MIL: Yes**<br>**Years:** 2007<br>**Models:** Jetta, New Beetle<br>**Engines:** 1.9L<br>**Transmissions**: All | **ECT Sensor Circuit Low Input Conditions:**<br>Engine started (cold), battery voltage must be 11.5, and all equipment must be off. The ECM detected the ECT sensor signal was less than the self-test minimum. This is a thermistor-type sensor with a variable resistance that changes when exposed to different temperatures<br>**Possible Causes:**<br>• ECT sensor signal circuit is grounded in the wiring harness<br>• ECT sensor doesn't react to changes in temperature<br>• ECT sensor is damaged or the ECM has failed |
| **DTC: P0118**<br>**2T, MIL: Yes**<br>**Years:** 2007<br>**Models:** Jetta, New Beetle<br>**Engines:** 1.9L<br>**Transmissions**: All | **ECT Sensor Circuit High Input Conditions:**<br>Engine started (cold), battery voltage must be 11.5, and all equipment must be off. The ECM detected the ECT sensor signal was more than the self-test maximum. This is a thermistor-type sensor with a variable resistance that changes when exposed to different temperatures<br>**Possible Causes:**<br>• ECT sensor signal circuit is open (inspect wiring & connector)<br>• ECT sensor signal circuit is shorted to ground<br>• ECT sensor is damaged or it has failed<br>• ECM has failed |
| **DTC: P0121**<br>**Years:** 2007<br>**Models:** Jetta, New Beetle<br>**Engines:** 1.9L<br>**Transmissions**: All | **TP Sensor Signal Range/Performance Conditions:**<br>Engine started; then immediately following a condition where the engine was running under at off-idle, the ECM detected the TP sensor signal indicated the throttle did not return to its previous closed position during the Rationality test.<br>**Possible Causes:**<br>• Throttle plate is binding, dirty or sticking<br>• Throttle valve is damaged or dirty<br>• Throttle valve control module is faulty<br>• TP sensor signal circuit open (inspect wiring & connector)<br>• TP sensor ground circuit open (inspect wiring & connector)<br>• TP sensor and/or control module is damaged or has failed<br>• MAF sensor signal is damaged, has failed or a short is present |

| DTC | Trouble Code Title, Conditions & Possible Causes |
|---|---|
| **DTC: P0123**<br>**Years:** 2007<br>**Models:** Jetta, New Beetle<br>**Engines:** 1.9L<br>**Transmissions:** All | **TP Sensor Circuit High Input Conditions:**<br>Engine started, at idle, the temperature must be at least 80 degrees Celsius. The ECM detected the TP sensor signal was more than the self-test maximum during testing.<br>**Possible Causes:**<br>• TP sensor not seated correctly in housing (may be damaged)<br>• TP sensor signal is circuit shorted to ground or system voltage<br>• TP sensor ground circuit is open (check the wiring harness)<br>• TP sensor and/or ECM has failed |
| **DTC: P0128**<br>**2T, MIL: Yes**<br>**Years:** 2007<br>**Models:** Jetta, New Beetle<br>**Engines:** 1.9L<br>**Transmissions:** All | **Intake Air Temperature Sensor 2 Circuit High Input Conditions:**<br>Engine started, vehicle driven for over 10 minutes, and the ECM detected the engine did not reach an engine operating temperature of 160°F after an additional runtime of 2 minutes.<br>**Possible Causes:**<br>• Check the operation of the thermostat (it may be stuck open)<br>• ECT sensor or CHT sensor is out-of-calibration, or has failed<br>• Inspect for low coolant level or an incorrect coolant mixture |
| **DTC: P0135**<br>**Years:** 2007<br>**Models:** Jetta, New Beetle<br>**Engines:** 1.9L<br>**Transmissions:** All | **HO2S (Bank 1 Sensor 1) Heater Circuit Malfunction Conditions:**<br>Engine running, battery voltage 11.5, all electrical components off, ground between engine and chassis well connected and the exhaust system must be properly sealed between catalytic converter and the cylinder head. The ECM detected an unexpected voltage condition, or it detected excessive current draw in the heater circuit during the CCM test.<br>**Possible Causes:**<br>• HO2S heater power circuit is open or heater ground circuit open<br>• HO2S signal tracking (due to oil or moisture in the connector)<br>• HO2S is damaged or has failed<br>• ECM has failed |
| **DTC: P0181**<br>**Years:** 2007<br>**Models:** Jetta, New Beetle<br>**Engines:** 1.9L<br>**Transmissions:** All | **Engine Fuel Temperature Sensor 'A' Circuit Range/Performance Conditions:**<br>Key on or engine running, all electrical components off and coolant temperature at least 80 degrees Celsius; and the ECM detected the Engine Fuel Temperature (EFT) Sensor 'A' signal was intermittent in the self-test.<br>**Possible Causes:**<br>• EFT sensor connector is damaged or shorted<br>• EFT sensor VREF circuit is open or shorted to ground<br>• EFT sensor circuit is shorted to chassis or to sensor ground<br>• EFT sensor is damaged or it has failed<br>• ECM has failed |
| **DTC: P0182**<br>**2T, MIL: Yes**<br>**Years:** 2007<br>**Models:** Jetta, New Beetle<br>**Engines:** 1.9L<br>**Transmissions:** All | **Engine Fuel Temperature Sensor 'A' Circuit Low Input Conditions:**<br>Key on or engine running; and the ECM detected the Engine Fuel Temperature (EFT) Sensor 'A' signal was under the required voltage in the self-test.<br>**Possible Causes:**<br>• EFT sensor connector is damaged or shorted<br>• EFT sensor VREF circuit is open or shorted to ground<br>• EFT sensor circuit is shorted to chassis or to sensor ground<br>• EFT sensor is damaged or it has failed<br>• ECM has failed |
| **DTC: P0183**<br>**2T, MIL: Yes**<br>**Years:** 2007<br>**Models:** Jetta, New Beetle<br>**Engines:** 1.9L<br>**Transmissions:** All | **Engine Fuel Temperature Sensor 'A' Circuit High Input Conditions:**<br>Key on or engine running, all electrical components off and coolant temperature at least 80 degrees Celsius; and the ECM detected the Engine Fuel Temperature (EFT) Sensor 'A' signal was intermittent in the self-test.<br>**Possible Causes:**<br>• EFT sensor connector is damaged, loose or open<br>• EFT sensor signal circuit is open or it is shorted to VREF (5v)<br>• EFT sensor is damaged or it has failed<br>• ECM has failed |
| **DTC: P0200**<br>**Years:** 2007<br>**Models:** Jetta, New Beetle<br>**Engines:** 1.9L<br>**Transmissions:** All | **Injector Circuit Open Conditions:**<br>Key on or engine running, all electrical components off and coolant temperature at least 80 degrees Celsius; and the ECM detected the fuel injector control circuit was not function within specs<br>**Possible Causes:**<br>• Injector connector is damaged, open or shorted<br>• Injector control circuit is open, shorted to ground or to power<br>• ECM has failed (the injector driver circuit may be damaged) |

| DTC | Trouble Code Title, Conditions & Possible Causes |
|---|---|
| **DTC: P0201**<br>**Years:** 2007<br>**Models:** Jetta, New Beetle<br>**Engines:** 1.9L<br>**Transmissions:** All | **Cylinder 1 Injector Circuit Malfunction Conditions:**<br>Engine started, and the ECM detected the fuel injector "1" control circuit was in a high state when it should have been low, or in a low state when it should have been high (wiring harness & injector okay).<br>**Possible Causes:**<br>• Injector 1 connector is damaged, open or shorted<br>• Injector 1 control circuit is open, shorted to ground or to power<br>• ECM has failed (the injector driver circuit may be damaged) |
| **DTC: P0202**<br>**Years:** 2007<br>**Models:** Jetta, New Beetle<br>**Engines:** 1.9L<br>**Transmissions:** All | **Cylinder 2 Injector Circuit Malfunction Conditions:**<br>Engine started, and the ECM detected the fuel injector "2" control circuit was in a high state when it should have been low, or in a low state when it should have been high (wiring harness & injector okay).<br>**Possible Causes:**<br>• Injector 2 connector is damaged, open or shorted<br>• Injector 2 control circuit is open, shorted to ground or to power<br>• ECM has failed (the injector driver circuit may be damaged) |
| **DTC: P0203**<br>**Years:** 2007<br>**Models:** Jetta, New Beetle<br>**Engines:** 1.9L<br>**Transmissions:** All | **Cylinder 3 Injector Circuit Malfunction Conditions:**<br>Engine started, and the ECM detected the fuel injector "3" control circuit was in a high state when it should have been low, or in a low state when it should have been high (wiring harness & injector okay).<br>**Possible Causes:**<br>• Injector 3 connector is damaged, open or shorted<br>• Injector 3 control circuit is open, shorted to ground or to power<br>• ECM has failed (the injector driver circuit may be damaged) |
| **DTC: P0204**<br>**Years:** 2007<br>**Models:** Jetta, New Beetle<br>**Engines:** 1.9L<br>**Transmissions:** All | **Cylinder 4 Injector Circuit Malfunction Conditions:**<br>Engine started, and the ECM detected the fuel injector "4" control circuit was in a high state when it should have been low, or in a low state when it should have been high (wiring harness & injector okay).<br>**Possible Causes:**<br>• Injector 4 connector is damaged, open or shorted<br>• Injector 4 control circuit is open, shorted to ground or to power<br>• ECM has failed (the injector driver circuit may be damaged) |
| **DTC: P0216**<br>**2T, MIL: Yes**<br>**Years:** 2007<br>**Models:** Jetta, New Beetle<br>**Engines:** 1.9L<br>**Transmissions:** All | **Injector/Injection Timing Control Malfunction Conditions:**<br>Engine started, and the ECM detected a malfunction in the injector timing control<br>**Possible Causes:**<br>• The cold start injector has failed<br>• Circuit wires have short circuited to each other ground or power<br>• The diesel direct fuel injection engine control module has failed |
| **DTC: P0219**<br>**Years:** 2007<br>**Models:** Passat<br>**Engines:** 2.0L<br>**Transmissions:** All | **Engine Over-Speed Condition Conditions:**<br>Engine started, and the ECM determined the vehicle had been driven in a manner that caused the engine to over-speed, and to exceed the engine speed calibration limit stored in memory.<br>**Possible Causes:**<br>• Engine operated in the wrong transmission gear position<br>• Excessive engine speed with gear selector in Neutral position<br>• Wheel slippage due to wet, muddy or snowing conditions |
| **DTC: P0225**<br>**2T, MIL: Yes**<br>**Years:** 2007<br>**Models:** Jetta, New Beetle<br>**Engines:** 1.9L<br>**Transmissions:** All | **Throttle Position Sensor 'C' Signal Voltage Supply Conditions:**<br>Engine started, battery voltage at least 11.5v, all electrical components off, ground connections between engine and chassis well connected, coolant temperature at least 80-degrees Celicius and the throttle valve must not be damaged or dirty; and the ECM detected the TP Sensor 'B' circuit was out of its normal operating range during a condition with the throttle wide open, or with it completely closed. The throttle valve activation occurs via an electric motor (throttle drive) in the throttle valve control module. It is activated by the ECM according to specifications of the two sensors, Throttle Position Sensor and Accelerator Pedal Position Sensor 2. Slowly depress accelerator pedal up to Wide Open Throttle (WOT) stop while observing the percentage display on the PID data function of the scan tool. The percentage display must increase uniformly.<br>**Possible Causes:**<br>• Throttle body is damaged<br>• Throttle linkage is binding or sticking<br>• ETC TP Sensor 'C' signal circuit to the ECM is open<br>• ETC TP Sensor 'C' ground circuit is open<br>• ETC TP Sensor 'C' is damaged or it has failed |

| DTC | Trouble Code Title, Conditions & Possible Causes |
|---|---|
| **DTC: P0226**<br>**2T, MIL: Yes**<br>**Years:** 2007<br>**Models:** Jetta, New Beetle<br>**Engines:** 1.9L<br>**Transmissions**: All | **Throttle Position Sensor 'C' Signal Performance Conditions:**<br>Engine started, battery voltage at least 11.5v, all electrical components off, ground connections between engine and chassis well connected, coolant temperature at least 80-degrees Celicius and the throttle valve must not be damaged or dirty; and the ECM detected the TP Sensor 'B' circuit was out of its normal operating range during a condition with the throttle wide open, or with it completely closed. The throttle valve activation occurs via an electric motor (throttle drive) in the throttle valve control module. It is activated by the ECM according to specifications of the two sensors, Throttle Position Sensor and Accelerator Pedal Position Sensor 2. Slowly depress accelerator pedal up to Wide Open Throttle (WOT) stop while observing the percentage display on the PID data function of the scan tool. The percentage display must increase uniformly..<br>**Possible Causes:**<br>• Throttle body is damaged<br>• Throttle linkage is binding or sticking<br>• ETC TP Sensor 'C' signal circuit to the ECM is open<br>• ETC TP Sensor 'C' ground circuit is open<br>• ETC TP Sensor 'C' is damaged or it has failed |
| **DTC: P0228**<br>**2T, MIL: Yes**<br>**Years:** 2007<br>**Models:** Jetta, New Beetle<br>**Engines:** 1.9L<br>**Transmissions**: All | **Throttle Position Sensor 'C' Circuit High Input Conditions:**<br>Engine started, battery voltage at least 11.5v, all electrical components off, ground connections between engine and chassis well connected, coolant temperature at least 80-degrees Celicius and the throttle valve must not be damaged or dirty; and the ECM detected the TP Sensor 'B' circuit was out of its normal operating range during a condition with the throttle wide open, or with it completely closed. The throttle valve activation occurs via an electric motor (throttle drive) in the throttle valve control module. It is activated by the ECM according to specifications of the two sensors, Throttle Position Sensor and Accelerator Pedal Position Sensor 2. Slowly depress accelerator pedal up to Wide Open Throttle (WOT) stop while observing the percentage display on the PID data function of the scan tool. The percentage display must increase uniformly.<br>**Possible Causes:**<br>• ETC TP Sensor 'C' connector is damaged or open<br>• ETC TP Sensor 'C' signal circuit is open<br>• ETC TP Sensor 'C' signal circuit is shorted to VREF (5v)<br>• ETC TP Sensor 'C' is damaged or it has failed |
| **DTC: P0230**<br>**Years:** 2007<br>**Models:** Jetta, New Beetle<br>**Engines:** 1.9L<br>**Transmissions**: All | **Fuel Pump Primary Circuit Malfunction Conditions:**<br>Engine started, battery voltage at least 11.5v, all electrical components off, ground connections between engine and chassis well connected, coolant temperature at least 80-degrees Celicius. The ECM detected high current in fuel pump or fuel shutoff valve (FSV) circuit, or it detected voltage with the valve off, or it did not detect voltage on the circuit. The circuit is used to energize the fuel pump relay at key on or while running. Fuel pressure value should be 3000 to 5000 kPa at idle.<br>**Possible Causes:**<br>• FP or FSV circuit is open or shorted<br>• Fuel pump relay VPWR circuit open<br>• Fuel pump relay is damaged or has failed<br>• Fuel pressure sensor has failed<br>• Fuel pump control module is faulty<br>• ECM has failed |
| **DTC: P0231**<br>**Years:** 2007<br>**Models:** Jetta, New Beetle<br>**Engines:** 1.9L<br>**Transmissions**: All | **Fuel Pump Secondary Circuit Low Conditions:**<br>Engine started, battery voltage at least 11.5v, all electrical components off, ground connections between engine and chassis well connected, coolant temperature at least 80-degrees Celicius. The ECM detected high current in fuel pump or fuel shutoff valve (FSV) circuit, or it detected voltage with the valve off, or it did not detect voltage on the circuit. The circuit is used to energize the fuel pump relay at key on or while running. Fuel pressure value should be 3000 to 5000 kPa at idle.<br>**Possible Causes:**<br>• FP or FSV circuit is open or shorted<br>• Fuel pump relay VPWR circuit open<br>• Fuel pump relay is damaged or has failed<br>• Fuel pressure sensor has failed<br>• Fuel pump control module is faulty<br>• ECM has failed |

| DTC | Trouble Code Title, Conditions & Possible Causes |
|---|---|
| **DTC: P0232**<br>**Years:** 2007<br>**Models:** Jetta, New Beetle<br>**Engines:** 1.9L<br>**Transmissions:** All | **Fuel Pump Secondary Circuit High Conditions:**<br>Engine started, battery voltage at least 11.5v, all electrical components off, ground connections between engine and chassis well connected, coolant temperature at least 80-degrees Celicius. The ECM detected high current in fuel pump or fuel shutoff valve (FSV) circuit, or it detected voltage with the valve off, or it did not detect voltage on the circuit. The circuit is used to energize the fuel pump relay at key on or while running. Fuel pressure value should be 3000 to 5000 kPa at idle.<br>**Possible Causes:**<br>• FP or FSV circuit is open or shorted<br>• Fuel pump relay VPWR circuit open<br>• Fuel pump relay is damaged or has failed<br>• Fuel pressure sensor has failed<br>• Fuel pump control module is faulty<br>• ECM has failed |
| **DTC: P0234**<br>**2T, MIL: Yes**<br>**Years:** 2007<br>**Models:** Jetta, New Beetle<br>**Engines:** 1.9L<br>**Transmissions:** All | **Turbo/Supercharger Overboost Condition Conditions:**<br>Engine started, battery voltage at least 11.5v, all electrical components off, ground connections between engine and chassis well connected, coolant temperature at least 80-degrees Celicius. The ECM detected an operating condition that could harm the engine or automatic transmission.<br>**Possible Causes:**<br>• Ignition misfire condition exceeds the calibrated threshold<br>• Knock sensor circuit has failed, or excessive knock detected<br>• Low speed fuel pump relay not switching properly<br>• Transmission oil temperature beyond the calibrated threshold<br>• Shaft bearing of charge pressure regulator valve in turbocharger is blocked |
| **DTC: P0236**<br>**Years:** 2007<br>**Models:** Jetta, New Beetle<br>**Engines:** 1.9L<br>**Transmissions:** All | **Turbocharger Boost Sensor (A) Circ Control Range/Performance Conditions:**<br>Engine started, battery voltage at least 11.5v, all electrical components off, ground connections between engine and chassis well connected, coolant temperature at least 80-degrees Celicius. The ECM detected an operating condition that could harm the engine or automatic transmission.<br>**Possible Causes:**<br>• Charge air pressure sensor is faulty<br>• Voltage supply to the charge air pressure sensor is open or shorted<br>• Check the charge air system for leaks<br>• Recirculating valve for turbocharger is faulty<br>• Turbocharging system is damaged or not functioning properly<br>• Turbocharger recirculating valve is faulty<br>• Vacuum diaphragm for turbocharger is out of adjustment<br>• Wastegate bypass regulator valve is faulty<br>• Boost sensor has failed<br>• ECM has failed |
| **DTC: P0237**<br>**Years:** 2007<br>**Models:** Jetta, New Beetle<br>**Engines:** 1.9L<br>**Transmissions:** All | **Turbocharger Boost Sensor (A) Circ Low Input Conditions:**<br>Engine started, battery voltage at least 11.5v, all electrical components off, ground connections between engine and chassis well connected, coolant temperature at least 80-degrees Celicius. The ECM detected an operating condition that could harm the engine or automatic transmission.<br>**Possible Causes:**<br>• Charge air pressure sensor is faulty<br>• Voltage supply to the charge air pressure sensor is open or shorted<br>• Check the charge air system for leaks<br>• Recirculating valve for turbocharger is faulty<br>• Turbocharging system is damaged or not functioning properly<br>• Turbocharger recirculating valve is faulty<br>• Vacuum diaphragm for turbocharger is out of adjustment<br>• Wastegate bypass regulator valve is faulty<br>• Boost sensor has failed<br>• ECM has failed |

| DTC | Trouble Code Title, Conditions & Possible Causes |
|---|---|
| **DTC: P0238**<br>**Years:** 2007<br>**Models:** Jetta, New Beetle<br>**Engines:** 1.9L<br>**Transmissions:** All | **Turbocharger Boost Sensor (A) Circ High Input Conditions:**<br>Engine started, battery voltage at least 11.5v, all electrical components off, ground connections between engine and chassis well connected, coolant temperature at least 80-degrees Celicius. The ECM detected an operating condition that could harm the engine or automatic transmission.<br>**Possible Causes:**<br>• Charge air pressure sensor is faulty<br>• Voltage supply to the charge air pressure sensor is open or shorted<br>• Check the charge air system for leaks<br>• Recirculating valve for turbocharger is faulty<br>• Turbocharging system is damaged or not functioning properly<br>• Turbocharger recirculating valve is faulty<br>• Vacuum diaphragm for turbocharger is out of adjustment<br>• Wastegate bypass regulator valve is faulty<br>• Boost sensor has failed<br>• ECM has failed |
| **DTC: P0243**<br>**Years:** 2007<br>**Models:** Jetta, New Beetle<br>**Engines:** 1.9L<br>**Transmissions:** All | **Turbocharger Boost Bypass Solenoid (A) Circuit Open/Short Circuit Conditions:**<br>Engine started, battery voltage at least 11.5v, all electrical components off, ground connections between engine and chassis well connected, coolant temperature at least 80-degrees Celicius. The ECM detected an unexpected voltage condition on the Bypass Solenoid control circuit<br>**Possible Causes:**<br>• Bypass solenoid power supply circuit is open<br>• Bypass solenoid control circuit is open, shorted to ground or system power<br>• Bypass solenoid assembly is damaged or has failed<br>• Charge air pressure sensor is faulty<br>• Voltage supply to the charge air pressure sensor is open or shorted<br>• Check the charge air system for leaks<br>• Recirculating valve for turbocharger is faulty<br>• Turbocharging system is damaged or not functioning properly<br>• Turbocharger recirculating valve is faulty<br>• Vacuum diaphragm for turbocharger is out of adjustment<br>• Wastegate bypass regulator valve is faulty<br>• Boost sensor has failed<br>• ECM has failed |
| **DTC: P0245**<br>**2T, MIL:** Yes<br>**Years:** 2007<br>**Models:** Jetta, New Beetle<br>**Engines:** 1.9L<br>**Transmissions:** All | **Turbocharger Boost Bypass Solenoid (A) Circuit Low Input/Short to Ground Conditions:**<br>Engine started, battery voltage at least 11.5v, all electrical components off, ground connections between engine and chassis well connected, coolant temperature at least 80-degrees Celicius. The ECM detected an unexpected voltage condition on the Bypass Solenoid control circuit<br>**Possible Causes:**<br>• Bypass solenoid power supply circuit is open<br>• Bypass solenoid control circuit is open, shorted to ground or system power<br>• Bypass solenoid assembly is damaged or has failed<br>• Charge air pressure sensor is faulty<br>• Voltage supply to the charge air pressure sensor is open or shorted<br>• Check the charge air system for leaks<br>• Recirculating valve for turbocharger is faulty<br>• Turbocharging system is damaged or not functioning properly<br>• Turbocharger recirculating valve is faulty<br>• Vacuum diaphragm for turbocharger is out of adjustment<br>• Wastegate bypass regulator valve is faulty<br>• Boost sensor has failed<br>• ECM has failed |

| DTC | Trouble Code Title, Conditions & Possible Causes |
|---|---|
| **DTC: P0246**<br>**2T, MIL: Yes**<br>**Years:** 2007<br>**Models:** Jetta, New Beetle<br>**Engines:** 1.9L<br>**Transmissions:** All | **Turbocharger Boost Bypass Solenoid (A) Circuit High Input/Short to B+ Conditions:**<br>Engine started, battery voltage at least 11.5v, all electrical components off, ground connections between engine and chassis well connected, coolant temperature at least 80-degrees Celicius. The ECM detected an unexpected voltage condition on the Bypass Solenoid control circuit<br>**Possible Causes:**<br>• Bypass solenoid power supply circuit is open<br>• Bypass solenoid control circuit is open, shorted to ground or system power<br>• Bypass solenoid assembly is damaged or has failed<br>• Charge air pressure sensor is faulty<br>• Voltage supply to the charge air pressure sensor is open or shorted<br>• Check the charge air system for leaks<br>• Recirculating valve for turbocharger is faulty<br>• Turbocharging system is damaged or not functioning properly<br>• Turbocharger recirculating valve is faulty<br>• Vacuum diaphragm for turbocharger is out of adjustment<br>• Wastegate bypass regulator valve is faulty<br>• Boost sensor has failed<br>• ECM has failed |
| **DTC: P0251**<br>**Years:** 2007<br>**Models:** Jetta, New Beetle<br>**Engines:** 1.9L<br>**Transmissions:** All | **Injection Pump Fuel Metering Control "A" (Cam/Rotor/Injector) Conditions:**<br>Engine running and the battery voltage must be 11.5v. The ECM detected a injection pump metering control problem.<br>**Note: The quantity adjuster is an electro-magnetic swiveling positioner which is controlled by the ECM via a directed duty cycle. The eccentric shaft on the quantity adjuster moves the modulating piston on the high pressure piston thereby regulating the quantity of fuel injected. The modulating piston displacement sensor informs the ECM of the position of Quantity Adjuster there determines the amount injected.**<br>**Possible Causes:**<br>• Circuit wires have short circuited to each other ground or power<br>• The diesel direct fuel injection engine control module has failed |
| **DTC: P0252**<br>**Years:** 2007<br>**Models:** Jetta, New Beetle<br>**Engines:** 1.9L<br>**Transmissions:** All | **Injection Pump Metering Control (A) Range/Performance Conditions:**<br>Engine running and the battery voltage must be 11.5v. The ECM detected a injection pump metering control problem.<br>**Note: The quantity adjuster is an electro-magnetic swiveling positioner which is controlled by the ECM via a directed duty cycle. The eccentric shaft on the quantity adjuster moves the modulating piston on the high pressure piston thereby regulating the quantity of fuel injected. The modulating piston displacement sensor informs the ECM of the position of Quantity Adjuster there determines the amount injected.**<br>**Possible Causes:**<br>• Circuit wires have short circuited to each other ground or power<br>• The diesel direct fuel injection engine control module has failed |
| **DTC: P0263**<br>**Years:** 2007<br>**Models:** Jetta, New Beetle<br>**Engines:** 1.9L<br>**Transmissions:** All | **Cylinder 1 Contribution/Balance Conditions:**<br>Key on or engine running; and the ECM detected an unexpected low or high voltage condition on the injector circuit<br>**Possible Causes:**<br>• Check activation of valve for pump/injector<br>• ECM has failed<br>• Fuel pump relay has failed<br>• Fuel injectors may have malfunctioned |
| **DTC: P0266**<br>**Years:** 2007<br>**Models:** Jetta, New Beetle<br>**Engines:** 1.9L<br>**Transmissions:** All | **Cylinder 2 Contribution/Balance Conditions:**<br>Key on or engine running; and the ECM detected an unexpected low or high voltage condition on the injector circuit<br>**Possible Causes:**<br>• Check activation of valve for pump/injector<br>• ECM has failed<br>• Fuel pump relay has failed<br>• Fuel injectors may have malfunctioned |
| **DTC: P0269**<br>**Years:** 2007<br>**Models:** Jetta, New Beetle<br>**Engines:** 1.9L<br>**Transmissions:** All | **Cylinder 3 Contribution/Balance Conditions:**<br>Key on or engine running; and the ECM detected an unexpected low or high voltage condition on the injector circuit<br>**Possible Causes:**<br>• Check activation of valve for pump/injector<br>• ECM has failed<br>• Fuel pump relay has failed<br>• Fuel injectors may have malfunctioned |

| DTC | Trouble Code Title, Conditions & Possible Causes |
|---|---|
| **DTC: P0272**<br>**Years:** 2007<br>**Models:** Jetta, New Beetle<br>**Engines:** 1.9L<br>**Transmissions:** All | **Cylinder 4 Contribution/Balance Conditions:**<br>Key on or engine running; and the ECM detected an unexpected low or high voltage condition on the injector circuit<br>**Possible Causes:**<br>• Check activation of valve for pump/injector<br>• ECM has failed<br>• Fuel pump relay has failed<br>• Fuel injectors may have malfunctioned |
| **DTC: P0299**<br>**Years:** 2007<br>**Models:** Jetta, New Beetle<br>**Engines:** 1.9L<br>**Transmissions:** All | **Turbocharger Underboost Conditions:**<br>Engine started, battery voltage at least 11.5v, all electrical components off, ground connections between engine and chassis well connected, coolant temperature at least 80-degrees Celcius. The ECM detected an operating condition that could harm the engine or automatic transmission.<br>**Possible Causes:**<br>• Charge air pressure sensor has failed<br>• Charge air system has leaks<br>• Recirculating valve for turbocharger is faulty<br>• Turbocharging system is faulty<br>• Vacuum diaphragm for turbocharger needs adjusting<br>• Wastegate bypass regulator valve is faulty |
| **DTC: P0300**<br>**2T, MIL: Yes**<br>**Years:** 2007<br>**Models:** Jetta, New Beetle<br>**Engines:** 1.9L<br>**Transmissions:** All | **Random Misfire Detected Conditions:**<br>Engine running under positive torque conditions, and the ECM detected a misfire or uneven engine running in two or more cylinders.<br>**Note: If the misfire is severe, the MIL will flash on/off on the 1st trip!**<br>**Possible Causes:**<br>• Base engine mechanical fault that affects two or more cylinders<br>• Fuel metering fault that affects two or more cylinders<br>• Fuel pressure too low or too high, fuel supply contaminated<br>• EVAP system problem or the EVAP canister is fuel saturated<br>• EGR valve is stuck open or the PCV system has a vacuum leak<br>• Ignition system fault (coil, plugs) affecting two or more cylinders<br>• MAF sensor contamination (it can cause a very lean condition)<br>• Vehicle driven while very low on fuel (less than 1/8 of a tank) |
| **DTC: P0301**<br>**2T, MIL: Yes**<br>**Years:** 2007<br>**Models:** Jetta, New Beetle<br>**Engines:** 1.9L<br>**Transmissions:** All | **Cylinder Number 1 Misfire Detected Conditions:**<br>Engine running under positive torque conditions, and the ECM detected a misfire or uneven engine function.<br>**Note: If the misfire is severe, the MIL will flash on/off on the 1st trip!**<br>**Possible Causes:**<br>• Air leak in the intake manifold, or in the EGR or ECM system<br>• Base engine mechanical problem<br>• Fuel delivery component problem (i.e., a contaminated, dirty or sticking fuel injector)<br>• Fuel pump relay defective<br>• Ignition coil fuses have failed<br>• Ignition system problem (dirty damaged coil or plug)<br>• Engine speed (RPM) sensor has failed<br>• Camshaft position sensors have failed<br>• Ignition coil is faulty<br>• Spark plugs are not working properly or are not gapped properly |
| **DTC: P0302**<br>**2T, MIL: Yes**<br>**Years:** 2007<br>**Models:** Jetta, New Beetle<br>**Engines:** 1.9L<br>**Transmissions:** All | **Cylinder Number 2 Misfire Detected Conditions:**<br>Engine running under positive torque conditions, and the ECM detected a misfire or uneven engine function.<br>**Note: If the misfire is severe, the MIL will flash on/off on the 1st trip!**<br>**Possible Causes:**<br>• Air leak in the intake manifold, or in the EGR or ECM system<br>• Base engine mechanical problem<br>• Fuel delivery component problem (i.e., a contaminated, dirty or sticking fuel injector)<br>• Fuel pump relay defective<br>• Ignition coil fuses have failed<br>• Ignition system problem (dirty damaged coil or plug)<br>• Engine speed (RPM) sensor has failed<br>• Camshaft position sensors have failed<br>• Ignition coil is faulty<br>• Spark plugs are not working properly or are not gapped properly |

| DTC | Trouble Code Title, Conditions & Possible Causes |
|---|---|
| **DTC: P0303**<br>**2T, MIL: Yes**<br>**Years:** 2007<br>**Models:** Jetta, New Beetle<br>**Engines:** 1.9L<br>**Transmissions:** All | **Cylinder Number 3 Misfire Detected Conditions:**<br>Engine running under positive torque conditions, and the ECM detected a misfire or uneven engine function.<br>**Note: If the misfire is severe, the MIL will flash on/off on the 1st trip!**<br>**Possible Causes:**<br>• Air leak in the intake manifold, or in the EGR or ECM system<br>• Base engine mechanical problem<br>• Fuel delivery component problem (i.e., a contaminated, dirty or sticking fuel injector)<br>• Fuel pump relay defective<br>• Ignition coil fuses have failed<br>• Ignition system problem (dirty damaged coil or plug)<br>• Engine speed (RPM) sensor has failed<br>• Camshaft position sensors have failed<br>• Ignition coil is faulty<br>• Spark plugs are not working properly or are not gapped properly |
| **DTC: P0304**<br>**2T, MIL: Yes**<br>**Years:** 2007<br>**Models:** Jetta, New Beetle<br>**Engines:** 1.9L<br>**Transmissions:** All | **Cylinder Number 4 Misfire Detected Conditions:**<br>Engine running under positive torque conditions, and the ECM detected a misfire or uneven engine function.<br>**Note: If the misfire is severe, the MIL will flash on/off on the 1st trip!**<br>**Possible Causes:**<br>• Air leak in the intake manifold, or in the EGR or ECM system<br>• Base engine mechanical problem<br>• Fuel delivery component problem (i.e., a contaminated, dirty or sticking fuel injector)<br>• Fuel pump relay defective<br>• Ignition coil fuses have failed<br>• Ignition system problem (dirty damaged coil or plug)<br>• Engine speed (RPM) sensor has failed<br>• Camshaft position sensors have failed<br>• Ignition coil is faulty<br>• Spark plugs are not working properly or are not gapped properly |
| **DTC: P0321**<br>**2T, MIL: Yes**<br>**Years:** 2007<br>**Models:** Jetta, New Beetle<br>**Engines:** 1.9L<br>**Transmissions:** All | **Ignition/Distributor Engine Speed Input Circuit Range/Performance Conditions:**<br>Engine started, vehicle driven, and the ECM detected the engine speed signal was more than the calibrated value.<br>**Note: The engine will not start if there is no speed signal. If the speed signal fails when the engine is running, it will cause the engine to stall immediately.**<br>**Possible Causes:**<br>• Engine speed sensor has failed or is damaged<br>• ECM has failed<br>• Sensor wheel is damaged or doesn't fit properly<br>• Sensor wheel spacer isn't seated properly |
| **DTC: P0401**<br>**2T, MIL: Yes**<br>**Years:** 2007<br>**Models:** Jetta, New Beetle<br>**Engines:** 1.9L<br>**Transmissions:** All | **Exhaust Gas Recirculation Malfunction (ESM System) Conditions:**<br>Engine started, battery voltage must be at least 11.5v, all electrical components must be off, parking brake must be engaged (to keep daytime driving lights off), automatic transmission selector must be in park and the ground between the engine and the chassis must be well connected. The ECM detected a problem in the EGR ESM system. Run the KOER self-test, make sure the EGR valve is set by switching the ignition on and then off, wait one minute, start the car and let it idle for at least one minute.<br>**Possible Causes:**<br>• EGR valve hoses are damaged, leaking or restricted<br>• EGR vacuum regulator solenoid valve has failed or is dirty<br>• EGR valve source hoses loose or connected wrong<br>• Potentiometer for the EGR is faulty<br>• EGR valve connector is damaged, loose or shorted<br>• EGR valve is damaged or it has failed<br>• ECM has failed |

| DTC | Trouble Code Title, Conditions & Possible Causes |
|---|---|
| **DTC: P0402**<br>**2T, MIL: Yes**<br>**Years:** 2007<br>**Models:** Jetta, New Beetle<br>**Engines:** 1.9L<br>**Transmissions**: All | **EGR Flow At Idle Speed Detected (ESM System) Conditions:**<br>Engine started, battery voltage must be at least 11.5v, all electrical components must be off, parking brake must be engaged (to keep daytime driving lights off), automatic transmission selector must be in park and the ground between the engine and the chassis must be well connected. The ECM detected a problem in the EGR ESM system. Run the KOER self-test, make sure the EGR valve is set by switching the ignition on and then off, wait one minute, start the car and let it idle for at least one minute.<br>**Possible Causes:**<br>• EGR valve hoses are damaged, leaking or restricted<br>• EGR vacuum regulator solenoid valve has failed or is dirty<br>• EGR valve source hoses loose or connected wrong<br>• Potentiometer for the EGR is faulty<br>• EGR valve connector is damaged, loose or shorted<br>• EGR valve is damaged or it has failed<br>• ECM has failed |
| **DTC: P0403**<br>**Years:** 2007<br>**Models:** Jetta, New Beetle<br>**Engines:** 1.9L<br>**Transmissions**: All | **EGR Solenoid Circuit Malfunction Conditions:**<br>Engine started, battery voltage must be at least 11.5v, all electrical components must be off, parking brake must be engaged (to keep daytime driving lights off), automatic transmission selector must be in park and the ground between the engine and the chassis must be well connected. The ECM detected a problem in the EGR ESM system. Run the KOER self-test, make sure the EGR valve is set by switching the ignition on and then off, wait one minute, start the car and let it idle for at least one minute.<br>**Possible Causes:**<br>• EGR valve hoses are damaged, leaking or restricted<br>• EGR vacuum regulator solenoid valve has failed or is dirty<br>• EGR valve source hoses loose or connected wrong<br>• Potentiometer for the EGR is faulty<br>• EGR valve connector is damaged, loose or shorted<br>• EGR valve is damaged or it has failed<br>• ECM has failed |
| **DTC: P0404**<br>**Years:** 2007<br>**Models:** Jetta, New Beetle<br>**Engines:** 1.9L<br>**Transmissions**: All | **EGR Control Circuit Range/Performance Conditions:**<br>Engine started, battery voltage must be at least 11.5v, all electrical components must be off, parking brake must be engaged (to keep daytime driving lights off), automatic transmission selector must be in park and the ground between the engine and the chassis must be well connected. The ECM detected a problem in the EGR ESM system. Run the KOER self-test, make sure the EGR valve is set by switching the ignition on and then off, wait one minute, start the car and let it idle for at least one minute.<br>**Possible Causes:**<br>• EGR valve hoses are damaged, leaking or restricted<br>• EGR vacuum regulator solenoid valve has failed or is dirty<br>• EGR valve source hoses loose or connected wrong<br>• Potentiometer for the EGR is faulty<br>• EGR valve connector is damaged, loose or shorted<br>• EGR valve is damaged or it has failed<br>• ECM has failed |
| **DTC: P0600**<br>**Years:** 2007<br>**Models:** Jetta, New Beetle<br>**Engines:** 1.9L<br>**Transmissions**: All | **Serial Communication Link (Data BUS) Message Missing Conditions:**<br>The Engine Control Module (ECM) communicates with all databus-capable control modules via a CAN databus. These databus-capable control modules are connected via two data bus wires which are twisted together (CAN_High and CAN_Low), and exchange information (messages). Missing information on the databus is recognized as a malfunction and stored. Trouble-free operation of the CAN-Bus requires that it have a terminal resistance. This central terminal resistor is located in the Engine Control Module (ECM).<br>**Possible Causes:**<br>• ECM has failed<br>• CAN data bus wires have short circuited to each other |
| **DTC: P0601**<br>**2T, MIL: Yes**<br>**Years:** 2007<br>**Models:** Jetta, New Beetle<br>**Engines:** 1.9L<br>**Transmissions**: All | **Internal Control Module Memory Check Sum Error Conditions:**<br>Key on, the ECM has detected a programming error<br>**Possible Causes:**<br>• Battery terminal corrosion, or loose battery connection<br>• Connection to the ECM interrupted, or the circuit has been opened<br>• Reprogramming error has occurred<br>• ECM has failed and needs replacement. Remember to check for Aftermarket Performance Products before replacing a ECM. |

| DTC | Trouble Code Title, Conditions & Possible Causes |
|---|---|
| **DTC: P0605**<br>**2T, MIL: Yes**<br>**Years:** 2007<br>**Models:** Jetta, New Beetle<br>**Engines:** 1.9L<br>**Transmissions:** All | **ECM Read Only Memory (ROM) Test Error Conditions:**<br>Key on, and the ECM detected a ROM test error (ROM inside ECM is corrupted). The ECM is normally replaced if this code has set.<br>**Possible Causes:**<br>• An attempt was made to change the module calibration, or a module programming error may have occurred<br>• Clear the trouble codes and then check for this trouble code. If it resets, the ECM has failed and needs replacement.<br>• Aftermarket performance products may have been installed.<br>• The Transmission Control Module (TCM) has failed. |
| **DTC: P0606**<br>**Years:** 2007<br>**Models:** Jetta, New Beetle<br>**Engines:** 1.9L<br>**Transmissions:** All | **ECM Internal Communication Error Conditions:**<br>Key on, and the ECM detected an internal communications register read back error during the initial key on check period.<br>**Possible Causes:**<br>• Clear the trouble codes and then check for this trouble code. If it resets, the ECM has failed and needs replacement.<br>• Remember to check for signs of Aftermarket Performance Products installation before replacing the ECM. |
| **DTC: P0607**<br>**Years:** 2007<br>**Models:** Jetta, New Beetle<br>**Engines:** 1.9L<br>**Transmissions:** All | **Control Module Performance Conditions:**<br>Key on, and the ECM detected an internal communications register read back error during the initial key on check period.<br>**Possible Causes:**<br>• Clear the trouble codes and then check for this trouble code. If it resets, the ECM has failed and needs replacement.<br>• Remember to check for signs of Aftermarket Performance Products installation before replacing the ECM. |
| **DTC: P0614**<br>**2T, MIL: Yes**<br>**Years:** 2007<br>**Models:** Jetta, New Beetle<br>**Engines:** 1.9L<br>**Transmissions:** All | **ECM / TCM Incompatible Conditions:**<br>Key on, and the ECM detected a communication error between the Transmission control module and the ECM<br>**Possible Causes:**<br>• TCM failed<br>• ECM failed<br>• Circuit shorting between ECM and TCM<br>• Replacement control module ID doesn't match old control module ID |
| **DTC: P0642**<br>**Years:** 2007<br>**Models:** Jetta, New Beetle<br>**Engines:** 1.9L<br>**Transmissions:** All | **Sensor Reference Voltage "A" Circuit Low Conditions:**<br>Engine started, battery voltage must be at least 11.5v, all electrical components must be off, parking brake must be engaged (to keep daytime driving lights off), automatic transmission selector must be in park, and the ground between the engine and the chassis must be well connected.<br>**Possible Causes:**<br>• Circuit harness connector contacts are corroded or ingressed of water<br>• Circuit wires have shorted to each other, to battery or ground<br>• Automatic Transmission Hydraulic Pressure Sensor 1 has failed<br>• Solenoid valves in valve body are faulty<br>• Transmission Control Module (TCM) needs replacing<br>• Transmission Input Speed (RPM) Sensor has failed<br>• Transmission Output Speed (RPM) Sensor has failed |
| **DTC: P0643**<br>**Years:** 2007<br>**Models:** Jetta, New Beetle<br>**Engines:** 1.9L<br>**Transmissions:** All | **Sensor Reference Voltage "A" Circuit High Conditions:**<br>Engine started, battery voltage must be at least 11.5v, all electrical components must be off, parking brake must be engaged (to keep daytime driving lights off), automatic transmission selector must be in park, and the ground between the engine and the chassis must be well connected.<br>**Possible Causes:**<br>• Circuit harness connector contacts are corroded or ingressed of water<br>• Circuit wires have shorted to each other, to battery or ground<br>• Automatic Transmission Hydraulic Pressure Sensor 1 has failed<br>• Solenoid valves in valve body are faulty<br>• Transmission Control Module (TCM) needs replacing<br>• Transmission Input Speed (RPM) Sensor has failed<br>• Transmission Output Speed (RPM) Sensor has failed |
| **DTC: P0652**<br>**Years:** 2007<br>**Models:** Jetta, New Beetle<br>**Engines:** 1.9L<br>**Transmissions:** All | **Sensor Reference Voltage "B" Circuit Low Conditions:**<br>Engine started, battery voltage must be at least 11.5v, all electrical components must be off, parking brake must be engaged (to keep daytime driving lights off), automatic transmission selector must be in park, and the ground between the engine and the chassis must be well connected.<br>**Possible Causes:**<br>• Circuit harness connector contacts are corroded or ingressed of water<br>• Circuit wires have shorted to each other, to battery or ground<br>• Automatic Transmission Hydraulic Pressure Sensor 1 has failed<br>• Solenoid valves in valve body are faulty<br>• Transmission Control Module (TCM) needs replacing<br>• Transmission Input Speed (RPM) Sensor has failed<br>• Transmission Output Speed (RPM) Sensor has failed |

| DTC | Trouble Code Title, Conditions & Possible Causes |
|---|---|
| **DTC: P0653**<br>**Years:** 2007<br>**Models:** Jetta, New Beetle<br>**Engines:** 1.9L<br>**Transmissions:** All | **Sensor Reference Voltage "B" Circuit High Conditions:**<br>Engine started, battery voltage must be at least 11.5v, all electrical components must be off, parking brake must be engaged (to keep daytime driving lights off), automatic transmission selector must be in park, and the ground between the engine and the chassis must be well connected.<br>**Possible Causes:**<br>• Circuit harness connector contacts are corroded or ingressed of water<br>• Circuit wires have shorted to each other, to battery or ground<br>• Automatic Transmission Hydraulic Pressure Sensor 1 has failed<br>• Solenoid valves in valve body are faulty<br>• Transmission Control Module (TCM) needs replacing<br>• Transmission Input Speed (RPM) Sensor has failed<br>• Transmission Output Speed (RPM) Sensor has failed |
| **DTC: P0670**<br>**2T, MIL:** Yes<br>**Years:** 2007<br>**Models:** Jetta, New Beetle<br>**Engines:** 1.9L<br>**Transmissions:** All | **Glow Plug Module Control Circuit Conditions:**<br>Key on, and the ECM detected an unexpected voltage condition on the Glow Plug Lamp circuit during the CCM test. The Glow Plug Lamp remains "on" for 1-12 seconds (depending on the Glow Plug relay on-time which can vary from 1 and 120 seconds).<br>**Possible Causes:**<br>• Glow plug lamp circuit is open or shorted to ground<br>• Glow plug relay control circuit is open or shorted to ground<br>• Glow plug relay power circuit is open (test the 12GA fuse link)<br>• Glow plug relay is damaged or it has failed |
| **DTC: P0671**<br>**2T, MIL:** Yes<br>**Years:** 2007<br>**Models:** Jetta, New Beetle<br>**Engines:** 1.9L<br>**Transmissions:** All | **Cylinder 1 Glow Plug Circuit Conditions:**<br>Key on, and the ECM detected an unexpected voltage condition on the Glow Plug Lamp circuit during the CCM test. The Glow Plug Lamp remains "on" for 1-12 seconds (depending on the Glow Plug relay on-time which can vary from 1 and 120 seconds).<br>**Possible Causes:**<br>• Glow plug lamp circuit is open or shorted to ground<br>• Glow plug relay control circuit is open or shorted to ground<br>• Glow plug relay power circuit is open (test the 12GA fuse link)<br>• Glow plug relay is damaged or it has failed |
| **DTC: P0672**<br>**2T, MIL:** Yes<br>**Years:** 2007<br>**Models:** Jetta, New Beetle<br>**Engines:** 1.9L<br>**Transmissions:** All | **Cylinder 2 Glow Plug Circuit Conditions:**<br>Key on, and the ECM detected an unexpected voltage condition on the Glow Plug Lamp circuit during the CCM test. The Glow Plug Lamp remains "on" for 1-12 seconds (depending on the Glow Plug relay on-time which can vary from 1 and 120 seconds).<br>**Possible Causes:**<br>• Glow plug lamp circuit is open or shorted to ground<br>• Glow plug relay control circuit is open or shorted to ground<br>• Glow plug relay power circuit is open (test the 12GA fuse link)<br>• Glow plug relay is damaged or it has failed |
| **DTC: P0673**<br>**2T, MIL:** Yes<br>**Years:** 2007<br>**Models:** Jetta, New Beetle<br>**Engines:** 1.9L<br>**Transmissions:** All | **Cylinder 3 Glow Plug Circuit Conditions:**<br>Key on, and the ECM detected an unexpected voltage condition on the Glow Plug Lamp circuit during the CCM test. The Glow Plug Lamp remains "on" for 1-12 seconds (depending on the Glow Plug relay on-time which can vary from 1 and 120 seconds).<br>**Possible Causes:**<br>• Glow plug lamp circuit is open or shorted to ground<br>• Glow plug relay control circuit is open or shorted to ground<br>• Glow plug relay power circuit is open (test the 12GA fuse link)<br>• Glow plug relay is damaged or it has failed |
| **DTC: P0674**<br>**2T, MIL:** Yes<br>**Years:** 2007<br>**Models:** Jetta, New Beetle<br>**Engines:** 1.9L<br>**Transmissions:** All | **Cylinder 4 Glow Plug Circuit Conditions:**<br>Key on, and the ECM detected an unexpected voltage condition on the Glow Plug Lamp circuit during the CCM test. The Glow Plug Lamp remains "on" for 1-12 seconds (depending on the Glow Plug relay on-time which can vary from 1 and 120 seconds).<br>**Possible Causes:**<br>• Glow plug lamp circuit is open or shorted to ground<br>• Glow plug relay control circuit is open or shorted to ground<br>• Glow plug relay power circuit is open (test the 12GA fuse link)<br>• Glow plug relay is damaged or it has failed |
| **DTC: P0684**<br>**2T, MIL:** Yes<br>**Years:** 2007<br>**Models:** Jetta, New Beetle<br>**Engines:** 1.9L<br>**Transmissions:** All | **Glow Plug Control Module to PCM Communication Circuit Range/Performance Conditions:**<br>Key on, and the ECM detected an unexpected voltage condition on the Glow Plug Lamp circuit during the CCM test. The Glow Plug Lamp remains "on" for 1-12 seconds (depending on the Glow Plug relay on-time which can vary from 1 and 120 seconds).<br>**Possible Causes:**<br>• Glow plug lamp circuit is open or shorted to ground<br>• Glow plug relay control circuit is open or shorted to ground<br>• Glow plug relay power circuit is open (test the 12GA fuse link)<br>• Glow plug relay is damaged or it has failed |

| DTC | Trouble Code Title, Conditions & Possible Causes |
|---|---|
| **DTC: P0700**<br>**2T, MIL: Yes**<br>**Years:** 2007<br>**Models:** Jetta, New Beetle<br>**Engines:** 1.9L<br>**Transmissions:** All | **Transmission Control System Malfunction Conditions:**<br>Engine started, battery voltage must be at least 11.5v, all electrical components must be off, parking brake must be engaged (to keep daytime driving lights off), automatic transmission selector must be in park, and the ground between the engine and the chassis must be well connected. The ECM detected a malfunction int the transmission control system.<br>**Possible Causes:**<br>• Circuit harness connector contacts are corroded or ingressed of water<br>• Circuit wires have shorted to each other, to battery or ground<br>• Automatic Transmission Hydraulic Pressure Sensor 1 has failed<br>• Solenoid valves in valve body are faulty<br>• Transmission Input Speed (RPM) Sensor has failed<br>• Transmission Output Speed (RPM) Sensor has failed<br>• Engine Control Module (ECM) is faulty<br>• Voltage supply for Engine Control Module (ECM) is faulty<br>• Transmission Control Module (TCM) is faulty |
| **DTC: P0705**<br>**2T, MIL: Yes**<br>**Years:** 2007<br>**Models:** Jetta, New Beetle<br>**Engines:** 1.9L<br>**Transmissions:** A/T | **TR Sensor Circuit Malfunction Conditions:**<br>Engine started, battery voltage must be at least 11.5v, all electrical components must be off, parking brake must be engaged (to keep daytime driving lights off), automatic transmission selector must be in park, and the ground between the engine and the chassis must be well connected. The ECM detected a voltage outside the normal performance range to allow the system to properly function.<br>**Possible Causes:**<br>• Circuit harness connector contacts are corroded or ingressed of water<br>• Circuit wires have shorted to each other, to battery or ground<br>• Automatic Transmission Hydraulic Pressure Sensor 1 has failed<br>• Solenoid valves in valve body are faulty<br>• Transmission Input Speed (RPM) Sensor has failed<br>• Transmission Output Speed (RPM) Sensor has failed<br>• Engine Control Module (ECM) is faulty<br>• Voltage supply for Engine Control Module (ECM) is faulty<br>• Transmission Control Module (TCM) is faulty |
| **DTC: P0710**<br>**Years:** 2007<br>**Models:** Jetta, New Beetle<br>**Engines:** 1.9L<br>**Transmissions:** A/T | **TFT Sensor Circuit Malfunction Conditions:**<br>Engine started, battery voltage must be at least 11.5v, all electrical components must be off, parking brake must be engaged (to keep daytime driving lights off), automatic transmission selector must be in park, and the ground between the engine and the chassis must be well connected. The ECM detected the Transmission Fluid Temperature (TFT) sensor value was not close its normal operating temperature.<br>**Possible Causes:**<br>• ATF is low, contaminated, dirty or burnt<br>• TFT sensor signal circuit has a high resistance condition<br>• TFT sensor is out-of-calibration ("skewed") or it has failed<br>• ECM has failed |
| **DTC: P0715**<br>**2T, MIL: Yes**<br>**Years:** 2007<br>**Models:** Jetta, New Beetle<br>**Engines:** 1.9L<br>**Transmissions:** A/T | **Input/Turbine Speed Sensor Circuit Malfunction Conditions:**<br>Engine started, vehicle driven with the vehicle speed sensor indicating more than 1 mph, and the ECM detected the Transmission Vehicle Speed Sensor signals were erratic, or that they were missing for a period of time.<br>**Possible Causes:**<br>• TVSS signal circuit is open<br>• TVSS signal is shorted to chassis ground<br>• TVSS signal is shorted to sensor ground<br>• TVSS assembly is damaged or it has failed<br>• ECM has failed |
| **DTC: P0717**<br>**Years:** 2007<br>**Models:** Jetta, New Beetle<br>**Engines:** 1.9L<br>**Transmissions:** A/T | **Transmission Speed Shaft Sensor Signal Intermittent Conditions:**<br>Engine started, vehicle speed sensor indicating over 1 mph, and the ECM detected an intermittent loss of TSS signals (i.e., the TSS signals were erratic, irregular or missing).<br>**Possible Causes:**<br>• TSS connector is damaged, loose or shorted<br>• TSS signal circuit has an intermittent open condition<br>• TSS signal circuit has an intermittent short to ground condition<br>• TSS assembly is damaged or is has failed<br>• ECM has failed |

| DTC | Trouble Code Title, Conditions & Possible Causes |
|---|---|
| **DTC: P0722**<br>**2T, MIL: Yes**<br>**Years:** 2007<br>**Models:** Jetta, New Beetle<br>**Engines:** 1.9L<br>**Transmissions:** A/T | **A/T Output Speed Sensor No Signal Conditions:**<br>Engine started, and the ECM did not detect any Vehicle Speed Sensor (VSS) sensor signals upon initial vehicle movement.<br>**Possible Causes:**<br>• After market add-on devices interfering with the VSS signal<br>• VSS sensor wiring is misrouted, damaged or shorting<br>• ECM and/or TCM has failed |
| **DTC: P0725**<br>**2T, MIL: Yes**<br>**Years:** 2007<br>**Models:** Jetta, New Beetle<br>**Engines:** 1.9L<br>**Transmissions:** A/T | **Engine Speed Input Circuit Malfunction Conditions:**<br>The Transmission Control Module (TCM) does not receive a signal from the Engine Control Module (ECM).<br>**Possible Causes:**<br>• The TCM circuit is shorting to ground, B+ or is open<br>• TCM has failed<br>• ECM has failed |
| **DTC: P0726**<br>**Years:** 2007<br>**Models:** Jetta, New Beetle<br>**Engines:** 1.9L<br>**Transmissions:** A/T | **Engine Speed Input Circuit Range/Performance Conditions:**<br>The Engine Speed (RPM) Sensor detects engine speed and reference marks. Without an engine speed signal, the engine will not start. If the engine speed signal fails while the engine is running, the engine will stop immediately.<br>**Note: There is a larger-sized gap on the sensor wheel. This gap is the reference mark and does not mean that the sensor wheel is damaged.**<br>**Possible Causes:**<br>• Engine speed sensor has failed<br>• Circuit is shorting to ground, B+ or is open<br>• Sensor wheel is damaged, run out or not properly secured<br>• ECM has failed |
| **DTC: P0727**<br>**Years:** 2007<br>**Models:** Jetta, New Beetle<br>**Engines:** 1.9L<br>**Transmissions:** A/T | **Engine Speed Input Circuit No Signal Conditions:**<br>The Engine Speed (RPM) Sensor detects engine speed and reference marks. Without an engine speed signal, the engine will not start. If the engine speed signal fails while the engine is running, the engine will stop immediately.<br>**Note: There is a larger-sized gap on the sensor wheel. This gap is the reference mark and does not mean that the sensor wheel is damaged.**<br>**Possible Causes:**<br>• Engine speed sensor has failed<br>• Circuit is shorting to ground, B+ or is open<br>• Sensor wheel is damaged, run out or not properly secured<br>• ECM has failed |
| **DTC: P0730**<br>**2T, MIL: Yes**<br>**Years:** 2007<br>**Models:** Jetta, New Beetle<br>**Engines:** 1.9L<br>**Transmissions:** A/T | **Gear Incorrect Ratio Conditions:**<br>Engine started, battery voltage must be at least 11.5v, all electrical components must be off, and the ground between the engine and the chassis must be well connected. The ECM detected an incorrect gear ratio.<br>**Possible Causes:**<br>• ATF level is low<br>• Circuit harness connector contacts are corroded or ingressed of water<br>• Circuit wires have shorted to each other, to battery or ground<br>• Automatic Transmission Hydraulic Pressure Sensor 1 has failed<br>• Solenoid valves in valve body are faulty<br>• Transmission Control Module (TCM) needs replacing<br>• Transmission Input Speed (RPM) Sensor has failed<br>• Transmission Output Speed (RPM) Sensor has failed |
| **DTC: P0731**<br>**2T, MIL: Yes**<br>**Years:** 2007<br>**Models:** Jetta, New Beetle<br>**Engines:** 1.9L<br>**Transmissions:** A/T | **Incorrect First Gear Ratio Conditions:**<br>Engine started, vehicle operating with 1st gear commanded "on", and the ECM detected an incorrect 1st gear ratio during the test.<br>**Possible Causes:**<br>• 1st Gear solenoid harness connector not properly seated<br>• 1st Gear solenoid signal shorted to ground, or open<br>• 1st Gear solenoid wiring harness connector is damaged<br>• 1st Gear solenoid is damaged or not properly installed<br>• ATF level is low<br>• Circuit harness connector contacts are corroded or ingressed of water<br>• Circuit wires have shorted to each other, to battery or ground<br>• Automatic Transmission Hydraulic Pressure Sensor 1 has failed<br>• Transmission Control Module (TCM) needs replacing<br>• Transmission Input Speed (RPM) Sensor has failed<br>• Transmission Output Speed (RPM) Sensor has failed |

| DTC | Trouble Code Title, Conditions & Possible Causes |
|---|---|
| **DTC: P0732**<br>**2T, MIL:** Yes<br>**Years:** 2007<br>**Models:** Jetta, New Beetle<br>**Engines:** 1.9L<br>**Transmissions:** A/T | **Incorrect Second Gear Ratio Conditions:**<br>Engine started, vehicle operating with 2nd Gear commanded "on", and the ECM detected an incorrect 2nd gear ratio during the test.<br>**Possible Causes:**<br>• 2nd Gear solenoid harness connector not properly seated<br>• 2nd Gear solenoid signal shorted to ground, or open<br>• 2nd Gear solenoid wring harness connector is damaged<br>• 2nd Gear solenoid is damaged or not properly installed<br>• ATF level is low<br>• Circuit harness connector contacts are corroded or ingressed of water<br>• Circuit wires have shorted to each other, to battery or ground<br>• Automatic Transmission Hydraulic Pressure Sensor 1 has failed<br>• Transmission Control Module (TCM) needs replacing<br>• Transmission Input Speed (RPM) Sensor has failed<br>• Transmission Output Speed (RPM) Sensor has failed |
| **DTC: P0733**<br>**2T, MIL:** Yes<br>**Years:** 2007<br>**Models:** Jetta, New Beetle<br>**Engines:** 1.9L<br>**Transmissions:** A/T | **Incorrect Third Gear Ratio Conditions:**<br>Engine started, vehicle operating with 3rd Gear commanded "on", and the ECM detected an incorrect 3rd gear ratio during the test.<br>**Possible Causes:**<br>• 3rd Gear solenoid harness connector not properly seated<br>• 3rd Gear solenoid signal shorted to ground, or open<br>• 3rd Gear solenoid wiring harness connector is damaged<br>• 3rd Gear solenoid is damaged or not properly installed<br>• ATF level is low<br>• Circuit harness connector contacts are corroded or ingressed of water<br>• Circuit wires have shorted to each other, to battery or ground<br>• Automatic Transmission Hydraulic Pressure Sensor 1 has failed<br>• Transmission Control Module (TCM) needs replacing<br>• Transmission Input Speed (RPM) Sensor has failed<br>• Transmission Output Speed (RPM) Sensor has failed |
| **DTC: P0734**<br>**2T, MIL:** Yes<br>**Years:** 2007<br>**Models:** Jetta, New Beetle<br>**Engines:** 1.9L<br>**Transmissions:** A/T | **Incorrect Fourth Gear Ratio Conditions:**<br>Engine started, vehicle operating with 4th Gear commanded "on", and the ECM detected an incorrect 4th gear ratio during the test.<br>**Possible Causes:**<br>• 4th Gear solenoid harness connector not properly seated<br>• 4th Gear solenoid signal shorted to ground, or open<br>• 4th Gear solenoid wiring harness connector is damaged<br>• 4th Gear solenoid is damaged or not properly installed<br>• ATF level is low<br>• Circuit harness connector contacts are corroded or ingressed of water<br>• Circuit wires have shorted to each other, to battery or ground<br>• Automatic Transmission Hydraulic Pressure Sensor 1 has failed<br>• Transmission Control Module (TCM) needs replacing<br>• Transmission Input Speed (RPM) Sensor has failed<br>• Transmission Output Speed (RPM) Sensor has failed |
| **DTC: P0740**<br>**2T, MIL:** Yes<br>**Years:** 2007<br>**Models:** Jetta, New Beetle<br>**Engines:** 1.9L<br>**Transmissions:** A/T | **TCC Solenoid Circuit Malfunction Conditions:**<br>Engine started, KOER Self-Test enabled, vehicle driven at cruise speed, and the ECM did not detect any voltage drop across the TCC solenoid circuit during the test period.<br>**Possible Causes:**<br>• TCC solenoid control circuit is open or shorted to ground<br>• TCC solenoid wiring harness connector is damaged<br>• TCC solenoid is damaged or has failed<br>• ECM has failed |
| **DTC: P0743**<br>**2T, MIL:** Yes<br>**Years:** 2007<br>**Models:** Jetta, New Beetle<br>**Engines:** 1.9L<br>**Transmissions:** A/T | **Torque Converter Clutch Circuit Electrical Malfunction Conditions:**<br>Engine started, KOER Self-Test enabled, vehicle driven at cruise speed, and the ECM did not detect any voltage drop across the TCC solenoid circuit during the test period.<br>**Possible Causes:**<br>• TCC solenoid control circuit is open or shorted to ground<br>• TCC solenoid wiring harness connector is damaged<br>• TCC solenoid is damaged or has failed<br>• ECM has failed |

| DTC | Trouble Code Title, Conditions & Possible Causes |
|---|---|
| **DTC: P0746**<br>**Years:** 2007<br>**Models:** Jetta, New Beetle<br>**Engines:** 1.9L<br>**Transmissions:** A/T | **Pressure Control Solenoid "A" Performance or Stuck Off Conditions:**<br>Engine started, battery voltage must be at least 11.5v, all electrical components must be off, and the ground between the engine and the chassis must be well connected. The ECM detected the pressure control solenoid was in the "stuck off" position.<br>**Possible Causes:**<br>• ATF level is low<br>• Circuit harness connector contacts are corroded or ingressed of water<br>• Circuit wires have shorted to each other, to battery or ground<br>• Automatic Transmission Hydraulic Pressure Sensor 1 has failed<br>• Solenoid valves in valve body are faulty<br>• Transmission Control Module (TCM) needs replacing<br>• Transmission Input Speed (RPM) Sensor has failed<br>• Transmission Output Speed (RPM) Sensor has failed |
| **DTC: P0748**<br>**2T, MIL: Yes**<br>**Years:** 2007<br>**Models:** Jetta, New Beetle<br>**Engines:** 1.9L<br>**Transmissions:** A/T | **Pressure Control Solenoid Electrical Conditions:**<br>The valve body solenoid valve is not receiving a signal.<br>**Possible Causes:**<br>• Pressure control solenoid circuit is shorting to ground<br>• Pressure control solenoid circuit is open<br>• Valve has failed<br>• TCM has failed<br>• ECM has failed |
| **DTC: P0753**<br>**2T, MIL: Yes**<br>**Years:** 2007<br>**Models:** Jetta, New Beetle<br>**Engines:** 1.9L<br>**Transmissions:** A/T | **A/T Shift Solenoid 1/A Circuit Malfunction Conditions:**<br>Engine started, vehicle driven with the solenoid applied, and the ECM detected an unexpected voltage condition on the SS1/A solenoid circuit was incorrect during the test.<br>**Possible Causes:**<br>• SS1/A solenoid control circuit is open<br>• SS1/A solenoid control circuit is shorted to ground<br>• SS1/A solenoid wiring harness connector is damaged<br>• SS1/A solenoid is damaged or has failed<br>• ECM has failed |
| **DTC: P0758**<br>**2T, MIL: Yes**<br>**Years:** 2007<br>**Models:** Jetta, New Beetle<br>**Engines:** 1.9L<br>**Transmissions:** A/T | **A/T Shift Solenoid 2/B Circuit Malfunction Conditions:**<br>Engine started, vehicle driven with the solenoid applied, and the ECM detected an unexpected voltage condition on the SS1/A solenoid circuit was incorrect during the test..<br>**Possible Causes:**<br>• Shift Solenoid 2/B connector is damaged, open or shorted<br>• Shift Solenoid 2/B control circuit is open<br>• Shift Solenoid 2/B control circuit is shorted to ground<br>• Shift Solenoid 2/B is damaged or it has failed<br>• ECM has failed |
| **DTC: P0763**<br>**2T, MIL: Yes**<br>**Years:** 2007<br>**Models:** Jetta, New Beetle<br>**Engines:** 1.9L<br>**Transmissions:** A/T | **A/T Shift Solenoid 3/C Electrical Conditions:**<br>Engine started, vehicle driven with the solenoid applied, and the ECM detected an unexpected voltage condition on the SS3/C solenoid circuit was incorrect during the test..<br>**Possible Causes:**<br>• Shift Solenoid 3/C connector is damaged, open or shorted<br>• Shift Solenoid 3/C control circuit is open<br>• Shift Solenoid 3/C control circuit is shorted to ground<br>• Shift Solenoid 3/C is damaged or it has failed<br>• ECM has failed |
| **DTC: P0773**<br>**2T, MIL: Yes**<br>**Years:** 2007<br>**Models:** Jetta, New Beetle<br>**Engines:** 1.9L<br>**Transmissions:** A/T | **A/T Shift Solenoid E Electrical Conditions:**<br>Engine started, vehicle driven with the solenoid applied, and the ECM detected an unexpected voltage condition on the SS3/D solenoid circuit was incorrect during the test..<br>**Possible Causes:**<br>• Shift Solenoid connector is damaged, open or shorted<br>• Shift Solenoid control circuit is open<br>• Shift Solenoid control circuit is shorted to ground<br>• Shift Solenoid is damaged or it has failed<br>• ECM has failed |

| DTC | Trouble Code Title, Conditions & Possible Causes |
|---|---|
| **DTC: P0778**<br>**Years:** 2007<br>**Models:** Jetta, New Beetle<br>**Engines:** 1.9L<br>**Transmissions:** A/T | **Pressure Control Solenoid "B" Electrical Conditions:**<br>Engine started, vehicle driven with the solenoid applied, and the ECM detected an unexpected voltage condition on the SS3/C solenoid circuit was incorrect during the test..<br>**Possible Causes:**<br>• Shift Solenoid connector is damaged, open or shorted<br>• Shift Solenoid control circuit is open<br>• Shift Solenoid control circuit is shorted to ground<br>• Shift Solenoid is damaged or it has failed<br>• ECM has failed |
| **DTC: P0785**<br>**Years:** 2007<br>**Models:** Jetta, New Beetle<br>**Engines:** 1.9L<br>**Transmissions:** A/T | **Shift/Timing Solenoid Conditions:**<br>Engine running and vehicle driven, the ECM detected a mechanical malfunction within the transmission<br>**Possible Causes:**<br>• Solenoid valves in valve body are faulty<br>• Solenoid circuit is shorting to ground<br>• Solenoid circuit is open<br>• TCM has failed or wiring is shorting<br>• ECM has failed<br>• Mechanical malfunction in transmission |
| **DTC: P0791**<br>**2T, MIL: Yes**<br>**Years:** 2007<br>**Models:** Jetta, New Beetle<br>**Engines:** 1.9L<br>**Transmissions:** A/T | **Intermediate Shaft Speed Sensor "A" Circuit Conditions:**<br>Engine running and vehicle driven, the ECM detected a mechanical malfunction within the transmission<br>**Possible Causes:**<br>• Solenoid valves in valve body are faulty<br>• Solenoid circuit is shorting to ground<br>• Solenoid circuit is open<br>• TCM has failed or wiring is shorting<br>• ECM has failed<br>• Mechanical malfunction in transmission |
| **DTC: P0811**<br>**Years:** 2007<br>**Models:** Jetta, New Beetle<br>**Engines:** 1.9L<br>**Transmissions:** A/T | **Excessive Clutch Slippage Conditions:**<br>Engine started, vehicle driven and the ECM and/or TCM has detected that the clutch has slipped multiple times in a given time frame.<br>**Possible Causes:**<br>• Solenoid valves in valve body are faulty<br>• Solenoid circuit is shorting to ground<br>• Solenoid circuit is open<br>• TCM has failed or wiring is shorting<br>• ECM has failed |
| **DTC: P0863**<br>**2T, MIL: Yes**<br>**Years:** 2007<br>**Models:** Jetta, New Beetle<br>**Engines:** 1.9L<br>**Transmissions:** A/T | **TCM Communication Circuit Conditions:**<br>The Transmission Control Module (ECM) communicates with all databus-capable control modules via a CAN databus. These databus-capable control modules are connected via two data bus wires which are twisted together (CAN_High and CAN_Low), and exchange information (messages). Missing information on the databus is recognized as a malfunction and stored. Trouble-free operation of the CAN-Bus requires that it have a terminal resistance.<br>**Possible Causes:**<br>• ECM has failed<br>• Terminal resistance for CAN-bus are faulty<br>• Can data bus wires have short circuited to each other<br>• TCM has failed |
| **DTC: P0864**<br>**2T, MIL: Yes**<br>**Years:** 2007<br>**Models:** Jetta, New Beetle<br>**Engines:** 1.9L<br>**Transmissions:** A/T | **TCM Communication Circuit Range/Performance Conditions:**<br>The Transmission Control Module (ECM) communicates with all databus-capable control modules via a CAN databus. These databus-capable control modules are connected via two data bus wires which are twisted together (CAN_High and CAN_Low), and exchange information (messages). Missing information on the databus is recognized as a malfunction and stored. Trouble-free operation of the CAN-Bus requires that it have a terminal resistance.<br>**Possible Causes:**<br>• ECM has failed<br>• Terminal resistance for CAN-bus are faulty<br>• Can data bus wires have short circuited to each other<br>• TCM has failed |

| DTC | Trouble Code Title, Conditions & Possible Causes |
|---|---|
| **DTC: P0865**<br>**2T, MIL: Yes**<br>**Years:** 2007<br>**Models:** Jetta, New Beetle<br>**Engines:** 1.9L<br>**Transmissions:** A/T | **TCM Communication Circuit Low Conditions:**<br>The Transmission Control Module (ECM) communicates with all databus-capable control modules via a CAN databus. These databus-capable control modules are connected via two data bus wires which are twisted together (CAN_High and CAN_Low), and exchange information (messages). Missing information on the databus is recognized as a malfunction and stored. Trouble-free operation of the CAN-Bus requires that it have a terminal resistance.<br>**Possible Causes:**<br>• ECM has failed<br>• Terminal resistance for CAN-bus are faulty<br>• Can data bus wires have short circuited to each other<br>• TCM has failed |

## Diesel Engine OBD II Trouble Code List (P1xxx Codes)

| DTC | Trouble Code Title, Conditions & Possible Causes |
|---|---|
| **DTC: P1026**<br>**Years:** 2007<br>**Models:** Jetta, New Beetle<br>**Engines:** 1.9L<br>**Transmissions:** A/T | **Activation Intake Manifold Flap for Air Stream Regulation Short Circuit to B+ Conditions:**<br>The ECM detected a voltage irregularity on the intake manifold flap.<br>**Possible Causes:**<br>• Check activation of intake manifold flap for air stream regulation |
| **DTC: P1027**<br>**Years:** 2007<br>**Models:** Jetta, New Beetle<br>**Engines:** 1.9L<br>**Transmissions:** A/T | **Activation Intake Manifold Flap for Air Stream Regulation Short Circuit to Ground Conditions:**<br>The ECM detected a voltage irregularity on the intake manifold flap.<br>**Possible Causes:**<br>• Check activation of intake manifold flap for air stream regulation |
| **DTC: P1028**<br>**Years:** 2007<br>**Models:** Jetta, New Beetle<br>**Engines:** 1.9L<br>**Transmissions:** A/T | **Activation Intake Manifold Flap for Air Stream Regulation Open Circuit Conditions:**<br>The ECM detected a voltage irregularity on the intake manifold flap.<br>**Possible Causes:**<br>• Check activation of intake manifold flap for air stream regulation |
| **DTC: P1778**<br>**2T, MIL: Yes**<br>**Years:** 2007<br>**Models:** Jetta, New Beetle<br>**Engines:** 1.9L<br>**Transmissions:** A/T | **Solenoid EV7 Electrical Malfunction Conditions:**<br>Engine started, battery voltage must be at least 11.5v, all electrical components must be off, the ground between the engine and the chassis must be well connected, the exhaust system must be properly sealed between the catalytic converter and the cylinder head, and the oxygen sensor heater for oxygen sensor before the catalytic converter must be properly functioning. The ECM detected a loss of communication between the TCM and the valve body solenoid valve.<br>**Possible Causes:**<br>• Valve body solenoid valve has a short<br>• The TCM has failed<br>• ECM has failed |
| **DTC: P1780**<br>**2T, MIL: Yes**<br>**Years:** 2007<br>**Models:** Jetta, New Beetle<br>**Engines:** 1.9L<br>**Transmissions:** A/T | **Engine Intervention Readable Conditions:**<br>Key on or engine started, Self-Test enabled, and the ECM detected the Transmission Control Switch (TCS) was out of range during the test and a signal was not received.<br>**Note: The seal on the ATF level plug must always be replaced if the ATF level is checked.**<br>**Possible Causes:**<br>• TCS circuit open or shorted in the wiring harness<br>• TCS not cycled during the self-test<br>• TCS is damaged or failed, or the ECM has failed<br>• Check ATF level<br>• Mechanical malfunction in transmission |
| **DTC: P1850**<br>**Years:** 2007<br>**Models:** Jetta, New Beetle<br>**Engines:** 1.9L<br>**Transmissions:** A/T | **Data BUS Powertrain Missing Message from Engine Control Conditions:**<br>The Engine Control Module (ECM) communicates with all databus-capable control modules via a CAN databus. These databus-capable control modules are connected via two data bus wires which are twisted together (CAN_High and CAN_Low), and exchange information (messages). Missing information on the databus is recognized as a malfunction and stored. Trouble-free operation of the CAN-bus requires that it have a terminal resistance. This central terminal resistor is located in the Engine Control Module (ECM).<br>**Possible Causes:**<br>• Check the Terminal resistance for CAN-bus<br>• Data-Bus wires have short<br>• Data-Bus components are malfunctioning<br>• ECM has failed |

| DTC | Trouble Code Title, Conditions & Possible Causes |
|---|---|
| **DTC: P1854**<br>**Years:** 2007<br>**Models:** Jetta, New Beetle<br>**Engines:** 1.9L<br>**Transmissions:** A/T | **Data BUS Powertrain Hardware Defective Conditions:**<br>The Engine Control Module (ECM) communicates with all databus-capable control modules via a CAN databus. These databus-capable control modules are connected via two data bus wires which are twisted together (CAN_High and CAN_Low), and exchange information (messages). Missing information on the databus is recognized as a malfunction and stored. Trouble-free operation of the CAN-bus requires that it have a terminal resistance. This central terminal resistor is located in the Engine Control Module (ECM).<br>**Possible Causes:**<br>• Check the Terminal resistance for CAN-bus<br>• Data-Bus wires have short<br>• Data-Bus components are malfunctioning<br>• ECM has failed |
| **DTC: P1855**<br>**Years:** 2007<br>**Models:** Jetta, New Beetle<br>**Engines:** 1.9L<br>**Transmissions:** A/T | **Data BUS Powertrain Software Version Control Conditions:**<br>The Engine Control Module (ECM) communicates with all databus-capable control modules via a CAN databus. These databus-capable control modules are connected via two data bus wires which are twisted together (CAN_High and CAN_Low), and exchange information (messages). Missing information on the databus is recognized as a malfunction and stored. Trouble-free operation of the CAN-bus requires that it have a terminal resistance. This central terminal resistor is located in the Engine Control Module (ECM).<br>**Possible Causes:**<br>• Check the Terminal resistance for CAN-bus<br>• Data-Bus wires have short<br>• Data-Bus components are malfunctioning<br>• ECM has failed |
| **DTC: P1857**<br>**Years:** 2007<br>**Models:** Jetta, New Beetle<br>**Engines:** 1.9L<br>**Transmissions:** A/T | **Load Signal Error Message From Engine Control Conditions:**<br>The Engine Control Module (ECM) communicates with all databus-capable control modules via a CAN databus. These databus-capable control modules are connected via two data bus wires which are twisted together (CAN_High and CAN_Low), and exchange information (messages). Missing information on the databus is recognized as a malfunction and stored. Trouble-free operation of the CAN-bus requires that it have a terminal resistance. This central terminal resistor is located in the Engine Control Module (ECM).<br>**Possible Causes:**<br>• Intake Manifold Runner Position Sensor is faulty<br>• Intake system has leaks (false air)<br>• Motor for intake flap is faulty<br>• Mass Air Flow (MAF) sensor has failed<br>• ECM has failed |
| **DTC: P1866**<br>**Years:** 2007<br>**Models:** Jetta, New Beetle<br>**Engines:** 1.9L<br>**Transmissions:** A/T | **Data Bus Powertrain Missing Messages Conditions:**<br>The Engine Control Module (ECM) communicates with all databus-capable control modules via a CAN databus. These databus-capable control modules are connected via two data bus wires which are twisted together (CAN_High and CAN_Low), and exchange information (messages). Missing information on the databus is recognized as a malfunction and stored. Trouble-free operation of the CAN-bus requires that it have a terminal resistance. This central terminal resistor is located in the Engine Control Module (ECM).<br>**Possible Causes:**<br>• Check the Terminal resistance for CAN-bus<br>• Data-Bus wires have short<br>• Data-Bus components are malfunctioning<br>• ECM has failed |

## Diesel Engine OBD II Trouble Code List (P2xxx Codes)

| DTC | Trouble Code Title, Conditions & Possible Causes |
|---|---|
| **DTC: P2100**<br>**Years:** 2007<br>**Models:** Jetta, New Beetle<br>**Engines:** 1.9L<br>**Transmissions:** A/T | **Throttle Actuator Control Motor Circuit Open Conditions:**<br>Engine started, battery voltage must be at least 11.5v, all electrical components must be off, parking brake must be engaged (to keep daytime driving lights off), automatic transmission selector must be in park, the exhaust system must be properly sealed between the catalytic converter and the cylinder head, coolant temperature must be at least 80 degrees Celsius. The ECM detected an unexpected low or high voltage condition on the Throttle Actuator Control Motor (TACM) circuit during the CCM test.<br>**Note: The throttle valve activation occurs via an electric motor (throttle drive) in the throttle valve control module. It is activated by the Engine Control Module (ECM) according to specifications of the two sensors, Throttle Position (TP) Sensor and Sender 2 for accelerator pedal position.**<br>**Possible Causes:**<br>• TACM wiring harness connector is damaged or open<br>• TACM wiring may be crossed in the wire harness assembly<br>• TACM (motor) circuit is open, or TACM assembly is damaged (possible open circuit)<br>• TACM or the Throttle Valve is dirty<br>• Throttle Position sensor has failed<br>• ECM has failed |

| DTC | Trouble Code Title, Conditions & Possible Causes |
|---|---|
| **DTC: P2102**<br>**Years:** 2007<br>**Models:** Jetta, New Beetle<br>**Engines:** 1.9L<br>**Transmissions:** A/T | **Throttle Actuator Control Motor Circuit Low Conditions:**<br>Engine started, battery voltage must be at least 11.5v, all electrical components must be off, parking brake must be engaged (to keep daytime driving lights off), automatic transmission selector must be in park, the exhaust system must be properly sealed between the catalytic converter and the cylinder head, coolant temperature must be at least 80 degrees Celsius. The ECM detected an unexpected low or high voltage condition on the Throttle Actuator Control Motor (TACM) circuit during the CCM test.<br>**Note: The throttle valve activation occurs via an electric motor (throttle drive) in the throttle valve control module. It is activated by the Engine Control Module (ECM) according to specifications of the two sensors, Throttle Position (TP) Sensor and Sender 2 for accelerator pedal position.**<br>**Possible Causes:**<br>• TACM wiring harness connector is damaged or open<br>• TACM wiring may be crossed in the wire harness assembly<br>• TACM (motor) circuit is open, or TACM assembly is damaged (possible open circuit)<br>• TACM or the Throttle Valve is dirty<br>• Throttle Position sensor has failed<br>• ECM has failed |
| **DTC: P2103**<br>**Years:** 2007<br>**Models:** Jetta, New Beetle<br>**Engines:** 1.9L<br>**Transmissions:** A/T | **Throttle Actuator Control Motor Circuit High Conditions:**<br>Engine started, battery voltage must be at least 11.5v, all electrical components must be off, parking brake must be engaged (to keep daytime driving lights off), automatic transmission selector must be in park, the exhaust system must be properly sealed between the catalytic converter and the cylinder head, coolant temperature must be at least 80 degrees Celsius. The ECM detected an unexpected low or high voltage condition on the Throttle Actuator Control Motor (TACM) circuit during the CCM test.<br>**Note: The throttle valve activation occurs via an electric motor (throttle drive) in the throttle valve control module. It is activated by the Engine Control Module (ECM) according to specifications of the two sensors, Throttle Position (TP) Sensor and Sender 2 for accelerator pedal position.**<br>**Possible Causes:**<br>• TACM wiring harness connector is damaged or open<br>• TACM wiring may be crossed in the wire harness assembly<br>• TACM (motor) circuit is open, or TACM assembly is damaged (possible open circuit)<br>• TACM or the Throttle Valve is dirty<br>• Throttle Position sensor has failed<br>• ECM has failed |
| **DTC: P2108**<br>**Years:** 2007<br>**Models:** Jetta, New Beetle<br>**Engines:** 1.9L<br>**Transmissions:** A/T | **Throttle Actuator Control Motor Performance Conditions:**<br>Engine started, battery voltage must be at least 11.5v, all electrical components must be off, parking brake must be engaged (to keep daytime driving lights off), automatic transmission selector must be in park, the exhaust system must be properly sealed between the catalytic converter and the cylinder head, coolant temperature must be at least 80 degrees Celsius. The ECM detected an unexpected low or high voltage condition on the Throttle Actuator Control Motor (TACM) circuit during the CCM test.<br>**Note: The throttle valve activation occurs via an electric motor (throttle drive) in the throttle valve control module. It is activated by the Engine Control Module (ECM) according to specifications of the two sensors, Throttle Position (TP) Sensor and Sender 2 for accelerator pedal position.**<br>**Possible Causes:**<br>• TACM wiring harness connector is damaged or open<br>• TACM wiring may be crossed in the wire harness assembly<br>• TACM (motor) circuit is open, or TACM assembly is damaged (possible open circuit)<br>• TACM or the Throttle Valve is dirty<br>• Throttle Position sensor has failed<br>• ECM has failed |
| **DTC: P2122**<br>**Years:** 2007<br>**Models:** Jetta, New Beetle<br>**Engines:** 1.9L<br>**Transmissions:** A/T | **Accelerator Pedal Position Sensor 'D' Circuit Low Input Conditions:**<br>Engine started, battery voltage at least 11.5v, all electrical components off, ground connections between engine and chassis well connected, the ECM detected that the accelerator pedal position sensor signal was outside the parameters to function normally.<br>**Note: Both the Throttle Position (TP) Sensor and Accelerator Pedal Position Sensor are located at the accelerator pedal module and communicate the driver's intentions to the ECM completely independently of each other. Both sensors are stored in one housing.**<br>**Possible Causes:**<br>• Ground between engine and chassis may be broken<br>• Throttle position sensor may have failed<br>• Accelerator Pedal Position Sensor has failed<br>• Throttle position sensor wiring may have shorted<br>• Throttle position sensor has failed<br>• Faulty voltage supply<br>• ECM has failed |

| DTC | Trouble Code Title, Conditions & Possible Causes |
|---|---|
| **DTC: P2195**<br>**Years:** 2007<br>**Models:** Jetta, New Beetle<br>**Engines:** 1.9L<br>**Transmissions:** A/T | **O2 Sensor Signal Stuck Lean Bank 1 Sensor 1 Conditions:**<br>Engine running in closed loop, and the ECM detected the O2S indicated a lean signal, or it could no longer control Fuel Trim because it was at lean limit.<br>**Possible Causes:**<br>• Engine oil level high<br>• Camshaft timing error<br>• Cylinder compression low<br>• Exhaust leaks in front of O2S<br>• EGR valve is stuck open<br>• EGR gasket is leaking<br>• EVR diaphragm is leaking<br>• Damaged fuel pressure regulator or extremely low fuel pressure<br>• O2S circuit is open or shorted in the wiring harness<br>• Oxygen sensor (before catalytic converter) is faulty<br>• Oxygen sensor (behind catalytic converter) is faulty<br>• Oxygen sensor heater (before catalytic converter) is faulty<br>• Oxygen sensor heater (behind catalytic converter) is faulty<br>• Air leaks after the MAF sensor<br>• PCV system leaks<br>• Dip stick not seated properly |
| **DTC: P2196**<br>**Years:** 2007<br>**Models:** Jetta, New Beetle<br>**Engines:** 1.9L<br>**Transmissions:** A/T | **O2 Sensor Signal Stuck Rich Bank 1 Sensor 1 Conditions:**<br>Engine running in closed loop, and the ECM detected the O2S indicated a rich signal, or it could no longer control Fuel Trim because it was at its rich limit.<br>**Possible Causes:**<br>• Engine oil level high<br>• Camshaft timing error<br>• Cylinder compression low<br>• Exhaust leaks in front of O2S<br>• EGR valve is stuck open<br>• EGR gasket is leaking<br>• EVR diaphragm is leaking<br>• Damaged fuel pressure regulator or extremely low fuel pressure<br>• O2S circuit is open or shorted in the wiring harness<br>• Oxygen sensor (before catalytic converter) is faulty<br>• Oxygen sensor (behind catalytic converter) is faulty<br>• Oxygen sensor heater (before catalytic converter) is faulty<br>• Oxygen sensor heater (behind catalytic converter) is faulty<br>• Air leaks after the MAF sensor<br>• PCV system leaks<br>• Dip stick not seated properly |
| **DTC: P2237**<br>**Years:** 2007<br>**Models:** Jetta, New Beetle<br>**Engines:** 1.9L<br>**Transmissions:** A/T | **O2 Sensor Positive Current Control Circuit/Open Bank 1 Sensor 1 Conditions:**<br>Engine started, battery voltage must be at least 11.5v, all electrical components must be off, parking brake must be engaged (to keep daytime driving lights off), automatic transmission selector must be in park. The ECM detected an unexpected voltage condition, or it detected an unexpected current draw in the sensor circuit during the CCM test.<br>**Note: Vehicle must be raised before connector for oxygen sensors is accessible.**<br>**Possible Causes:**<br>• Oxygen sensor (before catalytic converter) is faulty<br>• Oxygen sensor heater (before catalytic converter) is faulty<br>• Oxygen sensor heater (before catalytic converter) is faulty<br>• Oxygen sensor heater (behind catalytic converter) is faulty<br>• O2S circuit is open or shorted in the wiring harness<br>• ECM has failed |

| DTC | Trouble Code Title, Conditions & Possible Causes |
|---|---|
| **DTC: P2238**<br>**Years:** 2007<br>**Models:** Jetta, New Beetle<br>**Engines:** 1.9L<br>**Transmissions:** A/T | **O2 Sensor Positive Current Control Circuit Low Bank 1 Sensor 1 Conditions:**<br>Engine started, battery voltage must be at least 11.5v, all electrical components must be off, parking brake must be engaged (to keep daytime driving lights off), automatic transmission selector must be in park. The ECM detected an unexpected voltage condition, or it detected an unexpected current draw in the sensor circuit during the CCM test.<br>**Note: Vehicle must be raised before connector for oxygen sensors is accessible.**<br>**Possible Causes:**<br>• Oxygen sensor (before catalytic converter) is faulty<br>• Oxygen sensor heater (before catalytic converter) is faulty<br>• Oxygen sensor heater (before catalytic converter) is faulty<br>• Oxygen sensor heater (behind catalytic converter) is faulty<br>• O2S circuit is open or shorted in the wiring harness<br>• ECM has failed |
| **DTC: P2239**<br>**Years:** 2007<br>**Models:** Jetta, New Beetle<br>**Engines:** 1.9L<br>**Transmissions:** A/T | **O2 Sensor Positive Current Control Circuit High Bank 1 Sensor 1 Conditions:**<br>Engine started, battery voltage must be at least 11.5v, all electrical components must be off, parking brake must be engaged (to keep daytime driving lights off), automatic transmission selector must be in park. The ECM detected an unexpected voltage condition, or it detected an unexpected current draw in the sensor circuit during the CCM test.<br>**Note: Vehicle must be raised before connector for oxygen sensors is accessible.**<br>**Possible Causes:**<br>• Oxygen sensor (before catalytic converter) is faulty<br>• Oxygen sensor heater (before catalytic converter) is faulty<br>• Oxygen sensor heater (before catalytic converter) is faulty<br>• Oxygen sensor heater (behind catalytic converter) is faulty<br>• O2S circuit is open or shorted in the wiring harness<br>• ECM has failed |
| **DTC: P2243**<br>**Years:** 2007<br>**Models:** Jetta, New Beetle<br>**Engines:** 1.9L<br>**Transmissions:** A/T | **O2 Sensor Reference Voltage Circuit/Open Bank 1 Sensor 1 Conditions:**<br>Engine started, battery voltage must be at least 11.5v, all electrical components must be off, parking brake must be engaged (to keep daytime driving lights off), automatic transmission selector must be in park. The ECM detected an unexpected voltage condition, or it detected an unexpected current draw in the sensor circuit during the CCM test.<br>**Note: Vehicle must be raised before connector for oxygen sensors is accessible.**<br>**Possible Causes:**<br>• Oxygen sensor (before catalytic converter) is faulty<br>• Oxygen sensor heater (before catalytic converter) is faulty<br>• Oxygen sensor heater (before catalytic converter) is faulty<br>• Oxygen sensor heater (behind catalytic converter) is faulty<br>• O2S circuit is open or shorted in the wiring harness<br>• ECM has failed |
| **DTC: P2245**<br>**Years:** 2007<br>**Models:** Jetta, New Beetle<br>**Engines:** 1.9L<br>**Transmissions:** A/T | **O2 Sensor Reference Voltage Circuit Low Bank 1 Sensor 1 Conditions:**<br>Engine started, battery voltage must be at least 11.5v, all electrical components must be off, parking brake must be engaged (to keep daytime driving lights off), automatic transmission selector must be in park. The ECM detected an unexpected voltage condition, or it detected an unexpected current draw in the sensor circuit during the CCM test.<br>**Note: Vehicle must be raised before connector for oxygen sensors is accessible.**<br>**Possible Causes:**<br>• Oxygen sensor (before catalytic converter) is faulty<br>• Oxygen sensor heater (before catalytic converter) is faulty<br>• Oxygen sensor heater (before catalytic converter) is faulty<br>• Oxygen sensor heater (behind catalytic converter) is faulty<br>• O2S circuit is open or shorted in the wiring harness<br>• ECM has failed |
| **DTC: P2246**<br>**Years:** 2007<br>**Models:** Jetta, New Beetle<br>**Engines:** 1.9L<br>**Transmissions:** A/T | **O2 Sensor Reference Voltage Circuit High Bank 1 Sensor 1 Conditions:**<br>Engine started, battery voltage must be at least 11.5v, all electrical components must be off, parking brake must be engaged (to keep daytime driving lights off), automatic transmission selector must be in park. The ECM detected an unexpected voltage condition, or it detected an unexpected current draw in the sensor circuit during the CCM test.<br>**Note: Vehicle must be raised before connector for oxygen sensors is accessible.**<br>**Possible Causes:**<br>• Oxygen sensor (before catalytic converter) is faulty<br>• Oxygen sensor heater (before catalytic converter) is faulty<br>• Oxygen sensor heater (before catalytic converter) is faulty<br>• Oxygen sensor heater (behind catalytic converter) is faulty<br>• O2S circuit is open or shorted in the wiring harness<br>• ECM has failed |

| DTC | Trouble Code Title, Conditions & Possible Causes |
|---|---|
| **DTC: P2251**<br>**Years:** 2007<br>**Models:** Jetta, New Beetle<br>**Engines:** 1.9L<br>**Transmissions:** A/T | **O2 Sensor Negative Voltage Circuit/Open Bank 1 Sensor 1 Conditions:**<br>Engine started, battery voltage must be at least 11.5v, all electrical components must be off, parking brake must be engaged (to keep daytime driving lights off), automatic transmission selector must be in park. The ECM detected an unexpected voltage condition, or it detected an unexpected current draw in the sensor circuit during the CCM test.<br>**Note: Vehicle must be raised before connector for oxygen sensors is accessible.**<br>**Possible Causes:**<br>• Oxygen sensor (before catalytic converter) is faulty<br>• Oxygen sensor heater (before catalytic converter) is faulty<br>• Oxygen sensor heater (before catalytic converter) is faulty<br>• Oxygen sensor heater (behind catalytic converter) is faulty<br>• O2S circuit is open or shorted in the wiring harness<br>• ECM has failed |
| **DTC: P2252**<br>**Years:** 2007<br>**Models:** Jetta, New Beetle<br>**Engines:** 1.9L<br>**Transmissions:** A/T | **O2 Sensor Negative Voltage Circuit Low Bank 1 Sensor 1 Conditions:**<br>Engine started, battery voltage must be at least 11.5v, all electrical components must be off, parking brake must be engaged (to keep daytime driving lights off), automatic transmission selector must be in park. The ECM detected an unexpected voltage condition, or it detected an unexpected current draw in the sensor circuit during the CCM test.<br>**Note: Vehicle must be raised before connector for oxygen sensors is accessible.**<br>**Possible Causes:**<br>• Oxygen sensor (before catalytic converter) is faulty<br>• Oxygen sensor heater (before catalytic converter) is faulty<br>• Oxygen sensor heater (before catalytic converter) is faulty<br>• Oxygen sensor heater (behind catalytic converter) is faulty<br>• O2S circuit is open or shorted in the wiring harness<br>• ECM has failed |
| **DTC: P2253**<br>**Years:** 2007<br>**Models:** Jetta, New Beetle<br>**Engines:** 1.9L<br>**Transmissions:** A/T | **O2 Sensor Negative Voltage Circuit High Bank 1 Sensor 1 Conditions:**<br>Engine started, battery voltage must be at least 11.5v, all electrical components must be off, parking brake must be engaged (to keep daytime driving lights off), automatic transmission selector must be in park. The ECM detected an unexpected voltage condition, or it detected an unexpected current draw in the sensor circuit during the CCM test.<br>**Note: Vehicle must be raised before connector for oxygen sensors is accessible.**<br>**Possible Causes:**<br>• Oxygen sensor (before catalytic converter) is faulty<br>• Oxygen sensor heater (before catalytic converter) is faulty<br>• Oxygen sensor heater (before catalytic converter) is faulty<br>• Oxygen sensor heater (behind catalytic converter) is faulty<br>• O2S circuit is open or shorted in the wiring harness<br>• ECM has failed |
| **DTC: P2413**<br>**Years:** 2007<br>**Models:** Jetta, New Beetle<br>**Engines:** 1.9L<br>**Transmissions:** A/T | **Exhaust Gas Recirculation System Performance Conditions:**<br>Engine started, battery voltage must be at least 11.5v, all electrical components must be off, parking brake must be engaged (to keep daytime driving lights off), automatic transmission selector must be in park. The ECM detected a EGR fault that could affect the system's performance.<br>**Note: If the EGR Vacuum Regulator Solenoid Valve with Potentiometer for EGR was replaced, a basic setting must be performed.**<br>**Possible Causes:**<br>• EGR Vacuum Regulator Solenoid Valve is faulty<br>• EGR valve is malfunctioning<br>• Circuit wires have short circuited to each other, to ground or to power<br>• Potentiometer for EGR has failed |
| **DTC: P2425**<br>**Years:** 2007<br>**Models:** Jetta, New Beetle<br>**Engines:** 1.9L<br>**Transmissions:** All | **Exhaust Gas Recirculation Cooling Valve Control Circuit/Open Conditions:**<br>Engine started, battery voltage must be at least 11.5v. The ECM detected a problem with the EGR cooling valve circuit that could affect the performance of the vehicle.<br>**Possible Causes:**<br>• The exhaust gas recirculation cooler switch-over valve has failed<br>• Circuit wires have short circuited to each other, to ground or to power<br>• ECM has failed |
| **DTC: P2426**<br>**Years:** 2007<br>**Models:** Jetta, New Beetle<br>**Engines:** 1.9L<br>**Transmissions:** All | **Exhaust Gas Recirculation Cooling Valve Control Circuit Low Conditions:**<br>Engine started, battery voltage must be at least 11.5v. The ECM detected a problem with the EGR cooling valve circuit that could affect the performance of the vehicle.<br>**Possible Causes:**<br>• The exhaust gas recirculation cooler switch-over valve has failed<br>• Circuit wires have short circuited to each other, to ground or to power<br>• ECM has failed |

| DTC | Trouble Code Title, Conditions & Possible Causes |
|---|---|
| **DTC: P2427**<br>**Years:** 2007<br>**Models:** Jetta, New Beetle<br>**Engines:** 1.9L<br>**Transmissions:** All | **Exhaust Gas Recirculation Cooling Valve Control Circuit High Conditions:**<br>Engine started, battery voltage must be at least 11.5v. The ECM detected a problem with the EGR cooling valve circuit that could affect the performance of the vehicle.<br>**Possible Causes:**<br>• The exhaust gas recirculation cooler switch-over valve has failed<br>• Circuit wires have short circuited to each other, to ground or to power<br>• ECM has failed |
| **DTC: P2637**<br>**Years:** 2007<br>**Models:** Jetta, New Beetle<br>**Engines:** 1.9L<br>**Transmissions:** All | **Torque Management Feedback Signal "A" Conditions:**<br>Engine started, battery voltage must be at least 11.5v, all electrical components must be off, parking brake must be engaged (to keep daytime driving lights off), automatic transmission selector must be in park, the exhaust system must be properly sealed between the catalytic converter and the cylinder head, coolant temperature must be at least 80 degrees Celsius and oxygen sensor heaters for oxygen sensors before the catalytic converter must be functioning properly and the ground between the engine and the chassis must be well connected. The ECM detected a voltage value on the torque management circuits that doesn't fall within the desired parameters<br>**Possible Causes:**<br>• Engine Control Module (ECM) has failed<br>• Voltage supply for Engine Control Module (ECM) is damaged<br>• Engine Coolant Temperature (ECT) sensor has failed<br>• Intake Air Temperature (IAT) sensor has failed<br>• Intake Manifold Runner Position Sensor has failed<br>• Intake system has leaks (false air)<br>• Motor for intake flap has failed<br>• Mass Air Flow (MAF) sensor has failed |

## SPECIFICATIONS AND MAINTENANCE CHARTS

### ENGINE AND VEHICLE IDENTIFICATION CHART

| Engine | | | | | | | Model Year | |
|---|---|---|---|---|---|---|---|---|
| Code | Liters (cc) | Cu. in. | Cyl. | Fuel Sys. | Type | Eng. Mfg. | Code ① | Year |
| B5244S4 | 2.4 (2435) | 154 | 5 | EFI | DOHC | Volvo | 7 | 2007 |
| B5244S | 2.4 (2435) | 154 | 5 | EFI | DOHC | Volvo | 8 | 2008 |
| B5244TS | 2.4 (2435) | 154 | 5 | EFI | DOHC | Volvo | | |
| B5254T2 | 2.5 (2521) | 154 | 5 | EFI | DOHC | Volvo | | |
| B5254T3 | 2.5 (2521) | 154 | 5 | EFI | DOHC | Volvo | | |
| B5254T4 | 2.5 (2521) | 154 | 5 | EFI | DOHC | Volvo | | |
| B5254T7 | 2.5 (2521) | 154 | 5 | EFI | DOHC | Volvo | | |
| B5244S4 | 2.4 (2435) | 154 | 5 | EFI | DOHC | Volvo | | |
| B5244S7 | 2.4 (2435) | 154 | 5 | EFI | DOHC | Volvo | | |
| B5254S | 2.5 (2521) | 154 | 5 | EFI | DOHC | Volvo | | |
| B5254T | 2.5 (2521) | 154 | 5 | EFI | DOHC | Volvo | | |
| B5254T5 | 2.5 (2521) | 154 | 5 | EFI | DOHC | Volvo | | |
| B63042T | 3.0 (3000) | 183 | 5 | EFI | DOHC | Volvo | | |
| B6324S | 3.2 (3192) | 195 | 6 | EFI | DOHC | Volvo | | |
| B8444S | 4.4 (4400) | 268 | 8 | EFI | DOHC | Volvo | | |

① 10th Digit of VIN

EFI: Electronic Fuel Injection

DOHC: Double Overhead Camshafts

22250_VOLC_C0001

## GENERAL ENGINE SPECIFICATIONS

| Year | Model | Engine Displ. Liters | Engine ID | Net Horsepower @ rpm | Net Torque @ rpm (ft. lbs.) | Bore x Stroke (in.) | Compression Ratio | Oil Pressure @ rpm |
|------|-------|------|------|------|------|------|------|------|
| 2007 | C30 | 2.4 | B5244S4 | 227@5000 | 236@5000 | 3.27 x 3.67 | 9.0:1 | 51@4000 |
| | C30 | 2.5 | B5254T3 | 227@5000 | 236@5000 | 3.27 x 3.67 | 9.0:1 | 51@4000 |
| | C70 | 2.5 | B5244S4 | 227@5000 | 236@5000 | 3.27 x 3.67 | 9.0:1 | 51@4000 |
| | C70 | 2.5 | B5254T3 | 227@5000 | 236@5000 | 3.27 x 3.67 | 9.0:1 | 51@4000 |
| | S40 | 2.4 | B5244S4 | 168@6000 | 170@4400 | 3.27 x 3.54 | 10.3:1 | 51@4000 |
| | S40 | 2.5 | B5254T3 | 227@5000 | 236@4800 | 3.27 x 3.67 | 9.0:1 | 51@4000 |
| | S40 | 2.4 | B5244S7 | 227@5000 | 236@5000 | 3.27 x 3.67 | 9.0:1 | 51@4000 |
| | S40 | 2.5 | B5254T7 | 168@6000 | 170@4400 | 3.27 x 3.67 | 10.3:1 | 51@4000 |
| | S60 | 2.4 | B5244T5 | 208@5000 | 236@4500 | 3.27 x 3.67 | 8.5:1 | 51@4000 |
| | S60 T5 | 2.5 | B5254T5 | 257@5500 | 258@5000 | 3.19 x 3.67 | 8.5:1 | 51@4000 |
| | S80 T6 | 3.0 | B63042T | 281@5600 | 295@4800 | 3.23 x 3.67 | 9.3:1 | 51@4000 |
| | S80 3.2 | 3.2 | B6324S | 235@6200 | 236@3200 | 3.31 x 3.78 | 10.8:1 | 51@4000 |
| | S80 V8 | 4.4 | B8444S | 311@5950 | 325@3950 | 3.70 x 3.13 | 10.4:1 | ① |
| | V50 25i | 2.4 | B5244S7 | 168@6000 | 170@4400 | 3.27 x 3.67 | 10.3:1 | 51@4000 |
| | V50 T5 | 2.5 | B5254T3 | 227@5000 | 236@4800 | 3.27 x 3.67 | 9.0:1 | 51@4000 |
| | V70 | 3.2 | B6324S | 235@6200 | 236@3200 | 3.31 x 3.78 | 10.8:1 | 51@4000 |
| | XC70 | 2.5 | B5254T | 208@5000 | 236@1500 | 3.27 x 3.67 | 9.0:1 | 51@4000 |
| 2008 | C30 | 2.4 | B5244S4 | 227@5000 | 236@5000 | 3.27 x 3.67 | 9.0:1 | 51@4000 |
| | C30 | 2.5 | B5254T3 | 227@5000 | 236@5000 | 3.27 x 3.67 | 9.0:1 | 51@4000 |
| | C70 | 2.5 | B5244S4 | 227@5000 | 236@5000 | 3.27 x 3.67 | 9.0:1 | 51@4000 |
| | C70 | 2.5 | B5254T3 | 227@5000 | 236@5000 | 3.27 x 3.67 | 9.0:1 | 51@4000 |
| | S40 | 2.4 | B5244S4 | 168@6000 | 170@4400 | 3.27 x 3.54 | 10.3:1 | 51@4000 |
| | S40 | 2.5 | B5254T3 | 227@5000 | 236@4800 | 3.27 x 3.67 | 9.0:1 | 51@4000 |
| | S40 | 2.4 | B5244S7 | 227@5000 | 236@5000 | 3.27 x 3.67 | 9.0:1 | 51@4000 |
| | S40 | 2.5 | B5254T7 | 168@6000 | 170@4400 | 3.27 x 3.67 | 10.3:1 | 51@4000 |
| | S60 | 2.4 | B5244T5 | 208@5000 | 236@4500 | 3.27 x 3.67 | 8.5:1 | 51@4000 |
| | S60 T5 | 2.5 | B5254T5 | 257@5500 | 258@5000 | 3.19 x 3.67 | 8.5:1 | 51@4000 |
| | S80 T6 | 3.0 | B63042T | 281@5600 | 295@4800 | 3.23 x 3.67 | 9.3:1 | 51@4000 |
| | S80 3.2 | 3.2 | B6324S | 235@6200 | 236@3200 | 3.31 x 3.78 | 10.8:1 | 51@4000 |
| | S80 V8 | 4.4 | B8444S | 311@5950 | 325@3950 | 3.70 x 3.13 | 10.4:1 | ① |
| | V50 25i | 2.4 | B5244S7 | 168@6000 | 170@4400 | 3.27 x 3.67 | 10.3:1 | 51@4000 |
| | V50 T5 | 2.5 | B5254T3 | 227@5000 | 236@4800 | 3.27 x 3.67 | 9.0:1 | 51@4000 |
| | V70 | 3.2 | B6324S | 235@6200 | 236@3200 | 3.31 x 3.78 | 10.8:1 | 51@4000 |

① Oil pressure should be between 35 and 54 psi at 2000 rpm

22250_VOLC_C0002

## ENGINE TUNE-UP SPECIFICATIONS

| | Engine Displacement Liters | Engine ID | Spark Plug Gap (in.) ① | Ignition Timing (deg.) MT | AT | Fuel Pump (psi) | Idle Speed (rpm) MT | AT | Valve Clearance (in.) In. | Ex. |
|---|---|---|---|---|---|---|---|---|---|---|
| **2007** | 2.4 | B5244S4 | 0.028-0.032 | NA | NA | 55-58 | 720 | 720 | HYD | HYD |
| | 2.5 | B5254T3 | 0.028-0.032 | NA | NA | 55-58 | 720 | 720 | HYD | HYD |
| | 2.4 | B5244S4 | 0.028-0.032 | NA | NA | 55-58 | 720 | 720 | HYD | HYD |
| | 2.5 | B5254S4 | 0.028-0.032 | NA | NA | 55-58 | 720 | 720 | HYD | HYD |
| | 2.5 | B5244T5 | 0.028-0.032 | NA | NA | 55-58 | 670 | 670 | HYD | HYD |
| | 2.5 | B5254T5 | 0.028-0.032 | NA | NA | 55-58 | 650 | 650 | HYD | HYD |
| | 3.0 | B63042T | 0.028-0.032 | NA | NA | 55-58 | 670 | 670 | HYD | HYD |
| | 3.2 | B6324S | 0.028-0.032 | NA | NA | 55-58 | 650 | 650 | HYD | HYD |
| | 2.4 | B5244S7 | 0.028-0.032 | NA | NA | 55-58 | 720 | 720 | HYD | HYD |
| | 2.5 | B5254T3 | 0.028-0.032 | NA | NA | 55-58 | 670 | 670 | HYD | HYD |
| | 3.2 | B6324S | 0.028-0.032 | NA | NA | 55-58 | 650 | 650 | HYD | HYD |
| | 4.4 | B8444S | 0.028-0.032 | NA | NA | 55-58 | 675 | 675 | HYD | HYD |
| **2008** | 2.4 | B5244S4 | 0.028-0.032 | NA | NA | 55-58 | 720 | 720 | HYD | HYD |
| | 2.5 | B5254T3 | 0.028-0.032 | NA | NA | 55-58 | 720 | 720 | HYD | HYD |
| | 2.4 | B5244S4 | 0.028-0.032 | NA | NA | 55-58 | 720 | 720 | HYD | HYD |
| | 2.5 | B5254S4 | 0.028-0.032 | NA | NA | 55-58 | 720 | 720 | HYD | HYD |
| | 2.5 | B5244T5 | 0.028-0.032 | NA | NA | 55-58 | 670 | 670 | HYD | HYD |
| | 2.5 | B5254T5 | 0.028-0.032 | NA | NA | 55-58 | 650 | 650 | HYD | HYD |
| | 3.0 | B63042T | 0.028-0.032 | NA | NA | 55-58 | 670 | 670 | HYD | HYD |
| | 3.2 | B6324S | 0.028-0.032 | NA | NA | 55-58 | 650 | 650 | HYD | HYD |
| | 2.4 | B5244S7 | 0.028-0.032 | NA | NA | 55-58 | 720 | 720 | HYD | HYD |
| | 2.5 | B5254T3 | 0.028-0.032 | NA | NA | 55-58 | 670 | 670 | HYD | HYD |
| | 3.2 | B6324S | 0.028-0.032 | NA | NA | 55-58 | 650 | 650 | HYD | HYD |
| | 4.4 | B8444S | 0.028-0.032 | NA | NA | 55-58 | 675 | 675 | HYD | HYD |

HYD: Hydraulic lash adjusters

NA: Information not available

① The label figuires must be used if they differ from those in this chart.

22250_VOLC_C0003

## CAPACITIES

| Year | Model | Engine Displacement Liters | Engine ID | Engine Oil with Filter (qts.)① | Transmission (qts.) Man | Transmission (qts.) Auto. | Transfer Case (pts.) | Drive Axle Front (pts.) | Drive Axle Rear (pts.) | Fuel Tank (gal.) | Cooling System (qts.) |
|---|---|---|---|---|---|---|---|---|---|---|---|
| 2007 | C30 T5 | 2.4 | B5244S | 6.1 | ② | 8.5 | — | — | — | 15.9 | NA |
| | C70 T5 | 2.5 | B5254T3 | 5.8 | ② | 8.5 | — | — | — | 16.4 | NA |
| | S40 2.4i | 2.4 | B5244S | 5.8 | ② | 8.5 | — | — | — | 15.9 | NA |
| | S40 T5 | 2.5 | B5254T | 5.8 | ② | 8.5 | — | — | — | 15.9 | NA |
| | S60 2.5T | 2.5 | B5254T4 | 5.8 | ② | 8.5 | — | — | — | 18 | NA |
| | S60 T5 | 2.5 | B5254T | 5.8 | ② | 8.5 | — | — | — | 18 | NA |
| | S80 T6 | 3.0 | B63042T | 7.1 | ② | 8.5 | — | — | — | 18.5 | NA |
| | S80 3.2 | 3.2 | B6324S | 8.1 | ② | 8.5 | — | — | — | 18.5 | NA |
| | S80 V8 | 4.4 | B5244S | 7.4 | ② | 8.5 | — | — | — | 18.5 | NA |
| | V50 25i | 3.2 | B5254T3 | 5.8 | ② | 8.5 | — | — | — | 15.9 | NA |
| | V50 T5 | 2.4 | B5244S | 5.8 | ② | 8.5 | — | — | — | 15.9 | NA |
| | V70 | 2.5 | B5254T2 | 7.8 | ② | 8.5 | — | — | — | 18.5 | NA |
| | XC70 | 2.5 | B5254T | 5.8 | ② | 8.5 | — | — | — | 18 | NA |
| 2008 | C30 T5 | 2.4 | B5254T3 | 6.1 | ② | 8.5 | — | — | — | 15.9 | NA |
| | C70 T5 | 2.5 | B5244S | 5.8 | ② | 8.5 | — | — | — | 15.9 | NA |
| | S40 2.4i | 2.4 | B5254T3 | 5.8 | ② | 8.2 | — | — | — | 15.9 | NA |
| | S40 T5 | 2.5 | B5254T | 5.8 | ② | 8.5 | — | — | — | 18 | NA |
| | S60 2.5T | 2.5 | B5254T4 | 5.8 | ② | 8.5 | — | — | — | 18 | NA |
| | S60 T5 | 2.5 | B5244T | 5.8 | ② | 8.5 | — | — | — | 18 | NA |
| | S80 T6 | 3.0 | B63042T | 7.1 | ② | 8.5 | — | — | — | 18 | NA |
| | S80 3.2 | 3.2 | B6324S | 7.8 | ② | 8.2 | — | — | — | 15.9 | NA |
| | S80 V8 | 4.4 | B5254T3 | 5.8 | ② | 8.2 | — | — | — | 15.9 | NA |
| | V50 25i | 3.2 | B5244S | 5.8 | ② | 8.5 | — | — | — | 18.5 | NA |
| | V50 T5 | 2.4 | B5254T2 | 5.8 | ② | 8.5 | — | — | — | 18 | NA |
| | V70 | 2.5 | B5254T4 | 7.7 | ② | 8.5 | — | — | — | 18 | NA |

NOTE: All capacities are approximate. Add fluid gradualy and check to be sure a proper fluid level is obtained.

① On turbocharged engines, add 0.7 US qts. if the cooler is drained

② 3.5 to 4.5 qts

22250_VOLC_C0005

## FLUID SPECIFICATIONS

| Year | Model | Engine Size Liters | Engine ID | Engine Oil | Auto Transmission | Manual Transmission | Rear Axle | Power Steering Fluid | Engine Coolant | Brake Fluid |
|------|-------|------|------|------|------|------|------|------|------|------|
| **2007** | C30 T5 | 2.4 | B5244S4 | 5W-30 | Volvo | Volvo | Volvo | Volvo | Volvo | DOT 4+ |
| | C70 T5 | 2.5 | B5254T3 | 5W-30 | Volvo | Volvo | Volvo | Volvo | Volvo | DOT 4+ |
| | S40 2.4i | 2.4 | B5244S4 | 5W-30 | Volvo | Volvo | Volvo | Volvo | Volvo | DOT 4+ |
| | S40 T5 | 2.5 | B5254S4 | 5W-30 | Volvo | Volvo | Volvo | Volvo | Volvo | DOT 4+ |
| | S60 2.5T | 2.5 | B5254T5 | 5W-30 | Volvo | Volvo | Volvo | Volvo | Volvo | DOT 4+ |
| | S60 T5 | 2.5 | B5254T5 | 5W-30 | Volvo | Volvo | Volvo | Volvo | Volvo | DOT 4+ |
| | S80 T6 | 3.0 | B63042T | 5W-30 | Volvo | Volvo | Volvo | Volvo | Volvo | DOT 4+ |
| | S80 3.2 | 3.2 | B6324S | 5W-30 | Volvo | Volvo | Volvo | Volvo | Volvo | DOT 4+ |
| | S80 V8 | 4.4 | B8444S | 5W-30 | Volvo | Volvo | Volvo | Volvo | Volvo | DOT 4+ |
| | V50 25i | 3.2 | B5244S7 | 5W-30 | Volvo | Volvo | Volvo | Volvo | Volvo | DOT 4+ |
| | V50 T5 | 2.4 | B5254T3 | 5W-30 | Volvo | Volvo | Volvo | Volvo | Volvo | DOT 4+ |
| | V70 | 2.5 | B5254T2 | 5W-30 | Volvo | Volvo | Volvo | Volvo | Volvo | DOT 4+ |
| | XC70 | 2.5 | B5254T | 5W-30 | Volvo | Volvo | Volvo | Volvo | Volvo | DOT 4+ |
| **2008** | C30 T5 | 2.4 | B5244S4 | 5W-30 | Volvo | Volvo | Volvo | Volvo | Volvo | DOT 4+ |
| | C70 T5 | 2.5 | B5254T3 | 5W-30 | Volvo | Volvo | Volvo | Volvo | Volvo | DOT 4+ |
| | S40 2.4i | 2.4 | B5244S4 | 5W-30 | Volvo | Volvo | Volvo | Volvo | Volvo | DOT 4+ |
| | S40 T5 | 2.5 | B5254S4 | 5W-30 | Volvo | Volvo | Volvo | Volvo | Volvo | DOT 4+ |
| | S60 2.5T | 2.5 | B5254T5 | 5W-30 | Volvo | Volvo | Volvo | Volvo | Volvo | DOT 4+ |
| | S60 T5 | 2.5 | B5254T5 | 5W-30 | Volvo | Volvo | Volvo | Volvo | Volvo | DOT 4+ |
| | S80 T6 | 3.0 | B63042T | 5W-30 | Volvo | Volvo | Volvo | Volvo | Volvo | DOT 4+ |
| | S80 3.2 | 3.2 | B6324S | 5W-30 | Volvo | Volvo | Volvo | Volvo | Volvo | DOT 4+ |
| | S80 V8 | 4.4 | B8444S | 5W-30 | Volvo | Volvo | Volvo | Volvo | Volvo | DOT 4+ |
| | V50 25i | 3.2 | B5244S7 | 5W-30 | Volvo | Volvo | Volvo | Volvo | Volvo | DOT 4+ |
| | V50 T5 | 2.4 | B5254T3 | 5W-30 | Volvo | Volvo | Volvo | Volvo | Volvo | DOT 4+ |
| | V70 | 2.5 | B5254T2 | 5W-30 | Volvo | Volvo | Volvo | Volvo | Volvo | DOT 4+ |

DOT: Department Of Transpotation

## VALVE SPECIFICATIONS

| Year | Engine Displacement Liters | Engine ID | Seat Angle (deg.) | Face Angle (deg.) | Spring Test Pressure (lbs. @ in.) | Spring Installed Height (in.) | Stem-to-Guide Clearance (in.) Intake | Stem-to-Guide Clearance (in.) Exhaust | Stem Diameter (in.) Intake | Stem Diameter (in.) Exhaust |
|---|---|---|---|---|---|---|---|---|---|---|
| 2007 | 2.4 | B5244S4 | 45 | 44.5 | NA | NA | 0.0012-0.0024 | 0.0012-0.0024 | 0.2344-0.2350 | 0.2344-0.2350 |
| | 2.5 | B5254T3 | 45 | 44.5 | NA | NA | 0.0012-0.0024 | 0.0012-0.0024 | 0.2344-0.2350 | 0.2344-0.2350 |
| | 2.5 | B5254S4 | 45 | 44.5 | NA | NA | 0.0012-0.0024 | 0.0012-0.0024 | 0.2344-0.2350 | 0.2344-0.2350 |
| | 2.5 | B5254T3 | 45 | 44.5 | NA | NA | 0.0012-0.0024 | 0.0012-0.0024 | 0.2344-0.2350 | 0.2344-0.2350 |
| | 2.5 | B5254T2 | 45 | 44.5 | NA | NA | 0.0012-0.0024 | 0.0012-0.0024 | 0.2344-0.2350 | 0.2344-0.2350 |
| | 2.4 | B5244S7 | 45 | 44.5 | NA | NA | 0.0012-0.0024 | 0.0012-0.0024 | 0.2344-0.2350 | 0.2344-0.2350 |
| | 2.5 | B5254T7 | 45 | 44.5 | NA | NA | 0.0012-0.0024 | 0.0012-0.0024 | 0.2344-0.2350 | 0.2344-0.2350 |
| | 3.0 | B63042T | 45 | 44.5 | NA | NA | NA | NA | 0.2350-0.2370 | 0.2350-0.2370 |
| | 3.2 | B6324S | 45 | 44.5 | NA | NA | NA | NA | 0.2350-0.2370 | 0.2350-0.2370 |
| | 4.4 | B8444S | 45 | 44.5 | NA | NA | 0.0012-0.0024 | 0.0012-0.0024 | NA | NA |
| 2008 | 2.4 | B5244S4 | 45 | 44.5 | NA | NA | 0.0012-0.0024 | 0.0012-0.0024 | 0.2344-0.2350 | 0.2344-0.2350 |
| | 2.5 | B5254T3 | 45 | 44.5 | NA | NA | 0.0012-0.0024 | 0.0012-0.0024 | 0.2344-0.2350 | 0.2344-0.2350 |
| | 2.5 | B5254S4 | 45 | 44.5 | NA | NA | 0.0012-0.0024 | 0.0012-0.0024 | 0.2344-0.2350 | 0.2344-0.2350 |
| | 2.5 | B5254T3 | 45 | 44.5 | NA | NA | 0.0012-0.0024 | 0.0012-0.0024 | 0.2344-0.2350 | 0.2344-0.2350 |
| | 2.5 | B5254T2 | 45 | 44.5 | NA | NA | 0.0012-0.0024 | 0.0012-0.0024 | 0.2344-0.2350 | 0.2344-0.2350 |
| | 2.4 | B5244S7 | 45 | 44.5 | NA | NA | 0.0012-0.0024 | 0.0012-0.0024 | 0.2344-0.2350 | 0.2344-0.2350 |
| | 2.5 | B5254T7 | 45 | 44.5 | NA | NA | 0.0012-0.0024 | 0.0012-0.0024 | 0.2344-0.2350 | 0.2344-0.2350 |
| | 3.0 | B63042T | 45 | 44.5 | NA | NA | NA | NA | 0.2350-0.2370 | 0.2350-0.2370 |
| | 3.2 | B6324S | 45 | 44.5 | NA | NA | NA | NA | 0.2350-0.2370 | 0.2350-0.2370 |
| | 4.4 | B8444S | 45 | 44.5 | NA | NA | 0.0012-0.0024 | 0.0012-0.0024 | NA | NA |

NA: Information not available

## CRANKSHAFT AND CONNECTING ROD SPECIFICATIONS

All measurements are given in inches.

| Year | Engine Displacement Liters | Engine ID | Crankshaft | | | | Connecting Rod | | |
|---|---|---|---|---|---|---|---|---|---|
| | | | Main Brg. Journal Dia. | Main Brg. Oil Clearance | Shaft End-play | Thrust on No. | Journal Diameter | Oil Clearance | Side Clearance |
| **2007** | 2.4 | B5244S4 | 2.5584 - 2.5592 | NA | 0.003-0.007 | NA | NA | NA | NA |
| | 2.5 | B5254T3 | 2.5584 - 2.5592 | NA | 0.003-0.007 | NA | NA | NA | NA |
| | 2.5 | B5254S4 | 2.5584 - 2.5592 | NA | 0.003-0.007 | NA | NA | NA | NA |
| | 2.5 | B5254T3 | 2.5584 - 2.5592 | NA | 0.003-0.007 | NA | NA | NA | NA |
| | 2.5 | B5254T2 | 2.5584 - 2.5592 | NA | 0.003-0.007 | NA | NA | NA | NA |
| | 2.4 | B5244S7 | 2.5584 - 2.5592 | NA | 0.003-0.007 | NA | NA | NA | NA |
| | 2.5 | B5254T7 | 2.5584 - 2.5592 | NA | 0.003-0.007 | NA | NA | NA | NA |
| | 3.0 | B63042T | 2.5584 - 2.5592 | NA | 0.003-0.007 | NA | NA | NA | NA |
| | 3.2 | B6324S | 2.5584 - 2.5592 | NA | 0.003-0.007 | NA | NA | NA | NA |
| | 4.4 | B8444S | NA | NA | NA | NA | NA | NA | NA |
| **2008** | 2.4 | B5244S4 | 2.5584 - 2.5592 | NA | 0.003-0.007 | NA | NA | NA | NA |
| | 2.5 | B5254T3 | 2.5584 - 2.5592 | NA | 0.003-0.007 | NA | NA | NA | NA |
| | 2.5 | B5254S4 | 2.5584 - 2.5592 | NA | 0.003-0.007 | NA | NA | NA | NA |
| | 2.5 | B5254T3 | 2.5584 - 2.5592 | NA | 0.003-0.007 | NA | NA | NA | NA |
| | 2.5 | B5254T2 | 2.5584 - 2.5592 | NA | 0.003-0.007 | NA | NA | NA | NA |
| | 2.4 | B5244S7 | 2.5584 - 2.5592 | NA | 0.003-0.007 | NA | NA | NA | NA |
| | 2.5 | B5254T7 | 2.5584 - 2.5592 | NA | 0.003-0.007 | NA | NA | NA | NA |
| | 3.0 | B63042T | 2.5584 - 2.5592 | NA | 0.003-0.007 | NA | NA | NA | NA |
| | 3.2 | B6324S | 2.5584 - 2.5592 | NA | 0.003-0.007 | NA | NA | NA | NA |
| | 4.4 | B8444S | NA | NA | NA | NA | NA | NA | NA |

NA: Information not available

## PISTON AND RING SPECIFICATIONS

All measurements are given in inches.

| Year | Engine Displ. Liters | Engine ID | Piston Clearance | Ring Gap | | | Ring Side Clearance | | |
|---|---|---|---|---|---|---|---|---|---|
| | | | | Top Compression | Bottom Compression | Oil Control | Top Compression | Bottom Compression | Oil Control |
| 2007 | 2.4 | B5244S4 | 0.0004-0.0012 | 0.047 | 0.059 | NA | 0.0012-0.0028 | 0.0012-0.0028 | 0.0015-0.0056 |
| | 2.5 | B5254T3 | 0.0004-0.0012 | 0.047 | 0.059 | NA | 0.0012-0.0028 | 0.0012-0.0028 | 0.0015-0.0056 |
| | 2.5 | B5254S4 | 0.0004-0.0012 | 0.047 | 0.059 | NA | 0.0012-0.0028 | 0.0012-0.0028 | 0.0015-0.0056 |
| | 2.5 | B5254T3 | 0.0004-0.0012 | 0.047 | 0.059 | NA | 0.0012-0.0028 | 0.0012-0.0028 | 0.0015-0.0056 |
| | 2.5 | B5254T2 | 0.0004-0.0012 | 0.047 | 0.059 | NA | 0.0012-0.0028 | 0.0012-0.0028 | 0.0015-0.0056 |
| | 2.4 | B5244S7 | 0.0004-0.0012 | 0.047 | 0.059 | NA | 0.0012-0.0028 | 0.0012-0.0028 | 0.0015-0.0056 |
| | 2.5 | B5254T7 | 0.0004-0.0012 | 0.047 | 0.059 | NA | 0.0012-0.0028 | 0.0012-0.0028 | 0.0015-0.0056 |
| | 3.0 | B63042T | 0.0004-0.0012 | 0.047 | 0.059 | NA | 0.0012-0.0028 | 0.0012-0.0028 | 0.0015-0.0056 |
| | 3.2 | B6324S | 0.0004-0.0012 | 0.047 | 0.059 | NA | 0.0012-0.0028 | 0.0012-0.0028 | 0.0015-0.0056 |
| | 4.4 | B8444S | NA | NA | NA | NA | NA | NA | NA |
| 2008 | 2.4 | B5244S4 | 0.0004-0.0012 | 0.047 | 0.059 | NA | 0.0012-0.0028 | 0.0012-0.0028 | 0.0015-0.0056 |
| | 2.5 | B5254T3 | 0.0004-0.0012 | 0.047 | 0.059 | NA | 0.0012-0.0028 | 0.0012-0.0028 | 0.0015-0.0056 |
| | 2.5 | B5254S4 | 0.0004-0.0012 | 0.047 | 0.059 | NA | 0.0012-0.0028 | 0.0012-0.0028 | 0.0015-0.0056 |
| | 2.5 | B5254T3 | 0.0004-0.0012 | 0.047 | 0.059 | NA | 0.0012-0.0028 | 0.0012-0.0028 | 0.0015-0.0056 |
| | 2.5 | B5254T2 | 0.0004-0.0012 | 0.047 | 0.059 | NA | 0.0012-0.0028 | 0.0012-0.0028 | 0.0015-0.0056 |
| | 2.4 | B5244S7 | 0.0004-0.0012 | 0.047 | 0.059 | NA | 0.0012-0.0028 | 0.0012-0.0028 | 0.0015-0.0056 |
| | 2.5 | B5254T7 | 0.0004-0.0012 | 0.047 | 0.059 | NA | 0.0012-0.0028 | 0.0012-0.0028 | 0.0015-0.0056 |
| | 3.0 | B63042T | 0.0004-0.0012 | 0.047 | 0.059 | NA | 0.0012-0.0028 | 0.0012-0.0028 | 0.0015-0.0056 |
| | 3.2 | B6324S | 0.0004-0.0012 | 0.047 | 0.059 | NA | 0.0012-0.0028 | 0.0012-0.0028 | 0.0015-0.0056 |
| | 4.4 | B8444S | NA | NA | NA | NA | NA | NA | NA |

NA: Information not available

## TORQUE SPECIFICATIONS
All readings in ft. lbs.

| Year | Engine Displacement Liters | Engine ID | Cylinder Head Bolts | Main Bearing Bolts | Rod Bearing Bolts | Crankshaft Damper Bolts | Flywheel Bolts | Manifold | | Spark Plugs | Lug Nut |
|---|---|---|---|---|---|---|---|---|---|---|---|
| | | | | | | | | Intake | Exhaust | | |
| 2007 | 2.4 | B5244S4 | ① | ② | ③ | 132-134 | ④ | 14 | 18 | 22 | 103 |
| | 2.5 | B5254T3 | ① | ② | ③ | 132-134 | ④ | 14 | 18 | 22 | 103 |
| | 2.5 | B5254S4 | ① | ② | ③ | 132-134 | ④ | 14 | 18 | 22 | 103 |
| | 2.5 | B5254T3 | ① | ② | ③ | 132-134 | ④ | 14 | 18 | 22 | 103 |
| | 2.5 | B5254T2 | ① | ② | ③ | 132-134 | ④ | 14 | 18 | 22 | 103 |
| | 2.4 | B5244S7 | ① | ② | ③ | 220-222 | ④ | 14 | 18 | 22 | 103 |
| | 2.5 | B5254T7 | ① | ② | ③ | 220-222 | ④ | 14 | 18 | 22 | 103 |
| | 3.0 | B63042T | ⑤ | ⑥ | ⑦ | NA | ⑧ | 18 | 20 | 22 | 103 |
| | 3.2 | B6324S | ⑤ | ⑥ | ⑦ | NA | ⑧ | 18 | 20 | 22 | 103 |
| | 4.4 | B8444S | ⑨ | NA | NA | NA | 59 | 14 | 18 | 22 | 103 |
| 2008 | 2.4 | B5244S4 | ② | ③ | ④ | 132 | ⑤ | 14 | 18 | 22 | 103 |
| | 2.5 | B5254T3 | ② | ③ | ④ | 132 | ⑤ | 14 | 18 | 22 | 103 |
| | 2.5 | B5254S4 | ② | ③ | ④ | 132 | ⑤ | 14 | 18 | 22 | 103 |
| | 2.5 | B5254T3 | ② | ③ | ④ | 132 | ⑤ | 14 | 18 | 22 | 103 |
| | 2.5 | B5254T2 | ② | ③ | ④ | 132 | ⑤ | 14 | 18 | 22 | 103 |
| | 2.4 | B5244S7 | ② | ③ | ④ | 220-222 | ⑤ | 14 | 18 | 22 | 103 |
| | 2.5 | B5254T7 | ② | ③ | ④ | 220-222 | ⑤ | 14 | 18 | 22 | 103 |
| | 3.0 | B63042T | ⑤ | ⑥ | ⑦ | NA | ⑧ | 18 | 20 | 22 | 103 |
| | 3.2 | B6324S | ⑤ | ⑥ | ⑦ | NA | ⑧ | 18 | 20 | 22 | 103 |
| | 4.4 | B8444S | ⑨ | NA | NA | NA | 59 | 15 | 15 | 22 | 103 |

① Step 1: 15 ft. lbs.
Step 2: 44 ft. lbs.
Step 3: Plus 130 degrees

② Step 1: M10 bolts: 15 ft. lbs.
Step 2: M10 bolts: 33 ft. lbs.
Step 3: M8 bolts: 18 ft. lbs.
Step 4: M7 bolts: 13 ft. lbs.
Step 5: M10 bolts: Plus 90 degrees

③ Step 1: 15 ft. lbs.
Step 2: Plus 90 degrees

④ Step 1: 33 ft. lbs.
Step 2: Plus 65 degrees

⑤ Step 1: 33 ft. lbs.
Step 2: 33 ft. lbs.
Step 3: Plus 90 degrees
Step 4: Plus 180 degrees

⑥ Step 1: M10 bolts: 4 ft. lbs.
Step 2: M10 bolts: 23 ft. lbs.
Step 3: M8 bolts: 19 ft. lbs.
Step 4: M10 bolts: Plus 105 degrees
Step 5: M8 bolts: 19 ft. lbs.

⑦ Step 1: 12 ft. lbs.
Step 2: 15 ft. lbs.
Step 3: Plus 90 degrees

⑧ 59 ft. lbs.

⑨ Step 1: 30 ft. lbs.
Step 2: Plus 90 degrees

## WHEEL ALIGNMENT

| Year | Model | | Caster Range (+/-Deg.) | Caster Preferred Setting (Deg.) | Camber Range (+/-Deg.) | Camber Preferred Setting (Deg.) | Toe-in (in.) |
|------|-------|---|------|-------|------|-------|-------|
| 2007 | C30 | F | 1.50 | +3.60 | 0.70 | -0.60 | 0.20+/-0.10 |
|      |     | R | — | — | 1.00 | -1.50 | 0.30+/-0.10 |
|      | C70 | F | 1.50 | +3.60 | 0.70 | -0.60 | 0.20+/-0.10 |
|      |     | R | — | — | 1.00 | -1.50 | 0.30+/-0.10 |
|      | S40 | F | 1.50 | +3.60 | 0.70 | -0.60 | 0.20+/-0.10 |
|      |     | R | — | — | 1.00 | -1.50 | 0.30+/-0.20 |
|      | S60 | F | 1.00 | +4.00 | 0.90 | -0.30 | 0.10+/-0.10 |
|      |     | R | — | — | 1.00 | 0 | 0.20+/-0.20 |
|      | S80 | F | 1.00 | +4.00 | 0.90 | -0.30 | 0.10+/-0.10 |
|      |     | R | — | — | 1.00 | 0 | 0.20+/-0.20 |
|      | V50 | F | 1.50 | +3.60 | 0.70 | -0.60 | 0.20+/-0.10 |
|      |     | R | — | — | 1.00 | -1.50 | 0.30+/-0.20 |
|      | V70 | F | 1.00 | +4.00 | 0.90 | -0.30 | 0.10+/-0.10 |
|      |     | R | — | — | 1.00 | 0 | 0.20+/-0.20 |
|      | XC70 | F | 1.00 | +5.00 | 0.90 | -0.30 | 0.20+/-0.10 |
|      |      | R | — | — | 1.00 | +0.30 | 0.20+/-0.20 |
| 2008 | C30 | F | 1.50 | +3.60 | 0.70 | -0.60 | 0.20+/-0.10 |
|      |     | R | — | — | 1.00 | -1.50 | 0.30+/-0.10 |
|      | C70 | F | 1.50 | +3.60 | 0.70 | -0.60 | 0.20+/-0.10 |
|      |     | R | — | — | 1.00 | -1.50 | 0.30+/-0.10 |
|      | S40 | F | 1.50 | +3.60 | 0.70 | -0.60 | 0.20+/-0.10 |
|      |     | R | — | — | 1.00 | -1.50 | 0.30+/-0.20 |
|      | S60 | F | 1.00 | +4.00 | 0.90 | -0.30 | 0.10+/-0.10 |
|      |     | R | — | — | 1.00 | 0 | 0.20+/-0.20 |
|      | S80 | F | 1.00 | +4.00 | 0.90 | -0.30 | 0.10+/-0.10 |
|      |     | R | — | — | 1.00 | 0 | 0.20+/-0.20 |
|      | V50 | F | 1.50 | +3.60 | 0.70 | -0.60 | 0.20+/-0.10 |
|      |     | R | — | — | 1.00 | -1.50 | 0.30+/-0.20 |
|      | V70 | F | 1.00 | +4.00 | 0.90 | -0.30 | 0.10+/-0.10 |
|      |     | R | — | — | 1.00 | 0 | 0.20+/-0.20 |

F: Front

R: Rear

22250_VOLC_C0010

## TIRE, WHEEL AND BALL JOINT SPECIFICATIONS

| Year | Model | OEM Tires | | Tire Pressures (psi) | | Wheel Size | Ball Joint Inspection |
|------|-------|-----------|----------|-------|------|------|------|
|      |       | Standard  | Optional | Front | Rear |      |      |
| 2007 | C30 | 235/45R17<br>235/40R18 | 235/40ZR18 | ① | ① | 7.5J<br>8J | NA |
|      | C70 | 235/45R17<br>235/40R18 | 235/40ZR18 | ① | ① | 7.5J<br>8J | NA |
|      | S40 | 205/55R16 | 205/50R17 | ① | ① | 6.5J/7J | NA |
|      | S60 | 215/55R16<br>235/45R17 | 235/40ZR18 | ① | ① | 7.5J/7J<br>8J | NA |
|      | S80 | 205/50R17<br>245/40R18<br>245/40R17 | None | ① | ① | 7.5J/7J<br>8J | NA |
|      | V50 | 205/55R16 | 205/50R17 | ① | ① | 6.5J/7J | NA |
|      | V70 | 205/55R16<br>215/55R16<br>235/45ZR17 | 235/45ZR17 | ① | ① | 6.5J/7J<br>7.5J | NA |
|      | XC70 | 205/55R16<br>215/55R16<br>235/45ZR17 | 235/45ZR17 | ① | ① | 6.5J/7J<br>7.5J | NA |
| 2008 | C30 | 235/45R17<br>235/40R18 | 235/40ZR18 | ① | ① | 7.5J<br>8J | NA |
|      | C70 | 235/45R17<br>235/40R18 | 235/40ZR18 | ① | ① | 7.5J<br>8J | NA |
|      | S40 | 205/55R16 | 205/50R17 | ① | ① | 6.5J/7J | NA |
|      | S60 | 215/55R16<br>235/45R17 | 235/40ZR18 | ① | ① | 7.5J/7J<br>8J | NA |
|      | S80 | 205/50R17<br>245/40R18<br>245/40R17 | None | ① | ① | 7.5J/7J<br>8J | NA |
|      | V50 | 205/55R16 | 205/50R17 | ① | ① | 6.5J/7J | NA |
|      | V70 | 205/55R16<br>215/55R16<br>235/45ZR17 | 235/45ZR17 | ① | ① | 6.5J/7J<br>7.5J | NA |

OEM: Original Equipment Manufacturer

PSI: Pounds Per Square Inch

STD: Standard

NA: Information not available

OPT: Optional

① See the tire placard on the vehicle

# BRAKE SPECIFICATIONS

All measurements in inches unless noted

| Year | Model | | Brake Disc Original Thickness | Brake Disc Minimum Thickness | Brake Disc Maximum Runout | Minimum Lining Thickness | Brake Caliper Bracket bolts (ft. lbs.) | Brake Caliper Mounting bolts (ft. lbs.) |
|------|-------|---|---------------------|----------------------|-------------------|---------------|------------------|------------------|
| 2007 | C30 | F | NA | NA | 0.003 | 0.118 | 88 | 22 |
| | | R | NA | NA | 0.002 | 0.118 | 52 | 22 |
| | C70 | F | NA | NA | 0.003 | 0.118 | 88 | 22 |
| | | R | NA | NA | 0.002 | 0.118 | 52 | 22 |
| | S40 | F | 1.000 | 0.905 | 0.003 | 0.118 | 88 | 20 |
| | | R | 0.400 | 0.350 | 0.002 | 0.118 | 52 | 20 |
| | S60 | F | NA | NA | 0.001 | 0.118 | 74 | 22 |
| | | R | 0.394 | 0.421 | 0.002 | 0.118 | 44 | 22 |
| | S80 | F | NA | NA | 0.002 | 0.118 | 74 | 22 |
| | | R | 0.421 | 0.397 | 0.002 | 0.118 | 44 | 22 |
| | V50 | F | 0.937 | 0.906 | 0.001 | 0.118 | 88 | 20 |
| | | R | 0.385 | 0.354 | 0.002 | 0.118 | 52 | 20 |
| | V70 | F | NA | NA | 0.001 | 0.118 | 74 | 22 |
| | | R | 0.394 | 0.421 | 0.002 | 0.118 | 44 | 22 |
| | XC70 | F | NA | NA | 0.001 | 0.118 | 74 | 22 |
| | | R | 0.394 | 0.421 | 0.002 | 0.118 | 44 | 22 |
| 2008 | C30 | F | NA | NA | 0.003 | 0.118 | 88 | 22 |
| | | R | NA | NA | 0.002 | 0.118 | 52 | 22 |
| | C70 | F | NA | NA | 0.003 | 0.118 | 88 | 22 |
| | | R | NA | NA | 0.002 | 0.118 | 52 | 22 |
| | S40 | F | 1.000 | 0.905 | 0.003 | 0.118 | 88 | 20 |
| | | R | 0.400 | 0.350 | 0.002 | 0.118 | 52 | 20 |
| | S60 | F | NA | NA | 0.001 | 0.118 | 74 | 22 |
| | | R | 0.394 | 0.421 | 0.002 | 0.118 | 44 | 22 |
| | S80 | F | NA | NA | 0.002 | 0.118 | 74 | 22 |
| | | R | 0.421 | 0.397 | 0.002 | 0.118 | 44 | 22 |
| | V50 | F | 0.937 | 0.906 | 0.001 | 0.118 | 88 | 20 |
| | | R | 0.385 | 0.354 | 0.002 | 0.118 | 52 | 20 |
| | V70 | F | NA | NA | 0.001 | 0.118 | 74 | 22 |
| | | R | 0.394 | 0.421 | 0.002 | 0.118 | 44 | 22 |

NA: Information not available

F: Front

R: Rear

22250_VOLC_C0012

## SCHEDULED MAINTENANCE INTERVALS
### VOLVO CARS

| TO BE SERVICED | TYPE OF SERVICE | VEHICLE MILEAGE INTERVAL (x1000) | | | | | | | | | |
|---|---|---|---|---|---|---|---|---|---|---|---|
| | | 7.5 | 15 | 22.5 | 30 | 37.5 | 45 | 52.5 | 60 | 67.5 | 75 |
| Engine oil & filter | R | ✓ | ✓ | ✓ | ✓ | ✓ | ✓ | ✓ | ✓ | ✓ | ✓ |
| Automatic transmission fluid | S/I | ✓ | ✓ | ✓ | ✓ | ✓ | ✓ | ✓ | ✓ | ✓ | ✓ |
| Brake pads & parking brake | I | ✓ | ✓ | ✓ | ✓ | ✓ | ✓ | ✓ | ✓ | ✓ | ✓ |
| Cabin air filter | R | | ✓ | | ✓ | | ✓ | | ✓ | | ✓ |
| Engine coolant | S/I | | ✓ | | ✓ | | ✓ | | ✓ | | ✓ |
| Air cleaner filter | R | | | | | ✓ | | | | | ✓ |
| Spark plugs | R | | | | | | | ✓ | | | |
| Accessory drive belt | R | | | | | | | | | | ✓ |
| Fuel lines | I | | | | | | | | ✓ | ✓ | ✓ |
| Exhaust system | S/I | | | | | | | ✓ | | | |
| Check suspension | S/I | | | | | | | | ✓ | ✓ | ✓ |
| Brake fluid ① | R | | | | | | | | | | |

R: Replace          S/I: Service or Inspect

① Replace every 2 years or 30,000 miles, whichever comes first under normal conditions, more frequently in mountainous areas or moist climates.

### FREQUENT OPERATION MAINTENANCE (SEVERE SERVICE)

If a vehicle is operated under any of the following conditions it is considered severe service:

- Extremely dusty areas.
- 50% or more of the vehicle operation is in 90°F (32°C) or higher temperatures, or constant operation in temperatures below 32°F (0°C).
- Prolonged idling (vehicle operation in stop and go traffic).
- Frequent short running periods (engine does not warm to normal operating temperatures).
- Police, taxi, delivery usage or trailer towing usage.

Oil & oil filter: change every 5000 miles.

Air filter element: service or inspect every 15,000 miles.

22250_VOLC_C0013

## PRECAUTIONS

Before servicing any vehicle, please be sure to read all of the following precautions, which deal with personal safety, prevention of component damage, and important points to take into consideration when servicing a motor vehicle:

• Never open, service or drain the radiator or cooling system when the engine is hot; serious burns can occur from the steam and hot coolant.

• Observe all applicable safety precautions when working around fuel. Whenever servicing the fuel system, always work in a well-ventilated area. Do not allow fuel spray or vapors to come in contact with a spark, open flame, or excessive heat (a hot drop light, for example). Keep a dry chemical fire extinguisher near the work area. Always keep fuel in a container specifically designed for fuel storage; also, always properly seal fuel containers to avoid the possibility of fire or explosion. Refer to the additional fuel system precautions later in this section.

• Fuel injection systems often remain pressurized, even after the engine has been turned **OFF**. The fuel system pressure must be relieved before disconnecting any fuel lines. Failure to do so may result in fire and/or personal injury.

• Brake fluid often contains polyglycol ethers and polyglycols. Avoid contact with the eyes and wash your hands thoroughly after handling brake fluid. If you do get brake fluid in your eyes, flush your eyes with clean, running water for 15 minutes. If eye irritation persists, or if you have taken brake fluid internally, IMMEDIATELY seek medical assistance.

• The EPA warns that prolonged contact with used engine oil may cause a number of skin disorders, including cancer. You should make every effort to minimize your exposure to used engine oil. Protective gloves should be worn when changing oil. Wash your hands and any other exposed skin areas as soon as possible after exposure to used engine oil. Soap and water, or waterless hand cleaner should be used.

• All new vehicles are now equipped with an air bag system, often referred to as a Supplemental Restraint System (SRS) or Supplemental Inflatable Restraint (SIR) system. The system must be disabled before performing service on or around system components, steering column, instrument panel components, wiring and sensors. Failure to follow safety and disabling procedures could result in accidental air bag deployment, possible personal injury and unnecessary system repairs.

• Always wear safety goggles when working with, or around, the air bag system. When carrying a non-deployed air bag, be sure the bag and trim cover are pointed away from your body. When placing a non-deployed air bag on a work surface, always face the bag and trim cover upward, away from the surface. This will reduce the motion of the module if it is accidentally deployed. Refer to the additional air bag system precautions later in this section.

• Clean, high quality brake fluid from a sealed container is essential to the safe and proper operation of the brake system. You should always buy the correct type of brake fluid for your vehicle. If the brake fluid becomes contaminated, completely flush the system with new fluid. Never reuse any brake fluid. Any brake fluid that is removed from the system should be discarded. Also, do not allow any brake fluid to come in contact with a painted surface; it will damage the paint.

• Never operate the engine without the proper amount and type of engine oil; doing so WILL result in severe engine damage.

• Timing belt maintenance is extremely important. Many models utilize an interference-type, non-freewheeling engine. If the timing belt breaks, the valves in the cylinder head may strike the pistons, causing potentially serious (also time-consuming and expensive) engine damage. Refer to the maintenance interval charts for the recommended replacement interval for the timing belt, and to the timing belt section for belt replacement and inspection.

• Disconnecting the negative battery cable on some vehicles may interfere with the functions of the on-board computer system(s) and may require the computer to undergo a relearning process once the negative battery cable is reconnected.

• When servicing drum brakes, only disassemble and assemble one side at a time, leaving the remaining side intact for reference.

• Only an MVAC-trained, EPA-certified automotive technician should service the air conditioning system or its components.

## BRAKES | ANTI-LOCK BRAKE SYSTEM (ABS)

### GENERAL INFORMATION

*PRECAUTIONS*

• Certain components within the ABS system are not intended to be serviced or repaired individually.

• Do not use rubber hoses or other parts not specifically specified for and ABS system. When using repair kits, replace all parts included in the kit. Partial or incorrect repair may lead to functional problems and require the replacement of components.

• Lubricate rubber parts with clean, fresh brake fluid to ease assembly. Do not use shop air to clean parts; damage to rubber components may result.

• Use only DOT 4+ brake fluid from an unopened container.

• If any hydraulic component or line is removed or replaced, it may be necessary to bleed the entire system.

• A clean repair area is essential. Always clean the reservoir and cap thoroughly before removing the cap. The slightest amount of dirt in the fluid may plug an orifice and impair the system function. Perform repairs after components have been thoroughly cleaned; use only denatured alcohol to clean components. Do not allow ABS components to come into contact with any substance containing mineral oil; this includes used shop rags.

• The Anti-Lock control unit is a microprocessor similar to other computer units in the vehicle. Ensure that the ignition switch is **OFF** before removing or installing controller harnesses. Avoid static electricity discharge at or near the controller.

• If any arc welding is to be done on the vehicle, the control unit should be unplugged before welding operations begin.

**BRAKES**                                                                    **BLEEDING THE BRAKE SYSTEM**

## BLEEDING PROCEDURE

*BLEEDING PROCEDURE*

### ✲✲ CAUTION

Brake fluid contains polyglycol ethers and polyglycols. Avoid contact with the eyes and wash your hands thoroughly after handling brake fluid. If you do get brake fluid in your eyes, flush your eyes with clean, running water for 15 minutes. If eye irritation persists, or if you have taken brake fluid internally, IMMEDIATELY seek medical assistance.

### ✲✲ WARNING

Clean, high quality brake fluid is essential to the safe and proper operation of the brake system. You should always buy the highest quality brake fluid that is available. If the brake fluid becomes contaminated, drain and flush the system, then refill the master cylinder with new fluid. Never reuse any brake fluid. Any brake fluid that is removed from the system should be discarded. Also, do not allow any brake fluid to come in contact with a painted surface; it will damage the paint.

1.  Remove the reservoir cap and fill the brake reservoir with brake fluid.
2.  Pump the brake pedal several times, and then loosen the bleeder screw until fluid starts to run out without bubbles. Then close the bleeder screw.
3.  Repeat until there are no more bubbles in the fluid escaping.
4.  Tighten the bleeder screw to 80 inch lbs. (9 Nm).
5.  Check if there is still air in the system by pressing down the brake pedal with a jack at a pressure equivalent to heavy braking, 148 ft. lbs. (200 Nm).
6.  With the engine off and the brake pedal depressed 3–4 times, the pedal travel must not exceed 1 ½ inch (40mm).
7.  If the pedal travel exceeds this limit, bleed again and re-check the pedal travel.
8.  Check that there is no brake fluid leakage.
9.  Fill the brake reservoir with the proper amount of DOT 4+ brake fluid.

*BRAKE LINE BLEEDING*

### ✲✲ CAUTION

Brake fluid contains polyglycol ethers and polyglycols. Avoid contact with the eyes and wash your hands thoroughly after handling brake fluid. If you do get brake fluid in your eyes, flush your eyes with clean, running water for 15 minutes. If eye irritation persists, or if you have taken brake fluid internally, IMMEDIATELY seek medical assistance.

1.  Make sure the master cylinder is full and stays full during this procedure.
2.  You will need the help of another technician.
3.  As technician (A) applies brake pedal pressure from inside of the vehicle, technician (B) slowly opens the brake line.
4.  Repeat the procedure on any line that may have been replaced or is suspected of air contamination.

➡ Make sure the master cylinder stays full during the bleed procedure.

5.  Repeat the process until the brake pedal returns to a normal position.
6.  Refill the master cylinder with brake fluid.
7.  Test drive the vehicle for correct brake operation.

*BLEEDING THE ABS SYSTEM*

*See Figures 1 and 2.*

### ✲✲ CAUTION

Brake fluid contains polyglycol ethers and polyglycols. Avoid contact with the eyes and wash your hands thoroughly after handling brake fluid. If you do get brake fluid in your eyes, flush your eyes with clean, running water for 15 minutes. If eye irritation persists, or if you have taken brake fluid internally, IMMEDIATELY seek medical assistance.

### ✲✲ WARNING

Clean, high quality brake fluid is essential to the safe and proper operation of the brake system. You should always buy the highest quality brake fluid that is available. If the brake

fluid becomes contaminated, drain and flush the system, then refill the master cylinder with new fluid. Never reuse any brake fluid. Any brake fluid that is removed from the system should be discarded. Also, do not allow any brake fluid to come in contact with a painted surface; it will damage the paint.

1.  When reconditioning or replacing brake calipers bleed the relevant brake pipe as follows.

### ✲✲ WARNING

The brake pedal must be depressed throughout the operation. This is so that the brake system is not drained of brake fluid.

➡ If the braking system has been completely or partly drained, bleed the whole system. A bleeding unit that can pressurize the brake system to 0.2–0.3 Mpa must be used.

2.  Turn off the ignition switch.
3.  Clean all around the brake fluid reservoir filler cap.
4.  Take out the special brake fluid collection bottle.
5.  Connect bleeding unit to the brake fluid reservoir. Follow manufacturer's instructions for connecting and using bleeding unit.
6.  Pressurize the brake system. Check the brake fluid reservoir connector.
7.  Raise the car and remove the wheels.

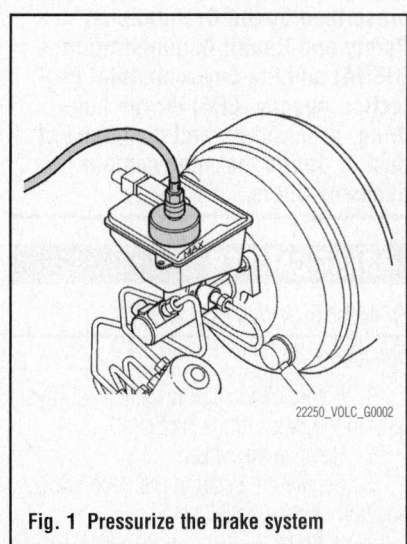

22250_VOLC_G0002

**Fig. 1 Pressurize the brake system**

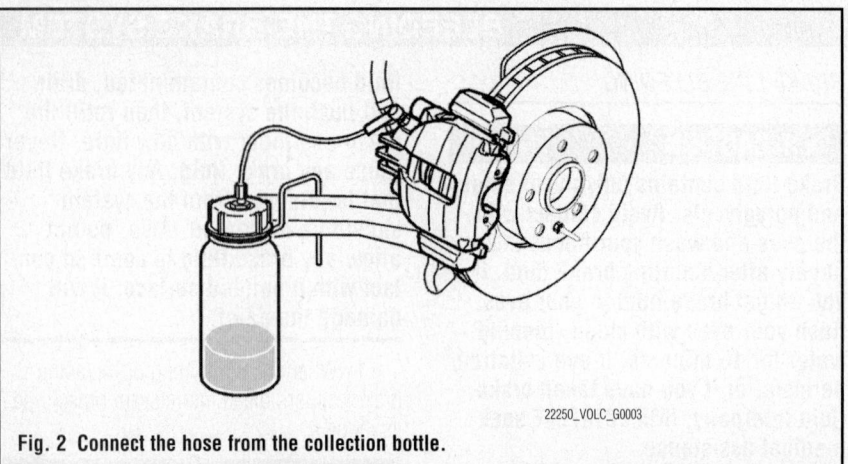

**Fig. 2 Connect the hose from the collection bottle.**

22250_VOLC_G0003

8. Remove the protective cap from the bleed nipple. Connect the hose from the collection bottle.

9. Open the bleed nipple. Close it when there are no more air bubbles in the fluid coming out of the hose.

10. Tighten the bleed nipple.

11. Continue to bleed the other wheels.

12. Check for leakage from the bleed nipples.

13. Depressurize the brake system.

14. Check for air in the brake system and for brake fluid leakage.

15. Fill the brake reservoir with the proper amount of DOT 4+ brake fluid.

## BRAKES

### ✳✳ CAUTION

**Dust and dirt accumulating on brake parts during normal use may contain asbestos fibers from production or aftermarket brake linings. Breathing excessive concentrations of asbestos fibers can cause serious bodily harm. Exercise care when servicing brake parts. Do not sand or grind brake lining unless equipment used is designed to contain the dust residue. Do not clean brake parts with compressed air or by dry brushing. Cleaning should be done by dampening the brake components with a fine mist of water, then wiping the brake components clean with a dampened cloth. Dispose of cloth and all residue containing asbestos fibers in an impermeable container with the appropriate label. Follow practices prescribed by the Occupational Safety and Health Administration (OSHA) and the Environmental Protection Agency (EPA) for the handling, processing, and disposing of dust or debris that may contain asbestos fibers.**

### BRAKE CALIPER

#### REMOVAL & INSTALLATION

*See Figure 3.*

1. Before servicing the vehicle, always refer to the precautions sections.

2. Remove the wheel.

3. Secure the pedal in the depressed position using a pedal jack.

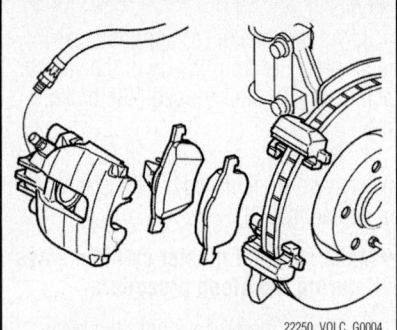

**Fig. 3 Exploded view of the front brake caliper**

22250_VOLC_G0004

4. Remove the protective cap from the bleed nipple, connect a hose and open the nipple.

5. Use a container to collect brake fluid.

6. Slacken off the brake hose half a turn.

7. Remove the following:
- The spring
- The protective caps
- The locating pins, use a 7 mm hex socket
- The brake pads
- The brake caliper (Unscrew the caliper from the brake hose)

**To install:**

8. Install the following.
- The brake hose in the brake caliper (Do not tighten the hose fully yet)
- The brake pads
- The locating pins, use a 7 mm hex socket and tighten to 22 ft. lbs. (30 Nm).

### FRONT DISC BRAKES

- The protective caps
- The spring

9. Tighten the brake hose to 17 ft. lbs. (23 Nm).

10. Bleed the brake system.

11. Install the wheel.

12. Depress the brake pedal a few times and check the brake fluid level.

### DISC BRAKE PADS

#### REMOVAL & INSTALLATION

1. Before servicing the vehicle, always refer to the precautions sections.

2. Remove the wheel.

3. Secure the pedal in the depressed position using a pedal jack.

4. Remove the protective cap from the bleed nipple, connect a hose and open the nipple.

5. Use a container to collect brake fluid.

6. Slacken off the brake hose half a turn.

7. Remove the following:
- The spring
- The protective caps
- The locating pins, use a 7 mm hex socket
- The brake pads

**To install:**

8. Install the following.
- The brake hose in the brake caliper (Do not tighten the hose fully yet)
- The brake pads
- The locating pins, use a 7 mm hex socket and tighten to 22 ft. lbs. (30 Nm).
- The protective caps
- The spring

**BRAKES**                                                          **REAR DISC BRAKES**

**✱✱ CAUTION**

Dust and dirt accumulating on brake parts during normal use may contain asbestos fibers from production or aftermarket brake linings. Breathing excessive concentrations of asbestos fibers can cause serious bodily harm. Exercise care when servicing brake parts. Do not sand or grind brake lining unless equipment used is designed to contain the dust residue. Do not clean brake parts with compressed air or by dry brushing. Cleaning should be done by dampening the brake components with a fine mist of water, then wiping the brake components clean with a dampened cloth. Dispose of cloth and all residue containing asbestos fibers in an impermeable container with the appropriate label. Follow practices prescribed by the Occupational Safety and Health Administration (OSHA) and the Environmental Protection Agency (EPA) for the handling, processing, and disposing of dust or debris that may contain asbestos fibers.

**BRAKE CALIPER**

*REMOVAL & INSTALLATION*

### C30, C70 and V50 Models

*See Figure 4.*

1. Before servicing the vehicle, always refer to the precautions sections.
2. Remove the wheel.
3. Secure the pedal in the depressed position using a pedal jack.
4. Remove the protective cap from the bleed nipple, connect a hose and open the nipple.
5. Use a container to collect brake fluid.
6. Slacken off the brake hose half a turn.
7. Remove the following:
   - The spring
   - The protective caps
   - The locating pins, use a 7 mm hex socket
   - The brake pads
   - The brake caliper (Unscrew the caliper from the brake hose)
8. Remove the brake pads.
9. Remove the brake caliper
10. Remove the parking brake cable.
11. Unscrew the caliper from the brake hose.

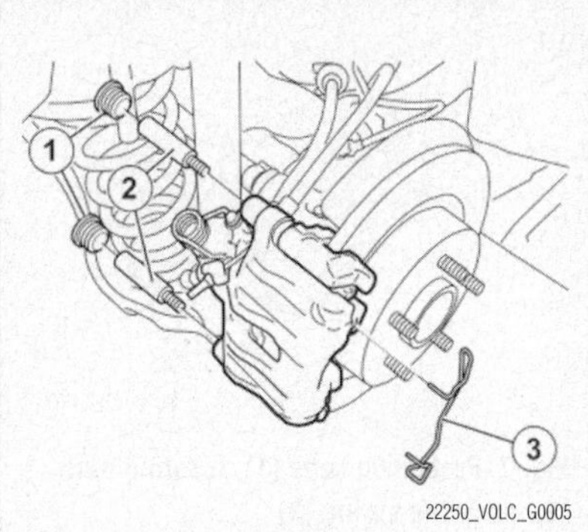

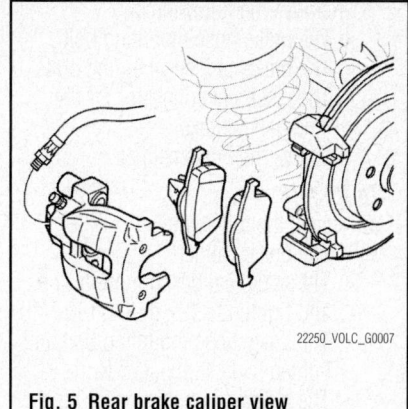

Fig. 4 Rear brake caliper view

*To install:*

12. Install the following:
   - The brake hose in the brake caliper (Do not tighten the hose fully yet)
   - The brake pads
   - The locating pins, use a 7 mm hex socket and tighten to 22 ft. lbs. (30 Nm).
   - The protective caps
   - The spring
13. Tighten the brake hose to 17 ft. lbs. (23 Nm).
14. Install the parking brake cable.
15. Bleed the brake system.
16. Install the wheel.
17. Depress the brake pedal a few times and check the brake fluid level.

### S40, S60 and V70 Models

*See Figure 5.*

1. Before servicing the vehicle, always refer to the precautions sections.
2. Remove the wheel.
3. Secure the pedal in the depressed position using a pedal jack.
4. Remove the protective cap from the bleed nipple, connect a hose and open the nipple.
5. Use a container to collect brake fluid.
6. Slacken off the brake hose half a turn.
7. Remove the following:
   - The spring
   - The protective caps
   - The locating pins, use a 7 mm hex socket
   - The brake pads
   - The brake caliper (Unscrew the caliper from the brake hose)
8. Remove the brake pads.
9. Remove the brake caliper
10. Unscrew the caliper from the brake hose.

Fig. 5 Rear brake caliper view

*To install:*

11. Install the following:
   - The brake hose in the brake caliper (Do not tighten the hose fully yet)
   - The brake pads
   - The locating pins, use a 7 mm hex socket and tighten to 22 ft. lbs. (30 Nm).
   - The protective caps
   - The spring
12. Tighten the brake hose to 17 ft. lbs. (23 Nm).
13. Bleed the brake system.
14. Install the wheel.
15. Depress the brake pedal a few times and check the brake fluid level.

### S80 Models With Electrical Emergency Brake System

*See Figure 6.*

1. Before servicing the vehicle, always refer to the precautions sections.
2. Remove the wheel.
3. Secure the pedal in the depressed position using a pedal jack.
4. Remove the protective cap from the bleed nipple, connect a hose and open the nipple.

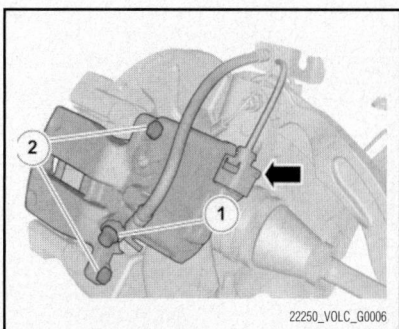

Fig. 6 Rear brake caliper view with electric emergency brake

5. Use a container to collect brake fluid.
6. Remove the following:
- The brake hose mounting bolt
- The brake caliper mounting bolts
- The electrical connector for the emergency brake
7. Remove the brake caliper.

**To install:**

8. Install the following:
- The brake caliper
- The brake caliper mounting bolts and tighten to (22 ft. lbs. (30 Nm)
- The brake hose mounting bolt and tighten to 17 ft. lbs. (23 Nm).
- The electrical connector for the emergency brake
9. Bleed the brake system.
10. Install the wheel.
11. Depress the brake pedal a few times and check the brake fluid level.

## DISC BRAKE PADS

### REMOVAL & INSTALLATION

#### C30, C70 and V50 Models

*See Figure 7.*

1. Remove the securing spring (3) carefully so that it does not deform
2. Remove the protective caps (1) from the two locating pins (2)
3. Remove the locating pins, use hex socket 7mm.
4. Remove the brake caliper from the holder.
5. Remove the brake pads.
6. Hang brake caliper from a steel wire from the spring so as not to damage brake hose.
7. Press the piston back into cylinder on brake caliper.
8. Check that the caliper dust boot is correctly positioned.

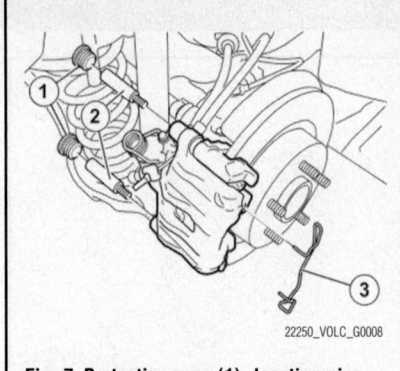

22250_VOLC_G0008

**Fig. 7 Protective caps (1), locating pins (2), securing spring (3)**

**To install:**

9. Install new brake pads
10. Install the brake caliper.
11. Check the rubber sleeves of the locating pins. Replace if necessary.
12. Lubricate the locating pins with silicone brake grease. Insert the locating pins (2) into the rubber sleeves. The pins should slide into the sleeves easily.
13. Tighten the locating pins. Tighten to 22 ft. lbs. (30 Nm). Install protective caps (1).
14. Install the retaining spring (3).
15. Depress the brake pedal a few times.
16. Add brake fluid if necessary.
17. Adjust the parking brake as required.

#### S40, S60 and V70 Models

1. Remove or disconnect the following:
- Wheel
- Caliper spring
- Protective caps
- Locating pins
- Brake caliper and hang in a suitable position
- Brake pads

➡**Do not depress the brake pedal while the brake pads are removed.**

**To install:**

2. Check the rubber sleeves on the locating pins. Replace if necessary. If necessary, lubricate the locating pins using silicon grease. Insert the locating pins in the rubber sleeves. The locating pins must slide into the sleeves easily.
3. Press the piston back into the cylinder using a C—clamp.
4. Install or connect the following:
- Brake pads
- Brake caliper
- Locating pins. Tighten to 22 ft. lbs. (30 Nm).
- Protective caps
- Caliper spring
- Wheel

#### S80 Models With Electrical Emergency Brake System

1. Before servicing the vehicle, always refer to the precautions sections.
2. Remove the wheel.
3. Remove the following:
- The brake caliper mounting bolts
- The electrical connector for the emergency brake
- The brake caliper
- The brake pads
4. Retract the brake caliper piston to except the new brake pads.

**To install:**

5. Install the following:
- The brake pads
- The brake caliper mounting bolts and tighten to (22 ft. lbs. (30 Nm)).
- The electrical connector for the emergency brake
6. Bleed the brake system if needed.
7. Adjust the parking brake system.
8. Install the wheel.
9. Depress the brake pedal a few times and check the brake fluid level.

## BRAKES

### PARKING BRAKE

### PARKING BRAKE SHOES

*REMOVAL & INSTALLATION*

**S40, S60 and V70 Models**
*See Figures 8 and 9.*

1. Before servicing the vehicle, always refer to the precautions sections.
2. Remove rear disc brake caliper assembly.

➡**Before removing the brake disc rotor, make a chalk marking by the bolts to aid in reassembly.**

3. Remove the rotor.
4. Remove the shoe hold spring by turning the pin to coincide with hole of spring cap.
5. Remove the return spring.
6. Disconnect the cable end from the trailing shoe.
7. Remove the parking brake shoes.

➡**Complete the removal and installation on one side of the vehicle at a time using the others side as a reference.**

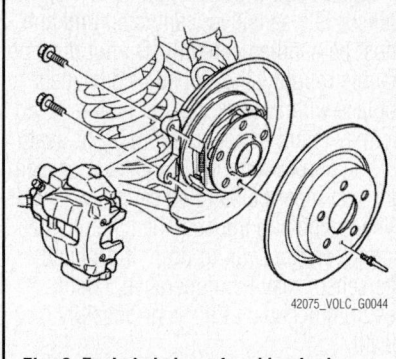

**Fig. 8 Exploded view of parking brake assembly**

*To install:*
8. Install the parking brake cable to the operating lever.
9. Apply brake grease to return spring and areas of movement on the mechanism, but do not apply grease to the brake shoe material.
10. Install the upper return spring and brake shoes.
11. Turn the adjuster in counter clockwise direction and install.

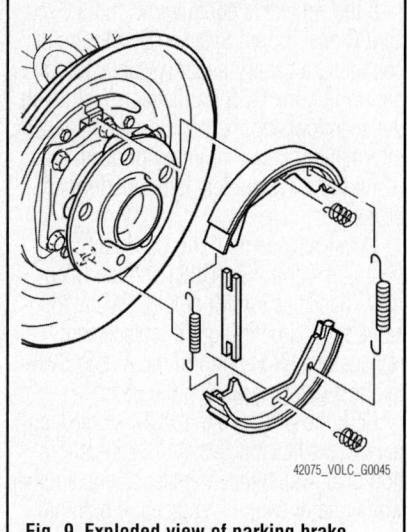

**Fig. 9 Exploded view of parking brake shoes and return springs**

12. Install the lower return spring.
13. Install the shoe hold spring with pliers.
14. Install the disc brake and then align the mark while tightening the screw.

## CHASSIS ELECTRICAL

## AIR BAG (SUPPLEMENTAL RESTRAINT SYSTEM)

### GENERAL INFORMATION

### ✴✴ CAUTION

**These vehicles are equipped with an air bag system. The system must be disarmed before performing service on, or around, system components, the steering column, instrument panel components, wiring and sensors. Failure to follow the safety precautions and the disarming procedure could result in accidental air bag deployment, possible injury and unnecessary system repairs.**

*SERVICE PRECAUTIONS*

Disconnect and isolate the battery negative cable before beginning any airbag system component diagnosis, testing, removal, or installation procedures. Allow system capacitor to discharge for two minutes before beginning any component service. This will disable the airbag system. Failure to disable the airbag system may result in accidental airbag deployment, personal injury, or death.

Do not place an intact undeployed airbag face down on a solid surface. The airbag will propel into the air if accidentally deployed and may result in personal injury or death.

When carrying or handling an undeployed airbag, the trim side (face) of the airbag should be pointing towards the body to minimize possibility of injury if accidental deployment occurs. Failure to do this may result in personal injury or death.

Replace airbag system components with OEM replacement parts. Substitute parts may appear interchangeable, but internal differences may result in inferior occupant protection. Failure to do so may result in occupant personal injury or death.

Wear safety glasses, rubber gloves, and long sleeved clothing when cleaning powder residue from vehicle after an airbag deployment. Powder residue emitted from a deployed airbag can cause skin irritation. Flush affected area with cool water if irritation is experienced. If nasal or throat irritation is experienced, exit the vehicle for fresh air until the irritation ceases. If irritation continues, see a physician.

Do not use a replacement airbag that is not in the original packaging. This may result in improper deployment, personal injury, or death.

The factory installed fasteners, screws and bolts used to fasten airbag components have a special coating and are specifically designed for the airbag system. Do not use substitute fasteners. Use only original equipment fasteners listed in the parts catalog when fastener replacement is required.

During, and following, any child restraint anchor service, due to impact event or vehicle repair, carefully inspect all mounting hardware, tether straps, and anchors for proper installation, operation, or damage. If a child restraint anchor is found damaged in any way, the anchor must be replaced. Failure to do this may result in personal injury or death.

Deployed and non-deployed airbags may or may not have live pyrotechnic material within the airbag inflator.

Do not dispose of driver/passenger/curtain airbags or seat belt tensioners unless you are sure of complete deployment. Refer to the Hazardous Substance Control System for proper disposal.

Dispose of deployed airbags and tensioners consistent with state, provincial, local, and federal regulations.

After any airbag component testing or service, do not connect the battery negative cable. Personal injury or death may result if the system test is not performed first.

If the vehicle is equipped with the Occupant Classification System (OCS), do not connect the battery negative cable before performing the OCS Verification Test using the scan tool and the appropriate diagnostic information. Personal injury or death may result if the system test is not performed properly.

Never replace both the Occupant Restraint Controller (ORC) and the Occupant Classification Module (OCM) at the same time. If both require replacement, replace one, then perform the Airbag System test before replacing the other.

Both the ORC and the OCM store Occupant Classification System (OCS) calibration data, which they transfer to one another when one of them is replaced. If both are replaced at the same time, an irreversible fault will be set in both modules and the

OCS may malfunction and cause personal injury or death.

If equipped with OCS, the Seat Weight Sensor is a sensitive, calibrated unit and must be handled carefully. Do not drop or handle roughly. If dropped or damaged, replace with another sensor. Failure to do so may result in occupant injury or death.

If equipped with OCS, the front passenger seat must be handled carefully as well. When removing the seat, be careful when setting on floor not to drop. If dropped, the sensor may be inoperative, could result in occupant injury, or possibly death.

If equipped with OCS, when the passenger front seat is on the floor, no one should sit in the front passenger seat. This uneven force may damage the sensing ability of the seat weight sensors. If sat on and damaged, the sensor may be inoperative, could result in occupant injury, or possibly death.

### DISARMING THE SYSTEM

1. Before servicing the vehicle, refer to the Precautions Section.
2. Disconnect and isolate the negative battery cable. Wait 3 minutes for the system capacitor to discharge before performing any service.

**❊❊ CAUTION**

**Wait at least 3 minutes before working on the vehicle. The air bag system is designed to retain enough power to deploy the air bag for a short time after the battery has been disconnected.**

### ARMING THE SYSTEM

1. After repairs are complete, connect the negative battery cable. Turn the ignition switch to the **ON** position and check the SRS light for proper operation.

## DRIVETRAIN

### AUTOMATIC TRANSAXLE ASSEMBLY

#### REMOVAL & INSTALLATION

**AW55-50/51SN Transmissions**

1. Before servicing the vehicle, refer to the Precautions Section.
2. Disconnect the negative battery cable.
3. Remove the box for the battery.
4. Slacken off the 2 x steering gear screws. Leave the screws attached by a few threads.
5. Remove the front wheels on both sides.
6. Remove the propeller shaft and transfer case. (AWD) models only.
7. Remove the splash guard under the engine.
8. Remove the drive shafts.
9. Remove the 2 x screws for the steering gear and sub-frame.
10. Slacken off the 2 x front upper screws on the sub-frame 3-4 turns.
11. Remove the anti-roll bar as follows:
   - Position the transmission jack together under the sub-frame.
   - Detach the rubber mountings for the exhaust pipe from the bracket.
   - Remove the screws at the rear edge of the sub-frame. Remove the

screw for the torque rod mounting at the front edge.
   - Lower the rear edge of the sub-frame a little. Remove the screws for the stabilizer bar that can be accessed from the top.
   - Remove the stabilizer bar.
12. Secure the steering gear using tie straps.
13. Remove the sub frame as follows:

➡**Note! Note the routing of the heated oxygen sensor (HO2S) wiring.**

   - Remove the wiring from the sub-frame and from the mounting on the transaxle.
   - Secure the sub-frame against the fixture plate. Use a retaining strap.
   - Remove the front screws for the sub-frame.
   - Lower and remove the frame.
14. Remove the transmission oil cooler lines and plug the line and the hole in the transaxle using plastic plugs.
15. Remove the ground cable from the transaxle.
16. Remove the torque rod mounting from the transaxle.
17. Remove the screws from the torque converter.
18. Remove the air cleaner module assembly.
19. Un-hook the gear selector cable and remove it from the bracket.

20. Remove the oil line from the transaxle cover and the oil cooler.
21. Plug the cover and the oil cooler using plastic plugs.
22. Remove the engine coolant hose bracket.
23. Remove the holder for the bracket.
24. Remove the transaxle bleed hose from the holder.
25. Secure the wiring harnesses and the hoses to expose the screwed joint for the transaxle and engine.
26. Remove the starter motor.
27. Remove the 4 engine and transmission screws.
28. Install an engine support system.
29. Lower the engine and the transmission as follows:
   - Remove the screw for the left-hand engine mount.
   - Lower the engine. Approximately 75 mm of the screw threads must remain. Also check that there are a few millimeters of play between the crankshaft generator pulley and the side member.

➡**Check carefully that no hoses or cables are stretched or trapped.**

30. Remove the mounting from the transaxle.
31. Install the fixture on the jack and transmission.
32. Apply the mobile jack lightly against the transmission.

33. Remove the remaining 6 screws for the transmission and engine.

34. Remove the transmission.

***To install:***

35. Install the transmission.

36. Install the 6 M10 screws to the transmission and engine.

37. First tighten alternately and cross-wise to light contact. At the same time, check that the whole of the transmission is in contact with the engine.

38. Tighten the transmission M10 screws to 37 ft. lbs. (50 Nm).

39. Remove the fixture from the transmission.

40. Tighten the transmission mounting as follows:
- Step 1: 22 ft. lbs. (30 Nm)
- Step 2: 180°

41. Lift the engine/transaxle. Align the mounting against the transaxle in the engine support insulator.

42. Install the M14 screw for the engine mount and tighten to 96 ft. lbs. (130 Nm).

43. Remove the engine support system.

44. Install the 4 x M10 screws for the transmission and engine. Tighten to 37 ft. lbs. (50 Nm).

45. Install the starter motor and tighten the screws to 30 ft. lbs. (40 Nm).

46. Install the electrical connector for the transmission.

47. Install the bracket holder.

48. Install the bracket and the coolant hose.

49. Install the oil line to the oil cooler and timing cover

50. Install the gear selector cable on the bracket and the lever.

51. Install the air cleaner assembly.

52. Align the torque converter against the carrier plate. Do this by turning the engine with the nut for the crankshaft pulley and turning the torque converter.

53. Install all 6 x screws loosely. Tighten the screws until the heads are almost in contact with the flexplate.

54. Tighten the torque converter screws to 23 ft. lbs. (30 Nm). Plus an additional 90°

55. Install the oil line. Use a new O-ring

56. Install the ground cable on the transmission.

57. Install the torque rod mounting to the transaxle, 3 x M10. Tighten to 37 ft. lbs. (50 Nm).

58. Install the sub-frame as follows:
- Secure the sub-frame against the fixture plate. Use a retaining strap.
- Lift the sub-frame. Align the sub-frame in the front mountings in the

side members. Install the screws loosely, 4–5 threads.
- Allow the sub-frame to hang down at the rear edge and to rest on the transmission jack.
- Remove the temporary mounting for the steering gear. Allow the steering gear to rest on the sub-frame.

59. Install the stabilizer bar.

60. Securing the sub-frame as follows:
- Install the adjustment tools 999 7089 (on both sides).
- Carefully raise the sub-frame so that the adjustment tools are lined up with the holes in the side members.
- Install the screws for the rear mounting of the sub-frame. Loosely install the new M16 screws, together with the brackets and the 4 x M8 screws.
- Tighten the M16 screws to 103 ft. lbs. (140 Nm). Plus an additional 120°

61. Tighten the screw for the lower torque rod mounting. Use a new screw.

62. Install the rubber mountings for the exhaust pipe mounting.

63. Install the catalyst monitor sensor wiring to the sub-frame and to the tab on the torque rod mounting in the transaxle.

64. Tighten the front sub-frame M12 screws to 59 ft. lbs (80 Nm).

65. Install the new screws loosely in the steering gear and sub-frame.

66. Tightening is performed later from the engine compartment.

67. Install the propeller shaft and transfer case. (AWD) models only.

68. Install the drive shafts.

69. Install the front wheels.

70. Tighten the 2 x M12 screws for the steering gear to 59 ft. lbs (80 Nm).

71. Install the battery box.

72. Connect the negative battery cable.

73. Install the transmission fluid.

74. Road test and check for leaks.

### TF-80SC Transmission—4.4L S80 Engine

*See Figure 10.*

1. Before servicing the vehicle, refer to the Precautions Section.

2. Remove the following items:
- Air cleaner housing
- Battery
- Battery tray
- Left fender splash guard
- Left drive shaft
- Left hand exhaust pipe

- Right hand exhaust pipe
- Propeller shaft
- Transfer case

3. Remove the shifter cable and the electrical connectors at transmission and starter.

4. Remove the starter motor.

5. Remove the top transmission bolts.

6. Remove the propeller shaft and transfer case. (AWD) models only.

7. Remove the transmission torque converter access plate bolts.

8. Remove all 6 torque converter screws.

9. Remove the transmission oil lines.

10. Remove the transmission ground cable.

11. Install an engine support and lower the left hand side of the engine/transaxle approximately 2.5 inches. (60 mm)

12. Install a transmission jack and support the transmission.

13. Remove the 7 remaining transmission bolts.

14. Remove the transmission assembly.

***To install:***

15. Adjust the height and incline using a transmission jack.

16. Align the transaxle with the engine.

17. Install the transmission assembly.

18. Install the 7 lower transmission bolts. Tighten the bolts to 37 ft. lbs. (50 Nm).

19. Remove the transmission jack.

20. Raise the engine/transaxle to the horizontal position.

21. Install the ground cable and transmission oil lines.

22. Install all 6 torque converter screws. Tighten to 45 ft. lbs. (60 Nm).

23. Install the transmission torque converter access plate bolts.

24. Remove the propeller shaft and transfer case. (AWD) models only.

25. Install the top transmission bolts.

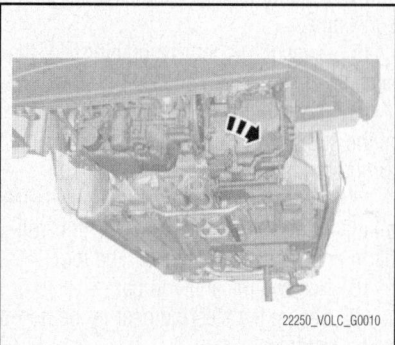

22250_VOLC_G0010

**Fig. 10 Remove the transmission assembly**

26. Install the starter motor.

27. Install the shifter cable and the electrical connectors at transmission and starter.

28. Remove the following items:
- Transfer case
- Propeller shaft
- Right hand exhaust pipe
- Left hand exhaust pipe
- Left drive shaft
- Left fender splash guard
- Battery tray
- Battery
- Air cleaner housing

## MANUAL TRANSAXLE ASSEMBLY

### REMOVAL & INSTALLATION

#### M56/M66 Transmissions

*See Figures 11 and 12.*

1. Before servicing the vehicle, refer to the Precautions Section.

2. Disconnect the negative battery cable.

3. Remove the battery box.

4. Slacken off the 2 steering gear screws. Leave the screws attached by a few threads.

5. Remove the front wheels.

6. Remove the splashguard under the engine.

7. Drain the transmission fluid.

8. Remove the propeller shaft and transfer case. (AWD) models.

9. Remove the drive shafts.

10. Remove the two screws between the steering gear and the sub-frame.

11. Slacken off the sub-frame front and upper screws (x2) at the front edge 3–4 turns on both sides.

12. Place a transmission jack under the sub-frame.

13. Detach the rubber mountings for the exhaust pipe from the bracket.

14. Install transmission jack under the sub-frame.

15. Detach the rubber mountings for the exhaust pipe from the bracket.

16. Remove the screws at the rear edge of the sub-frame. Remove the screw for the torque rod mounting at the front edge.

17. Lower the rear edge of the sub-frame a little. Remove the screws for the anti-roll bar that can be accessed from the top.

18. Remove the anti-roll bar.

19. Secure the steering gear in the hooks using tie straps.

20. Remove the cable harness from the sub-frame and from the mounting on the transmission.

21. Secure the sub-frame against the fixture plate using a retaining strap.

22. Remove the front screws for the sub-frame.

23. Lower and remove the frame.

24. Remove the ground lead from the transmission.

25. Remove the torque rod mounting from the transmission.

26. Remove the air cleaner module.

27. Remove the transmission shift cables.

28. Remove the engine coolant hose bracket.

29. Remove the holder for the bracket.

30. Secure the cable harnesses and the hoses to expose the screwed joint for the transmission and engine.

31. Connect the bleed equipment to the bleed valve. Open the bleed screw. Press the clutch pedal down using a pedal jack. Pull out the locking pin and remove the clutch line.

32. Tie up the clutch line.

33. Remove the bracket for the gear shift cables, 3 screws.

34. Install engine support system.

35. Lower the engine. Approximately 3 inches (75 mm) of the screw threads must remain. Also check that there are a few millimeters of play between the crankshaft pulley and the side member.

➡**Check carefully that no hoses/cables are stretched or trapped.**

36. Remove the bracket for the engine mounting, 3 screws.

37. Remove the starter motor and 3 screws.

38. Remove the 3 screws between the engine and transmission.

39. Remove the reverse light switch connector.

40. Install the transmission jack and support.

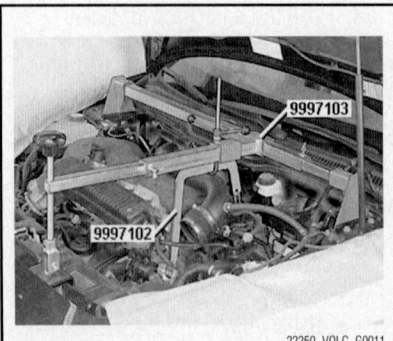

22250_VOLC_G0011

**Fig. 11 Engine support system installed**

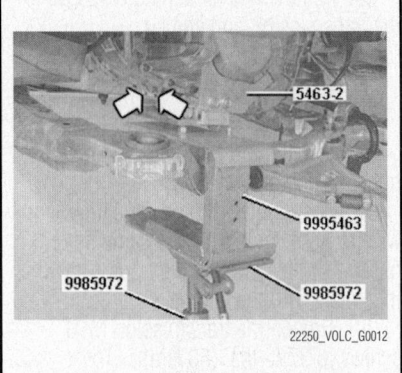

22250_VOLC_G0012

**Fig. 12 Engine support system installed**

41. Remove the remaining 7 screws for the transmission and engine.

42. Remove the transmission assembly.

*To install:*

43. Install the transmission assembly.

44. Locate the transmission on the engine. Ensure that the transmission goes straight in, in relation to the engine, and that no deviation occurs in the clutch driven plate center.

45. Install the 7 M10 screws securing the transmission to the engine. First tighten alternately and crosswise to light contact. At the same time, check that the whole of the transmission is in contact with the engine. Tighten the screws to 13 ft. lbs. (17 Nm).

46. Remove the fixture from the transmission.

47. Install the reverse light switch connector.

48. Install the 3 M12 screws for the engine mounting bracket and tighten to 59 ft. lbs. (80 Nm).

49. Install the starter motor.

50. Install the 3 x M10 screws between engine and transmission. Tighten to 36 ft. lbs. (50 Nm).

51. Lift the engine/transmission. Align the mounting against the transmission in the engine pad.

52. Install the M14 screw for the engine mounting. Tighten to 96 ft. lbs. (130 Nm).

53. Remove the engine support system.

54. Use a new gasket for the clutch line. Install the clutch line and lock using the locking pin.

55. Remove the pedal jack and lift the clutch pedal. Bleed the clutch system.

56. Install the bracket holder.

57. Install the bracket and the coolant hose.

58. Install the transmission cables on the bracket and the levers.

59. Make sure that the shift cable is locked in position before fitting the blue clip. Press the clip into place.

60. Install the air cleaner module.

61. Install the ground lead on the transmission.

- Install the torque rod mounting to the transmission, 3 x M10. Tighten to 36 ft. lbs. (50 Nm).

62. Secure the sub-frame against the fixture plate. Use a retaining strap. Lift the sub-frame.

63. Align the sub-frame in the front mountings in the side members. Install the screws loosely, 4–5 threads.

64. Allow the sub-frame to hang down at the rear edge and to rest on the mobile jack.

65. Remove the temporary mounting for the steering gear. Allow the steering gear to rest on the sub-frame.

66. Install the anti-roll bar. Use four new M10 screws and tighten to 36 ft. lbs. (50 Nm).

67. Carefully raise the sub-frame and align.

68. Install the screws for the rear mounting of the sub-frame. Loosely install the new M16 screws, together with the brackets and the 4x M8 screws.

69. Tighten the M16 screws.

70. Loosely tighten the 4 x M8 screws to the brackets.

71. Install and tighten the screw for the lower torque rod mounting.

72. Install the rubber mountings for the exhaust pipe mounting.

73. Install the heated oxygen sensor (HO2S) cable harness to the sub-frame and to the tab on the torque rod mounting in the transmission.

74. Tighten the front M12 screws to the sub-frame. Tighten to 59 ft. lbs. (80 Nm).

75. Install new screws for the steering gear and sub-frame loosely. Tightening is performed later from the engine compartment.

76. Install the drive shafts.

77. Install the propeller shaft and transfer case. (AWD) models.

78. Refill the transmission with proper fluid.

79. Install the front wheels.

80. Tighten the two M12 screws for steering gear to 59 ft. lbs. (80 Nm).

81. Install the battery box, battery, and cables.

82. Check the master cylinder fluid level.

83. Road test and check for proper operation.

## CLUTCH

### REMOVAL & INSTALLATION

*See Figures 13 and 14.*

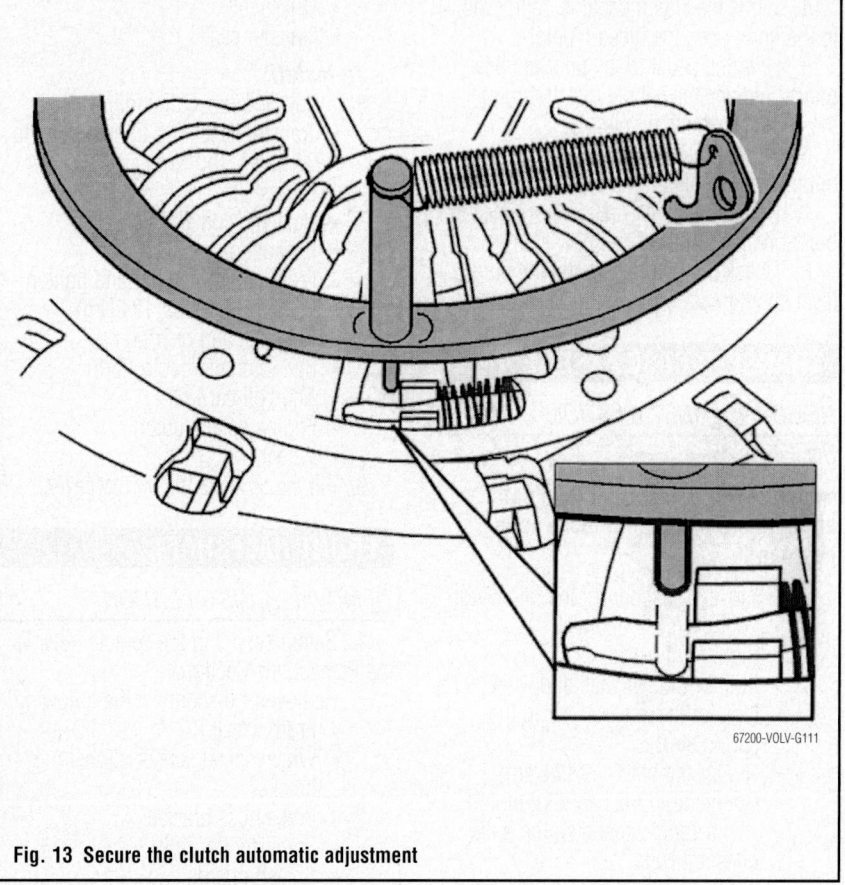

**Fig. 13  Secure the clutch automatic adjustment**

67200-VOLV-G110

**Fig. 14  Install the compression tool onto the clutch to release the load.**

1. Remove the transmission.

2. Block the flywheel using gear sector 999-5112.

3. Secure the clutch automatic adjustment with counterhold 999-5677.

4. Install the compression tool 999-5662 on the clutch. Center the tool and tighten the bolt to make light contact with the diaphragm spring.

5. Compress the clutch so the automatic adjuster is not under load. Screw in the compression tool until the diaphragm spring has pressed the pressure plate to a free position.

6. Remove the pressure plate and clutch disk.

### To install:

7. Install the compression tool 999-5662 on the clutch. Center the tool and tighten the bolt to make light contact against the diaphragm spring, and then compress the clutch.

8. Ensure the clutch automatic adjustment is secured with counterhold 999-5677.

9. Install the clutch, ensure the clutch is in contact with the flywheel all around its circumference, and tighten the six bolts crosswise and tighten to 18 ft. lbs. (25 Nm).

10. Remove the compression tool and gear sector.

11. Install the transmission.

### BLEEDING

**❊❊ CAUTION**

**Use only DOT 4 brake fluid.**

1. Before servicing the vehicle, refer to the Precautions Section.

2. Depress the clutch pedal a few times to purge the air bubbles in the master cylinder.

3. Connect a hose from a drain bottle to the nipple on the slave cylinder.

4. While the clutch pedal is depressed to the floor, open the bleed nipple.

5. Hold the pedal to the floor to allow brake fluid and air bubble to exit through the hose. Close the nipple.

6. Repeat this procedure until no air bubbles are visible in the escaping fluid.

7. Pump the clutch pedal a few times to build pressure in the system.

8. Check the fluid reservoir. The fluid level should not be above the MAX level.

## TRANSFER CASE ASSEMBLY

### REMOVAL & INSTALLATION

*See Figure 15.*

➡**Ensure the parking brake is not applied and the gear selector is in position P.**

1. Remove or disconnect the following:
   • Engine splash guard
   • Right front wheel
   • Bolt for the halfshaft and wheel hub
   • Link for the anti-roll bar in the spring strut
   • Nut for the ball joint/link arm. Counterhold the ball joint pinion with a Torx® wrench so the boot does not twist.

2. Position a jack under the right-hand side of the sub-frame. Remove the bolts from the support bracket for the right-hand side of the sub-frame.

3. Slacken off, but do not remove, the remaining bolts for the sub-frame.

4. Unhook the control arm from the ball joint.

5. Press out the wheel spindle.

6. Detach the halfshaft from the hub, remove the bolts from the bearing cap for the half-shaft. Pull the half-shaft straight out.

7. Remove or disconnect the following:
   • Front cross-member
   • Exhaust pipe
   • Catalytic converter

**Fig. 15 Remove the transfer case, and bolts.**

22250_VOLC_G0013

   • Driveshaft
   • Transfer case

### To install:

8. Install or connect the following:
   • Transfer case, and tighten bolts to 37 ft. lbs. (50 Nm)
   • Driveshaft
   • Catalytic converter
   • Exhaust pipe
   • Front cross-member and tighten bolts to 18 ft. lbs. (24 Nm)
   • Halfshaft and bearing cap.
   • Ball joint and control arm
   • Anti-roll bar link
   • Engine splash guard
   • Wheel

9. Fill the bevel to the correct level.

## FRONT HALFSHAFT

### REMOVAL & INSTALLATION

1. Before servicing the vehicle, refer to the Precautions Section.

2. Remove or disconnect the following:
   • Front wheel
   • Wheel speed sensor and wiring bracket
   • Brake hose bracket
   • Stabilizer bar link
   • Splash guards
   • Lower ball joint
   • Hub retainer nut

3. Pull the hub off of the stub shaft.

4. Pry the inner joint out of the transaxle and remove the axle halfshaft.

### To install:

➡**Use new fasteners for assembly.**

5. Install the axle halfshaft so that the circlip is felt to seat in the retainer groove.

6. Guide the stub shaft into the hub.

7. Install or connect the following:
   • Hub retainer nut and tighten the nut to 89 ft. lbs. (120 Nm) plus 60°
   • Lower ball joint
   • Splash guards
   • Stabilizer bar link
   • Brake hose bracket
   • Wheel speed sensor and wiring bracket
   • Front wheel

## REAR HALFSHAFT

### REMOVAL & INSTALLATION

*See Figure 16.*

1. Before servicing the vehicle, refer to the Precautions Section.

2. Remove the following items:

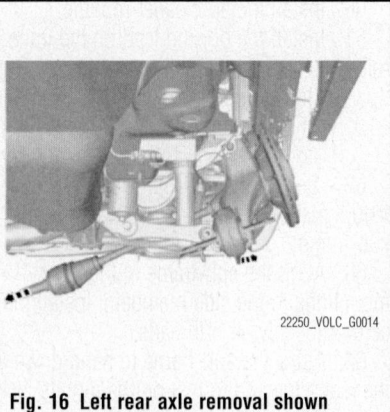

22250_VOLC_G0014

**Fig. 16 Left rear axle removal shown**

   • Wheel
   • Coil spring
   • Wheel knuckle rear to body
   • Driveshaft to wheel hub
   • Lower arm rear to sub-frame
   • Rear stabilizer bar link to lower stabilizer bar
   • Rear shock absorber to wheel knuckle

3. Remove the left rear axle.

### To install:

4. Install the left rear axle.

5. Install the following items:
   • Rear shock absorber to wheel knuckle
   • Rear stabilizer bar link to lower stabilizer bar
   • Lower arm rear to sub-frame
   • Driveshaft to wheel hub
   • Wheel knuckle rear to body
   • Coil spring
   • Wheel

## REAR PINION SEAL

### REMOVAL & INSTALLATION

*See Figures 17 and 18.*

1. Before servicing the vehicle, refer to the Precautions Section.

2. Remove or disconnect the following:
   • The exhaust system
   • The driveshaft
   • The Active on Demand Coupling (AOC), if equipped

➡**There are two parts to the pinion seal. Use extractor 999 7023 with care. The bolt and inner socket in the extractor must be unscrewed when tapping on the extractor to remove pinion seal.**

3. Tap the extractor tool 999 7023 onto the seal, so that the flat end is against the seal.

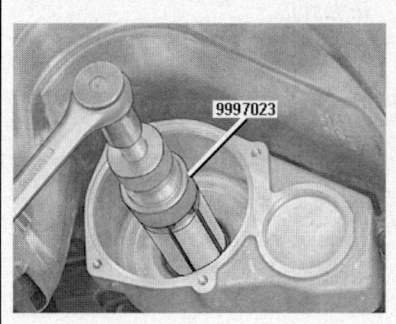

**Fig. 17 Using extractor 999 7023 to remove pinion seal**

4. Tighten the inner socket of tool 999 7023. Counter hold the outer part at the same time so that the tool expands against the seal.

5. Remove the seal using the bolt in the extractor.

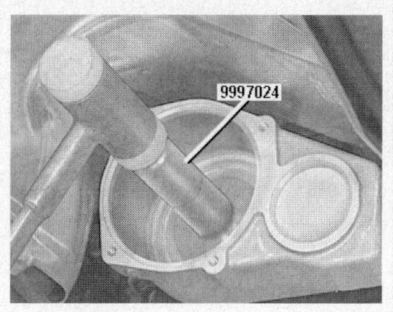

**Fig. 18 Using drift 999 7024 to install pinion seal**

➡ Check that no seal residue is left in the final drive. If the inner section of the seal remains in place, tap in the extractor again and pull out the inner section. Do not hammer on the extractor too hard. This can cause damage in

the extractor, which in turn can scratch the mating surface on the pinion shaft.

6. Remove the pinion seal.

*To install:*

➡ The seal consists of two sections but is installed as one unit.

7. Lubricate the edge of the seal.

8. Use the drift 999 7024 for installation. Position the section of the seal with the largest outer diameter against the drift.

9. Tap in the seal until the head of the drift is against the final drive housing.

10. Install or connect the following:
- The Active on Demand Coupling (AOC), if equipped
- The driveshaft
- The exhaust system

11. Refill the oil in the final drive and AOC, if equipped

# ENGINE COOLING

## THERMOSTAT

### REMOVAL & INSTALLATION

#### 2.4L and 2.5L Engines, Except S60

*See Figure 19.*

1. Before servicing the vehicle, refer to the Precautions Section.
2. Drain the cooling system.
3. Disconnect or remove the following:
- Hose clamp to the upper radiator hose
- Hose from the thermostat housing
- Connector from the temperature sensor
- Bolts for the thermostat housing
- Thermostat housing from cylinder head

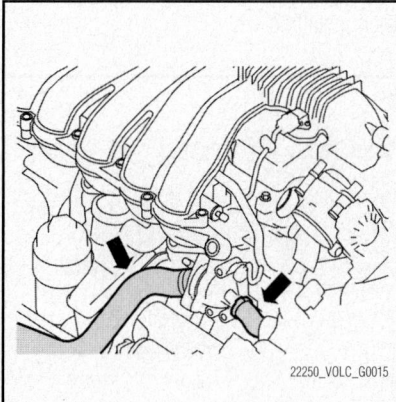

**Fig. 19 Location of thermostat housing— 5-Cylinder Engines**

*To install:*

4. Install or connect the following:
- Thermostat housing assembly to cylinder head using new gasket and new hose clamp
- Tighten thermostat housing to 16 ft. lbs. (22 Nm)
- Upper radiator hose and hose clamp
- Connector for the temperature sensor

5. Refill the engine coolant.

6. Test drive the engine until the thermostat has opened.

7. Check for leaks, and proper thermostat operation. Fill fluid as necessary.

#### S60 with 2.4L and 2.5L Engines

*See Figure 20.*

1. Before servicing the vehicle, refer to the Precautions Section.
2. Disconnect the negative battery cable.
3. Drain the cooling system.
4. Disconnect or remove the following:
- Control module cover
- Insert tool 999 5722-7 to remove control module
- Drive belt
- Upper timing belt cover
- Power steering pump
- Hose clamp to the upper radiator hose
- Hose from the thermostat housing
- Connector from the temperature sensor
- Bolts for the thermostat housing

**Fig. 20 Location of thermostat housing S60 with 5-Cylinder Engine**

- Water hose for heating crankcase ventilation
- Thermostat housing from cylinder head

*To install:*

5. Install or Reconnect the following:
- Thermostat housing assembly to cylinder head using new gasket and new hose clamp
- Tighten thermostat housing to 16 ft. lbs. (22 Nm)
- Water hose for heating crankcase ventilation
- Upper radiator hose and hose clamp
- Connector for the temperature sensor
- Power steering pump
- Upper timing belt cover
- Drive belt
- Insert tool 999 5722-7 to remove control module
- Control module cover

6. Connect the negative battery cable.

7. Refill engine with coolant and bleed the cooling system.

8. Test drive the engine until the thermostat has opened.

9. Check for coolant leaks and proper thermostat operation. Top off antifreeze as needed.

### 3.0L and 3.3L Engines

*See Figure 21.*

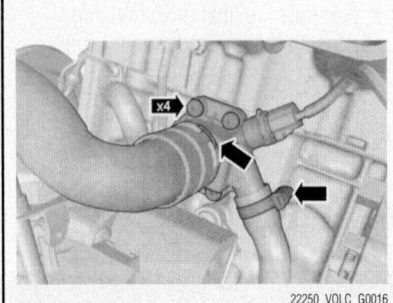

**Fig. 21 Location of thermostat housing–6–Cylinder Engines**

1. Before servicing the vehicle, refer to the Precautions Section.

2. Drain the engine coolant.

3. Disconnect or remove the following:
   • Engine cover
   • Engine under cover splash guard
   • Air intake hose
   • Electrical connectors at intake
   • Intake manifold bolts
   • Intake manifold
   • Thermostat bolts and assembly

*To install:*

4. Disconnect or remove the following:
   • Thermostat bolts and assembly
   • Intake manifold
   • Intake manifold bolts
   • Electrical connectors at intake
   • Air intake hose
   • Engine under cover splash guard
   • Engine cover

5. Install the engine coolant and bleed the cooling system.

### S80 with 4.4L Engine

*See Figure 22.*

1. Before servicing the vehicle, refer to the Precautions Section.

2. Raise the vehicle.

3. Drain the cooling system.

4. Remove the front engine cover.

5. Remove the thermostat housing bolts.

6. Remove the thermostat.

**Fig. 22 Location of thermostat housing–8–Cylinder 4.4L engine**

*To install:*

7. Install the thermostat

8. Tighten the thermostat housing bolts to 7 ft. lbs (10 Nm).

9. Install the front engine cover.

10. Install engine coolant and bleed the cooling system.

## WATER PUMP

### REMOVAL & INSTALLATION

### 2.4L and 2.5L Engines

*See Figure 23.*

1. Before servicing the vehicle, refer to the Precautions Section.

2. Drain the cooling system.

3. Remove or disconnect the following:
   • Negative battery cable
   • Timing belt (refer to timing belt)
   • Water pump

➡ **Raise the engine with a lifting beam to access the two upper screws, then lower the engine to access the remaining bolts from underneath.**

4. Remove the screws for the coolant pump.

5. Carefully tap the pump rotor with a plastic mallet and remove the pump.

**Fig. 23 View of water pump 5 cylinder engine**

*To install:*

6. Clean the engine surface.

7. Install a new gasket on water pump surface.

8. Tighten or reconnect the following:
   • Water pump
   • Timing belt (refer to timing belt)
   • Negative battery cable

9. Refill the engine coolant and bleed cooling system.

10. Check for any signs of coolant leaks.

### 3.0L and 3.2L Engines

*See Figures 24 and 25.*

1. Before servicing the vehicle, refer to the Precautions Section.

2. Drain the cooling system.

3. Remove the air hose connection at the air filter assembly.

4. Disconnect the Mass Air Flow sensor (MAF) and remove the air filter box if needed.

5. Remove the drive belt.

6. Remove the power steering pressure hose and nuts holding the hose retainer brackets.

7. Plug the hose connector to reduce fluid loss.

8. Remove the power steering pump and set aside.

9. Plug the pressure hose to reduce fluid loss.

10. Remove the bolts to the water pump.

11. Remove the water pump by carefully taping on pump shaft.

*To install:*

12. Clean the engine surface for water pump.

13. Install the water pump and tighten mounting bolts to 7 ft. lbs. (10 Nm).

14. Make sure that the coupling's spring (2) clicks into place on the bonded part of the water pump (1).

15. Always use a new guide pin (3) when installing.

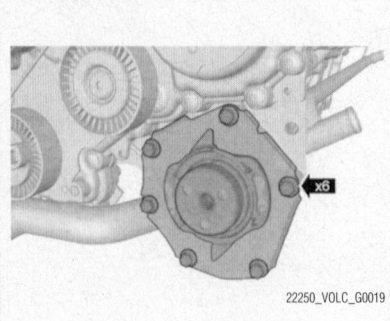

**Fig. 24 Water pump view 6 cylinder engines**

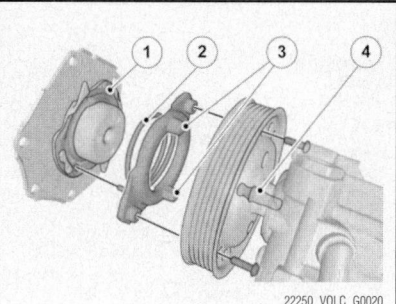

**Fig. 25 Bonded part of the water pump (1) coupling's spring (2) always use a new guide pin (3) pump guide pin (4)**

16. Install power steering pump making sure that alignment is correct. Tighten the mounting bolts to 28 ft. lbs. (24 Nm).

17. Install the power steering pressure hose and tighten to 22 ft. lbs. (30 Nm). Make sure the brackets are aligned and tighten the retaining nuts to 18 ft. lbs. (24 Nm).

18. Install the drive belt.

19. Install the air filter box and reconnect intake air hose.

20. Reconnect the MAF connector.

21. Install antifreeze and bleed the cooling system.

22. Top off the power steering fluid, bleed system.

23. Check for leaks on the coolant system, and power steering system.

### S80 with 4.4L Engine

*See Figures 26 and 27.*

1. Before servicing the vehicle, refer to the Precautions Section.

2. Raise the vehicle.

3. Drain the cooling system.

4. Loosen the water pump pulley bolts.

5. Remove the drive belt.

6. Completely remove the water pump pulley bolts.

7. Remove the water pump mounting bolts.

**Fig. 26 Remove the water pump pulley bolts**

**Fig. 27 4.4L Engine water pump view**

➡**Make sure that the mating surfaces are clean and free of foreign objects.**

*To install:*

8. Install the water pump and tighten the mounting bolts to 7 ft. lbs. (10 Nm).

9. Install the water pump pulley and bolts.

10. Install the drive belt.

11. Tighten the water pump pulley mounting bolts to 7 ft. lbs. (10 Nm).

12. Install fresh engine coolant and bleed the cooling system.

13. Start the engine and check cooling system for leaks.

---

## ENGINE ELECTRICAL

### ALTERNATOR

*REMOVAL & INSTALLATION*

#### 2.4L and 2.5L Engines (Except S60 Model)

*See Figure 28.*

1. Before servicing the vehicle, refer to the Precautions Section.

2. Disconnect the negative battery cable.

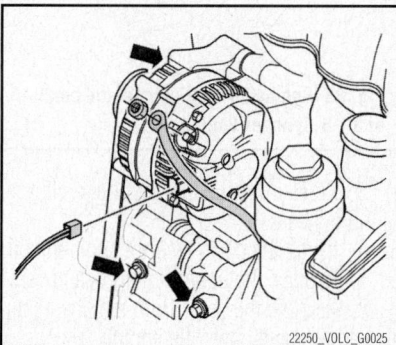

**Fig. 28 Alternator removal components shown**

3. Remove the air intake resonator box.

4. Remove the drive belt.

5. The retaining nut and battery lead.

6. The regulator connector.

7. Remove the mounting bolts and remove alternator.

*To install:*

8. Install the alternator and tighten the mounting bolts to 18 ft. lbs. (25 Nm)

9. Install battery lead and tighten the retaining nut to 11 ft. lbs. (15 Nm)

10. Install the drive belt.

11. Install the air intake resonator box.

12. Connect the negative battery cable.

#### S60 Model with 2.4L and 2.5L engines

*See Figure 29.*

1. Before servicing the vehicle, refer to the Precautions Section.

2. Drain the cooling system.

3. Remove or disconnect the following:
   • Negative battery cable
   • Accessory drive belt

## CHARGING SYSTEM

- Power steering pump
- Turbo hose from air conditioning assembly
- Radiator hose from radiator
- Mounting bolts common to alternator and compress
- Alternator connector and wiring

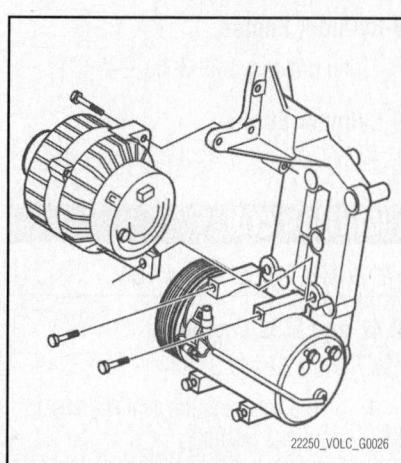

**Fig. 29 Alternator removal shown—S60 with 2.4L and 2.5L Engines**

- Alternator mounting bolts
- Alternator

## *To install:*

Install or connect the following:

- Alternator, and tighten mounting bolts and bolts common to the alternator and compressor to 18 ft. lbs. (25 Nm)
- Alternator connector and wiring
- Radiator hose to radiator
- Turbo hose to air conditioning assembly
- Power steering pump
- Accessory drive belt
- Negative battery cable

4. Refill with engine coolant and bleed the cooling system.

### 3.0L and 3.2L Engines

*See Figure 30.*

1. Before servicing the vehicle, refer to the Precautions Section.
2. Remove or disconnect the following:
   - Negative battery cable
   - Intake manifold
   - Alternator regulator connector
   - Alternator B+ terminal (1)
   - Alternator mounting bolts (2)
   - Remove alternator

## *To install:*

Install or connect the following:

- Alternator
- Alternator mounting bolts (2)

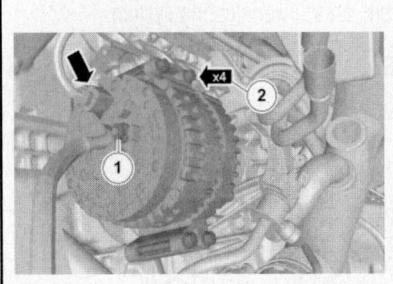

22250_VOLC_G0027

**Fig. 30 Alternator removal shown—3.0L and 3.2L Engines**

- Alternator B+ terminal (1)
- Alternator regulator connector
- Intake manifold

3. Connect the negative battery cable.

### S80 with 4.4L Engine

*See Figure 31.*

1. Before servicing the vehicle, refer to the Precautions Section.
2. Disconnect the negative battery cable.
3. Remove the drive belt.
4. Remove the rear engine cover.
5. Raise the vehicle.
6. Remove the alternator B+ terminal and regulator connector.

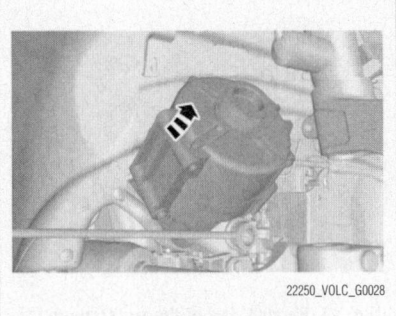

22250_VOLC_G0028

**Fig. 31 Alternator removal shown S80 4.4L engine**

7. Remove the alternator mounting bolts.
8. Remove the right-hand drive shaft.
9. Remove the right front stabilizer link from strut.
10. Remove the 2 bolts from the frame bracket.
11. Remove the alternator.

➡**Ensure that the components are not bent violently.**

12. To install, reverse the removal procedure and note the following:
   - Tighten the mounting bolts to 18 ft. lbs. (24 Nm)
   - Tighten the alternator B+ terminal to 11 ft lbs. (15 Nm)

---

# ENGINE ELECTRICAL

## FIRING ORDER

### 5-Cylinder Engine

Firing order is: 1–2–4–5–3

### 6-Cylinder Engine

Firing order is: 1–5–3–6–2–4

### 8-Cylinder Engine

Firing order is: 1–8–4–3–6–5–7–2

## IGNITION COIL

### *REMOVAL & INSTALLATION*

#### 2.4L and 2.5L Engines

*See Figures 32 and 33.*

1. Before servicing the vehicle, refer to the Precautions Section.
2. Remove or disconnect the following:
   - Charge air pipe over the engine. Seal the openings.
   - Upper timing belt cover

- Cover over the ignition coils
- Ignition coil connector
- Ignition coil mounting bolt
- Carefully pull coil up and out

## *To install:*

3. Align the coil and press it down.
4. Install ignition coil mounting bolt and tighten to 88 inch lbs. (10 Nm).

42075_VOLC_G0002

**Fig. 32 Charge Air Pipe Location—5-Cylinder Engine**

# IGNITION SYSTEM

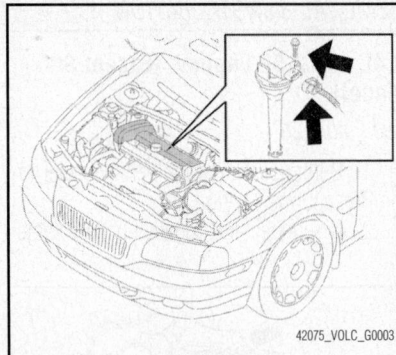

42075_VOLC_G0003

**Fig. 33 Removing ignition coil and small bolt—5-Cylinder Engine**

5. Press in the connector until a click sound is heard.
6. Install the cover over the ignition coil.
7. Replace the upper timing belt cover.
8. Remove the seals from the charge air pipe and reinstall over the engine.

### 3.0L and 3.2L Engines

*See Figures 34 and 35.*

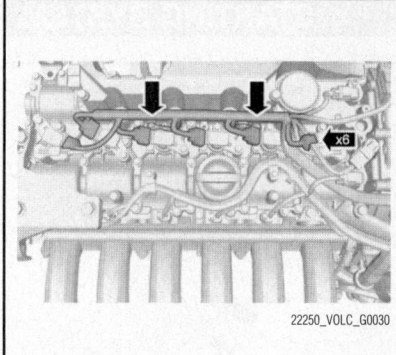

Fig. 34 Ignition coil electrical connectors

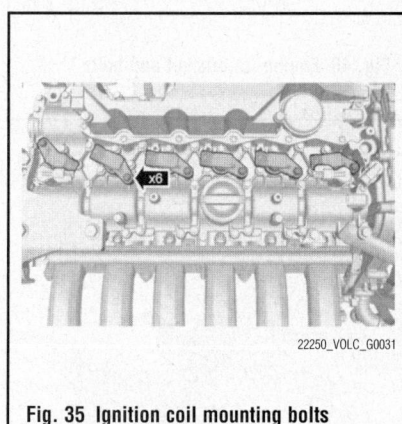

Fig. 35 Ignition coil mounting bolts

1. Before servicing the vehicle, refer to the Precautions Section.
2. Remove or disconnect the following:
   • Engine cover
   • Ignition coil electrical connector

➡ **Seal the open ends of pipes.**

   • Ignition coil mounting bolt
   • Carefully pull coil up and out

*To install:*
3. Align the coil and press it down.
4. Install ignition coil mounting bolt and tighten to 88 inch lbs. (10 Nm).
5. Press in the connector until a click sound is heard.
6. Install the engine cover.

### S80 with 4.4L Engine

*See Figure 36.*

1. Before servicing the vehicle, refer to the Precautions Section.
2. Remove the left and right engine covers.
3. Remove the mounting bolt for ignition coil.
4. Lift up the ignition coil.

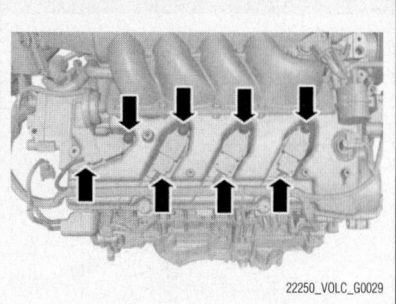

Fig. 36 Remove the mounting bolt for ignition coil

5. Remove the coil electrical connector.
6. Remove the ignition coil.

*To install:*
7. To install reverse the removal procedure and note the following:
   • Tighten the coil mounting bolt to 88 inch lbs. (10 Nm).

## IGNITION TIMING

### ADJUSTMENT

The ignition timing is controlled by the PCM.

## SPARK PLUGS

### REMOVAL & INSTALLATION

#### 2.4L and 2.5L Engines

*See Figure 37.*

1. Before servicing the vehicle, refer to the Precautions Section.
2. Remove or disconnect the following:
   • Charge air pipe over the engine. Seal the openings.
   • Upper timing belt cover
   • Cover over the ignition coils
   • Ignition coil connector
   • Ignition coil mounting bolt
   • Carefully pull coil up and out
   • Spark plugs

*To install:*
3. Install the spark plug and tighten to 22 ft. lbs. (28 Nm)
4. Align the coil and press it down.
5. Install ignition coil mounting bolt and tighten to 88 inch lbs. (10 Nm).
6. Press in the connector until a click sound is heard.
7. Install the cover over the ignition coil.
8. Replace the upper timing belt cover.

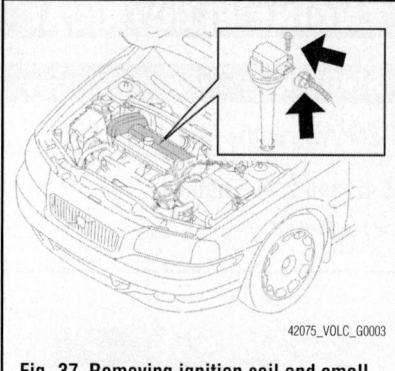

Fig. 37 Removing ignition coil and small bolt—5-Cylinder Engine

9. Remove the seals from the charge air pipe and reinstall over the engine.

#### 3.0L and 3.2L Engines

1. Before servicing the vehicle, refer to the Precautions Section.
2. Remove or disconnect the following:
   • Engine cover
   • Ignition coil electrical connector

➡ **Seal the open ends of pipes.**

   • Ignition coil mounting bolt
   • Carefully pull coil up and out
   • Spark plug

*To install:*
3. Install the spark plug and tighten to 22 ft. lbs. (28 Nm).
4. Align the coil and press it down.
5. Install ignition coil mounting bolt and tighten to 88 inch lbs. (10 Nm).
6. Press in the connector until a click sound is heard.
7. Install the engine cover.

#### S80 with 4.4L Engine

1. Before servicing the vehicle, refer to the Precautions Section.
2. Remove the left and right engine covers.
3. Remove the mounting bolt for ignition coil.
4. Lift up the ignition coil.
5. Remove the coil electrical connector.
6. Remove the ignition coil.
7. Remove the spark plug with a proper socket.

*To install:*
8. To install reverse the removal procedure and note the following:
   a. Install the spark plug and tighten to 22 ft. lbs. (28 Nm).
   b. Tighten the coil mounting bolt to 88 inch lbs. (10 Nm).

## STARTER

*REMOVAL & INSTALLATION*

### 2.4L and 2.5L Engines

*See Figure 38.*

Fig. 38 Starter motor and mounting bolt view 2.4L and 2.5L engines

1. Before servicing the vehicle, refer to the Precautions Section.
2. Remove or disconnect the following:
   - Negative battery cable
   - Air intake assembly
   - Battery positive cable to starter solenoid
   - Starter impulse lead tighten to 68 inch lbs. (8.5 Nm)
3. Remove the coolant hose from the bracket and slacken the nuts off for the hose bracket. Bend the bracket to one side to allow access for bolts to the starter motor.
4. Remove the start motor.

*To install:*

5. Install or connect the following:
   - Starter motor and tighten the bolts to 36 ft. lbs. (50 Nm).
   - Battery positive cable to starter solenoid
   - Starter impulse lead
   - Coolant hose bracket
   - Air intake assembly
   - Negative battery cable

### 3.0L and 3.2L Engines

*See Figure 39.*

1. Before servicing the vehicle, refer to the Precautions Section.

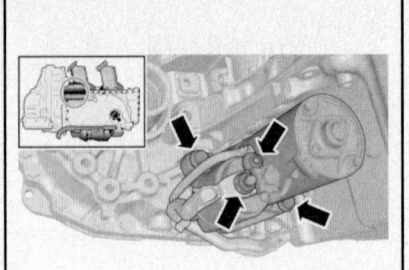

Fig. 39 Starter motor view for 3.0L and 3.2L Engines

2. Disconnect the negative battery cable.
3. Raise the vehicle and remove the undercover splash guard.
4. The battery positive cable to starter solenoid.
5. Remove the starter impulse lead.
6. Remove the starter mounting bolts and starter motor.

*To install:*

7. Install the starter mounting bolts and starter motor. Tighten the bolts to 36 ft. lbs. (50 Nm)
8. Install the starter impulse lead.
9. Install the battery positive cable to starter solenoid. Tighten the mounting nut to 8 ft. lbs. (11 Nm)
10. Install the vehicle under cover splash guard.
11. Lower the vehicle and connect the negative battery cable.
12. Check for proper starter motor operation.

### S80 with 4.4L Engine

*See Figures 40 and 41.*

1. Before servicing the vehicle, refer to the Precautions Section.
2. Remove the battery and battery box.
3. The battery positive cable to starter solenoid.
4. Remove the starter impulse lead.
5. Remove any components that are in the way of starter removal.
6. Remove the engine lift bracket and bolts.

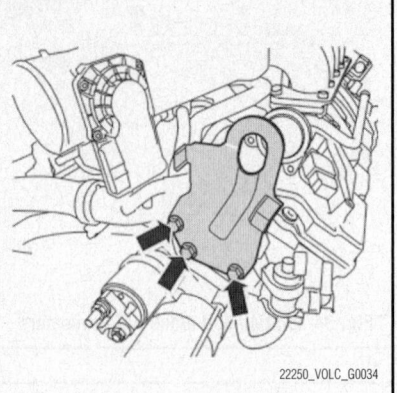

Fig. 40 Engine lift bracket and bolts shown

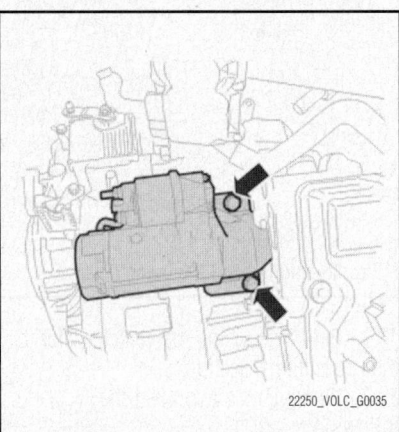

Fig. 41 Starter motor view—S80 4.4L engine

7. Remove the starter mounting bolts and starter motor.

*To install:*

8. Install the starter motor and mounting bolts. Tighten the mounting bolts to 37 ft. lbs. (50 Nm)
9. Install the engine lift bracket and bolts.
10. Remove any components that are in the way of starter removal.
11. Install the starter impulse lead.
12. Install the battery positive cable to starter solenoid. Tighten the cable mounting nut to 18 ft. lbs. (24 Nm).
13. Install the battery and battery box.
14. Install the battery cables.

## ENGINE MECHANICAL

➡Disconnecting the negative battery cable may interfere with the functions of the on board computer systems and may require the computer to undergo a relearning process, once the negative battery cable is reconnected.

### ACCESSORY DRIVE BELTS

#### ACCESSORY BELT ROUTING

*See Figures 42 through 44.*

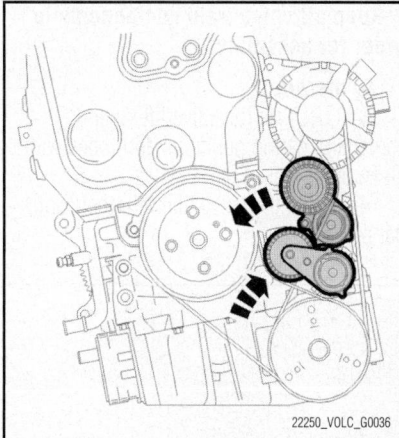

**Fig. 42 2.4L and 2.5L engines**

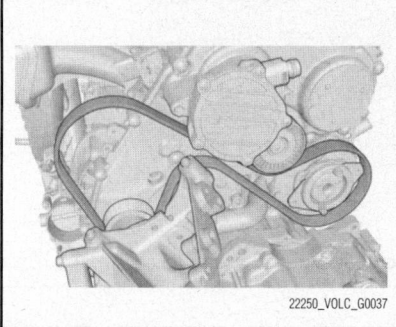

**Fig. 43 3.0L and 3.2L Engines shown with A/C compressor removed on the left**

#### INSPECTION

Inspect the drive belt for signs of glazing or cracking. A glazed belt will be perfectly smooth from slippage, while a good belt will have a slight texture of fabric visible. Cracks will usually start at the inner edge of the belt and run outward. All worn or damaged drive belts should be replaced immediately.

#### ADJUSTMENT

The accessory drive belt adjustment is maintained by an automatic tensioner.

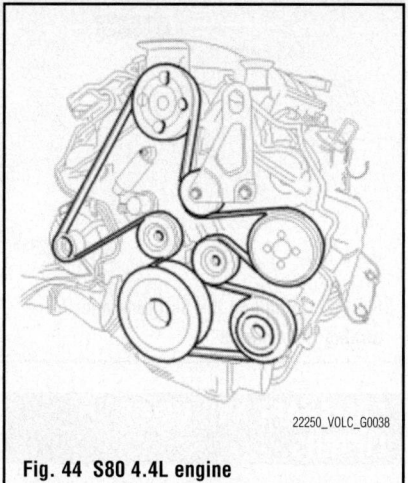

**Fig. 44 S80 4.4L engine**

#### REMOVAL & INSTALLATION

#### 2.4L and 2.5L Engines

*See Figure 42.*

1. Before servicing the vehicle, refer to the Precautions Section.
2. Raise the vehicle.
3. Remove or disconnect the following:
   - Right front wheel
   - Right-hand fender liner. Slacken off the screws at the front edge and fold back the fender liner.
4. Remove the outer drive belt as follows:
   - Relieve the load from the belt tensioner. Use: 999-7109. Turn the belt tensioner **clockwise** as far as it will go. Remove the belt. Remove the inner drive belt as follows:
   - Relieve the load from the belt tensioner using a Torx-wrench®. Turn the belt tensioner **counter clockwise**as far as it will go. Remove the belt.
5. Installation is in the reverse of the removal procedure.

#### 3.0L and 3.2L Engines

*See Figures 45 through 47.*

1. Before servicing the vehicle, refer to the Precautions Section.
2. Remove the engine cover.
3. Remove the battery and battery box.
4. Raise the vehicle and remove the engine under cover splash guard.
5. Remove the air intake clamp at the throttle body.
6. Lower the vehicle.
7. Remove the air intake at the air filter box.

8. Remove the air filter box.
9. Remove the power steering pressure line brackets at the A/C compressor.
10. Slacken the tensioner using a 19 mm fixed wrench. Lock the tensioner in the slackened position using a 3 mm pin.
11. Remove the engine support brackets at the A/C compressor.
12. Remove the Power steering pump and set aside
13. Remove the A/C compressor mounting bolts and line bracket retaining nuts. (do not remove lines)
14. Remove the A/C compressor and set aside.
15. Remove the belt tensioner bolt.
16. Remove the drive belt. If needed remove the belt tensioner bracket to remove the drive belt.

#### *To install:*

17. Install in the reverse of the removal procedure and note the following:
   a. Tighten the tensioner bolt to 18 ft. lbs. (24 Nm)
   b. Tighten the A/C mounting bolts and bracket nuts to 18 ft. lbs. (24 Nm)

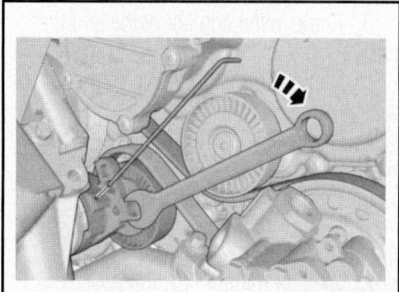

**Fig. 45 Slacken the tensioner using a 19 mm fixed wrench. Lock the tensioner in the slackened position using a 3 mm pin—3.0L and 3.2L engines**

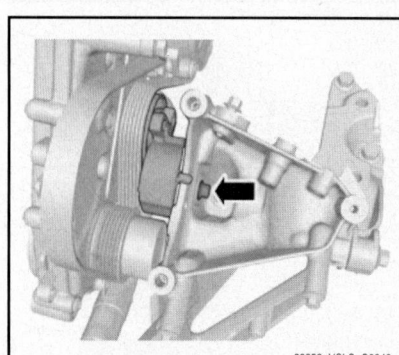

**Fig. 46 Tensioner and mounting bolt shown—3.0L and 3.2L engines**

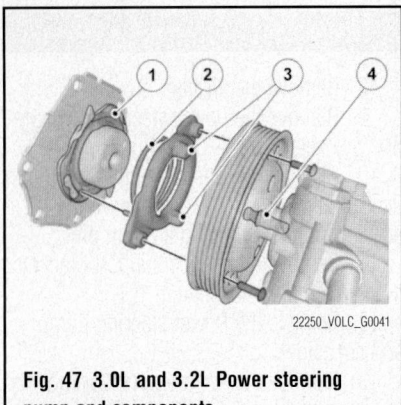

**Fig. 47 3.0L and 3.2L Power steering pump and components**

c. At installation, ensure that the coupling spring (2) clicks into position on the bonded section of the water pump (1) and that the lugs (3) meet the holes on the pulley.

Always use a new guide pin (4) when installing the power steering pump.

### S80 with 4.4L Engine

*See Figures 48 and 49.*

1. Before servicing the vehicle, refer to the Precautions Section.

2. Remove the upper drive belt cover.

3. Remove the coolant bottle and the torque rod.

4. Remove the 3 bolts and torque rod bracket.

5. Remove the bracket for torque rod bracket.

6. Remove the ground strap.

7. Remove the power steering pressure line and bracket retaining nut.

8. Relieve the load on the belt tensioner by turning it clockwise to max. 230 Nm and lock the tensioner with relevant tool.

9. Install in the reverse of the removal procedure.

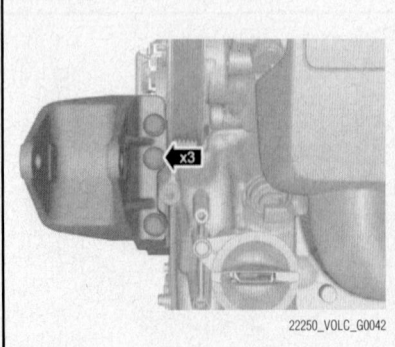

**Fig. 48 Remove the 3 bolts and torque rod bracket**

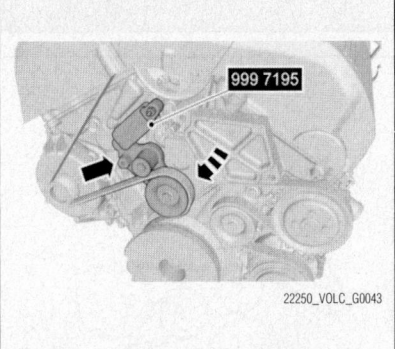

**Fig. 49 Unload on the belt tensioner by turning it clockwise**

## CAMSHAFT AND VALVE LIFTERS

### REMOVAL & INSTALLATION

#### 2.4L and 2.5L Engines

*See Figures 50 and 51.*

1. Before servicing the vehicle, refer to the Precautions Section.

2. Remove or disconnect the following:
- Negative battery cable
- Accessory drive belts
- Front cover
- Timing belt
- Ignition coil cover
- Switch holder and shield at left rear of the engine
- Ignition coils
- Camshaft sprockets
- Cylinder head cover
- Camshafts and lifters

➡ **Keep all valve train components in order for assembly.**

*To install:*

3. Lubricate the valve lifters and camshaft bearing positions and lobes with clean engine oil.

4. Install the valve lifters in their original positions.

**Fig. 50 Using Special Tool 951-2767 to apply the liquid gasket to the cylinder head cover.**

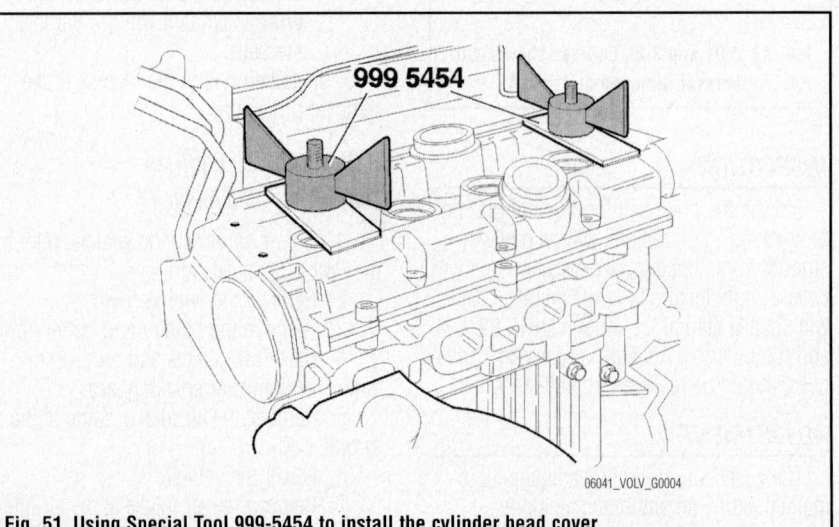

**Fig. 51 Using Special Tool 999-5454 to install the cylinder head cover**

5. Install the intake camshaft. Make sure that the groove in the rear edge of the camshaft extends across an imagined center line.

6. Install the exhaust camshaft. Make sure that the groove in the rear edge of the camshaft extends across an imagined center line.

7. Tighten the camshaft cover bolts alternately, keeping it parallel to the cylinder head using the press tools.

8. Using Special Tool 951-2767 or equivalent, apply liquid gasket to the mating surface of the cylinder head cover.

### ✳✳ WARNING

**Ensure no liquid gasket material gets in the oil ducts of the cylinder head cover.**

9. Position the Press Tool 999-5454 in the spark plug holes. Fit the two M6 bolts under each press tool to set the right height.

10. Crosswise tighten the cylinder head cover parallel to the cylinder head with the press tool. Install all the bolts and tighten from the middle out to the edges. Tighten the bolts to 13 ft. lbs. (17 Nm). Remove the press tool and bolts under the tool.
- Camshaft sprockets and tighten the bolts to 15 ft. lbs. (20 Nm)
- Ignition coils
- Switch holder and shield at left rear of the engine
- CMP sensor or distributor, as equipped
- Ignition coil cover
- Timing belt
- Front cover
- Accessory drive belts
- Negative battery cable

### 3.0L and 3.2L Engines

*See Figures 52 and 53.*

1. Before servicing the vehicle, refer to the Precautions Section.
2. Depressurize the fuel system.
3. Remove or disconnect the following:
- Negative battery cable
- Accessory drive belts
- Front cover
- Timing chain
- Ignition coil cover
- Ignition coils
- Camshaft sprockets
- Cylinder head cover
- Camshafts and lifters

➡**Mark the position of the components before removal.**

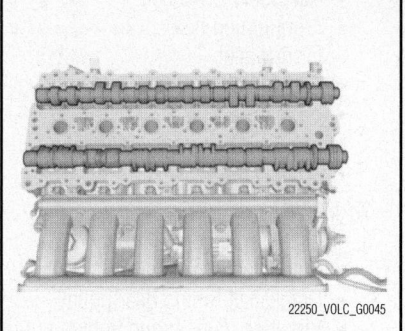

**Fig. 52 Camshafts shown in cylinder head**

*To install:*

4. Lubricate the valve lifters and camshaft bearing positions and lobes with clean engine oil.

5. Install the valve lifters in their original positions.

6. Install the intake camshaft. Make sure that the groove in the rear edge of the camshaft extends across an imagined center line.

7. Install the exhaust camshaft. Make sure that the groove in the rear edge of the camshaft extends across an imagined center line.

8. Tighten the camshaft cover bolts alternately, keeping it parallel to the cylinder head using the press tools.

9. Using Special Tool 951-2767 or equivalent, apply liquid gasket to the mating surface of the cylinder head cover.

### ✳✳ WARNING

**Ensure no liquid gasket material gets in the oil ducts of the cylinder head cover.**

10. Position the Press Tool 999-5454 in the spark plug holes. Fit the two M6 bolts under each press tool to set the right height.

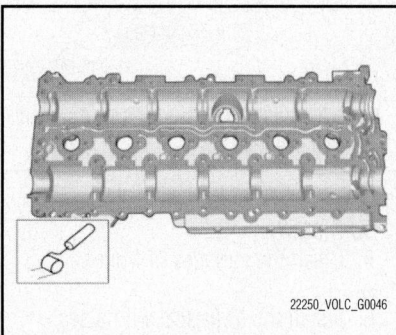

**Fig. 53 Apply liquid gasket to the mating surface**

11. Crosswise tighten the cylinder head cover parallel to the cylinder head with the press tool. Install all the bolts and tighten from the middle out to the edges. Tighten the bolts to 13 ft. lbs. (17 Nm). Remove the press tool and bolts under the tool.
- Camshaft sprockets and tighten the bolts to 15 ft. lbs. (20 Nm)
- Ignition coils
- Ignition coil cover
- Timing chain
- Front cover
- Accessory drive belts
- Negative battery cable

### S80 with 4.4L Engine

*See Figures 54 and 55.*

1. Before servicing the vehicle, refer to the Precautions Section.
2. Depressurize the fuel system.
3. Remove or disconnect the following:
- Negative battery cable
- Accessory drive belts
- Front cover
- Timing chain
- Ignition coil cover
- Ignition coils
- Camshaft sprockets

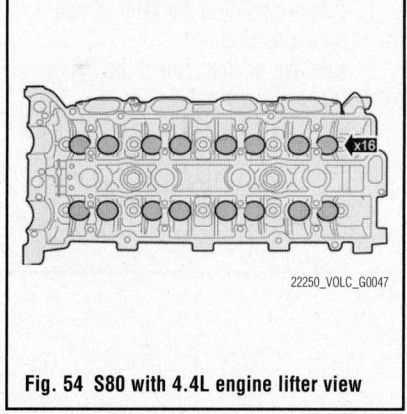

**Fig. 54 S80 with 4.4L engine lifter view**

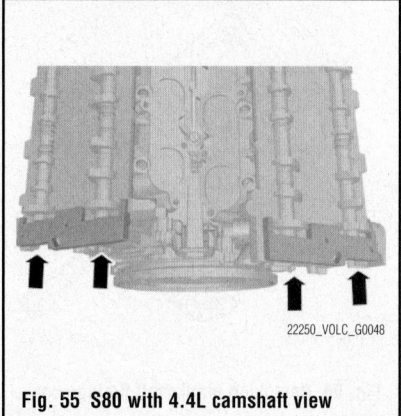

**Fig. 55 S80 with 4.4L camshaft view**

- Cylinder head cover
- Camshafts and lifters

➡**Mark the position of the components before removal.**

*To install:*

4. Lubricate the valve lifters and camshaft bearing positions and lobes with clean engine oil.

5. Install the valve lifters in their original positions.

6. Tighten the camshaft cover bolts alternately, keeping it parallel to the cylinder head using the press tools. Tighten to 7 ft. lbs. (10 Nm)

7. Using Special Tool 951-2767 or equivalent, apply liquid gasket to the mating surface of the cylinder head cover.

- Ignition coils
- Ignition coil cover
- Timing chain
- Front cover
- Accessory drive belts
- Negative battery cable

## CRANKSHAFT FRONT SEAL

*REMOVAL & INSTALLATION*

### 2.4L and 2.5L Engines

*See Figure 56.*

1. Before servicing the vehicle, refer to the Precautions Section.

2. Remove or disconnect the following:

- Engine stabilizer brace
- Upper timing belt cover
- Servo reservoir
- Expansion tank

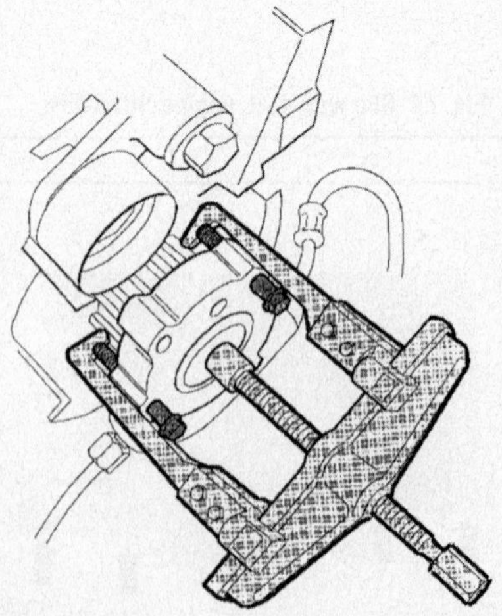

- Accessory drive belt
- Front timing belt cover
- Front wheel
- Timing belt
- Vibration damper
- Crankshaft timing gear pulley
- Front crankshaft seal

*To install:*

3. Install or connect the following:

- Front crankshaft seal
- Crankshaft timing gear pulley
- Vibration damper and tighten nut to 133 ft. lbs. (180 Nm)
- Timing belt
- Front wheel
- Front timing belt cover
- Upper timing belt cover
- Accessory drive belt
- Expansion tank
- Servo reservoir
- Engine stabilizer brace. Tighten the bolts at the suspension turrets to 37 ft. lbs. (50 Nm). Tighten the engine bracket bolt to 59 ft. lbs. (80 Nm).

4. Start the engine and check for leaks.

### 3.0L and 3.2L Engines

*See Figure 57.*

1. Before servicing the vehicle, refer to the Precautions Section.

2. Raise the vehicle.

3. Remove the right hand front wheel.

4. With flat head screw driver pry out the front seal.

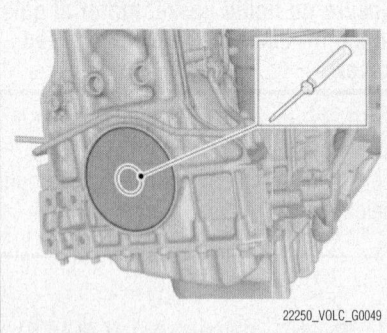

**Fig. 57 With flat head screw driver pry out the front seal**

*To install:*

5. Clean the surfaces of front seal area.

6. Install the front seal with a seal driver.

### S80 with 4.4L Engine

*See Figures 58 through 60.*

**Fig. 58 Install the counter hold tool 999-7196**

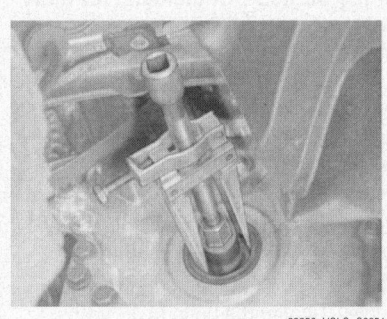

**Fig. 59 Install seal puller 999-5069 and remove the front crankshaft seal**

**Fig. 60 Install front crank seal and press in place with special seal driver tool 999-7197**

1. Remove the drive belt.

2. Remove the right hand front tire.

3. Remove the fender liner.

4. Install the counter hold tool 999-7196.

5. Remove the crank pulley mounting bolt.

6. Install the separator flange and remove the crank pulley.

7. Install seal puller 999-5069 and remove the front crankshaft seal.

**Fig. 56 Removing crankshaft timing gear pulley**

## ⚜ WARNING

**Cleaning must be carried out around the crankshaft journal, not along it. Make sure that all residues from the emery cloth and other foreign substances is completely removed before installing a new crankshaft seal.**

*To install:*

8. Install front crank seal and press in place with special seal driver tool 999-7197.

9. Install the counter hold tool 999-7196 with special tool 999-7198 and press the crankshaft pulley to 74 ft. lbs. (100 Nm)

10. Remove the special tool 999-7198 and just install the counter hold tool 999-7196.

11. The final crankshaft pulley tightening sequence is as follows:
- Stage 1: 89 ft. lbs. (120 Nm)
- Stage 2: Loosen the bolt 360°
- Stage 3: 37 ft. lbs. (50 Nm)
- Stage 4: Additional 90°

12. Remove the counter hold tool.

13. Install the drive belt.

14. Install the fender liner.

15. Install the right front tire.

## CYLINDER HEAD

*REMOVAL & INSTALLATION*

### 2.4L and 2.5L Engines

*See Figures 61 through 64.*

1. Before servicing the vehicle, refer to the Precautions Section.

2. Drain the cooling system.

3. Relieve the fuel pressure

4. Remove or disconnect the following:
- Negative battery cable
- Engine appearance cover
- Air cleaner housing

Fig. 61 Remove the valve cover

Fig. 62 Cylinder head loosening sequence—5-cylinder engine

Fig. 63 Cylinder head torque sequence—5-cylinder engines

- Ignition coils
- Intake manifold
- Fuel rail and injectors
- O2 sensor connector
- Front pipe at exhaust manifold
- Exhaust manifold
- Turbocharger, if equipped

5. Remove the screws for the bracket for the oil pipes for the power steering.

6. Remove the valve cover

## ⚜ WARNING

**Take extra care not to damage the mating faces.**

7. Remove the tensioned mounting yoke.

8. Remove or disconnect the following:
- Upper engine coolant hose
- Accessory drive belt
- Timing belt
- Camshafts
- Valve tappets (keep in their natural order)
- Cylinder head

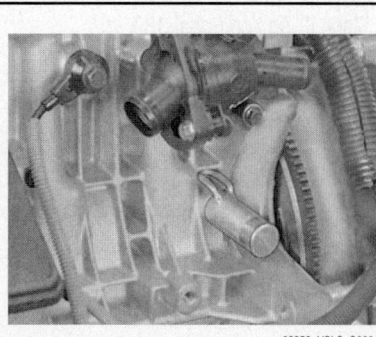

Fig. 64 Crankshaft adjustment tool 999-5451

*To install:*

9. Clean the mating surfaces of cylinder head and engine block.

10. Install the cylinder head with a new gasket. Tighten the bolts in the following sequence:

   a. Step 1: 15 ft. lbs. (20 Nm)

   b. Step 2: 44 ft. lbs. (60 Nm)

   c. Step 3: Plus 130°

11. Remove or disconnect the following:

- Camshafts (Lubricate all components)
- Valve tappets (in their natural order)
- Camshafts

12. Remove the cover plug. Turn the crankshaft clockwise slightly. Install the crankshaft adjustment tool.

13. Install or connect the following:

- Timing belt
- Accessory drive belt
- Intake manifold
- Upper engine coolant hose
- Valve cover (with new gasket)
- Steering pipe bracket bolts
- Turbocharger, if equipped
- Exhaust manifold
- Front pipe at exhaust manifold
- O2 sensor connector
- Fuel rail and injectors
- Intake manifold
- Ignition coils
- Air cleaner housing
- Engine appearance cover

14. Connect the negative battery cable.

15. Install new engine coolant and bleed the system.

16. Start the engine and check for leaks.

## 3.0L and 3.2L Engines

*See Figures 65 and 66.*

1. Before servicing the vehicle, refer to the Precautions Section.

2. Depressurize the fuel system.

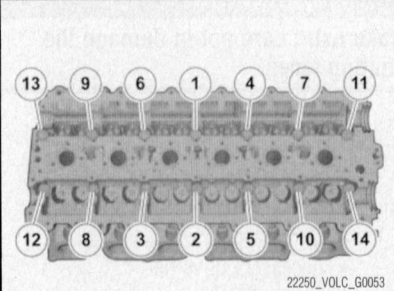

**Fig. 65 3.0L–3.2L Cylinder head torque sequence**

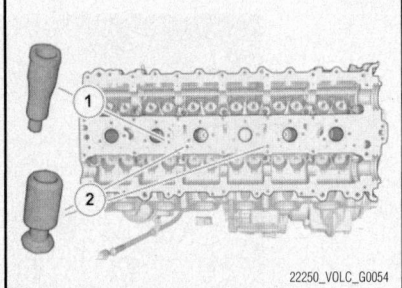

**Fig. 66 3.0L and 3.2L Choke valves (1) and (2)**

3. Disconnect the negative battery cable.

4. Remove or disconnect the following:

- Exhaust manifold
- Intake manifold
- Valve covers
- Injectors and fuel rail

➡**Mark the position of the components before removal.**

5. Lift out the valve lifters. Use a magnet or suction cup.

### ✳✳ WARNING

**Take extra care not to damage the mating faces.**

6. Remove the cylinder head bolts in sequence.

*To install:*

7. Make sure that the mating faces are clean and free of foreign material.

8. Make sure that the cylinder head screws' threaded holes in the engine block are free of fluid.

9. Check that the left—threaded sleeves go to the bottom.

10. Install new head gasket

11. Tighten the cylinder head as follows:

Stage 1: 33 ft. lbs. (45 Nm)

Stage 2: 33 ft. lbs. (45 Nm)

Stage 3: 90°

Stage 4: An additional 180°

12. Make sure that the choke valves (1), (2) are in place.

13. If only replacing the cylinder head gasket, reinstall the valve lifters in their original position.

➡**Valve clearance must be checked if the cylinder head is replaced.**

14. Install or reconnect the following:

- Injectors and fuel rail
- Valve covers
- Intake manifold
- Exhaust manifold

15. Connect the negative battery cable.

## S80 with 4.4L Engine

### *Right Cylinder Head*

*See Figure 67.*

1. Before servicing the vehicle, refer to the Precautions Section.

2. Disconnect the negative battery cable.

3. Drain the engine coolant.

4. Remove or disconnect the following:

- Drive belt
- Front cover
- Timing chain
- Valve cover
- Cam caps and bolts
- Intake camshaft
- Exhaust camshaft
- Catalytic converter
- Coolant hose at rear of cylinder head
- Cam sensors
- Vacuum lines
- Right cam sensor cover
- Converter support bracket and bolts
- Exhaust manifold

➡**Note the position of the component before removal.**

5. Lift out the valve lifters. Use a magnet or suction cup.

### ✳✳ WARNING

**Take extra care not to damage the mating faces.**

6. Remove the cylinder head bolts in sequence.

7. Remove the right cylinder head.

*To install:*

8. Clean the mating surfaces.

9. Install a new cylinder head gasket.

10. Install the right cylinder head.

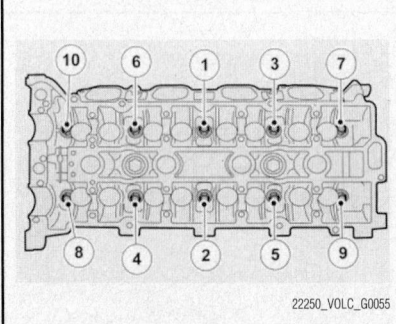

**Fig. 67 S80—Right cylinder head tightening sequence 4.4L engine**

11. Tighten the right cylinder head as follows:
- Stage 1: 30 ft. lbs. (40 Nm)
- Stage 2: 90°

12. Install or reconnect the following:
- Camshaft lifters
- Exhaust manifold and tighten to 18 ft. lbs. (24 Nm)
- Converter mounting bracket and tighten the bolts to 18 ft. lbs. (24 Nm)
- Alternator and tight bolts to18 ft. lbs. (24 Nm)
- Cam sensor bracket tighten the bolts to 7 ft. lbs. (10 Nm)
- Vacuum components
- Cam sensor electrical connectors
- Coolant hose at rear of cylinder head
- Catalytic converter
- Exhaust camshaft
- Intake camshaft
- Cam caps and bolts
- Valve cover
- Timing chain
- Front cover
- Drive belt

13. Change the oil and oil filter.

14. Install new engine coolant and bleed the cooling system.

### Left Cylinder Head

*See Figure 68.*

1. Before servicing the vehicle, refer to the Precautions Section.

2. Disconnect the negative battery cable.

3. Drain the engine coolant.

4. Remove or disconnect the following:
- Timing chain
- Valve cover
- Cam caps and bolts
- Intake camshaft
- Exhaust camshaft
- Coolant fan
- Temperature sensor

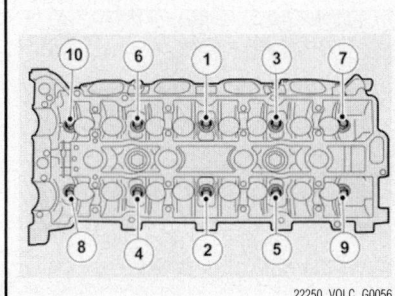

Fig. 68 Left cylinder head tightening sequence—S80 with 4.4L engine

22250_VOLC_G0056

- Cam sensor connectors
- Coolant lines
- Water pump assembly
- Coolant pipes mounted to cylinder head
- Exhaust manifold
- Lifters

➡ Note the position of the components before removal.

5. Lift out the valve lifters. Use a magnet or suction cup.

### ✳✳ WARNING

**Take extra care not to damage the mating faces.**

6. Remove the cylinder head bolts in sequence.

7. Remove the left cylinder head.

### *To install:*

8. Clean the mating surfaces.

9. Install a new cylinder head gasket.

10. Install the left cylinder head.

11. Tighten the left cylinder head as follows:
- Stage 1: 30 ft. lbs. (40 Nm)
- Stage 2: 90°

12. Install or reconnect the following:
- Camshaft lifters
- Exhaust manifold
- Coolant pipes mounted to cylinder head
- Water pump assembly
- Coolant lines
- Cam sensor connectors
- Temperature sensor
- Coolant fan
- Exhaust camshaft
- Intake camshaft
- Cam caps and bolts
- Valve cover
- Timing chain
- Exhaust manifold and tighten to 18 ft. lbs. (24 Nm)
- Converter mounting bracket and tighten the bolts to 18 ft. lbs. (24 Nm)

13. Change the engine oil and oil filter.

14. Install new engine coolant and bleed the cooling system.

### ENGINE ASSEMBLY

*REMOVAL & INSTALLATION*

#### 2.4L and 2.5L Engines

*See Figures 69 and 70.*

1. Before servicing the vehicle, refer to the Precautions Section.

2. Relieve the fuel system pressure.

3. Drain the cooling system.

4. Remove or disconnect the following:
- Negative battery cable
- Battery and battery box
- Air intake assembly
- Oil dipstick assembly
- Cable harness for fuse box on left hand fender liner and clamp
- Ground lead in left-hand suspension turret
- Positive cable on the starter and cable clamp
- Ground lead on the front of the transmission
- Steering shaft joint bolt, and release the steering shaft from the steering gear

5. Separate the quick-release fuel pressure line connector using special tool 999-5666

6. Separate the quick-release connector on the canister purge (CP) valve by pressing the connector together.

7. Disconnect the heated oxygen sensors (HO$_2$S) connectors

8. Lower the rear section of the engine.

9. Drain the cooling system.

10. Remove or disconnect the following:
- Halfshafts
- Lower torque rod
- Delivery and return lines from the power steering pump and seal the openings
- Power steering lines from the mountings on the engine and sub-frame
- Catalytic converter from the sub-frame
- Rear HO$_2$S connector and cable harness

11. Support the sub-frame with a jack with universal plate 999-5972 or equivalent.

12. Remove the following:
- Bolts for the front edge of the sub-frame
- Bolts for rear support brackets of the sub-frame
- Bolts for rear edge of the sub-frame

13. Carefully lower the sub-frame

14. Remove or disconnect the following:
- Side impact protection system (SIPS) member
- Catalytic converter
- Accessory drive belt
- Bolts for A/C compressor and hang the compressor on the radiator member
- Heater hoses
- Coolant hoses from radiator
- Gear selector cable using special tool 999-7077

**Fig. 69 Engine assembly shown removed from the vehicle**

- Clutch cable at the firewall, if equipped with manual transmission
- Oil hoses for transmission, if equipped with automatic transmission
- Transmission coolant hoses, if equipped with automatic transmission

15. Position a retaining strap over the engine and either side of the engine.

16. Raise the vehicle and position a lifting table against the oil pan and use post hoist support arms to support the engine and transmission.

17. Secure the engine and transmission in the lifting table with the retaining straps.

18. Open the cock for the hydraulic modulator for the lifting table and lower the vehicle at the same time as opening the lifting table compressed air unit. Stop the car immediately before the lifting table reaches its lower position.

19. Close the cock for the hydraulic modulator for the lifting table.

20. Using the lifting table, lift up the engine assembly slightly to take the slack of the engine mounts.

21. Remove or disconnect the following:
- Ground lead in the right-hand suspension turret
- Deflection limiter

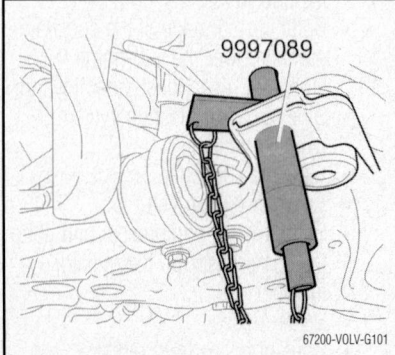

**Fig. 70 Sub-frame adjustment tool**

- Right hand engine mount
- Bolts for the engine mount on the transmission
- Raise the vehicle to free the engine assembly from the vehicle.

### To install:

22. Place the engine assembly on a lifting table and raise the vehicle enough to clear the engine assembly.

23. Lower the vehicle over the engine assembly and stop lowering the vehicle just before the engine assembly is in position. Raise the engine the last bit to put it in position to install the engine mounts.

24. Install or connect the following:
- Front engine mount
- Bolt for engine mount on transmission and tighten to 26 ft. lbs. (35 Nm) plus 60°
- Deflection limiter and right-hand engine mount and tighten to 59 ft. lbs. (80 Nm)

25. Disconnect the restraining straps around the engine and remove the lifting table.

26. Install or connect the following:
- Catalytic converter using new gaskets and nuts
- SIPS member
- Lower torque rod

27. Raise the sub-frame so adjustment tools 999-7089 are lined up with the holes in the side members.

28. Tighten the rear sub-frame bolts to 85 ft. lbs. (115 Nm).

29. Tighten the front sub-frame bolts to 85 ft. lbs. (115 Nm).

30. Tighten the support brackets for the rear sub-frame mounts.

31. Install or connect the following:
- Rear HO2S connector and cable harness
- Halfshafts
- A/C compressor
- Accessory drive belt
- Power steering lines and tighten as follows:
  a. Return line to 13 ft. lbs. (18 Nm)
  b. Delivery line to 22 ft. lbs. (30 Nm)
- Front wheels
- Engine splash guard, if equipped
- Coolant hoses to radiator
- Heater hoses
- Fuel pressure line
- EVAP canister hose
- Oil hoses for transmission, if equipped with automatic transmission
- Transmission coolant hoses, if equipped with automatic transmission

- Clutch cable, if equipped with manual transmission
- Gear selector cable
- Steering shaft on the steering gear and tighten the bolt to 18 ft. lbs. (24 Nm)
- Ground lead on transmission
- Positive cable on the starter and cable clamp
- Ground leads for each suspension turret
- Cable harness for fuse box and fuse box cover
- Oil dipstick assembly
- Air intake assembly
- Battery box and battery
- Battery cables

32. Fill the engine with coolant.
33. Fill the crankcase to correct level.
34. Start the engine and check for leaks.

### 3.0L and 3.2L Engines

*See Figure 71.*

1. Before servicing the vehicle, refer to the Precautions Section.

2. Relieve the fuel system pressure.

3. Drain the cooling system.

4. Disconnect the negative battery cable.

5. Remove or disconnect the following:
- Drive belt
- Engine Control Module (ECM)
- Exhaust manifolds
- Left driveshaft
- Right driveshaft

### ✳✳ WARNING

**Make sure that the steering lock is locked. Make sure that the road wheels are in the straight ahead position.**

6. The steering column to steering gear (steering shaft joint).

7. Remove the sub frame as follows:
- Control arms from the spindle

**Fig. 71 Check that the lifting table is correctly positioned**

- Torque rod
- Fuel lines
- Stabilizer bar link to stabilizer bar
- Exhaust system flex pipe
- Exhaust system hangers

8. Install the transmission jack with adapter to the sub-frame

9. Remove the sub-frame bolts.

10. Remove the steering gear mounting bolts.

11. Remove the transfer case and propeller shaft. (AWD models)

12. Remove all coolant lines.

13. Remove the under hood fuse and relay assembly. (Set aside)

14. Remove all the ground straps.

15. Remove all the electrical connectors as needed.

16. Install the power steering lines as needed.

17. Remove the transmission shift cable.

18. Remove the EVAP lines.

### ✲✲ WARNING

**Check and remove any components that may hinder the engine and trans mission removal.**

19. Ensure that the lifting table is centered under the engine/transmission for optimal weight distribution.

20. Check that the lifting table is correctly positioned.

21. Secure the engine / transmission in the lifting table with a retaining strap.

22. Open the cock for the hydraulic modulator for the lifting table. Ask a colleague to help. Lower the vehicle at the same time as opening the lifting table compressed air unit.

23. Stop the vehicle immediately before the lifting table reaches its lower position. Close the cock for the hydraulic modulator for the lifting table.

24. Remove the engine mounts.

### ✲✲ WARNING

**Make sure that no components catch when raising the car.**

25. Raise the car.

26. Separate the engine from the transaxle.

#### *To install:*

27. Install the engine to the transaxle.

28. Raise the vehicle so that the engine / transmission unit are free beneath the vehicle. Place the engine and transmission on the lifting table. Lower the lifting table to its lowest position.

29. Lower the vehicle over the engine/transmission unit, being sure to fit the unit between the frame members. Stop lowering the vehicle just before the engine/transmission unit is in position. Raise the engine/transmission unit.

### ✲✲ WARNING

**Make sure that no components catch when raising the car.**

30. Install the engine mounts. Tighten the engine mount retaining bolts to 59 ft. lbs. (80 Nm)

31. Install the EVAP lines.

32. Install the transmission shift cable.

33. Install all the electrical connectors as needed.

34. Install the power steering lines as needed.

35. Install all the ground straps.

36. Install the under hood fuse and relay assembly. (Set aside)

37. Install all coolant lines.

38. Install the transfer case and propeller shaft. (AWD models)

39. Install the steering gear mounting bolts.

40. Install the sub-frame bolts.

41. Install the transmission jack with adapter from the sub-frame

42. Install the sub frame as follows:
- Right driveshaft
- Left driveshaft
- Exhaust manifolds
- Engine Control Module (ECM)
- Drive belt

43. Connect the negative battery cable.

44. Install engine oil and filter.

45. Install new engine coolant and bleed the cooling system.

#### 2.4L and 2.5L Engines

*See Figures 72 through 74.*

1. Drain the cooling system.

2. Disconnect the negative battery cable.

3. Remove or disconnect the following:
- Air cleaner housing
- Battery
- Upper torque rod
- Drive belt
- Catalytic converter
- Front suspension position sensor
- Front wheels
- Engine Control Module (ECM)
- Air filter box bracket
- Coolant hoses
- Transmission lines
- Electrical connectors at starter and shifter cable

- Coolant hoses at radiator
- Engine covers
- Power steering reservoir
- Coolant tank
- Ball joints for the wheel knuckles.
- Left and right axles.
- Left and right tie-rod ends.
- Stabilizer links at strut
- Fuel lines
- Propeller shaft (AWD models)
- Lower steering shaft joint
- Electrical connector for the steering gear

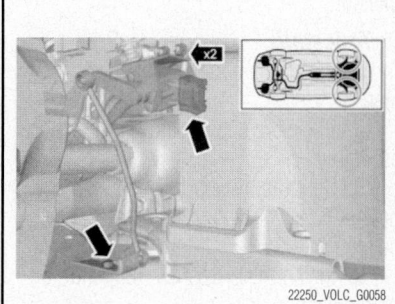

22250_VOLC_G0058

**Fig. 72 Front suspension position sensor shown**

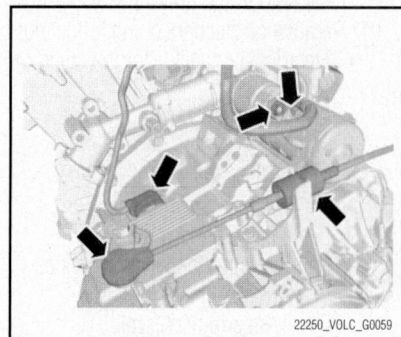

22250_VOLC_G0059

**Fig. 73 Electrical connectors and shifter cable**

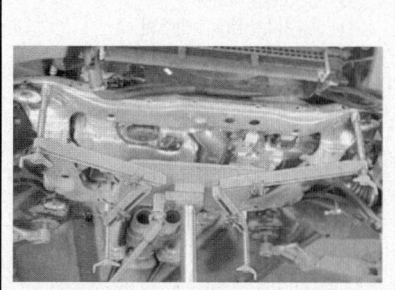

22250_VOLC_G0060

**Fig. 74 Lowering the engine and transmission unit**

## ✳✳ WARNING

**Ensure that the lifting table is properly centered under the engine. Make sure that no components catch when lowering the engine.**

- Lower the engine/transmission unit approximately 15 cm
- Sub-frame mounting bolts.
- Bend out the A/C compressor and suspend it in a suitable manner
- Continue lowering the engine and transmission unit.

4. Separate the engine from the transmission.

**To install:**

5. Install the engine to the transmission.

6. Raise the engine and transmission assembly.

7. Stop the lift approximately 15 cm before the sub-frame reaches the frame members.

8. Install the driveshaft to the hub on both sides

9. Install or connect the following:
- A/C compressor
- Raise the engine and transaxle.
- Guide the lower steering shaft into the hole in the firewall

10. Remove or disconnect the following:
- Electrical connector for the steering gear
- Propeller shaft (AWD models)
- Fuel lines
- Stabilizer links at strut
- Left and right tie-rod ends.
- Left and right axles.
- Ball joints for the wheel knuckles.
- Coolant tank
- Power steering reservoir
- Engine covers
- Coolant hoses at radiator
- Electrical connectors at starter and the shifter cable
- Transmission lines
- Coolant hoses
- Air filter box bracket
- Engine Control Module (ECM)
- Front wheels
- Front suspension position sensor
- Catalytic converter
- Drive belt
- Upper torque rod
- Battery
- Air cleaner housing

11. Refill the crankcase with oil and install a new oil filter.

12. Install new engine coolant and bleed the cooling system.

13. Start the engine and check for leaks.

## EXHAUST MANIFOLD

*REMOVAL & INSTALLATION*

### 2.4L and 2.5L Engines

*See Figure 75.*

1. Before servicing the vehicle, refer to the Precautions Section.

2. Remove or disconnect the following:
- Negative battery cable
- Exhaust manifold heat shield
- Exhaust front pipe
- Turbocharger
- Exhaust manifold

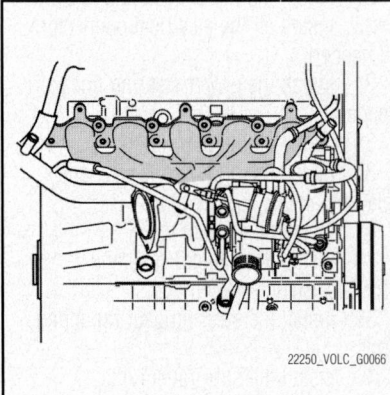

**Fig. 75 Exhaust manifold view**

22250_VOLC_G0066

**To install:**

3. Install or connect the following:
- Exhaust manifold and tighten the fasteners to 18 ft. lbs. (25 Nm)
- Turbocharger, if equipped
- Exhaust front pipe and tighten the fasteners to 44 ft. lbs. (60 Nm)
- Exhaust manifold heat shield and tighten the fasteners to 88 inch lbs. (10 Nm)

4. Loosen the joint at the catalytic converter and re-tighten to 18 ft. lbs. (25 Nm). This is necessary to prevent stresses in the system.

5. Connect the negative battery cable.

6. Start the engine and check for leaks.

### 3.0L and 3.2L Engines

*See Figure 76.*

1. Before servicing the vehicle, refer to the Precautions Section.

2. Remove or disconnect the following:
- The battery negative lead
- plastic charge air pipe and hose to charge air cooler
- The intake pipe between the air cleaner and the turbocharger

67200-XC90-G01

**Fig. 76 Intake manifold—6-cylinder engine**

3. Remove the intake pipe for the air cleaner housing from the connection above radiator. Twist the pipe up.

4. Drain the fuel line.

5. Remove:
- The protective cover over the injector connectors
- injector connectors
- The fuel rail mounting bolts

➡**Handle the injectors carefully to avoid damaging the nozzles and needles.**

6. Spray universal oil or similar around the injector terminals on the intake manifold.

7. Gently work the fuel rail and injectors loose.

8. Detach the fuel line from the nozzle pipe by pressing the quick-release connector.

9. Remove:
- The dip stick pipe mounting at the intake manifold
- The nipple for the crankcase ventilation under the intake

10. disconnect the throttle body connector. Remove the plastic charge air pipe from the throttle body.

11. Remove the upper bolts.

12. Remove the outer bolts from the lower row.

13. Slacken off the remaining bolts.

14. Lift up and remove the intake manifold.

15. Remove the remaining bolts and gasket.

**To install:**

16. Transfer the throttle body with a new gasket. Tighten to 88 inch lbs. (10 Nm).

17. Ensure that the gasket faces are clean.

18. Install a new gasket with the two centermost bolts in the lower row.

19. Lower the intake over the 2 bolts.

20. Install the remaining bolts.

21. Tighten all bolts. Tighten to 15 ft. lbs. (20 Nm).

22. Connect the throttle body connector.

23. Install or reconnect the following:
- The vacuum hoses according to the earlier markings
- The nipple for the crankcase ventilation under the intake manifold
- The dip stick mounting bolt to the intake manifold. Tighten to 18 ft. lbs. (25 Nm)
- The plastic charge air pipe between the throttle body and the charge air cooler

24. Lubricate the injector rings with petroleum jelly.

25. Hold the fuel rail. Press the fuel line quick-release connector together until it clicks.

26. Press down on the fuel rail. Check that all injectors are correctly positioned.

27. Tighten the fuel rail to 88 inch lbs. (10 Nm). Use new bolts.

28. Connect all the connectors to the injectors.

29. Install the protective cover over the connectors to the nozzles.

30. Install:
- intake manifold between air cleaner and turbocharger
- plastic charge air pipe and hose to charge air cooler
- The air cleaner intake pipe
- The battery negative lead

### S80 with 4.4L Engine

*See Figures 77 through 79.*

1. Before servicing the vehicle, refer to the Precautions Section.
2. Relieve the fuel system pressure.
3. Drain the cooling system.
4. Disconnect the negative battery cable.
5. Install the engine support system.

6. Remove the control arms on both sides.

7. Remove the front stabilizer bar link to stabilizer bar.

8. Remove the engine under cover.

9. Remove the catalytic converter.

10. Right-hand engine pad to sub-frame.

11. Remove the flex pipe heat shield near rack and pinion unit.

12. Remove the right lower torque mount.

13. Remove the right lower torque rod bracket.

14. Remove the rack and pinion mounting bolts.

15. Support the steering rack with tie straps or mechanics wire.

16. Collision member to sub-frame bracket.

17. Install a transmission jack with adapters and support the sub-frame.

18. Remove the sub-frame mounting bolts and carefully lower the sub-frame.

19. Remove the right driveshaft.

20. Disconnect the O2 sensor electrical connectors.

21. Remove the bolts holding the exhaust bracket to the oil pan.

22. Remove the right or left manifold as needed.

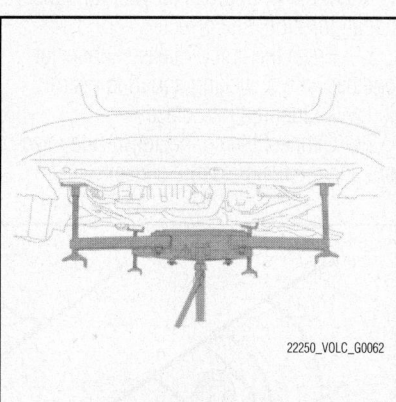

**Fig. 78 Transmission jack with adapters**

### *To install:*

23. Clean the mating surface and install new gasket.

24. Install the right or left manifold as needed. Tighten to 18 ft. lbs. (24 Nm)

25. Install the bolts holding the exhaust bracket to the oil pan. Tighten to 18 ft. lbs. (24 Nm)

26. Reconnect the O2 sensor electrical connectors.

27. Install the right driveshaft.

28. Install the sub-frame alignment tools 9997089 and carefully raise the sub-frame.

29. Install a transmission jack with adapters and support the sub-frame.

30. Install the sub frame mounting bolts and tighten in 2 stages:
- Stage 1: 111 ft. lbs. (150 Nm)
- Stage 2: Additional 90°

31. Install the collision member to sub-frame bracket

32. Install the steering rack and tighten mounting bolts

33. Install the right lower torque rod bracket and to 37 ft. lbs. (50 Nm)

34. Install the right lower torque mount and to 59 ft. lbs. (80 Nm)

35. Install the flex pipe heat shield near the rack and pinion unit.

36. Right-hand engine pad to sub-frame.

37. Install the catalytic converter.

38. Install the engine under cover.

39. Install the front stabilizer bar link to stabilizer bar.

40. Install the control arms on both sides.

41. Remove the engine support system.

42. Check for exhaust leaks.

### INTAKE MANIFOLD

*REMOVAL & INSTALLATION*

#### 2.4L and 2.5L Engines

*See Figures 80 and 81.*

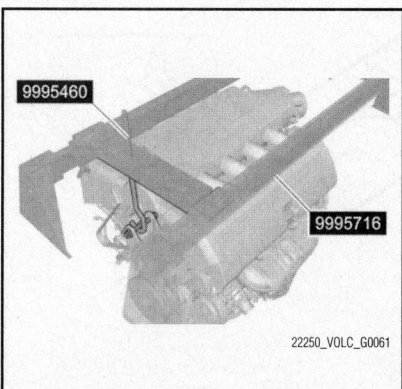

**Fig. 77 Engine support system shown**

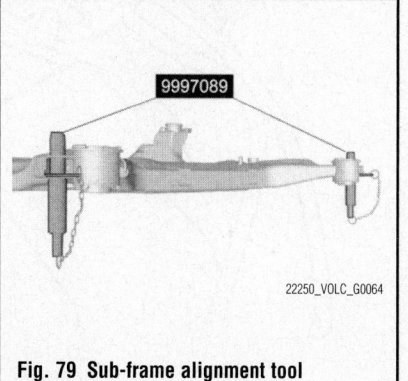

**Fig. 79 Sub-frame alignment tool 9997089**

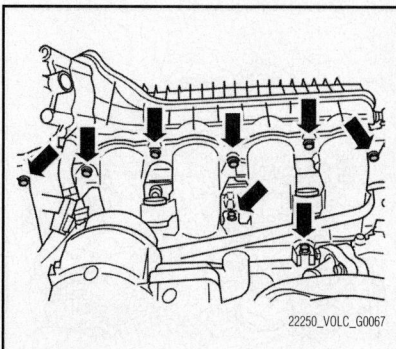

**Fig. 80 Remove the 6 screws for the upper section of the intake**

1. Before servicing the vehicle, refer to the Precautions Section.

2. Relieve the fuel system pressure.

3. Disconnect the negative battery cable.

4. Remove the brake vacuum hose.

5. Remove the hose for the throttle body.

6. Remove the intake manifold for the turbocharger. Remove the seals.

7. Remove the 6 screws for the upper section of the intake manifold.

8. Disconnect the vacuum hoses from the intake manifold. Put the pipe to one side.

9. Disconnect the connectors for the fuel pressure sensor and the connector for the injectors.

10. Remove the 2 fuel rail mounting screws.

11. Spray universal oil or similar around the injector nozzle at the terminal on the intake manifold.

12. Gently work the fuel rail and injector nozzles loose. Disconnect the fuel line to the fuel rail. Press in the two blue clips on the fuel rail. Pull the fuel line out of the fuel rail.

13. Place the fuel rail to one side.

14. Remove the hose from the oil trap to the timing belt cover at the terminal in the timing belt cover.

15. Remove the crankcase ventilation hoses from the intake manifold.

16. Remove the clamp for the fuel intake manifold.

17. Remove the lower screws for the intake manifold a few turns.

18. Remove the mounting screws in the upper row. Lift out the intake manifold.

19. Remove the screws in the lower row.

20. Remove the gasket for the intake manifold.

### To install:

21. Install a new gasket held in position by the lower screws for the intake manifold.

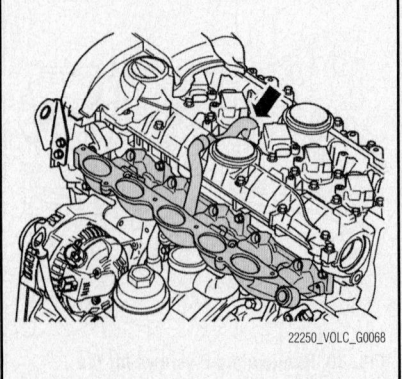

**Fig. 81 Lower intake manifold view**

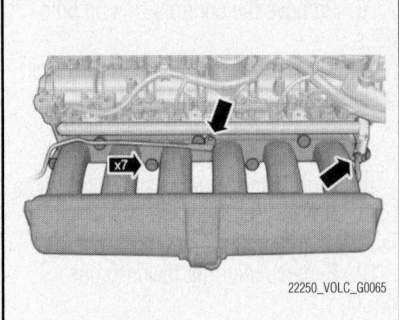

**Fig. 82 Intake manifold bolts and manifold shown**

➡ **Do not forget the crankcase ventilation hose. The hose must be inserted up through the gap between the second and third ducts.**

22. Install the three upper screws. Tighten all the screws starting from the center to 13 ft. lbs. (17 Nm)

23. Install the clamp for the fuel line on the intake manifold.

24. Install the crankcase ventilation hoses on the intake manifold.

25. Install the hose from the oil trap to the camshaft cover. Use a new hose clamp.

26. Lubricate the injector O-rings using petroleum jelly. Depress the fuel rail. Ensure that all injectors sit correctly.

27. Press the quick-release connector together until a clicking sound is heard. Screw the fuel rail into place.

28. Connect the connector for the injectors and the fuel pressure sensors

29. Install a new gasket for the intake manifold.

30. Install the vacuum hoses for the intake manifold

31. Install the intake manifold. Align the guides with the lower section of the intake manifold.

32. Install the screws for the intake manifold and tighten to 13 ft. lbs. (17 Nm)

33. Install the intake manifold for the turbocharger. Install the seals.

34. Install the hose for the throttle body.

35. Install the brake vacuum hose.

### 3.0L and 3.2L Engines

*See Figure 82.*

1. Before servicing the vehicle, refer to the Precautions Section.

2. Disconnect the negative battery cable.

3. Remove the engine cover.

4. Remove the engine under cover.

5. Remove the intake hose at the intake manifold and air filter housing.

6. Remove all the electrical connectors at the intake manifold.

7. Remove the intake manifold bolts and manifold.

8. To install, reverse the removal procedure and note the following:

9. Clean mating surfaces and install new gasket.

• Tighten the intake manifold bolts to 13 ft. lbs. (17 Nm)

### S80 with 4.4L Engine

*See Figures 83 through 85.*

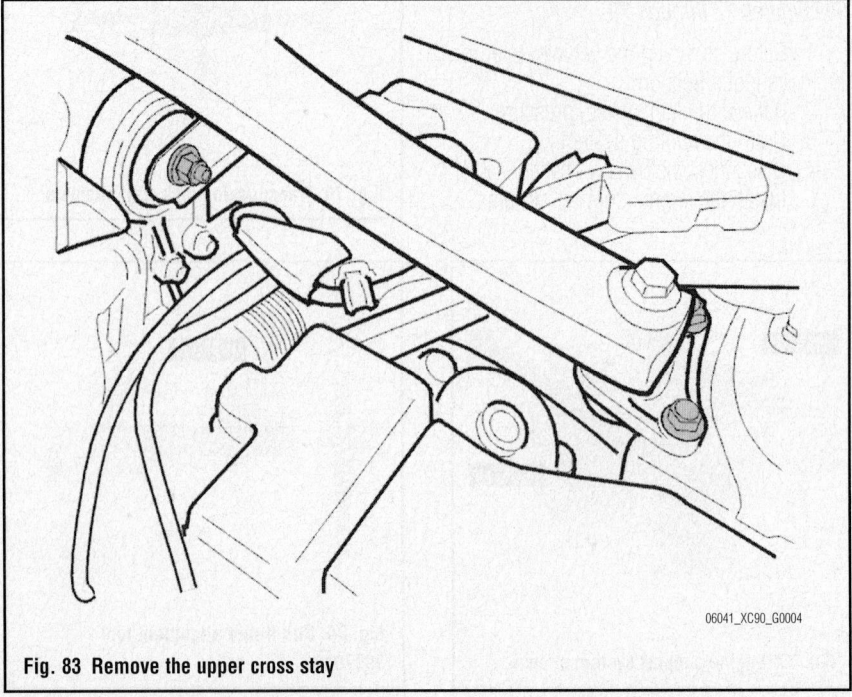

**Fig. 83 Remove the upper cross stay**

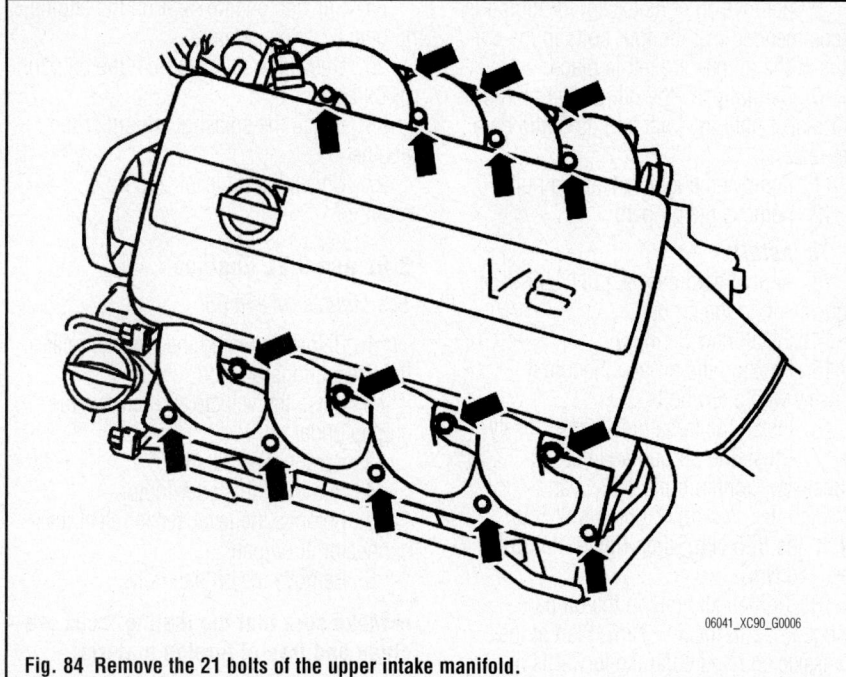

Fig. 84 Remove the 21 bolts of the upper intake manifold.

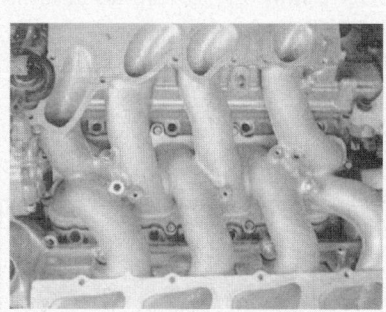

Fig. 85 Remove the 12 bolts and the lower intake manifold

1. Remove the 4 bolts at the suspension turrets
2. Remove the bolts of the upper engine mount on the engine.
3. Remove the upper cross stay.
4. Remove the air cleaner cover by pulling it straight up.
5. Remove the 2 connectors from the engine control module (ECM); clip the band clamp to the cable harness.
6. Remove the connector from the mass air flow sensor.
7. Remove the brake vacuum hose from the vacuum pump.
8. Remove the 2 bolts of the fresh air hose in the front plate.
9. Remove the fresh air hose.
10. Remove the air cleaner by pulling it straight up.
11. Remove the upper engine covers by pulling them straight up.

12. Remove the cover over the servo pump.
13. Detach the connector of the pressure and temperature sensor.
14. Remove the 2 vacuum hoses of the check valve of the distribution damper
15. Remove the 6 bolts.
16. Lift and slide the protective plate forward.
17. Remove the protective plate.
18. Open the cock on the lower left edge of the radiator.
19. Drain the coolant into a clean receptacle.
20. Close the cock.
21. Install the protective plate.
22. Install the 6 bolts.
23. Lower the vehicle.
24. Remove the 4 bolts and the upper engine mount.
25. Remove the brake vacuum hose from the intake manifold by releasing the clamp.
26. Remove the intake hose between the air cleaner and throttle body by releasing the 2 clamps.
27. Remove the 2 nuts on the throttle body.
28. Remove the 2 bolts on the throttle body.
29. Position aside the throttle body.
30. Detach the 2 coolant hoses, one from the throttle body and one from the coolant distribution pipe.
31. Remove the hose clip on the vacuum hose.
32. Remove the vacuum hose.

33. Remove the 2 M6 bolts on the non-return valve for crankcase ventilation
34. Remove the non-return valve for crankcase ventilation.
35. Remove the 21 bolts of the upper intake manifold.
36. Remove the upper intake manifold. Discard the gaskets.
37. Drain the fuel system.
38. Remove the crankcase ventilation hose between the cylinder heads by releasing the 2 clamps.
39. Remove the 3 bolts from the fuel rail.
40. Remove the 2 bolts on the fuel pressure sensor.
41. Lift the fuel rail assembly with injector straight up and position it aside.
42. Remove the 12 bolts and the lower intake manifold. Discard the gaskets

**To install:**
43. Clean the gasket surfaces.
44. Install new gaskets.
45. Install the lower intake manifold. Tighten the 12 bolts to 14 ft. lbs. (19 Nm).
46. Lubricate the O-rings on the injectors with Vaseline or the like.
47. Install the fuel rail assembly with injectors.
48. Install the 3 bolts of the fuel rail.
49. Install the 2 bolts on the fuel pressure sensor.
50. Install the crankcase ventilation hose and 2 clamps.
51. Connect the fuel supply line and secure the line in the clamp.
52. Install new gaskets.
53. Install the upper intake manifold. Tighten the 21 bolts to 14 ft. lbs. (19 Nm).
54. Install the hoses to the intake manifold.
55. Install the non-return valve for crankcase ventilation.
56. Install the connector to the pressure and temperature sensor.
57. Install the 2 vacuum hoses of the check valve of the distribution damper.
58. Install the upper engine mount. Tighten the 4 M10 bolts.
50. Install the cover over the servo pump.
60. Position the engine covers and press them into the retainers.
61. Install the upper cross stay between the suspension turrets. Tighten the 4 bolts.
62. Install the bolt of the cross stay mount on the engine.
63. Install the air cleaner.
64. Install the brake vacuum hose on the brake pump.
65. Install the fresh air hose.

66. Install the 2 bolts of the fresh air hose in the front plate.

67. Install the 2 connectors to the engine control module (ECM).

68. Install the connector to the mass air flow sensor.

69. Install the band clamp on the cable harness in the air cleaner housing.

70. Install the cover over the air cleaner.

71. Fill with coolant.

72. Start the engine and allow it to warm to normal operating temperature. Check for leaks and proper function. Fill up fluids as necessary

## OIL PAN

### REMOVAL & INSTALLATION

#### 2.4L and 2.5L Engines

*See Figure 86.*

1. Before servicing the vehicle, refer to the Precautions Section.

2. Remove the oil dipstick and its pipe.

3. Remove the splashguard under the engine.

4. Drain the engine oil and remove the oil filter.

5. Release the oil cooler from the oil pan. Hang up at the rear.

6. Remove the bolt from the bracket for the fuel line.

7. Removing the oil pan

8. Back off all bolts holding the oil pan.

9. Remove all bolts except for four. It is recommended that the four bolts in the corners of the oil pan are left in place.

10. Carefully tap the oil pan with a rubber mallet until the joint and its liquid gasket releases.

11. Remove the four remaining bolts.

12. Remove the oil pan.

#### To install:

13. Apply liquid gasket 1161 059-9, or equivalent, to the oil pan.

14. Install new O-rings.

15. Position the oil pan. Secure it loosely with a few bolts.

16. Install the remaining bolts loosely.

17. Press the oil pan against the transaxle. Tighten bolts 1, 2, 3 and 4 to 27 inch lbs. (3 Nm). Tighten bolt 5 to 18 ft. lbs. (25 Nm); then tighten to 35 ft. lbs. (48 Nm).

18. Tighten all bolts in the oil pan flange to 12 ft. lbs. (17 Nm). Start at the transaxle end and continue forwards in pairs.

19. Install the bolt for the bracket for the fuel line.

20. Connect the oil cooler to the oil pan. Use new O-rings. Reinstall the pipe on the sub-frame.

21. Install a new oil filter.

22. Install the oil drain plug with a new gasket.

23. Install the oil dipstick and its pipe. Use a new O-ring.

➡**Check that the O-ring is correctly positioned.**

24. Fill with engine oil. Run the engine to operating temperature.

25. Check for oil leaks from the oil pan or oil cooler.

26. Install the splashguard under the engine.

27. Check the oil level. Top up if required.

#### 3.0L and 3.2L Engines

*See Figures 87 and 88.*

1. Before servicing the vehicle, refer to the Precautions Section.

2. Raise the vehicle and remove the engine under cover.

3. Remove the dipstick.

4. Remove the starter motor.

5. Remove the level sensor electrical connector at oil pan.

6. Remove the oil pan bolts.

➡**Make sure that the mating faces are clean and free of foreign material.**

#### To install:

7. Apply (2.5 ± 0.5 mm) sealant to the engine block.

8. Install the oil pan.

9. To ensure sealing in the joint, all screws shall be tightened twice in a row to 13 ft. lbs. (17 Nm)

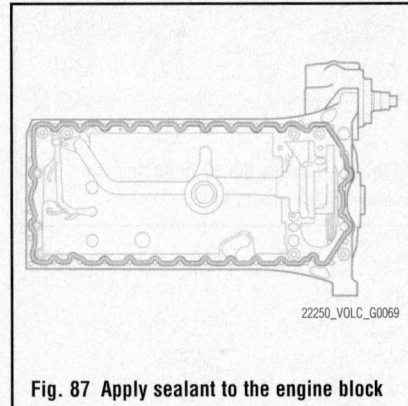

Fig. 87 Apply sealant to the engine block

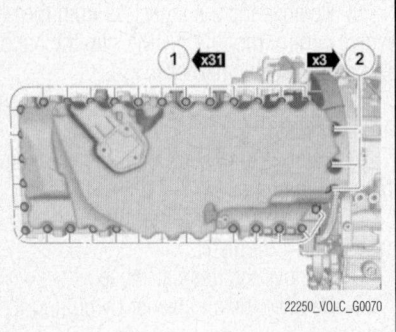

Fig. 88 Oil pan view—3.0L and 3.2L Engines

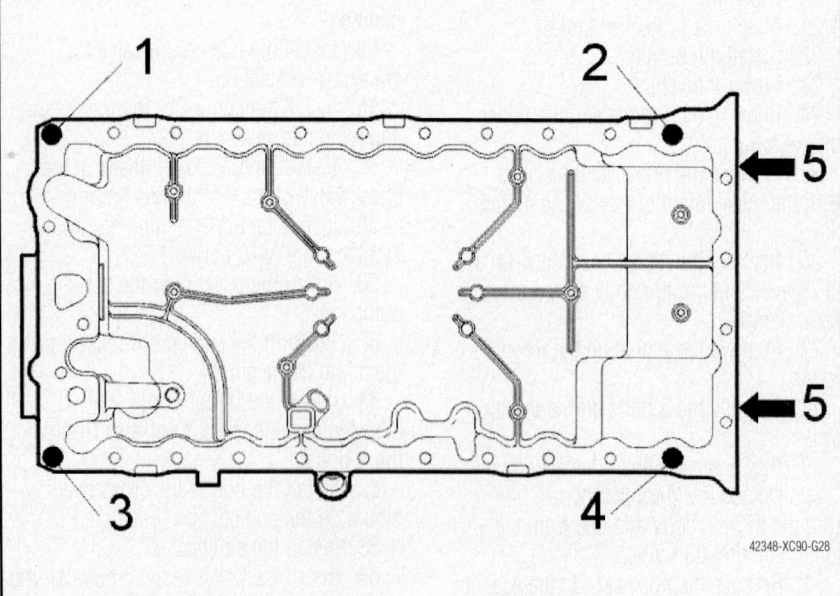

Fig. 86 5-cylinder oil pan

## S80 with 4.4L Engine

*See Figures 89 through 95.*

1.  Before servicing the vehicle, refer to the Precautions Section.

2.  Drain the engine oil, remove the oil filter.

3.  Remove the exhaust manifolds.

4.  Remove the right-hand engine mounting.

5.  Remove the 1 M10 bolts for the bracket.

6.  Remove the 4 M10 bolts for the transaxle.

7.  Remove the oil cooler. Discard the O-ring.

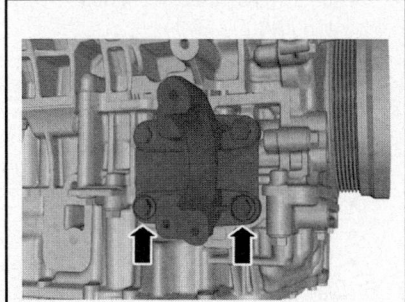

**Fig. 89 Right side engine mount—V8 engine**

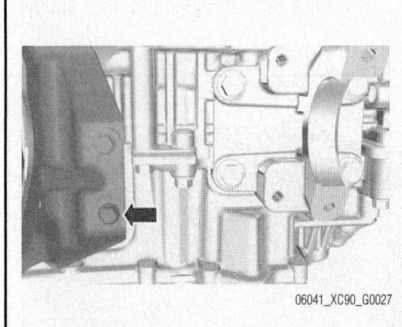

**Fig. 90 M10 bracket bolt—V8 engine**

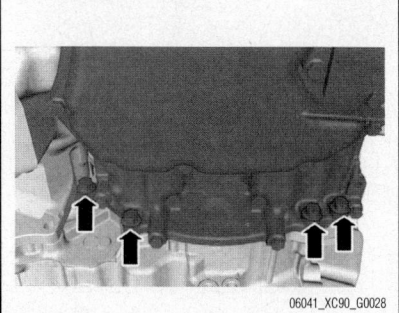

**Fig. 91 Transaxle bolts—V8 engine**

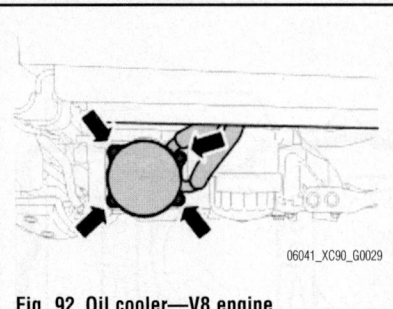

**Fig. 92 Oil cooler—V8 engine**

8.  Remove the 20 M6 bolts.

9.  Remove the oil level sensor.

10.  Remove the 17 x M8 bolts and the 1 x M6 bolt securing the intermediate section.

11.  Carefully tap the side of the oil pan using a rubber mallet. Lift out the oil pan.

12.  Discard the O-rings.

### To install:

➡ Clean the surface, using a razor blade or plastic scraper.

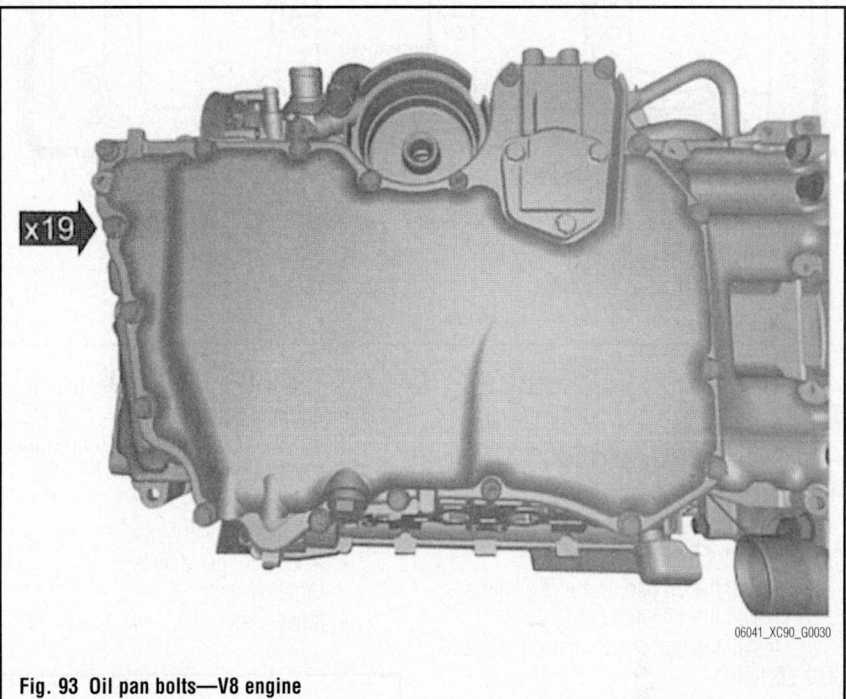

**Fig. 93 Oil pan bolts—V8 engine**

**Fig. 94 Intermediate section bolts—V8 engine**

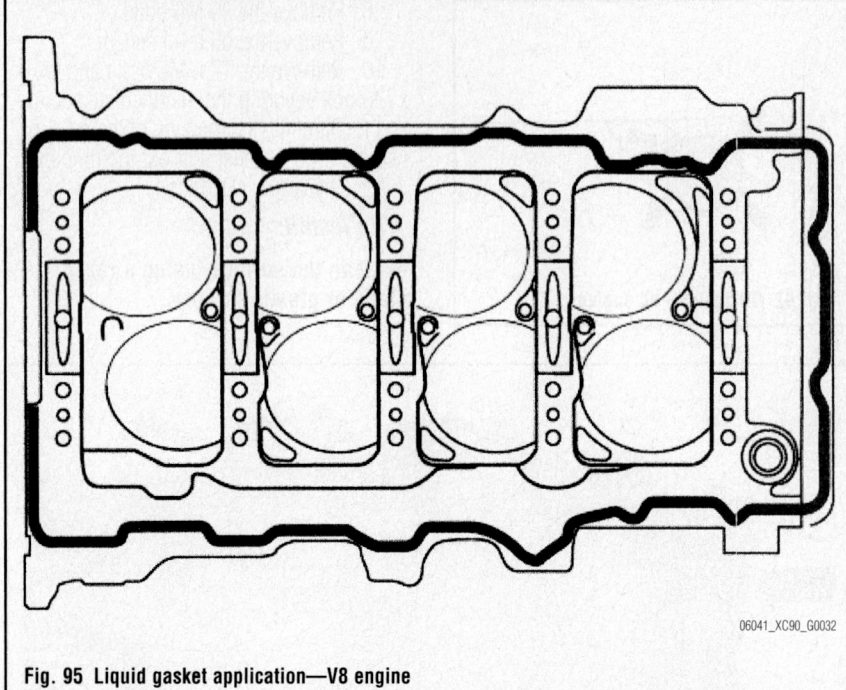

06041_XC90_G0032

**Fig. 95 Liquid gasket application—V8 engine**

13. Apply liquid gasket 307 57050 as illustrated.

**➡Minimum 0.039 inch (1mm) bead width.**

14. Install the oil pan within 5 minutes (curing time for liquid gasket).

15. Install the oil pan, use new O-rings

16. Install the 17 x M8 bolts.

17. Install the bolt securing the intermediate section.

**➡Clean the surface, using a razor blade or plastic scraper.**

18. Apply liquid gasket 307 57050 as illustrated.

**➡Minimum 0.039 inch (1mm) bead width.**

19. Install the oil pan within 5 minutes (cure time for liquid gasket)

20. Install the 20 M6 bolts.

21. Install the oil level sensor.

22. Install the oil filter, using a new O-ring.

23. Install the oil cooler. Use new O-ring

24. The remainder of installation is the reverse of removal.

25. Top up oil.

## OIL PUMP

### REMOVAL & INSTALLATION

**2.4L and 2.5L Engines**
*See Figure 96.*

1. Before servicing the vehicle, refer to the Precautions Section.
2. Remove or disconnect the following:
   • Negative battery cable
   • Fuel line clips
   • Coolant recovery tank
   • Accessory drive belts
   • Right front wheel
   • Inner fender liner

   • Front cover
   • Timing belt
   • Crankshaft timing sprocket
   • Front crankshaft seal
   • Oil pump

*To install:*

3. Install the oil pump using special tool 999-5455. Use the oil pump bolts to guide the pump. Use the crankshaft nut to press the pump in until it is seated fully. Tighten the oil pump bolts to 88 inch lbs. (10 Nm).

4. Remove the crankshaft nut and the press tool.

5. Install or connect the following:
   • Front crankshaft seal
   • Crankshaft timing sprocket and tighten the nut to 133 ft. lbs. (180 Nm)
   • Timing belt
   • Front cover
   • Inner fender liner
   • Right front wheel
   • Accessory drive belts
   • Coolant recovery tank
   • Fuel line clips
   • Negative battery cable

6. Start the engine and check for leaks.

### 3.0L and 3.2L Engines

1. Before servicing the vehicle, refer to the Precautions Section.
2. Remove the timing belt.
3. Check the tensioner and idler pulleys.

42348-XC90-G30

**Fig. 96 Oil pump installation**

4. Carefully pull crankshaft timing gear pulley free. Remove the four oil pump bolts.

5. Carefully pry upward diagonally between the stop lugs and the cylinder block.

6. Lift out the oil pump.

*To install:*

7. Install a new gasket.

8. Carefully insert the oil pump over the end of the crankshaft.

➡**The sealing ring in the oil pump is very easy to damage if not installed correctly.**

9. Install four new bolts as a guide.

10. Pull in the oil pump with tool 999 5455, or equivalent, and the crankshaft center nut.

11. Tighten the oil pump bolt. Tighten to 88 inch lbs. (10 Nm).

12. Install the timing gear pulley. Carefully tap alternately around the timing gear pulley until reaching its end position.

➡**The timing gear pulley can only be in one position on the crankshaft end splines.**

13. Install a new timing belt.

14. Install the pulley for the auxiliary belt. Locate the steering gear on the locating pin in the timing gear pulley.

15. Install the auxiliary belt.

16. Install counterhold 999 5433.

17. Tighten crankshaft center nut. See the torque chart at the beginning of this chapter.

18. Install the crankshaft damper. Use the crankshaft center nut as a counterhold.

19. Check the engine oil level.

**S80 with 4.4L Engine**

*See Figures 97 and 98.*

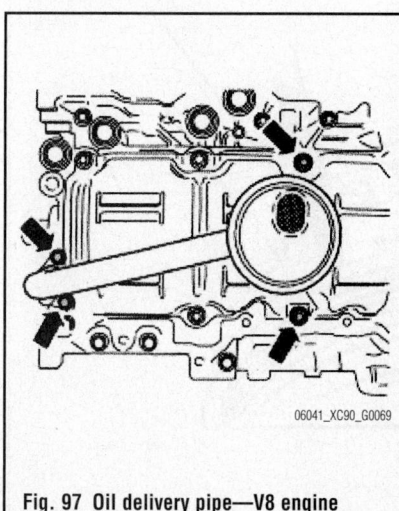

**Fig. 97 Oil delivery pipe—V8 engine**

06041_XC90_G0069

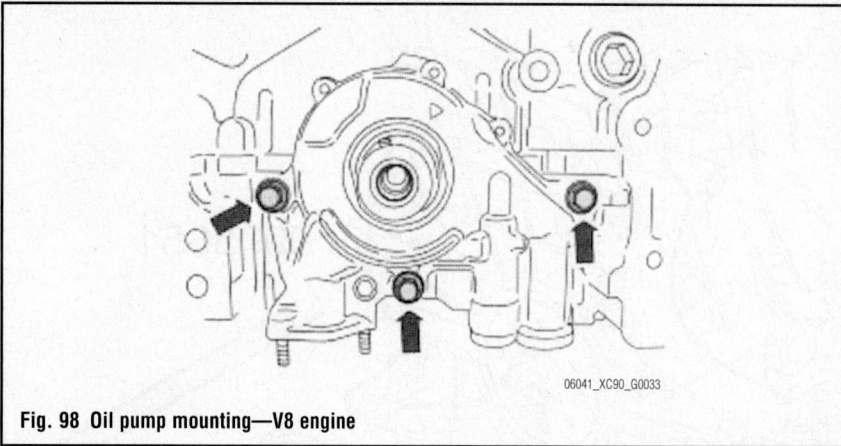

Fig. 98 Oil pump mounting—V8 engine

06041_XC90_G0033

1. Remove the timing chain.

2. Install the front engine bracket.

3. Remove the oil pan.

4. Remove the oil delivery pipe. Discard the gasket.

5. Remove the 3 M7 bolts and the oil pump. Discard the O-ring and the gasket for the oil delivery pipe.

*To install:*

6. Install a new O-ring and the oil pump.

7. Install the oil delivery pipe using a new gasket.

8. Install the oil pan.

9. Install the timing chain.

## PISTON AND RING

*POSITIONING*

*See Figures 99 and 100.*

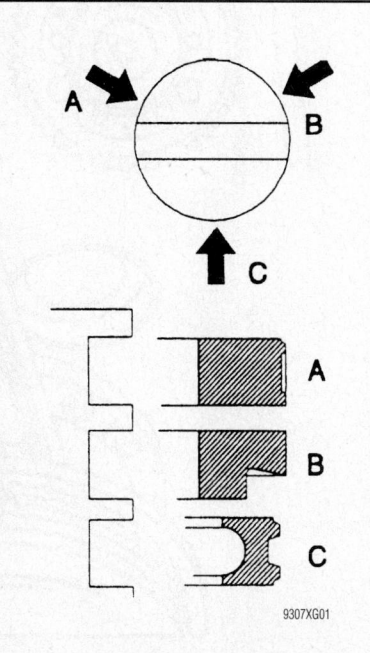

**Fig. 99 Ring identification and end-gap spacing—all engines**

9307XG01

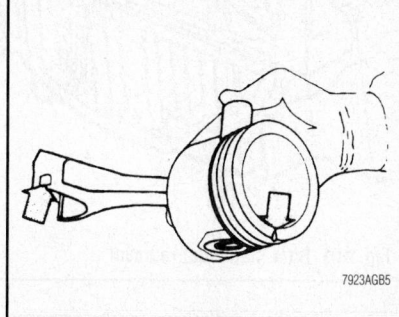

**Fig. 100 Piston and rod positioning—all engines. The notch on the piston crown faces the front of the engine**

7923AGB5

## REAR MAIN SEAL

*REMOVAL & INSTALLATION*

*See Figure 101.*

1. Before servicing the vehicle, refer to the Precautions Section.

2. Remove or disconnect the following:
   • Negative battery cable
   • Transmission
   • Clutch, if equipped
   • Flexplate or flywheel
   • Rear main seal

*To install:*

3. Install the rear main seal using Special Tool 999-7174 or equivalent, so that the seal is flush with the cylinder block.

➡**The felt side of the seal should face outward.**

4. Install or connect the following:
   • Flexplate or flywheel. Tighten the bolts to 33 ft. lbs. (45 Nm) plus 65°.
   • Clutch, if equipped
   • Transmission
   • Negative battery cable

5. Start the engine and check for leaks.

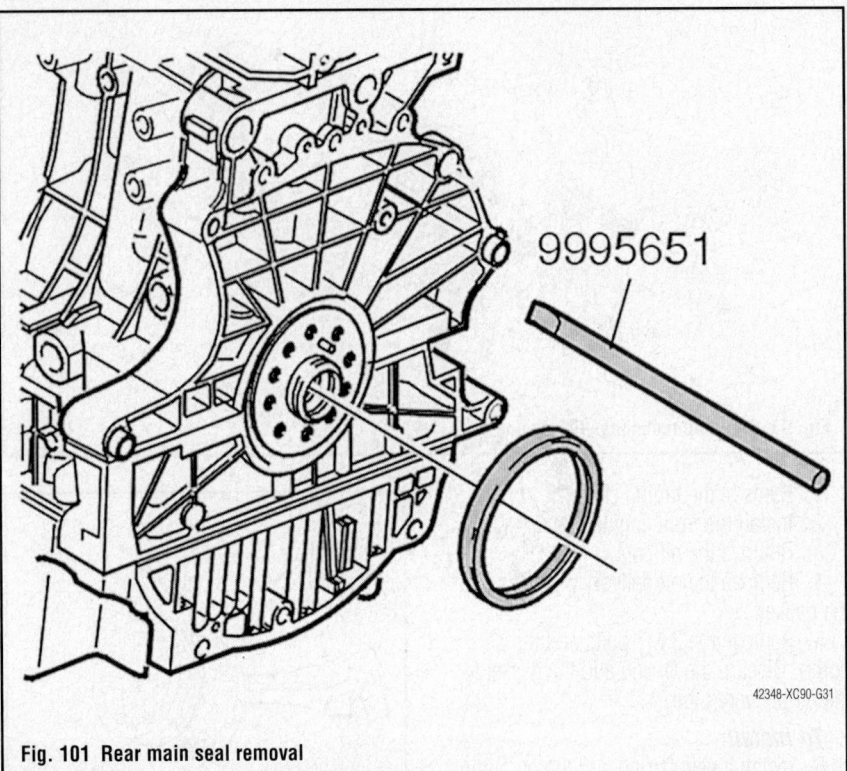

**Fig. 101 Rear main seal removal**

42348-XC90-G31

## TIMING CHAIN, SPROCKETS, FRONT COVER AND SEAL

### REMOVAL & INSTALLATION

### 2.4L and 2.5L Engines

*See Figures 102 and 103.*

1. Before servicing the vehicle, refer to the Precautions Section.

2. Lift the expansion tank and place it on top of the engine.

3. Install lifting beam 99-7103 or equivalent and tighten.

4. Remove or disconnect the following:

- Engine limiter
- Engine pad on the cylinder head
- Front transmission covers
- Right hand front wheel
- Inner fender liner
- Delivery line from the power steering pump
- Accessory drive belt
- Vibration damper
- Upper timing belt cover

5. Lower the engine slightly until the belt tensioner bolts can be accessed from beneath.

6. Turn the crankshaft clockwise until the markings on the crankshaft and camshaft pulleys correspond.

7. Slacken off the center bolt for the belt tensioner slightly and hold the center

bolt still. Turn the tensioner clockwise to 10 o'clock.

8. Remove the timing belt.

**To install:**

9. Install the timing belt in the following order:

a. Around the crankshaft

b. Around the idler pulley

c. Around the intake camshaft pulley

d. Around the exhaust camshaft pulley

e. Around the water pump

f. Onto the belt tensioner

10. Carefully turn the crankshaft clockwise until the timing belt is tensioned. The belt must be tensioned between the intake camshaft pulley, the idler pulley and the crankshaft.

11. Hold the belt tensioner center bolt secure. Turn the belt tensioner eccentric counter-clockwise until the tensioner indicator passes the marked position.

12. Then turn the belt tensioner eccentric back so that the indicator reaches the marked position in the center of the window and tighten the center bolt to 15 ft. lbs. (20 Nm).

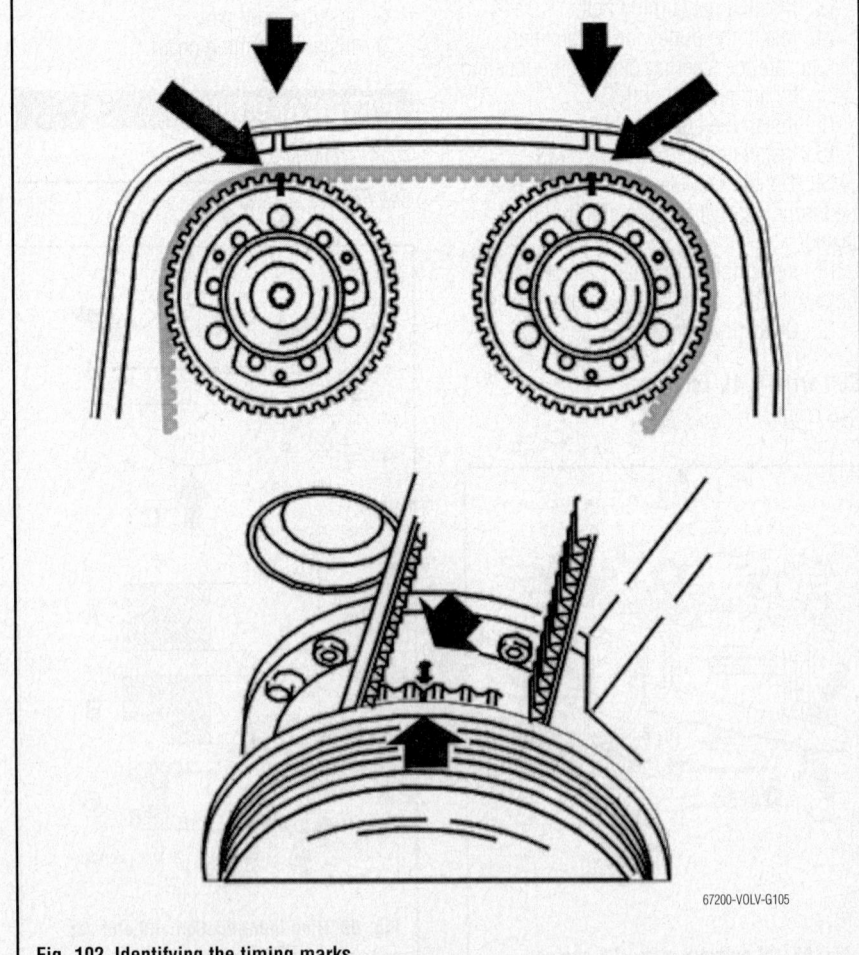

**Fig. 102 Identifying the timing marks**

67200-VOLV-G105

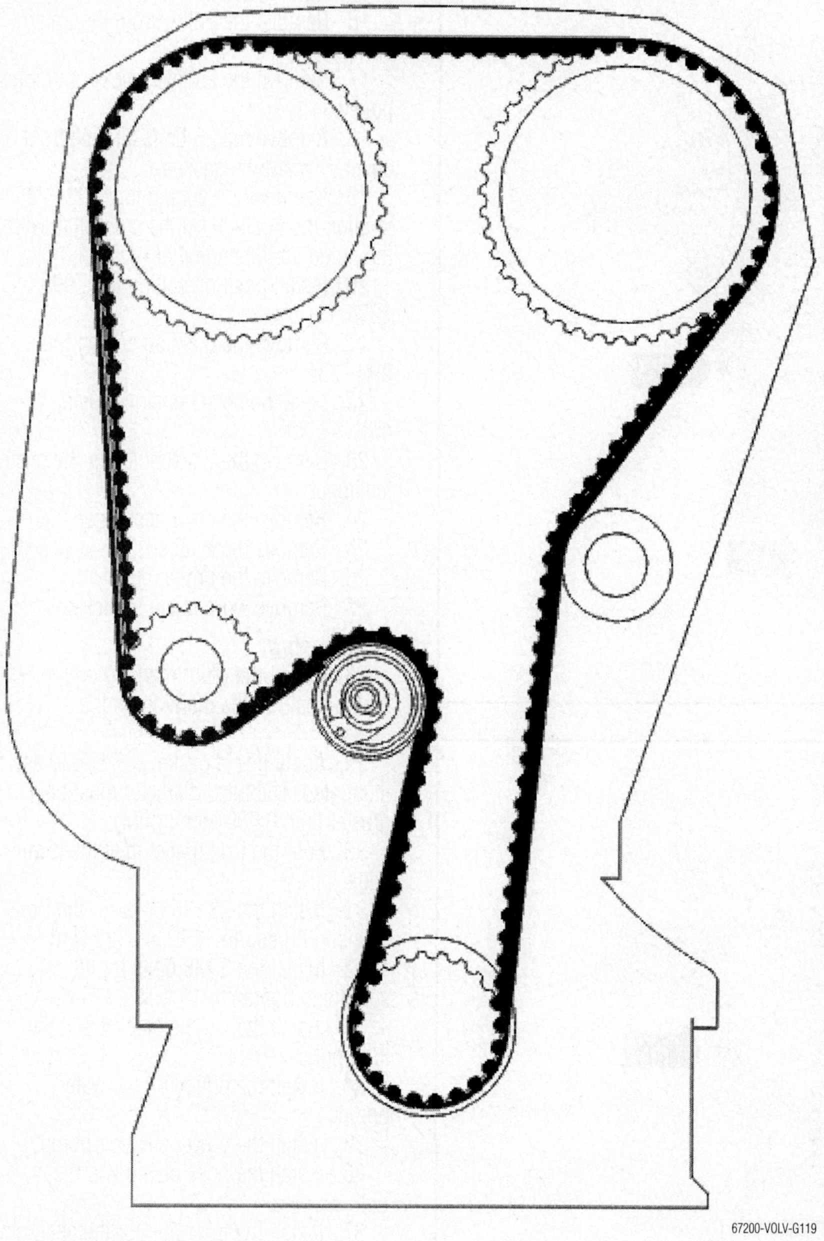

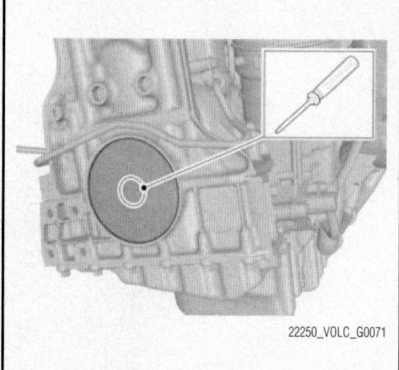

**Fig. 104 Remove the front crankshaft seal**

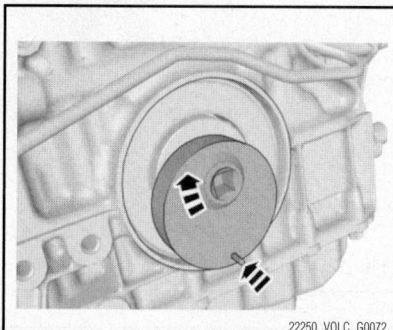

**Fig. 105 Rotate the engine to the zero position**

**Fig. 106 Lock cams in place with special tool 999 7261**

**Fig. 103 S40/V50/C70 timing belt routing**

13. Press the belt to check that the indicator on the tensioner moves easily.

14. Install or connect the following:
- Front timing belt cover
- Upper timing belt cover
- Vibration damper and tighten the center nut to 133 ft. lbs. (180 Nm) and 4 outer bolts to 18 ft. lbs. (25 Nm) plus 60°.
- Accessory drive belt
- Delivery hose for power steering pump
- Inner fender liner
- Front wheel
- Front transmission covers
- Engine pad on the cylinder head
- Engine limiter

15. Remove the lifting beam.

16. Install the expansion tank.

### 3.0L and 3.2L Engines

*See Figures 104 through 106.*

1. Before servicing the vehicle, refer to the Precautions Section.

2. Remove the front crankshaft seal.

3. Rotate the engine to the zero position.

4. Lock cams in place with the special tool 999 7261.

5. Remove the drive belt.

6. Remove the crank pulley cover.

7. Install the pulley removal tools 951 2926 bit Torx®50 and counter-hold 999 5760.

### S80 with 4.4L Engine

*See Figures 107 through 111.*

1. Before servicing the vehicle, refer to the Precautions Section.

2. Disconnect the battery negative cable.

3. Raise and support the vehicle safely.

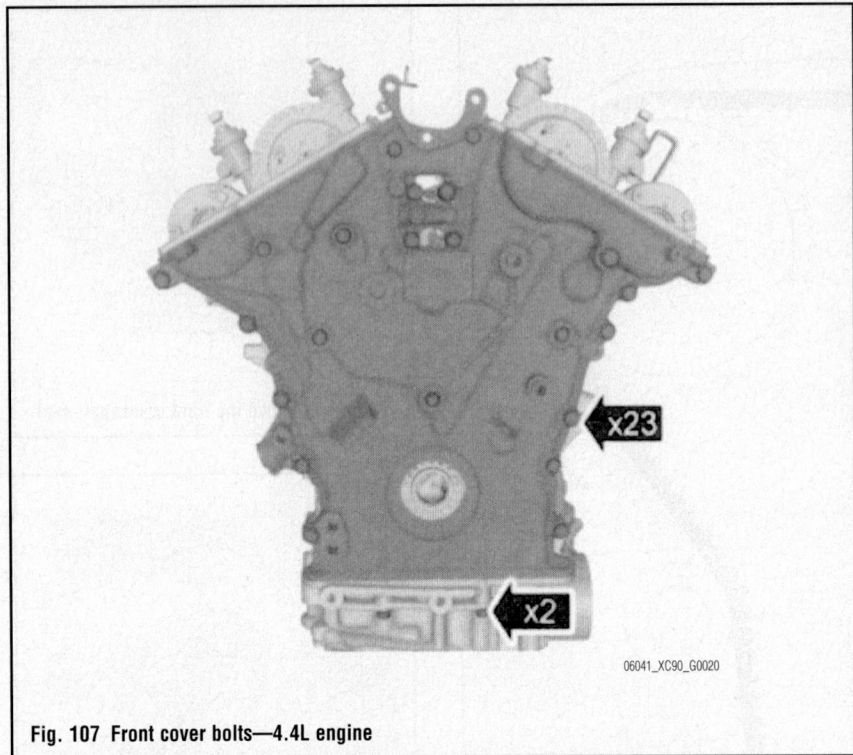

**Fig. 107 Front cover bolts—4.4L engine**

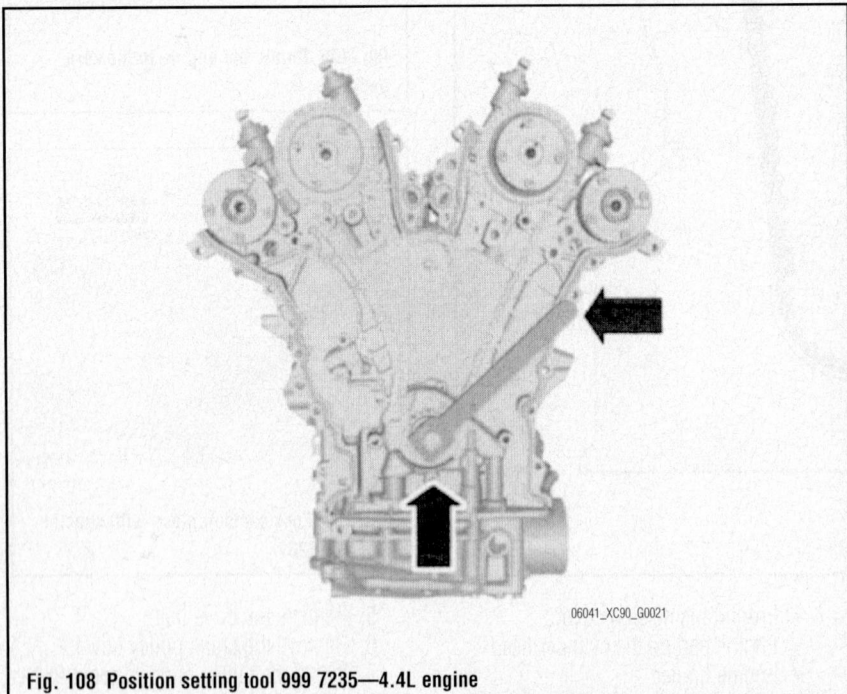

**Fig. 108 Position setting tool 999 7235—4.4L engine**

15. Remove the bolt and clamp for the power steering hose.

16. Remove the pulley from the water pump.

17. Remove the engine bracket and idler pulley.

18. Remove the 25 bolts and the front cover. Discard the gaskets.

19. Use position setting tool 999 7235 to align the position on the crankshaft with the guide on the engine block.

20. Install position setting tool: 999 7236

21. Remove the position setting tool: 999 7235

22. Lock the chain tensioner using a drift.

23. Remove the 2 M6 bolts for the chain tensioner.

24. Remove the chain tensioner.

25. Remove the 2 rails.

26. Remove the timing chain.

27. Remove worn chain sprocket.

### To install:

28. Adjust the color marking on the timing chain to the marking on the gears.

29. Apply liquid gasket 307 57050 as illustrated. The bead diameter must be greater than 0.039 inch (1mm).

30. Fold the front panel in at the lower edge.

31. Install the 23 M8 bolts for the front side and tighten to 18 ft. lbs. (24 Nm).

32. Install the 2 M8 bolts for the oil sump and tighten to 18 ft. lbs. (24 Nm).

33. Install the engine bracket and the idler pulley.

34. Install the pulley for the water pump.

35. Install the power steering pump.

36. Install the hose clamp and the bolt.

37. Install the front bolt for the alternator and tighten to 18 ft. lbs. (24 Nm).

38. Install lifting beam tool 999 5716, converted with the addition of 999 7070 and 999 5460.

39. Remove tool 999 5550.

40. Install the right-hand engine pad/brace

41. Clean the surface for the sealing ring on the crankshaft damper with cloth 951 10

42. Install tool 999 7196, drift: 999 7198 and press in the crankshaft damper to its limit. Position using the old center bolt.

43. Remove the center bolt.

44. Insert a new bolt for the crankshaft damper and tighten 133 ft. lbs. (180 Nm).

4. Remove the accessory drive belt, refer to Accessory Drive Belts removal and installation.

5. Remove the right front wheel.

6. Remove the crankshaft damper protective cover nuts.

7. Push the protection towards the crankshaft damper.

8. Remove the center bolt using special tool 999 7196.

9. Install 999 7198 in the center hole of the crankshaft.

10. Pull off the crankshaft damper.

11. Remove the alternator.

12. Remove right engine pad.

13. Remove the right-hand engine pad/bracket.

14. Disconnect the power steering pump. Place the power steering pump to one side.

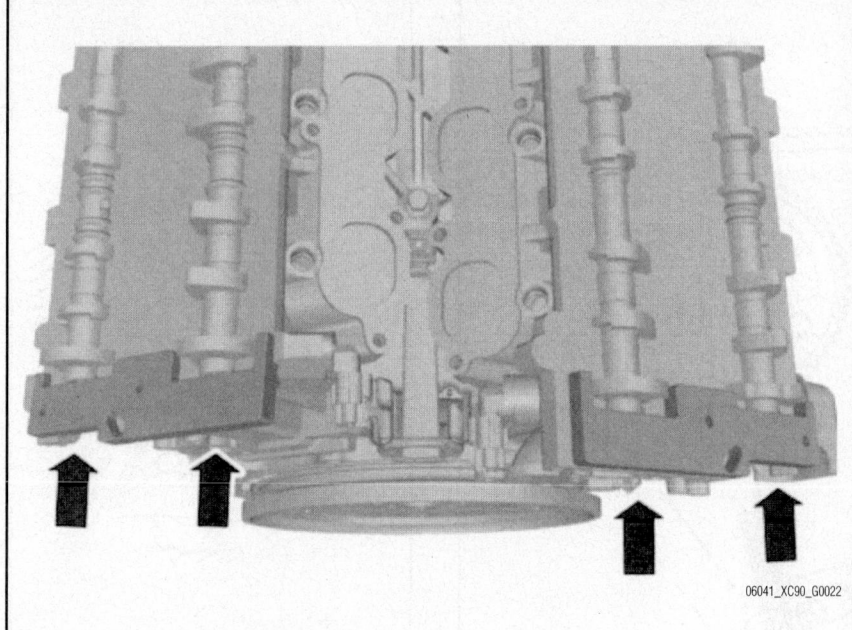

**Fig. 109 Position setting tool 999 7235—4.4L engine**

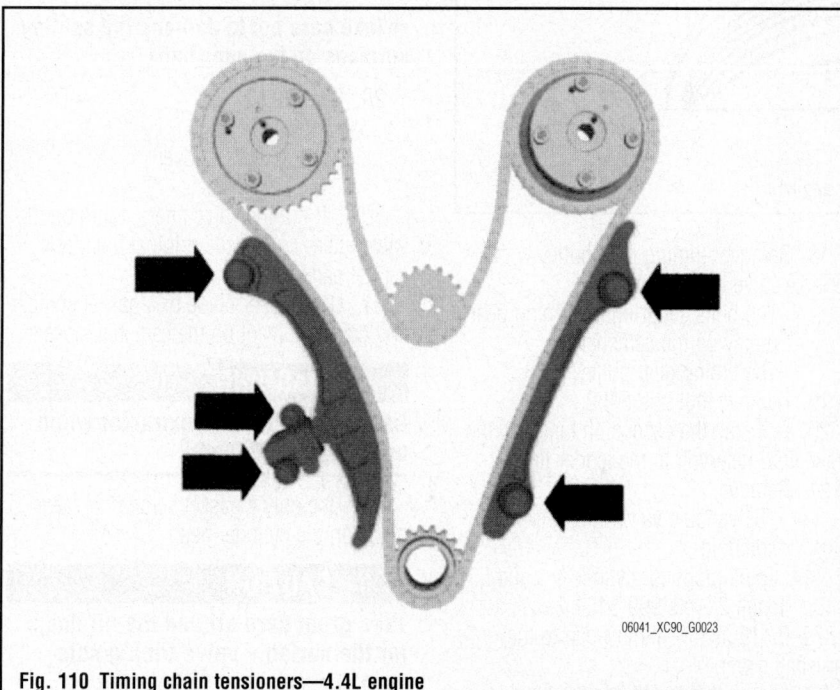

**Fig. 110 Timing chain tensioners—4.4L engine**

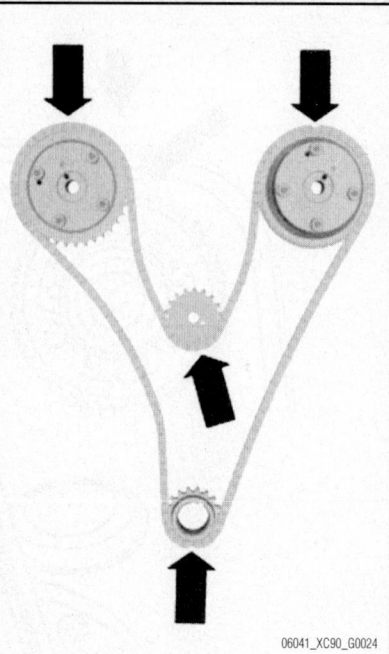

**Fig. 111 Timing chain alignment marks— 4.4L engine**

45. Remove the tool 999 7196.
46. Install the protective cover.
47. Install the camshaft covers.
48. Install the accessory drive belt tensioner.
49. Install the right front wheel and tighten lug nuts to 102 ft. lbs. (138 Nm).
50. Start the engine and run the engine to normal operating temperature.
51. Check for leakage. Fill fluids as necessary

## VALVE LASH

*ADJUSTMENT*

### 2.4L, 2.5L, 3.0L and 3.2L Engines

*See Figures 112 through 119.*

1. Remove the cable from the battery negative terminal.
2. Remove:
   • The cross stay between the suspension turrets

- The ground strip from the cylinder head
- The upper engine stabilizer brace
- The cover in the cylinder head at the rear of the exhaust camshaft
- The crankcase ventilation hose from the top of the camshaft cover
- The radiator breather tube from the expansion tank. Install lock grip pliers.

3. Lift up the brake fluid reservoir.
4. Disconnect the ABS sensor connector.
5. Place the brake fluid reservoir over the engine.

### ❊❊ WARNING

**Ensure that no fluid is spilled on the engine. It is extremely flammable!**

6. Disconnect the connector for the level sensor in the expansion tank.
7. Lift up and place the expansion tank on top of the engine.
8. Remove:
   • The accessory drive belt
   • The front timing cover.
9. Position the upper timing cover.
10. Turn the crankshaft clockwise until the markings on the crankshaft belt pulley and the timing belt pulley are aligned with the markings on the oil pump and the upper timing cover.
11. Remove the upper timing cover.

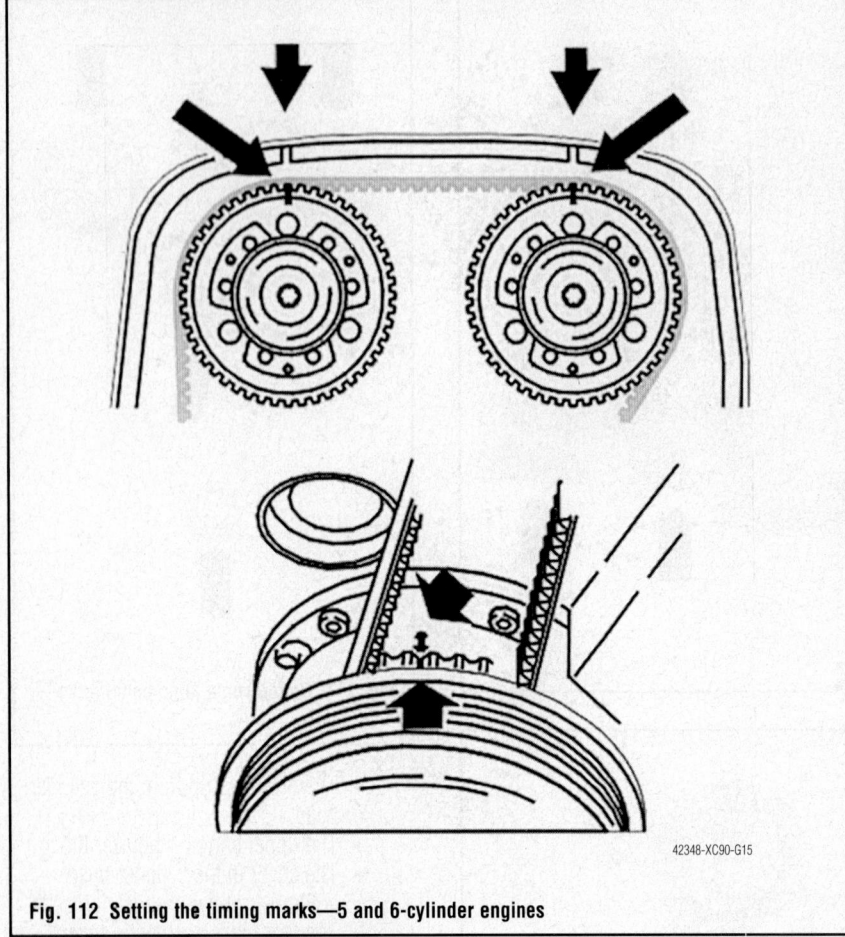

**Fig. 112 Setting the timing marks—5 and 6-cylinder engines**

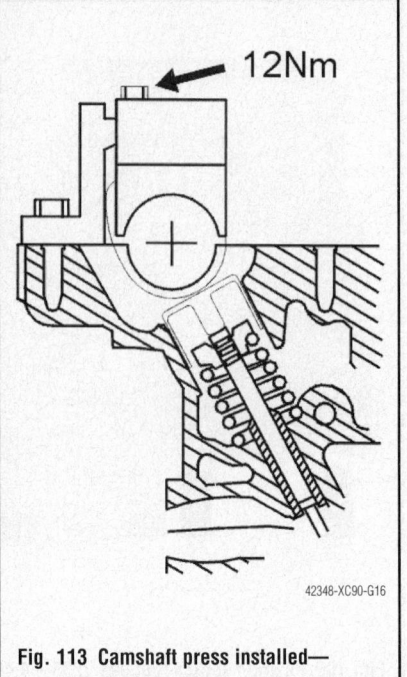

**Fig. 113 Camshaft press installed— 5 and 6-cylinder engines**

---

**※※ CAUTION**

**Crankshaft and camshafts must not be turned more than is stated in the method description. If the shafts are turned in any other way the valves may be damaged.**

12. Slacken off the center bolt for the belt tensioner slightly.

13. Hold the center bolt still. Turn the tensioner eccentric clockwise to 10 o'clock using a 6mm Allen key.

14. Remove the timing belt from the camshaft pulleys.

15. Install camshaft adjustment tool 999 5452 at the rear of the camshafts.

16. Check that the bolts securing the adjustment tool to the camshafts and the bolt holding the tool together are well tightened

17. Remove: (timing gear pulley with variable valve timing unit)
- The plug at the front of the variable valve timing unit
- The center bolt in the variable valve timing unit. Carefully pull out the variable valve timing unit with the timing gear pulley.

18. Remove: (timing gear pulley without variable valve timing unit)
- The bolts securing the timing gear pulley on the camshaft
- The timing gear pulley.

19. Remove tool 999 5452.

20. Reinstall the expansion tank and the brake fluid reservoir at the fender liner.

21. Remove:
- The variable valve timing (VVT) solenoid
- spark plugs for cylinders 1 and 5.

22. Install 2 tools 999 5454. Leave a 0.079–0.118 inch (2–3mm) gap to the camshaft cover.

23. Ensure that the bolt in the spark plug well is fully tightened.

24. Remove all the bolts securing the camshaft cover to the cylinder head.

25. Use pliers 999 5670 to lift the cover from the cylinder head.

26. Install the pliers at the stop lugs. Start with cylinder 1 and work alternately backward.

27. Slacken off the wing nuts approximately 2 turns. Repeat the procedure with the pliers.

28. Carefully press the camshaft seals free.

➡Take care not to damage the sealing surfaces on the camshafts.

29. Remove:
- Tool 999 5454
- The camshaft cover
- The camshafts.

30. Lift out the valve lifters. Mark up the valve lifters so that the original positions can be established.

31. Use a razor blade or a gasket scraper and gasket solvent on the camshaft cover.

**※※ WARNING**

**Use a fume hood or extractor when using gasket solvent!**

32. Use only a gasket scraper or razor blade on the cylinder head.

**※※ CAUTION**

**Take great care around the oil ducts for the variable valve timing sole-noid. This applies to both the camshaft cover and the cylinder head. The solenoid is extremely sen-sitive to contaminants.**

33. Dry and blow all surfaces clean.

34. Carefully tap the end of the valve stem to ensure that the valve is correctly located in the seat.

35. Use a plastic, aluminum or brass drift to protect the valve and the surface of the valve lifter.

36. The sound made by tapping reveals if the valve is correctly seated.

37. Install both the valve lifters for the inlet valves at cylinder 1.

38. Check the notes made earlier. Select new valve lifters if necessary.

➡**Only install two valve lifters. The valve lifters should be placed at the same cylinder!**

39. Other valve lifters are available as replacement part / replacement part kits.

40. The valve clearance on a cold engine (approximately 20°C) should be:
- Inlet valve: 0.008 plus or minus 0.0010 inch (0.20 plus or minus 0.03mm).
- Exhaust valve: 0.016 plus or minus 0.0010 inch (0.40 plus or minus 0.03mm).

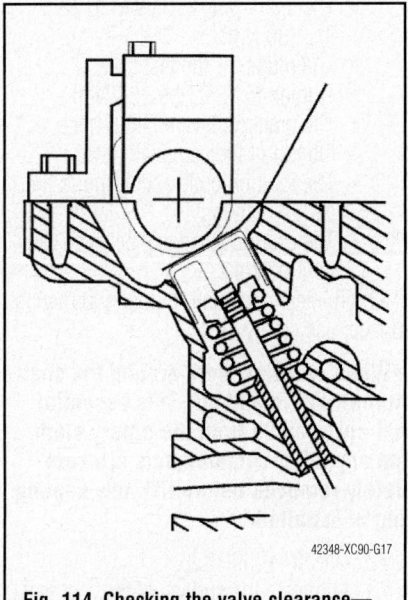

42348-XC90-G17

**Fig. 114 Checking the valve clearance— 5 and 6-cylinder engines**

➡**The tolerances are less at setting. When checking the valve clearance through the plug hole the tolerances are larger.**

41. Position the intake camshaft. Ensure that the lobes at cylinder 1 point upwards.

42. Apply a little oil to the cam lobe and the upper side of the valve lifter to facilitate later measurement.

43. Install the lower section of camshaft press 999 5765 at the inlet valves for cylinder 1.

44. Tighten the tool against the cylinder head. Tighten to 13 ft. lbs. (17 Nm).

45. Turn the camshaft until it stops against the camshaft press.

46. Install the upper section of the camshaft press.

47. Tighten the bolt which tensions the camshaft. Tighten to 106 inch lbs. (12 Nm).

➡**Measurements should only be taken on a cold engine. A suitable temperature is approximately 68°F (20°C).**

48. Using a feeler gauge, press with a finger so that the feeler gauge lies parallel to the upper side of the valve lifter.

49. Move the feeler gauge sideways when taking the reading in order to obtain as accurate a measurement as possible.

50. The valve clearance measured on cold engines (approximately 68°F (20°C) should be:
- Intake valve: 0.008 inch plus or minus 0.001 inch (0.20mm plus or minus 0.03mm).
- Exhaust valve: 0.016 inch plus or minus 0.001 inch (0.30mm plus or minus 0.03mm).

➡**The tolerances are less at setting! When checking the valve clearance through the plug hole the tolerances are larger.**

51. Differences in valve clearance for different engines/ambient temperatures:
- −0.0004 inch (− 0.01mm) at 59°F (15°C)
- +0.0004 inch (+ 0.01mm) at 77°F (25°C)
- +0.0008 inch (+ 0.02mm) at 86°F (30°C)
- +0.0012 inch (+ 0.03mm) at 95°F (35°C)
- +0.0016 inch (+ 0.04mm) at 113°F (45°C)

52. Correcting measured clearance:
a. Lift out the upper section of press tool.
b. Lift out the camshaft.
c. Adjust the play by replacing the valve lifters.
d. Other valve lifters are available as replacement part / replacement part kits.
e. Reinstall the camshaft and the upper section of the press tool. Tighten to 106 inch lbs. (12 Nm).
f. Take a new measurement.

53. When the correct valve clearance is reached, remove:
- The press tool 999 5765
- The camshaft
- The valve lifters.

54. Carefully mark the valve lifters so that exact reinstallation can be carried out. For example:
Intake side: I1, I2, I3 . . . I10.
Exhaust side: A1, A2, A3 . . . A10.

55. Repeat the procedure for measuring the valve clearance for all cylinders on both the intake and exhaust sides.

56. Lubricate the valve guide wells.

57. Install all the valve lifters.

58. Lubricate the camshaft bearing seats and the upper sides of the valve lifters.

59. Position the intake camshaft. Ensure that the groove at the rear edge of the camshaft is above an imaginary center line.

60. Position the exhaust camshaft. Ensure that the groove at the rear edge of the camshaft is below an imaginary center line.

61. Wipe the oil film off the mating surfaces on the camshaft cover and cylinder head.

62. Install new O-rings around the spark plug wells at the cylinder head.

63. Apply liquid gasket 1161 059 on the camshaft cover. Use roller 951 2767.

➡**The surface must be completely covered without any excess.**

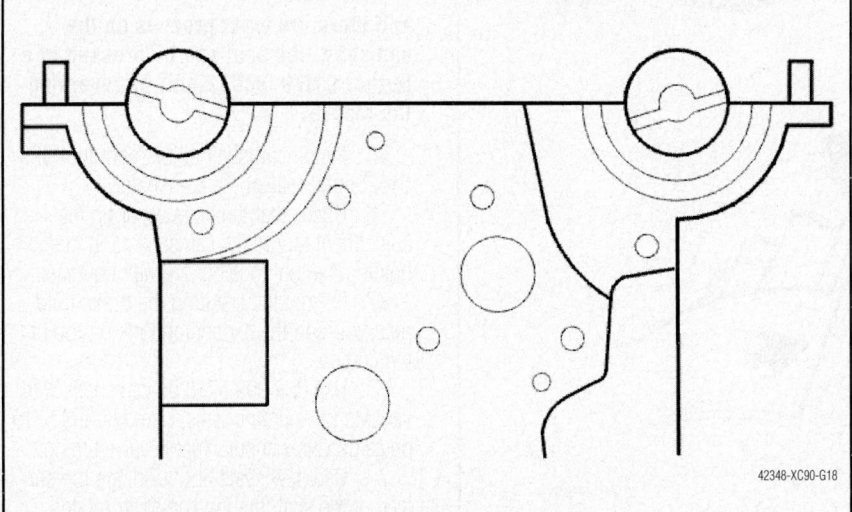

42348-XC90-G18

**Fig. 115 Camshafts properly positioned—5 and 6-cylinder engines**

Fig. 116 Front seal installation tools—5 and 6-cylinder engines

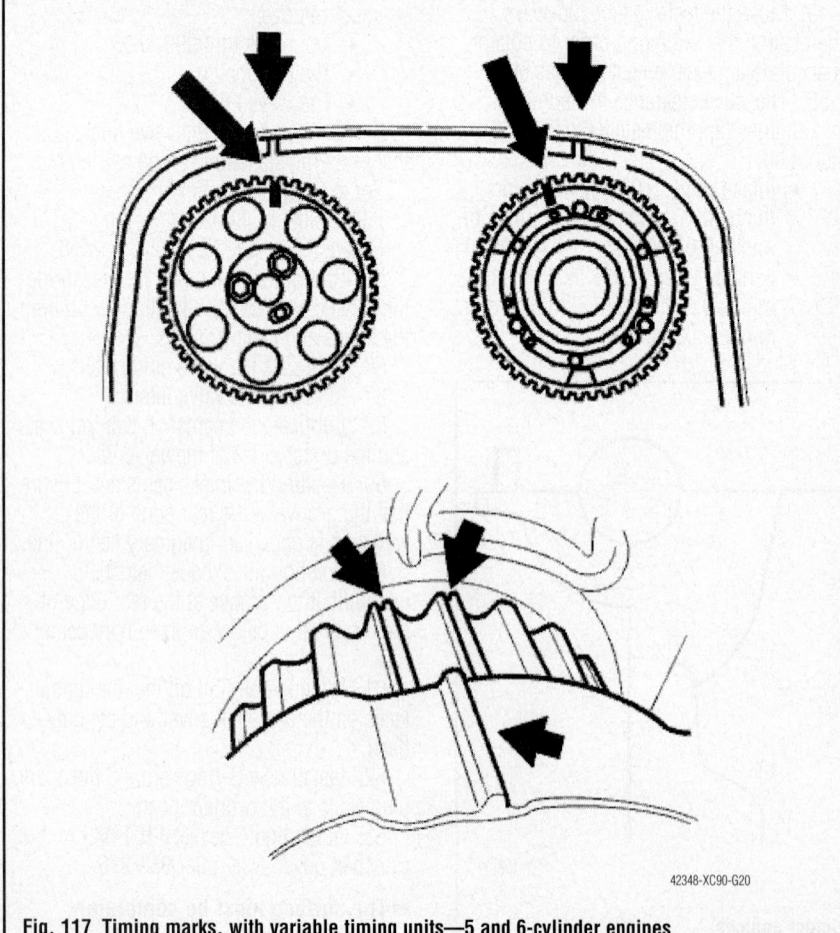

Fig. 117 Timing marks, with variable timing units—5 and 6-cylinder engines

### ✳✳ CAUTION

**Ensure that no liquid gasket gets in to the oil ducts.**

64. Lubricate the camshaft lobes, the camshaft bearing surfaces and the valve lifters.

65. Install the camshaft cover.

66. Install press tool 999 5454 (2x).

67. Tighten the camshaft cover bolts alternately, keeping it parallel to the cylinder head using the press tools.

68. Install all the bolts. Tighten the bolts from the middle and outwards.

69. Remove the press tool 999 5454

70. Install:
- The variable valve timing (VVT) solenoid. Use a new gasket
- The spark plugs. Tighten to 22 ft. lbs. (30 Nm)
- The plugs for the test holes. Tighten to 15 ft. lbs. (20 Nm)
- The crankcase ventilation hose to the top of the camshaft cover
- The ignition coils according to the earlier marking
- The ground terminals between the ignition coils.

71. To clean the shaft journal and mating surface, use emery cloth.

➡ **When cleaning work around the shaft journal, not in and out. It is essential that any residue from the emery cloth and any other contaminants are completely removed before the new sealing ring is installed.**

72. Use drift 999 5450.

73. Lubricate the surface of the seal that the camshaft rotates against.

74. Press in the seal until the drift bottoms out.

➡ **If there are wear grooves on the camshaft, the seal can be pressed in a further 0.079 inch (2mm) by reversing the sleeve.**

75. Install camshaft adjustment tool 999 5452 at the rear of the camshafts.

76. Check that the bolts securing the adjustment tool to the camshafts and the bolt holding the tool together are well tightened.

77. Lift up and position the brake fluid reservoir and the expansion tank on top of the engine.

78. Use drift 999 5718 on camshafts with variable valve timing units. Use drift 999 5719 on camshafts without variable valve timing.

79. Use new seals and lubricate the surface of the seal that the camshaft rotates against.

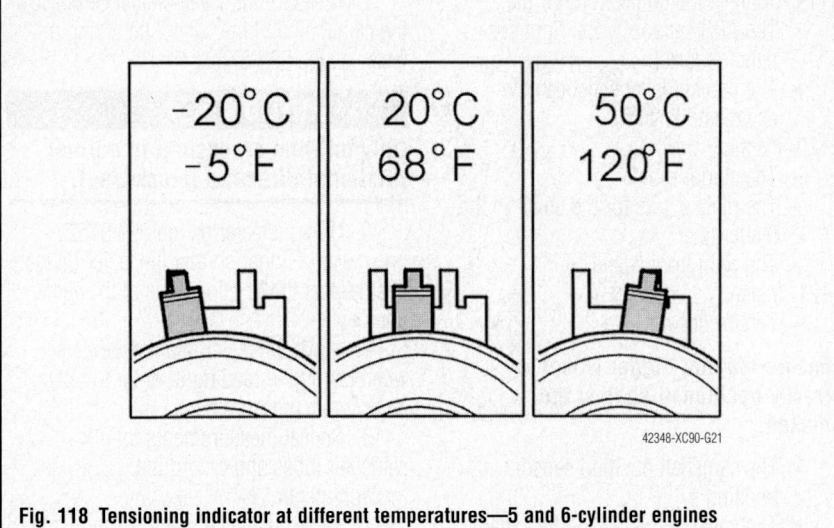

**Fig. 118 Tensioning indicator at different temperatures—5 and 6-cylinder engines**

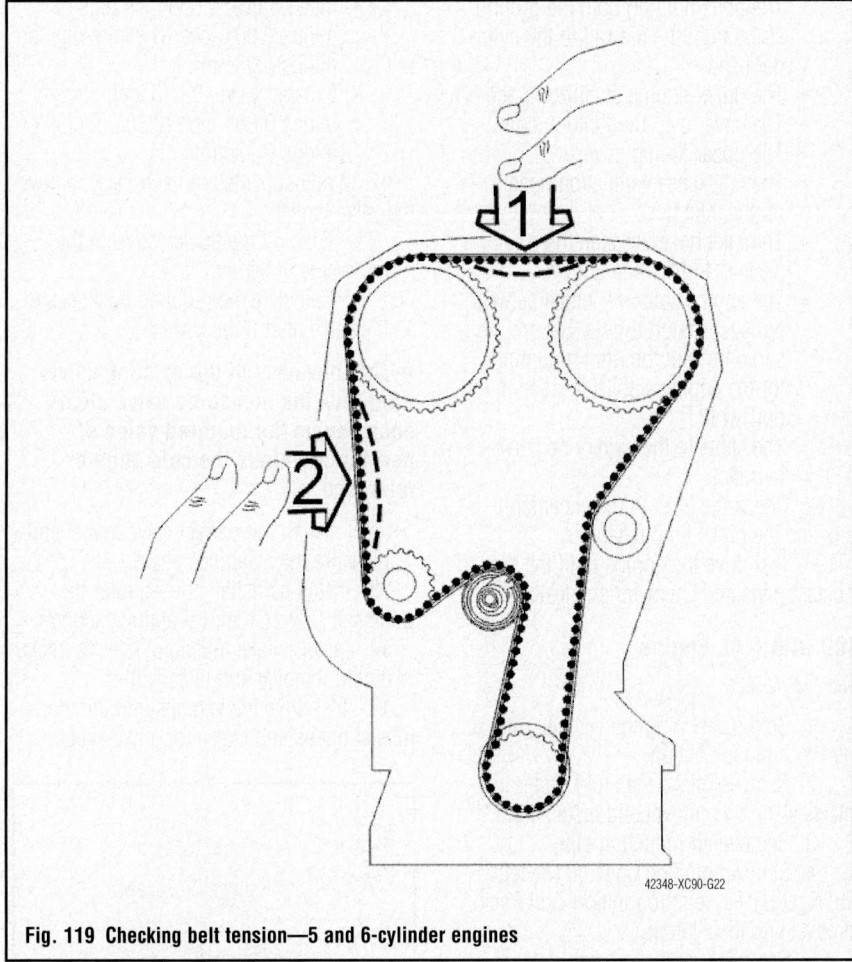

**Fig. 119 Checking belt tension—5 and 6-cylinder engines**

80. Use the variable valve timing unit/ timing belt pulley mounting bolts. Tighten the bolts until the drift bottoms out.

81. Remove the drift.

82. Remove:
• The mounting bolts for the starter motor. Pull off the starter motor. Place the starter motor to one side
• The blind cover plug and the seal washer.

83. Turn the crankshaft slightly clockwise.

84. Install the crankshaft stop 999 5451 Ensure that it bottoms out against the cylinder block.

85. Turn the crankshaft counter-clockwise until it stops against the crankshaft stop.

86. Check that the marking on the crankshaft timing gear pulley corresponds with the marking on the oil pump.

➡**The purpose of the section is to ensure that the VVT unit is correctly positioned and to reset the camshaft timing gear pulley to the correct position using the markings made at the factory. This is to ensure that the conditions are correct for any later fault-tracing.**

87. Slacken off, but do not remove the bolts which secure the timing gear pulley to the variable valve timing unit.

88. Press the variable valve timing unit/timing gear onto the camshaft.

89. Install the center bolt which secures the variable valve timing unit to the camshaft. Tighten slightly.

90. Turn the variable valve timing unit counter-clockwise as far as it will go.

91. Remove the center bolt.

92. Position the upper timing cover.

93. Turn the timing gear pulley clockwise until the bolts at the oval holes are in the limit position.

94. Continue turning clockwise until the timing gear pulley marking is 1 tooth before the marking on the upper timing cover.

**✳✳ CAUTION**

**Do not turn counter-clockwise during this procedure.**

95. Check that the timing gear pulley is still in its limit position at the oval holes.

96. Tighten the center bolt for the variable valve timing (VVT) unit. Tighten to 89 ft. lbs. (120 Nm). Check that the variable valve timing unit does not rotate when tightening.

97. Install the center bolt. Tighten to 26 ft. lbs. (35 Nm).

98. Install the timing gear pulley. Install the bolts.

99. Install two bolts without tightening. Allow the third bolt to protrude.

100. Adjust the timing gear pulleys so that the markings on the timing gear pulleys / upper timing cover correspond.

101. Tighten the center bolt on the belt tensioner. Tighten to 44 inch lbs. (5 Nm).

102. Turn the variable valve timing unit clockwise to the stop.

103. Hold it secure in the limit position.

104. Install the belt in the following order:
• Crankshaft
• The idler pulley
• Intake cam
• Exhaust cam
• Water pump
• Belt tensioner

➡Adjust the timing gear pulleys so that the bolts are not at a limit position in the oval holes. Also check that the markings correspond.

➡This adjustment is always carried out on a cold engine. A suitable temperature is approximately 68°F (20°C).

At higher temperatures (with the engine at operating temperature or a high outside temperature for example) the indicator is further to the right.

The illustration shows the position of the indicator when aligning the timing belt tensioner at different temperatures.

105. Hold the center bolt secure and turn the belt tensioner eccentric counterclockwise until the tensioner indicator passes the marked position.

➡Check that the variable valve timing unit is in its limit position.

106. Tighten the three bolts at the intake camshaft timing gear pulley. Tighten to 88 inch lbs. (10 Nm).

107. Tighten the three bolts at the exhaust camshaft timing gear pulley. Tighten to 15 ft. lbs. (20 Nm).

108. Turn the eccentric on the belt tensioner back so that the indicator reaches the marked position in the center of the window. Remember to hold the center bolt secure at the same time.

109. Hold the eccentric secure and tighten the center bolt. Tighten to 15 ft. lbs. (20 Nm).

110. Check that the indicator is in the correct position.

111. Remove camshaft adjustment tool 999 5451 and crankshaft stop 999 5452.

112. Install the plug with a new blind cover plug. Tighten to 30 ft. lbs. (40 Nm).

113. Press the timing belt to check that the indicator on the tensioner moves easily.

114. Turn the crankshaft two turns. Check that the markings on the crankshaft timing gear pulley and the camshaft timing gear pulley match up with the markings on the oil pump and upper timing cover respectively.

115. Check that the indicator on the belt tensioner is within the marked position.

116. Remove the upper timing cover.

117. Install:
- front timing cover
- The accessory drive belt
- The expansion tank
- The servo oil reservoir
- The bleed hose for the expansion tank

118. Close the clamp and check that the hoses lie correctly.

119. Connect the connectors for the:
- The ABS sensor by the right suspension turret
- The coolant level sensor in the expansion tank

120. Install:
- The starter motor
- The plastic nuts for the cover in the fender liner
- The right front wheel

121. Install:
- The trigger wheel.

➡Ensure that the trigger wheel is correctly positioned against the camshaft.

- The camshaft position sensor housing
- The cover at the rear of the exhaust camshaft
- The bolts holding both the ground strips from the firewall to the cylinder head
- The upper engine stabilizer brace
- The cover over the ignition coils
- The upper timing cover
- The crankcase ventilation hose on the inlet hose
- The inlet hose between the air cleaner and throttle body
- The engine stabilizer brace between the suspension turrets. Secure the servo hose at the right mounting for the engine stabilizer brace in the bodywork
- The cable to the battery negative terminal.

122. Check the level in the expansion tank and the brake fluid reservoir.

123. Test drive the engine until the thermostat opens and check for any leakage.

### S80 with 4.4L Engine

*See Figure 120.*

1. Before servicing the vehicle, refer to the Precautions Section.

2. Remove intake manifold. Refer to Intake Manifold removal and installation.

3. Remove air induction pipe.

4. Remove ignition coils on the left and right banks, refer to Ignition Coil Pack removal and installation.

5. Remove camshaft covers, refer to Camshaft Covers removal and installation.

➡Measurements should only be taken on a cold engine. A suitable temperature is approximately 68°F (20°C).

6. Turn the crankshaft until the cam lobe is 180° to the valve lifter.

7. Measure the valve clearance between the camshaft and the valve lifter using a feeler gauge 999 5752.

### ✳✳ CAUTION

**Only turn the crankshaft in normal rotational direction (clockwise).**

8. Using a feeler gauge 999 5752, press with a finger so that the feeler gauge lies parallel to the upper side of the valve lifter.

9. Move the feeler gauge sideways when taking the reading in order to obtain an accurate measurement.

10. Repeat measurements for all camshaft lobes and record the measurements.

11. Valve clearance readings on a cold engine, approximately 68°F (20°C):
- Intake valve: 0.008 inch plus or minus 0.001 inch (0.20mm plus or minus 0.03mm).
- Exhaust valve: 0.016 inch plus or minus 0.001 inch (0.30mm plus or minus 0.03mm).

12. If adjustments are necessary, remove the valve tappet:

a. Read off the thickness from the underside of tappet.

b. Calculate the required thickness of the valve lifter to be used.

➡The thickness of the existing valve lifter plus the measured valve clearance equals the required value of new tappet minus the cold engine tolerance.

c. Install the correct valve tappet and measure the clearance again.

13. Install camshaft covers, refer to Camshaft Covers removal and installation.

14. Install intake manifold, refer to Intake Manifold removal and installation.

15. Test drive the vehicle until the thermostat opens and check for any leakage.

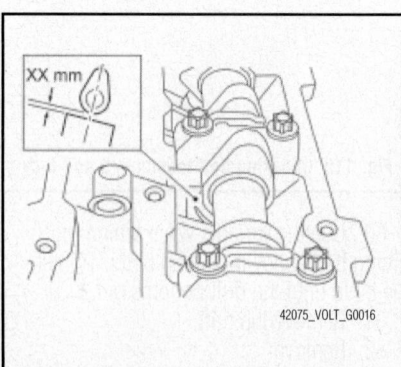

42075_VOLT_G0016

**Fig. 120 Using a feeler gauge to measure valve clearance**

# ENGINE PERFORMANCE & EMISSION CONTROL

## COMPONENT LOCATIONS

*See Figures 121 through 123.*

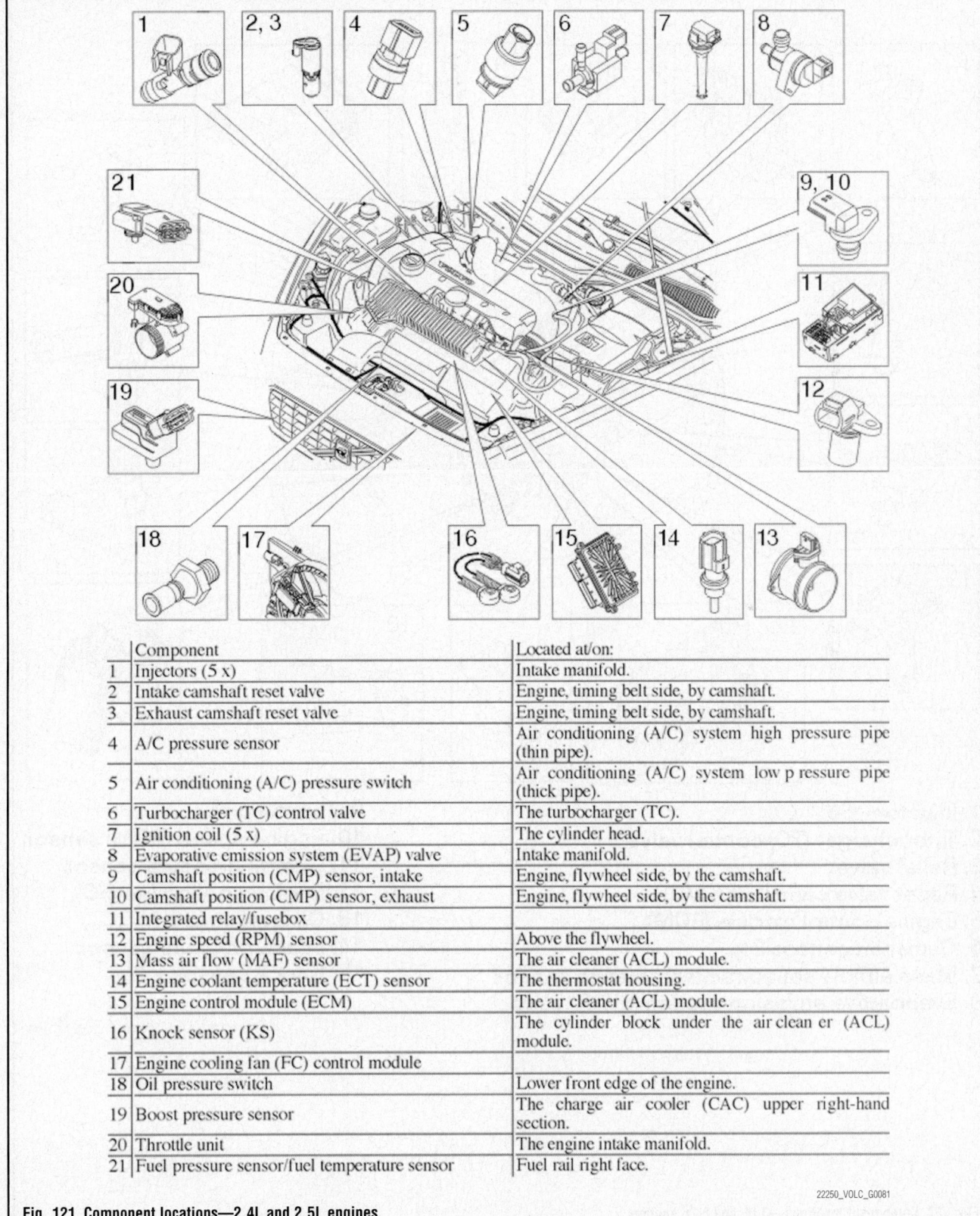

| | Component | Located at/on: |
|---|---|---|
| 1 | Injectors (5 x) | Intake manifold. |
| 2 | Intake camshaft reset valve | Engine, timing belt side, by camshaft. |
| 3 | Exhaust camshaft reset valve | Engine, timing belt side, by camshaft. |
| 4 | A/C pressure sensor | Air conditioning (A/C) system high pressure pipe (thin pipe). |
| 5 | Air conditioning (A/C) pressure switch | Air conditioning (A/C) system low pressure pipe (thick pipe). |
| 6 | Turbocharger (TC) control valve | The turbocharger (TC). |
| 7 | Ignition coil (5 x) | The cylinder head. |
| 8 | Evaporative emission system (EVAP) valve | Intake manifold. |
| 9 | Camshaft position (CMP) sensor, intake | Engine, flywheel side, by the camshaft. |
| 10 | Camshaft position (CMP) sensor, exhaust | Engine, flywheel side, by the camshaft. |
| 11 | Integrated relay/fusebox | |
| 12 | Engine speed (RPM) sensor | Above the flywheel. |
| 13 | Mass air flow (MAF) sensor | The air cleaner (ACL) module. |
| 14 | Engine coolant temperature (ECT) sensor | The thermostat housing. |
| 15 | Engine control module (ECM) | The air cleaner (ACL) module. |
| 16 | Knock sensor (KS) | The cylinder block under the air cleaner (ACL) module. |
| 17 | Engine cooling fan (FC) control module | |
| 18 | Oil pressure switch | Lower front edge of the engine. |
| 19 | Boost pressure sensor | The charge air cooler (CAC) upper right-hand section. |
| 20 | Throttle unit | The engine intake manifold. |
| 21 | Fuel pressure sensor/fuel temperature sensor | Fuel rail right face. |

22250_VOLC_G0081

**Fig. 121 Component locations—2.4L and 2.5L engines**

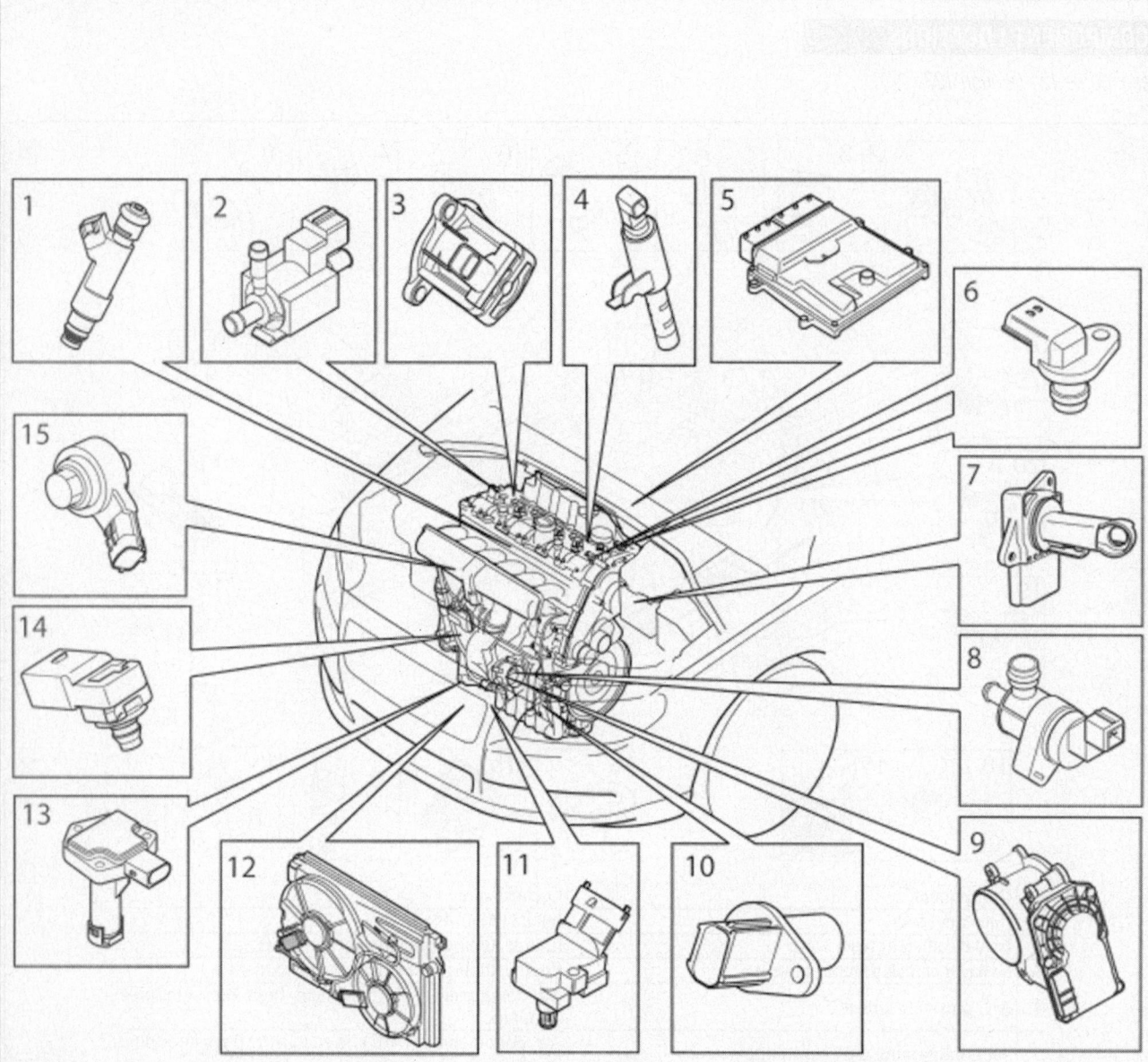

1. Injectors x 6
2. Turbocharger (TC) control valve
3. Relief valve.
4. Reset valve camshaft (CVVT)
5. Engine control module (ECM).
6. Camshaft sensor, 2 x.
7. Mass airflow sensor/air temperature sensor.
8. Evaporative emission system (EVAP) valve.

9. Throttle unit.
10. Engine speed (RPM) sensor.
11. Charge pressure sensor.
12. Engine cooling fan (FC).
13. Oil level sensor.
14. Intake pressure sensor.
15. Knock sensor, x 2

22250_VOLC_G0082

**Fig. 122 Component locations—3.0L and 3.2L engines**

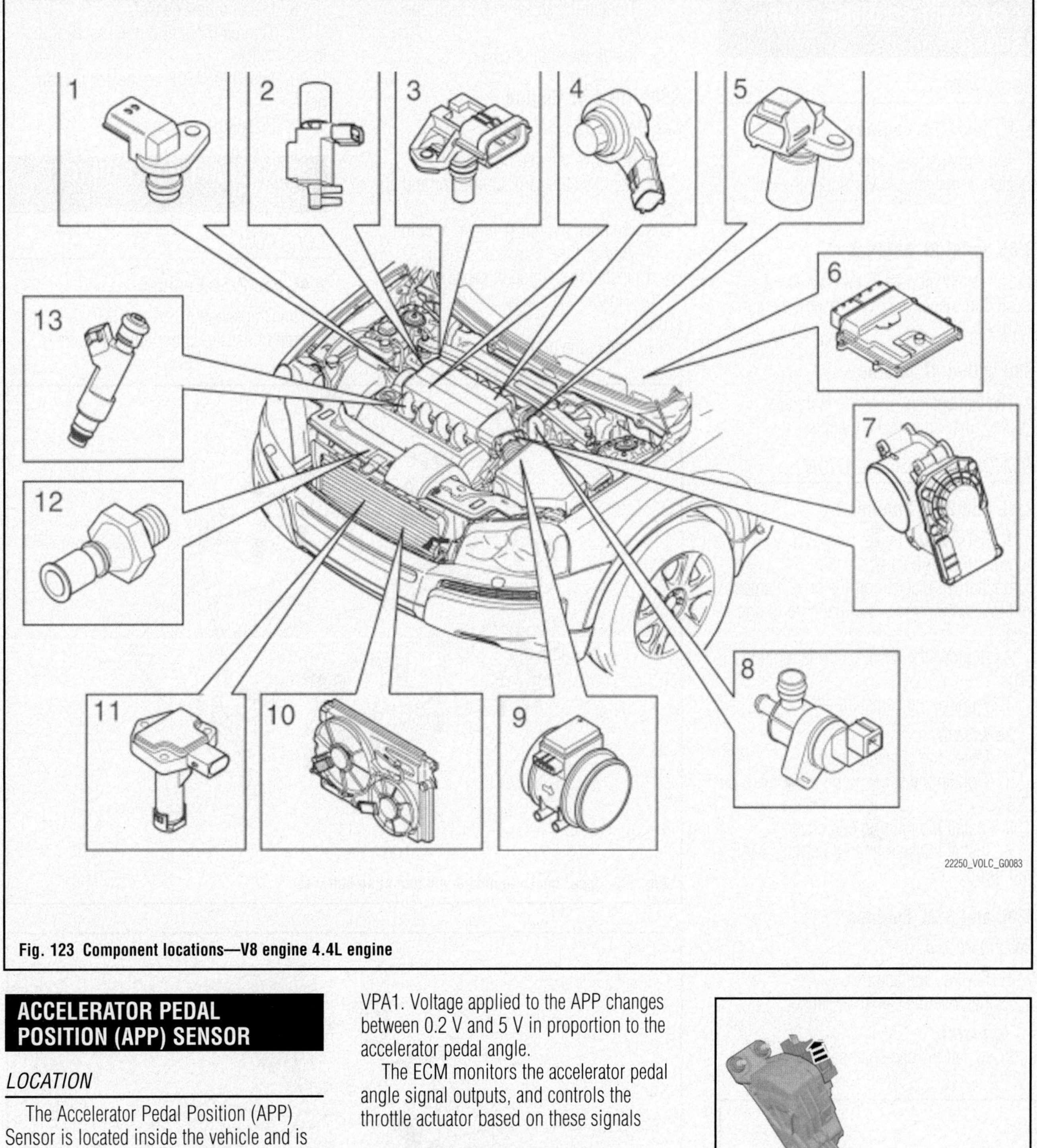

**Fig. 123 Component locations—V8 engine 4.4L engine**

22250_VOLC_G0083

## ACCELERATOR PEDAL POSITION (APP) SENSOR

### LOCATION

The Accelerator Pedal Position (APP) Sensor is located inside the vehicle and is integral to the accelerator pedal.

### OPERATION

A vehicle that is equipped with an APP does not have a throttle cable. The APP sensor is mounted on the accelerator pedal bracket. The APP sensor has 2 sensor elements/signal outputs: VPA1 and VPA2. VPA1 is used to detect the actual accelerator pedal angle (used for engine control) and VPA2 is used to detect malfunctions in VPA1. Voltage applied to the APP changes between 0.2 V and 5 V in proportion to the accelerator pedal angle.

The ECM monitors the accelerator pedal angle signal outputs, and controls the throttle actuator based on these signals

### REMOVAL & INSTALLATION

*See Figure 124.*

1. Remove the accelerator pedal position sensor connector.
2. Remove the 3 mounting nuts.
3. Remove the APP sensor.

**To install:**

4. Install the APP sensor and tighten the mounting nuts to 7 ft. lbs. (10 Nm).
5. Reconnect the APP sensor connector.

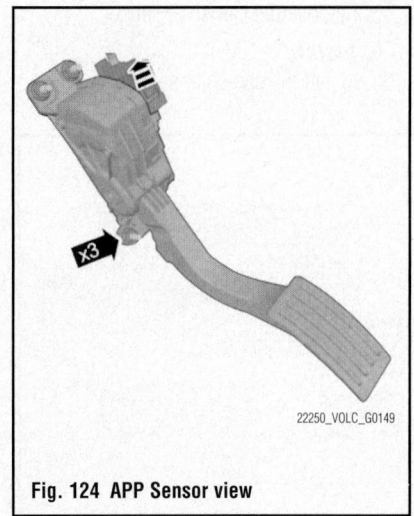

22250_VOLC_G0149

**Fig. 124 APP Sensor view**

## CAMSHAFT POSITION (CMP) SENSOR

### LOCATION

#### 2.4L and 2.5L Engines

The camshaft position sensor is located in the valve cover under the coil cover.

#### 3.0L and 3.2L Engines

The camshaft position sensor is located in the valve cover under the engine cover.

#### S80 with 4.4L Engine

The camshaft position sensors are located under the intake manifold.

### REMOVAL & INSTALLATION

#### 2.4L and 2.5L Engines

1. Remove the intake manifold between the MAF sensor and the Turbocharger. Slacken off the hose clamps and the screws. Place the manifold to one side.
2. Remove the cover over the ignition coils.
3. Remove the camshaft sensor.

**To install:**

4. Install the camshaft sensor.
5. Tighten the screw to 66 inch. lbs. (7.5 Nm)
6. Install the ignition coil cover.
7. Install intake manifold between MAF and turbo.

#### 3.0L and 3.2L Engines

See Figure 125.

1. Remove the engine cover.
2. Remove the camshaft sensor.

**To install:**

3. Install the camshaft sensor.

4. Tighten the screw to 66 inch. lbs. (7.5 Nm)
5. Install the engine cover.

#### S80 with 4.4L Engine

See Figures 126 and 127.

1. Remove the engine covers.
2. Remove the MAF connector and air hose.
3. Remove the upper intake manifold.
4. Remove the camshaft sensor.
5. Install the camshaft sensor.
6. Tighten the sensor to 7 ft. lbs. (10 Nm)
7. Install the upper intake manifold.

8. Install a new gasket.
9. Tighten mounting bolts to 18 ft. lbs.(24 Nm)
10. Install the MAF connector and air hose.
11. Install the engine covers.

## CRANKSHAFT POSITION (CKP) SENSOR

### LOCATION

#### 2.4L and 2.5L Engines

The Crankshaft Position (CKP) Sensor is located under the upper intake manifold.

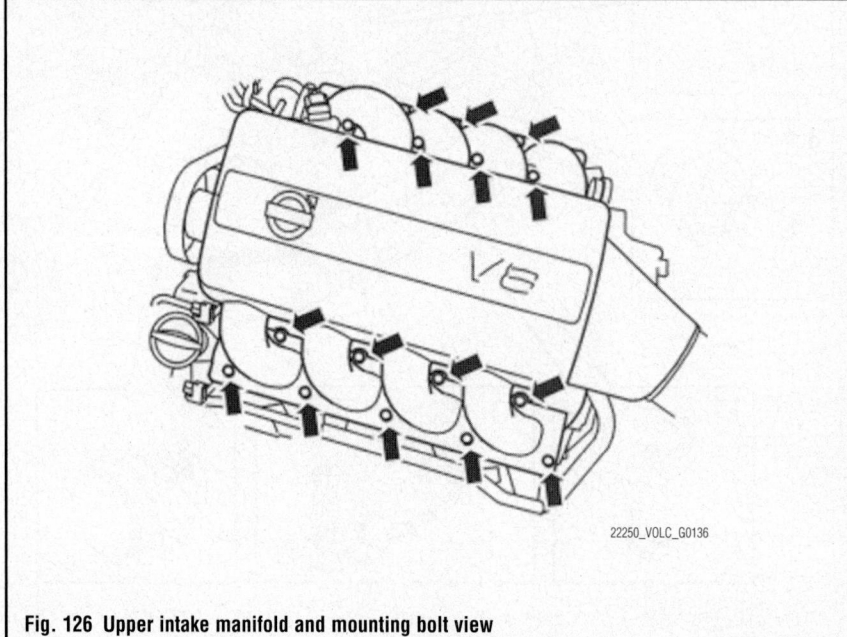

**Fig. 126 Upper intake manifold and mounting bolt view**

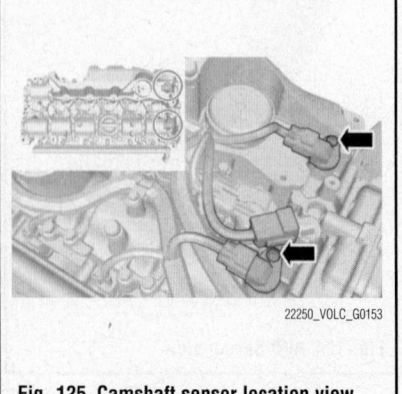

**Fig. 125 Camshaft sensor location view**

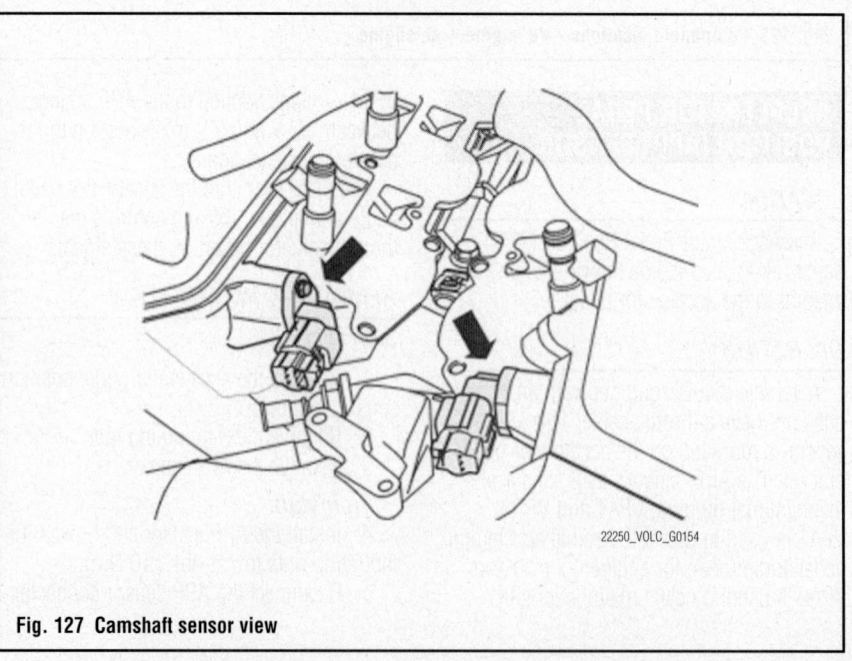

**Fig. 127 Camshaft sensor view**

### 3.0L and 3.2L Engines

The Crankshaft Position (CKP) Sensor is located at the front of the engine in the engine block.

### S80 with 4.4L Engine

The Crankshaft Position (CKP) Sensor is located at the back of the engine facing the firewall.

### OPERATION

The signal sent by the engine speed sensor to the engine control module (ECM) provides information on the number of teeth in the flywheel/number of holes in the carrier plate that has passed the engine speed sensor. The control module uses the signal to calculate engine speed. The diagnostic code is generated if via the engine speed sensor signal the control module detects an incorrect number of teeth on the flywheel/holes in the carrier plate for a certain number of crankshaft revolutions.

### REMOVAL & INSTALLATION

### 2.4L and 2.5L Engines

*See Figure 128.*

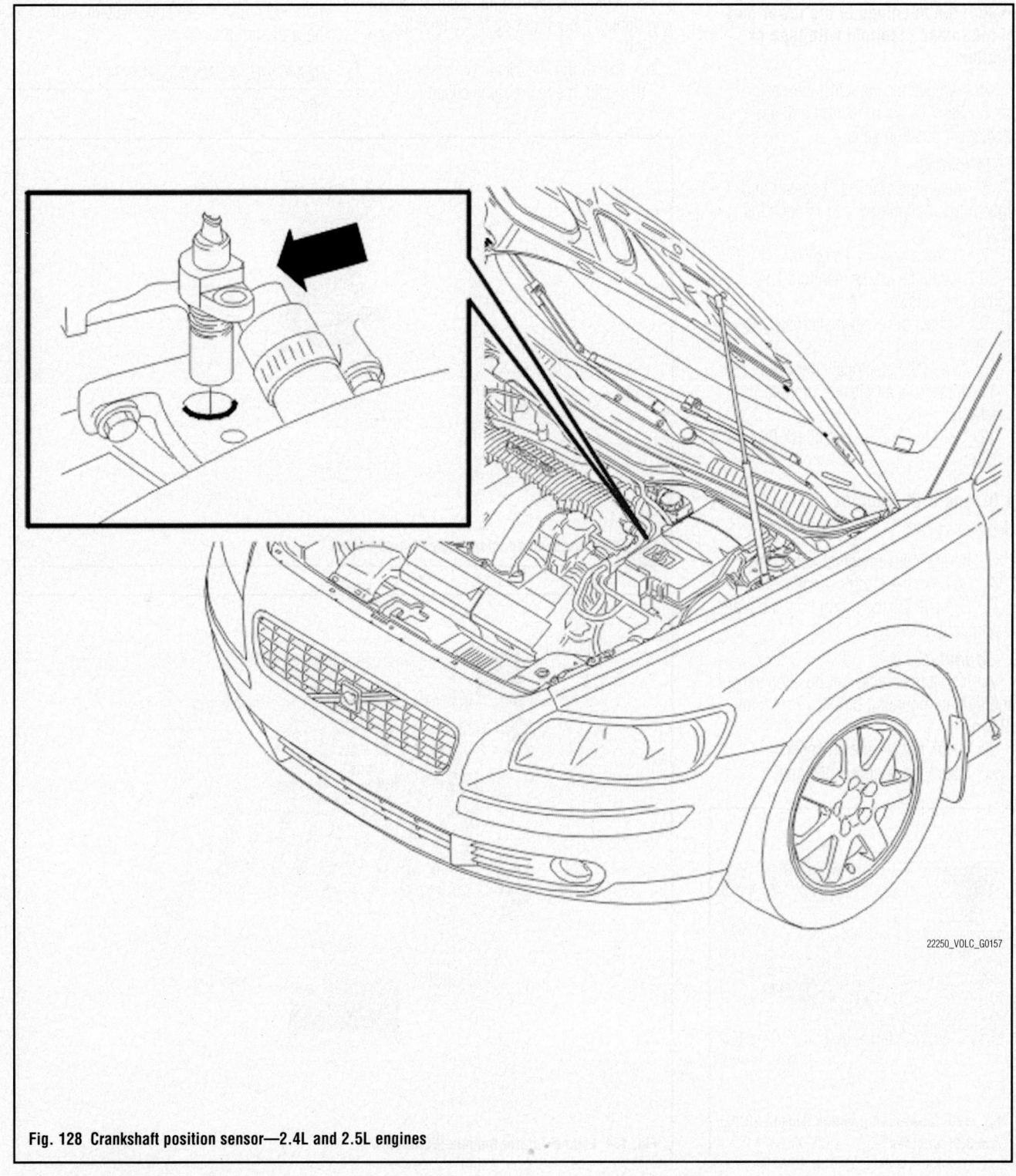

22250_VOLC_G0157

**Fig. 128 Crankshaft position sensor—2.4L and 2.5L engines**

1. Remove the air cleaner module assembly.

2. Remove the upper intake manifold.

3. Remove / slacken off the 6 screws for the upper intake manifold, the 2 screws closest to the front cover plate and the 2 screws against the firewall. Bend up the intake manifold and hold it to one side.

➡ **Seal the openings in the lower part of the intake manifold with tape or similar.**

4. Remove the electrical connector.

5. Remove the mounting bolt and crankshaft position sensor.

### To install:

6. Install the crankshaft sensor and tighten the mounting nut to 77 inch lbs. (2 Nm).

7. Install a new intake gasket.

8. Install the intake manifold and tighten the screws.

9. Tighten the screws alternately to 7 ft. lbs. (10 Nm).

10. Install the electrical connector.

11. Install the air cleaner module assembly.

12. Start the engine and check the function.

### 3.0L and 3.2L Engines

*See Figure 129.*

1. Remove the engine under cover.

2. Remove the electrical connector.

3. Remove the mounting bolt and crankshaft position sensor.

### To install:

4. Install the crankshaft position sensor and tighten mounting bolt to 77 inch lbs. (2 Nm).

5. Install the electrical connector.

6. Install the engine under cover.

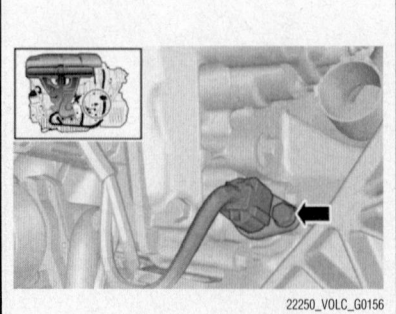

22250_VOLC_G0156

**Fig. 129 Crankshaft position sensor—3.0L and 3.2L engines**

### S80 with 4.4L Engine

*See Figure 130.*

1. Remove the battery.

2. Remove the air cleaner housing and holding bracket.

3. Remove the rear engine cover.

4. Remove the electrical connector.

5. Remove the crankshaft position sensor.

### To install:

6. Install the crankshaft position sensor and tighten mounting bolt to 77 inch lbs. (2 Nm).

7. Install the electrical connector.

8. Install the rear engine cover.

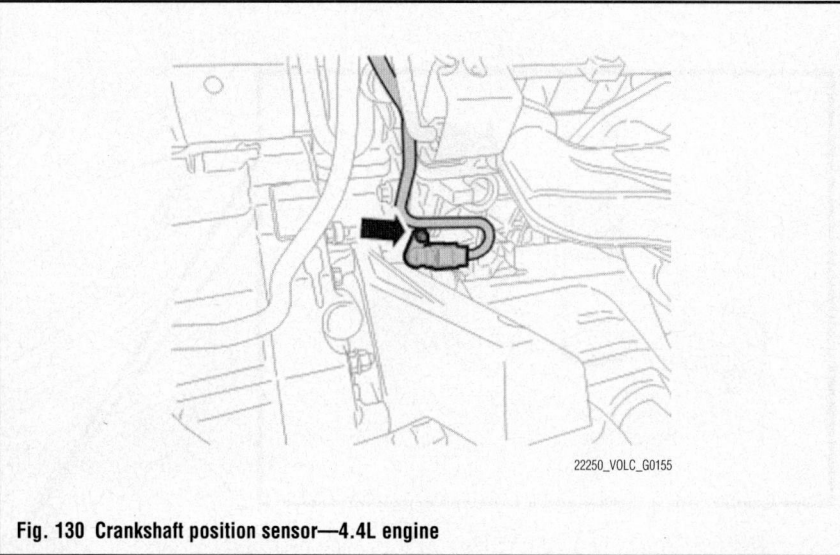

22250_VOLC_G0155

**Fig. 130 Crankshaft position sensor—4.4L engine**

9. Install the air cleaner housing and tighten the holding bracket to 7 ft. lbs. (10 Nm).

10. Install the battery.

## ELECTRONIC CONTROL MODULE (ECM)

### LOCATION

The Engine Control Module (ECM) is mounted in the intake cowl system under the wiper arms.

### REMOVAL & INSTALLATION

*See Figure 131.*

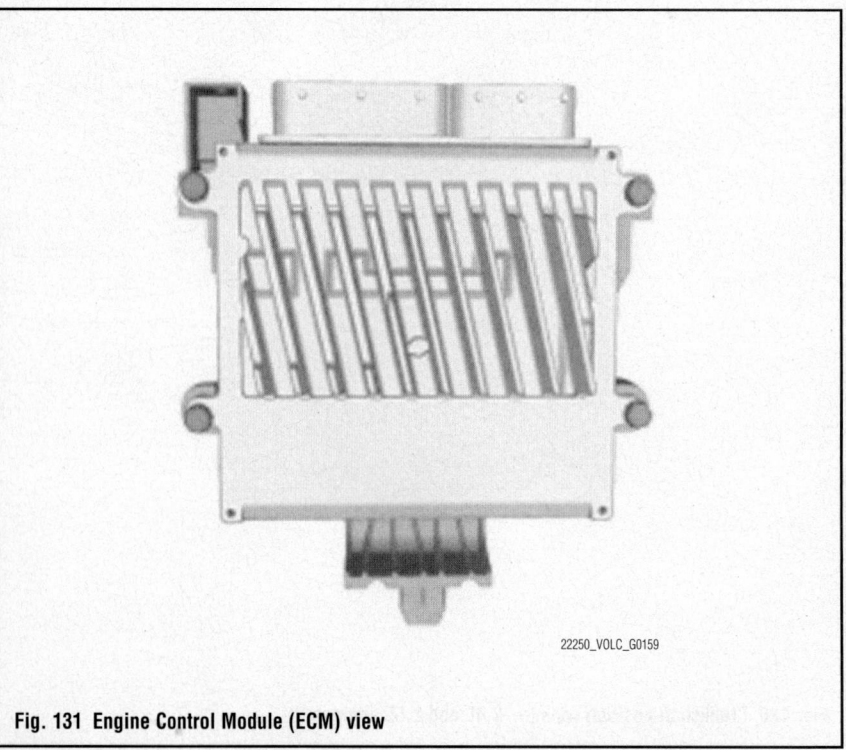

22250_VOLC_G0159

**Fig. 131 Engine Control Module (ECM) view**

1. Remove wiper arm assembly.
2. Remove the 4 retaining clips. To remove press the center pin down onto the clips.
3. Turn the 2 catches (1) anti-clockwise 1/4 turn
4. Lift out the cowl.
5. Remove the old protective cover from under the cowl using a screwdriver or similar.
6. Remove the remains from the protective cover's plastic welds using a small chisel.
7. Remove the ECM.

**To install:**
8. Install the ECM.
9. Install the retaining screws and tighten to 7 ft. lbs. (10 Nm).
10. Apply butyl tape to the protective cover. Press the tape out slightly before securing to the vehicle
11. Install the protective cover by sliding it towards and under the windshield (1). Press the front edge of the protective cover firmly so that the butyl tape adheres to the plate (2).
12. Install the cowl in the correct position. Ensure that the rubber seals against the fenders and the windshield are correctly located. Check that the guide clips underneath the cowl are in the correct position in relation to the windshield.
13. Turn the 2 catches clockwise 1/4 turn. Check that the marking on the relevant catch is turned toward the windshield.
14. Use a thin plastic film (transparent). Check that the cowl is correctly positioned against the windshield
15. Install the wiper arm assembly.

## ENGINE COOLANT TEMPERATURE (ECT) SENSOR

### LOCATION

#### 2.4L and 2.5L Engines

The ECT is located at the thermostat housing.

#### 3.0L and 3.2L Engines

The ECT is located behind the power steering reservoir in the coolant housing.

#### S80 with 4.4L Engine

The ECT is located at the rear of the engine above the transaxle, in the coolant housing.

### REMOVAL & INSTALLATION

#### 2.4L and 2.5L Engines
*See Figure 132.*

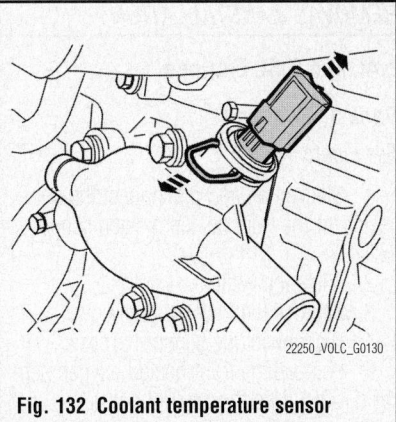

**Fig. 132 Coolant temperature sensor**

22250_VOLC_G0130

**✳✳ CAUTION**

**Never open, service or drain the radiator or cooling system when hot; serious burns can occur from the steam and hot coolant. Also, when draining engine coolant, keep in mind that cats and dogs are attracted to ethylene glycol antifreeze and could drink any that is left in an uncovered container or in puddles on the ground. This will prove fatal in sufficient quantities. Always drain coolant into a sealable container. Coolant should be reused unless it is contaminated or is several years old.**

1. Drain the engine coolant.
2. Disconnect the ground cable of battery.
3. Remove the electrical connector from the sensor.
4. Remove the clip from the coolant temperature sensor.

**To install:**
5. Install the sensor with a new O-ring and install clip.
6. Attach electrical connector to sensor.
7. Connect the ground cable of battery.
8. Refill the coolant.

#### 3.0L and 3.2L Engines
*See Figure 133.*

1. Before servicing the vehicle, refer to the Precautions Section.
2. Remove the power steering reservoir and set aside.
3. Remove the connector harness.
4. Remove the clip that retains the coolant temperature sensor.
5. Remove the coolant temperature sensor.

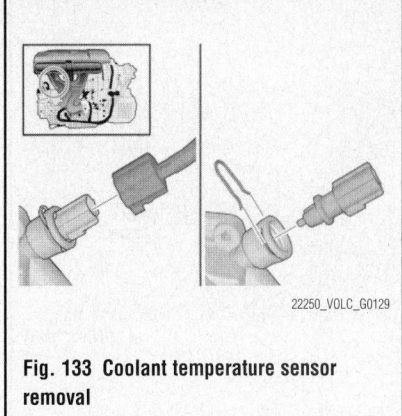

22250_VOLC_G0129

**Fig. 133 Coolant temperature sensor removal**

**To install:**
6. Install the coolant temperature sensor. Use new O-ring.
7. Install the clip that retains the coolant temperature sensor.
8. Install the connector harness.
9. Install the power steering reservoir and set aside.
10. Top off coolant.

#### S80 with 4.4L Engine
*See Figures 134 and 135.*

**✳✳ CAUTION**

**Never open, service or drain the radiator or cooling system when hot; serious burns can occur from the steam and hot coolant. Also, when draining engine coolant, keep in mind that cats and dogs are attracted to ethylene glycol antifreeze and could drink any that is left in an uncovered container or in puddles on the ground. This will prove fatal in sufficient quantities. Always drain coolant into a sealable container. Coolant should be reused unless it is contaminated or is several years old.**

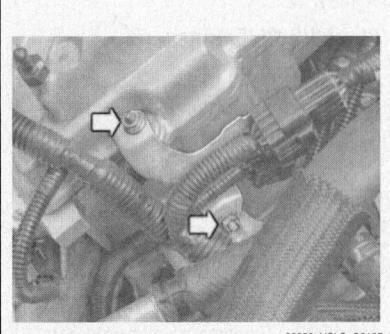

22250_VOLC_G0127

**Fig. 134 Engine harness and bracket shown**

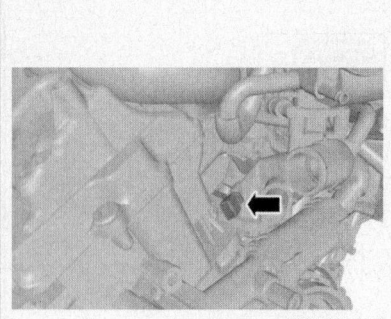

**Fig. 135 Coolant temperature sensor location view**

1. Before servicing the vehicle, refer to the Precautions Section.
2. Remove the front engine cover.
3. Remove the engine harness and bracket, set aside.
4. Remove the coolant sensor harness and sensor.

### To install:

5. Install the coolant temperature sensor and tighten to 16 ft. lbs. (22 Nm).
6. Reconnect the sensors connector.
7. Install the harness and tighten the bracket nuts to 7 ft. lbs. (10 Nm).
8. Top off cooling system.

## HEATED OXYGEN (HO2S) SENSOR

### LOCATION

#### 2.4L and 2.5L Engines

The front Heated Oxygen (HO2S) sensor is mounted in the exhaust manifold converter sub-assembly.

The rear Heated Oxygen (HO2S) sensor is mounted in the rear converter sub-assembly.

#### 3.0L and 3.2L Engines

The front Heated Oxygen (HO2S) sensor is mounted in the exhaust manifold converter sub-assembly.

The rear Heated Oxygen (HO2S) sensor is mounted in the rear converter sub-assembly.

#### S80 with 4.4L Engine

The front Heated Oxygen (HO2S) sensor is mounted in the exhaust manifold converter sub-assembly.

The rear Heated Oxygen (HO2S) sensor is mounted in the rear of exhaust manifold converter sub-assembly.

### REMOVAL & INSTALLATION

#### 2.4L and 2.5L Engines

##### Bank 1, Sensor 1

*See Figure 136.*

1. Slacken off and disconnect the connector for the front heated oxygen sensor (HO2S).
2. Raise the vehicle.
3. Remove the lower engine cover.
4. Disconnect the sensor harness.
5. Remove the front heated oxygen sensor (HO2S). Use Proper tool.

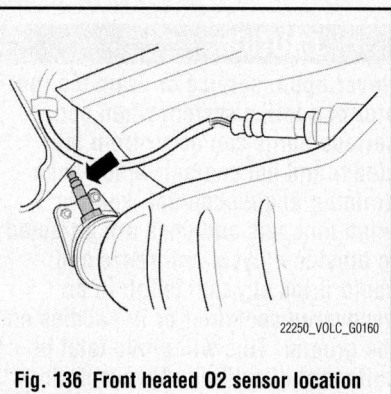

**Fig. 136 Front heated O2 sensor location view**

### To install:

6. Install the front heated oxygen sensor (HO2S). Tighten to 33 ft. lbs. (45 Nm).
7. Install the lower engine cover.
8. Reconnect the connector for the front heated oxygen sensor (HO2S).

##### Bank 1, Sensor 2

1. Raise the vehicle.
2. Remove the lower engine cover.
3. Raise the vehicle. Disconnect the heated oxygen sensor cable harness above the sub-frame.
4. Disconnect the cable harness from the SIPS member.
5. Remove the screw for the cable harness clamp in the three-way catalytic converter.
6. Remove the rear heated oxygen sensor. Use Proper tool.

### To install:

7. Install the rear heated oxygen sensor. Tighten to 33 ft. lbs. (45 Nm).
8. Reconnect the heated oxygen sensor cable harness above the sub-frame.
9. Reconnect the cable harness from the SIPS member.
10. Install the lower engine cover.

11. Reconnect the cable harness clamp in the three-way catalytic converter. Tighten the screw.

#### 3.0L and 3.2L Engines

##### Front Heated Oxygen Sensor

*See Figure 137.*

1. Remove the Engine cover.
2. Remove the clip and harness connector.
3. Remove the front heated oxygen sensor. Use Proper tool.

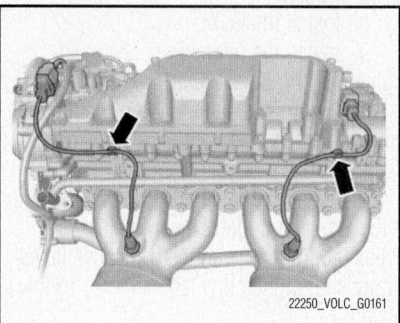

**Fig. 137 Front heated O2 sensor location view—3.0L–3.2L**

### To install:

4. Install the front heated oxygen sensor. Tighten to 33 ft. lbs. (45 Nm).
5. Reconnect the clip and harness connector.

##### Rear Heated Oxygen Sensor

*See Figure 138.*

1. Raise the vehicle.
2. Remove the clip and harness connector from the frame.
3. Remove the rear heated oxygen sensor. Use Proper tool.

### To install:

4. Install the rear Heated Oxygen Sensor (HO2S). Tighten to 33 ft. lbs. (45 Nm).

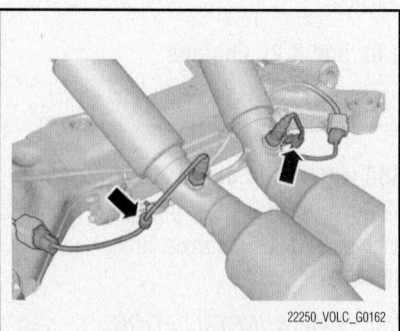

**Fig. 138 Rear heated O2 sensor location view**

5. Install the clip and harness connector from the frame.

6. Lower the vehicle.

### S80 with 4.4L Engine

**Bank 1, Sensor 1 And 2**

*See Figure 139.*

➡**Bank 1 (cylinder row nearest the front): cylinders 1, 3, 5 and 7.**

1. Raise the vehicle.
2. Remove the engine undercover.
3. Remove the retaining clip and harness connector.
4. Remove the O2 sensor.

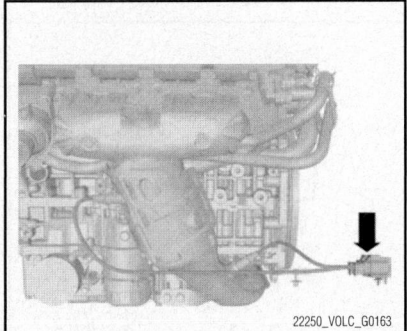

**Fig. 139 Bank 1 sensor 1–2 view**

**To install:**

5. Install the O2 sensor. Tighten to 33 ft. lbs. (45 Nm).
6. Install the retaining clip and harness connector.
7. Install the engine under cover.
8. Lower the vehicle.

**Bank 2, Sensor 1 And 2**

*See Figure 140.*

➡**Bank 2 (cylinder row furthest to the rear, nearest the passenger compartment): cylinders 2, 4, 6, and 8.**

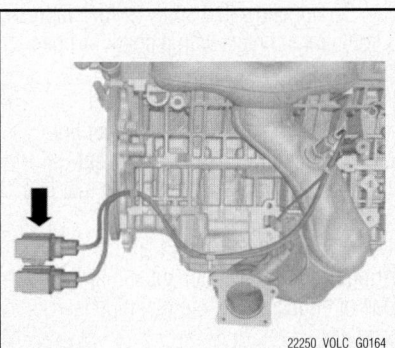

**Fig. 140 Bank 2 sensor 1–2 view**

1. Raise the vehicle.
2. Remove the engine undercover.
3. Remove the retaining clip and harness connector.
4. Remove the O2 sensor.

**To install:**

5. Install the O2 sensor. Tighten to 33 ft. lbs. (45 Nm).
6. Install the retaining clip and harness connector.
7. Install the engine under cover.
8. Lower the vehicle.

### KNOCK SENSOR (KS)

#### LOCATION

##### 2.4L and 2.5L Engines

The Knock Sensors (KS) are mounted on the left and the right front of the engine block.

##### 3.0L and 3.2L Engines

The Knock Sensors (KS) are mounted on the left and the right front of the engine block.

##### S80 with 4.4L Engine

The Knock Sensors (KS) are mounted under the intake manifold.

#### OPERATION

A flat type knock sensor has a structure that can detect vibrations over a wide band of frequencies.

The knock sensor contains a piezoelectric element which generates a voltage when it becomes deformed.

The voltage is generated when the engine block vibrates due to knocking. Any occurrence of engine knocking can be suppressed by delaying the ignition timing.

#### REMOVAL & INSTALLATION

##### 2.4L and 2.5L Engines

1. Remove the air filter module.
2. Remove the screws for the knock sensor.
3. Remove knock sensor.
4. Disconnect the knock sensor connector.

**To install:**

5. Reconnect the knock sensor connector.
6. Install the knock sensor.
7. Tighten the mounting bolt to 15 ft. lbs. (20 Nm

##### 3.0L and 3.2L Engines

*See Figures 141 and 142.*

1. Remove the engine cover.
2. Remove the intake manifold.

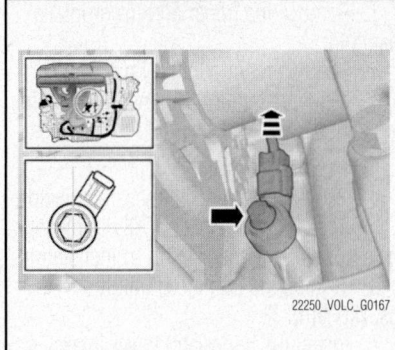

**Fig. 141 Right side knock sensor view**

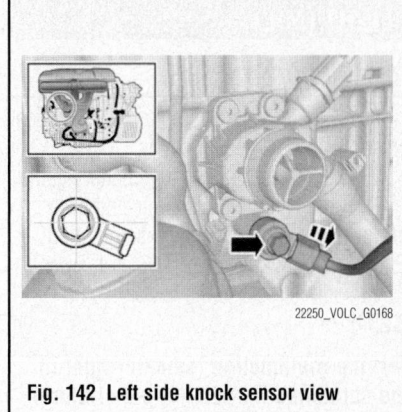

**Fig. 142 Left side knock sensor view**

3. Remove the mounting bolt and harness connector.
4. Remove the right knock sensor.
5. Remove the oil cooler to remove the left side knock sensor.
6. To install, reverse the removal procedure. Tighten the KS mounting bolts to 15 ft. lbs. (20 Nm)

##### S80 with 4.4L Engine

*See Figure 143.*

➡**Knock sensor 1 detects knocking on cylinders 5, 6, 7 and 8. Knock sensor 2 detects knocking on cylinders 1, 2, 3 and 4.**

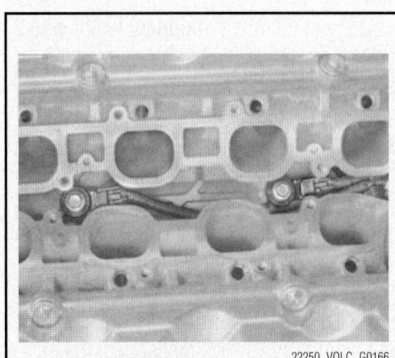

**Fig. 143 4.4L Engine KS sensor view**

1. Remove the upper and lower intake manifold.
2. Remove the mounting bolt.
3. Disconnect the KS harness connector and sensor.

### To install:

4. Reconnect the KS harness connector.
5. Install the sensor correctly in place, as to not damage the sensor on installation.
6. Tighten the mounting bolt to 15 ft. lbs. (20 Nm).
7. Install the upper and lower intake manifold.

## MALFUNCTION INDICATOR LIGHT (MIL)

### RESET PROCEDURES

1. Scan and check for any trouble codes present. Record present codes.
2. Disconnect the negative battery cable.
3. If the battery is disconnected or the power is cut and the vehicle is equipped with remote of the parking heater, the personal code is reset to the factory value (1234).

➡**If the malfunction indicator light in the combined instrument panel illuminates, any diagnostic trouble codes for the engine management system must be read out and remedied before the battery is disconnected.**

4. Initiate the central locking system and interior lighting by unlocking the vehicle with the remote control, key or via VIDA vehicle communication.
5. Initiate window positions. See: VIDA Car communication, Initiating the window position.

➡**If the battery was disconnected, the engine may need to run for a few minutes before it runs smoothly.**

## MASS AIR FLOW (MAF) SENSOR

### LOCATION

The Mass Air Flow (MAF) sensor is located in the air cleaner housing or the air tube.

### REMOVAL & INSTALLATION

See Figure 144.

1. Remove the Mass Air Flow (MAF) sensor harness.
2. Remove the MAF sensor.

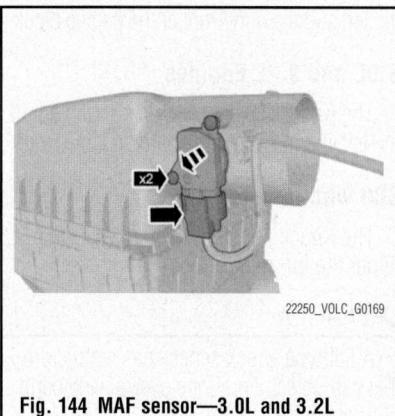

22250_VOLVO_G0169

**Fig. 144 MAF sensor—3.0L and 3.2L engines shown; other engines similar**

### To install:

3. Install the MAF sensor.
4. Install the Mass Air Flow (MAF) sensor harness.

## THROTTLE POSITION SENSOR (TPS)

### LOCATION

The Throttle Position Sensor (TPS) is located between the MAF and intake manifold.

### REMOVAL & INSTALLATION

See Figure 145.

1. Disconnect the negative battery cable.
2. Remove the air cleaner module
3. Disconnect the connector for the throttle body.
4. Remove the screws for the throttle body. Lift out the throttle body.

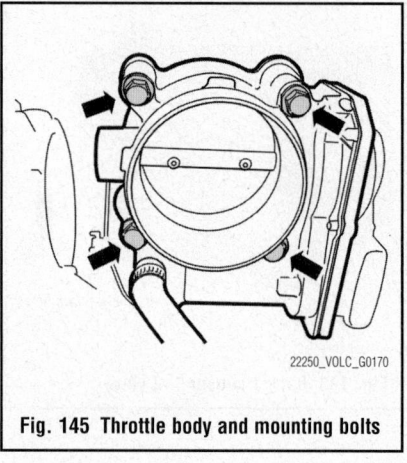

22250_VOLVO_G0170

**Fig. 145 Throttle body and mounting bolts**

### To install:

5. Install the throttle body with a new gasket.
6. Install new screws and tighten the screws to 6 ft. lbs. (8 Nm).
7. Connect the negative battery cable.
8. Install the air cleaner module.

# FUEL

## FUEL SYSTEM SERVICE PRECAUTIONS

Safety is the most important factor when performing not only fuel system maintenance but any type of maintenance. Failure to conduct maintenance and repairs in a safe manner may result in serious personal injury or death. Maintenance and testing of the vehicle's fuel system components can be accomplished safely and effectively by adhering to the following rules and guidelines.

• To avoid the possibility of fire and personal injury, always disconnect the negative battery cable unless the repair or test procedure requires that battery voltage be applied.

# GASOLINE FUEL INJECTION SYSTEM

• Always relieve the fuel system pressure prior to disconnecting any fuel system component (injector, fuel rail, pressure regulator, etc.), fitting or fuel line connection. Exercise extreme caution whenever relieving fuel system pressure to avoid exposing skin, face and eyes to fuel spray. Please be advised that fuel under pressure may penetrate the skin or any part of the body that it contacts.

• Always place a shop towel or cloth around the fitting or connection prior to loosening to absorb any excess fuel due to spillage. Ensure that all fuel spillage (should it occur) is quickly removed from engine surfaces. Ensure that all fuel soaked cloths or towels are deposited into a suitable waste container.

• Always keep a dry chemical (Class B) fire extinguisher near the work area.

• Do not allow fuel spray or fuel vapors to come into contact with a spark or open flame.

• Always use a back-up wrench when loosening and tightening fuel line connection fittings. This will prevent unnecessary stress and torsion to fuel line piping.

• Always replace worn fuel fitting O-rings with new Do not substitute fuel hose or equivalent where fuel pipe is installed.

Before servicing the vehicle, make sure to also refer to the precautions in the beginning of this section as well.

## RELIEVING FUEL SYSTEM PRESSURE

### 2.4L, 2.5L, 3.0L and 3.2L Engines

*See Figure 146.*

1. Before servicing the vehicle, refer to the Precautions Section.
2. Disconnect the negative battery cable.
3. Remove protective cap from the valve on the fuel rail.
4. Connect a hose to adapter 999-5484 and place the other end in a clean container.
5. Connect the adapter in the locked or closed position to the valve on the fuel rail.
6. Unlock or open the adapter valve.
7. After the fuel system pressure is relieved, remove the adapter and hose and replace the protective cap.
8. Connect the negative battery cable when repairs are complete.

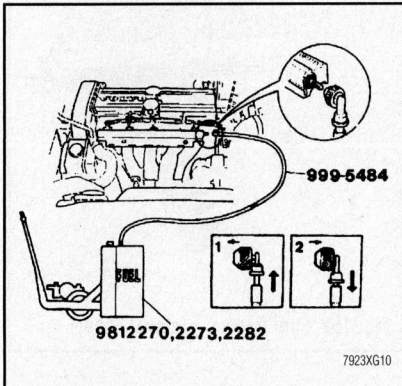

**Fig. 146 Connecting the adapter and fuel drainage unit—5-cylinder engine shown**

### S80 with 4.4L Engine

*See Figures 147 and 148.*

1. Remove the cover on the valve which is positioned on the fuel rail.
2. Connect adapter 999 5484 to fuel pressure relief unit 981 2270, 981 2273 and 981 2282.
3. Connect the adapter to the valve on the fuel rail in the locked position (figure 1, valve closed).
4. Unlock the adapter (illustration 2, valve open).
5. Raise the vehicle.
6. Remove the cover on the valve cap positioned at the fuel filter.
7. Connect venting hose 999 5480 to the valve prior to the fuel filter.

➥**It takes approximately 2 minutes to drain the system.**

➥**Do not forget to reinstall the covers over the valves.**

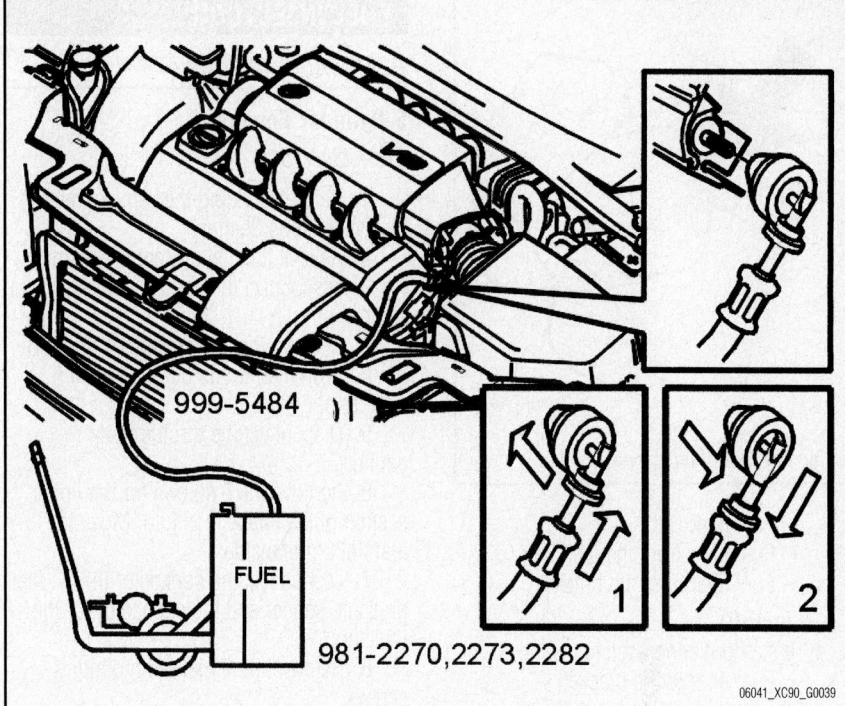

**Fig. 147 Connect adapter 999 5484 to fuel draining unit 981 2270, 981 2273 and 981 2282—V8 engine**

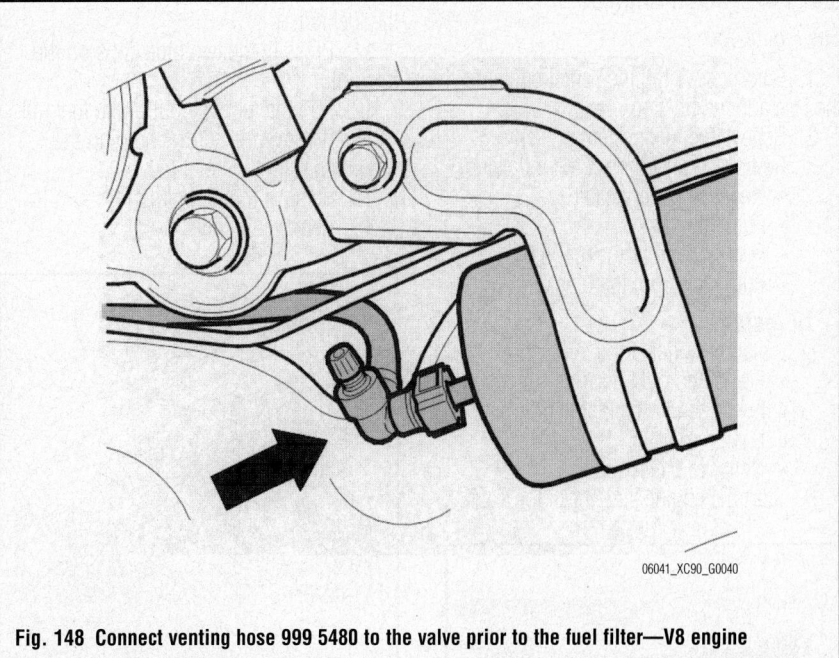

**Fig. 148 Connect venting hose 999 5480 to the valve prior to the fuel filter—V8 engine**

8. Reinstall the components in reverse order.

## FUEL FILTER

*REMOVAL & INSTALLATION*

### 5 Cylinder Engine

*See Figure 149.*

➥**The fuel filter is either on the left side of the firewall or next to the fuel pump near the left side of the fuel tank.**

1. Before servicing the vehicle, refer to the Precautions Section.
2. Relieve the fuel system pressure.
3. Remove or disconnect the following:
   • Negative battery cable

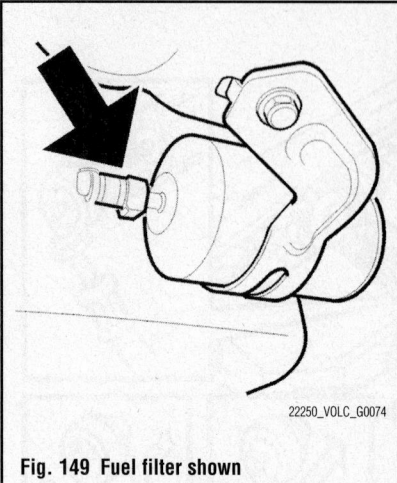

**Fig. 149 Fuel filter shown**

- Fuel filler cap
- Fuel lines from the fuel filter
- Fuel filter from the bracket

**To install:**

4. Install or connect the following:
   - Fuel filter to the bracket
   - Fuel lines to the fuel filter
   - Fuel filler cap
   - Negative battery cable
5. Start the engine and check for leaks.

### 6 and 8-Cylinder Engines

*See Figure 150.*

1. Before servicing the vehicle, refer to the Precautions Section.
2. Relieve the fuel system pressure.
3. Remove or disconnect the following:
   - Negative battery cable
   - Fuel filler cap
   - Fuel lines from the fuel filter
   - Fuel filter from the bracket

**To install:**

4. Install or connect the following:
   - Fuel filter to the bracket
   - Fuel lines to the fuel filter
   - Fuel filler cap
   - Negative battery cable
5. Start the engine and check for leaks.

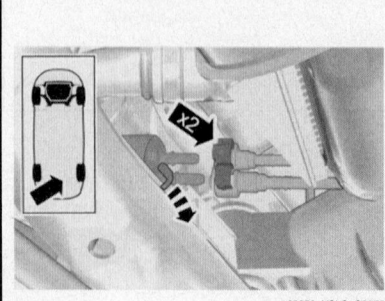

**Fig. 150 6 and 8 Cylinder fuel filter shown**

## FUEL INJECTORS

### REMOVAL & INSTALLATION

#### 5-Cylinder Engines

*See Figure 151.*

1. Before servicing the vehicle, refer to the Precautions Section.
2. Relieve the fuel system pressure.
3. Disconnect the negative battery cable.
4. Remove the hose clamp for the rubber hose on the throttle body. Move the hose to one side.
5. Disconnect the throttle body connector.
6. Remove the 6 screws for the upper section of the intake manifold. Move the manifold to one side.
7. Disconnect the connector for the fuel pressure sensor and the connector for the injectors.
8. Remove the 2 fuel rail mounting screws.
9. Spray with universal oil or similar around the injector nozzle at the terminal on the intake manifold.
10. Gently work the fuel rail and injector nozzles loose. Disconnect the fuel line to the fuel rail.
11. Press in the two blue clips on the fuel rail.
12. Pull the fuel line out of the fuel rail.
13. Remove the screws holding the mounting rail to the fuel rail.
14. Remove the mounting rail.
15. Remove the injector.

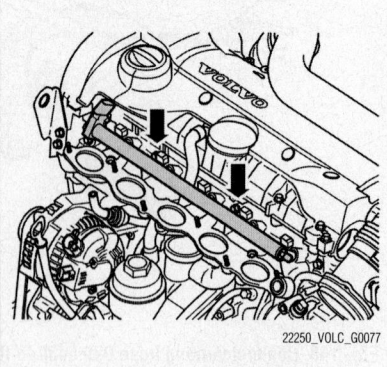

**Fig. 151 Fuel rail shown 2.4L and 2.5L engines**

**To install:**

16. Installation is in reverse of the removal procedure, note the following:
   - Tighten the fuel rail bolts to 88 inch lbs. (10 Nm)
17. Start the engine and check for leaks.

#### 6-Cylinder Engines

*See Figure 152.*

1. Before servicing the vehicle, refer to the Precautions Section.
2. Relieve the fuel system pressure.
3. Disconnect the negative battery cable.
4. Remove the engine cover.
5. Remove the engine torque mount and bracket.
6. Remove the fuel rail electrical connector.
7. Remove the fuel supply line.
8. Remove the fuel rail mounting bolts.
9. Carefully remove the fuel rail with injectors.
10. Remove the screws that retain the bracket for fuel injectors.

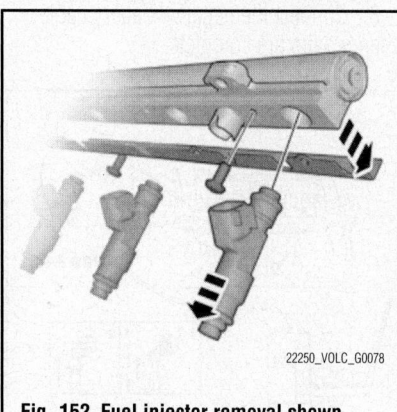

**Fig. 152 Fuel injector removal shown**

**To install:**

11. Installation is in reverse of the removal procedure, note the following:
   - Fuel injector retaining screws Tighten to 44 inch lbs. (5 Nm)
   - Tighten fuel rail mounting bolts to 44 inch lbs. (5 Nm)
   - Tighten engine torque mount and bracket to 59 ft. lbs. (80 Nm)
12. Connect the negative battery cable.
13. Start the engine and check for leaks.

#### 8-Cylinder Engines

*See Figures 153 and 154.*

1. Before servicing the vehicle, refer to the Precautions Section.
2. Relieve the fuel system pressure.
3. Disconnect the negative battery cable.
4. Remove the engine cover.
5. Remove the upper intake manifold. Refer to Intake Manifold removal and installation.
6. Remove the crankcase ventilation hose between the cylinder heads by releasing the 2 clamps.

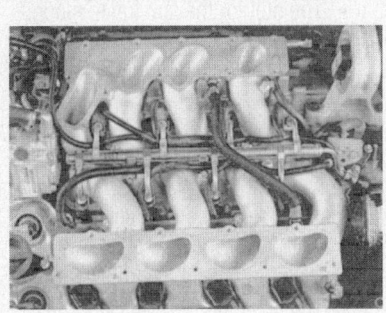

**Fig. 153 Fuel rail assembly 4.4L—V8 engine**

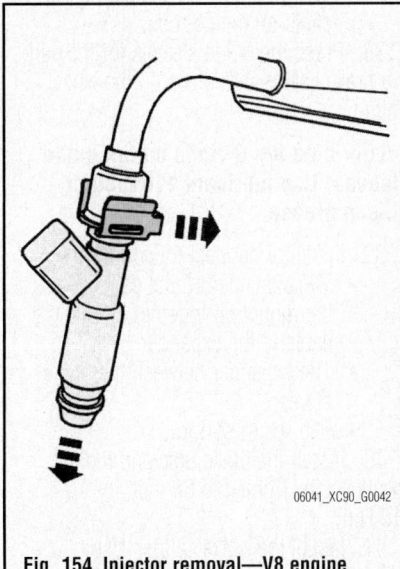

**Fig. 154 Injector removal—V8 engine**

7. Remove the 3 bolts from the fuel rail
8. Remove the 2 bolts on the fuel pressure sensor.
9. Lift the fuel rail assembly with injector straight up and bend it aside.
10. Unplug the connector of the injector.
11. Remove the circlip of the injector; pull the clip laterally.
12. Remove the injector; jiggle and pull it straight out.

*To install:*

➡**Lubricate the O-rings on the injectors with Vaseline or the like.**

13. Install the injector.
14. Install the circlip.
15. Plug in the connector to the fuel injector.
16. Install the fuel rail assembly with injector.
17. Install the 3 bolts of the fuel rail.
18. Install the 2 bolts on the fuel pressure sensor.
19. Install the crankcase ventilation hose and 2 clamps.

**✳✳ WARNING**

**Make sure there is an audible click during connection.**

20. Connect the fuel supply line and secure the line in the clamp.
21. Install the upper intake manifold.
22. Connect the negative battery cable.
23. Start the engine and check for leaks.
24. Install the engine cover.

## FUEL PUMP

*REMOVAL & INSTALLATION*

### 5 Cylinder Engines

1. Before servicing the vehicle, refer to the Precautions Section.
2. Relieve the fuel system pressure.
3. Remove or disconnect the following:
   • Negative battery cable
   • Upper Intake manifold
   • Injector connectors
   • Fuel rail with injectors attached
   • Injector mounting rail
   • Injectors

*To install:*

4. Install or connect the following:
   • Injectors to the injector mounting rail
   • Fuel rail
   • Injector connectors
   • Intake manifold
   • Negative battery cable
5. Start the engine and check for leaks.

### 6 and 8-Cylinder Engines

*See Figure 155.*

➡**The fuel pump for the 6–8 Cylinder Engine is external.**

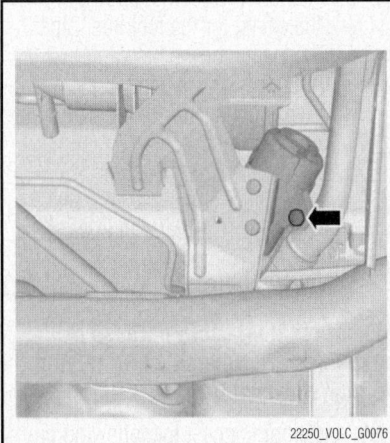

**Fig. 155 Fuel pump view 6–8 Cylinder Engines**

1. Before servicing the vehicle, refer to the Precautions Section.
2. Raise the vehicle and remove the fuel pump electrical connector and mounting bolt.
3. Remove the fuel filter.
4. Installation is in the reverse of the removal procedure.

## FUEL TANK

*REMOVAL & INSTALLATION*

*See Figures 156 and 157.*

**✳✳ CAUTION**

**Observe all applicable safety precautions when working around fuel. Whenever servicing the fuel system, always work in a well ventilated area. Do not allow fuel spray or vapors to come in contact with a spark or open flame. Keep a dry chemical fire extinguisher near the work area. Always keep fuel in a container specifically designed for fuel storage; also, always properly seal fuel containers to avoid the possibility of fire or explosion.**

1. Before servicing the vehicle, refer to the Precautions Section.
2. Relieve the fuel system pressure.
3. Drain the fuel tank.
4. Block the parking brake pedal in the neutral position.
5. Remove or disconnect the following on both sides:
   • Rear wheels
   • Rear brake calipers
   • Rear brake discs
   • Rear tension spring and expander
6. Lift the brake shoe slightly. Grip the expander. Pull the expander outwards.
7. Unhook the expander from the metal bracket.

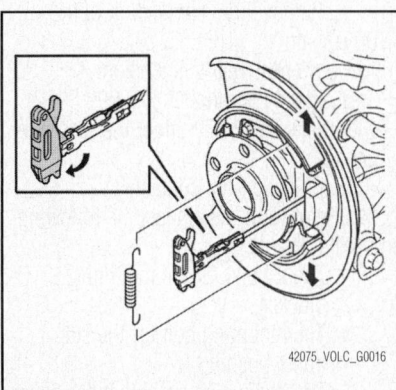

**Fig. 156 Removing expansion spring from rear brake**

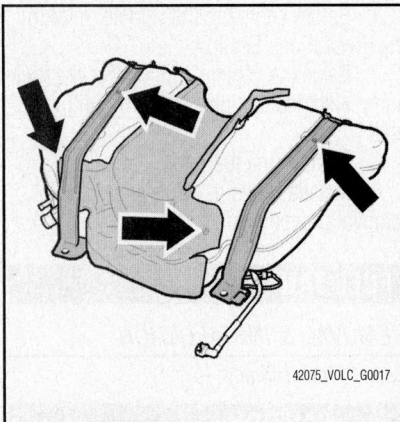

**Fig. 157 Removing fuel tank straps and heat shield**

8. Remove or disconnect the following on both sides:
- The guide sleeve for the cable sleeve. Apply a screwdriver between the guide sleeve lug and wheel arch. Pry apart slightly to remove the sleeve.
- The parking brake cable from the tie rod and the sub-frame

9. Remove 2 bolts from the cable clips and 2 bolts from the front mounting for the heat deflector plate.

10. Fold out the cables from the heat deflector plate. Hang the cables up at the front of the car.

11. Remove the exhaust system from the three-way catalytic converter and backwards.

➡ **Hang up the three-way catalytic converter to prevent damaging the flex boot.**

12. Remove the rear driveshaft:
 a. Mark up the position of the joint on the flange on the final drive and at the center bearing.
 b. Remove bolts from the joints. Use counterhold 999 7057.
 c. Remove the rear section of the driveshaft.

13. Lower the rear suspension:
 a. Place the mobile jack: 999 5972 with plate: 999 5972 under the rear suspension.
 b. Raise to light contact.

14. Remove or disconnect the following on both sides:
- The 2 bolts for the mounting bracket
- The mounting bolt for the sub frame bushing
- The brake pipe on both sides from the plastic clips on the bracket

15. Lower the rear suspension carefully.

16. Place a lifting table under the fuel tank.

17. Remove or disconnect the following:
- The upper quick-release connector plastic pipe on the carbon filter container
- The 4 bolts for the fuel tank securing strap
- Remove the clamp for the fuel filler pipe
- Remove the clamp for the bleed pipe

18. Lower the lifting table to access the plastic pipes on top of the tank.

19. Remove the plastic pipe.

20. Remove the ABS cable harness from the clips on both sides of the fuel tank.

21. Remove the fuel tank.

***To install:***

22. Place the fuel tank on the lifting table.

23. Raise the lifting table to install the plastic hoses on the right-hand level sensor and in the grooves in the tank.

24. Secure the hoses in place using silver tape.

25. Install the ABS cable harnesses in the clips on both sides of the tank.

26. Align the fuel filler pipe and the bleed pipe.

27. Insert the cable harnesses for the level sensor and the fuel tank pressure sensor through the inspection holes in the floor.

28. Raise the fuel tank to the installation position.

➡ **Ensure that the fuel filler pipe and bleed pipe slide into position while raising the fuel tank.**

29. Install the 4 bolts for the fuel tank securing strap. Tighten to 18 ft. lbs. (24 Nm).

30. Remove the lifting table.

31. Install or connect the following:
- The clamp for the fuel filler pipe
- The clamp for the bleed pipe
- The upper plastic pipe in the quick-release connector for the carbon filter container

32. Install the rear suspension:
 a. Place the mobile jack: 999 5972 and plate: 999 5972 under the rear suspension.
 b. Carefully raise the rear suspension.

➡ **Ensure that no cable harness or brake lines are trapped while raising the rear suspension.**

33. Install or connect the following on both sides:
- The 2 bolts for the mounting bracket. Tighten to 59 ft. lbs. (80 Nm)

- The bolt for the sub frame mounting. Tighten to 59 ft. lbs. (80 Nm)
- The brake line in the plastic clip and bracket

34. Install the brakes.

35. Install the parking brake cable along the heat shields.

36. Install the bolts in the center and front mountings for the heat shield.

37. Install or connect the following:
- The parking brake cables on the wheel units
- The parking brake cable mountings in the sub-frame. Tighten bolts to 19 ft. lbs. (25 Nm).
- The mounting for the parking brake cable on the tie rods.

38. Press the guide sleeves for the parking brake cables into place on the wheel units.

➡ **Lubricate the O-rings on the guide sleeves. Use lubricant 1161580 or silicon grease.**

39. Install or connect the following:
- The expander on the cable. Press the mounting together to lock the cable in the expander
- The expander between the brake shoes
- The return spring.

40. Install the brake disc and the locating pin. Tighten to 88 inch lbs. (10 Nm).

41. Install the brake caliper. Use new bolts. Tighten to 44 ft. lbs. (60 Nm).

➡ **Activate parking brake adjuster.**

42. Install the rear drive shaft.
 a. Position the propeller shaft as marked previously.
 b. Install new bolts for the joints. Use counterhold 999 7057. Tighten all bolts to 22 ft. lbs. (30 Nm).

43. Install the exhaust system. Use a new flange gasket. Tighten the bolts to 18 ft. lbs. (25 Nm).

44. Install the cover over the level sensor.

45. Connect the wiring and the connectors.

46. Install rear wheels and tighten lug nuts to 81–103 ft. lbs. (110–140 Nm). See torque specification in the following chart.

## IDLE SPEED

*ADJUSTMENT*

The idle speed is controlled by the ECM.

## THROTTLE BODY

*REMOVAL & INSTALLATION*

### 5-Cylinder Engines

*See Figure 158.*

1. Before servicing the vehicle, refer to the Precautions Section.
2. Remove or disconnect the following:
   - The negative battery terminal.
   - The air intake between the front cover plate and the air cleaner housing.
   - The charge air pipe between the charge air cooler and the electronic throttle body.
3. Remove the throttle body:
   a. Disconnect the connector for the throttle body.
   b. Remove the 4 bolts from the throttle body.
   c. Lift out the throttle body and discard the gasket.

42075_VOLC_G0015

**Fig. 158 Throttle body location— 5-Cylinder Engines**

*To install:*

➡Ensure that the surfaces between the throttle body and the intake manifold are clean.

4. Install electronic throttle body using a new gasket. Tighten the 4 bolts to 88 inch lbs. (10 Nm).
5. Install the charge air pipe between the electronic throttle body module and the charge air cooler.

➡**Heat the pipe at the connections to facilitate installation.**

6. Install the intake pipe between the front cover plate and the air cleaner housing.
7. Start the engine and check the function of the electronic throttle body module.

➡**After replacing the throttle unit, the throttle unit must be adapted using the vehicle communication input: Adaptation of the electronic throttle unit. Remedy as necessary.**

### 6-Cylinder Engines

1. Before servicing the vehicle, refer to the Precautions Section.
2. Remove or disconnect the following:
   - The negative battery terminal.
   - The charge air cooler.
   - The intake pipes between the front cover plate and air cleaner housing.
   - The hose between the turbocharger and the charge air cooler at its upper end. Push it to one side.
   - The pipe between the throttle body and the charge air cooler.
3. Removing the throttle body:
   a. Disconnect the throttle body connector.
   b. Remove the mounting bolts from the throttle body.
   c. Lift up and remove the throttle body. Discard the gasket.

*To install:*

4. Clean the mating surfaces thoroughly.

5. Install the throttle body using a new gasket. Tighten the 4 bolts to 88 inch lbs. (10 Nm).
6. Connect the throttle body connector.
7. Install or connect the following:
   - The pipe between the charge air cooler and the throttle body.
   - The hose between the charge air cooler and the turbocharger.
   - The intake pipes between the front cover plate and the air cleaner housing.
   - The battery negative lead.

➡**After replacing the throttle unit, the throttle unit must be adapted using the vehicle communication input: Adaptation of the electronic throttle unit. Remedy as necessary.**

### 8-Cylinder Engines

1. Before servicing the vehicle, refer to the Precautions Section.
2. Remove or disconnect the following:
   - Hose between throttle body and air filter housing
   - MAF connector
   - Throttle body coolant hose
   - 4 Throttle body mounting bolts
   - Throttle body

*To install:*

3. Install or connect the following:
   - Throttle body
   - 4 Throttle body mounting bolts Tighten to
   - Throttle body coolant hose
   - MAF connector
   - Hose between throttle body and air filter housing
4. Connect the negative battery cable.

➡**After replacing the throttle unit, the throttle unit must be adapted using the vehicle communication input: Adaptation of the electronic throttle unit. Remedy as necessary.**

## HEATING & AIR CONDITIONING SYSTEM

### BLOWER MOTOR

#### REMOVAL & INSTALLATION

*See Figure 159.*

1. Before servicing the vehicle, refer to the Precautions Section.
2. Remove or disconnect the following:
   - The negative battery cable
   - The rear side panel
   - The rubber seals and the cable terminals
   - The bolts for the blower motor
   - The blower motor

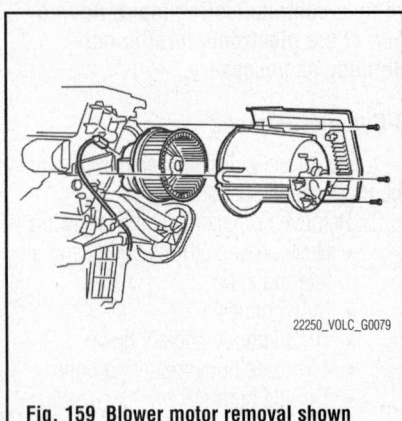

22250_VOLC_G0079

**Fig. 159 Blower motor removal shown**

**To install:**

3. Install or connect the following:
   - The blower motor using the M5 bolts. Tighten to 44 inch lbs. (5 Nm)
   - The cable terminals and the rubber seals
   - The rear side panel
   - The negative battery cable

### HEATER CORE

#### REMOVAL & INSTALLATION

#### C70 and V70 Models

*See Figures 160 through 164.*

1. Before servicing the vehicle, refer to the Precautions Section.
2. Disconnect the negative battery cable.
3. Drain the cooling system into a clean container for reuse.
4. Using tools No. 155 8957, disconnect the heater hoses from the heater core by pressing in the hose connections, squeezing the locking catches and pulling out the hoses. Discard the O-rings.
5. At the driver's side, remove the soundproofing panel—to—dashboard Torx® 20 screw and the soundproofing panel.

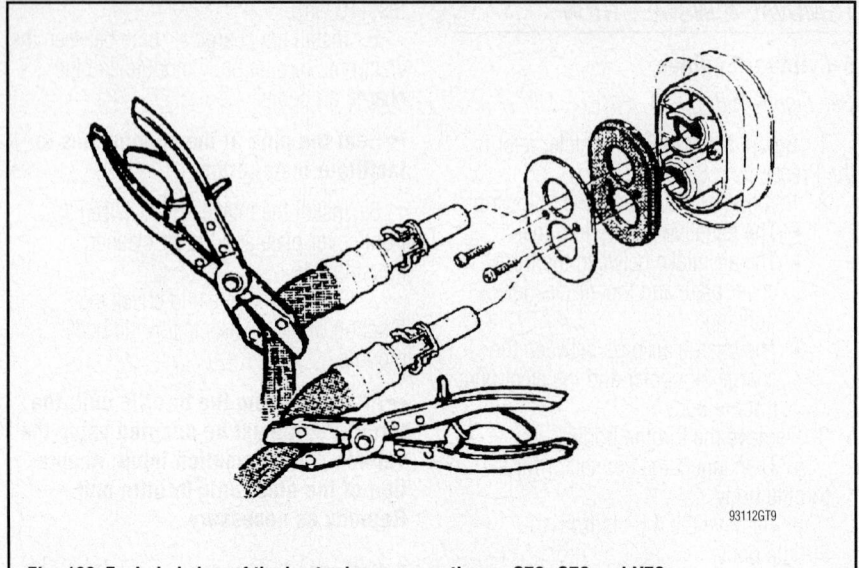

93112GT9

**Fig. 160 Exploded view of the heater hose connections—C70, S70 and V70**

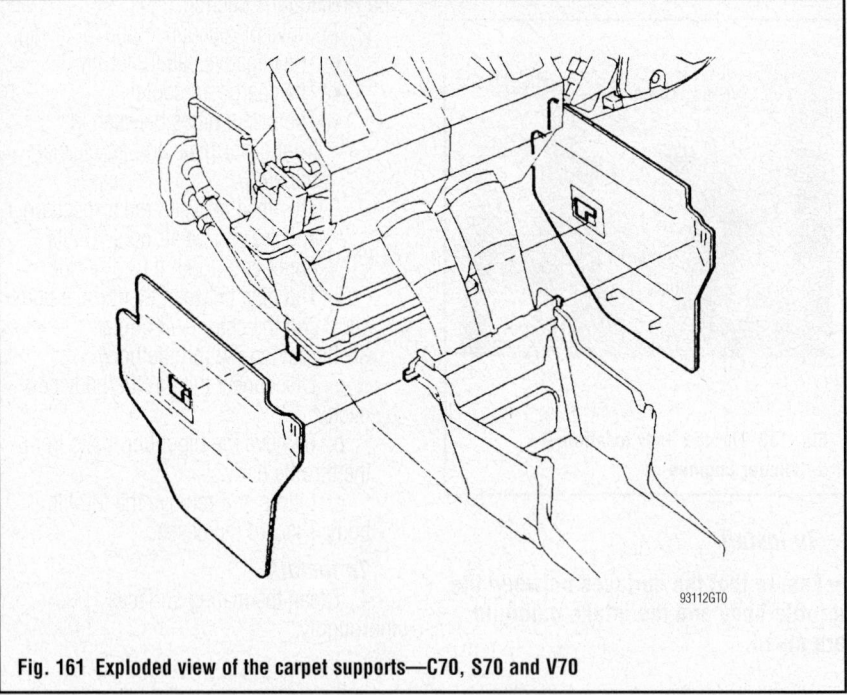

93112GT0

**Fig. 161 Exploded view of the carpet supports—C70, S70 and V70**

6. Open the glove box door; then, remove the 4 glove compartment—to—dashboard Torx® 20 screws and the glove compartment.
7. At the passenger's side, remove the 2 soundproofing panel—to—dashboard screws and the soundproofing panel.
8. At both the driver's side and passenger's side, remove the carpet supports.
9. If equipped, remove the Road Traffic Information (RTI) control module bracket or booster bracket.
10. If equipped, remove the knee bolster.

11. Disconnect the drain hose and fold it out of the way.
12. At both sides of the heater/air conditioning housing assembly, remove the heater core cover screws.
13. Remove the heater housing pipe flange screw.
14. Carefully, remove the heater core with the heater core cover.
15. Remove the heater core from the cover.

**To install:**

16. Install the heater core to the cover.

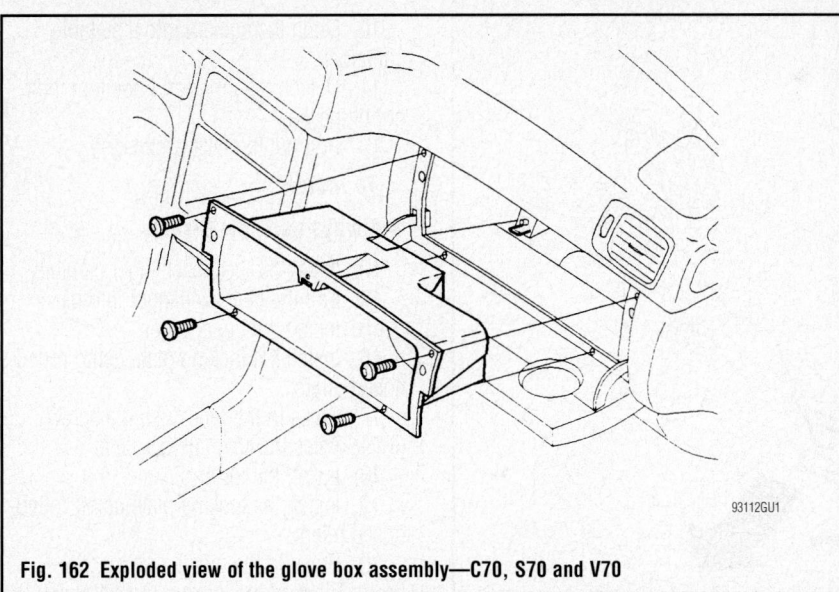

**Fig. 162 Exploded view of the glove box assembly—C70, S70 and V70**

17. Carefully, install the heater core with the heater core cover.

18. Install the heater housing pipe flange screw.

19. At both sides of the heater/air conditioning housing assembly, install the heater core cover screws.

20. Connect the drain hose.

21. If equipped, install the knee bolster.

22. If equipped, install the Road Traffic Information (RTI) control module bracket or booster bracket.

23. At both the driver's side and passenger's side, install the carpet supports.

24. At the passenger's side, install the soundproofing panel and the 2 soundproofing panel—to—dashboard screws.

25. Install the glove compartment and the 4 glove compartment—to—dashboard Torx® 20 screws.

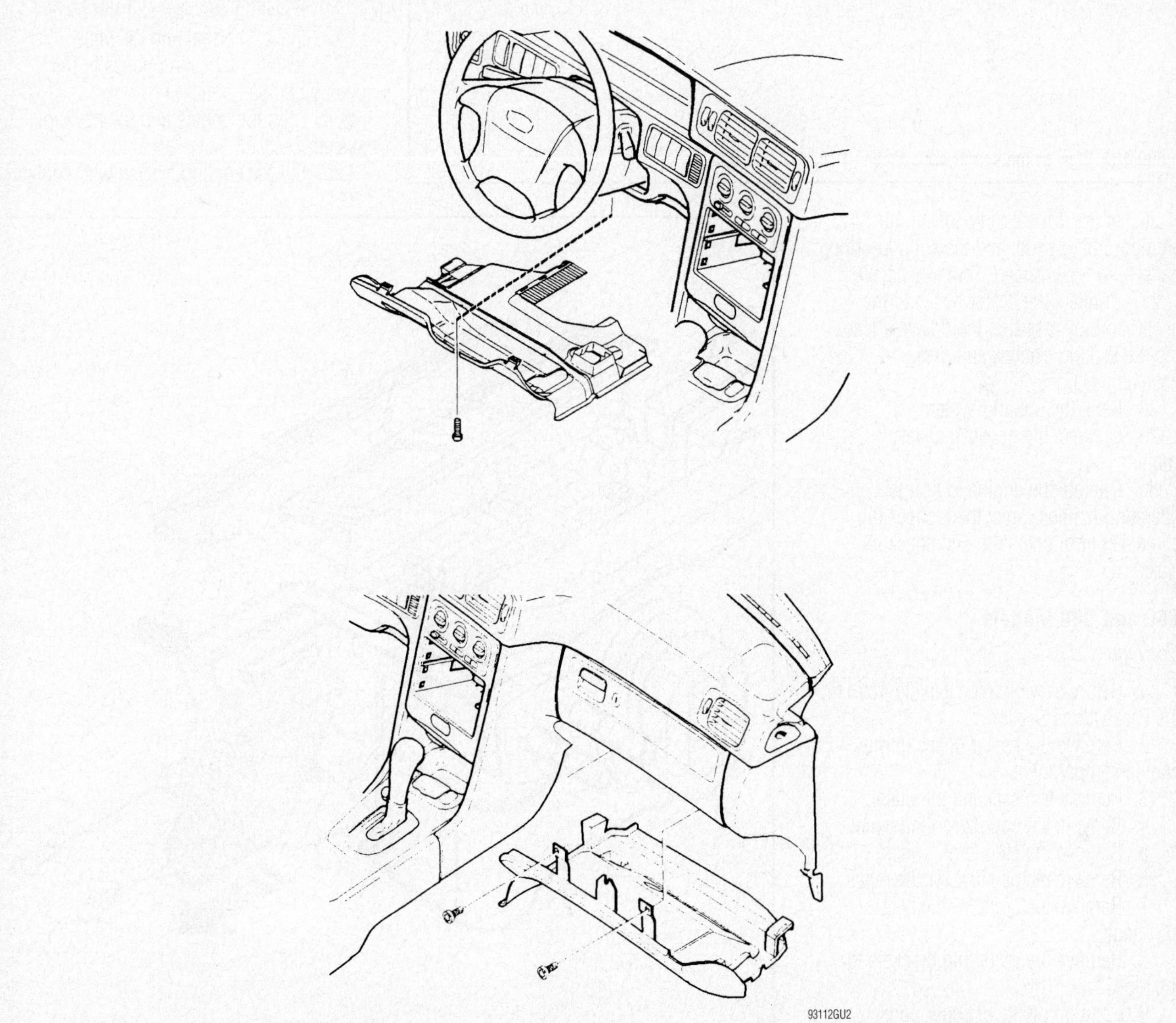

**Fig. 163 Exploded view of the driver's side and passenger's side soundproofing panels—C70, S70 and V70**

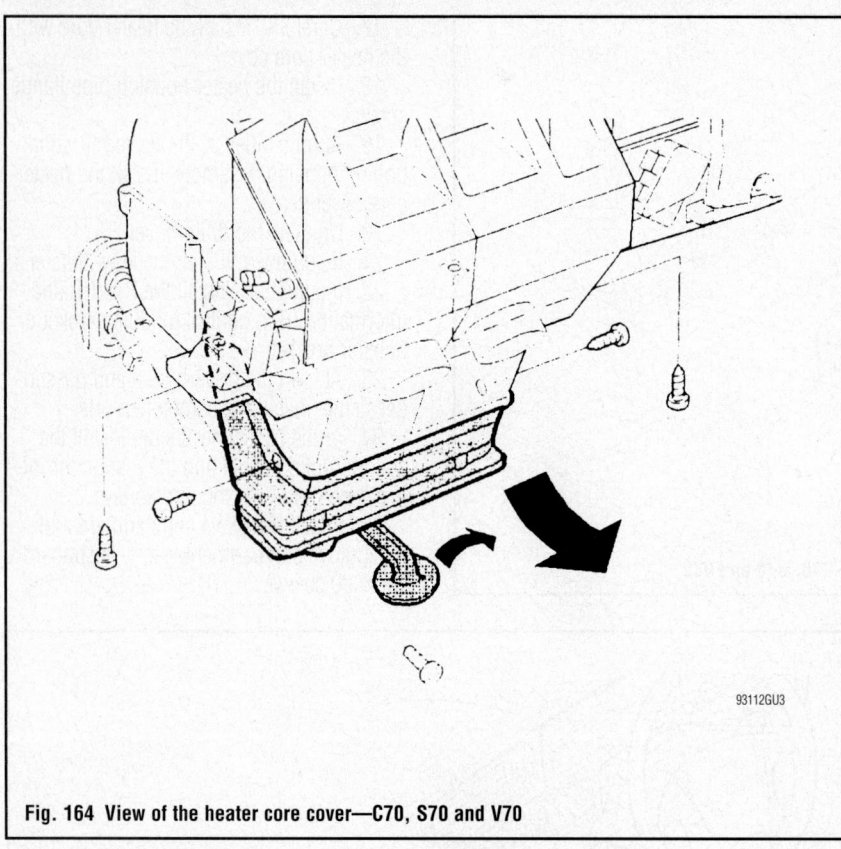

**Fig. 164 View of the heater core cover—C70, S70 and V70**

10. Drain the coolant into a suitable container.

11. Detach the pipes and position them out of the way

12. Pull out the heat exchanger.

***To install:***

➡**Always use new O-rings**

13. Install the heat exchanger carefully

14. Bolt the heat exchanger into place

15. Install the upper pipes

16. Install the locking plate using round-nosed pliers

17. Press in the pipes using a screwdriver. Press the catch using pliers

18. Install the lower pipes

19. Install the locking plate using round-nosed pliers

20. Press in the pipe using a screwdriver. Press in the locking bracket using round-nosed pliers

21. Install the tie strap for the pipes

22. Install the seal and the plate

23. Install the heating hoses in the engine compartment

24. Start the engine and the heating system

25. Fill the cooling system with coolant

26. At the driver's side, install the soundproofing panel and the soundproofing panel—to—dashboard Torx® 20 screw.

27. Connect the heater hoses to the heater core by inserting the hoses, squeezing the locking catches and pressing in the hose connections.

28. Refill the cooling system.

29. Connect the negative battery cable.

30. Operate the engine to normal operating temperatures; then, check the climate control operation and check for leaks.

## S60 and S80 Models

*See Figure 165.*

1. Before servicing the vehicle, refer to the Precautions Section.

2. Remove the heater hoses in the engine compartment.

3. Remove the seal and the plate.

4. Remove the soundproofing panel.

5. Remove the stop lamp switch.

6. Remove the tie strap for the pipes

7. Remove the bolts for the heat exchanger

8. Remove the mounting brackets for the pipes.

9. Position plenty of paper under the heat exchanger and around the pipes

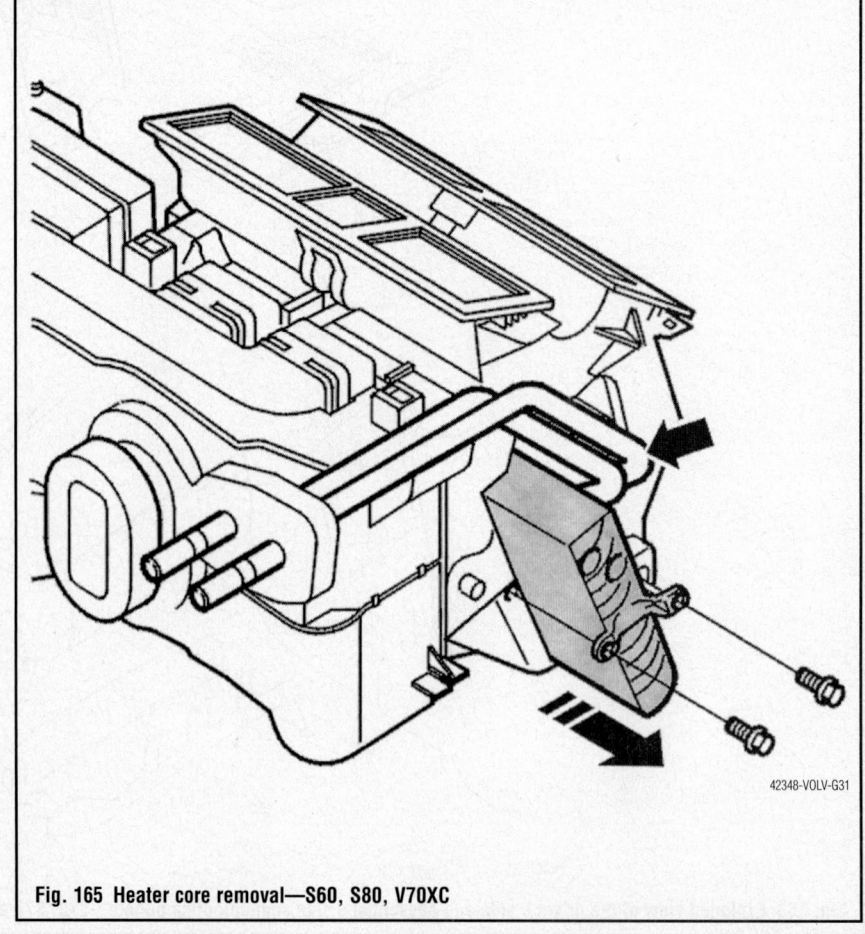

**Fig. 165 Heater core removal—S60, S80, V70XC**

26. Check that the pipe connectors do not leak
27. Install the soundproofing panel.

### S40 and V50 Models

*See Figure 166.*

1. Before servicing the vehicle, refer to the Precautions Section.
2. In the engine compartment:
   a. Remove the top splashguard.

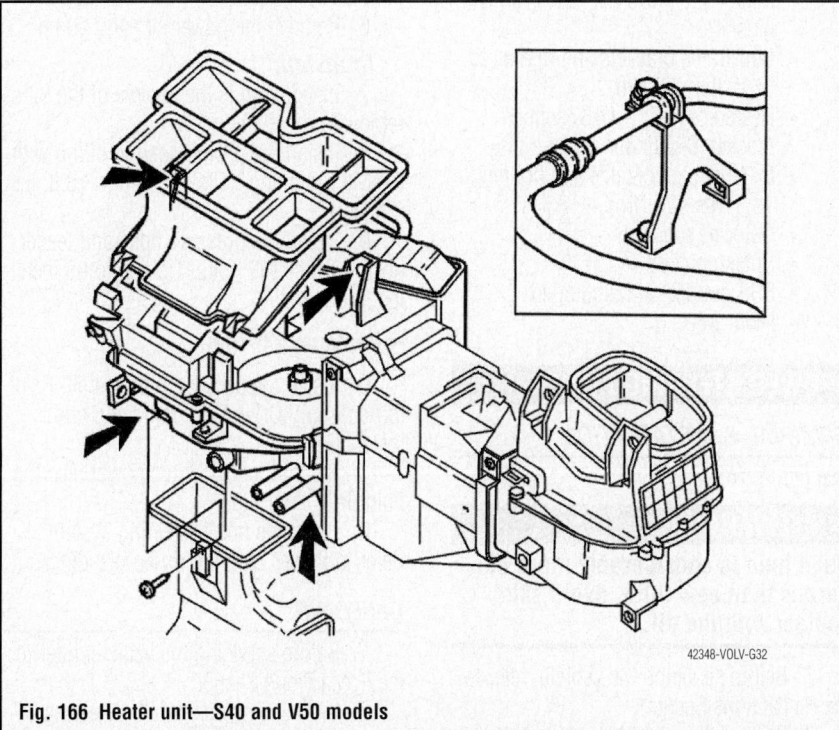

**Fig. 166 Heater unit—S40 and V50 models**

b. Install two hose clips on the hoses to the heat exchanger. Remove the hoses.
   c. Remove the air conditioning pipe in the engine compartment from the support.
3. In the passenger compartment:
   a. Disconnect the wiring and the connectors located on the center support.
   b. Remove the mounting bolts from the center support. Remove the support.

c. Remove the control panel.
   d. Disconnect all the connectors.
   e. Remove the power unit
4. Cars with air conditioning remove the air conditioning pipe in the engine compartment from the support.
5. Remove the bolts and the nut from the power unit. Pull the power unit backwards as far as possible
6. Remove the upper section from the housing.
7. Remove the section that connects with the floor heater.
8. Press the pipes down. Remove the lower section.
9. Remove the four mounting bolts.
10. Pull off the housing. Press the floor heater down. Remove the housing.
11. Remove the control lever on the valve side.
12. Remove the mounting and the clips around it.
13. Carefully separate the components.
14. Replace damaged components if necessary.

➡**Do not disassemble damper motors when adjusting.**

15. Replace the refrigerant temperature sensor.
16. Replace the heat exchanger.
17. Check the seals (also on the dampers).
18. Install the control mechanism last.
19. Install in reverse order.
20. Check the setting of the damper motors.

## STEERING

### POWER STEERING GEAR

*REMOVAL & INSTALLATION*

#### C30, C70, S40 and V50 Models

*See Figure 167.*

1. Before servicing the vehicle, refer to the precautions in the beginning of the section.

➡**Ensure the steering wheel is set to the 'straight ahead' position and remove the key so the steering wheel lock engages.**

2. Remove the left front wheel and measure and record the length of the tie rod in relation to the steering gear housing.
3. Remove or disconnect the following:
   • Left side soundproofing panel
   • Left front wheel
   • Steering shaft from the steering gear

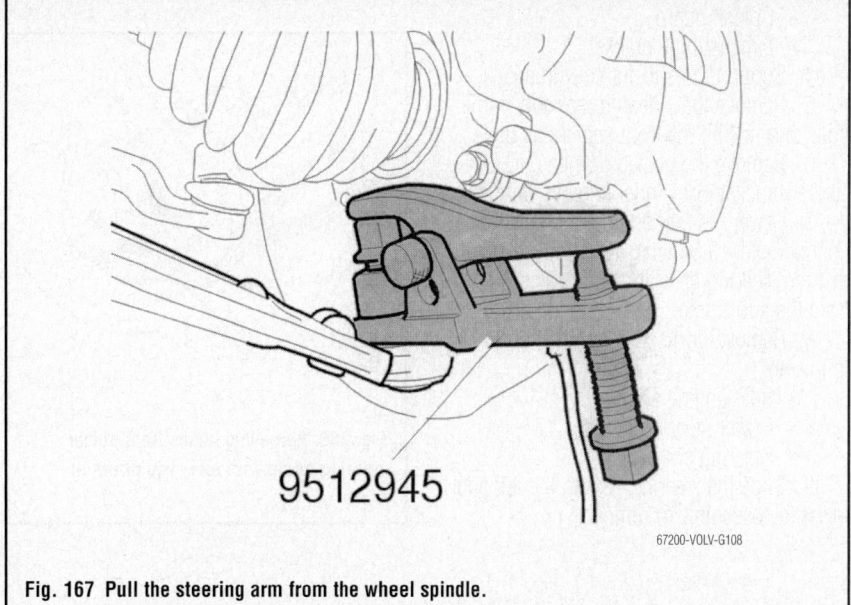

**Fig. 167 Pull the steering arm from the wheel spindle.**

- Sub-frame
- Powering steering hoses from the steering gear

4. Slacken off the nut for the outer steering arm. Press the steering arm out of the wheel spindle using special tool 951-2945 or equivalent.

5. Remove the steering gear.

***To install:***

6. Install or connect the following:
- Steering gear
- Outer steering arm, tighten nuts to 37 ft. lbs. (50 Nm)
- Power steering hoses using new O-rings
- Sub-frame
- Steering shaft to the steering gear
- Wheel
- Soundproofing panel

7. Fill and bleed the power steering system.

### S60, S80 and V70 Models

1. Before servicing the vehicle, refer to the Precautions Section.

➡**Do not turn the front wheels so that the steering gear position is changed. The SRS-system contact reel can be damaged.**

2. Raise and support the engine.

3. Remove or disconnect the following:
- Front wheels
- SRS connector, if equipped
Disconnect the connector cable from the clips along the ABS cable.
Push down the connector and cable to the steering gear.
- Outer tie rod ends
- Splash guard
- Exhaust pipe bracket

4. Support the sub-frame with a jack

5. Remove the side and rear sub-frame bolts and loosen the front sub-frame bolts.

6. Remove the heated oxygen (HO2S) sensor bracket, 6-cylinder engines only

7. Lower the rear edge of the sub-frame approximately 3.5 inches (90mm). Ensure that the steering gear bolts are released from the sub-frame.

8. Remove or disconnect the following:
- Rear engine mount
- Power steering hoses
- Steering shaft coupler

9. Slide the steering gear to the left and lower the assembly to remove.

***To install:***

10. Slide the steering gear in from the left hand side.

11. Install or connect the following:
- Power steering hoses using new O-rings.
- Steering shaft coupler and tighten to 15 ft. lbs. (20 Nm)
- Rear engine mount and tighten to 37 ft. lbs. (50 Nm)
- Sub-frame. Use new bolts and tighten them to 77 ft. lbs. (105 Nm) plus 120°.
- Sub-frame brackets and tighten to 37 ft. lbs. (50 Nm)
- Heated oxygen (HO2S) sensor bracket, 6-cylinder engines only
- Exhaust pipe bracket and tighten to 18 ft. lbs. (25 Nm)
- Outer tie rod ends
- Splash guards
- SRS connector, if equipped
- Front wheels

## POWER STEERING PUMP

*REMOVAL & INSTALLATION*

*See Figure 168.*

### ❋❋ WARNING

**Used fluid is considerably more dangerous than new fluid. Avoid skin contact with the oil.**

1. Before servicing the vehicle, refer to the Precautions Section.

2. Remove the accessory drive belt. Refer to Accessory Drive Belts removal and installation.

3. Drain the power steering pump oil:

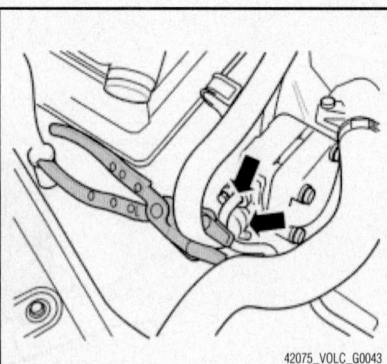

42075_VOLC_G0043

**Fig. 168 Removing hoses from power steering pump with lock-grip pliers in place**

a. Install a lock-grip pliers on the hose from the fluid reservoir as close to the power steering pump as possible.

b. Extract the oil.

➡**Place a covering over the alternator as protection against oil.**

4. Remove or disconnect the following:
- The pressure hose
- The feeder hose

5. Remove the mounting bolts from power steering pump.

6. Remove the power steering pump.

***To install:***

7. Installation is the reverse of the removal procedure.

8. Install the power steering pump with the M8 mounting bolts. Tighten to 18 ft. lbs. (24 Nm).

9. Install the pressure hose and feeder hose with the M6 bolts. Tighten to 88 inch lbs. (10 Nm).

➡**Use a new O-ring.**

10. Install the accessory drive belt. Refer to Accessory Drive Belts removing and installation.

11. Refill oil using Volvo power steering fluid, P/N 1161529.

12. Bleed the power steering system. Refer to Power Steering Pump bleeding.

*BLEEDING*

1. Before servicing the vehicle, refer to the Precautions Section

2. With engine off, turn the steering wheel fully to the right and left several times.

➡**Do not allow the fluid level in the reservoir tank to go below the MIN level line. Check and add fluid as needed.**

3. Run the engine at idle speed. Turn the steering wheel fully to the right and then fully to the left. Hold for about three seconds. Check for fluid leakage.

4. Repeat the above step several times at three second intervals.

➡**Do not hold the steering wheel in the locked position for more than ten seconds.**

5. Check for air bubbles or cloudy fluid. If found, repeat the bleeding procedure.

6. Stop the engine and check the fluid level. Fill as required.

**SUSPENSION**                                              **FRONT SUSPENSION**

## COIL SPRING

### REMOVAL & INSTALLATION

1. Before servicing the vehicle, refer to the Precautions Section.
2. Remove the strut assembly.
3. Install spring compressor 951-2911 or suitable equivalent on the coil spring.
4. Remove the shock absorber mounting nut using socket 999-5500 using a Torx® as a counterhold.
5. Remove or disconnect the following:
   - Shock absorber
   - Spring seat
   - Bump stop
   - Coil spring

### To install:

6. Install spring compressor 951-2911 or suitable equivalent on the coil spring.
7. Install or connect the following:
   - Bump stop
   - Coil spring
   - Spring seat
   - Shock absorber and tighten mounting nut to 52 ft. lbs. (70 Nm)
8. Remove the spring compressor.
9. Install the strut into the vehicle.
10. Check the wheel alignment and adjust as necessary.

## LOWER BALL JOINT

### REMOVAL & INSTALLATION

**S60, S80 and V70 Models**
*See Figures 169 through 175.*

1. Before servicing the vehicle, refer to the Precautions Section.
2. Remove the front wheels.
3. Remove the nut from the control arm. Use a Torx® wrench as a counterhold.

### ✳ WARNING

**Ensure that the tension strap is correctly secured in the control arms.**

4. Pull down the control arms (1) using a tension strap (2).
5. Release the spring strut from the control arm.
6. Remove the bolt from the halfshaft. Use a screwdriver as a counterhold on the brake disc.
7. Remove the halfshaft from the hub. Knock the halfshaft out using a brass drift.
8. Push the spring strut to one side.
9. Remove the bolts from the ball joint.
10. Remove the rubber seal (1).

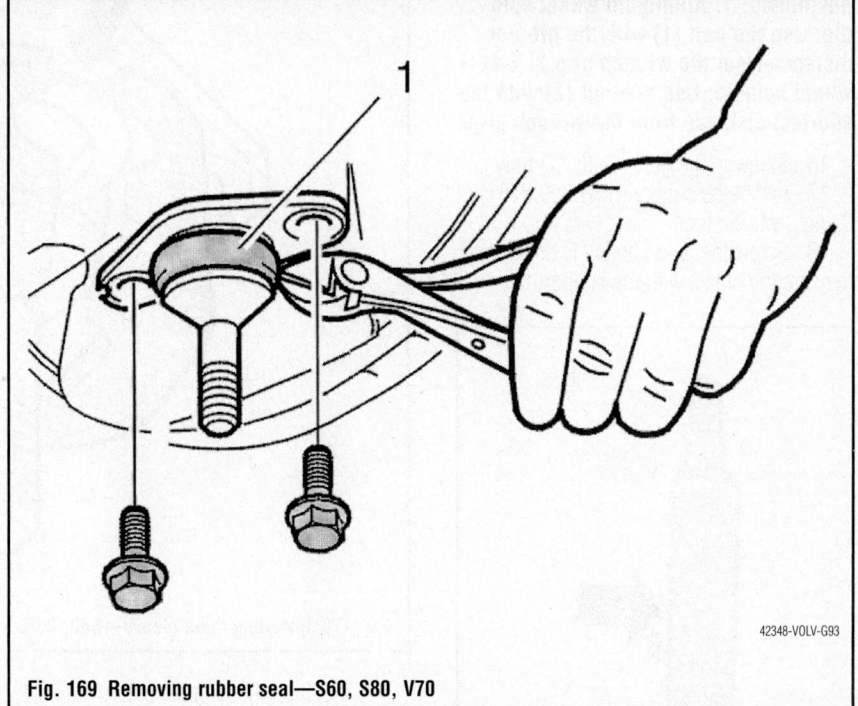

**Fig. 169 Removing rubber seal—S60, S80, V70**

42348-VOLV-G93

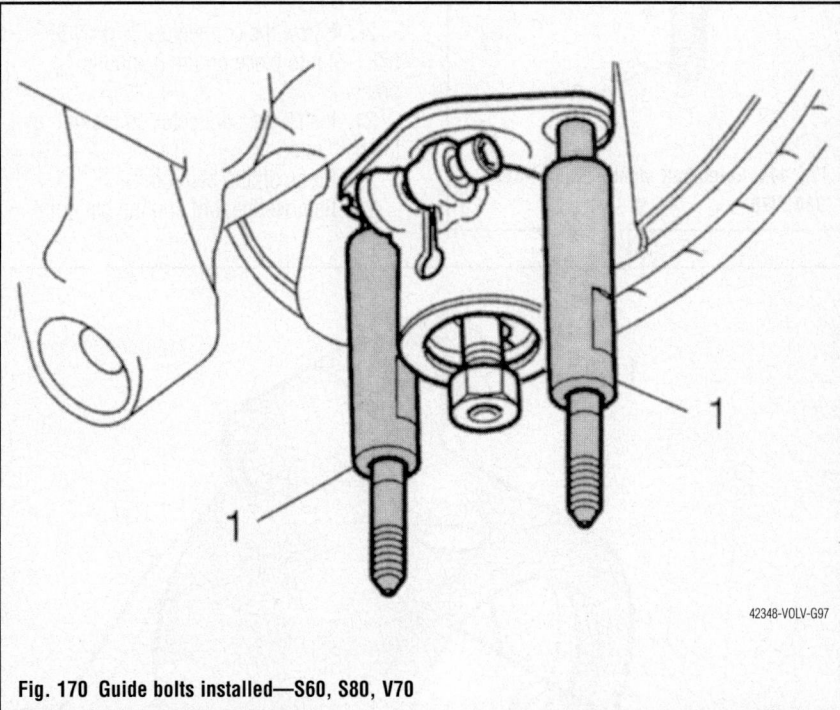

**Fig. 170 Guide bolts installed—S60, S80, V70**

42348-VOLV-G97

**Installing the inner sleeve (P/N 999 5781-1)**

11. Install the inner sleeve so that both halves hook securely in the flange (see arrow) on the ball joint.
12. Turn the inner sleeve so that the slit is at right angles to the bolt holes.

**Installing the outer sleeve (P/N 999 5781-2)**

13. Install the outer sleeve around the inner sleeve halves. Tighten the bolt (1).
14. Install an M12 nut on the ball joint bolt (2).
15. Screw on the nut so that the ball joint bolt is just protruding from the nut.

**Installing the guide bolts (P/N 999 5781-3)**

➡The guide bolts have a wrench grip (see arrow) which is asymmetrically positioned. 1. Aluminum wheel spindle: Use the bolt (1) with the greater distance from the wrench grip 2. Steel wheel spindle: Use the bolt (2) with the shortest distance from the wrench grip.

16. Screw in the guide bolts (1) fully.
17. Install the support (P/N 999 5781) using 3 wheel studs.
18. Screw the guide bolts (1) down until they are in contact with the support.

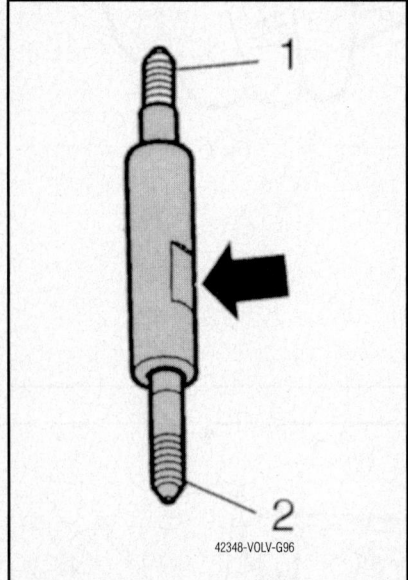

Fig. 171 Guide bolt identification—S60, S80, V70

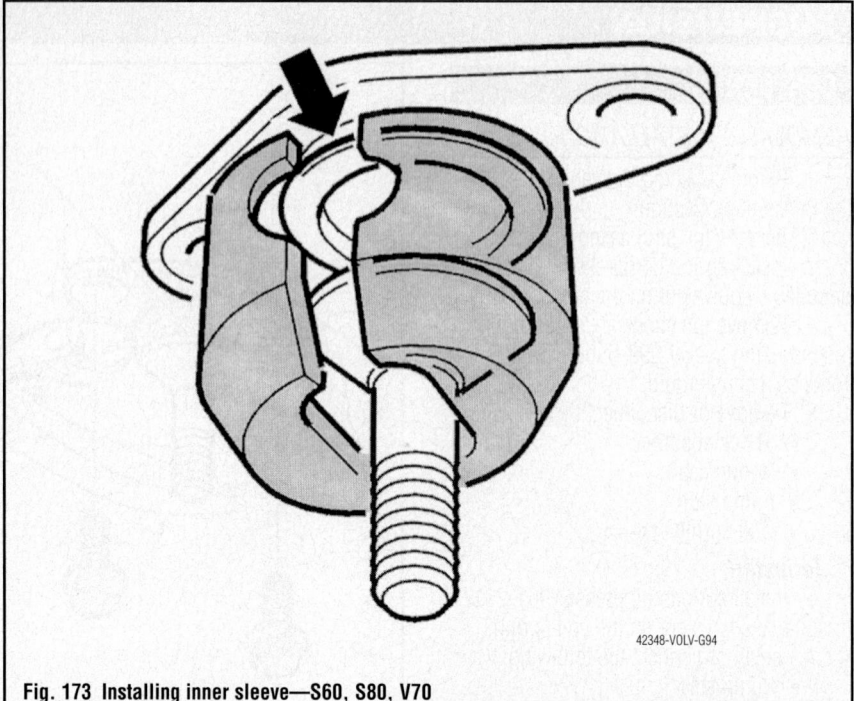

Fig. 173 Installing inner sleeve—S60, S80, V70

19. Position the supplementary support (2) (P/N 999 5781-4) on the hydraulic press (1).
20. Screw the connector (3) (P/N 999 5781-5) into place on the hydraulic press.
21. Install the connector on the nut on the ball joint.
22. Press off the ball joint.
23. Remove the tool and the ball joint.

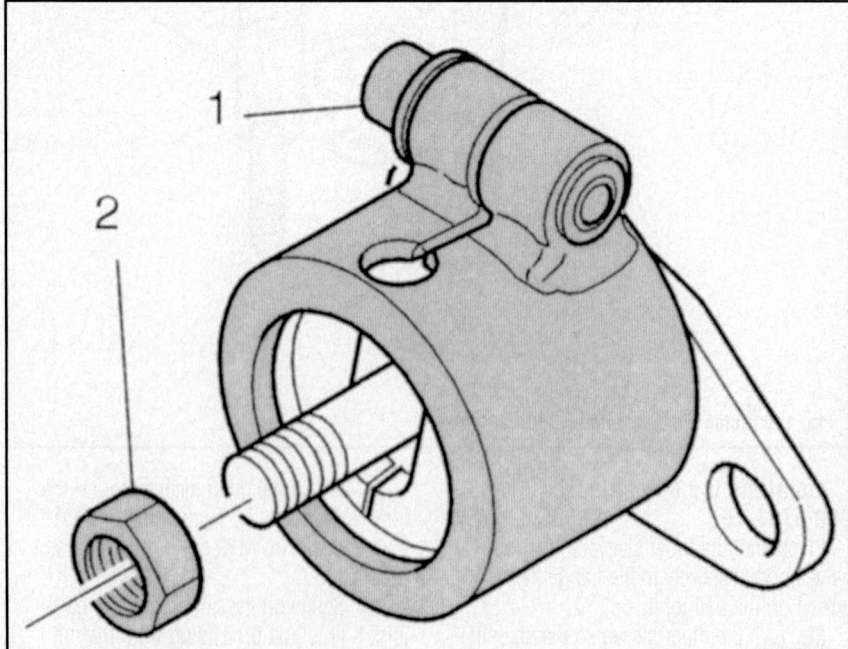

Fig. 172 Installing outer sleeve—S60, S80, V70

**✳✳ WARNING**

**Use protective goggles.**

24. Clean the ball joint seat and the mating surfaces of the ball joint to the wheel spindle. Use a rotary wire brush.
25. Lubricate the seat using wheel bearing grease.

***To install:***
26. Loosely install the new ball joint using the guide bolts.

**✳✳ WARNING**

**Leave the protective cap for the ball joint in position to prevent damaging the rubber seal.**

27. Press the ball joint up towards the seat using the impact drift (1) (P/N 999 5796). Check that the ball joint is centered in the seat.
28. Knock in the ball joint using a copper mallet.
29. Tighten the ball joint using new bolts. Tighten to 37 ft. lbs. (50 Nm)
30. Remove the protective cap (1).
31. Install the halfshaft in the hub. Install a new bolt.
32. Align the ball joints (1) in the control arm.

**✳✳ CAUTION**

**Take care when releasing the tension strap.**

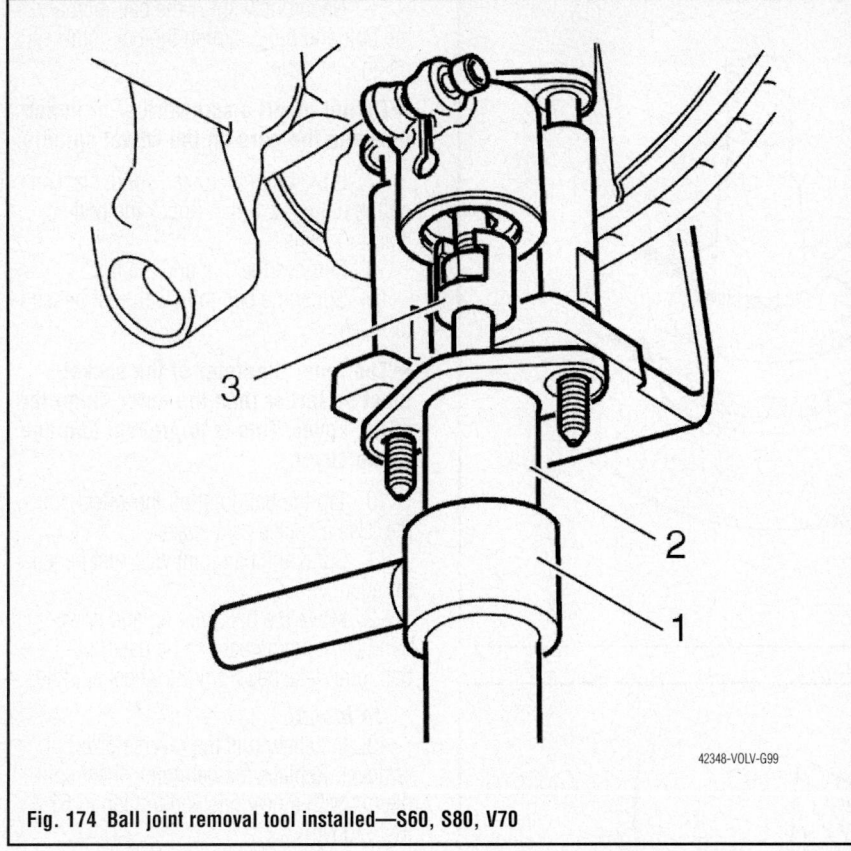

**Fig. 174 Ball joint removal tool installed—S60, S80, V70**

33. Slowly release the control arms (1) by releasing the tension strap (2).

34. Install a new nut.

→**The ball joint pinion must not rotate. Use a Torx® wrench as a counterhold so that the rubber boot is not damaged. Tighten to 59 ft. lbs. (80 Nm).**

35. Install the wheels. Tighten to 103 ft. lbs. (140 Nm).

**S40 and V50 Models**

*See Figures 176 and 177.*

1. Before servicing the vehicle, refer to the Precautions Section.

2. Raise the car.

3. Slacken off the ball joint nut as much as possible.

→**Clean the ball joint thread using a wire brush. Use rust penetrator. If the ball joint releases in the cone, position a transmission jack under the control arm. Carefully lift (not more than so the front spring is in the resting position) until the cone jams. Remove the nut.**

4. Remove the stud in the front edge

5. Remove the two bolts at the rear edge. Move the wheel to one side. Pull out the control arm from the cross member.

→**Ensure that the halfshaft is not pulled out of the transmission.**

6. Slacken off the nut positioned on the ball joint as much as possible. Press the rubber down.

**✳✳ WARNING**

**Secure the tool to prevent damage.**

7. Press out the ball joint.

8. Remove the nut and draw out the control arm.

*To install:*

**✳✳ CAUTION**

**Replace the ball joint rubber ring if it is worn or damaged.**

9. Clean the area around the ball joint.

10. Remove the old rubber ring.

11. Lubricate the new ring and install it.

12. Ensure that the locking springs are correctly installed.

13. Install the control arm with the ball joint on the wheel spindle.

14. Install the control arm in the cross member. Install the bolt and nut.

15. Install the two bolts in the rear bushing.

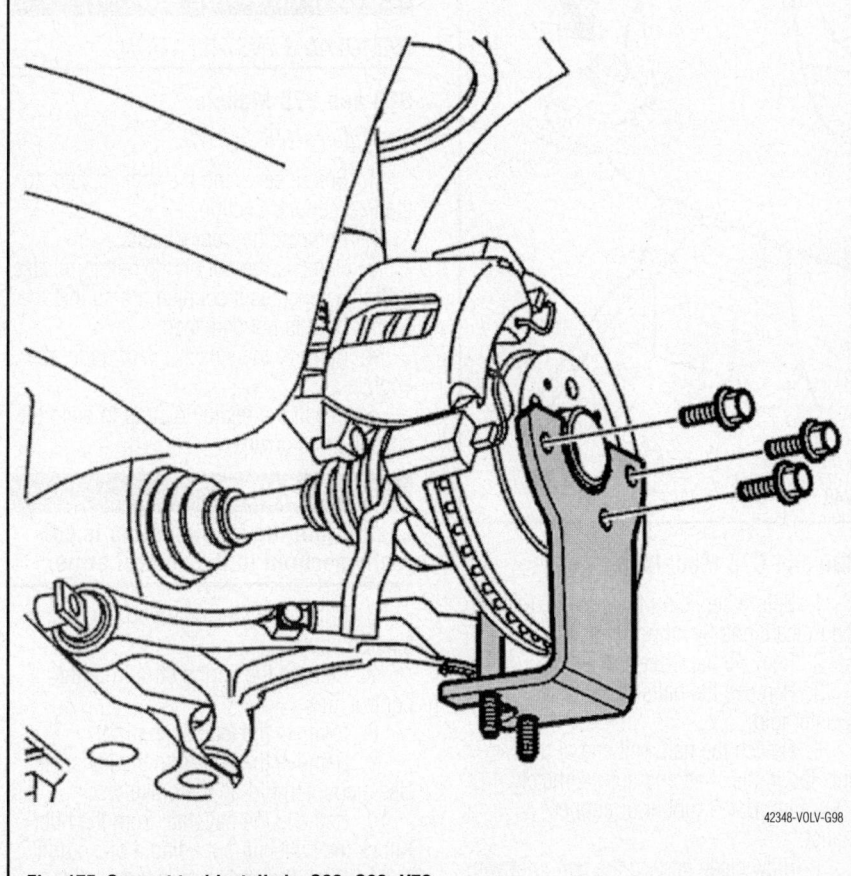

**Fig. 175 Support tool installed—S60, S80, V70**

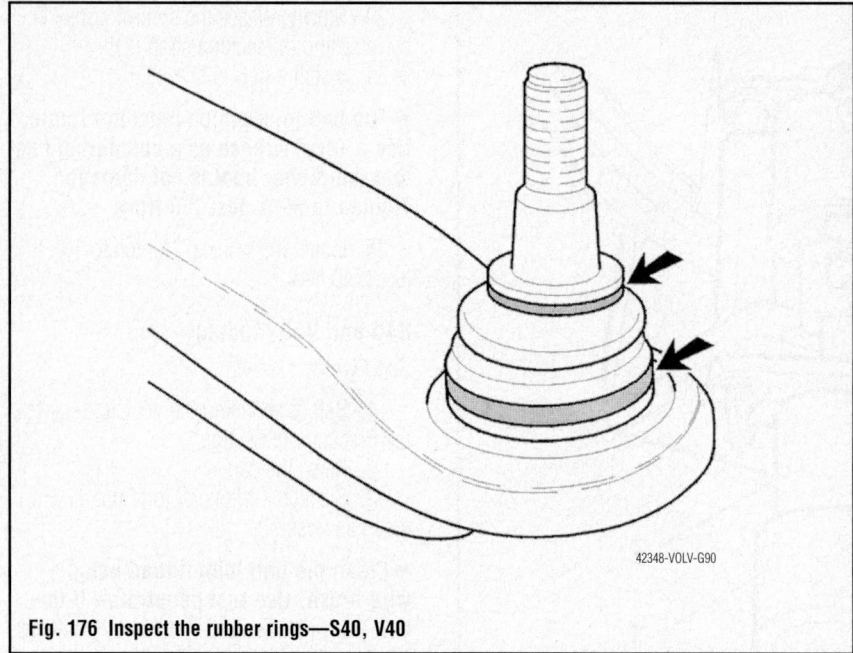

**Fig. 176 Inspect the rubber rings—S40, V40**

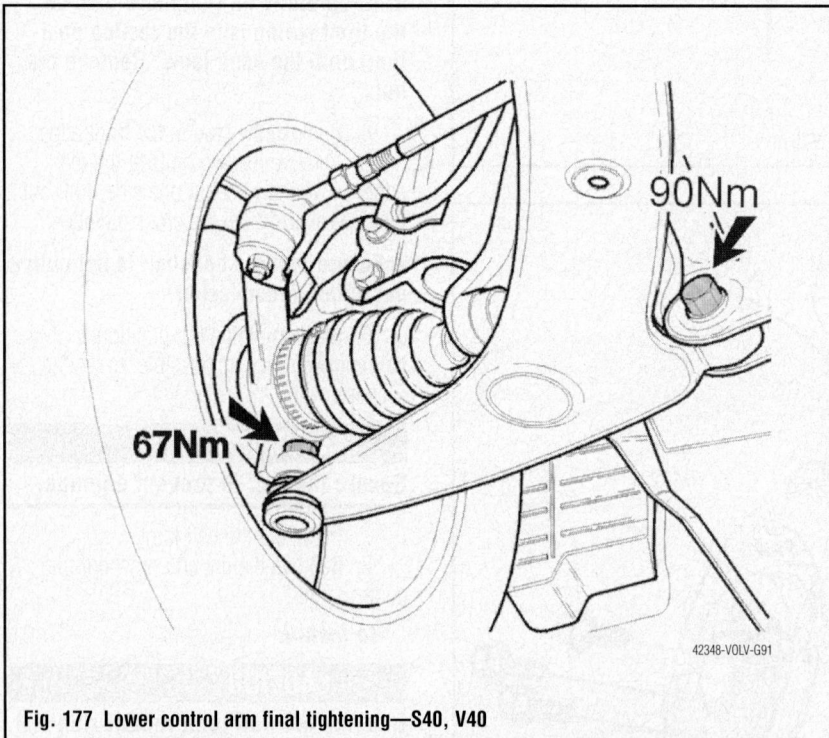

**Fig. 177 Lower control arm final tightening—S40, V40**

16. Tighten all bolts and nuts by hand.

17. Post-tighten the two rear bolts. Tighten to 68 ft. lbs. (90 Nm).

18. Press the ball joint all the way down. Tighten to 51 ft. lbs. (67 Nm).

19. Remove the support and rock the car a few times so that the bushings come into the correct position.

20. Tighten the nut. Tighten to 68 ft. lbs. (90 Nm) plus 60 deg.

21. Check the front mounting and wheel alignment.

### C30 and C70 Models

1. Before servicing the vehicle, refer to the Precautions Section.

2. Remove the wheel.

3. Remove the halfshaft bolt. Use a counterhold.

4. Detach the halfshaft end at the hub Tap in the shaft end approximately 10-15mm. Use a rubber or copper mallet.

5. Blow clean around the ball joint with compressed air.

6. Generously spray the ball joint and the bolt and nut retaining the ball joint using rust penetrator.

➡ **Do not insert a screwdriver or punch between the ears on the wheel spindle.**

7. If the bolt has rusted solid, slacken off the nut a few turns. Knock the bolt so that it loosens.

8. Remove the bolt and the nut.

9. Spray the ball joint with rust penetrator again.

➡ **The inner diameter of the socket must be larger than the outer diameter of the cover. This is to prevent damage to the cover.**

10. Tap the ball joint off the wheel spindle. Use a socket as a spacer.

11. Spray the ball joint with rust penetrator again.

12. Move the ball joint up and down by hand (or a jimmy bar can be used) until the ball joint detaches from the wheel spindle.

### To install:

13. Installation is the reverse order of removal. Replace the ball joint wheel spindle nut with a new one and tighten to 52 ft. lbs. (70 Nm).

## LOWER CONTROL ARM

### REMOVAL & INSTALLATION

#### S60 and V70 Models

*See Figures 178 and 179.*

1. Before servicing the vehicle, refer to the Precautions Section.

2. Remove the front wheels.

3. Remove the nut on the ball joint. Use a Torx® wrench as a counterhold so that the rubber boot is not damaged.

4. Remove the splash guard under the engine.

5. Install the retaining strap to bend the control arms down.

### ❄ WARNING

**Ensure that the tension strap is correctly secured in the control arms.**

6. Pull down the control arms using a tension strap.

7. Release the spring strut from the control arm

8. Remove the tensioner strap.

9. Remove the bolt from the halfshaft. Use a counterhold on the brake disc

10. Remove the halfshaft from the hub. Knock the halfshaft out using a brass drift

11. Push the spring strut to one side.

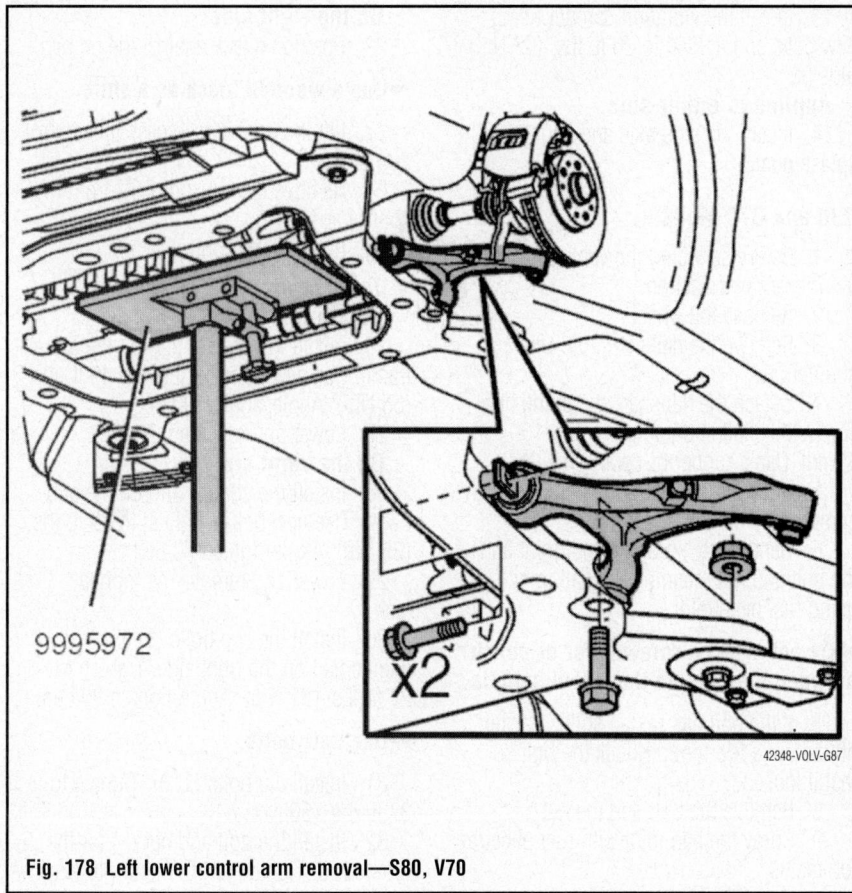

9995972

**Fig. 178 Left lower control arm removal—S80, V70**

42348-VOLV-G87

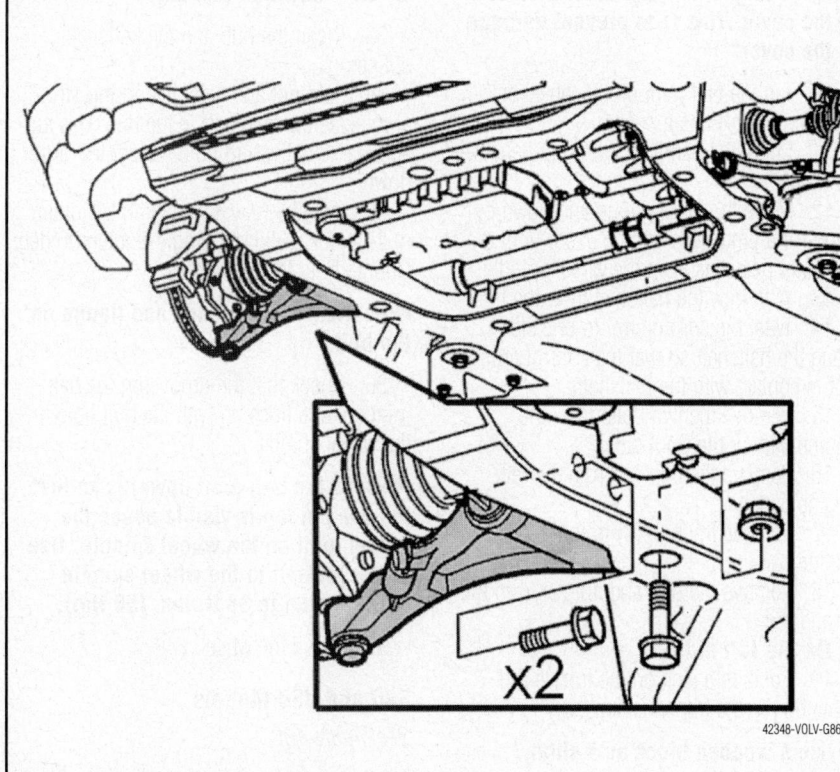

**Fig. 179 Right lower control arm removal—S80, V70**

42348-VOLV-G86

**On the right side:**

12. Remove the hose from the EVAP canister and the fuel line from the engine block heater (where applicable) from their mountings. Move the hose and fuel line to one side.

13. Lift the engine approximately 10mm using a mobile jack so that the front 2 bolts on the control arm can be removed

14. Position a mobile jack with a universal plate under the oil pan on the right side of the engine.

15. Remove the bolts and the nut from the control arm in the sub-frame

16. Remove the control arm.

**On the left side:**

17. Lift the engine approximately 15mm using a mobile jack so that the front bolts on the control arm can be removed.

18. Position a mobile jack with a universal plate under the oil pan on the left side of the engine.

19. Remove the bolts and the nuts from the control arm in the sub-frame

20. Remove the control arm.

*To install:*
**On the right side:**

21. For the right-hand control arm, lift the engine 10mm using a mobile jack.

22. Install the control arm on the sub-frame. Use new bolts and a new nut

23. Tighten the front bolts. Tighten to 49 ft. lbs. (65 Nm). Angle-tighten 90 deg.

24. Tighten the rear bolt and nut. Tighten to 77 ft. lbs. (105 Nm). Angle-tighten 90 deg.

25. Remove the jack.

26. Install the hose at the EVAP canister and the fuel line for the engine block heater (if applicable) in their mountings on the sub-frame.

27. Reinstall the splash guard under the engine.

**On the left side:**

28. Lift the engine 15mm using a mobile jack.

29. Install the control arm on the sub-frame. Use new bolts and new nuts

30. Tighten the front bolts. Tighten to 49 ft. lbs. (65 Nm). Angle-tighten 90 deg.

31. Tighten the rear bolt and nut. Tighten to 77 ft. lbs. (105 Nm). Angle-tighten 90 deg.

32. Remove the mobile jack

33. Reinstall the splash guard under the engine.

34. Turn in the wheel spindle and locate the halfshaft in the hub.

➡**Check that the seal is not worn or damaged. Replace if necessary. Ensure that the sealing ring is correctly positioned on the halfshaft.**

35. Clean the splines on the halfshaft.
36. Install the halfshaft in the hub.
37. Install a new bolt. Use a counterhold. Tighten to 27 ft. lbs. (35 Nm) plus 90 deg.
38. Install the retaining strap to bend the control arms down

➡**Ensure that the tension strap is correctly secured in the control arms.**

39. Align the ball joints (1) in the control arm.

➡**Take care when releasing the tension strap.**

40. Slowly release the control arms by releasing the tension strap.
41. Install a new nut

➡**The ball joint pinion must not rotate. Use a Torx® wrench as a counterhold so that the rubber boot is not damaged. Tighten to 59 ft. lbs. (80 Nm).**

42. Install the wheels.

## S80 Models

1. Before servicing the vehicle, refer to the Precautions Section.
2. Remove the stub axle and engine splash guard.

**On the left side:**
3. Remove the front engine mounting bolt.
4. Lift the left side of the engine approx. 1 inch (25mm) to expose the control arm bolt.
5. Remove the bolts for the control arm in the sub-frame and remove the control arm.

**On the right side:**
6. Remove the vibration damper, if equipped with a 6-cylinder engine.
7. Remove both right hand engine mounting bolts.
8. Lift up the ride side of the engine approx. 1 inch (25mm) to expose the control arm bolt.
9. Remove the bolts for the control arm in the sub-frame and remove the control arm.

*To install:*
10. Install the control arm on the sub-frame using new fasteners. Tighten the front bolts to 48 ft. lbs. (65 Nm) plus 90° and tighten the rear bolt to 77 ft. lbs. (105 Nm) plus 90°.

**On the left side:**
11. Lower the engine and tighten the front engine mounting bolt to 37 ft. lbs. (50 Nm).

**On the right side:**
12. Lower the engine and tighten the right engine mounting bolts to 26 ft. lbs. (35 Nm) plus 90°.

13. Install the vibration damper using new bolts and tighten to 26 ft. lbs. (35 Nm) plus 50°.

**Applies to either side:**
14. Install the stub axle and engine splash guard.

## C30 and C70 Models

1. Before servicing the vehicle, refer to the Precautions Section.
2. Remove the wheel.
3. Remove the halfshaft bolt. Use a counterhold.
4. Detach the halfshaft end at the hub Tap in the shaft end approximately 10-15mm. Use a rubber or copper mallet.
5. Blow clean around the ball joint with compressed air.
6. Generously spray the ball joint and the bolt and nut retaining the ball joint using rust penetrator.

➡**Do not insert a screwdriver or punch between the ears on the wheel spindle.**

7. If the bolt has rusted solid, slacken off the nut a few turns. Knock the bolt so that it loosens.
8. Remove the bolt and the nut.
9. Spray the ball joint with rust penetrator again.

➡**The inner diameter of the socket must be larger than the outer diameter of the cover. This is to prevent damage to the cover.**

10. Tap the ball joint off the wheel spindle. Use a socket as a spacer.
11. Spray the ball joint with rust penetrator again.
12. Move the ball joint up and down by hand (or a jimmy bar can be used) until the ball joint detaches from the wheel spindle.
13. Withdraw the halfshaft from the hub.
14. Twist the spring strut to one side. Hang the halfshaft so that the control arm is not in contact with the halfshafts.
15. Remove the two bolts from the engine pad on the right side.
16. Remove the bolt (1) from the rear engine pad
17. Remove the bolt (3) from the front engine pad
18. Remove the bolt and nut (2) from the torque rod.

**On the left side:**
19. Position a jack on the torque rod mounting in the transmission.

➡**Use a wooden block as a shim.**

20. Remove the four bolts from the control arm
21. Remove the control arm.

**On the right side:**
22. Position a jack against the oil pan.

➡**Use a wooden block as a shim.**

23. Lift the engine to access the control arm bolts.
24. Remove the four bolts (1) from the control arm
25. Remove the control arm.

*To install:*
**On the left side:**
26. Install the control arm on the sub-frame. Use new bolts. Tighten to 49 ft. lbs. (65 Nm). Angle-tighten 90 deg.
27. Lower and remove the jack.

**On the right side:**
28. Install the control arm on the sub-frame. Use new bolts. Tighten to 49 ft. lbs. (65 Nm). Angle-tighten 90 deg.
29. Lower and remove the mobile jack.
30. Install the two bolts (1) for the engine pad on the right side. Tighten to 27 ft. lbs. (35 Nm). Angle-tighten 90 deg.

➡**Use new bolts.**

31. Install the bolts (1, 3). Tighten to 37 ft. lbs. (50 Nm).
32. Install the bolt and nut (2) for the torque rod. Tighten to 30 ft. lbs. (40 Nm). Angle-tighten 40 deg.

➡**Use a new bolt and nut.**

33. Clean the hub and the halfshaft splines.
34. Fold out and twist the spring strut and locate the halfshaft in the hub. Use an adjustable wrench to hold the control arm down.
35. Install a *new* halfshaft bolt. Tighten to 27 ft. lbs. (35 Nm). Angle-tighten 90 deg. Counterhold the brake disc.

➡**Lubricate the threads and flange on the bolt.**

36. Check that the groove on the ball joint spindle lines up with the bolt hole in the wheel spindle.

➡**Press the ball joint upwards so that the conical top is visible above the bolted joint on the wheel spindle. Use a new locknut in the wheel spindle bolt. Tighten to 38 ft. lbs. (50 Nm).**

37. Install the wheel.

## S40 and V50 Models
*See Figure 180.*

1. Before servicing the vehicle, refer to the Precautions Section.
2. Raise the car.

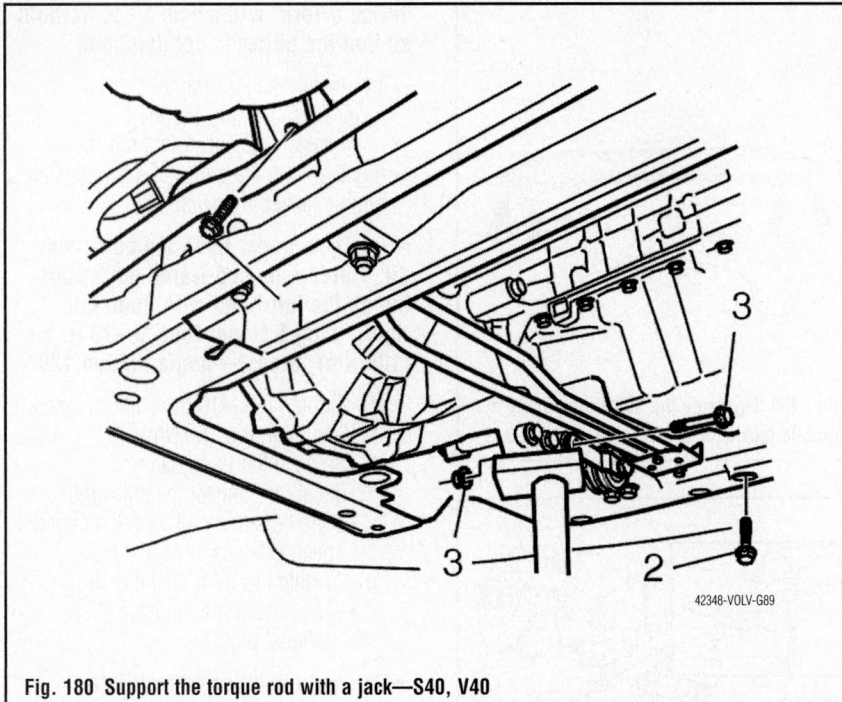

**Fig. 180 Support the torque rod with a jack—S40, V40**

3. Slacken off the ball joint nut as much as possible.

➡**Clean the ball joint thread using a wire brush. Use rust penetrator. If the ball joint releases in the cone, position a transmission jack under the control arm. Carefully lift (not more than so the front spring is in the resting position) until the cone jams. Remove the nut.**

4. Remove the stud in the front edge.
5. Remove the two bolts at the rear edge.
Move the wheel to one side. Pull out the control arm from the cross member.

➡**Ensure that the halfshaft is not pulled out of the transmission.**

6. Slacken off the nut positioned on the ball joint as much as possible. Press the rubber down.

### ✳✳ WARNING

**Secure the tool to prevent damage.**

7. Press out the ball joint.
8. Remove the nut and draw out the control arm.

*To install:*

### ✳✳ CAUTION

**Replace the ball joint rubber ring if it is worn or damaged.**

9. Clean the area around the ball joint.
10. Remove the old rubber ring.

11. Lubricate the new ring and install it.
12. Ensure that the locking springs are correctly installed.
13. Install the control arm with the ball joint on the wheel spindle.
14. Install the control arm in the cross member. Install the bolt and nut.
15. Install the two bolts in the rear bushing.
16. Tighten all bolts and nuts by hand.
17. Post-tighten the two rear bolts. Tighten to 68 ft. lbs. (90 Nm).
18. Press the ball joint all the way down. Tighten to 51 ft. lbs. (67 Nm).
19. Remove the support and rock the car a few times so that the bushings come into the correct position.
20. Tighten the nut. Tighten to 68 ft. lbs. (90 Nm) plus 60 deg.
21. Check the front mounting and wheel alignment.

### MACPHERSON STRUT

*REMOVAL & INSTALLATION*

#### C30, C70, S60, S80 and V70 Models

1. Before servicing the vehicle, refer to the Precautions Section.
2. Remove or disconnect the following:
   • Wheel
   • Anti-roll bar link
   • ABS sensor
   • Bolts holding the strut in the wheel spindle
   • Strut assembly

*To install:*

➡**Use new fasteners for safe assembly.**

3. Install or connect the following:
   • Strut assembly and tighten the upper mount nuts to 18 ft. lbs. (25 Nm)
   • Strut assembly onto the wheel spindle and tighten to 77 ft. lbs. (105 Nm) plus 90°
   • Anti-roll bar link and tighten to 37 ft. lbs. (50 Nm)
   • ABS sensor
   • Wheel

#### S40 and V50 Models

1. Before servicing the vehicle, refer to the Precautions Section.
2. Remove or disconnect the following:
   • Wheel
   • Halfshaft retainer bolt
   • Ball joint from the wheel spindle
   • Brake caliper
   • Halfshaft from the wheel hub
   • Anti-roll bar link
3. Remove the lower strut mounting bolt and carefully tap the wheel spindle down until it release from the strut assembly.
4. Remove the three top strut mounting bolts and remove the strut assembly.

*To install:*

5. Install or connect the following:
   • Strut assembly using new fasteners
   • Wheel spindle to strut assembly and tighten to 66 ft. lbs. (90 Nm)
   • Halfshaft into the wheel hub
   • Ball joint to the wheel spindle and tighten to 52 ft. lbs. (70 Nm)
   • Halfshaft retainer bolt
   • Brake caliper
   • Anti-roll bar link
   • Wheel

### STABILIZER BAR

*REMOVAL & INSTALLATION*

See Figures 181 through 184.

1. Before servicing the vehicle, refer to the Precautions Section.

➡**The anti-roll bar bushings and caps are vulcanized to the anti-roll bar and cannot be replaced separately.**

2. Install the lifting tool 999 5716, modified according to the tool bulletin for version C, and lifting hooks 999 5460.
3. Lift the engine to relieve the load on the engine pads.
4. Remove the front wheels.
5. Slacken off the right lower front part of the fender liner.

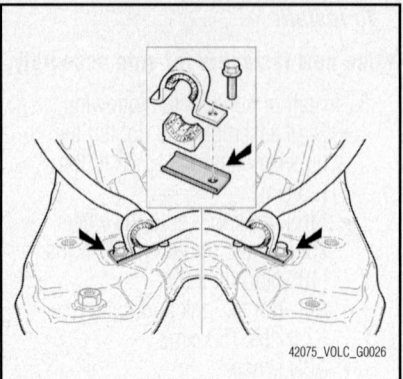

**Fig. 181 Installation of lifting tool 999 5716 and lifting hooks 999 5460**

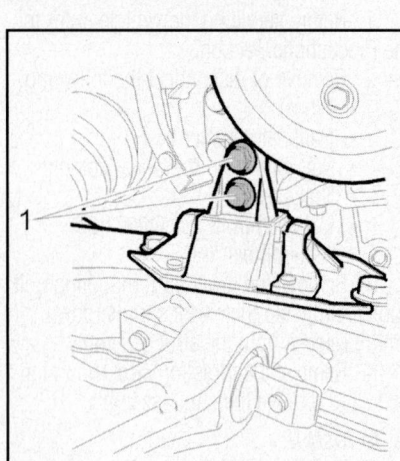

**Fig. 182 Location and removal of the SIPS member**

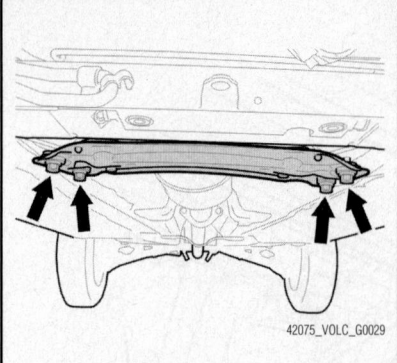

**Fig. 183 Lowering the sub-frame using a mobile jack with universal plate 999 5972**

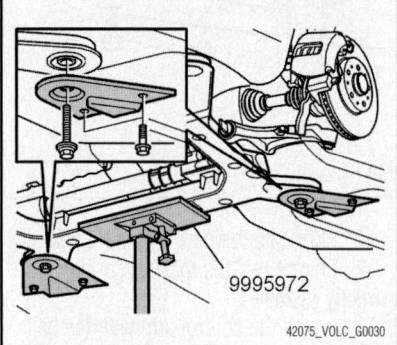

**Fig. 184 Lowering the sub-frame to the appropriate clearance**

6. Remove or disconnect the following:
   - The plastic nuts on the fender liner
   - The protective plate
   - Bend out the fender liner
   - Bolts for the right-hand engine mounting (1)
7. Remove or disconnect the following:
   - The splash guard under the engine
   - The skid guard, if equipped
   - The 4 mounting bolts for the Side Impact Protection System (SIPS) member
   - The brake pipe from the snap fastener on the member
   - The SIPS member
8. Remove or disconnect the following:
   - The rubber mountings for the exhaust pipe
   - The nuts and bolts holding the steering gear to the sub-frame
   - The power steering line from the clips on the sub-frame
   - The bracket for the heated oxygen sensor (HO2S) from the sub-frame
   - The heated oxygen sensor wiring from the right side of the sub-frame (only applies to 6-Cylinder engines)
9. Lower the sub-frame:
   a. Position a mobile jack with universal plate 999 5972 at the rear of the sub-frame.
   b. Remove the bolts holding the sub-frame bracket in the body.
   c. Remove the rear bolt for the sub-frame with the bracket and washer.
   d. Carefully lower the sub-frame until the clearance **A** between the sub-frame and the body is 4 ⅓ inches (110mm).
10. Remove the anti-roll bar on both sides. Use a Torx® wrench as a counterhold so that the boot is not damaged.
11. Remove the bolts for the anti-roll bar.
12. Remove the anti-roll bar.

**To install:**
13. Install or connect the following:
   - The anti-roll bar
   - The (M10) bolts for the anti-roll bar. Tighten to 37 ft. lbs. (50 Nm)
   - The nut for the anti-roll bar link on both sides. Tighten to 59 ft. lbs. (80 Nm)

➡Use a Torx® wrench as a counterhold so that the gaiter is not damaged.

14. Install the sub-frame:
   a. Raise the sub-frame.
   b. Install the sub-frame four bolts together with the sub-frame brackets and engine heater, if installed.

➡Use new bolts. Lube the bolts with oil. Tighten the sub-frame bolts starting on the left-hand side, then the right, using 2 steps: Step 1—78 ft. lbs. (105 Nm), Step 2—angle tighten 120°.

15. Tighten the M10 bolts for the brackets M10 to 37 ft. lbs. (50 Nm).
16. Remove the lifting table.
17. Install or connect the following:
   - The nuts and bolt for the steering gear. Use new nuts and bolt. Tighten to 37 ft. lbs. (50 Nm)
   - The rubber mountings for the exhaust pipe

➡Use soap solution to facilitate installing the rubber mountings.

18. Install the SIPS member:
   a. Install the brake pipe in the snap fastener on the member.
   b. Install the SIPS member. Tighten the M8 bolts to 18 ft. lbs. (24 Nm).
19. Install or connect the following:
   - The skid plate, if equipped, up into the bumper wraparound. Press forwards until the mounting brackets engage in the sub-frame
   - The M10 skid plate bolts. Tighten to 37 ft. lbs. (50 Nm)
   - The splash guard under the engine. Tighten the M8 bolts to 18 ft. lbs. (24 Nm)
20. Install the right-hand engine mounting:
   a. Install the bolts for the right-hand engine mounting. Use new bolts. Tighten in 2 steps:
   - Step 1—89 ft. lbs. (120 Nm)
   - Step 2—angle-tighten 40°
21. Fold the fender liner back into position.
22. Install the plastic nuts.
23. Install the front wheels and tighten lug nuts to 81–103 ft. lbs. (110–140 Nm). See torque specification in the following chart.
24. Remove the lifting tools 999 5716 and lifting hooks 999 5460

## STEERING KNUCKLE

### REMOVAL & INSTALLATION

*See Figures 185 through 187.*

1. Before servicing the vehicle, refer to the Precautions Section.

2. Raise and support the vehicle safely.

3. Remove front wheel.

4. Remove or disconnect the following:
- The anti-roll bar link from the spring strut. Use a fixed wrench as a counterhold
- The ABS sensor. Hang up the sensor using a piece of wire
- The ABS sensor cable from the spring strut
- The brake caliper mounting bolts. Hang the caliper up using a piece of wire
- The halfshaft bolt. Use a screwdriver as a counterhold on the brake disc
- The locating pin holding the brake disc
- The brake disc. Detach the end of the halfshaft in the hub by knocking the drive shaft into the hub approximately ⅓–½ inch (10–15mm). Use a rubber or copper mallet
- The tie rod ends from the steering arm

5. Measure the position of the steering knuckle and spring strut for installation purposes.

6. Remove the bolts retaining the spring strut and the steering knuckle.

7. Press out the steering knuckle. At the same time pull the halfshaft from the transaxle.

8. Suspend the drive shaft from a hook 999 5045.

➡**Take care not to damage the halfshaft boot.**

9. Disconnect the ball joint pinion from the control arm. Use tool 999 7062.

10. Install cap nut 7062-1 on the ball joint pinion.

11. Tighten the nut against the control arm.

12. Then turn back one turn to form a gap between the cap nut and the control arm.

13. Position the extractor on the control arm. Detach the ball joint from the control arm.

### To install:

14. Install the steering knuckle on the trailing arm.

15. Install a new nut on the ball joint. Tighten to 59 ft. lbs. (80 Nm). Use a Torx® wrench as counterhold.

➡**Make sure that the mating surfaces on the ball joint and link are clean.**

16. Clean the splines on the halfshaft.

17. Turn the steering knuckle and bring the halfshaft into the hub.

18. Install halfshaft bolt and finger-tighten. Lubricate the bolt.

19. Install bolts retaining the spring strut in the steering knuckle. Use new bolts and nuts. Tighten to 78 ft. lbs. (105 Nm) and then angle-tighten 60°.

20. Remove or disconnect the following:

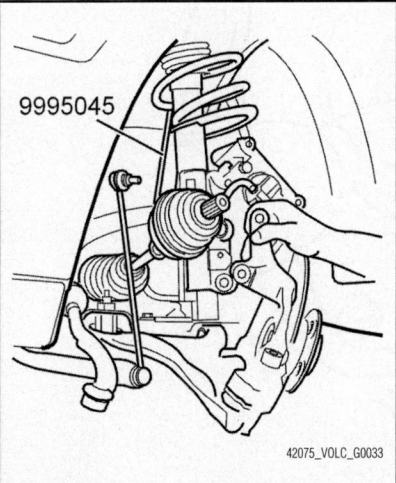

**Fig. 186 Suspending halfshaft from hook 999 5045**

- The anti-roll bar link to the spring strut. Use a new nut. Counterhold using a Torx® wrench so that the boot is not damaged. Tighten to 37 ft. lbs. (50 Nm)
- The ABS sensor cable
- The ABS sensor onto the steering knuckle. Tighten the ABS sensor to 88 inch lbs. (10 Nm)

➡**Ensure that the sensor seat in the steering knuckle is absolutely clean. Clean the ABS sensor with a soft brush.**

- The tie rod end onto the steering knuckle. Use a new nut. Tighten to 59 ft. lbs. (80 Nm)
- The brake disc. Tighten the brake disc locating pin to 88 inch lbs. (10 Nm)

➡**Ensure that the brake disc and wheel rim hub mating surfaces are clean.**

21. Tighten the halfshaft bolt. Use a screwdriver as a counterhold on the brake disc. Tighten to 26 ft. lbs. (35 Nm), plus 120°.

22. Install the brake caliper. Use new bolts

23. Install the wheel.

### WHEEL BEARINGS

#### REMOVAL & INSTALLATION

#### C30, C70, S60 and S80 Models w/Steel Spindle

*See Figures 188 and 189.*

1. Before servicing the vehicle, refer to the Precautions Section.

2. Remove the wheel

3. Remove the ABS sensor.

4. Detach the ABS sensor cable from the spring strut.

5. Remove both mounting bolts from the brake caliper.

6. Hang up the brake caliper, on the spring strut for example.

➡**Take care not to damage the brake hose while working.**

7. Remove the bolt holding the halfshaft. Counterhold the brake disc.

8. Remove the locating pin from the brake disc

9. Remove the brake disc.

10. Tap the end of the halfshaft into the hub approximately 10–15mm. Use a brass drift.

11. Remove the upper link from the anti-roll bar from the spring strut.

12. Remove the steering arm from the wheel spindle.

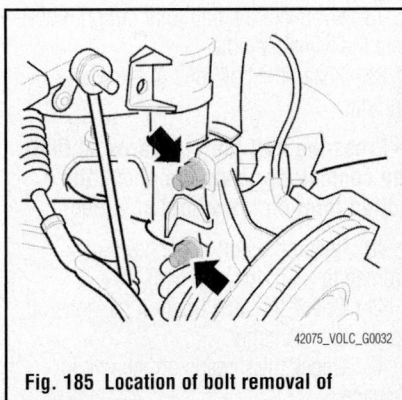

**Fig. 185 Location of bolt removal of steering knuckle**

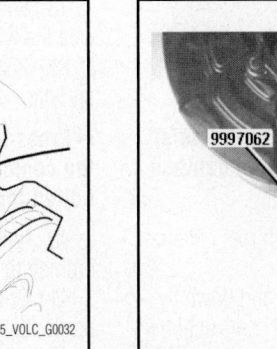

**Fig. 187 Using tool 999 7062 to extract ball joint**

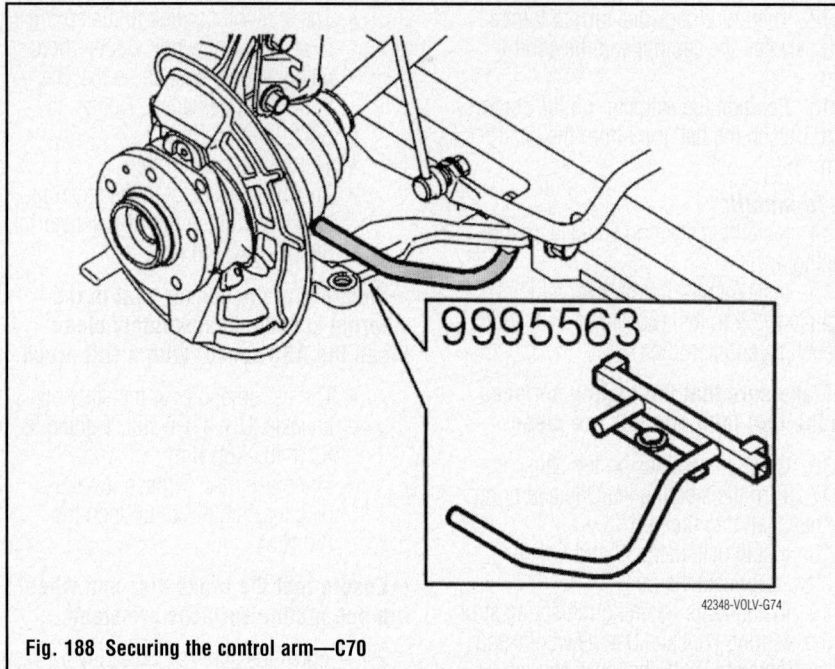

**Fig. 188 Securing the control arm—C70**

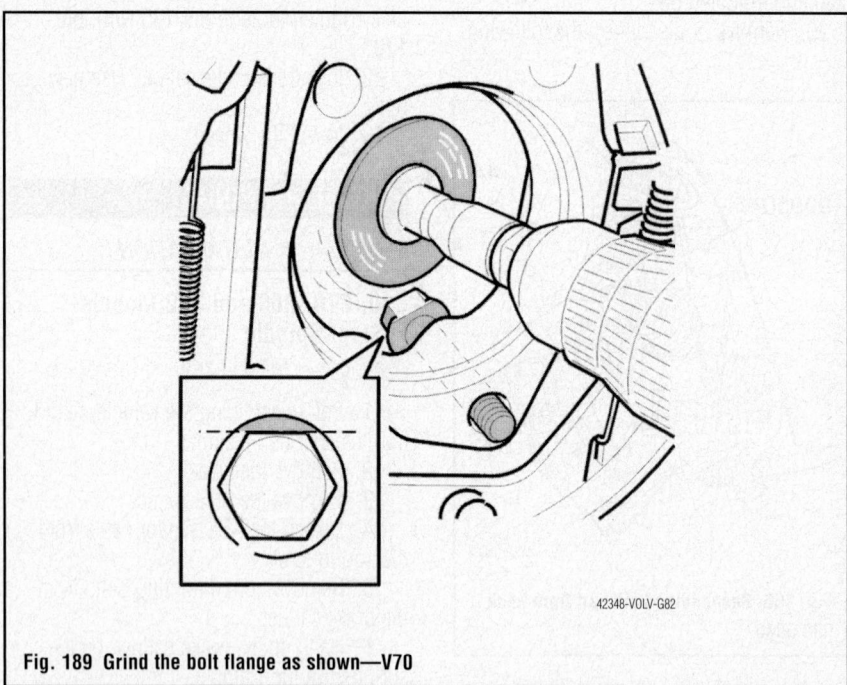

**Fig. 189 Grind the bolt flange as shown—V70**

13. Clean around the ball joint with compressed air.

14. Spray the area around the ball joint with plenty of rust penetrator. Spray the bolt and the nut holding the ball joint with rust penetrator. If the bolt has rusted solid, slacken of the nut a few turns. Knock the bolt so that it loosens.

15. Remove the nut and bolt.

16. Spray the ball joint with rust penetrator again.

**→Do not insert a punch between the ears on the wheel spindle.**

17. Tap the ball joint off the wheel spindle. Use a socket as a spacer.

**→The inner diameter of the sleeve must be larger than the outer diameter of the cover. This is to prevent damage to the cover.**

18. Spray the ball joint with rust penetrator again.

19. Move the ball joint up and down by hand (alternatively use a jimmy bar) until the ball joint detaches from the wheel spindle.

20. Install protective socket 999 5562 on the ball joint.

21. Secure the control arm in the depressed position using tool 999 5563.

22. Remove the halfshaft from the hub.

23. Turn the spring strut a half turn and lock it by positioning the steering arm on the control arm.

**→Take care not to damage the brake hose.**

24. Hang up the halfshaft on the spring strut.

25. Remove the wheel bearing unit.

26. Check the wheel spindle mating surfaces against the wheel bearing unit. Also check the bolt head mating surfaces on the wheel spindle.

*To install:*

27. Lubricate all the mating surfaces sparingly.

28. Install the wheel bearing unit. Use new bolts.

29. Lightly lubricate the outer threads and tighten crosswise. Tighten to 49 ft. lbs. (65 Nm). Angle-tighten the bolts for the wheel bearing unit crosswise. Tighten an additional 60 deg.

30. Check the halfshaft for damage to the mating surfaces with the wheel bearing.

31. Check the splines.

32. Clean and check the pulse wheel. Check the shape of the cogs.

33. Lubricate the splines and mating surfaces with the wheel bearing sparingly.

34. Remove the halfshaft from where it was hung earlier. Turn the spring strut a half turn to disconnect it from the control arm.

**→Take care not to damage the brake hose.**

35. Install the halfshaft in the hub by hand. It must be possible to move the halfshaft easily backward and forward when it is in position.

36. Lubricate a new halfshaft bolt. Screw it in a few turns by hand.

37. Clean off any rust from the ball joint and wheel spindle.

38. Lubricate the ball joint pinion. Lubricate the wheel spindle.

39. Install the ball joint in the wheel spindle

**→Press the ball joint upwards so that the conical top is visible above the bolted joint on the wheel spindle.**

40. Install the bolt and a new nut. Tighten to 38 ft. lbs. (50 Nm).

41. Check the dust boot on the anti-roll bar link for damage.

42. Check the threads on the link for damage.

43. Lubricate the threads.

44. Install the upper link of the anti-roll bar on the spring strut. Use a new lock nut. Tighten to 38 ft. lbs. (50 Nm).

45. Install the tie rod on the wheel spindle. Use a new lock nut. Tighten to 53 ft. lbs. (70 Nm).

46. Check that the brake disc mating surfaces with the wheel bearing unit are smooth and even.

47. Install the brake disc

48. Install the locating pin for the brake disc. Tighten to 72 inch lbs. (8 Nm).

49. Check the mating surfaces of the brake caliper and wheel spindle.

50. Clean and grease the surfaces of the steering limiter.

51. Install the brake caliper. Use new bolts. Tighten to 76 ft. lbs. (100 Nm).

52. Tighten the center bolt for the halfshaft to 27 ft. lbs. (35 Nm) plus 90°. Counterhold the brake disc.

53. Clean and check that the ABS sensor mating surface on the wheel spindle is true.

54. Apply a very small amount of grease to the mating surface.

55. Install the ABS sensor cable in the holder on the spring strut.

56. Install the ABS sensor. Lubricate the bolt and tighten it. Tighten to 88 inch lbs. (10 Nm).

57. Install the wheel

### S80 and V70 Models w/Aluminum Spindle

1. Before servicing the vehicle, refer to the Precautions Section.

2. Remove the wheel.

3. Remove the brake caliper and limiter bolts.

4. Hang the caliper up using a piece of wire.

5. Remove the halfshaft bolt. Use a screwdriver as a counterhold on the brake disc.

6. Remove the locating pin holding the brake disc.

7. Remove the brake disc.

8. Slacken off the end of the halfshaft in the hub by knocking the halfshaft into the hub approximately 10–15mm.

9. Use a rubber or copper mallet.

10. Remove the nut holding the control arm. Use a Torx® wrench as a counterhold.

### ✳✳ WARNING

**Ensure that the tension strap is correctly secured in the control arms.**

11. Pull down the control arm (1) using the tension strap (2)

12. Release the spring strut from the control arm.

13. Remove the halfshaft. Knock the halfshaft out using a brass drift. Hang the halfshaft as illustrated

14. Secure the spring strut in position

15. Remove the 4 bolts from the hub.

*To install:*

16. Install the hub. Use new bolts

17. Tighten crosswise to 15 ft. lbs. (20 Nm). Then tighten to 49 ft. lbs. (65 Nm). Angle-tighten 60 deg.

18. Turn in the spring strut and position the halfshaft in the hub.

➡**Check that the seal is not worn or damaged.**

19. Install the bolt loosely on the halfshaft. Use a new bolt

20. Align the ball joint in the control arm

➡**Be very careful when releasing the tension strap.**

21. Release the control arm slowly using the tension strap.

22. Install a new nut on the ball joint. Tighten to 43 ft. lbs. (80 Nm).

➡**The ball joint must not rotate. Use a Torx® wrench as a counterhold so that the rubber boot is not damaged.**

23. Install the brake disc.

24. Tighten the brake disc locating pin. Tighten to 72 inch lbs. (8 Nm).

➡**Ensure that the brake disc and wheel rim hub mating surfaces are clean.**

25. Tighten the new halfshaft bolt.

### S40 and V50 Models

*See Figure 190.*

1. Remove the wheel.

2. Slacken off the center nut from the halfshaft.

3. Remove the two bolts from the brake caliper holder and hang them up to prevent damage to the brake hose.

4. Remove the brake disc.

5. Remove the brake backing plate.

6. Pull the steering arm out of the wheel spindle.

7. Remove the nut from the halfshaft.

8. Remove the ABS sensor and hang in a suitable position.

9. Remove the tie rod ball joint.

10. Remove slacken off the wheel spindle nut as much as possible.

11. Remove the ball-joint from the wheel spindle.

12. Press the halfshaft out of the wheel spindle.

13. Remove the wheel spindle from the spring strut.

14. Secure the wheel spindle in a hydraulic press using socket 999-7090 and drift 999-1801 to press out the bearing.

*To install:*

15. Press both halves of 7090-1 between the bearing and the hub on the press.

16. Press the bearing into the wheel spindle.

17. Install the wheel spindle on the strut assembly.

18. Press the halfshaft into the wheel spindle.

19. Install the ball-joint onto the wheel spindle and tighten the nut to 52 ft. lbs. (70 Nm).

20. Install the center nut for the halfshaft.

21. Install or connect the following:
- ABS sensor
- Steering arm to the wheel spindle
- Brake backing plate
- Brake disc
- Brake caliper
- Wheel

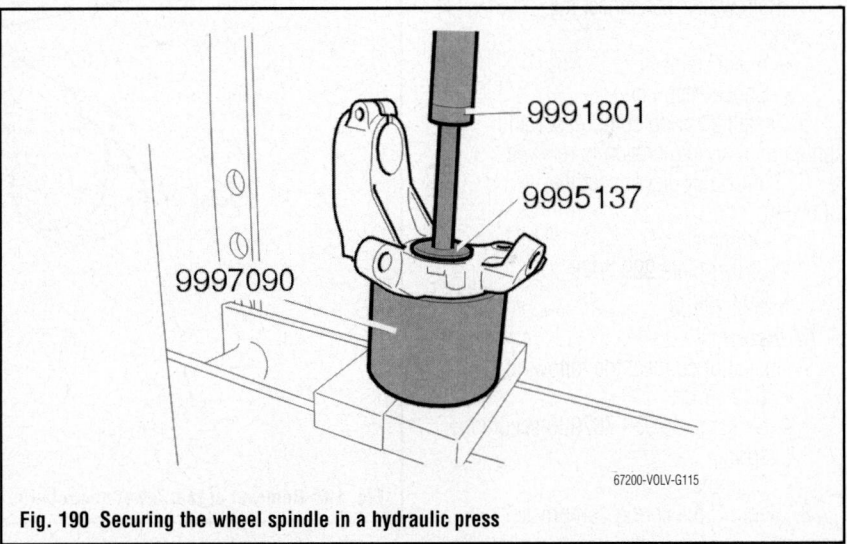

9991801
9995137
9997090

67200-VOLV-G115

**Fig. 190 Securing the wheel spindle in a hydraulic press**

## COIL SPRING

### REMOVAL & INSTALLATION

#### C30, C70 and V50 Models

*See Figure 191.*

1. Before servicing the vehicle, refer to the Precautions Section.
2. Remove the rear wheel
3. For right-hand side spring only:
   a. Place a transmission jack under the control arm to relieve the tension on the shock absorber.
   b. Remove the shock absorber lower mounting bolt.
4. Install spring compressor 951-2911 or suitable equivalent to compress the spring.
5. Remove the spring.

**To install:**

6. Install spring compressor 951-2911 or suitable equivalent to compress the spring.
7. Install the spring.

➡**Ensure that the spring engages in the control arm grooves.**

8. For right side spring only:
   a. Place a transmission jack under the control arm so the lower shock absorber mounting bolt can be installed.

#### S60, S80 and V70 Models

1. Before servicing the vehicle, refer to the Precautions Section.
2. Install plate 999 7079 to secure the spring.
3. Install a retaining strap between the control arms.
4. Remove or disconnect the following:
   • Brake caliper
   • Shock absorber
5. Install a spring compressor and tighten until spring tension is relieved.
6. Remove or disconnect the following:
   • Grommet
   • Striker plate 999 7079
   • Coil spring

**To install:**

7. Install or connect the following:
   • Coil spring
   • Striker plate 999 7079 to secure the spring
   • Grommet
8. Remove the spring compressor.

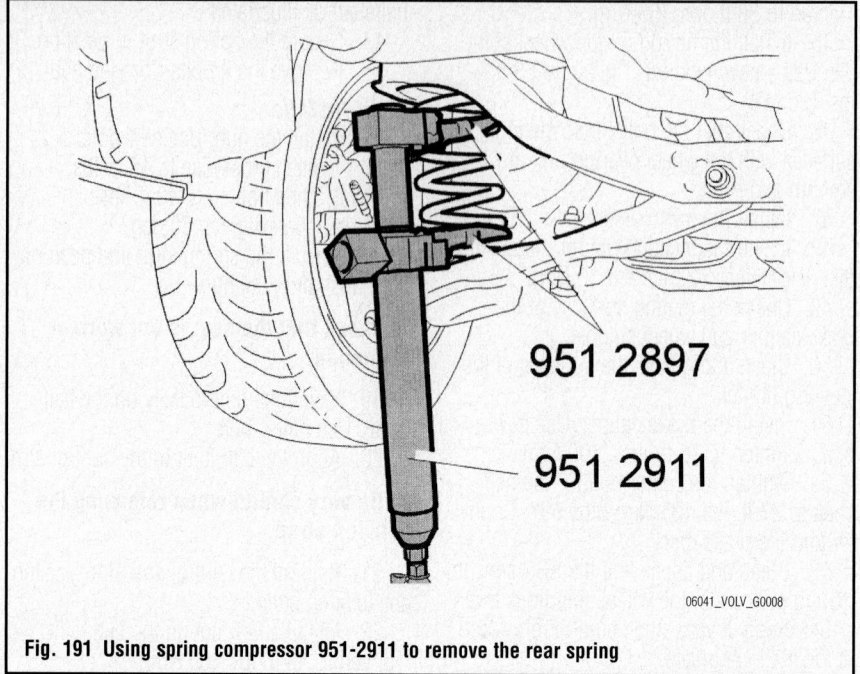

951 2897

951 2911

**Fig. 191 Using spring compressor 951-2911 to remove the rear spring**

06041_VOLV_G0008

9. Install or connect the following:
   • Shock absorber
   • Brake caliper
10. Remove striker plate 999 7079

## LOWER CONTROL ARM

### REMOVAL & INSTALLATION

*See Figure 192.*

1. Before servicing the vehicle, refer to the Precautions Section.
2. Raise the vehicle.

3. Remove or disconnect the following:
   • The wheel
   • The coil spring. Refer to Coil Spring removal and installation
   • The bolt in the sub-frame for the lower control arm
   • The control arm
   • The control arm rail (5740-1).

**To install:**

➡**Tighten all joints and rubber bushings with the rear suspension in the normal position.**

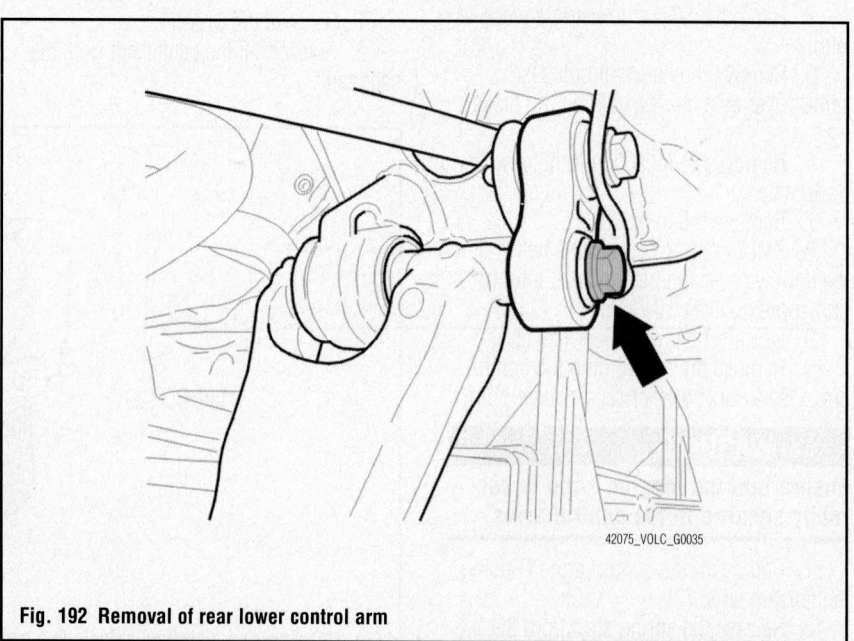

42075_VOLC_G0035

**Fig. 192 Removal of rear lower control arm**

4. Install or connect the following:
- The control arm rail (5742-1) in the control arm
- The control arm in the sub-frame. Do not tighten yet
- The coil spring.

5. Raise the control arms to their normal position using the tensioner 999 5659

6. Tighten the M12 bolt for the inner mounting for the control arm. Tighten to 59 ft. lbs. (80 Nm).

➡**If the vehicle is equipped with Bi-Xenon lamps, the position sensor must be calibrated. Carry out calibration according to VIDA Vehicle communication (Function area 3, Electrical system, Rear electronic module).**

## UPPER CONTROL ARM

### REMOVAL & INSTALLATION

*See Figures 193 through 195.*

1. Before servicing the vehicle, refer to the Precautions Section.
2. Raise the vehicle.
3. Remove or disconnect the following:
- The wheel
- The brake caliper. Hang the brake caliper on a hook in the sub-frame.

4. Slacken off the bolt for the outer control arm mounting by 3 turns.

5. Use tool: 951 2923, 999 7031 and 999 7030 together with threaded rod 999 7039 to press off the steering knuckle.

6. Remove the bolt for the drive shaft.

7. Press in the drive shaft and remove the spindle.

➡**Only press until the bushing releases from the splined area.**

8. Slacken off the bolt 3 turns for the rear control arm mounting. Use puller 998 5434 with counterhold 999 7074.

9. Remove or disconnect the following:
- The bolt for the inner rear control arm mounting
- The two bolts for the inner front control arm mounting
- The control arm

**To install:**

10. Install or connect the following:
- The bolts for the inner rear control arm mounting
- The bolts for the front inner control arm mounting
- The bolts for the outer control arm mounting

11. Install tensioner with tool 999 5659.

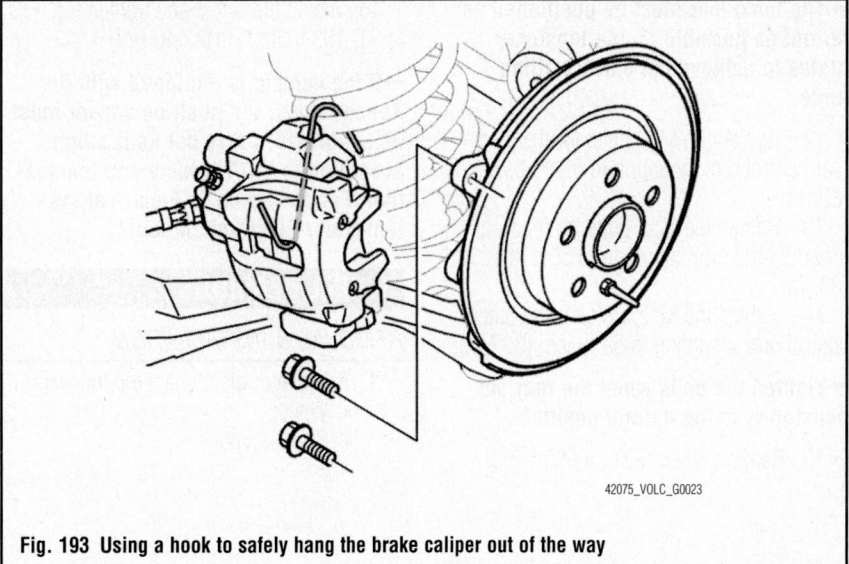

**Fig. 193 Using a hook to safely hang the brake caliper out of the way**

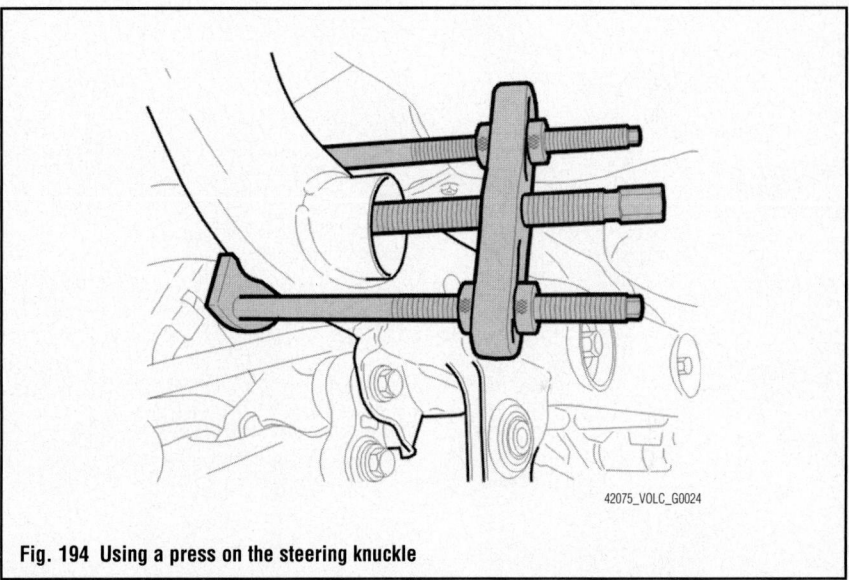

**Fig. 194 Using a press on the steering knuckle**

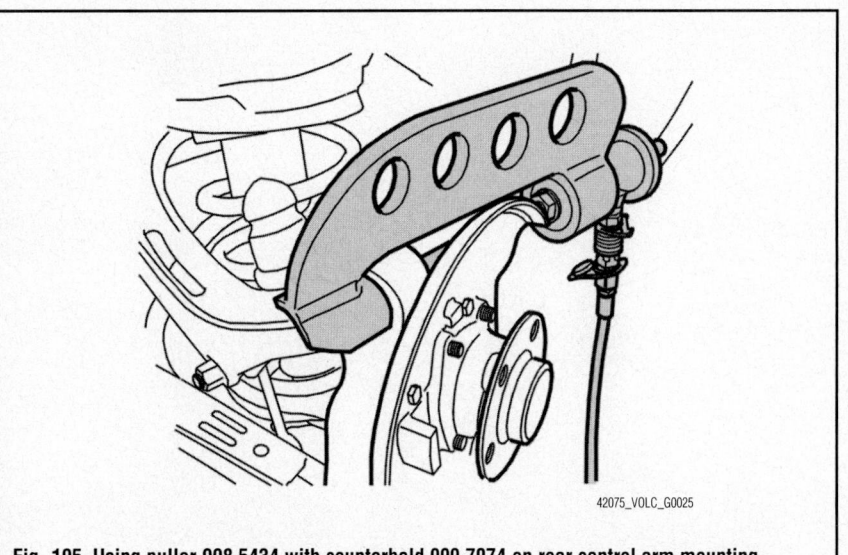

**Fig. 195 Using puller 998 5434 with counterhold 999 7074 on rear control arm mounting**

➡️**The tensioner must be positioned as far out as possible on the tensioner plates to achieve the correct lifting force.**

12. Tighten the M12 bolts for the inner rear control arm mounting to 59 ft. lbs. (80 Nm).

13. Tighten the M12 bolts for the front inner control arm mounting to 59 ft. lbs. (80 Nm).

14. Tighten the M12 bolts for the outer control arm mounting to 59 ft. lbs. (80 Nm).

➡️**Tighten the bolts when the rear suspension is in the normal position.**

15. Remove tensioner tool 999 5659.

16. Install the wheel and tighten lug nuts to 81–103 ft. lbs. (110–140 Nm).

➡️**If the vehicle is equipped with Bi-Xenon lamps, the position sensor must be calibrated. Carry out calibration according to VIDA vehicle communication (Function area 3, Electrical system, Rear Electronic Module).**

## WHEEL BEARINGS

### REMOVAL & INSTALLATION

1. Remove or disconnect the following:
   - Wheel
   - Brake caliper
   - Brake disc
   - ABS sensor
   - Wheel hub

*To install:*

2. Install or connect the following:
   - Wheel hub and tighten to 41 ft. lbs. (55 Nm)
   - ABS sensor and tighten to 42 inch lbs. (5 Nm)
   - Brake disc
   - Brake caliper
   - Wheel

## SPECIFICATIONS AND MAINTENANCE CHARTS

### ENGINE AND VEHICLE IDENTIFICATION

| | | | Engine | | | | | Model Year | |
|---|---|---|---|---|---|---|---|---|---|
| Code | Liters (cc) | Cu. In. | Cyl. | Fuel Sys. | Engine | Eng. Mfg. | | Code | Year |
| B5254T2 | 2.5 (2435) | 151 | 5 | MFI | DOHC | Volvo | | 6 | 2006 |
| B6294T | 2.9 (2922) | 178 | 6 | EFI | DOHC | Volvo | | 7 | 2007 |
| B6324S | 3.2 (3192) | 195 | 6 | EFI | DOHC | Volvo | | 8 | 2008 |
| B8444S | 4.4 (4414) | 269 | 8 | EFI | DOHC | Volvo | | | |

MFI: Multi-port Fuel Injection

EFI: Electronic Fuel Injection

DOHC: Double Overhead Camshafts

22140_VOLT_C0001

### GENERAL ENGINE SPECIFICATIONS

| Year | Model | Engine Displacement Liters | Engine ID | Net Horsepower @ rpm | Net Torque @ rpm (ft. lbs.) | Bore x Stroke (in.) | Compression Ratio | Oil Pressure @ rpm |
|---|---|---|---|---|---|---|---|---|
| 2006 | XC90 | 2.5 | B5254T2 | 208@5700 | 208@1500 | 3.27 x 3.54 | 8.5:1 | 43@2000 |
| | | 2.9 | B6294T | 268@5200 | 280@1800 | 3.27 x 3.54 | 10.0:1 | 50@4000 |
| | | 4.4 | B8444S | 315@5850 | 325@3900 | 3.70 x 3.13 | 10.4:1 | 68-109@2000 |
| 2007 | XC90 | 2.5 | B5254T2 | 208@5700 | 208@1500 | 3.27 x 3.54 | 8.5:1 | 43@2000 |
| | | 3.2 | B6324S | 235@6200 | 236@3200 | 3.30 x 3.78 | 10.8:1 | 50@4000 |
| | | 4.4 | B8444S | 315@5850 | 325@3900 | 3.70 x 3.13 | 10.4:1 | 68-109@2000 |
| 2008 | XC90 | 3.2 | B6324S | 235@6200 | 236@3200 | 3.30 x 3.78 | 10.8:1 | 50@4000 |
| | | 4.4 | B8444S | 315@5850 | 325@3900 | 3.70 x 3.13 | 10.4:1 | 68-109@2000 |

22140_VOLT_C0002

### GASOLINE ENGINE TUNE-UP SPECIFICATIONS

| Year | Engine Displacement Liters | Engine ID | Spark Plug Gap (in.) | Ignition Timing | Fuel Pump (psi) | Idle Speed RPM | Valve Clearance (in.) In. | Valve Clearance (in.) Ex. |
|---|---|---|---|---|---|---|---|---|
| 2006 | 2.5 | B5254T2 | 0.030 | 10B | 51 ① | 850 | HYD | HYD |
| | 2.9 | B6294T | 0.030 | NA | NA | 650 | HYD | HYD |
| | 4.4 | B8444S | 0.028 | NA | NA | 675 | HYD | HYD |
| 2007 | 2.5 | B5254T2 | 0.030 | 10B | 51 ① | 850 | HYD | HYD |
| | 3.2 | B6324S | 0.030 | NA | NA | 650 | HYD | HYD |
| | 4.4 | B8444S | 0.028 | NA | NA | 675 | HYD | HYD |
| 2008 | 3.2 | B6324S | 0.030 | NA | NA | 650 | HYD | HYD |
| | 4.4 | B8444S | 0.028 | NA | NA | 675 | HYD | HYD |

NA: Not available

NOTE: The Vehicle Emission Control Information label often reflects specification changes made during production. The label figures must be used if they differ from those in this chart.

B: Before top dead center

HYD: Hydraulic

① At idle

22140_VOLT_C0003

## CAPACITIES

| Year | Model | Engine Displacement Liters | Engine ID | Engine Oil with Filter (qts.) | Transmission (pts.) | Transfer Case (pts.) | Drive Axle Front (pts.) | Drive Axle Rear (pts.) | Fuel Tank (gal.) | Cooling System (qts.) |
|---|---|---|---|---|---|---|---|---|---|---|
| 2006 | XC90 | 2.5 | B5254T2 | 6.9 ① | 15.0 | ② | — | NA | 19.0 | NA |
| | | 2.9 | B6294T | 7.6 ① | NA | ② | — | NA | 19.0 | NA |
| | | 4.4 | B8444S | 7.0 | ③ | ② | — | NA | 21.1 | NA |
| 2007 | XC90 | 2.5 | B5254T2 | 6.9 ① | 15.0 | ② | — | NA | 19.0 | NA |
| | | 3.2 | B6324S | 7.6 ① | NA | ② | — | NA | 19.0 | NA |
| | | 4.4 | B8444S | 7.0 | ③ | ② | — | NA | 21.1 | NA |
| 2008 | XC90 | 3.2 | B6324S | 7.6 ① | NA | ② | — | NA | 19.0 | NA |
| | | 4.4 | B8444S | 7.0 | ③ | ② | — | NA | 21.1 | NA |

NA: Information not available

NOTE: All capacities are approximate. Add fluid gradually and check to be sure a proper fluid level is obtained.

① Amount to fill after engine reconditioning, including filter

② Included in transmission amount

③ AW55-51: drain & refill, 6.4 pts.
GM 4T65: drain & refill, 8.4 pts.
TF-80SC (AWD) drain & refill, 8.4 pts.

22140_VOLT_C0004

## VALVE SPECIFICATIONS

| Year | Engine Displacement Liters | Engine ID | Seat Angle (deg.) | Face Angle (deg.) | Spring Test Pressure (lbs. @ in.) | Spring Installed Height (in.) | Stem-to-Guide Clearance (in.) Intake | Stem-to-Guide Clearance (in.) Exhaust | Stem Diameter (in.) Intake | Stem Diameter (in.) Exhaust |
|---|---|---|---|---|---|---|---|---|---|---|
| 2006 | 2.5 | B5254T2 | 45 | 44.5 | NA | NA | 0.0012-0.0024 | 0.0015-0.0028 | 0.2738-0.2750 | 0.2734-0.2746 |
| | 2.9 | B6294T | 45 | 44.5 | NA | NA | 0.0012-0.0024 | 0.0015-0.0028 | 0.2738-0.2750 | 0.2734-0.2746 |
| | 4.4 | B8444S | NA | NA | NA | NA | NA | NA | NA | NA |
| 2007 | 2.5 | B5254T2 | 45 | 44.5 | NA | NA | 0.0012-0.0024 | 0.0015-0.0028 | 0.2738-0.2750 | 0.2734-0.2746 |
| | 3.2 | B6324S | 45 | 44.5 | NA | NA | 0.0012-0.0024 | 0.0015-0.0028 | 0.2344-0.2350 | 0.2343-0.2345 |
| | 4.4 | B8444S | NA | NA | NA | NA | NA | NA | NA | NA |
| 2008 | 3.2 | B6324S | 45 | 44.5 | NA | NA | 0.0012-0.0024 | 0.0015-0.0028 | 0.2344-0.2350 | 0.2343-0.2345 |
| | 4.4 | B8444S | NA | NA | NA | NA | NA | NA | NA | NA |

NA: Information not available

22140_VOLT_C0005

## CRANKSHAFT AND CONNECTING ROD SPECIFICATIONS

All measurements are given in inches.

| Year | Engine Displacement Liters | Engine ID | Crankshaft | | | | Connecting Rod | | |
|------|------|------|------|------|------|------|------|------|------|
| | | | Main Brg. Journal Dia. | Main Brg. Oil Clearance | Shaft End-play | Thrust on No. | Journal Diameter | Oil Clearance | Side Clearance |
| 2006 | 2.5 | B5254T2 | 2.5584-2.5592 | 0.0007-0.0017 | 0.003-0.007 | 5 | 1.9679-1.9685 | NA | 0.006-0.018 |
| | 2.9 | B6294T | 2.5584-2.5592 | 0.0007-0.0017 | 0.003-0.007 | 5 | 1.9679-1.9685 | NA | 0.006-0.018 |
| | 4.4 | B8444S | N/A | NA | NA | NA | NA | NA | NA |
| 2007 | 2.5 | B5254T2 | 2.5584-2.5592 | 0.0007-0.0017 | 0.003-0.007 | 5 | 1.9679-1.9685 | NA | 0.006-0.018 |
| | 3.2 | B6324S | 2.5584-2.5592 | 0.0007-0.0017 | 0.003-0.007 | 5 | 1.9679-1.9685 | NA | 0.006-0.018 |
| | 4.4 | B8444S | NA | NA | NA | NA | NA | NA | NA |
| 2008 | 3.2 | B6324S | 2.5584-2.5592 | 0.0007-0.0017 | 0.003-0.007 | 5 | 1.9679-1.9685 | NA | 0.006-0.018 |
| | 4.4 | B8444S | NA | NA | NA | NA | NA | NA | NA |

NA: Information not available

22140_VOLT_C0006

## PISTON AND RING SPECIFICATIONS

All measurements are given in inches.

| Year | Engine Displacement Liters | Engine ID | Piston Clearance | Ring Gap | | | Ring Side Clearance | | |
|------|------|------|------|------|------|------|------|------|------|
| | | | | Top Comp. | Bottom Comp. | Oil Control | Top Comp. | Bottom Comp. | Oil Control |
| 2006 | 2.5 | B5254T/2 | 0.0004-0.0012 | 0.047 | 0.069 | 0.118 | 0.0020-0.0033 | 0.0011-0.0026 | 0.0008-0.0022 |
| | 2.9 | B6249T | 0.0004-0.0012 | 0.047 | 0.069 | 0.118 | 0.0020-0.0033 | 0.0011-0.0026 | 0.0008-0.0022 |
| | 4.4 | B8444S | NA | NA | NA | NA | 0.0394-0.0787 | 0.0394-0.0787 | 0.0787-0.1969 |
| 2007 | 2.5 | B5254T/2 | 0.0004-0.0012 | 0.047 | 0.069 | 0.118 | 0.0020-0.0033 | 0.0011-0.0026 | 0.0008-0.0022 |
| | 3.2 | B6324S | 0.001-0.002 | N/A | N/A | N/A | 0.0012-0.0028 | 0.0012-0.0028 | 0.0008-0.0065 |
| | 4.4 | B8444S | NA | NA | NA | NA | 0.0394-0.0787 | 0.0394-0.0787 | 0.0787-0.1969 |
| 2008 | 3.2 | B6324S | 0.001-0.002 | N/A | N/A | N/A | 0.0012-0.0028 | 0.0012-0.0028 | 0.0008-0.0065 |
| | 4.4 | B8444S | NA | NA | NA | NA | 0.0394-0.0787 | 0.0394-0.0787 | 0.0787-0.1969 |

NA: Information not available

22140_VOLT_C0007

## TORQUE SPECIFICATIONS
All readings in ft. lbs.

| Year | Engine Displacement Liters | Engine ID | Cylinder Head Bolts | Main Bearing Bolts | Rod Bearing Bolts | Crankshaft Damper Bolts | Driveplate Bolts | Manifold Intake | Manifold Exhaust | Spark Plugs | Oil Pan Drain Plug |
|------|------|------|------|------|------|------|------|------|------|------|------|
| 2006 | 2.5 | B5254T/2 | ① | ② | ③ | 133 | ④ | 15 | 18 | 20 | 28 |
|      | 2.9 | B6249T | ① | ② | ⑤ | 222 | ④ | 15 | 18 | 22 | 28 |
|      | 4.4 | B8444S | NA | NA | NA | ⑥ | NA | 14 | 18 | 18 | 36 |
| 2007 | 2.5 | B5254T/2 | ① | ② | ③ | 133 | ④ | 15 | 18 | 20 | 28 |
|      | 3.2 | B6324S | ⑦ | ⑧ | ⑨ | NA | ⑩ | 13 | 18 | 22 | 28 |
|      | 4.4 | B8444S | NA | NA | NA | ⑥ | NA | 14 | 18 | 18 | 36 |
| 2008 | 3.2 | B6324S | ⑦ | ⑧ | ⑨ | NA | ⑩ | 13 | 18 | 22 | 28 |
|      | 4.4 | B8444S | NA | NA | NA | ⑥ | NA | 14 | 18 | 18 | 36 |

NA: Information not available

① Step 1: 14 ft. lbs.

Step 2: 43 ft. lbs.

Step 3: Plus 130 degrees

② Tighten cylinder block, intermediate section, in stages:

Step 1: M10 bolts: 15 ft. lbs. (20mm)

Step 2: M10 bolts: 33 ft. lbs. (45mm)

Step 3: M8 bolts: 18 ft. lbs. (25mm)

Step 4: M7 bolts: 13 ft. lbs. (17mm)

Step 5: M10 bolts: Plus 90 degrees

③ Connecting rod with treated toothed surface between the connecting rod and cap, screw with shoulder:

Step 1: 14 ft. lbs.

Step 2: plus 90 degrees

Connecting rod with fracture surface between the connecting rod and cap, fully threaded screw:

Step 1: 22 ft. lbs.

Step 2: Plus 90 degrees

④ Step 1: 33 ft. lbs.

Step 2: Plus 65 degrees

⑤ 15 ft. lbs., plus 90 degrees

⑥ Step 1: 89 ft. lbs.

Step 2: Back off at least 1 full turn

Step 3: 37 ft. lbs.

Step 4: Plus 90 degrees

⑦ Step 1: 33 ft. lbs.

Step 2: 33 ft. lbs.

Step 3: Plus 90 degrees

Step 4: Plus 180 degrees

⑧ Stage 1: 44 inch lbs.

Stage 2: 22 ft. lbs.

Step 2: 44 inch lbs.

Step 3: 18 ft. lbs.

Step 4: Plus 105 Degrees

Step 5: 18 ft. lbs.

⑨ Step 1: 106 inch lbs.

Step 2: 15 ft. lbs.

Step 3: Plus 90 degrees

⑩ 59 ft. lbs.

22140_VOLT_C0008

## TIRE, WHEEL AND BALL JOINT SPECIFICATIONS

| Year | Model | OEM Tires Standard | OEM Tires Optional | Tire Pressures (psi) Front | Tire Pressures (psi) Rear | Wheel Size | Ball Joint Inspection | Lug Nut Torque (ft. lbs.) |
|------|------|------|------|------|------|------|------|------|
| 2006 | XC90 | P225/70R16 | P235/65R17 | ① | ① | 7-JJ | NA | 102 |
| 2007 | XC90 | P235/60R18 | P255/50ZR19 | ① | ① | 7.5-JJ | NA | 102 |
| 2008 | XC90 | P235/65R18 | P255/50ZR19 | ① | ① | 7.5-JJ | NA | 102 |

OEM: Original Equipment Manufacturer

PSI: Pounds Per Square Inch

NA: Information not available

① See placard on vehicle

22140_VOLT_C0009

## BRAKE SPECIFICATIONS
All measurements in inches unless noted

| Year | Model | | Brake Disc Original Thickness | Brake Disc Minimum Thickness | Brake Disc Maximum Runout | Minimum Lining Thickness | Brake Caliper Bracket Bolts (ft. lbs.) | Brake Caliper Mounting Bolts (ft. lbs.) |
|------|-------|---|-----------|-----------|-----------|-----------|-----------|-----------|
| 2006 | XC90 | F | NA | 0.984 | 0.0015 | 0.12 | ① | ① |
|      |      | R | NA | 0.669 | 0.0019 | 0.12 | ② | ② |
| 2007 | XC90 | F | NA | 0.984 | 0.0015 | 0.12 | ① | ① |
|      |      | R | NA | 0.669 | 0.0019 | 0.12 | ② | ② |
| 2008 | XC90 | F | NA | 0.984 | 0.0015 | 0.12 | ① | ① |
|      |      | R | NA | 0.669 | 0.0019 | 0.12 | ② | ② |

NA: Not available

① Caliper holder (steel spindle): 77 ft. lbs. plus 60 degrees

   Caliper sliding pin: 20 ft. lbs.

② Caliper to wheel bearing housing: 26 ft. lbs. plus 60 degrees

③ See brake rotor for miminum thickness

22140_VOLT_C0010

## SCHEDULED MAINTENANCE INTERVALS
### VOLVO XC90

| TO BE SERVICED | SERVICE | 7.5 | 15 | 22.5 | 30 | 37.5 | 45 | 52.5 | 60 | 67.5 | 75 |
|----------------|---------|-----|----|------|----|------|----|------|----|------|----|
| Engine oil & filter | R | ✓ | ✓ | ✓ | ✓ | ✓ | ✓ | ✓ | ✓ | ✓ | ✓ |
| Automatic transmission fluid | S/I | ✓ | ✓ | ✓ | ✓ | ✓ | ✓ | ✓ | ✓ | ✓ | ✓ |
| Brake pads & parking brake | I | ✓ | ✓ | ✓ | ✓ | ✓ | ✓ | ✓ | ✓ | ✓ | ✓ |
| Cabin air filter | R | | ✓ | | ✓ | | ✓ | | ✓ | | ✓ |
| Engine coolant | S/I | | ✓ | | ✓ | | ✓ | | ✓ | | ✓ |
| Air cleaner filter | R | | | | | ✓ | | | | | ✓ |
| Spark plugs | R | | | | | | | | ✓ | | |
| Accessory drive belt | R | | | | | | | | | | ✓ |
| Fuel lines | I | | | | | | | | ✓ | ✓ | ✓ |
| Exhaust system | S/I | | | | | | | ✓ | | | |
| Check suspension | S/I | | | | | | | | ✓ | ✓ | ✓ |
| Brake fluid ① | R | | | | | | | | | | |

R: Replace       S/I: Service or Inspect

① Replace every 2 years or 30,000 miles, whichever comes first under normal conditions, more frequently in mountainous areas or moist climates.

### FREQUENT OPERATION MAINTENANCE (SEVERE SERVICE)

If a vehicle is operated under any of the following conditions it is considered severe service:

- Extremely dusty areas.

- 50% or more of the vehicle operation is in 32°C (90°F) or higher, or constant operation below 0°C (32°F).

- Prolonged idling (vehicle operation in stop and go traffic).

- Frequent short running periods (engine does not warm to normal operating temperatures).

- Police, taxi, delivery usage or trailer towing usage.

Oil & oil filter: change every 5000 miles.

Air filter element: service or inspect every 15,000 miles.

22140_VOLT_C0011

## PRECAUTIONS

Before servicing any vehicle, please be sure to read all of the following precautions, which deal with personal safety, prevention of component damage, and important points to take into consideration when servicing a motor vehicle:

• Never open, service or drain the radiator or cooling system when the engine is hot; serious burns can occur from the steam and hot coolant.

• Observe all applicable safety precautions when working around fuel. Whenever servicing the fuel system, always work in a well-ventilated area. Do not allow fuel spray or vapors to come in contact with a spark, open flame, or excessive heat (a hot drop light, for example). Keep a dry chemical fire extinguisher near the work area. Always keep fuel in a container specifically designed for fuel storage; also, always properly seal fuel containers to avoid the possibility of fire or explosion. Refer to the additional fuel system precautions later in this section.

• Fuel injection systems often remain pressurized, even after the engine has been turned **OFF**. The fuel system pressure must be relieved before disconnecting any fuel lines. Failure to do so may result in fire and/or personal injury.

• Brake fluid often contains polyglycol ethers and polyglycols. Avoid contact with the eyes and wash your hands thoroughly after handling brake fluid. If you do get brake fluid in your eyes, flush your eyes with clean, running water for 15 minutes. If eye irritation persists, or if you have taken

brake fluid internally, IMMEDIATELY seek medical assistance.

• The EPA warns that prolonged contact with used engine oil may cause a number of skin disorders, including cancer. You should make every effort to minimize your exposure to used engine oil. Protective gloves should be worn when changing oil. Wash your hands and any other exposed skin areas as soon as possible after exposure to used engine oil. Soap and water, or waterless hand cleaner should be used.

• All new vehicles are now equipped with an air bag system, often referred to as a Supplemental Restraint System (SRS) or Supplemental Inflatable Restraint (SIR) system. The system must be disabled before performing service on or around system components, steering column, instrument panel components, wiring and sensors. Failure to follow safety and disabling procedures could result in accidental air bag deployment, possible personal injury and unnecessary system repairs.

• Always wear safety goggles when working with, or around, the air bag system. When carrying a non-deployed air bag, be sure the bag and trim cover are pointed away from your body. When placing a non-deployed air bag on a work surface, always face the bag and trim cover upward, away from the surface. This will reduce the motion of the module if it is accidentally deployed. Refer to the additional air bag system precautions later in this section.

• Clean, high quality brake fluid from a sealed container is essential to the safe and

proper operation of the brake system. You should always buy the correct type of brake fluid for your vehicle. If the brake fluid becomes contaminated, completely flush the system with new fluid. Never reuse any brake fluid. Any brake fluid that is removed from the system should be discarded. Also, do not allow any brake fluid to come in contact with a painted surface; it will damage the paint.

• Never operate the engine without the proper amount and type of engine oil; doing so WILL result in severe engine damage.

• Timing belt maintenance is extremely important. Many models utilize an interference-type, non-freewheeling engine. If the timing belt breaks, the valves in the cylinder head may strike the pistons, causing potentially serious (also time-consuming and expensive) engine damage. Refer to the maintenance interval charts for the recommended replacement interval for the timing belt, and to the timing belt section for belt replacement and inspection.

• Disconnecting the negative battery cable on some vehicles may interfere with the functions of the on-board computer system(s) and may require the computer to undergo a relearning process once the negative battery cable is reconnected.

• When servicing drum brakes, only disassemble and assemble one side at a time, leaving the remaining side intact for reference.

• Only an MVAC-trained, EPA-certified automotive technician should service the air conditioning system or its components.

## BRAKES

### GENERAL INFORMATION

*PRECAUTIONS*

• Certain components within the ABS system are not intended to be serviced or repaired individually.

• Do not use rubber hoses or other parts not specifically specified for and ABS system. When using repair kits, replace all parts included in the kit. Partial or incorrect repair may lead to functional problems and require the replacement of components.

• Lubricate rubber parts with clean, fresh brake fluid to ease assembly. Do not

use shop air to clean parts; damage to rubber components may result.

• Use only DOT 3 brake fluid from an unopened container.

• If any hydraulic component or line is removed or replaced, it may be necessary to bleed the entire system.

• A clean repair area is essential. Always clean the reservoir and cap thoroughly before removing the cap. The slightest amount of dirt in the fluid may plug an orifice and impair the system function. Perform repairs after components have been thoroughly cleaned; use only denatured alcohol

## ANTI-LOCK BRAKE SYSTEM (ABS)

to clean components. Do not allow ABS components to come into contact with any substance containing mineral oil; this includes used shop rags.

• The Anti-Lock control unit is a microprocessor similar to other computer units in the vehicle. Ensure that the ignition switch is **OFF** before removing or installing controller harnesses. Avoid static electricity discharge at or near the controller.

• If any arc welding is to be done on the vehicle, the control unit should be unplugged before welding operations begin.

## BLEEDING PROCEDURE

*BLEEDING PROCEDURE*

### ✳✳ WARNING

Clean, high quality brake fluid is essential to the safe and proper operation of the brake system. You should always buy the highest quality brake fluid that is available. If the brake fluid becomes contaminated, drain and flush the system, then refill the master cylinder with new fluid. Never reuse any brake fluid. Any brake fluid that is removed from the system should be discarded. Also, do not allow any brake fluid to come in contact with a painted surface; it will damage the paint.

### ✳✳ CAUTION

Brake fluid contains polyglycol ethers and polyglycols. Avoid contact with the eyes and wash your hands thoroughly after handling brake fluid. If you do get brake fluid in your eyes, flush your eyes with clean, running water for 15 minutes. If eye irritation persists, or if you have taken brake fluid internally, IMMEDIATELY seek medical assistance.

1. Remove the reservoir cap and fill the brake reservoir with brake fluid.

2. Connect a vinyl tube to the wheel cylinder bleeder screw and insert the other end of the tube in a clear container.
3. Slowly depress the brake pedal several times.
4. While depressing the brake pedal fully, loosen the bleeder screw until fluid runs out. Then close the bleeder screw and release the brake pedal.
5. Repeat these steps until there are no more bubbles in the fluid escaping to the clear container.
6. Tighten the bleeder screw to 80 inch lbs. (9 Nm).
7. Repeat the above procedure for each wheel.

### ✳✳ CAUTION

Dust and dirt accumulating on brake parts during normal use may contain asbestos fibers from production or aftermarket brake linings. Breathing excessive concentrations of asbestos fibers can cause serious bodily harm. Exercise care when servicing brake parts. Do not sand or grind brake lining unless equipment used is designed to contain the dust residue. Do not clean brake parts with compressed air or by dry brushing. Cleaning should be done by dampening the brake components with a fine mist of water, then wiping the brake components clean with a dampened cloth. Dispose of cloth and all residue containing asbestos fibers in an impermeable container with the appropriate label. Follow practices prescribed by the Occupational Safety and Health Administration (OSHA) and the Environmental Protection Agency (EPA) for the handling, processing, and disposing of dust or debris that may contain asbestos fibers.

## BRAKE CALIPER

*REMOVAL & INSTALLATION*

See Figure 1.

1. Remove the wheel.
2. If the caliper is being replaced:
   a. Secure the pedal in the depressed position. Use a pedal jack.

   b. Clean the brake caliper thoroughly.
   c. Remove the protective cap from the bleed nipple.
   d. Install a plastic hose on the nipple.
   e. Open the bleed nipple. Collect the brake fluid in a container. Shut the bleed nipple.
   f. Slacken the off brake hose approximately half a turn.

3. Remove the retaining spring carefully so as not to deform it.
4. Remove:
   • The protective caps from the two locating pins
   • The locating pins
   • The brake caliper from the holder
   • The brake pads.

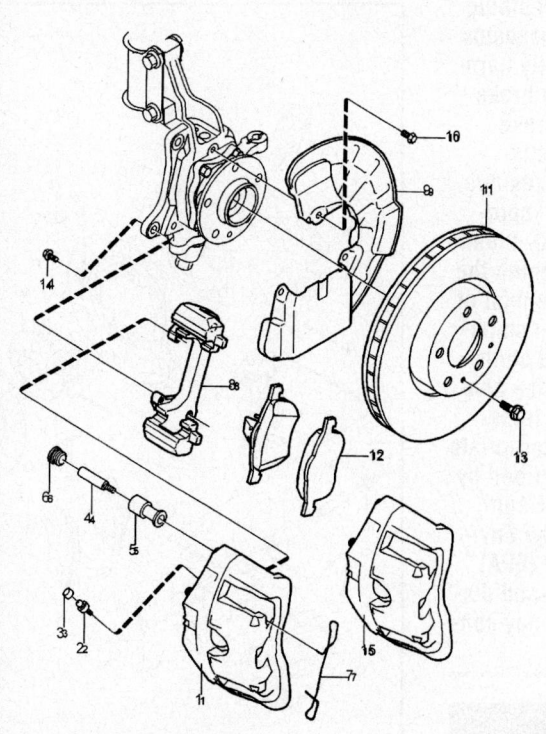

| | |
|---|---|
| 1 | Brake caliper, exch left |
| | Brake caliper, exch left |
| 2 | • Bleeder screw |
| 3 | • Protection |
| 4 | • Guide pin upper |
| | • Guide pin |
| 5 | • Rubber bellows upper |
| | • Bushing |
| 6 | • Protecting cover |
| 7 | • Spring |
| 8 | • Brace |
| | • Brace |
| 9 | Protecting plate |
| | Protecting plate |
| 10 | Flange screw |
| 11 | Brake disc |
| | Brake disc |
| 12 | Brake pad kit |
| | Brake pad kit |
| 13 | Hexagon screw |
| 14 | Flange screw |
| 15 | Brake caliper, exch left |
| | Housing left |

42348-XC90-G63

**Fig. 1 Front brake exploded view**

5. Unscrew the brake caliper from the brake hose.

6. Drain the remaining brake fluid.

7. Insert the brake caliper on brake hose (Do not tighten)

**To install:**

8. Install:
- brake pads
- brake caliper in the holder.

9. Lubricate the locating pins using grease caliper grease.

10. Insert the locating pins into the rubber sleeves. The pins should slide into the sleeves easily.

11. Tighten the locating pins to 20 ft. lbs. (28 Nm).

12. Install the protective caps.

13. Install the securing spring.

14. Tighten the brake hose to 13 ft. lbs. (18 Nm).

➡**The brake hose must not be twisted.**

15. Fill and bleed the brake system.

## DISC BRAKE PADS

### REMOVAL & INSTALLATION

1. Remove the wheels on both sides.

2. Remove the pad retaining spring carefully so as not to deform it.

3. Remove:
- The protective caps (1) from the two locating pins (2)
- The locating pins, use hex socket 7mm.
- The brake caliper from the holder
- Brake pads.

4. Hang brake caliper from a steel wire from the front spring so as not to damage brake hose.

➡**Do not depress the brake pedal while the brake pads are removed.**

5. Clean and check the brake caliper and dust cover.

6. Clean and check the brake pad mating surfaces in the brake caliper and caliper holder.

7. Check piston dust boot.

➡**If the dust boot is damaged dirt may have penetrated the cylinder. If this is the case the caliper must be replaced.**

8. Check brake disc friction surfaces.

➡**To install:**

9. Press piston back into cylinder on brake caliper.

10. Check that the dust cover is correctly positioned.

11. Install:
- New brake pads
- The brake caliper.

12. Lubricate the locating pins using caliper grease. Insert the locating pins into the rubber sleeves. The pins should slide into the sleeves easily.

13. Tighten the locating pins to 20 ft. lbs. (28 Nm).

14. Install the protective caps.

15. Install the securing spring.

16. Check the brake fluid level in the reservoir.

17. Depress the brake pedal a few times. Check the level of the brake fluid reservoir.

18. Install the wheels.

## BRAKES

### ✳✳ CAUTION

**Dust and dirt accumulating on brake parts during normal use may contain asbestos fibers from production or aftermarket brake linings. Breathing excessive concentrations of asbestos fibers can cause serious bodily harm. Exercise care when servicing brake parts. Do not sand or grind brake lining unless equipment used is designed to contain the dust residue. Do not clean brake parts with compressed air or by dry brushing. Cleaning should be done by dampening the brake components with a fine mist of water, then wiping the brake components clean with a dampened cloth. Dispose of cloth and all residue containing asbestos fibers in an impermeable container with the appropriate label. Follow practices prescribed by the Occupational Safety and Health Administration (OSHA) and the Environmental Protection Agency (EPA) for the handling, processing, and disposing of dust or debris that may contain asbestos fibers.**

## BRAKE CALIPER

### REMOVAL & INSTALLATION

*See Figure 2.*

1. Remove the wheel.

2. If the caliper is being replaced:
   a. Secure the pedal in the depressed position. Use a pedal jack.

## REAR DISC BRAKES

   b. Clean the brake caliper thoroughly.

   c. Remove the protective cap from the bleed nipple.

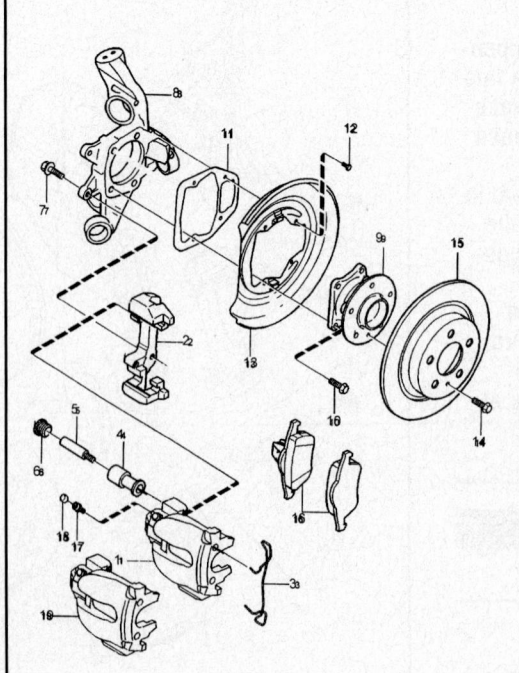

| | |
|---|---|
| 1 | Brake caliper, exch |
| 2 | • Brace |
| 3 | • Spring |
| 4 | • Seal |
| 5 | • Bolt |
| 6 | • Plug |
| 7 | Flange screw |
| 8 | Bearing housing, l.h. |
| 9 | Rear wheel hub |
| 10 | Flange screw |
| 11 | Gasket |
| 12 | Flange screw |
| 13 | Protecting plate |
| 14 | Screw |
| 15 | Brake disc |
| 16 | Brake pad kit |
| | **Service kits Rear wheel brake** |
| 17 | Bleeder screw |
| 18 | Protection |
| 19 | Brake caliper, exch left |

**Fig. 2 Rear brake exploded view**

42348-XC90-G64

d. Install a plastic hose on the nipple.

e. Open the bleed nipple. Collect the brake fluid in a container. Shut the bleed nipple.

f. Slacken the off brake hose approximately half a turn.

3. Remove the brake caliper and brake pads. See Brake Pads removal and installation.

4. Unscrew the brake caliper from the brake hose.

5. Drain the remaining brake fluid.

### *To install:*

6. Insert the brake caliper on the hose and hand tighten.

7. Clean and install the brake pads and brake calipers.

8. Tighten the locating pins to 20 ft. lbs. (28 Nm) for XC90 and 26 ft. lbs. (35 Nm) for XC70.

9. Tighten the brake hose to 13 ft. lbs. (18 Nm).

➡**The brake hose must not be twisted.**

10. Fill and bleed the brake system.

11. Depress the brake pedal a few times. Check the level of the brake fluid reservoir.

## DISC BRAKE PADS

### *REMOVAL & INSTALLATION*

1. Remove the wheels on both sides.

2. Remove the pad retaining spring carefully so as not to deform it.

3. Remove:
- The protective caps (1) from the two locating pins (2)
- The locating pins, use hex socket 7mm.
- The brake caliper from the holder
- Brake pads.

4. Hang brake caliper from a steel wire from the front spring so as not to damage brake hose.

➡**Do not depress the brake pedal while the brake pads are removed.**

5. Clean and check the brake caliper and dust cover.

6. Clean and check the brake pad mating surfaces in the brake caliper and caliper holder.

7. Check piston dust boot.

➡**If the dust boot is damaged dirt may have penetrated the cylinder. If**
this is the case the caliper must be replaced.

8. Check brake disc friction surfaces.

### *To install:*

9. Press piston back into cylinder on brake caliper.

10. Check that the dust cover is correctly positioned.

11. Install:
- New brake pads
- The brake caliper.

12. Lubricate the locating pins using caliper grease. Insert the locating pins into the rubber sleeves. The pins should slide into the sleeves easily.

13. Tighten the locating pins to 20 ft. lbs. (28 Nm) for XC90 and 26 ft. lbs. (35 Nm) for XC70.

14. Install the protective caps.

15. Install the securing spring.

16. Check the brake fluid level in the reservoir.

17. Depress the brake pedal a few times. Check the level of the brake fluid reservoir.

18. Install the wheels.

## BRAKES

PARKING BRAKE

## PARKING BRAKE SHOES

### *REMOVAL & INSTALLATION*

*See Figures 3 and 4.*

1. Remove rear disc brake caliper assembly

➡**Before removing the brake disc rotor, make a chalk marking by the bolts to aid in reassembly.**

2. Remove the rotor.

3. Remove the shoe hold spring by turning the pin to coincide with hole of spring cap.

4. Remove the return spring.

5. Disconnect the cable end from the trailing shoe.

6. Remove the parking brake shoes.

➡**Complete the removal and installation on one side of the vehicle at a time using the others side as a reference.**

### *To install:*

7. Install the parking brake cable to the operating lever.

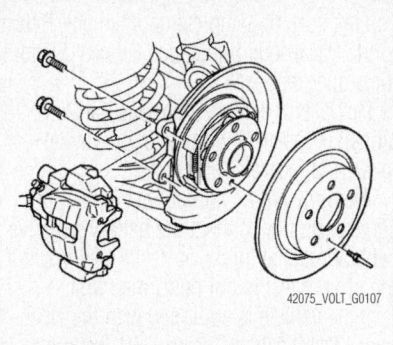

**Fig. 3 Exploded view of parking brake assembly**

8. Apply brake grease to return spring and areas of movement on the mechanism, but do not apply grease to the brake shoe material.

9. Install the upper return spring and brake shoes.

10. Turn the adjuster in counter clockwise direction and install.

11. Install the lower return spring.

12. Install the shoe hold spring with a pliers.

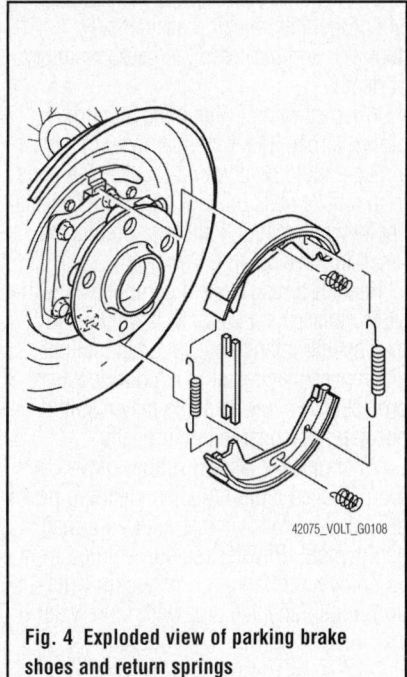

**Fig. 4 Exploded view of parking brake shoes and return springs**

13. Install the disc brake and then align the mark while tightening the screw.

## GENERAL INFORMATION

### ✳✳ CAUTION

**These vehicles are equipped with an air bag system. The system must be disarmed before performing service on, or around, system components, the steering column, instrument panel components, wiring and sensors. Failure to follow the safety precautions and the disarming procedure could result in accidental air bag deployment, possible injury and unnecessary system repairs.**

### SERVICE PRECAUTIONS

Disconnect and isolate the battery negative cable before beginning any airbag system component diagnosis, testing, removal, or installation procedures. Allow system capacitor to discharge for two minutes before beginning any component service. This will disable the airbag system. Failure to disable the airbag system may result in accidental airbag deployment, personal injury, or death.

Do not place an intact undeployed airbag face down on a solid surface. The airbag will propel into the air if accidentally deployed and may result in personal injury or death.

When carrying or handling an undeployed airbag, the trim side (face) of the airbag should be pointing towards the body to minimize possibility of injury if accidental deployment occurs. Failure to do this may result in personal injury or death.

Replace airbag system components with OEM replacement parts. Substitute parts may appear interchangeable, but internal differences may result in inferior occupant protection. Failure to do so may result in occupant personal injury or death.

Wear safety glasses, rubber gloves, and long sleeved clothing when cleaning powder residue from vehicle after an airbag deployment. Powder residue emitted from a deployed airbag can cause skin irritation. Flush affected area with cool water if irritation is experienced. If nasal or throat irritation is experienced, exit the vehicle for fresh air until the irritation ceases. If irritation continues, see a physician.

Do not use a replacement airbag that is not in the original packaging. This may result in improper deployment, personal injury, or death.

The factory installed fasteners, screws and bolts used to fasten airbag components have a special coating and are specifically designed for the airbag system. Do not use substitute fasteners. Use only original equipment fasteners listed in the parts catalog when fastener replacement is required.

During, and following, any child restraint anchor service, due to impact event or vehicle repair, carefully inspect all mounting hardware, tether straps, and anchors for proper installation, operation, or damage. If a child restraint anchor is found damaged in any way, the anchor must be replaced. Failure to do this may result in personal injury or death.

Deployed and non-deployed airbags may or may not have live pyrotechnic material within the airbag inflator.

Do not dispose of driver/passenger/curtain airbags or seat belt tensioners unless you are sure of complete deployment. Refer to the Hazardous Substance Control System for proper disposal.

Dispose of deployed airbags and tensioners consistent with state, provincial, local, and federal regulations.

After any airbag component testing or service, do not connect the battery negative cable. Personal injury or death may result if the system test is not performed first.

If the vehicle is equipped with the Occupant Classification System (OCS), do not connect the battery negative cable before performing the OCS Verification Test using the scan tool and the appropriate diagnostic information. Personal injury or death may result if the system test is not performed properly.

Never replace both the Occupant Restraint Controller (ORC) and the Occupant Classification Module (OCM) at the same time. If both require replacement, replace one, then perform the Airbag System test before replacing the other.

Both the ORC and the OCM store Occupant Classification System (OCS) calibration data, which they transfer to one another when one of them is replaced. If both are replaced at the same time, an irreversible fault will be set in both modules and the OCS may malfunction and cause personal injury or death.

If equipped with OCS, the Seat Weight Sensor is a sensitive, calibrated unit and must be handled carefully. Do not drop or handle roughly. If dropped or damaged, replace with another sensor. Failure to do so may result in occupant injury or death.

If equipped with OCS, the front passenger seat must be handled carefully as well. When removing the seat, be careful when setting on floor not to drop. If dropped, the sensor may be inoperative, could result in occupant injury, or possibly death.

If equipped with OCS, when the passenger front seat is on the floor, no one should sit in the front passenger seat. This uneven force may damage the sensing ability of the seat weight sensors. If sat on and damaged, the sensor may be inoperative, could result in occupant injury, or possibly death.

### DISARMING THE SYSTEM

1. Before servicing the vehicle, refer to the Precautions Section.

2. Disconnect and isolate the negative battery cable. Wait 3 minutes for the system capacitor to discharge before performing any service.

➡ **Wait at least 3 minutes before working on the vehicle. The air bag system is designed to retain enough power to deploy the air bag for a short time after the battery has been disconnected.**

### ARMING THE SYSTEM

1. After repairs are complete, connect the negative battery cable. Turn the ignition switch to the **ON** position and check the SRS light for proper operation.

# DRIVETRAIN

## AUTOMATIC TRANSAXLE ASSEMBLY

### REMOVAL & INSTALLATION

**5-Cylinder Engines**

*See Figure 5.*

1. Before servicing the vehicle, refer to the Precautions Section.
2. Drain the transaxle fluid.
3. Attach a support fixture to the engine lifting eyes.
4. Remove or disconnect the following:
   - Battery and tray
   - Air intake assembly
   - Turbo control valve
   - Turbocharger inlet tube
   - Gear select cable
   - Transaxle harness connector
   - Transaxle ground cable
   - Heated Oxygen (HO2S) sensor connector
   - Transaxle oil cooler lines
   - Transaxle dipstick tube
   - Exhaust Gas Recirculation (EGR) valve hoses, if equipped
   - Starter motor
   - Coolant recovery tank
   - Torque rod extension arm
   - Front wheels
   - Wheel speed sensors
   - Brake line brackets
   - Wheel speed sensor wiring brackets
   - Inner fender liners
   - Axle halfshafts
   - Transfer case, if equipped
   - Splash guards
   - Lower ball joints
   - Stabilizer bar links
   - Evaporative Emissions (EVAP) canister and hoses
   - Exhaust front pipe
   - Oil line bracket
   - Steering gear engine mount
   - Steering gear mounting nuts
   - Vehicle Speed (VSS) sensor connector
   - Transaxle mount
   - Torque converter
5. Loosen the 2 right side sub-frame-to-body bolts approximately ½ inch. Support the sub-frame with a jack and remove the sub-frame-to-body bolts on the left side.
6. Lower the jack and let the frame hang down from the right side bolts.
7. Tie the left side of the steering gear to the left side frame rail for support. Remove the steering gear engine mount.

8. Lower the engine and transaxle with the lifting hook.
9. Install transaxle fixture 5463 on the transaxle jack, using the torque rod mounting bolts to hold it in place. At the same time, fit tool 5463-1 support plate and raise the jack so that it is making light contact.
10. Remove the transaxle flange bolts.
11. Remove the transaxle.

***To install:***

➡**Use new fasteners where indicated.**

12. Install or connect the following:
   - Transaxle and tighten the bolts to 37 ft. lbs. (50 Nm)
   - Torque converter. Use new bolts and tighten them to 22 ft. lbs. (30 Nm).
   - Rear transaxle mount and torque the bolts to 37 ft. lbs. (50 Nm)
   - HO2S sensor connector
   - VSS sensor connector
13. Install the sub-frame using new bolts. Starting on the left side, lift the frame with a jack. Mount the support brackets on both sides. Tighten the frame bolts to 78 ft. lbs. (105 Nm) plus 120°. Tighten the bracket bolts to 37 ft. lbs. (50 Nm). Repeat the procedure for the right side. Remove the sub-frame jack.
14. Install or connect the following:
   - Steering gear. Use new mounting nuts and tighten them to 37 ft. lbs. (50 Nm).
   - Steering gear engine mount and tighten the bolts to 37 ft. lbs. (50 Nm)
   - Oil line bracket
   - Torque rod extension arm and tighten the bolts to 26 ft. lbs. (35 Nm) plus 40°
   - Exhaust front pipe
   - EVAP canister and hoses
   - Stabilizer bar links

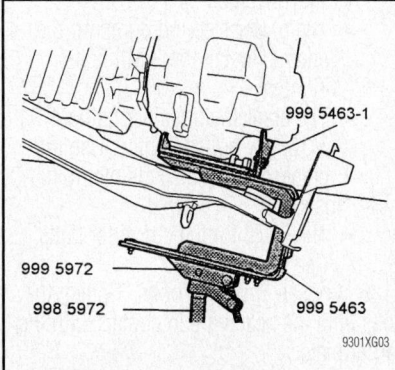

**Fig. 5  Using the transaxle fixture**

   - Lower ball joints
   - Splash guards
   - Transfer case, if equipped
   - Axle halfshafts
   - Inner fender liners
   - Wheel speed sensor wiring brackets
   - Brake line brackets
   - Wheel speed sensors
   - Front wheels
   - Coolant recovery tank
   - Starter motor
   - EGR valve hoses, if equipped
   - Transaxle dipstick tube
   - Transaxle oil cooler lines
   - Transaxle ground cable
   - Transaxle harness connector
   - Gear select cable
   - Turbocharger inlet tube
   - Turbo control valve
   - Air intake assembly
   - Battery and tray
15. Fill the transaxle to the correct level.

## 6-Cylinder Engines, 2006 Only

1. Before servicing the vehicle, refer to the Precautions Section.

➡**Remove the transaxle and engine from the engine compartment as one unit.**

2. Drain the transaxle fluid at a suitable time during work. Use an oil suction unit.
3. Remove:
   - The bolts for the lower torque rod mounting in the engine
   - The lower bolts for the engine transaxle cover.
4. Remove the transaxle from the engine. See the Engine removal instructions earlier in this chapter.
5. Remove:
   - The halfshafts. Install sealing plug on the transaxle. Install a plug on the chain housing
   - The bolt for the engine pad and steering gear
   - The cable harness from the bracket on the steering gear.
   - The bolt for the engine pad/sub-frame
   - The bolts for the cable duct/sub-frame.
6. Detach the starter motor from the transaxle cover and the engine to access the carrier plate/torque converter bolts.
7. Remove:
   - The torque converter bolts.
   - The crankshaft damper. Use the crankshaft nut as a counterhold

- The bolts between the engine mounting in the engine and the sub-frame
- The drive belt
- The power steering pump from the engine.

8. Move the power steering pump with hoses and fluid reservoir to one side.

9. Install lifting lug 999 7018, or equivalent, at the rear edge of the engine. Use lifting lugs 999 5185 and 999 5186 together with lifting lug 999 2810.

10. Attach the lifting yoke and adjust for optimum balance. Raise the engine/transaxle from the sub-frame.

11. Place the unit on a bench with a strong top surface and a stable base.

12. Place wooden or hard rubber blocks, approximately 1 inch (30mm) thick, between the bench top and the entire flat surface of the engine oil pan.

13. Place a tensioning strap between the engine intake manifold and the bench. Tighten the tensioning strap so that the engine oil pan lies flat against the blocks.

14. Remove:
- The bolt for dipstick pipe in the engine.

15. Detach the cable harness from the mountings on the transaxle.

16. Remove:
- The bracket between the engine and chain housing
- The bolts between the chain housing/brackets and transaxle
- The bevel gear/chain housing from the transaxle. Install sealing plug 999 5733
- The bracket from the end of the transaxle.

17. Install lifting lug 999 5464 on the engine mounting at the rear edge of the transaxle. Install lifting lug 999 5267 at the front edge.

18. Also use lifting hook 999 5186 and 999 5642 together with lifting yoke 999 5100.

19. Gently apply the lift. Remove all bolts for the transaxle/engine. Remove the transaxle. Ensure that the torque converter accompanies the transaxle during removal and does not fall out.

20. Place the transaxle on a workbench.

### To install:

21. Install lifting lug 999 5464 on the engine mounting at the rear edge of the transaxle. Install lifting lug 999 5267 at the front edge.

22. Also use lifting hook 999 5186 and 999 5642 together with lifting yoke 999 5100.

23. Ensure that the torque converter does not slide from its axle.

24. Adjust the transaxle against the guide sleeves on the engine. Install the 7 M10 bolts. Tighten alternately to a light contact. Tighten. (2 more bolts will be installed and tightened later.)

25. Remove the lifting tools from the transaxle.

26. Install:
- The bracket on the end of the transaxle. Tighten to 37 ft. lbs. (50 Nm)
- The bolt for the upper bracket. Tighten to 37 ft. lbs. (50 Nm)

27. Tighten the bolts for the bracket in the cylinder block: 4 M10 bolts. Tighten to 37 ft. lbs. (50 Nm). 1 M8 bolt. Tighten to 18 ft. lbs. (25 Nm).

28. Install:
- The bevel gear/chain housing on the transaxle. Lightly tighten the 2 M10 bolts alternately. Tighten
- The bracket between the engine and the chain housing. Lightly tighten the 4 M10 bolts to the cylinder block
- The 2 M10 bolts to the chain housing/bracket. Tighten.

29. Tighten the 4 M10 bolts for the bracket/cylinder block. Tighten.

30. Install:
- The 1 M8 bolt for the dip stick pipe in the engine. Tighten
- The cable harness to the mountings on the transaxle

31. Install lifting lug 999 7018 at the rear edge of the engine. Use lifting lugs 999 5185 and 999 5186 together with lifting lug 999 2810 .

32. Apply the lifting yoke and adjust for optimum balance. Raise the engine/transaxle to the sub-frame.

33. Install:
- The 2 M10 bolts between the engine mounting in the engine and the sub-frame. Use new bolts
- The drive belt
- The power steering pump on the engine. Install the 3 M8 bolts. Tighten
- The crankshaft damper using 4 M10 new bolts. Tighten. Use the crankshaft nut as a counterhold.

34. Install:
- The 6 M8 torque converter bolts. Use new bolts.

35. Loosely insert all bolts. Tighten the bolts until the heads are in contact with the carrier plate.

36. Use tool 999 5734 to turn the engine and as a counterhold.

37. Install:
- 1 M10 bolt for the engine pad/sub-frame.
- The bolt for the cable duct/sub-frame
- The starter motor on the transaxle cover and engine. Install the 2 M10 bolts in the transaxle cover. Install 1 M8 bolt in the bracket. Tighten.
- The halfshafts. Check that the snap rings for the halfshafts are in the grooves by pulling the inner constant velocity joint
- 1 M10 bolt in the steering gear engine pad. Tighten
- The cable harness to the bracket on the steering gear.

38. Check the oil level in the chain housing.

39. Tighten the level plug.

40. Check the oil level in the bevel gear.

41. Tighten the level plug.

42. Top up the oil in the transaxle via the dip stick pipe.

43. Install:
- The 4 M10 bolts for the lower torque rod mounting in the engine. Tighten
- The 2 M10 lower bolts for the engine transaxle cover. Tighten.

44. To install the engine/transaxle, see the Engine instructions earlier in this chapter.

45. Check the oil level.

### 8-Cylinder Engines

*See Figures 6 through 12.*

1. Before servicing the vehicle, refer to the Precautions Section.

2. Disconnect the negative battery cable.

3. Remove the expansion tank cap.

4. Remove protective skid plate:

➡**There is a guide sleeve on the upper mounting for the plate. This must be lifted slightly so that the panel can be slid forward.**

a. Remove the 6 bolts.

b. Lift and slide the skid plate forward.

c. Remove the skid plate.

5. Remove the 6 bolts of lower engine cover.

6. Remove the lower engine cover.

7. Open the cock on the lower left edge of the radiator and drain engine coolant into a clean receptacle.

8. Close the cock.

9. Remove the upper cross stay:

a. Remove the 4 bolts at the suspension turrets.

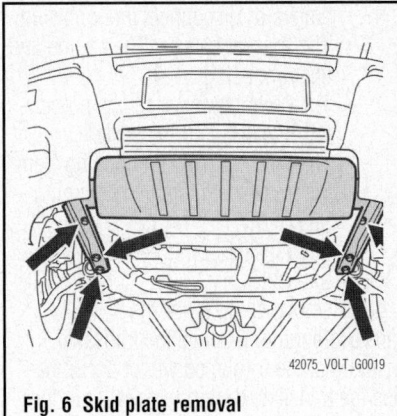

**Fig. 6  Skid plate removal**

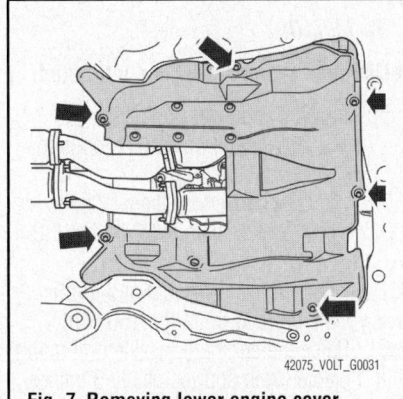

**Fig. 7  Removing lower engine cover**

b.  Remove the bolts of the upper engine mount on the engine

c.  Remove the upper cross stay.

10.  Remove the air cleaner cover by pulling it straight up.

11.  Disconnect the 2 connectors from the engine control module (ECM).

12.  Clip the band clamp to the cable harness.

13.  Disconnect the mass air flow sensor.

14.  Remove the air cleaner.

15.  Remove or disconnect the following:

- Brake vacuum hose from the vacuum pump
- The 2 bolts of the fresh air hose in the front plate
- The fresh air hose
- The air cleaner by pulling it straight up
- The positive lead (1) from the electrical distribution box
- The ground cable (2) between the engine and body
- Disconnect the connection at the electrical distribution box (3)

16.  Remove or disconnect the following:

- The brake vacuum valve
- Detach the vacuum hose from the valve that leads to the brake vacuum amplifier

**Fig. 8  Disconnecting from the electrical distribution box**

17.  Remove components from the transaxle:

a.  The selector lever cable using a break fork ( 999 7077 ) to break the cable loose from its ball.

b.  The transaxle control module (TCM) connector.

c.  Remove the oil cooler hoses.

➡**Mark the hoses for proper installation.**

d.  Remove the lower line of the transaxle oil cooler.

e.  Remove the upper line of the transaxle oil cooler.

➡**Seal the openings and collect the oil from the oil cooler in a receptacle. Do not reuse the drained oil.**

18.  Detach the radiator hoses at the quick-release coupling toward the firewall.

19.  Remove the upper radiator hose from the distributor housing.

20.  Remove the lower radiator hose from the thermostat housing.

21.  Remove or disconnect the following:

- Upper engine covers by pulling them straight up
- The cover over the servo pump
- The 2 bolts on the cover over the servo pump
- The bolt on the washer fluid filler pipe. Bend the pipe out of the way
- The return line for servo oil from the clips and clamps
- Lift up the servo oil reservoir

22.  Remove or disconnect the following:

- The 2 coolant hoses from the expansion tank
- The 3 ground cables from the body at the expansion tank
- The connector from the electric cooling fan
- Position the servo oil reservoir on the engine

➡**Be careful that no oil runs out.**

23.  Remove the accessory drive belt, refer to Accessory Drive Belts removal and installation.

24.  Remove the ground cable bolt between the engine and body.

➡**It is located beneath the expansion tank.**

25.  Raise the vehicle.

26.  Remove the front wheels.

27.  Remove the ball joint from the link arm on both sides of the vehicle:

a.  Remove the drive shaft bolt. Counterhold with a screwdriver in the brake disc cooling channels.

b.  Remove the rubber sleeve in the wheel hub.

c.  Remove the nut between the ball joint and link arm.

d.  Remove the nut of the anti-roll bar link in the anti-roll bar.

➡**Counterhold with a Torx® wrench so that the ball joint does not turn when the nuts are removed. Mark the position of the track rod.**

e.  Remove the track rod end from the guide arm. Use puller 999 5259.

f.  Detach the ball joint pin from the link arm using 999 7062.

g.  Fit the cap nut, 7062-1, on the ball joint pin.

h.  Tighten the nut toward the link arm. Then turn it back one rotation to create play between the cap nut and the link arm.

i.  Position the puller on the link arm and detach the ball joint from the link arm.

28.  Remove the nuts at intermediate exhaust pipe joints. Discard the gaskets.

29.  Remove front propeller shaft bolts.

➡**Mark the position of the carrier on the propeller shaft before removal.**

30.  Remove the propeller shaft bolts using the counterhold 999 7057.

**Fig. 9  Using puller 999 5259 to remove track rod end from guide arm**

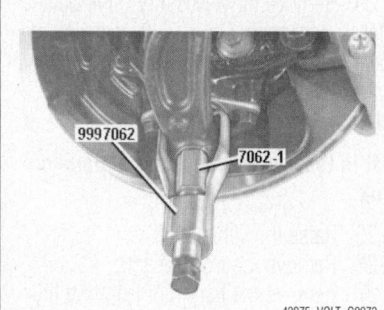

**Fig. 10 Using puller 999 7062 and 7062-1 to detach ball joint from link arm**

**Fig. 11 Intermediate exhaust pipe location**

**Fig. 12 Installation of counterhold 999 7057 on driveshaft**

31. Remove the lower steering shaft joint:

### ✸✸ CAUTION

**Do not move the steering wheel. The contact reel in the supplemental restraint system (SRS) could be damaged.**

 a. Set the front wheels in the straight-ahead position.
 b. Fold the floor carpet out of the way.
 c. Bend up the outer section of the steering shaft boot.
 d. Remove the retaining clip around the boot.

 e. Pull the boot up.
 f. Remove the ignition key and activate the steering wheel lock.
 g. Remove the bolt securing the mounting of the steering shaft in the joint.
 h. Remove the joint from the steering shaft by pressing the shaft upwards.

32. Detach the steering gear connector (blue) from the steering gear.

33. Relieve the pressure from the fuel system. Refer to Fuel System, Relieving Fuel System Pressure.

34. Drain the fuel line.

35. Detach the quick-release coupling of the fuel line at the right rear sub-frame mount. Use 999 5666 for the feed line.

36. Detach the fuel return line.

→**Seal the openings against dirt.**

37. Disconnect the connector from:
 • The oil level sensor in the oil sump
 • The oil pressure sensor
 • The 2 connections at the A/C compressor

38. Pull out the cable harness and bend it aside.

39. Position a lifting table under the sub-frame. Secure the table against the sub-frame.

### ✸✸ CAUTION

**Ensure that the lifting table is properly centered under the engine.**

40. Remove or disconnect the following:
 • The connector from the circulation pump.
 • The 4 M10 bolts in the rear edge of the sub-frame brackets
 • The 4 bolts in the sub-frame.

41. Lower the engine/transaxle unit approximately 6 inches (15cm) so that the drive shafts create a horizontal line between the hub and the transaxle.

→**Take great care that no hoses, wiring, or anything else catches in the engine/transaxle unit when it is being lowered.**

42. Tap the link arms off of the ball joint pin using 999 7076.

43. Release the driveshafts from the front wheel hubs by pressing the drive shafts into the hub.

44. Remove the 3 A/C compressor bolts.

45. Bend out the A/C compressor and suspend it in a suitable manner.

46. Continue lowering the engine/transaxle unit until it is clear of the vehicle.

47. Remove or disconnect the following:
 • The bracket between the engine and chain housing
 • The bolts between the chain housing/brackets and transaxle
 • The bevel gear/chain housing from the transaxle. Install sealing plug 999 5733
 • The bracket from the end of the transaxle.

48. Remove all bolts for the transaxle/engine. Remove the transaxle. Ensure that the torque converter accompanies the transaxle during removal and does not fall out.

49. Place the transaxle on a workbench.

 ***To install:***

→**Use new fasteners where indicated.**

50. Install or connect the following:
 • Transaxle and tighten the bolts to 37 ft. lbs. (50 Nm)
 • Torque converter. Use new bolts and tighten them to 22 ft. lbs. (30 Nm).
 • Rear transaxle mount and torque the bolts to 37 ft. lbs. (50 Nm)

51. Raise the engine/transaxle unit using a lifting table. Stop approximately 6 inches (15cm) before the sub-frame reaches the frame members.

### ✸✸ CAUTION

**Ensure that no hoses, wiring, or other components catch or become trapped during lifting.**

→**Make sure that the contact faces on the ball joints and link arms are clean.**

52. Install the lower radiator hose in the mount on the sub-frame

53. Install the A/C compressor and tighten the 3 bolts to 37 ft. lbs. (50 Nm).

54. Attach the connectors to:
 a. The A/C compressor.
 b. The oil pressure sensor
 c. The oil level sensor and secure the cable harness in the clips and clamps.

55. Install the driveshaft in the hub on both sides.

56. Continue raising the engine to approximately 2 inches (50mm) before it makes contact.

### ✸✸ CAUTION

**Guide the lower steering shaft into the hole in the firewall.**

57. Guide the ball joint to the link arm.

58. Raise the sub-frame toward the frame members until it makes contact.

59. Install the 4 sub-frame bolts together with the sub-frame brackets and engine heater, if installed.

➡**Use new bolts. Lube the bolts with oil. Tighten bolts first on the sub-frame left side, then on the right. Use: 951 2050.**

60. Tighten sub-frame bolts in two steps:
   a. Step 1—78 ft. lbs. (105 Nm).
   b. Step 2—angle-tighten 120°.

61. Tighten the M10 bracket bolts to 37 ft. lbs. (50 Nm).

62. Remove the lifting table.

63. Tighten the ball joint and link arm using a counterhold with Torx® 40 so that the ball joint gaiter does not rotate. Tighten to 59 ft. lbs. (80 Nm).

64. Install or connect the following:
   • The anti-roll bar links. Use new lock nuts
   • New center bolts and a rubber sleeve on the drive shaft
   • The track rods following the marks made during removal. Use new lock nuts
   • The fuel lines
   • Attach the connector to the steering gear. Use new mounting nuts and tighten them to 37 ft. lbs. (50 Nm)
   • The anti-roll bar
   • The intermediate pipes using new gaskets
   • The gear selector cable in the mount on the transaxle and press the cable into the ball
   • The transaxle control module (TCM) connector
   • The upper radiator hose on the distributor housing
   • The lower radiator hose on the thermostat housing
   • Connect the radiator hoses at the firewall
   • The lower line to the transaxle oil cooler as marked
   • The upper line to the transaxle oil cooler as marked
   • The positive lead (1) in the electrical distribution box
   • The ground cable (2) in the body (between the engine and body)
   • The connector at the electrical distribution box (3)
   • The connector to the brake vacuum valve and the vacuum hose on the valve leading to the brake vacuum amplifier.
   • The air cleaner and tighten the hose clip on the mass air flow sensor
   • The brake vacuum hose on the brake pump

   • The fresh air hose
   • The 2 bolts of the fresh air hose in the front plate
   • The 2 connectors to the engine control module (ECM)
   • The connector to the mass air flow sensor
   • The band clamp on the cable harness in the air cleaner housing
   • The cover over the air cleaner
   • The accessory drive belt.
   • The ground cable between the engine and body, located beneath the expansion tank
   • The servo oil reservoir
   • The return line for servo oil
   • The bolt on the washer fluid filler pipe
   • The 3 ground cables in the body at the expansion tank
   • The 2 coolant hoses to the expansion tank
   • The connector to the electric cooling fan.
   • The cover over the servo pump
   • Position the upper engine covers and press them into the retainers
   • The upper cross stay between the suspension turrets and tighten the 4 bolts to 37 ft. lbs. (50 Nm)
   • Tighten the engine bracket bolts and tighten to 59 ft. lbs. (80 Nm)
   • The lower engine cover 6 bolts and tighten to 37 ft. lbs. (50 Nm)
   • The skid plate and tighten the six M10 bolts to 37 ft. lbs. (50 Nm)
   • The front wheels and tighten the lug nuts to 102 ft. lbs. (138 Nm)
   • The battery cable

65. Check the oil level in the engine and transaxle.

66. Start the engine and allow it to warm to normal operating temperature.

67. Check for leaks and proper function. Fill fluids as necessary.

## TRANSFER CASE ASSEMBLY

*REMOVAL & INSTALLATION*

1. Before servicing the vehicle, refer to the Precautions Section.

2. Remove or disconnect the following:
   • Front right wheel and tire
   • Front right halfshaft
   • Engine splash guard
   • The mounting for the center bearing for the driveshaft

➡**Mark the driveshaft CV joint in relation to the transfer case assembly flange.**

   • Bolts on driveshaft using counterhold 999 7057
   • Press the driveshaft CV joints together and remove driveshaft
   • Remove bolts holding the transfer case and pull it straight out from the transaxle assembly

*To install:*

➡**Before installing the transfer case, lubricate the spline joint that joins to the transaxle. Make sure that the mating surfaces on the transaxle and transfer case are clean and not damaged.**

3. Use new bolts to attach transfer case. Tighten the bolts alternately to light contact. Then, tighten alternately to 37 ft. lbs. (50 Nm).

➡**Check carefully that the mating surfaces on the driveshaft CV joint and flange are clean.**

4. Install or connect the following:
   • The driveshaft according to the previous marking. Use new bolts and counterhold 999 7057 to tighten driveshaft bolts to 18 ft. lbs. (24 Nm).
   • The center bearing and member for the driveshaft. Use new bolts and tighten to 18 ft. lbs. (24 Nm).
   • Right front halfshaft to the wheel hub. Refer to Halfshafts removal and installation

5. Fill the transaxle with proper amount of oil.

6. Check the oil level in the transfer case. Fill with oil up to the filling plug.

7. Install engine splash guard.

8. Install right front wheel and tire. Tighten lug nuts to 102 ft. lbs. (138 Nm).

## FRONT HALFSHAFT

*REMOVAL & INSTALLATION*

1. Before servicing the vehicle, refer to the Precautions Section.

2. Drain transaxle oil.

3. Remove or disconnect the following:
   • Front wheel
   • Spindle nut on halfaxle

4. Remove stub of halfshaft from hub.

➡**If necessary, use a screwdriver in brake disc vents as a counterhold and tap the end of the halfshaft with a plastic hammer or brass drift until halfshaft moves into the hub approximately ¾–1 inch (20–25mm).**

5. Remove lower ball joint from the knuckle. Refer to Lower Ball Joint removal and installation.

6. Remove halfshaft from the hub.

7. If removing the left side halfshaft:

a. Use prybar 999 5462 B, or suitable prybar, to pry the inner joint out of the transaxle.

8. If removing the right side halfshaft:

a. 5- & 8-Cylinder transaxles: Remove the bearing cap for the halfshaft, hold down the spring strut, and pull out the halfshaft.

b. 6-Cylinder transaxles: Use prybar 999 5462 B, or suitable prybar, to pry the inner joint out of the transaxle.

9. Remove halfshaft.

### To install:

➡**Use new circlips, split pins, and self-locking nuts for assembly.**

10. Apply gear oil on the halfshaft splines.

11. When installing the halfshaft, set the opening side of the circlip so it faces downward.

12. Install the inner joint so that the circlip is felt to seat in the retaining groove.

13. Lubricate the halfshaft splines for the wheel hub with a small amount of oil.

14. Hold down the control arm as when removing. Align the halfshaft in the hub.

15. Align the ball joint in the control arm.

16. Install a new lock nut on the ball joint. Tighten to 59 ft. lbs. (80 Nm).

➡**Counterhold the end of the ball joint while tightening so that the boot does not twist. Use a Torx® wrench.**

17. Install a new bolt for the halfshaft/wheel hub. Tighten to 26 ft. lbs. (35 Nm), plus 120°.

18. Install front wheel and tighten lug nuts to 102 ft. lbs. (138 Nm).

19. Check and/or adjust the wheel alignment.

### REAR HALFSHAFT

### REMOVAL & INSTALLATION

*See Figures 13 and 14.*

1. Raise and support the vehicle.

2. Remove the wheel.

3. Push up the rear suspension with tool 999 5659 to a measurement of 20 inches (50cm) between the fender edge and the center of the wheel hub.

4. Remove or disconnect the following:

- The parking brake cable from the mounting in the tie rod
- The bolt for the halfshaft/wheel hub (1)
- The bolt for the tie rod mounting (2)

- The bolt for the control arm mounting (3)
- The nut for the anti-roll bar link mounting (4). Counterhold using a Torx® wrench so that the boot does not rotate
- The bolt for the lateral link mounting (5).

5. Install mobile jack 998 5972 under the knuckle and raise carefully.

➡**Pull out the halfshaft from the wheel hub while raising the steering knuckle. Angle the shaft past the knuckle.**

6. Release the tensioner so that the control arm hangs on the shock absorber.

7. Using a prybar, position the tip between the constant velocity joint housing and the final drive housing.

8. Tap firmly so that the snap ring for the halfshaft releases.

9. Pull out the halfshaft.

➡**Do not damage the halfshaft seal.**

### To install:

10. Lubricate the halfshaft mating surface using P/N 116 1329-6, or equivalent.

➡**Use new snap ring for installation.**

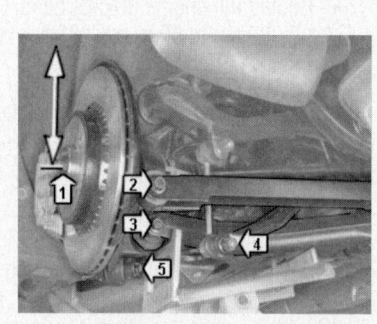

42075_VOLT_G0076

**Fig. 13 Components numbered for rear halfshaft removal and installation**

998 5972

42075_VOLT_G0077

**Fig. 14 Using mobile jack 998 5972 to raise rear knuckle**

11. Press the halfshaft into the final drive housing. Check that the snap ring is in the groove in the final drive.

12. Check by pulling the constant velocity joint on the halfshaft.

13. Lubricate the halfshaft splines using a small amount of oil.

14. Position the halfshaft splines in the wheel hub. Carefully lower the mobile jack.

15. Remove the mobile jack.

16. Press up the control arm and tensioner to the previous specified measurement so that the bolts can be installed.

➡**After installing the bolts, press tensioner 999 5659, or equivalent, up to the normal position. Tighten the bolts.**

17. Install or connect the following:

- One M12 bolt for the lateral link mounting (5). Tighten to 59 ft. lbs. (80 Nm)
- One M12 nut for the lateral link mounting (4). Tighten to 59 ft. lbs. (80 Nm). Counterhold using a Torx® wrench so that the boot does not rotate
- One M12 bolt for the lateral link mounting (3). Tighten to 59 ft. lbs. (80 Nm)
- One M12 bolt for the tie rod mounting (2). Tighten to 59 ft. lbs. (80 Nm)
- One M10 bolt to the halfshaft/wheel hub (1). Tighten to 26 ft. lbs. (35 Nm)
- The parking brake cable and mounting for the tie rod

18. Remove the tensioner.

19. Install the wheel and tighten the lug nuts to 102 ft. lbs. (138 Nm).

### REAR PINION SEAL

### REMOVAL & INSTALLATION

*See Figures 15 and 16.*

1. Before servicing the vehicle, refer to the Precautions Section.

2. Remove or disconnect the following:

- The exhaust system
- The driveshaft
- The Active on Demand Coupling (AOC)

➡**There are two parts to the pinion seal. Use extractor 999 7023 with care. The bolt and inner socket in the extractor must be unscrewed when tapping on the extractor to remove pinion seal.**

3. Tap extractor tool 999 7023 onto the seal so that the flat end is against the seal.

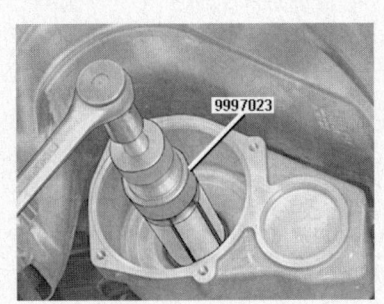

**Fig. 15 Using extractor 999 7023 to remove pinion seal**

4. Tighten the inner socket of tool 999 7023. Counter hold the outer part at the same time so that the tool expands against the seal.

5. Remove the seal using the screw in the extractor.

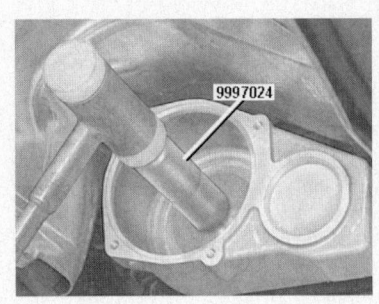

**Fig. 16 Using drift 999 7024 to install pinion seal**

➡Check that no seal residue is left in the final drive. If the inner section of the seal remains in place, tap in the extractor again and pull out the inner section. Do not hammer on the extractor too hard. This can cause damage in the extractor, which in turn can scratch the mating surface on the pinion shaft.

6. Remove the pinion seal.

*To install:*

➡The seal consists of two sections but is installed as one unit.

7. Lubricate the edge of the seal.

8. Use drift 999 7024 for installation. Position the section of the seal with the largest outer diameter against the drift.

9. Tap in the seal until the head of the drift is against the final drive housing.

10. Install or connect the following:
- The Active on Demand Coupling (AOC)
- The driveshaft
- The exhaust system

11. Refill the oil in the final drive and AOC

# ENGINE COOLING

## THERMOSTAT

*REMOVAL & INSTALLATION*

### ❊❊ CAUTION

**Never open, service or drain the radiator or cooling system when hot; serious burns can occur from the steam and hot coolant. Also, when draining engine coolant, keep in mind that cats and dogs are attracted to ethylene glycol antifreeze and could drink any that is left in an uncovered container or in puddles on the ground. This will prove fatal in sufficient quantities. Always drain coolant into a sealable container. Coolant should be reused unless it is contaminated or is several years old.**

### 5-Cylinder Engines

*See Figures 17 and 18.*

1. Before servicing the vehicle, refer to the Precautions Section.

2. Drain the cooling system:
   a. Remove cap from expansion tank
   b. Raise vehicle
   c. Remove the skid plate:
   - Loosen the six bolts and remove
   - Lift and slide skid plate forward

➡There is a guide sleeve on the upper mounting for the plate. This must be lifted slightly so that the panel can be slid forward.

d. Position a container under the engine drain cock and drain 2 quarts (2 liters) of engine coolant.
   e. Close the cock.

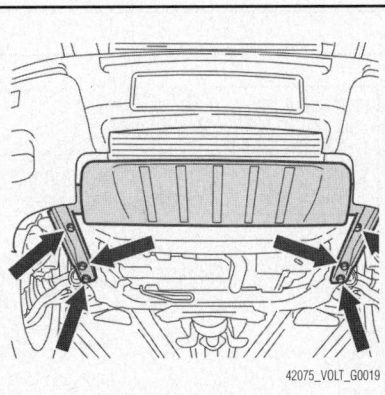

**Fig. 17 Skid plate removal**

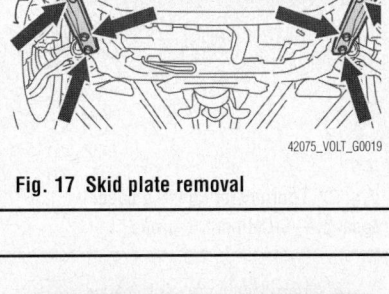

**Fig. 18 Location of thermostat housing— 5-Cylinder Engines**

3. Install the skid plate and tighten the six M10 bolts to 37 ft. lbs. (50 Nm).

4. Lower the vehicle.

5. Disconnect or remove:
- Hose clamp to the upper radiator hose
- Hose from the thermostat housing
- Connector from the temperature sensor
- The two bolts for the thermostat housing at the section against the cylinder head
- The thermostat housing from the cylinder head
- The hose clamp for the water hose for heating crankcase ventilation

*To install:*

6. Install or connect:
- The hose for the water heated crankcase ventilation
- Thermostat housing assembly to cylinder head using new gasket and new hose clamp
- Tighten thermostat housing to 16 ft. lbs. (22 Nm)
- Upper radiator hose and hose clamp
- Connector for the temperature sensor

7. Refill with coolant.

8. Install the cap on the expansion tank.

9. Test drive the engine until the thermostat has opened.

10. Check for leaks and proper function. Fill fluid as necessary.

### 6-Cylinder Engines, 2006 Only

*See Figures 19 through 22.*

1. Before servicing the vehicle, refer to the Precautions Section.
2. Disconnect or remove:
   - Clamp holding the return hose for the power steering
   - Bolt holding the engine stabilizer brace to the bracket on the engine
   - Bolts holding the engine stabilizer brace to the suspension turrets
   - Engine stabilizer brace
   - Expansion tank cap
   a. Raise vehicle
   b. Remove the skid plate:
   - Loosen the six bolts and remove
   - Lift and slide skid plate forward

➡ **There is a guide sleeve on the upper mounting for the plate. This must be lifted slightly so that the panel can be slid forward.**

   c. Position a container under the engine drain cock and drain engine coolant.
   d. Close the cock.
3. Install the skid plate and tighten the six M10 bolts to 37 ft. lbs. (50 Nm).
4. Lower the vehicle.
5. Disconnect or remove:
   - Plastic hoses between the turbocharger, charge air cooler, and air cleaner
   - Lift servo reservoir and place it on top of the engine

### ❊❊ WARNING

**Seal the cover for the servo reservoir and check that no oil leaks out. Servo oil is highly flammable.**

➡ **Ensure that the timing belt cover has cleared the nipple on the thermostat housing before the cover is pulled up.**

6. Remove:
   - Front cover
   - Timing belt cover. Use weatherstrip tool

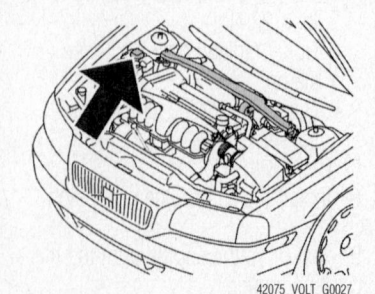
**Fig. 19 Location of engine stabilizer bar—6-Cylinder Engines**

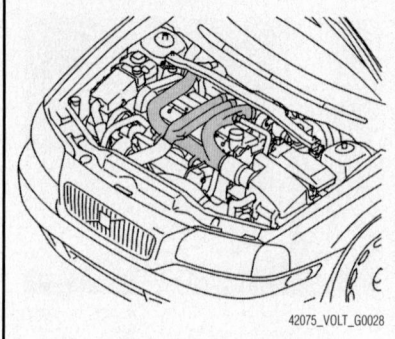
**Fig. 20 Location of plastic hoses for removal—6-Cylinder Engines**

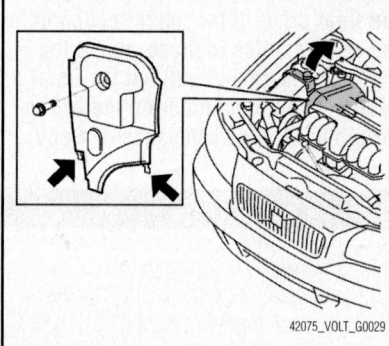
**Fig. 21 Location of servo reservoir for removal—6-Cylinder Engines**

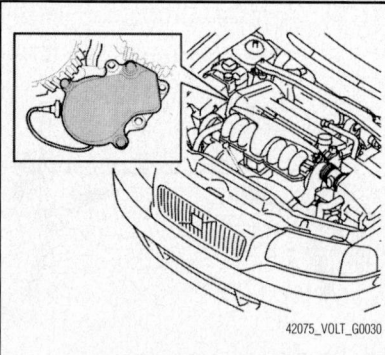
**Fig. 22 Thermostat housing cover location—6-Cylinder Engines**

   - Thermostat housing cover
   - Thermostat

*To install:*

7. Install or connect:
   - The hose for the water heated crankcase ventilation
   - Thermostat housing assembly to cylinder head using new gasket and new hose clamp
   - Tighten thermostat housing to 16 ft. lbs. (22 Nm)
   - Upper radiator hose and hose clamp
   - Connector for the temperature sensor

8. Refill with coolant.
9. Install the cap on the expansion tank.
10. Test drive the engine until the thermostat has opened.
11. Check for leaks and proper function. Fill fluid as necessary.

### 8-Cylinder Engines

*See Figures 23 and 24.*

1. Before servicing the vehicle, refer to the Precautions Section.
2. Drain the cooling system:
   a. Remove cap from expansion tank
   b. Raise vehicle
   c. Remove the skid plate:
   - Loosen the six bolts and remove
   - Lift and slide skid plate forward

➡ **There is a guide sleeve on the upper mounting for the plate. This must be lifted slightly so that the panel can be slid forward.**

   d. Position a container under the engine drain cock and drain engine coolant.
   e. Close the cock.
3. Install the skid plate and tighten the six M10 bolts to 37 ft. lbs. (50 Nm).

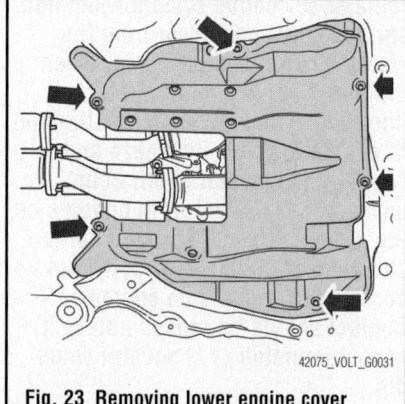
**Fig. 23 Removing lower engine cover**

**Fig. 24 Thermostat housing cover location—8-Cylinder Engines**

4. Remove the lower engine cover:
- Loosen and remove the six bolts
- Lower engine cover

5. Remove the two bolts from the thermostat housing cover.

6. Bend aside the radiator hose.

➡**The upper bolt can be loosened more easily from above.**

7. Remove the thermostat.

*To install:*

8. Install a new thermostat with a new gasket tightening bolts of housing to 16 ft. lbs. (22 Nm).

9. Install lower engine cover tightening the six bolts to 37 ft. lbs. (50 Nm).

10. Lower the vehicle.

11. Refill with coolant.

12. Install the cap on the expansion tank.

13. Test drive the engine until the thermostat has opened.

14. Check for leaks and proper function. Fill fluid as necessary.

## WATER PUMP

*REMOVAL & INSTALLATION*

### 5-Cylinder Engines

*See Figures 25 and 26.*

1. Before servicing the vehicle, refer to the Precautions Section.

2. Drain the cooling system.

3. Remove or disconnect the following:
- Negative battery cable
- Spark plug cover
- Fuel line clips
- Expansion tank
- Accessory drive belts
- Crankshaft damper guard
- Front wheel
- Inner fender panel
- Front timing cover
- Timing belt
- Water pump

*To install:*

4. Install or connect the following:
- Water pump with a new gasket. Tighten the bolts to 15 ft. lbs. (20 Nm).
- Timing belt
- Front timing cover
- Inner fender panel
- Front wheel
- Crankshaft damper guard
- Accessory drive belts
- Expansion tank
- Fuel line clips
- Spark plug cover
- Negative battery cable

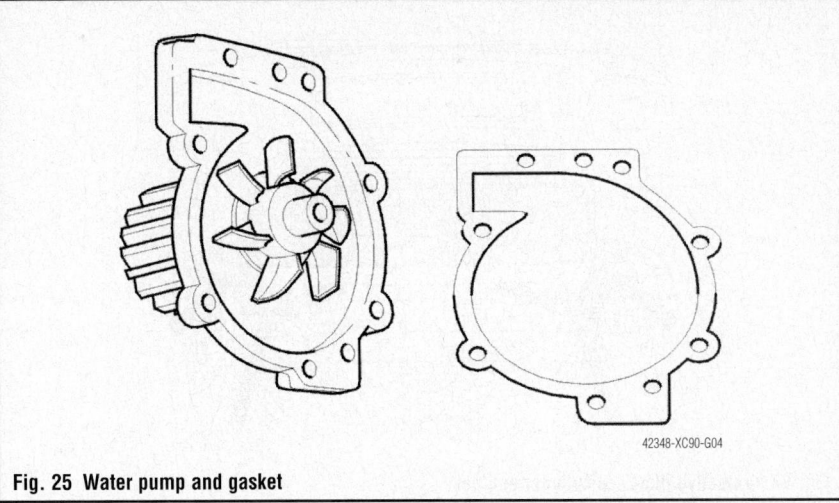

Fig. 25 Water pump and gasket

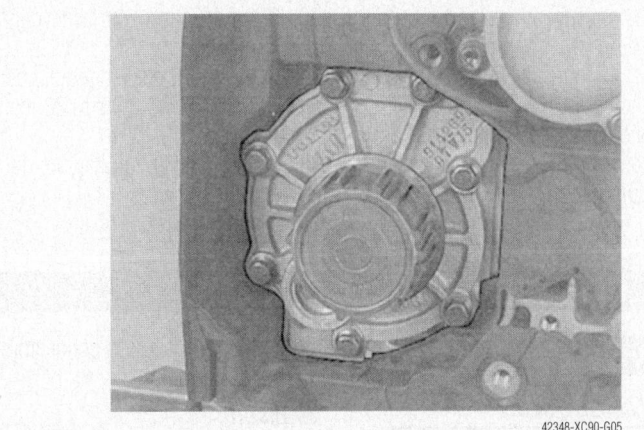

Fig. 26 Water pump installation

5. Fill the cooling system.

6. Start the engine and check for leaks.

### 6-Cylinder Engines, 2006 Only

1. Before servicing the vehicle, refer to the Precautions Section.

2. Drain the cooling system.

3. Raise and support the vehicle.

4. Remove the lower engine splash guard.

5. Open the engine nipple. Drain the coolant into a container. Close the nipple.

6. Remove the timing belt.

7. Remove the bolts holding the pump at the cylinder block.

8. Tap the pump wheel using the shaft of a hammer. Remove the pump.

*To install:*

9. Carefully clean the gasket face.

10. Install a new gasket and the coolant pump. Tighten crosswise. Tighten to 13 ft. lbs. (17 Nm).

11. Install the timing belt.

12. Remove the lock grip pliers from the lower coolant hose.

13. Install the lower engine splash guard.

14. Remove the lock grip pliers from the upper coolant hose.

15. Fill coolant and check the level.

16. Check the power steering fluid level.

### 8-Cylinder Engines

*See Figures 27 and 28.*

1. Remove the expansion tank cap.

2. Raise the vehicle.

3. Remove protective plate

➡**There is a guide sleeve on the upper mounting for the plate. This must be lifted slightly so that the panel can be slid forward.**

4. Remove the 6 bolts.

5. Lift and slide the protective plate forward.

6. Remove the protective plate.

7. Draining coolant

8. Open the cock on the lower left edge of the radiator.

9. Drain the coolant into a clean receptacle.

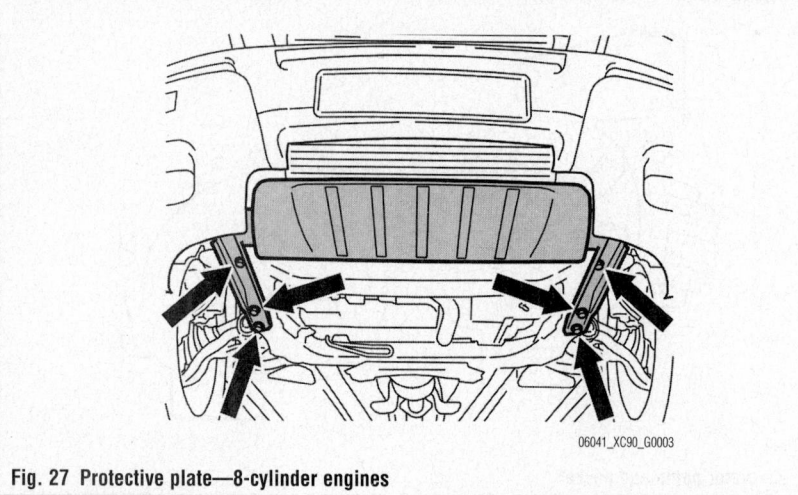

**Fig. 27 Protective plate—8-cylinder engines**

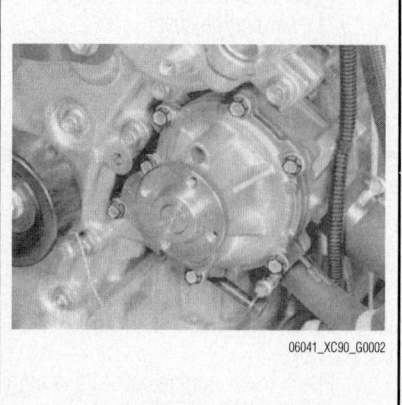

**Fig. 28 Water pump—8-cylinder engines**

10. Close the cock.
11. Install the protective plate.
12. Loosen the 4 bolts of the coolant pump belt pulley a few turns.
13. Remove the belt.
14. Remove the 4 bolts of the belt pulley.
15. Remove the 7 bolts of the coolant pump and the coolant pump. Discard the gasket.

*To install:*
16. Clean the gasket surface for the coolant pump.
17. Install a new gasket and the coolant pump. Tighten the M6 bolts to 88 inch lbs. (10 Nm).
18. Position the belt pulley but do not tighten the bolts.
19. Install the auxiliary belt.

20. Tighten the 4 M6 bolts of the belt pulley to 88 inch lbs. (10 Nm).
21. Fill with coolant.
22. Start the engine and allow it to warm to normal operating temperature. Check for leaks and proper function. Fill up fluid as necessary.

# ENGINE ELECTRICAL

## ALTERNATOR

### REMOVAL & INSTALLATION

#### 5-Cylinder Engines

*See Figure 29.*

1. Before servicing the vehicle, refer to the Precautions Section.
2. Remove or disconnect the following:
- Negative battery cable
- Accessory drive belt
- Power steering pump
- Turbocharger hose

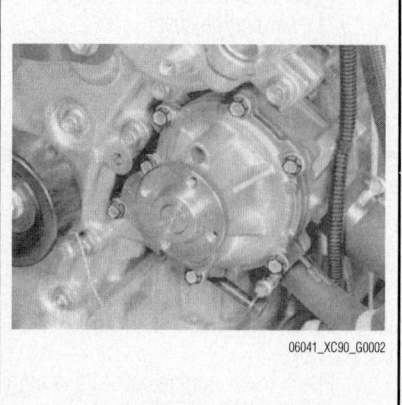

**Fig. 29 Alternator mounting**

- Alternator harness connectors
- Alternator

*To install:*
Install or connect the following:
- Alternator
- Alternator harness connectors
- Turbocharger hose
- Power steering pump
- Accessory drive belt
- Negative battery cable

#### 6-Cylinder Engines, 2006 Only

1. Disconnect the battery negative lead.
2. Raise and support the vehicle.
3. Remove the lower cover plate.
4. Drain the coolant.
5. Install the lower splash guard.
6. Lower the car.
7. Remove the upper engine coolant hose.
8. Remove the auxiliary belt.
9. Remove the power steering pump.
10. Disconnect the turbocharger hose from the cooling system. Push the hose to one side.
11. Remove the right-hand headlamp.
12. Detach the bumper cover on the right-hand side.
13. Remove the bolts from the front panel on the right-hand side as illustrated.

# CHARGING SYSTEM

14. Remove the bolt that runs through the upper mounting and the lower bolts.
15. With AC, slacken off the lower bolts for the compressor, to provide clearance.
16. Disconnect the battery positive lead.
17. Disconnect the connector.
18. Remove the alternator .
19. Turn the alternator into position.
20. Carefully pry the front panel forwards.
21. Pull the alternator upwards and out.

*To install:*
22. Position the alternator. Install the bolt through the upper mounting. Do not tighten.
23. Connect the cable and the connector.
24. Install the lower bolts for the alternator (upper bolts for the compressor).
25. With AC, tighten the 2 lower bolts for the compressor.
26. Tighten the 3 bolts for the alternator.
27. Install:
- The front panel
- The bumper cover
- The right-hand headlamp
- The turbocharger hose
- The power steering pump
- The accessory drive belt
- The engine coolant hose. Top up the coolant
- The battery negative lead

### 8-Cylinder Engines

*See Figure 30.*

1. Disconnect the battery negative cable.
2. Remove the auxiliary drive belt.
3. Detach the rear ignition coil cover on the right-hand side. Pull straight up and fold it aside.
4. Remove the right-hand halfshaft.
5. Remove the cover to the wheel arch
6. Remove the plastic nuts.
7. Removing the fuel line brackets
8. Remove the bolts and nuts.
9. Move the fuel lines to the side.
10. Lift the alternator out through the wheel arch.

#### To install:

11. Lift the alternator into position.
12. Install the bolts and connect the battery cable.
13. Install the fuel lines
14. Install the cover to the wheel arch
15. Install the right-hand halfshaft

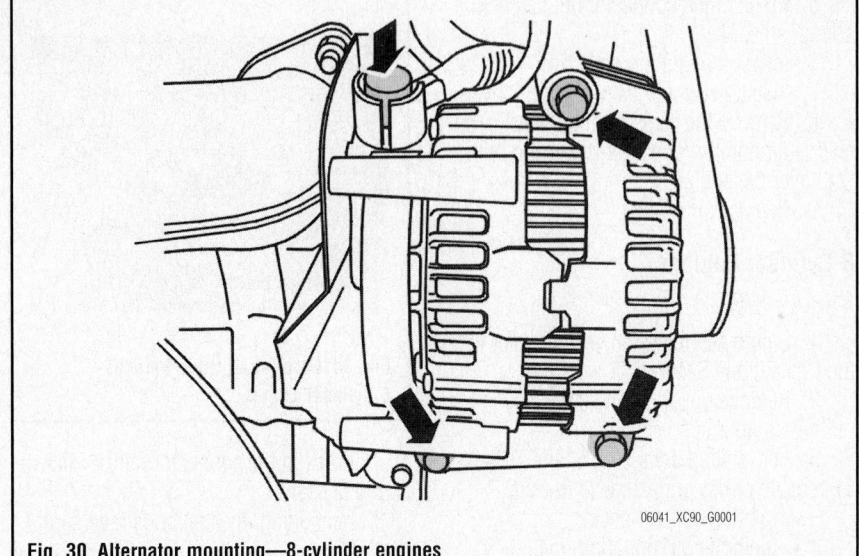

**Fig. 30 Alternator mounting—8-cylinder engines**

16. Install the fuel pipe
17. Press in the ignition coil cover.
18. Install the auxiliary belt.
19. Connect the battery negative cable.

## ENGINE ELECTRICAL

### FIRING ORDER

#### 5-Cylinder Engines

Firing order is: 1–2–4–5–3

#### 6-Cylinder Engines, 2006 Only

Firing order is: 1–5–3–6–2–4

#### 8-Cylinder Engines

Firing order is: 1–8–4–3–6–5–7–2

### IGNITION COIL

*REMOVAL & INSTALLATION*

#### 5-Cylinder Engines

*See Figures 31 and 32.*

1. Before servicing the vehicle, refer to the Precautions Section.
2. Remove or disconnect the following:
   - Charge air pipe over the engine. Seal the openings.
   - Upper timing belt cover
   - Cover over the ignition coils
   - Ignition coil connector
   - Ignition coil mounting bolt
   - Carefully pull coil up and out

#### To install:

3. Align the coil and press it down.
4. Install ignition coil mounting bolt and tighten to 88 inch lbs. (10 Nm).
5. Press in the connector until a click sound is heard.

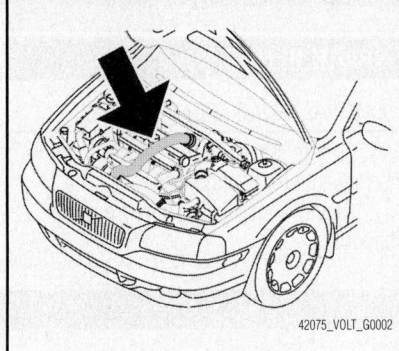

**Fig. 31 Charge Air Pipe Location— 5-Cylinder Engine**

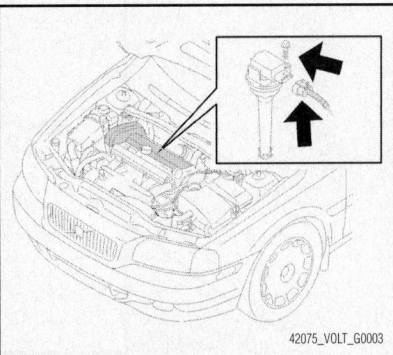

**Fig. 32 Removing ignition coil and small bolt—5-Cylinder Engine**

6. Install the cover over the ignition coil.
7. Replace the upper timing belt cover.
8. Remove the seals from the charge air pipe and reinstall over the engine.

## IGNITION SYSTEM

#### 6-Cylinder Engines, 2006 Only

*See Figure 33.*

1. Before servicing the vehicle, refer to the Precautions Section.
2. Remove or disconnect the following:
   - Plastic charge air pipes of turbocharger
   - Air cleaner.

➡ **Seal the open ends of pipes.**

   - Upper camshaft belt cover
   - Cover over the ignition coils
   - Ignition coil connector
   - Ignition coil mounting bolt
   - Carefully pull coil up and out

#### To install:

3. Align the coil and press it down.
4. Install ignition coil mounting bolt and tighten to 88 inch lbs. (10 Nm).

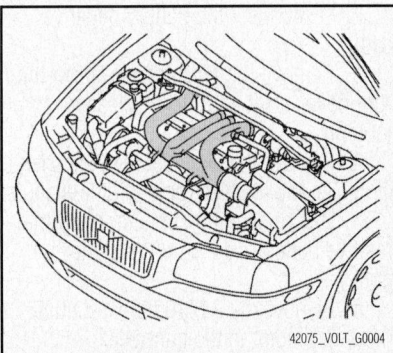

**Fig. 33 Charge air and turbocharger pipes location—6-Cylinder Engine**

5. Press in the connector until a click sound is heard.

6. Install the ignition coil cover.

7. Install the upper camshaft belt cover.

8. Remove the seals from the plastic charge air and turbocharger pipes and reinstall over the engine.

9. Replace air cleaner.

## 8-Cylinder Engines

*See Figure 34.*

1. Before servicing the vehicle, refer to the Precautions Section.

2. Remove upper engine covers by pulling straight up

3. For ignition coils 1, 3, 5, and 7 (Bank 1), remove or disconnect the following:
   - Ignition coil connector
   - Ignition coil mounting bolt
   - Carefully pull coil up and out

4. For ignition coils 2, 4, 6, and 8 (Bank 2), remove or disconnect the following:
   - Two bolts of the cable harness
   - Retaining bolt of the upper engine cover
   - Carefully bend aside the cable harness
   - Ignition coil connector
   - Ignition coil mounting bolt
   - Carefully pull coil up and out

### To install:

5. Align the coil and press it down.

6. Install ignition coil mounting bolt and tighten to 88 inch lbs. (10 Nm).

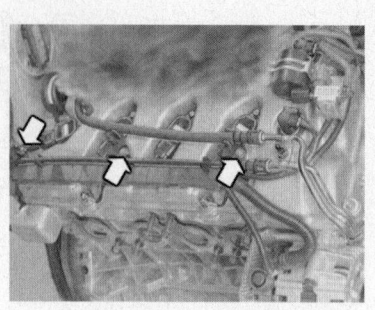

**Fig. 34 Location of cable harness— 8-Cylinder Engine**

7. Press in the connector until a click sound is heard.

8. Remaining installation is reverse of removal.

## IGNITION TIMING

### ADJUSTMENT

The ignition timing is controlled by the Powertrain Control Module (PCM). No adjustment is necessary or possible.

## SPARK PLUGS

### REMOVAL & INSTALLATION

1. Before servicing the vehicle, refer to the Precautions Section.

2. Remove spark plug wire from the spark plug.

➡**Pull on the spark plug wire boot when removing the spark plug wire to avoid damage to the wire.**

3. Clean loose debris away from area of spark plug to keep contaminants from entering engine when spark plug is removed.

4. Remove the spark plug using a spark plug socket and wrench.

### To install:

### 2006–07 5-Cylinder Engine and 2006 6-Cylinder Engine

1. Be sure the spark plug gap is set to 0.030 inch (0.763mm).

2. Carefully install the spark plug and tighten to 20–22 ft. lbs. (27–29 Nm).

### 8-Cylinder Engines

1. Be sure the spark plug gap is set to 0.028 in. (0.712mm).

2. Carefully install the spark plug and tighten 18 ft. lbs. (24 Nm).

### INSPECTION

1. Check spark plugs for the following:
   a. Broken insulator
   b. Worn electrode
   c. Carbon deposits
   d. Damaged or broken gasket
   e. Condition of the porcelain insulator at the tip of the spark plug

2. Replace as necessary

---

# ENGINE ELECTRICAL                    STARTING SYSTEM

## STARTER

### REMOVAL & INSTALLATION

#### 5-Cylinder Engines

*See Figure 35.*

1. Before servicing the vehicle, refer to the Precautions Section.

2. Disconnect the battery negative lead.

3. Remove the air intake between the front cover plate and the air cleaner housing

4. Remove the hose between the charge air cooler and the electronic throttle module.

5. Disconnect the connector and the positive battery lead from the starter motor.

6. Remove the 2 M10 bolts securing the starter motor to the transaxle.

7. Remove the starter motor.

8. Installation is the reverse of removal.

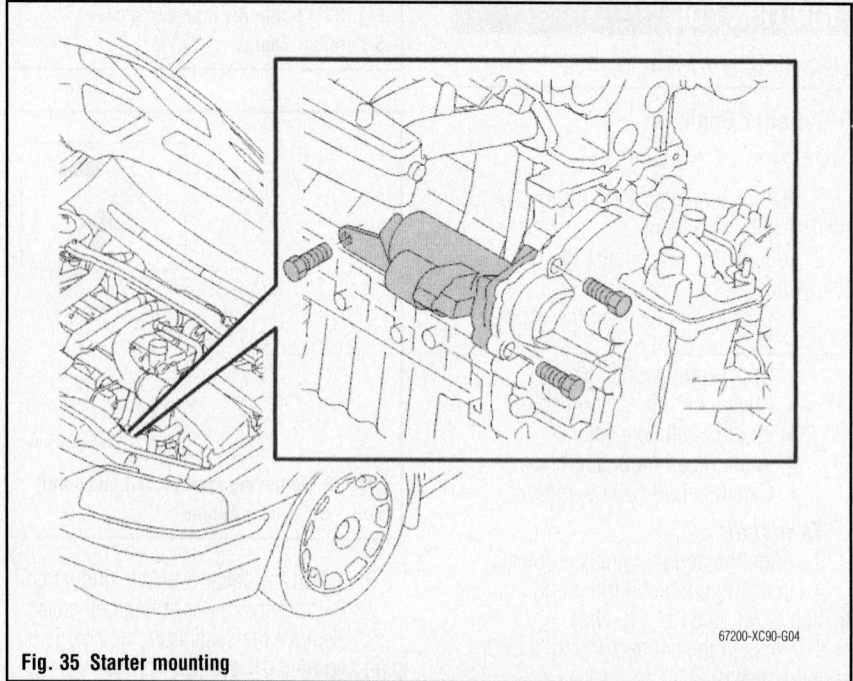

**Fig. 35 Starter mounting**

### 6-Cylinder Engines, 2006 Only

1. Before servicing the vehicle, refer to the Precautions Section.
2. Disconnect the battery negative lead.
3. Remove:
   - The intake pipe between the front cover plate and the air cleaner
   - The connector from the mass air flow sensor
   - The turbocharger control valve from the air cleaner housing
   - The brake vacuum hose from the air cleaner
   - The hose clamp between the fresh air intakes
   - Mass air flow sensor
   - The air cleaner by pulling it straight up.
   - The hose clamp on the charge air pipe on the throttle body. Pull the pipe off the throttle body
   - M8 bolt for the bracket in the cylinder block
   - M8 nuts in the bracket for the wiring and the engine coolant hose
   - The blade terminal connector on the solenoid
   - The nut for the positive lead
   - The M10 bolts for the mounting for the starter motor
   - The starter motor.

#### To install:

4. Install:
   - The starter motor
   - The M10 bolts in the transaxle cover
   - M8 bolt for the support bracket on the engine block. Tighten. The M6 nuts for the bracket on the starter motor
   - The bracket for the wiring. Use 2 M8 nuts.
   - The battery positive lead
   - The blade terminal connector on the solenoid
   - The charge air pipe on the throttle body
   - The air cleaner
   - The fresh air intake
   - The turbocharger control valve
   - The brake vacuum hose
   - The mass air flow sensor connector.

### 8-Cylinder Engines

*See Figure 36.*

1. Remove:
   a. The 4 bolts at the suspension turrets
   b. The bolts of the upper engine mount on the engine
   c. The upper cross stay.
   d. The battery negative cable.
   e. The expansion tank cap.

2. Raise the vehicle.
3. Remove the protective plate (skid plate).

➡**There is a guide sleeve on the upper mounting for the plate. This must be lifted slightly so that the panel can be slid forward.**

4. Draining coolant
5. Open the cock on the lower left edge of the radiator.
6. Drain the coolant into a clean receptacle.
7. Close the cock.
8. Installing the protective plate.
9. Install the protective plate.
10. Lower the vehicle.
11. Remove the air cleaner cover by pulling it straight up.
12. Remove:
    a. The 2 connectors from the engine control module (ECM); clip the band clamp to the cable harness.
    b. The connector from the mass air flow sensor.
13. Remove:
    a. The 2 bolts of the fresh air hose in the front plate.
    b. The fresh air hose.
    c. The hose clips of the intake manifold between the mass air flow sensor and the electronic throttle module (ETM).
    d. The air cleaner by pulling it straight up.
14. Remove:
    a. The connector to the brake vacuum valve.
    b. The vacuum hose from the vacuum pump by pressing in the quick-action lock.
    c. The vacuum hose from the brake vacuum valve.
    d. The vacuum hose from the intake manifold.
    e. The brake vacuum from the retainer.
15. Detach the radiator hoses at the quick-release coupling toward the firewall.
16. Detach the coolant hose from the throttle body by undoing the clamp.
17. Bend aside the coolant hoses.
18. Remove:
    a. The connector to the electronic throttle module (ETM) and the electrical distribution center.
    b. The protective cap from the starter motor solenoid.
    c. The nut of the positive leads on the starter motor solenoid.
    d. The positive leads.
    e. The control current lead from the starter motor solenoid.
19. Remove:
    a. The 4 bolts.
    b. The upper engine mount.
20. Remove:
    a. The EVAP valve from the bracket.
    b. The line from the EVAP valve to the tank by releasing the quick-action lock.
    c. The hose from the intake manifold to the EVAP valve by releasing the hose clip.
    d. The 2 bolts of the EVAP valve bracket.
    e. The EVAP valve bracket.
21. Remove the 3 bolts of the lifting eye and the lifting eye.
22. Open the hose clip and move it to the side.
23. Remove the 2 bolts and the connection pipe. Discard the gasket.
24. Remove the 2 bolts and the starter motor.

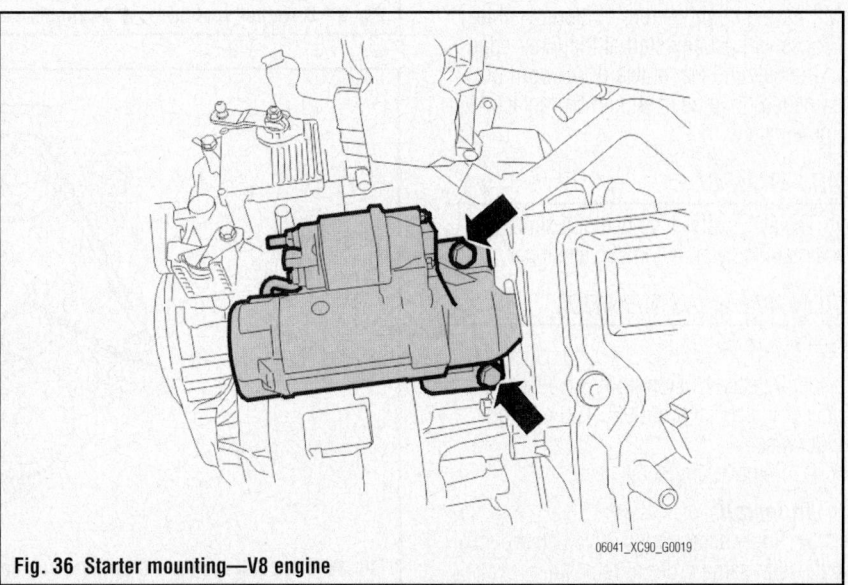

**Fig. 36 Starter mounting—V8 engine**

06041_XC90_G0019

**To install:**

25. Install the starter motor. Tighten the 2 bolts.

26. Install a new gasket and the connection pipe on the coolant hose.

27. Install the 2 bolts.

28. Install the clamp on the coolant hose.

29. Fit the clamp in its original position.

30. Install the lifting eye.

31. Install the EVAP valve bracket.

32. Install the hose from the intake manifold to the EVAP valve; secure with the clamp.

33. Install the line to the tank on the EVAP valve; secure with the quick-action lock.

34. Install the EVAP valve on the bracket.

35. Install the upper engine mount.

36. Install the connector to the electronic throttle module (ETM) and the electrical distribution center.

37. Install the positive leads, x 2, on the solenoid.

38. Install the nut.

39. Install the protective cover on the nut.

40. Install the control current lead on the solenoid.

41. Install the 2 coolant hoses on the connections at the firewall; secure with the quick-action lock.

42. Install the coolant hose on the throttle body; secure with the clamp.

43. Install the air cleaner on the left frame member.

44. Install the intake hose on the throttle body; secure with the clamp.

45. Install the connector to the mass air flow sensor.

46. Install the vacuum hose to the vacuum pump; secure with the quick-action lock.

47. Install the vacuum hose on the brake vacuum valve.

48. Install the vacuum hose on the intake manifold; secure with the clamp.

49. Install the brake vacuum valve in the retainer.

50. Install the connector to the brake vacuum valve.

51. Install the fresh air intake manifold.

52. Install the engine control module (ECM) connectors.

53. Install the cross stay between the suspension turrets.

54. Install the 4 bolts in the suspension turrets.

55. Install the bolt of the cross stay mount on the upper engine mount.

➡**Make sure the outlet hose of filling equipment is not blocked as this could damage the cooling system.**

56. Fill with coolant.

57. Start the engine and allow it to warm to normal operating temperature. Check for leaks and proper function. Fill up fluid as necessary.

## ENGINE MECHANICAL

➡**Disconnecting the negative battery cable may interfere with the functions of the on board computer systems and may require the computer to undergo a relearning process, once the negative battery cable is reconnected.**

### ACCESSORY DRIVE BELTS

*ACCESSORY BELT ROUTING*

*See Figures 37 and 38.*

*INSPECTION*

Inspect the drive belt for signs of glazing or cracking. A glazed belt will be perfectly smooth from slippage, while a good belt will have a slight texture of fabric visible. Cracks will usually start at the inner edge of the belt and run outward. All worn or damaged drive belts should be replaced immediately.

*ADJUSTMENT*

The accessory drive belt adjustment is maintained by an automatic tensioner.

*REMOVAL & INSTALLATION*

*See Figure 39.*

1. Release tension from drive belt by attaching special tool 999 7109 and turn clockwise.

2. Remove drive belt.

**To install:**

3. Route accessories drive belt around pulleys in same order of removal.

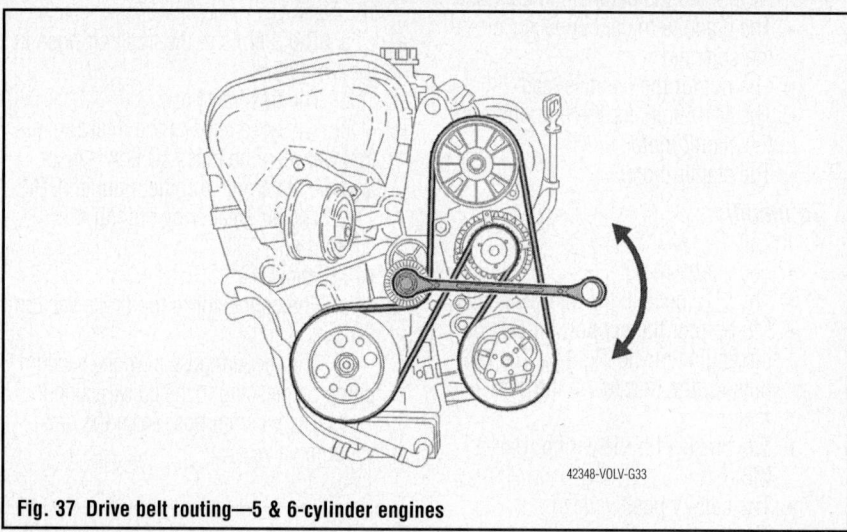

42348-VOLV-G33

**Fig. 37 Drive belt routing—5 & 6-cylinder engines**

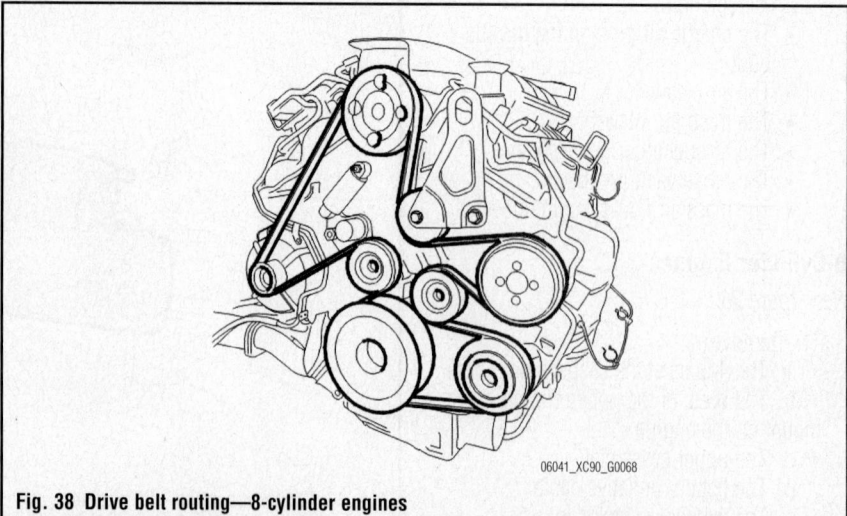

06041_XC90_G0068

**Fig. 38 Drive belt routing—8-cylinder engines**

**Fig. 39 Accessory drive belt with special tool 999 7109**

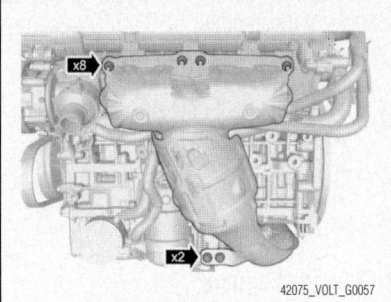

**Fig. 40 Exhaust manifold and catalytic converter removal—8-Cylinder Engine**

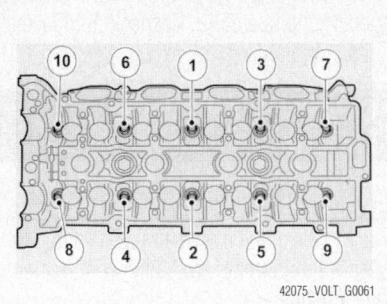

**Fig. 44 Cylinder head torque sequence— 8-Cylinder Engine**

4. Use special tool 999 7109 to back tensioner off by turning clockwise.
5. Release tensioner when drive belt is fully in place.
6. Be sure drive belt is securely installed around all pulleys.

## BALANCE SHAFT

### *REMOVAL & INSTALLATION*

**8-Cylinder Engines**
*See Figures 40 through 44.*

1. Before servicing the vehicle, refer to the Precautions Section.
2. Disconnect or remove the following:
   • Engine fan, refer to Engine Fan removal and installation
   • Engine coolant
   • Coolant hoses
   • Intake manifold
   • Water pump
   • Exhaust manifold on Bank 1 with catalytic converter
   • Timing chain. Refer to Timing Chain and Sprockets removal and installation
   • Variable valve timing (VVT) units on intake and exhaust camshafts, Bank 1. Refer to Camshaft and Valve Lifters removal and installation
3. Remove camshafts.
4. Lift out the valve lifters using a magnet or suction cup.

➡**Make a note of the positions of each part.**

5. Remove the balance shaft.

➡**Check that the engine block has a drainage hole by the rear balance shaft bearing. If there is no drainage hole, one will need to be drilled.**

6. To drill the drainage hole for rear balance shaft bearing:

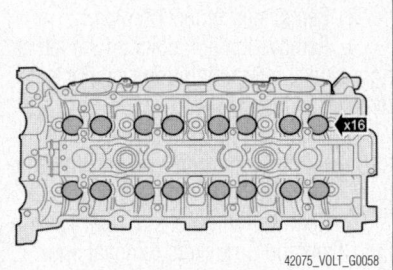

**Fig. 41 Valve lifters location—8-Cylinder Engine**

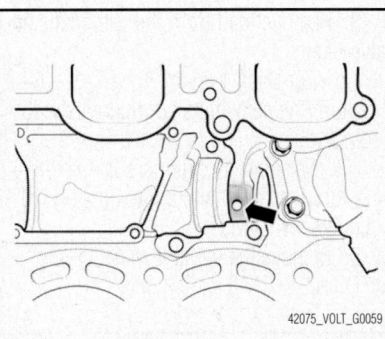

**Fig. 42 Location of drainage hole for rear balance shaft bearing—8-Cylinder Engine**

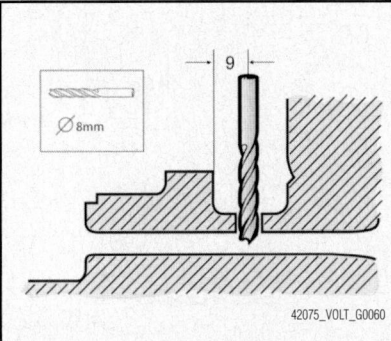

**Fig. 43 Drilling drainage hole for balance shaft bearing—8-Cylinder Engine**

• Drill down to the drainage duct using the correct drill bit 0.315 inch (8mm)
• Carefully blow the area clean after drilling

***To install:***
7. Install balance shaft into position.
8. Install cylinder head and tighten bolts in sequence from center outwards in two steps:
   a. Step 1—30 ft. lbs. (40 Nm).
   b. Step 2—angle-tighten 90°.
9. Install the valve lifters.

➡**If the cylinder head has been replaced, the valve clearance must be adjusted. Refer to Valve Lash adjustment.**

10. Lubricate and install valve lifters.
11. Install:
   • Exhaust manifold and tighten the eight M8 nuts to 18 ft. lbs. (24 Nm)
   • Two M8 bolts for the exhaust manifold bracket and tighten to 18 ft. lbs. (24 Nm)
   • The M8 bolt and two M8 nuts attaching coolant pipes and tighten to 18 ft. lbs. (24 Nm)
   • Two M8 upper bolts holding the water pump and tighten to 18 ft. lbs. (24 Nm)
   • Coolant hoses and clamps
   • Camshafts
   • Timing chain and sprockets
   • Camshaft cover
   • Ignition coils
   • Air induction pipe.
   • Intake manifold, refer to Intake Manifold removal and installation.
   • Engine fan, refer to Engine Fan removal and installation
12. Refill the cooling system using the proper grade and type of engine coolant.
13. Start the engine and check for leaks.

14. Run engine and allow it to reach full operating temperature. Recheck fluid levels. Fill as required.

## CAMSHAFT AND VALVE LIFTERS

### INSPECTION

1. Check the camshaft journals for wear. If the journals are badly worn, replace the camshaft.

2. Check the cam lobes for damage. If the lobe is damaged or excessively worn, replace the camshaft.

3. Measure the cam lift height. The specifications for cam lobe height:

   a. 5-Cylinder Engines—Intake 0.331 inch (8.40mm), Exhaust 0.356 inch (9.05mm)

   b. 6-Cylinder Engines—Intake 0.331 inch (8.40mm), Exhaust 0.356 inch (9.05mm)

4. Check the cam surface for abnormal wear or damage, and replace if necessary.

5. Check each bearing for damage. If the bearing surface is excessively damaged, replace the cylinder head assembly or camshaft bearing cap, as necessary.

### REMOVAL & INSTALLATION

#### 2006–07 5-Cylinder Engine and 2006 6-Cylinder Engine

1. Before servicing the vehicle, refer to the Precautions Section.

2. Remove or disconnect the following:
   • Negative battery cable
   • Accessory drive belt
   • Front cover
   • Timing belt
   • Ignition coil cover
   • Camshaft Position (CMP) sensor or distributor, as equipped
   • Switch holder and shield at left rear of the engine
   • Ignition coils
   • Camshaft sprockets
   • Cylinder head cover
   • Camshafts
   • Hydraulic lash adjusters

➡Keep all valvetrain components in order for assembly.

**To install:**

3. Install or connect the following:
   • Hydraulic lash adjusters
   • Camshafts
   • Cylinder head cover and tighten the bolts to 13 ft. lbs. (17 Nm)
   • Camshaft sprockets and tighten the bolts to 15 ft. lbs. (20 Nm)
   • Ignition coils

   • Switch holder and shield at left rear of the engine
   • CMP sensor or distributor, as equipped
   • Ignition coil cover
   • Timing belt
   • Front cover
   • Accessory drive belt
   • Negative battery cable

#### 8-Cylinder Engines

*See Figures 45 through 53.*

1. Before servicing the vehicle, refer to the Precautions Section.

2. Remove intake manifold. Refer to Intake Manifold removal and installation.

3. Remove air induction pipe.

4. Remove ignition coils on the left and right banks, refer to Ignition Coil Pack removal and installation.

5. Remove camshaft covers, refer to Camshaft Covers removal and installation.

6. Remove timing chain, refer to Timing Chain and Sprockets removal and installation.

7. Turn the crankshaft back approximately 45° so that the position setting tool is straight up.

➡The camshafts should rotate freely.

8. Remove load from the camshafts and valve lifters:

   a. For Bank 1, cylinders 1—3—5—7, turn the camshafts approximately 45° clockwise.

   b. For Bank 2, cylinders 2—4—6—8, turn the camshafts approximately 10° counter-clockwise.

9. Remove the variable valve timing (VVT) unit:

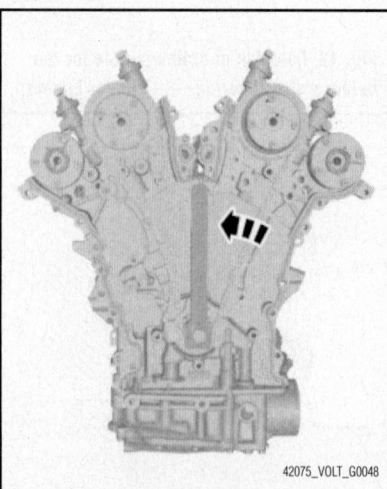

**Fig. 45 Position setting tool installed backed off 45°—8-Cylinder Engine**

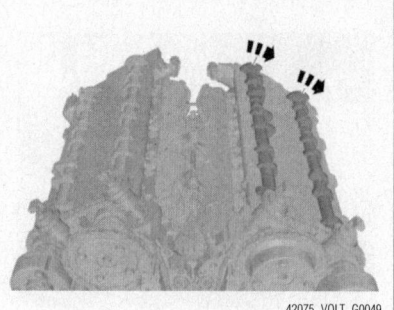

**Fig. 46 Removing load from Bank 1 camshafts—8-Cylinder Engine**

**Fig. 47 Removing load from Bank 2 camshafts—8-Cylinder Engine**

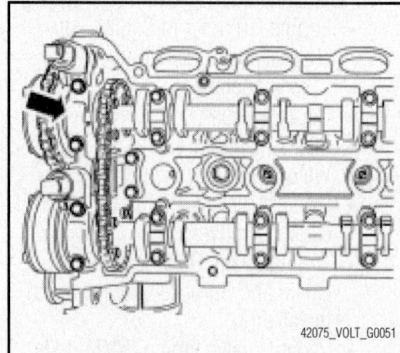

**Fig. 48 Removing the variable valve timing (VVT) unit—8-Cylinder Engine**

   • Remove the 4 M6 bolts and the cap
   • Remove the 2 M6 bolts for the chain tensioner
   • Discard the 2 filters

10. Remove the 8 M6 bolts holding the intake camshaft bearing caps.

➡Ensure that the camshaft is relieved from load against the valve lifters to prevent damaging the components. Note location of components.

11. Remove the chain tensioner.

12. Remove the camshaft.

13. To remove exhaust camshaft, remove the 8 M6 bolts holding the exhaust

camshaft bearing caps and remove exhaust camshaft.

14. Remove the valve lifters using a magnet or suction cup. Make note of the position of each lifter.

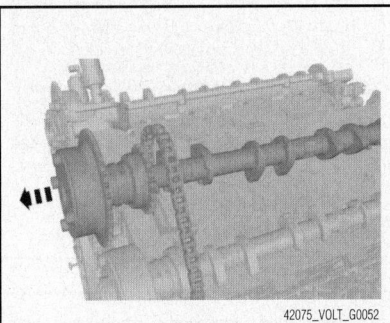

42075_VOLT_G0052

**Fig. 49 Removing camshaft and chain—8-Cylinder Engine**

*To install:*

15. Install valve lifters, refer to Valve Lifters adjustment.

16. Install VVT unit on the camshaft with 1 M10 bolt, but do not tighten.

17. Install the camshaft, chain, and chain tensioner.

➡Adjust the marking on the VVT unit to the color marking on the camshaft chain.

18. Install the camshaft bearing caps and tighten to 88 inch lbs. (10 Nm).

19. Install 2 new filters.

20. Install the 4 M6 bolts on the end cap of the camshafts and tighten to 18 inch lbs. (10 Nm).

21. Install the 2 M6 bolts for the chain tensioner and tighten to 18 inch lbs. (10 Nm).

22. Install tool 999 7236.

23. Minimize the slack of the camshaft chain.

   a. Use tool 999 7232.

   b. For cylinder head 1, 3, 5 and 7

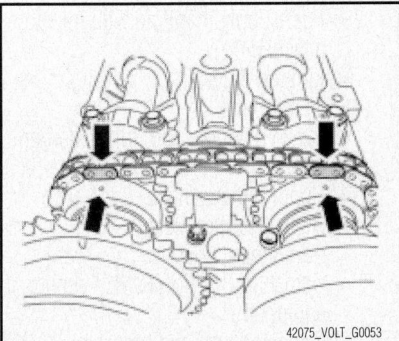

42075_VOLT_G0053

**Fig. 50 Aligning timing marks on VVT unit—8-Cylinder Engine**

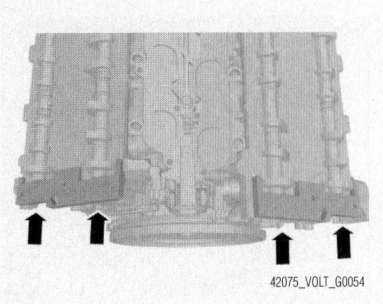

42075_VOLT_G0054

**Fig. 51 Installation of camshaft tool 999 7236—8-Cylinder Engine**

(Bank 1), turn the VVT unit on the exhaust camshaft counter-clockwise, until the chain is tensioned underneath.

   c. For cylinder head 2, 4, 6 and 8 (Bank 2), turn the VVT unit on the exhaust camshaft counter-clockwise, until the chain is tensioned on top.

   d. Tighten the bolt in the center of the intake camshaft VVT unit. Tighten to 15 ft. lbs. (20 Nm).

24. Install tool 999 7229 on the intake camshaft.

25. Tighten the center 1 M10 bolt on the camshaft to 37 ft. lbs. (50 Nm).

26. Remove the counterhold.

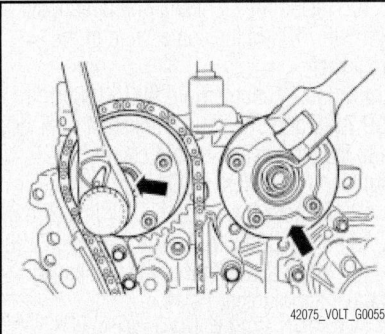

42075_VOLT_G0055

**Fig. 52 Adjusting camshaft chain tension—8-Cylinder Engine**

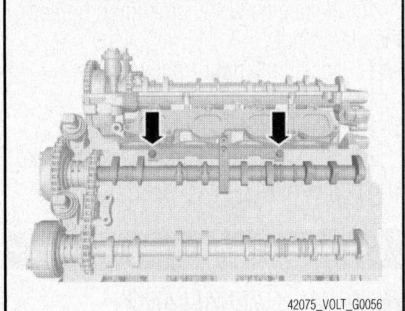

42075_VOLT_G0056

**Fig. 53 Installation of camshaft tool 999 7229—8-Cylinder Engine**

27. Install timing chain, refer to Timing Chain and Sprockets removal and installation.

28. Install camshaft covers, refer to Camshaft Covers removal and installation.

29. Install ignition coils on the left and right banks, refer to Ignition Coil Pack removal and installation.

30. Install air induction pipe.

31. Install intake manifold. Refer to Intake Manifold removal and installation.

## CRANKSHAFT FRONT SEAL

*REMOVAL & INSTALLATION*

### 2006–07 5-Cylinder Engine and 2006 6-Cylinder Engine

*See Figures 54 through 56.*

1. Before servicing the vehicle, refer to the Precautions Section.

2. Remove or disconnect the following:
- Negative battery cable
- Fuel line clips
- Coolant recovery tank
- Accessory drive belts
- Right front wheel
- Inner fender liner
- Front cover
- Timing belt
- Crankshaft timing sprocket
- Front crankshaft seal

*To install:*

3. Install or connect the following:
- Front crankshaft seal so that it is flush with the oil pump housing
- Crankshaft timing sprocket and tighten the nut to 133 ft. lbs. (180 Nm)
- Timing belt
- Front cover

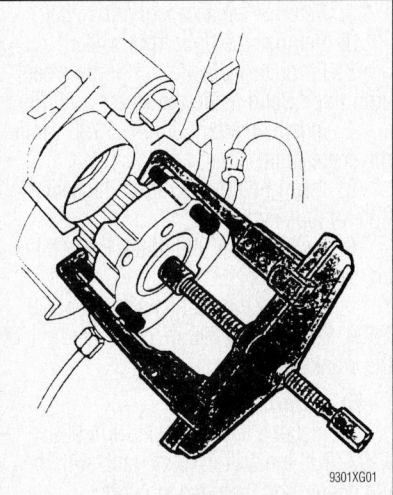

9301XG01

**Fig. 54 Removing the crankshaft timing belt sprocket—5 & 6-cylinder engines**

Fig. 55 Removing the front seal—5 & 6-cylinder engines

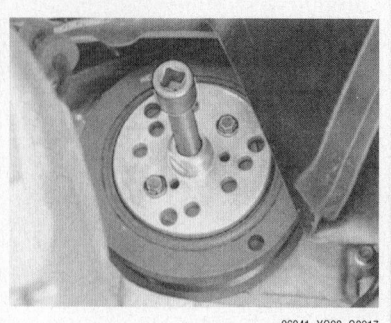

Fig. 57 Removing the damper—V8 engine

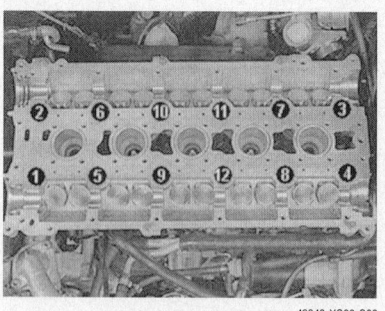

Fig. 59 Cylinder head loosening sequence—5-cylinder engine

Fig. 56 Installing the front seal—5 & 6-cylinder engines

Fig. 58 Removing the seal—V8 engine

Fig. 60 Cylinder head torque sequence—5-cylinder engines

- Inner fender liner
- Right front wheel
- Accessory drive belts
- Coolant recovery tank
- Fuel line clips
- Negative battery cable

4. Start the engine and check for leaks.

### 8-Cylinder Engines

*See Figures 57 and 58.*

1. Remove the accessory drive belt.
2. Remove the right front wheel.
3. Remove 1 plastic nut of the wheel arch liner. Bend aside the wheel arch liner.
4. Install counterhold 999 7196. Undo the center bolt.
5. Insert bolt 999 7198 in the center hole of the crankshaft.
6. Install puller 999 5304. Pull off the crankshaft damper.
7. Remove the seal using puller 999 5069. Screw 999 7198 in the center hole of the crankshaft.

#### *To install:*

8. Install a new seal with drift 999 7197. Pull the drift in to the stop with the old center bolt. Remove the drift.
9. Clean the surface for the sealing ring on the crankshaft damper with cloth 951 1024.

10. Install the crankshaft damper. Install bolt 999 7198 in the center hole of the crankshaft.
11. Install counterhold 999 7196, drift 999 7198 and press the crankshaft damper in to the stop using the old center bolt. Tighten to 74 ft. lbs. (100 Nm).
12. Remove the center bolt and discard it.
13. Install tools 999 7198 and 999 7198.
14. Insert a new bolt at the crankshaft damper and tighten:
    - Step 1: 89 ft. lbs. (120 Nm)
    - Step 2: back off at least 1 turn
    - Step 3: 37 ft. lbs. (50 Nm)
    - Step 4: plus 90°
15. Remove counterhold 999 7196.
16. Install the wheel arch liner; tighten the plastic nut.
17. Install the right front wheel.
18. Lower the vehicle.
19. Installing the auxiliary belt.
20. Start the engine and check for leaks and proper function.

### CYLINDER HEAD

#### *REMOVAL & INSTALLATION*

#### 5-Cylinder Engines

*See Figures 59 and 60.*

1. Before servicing the vehicle, refer to the Precautions Section.
2. Drain the cooling system.
3. Remove or disconnect the following:
    - Negative battery cable
    - Air intake assembly
    - Accessory drive belts
    - Front cover
    - Timing belt
    - Exhaust manifold
    - Fuel supply manifold covers
    - Fuel line retainers
4. Install fuel injector holders on the injectors.
    - Fuel pressure regulator vacuum hose
    - Fuel supply manifold with injectors attached
    - Ignition coils
    - Cooling fan
    - Intake manifold with turbocharger attached
    - Upper radiator hose
    - Camshaft sprockets
    - Rear timing cover
    - Camshaft Position (CMP) sensor, if equipped
    - Extension arm and brackets
    - Cylinder head cover
    - Camshafts

- Coolant pipe
- Cylinder head

***To install:***

5. Install the cylinder head with a new gasket. Tighten the bolts in sequence as follows:

  a. Step 1: 15 ft. lbs. (20 Nm)
  b. Step 2: 44 ft. lbs. (60 Nm)
  c. Step 3: Plus 130°

6. Install or connect the following:
- Coolant pipe
- Camshafts
- Upper cylinder head. Tighten the bolts to 13 ft. lbs. (17 Nm).
- Extension arm and brackets
- CMP sensor, if equipped
- Distributor, if equipped
- Rear timing cover
- Camshaft sprockets
- Upper radiator hose
- Intake manifold
- Cooling fan
- Fuel supply manifold with injectors attached and remove the injector holders
- Fuel pressure regulator vacuum hose
- Fuel line retainers
- Fuel supply manifold covers
- Exhaust manifold
- Timing belt
- Front cover
- Accessory drive belts
- Air intake assembly
- Negative battery cable

7. Fill the cooling system.
8. Start the engine and check for leaks.

### 6-Cylinder Engines, 2006 Only

*See Figures 61 through 66.*

1. Before servicing the vehicle, refer to the Precautions Section.
2. Disconnect the battery negative lead.
3. Remove the radiator breather tube at the expansion tank
4. Install lock grip pliers
5. Install the cover from the expansion tank.
6. Lift up the servo reservoir and place it on top of the engine.

➡**Ensure that the oil does not leak from the ventilation hole in the filler cap.**

7. Seal the hose between the expansion tank and the radiator.
8. Disconnect the hose at the tank. Lift up the expansion tank and place it on top of engine.
9. Remove the turbocharger.

10. Remove:
- The ground strip between the cylinder head and the engine
- The upper engine coolant hose between the engine and radiator
- The solenoid valve (turbocharger control system) from the air cleaner
- The air cleaner intake from the connection on the front cover plate
- Air cleaner assembly
- Torque rod for cross stay
- The upper timing cover
- The cover over the ignition coils
- The front timing cover.

11. Relieve the load from the belt tensioner. Remove the auxiliary belt.
12. Drain the fuel line.
13. Remove:
- The brake vacuum hose from the intake manifold
- The charge air pipe from the electronic throttle module
- The dip stick pipe from the intake manifold
- The pipe bolt for the crankcase ventilation connection at the intake manifold
- The protective cover over the nozzle connectors
- The fuel rail mounting bolts.

14. Spray penetrating oil around the injector connectors on the intake manifold. Gently work the fuel rail and injectors loose.

15. Separate the fuel line quick-release connector under the intake manifold. Use tool 951 2666, or equivalent. Carefully put the fuel rail to one side.

16. Remove the EVAP hose from the intake manifold.

17. Remove:
- The camshaft position sensor housing and trigger wheels
- The pipe bolt at the rear edge of the cylinder head for the water heated crankcase ventilation
- The clamps for the high tension wiring and positive cable.
- The bolt that holds the inlet pipe for the coolant pump on the cylinder block
- The two bolts for the connector between the coolant pipe and the bypass channel. Carefully turn the pipe to remove it from the cylinder block.

18. Remove the nuts from the cover on the fender liner.
19. Position the upper timing cover.
20. Turn the crankshaft clockwise until the markings on the crankshaft and camshafts correspond.

21. Remove the upper timing cover.
22. Remove:
- The variable valve timing solenoid connectors
- The connector for the coolant temperature sensor
- The bolts for the ignition coils and the two ground terminals.

➡**Mark up the ignition coils before removal.**

23. Lift up and place the ignition coils and wiring to one side.
24. Slacken off the center bolt for the belt tensioner slightly.
25. Hold the center bolt still. Turn the tensioner eccentric clockwise to 10 o'clock using a 6mm Allen key.
26. Remove the timing belt.
27. Install tool 999 5452, or equivalent, at the rear of the camshafts.
28. Remove the center plugs for the variable valve timing units.

➡**Collect any oil spills using paper under the plugs.**

29. Remove the variable valve timing unit center bolts.
30. Pull or work off the variable valve timing units and timing gear pulley.
31. Remove tool 999 5452.
32. Remove:
- The bolt for the thermostat housing cable duct
- bolts holding the thermostat housing to the cylinder head. Lift out the thermostat housing, gasket and cable duct
- The bolt holding the inner timing cover to the cylinder head.
- Spark plugs for cylinders 2 and 5. Install tool 999 5454, or equivalent. Leave a 0.079–0.118 inch (2–3mm) gap to the camshaft cover. Ensure that the bolt in the spark plug well is fully tightened

42348-XC90-G11

**Fig. 61 Tool 9995452 installed**

- Variable valve timing solenoid
- The bolts holding the camshaft cover to the cylinder head.

33. Use to lift the cover from the cylinder head. Install the pliers at the stop lugs.

34. Start with cylinder 1 and work alternately backward.

35. Slacken off the wing nuts approximately 2 turns.

36. Repeat the procedure with the pliers.

37. Carefully press out the front and rear camshaft seals.

38. Remove the seals.

39. Remove:
- Tools 999 5454
- The camshaft cover
- The camshafts.

### ❊❊ CAUTION

**Mark up and position the lifters so that their original positions can be established.**

40. Lift out the valve lifters. Use a suction cup.

41. Position the valve lifters on a piece of paper to drain.

42. Remove the bolts holding the cylinder head on the cylinder block.

43. Start at the sides and work alternately towards the center.

➡ **Do not damage the mating surfaces.**

44. Remove:
- The intake manifold and gasket
- The coolant outlet pipe.

### *To install:*

45. Clean the gasket surfaces of:
- The manifold. Check that the studs are tightened
- The coolant bypass channel
- The coolant outlet pipe
- The intake manifold
- The cylinder block
- Camshaft cover

46. Blow the oil ducts clean.

47. Remove the starter motor mounting bolts. Pull out the starter motor and place to one side.

48. Remove the blind cover plug.

49. Turn the crankshaft clockwise slightly. Install the crankshaft stop 999 5451, or equivalent. Ensure that it bottoms out against the cylinder block.

50. Turn the crankshaft counter-clockwise until it stops against the crankshaft stop.

51. Check that the marking on the crankshaft timing gear pulley corresponds with the marking on the oil pump.

**Fig. 62 Cylinder head loosening sequence—6-cylinder engine**

42348-XC90-G08

**Fig. 63 Cylinder head torque sequence—6-cylinder engines**

42348-XC90-G09

52. Install a new cylinder head gasket.

53. Install the cylinder head.

54. Use new bolts. Lubricate and install all the bolts.

55. Tighten the bolts in the order illustrated. See the torque chart at the beginning of this chapter.

56. When installing a new cylinder head, the valve clearance must be set.

### ❊❊ CAUTION

**Make sure that the lifters are in the same position as before.**

57. Lubricate the valve lifter wells and the lifters. Install all the valve lifters.

58. Lubricate the camshaft bearing positions.

59. Install the intake camshaft. Ensure that the groove at the back of the camshaft is above an imaginary center line.

60. Position the exhaust camshaft. Ensure that the groove at the back of the camshaft is below an imaginary center line.

➡ **Ensure that no liquid gasket gets into the oil ducts.**

61. Install new O-rings around the spark plug wells at the cylinder head.

62. Apply liquid gasket 11 61,059, or equivalent, to the camshaft cover. Use a roller. The surface must be completely covered without any excess.

63. Lubricate the camshaft lobes, the camshaft bearing surfaces and the valve lifters.

64. Install the camshaft cover. Install 2 press tools 999 5454, or equivalent.

65. Tighten the camshaft cover bolts alternately, keeping it parallel to the cylinder head using the press tools.

66. Install all the bolts and tighten.

67. Install the variable valve timing solenoid using a new gasket.

68. Remove the press tools.

69. Install the spark plugs.

70. Lubricate the surface of the seal that the camshaft rotates against.

71. Using a drift, press in the seal until the drift bottoms out.

➡If there are grooves worn into the camshaft, the seal can be pressed in a further 0.078 inch (2mm) by reversing the sleeve on the tool.

72. Using camshaft tool 999 5452, or equivalent, separate the camshaft adjustment tool into two units.

**✳✳ CAUTION**

**The camshafts must not be turned more than is listed in the text below. The valves may be damaged.**

73. Install the exhaust camshaft section of the adjustment tool. Screw into the rear edge of the camshaft.

74. Carefully turn the camshaft adjustment tool clockwise (as viewed from the back of the engine) until the camshaft adjustment tool is parallel with the join between the cylinder head and camshaft cover.

75. Install the intake camshaft section of the adjustment tool.

76. Screw into the rear edge of the camshaft.

77. Carefully turn the camshaft adjustment tool counter-clockwise until it is in contact with the exhaust camshaft section of the camshaft adjustment tool.

78. Screw the adjustment tool sections together into one.

79. Check that the bolts retaining the adjustment tool to the camshafts and the bolts holding the tool together are well tightened.

80. Lubricate the surface of the seal that the camshaft rotates against.

81. Use the mounting bolts for the variable valve timing unit.

82. Using drift 999 5718, or equivalent, tighten the bolts until the drift bottoms out.

83. Install:
- The bolt holding the inner timing cover to the cylinder head
- The thermostat housing and cable duct. Use a new gasket
- The bolt for the thermostat housing cable duct.

84. Spin the idler pulley and listen for noise. If replacing with a new idler pulley, tighten to 18 ft. lbs. (24 Nm).

85. Spin the tension pulley and listen for noise. When replacing, bolt the tension pulley into place using the center bolt. Screw in the center bolt by hand.

86. Ensure that the tensioner fork is centered over the cylinder block rib/bracket. Ensure that the Allen hole on the eccentric is at 10 o'clock.

87. Slacken off, but do not remove the bolts which secure the timing gear pulley to the VVT unit.

88. Press the variable valve timing unit and timing gear onto the camshaft.

89. Install the center bolt which secures the variable valve timing unit to the camshaft. Tighten slightly.

90. Turn the variable valve timing unit counter-clockwise as far as it will go. Slacken off the center bolt.

91. Position the upper timing cover.

92. Turn the timing gear pulley clockwise until the bolts at the oval holes are in the

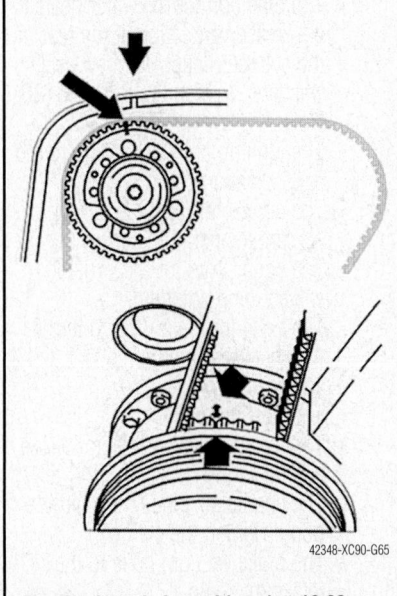

**Fig. 64 Allen hole positioned at 10:00 o'clock**

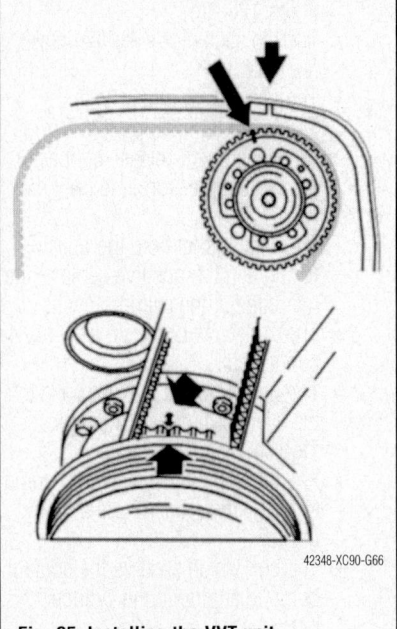

**Fig. 65 Installing the VVT unit**

limit position. Continue turning clockwise until the timing gear pulley marking is 1 cog before the marking on the upper timing cover.

➡**Do not turn counter clockwise during this procedure.**

93. Check that the timing gear pulley is still in the limit position in the oval holes.

94. Tighten the center bolt in the VVT unit to 89 ft. lbs. (120 Nm). Check that the variable valve timing unit does not rotate when tightening.

95. Install and tighten the center plug.

96. Turn the variable valve timing unit clockwise to the limit position. Turn the timing gear pulley so that the markings correspond.

97. Slacken off, but do not remove the bolts which secure the timing gear pulley to the VVT unit.

98. Press the variable valve timing unit/timing gear onto the camshaft.

99. Install the center bolt which secures the variable valve timing unit to the camshaft. Tighten slightly.

100. Turn the variable valve timing unit counter-clockwise as far as it will go.

101. Slacken off the center bolt.

102. Turn the timing gear pulley clockwise until the bolts at the oval holes are in the limit position. Continue turning clockwise until the timing gear pulley marking is 2 cogs before the marking on the upper timing cover.

➡**Do not turn counter clockwise during this procedure.**

103. Check that the timing gear pulley is still in the limit position in the oval holes.

104. Tighten the center bolt in the VVT unit to 89 ft. lbs. (120 Nm). Check that the variable valve timing unit does not rotate when tightening.

105. XII Install and tighten the center plug.

106. Turn the variable valve timing unit clockwise to the limit position.

107. Turn the timing gear pulley so that the markings correspond.

108. Tighten the timing belt tensioner center bolt. Tighten to 44 inch lbs. (5 Nm).

109. Install the timing belt in the following order:
- Crankshaft
- The idler pulley
- Intake cam
- Exhaust cam
- Water pump
- Tension pulley.

110. The variable valve timing unit does not have a return spring and is easily dislodged when reinstalling the timing belt. Check that the markings correspond.

➡**Adjust the timing gear pulleys so that the bolts are not at the limit positions in any of the oval holes.**

111. This adjustment is to be made with a cold engine. A suitable temperature is approximately 68°F (20°C).

112. At higher temperatures (with the engine at operating temperature or a high outside temperature for example) the indicator is further to the right.

113. The illustration shows the position of the indicator at different engine temperatures.

114. Hold the center bolt secure and turn the belt tensioner eccentric clockwise until the tensioner indicator passes the marked position.

115. Turn the eccentric back so that the indicator reaches the marked position in the center of the window.

116. Remember to hold the center bolt secure at the same time.

117. Hold the eccentric secure and tighten the center bolt. Tighten to 18 ft. lbs. (25 Nm).

118. Check that the indicator is in the correct position.

119. Check:
  • That the timing gear pulley is still in the limit position in the oval holes
  • The timing gear pulley so that the markings line up with the upper timing cover.

120. Tighten the bolts for the timing gear pulleys.

121. Remove:
  • Camshaft adjustment tools 999 5451
  • Crankshaft stop 999 5452

122. Install the plug with a new blind cover plug. Tighten to 30 ft. lbs. (40 Nm).

123. Press the timing belt to check that the indicator on the tensioner moves easily.

124. Position the upper timing cover.

125. Turn the crankshaft two turns. Check that the markings on the crankshaft timing gear pulley and the camshaft timing

42348-XC90-G12

**Fig. 66 Camshaft adjustment tools installed**

gear pulley match up with the markings on the oil pump and upper timing cover respectively.

126. Check that the indicator on the belt tensioner is within the marked position.

127. Remove the upper timing cover.

128. Install:
  • The trigger wheel.

---

**✳✳ CAUTION**

**The trigger wheel can only be installed in one way**

---

  • The camshaft position sensor housing
  • The pipe bolt for the water heated crankcase ventilation at the rear of the cylinder head. Use new seal washers. Tighten to 19 ft. lbs. (26 Nm).
  • The ignition coils according to the earlier markings
  • The connector for the engine coolant temperature sensor
  • The connectors for the variable valve timing solenoid
  • The bolts for the 2 ground terminals between ignition coils 1 and 2 and 5 and 6.
  • The starter motor
  • The throttle body connector to the engine wiring
  • The charge air pipe to the throttle body. Tighten the clamp
  • The brake vacuum hose to the intake manifold
  • The pipe bolts at the underside of the intake pipe for the water heated crankcase ventilation. Tighten to 19 ft. lbs. (26 Nm)
  • The dip stick pipe onto the intake manifold.
  • The fuel rail. Use new bolts. Tighten to 88 inch lbs. (10 Nm). Press the quick-release connector for the fuel pressure line until it clicks
  • The clamps that hold the injector wiring and the positive cable to the rear edge of the cylinder head
  • The protective cover over the nozzle connectors
  • The upper engine coolant hose to the cylinder head and radiator. Tighten the clamp.
  • The inlet pipe for the coolant pump to the cylinder head. Use a new gasket
  • The bolt which secures the coolant pump inlet pipe to the cylinder block
  • The turbocharger.

  • The bracket for the torque rod
  • The upper bolt securing the automatic transaxle dip stick pipe to the engine
  • The ground strip between the right frame member and the cylinder head
  • The cable duct on the inside of the right suspension turret
  • The front timing cover.
  • Expansion tank. Connect the connector to the coolant level sensor
  • The hose to the expansion tank from the radiator. Secure the clamp and remove the lock grip pliers.
  • Connect the connector to the ABS sensor
  • The servo reservoir. Ensure that the hoses are correctly positioned.
  • The upper timing cover
  • The cover over the ignition coils
  • The plastic intake pipe over the engine. Adjust and tighten the intake pipe at the turbocharger. Tighten the clamps and mounting bolts on the plastic pipe
  • The plastic charge air pipe over the engine. Tighten the clamps and mounting bolts
  • The hose between the charge air cooler and the plastic charge air pipe over the engine. Tighten the clamps
  • The crankcase ventilation hose to the PTC resistor in the plastic intake pipe over the engine. Use a new clamp
  • The connector for the PTC resistor
  • The protective cover over the PTC resistor
  • The EVAP hose to the plastic intake pipe
  • The air cleaner assembly
  • The plastic intake pipe to the mass air flow sensor. Tighten the clamp
  • The connector to the mass air flow sensor. Clamp the wiring at the front of the air cleaner
  • The turbocharger control valve to the mounting on the air cleaner
  • The blue marked hose and the turbocharger control system to the intake manifold
  • The red marked hose and turbocharger control system to the charge air pipe
  • The yellow marked hose and turbocharger control system to the pressure regulators
  • The cross stay between the suspension turrets

- The suspension turret covers. Clamp the servo hose to the right mounting on the engine stabilizer brace
- The battery negative lead.

129. Replace the oil and oil filter if necessary.

130. Check:
- The engine oil level
- The servo oil level
- The coolant level.
- The engine for leaks
- The oil level in the transaxle
- The coolant level

### 8-Cylinder Engines

1. Before servicing the vehicle, refer to the Precautions Section.

2. Remove intake manifold. Refer to Intake Manifold removal and installation.

3. Remove air induction pipe.

4. Remove ignition coils on the left and right banks, refer to Ignition Coil Pack removal and installation.

5. Remove camshaft covers, refer to Camshaft Covers removal and installation.

6. Remove timing chain and sprockets, refer to Timing Chain and Sprockets removal and installation.

7. Remove camshafts, refer to Camshaft and Valve Lifters removal and installation.

### Right Cylinder Head—8-Cylinder

*See Figures 67 through 72.*

This procedure will expose the right cylinder head, cylinders 1, 3, 5 and 7 (Bank 1).

1. Disconnect or remove:
- Hose clamp for the upper coolant hose
- Electrical connectors for the camshaft position sensors
- Coolant hoses and place them to one side

2. Remove the two M8 upper bolts holding the water pump

3. Remove the M8 bolt and two M8 nuts attaching coolant pipes illustrated in the graphic.

4. Remove exhaust manifold:
   a. Remove the two M8 bolts holding the bracket to the manifold.
   b. Remove the eight M8 nuts holding the flanged joint to the manifold.

➡**Discard the exhaust manifold gasket.**

When replacing or working on the cylinder head, lift out the lifters using a magnet or suction cup. Make note of the positions. Also, take care not to damage the mating surfaces when removing the cylinder head.

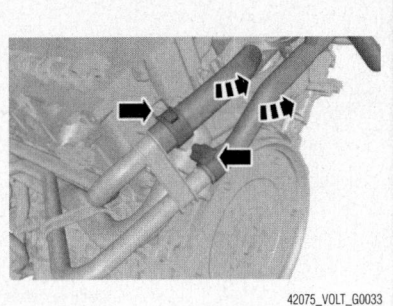
**Fig. 67 Coolant hose location—8-Cylinder Engine**

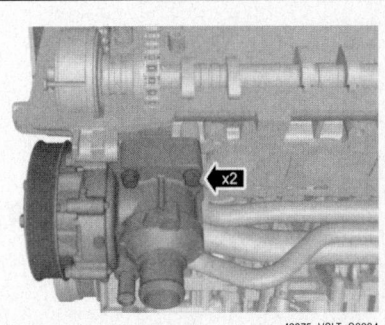
**Fig. 68 Water pump location—8-Cylinder Engine**

**Fig. 69 Coolant pipe removal—8-Cylinder Engine**

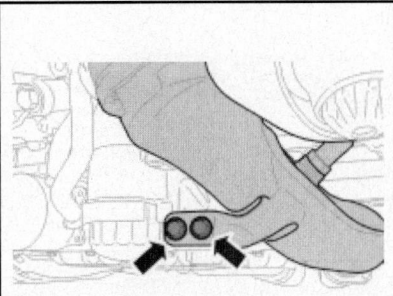
**Fig. 70 Bracket bolt removal on exhaust manifold—8-Cylinder Engine**

5. Remove the ten M12 bolts in the cylinder head.

6. Remove cylinder head.

7. Discard the cylinder head gasket.

***To install:***

8. Clean and check the gasket faces:
- The coolant bypass channel
- The intake manifold
- The cylinder block
- Camshaft cover

9. Carefully blow the oil ducts clean.

➡**Check that all studs at the exhaust ports are tightened.**

10. Install:
- New exhaust manifold gasket. Lubricate the studs using paste 116 1408
- New cylinder head gasket. Apply sealing compound 307 57050 to both sides
- The cylinder head
- New bolts in cylinder head. Lubricate the bolts

11. Tighten the cylinder head bolts in a circle from the inside and outward. Use protractor 951 2050. Tighten in two steps:
   a. Step 1 to 30 ft. lbs. (40 Nm).
   b. Step 2 angle tighten to 90°

12. Install the valve lifters.

**Fig. 71 Flanged joint removal—8-Cylinder Engine**

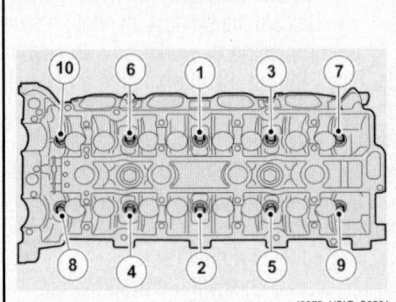
**Fig. 72 Cylinder head torque sequence—8-Cylinder Engine**

➡If the cylinder head has been replaced, the valve clearance must be adjusted. Refer to Valve Lash adjustment.

13. Lubricate and install valve lifters.
14. Install:
- Exhaust manifold flanged joint and tighten the eight M8 nuts to 18 ft. lbs. (24 Nm)
- Two M8 bolts for the exhaust manifold bracket and tighten to 18 ft. lbs. (24 Nm)
- The M8 bolt and two M8 nuts attaching coolant pipes and tighten to 18 ft. lbs. (24 Nm)
- Two M8 upper bolts holding the water pump and tighten to 18 ft. lbs. (24 Nm)
- Coolant hoses and clamps
- Camshafts
- Timing chain and sprockets
- Camshaft cover
- Ignition coils
- Air induction pipe.
- Intake manifold, refer to Intake Manifold removal and installation.

### Left Cylinder Head—8-Cylinder

*See Figures 73 through 78.*

This procedure will expose the left cylinder head, cylinders 2, 4, 6 and 8 (Bank 2).

1. Disconnect or remove:
- Cable harness from the engine bracket
- Connector for the camshaft position sensor
- Vacuum valve
- Hose clip for the coolant hose
2. Remove M6 bolt and heat shield around connector.
3. Remove the M8 upper mounting bolt from the alternator.
4. To remove the exhaust manifold:
- Remove the two M8 bracket bolts
- Remove the eight M8 nuts from the exhaust manifold
- Discard the exhaust manifold gasket
When replacing or working on the cylinder head, lift out the lifters using a magnet or suction cup. Make note of the positions. Also, take care not to damage the mating surfaces when removing the cylinder head.
5. Remove the ten M12 bolts in the cylinder head.
6. Remove cylinder head.
7. Discard the cylinder head gasket.

**To install:**

8. Clean and check the gasket faces:
- The coolant bypass channel
- The intake manifold

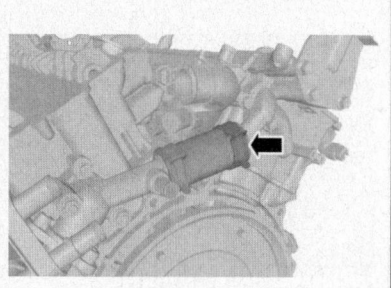

**Fig. 73 Hose clamp location for removal—8-Cylinder Engine**

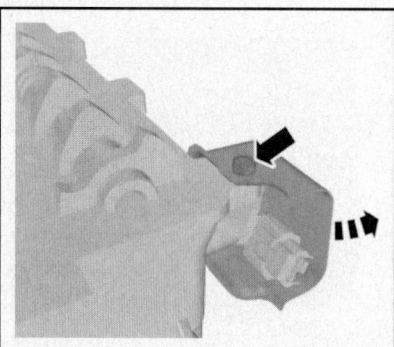

**Fig. 74 Heat shield location—8-Cylinder Engine**

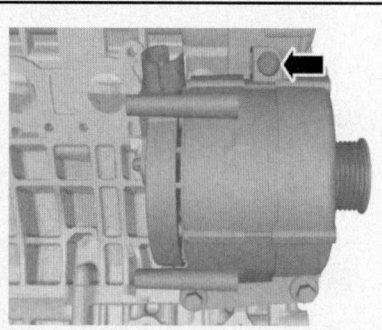

**Fig. 75 Alternator upper mounting bolt location—8-Cylinder Engine**

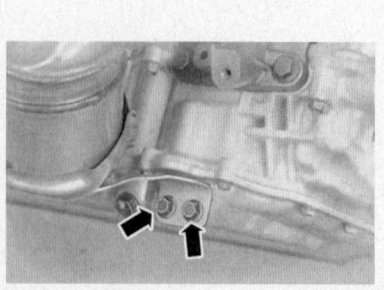

**Fig. 76 Exhaust bracket and bolt location for removal (Bank 2)—8-Cylinder Engine**

- The cylinder block
- Camshaft cover
9. Carefully blow the oil ducts clean.

➡Check that all studs at the exhaust ports are tightened.

10. Install:
- New exhaust manifold gasket. Lubricate the studs using paste 116 1408
- New cylinder head gasket. Apply sealing compound 307 57050 to both sides
- The cylinder head
- New bolts in cylinder head. Lubricate the bolts
11. Tighten the cylinder head bolts in a circle from the inside and outwards. Use protractor 951 2050. Tighten in two steps:
  a. Step 1 to 30 ft. lbs. (40 Nm).
  b. Step 2 angle tighten to 90°
12. Install the valve lifters.

➡If the cylinder head has been replaced, the valve clearance must be adjusted. Refer to Valve Lash adjustment.

13. Lubricate and install valve lifters.
14. Install or connect:
- Exhaust manifold and tighten the eight M8 nuts to 18 ft. lbs. (24 Nm)

**Fig. 77 Exhaust manifold bolt removal (Bank 2)—8-Cylinder Engine**

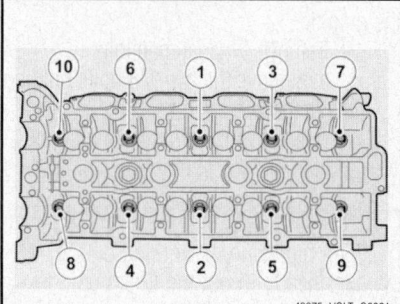

**Fig. 78 Cylinder head torque sequence—8-Cylinder Engine**

- Two M8 bolts for the exhaust manifold bracket and tighten to 18 ft. lbs. (24 Nm)
- The M8 upper mounting bolt from the alternator and tighten to 18 ft. lbs. (24 Nm)
- Heat shield and M6 bolt, tighten to 88 inch lbs. (10 Nm)
- Connector for camshaft position sensor
- Cable harness on the engine bracket
- Vacuum valve
- Hose clip for the coolant hose
- Camshafts
- Timing chain and sprockets
- Camshaft cover
- Ignition coils
- Intake manifold, refer to Intake Manifold removal and installation.

## ENGINE ASSEMBLY

### REMOVAL & INSTALLATION

#### 5-Cylinder Engines

*See Figure 79.*

1. Before servicing the vehicle, refer to the Precautions Section.
2. Drain the cooling system.
3. Remove or disconnect the following:
   - The battery negative lead
   - The cross stay between the suspension turrets
   - The turbocharger control valve from the air cleaner
   - Mass air flow sensor connector
   - The brake vacuum hose from the air cleaner
   - The air cleaner assembly. Seal the openings in the intake pipe for the turbocharger
   - The plastic charge air pipe above the engine. Seal the opening in the turbocharger
   - The ground strip at the top/rear of the cylinder head.
4. Position a container under the radiator drain cock. Drain the coolant. Close the cock.
5. Drain the fuel line.
6. Remove or disconnect the following:
   - The cover over the central electrical unit
   - The connector for the engine wiring from the central electrical unit
   - The positive lead for the starter motor from the terminal block
   - The ground lead at the front frame member
   - The bolts for the transaxle cable bracket at the transaxle. Tie up the cable

- Separate the transaxle wiring
- The oil cooler hoses for the transaxle. Seal the openings.
- The connectors for the temperature and pressure sensors in the charge air cooler
- The rubber hose at the intake for the charge air cooler
- The brake vacuum hose at the intake manifold
- The lower engine coolant hose from the engine
- The plastic pipe between the throttle body and the charge air cooler
- The dip stick pipe. Separate the wiring from the front heated oxygen sensor (HO2S). Remove the wiring from the clips and clamps.
- Connector for the level sensor in the expansion tank
- The hoses for the heater unit at the terminal in the firewall
- The bleed hose for the radiator at the expansion tank
- The ABS line and clips from the power steering hose. Lift up the expansion tank and servo oil reservoir. Place them on top of the engine.
- Control module unit cover
- The control modules
- The surround for the control module box.
- The red retaining clip from the connector using a screwdriver
- The connector by releasing the catches at the front edge using pliers. Pry out the connector using a screwdriver. Lift up the connector with the protection underneath and put it to one side.
- Remove the ground lead from the right fender liner. Pull out the wiring between the front cover plate and the air conditioning pipe.
- The right headlamp by pulling up the two lock facings. Disconnect the wiring and put the wiring to one side
- The connector on the air conditioning receiver drier
- The gray connector on the fan shroud. Remove the wiring from the clips and clamps on the fan shroud.
- Remove the upper radiator hose
- Unhook the accessory drive belt
- Disconnect the one-pin connector for the compressor magnetic clutch
- Remove the four bolts holding the air conditioning compressor to the

engine. Tie up the compressor in the front cover plate.

➡**Handle the compressor carefully. Ensure that there are no kinks in the air conditioning pipes.**

7. Set the front wheels so that they are straight
8. Remove the soundproofing panel under the steering wheel. Fold back the floor carpet
9. Bend up the outer section of the steering shaft boot
10. Remove the retaining clip around the boot. Pull the boot up
11. Turn the steering wheel to gain the best possible access to the steering shaft nut
12. Remove the ignition key. Activate the steering wheel lock.

### ✳✳ WARNING

**Do not move the steering wheel. The contact reel in the SRS (supplemental restraint system) could be damaged.**

13. Remove or disconnect the following:
    - The bolt securing the mounting for the steering shaft in the joint
    - The joint from the steering shaft by pressing the shaft upwards
14. Raise and support the vehicle.
15. Remove or disconnect the following:
    - The left and right front wheels
    - The splash guard under the engine.
    - The 6 bolts. Lift up and slide the plate forward
    - The protective plate.

➡**There is a guide sleeve on the upper mounting for the plate. This must be lifted slightly so that the panel can be slid forward.**

16. Remove the tie rod from the steering knuckle on the left and right sides.
17. On one side, measure the length of the tie rod in relation to the steering gear housing. Note the measurement.
18. Remove on the left and right sides:
    - The link rod from the anti-roll bar
    - The sound insulation cover for the halfshaft
    - The nut holding the control arm to the ball joint
    - The center bolt and the end of the halfshaft.

➡**If an engine block heater is installed, mark the position of the fuel lines.**

19. Separate the quick-release connectors for the fuel pressure line and engine block heater (option).

20. Pull apart the EVAP line at the rubber joint behind the rear right mounting for the sub-frame.

21. Remove the three-way catalytic converter.

22. Remove the propeller shaft.

23. Use a post hoist with a lifting table and stud guides. Position the lifting table against the sub-frame.

➡ **Ensure that the lifting table is centered under the sub-frame for optimal weight distribution. Check that the lifting table is correctly applied against the sub-frame.**

24. Remove or disconnect the following:
- The bumper cover on the left-hand side. Use a weatherstrip tool
- The M8 nut at the top of the left-hand fender liner for the engine coolant heater
- The connector for the engine coolant heater. Press up the engine coolant heater. Lift it out of its upper mounting.
- The bolts for the sub-frame brackets
- The bolts for the sub-frame. Lower the engine and transaxle approximately 6 inches (15cm) so that the halfshafts describe a horizontal line between the hub and transaxle.

### ✳✳ WARNING

**Ensure that no hoses, wiring or anything else catch in the engine and transaxle unit when it is lowered.**

- The control arms from the ball joint pinion

25. Hold the constant velocity joint. Pull the halfshaft off the hub

26. Let the halfshafts rest against the control arms

27. Turn the engine block heater out of its position

28. Check that the air conditioning compressor is hanging free. Continue lowering the engine and transaxle unit.

### To install:

29. Raise the engine and the transaxle unit. Use a lifting table with locating pins. Stop when the sub-frame is approximately 6 inches (15cm) below the frame members.

30. Install, on the left and right sides:
- The halfshaft to the hub. Install the center bolts. Do not tighten yet!
- The ball joint to the control arm. Install the nuts. Do not tighten yet!

➡ **Ensure that the mating surfaces on the ball joints and control arms are clean.**

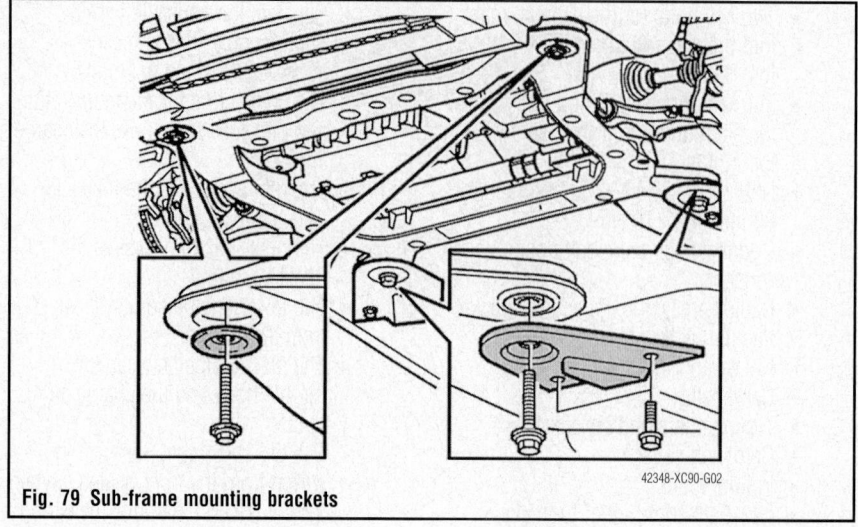

**Fig. 79 Sub-frame mounting brackets**

42348-XC90-G02

31. Twist the engine block heater (option) into position.

32. Check that the air conditioning compressor is correctly positioned.

33. Check that the tie rod is in the correct position. Check that the transaxle cable is routed between the steering shaft and the firewall.

34. Align the steering shaft joint with the hole in the firewall.

35. Lift the engine and transaxle unit up to the frame members.

➡ **Ensure that no hoses, wiring or anything else catches.**

36. Install or connect the following:
- New bolt in the sub-frame. Lubricate the bolts.
- The washers at the front edge and the support plates at the rear edge of the sub-frame
- Engine coolant heater left front (option).

37. Tighten the bolts for the sub-frame. Tighten to 77 ft. lbs. (105 Nm) plus 120°. Start on the left-hand side of the sub-frame. Continue with the right-hand side.

38. Tighten the bolts for the brackets at the rear of the sub-frame. Tighten to 37 ft. lbs. (50 Nm).

39. Remove the lifting table.

40. Connect the quick-release connectors and fuel line to the engine and engine block heater (option). Press the EVAP line together at the rubber joint.

41. Install and tighten the M8 nut in the upper mounting for the engine coolant heater. Connect the connector. Install the front bumper cover.

42. Install or connect the following:
- The propeller shaft
- The three-way catalytic converter (TWC)
- The tie rods. Use new nuts. Tighten to 52 ft. lbs. (70 Nm)
- The link rod to the anti-roll bar. Use new nuts. Tighten to 37 ft. lbs. (50 Nm). Counterhold using Torx® 27
- The bracket for the brake pipe to the body
- The nut for the ball joint and control arm. Tighten to 59 ft. lbs. (80 Nm). Counterhold using Torx® 40.
- The center bolt in the halfshaft. Tighten to 37 ft. lbs. (50 Nm).
- The front skid plate
- The lower engine splash guard
- The front wheels
- The steering shaft joint on the steering gear and steering shaft. Install a new bolt. Tighten to 18 ft. lbs. (25 Nm).
- The steering shaft boot and snapring. Fold back the carpet
- Soundproofing panel.
- The compressor

➡ **Handle the compressor carefully. Ensure that there are no kinks in the air conditioning (A/C) pipes. Connect the one-pin connector to the compressor magnetic clutch.**

- The connector on the air conditioning receiver/drier
- The gray connector for the fan shroud
- The wiring in the clips and clamps on the fan shroud
- The connectors for the pressure and temperature sensors on the charge air cooler
- The right-hand headlamp using two lock facings.
- The engine wiring to the control module box.

- The ground lead to the ground terminal at the right fender edge in the engine compartment
- The auxiliary belt
- The intermediate piece for the control module box
- The control modules
- The cover for the control module box
- Expansion tank. Connect the radiator breather tube to the expansion tank. Tighten the clamp. Connect the connector for the level sensor in the expansion tank
- Servo oil reservoir. Check that the hoses are correctly positioned.

43. Clamp the ABS cable to the power steering hose.

44. Install or connect the following:
- The transaxle cable
- The oil cooler hoses
- The lower and upper coolant hoses for the engine
- The rubber hose on the charge air pipe on the intake for the charge air cooler
- The dip stick pipe
- The brake vacuum hose to the intake
- The plastic charge air pipe between the throttle body and the charge air cooler. Tighten both the clamps.

➡️**Use a hot air gun to facilitate installation.**

- The connector for the engine wiring to the central electrical unit
- The positive lead for the starter motor to the terminal block
- The cover over the central electrical unit
- The ground lead between the engine and body in the front frame member.
- The ground strip at the top/rear of the cylinder head
- The plastic charge air pipe over the engine
- Air cleaner assembly
- The inlet hose for the turbocharger to the air cleaner
- Mass air flow sensor connector
- The turbocharger control valve on the air cleaner
- The brake vacuum hose in the clamp on the air cleaner housing
- The intake pipe between the front cover plate and the air cleaner
- The engine stabilizer brace between the suspension turrets
- The battery negative lead.
- Coolant
- Engine oil
- Power steering fluid.

45. Start the engine
46. Warm up the engine until the thermostat opens
47. Check for leakage
48. Check the oil and coolant levels. Adjust if necessary
49. Check the transaxle oil level.

### 6-Cylinder Engines, 2006 Only

1. Remove the cable from the battery negative terminal.
2. The expansion tank cap.
3. Drain the refrigerant.

➡️**There is a guide sleeve on the upper mounting for the plate. This must be lifted slightly so that the panel can be slid forward.**

4. Remove the 6 bolts.
5. Lift up and slide the plate forward.
6. Remove the protective plate.
7. Remove the splash guard under the engine
8. Disconnect the connectors for the heated oxygen sensors (HO2S). Remove the wiring from the clips and tie straps.
9. Connect a hose to the engine coolant drain cock.
10. Drain the coolant into a clean container.
11. Remove:
- The bolt for the engine stabilizer brace mounting on the engine
- The cross stay between the suspension turrets
- The plastic charge air pipe over the engine
- The charge air pipes which connect to the turbocharger. Seal the openings
- The air cleaner.

➡️**Note the position of the probes to make it easier to reinstall them. If they are mixed up, diagnostic trouble codes (DTCs) may be stored.**

12. Slide the wiring for the front heated oxygen sensors (HO2S) upwards, Unscrew the heated oxygen sensors (HO2S) from the three-way catalytic converter.
- Remove the heat deflector plates from the turbocharger
- Remove the nuts from the flange between the turbocharger and the three-way catalytic converter.

➡️**Rustproof the studs.**

13. Separate the connections. Unhook the three-way catalytic converter.
14. Remove the ground strip between the engine and bodywork which is positioned

above the turbocharger for cylinders 1, 2 and 3.

15. Remove both the hoses for the heater unit on the firewall.
16. Remove:
- The battery lead from the terminal block at the fuse box
- The connector in the fuse box.
- The rear radiator hose at the transaxle
- The gear selector cable from the transaxle
- The connector for the input speed sensor on the transaxle
- The connector on the transaxle
- The vacuum hose on the intake for the power brake booster
- The connector on the sensor in the charge air cooler
- The ground lead from the engine to the car body (detach in the car body).

17. Remove the charge air pipe on the intake between the throttle module and the charge air cooler.

➡️**If necessary use a hot air gun to loosen the pipe.**

18. Detach the upper radiator hoses on the engine.
19. Lift the power steering reservoir. Place it on top of the engine.

➡️**Ensure that no oil leaks out of the reservoir through the ventilation holes in the cover.**

20. Remove the coolant expansion tank. Slacken off the hose clamp on the hose to the engine. Disconnect the connector.
21. Remove the radiator breather tube by the expansion tank. Put the expansion tank to one side.
22. Remove:
- The cover from the distribution box
- The control module
- The surround from the distribution box.

23. Pry out the connector plate from the lower section of the control module box. Pull out the engine and transaxle wiring connectors from the plate.
24. Detach the ground lead on the fender liner.

➡️**Seal the openings.**

25. Separate the climate control unit pipe at the frame member inside the central electrical unit.
26. Remove the right-hand headlamp by pulling the 2 lock facings on the lamp straight up. Then carefully pull the headlamp

forward. Disconnect the connector and lift out the headlamp.

27. Disconnect the connector from the sensor on the accumulator. Pull the wiring through to the engine compartment.

28. Set the front wheels so that they are straight.

29. Fold back the floor carpet.

30. Bend up the outer section of the steering shaft boot.

31. Remove the retaining clip around the boot.

32. Pull the boot up.

33. Turn the steering wheel to gain the best possible access to the steering shaft nut.

34. Remove the ignition key. Activate the steering wheel lock.

### ❋❋ WARNING

**Do not move the steering wheel. The contact reel in the SRS (supplemental restraint system) could be damaged.**

35. Remove:
  • The bolt securing the mounting for the steering shaft in the joint
  • The joint from the steering shaft by pressing the shaft upwards.

➡**Clean and spray the exposed threads with rust-proofing agent before beginning removal.**

  • The front wheel
  • The tie rod ends from the steering arm.
  • The bolt holding the halfshaft to the hub
  • The nut for the ball joint / link arm
  • The nut for the anti-roll bar.

➡**Counterhold using Torx® 40 so that the ball joint boot does not turn when the nut is removed.**

### ❋❋ WARNING

**The steering shaft must not be turned. The contact reel in the SRS system could be damaged.**

36. On one side, measure the length of the tie rod in relation to the steering gear housing. Note the measurement.

37. Disconnect the brake pipe from the clips on the front SIPS member.

38. Remove the 4 bolts for the front SIPS member.

39. Remove the member.

40. Unhook the three-way catalytic converter (from the rubber mounting.

41. Remove:
  • The tie straps from the wiring for the rear probe. Separate the probe wiring
  • The nut for the flange between the three-way catalytic converter and exhaust system
  • The propeller shaft.

42. Drain the fuel line.

43. Disconnect the fuel line quick-release connector at the engine.

➡**Seal the openings against dirt.**

44. Place the cable to one side.

45. Separate the EVAP canister cable at the rear right sub-frame mounting.

46. Remove the transaxle oil cooling hoses from the transaxle. Seal the openings.

### ❋❋ CAUTION

**Ensure that the lifting table is properly centered under the engine.**

47. Position a lifting table under the sub-frame. Secure the table against the sub-frame.

### ❋❋ CAUTION

**Do not damage the brake pipe that runs across the transaxle tunnel.**

48. Remove:
  • The sub-frame brackets bolts
  • The bolts for the sub-frame. Lower the engine and transaxle approximately 6 inches (15cm) so that the halfshafts describe a horizontal line between the hub and transaxle.

➡**Take great care that no hoses, wiring or anything else catches in the engine/transaxle unit when it is lowered.**

49. Unhook the control arms from the ball joint pinion.

50. Remove the halfshafts from the wheel hubs.

➡**Do not pull the halfshafts. They do not have axial locks.**

### ❋❋ CAUTION

**Do not damage the ABS wiring. Do not damage the brake pipe.**

51. Let the halfshafts rest against the control arms. Lower the three-way catalytic converter.

52. Continue lowering until the hose for the compressor in the climate control unit is accessible. Detach the inner hose for the climate control unit. Seal the openings. Continue lowering the unit.

### *To install:*

53. Lift the engine and transaxle unit. Use a lifting table.

54. Stop the lift approximately 6 inches (15cm) before the sub-frame reaches the frame members.

55. Install (on both sides) the halfshaft on the hub.

56. Lift the three-way catalytic converter up into position. Connect the exhaust system to the rear flange. Use a new gasket. Install new nuts loosely.

➡**The three-way catalytic converter cannot be raised after the engine has been raised to its correct position.**

57. Continue raising the engine until it is only approximately 2 inches (50mm) from its correct position.

➡**Guide the lower steering shaft into the hole in the firewall.**

58. Align the ball joint in the control arm.

59. Install the inner pipe on the compressor in the climate control unit. Use a new O-ring.

60. Lift the sub-frame until it is against the frame members.

61. Install new bolts in the sub-frame. Lubricate the bolt threads.

62. Install the washers in the front edge and support plates at the rear edge of the sub-frame.

63. First tighten the bolts on the left side of the sub-frame. Tighten to 48 ft. lbs. (65 Nm) plus 60°.

64. Then tighten the bolts on the right-hand side of the sub-frame to the same values.

65. Tighten the brackets. Tighten to 37 ft. lbs. (50 Nm).

66. Remove the lifting table.

67. Tighten the ball joint and control arm. Tighten to 59 ft. lbs. (80 Nm). Counterhold using Torx® 40 so that the ball joint boot does not rotate.

68. Install:
  • The anti-roll bar links. Use new locknuts. Tighten to 37 ft. lbs. (50 Nm).
  • New center bolts for the halfshafts. Tighten to 37 ft. lbs. (50 Nm).

69. Install the tie rod using the marking made during removal. Use new locknuts. Tighten to 52 ft. lbs. (70 Nm).

70. Connect the fuel line to the engine.

71. Connect the evaporative emission system (EVAP) line at the right rear sub-frame mounting.

72. Install the oil cooler hoses on the transaxle.

73. Install the propeller shaft.

74. Lower the car.

75. Pull up and align the three-way catalytic converter (TWC) against the turbocharger (TC).

76. Lubricate the threads using copper paste 116 1408, or equivalent. Install new nuts loosely.

77. Lower the car.

78. Install the front SIPS member. Press the brake pipe into place in the member.

79. Hook the rubber mounting for the three-way catalytic converter (TWC) in the bracket on the sub-frame.

80. Connect the connector to the rear probe. Secure the wiring as before.

81. Tighten the screwed joint between the three-way catalytic converter (TWC) and the exhaust system. Tighten to 18 ft. lbs. (24 Nm).

82. Install:
- The steering shaft joint on the steering gear and steering shaft
- A new bolt. Tighten to 18 ft. lbs. (25 Nm)
- The steering shaft boot and snap-ring. Fold back the carpet.

83. Connect the engine and transaxle wiring connectors to the control module box base plate.

84. Press the base plate into place in the lower section of the control module box.
- The climate control unit pipe at the frame member inside the control unit box. Use a new O-ring
- The cable from the engine wiring for the sensor on the receiver to the right of the radiator
- The front right headlamp

85. Install the intermediate section. Align the ventilation pipe in the rear edge of the control module box.

86. Press the control modules down carefully. Lock into place using mounting bracket 999 5722, or equivalent.

87. Press the cover into place. Screw the ground lead into place on the fender liner.
- The connector for the expansion tank from the engine compartment wiring
- The hose from the engine to the expansion tank
- The breather tube from the radiator
- The expansion tank
- The servo reservoir
- The upper radiator hose.
- The plastic pipe between the charge air cooler and the throttle module.

➡ **Use a hot-air gun to facilitate installation of the pipe.**

- The connector for the sensor on the charge air cooler
- The ground lead from the engine to the car body
- The vacuum hose to the intake and power brake booster
- The connectors for the transaxle
- The mechanical transaxle cable to the transaxle
- The rear radiator hose
- The connector to the fuse box
- The battery lead to the terminal block on the fuse box.
- The radiator hoses on the heater unit. They are coded and will only connect to the correct sleeve connector
- The ground strip between the engine and the car body

88. Tighten the three-way catalytic converter to the turbocharger. Use new nuts. Tighten to 18 ft. lbs. (24 Nm).

89. Install the upper heated oxygen sensors using the marks made during removal. Lubricate the threads. Use copper paste 116 1408, or equivalent. Tighten to 33 ft. lbs. (45 Nm).

90. Lower the wiring down to the probes. Secure the wiring in the clips and clamps as before. Connect the wiring.

91. Install the heat deflector plates.

92. Install:
- The charge air pipes on the turbocharger
- The plastic charge air pipe over the engine
- Air cleaner assembly
- Mass air flow sensor connector
- The turbocharger control valve on the air cleaner
- The intake manifold to the air cleaner
- The engine stabilizer brace between the suspension turrets.

93. Tighten the bolts on the suspension turrets. Tighten to 37 ft. lbs. (50 Nm). Tighten the bolt in the torque bracket. Tighten to 59 ft. lbs. (80 Nm).

94. Install the battery negative lead.

95. Fill with coolant. The coolant volume is approximately 7.6 quarts (7.2 liters).

96. Check the engine oil.

97. Start the engine.

98. Warm up the engine until the thermostat opens.

99. Check for leakage.

100. Check the oil and coolant levels. Adjust if necessary.

101. Check the transaxle oil level.

102. Install:
- The splash guard under the engine

- The front skid plate on the sub-frame
- The front wheels

### 8-Cylinder Engines

1. Remove the battery negative lead.

2. Remove the expansion tank cap.

➡ **There is a guide sleeve on the upper mounting for the plate. This must be lifted slightly so that the panel can be slid forward.**

3. Remove the 6 bolts. Lift and slide the protective plate forward. Remove the protective plate.

4. Remove the 6 bolts and the lower engine cover.

5. Drain coolant.

6. Remove the 4 bolts at the suspension turrets, the bolts of the upper engine mount on the engine and the upper cross stay.

7. Remove the air cleaner cover

8. Remove the 2 connectors from the engine control module (ECM); clip the band clamp to the cable harness, remove the connector from the mass air flow sensor.

9. Remove the brake vacuum hose from the vacuum pump.

10. Remove the 2 bolts from the fresh air hose in the front plate.

11. Remove the fresh air hose.

12. Remove the air cleaner by pulling it straight up.

13. Remove the positive lead from the electrical distribution box.

14. Remove the ground cable between the engine and body (loosen in the body).

15. Remove the connector at the electrical distribution box from the brake vacuum valve. Detach the vacuum hose from the valve that leads to the brake vacuum amplifier.

16. Remove the selector lever cable using a break fork to break the cable loose from its ball.

17. Remove the transaxle control module (TCM) connector.

18. Remove the lower line of the transaxle oil cooler.

19. Remove the upper line of the transaxle oil cooler.

➡ **Seal the openings and collect the oil from the oil cooler in a receptacle. Do not reuse the drained oil.**

20. Detach the radiator hoses at the quick-release coupling toward the firewall.

21. Remove the upper radiator hose from the distributor housing.

22. Remove the lower radiator hose from the thermostat housing.

23. Remove the upper engine covers.

24. Remove the upper engine covers by pulling them straight up.

25. Remove the 2 bolts on the cover over the servo pump.

26. Remove the cover.

27. Remove the bolt on the washer fluid filler pipe; bend out the pipe.

28. Remove the return line for servo oil from the clips and clamps.

29. Lift up the servo oil reservoir.

30. Remove the 2 coolant hoses from the expansion tank.

31. Remove the 3 ground cables from the body at the expansion tank.

32. Remove the connector from the electric cooling fan.

33. Position the oil reservoir on the engine.

➡ **Make sure no oil runs out. Make sure that the oil return line is on the correct side of the upper coolant hose that runs to the engine.**

34. Remove the auxiliary drive belt.

35. Remove the ground cable bolt between the engine and body. It is located beneath the expansion tank.

36. Raise the vehicle.

37. Remove the front wheels.

38. Remove the ball joint from the link arm (on both sides of the vehicle)

  a. Remove the halfshaft bolt. Counterhold with a prybar in the brake disc cooling channels.

  b. Remove the rubber sleeve in the wheel hub.

  c. Remove the nut between the ball joint and link arm.

  d. Remove the nut of the anti-roll bar link in the anti-roll bar.

➡ **Counterhold with a Torx® wrench so that the ball joint gaiters do not turn when the nuts are removed.**

39. Mark the position of the track rod.

40. Remove the track rod end from the guide arm, using a puller.

41. Detach the ball joint pin from the link arm.

42. Fit the cap nut, 7062-1, on the ball joint pin.

43. Tighten the nut toward the link arm. Then turn it back one rotation to create play between the cap nut and the link arm. Position the puller on the link arm and detach the ball joint from the link arm.

44. Remove the nuts at intermediate pipe joints and the intermediate pipes; scrap the gaskets.

➡ **Mark the position of the carrier on the driveshaft.**

45. Remove the driveshaft bolts.

➡ **Do not move the steering wheel. The contact reel in the SRS (supplemental restraint system) could be damaged.**

46. Set the front wheels in the straight-ahead position.

47. Fold the floor carpet out of the way.

48. Bend up the outer section of the steering shaft boot.

49. Remove the retaining clip around the boot.

50. Pull the boot up.

51. Turn the steering wheel to gain the best possible access to the steering shaft nut.

52. Remove the ignition key and activate the steering wheel lock.

53. Remove the bolt securing the mounting for the steering shaft in the joint

54. Remove the joint from the steering shaft by pressing the shaft upwards.

55. Detach the connector (blue) from the steering gear.

56. Drain the fuel line.

57. Detach the quick-release coupling of the fuel line at the right rear sub-frame mount.

58. Detach the fuel return line.

➡ **Seal the openings against dirt.**

**Detach the connector from the oil level sensor in the oil sump and the oil pressure sensor.**

59. Remove the A/C compressor.

60. Pull out the cable harness and bend it aside.

➡ **Ensure that the lifting table is properly centered under the engine.**

61. Position a lifting table under the sub-frame. Secure the table against the sub-frame.

62. Remove the connector from the circulation pump.

63. Remove the 4 M10 bolts in the rear edge of the sub-frame brackets.

64. Remove the 4 bolts in the sub-frame. Lower the engine/transaxle unit approx. 6 inches (15cm) so that the halfshafts create a horizontal line between the hub and the transaxle.

➡ **Take great care that no hoses, wiring or anything else catches in the engine/transaxle unit when it is lowered.**

65. Tap the link arms off of the ball joint pin. Release the drive shafts from the wheel hubs by pressing the drive shafts into the hub.

66. Remove the 3 A/C compressor bolts.

67. Lift out the A/C compressor and suspend it in a suitable manner.

68. Continue lowering the engine/transaxle unit.

  *To install:*

### ✳✳ WARNING

**Ensure that no hoses, wiring or other components catch or become trapped during lifting.**

➡ **Make sure that the mating surfaces on the ball joints and control arms are clean.**

69. Raise the engine/transaxle unit using a lifting table. Stop approx. 6 inches (15cm) before the sub-frame reaches the frame members.

70. Install the lower radiator hose in the mount on the sub-frame.

71. Install the A/C compressor; tighten 3 bolts

72. Attach the connectors to the A/C compressor, the oil pressure sensor, and the oil level sensor; secure the cable harness in the clips and clamps.

73. Install the drive shaft in the hub on both sides.

74. Continue raising the engine to approximately 2 inches (50mm) before it makes contact.

➡ **Guide the lower steering shaft into the hole in the firewall.**

75. Guide the ball joint to the link arm.

76. Raise the sub-frame toward the frame members until it makes contact.

77. Install new bolts on the sub-frame. Oil the bolt threads.

78. Install the washers in the front edge and the support plates in the rear edge of the sub-frame.

79. First tighten the bolts on the left side of the sub-frame. Tighten the M10 bolts to 26 ft. lbs. (35 Nm) plus 60°. Tighten the M8 bolts to 15 ft. lbs. (20 Nm) plus 60°.

80. Then tighten the bolts on the right side of the sub-frame to the same values.

81. Tighten the bolts of the support plates.

82. Remove the lifting table.

83. Tighten the ball joint and link arm to 59 ft. lbs. (80 Nm). Counterhold with Torx® 40 so that the ball joint gaiter does not rotate.

84. Install the anti-roll bar links. Use new lock nuts.

85. Install new center bolts and a rubber sleeve on the drive shafts.

86. Install the track rods following the marks made during removal. Use new lock nuts.

➥**Remove the seals.**

87. Connect the fuel lines.
88. Attaching the steering gear connector
89. Attach the connector to the steering gear.
90. Install the anti-roll bar.
91. Install the driveshaft.
92. Install the intermediate pipes using new gaskets.
93. Install the nuts at intermediate pipe joints.
94. Install the gear selector cable in the mount on the transaxle and press the cable into the ball.
95. Install the transaxle control module (TCM) connector.
96. Install the upper radiator hose on the distributor housing; tighten the clamp.
97. Install the lower radiator hose on the thermostat housing; tighten the clamp.
98. Connect the radiator hoses at the firewall.
99. Remove the seals.
100. Install the lower line to the transaxle oil cooler as marked; secure with a quick-acting lock.
101. Install the upper line to the transaxle oil cooler as marked; secure with a quick-acting lock.
102. Install the positive lead in the electrical distribution box.
103. Install the ground cable in the body (between the engine and body)
104. Install the connector at the electrical distribution box.
105. Install the connector to the brake vacuum valve and the vacuum hose on the valve leading to the brake vacuum amplifier.
106. Install the auxiliary belt.
107. Installing the ground cable.
108. Install the servo oil reservoir.
109. Install the cover over the servo pump.
110. Position the upper engine covers and press them into the retainers.
111. Install the upper cross stay.
112. Install the lower engine cover.
113. Install the protective plate.
114. Install the front wheels.
115. Install the battery cable.

➥**Make sure the outlet hose of filling equipment is not blocked as this could damage the cooling system.**

116. Fill with coolant.
117. Check the oil level in the engine and transaxle. Start the engine and allow it to warm to normal operating temperature.

Check for leaks and proper function. Fill up fluids as necessary.

## EXHAUST MANIFOLD

### REMOVAL & INSTALLATION

#### 5-Cylinder Engines

1. Before servicing the vehicle, refer to the Precautions Section.
2. Remove or disconnect the following:
- Negative battery cable
- Exhaust manifold heat shield
- Exhaust front pipe
- Turbocharger
- Exhaust manifold

**To install:**
3. Install or connect the following:
- Exhaust manifold and tighten the fasteners to 18 ft. lbs. (25 Nm)
- Turbocharger, if equipped
- Exhaust front pipe and tighten the fasteners to 44 ft. lbs. (60 Nm)
- Exhaust manifold heat shield and tighten the fasteners to 88 inch lbs. (10 Nm)
4. Loosen the joint at the catalytic converter and re-tighten to 18 ft. lbs. (25 Nm). This is necessary to prevent stresses in the system.
5. Connect the negative battery cable.
6. Start the engine and check for leaks.

#### 6-Cylinder Engines, 2006 Only

See the procedures under Turbocharger Removal and Installation.

#### 8-Cylinder Engines

*See Figures 80 through 86.*

1. Remove the front wheels on both sides.
2. Remove the bolts to the torque rod.
3. Remove the oil filler cap.
4. Remove the upper front engine casing by pulling it straight up.

### ❋❋ WARNING

**When lifting the engine ensure that the lifting eye does not scratch the torque rod. Tape the torque rod and keep it away from your eyes.**

5. Install the oil filler cap.
6. Install the lifting tool: 999 5716 converted with kit 999 7070 and the lifting hooks: 999 5460 . Raise the engine slightly so that the engine pads are unloaded.
7. Remove the wing liner
8. Remove the bolts for the right-hand engine pad

➥**There is a guide sleeve on the upper mounting for the plate. This must be lifted slightly so that the panel can be slid forward.**

9. Raise and move the protective plate forward. Remove the protective plate.
10. Remove the protective cover under the engine
11. Remove the gearbox bolts to the lower torque rod.
12. Remove the exhaust pipes between front and rear catalytic converters
13. Remove the lower bolt to the engine pad.
14. Remove the anti-roll bar links from the anti-roll bar on both sides. Use: 999 5500 and a Torx® wrench as counterhold to avoid damaging the rubber gaiter.

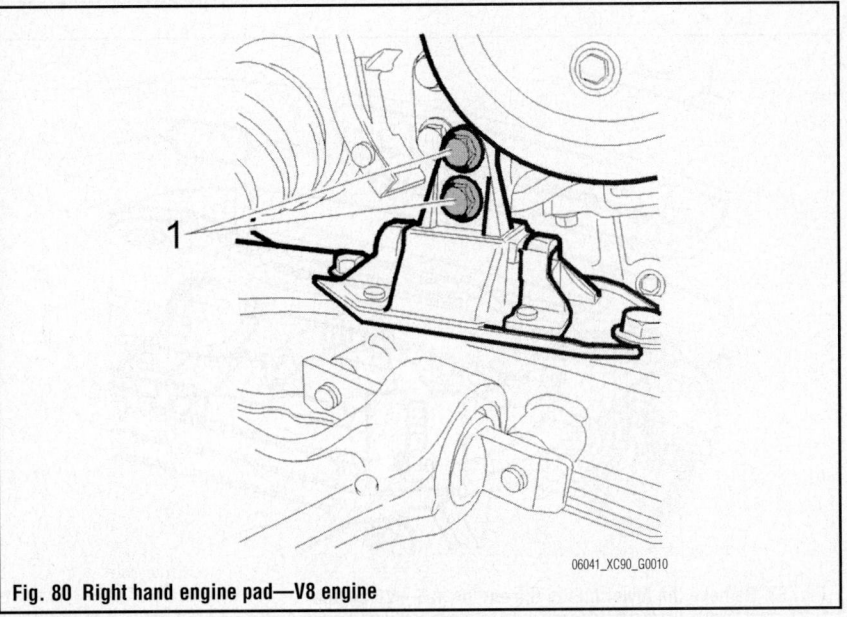

06041_XC90_G0010
**Fig. 80 Right hand engine pad—V8 engine**

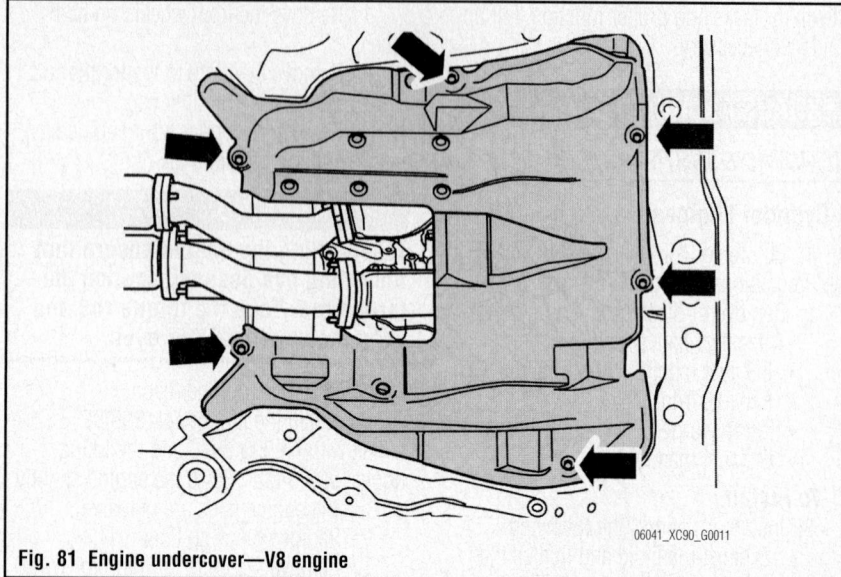

**Fig. 81 Engine undercover—V8 engine**

06041_XC90_G0011

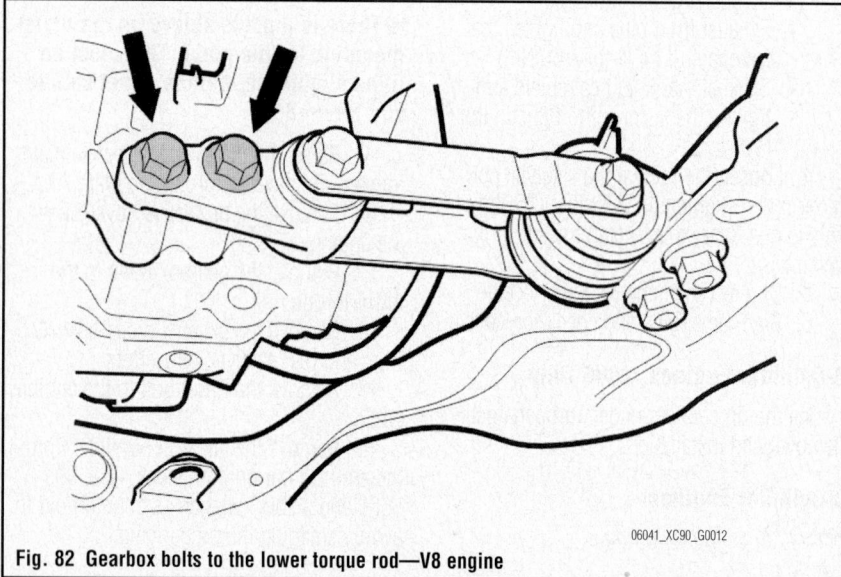

**Fig. 82 Gearbox bolts to the lower torque rod—V8 engine**

06041_XC90_G0012

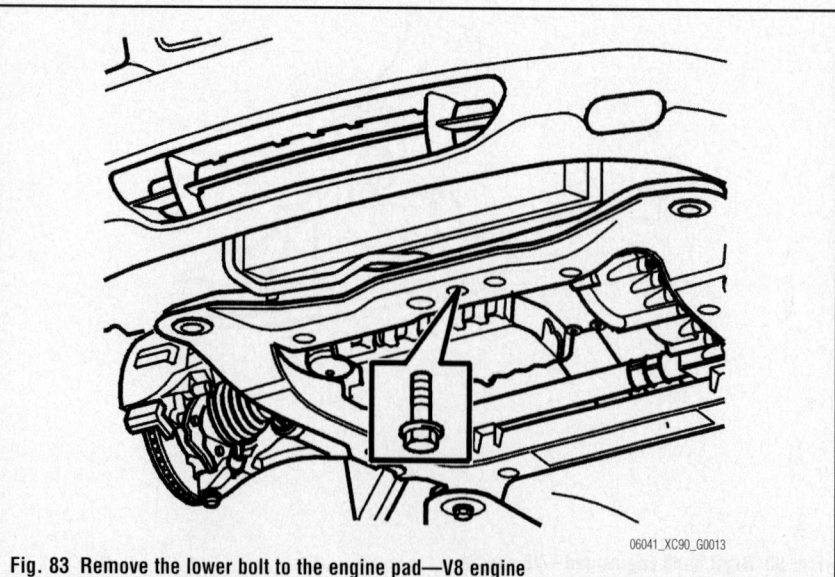

**Fig. 83 Remove the lower bolt to the engine pad—V8 engine**

06041_XC90_G0013

15. Remove the halfshaft

16. Remove the link arm from the steering knuckle on both sides

17. Detach the ball joint pin from the control arm.

18. Fit the cap nut, 7062-1, for the ball joint pin. Tighten the nut against the link arm. Then turn back one turn so that clearance is created between the cap nut and link arm.

19. Place the puller on the link arm and loosen the ball joint from the link arm.

20. Detach the link arm from the ball joint pin.

21. Pull the control arm down. Use: 999 7076 and move the steering knuckle to the side, and install the protective sleeve 999 5562 on the ball joint pin at the same time. Keep it in position using the ball joint nut.

22. Remove the steering gear from the sub-frame.

**※※ WARNING**

**Ensure that no pipes or cables are under load or are trapped when lowering the sub-frame.**

23. Release the 2 x heated oxygen sensor (HO2S) connectors from the rear edge of the sub-frame; the heated oxygen sensor (HO2S) connector from the front edge of the sub-frame.

24. Remove the bolt to the ground cable bracket on the front edge of the sub-frame.

25. Position a lifting table under the sub-frame.

26. Remove the bolts to the sub-frame's brackets at the rear edge.

27. Remove the bolts to the sub-frame in the front and rear edge.

28. Remove the sub-frame's brackets.

29. Release the heated oxygen sensor (HO2S) cables from the anti-roll bar.

30. Remove the lower coolant hose from the bracket on the sub-frame.

31. Lower the sub-frame carefully. Release the oil pipes to the steering gear from the clips on the sub-frame.

32. Remove the bolt to the bracket holding the pressure pipe to the steering gear on the sub-frame.

33. Remove the electric cooling fan.

34. Remove the front engine pad

35. Remove the connectors from the heated oxygen sensors and detach the lead from the clips and clamps.

36. Remove the 2 M6 bolts of the support bracket for the exhaust manifold in the oil sump.

37. Remove 8 M8 nuts at the exhaust manifold flange toward the cylinder head.

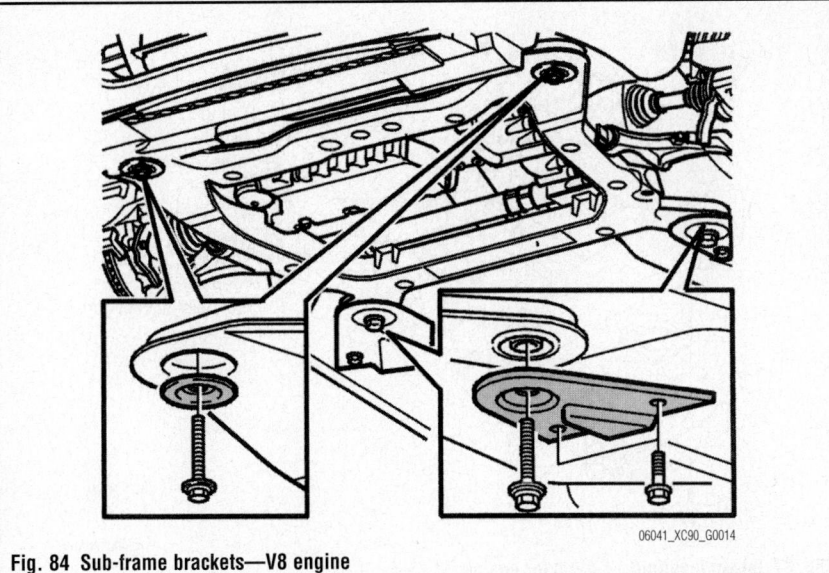

**Fig. 84 Sub-frame brackets—V8 engine**

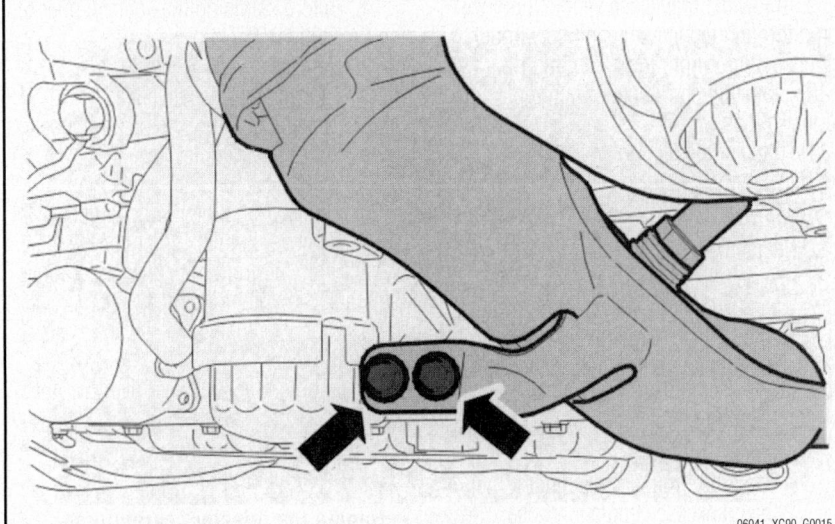

**Fig. 85 Exhaust manifold support bracket bolts—V8 engine**

**Fig. 86 Exhaust manifold nuts—V8 engine**

38. Lower the exhaust manifold.

39. Remove the exhaust manifold gasket. Discard the gasket.

➡**When replacing the exhaust manifold, transfer the heated oxygen sensors. Use 999 5543.**

*To install:*

40. Install a new gasket.

41. Install the exhaust manifold.

42. Install the 8 nuts. Tighten to 88 inch lbs. (10 Nm).

43. Install the 2 M6 bolts of the support bracket for the exhaust manifold in the oil sump.

44. Install the connectors to the heated oxygen sensors and the lead in the clips and clamps.

45. Install the front engine pad.

46. Install the engine cooling fan.

➡**Ensure that no pipes or cables are loaded or trapped when raising the sub-frame.**

47. Raise the sub-frame.

48. Install:

   a. The oil pipes to the steering gear in the clips on the sub-frame

   b. The bolt, M6, to the bracket holding the pressure pipe to the steering gear on the sub-frame

   c. The lower coolant hose in the bracket on the sub-frame

   d. The bolts and washers at the front edge and the brackets at the rear edge.

   e. The bolts to the brackets.

   f. The heated oxygen sensor (HO2S) cables on the anti-roll bar.

49. Remove the lifting table.

50. Install:

   a. The 2 x heated oxygen sensor (HO2S) connectors on the rear edge of the sub-frame

   b. The heated oxygen sensor (HO2S) connector on the front edge of the sub-frame

   c. The M6 bolt to the ground cable mounting on the front edge of the sub-frame.

51. Install the steering gear on the sub-frame.

52. Install the lower M10 bolt for the engine pad.

53. Install the gearbox bolts to the lower torque rod.

54. Install the exhaust pipes between front and rear catalytic converters.

55. Installing the link arm on both sides.

56. Install the halfshaft.

57. Install the protective cover under the engine.

58. Install the protective plate.

59. Move the skid plate up into the bumper cover. Press forward until the mounting brackets enter the sub-frame. Install the bolts.

60. Fit the bolts to the right-hand engine pad.

61. Install the wing liner.

62. Remove the lifting tool 999 5716, kit 999 7070 and lifting hooks 999 5460.

63. Fit the bolts to the torque rod.

64. Install the 2 x M10 bolts, by the suspension turret on both sides.

65. Install the front wheels on both sides.

66. Remove the oil filler cap.

67. Install the upper front engine casing and the oil filler cap. Carry out wheel alignment in accordance with.

68. Start the engine and check for leaks and proper function.

## INTAKE MANIFOLD

### REMOVAL & INSTALLATION

#### 5-Cylinder Engines

*See Figure 87.*

1. Disconnect the battery negative lead.
2. Remove the fuel injection system.
3. Drain the fuel injection system.
4. Remove or disconnect:
   - The charge air pipe over the engine
   - The cover plate over the connectors from the injectors
   - The fuel rail mounting bolts. Spray penetrating oil around the injector nozzle at the terminal on the intake manifold. Gently work the fuel rail and injector nozzles loose. Separate the fuel line quick-release connector.
   - Separate the throttle body connector
   - The plastic charge air pipe between the throttle body and the charge air cooler
   - Vacuum hoses from the intake manifold.
   - The oil filler cap
   - The cover over the ignition coils
   - The hose from the flame trap to the cam cover at the terminal in the cam cover.
   - The dip stick pipe
5. Slacken off the lower bolts for the intake manifold a few turns.
6. Remove the mounting bolts in the upper row and the outer bolts in the lower row.
7. Remove the intake manifold by lifting it approximately ¾ inch (20mm).
8. Remove the nipple for the water heated crankcase ventilation.

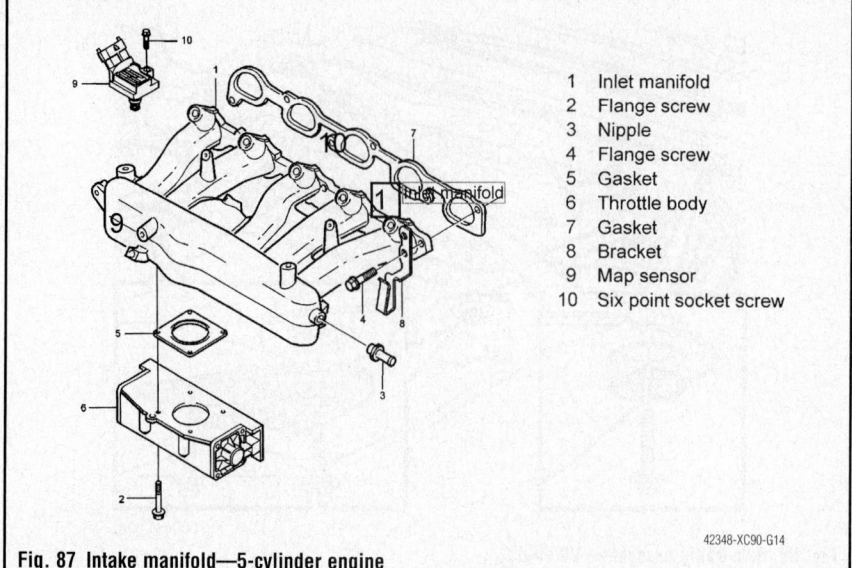

**Fig. 87 Intake manifold—5-cylinder engine**

1. Inlet manifold
2. Flange screw
3. Nipple
4. Flange screw
5. Gasket
6. Throttle body
7. Gasket
8. Bracket
9. Map sensor
10. Six point socket screw

42348-XC90-G14

9. Allow the crankcase ventilation hose to run through the intake manifold without disconnecting it from the flame trap.

10. Transfer components when replacing the intake manifold

11. Transfer the electronic throttle module. Use a new gasket.

**To install:**

12. Install or connect:
   - A new gasket held in position by the lower bolts for the intake manifold. Do not forget the clamp for the fuel line at the bolt in the lower row.
   - The nipple for the water heated crankcase ventilation
   - The intake manifold. Do not forget the crankcase ventilation hose. The hose must be inserted up through the gap between the second and third ducts
   - The 3 upper bolts. Tighten all the bolts starting from the center to 14 ft. lbs. (19 Nm).
   - The fuel rail. Press together the quick-release connector. Use new bolts.
   - Injector connectors
   - The protective cover over the fuel rail
   - The vacuum hoses according to the earlier markings
   - The plastic charge air pipe between the throttle body and the charge air cooler
   - The connectors to the throttle body
   - The dip stick in the intake manifold.

#### 6-Cylinder Engines, 2006 Only

*See Figure 88.*

1. Before servicing the vehicle, refer to the Precautions Section.

2. Remove:
   - The battery negative lead
   - plastic charge air pipe and hose to charge air cooler
   - The intake pipe between the air cleaner and the turbocharger

3. Remove the intake pipe for the air cleaner housing from the connection above radiator. Twist the pipe up.

4. Drain the fuel line.

5. Remove:
   - The protective cover over the injector connectors
   - injector connectors
   - The fuel rail mounting bolts

➡ **Handle the injectors carefully to avoid damaging the nozzles and needles.**

6. Spray universal oil or similar around the injector terminals on the intake manifold.

7. Gently work the fuel rail and injectors loose.

8. Detach the fuel line from the nozzle pipe by pressing the quick-release connector.

9. Remove:
   - The dip stick pipe mounting at the intake manifold
   - The nipple for the crankcase ventilation under the intake

10. disconnect the throttle body connector. Remove the plastic charge air pipe from the throttle body.

11. Remove the upper bolts.

12. Remove the outer bolts from the lower row.

13. Slacken off the remaining bolts.

**Fig. 88 Intake manifold—6-cylinder engine**

14. Lift up and remove the intake manifold.

15. Remove the remaining bolts and gasket.

### To install:

16. Transfer the throttle body with a new gasket. Tighten to 88 inch lbs. (10 Nm).

17. Ensure that the gasket faces are clean.

18. Install a new gasket with the two centermost bolts in the lower row.

19. Lower the intake over the 2 bolts.

20. Install the remaining bolts.

21. Tighten all bolts. Tighten to 15 ft. lbs. (20 Nm).

22. Connect the throttle body connector.

23. Install:
   • The vacuum hoses according to the earlier markings
   • The nipple for the crankcase ventilation under the intake manifold
   • The dip stick mounting bolt to the intake manifold. Tighten to 18 ft. lbs. (25 Nm)
   • The plastic charge air pipe between the throttle body and the charge air cooler

24. Lubricate the injector rings with petroleum jelly.

25. Hold the fuel rail. Press the fuel line quick-release connector together until it clicks.

26. Press down the fuel rail. Check that all injectors are correctly positioned.

27. Tighten the fuel rail to 88 inch lbs. (10 Nm). Use new bolts.

28. Connect all the connectors to the injectors.

29. Install the protective cover over the connectors to the nozzles.

30. Install:
   • intake manifold between air cleaner and turbocharger
   • plastic charge air pipe and hose to charge air cooler

   • The air cleaner intake pipe
   • The battery negative lead

### 8-Cylinder Engines

*See Figures 89 through 93.*

1. Remove the 4 bolts at the suspension turrets

2. Remove the bolts of the upper engine mount on the engine.

3. Remove the upper cross stay.

4. Remove the air cleaner cover by pulling it straight up.

5. Remove the 2 connectors from the engine control module (ECM); clip the band clamp to the cable harness.

6. Remove the connector from the mass air flow sensor.

7. Remove the brake vacuum hose from the vacuum pump.

8. Remove the 2 bolts of the fresh air hose in the front plate.

9. Remove the fresh air hose.

10. Remove the air cleaner by pulling it straight up.

11. Remove the upper engine covers by pulling them straight up.

12. Remove the cover over the servo pump.

13. Detach the connector of the pressure and temperature sensor.

14. Remove the 2 vacuum hoses of the check valve of the distribution damper.

➡**There is a guide sleeve on the upper mounting for the plate. This must be lifted slightly so that the panel can be slid forward.**

15. Remove the 6 bolts.

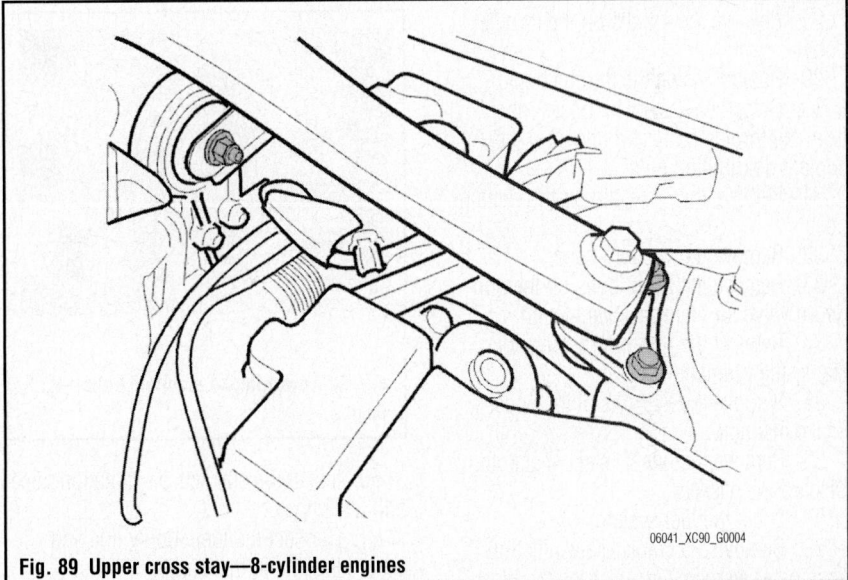

06041_XC90_G0004

**Fig. 89 Upper cross stay—8-cylinder engines**

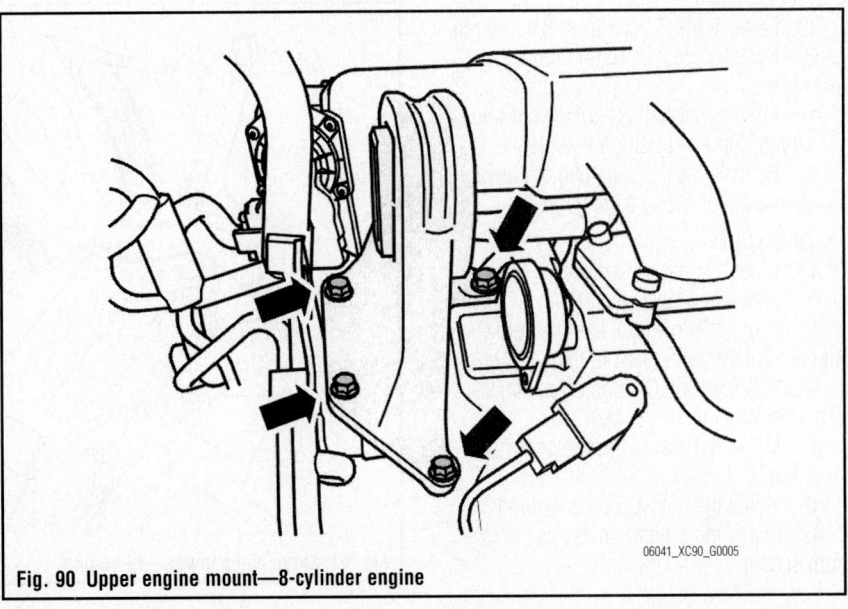

06041_XC90_G0005

**Fig. 90 Upper engine mount—8-cylinder engine**

16. Lift and slide the protective plate forward.

17. Remove the protective plate.

18. Open the cock on the lower left edge of the radiator.

19. Drain the coolant into a clean receptacle.

20. Close the cock.

21. Install the protective plate.

22. Install the 6 bolts.

23. Lower the vehicle.

24. Remove the 4 bolts and the upper engine mount.

25. Remove the brake vacuum hose from the intake manifold by releasing the clamp.

26. Remove the intake hose between the air cleaner and throttle body by releasing the 2 clamps.

27. Remove the 2 nuts on the throttle body.

28. Remove the 2 bolts on the throttle body.

29. Position aside the throttle body.

30. Detach the 2 coolant hoses, one from the throttle body and one from the coolant distribution pipe.

31. Remove the hose clip on the vacuum hose.

32. Remove the vacuum hose.

33. Remove the 2 M6 bolts on the non-return valve for crankcase ventilation

34. Remove the non-return valve for crankcase ventilation.

35. Remove the 21 bolts of the upper intake manifold.

36. Remove the upper intake manifold. Discard the gaskets.

37. Drain the fuel system.

38. Remove the crankcase ventilation hose between the cylinder heads by releasing the 2 clamps.

39. Remove the 3 bolts from the fuel rail.

40. Remove the 2 bolts on the fuel pressure sensor.

41. Lift the fuel rail assembly with injector straight up and position it aside.

42. Remove the 12 bolts and the lower intake manifold. Discard the gaskets.

**To install:**

43. Clean the gasket surfaces.

44. Install new gaskets.

45. Install the lower intake manifold. Tighten the 12 bolts to 14 ft. lbs. (19 Nm).

46. Lubricate the O-rings on the injectors with Vaseline or the like.

47. Install the fuel rail assembly with injectors.

48. Install the 3 bolts of the fuel rail.

49. Install the 2 bolts on the fuel pressure sensor.

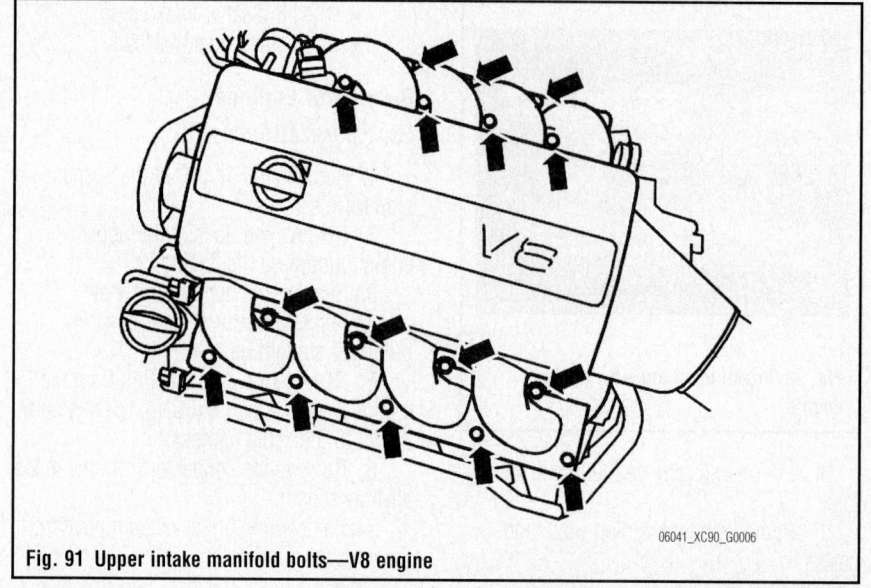

**Fig. 91 Upper intake manifold bolts—V8 engine**

06041_XC90_G0006

06041_XC90_G0007

**Fig. 92 Lower intake manifold bolts—V8 engine**

50. Install the crankcase ventilation hose and 2 clamps.

51. Connect the fuel supply line and secure the line in the clamp.

**✷✷ WARNING**

**Make sure there is an audible click during connection.**

52. Install new gaskets.

53. Install the upper intake manifold. Tighten the 21 bolts to 14 ft. lbs. (19 Nm).

54. Install the hoses to the intake manifold.

55. Install the non-return valve for crankcase ventilation.

56. Install the connector to the pressure and temperature sensor.

57. Install the 2 vacuum hoses of the check valve of the distribution damper.

58. Install the upper engine mount. Tighten the 4 M10 bolts.

59. Install the cover over the servo pump

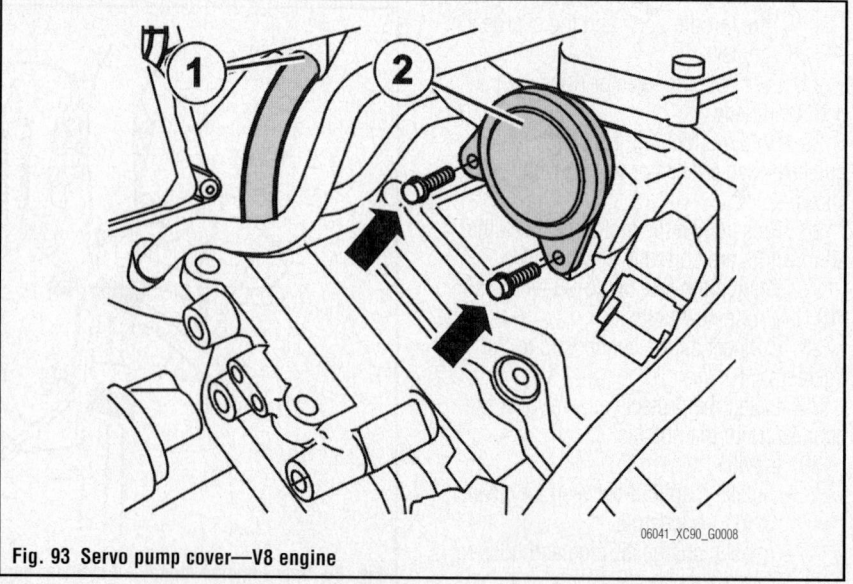

06041_XC90_G0008

**Fig. 93 Servo pump cover—V8 engine**

60. Position the engine covers and press them into the retainers.

61. Install the upper cross stay between the suspension turrets. Tighten the 4 bolts.

62. Install the bolt of the cross stay mount on the engine.

63. Install the air cleaner.

64. Install the brake vacuum hose on the brake pump.

65. Install the fresh air hose.

66. Install the 2 bolts of the fresh air hose in the front plate.

67. Install the 2 connectors to the engine control module (ECM).

68. Install the connector to the mass air flow sensor.

69. Install the band clamp on the cable harness in the air cleaner housing.

70. Install the cover over the air cleaner.

71. Fill with coolant.

72. Start the engine and allow it to warm to normal operating temperature. Check for leaks and proper function. Fill up fluids as necessary.

## OIL PAN

### *REMOVAL & INSTALLATION*

**5-Cylinder Engines**

*See Figure 94.*

1. Before servicing the vehicle, refer to the Precautions Section.

2. Remove the oil dipstick and its pipe.

3. Remove the splashguard under the engine.

4. Drain the engine oil and remove the oil filter.

5. Release the oil cooler from the oil pan. Hang up at the rear.

6. Remove the bolt from the bracket for the fuel line.

7. Removing the oil pan

8. Back off all bolts holding the oil pan.

9. Remove all bolts except for four. It is recommended that the four bolts in the corners of the oil pan are left in place.

10. Carefully tap the oil pan with a rubber mallet until the joint and its liquid gasket releases.

11. Remove the four remaining bolts.

12. Remove the oil pan.

**To install:**

13. Apply liquid gasket 1161 059-9, or equivalent, to the oil pan.

14. Install new O-rings.

15. Position the oil pan. Secure it loosely with a few bolts.

16. Install the remaining bolts loosely.

17. Press the oil pan against the transaxle. Tighten bolts 1, 2, 3 and 4 to

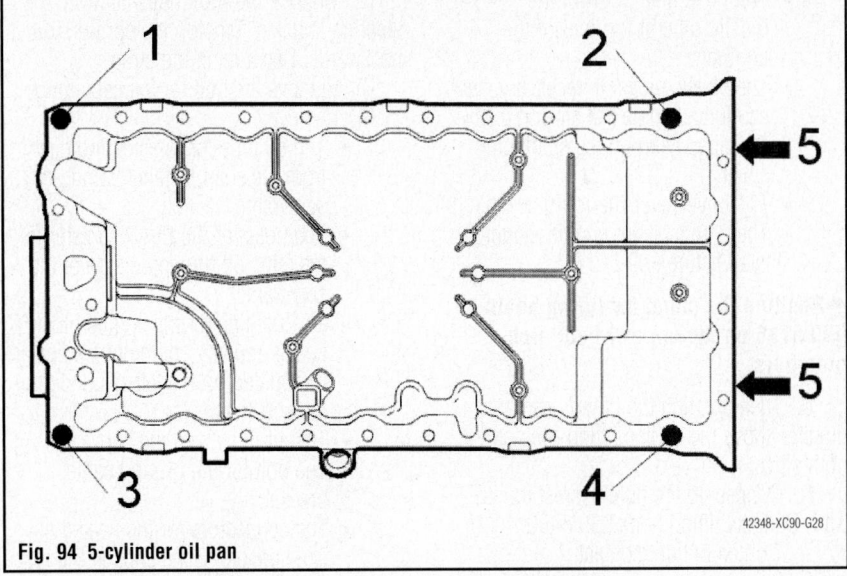

**Fig. 94 5-cylinder oil pan**

27 inch lbs. (3 Nm). Tighten bolt 5 to 18 ft. lbs. (25 Nm); then tighten to 35 ft. lbs. (48 Nm).

18. Tighten all bolts in the oil pan flange to 12 ft. lbs. (17 Nm). Start at the transaxle end and continue forwards in pairs.

19. Install the bolt for the bracket for the fuel line.

20. Connect the oil cooler to the oil pan. Use new O-rings. Reinstall the pipe on the sub-frame.

21. Install a new oil filter.

22. Install the oil drain plug with a new gasket.

23. Install the oil dipstick and its pipe. Use a new O-ring.

➡**Check that the O-ring is correctly positioned.**

24. Fill with engine oil. Run the engine to operating temperature.

25. Check for oil leaks from the oil pan or oil cooler.

26. Install the splashguard under the engine.

27. Check the oil level. Top up if required.

**6-Cylinder Engines, 2006 Only**

*See Figures 95 and 96.*

1. Before servicing the vehicle, refer to the Precautions Section.

2. Remove the oil dipstick and pipe.

3. Remove:
- The connector and hose from the crankcase ventilation terminal in the plastic intake pipe
- The hose from the plastic charge air pipe
- The bolts for the charge air pipe in the intake pipe

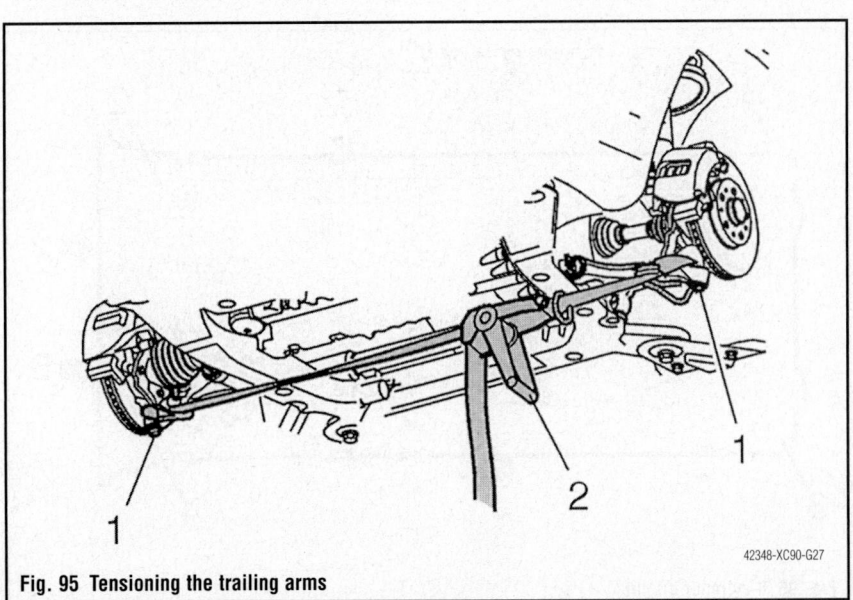

**Fig. 95 Tensioning the trailing arms**

- The hose clamps from the plastic pipe at the rear of the engine
- The plastic pipes. If necessary, heat carefully using a hot air gun. Leave the pipes lying on the spark plug cover
- The cover over the ignition coils
- The 2 bolts at the engine mounting

4. Install fixture 999 5717.

➡**Position the stand for lifting beam 999 5716 on the upper wheel arch members.**

5. Then position the lifting beam directly above the engine lifting eyes on both sides.

6. Connect to the hole nearest the firewall. Connect lifting hook 999 5460.

7. Tighten to light contact.

8. Remove:
- The lower engine cover
- The front air baffle

9. Drain the engine oil. Remove the oil filter.

10. Remove:
- The front wheel
- The nuts for the lower ball joints
- The anti-roll bar links from the anti-roll bar
- The center bolt for the right-hand halfshaft
- The bolt and nuts for the steering gear
- The bolt for the front engine mounting
- The torque rod including the mounting in the sub-frame

11. Tension the trailing arms together to release the spindles. Use a retaining strap.

12. Unhook the right halfshaft from the steering knuckle. Tension the spring strut backwards. Use a retaining strap.

13. Remove the bolt for the cable duct.

14. Remove:
- The oil pipe for the steering gear from the snap fasteners along the sub-frame
- The hose for the EVAP canister in the clips on the upper side of the sub-frame
- any fuel line for the engine block heater and any engine block heater if mounted on the left-hand side of the sub-frame
- The air duct from the clip
- The bolt for the ground cable bracket
- The connectors for the heated oxygen sensors in the clips at the rear of the sub-frame

15. Slacken off the sub-frame bolts. Lower the sub-frame slightly.

16. Remove the heated oxygen sensor bracket on the reverse of the sub-frame.

17. Unbolt the bolts for the sub-frame approximately 1 inch (3cm) on the right-hand side.

18. Fully remove the bolts on the left-hand side.

19. Angle the sub-frame down on the left-hand side.

20. Ensure that the bolts for the steering gear release from the frame.

21. Disconnect the oil cooler from the oil pan. Hang the oil cooler from a suitable place.

22. Remove the bolt from the bracket for the fuel line.

23. Remove the bolt for the bracket for the fuel line on the auxiliary bracket.

24. Slacken off all the bolts holding the oil pan.

25. Remove all but four bolts. It is recommended that the four bolts in the corners of the pan are left in place.

26. Carefully tap the pan using a rubber mallet until the joint and its liquid gasket releases.

27. Remove:
- The four remaining bolts
- The oil pan

28. Clean the gasket surfaces on the oil pan and cylinder block.

**✲✲ WARNING**

**Use a fume hood or extractor when using gasket solvent. Note! Also clean the gasket faces for the bolts for the torque rod.**

29. Apply liquid gasket 116 1059-9 to the oil pan.

Note! Also apply liquid gasket around the holes for the bolts for the torque rod.

***To install:***

30. Install:
- new O-rings
- The oil pan. Secure the oil pan loosely using a few bolts
- The remaining bolts loosely

31. Press the oil pan against the transaxle. Tighten the bolts (1), (2), (3) and (4). Tighten to 27 inch lbs. (3 Nm).

32. Tighten the bolts 5. First tighten to 18 ft. lbs. (25 Nm). Then tighten to 35 ft. lbs. (48 Nm).

33. Tighten all bolts in the oil pan flange. Tighten to 12 ft. lbs. (17 Nm). Start at the transaxle and continue forwards in pairs.

34. Connect the oil cooler to the oil pan. Use new O-rings.

35. Install the bolt for the bracket for the fuel line.

36. Install:
- A new oil filter
- The oil plug with a new gasket

37. Lift the sub-frame. Use a mobile jack.

38. Install:
- The bracket for the heated oxygen sensors in the rear edge of the sub-frame
- New bolt in the sub-frame. Lubricate the bolts
- The washers at the front edge and the support plates at the rear edge of the sub-frame

39. Tighten the bolts for the sub-frame. Tighten to 77 ft. lbs. (105 Nm). Angle-tighten 120°.

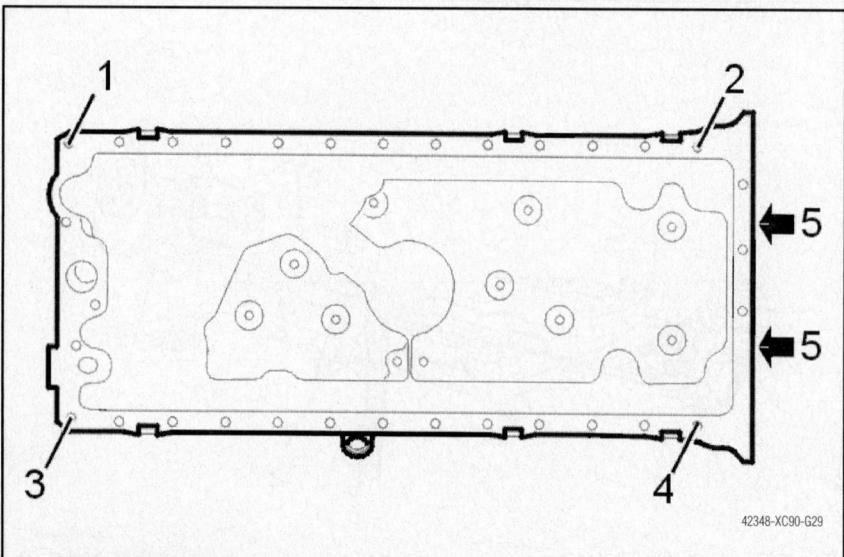

42348-XC90-G29

**Fig. 96 6-cylinder oil pan**

40. Start on the left-hand side of the sub-frame. Continue with the right-hand side.

41. Tighten the bolts for the brackets at the rear of the sub-frame. Tighten to 37 ft. lbs. (50 Nm).

42. Install:
- The bolt for the cable and air duct
- The oil pipe for the steering gear in the snap fasteners along the sub-frame
- The hose for the EVAP canister in the clip on the upper side of the sub-frame
- The engine block heater and the fuel line to the engine block heater if applicable
- The air duct in the clips
- The bolt for the ground cable bracket
- The connectors for the heated oxygen sensors in the clips at the rear of the sub-frame

43. Slacken off the retaining straps for the spring strut. Thread the halfshaft into the steering knuckle.

44. Carefully release the retaining strap for trailing arms. At the same time guide the lower ball joints into the trailing arms.

45. Install new nuts on the ball joints. Tighten to 59 ft. lbs. (80 Nm).

46. Counterhold using a Torx® wrench so that the ball joint boot is not damaged.

47. Install the anti-roll bar links. Use new nuts. Tighten to 37 ft. lbs. (50 Nm).

48. Counterhold using a Torx® wrench so that the boot is not damaged.

49. Install:
- The bolt and nuts for the steering gear. Use new nuts and new bolts. Tighten to 37 ft. lbs. (50 Nm)
- The front engine pad. Tighten to 37 ft. lbs. (50 Nm)
- The torque rod. Tighten the bolts through the oil pan. Tighten to 37 ft. lbs. (50 Nm). Tighten the nuts in the sub-frame. Tighten to 48 ft. lbs. (65 Nm)
- The center bolt for the halfshaft. Use a new bolt. Counterhold using a prybar in the brake disc vents. Tighten to 37 ft. lbs. (50 Nm)
- The front air baffle
- The front wheel
- The lifting beam
- The lifting hooks
- The lifting fixture from the engine. Install the bolts in the engine mounting
- The cover over the ignition coils
- The oil dipstick and pipe. Use a new O-ring. Check that the O-ring is correctly positioned

- The plastic pipes. Tighten the hose clamps
- The bolts for the charge air pipe in the intake pipe
- The connector and hose to the crankcase ventilation terminal in the plastic intake pipe
- The hose to the plastic charge air pipe

50. Fill with new engine oil.

51. Start the engine and check for oil leakage.

52. Stop the engine. Give the oil time to run down into the oil pan. Then check the oil level.

53. Top up if necessary but do not overfill.

54. Wipe the engine compartment clean.

55. Install the lower engine cover.

### 8-Cylinder Engines

*See Figures 97 through 103.*

1. Drain the engine oil, remove the oil filter.

2. Remove the exhaust manifolds.

3. Remove the right-hand engine mounting.

4. Remove the 1 M10 bolts for the bracket.

5. Remove the 4 M10 bolts for the transaxle.

6. Remove the oil cooler. Discard the O-ring.

7. Remove the 20 M6 bolts.

8. Remove the oil level sensor.

9. Remove the 17 x M8 bolts and the 1 x M6 bolt securing the intermediate section.

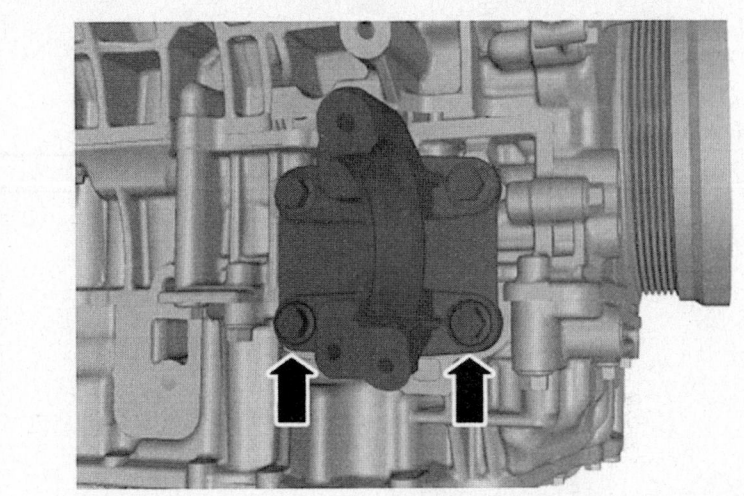

06041_XC90_G0026

**Fig. 97 Right side engine mount—V8**

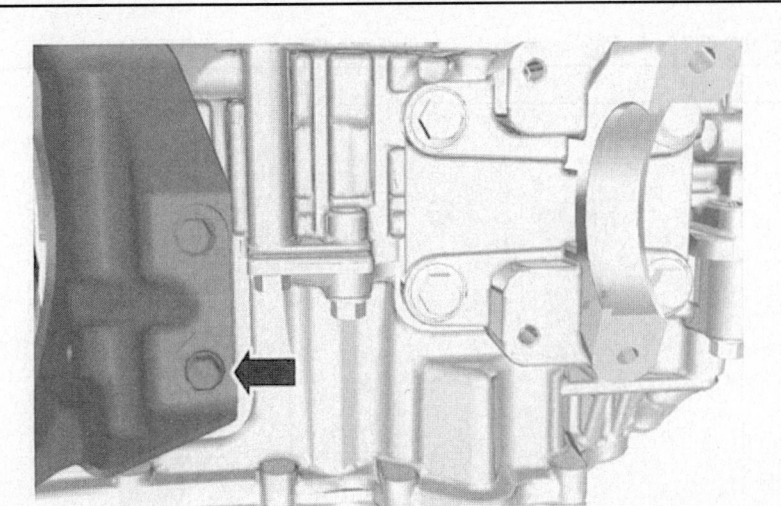

06041_XC90_G0027

**Fig. 98 M10 bracket bolt—V8**

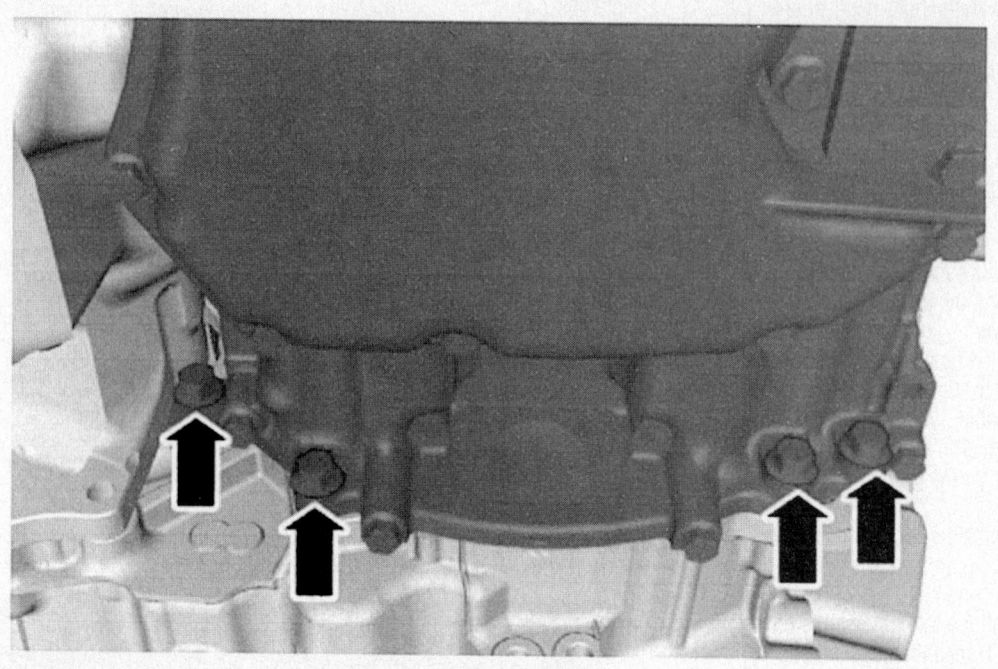

06041_XC90_G0028

**Fig. 99 Transaxle bolts—V8**

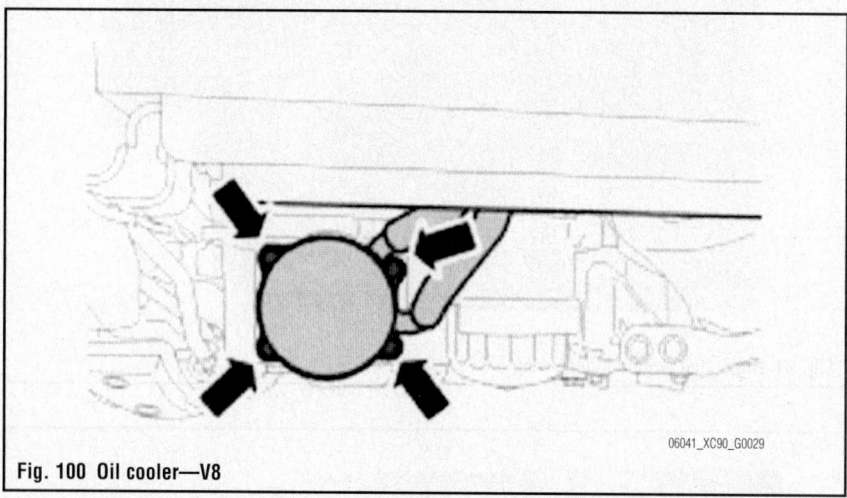

06041_XC90_G0029

**Fig. 100 Oil cooler—V8**

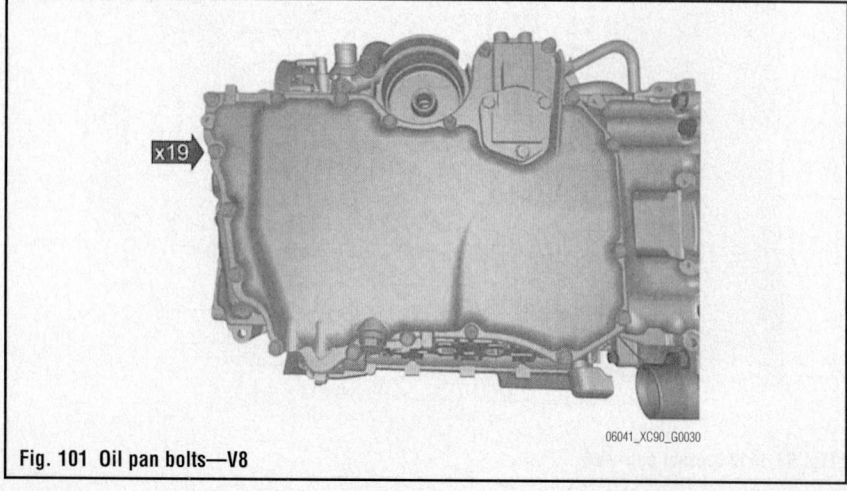

06041_XC90_G0030

**Fig. 101 Oil pan bolts—V8**

10. Carefully tap the side of the oil pan using a rubber mallet. Lift out the oil pan.

11. Discard the O-rings.

***To install:***

➡**Clean the surface, using a razor blade or plastic scraper.**

12. Apply liquid gasket 307 57050 as illustrated.

➡**Minimum 0.039 inch (1mm) bead width.**

13. Install the oil pan within 5 minutes (curing time for liquid gasket).

14. Install the oil pan, use new O-rings

15. Install the 17 x M8 bolts.

16. Install the bolt securing the intermediate section.

➡**Clean the surface, using a razor blade or plastic scraper.**

17. Apply liquid gasket 307 57050 as illustrated.

➡**Minimum 0.039 inch (1mm) bead width.**

18. Install the oil pan within 5 minutes (cure time for liquid gasket)

19. Install the 20 M6 bolts.

20. Install the oil level sensor.

21. Install the oil filter, using a new O-ring.

22. Install the oil cooler. Use new O-ring

**Fig. 102 Intermediate section bolts—V8**

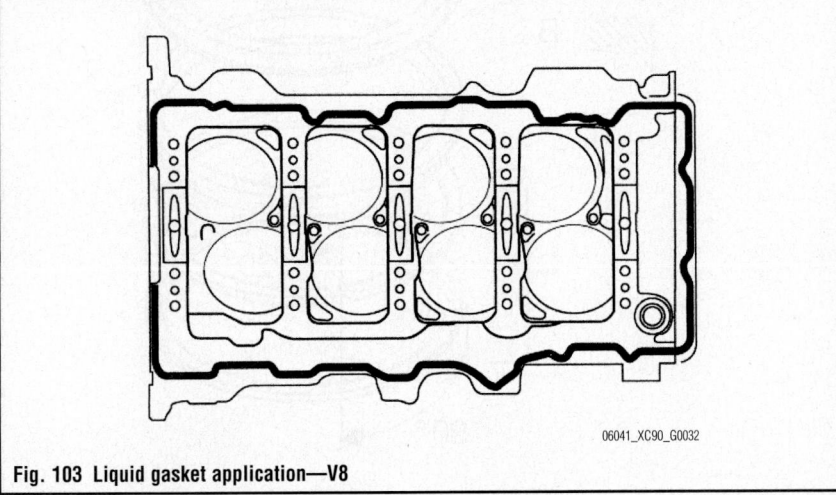

**Fig. 103 Liquid gasket application—V8**

23. The remainder of installation is the reverse of removal.
24. Top up oil.

## OIL PUMP

### REMOVAL & INSTALLATION

#### 5-Cylinder Engines

*See Figure 104.*

1. Before servicing the vehicle, refer to the Precautions Section.
2. Remove or disconnect the following:
   - Negative battery cable
   - Fuel line clips
   - Coolant recovery tank
   - Accessory drive belts
   - Right front wheel
   - Inner fender liner
   - Front cover
   - Timing belt
   - Crankshaft timing sprocket
   - Front crankshaft seal
   - Oil pump

*To install:*
3. Install the oil pump using special tool 999-5455. Use the oil pump bolts to guide the pump. Use the crankshaft nut to press the pump in until it is seated fully. Tighten the oil pump bolts to 88 inch lbs. (10 Nm).

4. Remove the crankshaft nut and the press tool.
5. Install or connect the following:
   - Front crankshaft seal
   - Crankshaft timing sprocket and tighten the nut to 133 ft. lbs. (180 Nm)
   - Timing belt
   - Front cover
   - Inner fender liner
   - Right front wheel
   - Accessory drive belts
   - Coolant recovery tank
   - Fuel line clips
   - Negative battery cable
6. Start the engine and check for leaks.

#### 6-Cylinder Engines, 2006 Only

1. Before servicing the vehicle, refer to the Precautions Section.
2. Remove the timing belt.
3. Check the tensioner and idler pulleys.
4. Carefully pull crankshaft timing gear pulley free. Remove the four oil pump bolts.
5. Carefully pry upward diagonally between the stop lugs and the cylinder block.
6. Lift out the oil pump.

*To install:*
7. Install a new gasket.
8. Carefully insert the oil pump over the end of the crankshaft.

➡ **The sealing ring in the oil pump is very easy to damage if not installed correctly.**

9. Install four new bolts as a guide.
10. Pull in the oil pump with tool 999 5455, or equivalent, and the crankshaft center nut.
11. Tighten the oil pump bolt. Tighten to 88 inch lbs. (10 Nm).

**Fig. 104 Oil pump installation**

12. Install the timing gear pulley. Carefully tap alternately around the timing gear pulley until reaching its end position.

➡ **The timing gear pulley can only be in one position on the crankshaft end splines.**

13. Install a new timing belt.
14. Install the pulley for the auxiliary belt. Locate the steering gear on the locating pin in the timing gear pulley.
15. Install the auxiliary belt.
16. Install counterhold 999 5433.
17. Tighten crankshaft center nut. See the torque chart at the beginning of this chapter.
18. Install the crankshaft damper. Use the crankshaft center nut as a counterhold.
19. Check the engine oil level.

### 8-Cylinder Engines

*See Figures 105 and 106.*

1. Remove the timing chain.
2. Install the front engine bracket.
3. Remove the oil pan.
4. Remove the oil delivery pipe. Discard the gasket.
5. Remove the 3 M7 bolts and the oil pump. Discard the O-ring and the gasket for the oil delivery pipe.

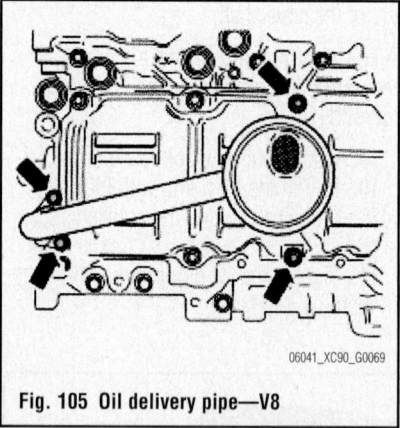

**Fig. 105 Oil delivery pipe—V8**

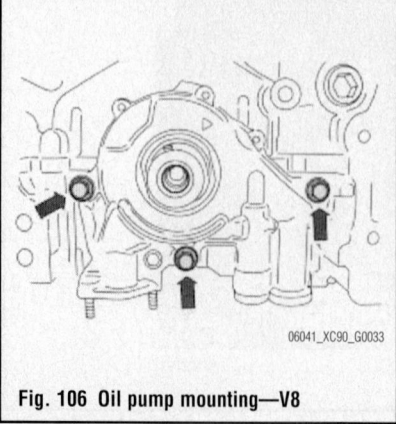

**Fig. 106 Oil pump mounting—V8**

### To install

6. Install a new O-ring and the oil pump.
7. Install the oil delivery pipe using a new gasket.
8. Install the oil pan.
9. Install the timing chain.

## PISTON AND RING

### POSITIONING

*See Figures 107 and 108.*

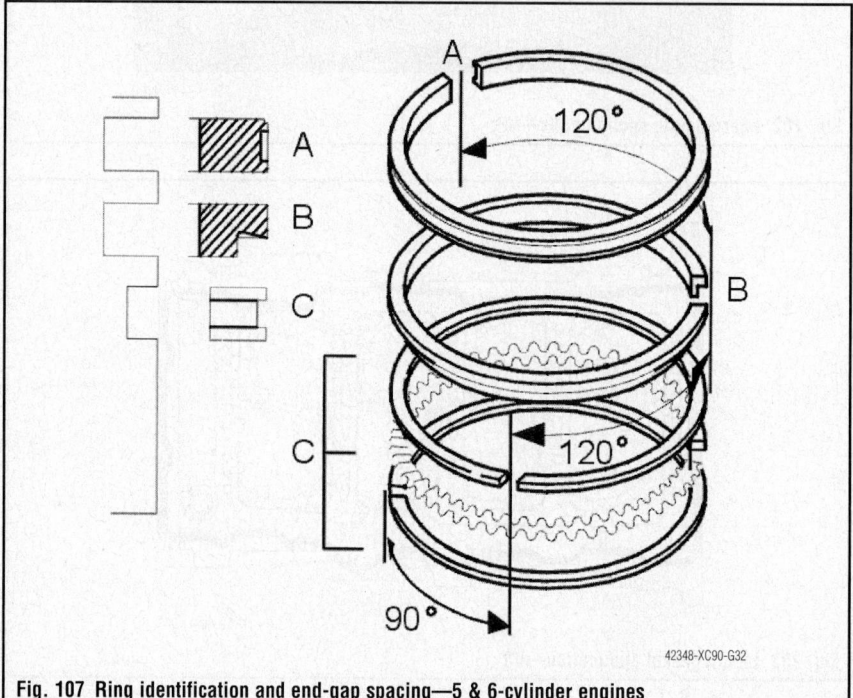

**Fig. 107 Ring identification and end-gap spacing—5 & 6-cylinder engines**

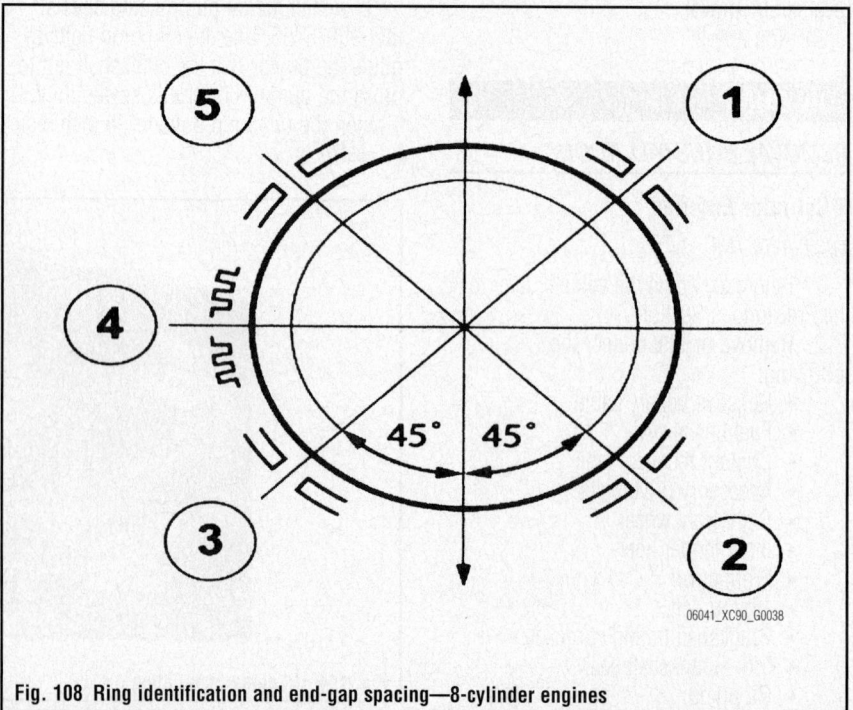

**Fig. 108 Ring identification and end-gap spacing—8-cylinder engines**

## REAR MAIN SEAL

### REMOVAL & INSTALLATION

#### 2006–07 5-Cylinder Engine and 2006 6-Cylinder Engine

*See Figure 109.*

1. Before servicing the vehicle, refer to the Precautions Section.
2. Remove or disconnect the following:

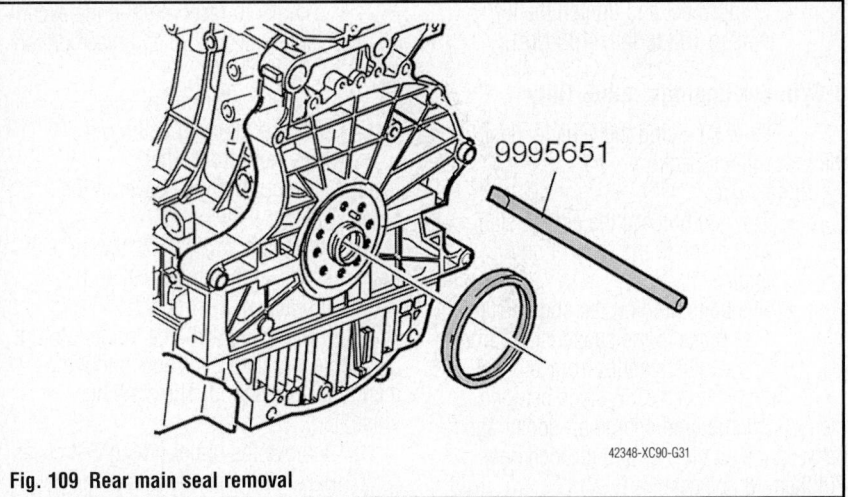

**Fig. 109 Rear main seal removal**

- Negative battery cable
- Transaxle
- Flexplate
- Rear main seal

***To install:***

3. Install the rear main seal so that it is flush with the cylinder block.

4. Install or connect the following:

- Flexplate. Tighten the bolts to 33 ft. lbs. (45 Nm) plus 65°.
- Transaxle
- Negative battery cable

5. Start the engine and check for leaks.

### 8-Cylinder Engines

*See Figures 110 through 113.*

1. Remove the transaxle according.

2. Lock the carrier plate using an M10 bolt.

3. Remove the carrier plate.

### ✳✳ WARNING

**Do not damage the mating surfaces of the crankshaft and cylinder block.**

4. Remove the old sealing ring using puller 999 5651.

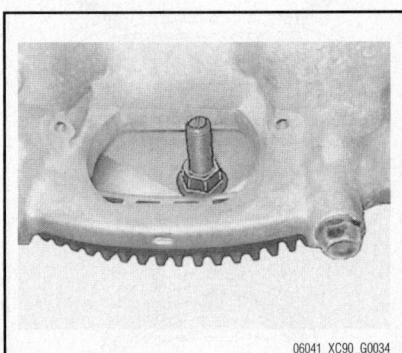

**Fig. 110 Lock the carrier plate using an M10 bolt—V8**

5. Clean around the crankshaft flange and bolt holes to remove any thread sealant residue. If necessary, wash and dry away any oil from the engine's rear engine block plane and the transaxle clutch housing. Blow clean with compressed air.

### ✳✳ WARNING

**When cleaning work around the flange, not in and out. No grinding tools or similar may be used. It is extremely important that there is no thread sealant residue or other dirt. This could cause leakage.**

➡ To clean the crankshaft, use emery cloth such as 9511024.

***To install:***

➡ Only Volvo approved engine oil may be used. Grease must not be used at all.

6. Lubricate the sealing ring mating surface against the cylinder block and the lips of the seal.

**Fig. 112 Fit a new sealing ring on the inner part of tool 999 7194 and screw the inner part onto the crankshaft using two old bolts from the carrier plate—V8**

7. Fit a new sealing ring on the inner part of tool 999 7194 and bolt the inner part onto the crankshaft using two old bolts from the carrier plate.

8. Press in the sealing ring with the outer part of tool 999 7194 with the center bolt to the stop.

**Fig. 113 Press in the sealing ring with the outer part of tool 999 7194 with the center bolt to the stop—V8**

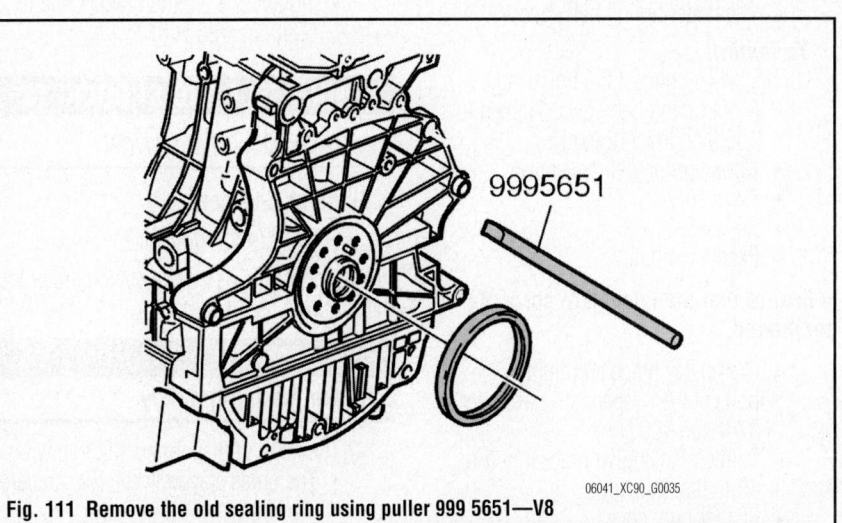

**Fig. 111 Remove the old sealing ring using puller 999 5651—V8**

9. Remove tool 999 7194. Discard the carrier plate bolts.

## ✻✻ WARNING

**Ensure that the mating surfaces on the crankshaft and carrier plate are completely clean. The carrier plate must lie against the crankshaft flange before the bolts are installed. There is a risk that excess locking fluid on the bolts will come between the mating surfaces.**

10. Brush off all the old locking fluid from the two old bolts. Use a wire brush. Mark the cleaned bolts using a marker pen.

11. Install the carrier plate, tighten it in towards the crankshaft with the two cleaned and marked bolts.

➡**Check that the locating pin on the crankshaft is positioned opposite the hole in the carrier plate.**

12. Remove the two old marked bolts.
13. Install the remaining new bolts.
14. Remove the locking bolt.
15. Install the transaxle.

## TIMING BELT FRONT COVER

### *REMOVAL & INSTALLATION*

#### 5-Cylinder Engines

1. Before servicing the vehicle, refer to the Precautions Section.

2. Remove or disconnect the following:
- The cross stay between the suspension turrets
- The servo reservoir and the expansion tank. Lift up and place on top of the engine.
- The accessory drive belt, refer to Accessory Drive Belts removal and installation
- The upper timing belt cover

#### *To install:*

3. Install or connect the following:
- Front timing belt cover. Tighten to 106 inch lbs. (12 Nm).
- Upper timing belt cover
- Accessory drive belt
- Servo reservoir
- Expansion tank

➡**Ensure that the hoses are correctly positioned.**

- Cross stay brace and tighten the bolts at the suspension turrets to 37 ft. lbs. (50 Nm)
- Tighten the engine bracket bolt to 59 ft. lbs. (80 Nm).
- Fender liner cover

- Front wheel and tighten the lug nuts to 102 ft. lbs. (138 Nm).

#### 6-Cylinder Engines, 2006 Only

1. Before servicing the vehicle, refer to the Precautions Section.

2. Remove:
- The bolt holding the engine stabilizer brace to the bracket on the engine
- The bolts holding the engine stabilizer brace to the suspension turrets
- The engine stabilizer brace.

3. Remove the plastic pipes between the turbocharger and charge air cooler and between the air cleaner and turbocharger. Put them to one side.

4. Remove the clamp from the intake manifold for the turbocharger for cylinders 1, 2 and 3. Turn the upper section of the pipe towards the firewall.

5. Relieve the load from the belt tensioner and remove the accessory drive belt.

6. Remove the upper timing belt cover.

7. Remove front timing belt cover.

#### *To install:*

8. Install or connect the following:
- Front timing belt cover. Tighten to 106 inch lbs. (12 Nm)
- Upper timing belt cover
- Accessory drive belt
- Twist the intake pipe for the turbocharger for cylinders 1, 2 and 3 into position. Tighten the clamp.
- Install the plastic hoses between the turbocharger and charge air cooler and between the air cleaner and turbocharger.
- Tighten the hose clamps.

9. Install:
- Engine stabilizer brace and tighten the bolts at the suspension turrets to 37 ft. lbs. (50 Nm)
- Tighten the engine bracket bolt to 59 ft. lbs. (80 Nm).

## TIMING BELT AND SPROCKETS

### *REMOVAL & INSTALLATION*

#### 5-Cylinder Engines

*See Figures 114 through 117.*

1. Before servicing the vehicle, refer to the Precautions Section.

#### ✻✻ CAUTION

**Remove the ignition key.**

2. Remove or disconnect the following:
- The cross stay between the suspension turrets

- The servo reservoir and the expansion tank. Lift up and place on top of the engine.
- The accessory drive belt
- The front timing belt cover.
- The right front wheel
- The nut from the cover in the fender liner

3. Turn the crankshaft clockwise until the markings on the crankshaft and camshaft pulley correspond

4. Turn the crankshaft a further ¼ turn clockwise and then back again until the markings correspond. The markings are illustrated.

5. Remove the upper timing belt cover.

6. Back off the center bolt of the belt tensioner slightly.

7. Hold the center bolt still. Turn the tensioner eccentric clockwise, using a 6mm Allen key, to 10 o'clock.

8. Remove the timing belt from the tension pulley, camshaft pulley, and water pump.

9. Remove the crankshaft damper, refer to Crankshaft Damper removal and installation.

10. Remove the timing belt.

11. Spin the idler pulley and listen for noise. If replacing with a new idler pulley, tighten to 18 ft. lbs. (24 Nm).

12. Spin the tension pulley and listen for noise. Replace if necessary

13. Ensure that the tensioner fork is centered over the cylinder block rib.

14. Ensure that the Allen hole on the eccentric is at "10 o'clock".

#### *To install:*

15. Install the timing belt over the pulley on the crankshaft.

16. Install the crankshaft damper. Tighten the center nut. Tighten to 133 ft. lbs. (180 Nm).

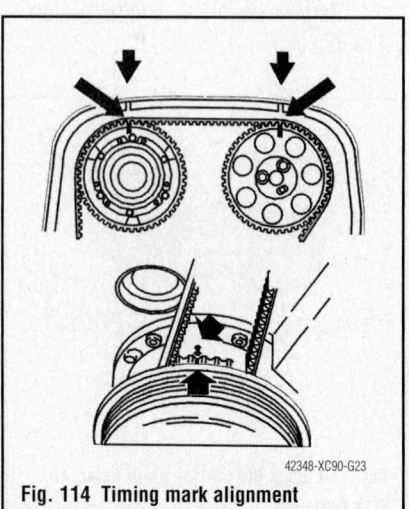

**Fig. 114 Timing mark alignment**

42348-XC90-G23

17. Remove the counterhold and install new bolts. Tighten to 18 ft. lbs. (25 Nm) and angle-tighten 30°.

18. Install or connect the following:
- Crankshaft pulley
- The idler pulley
- Intake camshaft pulley
- Exhaust camshaft pulley

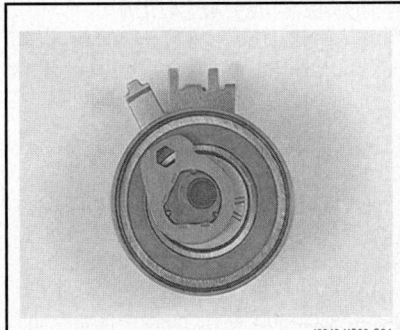

42348-XC90-G24

**Fig. 115 Belt tensioner—5-cylinder engine**

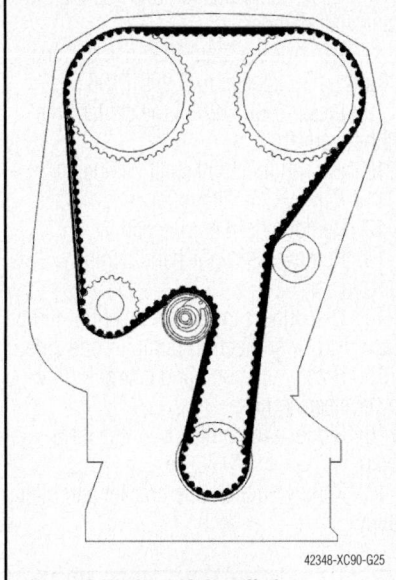

42348-XC90-G25

**Fig. 116 Timing belt installed**

- Water pump
- Belt tensioner.

➡️**The following adjustment is carried out on a cold engine. A suitable temperature is approximately 68°F. At higher temperatures, for example with the engine at operating temperature or at higher ambient temperature, the needle is further to the right. The illustration shows the position of the indicator when aligning the timing belt tensioner at different temperatures.**

19. Tension the timing belt as follows:
- Turn the crankshaft clockwise carefully until the timing belt is tensioned. The belt must be tensioned between the intake camshaft pulley, the idler pulley, and the crankshaft
- Hold the center bolt on the belt tensioner secure. Turn the belt tensioner eccentric counter-clockwise until the tensioner indicator passes the marked position. Then turn the eccentric back so that the indicator reaches the marked position in the center of the window
- Hold the eccentric secure and tighten the center bolt. Tighten to 15 ft. lbs. (20 Nm). Check that the indicator is in the correct position

20. To check the alignment:
- Press the belt to check that the indicator on the tensioner moves easily
- Install the upper timing belt cover
- Turn the crankshaft 2 turns. Check that the markings on the crankshaft and camshaft pulley correspond
- Check that the indicator on the belt tensioner is within the marked area

21. Install or connect the following:
- The front timing belt cover. Tighten to 106 inch lbs. (12 Nm)
- The upper timing belt cover
- Install the accessory drive belt

- The servo reservoir
- The expansion tank

➡️**Ensure that the hoses are correctly positioned.**

- The engine stabilizer brace. Tighten the bolts at the suspension turrets to 37 ft. lbs. (50 Nm)
- Tighten the engine bracket bolt to 59 ft. lbs. (80 Nm).
- The cover in the fender liner
- The front wheel and tighten the lug nuts to 102 ft. lbs. (138 Nm)

### 6-Cylinder Engines, 2006 Only

1. Before servicing the vehicle, refer to the Precautions Section.

**✳✳ CAUTION**

**Remove the ignition key.**

2. Remove:
- The bolt holding the engine stabilizer brace to the bracket on the engine
- The bolts holding the engine stabilizer brace to the suspension turrets
- The engine stabilizer brace

3. Remove the plastic pipes between the turbocharger and charge air cooler and between the air cleaner and turbocharger. Put them to one side.

4. Remove the clamp from the intake manifold for the turbocharger for cylinders 1, 2 and 3. Turn the upper section of the pipe towards the firewall.

5. Relieve the load from the belt tensioner. Remove the accessory drive belt.

6. Remove the upper timing belt cover.

7. Remove the front timing belt cover:
   a. The air duct, cover, surround, and control module box
   b. The hose between the thermostat housing and the expansion tank
   c. The front timing belt cover

8. Lift up the servo reservoir and place it on top of the engine.

➡️**Ensure that the oil does not leak from the ventilation hole in the filler cap.**

9. Seal the hose between the expansion tank and the radiator. Disconnect the hose at the tank.

10. Lift up the expansion tank and place it on top of engine.

11. Raise and support the vehicle.

12. Remove:
- The right front wheel
- The plastic nuts on the cover in the fender liner.

13. Install the upper timing belt cover.

-20°C
-5°F

20°C
68°F

50°C
120°F

42348-XC90-G26

**Fig. 117 Tension indicator at different temperatures**

14. Turn the crankshaft clockwise until the markings on the crankshaft and camshaft pulleys correspond.

15. Turn the crankshaft a further ¼ turn clockwise.

16. Then turn back counter-clockwise until the markings correspond.

17. Remove the upper timing belt cover.

18. Remove the 4 crankshaft damper bolts. Counterhold the crankshaft center nut.

19. Remove:
- The crankshaft damper
- The accessory drive belt
- The belt cover behind the crankshaft pulley for the accessory drive belt

20. Spray universal oil or similar around the rubber sleeve on the underside of the oil pump.

21. Remove the rubber sleeve.

22. Slacken off the center bolt for the belt tensioner slightly.

23. Hold the center bolt still. Turn the tensioner eccentric counter-clockwise using a 6mm Allen key. Turn to 10 o'clock.

24. Unhook and remove the timing belt.

### ✳✳ CAUTION

**Do not turn the camshafts or the crankshaft when the timing belt has been removed.**

25. Spin the idler pulley and listen for noise. If replacing with a new idler pulley, tighten to 17 ft. lbs. (24 Nm).

26. Spin the tension pulley and listen for noise. Replace if necessary.

### *To install:*

27. Ensure that the tensioner fork is centered over the cylinder block rib bracket.

28. Ensure that the Allen hole on the eccentric is at 10 o'clock.

29. Install the timing belt in this order:
- Crankshaft
- Idler pulley
- Intake cam
- Exhaust cam
- Water pump
- Tension pulley

➡**Belt tension adjustment is carried out on a cold engine. A suitable temperature is approximately 68°F (20°C).**

30. At higher temperatures (with the engine at operating temperature or a high outside temperature for example) the indicator is further to the right.

31. The illustration shows the position of the indicator when aligning the timing belt tensioner at different temperatures.

32. Carefully turn the crankshaft clockwise until the timing belt is tensioned. The belt must be in tension between the intake camshaft pulley, the idler pulley, and the crankshaft.

33. Hold the belt tensioner center bolt secure. Turn the belt tensioner eccentric clockwise until the tensioner indicator passes the marked position.

34. Then turn the eccentric back so that the indicator reaches the marked position in the center of the window.

35. Hold the eccentric secure. Tighten the center bolt to 18 ft. lbs. (25 Nm).

36. Check that the indicator is in the correct position.

37. Press the belt to check that the indicator on the tensioner moves easily.

38. Install the upper timing belt cover.

39. Turn the crankshaft 2 turns. Check that the markings on the crankshaft and camshaft pulley correspond.

40. Check that the indicator on the belt tensioner is within the marked area.

41. Install the accessory drive belt around the pulley on the crankshaft.

42. Install:
- The rubber sleeve on the underside of the oil pump
- The crankshaft damper. Use new bolts. Tighten to 133 ft. lbs. (180 Nm).
- The plastic nuts for the cover in the right-hand fender liner
- The right front wheel. Tighten lug nuts to 102 ft. lbs. (138 Nm)

43. Lower the car.

44. Remove the front timing belt cover.

45. Tension the accessory drive belt.

46. Install the front cover:
  a. The front timing belt cover
  b. The hose between the thermostat housing and the expansion tank
  c. The surround, cover and air ducts for the control module box

47. Install:
- The expansion tank
- The hose between the expansion tank and the radiator
- The servo reservoir
- The upper timing belt cover

48. Twist the intake pipe for the turbocharger for cylinders 1, 2 and 3 into position. Tighten the clamp.

49. Install the plastic hoses between the turbocharger and charge air cooler and between the air cleaner and turbocharger.

50. Tighten the hose clamps.

51. Install:
- The engine stabilizer brace. Tighten the bolts at the suspension turrets to 37 ft. lbs. (50 Nm)
- Tighten the engine bracket bolt to 59 ft. lbs. (80 Nm).

52. Check:
- Coolant level
- The servo fluid level

### TIMING CHAIN, SPROCKETS, FRONT COVER AND SEAL

#### *REMOVAL & INSTALLATION*

#### 8-Cylinder Engines Cover and Seal

*See Figures 118 and 119.*

1. Disconnect the battery negative cable.

2. Raise and support the vehicle safely.

3. Remove the camshaft covers, refer to Camshaft Covers removal and installation.

4. Remove the accessory drive belt, refer to Accessory Drive Belts removal and installation.

5. Remove the right front wheel.

6. Remove crankshaft damper protective cover nuts.

7. Push the protection towards the crankshaft damper.

8. Remove the crankshaft damper center bolt using special tool 999 7196.

9. Install 999 7198 in the center hole on the crankshaft.

10. Pull off the crankshaft damper.

11. Remove the alternator.

12. Remove right engine pad.

13. Remove the right-hand engine pad/bracket.

14. Disconnect the power steering pump. Place the power steering pump to one side.

15. Remove the bolt and clamp for the power steering hose.

16. Remove the pulley from the water pump.

17. Remove the engine bracket and idler pulley.

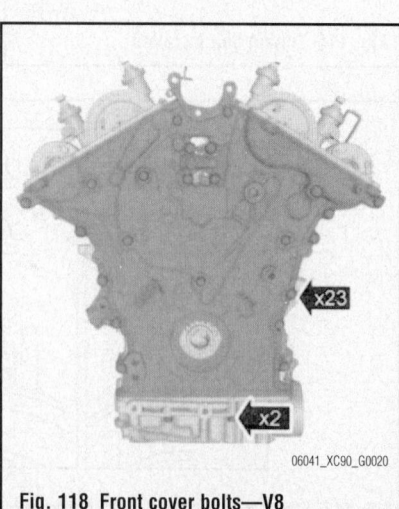

06041_XC90_G0020

**Fig. 118 Front cover bolts—V8**

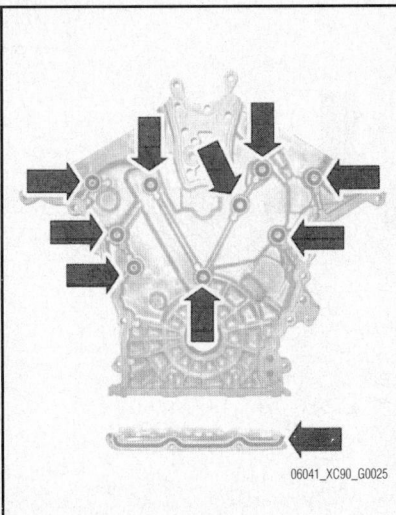

**Fig. 119 Liquid gasket application points—V8**

18. Remove the 25 bolts and the front cover. Discard the gaskets.

➡**Cover the hole in oil pan using a piece of paper.**

*To install:*

➡**The front cover must be cleaned before installing. Use a razor blade or plastic scraper.**

19. Apply liquid gasket 307 57050 as illustrated. The bead diameter must be greater than 0.039 inch (1mm).

➡**Install the front plate within 5 minutes (curing time for liquid gasket).**

20. Fold the front panel in at the lower edge.

21. Install the 23 M8 bolts for the front side and tighten to 18 ft. lbs. (24 Nm).

22. Install the 2 M8 bolts for the oil sump and tighten to 18 ft. lbs. (24 Nm).

23. Install the engine bracket and the idler pulley.

➡**The washer must be correctly installed.**

24. Install the pulley for the water pump.

25. Install the power steering pump.

26. Install the hose clamp and the bolt.

27. Install the front bolt for the alternator and tighten to 18 ft. lbs. (24 Nm).

28. Install lifting beam tool 999 5716, converted with the addition of 999 7070 and 999 5460.

29. Remove tool 999 5550.

30. Install the right-hand engine pad/bracket.

➡**The crankshaft flange must be cleaned.**

31. Clean the surface for the sealing ring on the crankshaft damper with cloth 951 1024.

➡**If necessary, replace the front crankshaft seal.**

32. Install tool 999 7196, drift: 999 7198 and press in the crankshaft damper to its limit. Position using the old center bolt.

33. Remove the center bolt.

34. Insert a new bolt for the crankshaft damper and tighten 133 ft. lbs. (180 Nm).

35. Remove tool 999 7196.

36. Install the protective cover.

37. Install the camshaft covers.

38. Install the accessory drive belt tensioner.

39. Install the right front wheel and tighten lug nuts to 102 ft. lbs. (138 Nm).

40. Start the engine and run the engine to normal operating temperature.

41. Check for leakage. Fill fluids as necessary.

## 8-Cylinder Engines Chain and Sprockets

*See Figures 120 through 125.*

1. Disconnect the battery negative cable.

**✳✳ CAUTION**
**Remove ignition key.**

2. Raise and support the vehicle safely.

3. Remove the accessory drive belt, refer to Accessory Drive Belts removal and installation.

4. Remove the right front wheel.

5. Remove crankshaft damper protective cover nuts.

6. Push the protection towards the crankshaft damper.

7. Remove the center bolt using special tool 999 7196.

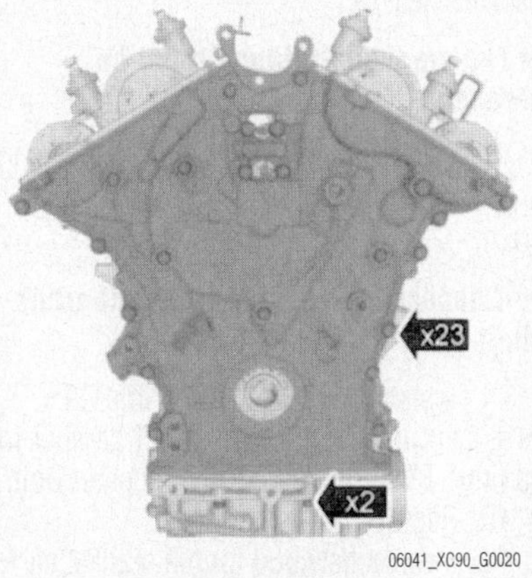

**Fig. 120 Front cover bolts—V8**

8. Install 999 7198 in the center hole of the crankshaft.

9. Pull off the crankshaft damper.

10. Remove the alternator.

11. Remove right engine pad

12. Remove the right-hand engine pad/bracket.

13. Disconnect the power steering pump. Place the power steering pump to one side.

14. Remove the bolt and clamp for the power steering hose.

15. Remove the pulley from the water pump.

16. Remove the engine bracket and idler pulley.

17. Remove the 25 bolts and the front cover. Discard the gaskets.

➡**Cover the hole in oil pan using a piece of paper.**

18. Use position setting tool 999 7235 to align the position on the crankshaft with the guide on the engine block.

19. Install position setting tool: 999 7236.

20. Remove the position setting tool: 999 7235.

21. Lock the chain tensioner using a drift.

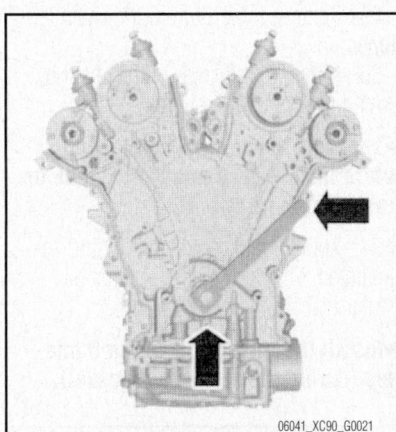

**Fig. 121 Position setting tool 999 7235—V8 engine**

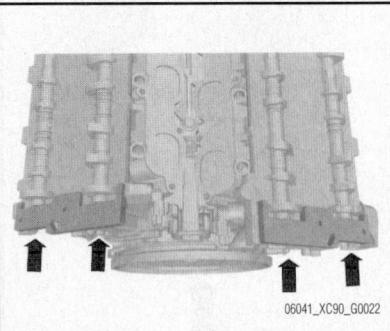

**Fig. 122 Position setting tool 999 7236—V8 engine**

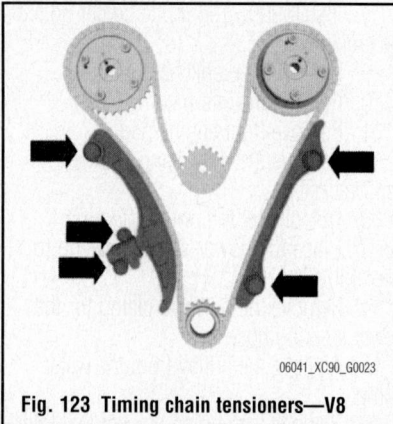

**Fig. 123 Timing chain tensioners—V8**

22. Remove the 2 M6 bolts for the chain tensioner.
23. Remove the chain tensioner.
24. Remove the 2 rails.
25. Remove the timing chain.
26. Remove worn chain sprockets.

### To install:

➥**Adjust the color marking on the timing chain to the marking on the gear.**

27. Install the timing chain, sprockets, and both rails.
28. Install the chain tensioner.
29. Remove the drift from the chain tensioner.
30. Minimize the slack of the chain by turning the crankshaft slightly.

➥**The front cover must be cleaned before installing. Use a razor blade or plastic scraper.**

31. Apply liquid gasket 307 57050 as illustrated. The bead diameter must be greater than 0.039 inch (1mm).

➥**Install the front plate within 5 minutes (curing time for liquid gasket).**

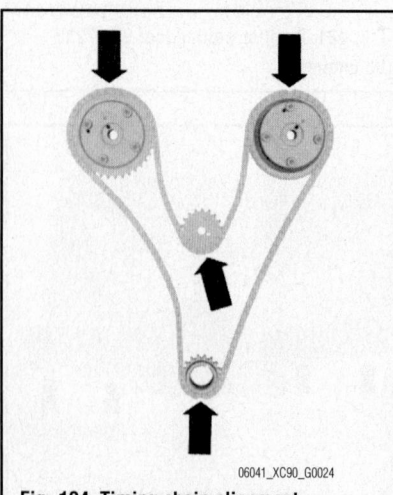

**Fig. 124 Timing chain alignment marks—V8**

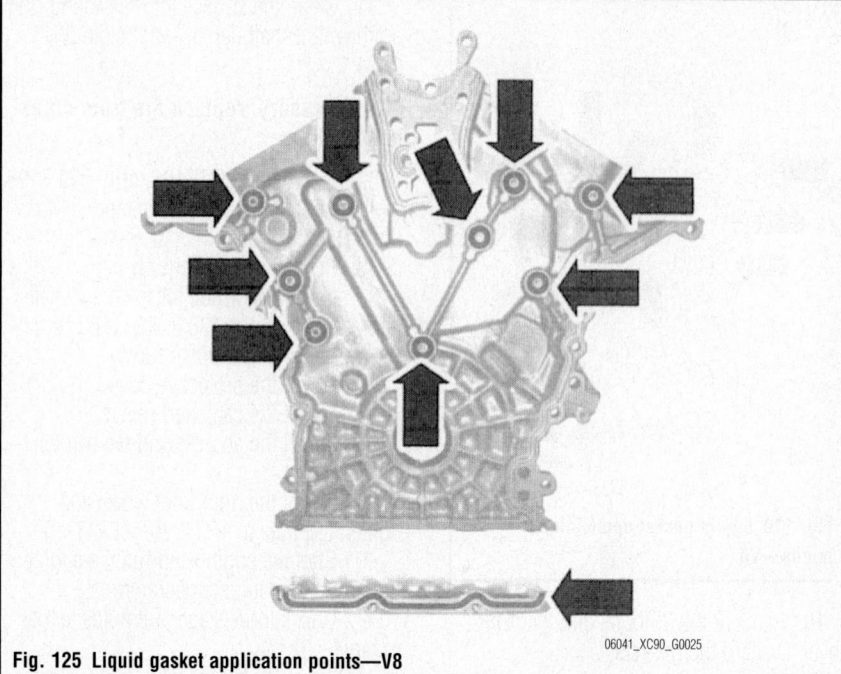

**Fig. 125 Liquid gasket application points—V8**

32. Fold the front panel in at the lower edge.
33. Install the 23 M8 bolts for the front side and tighten to 18 ft. lbs. (24 Nm).
34. Install the 2 M8 bolts for the oil sump and tighten to 18 ft. lbs. (24 Nm).
35. Install the engine bracket and the idler pulley.

➥**The washer must be correctly installed.**

36. Install the pulley for the water pump.
37. Install the power steering pump.
38. Install the hose clamp and the bolt.
39. Install the front bolt for the alternator and tighten to 18 ft. lbs. (24 Nm).
40. Install lifting beam tool 999 5716, converted with the addition of 999 7070 and 999 5460.
41. Remove tool 999 5550.
42. Install the right-hand engine pad/bracket.

➥**The crankshaft flange must be cleaned.**

43. Clean the surface for the sealing ring on the crankshaft damper with cloth 951 1024.

➥**If necessary, replace the front crankshaft seal.**

44. Install tool 999 7196, drift: 999 7198 and press in the crankshaft damper to its limit. Position using the old center bolt.
45. Remove the center bolt.
46. Insert a new bolt for the crankshaft damper and tighten 133 ft. lbs. (180 Nm).
47. Remove tool 999 7196.

48. Install the protective cover.
49. Install the camshaft covers.
50. Install the accessory drive belt tensioner.
51. Install the right front wheel and tighten lug nuts to 102 ft. lbs. (138 Nm).
52. Start the engine and run the engine to normal operating temperature.
53. Check for leakage. Fill fluids as necessary.

## TURBOCHARGER

### REMOVAL & INSTALLATION

#### 5-Cylinder Engines

*See Figure 126.*

1. Before servicing the vehicle, refer to the Precautions Section.
2. Drain the cooling system.
3. Remove or disconnect the following:
   - Negative battery cable
   - Exhaust manifold heat shield
   - Upper air charge pipe
   - Inner heat shield
   - Air intake hose
   - Turbo coolant hoses
   - Turbo oil lines
   - Exhaust front pipe and bracket
   - Red boost pressure hose
   - White bypass valve hose
   - Yellow pressure regulator hose
   - Turbocharger

### To install:

4. Install new pin bolts with thread locking compound. Tighten them to 15 ft. lbs. (20 Nm).

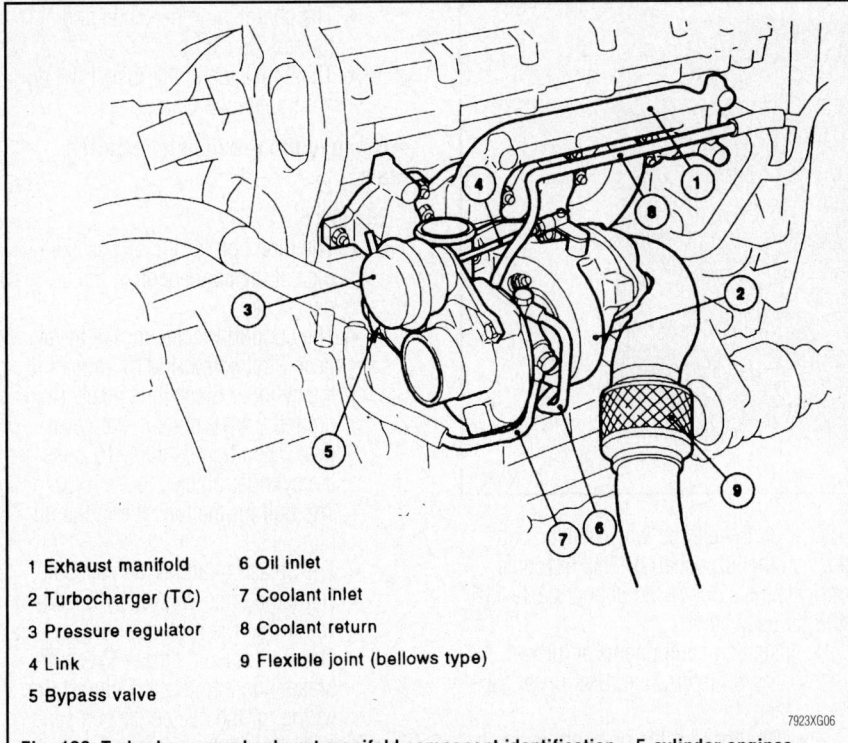

1 Exhaust manifold   6 Oil inlet
2 Turbocharger (TC)   7 Coolant inlet
3 Pressure regulator   8 Coolant return
4 Link   9 Flexible joint (bellows type)
5 Bypass valve

7923XG06

**Fig. 126 Turbocharger and exhaust manifold component identification—5-cylinder engines**

5. Install or connect the following:
   - Turbocharger and tighten the fasteners to 18 ft. lbs. (25 Nm) on models without a gasket; 27 ft. lbs. (37 Nm) on models with a gasket.
   - Yellow pressure regulator hose
   - White bypass valve hose
   - Red boost pressure hose
   - Exhaust front pipe and bracket
   - Turbo oil lines
   - Turbo coolant hoses
   - Air intake hose
   - Inner heat shield
   - Upper air charge pipe
   - Exhaust manifold heat shield
   - Negative battery cable
6. Fill the cooling system.
7. Start the engine and check for leaks.

### 6-Cylinder Engines, 2006 Only

*See Figure 127.*

➡ **The turbocharger for cylinders 1, 2 and 3 is designated turbo 1. The turbo for cylinders 4, 5 and 6 is designated turbocharger 2.**

1. Before servicing the vehicle, refer to the Precautions Section.
2. Remove:
   - The splash guard under the engine
   - The right front wheel and the mudguard at the halfshaft on the right-hand side
   - The bolt for the fuel line in the oil pan
   - The support bracket for the bevel gear in the cylinder block
   - The pipe bolt for the turbocharger oil pressure pipe by the cylinder block.
   - The connectors for the front heated oxygen sensors. Remove the wiring from the clips and clamps
   - The brake pipe from the clips on the front SIPS member
   - The front SIPS member
   - The three-way catalytic converter from the rubber mountings.
3. Connect a hose to the engine coolant drain cock.
4. Drain the coolant into a clean container. Close the cock.
5. Remove:
   - The bolt for the engine stabilizer brace mounting on the engine
   - The engine stabilizer brace between the suspension turrets
   - The plastic charge air pipe. Place to one side
   - The plastic intake pipe. Put it to one side.
6. Remove from turbo 1:
   - The intake and charge air pipes. Seal the openings
   - The heated oxygen sensor.

### ❊❊ CAUTION

**Note the location of the heated oxygen sensor**

   - The heat deflector plate
   - The 3 nuts for the three-way catalytic converter

➡**use rust solvent**

   - The pipe bolt for the oil pressure pipe
   - The oil pressure pipe
   - The pipe bolts for the coolant pipe. Move the upper coolant pipe and hose to one side

➡**Position a container underneath for the coolant.**

   - clamp the control pipe for the pressure regulator.
7. Remove from turbo 2:
   - The heated oxygen sensor
   - The heat deflector plate
   - The 3 nuts for the three-way catalytic converter

➡**Use rust solvent**

   - The upper bolt for the automatic transaxle dip stick pipe
   - The intake and charge air pipes. Seal the openings
   - The hose from the pressure regulator
   - The clamps between the coolant pump intake pipe and the turbocharger oil pressure pipe.
8. Raise and support the vehicle.
9. Remove:
   - The oil pressure pipe for turbo 1
   - The hose from the pressure regulator.
10. Lower the car.
11. Remove:
   - The pipe bolts for the coolant pipe at turbo 2. Place the lower pipe to one side
   - move the upper coolant pipe and hose from turbo 2 to one side
   - The air control pipe for the pressure regulators, 2 bolts
   - The nuts and washers for the manifold mounting at cylinders 1, 2 and 3.

➡**Use rust solvent or similar. Lift off the manifold/turbocharger.**

12. Remove from turbo 2:
   - The bolts for the oil return line
   - The pipe bolt for the oil pressure pipe
   - The nuts and washers where the manifold is connected to the cylinder head for cylinders 4, 5 and 6. Carefully lift off the turbocharger and manifold so that the oil pressure pipe is exposed

Fig. 127 Turbocharger mounting—6-cylinder engine

42348-XC90-G13

➡ **Do not damage the oil pressure pipe. Lift up and remove the turbocharger and manifold.**

*To install:*

13. Remove the old gaskets
14. Clean the gasket faces thoroughly
15. Ensure that all the studs are tightened in the cylinder head. Tighten to 15 ft. lbs. (20 Nm).
16. Installing the manifold and turbo 2
    - Install new gaskets
    - Lubricate the studs at the exhaust ports and at the flange on the turbocharger flange to the three-way catalytic converter using paste P/N 1161408, or equivalent.
    - Lower and install the exhaust manifold and turbocharger.
17. Install:
    - The washers and new nuts that hold the manifold onto the cylinder head.
    - The pipe bolt holding the oil pressure pipe to the turbocharger. Use new seal washers. Do not tighten
    - The bolts for the oil return line.
    - lower coolant pipe. Use new gaskets.
18. Installing the manifold and turbo 1
    - Install new gaskets
    - Lubricate the studs at the exhaust ports and at the flange on the turbocharger flange to the three-way catalytic converter using paste P/N 1161408, or equivalent.
19. Install:
    - The washers and new nuts that hold the manifold onto the cylinder head.
    - The pipe for pressure regulator control
    - The upper coolant pipe and hose to turbo 2. Use new seal washers.
    - The upper coolant pipe and hose to turbo 1. Use new seal washers.

20. Install the three-way catalytic converter on the turbocharger. Use new nuts. Lubricate the studs using copper paste 116 1408, or equivalent.
21. Installing components at turbo 1
    - The oil return line. Use a new gasket
    - The hose for the pressure regulator
    - The oil pressure pipe
    - The pipe bolt for the connection between the oil pressure pipes and the cylinder block. Use new seal washers. Do not tighten.
22. Installing components at turbo 2
23. Lower the car. Tighten the pipe bolt for the oil pressure pipe.
    - The charge air pipe. Align and tighten the clamp
    - The intake manifold. Align and tighten the clamp

➡ **Remove the previously installed seals**

- The upper bolt securing the automatic transaxle dip stick pipe to the engine. Tighten the lower bolt
- The heat deflector plate above the turbocharger
- The heated oxygen sensor
- The clamps between the coolant pump intake pipe and the turbocharger oil pressure pipe
24. Installing components at turbo 1
    - The pipe bolt for the oil pressure pipe. Use new seal washers
    - The heat deflector plate above the turbocharger
    - The heated oxygen sensor in its original position

❋❋ **CAUTION**

**Ensure that the probes are in the same position to prevent diagnostic trouble codes (DTCs) being stored**

- The charge air pipe. Align and tighten the clamp
- The intake manifold. Install but do not tighten the clamp

➡ **Remove the previously installed seals.**

25. Install:
    - The pipe bolt for the oil pressure pipe at the connection in the cylinder block
    - The support bracket for the bevel gear. First screw 4 M10 bolts in to the cylinder block to contact. Then tighten 2 M10 bolts in the bevel gear. Then tighten the M10 bolts in the cylinder block.
    - The bolt for the fuel line in the oil pan
    - The splash guard at the halfshaft
    - Hook the three-way catalytic converter into the rubber mounting
    - The front SIPS member. Press the brake pipe into place. Connect the wiring for the heated oxygen sensors. Secure the wiring as before
    - The right front wheel
    - The plastic intake pipe
    - The plastic charge air pipe
    - The vacuum hoses for the purge valve and turbocharger control valve
    - The engine stabilizer brace
26. Fill with coolant.
27. Warm up the engine until the thermostat opens.
28. Check:
    - The engine for leaks
    - The coolant level
29. Install the lower engine cover.

## VALVE LASH

*ADJUSTMENT*

### 2006–07 5-Cylinder Engine and 2006 6-Cylinder Engine

*See Figures 128 through 135.*

1. Remove the cable from the battery negative terminal.
2. Remove:
    - The cross stay between the suspension turrets
    - The ground strip from the cylinder head
    - The upper engine stabilizer brace
    - The cover in the cylinder head at the rear of the exhaust camshaft
    - The crankcase ventilation hose from the top of the camshaft cover
    - The radiator breather tube from the expansion tank. Install lock grip pliers.

3. Lift up the brake fluid reservoir.

4. Disconnect the ABS sensor connector.

5. Place the brake fluid reservoir over the engine.

### ✳✳ WARNING

**Ensure that no fluid is spilled on the engine. It is extremely flammable!**

6. Disconnect the connector for the level sensor in the expansion tank.

7. Lift up and place the expansion tank on top of the engine.

8. Remove:
- The accessory drive belt
- The front timing cover.

9. Position the upper timing cover.

10. Turn the crankshaft clockwise until the markings on the crankshaft belt pulley and the timing belt pulley are aligned with the markings on the oil pump and the upper timing cover.

11. Remove the upper timing cover.

### ✳✳ CAUTION

**Crankshaft and camshafts must not be turned more than is stated in the method description. If the shafts are turned in any other way the valves may be damaged.**

12. Slacken off the center bolt for the belt tensioner slightly.

13. Hold the center bolt still. Turn the tensioner eccentric clockwise to 10 o'clock using a 6mm Allen key.

14. Remove the timing belt from the camshaft pulleys.

15. Install camshaft adjustment tool 999 5452 at the rear of the camshafts.

16. Check that the bolts securing the adjustment tool to the camshafts and the bolt holding the tool together are well tightened

17. Remove: (timing gear pulley with variable valve timing unit)
- The plug at the front of the variable valve timing unit
- The center bolt in the variable valve timing unit. Carefully pull out the variable valve timing unit with the timing gear pulley.

18. Remove: (timing gear pulley without variable valve timing unit)
- The bolts securing the timing gear pulley on the camshaft
- The timing gear pulley.

19. Remove tool 999 5452.

20. Reinstall the expansion tank and the brake fluid reservoir at the fender liner.

21. Remove:
- The variable valve timing (VVT) solenoid
- spark plugs for cylinders 1 and 5.

22. Install 2 tools 999 5454. Leave a 0.079–0.118 inch (2–3mm) gap to the camshaft cover.

23. Ensure that the bolt in the spark plug well is fully tightened.

24. Remove all the bolts securing the camshaft cover to the cylinder head.

25. Use pliers 999 5670 to lift the cover from the cylinder head.

26. Install the pliers at the stop lugs. Start with cylinder 1 and work alternately backward.

27. Slacken off the wing nuts approximately 2 turns. Repeat the procedure with the pliers.

28. Carefully press the camshaft seals free.

➡ **Take care not to damage the sealing surfaces on the camshafts.**

29. Remove:
- Tool 999 5454
- The camshaft cover
- The camshafts.

30. Lift out the valve lifters. Mark up the valve lifters so that the original positions can be established.

31. Use a razor blade or a gasket scraper and gasket solvent on the camshaft cover.

### ✳✳ WARNING

**Use a fume hood or extractor when using gasket solvent!**

32. Use only a gasket scraper or razor blade on the cylinder head.

### ✳✳ CAUTION

**Take great care around the oil ducts for the variable valve timing solenoid. This applies to both the camshaft cover and the cylinder head. The solenoid is extremely sensitive to contaminants.**

33. Dry and blow all surfaces clean.

34. Carefully tap the end of the valve stem to ensure that the valve is correctly located in the seat.

35. Use a plastic, aluminum or brass drift to protect the valve and the surface of the valve lifter.

36. The sound made by tapping reveals if the valve is correctly seated.

37. Install both the valve lifters for the inlet valves at cylinder 1.

38. Check the notes made earlier. Select new valve lifters if necessary.

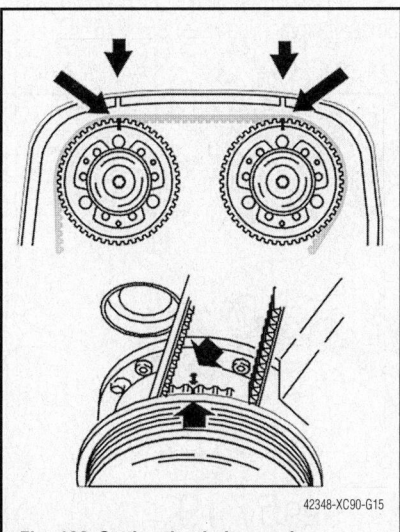

**Fig. 128 Setting the timing marks—
5 & 6-cylinder engines**

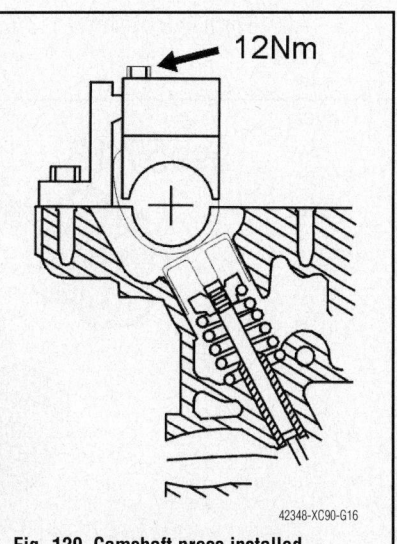

**Fig. 129 Camshaft press installed—
5 & 6-cylinder engines**

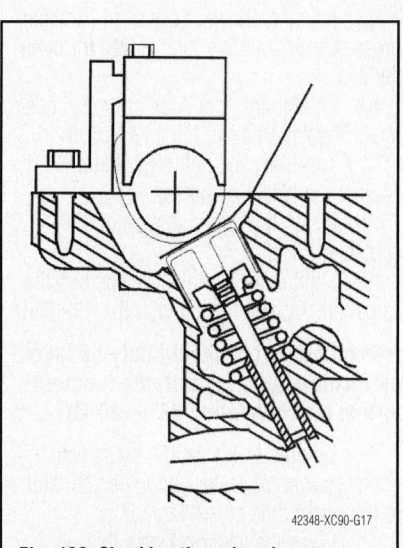

**Fig. 130 Checking the valve clearance—
5 & 6-cylinder engines**

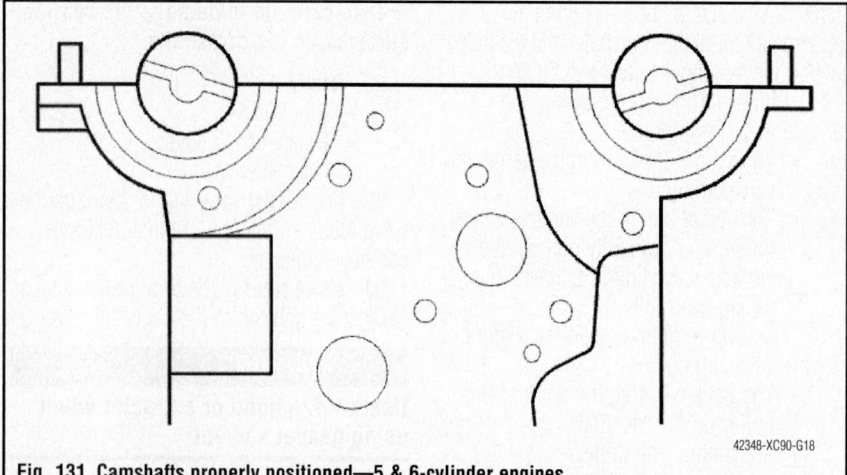

Fig. 131 Camshafts properly positioned—5 & 6-cylinder engines

➡ **Only install two valve lifters. The valve lifters should be placed at the same cylinder!**

39. Other valve lifters are available as replacement part / replacement part kits.

40. The valve clearance on a cold engine (approximately 20°C) should be:
- Inlet valve: 0.008 plus or minus 0.0010 inch (0.20 plus or minus 0.03mm).
- Exhaust valve: 0.016 plus or minus 0.0010 inch (0.40 plus or minus 0.03mm).

➡ **The tolerances are less at setting. When checking the valve clearance through the plug hole the tolerances are larger.**

41. Position the intake camshaft. Ensure that the lobes at cylinder 1 point upwards.

42. Apply a little oil to the cam lobe and the upper side of the valve lifter to facilitate later measurement.

43. Install the lower section of camshaft press 999 5765 at the inlet valves for cylinder 1.

44. Tighten the tool against the cylinder head. Tighten to 13 ft. lbs. (17 Nm).

45. Turn the camshaft until it stops against the camshaft press.

46. Install the upper section of the camshaft press.

47. Tighten the bolt which tensions the camshaft. Tighten to 106 inch lbs. (12 Nm).

➡ **Measurements should only be taken on a cold engine. A suitable temperature is approximately 68°F (20°C).**

48. Using a feeler gauge, press with a finger so that the feeler gauge lies parallel to the upper side of the valve lifter.

49. Move the feeler gauge sideways when taking the reading in order to obtain as accurate a measurement as possible.

50. The valve clearance measured on cold engines (approximately 68°F (20°C) should be:
- Intake valve: 0.008 inch plus or minus 0.001 inch (0.20mm plus or minus 0.03mm).
- Exhaust valve: 0.016 inch plus or minus 0.001 inch (0.30mm plus or minus 0.03mm).

➡ **The tolerances are less at setting! When checking the valve clearance through the plug hole the tolerances are larger.**

51. Differences in valve clearance for different engines/ambient temperatures:
- -0.0004 inch (- 0.01mm) at 59°F (15°C)
- +0.0004 inch (+ 0.01mm) at 77°F (25°C)
- +0.0008 inch (+ 0.02mm) at 86°F (30°C)
- +0.0012 inch (+ 0.03mm) at 95°F (35°C)
- +0.0016 inch (+ 0.04mm) at 113°F (45°C)

52. Correcting measured clearance:
   a. Lift out the upper section of the press tool.
   b. Lift out the camshaft.
   c. Adjust the play by replacing the valve lifters.
   d. Other valve lifters are available as replacement part / replacement part kits.
   e. Reinstall the camshaft and the upper section of the press tool. Tighten to 106 inch lbs. (12 Nm).
   f. Take a new measurement.

53. When the correct valve clearance is reached, remove:
- The press tool 999 5765
- The camshaft
- The valve lifters.

54. Carefully mark the valve lifters so that exact reinstallation can be carried out.
   For example:
   Intake side: I1, I2, I3......I10.
   Exhaust side: A1, A2, A3......A10.

55. Repeat the procedure for measuring the valve clearance for all cylinders on both the intake and exhaust sides.

56. Lubricate the valve guide wells.

57. Install all the valve lifters.

58. Lubricate the camshaft bearing seats and the upper sides of the valve lifters.

59. Position the intake camshaft. Ensure that the groove at the rear edge of the camshaft is above an imaginary center line.

60. Position the exhaust camshaft. Ensure that the groove at the rear edge of the camshaft is below an imaginary center line.

61. Wipe the oil film off the mating surfaces on the camshaft cover and cylinder head.

62. Install new O-rings around the spark plug wells at the cylinder head.

63. Apply liquid gasket 1161 059 on the camshaft cover. Use roller 951 2767.

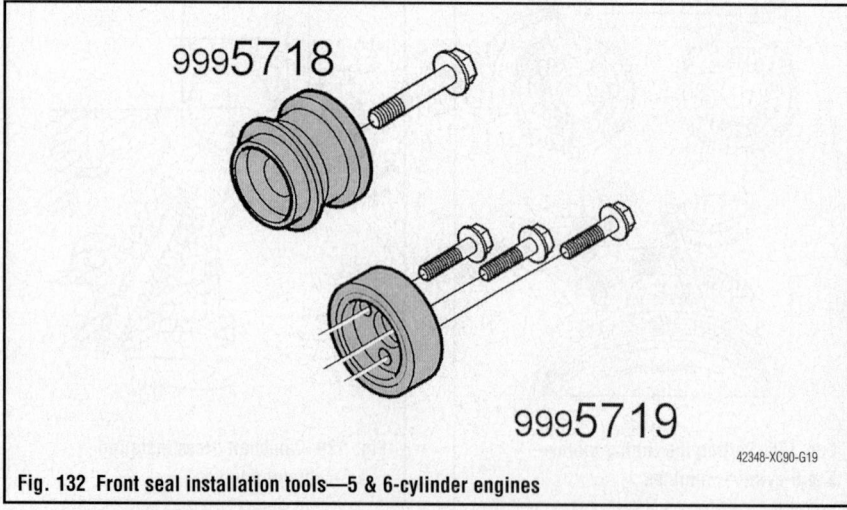

Fig. 132 Front seal installation tools—5 & 6-cylinder engines

➡The surface must be completely covered without any excess.

❉❉ **CAUTION**

Ensure that no liquid gasket gets in to the oil ducts.

64. Lubricate the camshaft lobes, the camshaft bearing surfaces and the valve lifters.
65. Install the camshaft cover.
66. Install press tool 999 5454 (2x).
67. Tighten the camshaft cover bolts alternately, keeping it parallel to the cylinder head using the press tools.
68. Install all the bolts. Tighten the bolts from the middle and outwards.
69. Remove the press tool 999 5454
70. Install:
   - The variable valve timing (VVT) solenoid. Use a new gasket
   - The spark plugs. Tighten to 22 ft. lbs. (30 Nm)
   - The plugs for the test holes. Tighten to 15 ft. lbs. (20 Nm)
   - The crankcase ventilation hose to the top of the camshaft cover
   - The ignition coils according to the earlier marking
   - The ground terminals between the ignition coils.
71. To clean the shaft journal and mating surface, use emery cloth.

➡When cleaning work around the shaft journal, not in and out. It is essential that any residue from the emery cloth and any other contaminants are completely removed before the new sealing ring is installed.

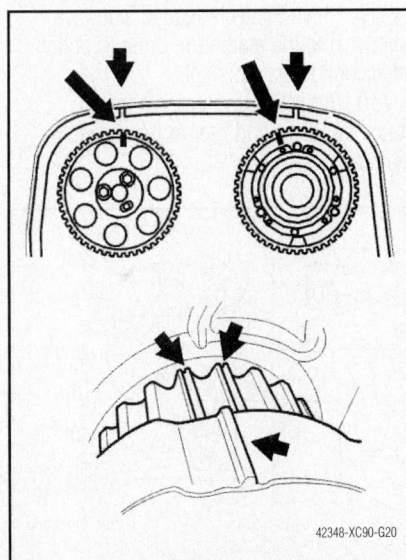

Fig. 133 Timing marks, with variable timing units—5 & 6-cylinder engines

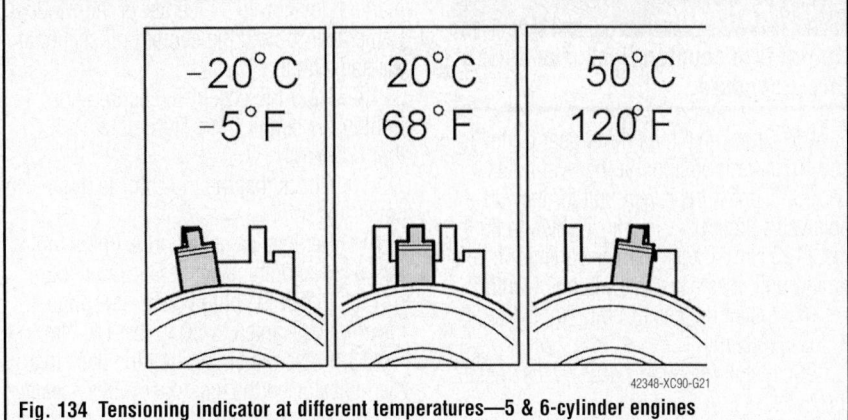

Fig. 134 Tensioning indicator at different temperatures—5 & 6-cylinder engines

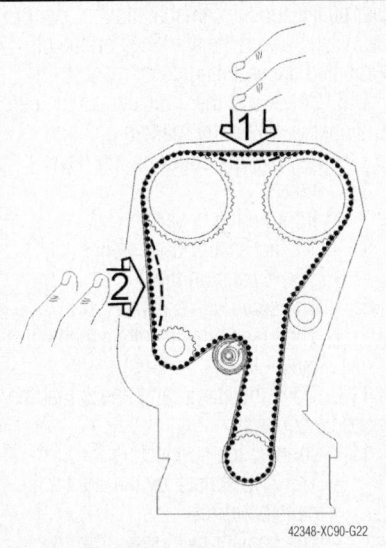

Fig. 135 Checking belt tension—5 & 6-cylinder engines

72. Use drift 999 5450.
73. Lubricate the surface of the seal that the camshaft rotates against.
74. Press in the seal until the drift bottoms out.

➡If there are wear grooves on the camshaft, the seal can be pressed in a further 0.079 inch (2mm) by reversing the sleeve.

75. Install camshaft adjustment tool 999 5452 at the rear of the camshafts.
76. Check that the bolts securing the adjustment tool to the
77. Lift up and position the brake fluid reservoir and the expansion tank on top of the engine.
78. Use drift 999 5718 on camshafts with variable valve timing units. Use drift 999 5719 on camshafts without variable valve timing.
79. Use new seals and lubricate the surface of the seal that the camshaft rotates against.

80. Use the variable valve timing unit/timing belt pulley mounting bolts. Tighten the bolts until the drift bottoms out.
81. Remove the drift.
82. Remove:
   - The mounting bolts for the starter motor. Pull off the starter motor. Place the starter motor to one side
   - The blind cover plug and the seal washer.
83. Turn the crankshaft slightly clockwise.
84. Install the crankshaft stop 999 5451 Ensure that it bottoms out against the cylinder block.
85. Turn the crankshaft counter-clockwise until it stops against the crankshaft stop.
86. Check that the marking on the crankshaft timing gear pulley corresponds with the marking on the oil pump.

➡The purpose of the section is to ensure that the VVT unit is correctly positioned and to reset the camshaft timing gear pulley to the correct position using the markings made at the factory. This is to ensure that the conditions are correct for any later fault-tracing.

87. Slacken off, but do not remove the bolts which secure the timing gear pulley to the variable valve timing unit.
88. Press the variable valve timing unit/timing gear onto the camshaft.
89. Install the center bolt which secures the variable valve timing unit to the camshaft. Tighten slightly.
90. Turn the variable valve timing unit counter-clockwise as far as it will go.
91. Remove the center bolt.
92. Position the upper timing cover.
93. Turn the timing gear pulley clockwise until the bolts at the oval holes are in the limit position.
94. Continue turning clockwise until the timing gear pulley marking is 1 tooth before the marking on the upper timing cover.

95. Check that the timing gear pulley is still in its limit position at the oval holes.

96. Tighten the center bolt for the variable valve timing (VVT) unit. Tighten to 89 ft. lbs. (120 Nm). Check that the variable valve timing unit does not rotate when tightening.

97. Install the center bolt. Tighten to 26 ft. lbs. (35 Nm).

98. Install the timing gear pulley. Install the bolts.

99. Install two bolts without tightening. Allow the third bolt to protrude.

100. Adjust the timing gear pulleys so that the markings on the timing gear pulleys / upper timing cover correspond.

101. Tighten the center bolt on the belt tensioner. Tighten to 44 inch lbs. (5 Nm).

102. Turn the variable valve timing unit clockwise to the stop.

103. Hold it secure in the limit position.

104. Install the belt in the following order:
- Crankshaft
- The idler pulley
- Intake cam
- Exhaust cam
- Water pump
- Belt tensioner

➡Adjust the timing gear pulleys so that the bolts are not at a limit position in the oval holes. Also check that the markings correspond.

➡This adjustment is always carried out on a cold engine. A suitable temperature is approximately 68°F (20°C).

At higher temperatures (with the engine at operating temperature or a high outside temperature for example) the indicator is further to the right.

The illustration shows the position of the indicator when aligning the timing belt tensioner at different temperatures.

105. Hold the center bolt secure and turn the belt tensioner eccentric counter-clockwise until the tensioner indicator passes the marked position.

➡Check that the variable valve timing unit is in its limit position.

106. Tighten the three bolts at the intake camshaft timing gear pulley. Tighten to 88 inch lbs. (10 Nm).

107. Tighten the three bolts at the exhaust camshaft timing gear pulley. Tighten to 15 ft. lbs. (20 Nm).

108. Turn the eccentric on the belt tensioner back so that the indicator reaches the marked position in the center of the window. Remember to hold the center bolt secure at the same time.

109. Hold the eccentric secure and tighten the center bolt. Tighten to 15 ft. lbs. (20 Nm).

110. Check that the indicator is in the correct position.

111. Remove camshaft adjustment tool 999 5451 and crankshaft stop 999 5452.

112. Install the plug with a new blind cover plug. Tighten to 30 ft. lbs. (40 Nm).

113. Press the timing belt to check that the indicator on the tensioner moves easily.

114. Turn the crankshaft two turns. Check that the markings on the crankshaft timing gear pulley and the camshaft timing gear pulley match up with the markings on the oil pump and upper timing cover respectively.

115. Check that the indicator on the belt tensioner is within the marked position.

116. Remove the upper timing cover.

117. Install:
- front timing cover
- The accessory drive belt
- The expansion tank
- The servo oil reservoir
- The bleed hose for the expansion tank

118. Close the clamp and check that the hoses lie correctly.

119. Connect the connectors for the:
- The ABS sensor by the right suspension turret
- The coolant level sensor in the expansion tank

120. Install:
- The starter motor
- The plastic nuts for the cover in the fender liner
- The right front wheel

121. Install:
- The trigger wheel.

➡Ensure that the trigger wheel is correctly positioned against the camshaft.

- The camshaft position sensor housing
- The cover at the rear of the exhaust camshaft
- The bolts holding both the ground strips from the firewall to the cylinder head
- The upper engine stabilizer brace
- The cover over the ignition coils
- The upper timing cover
- The crankcase ventilation hose on the inlet hose
- The inlet hose between the air cleaner and throttle body
- The engine stabilizer brace between the suspension turrets. Secure the

servo hose at the right mounting for the engine stabilizer brace in the bodywork
- The cable to the battery negative terminal.

122. Check the level in the expansion tank and the brake fluid reservoir.

123. Test drive the engine until the thermostat opens and check for any leakage.

### 8-Cylinder Engines

*See Figure 136.*

1. Before servicing the vehicle, refer to the Precautions Section.

2. Remove intake manifold. Refer to Intake Manifold removal and installation.

3. Remove air induction pipe.

4. Remove ignition coils on the left and right banks, refer to Ignition Coil Pack removal and installation.

5. Remove camshaft covers, refer to Camshaft Covers removal and installation.

➡**Measurements should only be taken on a cold engine. A suitable temperature is approximately 68°F (20°C).**

6. Turn the crankshaft until the cam lobe is 180° to the valve lifter.

7. Measure the valve clearance between the camshaft and the valve lifter using a feeler gauge 999 5752.

8. Using a feeler gauge 999 5752, press with a finger so that the feeler gauge lies parallel to the upper side of the valve lifter.

9. Move the feeler gauge sideways when taking the reading in order to obtain an accurate measurement.

10. Repeat measurements for all camshaft lobes and record the measurements.

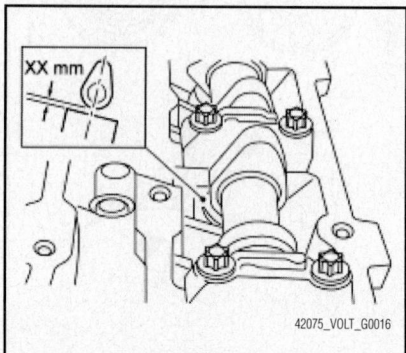

42075_VOLT_G0016

**Fig. 136 Using a feeler gauge to measure valve clearance**

11. Valve clearance readings on a cold engine, approximately 68°F (20°C):
- Intake valve: 0.008 inch plus or minus 0.001 inch (0.20mm plus or minus 0.03mm).
- Exhaust valve: 0.016 inch plus or minus 0.001 inch (0.30mm plus or minus 0.03mm).

12. If adjustments are necessary, remove the valve tappet:

a. Read off the thickness from the underside of tappet.

b. Calculate the required thickness of the valve lifter to be used.

➡ **The thickness of the existing valve lifter plus the measured valve clearance equals the required value of new tappet minus the cold engine tolerance.**

c. Install the correct valve tappet and measure the clearance again.

13. Install camshaft covers, refer to Camshaft Covers removal and installation.

14. Install intake manifold, refer to Intake Manifold removal and installation.

15. Test drive the vehicle until the thermostat opens and check for any leakage.

# ENGINE PERFORMANCE & EMISSION CONTROL

## ENGINE COOLANT TEMPERATURE (ECT) SENSOR

### REMOVAL & INSTALLATION
*See Figure 137.*

### ✳✳ CAUTION

**Never open, service or drain the radiator or cooling system when hot; serious burns can occur from the steam and hot coolant. Also, when draining engine coolant, keep in mind that cats and dogs are attracted to ethylene glycol antifreeze and could drink any that is left in an uncovered container or in puddles on the ground. This will prove fatal in sufficient quantities. Always drain coolant into a sealable container.**

Coolant should be reused unless it is contaminated or is several years old.

1. Drain the engine coolant.
2. Disconnect the ground cable of battery.
3. Remove the electrical connector from the sensor.
4. Remove the coolant temperature sensor.

### To install:
5. Apply sealant to sensor threads. Install the sensor and tighten to 15–29 ft. lbs. (20–39 Nm).
6. Attach electrical connector to sensor.
7. Connect the ground cable of battery.
8. Refill the coolant.

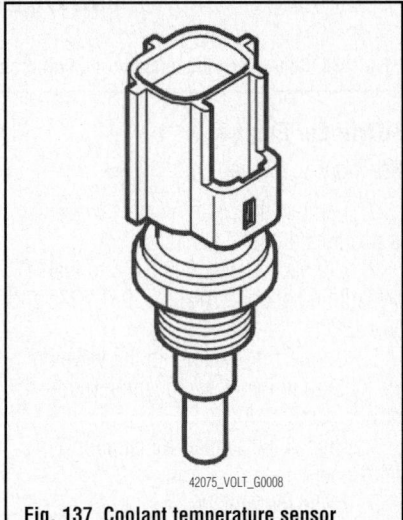

42075_VOLT_G0008

**Fig. 137 Coolant temperature sensor illustrated**

# FUEL                                                GASOLINE FUEL INJECTION SYSTEM

## FUEL SYSTEM SERVICE PRECAUTIONS

Safety is the most important factor when performing not only fuel system maintenance but any type of maintenance. Failure to conduct maintenance and repairs in a safe manner may result in serious personal injury or death. Maintenance and testing of the vehicle's fuel system components can be accomplished safely and effectively by adhering to the following rules and guidelines.

- To avoid the possibility of fire and personal injury, always disconnect the negative battery cable unless the repair or test procedure requires that battery voltage be applied.

- Always relieve the fuel system pressure prior to disconnecting any fuel system component (injector, fuel rail, pressure regulator, etc.), fitting or fuel line connection. Exercise extreme caution whenever relieving fuel system pressure to avoid exposing skin, face and eyes to fuel spray. Please be

advised that fuel under pressure may penetrate the skin or any part of the body that it contacts.

- Always place a shop towel or cloth around the fitting or connection prior to loosening to absorb any excess fuel due to spillage. Ensure that all fuel spillage (should it occur) is quickly removed from engine surfaces. Ensure that all fuel soaked cloths or towels are deposited into a suitable waste container.

- Always keep a dry chemical (Class B) fire extinguisher near the work area.

- Do not allow fuel spray or fuel vapors to come into contact with a spark or open flame.

- Always use a back-up wrench when loosening and tightening fuel line connection fittings. This will prevent unnecessary stress and torsion to fuel line piping.

- Always replace worn fuel fitting O-rings with new Do not substitute fuel hose or equivalent where fuel pipe is installed.

Before servicing the vehicle, make sure to also refer to the precautions in the beginning of this section as well.

## RELIEVING FUEL SYSTEM PRESSURE

### 2006–07 5-Cylinder Engine and 2006 6-Cylinder Engine
*See Figure 138.*

1. Before servicing the vehicle, refer to the Precautions Section.
2. Disconnect the negative battery cable.
3. Remove protective cap from the valve on the fuel rail.
4. Connect a hose to adapter 999-5484 and place the other end in a clean container.
5. Connect the adapter in the locked or closed position to the valve on the fuel rail.
6. Unlock or open the adapter valve.
7. After the fuel system pressure is relieved, remove the adapter and hose and replace the protective cap.
8. Connect the negative battery cable when repairs are complete.

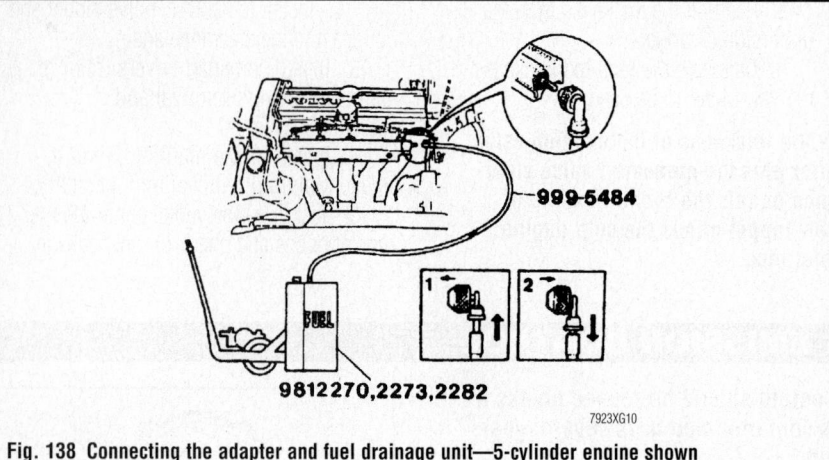

Fig. 138 Connecting the adapter and fuel drainage unit—5-cylinder engine shown

### 8-Cylinder Engines

*See Figures 139 and 140.*

1. Remove the cover on the valve which is positioned on the fuel rail.

2. Connect adapter 999 5484 to fuel pressure relief unit 981 2270, 981 2273 and 981 2282.

3. Connect the adapter to the valve on the fuel rail in the locked position (figure 1, valve closed).

4. Unlock the adapter (illustration 2, valve open).

5. Raise the vehicle.

6. Remove the cover on the valve cap positioned at the fuel filter.

7. Connect venting hose 999 5480 to the valve prior to the fuel filter.

➡ It takes approximately 2 minutes to drain the system.

➡ Do not forget to reinstall the covers over the valves.

8. Reinstall the components in reverse order.

## FUEL FILTER

### REMOVAL & INSTALLATION

*See Figure 141.*

➡ The fuel filter is either on the left side of the firewall or next to the fuel pump near the left side of the fuel tank.

1. Before servicing the vehicle, refer to the Precautions Section.

2. Relieve the fuel system pressure.

3. Remove or disconnect the following:

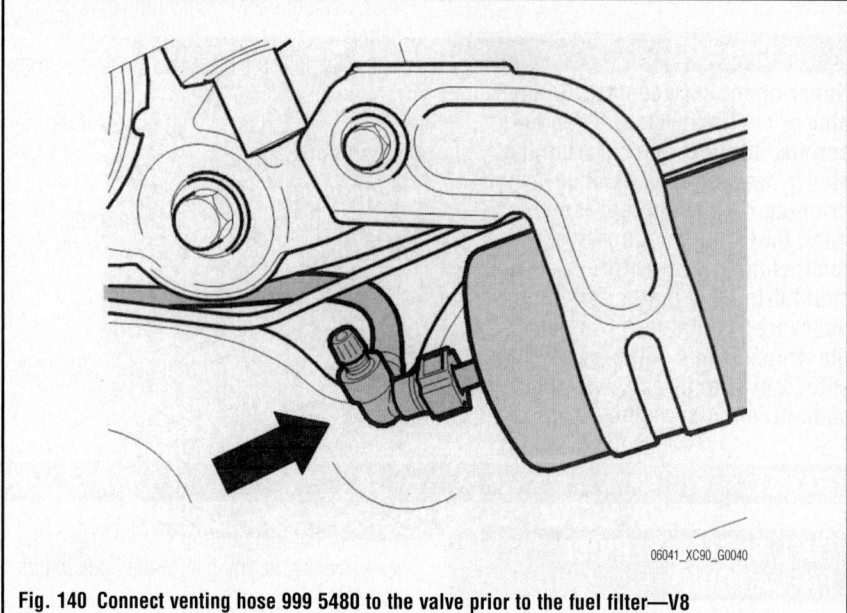

Fig. 140 Connect venting hose 999 5480 to the valve prior to the fuel filter—V8

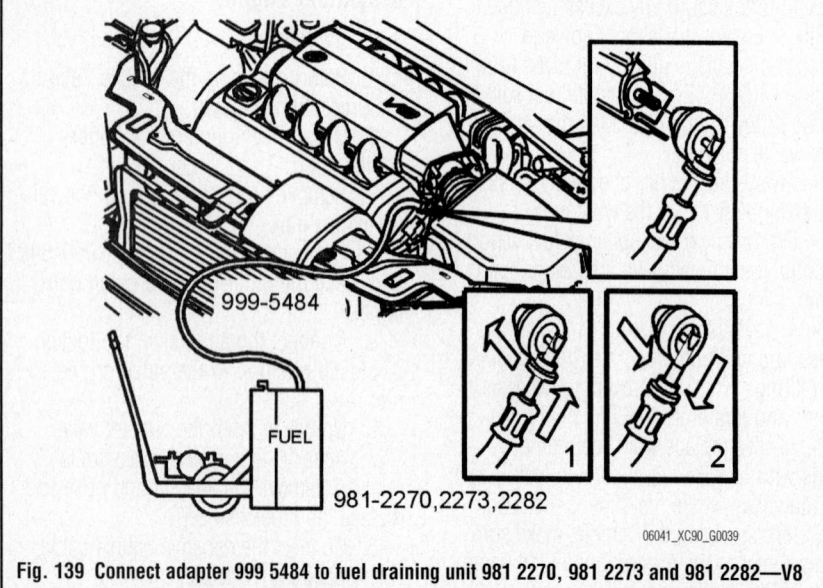

Fig. 139 Connect adapter 999 5484 to fuel draining unit 981 2270, 981 2273 and 981 2282—V8

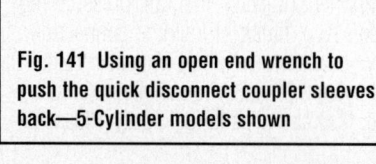

Fig. 141 Using an open end wrench to push the quick disconnect coupler sleeves back—5-Cylinder models shown

- Negative battery cable
- Fuel filler cap
- Fuel lines from the fuel filter
- Fuel filter from the bracket

### To install:

4. Install or connect the following:
- Fuel filter to the bracket
- Fuel lines to the fuel filter
- Fuel filler cap
- Negative battery cable

5. Start the engine and check for leaks.

## FUEL INJECTORS

### REMOVAL & INSTALLATION

#### 5-Cylinder Engines

*See Figure 142.*

1. Before servicing the vehicle, refer to the Precautions Section.

2. Remove or disconnect the following:
- Negative battery cable
- Injector cover
- Accelerator cable

3. Install fuel injector holders on the injectors.
- Fuel pressure regulator vacuum hose
- Fuel supply manifold with injectors attached

4. Remove the injector holders and separate the injectors from the fuel supply manifold.

### To install:

5. Install or connect the following:
- Injectors to the fuel supply manifold with new O-ring seals
- Injector holders
- Fuel supply manifold with injectors attached. Remove the injector holders.
- Fuel pressure regulator vacuum hose

- Accelerator cable
- Injector cover
- Negative battery cable

6. Start the engine and check for leaks.

#### 6-Cylinder Engines, 2006 Only

1. Drain the fuel injection system.

2. Remove the plastic hoses between the turbocharger and charge air cooler and the air cleaner and turbocharger. Place them to one side and seal the openings.

3. Remove the protective cover above the nozzles

4. Disconnect the connectors from the injectors

5. Remove the mounting bolts from the fuel rail

6. Spray universal oil or similar around the injector terminals on the intake manifold.

7. Gently work the fuel rail and injectors loose.

➡**Handle the injectors carefully to avoid damage to nozzles and needles.**

8. Remove:
- The bolts holding the mounting rail to the fuel rail
- The mounting rail
- The injector.

### To install:

9. Lubricate the O-ring for the new injector using petroleum jelly.

10. Install the injector.

11. Position the mounting rail. The injectors may need to be pulled away from the fuel rail a few millimeters to install the rail

12. Tighten the mounting rail to the fuel rail

13. Press down the fuel rail.

➡**Ensure all the injectors are correctly positioned.**

14. Screw the fuel rail into place. Use new bolts. Tighten to 88 inch lbs. (10 Nm)

15. Install all connectors on the injectors

16. Press the fuel line quick-release connector together until it clicks

17. Install the cover over the nozzles

18. Remove the seals. Install the plastic pipes between the turbocharger and charge air cooler and the air cleaner and turbocharger.

#### 8-Cylinder Engines

*See Figures 143 and 144.*

1. Remove the upper intake manifold. Refer to Intake Manifold removal and installation.

2. Remove the crankcase ventilation hose between the cylinder heads by releasing the 2 clamps.

3. Remove the 3 bolts from the fuel rail

4. Remove the 2 bolts on the fuel pressure sensor.

5. Lift the fuel rail assembly with injector straight up and bend it aside.

6. Unplug the connector of the injector.

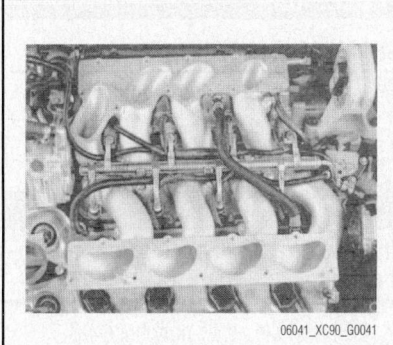

06041_XC90_G0041

**Fig. 143 Fuel rail assembly—V8**

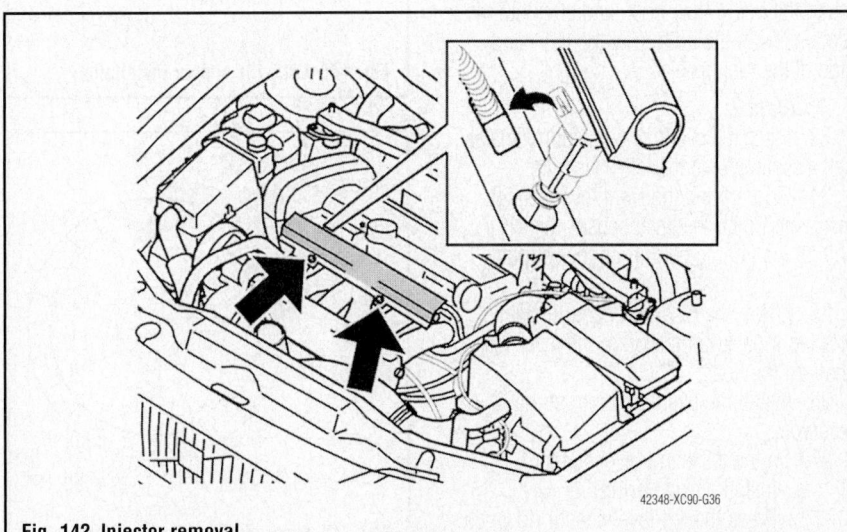

42348-XC90-G36

**Fig. 142 Injector removal**

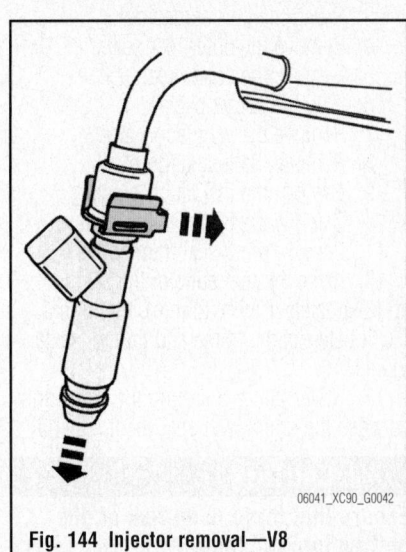

06041_XC90_G0042

**Fig. 144 Injector removal—V8**

7. Remove the circlip of the injector; pull the clip laterally.

8. Remove the injector; jiggle and pull it straight out.

***To install:***

→**Lubricate the O-rings on the injectors with Vaseline or the like.**

9. Install the injector.

10. Install the circlip.

11. Plug in the connector to the injector.

12. Install the fuel rail assembly with injector.

13. Install the 3 bolts of the fuel rail.

14. Install the 2 bolts on the fuel pressure sensor.

15. Install the crankcase ventilation hose and 2 clamps.

> ✳✳ **WARNING**
>
> **Make sure there is an audible click during connection.**

16. Connect the fuel supply line and secure the line in the clamp.

17. Install the upper intake manifold.

## FUEL PUMP

*REMOVAL & INSTALLATION*

*See Figures 145 through 147.*

1. Disconnect the battery negative terminal.

> ✳✳ **WARNING**
>
> **The use of a fresh air mask is recommended.**

2. Insert the heavy duty hose 1.3 meters into the fuel filler pipe, measured from the edge of the opening of the filler pipe.

3. Pump until air comes out.

4. Remove the outer rear seats.

5. Remove the center rear seat.

6. Fold the carpet back.

7. Remove the ventilation pipe

8. Remove the insulation block

9. Remove the insulation block

10. Remove the 6 seat frame bolts

11. Remove the 2 seat frame bolts

12. Raise the rear edge of the seat frame approximately 1 inch (25mm). Position a block between the frame and the transaxle tunnel.

13. Cover the area around the cover and between the doors with absorbent material.

> ✳✳ **CAUTION**
>
> **Ensure that there is no risk of dirt getting into the tank.**

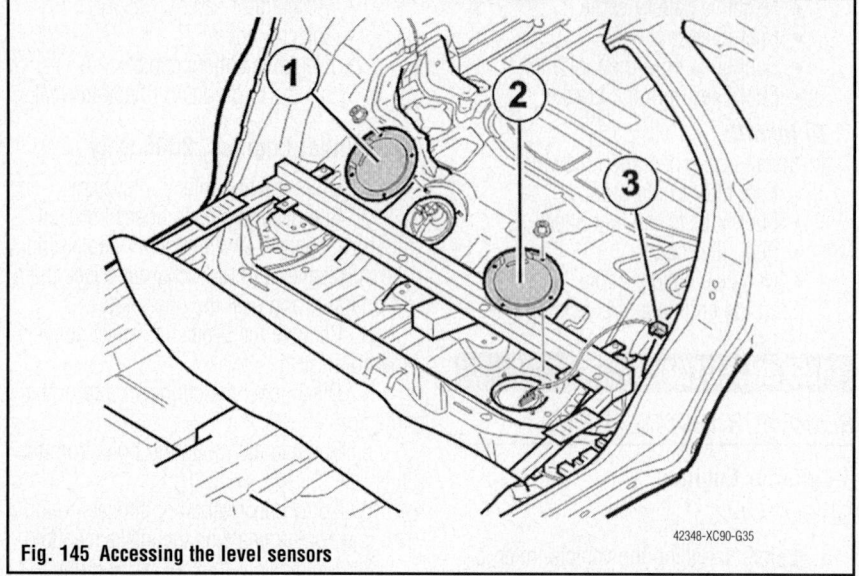

**Fig. 145 Accessing the level sensors**

14. Remove the covers (1 and 2) over the fuel tank units.

15. Disconnect the level sensor connector (3).

16. Remove the hoses from the right-hand level sensor.

17. Clean the area around the fuel tank unit thoroughly. Open the right-hand level sensor. Insert the heavy duty hose and suck out the remains from the bottom of the right-hand side of the fuel tank.

18. Connect hose 999 5721 to the heavy duty hose. Suck out the remains from the bottom of the level sensor reservoir.

19. Carefully remove the overflow pipe from the reservoir.

20. Remove the thin hose 999 5721 from the heavy duty hose. Connect the thin hose to the overflow pipe. Use a hose clamp.

21. Pump until a lot of air comes out.

22. Open the left-hand level sensor. Insert the heavy duty hose and suck out the remains from the bottom of the left-hand side of the fuel tank.

***To install:***

23. Place the overflow pipe approximately half a centimeter to the left-hand side.

24. Grip around the right level sensor reservoir. Pinch the level sensor reservoir so that the float is held in its lowest position.

25. Lower the reservoir carefully while rotating it backwards around the front-rear shaft on the float.

26. Install the overflow pipe on the reservoir.

27. On the left-hand level sensor:

a. Install a new O-ring.

b. Press the left level sensor down so that the row of protruding cables runs

along the car. Check that the O-ring is not trapped.

c. Tighten the bolt. Tighten to 44 ft. lbs. (60 Nm).

d. Position the wiring in the cut-out in the rubber ring

→**Check that the arrow on the cover is between the markings on the tank. If the arrow is outside the markings the**

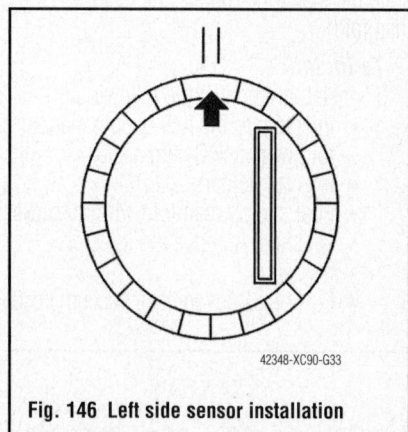

**Fig. 146 Left side sensor installation**

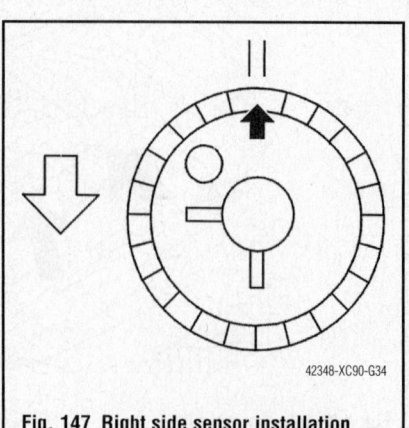

**Fig. 147 Right side sensor installation**

fuel gauge sensor will give an incorrect value and the float may catch against the inner wall of the tank. If not: Slacken off the bolt and adjust the position of the sensor. Tighten as above.

28. On the right-hand level sensor:
 a. Install a new O-ring.
 b. Press the right-hand level sensor down so that the fuel line connections are pointing forwards and to the right.
 c. Check that the O-ring is not trapped.
 d. Tighten the bolt. Tighten to 44 ft. lbs. (60 Nm).
 e. Install the hoses on the level sensor.
 f. Position the wiring in the cut-out in the rubber ring.

➡**Check that the arrow on the cover is between the markings on the tank. If the arrow is outside the markings the fuel gauge sensor will give an incorrect value and the float may catch against the inner wall of the tank. If not: Slacken off the bolt and adjust the position of the sensor. Tighten as above.**

29. Install the covers over the level sensors. Tighten the bolts.
30. Connect the wiring and the connectors.
31. Remove the block from under the seat frame
32. Install the 8 seat frame bolts. Tighten to 37 ft. lbs. (50 Nm).
33. Install the 2 seat frame bolts. Tighten to 18 ft. lbs. (24 Nm).
34. Fold the carpet down
35. Install the center rear seat.
36. Install the outer rear seat.

## FUEL TANK

*REMOVAL & INSTALLATION*

*See Figures 148 and 149.*

### ✳✳ CAUTION

**Observe all applicable safety precautions when working around fuel. Whenever servicing the fuel system, always work in a well ventilated area. Do not allow fuel spray or vapors to come in contact with a spark or open flame. Keep a dry chemical fire extinguisher near the work area. Always keep fuel in a container specifically designed for fuel storage; also, always properly seal**

fuel containers to avoid the possibility of fire or explosion.

1. Before servicing the vehicle, refer to the Precautions Section.
2. Relieve the fuel system pressure.
3. Drain the fuel tank.
4. Block the parking brake pedal in the neutral position.
5. Remove or disconnect the following on both sides:
 • Rear wheels
 • Rear brake calipers
 • Rear brake discs
 • Rear tension spring and expander
6. Lift the brake shoe slightly. Grip the expander. Pull the expander outward.
7. Unhook the expander from the metal bracket.
8. Remove or disconnect the following on both sides:
 • The guide sleeve for the cable sleeve. Apply a screwdriver between the guide sleeve lug and wheel arch. Pry apart slightly to remove the sleeve.
 • The parking brake cable from the tie rod and the sub-frame
9. Remove 2 bolts from the cable clips and 2 bolts from the front mounting for the heat deflector plate.
10. Fold out the cables from the heat deflector plate. Hang the cables up at the front of the car.
11. Remove the exhaust system from the three-way catalytic converter and backwards.

➡**Hang up the three-way catalytic converter to prevent damaging the flex boot.**

12. Remove the rear driveshaft:
 a. Mark up the position of the joint on the flange on the final drive and at the center bearing.

 b. Remove bolts from the joints. Use counterhold 999 7057.
 c. Remove the rear section of the driveshaft.
13. Lower the rear suspension:
 a. Place the mobile jack: 999 5972 with plate: 999 5972 under the rear suspension.
 b. Raise to light contact.
14. Remove or disconnect the following on both sides:
 • The 2 bolts for the mounting bracket
 • The mounting screw for the sub frame bushing
 • The brake pipe on both sides from the plastic clips on the bracket
15. Lower the rear suspension carefully.
16. Place a lifting table under the fuel tank.
17. Remove or disconnect the following:
 • The upper quick-release connector plastic pipe on the carbon filter container
 • The 4 bolts for the fuel tank securing strap
 • Remove the clamp for the fuel filler pipe
 • Remove the clamp for the bleed pipe
18. Lower the lifting table to access the plastic pipes on top of the tank.
19. Remove the plastic pipe.
20. Remove the ABS cable harness from the clips on both sides of the fuel tank.
21. Remove the fuel tank.

*To install:*
22. Place the fuel tank on the lifting table.
23. Raise the lifting table to install the plastic hoses on the right-hand level sensor and in the grooves in the tank.
24. Secure the hoses in place using silver tape.

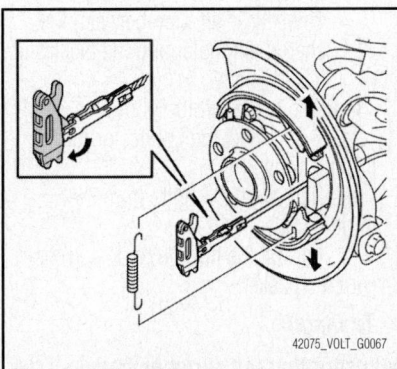
42075_VOLT_G0067
**Fig. 148 Removing expansion spring from rear brake**

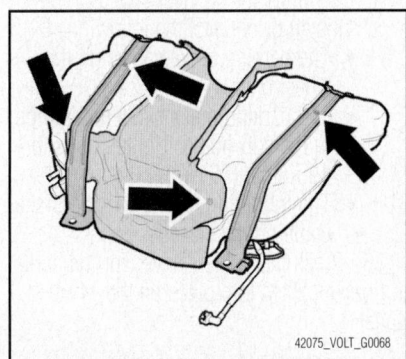
42075_VOLT_G0068
**Fig. 149 Removing fuel tank straps and heat shield**

25. Install the ABS cable harnesses in the clips on both sides of the tank.

26. Align the fuel filler pipe and the bleed pipe.

27. Insert the cable harnesses for the level sensor and the fuel tank pressure sensor through the inspection holes in the floor.

28. Raise the fuel tank to the installation position.

➡**Ensure that the fuel filler pipe and bleed pipe slide into position while raising the fuel tank.**

29. Install the 4 bolts for the fuel tank securing strap. Tighten to 18 ft. lbs. (24 Nm).

30. Remove the lifting table.

31. Install or connect the following:
- The clamp for the fuel filler pipe
- The clamp for the bleed pipe
- The upper plastic pipe in the quick-release connector for the carbon filter container

32. Install the rear suspension:
   a. Place the mobile jack: 999 5972 and plate: 999 5972 under the rear suspension.
   b. Carefully raise the rear suspension.

➡**Ensure that no cable harness or brake lines are trapped while raising the rear suspension.**

33. Install or connect the following on both sides:
- The 2 bolts for the mounting bracket. Tighten to 59 ft. lbs. (80 Nm)
- The bolt for the sub frame mounting. Tighten to 59 ft. lbs. (80 Nm)
- The brake line in the plastic clip and bracket

34. Install the brakes.

35. Install the parking brake cable along the heat shields.

36. Install the bolts in the center and front mountings for the heat shield.

37. Install or connect the following:
- The parking brake cables on the wheel units
- The parking brake cable mountings in the sub-frame. Tighten bolts to 19 ft. lbs. (25 Nm).
- The mounting for the parking brake cable on the tie rods.

38. Press the guide sleeves for the parking brake cables into place on the wheel units.

➡**Lubricate the O-rings on the guide sleeves. Use lubricant 1161580 or silicon grease.**

39. Install or connect the following:
- The expander on the cable. Press the mounting together to lock the cable in the expander
- The expander between the brake shoes
- The return spring.

40. Install the brake disc and the locating pin. Tighten to 88 inch lbs. (10 Nm).

41. Install the brake caliper. Use new bolts. Tighten to 44 ft. lbs. (60 Nm).

➡**Activate parking brake adjuster.**

42. Install the rear drive shaft.
   a. Position the propeller shaft as marked previously.
   b. Install new bolts for the joints. Use counterhold 999 7057. Tighten all bolts to 22 ft. lbs. (30 Nm).

43. Install the exhaust system. Use a new flange gasket. Tighten the bolts to 18 ft. lbs. (25 Nm).

44. Install the cover over the level sensor.

45. Connect the wiring and the connectors.

46. Install rear wheels and tighten lug nuts to 102 ft. lbs. (138 Nm).

## IDLE SPEED

### ADJUSTMENT

Idle speed is not adjustable.

## THROTTLE BODY

### REMOVAL & INSTALLATION

#### 5-Cylinder Engines

*See Figure 150.*

1. Before servicing the vehicle, refer to the Precautions Section.

2. Remove or disconnect the following:
- The negative battery terminal.
- The air intake between the front cover plate and the air cleaner housing.
- The charge air pipe between the charge air cooler and the electronic throttle body.

3. Remove the throttle body:
   a. Disconnect the connector for the throttle body.
   b. Remove the 4 bolts from the throttle body.
   c. Lift out the throttle body and discard the gasket.

#### To install:

➡**Ensure that the surfaces between the throttle body and the intake manifold are clean.**

Fig. 150 Throttle body location—5-Cylinder Engines

4. Install electronic throttle body using a new gasket. Tighten the 4 bolts to 88 inch lbs. (10 Nm).

5. Install the charge air pipe between the electronic throttle body module and the charge air cooler.

➡**Heat the pipe at the connections to facilitate installation.**

6. Install the intake pipe between the front cover plate and the air cleaner housing.

7. Start the engine and check the function of the electronic throttle body module.

➡**After replacing the throttle unit, the throttle unit must be adapted using the vehicle communication input: Adaptation of the electronic throttle unit. Remedy as necessary.**

#### 6-Cylinder Engines, 2006 Only

1. Before servicing the vehicle, refer to the Precautions Section.

2. Remove or disconnect the following:
- The negative battery terminal.
- The charge air cooler.
- The intake pipes between the front cover plate and air cleaner housing.
- The hose between the turbocharger and the charge air cooler at its upper end. Push it to one side.
- The pipe between the throttle body and the charge air cooler.

3. Removing the throttle body:
   a. Disconnect the throttle body connector.
   b. Remove the mounting bolts from the throttle body.
   c. Lift up and remove the throttle body. Discard the gasket.

#### To install:

4. Clean the mating surfaces thoroughly.

5. Install the throttle body using a new gasket. Tighten the 4 bolts to 88 inch lbs. (10 Nm).

6. Connect the throttle body connector.
7. Install or connect the following:
   - The pipe between the charge air cooler and the throttle body.
   - The hose between the charge air cooler and the turbocharger.
   - The intake pipes between the front cover plate and the air cleaner housing.
   - The battery negative lead.

➡**After replacing the throttle unit, the throttle unit must be adapted using the vehicle communication input: Adaptation of the electronic throttle unit. Remedy as necessary.**

### 8-Cylinder Engines

*See Figure 151.*

1. Before servicing the vehicle, refer to the Precautions Section.
2. Remove or disconnect the following:

- The negative battery terminal
- The intake hose by removing the 2 hose clips and bending aside the intake hose
- The connector from the electronic throttle body module
- The 2 coolant hoses attached to the throttle body and seal off the ends with a locking pliers
- The 2 nuts and the 2 bolts holding throttle body against intake manifold
- The throttle body

### *To install:*

3. Scrap the old gasket material to ensure a clean surface.
4. Install or connect the following:
   - A new throttle body gasket
   - The throttle body
   - The 2 nuts and 2 bolts. Tighten to 88 inch lbs. (10 Nm)
   - The 2 coolant hoses secure to throttle body with hose clips

- The connector to the electronic throttle body module
- The intake hose. Tighten the 2 hose clips.
- The negative battery terminal

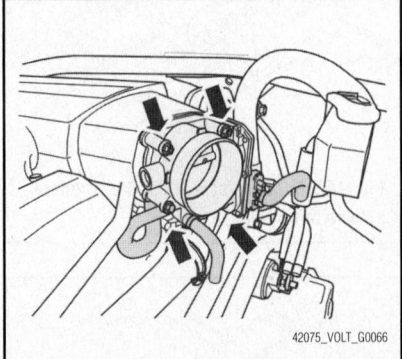

42075_VOLT_G0066

**Fig. 151 Location of throttle body with attached coolant hoses—8-Cylinder Engines**

## HEATING & AIR CONDITIONING SYSTEM

### BLOWER MOTOR

#### *REMOVAL & INSTALLATION*

*See Figures 152 through 157.*

➡**Before servicing, or working around, the SRS system, turn the ignition switch OFF, disconnect the battery and wait at least three minutes.**

1. Before servicing the vehicle, refer to the Precautions Section.

The dashboard will need to be removed in order to remove the blower motor.

2. Move the front seats back.
3. Remove or disconnect the following:
   - The negative battery lead.
   - The A-post panels of the dashboard
   - The end covers on both sides
   - The soundproofing panels

- The courtesy lighting
- The On-Board Diagnostic (OBD) connector
- The steering wheel
- The combined instrument panel
- Gear selector assembly and steering wheel module
- The center console
- The dashboard environment panel
- The bracket for the road traffic information loudspeaker
- The connector for the hazard warning signal flasher
- Lift up and disconnect the sun sensor
4. Remove or disconnect the following:
   - Unhook the cable for the parking brake at the cable terminal by the release handle.

- Disconnect the cable for the gear selector lever lock.
- Detach the gear selector assembly (four screws). Turn the gear selector assembly a quarter turn with the control downwards towards the driver's seat.
- The glove compartment
- The connector for the passenger air bag module
- The lower mounting for the passenger air bag module. Remove the 2 screws
- The 9 screws holding the dashboard
- Lift out dashboard.
5. Remove the fan motor from the climate control unit:

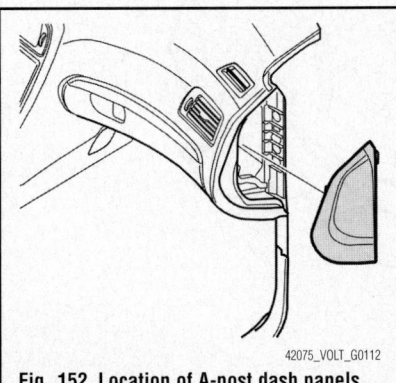

42075_VOLT_G0112

**Fig. 152 Location of A-post dash panels and end covers**

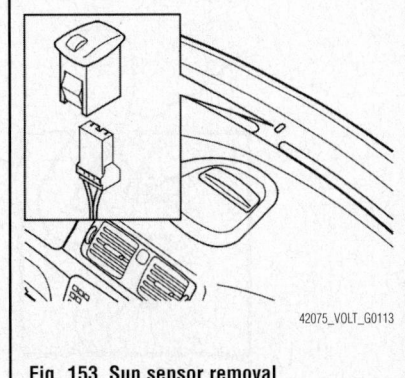

42075_VOLT_G0113

**Fig. 153 Sun sensor removal**

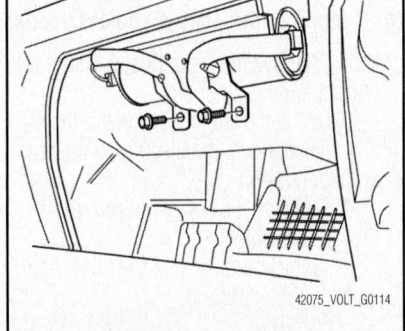

42075_VOLT_G0114

**Fig. 154 Location of lower mounting for the passenger air bag module**

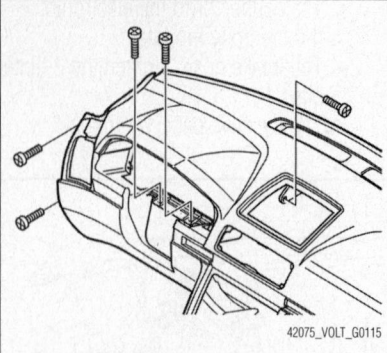

**Fig. 155 Location of 5 of the 9 dashboard screws to be removed**

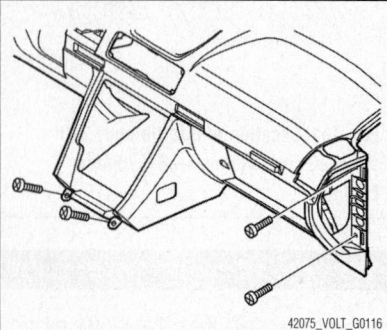

**Fig. 156 Location of 4 of the 9 dashboard screws to be removed**

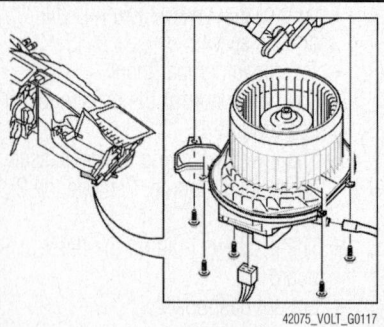

**Fig. 157 Removing the fan motor from the climate control unit**

➡**Do not detach the air conditioning (A/C) pipes or the engine coolant hoses.**

  a. Remove the drain hoses from the engine and filter housing.
  b. Move the drain hoses out of the way.
  c. Disconnect the electric connector at the fan motor.
  d. Remove the 5 screws from the blower fan motor.
  e. Detach and remove the blower fan motor and fan shroud.

**To install:**

6. Installation is the reverse of the removal procedure.

7. Install or connect the following:
- The blower fan motor and the cover on the climate control unit using the 5 screws
- The electrical connector for the blower motor
- The drain hose
- The climate control unit

8. Install or connect the following:
- Lift the dashboard
- Pull the connector for the sun sensor out through the hole in the dashboard.

➡**Check that no wiring is trapped at the cable ducts.**

- The dashboard. Tighten the 9 M6 screws to 88 inch lbs. (10 Nm)
- The lower mounting for the passenger air bag module. Tighten the 2 M6 screws to 88 inch lbs. (10 Nm)
- The sun sensor
- The connector for the hazard warning signal flasher
- The cable for the parking brake. Hook the cable into place at the release handle
- The On-Board Diagnostic (OBD) connector
- The combined instrument panel
- The gear selector assembly and steering wheel module
- The steering wheel
- The bracket for the road traffic information loudspeaker
- The dashboard environment panel

- The gear selector assembly. Tighten the 4 M8 bolts to 18 ft. lbs. (24 Nm)
- The mechanical cable for the gear selector lever knob
- The center console
- The courtesy lighting
- The end covers of the dashboard on both sides
- The A-post panels

## HEATER CORE

### REMOVAL & INSTALLATION

*See Figure 158.*

1. Before servicing the vehicle, refer to the Precautions Section.
2. Drain the cooling system.
3. Remove:
- The hoses from the heater core pipes in the engine compartment
- The seal and the plate
- The soundproofing panel
- The bolt for the heat exchanger
- The mounting brackets for the pipes
4. Position plenty of paper under the heat exchanger and around the pipes.
5. Drain the coolant into a suitable container.
6. Detach the pipes and move them out of the way.
7. Pull out the heat exchanger.

***To install:***

8. Install:
- The upper pipe
- The striker plate using round-nosed pliers.
- The catch using a pair of pliers.

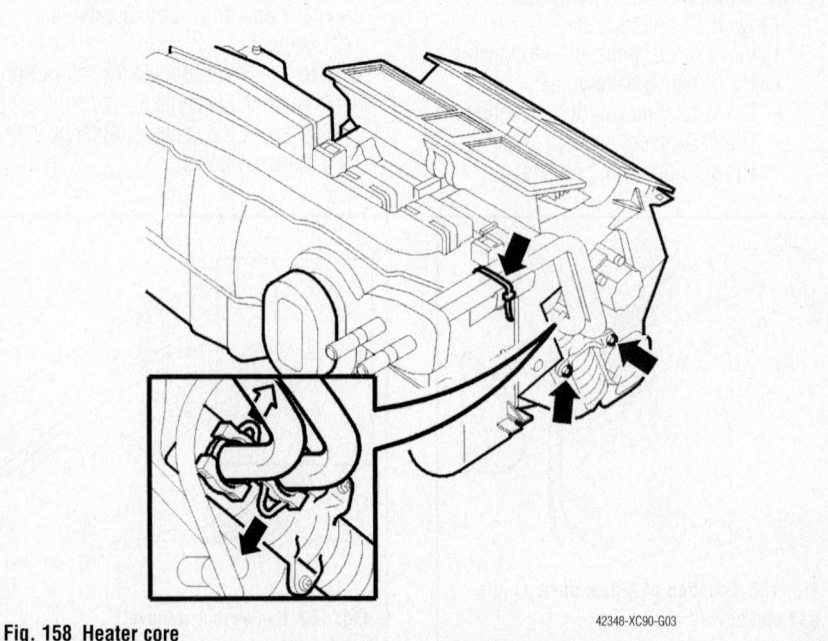

**Fig. 158 Heater core**

- The lower pipe
- The striker plate using round-nosed pliers.
- The striker plate using round-nosed pliers.

- The seal and the plate
- The heating loops in the engine compartment.

9. Start the engine and the heating system.

10. Fill the cooling system.
11. Check that the pipe connectors do not leak.
12. Install the soundproofing panel.

## SUSPENSION                    FRONT SUSPENSION

### LOWER BALL JOINT

#### REMOVAL & INSTALLATION

*See Figures 159 through 167.*

1. Remove the front wheels.
2. Remove the bolt for the drive shaft. Use a prybar in the brake disc as counter-hold.
3. Remove the drive shaft from the hub by tapping out the drive shaft with a brass drift.
4. Fold aside the spring strut.
5. Loosen the ball joint pin from the link arm. Use tool 999 7062.
6. Install the cap nut for the ball joint pinion. For earlier versions (deep thread pitch) use 999 7062 and for later versions (closer thread pitch) use 999 7231.
7. Tighten the nut against the link arm.
8. Then turn back one turn so that clearance is created between the cap nut and link arm.
9. Place the puller on the link arm and loosen the ball joint from the link arm.

➡**If the catch is placed by (2), the rubber gaiter on the ball joint may be damaged.**

10. Press the control arm down using lever 999 7076. Position the hook at (1).
11. Disconnect the spring strut from the link arm.
12. Remove the ball join bolts.
13. Install the nut on the ball joint. Use 999 7149 for earlier versions (deep thread pitch) and for later versions (closer thread pitch) use 999 7230.

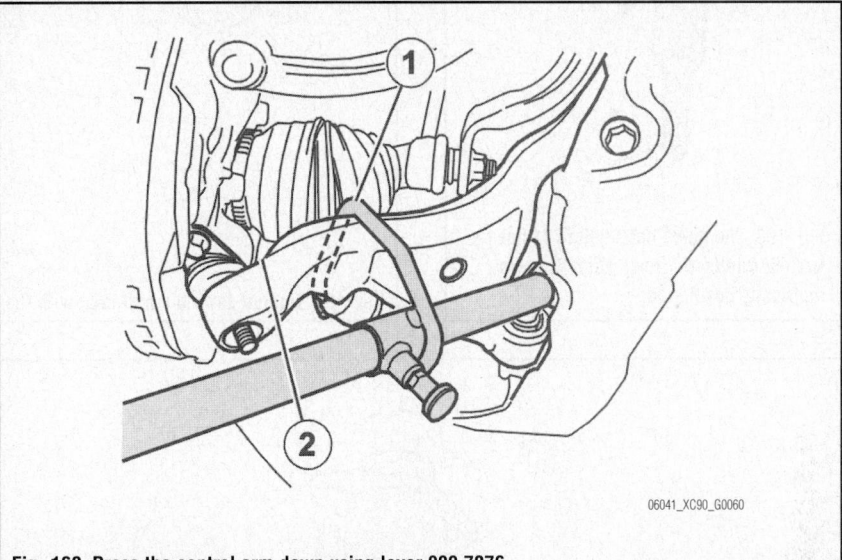

Fig. 160 Press the control arm down using lever 999 7076

14. Tap loose the ball joint.

### ✳✳ CAUTION

**Always use safety glasses. Clean both in the seat for the ball joint and the contact surfaces with a rotating steel brush. Grease the seat with wheel bearing grease.**

*To install:*

➡**The guide bolts 999 5781 have a wrench grip (see arrow) which is asymmetrically positioned.**

➡**Use the bolt (2) with the shortest distance from the wrench grip.**

➡**Leave the ball joint's protective cup on to avoid damaging the rubber gaiter.**

15. Install the new ball joint loosely using the guide bolts.
16. Press up the ball joint against the seat with press tool 999 5796.
17. Check that the ball joint is centered in the seat.
18. Drive in the ball joint with a copper hammer.

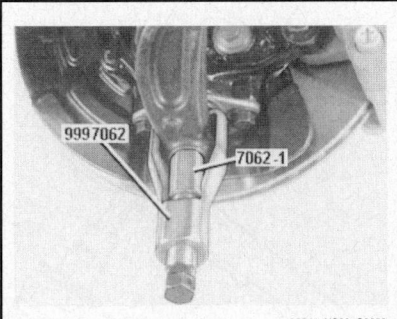

Fig. 159 Loosen the ball joint pin from the link arm

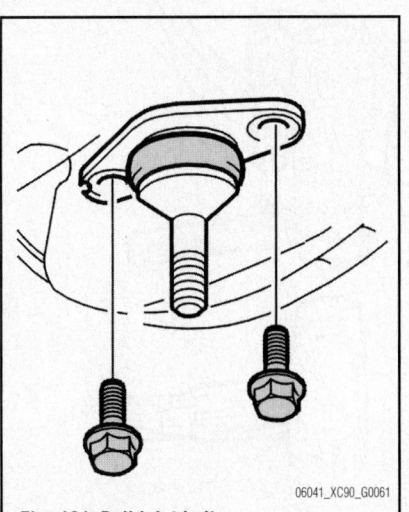

Fig. 161 Ball joint bolts

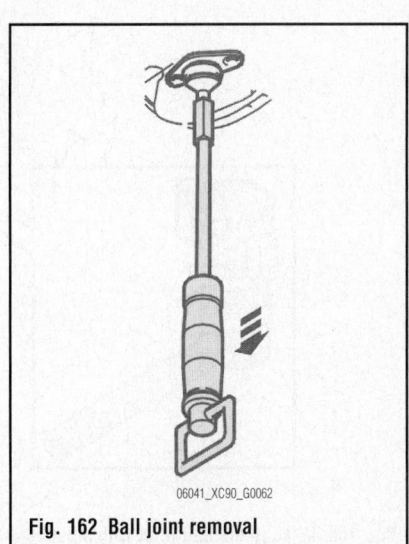

Fig. 162 Ball joint removal

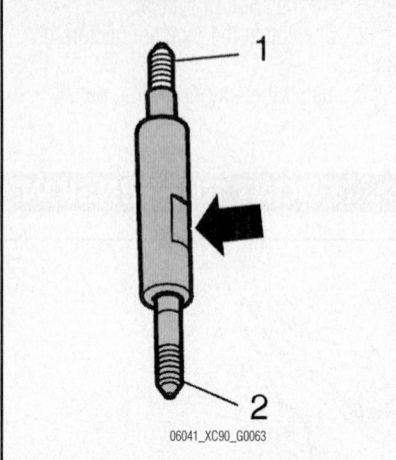

Fig. 163 The guide bolts 999 5781 have a wrench grip (see arrow) which is asymmetrically positioned

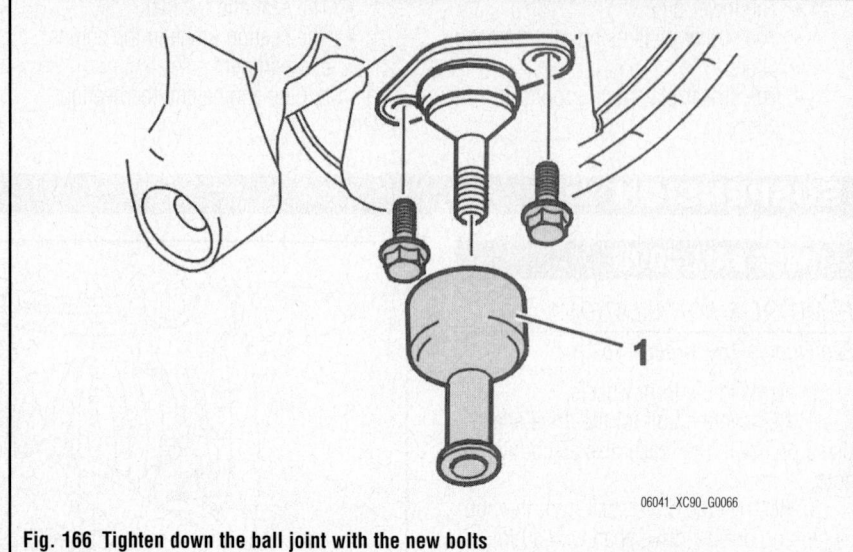

Fig. 166 Tighten down the ball joint with the new bolts

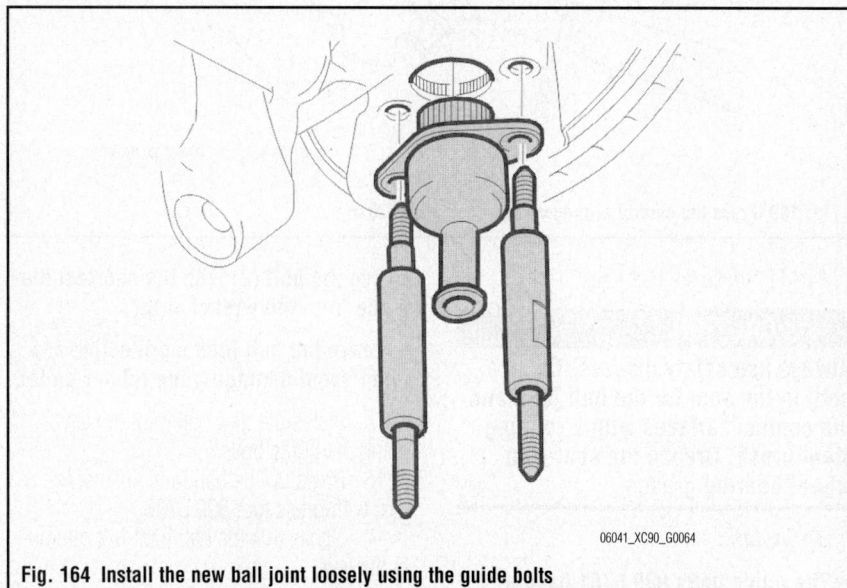

Fig. 164 Install the new ball joint loosely using the guide bolts

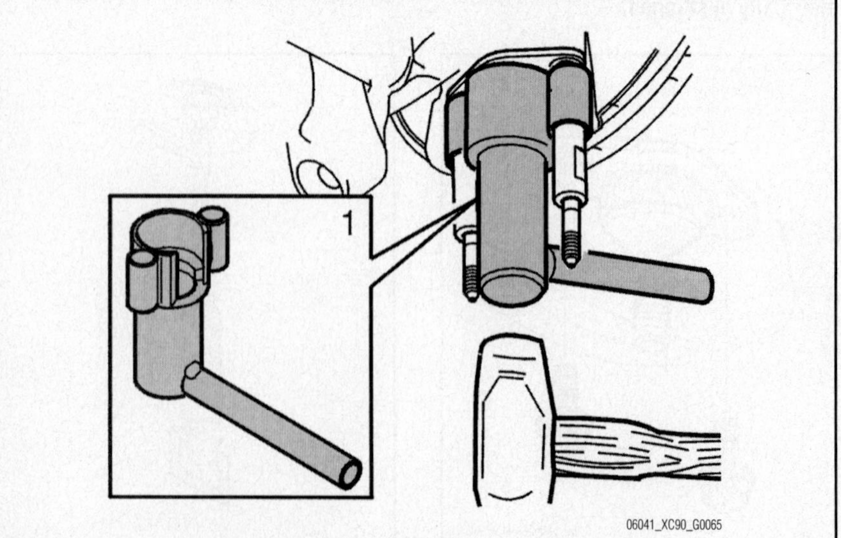

19. Remove the guide bolts.
20. Tighten down the ball joint with the new bolts. Tighten to 74 ft. lbs. (100 Nm).
21. Remove the protective cup.
22. Install the halfshaft.

**✳✳ WARNING**

**If the catch is placed by (2), the rubber gaiter on the ball joint may be damaged.**

23. Press down the link with lever 999 7076.
24. Place the catch by (1).
25. Install the spring strut in the link arm.

➡**The ball joint pin may not rotate. Use a Torx® bit as counterhold in order to prevent damaging the rubber gaiter.**

26. Use 999 5500 as adapter for the torque wrench.
27. Install the nut. Tighten according to 37 ft. lbs. (50 Nm) plus 35°.
28. Install the wheels.

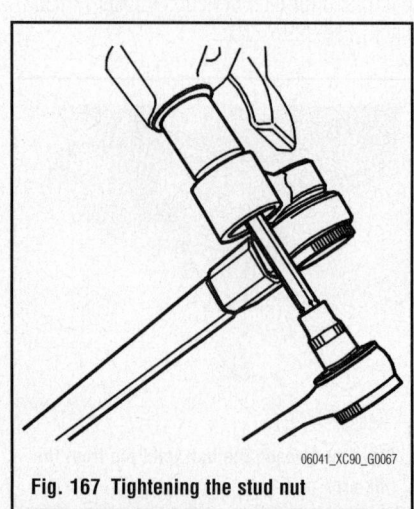

Fig. 167 Tightening the stud nut

Fig. 165 Drive in the ball joint with a copper hammer

## LOWER CONTROL ARM

### REMOVAL & INSTALLATION

*See Figure 168.*

1. Remove the wheel.
2. Disconnect the ball joint pinion from the control arm.
3. Install cap nut 7062-1 on the ball joint pinion.
4. Tighten the nut against the control arm.
5. Then turn back one turn to form a gap between the cap nut and the control arm.
6. Position the extractor on the control arm. Detach the ball joint from the control arm.
7. Pull down the control arms using a tension strap.
8. Release the spring strut from the control arm.

### ✳✳ WARNING

**Ensure that the tension strap is correctly secured in the control arms.**

9. Remove the bolt for the halfshaft. Use a prybar as a counterhold on the brake disc.
10. Remove the halfshaft from the hub. Knock the halfshaft out using a brass drift.
11. Push the spring strut one side. Remove the tensioning strap between the control arms.

### ✳✳ WARNING

**Ensure that the tensioning strap does not get trapped when releasing.**

12. Remove the splash guard under the engine.
13. Applies to the right-hand side only: Remove the hose for the EVAP canister and the fuel line for the engine block heater (where applicable) from their mountings. Move the hose and fuel line to one side.
14. Only applies to cars with 4T65EV transaxles when removing the left-hand control arm: Remove the bolt for the front engine pad. Lift the engine approximately ⅓ inch (10mm) using a mobile jack to remove the rear bolt on the front mounting for the control arm.

➡**Position the mobile jack so that it is raised towards the front right mounting bolt for the lower torque rod bracket.**

Applies to all models:
15. Remove:
   • The two bolts
   • The bolt and nut
   • The control arm

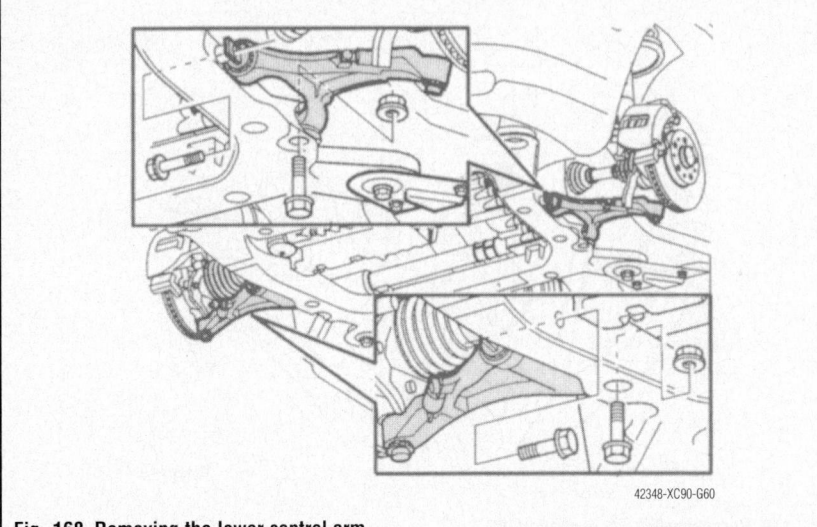
**Fig. 168 Removing the lower control arm**

42348-XC90-G60

### To install:

16. Install:
   • The control arm. Torque the front bolt to 48 ft. lbs. (65 Nm) plus 120°. Torque the rear bolt/nut to 78 ft. lbs. (105 Nm) plus 120°

➡**Use new bolts.**

17. Only applies to cars with 4T65EV transaxles when removing the left-hand control arm: Lower the engine. Install the bolt for the front engine pad. M10.
18. Applies to the right-hand side only: Install the hose for the EVAP canister and the fuel line for the engine block heater (if applicable) in their mountings.
Applies to all models:
19. Install the tensioning strap between the control arms and tighten.
20. Insert the halfshaft in the hub.
21. Position the spring strut over the ball joint. Release the tensioning strap so that the control arm meets the ball joint.

### ✳✳ WARNING

**Ensure that the tensioning strap does not get trapped when releasing.**

22. Install the halfshaft in the hub. Install a new bolt. Use a prybar as a counterhold.
23. Install the upper ball joint nut. Tighten to 59 ft. lbs. (80 Nm).
24. Install the splashguard under the engine.
25. Install the wheels.

## MACPHERSON STRUT

### REMOVAL & INSTALLATION

*See Figures 169 through 171.*

1. Remove:
   • The wheel
   • The anti-roll bar link from the spring strut.

➡**Use a Torx® wrench as a counterhold so that the boot is not damaged.**

   • The ABS line from the spring strut
   • The ABS sensor. Hang up the sensor using a piece of wire.
2. Measure over the steering knuckle and spring strut at the upper bolt as illustrated. Before measuring clean off any dirt from the measuring surfaces. Note the measurement.
3. Remove both the bolts retaining the spring strut in the steering knuckle.
4. Remove:
   • The three nuts holding the spring strut in the bodywork
   • The spring strut

### To install:

5. Install the spring strut in the bodywork using the new nuts. Tighten to 18 ft. lbs. (25 Nm).

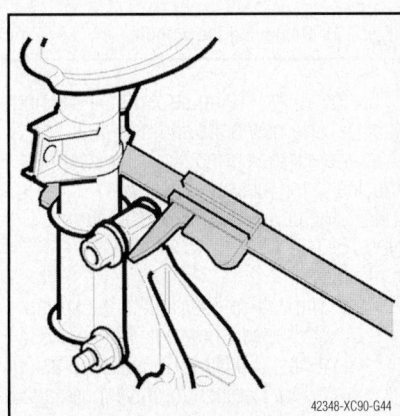
42348-XC90-G44
**Fig. 169 Measuring over the spindle and strut**

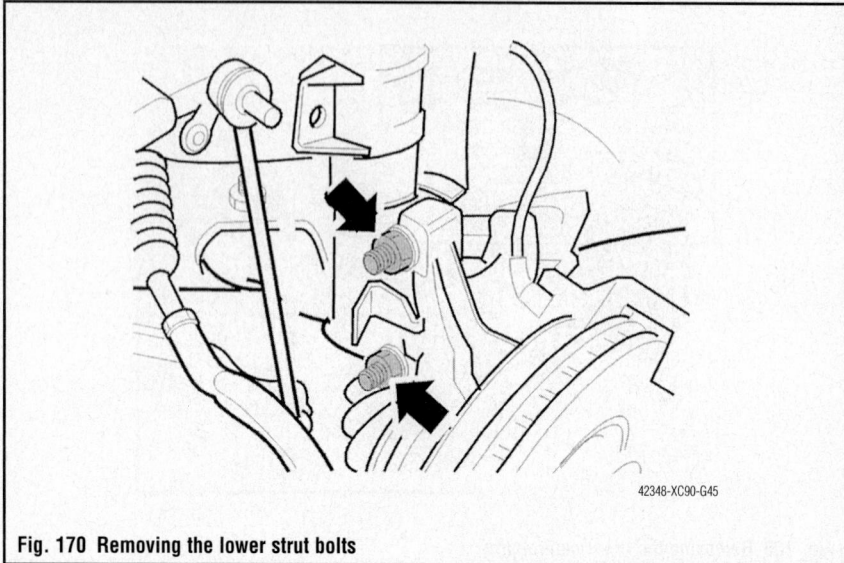

**Fig. 170 Removing the lower strut bolts**

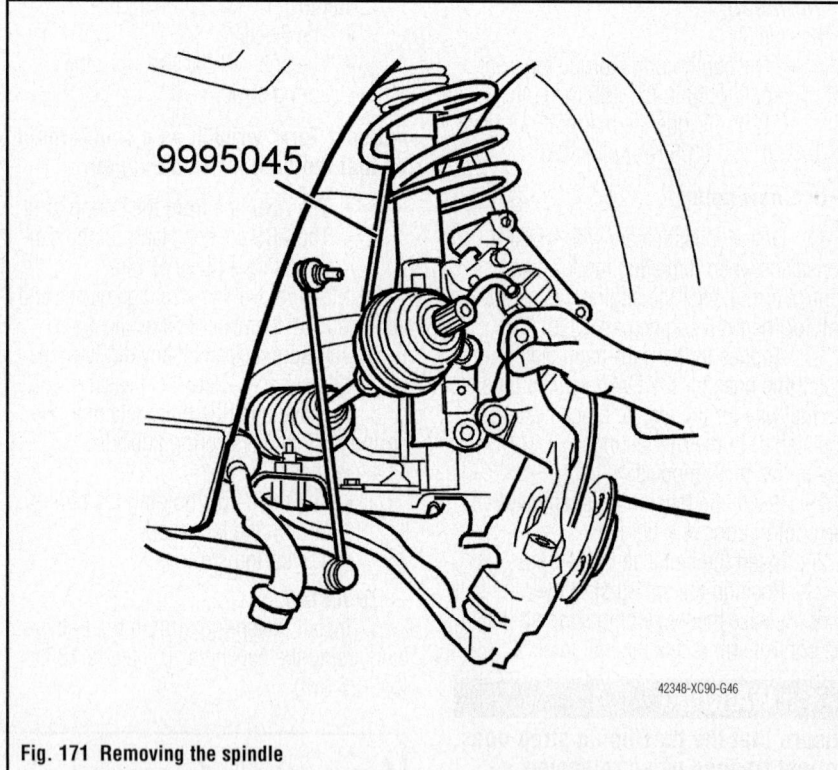

9995045

**Fig. 171 Removing the spindle**

6. Install the spring strut on the steering knuckle. Use new bolts and new nuts.

7. Adjust the spring strut and steering knuckle to the measured value.

8. Tighten to 77 ft. lbs. (105 Nm), plus 75°.

9. Install:
- The anti-roll bar link to the spring strut. Use a new nut. Tighten to 37 ft. lbs. (50 Nm). Counterhold using a Torx® wrench to prevent damage to the boot
- The ABS sensor. Tighten to 88 inch lbs. (10 Nm).

➡**Ensure that the ABS sensor seat in the steering knuckle is absolutely clean. Clean the ABS sensor with a soft brush.**

- The ABS sensor cable in the bracket
- The wheel and tighten to 102 ft. lbs. (138 Nm).

### STABILIZER BAR

*REMOVAL & INSTALLATION*
*See Figures 172 through 177.*

1. Before servicing the vehicle, refer to the Precautions Section.

➡**The anti-roll bar bushings and caps are vulcanized to the anti-roll bar and cannot be replaced separately.**

2. Install the lifting tool 999 5716, modified according to the tool bulletin for version C, and lifting hooks 999 5460.

3. Lift the engine to relieve the load on the engine pads.

4. Remove the front wheels.

5. Slacken off the right lower front part of the fender liner.

6. Remove or disconnect the following:
- The plastic nuts on the fender liner
- The protective plate
- Bend out the fender liner
- Bolts for the right-hand engine mounting (1)

7. Remove or disconnect the following:
- The splash guard under the ngine
- The 6 bolts of the skid guard
- The skid guard

➡**The skid guard upper mounting has a guide sleeve and therefore it must be lifted slightly so that the panel can be slid forward.**

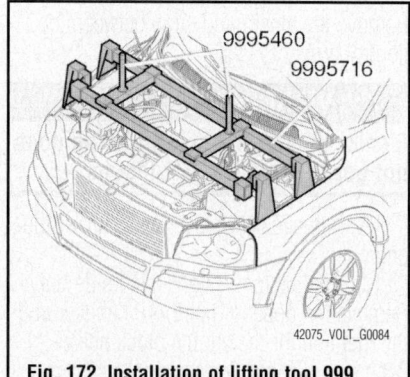

9995460
9995716

**Fig. 172 Installation of lifting tool 999 5716 and lifting hooks 999 5460**

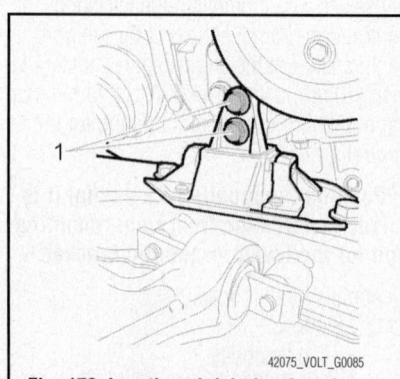

1

**Fig. 173 Location of right-hand engine mounting bolts**

- The 4 mounting bolts for the Side Impact Protection System (SIPS) member
- The brake pipe from the snap fastener on the member
- The SIPS member

8. Remove or disconnect the following:
- The rubber mountings for the exhaust pipe
- The nuts and bolts holding the steering gear to the sub-frame
- The power steering line from the clips on the sub-frame
- The bracket for the heated oxygen sensor (HO2S) from the sub-frame
- The heated oxygen sensor wiring from the right side of the sub-frame (only applies to 6-Cylinder engines)

9. Lower the sub-frame:
    a. Position a mobile jack with universal plate 999 5972 at the rear of the sub-frame.
    b. Remove the bolts holding the sub-frame bracket in the body.
    c. Remove the rear bolt for the sub-frame with the bracket and washer.
    d. Carefully lower the sub-frame until the clearance **A** between the sub-frame and the body is 4 ⅓ inches (110mm).

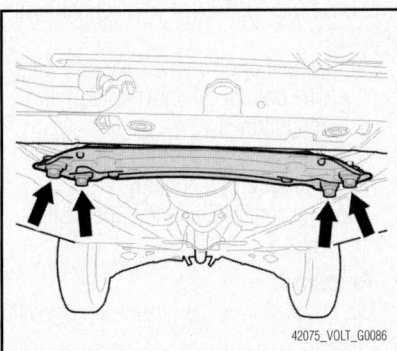

**Fig. 174 Location and removal of the SIPS member**

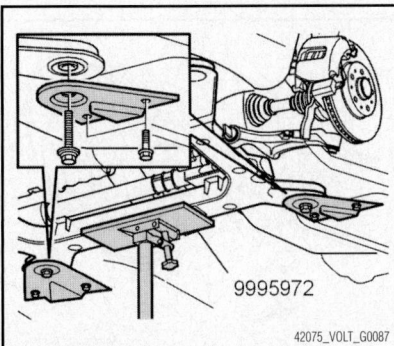

**Fig. 175 Lowering the sub-frame using a mobile jack with universal plate 999 5972**

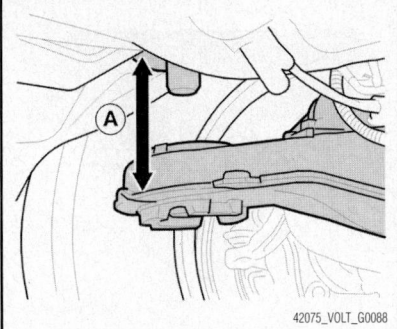

**Fig. 176 Lowering the sub-frame to the appropriate clearance**

10. Remove the anti-roll bar on both sides. Use a Torx® wrench as a counterhold so that the boot is not damaged.
11. Remove the bolts for the anti-roll bar.
12. Remove the anti-roll bar.

### To install:

13. Install or connect the following:
- The anti-roll bar
- The (M10) bolts for the anti-roll bar. Tighten to 37 ft. lbs. (50 Nm)
- The nut for the anti-roll bar link on both sides. Tighten to 59 ft. lbs. (80 Nm)

➡ **Use a Torx® wrench as a counterhold so that the gaiter is not damaged.**

14. Install the sub-frame:
    a. Raise the sub-frame.
    b. Install the sub-frame four bolts together with the sub-frame brackets and engine heater, if installed.

➡ **Use new bolts. Lube the bolts with oil. Tighten the sub-frame bolts starting on the left-hand side, then the right, using 2 steps: Step 1—78 ft. lbs. (105 Nm), Step 2—angle-tighten 120°.**

15. Tighten the M10 bolts for the brackets M10 to 37 ft. lbs. (50 Nm).

16. Remove the lifting table.
17. Install or connect the following:
- The nuts and bolt for the steering gear. Use new nuts and bolt. Tighten to 37 ft. lbs. (50 Nm)
- The rubber mountings for the exhaust pipe

➡ **Use soap solution to facilitate installing the rubber mountings.**

18. Install the SIPS member:
    a. Install the brake pipe in the snap fastener on the member.
    b. Install the SIPS member. Tighten the M8 bolts to 18 ft. lbs. (24 Nm).
19. Remove or disconnect the following:
- The skid plate up into the bumper wraparound. Press forwards until the mounting brackets engage in the sub-frame
- The 6 M10 skid plate bolts. Tighten to 37 ft. lbs. (50 Nm)
- The splash guard under the engine. Tighten the M8 bolts to 18 ft. lbs. (24 Nm)
20. Install the right-hand engine mounting:
    a. Install the bolts for the right-hand engine mounting. Use new bolts. Tighten in 2 steps:
- Step 1—89 ft. lbs. (120 Nm)
- Step 2—angle-tighten 40°
21. Fold the fender liner back into position.
22. Install the plastic nuts.
23. Install the front wheels and tighten lug nuts to 102 ft. lbs. (138 Nm).
24. Remove the lifting tools 999 5716 and lifting hooks 999 5460.

### STEERING KNUCKLE

#### REMOVAL & INSTALLATION

*See Figures 178 through 180.*

1. Before servicing the vehicle, refer to the Precautions Section.
2. Raise and support the vehicle safely.
3. Remove front wheel.
4. Remove or disconnect the following:
- The anti-roll bar link from the spring strut. Use a fixed wrench as a counterhold
- The ABS sensor. Hang up the sensor using a piece of wire
- The ABS sensor cable from the spring strut
- The brake caliper mounting bolts. Hang the caliper up using a piece of wire
- The halfshaft bolt. Use a screwdriver as a counterhold on the brake disc

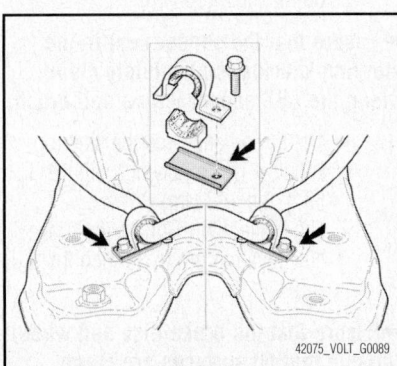

**Fig. 177 Location and removal of anti-roll bar bolts**

- The locating pin holding the brake disc
- The brake disc. Detach the end of the halfshaft in the hub by knocking the drive shaft into the hub approximately ⅓–½ inch (10–15mm). Use a rubber or copper mallet
- The tie rod ends from the steering arm

5. Measure the position of the steering knuckle and spring strut for installation purposes.

6. Remove the bolts retaining the spring strut and the steering knuckle.

7. Press out the steering knuckle. At the same time pull the halfshaft from the transaxle.

8. Suspend the drive shaft from a hook 999 5045.

➡**Take care not to damage the half-shaft boot.**

9. Disconnect the ball joint pinion from the control arm. Use tool 999 7062.

10. Install cap nut 7062-1 on the ball joint pinion.

11. Tighten the nut against the control arm.

12. Then turn back one turn to form a gap between the cap nut and the control arm.

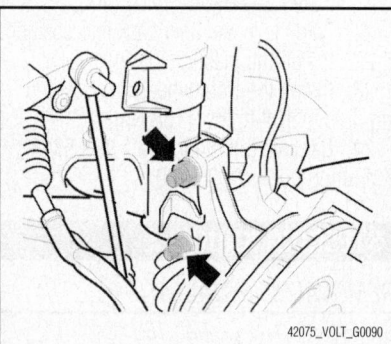

**Fig. 178 Location of bolt removal of steering knuckle**

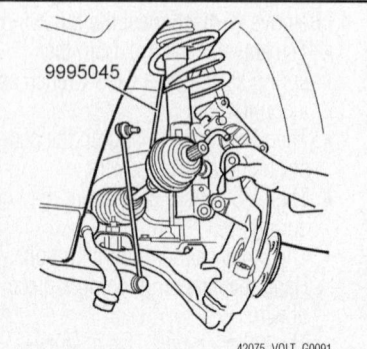

**Fig. 179 Suspending halfshaft from hook 999 5045**

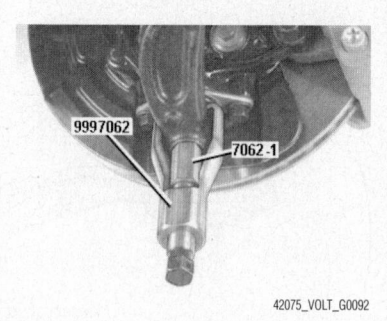

**Fig. 180 Using tool 999 7062 to extract ball joint**

13. Position the extractor on the control arm. Detach the ball joint from the control arm.

*To install:*

14. Install the steering knuckle on the trailing arm.

15. Install a new nut on the ball joint. Tighten to 59 ft. lbs. (80 Nm). Use a Torx® wrench as counterhold.

➡**Make sure that the mating surfaces on the ball joint and link are clean.**

16. Clean the splines on the halfshaft.

17. Turn the steering knuckle and bring the halfshaft into the hub.

18. Install halfshaft bolt and finger-tighten. Lubricate the bolt.

19. Install bolts retaining the spring strut in the steering knuckle. Use new bolts and nuts. Tighten to 78 ft. lbs. (105 Nm) and then angle-tighten 60°.

20. Remove or disconnect the following:
- The anti-roll bar link to the spring strut. Use a new nut. Counterhold using a Torx® wrench so that the boot is not damaged. Tighten to 37 ft. lbs. (50 Nm)
- The ABS sensor cable
- The ABS sensor onto the steering knuckle. Tighten the ABS sensor to 88 inch lbs. (10 Nm)

➡**Ensure that the sensor seat in the steering knuckle is absolutely clean. Clean the ABS sensor with a soft brush.**

- The tie rod end onto the steering knuckle. Use a new nut. Tighten to 59 ft. lbs. (80 Nm)
- The brake disc. Tighten the brake disc locating pin to 88 inch lbs. (10 Nm)

➡**Ensure that the brake disc and wheel rim hub mating surfaces are clean.**

21. Tighten the halfshaft bolt. Use a screwdriver as a counterhold on the brake

disc. Tighten to 26 ft. lbs. (35 Nm), plus 120°.

22. Install the brake caliper. Use new bolts

23. Install the wheel. Tighten to 102 ft. lbs. (138 Nm).

## UPPER BALL JOINT

### REMOVAL & INSTALLATION

*See Figure 181.*

1. Before servicing the vehicle, refer to the Precautions Section.

2. Raise the vehicle.

3. Remove or disconnect the following:
- The wheel
- The brake caliper. Hang the brake caliper on a hook in the sub-frame.

4. Loosen the nut for the control arm mounting but do not remove.

5. Use tool: 951 2923 , 999 7031 and 999 7030 together with threaded rod 999 7039 to press off knuckle.

6. Remove the nut for the drive shaft.

7. Press in the drive shaft to remove the spindle.

➡**Only press until the bushing releases from the splined area.**

8. Loosen the nut for the upper control arm. Use puller 998 5434 with counterhold 999 7074.

9. Remove or disconnect the following:
- The bolt for the inner control arm mounting
- The two bolts for the inner front control arm mounting
- The control arm and upper ball joint

*To install:*

10. Installation is the reverse of removal procedure.

11. Install or connect the following:

12. Tighten the M12 bolts to 59 ft. lbs. (80 Nm).

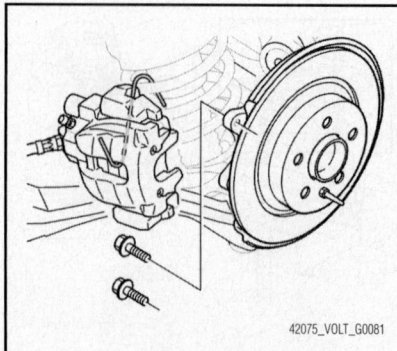

**Fig. 181 Using a hook to safely hang the brake caliper out of the way**

13. Install the wheel and tighten lug nuts to 102 ft. lbs. (138 Nm).

## UPPER CONTROL ARM

### REMOVAL & INSTALLATION

1. Before servicing the vehicle, refer to the Precautions Section.
2. Raise the vehicle.
3. Remove or disconnect the following:
   • The wheel
   • The brake caliper. Hang the brake caliper on a hook in the sub-frame.
4. Slacken off the screw for the outer control arm mounting by 3 turns.
5. Use tool: 951 2923 , 999 7031 and 999 7030 together with threaded rod 999 7039 to press off the steering knuckle.
6. Remove the bolt for the drive shaft.
7. Press in the drive shaft and remove the spindle.

➡**Only press until the bushing releases from the splined area.**

8. Slacken off the screw three turns for the rear control arm mounting. Use puller 998 5434 with counterhold 999 7074.
9. Remove or disconnect the following:
   • The bolt for the inner rear control arm mounting
   • The two bolts for the inner front control arm mounting
   • The control arm

### To install:
10. Install or connect the following:
   • The bolts for the inner rear control arm mounting
   • The bolts for the front inner control arm mounting
   • The bolts for the outer control arm mounting
11. Install tensioner 999 5659.

➡**The tensioner must be positioned as far out as possible on the tensioner plates to achieve the correct lifting force.**

12. Tighten the M12 bolts for the inner rear control arm mounting to 59 ft. lbs. (80 Nm).
13. Tighten the M12 bolts for the front inner control arm mounting to 59 ft. lbs. (80 Nm).
14. Tighten the M12 bolts for the outer control arm mounting to 59 ft. lbs. (80 Nm).

➡**Tighten the bolts when the rear suspension is in the normal position.**

15. Remove tensioner 999 5659.
16. Install the wheel and tighten lug nuts to 102 ft. lbs. (138 Nm).

➡**If the vehicle is equipped with Bi-Xenon lamps, the position sensor must be calibrated. Carry out calibration according to VIDA vehicle communication (Function area 3, Electrical system, Rear Electronic Module).**

## WHEEL BEARINGS

### REMOVAL & INSTALLATION

*See Figure 182.*

1. Remove the wheel.
2. Remove:
   • The two bolts
   • The brake caliper
3. Hang up the caliper in a suitable position.
4. Remove the halfshaft bolt. Use a prybar as a counterhold on the brake disc.
5. Detach the end of the halfshaft in the hub by knocking the halfshaft into the hub approximately ⅓–½ inch (10–15mm). Use a rubber or copper mallet.
6. Remove:
   • The bolt
   • The brake disc
7. Remove the nut.

➡**Counterhold using the wrench grip so that the bolt does not rotate and damage the rubber boot.**

8. Remove the nut. Pull the sway bar link out of its mounting.

➡**Counterhold using a Torx® wrench so that the boot for the link is not damaged.**

9. Remove the ABS cable harness from its brackets.
10. Remove the nut from the ball joint on both sides. Counterhold using a Torx® wrench so that the ball joint boot is not damaged.
11. Press out the ball joints.
12. Pull down the control arm using a tension strap.
13. Release the spring from the control arm and halfshaft.

### ✳✳ WARNING
**Ensure that the tension strap is correctly secured in the control arms.**

14. Remove:
   • The four bolts
   • The hub

### To install:
15. Install or connect the following:
   • The hub
   • The four bolts. For 5- and 6-cylinder models, torque to 15 ft. lbs. (20 Nm), then 44 ft. lbs. (60 Nm), then an additional 60°. For 8-cylinder models, torque to 15 ft. lbs. (20 Nm), then 33 ft. lbs. (45 Nm), then an additional 60°.
16. Install the halfshaft in the hub. Align the ball joint against the control arm.
17. Release the tensioning strap between the control arms.
18. Install a nut on the ball joint. Tighten to 59 ft. lbs. (80 Nm).

➡**Counterhold using the Torx® wrench so as not to damage the boot.**

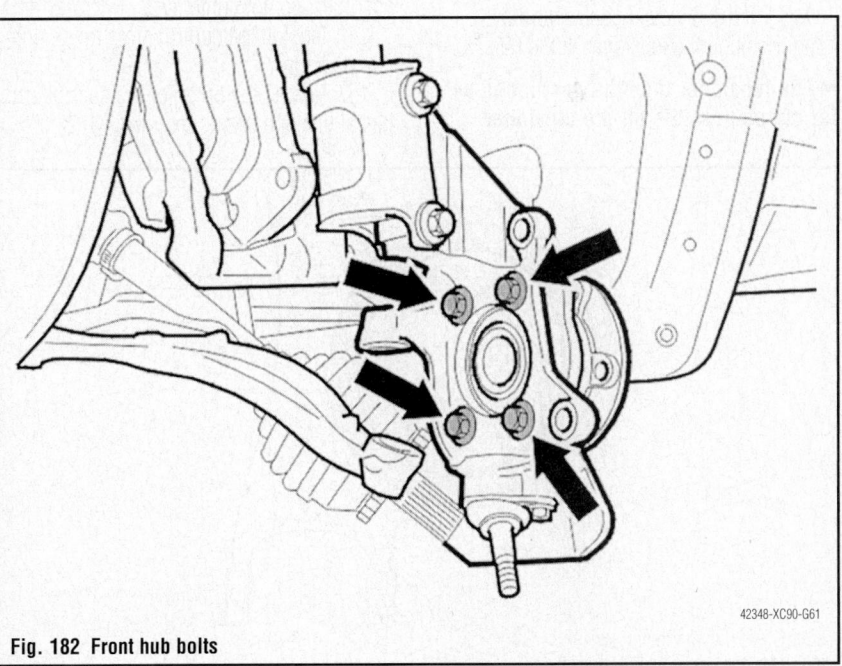
Fig. 182 Front hub bolts

19. Remove the tensioner strap.
20. Install the ABS cable harness in the brackets.
21. Install:
- The anti-roll bar link
- The nut (M12). Torque to 44 ft. lbs. (60 Nm)

- The steering arm
- The nut. Torque to 52 ft. lbs. (70 Nm)
- The brake disc
- The bolt.
22. Install a new halfshaft bolt.
23. Install:

- The brake caliper
- The two bolts
24. Install the wheel.

## ADJUSTMENT

If bearings are worn, replace with new sealed bearings.

---

# SUSPENSION

# REAR SUSPENSION

## COIL SPRING

### REMOVAL & INSTALLATION

*See Figures 183 and 184.*

1. Raise and support the vehicle.
2. Remove the wheels.
3. Install tensioner 999 5659. See the tensioner installation material under "Rear Shock Absorber".
4. Remove position sensor for Bi-Xenon lamps.
5. Remove:
- The bolt for the mounting for the lateral link
- The bolt holding the control arm in the steering knuckle.
6. Lower the control arm with the tensioner. The spring is now unloaded.
7. Lower the tensioner. Twist the tensioner so that it is positioned along the length of the car.
8. Press the control arm down by hand. Remove the spring.

### To install:

9. Install the new spring.
10. Check that the spring is correctly installed in the lower spring seat.
11. Lift the removed control arm by hand. Position the tensioner in the rails.

➡**The tensioner must be positioned as far out as possible on the tensioner**

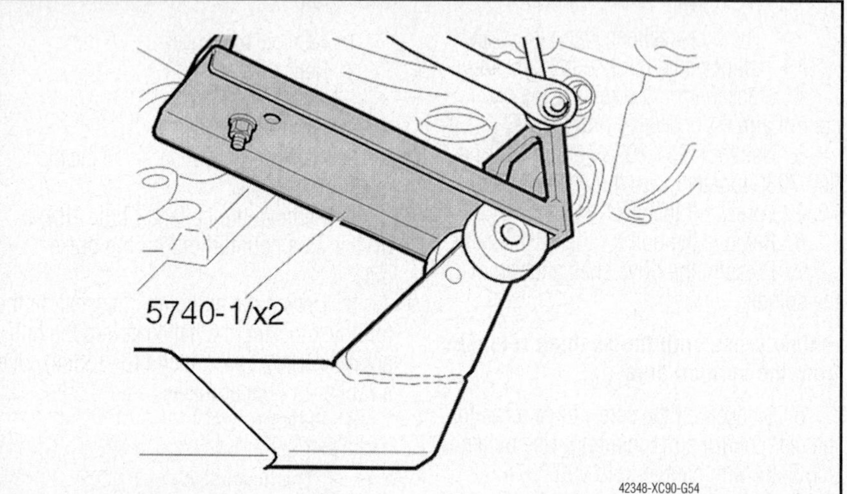

Fig. 184 Tensioner positioned on control arm

**plates to achieve the correct lifting force.**

12. Lift the control arms so that the bolt for the control arm and lateral link can be installed. Align the bolt. Tighten by hand. Do not tighten yet.
13. Lower the tensioner. Install the bolt in the lower mounting for the shock absorber. Do not tighten yet.
14. Tension the control arms to the normal position.
15. Tighten the bolts for the control arm, lateral link and shock absorber. M12

16. Install the position sensor for Bi-Xenon lamps. .
17. Remove the tensioner.
18. Install the wheel.

➡**If the car is equipped with Bi-Xenon lamps, the position sensor must be calibrated.**

## LOWER CONTROL ARM

### REMOVAL & INSTALLATION

*See Figure 185.*

1. Before servicing the vehicle, refer to the Precautions Section.
2. Raise the vehicle.

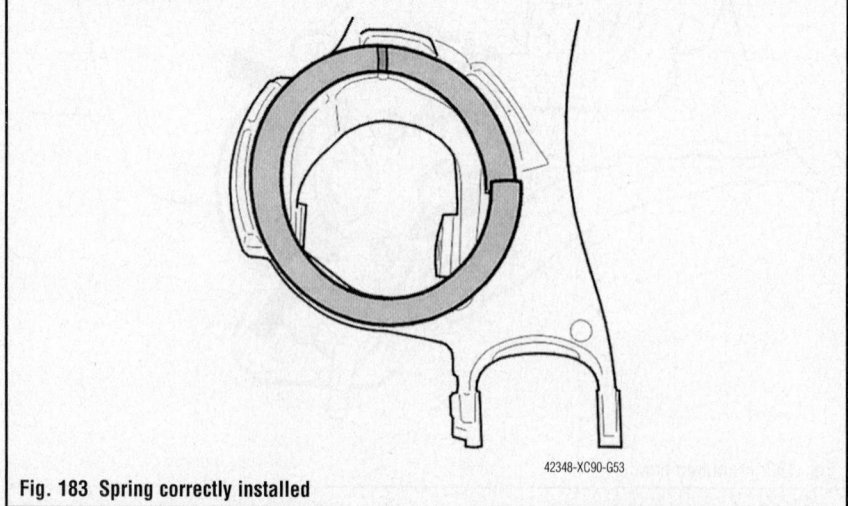

Fig. 183 Spring correctly installed

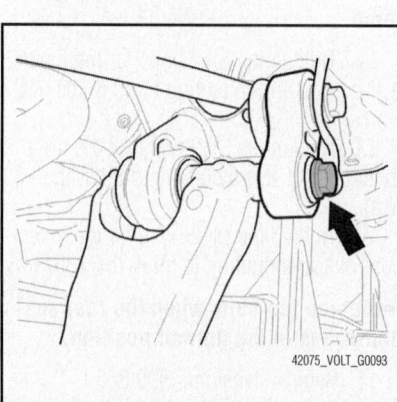

Fig. 185 Removal of rear lower control arm

3. Remove or disconnect the following:
- The wheel
- The coil spring. Refer to Coil Spring removal and installation
- The bolt in the sub-frame for the lower control arm
- The control arm
- The control arm rail (5740-1).

***To install:***

➡**Tighten all joints and rubber bushings with the rear suspension in the normal position.**

4. Install or connect the following:
- The control arm rail (5742-1) in the control arm
- The control arm in the sub-frame. Do not tighten yet
- The coil spring.

5. Raise the control arms to their normal position using the tensioner 999 5659

6. Tighten the M12 bolt for the inner mounting for the control arm. Tighten to 59 ft. lbs. (80 Nm).

➡**If the vehicle is equipped with Bi-Xenon lamps, the position sensor must be calibrated. Carry out calibration according to VIDA Vehicle communication (Function area 3, Electrical system, Rear electronic module).**

## SHOCK ABSORBER

*REMOVAL & INSTALLATION*

*See Figures 186 through 189.*

1. Raise the vehicle.
2. Remove the wheels.
3. Install tensioner 999 5659 as follows:

a. Remove the parking brake cable mountings from the sub-frame.
b. Install the tensioner.

➡**The tensioner must be positioned as far out as possible on the tensioner plates to provide the correct lifting force. It is vital that the screwed joints for the rubber bushings are tightened in the normal position (in the same position as when the car is on the ground and has three people in the car and a full fuel tank). The tensioner relieves the pressure on the components in the suspension when removing and installing.**

c. The tool is secured to the sub-frame with mounting 999 7061. This means that the lift stability is not affected during work.
d. Install the thread bar on the tensioner 999 5659

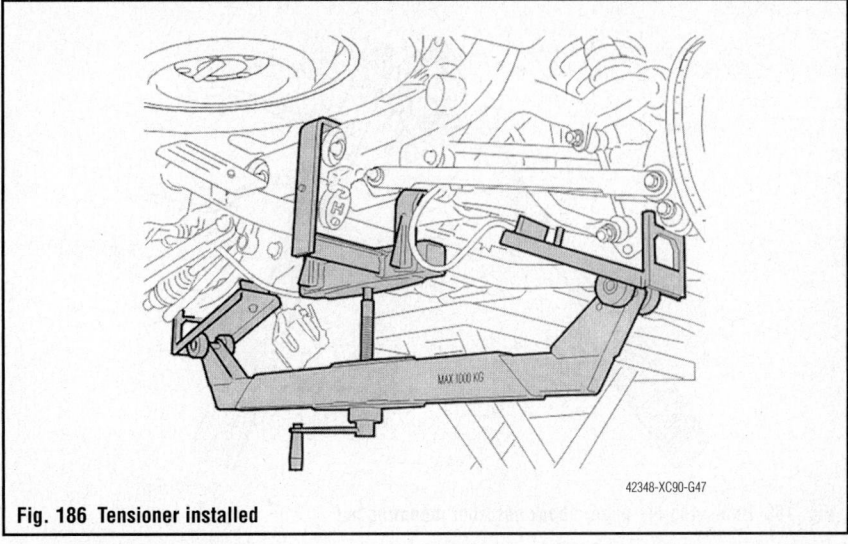

**Fig. 186 Tensioner installed**

e. Lift up the tensioner with the threaded bar 999 5659.

f. Insert the threaded bar in the mounting 999 7061 from underneath.

g. Insert the locking washer with the handle in from the side in mounting 999 7061 under the threaded bolt.

h. Lock the bolt in the locking washer. Align the locking washer in the mounting.

➡**The locking washer has a locating pin which must be aligned with the bolt.**

i. Install the 2 bolts (5740-36) from kit 998 9761 in the control arm holes.

j. Install the rails 5740-1 modified according to (WG-276) on the control arms.

k. Ensure that the rollers lie against the rails on both sides of the control arms.

➡**The tensioner must be positioned as far out as possible on the tensioner plates to provide the correct lifting force.**

l. Raise the control arms to their normal position using the tensioner. The normal position of the rear suspension is 18 inches (453mm) from the fender edge to the center of the wheel.

4. With 5 seats:
a. Remove:
- The front floor hatch
- The form-molded floor mat.

5. Fold the soundproofing over the shock absorber mounting out of the way.

6. With 7 seats:
a. Slide the seat cushion back on the rear row of seats.
b. Lift up the carpet.
c. Cut up the soundproofing as illustrated.

7. Fold up the soundproofing. Remove the cover on the splash guard.

8. Remove the shock absorber nut. Use socket 999 5500 and a Torx-wrench as a counterhold.

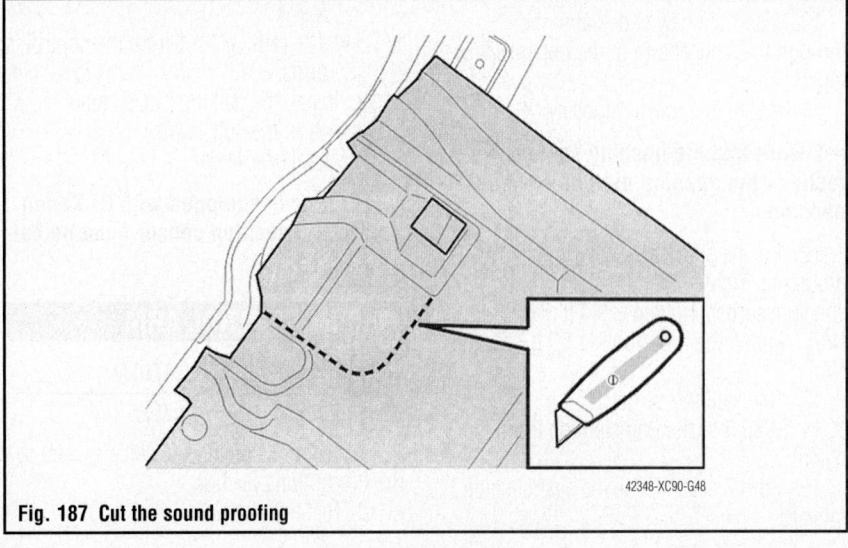

**Fig. 187 Cut the sound proofing**

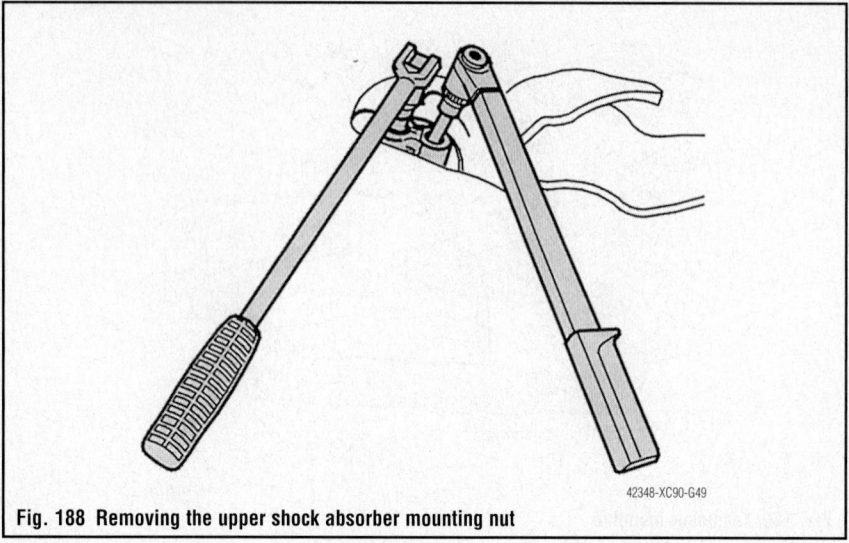

**Fig. 188 Removing the upper shock absorber mounting nut**

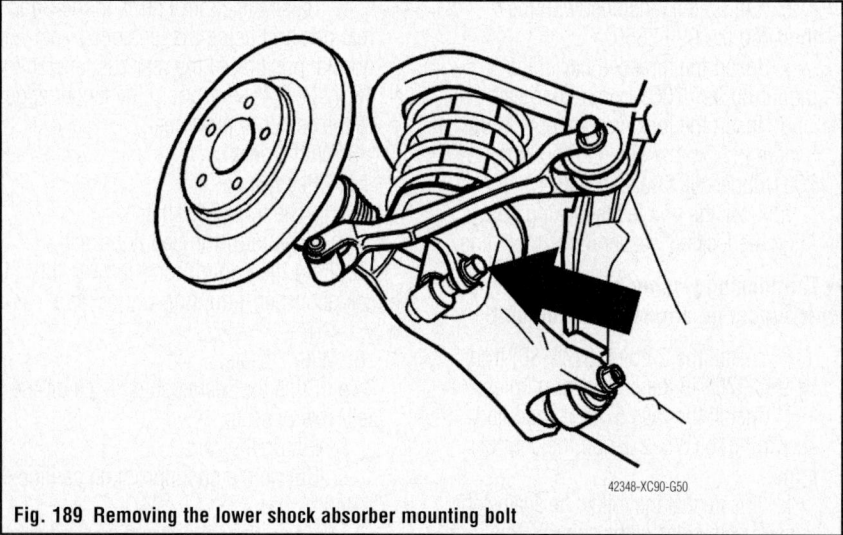

**Fig. 189 Removing the lower shock absorber mounting bolt**

9. Remove:
- The bolt in the lower mounting for the shock absorber
- The shock absorber.

### To install:

10. Insert the new shock absorber through the control arm to the upper mounting.

11. Install the new nut loosely.

➡**Ensure that the bushing seats correctly in the opening in the rear suspension.**

12. Fix the shock absorber in the lower mounting. Tighten the upper nut for the shock absorber. Tighten to 44 ft. lbs. (60 Nm). Tighten the lower bolt to 59 ft. lbs. (80 Nm).

13. Reinstall the carpet.

14. With 5 seats: Reinstall the front floor hatch.

15. With 7 seats: Slide the seat cushion forward.

16. Detach the tensioner, pull out the fork and lower the tensioner using the threaded bar.

17. Remove:
- Mounting 999 5659 from the sub-frame
- The rails 5740-1 from the control arms.

18. Install the parking brake cable mountings in the sub-frame.

19. Install the wheel.

➡**If the car is equipped with Bi-Xenon lamps, the position sensor must be calibrated.**

## UPPER CONTROL ARM

### REMOVAL & INSTALLATION

*See Figures 190 through 192.*

1. Before servicing the vehicle, refer to the Precautions Section.

2. Raise the vehicle.

3. Remove or disconnect the following:
- The wheel
- The brake caliper. Hang the brake caliper on a hook in the sub-frame.

4. Slacken off the screw for the outer control arm mounting by 3 turns.

5. Use tool: 951 2923, 999 7031 and 999 7030 together with threaded rod 999 7039 to press off the steering knuckle.

6. Remove the bolt for the drive shaft.

7. Press in the drive shaft and remove the spindle.

➡**Only press until the bushing releases from the splined area.**

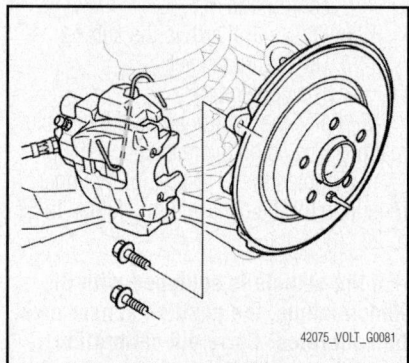

**Fig. 190 Using a hook to safely hang the brake caliper out of the way**

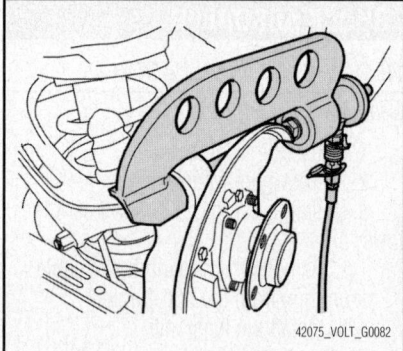

**Fig. 191 Using a press on the steering knuckle**

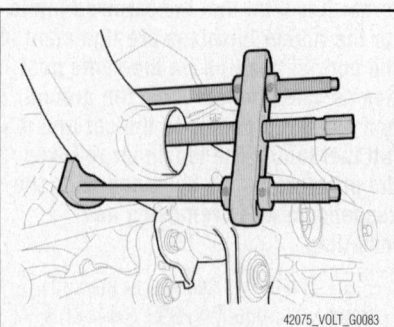

**Fig. 192 Using puller 998 5434 with counterhold 999 7074 on rear control arm mounting**

8. Slacken off the screw three turns for the rear control arm mounting. Use puller 998 5434 with counterhold 999 7074.

9. Remove or disconnect the following:
- The bolt for the inner rear control arm mounting
- The two bolts for the inner front control arm mounting
- The control arm

***To install:***

10. Install or connect the following:
- The bolts for the inner rear control arm mounting
- The bolts for the front inner control arm mounting
- The bolts for the outer control arm mounting

11. Install tensioner 999 5659.

➡**The tensioner must be positioned as far out as possible on the tensioner plates to achieve the correct lifting force.**

12. Tighten the M12 bolts for the inner rear control arm mounting to 59 ft. lbs. (80 Nm).

13. Tighten the M12 bolts for the front inner control arm mounting to 59 ft. lbs. (80 Nm).

14. Tighten the M12 bolts for the outer control arm mounting to 59 ft. lbs. (80 Nm).

➡**Tighten the bolts when the rear suspension is in the normal position.**

15. Remove tensioner 999 5659.

16. Install the wheel and tighten lug nuts to 102 ft. lbs. (138 Nm).

➡**If the vehicle is equipped with Bi-Xenon lamps, the position sensor must be calibrated. Carry out calibration according to VIDA vehicle communication (Function area 3, Electrical system, Rear Electronic Module).**

### WHEEL BEARINGS

*REMOVAL & INSTALLATION*

*See Figure 193.*

1. Lock the self-adjuster unit for the parking brake.

2. Remove:
- The rear wheel
- The bolt for the halfshaft.

3. Remove the wheel.

4. Remove:
- The brake caliper mounting bolts. Hang the caliper up using a piece of wire

- The brake disc locating pin/bolt
- The brake disc.

5. Install tensioner 999 5659. See the procedure under Shock Absorber removal and installation.

6. Support the rear suspension 20 inches (500mm) up between the fender edge and the center of the wheel.

7. Remove:
- The bolt in the mounting for the tie rod (1)
- The bolt in the mounting for the control arm (2)
- The nut in the mounting for the anti-roll bar link (3). Counterhold using a wrench so that the boot does not rotate
- The bolt for the mounting for the lateral link (4).

8. Position a transaxle jack under the steering knuckle and carefully raise.

➡**Pull the halfshaft from the hub whilst raising the steering knuckle.**

9. Release the tensioner so that the control arm hangs in the shock absorber.

10. Remove:
- The four bolts
- The hub.

***To install:***

11. Install:
- The hub
- The four bolts. Torque to 15 ft. lbs. (20 Nm), then 33 ft. lbs. (45 Nm) plus 60°

12. Carefully lower the transaxle jack. Install the halfshaft while lowering.

13. Remove the transaxle jack.

14. Support the tensioner to install the bolts for the control arm and lateral link.

15. Install:

➡**Tension the rear suspension to the normal position before tightening the bolts and mounting in the rubber bushing. (See normal position.)**

- The bolt for mounting the lateral link (4). (M12).
- The nut in the mounting for the anti-roll bar link (3). (M12) Counterhold using a wrench so that the boot does not rotate Torque to 59 ft. lbs. (80 Nm)

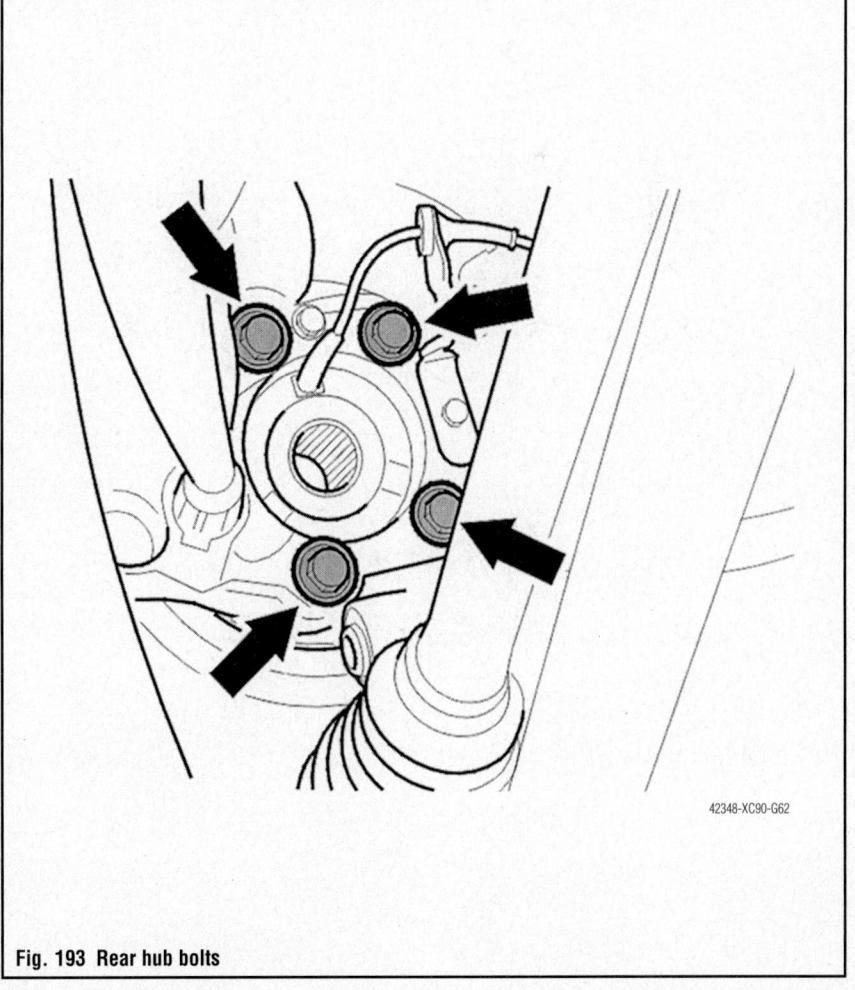

**Fig. 193 Rear hub bolts**

42348-XC90-G62

- The bolt for the mounting for the control arm (2) (M12)
- The bolt for the mounting for the tie rod (1). (M12).

16. Remove tensioner 999 5659.

17. Install:
- The brake disc
- The brake disc locating pin/bolt (M6)
- The brake caliper
- The brake caliper mounting bolts.
- The bolt for the halfshaft (M10)
- The rear wheel.

18. Activate the self adjuster for the parking brake.

## ADJUSTMENT

If bearings are worn, replace with new sealed bearings.

# GLOSSARY

**ABS:** Anti-lock braking system. An electro-mechanical braking system which is designed to minimize or prevent wheel lock-up during braking.

**ABSOLUTE PRESSURE:** Atmospheric (barometric) pressure plus the pressure gauge reading.

**ACCELERATOR PUMP:** A small pump located in the carburetor that feeds fuel into the air/fuel mixture during acceleration.

**ACCUMULATOR:** A device that controls shift quality by cushioning the shock of hydraulic oil pressure being applied to a clutch or band.

**ACTUATING MECHANISM:** The mechanical output devices of a hydraulic system, for example, clutch pistons and band servos.

**ACTUATOR:** The output component of a hydraulic or electronic system.

**ADVANCE:** Setting the ignition timing so that spark occurs earlier before the piston reaches top dead center (TDC).

**ADAPTIVE MEMORY (ADAPTIVE STRATEGY):** The learning ability of the TCM or PCM to redefine its decision-making process to provide optimum shift quality.

**AFTER TOP DEAD CENTER (ATDC):** The point after the piston reaches the top of its travel on the compression stroke.

**AIR BAG:** Device on the inside of the car designed to inflate on impact of crash, protecting the occupants of the car.

**AIR CHARGE TEMPERATURE (ACT) SENSOR:** The temperature of the airflow into the engine is measured by an ACT sensor, usually located in the lower intake manifold or air cleaner.

**AIR CLEANER:** An assembly consisting of a housing, filter and any connecting ductwork. The filter element is made up of a porous paper, sometimes with a wire mesh screening, and is designed to prevent airborne particles from entering the engine through the carburetor or throttle body.

**AIR INJECTION:** One method of reducing harmful exhaust emissions by injecting air into each of the exhaust ports of an engine. The fresh air entering the hot exhaust manifold causes any remaining fuel to be burned before it can exit the tailpipe.

**AIR PUMP:** An emission control device that supplies fresh air to the exhaust manifold to aid in more completely burning exhaust gases.

**AIR/FUEL RATIO:** The ratio of air-to-gasoline by weight in the fuel mixture drawn into the engine.

**ALDL (assembly line diagnostic link):** Electrical connector for scanning ECM/PCM/TCM input and output devices.

**ALIGNMENT RACK:** A special drive-on vehicle lift apparatus/measuring device used to adjust a vehicle's toe, caster and camber angles.

**ALL WHEEL DRIVE:** Term used to describe a full time four wheel drive system or any other vehicle drive system that continuously delivers power to all four wheels. This system is found primarily on station wagon vehicles and SUVs not utilized for significant off road use.

**ALTERNATING CURRENT (AC):** Electric current that flows first in one direction, then in the opposite direction, continually reversing flow.

**ALTERNATOR:** A device which produces AC (alternating current) which is converted to DC (direct current) to charge the car battery.

**AMMETER:** An instrument, calibrated in amperes, used to measure the flow of an electrical current in a circuit. Ammeters are always connected in series with the circuit being tested.

**AMPERAGE:** The total amount of current (amperes) flowing in a circuit.

**AMPLIFIER:** A device used in an electrical circuit to increase the voltage of an output signal.

**AMP/HR. RATING (BATTERY):** Measurement of the ability of a battery to deliver a stated amount of current for a stated period of time. The higher the amp/hr. rating, the better the battery.

**AMPERE:** The rate of flow of electrical current present when one volt of electrical pressure is applied against one ohm of electrical resistance.

**ANALOG COMPUTER:** Any microprocessor that uses similar (analogous) electrical signals to make its calculations.

**ANODIZED:** A special coating applied to the surface of aluminum valves for extended service life.

**ANTIFREEZE:** A substance (ethylene or propylene glycol) added to the coolant to prevent freezing in cold weather.

**ANTI-FOAM AGENTS:** Minimize fluid foaming from the whipping action encountered in the converter and planetary action.

**ANTI-WEAR AGENTS:** Zinc agents that control wear on the gears, bushings, and thrust washers.

**ANTI-LOCK BRAKING SYSTEM:** A supplementary system to the base hydraulic system that prevents sustained lock-up of the wheels during braking as well as automatically controlling wheel slip.

**ANTI-ROLL BAR:** See stabilizer bar.

**ARC:** A flow of electricity through the air between two electrodes or contact points that produces a spark.

**ARMATURE:** A laminated, soft iron core wrapped by a wire that converts electrical energy to mechanical energy as in a motor or relay. When rotated in a magnetic field, it changes mechanical energy into electrical energy as in a generator.

**ATDC:** After Top Dead Center.

**ATF:** Automatic transmission fluid.

**ATMOSPHERIC PRESSURE:** The pressure on the Earth's surface caused by the weight of the air in the atmosphere. At sea level, this pressure is 14.7 psi at 32°F (101 kPa at 0°C).

**ATOMIZATION:** The breaking down of a liquid into a fine mist that can be suspended in air.

**AUXILIARY ADD-ON COOLER:** A supplemental transmission fluid cooling device that is installed in series with the heat exchanger (cooler), located inside the radiator, to provide additional support to cool the hot fluid leaving the torque converter.

**AUXILIARY PRESSURE:** An added fluid pressure that is introduced into a regulator or balanced valve system to control valve movement. The auxiliary pressure itself can be either a fixed or a variable value. (See balanced valve; regulator valve.)

**AWD:** All wheel drive.

**AXIAL FORCE:** A side or end thrust force acting in or along the same plane as the power flow.

**AXIAL PLAY:** Movement parallel to a shaft or bearing bore.

**AXLE CAPACITY:** The maximum load-carrying capacity of the axle itself, as specified by the manufacturer. This is usually a higher number than the GAWR.

**AXLE RATIO:** This is a number (3.07:1, 4.56:1, for example) expressing the ratio between driveshaft revolutions and wheel revolutions. A low numerical ratio allows the engine to work easier because it doesn't have to turn as fast. A high numerical ratio means that the engine has to turn more rpm's to move the wheels through the same number of turns.

**BACKFIRE:** The sudden combustion of gases in the intake or exhaust system that results in a loud explosion.

**BACKLASH:** The clearance or play between two parts, such as meshed gears.

**BACKPRESSURE:** Restrictions in the exhaust system that slow the exit of exhaust gases from the combustion chamber.

**BAKELITE®:** A heat resistant, plastic insulator material commonly used in printed circuit boards and transistorized components.

**BALANCED VALVE:** A valve that is positioned by opposing auxiliary hydraulic pressures and/or spring force. Examples include mainline regulator, throttle, and governor valves. (See regulator valve.)

**BAND:** A flexible ring of steel with an inner lining of friction material. When tightened around the outside of a drum, a planetary member is held stationary to the transmission/transaxle case.

**BALL BEARING:** A bearing made up of hardened inner and outer races between which hardened steel balls roll.

**BALL JOINT:** A ball and matching socket connecting suspension components (steering knuckle to lower control arms). It permits rotating movement in any direction between the components that are joined.

**BARO (BAROMETRIC PRESSURE SENSOR):** Measures the change in the intake manifold pressure caused by changes in altitude.

**BAROMETRIC MANIFOLD ABSOLUTE PRESSURE (BMAP) SENSOR:** Operates similarly to a conventional MAP sensor; reads intake manifold pressure and is also responsible for determining altitude and barometric pressure prior to engine operation.

**BAROMETRIC PRESSURE:** (See atmospheric pressure.)

**BALLAST RESISTOR:** A resistor in the primary ignition circuit that lowers voltage after the engine is started to reduce wear on ignition components.

**BATTERY:** A direct current electrical storage unit, consisting of the basic active materials of lead and sulfuric acid, which converts chemical energy into electrical energy. Used to provide current for the operation of the starter as well as other equipment, such as the radio, lighting, etc.

**BEAD:** The portion of a tire that holds it on the rim.

**BEARING:** A friction reducing, supportive device usually located between a stationary part and a moving part.

**BEFORE TOP DEAD CENTER (BTDC):** The point just before the piston reaches the top of its travel on the compression stroke.

**BELTED TIRE:** Tire construction similar to bias-ply tires, but using two or more layers of reinforced belts between body plies and the tread.

**BEZEL:** Piece of metal surrounding radio, headlights, gauges or similar components; sometimes used to hold the glass face of a gauge in the dash.

**BIAS-PLY TIRE:** Tire construction, using body ply reinforcing cords which run at alternating angles to the center line of the tread.

**BI-METAL TEMPERATURE SENSOR:** Any sensor or switch made of two dissimilar types of metal that bend when heated or cooled due to the different expansion rates of the alloys. These types of sensors usually function as an on/off switch.

**BLOCK:** See Engine Block.

**BLOW-BY:** Combustion gases, composed of water vapor and unburned fuel, that leak past the piston rings into the crankcase during normal engine operation. These gases are removed by the PCV system to prevent the buildup of harmful acids in the crankcase.

**BOOK TIME:** See Labor Time.

**BOOK VALUE:** The average value of a car, widely used to determine trade-in and resale value.

**BOOST VALVE:** Used at the base of the regulator valve to increase mainline pressure.

**BORE:** Diameter of a cylinder.

**BRAKE CALIPER:** The housing that fits over the brake disc. The caliper holds the brake pads, which are pressed against the discs by the caliper pistons when the brake pedal is depressed.

**BRAKE HORSEPOWER (BHP):** The actual horsepower available at the engine flywheel as measured by a dynamometer.

**BRAKE FADE:** Loss of braking power, usually caused by excessive heat after repeated brake applications.

**BRAKE HORSEPOWER:** Usable horsepower of an engine measured at the crankshaft.

**BRAKE PAD:** A brake shoe and lining assembly used with disc brakes.

**BRAKE PROPORTIONING VALVE:** A valve on the master cylinder which restricts hydraulic brake pressure to the wheels to a specified amount, preventing wheel lock-up.

**BREAKAWAY:** Often used by Chrysler to identify first-gear operation in D and 2 ranges. In these ranges, first-gear operation depends on a one-way roller clutch that holds on acceleration and releases (breaks away) on deceleration, resulting in a freewheeling coast-down condition.

**BRAKE SHOE:** The backing for the brake lining. The term is, however, usually applied to the assembly of the brake backing and lining.

**BREAKER POINTS:** A set of points inside the distributor, operated by a cam, which make and break the ignition circuit.

**BRINNELLING:** A wear pattern identified by a series of indentations at regular intervals. This condition is caused by a lack of lube, overload situations, and/or vibrations.

**BTDC:** Before Top Dead Center.

**BUMP:** Sudden and forceful apply of a clutch or band.

**BUSHING:** A liner, usually removable, for a bearing; an anti-friction liner used in place of a bearing.

**CALIFORNIA ENGINE:** An engine certified by the EPA for use in California only; conforms to more stringent emission regulations than Federal engine.

**CALIPER:** A hydraulically activated device in a disc brake system,

which is mounted straddling the brake rotor (disc). The caliper contains at least one piston and two brake pads. Hydraulic pressure on the piston(s) forces the pads against the rotor.

**CAPACITY:** The quantity of electricity that can be delivered from a unit, as from a battery in ampere-hours, or output, as from a generator.

**CAMBER:** One of the factors of wheel alignment. Viewed from the front of the car, it is the inward or outward tilt of the wheel. The top of the tire will lean outward (positive camber) or inward (negative camber).

**CAMSHAFT:** A shaft in the engine on which are the lobes (cams) which operate the valves. The camshaft is driven by the crankshaft, via a belt, chain or gears, at one half the crankshaft speed.

**CAPACITOR:** A device which stores an electrical charge.

**CARBON MONOXIDE (CO):** A colorless, odorless gas given off as a normal byproduct of combustion. It is poisonous and extremely dangerous in confined areas, building up slowly to toxic levels without warning if adequate ventilation is not available.

**CARBURETOR:** A device, usually mounted on the intake manifold of an engine, which mixes the air and fuel in the proper proportion to allow even combustion.

**CASTER:** The forward or rearward tilt of an imaginary line drawn through the upper ball joint and the center of the wheel. Viewed from the sides, positive caster (forward tilt) lends directional stability, while negative caster (rearward tilt) produces instability.

**CATALYTIC CONVERTER:** A device installed in the exhaust system, like a muffler, that converts harmful byproducts of combustion into carbon dioxide and water vapor by means of a heat-producing chemical reaction.

**CENTRIFUGAL ADVANCE:** A mechanical method of advancing the spark timing by using flyweights in the distributor that react to centrifugal force generated by the distributor shaft rotation.

**CENTRIFUGAL FORCE:** The outward pull of a revolving object, away from the center of revolution. Centrifugal force increases with the speed of rotation.

**CETANE RATING:** A measure of the ignition value of diesel fuel. The higher the cetane rating, the better the fuel. Diesel fuel cetane rating is roughly comparable to gasoline octane rating.

**CHECK VALVE:** Any one-way valve installed to permit the flow of air, fuel or vacuum in one direction only.

**CHOKE:** The valve/plate that restricts the amount of air entering an engine on the induction stroke, thereby enriching the air/fuel ratio.

**CHUGGLE:** Bucking or jerking condition that may be engine related and may be most noticeable when converter clutch is engaged; similar to the feel of towing a trailer.

**CIRCLIP:** A split steel snapring that fits into a groove to hold various parts in place.

**CIRCUIT BREAKER:** A switch which protects an electrical circuit from overload by opening the circuit when the current flow exceeds a pre-determined level. Some circuit breakers must be reset manually, while most reset automatically.

**CIRCUIT:** Any unbroken path through which an electrical current can flow. Also used to describe fuel flow in some instances.

**CIRCUIT, BYPASS:** Another circuit in parallel with the major circuit through which power is diverted.

**CIRCUIT, CLOSED:** An electrical circuit in which there is no interruption of current flow.

**CIRCUIT, GROUND:** The non-insulated portion of a complete circuit used as a common potential point. In automotive circuits, the ground is composed of metal parts, such as the engine, body sheet metal, and frame and is usually a negative potential.

**CIRCUIT, HOT:** That portion of a circuit not at ground potential. The hot circuit is usually insulated and is connected to the positive side of the battery.

**CIRCUIT, OPEN:** A break or lack of contact in an electrical circuit, either intentional (switch) or unintentional (bad connection or broken wire).

**CIRCUIT, PARALLEL:** A circuit having two or more paths for current flow with common positive and negative tie points. The same voltage is applied to each load device or parallel branch.

**CIRCUIT, SERIES:** An electrical system in which separate parts are connected end to end, using one wire, to form a single path for current to flow.

**CIRCUIT, SHORT:** A circuit that is accidentally completed in an electrical path for which it was not intended.

**CLAMPING (ISOLATION) DIODES:** Diodes positioned in a circuit to prevent self-induction from damaging electronic components.

**CLEARCOAT:** A transparent layer which, when sprayed over a vehicle's paint job, adds gloss and depth as well as an additional protective coating to the finish.

**CLUTCH:** Part of the power train used to connect/disconnect power to the rear wheels.

**CLUTCH, FLUID:** The same as a fluid coupling. A fluid clutch or coupling performs the same function as a friction clutch by utilizing fluid friction and inertia as opposed to solid friction used by a friction clutch. (See fluid coupling.)

**CLUTCH, FRICTION:** A coupling device that provides a means of smooth and positive engagement and disengagement of engine torque to the vehicle powertrain. Transmission of power through the clutch is accomplished by bringing one or more rotating drive members into contact with complementing driven members.

**COAST:** Vehicle deceleration caused by engine braking conditions.

**COEFFICIENT OF FRICTION:** The amount of surface tension between two contacting surfaces; identified by a scientifically calculated number.

**COIL:** Part of the ignition system that boosts the relatively low voltage supplied by the car's electrical system to the high voltage required to fire the spark plugs.

**COMBINATION MANIFOLD:** An assembly which includes both the intake and exhaust manifolds in one casting.

**COMBINATION VALVE:** A device used in some fuel systems that routes fuel vapors to a charcoal storage canister instead of venting them into the atmosphere. The valve relieves fuel tank pressure and allows fresh air into the tank as the fuel level drops to prevent a vapor lock situation.

**COMBUSTION CHAMBER:** The part of the engine in the cylinder head where combustion takes place.

**COMPOUND GEAR:** A gear consisting of two or more simple gears with a common shaft.

**COMPOUND PLANETARY:** A gearset that has more than the three elements found in a simple gearset and is constructed by combining members of two planetary gearsets to create additional gear ratio possibilities.

**COMPRESSION CHECK:** A test involving removing each spark plug and inserting a gauge. When the engine is cranked, the gauge will record a pressure reading in the individual cylinder. General operating condition can be determined from a compression check.

**COMPRESSION RATIO:** The ratio of the volume between the piston and cylinder head when the piston is at the bottom of its stroke (bottom dead center) and when the piston is at the top of its stroke (top dead center).

**COMPUTER:** An electronic control module that correlates input data according to prearranged engineered instructions; used for the management of an actuator system or systems.

**CONDENSER:** An electrical device which acts to store an electrical charge, preventing voltage surges.

2. A radiator-like device in the air conditioning system in which refrigerant gas condenses into a liquid, giving off heat.

**CONDUCTOR:** Any material through which an electrical current can be transmitted easily.

**CONNECTING ROD:** The connecting link between the crankshaft and piston.

**CONSTANT VELOCITY JOINT:** Type of universal joint in a halfshaft assembly in which the output shaft turns at a constant angular velocity without variation, provided that the speed of the input shaft is constant.

**CONTINUITY:** Continuous or complete circuit. Can be checked with an ohmmeter.

**CONTROL ARM:** The upper or lower suspension components which are mounted on the frame and support the ball joints and steering knuckles.

**CONVENTIONAL IGNITION:** Ignition system which uses breaker points.

**CONVERTER:** (See torque converter.)

**CONVERTER LOCKUP:** The switching from hydrodynamic to direct mechanical drive, usually through the application of a friction element called the converter clutch.

**COOLANT:** Mixture of water and anti-freeze circulated through the engine to carry off heat produced by the engine.

**CORROSION INHIBITOR:** An inhibitor in ATF that prevents corrosion of bushings, thrust washers, and oil cooler brazed joints.

**COUNTERSHAFT:** An intermediate shaft which is rotated by a mainshaft and transmits, in turn, that rotation to a working part.

**COUPLING PHASE:** Occurs when the torque converter is operating at its greatest hydraulic efficiency. The speed differential between the impeller and the turbine is at its minimum. At this point, the stator freewheels, and there is no torque multiplication.

**CRANKCASE:** The lower part of an engine in which the crankshaft and related parts operate.

**CRANKSHAFT:** Engine component (connected to pistons by connecting rods) which converts the reciprocating (up and down) motion of pistons to rotary motion used to turn the driveshaft.

**CURB WEIGHT:** The weight of a vehicle without passengers or payload, but including all fluids (oil, gas, coolant, etc.) and other equipment specified as standard.

**CURRENT:** The flow (or rate) of electrons moving through a circuit. Current is measured in amperes (amp).

**CURRENT FLOW CONVENTIONAL:** Current flows through a circuit from the positive terminal of the source to the negative terminal (plus to minus).

**CURRENT FLOW, ELECTRON:** Current or electrons flow from the negative terminal of the source, through the circuit, to the positive terminal (minus to plus).

**CV-JOINT:** Constant velocity joint.

**CYCLIC VIBRATIONS:** The off-center movement of a rotating object that is affected by its initial balance, speed of rotation, and working angles.

**CYLINDER BLOCK:** See engine block.

**CYLINDER HEAD:** The detachable portion of the engine, usually fastened to the top of the cylinder block and containing all or most of the combustion chambers. On overhead valve engines, it contains the valves and their operating parts. On overhead cam engines, it contains the camshaft as well.

**CYLINDER:** In an engine, the round hole in the engine block in which the piston(s) ride.

**DATA LINK CONNECTOR (DLC):** Current acronym/term applied to the federally mandated, diagnostic junction connector that is used to monitor ECM/PC/TCM inputs, processing strategies, and outputs including diagnostic trouble codes (DTCs).

**DEAD CENTER:** The extreme top or bottom of the piston stroke.

**DECELERATION BUMP:** When referring to a torque converter clutch in the applied position, a sudden release of the accelerator pedal causes a forceful reversal of power through the drivetrain (engine braking), just prior to the apply plate actually being released.

**DELAYED (LATE OR EXTENDED):** Condition where shift is expected but does not occur for a period of time, for example, where clutch or band engagement does not occur as quickly as expected during part throttle or wide open throttle apply of accelerator or when manually downshifting to a lower range.

**DETENT:** A spring-loaded plunger, pin, ball, or pawl used as a holding device on a ratchet wheel or shaft. In automatic transmissions, a detent mechanism is used for locking the manual valve in place.

**DETENT DOWNSHIFT:** (See kickdown.)

**DETERGENT:** An additive in engine oil to improve its operating characteristics.

**DETONATION:** An unwanted explosion of the air/fuel mixture in the combustion chamber caused by excess heat and compression, advanced timing, or an overly lean mixture. Also referred to as "ping".

**DEXRON®:** A brand of automatic transmission fluid.

**DIAGNOSTIC TROUBLE CODES (DTCs):** A digital display from the control module memory that identifies the input, processor, or output device circuit that is related to the powertrain emission/driveability malfunction detected. Diagnostic trouble codes can be read by the MIL to flash any codes or by using a handheld scanner.

**DIAPHRAGM:** A thin, flexible wall separating two cavities, such as in a vacuum advance unit.

**DIESELING:** The engine continues to run after the car is shut off; caused by fuel continuing to be burned in the combustion chamber.

**DIFFERENTIAL:** A geared assembly which allows the transmission of motion between drive axles, giving one axle the ability to rotate faster than the other, as in cornering.

**DIFFERENTIAL AREAS:** When opposing faces of a spool valve are acted upon by the same pressure but their areas differ in size, the face with the larger area produces the differential force and valve movement. (See spool valve.)

**DIFFERENTIAL FORCE:** (See differential areas)

**DIGITAL READOUT:** A display of numbers or a combination of numbers and letters.

**DIGITAL VOLT OHMMETER:** An electronic diagnostic tool used to measure voltage, ohms and amps as well as several other functions, with the readings displayed on a digital screen in tenths, hundredths and thousandths.

**DIODE:** An electrical device that will allow current to flow in one direction only.

**DIRECT CURRENT (DC):** Electrical current that flows in one direction only.

**DIRECT DRIVE:** The gear ratio is 1:1, with no change occurring in the torque and speed input/output relationship.

**DISC BRAKE:** A hydraulic braking assembly consisting of a brake disc, or rotor, mounted on an axle shaft, and a caliper assembly containing, usually two brake pads which are activated by hydraulic pressure. The pads are forced against the sides of the disc, creating friction which slows the vehicle.

**DISPERSANTS:** Suspend dirt and prevent sludge buildup in a liquid, such as engine oil.

**DOUBLE BUMP (DOUBLE FEEL):** Two sudden and forceful applies of a clutch or band.

**DISPLACEMENT:** The total volume of air that is displaced by all pistons as the engine turns through one complete revolution.

**DISTRIBUTOR:** A mechanically driven device on an engine which is responsible for electrically firing the spark plug at a pre-determined point of the piston stroke.

**DOHC:** Double overhead camshaft.

**DOUBLE OVERHEAD CAMSHAFT:** The engine utilizes two camshafts mounted in one cylinder head. One camshaft operates the exhaust valves, while the other operates the intake valves.

**DOWEL PIN:** A pin, inserted in mating holes in two different parts allowing those parts to maintain a fixed relationship.

**DRIVELINE:** The drive connection between the transmission and the drive wheels.

**DRIVE TRAIN:** The components that transmit the flow of power from the engine to the wheels. The components include the clutch, transmission, driveshafts (or axle shafts in front wheel drive), U-joints and differential.

**DRUM BRAKE:** A braking system which consists of two brake shoes and one or two wheel cylinders, mounted on a fixed backing plate, and a brake drum, mounted on an axle, which revolves around the assembly.

**DRY CHARGED BATTERY:** Battery to which electrolyte is added when the battery is placed in service.

**DVOM:** Digital volt ohmmeter

**DWELL:** The rate, measured in degrees of shaft rotation, at which an electrical circuit cycles on and off.

**DYNAMIC:** An application in which there is rotating or reciprocating motion between the parts.

**EARLY:** Condition where shift occurs before vehicle has reached proper speed, which tends to labor engine after upshift.

**EBCM:** See Electronic Control Unit (ECU).

**ECM:** See Electronic Control Unit (ECU).

**ECU:** Electronic control unit.

**ELECTRODE:** Conductor (positive or negative) of electric current.

**ELECTROLYSIS:** A surface etching or bonding of current conducting transmission/transaxle components that may occur when grounding straps are missing or in poor condition.

**ELECTROLYTE:** A solution of water and sulfuric acid used to activate the battery. Electrolyte is extremely corrosive.

**ELECTROMAGNET:** A coil that produces a magnetic field when current flows through its windings.

**ELECTROMAGNETIC INDUCTION:** A method to create (generate) current flow through the use of magnetism.

**ELECTROMAGNETISM:** The effects surrounding the relationship between electricity and magnetism.

**ELECTROMOTIVE FORCE (EMF):** The force or pressure (voltage) that causes current movement in an electrical circuit.

**ELECTRONIC CONTROL UNIT:** A digital computer that controls engine (and sometimes transmission, brake or other vehicle system) functions based on data received from various sensors. Examples used by some manufacturers include Electronic Brake Control Module (EBCM), Engine Control Module (ECM), Powertrain Control Module (PCM) or Vehicle Control Module (VCM).

**ELECTRONIC IGNITION:** A system in which the timing and firing of the spark plugs is controlled by an electronic control unit, usually called a module. These systems have no points or condenser.

**ELECTRONIC PRESSURE CONTROL (EPC) SOLENOID:** A specially designed solenoid containing a spool valve and spring assembly to control fluid mainline pressure. A variable current flow, controlled by the ECM/PCM, varies the internal force of the solenoid on the spool valve and resulting mainline pressure. (See variable force solenoid.)

**ELECTRONICS:** Miniaturized electrical circuits utilizing semiconductors, solid-state devices, and printed circuits. Electronic circuits utilize small amounts of power.

**ELECTRONIFICATION:** The application of electronic circuitry to a mechanical device. Regarding automatic transmissions, electrification is incorporated into converter clutch lockup, shift scheduling, and line pressure control systems.

**ELECTROSTATIC DISCHARGE (ESD):** An unwanted, high-voltage electrical current released by an individual who has taken on a static charge of electricity. Electronic components can be easily damaged by ESD.

**ELEMENT:** A device within a hydrodynamic drive unit designed with a set of blades to direct fluid flow.

**ENAMEL:** Type of paint that dries to a smooth, glossy finish.

**END BUMP (END FEEL OR SLIP BUMP):** Firmer feel at end of shift when compared with feel at start of shift.

**END-PLAY:** The clearance/gap between two components that allows for expansion of the parts as they warm up, to prevent binding and to allow space for lubrication.

**ENERGY:** The ability or capacity to do work.

**ENGINE:** The primary motor or power apparatus of a vehicle, which converts liquid or gas fuel into mechanical energy.

**ENGINE BLOCK:** The basic engine casting containing the cylinders, the crankshaft main bearings, as well as machined surfaces for the mounting of other components such as the cylinder head, oil pan, transmission, etc.

**ENGINE BRAKING:** Use of engine to slow vehicle by manually downshifting during zero-throttle coast down.

**ENGINE CONTROL MODULE (ECM):** Manages the engine and incorporates output control over the torque converter clutch solenoid. (Note: Current designation for the ECM in late model vehicles is PCM.)

**ENGINE COOLANT TEMPERATURE (ECT) SENSOR:** Prevents converter clutch engagement with a cold engine; also used for shift timing and shift quality.

**EP LUBRICANT:** EP (extreme pressure) lubricants are specially formulated for use with gears involving heavy loads (transmissions, differentials, etc.).

**ETHYL:** A substance added to gasoline to improve its resistance to knock, by slowing down the rate of combustion.

**ETHYLENE GLYCOL:** The base substance of antifreeze.

**EXHAUST MANIFOLD:** A set of cast passages or pipes which conduct exhaust gases from the engine.

**FAIL-SAFE (BACKUP) CONTROL:** A substitute value used by the PCM/TCM to replace a faulty signal from an input sensor. The temporary value allows the vehicle to continue to be operated.

**FAST IDLE:** The speed of the engine when the choke is on. Fast idle speeds engine warm-up.

**FEDERAL ENGINE:** An engine certified by the EPA for use in any of the 49 states (except California).

**FEEDBACK:** A circuit malfunction whereby current can find another path to feed load devices.

**FEELER GAUGE:** A blade, usually metal, of precisely predetermined thickness, used to measure the clearance between two parts.

**FILAMENT:** The part of a bulb that glows; the filament creates high resistance to current flow and actually glows from the resulting heat.

**FINAL DRIVE:** An essential part of the axle drive assembly where final gear reduction takes place in the powertrain. In RWD applications and north-south FWD applications, it must also change the power flow direction to the axle shaft by ninety degrees. (Also see axle ratio).

**FIRING ORDER:** The order in which combustion occurs in the cylinders of an engine. Also the order in which spark is distributed to the plugs by the distributor.

**FIRM:** A noticeable quick apply of a clutch or band that is considered normal with medium to heavy throttle shift; should not be confused with harsh or rough.

**FLAME FRONT:** The term used to describe certain aspects of the fuel explosion in the cylinders. The flame front should move in a controlled pattern across the cylinder, rather than simply exploding immediately.

**FLARE (SLIPPING):** A quick increase in engine rpm accompanied by momentary loss of torque; generally occurs during shift.

**FLAT ENGINE:** Engine design in which the pistons are horizontally opposed. Porsche, Subaru and some old VW are common examples of flat engines.

**FLAT RATE:** A dealership term referring to the amount of money paid to a technician for a repair or diagnostic service based on that particular service versus dealership's labor time (NOT based on the actual time the technician spent on the job).

**FLAT SPOT:** A point during acceleration when the engine seems to lose power for an instant.

**FLOODING:** The presence of too much fuel in the intake manifold and combustion chamber which prevents the air/fuel mixture from firing, thereby causing a no-start situation.

**FLUID:** A fluid can be either liquid or gas. In hydraulics, a liquid is used for transmitting force or motion.

**FLUID COUPLING:** The simplest form of hydrodynamic drive, the fluid coupling consists of two look-alike members with straight radial varies referred to as the impeller (pump) and the turbine. Input torque is always equal to the output torque.

**FLUID DRIVE:** Either a fluid coupling or a fluid torque converter. (See hydrodynamic drive units.)

**FLUID TORQUE CONVERTER:** A hydrodynamic drive that has the ability to act both as a torque multiplier and fluid coupling. (See hydrodynamic drive units; torque converter.)

**FLUID VISCOSITY:** The resistance of a liquid to flow. A cold fluid (oil) has greater viscosity and flows more slowly than a hot fluid (oil).

**FLYWHEEL:** A heavy disc of metal attached to the rear of the crankshaft. It smoothes the firing impulses of the engine and keeps the crankshaft turning during periods when no firing takes place. The starter also engages the flywheel to start the engine.

**FOOT POUND (ft. lbs., lbs. ft. or sometimes, ft. lb.):** The amount of energy or work needed to raise an item weighing one pound, a distance of one foot.

**FREEZE PLUG:** A plug in the engine block which will be pushed out if the coolant freezes. Sometimes called expansion plugs, they protect the block from cracking should the coolant freeze.

**FRICTION:** The resistance that occurs between contacting surfaces. This relationship is expressed by a ratio called the coefficient of friction (CL).

**FRICTION, COEFFICIENT OF:** The amount of surface tension between two contacting surfaces; expressed by a scientifically calculated number.

**FRONT END ALIGNMENT:** A service to set caster, camber and toe-in to the correct specifications. This will ensure that the car steers and handles properly and that the tires wear properly.

**FRICTION MODIFIER:** Changes the coefficient of friction of the fluid between the mating steel and composition clutch/band surfaces during the engagement process and allows for a certain amount of intentional slipping for a good "shift-feel".

**FRONTAL AREA:** The total frontal area of a vehicle exposed to air flow.

**FUEL FILTER:** A component of the fuel system containing a porous paper element used to prevent any impurities from entering the engine through the fuel system. It usually takes the form of a canister-like housing, mounted in-line with the fuel hose, located anywhere on a vehicle between the fuel tank and engine.

**FUEL INJECTION:** A system replacing the carburetor that sprays fuel into the cylinder through nozzles. The amount of fuel can be more precisely controlled with fuel injection.

**FULL FLOATING AXLE:** An axle in which the axle housing extends through the wheel giving bearing support on the outside of the housing. The front axle of a four-wheel drive vehicle is usually a full floating axle, as are the rear axles of many larger (1 ton and over) pick-ups and vans.

**FULL-TIME FOUR-WHEEL DRIVE:** A four-wheel drive system that continuously delivers power to all four wheels. A differential between the front and rear driveshafts permits variations in axle speeds to control gear wind-up without damage.

**FULL THROTTLE DETENT DOWNSHIFT:** A quick apply of accelerator pedal to its full travel, forcing a downshift.

**FUSE:** A protective device in a circuit which prevents circuit overload by breaking the circuit when a specific amperage is present. The device is constructed around a strip or wire of a lower amperage rating than the circuit it is designed to protect. When an amperage higher than that stamped on the fuse is present in the circuit, the strip or wire melts, opening the circuit.

**FUSIBLE LINK:** A piece of wire in a wiring harness that performs the same job as a fuse. If overloaded, the fusible link will melt and interrupt the circuit.

**FWD:** Front wheel drive.

**GAWR:** (Gross axle weight rating) the total maximum weight an axle is designed to carry.

**GCW:** (Gross combined weight) total combined weight of a tow vehicle and trailer.

**GARAGE SHIFT:** initial engagement feel of transmission, neutral to reverse or neutral to a forward drive.

**GARAGE SHIFT FEEL:** A quick check of the engagement quality and responsiveness of reverse and forward gears. This test is done with the vehicle stationary.

**GEAR:** A toothed mechanical device that acts as a rotating lever to transmit power or turning effort from one shaft to another. (See gear ratio.)

**GEAR RATIO:** A ratio expressing the number of turns a smaller gear will make to turn a larger gear through one revolution. The ratio is found by dividing the number of teeth on the smaller gear into the number of teeth on the larger gear.

**GEARBOX:** Transmission

**GEAR REDUCTION:** Torque is multiplied and speed decreased by the factor of the gear ratio. For example, a 3:1 gear ratio changes an input torque of 180 ft. lbs. and an input speed of 2700 rpm to 540 Ft. lbs. and 900 rpm, respectively. (No account is taken of frictional losses, which are always present.)

**GEARTRAIN:** A succession of intermeshing gears that form an assembly and provide for one or more torque changes as the power input is transmitted to the power output.

**GEL COAT:** A thin coat of plastic resin covering fiberglass body panels.

**GENERATOR:** A device which produces direct current (DC) necessary to charge the battery.

**GOVERNOR:** A device that senses vehicle speed and generates a hydraulic oil pressure. As vehicle speed increases, governor oil pressure rises.

**GROUND CIRCUIT:** (See circuit, ground.)

**GROUND SIDE SWITCHING:** The electrical/electronic circuit control switch is located after the circuit load.

**GVWR:** (Gross vehicle weight rating) total maximum weight a vehicle is designed to carry including the weight of the vehicle, passengers, equipment, gas, oil, etc.

**HALOGEN:** A special type of lamp known for its quality of brilliant white light. Originally used for fog lights and driving lights.

**HARD CODES:** DTCs that are present at the time of testing; also called continuous or current codes.

**HARSH(ROUGH):** An apply of a clutch or band that is more noticeable than a firm one; considered undesirable at any throttle position.

**HEADER TANK:** An expansion tank for the radiator coolant. It can be located remotely or built into the radiator.

**HEAT RANGE:** A term used to describe the ability of a spark plug to carry away heat. Plugs with longer nosed insulators take longer to carry heat off effectively.

**HEAT RISER:** A flapper in the exhaust manifold that is closed when the engine is cold, causing hot exhaust gases to heat the intake manifold providing better cold engine operation. A thermostatic spring opens the flapper when the engine warms up.

**HEAVY THROTTLE:** Approximately three-fourths of accelerator pedal travel.

**HEMI:** A name given an engine using hemispherical combustion chambers.

**HERTZ (HZ):** The international unit of frequency equal to one cycle per second (10,000 Hertz equals 10,000 cycles per second).

**HIGH-IMPEDANCE DVOM (DIGITAL VOLT-OHMMETER):** This styled device provides a built-in resistance value and is capable of limiting circuit current flow to safe milliamp levels.

**HIGH RESISTANCE:** Often refers to a circuit where there is an excessive amount of opposition to normal current flow.

**HORSEPOWER:** A measurement of the amount of work; one horsepower is the amount of work necessary to lift 33,000 lbs. one foot in one minute. Brake horsepower (bhp) is the horsepower delivered by an engine on a dynamometer. Net horsepower is the power remaining (measured at the flywheel of the engine) that can be used to turn the wheels after power is consumed through friction and running the engine accessories (water pump, alternator, air pump, fan etc.)

**HOT CIRCUIT:** (See circuit, hot; hot lead.)

**HOT LEAD:** A wire or conductor in the power side of the circuit. (See circuit, hot.)

**HOT SIDE SWITCHING:** The electrical/electronic circuit control switch is located before the circuit load.

**HUB:** The center part of a wheel or gear.

**HUNTING (BUSYNESS):** Repeating quick series of up-shifts and downshifts that causes noticeable change in engine rpm, for example, as in a 4-3-4 shift pattern.

**HYDRAULICS:** The use of liquid under pressure to transfer force of motion.

**HYDROCARBON (HC):** Any chemical compound made up of hydrogen and carbon. A major pollutant formed by the engine as a by-product of combustion.

**HYDRODYNAMIC DRIVE UNITS:** Devices that transmit power solely by the action of a kinetic fluid flow in a closed recirculating path. An impeller energizes the fluid and discharges the high-speed jet stream into the turbine for power output.

**HYDROMETER:** An instrument used to measure the specific gravity of a solution.

**HYDROPLANING:** A phenomenon of driving when water builds up under the tire tread, causing it to lose contact with the road. Slowing down will usually restore normal tire contact with the road.

**HYPOID GEARSET:** The drive pinion gear may be placed below or above the centerline of the driven gear; often used as a final drive gearset.

**IDLE MIXTURE:** The mixture of air and fuel (usually about 14:1) being fed to the cylinders. The idle mixture screw(s) are sometimes adjusted as part of a tune-up.

**IDLER ARM:** Component of the steering linkage which is a geometric duplicate of the steering gear arm. It supports the right side of the center steering link.

**IMPELLER:** Often called a pump, the impeller is the power input (drive) member of a hydrodynamic drive. As part of the torque converter cover, it acts as a centrifugal pump and puts the fluid in motion.

**INCH POUND (inch lbs.; sometimes in. lb. or in. lbs.):** One twelfth of a foot pound.

**INDUCTANCE:** The force that produces voltage when a conductor is passed through a magnetic field.

**INDUCTION:** A means of transferring electrical energy in the form of a magnetic field. Principle used in the ignition coil to increase voltage.

**INITIAL FEEL:** A distinct firmer feel at start of shift when compared with feel at finish of shift.

**INJECTOR:** A device which receives metered fuel under relatively low pressure and is activated to inject the fuel into the engine under relatively high pressure at a predetermined time.

**INPUT:** In an automatic transmission, the source of power from the engine is absorbed by the torque converter, which provides the power input into the transmission. The turbine drives the input(turbine)shaft.

**INPUT SHAFT:** The shaft to which torque is applied, usually carrying the driving gear or gears.

**INTAKE MANIFOLD:** A casting of passages or pipes used to conduct air or a fuel/air mixture to the cylinders.

**INTERNAL GEAR:** The ring-like outer gear of a planetary gearset with the gear teeth cut on the inside of the ring to provide a mesh with the planet pinions.

**ISOLATION (CLAMPING) DIODES:** Diodes positioned in a circuit to prevent self-induction from damaging electronic components.

**IX ROTARY GEAR PUMP:** Contains two rotating members, one shaped with internal gear teeth and the other with external gear teeth. As the gears separate, the fluid fills the gaps between gear teeth, is pulled across a crescent-shaped divider, and then is forced to flow through the outlet as the gears mesh.

**IX ROTARY LOBE PUMP:** Sometimes referred to as a gerotor type pump. Two rotating members, one shaped with internal lobes and the other with external lobes, separate and then mesh to cause fluid to flow.

**JOURNAL:** The bearing surface within which a shaft operates.

**JUMPER CABLES:** Two heavy duty wires with large alligator clips used to provide power from a charged battery to a discharged battery mounted in a vehicle.

**JUMPSTART:** Utilizing the sufficiently charged battery of one vehicle to start the engine of another vehicle with a discharged battery by the use of jumper cables.

**KEY:** A small block usually fitted in a notch between a shaft and a hub to prevent slippage of the two parts.

**KICKDOWN:** Detent downshift system; either linkage, cable, or electrically controlled.

**KILO:** A prefix used in the metric system to indicate one thousand.

**KNOCK:** Noise which results from the spontaneous ignition of a portion of the air-fuel mixture in the engine cylinder caused by overly advanced ignition timing or use of incorrectly low octane fuel for that engine.

**KNOCK SENSOR:** An input device that responds to spark knock, caused by over advanced ignition timing.

**LABOR TIME:** A specific amount of time required to perform a certain repair or diagnostic service as defined by a vehicle or after-market manufacturer.

**LACQUER:** A quick-drying automotive paint.

**LATE:** Shift that occurs when engine is at higher than normal rpm for given amount of throttle.

**LIGHT-EMITTING DIODE (LED):** A semiconductor diode that emits light as electrical current flows through it; used in some electronic display devices to emit a red or other color light.

**LIGHT THROTTLE:** Approximately one-fourth of accelerator pedal travel.

**LIMITED SLIP:** A type of differential which transfers driving force to the wheel with the best traction.

**LIMP-IN MODE:** Electrical shutdown of the transmission/ transaxle output solenoids, allowing only forward and reverse gears that are hydraulically energized by the manual valve. This permits the vehicle to be driven to a service facility for repair.

**LIP SEAL:** Molded synthetic rubber seal designed with an outer sealing edge (lip) that points into the fluid containing area to be sealed. This type of seal is used where rotational and axial forces are present.

**LITHIUM-BASE GREASE:** Chassis and wheel bearing grease using lithium as a base. Not compatible with sodium-base grease.

**LOAD DEVICE:** A circuit's resistance that converts the electrical energy into light, sound, heat, or mechanical movement.

**LOAD RANGE:** Indicates the number of plies at which a tire is rated. Load range B equals four-ply rating; C equals six-ply rating; and, D equals an eight-ply rating.

**LOAD TORQUE:** The amount of output torque needed from the transmission/transaxle to overcome the vehicle load.

**LOCKING HUBS:** Accessories used on part-time four-wheel drive systems that allow the front wheels to be disengaged from the drive train when four-wheel drive is not being used. When four-wheel drive is desired, the hubs are engaged, locking the wheels to the drive train.

**LOCKUP CONVERTER:** A torque converter that operates hydraulically and mechanically. When an internal apply plate (lockup plate) clamps to the torque converter cover, hydraulic slippage is eliminated.

**LOCK RING:** See Circlip or Snapring

**MAGNET:** Any body with the property of attracting iron or steel.

**MAGNETIC FIELD:** The area surrounding the poles of a magnet that is affected by its attraction or repulsion forces.

**MAIN LINE PRESSURE:** Often called control pressure or line pressure, it refers to the pressure of the oil leaving the pump and is controlled by the pressure regulator valve.

**MALFUNCTION INDICATOR LAMP (MIL):** Previously known as a check engine light, the dash-mounted MIL illuminates and signals the driver that an emission or driveability problem with the powertrain has been detected by the ECM/PCM. When this occurs, at least one diagnostic trouble code (DTC) has been stored into the control module memory.

**MANIFOLD ABSOLUTE PRESSURE (MAP) SENSOR:** Reads the amount of air pressure (vacuum) in the engine's intake manifold system; its signal is used to analyze engine load conditions.

**MANIFOLD VACUUM:** Low pressure in an engine intake manifold formed just below the throttle plates. Manifold vacuum is highest at idle and drops under acceleration.

**MANIFOLD:** A casting of passages or set of pipes which connect the cylinders to an inlet or outlet source.

**MANUAL LEVER POSITION SWITCH (MLPS):** A mechanical switching unit that is typically mounted externally to the transmission/transaxle to inform the PCM/ECM which gear range the driver has selected.

**MANUAL VALVE:** Located inside the transmission/transaxle, it is directly connected to the driver's shift lever. The position of the manual valve determines which hydraulic circuits will be charged with oil pressure and the operating mode of the transmission.

**MANUAL VALVE LEVER POSITION SENSOR (MVLPS):** The input from this device tells the TCM what gear range was selected.

**MASS AIR FLOW (MAF) SENSOR:** Measures the airflow into the engine.

**MASTER CYLINDER:** The primary fluid pressurizing device in a hydraulic system. In automotive use, it is found in brake and hydraulic clutch systems and is pedal activated, either directly or, in a power brake system, through the power booster.

**MacPherson STRUT:** A suspension component combining a shock absorber and spring in one unit.

**MEDIUM THROTTLE:** Approximately one-half of accelerator pedal travel.

**MEGA:** A metric prefix indicating one million.

**MEMBER:** An independent component of a hydrodynamic unit such as an impeller, a stator, or a turbine. It may have one or more elements.

**MERCON:** A fluid developed by Ford Motor Company in 1988. It contains a friction modifier and closely resembles operating characteristics of Dexron.

**METAL SEALING RINGS:** Made from cast iron or aluminum, their primary application is with dynamic components involving pressure sealing circuits of rotating members. These rings are designed with either butt or hook lock end joints.

**METER (ANALOG):** A linear-style meter representing data as lengths; a needle-style instrument interfacing with logical numerical increments. This style of electrical meter uses relatively low impedance internal resistance and cannot be used for testing electronic circuitry.

**METER (DIGITAL):** Uses numbers as a direct readout to show values. Most meters of this style use high impedance internal resistance and must be used for testing low current electronic circuitry.

**MICRO:** A metric prefix indicating one-millionth (0.000001).

**MILLI:** A metric prefix indicating one-thousandth (0.001).

**MINIMUM THROTTLE:** The least amount of throttle opening required for upshift; normally close to zero throttle.

**MISFIRE:** Condition occurring when the fuel mixture in a cylinder fails to ignite, causing the engine to run roughly.

**MODULE:** Electronic control unit, amplifier or igniter of solid state or integrated design which controls the current flow in the ignition primary circuit based on input from the pick-up coil. When the module opens the primary circuit, high secondary voltage is induced in the coil.

**MODULATED:** In an electronic-hydraulic converter clutch system (or shift valve system), the term modulated refers to the pulsing of a solenoid, at a variable rate. This action controls the buildup of oil pressure in the hydraulic circuit to allow a controlled amount of clutch slippage.

**MODULATED CONVERTER CLUTCH CONTROL (MCCC):** A pulse width duty cycle valve that controls the converter lockup apply pressure and maximizes smoother transitions between lock and unlock conditions.

**MODULATOR PRESSURE (THROTTLE PRESSURE):** A hydraulic signal oil pressure relating to the amount of engine load, based on either the amount of throttle plate opening or engine vacuum.

**MODULATOR VALVE:** A regulator valve that is controlled by engine vacuum, providing a hydraulic pressure that varies in relation to engine torque. The hydraulic torque signal functions to delay the shift pattern and provide a line pressure boost. (See throttle valve.)

**MOTOR:** An electromagnetic device used to convert electrical energy into mechanical energy.

**MULTIPLE-DISC CLUTCH:** A grouping of steel and friction lined plates that, when compressed together by hydraulic pressure acting upon a piston, lock or unlock a planetary member.

**MULTI-WEIGHT:** Type of oil that provides adequate lubrication at both high and low temperatures.

needed to move one amp through a resistance of one ohm.

**MUSHY:** Same as soft; slow and drawn out clutch apply with very little shift feel.

**MUTUAL INDUCTION:** The generation of current from one wire circuit to another by movement of the magnetic field surrounding a current-carrying circuit as its ampere flow increases or decreases.

**NEEDLE BEARING:** A bearing which consists of a number (usually a large number) of long, thin rollers.

**NITROGEN OXIDE (NOx):** One of the three basic pollutants found in the exhaust emission of an internal combustion engine. The amount of NOx usually varies in an inverse proportion to the amount of HC and CO.

**NONPOSITIVE SEALING:** A sealing method that allows some minor leakage, which normally assists in lubrication.

**O2 SENSOR:** Located in the engine's exhaust system, it is an input device to the ECM/PCM for managing the fuel delivery and ignition system. A scanner can be used to observe the fluctuating voltage readings produced by an O2 sensor as the oxygen content of the exhaust is analyzed.

**O-RING SEAL:** Molded synthetic rubber seal designed with a circular cross-section. This type of seal is used primarily in static applications.

**OBD II (ON-BOARD DIAGNOSTICS, SECOND GENERATION):** Refers to the federal law mandating tighter control of 1996 and newer vehicle emissions, active monitoring of related devices, and standardization of terminology, data link connectors, and other technician concerns.

**OCTANE RATING:** A number, indicating the quality of gasoline based on its ability to resist knock. The higher the number, the better the quality. Higher compression engines require higher octane gas.

**OEM:** Original Equipment Manufactured. OEM equipment is that furnished standard by the manufacturer.

**OFFSET:** The distance between the vertical center of the wheel and the mounting surface at the lugs. Offset is positive if the center is outside the lug circle; negative offset puts the center line inside the lug circle.

**OHM'S LAW:** A law of electricity that states the relationship between voltage, current, and resistance. Volts = amperes x ohms

**OHM:** The unit used to measure the resistance of conductor-to-electrical flow. One ohm is the amount of resistance that limits current flow to one ampere in a circuit with one volt of pressure.

**OHMMETER:** An instrument used for measuring the resistance, in ohms, in an electrical circuit.

**ONE-WAY CLUTCH:** A mechanical clutch of roller or sprag design that resists torque or transmits power in one direction only. It is used to either hold or drive a planetary member.

**ONE-WAY ROLLER CLUTCH:** A mechanical device that transmits or holds torque in one direction only.

**OPEN CIRCUIT:** A break or lack of contact in an electrical circuit, either intentional (switch) or unintentional (bad connection or broken wire).

**ORIFICE:** Located in hydraulic oil circuits, it acts as a restriction. It slows down fluid flow to either create back pressure or delay pressure buildup downstream.

**OSCILLOSCOPE:** A piece of test equipment that shows electric impulses as a pattern on a screen. Engine performance can be analyzed by interpreting these patterns.

**OUTPUT SHAFT:** The shaft which transmits torque from a device, such as a transmission.

**OUTPUT SPEED SENSOR (OSS):** Identifies transmission/transaxle output shaft speed for shift timing and may be used to calculate TCC slip; often functions as the VSS (vehicle speed sensor).

**OVERDRIVE:** (1.) A device attached to or incorporated in a transmission/transaxle that allows the engine to turn less than one full revolution for every complete revolution of the wheels. The net effect is to reduce engine rpm, thereby using less fuel. A typical overdrive gear ratio would be .87:1, instead of the normal 1:1 in high gear. (2.) A gear assembly which produces more shaft revolutions than that transmitted to it.

**OVERDRIVE PLANETARY GEARSET:** A single planetary gearset designed to provide a direct drive and overdrive ratio. When coupled to a three-speed transmission/transaxle configuration, a four-speed/overdrive unit is present.

**OVERHEAD CAMSHAFT (OHC):** An engine configuration in which the camshaft is mounted on top of the cylinder head and operates the valve either directly or by means of rocker arms.

**OVERHEAD VALVE (OHV):** An engine configuration in which all of the valves are located in the cylinder head and the camshaft is located in the cylinder block. The camshaft operates the valves via lifters and pushrods.

**OVERRUNCLUTCH:** Another name for a one-way mechanical clutch. Applies to both roller and sprag designs.

**OVERSTEER:** The tendency of some vehicles, when steering into a turn, to over-respond or steer more than required, which could result in excessive slip of the rear wheels. Opposite of under-steer.

**OXIDATION STABILIZERS:** Absorb and dissipate heat. Automatic transmission fluid has high resistance to varnish and sludge buildup that occurs from excessive heat that is generated primarily in the torque converter. Local temperatures as high as 6000F (3150C) can occur at the clutch plates during engagement, and this heat must be absorbed and dissipated. If the fluid cannot withstand the heat, it burns or oxidizes, resulting in an almost immediate destruction of friction materials, clogged filter screen and hydraulic passages, and sticky valves.

**OXIDES OF NITROGEN:** See nitrogen oxide (NOx).

**OXYGEN SENSOR:** Used with a feedback system to sense the presence of oxygen in the exhaust gas and signal the computer which can use the voltage signal to determine engine operating efficiency and adjust the air/fuel ratio.

**PARALLEL CIRCUIT:** (See circuit, parallel.)

**PARTS WASHER:** A basin or tub, usually with a built-in pump mechanism and hose used for circulating chemical solvent for the purpose of cleaning greasy, oily and dirty components.

**PART-TIME FOUR WHEEL DRIVE:** A system that is normally in the two wheel drive mode and only runs in four-wheel drive when the system is manually engaged because more traction is desired. Two or four wheel drive is normally selected by a lever to engage the front axle, but if locking hubs are used, these must be manually engaged in the Lock position. Otherwise, the front axle will not drive the front wheels.

**PASSIVE RESTRAINT:** Safety systems such as air bags or automatic seat belts which operate with no action required on the part of the driver or passenger. Mandated by Federal regulations on all vehicles sold in the U.S. after 1990.

**PAYLOAD:** The weight the vehicle is capable of carrying in addition to its own weight. Payload includes weight of the driver, passengers and cargo, but not coolant, fuel, lubricant, spare tire, etc.

**PCM:** Powertrain control module.

**PCV VALVE:** A valve usually located in the rocker cover that vents crankcase vapors back into the engine to be reburned.

**PERCOLATION:** A condition in which the fuel actually "boils," due to excessive heat. Percolation prevents proper atomization of the fuel causing rough running.

**PICK-UP COIL:** The coil in which voltage is induced in an electronic ignition.

**PING:** A metallic rattling sound produced by the engine during acceleration. It is usually due to incorrect ignition timing or a poor grade of gasoline.

**PINION:** The smaller of two gears. The rear axle pinion drives the ring gear which transmits motion to the axle shafts.

**PINION GEAR:** The smallest gear in a drive gear assembly.

**PISTON:** A disc or cup that fits in a cylinder bore and is free to move. In hydraulics, it provides the means of converting hydraulic pressure into a usable force. Examples of piston applications are found in servo, clutch, and accumulator units.

**PISTON RING:** An open-ended ring which fits into a groove on the outer diameter of the piston. Its chief function is to form a seal between the piston and cylinder wall. Most automotive pistons have three rings: two for compression sealing; one for oil sealing.

**PITMAN ARM:** A lever which transmits steering force from the steering gear to the steering linkage.

**PLANET CARRIER:** A basic member of a planetary gear assembly that carries the pinion gears.

**PLANET PINIONS:** Gears housed in a planet carrier that are in constant mesh with the sun gear and internal gear. Because they have their own independent rotating centers, the pinions are capable of rotating around the sun gear or the inside of the internal gear.

**PLANETARY GEAR RATIO:** The reduction or overdrive ratio developed by a planetary gearset.

**PLANETARY GEARSET:** In its simplest form, it is made up of a basic assembly group containing a sun gear, internal gear, and planet carrier. The gears are always in constant mesh and offer a wide range of gear ratio possibilities.

**PLANETARY GEARSET (COMPOUND):** Two planetary gearsets combined together.

**PLANETARY GEARSET (SIMPLE):** An assembly of gears in constant mesh consisting of a sun gear, several pinion gears mounted in a carrier, and a ring gear. It provides gear ratio and direction changes, in addition to a direct drive and a neutral.

**PLY RATING:** A. rating given a tire which indicates strength (but not necessarily actual plies). A two-ply/four-ply rating has only two plies, but the strength of a four-ply tire.

**POLARITY:** Indication (positive or negative) of the two poles of a battery.

**PORT:** An opening for fluid intake or exhaust.

**POSITIVE SEALING:** A sealing method that completely prevents leakage.

**POTENTIAL:** Electrical force measured in volts; sometimes used interchangeably with voltage.

**POWER:** The ability to do work per unit of time, as expressed in horsepower; one horsepower equals 33,000 ft. lbs. of work per minute, or 550 ft. lbs. of work per second.

**POWER FLOW:** The systematic flow or transmission of power through the gears, from the input shaft to the output shaft.

**POWER-TO-WEIGHT RATIO:** Ratio of horsepower to weight of car.

**POWERTRAIN:** See Drivetrain.

**POWERTRAIN CONTROL MODULE (PCM):** Current designation for the engine control module (ECM). In many cases, late model vehicle control units manage the engine as well as the transmission. In other settings, the PCM controls the engine and is interfaced with a TCM to control transmission functions.

**Ppm:** Parts per million; unit used to measure exhaust emissions.

**PREIGNITION:** Early ignition of fuel in the cylinder, sometimes due to glowing carbon deposits in the combustion chamber. Preignition can be damaging since combustion takes place prematurely.

**PRELOAD:** A predetermined load placed on a bearing during assembly or by adjustment.

**PRESS FIT:** The mating of two parts under pressure, due to the inner diameter of one being smaller than the outer diameter of the other, or vice versa; an interference fit.

**PRESSURE:** The amount of force exerted upon a surface area.

**PRESSURE CONTROL SOLENOID (PCS):** An output device that provides a boost oil pressure to the mainline regulator valve to control line pressure. Its operation is determined by the amount of current sent from the PCM.

**PRESSURE GAUGE:** An instrument used for measuring the fluid pressure in a hydraulic circuit.

**PRESSURE REGULATOR VALVE:** In automatic transmissions, its purpose is to regulate the pressure of the pump output and supply the basic fluid pressure necessary to operate the transmission. The regulated fluid pressure may be referred to as mainline pressure, line pressure, or control pressure.

**PRESSURE SWITCH ASSEMBLY (PSA):** Mounted inside the transmission, it is a grouping of oil pressure switches that inputs to the PCM when certain hydraulic passages are charged with oil pressure.

**PRESSURE PLATE:** A spring-loaded plate (part of the clutch) that transmits power to the driven (friction) plate when the clutch is engaged.

**PRIMARY CIRCUIT:** The low voltage side of the ignition system which consists of the ignition switch, ballast resistor or resistance wire, bypass, coil, electronic control unit and pick-up coil as well as the connecting wires and harnesses.

**PROFILE:** Term used for tire measurement (tire series), which is the ratio of tire height to tread width.

**PROM (PROGRAMMABLE READ-ONLY MEMORY):** The heart of the computer that compares input data and makes the engineered program or strategy decisions about when to trigger the appropriate output based on stored computer instructions.

**PULSE GENERATOR:** A two-wire pickup sensor used to produce a fluctuating electrical signal. This changing signal is read by the controller to determine the speed of the object and can be used to measure transmission/transaxle input speed, output speed, and vehicle speed.

**PSI:** Pounds per square inch; a measurement of pressure.

**PULSE WIDTH DUTY CYCLE SOLENOID (PULSE WIDTH MODULATED SOLENOID):** A computer-controlled solenoid that turns on and off at a variable rate producing a modulated oil pressure; often referred to as a pulse width modulated (PWM) solenoid. Employed in many electronic automatic transmissions and transaxles, these solenoids are used to manage shift control and converter clutch hydraulic circuits.

**PUSHROD:** A steel rod between the hydraulic valve lifter and the valve rocker arm in overhead valve (OHV) engines.

**PUMP:** A mechanical device designed to create fluid flow and pressure buildup in a hydraulic system.

**QUARTER PANEL:** General term used to refer to a rear fender. Quarter panel is the area from the rear door opening to the tail light area and from rear wheel well to the base of the trunk and roof-line.

**RACE:** The surface on the inner or outer ring of a bearing on which the balls, needles or rollers move.

**RACK AND PINION:** A type of automotive steering system using a pinion gear attached to the end of the steering shaft. The pinion meshes with a long rack attached to the steering linkage.

**RADIAL TIRE:** Tire design which uses body cords running at right angles to the center line of the tire. Two or more belts are used to give tread strength. Radials can be identified by their characteristic sidewall bulge.

**RADIATOR:** Part of the cooling system for a water-cooled engine, mounted in the front of the vehicle and connected to the engine with rubber hoses. Through the radiator, excess combustion heat is dissipated into the atmosphere through forced convection using a water and glycol based mixture that circulates through, and cools, the engine.

**RANGE REFERENCE AND CLUTCH/BAND APPLY CHART:** A guide that shows the application of clutches and bands for each gear, within the selector range positions. These charts are extremely useful for understanding how the unit operates and for diagnosing malfunctions.

**RAVIGNEAUX GEARSET:** A compound planetary gearset that features matched dual planetary pinions (sets of two) mounted in a single planet carrier. Two sun gears and one ring mesh with the carrier pinions.

**REACTION MEMBER:** The stationary planetary member, in a planetary gearset, that is grounded to the transmission/transaxle case through the use of friction and wedging devices known as bands, disc clutches, and one-way clutches.

**REACTION PRESSURE:** The fluid pressure that moves a spool valve against an opposing force or forces; the area on which the opposing force acts. The opposing force can be a spring or a combination of spring force and auxiliary hydraulic force.

**REACTOR, TORQUE CONVERTER:** The reaction member of a fluid torque converter, more commonly called a stator. (See stator.)

**REAR MAIN OIL SEAL:** A synthetic or rope-type seal that prevents oil from leaking out of the engine past the rear main crankshaft bearing.

**RECIRCULATING BALL:** Type of steering system in which recirculating steel balls occupy the area between the nut and worm wheel, causing a reduction in friction.

**RECTIFIER:** A device (used primarily in alternators) that permits electrical current to flow in one direction only.

**REDUCTION:** (See gear reduction.)

**REGULATOR VALVE:** A valve that changes the pressure of the oil in a hydraulic circuit as the oil passes through the valve by bleeding off (or exhausting) some of the volume of oil supplied to the valve.

**REFRIGERANT 12 (R-12) or 134 (R-134):** The generic name of the refrigerant used in automotive air conditioning systems.

**REGULATOR:** A device which maintains the amperage and/or voltage levels of a circuit at predetermined values.

**RELAY:** A switch which automatically opens and/or closes a circuit.

**RELAY VALVE:** A valve that directs flow and pressure. Relay valves simply connect or disconnect interrelated passages without restricting the fluid flow or changing the pressure.

**RELIEF VALVE:** A spring-loaded, pressure-operated valve that limits oil pressure buildup in a hydraulic circuit to a predetermined maximum value.

**RELUCTOR:** A wheel that rotates inside the distributor and triggers the release of voltage in an electronic ignition.

**RESERVOIR:** The storage area for fluid in a hydraulic system; often called a sump.

**RESIN:** A liquid plastic used in body work.

**RESIDUAL MAGNETISM:** The magnetic strength stored in a material after a magnetizing field has been removed.

**RESISTANCE:** The opposition to the flow of current through a circuit or electrical device, and is measured in ohms. Resistance is equal to the voltage divided by the amperage.

**RESISTOR SPARK PLUG:** A spark plug using a resistor to shorten the spark duration. This suppresses radio interference and lengthens plug life.

**RESISTOR:** A device, usually made of wire, which offers a preset amount of resistance in an electrical circuit.

**RESULTANT FORCE:** The single effective directional thrust of the fluid force on the turbine produced by the vortex and rotary forces acting in different planes.

**RETARD:** Set the ignition timing so that spark occurs later (fewer degrees before TDC).

**RHEOSTAT:** A device for regulating a current by means of a variable resistance.

**RING GEAR:** The name given to a ring-shaped gear attached to a differential case, or affixed to a flywheel or as part of a planetary gear set.

**ROADLOAD:** grade.

**ROCKER ARM:** A lever which rotates around a shaft pushing down (opening) the valve with an end when the other end is pushed up by the pushrod. Spring pressure will later close the valve.

**ROCKER PANEL:** The body panel below the doors between the wheel opening.

**ROLLER BEARING:** A bearing made up of hardened inner and outer races between which hardened steel rollers move.

**ROLLER CLUTCH:** A type of one-way clutch design using rollers and springs mounted within an inner and outer cam race assembly.

**ROTARY FLOW:** The path of the fluid trapped between the blades of the members as they revolve with the rotation of the torque converter cover (rotational inertia).

**ROTOR:** (1.) The disc-shaped part of a disc brake assembly, upon which the brake pads bear; also called, brake disc. (2.) The device mounted atop the distributor shaft, which passes current to the distributor cap tower contacts.

**ROTARY ENGINE:** See Wankel engine.

**RPM:** Revolutions per minute (usually indicates engine speed).

**RTV:** A gasket making compound that cures as it is exposed to the atmosphere. It is used between surfaces that are not perfectly machined to one another, leaving a slight gap that the RTV fills and in which it hardens. The letters RTV represent room temperature vulcanizing.

**RUN-ON:** Condition when the engine continues to run, even when the key is turned off. See dieseling.

**SEALED BEAM:** A automotive headlight. The lens, reflector and filament from a single unit.

**SEATBELT INTERLOCK:** A system whereby the car cannot be started unless the seatbelt is buckled.

**SECONDARY CIRCUIT:** The high voltage side of the ignition system, usually above 20,000 volts. The secondary includes the ignition coil, coil wire, distributor cap and rotor, spark plug wires and spark plugs.

**SELF-INDUCTION:** The generation of voltage in a current-carrying wire by changing the amount of current flowing within that wire.

**SEMI-CONDUCTOR:** A material (silicon or germanium) that is neither a good conductor nor an insulator; used in diodes and transistors.

**SEMI-FLOATING AXLE:** In this design, a wheel is attached to the axle shaft, which takes both drive and cornering loads. Almost all solid axle passenger cars and light trucks use this design.

**SENDING UNIT:** A mechanical, electrical, hydraulic or electromagnetic device which transmits information to a gauge.

**SENSOR:** Any device designed to measure engine operating conditions or ambient pressures and temperatures. Usually electronic in nature and designed to send a voltage signal to an on-board computer, some sensors may operate as a simple on/off switch or they may provide a variable voltage signal (like a potentiometer) as conditions or measured parameters change.

**SERIES CIRCUIT:** (See circuit, series.)

**SERPENTINE BELT:** An accessory drive belt, with small multiple v-ribs, routed around most or all of the engine-powered accessories such as the alternator and power steering pump. Usually both the front and the back side of the belt comes into contact with various pulleys.

**SERVO:** In an automatic transmission, it is a piston in a cylinder assembly that converts hydraulic pressure into mechanical force and movement; used for the application of the bands and clutches.

**SHIFT BUSYNESS:** When referring to a torque converter clutch, it is the frequent apply and release of the clutch plate due to uncommon driving conditions.

**SHIFT VALVE:** Classified as a relay valve, it triggers the automatic shift in response to a governor and a throttle signal by directing fluid to the appropriate band and clutch apply combination to cause the shift to occur.

**SHIM:** Spacers of precise, predetermined thickness used between parts to establish a proper working relationship.

**SHIMMY:** Vibration (sometimes violent) in the front end caused by misaligned front end, out of balance tires or worn suspension components.

**SHORT CIRCUIT:** An electrical malfunction where current takes the path of least resistance to ground (usually through damaged insulation). Current flow is excessive from low resistance resulting in a blown fuse.

**SHUDDER:** Repeated jerking or stick-slip sensation, similar to chuggle but more severe and rapid in nature, that may be most noticeable during certain ranges of vehicle speed; also used to define condition after converter clutch engagement.

**SIMPSON GEARSET:** A compound planetary gear train that integrates two simple planetary gearsets referred to as the front planetary and the rear planetary.

**SINGLE OVERHEAD CAMSHAFT:** See overhead camshaft.

**SKIDPLATE:** A metal plate attached to the underside of the body to protect the fuel tank, transfer case or other vulnerable parts from damage.

**SLAVE CYLINDER:** In automotive use, a device in the hydraulic clutch system which is activated by hydraulic force, disengaging the clutch.

**SLIPPING:** Noticeable increase in engine rpm without vehicle speed increase; usually occurs during or after initial clutch or band engagement.

**SLUDGE:** Thick, black deposits in engine formed from dirt, oil, water, etc. It is usually formed in engines when oil changes are neglected.

**SNAP RING:** A circular retaining clip used inside or outside a shaft or part to secure a shaft, such as a floating wrist pin.

**SOFT:** Slow, almost unnoticeable clutch apply with very little shift feel.

**SOFTCODES:** DTCs that have been set into the PCM memory but are not present at the time of testing; often referred to as history or intermittent codes.

**SOHC:** Single overhead camshaft.

**SOLENOID:** An electrically operated, magnetic switching device.

**SPALLING:** A wear pattern identified by metal chips flaking off the hardened surface. This condition is caused by foreign particles, overloading situations, and/or normal wear.

**SPARK PLUG:** A device screwed into the combustion chamber of a spark ignition engine. The basic construction is a conductive core inside of a ceramic insulator, mounted in an outer conductive base. An electrical charge from the spark plug wire travels along the conductive core and jumps a preset air gap to a grounding point or points at the end of the conductive base. The resultant spark ignites the fuel/air mixture in the combustion chamber.

**SPECIFIC GRAVITY (BATTERY):** The relative weight of liquid (battery electrolyte) as compared to the weight of an equal volume of water.

**SPLINES:** Ridges machined or cast onto the outer diameter of a shaft or inner diameter of a bore to enable parts to mate without rotation.

**SPLIT TORQUE DRIVE:** In a torque converter, it refers to parallel paths of torque transmission, one of which is mechanical and the other hydraulic.

**SPONGY PEDAL:** A soft or spongy feeling when the brake pedal is depressed. It is usually due to air in the brake lines.

**SPOOLVALVE:** A precision-machined, cylindrically shaped valve made up of lands and grooves. Depending on its position in the valve bore, various interconnecting hydraulic circuit passages are either opened or closed.

**SPRAG CLUTCH:** A type of one-way clutch design using cams or contoured-shaped sprags between inner and outer races. (See one-way clutch.)

**SPRUNG WEIGHT:** The weight of a car supported by the springs.

**SQUARE-CUT SEAL:** Molded synthetic rubber seal designed with a square- or rectangular-shaped cross-section. This type of seal is used for both dynamic and static applications.

**SRS:** Supplemental restraint system

**STABILIZER (SWAY) BAR:** A bar linking both sides of the suspension. It resists sway on turns by taking some of added load from one wheel and putting it on the other.

**STAGE:** The number of turbine sets separated by a stator. A turbine set may be made up of one or more turbine members. A three-element converter is classified as a single stage.

**STALL:** In fluid drive transmission/transaxle applications, stall refers to engine rpm with the transmission/transaxle engaged and the vehicle stationary; throttle valve can be in any position between closed and wide open.

**STALL SPEED:** In fluid drive transmission/transaxle applications, stall speed refers to the maximum engine rpm with the transmission/transaxle engaged and vehicle stationary, when the throttle valve is wide open. (See stall; stall test.)

**STALL TEST:** A procedure recommended by many manufacturers to help determine the integrity of an engine, the torque converter stator, and certain clutch and band combinations. With the shift lever in each of the forward and reverse positions and with the brakes firmly applied, the accelerator pedal is momentarily pressed to the wide open throttle (WOT) position. The engine rpm reading at full throttle can provide clues for diagnosing the condition of the items listed above.

**STALL TORQUE:** The maximum design or engineered torque ratio of a fluid torque converter, produced under stall speed conditions. (See stall speed.)

**STARTER:** A high-torque electric motor used for the purpose of starting the engine, typically through a high ratio geared drive connected to the flywheel ring gear.

**STATIC:** A sealing application in which the parts being sealed do not move in relation to each other.

**STATOR (REACTOR):** The reaction member of a fluid torque converter that changes the direction of the fluid as it leaves the turbine to enter the impeller vanes. During the torque multiplication phase, this action assists the impeller's rotary force and results in an increase in torque.

**STEERING GEOMETRY:** Combination of various angles of suspension components (caster, camber, toe-in); roughly equivalent to front end alignment.

**STRAIGHT WEIGHT:** Term designating motor oil as suitable for use within a narrow range of temperatures. Outside the narrow temperature range its flow characteristics will not adequately lubricate.

**STROKE:** The distance the piston travels from bottom dead center to top dead center.

**SUBSTITUTION:** Replacing one part suspected of a defect with a like part of known quality.

**SUMP:** The storage vessel or reservoir that provides a ready source of fluid to the pump. In an automatic transmission, the sump is the oil pan. All fluid eventually returns to the sump for recycling into the hydraulic system.

**SUN GEAR:** In a planetary gearset, it is the center gear that meshes with a cluster of planet pinions.

**SUPERCHARGER:** An air pump driven mechanically by the engine through belts, chains, shafts or gears from the crankshaft. Two general types of supercharger are the positive displacement and centrifugal type, which pump air in direct relationship to the speed of the engine.

**SUPPLEMENTAL RESTRAINT SYSTEM:** See air bag.

**SURGE:** Repeating engine-related feeling of acceleration and deceleration that is less intense than chuggle.

**SWITCH:** A device used to open, close, or redirect the current in an electrical circuit.

**SYNCHROMESH:** A manual transmission/transaxle that is equipped with devices (synchronizers) that match the gear speeds so that the transmission/transaxle can be downshifted without clashing gears.

**SYNTHETIC OIL:** Non-petroleum based oil.

**TACHOMETER:** A device used to measure the rotary speed of an engine, shaft, gear, etc., usually in rotations per minute.

**TDC:** Top dead center. The exact top of the piston's stroke.

**TEFLON SEALING RINGS:** Teflon is a soft, durable, plastic-like material that is resistant to heat and provides excellent sealing. These rings are designed with either scarf-cut joints or as one-piece rings. Teflon sealing rings have replaced many metal ring applications.

**TERMINAL:** A device attached to the end of a wire or cable to make an electrical connection.

**TEST LIGHT, CIRCUIT-POWERED:** Uses available circuit voltage to test circuit continuity.

**TEST LIGHT, SELF-POWERED:** Uses its own battery source to test circuit continuity.

**THERMISTOR:** A special resistor used to measure fluid temperature; it decreases its resistance with increases in temperature.

**THERMOSTAT:** A valve, located in the cooling system of an engine, which is closed when cold and opens gradually in response to engine heating, controlling the temperature of the coolant and rate of coolant flow.

**THERMOSTATIC ELEMENT:** A heat-sensitive, spring-type device that controls a drain port from the upper sump area to the lower sump. When the transaxle fluid reaches operating temperature, the port is closed and the upper sump fills, thus reducing the fluid level in the lower sump.

**THROTTLE POSITION (TP) SENSOR:** Reads the degree of throttle opening; its signal is used to analyze engine load conditions. The ECM/PCM decides to apply the TCC, or to disengage it for coast or load conditions that need a converter torque boost.

**THROTTLE PRESSURE/MODULATOR PRESSURE:** A hydraulic signal oil pressure relating to the amount of engine load, based on either the amount of throttle plate opening or engine vacuum.

**THROTTLE VALVE:** A regulating or balanced valve that is controlled mechanically by throttle linkage or engine vacuum. It sends a hydraulic signal to the shift valve body to control shift timing and shift quality. (See balanced valve; modulator valve.)

**THROW-OUT BEARING:** As the clutch pedal is depressed, the throwout bearing moves against the spring fingers of the pressure plate, forcing the pressure plate to disengage from the driven disc.

**TIE ROD:** A rod connecting the steering arms. Tie rods have threaded ends that are used to adjust toe-in.

**TIE-UP:** Condition where two opposing clutches are attempting to apply at same time, causing engine to labor with noticeable loss of engine rpm.

**TIMING BELT:** A square-toothed, reinforced rubber belt that is driven by the crankshaft and operates the camshaft.

**TIMING CHAIN:** A roller chain that is driven by the crankshaft and operates the camshaft.

**TIRE ROTATION:** Moving the tires from one position to another to make the tires wear evenly.

**TOE-IN (OUT):** A term comparing the extreme front and rear of the front tires. Closer together at the front is toe-in; farther apart at the front is toe-out.

**TOP DEAD CENTER (TDC):** The point at which the piston reaches the top of its travel on the compression stroke.

**TORQUE:** Measurement of turning or twisting force, expressed as foot-pounds or inch-pounds.

**TORQUE CONVERTER:** A turbine used to transmit power from a driving member to a driven member via hydraulic action, providing changes in drive ratio and torque. In automotive use, it links the driveplate at the rear of the engine to the automatic transmission.

**TORQUE CONVERTER CLUTCH:** The apply plate (lockup plate) assembly used for mechanical power flow through the converter.

**TORQUE PHASE:** Sometimes referred to as slip phase or stall phase, torque multiplication occurs when the turbine is turning at a slower speed than the impeller, and the stator is reactionary (stationary). This sequence generates a boost in output torque.

**TORQUE RATING (STALL TORQUE):** The maximum torque multiplication that occurs during stall conditions, with the engine at wide open throttle (WOT) and zero turbine speed.

**TORQUE RATIO:** An expression of the gear ratio factor on torque effect. A 3:1 gear ratio or 3:1 torque ratio increases the torque input by the ratio factor of 3. Input torque (100 ft. lbs.) x 3 = output torque (300 ft. lbs.)

**TRACTION:** The amount of usable tractive effort before the drive wheels slip on the road contact surface.

**TORSION BAR SUSPENSION:** Long rods of spring steel which take the place of springs. One end of the bar is anchored and the other arm (attached to the suspension) is free to twist. The bars' resistance to twisting causes springing action.

**TRACK:** Distance between the centers of the tires where they contact the ground.

**TRACTION CONTROL:** A control system that prevents the spinning of a vehicle's drive wheels when excess power is applied.

**TRACTIVE EFFORT:** The amount of force available to the drive wheels, to move the vehicle.

**TRANSAXLE:** A single housing containing the transmission and differential. Transaxles are usually found on front engine/front wheel drive or rear engine/rear wheel drive cars.

**TRANSDUCER:** A device that changes energy from one form to another. For example, a transducer in a microphone changes sound energy to electrical energy. In automotive air-conditioning controls used in automatic temperature systems, a transducer changes an electrical signal to a vacuum signal, which operates mechanical doors.

**TRANSMISSION:** A powertrain component designed to modify torque and speed developed by the engine; also provides direct drive, reverse, and neutral.

**TRANSMISSION CONTROL MODULE (TCM):** Manages transmission functions. These vary according to the manufacturer's product design but may include converter clutch operation, electronic shift scheduling, and mainline pressure.

**TRANSMISSION FLUID TEMPERATURE (TFT) SENSOR:** Originally called a transmission oil temperature (TOT) sensor, this input device to the ECM/PCM senses the fluid temperature and provides a resistance value. It operates on the thermistor principle.

**TRANSMISSION INPUT SPEED (TIS) SENSOR:** Measures turbine shaft (input shaft) rpm's and compares to engine rpm's to determine torque converter slip. When compared to the transmission output speed sensor or VSS, gear ratio and clutch engagement timing can be determined.

**TRANSMISSION OIL TEMPERATURE (TOT) SENSOR:** (See transmission fluid temperature (TFT) sensor.)

**TRANSMISSION RANGE SELECTOR (TRS) SWITCH:** Tells the module which gear shift position the driver has chosen.

**TRANSFER CASE:** A gearbox driven from the transmission that delivers power to both front and rear driveshafts in a four-wheel drive system. Transfer cases usually have a high and low range set of gears, used depending on how much pulling power is needed.

**TRANSISTOR:** A semi-conductor component which can be actuated by a small voltage to perform an electrical switching function.

**TREAD WEAR INDICATOR:** Bars molded into the tire at right angles to the tread that appear as horizontal bars when 1/16 in. of tread remains.

**TREAD WEAR PATTERN:** The pattern of wear on tires which can be "read" to diagnose problems in the front suspension.

**TUNE-UP:** A regular maintenance function, usually associated with the replacement and adjustment of parts and components in the electrical and fuel systems of a vehicle for the purpose of attaining optimum performance.

**TURBINE:** The output (driven) member of a fluid coupling or fluid torque converter. It is splined to the input (turbine) shaft of the transmission.

**TURBOCHARGER:** An exhaust driven pump which compresses intake air and forces it into the combustion chambers at higher than atmospheric pressures. The increased air pressure allows more fuel to be burned and results in increased horsepower being produced.

**TURBULENCE:** The interference of molecules of a fluid (or vapor) with each other in a fluid flow.

**TYPE F:** Transmission fluid developed and used by Ford Motor Company up to 1982. This fluid type provides a high coefficient of friction.

**TYPE 7176:** The preferred choice of transmission fluid for Chrysler automatic transmissions and transaxles. Developed in 1986, it closely resembles Dexron and Mercon. Type 7176 is the recommended service fill fluid for all Chrysler products utilizing a lockup torque converter dating back to 1978.

**U-JOINT (UNIVERSAL JOINT):** A flexible coupling in the drive train that allows the driveshafts or axle shafts to operate at different angles and still transmit rotary power.

**UNDERSTEER:** The tendency of a car to continue straight ahead while negotiating a turn.

**UNIT BODY:** Design in which the car body acts as the frame.

**UNLEADED FUEL:** Fuel which contains no lead (a common gasoline additive). The presence of lead in fuel will destroy the functioning elements of a catalytic converter, making it useless.

**UNSPRUNG WEIGHT:** The weight of car components not supported by the springs (wheels, tires, brakes, rear axle, control arms, etc.).

**UPSHIFT:** A shift that results in a decrease in torque ratio and an increase in speed.

**VACUUM:** A negative pressure; any pressure less than atmospheric pressure.

**VACUUM ADVANCE:** A device which advances the ignition timing in response to increased engine vacuum.

**VACUUM GAUGE:** An instrument used for measuring the existing vacuum in a vacuum circuit or chamber. The unit of measure is inches (of mercury in a barometer).

**VACUUM MODULATOR:** Generates a hydraulic oil pressure in response to the amount of engine vacuum.

**VALVES:** Devices that can open or close fluid passages in a hydraulic system and are used for directing fluid flow and controlling pressure.

**VALVE BODY ASSEMBLY:** The main hydraulic control assembly of the transmission/transaxle that contains numerous valves, check balls, and other components to control the distribution of pressurized oil throughout the transmission.

**VALVE CLEARANCE:** The measured gap between the end of the valve stem and the rocker arm, cam lobe or follower that activates the valve.

**VALVE GUIDES:** The guide through which the stem of the valve passes.

The guide is designed to keep the valve in proper alignment.

**VALVE LASH (clearance):** The operating clearance in the valve train.

**VALVE TRAIN:** The system that operates intake and exhaust valves, consisting of camshaft, valves and springs, lifters, pushrods and rocker arms.

**VAPOR LOCK:** Boiling of the fuel in the fuel lines due to excess heat. This will interfere with the flow of fuel in the lines and can completely stop the flow. Vapor lock normally only occurs in hot weather.

**VARIABLE DISPLACEMENT (VARIABLE CAPACITY) VANE PUMP:** Slipper-type vanes, mounted in a revolving rotor and contained within the bore of a movable slide, capture and then force fluid to flow. Movement of the slide to various positions changes the size of the vane chambers and the amount of fluid flow. **Note:** GM refers to this pump design as variable displacement, and Ford terms it variable capacity.

**VARIABLE FORCE SOLENOID (VFS):** Commonly referred to as the electronic pressure control (EPC) solenoid, it replaces the cable/linkage style of TV system control and is integrated with a spool valve and spring assembly to control pressure. A variable computer-controlled current flow varies the internal force of the solenoid on the spool valve and resulting control pressure.

**VARIABLE ORIFICE THERMAL VALVE:** Temperature-sensitive hydraulic oil control device that adjusts the size of a circuit path opening. By altering the size of the opening, the oil flow rate is adapted for cold to hot oil viscosity changes.

**VARNISH:** Term applied to the residue formed when gasoline gets old and stale.

**VCM:** See Electronic Control Unit (ECU).

**VEHICLE SPEED SENSOR (VSS):** Provides an electrical signal to the computer module, measuring vehicle speed, and affects the torque converter clutch engagement and release.

**VESPEL SEALING RINGS:** Hard plastic material that produces excellent sealing in dynamic settings. These rings are found in late versions of the 4T60 and in all 4T60-E and 4T80-E transaxles.

**VISCOSITY:** The ability of a fluid to flow. The lower the viscosity rating, the easier the fluid will flow. 10 weight motor oil will flow much easier than 40 weight motor oil.

**VISCOSITY INDEX IMPROVERS:** Keeps the viscosity nearly constant with changes in temperature. This is especially important at low temperatures, when the oil needs to be thin to aid in shifting and for cold-weather starting. Yet it must not be so thin that at high temperatures it will cause excessive hydraulic leakage so that pumps are unable to maintain the proper pressures.

**VISCOUS CLUTCH:** A specially designed torque converter clutch apply plate that, through the use of a silicon fluid, clamps smoothly and absorbs torsional vibrations.

**VOLT:** Unit used to measure the force or pressure of electricity. It is defined as the pressure needed to move one amp through the resistance of one ohm.

**VOLTAGE:** The electrical pressure that causes current to flow. Voltage is measured in volts (V).

**VOLTAGE, APPLIED:** The actual voltage read at a given point in a circuit. It equals the available voltage of the power supply minus the losses in the circuit up to that point.

**VOLTAGE DROP:** The voltage lost or used in a circuit by normal loads such as a motor or lamp or by abnormal loads such as a poor (high-resistance) lead or terminal connection.

**VOLTAGE REGULATOR:** A device that controls the current output of the alternator or generator.

**VOLTMETER:** An instrument used for measuring electrical force in units called volts. Voltmeters are always connected parallel with the circuit being tested.

**VORTEX FLOW:** The crosswise or circulatory flow of oil between the blades of the members caused by the centrifugal pumping action of the impeller.

**WANKEL ENGINE:** An engine which uses no pistons. In place of pistons, triangular-shaped rotors revolve in specially shaped housings.

**WATER PUMP:** A belt driven component of the cooling system that mounts on the engine, circulating the coolant under pressure.

**WATT:** The unit for measuring electrical power. One watt is the product of one ampere and one volt (watts equals amps times volts). Wattage is the horsepower of electricity (746 watts equal one horsepower).

**WHEEL ALIGNMENT:** Inclusive term to describe the front end geometry (caster, camber, toe-in/out).

**WHEEL CYLINDER:** Found in the automotive drum brake assembly, it is a device, actuated by hydraulic pressure, which, through internal pistons, pushes the brake shoes outward against the drums.

**WHEEL WEIGHT:** Small weights attached to the wheel to balance the wheel and tire assembly. Out-of-balance tires quickly wear out and also give erratic handling when installed on the front.

**WHEELBASE:** Distance between the center of front wheels and the center of rear wheels.

**WIDE OPEN THROTTLE (WOT):** Full travel of accelerator pedal.

**WORK:** The force exerted to move a mass or object. Work involves motion; if a force is exerted and no motion takes place, no work is done. Work per unit of time is called power. Work = force x distance = ft. lbs. 33,000 ft. lbs. in one minute = 1 horsepower

**ZERO-THROTTLE COAST DOWN:** A full release of accelerator pedal while vehicle is in motion and in drive range.

## Commonly Used Abbreviations

### 2

| | |
|---|---|
| 2WD | Two Wheel Drive |

### 4

| | |
|---|---|
| 4WD | Four Wheel Drive |

### A

| | |
|---|---|
| A/C | Air Conditioning |
| ABDC | After Bottom Dead Center |
| ABS | Anti-lock Brakes |
| AC | Alternating Current |
| ACL | Air cleaner |
| ACT | Air Charge Temperature |
| AIR | Secondary Air Injection |
| ALCL | Assembly Line Communications Link |
| ALDL | Assembly Line Diagnostic Link |
| AT | Automatic Transaxle/Transmission |
| ATDC | After Top Dead Center |
| ATF | Automatic Transmission Fluid |
| ATS | Air Temperature Sensor |
| AWD | All Wheel Drive |

### B

| | |
|---|---|
| BAP | Barometric Absolute Pressure |
| BARO | Barometric Pressure |
| BBDC | Before Bottom Dead Center |
| BCM | Body Control Module |
| BDC | Bottom Dead Center |
| BPT | Backpressure Transducer |
| BTDC | Before Top Dead Center |
| BVSV | Bimetallic Vacuum Switching Valve |

### C

| | |
|---|---|
| CAC | Charge Air Cooler |
| CARB | California Air Resources Board |
| CAT | Catalytic Converter |
| CCC | Computer Command Control |
| CCCC | Computer Controlled Catalytic Converter |
| CCCI | Computer Controlled Coil Ignition |
| CCD | Computer Controlled Dwell |
| CDI | Capacitor Discharge Ignition |
| CEC | Computerized Engine Control |
| CFI | Continuous Fuel Injection |
| CIS | Continuous Injection System |
| CIS-E | Continuous Injection System - Electronic |
| CKP | Crankshaft Position |
| CL | Closed Loop |
| CMP | Camshaft Position |
| CPP | Clutch Pedal Position |
| CTOX | Continuous Trap Oxidizer System |
| CTP | Closed Throttle Position |
| CVC | Constant Vacuum Control |
| CYL | Cylinder |

### D

| | |
|---|---|
| DBC | Dual Bed Catalyst |
| DC | Direct Current |
| DFI | Direct Fuel Injection |
| DIS | Distributorless Ignition System |
| DLC | Data Link Connector |
| DMM | Digital Multimeter |
| DOHC | Double Overhead Camshaft |
| DRB | Diagnostic Readout Box |
| DTC | Diagnostic Trouble Code |
| DTM | Diagnostic Test Mode |
| DVOM | Digital Volt/Ohmmeter |

### E

| | |
|---|---|
| EBCM | Electronic Brake Control Module |
| ECM | Engine Control Module |
| ECT | Engine Coolant Temperature |
| ECU | Engine Control Unit or Electronic Control Unit |
| EDIS | Electronic Distributorless Ignition System |
| EEC | Electronic Engine Control |
| EEPROM | Electrically Erasable Programmable Read Only Memory |
| EFE | Early Fuel Evaporation |
| EGR | Exhaust Gas Recirculation |
| EGRT | Exhaust Gas Recirculation Temperature |
| EGRVC | EGR Valve Control |
| EPROM | Erasable Programmable Read Only Memory |
| EVAP | Evaporative Emissions |
| EVP | EGR Valve Position |

### F

| | |
|---|---|
| FBC | Feedback Carburetor |
| FEEPROM | Flash Electrically Erasable Programmable Read Only Memory |
| FF | Flexible Fuel |
| FI | Fuel Injection |
| FT | Fuel Trim |
| FWD | Front Wheel Drive |

### G

| | |
|---|---|
| GND | Ground |

### H

| | |
|---|---|
| HAC | High Altitude Compensation |
| HEGO | Heated Exhaust Gas Oxygen sensor |
| HEI | High Energy Ignition |
| HO2 Sensor | Heated Oxygen Sensor |

### I

| | |
|---|---|
| IAC | Idle Air Control |
| IAT | Intake Air Temperature |
| ICM | Ignition Control Module |
| IFI | Indirect Fuel Injection |
| IFS | Inertia Fuel Shutoff |
| ISC | Idle Speed Control |
| IVSV | Idle Vacuum Switching Valve |

# Commonly Used Abbreviations

## K

| | |
|---|---|
| KOEO | Key On, Engine Off |
| KOER | Key ON, Engine Running |
| KS | Knock Sensor |

## M

| | |
|---|---|
| MAF | Mass Air Flow |
| MAP | Manifold Absolute Pressure |
| MAT | Manifold Air Temperature |
| MC | Mixture Control |
| MDP | Manifold Differential Pressure |
| MFI | Multiport Fuel Injection |
| MIL | Malfunction Indicator Lamp or Maintenance |
| MST | Manifold Surface Temperature |
| MVZ | Manifold Vacuum Zone |

## N

| | |
|---|---|
| NVRAM | Nonvolatile Random Access Memory |

## O

| | |
|---|---|
| O2 Sensor | Oxygen Sensor |
| OBD | On-Board Diagnostic |
| OC | Oxidation Catalyst |
| OHC | Overhead Camshaft |
| OL | Open Loop |

## P

| | |
|---|---|
| P/S | Power Steering |
| PAIR | Pulsed Secondary Air Injection |
| PCM | Powertrain Control Module |
| PCS | Purge Control Solenoid |
| PCV | Positive Crankcase Ventilation |
| PIP | Profile Ignition Pick-up |
| PNP | Park/Neutral Position |
| PROM | Programmable Read Only Memory |
| PSP | Power Steering Pressure |
| PTO | Power Take-Off |
| PTOX | Periodic Trap Oxidizer System |

## R

| | |
|---|---|
| RABS | Rear Anti-lock Brake System |
| RAM | Random Access Memory |
| ROM | Read Only Memory |
| RPM | Revolutions Per Minute |
| RWAL | Rear Wheel Anti-lock Brakes |
| RWD | Rear Wheel Drive |

## S

| | |
|---|---|
| SBC | Single Bed Converter |
| SBEC | Single Board Engine Controller |
| SC | Supercharger |
| SCB | Supercharger Bypass |
| SFI | Sequential Multiport Fuel Injection |
| SIR | Supplemental Inflatable Restraint |
| SOHC | Single Overhead Camshaft |
| SPL | Smoke Puff Limiter |
| SPOUT | Spark Output |
| SRI | Service Reminder Indicator |
| SRS | Supplemental Restraint System |
| SRT | System Readiness Test |
| SSI | Solid State Ignition |
| ST | Scan Tool |
| STO | Self-Test Output |

## T

| | |
|---|---|
| TAC | Thermostatic Air Clearner |
| TBI | Throttle Body Fuel Injection |
| TC | Turbocharger |
| TCC | Torque Converter Clutch |
| TCM | Transmission Control Module |
| TDC | Top Dead Center |
| TFI | Thick Film Ignition |
| TP | Throttle Position |
| TR Sensor | Transaxle/Transmission Range Sensor |
| TVV | Thermal Vacuum Valve |
| TWC | Three-way Catalytic Converter |

## V

| | |
|---|---|
| VAF | Volume Air Flow, or Vane Air Flow |
| VAPS | Variable Assist Power Steering |
| VRV | Vacuum Regulator Valve |
| VSS | Vehicle Speed Sensor |
| VSV | Vacuum Switching Valve |

## W

| | |
|---|---|
| WOT | Wide Open Throttle |
| WU-TWC | Warm Up Three-way Catalytic Converter |

## ENGLISH TO METRIC CONVERSION: TORQUE

To convert foot-pounds (ft. lbs.) to Newton-meters (Nm), multiply the number of ft. lbs. by 1.36
To convert Newton-meters (Nm) to foot-pounds (ft. lbs.), multiply the number of Nm by 0.7376

| ft. lbs. | Nm | ft. lbs. | Nm | ft. lbs. | Nm | ft. lbs. | Nm |
|---|---|---|---|---|---|---|---|
| 0.1 | 0.1 | 34 | 46.2 | 76 | 103.4 | 118 | 160.5 |
| 0.2 | 0.3 | 35 | 47.6 | 77 | 104.7 | 119 | 161.8 |
| 0.3 | 0.4 | 36 | 49.0 | 78 | 106.1 | 120 | 163.2 |
| 0.4 | 0.5 | 37 | 50.3 | 79 | 107.4 | 121 | 164.6 |
| 0.5 | 0.7 | 38 | 51.7 | 80 | 108.8 | 122 | 165.9 |
| 0.6 | 0.8 | 39 | 53.0 | 81 | 110.2 | 123 | 167.3 |
| 0.7 | 1.0 | 40 | 54.4 | 82 | 111.5 | 124 | 168.6 |
| 0.8 | 1.1 | 41 | 55.8 | 83 | 112.9 | 125 | 170.0 |
| 0.9 | 1.2 | 42 | 57.1 | 84 | 114.2 | 126 | 171.4 |
| 1 | 1.4 | 43 | 58.5 | 85 | 115.6 | 127 | 172.7 |
| 2 | 2.7 | 44 | 59.8 | 86 | 117.0 | 128 | 174.1 |
| 3 | 4.1 | 45 | 61.2 | 87 | 118.3 | 129 | 175.4 |
| 4 | 5.4 | 46 | 62.6 | 88 | 119.7 | 130 | 176.8 |
| 5 | 6.8 | 47 | 63.9 | 89 | 121.0 | 131 | 178.2 |
| 6 | 8.2 | 48 | 65.3 | 90 | 122.4 | 132 | 179.5 |
| 7 | 9.5 | 49 | 66.6 | 91 | 123.8 | 133 | 180.9 |
| 8 | 10.9 | 50 | 68.0 | 92 | 125.1 | 134 | 182.2 |
| 9 | 12.2 | 51 | 69.4 | 93 | 126.5 | 135 | 183.6 |
| 10 | 13.6 | 52 | 70.7 | 94 | 127.8 | 136 | 185.0 |
| 11 | 15.0 | 53 | 72.1 | 95 | 129.2 | 137 | 186.3 |
| 12 | 16.3 | 54 | 73.4 | 96 | 130.6 | 138 | 187.7 |
| 13 | 17.7 | 55 | 74.8 | 97 | 131.9 | 139 | 189.0 |
| 14 | 19.0 | 56 | 76.2 | 98 | 133.3 | 140 | 190.4 |
| 15 | 20.4 | 57 | 77.5 | 99 | 134.6 | 141 | 191.8 |
| 16 | 21.8 | 58 | 78.9 | 100 | 136.0 | 142 | 193.1 |
| 17 | 23.1 | 59 | 80.2 | 101 | 137.4 | 143 | 194.5 |
| 18 | 24.5 | 60 | 81.6 | 102 | 138.7 | 144 | 195.8 |
| 19 | 25.8 | 61 | 83.0 | 103 | 140.1 | 145 | 197.2 |
| 20 | 27.2 | 62 | 84.3 | 104 | 141.4 | 146 | 198.6 |
| 21 | 28.6 | 63 | 85.7 | 105 | 142.8 | 147 | 199.9 |
| 22 | 29.9 | 64 | 87.0 | 106 | 144.2 | 148 | 201.3 |
| 23 | 31.3 | 65 | 88.4 | 107 | 145.5 | 149 | 202.6 |
| 24 | 32.6 | 66 | 89.8 | 108 | 146.9 | 150 | 204.0 |
| 25 | 34.0 | 67 | 91.1 | 109 | 148.2 | 151 | 205.4 |
| 26 | 35.4 | 68 | 92.5 | 110 | 149.6 | 152 | 206.7 |
| 27 | 36.7 | 69 | 93.8 | 111 | 151.0 | 153 | 208.1 |
| 28 | 38.1 | 70 | 95.2 | 112 | 152.3 | 154 | 209.4 |
| 29 | 39.4 | 71 | 96.6 | 113 | 153.7 | 155 | 210.8 |
| 30 | 40.8 | 72 | 97.9 | 114 | 155.0 | 156 | 212.2 |
| 31 | 42.2 | 73 | 99.3 | 115 | 156.4 | 157 | 213.5 |
| 32 | 43.5 | 74 | 100.6 | 116 | 157.8 | 158 | 214.9 |
| 33 | 44.9 | 75 | 102.0 | 117 | 159.1 | 159 | 216.2 |

## METRIC TO ENGLISH CONVERSION: TORQUE

To convert foot-pounds (ft. lbs.) to Newton-meters (Nm), multiply the number of ft. lbs. by 1.36

To convert Newton-meters (Nm) to foot-pounds (ft. lbs.), multiply the number of Nm by 0.7376

| Nm | ft. lbs. | Nm | ft. lbs. | Nm | ft. lbs. | Nm | ft. lbs. | Nm | ft. lbs. |
|----|----------|----|----------|----|----------|----|----------|----|----------|
| 0.1 | 0.1 | 34 | 25.0 | 76 | 55.9 | 118 | 86.8 | 160 | 117.6 |
| 0.2 | 0.1 | 35 | 25.7 | 77 | 56.6 | 119 | 87.5 | 161 | 118.4 |
| 0.3 | 0.2 | 36 | 26.5 | 78 | 57.4 | 120 | 88.2 | 162 | 119.1 |
| 0.4 | 0.3 | 37 | 27.2 | 79 | 58.1 | 121 | 89.0 | 163 | 119.9 |
| 0.5 | 0.4 | 38 | 27.9 | 80 | 58.8 | 122 | 89.7 | 164 | 120.6 |
| 0.6 | 0.4 | 39 | 28.7 | 81 | 59.6 | 123 | 90.4 | 165 | 121.3 |
| 0.7 | 0.5 | 40 | 29.4 | 82 | 60.3 | 124 | 91.2 | 166 | 122.1 |
| 0.8 | 0.6 | 41 | 30.1 | 83 | 61.0 | 125 | 91.9 | 167 | 122.8 |
| 0.9 | 0.7 | 42 | 30.9 | 84 | 61.8 | 126 | 92.6 | 168 | 123.5 |
| 1 | 0.7 | 43 | 31.6 | 85 | 62.5 | 127 | 93.4 | 169 | 124.3 |
| 2 | 1.5 | 44 | 32.4 | 86 | 63.2 | 128 | 94.1 | 170 | 125.0 |
| 3 | 2.2 | 45 | 33.1 | 87 | 64.0 | 129 | 94.9 | 171 | 125.7 |
| 4 | 2.9 | 46 | 33.8 | 88 | 64.7 | 130 | 95.6 | 172 | 126.5 |
| 5 | 3.7 | 47 | 34.6 | 89 | 65.4 | 131 | 96.3 | 173 | 127.2 |
| 6 | 4.4 | 48 | 35.3 | 90 | 66.2 | 132 | 97.1 | 174 | 127.9 |
| 7 | 5.1 | 49 | 36.0 | 91 | 66.9 | 133 | 97.8 | 175 | 128.7 |
| 8 | 5.9 | 50 | 36.8 | 92 | 67.6 | 134 | 98.5 | 176 | 129.4 |
| 9 | 6.6 | 51 | 37.5 | 93 | 68.4 | 135 | 99.3 | 177 | 130.1 |
| 10 | 7.4 | 52 | 38.2 | 94 | 69.1 | 136 | 100.0 | 178 | 130.9 |
| 11 | 8.1 | 53 | 39.0 | 95 | 69.9 | 137 | 100.7 | 179 | 131.6 |
| 12 | 8.8 | 54 | 39.7 | 96 | 70.6 | 138 | 101.5 | 180 | 132.4 |
| 13 | 9.6 | 55 | 40.4 | 97 | 71.3 | 139 | 102.2 | 181 | 133.1 |
| 14 | 10.3 | 56 | 41.2 | 98 | 72.1 | 140 | 102.9 | 182 | 133.8 |
| 15 | 11.0 | 57 | 41.9 | 99 | 72.8 | 141 | 103.7 | 183 | 134.6 |
| 16 | 11.8 | 58 | 42.6 | 100 | 73.5 | 142 | 104.4 | 184 | 135.3 |
| 17 | 12.5 | 59 | 43.4 | 101 | 74.3 | 143 | 105.1 | 185 | 136.0 |
| 18 | 13.2 | 60 | 44.1 | 102 | 75.0 | 144 | 105.9 | 186 | 136.8 |
| 19 | 14.0 | 61 | 44.9 | 103 | 75.7 | 145 | 106.6 | 187 | 137.5 |
| 20 | 14.7 | 62 | 45.6 | 104 | 76.5 | 146 | 107.4 | 188 | 138.2 |
| 21 | 15.4 | 63 | 46.3 | 105 | 77.2 | 147 | 108.1 | 189 | 139.0 |
| 22 | 16.2 | 64 | 47.1 | 106 | 77.9 | 148 | 108.8 | 190 | 139.7 |
| 23 | 16.9 | 65 | 47.8 | 107 | 78.7 | 149 | 109.6 | 191 | 140.4 |
| 24 | 17.6 | 66 | 48.5 | 108 | 79.4 | 150 | 110.3 | 192 | 141.2 |
| 25 | 18.4 | 67 | 49.3 | 109 | 80.1 | 151 | 111.0 | 193 | 141.9 |
| 26 | 19.1 | 68 | 50.0 | 110 | 80.9 | 152 | 111.8 | 194 | 142.6 |
| 27 | 19.9 | 69 | 50.7 | 111 | 81.6 | 153 | 112.5 | 195 | 143.4 |
| 28 | 20.6 | 70 | 51.5 | 112 | 82.4 | 154 | 113.2 | 196 | 144.1 |
| 29 | 21.3 | 71 | 52.2 | 113 | 83.1 | 155 | 114.0 | 197 | 144.9 |
| 30 | 22.1 | 72 | 52.9 | 114 | 83.8 | 156 | 114.7 | 198 | 145.6 |
| 31 | 22.8 | 73 | 53.7 | 115 | 84.6 | 157 | 115.4 | 199 | 146.3 |
| 32 | 23.5 | 74 | 54.4 | 116 | 85.3 | 158 | 116.2 | 200 | 147.1 |
| 33 | 24.3 | 75 | 55.1 | 117 | 86.0 | 159 | 116.9 | 201 | 147.8 |

## ENGLISH/METRIC CONVERSION: TEMPERATURE

To convert Fahrenheit (F°) to Celsius (C°), take F° temperature and subtract 32, multiply the result by 5 and divide the result by 9
To convert Celsius (C°) to Fahrenheit (F°), take C° temperature and multiply it by 9, divide the result by 5 and add 32

| F° | C° | F° | C° | C° | F° | C° | F° |
|---|---|---|---|---|---|---|---|
| -40 | -40.0 | 150 | 65.6 | -38 | -36.4 | 46 | 114.8 |
| -35 | -37.2 | 155 | 68.3 | -36 | -32.8 | 48 | 118.4 |
| -30 | -34.4 | 160 | 71.1 | -34 | -29.2 | 50 | 122 |
| -25 | -31.7 | 165 | 73.9 | -32 | -25.6 | 52 | 125.6 |
| -20 | -28.9 | 170 | 76.7 | -30 | -22 | 54 | 129.2 |
| -15 | -26.1 | 175 | 79.4 | -28 | -18.4 | 56 | 132.8 |
| -10 | -23.3 | 180 | 82.2 | -26 | -14.8 | 58 | 136.4 |
| -5 | -20.6 | 185 | 85.0 | -24 | -11.2 | 60 | 140 |
| 0 | -17.8 | 190 | 87.8 | -22 | -7.6 | 62 | 143.6 |
| 1 | -17.2 | 195 | 90.6 | -20 | -4 | 64 | 147.2 |
| 2 | -16.7 | 200 | 93.3 | -18 | -0.4 | 66 | 150.8 |
| 3 | -16.1 | 205 | 96.1 | -16 | 3.2 | 68 | 154.4 |
| 4 | -15.6 | 210 | 98.9 | -14 | 6.8 | 70 | 158 |
| 5 | -15.0 | 212 | 100.0 | -12 | 10.4 | 72 | 161.6 |
| 10 | -12.2 | 215 | 101.7 | -10 | 14 | 74 | 165.2 |
| 15 | -9.4 | 220 | 104.4 | -8 | 17.6 | 76 | 168.8 |
| 20 | -6.7 | 225 | 107.2 | -6 | 21.2 | 78 | 172.4 |
| 25 | -3.9 | 230 | 110.0 | -4 | 24.8 | 80 | 176 |
| 30 | -1.1 | 235 | 112.8 | -2 | 28.4 | 82 | 179.6 |
| 35 | 1.7 | 240 | 115.6 | 0 | 32 | 84 | 183.2 |
| 40 | 4.4 | 245 | 118.3 | 2 | 35.6 | 86 | 186.8 |
| 45 | 7.2 | 250 | 121.1 | 4 | 39.2 | 88 | 190.4 |
| 50 | 10.0 | 255 | 123.9 | 6 | 42.8 | 90 | 194 |
| 55 | 12.8 | 260 | 126.7 | 8 | 46.4 | 92 | 197.6 |
| 60 | 15.6 | 265 | 129.4 | 10 | 50 | 94 | 201.2 |
| 65 | 18.3 | 270 | 132.2 | 12 | 53.6 | 96 | 204.8 |
| 70 | 21.1 | 275 | 135.0 | 14 | 57.2 | 98 | 208.4 |
| 75 | 23.9 | 280 | 137.8 | 16 | 60.8 | 100 | 212 |
| 80 | 26.7 | 285 | 140.6 | 18 | 64.4 | 102 | 215.6 |
| 85 | 29.4 | 290 | 143.3 | 20 | 68 | 104 | 219.2 |
| 90 | 32.2 | 295 | 146.1 | 22 | 71.6 | 106 | 222.8 |
| 95 | 35.0 | 300 | 148.9 | 24 | 75.2 | 108 | 226.4 |
| 100 | 37.8 | 305 | 151.7 | 26 | 78.8 | 110 | 230 |
| 105 | 40.6 | 310 | 154.4 | 28 | 82.4 | 112 | 233.6 |
| 110 | 43.3 | 315 | 157.2 | 30 | 86 | 114 | 237.2 |
| 115 | 46.1 | 320 | 160.0 | 32 | 89.6 | 116 | 240.8 |
| 120 | 48.9 | 325 | 162.8 | 34 | 93.2 | 118 | 244.4 |
| 125 | 51.7 | 330 | 165.6 | 36 | 96.8 | 120 | 248 |
| 130 | 54.4 | 335 | 168.3 | 38 | 100.4 | 122 | 251.6 |
| 135 | 57.2 | 340 | 171.1 | 40 | 104 | 124 | 255.2 |
| 140 | 60.0 | 345 | 173.9 | 42 | 107.6 | 126 | 258.8 |
| 145 | 62.8 | 350 | 176.7 | 44 | 111.2 | 128 | 262.4 |

## LENGTH CONVERSION

To convert inches (in.) to millimeters (mm), multiply the number of inches by 25.4
To convert millimeters (mm) to inches (in.), multiply the number of millimeters by 0.04

| Inches | Millimeters | Inches | Millimeters | Inches | Millimeters | Inches | Millimeters |
|--------|-------------|--------|-------------|--------|-------------|--------|-------------|
| 0.0001 | 0.00254 | 0.005 | 0.1270 | 0.09 | 2.286 | 4 | 101.6 |
| 0.0002 | 0.00508 | 0.006 | 0.1524 | 0.1 | 2.54 | 5 | 127.0 |
| 0.0003 | 0.00762 | 0.007 | 0.1778 | 0.2 | 5.08 | 6 | 152.4 |
| 0.0004 | 0.01016 | 0.008 | 0.2032 | 0.3 | 7.62 | 7 | 177.8 |
| 0.0005 | 0.01270 | 0.009 | 0.2286 | 0.4 | 10.16 | 8 | 203.2 |
| 0.0006 | 0.01524 | 0.01 | 0.254 | 0.5 | 12.70 | 9 | 228.6 |
| 0.0007 | 0.01778 | 0.02 | 0.508 | 0.6 | 15.24 | 10 | 254.0 |
| 0.0008 | 0.02032 | 0.03 | 0.762 | 0.7 | 17.78 | 11 | 279.4 |
| 0.0009 | 0.02286 | 0.04 | 1.016 | 0.8 | 20.32 | 12 | 304.8 |
| 0.001 | 0.0254 | 0.05 | 1.270 | 0.9 | 22.86 | 13 | 330.2 |
| 0.002 | 0.0508 | 0.06 | 1.524 | 1 | 25.4 | 14 | 355.6 |
| 0.003 | 0.0762 | 0.07 | 1.778 | 2 | 50.8 | 15 | 381.0 |
| 0.004 | 0.1016 | 0.08 | 2.032 | 3 | 76.2 | 16 | 406.4 |

## ENGLISH/METRIC CONVERSION: LENGTH

To convert inches (in.) to millimeters (mm), multiply the number of inches by 25.4
To convert millimeters (mm) to inches (in.), multiply the number of millimeters by 0.04

| Inches | | Millimeters | Inches | | Millimeters | Inches | | Millimeters |
|--------|---------|-------------|--------|---------|-------------|--------|---------|-------------|
| Fraction | Decimal | Decimal | Fraction | Decimal | Decimal | Fraction | Decimal | Decimal |
| 1/64 | 0.016 | 0.397 | 11/32 | 0.344 | 8.731 | 11/16 | 0.688 | 17.463 |
| 1/32 | 0.031 | 0.794 | 23/64 | 0.359 | 9.128 | 45/64 | 0.703 | 17.859 |
| 3/64 | 0.047 | 1.191 | 3/8 | 0.375 | 9.525 | 23/32 | 0.719 | 18.256 |
| 1/16 | 0.063 | 1.588 | 25/64 | 0.391 | 9.922 | 47/64 | 0.734 | 18.653 |
| 5/64 | 0.078 | 1.984 | 13/32 | 0.406 | 10.319 | 3/4 | 0.750 | 19.050 |
| 3/32 | 0.094 | 2.381 | 27/64 | 0.422 | 10.716 | 49/64 | 0.766 | 19.447 |
| 7/64 | 0.109 | 2.778 | 7/16 | 0.438 | 11.113 | 25/32 | 0.781 | 19.844 |
| 1/8 | 0.125 | 3.175 | 29/64 | 0.453 | 11.509 | 51/64 | 0.797 | 20.241 |
| 9/64 | 0.141 | 3.572 | 15/32 | 0.469 | 11.906 | 13/16 | 0.813 | 20.638 |
| 5/32 | 0.156 | 3.969 | 31/64 | 0.484 | 12.303 | 53/64 | 0.828 | 21.034 |
| 11/64 | 0.172 | 4.366 | 1/2 | 0.500 | 12.700 | 27/32 | 0.844 | 21.431 |
| 3/16 | 0.188 | 4.763 | 33/64 | 0.516 | 13.097 | 55/64 | 0.859 | 21.828 |
| 13/64 | 0.203 | 5.159 | 17/32 | 0.531 | 13.494 | 7/8 | 0.875 | 22.225 |
| 7/32 | 0.219 | 5.556 | 35/64 | 0.547 | 13.891 | 57/64 | 0.891 | 22.622 |
| 15/64 | 0.234 | 5.953 | 9/16 | 0.563 | 14.288 | 29/32 | 0.906 | 23.019 |
| 1/4 | 0.250 | 6.350 | 37/64 | 0.578 | 14.684 | 59/64 | 0.922 | 23.416 |
| 17/64 | 0.266 | 6.747 | 19/32 | 0.594 | 15.081 | 15/16 | 0.938 | 23.813 |
| 9/32 | 0.281 | 7.144 | 39/64 | 0.609 | 15.478 | 61/64 | 0.953 | 24.209 |
| 19/64 | 0.297 | 7.541 | 5/8 | 0.625 | 15.875 | 31/32 | 0.969 | 24.606 |
| 5/16 | 0.313 | 7.938 | 41/64 | 0.641 | 16.272 | 63/64 | 0.984 | 25.003 |
| 21/64 | 0.328 | 8.334 | 21/32 | 0.656 | 16.669 | 1/1 | 1.000 | 25.400 |
| | | | 43/64 | 0.672 | 17.066 | | | |